PROBLEM-SOLVING STRATEGY BOXES

For users of the three-volume edition, pages 1–791 (Chapters 1–20) are in Volume 1; pages 792–1444 (Chapters 21–37) are in Volume 2; and pages 1403–1714 (Chapters 37–44) are in Volume 3.

Pages 1445–1714 (Chapters 38–44) are not in the Standard Edition of this textbook.

HE MOST TRUSTED AND ENDURING PHYSICS TEXTBOOK COMES WITH THE MOST EFFECTIVE AND VALUABLE MULTIMEDIA AVAILABLE TO HELP YOU SUCCEED IN YOUR COURSE.

ActivPhysics OnLine

ActivPhysics OnLine utilizes visualization, simulation and multiple representations to help you better understand key physical processes, experiment quantitatively, and develop your critical-thinking skills. This library of online interactive simulations are coupled with thought-provoking questions and activities to guide your understanding of physics.

Website: www.aw.com/young11

Minimum System Requirements for book specific web site:
Windows: 250 MHz; OS 98, NT, ME, 2000, XP
Macintosh: 233 MHz; OS 9.2, 10
Both:

- 64 RAM installed
- 1024 x 768 screen resolution
- Browsers: Internet Explorer 5.5, 6.0; Netscape 6.2.3, 7.0
- Plug Ins: Macromedia's Flash 6.0

Mastering PHYSICS

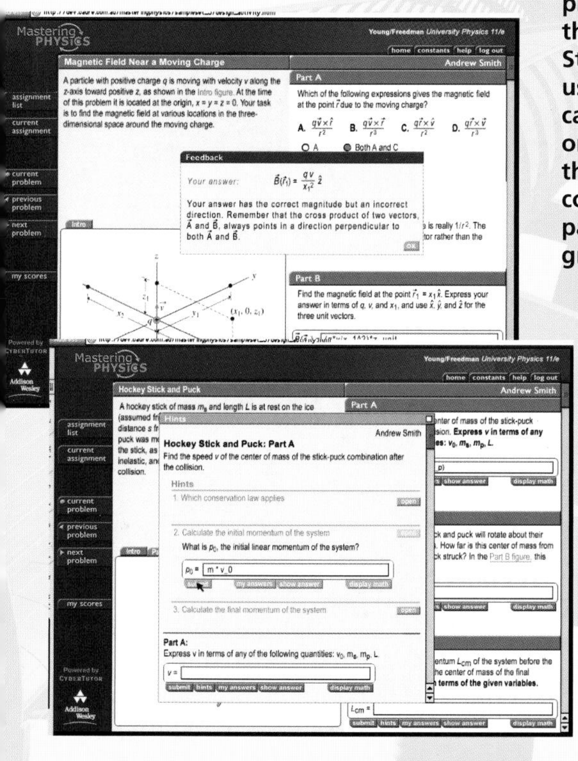

MasteringPhysics™ is the first Socratic tutoring system developed specifically for physics students like you. It is the result of years of detailed studies of how students work physics problems, and where they get stuck and need help. Studies show students who used MasteringPhysics significantly improved their scores on traditional final exams and the Force Concept Inventory (a conceptual test) when compared with traditional hand-graded homework.

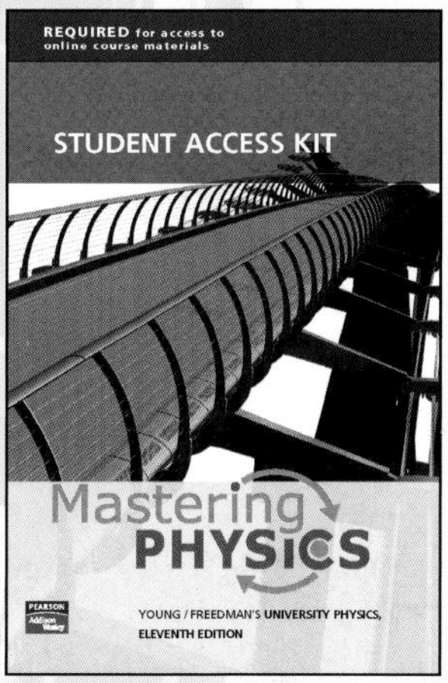

With your purchase of a new copy of Young/Freedman's ***University Physics*, Eleventh Edition,** you should have received a Student Access Kit for **MasteringPhysics™** if your professor required it as a component of your course. The kit contains instructions and a code for you to access MasteringPhysics.

If you did not purchase a new textbook and your professor requires you to enroll in the **MasteringPhysics** online homework and tutorial program, you may purchase an online subscription with a major credit card. Go to www.masteringphysics.com and follow the links to purchasing online.

Minimum System Requirements for MasteringPhysics:
Windows: 250 MHz; OS 98, NT, ME, 2000, XP
Macintosh: 233 MHz; OS 9.2, 10
RedHat Linux 8.0
All:

- 64 RAM installed
- 1024 x 768 screen resolution
- Browsers: Internet Explorer 5.0, 5.5, 6.0; Netscape 6.2.3, 7.0; Mozilla 1.2, 1.3

MasteringPhysics™ is powered by MyCyberTutor by Effective Educational Technologies

BC001AP/MP

Sears and Zemansky's

UNIVERSITY PHYSICS

11th Edition

with Modern Physics

HUGH D. YOUNG
CARNEGIE MELLON UNIVERSITY

ROGER A. FREEDMAN
UNIVERSITY OF CALIFORNIA, SANTA BARBARA

CONTRIBUTING AUTHOR
A. LEWIS FORD
TEXAS A&M UNIVERSITY

PEARSON
Addison Wesley

San Francisco Boston New York
Cape Town Hong Kong London Madrid Mexico City
Montreal Munich Paris Singapore Sydney Tokyo Toronto

Executive Editor: Adam Black, Ph.D.
Development Editors: David Chelton, Alice Houston
Project Editors: Nancy Benton, Richard Dal Porto
Production Editor, Art Coordinator, Copyeditor, Illustrator, Photo Editor, Permissions Editor, Proofreader, Compositor: Thompson Steele, Inc.
Media Producer: Claire Masson
Marketing Manager: Christy Lawrence
Director of Marketing: Stacy Treco
Art Director: Blake Kim
Manufacturing Supervisor: Vivian McDougal
Text and Cover Designer: Mark Ong, Side by Side Studios
Cover Printer: Phoenix Color Corporation
Printer and Binder: R.R. Donnelley & Sons

Cover Photo: Image of London's Millennium Bridge
created by Cimex Media, Ltd. http://www.cimexmedia.com
Photo Credits: see page A21

Library of Congress Cataloging-in-Publication Data

Young, Hugh D.
Sears and Zemansky's university physics: with modern physics. — 11th ed. / Hugh D. Young, Roger A. Freedman; contributing author, A. Lewis Ford.
p. cm.
Includes bibliographical references and index.
ISBN 0-8053-8684-X (alk, paper)
1. Physics. I. Title: University physics. II. Freedman, Roger A. III. Sears, Francis Weston, 1898–1975. University physics. IV. Title.

QC21.3.Y68 2004
530—dc21

2003049555

0-8053-8684-X Extended with Mastering Physics
0-8053-9185-1 Extended without Mastering Physics

45678910 — DOW — 06 05 04
www.aw.com/bc

PREFACE

This book is the product of more than half a century of innovation in physics education. When the first edition of *University Physics* by Francis W. Sears and Mark W. Zemansky appeared in 1949, it was revolutionary among calculus-based physics textbooks in its emphasis on the fundamental principles of physics and how to apply them. The success of *University Physics* with generations of students and educators throughout the world is a testament to the merits of this approach. In preparing this new Eleventh Edition, we have further enhanced and developed *University Physics* to emphasize two key objectives: helping students develop conceptual understanding, and helping them build strong problem-solving skills.

New to This Edition

- A **new arrangement of topics** places the chapters on mechanical waves (15 and 16) immediately after the chapters on oscillations and fluids and before the chapters on thermodynamics. The wave chapters have been rearranged to eliminate repetition of topics. The discussion of displacement current has been moved to Chapter 29 to follow the presentation of electromagnetic induction. The chapters on geometric optics and optical instruments have been streamlined into a single Chapter 34. Chapter 39 has an expanded discussion of the Schrödinger equation. Some of the optional sections from the Tenth Edition of *University Physics* have been moved to the companion web site.

- A totally revised **art program** helps students better visualize key concepts. More discussion has been placed *on* the figures themselves, and consistent colors for each kind of vector quantity are used throughout the book. In the discussion of mechanics, color and design have been streamlined to help students identify the object of interest. New photographs throughout the book illustrate the applications of physics to the natural world and to today's technology.

- A research-based, **step-by-step approach to problem solving** is used throughout the book. Called **ISEE**—an acronym for *Identify, Set up, Execute,* and *Evaluate*—this approach gives students a global strategy for solving physics problems of all kinds. The ISEE approach is used in *all* Problem-Solving Strategy boxes and in *all* worked examples. Educational research indicates that this consistent approach helps students build their confidence in attacking new problems.

- To help students check their comprehension of the material, there are **Chapter Opening Questions** at the beginning of each chapter and **Test Your Understanding Questions** at the end of almost every chapter section. The answers to the questions are found at the end of the chapter.

- New **visual chapter summaries** reinforce the key ideas in each chapter. They help students consolidate and connect important concepts by matching a descriptive explanation, a mathematical representation, and a thumbnail sketch in an easy-to-review grid.

■ **Careful use of subscripts** in kinematics, dynamics, and elsewhere makes it easier for students to distinguish between signed quantities such as velocity v_x and magnitudes such as speed v.

■ There are over **600 new end-of-chapter problems,** with an emphasis on context-rich problems, bringing the total problem count to more than 3700.

■ Accompanying *University Physics* are **two new web-based tools** for teaching and learning physics. **MasteringPhysics™** is a research-proven online tutorial and homework assignment system. MasteringPhysics™ includes an extensive library of entirely new and carefully tested problems that can be used for student assignments. These problems include *Skill Builders* (detailed worked examples that focus on conceptual understanding or key skills, and include multiple hint-giving options) and *Tutoring Problems* (standard multi-step homework problems that give on-demand hints and individualized feedback based on the student's particular answers). *End-of-Chapter Problems* (the majority of the end-of-chapter problems from *University Physics*) are also provided in the system with algorithmically generated variables and fewer hints to more accurately replicate a testing environment. **ActivPhysics OnLine** provides a comprehensive library of more than 220 applets and accompanying interactive activities that cover topics from mechanics through modern physics. It includes interactive conceptual exercises that encourage students to confront misconceptions, reason qualitatively, and think critically.

Key Features of *University Physics*

Standard, Extended, and Split Versions: This edition is available in three versions. The Standard version (ISBN 0-8053-8768-4) includes 37 chapters, ending with the special theory of relativity. The Extended version (ISBN 0-8053-8684-X) adds seven chapters on modern physics, including the physics of atoms, molecules, condensed matter, nuclei, and elementary particles. The Split version includes all 44 chapters in three volumes. Volume 1 (ISBN 0-8053-8767-6, Chapters 1–20) covers mechanics, waves, and thermodynamics; Volume 2 (ISBN 0-8053-8766-8, Chapters 21–37) covers electricity and magnetism, optics, and relativity; and Volume 3 (ISBN 0-8053-8765-X, Chapters 37–44) covers modern physics.

A Guide for the Student: Many physics students experience difficulty simply because they don't know how to use their textbook. A section entitled "How to Succeed in Physics by Really Trying," which follows this preface, is a "user's manual" to all the features of this book. This section, written by Professor Mark Hollabaugh (Normandale Community College), also gives a number of helpful study hints. *Every* student should read this section!

Chapter Organization: The first section of each chapter is an *Introduction* that gives specific examples of the chapter's content and connects it with what has come before. There is also a *Chapter Opening Question* to make the reader think about the subject matter of the chapter ahead. (To find the answer to the question, look for the ? icon.) Most sections end with a *Test Your Understanding Ques-*

tion, which can be conceptual or quantitative in nature. At the end of the last section of the chapter is a *Visual Chapter Summary* of the most important principles in the chapter, as well as a list of *Key Terms* with reference to the page number where each term is introduced. The answers to the Chapter Opening Question and Test Your Understanding Questions are on the page following the Key Terms.

Questions and Problems: At the end of each chapter is a collection of *Discussion Questions* that probe and extend the student's conceptual understanding. Following these are *Exercises,* which are single-concept problems keyed to specific sections of the text; *Problems,* usually requiring one or two nontrivial steps; and *Challenge Problems,* intended to challenge the strongest students. The problems include applications to such diverse fields as astrophysics, biology, and aerodynamics. Many problems have a conceptual part in which students must discuss and explain their results. The new questions, exercises, and problems for this edition were created and organized by Professor A. Lewis Ford (Texas A&M University) with contributions from Wayne Anderson (Sacramento City College) and Rich Gottfried (Frederick Community College).

Problem-Solving Strategies and Worked Examples: Throughout the book, *Problem-Solving Strategy* boxes provide students with specific tactics for solving particular types of problems. They address the needs of any student who's ever felt that they "understand the concepts but can't do the problems."

All Problem-Solving Strategy boxes follow the ISEE approach (Identify, Set up, Execute, and Evaluate) to solving problems. This approach helps students see how to begin with a seemingly complex situation, identify the relevant physical concepts, decide what tools are needed to solve the problem, carry out the solution, and then evaluate whether the result makes sense.

Each Problem-Solving Strategy box is followed by one or more worked-out *Examples* that illustrate the strategy. Many other worked-out Examples are found in each chapter. Like the Problem-Solving Strategy boxes, all of the quantitative Examples use the ISEE approach. Several of the examples are purely qualitative, and are labeled as *Conceptual Examples;* see, for instance, Conceptual Examples 6.6 (Comparing kinetic energies, p. 218), 8.1 (Momentum versus kinetic energy, p. 286) and 20.7 (A reversible adiabatic process, p. 775).

"Caution" paragraphs: Two decades of physics education research have revealed a number of conceptual pitfalls that commonly plague beginning physics students. These include the ideas that force is required for motion, that electric current is "used up" as it goes around a circuit, and that the product of an object's mass and its acceleration is itself a force. The "Caution" paragraphs alert students to these and other pitfalls, and explain why the wrong way to think about a certain situation (which may have occurred to the student first) is indeed wrong. (See, for example, pp. 132, 182, and 629.)

Notation and units: Students often have a hard time keeping track of which quantities are vectors and which are not. We use boldface italic symbols with an arrow on top for vector quantities, such as $\vec{\boldsymbol{v}}$, $\vec{\boldsymbol{a}}$, and $\vec{\boldsymbol{F}}$; unit vectors such as $\hat{\boldsymbol{\imath}}$, have a caret on top. Boldface $\mathbf{+}$, $\mathbf{-}$, $\boldsymbol{\times}$, and $\mathbf{=}$ signs are used in vector equations to emphasize the distinction between vector and scalar mathematical operations.

SI units are used exclusively (English unit conversions are included where appropriate). The joule is used as the standard unit of energy of all forms, including heat.

Flexibility: The book is adaptable to a wide variety of course outlines. There is plenty of material for a three-semester or a five-quarter course. Most instructors will find that there is too much material for a one-year course, but it is easy to tailor the book to a variety of one-year course plans by omitting certain chapters or sections. For example, any or all of the chapters on fluid mechanics, sound and hearing, electromagnetic waves, or relativity can be omitted without loss of continuity. In any case, no instructor should feel constrained to work straight through the entire book.

Online Supplements for Students and Instructors

MasteringPhysics™ (www.masteringphysics.com) is a sophisticated online tutoring and homework system developed specially for courses using *University Physics*. Originally developed by Professor David Pritchard of Effective Educational Technologies, MasteringPhysics™ provides students with individualized online tutoring by responding to their wrong answers and providing hints for solving multi-step problems when they get stuck. It gives them immediate and up-to-date assessment of their progress, and shows where they need to practice more. MasteringPhysics™ provides instructors with a fast and effective way to assign tried-and-tested online homework assignments that comprise a range of problem types. The powerful post-assignment diagnostics allow instructors to assess the progress of their class as a whole and individual students, and quickly identify areas of difficulty.

ActivPhysics OnLine

ActivPhysics OnLine (www.aw.com/young11) was developed by Professors Alan Van Heuvelen and Paul D'Alessandris. It provides simulations and accompanying conceptual exercises from mechanics, waves, thermodynamics, electricity and magnetism, and modern physics. Icons throughout the book indicate which ActivPhysics simulations and exercises correspond to various sections of *University Physics.*

The Addison Wesley Tutor Center (www.aw.com/tutorcenter) provides one-on-one tutoring via telephone, fax, email, or interactive web site during evening hours and on weekends. Qualified college instructors answer questions and provide instruction regarding MasteringPhysics™ as well as examples, exercises, and other content from *University Physics.*

The Companion Web Site (www.aw.com/young11) includes additional study material as well as in-depth discussions of physics topics not included in *University Physics.*

Additional Student Supplements

The **Study Guide,** prepared by Professor James R. Gaines and William F. Palmer, reinforces the text's emphasis on problem-solving strategies and student misconceptions. *The Study Guide for Volume 1* (ISBN 0-8053-8778-1) covers Chapters 1–20, and the *Study Guide for Volumes 2 and 3* (ISBN 0-8053-8743-9) covers Chapters 21–44.

The **Student Solutions Manual,** prepared by Professor A. Lewis Ford, includes careful, step-by-step solutions for most of the odd-numbered exercises and problems in *University Physics,* following the same ISEE approach used in the Problem-Solving Strategy boxes and in the Examples. (The answers for all odd-numbered problems are found in *University Physics* following the Appendices.) The *Student Solutions Manual for Volume 1* (ISBN 0-8053-8777-3) covers Chapters 1–20, and the *Student Solutions Manual for Volumes 2 and 3* (ISBN 0-8053-8696-3) covers Chapters 21–44.

Additional Instructor Supplements

The **Instructor's Solutions Manual,** prepared by Professor A. Lewis Ford, contains worked-out solutions to all exercises, problems, and challenge problems. (It is available to qualifying instructors only.) The *Instructor's Solutions Manual for Volume 1* (ISBN 0-8053-8775-7) covers Chapters 1–20, and the *Instructor's Solutions Manual for Volumes 2 and 3* (ISBN 0-8053-8776-5) covers Chapters 21–44. The complete text of both printed manuals is also available in Word format on the cross-platform *Instructor's Solutions Manual* CD-ROM (ISBN 0-8053-8774-9).

The **Test Bank** is available in both print and electronic formats. The *Printed Test Bank* (ISBN 0-8053-8772-2) includes more than 1500 problems. All of these problems are included in the cross-platform *Computerized Test Bank* CD-ROM (ISBN 0-8053-8771-4); more than half of the problems on the CD-ROM now offer algorithmically generated functionality thanks to Benjamin Grinstein (University of California, San Diego).

Five Easy Lessons: Strategies for Successful Physics Teaching (ISBN 0-8053-8702-1), written by Professor Randall Knight (California Polytechnic State University, San Luis Obispo), is packed with creative ideas on how to enhance any physics course. It is an invaluable companion for both novice and veteran physics instructors.

The **Simulation and Image Presentation CD-ROM** (ISBN 0-8053-8769-2) includes a library of more than 220 simulation applets from ActivPhysics OnLine. These make it easy to illustrate challenging physics concepts in class in a dynamic way. The CD-ROM also includes all of the illustrations (excluding photographs) from *University Physics,* all chapter summaries, Problem-Solving Strategy boxes, and key equations. All these items can easily be exported into lecture-presentation programs such as PowerPoint®, or posted online. A set of **Transparency Acetates** (ISBN 0-8053-8773-0) with more than 200 illustrations from *University Physics* is also available.

Acknowledgments

We would like to thank the hundreds of reviewers and colleagues who have offered valuable comments and suggestions over the life of this textbook. The continuing success of *University Physics* is due in large measure to their contributions.

Edward Adelson (Ohio State University), Ralph Alexander (University of Missouri at Rolla), J. G. Anderson, R. S. Anderson, Wayne Anderson (Sacramento City College), Alex Azima (Lansing Community College), Dilip Balamore (Nassau Community College), Harold Bale (University of North Dakota), Arun Bansil (Northeastern University), John Barach (Vanderbilt University), J. D. Barnett, H. H. Barschall, Albert Bartlett (University of Colorado), Paul Baum (CUNY, Queens College), Frederick Becchetti (University of Michigan), B. Bederson, Lev I. Berger (San Diego State University),

Robert Boeke (William Rainey Harper College), S. Borowitz, A. C. Braden, James Brooks (Boston University), Nicholas E. Brown (California Polytechnic State University, San Luis Obispo), Tony Buffa (California Polytechnic State University, San Luis Obispo), A. Capecelatro, Michael Cardamone (Pennsylvania State University), Duane Carmony (Purdue University), Troy Carter (UCLA), P. Catranides, John Cerne (SUNY at Buffalo), Roger Clapp (University of South Florida), William M. Cloud (Eastern Illinois University), Leonard Cohen (Drexel University), W. R. Coker (University of Texas, Austin), Malcolm D. Cole (University of Missouri at Rolla), H. Conrad, David Cook (Lawrence University), Gayl Cook (University of Colorado), Hans Courant (University of Minnesota), Bruce A. Craver (University of Dayton), Larry Curtis (University of Toledo), Jai Dahiya (Southeast Missouri State University), Steve Detweiler (University of Florida), George Dixon (Oklahoma State University), Donald S. Duncan, Boyd Edwards (West Virginia University), Robert Eisenstein (Carnegie Mellon University), Amy Emerson Missourn (Virginia Institute of Technology), William Faissler (Northeastern University), William Fasnacht (U.S. Naval Academy), Paul Feldker (St. Louis Community College), Carlos Figueroa (Cabrillo College), L. H. Fisher, Neil Fletcher (Florida State University), Robert Folk, Peter Fong (Emory University), A. Lewis Ford (Texas A&M University), D. Frantszog, James R. Gaines (Ohio State University), Solomon Gartenhaus (Purdue University), Ron Gautreau (New Jersey Institute of Technology), J. David Gavenda (University of Texas, Austin), Dennis Gay (University of North Florida), James Gerhart (University of Washington), N. S. Gingrich, J. L. Glathart, S. Goodwin, Rich Gottfried (Frederick Community College), Walter S. Gray (University of Michigan), Paul Gresser (University of Maryland), Benjamin Grinstein (UC San Diego), Howard Grotch (Pennsylvania State University), John Gruber (San Jose State University), Graham D. Gutsche (U.S. Naval Academy), Michael J. Harrison (Michigan State University), Harold Hart (Western Illinois University), Howard Hayden (University of Connecticut), Carl Helrich (Goshen College), Laurent Hodges (Iowa State University), C. D. Hodgman, Michael Hones (Villanova University), Keith Honey (West Virginia Institute of Technology), Gregory Hood (Tidewater Community College), John Hubisz (North Carolina State University), M. Iona, John Jaszczak (Michigan Technical University), Alvin Jenkins (North Carolina State University), Robert P. Johnson (UC Santa Cruz), Lorella Jones (University of Illinois), John Karchek (GMI Engineering & Management Institute), Thomas Keil (Worcester Polytechnic Institute), Robert Kraemer (Carnegie Mellon University), Jean P. Krisch (University of Michigan), Robert A. Kromhout, Charles Lane (Berry College), Robert J. Lee, Alfred Leitner (Rensselaer Polytechnic University), Gerald P. Lietz (De Paul University), Gordon Lind (Utah State University), S. Livingston, Elihu Lubkin (University of Wisconsin, Milwaukee), Robert Luke (Boise State University), David Lynch (Iowa State University), Michael Lysak (San Bernardino Valley College), Jeffrey Mallow (Loyola University), Robert Mania (Kentucky State University), Robert Marchina (University of Memphis), David Markowitz (University of Connecticut), R. J. Maurer, Oren Maxwell (Florida International University), Joseph L. McCauley (University of Houston), T. K. McCubbin, Jr. (Pennsylvania State University), Charles McFarland (University of Missouri at Rolla), James Mcguire (Tulane University), Lawrence McIntyre (University of Arizona), Fredric Messing (Carnegie-Mellon University), Thomas Meyer (Texas A&M University), Andre Mirabelli (St. Peter's College, New Jersey), Herbert Muether (S.U.N.Y., Stony Brook), Jack Munsee (California State University, Long Beach), Lorenzo Narducci (Drexel University), Van E. Neie (Purdue University), David A. Nordling (U. S. Naval Academy), Benedict Oh (Pennsylvania State University), L. O. Olsen, Jim Pannell (DeVry Institute of Technology), W. F. Parks (University of Missouri), Jerry Peacher (University of Missouri at Rolla), Arnold Perlmutter (University of Miami), Lennart Peterson (University of Florida), R. J. Peterson (University of Colorado, Boulder), R. Pinkston, Ronald Poling (University of Minnesota), J. G. Potter, C. W. Price (Millersville University), Francis Prosser (University of Kansas), Shelden H. Radin, Michael Rapport (Anne Arundel Community College), R. Resnick, James A. Richards, Jr., John S. Risley (North Carolina State University), Francesc Roig (University of California, Santa Barbara), T. L. Rokoske, Richard Roth (Eastern Michigan University), Carl Rotter (University of West Virginia), S. Clark Rowland (Andrews University), Rajarshi Roy (Georgia Institute of Technology), Russell A. Roy (Santa Fe Community College), Bruce Schumm (UC Santa Cruz), Melvin Schwartz (St. John's University), F. A. Scott, L. W. Seagondollar, Stan Shepherd (Pennsylvania State University), Douglas Sherman (San Jose State), Bruce Sherwood (Carnegie Mellon University), Hugh Siefkin (Greenville College), C. P. Slichter, Charles W. Smith (University of Maine, Orono), Malcolm Smith (University of Lowell), Ross Spencer (Brigham Young University), Julien Sprott (University of Wisconsin), Victor Stanionis (Iona College), James Stith (American Institute of Physics), Chuck Stone (North Carolina A&T State University), Edward Strother (Florida Institute of Technology), Conley Stutz (Bradley University), Albert Stwertka (U.S. Merchant Marine Academy), Martin Tiersten (CUNY, City College), David Toot (Alfred University), Somdev Tyagi (Drexel University), F. Verbrugge, Helmut Vogel (Carnegie Mellon University), Robert Webb (Texas A & M), Thomas Weber (Iowa State University), M. Russell Wehr, (Pennsylvania State University), Robert Weidman (Michigan Technical University), Dan Whalen (UC San Diego), Lester V. Whitney, Thomas Wiggins (Pennsylvania State University), David Willey (University of Pittsburgh, Johnstown), George Williams (University of Utah), John Williams (Auburn University), Stanley Williams (Iowa State University), Jack Willis, Suzanne Willis (Northern Illinois University), Robert Wilson (San Bernardino Valley College), L. Wolfenstein, James Wood (Palm Beach Junior College), Lowell Wood (University of Houston), R. E. Worley, D. H. Ziebell (Manatee Community College), George O. Zimmerman (Boston University)

In addition, we both have individual acknowledgments we would like to make.

I want to extend my heartfelt thanks to my colleagues at Carnegie Mellon, especially Professors Robert Kraemer, Bruce Sherwood, Ruth Chabay, Helmut Vogel, and Brian Quinn, for many stimulating discussions about physics pedagogy and for their support and encouragement during the writing of several successive editions of this book. I am equally indebted to the many generations of Carnegie Mellon students who have helped me learn what good teaching and good writing are, by showing me what works and what doesn't. It is always a joy and a privilege to express my gratitude to my wife Alice and our children Gretchen and Rebecca for their love, support, and emotional sustenance during the writing of several successive editions of this book. May all men and women be blessed with love such as theirs. — H. D. Y.

I would like to thank my past and present colleagues at UCSB, including Francesc Roig, Elisabeth Nicol, Al Nash, and Carl Gwinn, for their wholehearted support and for many helpful discussions. I owe a special debt of gratitude to my early teachers Willa Ramsay, Peter Zimmerman, William Little, Alan Schwettman, and Dirk Walecka for showing me what clear and engaging physics teaching is all about, and to Stuart Johnson for inviting me to become a co-author of *University Physics* beginning with the 9th edition. I want to express special thanks to Adam Black and Nancy Benton at Addison Wesley for their superb editorial guidance and vision, to David Chelton for aiding in the development of this edition through his many valuable suggestions and constant good humor, and to Nancy Freihofer of Thompson Steele for shepherding this book through the Byzantine production process. I want to thank my father for his continued love and support and for keeping a space open on his bookshelf for this book. Most of all, I want to express my gratitude and love to my wife Caroline, to whom I dedicate my contributions to this book. Hey, Caroline, the new edition's done at last — let's go flying! — R. A. F.

PLEASE TELL US WHAT YOU THINK!

We welcome communications from students and professors, especially concerning errors or deficiencies that you find in this edition. We have devoted a lot of time and effort to writing the best book we know how to write, and we hope it will help you to teach and learn physics. In turn, you can help us by letting us know what still needs to be improved! Please feel free to contact us either electronically or by ordinary mail. Your comments will be greatly appreciated.

April 2003

Hugh D. Young
Department of Physics
Carnegie Mellon University
Pittsburgh, Pennsylvania 15213
hdy@andrew.cmu.edu

Roger A. Freedman
Department of Physics
University of California, Santa Barbara
Santa Barbara, California 93106-9530
airboy@physics.ucsb.edu
http://www.physics.ucsb.edu/~airboy/

HOW TO SUCCEED IN PHYSICS BY REALLY TRYING

Mark Hollabaugh, Normandale Community College

Physics encompasses the large and the small, the old and the new. From the atom to galaxies, from electrical circuitry to aerodynamics, physics is very much a part of the world around us. You probably are taking this introductory course in calculus-based physics because it is required for subsequent courses you plan to take in preparation for a career in science or engineering. Your professor wants you to learn physics and to enjoy the experience. He or she is very interested in helping you learn this fascinating subject. That is part of the reason your professor chose this textbook for your course. That is also the reason why Drs. Young and Freedman asked me to write this introductory section. We want you to succeed!

The purpose of this section of *University Physics* is to give you some ideas that will assist your learning. Specific suggestions on how to use the textbook will follow a brief discussion of general study habits and strategies.

Preparation for This Course

If you had high school physics, you will probably learn concepts faster than those who have not because you will be familiar with the language of physics. If English is a second language for you, keep a glossary of new terms that you encounter and make sure you understand how they are used in physics. Likewise, if you are farther along in your mathematics courses, you will pick up the mathematical aspects of physics faster. Even if your mathematics is adequate, you may find a book such as Arnold D. Pickar's *Preparing for General Physics: Math Skill Drills and Other Useful Help (Calculus Version)* to be useful. Your professor may actually assign sections of this math review to assist your learning.

Learning to Learn

Each of us has a different learning style and a preferred means of learning. Understanding your own learning style will help you to focus on aspects of physics that may give you difficulty and to use those components of your course that will help you overcome the difficulty. Obviously you will want to spend more time on those aspects that give you the most trouble. If you learn by hearing, lectures will be very important. If you learn by explaining, then working with other students will be useful to you. If solving problems is difficult for you, spend more time learning how to solve problems. Also, it is important to understand and develop good study habits. Perhaps the most important thing you can do for yourself is to set aside adequate, regularly scheduled, study time in a distraction-free environment.

Answer the following questions for yourself:

- Am I able to use fundamental mathematical concepts from algebra, geometry and trigonometry? (If not, plan a program of review with help from your professor.)
- In similar courses, what activity has given me the most trouble? (Spend more time on this.) What has been the easiest for me? (Do this first; it will help to build your confidence.)
- Do I understand the material better if I read the book before or after the lecture? (You may learn best by skimming the material, going to lecture, and then undertaking an in-depth reading.)

- Do I spend adequate time in studying physics? (A rule of thumb for a class like this is to devote, on the average, 2.5 hours out of class for each hour in class. For a course meeting 5 hours each week, that means you should spend about 10 to 15 hours per week studying physics.)
- Do I study physics every day? (Spread that 10 to 15 hours out over an entire week!) At what time of the day am I at my best for studying physics? (Pick a specific time of the day and stick to it.)
- Do I work in a quiet place where I can maintain my focus? (Distractions will break your routine and cause you to miss important points.)

Working with Others

Scientists or engineers seldom work in isolation from one another but rather work cooperatively. You will learn more physics and have more fun doing it if you work with other students. Some professors may formalize the use of cooperative learning or facilitate the formation of study groups. You may wish to form your own informal study group with members of your class who live in your neighborhood or dorm. If you have access to e-mail, use it to keep in touch with one another. Your study group is an excellent resource when reviewing for exams.

Lectures and Taking Notes

An important component of any college course is the lecture. In physics this is especially important because your professor will frequently do demonstrations of physical principles, run computer simulations, or show video clips. All of these are learning activities that will help you to understand the basic principles of physics. Don't miss lectures, and if for some reason you do, ask a friend or member of your study group to provide you with notes and let you know what happened.

Take your class notes in outline form, and fill in the details later. It can be very difficult to take word for word notes, so just write down key ideas. Your professor may use a diagram from the textbook. Leave a space in your notes and just add the diagram later. After class, edit your notes, filling in any gaps or omissions and noting things you need to study further. Make references to the textbook by page, equation number, or section number.

Make sure you ask questions in class, or see your professor during office hours. Remember the only "dumb" question is the one that is not asked. Your college may also have teaching assistants or peer tutors who are available to help you with difficulties you may have.

Examinations

Taking an examination is stressful. But if you feel adequately prepared and are well-rested, your stress will be lessened. Preparing for an exam is a continual process; it begins the moment the last exam is over. You should immediately go over the exam and understand any mistakes you made. If you worked a problem and made substantial errors, try this: Take a piece of paper and divide it down the middle with a line from top to bottom. In one column, write the proper solution to the problem. In the other column, write what you did and why, if you know, and why your solution was incorrect. If you are uncertain why you made your mistake, or how to avoid making it again, talk with your professor. Physics continually builds on fundamental ideas and it is important to correct any misunderstandings immediately. Warning: While cramming at the last minute may get you through the *present* exam, you will not adequately retain the concepts for use on the *next* exam.

Using Your Textbook

Now let's take a look at specific features of *University Physics* that will help you understand the concepts of physics. At its heart, physics is not equations and numbers. Physics is a way of looking at the universe and understanding how the universe works and how its various parts relate to each other. And although solving quantitative problems is an important part of physics, it is equally important for you to understand concepts qualitatively. Your textbook will help you in both areas.

First of all, don't be afraid to write in your book. It is more important for you to learn the concepts of physics than to keep your book in pristine condition. Write in the margins, make cross references. Take notes in your notebook as you read. *University Physics* is your primary "reference book" for this course. Refer to it often to help you understand the concepts you hear in lecture. Become familiar with the contents of the appendices and end papers.

Chapter Opening Questions and Test Your Understanding Questions

Each chapter of *University Physics* opens with a photograph, and next to that photograph is a question. As you read the chapter, try to figure out the answer to this question. You'll also find a *Test Your Understanding* question at the end of most sections of the book. These are designed to help you check if you've understood a key concept that's just been presented. If you can't see how to answer the question, you may need to read the section again with more care. (You'll find the answers to both kinds of questions toward the end of the chapter, but you'll get the most benefit if you figure out the answers yourself!)

Test Your Understanding

You are driving your car on a country road when a mosquito splatters itself on the windshield. Which has the greater magnitude, the force that the car exerted on the mosquito or the force that the mosquito exerted on the car? Or are the magnitudes the same? If they are different, how can you reconcile this fact with Newton's third law? If they are equal, why is the mosquito splattered while the car is undamaged?

Worked Examples

Your professor will work example problems in class to illustrate the application of the concepts of physics to real-world problems. You should work through all the examples in the textbook, filling in any missing steps, and making note of things you don't understand. Get help with the concepts that confuse you!

Example 5.8 **Straight-line motion with friction**

Suppose the wind is once again blowing steadily in the $+x$-direction as in Example 5.6, so that the iceboat has a constant acceleration $a_x = 1.5\ \text{m/s}^2$. Now, however, there is a constant horizontal friction force with magnitude 100 N that opposes the motion of the iceboat. In this case, what force F_W must the wind exert on the iceboat?

SOLUTION

IDENTIFY: Once again the target variable is F_W. We are given the acceleration, so to find F_W all we need is Newton's second law.

SET UP: Figure 5.9 shows the new free-body diagram. The only difference from Fig. 5.7b is the addition of the friction force $\vec{f}$, which points opposite to the motion. (Note that its *magnitude*, $f = 100\ \text{N}$, is a positive quantity but that its *component* in the x-direction is negative, equal to $-f$ or $-100\ \text{N}$.)

EXECUTE: There are now two forces (the force of the wind and the friction force) that have x-components. The x-component of Newton's second law gives

$$\sum F_x = F_W + (-f) = ma_x$$

$$F_W = ma_x + f = (200\ \text{kg})(1.5\ \text{m/s}^2) + (100\ \text{N}) = 400\ \text{N}$$

EVALUATE: Because there is friction, a greater force F_W is needed than in Example 5.6. We need 100 N to overcome friction and 300 N more to give the iceboat the necessary acceleration.

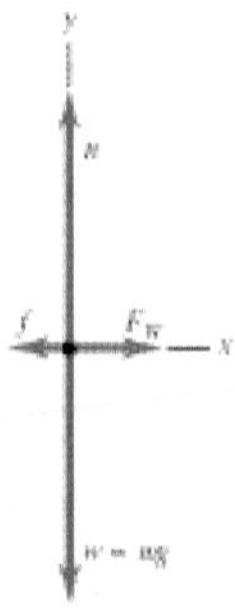

5.9 Free-body diagram for the iceboat and rider with a frictional force $\vec{f}$ opposing the motion.

Problem-Solving Strategies

One of the features of *University Physics* that first caught my eye as a teacher were the *Problem-Solving Strategy* boxes. This is the advice I would give to a student who came to me for help with a physics problem. Physics teachers approach a problem in a very systematic and logical manner. These boxes will help you as a beginning problem solver to do the same by following a systematic problem-solving strategy called ISEE—Identify, Set up, Execute, and Evaluate. Study these suggestions in great detail and implement them. In many cases these strategy boxes will tell you how to visualize an abstract concept. Finally, use the ISEE approach consistently. It will help you build confidence in tackling new problems.

Problem-Solving Strategy **Conservation of Momentum**

IDENTIFY *the relevant concepts:* Before applying conservation of momentum to a problem, you must first decide whether momentum *is* conserved! This will be true *only* if the vector sum of the external forces acting on the system of particles is zero. If this is not the case, you can't use conservation of momentum.

SET UP *the problem using the following steps:*

1. Define a coordinate system. Make a sketch showing the coordinate axes, including the positive direction for each. Often it is easiest to choose the x-axis to have the direction of one of the initial velocities. Make sure you are using an inertial frame of reference. Most of the problems in this chapter deal with two-dimensional situations, in which the vectors have only x- and y-components; all of the following statements can be generalized to include z-components when necessary.
2. Treat each body as a particle. Draw "before" and "after" sketches, and include vectors on each to represent all known velocities. Label the vectors with magnitudes, angles, components, or whatever information is given, and give each unknown magnitude, angle, or component an algebraic symbol. You may find it helpful to use the subscripts 1 and 2 for velocities before and after the interaction, respectively; if you use these subscripts, use letters (not numbers) to label each particle.
3. As always, identify the target variable(s) from among the unknowns.

EXECUTE *the solution as follows:*

1. Write an equation in terms of symbols equating the total *initial* x-component of momentum (that is, before the interaction) to the total *final* x-component of momentum (that is, after the interaction), using $p_x = mv_x$ for each particle. Write another equation for the y-components, using $p_y = mv_y$ for each particle. Remember that the x- and y-components of velocity or momentum are *never* added together in the same equation! Even when all the velocities lie along a line (such as the x-axis), the components of velocity along this line can be positive or negative; be careful with signs!
2. Solve these equations to determine whatever results are required. In some problems you will have to convert from the x- and y-components of a velocity to its magnitude and direction, or the reverse.
3. In some problems, energy considerations give additional relationships among the various velocities, as we will see later in this chapter.

EVALUATE *your answer:* Does your answer make physical sense? If your target variable is a certain body's momentum, check that the direction of the momentum is reasonable.

Caution!

Educational research has identified misconceptions or misunderstandings that students frequently have when they study physics. Dr. Freedman has added Caution! paragraphs to warn you about these potential pitfalls and help you avoid them.

CAUTION Please note that the quantity $m\vec{a}$ is *not* a force. All that Eqs. (4.7) and (4.8) say is that the vector $m\vec{a}$ is equal in magnitude and direction to the vector sum $\sum\vec{F}$ of all the forces acting on the body. It's incorrect to think of acceleration as a force; rather, acceleration is a result of a nonzero net force. It's "common sense" to think that there is a "force of acceleration" that pushes you back into your seat when your car accelerates forward from rest. But *there is no such force;* instead, your inertia causes you to tend to stay at rest relative to the earth, and the car accelerates around you. The "common sense" confusion arises from trying to apply Newton's second law in a frame of reference where it isn't valid, like the non-inertial reference frame of an accelerating car. We will always examine motion relative to *inertial* frames of reference only.

Key Terms, Summary, and Problems

The most important concepts are listed in the *Key Terms.* Keep a glossary of terms in your notebook. Your professor may indicate through the use of course objectives which terms are important for you to know. The *Summary* will give you a quick review of the chapter's main ideas and how they can be represented descriptively (in text), visually (in a figure), and mathematically (with an equation). Use this to make sure you understand how the same concept can be represented in these different ways. Everything else can be derived from the general equations provided in the summary.

If your professor assigns *Problems* at the end of the chapter, make sure you work them carefully with other students. If solutions are available, do not look at the answer until you have struggled with the problem and compared your answer with someone else's. If the two of you agree on the answer, then look at the solution. If you have made a mistake, go back and rework the problem. You will note that the *Exercises* are keyed to specific sections of the chapter and are easier. Work on these before you attempt the *Problems* or *Challenge Problems* which typically use multiple concepts.

Well, there you have it. We hope these suggestions will benefit your study of physics. Strive for understanding and excellence, and be persistent in your learning.

CONTENTS

MECHANICS

OPTICS

33 THE NATURE AND PROPAGATION OF LIGHT 1247

34 GEOMETRIC OPTICS AND OPTICAL INSTRUMENTS 1285

35 INTERFERENCE 1338

36 DIFFRACTION 1367

MODERN PHYSICS

37 RELATIVITY 1403

38 PHOTONS, ELECTRONS, AND ATOMS 1445

ABOUT THE AUTHORS

Hugh D. Young is Professor of Physics at Carnegie Mellon University in Pittsburgh, PA. He attended Carnegie Mellon for both undergraduate and graduate study and earned his Ph.D. in fundamental particle theory under the direction of the late Richard Cutkosky. He joined the faculty of Carnegie Mellon in 1956, and has also spent two years as a Visiting Professor at the University of California at Berkeley.

Prof. Young's career has centered entirely around undergraduate education. He has written several undergraduate-level textbooks, and in 1973 he became a coauthor with Francis Sears and Mark Zemansky for their well-known introductory texts. With their deaths, he assumed full responsibility for new editions of these books until joined by Prof. Freedman for *University Physics*.

Prof. Young is an enthusiastic skier, climber, and hiker. He also served for several years as Associate Organist at St. Paul's Cathedral in Pittsburgh, and has played numerous organ recitals in the Pittsburgh area. Prof. Young and his wife Alice usually travel extensively in the summer, especially in Europe and in the desert canyon country of southern Utah.

Roger A. Freedman is a Lecturer in Physics at the University of California, Santa Barbara. Dr. Freedman was an undergraduate at the University of California campuses in San Diego and Los Angeles, and did his doctoral research in nuclear theory at Stanford University under the direction of Professor J. Dirk Walecka. He came to UCSB in 1981 after three years teaching and doing research at the University of Washington.

At UCSB, Dr. Freedman has taught in both the Department of Physics and the College of Creative Studies, a branch of the university intended for highly gifted and motivated undergraduates. He has published research in nuclear physics, elementary particle physics, and laser physics. In recent years, he has helped to develop computer-based tools for learning introductory physics and astronomy.

When not in the classroom or slaving over a computer, Dr. Freedman can be found either flying (he holds a commercial pilot's license) or driving with his wife, Caroline, in their 1960 Nash Metropolitan convertible.

A. Lewis Ford is Professor of Physics at Texas A&M University. He received a B.A. from Rice University in 1968 and a Ph.D. in chemical physics from the University of Texas at Austin in 1972. After a one-year postdoc at Harvard University, he joined the Texas A&M physics faculty in 1973 and has been there ever since. Professor Ford's research area is theoretical atomic physics, with a specialization in atomic collisions. At Texas A&M he has taught a variety of undergraduate and graduate courses, but primarily introductory physics.

UNITS, PHYSICAL QUANTITIES, AND VECTORS

1

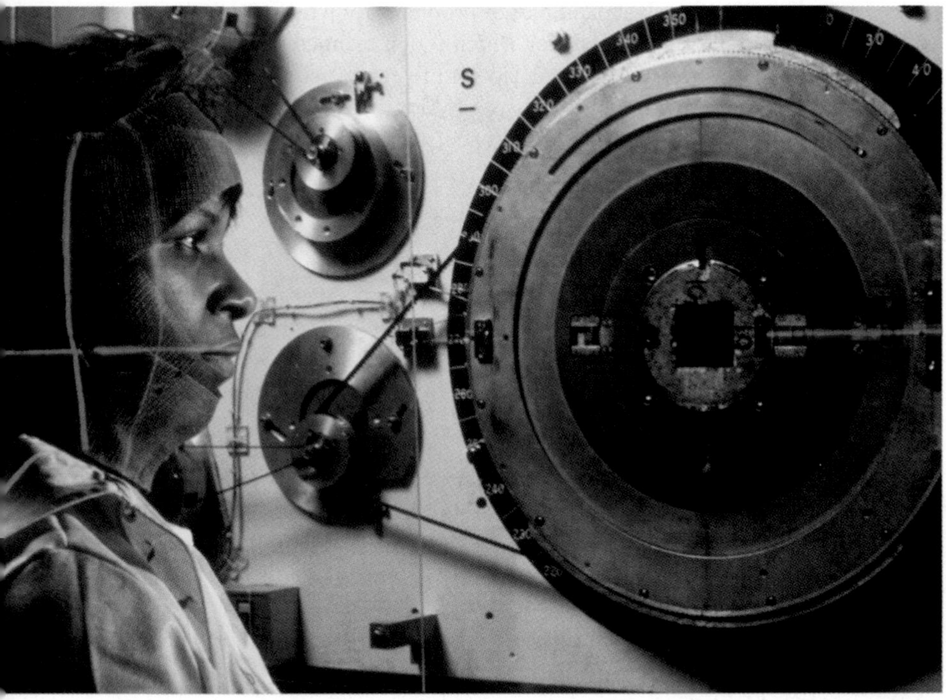

Accurate measurement is essential in medical applications of physics. The laser beams shining on this cancer patient form a cross-hairs on the site of a tumor, which is then bombarded by a stream of high-energy neutrons coming from the square aperture on the right. The neutrons deposit their energy in the tumor, stopping its growth and, ideally, destroying it completely. Because the narrow neutron beam is very accurately targeted, little damage occurs to the healthy tissue surrounding the tumor.

? Subatomic particles used in cancer therapy can be aimed at a tumor with an accuracy of 100 micrometers. How many human blood cells laid side by side would span this distance?

Why study physics? For two reasons. First, physics is one of the most fundamental of the sciences. Scientists of all disciplines make use of the ideas of physics, from chemists who study the structure of molecules to paleontologists who try to reconstruct how dinosaurs walked. The principles of physics play an essential role in the scientific quest to understand how human activities affect the atmosphere and oceans, and in the search for alternative sources of energy. Physics is also the foundation of all engineering and technology. No engineer could design any kind of practical device without first understanding the basic principles involved. No engineer could design a DVD player, a flat-screen TV, an interplanetary spacecraft, or even a better mousetrap without first understanding the basic laws of physics.

But there's another reason. The study of physics is an adventure. You will find it challenging, sometimes frustrating, occasionally painful, and often richly rewarding and satisfying. It will appeal to your sense of beauty as well as to your rational intelligence. Our present understanding of the physical world has been built on the foundations laid by scientific giants such as Galileo, Newton, Maxwell, and Einstein, and their influence has extended far beyond science to affect profoundly the ways in which we live and think. You can share some of the excitement of their discoveries when you learn to use physics to solve practical problems and to gain insight into everyday phenomena. If you've ever wondered why the sky is blue, how radio waves can travel through empty space, or how a satellite stays in orbit, you can find the answers by using fundamental physics. Above all, you will come to see physics as a towering achievement of the human intellect in its quest to understand our world and ourselves.

In this opening chapter, we'll go over some important preliminaries that we'll need throughout our study. We'll discuss the nature of physical theory and the use of idealized models to represent physical systems. We'll introduce the systems of units used to describe physical quantities and discuss ways to describe the accuracy of a number. We'll look at examples of problems for which we can't (or don't want to) find a precise answer, but for which rough estimates can be useful and interesting. Finally, we'll study several aspects of vectors and vector algebra. Vectors will be needed throughout our study of physics to describe and analyze physical quantities, such as velocity and force, that have direction as well as magnitude.

1.1 | The Nature of Physics

Physics is an *experimental* science. Physicists observe the phenomena of nature and try to find patterns and principles that relate these phenomena. These patterns are called physical theories or, when they are very well established and of broad use, physical laws or principles.

CAUTION Calling an idea a theory does *not* mean that it's just a random thought or an unproven concept. Rather, a theory is an explanation of natural phenomena based on observation and accepted fundamental principles. An example is the theory of biological evolution, which is the result of extensive research and observation by generations of biologists.

The development of physical theory requires creativity at every stage. The physicist has to learn to ask appropriate questions, design experiments to try to answer the questions, and draw appropriate conclusions from the results. Figure 1.1 shows two famous experimental facilities.

(a) (b)

1.1 Two research laboratories. (a) The Leaning Tower in Pisa, Italy. According to legend, Galileo studied the motion of freely falling bodies by dropping them from the tower. He is also said to have studied pendulum motion by observing the swinging of the chandelier in the cathedral behind the tower. (b) The Hubble Space Telescope is the first major telescope to operate outside the earth's atmosphere. Its sensitive instruments measure visible, ultraviolet, and near-infrared light from astronomical objects. The Space Telescope has been used to study phenomena ranging from eruptions on the moons of Jupiter to the cores of distant galaxies. It is shown here in March 2002 while being repaired in orbit by the crew of the Space Shuttle *Columbia*.

According to legend, Galileo Galilei (1564–1642) dropped light and heavy objects from the top of the Leaning Tower of Pisa (Fig. 1.1a) to find out whether their rates of fall were the same or different. Galileo recognized that only experimental investigation could answer this question. From examining the results of his experiments (which were actually much more sophisticated than in the legend), he made the inductive leap to the principle, or theory, that the acceleration of a falling body is independent of its weight.

The development of physical theories such as Galileo's is always a two-way process that starts and ends with observations or experiments. This development often takes an indirect path, with blind alleys, wrong guesses, and the discarding of unsuccessful theories in favor of more promising ones. Physics is not simply a collection of facts and principles; it is also the *process* by which we arrive at general principles that describe how the physical universe behaves.

No theory is ever regarded as the final or ultimate truth. The possibility always exists that new observations will require that a theory be revised or discarded. It is in the nature of physical theory that we can disprove a theory by finding behavior that is inconsistent with it, but we can never prove that a theory is always correct.

Getting back to Galileo, suppose we drop a feather and a cannonball. They certainly do *not* fall at the same rate. This does not mean that Galileo was wrong; it means that his theory was incomplete. If we drop the feather and the cannonball *in a vacuum* to eliminate the effects of the air, then they do fall at the same rate. Galileo's theory has a **range of validity:** it applies only to objects for which the force exerted by the air (due to air resistance and buoyancy) is much less than the weight. Objects like feathers or parachutes are clearly outside this range.

Every physical theory has a range of validity outside of which it is not applicable. Often a new development in physics extends a principle's range of validity. Galileo's analysis of falling bodies was greatly extended half a century later by Newton's laws of motion and law of gravitation.

1.2 | Solving Physics Problems

At some point in their studies, almost all physics students find themselves thinking, "I understand the concepts, but I just can't solve the problems." But in physics, truly understanding a concept or principle is the same thing as being able to apply it to a variety of practical problems. Learning how to solve problems is absolutely essential; you don't *know* physics unless you can *do* physics.

How do you learn to solve physics problems? In every chapter of this book you will find *Problem-Solving Strategies* that offer techniques for setting up and solving problems efficiently and accurately. Following each *Problem-Solving Strategy* are one or more worked *Examples* that show these techniques in action. (The *Problem-Solving Strategies* will also help steer you away from some *incorrect* techniques that you may be tempted to use.) You'll also find additional examples that aren't associated with a particular *Problem-Solving Strategy*. Study these strategies and examples carefully, and work through each example for yourself on a piece of paper.

Different techniques are useful for solving different kinds of physics problems, which is why this book offers dozens of *Problem-Solving Strategies*. No matter what kind of problem you're dealing with, however, there are certain key steps that you'll always follow. (These same steps are equally useful for problems in math, engineering, chemistry, and many other fields.) In this book we've organized these steps into four stages of solving a problem.

Problem-Solving Strategy

Solving Physics Problems

IDENTIFY *the relevant concepts:* First, decide which physics ideas are relevant to the problem. Although this step doesn't involve any calculations, it's sometimes the most challenging part of solving the problem. Don't skip over this step, though; choosing the wrong approach at the beginning can make the problem more difficult than it has to be, or even lead you to an incorrect answer.

At this stage you'll also want to identify the **target variable** of the problem. The target variable is the quantity whose value you're trying to find. It could be the speed at which a projectile hits the ground, the intensity of a sound made by a siren, or the strength of a magnetic field generated by an electromagnet. (Sometimes the goal will be to find a mathematical expression for the target variable rather than a numerical value. Sometimes, too, the problem will have more than one target variable.) The target variable is the goal of the problem—solving process; make sure you don't lose sight of this goal as you work through the mathematics of the solution.

SET UP *the problem:* If appropriate, draw a sketch of the situation described in the problem. Based on the concepts you selected in the *Identify* step, choose the equations that you'll use to solve the problem and decide how you'll use them.

EXECUTE *the solution:* In this step, you "do the math." Before you launch into a flurry of calculations, make a list of all known and unknown quantities, and note which are the target variable or variables. Then solve the equations for the unknowns.

EVALUATE *your answer:* The goal of physics problem solving isn't just to get a number or a formula; it's to achieve better understanding. That means you must examine your answer to see what it's telling you. In particular, ask yourself, "Does this answer make sense?" If your target variable was the radius of the Earth and your answer is 6.38 centimeters (or if your answer turned out to be a negative number!), something went wrong in your problem-solving process. Go back and check your work, and revise your solution as necessary.

All of the *Problem-Solving Strategies* and *Examples* in this book will follow these four steps. (In some cases we will combine the first two or three steps.) We encourage you to follow these same steps when you solve problems yourself. You may find it useful to remember the acronym ***I SEE***—short for *Identify, Set up, Execute,* and *Evaluate.*

Idealized Models

In everyday conversation we often use the word "model" to mean either a small-scale replica, such as a model railroad, or a person who displays articles of clothing (or the absence there of). In physics a **model** is a simplified version of a physical system that would be too complicated to analyze in full detail.

For example, suppose we want to analyze the motion of a baseball thrown through the air. How complicated is this problem? The ball is neither perfectly spherical nor perfectly rigid; it has raised seams, and it spins as it moves through the air. Wind and air resistance influence its motion, the earth rotates beneath it, the ball's weight varies a little as its distance from the center of the earth changes, and so on. If we try to include all these things, the analysis gets hopelessly complicated. Instead, we invent a simplified version of the problem. We neglect the size and shape of the ball by representing it as a point object, or **particle.** We neglect air resistance by making the ball move in a vacuum, we ignore the earth's rotation, and we make the weight constant. Now we have a problem that is simple enough to deal with. We will analyze this model in detail in Chapter 3.

To make an idealized model of the system, we have to overlook quite a few minor effects to concentrate on the most important features of the system. Of course, we have to be careful not to neglect too much. If we ignore the effects of gravity completely, then our model predicts that when we throw the ball up, it goes in a straight line and disappears into space. We need to use some judgment and creativity to construct a model that simplifies a problem enough to make it manageable, yet keeps its essential features.

When we use a model to predict how a system will behave, the validity of our predictions is limited by the validity of the model. Going back to Galileo once more, we see that his prediction about falling bodies corresponds to an idealized model that does not include the effects of air resistance. This model works fairly well for a dropped cannonball, but not so well for a feather.

The concept of idealized models is extremely important in all physical science and technology. When we apply physical principles to complex systems, we always use idealized models, and we have to be aware of the assumptions we are making. In fact, the principles of physics themselves are stated in terms of idealized models; we speak about point masses, rigid bodies, ideal insulators, and so on. Idealized models play a crucial role throughout this book. Watch for them in discussions of physical theories and their applications to specific problems.

1.3 | Standards and Units

As we learned in Section 1.1, physics is an experimental science. Experiments require measurements, and we generally use numbers to describe the results of measurements. Any number that is used to describe a physical phenomenon quantitatively is called a **physical quantity.** For example, two physical quantities that describe you are your weight and your height. Some physical quantities are so fundamental that we can define them only by describing how to measure them. Such a definition is called an **operational definition.** Some examples are measuring a distance by using a ruler, and measuring a time interval by using a stopwatch. In other cases we define a physical quantity by describing how to calculate it from other quantities that we *can* measure. Thus we might define the average speed of a moving object as the distance traveled (measured with a ruler) divided by the time of travel (measured with a stopwatch).

When we measure a quantity, we always compare it with some reference standard. When we say that a Porsche Carrera GT is 4.56 meters long, we mean that it is 4.56 times as long as a meter stick, which we define to be 1 meter long. Such a standard defines a **unit** of the quantity. The meter is a unit of distance, and the second is a unit of time. When we use a number to describe a physical quantity, we must always specify the unit that we are using; to describe a distance simply as "4.56" wouldn't mean anything.

To make accurate, reliable measurements, we need units of measurement that do not change and that can be duplicated by observers in various locations. The system of units used by scientists and engineers around the world is commonly called "the metric system," but since 1960 it has been known officially as the **International System,** or **SI** (the abbreviation for its French name, *Système International*). A list of all SI units is given in Appendix A, as are definitions of the most fundamental units.

The definitions of the basic units of the metric system have evolved over the years. When the metric system was established in 1791 by the French Academy of Sciences, the meter was defined as one ten-millionth of the distance from the North Pole to the equator (Fig. 1.2). The second was defined as the time required

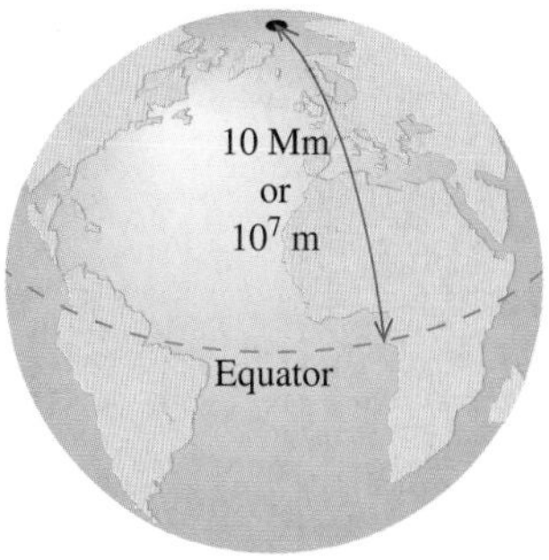

1.2 In 1791 the distance from the North Pole to the equator was defined to be exactly 10^7 m. With the modern definition of the meter this distance is about 0.02% more than 10^7 m.

for a pendulum one meter long to swing from one side to the other. These definitions were cumbersome and hard to duplicate precisely, and by international agreement they have been replaced with more refined definitions.

Time

From 1889 until 1967, the unit of time was defined as a certain fraction of the mean solar day, the average time between successive arrivals of the sun at its highest point in the sky. The present standard, adopted in 1967, is much more precise. It is based on an atomic clock, which uses the energy difference between the two lowest energy states of the cesium atom. When bombarded by microwaves of precisely the proper frequency, cesium atoms undergo a transition from one of these states to the other. One **second** is defined as the time required for 9,192,631,770 cycles of this microwave radiation.

Length

In 1960 an atomic standard for the meter was also established, using the wavelength of the orange-red light emitted by atoms of krypton (^{86}Kr) in a glow discharge tube. Using this length standard, the speed of light in a vacuum was measured to be 299,792,458 m/s. In November 1983, the length standard was changed again so that the speed of light in a vacuum was *defined* to be precisely 299,792,458 m/s. The meter is defined to be consistent with this number and with the above definition of the second. Hence the new definition of the **meter** is the distance that light travels in a vacuum in 1/299,792,458 second. This provides a much more precise standard of length than the one based on a wavelength of light.

Mass

The standard of mass, the **kilogram,** is defined to be the mass of a particular cylinder of platinum-iridium alloy. That cylinder is kept at the International Bureau of Weights and Measures at Sèvres, near Paris. An atomic standard of mass would be more fundamental, but at present we cannot measure masses on an atomic scale with as much accuracy as on a macroscopic scale. The *gram* (which is not a fundamental unit) is 0.001 kilogram.

Unit Prefixes

Once we have defined the fundamental units, it is easy to introduce larger and smaller units for the same physical quantities. In the metric system these other units are related to the fundamental units (or, in the case of mass, to the gram) by multiples of 10 or $\frac{1}{10}$. Thus one kilometer (1 km) is 1000 meters, and one centimeter (1 cm) is $\frac{1}{100}$ meter. We usually express multiples of 10 or $\frac{1}{10}$ in exponential notation: $1000 = 10^3$, $\frac{1}{1000} = 10^{-3}$, and so on. With this notation, $1\text{ km} = 10^3\text{ m}$ and $1\text{ cm} = 10^{-2}\text{ m}$.

The names of the additional units are derived by adding a **prefix** to the name of the fundamental unit. For example, the prefix "kilo-," abbreviated k, always means a unit larger by a factor of 1000; thus

$$\begin{aligned} 1\text{ kilometer} &= 1\text{ km} = 10^3\text{ meters} = 10^3\text{ m} \\ 1\text{ kilogram} &= 1\text{ kg} = 10^3\text{ grams} = 10^3\text{ g} \\ 1\text{ kilowatt} &= 1\text{ kW} = 10^3\text{ watts} = 10^3\text{ W} \end{aligned}$$

(a) 10^{26} m

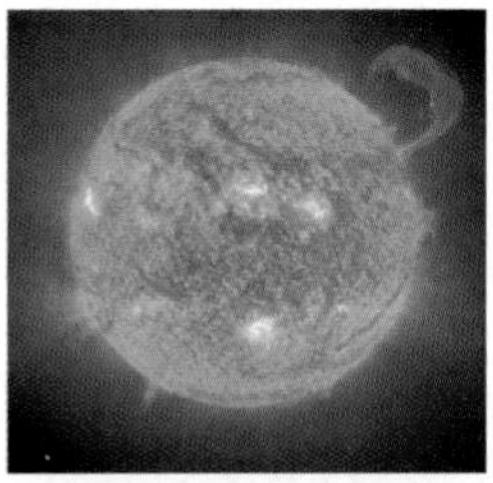

(b) 10^{11} m

(c) 10^{7} m

(d) 1 m

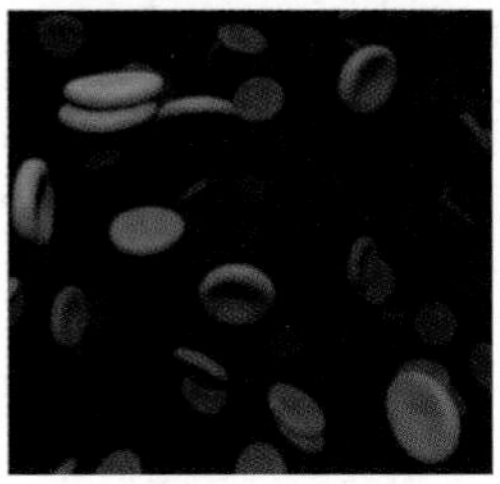

(e) 10^{-6} m

(f) 10^{-10} m

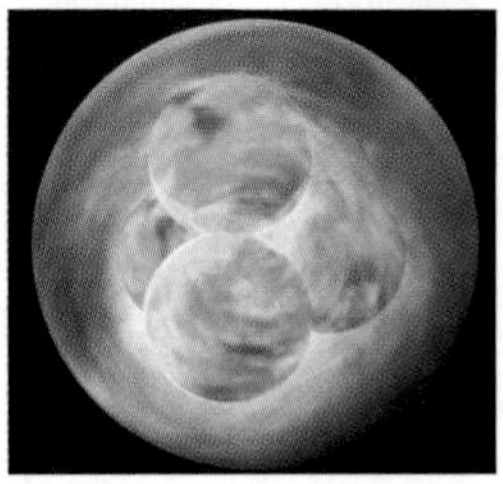

(g) 10^{-14} m

1.3 Some typical lengths in the universe. (a) The most distant galaxies are about 10^{26} m, or 10^{23} km, away. (b) The sun is 1.50×10^{11} m, or 1.50×10^{8} km, from our earth. (c) The diameter of the earth is 1.28×10^{7} m, or 12,800 km. (d) A typical human is about 1.7 m, or 170 cm, tall. (e) Human red blood cells are about 8×10^{-6} m (0.008 mm, or 8 μm) in diameter. (f) These oxygen atoms, shown arrayed on the surface of a crystal, are about 10^{-10} m, or 10^{-4} μm, in radius. (g) Typical atomic nuclei (shown in an artist's impression) have radii of about 10^{-14} m, or 10^{-5} nm.

A table on the inside back cover of this book lists the standard SI prefixes, with their meanings and abbreviations.

Here are several examples of the use of multiples of 10 and their prefixes with the units of length, mass, and time. Figure 1.3 shows how these prefixes help describe both large and small distances.

Length

1 nanometer = 1 nm = 10^{-9} m (a few times the size of the largest atom)
1 micrometer = 1 μm = 10^{-6} m (size of some bacteria and living cells)
1 millimeter = 1 mm = 10^{-3} m (diameter of the point of a ballpoint pen)
1 centimeter = 1 cm = 10^{-2} m (diameter of your little finger)
1 kilometer = 1 km = 10^{3} m (a 10-minute walk)

Mass

1 microgram = 1 μg = 10^{-6} g = 10^{-9} kg (mass of a very small dust particle)
1 milligram = 1 mg = 10^{-3} g = 10^{-6} kg (mass of a grain of salt)
1 gram = 1 g = 10^{-3} kg (mass of a paper clip)

Time

$$
\begin{aligned}
1 \text{ nanosecond} &= 1 \text{ ns} = 10^{-9} \text{ s} \ (\text{time for light to travel } 0.3 \text{ m}) \\
1 \text{ microsecond} &= 1\ \mu\text{s} = 10^{-6} \text{ s} \ (\text{time for an orbiting space shuttle to travel } 8 \text{ mm}) \\
1 \text{ millisecond} &= 1 \text{ ms} = 10^{-3} \text{ s} \ (\text{time for sound to travel } 0.35 \text{ m})
\end{aligned}
$$

1.4 Many everyday items make use of both SI and British units. An example is this speedometer from a U.S.-built automobile, which shows the speed in both kilometers per hour and miles per hour.

The British System

Finally, we mention the British system of units. These units are used only in the United States and a few other countries, and in most of these they are being replaced by SI units. British units are now officially defined in terms of SI units, as follows:

$$
\begin{aligned}
\textit{Length}: 1 \text{ inch} &= 2.54 \text{ cm (exactly)} \\
\textit{Force}: 1 \text{ pound} &= 4.448221615260 \text{ newtons (exactly)}
\end{aligned}
$$

The newton, abbreviated N, is the SI unit of force. The British unit of time is the second, defined the same way as in SI. In physics, British units are used only in mechanics and thermodynamics; there is no British system of electrical units.

In this book we use SI units for all examples and problems, but we occasionally give approximate equivalents in British units. As you do problems using SI units, you may also wish to convert to the approximate British equivalents if they are more familiar to you (see Fig. 1.4). But you should try to think directly in SI as much as you can.

1.4 | Unit Consistency and Conversions

We use equations to express relationships among physical quantities, represented by algebraic symbols. Each algebraic symbol always denotes both a number and a unit. For example, d might represent a distance of 10 m, t a time of 5 s, and v a speed of 2 m/s.

An equation must always be **dimensionally consistent.** You can't add apples and automobiles; two terms may be added or equated only if they have the same units. For example, if a body moving with constant speed v travels a distance d in a time t, these quantities are related by the equation

$$d = vt \tag{1.1}$$

If d is measured in meters, then the product vt must also be expressed in meters. Using the above numbers as an example, we may write

$$10 \text{ m} = \left(2 \frac{\text{m}}{\not{\text{s}}}\right)(5 \not{\text{s}})$$

Because the unit 1/s on the right side of the equation cancels the unit s, the product vt has units of meters, as it must. In calculations, units are treated just like algebraic symbols with respect to multiplication and division.

CAUTION When a problem requires calculations using numbers with units, *always* write the numbers with the correct units and carry the units through the calculation as in the example above. This provides a very useful check for calculations. If at some stage in a calculation you find that an equation or an expression has inconsistent units, you know you have made an error somewhere. In this book we will always carry units through all calculations, and we strongly urge you to follow this practice when you solve problems.

Problem-Solving Strategy: Unit Conversions

IDENTIFY *the relevant concepts:* Unit conversion is important, but it's also important to recognize when it's needed. In most cases, you're best off using the fundamental SI units (lengths in meters, masses in kilograms, and time in seconds) within a problem. If you need the answer to be in a different set of units (such as kilometers, grams, or hours), wait until the end of the problem to make the conversion. In the following examples, we'll concentrate on unit conversion alone, so we'll skip the *Identify* step.

SET UP *the problem and* **EXECUTE** *the solution:* Units are multiplied and divided just like ordinary algebraic symbols. This gives us an easy way to convert a quantity from one set of units to another. The key idea is to express the same physical quantity in two different units and form an equality.

For example, when we say that 1 min = 60 s, we don't mean that the number 1 is equal to the number 60; rather, we mean that 1 min represents the same physical time interval as 60 s. For this reason, the ratio (1 min)/(60 s) equals 1, as does its reciprocal (60 s)/(1 min). We may multiply a quantity by either of these factors without changing that quantity's physical meaning. For example, to find the number of seconds in 3 min, we write

$$3\text{ min} = (3\,\cancel{\text{min}})\left(\frac{60\text{ s}}{1\,\cancel{\text{min}}}\right) = 180\text{ s}$$

EVALUATE *your answer:* If you do your unit conversions correctly, unwanted units will cancel, as in the example above. If instead you had multiplied 3 min by (1 min)/(60 s), your result would have been $\frac{1}{20}\text{ min}^2/\text{s}$, which is a rather odd way of measuring time. To be sure you convert units properly, you must write down the units at *all* stages of the calculation.

Finally, check whether your answer is reasonable. Is the result 3 min = 180 s reasonable? The answer is yes; the second is a smaller unit than the minute, so there are more seconds than minutes in the same time interval.

Example 1.1 Converting speed units

The official world land speed record is 1228.0 km/h, set on October 15, 1997, by Andy Green in the jet engine car *Thrust SSC*. Express this speed in meters per second.

SOLUTION

IDENTIFY, SET UP and EXECUTE: The prefix k means 10^3, so the speed 1228.0 km/h = 1228.0×10^3 m/h. We also know that there are 3600 s in 1 h. So we must combine the speed of 1228.0×10^3 m/h and a factor of 3600. But should we multiply or divide by this factor? If we treat the factor as a pure number without units, we're forced to guess how to proceed.

The correct approach is to carry the units with each factor. We then arrange the factor so that the hour unit cancels:

$$1228.0\text{ km/h} = \left(1228.0 \times 10^3\,\frac{\text{m}}{\cancel{\text{h}}}\right)\left(\frac{1\,\cancel{\text{h}}}{3600\text{ s}}\right) = 341.11\text{ m/s}$$

If you multiplied by (3600 s)/(1 h) instead of (1 h)/(3600 s), the hour unit wouldn't cancel, and you would be able to easily recognize your error. Again, the *only* way to be sure that you correctly convert units is to carry the units throughout the calculation.

EVALUATE: While you probably have a good intuition for speeds in kilometers per hour or miles per hour, speeds in meters per second are likely to be a bit more mysterious. Notice that the length of an average person's stride is about one meter, and a good walking pace is about one stride per second. Thus a typical walking speed is about 1 m/s. By comparison, a speed of 341.11 m/s is rapid indeed!

Example 1.2 Converting volume units

The world's largest cut diamond is the First Star of Africa (mounted in the British Royal Sceptre and kept in the Tower of London). Its volume is 1.84 cubic inches. What is its volume in cubic centimeters? In cubic meters?

SOLUTION

IDENTIFY, SET UP and EXECUTE: To convert cubic inches to cubic centimeters, we multiply by $[(2.54\text{ cm})/(1\text{ in.})]^3$, not just (2.54 cm)/(1 in.). We find

$$1.84\text{ in.}^3 = (1.84\text{ in.}^3)\left(\frac{2.54\text{ cm}}{1\text{ in.}}\right)^3$$

$$= (1.84)(2.54)^3\frac{\cancel{\text{in.}^3}\text{ cm}^3}{\cancel{\text{in.}^3}} = 30.2\text{ cm}^3$$

Also, $1\text{ cm} = 10^{-2}\text{ m}$, and

$$30.2\text{ cm}^3 = (30.2\text{ cm}^3)\left(\frac{10^{-2}\text{ m}}{1\text{ cm}}\right)^3$$

$$= (30.2)(10^{-2})^3\frac{\cancel{\text{cm}^3}\text{ m}^3}{\cancel{\text{cm}^3}} = 30.2\times10^{-6}\text{ m}^3$$

EVALUATE: Our answer shows that while 1 centimeter is 10^{-2} of a meter (that is, $1\text{ cm} = 10^{-2}\text{ m}$), a cubic centimeter (1 cm^3) is *not* 10^{-2} of a cubic meter. Rather, it is the volume of a cube whose sides are 1 cm long. So $1\text{ cm}^3 = (1\text{ cm})^3 = (10^{-2}\text{ m})^3 = (10^{-2})^3\text{ m}^3$, or $1\text{ cm}^3 = 10^{-6}\text{ m}^3$.

1.5 | Uncertainty and Significant Figures

Measurements always have uncertainties. If you measure the thickness of the cover of this book using an ordinary ruler, your measurement is reliable only to the nearest millimeter, and your result will be 3 mm. It would be *wrong* to state this result as 3.00 mm; given the limitations of the measuring device, you can't tell whether the actual thickness is 3.00 mm, 2.85 mm, or 3.11 mm. But if you use a micrometer caliper, a device that measures distances reliably to the nearest 0.01 mm, the result will be 2.91 mm. The distinction between these two measurements is in their **uncertainty.** The measurement using the micrometer caliper has a smaller uncertainty; it's a more accurate measurement. The uncertainty is also called the **error,** because it indicates the maximum difference there is likely to be between the measured value and the true value. The uncertainty or error of a measured value depends on the measurement technique used.

We often indicate the **accuracy** of a measured value—that is, how close it is likely to be to the true value—by writing the number, the symbol $\pm$, and a second number indicating the uncertainty of the measurement. If the diameter of a steel rod is given as 56.47 ± 0.02 mm, this means that the true value is unlikely to be less than 56.45 mm or greater than 56.49 mm. In a commonly used shorthand notation, the number 1.6454(21) means 1.6454 ± 0.0021. The numbers in parentheses show the uncertainty in the final digits of the main number.

We can also express accuracy in terms of the maximum likely **fractional error** or **percent error** (also called *fractional uncertainty* and *percent uncertainty*). A resistor labeled "47 ohms $\pm$ 10%" probably has a true resistance differing from 47 ohms by no more than 10% of 47 ohms—that is, about 5 ohms. The resistance is probably between 42 and 52 ohms. For the diameter of the steel rod given above, the fractional error is (0.02 mm)/(56.47 mm), or about 0.0004; the percent error is (0.0004) (100%), or about 0.04%. Even small percent errors can sometimes be very significant (Fig. 1.5).

1.5 This spectacular mishap was the result of a very small percent error—traveling a few meters too far in a journey of hundreds of thousands of meters.

In many cases the uncertainty of a number is not stated explicitly. Instead, the uncertainty is indicated by the number of meaningful digits, or **significant figures,** in the measured value. We gave the thickness of the cover of this book as 2.91 mm, which has three significant figures. By this we mean that the first two digits are known to be correct, while the third digit is uncertain. The last digit is in the hundredths place, so the uncertainty is about 0.01 mm. Two values with the *same* number of significant figures may have *different* uncertainties; a distance given as 137 km also has three significant figures, but the uncertainty is about 1 km.

When you use numbers having uncertainties to compute other numbers, the computed numbers are also uncertain. It is especially important to understand this when you compare a number obtained from measurements with a value obtained from a theoretical prediction. Suppose you want to verify the value of π, the ratio of the circumference of a circle to its diameter. The true value of this ratio to ten digits is 3.141592654. To make your own calculation, you draw a large circle and measure its diameter and circumference to the nearest millimeter, obtaining the values 135 mm and 424 mm. You punch these into your calculator and obtain the quotient 3.140740741. Does this agree with the true value or not?

First, the last seven digits in this answer are meaningless; they imply a smaller uncertainty than is possible with your measurements. When numbers are multiplied or divided, the number of significant figures in the result can be no greater than in the factor with the fewest significant figures. For example, $3.1416 \times 2.34 \times 0.58 = 4.3$. Your measurements each have three significant figures, so your measured value of π, equal to (424 mm)/(135 mm), can have only three significant figures. It should be stated simply as 3.14. Within the limit of three significant figures, your value does agree with the true value.

When we add and subtract numbers, it's the location of the decimal point that matters, not the number of significant figures. For example, $123.62 + 8.9 = 132.5$. Although 123.62 has an uncertainty of about 0.01, 8.9 has an uncertainty of about 0.1. So their sum has an uncertainty of about 0.1 and should be written as 132.5, not 132.52.

In the examples and problems in this book we usually give numerical values with three significant figures, so your answers should usually have no more than three significant figures. (Many numbers that you encounter in the real world have even less accuracy. An automobile speedometer, for example, usually gives only two significant figures.) You might do the arithmetic with a calculator having a display with ten digits. To give a ten-digit answer is not merely unnecessary, but genuinely wrong, because it misrepresents the accuracy of the results. Always round your final answer to keep only the correct number of significant figures or, in doubtful cases, one more at most. In Example 1.1 it would have been wrong to state the answer as 341.11111 m/s. Note that when you reduce such an answer to the appropriate number of significant figures, you must *round,* not *truncate.* Your calculator will tell you that the ratio of 525 m to 311 m is 1.688102894; to three significant figures, this is 1.69, not 1.68.

When we calculate with very large or very small numbers, we can show significant figures much more easily by using **scientific notation,** sometimes called **powers-of-10 notation.** The distance from the earth to the moon is about 384,000,000 m, but writing the number in this form gives no indication of the number of significant figures. Instead, we move the decimal point eight places to the left (corresponding to dividing by 10^8) and multiply by 10^8. That is,

$$384{,}000{,}000 \text{ m} = 3.84 \times 10^8 \text{ m}$$

In this form, it is clear that we have three significant figures. The number 4.00×10^{-7} also has three significant figures, even though two of them are zeros.

Note that in scientific notation the usual practice is to express the quantity as a number between 1 and 10 multiplied by the appropriate power of 10. Table 1.1 summarizes the rules for significant figures.

Table 1.1 Using Significant Figures

Mathematical operation	Significant figures in result
Multiplication or division	No more than in the number with the fewest significant figures *Example:* $(0.745 \times 2.2)/3.885 = 0.42$ *Example:* $(1.32578 \times 10^7) \times (4.11 \times 10^{-3}) = 5.45 \times 10^4$
Addition or subtraction	Determined by the number with the smallest uncertainty (i.e., the fewest digits to the right of the decimal point) *Example:* $27.153 + 138.2 - 11.74 = 153.6$

Note: In this book we will usually give numerical values with three significant figures.

When an integer or a fraction occurs in a general equation, we treat that number as having no uncertainty at all. For example, in the equation $v_x^2 = v_{0x}^2 + 2a_x(x - x_0)$, which is Eq. (2.13) in Chapter 2, the coefficient 2 is *exactly* 2. We can consider this coefficient as having an infinite number of significant figures (2.000000 . . .). The same is true of the exponent 2 in v_x^2 and v_{0x}^2.

Finally, let's note that **precision** is not the same as *accuracy.* A cheap digital watch that says the time is 10:35:17 a.m. is very *precise* (the time is given to the second), but if the watch runs several minutes slow, then this value isn't very *accurate.* On the other hand, a grandfather clock might be very accurate (that is, display the correct time), but if the clock has no second hand, it isn't very precise. A high-quality measurement, like those used to define standards (Section 1.3), is both precise *and* accurate.

Example 1.3 Significant figures in multiplication

The rest energy E of an object with rest mass m is given by Einstein's equation

$$E = mc^2$$

where c is the speed of light in a vacuum. Find E for an object with $m = 9.11 \times 10^{-31}$ kg (to three significant figures, the mass of an electron). The SI unit for E is the joule (J); $1\text{ J} = 1\text{ kg}\cdot\text{m}^2/\text{s}^2$.

SOLUTION

IDENTIFY and SET UP: Our target variable is the energy E. We are given the equation to use and the value of the mass m; from Section 1.3 the exact value of the speed of light is $c = 299{,}792{,}458$ m/s $= 2.99792458 \times 10^8$ m/s.

EXECUTE Substituting the values of m and c into Einstein's equation, we find

$$\begin{aligned} E &= (9.11 \times 10^{-31}\text{ kg})(2.99792458 \times 10^8\text{ m/s})^2 \\ &= (9.11)(2.99792458)^2(10^{-31})(10^8)^2\text{ kg}\cdot\text{m}^2/\text{s}^2 \\ &= (81.87659678)(10^{[-31+(2\times 8)]})\text{ kg}\cdot\text{m}^2/\text{s}^2 \\ &= 8.187659678 \times 10^{-14}\text{ kg}\cdot\text{m}^2/\text{s}^2 \end{aligned}$$

Since the value of m was given to only three significant figures, we must round this to

$$E = 8.19 \times 10^{-14}\text{ kg}\cdot\text{m}^2/\text{s}^2 = 8.19 \times 10^{-14}\text{ J}$$

Most calculators use scientific notation and add exponents automatically, but you should be able to do such calculations by hand when necessary.

EVALUATE: While the rest energy contained in an electron may seem ridiculously small, on the atomic scale it is tremendous. Compare our answer to 10^{-19} J, the energy gained or lost by a single atom during a typical chemical reaction; the rest energy of an electron is about 1,000,000 times larger! (We will discuss the significance of rest energy in Chapter 37.)

Test Your Understanding

The density of a material is equal to its mass divided by its volume. What is the density (in kg/m^3) of a rock of mass 1.80 kg and volume 6.0×10^{-4} m^3? Make sure your answer has the correct number of significant figures.

1.6 | Estimates and Orders of Magnitude

We have stressed the importance of knowing the accuracy of numbers that represent physical quantities. But even a very crude estimate of a quantity often gives us useful information. Sometimes we know how to calculate a certain quantity but have to guess at the data we need for the calculation. Or the calculation might be too complicated to carry out exactly, so we make some rough approximations. In either case our result is also a guess, but such a guess can be useful even if it is uncertain by a factor of two, ten, or more. Such calculations are often called **order-of-magnitude estimates.** The great Italian American nuclear physicist Enrico Fermi (1901–1954) called them "back-of-the-envelope calculations."

Exercises 1.18 through 1.29 at the end of this chapter are of the estimating, or "order-of-magnitude," variety. Some are silly, and most require guesswork for the needed input data. Don't try to look up a lot of data; make the best guesses you can. Even when they are off by a factor of ten, the results can be useful and interesting.

Example 1.4 An order-of-magnitude estimate

You are writing an adventure novel in which the hero escapes across the border with a billion dollars worth of gold in his suitcase. Is this possible? Would that amount of gold fit in a suitcase? Would it be too heavy to carry?

SOLUTION

IDENTIFY, SET UP, and EXECUTE Gold sells for around \$400 an ounce. On a particular day the price might be \$200 or \$600, but never mind. An ounce is about 30 grams. Actually, an ordinary (avoirdupois) ounce is 28.35 g; an ounce of gold is a troy ounce, which is 9.45% more. Again, never mind. Ten dollars' worth of gold has a mass somewhere around one gram, so a billion (10^9) dollars worth of gold is a hundred million (10^8) grams, or a hundred thousand (10^5) kilograms. This corresponds to a weight in British units of around 200,000 lb, or 100 tons. Whether the precise number is closer to 50 tons or 200 tons doesn't matter. Either way, the hero is not about to carry it across the border in a suitcase.

We can also estimate the *volume* of this gold. If its density were the same as that of water $(1\ g/cm^3)$, the volume would be 10^8 cm^3, or 100 m^3. But gold is a heavy metal; we might guess its density to be ten times that of water. Gold is actually 19.3 times as dense as water. But by guessing ten, we find a volume of 10 m^3. Visualize ten cubical stacks of gold bricks, each 1 meter on a side, and ask yourself whether they would fit in a suitcase!

EVALUATE: Clearly, your novel needs rewriting. Try the calculation again with a suitcase full of five-carat (1-gram) diamonds, each worth \$100,000. Would this work?

Test Your Understanding

Can you estimate the total number of teeth in all the mouths of everyone (students, staff, and faculty) on your campus? (*Hint:* How many teeth are in your mouth? Count them!)

1.7 | Vectors and Vector Addition

Some physical quantities, such as time, temperature, mass, density, and electric charge, can be described completely by a single number with a unit. But many other important quantities have a *direction* associated with them and cannot be described by a single number. Such quantities play an essential role in many of the central topics of physics, including motion and its causes and the phenomena of electricity and magnetism. A simple example of a quantity with direction is the motion of an airplane. To describe this motion completely, we must say not only how fast the plane is moving, but also in what direction. To fly from Chicago to New York, a plane has to head east, not south. The speed of the airplane combined with its direction of motion together constitute a quantity called *velocity*. Another example is *force*, which in physics means a push or pull exerted on a body. Giving a complete description of a force means describing both how hard the force pushes or pulls on the body and the direction of the push or pull.

When a physical quantity is described by a single number, we call it a **scalar quantity.** In contrast, a **vector quantity** has both a **magnitude** (the "how much" or "how big" part) and a direction in space. Calculations with scalar quantities use the operations of ordinary arithmetic. For example, 6 kg + 3 kg = 9 kg, or 4 × 2 s = 8 s. However, combining vectors requires a different set of operations.

To understand more about vectors and how they combine, we start with the simplest vector quantity, **displacement.** Displacement is simply a change in position of a point. (The point may represent a particle or a small body.) In Fig. 1.6a we represent the change of position from point P_1 to point P_2 by a line from P_1 to P_2, with an arrowhead at P_2 to represent the direction of motion. Displacement is a vector quantity because we must state not only how far the particle moves, but also in what direction. Walking 3 km north from your front door doesn't get you to the same place as walking 3 km southeast; these two displacements have the same magnitude, but different directions.

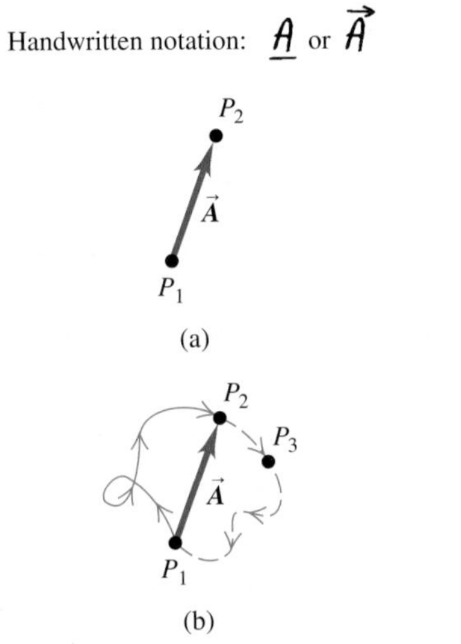

1.6 (a) Vector $\vec{A}$ is the displacement from point P_1 to point P_2. (b) A displacement is always a straight-line segment directed from the starting point to the end point, even if the actual path is curved. When a path ends at the same place where it started, the displacement is zero.

We usually represent a vector quantity such as displacement by a single letter, such as $\vec{A}$ in Fig. 1.6a. In this book we always print vector symbols in ***boldface italic type with an arrow above them.*** We do this to remind you that vector quantities have different properties from scalar quantities; the arrow is a reminder that vectors have direction. In handwriting, vector symbols are usually underlined or written with an arrow above them (Fig. 1.6). When you write a symbol for a vector, *always* write it in one of these ways. If you don't distinguish between scalar and vector quantities in your notation, you probably won't make the distinction in your thinking either, and hopeless confusion will result.

When drawing any vector, we always draw a line with an arrowhead at its tip. The length of the line shows the vector's magnitude, and the direction of the line shows the vector's direction. Displacement is always a straight-line segment, directed from the starting point to the end point, even though the actual path of the particle may be curved. In Fig. 1.6b the particle moves along the curved path shown from P_1 to P_2, but the displacement is still the vector $\vec{A}$. Note that displacement is not related directly to the total *distance* traveled. If the particle were to continue on to P_3 and then return to P_1, the displacement for the entire trip would be *zero*.

If two vectors have the same direction, they are **parallel.** If they have the same magnitude *and* the same direction, they are *equal,* no matter where they are located in space. The vector $\vec{A}'$ from point P_3 to point P_4 in Fig. 1.7 has the same length and direction as the vector $\vec{A}$ from P_1 to P_2. These two displacements are equal, even though they start at different points. We write this as $\vec{A} = \vec{A}'$ in Fig. 1.7, using a boldface equals sign to emphasize that equality of two vector quantities is not the same relationship as equality of two scalar quantities. Two vector quantities are equal only when they have the same magnitude *and* the same direction.

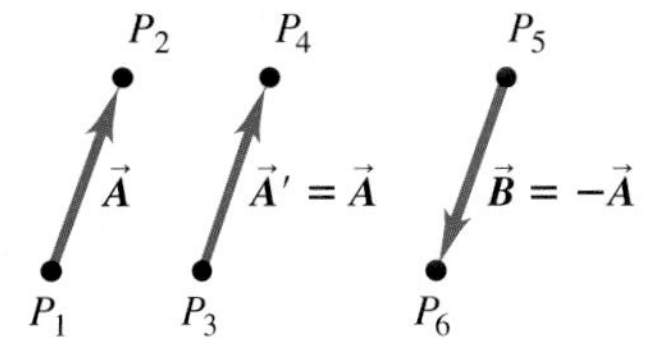

1.7 The displacement from P_3 to P_4 is equal to that from P_1 to P_2. The displacement $\vec{B}$ from P_5 to P_6 has the same magnitude as $\vec{A}$ and $\vec{A}'$ but opposite direction; displacement $\vec{B}$ is the negative of displacement $\vec{A}$.

The vector $\vec{B}$ in Fig. 1.7, however, is not equal to $\vec{A}$ because its direction is *opposite* to that of $\vec{A}$. We define the **negative of a vector** as a vector having the same magnitude as the original vector but the *opposite* direction. The negative of vector quantity $\vec{A}$. is denoted as $-\vec{A}$, and we use a boldface minus sign to emphasize the vector nature of the quantities. If $\vec{A}$ is 87 m south, then $-\vec{A}$ is 87 m north. Thus the relation between $\vec{A}$ and $\vec{B}$ of Fig. 1.7 may be written as $\vec{A} = -\vec{B}$ or $\vec{B} = -\vec{A}$. When two vectors $\vec{A}$ and $\vec{B}$ have opposite directions, whether their magnitudes are the same or not, we say that they are **antiparallel.**

We usually represent the *magnitude* of a vector quantity (its length in the case of a displacement vector) by the same letter used for the vector, but in *light italic type* with *no* arrow on top, rather than boldface italic with an arrow (which is reserved for vectors). An alternative notation is the vector symbol with vertical bars on both sides:

$$(\text{Magnitude of } \vec{A}) = A = |\vec{A}| \tag{1.2}$$

By definition the magnitude of a vector quantity is a scalar quantity (a number) and is *always positive.* We also note that a vector can never be equal to a scalar because they are different kinds of quantities. The expression "$\vec{A} = 6$ m" is just as wrong as "2 oranges = 3 apples" or "6 lb = 7 km"!

When drawing diagrams with vectors, we'll generally use a scale similar to those used for maps, in which the distance on the diagram is proportional to the magnitude of the vector. For example, a displacement of 5 km might be represented in a diagram by a vector 1 cm long, since an actual-size diagram wouldn't be practical. When we work with vector quantities with units other than displacement, such as force or velocity, we *must* use a scale. In a diagram for force vectors we might use a scale in which a vector that is 1 cm long represents a force of magnitude 5 N. A 20-N force would then be represented by a vector 4 cm long, with the appropriate direction.

Vector Addition

Now suppose a particle undergoes a displacement $\vec{A}$, followed by a second displacement $\vec{B}$ (Fig. 1.8a). The final result is the same as if the particle had started at the same initial point and undergone a single displacement $\vec{C}$, as shown. We call displacement $\vec{C}$ the **vector sum,** or **resultant,** of displacements $\vec{A}$ and $\vec{B}$. We express this relationship symbolically as

$$\vec{C} = \vec{A} + \vec{B} \tag{1.3}$$

The boldface plus sign emphasizes that adding two vector quantities requires a geometrical process and is not the same operation as adding two scalar quantities such as $2 + 3 = 5$. In vector addition we usually place the *tail* of the *second* vector at the *head,* or tip, of the *first* vector (Fig. 1.8a).

If we make the displacements $\vec{A}$ and $\vec{B}$ in reverse order, with $\vec{B}$ first and $\vec{A}$ second, the result is the same (Fig. 1.8b). Thus

$$\vec{C} = \vec{B} + \vec{A} \quad \text{and} \quad \vec{A} + \vec{B} = \vec{B} + \vec{A} \tag{1.4}$$

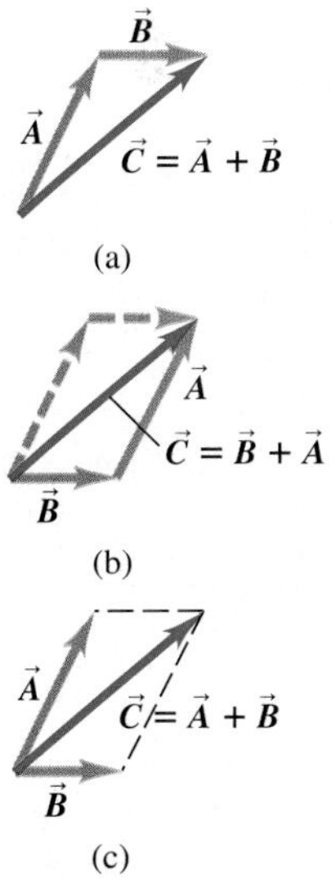

1.8 Vector $\vec{C}$ is the vector sum of vectors $\vec{A}$ and $\vec{B}$. The order in vector addition doesn't matter; vector addition is commutative.

This shows that the order of terms in a vector sum doesn't matter. In other words, vector addition obeys the commutative law.

Figure 1.8c shows an alternative representation of the vector sum. When vectors $\vec{A}$ and $\vec{B}$ are both drawn with their tails at the same point, vector $\vec{C}$ is the diagonal of a parallelogram constructed with $\vec{A}$ and $\vec{B}$ as two adjacent sides.

CAUTION A word of warning is in order about vector addition. It's a common error to conclude that if $\vec{C} = \vec{A} + \vec{B}$, the magnitude C should just equal the magnitude A plus the magnitude B. Figure 1.8 shows that in general, this conclusion is *wrong;* you can see from the figure that $C < A + B$. The magnitude of the vector sum $\vec{A} + \vec{B}$ depends on the magnitudes of $\vec{A}$ and of $\vec{B}$ *and* on the angle between $\vec{A}$ and $\vec{B}$ (see Problem 1.88). Only in the special case in which $\vec{A}$ and $\vec{B}$ are *parallel* is the magnitude of $\vec{C} = \vec{A} + \vec{B}$ equal to the sum of the magnitudes of $\vec{A}$ and $\vec{B}$ (Fig. 1.9a). By contrast, when the vectors are *antiparallel* (Fig. 1.9b) the magnitude of $\vec{C}$ equals the *difference* of the magnitudes of $\vec{A}$ and $\vec{B}$. Students who aren't careful about distinguishing between scalar and vector quantities frequently make errors about the magnitude of a vector sum. Don't let this happen to you!

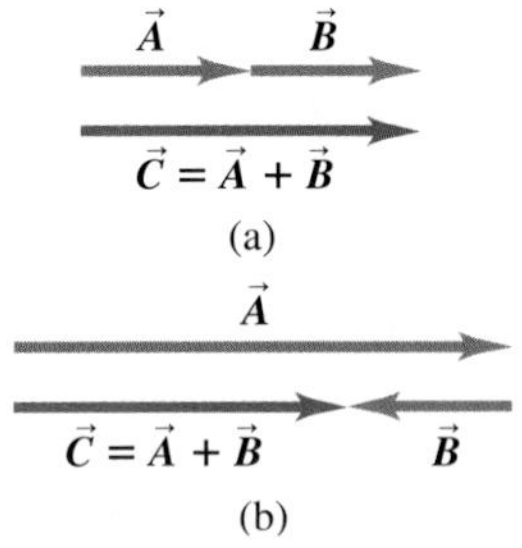

1.9 (a) In the special case when vectors $\vec{A}$ and $\vec{B}$ are parallel, and *only* in this case, the magnitude of their sum equals the sum of their magnitudes: $C = A + B$. (b) When $\vec{A}$ and $\vec{B}$ are antiparallel, the magnitude of their sum equals the *difference* of their magnitudes: $C = |A - B|$. Note that the vectors $\vec{A}$, $\vec{B}$, and $\vec{C}$ in part (a) are not the same as the vectors $\vec{A}$, $\vec{B}$, and $\vec{C}$ in part (b).

When we need to add more than two vectors, we may first find the vector sum of any two, add this vectorially to the third, and so on. Figure 1.10a shows three vectors $\vec{A}$, $\vec{B}$, and $\vec{C}$. In Fig. 1.10b, vectors $\vec{A}$ and $\vec{B}$ are added first, giving a vector sum $\vec{D}$; vectors $\vec{C}$ and $\vec{D}$ are then added by the same process to obtain the vector sum $\vec{R}$:

$$\vec{R} = (\vec{A} + \vec{B}) + \vec{C} = \vec{D} + \vec{C}$$

Alternatively, we can first add $\vec{B}$ and $\vec{C}$ (Fig. 1.10c) to obtain vector $\vec{E}$, and then add $\vec{A}$ and $\vec{E}$ to obtain $\vec{R}$:

$$\vec{R} = \vec{A} + (\vec{B} + \vec{C}) = \vec{A} + \vec{E}$$

We don't even need to draw vectors $\vec{D}$ and $\vec{E}$; all we need to do is draw the given vectors in succession, with the tail of each at the head of the one preceding it, and complete the polygon by a vector $\vec{R}$ from the tail of the first vector to the head of the last vector (Fig. 1.10d). The order makes no difference; Fig. 1.10e shows a different order, and we invite you to try others. We see that vector addition obeys the associative law.

We mentioned above that $-\vec{A}$ is a vector having the same magnitude as $\vec{A}$ but the opposite direction. This provides the basis for defining vector subtraction. We define the difference $\vec{A} - \vec{B}$ of two vectors $\vec{A}$ and $\vec{B}$ to be the vector sum of $\vec{A}$ and $-\vec{B}$:

$$\vec{A} - \vec{B} = \vec{A} + (-\vec{B}) \tag{1.5}$$

Figure 1.11 shows an example of vector subtraction. To construct the vector difference $\vec{A} - \vec{B}$, the tail of $-\vec{B}$ is placed at the head of $\vec{A}$.

A vector quantity such as a displacement can be multiplied by a scalar quantity (an ordinary number). The displacement $2\vec{A}$ is a displacement (vector quantity) in the same direction as the vector $\vec{A}$ but twice as long; this is the same as adding $\vec{A}$ to itself. In general, when a vector $\vec{A}$ is multiplied by a scalar c, the result $c\vec{A}$ has magnitude $|c|A$ (the absolute value of c multiplied by the magnitude of the vector

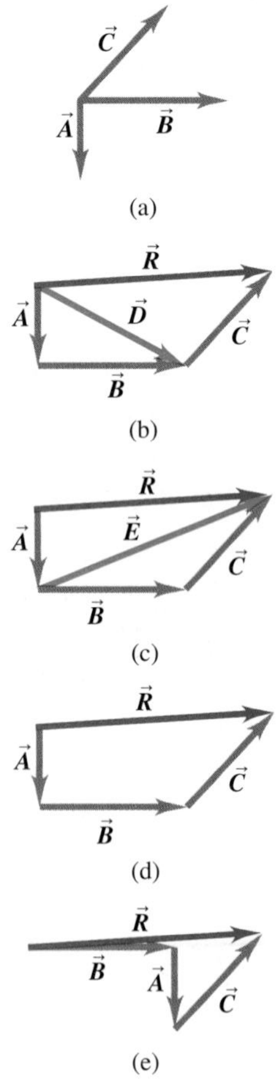

1.10 Several constructions for finding the vector sum $\vec{A} + \vec{B} + \vec{C}$.

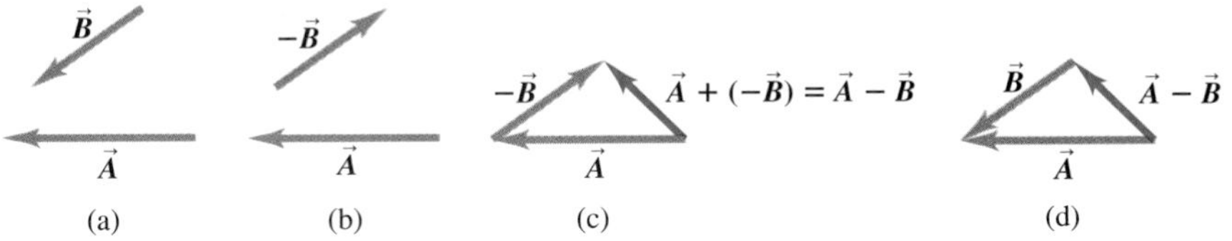

1.11 (a) Vector $\vec{A}$ and vector $\vec{B}$. (b) Vector $\vec{A}$ and vector $-\vec{B}$. (c) The vector difference $\vec{A} - \vec{B}$ is the sum of vectors $\vec{A}$ and $-\vec{B}$. The tail of $-\vec{B}$ is placed at the head of $\vec{A}$. (d) To check: $(\vec{A} - \vec{B}) + \vec{B} = \vec{A}$.

$\vec{A}$). If c is positive, $c\vec{A}$ is in the same direction as $\vec{A}$; if c is negative, $c\vec{A}$ is in the direction opposite to $\vec{A}$. Thus $5\vec{A}$ is parallel to $\vec{A}$, while $-5\vec{A}$ is antiparallel to $\vec{A}$.

The scalar quantity used to multiply a vector may also be a physical quantity having units. For example, you may be familiar with the relationship $\vec{F} = m\vec{a}$; the net force $\vec{F}$ (a vector quantity) that acts on a body is equal to the product of the body's mass m (a positive scalar quantity) and its acceleration $\vec{a}$ (a vector quantity). The magnitude of the net force is equal to the mass (which is positive and equals its own absolute value) multiplied by the magnitude of the acceleration. The unit of the magnitude of force is the unit of mass multiplied by the unit of the magnitude of acceleration. The direction of $\vec{F}$ is the same as that of $\vec{a}$ because m is positive.

Example 1.5 Vector addition

A cross-country skier skis 1.00 km north and then 2.00 km east on a horizontal snow field. a) How far and in what direction is she from the starting point? b) What are the magnitude and direction of her resultant displacement?

SOLUTION

IDENTIFY: The problem involves combining displacements, so we can solve it using vector addition. The target variables in part (a) are the skier's total distance and direction from her starting point. Note that the target variables in part (b) are the *same* as those in part (a): the "magnitude of her resultant displacement" is just her final distance from the point of origin, and the "direction of her resultant displacement" is just the direction from the point of origin to where she stops.

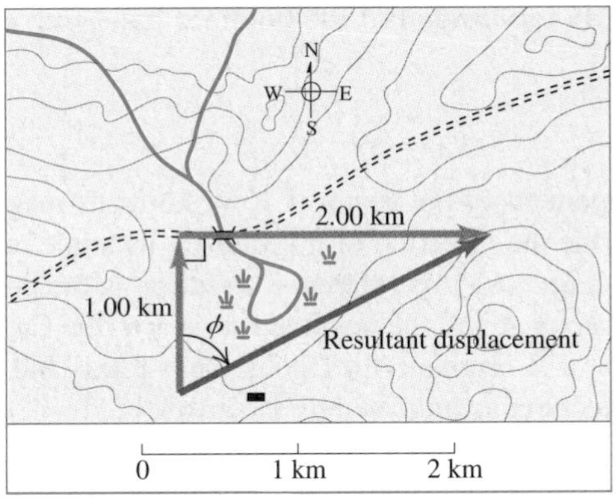

1.12 The vector diagram, drawn to scale, for a cross-country ski trip.

SET UP: Figure 1.12 is a scale diagram of the skier's displacements. We describe the direction from the starting point by the angle ϕ (the Greek letter "phi"). By careful measurement we find that the distance from the starting point is about 2.2 km and that ϕ is about 63°. But we can *calculate* a much more accurate result by adding the 1.00-km and 2.00-km displacement vectors.

EXECUTE: a) The vectors in the diagram form a right triangle; the distance from the starting point is equal to the length of the hypotenuse. We find this length by using the Pythagorean theorem:

$$\sqrt{(1.00\ \text{km})^2 + (2.00\ \text{km})^2} = 2.24\ \text{km}$$

The angle ϕ can be found with a little simple trigonometry. If you need a review, the trigonometric functions and identities are summarized in Appendix B, along with other useful mathematical and geometrical relations. By the definition of the tangent function,

$$\tan\phi = \frac{\text{opposite side}}{\text{adjacent side}} = \frac{2.00\ \text{km}}{1.00\ \text{km}}$$

$$\phi = 63.4°$$

b) The magnitude of the resultant displacement is just the distance that we found in part (a), 2.24 km. We can describe the direction as 63.4° east of north or 90° − 63.4° = 26.6° north of east. Take your choice!

EVALUATE: It's good practice to check the results of a vector-addition problem by making measurements on a drawing of the situation. Happily, the answers we found by calculation (2.24 km and ϕ = 63.4°) are very close to the cruder results we found by measurement (about 2.2 km and about 63°). If they were substantially different, we would have to go back and check for errors.

Test Your Understanding

If the skier had first gone 2.00 km east from her starting point, then 1.00 km north, what would have been the magnitude and direction of her resulting displacement?

1.8 | Components of Vectors

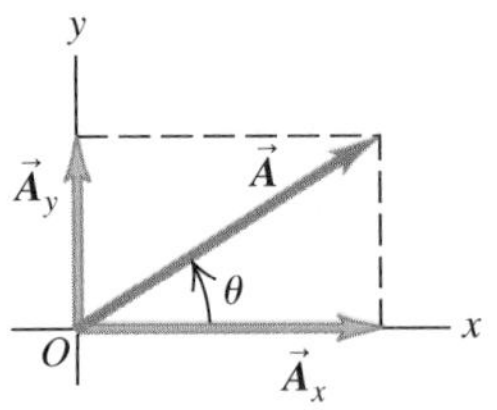

1.13 Vectors $\vec{A}_x$ and $\vec{A}_y$ are the rectangular component vectors of $\vec{A}$ in the directions of the x- and y-axes. For the vector $\vec{A}$ shown here, the components A_x and A_y are both positive.

In Section 1.7 we added vectors by using a scale diagram and by using properties of right triangles. Measuring a diagram offers only very limited accuracy, and calculations with right triangles work only when the two vectors are perpendicular. So we need a simple but general method for adding vectors. This is called the method of *components.*

To define what we mean by the components of a vector, we begin with a rectangular (Cartesian) coordinate system of axes (Fig. 1.13). We then draw the vector we're considering with its tail at O, the origin of the coordinate system. We can represent any vector lying in the xy-plane as the sum of a vector parallel to the x-axis and a vector parallel to the y-axis. These two vectors are labeled $\vec{A}_x$ and $\vec{A}_y$ in the figure; they are called the **component vectors** of vector $\vec{A}$, and their vector sum is equal to $\vec{A}$. In symbols,

$$\vec{A} = \vec{A}_x + \vec{A}_y \tag{1.6}$$

By definition, each component vector lies along a coordinate-axis direction. Thus we need only a single number to describe each one. When the component vector $\vec{A}_x$ points in the positive x-direction, we define the number A_x to be equal to the magnitude of $\vec{A}_x$. When the component vector $\vec{A}_x$ points in the negative x-direction, we define the number $\vec{A}_x$ to be equal to the negative of that magnitude, keeping in mind that the magnitude of a vector quantity is itself never negative. We define the number A_y in the same way. The two numbers A_x and A_y are called the **components** of $\vec{A}$.

CAUTION The components A_x and A_y of a vector $\vec{A}$ are just numbers; they are *not* vectors themselves. This is why we print the symbols for components in light italic type with no arrow on top instead of the boldface italic with an arrow reserved for vectors.

We can calculate the components of the vector $\vec{A}$ if we know its magnitude A and its direction. We'll describe the direction of a vector by its angle relative to some reference direction. In Fig. 1.13 this reference direction is the positive x-axis, and the angle between vector $\vec{A}$ and the positive x-axis is θ (the Greek letter "theta"). Imagine that the vector $\vec{A}$ originally lies along the $+x$-axis and that you then rotate it to its correct direction, as indicated by the arrow in Fig. 1.13 on the angle θ. If this rotation is from the $+x$-axis toward the $+y$-axis, as shown in Fig. 1.13, then θ is *positive;* if the rotation is from the $+x$-axis toward the $-y$-axis, θ is *negative.* Thus the $+y$-axis is at an angle of 90°, the $-x$-axis at 180°, and the $-y$-axis at 270° (or $-90°$). If θ is measured in this way, then from the definition of the trigonometric functions,

$$\frac{A_x}{A} = \cos\theta \quad \text{and} \quad \frac{A_y}{A} = \sin\theta$$

$$A_x = A\cos\theta \quad \text{and} \quad A_y = A\sin\theta \tag{1.7}$$

(θ measured from the $+x$-axis, rotating toward the $+y$-axis)

CAUTION Equations (1.7) are correct *only* when the angle θ is measured from the positive x-axis as described above. If the angle of the vector is given from a different reference direction or using a different sense of rotation, the relationships are different. Be careful! Example 1.6 illustrates this point.

In Fig. 1.13, A_x is positive because its direction is along the positive x-axis and A_y is positive because its direction is along the positive y-axis. This is consistent with Eqs. (1.7); θ is in the first quadrant (between 0° and 90°), and both the cosine and the sine of an angle in this quadrant are positive. But in Fig. 1.14a the component B_x is negative; its direction is opposite to that of the positive x-axis. Again, this agrees with Eqs. (1.7); the cosine of an angle in the second quadrant is negative. The component B_y is positive ($\sin\theta$ is positive in the second quadrant). In Fig. 1.14b, both C_x and C_y are negative (both $\cos\theta$ and $\sin\theta$ are negative in the third quadrant).

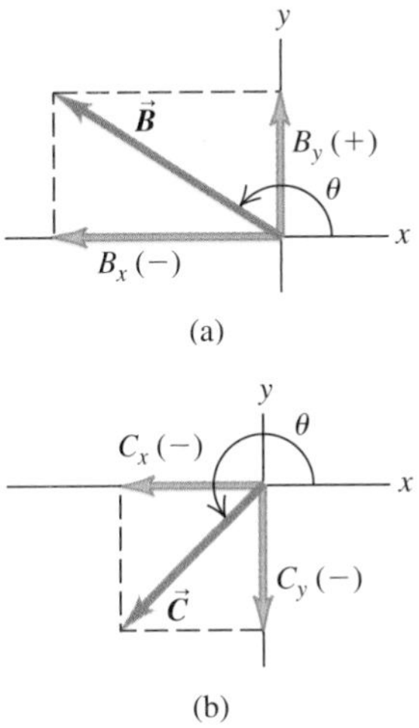

1.14 The components of a vector may be positive or negative numbers.

Example 1.6 Finding components

a) What are the x- and y-components of vector $\vec{D}$ in Fig. 1.15a? The magnitude of the vector is $D = 3.00$ m and the angle $\alpha = 45°$.
b) What are the x- and y-components of vector $\vec{E}$ in Fig. 1.15b? The magnitude of the vector is $E = 4.50$ m and the angle $\beta = 37.0°$.

SOLUTION

IDENTIFY and SET UP: The problem involves finding components, so it would seem that all we need is Eqs. (1.7). However, we need to be careful because the angles in Fig. 1.15 are *not* measured from the $+x$-axis toward the $+y$-axis.

EXECUTE: a) The angle between $\vec{D}$ and the positive x-axis is α (the Greek letter "alpha"), but this angle is measured toward the *negative* y-axis. So the angle we must use in Eqs. (1.7) is $\theta = -\alpha = -45°$. We find

$$D_x = D\cos\theta = (3.00\text{ m})(\cos(-45°)) = +2.1\text{ m}$$

$$D_y = D\sin\theta = (3.00\text{ m})(\sin(-45°)) = -2.1\text{ m}$$

The vector has a positive x-component and a negative y-component, as shown in the figure. Had you been careless and substituted $+45°$ for θ in Eqs. (1.7), you would have gotten the wrong sign for D_y.
b) The x-axis isn't horizontal in Fig. 1.15b, nor is the y-axis vertical. In general, *any* orientation of the x- and y-axes is permissible, just so the axes are mutually perpendicular. (Later we'll use axes like these to study the motion of an object sliding on a ramp or incline; one axis will lie along the ramp and the other will be perpendicular to the ramp.)

Here the angle β (the Greek letter "beta") is the angle between $\vec{E}$ and the positive y-axis, *not* the positive x-axis, so we *cannot* use this angle in Eqs. (1.7). Instead, note that $\vec{E}$ defines the hypotenuse of a right triangle; the other two sides of the triangle are the magnitudes of E_x and E_y, the x- and y-components of $\vec{E}$. The sine of β is the opposite side (the magnitude of E_x) divided by the hypotenuse (the magnitude E), and the cosine of β is the adjacent side (the magnitude of E_y) divided by

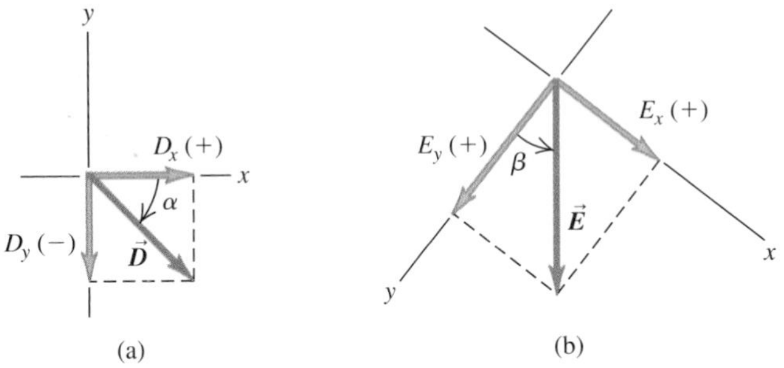

1.15 Calculating the x- and y-components of vectors.

the hypotenuse (again, the magnitude E). Both components of $\vec{E}$ are positive, so

$$E_x = E\sin\beta = (4.50\text{ m})(\sin 37.0°) = +2.71\text{ m}$$
$$E_y = E\cos\beta = (4.50\text{ m})(\cos 37.0°) = +3.59\text{ m}$$

Had you used Eqs. (1.7) directly and written $E_x = E\cos 37.0°$ and $E_y = E\sin 37.0°$, your answers for E_x and E_y would have been reversed!

If you insist on using Eqs. (1.7), you must first find the angle between $\vec{E}$ and the positive x-axis, measured toward the positive y-axis; this is $\theta = 90.0° - \beta = 90.0° - 37.0° = 53.0°$. Then $E_x = E\cos\theta$ and $E_y = E\sin\theta$. You can substitute the values of E and θ into Eqs. (1.7) to show that the results for E_x and E_y are the same as given above.

EVALUATE: Notice that the answers to part (b) have three significant figures, but the answers to part (a) have only two. Can you see why?

Using Components to Add Vectors

We can describe a vector completely by giving either its magnitude and direction or its x- and y-components. Equations (1.7) show how to find the components if we know the magnitude and direction. We can also reverse the process: we can find the magnitude and direction if we know the components. By applying the Pythagorean theorem to Fig. 1.13, we find that the magnitude of a vector $\vec{A}$ is

$$A = \sqrt{A_x^2 + A_y^2} \tag{1.8}$$

where we always take the positive root. Equation (1.8) is valid for any choice of x-axis and y-axis, as long as they are mutually perpendicular. The expression for the vector direction comes from the definition of the tangent of an angle. If θ is measured from the positive x-axis, and a positive angle is measured toward the positive y-axis (as in Fig. 1.13), then

$$\tan\theta = \frac{A_y}{A_x} \quad \text{and} \quad \theta = \arctan\frac{A_y}{A_x} \tag{1.9}$$

We will always use the notation arctan for the inverse tangent function. The notation $\tan^{-1}$ is also commonly used, and your calculator may have an INV button to be used with the TAN button. Microsoft Excel uses ATAN.

CAUTION There is one slight complication in using Eqs. (1.9) to find θ. Suppose $A_x = 2\text{ m}$ and $A_y = -2\text{ m}$; then $\tan\theta = -1$. But there are two angles having tangents of -1, namely, 135° and 315° (or −45°). In general, any two angles that differ by 180° have the same tangent. To decide which is correct, we have to look at the individual components. Because A_x is positive and A_y is negative, the angle must be in the fourth quadrant; thus $\theta = 315°$ (or $-45°$) is the correct value. Most pocket calculators give $\arctan(-1) = -45°$. In this case that is correct; but if instead we have $A_x = -2\text{ m}$ and $A_y = 2\text{ m}$, then the correct angle is 135°. Similarly, when A_x and A_y are both negative, the tangent is positive, but the angle is in the third quadrant. You should *always* draw a sketch to check which of the two possibilities is the correct one.

Here's how we use components to calculate the vector sum (resultant) of two or more vectors. Figure 1.16 shows two vectors $\vec{A}$ and $\vec{B}$ and their vector sum $\vec{R}$, along with the x- and y-components of all three vectors. You can see from the diagram that the x-component R_x of the vector sum is simply the sum $(A_x + B_x)$ of the x-components of the vectors being added. The same is true for the y-components. In symbols,

$$R_x = A_x + B_x, \quad R_y = A_y + B_y \quad (\text{components of } \vec{R} = \vec{A} + \vec{B}) \tag{1.10}$$

Figure 1.16 shows this result for the case in which the components A_x, A_y, B_x, and B_y are all positive. You should draw additional diagrams to verify for yourself that Eqs. (1.10) are valid for *any* signs of the components of $\vec{A}$ and $\vec{B}$.

If we know the components of any two vectors $\vec{A}$ and $\vec{B}$, perhaps by using Eqs. (1.7), we can compute the components of the vector sum $\vec{R}$. Then if we need the magnitude and direction of $\vec{R}$, we can obtain them from Eqs. (1.8) and (1.9) with the A's replaced by R's.

This procedure for finding the sum of two vectors can be extended to any number of vectors. Let $\vec{R}$ be the vector sum of $\vec{A}, \vec{B}, \vec{C}, \vec{D}, \vec{E}, \ldots$. Then the components of $\vec{R}$ are

$$R_x = A_x + B_x + C_x + D_x + E_x + \cdots$$
$$R_y = A_y + B_y + C_y + D_y + E_y + \cdots \qquad (1.11)$$

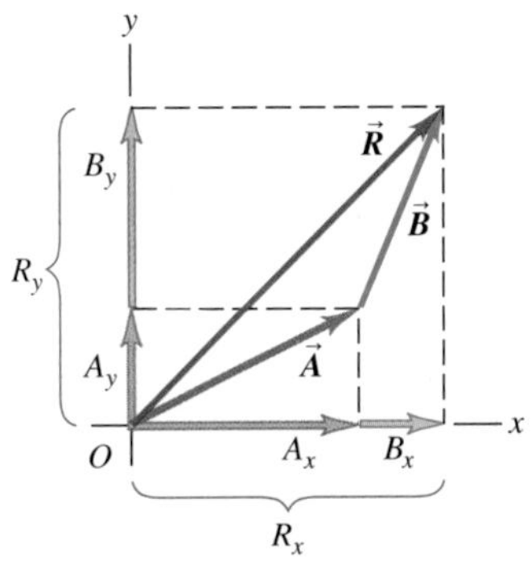

1.16 Vector $\vec{R}$ is the vector sum (resultant) of $\vec{A}$ and $\vec{B}$. Its x-component, R_x, equals the sum of the x-components of $\vec{A}$ and $\vec{B}$. The y-components are similarly related.

We have talked only about vectors that lie in the xy-plane, but the component method works just as well for vectors having any direction in space. We introduce a z-axis perpendicular to the xy-plane; then in general a vector $\vec{A}$ has components A_x, A_y, and A_z in the three coordinate directions. The magnitude A is given by

$$A = \sqrt{A_x^2 + A_y^2 + A_z^2} \qquad (1.12)$$

Again, we always take the positive root. Also, Eqs. (1.11) for the components of the vector sum $\vec{R}$ have an additional member:

$$R_z = A_z + B_z + C_z + D_z + E_z + \cdots$$

Finally, while our discussion of vector addition has centered on combining *displacement* vectors, the method is applicable to all other vector quantities as well. When we study the concept of force in Chapter 4, we'll find that forces are vectors that obey the same rules of vector addition that we've used with displacement. Other vector quantities will make their appearance in later chapters.

Problem-Solving Strategy

Vector Addition

IDENTIFY *the relevant concepts and* **SET UP** *the problem:* Decide what your target variable is. It may be the magnitude of the vector sum, the direction, or both. Then draw the individual vectors being summed and the coordinate axes being used. In your drawing, place the tail of the first vector at the origin of coordinates; place the tail of the second vector at the head of the first vector; and so on. Draw the vector sum $\vec{R}$ from the tail of the first vector to the head of the last vector. By examining your drawing, make a rough estimate of the magnitude and direction of $\vec{R}$; you'll use these estimates later to check your calculations.

EXECUTE *the solution as follows:*

1. Find the x- and y-components of each individual vector and record your results in a table. If a vector is described by its magnitude A and its angle θ, measured from the $+x$-axis toward the $+y$-axis, then the components are given by

$$A_x = A\cos\theta \qquad A_y = A\sin\theta$$

Some components may be positive and some may be negative, depending on how the vector is oriented (that is, what quadrant θ lies in). You can use this sign table as a check:

Quadrant	I	II	III	IV
A_x	+	−	−	+
A_y	+	+	−	−

If the angles of the vectors are given in some other way, perhaps using a different reference direction, convert them to angles measured from the $+x$-axis as described above. Be particularly careful with signs.

2. Add the individual x-components algebraically, including signs, to find R_x, the x-component of the vector sum. Do the same for the y-components to find R_y.
3. Then the magnitude R and direction θ of the vector sum are given by

$$R = \sqrt{R_x^2 + R_y^2}$$

$$\theta = \arctan \frac{R_y}{R_x}$$

EVALUATE *your answer:* Check your results for the magnitude and direction of the vector sum by comparing them with the rough estimates you made from your drawing. Remember that the magnitude R is *always* positive and that θ is measured from the positive x-axis. The value of θ that you find with a calculator may be the correct one, or it may be off by 180°. You can decide by examining your drawing.

If your calculations disagree totally with the estimates from your drawing, check whether your calculator is set in "radians" or "degrees" mode. If it's in "radians" mode, entering angles in degrees will give nonsensical answers. Watch out for this problem if you use Microsoft Excel, in which all trigonometric functions use units of radians rather than degrees. To convert from one to the other, remember that 360 degrees is the same as 2π radians.

Example 1.7 Adding vectors with components

The three finalists in a contest are brought to the center of a large, flat field. Each is given a meter stick, a compass, a calculator, a shovel, and (in a different order for each contestant) the following three displacements:

72.4 m, 32.0° east of north;
57.3 m, 36.0° south of west;
17.8 m straight south.

The three displacements lead to the point where the keys to a new Porsche are buried. Two contestants start measuring immediately, but the winner first *calculates* where to go. What does she calculate?

SOLUTION

IDENTIFY and SET UP: The situation is shown in Fig. 1.17. We have chosen the $+x$-axis as east and the $+y$-axis as north, the usual choice for maps. Let $\vec{A}$ be the first displacement, $\vec{B}$ the second, and $\vec{C}$ the third. We can estimate from the diagram that the vector sum $\vec{R}$ is about 10 m, 40° west of north.

EXECUTE: The angles of the vectors, measured from the $+x$-axis toward the $+y$-axis, are $(90.0° - 32.0°) = 58.0°$, $(180.0° + 36.0°) = 216.0°$, and 270.0°. We have to find the components of each. Because of our choice of axes, we may use Eqs. (1.7), and so the components of $\vec{A}$ are

$$A_x = A\cos\theta_A = (72.4\text{ m})(\cos 58.0°) = 38.37\text{ m}$$
$$A_y = A\sin\theta_A = (72.4\text{ m})(\sin 58.0°) = 61.40\text{ m}$$

Note that we have kept one too many significant figures in the components; we will wait until the end to round to the correct number of significant figures. The table below shows the components of all the displacements, the addition of components, and the other calculations. Always arrange your component calculations systematically like this.

Distance	Angle	x-component	y-component
$A = 72.4$ m	58.0°	38.37 m	61.40 m
$B = 57.3$ m	216.0°	−46.36 m	−33.68 m
$C = 17.8$ m	270.0°	0.00 m	−17.80 m
		$R_x = -7.99$ m	$R_y = 9.92$ m

$$R = \sqrt{(-7.99\text{ m})^2 + (9.92\text{ m})^2} = 12.7\text{ m}$$

$$\theta = \arctan \frac{9.92\text{ m}}{-7.99\text{ m}} = 129° = 39° \text{ west of north}$$

The losers try to measure three angles and three distances totaling 147.5 m, one meter at a time. The winner measured only one angle and one much shorter distance.

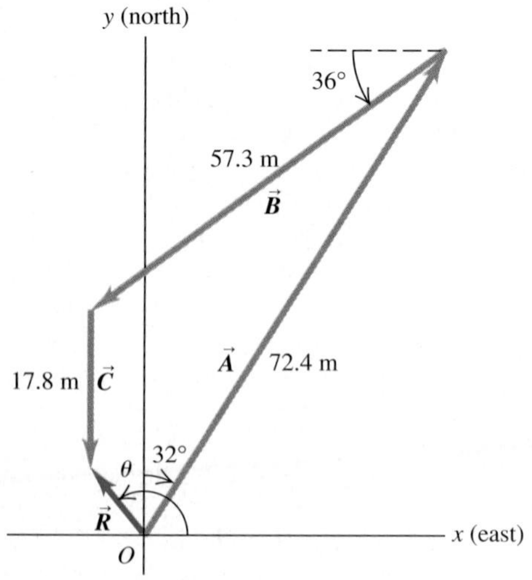

1.17 Three successive displacements $\vec{A}$, $\vec{B}$, and $\vec{C}$ and the resultant (vector sum) displacement $\vec{R} = \vec{A} + \vec{B} + \vec{C}$.

EVALUATE: Our calculated answers for R and θ are not too different from our estimates of 10 m and 40° west of north; that's good! Notice that $\theta = -51°$, or 51° south of east, also satisfies the equation for θ. But since the winner has made a drawing of the displacement vectors (Fig. 1.17), she knows that $\theta = 129°$ is the only correct solution for the angle.

Example 1.8 A vector in three dimensions

After an airplane takes off, it travels 10.4 km west, 8.7 km north, and 2.1 km up. How far is it from the takeoff point?

SOLUTION

Let the $+x$-axis be east, the $+y$-axis north, and the $+z$-axis up. Then $A_x = -10.4$ km, $A_y = 8.7$ km, and $A_z = 2.1$ km; Eq. (1.12) gives

$$A = \sqrt{(-10.4\text{ km})^2 + (8.7\text{ km})^2 + (2.1\text{ km})^2} = 13.7\text{ km}$$

Test Your Understanding

How far are you from your starting point if you first traveled 4.00 km due west, then 4.00 km due south? What is the direction from where you started to where you ended up?

1.9 | Unit Vectors

A **unit vector** is a vector that has a magnitude of 1, with no units. Its only purpose is to *point;* that is, to describe a direction in space. Unit vectors provide a convenient notation for many expressions involving components of vectors. We will always include a caret or "hat" (^) in the symbol for a unit vector to distinguish it from ordinary vectors whose magnitude may or may not be equal to 1.

In an x-y coordinate system we can define a unit vector $\hat{\imath}$ that points in the direction of the positive x-axis and a unit vector $\hat{\jmath}$ that points in the direction of the positive y-axis. Then we can express the relationship between component vectors and components, described at the beginning of Section 1.8, as follows:

$$\vec{A}_x = A_x\hat{\imath}$$
$$\vec{A}_y = A_y\hat{\jmath} \qquad (1.13)$$

Similarly, we can write a vector $\vec{A}$ in terms of its components as

$$\vec{A} = A_x\hat{\imath} + A_y\hat{\jmath} \qquad (1.14)$$

Equations (1.13) and (1.14) are vector equations; each term, such as $A_x\hat{\imath}$, is a vector quantity (Fig. 1.18). The boldface equals and plus signs denote vector equality and addition.

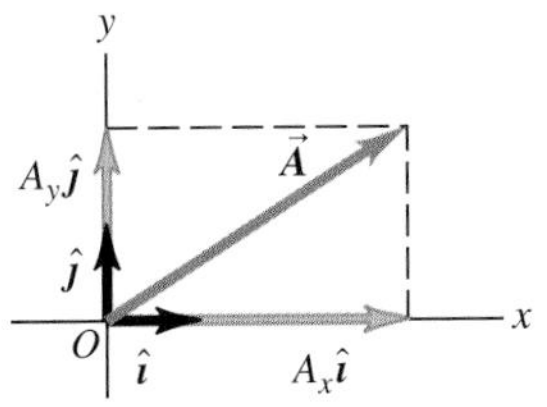

1.18 Using unit vectors, we can express a vector $\vec{A}$ in terms of its components A_x and A_y as $\vec{A} = A_x\hat{\imath} + A_y\hat{\jmath}$.

When two vectors $\vec{A}$ and $\vec{B}$ are represented in terms of their components, we can express the vector sum $\vec{R}$ using unit vectors as follows:

$$\vec{A} = A_x\hat{\imath} + A_y\hat{\jmath}$$
$$\vec{B} = B_x\hat{\imath} + B_y\hat{\jmath}$$

$$\begin{aligned}\vec{R} &= \vec{A} + \vec{B} \\ &= (A_x\hat{\imath} + A_y\hat{\jmath}) + (B_x\hat{\imath} + B_y\hat{\jmath}) \\ &= (A_x + B_x)\hat{\imath} + (A_y + B_y)\hat{\jmath} \\ &= R_x\hat{\imath} + R_y\hat{\jmath} \qquad (1.15)\end{aligned}$$

Equation (1.15) restates the content of Eqs. (1.10) in the form of a single vector equation rather than two component equations.

If the vectors do not all lie in the xy-plane, then we need a third component. We introduce a third unit vector $\hat{k}$ that points in the direction of the positive z-axis. The generalized forms of Eqs. (1.14) and (1.15) are

$$\begin{aligned}\vec{A} &= A_x\hat{\imath} + A_y\hat{\jmath} + A_z\hat{k} \\ \vec{B} &= B_x\hat{\imath} + B_y\hat{\jmath} + B_z\hat{k} \qquad (1.16)\end{aligned}$$

$$\begin{aligned}\vec{R} &= (A_x + B_x)\hat{\imath} + (A_y + B_y)\hat{\jmath} + (A_z + B_z)\hat{k} \\ &= R_x\hat{\imath} + R_y\hat{\jmath} + R_z\hat{k} \qquad (1.17)\end{aligned}$$

Example 1.9 Using unit vectors

Given the two displacements

$$\vec{D} = (6\hat{\imath} + 3\hat{\jmath} - \hat{k})\text{m} \quad \text{and} \quad \vec{E} = (4\hat{\imath} - 5\hat{\jmath} + 8\hat{k})\text{m}$$

find the magnitude of the displacement $2\vec{D} - \vec{E}$.

SOLUTION

IDENTIFY, SET UP, and EXECUTE: Letting $\vec{F} = 2\vec{D} - \vec{E}$, we have

$$\begin{aligned}\vec{F} &= 2(6\hat{\imath} + 3\hat{\jmath} - \hat{k})\text{m} - (4\hat{\imath} - 5\hat{\jmath} + 8\hat{k})\text{m} \\ &= [(12 - 4)\hat{\imath} + (6 + 5)\hat{\jmath} + (-2 - 8)\hat{k}]\text{m} \\ &= (8\hat{\imath} + 11\hat{\jmath} - 10\hat{k})\text{m}\end{aligned}$$

The units of the vectors $\vec{D}$, $\vec{E}$, and $\vec{F}$ are meters, so the components of these vectors are also in meters. From Eq. (1.12),

$$\begin{aligned}F &= \sqrt{F_x^2 + F_y^2 + F_z^2} \\ &= \sqrt{(8\text{ m})^2 + (11\text{ m})^2 + (-10\text{ m})^2} = 17\text{ m}\end{aligned}$$

EVALUATE: Working with unit vectors makes vector addition and subtraction no more complicated than adding and subtracting ordinary numbers. Still, be sure to check for simple arithmetic errors.

Test Your Understanding

Express each of the vectors $\vec{A}$, $\vec{B}$, and $\vec{C}$ in Example 1.7 (Section 1.8) in terms of unit vectors.

1.10 | Products of Vectors

We have seen how addition of vectors develops naturally from the problem of combining displacements, and we will use vector addition for calculating many other vector quantities later. We can also express many physical relationships concisely by using *products* of vectors. Vectors are not ordinary numbers, so ordinary multiplication is not directly applicable to vectors. We will define two different kinds of products of vectors. The first, called the scalar product, yields a result that is a scalar quantity. The second, the vector product, yields another vector.

Scalar Product

The **scalar product** of two vectors $\vec{A}$ and $\vec{B}$ is denoted by $\vec{A}\cdot\vec{B}$. Because of this notation, the scalar product is also called the **dot product.**

To define the scalar product $\vec{A}\cdot\vec{B}$ of two vectors $\vec{A}$ and $\vec{B}$, we draw the two vectors with their tails at the same point (Fig. 1.19a). The angle between their directions is ϕ as shown; the angle ϕ always lies between 0° and 180°. (As usual, we use Greek letters for angles.) Figure 1.19b shows the projection of the vector $\vec{B}$ onto the direction of $\vec{A}$; this projection is the component of $\vec{B}$ parallel to $\vec{A}$ and is equal to $B\cos\phi$. (We can take components along any direction that's convenient, not just the x- and y-axes.) We define $\vec{A}\cdot\vec{B}$ to be the magnitude of $\vec{A}$ multiplied by the component of $\vec{B}$ parallel to $\vec{A}$. Expressed as an equation,

$$\vec{A}\cdot\vec{B} = AB\cos\phi = |\vec{A}||\vec{B}|\cos\phi \tag{1.18}$$

(definition of the scalar (dot) product)

where ϕ ranges from 0° to 180°.

Alternatively, we can define $\vec{A}\cdot\vec{B}$ to be the magnitude of $\vec{B}$ multiplied by the component of $\vec{A}$ parallel to $\vec{B}$, as in Fig. 1.19c. Hence $\vec{A}\cdot\vec{B} = B(A\cos\phi) = AB\cos\phi$, which is the same as Eq. (1.18).

The scalar product is a scalar quantity, not a vector, and it may be positive, negative, or zero. When ϕ is between 0° and 90°, the scalar product is positive. When ϕ is between 90° and 180°, it is negative. You should draw a diagram like Fig. 1.19, but with ϕ between 90° and 180°, to confirm for yourself that the component of $\vec{B}$ parallel to $\vec{A}$ is negative in this case, as is the component of $\vec{A}$ parallel to $\vec{B}$. Finally, when $\phi = 90°$, $\vec{A}\cdot\vec{B} = 0$. *The scalar product of two perpendicular vectors is always zero.*

For any two vectors $\vec{A}$ and $\vec{B}$, $AB\cos\phi = BA\cos\phi$. This means that $\vec{A}\cdot\vec{B} = \vec{B}\cdot\vec{A}$. The scalar product obeys the commutative law of multiplication; the order of the two vectors does not matter.

We will use the scalar product in Chapter 6 to describe work done by a force. When a constant force $\vec{F}$ is applied to a body that undergoes a displacement $\vec{s}$, the work W (a scalar quantity) done by the force is given by

$$W = \vec{F}\cdot\vec{s}$$

The work done by the force is positive if the angle between $\vec{F}$ and $\vec{s}$ is between 0° and 90°, negative if this angle is between 90° and 180°, and zero if $\vec{F}$ and $\vec{s}$ are perpendicular. (This is another example of a term that has a special meaning in physics; in everyday language, "work" isn't something that can be positive or negative.) In later chapters we'll use the scalar product for a variety of purposes, from calculating electric potential to determining the effects that varying magnetic fields have on electric circuits.

We can calculate the scalar product $\vec{A}\cdot\vec{B}$ directly if we know the x-, y-, and z-components of $\vec{A}$ and $\vec{B}$. To see how this is done, let's first work out the scalar products of the unit vectors. This is easy, since $\hat{\imath}$, $\hat{\jmath}$, and $\hat{k}$ are all perpendicular to each other. Using Eq. (1.18), we find

$$\hat{\imath}\cdot\hat{\imath} = \hat{\jmath}\cdot\hat{\jmath} = \hat{k}\cdot\hat{k} = (1)(1)\cos 0 = 1$$
$$\hat{\imath}\cdot\hat{\jmath} = \hat{\imath}\cdot\hat{k} = \hat{\jmath}\cdot\hat{k} = (1)(1)\cos 90° = 0 \tag{1.19}$$

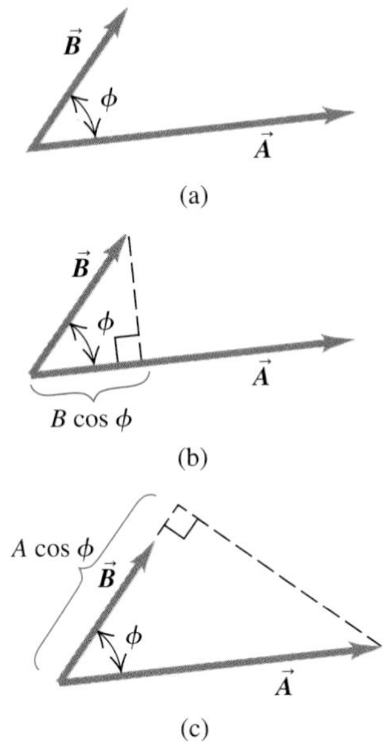

1.19 (a) Two vectors drawn from a common starting point to define their scalar product $\vec{A}\cdot\vec{B} = AB\cos\phi$. (b) $B\cos\phi$ is the component of $\vec{B}$ in the direction of $\vec{A}$, and $\vec{A}\cdot\vec{B}$ is the product of the magnitude of $\vec{A}$ and this component. (c) $\vec{A}\cdot\vec{B}$ is also the product of the magnitude of $\vec{B}$ and the component of $\vec{A}$ in the direction of $\vec{B}$.

Now we express $\vec{A}$ and $\vec{B}$ in terms of their components, expand the product, and use these products of unit vectors:

$$\begin{aligned}\vec{A}\cdot\vec{B} &= (A_x\hat{\imath} + A_y\hat{\jmath} + A_z\hat{k})\cdot(B_x\hat{\imath} + B_y\hat{\jmath} + B_z\hat{k}) \\ &= A_x\hat{\imath}\cdot B_x\hat{\imath} + A_x\hat{\imath}\cdot B_y\hat{\jmath} + A_x\hat{\imath}\cdot B_z\hat{k} \\ &\quad + A_y\hat{\jmath}\cdot B_x\hat{\imath} + A_y\hat{\jmath}\cdot B_y\hat{\jmath} + A_y\hat{\jmath}\cdot B_z\hat{k} \\ &\quad + A_z\hat{k}\cdot B_x\hat{\imath} + A_z\hat{k}\cdot B_y\hat{\jmath} + A_z\hat{k}\cdot B_z\hat{k} \\ &= A_xB_x\hat{\imath}\cdot\hat{\imath} + A_xB_y\hat{\imath}\cdot\hat{\jmath} + A_xB_z\hat{\imath}\cdot\hat{k} \\ &\quad + A_yB_x\hat{\jmath}\cdot\hat{\imath} + A_yB_y\hat{\jmath}\cdot\hat{\jmath} + A_yB_z\hat{\jmath}\cdot\hat{k} \\ &\quad + A_zB_x\hat{k}\cdot\hat{\imath} + A_zB_y\hat{k}\cdot\hat{\jmath} + A_zB_z\hat{k}\cdot\hat{k} \qquad (1.20)\end{aligned}$$

From Eqs. (1.19) we see that six of these nine terms are zero, and the three that survive give simply

$$\vec{A}\cdot\vec{B} = A_xB_x + A_yB_y + A_zB_z \qquad (1.21)$$

(scalar (dot) product in terms of components)

Thus the scalar product of two vectors is the sum of the products of their respective components.

The scalar product gives a straightforward way to find the angle ϕ between any two vectors $\vec{A}$ and $\vec{B}$ whose components are known. In this case, Eq. (1.21) can be used to find the scalar product of $\vec{A}$ and $\vec{B}$. From Eq. (1.18) the scalar product is also equal to $AB\cos\phi$. The vector magnitudes A and B can be found from the vector components with Eq. (1.12), so $\cos\phi$ and hence the angle ϕ can be determined (see Example 1.11).

Example 1.10 Calculating a scalar product

Find the scalar product $\vec{A}\cdot\vec{B}$ of the two vectors in Fig. 1.20. The magnitudes of the vectors are $A = 4.00$ and $B = 5.00$.

SOLUTION

IDENTIFY and SET UP: There are two ways to calculate the scalar product. The first way uses the magnitudes of the vectors and the angle between them (Eq. 1.18), and the second uses the components of the two vectors (Eq. 1.21).

EXECUTE: Using the first approach, the angle between the two vectors is $\phi = 130.0° - 53.0° = 77.0°$, so

$$\vec{A}\cdot\vec{B} = AB\cos\phi = (4.00)(5.00)\cos 77.0° = 4.50$$

This is positive because the angle between $\vec{A}$ and $\vec{B}$ is between 0° and 90°.

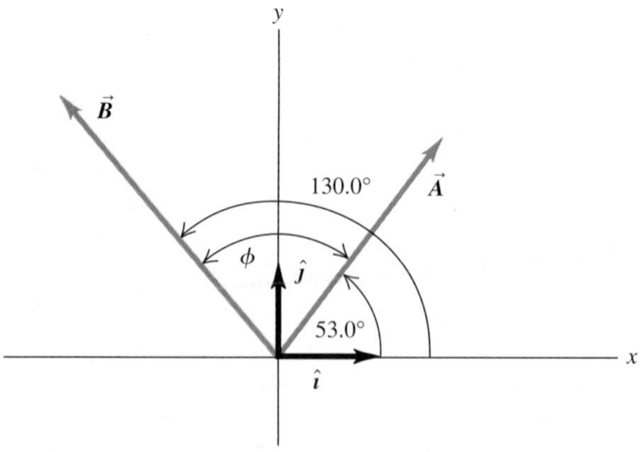

1.20 Two vectors in two dimensions.

To use the second approach, we first need to find the components of the two vectors. Since the angles of $\vec{A}$ and $\vec{B}$ are given with respect to the $+x$-axis, and these angles are measured in the sense from the $+x$-axis to the $+y$-axis, we can use Eqs. (1.7):

$$A_x = (4.00)\cos 53.0° = 2.407$$
$$A_y = (4.00)\sin 53.0° = 3.195$$
$$A_z = 0$$
$$B_x = (5.00)\cos 130.0° = -3.214$$
$$B_y = (5.00)\sin 130.0° = 3.830$$
$$B_z = 0$$

The z-components are zero because both vectors lie in the xy-plane. As in Example 1.7, we are keeping one too many significant figures in the components; we'll round to the correct number at the end. From Eq. (1.21) the scalar product is

$$\begin{aligned}\vec{A}\cdot\vec{B} &= A_xB_x + A_yB_y + A_zB_z\\ &= (2.407)(-3.214) + (3.195)(3.830)\\ &\quad + (0)(0) = 4.50\end{aligned}$$

EVALUATE: We get the same result for the scalar product with both methods, as we should.

Example 1.11 Finding angles with the scalar product

Find the angle between the two vectors

$$\vec{A} = 2\hat{\imath} + 3\hat{\jmath} + \hat{k} \quad \text{and} \quad \vec{B} = -4\hat{\imath} + 2\hat{\jmath} - \hat{k}$$

SOLUTION

IDENTIFY: The scalar product of two vectors $\vec{A}$ and $\vec{B}$ is related to the angle ϕ between them and to the magnitudes A and B. The scalar product is also related to the components of the two vectors. If we are given the components of the vectors (as we are in this example), we first determine the scalar product $\vec{A}\cdot\vec{B}$ and the values of A and B, and then determine the target variable ϕ.

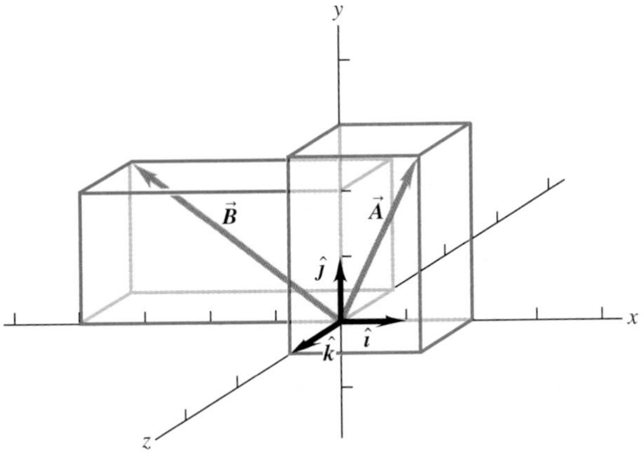

1.21 Two vectors in three dimensions.

SET UP and EXECUTE: The vectors are shown in Fig. 1.21. The scalar product of two vectors is given by either Eq. (1.18) or Eq. (1.21). Equating these two and rearranging, we obtain

$$\cos\phi = \frac{A_xB_x + A_yB_y + A_zB_z}{AB}$$

This formula can be used to find the angle between any two vectors $\vec{A}$ and $\vec{B}$. For our example the components of $\vec{A}$ are $A_x = 2$, $A_y = 3$, and $A_z = 1$, and the components of $\vec{B}$ are $B_x = -4$, $B_y = 2$, and $B_z = -1$. Thus

$$\begin{aligned}\vec{A}\cdot\vec{B} &= A_xB_x + A_yB_y + A_zB_z\\ &= (2)(-4) + (3)(2) + (1)(-1) = -3\\ A &= \sqrt{A_x^2 + A_y^2 + A_z^2} = \sqrt{2^2 + 3^2 + 1^2} = \sqrt{14}\\ B &= \sqrt{B_x^2 + B_y^2 + B_z^2} = \sqrt{(-4^2) + 2^2 + (-1)^2} = \sqrt{21}\\ \cos\phi &= \frac{A_xB_x + A_yB_y + A_zB_z}{AB} = \frac{-3}{\sqrt{14}\sqrt{21}} = -0.175\\ \phi &= 100°\end{aligned}$$

EVALUATE: As a check on this result, note that the scalar product $\vec{A}\cdot\vec{B}$ is negative. This means that ϕ is between 90° and 180°, in agreement with our answer.

Vector Product

The **vector product** of two vectors $\vec{A}$ and $\vec{B}$, also called the **cross product,** is denoted by $\vec{A}\times\vec{B}$. We will use this product in Chapter 10 to describe torque

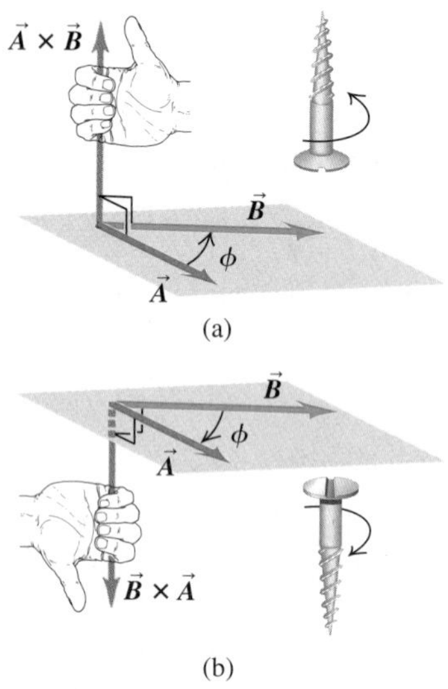

1.22 (a) Vectors $\vec{A}$ and $\vec{B}$ lie in a plane; the vector product $\vec{A} \times \vec{B}$ is perpendicular to this plane in a direction determined by the right-hand rule. (b) $\vec{B} \times \vec{A} = -\vec{A} \times \vec{B}$; the vector product is anticommutative.

and angular momentum. Later we will also use it extensively for magnetic fields, where it will help us describe the relationships among the directions of several vector quantities.

To define the vector product $\vec{A} \times \vec{B}$ of two vectors $\vec{A}$ and $\vec{B}$, we again draw the two vectors with their tails at the same point (Fig. 1.22a). The two vectors then lie in a plane. We define the vector product to be a vector quantity with a direction perpendicular to this plane (that is, perpendicular to both $\vec{A}$ and $\vec{B}$) and a magnitude equal to $AB \sin \phi$. That is, if $\vec{C} = \vec{A} \times \vec{B}$, then

$$C = AB \sin \phi \qquad (1.22)$$

(magnitude of the vector (cross) product of $\vec{A}$ and $\vec{B}$)

We measure the angle ϕ from $\vec{A}$ toward $\vec{B}$ and take it to be the smaller of the two possible angles, so ϕ ranges from 0° to 180°. Thus C in Eq. (1.22) is always positive, as a vector magnitude should be. Note also that when $\vec{A}$ and $\vec{B}$ are parallel or antiparallel, $\phi = 0$ or 180° and $C = 0$. That is, *the vector product of two parallel or antiparallel vectors is always zero.* In particular, *the vector product of any vector with itself is zero.* To see the contrast between the scalar product and the magnitude of the vector product, imagine that we vary the angle between $\vec{A}$ and $\vec{B}$ while keeping their magnitudes constant. When $\vec{A}$ and $\vec{B}$ are parallel, the scalar product will be maximum and the magnitude of the vector product will be zero. When $\vec{A}$ and $\vec{B}$ are perpendicular, the scalar product will be zero and the magnitude of the vector product will be maximum.

There are always *two* directions perpendicular to a given plane, one on each side of the plane. We choose which of these is the direction of $\vec{A} \times \vec{B}$ as follows. Imagine rotating vector $\vec{A}$ about the perpendicular line until it is aligned with $\vec{B}$, choosing the smaller of the two possible angles between $\vec{A}$ and $\vec{B}$. Curl the fingers of your right hand around the perpendicular line so that the fingertips point in the direction of rotation; your thumb will then point in the direction of $\vec{A} \times \vec{B}$. This **right-hand rule** is shown in Fig. 1.22a. The direction of the vector product is also the direction in which a right-hand screw advances if turned from $\vec{A}$ toward $\vec{B}$, as shown.

Similarly, we determine the direction of $\vec{B} \times \vec{A}$ by rotating $\vec{B}$ into $\vec{A}$ in Fig. 1.22b. The result is a vector that is *opposite* to the vector $\vec{A} \times \vec{B}$. The vector product is not commutative! In fact, for any two vectors $\vec{A}$ and $\vec{B}$,

$$\vec{A} \times \vec{B} = -\vec{B} \times \vec{A} \qquad (1.23)$$

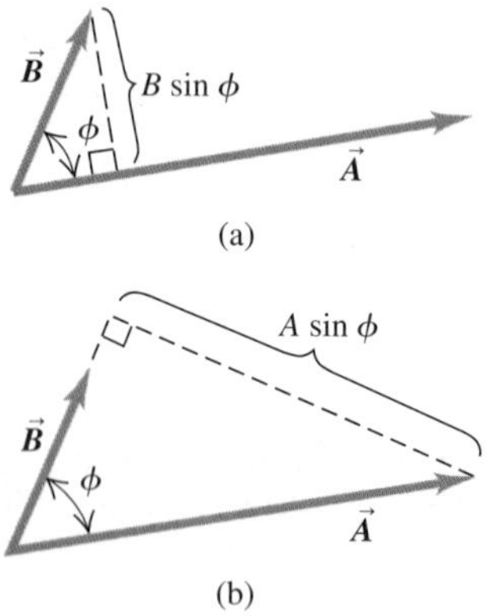

1.23 (a) $B \sin \phi$ is the component of $\vec{B}$ perpendicular to the direction of $\vec{A}$, and the magnitude of $\vec{A} \times \vec{B}$ is the product of the magnitude of $\vec{A}$ and this component. (b) The magnitude of $\vec{A} \times \vec{B}$ is also the product of the magnitude of $\vec{B}$ and the component of $\vec{A}$ perpendicular to $\vec{B}$.

Just as we did for the scalar product, we can give a geometrical interpretation of the magnitude of the vector product. In Fig. 1.23a, $B \sin \phi$ is the component of vector $\vec{B}$ that is *perpendicular* to the direction of vector $\vec{A}$. From Eq. (1.22) the magnitude of $\vec{A} \times \vec{B}$ equals the magnitude of $\vec{A}$ multiplied by the component of $\vec{B}$ perpendicular to $\vec{A}$. Figure 1.23b shows that the magnitude of $\vec{A} \times \vec{B}$ also equals the magnitude of $\vec{B}$ multiplied by the component of $\vec{A}$ perpendicular to $\vec{B}$. Note that Fig. 1.23 shows the case in which ϕ is between 0° and 90°; you should draw a similar diagram for ϕ between 90° and 180° to show that the same geometrical interpretation of the magnitude of $\vec{A} \times \vec{B}$ still applies.

If we know the components of $\vec{A}$ and $\vec{B}$, we can calculate the components of the vector product, using a procedure similar to that for the scalar product. First we work out the multiplication table for the unit vectors $\hat{\imath}$, $\hat{\jmath}$, and $\hat{k}$. The vector product of any vector with itself is zero, so

$$\hat{\imath} \times \hat{\imath} = \hat{\jmath} \times \hat{\jmath} = \hat{k} \times \hat{k} = \mathbf{0}$$

The boldface zero is a reminder that each product is a zero *vector*—that is, one with all components equal to zero and an undefined direction. Using Eqs. (1.22) and (1.23) and the right-hand rule, we find

$$\begin{aligned} \hat{\imath} \times \hat{\jmath} &= -\hat{\jmath} \times \hat{\imath} = \hat{k} \\ \hat{\jmath} \times \hat{k} &= -\hat{k} \times \hat{\jmath} = \hat{\imath} \\ \hat{k} \times \hat{\imath} &= -\hat{\imath} \times \hat{k} = \hat{\jmath} \end{aligned} \tag{1.24}$$

Next we express $\vec{A}$ and $\vec{B}$ in terms of their components and the corresponding unit vectors, and we expand the expression for the vector product:

$$\begin{aligned} \vec{A} \times \vec{B} &= (A_x\hat{\imath} + A_y\hat{\jmath} + A_z\hat{k}) \times (B_x\hat{\imath} + B_y\hat{\jmath} + B_z\hat{k}) \\ &= A_x\hat{\imath} \times B_x\hat{\imath} + A_x\hat{\imath} \times B_y\hat{\jmath} + A_x\hat{\imath} \times B_z\hat{k} \\ &\quad + A_y\hat{\jmath} \times B_x\hat{\imath} + A_y\hat{\jmath} \times B_y\hat{\jmath} + A_y\hat{\jmath} \times B_z\hat{k} \\ &\quad + A_z\hat{k} \times B_x\hat{\imath} + A_z\hat{k} \times B_y\hat{\jmath} + A_z\hat{k} \times B_z\hat{k} \end{aligned} \tag{1.25}$$

We can also rewrite the individual terms as $A_x\hat{\imath} \times B_y\hat{\jmath} = (A_xB_y)\hat{\imath} \times \hat{\jmath}$, and so on. Evaluating these by using the multiplication table for the unit vectors and then grouping the terms, we find

$$\vec{A} \times \vec{B} = (A_yB_z - A_zB_y)\hat{\imath} + (A_zB_x - A_xB_z)\hat{\jmath} + (A_xB_y - A_yB_x)\hat{k} \tag{1.26}$$

Thus the components of $\vec{C} = \vec{A} \times \vec{B}$ are given by

$$C_x = A_yB_z - A_zB_y \qquad C_y = A_zB_x - A_xB_z \qquad C_z = A_xB_y - A_yB_x$$
$$\text{(components of } \vec{C} = \vec{A} \times \vec{B}\text{)} \tag{1.27}$$

The vector product can also be expressed in determinant form as

$$\vec{A} \times \vec{B} = \begin{vmatrix} \hat{\imath} & \hat{\jmath} & \hat{k} \\ A_x & A_y & A_z \\ B_x & B_y & B_z \end{vmatrix}$$

If you aren't familiar with determinants, don't worry about this form.

With the axis system of Fig. 1.24a, if we reverse the direction of the z-axis, we get the system shown in Fig. 1.24b. Then, as you may verify, the definition of the vector product gives $\hat{\imath} \times \hat{\jmath} = -\hat{k}$ instead of $\hat{\imath} \times \hat{\jmath} = \hat{k}$. In fact, all vector products of the unit vectors $\hat{\imath}$, $\hat{\jmath}$, and $\hat{k}$ would have signs opposite to those in Eqs. (1.24). We see that there are two kinds of coordinate systems, differing in the signs of the vector products of unit vectors. An axis system in which $\hat{\imath} \times \hat{\jmath} = \hat{k}$, as in Fig. 1.24a, is called a **right-handed system.** The usual practice is to use *only* right-handed systems, and we will follow that practice throughout this book.

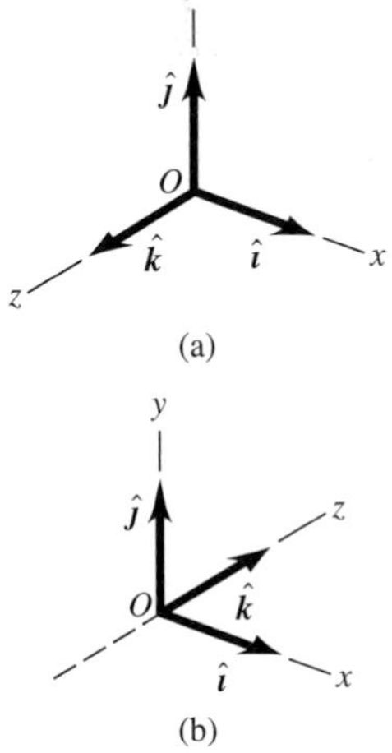

1.24 (a) A right-handed coordinate system, in which $\hat{\imath} \times \hat{\jmath} = \hat{k}$, $\hat{\jmath} \times \hat{k} = \hat{\imath}$, and $\hat{k} \times \hat{\imath} = \hat{\jmath}$. (b) A left-handed coordinate system, in which $\hat{\imath} \times \hat{\jmath} = -\hat{k}$, and so on. We'll use only right-handed systems.

Example 1.12 Calculating a vector product

Vector $\vec{A}$ has magnitude 6 units and is in the direction of the $+x$-axis. Vector $\vec{B}$ has magnitude 4 units and lies in the xy-plane, making an angle of 30° with the $+x$-axis (Fig. 1.25). Find the vector product $\vec{A} \times \vec{B}$.

SOLUTION

IDENTIFY and SET UP: We can find the vector product in one of two ways. The first way is to use Eq. (1.22) to determine the magnitude of $\vec{A} \times \vec{B}$ and then use the right-hand rule to find the direction of the vector product. The second way is to use the components of $\vec{A}$ and $\vec{B}$ to find the components of the vector product $\vec{C} = \vec{A} \times \vec{B}$ using Eqs. (1.27).

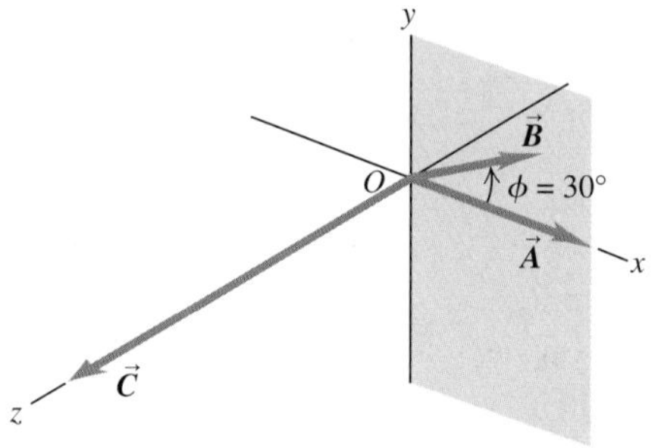

1.25 Vectors $\vec{A}$ and $\vec{B}$ and their vector product $\vec{C} = \vec{A} \times \vec{B}$. The vector $\vec{B}$ lies in the xy-plane.

EXECUTE: With the first approach, from Eq. (1.22) the magnitude of the vector product is

$$AB \sin \phi = (6)(4)(\sin 30°) = 12$$

From the right-hand rule, the direction of $\vec{A} \times \vec{B}$ is along the $+z$-axis, so we have $\vec{A} \times \vec{B} = 12\hat{k}$.

To use the second approach, we first write the components of $\vec{A}$ and $\vec{B}$:

$$A_x = 6 \qquad A_y = 0 \qquad A_z = 0$$
$$B_x = 4\cos 30° = 2\sqrt{3} \qquad B_y = 4\sin 30° = 2 \qquad B_z = 0$$

Defining $\vec{C} = \vec{A} \times \vec{B}$, we have from Eqs. (1.27) that

$$C_x = (0)(0) - (0)(2) = 0$$
$$C_y = (0)(2\sqrt{3}) - (6)(0) = 0$$
$$C_z = (6)(2) - (0)(2\sqrt{3}) = 12$$

The vector product $\vec{C}$ has only a z-component, and it lies along the $+z$-axis. The magnitude agrees with the result we obtained with the first approach, as it should.

EVALUATE: For this example the first approach was more direct because we knew the magnitudes of each vector and the angle between them, and furthermore, both vectors lay in one of the planes of the coordinate system. But often you will need to find the vector product of two vectors that are not so conveniently oriented or for which only the components are given. In such a case the second approach, using components, is the more direct one.

Test Your Understanding

For the two vectors $\vec{A} = 3\hat{\imath} + 2\hat{\jmath}$ and $\vec{B} = 4\hat{\imath} + 5\hat{k}$, find the scalar product $\vec{A} \cdot \vec{B}$ and the vector product $\vec{A} \times \vec{B}$.

SUMMARY

The fundamental physical quantities of mechanics are mass, length, and time. The corresponding basic SI units are the kilogram, the meter, and the second. Other units for these quantities, related by powers of 10, are identified by adding prefixes to the basic unit. Derived units for other physical quantities are products or quotients of the basic units. Equations must be dimensionally consistent; two terms can be added only when they have the same units. (See Examples 1.1 and 1.2)

The accuracy of a measurement can be indicated by the number of significant figures or by a stated uncertainty. The result of a calculation usually has no more significant figures than the input data. When only crude estimates are available for input data, we can often make useful order-of-magnitude estimates. (See Examples 1.3 and 1.4)

Scalar quantities are numbers, and combine with the usual rules of arithmetic. Vector quantities have direction as well as magnitude, and combine according to the rules of vector addition. Graphically, two vectors $\vec{A}$ and $\vec{B}$ are added by placing the tail of $\vec{B}$ at the head, or tip, of $\vec{A}$. The vector sum $\vec{A} + \vec{B}$ then extends from the tail of $\vec{A}$ to the head of $\vec{B}$. (See Example 1.5)

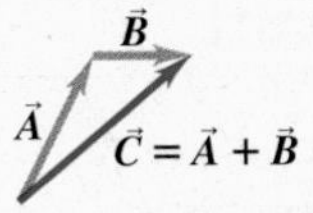

Vector addition can be carried out using components of vectors. The x-component of $\vec{R} = \vec{A} + \vec{B}$ is the sum of the x-components of $\vec{A}$ and $\vec{B}$, and likewise for the y- and z-components. (See Examples 1.6 and 1.7)

$$R_x = A_x + B_x$$
$$R_y = A_y + B_y$$
$$R_z = A_z + B_z \qquad (1.10)$$

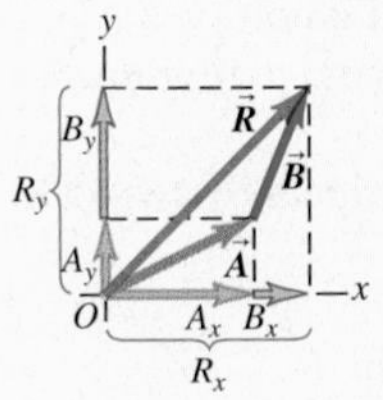

Unit vectors describe directions in space. A unit vector has a magnitude of one, with no units. The unit vectors $\hat{\imath}$, $\hat{\jmath}$, and $\hat{k}$, aligned with the x-, y-, and z-axes of a rectangular coordinate system, are especially useful. (See Example 1.9)

$$\vec{A} = A_x\hat{\imath} + A_y\hat{\jmath} + A_z\hat{k} \qquad (1.16)$$

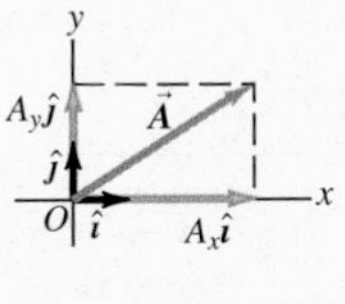

The scalar product $C = \vec{A} \cdot \vec{B}$ of two vectors $\vec{A}$ and $\vec{B}$ is a scalar quantity. It can be expressed in two ways: in terms of the magnitudes of $\vec{A}$ and $\vec{B}$ and the angle ϕ between the two vectors, or in terms of the components of $\vec{A}$ and $\vec{B}$. The scalar product is commutative; for any two vectors $\vec{A}$ and $\vec{B}$, $\vec{A} \cdot \vec{B} = \vec{B} \cdot \vec{A}$. The scalar product of two perpendicular vectors is zero. (See Examples 1.10 and 1.11)

$$\vec{A} \cdot \vec{B} = AB\cos\phi = |\vec{A}||\vec{B}|\cos\phi \qquad (1.18)$$

$$\vec{A} \cdot \vec{B} = A_xB_x + A_yB_y + A_zB_z \qquad (1.21)$$

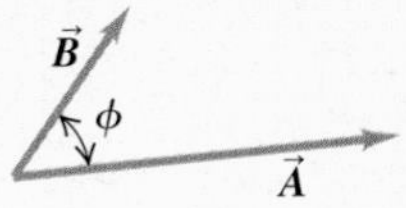

The vector product $\vec{C} = \vec{A} \times \vec{B}$ of two vectors $\vec{A}$ and $\vec{B}$ is another vector $\vec{C}$. The magnitude of $\vec{A} \times \vec{B}$ depends on the magnitudes of $\vec{A}$ and $\vec{B}$ and the angle ϕ between the two vectors. The direction of $\vec{A} \times \vec{B}$ is perpendicular to the plane of the two vectors being multiplied, as given by the right-hand rule. The components of $\vec{C} = \vec{A} \times \vec{B}$ can be expressed in terms of the components of $\vec{A}$ and $\vec{B}$. The vector product is not commutative; for any two vectors $\vec{A}$ and $\vec{B}$, $\vec{A} \times \vec{B} = -\vec{B} \times \vec{A}$. The vector product of two parallel or antiparallel vectors is zero. (See Example 1.12)

$$C = AB \sin \phi \qquad (1.22)$$

$$C_x = A_yB_z - A_zB_y$$
$$C_y = A_zB_x - A_xB_z$$
$$C_z = A_xB_y - A_yB_x \qquad (1.27)$$

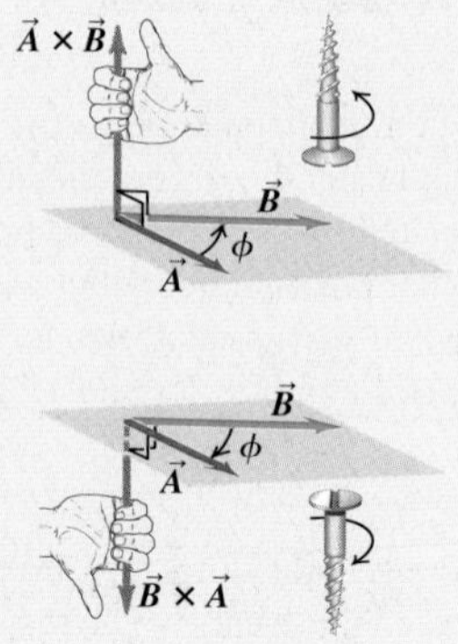

Key Terms

range of validity, 3
target variable, 4
model, 4
particle, 4
physical quantity, 5
operational definition, 5
unit, 5
International System (SI), 5
second, 6
meter, 6
kilogram, 6
prefix, 6
dimensionally consistent, 8
uncertainty (error), 10
accuracy, 10
fractional error, 10
percent error, 10
significant figures, 10
scientific (powers-of-10) notation, 11
precision, 12
order-of-magnitude estimates, 13
scalar quantity, 14
vector quantity, 14
magnitude of a vector, 14
displacement, 14
parallel vectors, 14
negative of a vector, 15
antiparallel vectors, 15
vector sum (resultant), 15
component vectors, 18
components, 18
unit vector, 23
scalar (dot) product, 25
vector (cross) product, 27
right-hand rule, 28
right-handed system, 29

Your Notes

Answer to Chapter Opening Question ?

Figure 1.3e shows that a typical human blood cell is about 8 μm in diameter. Twelve or thirteen such cells laid side by side would span a distance of 100 μm $= 10^{-4}$ m.

Answers to Test Your Understanding Questions

Section 1.5 Density $= (1.80\ \text{kg})/(6.0 \times 10^{-4}\ \text{m}^3) = 3.0 \times 10^3\ \text{kg/m}^3$. When multiplying or dividing, the number with the fewest significant figures controls the number of significant figures in the result.

Section 1.6 The answer depends on how many students are enrolled at your campus.

Section 1.7 When adding two vectors together, the order of the vectors doesn't matter. Hence the resultant displacement would be the same as in Example 1.5 (magnitude 2.24 km, direction 63.4° east of north).

Section 1.8 You are 5.66 km from your starting point in a direction 45.0° south of west.

Section 1.9
$\vec{A} = (38.37\hat{\imath} + 61.40\hat{\jmath})\text{m}$, $\vec{B} = (-46.36\hat{\imath} - 33.68\hat{\jmath})\text{m}$, $\vec{C} = (-17.80\hat{\jmath})\text{m}$

Section 1.10 $\vec{A}\cdot\vec{B} = (3)(4) + (2)(0) + (0)(5) = 12$

$$\begin{aligned}\vec{A}\times\vec{B} &= [(2)(5) - (0)(0)]\hat{\imath} + [(0)(4) - (3)(5)]\hat{\jmath}\\ &\quad + [(3)(0) - (2)(4)]\hat{k}\\ &= 10\hat{\imath} - 15\hat{\jmath} - 8\hat{k}\end{aligned}$$

Discussion Questions

Q1.1 How many correct experiments do we need to disprove a theory? How many to prove a theory? Explain.

Q1.2 A guidebook describes the rate of climb of a mountain trail as 120 meters per kilometer. How can you express this as a number with no units?

Q1.3 Suppose you are asked to compute the tangent of 5.00 meters. Is this possible? Why or why not?

Q1.4 A highway contractor stated that in building a bridge deck he poured 250 yards of concrete. What do you think he meant?

Q1.5 What is your height in centimeters? What is your weight in newtons?

Q1.6 The U.S. National Institute of Science and Technology (NIST) maintains several accurate copies of the international standard kilogram. Even after careful cleaning, these national standard kilograms are gaining mass at an average rate of about 1 μg/y (1 y = 1 year) when compared every ten years or so to the standard international kilogram. Does this apparent change have any importance? Explain.

Q1.7 What physical phenomena (other than a pendulum or cesium clock) could you use to define a time standard?

Q1.8 Describe how you could measure the thickness of a sheet of paper with an ordinary ruler.

Q1.9 The quantity $\pi = 3.14159\ldots$ is a number with no dimensions, since it is a ratio of two lengths. Describe two or three other geometrical or physical quantities that are dimensionless.

Q1.10 What are the units of volume? Suppose another student tells you that a cylinder of radius r and height h has volume given by $\pi r^3 h$. Explain why this cannot be right.

Q1.11 Three archers each fire four arrows at a target. Joe's four arrows hit at points 10 cm above, 10 cm below, 10 cm to the left, and 10 cm to the right of the center of the target. All four of Moe's arrows hit within 1 cm of a point 20 cm from the center, and Flo's four arrows all hit within 1 cm of the center. The contest judge says that one of the archers is precise but not accurate, another archer is accurate but not precise, and the third archer is both accurate and precise. Which description goes with which archer? Explain your reasoning.

Q1.12 A circular racetrack has a radius of 500 m. What is the displacement of a bicyclist when she travels around the track from the north side to the south side? When she makes one complete circle around the track? Explain your reasoning.

Q1.13 Can you find two vectors with different lengths that have a vector sum of zero? What length restrictions are required for three vectors to have a vector sum of zero? Explain your reasoning.

Q1.14 One sometimes speaks of the "direction of time," evolving from past to future. Does this mean that time is a vector quantity? Explain your reasoning.

Q1.15 Air traffic controllers give instructions to airline pilots telling them in which direction they are to fly. These instructions are called "vectors." If these are the only instructions given, is the name "vector" used correctly? Why or why not?

Q1.16 Can you find a vector quantity that has a magnitude of zero, but components that are different from zero? Explain. Can the magnitude of a vector be less than the magnitude of any of its components? Explain.

Q1.17 (a) Does it make sense to say that a vector is *negative*? Why? (b) Does it make sense to say that one vector is the negative of another? Why? Does your answer here contradict what you said in part (b)?

Q1.18 If $\vec{C}$ is the vector sum of $\vec{A}$ and $\vec{B}$, $\vec{C} = \vec{A} + \vec{B}$, what must be true if $C = A + B$? What must be true if $C = 0$?

Q1.19 If $\vec{A}$ and $\vec{B}$ are nonzero vectors, is it possible for $\vec{A}\cdot\vec{B}$ and $\vec{A}\times\vec{B}$ *both* to be zero? Explain.

Q1.20 What does $\vec{A}\cdot\vec{A}$, the scalar product of a vector with itself, give? What about $\vec{A}\times\vec{A}$, the vector product of a vector with itself?

Q1.21 Let $\vec{A}$ represent any nonzero vector. Why is $\vec{A}/A$ a unit vector and what is its direction? If θ is the angle that $\vec{A}$ makes with the $+x$-axis, explain why $(\vec{A}/A)\cdot\hat{\imath}$ is called the *direction cosine* for that axis.

Q1.22 Which of the following are legitimate mathematical operations: a) $\vec{A}\cdot(\vec{B} - \vec{C})$; b) $(\vec{A} - \vec{B})\times\vec{C}$; c) $\vec{A}\cdot(\vec{B}\times\vec{C})$; d) $\vec{A}\times(\vec{B}\times\vec{C})$; e) $\vec{A}\times(\vec{B}\cdot\vec{C})$? In each case, give the reason for your answer.

Q1.23 Consider the two repeated vector products $\vec{A}\times(\vec{B}\times\vec{C})$ and $(\vec{A}\times\vec{B})\times\vec{C}$. Give an example that illustrates the general rule that these two vector products do not have the same magnitude or direction. Can you choose the vectors $\vec{A}$, $\vec{B}$, and $\vec{C}$ such that these two vector products *are* equal to each other? If so, give an example.

Exercises

Section 1.3 Standards and Units
Section 1.4 Unit Consistency and Conversions

1.1 Starting with the definition 1 in. = 2.54 cm, find the number of kilometers in 1.00 mile.

1.2 According to the label on a bottle of salad dressing, the volume of the contents is 0.473 liter (L). Using only the conversions 1 L = 1000 cm^3 and 1 in. = 2.54 cm, express this volume in cubic inches.

1.3 How many nanoseconds does it take light to travel 1.00 km in vacuum?

1.4 The density of lead is 11.3 g/cm^3. What is this value in kilograms per cubic meter?

1.5 The most powerful engine available for the classic 1963 Chevrolet Corvette Sting Ray developed 360 horsepower and had a displacement of 327 cubic inches. Express this displacement in liters (L) by using only the conversions 1 L = 1000 cm^3 and 1 in. = 2.54 cm.

1.6 Told that he needs to set goals for himself, Billy Joe Bob decides to drink one cubic meter of his favorite beverage during the coming year. How many 16-fluid-ounce bottles should he drink each day? (Use Appendix E. A fluid ounce is a volume unit; 128 fluid ounces equals one gallon.)

1.7 The Concorde is the fastest airliner used for commercial service. It cruises at 1450 mi/h (about two times the speed of sound, or Mach 2). a) What is the cruise speed of the Concorde in km/h? b) What is it in m/s?

1.8 While driving in an exotic foreign land you see a speed limit sign on a highway that reads 180,000 furlongs per fortnight. How many miles per hour is this? (One furlong is $\frac{1}{8}$ mile, and a fortnight is 14 days. A furlong originally referred to the length of a plowed furrow.)

1.9 The gasoline consumption of a small car is advertised as 15.0 km/L (1 L = 1 liter). How many miles per gallon is this? Use the conversion factors in Appendix E.

1.10 The following conversions occur frequently in physics and are very useful. a) Use 1 mi = 5280 ft and 1 h = 3600 s to convert 60 mph to units of ft/s. b) The acceleration of a freely falling object is 32 ft/s^2. Use 1 ft = 30.48 cm to express this acceleration in units of m/s^2. c) The density of water is 1.0 g/cm^3. Convert this density to units of kg/m^3.

1.11 Neptunium. In the fall of 2002, a group of scientists at Los Alamos National Laboratory determined that the critical mass of neptunium–237 is about 60 kg. The critical mass of a fissionable material is the minimum amount that must be brought together to start a chain reaction. This element has a density of 19.5 g/cm^3. What would be the radius of a sphere of this material that has a critical mass?

Section 1.5 Uncertainty and Significant Figures

1.12 A useful and easy-to-remember approximate value for the number of seconds in a year is $\pi \times 10^7$. Determine the percent error in this approximate value. (There are 365.24 days in one year.)

1.13 Figure 1.5 shows the result of unacceptable error in the stopping position of a train. a) If a train travels 890 km from Berlin to Paris and then overshoots the end of the track by 10 m, what is the percentage error in the total distance covered? b) Would it be correct to write the total distance covered by the train as 890,010 m? Explain.

1.14 With a wooden ruler you measure the length of a rectangular piece of sheet metal to be 12 mm. You use micrometer calipers to measure the width of the rectangle and obtain the value 5.98 mm. Give your answers to the following questions to the correct number of significant figures. a) What is the area of the rectangle? b) What is the ratio of the rectangle's width to its length? c) What is the perimeter of the rectangle? d) What is the difference between the length and width? e) What is the ratio of the length to the width?

1.15 Estimate the percent error in measuring a) a distance of about 75 cm with a meter stick; b) a mass of about 12 g with a chemical balance; c) a time interval of about 6 min with a stopwatch.

1.16 A rectangular piece of aluminum is 5.10 ± 0.01 cm long and 1.90 ± 0.01 cm wide. a) Find the area of the rectangle and the uncertainty in the area. b) Verify that the fractional uncertainty in the area is equal to the sum of the fractional uncertainties in the length and in the width. (This is a general result; see Challenge Problem 1.94.)

1.17 As you eat your way through a bag of chocolate chip cookies, you observe that each cookie is a circular disk with a diameter of 8.50 ± 0.02 cm and a thickness of 0.050 ± 0.005 cm. a) Find the average volume of a cookie and the uncertainty in the volume. b) Find the ratio of the diameter to the thickness and the uncertainty in this ratio.

Section 1.6 Estimates and Orders of Magnitude

1.18 How many gallons of gasoline are used in the United States in one day?

1.19 A box of typewriter paper has dimensions 11 in. × 17 in. × 9 in.; it is marked "10 M." Does that mean it contains 10,000 sheets or 10 million?

1.20 How many kernels of corn does it take to fill a 2-L soft drink bottle?

1.21 How many words are there in this book?

1.22 What total volume of air does a person breathe in a lifetime? How does that compare with the volume of the Houston Astrodome? (Estimate that a person breathes about 500 cm^3 of air with each breath.)

1.23 How many hairs do you have on your head?

1.24 How many times does a human heart beat during a lifetime? How many gallons of blood does it pump? (Estimate that the heart pumps 50 cm^3 of blood with each beat.)

1.25 In Wagner's opera *Ring of the Nibelung*, the goddess Freia is ransomed for a pile of gold just tall enough and wide enough to hide her from sight. Estimate the monetary value of this pile. (Refer to Example 1.4 for information on the price per ounce and density of gold.)

1.26 How many drops of water are in all the oceans on earth?

1.27 How many pizzas are consumed each academic year by students at your school?

1.28 How many dollar bills would you have to stack to reach the moon? Would that be cheaper than building and launching a spacecraft?

1.29 How much would it cost to paper the entire United States (including Alaska and Hawaii) with dollar bills? What would be the cost to each person in the United States?

Section 1.7 Vectors and Vector Addition

1.30 Hearing rattles from a snake, you make two rapid displacements of magnitude 1.8 m and 2.4 m. In sketches (roughly to scale), show how your two displacements might add up to give a resultant of magnitude a) 4.2 m; b) 0.6 m; c) 3.0 m.

1.31 A postal employee drives a delivery truck along the route shown in Fig. 1.26. Determine the magnitude and direction of the resultant displacement by drawing a scale diagram. (See also Exercise 1.38 for a different approach to this same problem.)

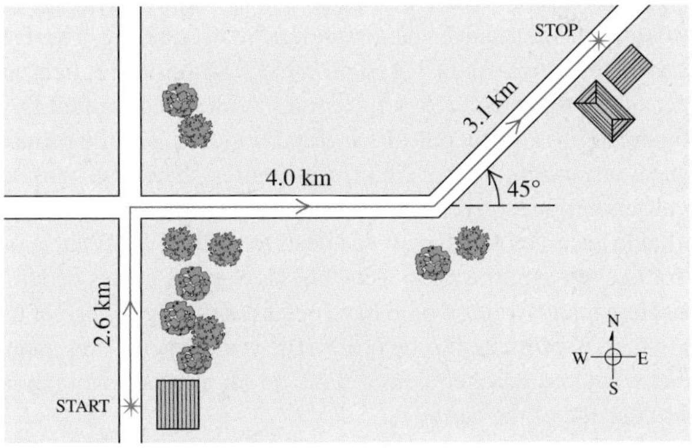

Figure 1.26 Exercises 1.31 and 1.38.

1.32 For the vectors $\vec{A}$ and $\vec{B}$ in Fig. 1.27, use a scale drawing to find the magnitude and direction of a) the vector sum $\vec{A} + \vec{B}$; b) the vector difference $\vec{A} - \vec{B}$. From your answers to parts (a) and (b), find the magnitude and direction of c) $-\vec{A} - \vec{B}$; d) $\vec{B} - \vec{A}$. (See also Exercise 1.39 for a different approach to this problem.)

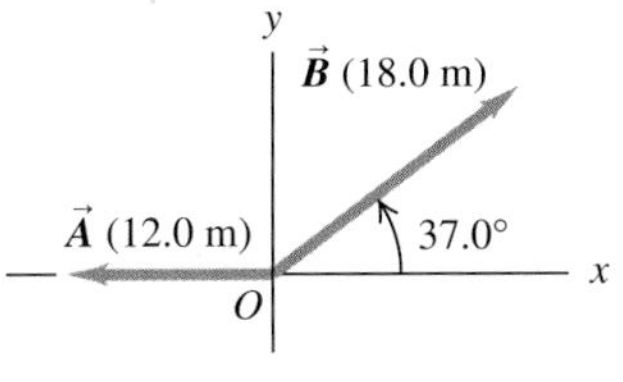

Figure 1.27 Exercises 1.32, 1.39, 1.44, and 1.54.

1.33 A spelunker is surveying a cave. She follows a passage 180 m straight west, then 210 m in a direction 45° east of south, and then 280 m at 30° east of north. After a fourth unmeasured displacement she finds herself back where she started. Use a scale drawing to determine the magnitude and direction of the fourth displacement. (See also Problem 1.69 for a different approach to this problem.)

Section 1.8 Components of Vectors

1.34 Use a scale drawing to find the x- and y-components of the following vectors. For each vector the numbers given are i) the magnitude of the vector and ii) the angle, measured in the sense from the $+x$-axis toward the $+y$-axis, that it makes with the $+x$-axis. Find a) magnitude 9.30 m, angle 60.0°; b) magnitude 22.0 km, angle 135°; c) magnitude 6.35 cm, angle 307°.

1.35 Compute the x- and y-components of the vectors $\vec{A}$, $\vec{B}$, and $\vec{C}$ in Fig. 1.28.

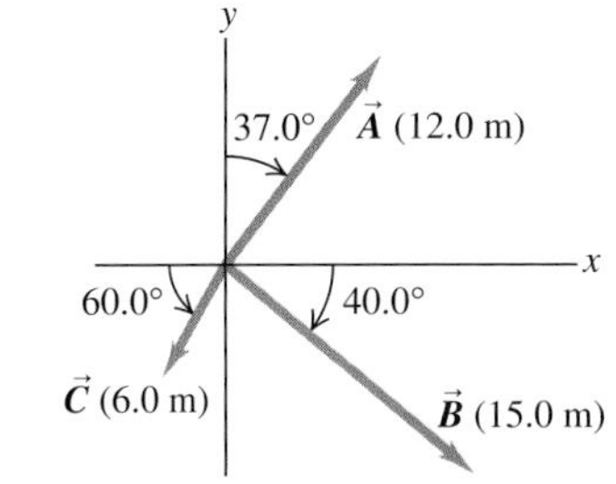

Figure 1.28 Exercises 1.35, 1.45, and 1.50, and Problem 1.68.

1.36 Let the angle θ be the angle that the vector $\vec{A}$ makes with the $+x$-axis, measured counterclockwise from that axis. Find the angle θ for a vector that has the following components: a) $A_x = 2.00$ m, $A_y = -1.00$ m; b) $A_x = 2.00$ m, $A_y = 1.00$ m; c) $A_x = -2.00$ m, $A_y = 1.00$ m; d) $A_x = -2.00$ m, $A_y = -1.00$ m.

1.37 A rocket fires two engines simultaneously. One produces a thrust of 725 N directly forward while the other gives a 513 N thrust at 32.4° above the forward direction. Find the magnitude and direction (relative to the forward direction) of the resultant force which these engines exert on the rocket.

1.38 A postal employee drives a delivery truck over the route shown in Fig. 1.26. Use the method of components to determine the magnitude and direction of her resultant displacement. In a vector addition diagram (roughly to scale), show that the resultant displacement found from your diagram is in qualitative agreement with the result you obtained using the method of components.

1.39 For the vectors $\vec{A}$ and $\vec{B}$ in Fig. 1.27, use the method of components to find the magnitude and direction of a) the vector sum $\vec{A} + \vec{B}$; b) the vector sum $\vec{B} + \vec{A}$; c) the vector difference $\vec{A} - \vec{B}$; d) the vector difference $\vec{B} - \vec{A}$.

1.40 Find the magnitude and direction of the vector represented by the following pairs of components: a) $A_x = -8.60$ cm, $A_y = 5.20$ cm; b) $A_x = -9.70$ m, $A_y = -2.45$ m; c) $A_x = 7.75$ km, $A_y = -2.70$ km.

1.41 A disoriented physics professor drives 3.25 km north, then 4.75 km west, and then 1.50 km south. Find the magnitude and direction of the resultant displacement, using the method of components. In a vector addition diagram (roughly to scale), show that the resultant displacement found from your diagram is in qualitative agreement with the result you obtained using the method of components.

1.42 Vector $\vec{A}$ has components $A_x = 1.30$ cm, $A_y = 2.25$ cm; vector $\vec{B}$ has components $B_x = 4.10$ cm, $B_y = -3.75$ cm. Find a) the components of the vector sum $\vec{A} + \vec{B}$; b) the magnitude and direction of $\vec{A} + \vec{B}$; c) the components of the vector difference $\vec{B} - \vec{A}$; d) the magnitude and direction of $\vec{B} - \vec{A}$.

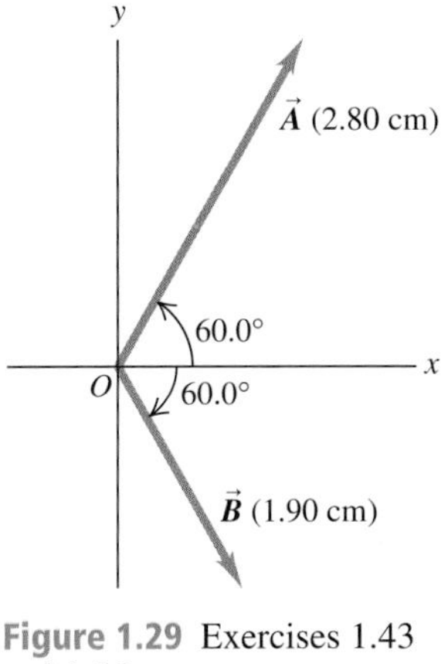

Figure 1.29 Exercises 1.43 and 1.56.

1.43 Vector $\vec{A}$ is 2.80 cm long and is 60.0° above the x-axis in the first quadrant. Vector $\vec{B}$ is 1.90 cm long and is 60.0° below the x-axis in the fourth quadrant (Fig. 1.29). Find the magnitude and direction of a) $\vec{A} + \vec{B}$; b) $\vec{A} - \vec{B}$; c) $\vec{B} - \vec{A}$. In each case, sketch the vector addition or subtraction and show that your numerical answers are in qualitative agreement with your sketch.

Section 1.9 Unit Vectors

1.44 Write each vector in Fig. 1.27 in terms of the unit vectors $\hat{\imath}$ and $\hat{\jmath}$.

1.45 Write each vector in Fig. 1.28 in terms of the unit vectors $\hat{\imath}$ and $\hat{\jmath}$.

1.46 a) Write each vector in Fig. 1.30 in terms of the unit vectors $\hat{\imath}$ and $\hat{\jmath}$. b) Use unit vectors to express the vector $\vec{C}$, where $\vec{C} = 3.00\vec{A} - 4.00\vec{B}$. c) Find the magnitude and direction of $\vec{C}$.

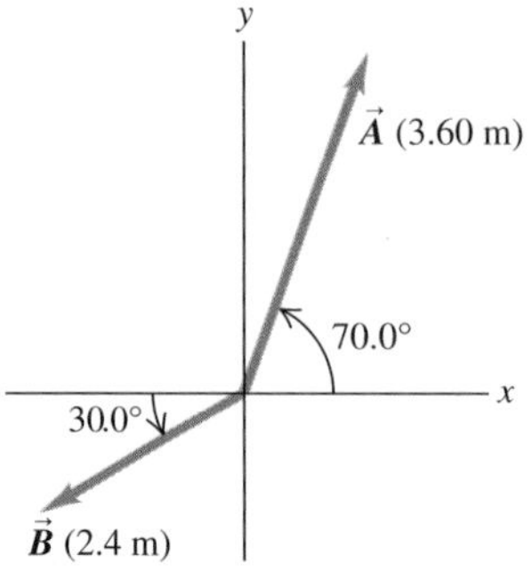

Figure 1.30 Exercise 1.46 and Problem 1.82.

1.47 Given two vectors $\vec{A} = 4.00\hat{\imath} + 3.00\hat{\jmath}$ and $\vec{B} = 5.00\hat{\imath} - 2.00\hat{\jmath}$, a) find the magnitude of each vector; b) write an expression for the vector difference $\vec{A} - \vec{B}$ using unit vectors; c) find the magnitude and direction of the vector difference $\vec{A} - \vec{B}$. d) In a vector diagram show $\vec{A}$, $\vec{B}$, and $\vec{A} - \vec{B}$, and also show that your diagram agrees qualitatively with your answer in part (c).

1.48 a) Is the vector $(\hat{\imath} + \hat{\jmath} + \hat{k})$ a unit vector? Justify your answer. b) Can a unit vector have any components with magnitude greater than unity? Can it have any negative components? In each case justify your answer. c) If $\vec{A} = a(3.0\hat{\imath} + 4.0\hat{\jmath})$, where a is a constant, determine the value of a that makes $\vec{A}$ a unit vector.

1.49 a) Use vector components to prove that two vectors commute for both addition and the scalar product. b) Use vector components to prove that two vectors anticommute for the vector product. That is, prove that $\vec{A} \times \vec{B} = -\vec{B} \times \vec{A}$.

Section 1.10 Products of Vectors

1.50 For the vectors $\vec{A}$, $\vec{B}$, and $\vec{C}$ in Fig. 1.28, find the scalar products a) $\vec{A} \cdot \vec{B}$; b) $\vec{B} \cdot \vec{C}$; c) $\vec{A} \cdot \vec{C}$.

1.51 a) Find the scalar product of the two vectors $\vec{A}$ and $\vec{B}$ given in Exercise 1.47. b) Find the angle between these two vectors.

1.52 Find the angle between each of the following pairs of vectors:

a) $\vec{A} = -2.00\hat{\imath} + 6.00\hat{\jmath}$ and $\vec{B} = 2.00\hat{\imath} - 3.00\hat{\jmath}$
b) $\vec{A} = 3.00\hat{\imath} + 5.00\hat{\jmath}$ and $\vec{B} = 10.00\hat{\imath} + 6.00\hat{\jmath}$
c) $\vec{A} = -4.00\hat{\imath} + 2.00\hat{\jmath}$ and $\vec{B} = 7.00\hat{\imath} + 14.00\hat{\jmath}$

1.53 Assuming a right-handed coordinate system, find the direction of the $+z$-axis in a) Fig. 1.15a; b) Fig. 1.15b.

1.54 For the two vectors in Fig. 1.27, a) find the magnitude and direction of the vector product $\vec{A} \times \vec{B}$; b) find the magnitude and direction of $\vec{B} \times \vec{A}$.

1.55 Find the vector product $\vec{A} \times \vec{B}$ (expressed in unit vectors) of the two vectors given in Exercise 1.47. What is the magnitude of the vector product?

1.56 For the two vectors in Fig. 1.29, a) find the magnitude and direction of the vector product $\vec{A} \times \vec{B}$; b) find the magnitude and direction of $\vec{B} \times \vec{A}$.

Problems

1.57 An acre, a unit of land measurement still in wide use, has a length of one furlong ($\frac{1}{8}$ mi) and a width one-tenth of its length. a) How many acres are in a square mile? b) How many square feet are in an acre? See Appendix E. c) An acre-foot is the volume of water that would cover one acre of flat land to a depth of one foot. How many gallons are in an acre-foot?

1.58 An estate on the California coast was offered for sale for \$4,950,000. The total area of the estate was 102 acres (see Problem 1.57). a) Considering the price of the estate to be proportional to its area, what was the cost of one square meter of the estate? b) What would be the price of a portion of the estate the size of a postage stamp ($\frac{7}{8}$ in. by 1.0 in.)?

1.59 The Hydrogen Maser. You can use the radio waves generated by a hydrogen maser as a standard of frequency. The frequency of these waves is 1,420,405,751.786 hertz. (A hertz is another name for one cycle per second.) A clock controlled by a hydrogen maser is off by only 1 s in 100,000 years. For the following questions, use only three significant figures. (The large number of significant figures given for the frequency simply illustrates the remarkable accuracy to which it has been measured.) a) What is the time for one cycle of the radio wave? b) How many cycles occur in 1 h? c) How many cycles would have occurred during the age of the earth, which is estimated to be 4.6×10^9 years? d) By how many seconds would a hydrogen maser clock be off after a time interval equal to the age of the earth?

1.60 Estimate the number of atoms in your body. (*Hint:* Based on what you know about biology and chemistry, what are the most common types of atom in your body? What is the mass of each type of atom? Appendix D gives the atomic masses for different elements, measured in atomic mass units; you can find the value of an atomic mass unit, or 1 u, in Appendix F.)

1.61 Biological tissues are typically made up of 98% water. Given that the density of water is 1.0×10^3 kg/m^3, estimate the mass of a) the heart of an adult human; b) a cell with a diameter of 0.5 μm; c) a honey bee.

1.62 Iron has a property such that a 1.00 m^3 volume has a mass of 7.86×10^3 kg (density equals 7.86×10^3 kg/m^3). You want to manufacture iron into cubes and spheres. Find a) the length of the side of a cube of iron that has a mass of 200.0 g; b) the radius of a solid sphere of iron that has a mass of 200.0 g.

1.63 a) Estimate the number of dentists in your city. You will need to consider the number of people in your city, how often they need to go to the dentist, how often they actually go, how many hours a typical dental procedure (filling, root canal, and so on) takes, and how many hours a dentist works in a week. b) Using your local telephone directory, check to see whether your estimate was roughly correct.

1.64 Physicists, mathematicians, and others often deal with large numbers. The number 10^{100} has been given the whimsical name *googol* by mathematicians. Let us compare some large numbers in physics with the googol. (*Note:* This problem requires numerical values that you can find in the appendices of the book, with which you should become familiar.) a) Approximately how many atoms make up our planet? For simplicity, assume the average atomic mass of the atoms to be 14 g/mol. Avogadro's number gives the number of atoms in a mole. b) Approximately how many neutrons are in a neutron star? Neutron stars are composed almost entirely of neutrons and have approximately twice the mass of the sun. c) In the leading theory of the origin of the universe, the entire universe that we can now observe occupied, at a very early time, a sphere whose radius was approximately equal to the present distance of the earth to the sun. At that time the universe had a density (mass divided by volume) of 10^{15} g/cm^3. Assuming that $\frac{1}{3}$ of the particles were protons, $\frac{1}{3}$ of the particles were neutrons, and the remaining $\frac{1}{3}$ were electrons, how many particles then made up the universe?

1.65 Three horizontal ropes pull on a large stone stuck in the ground, producing the vector forces $\vec{A}$, $\vec{B}$ and $\vec{C}$ shown in Fig. 1.31. Find the magnitude and direction of a fourth force on the stone that will make the vector sum of the four forces zero.

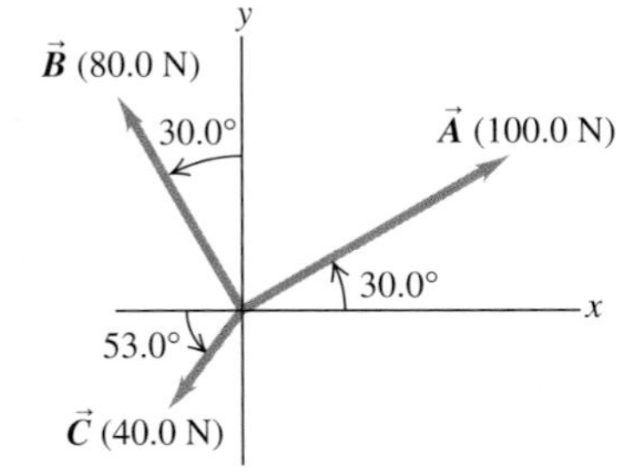

Figure 1.31 Problem 1.65.

1.66 Emergency Landing. A plane leaves the airport in Galisto and flies 170 km at 68° east of north and then changes direction to fly 230 km at 48° south of east, after which it makes an immediate emergency landing in a pasture. When the airport sends out a rescue crew, in which direction and how far should this crew fly to go directly to this plane?

1.67 You are to program a robotic arm on an assembly line to move in the xy-plane. Its first displacement is $\vec{A}$; its second displacement is $\vec{B}$, of magnitude 6.40 cm and direction 63.0° measured in the sense from the $+x$-axis toward the $-y$-axis. The resultant $\vec{C} = \vec{A} + \vec{B}$ of the two displacements should also have a magnitude of 6.40 cm, but a direction 22.0° measured in the sense from the $+x$-axis toward the $+y$-axis. a) Draw the vector addition diagram for these vectors, roughly to scale. b) Find the components of $\vec{A}$. c) Find the magnitude and direction of $\vec{A}$.

1.68 a) Find the magnitude and direction of the vector $\vec{R}$ that is the sum of the three vectors $\vec{A}$, $\vec{B}$, and $\vec{C}$ in Fig. 1.28. In a diagram, show how $\vec{R}$ is formed from the three vectors in Fig. 1.28. b) Find the magnitude and direction of the vector $\vec{S} = \vec{C} - \vec{A} - \vec{B}$. In a diagram, show how $\vec{S}$ is formed from the three vectors in Fig. 1.28.

1.69 As noted in Exercise 1.33, a spelunker is surveying a cave. She follows a passage 180 m straight west, then 210 m in a direction 45° east of south, then 280 m at 30° east of north. After a fourth unmeasured displacement she finds herself back where she started. Use the method of components to determine the magnitude and direction of the fourth displacement. Draw the vector addition diagram and show that it is in qualitative agreement with your numerical solution.

1.70 A sailor in a small sailboat encounters shifting winds. She sails 2.00 km east, then 3.50 km southeast, and then an additional distance in an unknown direction. Her final position is 5.80 km directly east of the starting point (Fig. 1.32). Find the magnitude and direction of the third leg of the journey. Draw the vector addition diagram and show that it is in qualitative agreement with your numerical solution.

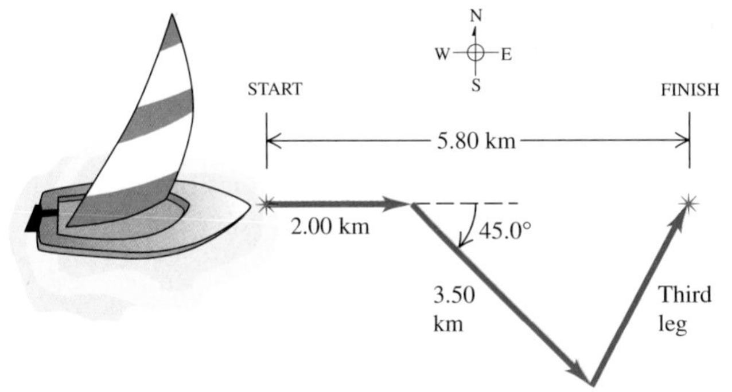

Figure 1.32 Problem 1.70.

1.71 A cross-country skier skis 2.80 km in the direction 45.0° west of south, then 7.40 km in the direction 30.0° north of east, and finally 3.30 km in the direction 22.0° south of west. a) Show these displacements in a diagram. b) How far is the skier from the starting point?

1.72 On a training flight, a student pilot flies from Lincoln, Nebraska to Clarinda, Iowa, then to St. Joseph, Missouri, and then to Manhattan, Kansas (Fig. 1.33). The directions are shown relative to north: 0° is north, 90° is east, 180° is south, and 270° is west. Use the method of components to find a) the distance she has to fly from Manhattan to get back to Lincoln; b) the direction (relative to north) she must fly to get there. Illustrate your solution with a vector diagram.

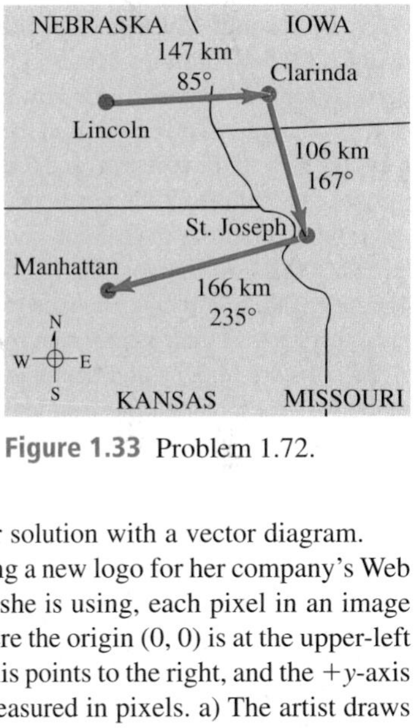

Figure 1.33 Problem 1.72.

1.73 A graphic artist is creating a new logo for her company's Web site. In the graphics program she is using, each pixel in an image file has coordinates (x, y), where the origin $(0, 0)$ is at the upper-left corner of the image, the $+x$-axis points to the right, and the $+y$-axis points down. Distances are measured in pixels. a) The artist draws a line from the pixel location (10, 20) to the location (210, 200). She wishes to draw a second line that starts at (10, 20), is 250 pixels long, and is at angle of 30° measured clockwise from the first line. At which pixel location should this second line end? Give your answer to the nearest pixel. b) The artist now draws an arrow that connects the lower-right end of the first line to the lower right end of the second line. Find the length and direction of this arrow. Draw a diagram showing all three lines.

1.74 Getting Back. An explorer in the dense jungles of equatorial Africa leaves his hut. He takes 40 steps northeast, then 80 steps 60° north of west, then 50 steps due south. Assume his steps all have equal length. a) Sketch, roughly to scale, the three vectors and their resultant. b) Save him from becoming hopelessly lost in the jungle by giving him the displacement, calculated using the method of components, that will return him to his hut.

1.75 A ship leaves the island of Guam and sails 285 km at 40.0° north of west. In which direction must it now head and how far must it sail so that its resultant displacement will be 115 km directly east of Guam?

1.76 A boulder of weight w rests on a hillside that rises at a constant angle α above the horizontal, as shown in Fig. 1.34. Its weight is a force on the boulder that has direction vertically downward. a) In terms of α and w, what is the component of the weight of the boulder in the direction parallel to the surface of the hill? b) What is the component of the weight in the direction perpendicular to the surface of the hill? c) An air conditioner unit is fastened to a roof that slopes upward at an angle of 35.0°. In order that the unit not slide down the roof, the component of the unit's weight parallel to the roof cannot exceed 550 N. What is the maximum allowed weight of the unit?

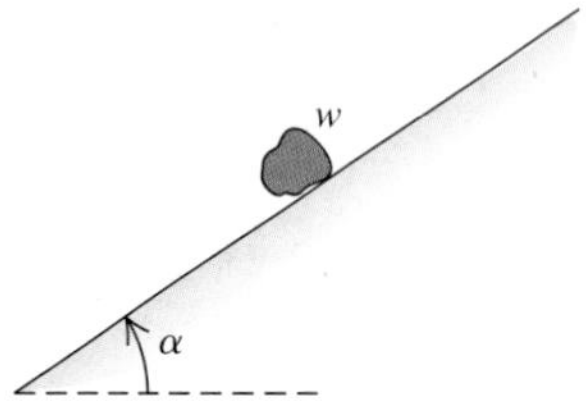

Figure 1.34 Problem 1.76.

1.77 Bones and Muscles. A patient in therapy has a forearm that weighs 20.5 N and that lifts a 112.0-N weight. These two forces have direction vertically downward. The only other significant forces on his forearm come from the biceps muscle (which acts perpendicularly to the forearm) and the force at the elbow. If the biceps produces a pull of 232 N when the forearm is raised 43° above the horizontal, find the magnitude and direction of the force that the elbow exerts on the forearm. (The sum of the elbow force and the biceps force must balance the weight of the arm and the weight it is carrying, so their vector sum must be 132.5 N, upward.)

1.78 You are hungry and decide to go to your favorite neighborhood fast-food restaurant. You leave your apartment and take the elevator 10 flights down (each flight is 3.0 m) and then go 15 m south to the apartment exit. You then proceed 0.2 km east, turn north, and go 0.1 km to the entrance of the restaurant. a) Determine the displacement from your apartment to the restaurant. Use unit-vector notation for your answer, being sure to make clear your choice of coordinates. b) How far did you travel along the path you took from your apartment to the restaurant and what is the magnitude of the displacement you calculated in part (a)?

1.79 You are canoeing on a lake. Starting at your camp on the shore, you travel 240 m in the direction 32° south of east to reach a store to purchase supplies. You know the distance because you have located both your camp and the store on a map of the lake. On the return trip you travel distance B in the direction 48° north of west, distance C in the direction 62° south of west, and then you are back at your camp. You measure the directions of travel with your compass, but you don't know the distances. Since you are curious to know the total distance you rowed, use vector methods to calculate the distances B and C.

1.80 You are camping with two friends, Joe and Karl. Since all three of you like your privacy, you don't pitch your tents close together. Joe's tent is 21.0 m from yours, in the direction 23.0° south of east. Karl's tent is 32.0 m from yours, in the direction 37.0° north of east. What is the distance between Karl's tent and Joe's tent?

1.81 Vectors $\vec{A}$ and $\vec{B}$ are drawn from a common point. Vector $\vec{A}$ has magnitude A and angle θ_A measured in the sense from the $+x$-axis to the $+y$-axis. The corresponding quantities for vector $\vec{B}$ are B and θ_B. Then $\vec{A} = A\cos\theta_A\hat{\imath} + A\sin\theta_A\hat{\jmath}$, $\vec{B} = B\cos\theta_B\hat{\imath} + B\sin\theta_B\hat{\jmath}$, and $\phi = |\theta_B - \theta_A|$ is the angle between $\vec{A}$ and $\vec{B}$. a) Derive Eq. (1.18) from Eq. (1.21). b) Derive Eq. (1.22) from Eq. (1.27).

1.82 For the two vectors $\vec{A}$ and $\vec{B}$ in Fig. 1.30, a) find the scalar product $\vec{A}\cdot\vec{B}$; b) find the magnitude and direction of the vector product $\vec{A}\times\vec{B}$.

1.83 Figure 1.8c shows a parallelogram based on the two vectors $\vec{A}$ and $\vec{B}$. a) Show that the magnitude of the cross product of these two vectors is equal to the area of the parallelogram. (*Hint:* area = base × height.) b) What is the angle between the cross product and the plane of the parallelogram?

1.84 The vector $\vec{A}$ is 3.50 cm long and is directed into this page. Vector $\vec{B}$ points from the lower-right corner of this page to the upper-left corner of this page. Define an appropriate right-handed coordinate system and find the three components of the vector product $\vec{A}\times\vec{B}$, measured in cm^2. In a diagram, show your coordinate system and the vectors $\vec{A}$, $\vec{B}$, and $\vec{A}\times\vec{B}$.

1.85 Given two vectors $\vec{A} = -2.00\hat{\imath} + 3.00\hat{\jmath} + 4.00\hat{k}$ and $\vec{B} = 3.00\hat{\imath} + 1.00\hat{\jmath} - 3.00\hat{k}$, do the following. a) Find the magnitude of each vector. b) Write an expression for the vector difference $\vec{A} - \vec{B}$, using unit vectors. c) Find the magnitude of the vector difference $\vec{A} - \vec{B}$. Is this the same as the magnitude of $\vec{B} - \vec{A}$? Explain.

1.86 Bond Angle in Methane. In the methane molecule, CH_4, each hydrogen atom is at a corner of a regular tetrahedron with the carbon atom at the center. In coordinates where one of the C—H bonds is in the direction of $\hat{\imath} + \hat{\jmath} + \hat{k}$, an adjacent C—H bond is in the $\hat{\imath} - \hat{\jmath} - \hat{k}$ direction. Calculate the angle between these two bonds.

1.87 The two vectors $\vec{A}$ and $\vec{B}$ are drawn from a common point, and $\vec{C} = \vec{A} + \vec{B}$. a) Show that if $C^2 = A^2 + B^2$, the angle between the vectors $\vec{A}$ and $\vec{B}$ is 90°. b) Show that if $C^2 < A^2 + B^2$, the angle between the vectors $\vec{A}$ and $\vec{B}$ is greater than 90°. c) Show that if $C^2 > A^2 + B^2$, the angle between the vectors $\vec{A}$ and $\vec{B}$ is between 0° and 90°.

1.88 When two vectors $\vec{A}$ and $\vec{B}$ are drawn from a common point, the angle between them is ϕ. a) Using vector techniques, show that the magnitude of their vector sum is given by

$$\sqrt{A^2 + B^2 + 2AB\cos\phi}$$

b) If $\vec{A}$ and $\vec{B}$ have the same magnitude, for which value of ϕ will their vector sum have the same magnitude as $\vec{A}$ or $\vec{B}$? c) Derive a result analogous to that in part (a) for the magnitude of the vector difference $\vec{A} - \vec{B}$. d) If $\vec{A}$ and $\vec{B}$ have the same magnitude, for what value of ϕ will $\vec{A} - \vec{B}$ have this same magnitude?

1.89 A cube is placed so that one corner is at the origin and three edges are along the x-, y-, and z-axes of a coordinate system (Fig. 1.35). Use vectors to compute a) the angle between the edge along the z-axis (line ab) and the diagonal from the origin to the opposite corner (line ad); b) the angle between line ac (the diagonal of a face) and line ad.

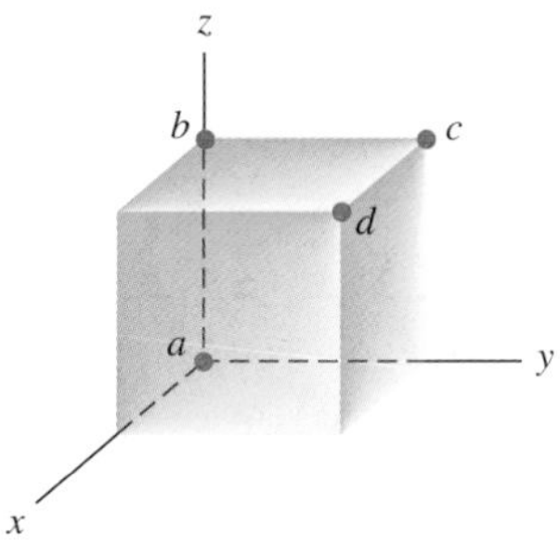

Figure 1.35 Problem 1.89.

1.90 Obtain a *unit vector* perpendicular to the two vectors given in Problem 1.85.

1.91 You are given vectors $\vec{A} = 5.0\hat{\imath} - 6.5\hat{\jmath}$ and $\vec{B} = -3.5\hat{\imath} + 7.0\hat{\jmath}$. A third vector $\vec{C}$ lies in the xy-plane. Vector $\vec{C}$ is perpendicular to vector $\vec{A}$, and the scalar product of $\vec{C}$ with $\vec{B}$ is 15.0. From this information, find the components of vector $\vec{C}$.

1.92 Two vectors $\vec{A}$ and $\vec{B}$ have magnitude $A = 3.00$ and $B = 3.00$. Their vector product is $\vec{A} \times \vec{B} = -5.00\hat{k} + 2.00\hat{\imath}$. What is the angle between $\vec{A}$ and $\vec{B}$?

1.93 Later in our study of physics we will encounter quantities represented by $(\vec{A} \times \vec{B}) \cdot \vec{C}$. a) Prove that for any three vectors $\vec{A}$, $\vec{B}$, and $\vec{C}$, $\vec{A} \cdot (\vec{B} \times \vec{C}) = (\vec{A} \times \vec{B}) \cdot \vec{C}$. b) Calculate $(\vec{A} \times \vec{B}) \cdot \vec{C}$ for the three vectors $\vec{A}$ with magnitude $A = 5.00$ and angle $\theta_A = 26.0°$ measured in the sense from the $+x$-axis toward the $+y$-axis, $\vec{B}$ with $B = 4.00$ and $\theta_B = 63.0°$, and $\vec{C}$ with magnitude 6.00 and in the $+z$-direction. Vectors $\vec{A}$ and $\vec{B}$ are in the xy-plane.

Challenge Problems

1.94 The length of a rectangle is given as $L \pm l$, and its width as $W \pm w$. a) Show that the uncertainty in its area A is $a = Lw + lW$. Assume that the uncertainties l and w are small, so that the product lw is very small and you can ignore it. b) Show that the fractional uncertainty in the area is equal to the sum of the fractional uncertainty in length and the fractional uncertainty in width. c) A rectangular solid has dimensions $L \pm l$, $W \pm w$, *and* $H \pm h$. Find the fractional uncertainty in the volume, and show that it equals the sum of the fractional uncertainties in the length, width, and height.

1.95 Completed Pass. At Enormous State University (ESU), the football team records its plays using vector displacements, with the origin taken to be the position of the ball before the play starts. In a certain pass play, the receiver starts at $+1.0\hat{\imath} - 5.0\hat{\jmath}$, where the units are yards, $\hat{\imath}$ is to the right, and $\hat{\jmath}$ is downfield. Subsequent displacements of the receiver are $+9.0\hat{\imath}$ (in motion before the snap), $+11.0\hat{\jmath}$ (breaks downfield), $-6.0\hat{\imath} + 4.0\hat{\jmath}$ (zigs), and $+12.0\hat{\imath} + 18.0\hat{\jmath}$ (zags). Meanwhile, the quarterback has dropped straight back to a position $-7.0\hat{\jmath}$. How far and in which direction must the quarterback throw the ball? (Like the coach, you will be well advised to diagram the situation before solving it numerically.)

1.96 Navigating in the Solar System. The *Mars Polar Lander* spacecraft was launched on January 3, 1999. On December 3, 1999, the day that *Mars Polar Lander* touched down on the Martian surface, the positions of the earth and Mars were given by these coordinates:

	x	y	z
Earth	0.3182 AU	0.9329 AU	0.0000 AU
Mars	1.3087 AU	−0.4423 AU	−0.0414 AU

In these coordinates, the sun is at the origin and the plane of the earth's orbit is the xy-plane. The earth passes through the $+x$-axis once a year on the autumnal equinox, the first day of autumn in the northern hemisphere (on or about September 22). One AU, or *astronomical unit*, is equal to 1.496×10^8 km, the average distance from the earth to the sun. a) In a diagram, show the positions of the sun, the earth, and Mars on December 3, 1999. b) Find the following distances in AU on December 3, 1999: i) from the sun to the earth; ii) from the sun to Mars; iii) from the earth to Mars. c) As seen from the earth, what was the angle between the direction to the sun and the direction to Mars on December 3, 1999? d) Explain whether Mars was visible from your location at midnight on December 3, 1999. (When it is midnight at your location, the sun is on the opposite side of the earth from you.)

1.97 Navigating in the Big Dipper. All the stars of the Big Dipper (part of the constellation Ursa Major) may appear to be the same distance from the earth, but in fact they are very far from each other. Figure 1.36 shows the distances from the earth to each of these stars. The distances are given in light years (ly), the distance that light travels in one year. One light year equals 9.461×10^{15} m. a) Alkaid and Merak are 25.6° apart in the earth's sky. In a diagram, show the relative positions of Alkaid, Merak, and our sun. Find the distance in light years from Alkaid to Merak. b) To an inhabitant of a planet orbiting Merak, how many degrees apart in the sky would Alkaid and our sun be?

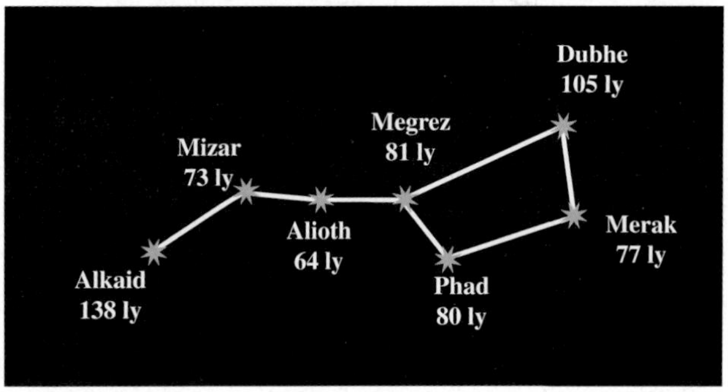

Figure 1.36 Challenge Problem 1.97.

1.98 The vector $\vec{r} = x\hat{\imath} + y\hat{\jmath} + z\hat{k}$, called the *position vector*, points from the origin (0, 0, 0) to an arbitrary point in space with coordinates (x, y, z). Use what you know about vectors to prove the following: All points (x, y, z) that satisfy the equation $Ax + By + Cz = 0$, where A, B, and C are constants, lie in a plane that passes through the origin and that is perpendicular to the vector $A\hat{\imath} + B\hat{\jmath} + C\hat{k}$. Sketch this vector and the plane.

2 MOTION ALONG A STRAIGHT LINE

Every successful space shuttle mission ends with a brief period of straight-line motion as it comes to a stop on the runway. No larger than a typical airliner, the space shuttle orbiter lands at more than 350 km/h (220 mi/h). Even with a drag parachute for braking, the fast-moving orbiter requires up to 3 km (about 2 mi) to stop.

? **Is it accurate to say that the space shuttle orbiter is *accelerating* as it slows to a stop?**

How do you describe the motion of a jet fighter being catapulted down the deck of an aircraft carrier? When you throw a baseball straight up in the air, how high does it go? When a glass slips from your hand, how much time do you have to catch it before it hits the floor? These are the kinds of questions you will learn to answer in this chapter. We are beginning our study of physics with *mechanics,* the study of the relationships among force, matter, and motion. The goal of this chapter and the next is to develop general methods for describing motion. The part of mechanics that deals with the description of motion is called *kinematics.* Later we will study *dynamics,* the relation of motion to its causes.

In this chapter we study the simplest kind of motion: a single particle moving along a straight line. We often use a particle as a model for a moving body when effects such as rotation or change of shape are not important. To describe the motion of a particle, we introduce the physical quantities *velocity* and *acceleration.* These quantities have simple definitions in physics; however, those definitions are more precise and slightly different than the ones used in everyday language. Paying careful attention to how velocity and acceleration are defined will help you work with these and other important physical quantities.

An important part of how a physicist defines velocity and acceleration is that these quantities are *vectors.* As you learned in Chapter 1, this means that they have both magnitude and direction. Our concern in this chapter is with motion along a straight line only, so we won't need the full mathematics of vectors just yet. But in Chapter 3 we'll extend our discussion to include motion in two or three dimensions, and using vectors will be essential.

An important special case of straight-line motion is when the acceleration is constant, a situation that we will encounter many times in our study of physics. An example is the motion of a freely falling body. We'll develop simple equations to describe motion with constant acceleration. We'll also consider situations in which the acceleration varies during the motion; in this case, it's necessary to use integration to describe the motion. (If you haven't studied integration yet, this section is optional.)

2.1 | Displacement, Time, and Average Velocity

Suppose a drag racer drives her AA-fuel dragster along a straight track. To study this motion, we need a coordinate system to describe the dragster's position. We choose the x-axis of our coordinate system to lie along the dragster's straight-line path, with the origin O at the starting line (Fig. 2.1). We will describe the dragster's position in terms of the position of a representative point, such as its front end. By doing this, we represent the entire dragster by that point and so treat the dragster as a **particle.**

A useful way to describe the motion of the front of the dragster—that is, the motion of the particle—is in terms of the change in the particle's position (in other words, the change in its coordinate x) over a time interval. Suppose that 1.0 s after the start the front of the dragster is at point P_1, 19 m from the origin, and 4.0 s after the start it is at point P_2, 277 m from the origin. The *displacement* of the particle is a vector that points from P_1 to P_2 (see Section 1.7). Figure 2.1 shows that this vector points along the x-axis. The x-component of the displacement is just the change in the value of x, $(277\ \text{m} - 19\ \text{m}) = 258\ \text{m}$, that took place during the time interval of $(4.0\ \text{s} - 1.0\ \text{s}) = 3.0\ \text{s}$. We define the dragster's **average velocity** during this time interval as a *vector* quantity whose x-component is the change in x divided by the time interval: $(258\ \text{m})/(3.0\ \text{s}) = 86\ \text{m/s}$. In general, the average velocity depends on the particular time interval chosen. For a 3.0-s time interval *before* the start of the race, the average velocity would be zero, because the dragster would be at rest at the starting line and would have zero displacement.

Let's generalize the concept of average velocity. At time t_1 the dragster is at point P_1, with coordinate x_1, and at time t_2 it is at point P_2, with coordinate x_2. The displacement of the dragster during the time interval from t_1 to t_2 is the vector from P_1 to P_2, with x-component $(x_2 - x_1)$ and with y- and z-components equal to zero. The x-component of the dragster's displacement is just the change in the coordinate x, which we write in a more compact way as

$$\Delta x = x_2 - x_1 \tag{2.1}$$

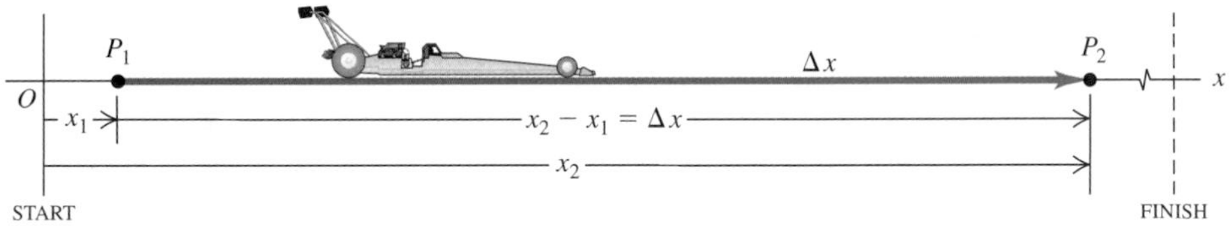

2.1 Positions of an AA-fuel dragster at two times during its run.

CAUTION Note that Δx is *not* the product of Δ and x; it is a single symbol that means "the change in the quantity x." We will always use the Greek capital letter Δ ("delta") to represent a *change* in a quantity, calculated by taking the *final* value of the quantity minus the *initial* value—never the reverse. Likewise, the time interval from t_1 to t_2 is Δt, the change in the quantity t: $\Delta t = t_2 - t_1$ (final time minus initial time).

We can now define the x-component of average velocity more precisely: it is the x-component of displacement, Δx, divided by the time interval Δt during which the displacement occurs. We represent this quantity by the symbol $v_{\text{av-}x}$, where the subscript "av" signifies average value and the subscript x indicates that this is the x-component:

$$v_{\text{av-}x} = \frac{x_2 - x_1}{t_2 - t_1} = \frac{\Delta x}{\Delta t} \quad \text{(average velocity, straight-line motion)} \quad (2.2)$$

For the above example we had $x_1 = 19$ m, $x_2 = 277$ m, $t_1 = 1.0$ s, and $t_2 = 4.0$ s, so Eq. (2.2) gives

$$v_{\text{av-}x} = \frac{277\ \text{m} - 19\ \text{m}}{4.0\ \text{s} - 1.0\ \text{s}} = \frac{258\ \text{m}}{3.0\ \text{s}} = 86\ \text{m/s}$$

The average velocity of the dragster is positive. This means that during the time interval, the coordinate x increased and the dragster moved in the positive x-direction (to the right in Fig. 2.1). If a particle moves in the *negative* x-direction during a time interval, its average velocity for that time interval is negative. For example, suppose an official's truck moves to the left along the track (Fig. 2.2). The truck is at $x_1 = 277$ m at $t_1 = 16.0$ s and is at $x_2 = 19$ m at $t_2 = 25.0$ s. Then $\Delta x = (19\ \text{m} - 277\ \text{m}) = -258$ m and $\Delta t = (25.0\ \text{s} - 16.0\ \text{s}) = 9.0$ s. The x-component of average velocity is $v_{\text{av-}x} = \Delta x/\Delta t = (-258\ \text{m})/(9.0\ \text{s}) = -29$ m/s. Whenever x is positive and increasing or is negative and becoming less negative, the particle is moving in the $+x$-direction and $v_{\text{av-}x}$ is positive (Fig. 2.1). Whenever x is positive and decreasing or is negative and becoming more negative, the particle is moving in the $-x$-direction and $v_{\text{av-}x}$ is negative (Fig. 2.2).

CAUTION You might be tempted to conclude that positive average velocity necessarily implies motion to the right, as in Fig. 2.1, and that negative average velocity necessarily implies motion to the left, as in Fig. 2.2. But these conclusions are correct *only* if the positive x-direction is to the right, as we have cho-

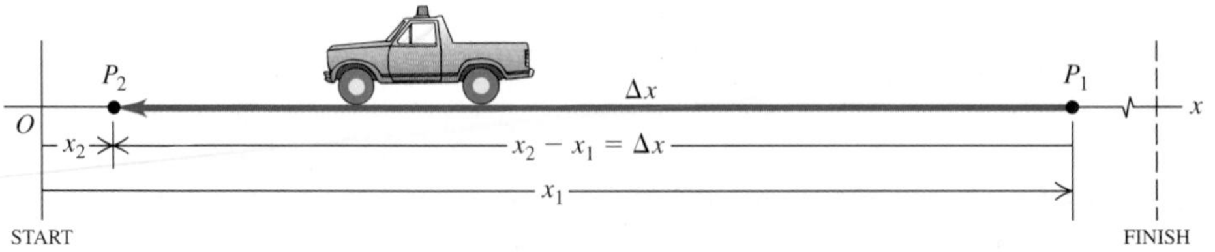

2.2 Positions of an official's truck at two times during its motion. The points P_1 and P_2 now refer to the motion of the truck, and so are different than in Fig. 2.1. The x-component of the truck's displacement is negative, so $v_{\text{av-}x}$ is negative.

sen it to be in Figs. 2.1 and 2.2. We could just as well have chosen the positive x-direction to be to the left, with the origin at the finish line. Then the dragster would have negative average velocity and the official's truck would have positive average velocity. In most problems the direction of the coordinate axis will be yours to choose. Once you've made your choice, you *must* take it into account when interpreting the signs of $v_{av\text{-}x}$ and other quantities that describe motion!

With straight-line motion we will usually call Δx simply the displacement and $v_{av\text{-}x}$ simply the average velocity. But be sure to remember that these are really the x-components of vector quantities that, in this special case, have *only* x-components. In Chapter 3, displacement, velocity, and acceleration vectors will have two or three nonzero components.

Figure 2.3 is a graph of the dragster's position as a function of time, that is, an ***x-t* graph.** The curve in the figure *does not* represent the dragster's path in space; as Fig. 2.1 shows, the path is a straight line. Rather, the graph is a pictorial way to represent how the dragster's position changes with time. The points on the graph labeled p_1 and p_2 correspond to the points P_1 and P_2 along the dragster's path. Line p_1p_2 is the hypotenuse of a right triangle with vertical side $\Delta x = x_2 - x_1$ and horizontal side $\Delta t = t_2 - t_1$. The average velocity $v_{av\text{-}x} = \Delta x/\Delta t$ of the dragster is then equal to the *slope* of the line p_1p_2, that is, the ratio of the vertical side Δx of the triangle to the horizontal side Δt.

Average velocity depends only on the total displacement $\Delta x = x_2 - x_1$ that occurs during the time interval $\Delta t = t_2 - t_1$, not on the details of what happens during the time interval. A second dragster might have passed through point P_1 in Fig. 2.1 at the same time t_1 as the first dragster, raced past the first dragster, then blown its engine and slowed down to pass through point P_2 at the same time t_2 as the first dragster. Both dragsters have the same displacement during the same time interval and so have the same average velocity.

If we express distance in meters and time in seconds, average velocity is measured in meters per second (m/s). Other common units of velocity are kilometers per hour (km/h), feet per second (ft/s), miles per hour (mi/h), and knots (1 knot = 1 nautical mile/h = 6080 ft/h). Table 2.1 shows a few typical velocity magnitudes.

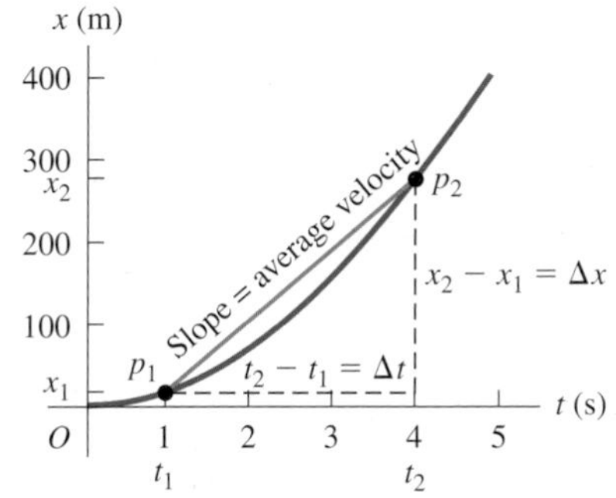

2.3 The position of an AA-fuel dragster as a function of time. The average velocity $v_{av\text{-}x}$ between points P_1 and P_2 in Fig. 2.1 is the slope of the line p_1p_2. This line slopes upward to the right, so the slope is positive and $v_{av\text{-}x}$ is positive.

Table 2.1 Typical Velocity Magnitudes

A snail's pace	10^{-3} m/s	Random motion of air molecules	500 m/s
A brisk walk	2 m/s	Fastest airplane	1000 m/s
Fastest human	11 m/s	Orbiting communications satellite	3000 m/s
Running cheetah	35 m/s	Electron orbiting in a hydrogen atom	2×10^6 m/s
Fastest car	341 m/s	Light traveling in a vacuum	3×10^8 m/s

Test Your Understanding

A truck drives due west from point A to point B, a distance of 60 km. A car drives due east from point A to point C, a distance of 30 km, then turns around and drives west to point B. The truck and car leave point A simultaneously and arrive at point B simultaneously. Explain why the car and truck have the *same* average velocity.

2.4 The winner of a 50-m swimming race is the swimmer whose average velocity has the greatest magnitude—that is, the swimmer who traverses a displacement Δx of 50 m in the shortest elapsed time Δt.

2.2 | Instantaneous Velocity

Sometimes the average velocity is all you need to know about a particle's motion. For example, a race along a straight line is really a competition to see whose average velocity, $v_{\text{av-}x}$, has the greatest magnitude. The prize goes to the competitor who can travel the displacement Δx from the start to the finish line in the briefest time interval, Δt (Fig. 2.4).

But the average velocity of a particle during a time interval can't tell us how fast, or in what direction, the particle was moving at any given time during the interval. To describe the motion in greater detail, we need to define the velocity at any specific instant of time or specific point along the path. Such a velocity is called **instantaneous velocity,** and it needs to be defined carefully.

Note that the word *instant* has a somewhat different definition in physics than in everyday language. You might use the phrase "It lasted just an instant" to refer to something that lasted for a very short time interval. But in physics an instant has *no* duration at all; it refers to a single value of time.

To find the instantaneous velocity of the dragster in Fig. 2.1 at the point P_1, we imagine moving the second point P_2 closer and closer to the first point P_1. We compute the average velocity $v_{\text{av-}x} = \Delta x/\Delta t$ over these shorter and shorter displacements and time intervals. Both Δx and Δt become very small, but their ratio does not necessarily become small. In the language of calculus, the limit of $\Delta x/\Delta t$ as Δt approaches zero is called the **derivative** of x with respect to t and is written *dx/dt. The instantaneous velocity is the limit of the average velocity as the time interval approaches zero; it equals the instantaneous rate of change of position with time.* We use the symbol v_x, with no "av" subscript, for the instantaneous velocity along the x-axis:

$$v_x = \lim_{\Delta t \to 0} \frac{\Delta x}{\Delta t} = \frac{dx}{dt} \qquad \text{(instantaneous velocity, straight-line motion)} \qquad (2.3)$$

We always assume that the time interval Δt is positive so that v_x has the same algebraic sign as Δx. If the positive x-axis points to the right, as in Fig. 2.1, a positive value of v_x means that x is increasing and the motion is toward the right; a negative value of v_x means that x is decreasing and the motion is toward the left. A body can have positive x and negative v_x, or the reverse; x tells us where the body is, while v_x tells us how it's moving (Fig 2.5).

2.5 Even when moving forward, this cyclist's instantaneous velocity can be negative—if he's traveling in the negative x-direction. In any problem, the choice of which direction is positive and which is negative is entirely up to you.

Instantaneous velocity, like average velocity, is a vector quantity. Equation (2.3) defines its x-component, which can be positive or negative. In straight-line motion, all other components of instantaneous velocity are zero, and in this case we will often call v_x simply the instantaneous velocity. (In Chapter 3 we'll deal with the general case in which the instantaneous velocity can have nonzero x-, y-, and z-components.) When we use the term "velocity," we will always mean instantaneous rather than average velocity, unless we state otherwise.

The terms "velocity" and "speed" are used interchangeably in everyday language, but they have distinct definitions in physics. We use the term **speed** to denote distance traveled divided by time, on either an average or an instantaneous basis. We use the symbol v with *no* subscripts to denote instantaneous speed. Instantaneous *speed* measures how fast a particle is moving; instantaneous *velocity* measures how fast *and* in what direction it's moving. For example, a particle with instantaneous velocity $v_x = 25$ m/s and a second particle with $v_x = -25$ m/s are moving in opposite directions at the same instantaneous speed 25 m/s. Instantaneous speed is the magnitude of instantaneous velocity, and so instantaneous speed can never be negative.

CAUTION Average speed, however, is *not* the magnitude of average velocity. When Alexander Popov set a world record in 1994 by swimming 100.0 m in 46.74 s, his average speed was $(100.0\text{ m})/(46.74\text{ s}) = 2.139\text{ m/s}$. But because he swam two lengths in a 50-m pool, he started and ended at the same point, giving him zero total displacement and zero average *velocity* for his effort! Both average speed and instantaneous speed are scalars, not vectors, because these quantities contain no information about direction.

Example 2.1 Average and instantaneous velocities

A cheetah is crouched in ambush 20 m to the east of an observer's blind (Fig. 2.6). At time $t = 0$ the cheetah charges an antelope in a clearing 50 m east of the observer. The cheetah runs along a straight line. Later analysis of a videotape shows that during the first 2.0 s of the attack, the cheetah's coordinate x varies with time according to the equation $x = 20\text{ m} + (5.0\text{ m/s}^2)t^2$. (Note that the units for the numbers 20 and 5.0 *must* be as shown to make the expression dimensionally consistent.) a) Find the displacement of the cheetah during the interval between $t_1 = 1.0\text{ s}$ and $t_2 = 2.0\text{ s}$. b) Find the average velocity during the same time interval. c) Find the instantaneous velocity at time $t_1 = 1.0\text{ s}$ by taking $\Delta t = 0.1$ s, then $\Delta t = 0.01$ s, then $\Delta t = 0.001$ s. d) Derive a general expression for the instantaneous velocity as a function of time, and from it find v_x at $t = 1.0$ s and $= 2.0$ s.

SOLUTION

IDENTIFY: This problem requires us to use the definitions of displacement, average velocity, and instantaneous velocity. Using the first two of these involves algebra; the last one requires us to use calculus to take a derivative.

SET UP: Figure 2.6 depicts the cheetah's motion. To analyze this problem we use Eq. (2.1) for displacement, Eq. (2.2) for average velocity, and Eq. (2.3) for instantaneous velocity.

EXECUTE: a) At time $t_1 = 1.0$ s the cheetah's position x_1 is

$$x_1 = 20\text{ m} + (5.0\text{ m/s}^2)(1.0\text{ s})^2 = 25\text{ m}$$

At time $t_2 = 2.0$ s its position x_2 is

$$x_2 = 20\text{ m} + (5.0\text{ m/s}^2)(2.0\text{ s})^2 = 40\text{ m}$$

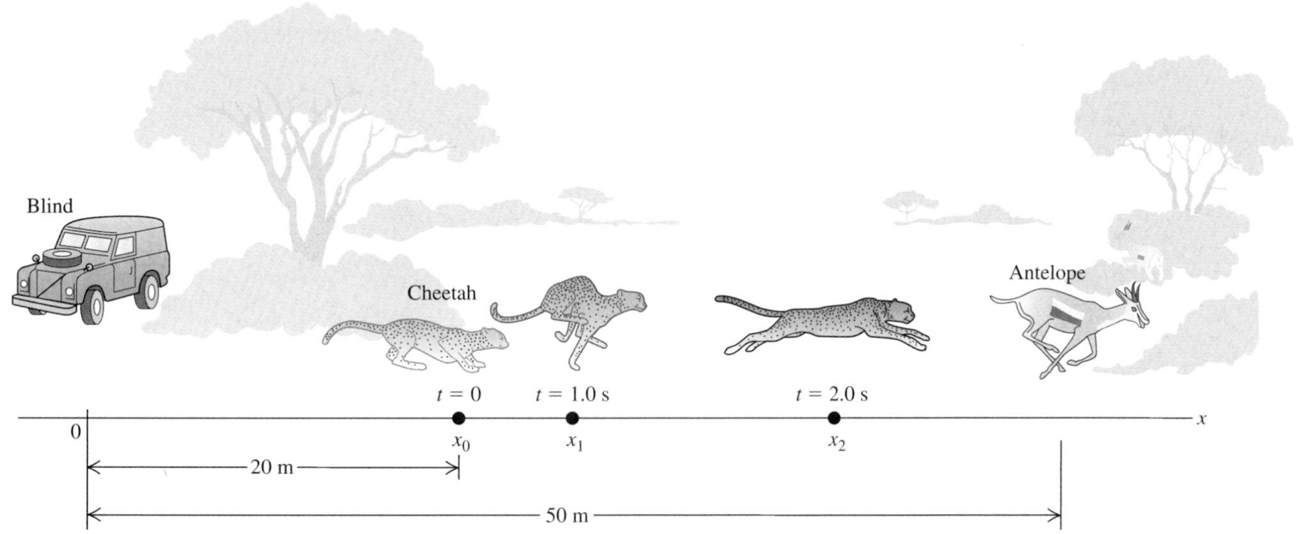

2.6 A cheetah attacking an antelope from ambush. The animals are not drawn to the same scale as the axis.

The displacement during this interval is

$$\Delta x = x_2 - x_1 = 40\text{ m} - 25\text{ m} = 15\text{ m}$$

b) The average velocity during this time interval is

$$v_{\text{av-}x} = \frac{x_2 - x_1}{t_2 - t_1} = \frac{40\text{ m} - 25\text{ m}}{2.0\text{ s} - 1.0\text{ s}} = \frac{15\text{ m}}{1.0\text{ s}} = 15\text{ m/s}$$

c) With $\Delta t = 0.1$ s, the time interval is from $t_1 = 1.0$ s to $t_2 = 1.1$ s. At time t_2, the position is

$$x_2 = 20\text{ m} + (5.0\text{ m/s}^2)(1.1\text{ s})^2 = 26.05\text{ m}$$

The average velocity during this interval is

$$v_{\text{av-}x} = \frac{26.05\text{ m} - 25\text{ m}}{1.1\text{ s} - 1.0\text{ s}} = 10.5\text{ m/s}$$

You should follow this same pattern to work out the average velocities for the 0.01-s and 0.001-s intervals. The results are 10.05 m/s and 10.005 m/s. As Δt gets smaller, the average velocity gets closer to 10.0 m/s, so we conclude that the instantaneous velocity at time $t = 1.0$ s is 10.0 m/s.

d) We find the instantaneous velocity as a function of time by taking the derivative of the expression for x with respect to t. For any n the derivative of t^n is nt^{n-1}, so the derivative of t^2 is $2t$. Therefore

$$v_x = \frac{dx}{dt} = (5.0\text{ m/s}^2)(2t) = (10\text{ m/s}^2)t$$

At time $t = 1.0$ s, $v_x = 10$ m/s as we found in part (c). At time $t = 2.0$ s, $v_x = 20$ m/s.

EVALUATE: Our results show that the cheetah picked up speed from $t = 0$ (when it was at rest) to $t = 1.0$ s ($v_x = 10$ m/s) to $t = 2.0$ s ($v_x = 20$ m/s). This makes sense; the cheetah covered only 5 m during the interval $t = 0$ to $t = 1.0$ s, but covered 15 m during the interval $t = 1.0$ s to $t = 2.0$ s.

ActivPhysics ONLINE

1.1 Analyzing Motion Using Diagrams

Finding Velocity on an *x-t* Graph

The velocity of a particle can also be found from the graph of the particle's position as a function of time. Suppose we want to find the velocity of the dragster in Fig. 2.1 at point P_1. As point P_2 in Fig. 2.1 approaches point P_1, point p_2 in the x-t graph of Fig. 2.3 approaches point p_1. This is shown in Figs. 2.7a and 2.7b, in which the average velocity is calculated over shorter time intervals Δt. In the limit that $\Delta t \to 0$, shown in Fig. 2.7c, the slope of the line p_1p_2 equals the slope of the line tangent to the curve at point p_1. *On a graph of position as a function of time for straight-line motion, the instantaneous velocity at any point is equal to the slope of the tangent to the curve at that point.*

If the tangent to the x-t curve slopes upward to the right, as in Fig. 2.7c, then its slope is positive, the velocity is positive, and the motion is in the positive x-direction. If the tangent slopes downward to the right, the slope and velocity are negative and the motion is in the negative x-direction. When the tangent is horizontal, the slope is zero and the velocity is zero. Figure 2.8 illustrates these three possibilities.

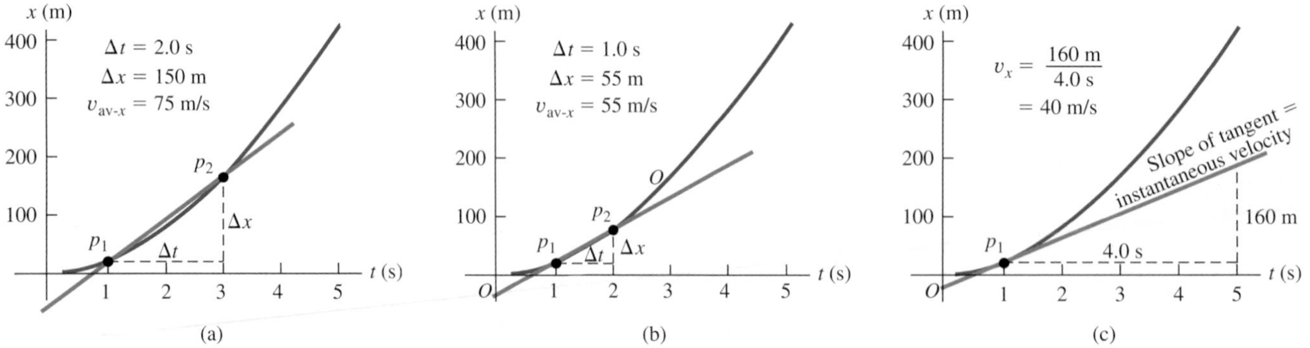

2.7 (a) and (b) As the average velocity $v_{\text{av-}x}$ is calculated over shorter and shorter time intervals, its value approaches the instantaneous velocity. (c) The instantaneous velocity v_x at a certain time equals the slope of the tangent to the x-t curve at that time. We find this slope by dividing any vertical interval (with distance units) along the tangent by the corresponding horizontal interval (with time units).

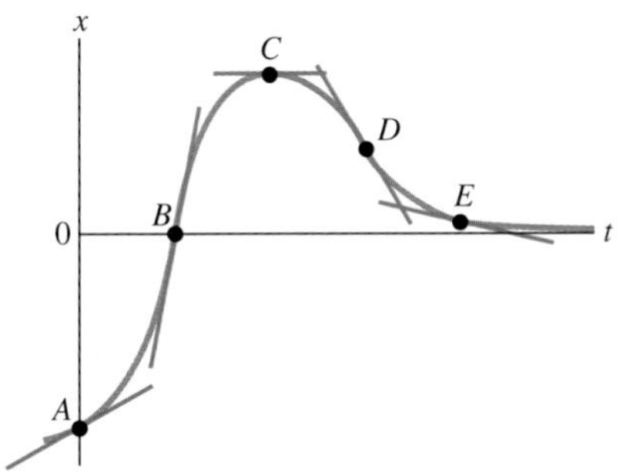

	x-t graph	Motion of particle
A	positive slope, so $v_x > 0$	moving in $+x$-direction
B	larger positive slope, so $v_x > 0$	moving in $+x$-direction faster than at A
C	zero slope, so $v_x = 0$	instantaneously at rest
D	negative slope, so $v_x < 0$	moving in $-x$-direction
E	smaller negative slope, so $v_x < 0$	moving in $-x$-direction more slowly than at D

(a)

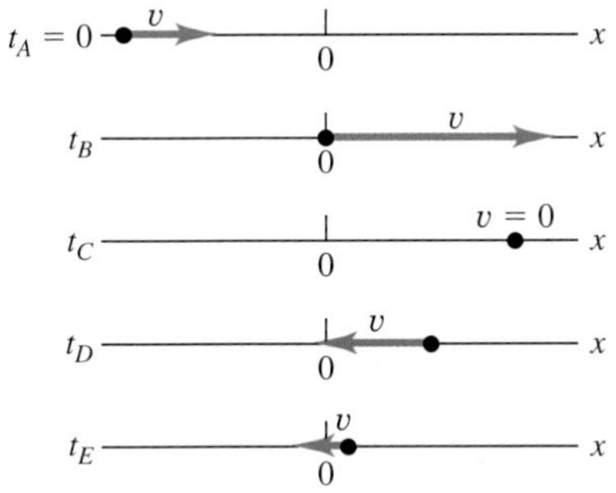

(b)

2.8 (a) The x-t graph of the motion of a particular particle. The slope of the tangent at any point equals the velocity at that point. (b) A motion diagram showing the position and velocity of the particle at each of the five times labeled on the x-t graph. The particle speeds up from A to B, then slows down from B to C, where it comes to a momentary halt. It then moves in the negative x-direction, speeding up from C to D and slowing down from D to E.

Note that Fig. 2.8 depicts the motion of a particle in two ways. Figure 2.8a is an x-t graph, and Fig. 2.8b is an example of a **motion diagram.** A motion diagram shows the position of the particle at various times during the motion, like frames from a movie or video of the particle's motion, as well as arrows to represent the particle's velocity at each instant. Both x-t graphs and motion diagrams are helpful aids to understanding motion, and we will use both frequently in this chapter. You will find it worth your while to draw *both* an x-t graph and a motion diagram as part of solving any problem involving motion.

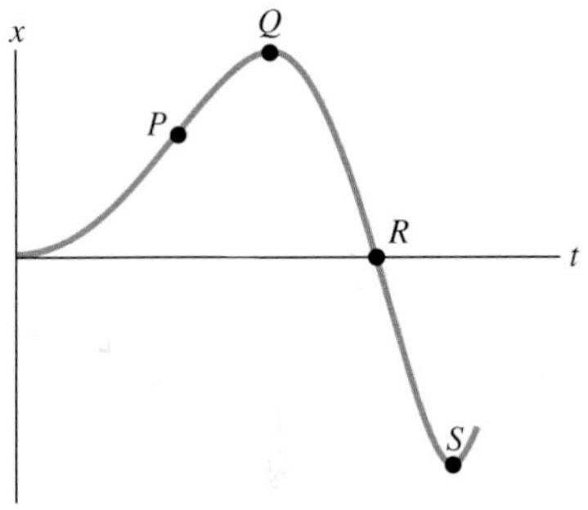

2.9 An x-t graph for a particle.

Test Your Understanding

Figure 2.9 is an x-t graph of the motion of a particle. At which of the points P, Q, R, and S is the velocity v_x positive? At which points is the velocity negative? At which points is it zero? At which point is the *speed* greatest?

2.3 | Average and Instantaneous Acceleration

When the velocity of a moving body changes with time, we say that the body has an *acceleration.* Just as velocity describes the rate of change of position with time, acceleration describes the rate of change of velocity with time. Like velocity, acceleration is a vector quantity. In straight-line motion its only nonzero component is along the axis along which the motion takes place.

Average Acceleration

Let's consider again the motion of a particle along the x-axis. Suppose that at time t_1 the particle is at point P_1 and has x-component of (instantaneous) velocity v_{1x}, and at a later time t_2 it is at point P_2 and has x-component of velocity v_{2x}. So the x-component of velocity changes by an amount $\Delta v_x = v_{2x} - v_{1x}$ during the time interval $\Delta t = t_2 - t_1$.

We define the **average acceleration** $a_{av\text{-}x}$ of the particle as it moves from P_1 to P_2 to be a vector quantity whose x-component is Δv_x, the change in the x-component of velocity, divided by the time interval Δt:

$$a_{av\text{-}x} = \frac{v_{2x} - v_{1x}}{t_2 - t_1} = \frac{\Delta v_x}{\Delta t} \qquad (2.4)$$

(average acceleration, straight-line motion)

For straight-line motion we will usually call $a_{av\text{-}x}$ simply the average acceleration, remembering that in fact it is the x-component of the average acceleration vector. (We'll encounter the other components of the average acceleration vector in Chapter 3.)

If we express velocity in meters per second and time in seconds, then average acceleration is in meters per second per second, or (m/s)/s. This is usually written as m/s^2 and is read "meters per second squared."

CAUTION Be very careful not to confuse acceleration with velocity! Velocity describes how an object's position changes with time; it tells us how fast and in what direction the object moves. Acceleration describes how the velocity changes with time; it tells us how the speed and direction of motion are changing. It may help to remember the phrase "acceleration is to velocity as velocity is to position."

Example 2.2 Average acceleration

An astronaut has left an orbiting space shuttle to test a new personal maneuvering unit. As she moves along a straight line, her partner on board the shuttle measures her velocity every 2.0 s, starting at time $t = 1.0$ s, as follows:

t	v_x	t	v_x
1.0 s	0.8 m/s	9.0 s	−0.4 m/s
3.0 s	1.2 m/s	11.0 s	−1.0 m/s
5.0 s	1.6 m/s	13.0 s	−1.6 m/s
7.0 s	1.2 m/s	15.0 s	−0.8 m/s

Find the average acceleration, and describe whether the speed of the astronaut increases or decreases, for each of the following time intervals: a) $t_1 = 1.0$ s to $t_2 = 3.0$ s; b) $t_1 = 5.0$ s to $t_2 = 7.0$ s; c) $t_1 = 9.0$ s to $t_2 = 11.0$ s; d) $t_1 = 13.0$ s to $t_2 = 15.0$ s.

SOLUTION

IDENTIFY and SET UP: We use the definition of average acceleration, Eq. (2.4), to determine the value of $a_{av\text{-}x}$ from the change in *velocity* for each time interval. We determine the change in *speed* for each time interval by remembering that speed v is the magnitude of the instantaneous velocity v_x.

EXECUTE: The upper part of Fig. 2.10 graphs the velocity as a function of time. On this graph, the slope of the line connecting the points at the beginning and end of each interval equals the average acceleration $a_{av\text{-}x} = \Delta v_x / \Delta t$ for that interval. The values of $a_{av\text{-}x}$ are

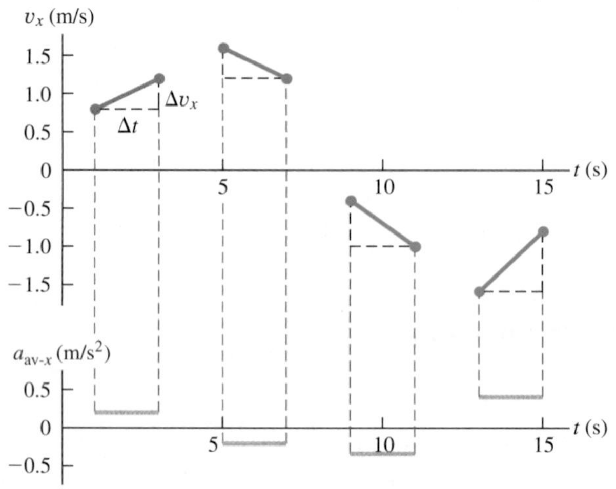

2.10 The slope of the line connecting two points on a graph of velocity versus time (upper graph) equals the average acceleration between those two points (lower graph).

graphed in the lower part of Fig. 2.10. For each time interval, we have

a) $a_{\text{av-}x} = (1.2\text{ m/s} - 0.8\text{ m/s})/(3.0\text{ s} - 1.0\text{ s}) = 0.2\text{ m/s}^2$. The speed (magnitude of instantaneous velocity) increases from 0.8 m/s to 1.2 m/s.

b) $a_{\text{av-}x} = (1.2\text{ m/s} - 1.6\text{ m/s})/(7.0\text{ s} - 5.0\text{ s}) = -0.2\text{ m/s}^2$. The speed decreases from 1.6 m/s to 1.2 m/s.

c) $a_{\text{av-}x} = [-1.0\text{ m/s} - (-0.4\text{ m/s})]/[11.0\text{ s} - 9.0\text{ s}] = -0.3\text{ m/s}^2$. The speed increases from 0.4 m/s to 1.0 m/s.

d) $a_{\text{av-}x} = [-0.8\text{ m/s} - (-1.6\text{ m/s})]/[15.0\text{ s} - 13.0\text{ s}] = 0.4\text{ m/s}^2$. The speed decreases from 1.6 m/s to 0.8 m/s.

EVALUATE: When the acceleration has the *same* direction (same algebraic sign) as the initial velocity, as in intervals a and c, the astronaut goes faster; when it has the *opposite* direction (opposite algebraic sign), as in intervals b and d, she slows down. When she moves in the negative direction with increasing speed (interval c), her velocity is algebraically decreasing (becoming more negative), and her acceleration is negative. But when she moves in the negative direction with decreasing speed (interval d), her velocity is algebraically increasing (becoming less negative) and her acceleration is positive.

Instantaneous Acceleration

We can now define **instantaneous acceleration,** following the same procedure that we used to define instantaneous velocity. Consider this situation: A race car driver has just entered the final straightaway at the Grand Prix. He reaches point P_1 at time t_1, moving with velocity v_{1x}. He passes point P_2, closer to the finish line, at time t_2 with velocity v_{2x} (Fig. 2.11).

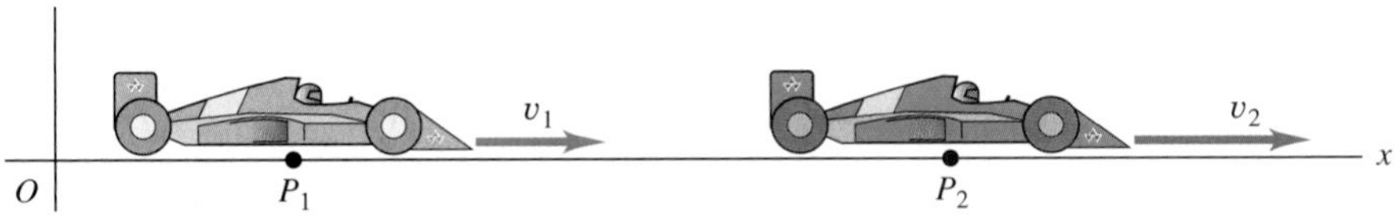

2.11 A Grand Prix car at two points on the straightaway.

To define the instantaneous acceleration at point P_1, we take the second point P_2 in Fig. 2.11 to be closer and closer to the first point P_1 so that the average acceleration is computed over shorter and shorter time intervals. *The instantaneous acceleration is the limit of the average acceleration as the time interval approaches zero.* In the language of calculus, *instantaneous acceleration equals the instantaneous rate of change of velocity with time.* Thus

$$a_x = \lim_{\Delta t \to 0} \frac{\Delta v_x}{\Delta t} = \frac{dv_x}{dt} \qquad (2.5)$$

(instantaneous acceleration, straight-line motion)

Note that Eq. (2.5) is really the definition of the x-component of the acceleration vector; in straight-line motion, all other components of this vector are zero. Instantaneous acceleration plays an essential role in the laws of mechanics. From now on, when we use the term "acceleration," we will always mean instantaneous acceleration, not average acceleration.

Example 2.3 Average and instantaneous accelerations

Suppose the velocity v_x of the car in Fig. 2.11 at any time t is given by the equation

$$v_x = 60\text{ m/s} + (0.50\text{ m/s}^3)t^2$$

(Note that the units for the numbers 60 and 0.50 *must* be as shown to make the expression dimensionally consistent.) a) Find the change in velocity of the car in the time interval between $t_1 = 1.0$ s and $t_2 = 3.0$ s. b) Find the average acceleration in this time interval.

c) Find the instantaneous acceleration at time $t_1 = 1.0$ s by taking Δt to be first 0.1 s, then 0.01 s, then 0.001 s. d) Derive an expression for the instantaneous acceleration at any time, and use it to find the acceleration at $t = 1.0$ s and $t = 3.0$ s.

SOLUTION

IDENTIFY: This example is analogous to Example 2.1 in Section 2.2. (Now would be a good time to review that example.) There we determined the average velocity over shorter and shorter time intervals by considering the change in displacement, and we determined the instantaneous velocity by differentiating the position as a function of time. In this example, we determine the average *acceleration* by considering changes in *velocity* over a time interval. Likewise, we find the *instantaneous* acceleration by differentiating the velocity as a function of time.

SET UP: We'll use Eq. (2.4) for average acceleration and Eq. (2.5) for instantaneous acceleration.

EXECUTE: a) We first find the velocity at each time by substituting each value of t into the equation. At time $t_1 = 1.0$ s,

$$v_{1x} = 60\text{ m/s} + (0.50\text{ m/s}^3)(1.0\text{ s})^2 = 60.5\text{ m/s}$$

At time $t_2 = 3.0$ s,

$$v_{2x} = 60\text{ m/s} + (0.50\text{ m/s}^3(3.0\text{ s})^2 = 64.5\text{ m/s}$$

The change in velocity Δv_x is

$$\Delta v_x = v_{2x} - v_{1x} = 64.5\text{ m/s} - 60.5\text{ m/s} = 4.0\text{ m/s}$$

The time interval is $\Delta t = 3.0\text{ s} - 1.0\text{ s} = 2.0$ s.

b) The average acceleration during this time interval is

$$a_{\text{av-}x} = \frac{v_{2x} - v_{1x}}{t_2 - t_1} = \frac{4.0\text{ m/s}}{2.0\text{ s}} = 2.0\text{ m/s}^2$$

During the time interval from $t_1 = 1.0$ s to $t_2 = 3.0$ s, the velocity and average acceleration have the same algebraic sign (in this case, positive), and the car speeds up.

c) When $\Delta t = 0.1$ s, $t_2 = 1.1$ s and

$$v_{2x} = 60\text{ m/s} + (0.50\text{ m/s}^3)(1.1\text{ s})^2 = 60.605\text{ m/s}$$

$$\Delta v_x = 0.105\text{ m/s}$$

$$a_{\text{av-}x} = \frac{\Delta v_x}{\Delta t} = \frac{0.105\text{ m/s}}{0.1\text{ s}} = 1.05\text{ m/s}^2$$

We invite you to repeat this pattern for $\Delta t = 0.01$ s and $\Delta t = 0.001$ s; the results are $a_{\text{av-}x} = 1.005\text{ m/s}^2$ and $a_{\text{av-}x} = 1.0005\text{ m/s}^2$, respectively. As Δt gets smaller, the average acceleration gets closer to 1.0 m/s². We conclude that the instantaneous acceleration at $t = 1.0$ s is 1.0 m/s².

d) The instantaneous acceleration is $a_x = dv_x/dt$, the derivative of a constant is zero, and the derivative of t^2 is $2t$. Using these formulas, we obtain

$$a_x = \frac{dv_x}{dt} = \frac{d}{dt}[60\text{ m/s} + (0.50\text{ m/s}^3)t^2]$$
$$= (0.50\text{ m/s}^3)(2t) = (1.0\text{ m/s}^3)t$$

When $t = 1.0$ s,

$$a_x = (1.0\text{ m/s}^3)(1.0\text{ s}) = 1.0\text{ m/s}^2$$

When $t = 3.0$ s,

$$a_x = (1.0\text{ m/s}^3)(3.0\text{ s}) = 3.0\text{ m/s}^2$$

EVALUATE: Note that neither of the values we found in part (d) is equal to the average acceleration found in part (b). The instantaneous acceleration of this car varies with time. Automotive engineers sometimes call the rate of change of acceleration with time the "jerk."

Finding Acceleration on a v_x-t Graph or an x-t Graph

We interpreted average and instantaneous velocity in terms of the slope of a graph of position versus time. In the same way, we can get additional insight into the concepts of average and instantaneous acceleration from a graph with instantaneous velocity v on the vertical axis and time t on the horizontal axis; that is, a **v_x-t graph** (Fig. 2.12). The points on the graph labeled p_1 and p_2 correspond to points P_1 and P_2 in Fig. 2.11. The average acceleration $a_{\text{av-}x} = \Delta v_x/\Delta t$ during this interval is the slope of the line p_1p_2. As point P_2 in Fig. 2.11 approaches point P_1, point p_2 in the v_x-t graph of Fig. 2.12 approaches point p_1, and the slope of the line p_1p_2 approaches the slope of the line tangent to the curve at point p_1. Thus, *on a graph of velocity as a function of time, the instantaneous acceleration at any point is equal to the slope of the tangent to the curve at that point.* In Fig. 2.12 the instantaneous acceleration varies with time.

?

Note that by itself, the algebraic sign of the acceleration doesn't tell you whether a body is speeding up or slowing down. You must compare the signs of the velocity and the acceleration. When v_x and a_x have the *same* sign, the body is

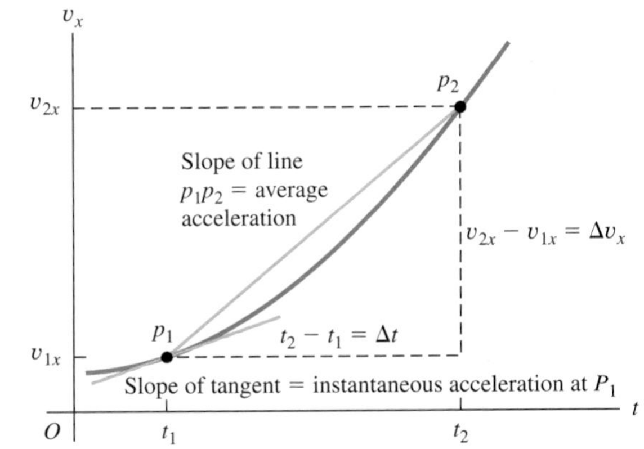

2.12 A v_x-t graph of the motion in Fig. 2.11. The average acceleration between t_1 and t_2 equals the slope of the line p_1p_2. The instantaneous acceleration at P_1 equals the slope of the tangent at p_1.

speeding up. If both are positive, the body is moving in the positive direction with increasing speed. If both are negative, the body is moving in the negative direction with a velocity that is becoming more and more negative, and again the speed is increasing. When v_x and a_x have *opposite* signs, the body is slowing down. If v_x is positive and a_x is negative, the body is moving in the positive direction with decreasing speed; if v_x is negative and a_x is positive, the body is moving in the negative direction with a velocity that is becoming less negative, and again the body is slowing down. Figure 2.13 illustrates these various possibilities.

The term *deceleration* is sometimes used for a decrease in speed. Because it may mean positive or negative a_x, depending on the sign of v_x, we avoid this term.

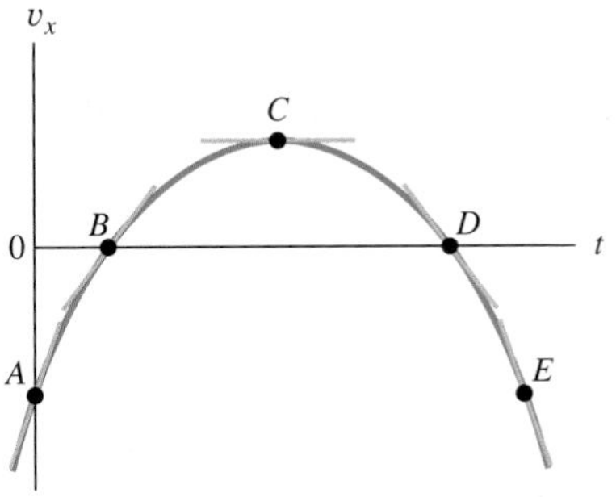

	v_x-t graph	Motion of particle
A	$v_x < 0$; positive slope, so $a_x > 0$	moving in $-x$-direction, slowing down
B	$v_x = 0$; positive slope, so $a_x > 0$	instantaneously at rest, about to move in $+x$-direction
C	$v_x > 0$; zero slope, so $a_x = 0$	moving in $+x$-direction at maximum speed
D	$v_x = 0$; negative slope, so $a_x < 0$	instantaneously at rest, about to move in $-x$-direction
E	$v_x < 0$; negative slope, so $a_x < 0$	moving in $-x$-direction, speeding up

(a)

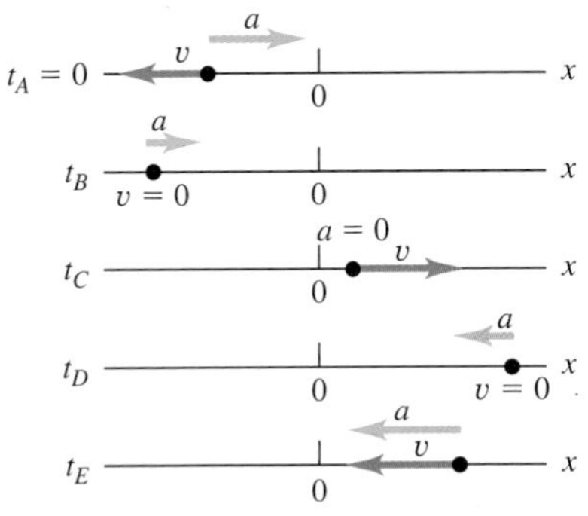

(b)

2.13 (a) A v_x-t graph of the motion of a particle (a different motion than in Fig. 2.8). The slope of the tangent at any point equals the acceleration at that point. (b) A motion diagram showing the position, velocity, and acceleration of the particle at each of the times labeled on the v_x-t graph. The positions are consistent with the v_x-t graph; for instance, from t_A to t_B the velocity is negative, so at t_B the particle is at a more negative value of x than at t_A.

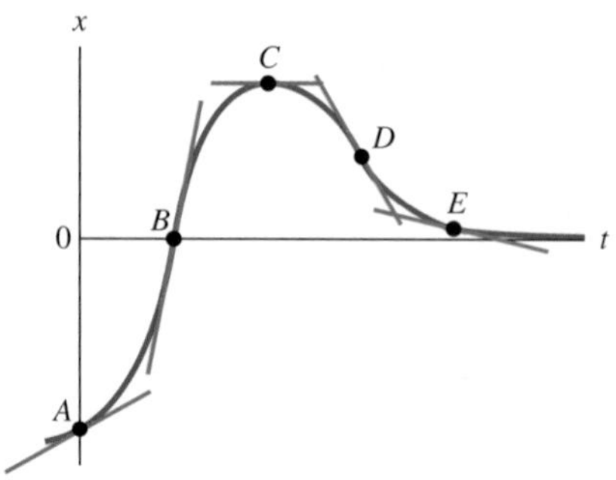

	x-t graph	Motion of particle
A	positive slope, upward curvature, so $v_x > 0$, $a_x > 0$	moving in $+x$-direction, speeding up
B	positive slope, zero curvature, so $v_x > 0$, $a_x = 0$	moving in $+x$-direction, speed not changing
C	zero slope, downward curvature, so $v_x = 0$, $a_x < 0$	instantaneously at rest, velocity changing from + to −
D	negative slope, zero curvature, so $v_x < 0$, $a_x = 0$	moving in $-x$-direction, speed not changing
E	negative slope, upward curvature, so $v_x < 0$, $a_x > 0$	moving in $-x$-direction, slowing down

(a)

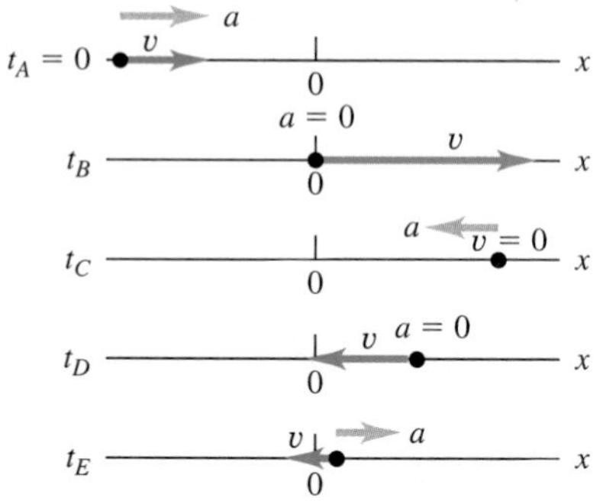

(b)

2.14 (a) The same x-t graph as shown in Fig. 2.8a. The velocity is equal to the *slope* of the graph, and the acceleration is given by the *concavity* or *curvature* of the graph. (b) A motion diagram showing the position, velocity, and acceleration of the particle at each of the times labeled on the x-t graph.

We can also learn about the acceleration of a body from a graph of its *position* versus time. Because $a_x = dv_x/dt$ and $v_x = dx/dt$, we can write

$$a_x = \frac{dv_x}{dt} = \frac{d}{dt}\left(\frac{dx}{dt}\right) = \frac{d^2x}{dt^2} \tag{2.6}$$

That is, a_x is the second derivative of x with respect to t. The second derivative of any function is directly related to the *concavity* or *curvature* of the graph of that function. At a point where the x-t graph is concave up (curved upward), the acceleration is positive and v_x is increasing; at a point where the x-t graph is concave down (curved downward), the acceleration is negative and v is decreasing. At a point where the x-t graph has no curvature, such as an inflection point, the acceleration is zero and the velocity is constant. All three of these possibilities are shown in Fig. 2.14. Examining the curvature of an x-t graph is an easy way to decide what the *sign* of acceleration is. This technique is less helpful for determining numerical values of acceleration, because the curvature of a graph is hard to measure accurately.

Test Your Understanding

Look again at the x-t graph in Fig. 2.9 at the end of Section 2.2. At which of the points P, Q, R, and S is the acceleration a_x positive? At which points is the acceleration negative? At which points does the acceleration appear to be zero? Compare the acceleration at each point to the velocity v_x at that point, and decide whether the speed is increasing, decreasing, or not changing at that point.

2.4 | Motion with Constant Acceleration

The simplest accelerated motion is straight-line motion with *constant* acceleration. In this case the velocity changes at the same rate throughout the motion. This is a very special situation, yet one that occurs often in nature. As we will discuss in the next section, a falling body has a constant acceleration if the effects of the air are not important. The same is true for a body sliding on an incline or along a rough horizontal surface. Straight-line motion with nearly constant acceleration also occurs in technology, such as a jet fighter being catapulted from the deck of an aircraft carrier.

In this section we'll derive key equations for straight-line motion with constant acceleration. These equations will enable us to solve a wide range of problems.

Figure 2.15 is a motion diagram showing the position, velocity, and acceleration at five different times for a particle moving with constant acceleration. Figures 2.16 and 2.17 depict this same motion in the form of graphs. Since the acceleration a_x is constant, the **a_x-t graph** (graph of acceleration versus time) in Fig. 2.16 is a horizontal line. The graph of velocity versus time has a constant *slope* because the acceleration is constant, and so the v_x-t graph is a straight line (Fig. 2.17).

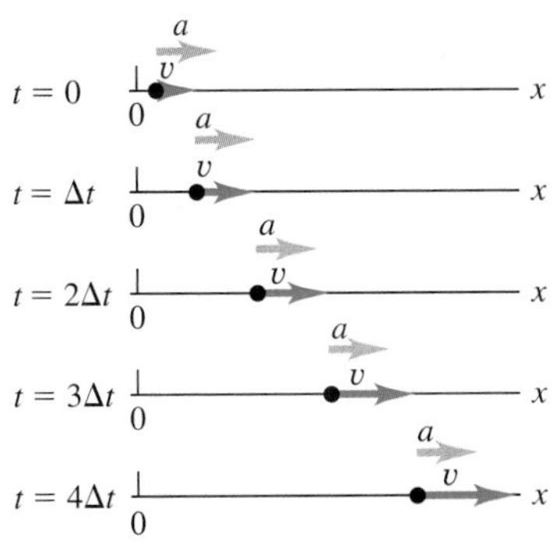

2.15 A motion diagram for a particle moving in a straight line in the positive x-direction with constant positive acceleration a_x. The position, velocity, and acceleration are shown at five equally spaced times. The velocity changes by equal amounts in equal time intervals because the acceleration is constant. The position changes by *different* amounts in equal time intervals because the velocity is changing.

When the acceleration is constant, it's easy to derive equations for position x and velocity v as functions of time. Let's start with velocity. In Eq. (2.4) we can replace the average acceleration $a_{\text{av-}x}$ by the constant (instantaneous) acceleration a_x. We then have

$$a_x = \frac{v_{2x} - v_{1x}}{t_2 - t_1} \qquad (2.7)$$

Now we let $t_1 = 0$ and let t_2 be any arbitrary later time t. We use the symbol v_{0x} for the x-component of velocity at the initial time $t = 0$; the x-component of velocity at the later time t is v_x. Then Eq. (2.7) becomes

$$a_x = \frac{v_x - v_{0x}}{t - 0} \quad \text{or}$$

$$v_x = v_{0x} + a_x t \qquad \text{(constant acceleration only)} \qquad (2.8)$$

We can interpret this equation as follows. The acceleration a_x is the constant rate of change of velocity; that is, the change in velocity per unit time. The term $a_x t$ is the product of the change in velocity per unit time, a_x, and the time interval t. Therefore it equals the *total* change in velocity from the initial time $t = 0$ to the later time t. The velocity v_x at any time t then equals the initial velocity v_{0x} (at $t = 0$) plus the change in velocity $a_x t$. Graphically, we can consider the height v_x of the graph in Fig. 2.17 at any time t to be the sum of two segments: one with length v_{0x} equal to the initial velocity, the other with length $a_x t$ equal to the change in velocity during time t. The graph of velocity as a function of time is a straight line with slope a_x that intercepts the vertical axis (the v-axis) at v_{0x}.

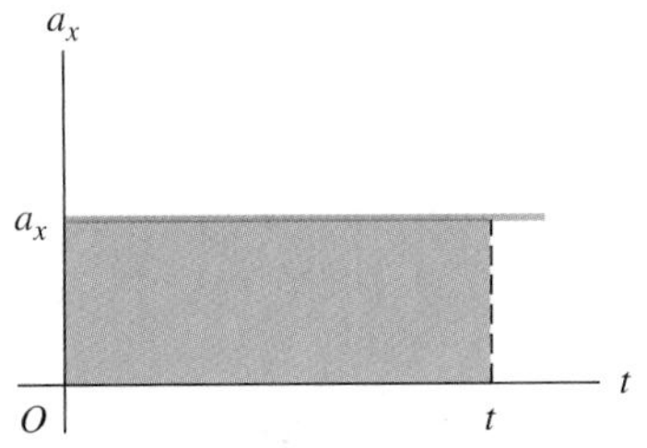

2.16 An acceleration-time (a_x-t) graph for straight-line motion with constant positive acceleration a_x.

Another interpretation of Eq. (2.8) is that the change in velocity $v_x - v_{0x}$ of the particle between $t = 0$ and any later time t equals the *area* under the a_x-t graph between those two times. In Fig. 2.16, the area under the graph of acceleration versus time is shown as a red rectangle of vertical side a_x and horizontal side t. The area of this rectangle is equal to $a_x t$, which from Eq. (2.8) is just equal to the change in velocity $v_x - v_{0x}$. In Section 2.6 we'll show that even if the acceleration is not constant, the change in velocity during a time interval is still equal to the area under the a_x-t curve, although in that case Eq. (2.8) does not apply.

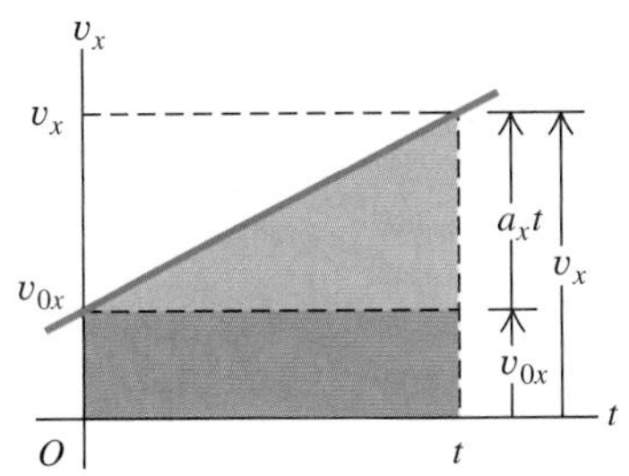

2.17 A velocity-time (v_x-t) graph for straight-line motion with constant positive acceleration a_x. The initial velocity v_{0x} is also positive.

Next we want to derive an equation for the position x of a particle moving with constant acceleration. To do this, we make use of two different expressions for the average velocity $v_{\text{av-}x}$ during the interval from $t = 0$ to any later time t. The first expression comes from the definition of $v_{\text{av-}x}$, Eq. (2.2), which holds true whether or not the acceleration is constant. We call the position at time $t = 0$ the *initial position*,

ActivPhysics ONLINE

1.1 Analyzing Motion Using Diagrams
1.2 Analyzing Motion Using Graphs
1.3 Predicting Motion from Graphs
1.4 Predicting Motion from Equations
1.5 Problem–Solving Strategies for Kinematics
1.6 Skier Races Downhill

denoted by x_0. The position at the later time t is simply x. Thus for the time interval $\Delta t = t - 0$ and the corresponding displacement $\Delta x = x - x_0$, Eq. (2.2) gives

$$v_{\text{av-}x} = \frac{x - x_0}{t} \tag{2.9}$$

We can also get a second expression for $v_{\text{av-}x}$ that is valid only when the acceleration is constant, so that the v_x-t graph is a straight line (as in Fig. 2.17) and the velocity changes at a constant rate. In this case the average velocity during any time interval is simply the arithmetic average of the velocities at the beginning and end of the interval. For the time interval 0 to t,

$$v_{\text{av-}x} = \frac{v_{0x} + v_x}{2} \qquad \text{(constant acceleration only)} \tag{2.10}$$

(This equation is *not* true if the acceleration varies and the v_x-t graph is a curve, as in Fig. 2.13.) We also know that with constant acceleration the velocity v_x at any time t is given by Eq. (2.8). Substituting that expression for v_x into Eq. (2.10), we find

$$v_{\text{av-}x} = \frac{1}{2}(v_{0x} + v_{0x} + a_x t)$$

$$= v_{0x} + \frac{1}{2}a_x t \qquad \text{(constant acceleration only)} \tag{2.11}$$

Finally, we equate Eqs. (2.9) and (2.11) and simplify the result:

$$v_{0x} + \frac{1}{2}a_x t = \frac{x - x_0}{t} \qquad \text{or}$$

$$x = x_0 + v_{0x}t + \frac{1}{2}a_x t^2 \qquad \text{(constant acceleration only)} \tag{2.12}$$

This equation states that if at the initial time $t = 0$, a particle is at position x_0 and has velocity v_{0x}, its new position x at any later time t is the sum of three terms: its initial position x_0, plus the distance $v_{0x}t$ that it would move if its velocity were constant, plus an additional distance $\frac{1}{2}a_x t^2$ caused by the change in velocity.

Just as the change in velocity of the particle equals the area under the a_x-t graph, the displacement—that is, the change in position—equals the area under the v_x-t graph. To be specific, the displacement $x - x_0$ of the particle between $t = 0$ and any later time t equals the area under the v_x-t graph between those two times. In Fig. 2.17 the area under the graph is divided into a dark-colored rectangle of vertical side v_{0x} and horizontal side t and a light-colored right triangle of vertical side $a_x t$ and horizontal side t. The area of the rectangle is $v_{0x}t$ and the area of the triangle is $\frac{1}{2}(a_x t)(t) = \frac{1}{2}a_x t^2$, so the total area under the curve is

$$x - x_0 = v_{0x}t + \frac{1}{2}a_x t^2$$

in agreement with Eq. (2.12).

The displacement during a time interval can always be found from the area under the v_x-t curve. This is true even if the acceleration is *not* constant, although in that case Eq. (2.12) does not apply. (We'll show this in Section 2.6.)

We can check whether Eqs. (2.8) and (2.12) are consistent with the assumption of constant acceleration by taking the derivative of Eq. (2.12). We find

$$v_x = \frac{dx}{dt} = v_{0x} + a_x t$$

which is Eq. (2.8). Differentiating again, we find simply

$$\frac{dv_x}{dt} = a_x$$

as we should expect.

In many problems, it's useful to have a relationship between position, velocity, and acceleration that does not involve the time. To obtain this, we first solve Eq. (2.8) for t, then substitute the resulting expression into Eq. (2.12) and simplify:

$$t = \frac{v_x - v_{0x}}{a_x}$$

$$x = x_0 + v_{0x}\left(\frac{v_x - v_{0x}}{a_x}\right) + \frac{1}{2}a_x\left(\frac{v_x - v_{0x}}{a_x}\right)^2$$

We transfer the term x_0 to the left side and multiply through by $2a_x$:

$$2a_x(x - x_0) = 2v_{0x}v_x - 2v_{0x}^{\ 2} + v_x^2 - 2v_{0x}v_x + v_{0x}^{\ 2}$$

Finally, simplifying gives us

$$v_x^2 = v_{0x}^{\ 2} + 2a_x(x - x_0) \qquad \text{(constant acceleration only)} \qquad (2.13)$$

We can get one more useful relationship by equating the two expressions for $v_{\text{av-}x}$, Eqs. (2.9) and (2.10), and multiplying through by t. Doing this, we obtain

$$x - x_0 = \left(\frac{v_{0x} + v_x}{2}\right)t \qquad \text{(constant acceleration only)} \qquad (2.14)$$

Note that Eq. (2.14) does not contain the acceleration a_x. This is sometimes a useful equation when a_x is constant but its value is unknown.

Equations (2.8), (2.12), (2.13), and (2.14) are the *equations of motion with constant acceleration.* By using these equations, we can solve *any* kinematics problem involving straight-line motion of a particle with constant acceleration.

Figure 2.18 shows a graph of the coordinate x as a function of time for motion with constant acceleration. That is, it is a graph of Eq. (2.12); the x-t graph for constant acceleration is always a *parabola.* The curve intercepts the vertical axis (x-axis) at x_0, the position at $t = 0$. The slope of the tangent at $t = 0$ equals v_{0x}, the initial velocity, and the slope of the tangent at any time t equals the velocity v_x at that time. The graph in Fig. 2.18 is concave up (it curves upward). The slope and velocity are continuously increasing, so the acceleration a_x is positive. If a_x is negative, the x-t graph is a parabola that is concave down (has a downward curvature).

For the particular case of motion with constant acceleration depicted in Fig. 2.15 and graphed in Figs. 2.16, 2.17, and 2.18, the values of x_0, v_{0x}, and a_x are all positive. We invite you to redraw these figures for cases in which one, two, or all three of these quantities are negative.

A special case of motion with constant acceleration occurs when the acceleration is *zero.* The velocity is then constant, and the equations of motion become simply

$$v_x = v_{0x} = \text{constant}$$
$$x = x_0 + v_x t$$

ActivPhysics Online

1.8 Seat Belts Save Lives
1.9 Screeching to a Halt
1.11 Car Starts, Then Stops
1.12 Solving Two-Vehicle Problems
1.13 Car Catches Truck
1.14 Avoiding a Rear-End Collision

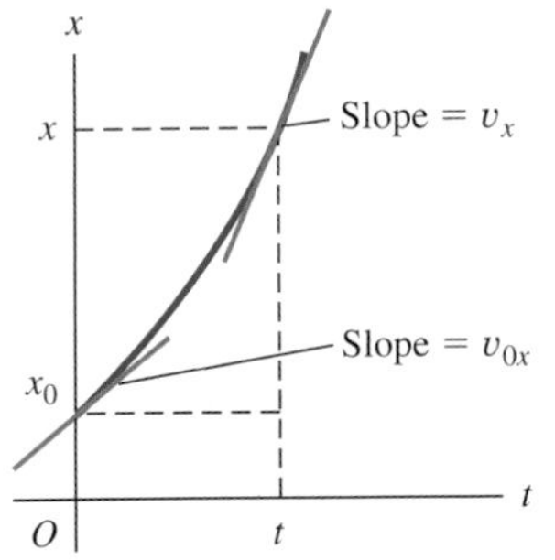

2.18 A position-time (x-t) graph for straight-line motion with constant acceleration. This graph refers to the same motion as is shown in Figs. 2.15, 2.16, and 2.17. For this motion the initial position x_0, the initial velocity v_{0x}, and the acceleration a_x are all positive.

Problem-Solving Strategy: Motion with Constant Acceleration

IDENTIFY *the relevant concepts:* In most straight-line motion problems, you can use the constant-acceleration equations. Occasionally, however, you will encounter a situation in which the acceleration *isn't* constant. In such a case, you'll need a different approach (see Section 2.6).

SET UP *the problem using the following steps:*

1. You *must* decide at the beginning of a problem where the origin of coordinates is and which axis direction is positive. These choices are usually a matter of convenience. It is often easiest to place the particle at the origin at time $t = 0$; then $x_0 = 0$. It is always helpful to make a motion diagram showing these choices and some later positions of the particle.
2. Remember that your choice of the positive axis direction automatically determines the positive directions for velocity and acceleration. If x is positive to the right of the origin, then v_x and a_x are also positive toward the right.
3. Restate the problem in words first, and then translate this description into symbols and equations. *When* does the particle arrive at a certain point (that is, what is the value of t)? *Where* is the particle when its velocity has a specified value (that is, what is the value of x when v_x has the specified value)? Example 2.4 asks, "Where is the motorcyclist when his velocity is 25 m/s?" Translated into symbols, this becomes "What is the value of x when $v_x = 25$ m/s?"
4. Make a list of quantities such as x, x_0, v_x, v_{0x}, a_x, and t. In general, some of them will be known and some will be unknown. Write down the values of the known quantities, and decide which of the unknowns are the target variables. Be on the lookout for implicit information. For example, "A car sits at a stoplight" usually means $v_{0x} = 0$.

EXECUTE *the solution:* Choose an equation from Eqs. (2.8), (2.12), (2.13), and (2.14) that contains only one of the target variables. Solve this equation for the target variable, using symbols only. Then substitute the known values and compute the value of the target variable. Sometimes you will have to solve two simultaneous equations for two unknown quantities.

EVALUATE *your answer:* Take a hard look at your results to see whether they make sense. Are they within the general range of values you expected?

Example 2.4: Constant-acceleration calculations

A motorcyclist heading east through a small Iowa city accelerates after he passes the signpost marking the city limits (Fig. 2.19). His acceleration is a constant 4.0 m/s². At time $t = 0$ he is 5.0 m east of the signpost, moving east at 15 m/s. a) Find his position and velocity at time $t = 2.0$ s. b) Where is the motorcyclist when his velocity is 25 m/s?

SOLUTION

IDENTIFY: The statement of the problem tells us explicitly that the acceleration is constant, so we can use the constant-acceleration equations.

SET UP: We take the signpost as the origin of coordinates ($x = 0$), and choose the positive x-axis to point east (Fig. 2.19). At the initial time $t = 0$, the initial position is $x_0 = 5.0$ m and the initial velocity is $v_{0x} = 15$ m/s. The constant acceleration is $a_x = 4.0\ \text{m/s}^2$. The unknown target variables in part (a) are the values of the position x and the velocity v_x at the later time $t = 2.0$ s; the target variable in part (b) is the value of x when $v_x = 25$ m/s.

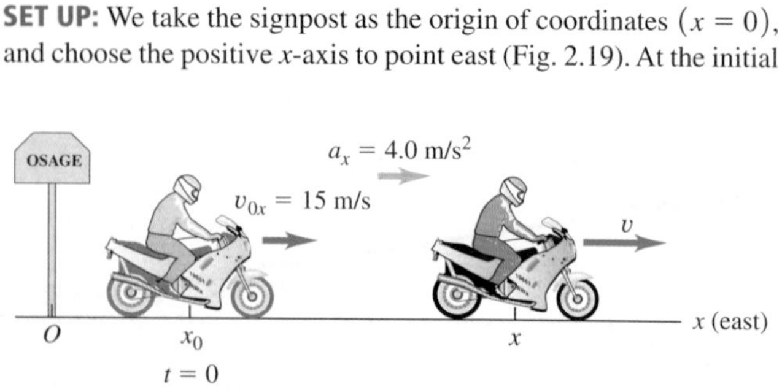

2.19 A motorcyclist traveling with constant acceleration.

EXECUTE: a) We can find the position at $t = 2.0$ s by using Eq. (2.12), which gives position x as a function of time t:

$$x = x_0 + v_{0x} + \frac{1}{2}a_x t^2$$

$$= 5.0\ \text{m} + (15\ \text{m/s})(2.0\ \text{s}) + \frac{1}{2}(4.0\ \text{m/s}^2)(2.0\ \text{s})^2$$

$$= 43\ \text{m}$$

We can find the velocity at this same time by using Eq. (2.8), which gives velocity v_x as a function of time t:

$$v_x = v_{0x} + a_x t$$

$$= 15\ \text{m/s} + (4.0\ \text{m/s}^2)(2.0\ \text{s}) = 23\ \text{m/s}$$

b) From our solution to part (a), we can see that the velocity is $v_x = 25$ m/s at a time later than 2.0 s and at a point farther than 43 m from the signpost. From Eq. (2.13) we have

$$v_x^2 = v_{0x}^2 + 2a_x(x - x_0)$$

Solving for x and substituting in the known values, we find

$$x = x_0 + \frac{v_x^2 - v_{0x}^2}{2a_x}$$

$$= 5.0\ \text{m} + \frac{(25\ \text{m/s})^2 - (15\ \text{m/s})^2}{2(4.0\ \text{m/s}^2)}$$

$$= 55\ \text{m}$$

Alternatively, we may use Eq. (2.8) to first find the time when $v_x = 25$ m/s:

$$v_x = v_{0x} + a_x t \quad \text{so}$$

$$t = \frac{v_x - v_{0x}}{a_x} = \frac{25\ \text{m/s} - 15\ \text{m/s}}{4.0\ \text{m/s}^2}$$

$$= 2.5\ \text{s}$$

Then from Eq. (2.12) we have

$$x = x_0 + v_{0x}t + \frac{1}{2}a_x t^2$$

$$= 5.0\ \text{m} + (15\ \text{m/s})(2.5\ \text{s}) + \frac{1}{2}(4.0\ \text{m/s}^2)(2.5\ \text{s})^2$$

$$= 55\ \text{m}$$

EVALUATE: Do these results make sense? According to our results in part (a), the motorcyclist accelerates from 15 m/s (about 34 mi/h, or 54 km/h) to 23 m/s (about 51 mi/h, or 83 km/h) in 2.0 s while traveling a distance of 38 m (about 125 ft). Our results in part (b) tell us that after an additional 0.5 s, the motorcyclist has moved an additional 12 m (about 39 ft) and has accelerated to 25 m/s (56 mi/h, or 90 km/h). This is pretty brisk acceleration, but well within the capabilities of a high-performance bike.

Example 2.5 Two bodies with different accelerations

A motorist traveling with constant velocity of 15 m/s (about 34 mi/h) passes a school-crossing corner, where the speed limit is 10 m/s (about 22 mi/h). Just as the motorist passes, a police officer on a motorcycle stopped at the corner starts off in pursuit with constant acceleration of 3.0 m/s² (Fig. 2.20a). a) How much time elapses before the officer catches up with the motorist? b) What is the officer's speed at that point? c) What is the total distance each vehicle has traveled at that point?

SOLUTION

IDENTIFY: The police officer and the motorist both move with constant acceleration (equal to zero for the motorist), so we can use the formulas we have developed.

SET UP: We take the origin at the corner, so $x_0 = 0$ for both, and take the positive direction to the right. Let x_P (for police) be the officer's position and x_M (for motorist) be the motorist's position at any time. The initial velocities are $v_{P0x} = 0$ for the officer and $v_{M0x} = 15$ m/s for the motorist; the constant accelerations are $a_{Px} = 3.0\ \text{m/s}^2$ for the officer and $a_{Mx} = 0$ for the motorist. Our target variable in part (a) is the time when the officer catches the motorist—that is, when the two vehicles are at the same position. In part (b) we're looking for the officer's speed v (the magnitude of his velocity) at the time found in part (a). In part (c) we want to find the position of either vehicle at this same time. Hence we use Eq. (2.12) (which relates position and time) in parts (a) and (c), and Eq. (2.8) (which relates velocity and time) in part (b).

EXECUTE: a) We want to find the value of the time t when the motorist and the police officer are at the same position: $x_M = x_P$. Applying Eq. (2.12), $x = x_0 + v_{0x}t + (1/2)a_x t^2$, to each vehicle, we get

$$x_M = 0 + v_{M0x}t + \frac{1}{2}(0)t^2 = v_{M0x}t$$

$$x_P = 0 + (0)t + \frac{1}{2}a_{Px}t^2 = \frac{1}{2}a_{Px}t^2$$

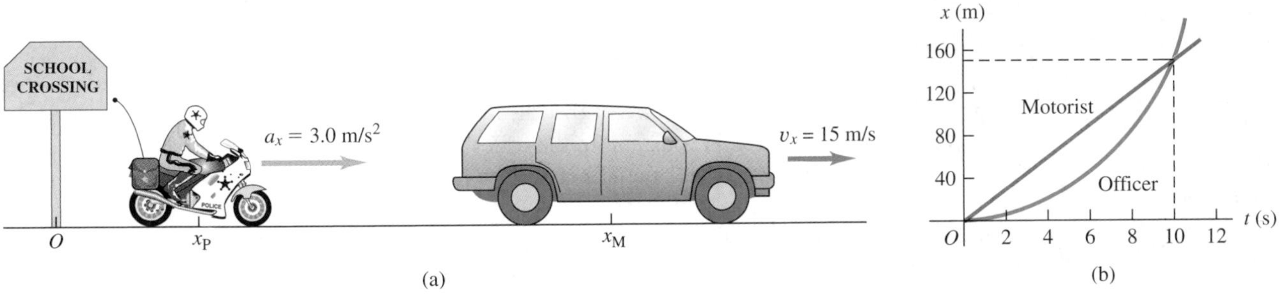

2.20 (a) Motion with constant acceleration overtaking motion with constant velocity. (b) Graph of x versus t for each vehicle.

Since $x_M = x_P$ at time t, we set these two expressions equal to each other and solve for t:

$$v_{M0x}t = \frac{1}{2}a_{Px}t^2$$

$$t = 0 \quad \text{or} \quad t = \frac{2v_{M0x}}{a_{Px}} = \frac{2(15\ \text{m/s})}{3.0\ \text{m/s}^2} = 10\ \text{s}$$

There are *two* times when both the vehicles have the same x-coordinate. The first, $t = 0$, is the time when the motorist passes the parked motorcycle at the corner. The second, $t = 10$ s, is the time when the officer catches up with the motorist.

b) We want the magnitude of the officer's velocity v_{Px} at the time t found in part (a). Her velocity at any time is given by Eq. (2.8):

$$v_{Px} = v_{P0x} + a_{Px}t = 0 + (3.0\ \text{m/s}^2)t$$

so when $t = 10$ s, we find $v_{Px} = 30$ m/s. When the officer overtakes the motorist, she is traveling twice as fast as the motorist is.

c) In 10 s the distance the motorist travels is

$$x_M = v_{M0x}t = (15\ \text{m/s})(10\ \text{s}) = 150\ \text{m}$$

and the distance the officer travels is

$$x_P = \frac{1}{2}a_{Px}t^2 = \frac{1}{2}(3.0\ \text{m/s}^2)(10\ \text{s})^2 = 150\ \text{m}$$

This verifies that at the time the officer catches the motorist, they have gone equal distances.

EVALUATE: Figure 2.20b shows graphs of x versus t for each vehicle. We see again that there are two times when the two positions are the same (where the two graphs cross). At neither of these times do the two vehicles have the same velocity (i.e., where the two graphs cross, their slopes are different). At $t = 0$, the officer is at rest; at $t = 10$ s, the officer has twice the speed of the motorist.

In a real pursuit the officer would accelerate to a speed faster than that of the motorist, then slow down in order to have the same velocity as the motorist when she catches him. We haven't treated this case here because it involves a changing acceleration (but see Problem 2.68).

Test Your Understanding

Figure 2.20b is an x-t graph for each vehicle in Example 2.5. Draw a v_x-t graph for each vehicle. Do the motorist and the police officer ever have the same velocity? At what time does this happen?

2.5 | Freely Falling Bodies

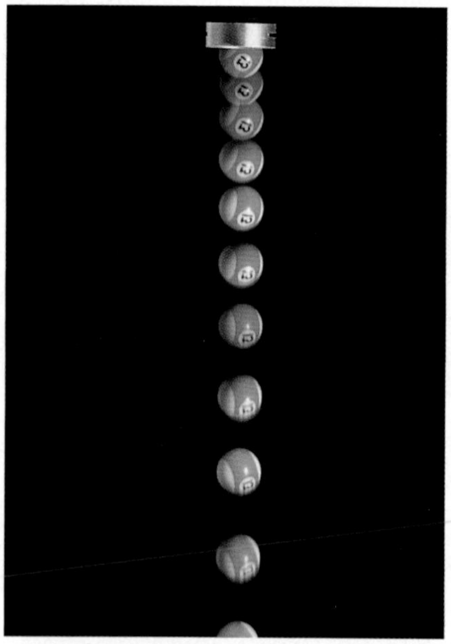

2.21 Multiflash photo of a freely falling ball.

The most familiar example of motion with (nearly) constant acceleration is that of a body falling under the influence of the earth's gravitational attraction. Such motion has held the attention of philosophers and scientists since ancient times. In the 4th century B.C., Aristotle thought (erroneously) that heavy objects fall faster than light objects, in proportion to their weight. Nineteen centuries later, Galileo argued that a body should fall with a downward acceleration that is constant and independent of its weight. We mentioned in Section 1.1 that, according to legend, Galileo experimented by dropping bullets and cannonballs from the Leaning Tower of Pisa.

The motion of falling bodies has since been studied with great precision. When the effects of the air can be neglected, Galileo is right; all bodies at a particular location fall with the same downward acceleration, regardless of their size or weight. If the distance of the fall is small compared to the radius of the earth, the acceleration is constant. In the following discussion we use an idealized model in which we neglect the effects of the air, the earth's rotation, and the decrease of acceleration with increasing altitude. We call this idealized motion **free fall,** although it includes rising as well as falling motion. (In Chapter 3 we will extend the discussion of free fall to include the motion of projectiles, which move both vertically and horizontally.)

Figure 2.21 is a photograph of a falling ball taken with a stroboscopic light source that produces a series of intense flashes at equal time intervals. Each flash is so short (a few millionths of a second) that there is little blur in the image of even a rapidly moving body. As each flash occurs, the position of the ball at that instant is recorded on the film. Because of the equal time intervals between flashes, the average velocity of the ball between any two flashes is proportional to the distance between corresponding images in the photograph. The increasing distances between images show that the velocity is continuously changing; the ball is accelerating downward. Careful measurement shows that the velocity change is the same in each time interval, so the acceleration of the freely falling ball is constant.

ActivPhysics ONLINE

1.7 Balloonist Drops Lemonade
1.10 Pole-Vaulter Lands

The constant acceleration of a freely falling body is called the **acceleration due to gravity,** and we denote its magnitude with the letter g. We will frequently use the approximate value of g at or near the earth's surface:

$$g = 9.8\ \text{m/s}^2 = 980\ \text{cm/s}^2$$
$$= 32\ \text{ft/s}^2 \qquad \text{(approximate value near the earth's surface)}$$

The exact value varies with location, so we will often give the value of g at the earth's surface to only two significant figures. Because g is the magnitude of a vector quantity, it is always a *positive* number. On the surface of the moon the acceleration due to gravity is caused by the attractive force of the moon rather than the earth, and $g = 1.6\ \text{m/s}^2$. Near the surface of the sun, $g = 270\ \text{m/s}^2$.

In the following examples we use the constant-acceleration equations developed in Section 2.4. We suggest that you review the problem-solving strategies discussed in that section before you study the examples below.

Example 2.6 A freely-falling coin

A one-euro coin is dropped from the Leaning Tower of Pisa. It starts from rest and falls freely. Compute its position and velocity after 1.0, 2.0, and 3.0 s.

SOLUTION

IDENTIFY: "Falls freely" means "has a constant acceleration due to gravity," so we can use the constant-acceleration equations to determine our target variables.

SET UP: We take the origin O at the starting point, the coordinate axis as vertical, and the upward direction as positive (Fig. 2.22). Because the coordinate axis is vertical, let's call the coordinate y instead of x. We replace all the x's in the constant-acceleration equations by y's. The initial coordinate y_0 and the initial velocity v_{0y} are both zero. The acceleration is downward (in the negative y-direction), so $a_y = -g = -9.8\ \text{m/s}^2$. (Remember that, by definition, g itself is *always* positive.) Hence our target variables are the values of y and v_y at the three given times. To find these, we use Eqs. (2.12) and (2.8) with x replaced by y.

EXECUTE: At an arbitrary time t, the position and velocity are

$$y = v_{0y}t + \frac{1}{2}a_y t^2 = 0 + \frac{1}{2}(-g)t^2 = (-4.9\ \text{m/s}^2)t^2$$

$$v_y = v_{0y} + a_y t = 0 + (-g)t = (-9.8\ \text{m/s}^2)t$$

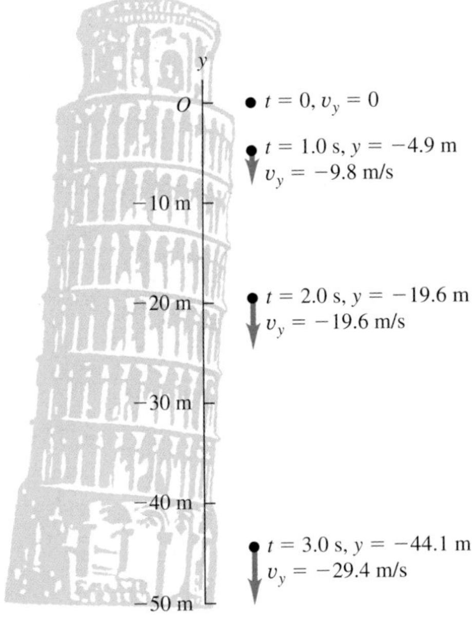

2.22 Position and velocity of a coin freely falling from rest.

When $t = 1.0$ s, $y = (-4.9\text{ m/s}^2)(1.0\text{ s})^2 = -4.9$ m and $v_y = (-9.8\text{ m/s}^2)(1.0\text{ s}) = -9.8$ m/s; after one second, the coin is 4.9 m below the origin (y is negative) and has a downward velocity (v_y is negative) with magnitude 9.8 m/s.

The position and velocity at 2.0 and 3.0 s are found in the same way. The results are shown in Fig. 2.22; check the numerical values for yourself.

EVALUATE: All our answers for v_y are negative because we chose the positive y-axis to point upward. But we could just as well have chosen the positive y-axis to point downward. In that case the acceleration would have been $a_y = +g$ and all our answers for v_y would have been positive. Either choice of axis is fine; just make sure that you state your choice explicitly in your solution and confirm that the acceleration has the correct sign.

Example 2.7 Up and down motion in free fall

You throw a ball vertically upward from the roof of a tall building. The ball leaves your hand at a point even with the roof railing with an upward speed of 15.0 m/s; the ball is then in free fall. On its way back down, it just misses the railing. At the location of the building, $g = 9.80\text{ m/s}^2$. Find a) the position and velocity of the ball 1.00 s and 4.00 s after leaving your hand; b) the velocity when the ball is 5.00 m above the railing; c) the maximum height reached and the time at which it is reached; and d) the acceleration of the ball when it is at its maximum height.

SOLUTION

IDENTIFY: The words "free fall" in the statement of the problem mean that the acceleration is constant and due to gravity. Our target variables are position [in parts (a) and (c)], velocity [in parts (a) and (b)], and acceleration [in part (d)].

SET UP: In Fig. 2.23 the downward path is displaced a little to the right of its actual position for clarity. Take the origin at the roof railing, at the point where the ball leaves your hand, and take the positive direction to be upward. First, let's collect our data. The initial position y_0 is zero. The initial velocity v_{0y} is $+15.0$ m/s, and the acceleration is $a_y = -g = -9.80\text{ m/s}^2$. We'll again use Eqs. (2.12) and (2.8) to find the position and velocity, respectively, as functions of time. In part (b) we're asked to find the velocity at a certain *position* rather than at a certain *time,* so it will also be convenient to use Eq. (2.13) for that part.

EXECUTE: a) The position y and velocity v_y at any time t after the ball leaves your hand are given by Eqs. (2.12) and (2.8) with x's replaced by y's, so

$$y = y_0 = v_{0y}t + \frac{1}{2}a_y t^2 = y_0 + v_{0y}t + \frac{1}{2}(-g)t^2$$

$$= (0) + (15.0\text{ m/s})t + \frac{1}{2}(-9.80\text{ m/s}^2)t^2$$

$$v_y = v_{0y} + a_y t = v_{0y} + (-g)t$$

$$= 15.0\text{ m/s} + (-9.80\text{ m/s}^2)t$$

When $t = 1.00$ s, these equations give

$$y = +10.1\text{ m} \qquad v_y = +5.2\text{ m/s}$$

The ball is 10.1 m above the origin (y is positive), and it is moving upward (v_y is positive) with a speed of 5.2 m/s. This is less than the initial speed of 15.0 m/s, as expected.

When $t = 4.00$ s, the equations for y and v_y as functions of time t give

$$y = -18.4\text{ m} \qquad v_y = -24.2\text{ m/s}$$

The ball has passed its highest point and is 18.4 m *below* the origin (y is negative). It has a *downward* velocity (v_y is negative) with magnitude 24.2 m/s. This speed is greater than the initial speed, as we should expect for points below the ball's launching point. Note that to get these results, we don't need to find the highest point reached or the time at which it was reached. The equations of motion give the position and velocity at any time, whether the ball is on the way up or on the way down.

b) The velocity v_y at any position y is given by Eq. (2.13) with x's replaced by y's:

$$v_y^2 = v_{0y}^2 + 2a_y(y - y_0) = v_{0y}^2 + 2(-g)(y - 0)$$

$$= (15.0\text{ m/s})^2 + 2(-9.80\text{ m/s}^2)y$$

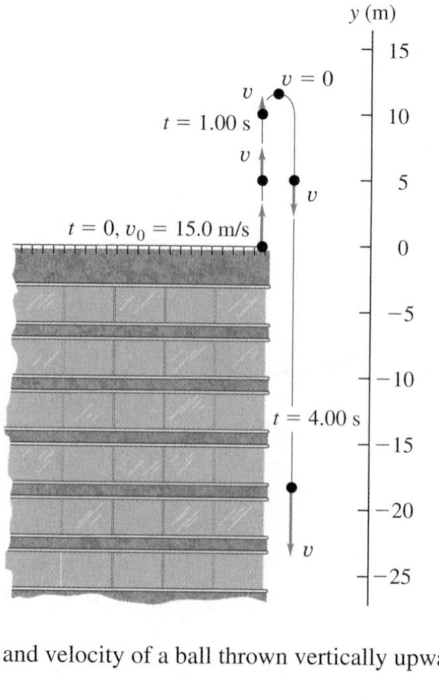

2.23 Position and velocity of a ball thrown vertically upward.

When the ball is 5.00 m above the origin, $y = +5.00$ m, so

$$v_y^2 = (15.0 \text{ m/s})^2 + 2(-9.80 \text{ m/s}^2)(5.00 \text{ m}) = 127 \text{ m}^2/\text{s}^2$$
$$v_y = \pm 11.3 \text{ m/s}$$

We get *two* values of v_y, one positive and one negative. As shown in Fig. 2.23, the ball passes this point twice, once on the way up and again on the way down. The velocity on the way up is +11.3 m/s, and on the way down it is −11.3 m/s.

c) At the highest point, the ball stops going up (positive v_y) and starts going down (negative v_y). Just at the instant when it reaches the highest point, $v_y = 0$. The maximum height y_1 can then be found in two ways. The first way is to use Eq. (2.13) and substitute $v_y = 0$, $y_0 = 0$, and $a_y = -g$:

$$0 = v_{0y}^2 + 2(-g)(y_1 - 0)$$
$$y_1 = \frac{v_{0y}^2}{2g} = \frac{(15.0 \text{ m/s})^2}{2(9.80 \text{ m/s}^2)} = +11.5 \text{ m}$$

The second way is find the time at which $v = 0$ using Eq. (2.8), $v_y = v_{0y} + a_y t$, and then substitute this value of t into Eq. (2.12) to find the position at this time. From Eq. (2.8), the time t_1 when the ball reaches the highest point is given by

$$v_y = 0 = v_{0y} + (-g)t_1$$
$$t_1 = \frac{v_{0y}}{g} = \frac{15.0 \text{ m/s}}{9.80 \text{ m/s}^2} = 1.53 \text{ s}$$

Substituting this value of t into Eq. (2.12), we find

$$y = y_0 + v_{0y}t + \frac{1}{2}a_y t^2 = (0) + (15 \text{ m/s})(1.53 \text{ s})$$
$$+ \frac{1}{2}(-9.8 \text{ m/s}^2)(1.53 \text{ s})^2 = +11.5 \text{ m}$$

Notice that with the first way of finding the maximum height, it's not necessary to find the time first.

d) **CAUTION** It's a common misconception that at the highest point of the motion the velocity is zero *and* the acceleration is zero. If this were so, once the ball reached the highest point it would hang there in midair forever! To see why, remember that acceleration is the rate of change of velocity. If the acceleration were zero at the highest point, the ball's velocity would no longer change, and once the ball was instantaneously at rest it would remain at rest forever.

The truth of the matter is that at the highest point, the acceleration is still $a_y = -g = -9.80$ m/s², the same value as when the ball is moving up and when it's moving down. The ball stops for an instant at the highest point, but its velocity is continuously changing, from positive values through zero to negative values.

EVALUATE: A useful way to check any motion problem is to draw the graphs of position versus time and velocity versus time. Figure 2.24 shows these graphs for the situation in this problem. Note that the v_y-t graph has a constant negative slope. This means that the acceleration is negative (downward) on the way up, at the highest point, and on the way down.

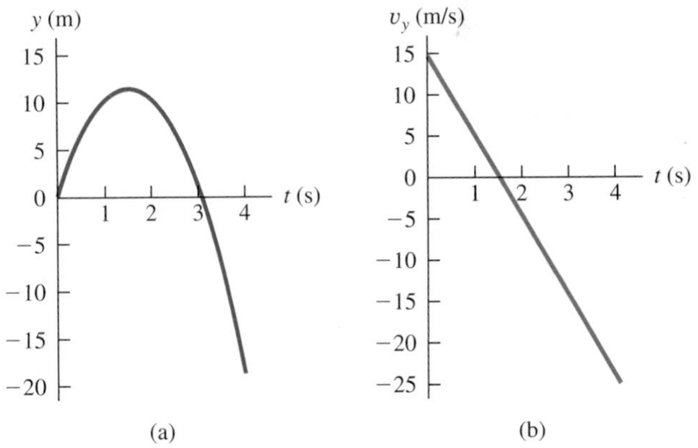

2.24 (a) Position and (b) velocity as functions of time for a ball thrown upward with an initial velocity of 15 m/s.

Example 2.8 Two solutions or one?

Find the time when the ball in Example 2.7 is 5.00 m below the roof railing.

SOLUTION

IDENTIFY AND SET UP: We again choose the y-axis as in Fig. 2.23, so y_0, v_{0y}, and $a_y = -g$ have the same values as in Example 2.7. The position y as a function of time t is again given by Eq. (2.12):

$$y = y_0 + v_{0y}t + \frac{1}{2}a_y t^2 = y_0 + v_{0y}t + \frac{1}{2}(-g)t^2$$

We want to solve this for the value of t when $y = -5.00$ m. Since this equation involves t^2, it is a *quadratic* equation for t.

EXECUTE: We first rearrange our equation into the same form as the standard quadratic equation for an unknown x, $Ax^2 + Bx + C = 0$:

$$\left(\frac{1}{2}g\right)t^2 + (-v_{0y})t + (y - y_0) = At^2 + Bt + C = 0$$

so $A = g/2$, $B = -v_{0y}$, and $C = y - y_0$. Using the quadratic formula (Appendix B), we find that this equation has *two* solutions:

$$t = \frac{-B \pm \sqrt{B^2 - 4AC}}{2A}$$

$$= \frac{-(-v_{0y}) \pm \sqrt{(-v_{0y})^2 - 4(g/2)(y - y_0)}}{2(g/2)}$$

$$t = \frac{v_{0y} \pm \sqrt{v_{0y}^2 - 2g(y - y_0)}}{g}$$

Substituting the values $y_0 = 0$, $v_{0y} = +15.0$ m/s, $g = 9.80$ m/s^2, and $y = -5.00$ m, we find

$$t = \frac{(15.0\ \text{m/s}) \pm \sqrt{(15.0\ \text{m/s})^2 - 2(9.80\ \text{m/s}^2)(-5.00\ \text{m} - 0)}}{9.80\ \text{m/s}^2}$$

$t = +3.36$ s or $t = -0.30$ s

To decide which of these is the right answer, the key question to ask is, "Are these answers reasonable?" The second answer, $t = -0.30$ s, is simply not reasonable; it refers to a time 0.30 s *before* the ball left your hand! The correct answer is $t = +3.36$ s. The ball is 5.00 m below the railing 3.36 s *after* it leaves your hand.

EVALUATE: Where did the erroneous "solution" $t = -0.30$ s come from? Remember that we began with Eq. (2.12) with $a_y = -g$, that is, $y = y_0 + v_{0y}t + \frac{1}{2}(-g)t^2$. This equation has built into it the assumption that the acceleration is constant for *all* values of t, whether positive, negative, or zero. Taken at face value, this equation would tell us that the ball had been moving upward in free fall ever since the dawn of time; it eventually passes your hand at $y = 0$ at the special instant we chose to call $t = 0$, then continues in free fall. But anything that this equation describes happening before $t = 0$ is pure fiction, since the ball went into free fall only after leaving your hand at $t = 0$; the "solution" $t = -0.30$ s is part of this fiction.

You should repeat these calculations to find the times at which the ball is 5.00 m *above* the origin ($y = +5.00$ m). The two answers are $t = +0.38$ s and $t = +2.68$ s; these are both positive values of t, and both refer to the real motion of the ball after leaving your hand. The earlier time is when the ball passes through $y = +5.00$ m moving upward, and the later time is when the ball passes through this point moving downward. (Compare this with part (b) of Example 2.7.) You should also solve for the times at which $y = +15.0$ m. In this case, both solutions involve the square root of a negative number, so there are *no* real solutions. This makes sense; we found in part (c) of Example 2.7 that the ball's maximum height is only $y = +11.5$ m, so it never reaches $y = +15.0$ m. While a quadratic equation such as Eq. (2.12) always has two solutions, in some situations one or both of the solutions will not be physically reasonable.

Test Your Understanding

If you toss a ball upward with a certain velocity, it falls freely and reaches a maximum height h. What maximum height does the ball reach if you throw it upward with double the initial velocity?

*2.6 | Velocity and Position by Integration

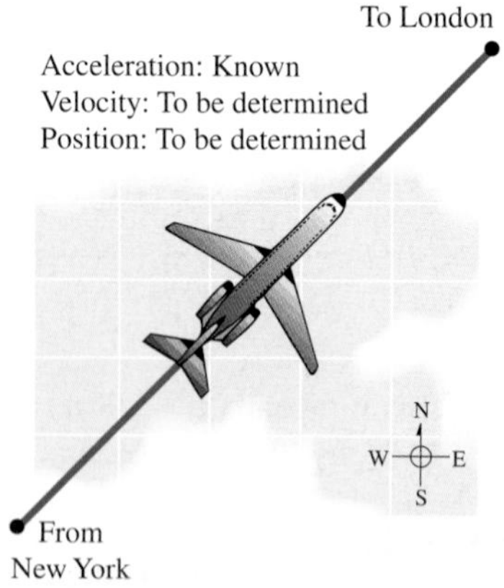

2.25 The position and velocity of an airliner crossing the Atlantic are found by integrating its acceleration with respect to time.

This optional section is intended for students who have already learned a little integral calculus. In Section 2.4 we analyzed the special case of straight-line motion with constant acceleration. When a_x is not constant, as is frequently the case, the equations that we derived in that section are no longer valid. But even when a_x varies with time, we can still use the relation $v_x = dx/dt$ to find the velocity v_x as a function of time if the position x is a known function of time. And we can still use $a_x = dv_x/dt$ to find the acceleration a_x as a function of time if the velocity v_x is a known function of time.

In many physical situations, however, position and velocity are not known as functions of time, while the acceleration is. How can we find the position and velocity from the acceleration function $a_x(t)$? This problem arises in navigating an airliner between North America and Europe (Fig. 2.25). The crew of the airliner must know their position precisely at all times, because the airspace over the North Atlantic is very congested. But over the ocean an airliner is usually out of range of both radio navigation beacons on land and air traffic controllers' radar. To determine their position, airliners carry a device called an inertial navigation system (INS), which measures the acceleration of the moving airliner. This is done in much the same way that you can sense changes in the velocity of a car in which you're riding, even when your eyes are closed. (In Chapter 4 we'll discuss how your body detects acceleration.) Given this information, along with the airliner's initial position (say, a particular gate at Kennedy International Airport) and its initial velocity (zero when parked at the gate), the INS calculates and displays for the crew the current velocity and position of the airliner at all times during the flight. Our goal in the remainder of this section is to see how these calculations are done for the simpler case of motion in a straight line with time-varying acceleration. Figure 2.26 shows an everyday example of straight-line motion with varying acceleration.

2.26 When you push your car's accelerator pedal to the floorboard, the resulting acceleration is *not* constant: the greater the car's speed, the more slowly it gains additional speed. A typical car takes twice as long to accelerate from 50 km/h to 100 km/h as it does to accelerate from 0 to 50 km/h.

We first consider a graphical approach. Figure 2.27 is a graph of acceleration versus time for a body whose acceleration is not constant but increases with time. We can divide the time interval between times t_1 and t_2 into many smaller intervals, calling a typical one Δt. Let the average acceleration during Δt be $a_{\text{av-}x}$. From Eq. (2.4) the change in velocity Δv_x during Δt is

$$\Delta v_x = a_{\text{av-}x}\,\Delta t$$

Graphically, Δv_x equals the area of the shaded strip with height $a_{\text{av-}x}$ and width Δt; that is, the area under the curve between the left and right sides of Δt. The total velocity change during any interval (say, t_1 to t_2) is the sum of the velocity changes Δv_x in the small subintervals. So the total velocity change is represented graphically by the *total* area under the a_x-t curve between the vertical lines t_1 and t_2. (In Section 2.4 we showed this to be true in the special case in which the acceleration is constant.)

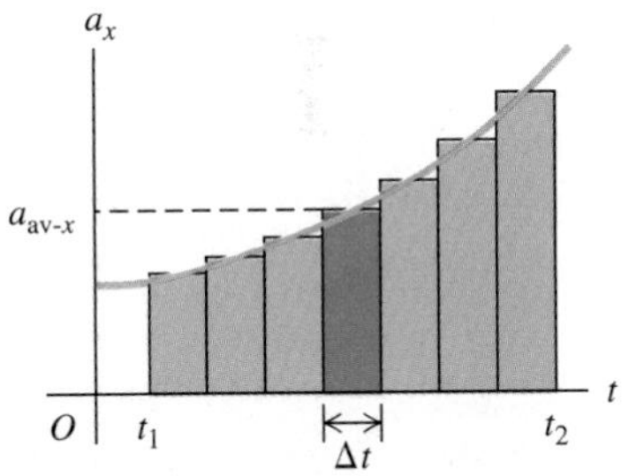

2.27 The area under an a_x-t graph between times t_1 and t_2 equals the change in velocity, $v_{2x} - v_{1x}$, that occurs between those times.

In the limit that all the Δt's become very small and their number very large, the value of $a_{\text{av-}x}$ for the interval from any time t to $t + \Delta t$ approaches the instantaneous acceleration a_x at time t. In this limit, the area under the a_x-t curve is the *integral* of a_x (which is in general a function of t) from t_1 to t_2. If v_{1x} is the velocity of the body at time t_1 and v_{2x} is the velocity at time t_2, then

$$v_{2x} - v_{1x} = \int_{v_{1x}}^{v_{2x}} dv_x = \int_{t_1}^{t_2} a_x\,dt \tag{2.15}$$

The change in velocity v_x is the integral of acceleration a_x with respect to time.

We can carry out exactly the same procedure with the curve of velocity versus time, where v_x is in general a function of t. If x_1 is a body's position at time t_1 and x_2 is its position at time t_2, from Eq. (2.2) the displacement Δx during a small time interval Δt is equal to $v_{\text{av-}x}\,\Delta t$, where $v_{\text{av-}x}$ is the average velocity during Δt. The total displacement $x_2 - x_1$ during the interval $t_2 - t_1$ is given by

$$x_2 - x_1 = \int_{x_1}^{x_2} dx = \int_{t_1}^{t_2} v_x\,dt \tag{2.16}$$

The change in position x—that is, the displacement—is the time integral of velocity v_x. Graphically, the displacement between times t_1 and t_2 is the area under the v_x-t curve between those two times. (This is the same result that we obtained in Section 2.4 for the special case in which v_x is given by Eq. (2.8).)

If $t_1 = 0$ and t_2 is any later time t, and if x_0 and v_{0x} are the position and velocity, respectively, at time $t = 0$, then we can rewrite Eqs. (2.15) and (2.16) as follows:

$$v_x = v_{0x} + \int_0^t a_x \, dt \qquad (2.17)$$

$$x = x_0 + \int_0^t v_x \, dt \qquad (2.18)$$

Here x and v_x are the position and velocity at time t. If we know the acceleration a_x as a function of time and we know the initial velocity v_{0x} we can use Eq. (2.17) to find the velocity v_x at any time; in other words, we can find v_x as a function of time. Once we know this function, and given the initial position x_0, we can use Eq. (2.18) to find the position x at any time.

Example 2.9 Motion with changing acceleration

Sally is driving along a straight highway in her classic 1965 Mustang. At time $t = 0$, when Sally is moving at 10 m/s in the positive x-direction, she passes a signpost at $x = 50$ m. Her acceleration is a function of time:

$$a_x = 2.0 \text{ m/s}^2 - (0.10 \text{ m/s}^3)t$$

a) Derive expressions for her velocity and position as functions of time. b) At what time is her velocity greatest? c) What is the maximum velocity? d) Where is the car when it reaches maximum velocity?

SOLUTION

IDENTIFY AND SET UP: The acceleration is a function of time, so we cannot use the constant-acceleration formulas of Section 2.4. Instead we must use Eqs. (2.17) and (2.18) to find the velocity and position, respectively, as functions of time. Once we have those functions, we'll be able to answer a variety of questions about the motion.

EXECUTE: a) At time $t = 0$, Sally's position is $x_0 = 50$ m and her velocity is $v_{0x} = 10$ m/s. Since we are given the acceleration a_x as a function of time, we first use Eq. (2.17) to find the velocity v_x as a function of time t. The integral of t^n is $\int t^n dt = \frac{1}{n+1}t^{n+1}$ for $n \neq -1$, so

$$v_x = 10 \text{ m/s} + \int_0^t [2.0 \text{ m/s}^2 - (0.10 \text{ m/s}^3)t]dt$$

$$= 10 \text{ m/s} + (2.0 \text{ m/s}^2)t - \frac{1}{2}(0.10 \text{ m/s}^3)t^2$$

Then we use Eq. (2.18) to find x as a function of t:

$$x = 50 \text{ m} + \int_0^t \left[10 \text{ m/s} + (2.0 \text{ m/s}^2)t - \frac{1}{2}(0.10 \text{ m/s}^3)t^2\right]dt$$

$$= 50 \text{ m} + (10 \text{ m/s})t + \frac{1}{2}(2.0 \text{ m/s}^2)t^2 - \frac{1}{6}(0.10 \text{ m/s}^3)t^3$$

Figure 2.28 shows graphs of x, v_x, and a as functions of time. Note that for any time t the slope of the x-t graph equals the value of v_x, and the slope of the v_x-t graph equals the value of a_x.

b) The maximum value of v_x occurs when v_x stops increasing and begins to decrease. At this instant, $dv_x/dt = a_x = 0$. Setting the expression for acceleration equal to zero, we obtain

$$0 = 2.0 \text{ m/s}^2 - (0.10 \text{ m/s}^3)t$$

$$t = \frac{2.0 \text{ m/s}^2}{0.10 \text{ m/s}^3} = 20 \text{ s}$$

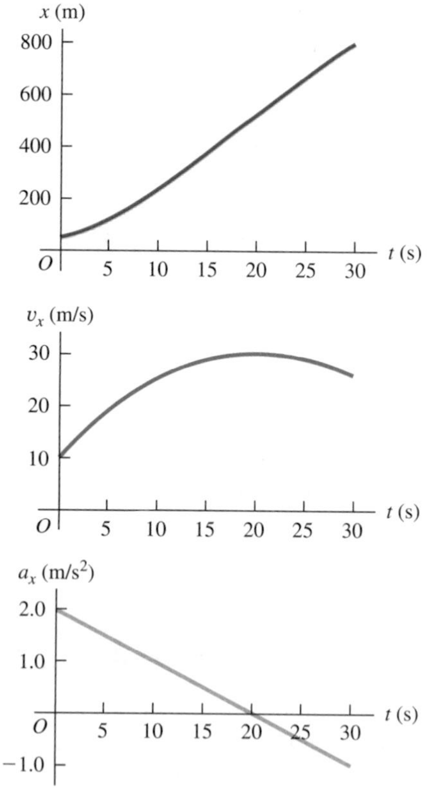

2.28 The position, velocity, and acceleration of the car in Example 2.9 as functions of time. Can you show that if this motion continues, the car will stop at $t = 44.5$ s?

c) We find the maximum velocity by substituting $t = 20$ s (when velocity is maximum) into the general velocity equation:

$$v_{\text{max-}x} = 10 \text{ m/s} + (2.0 \text{ m/s}^2)(20 \text{ s}) - \frac{1}{2}(0.10 \text{ m/s}^3)(20 \text{ s})^2$$
$$= 30 \text{ m/s}$$

d) The maximum value of v_x occurs at time $t = 20$ s. We obtain the position of the car (that is, the value of x) at that time by substituting $t = 20$ s into the general expression for x:

$$x = 50 \text{ m} + (10 \text{ m/s})(20 \text{ s}) + \frac{1}{2}(2.0 \text{ m/s}^2)(20 \text{ s})^2$$
$$- \frac{1}{6}(0.10 \text{ m/s}^3)(20 \text{ s})^3$$
$$= 517 \text{ m}$$

EVALUATE: We can use Fig. 2.28 to help us interpret our results. The bottom graph in this figure shows that a_x is positive between $t = 0$ and $t = 20$ s and negative after that. It is zero at $t = 20$ s, the time at which v_x is maximum (the high point in the middle graph). The car speeds up until $t = 20$ s (because v_x and a_x have the same sign) and slows down after $t = 20$ s (because v_x and a_x have opposite signs).

Since v_x is maximum at $t = 20$ s, the x-t graph (the top graph in Fig. 2.28) has its maximum positive slope at this time. Note that the x-t graph is concave up (curved upward) from $t = 0$ to $t = 20$ s, when a_x is positive. The graph is concave down (curved downward) after $t = 20$ s, when a_x is negative.

Example 2.10 Constant-acceleration formulas via integration

Use Eqs. (2.17) and (2.18) to find v_x and x as functions of time in the case in which the acceleration is constant. Compare the results to the constant-acceleration formulas $v_x = v_{0x} + a_x t$ (Eq. 2.8) and $x = x_0 + v_{0x}t + \frac{1}{2}a_x t^2$ (Eq. 2.12), which were derived in Section 2.4 without using integration.

SOLUTION

EXECUTE: From Eq. (2.17) the velocity is given by

$$v_x = v_{0x} + \int_0^t a_x\, dt = v_{0x} + a_x \int_0^t dt = v_{0x} + a_x t$$

We were able to take a_x outside the integral because it is constant. Our result is identical to Eq. (2.8), as it should be. Substituting this expression for v_x into Eq. (2.18), we get

$$x = x_0 + \int_0^t v_x\, dt = x_0 + \int_0^t (v_{0x} + a_x t)\, dt$$

Since v_{0x} and a_x are constants, they may be taken outside the integral:

$$x = x_0 + v_{0x} \int_0^t dt + a_x \int_0^t t\, dt = x_0 + v_{0x}t + \frac{1}{2}a_x t^2$$

EVALUATE: This result is the same as Eq. (2.12). Our expressions for v_x and x, Eqs. (2.17) and (2.18), which we developed to deal with cases in which acceleration depends on time, can be used just as well when the acceleration is constant.

Test Your Understanding

If the acceleration a_x is increasing with time, will the v_x-t graph be a straight line, concave up, or concave down?

SUMMARY

When a particle moves along a straight line, we describe its position with respect to an origin O by means of a coordinate such as x. The particle's average velocity $v_{\text{av-}x}$ during a time interval $\Delta t = t_2 - t_1$ is equal to its displacement $\Delta x = x_2 - x_1$ divided by Δt. (See Example 2.1)

$$v_{\text{av-}x} = \frac{x_2 - x_1}{t_2 - t_1} = \frac{\Delta x}{\Delta t} \qquad (2.2)$$

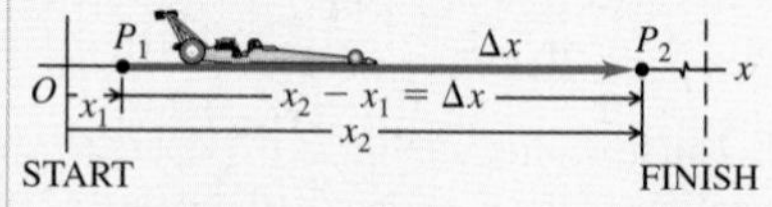

The instantaneous velocity v_x at any time t is equal to the average velocity for the time interval from t to $t + \Delta t$ in the limit that Δt goes to zero. Equivalently, v_x is the derivative of the position function with respect to time. (See Example 2.1)

$$v_x = \lim_{\Delta t \to 0} \frac{\Delta x}{\Delta t} = \frac{dx}{dt} \qquad (2.3)$$

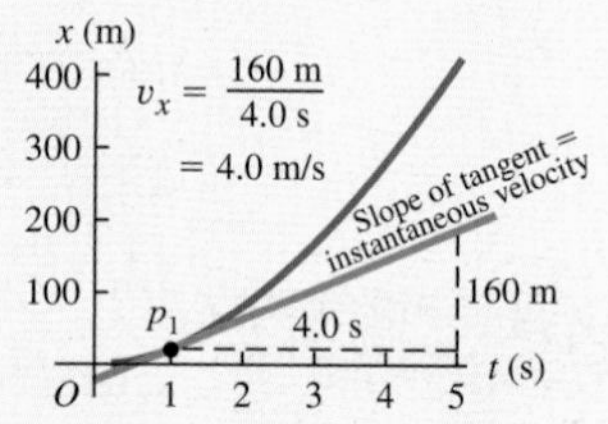

The average acceleration $a_{\text{av-}x}$ during a time interval Δt is equal to the change in velocity $\Delta v_x = v_{2x} - v_{1x}$ during that time interval divided by Δt. The instantaneous acceleration a_x is the limit of $a_{\text{av-}x}$ as Δt goes to zero, or the derivative of v_x with respect to t. (See Examples 2.2 and 2.3)

$$a_{\text{av-}x} = \frac{v_{2x} - v_{1x}}{t_2 - t_1} = \frac{\Delta v_x}{\Delta t} \qquad (2.4)$$

$$a_x = \lim_{\Delta t \to 0} \frac{\Delta v_x}{\Delta t} = \frac{dv_x}{dt} \qquad (2.5)$$

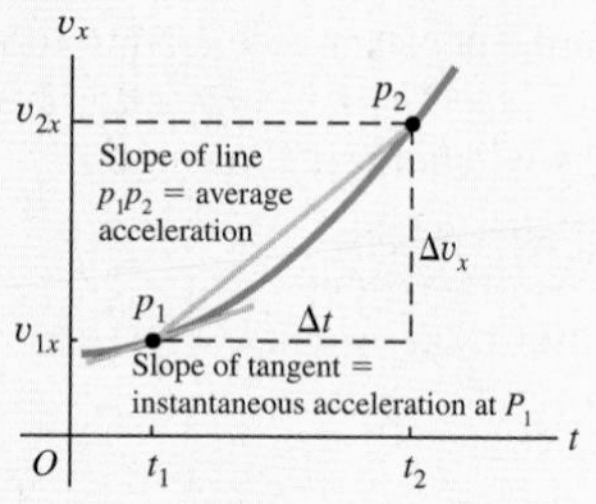

When the acceleration is constant, four equations relate the position x and velocity v_x at any time t to the initial position x_0, the initial velocity v_{0x} (both measured at time $t = 0$), and the acceleration a_x. (See Examples 2.4 and 2.5)

constant acceleration only:

$$v_x = v_{0x} + a_x t \qquad (2.8)$$

$$x = x_0 + v_{0x} t + \frac{1}{2} a_x t^2 \qquad (2.12)$$

$$v_x^2 = v_{0x}^2 + 2a_x(x - x_0) \qquad (2.13)$$

$$x - x_0 = \left(\frac{v_{0x} + v_x}{2}\right) t \qquad (2.14)$$

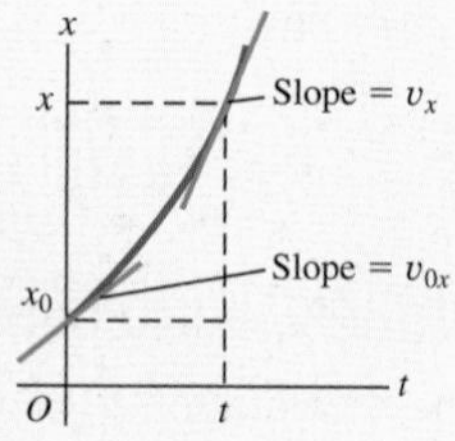

Free fall is a case of motion with constant acceleration. The magnitude of the acceleration due to gravity is a positive quantity, g. The acceleration of a body in free fall is always downward. (See Examples 2.6 through 2.8)

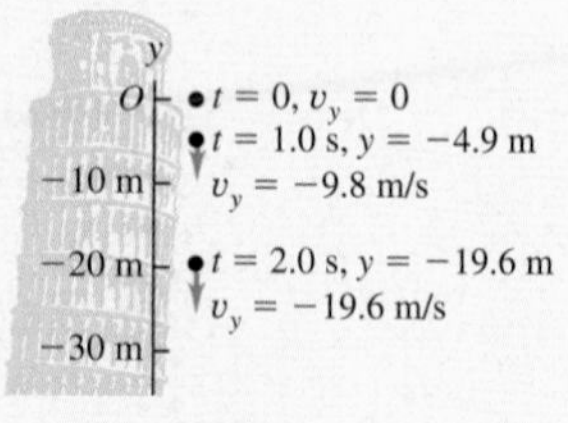

When the acceleration is not constant but is a known function of time, we can find the velocity and position as functions of time by integrating the acceleration function. (See Examples 2.9 and 2.10)	$v_x = v_{0x} + \int_0^t a_x\, dt$ (2.17) $x = x_0 + \int_0^t v_x\, dt$ (2.18)	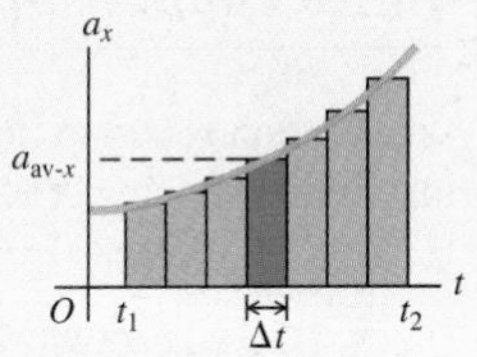

Key Terms

particle, 41
average velocity, 41
***x-t* graph, 43**
instantaneous velocity, 44
derivative, 44
speed, 44
motion diagram, 47
average acceleration, 48
instantaneous acceleration, 49
v_x-*t* graph, 50
a_x-*t* graph, 53
free fall, 58
acceleration due to gravity, 59

Your Notes

Answer to Chapter Opening Question

Yes. Acceleration refers to *any* change in velocity, including both speeding up and slowing down.

Answers to Test Your Understanding Questions

Section 2.1 Both the truck and the car undergo the same total displacement Δx (from point A to point B) during the same time interval Δt. Hence they have the same average velocity $v_{\text{av-}x} = \Delta x/\Delta t$. The details of what happens during the time interval don't matter.
Section 2.2 The velocity is positive when the slope of the x-t graph is positive (point P), negative when the slope is negative (point R), and zero when the slope is zero (points Q and S). The speed is greatest at times for which the slope of the x-t graph is steepest (either positive or negative), which is at point R.
Section 2.3 The acceleration is positive when the x-t graph is curved upward (concave up), as at point S, and negative when it is curved downward (concave down), as at point Q. The acceleration is zero when the x-t graph is not curved but is a straight line, as at points P and R. At P, $v_x > 0$ and $a_x = 0$ (speed is not changing); at Q, $v_x > 0$ and $a_x < 0$ (speed is decreasing); at R, $v_x < 0$ and $a_x = 0$ (speed is not changing); and at S, $v_x < 0$ and $a_x > 0$ (speed is decreasing).
Section 2.4 The motorist and the officer have the same velocity at $t = 5.0$ s.

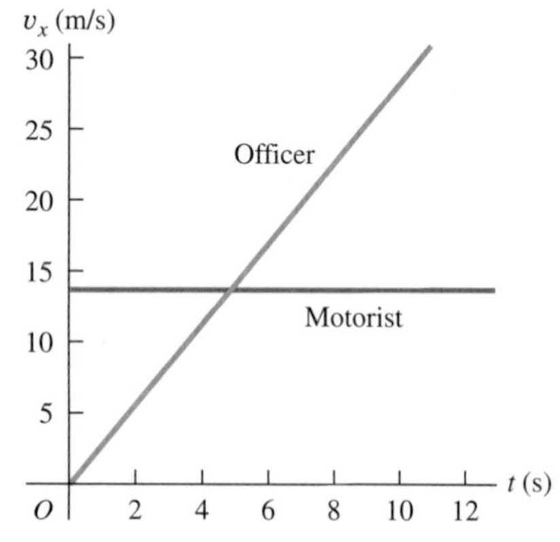

Section 2.5 Use Eq. (2.13) with x replaced by y and $a_y = g$; $v_y^2 = v_{0y}^2 - 2g(y - y_0)$. The starting height is $y_0 = 0$ and the velocity at the maximum height $y = h$ is $v_y = 0$, so $0 = v_{0y}^2 - 2gh$ and $h = v_{0y}^2/2g$. If the initial velocity is increased by a factor of 2, the maximum height increases by a factor of $2^2 = 4$ and the ball goes to height $4h$.
Section 2.6 The acceleration a_x is equal to the slope of the v_x-t graph. If a_x is increasing, the slope of the v_x-t graph is also increasing and the graph is concave up.

Discussion Questions

Q2.1 Does the speedometer of a car measure speed or velocity? Explain.
Q2.2 In the same time interval, a car accelerates from 15 m/s to 20 m/s while a truck accelerates from 36 m/s to 40 m/s. Which vehicle has the greater average acceleration? Explain.
Q2.3 Can an object with constant acceleration reverse its direction of travel? Explain.
Q2.4 Under what conditions is average velocity equal to instantaneous velocity?
Q2.5 In a given time interval, is the total displacement of a particle equal to the product of the average velocity and the time interval, even when the velocity is not constant? Explain.
Q2.6 Under what conditions does the magnitude of the average velocity equal the average speed?
Q2.7 When a Dodge Viper is at Elwood's Car Wash, a BMW Z3 is at Elm and Main. Later, when the Dodge reaches Elm and Main, the BMW reaches Elwood's Car Wash. How are the cars' average velocities between these two times related?
Q2.8 A driver in Massachusetts was sent to traffic court for speeding. The evidence against the driver was that a policewoman observed the driver's car alongside a second car at a certain moment, and that the policewoman had already clocked the second car as going faster than the speed limit. The driver argued that "the second car was passing me. I was not speeding." The judge ruled against the driver because, in the judge's words, "if two cars were side by side, you were both speeding." If you were a lawyer representing the accused driver, how would you argue this case?
Q2.9 Can you have a zero displacement and a nonzero average velocity? A nonzero velocity? Illustrate your answers on an x-t graph.
Q2.10 Can you have zero acceleration and nonzero velocity? Explain using a v_x-t graph.
Q2.11 Can you have zero velocity and nonzero average acceleration? Zero velocity and nonzero acceleration? Explain using a v_x-t graph and give an example of such motion.
Q2.12 An automobile is traveling west. Can it have a velocity toward the west and at the same time have an acceleration toward the east? Under what circumstances?
Q2.13 The official's truck in Fig. 2.2 is at $x_1 = 277$ m at $t_1 = 16.0$ s, and is at $x_2 = 19$ m at $t_2 = 25.0$ s. a) Sketch *two* different possible x-t graphs for the motion of the truck. b) Does the average velocity $v_{\text{av-}x}$ during the time interval from t_1 to t_2 have the same value for both of your graphs? Why or why not?
Q2.14 Under constant acceleration the average velocity of a particle is half the sum of its initial and final velocities. Is this still true if the acceleration is *not* constant? Explain.
Q2.15 You throw a baseball straight up in the air so that it rises to a maximum height much greater than your height. Is the magnitude of the acceleration greater while it is being thrown or after it leaves your hand? Explain.
Q2.16 Prove these statements: i) As long as you can neglect the effects of the air, if you throw anything vertically upward, it will have the same speed when it returns to the release point as when it was released. ii) The time of flight will be twice the time it takes to get to its highest point.
Q2.17 In Example 2.7, substituting $y = -18.4$ m into Eq. (2.13) gives $v_y = \pm 24.2$ m/s. The negative root is the velocity at $t = 4.00$ s. Explain the meaning of the positive root.

Q2.18 If the initial position and initial velocity of a vehicle are known and a record is kept of the acceleration at each instant, can you compute the vehicle's position after a certain time from these data? If so, explain how this might be done.

Q2.19 You and your friend are leaning over the edge of the roof of a tall building. At the same instant you throw a baseball straight up with an initial speed v_0 and your friend throws a marble straight down with the same initial speed v_0. Assume that you can ignore air resistance. Which object will reach the ground first, the baseball or the marble? How will their speeds compare just before hitting the ground?

Q2.20 A ball is dropped from rest from the top of a building of height h. At the same instant, a second ball is projected vertically upward from ground level, such that it has zero speed when it reaches the top of the building. When the two balls pass each other, which ball has the greater speed, or do they have the same speed? Explain. Where will the two balls be when they are alongside each other: at height $h/2$ above the ground, below this height, or above this height? Explain.

Exercises

Section 2.1 Displacement, Time, and Average Velocity

2.1 A rocket carrying a satellite is accelerating straight up from the earth's surface. At 1.15 s after liftoff, the rocket clears the top of its launch platform, 63 m above the ground. After an additional 4.75 s it is 1.00 km above the ground. Calculate the magnitude of the average velocity of the rocket for a) the 4.75-s part of its flight; b) the first 5.90 s of its flight.

2.2 In an experiment, a shearwater (a seabird) was taken from its nest, flown 5150 km away, and released. It found its way back to its nest 13.5 days after release. If we place the origin in the nest and extend the $+x$-axis to the release point, what was the bird's average velocity in m/s a) for the return flight? b) For the whole episode, from leaving the nest to returning?

2.3 Trip Home. You normally drive on the freeway between San Diego and Los Angeles at an average speed of 105 km/h (65 mi/h), and the trip takes 2 h and 20 min. On a Friday afternoon, however, heavy traffic slows you down and you drive the same distance at an average speed of only 70 km/h (43 mi/h). How much longer does the trip take?

2.4 From Pillar to Post. Starting from a pillar, you run 200 m east (the $+x$-direction) at an average speed of 5.0 m/s, and then run 280 m west at an average speed of 4.0 m/s to a post. Calculate a) your average speed from pillar to post; b) your average velocity from pillar to post.

2.5 Two runners start simultaneously from the same point on a circular 200-m track and run in the same direction. One runs at a constant speed of 6.20 m/s and the other at a constant speed of 5.50 m/s. When will the fast one overtake ("lap") the slower one and how far from the starting point will each have run?

2.6 Geology. Earthquakes produce several types of shock waves. The most well-known are the P-waves (P for *primary* or *pressure*) and the S-waves (S for *secondary* or *shear*). In the earth's crust, the P-waves travel at around 6.5 km/s while the S-waves move at about 3.5 km/s. The actual speeds vary depending on the type of material they are going through. The time delay between the arrival of these two waves at a seismic recording station tells geologists how far away the earthquake occurred. If the time delay is 33 s, how far from the seismic station did the earthquake occur?

2.7 a) Your old VW van sputters along at an average speed of 8.0 m/s for 60 s, and then catches hold and zips along at an average speed of 20.0 m/s for another 60 s. Calculate your average speed for the entire 120 s. b) Suppose the speed of 8.0 m/s was maintained while you traveled 240 m, followed by the average speed of 20.0 m/s for another 240 m. Calculate the van's average speed for the entire distance. c) In which case is the average speed for the entire motion the numerical average of the two speeds?

2.8 A Honda Civic travels in a straight line along a road. Its distance x from a stop sign is given as a function of time t by the equation $x(t) = \alpha t^2 - \beta t^3$, where $\alpha = 1.50\ \text{m/s}^2$ and $\beta = 0.0500\ \text{m/s}^3$. Calculate the average velocity of the car for the following time intervals: a) $t = 0$ to $t = 2.00$ s; b) $t = 0$ to $t = 4.00$ s; c) $t = 2.00$ s to $t = 4.00$ s.

Section 2.2 Instantaneous Velocity

2.9 A car is stopped at a traffic light. It then travels along a straight road so that its distance from the light is given by $x(t) = bt^2 - ct^3$, where $b = 2.40\ \text{m/s}^2$ and $c = 0.120\ \text{m/s}^3$. a) Calculate the average velocity of the car for the time interval $t = 0$ to $t = 10.0$ s. b) Calculate the instantaneous velocity of the car at i) $t = 0$; ii) $t = 5.0$ s; iii) $t = 10.0$ s. c) How long after starting from rest is the car again at rest?

2.10 A physics professor leaves her house and walks along the sidewalk toward campus. After 5 min it starts to rain and she returns home. Her distance from her house as a function of time is shown in Fig. 2.29. At which of the labeled points is her velocity a) zero? b) constant and positive? c) constant and negative? d) increasing in magnitude? e) decreasing in magnitude?

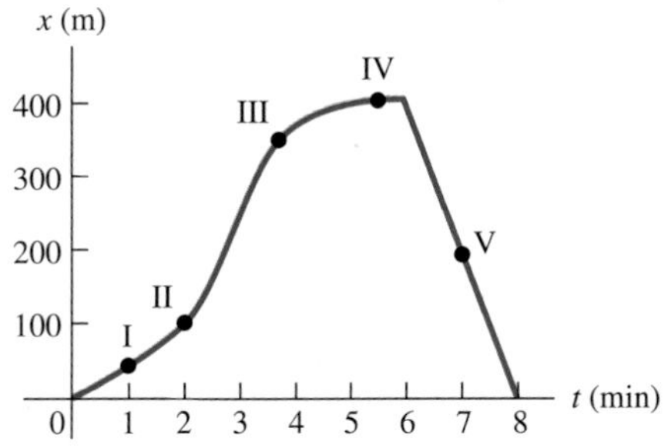

Figure 2.29 Exercise 2.10.

Section 2.3 Average and Instantaneous Acceleration

2.11 A test driver at Incredible Motors, Inc. is testing a new model car with a speedometer calibrated to read m/s rather than mi/h. The following series of speedometer readings was obtained during a test run along a long, straight road:

Time (s)	0	2	4	6	8	10	12	14	16
Speed (m/s)	0	0	2	6	10	16	19	22	22

a) Compute the average acceleration during each 2-s interval. Is the acceleration constant? Is it constant during any part of the test run? b) Make a v_x-t graph of the data above, using scales of 1 cm = 1 s horizontally and 1 cm = 2 m/s vertically. Draw a smooth curve through the plotted points. By measuring the slope of your curve, find the instantaneous acceleration at $t = 9$ s, 13 s, and 15 s.

2.12 Figure 2.30 shows the velocity of a solar-powered car as a function of time. The driver accelerates from a stop sign, cruises for 20 s at a constant speed of 60 km/h, and then brakes to come to a stop 40 s after leaving the stop sign. Compute the average acceleration for the following time intervals: a) $t = 0$ to $t = 10$ s; b) $t = 30$ s to $t = 40$ s; c) $t = 10$ s to $t = 30$ s; d) $t = 0$ to $t = 40$ s.

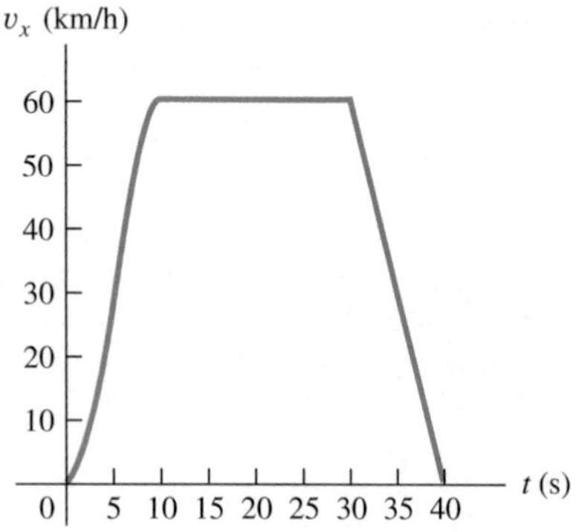

Figure 2.30 Exercises 2.12 and 2.13.

2.13 Refer to Exercise 2.12 and Fig. 2.30. a) During what time interval does the instantaneous acceleration a_x have its most positive value? b) During what time interval does the instantaneous acceleration a_x have its most negative value? c) What is the instantaneous acceleration at $t = 20$ s? d) What is the instantaneous acceleration at $t = 35$ s? e) In a motion diagram (like Fig. 2.13b or Fig. 2.14b), show the position, velocity, and acceleration of the car at the four times $t = 5$ s, $t = 15$ s, $t = 25$ s, and $t = 35$ s.

2.14 A person looking out the window of a tall office building observes what he suspects is a UFO. The person records the position of the object as a function of time and finds that it is given by $\vec{r}(t) = -(5.0\text{ m/s})t\hat{\imath} + (10.0\text{ m/s})t\hat{\jmath} + [(7.0\text{ m/s})t - (3.0\text{ m/s}^2)t^2]\hat{k}$. a) Find the displacement, velocity, and acceleration vectors for the object at $t = 5.0$ s. b) Is there a time when the object has a velocity of zero? c) Is the acceleration of the object constant, or does it change in time?

2.15 A turtle crawls along a straight line, which we will call the x-axis with the positive direction to the right. The equation for the turtle's position as a function of time is $x(t) = 50.0\text{ cm} + (2.00\text{ cm/s})t - (0.0625\text{ cm/s}^2)t^2$. a) Find the turtle's initial velocity, initial position, and initial acceleration. b) At what time t is the velocity of the turtle zero? c) How long after starting does it take the turtle to return to its starting point? d) At what times t is the turtle a distance of 10.0 cm from its starting point? What is the velocity (magnitude and direction) of the turtle at each of these times? e) Sketch graphs of x versus t, v_x versus t, and a_x versus t, for the time interval $t = 0$ to $t = 40$ s.

2.16 An astronaut has left the International Space Station to test a new space scooter. Her partner measures the following velocity changes, each taking place in a 10-s interval. What are the magnitude, the algebraic sign, and the direction of the average acceleration in each interval? Assume that the positive direction is to the right. a) At the beginning of the interval the astronaut is moving toward the right along the x-axis at 15.0 m/s, and at the end of the interval she is moving toward the right at 5.0 m/s. b) At the beginning she is moving toward the left at 5.0 m/s, and at the end she is moving toward the left at 15.0 m/s. c) At the beginning she is moving toward the right at 15.0 m/s, and at the end she is moving toward the left at 15.0 m/s.

2.17 Auto Acceleration. a) Based on your experiences of riding in automobiles, estimate the magnitude of a car's average acceleration when it brakes from highway speeds to a sudden stop. b) Explain why this average acceleration could be regarded as either positive or negative.

2.18 A car's velocity as a function of time is given by $v_x(t) = \alpha + \beta t^2$, where $\alpha = 3.00$ m/s and $\beta = 0.100\text{ m/s}^3$. a) Calculate the average acceleration for the time interval $t = 0$ to $t = 5.00$ s. b) Calculate the instantaneous acceleration for i) $t = 0$; ii) $t = 5.00$ s. c) Draw accurate v_x-t and a_x-t graphs for the car's motion between $t = 0$ and $t = 5.00$ s.

2.19 Figure 2.31 is a graph of the coordinate of a spider crawling along the x-axis. a) Graph its velocity and acceleration as functions of time. b) In a motion diagram (like Fig. 2.13b or Fig. 2.14b), show the position, velocity, and acceleration of the spider at the five times $t = 2.5$ s, $t = 10$ s, $t = 20$ s, $t = 30$ s, and $t = 37.5$ s.

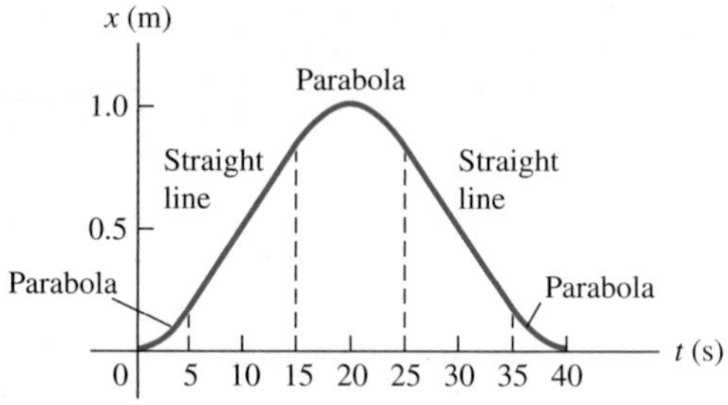

Figure 2.31 Exercise 2.19.

2.20 The position of the front bumper of a test car under microprocessor control is given by $x(t) = 2.17\text{ m} + (4.80\text{ m/s}^2)t^2 - (0.100\text{ m/s}^6)t^6$. Find its a) position and acceleration at the instants when the car has zero velocity. b) Draw x-t, v_x-t, and a_x-t graphs for the motion of the bumper between $t = 0$ and $t = 2.00$ s.

Section 2.4 Motion with Constant Acceleration

2.21 An antelope moving with constant acceleration covers the distance between two points 70.0 m apart in 7.00 s. Its speed as it passes the second point is 15.0 m/s. a) What is its speed at the first point? b) What is the acceleration?

2.22 The catapult of the aircraft carrier *USS Abraham Lincoln* accelerates an F/A-18 Hornet jet fighter from rest to a takeoff speed of 173 mi/h in a distance of 307 ft. Assume constant acceleration. a) Calculate the acceleration of the fighter in m/s^2. b) Calculate the time required for the fighter to accelerate to takeoff speed.

2.23 Automobile Airbags. The human body can survive a negative acceleration trauma incident (sudden stop) if the magnitude of the acceleration is less than 250 m/s^2 (approximately $25g$). If you are in an automobile accident with an initial speed of 105 km/h (65 mi/h) and are stopped by an airbag that inflates from the dashboard, over what distance must the airbag stop you for you to survive the crash?

2.24 An airplane travels 280 m down the runway before taking off. If it starts from rest, moves with constant acceleration, and becomes airborne in 8.00 s, what is its speed, in m/s, when it takes off?

2.25 Entering the Freeway. A car sits in an entrance ramp to a freeway, waiting for a break in the traffic. The driver sees a small gap between a van and an 18-wheel truck and accelerates with constant acceleration along the ramp and onto the freeway. The car starts from rest, moves in a straight line, and has a speed of 20 m/s (45 mi/h) when it reaches the end of the 120-m-long ramp. a) What is the acceleration of the car? b) How much time does it take the car to travel the length of the ramp? c) The traffic on the freeway is moving at a constant speed of 20 m/s. What distance does the traffic travel while the car is moving the length of the ramp?

2.26 Figures 2.15, 2.16, 2.17, and 2.18 were drawn for motion with constant acceleration with positive values of x_0, v_{0x}, and a_x. Redraw these four figures for the following cases: a) $x_0 < 0$, $v_{0x} < 0$, $a_x < 0$; b) $x_0 > 0$, $v_{0x} < 0$, $a_x > 0$; c) $x_0 > 0$, $v_{0x} > 0$, $a_x < 0$.

2.27 According to 1994 test data, a Ford Aspire automobile travels 0.250 miles in 19.9 s, starting from rest. The same car, when braking from 60.0 mph on dry pavement, stops in 146 ft. Assume constant acceleration in each part of the motion, but not necessarily the same acceleration when slowing down as when speeding up. a) Find the acceleration of this car when speeding up and when braking. b) If its acceleration is constant, how fast (in mph) should this car be traveling after 0.250 miles of acceleration? The actual measured speed is 70.0 mph; what does this tell you about the motion? c) How long does it take this car to stop while braking from 60.0 mph?

2.28 A cat walks in a straight line, which we shall call the x-axis with the positive direction to the right. As an observant physicist, you make measurements of this cat's motion and construct a graph of the feline's velocity as a function of time (Fig. 2.32). a) Find the cat's velocity at $t = 4.0$ s; at $t = 7.0$ s. b) What is the cat's acceleration at $t = 3.0$ s? At $t = 6.0$ s? At $t = 7.0$ s? c) What distance does the cat move during the first 4.5 s? From $t = 0$ to $t = 7.5$ s? d) Sketch clear graphs of the cat's acceleration and position as functions of time, assuming that the cat started at the origin.

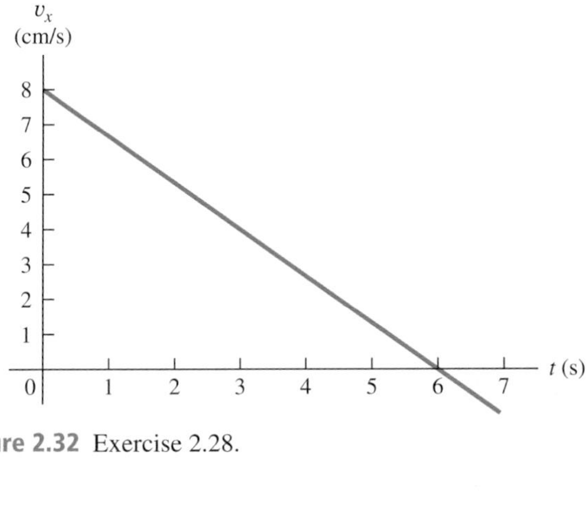

Figure 2.32 Exercise 2.28.

2.29 At time $t = 0$ a Corvette is traveling on a long, straight stretch of highway in Arizona at a constant speed of 30 m/s. This motion continues for 20 s. Then the driver, worried that she is behind schedule, accelerates at a constant rate for 5 s to bring the car's speed to 40 m/s. The car travels at this new speed for 10 s. But then the driver spots a police motorcycle parked behind a large cactus up ahead and slows with a constant acceleration of magnitude 4.0 m/s^2 until the car's speed is reduced once again to the legal limit of 30 m/s. She then maintains this speed, and waves to the officer as she passes him 5 s later. a) For the motion of the car from $t = 0$ until the moment the car passes the parked motorcycle, draw accurate a_x-t, v_x-t, and x-t graphs. b) In a motion diagram (like Fig. 2.13b or 2.14b), show the position, velocity, and acceleration of the car.

2.30 At $t = 0$ a car is stopped at a traffic light. When the light turns green, the car starts to speed up, and gains speed at a constant rate until it reaches a speed of 20 m/s 8 seconds after the light turns green. The car continues at a constant speed for 60 m. Then the driver sees a red light up ahead at the next intersection, and starts slowing down at a constant rate. The car stops at the red light, 180 m from where it was at $t = 0$. a) Draw accurate x-t, v_x-t, and a_x-t graphs for the motion of the car. b) In a motion diagram (like Fig. 2.13b or 2.14b), show the position, velocity, and acceleration of the car.

2.31 The graph in Fig. 2.33 shows the velocity of a motorcycle police officer plotted as a function of time. a) Find the instantaneous acceleration at $t = 3$ s, at $t = 7$ s, and at $t = 11$ s. b) How far does the officer go in the first 5 s? The first 9 s? The first 13 s?

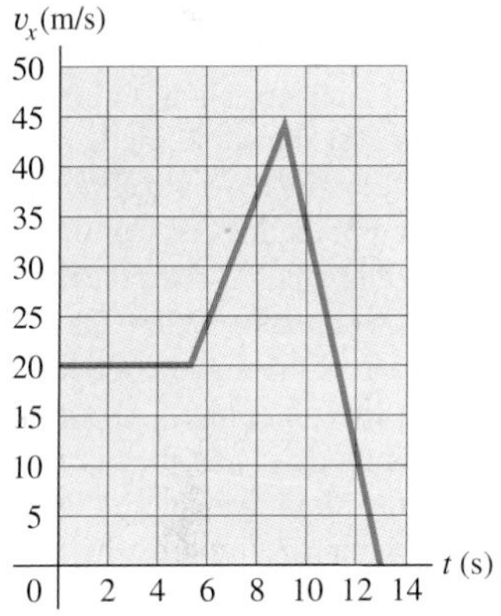

Figure 2.33 Exercise 2.31.

2.32 Figure 2.34 is a graph of the acceleration of a model railroad locomotive moving on the x-axis. Graph its velocity and coordinate as functions of time if $x = 0$ and $v_x = 0$ at $t = 0$.

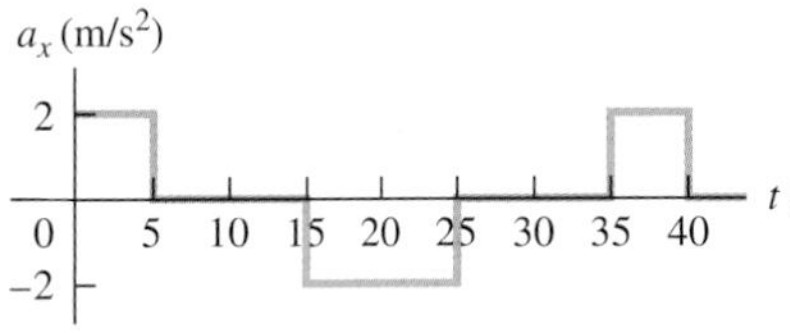

Figure 2.34 Exercise 2.32.

2.33 A spaceship ferrying workers to Moon Base I takes a straight-line path from the earth to the moon, a distance of 384,000 km. Suppose it accelerates at 20.0 m/s^2 for the first 15.0 min of the trip, then travels at constant speed until the last 15.0 min, when it accelerates at $-20.0\ m/s^2$, just coming to rest as it reaches the moon. a) What is the maximum speed attained? b) What fraction of the total distance is traveled at constant speed? c) What total time is required for the trip?

2.34 A subway train starts from rest at a station and accelerates at a rate of $1.60\ m/s^2$ for 14.0 s. It runs at constant speed for 70.0 s and slows down at a rate of $3.50\ m/s^2$ until it stops at the next station. Find the *total* distance covered.

2.35 Two cars, *A* and *B*, move along the *x*-axis. Figure 2.35 is a graph of the positions of *A* and *B* versus time. a) In motion diagrams (like Figure 2.13b or Figure 2.14b), show the position, velocity, and acceleration of each of the two cars at $t = 0$, $t = 1$ s, and $t = 3$ s. b) At what time(s), if any, do *A* and *B* have the same position? c) Graph velocity versus time for both *A* and *B*. d) At what time(s), if any, do *A* and *B* have the same velocity? e) At what time(s), if any, does car *A* pass car *B*? f) At what time(s), if any, does car *B* pass car *A*?

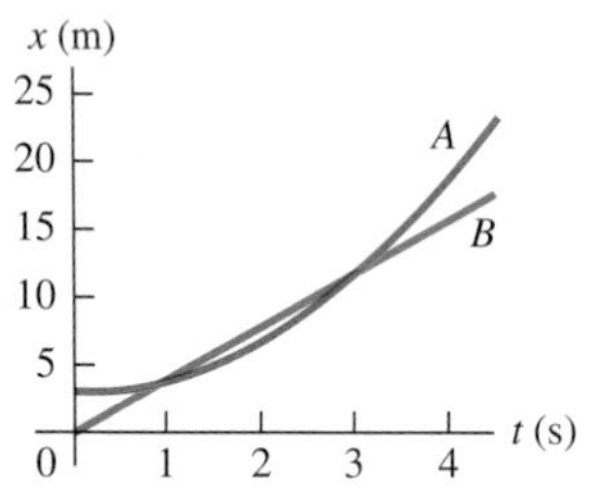

Figure 2.35 Exercise 2.35.

2.36 At the instant the traffic light turns green, a car that has been waiting at an intersection starts ahead with a constant acceleration of $3.20\ m/s^2$. At the same instant a truck, traveling with a constant speed of 20.0 m/s, overtakes and passes the car. a) How far beyond its starting point does the car overtake the truck? b) How fast is the car traveling when it overtakes the truck? c) Sketch an *x-t* graph of the motion of both vehicles. Take $x = 0$ at the intersection. d) Sketch a v_x-*t* graph of the motion of both vehicles.

2.37 As in Example 2.5, a car is traveling at a constant velocity with magnitude v_C. At the instant that the car passes a motorcycle officer, the motorcycle accelerates from rest with acceleration a_{Mx}. a) Sketch an *x-t* graph of the motion of both objects. Show that when the motorcycle overtakes the car, the motorcycle has a speed twice that of the car, no matter what the value of a_{Mx}. b) Let *d* be the distance the motorcycle travels before catching up with the car. In terms of *d*, how far has the motorcycle traveled when its velocity equals the velocity of the car?

Section 2.5 Freely Falling Bodies

2.38 Raindrops. If the effects of the air acting on falling raindrops is ignored, then we can treat raindrops as freely falling objects. a) Rain clouds are typically a few hundred meters above the ground. Estimate the speed with which raindrops would strike the ground if they were freely falling objects. Give your estimate in m/s, km/h, and mi/h. b) Estimate (from your own personal observations of rain) the speed with which raindrops actually strike the ground. c) Based on your answers to parts (a) and (b), is it a good approximation to neglect the effects of the air on falling raindrops? Explain.

2.39 a) If a flea can jump straight up to a height of 0.440 m, what is its initial speed as it leaves the ground? b) How long is it in the air?

2.40 Touchdown on the Moon. A lunar lander is making its descent to Moon Base I (Fig. 2.36). The lander descends slowly under the retro-thrust of its descent engine. The engine is cut off when the lander is 5.0 m above the surface and has a downward speed of 0.8 m/s. With the engine off, the lander is in free fall. What is the speed of the lander just before it touches the surface? The acceleration due to gravity on the moon is $1.6\ m/s^2$.

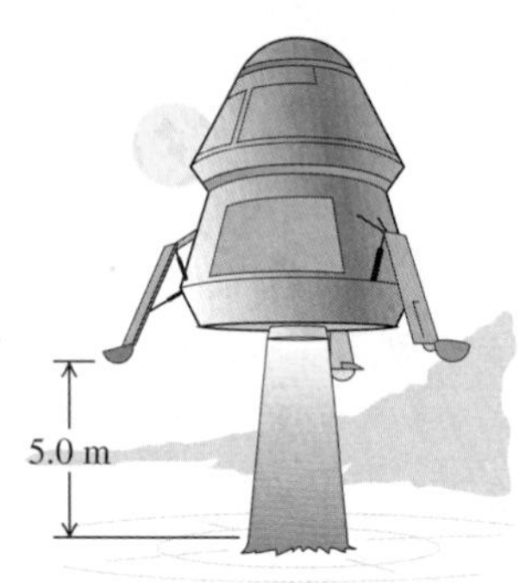

Figure 2.36 Exercise 2.40.

2.41 A Simple Reaction-Time Test. A meter stick is held vertically above your hand, with the lower end between your thumb and first finger. On seeing the meter stick released, you grab it with these two fingers. You can calculate your reaction time from the distance the meter stick falls, read directly from the point where your fingers grabbed it. a) Derive a relationship for your reaction time in terms of this measured distance, *d*. b) If the measured distance is 17.6 cm, what is the reaction time?

2.42 A brick is dropped (zero initial speed) from the roof of a building. The brick strikes the ground in 2.50 s. You may ignore air resistance, so the brick is in free fall. a) How tall, in meters, is the building? b) What is the magnitude of the brick's velocity just before it reaches the ground? c) Sketch a_y-*t*, v_y-*t*, and *y-t* graphs for the motion of the brick.

2.43 Jillian angrily throws her engagement ring straight up from the roof of a building, 12.0 m above the ground, with an initial speed of 5.00 m/s. You may ignore air resistance. For the motion from her hand to the ground, what are the magnitude and direction of a) the average velocity of the ring? b) the average acceleration of the ring? c) In how many seconds after being thrown does the ring strike the ground? d) What is the speed of the ring just before it strikes the ground? e) Sketch a_y-*t*, v_y-*t*, and *y-t* graphs for the motion.

2.44 A hot-air balloonist, rising vertically with a constant velocity of magnitude 5.00 m/s, releases a sandbag at an instant when the balloon is 40.0 m above the ground (Fig. 2.37). After it is released, the sandbag is in free fall. a) Compute the position and velocity of the sandbag at 0.250 s and 1.00 s after its release. b) How many seconds after its release will the bag strike the ground? c) With what magnitude of velocity does it strike? d) What is the greatest height above the ground that the sandbag reaches? e) Sketch a_y-*t*, v_y-*t*, and *y-t* graphs for the motion.

Figure 2.37 Exercise 2.44.

2.45 A student throws a water balloon vertically downward from the top of a building. The balloon leaves the thrower's hand with a speed of 6.00 m/s. Air resistance may be ignored, so the water balloon is in free fall after it leaves the thrower's hand. a) What is its speed after falling for 2.00 s? b) How far does it fall in 2.00 s? c) What is the magnitude of its velocity after falling 10.0 m? d) Sketch a_y-t, v_y-t, and y-t graphs for the motion.

2.46 An egg is thrown nearly vertically upward from a point near the cornice of a tall building. It just misses the cornice on the way down and passes a point 50.0 m below its starting point 5.00 s after it leaves the thrower's hand. Air resistance may be ignored. a) What is the initial speed of the egg? b) How high does it rise above its starting point? c) What is the magnitude of its velocity at the highest point? d) What are the magnitude and direction of its acceleration at the highest point? e) Sketch a_y-t, v_y-t, and y-t graphs for the motion of the egg.

2.47 The rocket-driven sled *Sonic Wind No. 2,* used for investigating the physiological effects of large accelerations, runs on a straight, level track 1070 m (3500 ft) long. Starting from rest, it can reach a speed of 224 m/s (500 mi/h) in 0.900 s. a) Compute the acceleration in m/s^2, assuming that it is constant. b) What is the ratio of this acceleration to that of a freely falling body (g)? c) What is the distance covered in 0.900 s? d) A magazine article states that at the end of a certain run, the speed of the sled decreased from 283 m/s (632 mi/h) to zero in 1.40 s and that during this time the magnitude of the acceleration was more than 40g. Are these figures consistent?

2.48 A large boulder is ejected vertically upward from a volcano with an initial speed of 40.0 m/s. Air resistance may be ignored. a) At what time after being ejected is the boulder moving at 20.0 m/s upward? b) At what time is it moving at 20.0 m/s downward? c) When is the displacement of the boulder from its initial position zero? d) When is the velocity of the boulder zero? e) What are the magnitude and direction of the acceleration while the boulder is i) moving upward? ii) moving downward? iii) at the highest point? f) Sketch a_y-t, v_y-t, and y-t graphs for the motion.

2.49 Suppose the acceleration of gravity were only 0.98 m/s^2 instead of 9.8 m/s^2, but the initial velocities with which you can jump upward or throw a ball are unchanged. a) Calculate the height to which you could jump vertically from a standing start if you can jump to 0.75 m when $g = 9.8$ m/s^2. b) How high could you throw a baseball if you can throw it 18 m up when $g = 9.8$ m/s^2? c) Calculate the maximum height of a window from which you would care to jump to a concrete sidewalk below if for $g = 9.8$ m/s^2 you are willing to jump from a height of 2.0 m, which is the typical height of a first-story window.

*Section 2.6 Velocity and Position by Integration

***2.50** The acceleration of a bus is given by $a_x(t) = \alpha t$, where $\alpha = 1.2$ m/s^3. a) If the bus's velocity at time $t = 1.0$ s is 5.0 m/s, what is its velocity at time $t = 2.0$ s? b) If the bus's position at time $t = 1.0$ s is 6.0 m, what is its position at time $t = 2.0$ s? c) Sketch a_x-t, v_x-t, and x-t graphs for the motion.

***2.51** The acceleration of a motorcycle is given by $a_x(t) = At - Bt^2$, where $A = 1.50$ m/s^3 and $B = 0.120$ m/s^4. The motorcycle is at rest at the origin at time $t = 0$. a) Find its position and velocity as functions of time. b) Calculate the maximum velocity it attains.

***2.52 Flying Leap of the Flea.** High speed motion pictures (3500 frames/second) of a jumping, 210-μg flea yielded the data used to plot the graph given in Fig. 2.38. (See "The Flying Leap of the Flea", by M. Rothschild, Y. Schlein, K. Parker, C. Neville, and S. Sternberg in the Nov. 1973 *Scientific American.*) This flea was about 2 mm long and jumped at a nearly vertical take-off angle. Use the graph to answer the following questions. a) Is the acceleration of the flea ever zero? If so, when? Justify your answer. b) Find the maximum height the flea reached in the first 2.5 ms. c) Find the flea's acceleration at 0.5 ms, 1.0 ms, and 1.5 ms. d) Find the flea's height at 0.5 ms, 1.0 ms, and 1.5 ms.

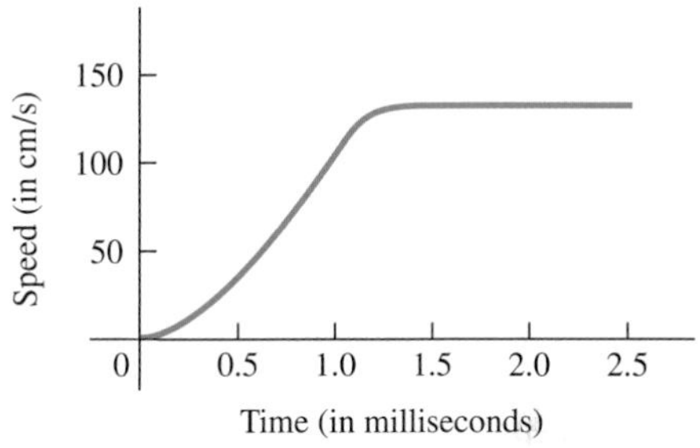

Figure 2.38 Exercise 2.52.

***2.53** The graph in Fig. 2.39 describes the acceleration as a function of time for a stone rolling down a hill starting from rest. a) Find the change in the stone's velocity between $t = 2.5$ s and $t = 7.5$ s. b) Sketch a graph of the stone's velocity as a function of time.

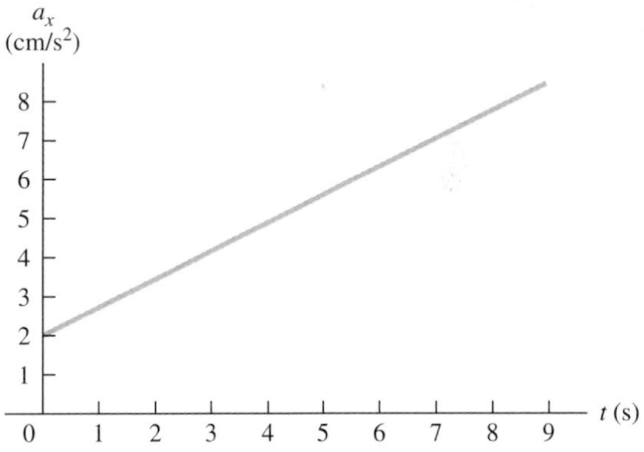

Figure 2.39 Exercise 2.53.

Problems

2.54 On a 20-mile bike ride, you ride the first 10 miles at an average speed of 8 mi/h. What must your average speed over the next 10 miles be to have your average speed for the total 20 miles be a) 4 mi/h? b) 12 mi/h? c) Given this average speed for the first 10 miles, can you possibly attain an average speed of 16 mi/h for the total 20-mile ride? Explain.

2.55 The position of a particle between $t = 0$ and $t = 2.00$ s is given by $x(t) = (3.00\ \text{m/s}^3)t^3 - (10.0\ \text{m/s}^2)t^2 + (9.00\ \text{m/s})t$. a) Draw the x-t, v_x-t, and a_x-t graphs of this particle. b) At what time(s) between $t = 0$ and $t = 2.00$ s is the particle instantaneously at rest? Does your numerical result agree with the v_x-t graph in part (a)? c) At each time calculated in part (b) is the acceleration of the particle positive or negative? Show that in each case the same answer is deduced from $a_x(t)$ and from the v_x-t graph. d) At what time(s) between $t = 0$ and $t = 2.00$ s is the velocity of the particle instantaneously not changing? Locate this point on the v_x-t and a_x-t graphs of part (a). e) What is the particle's greatest distance from the origin ($x = 0$) between $t = 0$ and $t = 2.00$ s? f) At what time(s) between $t = 0$ and $t = 2.00$ s is the particle *speeding up* at the greatest rate? At what time(s) between $t = 0$ and $t = 2.00$ s is the particle *slowing down* at the greatest rate? Locate these points on the v_x-t and a_x-t graphs of part (a).

2.56 Relay Race. In a relay race, each contestant runs 25.0 m while carrying an egg balanced on a spoon, turns around, and comes back to the starting point. Edith runs the first 25.0 m in 20.0 s. On the return trip she is more confident and takes only 15.0 s. What is the magnitude of her average velocity for a) the first 25.0 m? b) the return trip? c) What is her average velocity for the entire round trip? d) What is her average speed for the round trip?

2.57 Dan gets on Interstate Highway I–80 at Seward, Nebraska, and drives due west in a straight line and at an average velocity of magnitude 88 km/h. After traveling 76 km, he reaches the Aurora exit (Fig. 2.40). Realizing he has gone too far, he turns around and drives due east 34 km back to the York exit at an average velocity of magnitude 72 km/h. For his whole trip from Seward to the York exit, what is a) his average speed? b) the magnitude of his average velocity?

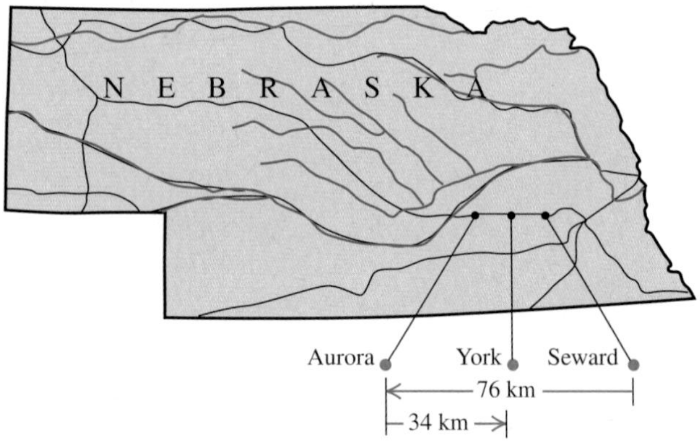

Figure 2.40 Problem 2.57.

2.58 Freeway Traffic. According to a *Scientific American* article (May, 1990), current freeways can sustain about 2400 vehicles per lane per hour in smooth traffic flow at 96 km/h (60 mi/h). Above that figure the traffic flow becomes "turbulent" (stop-and-go). a) If a vehicle is 4.6 m (15 ft) long on the average, what is the average spacing between vehicles at the above traffic density? b) Collision-avoidance automated control systems, which operate by bouncing radar or sonar signals off surrounding vehicles and then accelerate or brake the car when necessary, could greatly reduce the required spacing between vehicles. If the average spacing is 9.2 m (two car lengths), how many vehicles per hour can a lane of traffic carry at 96 km/h?

2.59 A world-class sprinter accelerates to his maximum speed in 4.0 s. He then maintains this speed for the remainder of a 100-m race, finishing with a total time of 9.1 s. a) What is the runner's average acceleration during the first 4.0 s? b) What is his average acceleration during the last 5.1 s? c) What is his average acceleration for the entire race? d) Explain why your answer to part (c) is not the average of the answers to parts (a) and (b).

2.60 A sled starts from rest at the top of a hill and slides down with a constant acceleration. At some later time it is 14.4 m from the top; 2.00 s after that it is 25.6 m from the top, 2.00 s later 40.0 m from the top, and 2.00 s later it is 57.6 m from the top. a) What is the magnitude of the average velocity of the sled during each of the 2.00-s intervals after passing the 14.4-m point? b) What is the acceleration of the sled? c) What is the speed of the sled when it passes the 14.4-m point? d) How much time did it take to go from the top to the 14.4-m point? e) How far did the sled go during the first second after passing the 14.4-m point?

2.61 A car 3.5 m in length approaches an intersection that is 20 m wide (Fig. 2.41). The light turns yellow when the car's speed is 20 m/s and the front of the car is 50 m from the beginning of the intersection. If the driver steps on the brake, the car slows at $-3.8\ \text{m/s}^2$. If he instead steps on the gas pedal, the car accelerates at $2.3\ \text{m/s}^2$. The light stays yellow for 3.0 s. Ignore the driver's reaction time. To avoid being in the intersection at the instant that the light turns red, should the driver step on the brake or the gas? Will the car remain clear of the intersection?

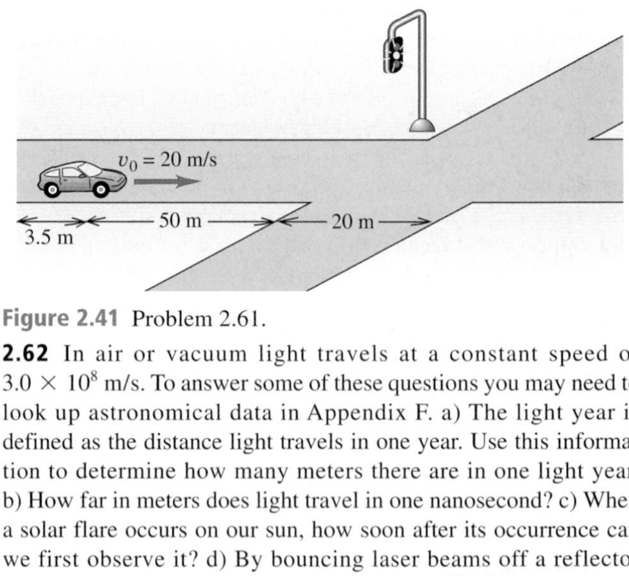

Figure 2.41 Problem 2.61.

2.62 In air or vacuum light travels at a constant speed of 3.0×10^8 m/s. To answer some of these questions you may need to look up astronomical data in Appendix F. a) The light year is defined as the distance light travels in one year. Use this information to determine how many meters there are in one light year. b) How far in meters does light travel in one nanosecond? c) When a solar flare occurs on our sun, how soon after its occurrence can we first observe it? d) By bouncing laser beams off a reflector placed on our moon by the Apollo astronauts, astronomers can make very accurate measurements of the earth-moon distance. How long after it is sent does it take such a laser beam (which is just a light beam) to return to earth? e) The Voyager probe, which passed by Neptune in Aug. 1989, was about 3.0 billion miles from earth at that time. Photographs and other information were sent to earth by radio waves, which travel at the speed of light. How long did it take these waves to reach earth from Voyager?

2.63 Use the information in Appendix F to answer each of the following questions. a) What is the speed of the Galapagos Islands, on the earth's equator, due to our planet's spin on its axis? b) What is the earth's speed due to its rotation around the sun? c) If light would bend around the curvature of the earth (which it does not), how many times would a light beam go around the equator in one second?

2.64 In a 350-m race, runner A starts from rest and accelerates at 1.6 m/s^2 for the first 30 m and then runs at constant speed. Runner B starts from rest and accelerates at 2.0 m/s^2 for the first 30 m and then runs at constant speed. Runner A begins running as soon as the race begins but B first takes a nap to rest up. What is the longest nap that B can take and still not lose the race?

2.65 A ball starts from rest and rolls down a hill with uniform acceleration, traveling 150 m during the second 5.0 s of its motion. How far did it roll during the first 5.0 s of motion?

2.66 Collision. The engineer of a passenger train traveling at 25.0 m/s sights a freight train whose caboose is 200 m ahead on the same track (Fig. 2.42). The freight train is traveling at 15.0 m/s in the same direction as the passenger train. The engineer of the passenger train immediately applies the brakes, causing a constant acceleration of -0.100 m/s^2, while the freight train continues with constant speed. Take $x = 0$ at the location of the front of the passenger train when the engineer applies the brakes. a) Will the cows nearby witness a collision? b) If so, where will it take place? c) On a single graph, sketch the positions of the front of the passenger train and the back of the freight train.

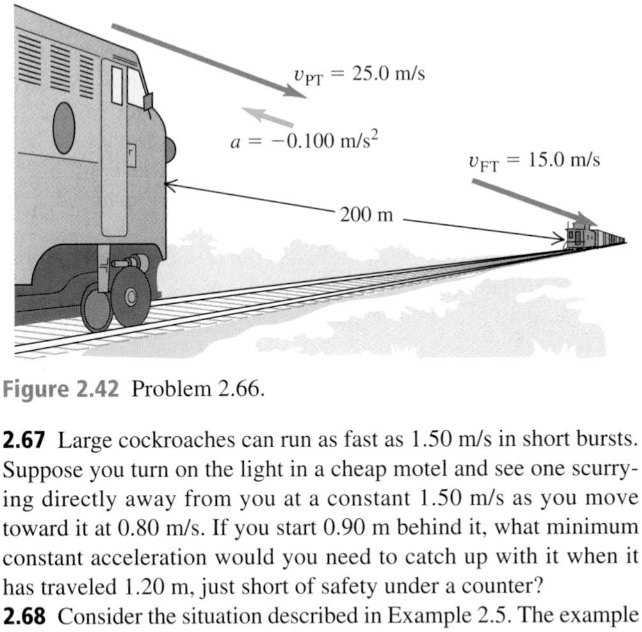

Figure 2.42 Problem 2.66.

2.67 Large cockroaches can run as fast as 1.50 m/s in short bursts. Suppose you turn on the light in a cheap motel and see one scurrying directly away from you at a constant 1.50 m/s as you move toward it at 0.80 m/s. If you start 0.90 m behind it, what minimum constant acceleration would you need to catch up with it when it has traveled 1.20 m, just short of safety under a counter?

2.68 Consider the situation described in Example 2.5. The example is somewhat unrealistic, because the constantly accelerating officer continues past the car. In a real pursuit the officer would accelerate to a speed faster than that of the car, and then slow down in order to have the same velocity as the car when she catches up with it. Suppose that the officer in Example 2.5 accelerates from rest at $a_x = 2.5$ m/s^2 until her speed is 20 m/s. She then slows down at a constant rate until she comes alongside the car at $x = 360$ m, traveling with the same speed of 15 m/s as the car. a) At what time does the officer catch up with the car? b) At what time does the officer change from speeding up to slowing down? At this time, how far is the officer from the stop sign? How far behind the car is she? c) Find the acceleration of the officer while she is slowing down. d) Draw an x-t graph for the two vehicles. e) Draw a v_x-t graph for the two vehicles.

2.69 An automobile and a truck start from rest at the same instant, with the automobile initially at some distance behind the truck. The truck has a constant acceleration of 2.10 m/s^2, and the automobile an acceleration of 3.40 m/s^2. The automobile overtakes the truck after the truck has moved 40.0 m. a) How much time does it take the automobile to overtake the truck? b) How far was the automobile behind the truck initially? c) What is the speed of each when they are abreast? d) On a single graph, sketch the position of each vehicle as a function of time. Take $x = 0$ at the initial location of the truck.

2.70 Two stunt drivers drive directly toward each other. At time $t = 0$ the two cars are a distance D apart, car 1 is at rest, and car 2 is moving to the left with speed v_0. Car 1 begins to move at $t = 0$, speeding up with a constant acceleration a_x. Car 2 continues to move with a constant velocity. a) At what time do the two cars collide? b) Find the speed of car 1 just before it collides with car 2. c) Sketch x-t and v_x-t graphs for car 1 and car 2. Draw the curves for each car using the same axes.

2.71 In his Mustang, Juan rounds a curve onto a straight section of country road while traveling at 20 m/s (45 mi/h), only to find a fully loaded manure spreader completely blocking the road 37 m ahead of him. Startled, he travels at a constant velocity for 0.80 s of reaction time, then hits the brakes and slows at constant acceleration so that he manages to stop just short of the spreader. With the same reaction time and acceleration, if he had roared around the corner at 25 m/s (56 mi/h) instead of 20 m/s, a) how fast would he have been going when he hit the spreader? b) How much time would he have had for his life to flash before his eyes from the time he first sees the spreader until he hits it?

2.72 A police car is traveling in a straight line with constant speed v_p. A truck traveling in the same direction with speed $\frac{3}{2}v_p$ passes the police car. The truck driver realizes that she is speeding, and immediately begins to slow down at a constant rate until she comes to a stop. This is her lucky day, however, and the police car (still moving with the same constant speed) passes the truck driver without giving her a ticket. a) Show that the truck's speed at the instant that the police car passes the truck does *not* depend on the magnitude of the truck's acceleration as it slows down, and find the value of that speed. b) Sketch the x-t graph for the two vehicles.

2.73 Passing. The driver of a car wishes to pass a truck that is traveling at a constant speed of 20.0 m/s (about 45 mi/h). Initially, the car is also traveling at 20.0 m/s and its front bumper is 24.0 m behind the truck's rear bumper. The car accelerates at a constant 0.600 m/s^2, then pulls back into the truck's lane when the rear of the car is 26.0 m ahead of the front of the truck. The car is 4.5 m long and the truck is 21.0 m long. a) How much time is required for the car to pass the truck? b) What distance does the car travel during this time? c) What is the final speed of the car?

***2.74** An object's velocity is measured to be $v_x(t) = \alpha - \beta t^2$, where $\alpha = 4.00$ m/s and $\beta = 2.00$ m/s^3. At $t = 0$ the object is at $x = 0$. a) Calculate the object's position and acceleration as functions of time. b) What is the object's maximum *positive* displacement from the origin?

***2.75** The acceleration of a particle is given by $a_x(t) = -2.00\ \text{m/s}^2 + (3.00\ \text{m/s}^3)t$. a) Find the initial velocity v_{0x} such that the particle will have the same x-coordinate at $t = 4.00$ s as it had at $t = 0$. b) What will be the velocity at $t = 4.00$ s?

2.76 Egg Drop. You are on the roof of the physics building, 46.0 m above the ground (Fig. 2.43). Your physics professor, who is 1.80 m tall, is walking alongside the building at a constant speed of 1.20 m/s. If you wish to drop an egg on your professor's head, where should the professor be when you release the egg? Assume that the egg is in free fall.

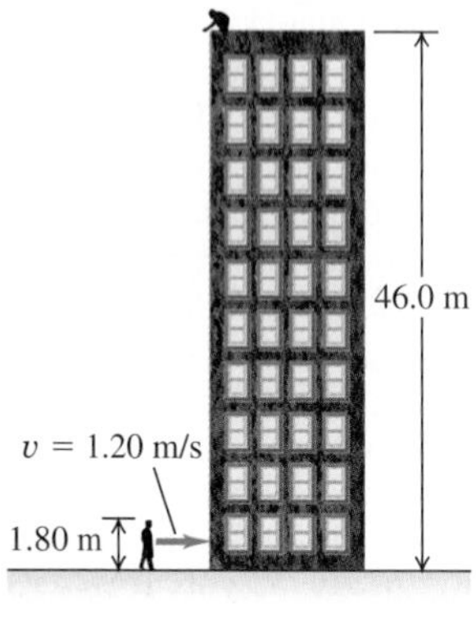

Figure 2.43 Problem 2.76.

2.77 A physics student with too much free time drops a watermelon from the roof of a building. He hears the sound of the watermelon going "splat" 2.50 s later. How high is the building? The speed of sound is 340 m/s. You may ignore air resistance.

2.78 Elevators. Estimate the maximum speed and the magnitude of acceleration of an elevator. You will need to use your observations of roughly how long it takes the elevator to go from one floor to another, the approximate vertical distance from one floor to the next, and the distance over which an elevator accelerates to its maximum speed or slows to a stop.

2.79 Visitors at an amusement park watch divers step off a platform 21.3 m (70 ft) above a pool of water. According to the announcer, the divers enter the water at a speed of 56 mi/h (25 m/s). Air resistance may be ignored. a) Is the announcer correct in this claim? b) Is it possible for a diver to leap directly upward off the board so that, missing the board on the way down, she enters the water at 25.0 m/s? If so, what initial upward speed is required? Is the required initial speed physically attainable?

2.80 A flowerpot falls off a windowsill and falls past the window below. You may ignore air resistance. It takes 0.420 s to pass this window, which is 1.90 m high. How far is the top of the window below the windowsill from which the flowerpot fell?

2.81 A football is kicked vertically upward from the ground and a student gazing out of the window sees it moving upward past her at 5.00 m/s. The window is 12.0 m above the ground. You may ignore air resistance. a) How high does the football go above ground? b) How much time does it take to go from the ground to its highest point?

2.82 A model rocket has a constant upward acceleration of $40.0\ \text{m/s}^2$ while its engine is running. The rocket is fired vertically, and the engine runs for 2.50 s before it uses up the fuel. After the engine stops, the rocket is in free fall. The motion of the rocket is purely up and down. a) Sketch the a_y-t, v_y-t, and y-t graphs for the rocket. b) What is the maximum height that the rocket reaches? c) What will be the speed of the rocket just before it hits the ground? d) Is the total time of flight equal to twice the time the rocket takes to reach its highest point? Why or why not? (See Question Q2.16.)

2.83 Look Out Below. Sam heaves a 16-lb shot straight upward, giving it a constant upward acceleration from rest of $45.0\ \text{m/s}^2$ for 64.0 cm. He releases it 2.20 m above the ground. You may ignore air resistance. a) What is the speed of the shot when he releases it? b) How high above the ground does it go? c) How much time does he have to get out of its way before it returns to the height of the top of his head, 1.83 m above the ground?

2.84 A physics teacher performing an outdoor demonstration suddenly falls from rest off a high cliff and simultaneously shouts "Help." When she has fallen for 3.0 s, she hears the echo of her shout from the valley floor below. The speed of sound is 340 m/s. a) How tall is the cliff? b) If air resistance is neglected, how fast will she be moving just before she hits the ground? (Her actual speed will be less than this, due to air resistance.)

2.85 Juggling Act. A juggler performs in a room whose ceiling is 3.0 m above the level of his hands. He throws a ball upward so that it just reaches the ceiling. a) What is the initial velocity of the ball? b) What is the time required for the ball to reach the ceiling? At the instant when the first ball is at the ceiling, he throws a second ball upward with two-thirds the initial velocity of the first. c) How long after the second ball is thrown did the two balls pass each other? d) At what distance above the juggler's hand do they pass each other?

2.86 A helicopter carrying Dr. Evil takes off with a constant upward acceleration of $5.0\ \text{m/s}^2$. Secret agent Austin Powers jumps on just as the helicopter lifts off the ground. After the two men struggle for 10.0 s, Powers shuts off the engine and steps out of the helicopter. Assume that the helicopter is in free fall after its engine is shut off and ignore effects of air resistance. a) What is the maximum height above ground reached by the helicopter? b) Powers deploys a jet pack strapped on his back 7.0 s after leaving the helicopter, and then he has a constant downward acceleration with magnitude $2.0\ \text{m/s}^2$. How far is Powers above the ground when the helicopter crashes into the ground?

2.87 Building Height. Spiderman steps from the top of a tall building. He falls freely from rest to the ground a distance of h. He falls a distance of $h/4$ in the last 1.0 s of his fall. What is the height h of the building?

2.88 Cliff Height. You are climbing in the High Sierra where you suddenly find yourself at the edge of a fog-shrouded cliff. To find the height of this cliff, you drop a rock from the top and 10.0 s later hear the sound of it hitting the ground at the foot of the cliff. a) Ignoring air resistance, how high is the cliff if the speed of sound is 330 m/s? b) Suppose you had ignored the time it takes the sound to reach you. In that case, would you have overestimated or underestimated the height of the cliff? Explain your reasoning.

2.89 Falling Can. A painter is standing on scaffolding that is raised at constant speed. As he travels upward, he accidentally nudges a paint can off the scaffolding and it falls 15.0 m to the ground. You are watching, and measure with your stopwatch that it takes 3.25 s for the can to reach the ground. Ignore air resistance. a) What is the speed of the can just before it hits the ground? b) Another painter is standing on a ledge, with his hands 4.00 m above the can when it falls off. He has lightning-fast reflexes and if the can passes in front of him, he can catch it. Does he get the chance?

2.90 Determined to test the law of gravity for himself, a student walks off a skyscraper 180 m high, stopwatch in hand, and starts his free fall (zero initial velocity). Five seconds later, Superman arrives at the scene and dives off the roof to save the student. a) Superman leaves the roof with an initial speed v_0 that he produces by pushing himself downward from the edge of the roof with his legs of steel. He then falls with the same acceleration as any freely falling body. What must the value of v_0 be so that Superman catches the student just before they reach the ground? b) On the same graph, sketch the positions of the student and of Superman as functions of time. Take Superman's initial speed to have the value calculated in part (a). c) If the height of the skyscraper is less than some minimum value, even Superman can't reach the student before he hits the ground. What is this minimum height?

2.91 Another determined student (see Problem 2.90) walks off the top of the CN Tower in Toronto, which is 553 m high, and falls freely. His initial velocity is zero. The Rocketeer arrives at the scene 5.00 s later and dives off the top of the tower to save the student. The Rocketeer leaves the roof with an initial downward speed v_0. In order both to catch the student and to prevent injury to him, the Rocketeer should catch the student at a sufficiently great height above ground so that the Rocketeer and the student slow down and arrive at the ground with zero velocity. The upward acceleration that accomplishes this is provided by the Rocketeer's jet pack, which he turns on just as he catches the student; before then the Rocketeer is in free fall. To prevent discomfort to the student, the magnitude of the acceleration of the Rocketeer and the student as they move downward together should be no more than five times g. a) What is the minimum height above the ground at which the Rocketeer should catch the student? b) What must the Rocketeer's initial downward speed be so that he catches the student at the minimum height found in part (a)? c) Sketch the y-t, v_y-t, and a_y-t graphs for the student and for the Rocketeer. On each graph, draw the curves for the Rocketeer and for the student using the same axes.

2.92 A ball is thrown straight up from the ground with speed v_0. At the same instant, a second ball is dropped from rest from a height H, directly above the point where the first ball was thrown upward. There is no air resistance. a) Find the time at which the two balls collide. b) Find the value of H in terms of v_0 and g so that at the instant when the balls collide, the first ball is at the highest point of its motion.

2.93 Two cars, A and B, travel in a straight line. The distance of A from the starting point is given as a function of time by $x_A(t) = \alpha t + \beta t^2$, with $\alpha = 2.60$ m/s and $\beta = 1.20$ m/s^2. The distance of B from the starting point is $x_B(t) = \gamma t^2 - \delta t^3$, with $\gamma = 2.80$ m/s^2 and $\delta = 0.20$ m/s^3. a) Which car is ahead just after they leave the starting point? b) At what time(s) are the cars at the same point? c) At what time(s) is the distance from A to B neither increasing nor decreasing? d) At what time(s) do A and B have the same acceleration?

2.94 An apple drops from the tree and falls freely. The apple is originally at rest a height H above the top of the grass of a thick lawn, which is made of blades of grass of height h. When the apple enters the grass, it slows down at a constant rate so that its speed is zero when it reaches ground level. a) Find the speed of the apple just before it enters the grass. b) Find the acceleration of the apple while it is in the grass. c) Sketch the y-t, v_y-t, and a_y-t graphs for the apple's motion.

Challenge Problems

2.95 Catching the Bus. A student is running at her top speed of 5.0 m/s to catch a bus, which is stopped at the bus stop. When the student is still 40.0 m from the bus, it starts to pull away, moving with a constant acceleration of 0.170 m/s^2. a) For how much time and what distance does the student have to run at 5.0 m/s before she overtakes the bus? b) When she reaches the bus, how fast is the bus traveling? c) Sketch an x-t graph for both the student and the bus. Take $x = 0$ at the initial position of the student. d) The equations you used in part (a) to find the time have a second solution, corresponding to a later time for which the student and bus are again at the same place if they continue their specified motions. Explain the significance of this second solution. How fast is the bus traveling at this point? e) If the student's top speed is 3.5 m/s, will she catch the bus? f) What is the *minimum* speed the student must have to just catch up with the bus? For what time and what distance does she have to run in that case?

2.96 In the vertical jump, an athlete starts from a crouch and jumps upward to reach as high as possible. Even the best athletes spend little more than 1.00 s in the air (their "hang time"). Treat the athlete as a particle and let y_{max} be his maximum height above the floor. To explain why he seems to hang in the air, calculate the ratio of the time he is above $y_{max}/2$ to the time it takes him to go from the floor to that height. You may ignore air resistance.

2.97 A ball is thrown straight up from the edge of the roof of a building. A second ball is dropped from the roof 1.00 s later. You may ignore air resistance. a) If the height of the building is 20.0 m, what must the initial speed be of the first ball if both are to hit the ground at the same time? On the same graph, sketch the position of each ball as a function of time, measured from when the first ball is thrown. Consider the same situation, but now let the initial speed v_0 of the first ball be given and treat the height h of the building as an unknown. b) What must the height of the building be for both balls to reach the ground at the same time for each of the following values of v_0: i) 6.0 m/s; ii) 9.5 m/s? c) If v_0 is greater than some value v_{max}, a value of h does not exist that allows both balls to hit the ground at the same time. Solve for v_{max}. The value v_{max} has a simple physical interpretation. What is it? d) If v_0 is less than some value v_{min}, a value of h does not exist that allows both balls to hit the ground at the same time. Solve for v_{min}. The value v_{min} also has a simple physical interpretation. What is it?

2.98 An alert hiker sees a boulder fall from the top of a distant cliff and notes that it takes 1.30 s for the boulder to fall the last third of the way to the ground. You may ignore air resistance. a) What is the height of the cliff in meters? b) If in part (a) you get two solutions of a quadratic equation and use one for your answer, what does the other solution represent?

3 MOTION IN TWO OR THREE DIMENSIONS

A spectacular example of motion occurs when a volcano ejects rock fragments into the air. Such eruptions are caused by the sudden release of high-pressure gas from within the magma (molten rock beneath the surface). The ejected fragments, called tephra, can be as large as a house and can fly out of the volcano at speeds of hundreds of meters per second.

? If all the fragments are ejected at the same speed, for which ejection angle do the fragments land farthest from the volcano?

When a baseball is hit by a bat, what determines where the baseball lands? How do you describe the motion of a roller coaster car along a curved track or the flight of a hawk circling over an open field? If you throw a water balloon horizontally from your window, will it take the same amount of time to hit the ground as a balloon that you simply drop?

We can't answer these kinds of questions using the techniques of Chapter 2, in which particles moved only along a straight line. Instead, we have to confront the reality that our world is three-dimensional. To understand the curved flight of a baseball, the orbital motion of a satellite, or the trajectory of a thrown projectile, we need to extend our descriptions of motion to two- and three-dimensional situations. We'll still use the vector quantities displacement, velocity, and acceleration, but now these quantities will have two or three components and will no longer lie along a single line. We'll find that several important and interesting kinds of motion take place in two dimensions only, that is, in a *plane*. These motions can be described with two coordinates and two components of velocity and acceleration.

We also need to consider how the motion of a particle is described by different observers who are moving relative to each other. The concept of *relative velocity* will play an important role later in the book when we study collisions, when we explore electromagnetic phenomena, and when we introduce Einstein's special theory of relativity.

This chapter merges the vector language we learned in Chapter 1 with the kinematic language of Chapter 2. As before, we are concerned with describing motion, not with analyzing its causes. But the language you learn here will be an essential tool in later chapters when you use Newton's laws of motion to study the relation between force and motion.

3.1 | Position and Velocity Vectors

To describe the *motion* of a particle in space, we first need to be able to describe the *position* of the particle. Consider a particle that is at a point P at a certain instant. The **position vector** $\vec{r}$ of the particle at this instant is a vector that goes from the origin of the coordinate system to the point P (Fig. 3.1). The figure also shows that the Cartesian coordinates x, y, and z of point P are the x-, y-, and z-components of vector $\vec{r}$. Using the unit vectors we introduced in Section 1.9, we can write

$$\vec{r} = x\hat{\imath} + y\hat{\jmath} + z\hat{k} \tag{3.1}$$

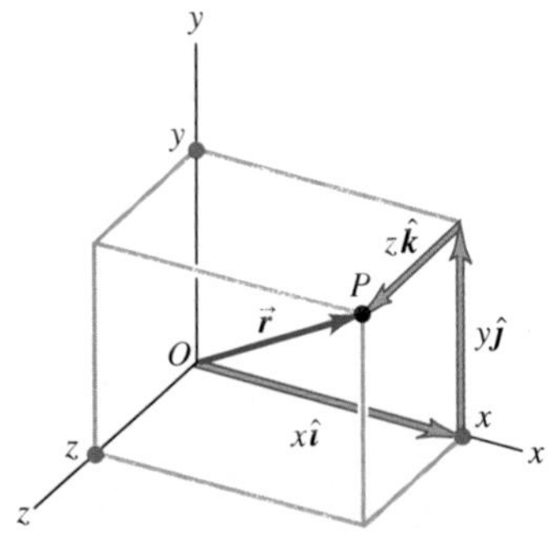

3.1 The position vector $\vec{r}$ from the origin to point P has components x, y, and z.

As the particle moves through space, the path that it follows is in general a curve (Fig. 3.2). During a time interval Δt the particle moves from P_1, where its position vector is $\vec{r}_1$, to P_2, where its position vector is $\vec{r}_2$. The change in position (the displacement) during this interval is $\Delta\vec{r} = \vec{r}_2 - \vec{r}_1 = (x_2 - x_1)\hat{\imath} + (y_2 - y_1)\hat{\jmath} + (z_2 - z_1)\hat{k}$. We define the **average velocity** $\vec{v}_{av}$ during this interval in the same way we did in Chapter 2 for straight-line motion, as the displacement divided by the time interval:

$$\vec{v}_{av} = \frac{\vec{r}_2 - \vec{r}_1}{t_2 - t_1} = \frac{\Delta\vec{r}}{\Delta t} \quad \text{(average velocity vector)} \tag{3.2}$$

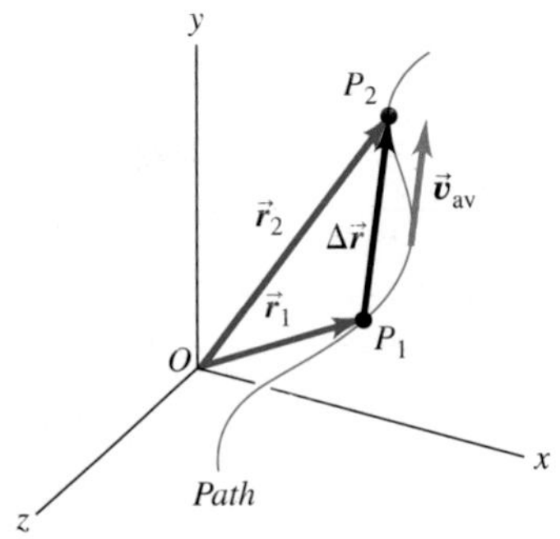

3.2 The average velocity $\vec{v}_{av}$ between points P_1 and P_2 has the same direction as the displacement $\Delta\vec{r}$.

Note that *dividing* a vector by a scalar is really a special case of *multiplying* a vector by a scalar, described in Section 1.7; the average velocity $\vec{v}_{av}$ is equal to the displacement vector $\Delta\vec{r}$ multiplied by $1/\Delta t$, the reciprocal of the time interval. Note also that the x-component of Eq. (3.2) is $v_{av\text{-}x} = (x_2 - x_1)/(t_2 - t_1) = \Delta x/\Delta t$. This is just Eq. (2.2), the expression for average velocity in one dimension that we found in Section 2.1.

We now define **instantaneous velocity** just as we did in Chapter 2; it is the limit of the average velocity as the time interval approaches zero, and it equals the instantaneous rate of change of position with time. The key difference is that position $\vec{r}$ and instantaneous velocity $\vec{v}$ are now both vectors:

$$\vec{v} = \lim_{\Delta t \to 0} \frac{\Delta\vec{r}}{\Delta t} = \frac{d\vec{r}}{dt} \quad \text{(instantaneous velocity vector)} \tag{3.3}$$

The *magnitude* of the vector $\vec{v}$ at any instant is the *speed* v of the particle at that instant. The *direction* of $\vec{v}$ at any instant is the same as the direction in which the particle is moving at that instant.

Note that as $\Delta t \to 0$, points P_1 and P_2 in Fig. 3.2 move closer and closer together. In this limit, the vector $\Delta\vec{r}$ becomes tangent to the curve. The direction of $\Delta\vec{r}$ in the limit is also the direction of the instantaneous velocity $\vec{v}$. This leads to an important conclusion: *at every point along the path, the instantaneous velocity vector is tangent to the path at that point* (Fig. 3.3).

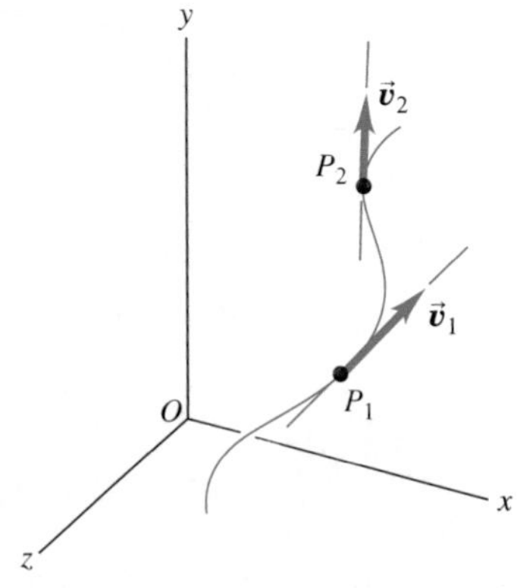

3.3 The instantaneous velocity $\vec{v}$ is tangent to the path at each point. Here $\vec{v}_1$ and $\vec{v}_2$ are the instantaneous velocities at the points P_1 and P_2 shown in Fig. 3.2.

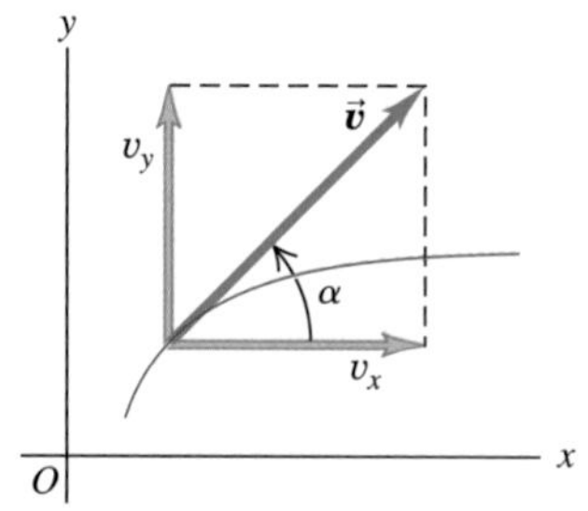

3.4 The two velocity components for motion in the xy-plane.

It's often easiest to calculate the instantaneous velocity vector using components. During any displacement $\Delta\vec{r}$, the changes Δx, Δy, and Δz in the three coordinates of the particle are the *components* of $\Delta\vec{r}$. It follows that the components v_x, v_y, and v_z of the instantaneous velocity $\vec{v}$ are simply the time derivatives of the coordinates x, y, and z. That is,

$$v_x = \frac{dx}{dt} \qquad v_y = \frac{dy}{dt} \qquad v_z = \frac{dz}{dt} \tag{3.4}$$

(components of instantaneous velocity)

The x-component of $\vec{v}$ is $v_x = dx/dt$, which is the same as Eq. (2.3)—the expression for instantaneous velocity for straight-line motion that we obtained in Section 2.2. Hence Eq. (3.4) is a direct extension of the idea of instantaneous velocity to motion in three dimensions.

We can also get this result by taking the derivative of Eq. (3.1). The unit vectors $\hat{\imath}$, $\hat{\jmath}$, and $\hat{k}$ are constant in magnitude and direction, so their derivatives are zero, and we find

$$\vec{v} = \frac{d\vec{r}}{dt} = \frac{dx}{dt}\hat{\imath} + \frac{dy}{dt}\hat{\jmath} + \frac{dz}{dt}\hat{k} \tag{3.5}$$

This shows again that the components of $\vec{v}$ are dx/dt, dy/dt, and dz/dt.

The magnitude of the instantaneous velocity vector $\vec{v}$—that is, the speed—is given in terms of the components v_x, v_y, and v_z by the Pythagorean relation

$$|\vec{v}| = v = \sqrt{v_x^2 + v_y^2 + v_z^2} \tag{3.6}$$

Figure 3.4 shows the situation when the particle moves in the xy-plane. In this case, z and v_z are zero. Then the speed (the magnitude of $\vec{v}$) is

$$v = \sqrt{v_x^2 + v_y^2}$$

and the direction of the instantaneous velocity $\vec{v}$ is given by the angle α in the figure. We see that

$$\tan\alpha = \frac{v_y}{v_x} \tag{3.7}$$

(We always use Greek letters for angles. We use α for the direction of the instantaneous velocity vector to avoid confusion with the direction θ of the *position* vector of the particle.)

The instantaneous velocity vector is usually more interesting and useful than the average velocity vector. From now on, when we use the word "velocity," we will always mean the instantaneous velocity vector $\vec{v}$ (rather than the average velocity vector). Usually, we won't even bother to call $\vec{v}$ a vector; it's up to you to remember that velocity is a vector quantity with both magnitude and direction.

Example 3.1 Calculating average and instantaneous velocity

A robotic vehicle, or rover, is being used to explore the surface of Mars. The landing craft is the origin of coordinates and the surrounding Martian surface lies in the xy-plane. The rover, which we represent as a point, has x- and y-coordinates that vary with time according to

$$x = 2.0\ \text{m} - (0.25\ \text{m/s}^2)t^2$$
$$y = (1.0\ \text{m/s})t + (0.025\ \text{m/s}^3)t^3$$

a) Find the rover's coordinates and its distance from the lander at time $t = 2.0$ s. b) Find the rover's displacement and average velocity vectors during the interval from $t = 0.0$ s to $t = 2.0$ s. c) Derive a general expression for the rover's instantaneous velocity vector, and find the instantaneous velocity at $t = 2.0$ s. Express the instantaneous velocity in component form and also in terms of magnitude and direction.

SOLUTION

IDENTIFY: This problem involves motion in a two-dimensional trajectory (i.e., in a plane). Hence we must use the expressions for the displacement, average velocity, and instantaneous velocity *vectors* obtained in this section. (The simpler expressions in Sections 2.1 and 2.2 don't involve vectors, but they apply only to motion along a straight line.)

SET UP: The rover's path is shown in Fig. 3.5. We'll use Eq. (3.1) for position $\vec{r}$, the expression $\Delta\vec{r} = \vec{r}_2 - \vec{r}_1$ for displacement, Eq. (3.2) for average velocity, and Eqs. (3.5) and (3.6) for instantaneous velocity and its direction. The target variables are as stated in the problem.

EXECUTE: a) At time $t = 2.0$ s the rover's coordinates are

$$x = 2.0\text{ m} - (0.25\text{ m/s}^2)(2.0\text{ s})^2 = 1.0\text{ m}$$

$$y = (1.0\text{ m/s})(2.0\text{ s}) + (0.025\text{ m/s}^3)(2.0\text{ s})^3 = 2.2\text{ m}$$

The rover's distance from the origin at this time is

$$r = \sqrt{x^2 + y^2} = \sqrt{(1.0\text{ m})^2 + (2.2\text{ m})^2} = 2.4\text{ m}$$

b) To find the displacement and average velocity, we express the position vector $\vec{r}$ as a function of time t. From Eq. (3.1), this is

$$\begin{aligned}\vec{r} &= x\hat{\imath} + y\hat{\jmath}\\ &= [2.0\text{ m} - (0.25\text{ m/s}^2)t^2]\hat{\imath}\\ &\quad + [(1.0\text{ m/s})t + (0.025\text{ m/s}^3)t^3]\hat{\jmath}\end{aligned}$$

At time $t = 0.0$ s the position vector is

$$\vec{r}_0 = (2.0\text{ m})\hat{\imath} + (0.0\text{ m})\hat{\jmath}$$

From part (a) we find that at time $t = 2.0$ s the position vector $\vec{r}_2$ is

$$\vec{r}_2 = (1.0\text{ m})\hat{\imath} + (2.2\text{ m})\hat{\jmath}$$

Therefore the displacement from $t = 0.0$ s to $t = 2.0$ s is

$$\begin{aligned}\Delta\vec{r} = \vec{r}_2 - \vec{r}_0 &= (1.0\text{ m})\hat{\imath} + (2.2\text{ m})\hat{\jmath} - (2.0\text{ m})\hat{\imath}\\ &= (-1.0\text{ m})\hat{\imath} + (2.2\text{ m})\hat{\jmath}\end{aligned}$$

During the time interval the rover has moved 1.0 m in the negative x-direction and 2.2 m in the positive y-direction. The average velocity during the interval from $t = 0.0$ s to $t = 2.0$ s is this displacement divided by the elapsed time, as in Eq. (3.2):

$$\begin{aligned}\vec{v}_{av} = \frac{\Delta\vec{r}}{\Delta t} &= \frac{(-1.0\text{ m})\hat{\imath} + (2.2\text{ m})\hat{\jmath}}{2.0\text{ s} - 0.0\text{ s}}\\ &= (-0.50\text{ m/s})\hat{\imath} + (1.1\text{ m/s})\hat{\jmath}\end{aligned}$$

The components of this average velocity are

$$v_{av\text{-}x} = -0.50\text{ m/s} \qquad v_{av\text{-}y} = 1.1\text{ m/s}$$

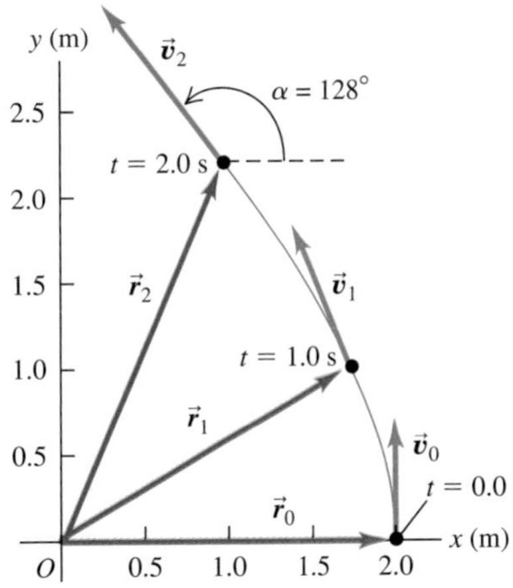

3.5 Path of a radio-controlled robotic vehicle. At $t = 0$ the rover has position vector $\vec{r}_0$ and instantaneous velocity vector $\vec{v}_0$. Likewise, $\vec{r}_1$ and $\vec{v}_1$ are the vectors at $t = 1.0$ s; $\vec{r}_2$ and $\vec{v}_2$ are the vectors at $t = 2.0$ s.

c) From Eq. (3.4), the components of instantaneous velocity are the time derivatives of the coordinates:

$$v_x = \frac{dx}{dt} = (-0.25\text{ m/s}^2)(2t)$$

$$v_y = \frac{dy}{dt} = 1.0\text{ m/s} + (0.025\text{ m/s}^3)(3t^2)$$

Then we can write the instantaneous velocity vector $\vec{v}$ as

$$\begin{aligned}\vec{v} = v_x\hat{\imath} + v_y\hat{\jmath} &= (-0.50\text{ m/s}^2)t\hat{\imath}\\ &\quad + [1.0\text{ m/s} + (0.075\text{ m/s}^3)t^2]\hat{\jmath}\end{aligned}$$

At time $t = 2.0$ s, the components of instantaneous velocity are

$$v_x = (-0.50\text{ m/s}^2)(2.0\text{ s}) = -1.0\text{ m/s}$$

$$v_y = 1.0\text{ m/s} + (0.075\text{ m/s}^3)(2.0\text{ s})^2 = 1.3\text{ m/s}$$

The magnitude of the instantaneous velocity (that is, the speed) at $t = 2.0$ s is

$$\begin{aligned}v = \sqrt{v_x^2 + v_y^2} &= \sqrt{(-1.0\text{ m/s})^2 + (1.3\text{ m/s})^2}\\ &= 1.6\text{ m/s}\end{aligned}$$

Its direction with respect to the positive x-axis is given by the angle α, where, from Eq. (3.7),

$$\tan\alpha = \frac{v_y}{v_x} = \frac{1.3\text{ m/s}}{-1.0\text{ m/s}} = -1.3 \quad \text{so} \quad \alpha = 128°$$

Your calculator will tell you that the inverse tangent of -1.3 is $-52°$. But as we learned in Section 1.8, you have to examine a sketch of a vector (Fig. 3.5) to decide on its direction. The correct answer for α is $-52° + 180° = 128°$.

EVALUATE: Take a moment to compare the components of *average* velocity that we found in part (b) for the interval from $t = 0.0$ s to $t = 2.0$ s $(v_{av\text{-}x} = -0.50\text{ m/s}, v_{av\text{-}y} = 1.1\text{ m/s})$ to the components of *instantaneous* velocity at $t = 2.0$ s that we found in part (c) $(v_x = -1.0 \text{ m/s}, v_y = 1.3 \text{ m/s})$. The comparison shows that, just as in one dimension, the average velocity vector $\vec{v}_{av}$ over an interval is in general *not* equal to the instantaneous velocity $\vec{v}$ at the end of the interval (see Example 2.1).

We invite you to calculate the position, instantaneous velocity vector, speed, and direction of motion at $t = 0.0$ s and $t = 1.0$ s. The position vectors and instantaneous velocity vectors at $t = 0.0$ s, 1.0 s, and 2.0 s are shown in Fig. 3.5. Notice that at every point, the instantaneous velocity vector $\vec{v}$ is tangent to the path. The magnitude of $\vec{v}$ increases as the rover moves, which shows that the rover's speed is increasing.

Test Your Understanding

Give an example of a situation in which the average velocity vector $\vec{v}_{av}$ over an interval *would* be equal to the instantaneous velocity $\vec{v}$ at the end of the interval.

3.2 | The Acceleration Vector

Now let's consider the *acceleration* of a particle moving in space. Just as for motion in a straight line, acceleration describes how the velocity of the particle changes. But now we'll generalize acceleration to describe changes in the magnitude of the velocity (that is, the speed) *and* changes in the direction of the velocity (that is, the direction in which the particle is moving in space).

In Fig. 3.6a, a particle is moving along a curved path. The vectors $\vec{v}_1$ and $\vec{v}_2$ represent the particle's instantaneous velocities at time t_1, when the particle is at point P_1, and at time t_2, when the particle is at point P_2. The two velocities may differ in both magnitude and direction. We define the **average acceleration** $\vec{a}_{av}$ of the particle as it moves from P_1 to P_2 as the *vector change in velocity,* $\vec{v}_2 - \vec{v}_1 = \Delta\vec{v}$, divided by the time interval $t_2 - t_1 = \Delta t$:

$$\vec{a}_{av} = \frac{\vec{v}_2 - \vec{v}_1}{t_2 - t_1} = \frac{\Delta\vec{v}}{\Delta t} \qquad \text{(average acceleration vector)} \qquad (3.8)$$

Average acceleration is a *vector* quantity in the same direction as the vector $\Delta\vec{v}$ (Fig. 3.6a). Note that $\vec{v}_2$ is the vector sum of the original velocity $\vec{v}_1$ and the change $\Delta\vec{v}$ (Fig. 3.6b). The x-component of Eq. (3.8) is $a_{av\text{-}x} = (v_{2x} - v_{1x})/(t_2 - t_1) = \Delta v_x/\Delta t$, which is just Eq. (2.4) for the average acceleration in straight-line motion.

As in Chapter 2, we define the **instantaneous acceleration** $\vec{a}$ at point P_1 as the limit of the average acceleration when point P_2 approaches point P_1 and $\Delta\vec{v}$ and Δt both approach zero; the instantaneous acceleration is also equal to the instantaneous rate of change of velocity with time. Because we are not restricted to straight-line motion, instantaneous acceleration is now a vector:

$$\vec{a} = \lim_{\Delta t \to 0} \frac{\Delta\vec{v}}{\Delta t} = \frac{d\vec{v}}{dt} \qquad \text{(instantaneous acceleration vector)} \qquad (3.9)$$

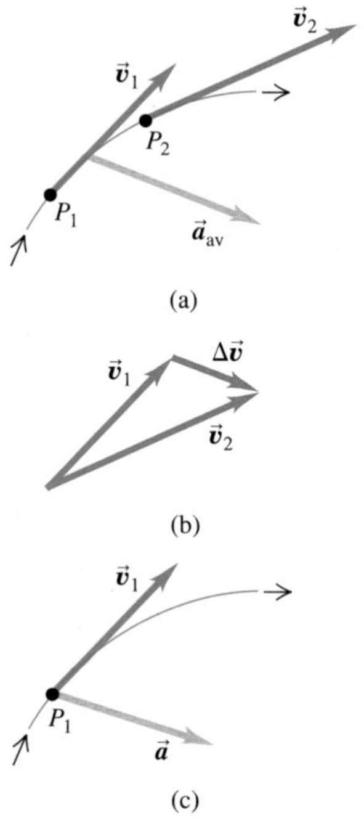

3.6 (a) The vector $\vec{a}_{av} = \Delta\vec{v}/\Delta t$ represents the average acceleration between P_1 and P_2. (b) Construction for obtaining $\Delta\vec{v} = \vec{v}_2 - \vec{v}_1$. (c) Instantaneous acceleration $\vec{a}$ at point P_1. Vector $\vec{v}$ is tangent to the path; vector $\vec{a}$ points toward the concave side of the path.

The velocity vector $\vec{v}$, as we have seen, is tangent to the path of the particle. But the construction in Fig. 3.6c shows that the instantaneous acceleration vector $\vec{a}$ of a moving particle always points toward the concave side of a curved path—that is, toward the inside of any turn that the particle is making. We can also see that when a particle is moving in a curved path, it *always* has nonzero acceleration, even when it moves with constant speed. This conclusion may seem contrary to your intuition, but it's really just contrary to the everyday use of the word "acceleration" to mean that speed is increasing. The more precise definition given in Eq. (3.9) shows that there is a nonzero acceleration whenever the velocity vector changes in *any* way, whether there is a change of speed, direction, or both.

To convince yourself that a particle has a nonzero acceleration when moving on a curved path with constant speed, think of your sensations when you ride in a car. When the car accelerates, you tend to move within the car in a direction *opposite* to the car's acceleration. (We'll discover the reason for this behavior in Chapter 4.) Thus you tend to slide toward the back of the car when it accelerates forward (speeds up) and toward the front of the car when it accelerates backward (slows down). If the car makes a turn on a level road, you tend to slide toward the outside of the turn; hence the car has an acceleration toward the inside of the turn.

We will usually be interested in the instantaneous acceleration, not the average acceleration. From now on, we will use the term "acceleration" to mean the instantaneous acceleration vector $\vec{a}$.

Each component of the acceleration vector is the derivative of the corresponding component of velocity:

$$a_x = \frac{dv_x}{dt} \qquad a_y = \frac{dv_y}{dt} \qquad a_z = \frac{dv_z}{dt} \tag{3.10}$$

(components of instantaneous acceleration)

In terms of unit vectors,

$$\vec{a} = \frac{dv_x}{dt}\hat{\imath} + \frac{dv_y}{dt}\hat{\jmath} + \frac{dv_z}{dt}\hat{k} \tag{3.11}$$

The x-component of Eqs. (3.10) and (3.11), $a_x = dv_x/dt$, is a familiar result: it's the expression from Section 2.3 for instantaneous acceleration in one dimension, Eq. (2.5). Figure 3.7 shows an example of an acceleration vector that has both x- and y-components.

Also, because each component of velocity is the derivative of the corresponding coordinate, we can express the components a_x, a_y, and a_z of the acceleration vector $\vec{a}$ as

$$a_x = \frac{d^2x}{dt^2} \qquad a_y = \frac{d^2y}{dt^2} \qquad a_z = \frac{d^2z}{dt^2} \tag{3.12}$$

and the acceleration vector $\vec{a}$ as

$$\vec{a} = \frac{d^2x}{dt^2}\hat{\imath} + \frac{d^2y}{dt^2}\hat{\jmath} + \frac{d^2z}{dt^2}\hat{k} \tag{3.13}$$

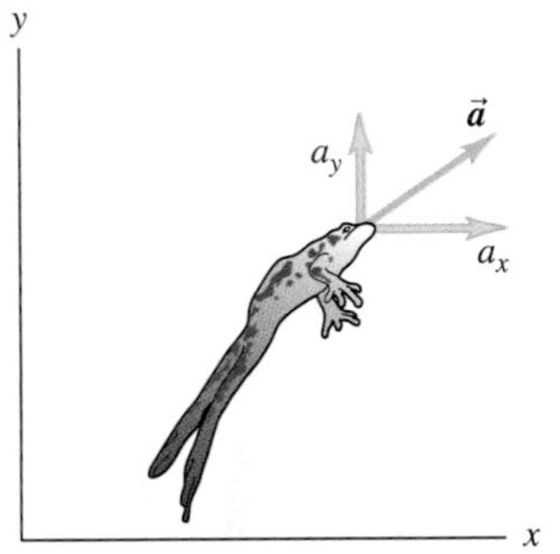

3.7 When a frog leaps, it accelerates both forward and upward. Thus its acceleration vector has both a horizontal component (a_x) and a vertical component (a_y).

Example 3.2 Calculating average and instantaneous acceleration

Let's look again at the motions of the robotic rover in Example 3.1. We found that the components of instantaneous velocity at any time t are

$$v_x = \frac{dx}{dt} = (-0.25\ \text{m/s}^2)(2t)$$

$$v_y = \frac{dy}{dt} = 1.0\ \text{m/s} + (0.025\ \text{m/s}^3)(3t^2)$$

and that the velocity vector is

$$\vec{v} = v_x\hat{\imath} + v_y\hat{\jmath} = (-0.50\ \text{m/s}^2)t\hat{\imath} + [1.0\ \text{m/s} + (0.075\ \text{m/s}^3)t^2]\hat{\jmath}$$

a) Find the components of the average acceleration in the interval from $t = 0.0$ s to $t = 2.0$ s. b) Find the instantaneous acceleration at $t = 2.0$ s.

SOLUTION

IDENTIFY and SET UP: In part (a) we use Eq. (3.8) to calculate the components of the average acceleration. To do so, we first have to use the expressions given above to determine the values of v_x and v_y at the beginning and end of the interval. In part (b) we determine the instantaneous acceleration components at *any* time t by taking the time derivatives of the velocity components as in Eq. (3.10).

EXECUTE: a) If we substitute $t = 0.0$ s or $t = 2.0$ s into the above expressions for v_x and v_y, we find that at the beginning of the interval $(t = 0.0\text{ s})$ the velocity components are

$$v_x = 0.0\text{ m/s} \qquad v_y = 1.0\text{ m/s}$$

and that at the end of the interval $(t = 2.0\text{ s})$ the components are

$$v_x = -2.0\text{ m/s} \qquad v_y = 1.3\text{ m/s}$$

(Note that the values at $t = 2.0$ s are the same as those we found in Example 3.1.) Thus the components of average acceleration in this interval are

$$a_{\text{av-}x} = \frac{\Delta v_x}{\Delta t} = \frac{-1.0\text{ m/s} - 0.0\text{ m/s}}{2.0\text{ s} - 0.0\text{ s}} = -0.5\text{ m/s}^2$$

$$a_{\text{av-}y} = \frac{\Delta v_y}{\Delta t} = \frac{1.3\text{ m/s} - 1.0\text{ m/s}}{2.0\text{ s} - 0.0\text{ s}} = 0.15\text{ m/s}^2$$

b) Using Eq. (3.10), we find

$$a_x = \frac{dv_x}{dt} = -0.50\text{ m/s}^2 \qquad a_y = \frac{dv_y}{dt} = (0.075\text{ m/s}^3)(2t)$$

We can write the instantaneous acceleration vector $\vec{a}$ as

$$\vec{a} = a_x\hat{\imath} + a_y\hat{\jmath} = (-0.50\text{ m/s}^2)\hat{\imath} + (0.15\text{ m/s}^3)t\hat{\jmath}$$

At time $t = 2.0$ s, the components of instantaneous acceleration are

$$a_x = -0.50\text{ m/s}^2 \qquad a_y = (0.15\text{ m/s}^3)(2.0\text{ s}) = 0.30\text{ m/s}^2$$

The acceleration vector at this time is

$$\vec{a} = (-0.50\text{ m/s}^2)\hat{\imath} + (0.30\text{ m/s}^2)\hat{\jmath}$$

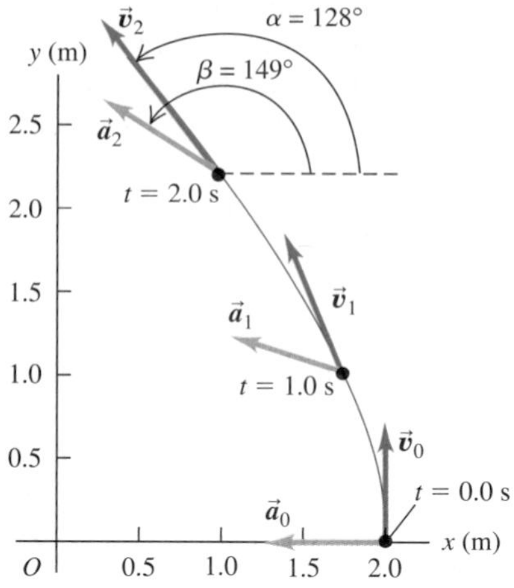

3.8 The path of the radio-controlled rover, showing the velocity and acceleration at $t = 0.0$ s ($\vec{v}_0$ and $\vec{a}_0$), $t = 1.0$ s ($\vec{v}_1$ and $\vec{a}_1$), and $t = 2.0$ s ($\vec{v}_2$ and $\vec{a}_2$).

The magnitude of acceleration at this time is

$$a = \sqrt{a_x^2 + a_y^2}$$
$$= \sqrt{(-0.50\text{ m/s}^2)^2 + (0.30\text{ m/s}^2)^2} = 0.58\text{ m/s}^2$$

The direction of $\vec{a}$ with respect to the positive x-axis is given by the angle β, where

$$\tan\beta = \frac{a_y}{a_x} = \frac{0.30\text{ m/s}^2}{-0.50\text{ m/s}^2} = -0.60$$
$$\beta = 180° - 31° = 149°$$

EVALUATE: We invite you to calculate the instantaneous acceleration at $t = 0.0$ s and $t = 1.0$ s. Figure 3.8 shows the path of the car and the velocity and acceleration vectors at $t = 0.0$ s, 1.0 s, and 2.0 s. Note that the two vectors $\vec{v}$ and $\vec{a}$ are *not* in the same direction at any of these times. The velocity vector is tangent to the path at each point, and the acceleration vector points toward the concave side of the path.

Example 3.3 Determining average acceleration graphically

For the rover in Examples 3.1 and 3.2, use vector subtraction to find the approximate direction of the average acceleration in the interval from $t = 0.0$ s to 2.0 s.

SOLUTION

IDENTIFY: Figure 3.6 shows that the average acceleration for the interval from t_1 to t_2 is in the direction of the velocity change $\Delta\vec{v}$. Given the velocity vectors at the beginning and end of the interval, we can find the direction of $\Delta\vec{v}$, and hence of $\vec{a}_{\text{av}}$, graphically.

SET UP and EXECUTE: From Fig. 3.8, the velocity $\vec{v}_0$ at $t = 0.0$ s points in the positive y-direction and the velocity $\vec{v}_2$ at $t = 2.0$ s points at an angle between the negative x-direction and the positive y-direction. Furthermore, $\vec{v}_2$ has a larger magnitude than $\vec{v}_0$.

Figure 3.9 shows how to find the direction of $\Delta\vec{v} = \vec{v}_2 - \vec{v}_0$ by subtracting the two vectors graphically. The result points at a small angle above the negative x-direction; the average acceleration points in the same direction as $\Delta\vec{v}$.

EVALUATE: The direction of $\Delta\vec{v}$ in Fig. 3.9 is very close to the direction of $\vec{a}_1$ (the instantaneous acceleration at the midpoint of the interval) shown in Fig. 3.8. It's also intermediate between the directions of $\vec{a}_0$ and $\vec{a}_2$, the instantaneous acceleration at the beginning and end of the interval, respectively. This is just how we would expect the *average* acceleration to behave.

This graphical procedure usually won't help you determine numerical values for the acceleration; for that, you need to do calculations like those in Example 3.2. But the graphical approach does offer a way to check whether your calculations are reasonable. It also provides important insight into the character of the motion, as we describe below.

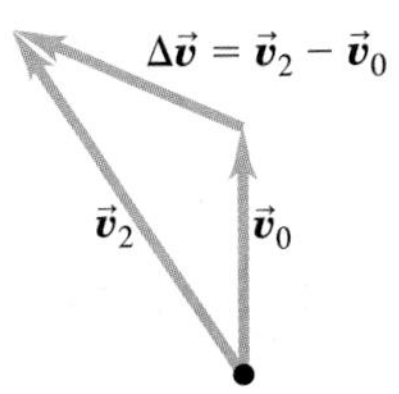

3.9 Using graphical vector subtraction to estimate the direction of average acceleration.

Parallel and Perpendicular Components of Acceleration

It's often useful to describe the acceleration of a particle moving along a curved path in terms of components parallel and perpendicular to the velocity at each point (Fig. 3.10). In Fig. 3.10 we label these components $a_\perp$ and $a_\parallel$. To see why these components are useful, we'll use the graphical interpretation of acceleration (as in Example 3.3) and consider two special cases.

In Fig. 3.11a the acceleration vector is *parallel* to the velocity $\vec{v}_1$. The change in $\vec{v}$ during a small time interval Δt is a vector $\Delta\vec{v}$ having the same direction as $\vec{a}$ and hence the same direction as $\vec{v}_1$. The velocity $\vec{v}_2$ at the end of Δt, given by $\vec{v}_2 = \vec{v}_1 + \Delta\vec{v}$, is a vector having the same direction as $\vec{v}_1$ but greater magnitude. In other words, during the time interval Δt the particle in Fig. 3.11a moved in a straight line with increasing speed.

In Fig. 3.11b the acceleration $\vec{a}$ is *perpendicular* to the velocity $\vec{v}$. In a small time interval Δt, the change $\Delta\vec{v}$ is a vector very nearly perpendicular to $\vec{v}_1$, as shown. Again $\vec{v}_2 = \vec{v}_1 + \Delta\vec{v}$, but in this case $\vec{v}_1$ and $\vec{v}_2$ have different directions. As the time interval Δt approaches zero, the angle ϕ in the figure also approaches zero, $\Delta\vec{v}$ becomes perpendicular to *both* $\vec{v}_1$ and $\vec{v}_2$, and $\vec{v}_1$ and $\vec{v}_2$ have the same magnitude. In other words, the speed of the particle stays the same, but the path of the particle curves.

Thus, when $\vec{a}$ is *parallel* (or antiparallel) to $\vec{v}$, its effect is to change the magnitude of $\vec{v}$ but not its direction; when $\vec{a}$ is *perpendicular* to $\vec{v}$, its effect is to change the direction of $\vec{v}$ but not its magnitude. In general, $\vec{a}$ may have components *both* parallel and perpendicular to $\vec{v}$, but the above statements are still valid for the individual components. In particular, when a particle travels along a curved path with constant speed, its acceleration is always perpendicular to $\vec{v}$ at each point.

Figure 3.12 shows a particle moving along a curved path for three different situations: constant speed, increasing speed, and decreasing speed. If the speed is constant, $\vec{a}$ is perpendicular, or *normal,* to the path and to $\vec{v}$ and points toward the concave side of the path (Fig. 3.12a). If the speed is increasing, there is still a perpendicular component of $\vec{a}$, but there is also a parallel component having the same direction as $\vec{v}$ (Fig. 3.12b). Then $\vec{a}$ points ahead of the normal to the path. (This was the case in Example 3.2.) If the speed is decreasing, the parallel component has the direction opposite to $\vec{v}$, and $\vec{a}$ points behind the normal to the path (Fig. 3.12c). These ideas help us describe the acceleration of a car going around a turn (Fig. 3.13). We will use these ideas again in Section 3.4 when we study the special case of motion in a circle.

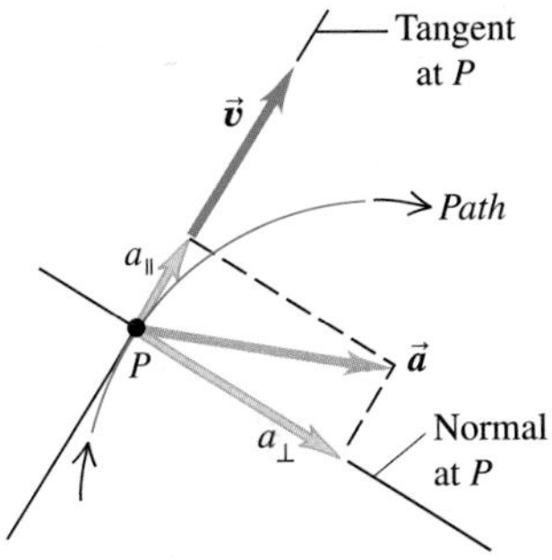

3.10 The acceleration can be resolved into a component $a_\parallel$ parallel to the path (and to the velocity) and a component $a_\perp$ perpendicular to the path (that is, along the normal to the path).

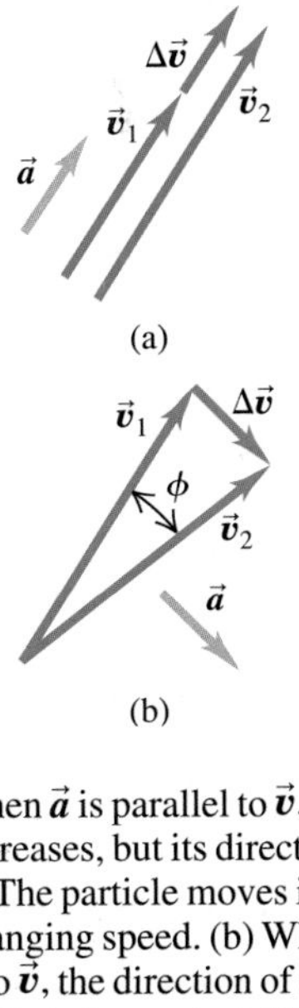

3.11 (a) When $\vec{a}$ is parallel to $\vec{v}$, the magnitude of $\vec{v}$ increases, but its direction does not change. The particle moves in a straight line with changing speed. (b) When $\vec{a}$ is perpendicular to $\vec{v}$, the direction of $\vec{v}$ changes, but its magnitude is constant. The particle moves along a curve with constant speed.

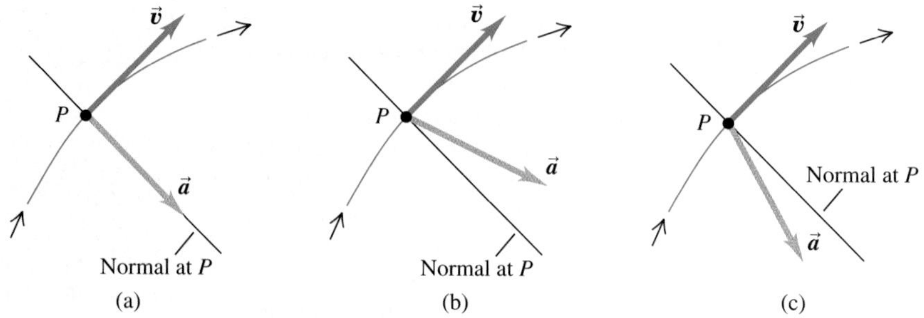

3.12 Velocity and acceleration vectors for a particle moving through a point P on a curved path with (a) constant speed, (b) increasing speed, and (c) decreasing speed.

3.13 A race car typically slows down when it enters a turn and speeds up as it exits the turn. Thus the car has a component of acceleration parallel or antiparallel to the direction of motion (which describes the change in speed) and a component perpendicular to the direction of motion (which describes the change in direction).

CAUTION Note that the two quantities

$$\frac{d|\vec{v}|}{dt} \quad \text{and} \quad \left|\frac{d\vec{v}}{dt}\right|$$

are not the same. The first is the rate of change of speed; it is zero whenever a particle moves with constant speed, even when its direction of motion changes. The second is the magnitude of the vector acceleration; it is zero only when the particle's acceleration is zero—that is, when the particle moves in a straight line with constant speed.

Example 3.4 Calculating parallel and perpendicular components of acceleration

For the rover of Examples 3.1 and 3.2, find the parallel and perpendicular components of the acceleration at $t = 2.0$ s.

SOLUTION

IDENTIFY and SET UP: To find the parallel and perpendicular components, we need to know the angle between the acceleration vector $\vec{a}$ and the velocity vector $\vec{v}$. We'll be able to determine this because we found the directions of $\vec{a}$ and $\vec{v}$ in Examples 3.2 and 3.1, respectively.

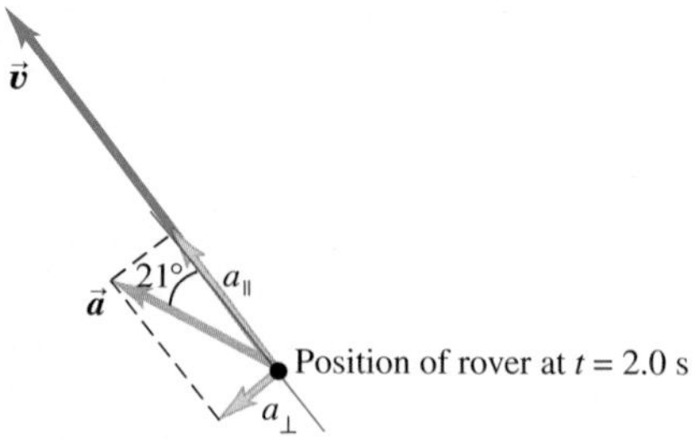

3.14 The parallel and perpendicular components of the acceleration of the rover at $t = 2.0$ s.

EXECUTE: In Example 3.2 we found that at $t = 2.0$ s the particle has an acceleration of magnitude $0.58\ \text{m/s}^2$ at an angle of 149° with respect to the positive x-axis. From Example 3.1, at this same time the velocity vector is at an angle of 128° with respect to the positive x-axis. So Fig. 3.8 shows that the angle between $\vec{a}$ and $\vec{v}$ is $149° - 128° = 21°$ (Fig. 3.14). The parallel and perpendicular components of acceleration are then

$$a_{\parallel} = a\cos 21° = (0.58\ \text{m/s}^2)\cos 21° = 0.54\ \text{m/s}^2$$
$$a_{\perp} = a\sin 21° = (0.58\ \text{m/s}^2)\sin 21° = 0.21\ \text{m/s}^2$$

EVALUATE: The parallel component $a_{\parallel}$ is in the same direction as $\vec{v}$, which means that the speed is increasing at this instant. Because the perpendicular component $a_{\perp}$ is not zero, it follows that the rover's path is curved at this instant; in other words, the rover is turning.

Conceptual Example 3.5 Acceleration of a skier

A skier moves along a ski-jump ramp as shown in Fig. 3.15. The ramp is straight from point A to point C and curved from point C onward. The skier picks up speed as she moves downhill from point A to point E, where her speed is maximum. She slows down after passing point E. Draw the direction of the acceleration vector at points B, D, E, and F.

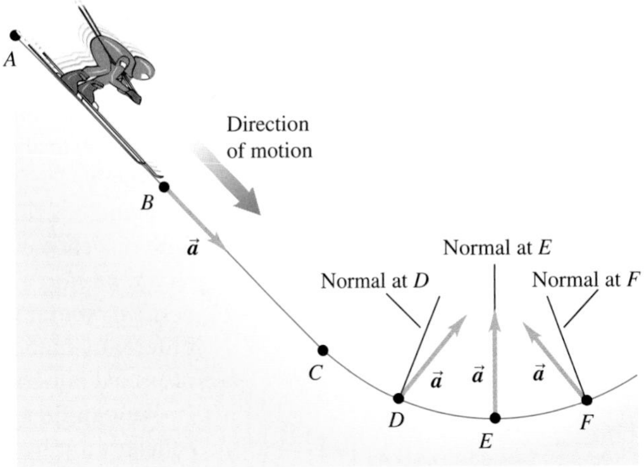

3.15 The direction of a skier's acceleration is different at different points along the path.

SOLUTION

At point B the skier is moving in a straight line with increasing speed, so her acceleration points downhill, in the same direction as her velocity.

At point D the skier is moving along a curved path, so her acceleration has a component perpendicular to the path. There is also a component in the direction of her motion because she is still speeding up at this point. So the acceleration vector points *ahead* of the normal to her path at point D.

The skier's speed is instantaneously not changing at point E; the speed is maximum at this point, so its derivative is zero. So there is no parallel component of $\vec{a}$, and the acceleration is perpendicular to her motion.

Finally, at point F the acceleration has a perpendicular component (because her path is curved at this point) and a parallel component *opposite* to the direction of her motion (because she's slowing down). So at this point, the acceleration vector points *behind* the normal to her path.

In the next section we'll examine the skier's acceleration after she flies off the ramp.

Test Your Understanding

Consider the volcanic rock fragments shown in the photograph that opens this chapter. Each fragment slows down as it climbs toward the highest point of its path, then speeds up as it descends from this point. What is the direction of the fragment's acceleration at the highest point of its path?

3.3 | Projectile Motion

A **projectile** is any body that is given an initial velocity and then follows a path determined entirely by the effects of gravitational acceleration and air resistance. A batted baseball, a thrown football, a package dropped from an airplane, and a bullet shot from a rifle are all projectiles. The path followed by a projectile is called its **trajectory.**

To analyze this common type of motion, we'll start with an idealized model, representing the projectile as a single particle with an acceleration (due to gravity) that is constant in both magnitude and direction. We'll neglect the effects of air resistance and the curvature and rotation of the earth. Like all models, this one has limitations. Curvature of the earth has to be considered in the flight of long-range missiles, and air resistance is of crucial importance to a sky diver. Nevertheless, we can learn a lot from analysis of this simple model. For the remainder of this chapter

ActivPhysics ONLINE

3.1 Solving Projectile Motion Problems
3.2 Two Balls Falling
3.3 Changing the x-Velocity
3.4 Projectile x- and y-Accelerations
3.5 Initial Velocity Components
3.6 Target Practice I

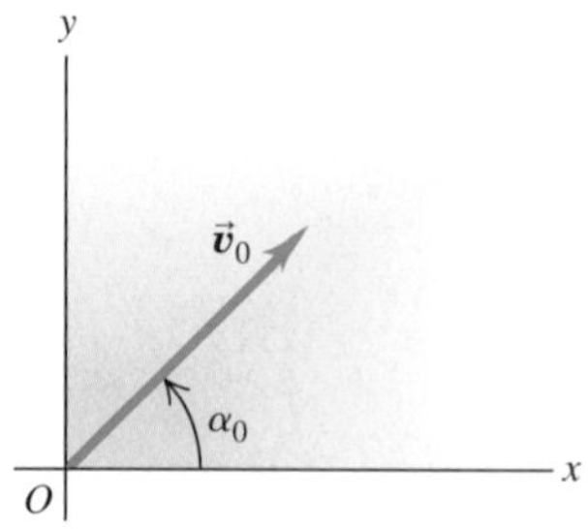

3.16 All projectile motion occurs in the vertical plane containing the initial velocity vector $\vec{v}_0$.

the phrase "projectile motion" will imply that we're ignoring air resistance. In Chapter 5 we will see what happens when air resistance cannot be ignored.

We first notice that projectile motion is always confined to a vertical plane determined by the direction of the initial velocity (Fig. 3.16). This is because the acceleration due to gravity is purely vertical; gravity can't move the projectile sideways. Thus projectile motion is *two-dimensional.* We will call the plane of motion the xy-coordinate plane, with the x-axis horizontal and the y-axis vertically upward.

The key to analyzing projectile motion is that we can treat the x- and y-coordinates separately. The x-component of acceleration is zero, and the y-component is constant and equal to $-g$. (Remember that by definition, g is always positive, while with our choice of coordinate directions, a_y is negative.) So *we can analyze projectile motion as a combination of horizontal motion with constant velocity and vertical motion with constant acceleration.* Figure 3.17 shows two projectiles with different x-motion but identical y-motion; one is dropped from rest and the other is projected horizontally, but both projectiles fall the same distance in the same time.

We can then express all the vector relationships for position, velocity, and acceleration by separate equations for the horizontal and vertical components. The actual motion of the projectile is the superposition of these separate motions. The components of $\vec{a}$ are

$$a_x = 0 \qquad a_y = -g \qquad \text{(projectile motion, no air resistance)} \qquad (3.14)$$

We will usually use $g = 9.8\ \text{m/s}^2$. Since the x-acceleration and y-acceleration are both constant, we can use Eqs. (2.8), (2.12), (2.13), and (2.14) directly. For example, suppose that at time $t = 0$ our particle is at the point (x_0, y_0) and that at this time its velocity components have the initial values v_{0x} and v_{0y}. The components of acceleration are $a_x = 0$, $a_y = -g$. Considering the x-motion first, we substitute 0 for a_x in Eqs. (2.8) and (2.12). We find

$$v_x = v_{0x} \qquad (3.15)$$

$$x = x_0 + v_{0x}t \qquad (3.16)$$

For the y-motion we substitute y for x, v_y for v_x, v_{0y} for v_{0x}, and $a_y = -g$ for a_x:

$$v_y = v_{0y} - gt \qquad (3.17)$$

$$y = y_0 + v_{0y}t - \frac{1}{2}gt^2 \qquad (3.18)$$

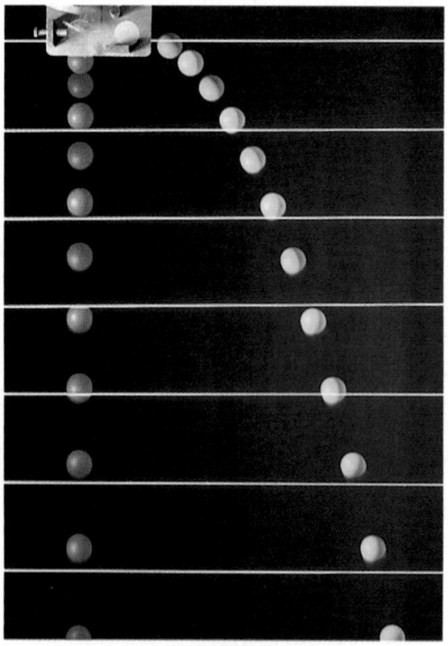

3.17 Independence of horizontal and vertical motion. The red ball is dropped from rest, and the yellow ball is simultaneously projected horizontally; successive images in this stroboscopic photograph are separated by equal time intervals. At any given time, both balls have the same y-position, y-velocity, and y-acceleration, despite having different x-positions and x-velocities.

It's usually simplest to take the initial position (at time $t = 0$) as the origin; in this case, $x_0 = y_0 = 0$. This point might be, for example, the position of a ball at the instant it leaves the thrower's hand or the position of a bullet at the instant it leaves the gun barrel.

Figure 3.18 shows the path of a projectile that starts at (or passes through) the origin at time $t = 0$. The position, velocity, velocity components, and acceleration are shown at a series of times separated by equal intervals. The x-component of acceleration is zero, so v_x is constant. The y-component of acceleration is constant and not zero, so v_y changes by equal amounts in equal times. At the highest point in the trajectory, $v_y = 0$.

We can also represent the initial velocity $\vec{v}_0$ by its magnitude v_0 (the initial speed) and its angle α_0 with the positive x-axis. In terms of these quantities, the components v_{0x} and v_{0y} of the initial velocity are

$$v_{0x} = v_0 \cos\alpha_0 \qquad v_{0y} = v_0 \sin\alpha_0 \qquad (3.19)$$

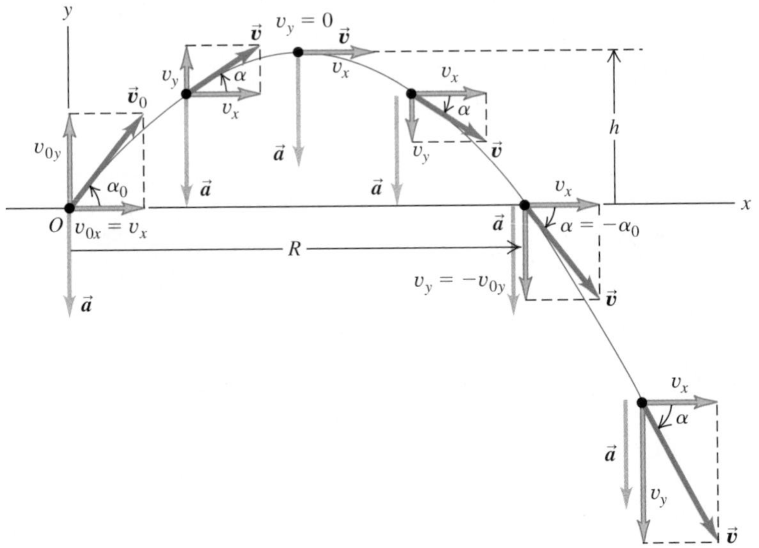

3.18 The trajectory of a body projected with an initial velocity $\vec{v}_0$ at an angle α_0 above the horizontal with negligible air resistance. The distance R is the horizontal range, and h is the maximum height.

ActivPhysics Online

3.7 Target Practice II

Using these relations in Eqs. (3.15) through (3.18) and setting $x_0 = y_0 = 0$, we find

$$x = (v_0 \cos \alpha_0)t \quad \text{(projectile motion)} \tag{3.20}$$

$$y = (v_0 \sin \alpha_0)t - \frac{1}{2}gt^2 \quad \text{(projectile motion)} \tag{3.21}$$

$$v_x = v_0 \cos \alpha_0 \quad \text{(projectile motion)} \tag{3.22}$$

$$v_y = v_0 \sin \alpha_0 - gt \quad \text{(projectile motion)} \tag{3.23}$$

These equations describe the position and velocity of the projectile in Fig. 3.18 at any time t.

We can get a lot of information from these equations. For example, at any time the distance r of the projectile from the origin (the magnitude of the position vector $\vec{r}$) is given by

$$r = \sqrt{x^2 + y^2} \tag{3.24}$$

The projectile's speed (the magnitude of its velocity) at any time is

$$v = \sqrt{v_x^2 + v_y^2} \tag{3.25}$$

The *direction* of the velocity, in terms of the angle α it makes with the positive x-axis, is given by

$$\tan \alpha = \frac{v_y}{v_x} \tag{3.26}$$

The velocity vector $\vec{v}$ is tangent to the trajectory at each point.

We can derive an equation for the trajectory's shape in terms of x and y by eliminating t. From Eqs. (3.20) and (3.21), which assume $x_0 = y_0 = 0$, we find $t = x/(v_0 \cos \alpha_0)$ and

$$y = (\tan \alpha_0)x - \frac{g}{2v_0^2 \cos^2 \alpha_0}x^2 \tag{3.27}$$

3.19 Stroboscopic photograph of a bouncing ball, showing parabolic trajectories after each bounce. Successive images are separated by equal time intervals, as in Fig. 3.17. Each peak in the trajectory is lower than the preceding one because of energy loss during the "bounce," or collision with the horizontal surface.

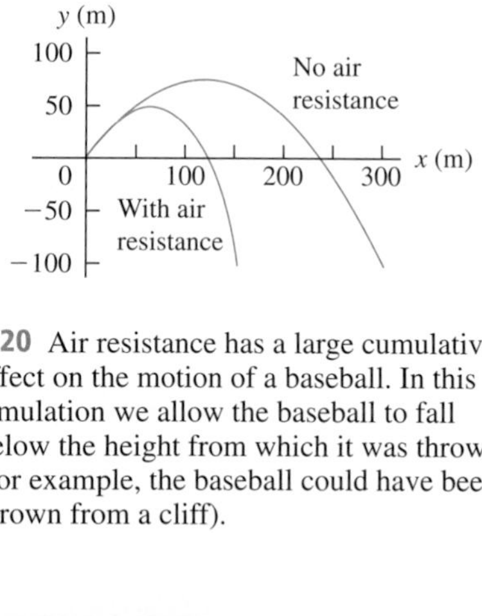

3.20 Air resistance has a large cumulative effect on the motion of a baseball. In this simulation we allow the baseball to fall below the height from which it was thrown (for example, the baseball could have been thrown from a cliff).

Don't worry about the details of this equation; the important point is its general form. The quantities v_0, $\tan\alpha_0$, $\cos\alpha_0$, and g are constants, so the equation has the form

$$y = bx - cx^2$$

where b and c are constants. This is the equation of a *parabola.* In projectile motion, with our simple model, the trajectory is always a parabola (Fig. 3.19).

We mentioned at the beginning of this section that air resistance isn't always negligible. When it has to be included, the calculations become a lot more complicated; the effects of air resistance depend on velocity, so the acceleration is no longer constant. Figure 3.20 shows a computer simulation of the trajectory of a baseball with $v_0 = 50$ m/s and $\alpha_0 = 53.1°$, both without air resistance and with air resistance proportional to the square of the baseball's speed. We see that air resistance has a very large effect; the maximum height and range both decrease, and the trajectory is no longer a parabola.

Conceptual Example 3.6 Acceleration of a skier, continued

Let's consider again the skier in Conceptual Example 3.5. What is her acceleration at points G, H, and I in Fig. 3.21 *after* she flies off the ramp? Neglect air resistance.

SOLUTION

The skier's acceleration changed from point to point while she was on the ramp. But as soon as she leaves the ramp, she becomes a projectile. So at points G, H, and I, and indeed at *all* points after she leaves the ramp, the skier's acceleration points vertically downward and has magnitude g. No matter how complicated the acceleration of a particle before it becomes a projectile, its acceleration as a projectile is given by $a_x = 0$, $a_y = -g$.

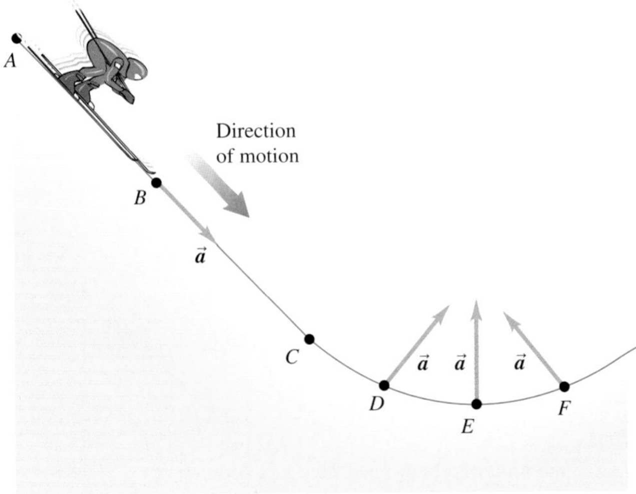

3.21 Once the skier is in projectile motion, her acceleration is constant and downward.

Problem-Solving Strategy

Projectile Motion

NOTE: The strategies we used in Sections 2.4 and 2.5 for straight-line, constant-acceleration problems are also useful here.

IDENTIFY *the relevant concepts:* The key concept to remember is that throughout projectile motion, the acceleration is downward and has a constant magnitude g. Be on the lookout for aspects of the problem that do *not* involve projectile motion. For example, the projectile-motion equations don't apply to throwing a ball, because during the throw the ball is acted on by both the thrower's hand and gravity. These equations come into play only after the ball leaves the thrower's hand.

SET UP *the problem using the following steps:*

1. Define your coordinate system and make a sketch showing your axes. Usually it's easiest to place the origin at the initial $(t = 0)$ position of the projectile. (If the projectile is a thrown ball or a dart shot from a gun, the initial position is where the projectile leaves the thrower's hand or exits the muzzle of the gun.) Also, it's usually best to take the x-axis as being horizontal and the y-axis as being upward. Then the initial position is $x_0 = 0$ and $y_0 = 0$, and the components of the (constant) acceleration are $a_x = 0$, $a_y = -g$.
2. List the unknown and known quantities, and decide which unknowns are your target variables. In some problems you'll be given the initial velocity (either in terms of components or in terms of magnitude and direction) and asked to find the coordinates and velocity components at some later time. In other problems you might be given two points on the trajectory and asked to find the initial velocity. In any case, you'll be using Eqs. (3.20) through (3.23). Make sure that you have as many equations as there are target variables to be found.
3. It often helps to state the problem in words and then translate those words into symbols. For example, *when* does the particle arrive at a certain point? (That is, at what value of t?) *Where* is the particle when its velocity has a certain value? (That is, what are the values of x and y when v_x or v_y has the specified value?) At the highest point in a trajectory, $v_y = 0$. So the question "When does the particle reach its highest point?" translates into "What is the value of t when $v_y = 0$?" Similarly, "When does the projectile return to its initial elevation?" translates into "What is the value of t when $y = y_0$?"

EXECUTE *the solution:* Use Eqs. (3.20) through (3.23) to find the target variables. (Certain other equations given in Section 3.3 may be useful as well.) As you do so, resist the temptation to break the trajectory into segments and analyze each segment separately. You don't have to start all over, with a new axis and a new time scale, when the projectile reaches its highest point! It's almost always easier to set up Eqs. (3.20) through (3.23) at the start and continue to use the same axes and time scale throughout the problem.

EVALUATE *your answer:* As always, look at your results to see whether they make sense and whether the numerical values seem reasonable.

Example 3.7 A body projected horizontally

A motorcycle stunt rider rides off the edge of a cliff. Just at the edge his velocity is horizontal, with magnitude 9.0 m/s. Find the motorcycle's position, distance from the edge of the cliff, and velocity after 0.50 s.

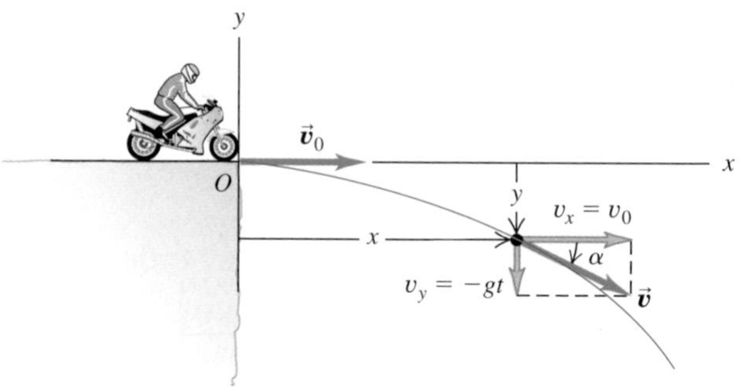

3.22 Trajectory of a body projected horizontally.

SOLUTION

IDENTIFY: Once the rider leaves the cliff, he is in projectile motion. His velocity at the edge of the cliff is therefore his initial velocity.

SET UP: The coordinate system is shown in Fig. 3.22. We choose the origin at the edge of the cliff, where the motorcycle first becomes a projectile, so $x_0 = 0$ and $y_0 = 0$. The initial velocity is purely horizontal (that is, $\alpha_0 = 0$), so the initial velocity components are $v_{0x} = v_0 \cos\alpha_0 = 9.0$ m/s and $v_{0y} = v_0 \sin\alpha_0 = 0$. To find the motorcycle's position at time $t = 0.50$ s, we use Eqs. (3.20) and (3.21), which give x and y as functions of time. Given these values, we find the distance from the origin using Eq. (3.24). Finally, we determine the velocity at $t = 0.50$ s with Eqs. (3.22) and (3.23), which give v_x and v_y as functions of time.

EXECUTE: Where is the motorcycle at $t = 0.50$ s? From Eqs. (3.20) and (3.21), the x- and y-coordinates are

$$x = v_{0x}t = (9.0\text{ m/s})(0.50\text{ s}) = 4.5\text{ m}$$

$$y = -\frac{1}{2}gt^2 = -\frac{1}{2}(9.8\text{ m/s}^2)(0.50\text{ s})^2 = -1.2\text{ m}$$

The negative value of y shows that at this time the motorcycle is below its starting point.

What is the motorcycle's distance from the origin at this time? From Eq. (3.24),

$$r = \sqrt{x^2 + y^2} = \sqrt{(4.5\text{ m})^2 + (-1.2\text{ m})^2} = 4.7\text{ m}$$

What is the velocity at time $t = 0.50$ s? From Eqs. (3.22) and (3.23), the components of velocity at this time are

$$v_x = v_{0x} = 9.0\text{ m/s}$$

$$v_y = -gt = (-9.8\text{ m/s}^2)(0.50\text{ s}) = -4.9\text{ m/s}$$

The motorcycle has the same horizontal velocity v_x as when it left the cliff at $t = 0$, but in addition there is a vertical velocity v_y directed downward (in the negative y-direction). If we use unit vectors, the velocity at $t = 0.50$ s is

$$\vec{v} = v_x\hat{\imath} + v_y\hat{\jmath} = (9.0\text{ m/s})\hat{\imath} + (-4.9\text{ m/s})\hat{\jmath}$$

We can also express the velocity in terms of magnitude and direction. From Eq. (3.25), the speed (magnitude of the velocity) at this time is

$$v = \sqrt{v_x^2 + v_y^2}$$
$$= \sqrt{(9.0\text{ m/s})^2 + (-4.9\text{ m/s})^2} = 10.2\text{ m/s}$$

From Eq. (3.26), the angle α of the velocity vector is

$$\alpha = \arctan\frac{v_y}{v_x}$$
$$= \arctan\left(\frac{-4.9\text{ m/s}}{9.0\text{ m/s}}\right) = -29^\circ$$

At this time the velocity is 29° below the horizontal.

EVALUATE: Our answers illustrate that the horizontal aspect of the motion is unchanged by gravity; the motorcycle continues to move horizontally at 9.0 m/s, covering 4.5 m in 0.50 s. As the motorcycle initially has zero vertical velocity, it falls vertically just like an object released from rest, descending a distance $\frac{1}{2}gt^2 = 1.2$ m in 0.50 s.

Example 3.8 Height and range of a baseball

A batter hits a baseball so that it leaves the bat with an initial speed $v_0 = 37.0$ m/s at an initial angle $\alpha_0 = 53.1^\circ$, at a location where $g = 9.80\text{ m/s}^2$ (Fig. 3.23). a) Find the position of the ball, and the magnitude and direction of its velocity, when $t = 2.00$ s. b) Find the time when the ball reaches the highest point of its flight and find its height h at this point. c) Find the *horizontal range R*—that is, the horizontal distance from the starting point to the point at which the ball hits the ground. For each part, treat the baseball as a projectile.

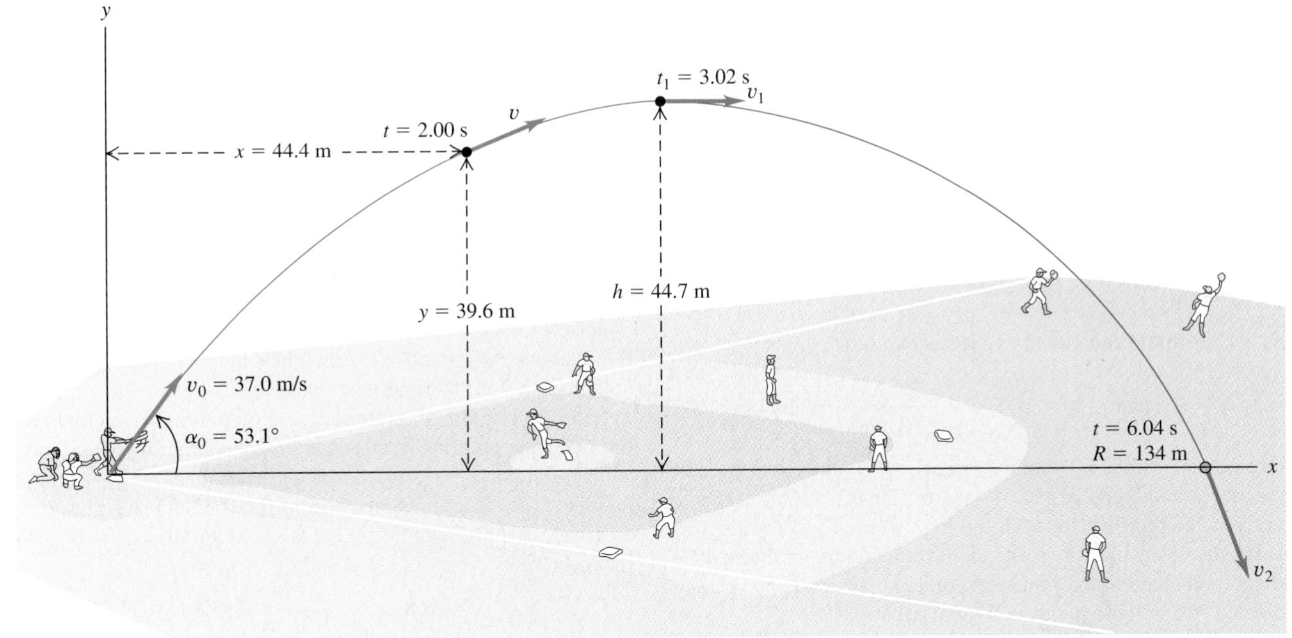

3.23 Trajectory of a batted baseball, treated as a projectile.

SOLUTION

IDENTIFY: As Fig. 3.20 shows, the effects of air resistance on the motion of a baseball aren't really negligible. For the sake of simplicity, however, we'll ignore air resistance for this example. Instead, we'll use the projectile-motion equations to describe the motion.

SET UP: We use the same coordinate system as in Fig. 3.18. Then we can use Eqs. (3.20) through (3.23) without any modifications. Our target variables are (1) the position and velocity of the ball 2.00 s after it leaves the bat, (2) the elapsed time after leaving the bat when the ball is at its maximum height—that is, when $v_y = 0$—and the y-coordinate at this time, and (3) the x-coordinate at the time when the y-coordinate is equal to the initial value y_0.

The ball is probably struck a meter or so above ground level, but we neglect this distance and assume that it starts at ground level $(y_0 = 0)$. The initial velocity of the ball has components

$$v_{0x} = v_0 \cos \alpha_0 = (37.0 \text{ m/s}) \cos 53.1° = 22.2 \text{ m/s}$$
$$v_{0y} = v_0 \sin \alpha_0 = (37.0 \text{ m/s}) \sin 53.1° = 29.6 \text{ m/s}$$

EXECUTE: a) We want to find x, y, v_x, and v_y at time $t = 2.00$ s. From Eqs. (3.20) through (3.23),

$$x = v_{0x}t = (22.2 \text{ m/s})(2.00 \text{ s}) = 44.4 \text{ m}$$
$$y = v_{0y}t - \frac{1}{2}gt^2$$
$$= (29.6 \text{ m/s})(2.00 \text{ s}) - \frac{1}{2}(9.80 \text{ m/s}^2)(2.00 \text{ s})^2$$
$$= 39.6 \text{ m}$$

$$v_x = v_{0x} = 22.2 \text{ m/s}$$
$$v_y = v_{0y} - gt = 29.6 \text{ m/s} - (9.80 \text{ m/s}^2)(2.00 \text{ s})$$
$$= 10.0 \text{ m/s}$$

The y-component of velocity is positive, which means that the ball is still moving upward at this time (Fig. 3.23). The magnitude and direction of the velocity are found from Eqs. (3.25) and (3.26):

$$v = \sqrt{v_x^2 + v_y^2} = \sqrt{(22.2 \text{ m/s})^2 + (10.0 \text{ m/s})^2}$$
$$= 24.3 \text{ m/s}$$
$$\alpha = \arctan\left(\frac{10.0 \text{ m/s}}{22.2 \text{ m/s}}\right) = \arctan 0.450 = 24.2°$$

The velocity is 24.2° above the horizontal.

b) At the highest point, the vertical velocity v_y is zero. When does this happen? Call the time t_1; then

$$v_y = 0 = v_{0y} - gt_1$$
$$t_1 = \frac{v_{0y}}{g} = \frac{29.6 \text{ m/s}}{9.80 \text{ m/s}^2} = 3.02 \text{ s}$$

The height h at this time is the value of y when $t = t_1 = 3.02$ s:

$$h = v_{0y}t_1 - \frac{1}{2}gt_1^2$$
$$= (29.6 \text{ m/s})(3.02 \text{ s}) - \frac{1}{2}(9.80 \text{ m/s}^2)(3.02 \text{ s})^2$$
$$= 44.7 \text{ m}$$

Alternatively, we can apply the constant-acceleration formula Eq. (2.13) to the y-motion:

$$v_y^2 = v_{0y}^2 + 2a_y(y - y_0) = v_{0y}^2 - 2g(y - y_0)$$

At the highest point, $v_y = 0$ and $y = h$. Substituting these, along with $y_0 = 0$, we find

$$0 = v_{0y}^2 - 2gh$$

$$h = \frac{v_{0y}^2}{2g} = \frac{(29.6 \text{ m/s})^2}{2(9.80 \text{ m/s}^2)} = 44.7 \text{ m}$$

c) We'll find the horizontal range in two steps. First, *when* does the ball hit the ground? This occurs when $y = 0$. Call this time t_2; then

$$y = 0 = v_{0y}t_2 - \frac{1}{2}gt_2^2 = t_2\left(v_{0y} - \frac{1}{2}gt_2\right)$$

This is a quadratic equation for t_2. It has two roots,

$$t_2 = 0 \quad \text{and} \quad t_2 = \frac{2v_{0y}}{g} = \frac{2(29.6 \text{ m/s})}{9.80 \text{ m/s}^2} = 6.04 \text{ s}$$

There are two times at which $y = 0$; $t_2 = 0$ is the time the ball *leaves* the ground, and $t_2 = 6.04$ s is the time of its return. This is exactly twice the time to reach the highest point, so the time of descent equals the time of ascent. (This is *always* true if the starting and end points are at the same elevation and air resistance can be neglected. We will prove this in Example 3.10.)

The horizontal range R is the value of x when the ball returns to the ground, that is, at $t = 6.04$ s:

$$R = v_{0x}t_2 = (22.2 \text{ m/s})(6.04 \text{ s}) = 134 \text{ m}$$

The vertical component of velocity when the ball hits the ground is

$$v_y = v_{0y} - gt_2 = 29.6 \text{ m/s} - (9.80 \text{ m/s}^2)(6.04 \text{ s})$$
$$= -29.6 \text{ m/s}$$

That is, v_y has the same magnitude as the initial vertical velocity v_{0y} but the opposite direction (down). Since v_x is constant, the angle $\alpha = -53.1°$ (below the horizontal) at this point is the negative of the initial angle $\alpha_0 = 53.1°$.

EVALUATE: Do our numerical results make sense? The maximum height $h = 44.7$ m in part (b) is roughly half the height above the playing field of the roof of Skydome in Toronto. The horizontal range $R = 134$ m in part (c) is greater than the 93.6-m distance from home plate to the right-field fence at Pacific Bell Park in San Francisco. Can you show that when the ball reaches this fence, it is 37.7 m above the ground? (This height is more than enough to clear the fence, so this ball is a home run.)

If the ball could continue to travel on *below* its original level (through an appropriately shaped hole in the ground), then negative values of y, corresponding to times greater than 6.04 s, would be possible. Can you compute the position and velocity at a time 7.55 s after the start, corresponding to the last position shown in Fig. 3.18? The results are

$$x = 168 \text{ m} \qquad y = -55.8 \text{ m}$$
$$v_x = 22.2 \text{ m/s} \qquad v_y = -44.4 \text{ m/s}$$

It's worth pointing out that in real life, a batted ball with the initial speed and angle we've used here won't go as high or as far as we've calculated. (If it did, home runs would be far more common and baseball would be a far less interesting game.) The reason is that air resistance, which we neglected in this example, is actually an important factor at the typical speeds of pitched and batted balls.

Example 3.9 The zoo keeper and the monkey

A clever monkey escapes from the zoo. The zoo keeper finds him in a tree. After failing to entice the monkey down, the zoo keeper points her tranquilizer gun directly at the monkey and shoots (Fig. 3.24). The clever monkey lets go at the same instant the dart leaves the gun barrel, intending to land on the ground and escape. Show that the dart *always* hits the monkey, regardless of the dart's muzzle velocity (provided that it gets to the monkey before he hits the ground).

SOLUTION

IDENTIFY: In this example we have *two* objects in projectile motion: the tranquilizer dart and the monkey. The dart and the monkey have different initial positions and initial velocities, but they go into projectile motion at the same time. To show that the dart hits the monkey, we have to prove that at some time the monkey and the dart have the same x-coordinate and the same y-coordinate.

SET UP: We make the usual choice for the x- and y-directions, and place the origin of coordinates at the end of the barrel of the tranquilizer gun (Fig. 3.24). We'll first use Eq. (3.20) to find the time t when the x-coordinates x_{monkey} and x_{dart} are the same. We'll then use Eq. (3.21) to check whether y_{monkey} and y_{dart} are also equal at this time; if they are, the dart hits the monkey.

EXECUTE: The monkey drops straight down, so $x_{\text{monkey}} = d$ at *all* times. For the dart, Eq. (3.20) tells us that $x_{\text{dart}} = (v_0 \cos \alpha_0)t$. When these x-coordinates are equal, $d = (v_0 \cos \alpha_0)t$, or

$$t = \frac{d}{v_0 \cos \alpha_0}$$

To have the dart hit the monkey, it must be true that $y_{\text{monkey}} = y_{\text{dart}}$ at this same time. The monkey is in one-dimensional free fall; his position at any time is given by Eq. (2.12), with appropriate symbol changes. Figure 3.24 shows that the monkey's initial height is $d \tan \alpha_0$ (the opposite side of a right triangle with angle α_0 and adjacent side d), and we find

$$y_{\text{monkey}} = d \tan \alpha_0 - \frac{1}{2}gt^2$$

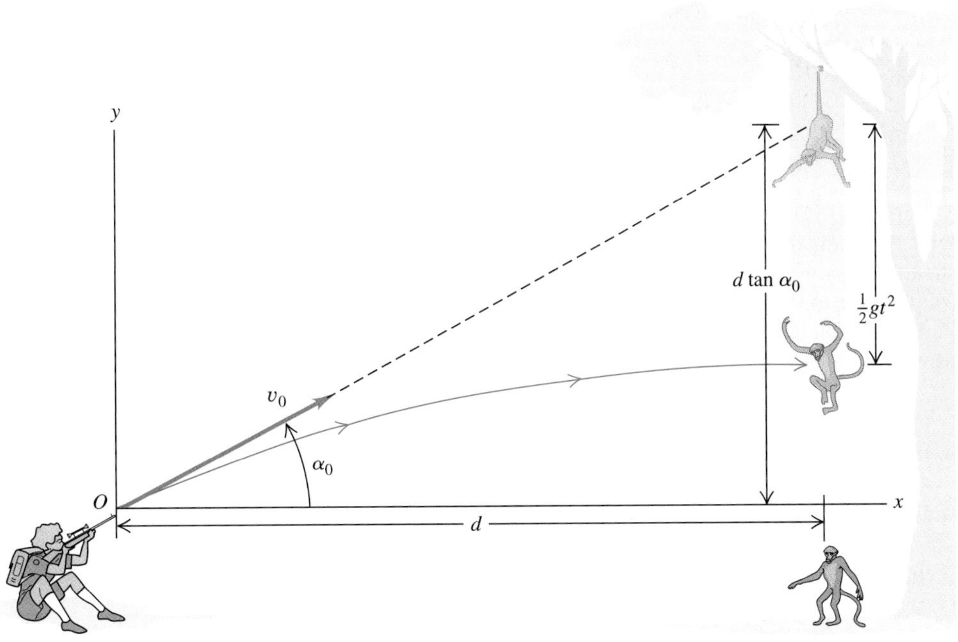

3.24 The tranquilizer dart hits the falling monkey.

For the dart we use Eq. (3.21):

$$y_{\text{dart}} = (v_0 \sin \alpha_0)t - \frac{1}{2}gt^2$$

So we see that if $d \tan \alpha_0 = (v_0 \sin \alpha_0)t$ at the time when the two x-coordinates are equal, then $y_{\text{monkey}} = y_{\text{dart}}$, and we have a hit. To prove that this happens, we replace t with $d/(v_0 \cos \alpha_0)$, the time when $x_{\text{monkey}} = x_{\text{dart}}$; then

$$(v_0 \sin \alpha_0)t = (v_0 \sin \alpha_0)\frac{d}{v_0 \cos \alpha_0} = d \tan \alpha_0$$

EVALUATE: We have proved that at the time the x-coordinates are equal, the y-coordinates are also equal; a dart aimed at the initial position of the monkey always hits it, no matter what v_0 is. This result is also independent of the value of g, the acceleration due to gravity. With no gravity $(g = 0)$, the monkey would remain motionless, and the dart would travel in a straight line to hit him. With gravity, both "fall" the same distance $(\frac{1}{2}gt^2)$ below their $g = 0$ positions, and the dart still hits the monkey.

Example 3.10 Height and range of a projectile

For a projectile launched with speed v_0 at initial angle α_0 (between 0 and 90°), derive general expressions for the maximum height h and horizontal range R (Fig. 3.18). For a given v_0, what value of α_0 gives maximum height? What value gives maximum horizontal range?

SOLUTION

IDENTIFY: This is really the same exercise as parts (b) and (c) of Example 3.8. The difference is that we are looking for general expressions for h and R. We'll also be looking for the values of α_0 that give the maximum values of h and R.

SET UP: As in Example 3.8, we use the coordinate system shown in Fig. 3.18. For convenience we choose the origin to be at the initial position of the projectile. To determine the height h at the high point of the trajectory, we use Eq. (3.23) to find the time t_1 at which $v_y = 0$, then find the y-coordinate at this time with Eq. (3.21). To determine R, we first use Eq. (3.21) to find the time t_2 at which $y = 0$ (i. e., when the projectile returns to its initial height), then substitute t_2 into Eq. (3.20) to determine the x-coordinate at t_2.

EXECUTE: From Eq. (3.23), the time t_1 when $v_y = 0$ is given by the equation

$$v_y = v_0 \sin \alpha_0 - gt_1 = 0 \qquad t_1 = \frac{v_0 \sin \alpha_0}{g}$$

Then, from Eq. (3.21), the height at this time is

$$h = v_0 \sin\alpha_0\left(\frac{v_0 \sin\alpha_0}{g}\right) - \frac{1}{2}g\left(\frac{v_0 \sin\alpha_0}{g}\right)^2$$
$$= \frac{v_0^2 \sin^2\alpha_0}{2g}$$

If we vary α_0, the maximum value of h occurs when $\sin\alpha_0 = 1$ and $\alpha_0 = 90°$; that is, when the projectile is launched straight up. That's what we should expect. If it is launched horizontally, as in Example 3.7, $\alpha_0 = 0$ and the maximum height is zero!

To find a general expression for the horizontal range R, we first find the time t_2 at which $y = 0$. From Eq. (3.21), t_2 satisfies the equation

$$(v_0 \sin\alpha_0)t_2 - \frac{1}{2}gt_2^2 = t_2\left(v_0 \sin\alpha_0 - \frac{1}{2}gt_2\right) = 0$$

The two roots of this quadratic equation for t_2 are $t_2 = 0$ (the time of launch) and $t_2 = (2v_0 \sin\alpha_0)/g$. The horizontal range R is the value of x at the second time. From Eq. (3.20),

$$R = (v_0 \cos\alpha_0)\frac{2v_0 \sin\alpha_0}{g}$$

We can now use the trigonometric identity $2\sin\alpha_0\cos\alpha_0 = \sin 2\alpha_0$ to rewrite this as

$$R = \frac{v_0^2 \sin 2\alpha_0}{g}$$

The maximum value of $\sin 2\alpha_0$ is 1; this occurs when $2\alpha_0 = 90°$, or $\alpha_0 = 45°$. This angle gives the maximum range for a given initial speed.

EVALUATE: Figure 3.25 is based on a composite photograph of three trajectories of a ball projected from a spring gun at angles of 30°, 45°, and 60°. The initial speed v_0 is approximately the same in all three cases. The horizontal ranges are nearly the same for the 30° and 60° angles, and the range for 45° is greater than either. Can you prove that for a given value of v_0 the range is the same for an initial angle α_0 as for an initial angle $90° - \alpha_0$?

Comparing the expressions for t_1 and t_2, we see that $t_2 = 2t_1$; that is, the total flight time is twice the time up to the highest point. It follows that the time to reach the highest point equals the time to fall from there back to the initial elevation, as we asserted in Example 3.8.

CAUTION We don't recommend memorizing the above expressions for h and R. They are applicable only in the special circumstances we have described. In particular, the expression for the range R can be used *only* when launch and landing heights are equal. There are many end-of-chapter problems to which these equations are *not* applicable.

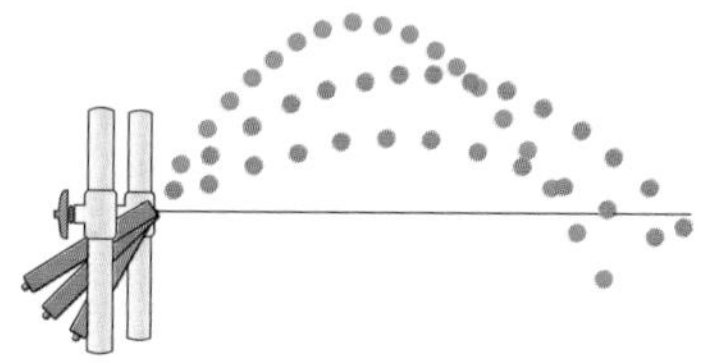

3.25 A firing angle of 45° (shown in purple) gives the maximum horizontal range. The range is less with firing angles of 30° (blue) or 60° (green).

Example 3.11 Different initial and final heights

You toss a ball from your window 8.0 m above the ground. When the ball leaves your hand, it is moving at 10.0 m/s at an angle of 20° below the horizontal. How far horizontally from your window will the ball hit the ground? Ignore air resistance.

SOLUTION

IDENTIFY: As in our calculation of the horizontal range in Examples 3.8 and 3.10, we are trying to find the horizontal coordinate of a projectile when it is at a given value of y. The difference here is that this value of y is *not* equal to the initial y-coordinate.

SET UP: Once again we choose the x-axis to be horizontal and the y-axis to be upward. We place the origin of coordinates at the point where the ball leaves your hand, so that the ball's initial coordinates are $x_0 = 0$, $y_0 = 0$ (Fig. 3.26). We have $v_0 = 10.0$ m/s and $\alpha_0 = -20°$; the angle is negative because the initial velocity is below the horizontal. Our target variable is the value of x at the

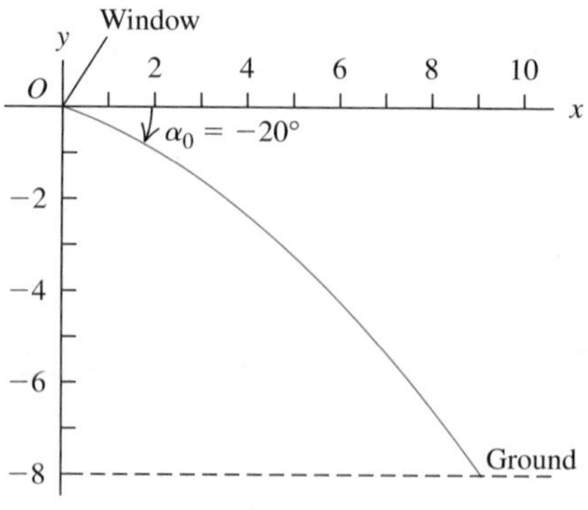

3.26 The trajectory of a ball launched at an angle below the horizontal.

point where the ball reaches the ground—that is, when $y = -8.0$ m. Because the initial and final heights of the ball are different, we can't simply use the expression for the horizontal range found in Example 3.10. Instead, we first use Eq. (3.21) to find the time t when the ball reaches $y = -8.0$ m, then calculate the value of x at this time, using Eq. (3.20).

EXECUTE: To determine t, we rewrite Eq. (3.21) in the standard form for a quadratic equation for t:

$$\frac{1}{2}gt^2 - (v_0 \sin \alpha_0)t + y = 0$$

The roots of this equation are

$$t = \frac{v_0 \sin \alpha_0 \pm \sqrt{(-v_0 \sin \alpha_0)^2 - 4\left(\frac{1}{2}g\right)y}}{2\left(\frac{1}{2}g\right)}$$

$$= \frac{v_0 \sin \alpha_0 \pm \sqrt{v_0^2 \sin^2 \alpha_0 - 2gy}}{g}$$

$$= \frac{\left[\begin{array}{l}(10.0\text{ m/s})\sin(-20°) \\ \pm\sqrt{(10.0\text{ m/s})^2 \sin^2(-20°) - 2(9.8\text{ m/s}^2)(-8.0\text{ m})}\end{array}\right]}{9.8\text{ m/s}^2}$$

We can discard the negative root, since it refers to a time before the ball left your hand. The positive root tells us that the ball takes 0.98 s to reach the ground. From Eq. (3.20), the ball's x-coordinate at that time is

$$x = (v_0 \cos \alpha_0)t = (10.0\text{ m/s})[\cos(-20°)](0.98\text{ s})$$
$$= 9.2\text{ m}$$

The ball hits the ground a horizontal distance of 9.2 m from your window.

EVALUATE: The root $t = -1.7$ s is an example of a "fictional" solution to a quadratic equation. We discussed these in Example 2.8 in Section 2.5; you should review that discussion.

With our choice of origin we had initial and final heights $y_0 = 0$ and $y = -8.0$ m. Can you use Eqs. (3.16) and (3.18) to show that you get the same answers for t and x if you choose the origin to be at the point on the ground directly below where the ball leaves your hand?

Test Your Understanding

In Example 3.9, suppose the tranquilizer dart has a relatively low muzzle velocity so that the dart reaches a maximum height at a point P before striking the monkey (see Fig. 3.27). When the dart is at point P, will the monkey be at point A (higher than P), point B (at the same height as P), or at point C (lower than P)? Ignore air resistance.

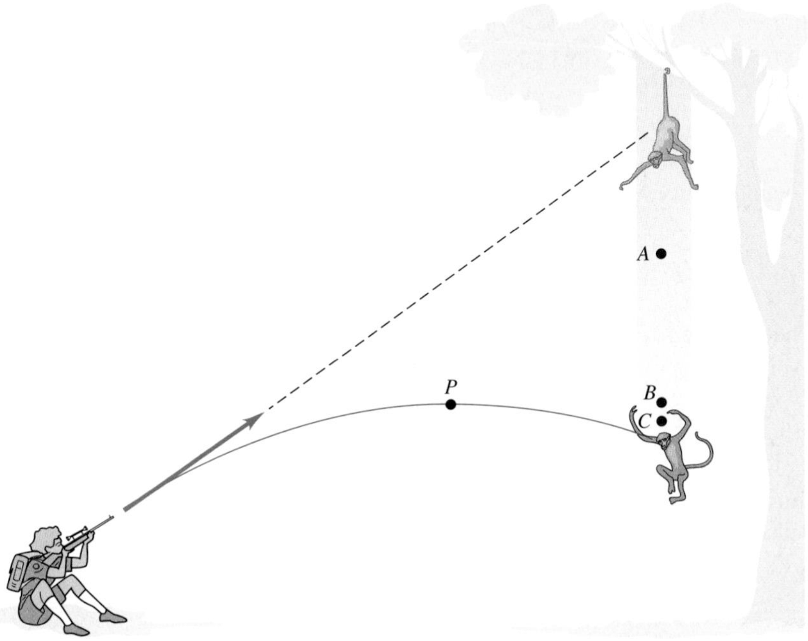

3.27 The same situation as in Example 3.9, but with a low muzzle velocity.

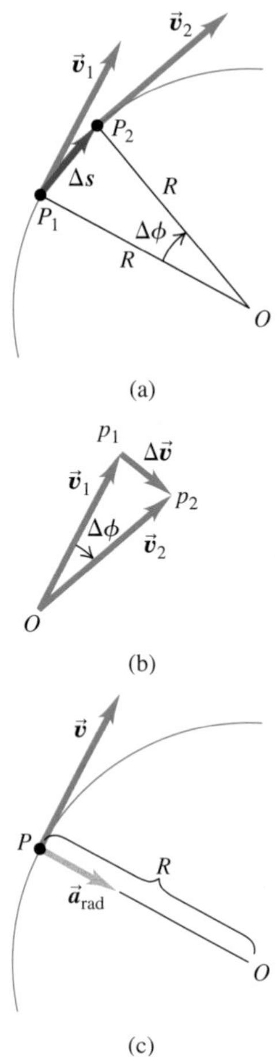

3.28 Finding the change in velocity, $\Delta\vec{v}$, of a particle moving in a circle with constant speed.

3.4 | Motion in a Circle

When a particle moves along a curved path, the direction of its velocity changes. As we saw in Section 3.2, this means that the particle *must* have a component of acceleration perpendicular to the path, even if its speed is constant. In this section we'll calculate the acceleration for the important special case of motion in a circle.

Uniform Circular Motion

When a particle moves in a circle with *constant speed,* the motion is called **uniform circular motion.** A car rounding a curve with constant radius at constant speed, a satellite moving in a circular orbit, and an ice skater skating in a circle with constant speed are all examples of uniform circular motion. There is no component of acceleration parallel (tangent) to the path; otherwise, the speed would change. The component of acceleration perpendicular (normal) to the path, which causes the direction of the velocity to change, is related in a simple way to the speed of the particle and the radius of the circle. Our next project is to derive this relation.

First we note that this is a different problem from the projectile motion considered in Section 3.3, in which the acceleration was constant in both its magnitude *(g)* and its direction (straight down). In uniform circular motion the acceleration is perpendicular to the velocity at each instant; as the direction of the velocity changes, the direction of the acceleration also changes. As we will see, the acceleration vector at each point in the circular path is directed toward the *center* of the circle.

Figure 3.28a shows a particle moving with constant speed in a circular path of radius R with center at O. The particle moves from P_1 to P_2 in a time Δt. The vector change in velocity $\Delta\vec{v}$ during this time is shown in Fig. 3.28b.

The angles labeled $\Delta\phi$ in Figs. 3.28a and 3.28b are the same because $\vec{v}_1$ is perpendicular to the line OP_1 and $\vec{v}_2$ is perpendicular to the line OP_2. Hence the triangles OP_1P_2 (Fig. 3.28a) and Op_1p_2 (Fig. 3.28b) are *similar.* Ratios of corresponding sides are equal, so

$$\frac{|\Delta\vec{v}|}{v_1} = \frac{\Delta s}{R} \quad \text{or} \quad |\Delta\vec{v}| = \frac{v_1}{R}\Delta s$$

The magnitude a_{av} of the average acceleration during Δt is therefore

$$a_{av} = \frac{|\Delta\vec{v}|}{\Delta t} = \frac{v_1}{R}\frac{\Delta s}{\Delta t}$$

The magnitude a of the *instantaneous* acceleration $\vec{a}$ at point P_1 is the limit of this expression as we take point P_2 closer and closer to point P_1:

$$a = \lim_{\Delta t \to 0} \frac{v_1}{R}\frac{\Delta s}{\Delta t} = \frac{v_1}{R} \lim_{\Delta t \to 0} \frac{\Delta s}{\Delta t}$$

But the limit of $\Delta s/\Delta t$ is the speed v_1 at point P_1. Also, P_1 can be any point on the path, so we can drop the subscript and let v represent the speed at any point. Then

$$a_{rad} = \frac{v^2}{R} \quad \text{(uniform circular motion)} \qquad (3.28)$$

ActivPhysics Online

4.1 Magnitude of Centripetal Acceleration

We have added the subscript "rad" as a reminder that the direction of the instantaneous acceleration at each point is always along a radius of the circle, toward its center. This is consistent with our discussion in Section 3.2: the acceleration vector points toward the *concave* side of the particle's circular path, that is, toward the *inside* of the circle (never the outside). Because the speed is constant, the acceleration is always perpendicular to the instantaneous velocity. This is shown in Fig. 3.28c; compare with Fig. 3.12a.

We conclude: *In uniform circular motion, the magnitude a of the instantaneous acceleration is equal to the square of the speed v divided by the radius R of the circle. Its direction is perpendicular to $\vec{v}$ and inward along the radius.* Because the acceleration is always directed toward the center of the circle, it is sometimes called **centripetal acceleration.** The word *centripetal* is derived from two Greek words meaning "seeking the center." Figure 3.29 shows the directions of the velocity and acceleration vectors at several points for a particle moving with uniform circular motion. Compare this with the projectile motion shown in Fig. 3.18, in which the acceleration is always directed straight down and is *not* perpendicular to the path, except at one point.

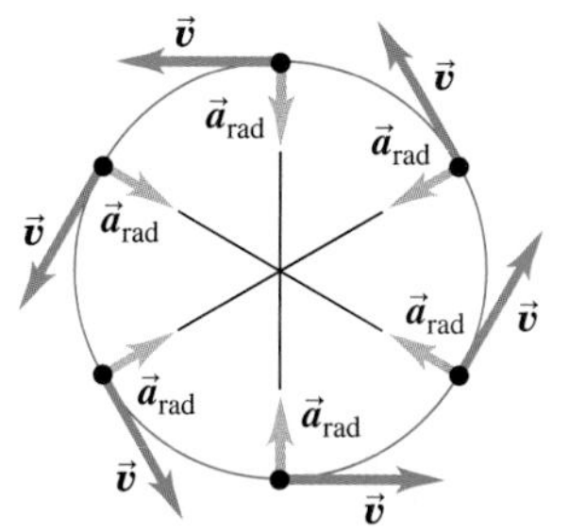

3.29 For a particle in uniform circular motion, the velocity at each point is tangent to the circle and the acceleration is directed toward the center.

We can also express the magnitude of the acceleration in uniform circular motion in terms of the **period** T of the motion, the time for one revolution (one complete trip around the circle). In a time T the particle travels a distance equal to the circumference $2\pi R$ of the circle, so its speed is

$$v = \frac{2\pi R}{T} \tag{3.29}$$

When we substitute this into Eq. (3.28), we obtain the alternative expression

$$a_{rad} = \frac{4\pi^2 R}{T^2} \quad \text{(uniform circular motion)} \tag{3.30}$$

Example 3.12 Centripetal acceleration on a curved road

A BMW Z4 roadster has a "lateral acceleration" of $0.87g$, which is $(0.87)(9.8\text{ m/s}^2) = 8.5\text{ m/s}^2$. This represents the maximum centripetal acceleration that can be attained without skidding out of the circular path. If the car is traveling at a constant 40 m/s (about 89 mi/h, or 144 km/h), what is the minimum radius of curve it can negotiate? (Assume that the curve is unbanked.)

SOLUTION

IDENTIFY and SET UP: Because the car is moving along a curve—that is, a segment of a circle—at a constant speed, we can apply the ideas of uniform circular motion. In particular, we can use Eq. (3.28) to find the target variable R (the radius of the curve) in terms of the given centripetal acceleration a_{rad} and speed v.

EXECUTE: We are given a_{rad} and v, so we solve Eq. (3.28) for R:

$$R = \frac{v^2}{a_{rad}} = \frac{(40\text{ m/s})^2}{8.5\text{ m/s}^2} = 190\text{ m} \qquad \text{(about 600 ft)}$$

EVALUATE: Our result shows that the required turning radius R is proportional to the *square* of the speed. Hence even a small reduction in speed can make R substantially smaller. For example, reducing v by 20% (from 40 to 32 m/s) would decrease R by 36% (from 190 m to 120 m).

Another way to make the required turning radius smaller is to *bank* the curve. We will investigate this option in Chapter 5.

Example 3.13

Centripetal acceleration on a carnival ride

In a carnival ride, the passengers travel at constant speed in a circle of radius 5.0 m. They make one complete circle in 4.0 s. What is their acceleration?

SOLUTION

IDENTIFY and SET UP: The speed is constant, so this is a problem involving uniform circular motion. We are given the radius $R = 5.0$ m and the period $T = 4.0$ s, so we can use Eq. (3.30) to calculate the acceleration. Alternatively, we can first calculate the speed v using Eq. (3.29), then find the acceleration using Eq. (3.28).

EXECUTE: From Eq. (3.30),

$$a_{rad} = \frac{4\pi^2(5.0\text{ m})}{(4.0\text{ s})^2} = 12\text{ m/s}^2$$

We'll check this answer by using Eq. (3.28) after first determining the speed v. From Eq. (3.29), the speed is the circumference of the circle divided by the period T:

$$v = \frac{2\pi R}{T} = \frac{2\pi(5.0\text{ m})}{4.0\text{ s}} = 7.9\text{ m/s}$$

The centripetal acceleration is then

$$a_{rad} = \frac{v^2}{R} = \frac{(7.9\text{ m/s})^2}{5.0\text{ m}} = 12\text{ m/s}^2$$

Happily, we get the same answer for a_{rad} with both approaches.

EVALUATE: As in the preceding example, the direction of $\vec{a}$ is always toward the center of the circle. The magnitude of $\vec{a}$ is greater than g, the acceleration due to gravity, so this is not a ride for the faint-hearted. (Some roller coasters subject their passengers to accelerations as great as $4g$.)

Non-Uniform Circular Motion

We have assumed throughout this section that the particle's speed is constant. If the speed varies, we call the motion **non-uniform circular motion.** An example is a roller coaster car that slows down and speeds up as it moves around a vertical loop. In non-uniform circular motion, Eq. (3.28) still gives the *radial* component of acceleration $a_{rad} = v^2/R$, which is always *perpendicular* to the instantaneous velocity and directed toward the center of the circle. But since the speed v has different values at different points in the motion, the value of a_{rad} is not constant. The radial (centripetal) acceleration is greatest at the point in the circle where the speed is greatest.

In non-uniform circular motion there is also a component of acceleration that is *parallel* to the instantaneous velocity. This is the component $a_{\parallel}$ that we discussed in Section 3.2; here we call this component a_{tan} to emphasize that it is *tangent* to the circle. From the discussion at the end of Section 3.2 we see that the tangential component of acceleration a_{tan} is equal to the rate of change of *speed.* Thus

$$a_{rad} = \frac{v^2}{R} \quad \text{and} \quad a_{tan} = \frac{d|\vec{v}|}{dt} \qquad \text{(non-uniform circular motion)} \quad (3.31)$$

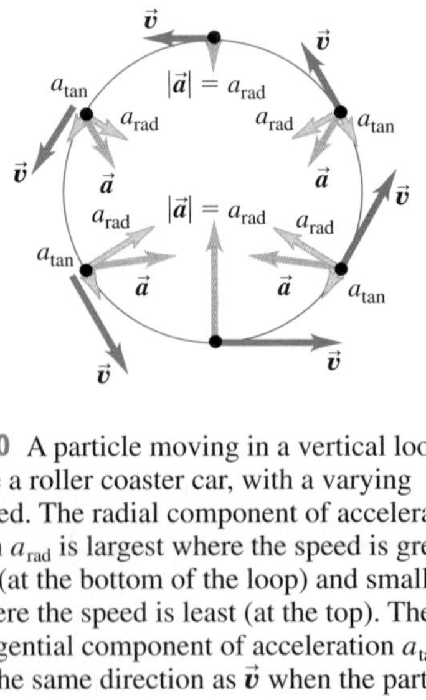

3.30 A particle moving in a vertical loop, like a roller coaster car, with a varying speed. The radial component of acceleration a_{rad} is largest where the speed is greatest (at the bottom of the loop) and smallest where the speed is least (at the top). The tangential component of acceleration a_{tan} is in the same direction as $\vec{v}$ when the particle is speeding up (going downhill) and opposite $\vec{v}$ when it is slowing down (going uphill).

The vector acceleration of a particle moving in a circle with varying speed is the vector sum of the radial and tangential components of accelerations. The tangential component is in the same direction as the velocity if the particle is speeding up, and is in the opposite direction if the particle is slowing down (Fig. 3.30).

In *uniform* circular motion there is no tangential component of acceleration, but the radial component is the magnitude of $d\vec{v}/dt$. We have mentioned before that the two quantities $|d\vec{v}/dt|$ and $d|\vec{v}|/dt$ are in general *not* equal. In uniform circular motion the first is constant and equal to v^2/R; the second is zero.

Test Your Understanding

Suppose that the roller coaster car in Fig. 3.30 experiences four times the acceleration at the bottom of the loop as it does at the top of the loop. How do the speeds at the top and bottom of the loop compare?

3.5 | Relative Velocity

3.31 Airshow pilots face a complicated problem involving relative velocities. They must keep track of their velocity relative to the air (to maintain enough airflow over the wings to sustain lift), their velocity relative to each other (to keep a tight formation without colliding), and their velocity relative to their audience (to remain in sight of the spectators).

You've no doubt observed how a car that is moving slowly forward appears to be moving backward when you pass it. In general, when two observers measure the velocity of a moving body, they get different results if one observer is moving relative to the other. The velocity seen by a particular observer is called the velocity *relative* to that observer, or simply **relative velocity.** Figure 3.31 shows a situation in which in which understanding relative velocity is extremely important.

We'll first consider relative velocity along a straight line, then generalize to relative velocity in a plane. Recall that for straight-line (one-dimensional) motion we use the term *velocity* to mean the component of the velocity vector along the line of motion; this can be positive, negative, or zero.

Relative Velocity in One Dimension

A woman walks with a velocity of 1.0 m/s along the aisle of a train that is moving with a velocity of 3.0 m/s (Fig. 3.32a). What is the woman's velocity? It's a simple enough question, but it has no single answer. As seen by a passenger sitting in the train, she is moving at 1.0 m/s. A person on a bicycle standing beside the train sees the woman moving at 1.0 m/s + 3.0 m/s = 4.0 m/s. An observer in another train going in the opposite direction would give still another answer. We have to specify which observer we mean, and we speak of the velocity *relative* to a particular observer. The woman's velocity relative to the train is 1.0 m/s, her velocity

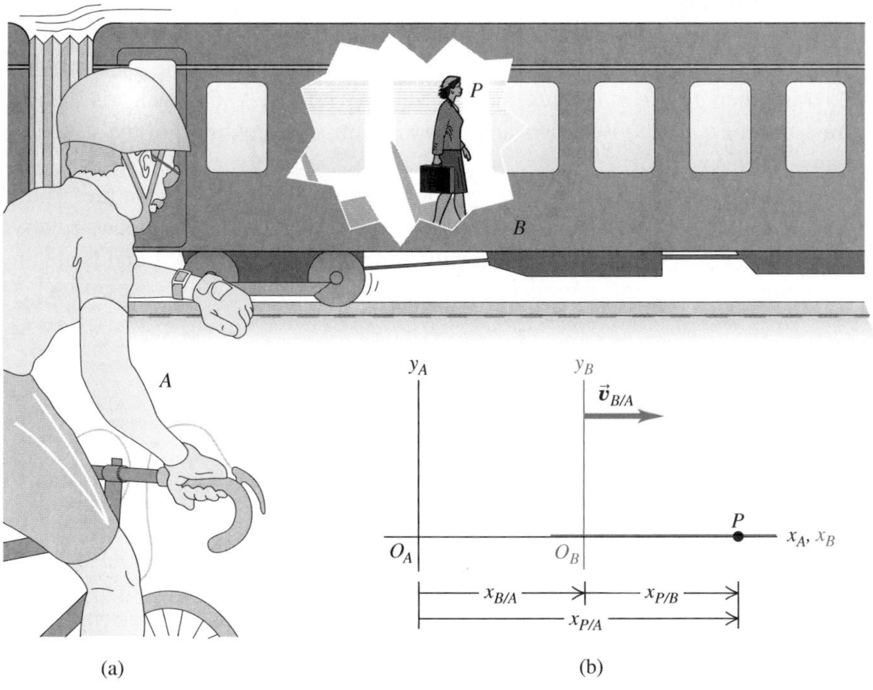

3.32 (a) A woman walking in a train. (b) At the instant shown, the position of the woman (particle P) relative to frame of reference A is different from her position relative to frame of reference B.

relative to the cyclist is 4.0 m/s, and so on. Each observer, equipped in principle with a meter stick and a stopwatch, forms what we call a **frame of reference.** Thus a frame of reference is a coordinate system plus a time scale.

Let's call the cyclist's frame of reference (at rest with respect to the ground) A and the frame of reference of the moving train B (Fig. 3.32b). In straight-line motion the position of a point P relative to frame of reference A is given by the distance $x_{P/A}$ (the position of P with respect to A), and the position relative to frame B is given by $x_{P/B}$. The distance from the origin of A to the origin of B (position of B with respect to A) is $x_{B/A}$. We can see from the figure that

$$x_{P/A} = x_{P/B} + x_{B/A} \qquad (3.32)$$

This says that the total distance from the origin of A to point P is the distance from the origin of B to point P plus the distance from the origin of A to the origin of B.

The velocity of P relative to frame A, denoted by $v_{P/A}$, is the derivative of $x_{P/A}$ with respect to time. The other velocities are similarly obtained. So the time derivative of Eq. (3.32) gives us a relationship among the various velocities:

$$\frac{dx_{P/A}}{dt} = \frac{dx_{P/B}}{dt} + \frac{dx_{B/A}}{dt}$$

or

$$v_{P/A} = v_{P/B} + v_{B/A} \quad \text{(relative velocity along a line)} \qquad (3.33)$$

Getting back to the woman on the train in Figure 3.32, A is the cyclist's frame of reference, B is the frame of reference of the train, and point P represents the woman. Using the above notation, we have

$$v_{P/B} = 1.0\text{ m/s} \qquad v_{B/A} = 3.0\text{ m/s}$$

From Eq. (3.33) the woman's velocity $v_{P/A}$ relative to the cyclist is

$$v_{P/A} = 1.0\text{ m/s} + 3.0\text{ m/s} = 4.0\text{ m/s}$$

as we already knew.

In this example, both velocities are toward the right, and we have implicitly taken this as the positive direction. If the woman walks toward the *left* relative to the train, then $v_{P/B} = -1.0$ m/s, and her velocity relative to the cyclist is 2.0 m/s. The sum in Eq. (3.33) is always an algebraic sum, and any or all of the velocities may be negative.

When the woman looks out the window, the stationary cyclist on the ground appears to her to be moving backward; we can call the cyclist's velocity relative to her $v_{A/P}$. Clearly, this is just the negative of $v_{P/A}$. In general, if A and B are any two points or frames of reference,

$$v_{A/B} = -v_{B/A} \qquad (3.34)$$

Problem-Solving Strategy

Relative Velocity

IDENTIFY *the relevant concepts:* Whenever you see the phrase "velocity relative to" or "velocity with respect to," it's likely that the concepts of relative velocity will be helpful.

SET UP *the problem:* Label each frame of reference in the problem. Each moving object has its own frame of reference; in addition, you'll almost always have to include the frame of reference of the earth's surface. (Statements such as "The car is traveling north at 90 km/h" implicitly refer to the car's velocity relative to the surface of the earth.) Use the labels to help identify the target variable. For example, if you want to find the velocity of a car (C) with respect to a bus (B), your target variable is $v_{C/B}$.

EXECUTE *the solution:* Solve for the target variable using Eq. (3.33). (If the velocities are not along the same direction, you'll need to use the vector form of this equation, derived later in this section.) It's important to note the order of the double subscripts in Eq. (3.33): $v_{A/B}$ always means "velocity of A relative to B." These subscripts obey an interesting kind of algebra, as Eq. (3.33) shows. If we regard each one as a fraction, then the fraction on the left side is the *product* of the fractions on the right sides: $P/A = (P/B)(B/A)$. This is a handy rule you can use when applying Eq. (3.33) to any number of frames of reference. For example, if there are three different frames of reference A, B, and C, we can write immediately

$$v_{P/A} = v_{P/C} + v_{C/B} + v_{B/A}$$

EVALUATE *your answer:* Be on the lookout for stray minus signs in your answer. If the target variable is the velocity of a car relative to a bus $(v_{C/B})$, make sure that you haven't accidentally calculated the velocity of the *bus* relative to the *car* $(v_{B/C})$. If you have made this mistake, you can recover using Eq. (3.34).

Example 3.14 Relative velocity on a straight road

You are driving north on a straight two-lane road at a constant 88 km/h. A truck traveling at a constant 104 km/h approaches you (in the other lane, fortunately). a) What is the truck's velocity relative to you? b) What is your velocity with respect to the truck? c) How do the relative velocities change after you and the truck have passed each other?

SOLUTION

IDENTIFY and SET UP: Let you be Y, the truck be T, and the earth's surface be E, and let the positive direction be north (Fig. 3.33). Then your velocity relative to the earth is $v_{Y/E} = +88$ km/h. As the truck is initially approaching you, it must be moving south and its velocity with respect to the earth is $v_{T/E} = -104$ km/h. The target variable in part (a) is $v_{T/Y}$; the target variable in part (b) is $v_{Y/T}$. We'll find the answers for both parts of the problem by using the equation for relative velocity, Eq. (3.33).

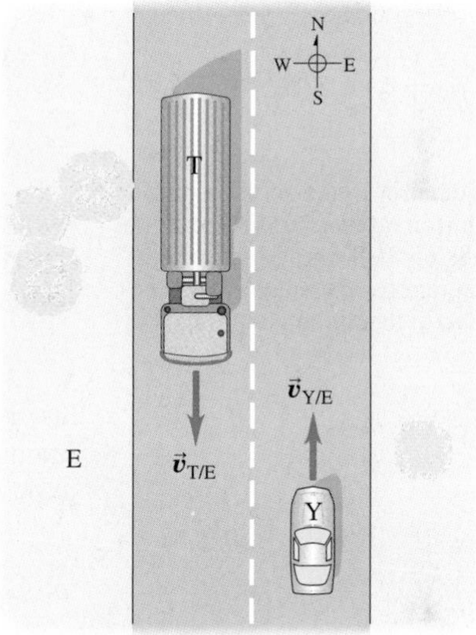

3.33 Reference frames for you and the truck.

EXECUTE: a) To find $v_{T/Y}$, we first write Eq. (3.33) for the three frames Y, T, and E, and then rearrange:

$$v_{T/E} = v_{T/Y} + v_{Y/E}$$
$$v_{T/Y} = v_{T/E} - v_{Y/E}$$
$$= -104 \text{ km/h} - 88 \text{ km/h} = -192 \text{ km/h}$$

The truck is moving 192 km/h south relative to you.

b) From Eq. (3.34),

$$v_{Y/T} = -v_{T/Y} = -(-192 \text{ km/h}) = +192 \text{ km/h}$$

You are moving 192 km/h north relative to the truck.

c) The relative velocities don't change at all after you and the truck have passed each other. The relative positions of the bodies don't matter. The velocity of the truck relative to you is still -192 km/h, but it is now moving away from you instead of toward you.

EVALUATE: As a check on your answer in part (b), try using Eq. (3.33) directly in the form $v_{Y/T} = v_{Y/E} + v_{E/T}$. (Remember that the velocity of the earth with respect to the truck is the opposite of the velocity of the truck with respect to the earth: $v_{E/T} = -v_{T/E}$.) Do you get the same result?

Relative Velocity in Two or Three Dimensions

We can extend the concept of relative velocity to include motion in a plane or in space by using vector addition to combine velocities. Suppose that the woman in Fig. 3.32a is walking not down the aisle of the railroad car but from one side of the car to the other, with a speed of 1.0 m/s (Fig. 3.34a). We can again describe the

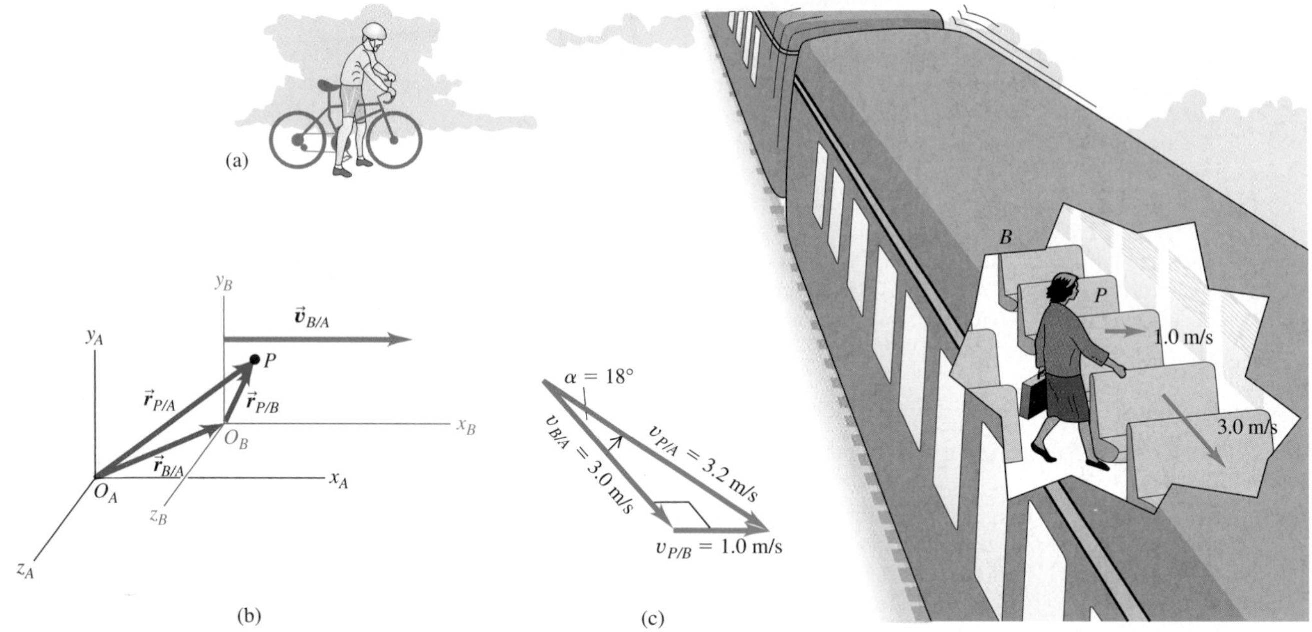

3.34 (a) A woman walking across a railroad car, seen from above. (b) The position vector depends on the frame of reference. (c) Vector diagram for the velocity of the woman relative to the ground.

woman's position P in two different frames of reference, A for the stationary ground observer and B for the moving train. But instead of coordinates x, we use position vectors $\vec{r}$ because the problem is now two-dimensional. Then, as Fig. 3.34b shows,

$$\vec{r}_{P/A} = \vec{r}_{P/B} + \vec{r}_{B/A} \tag{3.35}$$

Just as we did before, we take the time derivative of this equation to get a relation among the various velocities; the velocity of P relative to A is $\vec{v}_{P/A} = d\vec{r}_{P/A}/dt$, and so on for the other velocities. We get

$$\vec{v}_{P/A} = \vec{v}_{P/B} + \vec{v}_{B/A} \quad \text{(relative velocity in space)} \tag{3.36}$$

If all three of these velocities lie along the same line, then Eq. (3.36) reduces to Eq. (3.33) for the components of the velocities along that line.

If the train's velocity relative to the ground has magnitude $v_{B/A} = 3.0$ m/s and the woman's velocity relative to the railroad car has magnitude $v_{P/B} = 1.0$ m/s, then her velocity vector $\vec{v}_{P/A}$ relative to the ground is as shown in the vector diagram of Fig. 3.34c. The Pythagorean theorem then gives us

$$v_{P/A} = \sqrt{(3.0\ \text{m/s})^2 + (1.0\ \text{m/s})^2} = \sqrt{10\ \text{m}^2/\text{s}^2} = 3.2\ \text{m/s}$$

We can also see from the diagram that the *direction* of her velocity vector relative to the ground makes an angle α with the train's velocity vector $\vec{v}_{B/A}$, where

$$\tan\alpha = \frac{v_{P/B}}{v_{B/A}} = \frac{1.0\ \text{m/s}}{3.0\ \text{m/s}} \quad \text{and} \quad \alpha = 18°$$

As in the case of motion along a straight line, we have the general rule that if A and B are *any* two points or frames of reference,

$$\vec{v}_{A/B} = -\vec{v}_{B/A} \tag{3.37}$$

The velocity of the woman relative to the train is the negative of the velocity of the train relative to the woman, and so on.

Equation (3.36) is known as the *Galilean velocity transformation.* It shows that the velocity of an object P with respect to frame A and its velocity with respect to frame B ($\vec{v}_{P/A}$ and $\vec{v}_{P/B}$ respectively) are related by the velocity of frame B with respect to frame A ($\vec{v}_{P/B}$).

Example 3.15 Flying in a crosswind

The compass of an airplane indicates that it is headed due north, and its airspeed indicator shows that it is moving through the air at 240 km/h. If there is a wind of 100 km/h from west to east, what is the velocity of the airplane relative to the earth?

SOLUTION

IDENTIFY and SET UP: This is clearly a relative velocity problem, and we will use Eq. (3.36) to find the velocity of the plane (P) relative to the earth (E). We are given the magnitude and direction of the velocity of the plane relative to the air (A). We are also given the magnitude and direction of the wind velocity, which is the velocity of the air with respect to the earth:

$$\vec{v}_{P/A} = 240 \text{ km/h} \qquad \text{due north}$$

$$\vec{v}_{A/E} = 100 \text{ km/h} \qquad \text{due east}$$

Our target variables are the magnitude and direction of $\vec{v}_{P/E}$.

EXECUTE: Using Eq. (3.36), we have

$$\vec{v}_{P/E} = \vec{v}_{P/A} + \vec{v}_{A/E}$$

The three relative velocities and their relationship are shown in Fig. 3.35; the unknowns are the speed $v_{P/E}$ and the angle ϕ. From this diagram we find

$$v_{P/E} = \sqrt{(240 \text{ km/h})^2 + (100 \text{ km/h})^2} = 260 \text{ km/h}$$

$$\phi = \arctan\left(\frac{100 \text{ km/h}}{240 \text{ km/h}}\right) = 23° \text{ E of N}$$

EVALUATE: The crosswind increases the speed of the airplane relative to the earth, but at the price of pushing the airplane off course.

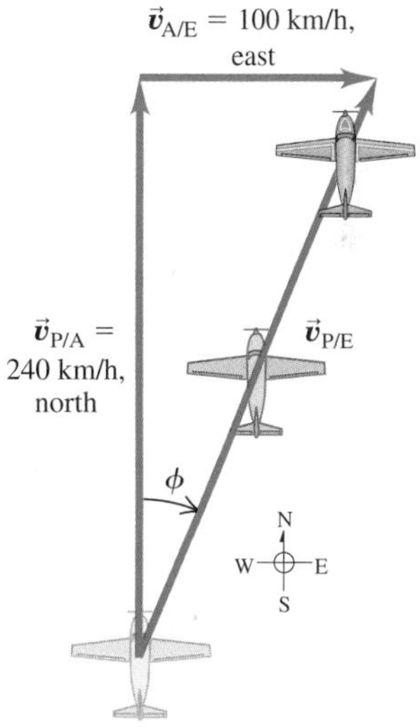

3.35 The plane is pointed north, but the wind blows east, giving the resultant velocity $\vec{v}_{P/E}$ relative to the earth.

Example 3.16 Correcting for a crosswind

In Example 3.15, in what direction should the pilot head to travel due north? What will then be her velocity relative to the earth? (Assume that her airspeed and the velocity of the wind are the same as in Example 3.15.)

SOLUTION

IDENTIFY: Figure 3.36 illustrates the situation. The vectors in Fig. 3.36 are arranged in accordance with the vector relative-velocity equation, Eq. (3.36):

$$\vec{v}_{P/E} = \vec{v}_{P/A} + \vec{v}_{A/E}$$

As Fig. 3.36 shows, the pilot points the nose of the airplane at an angle ϕ into the wind to compensate for the crosswind. This angle, which tells us the direction of the vector $\vec{v}_{P/A}$ (the velocity of the airplane relative to the air), is one of our target variables. The other target variable is the speed of the airplane over the ground, which is the magnitude of the vector $\vec{v}_{P/E}$ (the velocity of the airplane relative to the earth). Here are the known and unknown quantities:

$$\vec{v}_{P/E} = \text{magnitude unknown} \qquad \text{due north}$$

$$\vec{v}_{P/A} = 240 \text{ km/h} \qquad \text{direction unknown}$$

$$\vec{v}_{A/E} = 100 \text{ km/h} \qquad \text{due east}$$

We can solve for the unknown target variables using Fig. 3.36 and trigonometry.

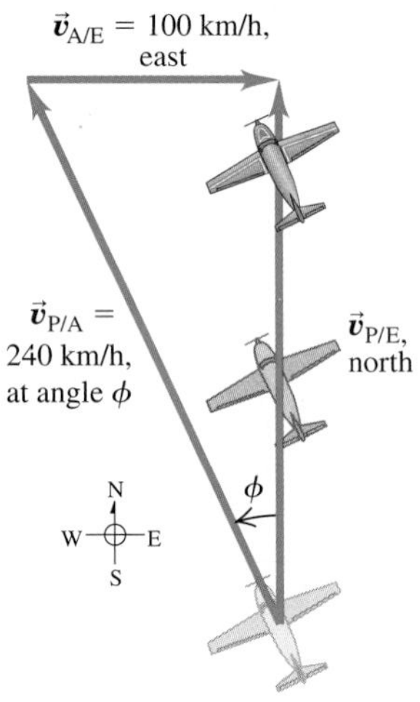

3.36 The pilot must point the plane in the direction of the vector $\vec{v}_{P/A}$ to travel due north relative to the earth.

EXECUTE: From the diagram, the speed $v_{P/E}$ and the angle ϕ are given by

$$v_{P/E} = \sqrt{(240\text{ km/h})^2 - (100\text{ km/h})^2} = 218\text{ km/h}^2$$

$$\phi = \arcsin\left(\frac{100\text{ km/h}}{240\text{ km/h}}\right) = 25^\circ$$

The pilot should point the airplane 25° west of north, and her ground speed is then 218 km/h.

EVALUATE: Note that there were two target variables—the magnitude of a vector and the direction of a vector—in both this example and Example 3.15. The difference is that in Example 3.15, the magnitude and direction referred to the *same* vector ($\vec{v}_{P/E}$), whereas in this example they referred to different vectors ($\vec{v}_{P/E}$ and $\vec{v}_{P/A}$).

It's no surprise that a headwind will reduce an airplane's speed relative to the ground. What this example shows is that a *crosswind* also slows an airplane down—an unfortunate fact of aeronautical life.

When we derived the relative-velocity relations, we assumed that all the observers use the same time scale. This is precisely the point at which Einstein's special theory of relativity departs from the physics of Galileo and Newton. When speeds approach the speed of light, denoted by c, the velocity-addition equation has to be modified. It turns out that if the woman in Fig. 3.32a could walk down the aisle at $0.30c$ and the train could move at $0.90c$, then her speed relative to the ground would be not $1.20c$ but $0.94c$; nothing can travel faster than light! We'll return to the special theory of relativity in Chapter 37.

Test Your Understanding

Suppose the wind in Example 3.16 dropped suddenly to zero, but the pilot kept the nose of the airplane pointed 25° to the west of north. How far off her desired south-to-north course would she be after one hour?

SUMMARY

The position vector $\vec{r}$ of a point P in space is the vector from the origin to P. Its components are the coordinates x, y, and z.

$$\vec{r} = x\hat{\imath} + y\hat{\jmath} + z\hat{k} \tag{3.1}$$

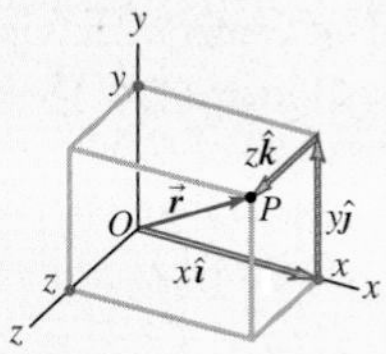

The average velocity vector $\vec{v}_{av}$ during the time interval Δt is the displacement $\Delta\vec{r}$ (the change in the position vector $\vec{r}$) divided by Δt. The instantaneous velocity vector $\vec{v}$ is the time derivative of $\vec{r}$, and its components are the time derivatives of x, y, and z. The instantaneous speed is the magnitude of $\vec{v}$. The velocity $\vec{v}$ of a particle is always tangent to the particle's path. (See Example 3.1)

$$\vec{v}_{av} = \frac{\vec{r}_2 - \vec{r}_1}{t_2 - t_1} = \frac{\Delta\vec{r}}{\Delta t} \tag{3.2}$$

$$\vec{v} = \lim_{\Delta t \to 0} \frac{\Delta\vec{r}}{\Delta t} = \frac{d\vec{r}}{dt} \tag{3.3}$$

$$v_x = \frac{dx}{dt} \qquad v_y = \frac{dy}{dt} \qquad v_z = \frac{dz}{dt} \tag{3.4}$$

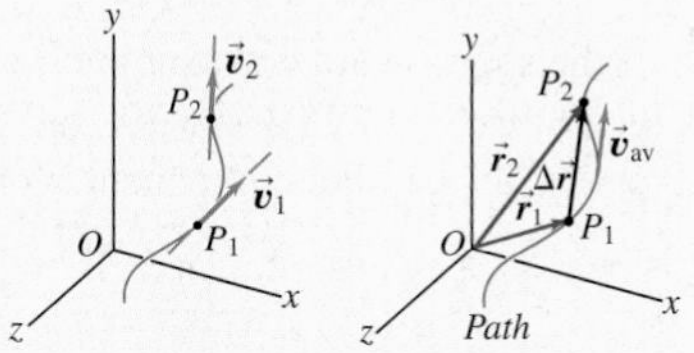

The average acceleration vector $\vec{a}_{av}$ during the time interval Δt equals $\Delta\vec{v}$ (the change in the velocity vector $\vec{v}$) divided by Δt. The instantaneous acceleration vector $\vec{a}$ is the time derivative of $\vec{v}$, and its components are the time derivatives of v_x, v_y, and v_z.
(See Examples 3.2 and 3.3)

$$\vec{a}_{av} = \frac{\vec{v}_2 - \vec{v}_1}{t_2 - t_1} = \frac{\Delta\vec{v}}{\Delta t} \tag{3.8}$$

$$\vec{a} = \lim_{\Delta t \to 0} \frac{\Delta\vec{v}}{\Delta t} = \frac{d\vec{v}}{dt} \tag{3.9}$$

$$a_x = \frac{dv_x}{dt} \qquad a_y = \frac{dv_y}{dt} \qquad a_z = \frac{dv_z}{dt} \tag{3.10}$$

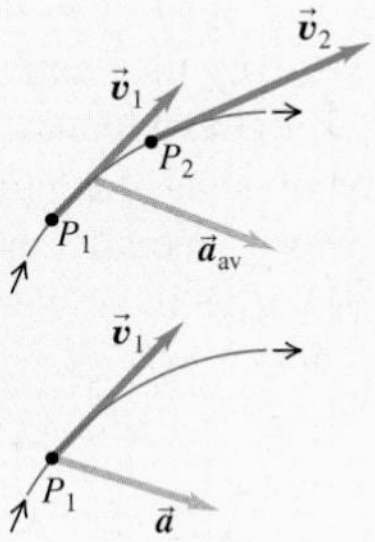

Acceleration can also be represented in terms of its components parallel and perpendicular to the direction of the instantaneous velocity. The parallel component $\vec{a}$ affects the speed, while the perpendicular component of $\vec{a}$ affects the direction of motion. (See Examples 3.4 and 3.5)

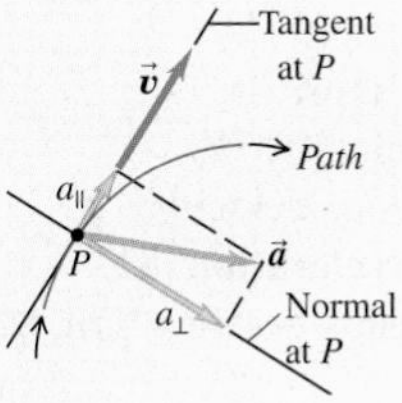

In projectile motion with no air resistance, $a_x = 0$ and $a_y = -g$. The coordinates and velocity components are simple functions of time, and the shape of the path is always a parabola. We conventionally choose the origin to be at the initial position of the projectile. (See Examples 3.6 through 3.11)

$$x = (v_0 \cos\alpha_0)t \tag{3.20}$$

$$y = (v_0 \sin\alpha_0)t - \frac{1}{2}gt^2 \tag{3.21}$$

$$v_x = v_0 \cos\alpha_0 \tag{3.22}$$

$$v_y = v_0 \sin\alpha_0 - gt \tag{3.23}$$

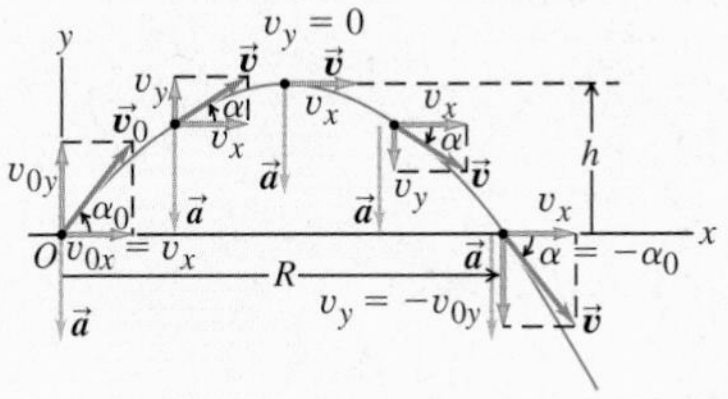

When a particle moves in a circular path of radius R with constant speed v, its acceleration $\vec{a}$ is directed toward the center of the circle and perpendicular to $\vec{v}$. The magnitude a_{rad} of the acceleration can be expressed in terms of v and R or in terms of R and the period T (the time for one revolution), where $v = 2\pi R/T$. (See Examples 3.12 and 3.13)

$$a_{\text{rad}} = \frac{v^2}{R} \quad (3.28)$$

$$a_{\text{rad}} = \frac{4\pi^2 R}{T^2} \quad (3.30)$$

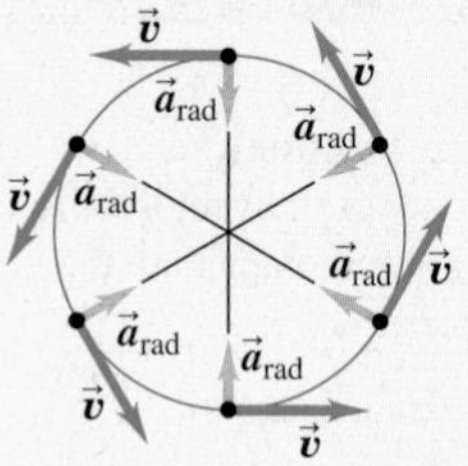

If the speed is not constant in circular motion, there is still a radial component of $\vec{a}$ given by Eq. (3.28) or (3.30), but there is also a component of $\vec{a}$ parallel to the path. This parallel component is equal to the rate of change of speed, dv/dt.

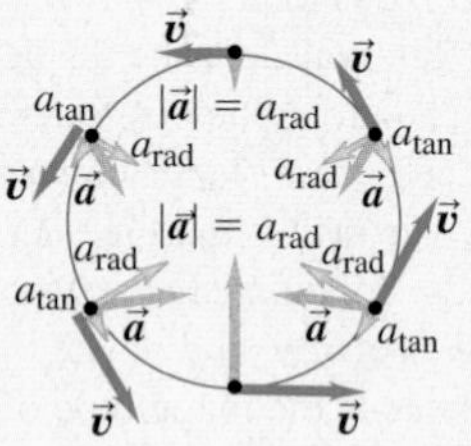

When a body P moves relative to a body (or reference frame) B, and B moves relative to A, we denote the velocity of P relative to B by $\vec{v}_{P/B}$, the velocity of P relative to A by $\vec{v}_{P/A}$, and the velocity of B relative to A by $\vec{v}_{B/A}$. If these velocities are all along the same line, their components along that line are related by Eq. (3.33). More generally, these velocities are related by Eq. (3.36). (See Examples 3.14 through 3.16)

$$v_{P/A} = v_{P/B} + v_{B/A} \quad (3.33)$$

(relative velocity along a line)

$$\vec{v}_{P/A} = \vec{v}_{P/B} + \vec{v}_{B/A} \quad (3.36)$$

(relative velocity in space)

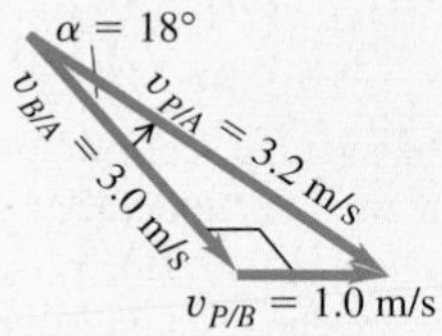

Key Terms

position vector, 79
average velocity, 79
instantaneous velocity, 79
average acceleration, 82
instantaneous acceleration, 82
projectile, 87
trajectory, 87
uniform circular motion, 98
centripetal acceleration, 99
period, 99
non-uniform circular motion, 100
relative velocity, 101
frame of reference, 102

Your Notes

Answer to Chapter Opening Question ?

We show in Example 3.10 (Section 3.3) that a projectile has the maximum horizontal range when it is launched at a 45° angle.

Answers to Test Your Understanding Questions

Section 3.1 The average velocity vector $\vec{v}_{av}$ over an interval and the instantaneous velocity $\vec{v}$ at the end of the interval are equal if the velocity is constant over that interval.
Section 3.2 At the high point of the fragment's path, the speed is minimum. At that point the speed is neither increasing nor decreasing, and the parallel component of the acceleration (that is, the horizontal component) is zero. The acceleration has only a perpendicular component toward the inside of the fragment's curved path. In other words, the acceleration is downward. (We discuss the acceleration of projectiles in more depth in Section 3.3.)
Section 3.3 If there were no gravity $(g = 0)$, the monkey would not fall and the dart would follow a straight-line path (shown as a dashed line in Fig. 3.27). The effect of gravity is to make the monkey and dart both fall the same distance $\frac{1}{2}gt^2$ below their $g = 0$ positions. In Fig. 3.27, point A is the same distance below the monkey's initial position as point P is below the dashed straight line, so point A is where we would find the monkey at the time in question.
Section 3.4 At both the top and bottom of the loop, the acceleration is purely radial and is given by Eq. (3.28). The radius R is the same at both points, so the difference in acceleration is due purely to differences in speed. Since a_{rad} is proportional to the square of v, the speed must be twice as great at the bottom of the loop as at the top.
Section 3.5 By pointing the nose of the airplane to the west, the pilot has given the airplane's velocity relative to the air $(\vec{v}_{P/A})$ a westward component of 100 km/h—just enough to cancel the eastward wind. With no wind present, the airplane will drift westward by 100 km in an hour.

Discussion Questions

Q3.1 A simple pendulum (a mass swinging at the end of a string) swings back and forth in a circular arc. What is the direction of its acceleration at the ends of the swing? At the midpoint? In each case, explain how you obtain your answer.
Q3.2 Redraw Fig. 3.11a if $\vec{a}$ is antiparallel to $\vec{v}_1$. Does the particle move in a straight line? What happens to the speed?
Q3.3 A projectile moves in a parabolic path without air resistance. Is there any point at which $\vec{a}$ is parallel to $\vec{v}$? Perpendicular to $\vec{v}$? Explain.
Q3.4 When a rifle is fired at a distant target, the barrel is not lined up exactly on the target. Why not? Does the angle of correction depend on the distance of the target?
Q3.5 At the same instant that you fire a bullet horizontally from a gun, you drop a bullet from the height of the barrel. If there is no air resistance, which bullet hits the ground first? Explain.
Q3.6 A package falls out of an airplane that is flying in a straight line at a constant altitude and speed. If you could ignore air resistance, what would be the path of the package as observed by the pilot? As observed by a person on the ground?
Q3.7 Sketch the six graphs of the x- and y-components of position, velocity, and acceleration versus time for projectile motion with $x_0 = y_0 = 0$ and $0 < \alpha_0 < 90°$.
Q3.8 If $y_0 = 0$ and α_0 is negative, y can never be positive for a projectile. However, the expression for h in Example 3.10 seems to give a positive maximum height for a negative α_0. Explain this apparent contradiction.
Q3.9 If a jumping frog can give itself the same initial speed regardless of the direction in which it jumps (forward or straight up), how is the maximum vertical height to which it can jump related to its maximum horizontal range $R_{max} = v_0^2/g$?
Q3.10 When a projectile reaches its maximum height, is its speed zero? Explain.
Q3.11 In uniform circular motion, what are the *average* velocity and *average* acceleration for one revolution? Explain.
Q3.12 In uniform circular motion, how does the acceleration change when the speed is increased by a factor of 3? When the radius is decreased by a factor of 2?
Q3.13 In uniform circular motion, the acceleration is perpendicular to the velocity at every instant. Is this still true when the motion is not uniform, that is, when the speed is not constant?
Q3.14 Raindrops hitting the side windows of a car in motion often leave diagonal streaks. Why? Is the explanation the same or different for diagonal streaks on the windshield?
Q3.15 In a rainstorm with a strong wind, what determines the best position in which to hold an umbrella?
Q3.16 You are on the west bank of a river that is flowing north with a speed of 1.2 m/s. Your swimming speed relative to the water is 1.5 m/s and the river is 60 m wide. What is your path relative to earth that allows you to cross the river in the shortest time? Explain your reasoning.

Exercises

Section 3.1 Position and Velocity Vectors

3.1 A squirrel has x- and y-coordinates (1.1 m, 3.4 m) at time $t_1 = 0$ and coordinates (5.3 m, −0.5 m) at time $t_2 = 3.0$ s. For this time interval, find a) the components of the average velocity; b) the magnitude and direction of the average velocity.
3.2 A rhinoceros is at the origin of coordinates at time $t_1 = 0$. For the time interval from $t_1 = 0$ to $t_2 = 12.0$ s, the rhino's average velocity has x-component −3.8 m/s and y-component 4.9 m/s. At time $t_2 = 12.0$ s, a) what are the x- and y-coordinates of the rhino? b) How far is the rhino from the origin?
3.3 A web page designer creates an animation in which a dot on a computer screen has a position of $\vec{r} = [4.0\text{ cm} + (2.5\text{ cm/s}^2)t^2]\hat{\imath} + (5.0\text{ cm/s})t\hat{\jmath}$. a) Find the magnitude and direction of the dot's average velocity between $t = 0$ and $t = 2.0$ s. b) Find the magnitude and direction of the instantaneous velocity at $t = 0$, $t = 1.0$ s, and $t = 2.0$ s. c) Sketch the dot's trajectory from $t = 0$ to $t = 2.0$ s, and show the velocities calculated in part (b).

3.4 If $\vec{r} = bt^2\hat{\imath} + ct^3\hat{\jmath}$, where b and c are positive constants, when does the velocity vector make an angle of 45.0° with the x- and y-axes?

Section 3.2 The Acceleration Vector

3.5 A jet plane is flying at a constant altitude. At time $t_1 = 0$ it has components of velocity $v_x = 90$ m/s, $v_y = 110$ m/s. At time $t_2 = 30.0$ s the components are $v_x = -170$ m/s, $v_y = 40$ m/s. a) Sketch the velocity vectors at t_1 and t_2. How do these two vectors differ? For this time interval calculate b) the components of the average acceleration; c) the magnitude and direction of the average acceleration.

3.6 A dog running in an open field has components of velocity $v_x = 2.6$ m/s and $v_y = -1.8$ m/s at $t_1 = 10.0$ s. For the time interval from $t_1 = 10.0$ s to $t_2 = 20.0$ s, the average acceleration of the dog has magnitude 0.45 m/s^2 and direction 31.0° measured from the $+x$-axis toward the $+y$-axis. At $t_2 = 20.0$ s, a) what are the x- and y-components of the dog's velocity? b) What are the magnitude and direction of the dog's velocity? c) Sketch the velocity vectors at t_1 and t_2. How do these two vectors differ?

3.7 The coordinates of a bird flying in the xy-plane are given by $x(t) = \alpha t$ and $y(t) = 3.0\text{ m} - \beta t^2$, where $\alpha = 2.4$ m/s and $\beta = 1.2$ m/s^2. a) Sketch the path of the bird between $t = 0$ and $t = 2.0$ s. b) Calculate the velocity and acceleration vectors of the bird as functions of time. c) Calculate the magnitude and direction of the bird's velocity and acceleration at $t = 2.0$ s. d) Sketch the velocity and acceleration vectors at $t = 2.0$ s. At this instant, is the bird speeding up, is it slowing down, or is its speed instantaneously not changing? Is the bird turning? If so, in what direction?

3.8 A particle moves along a path as shown in Fig. 3.37. Between points B and D, the path is a straight line. Sketch the acceleration vectors at A, C, and E in the cases in which a) the particle moves with a constant speed; b) the particle moves with a steadily increasing speed; c) the particle moves with a steadily decreasing speed.

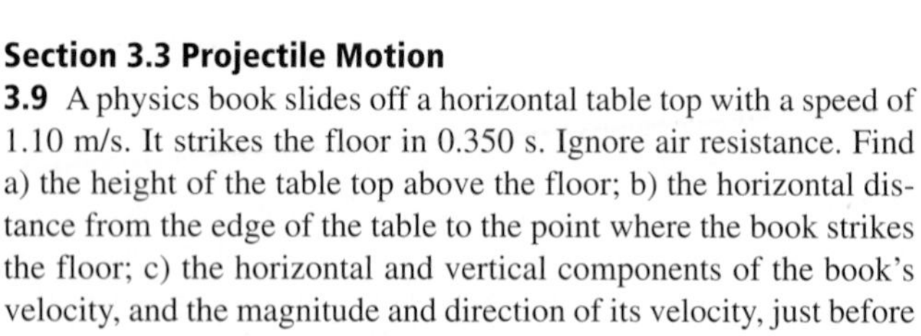

Figure 3.37 Exercise 3.8.

Section 3.3 Projectile Motion

3.9 A physics book slides off a horizontal table top with a speed of 1.10 m/s. It strikes the floor in 0.350 s. Ignore air resistance. Find a) the height of the table top above the floor; b) the horizontal distance from the edge of the table to the point where the book strikes the floor; c) the horizontal and vertical components of the book's velocity, and the magnitude and direction of its velocity, just before the book reaches the floor. d) Draw x-t, y-t, v_x-t, and v_y-t graphs for the motion.

3.10 A military helicopter on a training mission is flying horizontally at a speed of 60.0 m/s and accidentally drops a bomb (fortunately not armed) at an elevation of 300 m. You can ignore air resistance. a) How much time is required for the bomb to reach the earth? b) How far does it travel horizontally while falling? c) Find the horizontal and vertical components of its velocity just before it strikes the earth. d) Draw x-t, y-t, v_x-t, and v_y-t graphs for the bomb's motion. e) If the velocity of the helicopter remains constant, where is the helicopter when the bomb hits the ground?

3.11 Two crickets, Chirpy and Milada, jump from the top of a vertical cliff. Chirpy jumps horizontally and reaches the ground in 3.50 s. Milada jumps with an initial velocity of 95.0 cm/s at an angle of 32.0° above the horizontal. How far from the base of the cliff will Milada hit the ground?

3.12 A daring 510-N swimmer dives off a cliff with a running horizontal leap, as shown in Fig. 3.38. What must her minimum speed be just as she leaves the top of the cliff so that she will miss the ledge at the bottom, which is 1.75 m wide and 9.00 m below the top of the cliff?

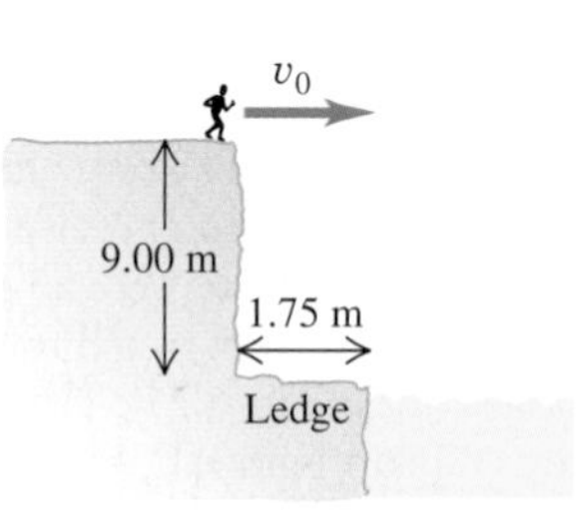

Figure 3.38 Exercise 3.12

3.13 Leaping the River I. A 10,000-N car comes to a bridge during a storm and finds the bridge washed out. The 650-N driver must get to the other side, so he decides to try leaping it with his car. The side the car is on is 21.3 m above the river, while the opposite side is a mere 1.8 m above the river. The river itself is a raging torrent 61.0 m wide. a) How fast should the car be traveling just as it leaves the cliff in order just to clear the river and land safely on the opposite side? b) What is the speed of the car just before it lands safely on the other side?

3.14 A student makes a bet with another student that she can roll a marble off the top of a table and have it land in a cup placed on the floor. She claims she can do this without letting the marble roll off the table before the actual attempt. If you were the student making the bet, what would you do to determine where to place the cup? Be sure to explain what measurements you need to make and how you would use them to determine the placement of the cup.

3.15 You throw a football with an initial speed $v_0 = 15.0$ m/s at an initial angle $\alpha_0 = 45.0°$. a) Find the time T when the football is at its maximum height. b) At the three times $t_1 = T - 0.50$ s, $t_2 = T$, and $t_3 = T + 0.50$ s, find the x- and y-components of the position vector. c) At the three times t_1, t_2, and t_3, find the magnitude and direction of the velocity vector. d) At the three times t_1, t_2, and t_3, find the component of the acceleration vector that is parallel (or antiparallel) to the velocity, and find the component of the acceleration vector that is perpendicular to the velocity. e) Sketch the trajectory of the football. On your sketch, label the position of the football at the three times t_1, t_2, and t_3. At each of these positions, draw the velocity vector and the parallel and perpendicular components of the acceleration vector. f) Discuss how the speed and the direction of motion of the football are changing at the three times t_1, t_2, and t_3, and explain how the vectors in your sketch describe these changes.

3.16 A tennis ball rolls off the edge of a tabletop 0.750 m above the floor and strikes the floor at a point 1.40 m horizontally from the edge of the table. You can ignore air resistance. a) Find the time of flight. b) Find the magnitude of the initial velocity. c) Find the magnitude and direction of the velocity of the ball just before it strikes the floor. d) Draw x-t, y-t, v_x-t, and v_y-t graphs for the motion.

3.17 A pistol that fires a signal flare gives the flare an initial speed (muzzle speed) of 120 m/s. a) If the flare is fired at an angle of 55° above the horizontal on the level salt flats of Utah, what is its horizontal range? You can ignore air resistance. b) If the flare is fired at the same angle over the flat Sea of Tranquility on the moon, where $g = 1.6\ \text{m/s}^2$, what is its horizontal range?

3.18 A rookie quarterback throws a football with an initial upward velocity component of 16.0 m/s and a horizontal velocity component of 20.0 m/s. Ignore air resistance. a) How much time is required for the football to reach the highest point of the trajectory? b) How high is this point? c) How much time (after it is thrown) is required for the football to return to its original level? How does this compare with the time calculated in part (a)? d) How far has it traveled horizontally during this time? e) Draw x-t, y-t, v_x-t, and v_y-t graphs for the motion.

3.19 A major leaguer hits a baseball so that it leaves the bat at a speed of 30.0 m/s and at an angle of 36.9° above the horizontal. You can ignore air resistance. a) At what *two* times is the baseball at a height of 10.0 m above the point at which it left the bat? b) Calculate the horizontal and vertical components of the baseball's velocity at each of the two times calculated in part (a). c) What is the magnitude and direction of the baseball's velocity when it returns to the level at which it left the bat?

3.20 A shot putter releases the shot some distance above the level ground with a velocity of 12.0 m/s, 51.0° above the horizontal. The shot hits the ground 2.08 s later. You can ignore air resistance. a) What are the components of the shot's acceleration while in flight? b) What are the components of the shot's velocity at the beginning and at the end of its trajectory? c) How far did she throw the shot horizontally? d) Why does the expression for R in Example 3.10 *not* give the correct answer for part (c)? e) How high was the shot above the ground when she released it? f) Draw x-t, y-t, v_x-t, and v_y-t graphs for the motion.

3.21 Win the Prize. In a carnival booth, you win a stuffed giraffe if you toss a quarter into a small dish. The dish is on a shelf above the point where the quarter leaves your hand and is a horizontal distance of 2.1 m from this point (Fig. 3.39). If you toss the coin with a velocity of 6.4 m/s at an angle of 60° above the horizontal, the coin lands in the dish. You can ignore air resistance. a) What is the height of the shelf above the point where the quarter leaves your hand? b) What is the vertical component of the velocity of the quarter just before it lands in the dish?

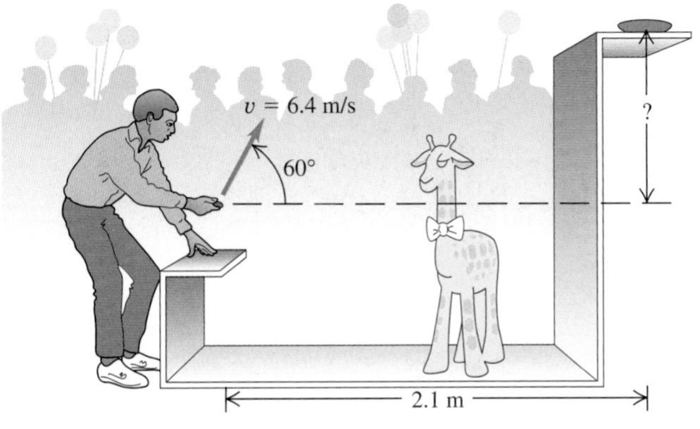

Figure 3.39 Exercise 3.21.

3.22 Suppose the departure angle α_0 in Fig. 3.24 is 42.0° and the distance d is 3.00 m. Where will the dart and monkey meet if the initial speed of the dart is a) 12.0 m/s? b) 8.0 m/s? c) What will happen if the initial speed of the dart is 4.0 m/s? Sketch the trajectory in each case.

3.23 A man stands on the roof of a 15.0-m-tall building and throws a rock with a velocity of magnitude 30.0 m/s at an angle of 33.0° above the horizontal. You can ignore air resistance. Calculate a) the maximum height above the roof reached by the rock; b) the magnitude of the velocity of the rock just before it strikes the ground; c) the horizontal distance from the base of the building to the point where the rock strikes the ground. d) Draw x-t, y-t, v_x-t, and v_y-t graphs for the motion.

3.24 Firemen are shooting a stream of water at a burning building using a high-pressure hose that shoots out the water with a speed of 25.0 m/s as it leaves the end of the hose. Once it leaves the hose, the water moves in projectile motion. The firemen adjust the angle of elevation α of the hose until the water takes 3.00 s to reach a building 45.0 m away. You can ignore air resistance; assume that the end of the hose is at ground level. a) Find the angle of elevation α. b) Find the speed and acceleration of the water at the highest point in its trajectory. c) How high above the ground does the water strike the building, and how fast is it moving just before it hits the building?

3.25 A 124-kg balloon carrying a 22-kg basket is descending with a constant downward velocity of 20.0 m/s. A 1.0-kg stone is thrown from the basket with an initial velocity of 15.0 m/s perpendicular to the path of the descending balloon, as measured relative to a person at rest in the basket. The person in the basket sees the stone hit the ground 6.00 s after being thrown. Assume that the balloon continues its downward descent with the same constant speed of 20.0 m/s. a) How high was the balloon when the rock was thrown out? b) How high is the balloon when the rock hits the ground? c) At the instant the rock hits the ground, how far is it from the basket? d) Just before the rock hits the ground, find its horizontal and vertical velocity components as measured by an observer: i) at rest in the basket; ii) at rest on the ground.

3.26 A cannon, located 60.0 m from the base of a vertical 25.0-m-tall cliff, shoots a 15-kg shell at 43.0° above the horizontal toward the cliff. a) What must the minimum muzzle velocity be for the shell to clear the top of the cliff? b) The ground at the top of the cliff is level, with a constant elevation of 25.0 m above the cannon. Under the conditions of part (a), how far does the shell land past the edge of the cliff?

3.27 An airplane is flying with a velocity of 90.0 m/s at an angle of 23.0° above the horizontal. When the plane is 114 m directly above a dog that is standing on level ground, a suitcase drops out of the luggage compartment. How far from the dog will the suitcase land? You can ignore air resistance.

Section 3.4 Motion in a Circle

3.28 On your first day at work for an appliance manufacturer, you are told to figure out what to do to the period of rotation during a washer spin cycle to triple the centripetal acceleration. You impress your boss by answering immediately. What do you tell her?

3.29 The earth has a radius of 6380 km and turns around once on its axis in 24 h. a) What is the radial acceleration of an object at the earth's equator? Give your answer in m/s^2 and as a fraction of g. b) If a_{rad} at the equator is greater than g, objects would fly off the earth's surface and into space. (We will see the reason for this in Chapter 5.) What would the period of the earth's rotation have to be for this to occur?

3.30 A model of a helicopter rotor has four blades, each 3.40 m in length from the central shaft to the blade tip. The model is rotated in a wind tunnel at 550 rev/min. a) What is the linear speed of the blade tip, in m/s? b) What is the radial acceleration of the blade tip expressed as a multiple of the acceleration of gravity, g?

3.31 In a test of a "g-suit," a volunteer is rotated in a horizontal circle of radius 7.0 m. What is the period of rotation at which the centripetal acceleration has a magnitude of a) $3.0g$? b) $10g$?

3.32 The radius of the earth's orbit around the sun (assumed to be circular) is 1.50×10^8 km, and the earth travels around this orbit in 365 days. a) What is the magnitude of the orbital velocity of the earth in m/s? b) What is the radial acceleration of the earth toward the sun in m/s^2? c) Repeat parts (a) and (b) for the motion of the planet Mercury (orbit radius $= 5.79 \times 10^7$ km, orbital period $= 88.0$ days).

3.33 A Ferris wheel with radius 14.0 m is turning about a horizontal axis through its center (Fig. 3.40). The linear speed of a passenger on the rim is constant and equal to 7.00 m/s. What are the magnitude and direction of the passenger's acceleration as she passes through a) the lowest point in her circular motion? b) the highest point in her circular motion? c) How much time does it take the Ferris wheel to make one revolution?

Figure 3.40 Exercises 3.33 and 3.34.

3.34 The Ferris wheel in Fig. 3.40, which rotates counterclockwise, is just starting up. At a given instant, a passenger on the rim of the wheel and passing through the lowest point of his circular motion is moving at 3.00 m/s and is gaining speed at a rate of 0.500 m/s^2. a) Find the magnitude and the direction of the passenger's acceleration at this instant. b) Sketch the Ferris wheel and passenger showing his velocity and acceleration vectors.

3.35 Elliptical Path. A level racetrack has the shape of an ellipse. (See your math textbook, a dictionary, or an encyclopedia for an illustration of an ellipse.) A car drives around this track at a constant speed. a) Draw a picture of the track showing the car's velocity vector and acceleration vector at five or more different places around the track. b) Does the car's acceleration vector always point toward the geometrical center of the ellipse? Explain. c) At what point(s) on the ellipse does the car's acceleration have the greatest magnitude? Explain.

Section 3.5 Relative Velocity

3.36 A railroad flatcar is traveling to the right at a speed of 13.0 m/s relative to an observer standing on the ground. Someone is riding a motor scooter on the flatcar (Fig. 3.41). What is the velocity (magnitude and direction) of the motor scooter relative to the flatcar if its velocity relative to the observer on the ground is a) 18.0 m/s to the right? b) 3.0 m/s to the left? c) zero?

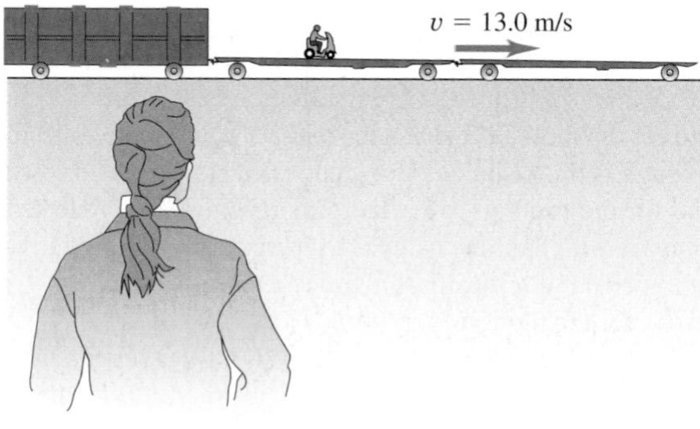

Figure 3.41 Exercise 3.36.

3.37 A "moving sidewalk" in an airport terminal building moves at 1.0 m/s and is 35.0 m long. If a woman steps on at one end and walks at 1.5 m/s relative to the moving sidewalk, how much time does she require to reach the opposite end if she walks a) in the same direction the sidewalk is moving? b) in the opposite direction?

3.38 Two piers, A and B, are located on a river: B is 1500 m downstream from A (Fig. 3.42). Two friends must make round trips from pier A to pier B and return. One rows a boat at a constant speed of 4.00 km/h relative to the water; the other walks on the shore at a constant speed of 4.00 km/h. The velocity of the river is 2.80 km/h in the direction from A to B. How much time does it take each person to make the round trip?

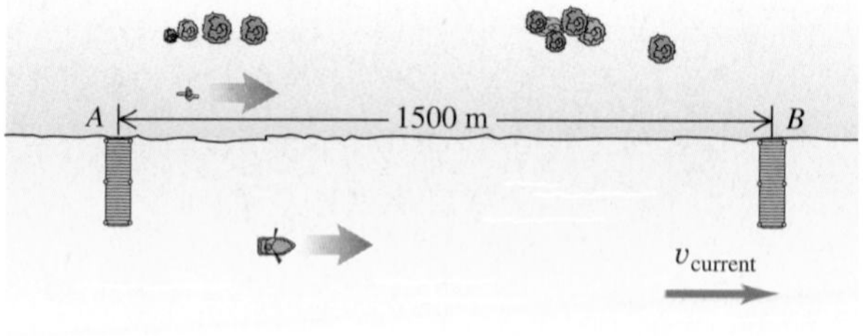

Figure 3.42 Exercise 3.38.

3.39 A canoe has a velocity of 0.40 m/s southeast relative to the earth. The canoe is on a river that is flowing 0.50 m/s east relative to the earth. Find the velocity (magnitude and direction) of the canoe relative to the river.

3.40 An airplane pilot wishes to fly due west. A wind of 80.0 km/h (about 50 mi/h) is blowing toward the south. a) If the airspeed of the plane (its speed in still air) is 320.0 km/h (about 200 mi/h), in which direction should the pilot head? b) What is the speed of the plane over the ground? Illustrate with a vector diagram.

3.41 Crossing the River I. A river flows due south with a speed of 2.0 m/s. A man steers a motorboat across the river; his velocity relative to the water is 4.2 m/s due east. The river is 800 m wide. a) What is his velocity (magnitude and direction) relative to the earth? b) How much time is required to cross the river? c) How far south of his starting point will he reach the opposite bank?

3.42 Crossing the River II. a) In which direction should the motorboat in Exercise 3.41 head in order to reach a point on the opposite bank directly east from the starting point? (The boat's speed relative to the water remains 4.2 m/s.) b) What is the velocity of the boat relative to the earth? c) How much time is required to cross the river?

3.43 The nose of an ultralight plane is pointed south, and its air speed indicator shows 35 m/s. The plane is in a 10 m/s wind blowing toward the southwest relative to the earth. a) In a vector addition diagram, show the relation of $\vec{v}_{P/E}$ (the velocity of the plane relative to the earth) to the two given vectors. b) Letting x be east and y be north, find the components of $\vec{v}_{P/E}$. c) Find the magnitude and direction of $\vec{v}_{P/E}$.

Problems

3.44 A faulty model rocket moves in the xy-plane (the positive y-direction is vertically upward). The rocket's acceleration has components $a_x(t) = \alpha t^2$ and $a_y(t) = \beta - \gamma t$, where $\alpha = 2.50\text{ m/s}^4$, $\beta = 9.00\text{ m/s}^2$, and $\gamma = 1.40\text{ m/s}^3$. At $t = 0$ the rocket is at the origin and has velocity $\vec{v}_0 = v_{0x}\hat{\imath} + v_{0y}\hat{\jmath}$ with $v_{0x} = 1.00$ m/s and $v_{0y} = 7.00$ m/s. a) Calculate the velocity and position vectors as functions of time. b) What is the maximum height reached by the rocket? c) Sketch the path of the rocket. d) What is the horizontal displacement of the rocket when it returns to $y = 0$?

3.45 A student is moving in the xy-plane in a darkened room, trying to locate a lost \$20 bill. The student's coordinates are given as functions of time by $x(t) = \alpha t$ and $y(t) = 15.0\text{ m} - \beta t^2$, where $\alpha = 1.20$ m/s and $\beta = 0.500\text{ m/s}^2$. Unknown to the student, the \$20 bill is at the origin. a) At which times is the student's velocity perpendicular to his acceleration? b) At which times is the student's speed instantaneously not changing? c) At which times is the student's velocity perpendicular to his position vector? What are the locations of the student at these times? d) What is the minimum distance from the student to the \$20 bill? At what time does this minimum occur? e) Sketch the path of the hapless student.

3.46 A bird flies in the xy-plane with a velocity vector given by $\vec{v} = (\alpha - \beta t^2)\hat{\imath} + \gamma t\hat{\jmath}$, with $\alpha = 2.4$ m/s, $\beta = 1.6\text{ m/s}^3$, and $\gamma = 4.0\text{ m/s}^2$. The positive y-direction is vertically upward. At $t = 0$ the bird is at the origin. a) Calculate the position and acceleration vectors of the bird as functions of time. b) What is the bird's altitude (y-coordinate) as it flies over $x = 0$ for the first time after $t = 0$?

3.47 A Piper Warrior, a small four-seat airplane, requires 300 m of runway to take off. Its liftoff speed is 88 km/h. It then climbs with a constant speed of 88 km/h along a straight-line path, just clearing a power line 15 m high at a horizontal distance of 460 m from where the Piper began its takeoff roll. a) What was the acceleration of the Piper (assumed constant) during its takeoff roll? b) After the Piper takes off, what is the angle of its flight path above the horizontal? c) What is the rate of climb (in m/s) of the Piper? d) What is the elapsed time from the beginning of the takeoff roll to when the Piper just clears the power line?

3.48 A track and field coach (who is also a physics teacher) trains an athlete to throw a javelin so that it leaves the athlete's hand at a height h above the ground, moving at a speed of $\sqrt{25gh/8}$ and at an angle of 36.9° above the horizontal. The javelin continues in flight until it strikes the ground. The ground around the athlete is level, and air resistance is negligible. a) Graph the javelin's horizontal velocity versus time and the javelin's vertical velocity versus time. b) Find the maximum height reached by the javelin. c) Find the horizontal distance that the javelin travels from where it leaves the athlete's hand to where it strikes the ground.

3.49 Dynamite! A demolition crew uses dynamite to blow an old building apart. Debris from the explosion flies off in all directions, and is later found at distances up to 50 m from the explosion. Estimate the maximum speed at which debris was blown outward by the explosion. Describe any assumptions that you make.

3.50 Spiraling Up. It is common to see birds of prey rising upward on thermals. The paths they take may be spiral-like. You can model the spiral motion as uniform circular motion combined with a constant upward velocity. Assume a bird completes a circle of radius 8.00 m every 5.00 s and rises vertically at a rate of 3.00 m/s. Determine the following: a) the speed of the bird relative to the ground; b) the bird's acceleration (magnitude and direction); c) the angle between the bird's velocity vector and the horizontal.

3.51 A jungle veterinarian with a blow-gun loaded with a tranquilizer dart and a sly 1.5-kg monkey are each 25 m above the ground in trees 90 m apart. Just as the hunter shoots horizontally at the monkey, the monkey drops from the tree in a vain attempt to escape being hit. What must the minimum muzzle velocity of the dart have been for the hunter to hit the monkey before it reached the ground?

3.52 A movie stuntwoman drops from a helicopter that is 30.0 m above the ground and moving with a constant velocity whose components are 10.0 m/s upward and 15.0 m/s horizontal and toward the south. You can ignore air resistance. a) Where on the ground (relative to the position of the helicopter when she drops) should the stuntwoman have placed the foam mats that break her fall? b) Draw x-t, y-t, v_x-t, and v_y-t graphs of the motion.

3.53 In fighting forest fires, airplanes work in support of ground crews by dropping water on the fires. A pilot is practicing by dropping a canister of red dye, hoping to hit a target on the ground below. If the plane is flying in a horizontal path 90.0 m above the ground and with a speed of 64.0 m/s (143 mi/h), at what horizontal distance from the target should the pilot release the canister? Ignore air resistance.

3.54 Don't Do This I. A girl throws a water-filled balloon at an angle of 50.0° above the horizontal with a speed of 12.0 m/s. The

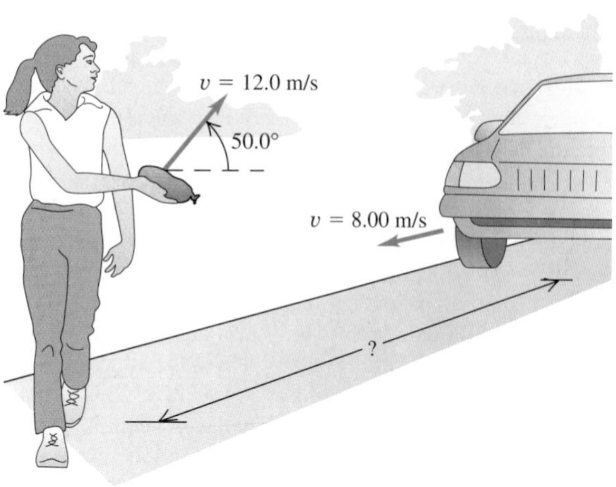

Figure 3.43 Problem 3.54.

horizontal component of the balloon's velocity is directed toward a car that is approaching the girl at a constant speed of 8.00 m/s (Fig. 3.43). If the balloon is to hit the car at the same height at which it leaves her hand, what is the maximum distance the car can be from the girl when the balloon is thrown? You can ignore air resistance.

3.55 The Longest Home Run. According to the *Guinness Book of World Records,* the longest home run ever measured was hit by Roy "Dizzy" Carlyle in a minor league game. The ball traveled 188 m (618 ft) before landing on the ground outside the ballpark. a) Assuming the ball's initial velocity was 45° above the horizontal and ignoring air resistance, what did the initial speed of the ball need to be to produce such a home run if the ball was hit at a point 0.9 m (3.0 ft) above ground level? Assume that the ground was perfectly flat. b) How far would the ball be above a fence 3.0 m (10 ft) in height if the fence was 116 m (380 ft) from home plate?

3.56 Don't Do This II. On the day after graduation, you decide to toss a burning match into the top of a cylindrical wastebasket (diameter D and height $2D$) full of old homework papers. To make this event more sporting, the bottom of the wastebasket is at the same height as the point where the match leaves your hand, and the near side of the wastebasket is a horizontal distance $6D$ from the point where the match leaves your hand. You launch the match at a 45.0° angle above the horizontal. Find the *minimum* and *maximum* values of the launch speed that will make the match enter the top of the wastebasket. You can ignore air resistance and give your answers in terms of g and D.

3.57 Hallway Catch. You are playing catch with a friend in the hallway of your dormitory. The distance from floor to ceiling is D, and you throw the ball with an initial speed $v_0 = \sqrt{6gD}$. What is the maximum horizontal distance (in terms of D) that the ball can travel without bouncing? (Assume that the ball is launched from the floor.)

3.58 A baseball thrown at an angle of 60.0° above the horizontal strikes a building 18.0 m away at a point 8.00 m above the point from which it is thrown. Ignore air resistance. a) Find the magnitude of the initial velocity of the baseball (the velocity with which the baseball is thrown). b) Find the magnitude and direction of the velocity of the baseball just before it strikes the building. c) Draw x-t, y-t, v_x-t, and v_y-t graphs of the motion.

3.59 A projectile is launched with speed v_0 at an angle α_0 above the horizontal. The launch point is a height h above the ground. a) Show that if air resistance is ignored, the horizontal distance that the projectile travels before striking the ground is

$$x = \frac{v_0 \cos \alpha_0}{g}\left(v_0 \sin \alpha_0 + \sqrt{v_0^2 \sin^2 \alpha_0 + 2gh}\right)$$

Verify that if the launch point is at ground level so that $h = 0$, this is equal to the horizontal range R found in Example 3.10. b) For the case where $v_0 = 10$ m/s and $h = 5.0$ m, graph x as a function of launch angle α_0 for values of α_0 from 0° to 90°. Your graph should show that x is zero if $\alpha_0 = 90°$, but x is nonzero if $\alpha_0 = 0$; explain why this is so. c) We saw in Example 3.10 that for a projectile that lands at the same height from which it is launched, the horizontal range is maximum for $\alpha_0 = 45°$. For the case graphed in part (b), is the angle for maximum horizontal distance equal to, less than, or greater than 45°? (This is a general result for the situation where a projectile is launched from a point higher than where it lands.)

3.60 Look Out! A snowball rolls off a barn roof that slopes downward at an angle of 40° (Fig. 3.44). The edge of the roof is 14.0 m above the ground, and the snowball has a speed of 7.00 m/s as it rolls off the roof. Ignore air resistance. a) How far from the edge of the barn does the snowball strike the ground if it doesn't strike anything else while falling? b) Draw x-t, y-t, v_x-t, and v_y-t graphs for the motion in part (a). c) A man 1.9 m tall is standing 4.0 m from the edge of the barn. Will he be hit by the snowball?

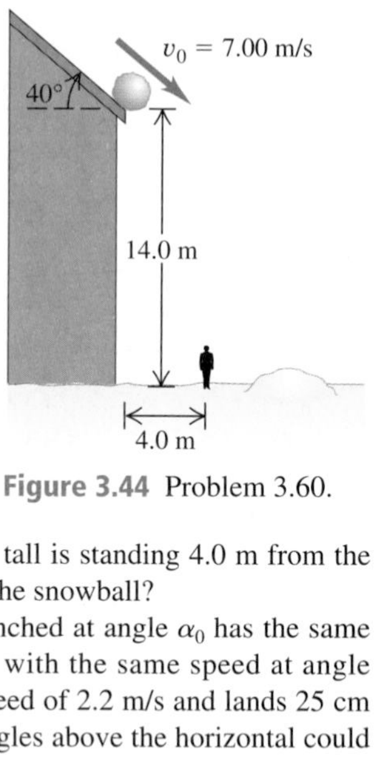

Figure 3.44 Problem 3.60.

3.61 a) Prove that a projectile launched at angle α_0 has the same horizontal range as one launched with the same speed at angle $(90° - \alpha_0)$. b) A frog jumps at a speed of 2.2 m/s and lands 25 cm from its starting point. At which angles above the horizontal could it have jumped?

3.62 On the Flying Trapeze. A new circus act is called the Texas Tumblers. Lovely Mary Belle swings from a trapeze, projects herself at an angle of 53°, and is supposed to be caught by Joe Bob, whose hands are 6.1 m above and 8.2 m horizontally from her launch point (Fig. 3.45). You can ignore air resistance. a) What initial speed v_0 must Mary Belle have just to reach Joe Bob? b) For the initial

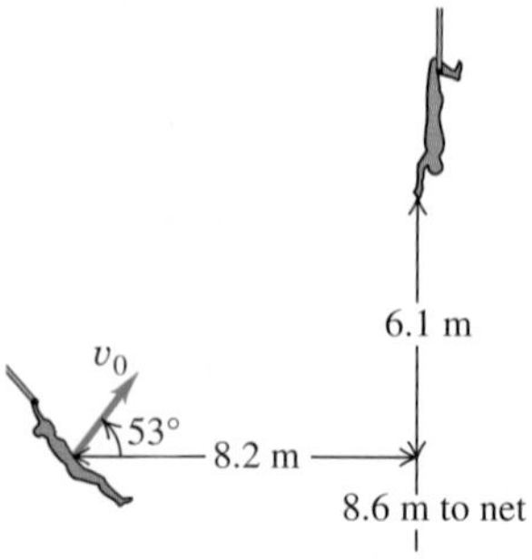

Figure 3.45 Problem 3.62.

speed calculated in part (a), what are the magnitude and direction of her velocity when Mary Belle reaches Joe Bob? c) Assuming that Mary Belle has the initial speed calculated in part (a), draw x-t, y-t, v_x-t, and v_y-t graphs showing the motion of both tumblers. Your graphs should show the motion up until the point where Mary Belle reaches Joe Bob. d) The night of their debut performance, Joe Bob misses her completely as she flies past. How far horizontally does Mary Belle travel, from her initial launch point, before landing in the safety net 8.6 m below her starting point?

3.63 Leaping the River II. A physics professor did daredevil stunts in his spare time. His last stunt was an attempt to jump across a river on a motorcycle (Fig. 3.46). The takeoff ramp was inclined at 53.0°, the river was 40.0 m wide, and the far bank was 15.0 m lower than the top of the ramp. The river itself was 100 m below the ramp. You can ignore air resistance. a) What should his speed have been at the top of the ramp to have just made it to the edge of the far bank? b) If his speed was only half the value found in (a), where did he land?

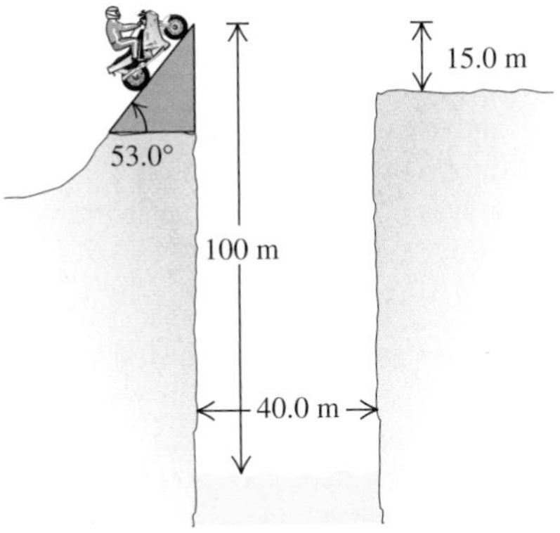

Figure 3.46 Problem 3.63.

3.64 A rock is thrown from the roof of a building with a velocity v_0 at an angle of α_0 from the horizontal. The building has height h. You can ignore air resistance. Calculate the magnitude of the velocity of the rock just before it strikes the ground, and show that this speed is independent of α_0.

3.65 A 5500-kg cart carrying a vertical rocket launcher moves to the right at a constant speed of 30.0 m/s along a horizontal track. It launches a 45.0-kg rocket vertically upward with an initial speed of 40.0 m/s relative to the cart. a) How high will the rocket go? b) Where, relative to the cart, will the rocket land? c) How far does the cart move while the rocket is in the air? d) At what angle, relative to the horizontal, is the rocket traveling just as it leaves the cart, as measured by an observer at rest on the ground? e) Sketch the rocket's trajectory as seen by an observer: i) stationary on the cart; ii) stationary on the ground.

3.66 A 2.7-kg ball is thrown upward with an initial speed of 20.0 m/s from the edge of a 45.0 m high cliff. At the instant the ball is thrown, a woman starts running away from the base of the cliff with a constant speed of 6.00 m/s. The woman runs in a straight line on level ground, and air resistance acting on the ball can be ignored. a) At what angle above the horizontal should the ball be thrown so that the runner will catch it just before it hits the ground, and how far does the woman run before she catches the ball? b) Carefully sketch the ball's trajectory as viewed by: i) a person at rest on the ground; ii) the runner.

3.67 A 76.0-kg boulder is rolling horizontally at the top of a vertical cliff that is 20 m above the surface of a lake, as shown in Fig. 3.47. The top of the vertical face of a dam is located 100 m from the foot of the cliff, with the top of the dam level with the surface of the water in the lake. A level plain is 25 m below the top of the dam. a) What must the minimum speed of the rock be just as it leaves the cliff so it will travel to the plain without striking the dam? b) How far from the foot of the dam does the rock hit the plain?

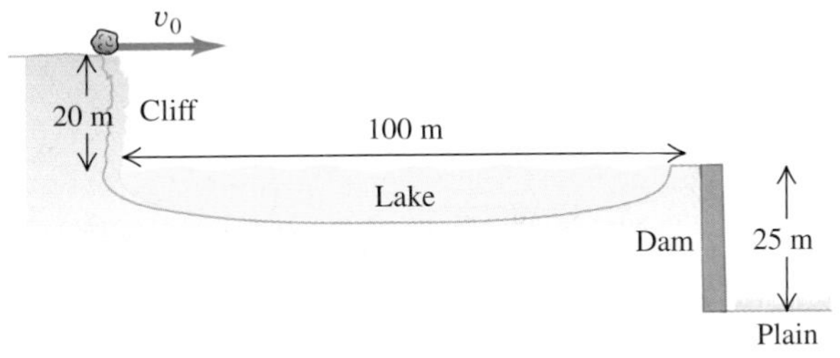

Figure 3.47 Problem 3.67.

3.68 Tossing Your Lunch. Henrietta is going off to her physics class, jogging down the sidewalk at 3.05 m/s. Her husband Bruce suddenly realizes that she left in such a hurry that she forgot her lunch of bagels, so he runs to the window of their apartment, which is 43.9 m above the street level and directly above the sidewalk, to throw them to her. Bruce throws them horizontally 9.00 s after Henrietta has passed below the window, and she catches them on the run. You can ignore air resistance. a) With what initial speed must Bruce throw the bagels so Henrietta can catch them just before they hit the ground? b) Where is Henrietta when she catches the bagels?

3.69 Two tanks are engaged in a training exercise on level ground. The first tank fires a paint-filled training round with a muzzle speed of 250 m/s at 10.0° above the horizontal while advancing toward the second tank with a speed of 15.0 m/s relative to the ground. The second tank is retreating at 35.0 m/s relative to the ground, but is hit by the shell. You can ignore air resistance and assume the shell hits at the same height above ground from which it was fired. Find the distance between the tanks a) when the round was first fired; b) at the time of impact.

3.70 Bang! A student sits atop a platform a distance h above the ground. He throws a large firecracker horizontally with a speed v. However, a wind blowing parallel to the ground gives the firecracker a constant horizontal acceleration with magnitude a. This results in the firecracker reaching the ground directly under the student. Determine the height h in terms of v, a, and g. You can ignore the effect of air resistance on the vertical motion.

3.71 Free Throw. A basketball player is fouled and knocked to the floor during a layup attempt. The player is awarded two free throws. The center of the basket is a horizontal distance of 4.21 m (13.8 ft) from the foul line and is a height of 3.05 m (10.0 ft) above the floor (Fig. 3.48). On the first free-throw attempt, he shoots the ball at an angle of 35° above the horizontal and with a speed of

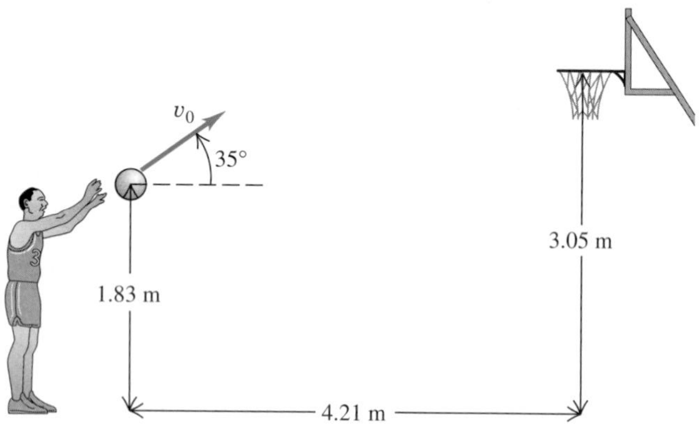

Figure 3.48 Problem 3.71.

$v_0 = 4.88$ m/s (16.0 ft/s). The ball is released 1.83 m (6.0 ft) above the floor. This shot misses badly. You can ignore air resistance. a) What is the maximum height reached by the ball? b) At what distance along the floor from the free-throw line does the ball land? c) For the second throw, the ball goes through the center of the basket. For this second free throw, the player again shoots the ball at 35° above the horizontal and releases it 1.83 m above the floor. What initial speed does the player give the ball on this second attempt? d) For the second throw, what is the maximum height reached by the ball? At this point, how far horizontally is the ball from the basket?

3.72 Romeo tosses a pebble at Juliet's window to wake her. Unfortunately, he throws too large a pebble too fast. Just before crashing through the glass, the pebble is moving horizontally, having traveled a horizontal distance x and a vertical distance y as a projectile. Find the magnitude and direction of the pebble's velocity as it leaves Romeo's hand.

3.73 A rocket is initially at rest on the ground. When its engines fire, the rocket flies off in a straight line at an angle of 53.1° above the horizontal with a constant acceleration of magnitude g. The engines stop at a time T after launch, after which the rocket is in projectile motion. You can ignore air resistance and assume g is independent of altitude. a) Draw the trajectory of the rocket from when its engines first fire until the rocket hits the ground. Indicate the direction of the velocity and acceleration vectors at various points along the trajectory. b) Sketch v_x-t and v_y-t graphs for the motion of the rocket from when its engines first fire until the rocket hits the ground. c) Find the maximum altitude reached by the rocket. Your answer will be in terms of g and T. d) Find the horizontal distance from the launch point to where the rocket hits the ground (that is, the range). Your answer will be in terms of g and T.

3.74 In an action-adventure film, the hero is supposed to throw a grenade from his car, which is going 90.0 km/h, to his enemy's car, which is going 110 km/h. The enemy's car is 15.8 m in front of the hero's when he lets go of the grenade. If the hero throws the grenade so its initial velocity relative to him is at an angle of 45° above the horizontal, what should the magnitude of the initial velocity be? The cars are both traveling in the same direction on a level road. You can ignore air resistance. Find the magnitude of the velocity both relative to the hero and relative to the earth.

3.75 A rock tied to a rope moves in the xy-plane. Its coordinates are given as functions of time by

$$x(t) = R\cos\omega t \qquad y(t) = R\sin\omega t$$

where R and ω are constants. a) Show that the rock's distance from the origin is constant and equal to R, that is, that its path is a circle of radius R. b) Show that at every point the rock's velocity is perpendicular to its position vector. c) Show that the rock's acceleration is always opposite in direction to its position vector and has magnitude $\omega^2 R$. d) Show that the magnitude of the rock's velocity is constant and equal to ωR. e) Combine the results of parts (c) and (d) to show that the rock's acceleration has constant magnitude v^2/R.

3.76 The speed of a particle moving in a plane is equal to the magnitude of its instantaneous velocity, $v = |\vec{v}| = \sqrt{v_x^2 + v_y^2}$. a) Show that the rate of change of the speed is $dv/dt = (v_x a_x + v_y a_y)/\sqrt{v_x^2 + v_y^2}$. b) Use this expression to find the value of dv/dt at time $t = 2.0$ s for the rover in Examples 3.1 through 3.4. Compare your answer to the components of acceleration found in Example 3.4. Explain why your answer is *not* equal to the magnitude of the acceleration found in part (b) of Example 3.2. c) Show that the rate of change of speed can be expressed as $dv/dt = \vec{v}\cdot\vec{a}/v$, and use this result to explain why dv/dt is equal to $a_{\parallel}$, the component of $\vec{a}$ that is parallel to $\vec{v}$.

3.77 A particle moves in the xy-plane. Its coordinates are given as functions of time by

$$x(t) = R(\omega t - \sin\omega t) \qquad y(t) = R(1 - \cos\omega t)$$

where R and ω are constants. a) Sketch the trajectory of the particle. (This is the trajectory of a point on the rim of a wheel that is rolling at a constant speed on a horizontal surface. The curve traced out by such a point as it moves through space is called a *cycloid.*) b) Determine the velocity components and the acceleration components of the particle at any time t. c) At which times is the particle momentarily at rest? What are the coordinates of the particle at these times? What are the magnitude and direction of the acceleration at these times? d) Does the magnitude of the acceleration depend on time? Compare to uniform circular motion.

3.78 You are flying in a light airplane spotting traffic for a radio station. Your flight carries you due east above a highway. Landmarks below tell you that your speed is 50 m/s relative to the ground and your air speed indicator also reads 50 m/s. However, the nose of your airplane is pointed somewhat south of east and the station's weather person tells you that a 20 m/s wind is blowing. In which direction is the wind blowing?

3.79 The Carrier Pigeon Problem. Larry is driving east at 40 km/h. His twin brother Harry is driving west at 30 km/h, toward Larry in an identical car on the same straight road. When they are 42 km apart, Larry sends out a carrier pigeon, which flies at a constant speed of 50 km/h. (All speeds are relative to the earth.) The pigeon flies to Harry, becomes confused and immediately returns to Larry, becomes more confused, and immediately flies back to Harry. This continues until the twins meet, at which time the dazed pigeon drops to the ground in exhaustion. Ignoring turnaround time, how far did the pigeon fly?

3.80 Raindrops. When a train's velocity is 12.0 m/s eastward, raindrops that are falling vertically with respect to the earth make traces that are inclined 30.0° to the vertical on the windows of the train. a) What is the horizontal component of a drop's velocity with respect to the earth? With respect to the train? b) What is the magnitude of the velocity of the raindrop with respect to the earth? With respect to the train?

3.81 An airplane pilot sets a compass course due west and maintains an airspeed of 220 km/h. After flying for 0.500 h, she finds herself over a town 120 km west and 20 km south of her starting point. a) Find the wind velocity (magnitude and direction). b) If the wind velocity is 40 km/h due south, in what direction should the pilot set her course to travel due west? Use the same airspeed of 220 km/h.

3.82 In an air race, an airplane flies from a point directly over Metropolis to a point directly over Gotham City, and then turns around and flies back to the starting point. The airspeed (that is, speed of the airplane relative to the air) is constant throughout the flight and equal to v. Gotham City lies a distance D due east of Metropolis. a) If there is no wind, how much time is required for the round trip? How much time is required for the round trip if a steady wind of speed w is blowing b) toward the east? c) toward the south? d) If $D = 3.00 \times 10^2$ km, $v = 4.00 \times 10^2$ km/h, and $w = 1.00 \times 10^2$ km/h, calculate the round-trip flight times for parts (a), (b), and (c). Which case gives the slowest round-trip?

3.83 An elevator is moving upward at a constant speed of 2.50 m/s. A bolt in the elevator ceiling 3.00 m above the elevator floor works loose and falls. a) How long does it take for the bolt to fall to the elevator floor? What is the speed of the bolt just as it hits the elevator floor b) according to an observer in the elevator? c) according to an observer standing on one of the floor landings of the building? d) According to the observer in part (c), what distance did the bolt travel between the ceiling and the floor of the elevator?

3.84 City A lies directly west of city B. When there is no wind, an airliner makes the 5310 km round trip flight between them in 6.60 h of flying time while traveling at the same speed in both directions. When a strong, steady wind is blowing from west to east and the airliner has the same air speed as before, the round trip takes 6.70 h of flight time. How fast is the wind blowing?

3.85 In a World Cup soccer match, Juan is running due north toward the goal with a speed of 8.00 m/s relative to the ground. A teammate passes the ball to him. The ball has a speed of 12.0 m/s and is moving in a direction of 37.0° east of north, relative to the ground. What are the magnitude and direction of the ball's velocity relative to Juan?

Challenge Problems

3.86 A man is riding on a flatcar traveling at a constant speed of 9.10 m/s (Fig. 3.49). He wishes to throw a ball through a stationary hoop 4.90 m above the height of his hands in such a manner that the ball will move horizontally as it passes through the hoop. He throws the ball with a speed of 10.8 m/s with respect to himself. a) What must the vertical component of the initial velocity of the ball be? b) How many seconds after he releases the ball will it pass through the hoop? c) At what horizontal distance in front of the hoop must he release the ball? d) When the ball leaves the man's hands, what is the direction of its velocity relative to the frame of reference of the flatcar? Relative to the frame of reference of an observer standing on the ground?

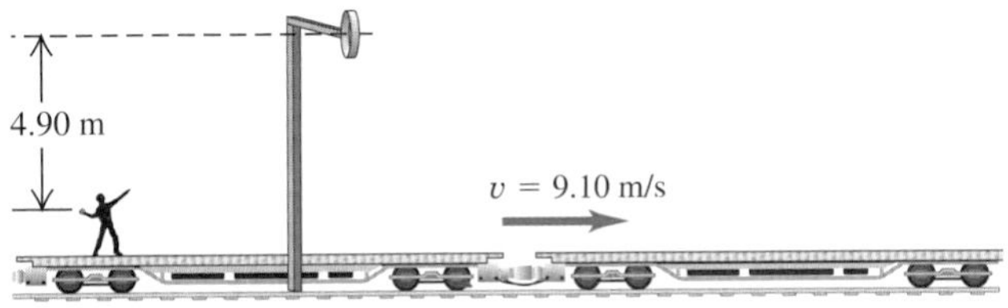

Figure 3.49 Challenge Problem 3.86.

3.87 A shotgun fires a large number of pellets upward, with some pellets traveling very nearly vertically and others as much as 1.0° from the vertical. Assume that the initial speed of the pellets is uniformly 150 m/s, and ignore air resistance. a) Within what radius from the point of firing will the pellets land? b) If there are 1000 pellets, and they fall in a uniform distribution over a circle with the radius calculated in part (a), what is the probability that at least one pellet will fall on the head of the person who fires the shotgun? Assume that his head has a radius of 10 cm. c) Air resistance, in fact, has several effects. It slows down the rising pellets, decreases their horizontal component of velocity, and limits the speed with which they fall. Which of these effects will tend to make the radius greater than calculated in part (a), and which will tend to make it less? What do you think the overall effect of air resistance will be? (The effect of air resistance on a velocity component increases as the magnitude of the component increases.)

3.88 A projectile is thrown from a point P. It moves in such a way that its distance from P is always increasing. Find the maximum angle above the horizontal with which the projectile could have been thrown. You can ignore air resistance.

3.89 Projectile Motion on an Incline I. A baseball is given an initial velocity with magnitude v_0 at an angle ϕ above the surface of an incline, which is in turn inclined at an angle θ above the horizontal (Fig. 3.50). a) Calculate the distance, measured along the incline, from the launch point to where the baseball strikes the incline. Your answer will be in terms of v_0, g, θ, and ϕ. b) What angle ϕ gives the maximum range, measured along the incline? (*Note:* You might be interested in the three different methods of solution presented by I. R. Lapidus in *Amer. Jour. of Phys.*, Vol. 51 (1983), pp. 806 and 847. See also H. A. Buckmaster in *Amer. Jour. of Phys.*, Vol. 53 (1985), pp. 638–641, for a thorough study of this and some similar problems.)

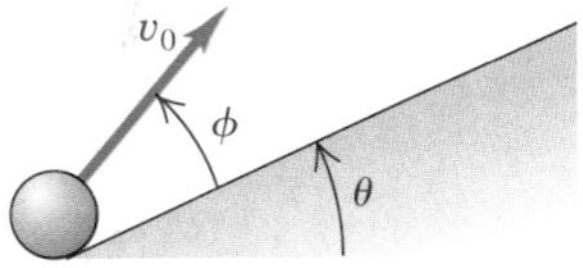

Figure 3.50 Challenge Problem 3.89.

3.90 Projectile Motion on an Incline II. Refer to Challenge Problem 3.89. a) An archer on ground that has a constant upward slope of 30.0° aims at a target 60.0 m farther up the incline. The arrow in the bow and the bull's-eye at the center of the target are each 1.50 m above the ground. The initial velocity of the arrow just after it leaves the bow has magnitude 32.0 m/s. At what angle above the *horizontal* should the archer aim to hit the bull's-eye? If there are two such angles, calculate the smaller of the two. You might have to solve the equation for the angle by iteration, that is, by trial and error. How does the angle compare to that required when the ground is level, with zero slope? b) Repeat the above for ground that has a constant *downward* slope of 30.0°.

3.91 For no apparent reason, a poodle is running at a constant speed of $v = 5.00$ m/s in a circle with radius $R = 2.50$ m. Let $\vec{v}_1$ be the velocity vector at time t_1, and let $\vec{v}_2$ be the velocity vector at time t_2. Consider $\Delta\vec{v} = \vec{v}_2 - \vec{v}_1$ and $\Delta t = t_2 - t_1$. Recall that $\vec{a}_{av} = \Delta\vec{v}/\Delta t$. For $\Delta t = 0.5$ s, 0.1 s, and 0.05 s, calculate the magnitude (to four significant figures) and direction (relative to $\vec{v}_1$) of the average acceleration $\vec{a}_{av}$. Compare your results to the general expression for the instantaneous acceleration $\vec{a}$ for uniform circular motion that is derived in the text.

3.92 A rocket designed to place small payloads into orbit is carried to an altitude of 12.0 km above sea level by a converted airliner. When the airliner is flying in a straight line at a constant speed of 850 km/h, the rocket is dropped. After the drop, the airliner maintains the same altitude and speed and continues to fly in a straight line. The rocket falls for a brief time, after which its rocket motor turns on. Once its rocket motor is on, the combined effects of thrust and gravity give the rocket a constant acceleration of magnitude $3.00g$ directed at an angle of 30.0° above the horizontal. For reasons of safety, the rocket should be at least 1.00 km in front of the airliner when it climbs through the airliner's altitude. Your job is to determine the minimum time that the rocket must fall before its engine starts. You can ignore air resistance. Your answer should include i) a diagram showing the flight paths of both the rocket and the airliner, labeled at several points with vectors for their velocities and accelerations; ii) an x-t graph showing the motions of both the rocket and the airliner; and iii) a y-t graph showing the motions of both the rocket and the airliner. In the diagram and the graphs, indicate when the rocket is dropped, when the rocket motor turns on, and when the rocket climbs through the altitude of the airliner.

3.93 Two students are canoeing on a river. While heading upstream they accidentally drop an empty bottle overboard. They then continue paddling for 60 minutes, reaching a point 2.0 km further upstream. At this point they realize that the bottle is missing and, driven by ecological awareness, they turn around and head downstream. They catch up with and retrieve the bottle (which has been moving along with the current) 5.0 km downstream from the turn-around point. a) Assuming a constant paddling effort throughout, how fast is the river flowing? b) What would be the canoe speed in a still lake for the same paddling effort?

NEWTON'S LAWS OF MOTION

4

A man pulling his son's sled is illustrating Newton's third law—the law of action and reaction. To propel himself forward, the man pushes backward on the ground with his feet (the "action"), knowing that the ground will respond by pushing forward on him with an equal amount of force (the "reaction"). Similarly, the force that he exerts on the sled to pull it forward ("action") results in him feeling an equally strong force pulling him back ("reaction").

? **If the man pulls forward on the sled, and the sled pulls backward on the man with an equal amount of force, how can the man and sled move at all?**

We've seen in the last two chapters how to describe motion in one, two, or three dimensions. But what are the underlying *causes* of motion? For example, how can a tugboat push a cruise ship that's much heavier than the tug? Why does it take a long distance to stop the ship once it is in motion? Why is it harder to control a car on wet ice than on dry concrete? The answers to these and similar questions take us into the subject of **dynamics,** the relationship of motion to the forces that cause it. In the two preceding chapters we studied *kinematics,* the language for *describing* motion. Now we are ready to think about what makes bodies move the way they do.

In this chapter we will use the kinematic quantities displacement, velocity, and acceleration along with two new concepts, *force* and *mass,* to analyze the principles of dynamics. These principles can be wrapped up in a neat package of three statements called **Newton's laws of motion.** The first law states that when the net force on a body is zero, its motion doesn't change. The second law relates force to acceleration when the net force is *not* zero. The third law is a relation between the forces that two interacting bodies exert on each other.

Newton's laws are not the product of mathematical derivations, but rather a synthesis of what physicists have learned from a multitude of *experiments* about how objects move. These laws are truly fundamental, for they cannot be deduced or proved from other principles. The overarching importance of Newton's laws is that they make it possible to understand most familiar kinds of motion; they are the foundation of **classical mechanics** (also called **Newtonian mechanics**). Newton's laws are not universal, however; they require modification at very high speeds (near the speed of light) and for very small sizes (such as within the atom).

The laws of motion were clearly stated for the first time by Sir Isaac Newton (1642–1727), who published them in 1687 in his *Philosophiae Naturalis Principia Mathematica* ("Mathematical Principles of Natural Philosophy"). Many other scientists before Newton contributed to the foundations of mechanics, including Copernicus, Brahe, Kepler, and especially Galileo Galilei (1564–1642), who died the same year Newton was born. Indeed, Newton himself said, "If I have been able to see a little farther than other men, it is because I have stood on the shoulders of giants." Now it's your turn to stand on the shoulders of Newton and use his laws to understand how the physical world works.

Newton's laws are very simple to state, yet many students find these laws difficult to grasp and to work with. The reason is that before studying physics, you've spent years walking, throwing balls, pushing boxes, and doing dozens of things that involve motion. Along the way, you've developed a set of "common sense" ideas about motion and its causes. But many of these "common sense" ideas don't stand up to logical analysis. A big part of the job of this chapter—and of the rest of our study of physics—is helping you to recognize how "common sense" ideas can sometimes lead you astray, and how to adjust your understanding of the physical world to make it consistent with what experiments tell us.

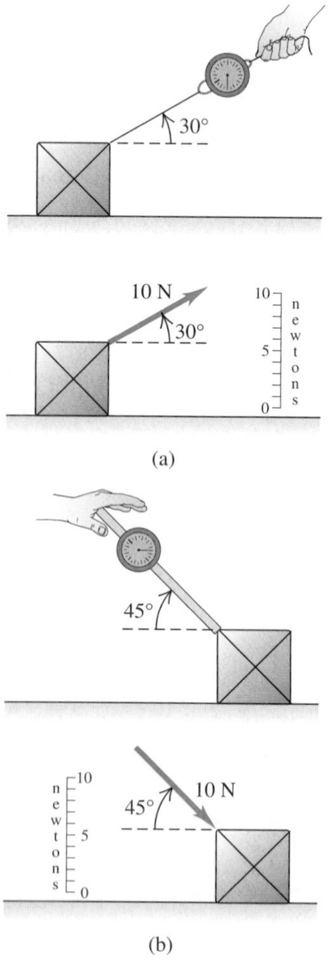

4.1 A force may be exerted on the box by either (a) pulling it or (b) pushing it. A force diagram illustrates each case.

4.1 | Force and Interactions

In everyday language, a **force** is a push or pull. The concept of force gives us a quantitative description of the interaction between two bodies or between a body and its environment. When you push on a car that is stuck in the snow, you exert a force on it. A locomotive exerts a force on the train it is pulling or pushing, a steel cable exerts a force on the beam it is hoisting at a construction site, and so on.

When a force involves direct contact between two bodies, we call it a **contact force.** Contact forces include the pushes or pulls you exert with your hand, the force of a rope pulling on a block to which it is tied, and the friction force that the ground exerts on a ball player sliding into home. There are also forces, called **long-range forces,** that act even when the bodies are separated by empty space. You've experienced long-range forces if you've ever played with a pair of magnets. Gravity, too, is a long-range force; the sun exerts a gravitational pull on the earth, even over a distance of 150 million kilometers, that keeps the earth in orbit. The force of gravitational attraction that the earth exerts on your body is called your **weight.**

Force is a vector quantity; you can push or pull a body in different directions. Thus, to describe a force, we need to describe the *direction* in which it acts as well as its *magnitude,* the quantity that describes "how much" or "how hard" the force pushes or pulls. The SI unit of the magnitude of force is the *newton,* abbreviated N. (We'll give a precise definition of the newton in Section 4.3.) Table 4.1 lists some typical force magnitudes.

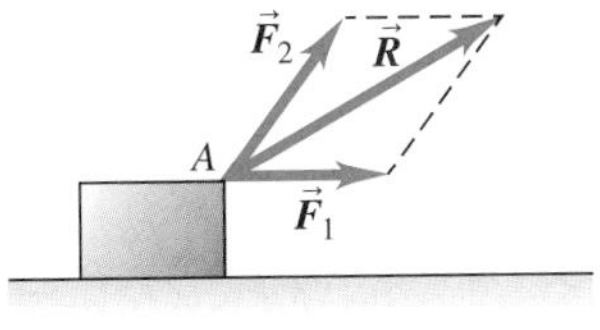

4.2 Two forces $\vec{F}_1$ and $\vec{F}_2$ acting simultaneously on a body have the same effect as a single force $\vec{R}$ equal to the vector sum (resultant) of $\vec{F}_1$ and $\vec{F}_2$.

Table 4.1 Typical Force Magnitudes

Sun's gravitational force on the earth	3.5×10^{22} N
Thrust of a space shuttle during launch	3.1×10^{7} N
Weight of a large blue whale	1.9×10^{6} N
Maximum pulling force of a locomotive	8.9×10^{5} N
Weight of a 250-lb linebacker	1.1×10^{3} N
Weight of a medium apple	1 N
Weight of smallest insect eggs	2×10^{-6} N
Electric attraction between the proton and the electron in a hydrogen atom	8.2×10^{-8} N
Weight of a very small bacterium	1×10^{-18} N
Weight of a hydrogen atom	1.6×10^{-26} N
Weight of an electron	8.9×10^{-30} N
Gravitational attraction between the proton and the electron in a hydrogen atom	3.6×10^{-47} N

A common instrument for measuring forces is the *spring balance.* It consists of a coil spring, enclosed in a case for protection, with a pointer attached to one end. When forces are applied to the ends of the spring, it stretches; the amount of stretch depends on the force. We can make a scale for the pointer and calibrate it by using a number of identical bodies with weights of exactly 1 N each. When two, three, or more of these are suspended simultaneously from the balance, the total force stretching the spring is 2 N, 3 N, and so on, and we can label the corresponding positions of the pointer 2 N, 3 N, and so on. Then we can use this instrument to measure the magnitude of an unknown force. We can also make a similar instrument that measures pushes instead of pulls.

Suppose we slide a box along the floor, applying a force to it by pulling it with a string or pushing it with a stick (Fig. 4.1). In each case we draw a vector to represent the force applied. The labels indicate the magnitude and direction of the force, and the length of the arrow (drawn to some scale, such as 1 cm = 10 N) also shows the magnitude.

When *two* forces $\vec{F}_1$ and $\vec{F}_2$ act at the same time at a point *A* of a body (Fig. 4.2), experiment shows that the effect on the body's motion is the same as the effect of a single force $\vec{R}$ equal to the *vector sum* of the original forces: $\vec{R} = \vec{F}_1 + \vec{F}_2$. More generally, the effect of any number of forces applied at a point on a body is the same as the effect of a single force equal to the vector sum of the forces. This important principle goes by the name **superposition of forces.**

The experimental discovery that forces combine according to vector addition is of the utmost importance. We will use this fact many times throughout our study of physics. It allows us to replace a force by its component vectors, as we did with displacements in Section 1.8. For example, in Fig. 4.3a, force $\vec{F}$ acts on a body at point *O*. The component vectors of $\vec{F}$ in the directions *Ox* and *Oy* are $\vec{F}_x$ and $\vec{F}_y$. When $\vec{F}_x$ and $\vec{F}_y$ are applied simultaneously, as in Fig. 4.3b, the effect is exactly the same as the effect of the original force $\vec{F}$. *Any force can be replaced by its component vectors, acting at the same point.*

It's frequently more convenient to describe a force $\vec{F}$ in terms of its *x*- and *y*-components F_x and F_y rather than by its component vectors (recall from

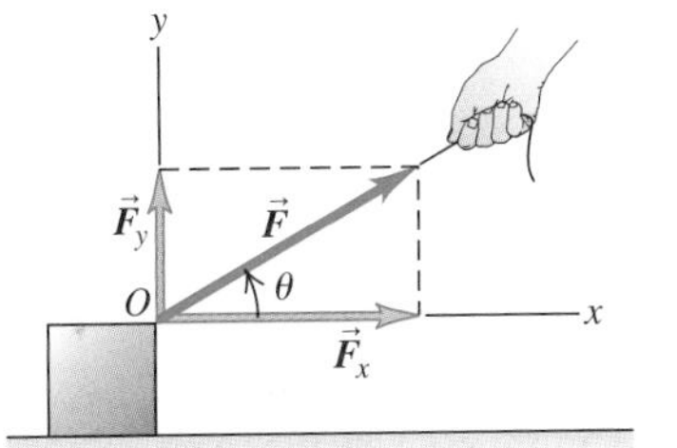

Component vectors: $\vec{F}_x$ and $\vec{F}_y$
Components: $F_x = F\cos\theta$ and $F_y = F\sin\theta$

(a)

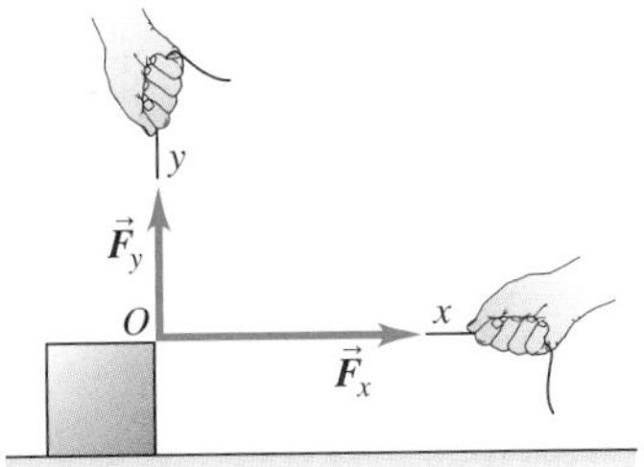

Component vectors $\vec{F}_x$ and $\vec{F}_y$ together have the same effect as original force $\vec{F}$.

(b)

4.3 The force $\vec{F}$, which acts at an angle θ from the *x*-axis, may be replaced by its rectangular component vectors $\vec{F}_x$ and $\vec{F}_y$.

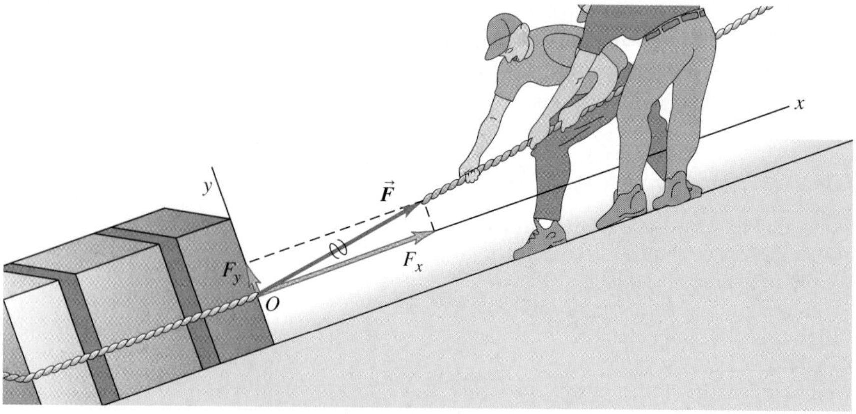

4.4 F_x and F_y are the components of $\vec{F}$ parallel and perpendicular to the sloping surface of the inclined plane.

Section 1.8 that *component vectors* are vectors, but *components* are just numbers). For the case shown in Fig. 4.3, both F_x and F_y are positive; for other orientations of the force $\vec{F}$, either F_x or F_y can be negative or zero.

There is no law that says our coordinate axes have to be vertical and horizontal. Figure 4.4 shows a crate being pulled up a ramp by a force $\vec{F}$, represented by its components F_x and F_y parallel and perpendicular to the sloping surface of the ramp.

CAUTION In Fig. 4.4 we draw a wiggly line through the force vector $\vec{F}$ to show that we have replaced it by its x- and y-components. Otherwise, the diagram would include the same force twice. We will draw such a wiggly line in any force diagram where a force is replaced by its components. Look for this wiggly line in other figures in this and subsequent chapters.

We will often need to find the vector sum (resultant) of *all* the forces acting on a body. We will call this the **net force** acting on the body. We will use the Greek letter Σ (capital "sigma," equivalent to the Roman S) as a shorthand notation for a sum. If the forces are labeled $\vec{F}_1$, $\vec{F}_2$, $\vec{F}_3$, and so on, we abbreviate the sum as

$$\vec{R} = \vec{F}_1 + \vec{F}_2 + \vec{F}_3 + \cdots = \sum \vec{F} \qquad (4.1)$$

where $\sum \vec{F}$ is read as "the vector sum of the forces" or "the net force." The component version of Eq. (4.1) is the pair of component equations

$$R_x = \sum F_x \qquad R_y = \sum F_y \qquad (4.2)$$

where $\sum F_x$ is the sum of the x-components, and so on. Each component may be positive or negative, so be careful with signs when you evaluate the sums in Eq. (4.2).

Once we have R_x and R_y, we can find the magnitude and direction of the net force $\vec{R} = \sum \vec{F}$ acting on the body. The magnitude is

$$R = \sqrt{R_x^2 + R_y^2}$$

and the angle θ between $\vec{R}$ and the $+x$-axis can be found from the relation $\tan\theta = R_y/R_x$. The components R_x and R_y may be positive, negative, or zero, and the angle θ may be in any of the four quadrants.

In three-dimensional problems, forces may also have z-components; then we add the equation $R_z = \sum F_z$ to Eq. (4.2). The magnitude of the net force is then

$$R = \sqrt{R_x^2 + R_y^2 + R_z^2}$$

Example 4.1 Superposition of forces

Three professional wrestlers are fighting over the same champion's belt. As viewed from above, they apply the three horizontal forces to the belt that are shown in Fig. 4.5a, where the belt is located at the origin. The magnitudes of the three forces are $F_1 = 250$ N, $F_2 = 50$ N, and $F_3 = 120$ N. Find the x- and y-components of the net force on the belt, and find the magnitude and direction of the net force.

SOLUTION

IDENTIFY AND SET UP: This example is just a problem in vector addition. We are asked to find the components of the net force, so we'll attack this problem using the component method. Our target variables are the magnitude and direction of the net force $\vec{R}$ as well as its x- and y-components; we'll find these using Eq. (4.2).

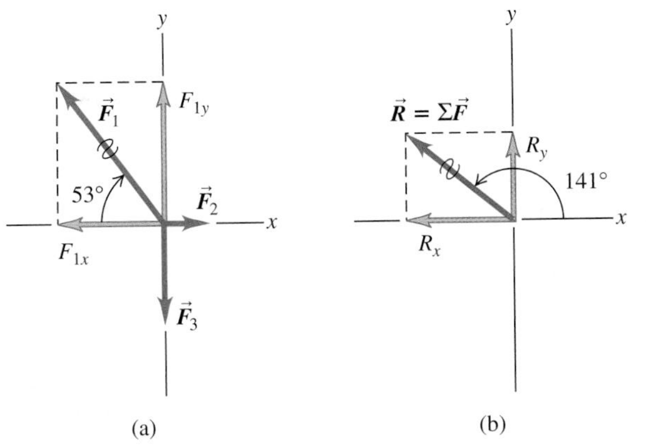

4.5 (a) Three forces acting on a belt. The x- and y-components of $\vec{F}_1$ are shown; $\vec{F}_2$ has zero y-component and $\vec{F}_3$ has zero x-component. (b) The net force $\vec{R}$ and its components.

EXECUTE: From Fig. 4.5a, the angles between the three forces $\vec{F}_1$, $\vec{F}_2$, and $\vec{F}_3$ and the $+x$-axis are $\theta_1 = 180° - 53° = 127°$, $\theta_2 = 0°$, and $\theta_3 = 270°$. The x- and y-components of the three forces are

$$F_{1x} = (250\text{ N})\cos 127° = -150\text{ N}$$
$$F_{1y} = (250\text{ N})\sin 127° = 200\text{ N}$$
$$F_{2x} = (50\text{ N})\cos 0° = 50\text{ N}$$
$$F_{2y} = (50\text{ N})\sin 0° = 0\text{ N}$$
$$F_{3x} = (120\text{ N})\cos 270° = 0\text{ N}$$
$$F_{3y} = (120\text{ N})\sin 270° = -120\text{ N}$$

From Eq. (4.2) the net force $\vec{R} = \sum\vec{F}$ has components

$$R_x = F_{1x} + F_{2x} + F_{3x} = (-150\text{ N}) + 50\text{ N} + 0\text{ N} = -100\text{ N}$$
$$R_y = F_{1y} + F_{2y} + F_{3y} = 200\text{ N} + 0\text{ N} + (-120\text{ N}) = 80\text{ N}$$

The net force has a negative x-component and a positive y-component, so it points to the left and toward the top of the page in Fig. 4.5b (that is, in the second quadrant).

The magnitude of the net force $\vec{R} = \sum\vec{F}$ is

$$R = \sqrt{R_x^2 + R_y^2} = \sqrt{(-100\text{ N})^2 + (80\text{ N})^2} = 128\text{ N}$$

To find the angle between the net force and the $+x$-axis, we use the relation $\tan\theta = R_y/R_x$, or

$$\theta = \arctan\frac{R_y}{R_x} = \arctan\left(\frac{80\text{ N}}{-100\text{ N}}\right) = \arctan(-0.80)$$

The two possible solutions are $\theta = -39°$ or $\theta = -39° + 180° = 141°$. Since the net force lies in the second quadrant, as mentioned earlier, the correct answer is 141° (see Fig. 4.5b).

EVALUATE: In this situation the net force is *not* zero, and you can see intuitively that wrestler 1 (who exerts the largest force, $\vec{F}_1$, on the belt) is likely to walk away with the belt at the end of the struggle. In Section 4.2 we will explore in detail what happens in situations in which the net force *is* zero.

Test Your Understanding

Suppose that in Figure 4.5a the forces $\vec{F}_1$ and $\vec{F}_2$ remain unchanged, but the magnitude and direction of $\vec{F}_3$ are changed so that the net force on the belt is zero. What are the x- and y-components of the new force $\vec{F}_3$? What are its magnitude and direction?

4.2 | Newton's First Law

We have discussed some of the properties of forces, but so far have said nothing about how forces affect motion. To begin with, let's consider what happens when the net force on a body is *zero*. You would almost certainly agree that if a body is at rest, and if no net force acts on it (that is, no net push or pull), that body will remain at rest. But what if there is zero net force acting on a body in *motion?*

To see what happens in this case, suppose you slide a hockey puck along a horizontal table top, applying a horizontal force to it with your hand (Fig. 4.6a). After you stop pushing, the puck *does not* continue to move indefinitely; it slows down and stops. To keep it moving, you have to keep pushing (that is, applying a force). You might come to the "common sense" conclusions that bodies in motion naturally come to rest and that a force is required to sustain motion.

But now imagine pushing the puck across the smooth surface of a freshly waxed floor (Fig. 4.6b). After you quit pushing, the puck will slide a lot farther before it stops. Put it on an air-hockey table, where it floats on a thin cushion of air, and it moves still farther (Fig. 4.6c). In each case, what slows the puck down is *friction,* an interaction between the lower surface of the puck and the surface on which it slides. Each surface exerts a frictional force on the puck that resists the puck's motion; the difference between the three cases is the magnitude of the frictional force. The slippery waxed floor exerts less friction than the table top, so the puck travels farther. The gas molecules of the air-hockey table exert the least friction of all. If we could eliminate friction completely, the puck would never slow down, and we would need no force at all to keep the puck moving once it had been started. Thus the "common sense" idea that a force is required to sustain motion is *incorrect.*

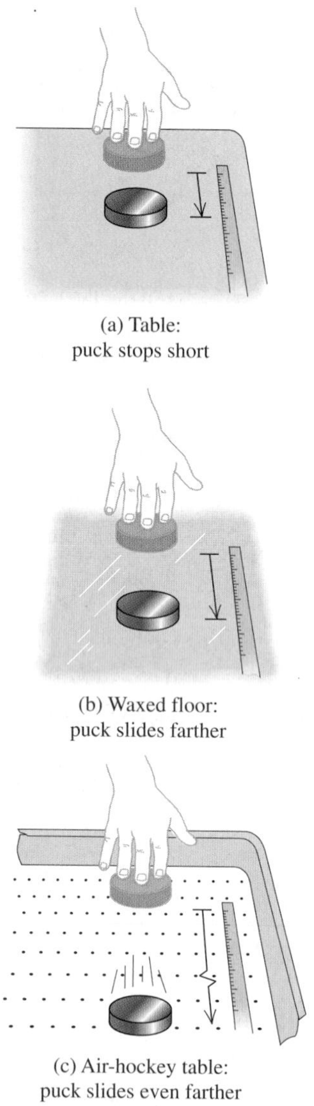

4.6 A hockey puck is given an initial velocity. (a) It stops in a short distance on a table top. (b) A freshly waxed surface decreases the friction force, and the puck slides farther. (c) On an air-hockey table the friction force is practically zero, so the puck continues with almost constant velocity.

Experiments like the ones we've just described show that when no net force acts on a body, the body either remains at rest or moves with constant velocity in a straight line. Once a body has been set in motion, no net force is needed to keep it moving. In other words:

A body acted on by no net force moves with constant velocity (which may be zero) and zero acceleration.

This is **Newton's first law of motion.**

The tendency of a body to keep moving once it is set in motion results from a property called **inertia.** You use this when you try to get ketchup out of a bottle by shaking it. First you start the bottle (and the ketchup inside) moving forward; when you jerk the bottle back, the ketchup tends to keep moving forward and, you hope, ends up on your burger. The tendency of a body at rest to remain at rest is also due to inertia. You may have seen a tablecloth yanked out from under the china without breaking anything. The force on the china isn't great enough to make it move appreciably during the short time it takes to pull the tablecloth away.

It's important to note that the *net* force is what matters in Newton's first law. For example, a physics book at rest on a horizontal table top has two forces acting on it: the downward force of the earth's gravitational attraction (a long-range force that acts even if the table top is elevated above the ground) and an upward supporting force exerted by the table top (a contact force). The upward push of the surface is just as great as the downward pull of gravity, so the *net* force acting on the book (that is, the vector sum of the two forces) is zero. In agreement with Newton's first law, if the book is at rest on the table top, it remains at rest.

The same principle applies to a hockey puck sliding on a horizontal, frictionless surface: The vector sum of the upward push of the surface and the downward pull of gravity is zero. Once the puck is in motion, it continues to move with constant velocity because the *net* force acting on it is zero. (The upward supporting force of the surface is called a **normal force** because it is *normal,* or perpendicular, to the surface of contact. We'll discuss the normal force in more detail in Chapter 5.)

Here's another example. Suppose a hockey puck rests on a horizontal surface with negligible friction, such as an air-hockey table or a slab of wet ice. If the puck is initially at rest and a single horizontal force $\vec{F}_1$ acts on it (Fig. 4.7a), the puck starts to move. If the puck is in motion to begin with, the force changes its speed, its direction, or both, depending on the direction of the force. In this case the net force is equal to $\vec{F}_1$, which is *not* zero. (There are also two vertical forces: the earth's gravitational attraction and the upward contact force exerted by the surface. But as we mentioned earlier, these two forces cancel.)

Now suppose we apply a second force $\vec{F}_2$ (Fig. 4.7b), equal in magnitude to $\vec{F}_1$ but opposite in direction. The two forces are negatives of each other, $\vec{F}_2 = -\vec{F}_1$, and their vector sum is zero:

$$\sum \vec{F} = \vec{F}_1 + \vec{F}_2 = \vec{F}_1 + (-\vec{F}_1) = \mathbf{0}$$

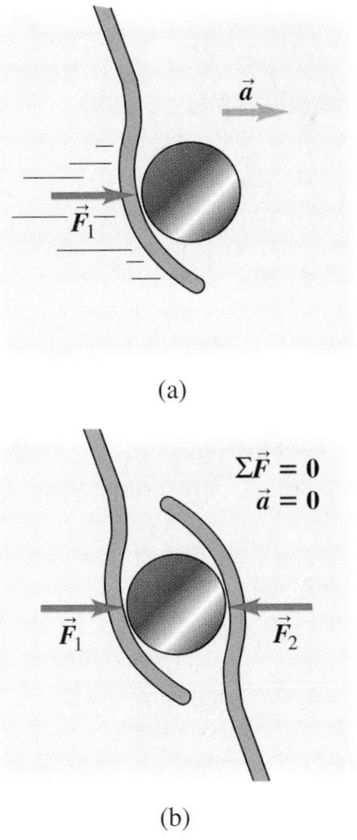

4.7 (a) A hockey puck accelerates in the direction of a net applied force $\vec{F}_1$. (b) When the net force is zero, the acceleration is zero, and the puck is in equilibrium.

Again, we find that if the body is at rest at the start, it remains at rest; if it is initially moving, it continues to move in the same direction with constant speed. These results show that in Newton's first law, *zero net force is equivalent to no force at all.* This is just the principle of superposition of forces that we saw in Section 4.1.

When a body is acted on by no forces, or by several forces such that their vector sum (resultant) is zero, we say that the body is in **equilibrium.** In equilibrium, a body is either at rest or moving in a straight line with constant velocity. For a body in equilibrium, the net force is zero:

$$\sum \vec{F} = \mathbf{0} \quad \text{(body in equilibrium)} \tag{4.3}$$

For this to be true, each component of the net force must be zero, so

$$\sum F_x = 0 \qquad \sum F_y = 0 \qquad \text{(body in equilibrium)} \tag{4.4}$$

When Eqs. (4.4) are satisfied, the body is in equilibrium.

We are assuming that the body can be represented adequately as a point particle. When the body has finite size, we also have to consider *where* on the body the forces are applied. We will return to this point in Chapter 11.

Conceptual Example 4.2 Zero net force means constant velocity

In the classic 1950 science fiction film *Rocketship X-M,* a spaceship is moving in the vacuum of outer space, far from any planet, when its engine dies. As a result, the spaceship slows down and stops. What does Newton's first law say about this event?

SOLUTION

In this situation there are no forces acting on the spaceship, so according to Newton's first law, it will *not* stop. It continues to move in a straight line with constant speed. Some science fiction movies have made use of very accurate science; this was not one of them.

Conceptual Example 4.3 Constant velocity means zero net force

You are driving a Porsche 911 Carrera on a straight testing track at a constant speed of 150 km/h. You pass a 1971 Volkswagen Beetle doing a constant 75 km/h. For which car is the net force greater?

SOLUTION

The key word in this question is "net." Both cars are in equilibrium because their velocities are both constant; therefore the *net* force on each car is *zero*.

This conclusion seems to contradict the "common sense" idea that the faster car must have a greater force pushing it. It's true that there is a forward force on both cars, and it's true that the forward force on your Porsche is much greater than that on the Volkswagen (thanks to your Porsche's high-power engine). But there is also a *backward* force acting on each car due to friction with the road and air resistance. The only reason these cars need their engines running at all is to counteract this backward force so that the vector sum of the forward and backward forces will be zero and the car will travel with constant velocity. The backward force on your Porsche is greater because of its greater speed, which is why your engine has to be more powerful than the Volkswagen's.

Inertial Frames of Reference

In discussing relative velocity in Section 3.5, we introduced the concept of *frame of reference.* This concept is central to Newton's laws of motion. Suppose you are in an airplane that is accelerating down the runway during takeoff. If you could stand in the aisle on roller skates, you would start moving *backward* relative to the plane as the pilot opens the throttle. If instead the airplane was landing, you would start moving forward down the aisle as the airplane slowed to a stop. In either case, it looks as though Newton's first law is not obeyed; there is no net force acting on you, yet your velocity changes. What's wrong?

The point is that the airplane is accelerating with respect to the earth and is *not* a suitable frame of reference for Newton's first law. This law is valid in some frames of reference, and is not valid in others. A frame of reference in which Newton's first law *is* valid is called an **inertial frame of reference.** The earth is at least approximately an inertial frame of reference, but the airplane is not. (The earth is not a completely inertial frame, owing to the acceleration associated with its rotation and its motion around the sun. These effects are quite small, however; see Exercises 3.29 and 3.32.) Because Newton's first law is used to define what we mean by an inertial frame of reference, it is sometimes called the *law of inertia.*

Figure 4.8 shows how we can use Newton's first law to understand what you experience when riding in a vehicle that's accelerating. In Fig. 4.8a a vehicle is initially at rest, then begins to accelerate to the right. A passenger on roller skates has virtually no net force acting on her, since the skates nearly eliminate the effects of friction; hence she tends to remain at rest relative to the inertial frame of the earth, in accordance with Newton's first law. As the vehicle accelerates around her, she moves backward relative to the vehicle. In the same way, a passenger in a vehicle that is slowing down tends to continue moving with constant velocity relative to the earth (Fig. 4.8b). This passenger moves forward relative to the vehicle. A vehicle is also accelerating if it moves at a constant speed but is turning (Fig. 4.8c). In this case a passenger tends to continue moving relative to the earth at constant speed in a straight line; relative to the vehicle, the passenger moves to the side of the vehicle on the outside of the turn.

In each case shown in Fig. 4.8, an observer in the vehicle's frame of reference might be tempted to conclude that there *is* a net force acting on the passenger, since the passenger's velocity *relative to the vehicle* changes in each case. This

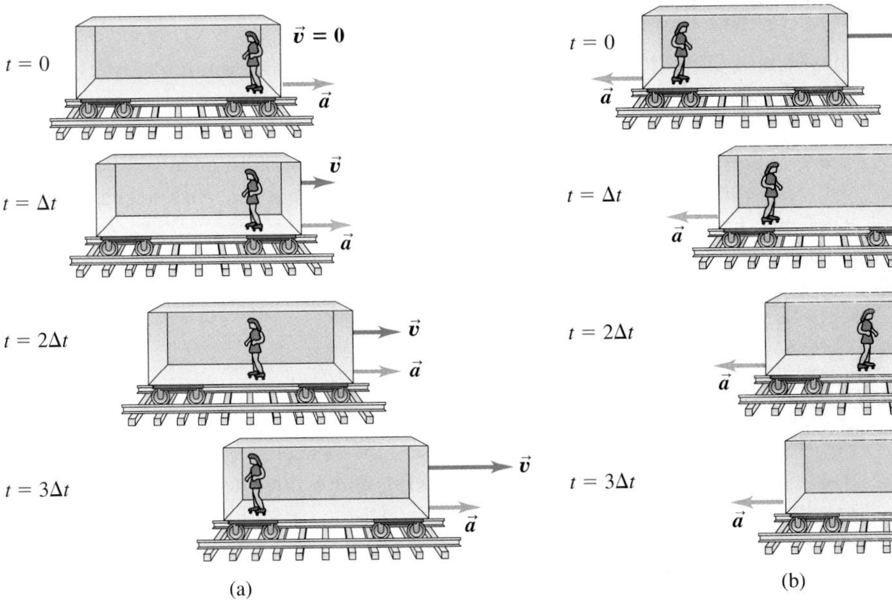

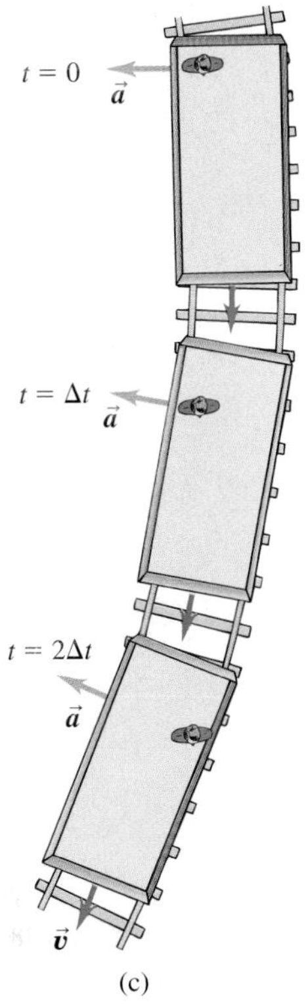

4.8 Riding in an accelerating vehicle. (a) If you and the vehicle are initially at rest, you tend to remain at rest as the vehicle accelerates around you. (b) If you and the vehicle are initially moving, you tend to continue moving with constant velocity as the vehicle slows down. (c) You tend to continue moving in a straight line as the vehicle turns.

conclusion is simply wrong; the net force on the passenger is indeed zero. The vehicle observer's mistake is in trying to apply Newton's first law in the vehicle's frame of reference, which is *not* an inertial frame and in which Newton's first law isn't valid (Fig. 4.9). In this book we will use *only* inertial frames of reference.

We've mentioned only one (approximately) inertial frame of reference: the earth's surface. But there are many inertial frames. If we have an inertial frame of reference A, in which Newton's first law is obeyed, then any second frame of reference B will also be inertial if it moves relative to A with constant velocity $\vec{v}_{B/A}$. To prove this, we use the relative velocity relation Eq. (3.36) from Section 3.5:

$$\vec{v}_{P/A} = \vec{v}_{P/B} + \vec{v}_{B/A}$$

Suppose that P is a body that moves with constant velocity $\vec{v}_{P/A}$ with respect to an inertial frame A. By Newton's first law the net force on this body is zero. The velocity of P relative to another frame B has a different value, $\vec{v}_{P/B} = \vec{v}_{P/A} - \vec{v}_{B/A}$. But if the relative velocity $\vec{v}_{B/A}$ of the two frames is constant, then $\vec{v}_{P/B}$ is constant as well. Thus B is also an inertial frame; the velocity of P in this frame is constant, and the net force on P is zero, so Newton's first law is obeyed in B. Observers in frames A and B will disagree about the velocity of P, but they will agree that P has a constant velocity (zero acceleration) and has zero net force acting on it.

There is no single inertial frame of reference that is preferred over all others for formulating Newton's laws. If one frame is inertial, then every other frame moving relative to it with constant velocity is also inertial. Viewed in this light, the state of rest and the state of motion with constant velocity are not very different; both occur when the vector sum of forces acting on the body is zero.

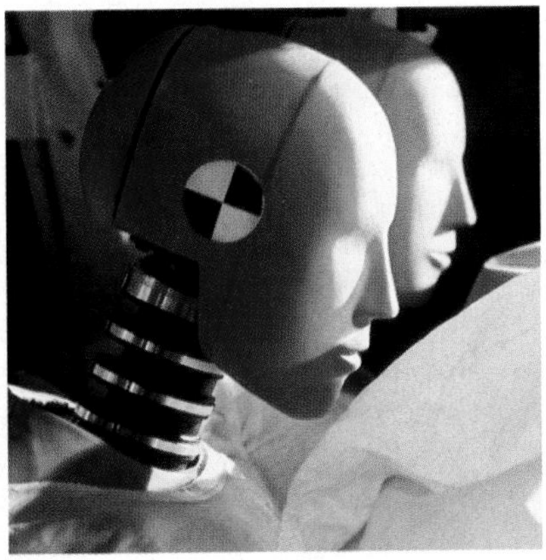

4.9 From the frame of reference of this car, it seems as though a force is pushing the crash test dummies forward as the car comes to a sudden stop. But there is really no such force: as the car stops, the dummies keep moving forward as a consequence of Newton's first law.

Test Your Understanding

In which of the following situations is there zero net force on the body: (i) an airplane flying due north at a steady 120 m/s and at a constant altitude; (ii) a car driving straight up a hill with a 3° slope at a constant 90 km/h; (iii) a hawk circling at a constant 20 km/h at a constant height of 15 m above an open field; (iv) a box with slick, frictionless surfaces in the back of a truck as the truck accelerates forward on a level road at 5 m/s^2?

4.3 | Newton's Second Law

In discussing Newton's first law, we have seen that when a body is acted on by no force or zero net force, it moves with constant velocity and zero acceleration. In Fig. 4.10a a hockey puck is sliding to the right on wet ice, so there is negligible friction. There are no horizontal forces acting on the puck; the downward force of gravity and the upward contact force exerted by the ice surface sum to zero. So the net force $\sum\vec{F}$ acting on the puck is zero, the puck has zero acceleration, and its velocity is constant.

But what happens when the net force is *not* zero? In Fig. 4.10b we apply a constant horizontal force to a sliding puck in the same direction that the puck is moving. Then $\sum\vec{F}$ is constant and in the same horizontal direction as $\vec{v}$. We find that during the time the force is acting, the velocity of the puck changes at a constant rate; that is, the puck moves with constant acceleration. The speed of the puck increases, so the acceleration $\vec{a}$ is in the same direction as $\vec{v}$ and $\sum\vec{F}$.

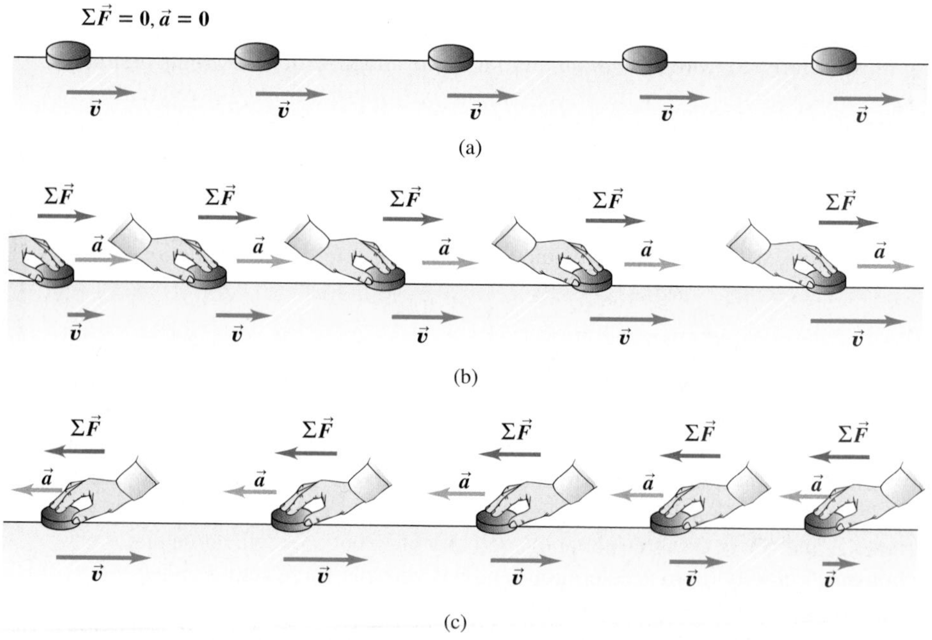

4.10 The acceleration of a body is in the same direction as the net force acting on the body (in this case, a hockey puck on a frictionless surface). (a) If $\sum\vec{F} = \mathbf{0}$, the puck is in equilibrium; the velocity is constant and the acceleration is zero. (b) If $\sum\vec{F}$ is to the right, the acceleration is to the right. (c) If $\sum\vec{F}$ is to the left, the acceleration is to the left.

Figure 4.10c shows another experiment, in which we reverse the direction of the force on the puck so that $\sum\vec{F}$ acts in the direction opposite to $\vec{v}$. In this case as well, the puck has an acceleration; the puck moves more and more slowly to the right. If the leftward force continues to act, the puck eventually stops and begins to move more and more rapidly to the left. The acceleration $\vec{a}$ in this experiment is to the left, in the same direction as $\sum\vec{F}$. As in the previous case, experiment shows that the acceleration is constant if $\sum\vec{F}$ is constant.

We conclude that the presence of a net force acting on a body causes the body to accelerate. The direction of the acceleration is the same as that of the net force. If the magnitude of the net force is constant, as in Figs. 4.10b and 4.10c, then so is the magnitude of the acceleration.

These conclusions about net force and acceleration also apply to a body moving along a curved path. For example, Fig. 4.11 shows a hockey puck moving in a horizontal circle on an ice surface of negligible friction. A string attaching the puck to the ice exerts a force of constant magnitude toward the center of the circle. The result is acceleration that is constant in magnitude and directed toward the center of the circle. The speed of the puck is constant, so this is uniform circular motion, as discussed in Section 3.4.

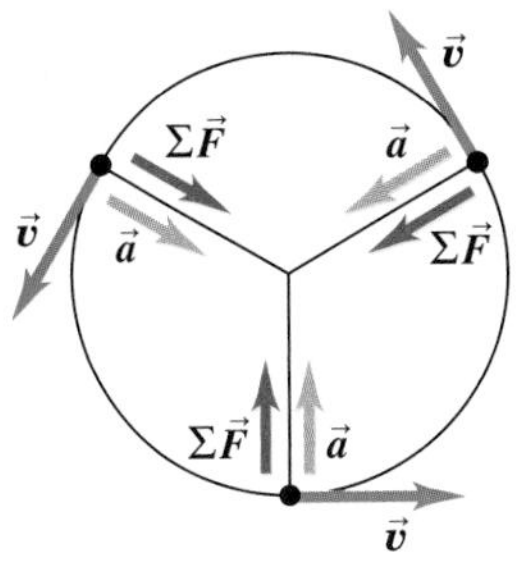

4.11 A top view of a hockey puck in uniform circular motion on a frictionless horizontal surface. At any point in the motion the acceleration $\vec{a}$ and the net force $\sum\vec{F}$ are in the same direction, toward the center of the circle.

Figure 4.12a shows another experiment to explore the relationship between the acceleration of a body and the net force acting on that body. We apply a constant horizontal force to a puck on a frictionless horizontal surface, using the spring balance described in Section 4.1 with the spring stretched a constant amount. As in Figs. 4.10b and 4.10c, this horizontal force equals the net force on the puck. If we change the magnitude of the net force, the acceleration changes in the same proportion. Doubling the net force doubles the acceleration (Fig. 4.12b), halving the force halves the acceleration (Fig. 4.12c), and so on. Many such experiments show that for any given body the magnitude of the acceleration is directly proportional to the magnitude of the net force acting on the body.

For a given body the ratio of the magnitude $|\sum\vec{F}|$ of the net force to the magnitude $a = |\vec{a}|$ of the acceleration is constant, regardless of the magnitude of the net force. We call this ratio the inertial mass, or simply the **mass,** of the body and denote it by m. That is, $m = |\sum\vec{F}|/a$, or

$$\left|\sum\vec{F}\right| = ma \qquad (4.5)$$

This equation *defines* what we mean by mass, and it also tells us how to measure mass through the relation $m = |\sum\vec{F}|/a$. When you hold a piece of fruit in your hand at the supermarket and move it slightly up and down to estimate its heft, you're using this relation to estimate the fruit's mass: you're seeing how much force you must exert with your hand to make the fruit accelerate up and down.

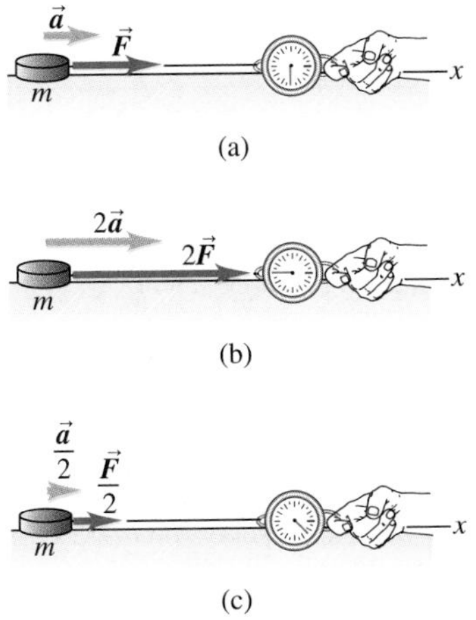

4.12 (a) Acceleration $\vec{a}$ is proportional to the net force $\sum\vec{F}$. (b) Doubling the net force doubles the acceleration. (c) Halving the net force halves the acceleration.

Mass and Force

Mass is a quantitative measure of inertia, which we discussed in Section 4.2. The greater its mass, the more a body "resists" being accelerated. The concept is easy to relate to everyday experience. If you hit a table-tennis ball and then a basketball with the same force, the basketball has much smaller acceleration because it has much greater mass. If a force causes a large acceleration, the mass of the body is small; if the same force causes only a small acceleration, the mass of the body is large.

The SI unit of mass is the **kilogram.** We mentioned in Section 1.3 that the kilogram is officially defined to be the mass of a chunk of platinum-iridium alloy kept in a vault near Paris. We can use this standard kilogram, along with Eq. (4.5), to define the **newton:**

One newton is the amount of net force that gives an acceleration of one meter per second squared to a body with a mass of one kilogram.

We can use this definition to calibrate the spring balances and other instruments used to measure forces. Because of the way we have defined the newton, it is related to the units of mass, length, and time. For Eq. (4.5) to be dimensionally consistent, it must be true that

$$1 \text{ newton} = (1 \text{ kilogram})(1 \text{ meter per second squared})$$

or

$$1\text{ N} = 1\text{ kg}\cdot\text{m/s}^2$$

We will use this relation many times in the next few chapters, so keep it in mind.

We can also use Eq. (4.5) to compare a mass with the standard mass and thus to *measure* masses. Suppose we apply a constant net force F to a body having a known mass m_1 and we find an acceleration of magnitude a_1. We then apply the same force to another body having an unknown mass m_2, and we find an acceleration of magnitude a_2. Then, according to Eq. (4.5),

$$m_1 a_1 = m_2 a_2$$

$$\frac{m_2}{m_1} = \frac{a_1}{a_2} \quad \text{(same net force)} \tag{4.6}$$

The ratio of the masses is the inverse of the ratio of the accelerations. Figure 4.13 shows the inverse proportionality between mass and acceleration. In principle we could use Eq. (4.6) to measure an unknown mass m_2, but it is usually easier to determine mass indirectly by measuring the body's *weight.* We'll return to this point in Section 4.4.

When two bodies with masses m_1 and m_2 are fastened together, we find that the mass of the composite body is always $m_1 + m_2$ (Fig. 4.13c). This additive property of mass may seem obvious, but it has to be verified experimentally. Ultimately, the mass of a body is related to the number of protons, electrons, and neutrons it contains. This wouldn't be a good way to *define* mass because there is no practical way to count these particles. But the concept of mass is the most fundamental way to characterize the quantity of matter in a body.

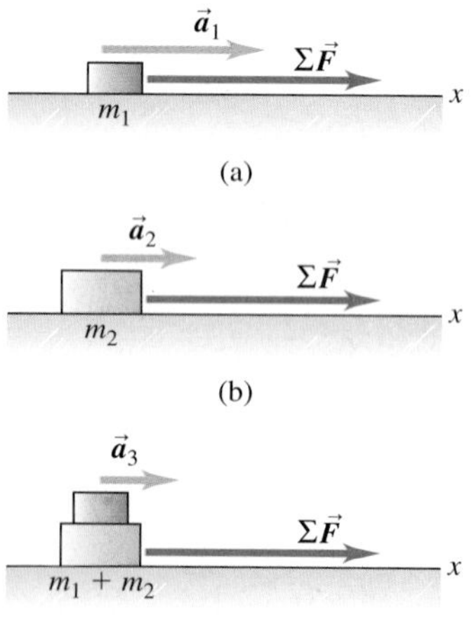

4.13 For a constant net force $\Sigma\vec{F}$, acceleration is inversely proportional to the mass of a body. Masses add like ordinary scalars. (a) $\vec{a}_1 = \Sigma\vec{F}/m_1$. (b) With a greater mass m_2, the acceleration has smaller magnitude: $\vec{a}_2 = \Sigma\vec{F}/m_2$. (c) If the two masses are combined, the acceleration is $\vec{a}_3 = \Sigma\vec{F}/(m_1 + m_2)$.

Newton's Second Law

We've been careful to state that the *net* force on a body is what causes that body to accelerate. Experiment shows that if a combination of forces $\vec{F}_1, \vec{F}_2, \vec{F}_3, \ldots$,is applied to a body, the body will have the same acceleration (magnitude and direction) as when only a single force is applied, if that single force is equal to the vector sum $\vec{F}_1 + \vec{F}_2 + \vec{F}_3 + \ldots$. In other words, the principle of superposition of forces also holds true when the net force is not zero and the body is accelerating.

Equation (4.5) relates the magnitude of the net force on a body to the magnitude of the acceleration that it produces. We have also seen that the direction of the net force is the same as the direction of the acceleration, whether the body's path is straight or curved. Newton wrapped up all these relationships and experimental results in a single concise statement that we now call **Newton's second law of motion:**

If a net external force acts on a body, the body accelerates. The direction of acceleration is the same as the direction of the net force. The net force vector is equal to the mass of the body times the acceleration of the body.

In symbols,

$$\sum \vec{F} = m\vec{a} \qquad \text{(Newton's second law)} \qquad (4.7)$$

An alternative statement is that the acceleration of a body (the rate of change of its velocity) is equal to the vector sum (resultant) of all forces acting on the body, divided by its mass:

$$\vec{a} = \frac{\sum \vec{F}}{m}$$

The acceleration is in the same direction as the net force.

Newton's second law is a fundamental law of nature, the basic relation between force and motion. Most of the remainder of this chapter and all of the next are devoted to learning how to apply this principle in various situations.

Equation (4.7) has many practical applications (Fig. 4.14). In fact, you've actually been using it all your life to measure your body's acceleration. In your inner ear, microscopic hair cells sense the magnitude and direction of the force that they must exert to cause small membranes to accelerate along with the rest of your body. By Newton's second law, the acceleration of the membranes—and hence that of your body as a whole—is proportional to this force and has the same direction. In this way, you can sense the magnitude and direction of your acceleration even with your eyes closed!

There are at least four aspects of Newton's second law that deserve special attention. First, Eq. (4.7) is a *vector* equation. Usually, we will use it in component form, with a separate equation for each component of force and the corresponding acceleration:

$$\sum F_x = ma_x \qquad \sum F_y = ma_y \qquad \sum F_z = ma_z \qquad (4.8)$$
$$\text{(Newton's Second Law)}$$

This set of component equations is equivalent to the single vector equation (4.7). Each component of total force equals the mass times the corresponding component of acceleration.

Second, the statement of Newton's second law refers to *external* forces. By this we mean forces exerted on the body by other bodies in its environment. It's impossible for a body to affect its own motion by exerting a force on itself; if it were possible, you could lift yourself to the ceiling by pulling up on your belt! That's why only external forces are included in the sum $\sum \vec{F}$ in Eqs. (4.7) and (4.8).

Third, Eqs. (4.7) and (4.8) are valid only when the mass m is *constant.* It's easy to think of systems whose masses change, such as a leaking tank truck, a rocket ship, or a moving railroad car being loaded with coal. But such systems are better handled by using the concept of momentum; we'll get to that in Chapter 8.

Finally, Newton's second law is valid only in inertial frames of reference, just like the first law. Thus it is not valid in the reference frame of any of the accelerating

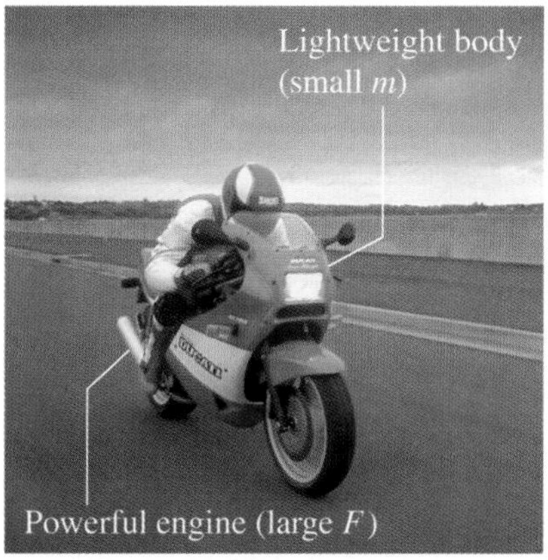

4.14 The design of high-performance motorcycles depends fundamentally on Newton's second law. To maximize the forward acceleration, the designer makes the motorcycle as light as possible (that is, minimizes the mass) and uses the most powerful engine possible (thus maximizing the forward force).

ActivPhysics ONLINE

2.1.3 Tension Change

2.1.4 Sliding on an Incline

vehicles in Fig. 4.8; relative to any of these frames, the passenger accelerates even though the net force on the passenger is zero. We will usually assume that the earth is an adequate approximation to an inertial frame, although because of its rotation and orbital motion it is not precisely inertial.

CAUTION Please note that the quantity $m\vec{a}$ is *not* a force. All that Eqs. (4.7) and (4.8) say is that the vector $m\vec{a}$ is equal in magnitude and direction to the vector sum $\sum\vec{F}$ of all the forces acting on the body. It's incorrect to think of acceleration as a force; rather, acceleration is a result of a nonzero net force. It's "common sense" to think that there is a "force of acceleration" that pushes you back into your seat when your car accelerates forward from rest. But *there is no such force;* instead, your inertia causes you to tend to stay at rest relative to the earth, and the car accelerates around you. The "common sense" confusion arises from trying to apply Newton's second law in a frame of reference where it isn't valid, like the non-inertial reference frame of an accelerating car. We will always examine motion relative to *inertial* frames of reference only.

In learning how to use Newton's second law, we will begin in this chapter with examples of straight-line motion. Then in Chapter 5 we will consider more general cases and develop more detailed problem-solving strategies for applying Newton's laws of motion.

Example 4.4 Determining acceleration from force

A worker applies a constant horizontal force with magnitude 20 N to a box with mass 40 kg resting on a level floor with negligible friction. What is the acceleration of the box?

SOLUTION

IDENTIFY: This problem involves force and acceleration. Whenever you encounter a problem of this kind, you should approach it using Newton's second law.

SET UP: In *any* problem involving forces, the first steps are to (i) choose a coordinate system and (ii) identify all of the forces acting on the body in question.

It's usually convenient to take one axis either along or opposite to the direction of the body's acceleration, which in this case is horizontal (see Fig. 4.15a). Hence we take the $+x$-axis to be in the direction of the applied horizontal force (that is, the direction in

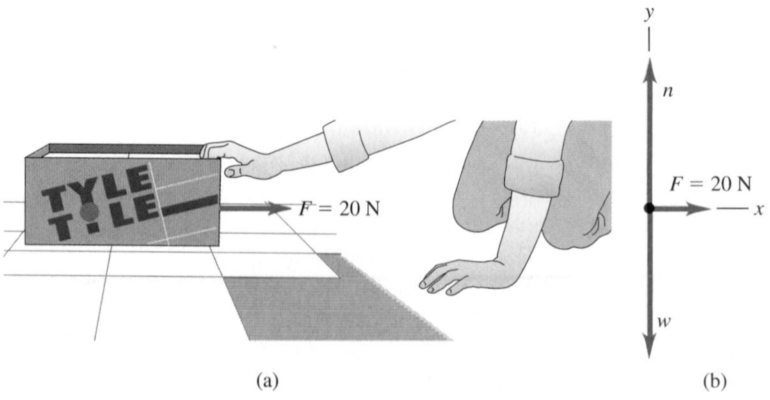

4.15 (a) A 20-N force accelerates the box to the right. (b) Vector diagram of forces acting on the box, considered as a particle. The tiles under the box are freshly waxed, so friction is negligible.

which the box accelerates) and the $+y$-axis to be upward (Fig. 4.15b). In most force problems that you'll encounter (including this one), the force vectors all lie in a plane, so the z-axis isn't used.

The forces acting on the box are (i) the horizontal force $\vec{F}$ exerted by the worker; (ii) the weight $\vec{w}$ of the box, that is, the downward force of the earth's gravitational attraction; and (iii) the upward supporting force $\vec{n}$ exerted by the floor. As in Section 4.2, we call the force $\vec{n}$ a *normal* force because it is normal (perpendicular) to the surface of contact. (We use an italic letter n to avoid confusion with the abbreviation N for newton.) We are told that friction is negligible, so no friction force is present.

Since the box doesn't move vertically at all, the y-acceleration is zero: $a_y = 0$. Our target variable is the x-component of acceleration, a_x. We'll find it using Newton's second law in component form as given by Eq. (4.8).

EXECUTE: From Fig. 4.15b, only the 20-N force has a nonzero x-component. Hence the first relation in Eqs. (4.8) tells us that

$$\sum F_x = F = 20\text{ N}$$

Hence the x-component of acceleration is

$$a_x = \frac{\sum F_x}{m} = \frac{20\text{ N}}{40\text{ kg}} = \frac{20\text{ kg}\cdot\text{m/s}^2}{40\text{ kg}} = 0.50\text{ m/s}^2$$

EVALUATE: The acceleration is in the $+x$-direction, the same direction as the net force. The net force is constant, so the acceleration is also constant. If we are given the initial position and velocity of the box, we can find the position and velocity at any later time from the equations of motion with constant acceleration we derived in Chapter 2.

Notice that to determine a_x, we didn't have to use the y-component of Newton's second law from Eq. (4.8), $\sum F_y = ma_y$. By using this equation, can you show that the magnitude n of the normal force in this situation is equal to the weight of the box?

Example 4.5 Determining force from acceleration

A waitress shoves a ketchup bottle with mass 0.45 kg toward the right along a smooth, level lunch counter. As the bottle leaves her hand, it has an initial velocity of 2.8 m/s. As it slides, it slows down because of the constant horizontal friction force exerted on it by the counter top. It slides a distance of 1.0 m before coming to rest. What are the magnitude and direction of the friction force acting on it?

SOLUTION

IDENTIFY: Like the previous example, this problem involves forces and acceleration (the slowing of the ketchup bottle), so we'll use Newton's second law to solve it.

SET UP: As in Example 4.4, our first steps are to choose a coordinate system and to identify the forces acting on the body (in this case, the ketchup bottle). We choose the $+x$-axis to be in the direction that the bottle slides, and we take the origin to be the point at which the bottle leaves the waitress's hand with the given initial velocity (Fig. 4.16a). The forces acting on the bottle are shown in Fig. 4.16b. The friction force $\vec{f}$ acts to slow the bottle down, so its direction must be opposite the direction of velocity (see Fig. 4.10c).

Our target variable is the magnitude f of the friction force. We'll find it using the x-component of Newton's second law from

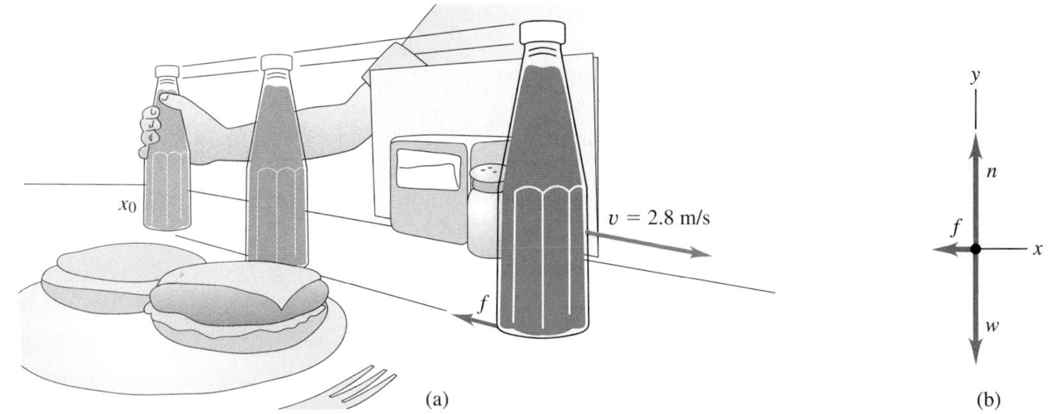

4.16 (a) As the bottle slides to the right, the friction force slows it down. (b) Vector diagram of forces acting on the bottle, considered as a particle.

Eq. (4.8). To do so, we'll first need to know the x-component of the bottle's acceleration, a_x. We aren't told the value of a_x in the problem, but we are told that the friction force is constant. Hence the acceleration is constant as well, and we can calculate a_x by using one of the constant-acceleration formulas from Section 2.4. Since we know the bottle's initial x-coordinate and x-velocity ($x_0 = 0$, $v_{0x} = 2.8\ \text{m/s}$) as well as its final x-coordinate and x-velocity ($x = 1.0\ \text{m}$, $v_x = 0$), the easiest equation to use to determine a_x is Eq. (2.13).

EXECUTE: From Eq. (2.13),

$$v_x^2 = v_{0x}^2 + 2a_x(x - x_0)$$

$$a_x = \frac{v_x^2 - v_{0x}^2}{2(x - x_0)} = \frac{(0\ \text{m/s})^2 - (2.8\ \text{m/s})^2}{2(1.0\ \text{m} - 0\ \text{m})} = -3.9\ \text{m/s}^2$$

The negative sign means that the acceleration is toward the *left;* the velocity is in the opposite direction to acceleration, as it must be, since the bottle is slowing down. The net force in the x-direction is the x-component $-f$ of the friction force, so

$$\sum F_x = -f = ma_x = (0.45\ \text{kg})(-3.9\ \text{m/s}^2)$$
$$= -1.8\ \text{kg} \cdot \text{m/s}^2 = -1.8\ \text{N}$$

Again the negative sign shows that the force on the bottle is directed toward the left. The magnitude of the friction force is $f = 1.8\ \text{N}$. Remember that magnitudes are *always* positive!

EVALUATE: We chose the $+x$-axis to be in the direction of the bottle's motion, so that a_x was negative. As a check on the result, we invite you to repeat the calculation with the $+x$-axis directed *opposite* to the motion (to the left in Fig. 4.16b), so that a_x is positive. In this case you should find that $\sum F_x$ is equal to $+f$ (because the friction force is now in the $+x$-direction), which in turn is equal to $+1.8\ \text{N}$. Your answers for the magnitudes of forces (which are always positive numbers) should never depend on your choice of coordinate axes!

Some Notes on Units

A few words about units are in order. In the cgs metric system (not used in this book), the unit of mass is the gram, equal to 10^{-3} kg, and the unit of distance is the centimeter, equal to 10^{-2} m. The corresponding unit of force is called the *dyne:*

$$1\ \text{dyne} = 1\ \text{g} \cdot \text{cm/s}^2$$

One dyne is equal to 10^{-5} N. In the British system, the unit of force is the *pound* (or pound-force) and the unit of mass is the *slug* (Fig. 4.17). The unit of acceleration is one foot per second squared, so

$$1\ \text{pound} = 1\ \text{slug} \cdot \text{ft/s}^2$$

The official definition of the pound is

$$1\ \text{pound} = 4.448221615260\ \text{newtons}$$

It is handy to remember that a pound is about 4.4 N and a newton is about 0.22 pound. Next time you want to order a "quarter-pounder," try asking for a "one-newtoner" and see what happens. Another useful fact: A body with a mass of 1 kg has a weight of about 2.2 lb at the earth's surface.

Table 4.2 summarizes the units of force, mass, and acceleration in the three systems.

4.17 Despite its name, the English unit of mass has nothing to do with the type of slug shown here. A common garden slug has a mass of about 15 grams, or about 10^{-3} slug.

Table 4.2 Units of Force, Mass, and Acceleration

System of Units	Force	Mass	Acceleration
SI	newton (N)	kilogram (kg)	m/s^2
cgs	dyne (dyn)	gram (g)	cm/s^2
British	pound (lb)	slug	ft/s^2

Test Your Understanding

Suppose the championship belt in Example 4.1 (Section 4.1) has a mass of 8.0 kg. What are the magnitude and direction of the belt's acceleration? If the belt was initially at rest, how far has it moved after 0.50 s?

4.4 | Mass and Weight

The *weight* of a body is a familiar force. It is the force of the earth's gravitational attraction for the body. We will study gravitational interactions in detail in Chapter 12, but we need some preliminary discussion now. The terms *mass* and *weight* are often misused and interchanged in everyday conversation. It is absolutely essential for you to understand clearly the distinctions between these two physical quantities.

Mass characterizes the *inertial* properties of a body. Mass is what keeps the china on the table when you yank the tablecloth out from under it. The greater the mass, the greater the force needed to cause a given acceleration; this is reflected in Newton's second law, $\sum\vec{F} = m\vec{a}$. Weight, on the other hand, is a *force* exerted on a body by the pull of the earth or some other large body. Everyday experience shows us that bodies having large mass also have large weight. A large stone is hard to throw because of its large *mass,* and hard to lift off the ground because of its large *weight.* On the moon the stone would be just as hard to throw horizontally, but it would be easier to lift. So what exactly *is* the relationship between mass and weight?

The answer to this question, according to legend, came to Newton as he sat under an apple tree watching the apples fall. A freely falling body has an acceleration equal to g, and because of Newton's second law, a force must act to produce this acceleration. If a 1-kg body falls with an acceleration of 9.8 m/s^2, the required force has magnitude

ActivPhysics ONLINE 2.9 Pole-Vaulter Vaults

$$F = ma = (1\text{ kg})(9.8\text{ m/s}^2) = 9.8\text{ kg}\cdot\text{m/s}^2 = 9.8\text{ N}$$

But the force that makes the body accelerate downward is the gravitational pull of the earth, that is, the *weight* of the body. Any body near the surface of the earth that has a mass of 1 kg *must* have a weight of 9.8 N to give it the acceleration we observe when it is in free fall. More generally, a body with mass m must have weight with magnitude w given by

$$w = mg \qquad \text{(magnitude of the weight of a body of mass } m) \qquad (4.9)$$

The weight of a body is a force, a vector quantity, and we can write Eq. (4.9) as a vector equation:

$$\vec{w} = m\vec{g} \qquad (4.10)$$

Remember that g is the *magnitude* of $\vec{g}$, the acceleration due to gravity, so g is always a positive number, by definition. Thus w, given by Eq. (4.9), is the *magnitude* of the weight and is also always positive.

CAUTION It is important to understand that the weight of a body acts on the body *all the time,* whether it is in free fall or not. When a 10-kg flowerpot hangs suspended from a chain, it is in equilibrium, and its acceleration is zero. But its weight, given by Eq. (4.10), is still pulling down on it. In this case the chain pulls up on the pot, applying an upward force. The *vector sum* of the forces is zero, and the pot is in equilibrium.

Conceptual Example 4.6 Net force and acceleration in free fall

In Example 2.6 (Section 2.5) a one-euro coin was dropped from rest from the Leaning Tower of Pisa. If the coin falls freely, so that the effects of the air are negligible, how does the net force on the coin vary as it falls?

SOLUTION

In free fall, the acceleration $\vec{a}$ of the coin is constant and equal to $\vec{g}$. Hence by Newton's second law the net force $\sum\vec{F} = m\vec{a}$ is also constant and equal to $m\vec{g}$, which is the weight $\vec{w}$ of the coin (Fig. 4.18). The velocity of the coin changes as it falls, but the net force acting on it remains constant. If this result surprises you, it's because you still believe in the erroneous "common sense" idea that greater speed implies greater force. If so, you should reread Conceptual Example 4.3.

The net force on a freely falling coin is constant even if you initially toss it upward. The force that your hand exerts on the coin is a contact force, and it disappears the instant that the coin loses contact with your hand. From then on, the only force acting on the coin is its weight $\vec{w}$.

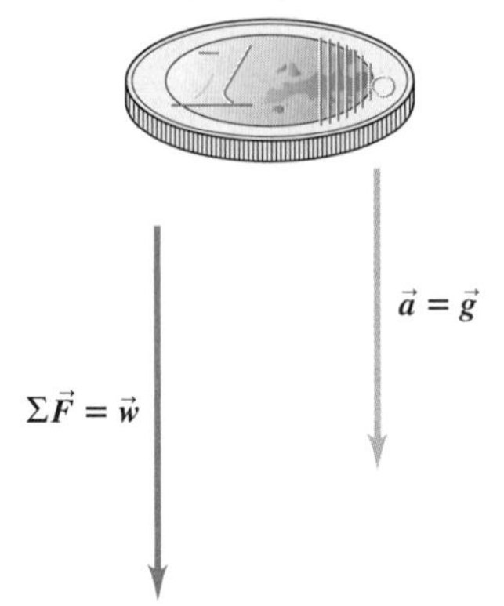

4.18 The acceleration of a freely falling object is constant, and so is the net force acting on the object.

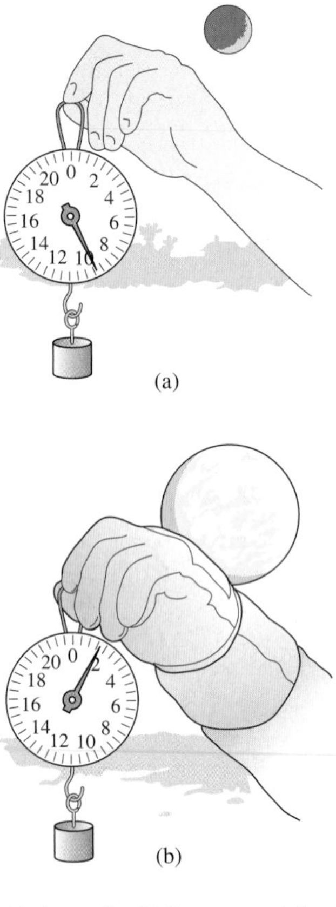

4.19 (a) A standard kilogram weighs about 9.8 N on earth. (b) The same kilogram weighs only about 1.6 N on the moon.

Variation of *g* with Location

We will use $g = 9.80\ \text{m/s}^2$ for problems on the earth (or, if the other data in the problem are given to only two significant figures, $g = 9.8\ \text{m/s}^2$). In fact, the value of g varies somewhat from point to point on the earth's surface, from about 9.78 to $9.82\ \text{m/s}^2$, because the earth is not perfectly spherical and because of effects due to its rotation and orbital motion. At a point where $g = 9.80\ \text{m/s}^2$, the weight of a standard kilogram is $w = 9.80\ \text{N}$. At a different point, where $g = 9.78\ \text{m/s}^2$, the weight is $w = 9.78\ \text{N}$ but the mass is still 1 kg. The weight of a body varies from one location to another; the mass does not. If we take a standard kilogram to the surface of the moon, where the acceleration of free fall (equal to the value of g at the moon's surface) is $1.62\ \text{m/s}^2$, its weight is 1.62 N, but its mass is still 1 kg (Fig. 4.19). An 80.0-kg astronaut has a weight on earth of $(80.0\ \text{kg})(9.80\ \text{m/s}^2) = 784\ \text{N}$, but on the moon the astronaut's weight would be only $(80.0\ \text{kg})(1.62\ \text{m/s}^2) = 130\ \text{N}$.

The following excerpt from Astronaut Buzz Aldrin's book *Men from Earth,* describing his experiences during the Apollo 11 mission to the moon in 1969, offers some interesting insights into the distinction between mass and weight:

> Our portable life-support system backpacks looked simple, but they were hard to put on and tricky to operate. On Earth the portable life-support system and space suit combination weighed 190 pounds, but on the moon it was only 30. Combined with my own body weight, that brought me to a total lunar-gravity weight of around 60 pounds.
>
> One of my tests was to jog away from the lunar module to see how maneuverable an astronaut was on the surface. I remembered what Isaac Newton had taught us two centuries before: Mass and weight are not the same. I weighed only 60 pounds, but my *mass* was the same as it was on Earth. Inertia was a problem. I had to plan ahead several steps to bring myself to a stop or to turn without falling.

Measuring Mass and Weight

In Section 4.3 we described a way to compare masses by comparing their accelerations when subjected to the same net force. Usually, however, the easiest way to measure the mass of a body is to measure its weight, often by comparing with a standard. Because of Eq. (4.9), two bodies that have the same weight at a particular location also have the same mass. We can compare weights very precisely; the familiar equal-arm balance (Fig. 4.20) can determine with great precision (up to 1 part in 10^6) when the weights of two bodies are equal and hence when their masses are equal. This method doesn't work in the apparent "zero-gravity" environment of outer space. Instead, we have to use Newton's second law directly. We apply a known force to the body, measure its acceleration, and compute the mass as the ratio of force to acceleration. This method, or a variation of it, is used to measure the masses of astronauts in orbiting space stations as well as the masses of atomic and subatomic particles.

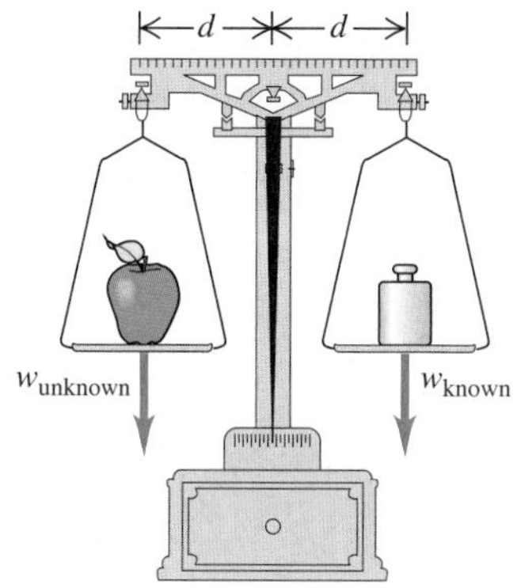

4.20 An equal-arm balance determines the mass of a body by comparing its weight to a known weight.

The concept of mass plays two rather different roles in mechanics. The weight of a body (the gravitational force acting on it) is proportional to its mass; we may call the property related to gravitational interactions *gravitational mass.* On the other hand, we can call the inertial property that appears in Newton's second law the *inertial mass.* If these two quantities were different, the acceleration due to gravity might well be different for different bodies. However, extraordinarily precise experiments have established with a precision of better than one part in 10^{12} that in fact the two *are* the same.

CAUTION The SI units for mass and weight are often misused in everyday life. Incorrect expressions such as "This box weighs 6 kg" are nearly universal. What is meant is that the *mass* of the box, probably determined indirectly by *weighing,* is 6 kg. This usage is so common that there is probably no hope of straightening things out, but be sure you recognize that the term *weight* is often used when *mass* is meant. Be careful to avoid this kind of mistake in your own work! In SI units, weight (a force) is measured in newtons, while mass is measured in kilograms.

Example 4.7 Mass and weight

A 1.96×10^4 N Lincoln Town Car traveling in the $+x$-direction makes a fast stop; the x-component of the net force acting on it is -1.50×10^4 N. What is its acceleration?

SOLUTION

IDENTIFY: Again we will use Newton's second law to relate force and acceleration. To use this relationship, we need to know the car's mass. However, because the newton is a unit for force, we know that 1.96×10^4 N is the car's *weight,* not its mass. So we'll also have to use the relationship between a body's mass and its weight.

SET UP: Our target variable is the x-component of acceleration of the car, a_x. (The motion is purely in the x-direction.) We use Eq. (4.9) to determine the car's mass from its weight, then use the x-component of Newton's second law from Eq. (4.8) to determine a_x.

EXECUTE: The mass m of the car is

$$m = \frac{w}{g} = \frac{1.96 \times 10^4\ \text{N}}{9.80\ \text{m/s}^2} = \frac{1.96 \times 10^4\ \text{kg} \cdot \text{m/s}^2}{9.80\ \text{m/s}^2} = 2000\ \text{kg}$$

Then $\sum F_x = ma_x$ gives

$$a_x = \frac{\sum F_x}{m} = \frac{-1.50 \times 10^4\ \text{N}}{2000\ \text{kg}} = \frac{-1.50 \times 10^4\ \text{kg} \cdot \text{m/s}^2}{2000\ \text{kg}} = -7.5\ \text{m/s}^2$$

EVALUATE: The negative sign means that the acceleration vector points in the negative x-direction. This makes sense: the car is moving in the positive x-direction and is slowing down.

Note that the acceleration can alternatively be written as $-0.77g$. It's of interest that -0.77 is also the ratio of -1.50×10^4 N (the x-component of the net force) to 1.96×10^4 N (the weight). Indeed, the acceleration of a body expressed as a multiple of g is always equal to the ratio of the net force on the body to its weight. Can you see why?

Test Your Understanding

Suppose an astronaut landed on a planet where $g = 19.6 \text{ m/s}^2$. Compared to earth, would it be easier, harder, or just as easy for her to walk around? Would it be easier, harder, or just as easy for her to catch a ball that is moving horizontally at 12 m/s? (Assume that the astronaut's spacesuit is a lightweight model that doesn't impair her movements in any way.)

4.5 | Newton's Third Law

A force acting on a body is always the result of its interaction with another body, so forces always come in pairs. You can't pull on a doorknob without the doorknob pulling back on you. When you kick a football, the forward force that your foot exerts on the ball launches it into its trajectory, but you also feel the force the ball exerts back on your foot. If you kick a boulder, the pain you feel is due to the force that the boulder exerts on your foot.

In each of these cases, the force that you exert on the other body is in the opposite direction to the force that body exerts on you. Experiments show that whenever two bodies interact, the two forces that they exert on each other are always *equal in magnitude* and *opposite in direction.* This fact is called **Newton's third law of motion.** In Fig. 4.21, $\vec{F}_{A \text{ on } B}$ is the force applied *by* body A (first subscript) *on* body B (second subscript), and $\vec{F}_{B \text{ on } A}$ is the force applied *by* body B (first subscript) *on* body A (second subscript). The mathematical statement of Newton's third law is

$$\vec{F}_{A \text{ on } B} = -\vec{F}_{B \text{ on } A} \tag{4.11}$$

Expressed in words,

If body A exerts a force on body B (an "action"), then body B exerts a force on body A (a "reaction"). These two forces have the same magnitude but are opposite in direction. These two forces act on *different* bodies.

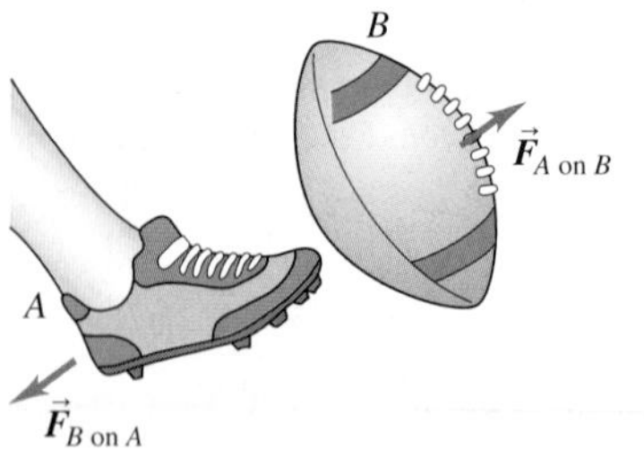

4.21 If body A exerts a force $\vec{F}_{A \text{ on } B}$ on body B, then body B exerts a force $\vec{F}_{B \text{ on } A}$ on body A that is equal in magnitude and opposite in direction: $\vec{F}_{A \text{ on } B} = -\vec{F}_{B \text{ on } A}$.

In this statement, "action" and "reaction" are the two opposite forces; we sometimes refer to them as an **action-reaction pair.** This is not meant to imply any cause-and-effect relationship; we can consider either force as the "action" and the other as the "reaction." We often say simply that the forces are "equal and opposite," meaning that they have equal magnitudes and opposite directions.

We stress that the two forces described in Newton's third law act on *different* bodies. This is important in problems involving Newton's first or second law, which involve the forces that act *on* a body. For instance, the net force acting on the football in Fig. 4.21 is the vector sum of the weight of the football and the force $\vec{F}_{A \text{ on } B}$ exerted by the kicker. You would not include the force $\vec{F}_{B \text{ on } A}$ because this force acts on the *kicker,* not on the football.

In Fig. 4.21 the action and reaction forces are *contact* forces that are present only when the two bodies are touching. But Newton's third law also applies to *long-range* forces that do not require physical contact, such as the force of gravitational attraction. A table-tennis ball exerts an upward gravitational force on the earth that's equal in magnitude to the downward gravitational force the earth exerts on the ball. When you drop the ball, both the ball and the earth accelerate toward each other. The net force on each body has the same magnitude, but the earth's acceleration is microscopically small because its mass is so great. Nevertheless, it does move!

Conceptual Example 4.8 Which force is greater?

After your sports car breaks down, you start to push it to the nearest repair shop. While the car is starting to move, how does the force you exert on the car compare to the force the car exerts on you? How do these forces compare when you are pushing the car along at a constant speed?

SOLUTION

In *both* cases, the force you exert on the car is equal in magnitude and opposite in direction to the force the car exerts on you. It's true that you have to push harder to get the car going than to keep it going. But no matter how hard you push on the car, the car pushes just as hard back on you. Newton's third law gives the same result whether the two bodies are at rest, moving with constant velocity, or accelerating.

You may wonder how the car "knows" to push back on you with the same magnitude of force that you exert on it. It may help to remember that the forces you and the car exert on each other are really interactions between the atoms at the surface of your hand and the atoms at the surface of the car. These interactions are analogous to miniature springs between adjacent atoms, and a compressed spring exerts equally strong forces on either of its ends.

Fundamentally, though, the reason why we know that objects of different masses exert equally strong forces on each other is that *experiment tells us so.* Never forget that physics isn't merely a collection of rules and equations; rather, it's a systematic description of the natural world based on experiment and observation.

Conceptual Example 4.9 Applying Newton's third law: Objects at rest

An apple sits on a table in equilibrium. What forces act on it? What is the reaction force to each of the forces acting on the apple? What are the action-reaction pairs?

SOLUTION

Figure 4.22a shows the apple on the table, and Fig. 4.22b shows the forces acting on the apple. The apple is A, the table T, and the earth E. In the diagram, $\vec{F}_{\text{E on A}}$ is the weight of the apple; that is, the downward gravitational force exerted *by* the earth E (first subscript) *on* the apple A (second subscript). Similarly, $\vec{F}_{\text{T on A}}$ is the upward force exerted *by* the table T (first subscript) *on* the apple A (second subscript).

As the earth pulls down on the apple, the apple exerts an equally strong upward pull $\vec{F}_{\text{A on E}}$ on the earth, as shown in Fig. 4.22d. $\vec{F}_{\text{A on E}}$ and $\vec{F}_{\text{E on A}}$ are an action-reaction pair, representing the mutual interaction of the apple and the earth, so

$$\vec{F}_{\text{A on E}} = -\vec{F}_{\text{E on A}}$$

Also, as the table pushes up on the apple with force $\vec{F}_{\text{T on A}}$, the corresponding reaction is the downward force $\vec{F}_{\text{A on T}}$ exerted by the apple on the table (Fig. 4.22c). So we have

$$\vec{F}_{\text{A on T}} = -\vec{F}_{\text{T on A}}$$

The two forces acting on the apple are $\vec{F}_{\text{T on A}}$ and $\vec{F}_{\text{E on A}}$. Are they an action-reaction pair? No, they aren't, despite being equal and opposite. They do not represent the mutual interaction of two bodies; they are two different forces acting on the *same* body. *The two forces in an action-reaction pair* ***never*** *act on the same body.* Here's another way to look at it. Suppose we suddenly yank the table out from under the apple (Fig. 4.22e). The two forces $\vec{F}_{\text{A on T}}$ and $\vec{F}_{\text{T on A}}$ then become zero, but $\vec{F}_{\text{A on E}}$ and $\vec{F}_{\text{E on A}}$ are still there (the gravitational interaction is still present). Since $\vec{F}_{\text{T on A}}$ is now zero, it can't be the negative of $\vec{F}_{\text{E on A}}$, and these two forces can't be an action-reaction pair.

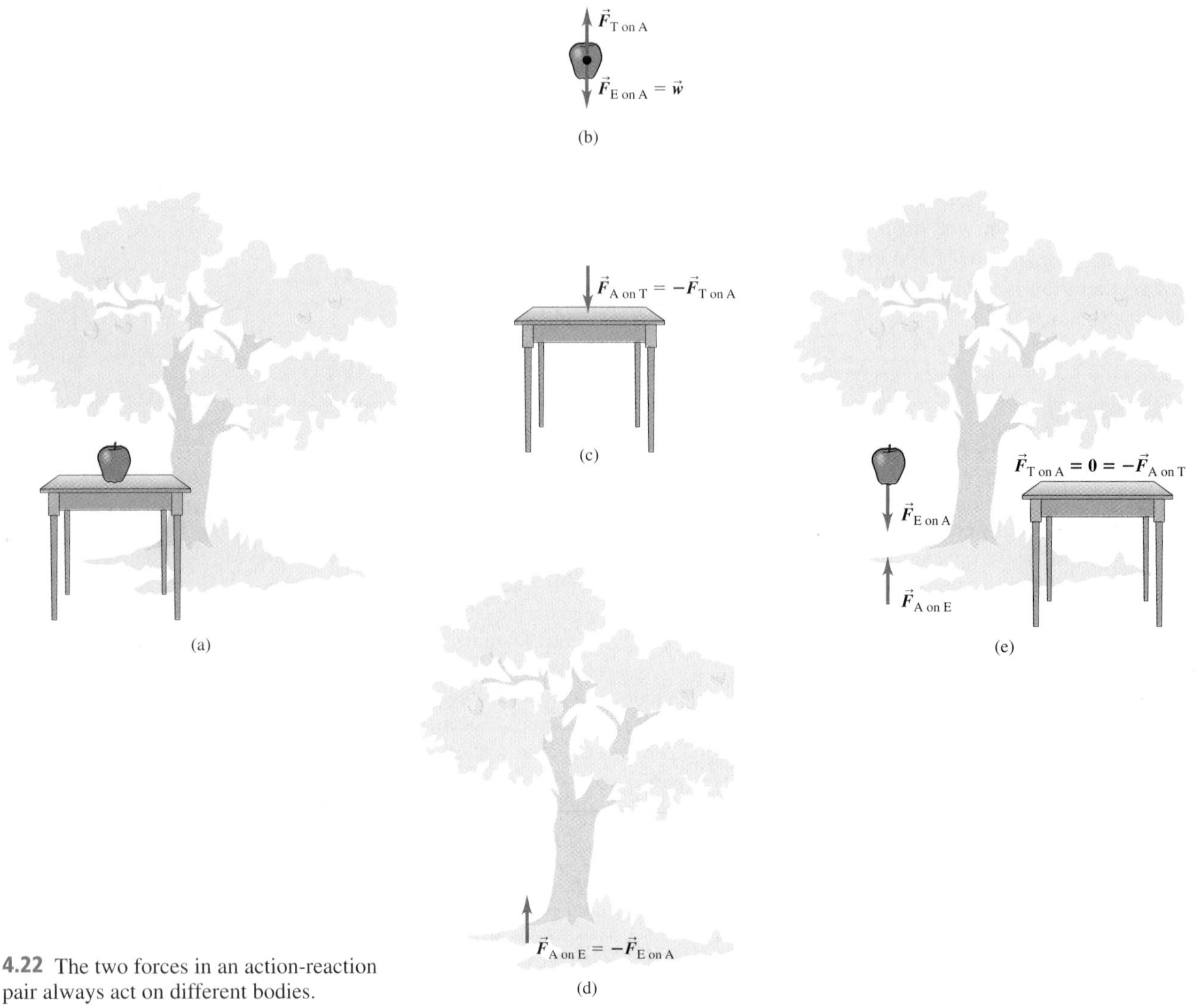

4.22 The two forces in an action-reaction pair always act on different bodies.

Conceptual Example 4.10 Applying Newton's third law: Objects in motion

A stonemason drags a marble block across a floor by pulling on a rope attached to the block (Fig. 4.23a). The block may or may not be in equilibrium. How are the various forces related? What are the action-reaction pairs?

SOLUTION

Figure 4.23b shows the horizontal forces acting on each body: the block (B), the rope (R), and the mason (M). We'll use subscripts on all the forces to help explain things. Vector $\vec{F}_{\text{M on R}}$ represents the force exerted *by* the mason *on* the rope. Its reaction is the equal and opposite force $\vec{F}_{\text{R on M}}$ exerted *by* the rope *on* the mason. Vector $\vec{F}_{\text{R on B}}$ represents the force exerted by the rope on the block. The reaction to it is the equal and opposite force $\vec{F}_{\text{B on R}}$ exerted by the block on the rope:

$$\vec{F}_{\text{R on M}} = -\vec{F}_{\text{M on R}} \quad \text{and} \quad \vec{F}_{\text{B on R}} = -\vec{F}_{\text{R on B}}$$

Be sure you understand that the forces $\vec{F}_{\text{M on R}}$ and $\vec{F}_{\text{B on R}}$ are *not* an action-reaction pair. Both of these forces act on the *same* body (the rope); an action and its reaction *must* always act on *different* bodies. Furthermore, the forces $\vec{F}_{\text{M on R}}$ and $\vec{F}_{\text{B on R}}$ are not necessarily equal in magnitude. Applying Newton's second law to the rope, we get

$$\sum \vec{F} = \vec{F}_{\text{M on R}} + \vec{F}_{\text{B on R}} = m_{\text{rope}}\vec{a}_{\text{rope}}$$

If the block and rope are accelerating (that is, speeding up or slowing down), the rope is not in equilibrium, and $\vec{F}_{\text{M on R}}$ must have a different magnitude than $\vec{F}_{\text{B on R}}$. By contrast, the action-reaction forces $\vec{F}_{\text{M on R}}$ and $\vec{F}_{\text{R on M}}$ are always equal in magnitude, as are

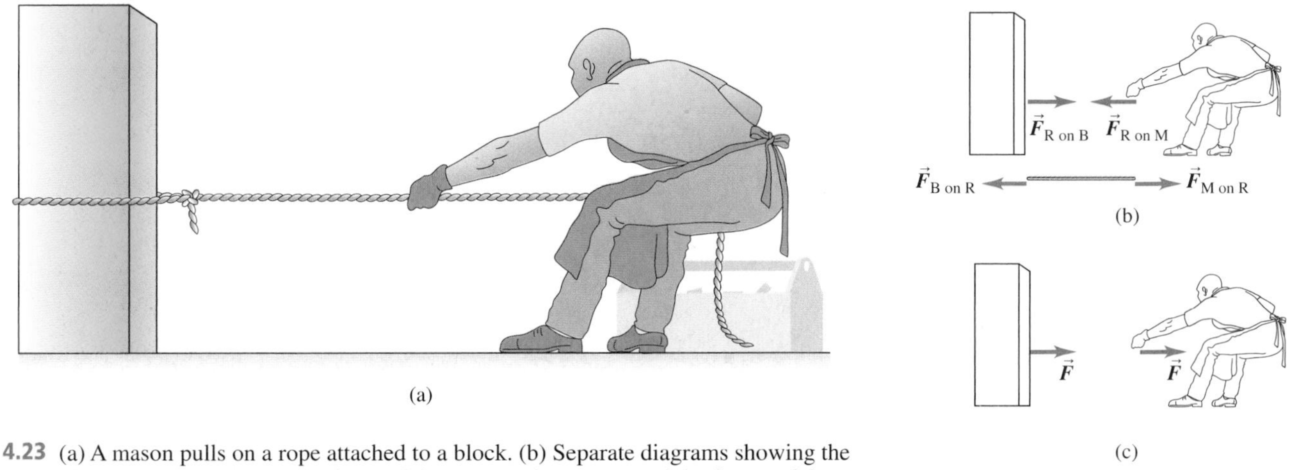

4.23 (a) A mason pulls on a rope attached to a block. (b) Separate diagrams showing the force of the rope on the block, the force of the rope on the mason, and the forces of the block and the mason on the rope. (c) If the rope is not accelerating or if its mass can be neglected, it can be considered to transmit an undiminished force from the mason to the block, and vice versa.

$\vec{F}_{R\ on\ B}$ and $\vec{F}_{B\ on\ R}$. Newton's third law holds whether or not the bodies are accelerating.

In the special case in which the rope is in equilibrium, the forces $\vec{F}_{M\ on\ R}$ and $\vec{F}_{B\ on\ R}$ are equal in magnitude. But this is an example of Newton's *first* law, not his *third.* Another way to look at this is that in equilibrium, $\vec{a}_{rope} = \mathbf{0}$ in the equation on the previous page. Then $\vec{F}_{B\ on\ R} = -\vec{F}_{M\ on\ R}$ because of Newton's first or second law.

This is also true if the rope is accelerating but has negligibly small mass compared to the block or the mason. In this case, $m_{rope} = 0$ in the equation on the previous page, so again $\vec{F}_{B\ on\ R} = -\vec{F}_{M\ on\ R}$. Since $\vec{F}_{B\ on\ R}$ *always* equals $-\vec{F}_{R\ on\ B}$ by Newton's third law (they are an action-reaction pair), in these same special cases, $\vec{F}_{R\ on\ B}$ also equals $\vec{F}_{M\ on\ R}$. In other words, in these cases the force of the rope on the block equals the force of the mason on the rope, and we can then think of the rope as "transmitting" to the block, without change, the force the person exerts on the rope (Fig. 4.23c). This is a useful point of view, but you have to remember that it is valid *only* when the rope has negligibly small mass or is in equilibrium.

If you feel as though you're drowning in subscripts at this point, take heart. Go over this discussion again, comparing the symbols with the vector diagrams, until you're sure you see what's going on.

Conceptual Example 4.11 A Newton's third law paradox?

We saw in Conceptual Example 4.10 that the stonemason pulls as hard on the rope–block combination as that combination pulls back on him. Why, then, does the block move while the stonemason remains stationary?

SOLUTION

The way out of this seeming conundrum is to keep in mind the difference between Newton's *second* law and his *third* law. The only forces involved in Newton's second law are those that act *on* that body. The vector sum of these forces determines how the body accelerates (and whether it accelerates at all). By contrast, Newton's third law relates the forces that two *different* bodies exert on each other. The third law alone tells you nothing about the motion of either body.

The reason why the stonemason doesn't move is that the net force acting on him is zero. It is the vector sum of the upward normal force from the floor, his weight acting downward, the rope pulling to the left, and the friction force of the floor pushing him to the right (Fig. 4.24). Because the stonemason has shoes with nonskid soles that don't slip on the floor, the friction force is strong enough to exactly balance the pull of the rope. If the floor were freshly waxed, so that there was little friction between the floor and the stonemason's shoes, he would indeed start sliding to the left.

Four forces also act on the block–rope combination: normal force, weight, friction, and the force of the stonemason pulling to the right (see Fig. 4.24). If the block is initially at rest, it begins to slide if the stonemason's force is *greater* than the friction force that the floor exerts on the block. (The marble block has a smooth underside, which helps to minimize friction.) Then the net force on the block is not zero, and the block accelerates to the right. Once the block is moving, the stonemason doesn't need to pull quite so hard; he need exert only enough force to exactly balance the friction force on the block. Then the net force on the moving block is zero, and the block continues to move at a constant velocity in accordance with Newton's first law.

The moral of this example is that when analyzing the motion of a body, remember that only forces acting *on* the body determine its motion. From this perspective, Newton's third law is merely a tool that can help you determine what those forces are.

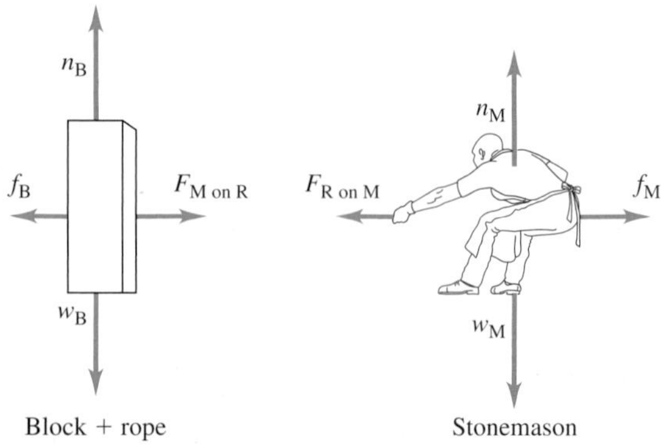

4.24 Force diagrams for the stonemason (right) and the block–rope combination (left) in Conceptual Example 4.10. Both the block and the stonemason are acted on by normal forces exerted by the floor (n_B and n_M, respectively), weight forces (w_B and w_M, respectively), and friction forces exerted by the floor (f_B and f_M, respectively). (We ignore the relatively small mass of the rope.) The force of the rope on the stonemason has the same magnitude as the force of the stonemason on the rope ($F_{R\text{ on }M} = F_{M\text{ on }R}$).

A body, such as the rope in Fig. 4.23, that has pulling forces applied at its ends is said to be in **tension.** The tension at any point is the magnitude of force acting at that point. In Fig. 4.23b the tension at the right-hand end of the rope is the magnitude of $\vec{F}_{M\text{ on }R}$ (or of $\vec{F}_{R\text{ on }M}$), and the tension at the left-hand end equals the magnitude of $\vec{F}_{B\text{ on }R}$ (or of $\vec{F}_{R\text{ on }B}$). If the rope is in equilibrium and if no forces act except at its ends, the tension is the same at both ends and throughout the rope. Thus, if the magnitudes of $\vec{F}_{B\text{ on }R}$ and $\vec{F}_{M\text{ on }R}$ are 50 N each, the tension in the rope is 50 N (*not* 100 N). The *total* force vector $\vec{F}_{B\text{ on }R} + \vec{F}_{M\text{ on }R}$ acting on the rope in this case is zero!

CAUTION We emphasize once more a fundamental truth: The two forces in an action-reaction pair *never* act on the same body. Remembering this simple fact can often help you avoid confusion about action-reaction pairs and Newton's third law.

Test Your Understanding

You are driving your car on a country road when a mosquito splatters itself on the windshield. Which has the greater magnitude, the force that the car exerted on the mosquito or the force that the mosquito exerted on the car? Or are the magnitudes the same? If they are different, how can you reconcile this fact with Newton's third law? If they are equal, why is the mosquito splattered while the car is undamaged?

4.6 | Free-Body Diagrams

Newton's three laws of motion contain all the basic principles we need to solve a wide variety of problems in mechanics. These laws are very simple in form, but the process of applying them to specific situations can pose real challenges. In this brief section we'll point out some key ideas and techniques to use in any problems involving Newton's laws. You'll learn others in Chapter 5, which also extends the use of Newton's laws to cover more complex situations.

Newton's first and second laws apply to a specific body. Whenever you use Newton's first law, $\sum \vec{F} = \mathbf{0}$, for an equilibrium situation or Newton's second law, $\sum \vec{F} = m\vec{a}$, for a non-equilibrium situation, you must decide at the beginning to which body you are referring. This decision may sound trivial, but it isn't.

Only forces acting on the body matter. The sum $\sum \vec{F}$ includes all the forces that act *on* the body in question. Hence, once you've chosen the body to analyze, you have to identify all the forces acting on it. Don't get confused between the forces acting on a body and the forces exerted by that body on some other body. For example, to analyze a person walking, you would include in $\sum \vec{F}$ the force that the ground exerts on the person as he walks, but *not* the force that the person exerts on the ground (Fig. 4.25). These forces form an action-reaction pair and are related by Newton's third law, but only the member of the pair that acts on the body you're working with goes into $\sum \vec{F}$.

4.25 The simple act of walking depends crucially on Newton's third law. To start moving forward, you push backward on the ground with your foot. As a reaction, the ground pushes forward on your foot (and hence on your body as a whole) with a force of the same magnitude. This *external* force provided by the ground is what accelerates your body forward.

Free-body diagrams are essential to help identify the relevant forces. A **free-body diagram** is a diagram showing the chosen body by itself, "free" of its surroundings, with vectors drawn to show the magnitudes and directions of all the forces applied to the body by the various other bodies that interact with it. We have already shown some free-body diagrams in Figs. 4.15, 4.16, and 4.22. Be careful to include all the forces acting *on* the body, but be equally careful *not* to include any forces that the body exerts on any other body. In particular, the two forces in an action-reaction pair must *never* appear in the same free-body diagram because they never act on the same body. Furthermore, forces that a body exerts on itself are never included, since these can't affect the body's motion.

CAUTION When you have a complete free-body diagram, you *must* be able to answer for each force the question, "What other body is applying this force?" If you can't answer that question, you may be dealing with a nonexistent force. Be especially on your guard to avoid nonexistent forces such as "the force of acceleration" or "the $m\vec{a}$ force," discussed in Section 4.3.

ActivPhysics Online

2.1.1 Force Magnitudes

When a problem involves more than one body, you have to take the problem apart and draw a separate free-body diagram for each body. For example, Fig. 4.23b shows a separate free-body diagram for the rope in the case in which the rope is considered massless (so that no gravitational force acts on it). This figure also shows diagrams for the block and the mason, but these are *not* complete free-body diagrams because they don't show all the forces acting on each body. Which forces are missing in these diagrams? (Compare the complete free-body diagrams in Fig. 4.24.)

On the next page we present some real-life situations and the corresponding free-body diagrams. Note that while the situations depicted in the following figures occur in very different surroundings, the free-body diagrams are remarkably similar.

A sprinter gains a large forward acceleration at the start of a race by kicking back hard against the angled starting blocks. The blocks exert an equally large reaction force $\vec{n}$ on the sprinter. This force has a large forward component n_x that springs her into motion, as well as a smaller upward component n_y. If this upward component has the same magnitude as the sprinter's weight w, the net vertical force acting on her is zero and she does not accelerate vertically.

Immersed in water, a person's body experiences an upward force $\vec{B}$ due to buoyancy. This is balanced by the downward force of the diver's weight $\vec{w}$. In this situation the diver's motion depends on the force with which the water pushes on her, due either to the water currents or to reaction forces to the swimmer's arm and leg movements.

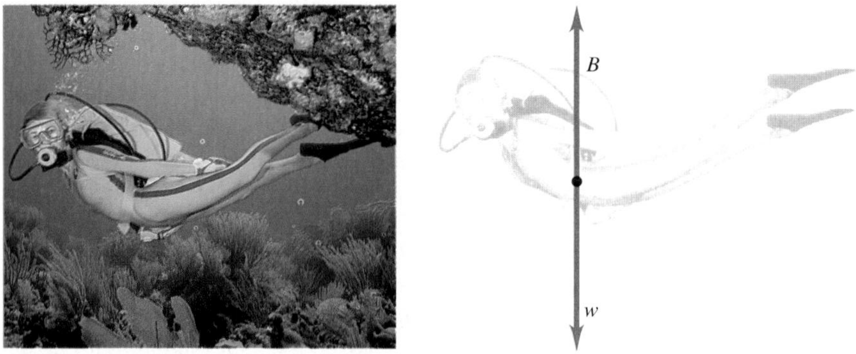

A basketball player jumps by pushing with his feet against the floor. The forces acting on him are the reaction force with which the floor pushes back up and his weight $\vec{w}$. Once the player is in the air, the only force acting on him is his weight; his acceleration is downward, even as he rises up. His stationary opponent is acted on by his own weight and by the upward normal force $\vec{n}$ exerted on him by the floor.

SUMMARY

Force is a quantitative measure of the interaction between two bodies. It is a vector quantity. When several forces act on a body, the effect on its motion is the same as when a single force, equal to the vector sum (resultant) of the forces, acts on the body. (See Example 4.1)

$$\vec{R} = \vec{F}_1 + \vec{F}_2 + \vec{F}_3 + \cdots = \sum \vec{F} \quad (4.1)$$

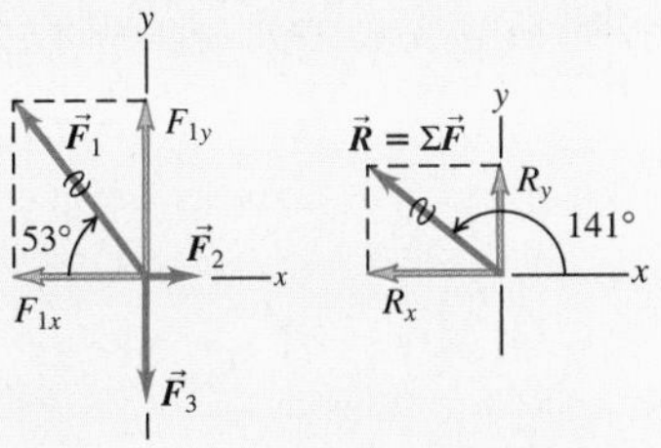

Newton's first law states that when the vector sum of all forces acting on a body (the *net force*) is zero, the body is in equilibrium. If the body is initially at rest, it remains at rest; if it is initially in motion, it continues to move with constant velocity. This law is valid only in inertial frames of reference. (See Examples 4.2 and 4.3)

$$\sum \vec{F} = \mathbf{0} \quad (4.3)$$

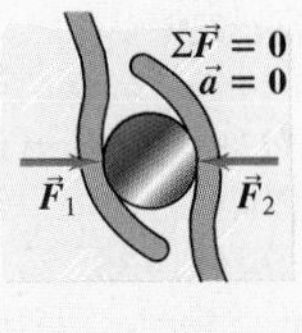

The inertial properties of a body are characterized by its *mass.* The acceleration of a body under the action of a given set of forces is directly proportional to the vector sum of the forces (the *net force*) and inversely proportional to the mass of the body. This relationship is Newton's second law. Like Newton's first law, this law is valid only in inertial frames of reference. (See Examples 4.4 and 4.5)

$$\sum \vec{F} = m\vec{a} \quad (4.7)$$

$$\sum F_x = ma_x$$
$$\sum F_y = ma_y$$
$$\sum F_z = ma_z \quad (4.8)$$

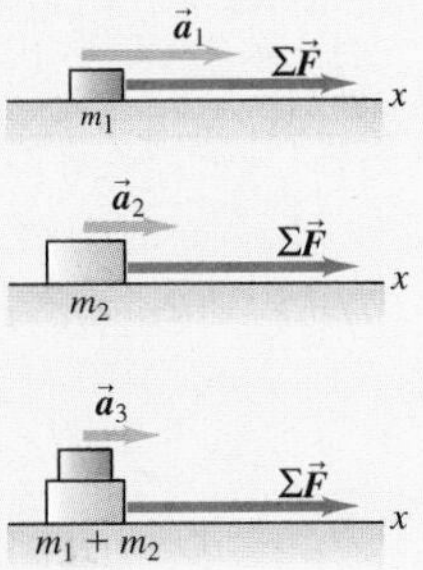

The unit of force is defined in terms of the units of mass and acceleration. In SI units, the unit of force is the newton (N), equal to 1 kg · m/s^2.

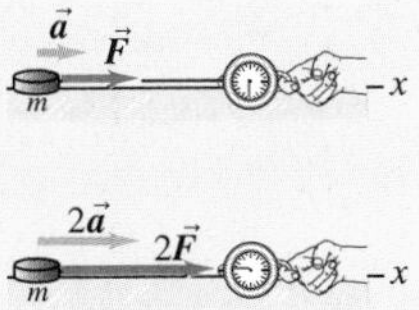

The weight of a body is the gravitational force exerted on it by the earth (or whatever other body exerts the gravitational force). Weight is a force and is therefore a vector quantity. The magnitude of the weight of a body at any specific location is equal to the product of its mass m and the magnitude of the acceleration due to gravity g at that location. While the weight of a body depends on its location, the mass is independent of location. (See Examples 4.6 and 4.7)

$$w = mg \quad (4.9)$$

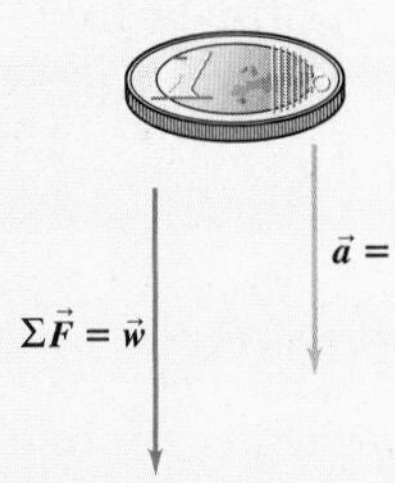

Newton's third law states that "action equals reaction." That is, when two bodies interact, they exert forces on each other that at each instant are equal in magnitude and opposite in direction. Each force in an action–reaction pair acts on only one of the two bodies; the action and reaction forces never act on the same body. (See Examples 4.8 through 4.11)

$$\vec{F}_{A \text{ on } B} = -\vec{F}_{B \text{ on } A} \quad (4.11)$$

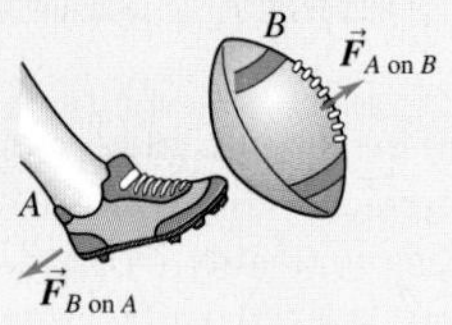

Key Terms

dynamics, 119
Newton's laws of motion, 120
classical (Newtonian) mechanics, 120
force, 120
contact force, 120
long-range forces, 120
weight, 120
superposition of forces, 121
net force, 122
Newton's first law of motion, 124
inertia, 124
normal force, 125
equilibrium, 125
inertial frame of reference, 126
mass, 129
kilogram, 129
newton, 130
Newton's second law of motion, 131
Newton's third law of motion, 138
action-reaction pair, 138
tension, 142
free-body diagram, 143

Your Notes

Answer to Chapter Opening Question

What determines the motion of the man and sled together are the forces that *other* bodies exert on them. In particular, the man pushes backward on the ground with his feet, and the ground pushes forward on him with a force of equal magnitude (in accordance with Newton's third law). The ground also exerts a backward friction force on the sled. Thus, to pull the sled forward, the force that the man exerts on the ground must be at least as great as the friction force on the sled.

Answers to Test Your Understanding Questions

Section 4.1 We demand that the net force be zero, so

$$F_{1x} + F_{2x} + F_{3x} = 0 \quad \text{and} \quad F_{1y} + F_{2y} + F_{3y} = 0$$

Hence

$$F_{3x} = -(F_{1x} + F_{2x}) = -[(-150\text{ N}) + 50\text{ N}] = 100\text{ N}$$

$$F_{3y} = -(F_{1y} + F_{2y}) = -(200\text{ N} + 0\text{ N}) = -200\text{ N}$$

The magnitude is

$$F_3 = \sqrt{F_{3x}^2 + F_{3y}^2} = \sqrt{(100\text{ N})^2 + (-200\text{ N})^2} = 224\text{ N}$$

and the direction is

$$\theta = \arctan\left(\frac{F_{3y}}{F_{3x}}\right) = \arctan\left(\frac{100\text{ N}}{-200\text{ N}}\right) = \arctan\left(-\frac{1}{2}\right) = -27°$$

Section 4.2 There is zero net force in situations (i), (ii), and (iv). In situation (iii), the hawk is moving in a circle; hence it is accelerating and is not in equilibrium. In situation (iv), the box remains stationary (as seen in the inertial reference frame of the ground) as the truck accelerates forward, like the skater in Fig. 4.8a.

Section 4.3 The acceleration is equal to the net force divided by the mass. Hence the magnitude of the acceleration is $a = (128\text{ N})/(8.0\text{ kg}) = 16\text{ m/s}^2$. The acceleration is in the same direction as the net force, $\theta = 141°$. From Eq. (2.12), after 0.50 s the belt will have moved a distance $\frac{1}{2}at^2 = (1/2)(16\text{ m/s}^2)(0.50\text{ s})^2 = 2.0\text{ m}$.

Section 4.4 It would take twice the effort for the astronaut to walk around because her weight on the planet would be twice as much as on the earth. But it would be just as easy to catch the ball. The ball's *mass* remains the same, so the horizontal force the astronaut would have to exert to bring it to a stop (i.e., to give it the same acceleration) would be the same as on earth.

Section 4.5 By Newton's third law, the two forces have equal magnitudes. Because the car has much greater mass than the mosquito, it undergoes only a tiny, imperceptible acceleration in response to the force of the impact. By contrast, the mosquito, with its miniscule mass, undergoes a catastrophically large acceleration.

Discussion Questions

Q4.1 Can a body be in equilibrium when only one force acts on it? Explain.

Q4.2 A ball thrown straight up has zero velocity at its highest point. Is the ball in equilibrium at this point? Why or why not?

Q4.3 A helium balloon hovers in midair, neither ascending nor descending. Is it in equilibrium? What forces act on it?

Q4.4 When you fly in an airplane at night in smooth air, there is no sensation of motion, even though the plane may be moving at 800 km/h (500 mi/h). Why is this?

Q4.5 If the two ends of a rope in equilibrium are pulled with forces of equal magnitude and opposite direction, why is the total tension in the rope not zero?

Q4.6 You tie a brick to the end of a rope and whirl the brick around you in a horizontal circle. Describe the path of the brick after you suddenly let go of the rope.

Q4.7 When a car stops suddenly, the passengers tend to move forward relative to their seats. Why? When a car makes a sharp turn, the passengers tend to slide to one side of the car. Why?

Q4.8 Some people say that the "force of inertia" (or "force of momentum") throws the passengers forward when a car brakes sharply. What is wrong with this explanation?

Q4.9 A passenger in a moving bus with no windows notices that a ball that has been at rest in the aisle suddenly starts to move toward the rear of the bus. Think of two different possible explanations, and devise a way to decide which is correct.

Q4.10 Suppose you chose the fundamental SI units to be force, length, and time instead of mass, length, and time. What would be the units of mass in terms of those fundamental units?

Q4.11 Inertia is *not* a force that keeps things in place or that keeps them moving. How do we know this?

Q4.12 Why is the earth only approximately an inertial reference frame?

Q4.13 Does Newton's second law hold true for an observer in a van as it speeds up, slows down, or rounds a corner? Explain.

Q4.14 Some students refer to the quantity $m\vec{a}$ as "the force of acceleration." Is it correct to refer to this quantity as a force? If so, what exerts this force? If not, what is a better description of this quantity?

Q4.15 The acceleration of a falling body is measured in an elevator traveling upward at a constant speed of 9.8 m/s. What result is obtained?

Q4.16 You can play catch with a softball in a bus moving with constant speed on a straight road, just as though the bus were at rest. Is this still possible when the bus is making a turn at constant speed on a level road? Why or why not?

Q4.17 Is the belt of Example 4.1 in equilibrium? Explain.

Q4.18 The head of a hammer begins to come loose from its wooden handle. How should you strike the handle on a concrete sidewalk to reset the head? Why does this work?

Q4.19 Why can it hurt your foot more to kick a big rock than a small pebble? *Must* the big rock hurt more? Explain.

Q4.20 "It's not the fall that hurts you; it's the sudden stop at the bottom." Translate this saying into the language of Newton's laws of motion.

Q4.21 A person can dive into water from a height of 10 m without injury, but a person who jumps off the roof of a 10-m-tall building and lands on a concrete street is likely to be seriously injured. Why is there a difference?

Q4.22 Why are cars designed to crumple up in front and back for safety? Why not for side collisions and rollovers?

Q4.23 When a bullet is fired from a gun, what is the origin of the force that accelerates the bullet?

Q4.24 When a string barely strong enough lifts a heavy weight, it can lift the weight by a steady pull; but if you jerk the string, it will break. Explain in terms of Newton's laws of motion.

Q4.25 A large crate is suspended from the end of a vertical rope. Is the tension in the rope larger when the crate is at rest or when it is moving upward at constant speed? If the crate is traveling upward, is the tension in the rope greatest when the crate is speeding up or when it is slowing down? In each case explain in terms of Newton's laws of motion.

Q4.26 Which feels a greater pull due to the earth's gravity, a 10-kg stone or a 20-kg stone? If you drop them, why does the 20-kg stone not fall with twice the acceleration of the 10-kg stone? Explain your reasoning.

Q4.27 Why is it incorrect to say that 1.0 kg *equals* 2.2 lb?

Q4.28 A horse is hitched to a wagon. Since the wagon pulls back on the horse just as hard as the horse pulls on the wagon, why does the wagon not remain in equilibrium, no matter how hard the horse pulls?

Q4.29 A 450-N girl from the south slaps a 800-N boy from the north. Her fingers exert a force of 30 N to the west on his cheek. There may be other reactions, but by Newton's third law, what is the reaction *force* to the slap?

Q4.30 A large truck and a small compact car have a head-on collision. During the collision the truck exerts a force $\vec{F}_{\text{T on C}}$ on the car, and the car exerts a force $\vec{F}_{\text{C on T}}$ on the truck. Which force has the larger magnitude, or are they the same? Does your answer depend on how fast each vehicle was moving before the collision? Why or why not?

Q4.31 If you ask most people what force makes a car accelerate forward, they will say "the force of the engine." But what force is *directly* responsible for making the car accelerate?

Q4.32 A small compact car is pushing a large van that has broken down, and they travel along the road with equal velocities and accelerations. While the car is speeding up, is the force it exerts on the van larger than, smaller than, or the same magnitude as the force the van exerts on it? Which object, the car or the van, has the largest net force on it, or are the net forces the same? Explain.

Q4.33 Consider a tug-of-war between two people who pull in opposite directions on the ends of a rope. By Newton's third law, the force that A exerts on B is just as great as the force that B exerts on A. So what determines who wins? (*Hint:* Draw a free-body diagram showing all the forces that act on each person.)

Q4.34 On the moon, $g = 1.62\ \text{m/s}^2$. If a 2-kg brick drops on your foot from a height of 2 m, will this hurt more, or less, or the same if it happens on the moon instead of on the earth? Explain. If a 2-kg brick is thrown and hits you when it is moving horizontally at 6 m/s, will this hurt more, less, or the same if it happens on the moon instead of on the earth? Explain. (On the moon, assume that you are inside a pressurized structure, so you are not wearing a spacesuit.)

Q4.35 A manual for student pilots contains the following passage: "When an airplane flies at a steady altitude, neither climbing nor descending, the upward lift force from the wings equals the airplane's weight. When the airplane is climbing at a steady rate, the upward lift is greater than the weight; when the airplane is descending at a steady rate, the upward lift is less than the weight." Are these statements correct? Explain.

Q4.36 If your hands are wet and no towel is handy, you can remove some of the excess water by shaking them. Why does this get rid of the water?

Q4.37 If you are squatting down (such as when you are examining the books on the bottom shelf in a library or bookstore) and suddenly get up, you can temporarily feel light-headed. What do Newton's laws of motion have to say about why this happens?

Q4.38 When a car is hit from behind, the passengers can receive a whiplash. Use Newton's laws of motion to explain what causes this to occur.

Q4.39 In a head-on auto collision, passengers not wearing seat belts can be thrown through the windshield. Use Newton's laws of motion to explain why this happens.

Q4.40 In a head-on collision between a compact 1000-kg car and a large 2500-kg car, which one experiences the greater force? Explain. Which one experiences the greater acceleration? Explain why. Now explain why passengers in the smaller car are more likely to be injured than those in the large car, even if the bodies of both cars are equally strong.

Q4.41 Suppose you are in a rocket with no windows, traveling in deep space far from any other objects. Without looking outside the rocket or making any contact with the outside world, explain how you could determine if the rocket is i) moving forward at a constant 80% of the speed of light; ii) accelerating in the forward direction.

Exercises

Section 4.1 Force and Interactions

4.1 Two forces have the same magnitude F. What is the angle between the two vectors if their sum has a magnitude of a) $2F$? b) $\sqrt{2}F$? c) zero? Sketch the three vectors in each case.

4.2 Instead of using the x- and y-axes of Fig. 4.5 to analyze the situation of Example 4.1, use axes rotated 37.0° counterclockwise, so the y-axis is parallel to the 250-N force. a) For these axes find the x- and y-components of the net force on the belt. b) From the components computed in part (a) find the magnitude and direction of the net force. Compare your results to Example 4.1.

4.3 A warehouse worker pushes a crate along the floor, as shown in Fig. 4.1b, by a force of 10 N that points downward at an angle of 45° below the horizontal. Find the horizontal and vertical components of the force.

4.4 A man is dragging a trunk up the loading ramp of a mover's truck. The ramp has a slope angle of 20.0°, and the man pulls upward with a force $\vec{F}$ whose direction makes an angle of 30.0° with the ramp (Fig. 4.26). a) How large a force $\vec{F}$ is necessary for the component F_x parallel to the ramp to be 60.0

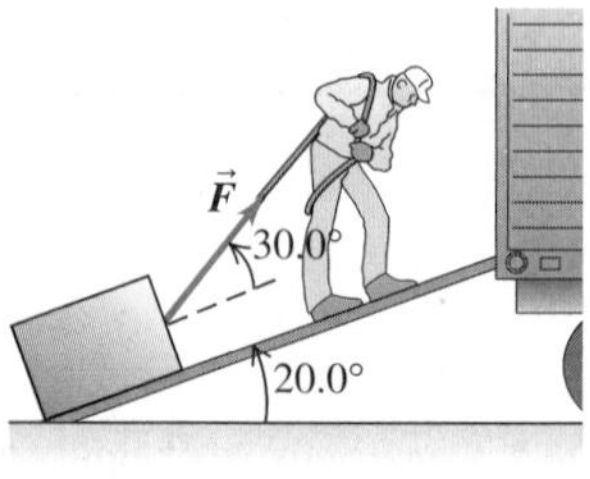

Figure 4.26 Exercise 4.4.

N? b) How large will the component F_y perpendicular to the ramp then be?

4.5 Two dogs pull horizontally on ropes attached to a post; the angle between the ropes is 60.0°. If dog *A* exerts a force of 270 N and dog *B* exerts a force of 300 N, find the magnitude of the resultant force and the angle it makes with dog *A*'s rope.

4.6 Two forces, $\vec{F}_1$ and $\vec{F}_2$, act at a point. The magnitude of $\vec{F}_1$ is 9.00 N, and its direction is 60.0° above the x-axis in the second quadrant. The magnitude of $\vec{F}_2$ is 6.00 N, and its direction is 53.1° below the x-axis in the third quadrant. a) What are the x- and y-components of the resultant force? b) What is the magnitude of the resultant force?

Section 4.3 Newton's Second Law

4.7 If a net horizontal force of 132 N is applied to a person with mass 60 kg who is resting on the edge of a swimming pool, what horizontal acceleration is produced?

4.8 What magnitude of net force is required to give a 135-kg refrigerator an acceleration of magnitude 1.40 m/s^2?

4.9 A box rests on a frozen pond, which serves as a frictionless horizontal surface. If a fisherman applies a horizontal force with magnitude 48.0 N to the box and produces an acceleration of magnitude 3.00 m/s^2, what is the mass of the box?

4.10 A dockworker applies a constant horizontal force of 80.0 N to a block of ice on a smooth horizontal floor. The frictional force is negligible. The block starts from rest and moves 11.0 m in 5.00 s. a) What is the mass of the block of ice? b) If the worker stops pushing at the end of 5.00 s, how far does the block move in the next 5.00 s?

4.11 A hockey puck with mass 0.160 kg is at rest at the origin ($x = 0$) on the horizontal, frictionless surface of the rink. At time $t = 0$ a player applies a force of 0.250 N to the puck, parallel to the x-axis; he continues to apply this force until $t = 2.00$ s. a) What are the position and speed of the puck at $t = 2.00$ s? b) If the same force is again applied at $t = 5.00$ s, what are the position and speed of the puck at $t = 7.00$ s?

4.12 A crate with mass 32.5 kg initially at rest on a warehouse floor is acted on by a net horizontal force of 140 N. a) What acceleration is produced? b) How far does the crate travel in 10.0 s? c) What is its speed at the end of 10.0 s?

4.13 A hockey puck moves from point *A* to point *B* at constant velocity while under the influence of several forces. a) What can you say about the forces? b) Graph the puck's path from *A* to *B*. c) On the graph, continue the path to point *C* if a new constant force is added to the puck at point *B*, where the new force is perpendicular to the puck's velocity at point *B*. d) Continue the path to point *D* if at point *C* the constant force added at *B* is replaced by one of constant magnitude but in a direction always perpendicular to the path of the puck.

4.14 An electron (mass $= 9.11 \times 10^{-31}$ kg) leaves one end of a TV picture tube with zero initial speed and travels in a straight line to the accelerating grid, which is 1.80 cm away. It reaches the grid with a speed of 3.00×10^6 m/s. If the accelerating force is constant, compute a) the acceleration; b) the time to reach the grid; c) the net force, in newtons. (You can ignore the gravitational force on the electron.)

Section 4.4 Mass and Weight

4.15 Superman throws a 2400-N boulder at an adversary. What horizontal force must Superman apply to the boulder to give it a horizontal acceleration of 12.0 m/s^2?

4.16 A bowling ball weighs 71.2 N (16.0 lb). The bowler applies a horizontal force of 160 N (36.0 lb) to the ball. What is the magnitude of the horizontal acceleration of the ball?

4.17 At the surface of Jupiter's moon Io, the acceleration due to gravity is $g = 1.81$ m/s^2. A watermelon weighs 44.0 N at the surface of the earth. a) What is its mass on the earth's surface? b) What are its mass and weight on the surface of Io?

4.18 a) What is the mass of a book that weighs 3.20 N at a point where $g = 9.80$ m/s^2? b) At the same location, what is the weight of a dog whose mass is 14.0 kg?

Section 4.5 Newton's Third Law

4.19 World-class sprinters can accelerate out of the starting blocks with an acceleration that is nearly horizontal and has magnitude 15 m/s^2. How much horizontal force must a 55-kg sprinter exert on the starting blocks during a start to produce this acceleration? Which body exerts the force that propels the sprinter: the blocks or the sprinter herself?

4.20 Imagine that you are holding a book weighing 4 N at rest on the palm of your hand. Complete the following sentences: a) A downward force of magnitude 4 N is exerted on the book by __________. b) An upward force of magnitude __________ is exerted on __________ by the hand. c) Is the upward force in part (b) the reaction to the downward force in part (a)? d) The reaction to the force in part (a) is a force of magnitude __________, exerted on __________ by __________. Its direction is __________. e) The reaction to the force in part (b) is a force of magnitude __________, exerted on __________ by __________. Its direction is __________. f) The forces in parts (a) and (b) are equal and opposite because of Newton's __________ law. g) The forces in parts (b) and (e) are equal and opposite because of Newton's __________ law. Now suppose that you exert an upward force of magnitude 5 N on the book. h) Does the book remain in equilibrium? i) Is the force exerted on the book by your hand equal and opposite to the force exerted on the book by the earth? j) Is the force exerted on the book by the earth equal and opposite to the force exerted on the earth by the book? k) Is the force exerted on the book by your hand equal and opposite to the force exerted on your hand by the book? Finally, suppose that you snatch your hand away while the book is moving upward. l) How many forces then act on the book? m) Is the book in equilibrium?

4.21 A bottle is given a push along a tabletop and slides off the edge of the table. Do *not* ignore air resistance. a) What forces are exerted on the bottle while it is falling from the table to the floor? b) What is the reaction to each force; that is, on which body and by which body is the reaction exerted?

4.22 The upward normal force exerted by the floor is 620 N on an elevator passenger who weighs 650 N. What are the reaction forces to these two forces? Is the passenger accelerating? If so, what are the magnitude and direction of the acceleration?

4.23 A student with mass 45 kg jumps off a high diving board. Using 6.0×10^{24} kg for the mass of the earth, what is the acceleration of the earth toward her as she accelerates toward the earth with an acceleration of 9.8 m/s^2? Assume that the net force on the earth is the force of gravity she exerts on it.

Section 4.6 Free-Body Diagrams

4.24 Two crates, A and B, sit at rest side-by-side on a frictionless horizontal surface, as shown in Fig. 4.27. The crates have masses m_A and m_B. A horizontal force $\vec{F}$ is applied to crate A and the two crates move off to the right. a) Draw clearly labeled free-body diagrams for crate A and for crate B. Indicate which pairs of forces, if any, are third-law action-reaction pairs. b) If the magnitude of force $\vec{F}$ is less than the total weight of the two crates, will it cause the crates to move? Explain.

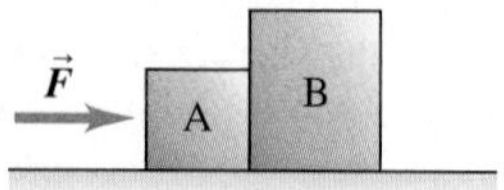

Figure 4.27 Exercise 4.24.

4.25 A ball is hanging from a long string that is tied to the ceiling of a train car that is traveling eastward on horizontal tracks and speeding up. An observer inside the train car sees the ball hang motionless. Draw a clearly labeled free-body diagram for the ball. Is the net force on the ball zero? Explain.

4.26 A large box containing your new computer sits on the bed of your pickup truck. You are stopped at a red light. The light turns green and you stomp on the gas and the truck accelerates. To your horror, the box starts to slide toward the back of the truck. Draw a clearly labeled free-body diagram for the truck and for the box. Indicate pairs of forces, if any, that are third-law action-reaction pairs.

4.27 A chair of mass 12.0 kg is sitting on the horizontal floor; the floor is not frictionless. You push on the chair with a force $F = 40.0$ N that is directed at an angle of 37.0° below the horizontal and the chair slides along the floor. a) Draw a clearly labeled free-body diagram for the chair. b) Use your diagram and Newton's laws to calculate the normal force that the floor exerts on the chair.

4.28 A skier of mass 65.0 kg is pulled up a snow-covered slope at constant speed by a tow rope that is parallel to the ground. The ground slopes upward at a constant angle of 26.0° above the horizontal and you can ignore friction. a) Draw a clearly labeled free-body diagram for the skier. b) Calculate the tension in the tow rope.

4.29 Your nephew Frank is riding his tricycle. The tricycle is connected by a light horizontal rope to a wagon in which his dog Spot is sitting. Treat the tricycle and Frank as one object and Spot and the wagon as one object. Draw a clearly labeled free-body diagram for each object. Indicate which pairs of forces, if any, form third-law action-reaction pairs.

Problems

4.30 A .22 rifle bullet, traveling at 350 m/s, strikes a block of soft wood, which it penetrates to a depth of 0.130 m. The block of wood is clamped in place and doesn't move. The mass of the bullet is 1.80 g. Assume a constant retarding force. a) How much time is required for the bullet to stop? b) What force, in newtons, does the wood exert on the bullet?

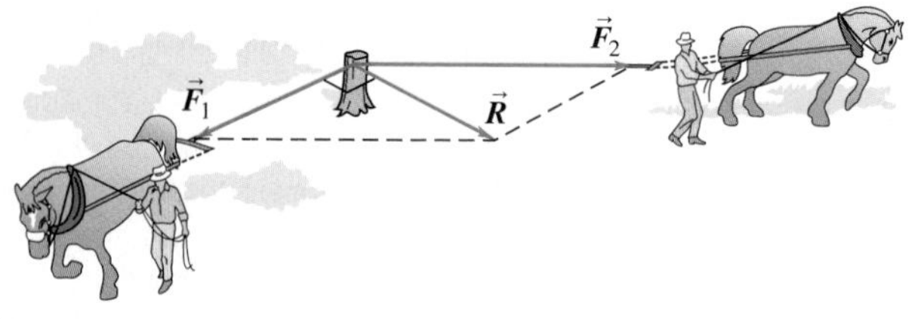

Figure 4.28 Problem 4.31.

4.31 Two horses pull horizontally on ropes attached to a stump. The two forces $\vec{F}_1$ and $\vec{F}_2$ that they apply to the stump are such that the net (resultant) force $\vec{R}$ has a magnitude equal to that of $\vec{F}_1$ and makes an angle of 90° with $\vec{F}_1$ (Fig. 4.28). Let $F_1 = 1300$ N and $R = 1300$ N also. Find the magnitude of $\vec{F}_2$ and its direction (relative to $\vec{F}_1$).

4.32 You have just landed on Planet X. You take out a 100-g ball, release it from rest from a height of 10.0 m and measure that it takes 2.2 s to reach the ground. You can ignore any force on the ball from the atmosphere of the planet. How much does the 100-g ball weigh on the surface of Planet X?

4.33 Two adults and a child want to push a wheeled cart in the direction marked x in Fig. 4.29. The two adults push with horizontal forces $\vec{F}_1$ and $\vec{F}_2$ as shown in the figure. a) Find the magnitude and direction of the *smallest* force that the child should exert. You can ignore the effects of friction. b) If the child exerts the minimum force found in part (a), the cart accelerates at 2.0 m/s^2 in the $+x$-direction. What is the weight of the cart?

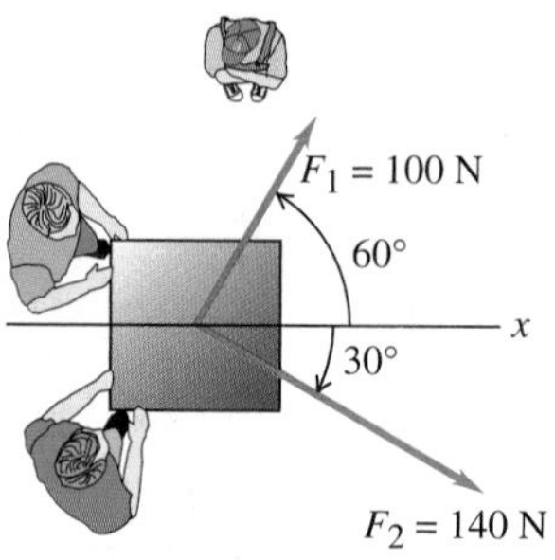

Figure 4.29 Problem 4.33.

4.34 An oil tanker's engines have broken down, and the wind is blowing the tanker straight toward a reef at a constant speed of 1.5 m/s (Fig. 4.30). When the tanker is 500 m from the reef, the wind dies down just as the engineer gets the engines going again. The rudder is stuck, so the only choice is to try to accelerate straight backward away from the reef. The mass of the tanker and cargo is 3.6×10^7 kg, and the engines produce a net horizontal force of 8.0×10^4 N on the tanker. Will the ship hit the reef? If it does, will the oil be safe? The hull can withstand an impact at a speed of 0.2 m/s or less. You can ignore the retarding force of the water on the tanker's hull.

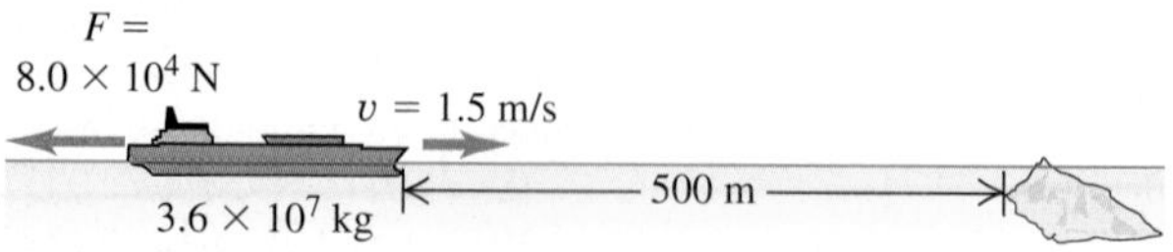

Figure 4.30 Problem 4.34.

4.35 A Standing Vertical Jump. Basketball player Darrell Griffith is on record as attaining a standing vertical jump of 1.2 m (4 ft). (This means that he moved upward by 1.2 m after his feet left the floor.) Griffith weighed 890 N (200 lb). a) What is his speed as he leaves the floor? b) If the time of the part of the jump before his feet left the floor was 0.300 s, what was his average acceleration (magnitude and direction) while he was pushing against the floor? c) Draw his free-body diagram (see Section 4.6). In terms of the forces on the diagram, what is the net force on him? Use Newton's laws and the results of part (b) to calculate the average force he applied to the ground.

4.36 An advertisement claims that a particular automobile can "stop on a dime." What net force would actually be necessary to stop a 850-kg automobile traveling initially at 45.0 km/h in a distance equal to the diameter of a dime, which is 1.8 cm?

4.37 A 4.80-kg bucket of water is accelerated upward by a cord of negligible mass whose breaking strength is 75.0 N. a) Draw the free-body force diagram for the bucket. In terms of the forces on your diagram, what is the net force on the bucket? b) Apply Newton's second law to the bucket and find the maximum upward acceleration that can be given to the bucket without breaking the cord.

4.38 A parachutist relies on air resistance (mainly on her parachute) to decrease her downward velocity. She and her parachute have a mass of 55.0 kg and air resistance exerts a total upward force of 620 N on her and her parachute. a) What is the weight of the parachutist? b) Draw a free-body diagram for the parachutist (see Section 4.6). Use that diagram to calculate the net force on the parachutist. Is the net force upward or downward? c) What is the acceleration (magnitude and direction) of the parachutist?

4.39 Two crates, one with mass 4.00 kg and the other with mass 6.00 kg, sit on the frictionless surface of a frozen pond, connected by a light rope (Fig. 4.31). A woman wearing golf shoes (so she can get traction on the ice) pulls horizontally on the 6.00-kg crate with a force F that gives the crate an acceleration of 2.50 m/s^2. a) What is the acceleration of the 4.00-kg crate? b) Draw a free-body diagram for the 4.00-kg crate. Use that diagram and Newton's second law to find the tension T in the rope that connects the two crates. c) Draw a free-body diagram for the 6.00-kg crate. What is the direction of the net force on the 6.00-kg crate? Which is larger in magnitude, force T or force F? d) Use part (c) and Newton's second law to calculate the magnitude of the force F.

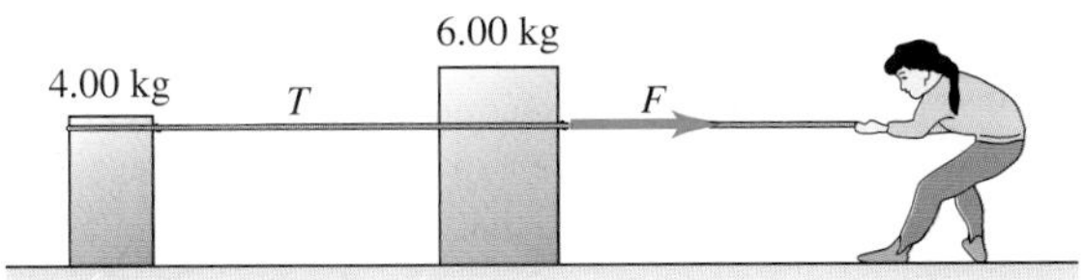

Figure 4.31 Problem 4.39.

4.40 An astronaut is tethered by a strong cable to a spacecraft. The astronaut and her spacesuit have a total mass of 105 kg, while the mass of the cable is negligible. The mass of the spacecraft is 9.05×10^4 kg. The spacecraft is far from any large astronomical bodies, so we can ignore the gravitational forces on it and the astronaut. We also assume that both the spacecraft and the astronaut are initially at rest in an inertial reference frame. The astronaut then pulls on the cable with a force of 80.0 N. a) What force does the cable exert on the astronaut? b) Since $\sum \vec{F} = m\vec{a}$, how can a "massless" $(m = 0)$ cable exert a force? c) What is the astronaut's acceleration? d) What force does the cable exert on the spacecraft? e) What is the acceleration of the spacecraft?

4.41 To study damage to aircraft that collide with large birds, you design a test gun that will accelerate chicken-sized objects so that their displacement along the gun barrel is given by $x = (9.0 \times 10^3\ \text{m/s}^2)t^2 - (8.0 \times 10^4\ \text{m/s}^3)t^3$. The object leaves the end of the barrel at $t = 0.025$ s. a) How long must the gun barrel be? b) What will be the speed of the objects as they leave the end of the barrel? c) What net force must be exerted on a 1.50-kg object at i) $t = 0$? ii) $t = 0.025$ s?

4.42 A spacecraft descends vertically near the surface of Planet X. An upward thrust of 25.0 kN from its engines slows it down at a rate of 1.20 m/s^2, but it speeds up at a rate of 0.80 m/s^2 with an upward thrust of 10.0 kN. a) In each case, what is the direction of the acceleration of the spacecraft? b) Draw a free-body diagram for the spacecraft. In each case, speeding up or slowing down, what is the direction of the net force on the spacecraft? c) Apply Newton's second law to each case, slowing down or speeding up, and use this to find the spacecraft's weight near the surface of Planet X.

4.43 A train (an engine plus four cars) running on level tracks is speeding up with an acceleration of magnitude $|\vec{a}|$. If each car has a mass m and if each car has negligible frictional forces acting on it, what is a) the force of the engine on the first car? b) the force of the first car on the second car? c) the force of the second car on the third car? d) the force of the fourth car on the third car? e) What would these same four forces be if the train were slowing down with an acceleration of magnitude $|\vec{a}|$? Your answer to each question must include a clearly labeled free-body diagram. In each case indicate what body you are considering.

4.44 A gymnast of mass m climbs a vertical rope attached to the ceiling. You can ignore the weight of the rope. Draw a free-body diagram for the gymnast. Calculate the tension in the rope if the gymnast a) climbs at a constant rate; b) hangs motionless on the rope; c) accelerates up the rope with an acceleration of magnitude $|\vec{a}|$; d) slides down the rope with a downward acceleration of magnitude $|\vec{a}|$.

4.45 A loaded elevator with very worn cables has a total mass of 2200 kg, and the cables can withstand a maximum tension of 28,000 N. a) Draw the free-body force diagram for the elevator. In terms of the forces on your diagram, what is the net force on the elevator? Apply Newton's second law to the elevator and find the maximum upward acceleration for the elevator if the cables are not to break? b) What would be the answer to part (a) if the elevator were on the moon, where $g = 1.62$ m/s^2?

4.46 Jumping to the Ground. A 75.0-kg man steps off a platform 3.10 m above the ground. He keeps his legs straight as he falls, but at the moment his feet touch the ground his knees begin to bend, and, treated as a particle, he moves an additional 0.60 m before coming to rest. a) What is his speed at the instant his feet touch the ground? b) Treating him as a particle, what is his acceleration (magnitude and direction) as he slows down, if the acceleration is assumed to be constant? c) Draw his free-body diagram (see Section 4.6). In terms of the forces on the diagram, what is the net force on him? Use Newton's laws and the results of part (b) to calculate the average force his feet exert on the ground while he slows down. Express this force in newtons and also as a multiple of his weight.

4.47 A 4.9-N hammer head is stopped from an initial downward velocity of 3.2 m/s in a distance of 0.45 cm by a nail in a pine board. In addition to its weight, there is a 15-N downward force on the hammer head applied by the person using the hammer. Assume that the acceleration of the hammer head is constant while it is in contact with the nail and moving downward. a) Draw a free-body diagram for the hammer head. Identify the reaction force to each force in the diagram. b) Calculate the downward force $\vec{F}$ exerted by the hammer head on the nail while the hammer head is in contact with the nail and moving downward. c) Suppose the nail is in hardwood and the distance the hammer head travels in coming to rest is only 0.12 cm. The downward forces on the hammer head are the same as on part (b). What then is the force $\vec{F}$ exerted by the hammer head on the nail while the hammer head is in contact with the nail and moving downward?

4.48 A uniform cable of weight w hangs vertically downward, supported by an upward force of magnitude w at its top end. What is the tension in the cable a) at its top end? b) at its bottom end? c) at its middle? Your answer to each part must include a free-body diagram. (*Hint:* For each question choose the body to analyze to be a section of the cable or a point along the cable.) d) Graph the tension in the rope versus the distance from its top end.

4.49 The two blocks in Fig. 4.32 are connected by a heavy uniform rope with a mass of 4.00 kg. An upward force of 200 N is applied as shown. a) Draw three free-body diagrams, one for the 6.00-kg block, one for the 4.00-kg rope, and another one for the 5.00-kg block. For each force, indicate what body exerts that force. b) What is the acceleration of the system? c) What is the tension at the top of the heavy rope? d) What is the tension at the midpoint of the rope?

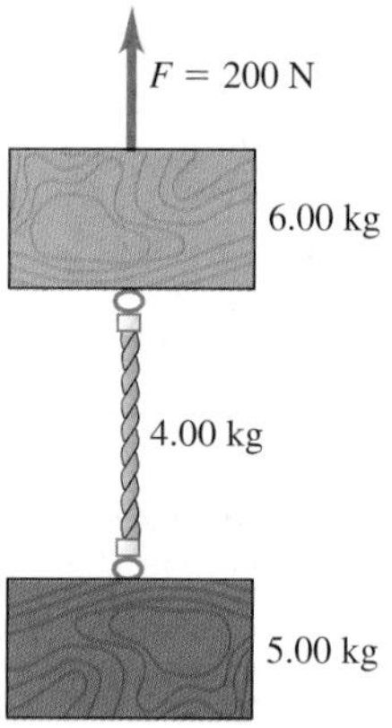

Figure 4.32 Problem 4.49.

4.50 An athlete whose mass is 90.0 kg is performing weight-lifting exercises. Starting from the rest position, he lifts, with constant acceleration, a barbell that weighs 490 N. He lifts the barbell a distance of 0.60 m in 1.6 s. a) Draw a clearly labeled free-body force diagram for the barbell and for the athlete. b) Use the diagrams in part (a) and Newton's laws to find the total force that his feet exert on the ground as he lifts the barbell.

4.51 A hot-air balloon consists of a basket, one passenger, and some cargo. Let the total mass be M. Even though there is an upward lift force on the balloon, the balloon is initially accelerating downward at a rate of $g/3$. a) Draw a free-body diagram for the descending balloon. b) Find the upward lift force in terms of the initial total weight Mg. c) The passenger notices that he is heading straight for a waterfall and decides he needs to go up. What fraction of the total weight must he drop overboard so that the balloon accelerates *upward* at a rate of $g/2$? Assume that the upward lift force remains the same.

4.52 A student tries to raise a chain consisting of three identical links. Each link has a mass of 300 g. The three-piece chain is connected to a string and then suspended vertically, with the student holding the upper end of the string and pulling upward. Because of the student's pull, an upward force of 12 N is applied to the chain by the string. a) Draw a free-body diagram for *each* of the links in the chain and also for the entire chain considered as a single body. b) Use the results of part (a) and Newton's laws to find (i) the acceleration of the chain; and (ii) the force exerted by the top link on the middle link.

4.53 The position of a 2.75×10^5 N training helicopter under test is given by $\vec{r} = (0.020\ \text{m/s}^3)t^3\,\hat{\imath} + (2.2\ \text{m/s})t\hat{\jmath} - (0.060\ \text{m/s}^2)t^2\hat{k}$. Find the net force on the helicopter at $t = 5.0$ s.

4.54 An object with mass m moves along the x-axis. Its position as a function of time is given by $x(t) = At - Bt^3$, where A and B are constants. Calculate the net force on the object as a function of time.

***4.55** An object with mass m initially at rest is acted on by a force $\vec{F} = k_1\hat{\imath} + k_2t^3\hat{\jmath}$, where k_1 and k_2 are constants. Calculate the velocity $\vec{v}(t)$ of the object as a function of time.

Challenge Problems

***4.56** If we know $F(t)$, the force as a function of time, for straight-line motion, Newton's second law gives us $a(t)$, the acceleration as a function of time. We can then integrate $a(t)$ to find $v(t)$ and $x(t)$. However, suppose we know $F(v)$ instead. a) The net force on a body moving along the x-axis equals $-Cv^2$. Use Newton's second law written as $\sum F = m\,dv/dt$ and two integrations to show that $x - x_0 = (m/C)\ln(v_0/v)$. b) Show that Newton's second law can be written as $\sum F = mv\,dv/dx$. Derive the same expression as in part (a) using this form of the second law and one integration.

***4.57** An object of mass m is at rest in equilibrium at the origin. At $t = 0$ a new force $\vec{F}(t)$ is applied that has components

$$F_x(t) = k_1 + k_2y \qquad F_y(t) = k_3t$$

where k_1, k_2, and k_3 are constants. Calculate the position $\vec{r}(t)$ and velocity $\vec{v}(t)$ vectors as functions of time.

5 APPLYING NEWTON'S LAWS

A bird glides by making use of Newton's first and third laws. The wings push downward on the air as it flows past the bird, and the air reacts by pushing upward on the wings with a force—called *lift*—of equal magnitude and opposite direction. In a steady glide the aerodynamic forces on the bird exactly balance the downward force of gravity, so that the net external force on the bird is zero.

? **Suppose the bird is caught in an updraft so that it is ascending at a steady rate. In this situation, which has a greater magnitude: the force of gravity or the aerodynamic forces?**

We saw in Chapter 4 that Newton's three laws of motion, the foundation of classical mechanics, can be stated very simply. But *applying* these laws to situations such as an iceboat skating across a frozen lake, a toboggan sliding down a hill, or a jet airplane making a steep turn requires analytical skills and problem-solving technique. In this chapter we'll help you extend the problem-solving skills you began to develop in Chapter 4.

We begin with equilibrium problems, in which a body is at rest or moving with constant velocity. Then we generalize our problem-solving techniques to include bodies that are not in equilibrium, for which we need to deal precisely with the relationships between forces and motion. We will learn how to describe and analyze the contact force acting on a body when it rests or slides on a surface. Finally, we study the important case of uniform circular motion, in which a body moves in a circle with constant speed.

All these situations involve the concept of force, a concept we'll use throughout our study of physics. We close the chapter with a brief look at the fundamental nature of force and the classes of forces found in our physical universe.

5.1 | Using Newton's First Law: Particles in Equilibrium

?

We learned in Chapter 4 that a body is in *equilibrium* when it is at rest or moving with constant velocity in an inertial frame of reference. A hanging lamp, a suspension bridge, an airplane flying straight and level at a constant speed—all are examples of equilibrium situations. In this section we consider only equilibrium of a body that can be modeled as a particle. (In Chapter 11, we'll consider the additional principles needed when the body can't be represented adequately as a particle.) The essential physical principle is Newton's first law: When a particle is at rest or is moving with constant velocity in an inertial frame of reference, the net force acting on it—that is, the vector sum of all the forces acting on it—must be zero:

$$\sum \vec{F} = \mathbf{0} \qquad \text{(particle in equilibrium, vector form)} \qquad (5.1)$$

We most often use this equation in component form:

$$\sum F_x = 0 \qquad \sum F_y = 0 \qquad (5.2)$$

$$\text{(particle in equilibrium, component form)}$$

This section is about using Newton's first law to solve problems dealing with bodies in equilibrium. Some of these problems may seem complicated, but the important thing to remember is that *all* problems involving particles in equilibrium are done in the same way. The following strategy details the steps you need to follow for any and all such problems. (Like all Problem-Solving Strategies, this one follows the "I SEE" format—Identify, Set up, Execute, and Evaluate—that we introduced in Section 1.2.) Study this strategy carefully, look at how it's applied in the worked-out examples, and try to apply it yourself when you solve assigned problems.

Problem-Solving Strategy

Newton's First Law: Equilibrium of a Particle

IDENTIFY *the relevant concepts:* You must use Newton's first law for any problem that involves forces acting on a body in equilibrium. Remember that "equilibrium" means that the body either remains at rest or moves with constant velocity. For example, a car is in equilibrium when it's parked, but also when it's driving down a straight road at a steady speed.

If the problem involves more than one body, and the bodies interact with each other, you'll also need to use Newton's *third* law. This law allows you to relate the force that one body exerts on a second body to the force that the second body exerts on the first one.

Be certain that you identify the target variable(s). Common target variables in equilibrium problems include the magnitude of one of the forces, the components of a force, or the direction of a force.

SET UP *the problem using the following steps:*

1. Draw a very simple sketch of the physical situation, showing dimensions and angles. You don't have to be an artist!
2. For each body that is in equilibrium, draw a free-body diagram of this body. For the present, we consider the body as a particle, so a large dot will do to represent it. In your free-body diagram, *do not* include the other bodies that interact with it, such as a surface it may be resting on, or a rope pulling on it.
3. Now ask yourself what is interacting with the body by touching it or in any other way. On your free-body diagram, draw a force vector for each interaction. If you know the angle at which a force is directed, draw the angle accurately and label it. A surface in contact with the body exerts a normal force perpendicular to the surface and possibly a friction force parallel to the surface. Remember that a rope or chain can't push on a body, but can only pull in a direction along its length. Be sure to include the body's weight, except in cases where the body has negligible mass (and hence negligible weight). If the mass is given, use $w = mg$ to find the weight. Label each force with a symbol representing the *magnitude* of the force.

4. *Do not* show in the free-body diagram any forces exerted *by* the body on any other body. The sums in Eqs. (5.1) and (5.2) include only forces that act *on* the body. Make sure you can answer the question "What other body causes that force?" for each force. If you can't answer that question, you may be imagining a force that isn't there.
5. Choose a set of coordinate axes and include them in your free-body diagram. (If there is more than one body in the problem, you'll need to choose axes for each body separately.) Make sure you label the positive direction for each axis. This will be crucially important when you take components of the force vectors as part of your solution. Often you can simplify the problem by your choice of coordinate axes. For example, when a body rests or slides on a plane surface, it's usually simplest to take the axes in the directions parallel and perpendicular to this surface, even when the plane is tilted.

EXECUTE *the solution as follows:*

1. Find the components of each force along each of the body's coordinate axes. Draw a wiggly line through each force vector that has been replaced by its components, so you don't count it twice. Keep in mind that while the *magnitude* of a force is always positive, the *component* of a force along a particular direction may be positive or negative.
2. Set the algebraic sum of all x-components of force equal to zero. In a separate equation, set the algebraic sum of all y-components equal to zero. (*Never* add x- and y-components in a single equation.) You can then solve these equations for up to two unknown quantities, which may be force magnitudes, components, or angles.
3. If there are two or more bodies, repeat all of the above steps for each body. If the bodies interact with each other, use Newton's third law to relate the forces they exert on each other.
4. Make sure that you have as many independent equations as the number of unknown quantities. Then solve these equations to obtain the target variables. This part is algebra, not physics, but it's an essential step.

EVALUATE *your answer:* Look at your results and ask whether they make sense. When the result is a symbolic expression or formula, try to think of special cases (particular values or extreme cases for the various quantities) for which you can guess what the results ought to be. Check to see that your formula works in these particular cases.

Example 5.1 One-dimensional equilibrium: Tension in a massless rope

A gymnast with mass $m_G = 50.0$ kg suspends herself from the lower end of a hanging rope. The upper end of the rope is attached to the gymnasium ceiling (Fig. 5.1a). What is the gymnast's weight? What force (magnitude and direction) does the rope exert on her? What is the tension at the top of the rope? Assume that the mass of the rope itself is negligible.

SOLUTION

IDENTIFY: The gymnast and the rope are in equilibrium, so we can apply Newton's first law to both bodies. The gymnast and the rope exert forces on each other—that is, they interact—so we'll also use Newton's third law to relate these forces. The target variables are the weight of the gymnast, w_G; the force that the rope exerts on the gymnast (call it $T_{R\text{ on }G}$); and the tension that the ceiling exerts on the top of the rope (call it $T_{C\text{ on }R}$).

SET UP: We draw separate free-body diagrams for the gymnast (Fig. 5.1b) and the rope (Fig. 5.1c). We take the positive y-axis to be upward, as shown. Each force acts in the vertical direction and so has only a y-component.

The two forces $T_{R\text{ on }G}$ and $T_{G\text{ on }R}$ are the upward force of the rope on the gymnast (in Fig. 5.1b) and the downward force of the gymnast on the rope (in Fig. 5.1c). These forces form an action-reaction pair, so they must have the same magnitude.

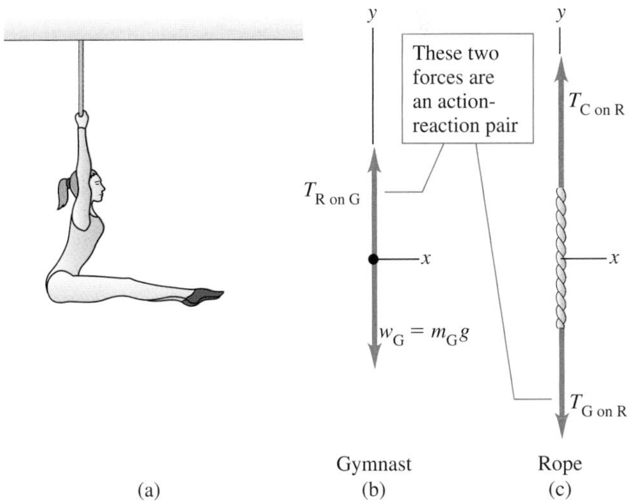

5.1 (a) A gymnast hanging at rest from the end of a vertical rope. (b) Free-body diagram for the gymnast. (c) Free-body diagram for the rope, assuming its weight to be negligible. The upward force of the rope on the gymnast and the downward force of the gymnast on the rope are an action-reaction pair.

Note also that the gymnast's weight w_G is the attractive (downward) force exerted on the gymnast by the earth. Its reaction force is the equal and opposite (upward) attractive force exerted on the *earth* by the *gymnast.* This force acts on the earth, not on the gymnast, so it doesn't appear in her free-body diagram. Compare the discussion of the apple in Conceptual Example 4.9 (Section 4.5).

EXECUTE: The magnitude of the weight of any object is the product of the object's mass and the acceleration due to gravity, g. For the gymnast this is

$$w_G = m_G g = (50.0\text{ kg})(9.80\text{ m/s}^2) = 490\text{ N}$$

This force points in the negative y-direction, so its y-component is -490 N. The upward force exerted by the rope has unknown magnitude $T_{R\text{ on }G}$. From Eq. (5.2), because the gymnast is in equilibrium, the sum of the y-components of force acting on her must be zero:

$$\text{Gymnast: } \sum F_y = T_{R\text{ on }G} + (-w_G) = 0 \quad \text{so}$$
$$T_{R\text{ on }G} = w_G = 490\text{ N}$$

The rope pulls *up* on the gymnast with a force $T_{R\text{ on }G}$ of magnitude 490 N. By Newton's third law, the gymnast pulls *down* on the rope with a force of the same magnitude, $T_{G\text{ on }R} = 490$ N.

The rope is also in equilibrium. We have assumed that it is weightless, so the upward force of magnitude $T_{C\text{ on }R}$ that the ceiling exerts on its top end must make the *net* vertical force on the rope equal to zero. Expressed as an equation, this says

$$\text{Rope: } \sum F_y = T_{C\text{ on }R} + (-T_{G\text{ on }R}) = 0 \quad \text{so}$$
$$T_{C\text{ on }R} = T_{G\text{ on }R} = 490\text{ N}$$

EVALUATE: The *tension* at any point in the rope is the force that acts at that point. For this weightless rope, the tension $T_{G\text{ on }R}$ at the lower end has the same value as the tension $T_{C\text{ on }R}$ at the upper end. Indeed, for an ideal weightless rope, the tension has the same value at any point along the rope's length. (Compare the discussion of Conceptual Example 4.10 in Section 4.5.)

Note that we have defined tension to be the *magnitude* of a force, so it is always positive. But the y-component of force acting on the rope at its lower end is $-T_{G\text{ on }R} = -490$ N.

Example 5.2 One-dimensional equilibrium: Tension in a rope with mass

Suppose that in Example 5.1, the weight of the rope is not negligible but is 120 N. Find the tension at each end of the rope.

SOLUTION

IDENTIFY: As in the previous example, we will apply Newton's first law to each of the two bodies in the problem (the gymnast and the rope) and use Newton's third law to relate the forces that the gymnast and rope exert on each other. The target variables are the magnitudes $T_{G\text{ on }R}$ and $T_{C\text{ on }R}$ of the forces that act at the bottom and top of the rope, respectively.

SET UP: Again we draw separate free-body diagrams for the gymnast (Fig. 5.2a) and the rope (Fig. 5.2b). The only difference in the diagrams from those in Example 5.1 is that there are now three forces acting on the rope: the downward force exerted by the gymnast ($T_{G\text{ on }R}$), the upward force exerted by the ceiling ($T_{C\text{ on }R}$), and the downward force of gravity (the weight of the rope, of magnitude $w_R = 120$ N).

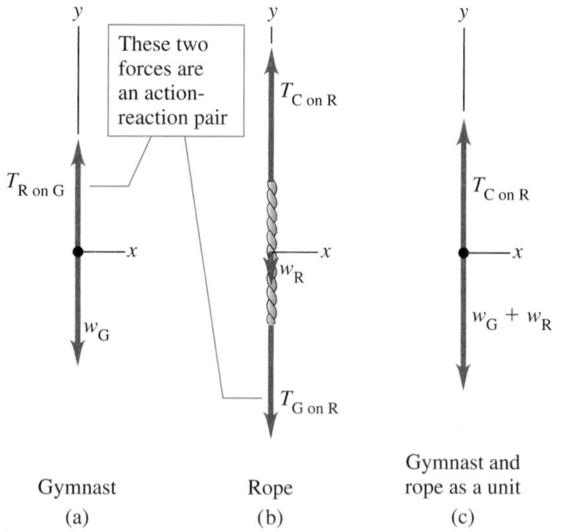

5.2 Free-body diagrams (a) for the gymnast (weight w_G) and (b) for the rope (weight w_R). (Compare Fig. 5.1.) (c) Free-body diagram for the gymnast and rope, considered as a single composite body.

EXECUTE: The gymnast's free-body diagram is unchanged from Example 5.1, so her equilibrium condition is also unchanged. From Newton's third law, $T_{R\text{ on }G} = T_{G\text{ on }R}$, and we have

$$\text{Gymnast: } \sum F_y = T_{R\text{ on }G} + (-w_G) = 0 \quad \text{so}$$
$$T_{R\text{ on }G} = T_{G\text{ on }R} = w_G = 490\text{ N}$$

The equilibrium condition $\sum F_y = 0$ for the rope is

$$\text{Rope: } \sum F_y = T_{C\text{ on }R} + (-T_{G\text{ on }R}) + (-w_R) = 0$$

Note that the y-component of $T_{C\text{ on }R}$ is positive because it points in the $+y$-direction, but the y-components of both $T_{G\text{ on }R}$ and w_R are negative. When we solve for $T_{C\text{ on }R}$ and substitute the values $T_{G\text{ on }R} = T_{R\text{ on }G} = 490$ N and $w_R = 120$ N, we find

$$T_{C\text{ on }R} = T_{G\text{ on }R} + w_R = 490\text{ N} + 120\text{ N} = 610\text{ N}$$

EVALUATE: When we include the weight of the rope, we find that the tension is *different* at the rope's two ends. This makes sense: the force $T_{\text{C on R}}$ exerted by the ceiling has to hold up both the 490-N weight of the gymnast and the 120-N weight of the rope, so $T_{\text{C on R}} = 610\text{ N}$.

To see this more explicitly, draw a free-body diagram for a composite body consisting of the gymnast and rope considered as a unit (Fig. 5.2c). There are only two external forces acting on this composite body: the force $T_{\text{C on R}}$ exerted by the ceiling and the total weight $w_G + w_R = 490\text{ N} + 120\text{ N} = 610\text{ N}$. (The forces $T_{\text{G on R}}$ and $T_{\text{R on G}}$ are *internal* to the composite body. Since Newton's first law involves only *external* forces, the internal forces play no role.) Hence Newton's first law applied to this composite body is

$$\text{Composite body: } \sum F_y = T_{\text{C on R}} + [-(w_G + w_R)] = 0$$

and so $T_{\text{C on R}} = w_G + w_R = 610\text{ N}$.

This method of treating the gymnast and rope as a composite body seems a lot simpler, and you may be wondering why we didn't use it first. The answer is that we can't find the tension $T_{\text{G on R}}$ at the bottom of the rope by this method. The moral is that whenever you have more than one body in a problem involving Newton's laws, the safest approach is to treat each body separately.

Example 5.3 Two-dimensional equilibrium

In Fig. 5.3a a car engine with weight w hangs from a chain that is linked at ring O to two other chains, one fastened to the ceiling and the other to the wall. Find the tensions in each of these three chains, assuming that w is given and the weights of the ring and chains are negligible.

SOLUTION

IDENTIFY: The target variables are the tensions T_1, T_2, and T_3 in the three chains (Fig. 5.3a). It may seem strange that we neglect the weight of the chains and ring in this example, while in Example 5.2 we did *not* neglect the weight of a mere rope. The reason is that the weight of the chains or ring is very small compared to the weight of the massive engine. By contrast, in Example 5.2 the rope weighed a reasonable fraction of the gymnast's weight (100 N compared to 490 N).

All the bodies in the example are in equilibrium, so we'll use Newton's first law to determine T_1, T_2, and T_3. We need three simultaneous equations, one for each target variable. However, applying Newton's first law to just one body gives us just *two* equations, as in Eq. (5.2). So to solve the problem, we'll have to consider more than one body in equilibrium. We'll look at the engine (which is acted on by T_1) and the ring (which is connected to all three chains and so is acted on by all three tensions).

SET UP: Figures 5.3b and 5.3c are free-body diagrams for the engine and the ring respectively. As in Examples 5.1 and 5.2, we've included an x-y coordinate system in each diagram.

The two forces acting on the engine are its weight w and the upward force T_1 exerted by the vertical chain; the three forces acting on the ring are the tensions from the vertical chain (T_1), the

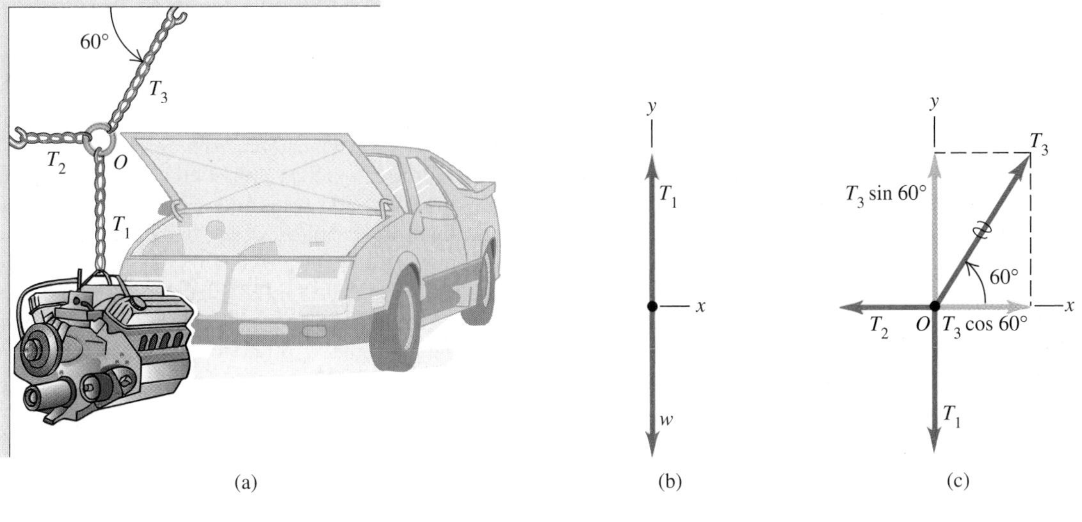

5.3 (a) A car engine with weight w is suspended from a chain linked at O to two other chains. The chains are considered to be massless. (b) Free-body diagram for the engine. (c) Free-body diagram for the ring with $\vec{T}_3$ replaced by its components.

horizontal chain (T_2), and the slanted chain (T_3). Note that the vertical chain exerts forces of the same magnitude T_1 at both of its ends: upward on the engine in Fig. 5.3b and downward on the ring in Fig. 5.3c. This is because the weight of the chain is negligible (see Example 5.1). If the weight were not negligible, these two forces would have different magnitudes, as was the case for the rope in Example 5.2. Remember that we're also neglecting the weight of the ring, which is why it isn't included in the forces in Fig. 5.3c.

EXECUTE: The forces acting on the engine are along the y-axis only, so Newton's first law says

$$\text{Engine: } \sum F_y = T_1 + (-w) = 0 \quad \text{and} \quad T_1 = w$$

Note that the horizontal and slanted chains do not exert forces on the engine itself, because they are not attached to it. These forces do appear in the application of Newton's first law to the ring, however.

In the free-body diagram for the ring (Fig. 5.3c), remember that T_1, T_2, and T_3 are the *magnitudes* of the forces; their directions are shown by the vectors on the diagram. We first resolve the force with magnitude T_3 into its x- and y-components. We can then write the equilibrium conditions for the ring, writing separate equations for the x- and y-components. (Remember from the Problem-Solving Strategy that x- and y-components are *never* added together in a single equation.) We find

$$\text{Ring: } \sum F_x = T_3 \cos 60° + (-T_2) = 0$$
$$\text{Ring: } \sum F_y = T_3 \sin 60° + (-T_1) = 0$$

Because $T_1 = w$ (from the engine equation), we can rewrite the second ring equation as

$$T_3 = \frac{T_1}{\sin 60°} = \frac{w}{\sin 60°} = 1.155w$$

We can now use this result in the first ring equation:

$$T_2 = T_3 \cos 60° = w\frac{\cos 60°}{\sin 60°} = 0.577w$$

So we can express all three tensions as multiples of the weight w of the engine, which we assume is known. To summarize,

$$T_1 = w$$
$$T_2 = 0.577w$$
$$T_3 = 1.155w$$

EVALUATE: Our results show that the chain attached to the ceiling exerts a force on the ring of magnitude T_3, which is *greater* than the weight of the engine. If this seems strange, note that the vertical component of this force is equal to T_1, which in turn is equal to w. But since this force also has a horizontal component, its magnitude T_3 must be somewhat larger than w. Hence the chain attached to the ceiling is under the greatest tension, and is the one most susceptible to breaking.

It probably seemed to you at first glance that the most important body in this problem was the engine. But in order to get enough equations to solve the problem, we also had to consider the forces acting on a second body (in this case, the ring connecting the chains). Situations like this are fairly common in equilibrium problems, so keep this technique in mind.

Example 5.4 An inclined plane

A car rests on the slanted tracks of a ramp leading to a car-transporter trailer (Fig. 5.4a). Only a cable attached to the car and to the frame of the trailer prevents the car from rolling backward off the trailer. (The car's brakes and transmission lock are both released.) If the weight of the car is w, find the tension in the cable and the force with which the tracks push on the car's tires.

SOLUTION

IDENTIFY: The car is in equilibrium, so once again we use Newton's first law. One complication is that the ramp exerts *four* forces on the car, one on each of the car's tires. For simplicity, we lump all of these forces together into a single force. For a further simplification, we'll ignore the component of this force that acts *parallel* to the tracks. This is equivalent to saying that there's very little friction on the car. (If there were substantial friction, the tracks would exert a *friction force* acting parallel to the ramp that would tend to prevent the car from rolling down the ramp. We'll ignore this effect here, but will return to it in Section 5.3.) Hence we can say that the ramp only exerts a force on the car that is *perpendicular* to the tracks. This force appears because the atoms on the surface of the track resist having the atoms of the tires squeezed into them. As in Section 4.2, we call this force *the normal force.*

The forces on the car do not all act along the same line—the force of gravity acts straight downward, but the normal force is perpendicular to the tilted surface of the tracks. Hence we have two separate equations for Newton's first law, one for the x-components of force and one for the y-components [see Eq. (5.2)]. This is just the right number of equations to solve for the two target variables, the magnitude T of the tension in the cable and the magnitude n of the normal force.

SET UP: Figure 5.4b shows a free-body diagram for the car. The three forces acting on the car are its weight (magnitude w), the tension in the cable (magnitude T), and the normal force (magnitude n). Note that the normal force acts up and to the left because it's preventing the car from penetrating into the solid tracks.

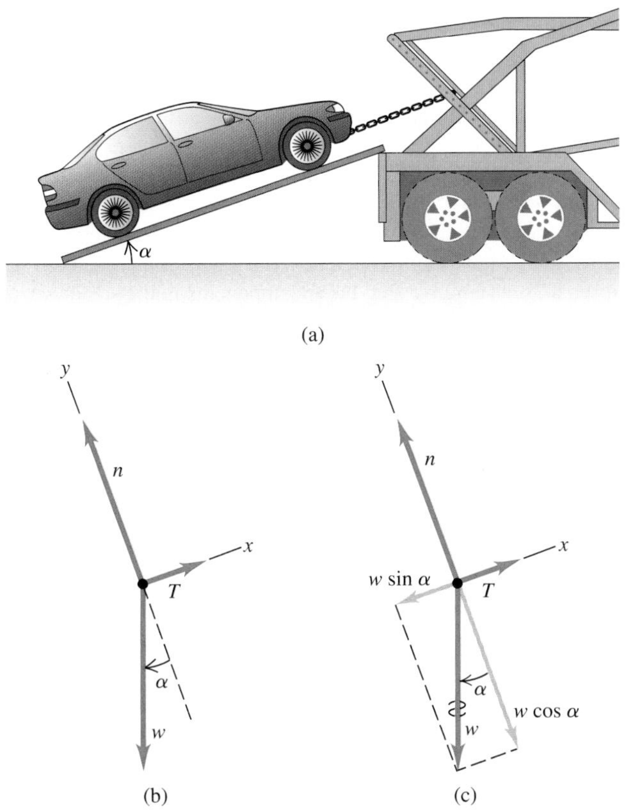

5.4 (a) A cable holds the car on the ramp. (b) Free-body diagram for the car. (c) Free-body diagram with the car's weight replaced by its components ($w \sin \alpha$ down the ramp, $w \cos \alpha$ perpendicular to the ramp).

We choose the x- and y-axes to be parallel and perpendicular to the ramp as shown. This choice makes the problem easier to analyze because only the weight force has both an x- and y-component. If we were to choose axes that were horizontal and vertical, our job would be harder because we'd need to find x- and y-components for *two* forces (the normal force and the tension).

Note that the angle α between the ramp and the horizontal is equal to the angle α between the weight vector $\vec{w}$ and the normal to the plane of the ramp.

EXECUTE: To write down the x- and y-components of Newton's first law, we need to find the components of the weight. One complication is that the angle α in Fig. 5.4b is *not* measured from the $+x$-axis toward the $+y$-axis. Hence we *cannot* use Eqs. (1.7) directly to find the components. (You may want to review Section 1.8 to make sure that you understand this important point.)

One approach to finding the components of $\vec{w}$ is to consider the right triangles in Fig. 5.4c. The sine of α is the magnitude of the x-component of $\vec{w}$ (that is, the side of the triangle opposite α) divided by the magnitude w (the hypotenuse of the triangle). Similarly, the cosine of α is the magnitude of the y-component (the side of the triangle adjacent to α) divided by w. Both components are negative, so $w_x = -w \sin \alpha$ and $w_y = -w \cos \alpha$.

Another approach is to recognize that one component of $\vec{w}$ must involve $\sin \alpha$ while the other component involves $\cos \alpha$. To decide which is which, it helps to draw the free-body diagram so that the angle α is noticeably less than or greater than 45°. (You're free to do this, since the value of α isn't specified. You'll have to fight the natural tendency to draw such angles as being close to 45°.) We've drawn Figs. 5.4b and 5.4c so that α is less than 45°, which means that $\sin \alpha$ is less than $\cos \alpha$. Figure 5.4c shows that the x-component of $\vec{w}$ is smaller than the y-component, so the x-component must involve $\sin \alpha$ and the y-component must involve $\cos \alpha$. Since both components are negative, we again find $w_x = -w \sin \alpha$ and $w_y = -w \cos \alpha$.

We draw a wiggly line through the original vector representing the weight to remind us not to count it twice. The equilibrium conditions then give us

$$\sum F_x = T + (-w \sin \alpha) = 0$$
$$\sum F_y = n + (-w \cos \alpha) = 0$$

Be sure you understand how these signs are related to our choice of coordinates. Remember that, by definition, T, w, and n are all *magnitudes* of vectors and are therefore all positive.

Solving these equations for T and n, we find

$$T = w \sin \alpha$$
$$n = w \cos \alpha$$

EVALUATE: Our answers for T and n depend on the value of α. To check these answers to see whether they're reasonable, let's look at some special cases. If the angle α is zero, then $\sin \alpha = 0$ and $\cos \alpha = 1$. In this case, the tracks are horizontal; our answers tell us that no cable tension T is needed to hold the car, and the total normal force n is equal in magnitude to the weight. If the angle is 90°, then $\sin \alpha = 1$ and $\cos \alpha = 0$. Then the cable tension T equals the weight w, and the normal force n is zero. Are these the results you would expect for these particular cases?

CAUTION It's a common error to automatically assume that the magnitude n of the normal force is equal to the weight w. But our result shows that this is *not* true in general. It's always best to treat n as a variable and solve for its value, as we have done here.

As a final note, ask yourself how the answers for T and n would be affected if the car were not stationary but were being pulled up the ramp at a constant speed. You should immediately realize that this, too, is an equilibrium situation, since the car's velocity is constant. So the calculation is exactly the same, and T and n have the same values as when the car is at rest. (It's true that T must be greater than $w \sin \alpha$ to *start* the car moving up the ramp, but that's not what we asked.)

Example 5.5 Tension over a frictionless pulley

Blocks of granite are being hauled up a 15° slope out of a quarry. For environmental reasons, dirt is also being dumped into the quarry to fill up old holes. You have been asked to find a way to use this dirt to move the granite out more easily. You design a system in which a granite block on a cart with steel wheels (weight w_1, including both the block and cart) is pulled uphill on steel rails by a dirt-filled bucket (weight w_2, including both the dirt and the bucket) dropping vertically into the quarry (Fig. 5.5a). Ignoring friction in the pulley and wheels and the weight of the cable, determine how the weights w_1 and w_2 must be related in order for the system to move with constant speed.

SOLUTION

IDENTIFY: The cart and bucket each move with a constant velocity (that is, in a straight line at constant speed). Hence each body is in equilibrium and we can apply Newton's first law to each.

Our two target variables are the weights w_1 and w_2. The forces that act on the bucket are its weight w_2 and an upward tension exerted by the cable; both these forces are purely vertical, so Newton's first law applied to the bucket gives us just one equation. The cart has *three* forces acting on it: its weight w_1, a normal force of magnitude n exerted by the rails, and a tension force from the cable. (We're ignoring friction, so we're assuming that the rails exert no force parallel to the incline.) This is exactly like the situation for the car on the ramp in Example 5.4. As in that example, the forces on the cart are not all along the same direction, so we'll need to use both components of Newton's first law in Eq. (5.2).

We're assuming that the cable has negligible weight, so the tension forces that the rope exerts on the cart and on the bucket have the same magnitude T.

SET UP: Figure 5.5b is our idealized model for the system. Figure 5.5c is the free-body diagram for the dirt and bucket and Fig. 5.5d is the free-body diagram for the granite block and cart. Note that we are at liberty to orient the axes differently for each body; the choices shown are the most convenient ones. We have represented the weight of the granite block in terms of its components in the chosen axis system; we find these components in the same way as in Example 5.4.

EXECUTE: Applying $\sum F_y = 0$ to the dirt-filled bucket in Fig. 5.5c, we find

$$\sum F_y = T + (-w_2) = 0 \quad \text{so} \quad T = w_2$$

Applying $\sum F_x = 0$ to the block and cart in Fig. 5.5d, we get

$$\sum F_x = T + (-w_1 \sin 15°) = 0 \quad \text{so} \quad T = w_1 \sin 15°$$

Equating the two expressions for T, we find

$$w_2 = w_1 \sin 15° = 0.26 w_1$$

EVALUATE: If the weight of dirt and bucket totals 26% of the weight of the granite block and cart, the system can move with constant speed in *either* direction. This is because our analysis doesn't depend on the direction of motion, only on the velocity being constant. Can you see what would happen if w_2 were greater than $0.26w_1$? If it were less than $0.26w_1$?

Notice that we didn't need to apply the equation $\sum F_y = 0$ to the cart and block; this would be useful only if we wanted to find the value of n. Can you show that $n = w_1 \cos 15°$?

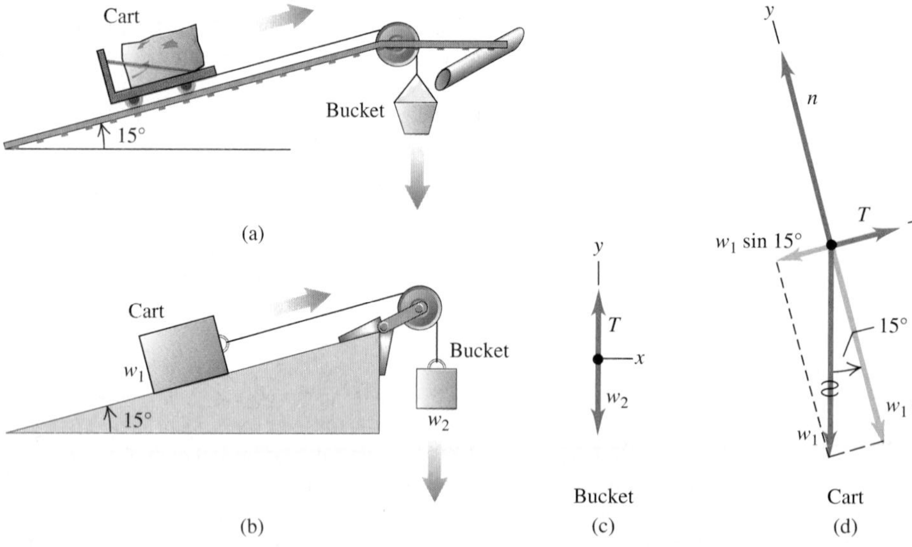

5.5 (a) A bucket of dirt pulls up a cart carrying a granite block. (b) Idealized model of the system. (c) Free-body diagram for the bucket of dirt. (d) Free-body diagram for the cart with block.

Test Your Understanding

A traffic light of mass m hangs from two lightweight cables, one on each side of the light. Each cable hangs at a 45° angle from the horizontal. What is the tension in each cable?

5.2 | Using Newton's Second Law: Dynamics of Particles

We are now ready to discuss *dynamics* problems. In these problems, we apply Newton's second law to bodies that are accelerating and hence are *not* in equilibrium. In this case the net force on the body is not zero, but is equal to the mass of the body times its acceleration:

$$\sum \vec{F} = m\vec{a} \qquad \text{(Newton's second law, vector form)} \qquad (5.3)$$

We most often use this relation in component form:

$$\sum F_x = ma_x \qquad \sum F_y = ma_y \qquad (5.4)$$

(Newton's second law, component form)

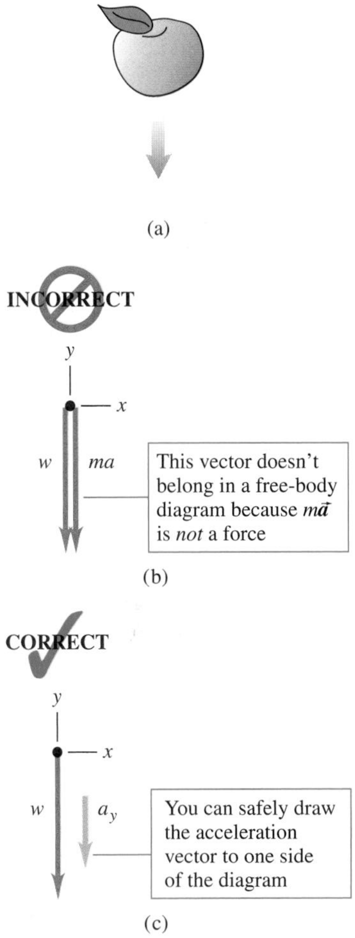

5.6 (a) The only force acting on this falling piece of fruit is the earth's gravitational pull. (b) An incorrect free-body diagram. (c) A correct free-body diagram.

The following problem-solving strategy is very similar to our strategy for equilibrium problems in Section 5.1. We urge you to study it carefully, watch how we apply it in our examples, and use it when you tackle the end-of-chapter problems. Remember that you can solve *any* dynamics problem using this strategy.

CAUTION We emphasize again that the quantity $m\vec{a}$ is *not* a force; it's not a push or a pull exerted by anything in the body's environment. All that Eqs. (5.3) and (5.4) say is that the acceleration $\vec{a}$ is proportional to the net force $\sum \vec{F}$. When you draw the free-body diagram for an accelerating body (like the fruit in Fig. 5.6a), make sure you *never* include "the $m\vec{a}$ force," because *there is no such force* (Fig. 5.6b). You should review Section 4.3 if you're not clear on this point. Sometimes we will draw the acceleration vector $\vec{a}$ *alongside* a free-body diagram, as in Fig. 5.6c. But we will *never* draw the acceleration vector with its tail touching the body (a position reserved exclusively for the forces that act on the body).

Problem-Solving Strategy — Newton's Second Law: Dynamics of Particles

IDENTIFY *the relevant concepts:* You have to use Newton's second law for *any* problem that involves forces acting on an accelerating body.

As with any problem, identify the target variable—usually an acceleration or a force. If the target variable is something else, you'll need to identify another concept to use. For example, suppose the problem asks you to find how fast a sled is moving when it reaches the bottom of a hill. This means your target variable is the sled's final velocity. To find this, you'll first need to use Newton's second law to find the sled's acceleration. Then you'll also have to use the constant-acceleration relationships from Section 2.4 to find velocity from acceleration.

SET UP *the problem using the following steps:*

1. Draw a simple sketch of the situation. Identify one or more moving bodies to which you'll apply Newton's second law.
2. For each body you identified, draw a free-body diagram that shows all the forces acting *on* the body. (Don't try to be fancy—just represent the object by a point.) Be careful

not to include any forces exerted *by* the object on some other object. Remember, the acceleration of a body is determined by the forces that act on it, not by the forces that it exerts on anything else. Make sure you can answer the question "What other body is applying this force?" for each force in your diagram. Furthermore, never include the quantity $m\vec{a}$ in your free-body diagram; it's not a force!

3. Label each force with an algebraic symbol for the force's *magnitude,* as well as a numerical value of the magnitude if it's given in the problem. (Remember that magnitudes are always positive. Minus signs show up later when you take components of the forces.) Usually, one of the forces will be the body's weight; it's usually best to label this as $w = mg$. If a numerical value of mass is given, you can compute the corresponding weight.
4. Choose your x- and y-coordinate axes for each object, and show them explicitly in each free-body diagram. Be sure to indicate which is the positive direction for each axis. If you know the direction of the acceleration, it usually simplifies things to take one positive axis along that direction. Note that if your problem involves more than one object and the objects accelerate in different directions, you can use a different set of axes for each object.
5. In addition to Newton's second law, $\sum\vec{F} = m\vec{a}$, identify any other equations you might need. (You need one equation for each target variable.) For example, you might need one or more of the equations for motion with constant acceleration. If more than one body is involved, there may be relationships among their motions; for example, they may be connected by a rope. Express any such relationships as equations relating the accelerations of the various bodies.

EXECUTE *the solution as follows:*

1. For each object, determine the components of the forces along each of the object's coordinate axes. When you represent a force in terms of its components, draw a wiggly line through the original force vector to remind you not to include it twice.
2. For each object, write a separate equation for each component of Newton's second law, as in Eq. (5.4).
3. Make a list of all the known and unknown quantities. In your list, identify the target variable or variables.
4. Check that you have as many equations as there are unknowns. If you have too few equations, go back to step 5 of "Set up the problem." If you have too many equations, perhaps there is an unknown quantity that you haven't identified as such.
5. Do the easy part—the math! Solve the equations to find the target variable(s).

EVALUATE *your answer:* Does your answer have the correct units? (When appropriate, use the conversion $1\ \text{N} = 1\ \text{kg}\cdot\text{m/s}^2$.) Does it have the correct algebraic sign? (If the problem is about a sled sliding downhill, you probably took the positive x-axis to point down the hill. If you then find that the sled has a negative acceleration—that is, the acceleration is uphill—then something went wrong in your calculations.) When possible, consider particular values or extreme cases of quantities and compare the results with your intuitive expectations. Ask, "Does this result make sense?"

Example 5.6 Straight-line motion with a constant force

An iceboat is at rest on a perfectly frictionless horizontal surface (Fig. 5.7a). A steady wind is blowing (along the direction of the runners) so that 4.0 s after the iceboat is released, it attains a velocity of 6.0 m/s (about 22 km/h, or 13 mi/h). What constant horizontal force F_W does the wind exert on the iceboat? The mass of iceboat and rider is 200 kg.

SOLUTION

IDENTIFY: Our target variable is one of the forces (F_W) acting on the iceboat, so we'll need to use Newton's second law. This law involves forces and acceleration, but the acceleration isn't given; hence we'll need to find it. The wind is assumed to be steady, so the forces don't change with time and the resulting acceleration is constant. That means we'll be able to use one of the constant-acceleration formulas.

SET UP: Figure 5.7b shows the free-body diagram for the iceboat and rider considered as a unit. The forces acting on this object are the weight w, the normal force n exerted by the surface, and the horizontal force F_W (our target variable). Note that the net force and hence the acceleration are to the right, so we chose the positive x-axis in this direction.

Since the acceleration isn't given, we'll need to find it from the other information given in the problem—namely, the final velocity $v_x = 6.0$ m/s and elapsed time $t = 4.0$ s. The iceboat starts at rest, so the initial velocity is $v_{0x} = 0$. From Section 2.4, an equation we can use to relate the x-acceleration a_x to these quantities is Eq. (2.8), $v_x = v_{0x} + a_x t$.

EXECUTE: The force F_W is in the positive x-direction, while the forces n and mg are in the positive and negative y-directions respectively. Hence the x- and y-equations for Newton's second law are

$$\sum F_x = F_W = ma_x$$
$$\sum F_y = n + (-mg) = 0$$

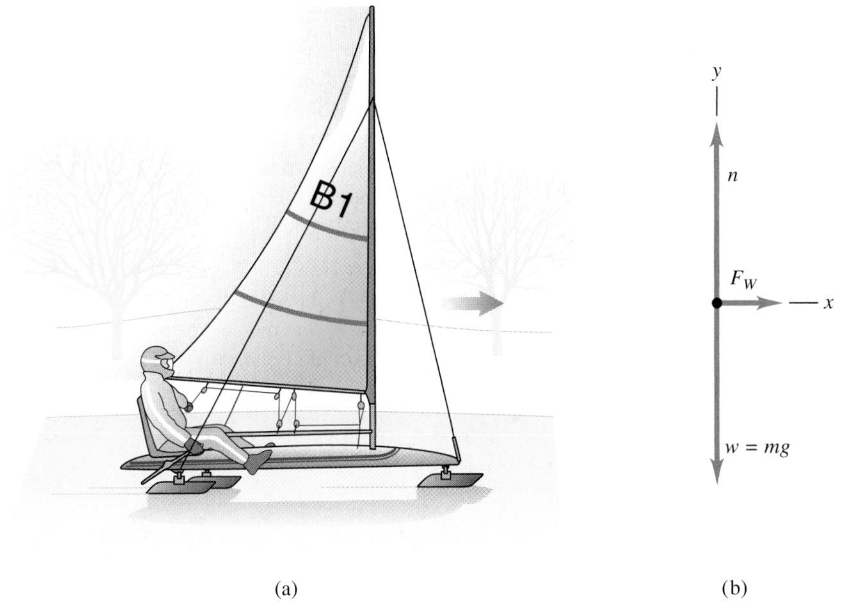

5.7 (a) An iceboat starting from rest. (b) Free-body diagram for the iceboat and rider with no friction.

(Note that there is zero acceleration in the y-direction; the iceboat doesn't accelerate up or down.) In addition, we have the constant-acceleration equation

$$v_x = v_{0x} + a_x t$$

The *known* quantities are the mass $m = 200$ kg, the final velocity $v_x = 6.0$ m/s, the initial velocity $v_{0x} = 0$, and the elapsed time $t = 4.0$ s. The *unknown* quantities are the acceleration a_x, the normal force n, and the horizontal force F_W (the target variable). There are as many unknowns (three) as equations, so we're in business.

To find F_W, first solve the constant-acceleration equation for a_x, then substitute this into the $\sum F_x$ equation:

$$a_x = \frac{v_x - v_{0x}}{t} = \frac{6.0\ \text{m/s} - 0\ \text{m/s}}{4.0\ \text{s}} = 1.5\ \text{m/s}^2$$

$$F_W = ma_x = (200\ \text{kg})(1.5\ \text{m/s}^2) = 300\ \text{kg}\cdot\text{m/s}^2$$

One kg · m/s² is the same as one newton (N), so the final answer is

$$F_W = 300\ \text{N}\ (\text{about } 67\ \text{lb})$$

Note that we did not need the $\sum F_y$ equation at all to find F_W. We would need this equation if we wanted to find the normal force n:

$$n - mg = 0$$

$$n - mg = (200\ \text{kg})(9.8\ \text{m/s}^2)$$

$$= 2.0 \times 10^3\ \text{N} \qquad (\text{about } 440\ \text{lb})$$

The magnitude n of the normal force is equal to the combined weight of the iceboat and rider, because the surface is horizontal and these are the only vertical forces acting.

EVALUATE: Our answers for F_W and n have the correct units for a force, as they should. Does it seem reasonable that the force F_W is substantially less than the combined weight of the iceboat and rider?

Example 5.7 Straight-line motion with a time-varying force

Let's again consider the iceboat moving on a frictionless surface, as in Example 5.6. But now let's suppose that once the iceboat starts to move, its position as a function of time is

$$x = (1.2\ \text{m/s}^2)t^2 - (0.20\ \text{m/s}^3)t^3$$

Find the force F_W exerted by the wind as a function of time in this case. What is this force at time $t = 3.0$ s? For what times is the force zero? Positive? Negative?

SOLUTION

IDENTIFY: Just as in Example 5.6, F_W is our target variable, so we must once again use Newton's second law. We aren't given the acceleration, but we do know the position x as a function of time t; hence we can determine the acceleration by taking the second derivative of x with respect to t, as in Section 2.3.

Note that our expression for x includes a t^3 term, which doesn't appear in any of the constant-acceleration formulas. This tells us that the acceleration a_x is not constant, and hence neither is the force F_W. Hence the wind force is varying in this situation.

SET UP: The free-body diagram is exactly the same as in Fig. 5.7; the diagram is equally valid whether F_W is constant or not. To determine the acceleration a_x from the position x as a function of time, we use Eq. (2.6):

$$a_x = \frac{dv_x}{dt} = \frac{d^2x}{dt^2}$$

EXECUTE: Because the free-body diagram is unchanged from Example 5.6, the equation for the x-component of Newton's second law is also unchanged:

$$\sum F_x = F_W = ma_x.$$

To determine F_W, all that's left is to find a_x as a function of time. The second derivative of t^2 is 2, and the second derivative of t^3 is $6t$, so

$$\begin{aligned} a_x &= \frac{d^2x}{dt^2} = \frac{d^2}{dt^2}[(1.2\ \text{m/s}^2)t^2 - (0.20\ \text{m/s}^3)t^3] \\ &= 2.4\ \text{m/s}^2 - (1.2\ \text{m/s}^3)t \end{aligned}$$

Then the force of the wind as a function of time is

$$\begin{aligned} F_W = ma_x &= (200\ \text{kg})[2.4\ \text{m/s}^2 - (1.2\ \text{m/s}^3)t] \\ &= 480\ \text{N} - (240\ \text{N/s})t \end{aligned}$$

At time $t = 3.0$ s the value of F_W is $480\ \text{N} - (240\ \text{N/s})(3.0\ \text{s}) = -240$ N. The minus sign means that the direction of the wind force is actually opposite to what we assumed in Fig. 5.7. The wind has shifted and is now opposing the iceboat's motion! The force is zero when $480\ \text{N} - (240\ \text{N/s})t = 0$; this happens when $t = 2.0$ s, which is when the wind has momentarily died down. When $t < 2.0$ s, F_W is positive and the wind is pushing to the right in Fig. 5.7 (in the $+x$-direction). When $t > 2.0$ s, F_W is negative and the wind is pushing to the left.

EVALUATE: Figure 5.8 shows graphs of F_W and a_x as functions of time. Note that in this case F_W is the net horizontal force on the iceboat. You shouldn't be surprised that the net force and acceleration are directly proportional; according to Newton's second law, this is *always* true.

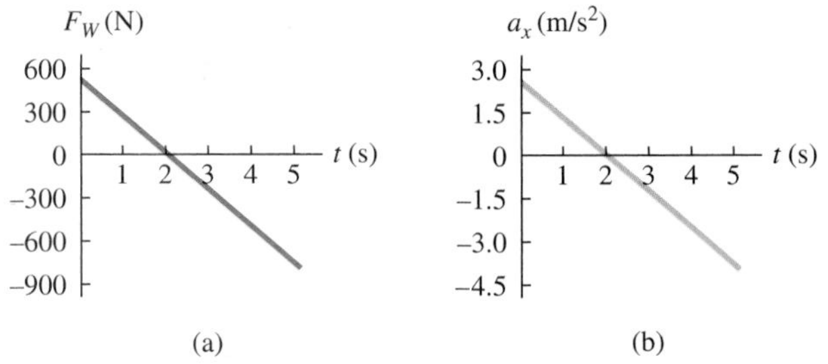

5.8 (a) The net force on the iceboat is directly proportional to (b) its acceleration.

Example 5.8 Straight-line motion with friction

Suppose the wind is once again blowing steadily in the $+x$-direction as in Example 5.6, so that the iceboat has a constant acceleration $a_x = 1.5\ \text{m/s}^2$. Now, however, there is a constant horizontal friction force with magnitude 100 N that opposes the motion of the iceboat. In this case, what force F_W must the wind exert on the iceboat?

SOLUTION

IDENTIFY: Once again the target variable is F_W. We are given the acceleration, so to find F_W all we need is Newton's second law.

SET UP: Figure 5.9 shows the new free-body diagram. The only difference from Fig. 5.7b is the addition of the friction force $\vec{f}$, which points opposite to the motion. (Note that its *magnitude*, $f = 100$ N, is a positive quantity but that its *component* in the x-direction is negative, equal to $-f$ or -100 N.)

EXECUTE: There are now two forces (the force of the wind and the friction force) that have x-components. The x-component of Newton's second law gives

$$\begin{aligned} \sum F_x &= F_W + (-f) = ma_x \\ F_W &= ma_x + f = (200\ \text{kg})(1.5\ \text{m/s}^2) + (100\ \text{N}) = 400\ \text{N} \end{aligned}$$

EVALUATE: Because there is friction, a greater force F_W is needed than in Example 5.6. We need 100 N to overcome friction and 300 N more to give the iceboat the necessary acceleration.

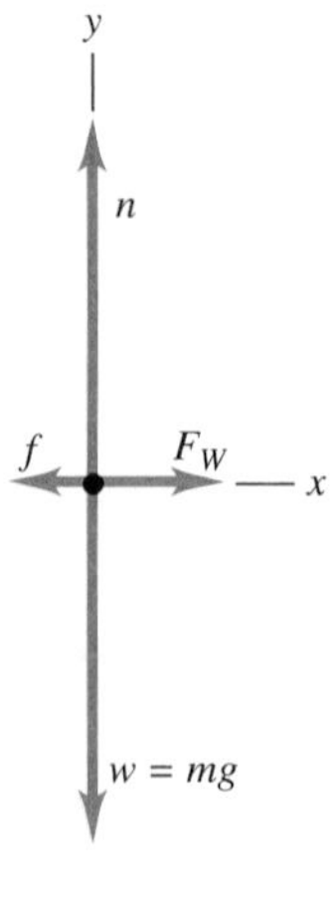

5.9 Free-body diagram for the iceboat and rider with a frictional force $\vec{f}$ opposing the motion.

Example 5.9 Tension in an elevator cable

An elevator and its load have a total mass of 800 kg (Fig. 5.10a). The elevator is originally moving downward at 10.0 m/s; it slows to a stop with constant acceleration in a distance of 25.0 m. Find the tension T in the supporting cable while the elevator is being brought to rest.

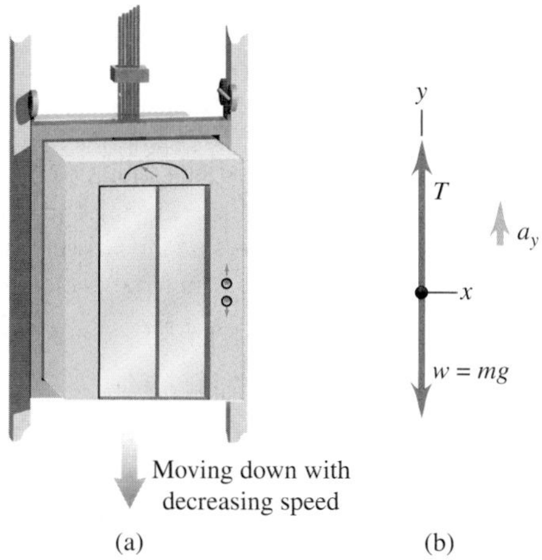

5.10 (a) A loaded elevator moving downward and coming to rest. (b) Free-body diagram for the elevator. As in Fig. 5.6, we draw the acceleration vector to one side of the free-body diagram because acceleration is not a force.

SOLUTION

IDENTIFY: The target variable is the tension T, which we will find using Newton's second law. As in Example 5.6, we'll have to determine the acceleration using the constant-acceleration formulas.

SET UP: The free-body diagram in Figure 5.10b shows the only forces acting on the elevator: its weight w and the tension force T of the cable. The elevator is moving downward with decreasing speed, so its acceleration is upward; we choose the positive y-axis to be in this direction.

The elevator is moving downward, in the negative y-direction. Hence its initial velocity v_{0y} and its displacement $y - y_0$ are both negative: $v_{0y} = -10.0\text{ m/s}$ and $y - y_0 = -25.0\text{ m}$. The final velocity is $v_y = 0$. To find the acceleration a_y from this information, we'll use Eq. (2.13) in the form $v_y^2 = v_{0y}^2 + 2a_y(y - y_0)$. Once we have a_y, we'll substitute it into the y-component of Newton's second law from Eq. (5.4).

EXECUTE: First let's write out Newton's second law. The tension force acts upward and the weight acts downward, so

$$\sum F_y = T + (-w) = ma_y$$

We solve for the target variable T:

$$T = w + ma_y = mg + ma_y = m(g + a_y)$$

To determine a_y, we rewrite the constant-acceleration equation $v_y^2 = v_{0y}^2 + 2a_y(y - y_0)$:

$$a_y = \frac{v_y^2 - v_{0y}^2}{2(y - y_0)} = \frac{(0)^2 - (-10.0\text{ m/s})^2}{2(-25.0\text{ m})} = +2.00\text{ m/s}^2$$

The acceleration is upward (positive), just as it should be for downward motion with decreasing speed.

Now we can substitute the acceleration into the equation for the tension:

$$T = m(g + a_y) = (800\text{ kg})(9.80\text{ m/s}^2 + 2.00\text{ m/s}^2)$$
$$= 9440\text{ N}$$

EVALUATE: Our answer shows that the tension is 1600 N *greater* than the weight ($w = mg = 7840\text{ N}$). This makes sense: there must be an upward net force to provide the upward acceleration that brings the elevator to a halt. Can you see that we would get the same answer for T if the elevator were moving *upward* and *gaining* speed at a rate of 2.00 m/s^2?

Example 5.10 Apparent weight in an accelerating elevator

A 50.0-kg woman stands on a bathroom scale while riding in the elevator in Example 5.9 (Fig. 5.11a). What is the reading on the scale?

SOLUTION

IDENTIFY: The scale reads the magnitude of the downward force exerted *by* the woman *on* the scale. By Newton's third law, this equals the magnitude of the upward normal force exerted *by* the scale *on* the woman. Hence our target variable is the magnitude n of the normal force.

We'll find n by applying Newton's second law. Happily, we already know the acceleration of the woman; it's the same as the acceleration of the elevator, which we calculated in Example 5.9.

SET UP: Figure 5.11b shows a free-body diagram for the woman. The forces acting on her are the normal force n exerted by the scale

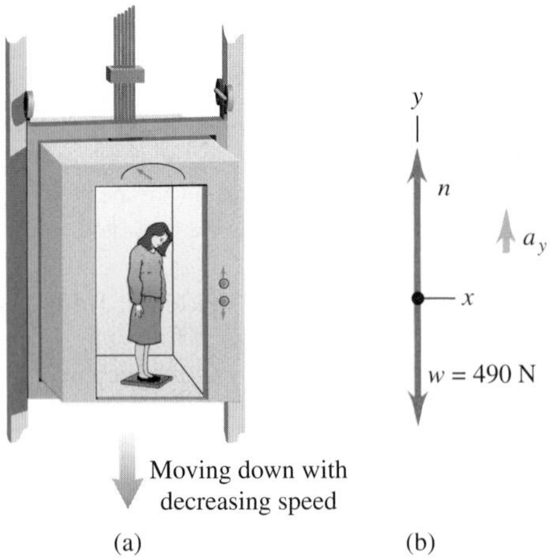

5.11 (a) A woman in the elevator as it slows down. (b) Free-body diagram for the woman.

and her weight $w = mg = (50.0 \text{ kg})(9.80 \text{ m/s}^2) = 490 \text{ N}$. (The tension force, which played a major role in Example 5.9, doesn't appear here. That's because the tension doesn't act directly on the woman. What she feels pushing upward on her feet is the scale, *not* the elevator cable.) From Example 5.9, the acceleration of the elevator and of the woman is $a_y = +2.00 \text{ m/s}^2$.

EXECUTE: Newton's second law gives

$$\sum F_y = n + (-mg) = ma_y$$
$$n = mg + ma_y = m(g + a_y)$$
$$= (50.0 \text{ kg})(9.80 \text{ m/s}^2 + 2.00 \text{ m/s}^2) = 590 \text{ N}$$

EVALUATE: Our answer for n means that while the elevator is stopping, the scale pushes up on the woman with a force of 590 N. By Newton's third law, she pushes down on the scale with the same force; so the scale reads 590 N, which is 100 N more than her actual weight. The scale reading is called the passenger's **apparent weight.** The woman *feels* the floor pushing up harder on her feet than when the elevator is stationary or moving with constant velocity.

What would the woman feel if the elevator were accelerating *downward* so that $a_y = -2.00 \text{ m/s}^2$? This would be the case if the elevator were moving upward with decreasing speed or moving downward with increasing speed. To find the answer for this situation, we just insert the new value of a_y in our equation for n:

$$n = m(g + a_y) = (50.0 \text{ kg})[9.80 \text{ m/s}^2 + (-2.00 \text{ m/s}^2)]$$
$$= 390 \text{ N}$$

Now the woman feels as though she weighs only 390 N, or 100 N *less* than her actual weight.

You can feel these effects yourself; try taking a few steps in an elevator that is coming to a stop after descending (when your apparent weight is greater than your true weight w) or coming to a stop after ascending (when your apparent weight is less than w).

ActivPhysics OnLine

2.1.5 Car Race

2.2 Lifting a Crate

2.3 Lowering a Crate

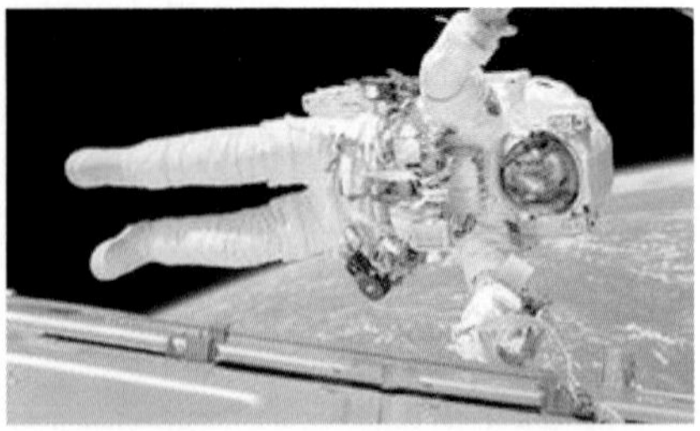

5.12 Astronauts in orbit feel "weightless" because they have the same acceleration as their spacecraft—*not* because they are "outside the pull of the earth's gravity." (If they were, the astronauts and their spacecraft wouldn't remain in orbit, but would fly off into deep space.)

Let's generalize the result of Example 5.10. When a passenger with mass m rides in an elevator with y-acceleration a_y, a scale shows the passenger's apparent weight to be

$$n = m(g + a_y)$$

When a_y is positive, the elevator is accelerating upward (moving upward with increasing speed, or moving downward with decreasing speed) and n is greater than the passenger's weight $w = mg$. When the elevator is accelerating downward (moving upward with decreasing speed, or moving downward with increasing speed), a_y is negative and n is less than the weight. If the passenger doesn't know the elevator is accelerating, she may feel as though her weight is changing; indeed, this is just what the scale shows.

The extreme case occurs when the elevator has a downward acceleration $a_y = -g$, that is, when it is in free fall. In that case, $n = 0$ and the passenger *seems* to be weightless. Similarly, an astronaut orbiting the Earth in a spacecraft experiences apparent weightlessness (Fig. 5.12). In each case, the person is not truly weightless, because there is still a gravitational force acting. But the effect of this free-fall condition is exactly the same as though the body were in outer space with no gravitational force at all. In both cases the person and vehicle (elevator or spacecraft) are falling together with the same acceleration g, so nothing pushes the person against the floor or walls of the vehicle.

Prolonged apparent weightlessness can have major physiological consequences for astronauts. The fluids of the body redistribute themselves, causing congested sinuses and shrinkage of the legs. Muscles and bones that normally carry weight on earth tend to deteriorate, and the resulting bone loss can cause kidney stones. Medical researchers are investigating ways to confront these and other effects of long-duration space flight.

ActivPhysics ONLINE

2.4 Rocket Blasts Off

2.11 Modified Atwood Machine

Example 5.11 Acceleration down a hill

A toboggan loaded with vacationing students (total weight w) slides down a long, snow-covered slope (Fig. 5.13a). The hill slopes at a constant angle α and the toboggan is so well waxed that there is virtually no friction. What is its acceleration?

SOLUTION

IDENTIFY: Our target variable is the acceleration, which we'll find using Newton's second law. There is no friction, so just two forces act on the toboggan: its weight w and the normal force n exerted by the hill. The new feature in this example is that the normal force points at an angle from the vertical and is not opposite to the weight. Hence we must use both components of $\sum \vec{F} = m\vec{a}$ in Eq. (5.4).

SET UP: Figure 5.13b shows the free-body diagram. We take axes parallel and perpendicular to the surface of the hill, so that the acceleration (which is parallel to the hill) is along the positive x-direction.

EXECUTE: The normal force has only a y-component, but the weight has both x- and y-components: $w_x = w \sin\alpha$ and $w_y = -w\cos\alpha$. (Compare to Example 5.4, in which the x-component of weight was $-w\sin\alpha$. The difference is that the positive x-axis was uphill in Example 5.4, while in Fig. 5.13b it is downhill.) The wiggly line in Fig. 5.13b reminds us that we have resolved the weight into its components.

The acceleration is purely in the $+x$-direction, so $a_y = 0$. Newton's second law in component form then tells us that

$$\sum F_x = w \sin\alpha = ma_x$$

$$\sum F_y = n - w\cos\alpha = ma_y = 0$$

Since $w = mg$, the x-component equation tells us that $mg \sin\alpha = ma_x$, or

$$a_x = g \sin\alpha$$

Note that we didn't need the y-component equation to find the acceleration. That's the beauty of choosing the x-axis to lie along the acceleration direction! What the y-components tell us is the magnitude of the normal force that the surface of the hill exerts on the toboggan:

$$n = w\cos\alpha = mg\cos\alpha$$

EVALUATE: Notice that the mass does not appear in our answer for the acceleration. This means that *any* toboggan, regardless of its mass or number of passengers, slides down a frictionless hill with an acceleration of $g \sin\alpha$. In particular, if the plane is horizontal, $\alpha = 0$ and $a_x = 0$ (the toboggan does not accelerate); if the plane is vertical, $\alpha = 90°$ and $a_x = g$ (the toboggan is in free fall).

(a) (b)

5.13 (a) A loaded toboggan slides down a frictionless hill. (b) The free-body diagram shows that the x-component of the weight, $w \sin\alpha$, accelerates the toboggan down the hill.

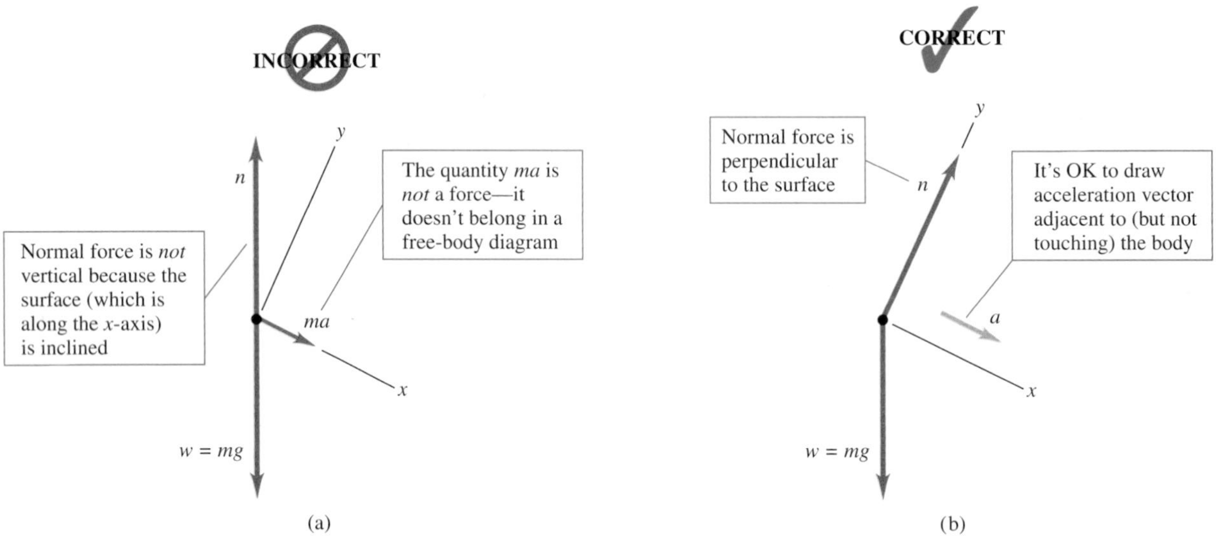

5.14 (a) An incorrect free-body diagram for the toboggan. (b) A correct free-body diagram.

Notice also that the normal force n is not equal to the toboggan's weight (compare Example 5.4 in Section 5.1). We don't need this result here, but it will be useful in a later example.

CAUTION Figure 5.14a shows a common *incorrect* way to draw the free-body diagram for the toboggan. This diagram is wrong for two reasons: the normal force must be drawn perpendicular to the surface, and it's completely incorrect to include an "$m\vec{a}$ force." If you remember that "normal" means "perpendicular" and that the mass times the acceleration is not itself a force, you'll be well on your way to always drawing correct free-body diagrams (Fig. 5.14b)

Example 5.12 Two bodies with the same acceleration

You are pushing a 1.00-kg food tray through the cafeteria line with a constant 9.0-N force. As the tray moves, it pushes in turn on a 0.50-kg carton of milk (Fig. 5.15a). The tray and carton slide on a horizontal surface that is so greasy that friction can be neglected. Find the acceleration of the system and the horizontal force that the tray exerts on the carton.

SOLUTION

IDENTIFY: In this example we have *two* target variables, the acceleration a_x of the system of tray and carton and the force $F_{\text{T on C}}$ of the tray on the carton. Again we will use Newton's second law, but we'll have to apply it to two different bodies to get two equations (one for each target variable).

SET UP: We can set up the problem in either of two ways.

Method 1: We can treat the tray (mass m_T) and milk carton (mass m_C) as separate bodies, each with its own free-body diagram (Fig. 5.15b). Note that the force F that you exert on the tray doesn't appear in the free-body diagram for the milk carton. Instead, what's making the carton accelerate is the force of magnitude $F_{\text{T on C}}$ exerted on it by the tray. By Newton's third law, the carton exerts a force of equal magnitude on the tray: $F_{\text{C on T}} = F_{\text{T on C}}$. We take the acceleration to be in the positive x-direction; both the tray and milk carton move with the same acceleration a_x.

Method 2: We can treat the tray and milk carton as a composite body of mass $m = m_T + m_C = 1.50$ kg (Fig. 5.15c). The only horizontal force acting on this composite body is the force F that you exert. The forces $F_{\text{T on C}}$ and $F_{\text{C on T}}$ don't come into play because they're *internal* to this composite body, and Newton's second law tells us that only *external* forces affect a body's acceleration (see Section 4.3). That means we'll need an additional equation to find the magnitude $F_{\text{T on C}}$ using this method; we'll get this equation by also applying Newton's second law to the milk carton, as in Method 1.

EXECUTE: *Method 1:* The x-component equations of Newton's second law for the tray and for the carton are

$$\text{Tray:} \quad \sum F_x = F - F_{\text{C on T}} = F - F_{\text{T on C}} = m_T a_x$$
$$\text{Carton:} \quad \sum F_x = F_{\text{T on C}} = m_C a_x$$

These are two simultaneous equations for the two target variables a_x and $F_{\text{T on C}}$. (Two equations are all we need, which means that the y-components don't play a role in this example.) An easy way

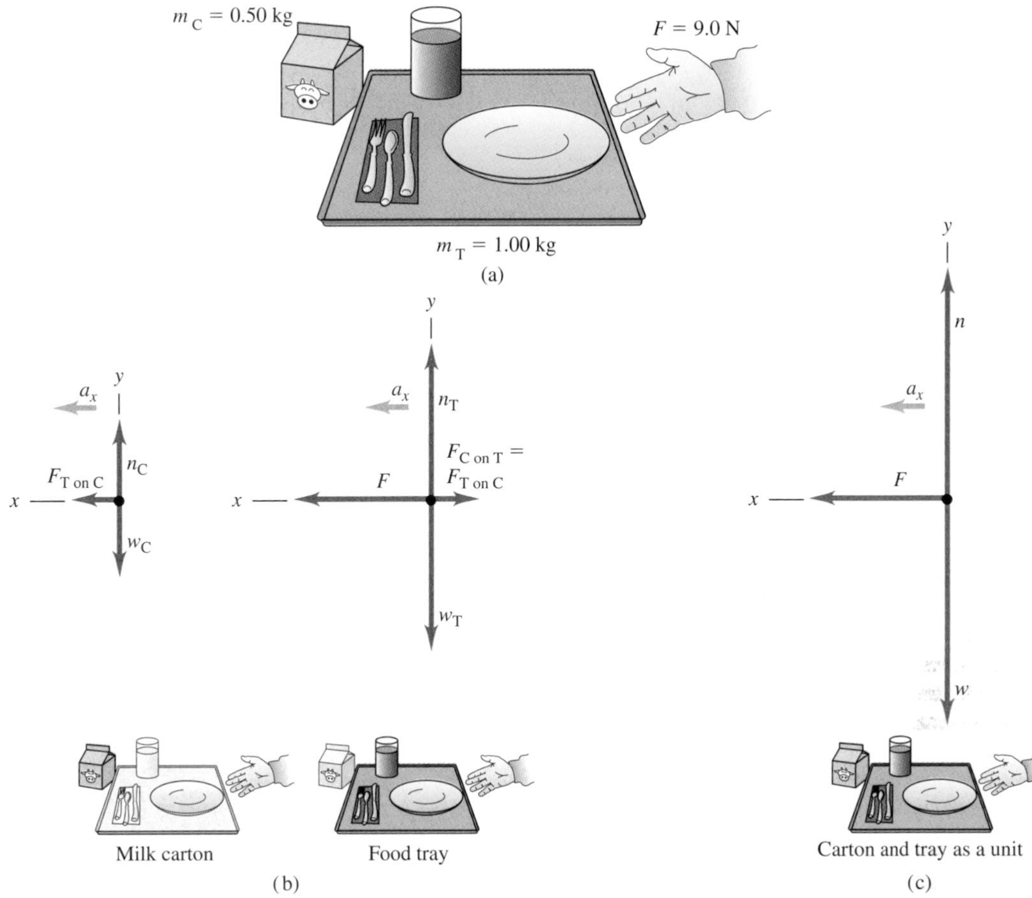

5.15 (a) Pushing a food tray and milk carton in the cafeteria line. (b) Individual free-body diagrams for the carton (left) and the tray (right). (c) Free-body diagram for the carton and the tray as a composite body.

to solve the two equations is to add them; this eliminates $F_{T \text{ on } C}$, giving

$$F = m_T a_x + m_C a_x = (m_T + m_C)a_x$$

and

$$a_x = \frac{F}{m_T + m_C} = \frac{9.0\text{ N}}{1.00\text{ kg} + 0.50\text{ kg}} = 6.0\text{ m/s}^2$$

Substituting this back into the second equation (for the carton) gives

$$F_{T \text{ on } C} = m_C a_x = (0.50\text{ kg})(6.0\text{ m/s}^2) = 3.0\text{ N}$$

Method 2: The x-component of Newton's second law for the composite body of mass m is

$$\sum F_x = F = ma_x$$

and the acceleration of this composite body is

$$a_x = \frac{F}{m} = \frac{9.0\text{ N}}{1.50\text{ kg}} = 6.0\text{ m/s}^2$$

Then, looking at the milk carton by itself, we see that to give it an acceleration of 6.0 m/s^2 requires that the tray exert a force

$$F_{T \text{ on } C} = m_C a_x = (0.50\text{ kg})(6.0\text{ m/s}^2) = 3.0\text{ N}$$

EVALUATE: The answers are the same with either method, as they should be. To check the answers, note that there are different forces on the two sides of the tray: $F = 9.0$ N on the right and $F_{C \text{ on } T} = 3.0$ N on the left. Hence the net horizontal force on the tray is $F - F_{C \text{ on } T} = 6.0$ N, exactly enough to accelerate a 1.00-kg tray at 6.0 m/s^2.

The method of treating the two bodies as a single composite body works *only* if the two bodies have the same magnitude *and* direction of acceleration. If the accelerations are different, we must treat the two bodies separately, as in the following example.

Example 5.13 Two bodies with the same magnitude of acceleration

Figure 5.16a shows an air-track glider with mass m_1 moving on a level, frictionless air track in the physics lab. The glider is connected to a lab weight with mass m_2 by a light, flexible, nonstretching string that passes over a small frictionless pulley. Find the acceleration of each body and the tension in the string.

SOLUTION

IDENTIFY: We have objects that are accelerating, so again we must use Newton's second law. In this example there are *three* target variables: the tension T in the string and the accelerations of the two bodies. Hence we'll need to find three simultaneous equations that involve these variables.

SET UP: The two bodies have different motions—one horizontal, one vertical—so we can't consider them together as we did the bodies in Example 5.12. We need a separate free-body diagram for each. Figures 5.16b and 5.16c show a free-body diagram and a coordinate system for each body. It's convenient to have both bodies accelerate in positive axis directions, so we choose the positive y-direction for the lab weight to be downward. (It's perfectly all right to use different coordinate axes for the two bodies.)

There is no friction in the pulley and we consider the string to be massless, so the tension T in the string is the same throughout; it applies a force of the same magnitude T to each body. (You may want to review Conceptual Example 4.10 in Section 4.5, where we discussed the tension force exerted by a massless string.) The weights are m_1g and m_2g.

While the *directions* of the two accelerations are different, their *magnitudes* are the same. That's because the string doesn't stretch. Hence the two bodies must move equal distances in equal times, and so their speeds at any instant must be equal. When the speeds change, they change by equal amounts in a given time, so the accelerations of the two bodies must have the same magnitude a. We can express this relation as

$$a_{1x} = a_{2y} = a$$

Thanks to this relationship, we actually have only *two* target variables: a and the tension T.

EXECUTE: For the glider on the track, Newton's second law gives

$$\text{Glider: } \sum F_x = T = m_1 a_{1x} = m_1 a$$
$$\text{Glider: } \sum F_y = n + (-m_1 g) = m_1 a_{1y} = 0$$

For the lab weight, the only forces are in the y-direction, and

$$\text{Lab weight: } \sum F_y = m_2 g + (-T) = m_2 a_{2y} = m_2 a$$

In these equations we've used the relationships $a_{1y} = 0$ (the glider doesn't accelerate vertically) and $a_{1x} = a_{2y} = a$ (the two objects have the same magnitude of acceleration).

The x-equation for the glider and the equation for the lab weight give us two simultaneous equations for the target variables T and a:

$$\text{Glider: } T = m_1 a$$
$$\text{Lab weight: } m_2 g - T = m_2 a$$

We add the two equations to eliminate T, giving

$$m_2 g = m_1 a + m_2 a = (m_1 + m_2)a$$

and so the magnitude of each body's acceleration is

$$a = \frac{m_2}{m_1 + m_2} g$$

Substituting this back into the first equation (for the glider), we get

$$T = \frac{m_1 m_2}{m_1 + m_2} g$$

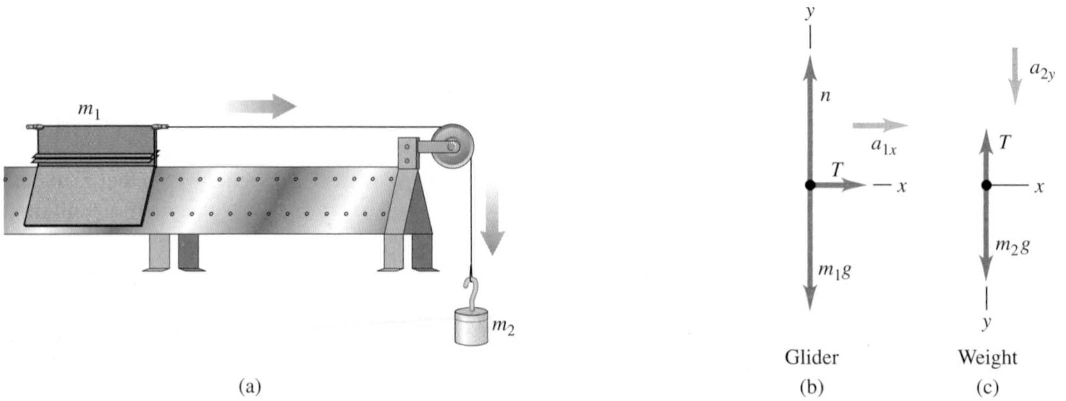

5.16 (a) A lab weight accelerates a glider along an air track. Free-body diagrams for (b) the glider (mass m_1) and for (c) the lab weight (mass m_2).

EVALUATE: The acceleration is less than g, as you might expect; the lab weight accelerates more slowly because the string tension pulls it back.

The tension T is *not* equal to the weight m_2g of the lab weight, but is *less* by a factor of $m_1/(m_1 + m_2)$. If T *were* equal to m_2g, then the lab weight would be in equilibrium, and it isn't.

CAUTION It's a common mistake to assume that if an object is attached to a vertical string, the string tension must be equal to the object's weight. That was the case in Example 5.5, where the acceleration was zero, but it would certainly be wrong in this example! The only safe approach is to *always* treat the tension as a variable, as we have done here.

Finally, let's check some special cases. If $m_1 = 0$, then the lab weight would fall freely and there would be no tension in the string. The equations do give $T = 0$ and $a = g$ when $m_1 = 0$. Also, if $m_2 = 0$, we expect no tension and no acceleration; for this case the equations give $T = 0$ and $a = 0$. Thus in these two special cases the results agree with our intuitive expectations.

Test Your Understanding

You are riding a sled with frictionless runners down a hill with a 3.0° slope. The combined mass of you and the sled is 90 kg. As the sled moves, you drag your heels in the snow to keep your speed under control. What force must you exert parallel to the slope to make the sled slow down at a rate of 2.0 m/s^2?

5.3 | Frictional Forces

We have seen several problems where a body rests or slides on a surface that exerts forces on the body, and we have used the terms *normal force* and *friction force* to describe these forces. Whenever two bodies interact by direct contact (touching) of their surfaces, we call the interaction forces *contact forces.* Normal forces and friction forces are both contact forces.

Our principal concern in this section is with friction, an important force in many aspects of everyday life. The oil in a car engine minimizes friction between moving parts, but without friction between the tires and the road we couldn't drive or turn the car. Air drag—the frictional force exerted by the air on a body moving through it—decreases automotive fuel economy but makes parachutes work. Without friction, nails would pull out, light bulbs and bottle caps would unscrew effortlessly, and sports such as bicycling and ice hockey would be hopeless (Fig. 5.17).

5.17 The sport of ice hockey depends critically on having just the right amount of friction between a player's skates and the ice. If there were too much friction, the players would move too slowly; if there were too little friction, they could hardly keep from falling over.

Kinetic and Static Friction

Let's consider a body sliding across a surface. When you try to slide a heavy box of books across the floor, the box doesn't move at all unless you push with a certain minimum force. Then the box starts moving, and you can usually keep it moving with less force than you needed to get it started. If you take some of the books out, you need less force than before to get it started or keep it moving. What general statements can we make about this behavior?

First, when a body rests or slides on a surface, we can always represent the contact force exerted by the surface on the body in terms of components of force perpendicular and parallel to the surface. As before, we call the perpendicular component vector the normal force, denoted by $\vec{n}$. The component vector parallel to the surface is the **friction force,** denoted by $\vec{f}$. By definition, $\vec{n}$ and $\vec{f}$ are always perpendicular to each other. If the surface is frictionless, then the contact force has only a normal component and $\vec{f}$ is zero. (Frictionless surfaces are an unattainable idealization, like a massless rope. But we can approximate a surface

ActivPhysics Online

2.5 Truck Pulls Crate
2.6 Pushing a Crate Up a Wall
2.7 Skier Goes Down a Slope
2.8 Skier and Rope Tow
2.10 Truck Pulls Two Crates

as frictionless if the effects of friction are negligibly small.) The direction of the friction force is always such as to oppose relative motion of the two surfaces.

The kind of friction that acts when a body slides over a surface is called a **kinetic friction force** $\vec{f}_k$. The adjective "kinetic" and the subscript "k" remind us that the two surfaces are moving relative to each other. The *magnitude* of the kinetic friction force usually increases when the normal force increases. This is why it takes more force to slide a box full of books across the floor than to slide the same box when it is empty. This principle is also used in automotive braking systems: the harder the brake pads are squeezed against the rotating brake disks, the greater the braking effect. In many cases the magnitude of the kinetic friction force f_k is found experimentally to be approximately *proportional* to the magnitude n of the normal force. In such cases we represent the relation by the equation

$$f_k = \mu_k n \qquad \text{(magnitude of kinetic friction force)} \qquad (5.5)$$

where μ_k (pronounced "mu-sub-k") is a constant called the **coefficient of kinetic friction.** The more slippery the surface, the smaller the coefficient of friction. Because it is a quotient of two force magnitudes, μ_k is a pure number, without units.

CAUTION Remember, the friction force and the normal force are always perpendicular. Equation (5.5) is not a vector equation, but a scalar relation between the magnitudes of the two perpendicular forces.

Equation (5.5) is only an approximate representation of a complex phenomenon. On a microscopic level, friction and normal forces result from the intermolecular forces (fundamentally electrical in nature) between two rough surfaces at points where they come into contact (Fig. 5.18). The actual area of contact is usually much smaller than the total surface area. As a box slides over the floor, bonds between the two surfaces form and break, and the total number of such bonds varies; hence the kinetic friction force is not perfectly constant. Smoothing the surfaces can actually increase friction, since more molecules are able to interact and bond; bringing two smooth surfaces of the same metal together can cause a "cold weld." Lubricating oils work because an oil film between two surfaces (such as the pistons and cylinder walls in a car engine) prevents them from coming into actual contact.

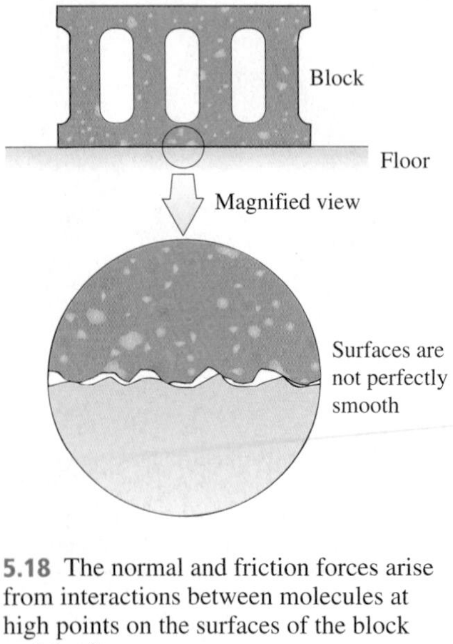

5.18 The normal and friction forces arise from interactions between molecules at high points on the surfaces of the block and the floor.

Table 5.1 shows a few representative values of μ_k. Although these values are given with two significant figures, they are only approximate since friction forces can also depend on the speed of the body relative to the surface. For now we'll ignore this effect and assume that μ_k and f_k are independent of speed, in order to concentrate on the simplest cases. Table 5.1 also lists coefficients of static friction; we'll define these shortly.

Friction forces may also act when there is *no* relative motion. If you try to slide that box of books across the floor, the box may not move at all because the floor exerts an equal and opposite friction force on the box. This is called a **static friction force** $\vec{f}_s$. In Fig. 5.19a, the box is at rest, in equilibrium, under the action of its weight $\vec{w}$ and the upward normal force $\vec{n}$. The normal force is equal in magnitude to the weight $(n = w)$ and is exerted on the box by the floor. Now we tie a rope to the box (Fig. 5.19b) and gradually increase the tension T in the rope. At first the box remains at rest because, as T increases, the force of static friction f_s also increases (staying equal in magnitude to T).

Table 5.1 Approximate Coefficients of Friction

Materials	Coefficient of static friction, μ_s	Coefficient of kinetic friction, μ_k
Steel on steel	0.74	0.57
Aluminum on steel	0.61	0.47
Copper on steel	0.53	0.36
Brass on steel	0.51	0.44
Zinc on cast iron	0.85	0.21
Copper on cast iron	1.05	0.29
Glass on glass	0.94	0.40
Copper on glass	0.68	0.53
Teflon on Teflon	0.04	0.04
Teflon on steel	0.04	0.04
Rubber on concrete (dry)	1.0	0.8
Rubber on concrete (wet)	0.30	0.25

At some point T becomes greater than the maximum static friction force f_s the surface can exert. Then the box "breaks loose" (the tension T is able to break the bonds between molecules in the surfaces of the box and floor) and starts to slide. Figure 5.19c is the force diagram when T is at this critical value. If T exceeds this value, the box is no longer in equilibrium. For a given pair of surfaces the maximum value of f_s depends on the normal force. Experiment shows that in many cases this maximum value, called $(f_s)_{max}$, is approximately *proportional* to n; we call the proportionality factor μ_s the **coefficient of static friction.** Table 5.1 lists some representative values of μ_s. In a particular situation, the actual force of static friction can have any magnitude between zero (when there is no other force parallel to the surface) and a maximum value given by $\mu_s n$. In symbols,

$$f_s \le \mu_s n \qquad \text{(magnitude of static friction force)} \tag{5.6}$$

Like Eq. (5.5), this is a relation between magnitudes, *not* a vector relation. The equality sign holds only when the applied force T, parallel to the surface, has reached the critical value at which motion is about to start (Fig. 5.19c). When T is less than this value (Fig. 5.19b), the inequality sign holds. In that case we have to use the equilibrium conditions $(\sum \vec{F} = \mathbf{0})$ to find f_s. If there is no applied force $(T = 0)$ as in Fig. 5.19a, then there is no static friction force either $(f_s = 0)$.

As soon as sliding starts (Fig. 5.19d), the friction force usually *decreases;* in other words, it's easier to keep the box moving than to start it moving. Hence the coefficient of kinetic friction is usually *less* than the coefficient of static friction for any given pair of surfaces, as shown in Table 5.1. If we start with no applied force $(T = 0)$ at time $t = 0$ and gradually increase the force, the friction force varies somewhat as shown in Fig. 5.20.

In some situations the surfaces will alternately stick (static friction) and slip (kinetic friction). This is what causes the horrible squeak made by chalk held at the wrong angle while writing on the blackboard. Another stick-slip phenomenon is the squeaky noise your windshield-wiper blades make when the glass is nearly dry; still another is the outraged shriek of tires sliding on asphalt pavement. A more positive example is the motion of a violin bow against the string.

When a body slides on a layer of gas, friction can be made very small. In the linear air track used in physics laboratories, the gliders are supported on a layer of air. The frictional force is velocity-dependent, but at typical speeds the effective coefficient of friction is of the order of 0.001. A similar device is the air hockey table, where the pucks are supported by an array of small air jets about 2 cm apart.

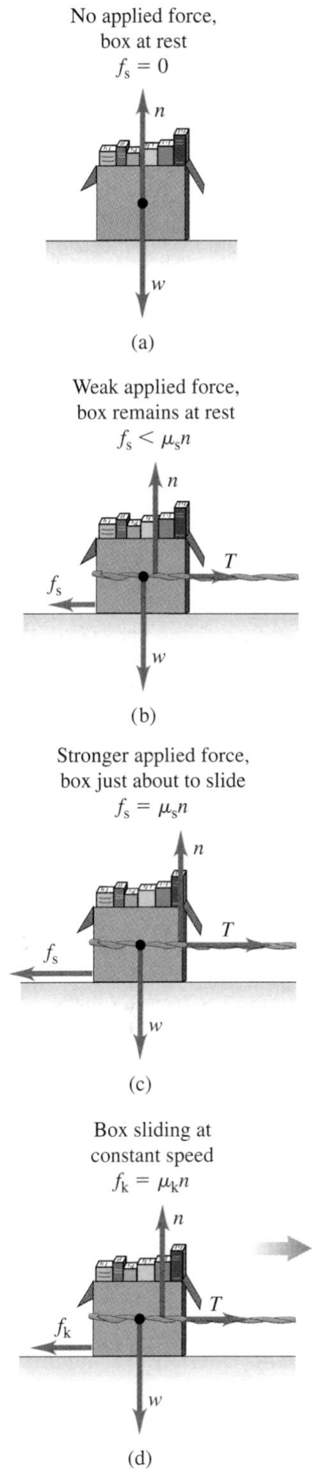

5.19 (a), (b), (c) When there is no relative motion of the surfaces, the magnitude of the static friction force f_s is less than or equal to $\mu_s n$. (d) When there is relative motion, the magnitude of the kinetic friction force f_k equals $\mu_k n$.

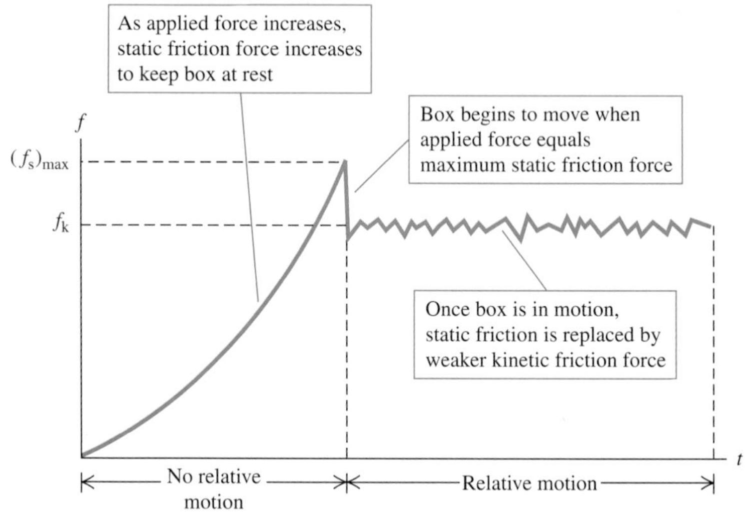

5.20 In response to an increasing externally applied force, the friction force increases to $(f_s)_{max}$. Then the surfaces begin to slide across one another, and the frictional force drops back to a nearly constant value f_k. The kinetic friction force varies somewhat as intermolecular bonds form and break.

Example 5.14 Friction in horizontal motion

A delivery company has just unloaded a 500-N crate full of home exercise equipment in your driveway (Fig. 5.21a). You find that to get it started moving toward your garage, you have to pull with a horizontal force of magnitude 230 N. Once it "breaks loose" and starts to move, you can keep it moving at constant velocity with only 200 N. What are the coefficients of static and kinetic friction?

SOLUTION

IDENTIFY: The state of rest and the state of motion with constant velocity are both equilibrium conditions, so we use Newton's first law as expressed by Eq. (5.2). In either situation there are four forces acting on the crate: the downward weight force (magnitude $w = 500$ N), the upward normal force (magnitude n) exerted by the ground, a tension force (magnitude T) to the right exerted by the rope, and a friction force to the left exerted by the ground. The target variables are μ_s and μ_k; to find these we'll need Eqs. (5.5) and (5.6), which involve the friction coefficients.

SET UP: Figure 5.21b shows the free-body diagram an instant before the crate starts to move, when the static friction force has its maximum possible value $(f_s)_{max} = \mu_s n$. Once the crate is moving to the right at constant velocity, the friction force changes to its

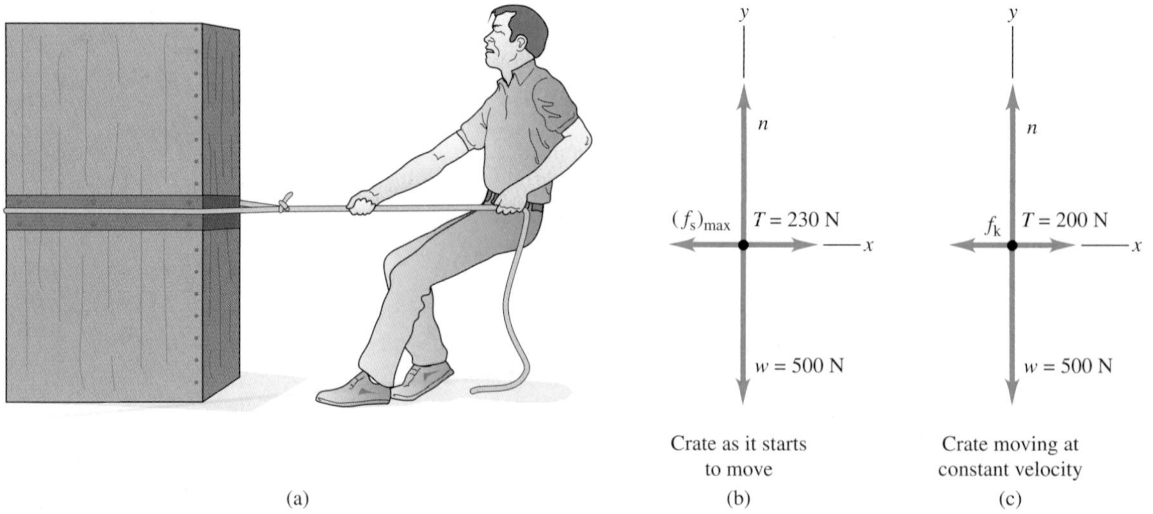

5.21 (a) Pulling a crate with a horizontal force. (b) Free-body diagram for the crate as it starts to move. (c) Free-body diagram for the crate moving at constant velocity.

kinetic form (Fig. 5.21c). Because the rope in Fig. 5.21a is in equilibrium, the tension is the same at both ends. Hence the tension force that the rope exerts on the crate has the same magnitude as the force you exert on the rope.

EXECUTE: Just before the crate starts to move, we have

$$\sum F_x = T + (-(f_s)_{max}) = 0 \quad \text{so} \quad (f_s)_{max} = T = 230 \text{ N}$$
$$\sum F_y = n + (-w) = 0 \quad \text{so} \quad n = w = 500 \text{ N}$$

To find the value of μ_s, we use the relationship $(f_s)_{max} = \mu_s n$. Hence

$$\mu_s = \frac{(f_s)_{max}}{n} = \frac{230 \text{ N}}{500 \text{ N}} = 0.46$$

After the crate starts to move, the forces are as shown in Fig. 5.21c, and we have

$$\sum F_x = T + (-f_k) = 0 \quad \text{so} \quad f_k = T = 200 \text{ N}$$
$$\sum F_y = n + (-w) = 0 \quad \text{so} \quad n = w = 500 \text{ N}$$

Then, using $f_k = \mu_k n$ from Eq. (5.5),

$$\mu_k = \frac{f_k}{n} = \frac{200 \text{ N}}{500 \text{ N}} = 0.40$$

EVALUATE: It's easier to keep the crate moving than to start it moving, and so the coefficient of kinetic friction is less than the coefficient of static friction.

Example 5.15 Static friction can be less than the maximum

In Example 5.14, what is the friction force if the crate is at rest on the surface and a horizontal force of 50 N is applied to it?

SOLUTION

IDENTIFY: The applied force is less than the maximum force of static friction, $(f_s)_{max} = 230$ N. Hence the crate remains at rest and the net force acting on it is zero. The target variable is the magnitude f_s of the friction force.

SET UP: The free-body diagram is the same as in Fig. 5.21b, but with $(f_s)_{max}$ replaced by f_s and $T = 230$ N replaced by $T = 50$ N.

EXECUTE: From the equilibrium conditions, Eq. (5.2), we have

$$\sum F_x = T + (-f_s) = 0 \quad \text{so} \quad f_s = T = 50 \text{ N}$$

EVALUATE: In this case, f_s is less than the maximum value $(f_s)_{max} = \mu_s n$. The frictional force can prevent motion for any horizontal applied force up to 230 N.

Example 5.16 Minimizing kinetic friction

In Example 5.14, suppose you try to move your crate full of exercise equipment by tying a rope around it and pulling upward on the rope at an angle of 30° above the horizontal (Fig. 5.22a). How hard do you have to pull to keep the crate moving with constant velocity? Is this easier or harder than pulling horizontally? Assume $w = 500$ N and $\mu_k = 0.40$.

SOLUTION

IDENTIFY: The crate is in equilibrium since its velocity is constant, so we again apply Newton's first law. Since the crate is in motion, the ground exerts a *kinetic* friction force. The target variable is the magnitude T of the tension force.

SET UP: Figure 5.22b is a free-body diagram showing the forces on the crate. The kinetic friction force f_k is still equal to $\mu_k n$, but now the normal force n is *not* equal in magnitude to the weight of the crate. The force exerted by the rope has an additional vertical component that tends to lift the crate off the floor.

EXECUTE: From the equilibrium conditions and the equation $f_k = \mu_k n$, we have

$$\sum F_x = T\cos 30° + (-f_k) = 0 \quad \text{so} \quad T\cos 30° = \mu_k n$$
$$\sum F_y = T\sin 30° + n + (-w) = 0 \quad \text{so} \quad n = w - T\sin 30°$$

These are two simultaneous equations for the two unknown quantities T and n. To solve them, we can eliminate one unknown and solve for the other. There are many ways to do this; one way is to substitute the expression for n in the second equation back into the first equation:

$$T\cos 30° = \mu_k(w - T\sin 30°)$$

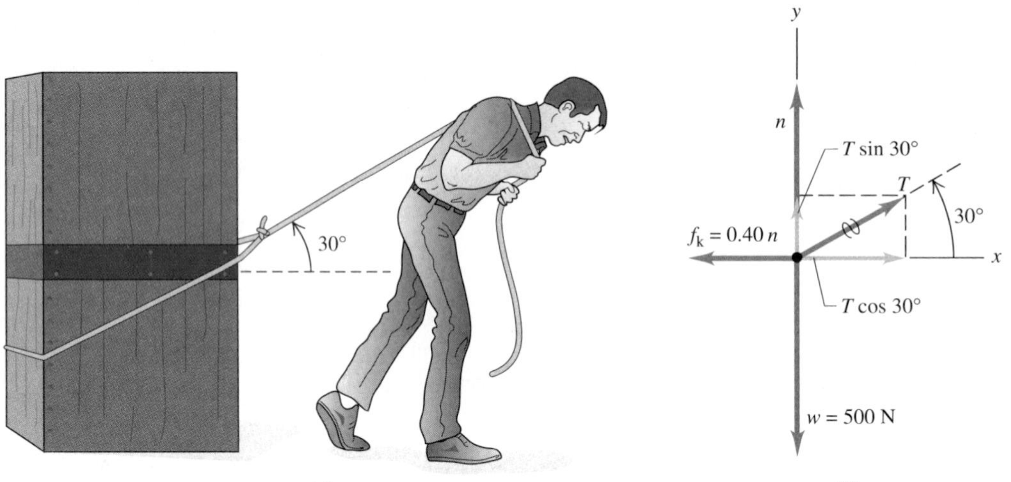

(a) (b)

5.22 (a) Pulling a crate with a force applied at an upward angle. (b) Free-body diagram for the crate moving at constant velocity.

Then solve this equation for T, with the result

$$T = \frac{\mu_k w}{\cos 30° + \mu_k \sin 30°} = 188\ \text{N}$$

Finally, substitute this result back into either of the original equations to obtain n. If we use the second equation to do this, we get

$$n = w - T \sin 30° = (500\ \text{N}) - (188\ \text{N}) \sin 30° = 406\ \text{N}$$

EVALUATE: The normal force is *less* than the weight of the box $(w = 500\ \text{N})$ because the vertical component of tension pulls upward on the crate. Despite this, the tension required is a little less than the 200-N force needed when you pulled horizontally in Example 5.14. Try pulling at 22°; you'll find you need even less force (see Challenge Problem 5.123).

Example 5.17 Toboggan ride with friction I

Let's go back to the toboggan we studied in Example 5.11 (Section 5.2). The wax has worn off and there is now a coefficient of kinetic friction μ_k. The slope has just the right angle to make the toboggan slide with constant speed. Derive an expression for the slope angle in terms of w and μ_k.

SOLUTION

IDENTIFY: Our target variable is the slope angle α. The toboggan is in equilibrium because its velocity is constant, so once again we use Newton's first law. There are three forces acting on the toboggan: its weight, the normal force, and the kinetic friction force. Since the motion is downhill, the kinetic friction force (which opposes the motion of the toboggan over the hill) is directed uphill. The magnitude of the friction force is given by Eq. (5.5), $f_k = \mu_k n$.

SET UP: Figure 5.23 shows the free-body diagram. We take axes perpendicular and parallel to the surface and represent the weight in terms of its components in these two directions, as shown. (Compare Fig. 5.13b in Example 5.11.)

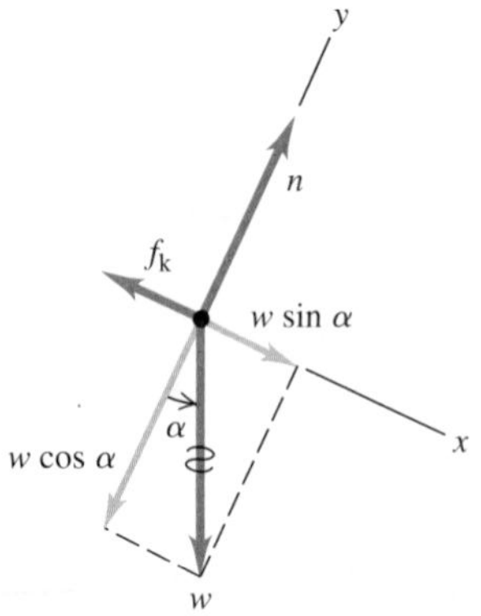

5.23 Free-body diagram for the toboggan with friction.

EXECUTE: The equilibrium conditions are

$$\sum F_x = w \sin \alpha + (-f_k) = w \sin \alpha - \mu_k n = 0$$
$$\sum F_y = n + (-w \cos \alpha) = 0$$

(We used the relationship $f_k = \mu_k n$ in the equation for the x-components.) Rearranging, we get

$$\mu_k n = w \sin \alpha \quad \text{and} \quad n = w \cos \alpha$$

Just as in Example 5.11, the normal force n is *not* equal to the weight w. When we divide the first of these equations by the second, we find

$$\mu_k = \frac{\sin \alpha}{\cos \alpha} = \tan \alpha \quad \text{so} \quad \alpha = \arctan \mu_k$$

EVALUATE: The weight w doesn't appear in this expression. *Any* toboggan, regardless of its weight, slides down an incline with constant speed if the coefficient of kinetic friction equals the tangent of the slope angle of the incline. The greater the coefficient of friction, the steeper the slope has to be for the toboggan to slide with constant velocity. This is just what we should expect.

Example 5.18 Toboggan ride with friction II

What if the same toboggan with the same coefficient of friction is sliding downhill as in Example 5.17, but on a steeper hill? This time the toboggan accelerates, although not as much as in Example 5.11, when there was no friction. Derive an expression for the acceleration in terms of g, α, μ_k, and w.

SOLUTION

IDENTIFY: The toboggan is no longer in equilibrium, but accelerates as it slides down the hill. Hence we must use Newton's second law, $\sum \vec{F} = m\vec{a}$, in its component form as given in Eq. (5.4). Our target variable is the downhill acceleration.

SET UP: The free-body diagram (Fig. 5.24) is almost the same as for Example 5.17. The toboggan's y-component of acceleration a_y is still zero, but the x-component a_x is not.

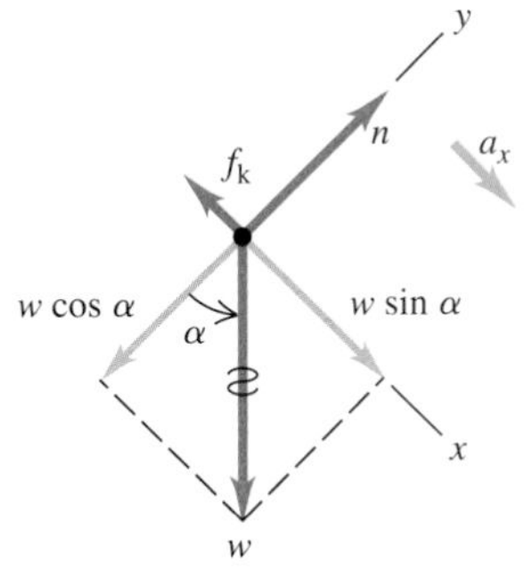

5.24 Free-body diagram for the toboggan with friction, going down a steeper hill.

EXECUTE: It's convenient to express the weight as $w = mg$. Then, using Newton's second law for the x- and y-components of force, we have

$$\sum F_x = mg \sin \alpha + (-f_k) = ma_x$$
$$\sum F_y = n + (-mg \cos \alpha) = 0$$

From the second equation and Eq. (5.5) we get an expression for f_k:

$$n = mg \cos \alpha$$
$$f_k = \mu_k n = \mu_k mg \cos \alpha$$

We substitute this back into the x-component equation. The result is

$$mg \sin \alpha + (-\mu_k mg \cos \alpha) = ma_x$$
$$a_x = g(\sin \alpha - \mu_k \cos \alpha)$$

EVALUATE: Does this result make sense? Here are some special cases we can check. First, if the hill is vertical, $\alpha = 90°$; then $\sin \alpha = 1$, $\cos \alpha = 0$, and $a_x = g$. This is free fall, just what we would expect. Second, on a hill at angle α with no friction, $\mu_k = 0$. Then $a_x = g \sin \alpha$. The situation is the same as in Example 5.11, and we get the same result; that's encouraging! Next, suppose that there is just enough friction to make the toboggan move with constant velocity. In that case, $a_x = 0$ and our result gives

$$\sin \alpha = \mu_k \cos \alpha \quad \text{and} \quad \mu_k = \tan \alpha$$

This agrees with our result from Example 5.17; good! Finally, note that there may be so much friction that $\mu_k \cos \alpha$ is actually greater than $\sin \alpha$. In that case, a_x is negative; if we give the toboggan an initial downhill push to start it moving, it will slow down and eventually stop.

We have pretty much beaten the toboggan problem to death, but there is an important lesson to be learned. We started out with a simple problem and then extended it to more and more general situations. The general result we found in this example includes *all* the previous ones as special cases, and that's a nice, neat package! Don't memorize this package; it is useful only for this one set of problems. But do try to understand how we obtained it and what it means.

One final variation that you may want to try out is the case in which we give the toboggan an initial push up the hill. The direction of the kinetic friction force is now reversed, so the acceleration is different from the downhill value. It turns out that the expression for a_x is the same as for downhill motion except that the minus sign becomes plus. Can you prove this?

Rolling Friction

It's a lot easier to move a loaded filing cabinet across a horizontal floor using a cart with wheels than to slide it. How much easier? We can define a **coefficient of rolling friction** μ_r, which is the horizontal force needed for constant speed on a flat surface divided by the upward normal force exerted by the surface. Transportation engineers call μ_r the *tractive resistance.* Typical values of μ_r are 0.002 to 0.003 for steel wheels on steel rails and 0.01 to 0.02 for rubber tires on concrete. These values show one reason why railroad trains are generally much more fuel-efficient than highway trucks.

Example 5.19 Motion with rolling friction

A typical car weighs about 12,000 N (about 2700 lb). If the coefficient of rolling friction is $\mu_r = 0.015$, what horizontal force is needed to make the car move with constant speed on a level road? Neglect air resistance.

SOLUTION

IDENTIFY: The car is moving with constant velocity, so this is an equilibrium problem that uses Newton's first law. The four forces on the car are the weight, the upward normal force, the backward force of rolling friction, and the unknown forward force F (our target variable).

SET UP: The free-body diagram is much like the one in Fig. 5.21c of Example 5.14, but with the kinetic friction force replaced by the rolling friction force f_r and with the tension force replaced by the unknown force F.

EXECUTE: As in Example 5.14, Newton's first law for the *vertical* components tells us that the normal force is equal in magnitude to the car's weight. Hence, from the definition of μ_r, the rolling friction force f_r is

$$f_r = \mu_r n = (0.015)(12{,}000\ \text{N}) = 180\ \text{N} \quad (\text{about 41 lb})$$

Newton's first law for the *horizontal* components tells us that a forward force with this magnitude is needed to keep the car moving with constant speed.

EVALUATE: The required force is rather small, which is why it's possible to push a car by hand. (As in the case of sliding, it's easier to keep a car rolling than it is to start it rolling.) We've ignored the effects of air resistance, which is a pretty good approximation if the car is moving slowly. But at typical highway speeds air resistance is a larger effect than rolling friction.

Try applying this analysis to your crate of exercise equipment (Example 5.14). If the delivery company brings it on a rubber-wheel dolly with $\mu_r = 0.02$, only a 10-N force is needed to keep it moving at constant velocity. Can you verify this?

Fluid Resistance and Terminal Speed

Sticking your hand out the window of a fast-moving car will convince you of the existence of **fluid resistance,** the force that a fluid (a gas or liquid) exerts on a body moving through it. The moving body exerts a force on the fluid to push it out of the way. By Newton's third law, the fluid pushes back on the body with an equal and opposite force.

The *direction* of the fluid resistance force acting on a body is always opposite the direction of the body's velocity relative to the fluid. The *magnitude* of the fluid resistance force usually increases with the speed of the body through the fluid; contrast this behavior with that of the kinetic friction force between two surfaces in contact, which we can usually regard as independent of speed. For low speeds, the magnitude f of the resisting force of the fluid is approximately proportional to the body's speed v:

$$f = kv \qquad (\text{fluid resistance at low speed}) \tag{5.7}$$

where k is a proportionality constant that depends on the shape and size of the body and the properties of the fluid. In motion through air at the speed of a tossed tennis ball or faster, the resisting force is approximately proportional to v^2 rather than to v. It is then called **air drag** or simply *drag*. Airplanes, falling raindrops, and cars moving at high speed all experience air drag. In this case we replace Eq. (5.7) by

$$f = Dv^2 \qquad \text{(fluid resistance at high speed)} \qquad (5.8)$$

Because of the v^2 dependence, air drag increases rapidly with increasing speed. The air drag on a typical car is negligible at low speeds but comparable to or greater than tractive resistance at highway speeds. The value of D depends on the shape and size of the body and on the density of the air. You should verify that the units of the constant k in Eq. (5.7) are $\text{N} \cdot \text{s/m}$ or kg/s, and that the units of the constant D in Eq. (5.8) are $\text{N} \cdot \text{s}^2/\text{m}^2$ or kg/m.

Because of the effects of fluid resistance, an object falling in a fluid does *not* have a constant acceleration. To describe its motion, we can't use the constant-acceleration relationships from Chapter 2; instead, we have to start over using Newton's second law. Let's consider the following situation. You release a rock at the surface of a deep pond and it falls to the bottom (Fig. 5.25a). The fluid resistance force in this situation is given by Eq. (5.7). What are the acceleration, velocity, and position of the rock as functions of time?

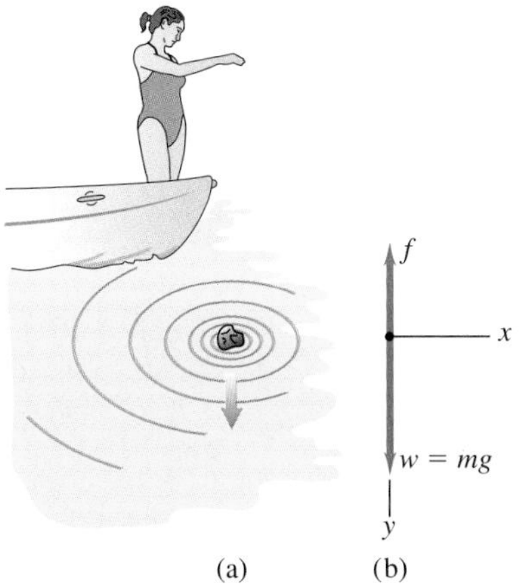

5.25 (a) A rock falling downward through a fluid (water). (b) Free-body diagram for the rock.

Figure 5.25b shows the free-body diagram. We take the positive y-direction to be downward and neglect any force associated with buoyancy in the water. Since the rock is moving downward, the speed v is equal to the y-velocity v_y and the fluid resistance force is in the $-y$-direction. There are no x-components, so Newton's second law gives

$$\sum F_y = mg + (-kv_y) = ma_y$$

When the rock first starts to move, $v_y = 0$, the resisting force is zero, and the initial acceleration is $a_y = g$. As the speed increases, the resisting force also increases, until finally it is equal in magnitude to the weight. At this time $mg - kv_y = 0$, the acceleration becomes zero, and there is no further increase in speed. The final speed v_t, called the **terminal speed,** is given by $mg - kv_t = 0$, or

$$v_t = \frac{mg}{k} \qquad \text{(terminal speed, fluid resistance } f = kv\text{)} \qquad (5.9)$$

Figure 5.26 shows how the acceleration, velocity, and position vary with time. As time goes by, the acceleration approaches zero and the velocity approaches v_t (remember that we chose the positive y-direction to be down). The slope of the graph of y versus t becomes constant as the velocity becomes constant.

To see how the graphs in Fig. 5.26 are derived, we must find the relation between speed and time during the interval before the terminal speed is reached. We go back to Newton's second law, which we rewrite as

$$m\frac{dv_y}{dt} = mg - kv_y$$

After rearranging terms and replacing mg/k by v_t, we integrate both sides, noting that $v_y = 0$ when $t = 0$:

$$\int_0^v \frac{dv_y}{v_y - v_t} = -\frac{k}{m}\int_0^t dt$$

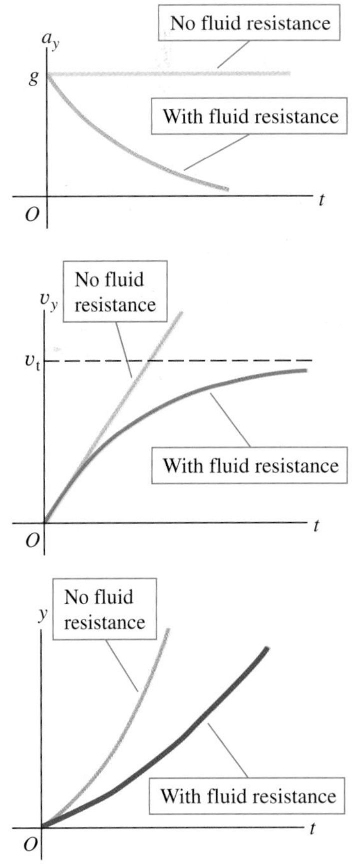

5.26 Graphs of acceleration, velocity, and position versus time for a body falling with fluid resistance proportional to the speed (dark color curves) and without fluid resistance (light color curves).

which integrates to

$$\ln\frac{v_t - v_y}{v_t} = -\frac{k}{m}t \quad \text{or} \quad 1 - \frac{v_y}{v_t} = e^{-(k/m)t}$$

and finally

$$v_y = v_t\left[1 - e^{-(k/m)t}\right] \tag{5.10}$$

ActivPhysics Online

2.1.2 Skydiver

Note that v_y becomes equal to the terminal speed v_t only in the limit that $t \to \infty$; the rock cannot attain terminal speed in any finite length of time.

The derivative of v_y gives a_y as a function of time, and the integral of v_y gives y as a function of time. We leave the derivations for you to complete (see Exercise 5.40); the results are

$$a_y = ge^{-(k/m)t} \tag{5.11}$$

$$y = v_t\left[t - \frac{m}{k}\left(1 - e^{-(k/m)t}\right)\right] \tag{5.12}$$

Now look again at Fig. 5.26, which shows graphs of these three relations.

In deriving the terminal speed in Eq. (5.9), we assumed that the fluid resistance force is proportional to the speed. For an object falling through the air at high speeds, so that the fluid resistance is proportional to v^2 as in Eq. (5.8), you can show that the terminal speed v_t is given by

$$v_t = \sqrt{\frac{mg}{D}} \qquad \text{(terminal speed, fluid resistance } f = Dv^2\text{)} \tag{5.13}$$

5.27 By changing the positions of their arms and legs while falling, skydivers can change the value of the constant D in Eq. (5.8) and hence adjust the terminal speed of their fall [Eq. (5.13)].

This expression for terminal speed explains why heavy objects in air tend to fall faster than light objects. Two objects with the same physical size but different mass (say, a table-tennis ball and a lead ball with the same radius) have the same value of D but different values of m. The more massive object has a larger terminal speed and falls faster. The same idea explains why a sheet of paper falls faster if you first crumple it into a ball; the mass m is the same, but the smaller size makes D smaller (less air drag for a given speed) and v_t larger. Skydivers use the same principle to control their descent (Fig. 5.27).

Figure 5.28 shows the trajectory of a baseball with and without air drag, assuming a coefficient $D = 1.3 \times 10^{-3}$ kg/m (appropriate for a batted ball or

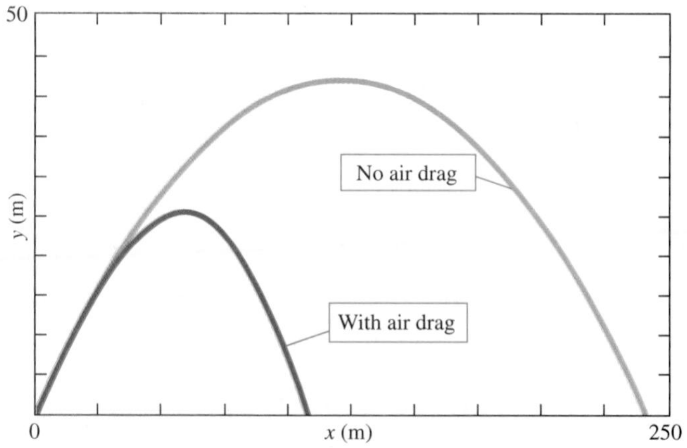

5.28 Computer-generated trajectories of a baseball (mass 0.145 kg, radius 0.0366 m) launched at a 35° angle above the horizontal at a speed of 50 m/s. The two curves show the trajectories with and without air drag. Note that the scales are different on the horizontal and vertical axes.

pitched fastball and a ballpark at sea level). In this example the baseball was given an initial velocity of 50 m/s at an angle of 35° above the $+x$-axis. You can see that both the range of the baseball and the maximum height reached are substantially less than the zero-drag calculation would lead you to believe. Hence the baseball trajectory we calculated in Example 3.8 (Section 3.3) by ignoring air drag is quite unrealistic. Air drag is an important part of the game of baseball!

Example 5.20 Terminal speed of a skydiver

For a human body falling though air in a spread-eagle position (Fig. 5.27), the numerical value of the constant D in Eq. (5.8) is about 0.25 kg/m. Find the terminal speed for an 80-kg skydiver.

SOLUTION

IDENTIFY AND SET UP: We use Eq. (5.13) to find the target variable v_t.

EXECUTE: We find

$$v_t = \sqrt{\frac{mg}{D}} = \sqrt{\frac{(80\text{ kg})(9.8\text{ m/s}^2)}{0.25\text{ kg/m}}}$$

$$= 56\text{ m/s (about 200 km/h, or 125 mi/h)}$$

EVALUATE: Because the terminal speed is quite high, skydives don't last very long. A drop from 2800 m (9200 ft) to the surface takes only $(2800\text{ m})/(56\text{ m/s}) = 50$ s.

When the sky diver deploys the parachute, the value of D increases greatly. Hence the terminal speed of the sky diver and parachute decreases dramatically to much less than 56 m/s.

Test Your Understanding

Once again you are riding a sled down a hill with a 3.0° slope. The combined mass of you and the sled is 90 kg. This time you don't drag your heels, and the sled speeds up at a rate of 0.25 m/s^2. What is the coefficient of kinetic friction between the snow and the sled's runners?

5.4 | Dynamics of Circular Motion

ActivPhysics Online

4.2 Circular Motion Problem Solving
4.3 Cart Goes Over Circular Path
4.4 Ball Swings on a String
4.5 Car Circles a Track

We talked about uniform circular motion in Section 3.4. We showed that when a particle moves in a circular path with constant speed, the particle's acceleration is always directed toward the center of the circle (perpendicular to the instantaneous velocity). The magnitude a_{rad} of the acceleration is constant and is given in terms of the speed v and the radius R of the circle by

$$a_{rad} = \frac{v^2}{R} \quad (5.14)$$

The subscript "rad" is a reminder that at each point the acceleration is radially inward toward the center of the circle, perpendicular to the instantaneous velocity. We explained in Section 3.4 why this acceleration is often called *centripetal acceleration.*

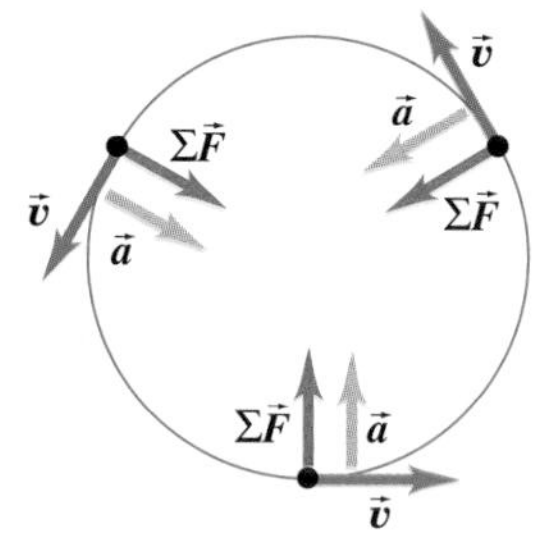

5.29 In uniform circular motion, both the acceleration and the net force are directed toward the center of the circle.

We can also express the centripetal acceleration a_{rad} in terms of the *period* T, the time for one revolution:

$$T = \frac{2\pi R}{v} \tag{5.15}$$

In terms of the period, a_{rad} is

$$a_{\mathrm{rad}} = \frac{4\pi^2 R}{T^2} \tag{5.16}$$

Uniform circular motion, like all other motion of a particle, is governed by Newton's second law. The particle's acceleration toward the center of the circle must be caused by a force, or several forces, such that the vector sum $\sum\vec{F}$ is a vector that is always directed toward the center (Fig. 5.29). The magnitude of the acceleration is constant, so the magnitude F_{net} of the net inward radial force must also be constant. If an ice skater whirls his partner around in a circle on the ice, he must constantly pull on her inward toward the center of the circle. If he lets go, then the inward force no longer acts and his partner flies off in a straight line tangent to the circle (Fig. 5.30).

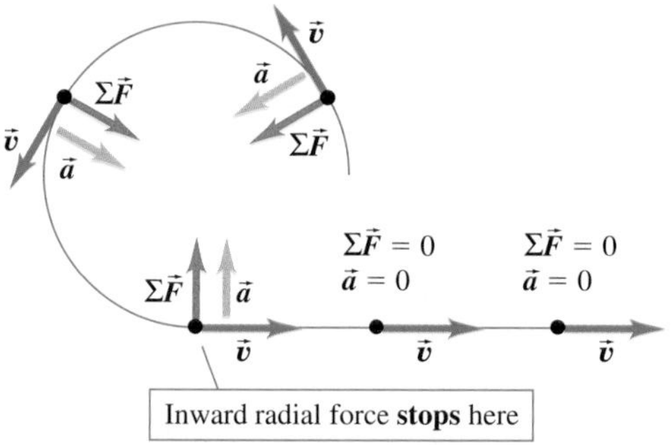

5.30 If the inward radial force suddenly ceases to act on a body in circular motion, the body flies off in a straight line with constant velocity (as it must, since the net force acting is zero).

The magnitude of the radial acceleration is given by $a_{\mathrm{rad}} = v^2/R$, so the magnitude F_{net} of the net inward radial force on a particle with mass m must be

$$F_{\mathrm{net}} = ma_{\mathrm{rad}} = m\frac{v^2}{R} \tag{5.17}$$

Uniform circular motion can result from *any* combination of forces, just so the net force $\sum\vec{F}$ is always directed toward the center of the circle and has a constant magnitude.

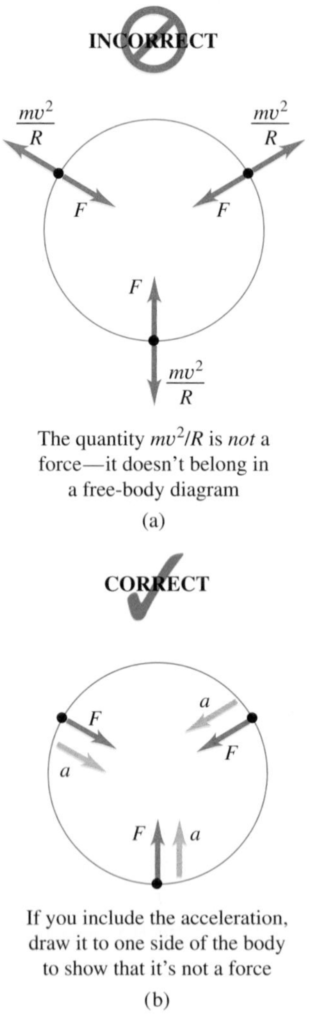

5.31 (a) Incorrect and (b) correct free-body diagrams for a body in uniform circular motion.

CAUTION In doing problems involving uniform circular motion, you may be tempted to include an extra outward force of magnitude $m(v^2/R)$ to "keep the body out there" or to "keep it in equilibrium"; this outward force is usually called *"centrifugal force."* ("Centrifugal" means "fleeing from the center.") Resist this temptation, because this approach is simply wrong! (See Fig. 5.31a.) Here's why: First, the body does *not* "stay out there." It is in constant motion around its circular path. Because its velocity is constantly changing in direction, the body accelerates and is *not* in equilibrium. Second, if there *were* an additional outward ("centrifugal") force to balance the inward force, there would then be no net inward force to cause the circular motion, and the body would move in a straight line, not a circle (Fig. 5.30). And third, the quantity $m(v^2/R)$ is *not* a force; it corresponds to the $m\vec{a}$ side of $\sum\vec{F} = m\vec{a}$, and does not appear in $\sum\vec{F}$ (Fig. 5.31b). It's certainly true that a passenger riding in a car going around a circular path on a level road tends to slide to the outside of the turn, as though responding to a "centrifugal force." But such a passenger is in an accelerating, noninertial frame of reference in which Newton's first and second laws don't apply. As we discussed in Section 4.2, what really happens is that the passenger tends to keep moving in a straight line, and the outer side of the car "runs into" the passenger as the car turns (Fig. 4.8c). *In an inertial frame of reference there is no such thing as "centrifugal force."* We promise not to mention this term again, and we strongly advise you to avoid using it as well.

Example 5.21 Force in uniform circular motion

A sled with a mass of 25.0 kg rests on a horizontal sheet of essentially frictionless ice. It is attached by a 5.00-m rope to a post set in the ice. Once given a push, the sled revolves uniformly in a circle around the post (Fig. 5.32a). If the sled makes five complete revolutions every minute, find the force F exerted on it by the cord.

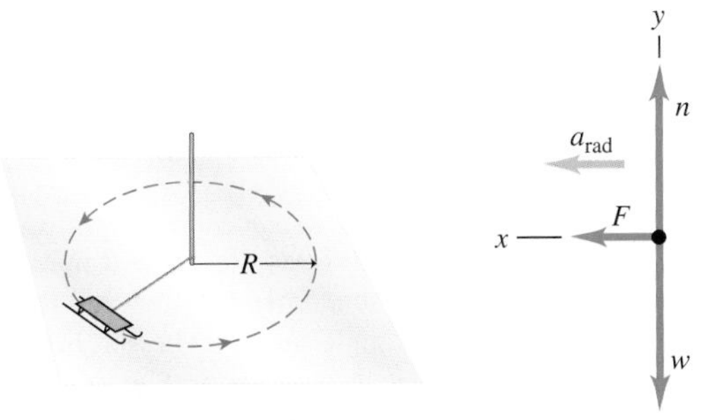

(a) (b)

5.32 (a) A sled moving in uniform circular motion on a horizontal frictionless surface. (b) Free-body diagram for the sled. The $+x$-direction points toward the center of the circle.

SOLUTION

IDENTIFY: The sled is moving in uniform circular motion, so it has a radial acceleration. We will apply the component form of Newton's second law to the sled to determine the magnitude F of the force exerted by the rope, which is our target variable.

SET UP: Figure 5.32b shows the free-body diagram for the sled. The acceleration has only an x-component; this is toward the center of the circle, so we denote it as a_{rad}. The acceleration isn't given, so we'll need to determine its value using either Eq. (5.14) or Eq. (5.16).

EXECUTE: There is no acceleration in the y-direction, so the net force in that direction (the sum of the normal force and the weight) is zero. For the x-direction, Newton's second law gives

$$\sum F_x = F = ma_{rad}$$

We can find the centripetal acceleration a_{rad} using Eq. (5.16). The sled moves in a circle of radius $R = 5.00\text{ m}$ with a period $T = (60.0\text{ s})/(5\text{ rev}) = 12.0\text{ s}$, so

$$a_{rad} = \frac{4\pi^2 R}{T^2} = \frac{4\pi^2(5.00\text{ m})}{(12.0\text{ s})^2} = 1.37\text{ m/s}^2$$

Alternatively, we can first use Eq. (5.15) to find the speed v:

$$v = \frac{2\pi R}{T} = \frac{2\pi(5.00\text{ m})}{12.0\text{ s}} = 2.62\text{ m/s}$$

Then, using Eq. (5.14),

$$a_{rad} = \frac{v^2}{R} = \frac{(2.62\text{ m/s})^2}{5.00\text{ m}} = 1.37\text{ m/s}^2$$

Hence the magnitude F of the force exerted by the rope is

$$F = ma_{rad} = (25.0\text{ kg})(1.37\text{ m/s}^2)$$
$$= 34.3\text{ kg}\cdot\text{m/s}^2 = 34.3\text{ N}$$

EVALUATE: A greater force would be needed if the sled moved around the circle at a higher speed v. In fact, if v were doubled while R remained the same, F would be four times greater. Can you show this? How would F change if v remained the same but the radius R were doubled?

Example 5.22 The conical pendulum

An inventor who dares to be different proposes to make a pendulum clock using a pendulum bob with mass m at the end of a thin wire of length L. Instead of swinging back and forth, the bob moves in a horizontal circle with constant speed v, with the wire making a constant angle β with the vertical direction (Fig. 5.33a). Assuming that the angle β is known, find the tension F in the wire and the period T (the time for one revolution of the bob).

SOLUTION

IDENTIFY: To find our two target variables—the tension F and period T—we need two simultaneous equations. These will be the horizontal and vertical components of Newton's second law applied to the bob. We'll find the acceleration of the bob toward the center of the circle using one of the circular motion equations.

SET UP: Figure 5.33b shows a free-body diagram for the bob as well as a coordinate system. The forces on the bob in the position shown are the weight mg and the tension F in the wire. Note that the center of the circular path is at the center of the shaded area, *not* at the top end of the wire. The horizontal component of tension is the force that produces the horizontal acceleration (toward the center of the circle).

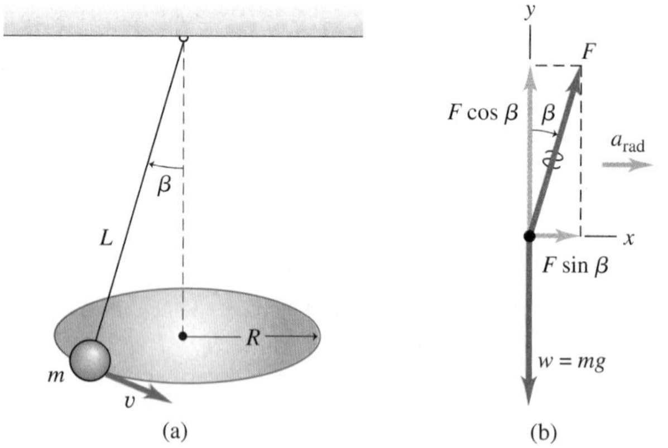

5.33 (a) The bob on the end of the wire moves in uniform circular motion. (b) Free-body diagram for the bob. The $+x$-direction points toward the center of the circle.

EXECUTE: The tension has a horizontal component $F \sin \beta$ and a vertical component $F \cos \beta$. The system has zero vertical acceleration; the horizontal acceleration is toward the center of the circle, so we use the symbol a_{rad}. The $\sum \vec{F} = m\vec{a}$ equations are

$$\sum F_x = F \sin \beta = ma_{rad}$$
$$\sum F_y = F \cos \beta + (-mg) = 0$$

These are two simultaneous equations for the unknowns F and β. The equation for $\sum F_y$ gives $F = mg/\cos \beta$; substituting this result into the equation for $\sum F_x$ and using $(\sin \beta)/(\cos \beta) = \tan \beta$, we find

$$\tan \beta = \frac{a_{rad}}{g}$$

To relate β to the period T, we use Eq. (5.16) for a_{rad}. The radius of the circle is $R = L \sin \beta$, so

$$a_{rad} = \frac{4\pi^2 R}{T^2} = \frac{4\pi^2 L \sin \beta}{T^2}$$

Substituting this into $\tan \beta = a_{rad}/g$, we obtain

$$\tan \beta = \frac{4\pi^2 L \sin \beta}{gT^2}$$

which we can rewrite as

$$T = 2\pi\sqrt{\frac{L \cos \beta}{g}}$$

EVALUATE: For a given length L, as the angle β increases, the $\cos \beta$ decreases, the period T becomes smaller, and the tension $F = mg/\cos \beta$ increases. The angle can never be 90°, however; this would require that $T = 0$, $F = \infty$, and $v = \infty$. This system is called a *conical pendulum* because the suspending wire traces out a cone. It would not make a very good clock because the period depends on the angle β in such a direct way.

Example 5.23 Rounding a flat curve

The BMW Z4 roadster in Example 3.12 (Section 3.4) is rounding a flat, unbanked curve with radius R (Fig. 5.34a). If the coefficient of friction between tires and road is μ_s, what is the maximum speed v_{max} at which the driver can take the curve without sliding?

SOLUTION

IDENTIFY: The car's acceleration as it rounds the curve has magnitude $a_{rad} = v^2/R$, so the maximum speed corresponds to the maximum acceleration a_{rad}. This acceleration corresponds to the maximum horizontal force on the car toward the center of its circular path. The only horizontal force acting on the car is the friction force exerted by the road. So we'll approach this problem using Newton's second law and our knowledge of the friction force from Section 5.3.

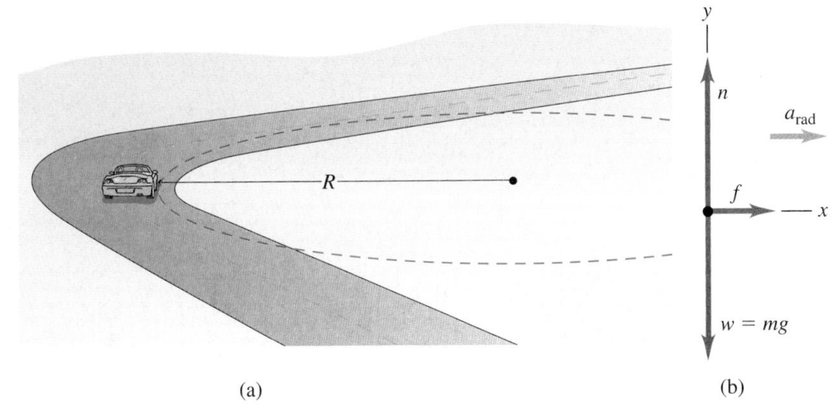

5.34 (a) A sports car rounding a curve on a level road. (b) Free-body diagram for the car. The $+x$-direction is toward the center of the car's circular path.

SET UP: The free-body diagram in Fig. 5.34b includes the car's weight $w = mg$ and two contact forces exerted by the road: the normal force (magnitude n) and the horizontal friction force (magnitude f). The friction force must point in the positive x-direction (toward the center of the circular path) in order to cause the radial acceleration. Since the car doesn't move in the radial direction (that is, it doesn't slide toward or away from the center of the circle), the friction force is *static* friction with a maximum magnitude $f_{max} = \mu_s n$ [see Eq. (5.6)].

EXECUTE: The acceleration toward the center of the circular path is $a_{rad} = v^2/R$ and there is no vertical acceleration. Thus we have

$$\sum F_x = f = ma_{rad} = m\frac{v^2}{R}$$
$$\sum F_y = n + (-mg) = 0$$

The second equation shows that $n = mg$. The first equation shows that the friction force *needed* to keep the car moving in its circular path increases with the car's speed. But the maximum friction force *available* is $f_{max} = \mu_s n = \mu_s mg$, and this determines the car's maximum speed. Substituting f_{max} for f and v_{max} for v in the $\sum F_x$ equation, we find

$$\mu_s mg = m\frac{v_{max}^2}{R}$$

so the maximum speed is

$$v_{max} = \sqrt{\mu_s gR}$$

As an example, if $\mu_s = 0.87$ and $R = 230$ m, then

$$v_{max} = \sqrt{(0.87)(9.8\ \text{m/s}^2)(230\ \text{m})} = 44\ \text{m/s}$$

or about 160 km/h (100 mi/h). This is the maximum speed for this radius.

EVALUATE: If the car's speed is slower than $\sqrt{\mu_s gR}$, the required friction force is less than the maximum possible value $f_{max} = \mu_s mg$ and the car can easily make the curve. If we try to take the curve going *faster* than the maximum speed, the car can still go in a circle without skidding, but the radius will be larger and the car will run off the road.

Note that the maximum centripetal acceleration (called the "lateral acceleration" in Example 3.12) is equal to $\mu_s g$. If the coefficient of friction is reduced, the maximum centripetal acceleration and v_{max} are also reduced. That's why it's best to take curves at a lower speed if the road is wet or icy (either of which can reduce the value of μ_s).

Example 5.24 Rounding a banked curve

For a car traveling at a certain speed, it is possible to bank a curve at just the right angle so that no friction at all is needed to maintain the car's turning radius. Then a car can round the curve even on wet ice with Teflon tires. Bobsled racing depends on this same idea. An engineer proposes to rebuild the curve in Example 5.23 so that a car moving at speed v can safely make the turn even if there is no friction (Fig. 5.35a). At what angle β should the curve be banked?

SOLUTION

IDENTIFY: With no friction, the only two forces acting on the car are its weight and the normal force. Because the road is banked, the normal force (which acts perpendicular to the road surface) has a horizontal component. This component causes the car's centripetal acceleration (which is in the horizontal direction, toward the center of the car's circular path). Since forces and acceleration are involved, we'll use Newton's second law to find the target variable β.

SET UP: The free-body diagram (Fig. 5.35b) is very similar to the diagram for the conical pendulum in Example 5.22 (Fig. 5.33b). The normal force acting on the car plays the role of the tension acting on the pendulum bob.

EXECUTE: The normal force $\vec{n}$ is perpendicular to the roadway at an angle β with the vertical. Thus it has a vertical component $n \cos \beta$ and a horizontal component $n \sin \beta$, as Fig. 5.35b shows. The acceleration in the x-direction is the centripetal acceleration,

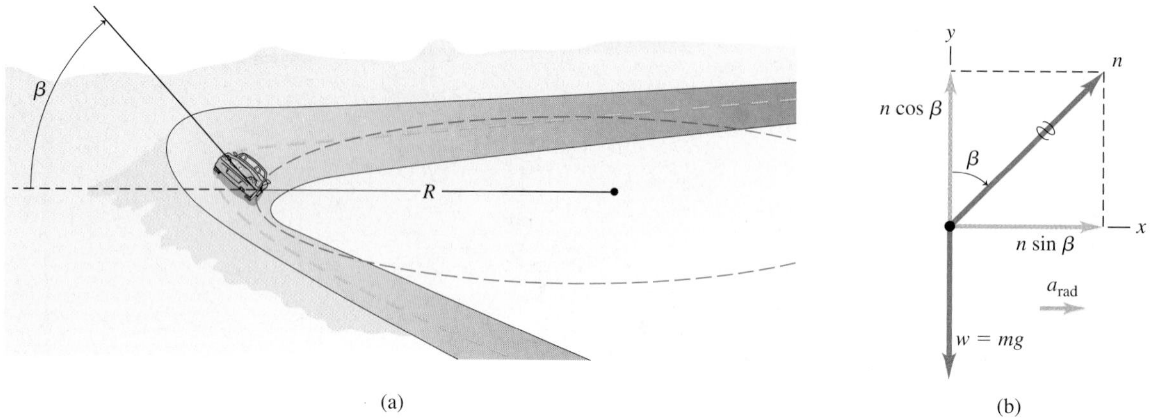

5.35 (a) A sports car rounding a banked curve. If the curved is banked at the correct angle β, no friction is needed to round the curve at speed v, (b) Free-body diagram for the car. The $+x$-direction points toward the center of the car's circular path.

there is no acceleration in the y-direction. Thus the equations of Newton's second law are

$$\sum F_x = n \sin \beta = ma_{rad}$$

$$\sum F_y = n \cos \beta + (-mg) = 0$$

From the $\sum F_y$ equation, $n = mg/\cos \beta$. Substituting this into the $\sum F_x$ equation gives an expression for the banking angle:

$$\tan \beta = \frac{a_{rad}}{g}$$

which is the same expression we found in Example 5.23. Finally, substituting the expression $a_{rad} = v^2/R$, we have

$$\tan \beta = \frac{v^2}{gR}$$

EVALUATE: The banking angle depends on the speed and the radius. For a given radius, no one angle is correct for all speeds. In the design of highways and railroads, curves are often banked for the average speed of the traffic over them. If $R = 230$ m and $v = 25$ m/s (equal to a highway speed of 88 km/h, or 55 mi/h), then

$$\beta = \arctan \frac{(25\text{ m/s})^2}{(9.8\text{ m/s}^2)(230\text{ m})} = 15°$$

This is within the range of banking angles actually used in highways. With the same radius and $v = 44$ m/s, as in Example 5.23, $\beta = 41°$; such steeply banked curves are found at automobile raceways.

The results of Example 5.24 also apply to an airplane when it makes a turn in level flight (Fig 5.36a). When an airplane is flying in a straight line at a constant speed and at a steady altitude, the airplane's weight is exactly balanced by the lift force $\vec{L}$ due to the wings. (The lift force is actually exerted by the air; as the wings move through the air, they push air downward, and so by Newton's third law, the air pushes upward on the wings.) To make the airplane turn, the pilot banks the airplane to one side (Fig. 5.36a) so that the lift force has a horizontal component as in Fig. 5.36b. (The pilot also changes the angle at which the wings "bite" into the air so that the vertical component of lift continues to balance the weight.) The bank angle is related to the airplane's speed v and the radius R of the turn by the same expression as in Example 5.24: $\tan \beta = v^2/gR$. For an airplane to make a tight turn (small R) at high speed (large v), $\tan \beta$ must be large and the required bank angle β must approach 90°.

We can also apply the results of Example 5.24 to the *pilot* of an airplane. The free-body diagram for the pilot of the airplane is exactly as shown in Fig. 5.35b; the normal force $n = mg/ \cos \beta$ is exerted on the pilot by the seat. As in Example

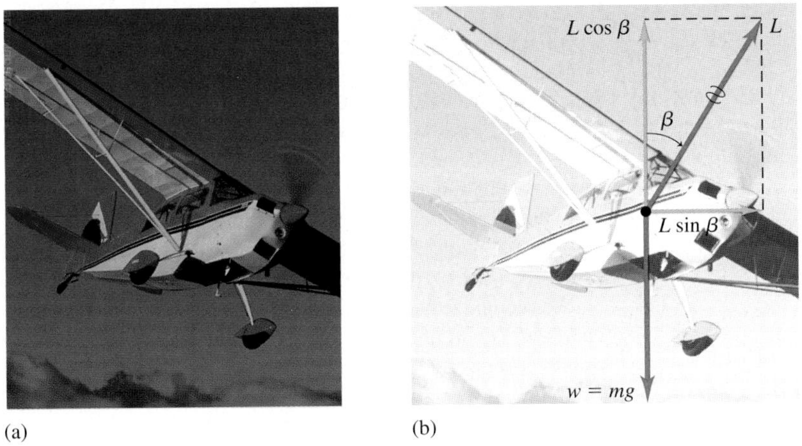

5.36 (a) An airplane banks to one side in order to turn in that direction. (b) Free-body diagram for the airplane. The vertical component of the lift force $\vec{L}$ (provided by the air pushing on the wing) balances the force of gravity; the horizontal component of $\vec{L}$ causes the acceleration v^2/R.

5.10, n is equal to the apparent weight of the pilot, which is greater than the pilot's true weight mg. In a tight turn with a large bank angle β, the pilot's apparent weight can be tremendous: $n = 5.8mg$ at $\beta = 80°$ and $n = 9.6mg$ at $\beta = 84°$. Pilots black out in such tight turns because the apparent weight of their blood increases by the same factor, and the human heart isn't strong enough to pump such apparently "heavy" blood to the brain.

Motion in a Vertical Circle

In all the preceding examples the body moved in a horizontal circle. Motion in a vertical circle is no different in principle, but the weight of the body has to be treated carefully. The following example shows what we mean.

Example 5.25 Uniform circular motion in a vertical circle

A passenger on a carnival Ferris wheel moves in a vertical circle of radius R with constant speed v. Assuming that the seat remains upright during the motion, derive expressions for the force the seat exerts on the passenger at the top of the circle and at the bottom.

SOLUTION

IDENTIFY: At both the top and bottom of the circle, the target variable is the magnitude n of the normal force that the seat exerts on the passenger. We'll find this force at each position using Newton's second law and the equations of uniform circular motion.

SET UP: The diagram in Fig. 5.37a shows the passenger's velocity and acceleration at the two positions. Note that the acceleration points downward at the top of the circle but upward at the bottom of the circle. At each position the only forces acting are vertical: the

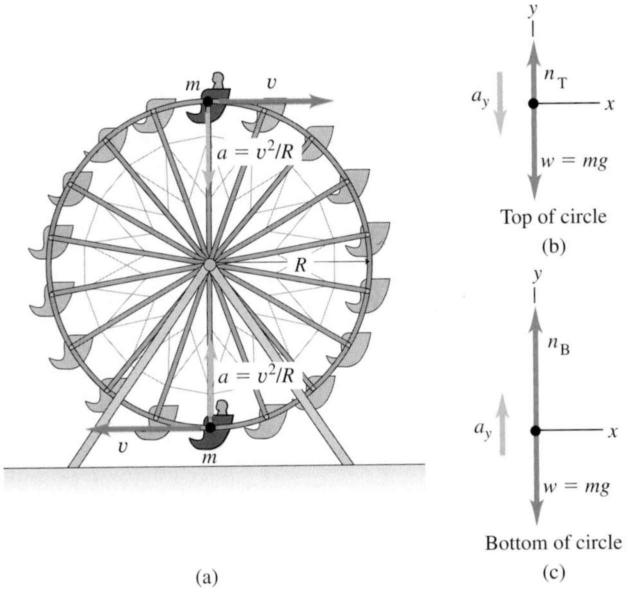

5.37 (a) The passenger's acceleration vector has the same magnitude at the top and bottom of the Ferris wheel, and in both cases points toward the center of the circle. (b) Free-body diagram for a passenger at the top of the circle, and (c) free-body diagram for a passenger at the bottom of the circle.

upward normal force and the downward force of gravity. Hence we only need the vertical component of Newton's second law.

EXECUTE: Figures 5.37b and 5.37c show free-body diagrams for the two positions. We take the positive y-direction as upward in both cases. Let n_T be the upward normal force the seat applies to the passenger at the top of the circle, and let n_B be the normal force at the bottom. At the top the acceleration has magnitude v^2/R, but its vertical component is negative because its direction is downward. Hence $a_y = -v^2/R$, and Newton's second law at the top tells us that

$$\text{Top: } \sum F_y = n_T + (-mg) = -m\frac{v^2}{R} \quad \text{or}$$

$$n_T = m\left(g - \frac{v^2}{R}\right)$$

At the bottom the acceleration is upward, so $a_y = +v^2/R$ and Newton's second law is

$$\text{Bottom: } \sum F_y = n_B + (-mg) = +m\frac{v^2}{R} \quad \text{or}$$

$$n_B = m\left(g + \frac{v^2}{R}\right)$$

EVALUATE: Our result for n_T tells us that at the top of the Ferris wheel, the upward force the seat applies to the passenger is *smaller* in magnitude than the passenger's weight. If the ride goes fast enough that $g - v^2/R$ becomes zero, the seat applies *no* force, and the passenger is about to become airborne. If v becomes still larger, n_T becomes negative; this means that a *downward* force (such as from a seat belt) is needed to keep the passenger in the seat. By contrast, the normal force n_B at the bottom is always *greater than* the passenger's weight. You feel the seat pushing up on you more firmly than when you are at rest.

We recognize that n_T and n_B are the values of the passenger's *apparent weight* at the top and bottom of the circle. As a specific example, suppose the passenger's mass is 60.0 kg, the radius of the circle is $R = 8.00$ m, and the wheel makes one revolution in 10.0 s. From Eqs. (5.14) and (5.16),

$$a_{rad} = \frac{v^2}{R} = \frac{4\pi^2 R}{T^2} = \frac{4\pi^2(8.00\text{ m})}{(10.0\text{ s})^2} = 3.16\text{ m/s}^2$$

The two values of the apparent weight are

$$n_T = (60.0\text{ kg})(9.80\text{ m/s}^2 - 3.16\text{ m/s}^2) = 398\text{ N}$$
$$n_B = (60.0\text{ kg})(9.80\text{ m/s}^2 + 3.16\text{ m/s}^2) = 778\text{ N}$$

The passenger's weight is 588 N, and n_T and n_B are about 30% less than and greater than the weight, respectively.

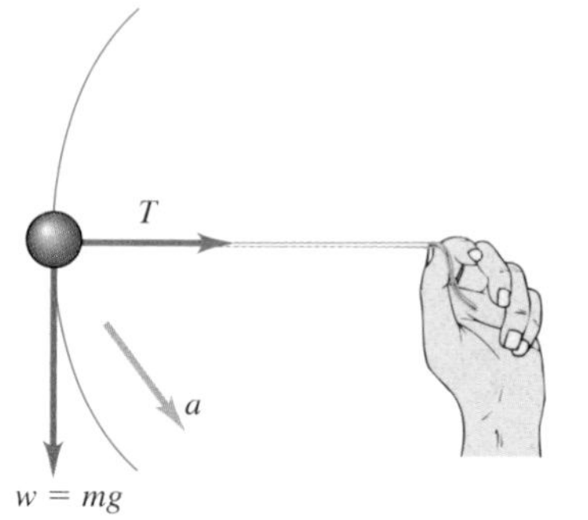

5.38 A ball moving in a vertical circle. At most points around the circle, a component of the net force (and hence a component of acceleration) is tangent to the circle. Hence the ball is in nonuniform circular motion.

When we tie a string to an object and whirl it in a vertical circle, the analysis in Example 5.25 isn't directly applicable. The reason is that v is *not* constant in this case; at every point in the circle except the top and bottom, the net force (and hence the acceleration) does *not* point toward the center of the circle (Fig. 5.38). So both $\sum\vec{F}$ and $\vec{a}$ have a component tangent to the circle, and this is a case of *non-uniform* circular motion (see Section 3.5). Even worse, we can't use the constant-acceleration formulas to relate the speeds at various points because *neither* the magnitude nor the direction of the acceleration is constant. The speed relations we need are best obtained by using the concept of energy. We'll consider such problems in Chapter 7.

Test Your Understanding

Satellites are held in orbit by the force of our planet's gravitational attraction. A satellite in a small-radius orbit moves at a higher speed than a satellite in an orbit of large radius. Based on this information, explain how you can conclude that the Earth's gravitational attraction decreases with increasing distance from the planet.

*5.5 | The Fundamental Forces of Nature

We have discussed several kinds of forces—including weight, tension, friction, fluid resistance, and the normal force—and we will encounter others as we continue our study of physics. But just how many kinds of forces are there? Our current understanding is that all forces are expressions of just four distinct classes of *fundamental* forces, or interactions between particles (Fig. 5.39). Two are familiar in everyday experience. The other two involve interactions between subatomic particles that we cannot observe with the unaided senses.

Of the two familiar classes, **gravitational interactions** were the first to be studied in detail. The *weight* of a body results from the Earth's gravitational attraction acting on it. The sun's gravitational attraction for the Earth keeps the Earth in its nearly circular orbit around the sun. Newton recognized that the motions of the planets around the sun and the free fall of objects on Earth both result from gravitational forces. In Chapter 12 we will study gravitational interactions in greater detail, and we will analyze their vital role in the motions of planets and satellites.

The second familiar class of forces, **electromagnetic interactions,** includes electric and magnetic forces. If you run a comb through your hair, you can then use it to pick up bits of paper or fluff; this interaction is the result of electric charge on the comb. All atoms contain positive and negative electric charge, so atoms and molecules can exert electric forces on each other. Contact forces, including the normal force, friction, and fluid resistance, are the combination of all such forces exerted on the atoms of a body by atoms in its surroundings. *Magnetic* forces occur in interactions between magnets or between a magnet and a piece of iron. These may seem to form a different category, but magnetic interactions are actually the result of electric charges in motion. In an electromagnet an electric current in a coil of wire causes magnetic interactions. We will study electric and magnetic interactions in detail in the second half of this book.

These two interactions differ enormously in their strength. The electrical repulsion between two protons at a given distance is stronger than their gravitational attraction by a factor of the order of 10^{35}. Gravitational forces play no significant role in atomic or molecular structure. But in bodies of astronomical size, positive and negative charge are usually present in nearly equal amounts, and the resulting electrical interactions nearly cancel out. Gravitational interactions are thus the dominant influence in the motion of planets and in the internal structure of stars.

The other two classes of interactions are less familiar. One, the **strong interaction,** is responsible for holding the nucleus of an atom together. Nuclei contain electrically neutral neutrons and positively charged protons. The charged protons repel each other, and a nucleus could not be stable if it were not for the presence of an attractive force of a different kind that counteracts the repulsive electrical interactions. In this context the strong interaction is also called the *nuclear force.* It has much shorter range than electrical interactions, but within its range it is much stronger. The strong interaction is also responsible for the creation of unstable particles in high-energy particle collisions.

Finally, there is the **weak interaction.** It plays no direct role in the behavior of ordinary matter, but it is of vital importance in interactions among fundamental particles. The weak interaction is responsible for a common form of radioactivity called beta decay, in which a neutron in a radioactive nucleus is transformed into a proton while ejecting an electron and an essentially massless particle called an antineutrino. The weak interaction between the antineutrino and ordinary matter is so feeble that an antineutrino could easily penetrate a wall of lead a million kilometers thick!

During the past several decades a unified theory of the electromagnetic and weak interactions has been developed. We now speak of the *electroweak* interaction, and in a sense this reduces the number of classes of interactions from four to three. Similar attempts have been made to understand strong, electromagnetic, and weak interactions on the basis of a single unified theory called a *grand unified theory* (GUT), and the first tentative steps have been taken toward a possible unification of all interactions into a *theory of everything* (TOE). Such theories are still speculative, and there are many unanswered questions in this very active field of current research.

(a)

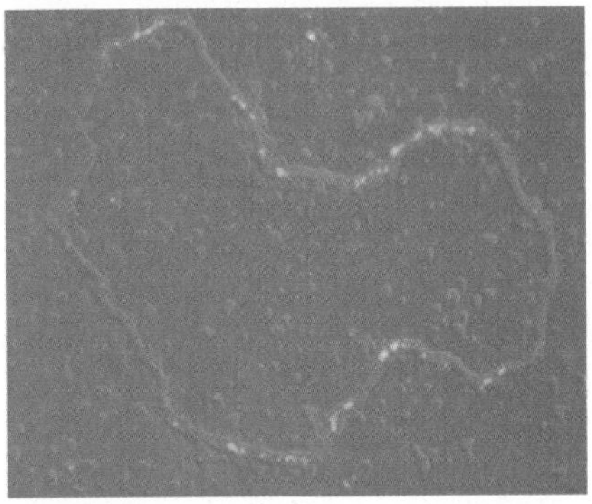

(b)

(c)

(d)

5.39 Examples of the fundamental interactions in nature. (a) Gravitational forces cause the motion of the moon around the earth. (b) Electromagnetic forces form atoms into molecules. This atomic force microscope image shows a molecule of bacterial plasmid DNA. (c) Strong forces between nuclear particles are responsible for thermonuclear reactions in the center of the sun. The released energy reaches us as sunlight. (d) Weak forces within atomic nuclei play a crucial role when a star (shown on the left by an arrow) explodes into a supernova (shown on the right).

SUMMARY

When a body is in equilibrium in an inertial frame of reference, the vector sum of forces acting on it must be zero. Free-body diagrams are essential in identifying the forces that act on the body being considered.

Newton's third law is also frequently needed in equilibrium problems. The two forces in an action-reaction pair *never* act on the same body. (See Examples 5.1 through 5.5)

Vector form:

$$\sum \vec{F} = \mathbf{0} \quad (5.1)$$

Component form:

$$\sum F_x = 0$$
$$\sum F_y = 0 \quad (5.2)$$

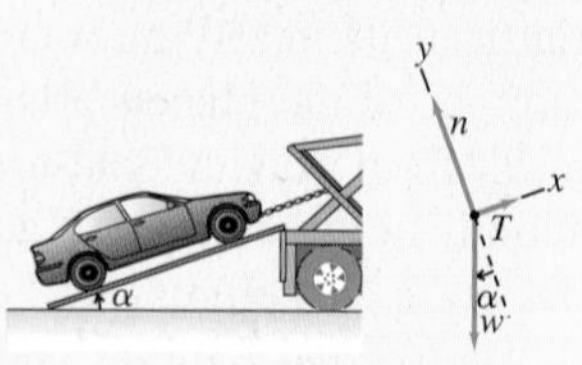

If the vector sum of forces on a body is *not* zero, the body accelerates. Its acceleration is given by Newton's second law.

As they are for equilibrium problems, free-body diagrams are essential for solving problems involving Newton's second law. (See Examples 5.6 through 5.13)

Vector form:

$$\sum \vec{F} = m\vec{a} \quad (5.3)$$

Component form:

$$\sum F_x = ma_x$$
$$\sum F_y = ma_y \quad (5.4)$$

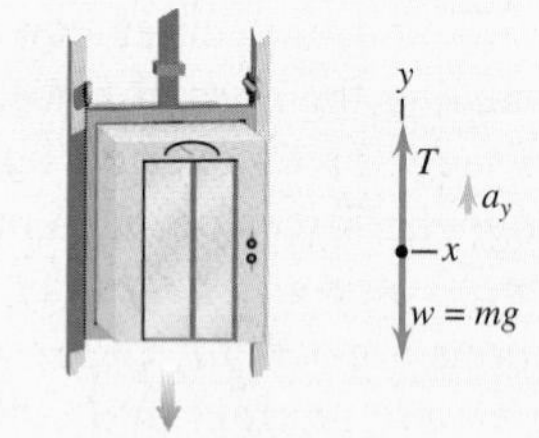

The contact force between two bodies can always be represented in terms of a normal force $\vec{n}$ perpendicular to the surface of contact and a friction force $\vec{f}$ parallel to the surface. The normal force exerted on a body by a surface is *not* always equal to the body's weight. (See Examples 5.3, 5.10, and 5.11)

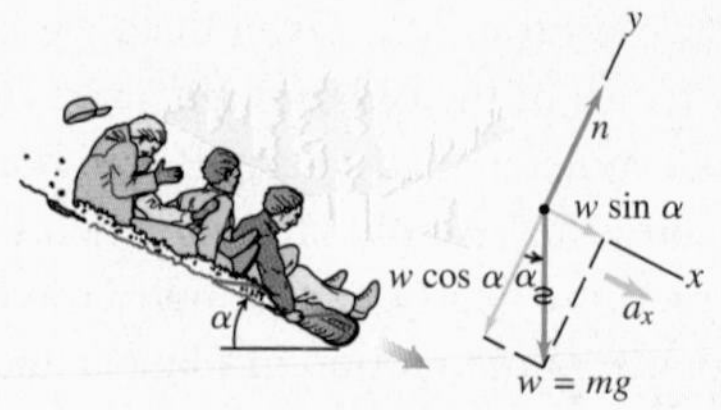

When a body is sliding over the surface, the friction force is called kinetic friction. Its magnitude f_k is approximately equal to the normal force magnitude n multiplied by the coefficient of kinetic friction μ_k. (See Examples 5.14 and 5.16 through 5.18)

Magnitude of kinetic friction force:

$$f_k = \mu_k n \quad (5.5)$$

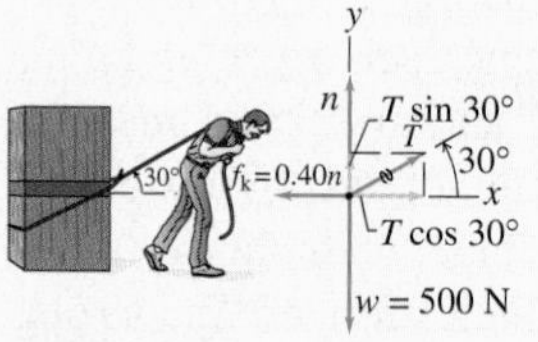

When a body is not moving relative to a surface, the friction force is called static friction. The *maximum* possible static friction force is approximately equal to the magnitude n of the normal force multiplied by the coefficient of static friction μ_s. The *actual* static friction force may be anything from zero to this maximum value, depending on the situation. Usually μ_s is greater than μ_k for a given pair of surfaces in contact. (See Examples 5.14 and 5.15)

Magnitude of static friction force:

$$f_s \le \mu_s n \quad (5.6)$$

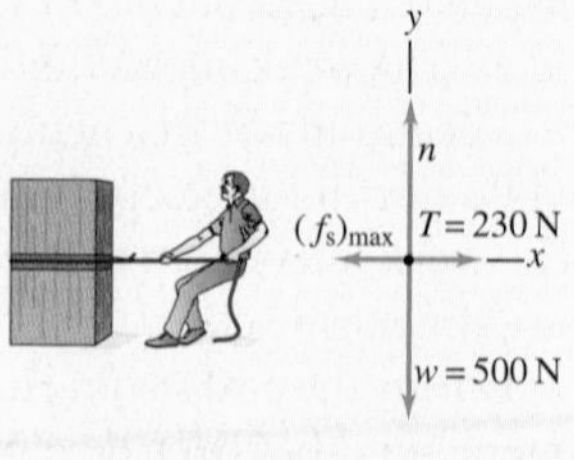

In uniform circular motion, the acceleration vector is directed toward the center of the circle. Just as in any other dynamics problem, the motion is governed by Newton's second law, $\sum\vec{F} = m\vec{a}$. (See Examples 5.21 through 5.25)

Acceleration in uniform circular motion:

$$a_{\text{rad}} = \frac{v^2}{R} = \frac{4\pi^2 R}{T^2} \qquad (5.14), (5.16)$$

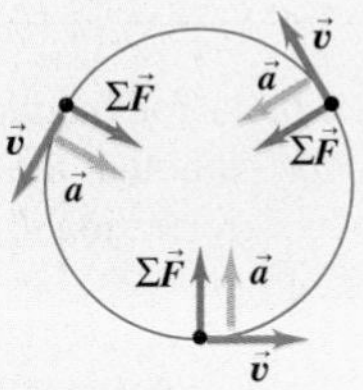

Key Terms

apparent weight, 166
friction force, 171
kinetic friction force, 172
coefficient of kinetic friction, 172
static friction force, 172
coefficient of static friction, 173
coefficient of rolling friction, 178
fluid resistance, 178
air drag, 179
terminal speed, 179
gravitational interaction, 189
electromagnetic interaction, 189
strong interaction, 189
weak interaction, 189

Your Notes

Answer to Chapter Opening Question

Neither: the aerodynamic forces are still in balance with the force of gravity. Although the bird is ascending, its velocity is constant and so its acceleration is zero. Hence the net force on the bird must also be zero.

Answers to Test Your Understanding Questions

Section 5.1 The two cables are arranged symmetrically, so the tension in each cable has the same magnitude T. The vertical component of the tension from each cable is $T \sin 45°$ (or, equivalently, $T \cos 45°$), so Newton's first law applied to the vertical forces tells us that $2T \sin 45° - mg = 0$. Hence $T = mg/(2 \sin 45°) = mg/\sqrt{2} = 0.71mg$.
Section 5.2 The situation is the same as in Example 5.11, but with an additional force of unknown magnitude F_{heels} acting in the negative x-direction (uphill). Hence the x-component equation for Newton's second law is $w \sin \alpha - F_{\text{heels}} = ma_x$, and so $F_{\text{heels}} = w \sin \alpha - ma_x$. With $m = 90$ kg, $w = mg = (90\text{ kg})(9.8\text{ m/s}^2)$, $\alpha = 3.0°$, and $a_x = -2.0\text{ m/s}^2$ (the acceleration is uphill, opposite the motion), we get $F_{\text{heels}} = 2.3 \times 10^2$ N.
Section 5.3 This situation is the same as in Example 5.18. We found in that example that $a_x = g(\sin \alpha - \mu_k \cos \alpha)$; solving for μ_k, we get $\mu_k = (g \sin \alpha - a_x)/(g \cos \alpha) = [(9.8\text{ m/s}^2)\sin 3.0° - (0.25\text{ m/s}^2)]/[(9.8\text{ m/s}^2)\cos 3.0°] = 0.27$.
Section 5.4 A satellite of mass m orbiting the earth at speed v in an orbit of radius r has an acceleration of magnitude v^2/r, so the net force acting on it from the earth's gravity has magnitude $F = mv^2/r$. The farther the satellite is from earth, the greater the value of r, the smaller the value of v, and hence the smaller the values of v^2/r and of F. In other words, the earth's gravitational force decreases with increasing distance.

Discussion Questions

Q5.1 A man sits in a seat that is suspended from a rope. The rope passes over a pulley suspended from the ceiling, and the man holds the other end of the rope in his hands. What is the tension in the rope, and what force does the seat exert on the man? Draw a free-body force diagram for the man.
Q5.2 "In general, the normal force is not equal to the weight." Give an example where the two forces are equal in magnitude, and at least two examples where they are not.
Q5.3 A clothesline hangs between two poles. No matter how tightly the line is stretched, it always sags a little at the center. Explain why.
Q5.4 A car is driven up a steep hill at constant speed. Discuss all the forces acting on the car. What pushes it up the hill?
Q5.5 For medical reasons it is important for astronauts in outer space to determine their body mass at regular intervals. Devise a scheme for measuring body mass in an apparently weightless environment.
Q5.6 To push a box up a ramp, is the force required smaller if you push horizontally or if you push parallel to the ramp? Why?
Q5.7 A woman in an elevator lets go of her briefcase but it does not fall to the floor. How is the elevator moving?
Q5.8 You can classify scales for weighing objects as those that use springs and those that use standard masses to balance unknown masses. Which group would be more accurate when you use it in an accelerating elevator? When you use it on the moon? Does it matter whether you are trying to determine mass or weight?
Q5.9 When you tighten a nut on a bolt, how are you increasing the frictional force? How does a lock washer work?
Q5.10 A block rests on an inclined plane with enough friction to prevent it from sliding down. To start the block moving, is it easier to push it up the plane, down the plane, or sideways? Why?
Q5.11 A crate of books rests on a level floor. To move it along the floor at a constant velocity, why do you exert a smaller force if you pull it at an angle θ above the horizontal than if you push it at the same angle below the horizontal?
Q5.12 To stop a car on an icy road, is it better to push the brake pedal hard enough to "lock" the wheels and make them slide or to push gently so that the wheels continue to roll? Why?
Q5.13 Can the force of kinetic friction acting on an object ever cause that object to *speed up?* If it cannot, explain why not. If it can, give at least one example. Repeat for the force of *static* friction.
Q5.14 When you stand with bare feet in a wet bathtub, the grip feels fairly secure, and yet a catastrophic slip is quite possible. Explain this in terms of the two coefficients of friction.
Q5.15 You are pushing a large crate from the back of a freight elevator to the front as the elevator is moving to the next floor. In which situation is the force you must apply to move the crate the smallest and in which is it the largest: when the elevator is accelerating upward, when it is accelerating downward, or when it is traveling at constant speed? Explain.
Q5.16 Roller Coaster Road near Tucson, Arizona, was so named because it resembled an amusement park ride. If you drove on it at constant speed, where was the normal force the greatest and the least? Why would it have been especially poor road design to put an unbanked right or left curve at the top of one of the hills?
Q5.17 An automotive magazine calls decreasing-radius curves "the bane of the Sunday driver." Explain.
Q5.18 If you hang fuzzy dice from your rear-view mirror and drive through a banked curve, how can you tell whether you are traveling less than, equal to, or greater than the speed used to calculate the banking angle?
Q5.19 If there is a net force on a particle in uniform circular motion, why does the particle's speed not change?
Q5.20 A curve in a road has the banking angle calculated for 80 km/h. However, the road is covered with ice, and you plan to creep around the highest lane at 20 km/h. What may happen to your car? Why?
Q5.21 If you swing a ball on the end of a lightweight string in a horizontal circle at constant speed, the string at any instant does not lie exactly along the radius vector from the center of the path to the ball's location. Does the string lie above or below the horizontal? In relation to the direction the ball is moving, is the string in front of or behind the radius vector? Draw a free-body force diagram for the

ball and use it to explain your answers. (Note that air resistance may be a factor.)

Q5.22 The centrifugal force is not included in the free-body diagrams of Figs. 5.34b and 5.35b. Explain why not.

Q5.23 A professor swings a rubber stopper in a horizontal circle on the end of a string in front of his class. He tells Caroline, in the first row, that he is going to let the string go when the stopper is directly in front of her face. Should Caroline worry?

Q5.24 To keep the forces on the riders within allowable limits, loop-the-loop roller coaster rides are often designed so the loop, rather than being a perfect circle, has a larger radius of curvature at the bottom than at the top. Explain.

Q5.25 A tennis ball drops from rest at the top of a tall glass cylinder, first with the air pumped out of the cylinder so there is no air resistance, and then a second time after the air has been readmitted to the cylinder. You examine multiflash photographs (like Fig. 2.21) of the two drops. From these photos how can you tell which one is which, or can you?

Q5.26 If you throw a baseball straight upward with speed v_0, how does its speed, when it returns to the point from where you threw it, compare to v_0 a) in the absence of air resistance? b) in the presence of air resistance? Explain.

Q5.27 You take two identical tennis balls and fill one with water. You release both balls simultaneously from the top of a tall building. If air resistance is negligible, which ball strikes the ground first? Explain. What is the answer if air resistance is *not* negligible?

Q5.28 A sheet of paper is falling through the air at its terminal speed. An identical sheet of paper is crumpled tightly into a ball and then dropped from the top of a tall ladder. Which piece of paper, the flat sheet or the crumpled ball, falls with the greater terminal speed? Explain. How does the air resistance force on the flat sheet of paper compare with the air resistance force on the crumpled ball when each is falling at its terminal speed? Explain.

Q5.29 You throw a baseball straight upward. If air resistance is *not* ignored, how does the time required for the ball to go from the height at which it was thrown up to its maximum height compare to the time required for it to fall from its maximum height back down to the height from which it was thrown? Explain your answer.

Q5.30 When does a baseball in flight have an acceleration with a positive upward component? Explain in terms of the forces on the ball and also in terms of the velocity components compared to the terminal speed. Do *not* ignore air resistance.

Q5.31 When a batted baseball moves with air drag, does it spend more time climbing to its maximum height or descending from its maximum height back to the ground? Or is the time the same for both? Explain in terms of the forces acting on the ball.

Q5.32 When a batted baseball moves with air drag, does it travel a greater horizontal distance while climbing to its maximum height or while descending from its maximum height back to the ground? Or is the horizontal distance traveled the same for both? Explain in terms of the forces acting on the ball.

Q5.33 "A ball is thrown from the edge of a high cliff. No matter what the angle at which it is thrown, due to air resistance, the ball will eventually end up moving vertically downward." Justify this statement.

Exercises

Section 5.1 Using Newton's First Law: Particles In Equilibrium

5.1 Two 25.0-N weights are suspended at opposite ends of a rope that passes over a light, frictionless pulley. The pulley is attached to a chain that goes to the ceiling. a) What is the tension in the rope? b) What is the tension in the chain?

5.2 In Fig. 5.40 each of the suspended blocks has weight w. The pulleys are frictionless and the ropes have negligible weight. Calculate, in each case, the tension T in the rope in terms of the weight w. In each case, include the free-body diagram or diagrams you used to determine the answer.

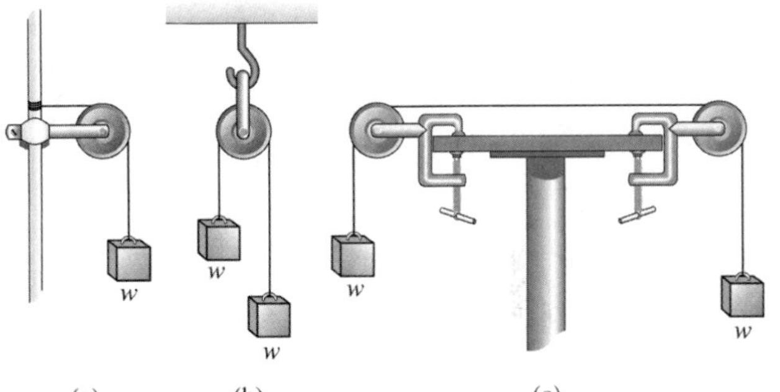

Figure 5.40 Exercise 5.2.

5.3 An adventurous archaeologist crosses between two rock cliffs by slowly going hand-over-hand along a rope stretched between the cliffs. He stops to rest at the middle of the rope (Fig. 5.41). The rope will break if the tension in it exceeds 2.50×10^4 N, and our hero's mass is 90.0 kg. a) If the angle θ is 10.0°, find the tension in the rope. b) What is the smallest value the angle θ can have if the rope is not to break?

Figure 5.41 Exercise 5.3.

5.4 A picture frame hung against a wall is suspended by two wires attached to its upper corners. If the two wires make the same angle with the vertical, what must this angle be if the tension in each wire is equal to 0.75 of the weight of the frame? (Ignore any friction between the wall and the picture frame.)

5.5 Solve the problem in Example 5.4 using coordinate axes where the y-axis is vertical and the x-axis is horizontal. Do you get the same answers as obtained in the example using this different set of axes?

5.6 Certain streets in San Francisco make an angle of 17.5° with the horizontal. What force parallel to the street surface is required to keep a loaded 1967 Corvette of mass 1390 kg from rolling down such a street?

5.7 A large wrecking ball is held in place by two light steel cables (Fig. 5.42). If the mass m of the wrecking ball is 4090 kg, what is a) the tension T_B in the cable that makes an angle of 40° with the vertical? b) the tension T_A in the horizontal cable?

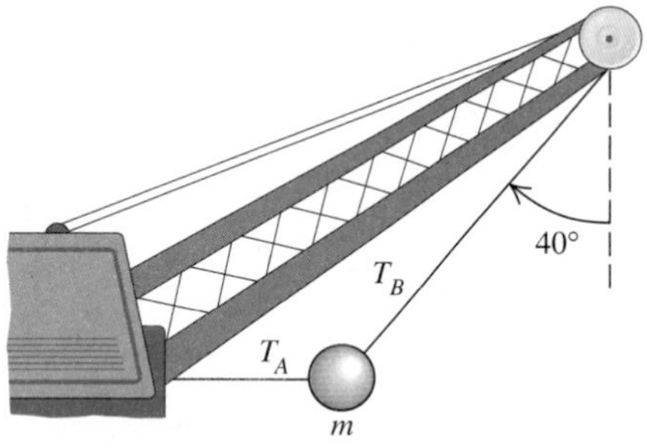

Figure 5.42 Exercise 5.7.

5.8 Find the tension in each cord in Fig. 5.43 if the weight of the suspended object is w.

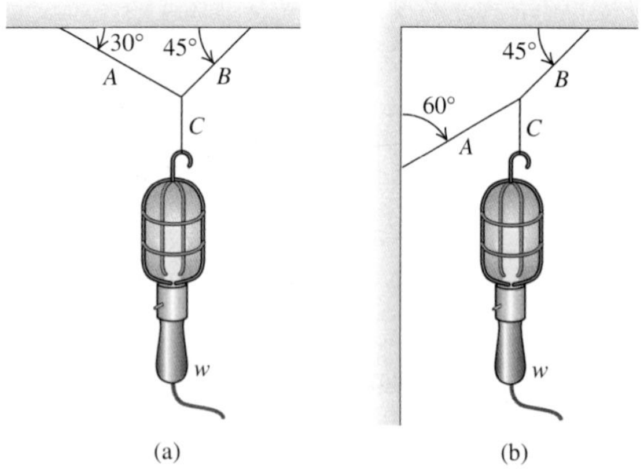

Figure 5.43 Exercise 5.8.

5.9 At a certain part of your drive from home to school, your car (mass 1600 kg) will coast in neutral at a constant speed of 72 km/h if there is no wind. Examination of a topographical map shows that for this stretch of straight highway the elevation decreases 200 m for each 6000 m of road. What is the total resistive force (friction plus air resistance) that acts on your car when it is traveling at 72 km/h?

5.10 A man pushes on a piano with mass 180 kg so that it slides at constant velocity down a ramp that is inclined at 11.0° above the horizontal. Neglect friction acting on the piano. If the force applied by the man is parallel to the incline, calculate the magnitude of this force.

5.11 In Fig. 5.44 the weight w is 60.0 N. a) What is the tension in the diagonal string? b) Find the magnitudes of the horizontal forces $\vec{F}_1$ and $\vec{F}_2$ that must be applied to hold the system in the position shown.

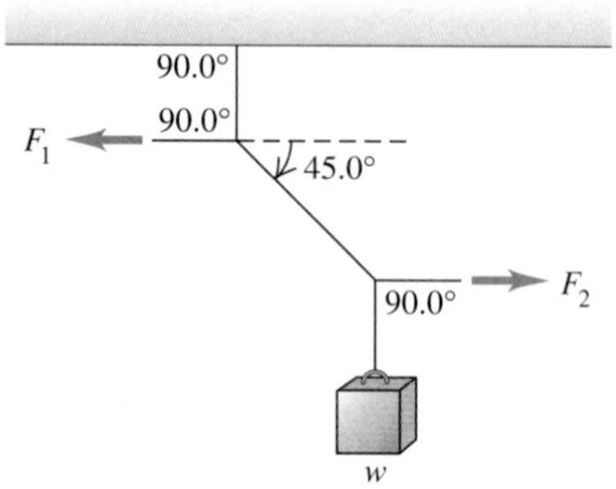

Figure 5.44 Exercise 5.11.

5.12 A tether ball leans against the post to which it is attached (Fig. 5.45). If the string to which the ball is attached is 1.40 m long and the ball has a radius of 0.110 m with mass 0.270 kg, what is the tension in the rope and what is the force the pole exerts on the ball? You can ignore any friction between the ball and the pole. (The string is attached to the ball such that a line along the string passes through the center of the ball.)

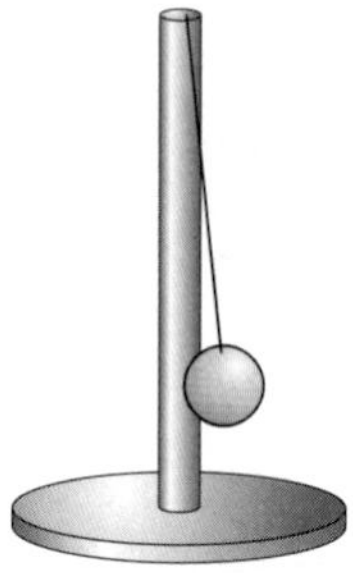

Figure 5.45 Exercise 5.12

5.13 Two blocks, each with weight w, are held in place on a frictionless incline (Fig. 5.46). In terms of w and the angle α of the incline, calculate the tension in a) the rope connecting the blocks; b) the rope that connects block A to the wall. c) Calculate the magnitude of the force that the incline exerts on each block. d) Interpret your answers for the cases $\alpha = 0$ and $\alpha = 90°$.

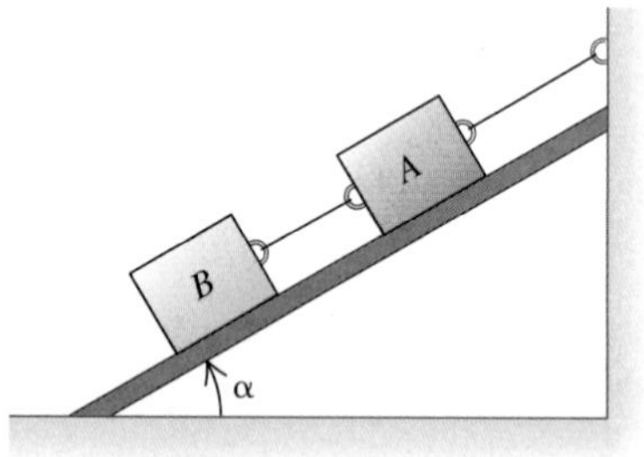

Figure 5.46 Exercise 5.13.

5.14 An airplane is in level flight at a constant speed. There are four forces acting on it: its weight $w = mg$; a forward force F provided by the engine (the *thrust*); the air resistance or *drag* force f, which acts opposite to the direction of flight; and an upward lift force L provided by the wings, which acts perpendicular to the plane of the wings and to the direction of flight. The drag force f is proportional to the square of the speed. a) Show that $F = f$ and $L = w$. b) Suppose the pilot pushes the throttle forward, doubling the value of F, while maintaining a constant altitude. The airplane will eventually attain a new, faster, constant speed. At this faster speed, how does the new value of f compare to the previous value? c) How much faster than the original speed is the airplane going now?

Section 5.2 Using Newton's Second Law: Dynamics of Particles

5.15 Atwood's Machine. A 15.0-kg load of bricks hangs from one end of a rope that passes over a small, frictionless pulley. A 28.0-kg counterweight is suspended from the other end of the rope, as shown in Fig. 5.47. The system is released from rest. a) Draw two free-body diagrams, one for the load of bricks and one for the counterweight. b) What is the magnitude of the upward acceleration of the load of bricks? c) What is the tension in the rope while the load is moving? How does the tension compare to the weight of the load of bricks? To the weight of the counterweight?

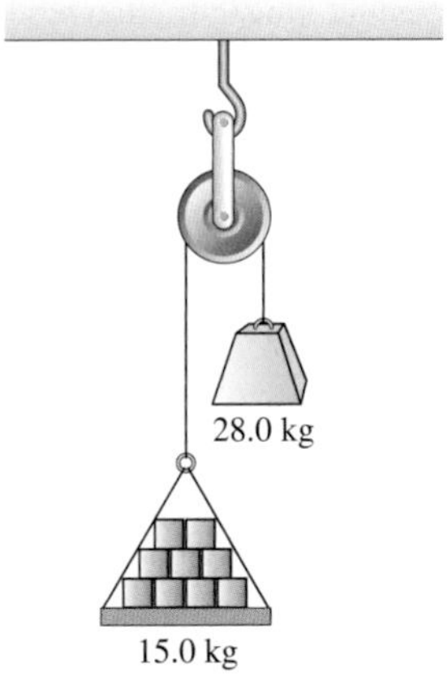

Figure 5.47 Exercise 5.15.

5.16 A 8.00-kg block of ice, released from rest at the top of a 1.50-m-long frictionless ramp, slides downhill, reaching a speed of 2.50 m/s at the bottom. What is the angle between the ramp and the horizontal?

5.17 A light rope is attached to a block with mass 4.00 kg that rests on a frictionless, horizontal surface. The horizontal rope passes over a frictionless, massless pulley and a block with mass m is suspended from the other end. When the blocks are released, the tension in the rope is 10.0 N. a) Draw two free-body diagrams, one for the 4.00-kg block and one for the block with mass m. What is b) the acceleration of either block? c) the mass m of the hanging block? d) How does the tension compare to the weight of the hanging block?

5.18 Runway Design. A transport plane takes off from a level landing field with two gliders in tow, one behind the other. The mass of each glider is 700 kg, and the total resistance (air drag plus friction with the runway) on each may be assumed constant and equal to 2500 N. The tension in the towrope between the transport plane and the first glider is not to exceed 12,000 N. a) If a speed of 40 m/s is required for takeoff, what minimum length of runway is needed? b) What is the tension in the towrope between the two gliders while they are accelerating for the takeoff?

5.19 Apparent Weight. A 550-N physics student stands on a bathroom scale in an elevator. As the elevator starts moving, the scale reads 450 N. a) Find the acceleration of the elevator (magnitude and direction). b) What is the acceleration if the scale reads 670 N? c) If the scale reads zero, should the student worry? Explain.

5.20 A physics student playing with an air hockey table (a frictionless surface) finds that if she gives the puck a velocity of 3.80 m/s along the length (1.75 m) of the table at one end, by the time it has reached the other end the puck has drifted 2.50 cm to the right but still has a velocity component along the length of 3.80 m/s. She correctly concludes that the table is not level and correctly calculates its inclination from the above information. What is the angle of inclination?

Section 5.3 Frictional Forces

5.21 Free-Body Diagrams. The first two steps in the solution of Newton's second law problems are to select an object for analysis and then to draw free-body diagrams for that object. Draw free-body diagrams for the following situations: a) a mass M sliding down a frictionless inclined plane of angle α; b) a mass M sliding up a frictionless inclined plane of angle α; c) a mass M sliding up an inclined plane of angle α with kinetic friction present; d) masses M and m sliding down an inclined plane of angle α with friction present, as shown in Fig. 5.48a. Here, draw free-body diagrams for both m and M. Identify the forces that are action-reaction pairs. e) Draw free-body diagrams for masses m and M shown in Fig. 5.48b. Identify all action-reaction pairs. There is a frictional force between all surfaces in contact. The pulley is frictionless and massless. In all cases be sure you have the correct direction of the forces and are completely clear on what object is causing each force in your free-body diagram.

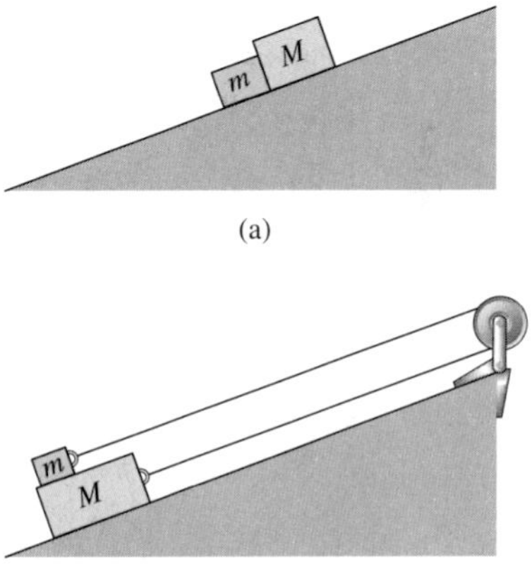

(a)

(b)

Figure 5.48 Exercise 5.21.

5.22 a) A large rock rests upon a rough horizontal surface. A bulldozer pushes on the rock with a horizontal force T that is slowly increased, starting from zero. In a graph with T along the x-axis and the friction force f along the y-axis, start at $T = 0$ and show the region of no motion, the point at which the rock is ready to move, and the region where the rock is moving. b) A block with weight w rests on a rough horizontal plank. The slope angle of the plank θ is gradually increased until the block starts to slip. Draw two graphs, both with θ along the x-axis. In one graph, show the ratio of the normal force to the weight, n/w, as a function of θ. In the second graph, show the ratio of the friction force to the weight, f/w. Indicate the region of no motion, the point at which the block is ready to move, and the region where the block is moving.

5.23 A stockroom worker pushes a box with mass 11.2 kg on a horizontal surface with a constant speed of 3.50 m/s. The coefficient of kinetic friction between the box and the surface is 0.20. a) What horizontal force must the worker apply to maintain the motion? b) If the force calculated in part (a) is removed, how far does the box slide before coming to rest?

5.24 A box of bananas weighing 40.0 N rests on a horizontal surface. The coefficient of static friction between the box and the surface is 0.40 and the coefficient of kinetic friction is 0.20. a) If no horizontal force is applied to the box and the box is at rest, how large is the friction force exerted on the box? b) What is the magnitude of the friction force if a monkey applies a horizontal force of 6.0 N to the box and the box is initially at rest? c) What minimum horizontal force must the monkey apply to start the box in motion? d) What minimum horizontal force must the monkey apply to keep the box moving at constant velocity once it has been started? e) If the monkey applies a horizontal force of 18.0 N, what is the magnitude of the friction force and what is the box's acceleration?

5.25 In a physics lab experiment, a 6.00-kg box is pushed across a flat table by a horizontal force $\vec{F}$. a) If the box is moving at a constant speed of 0.350 m/s and the coefficient of kinetic friction is 0.12, what is the magnitude of $\vec{F}$? b) What is the magnitude of $\vec{F}$ if the box is speeding up with a constant acceleration of 0.180 m/s^2? c) How would your answers to parts (a) and (b) change if the experiments were performed on the moon, where $g = 1.62$ m/s^2?

5.26 An 85-N box of oranges is being pushed across a horizontal floor. As it moves, it is slowing at a constant rate of 0.90 m/s each second. The push force has a horizontal component of 20 N and a vertical component of 25 N downward. Calculate the coefficient of kinetic friction between the box and floor.

5.27 You are to lower a safe with mass 260 kg at constant speed down skids 20.0 m long, from a truck 2.00 m high. a) If the coefficient of kinetic friction between safe and skids is 0.25, do you need to pull down or hold back the safe? b) How great a force parallel to the skids is needed?

5.28 Stopping Distance. a) If the coefficient of kinetic friction between tires and dry pavement is 0.80, what is the shortest distance in which you can stop an automobile by locking the brakes when traveling at 28.7 m/s (about 65 mi/h)? b) On wet pavement the coefficient of kinetic friction may be only 0.25. How fast should you drive on wet pavement in order to be able to stop in the same distance as in part (a)? (*Note:* Locking the brakes is *not* the safest way to stop.)

5.29 Coefficient of Friction. A clean brass washer slides along a horizontal clean steel surface until it stops. Using the values from Table 5.1, how many times farther would it slide with the same initial speed if the washer were Teflon-coated?

5.30 Consider the system shown in Fig. 5.49. Block A has weight w_A and block B has weight w_B. Once block B is set into downward motion, it descends at a constant speed. a) Calculate the coefficient of kinetic friction between block A and the tabletop. b) A cat, also of weight w_A, falls asleep on top of block A. If block B is now set into downward motion, what is its acceleration (magnitude and direction)?

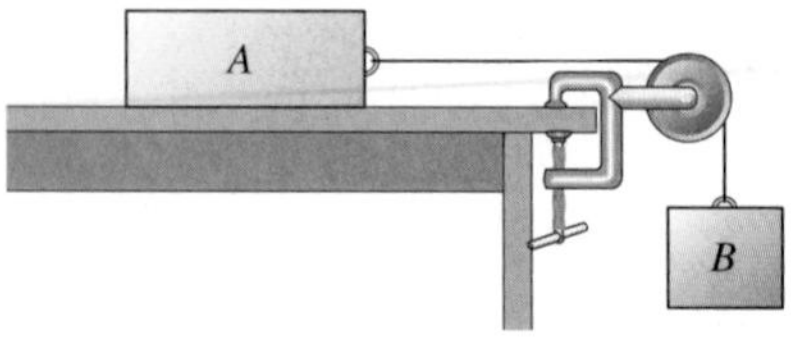

Figure 5.49 Exercises 5.30, 5.35, and Problem 5.72.

5.31 Two crates connected by a rope lie on a horizontal surface (Fig. 5.50). Crate A has mass m_A and crate B has mass m_B. The coefficient of kinetic friction between each crate and the surface is μ_k. The crates are pulled to the right at constant velocity by a horizontal force $\vec{F}$. In terms of m_A, m_B, and μ_k, calculate a) the magnitude of the force $\vec{F}$ and b) the tension in the rope connecting the blocks. Include the free-body diagram or diagrams you used to determine each answer.

Figure 5.50 Exercise 5.31.

5.32 Rolling Friction. Two bicycle tires are set rolling with the same initial speed of 3.50 m/s on a long, straight road, and the distance each travels before its speed is reduced by half is measured. One tire is inflated to a pressure of 40 psi and goes 18.1 m; the other is at 105 psi and goes 92.9 m. What is the coefficient of rolling friction μ_r for each? Assume that the net horizontal force is due to rolling friction only.

5.33 Wheels. You find that it takes a horizontal force of 160 N to slide a box along the surface of a level floor at constant speed. The coefficient of static friction is 0.52 and the coefficient of kinetic friction is 0.47. If you place the box on a dolly of mass 5.3 kg and with coefficient of rolling friction 0.018, what horizontal acceleration would that 160-N force provide?

5.34 You find it takes 200 N of horizontal force to move an empty pickup truck along a level road at a speed of 2.4 m/s. You then load up the pickup and pump up its tires so that its total weight increases by 42% while the coefficient of rolling friction decreases by 19%. Now what horizontal force will you need to move the pickup along the same road at the same speed? The speed is low enough that you can ignore air resistance.

5.35 As shown in Fig. 5.49, block A (mass 2.25 kg) rests on a tabletop. It is connected by a horizontal cord passing over a light, frictionless pulley to a hanging block B (mass 1.30 kg). The coefficient of kinetic friction between block A and the table top is 0.450. After the blocks are released from rest, find a) the speed of each block after moving 3.00 cm and b) the tension in the cord. Include the free-body diagram or diagrams you used to determine the answers.

5.36 A 25.0-kg box of textbooks rests on a loading ramp that makes an angle α with the horizontal. The coefficient of kinetic friction is 0.25 and the coefficient of static friction is 0.35. a) As the angle α is increased, find the minimum angle at which the box starts to slip. b) At this angle, find the acceleration once the box has begun to move. c) At this angle, how fast will the box be moving after it has slid 5.0 m along the loading ramp?

5.37 A large crate with mass m rests on a horizontal floor. The coefficients of friction between the crate and the floor are μ_s and μ_k. A woman pushes downward at an angle θ below the horizontal on the crate with a force $\vec{F}$. a) What magnitude of force $\vec{F}$ is required to keep the crate moving at constant velocity? b) If μ_s is larger than some critical value, the woman cannot start the crate moving no matter how hard she pushes. Calculate this critical value of μ_s.

5.38 A box with mass m is dragged across a level floor having a coefficient of kinetic friction μ_k by a rope that is pulled upward at an angle θ above the horizontal with a force of magnitude F. a) In terms of m, μ_k, θ, and g, obtain an expression for the magnitude of force required to move the box with constant speed. b) Knowing that you are studying physics, a CPR instructor asks you how much force it would take to slide a 90-kg patient across a floor at constant speed by pulling on him at an angle of 25° above the horizontal. By dragging some weights wrapped in an old pair of pants down the hall with a spring balance, you find that $\mu_k = 0.35$. Use the result of part (a) to answer the instructor's question.

5.39 Blocks A, B, and C are placed as in Fig. 5.51 and connected by ropes of negligible mass. Both A and B weigh 25.0 N each, and the coefficient of kinetic friction between each block and the surface is 0.35. Block C descends with constant velocity. a) Draw two separate free-body diagrams showing the forces acting on A and on B. b) Find the tension in the rope connecting blocks A and B. c) What is the weight of block C? d) If the rope connecting A and B were cut, what would be the acceleration of C?

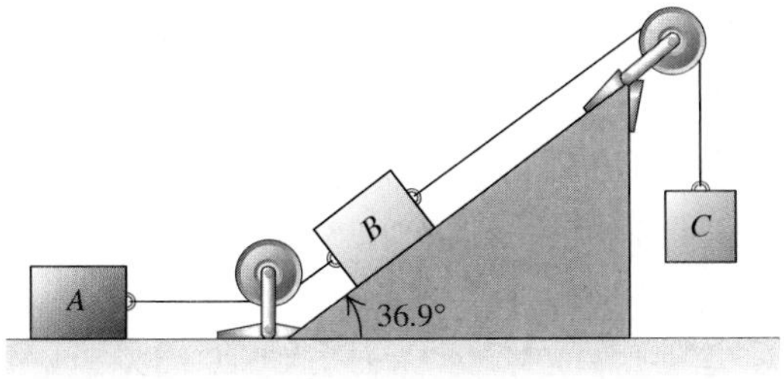

Figure 5.51 Exercise 5.39.

5.40 Starting from Eq. (5.10), derive Eqs. (5.11) and (5.12).

5.41 a) In Example 5.20, what value of D is required to make $v_t = 42$ m/s for this sky diver? b) If the sky diver's daughter, whose mass is 45 kg, is falling through the air and has the same D (0.25 kg/m) as her father, what is the daughter's terminal speed?

5.42 You throw a baseball straight up. The drag force is proportional to v^2. In terms of g, what is the y-component of the ball's acceleration when its speed is half its terminal speed and a) it is moving up? b) it is moving back down?

Section 5.4 Dynamics of Circular Motion

5.43 A stone with mass 0.80 kg is attached to one end of a string 0.90 m long. The string will break if its tension exceeds 600 N. (This is called the *breaking strength* of the string.) The stone is whirled in a horizontal circle on a frictionless tabletop; the other end of the string remains fixed. Find the maximum speed the stone can attain without breaking the string.

5.44 A flat (unbanked) curve on a highway has a radius of 220 m. A car rounds the curve at a speed of 25.0 m/s. What is the minimum coefficient of friction that will prevent sliding?

5.45 Aircraft experience a lift force (due to the air) that is perpendicular to the plane of the wings and to the direction of flight. Light airplanes are designed so that their wings can safely provide a lift force of 3.8 times the weight of the airplane. A greater force can damage the structure of the wing. (Aerobatic airplanes and jet fighters are designed with much higher limits.) a) What is the maximum banking angle that a pilot can safely maintain in level flight without threatening the safety of the airplane (and her own safety)? b) Does your answer to part (a) depend on the speed of the airplane? Explain why or why not.

5.46 The "Giant Swing" at a county fair consists of a vertical central shaft with a number of horizontal arms attached at its upper end (Fig. 5.52). Each arm supports a seat suspended from a cable 5.00 m long, the upper end of the cable being fastened to the arm at a point 3.00 m from the central shaft. a) Find the time of one revolution of the swing if the cable supporting a seat makes an angle of 30.0° with the vertical. b) Does the angle depend on the weight of the passenger for a given rate of revolution?

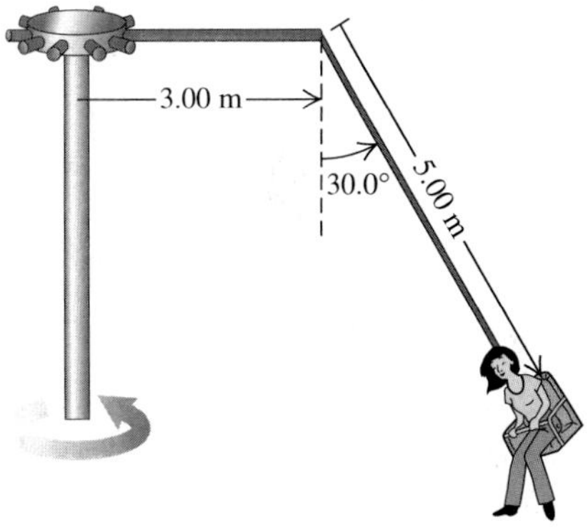

Figure 5.52 Exercise 5.46.

5.47 Aircraft experience a lift force (due to the air) that is perpendicular to the plane of the wings and to the direction of flight. A small airplane is flying at a constant speed of 240 km/h. At what angle from the horizontal must the wings of the airplane be tilted for the plane to execute a horizontal turn from east to north with a turning radius of 1200 m?

5.48 A small button placed on a horizontal rotating platform with diameter 0.320 m will revolve with the platform when it is brought up to a speed of 40.0 rev/min, provided the button is no more than 0.150 m from the axis. a) What is the coefficient of static friction between the button and the platform? b) How far from the axis can the button be placed, without slipping, if the platform rotates at 60.0 rev/min?

5.49 Rotating Space Stations. One problem for humans living in outer space is that they are apparently weightless. One way around this problem is to design a space station that spins about its center at a constant rate. This creates "artificial gravity" at the outside rim of the station. a) If the diameter of the space station is 800 m, how many revolutions per minute are needed in order for the "artificial gravity" acceleration to be 9.80 m/s^2? b) If the space station is a waiting area for travelers going to Mars, it might be desirable to simulate the acceleration due to gravity on the Martian surface (3.70 m/s^2). How many revolutions per minute are needed in this case?

5.50 The Cosmoclock 21 Ferris wheel in Yokohama City, Japan, has a diameter of 100 m. Its name comes from its 60 arms, each of which can function as a second hand (so that it makes one revolution every 60.0 s). a) Find the speed of the passengers when the Ferris wheel is rotating at this rate. b) A passenger weighs 882 N at the weight-guessing booth on the ground. What is his apparent weight at the highest and at the lowest point on the Ferris wheel? c) What would be the time for one revolution if the passenger's apparent weight at the highest point were zero? d) What then would be the passenger's apparent weight at the lowest point?

5.51 An airplane flies in a loop (a circular path in a vertical plane) of radius 150 m. The pilot's head always points toward the center of the loop. The speed of the airplane is not constant; the airplane goes slowest at the top of the loop and fastest at the bottom. a) At the top of the loop, the pilot feels weightless. What is the speed of the airplane at this point? b) At the bottom of the loop, the speed of the airplane is 280 km/h. What is the apparent weight of the pilot at this point? His true weight is 700 N.

5.52 A 50.0-kg stunt pilot who has been diving her airplane vertically pulls out of the dive by changing her course to a circle in a vertical plane. a) If the plane's speed at the lowest point of the circle is 95.0 m/s, what is the minimum radius of the circle for the acceleration at this point not to exceed $4.00g$? b) What is the apparent weight of the pilot at the lowest point of the pullout?

5.53 Stay Dry! You tie a cord to a pail of water, and you swing the pail in a vertical circle of radius 0.600 m. What minimum speed must you give the pail at the highest point of the circle if no water is to spill from it?

5.54 A bowling ball weighing 71.2 N (16.0 lb) is attached to the ceiling by a 3.80-m rope. The ball is pulled to one side and released; it then swings back and forth as a pendulum. As the rope swings through the vertical, the speed of the bowling ball is 4.20 m/s. a) What is the acceleration of the bowling ball, in magnitude and direction, at this instant? b) What is the tension in the rope at this instant?

Problems

5.55 Two ropes are connected to a steel cable that supports a hanging weight as shown in Fig. 5.53. a) Draw a free-body diagram showing all of the forces acting at the knot connecting the two ropes to the steel cable. Based on your force diagram, which of the two ropes will have the greater tension? b) If the maximum tension either rope can sustain without breaking is 5000 N, determine the maximum value of the hanging weight that these ropes can safely support. You can ignore the weight of the ropes and the steel cable.

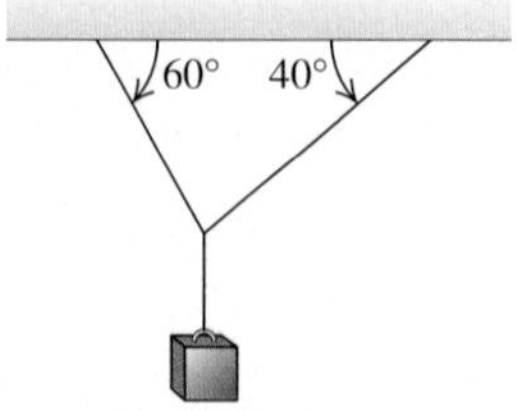

Figure 5.53 Problem 5.55.

5.56 In Fig. 5.54 a worker lifts a weight w by pulling down on a rope with a force $\vec{F}$. The upper pulley is attached to the ceiling by a chain, and the lower pulley is attached to the weight by another chain. In terms of w, find the tension in each chain and the magnitude of the force $\vec{F}$ if the weight is lifted at constant speed. Include the free-body diagram or diagrams you used to determine your answers. Assume that the rope, pulleys, and chains all have negligible weights.

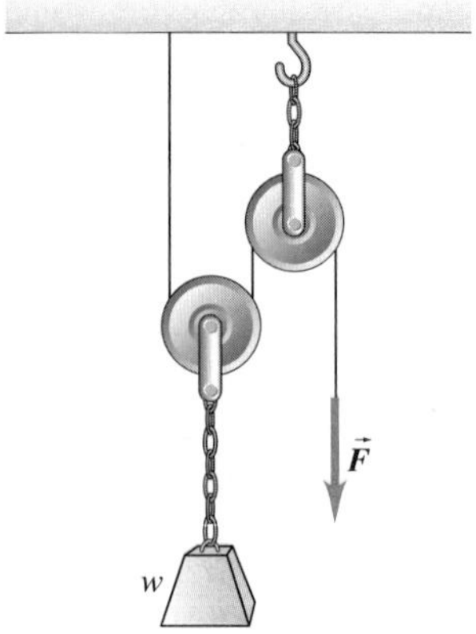

Figure 5.54 Problem 5.56.

5.57 A man is pushing a refrigerator up a ramp at constant speed. The ramp is inclined at an angle α above the horizontal, but the man applies a *horizontal* force $\vec{F}$. Calculate the magnitude of $\vec{F}$ in terms of α and the refrigerator mass m. You can ignore friction on the refrigerator.

5.58 A Rope with Mass. In most problems in this book, the ropes, cords, or cables have so little mass compared to other objects in the problem that you can safely ignore their mass. But if the rope is the *only* object in the problem, then clearly you cannot ignore its mass. For example, suppose we have a clothesline attached to two poles (Fig. 5.55). The clothesline has a mass M, and each end makes an angle θ with the horizontal. What is a) the tension at the ends of the clothesline? b) the tension at the lowest point? c) Why can we not have $\theta = 0$? (See Discussion Question Q5.3.) d) Discuss your results for parts (a) and (b) in the limit that $\theta \to 90°$. The curve of the clothesline, or of any flexible cable hanging under its own weight, is called a *catenary.* For a more advanced treatment of this curve, see K. R. Symon, *Mechanics,* 3rd ed. (Addison–Wesley, Reading, MA, 1971), pp. 237–241.

Figure 5.55 Problem 5.58.

5.59 Another Rope with Mass. A block with mass M is attached to the lower end of a vertical, uniform rope with mass m and length L. A constant upward force $\vec{F}$ is applied to the top of the rope, causing the rope and block to accelerate upward. Find the tension in the rope at a distance x from the top end of the rope, where x can have any value from 0 to L.

5.60 A block with mass m_1 is placed on an inclined plane with slope angle α and is connected to a second hanging block with mass m_2 by a cord passing over a small, frictionless pulley (Fig. 5.56). The coefficient of static friction is μ_s and the coefficient of kinetic friction is μ_k. a) Find the mass m_2 for which block m_1 moves up the plane at constant speed once it is set in motion. b) Find the mass m_2 for which block m_1 moves down the plane at constant speed once it is set in motion. c) For what range of values of m_2 will the blocks remain at rest if they are released from rest?

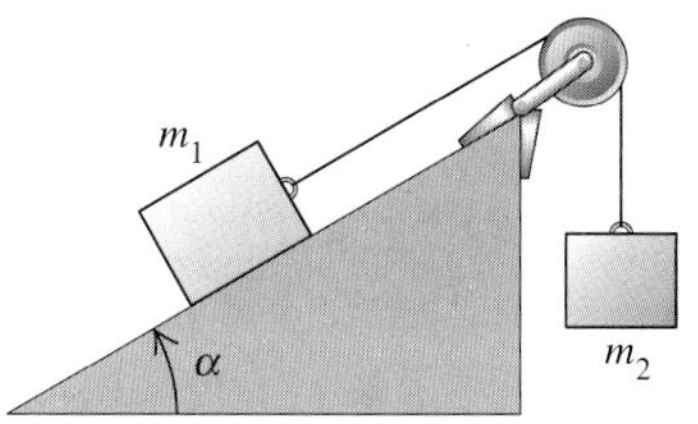

Figure 5.56 Problem 5.60.

5.61 a) Block A in Fig. 5.57 weighs 60.0 N. The coefficient of static friction between the block and the surface on which it rests is 0.25. The weight w is 12.0 N and the system is in equilibrium. Find the friction force exerted on block A. b) Find the maximum weight w for which the system will remain in equilibrium.

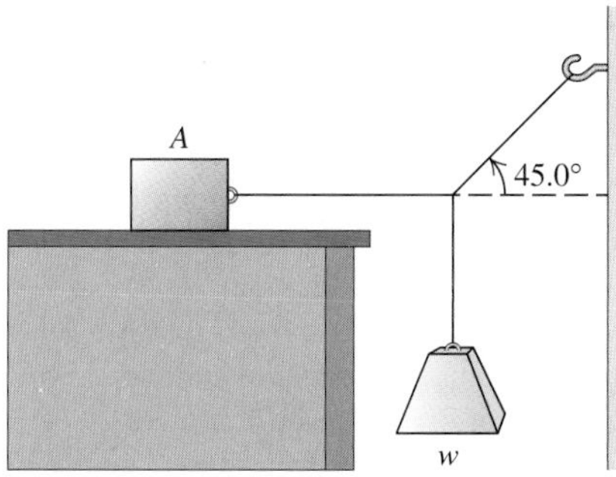

Figure 5.57 Problem 5.61.

5.62 Block A in Fig. 5.58 weighs 1.20 N and block B weighs 3.60 N. The coefficient of kinetic friction between all surfaces is 0.300. Find the magnitude of the horizontal force $\vec{F}$ necessary to drag block B to the left at constant speed a) if A rests on B and moves with it (Fig. 5.58a); b) if A is held at rest (Fig. 5.58b).

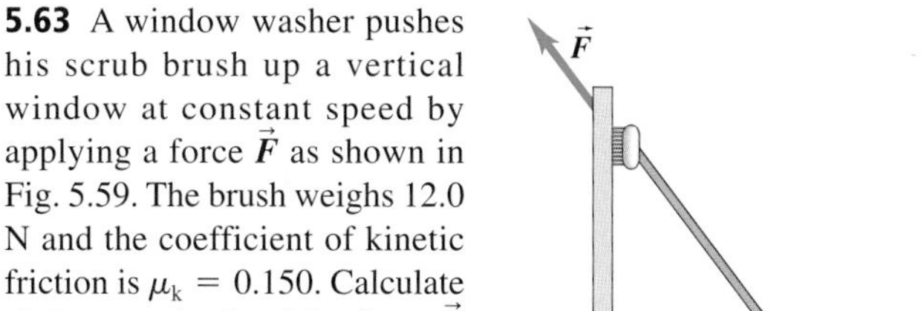

Figure 5.58 Problem 5.62.

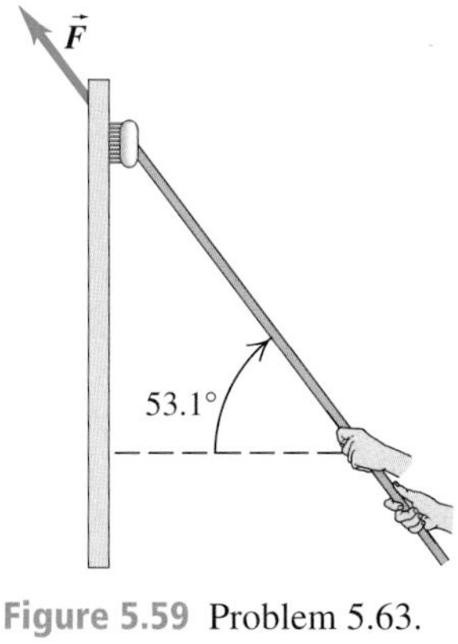

Figure 5.59 Problem 5.63.

5.63 A window washer pushes his scrub brush up a vertical window at constant speed by applying a force $\vec{F}$ as shown in Fig. 5.59. The brush weighs 12.0 N and the coefficient of kinetic friction is $\mu_k = 0.150$. Calculate a) the magnitude of the force $\vec{F}$ and b) the normal force exerted by the window on the brush.

5.64 The Flying Leap of a Flea. High-speed motion pictures (3500 frames/second) of a jumping 210 μg flea yielded the data to plot the flea's acceleration as a function of time as shown in Fig. 5.60. (See "The Flying Leap of the Flea," by M. Rothschild et al. in the Nov. 1973 *Scientific American*.) This flea was about 2 mm long and jumped at a nearly vertical takeoff angle. Use measurements made on the graph to answer the following questions. a) Find the *initial* net external force on the flea. How does it compare to the flea's weight? b) Find the *maximum* net external force on this jumping flea. When does this maximum force occur? c) Use the graph to find the flea's maximum speed.

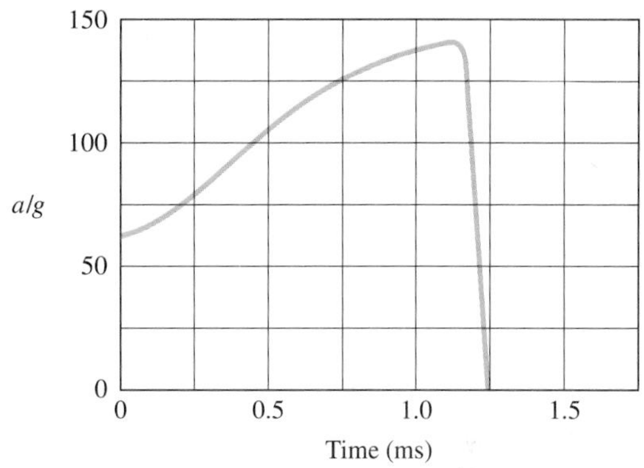

Figure 5.60 Problem 5.64.

5.65 A 25,000-kg rocket blasts off vertically from the earth's surface with a constant acceleration. During the motion considered in the problem, assume that g remains constant (see Chapter 12). Inside the rocket, a 15.0-N instrument hangs from a wire that can support a maximum tension of 35.0 N. a) Find the minimum time for this rocket to reach the sound barrier (330 m/s) without breaking the inside wire and the maximum vertical thrust of the rocket engines under these conditions. b) How far is the rocket above the earth's surface when it breaks the sound barrier?

5.66 You are standing on a bathroom scale in an elevator in a tall building. Your mass is 72 kg. The elevator starts from rest and travels upward with a speed that varies with time according to $v(t) = (3.0\ \text{m/s}^2)t + (0.20\ \text{m/s}^3)t^2$. When $t = 4.0$ s, what is the reading of the bathroom scale?

5.67 Elevator Design. You are designing an elevator for a hospital. The force exerted on a passenger by the floor of the elevator is not to exceed 1.60 times the passenger's weight. The elevator accelerates upward with constant acceleration for a distance of 3.0 m and then starts to slow down. What is the maximum speed of the elevator?

5.68 You are working for a shipping company. Your job is to stand at the bottom of a 8.0-m-long ramp that is inclined at 37° above the horizontal. You grab packages off a conveyor belt and propel them up the ramp. The coefficient of kinetic friction between the packages and the ramp is $\mu_k = 0.30$. a) What speed do you need to give a package at the bottom of the ramp so that it has zero speed at the top of the ramp? b) Your coworker is supposed to grab the packages as they arrive at the top of the ramp, but she misses one and it slides back down. What is its speed when it returns to you?

5.69 A hammer is hanging by a light rope from the ceiling of a bus. The ceiling of the bus is parallel to the roadway. The bus is traveling in a straight line on a horizontal street. You observe that the hammer hangs at rest with respect to the bus when the angle between the rope and the ceiling of the bus is 74°. What is the acceleration of the bus?

5.70 A steel washer is suspended inside an empty shipping crate from a light string attached to the top of the crate. The crate slides down a long ramp that is inclined at an angle of 37° above the horizontal. The crate has mass 180 kg. You are sitting inside the crate (with a flashlight); your mass is 55 kg. As the crate is sliding down the ramp, you find the washer is at rest with respect to the crate when the string makes an angle of 68° with the top of the crate. What is the coefficient of kinetic friction between the ramp and the crate?

5.71 Lunch Time! You are riding your motorcycle one day down a wet street that slopes downward at an angle of 20° below the horizontal. As you start to ride down the hill, you notice a construction crew has dug a deep hole in the street at the bottom of the hill. A Siberian tiger, escaped from the City Zoo, has taken up residence in the hole. a) You apply the brakes and lock your wheels at the top of the hill, where you are moving with a speed of 20 m/s. The inclined street in front of you is 40 m long. a) Will you plunge into the hole and become the tiger's lunch, or do you skid to a stop before you reach the hole? (The coefficients of friction between your motorcycle tires and the wet pavement are $\mu_s = 0.90$ and $\mu_k = 0.70$.) b) What must your initial speed be if you are to stop just before reaching the hole?

5.72 In the system shown in Fig. 5.49, block A has mass m_A, block B has mass m_B, and the rope connecting them has a *nonzero* mass m_{rope}. The rope has a total length L and the pulley has a very small radius. You can ignore any sag in the horizontal part of the rope. a) If there is no friction between block A and the tabletop, find the acceleration of the blocks at an instant when a length d of rope hangs vertically between the pulley and block B. As block B falls, will the magnitude of the acceleration of the system increase, decrease, or remain constant? Explain. b) Let $m_A = 2.00$ kg, $m_B = 0.400$ kg, $m_{\text{rope}} = 0.160$ kg, and $L = 1.00$ m. If there is friction between block A and the tabletop, with $\mu_k = 0.200$ and $\mu_s = 0.250$, find the minimum value of the distance d such that the blocks will start to move if they are initially at rest. c) Repeat part (b) for the case $m_{\text{rope}} = 0.040$ kg. Will the blocks move in this case?

5.73 If the coefficient of static friction between a table and a uniform massive rope is μ_s, what fraction of the rope can hang over the edge of the table without the rope sliding?

5.74 A woman attempts to push a box of books that has mass m up a ramp inclined at an angle α above the horizontal. The coefficients of friction between the ramp and the box are μ_s and μ_k. The force $\vec{F}$ applied by the woman is *horizontal.* a) If μ_s is greater than some critical value, the woman cannot start the box moving up the ramp no matter how hard she pushes. Calculate this critical value of μ_s. b) Assume that μ_s is less than this critical value. What magnitude of force must the woman apply to keep the box moving up the ramp at constant speed?

5.75 A 30.0-kg packing case is initially at rest on the floor of a 1500-kg pickup truck. The coefficient of static friction between the case and the truck floor is 0.30 and the coefficient of kinetic friction is 0.20. Before each acceleration given below, the truck is traveling due north at constant speed. Find the magnitude and direction of the friction force acting on the case a) when the truck accelerates at 2.20 m/s^2 northward and b) when it accelerates at 3.40 m/s^2 southward.

5.76 Traffic Court. The truck in Problem 5.75 is traveling at a constant speed on a country road where the speed limit is 72 km/h (45 mi/h). Spotting a stop sign ahead, the driver applies the brakes and comes to a halt in a distance of 47.0 m. Without warning a policeman emerges from behind a bush and hands the driver a speeding ticket. When the driver protests that he was going less than the speed limit, the policeman replies that "I saw that packing case slide from the back to the front of your truck's cargo area. To make it slide like that, you must have had to brake very hard, which means you had to have been speeding." Will the policeman's argument hold up in traffic court? (Assume that the judge, like you, knows physics.)

5.77 Troy uses a frayed rope to pull a crate across a level floor. The maximum tension in the rope without breaking is T_{max} and the coefficient of kinetic friction is μ_k. a) Show that the maximum weight that he can pull at constant speed is $T_{\text{max}}/(\sin\theta)$, where $\theta = \arctan\mu_k$ is the angle of the rope above the horizontal. b) The answer to part (a) suggests that Troy can pull a weight that approaches infinity with a bit of spiderweb as the coefficient of kinetic friction approaches zero. Explain.

5.78 The airplane in Exercise 5.14 has an engine failure (so $F = 0$), and the airplane glides at constant speed toward a safe landing. The direction of flight is a constant angle α (called the *glide angle*) below the horizontal (Fig. 5.61). a) Find the magnitude of the lift force L (which acts perpendicular to the plane of the wings and to the direction of flight) and of the drag force f in terms of w and α. b) Show that $\alpha = \arctan(f/L)$. c) A fully loaded Cessna 182 (a single-engine airplane) weighs 12,900 N and has 1300 N of drag at a speed of 130 km/h. If this airplane has an engine failure at an altitude of 2500 m, what maximum distance over the ground can it glide while searching for a safe landing spot? d) Justify the statement that "it's drag, not gravity, that makes the airplane go down."

Figure 5.61 Problem 5.78.

5.79 Thrust. The pilot of the Cessna 182 airplane in part (c) of Problem 5.78 manages to get the engine restarted. He applies full throttle and the airplane climbs along a straight line angled above the horizontal. The airplane is flying at a constant speed of 130 km/h, weighs 12,900 N, and has 1300 N of drag. The rate-of-climb indicator on the instrument panel shows that the airplane is gaining altitude at a steady rate of 5.00 m/s (300 m/min, or about 980 feet/min). Determine the magnitude of the thrust (the forward force produced by the engine). (*Hint:* The thrust acts in the direction that the airplane is moving.)

5.80 Losing Cargo. A 12.0-kg box rests on the flat floor of a truck. The coefficients of friction between the box and floor are $\mu_s = 0.19$ and $\mu_k = 0.15$. The truck stops at a stop sign and then starts to move with an acceleration of 2.20 m/s^2. If the box is 1.80 m from the rear of the truck when the truck starts, how much time elapses before the box falls off the truck? How far does the truck travel in this time?

5.81 Block *A* in Fig. 5.62 weighs 1.40 N, and block *B* weighs 4.20 N. The coefficient of kinetic friction between all surfaces is 0.30. Find the magnitude of the horizontal force $\vec{F}$ necessary to drag block *B* to the left at constant speed if *A* and *B* are connected by a light, flexible cord passing around a fixed, frictionless pulley.

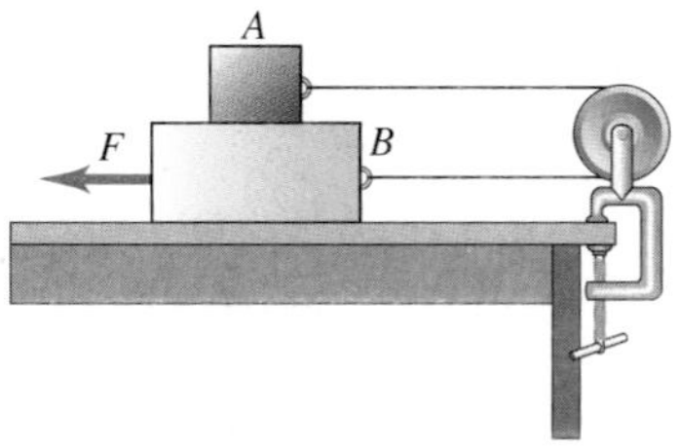

Figure 5.62 Problem 5.81.

5.82 You are part of a design team for future exploration of the planet Mars, where $g = 3.7$ m/s^2. An explorer is to step out of a survey vehicle traveling horizontally at 33 m/s when it is 1200 m above the surface and then fall freely for 20 s. At that time, a portable advanced propulsion system (PAPS) is to exert a constant force that will decrease the explorer's speed to zero at the instant she touches the surface. The total mass (explorer, suit, equipment, and PAPS) is 150 kg. Assume the change in mass of the PAPS to be negligible. Find the horizontal and vertical components of the force the PAPS must exert, and for what interval of time the PAPS must exert it. You can ignore air resistance.

5.83 Block *A* in Fig. 5.63 has a mass of 4.00 kg and block *B* has mass 12.0 kg. The coefficient of kinetic friction between block *B* and the horizontal surface is 0.25. a) What is the mass of block *C* if block *B* is moving to the right and speeding up with an acceleration 2.00 m/s^2? b) What is the tension in each cord when block *B* has this acceleration?

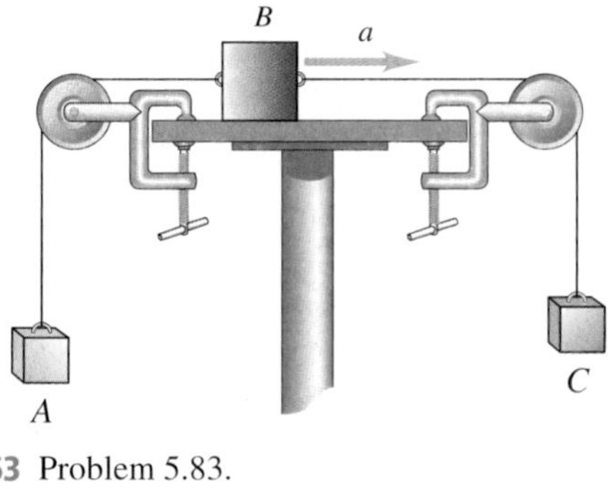

Figure 5.63 Problem 5.83.

5.84 Two blocks connected by a cord passing over a small, frictionless pulley rest on frictionless planes (Fig. 5.64). a) Which way will the system move when the blocks are released from rest? b) What is the acceleration of the blocks? c) What is the tension in the cord?

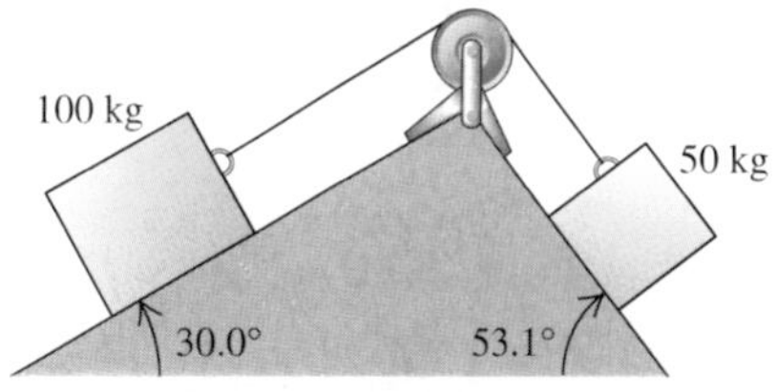

Figure 5.64 Problem 5.84.

5.85 In terms of m_1, m_2, and g, find the accelerations of each block in Fig. 5.65. There is no friction anywhere in the system.

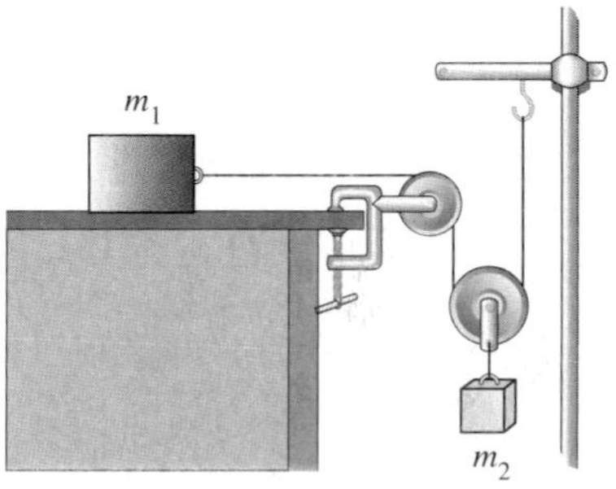

Figure 5.65 Problem 5.85.

5.86 Block *B*, with mass m_B, rests on block *A*, with mass m_A, which in turn is on a horizontal tabletop (Fig. 5.66). The coefficient of kinetic friction between block *A* and the tabletop is μ_k, and the coefficient of static friction between block *A* and block *B* is μ_s. A light string attached to block *A* passes over a frictionless, massless pulley and block *C* is suspended from the other end of the string. What is the largest mass m_C that block *C* can have so that blocks *A* and *B* still slide together when the system is released from rest?

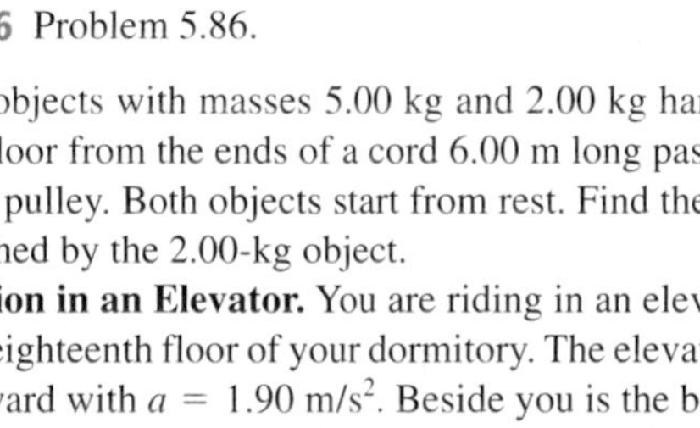

Figure 5.66 Problem 5.86.

5.87 Two objects with masses 5.00 kg and 2.00 kg hang 0.600 m above the floor from the ends of a cord 6.00 m long passing over a frictionless pulley. Both objects start from rest. Find the maximum height reached by the 2.00-kg object.

5.88 Friction in an Elevator. You are riding in an elevator on the way to the eighteenth floor of your dormitory. The elevator is accelerating upward with $a = 1.90$ m/s^2. Beside you is the box containing your new computer; box and contents have a total mass 28.0 kg. While the elevator is accelerating upward, you push horizontally on the box to slide it at constant speed toward the elevator door. If the coefficient of kinetic friction between the box and elevator floor is $\mu_k = 0.32$, what magnitude of force must you apply?

5.89 A block is placed against the vertical front of a cart as shown in Fig. 5.67. What acceleration must the cart have in order that block A does not fall? The coefficient of static friction between the block and the cart is μ_s. How would an observer on the cart describe the behavior of the block?

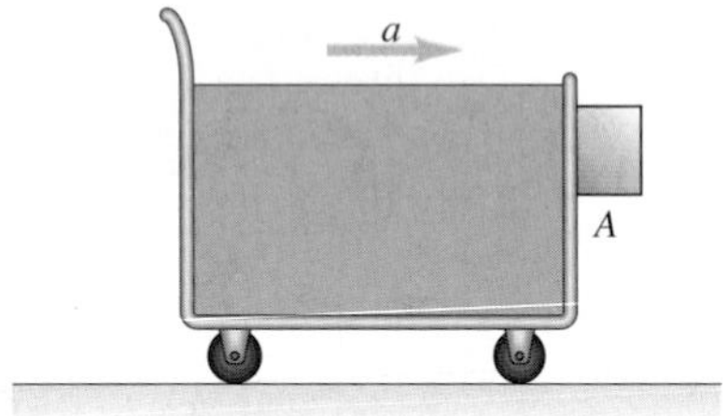

Figure 5.67 Problem 5.89.

5.90 Two blocks with masses 4.00 kg and 8.00 kg are connected by a string and slide down a 30.0° inclined plane (Fig. 5.68). The coefficient of kinetic friction between the 4.00-kg block and the plane is 0.25; that between the 8.00-kg block and the plane is 0.35. a) Calculate the acceleration of each block. b) Calculate the tension in the string. c) What happens if the positions of the blocks are reversed, so the 4.00-kg block is above the 8.00-kg block?

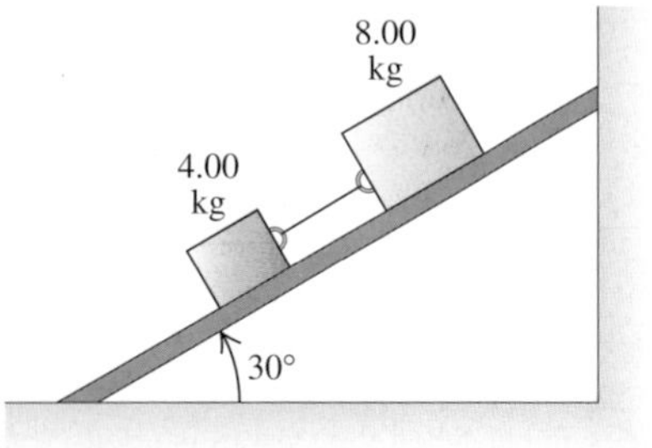

Figure 5.68 Problem 5.90.

5.91 Block A, with weight $3w$, slides down an inclined plane S of slope angle 36.9° at a constant speed while plank B, with weight w, rests on top of A. The plank is attached by a cord to the wall (Fig. 5.69). a) Draw a diagram of all the forces acting on block A. b) If the coefficient of kinetic friction is the same between A and B and between S and A, determine its value.

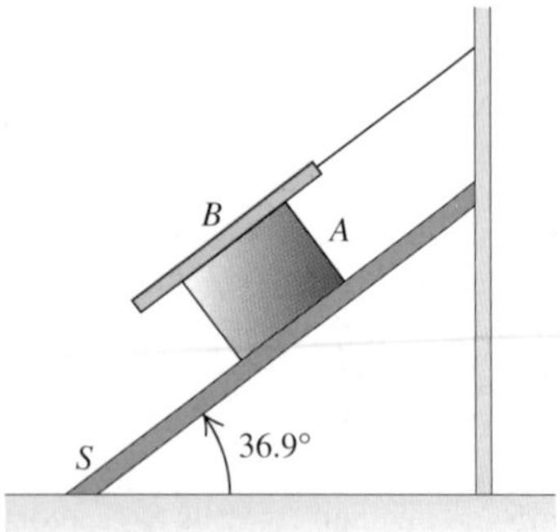

Figure 5.69 Problem 5.91.

5.92 A man with mass 70.0 kg stands on a platform with mass 25.0 kg. He pulls on the free end of a rope that runs over a pulley on the ceiling and has its other end fastened to the platform. You can ignore the mass of the rope and the mass of the pulley, and the pulley is frictionless. The rope is vertical on either side of the pulley. a) With what force does he have to pull so that he and the platform have an upward acceleration of 1.80 m/s^2? b) What is the acceleration of the rope relative to him?

5.93 Two blocks, with masses m_1 and m_2, are stacked as shown in Fig. 5.70 and placed on a frictionless horizontal surface. There is friction between the two blocks. An external force of magnitude F is applied to the top block at an angle α below the horizontal. a) If the two blocks move together, find their acceleration. b) Show that the two blocks will move together only if

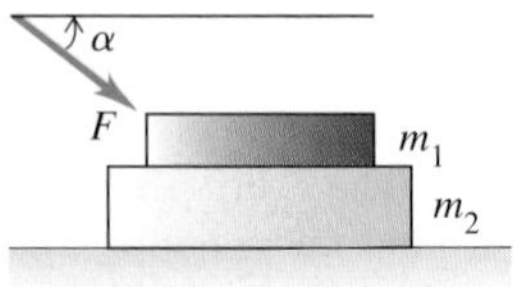

Figure 5.70 Problem 5.93.

$$F \le \frac{\mu_s m_1(m_1 + m_2)g}{m_2 \cos\alpha - \mu_s(m_1 + m_2)\sin\alpha}$$

where μ_s is the coefficient of static friction between the two blocks.

5.94 Banked Curve I. A curve with a 120-m radius on a level road is banked at the correct angle for a speed of 20 m/s. If an automobile rounds this curve at 30 m/s, what is the minimum coefficient of static friction needed between tires and road to prevent skidding?

5.95 Banked Curve II. Consider a wet roadway banked as in Example 5.24 (Section 5.4), where there is a coefficient of static friction of 0.30 and a coefficient of kinetic friction of 0.25 between the tires and the roadway. The radius of the curve is $R = 50$ m. a) If the banking angle is $\beta = 25°$, what is the *maximum* speed the automobile can have before sliding *up* the banking? b) What is the *minimum* speed the automobile can have before sliding *down* the banking?

5.96 Maximum Safe Speed. As you travel every day to campus, the road makes a large turn that is approximately an arc of a circle. You notice the warning sign at the start of the turn, asking for a maximum speed of 55 mi/h. You also notice that in the curved portion the road is level; that is, not banked at all. On a dry day with very little traffic, you enter the turn at a constant speed of 80 mi/h and feel the car may skid if you do not slow down quickly. You conclude that your speed is at the limit of safety for this curve and you slow down. However, you remember reading that on dry pavement new tires have an average coefficient of static friction of about 0.76, while under the worst winter driving conditions, you may encounter wet ice for which the coefficient of static friction can be as low as 0.20. Wet ice is not unheard of on this road, so you ask yourself whether the speed limit for the turn on the roadside warning sign is for the worst-case scenario. a) Estimate the radius of the curve from your 80 mi/h experience in the dry turn. b) Use this estimate to find the maximum speed limit in the turn under the worst wet-ice conditions. How does this compare with the speed limit on the sign? Is the sign misleading drivers?

c) On a rainy day, the coefficient of static friction would be about 0.37. What is the maximum safe speed for the turn when the road is wet? Does your answer help you understand the maximum-speed sign?

5.97 On a rainy day, you drive on a wet one-way road. The road is level with two lanes and you drive in the second lane. You are on the straight part of the road, but you are aware that a turn having the shape of a circular arc will start in 0.80 km (about half a mile) from your current position. Your speed is 27 m/s (60 mi/h), and you know from past experience that on dry pavement this is the maximum speed at which you can round the coming curve safely. However, the road is wet and you also know that the rain makes the road slippery, reducing the coefficient of static friction to half its value for a dry road. Suddenly you notice a car 0.50 km (about one-third of a mile) behind you in the other lane and moving very fast, in your estimation at a speed of 36 m/s (80 mi/h). The driver of this car apparently does not notice the warning sign about the coming curve and continues to travel at constant speed. You realize that this car may catch up with you at the first section of the turn, skid in the turn, and involve you in a bad accident. a) On the wet road, what is the maximum safe speed for the turn? b) If you slow down with constant acceleration so that you have the speed calculated in (a), where will the second car be as you enter the turn? Is a collision likely?

5.98 You are riding in a school bus. As the bus rounds a flat curve at constant speed, a lunch box with mass 0.500 kg, suspended from the ceiling of the bus by a string 1.80 m long, is found to hang at rest relative to the bus when the string makes an angle of 30.0° with the vertical. In this position the lunch box is 50.0 m from the center of curvature of the curve. What is the speed v of the bus?

5.99 The Monkey and Bananas Problem. A 20-kg monkey has a firm hold on a light rope that passes over a frictionless pulley and is attached to a 20-kg bunch of bananas (Fig. 5.71). The monkey looks up, sees the bananas, and starts to climb the rope to get them. a) As the monkey climbs, do the bananas move up, down, or remain at rest? b) As the monkey climbs, does the distance between the monkey and the bananas decrease, increase, or remain constant? c) The monkey releases her hold on the rope. What happens to the distance between the monkey and the bananas while she is falling? d) Before reaching the ground, the monkey grabs the rope to stop her fall. What do the bananas do?

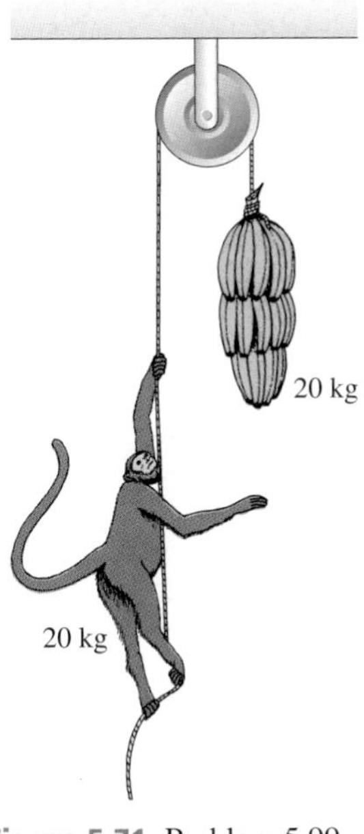

Figure 5.71 Problem 5.99.

5.100 You throw a rock downward into water with a speed of $3mg/k$, where k is the coefficient in Eq. (5.7). Assume that the relation between fluid resistance and speed is as given in Eq. (5.7) and calculate the speed of the rock as a function of time.

5.101 A rock with mass $m = 3.00$ kg falls from rest in a viscous medium. The rock is acted on by a net constant downward force of 18.0 N (a combination of gravity and the buoyant force exerted by the medium) and by a fluid resistance force $f = kv$, where v is the speed in m/s and $k = 2.20$ N·s/m. (See Section 5.3.) a) Find the initial acceleration, a_0. b) Find the acceleration when the speed is 3.00 m/s. c) Find the speed when the acceleration equals $0.1a_0$. d) Find the terminal speed, v_t. e) Find the coordinate, speed, and acceleration 2.00 s after the start of the motion. f) Find the time required to reach a speed $0.9v_t$.

5.102 A rock with mass m slides with initial velocity v_0 on a horizontal surface. A retarding force F_R that the surface exerts on the rock is proportional to the square root of the instantaneous velocity of the rock $\left(F_R = -kv^{1/2}\right)$. a) Find expressions for the velocity and position of the rock as a function of time. b) In terms of m, k, and v_0, at what time will the rock come to rest? c) In terms of m, k, and v_0, what is the distance of the rock from its starting point when it comes to rest?

5.103 A fluid exerts an upward buoyancy force on an object immersed in it. In the derivation of Eq. (5.9) the buoyancy force exerted on an object by the fluid was ignored. But in some situations, where the density of the object is not much greater than the density of the fluid, you cannot ignore the buoyancy force. For a plastic sphere falling in water, you calculate the terminal speed to be 0.36 m/s when you ignore buoyancy, but you measure it to be 0.24 m/s. The buoyancy force is what fraction of the weight?

5.104 The 4.00-kg block in Fig. 5.72 is attached to a vertical rod by means of two strings. When the system rotates about the axis of the rod, the strings are extended as shown in the diagram and the tension in the upper string is 80.0 N. a) What is the tension in the lower cord? b) How many revolutions per minute does the system make? c) Find the number of revolutions per minute at which the lower cord just goes slack. d) Explain what happens if the number of revolutions per minute is less than in part (c).

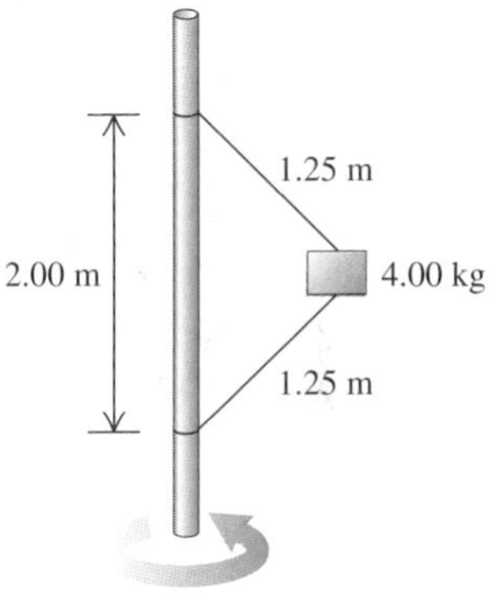

Figure 5.72 Problem 5.104.

5.105 Equation (5.10) applies to the case where the initial velocity is zero. a) Derive the corresponding equation for $v_y(t)$ when the falling object has an initial downward velocity with magnitude v_0. b) For the case where $v_0 < v_t$, sketch a graph of v_y as a function of t and label v_t on your graph. c) Repeat part (b) for the case where $v_0 > v_t$. d) Discuss what your result says about $v_y(t)$ when $v_0 = v_t$.

5.106 A small rock moves in water and the force exerted on it by the water is given by Eq. (5.7). The terminal speed of the rock is measured and found to be 2.0 m/s. The rock is projected *upward* at an initial speed of 6.0 m/s. You can ignore the buoyancy force on the rock. a) In the absence of fluid resistance, how high will the rock rise and how long will it take to reach this maximum height? b) When the effects of fluid resistance are included, what are the answers to the questions in part (a)?

5.107 You observe a 1350-kg sports car rolling along flat pavement in a straight line. The only horizontal forces acting on it are a constant rolling friction and air resistance (proportional to the square of its speed). You take the following data during a time interval of 25 s: When its speed is 32 m/s, it slows down at a rate of $-0.42\ \text{m/s}^2$, and when its speed is decreased to 24 m/s, it slows down at $-0.30\ \text{m/s}^2$. a) Find the coefficient of rolling friction and the air drag constant D. b) At what constant speed will this car move down an incline that makes a 2.2° angle with the horizontal? c) How is the constant speed for an incline of angle β related to the terminal speed of this sports car if the car drops off a high cliff? Assume that in both cases the air resistance force is proportional to the square of the speed, and the air drag constant is the same.

5.108 A 70-kg person rides in a 30-kg cart moving at 12 m/s at the top of a hill that is in the shape of an arc of a circle with radius of 40 m. a) What is the apparent weight of the person as the cart passes over the top of the hill? b) Determine the maximum speed that the cart may travel at the top of the hill without losing contact with the surface. Does your answer depend on the mass of the cart or that of the person? Explain.

5.109 Merry-Go-Round. One December identical twins Jena and Jackie are playing on a large merry-go-round (a disk mounted parallel to the ground, on a vertical axle through its center) in their school playground in northern Minnesota. Each twin has mass 30.0 kg. The icy coating on the merry-go-round surface makes it frictionless. The merry-go-round revolves at a constant rate as the twins ride on it. Jena, sitting 1.80 m from the center of the merry-go-round, must hold on to one of the metal posts attached to the merry-go-round with a horizontal force of 60.0 N to keep from sliding off. Jackie is sitting at the edge, 3.60 m from the center. a) With what horizontal force must Jackie hold on to keep from falling off? b) If Jackie falls off, what will be her horizontal velocity when she becomes airborne?

5.110 A passenger with mass 85 kg rides in a Ferris wheel like that in Example 5.25. The seats travel in a circle of radius 35 m. The Ferris wheel rotates at constant speed and makes one complete revolution every 25 s. Calculate the magnitude and direction of the net force exerted on the passenger by the seat when she is a) one-quarter revolution past her lowest point and b) one-quarter revolution past her highest point.

5.111 On the ride "Spindletop" at the amusement park Six Flags Over Texas, people stood against the inner wall of a hollow vertical cylinder with radius 2.5 m. The cylinder started to rotate, and when it reached a constant rotation rate of 0.60 rev/s, the floor on which people were standing dropped about 0.5 m. The people remained pinned against the wall. a) Draw a force diagram for a person on this ride, after the floor has dropped. b) What minimum coefficient of static friction is required if the person on the ride is not to slide downward to the new position of the floor? c) Does your answer in part (b) depend on the mass of the passenger? (*Note:* When the ride is over, the cylinder is slowly brought to rest. As it slows down, people slide down the walls to the floor.)

5.112 A physics major is working to pay his college tuition by performing in a traveling carnival. He rides a motorcycle inside a hollow transparent plastic sphere. After gaining sufficient speed, he travels in a vertical circle with a radius of 13.0 m. The physics major has mass 70.0 kg, and his motorcycle has mass 40.0 kg. a) What minimum speed must he have at the top of the circle if the tires of the motorcycle are not to lose contact with the sphere? b) At the bottom of the circle, his speed is twice the value calculated in part (a). What is the magnitude of the normal force exerted on the motorcycle by the sphere at this point?

5.113 Ulterior Motives. You are driving a classic 1954 Nash Ambassador with a friend who is sitting to your right on the passenger side of the front seat. The Ambassador has flat bench seats. You would like to be closer to your friend and decide to use physics to achieve your romantic goal by making a quick turn. a) Which way (to the left or to the right) should you turn the car to get your friend to slide closer to you? b) If the coefficient of static friction between your friend and the car seat is 0.35, and you keep driving at a constant speed of 20 m/s, what is the maximum radius you could make your turn and still have your friend slide your way?

5.114 A small block with mass m rests on a frictionless horizontal tabletop a distance r from a hole in the center of the table (Fig. 5.73). A string tied to the small block passes down through the hole, and a larger block with mass M is suspended from the free end of the string. The small block is set into uniform circular motion with radius r and speed v. What must v be if the large block is to remain motionless when released?

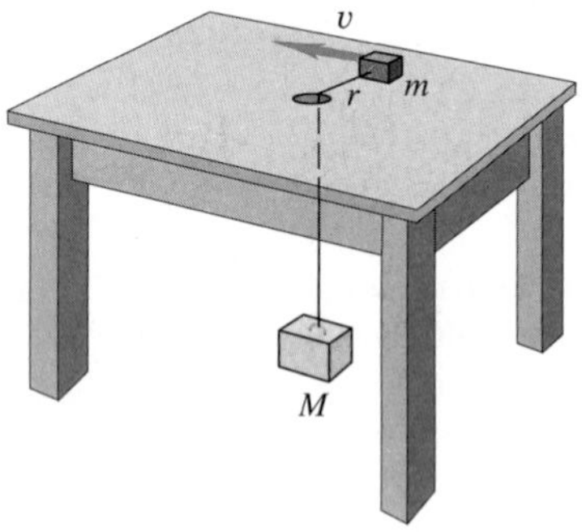

Figure 5.73 Problem 5.114.

5.115 A small bead can slide without friction on a circular hoop that is in a vertical plane and has a radius of 0.100 m. The hoop rotates at a constant rate of 4.00 rev/s about a vertical diameter (Fig. 5.74). a) Find the angle β at which the bead is in vertical equilibrium. (Of course, it has a radial acceleration toward the axis.) b) Is it possible for the bead to "ride" at the same elevation as the center of the hoop? c) What will happen if the hoop rotates at 1.00 rev/s?

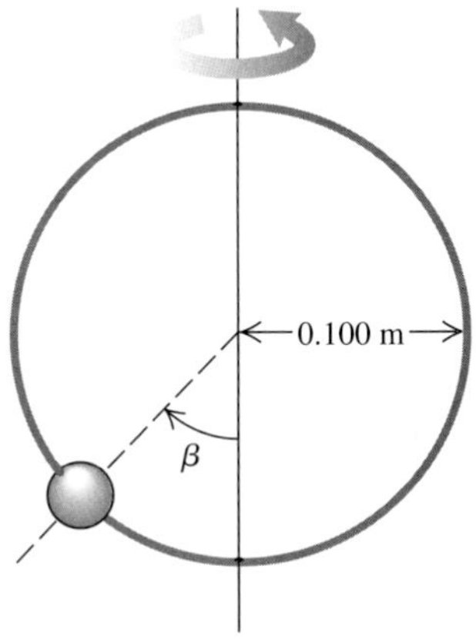

Figure 5.74 Problem 5.115.

5.116 A model airplane with mass 2.20 kg moves in the xy-plane such that its x- and y-coordinates vary in time according to $x(t) = \alpha - \beta t^3$ and $y(t) = \gamma t - \delta t^2$, where $\alpha = 1.50$ m, $\beta = 0.120$ m/s^3, $\gamma = 3.00$ m/s, and $\delta = 1.00$ m/s^2. a) Calculate the x- and y-components of the net force on the plane as functions of time. b) Sketch the trajectory of the airplane between $t = 0$ and $t = 3.00$ s, and draw on your sketch vectors showing the net force on the airplane at $t = 0$, $t = 1.00$ s, $t = 2.00$ s, and $t = 3.00$ s. For each of these times, relate the direction of the net force to the direction that the airplane is turning, and to whether the airplane is speeding up or slowing down (or neither). c) What are the magnitude and direction of the net force at $t = 3.00$ s?

5.117 A particle moves on a frictionless surface along a path as shown in Fig. 5.75. (The figure gives a view looking down on the surface.) The particle is initially at rest at point A, and then begins to move toward B as it gains speed at a constant rate. From B to C, the particle moves along a circular path at a constant speed. The speed remains constant along the straight-line path from C to D. From D to E, the particle moves along a circular path, but now its speed is decreasing at a constant rate. The speed continues to decrease at a constant rate as the particle moves from E to F; the particle comes to a halt at F. (The time intervals between the marked points are not equal.) At each point marked with a dot, draw arrows to represent the velocity, the acceleration, and the net force acting on the particle. Draw a longer or shorter arrow to represent vectors of larger or smaller magnitude.

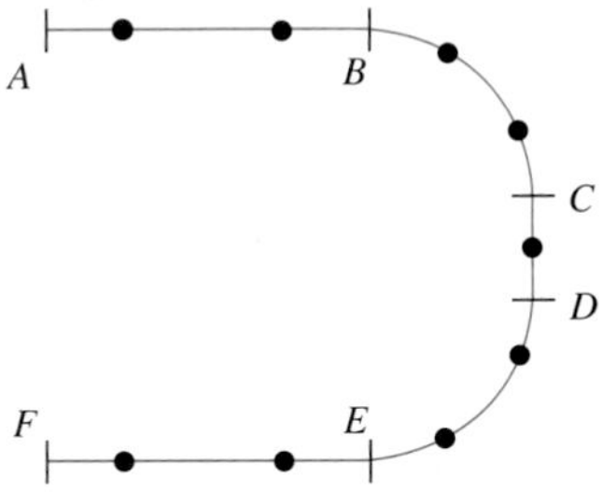

Figure 5.75 Problem 5.117.

5.118 A small remote-control car with mass 1.60 kg moves at a constant speed of $v = 12.0$ m/s in a vertical circle inside a hollow metal cylinder that has a radius of 5.00 m (Fig. 5.76). What is the magnitude of the normal force exerted on the car by the walls of the cylinder at a) point A (at the bottom of the vertical circle)? b) point B (at the top of the vertical circle)?

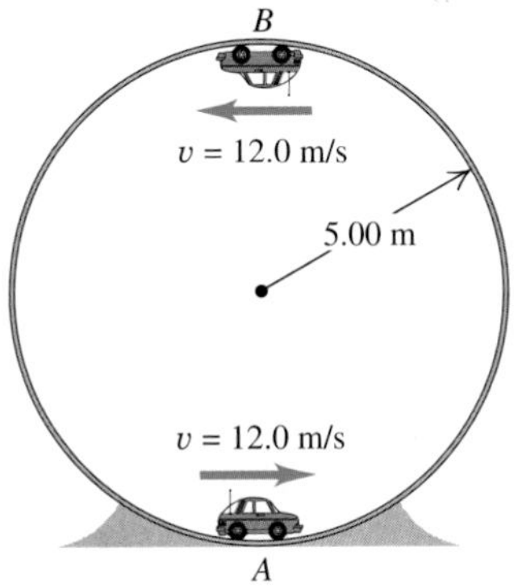

Figure 5.76 Problem 5.118.

5.119 A small block with mass m is placed inside an inverted cone that is rotating about a vertical axis such that the time for one revolution of the cone is T (Fig. 5.77). The walls of the cone make an angle β with the vertical. The coefficient of static friction between the block and the cone is μ_s. If the block is to remain at a constant height h above the apex of the cone, what are the maximum and minimum values of T?

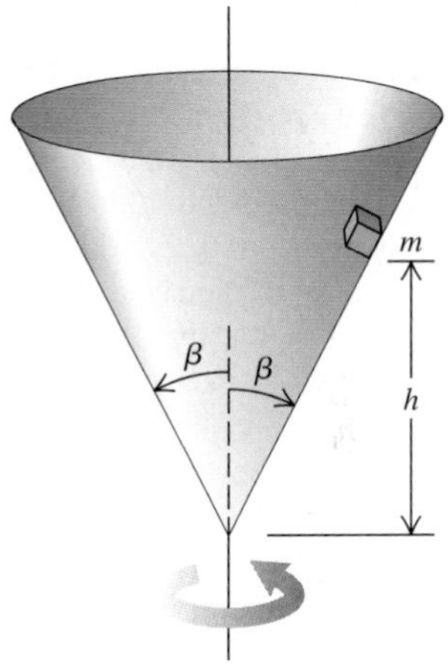

Figure 5.77 Problem 5.119.

Challenge Problems

5.120 Moving Wedge. A wedge with mass M rests on a frictionless, horizontal table top. A block with mass m is placed on the wedge (Fig. 5.78a). There is no friction between the block and the wedge. The system is released from rest. a) Calculate the acceleration of the wedge and the horizontal and vertical components of the acceleration of the block. b) Do your answers to part (a) reduce to the correct results when M is very large? c) As seen by a stationary observer, what is the shape of the trajectory of the block?

5.121 A wedge with mass M rests on a frictionless horizontal tabletop. A block with mass m is placed on the wedge and a horizontal force $\vec{F}$ is applied to the wedge (Fig. 5.78b). What must the magnitude of $\vec{F}$ be if the block is to remain at a constant height above the tabletop?

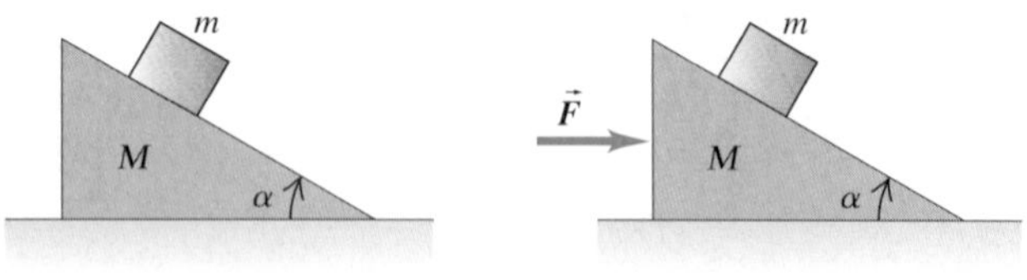

(a) (b)

Figure 5.78 Challenge Problems 5.120 and 5.121.

5.122 A box of weight w is accelerated up a ramp by a rope that exerts a tension T. The ramp makes an angle α with the horizontal and the rope makes an angle θ above the ramp. The coefficient of kinetic friction between the box and the ramp is μ_k. Show that no matter what the value of α, the acceleration is maximum if $\theta = \arctan \mu_k$ (as long as the box remains in contact with the ramp).

5.123 Angle for Minimum Force. A box with weight w is pulled at constant speed along a level floor by a force $\vec{F}$ that is at an angle θ above the horizontal. The coefficient of kinetic friction between the floor and box is μ_k. a) In terms of θ, μ_k, and w, calculate F. b) For $w = 400$ N and $\mu_k = 0.25$, calculate F for θ ranging from 0° to 90° in increments of 10°. Graph F versus θ. c) From the general expression in part (a), calculate the value of θ for which the value of F, required to maintain constant speed, is a minimum. (*Hint:* At a point where a function is minimum, what are the first and second derivatives of the function? Here F is a function of θ.) For the special case of $w = 400$ N and $\mu_k = 0.25$, evaluate this optimal θ and compare your result to the graph you constructed in part (b).

5.124 Falling Baseball. You drop a baseball from the roof of a tall building. As the ball falls, the air exerts a drag force proportional to the square of the ball's speed $(f = Dv^2)$. a) In a diagram, show the direction of motion and indicate, with the aid of vectors, all of the forces acting on the ball. b) Apply Newton's second law and infer from the resulting equation the general properties of the motion. c) Show that the ball acquires a terminal speed that is as given in Eq. (5.13). d) Derive the equation for the speed at any time. (*Note:*

$$\int \frac{dx}{a^2 - x^2} = \frac{1}{a}\,\text{arctanh}\left(\frac{x}{a}\right)$$

where

$$\tanh(x) = \frac{e^x - e^{-x}}{e^x + e^{-x}} = \frac{e^{2x} - 1}{e^{2x} + 1}$$

defines the hyperbolic tangent.)

5.125 Double Atwood's Machine. In Fig. 5.79 masses m_1 and m_2 are connected by a light string A over a light frictionless pulley B. The axle of pulley B is connected by a second light string C over a second light, frictionless pulley D to a mass m_3. Pulley D is suspended from the ceiling by an attachment to its axle. The system is released from rest. In terms of m_1, m_2, m_3, and g, what is a) the acceleration of block m_3? b) the acceleration of pulley B? c) the acceleration of block m_1? d) the acceleration of block m_2? e) the tension in string A? f) the tension in string C? g) What do your expressions give for the special case of $m_1 = m_2$ and $m_3 = m_1 + m_2$? Is this sensible?

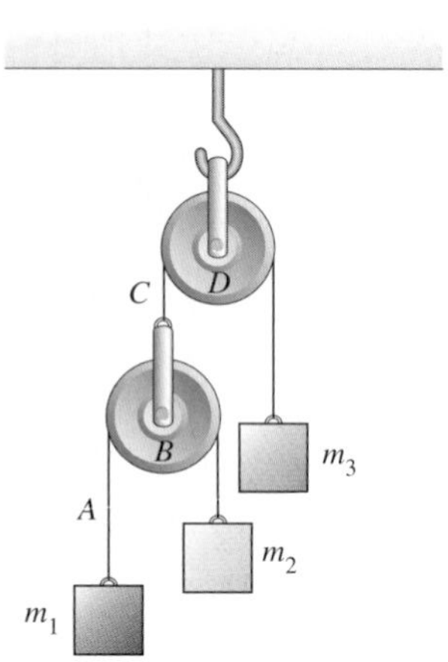

Figure 5.79 Challenge Problem 5.125.

5.126 The masses of blocks A and B in Fig. 5.80 are 20.0 kg and 10.0 kg, respectively. The blocks are initially at rest on the floor and are connected by a massless string passing over a massless and frictionless pulley. An upward force $\vec{F}$ is applied to the pulley. Find the accelerations $\vec{a}_A$ of block A and $\vec{a}_B$ of block B when F is a) 124 N; b) 294 N; c) 424 N.

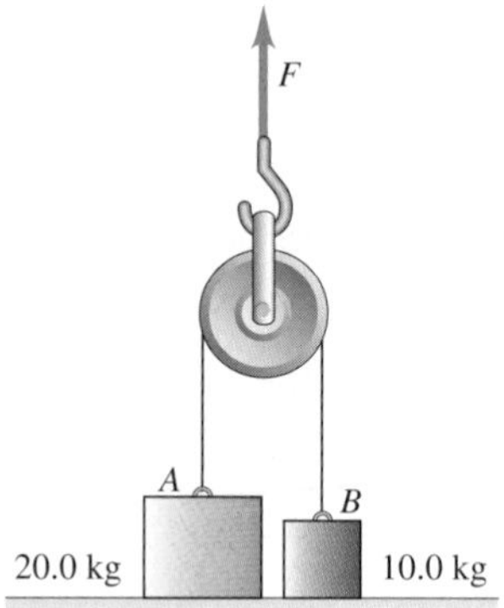

Figure 5.80 Challenge Problem 5.126.

5.127 A ball is held at rest at position A in Fig. 5.81 by two light strings. The horizontal string is cut and the ball starts swinging as a pendulum. Point B is the farthest to the right the ball goes as it swings back and forth. What is the ratio of the tension in the supporting string in position B to its value at A before the horizontal string was cut?

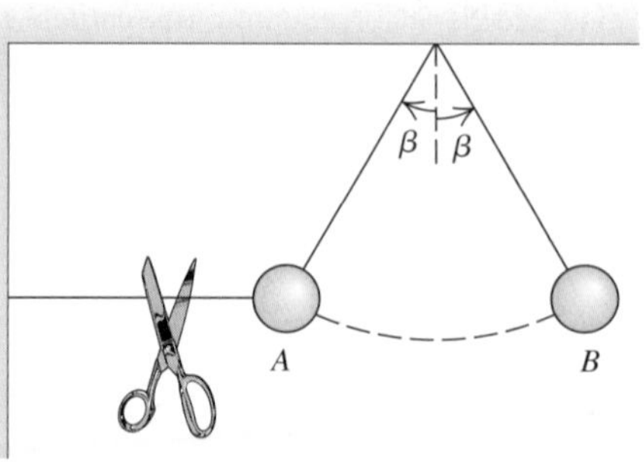

Figure 5.81 Challenge Problem 5.127.

WORK AND KINETIC ENERGY

6

When an athlete throws the javelin in a track and field competition, she does work on the javelin—that is, she exerts a force over a distance. As a result the javelin gains kinetic energy, or energy of motion. At the end of the javelin's flight, this kinetic energy, in turn, does work on the ground as it penetrates into the surface.

? **According to Newton's third law, the javelin exerts as much force on the athlete as the athlete exerts on the javelin. Would it be correct to say that the *javelin* does work on the *athlete*?**

Some problems are harder than they look. For example, suppose you try to find the speed of an arrow that has been shot from a bow. You apply Newton's laws and all the problem-solving techniques that we've learned, but you run across a major stumbling block: After the archer releases the arrow, the bow string exerts a *varying* force that depends on the arrow's position. As a result, the simple methods that we've learned aren't enough to calculate the speed. Never fear; we aren't by any means finished with mechanics, and there are other methods for dealing with such problems.

The new method that we're about to introduce uses the ideas of *work* and *energy*. The importance of the energy idea stems from the *principle of conservation of energy*: Energy is a quantity that can be converted from one form to another but cannot be created or destroyed. In an automobile engine, chemical energy stored in the fuel is converted partially to the energy of the automobile's motion and partially to thermal energy. In a microwave oven, electromagnetic energy obtained from your power company is converted to thermal energy of the food being cooked. In these and all other processes, the *total* energy—the sum of all energy present in all different form—remains the same. No exception has ever been found.

We'll use the energy idea throughout the rest of this book to study a tremendous range of physical phenomena. This idea will help you understand why a sweater keeps you warm, how a camera's flash unit can produce a short burst of light, and the meaning of Einstein's famous equation $E = mc^2$.

In this chapter, though, our concentration will be on mechanics. We'll learn about one important form of energy called *kinetic energy,* or energy of motion, and how it relates to the concept of *work*. We'll also consider *power,* which is the time rate of doing work. In Chapter 7 we'll expand the ideas of work and kinetic energy into a deeper understanding of the concepts of energy and the conservation of energy.

6.1 | Work

You'd probably agree that it's hard work to pull a heavy sofa across the room, to lift a stack of encyclopedias from the floor to a high shelf, or to push a stalled car off the road. Indeed, all of these examples agree with the everyday meaning of *work*—any activity that requires muscular or mental effort.

In physics, work has a much more precise definition. By making use of this definition we'll find that in any motion, no matter how complicated, the total work done on a particle by all forces that act on it equals the change in its *kinetic energy*—a quantity that's related to the particle's speed. This relationship holds even when the forces acting on the particle aren't constant, a situation that can be difficult or impossible to handle with the techniques you learned in Chapters 4 and 5. The ideas of work and kinetic energy enable us to solve problems in mechanics that we could not have attempted before.

We'll develop the relationship between work and kinetic energy in Section 6.2, and see what to do with nonconstant forces in Section 6.3. Meanwhile, let's see how work is defined and how to calculate work in a variety of situations involving *constant* forces. Even though we already know how to solve problems in which the forces are constant, the idea of work is still useful in such problems.

The three examples of work described above—pulling a sofa, lifting encyclopedias, and pushing a car—have something in common. In each case you do work by exerting a *force* on a body while that body *moves* from one place to another, that is, undergoes a *displacement*. You do more work if the force is greater (you pull harder on the sofa) or if the displacement is greater (you pull the sofa farther across the room).

The physicist's definition of work is based on these observations. Consider a body that undergoes a displacement of magnitude s along a straight line. (For now, we'll assume that any body we discuss can be treated as a particle so that we can ignore any rotation or changes in shape of the body.) While the body moves, a constant force $\vec{F}$ acts on it in the same direction as the displacement $\vec{s}$ (Fig. 6.1). We define the **work** W done by this constant force under these circumstances as the product of the force magnitude F and the displacement magnitude s:

$$W = Fs \quad \text{(constant force in direction of straight-line displacement)} \qquad (6.1)$$

The work done on the body is greater if either the force F or the displacement s is greater, in agreement with our observations above.

CAUTION Don't confuse W (work) with w (weight). Though the symbols may be almost the same, work and weight are different quantities.

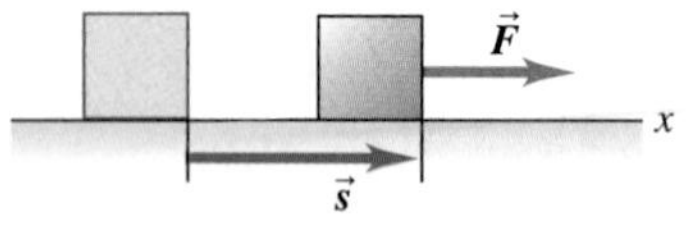

6.1 When a constant force $\vec{F}$ acts in the same direction as the displacement $\vec{s}$, the work done by the force is $W = Fs$.

The SI unit of work is the **joule** (abbreviated J, pronounced "jewel," and named in honor of the 19th-century English physicist James Prescott Joule). From

Eq. (6.1) we see that in any system of units, the unit of work is the unit of force multiplied by the unit of distance. In SI units the unit of force is the newton and the unit of distance is the meter, so one joule is equivalent to one *newton-meter* $(\mathrm{N \cdot m})$:

$$1 \text{ joule} = (1 \text{ newton})(1 \text{ meter}) \quad \text{or} \quad 1\ \mathrm{J} = 1\ \mathrm{N \cdot m}$$

In the British system the unit of force is the pound (lb), the unit of distance is the foot, and the unit of work is the *foot-pound* $(\mathrm{ft \cdot lb})$. The following conversions are useful:

$$1\ \mathrm{J} = 0.7376\ \mathrm{ft \cdot lb} \qquad 1\ \mathrm{ft \cdot lb} = 1.356\ \mathrm{J}$$

As an illustration of Eq. (6.1), think of a person pushing a stalled car. If he pushes the car through a displacement $\vec{s}$ with a constant force $\vec{F}$ in the direction of motion, the amount of work he does on the car is given by Eq. (6.1): $W = Fs$. But what if the person pushes at an angle ϕ with the car's displacement (Fig. 6.2)? Only the component of force in the direction of the car's motion would be effective in moving the car. (Other forces must act on the car so that it moves along $\vec{s}$, not in the direction of $\vec{F}$. We're interested only in the work that the person does, however, so we'll consider only the force he exerts.) When the force $\vec{F}$ and the displacement $\vec{s}$ have different directions, we take the component of $\vec{F}$ in the direction of the displacement $\vec{s}$, and we define the work as the product of this component and the magnitude of the displacement. The component of $\vec{F}$ in the direction of $\vec{s}$ is $F \cos \phi$, so

$$W = Fs \cos \phi \qquad \text{(constant force, straight-line displacement)} \qquad (6.2)$$

We are assuming that F and ϕ are constant during the displacement. If $\phi = 0$, so that $\vec{F}$ and $\vec{s}$ are in the same direction, then $\cos \phi = 1$ and we are back to Eq. (6.1).

Equation (6.2) has the form of the *scalar product* of two vectors, introduced in Section 1.10: $\vec{A} \cdot \vec{B} = AB \cos \phi$. You may want to review that definition. Using this, we can write Eq. (6.2) more compactly as

$$W = \vec{F} \cdot \vec{s} \qquad \text{(constant force, straight-line displacement)} \qquad (6.3)$$

CAUTION Here's an essential point: work is a *scalar* quantity, even though it's calculated by using two vector quantities (force and displacement). A 5-N force toward the east acting on a body that moves 6 m to the east does exactly the same work as a 5-N force toward the north acting on a body that moves 6 m to the north.

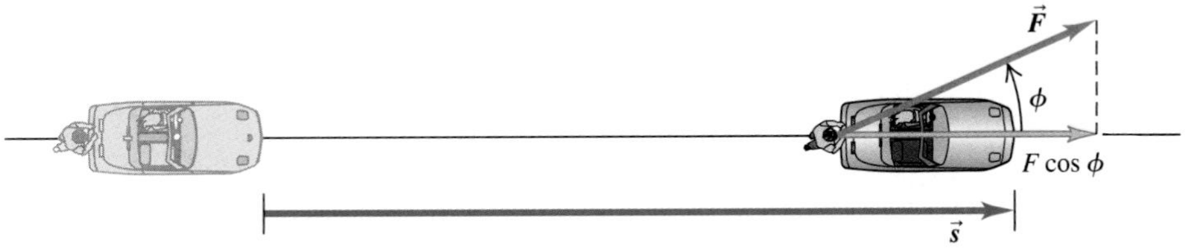

6.2 When a constant force $\vec{F}$ acts at an angle ϕ to the displacement $\vec{s}$, the work done by the force is $(F \cos \phi)\, s = Fs \cos \phi$.

Example 6.1 Work done by a constant force

a) Steve exerts a steady force of magnitude 210 N (about 47 lb) on the stalled car in Fig. 6.2 as he pushes it a distance of 18 m. The car also has a flat tire, so to make the car track straight Steve must push at an angle of 30° to the direction of motion. How much work does Steve do? b) In a helpful mood, Steve pushes a second stalled car with a steady force $\vec{F} = (160\text{ N})\hat{\imath} - (40\text{ N})\hat{\jmath}$. The displacement of the car is $\vec{s} = (14\text{ m})\hat{\imath} + (11\text{ m})\hat{\jmath}$. How much work does Steve do in this case?

SOLUTION

IDENTIFY and SET UP: In both parts (a) and (b), the target variable is the work W done by Steve. In each case the force is constant and the displacement is along a straight line, so we can use Eq. (6.2) or (6.3). The angle between $\vec{F}$ and $\vec{s}$ is given explicitly in part (a), so we can apply Eq. (6.2) directly. In part (b) the angle isn't given, so we're better off calculating the scalar product in Eq. (6.3) from the components of $\vec{F}$ and $\vec{s}$, as in Eq. (1.21).

EXECUTE: a) From Eq. (6.2),

$$W = Fs\cos\phi = (210\text{ N})(18\text{ m})\cos 30° = 3.3 \times 10^3\text{ J}$$

b) The components of $\vec{F}$ are $F_x = 160$ N and $F_y = -40$ N, and the components of $\vec{s}$ are $x = 14$ m and $y = 11$ m. Hence, using Eqs. (1.21) and (6.3),

$$\begin{aligned} W &= \vec{F}\cdot\vec{s} = F_x x + F_y y \\ &= (160\text{ N})(14\text{ m}) + (-40\text{ N})(11\text{ m}) \\ &= 1.8 \times 10^3\text{ J} \end{aligned}$$

EVALUATE: Our results show that 1 joule is a rather small amount of work.

In Example 6.1 the work done in pushing the cars was positive. But it's important to understand that work can also be negative or zero. This is the essential way in which work as defined in physics differs from the "everyday" definition of work. When the force has a component in the *same direction* as the displacement (ϕ between zero and 90°), $\cos\phi$ in Eq. (6.2) is positive and the work W is *positive* (Fig. 6.3a). When the force has a component *opposite* to the displacement (ϕ between 90° and 180°), $\cos\phi$ is negative and the work is *negative* (Fig. 6.3b). When the force is *perpendicular* to the displacement, $\phi = 90°$ and the work done by the force is *zero* (Fig. 6.3c). The cases of zero work and negative work bear closer examination, so let's look at some examples.

There are many situations in which forces act but do zero work. You might think it's "hard work" to hold a barbell motionless in the air for five minutes (Fig. 6.4). But in fact, you aren't doing any work at all on the barbell because there is no displacement. You get tired because the muscle fibers in your arm do work as they continually contract and relax. This is work done by one part of the arm exerting force on another part, however, *not* on the barbell. (We'll say more in Section 6.2 about work done by one part of a body on another part.) Even when

6.3 A constant force $\vec{F}$ can do positive, negative, or zero work depending on the angle between $\vec{F}$ and the displacement $\vec{s}$.

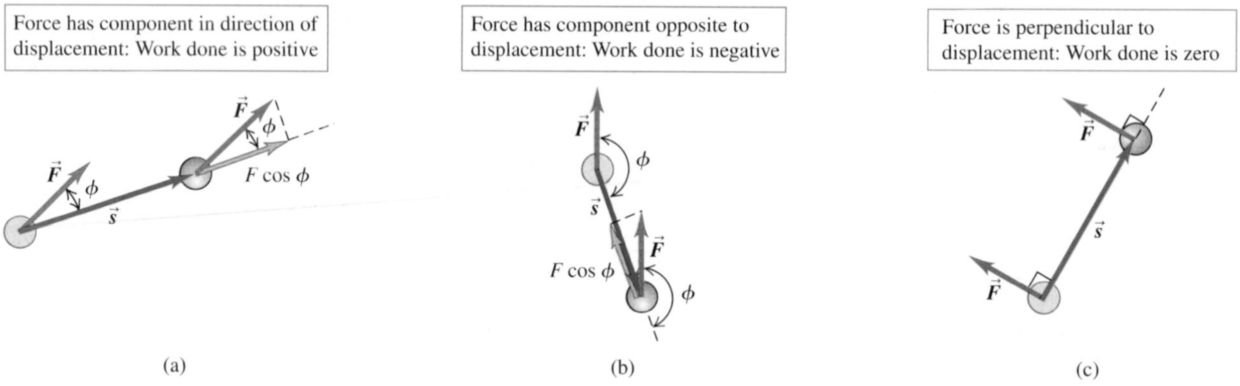

(a) Positive work (b) Zero work (c) Negative work

6.4 This weight lifter exerts an upward force $\vec{F}$ on the barbell at all times. (a) He does *positive* work on the barbell only when he is lifting it, so that the force is in the same direction as the displacement $\vec{s}$. (b) He does *zero* work while the barbell is stationary (the displacement is zero). (c) He does *negative* work while lowering the barbell ($\vec{F}$ is opposite to $\vec{s}$).

you walk with constant velocity on a level floor while carrying a book, you still do no work on it. The book has a displacement, but the (vertical) supporting force that you exert on the book has no component in the direction of the (horizontal) motion. Then $\phi = 90°$ in Eq. (6.2), and $\cos\phi = 0$. When a body slides along a surface, the work done on the body by the normal force is zero; and when a ball on a string moves in uniform circular motion, the work done on the ball by the tension in the string is also zero. In both cases the work is zero because the force has no component in the direction of motion.

What does it really mean to do *negative* work? The answer comes from Newton's third law of motion. When you catch a ball as in Fig. 6.5a, your hand and the ball move together with the same displacement $\vec{s}$ (Fig. 6.5b). The ball exerts a force $\vec{F}_{\text{B on H}}$ on your hand in the same direction as the hand's displacement, so the work done by the *ball* on your *hand* is positive. But by Newton's third law your hand exerts an equal and opposite force $\vec{F}_{\text{H on B}} = -\vec{F}_{\text{B on H}}$ on the ball (Fig. 6.5c). This force, which slows the ball to a stop, acts opposite to the ball's displacement. Thus the work done by your *hand* on the *ball* is negative. Because your hand and the ball have the same displacement, the work that your hand does on the ball is just the negative of the work that the ball does on your hand. In general, when one body does negative work on a second body, the second body does an equal amount of *positive* work on the first body.

CAUTION We always speak of work done *on* a particular body *by* a specific force. Always be sure to specify exactly what force is doing the work you are talking about. When you lift a book, you exert an upward force on the book and the book's displacement is upward, so the work done by the lifting force on the book is positive. But the work done by the *gravitational* force (weight) on a book being lifted is *negative* because the downward gravitational force is opposite to the upward displacement.

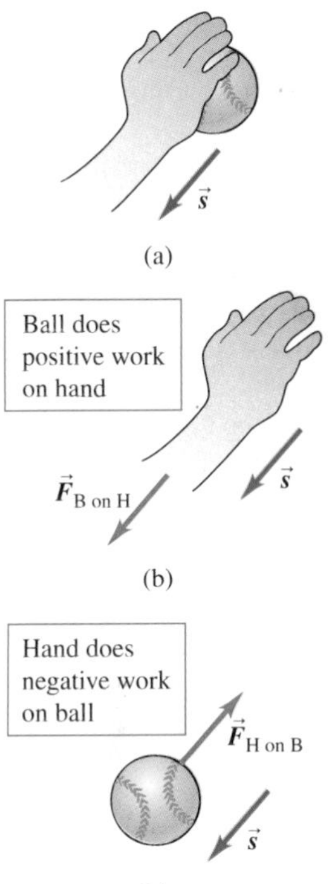

6.5 (a) When you catch a baseball, your hand and the ball have the same displacement $\vec{s}$. (b) The ball exerts a force $\vec{F}_{\text{B on H}}$ on your hand in the same direction as $\vec{s}$. (c) Your hand exerts an equal and opposite force $\vec{F}_{\text{H on B}} = -\vec{F}_{\text{B on H}}$ on the ball in the direction opposite to $\vec{s}$.

How do we calculate work when several forces act on a body? One way is to use Eq. (6.2) or (6.3) to compute the work done by each separate force. Then, because work is a scalar quantity, the *total* work W_{tot} done on the body by all the forces is the algebraic sum of the quantities of work done by the individual forces. An alternative way to find the total work W_{tot} is to compute the vector sum of the forces (that is, the net force) and then use this vector sum as $\vec{F}$ in Eq. (6.2) or (6.3).

ActivPhysics Online
5.1 Work Calculations

Example 6.2 Work done by several forces

Farmer Benton hitches his tractor to a sled loaded with firewood and pulls it a distance of 20 m along level ground (Fig. 6.6a). The total weight of sled and load is 14,700 N. The tractor exerts a constant 5000-N force at an angle of 36.9° above the horizontal, as shown in Fig. 6.6b. There is a 3500-N friction force opposing the sled's motion. Find the work done by each force acting on the sled and the total work done by all the forces.

SOLUTION

IDENTIFY: All of the forces are constant and the displacement is along a straight line, so we can calculate the work using the formulas given in this section. We'll find the net work in two ways: (1) by adding together the work done on the sled by each force and (2) by finding the amount of work done by the net force on the sled. We'll find the net force using the techniques described in Chapter 5.

SET UP: As we are working with forces, our first steps are to draw a free-body diagram showing all of the forces acting on the sled and to choose a coordinate system (Fig. 6.6b). For each of the four forces—weight, normal force, force of the tractor, and friction force—we know the angle between the displacement (which is in the positive x-direction) and the force. Hence we can calculate the work done by each force using Eq. (6.2).

To find the net force, we'll add the components of the four forces. Newton's second law tells us that because the sled's motion is purely horizontal, the net force will have only a horizontal component.

EXECUTE: The work W_w done by the weight is zero because its direction is perpendicular to the displacement. (The angle between the force of gravity and the displacement is 90°, and the cosine of the angle is zero.) For the same reason, the work W_n done by the normal force is also zero. So $W_w = W_n = 0$. (Incidentally, the normal force is *not* equal in magnitude to the weight; see Example 5.16 of Section 5.3, in which the free-body diagram is very similar to Fig. 6.6b.)

That leaves the force F_T exerted by the tractor and the friction force f. From Eq. (6.2) the work W_T done by the tractor is

$$W_T = F_T s \cos\phi = (5000\text{ N})(20\text{ m})(0.800) = 80{,}000\text{ N}\cdot\text{m} \\ = 80\text{ kJ}$$

The friction force $\vec{f}$ is opposite to the displacement, so for this force $\phi = 180°$ and $\cos\phi = -1$. The work W_f done by the friction force is

$$W_f = fs\cos 180° = (3500\text{ N})(20\text{ m})(-1) = -70{,}000\text{ N}\cdot\text{m} \\ = -70\text{ kJ}$$

The total work W_{tot} done on the sled by all forces is the *algebraic* sum of the work done by the individual forces:

$$W_{tot} = W_w + W_n + W_T + W_f = 0 + 0 + 80\text{ kJ} + (-70\text{ kJ}) \\ = 10\text{ kJ}$$

In the alternative approach, we first find the *vector* sum of all the forces (the net force) and then use it to compute the total work. The vector sum is best found by using components. From Fig. 6.6b,

$$\sum F_x = F_T\cos\phi + (-f) = (5000\text{ N})\cos 36.9° - 3500\text{ N} \\ = 500\text{ N}$$

$$\sum F_y = F_T\sin\phi + n + (-w) \\ = (5000\text{ N})\sin 36.9° + n - 14{,}700\text{ N}$$

We don't really need the second equation; we know that the y-component of force is perpendicular to the displacement, so it

6.6 (a) A tractor pulling a sled of firewood. (b) Free-body diagram for the sled and its load, treated as a particle.

does no work. Besides, there is no y-component of acceleration, so $\sum F_y$ has to be zero anyway. The total work is therefore the work done by the total x-component:

$$W_{tot} = (\sum \vec{F}) \cdot \vec{s} = (\sum F_x)s = (500\text{ N})(20\text{ m}) = 10{,}000\text{ J}$$
$$= 10\text{ kJ}$$

EVALUATE: We get the same result for W_{tot} with either method, as we should.

Note that the net force in the x-direction is *not* zero, and so the sled must accelerate as it moves. In Section 6.2 we'll return to this example and see how to use the concept of work to explore the sled's motion.

Conceptual Example 6.3 Total work when velocity is constant

An electron moves in a straight line toward the east with a constant speed of 8×10^7 m/s. It has electric, magnetic, and gravitational forces acting on it. Calculate the total work done on the electron during a 1-m displacement.

SOLUTION

The electron moves with constant velocity, so its acceleration is zero. By Newton's second law, the vector sum of forces is also zero. Therefore the total work done by all the forces (equal to the work done by the vector sum of all the forces) must be *zero*. Individual forces may do nonzero work, but that's not what the problem asks for.

Test Your Understanding

Suppose the barbell in Fig. 6.4 weighs 500 N. Calculate how much work the weight lifter does on the barbell (a) to raise it at constant speed a distance of 1.5 m; (b) to hold it stationary for 3.0 s at a height 1.5 m above the ground; (c) to lower it at constant speed a distance of 1.5 m.

6.2 | Work and Kinetic Energy

The total work done on a body by external forces is related to the body's displacement—that is, to changes in its position. But the total work is also related to changes in the *speed* of the body. To see this, consider Fig. 6.7, which shows several exam-

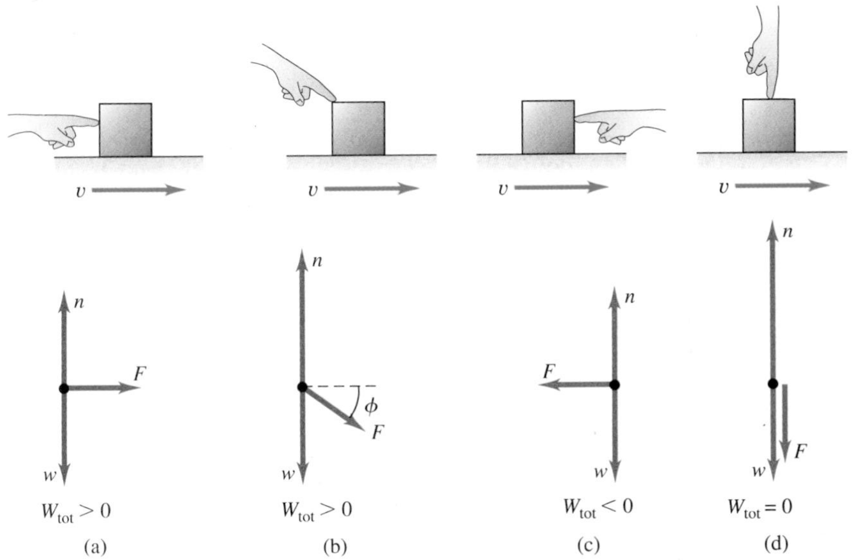

6.7 A block sliding on a frictionless table. (a) The net force causes the speed to increase and does positive work. (b) Again the net force causes the speed to increase and does positive work. (c) The net force opposes the displacement, causes the speed to decrease, and does negative work. (d) The net force is zero and does no work, and the speed is constant.

ples of a block sliding on a frictionless table. The forces acting on the block are its weight $\vec{w}$, the normal force $\vec{n}$, and the force $\vec{F}$ exerted on it by the hand.

In Fig. 6.7a the net force on the block is in the direction of its motion. From Newton's second law, this means that the block speeds up; from Eq. (6.1), this also means that the total work W_{tot} done on the block is positive. The total work is also positive in Fig. 6.7b, but only the component $F \cos \phi$ contributes to W_{tot}. The block again speeds up, and this same component $F \cos \phi$ is what causes the acceleration. The total work is *negative* in Fig. 6.7c because the net force opposes the displacement; in this case the block slows down. The net force is zero in Fig. 6.7d, so the speed of the block stays the same and the total work done on the block is zero. We can conclude that when a particle undergoes a displacement, it speeds up if $W_{tot} > 0$, slows down if $W_{tot} < 0$, and maintains the same speed if $W_{tot} = 0$.

Let's make these observations more quantitative. Consider a particle with mass m moving along the x-axis under the action of a constant net force with magnitude F directed along the positive x-axis (Fig. 6.1). The particle's acceleration is constant and given by Newton's second law, $F = ma_x$. Suppose the speed changes from v_1 to v_2 while the particle undergoes a displacement $s = x_2 - x_1$ from point x_1 to x_2. Using a constant-acceleration equation, Eq. (2.13), and replacing v_{0x} by v_1, v_x by v_2, and $(x - x_0)$ by s, we have

$$v_2^2 = v_1^2 + 2a_x s$$

$$a_x = \frac{v_2^2 - v_1^2}{2s}$$

When we multiply this equation by m and equate ma_x to the net force F, we find

$$F = ma_x = m\frac{v_2^2 - v_1^2}{2s}$$

and

$$Fs = \frac{1}{2}mv_2^2 - \frac{1}{2}mv_1^2 \tag{6.4}$$

The product Fs is the work done by the net force F and thus is equal to the total work W_{tot} done by all the forces acting on the particle. The quantity $\frac{1}{2}mv^2$ is called the **kinetic energy** K of the particle:

$$K = \frac{1}{2}mv^2 \quad \text{(definition of kinetic energy)} \tag{6.5}$$

Like work, the kinetic energy of a particle is a scalar quantity; it depends only on the particle's mass and speed, not its direction of motion. A car (viewed as a particle) has the same kinetic energy when going north at 10 m/s as when going east at 10 m/s. Kinetic energy can never be negative, and it is zero only when the particle is at rest.

We can now interpret Eq. (6.4) in terms of work and kinetic energy. The first term on the right side of Eq. (6.4) is $K_2 = \frac{1}{2}mv_2^2$, the final kinetic energy of the particle (that is, after the displacement). The second term is the initial kinetic energy, $K_1 = \frac{1}{2}mv_1^2$, and the difference between these terms is the *change* in kinetic energy. So Eq. (6.4) says that **the work done by the net force on a particle equals the change in the particle's kinetic energy:**

$$W_{tot} = K_2 - K_1 = \Delta K \quad \text{(work-energy theorem)} \tag{6.6}$$

This result is the **work-energy theorem.**

The work-energy theorem agrees with our observations about the block in Fig. 6.7. When W_{tot} is *positive;* K_2 is greater than K_1, the kinetic energy *increases,* and the particle is going faster at the end of the displacement than at the beginning. When W_{tot} is *negative,* the kinetic energy *decreases* and the speed is less after the displacement. When $W_{tot} = 0$, the initial and final kinetic energies K_1 and K_2 are the same and the speed is unchanged. We stress that the work-energy theorem by itself tells us only about changes in *speed,* not velocity, since the kinetic energy carries no information about the direction of motion.

From Eq. (6.4) or (6.6), kinetic energy and work must have the same units. Hence the joule is the SI unit of both work and kinetic energy (and, as we will see later, of all kinds of energy). To verify this, note that in SI units the quantity $K = \frac{1}{2}mv^2$ has units $\text{kg} \cdot (\text{m/s})^2$ or $\text{kg} \cdot \text{m}^2/\text{s}^2$; we recall that $1\ \text{N} = 1\ \text{kg} \cdot \text{m/s}^2$, so

$$1\ \text{J} = 1\ \text{N} \cdot \text{m} = 1\ (\text{kg} \cdot \text{m/s}^2) \cdot \text{m} = 1\ \text{kg} \cdot \text{m}^2/\text{s}^2$$

In the British system the unit of kinetic energy and of work is

$$1\ \text{ft} \cdot \text{lb} = 1\ \text{ft} \cdot \text{slug} \cdot \text{ft/s}^2 = 1\ \text{slug} \cdot \text{ft}^2/\text{s}^2$$

Because we used Newton's laws in deriving the work-energy theorem, we can use it only in an inertial frame of reference. The speeds that we use to compute the kinetic energies and the distances that we use to compute work *must* be measured in an inertial frame. Note also that the work-energy theorem is valid in *any* inertial frame, but the values of W_{tot} and $K_2 - K_1$ may differ from one inertial frame to another (because the displacement and speed of a body may be different in different frames).

We have derived the work-energy theorem for the special case of straight-line motion with constant forces, and in the following examples we'll apply it to this special case only. We'll find in the next section that the theorem is valid in general, even when the forces are not constant and the particle's trajectory is curved.

Problem Solving Strategy — Work and Kinetic Energy

IDENTIFY *the relevant concepts:* The work-energy theorem is extremely useful in situations where you want to relate a body's speed v_1 at one point in its motion to its speed v_2 at a different point. This approach is less useful for problems that involve time, such as finding the time it takes a body to go from point 1 to point 2. The reason is that the work-energy theorem, $W_{tot} = K_2 - K_1$, doesn't involve time at all. For problems that involve time, it's usually best to use the relationships among time, position, velocity, and acceleration described in Chapters 2 and 3.

SET UP *the problem using the following steps:*

1. Choose the initial and final positions of the body, and draw a free-body diagram showing all the forces that act on the body.
2. Choose a coordinate system. (If the motion is along a straight line, it's usually easiest to have both the initial and final positions lie along the x-axis.)
3. List the unknown and known quantities, and decide which unknowns are your target variables. In some cases the target variable will be the body's initial or final speed; in other cases it will be the magnitude of one of the forces acting on the body.

EXECUTE *the solution:* Calculate the work done by each force. If the force is constant and the displacement is a straight line, you can use Eq. (6.2) or (6.3). (Later in this chapter we'll see how to handle varying forces and curved trajectories.) Be sure to check the sign of the work for each force; it must be positive if the force has a component in the direction of the displacement, negative if it has a component opposite the displacement, and zero if the force and displacement are perpendicular.

Add the amounts of work done by each force to find the total work W_{tot}. Be careful with signs! Sometimes it may be easier to calculate the vector sum of the forces (the net force), then find the work done by the net force; this value is also equal to W_{tot}.

Write expressions for the initial and final kinetic energies, K_1 and K_2. Note that kinetic energy involves *mass,* not *weight;* if you are given the body's weight, you'll need to use the relationship $w = mg$ to find the mass.

Finally, use the relationship $W_{tot} = K_2 - K_1$ to solve for the target variable. Remember that the right-hand side of this equation is the *final* kinetic energy minus the *initial* kinetic energy, never the other way around.

EVALUATE *your answer:* Check whether your answer makes physical sense. A key item to remember is that kinetic energy $K = \frac{1}{2}mv^2$ can never be negative. If you come up with a negative value of K, you've made a mistake. Perhaps you interchanged the initial and final kinetic energies in $W_{tot} = K_2 - K_1$ or made a sign error in one of the work calculations.

Example 6.4 Using work and energy to calculate speed

Let's look again at the sled in Fig. 6.6 and the numbers at the end of Example 6.2. Suppose the initial speed v_1 is 2.0 m/s. What is the speed of the sled after it moves 20 m?

SOLUTION

IDENTIFY and SET UP: We'll use the work-energy theorem, since we are given the initial speed $v_1 = 2.0$ m/s and want to find the final speed v_2. Figure 6.8 shows the free-body diagram from Example 6.2. The motion is in the positive x-direction.

EXECUTE: We already calculated the total work done by all the forces in Example 6.2, where we found $W_{tot} = 10$ kJ. Hence the kinetic energy of the sled and its load must increase by 10 kJ.

To write expressions for the initial and final kinetic energies, we need the mass of the sled and load. We are given that the *weight* is 14,700 N, so the mass is

$$m = w/g = (14{,}700\ \text{N})/(9.8\ \text{m/s}^2) = 1500\ \text{kg}$$

Then the initial kinetic energy K_1 is

$$K_1 = \frac{1}{2}mv_1^2 = \frac{1}{2}(1500\ \text{kg})(2.0\ \text{m/s})^2 = 3000\ \text{kg}\cdot\text{m}^2/\text{s}^2$$
$$= 3000\ \text{J}$$

The final kinetic energy K_2 is

$$K_2 = \frac{1}{2}mv_2^2 = \frac{1}{2}(1500\ \text{kg})v_2^2$$

where v_2 is the unknown speed we want to find. Equation (6.6) gives

$$K_2 = K_1 + W_{tot} = 3000\ \text{J} + 10{,}000\ \text{J} = 13{,}000\ \text{J}$$

Setting these two expressions for K_2 equal, substituting $1\ \text{J} = 1\ \text{kg}\cdot\text{m}^2/\text{s}^2$, and solving for v_2, we find

$$v_2 = 4.2\ \text{m/s}$$

EVALUATE: The total work is positive, so the kinetic energy increases $(K_2 > K_1)$ and the speed increases $(v_2 > v_1)$.

This problem can also be done without the work-energy theorem. We can find the acceleration from $\sum\vec{F} = m\vec{a}$ and then use the equations of motion for constant acceleration to find v_2. Since the acceleration is along the x-axis,

$$a = a_x = \frac{\sum F_x}{m} = \frac{(5000\ \text{N})\cos 36.9° - 3500\ \text{N}}{1500\ \text{kg}}$$
$$= 0.333\ \text{m/s}^2$$

then

$$v_2^2 = v_1^2 + 2as = (2.0\ \text{m/s})^2 + 2(0.333\ \text{m/s}^2)(20\ \text{m})$$
$$= 17.3\ \text{m}^2/\text{s}^2$$
$$v_2 = 4.2\ \text{m/s}$$

This is the same result that we obtained with the work-energy approach, but there we avoided the intermediate step of finding the acceleration. You will find several other examples in this chapter and the next that *can* be done without using energy considerations but that are easier when energy methods are used. When a problem can be done by two different methods, doing it by both methods (as we did in this example) is a very good way to check your work.

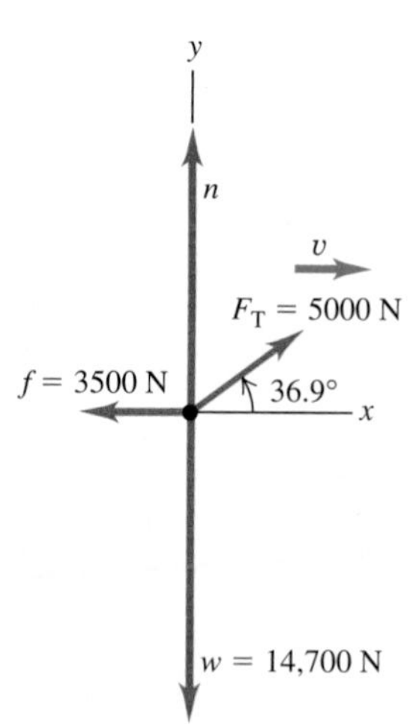

6.8 Free-body diagram for the sled and its load in Example 6.2.

Example 6.5 Forces on a hammerhead

In a pile driver, a steel hammerhead with mass 200 kg is lifted 3.00 m above the top of a vertical I-beam being driven into the ground (Fig. 6.9a). The hammer is then dropped, driving the I-beam 7.4 cm farther into the ground. The vertical rails that guide the hammerhead exert a constant 60-N friction force on the hammerhead. Use the work-energy theorem to find a) the speed of the hammerhead just as it hits the I-beam and b) the average force the hammerhead exerts on the I-beam. Ignore the effects of the air.

SOLUTION

IDENTIFY: This problem is an ideal candidate for the work-energy theorem, as it relates a body's speed at different locations and the forces acting on the body. In this problem there are *three* positions of interest: point 1, where the hammerhead starts from rest; point 2, where it first contacts the I-beam; and point 3, where the hammerhead comes to a halt (see Fig. 6.9a). We have two unknowns—the hammerhead's speed at point 2, and the force the I-beam exerts between points 2 and 3—we'll apply the work-energy theorem twice: once for the motion from 1 to 2 and once for the motion from 2 to 3.

SET UP: Figure 6.9b is a free-body diagram showing the vertical forces on the hammerhead as it falls from point 1 to point 2. (Because the displacement is vertical, we ignore any horizontal forces that may be present because they do no work.) For this part of the motion, our target variable is the hammerhead's speed v_2.

The free-body diagram in Fig. 6.9c shows the vertical forces on the hammerhead during the motion from point 2 to point 3. In addition to the forces shown in Fig. 6.9b, the I-beam exerts an upward normal force of magnitude n on the hammerhead. This force actually varies as the hammerhead comes to a halt, but for simplicity we'll treat n as a constant. Hence n represents the *average* value of this upward force during the motion. Our target variable for this part of the motion is the force that the *hammerhead* exerts on the I-beam; it is the reaction force to the normal force exerted by the I-beam, so by Newton's third law its magnitude is also n.

EXECUTE: a) From point 1 to point 2, the vertical forces are the downward weight $w = mg = (200\text{ kg})(9.8\text{ m/s}^2) = 1960\text{ N}$ and the upward friction force $f = 60\text{ N}$. Thus the net downward force is $w - f = 1900\text{ N}$. The displacement of the hammerhead from

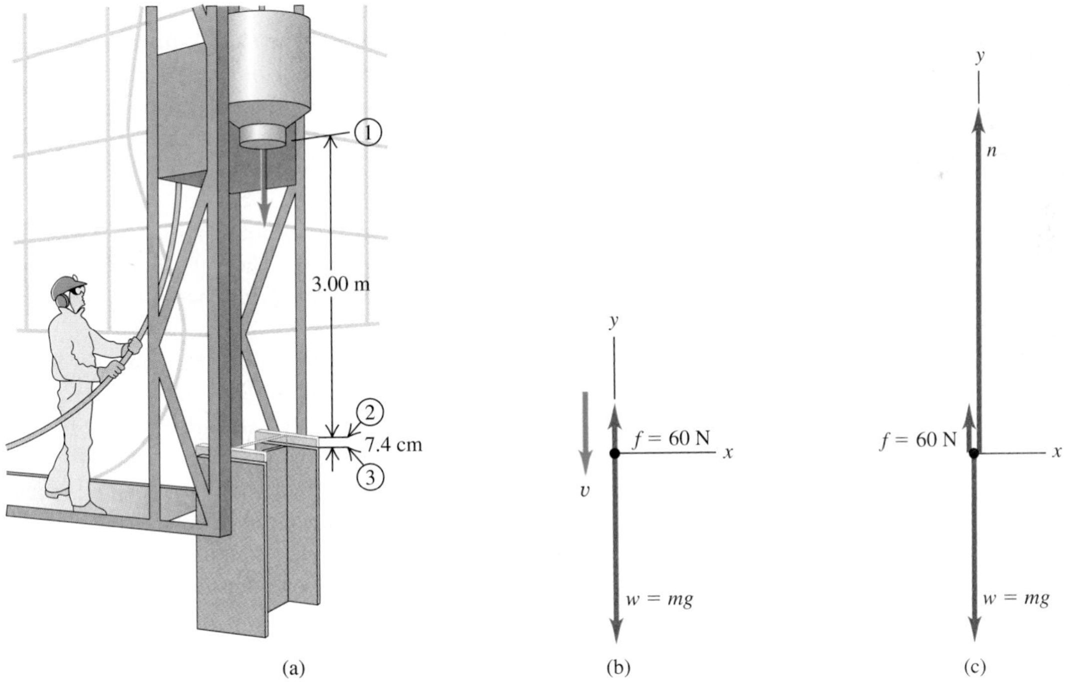

6.9 (a) A pile driver pounds an I-beam into the ground. (b) Free-body diagram for the hammerhead while falling. (c) Free-body diagram for the hammerhead while driving the I-beam. Vector lengths are not to scale.

point 1 to point 2 is downward and equal to $s_{12} = 3.00$ m. The total work done on the hammerhead as it moves from point 1 to point 2 is then

$$W_{tot} = (w - f)s_{12} = (1900\ \text{N})(3.00\ \text{m}) = 5700\ \text{J}$$

At point 1 the hammerhead is at rest, so its initial kinetic energy K_1 is zero. Equation (6.6) gives

$$W_{tot} = K_2 - K_1 = \frac{1}{2}mv_2^2 - 0$$

$$v_2 = \sqrt{\frac{2W_{tot}}{m}} = \sqrt{\frac{2(5700\ \text{J})}{200\ \text{kg}}} = 7.55\ \text{m/s}$$

This is the hammerhead's speed at point 2, just as it hits the beam.

b) As the hammerhead moves downward between points 2 and 3, the net downward force acting on it is $w - f - n$ (see Fig. 6.9c). The total work done on the hammerhead during this displacement is

$$W_{tot} = (w - f - n)s_{23}$$

The initial kinetic energy for this part of the motion is K_2, which from part (a) equals 5700 J (the total work done on the hammerhead as it moved from point 1 to point 2). The final kinetic energy is $K_3 = 0$, since the hammerhead ends at rest. Then, from the work-energy theorem,

$$W_{tot} = (w - f - n)s_{23} = K_3 - K_2$$

$$n = w - f - \frac{(K_3 - K_2)}{s_{23}}$$

$$= 1960\ \text{N} - 60\ \text{N} - \frac{(0\ \text{J} - 5700\ \text{J})}{0.074\ \text{m}}$$

$$= 79{,}000\ \text{N}$$

The downward force that the hammerhead exerts on the I-beam has this same magnitude, 79,000 N (about 9 tons)—more than 40 times the weight of the hammerhead.

EVALUATE: The total change in the hammerhead's kinetic energy during the whole process is zero; a relatively small net force does positive work over a large distance, and then a much larger net force does negative work over a much smaller distance. The same thing happens if you speed up your car gradually and then drive it into a brick wall. The very large force needed to reduce the kinetic energy to zero over a short distance is what does the damage to your car—and possibly to you.

The Meaning Of Kinetic Energy

Example 6.5 gives insight into the physical meaning of kinetic energy. The hammerhead is dropped from rest, and its kinetic energy when it hits the I-beam equals the total work done on it up to that point by the net force. This result is true in general: to accelerate a particle of mass m from rest (zero kinetic energy) up to a speed v, the total work done on it must equal the change in kinetic energy from zero to $K = \frac{1}{2}mv^2$;

$$W_{tot} = K - 0 = K$$

So *the kinetic energy of a particle is equal to the total work that was done to accelerate it from rest to its present speed* (Fig. 6.10). The definition $K = \frac{1}{2}mv^2$, Eq. (6.5), wasn't chosen at random; it's the *only* definition that agrees with this interpretation of kinetic energy.

In the second part of Example 6.5 the kinetic energy of the hammerhead did work on the I-beam and drove it into the ground. This gives us another interpretation of kinetic energy: *the kinetic energy of a particle is equal to the total work that particle can do in the process of being brought to rest.* This is why you pull your hand and arm backward when you catch a ball (Fig. 6.5). As the ball comes to rest, it does an amount of work (force times distance) on your hand equal to the ball's initial kinetic energy. By pulling your hand back, you maximize the distance over which the force acts and so minimize the force on your hand.

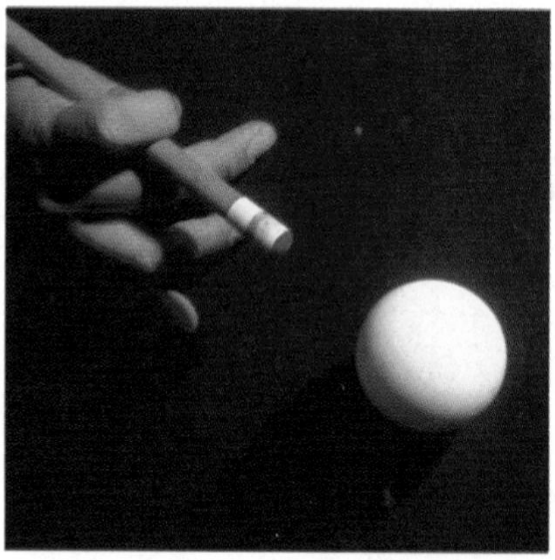

6.10 When a billiards player hits a cue ball at rest, the ball's kinetic energy after being hit is equal to the work that was done on it by the cue. The greater the force exerted by the cue and the greater the distance the ball moves while in contact with it, the greater the ball's kinetic energy.

Conceptual Example 6.6

Comparing kinetic energies

Two iceboats like the one in Example 5.6 (Section 5.2) hold a race on a frictionless horizontal lake (Fig. 6.11). The two iceboats have masses m and $2m$, respectively. Each iceboat has an identical sail, so the wind exerts the same constant force $\vec{F}$ on each iceboat. The two iceboats start from rest and cross the finish line a distance s away. Which iceboat crosses the finish line with greater kinetic energy?

SOLUTION

If you simply use the mathematical definition of kinetic energy, $K = \frac{1}{2}mv^2$, Eq. (6.5), the answer to this problem isn't immediately obvious. The iceboat of mass $2m$ has greater mass, so you might guess that the larger iceboat attains a greater kinetic energy at the finish line. But the smaller iceboat, of mass m, crosses the finish line with a greater speed, and you might guess that *this* iceboat has the greater kinetic energy. How can we decide?

The correct way to approach this problem is to remember that *the kinetic energy of a particle is just equal to the total work done to accelerate it from rest.* Both iceboats travel the same distance s, and only the horizontal force F in the direction of motion does work on either iceboat. Hence the total work done between the starting line and the finish line is the *same* for each iceboat, $W_{tot} = Fs$. At the finish line, each iceboat has a kinetic energy equal to the work W_{tot} done on it, because each iceboat started from rest. So both iceboats have the *same* kinetic energy at the finish line!

You might think this is a "trick" question, but it isn't. If you really understand the physical meanings of quantities such as kinetic energy, you can solve problems more easily and with better insight into the physics.

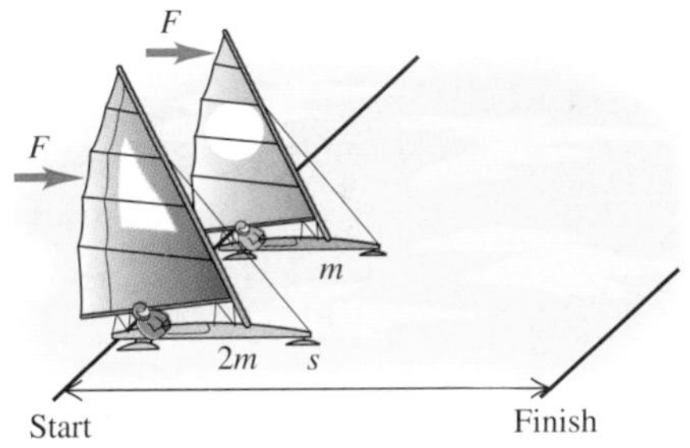

6.11 A race between iceboats.

Notice that we didn't need to say anything about how much time each iceboat took to reach the finish line. This is because the work-energy theorem makes no direct reference to time, only to displacement. In fact the iceboat of mass m takes less time to reach the finish line than does the larger iceboat of mass $2m$; we leave the calculation to you (Exercise 6.18).

Work and Kinetic Energy in Composite Systems

You may have noticed that in this section we've been careful to apply the work-energy theorem only to bodies that we can represent as *particles*—that is, as moving point masses. The reason is that new subtleties appear for more complex systems that have to be represented in terms of many particles with different motions. We can't go into these subtleties in detail in this chapter, but here's an example.

Consider a man standing on frictionless roller skates on a level surface, facing a rigid wall (Fig. 6.12). He pushes against the wall, setting himself in motion to the right. The forces acting on him are his weight $\vec{w}$, the upward normal forces $\vec{n}_1$ and $\vec{n}_2$ exerted by the ground on his skates, and the horizontal force $\vec{F}$ exerted on him by the wall. There is no vertical displacement, so $\vec{w}$, $\vec{n}_1$, and $\vec{n}_2$ do no work. The force $\vec{F}$ is the horizontal force that accelerates him to the right, but the parts of his body where that force is applied (the man's hands) do not move. Thus the force $\vec{F}$ also does no work. So where does the man's kinetic energy come from?

The difficulty is that it's simply not adequate to represent the man as a single point mass. For the motion to occur as we've described it, different parts of the man's body must have different motions; his hands are stationary against the wall while his torso is moving away from the wall. The various parts of his body interact with each other, and one part can exert forces and do work on another part. Therefore the *total* kinetic energy of this composite system of body parts can change, even though no work is done by forces applied by bodies (such as the wall) that are outside the system. This would not be possible with a system that can be represented as a single point particle. In Chapter 8 we'll consider further the motion of a collection of particles that interact with each other. We'll discover that just as for the man in this example, the total kinetic energy of such a system can change even when no work is done on any part of the system by anything outside it.

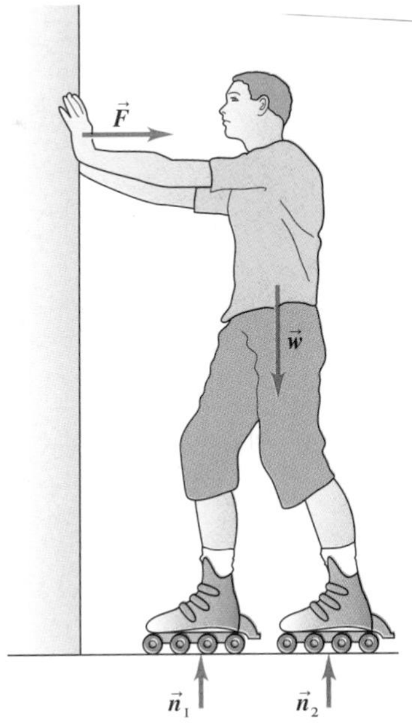

6.12 The external forces acting on a skater pushing off a wall. The work done by these forces is zero, but his kinetic energy changes nonetheless.

Test Your Understanding

A baseball (mass 0.145 kg) leaves the pitcher's hand moving horizontally at a speed of 40 m/s (144 km/h, or 89 mi/h). If his throwing motion moves the ball forward over a distance of 2.0 m, what average horizontal force did the pitcher exert on the ball during the throw?

6.3 | Work and Energy with Varying Forces

So far in this chapter we've considered work done by *constant forces* only. But what happens when you stretch a spring? The more you stretch it, the harder you have to pull, so the force you exert is *not* constant as the spring is stretched. We've also restricted our discussion to *straight-line* motion. You can think of many situations in which a force that varies in magnitude, direction, or both acts on a body moving along a curved path. We need to be able to compute the work done by the force in these more general cases. Fortunately, we'll find that the work-energy theorem holds true even when varying forces are considered and when the body's path is not straight.

To add only one complication at a time, let's consider straight-line motion with a force that is directed along the line but with an x-component F_x that may change as the body moves. For example, imagine a train moving on a straight track with the engineer constantly changing the locomotive's throttle setting or applying the brakes. Suppose a particle moves along the x-axis from point x_1 to x_2. Figure 6.13a is a graph of the x-component of force as a function of the particle's coordinate x. To find the work done by this force, we divide the total displacement into small segments Δx_a, Δx_b, and so on (Fig. 6.13b). We approximate the work done by the force during segment Δx_a as the average force F_a in that segment multiplied by the displacement Δx_a. We do this for each segment and then add the results for all the segments. The work done by the force in the total displacement from x_1 to x_2 is approximately

$$W = F_a\Delta x_a + F_b\Delta x_b + \cdots$$

As the number of segments becomes very large and the width of each becomes very small, this sum becomes (in the limit) the *integral* of F_x from x_1 to x_2:

$$W = \int_{x_1}^{x_2} F_x\, dx \qquad (6.7)$$

(varying x-component of force, straight-line displacement)

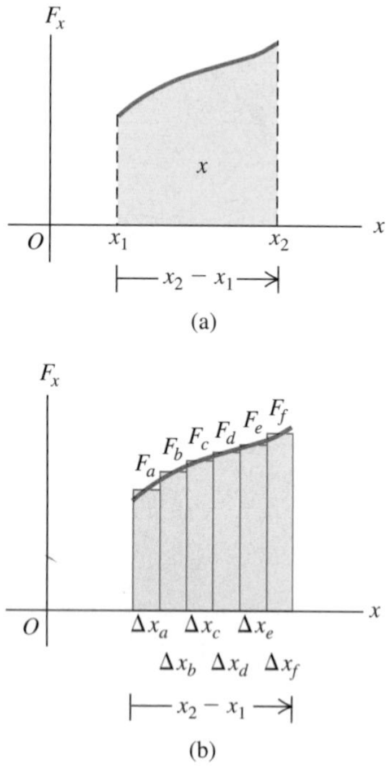

6.13 (a) Curve showing how a certain force F_x varies with x. (b) If the area is partitioned into small rectangles, the sum of their areas approximates the total work done during the displacement. The total area under the curve equals the work done by the force as the particle moves from x_1 to x_2.

Note that $F_a\,\Delta x_a$ represents the *area* of the first vertical strip in Fig. 6.13b and that the integral in Eq. (6.7) represents the area under the curve of Fig. 6.13a between x_1 and x_2. *On a graph of force as a function of position, the total work done by the force is represented by the area under the curve between the initial and final positions.* An alternative interpretation of Eq. (6.7) is that the work W equals the average force that acts over the entire displacement, multiplied by the displacement.

Equation (6.7) also applies if F_x, the x-component of the force, is constant. In that case, F_x may be taken outside the integral:

$$W = \int_{x_1}^{x_2} F_x\, dx = F_x \int_{x_1}^{x_2} dx = F_x(x_2 - x_1)$$

But $x_2 - x_1 = s$, the total displacement of the particle. So in the case of a constant force F, Eq. (6.7) says that $W = Fs$, in agreement with Eq. (6.1). The interpretation of work as the area under the curve of F_x as a function of x also holds for a constant force; $W = Fs$ is the area of a rectangle of height F and width s (Fig. 6.14).

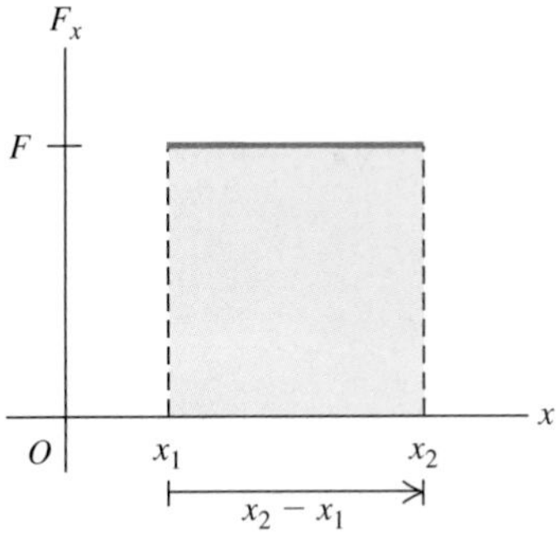

6.14 The work done by a constant force F in the x-direction as a particle moves from x_1 to x_2 is equal to the rectangular area under the graph of force versus displacement.

Now let's apply what we've learned to the stretched spring. To keep a spring stretched beyond its unstretched length by an amount x, we have to apply a force with magnitude F at each end (Fig. 6.15). If the elongation x is not too great, we find that the force we apply to the right-hand end has an x-component directly proportional to x:

$$F_x = kx \quad \text{(force required to stretch a spring)} \tag{6.8}$$

where k is a constant called the **force constant** (or spring constant) of the spring. Equation (6.8) shows that the units of k are force divided by distance, N/m in SI units and lb/ft in British units. A floppy toy spring such as a Slinky™ has a force constant of about 1 N/m; for the much stiffer springs in an automobile's suspension, k is about 10^5 N/m. The observation that elongation is directly proportional to force for elongations that are not too great was made by Robert Hooke in 1678 and is known as **Hooke's law.** It really shouldn't be called a "law," since it's a statement about a specific device and not a fundamental law of nature. Real springs don't always obey Eq. (6.8) precisely, but it's still a useful idealized model. We'll discuss Hooke's law more fully in Chapter 11.

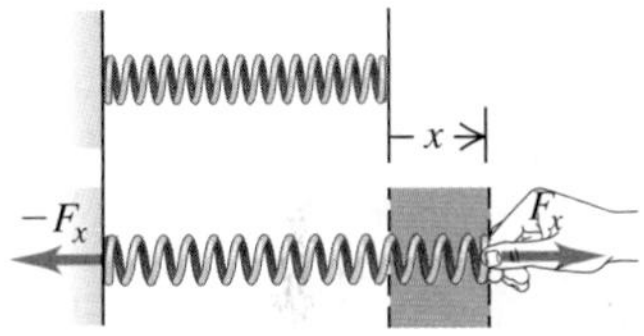

6.15 The force needed to stretch an ideal spring is proportional to its elongation: $F_x = kx$.

To stretch a spring, we must do work. We apply equal and opposite forces to the ends of the spring and gradually increase the forces. We hold the left end stationary, so the force we apply at this end does no work. The force at the moving end *does* do work. Figure 6.16 is a graph of F_x as a function of x, the elongation of the spring. The work done by this force when the elongation goes from zero to a maximum value X is

$$W = \int_0^X F_x\,dx = \int_0^X kx\,dx = \frac{1}{2}kX^2 \tag{6.9}$$

We can also obtain this result graphically. The area of the shaded triangle in Fig. 6.16, representing the total work done by the force, is equal to half the product of the base and altitude, or

$$W = \frac{1}{2}(X)(kX) = \frac{1}{2}kX^2$$

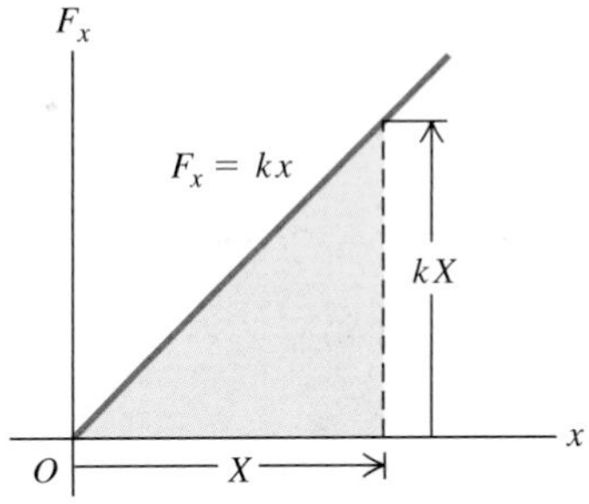

6.16 The work done to stretch a spring by a length X is equal to the triangular area under the graph of force versus displacement.

This equation also says that the work is the *average* force $kX/2$ multiplied by the total displacement X. We see that the total work is proportional to the *square* of the final elongation X. To stretch an ideal spring by 2 cm, you must do four times as much work as is needed to stretch it by 1 cm.

Equation (6.9) assumes that the spring was originally unstretched. If initially the spring is already stretched a distance x_1, the work we must do to stretch it to a greater elongation x_2 is

$$W = \int_{x_1}^{x_2} F_x\,dx = \int_{x_1}^{x_2} kx\,dx = \frac{1}{2}kx_2^2 - \frac{1}{2}kx_1^2 \tag{6.10}$$

If the spring has spaces between the coils when it is unstretched, then it can also be compressed, and Hooke's law holds for compression as well as stretching. In this case the force F and displacement x are in the opposite directions from those shown in Fig. 6.15, and so F_x and x in Eq. (6.8) are both negative. Since both F_x and x are reversed, the force again is in the same direction as the displacement, and the work done by F_x is again positive. So the total work is still given by Eq. (6.9) or (6.10), even when X is negative or either or both of x_1 and x_2 are negative.

CAUTION Note that the work given by Eq. (6.10) is the work that *you* must do *on* a spring to change its length. For example, if you stretch a spring that's originally relaxed, then $x_1 = 0$, $x_2 > 0$, and $W > 0$. That's because the force you apply to one end of the spring is in the same direction as the displacement and the work you do is positive. By contrast, the work that the *spring* does on whatever it's attached to is given by the *negative* of Eq. (6.10). Thus, as you pull on the spring, the spring does negative work on you. Paying careful attention to the sign of work will eliminate confusion later on!

Example 6.7 Work done on a spring scale

A woman weighing 600 N steps on a bathroom scale containing a stiff spring (Fig. 6.17). In equilibrium the spring is compressed 1.0 cm under her weight. Find the force constant of the spring and the total work done on it during the compression.

SOLUTION

IDENTIFY and SET UP: In equilibrium the upward force exerted by the spring balances the downward force of the woman's weight. We'll use this principle and Eq. (6.8) to determine the force constant k, and we'll use Eq. (6.10) to calculate the work W that the woman does on the spring to compress it. We take positive values of x to correspond to elongation, so that the displacement of the spring and the force that the woman exerts on it are both negative.

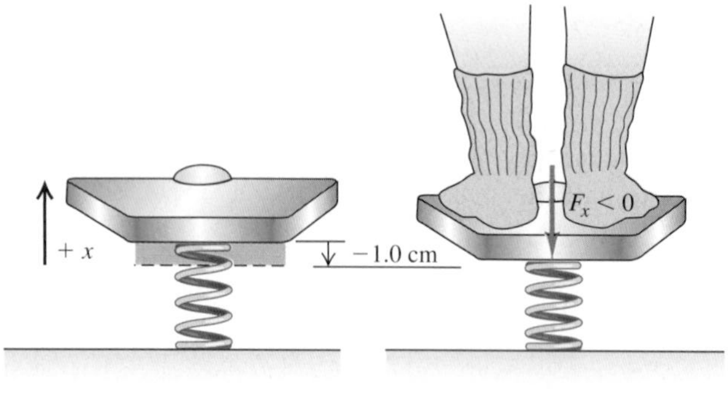

6.17 Compressing a spring in a bathroom scale. A negative force causes a negative displacement but does a positive amount of work on the spring.

EXECUTE: With our choice of coordinates, the top of the spring is displaced by $x = -1.0\text{ cm} = -0.010\text{ m}$ and the force the woman exerts on the spring is $F_x = -600\text{ N}$. From Eq. (6.8) the force constant is

$$k = \frac{F_x}{x} = \frac{-600\text{ N}}{-0.010\text{ m}} = 6.0 \times 10^4\text{ N/m}$$

Then, using $x_1 = 0$ and $x_2 = -0.010$ m in Eq. (6.10),

$$W = \frac{1}{2}kx_2^2 - \frac{1}{2}kx_1^2$$
$$= \frac{1}{2}(6.0 \times 10^4\text{ N/m})(-0.010\text{ m})^2 - 0 = 3.0\text{ J}$$

EVALUATE: The applied force and the displacement of the end of the spring were in the same direction, so the work done must have been positive—which is just what we found by calculation. Our arbitrary choice of the positive direction has no effect on the answer for W. (You can test this by taking the positive x-direction to correspond to compression. You'll get the same values for k and for W.)

Work-Energy Theorem for Straight-Line Motion, Varying Forces

In Section 6.2 we derived the work-energy theorem, $W_{tot} = K_2 - K_1$, for the special case of straight-line motion with a constant net force. We can now prove that this theorem is true even when the force varies with position. As in Section 6.2, let's consider a particle that undergoes a displacement x while being acted on by a net force with x-component F, which we now allow to vary. Just as in Fig. 6.13, we divide the total displacement x into a large number of small segments Δx. We can apply the work-energy theorem, Eq. (6.6), to each segment because the value of F in each small segment is approximately constant. The change in kinetic energy in segment Δx_a is equal to the work $F_a \Delta x_a$, and so on. The total change of kinetic energy is the sum of the changes in the individual segments, and thus is equal to the total work done on the particle during the entire displacement. So $W_{tot} = \Delta K$ holds for varying forces as well as for constant ones.

Here's an alternative derivation of the work-energy theorem for a force that may vary with position. It involves making a change of variable from x to v_x in the work integral. As a preliminary, we note that the acceleration a of the particle can be expressed in various ways, using $a_x = dv_x/dt$, $v_x = dx/dt$, and the chain rule for derivatives:

$$a_x \frac{dv_x}{dt} = \frac{dv_x}{dx}\frac{dx}{dt} = v_x \frac{dv_x}{dx} \tag{6.11}$$

Using this result, Eq. (6.7) tells us that the total work done by the *net* force F_x is

$$W_{tot} = \int_{x_1}^{x_2} F_x\,dx = \int_{x_1}^{x_2} ma_x\,dx = \int_{x_1}^{x_2} mv_x \frac{dv_x}{dx}\,dx \tag{6.12}$$

Now $(dv_x/dx)\,dx$ is the change in velocity dv_x during the displacement dx, so in Eq. (6.12) we can substitute dv_x for $(dv_x/dx)\,dx$. This changes the integration variable from x to v_x, so we change the limits from x_1 and x_2 to the corresponding velocities v_1 and v_2 at these points. This gives us

$$W_{tot} = \int_{v_1}^{v_2} mv_x\,dv_x$$

The integral of $v_x\,dv_x$ is just $v_x^2/2$. Substituting the upper and lower limits, we finally find

$$W_{tot} = \frac{1}{2}mv_2^2 - \frac{1}{2}mv_1^2 \tag{6.13}$$

This is the same result as Eq. (6.6) but without the assumption that the net force F is constant. So the work-energy theorem is valid even when F varies during the displacement.

Example 6.8 Motion with a varying force

An air-track glider of mass 0.100 kg is attached to the end of a horizontal air track by a spring with force constant 20.0 N/m (Fig. 6.18a). Initially the spring is unstretched and the glider is moving at 1.50 m/s to the right. Find the maximum distance d that the glider moves to the right a) if the air track is turned on so that there is no friction, and b) if the air is turned off so that there is kinetic friction with coefficient $\mu_k = 0.47$.

SOLUTION

IDENTIFY: The force exerted by the spring is not constant, so we *cannot* use the constant-acceleration formulas of Chapter 2 to solve this problem. Instead, we'll use the work-energy theorem, which involves the distance moved (our target variable) through the formula for work.

SET UP: Figures 6.18b and 6.18c show the free-body diagrams for the glider without friction and with friction, respectively. We choose the positive x-direction to be to the right (in the direction of the glider's motion), with $x = 0$ at the glider's initial position (where the spring is relaxed) and $x = d$ (the target variable) at the position where the glider stops. In both cases the motion is purely horizontal, so only the horizontal forces do work. Note that

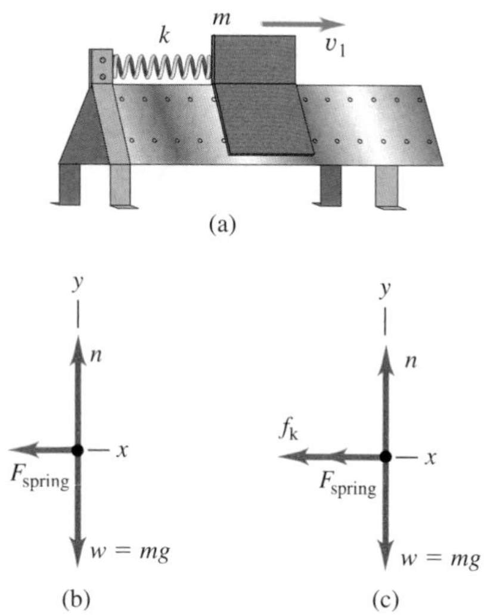

6.18 (a) A glider attached to an air track by a spring. (b) Free-body diagram for the glider without friction. (c) Free-body diagram for the glider with kinetic friction.

Eq. (6.10) gives the work done *on* the spring as it stretches, but to use the work-energy theorem we need the work done *by* the spring *on* the glider—which is the negative of Eq. (6.10).

EXECUTE: a) As the glider moves from $x_1 = 0$ to $x_2 = d$, it does an amount of work on the spring given by Eq. (6.10): $W = \frac{1}{2}kd^2 - \frac{1}{2}k(0)^2 = \frac{1}{2}kd^2$. The amount of work that the spring does on the glider is the negative of this value, or $-\frac{1}{2}kd^2$. The spring stretches until the glider comes instantaneously to rest, so the final kinetic energy K_2 is zero. The initial kinetic energy is $\frac{1}{2}mv_1^2$, where $v_1 = 1.50$ m/s is the glider's initial speed. Using the work-energy theorem, we find

$$-\frac{1}{2}kd^2 = 0 - \frac{1}{2}mv_1^2$$

so the distance the glider moves is

$$d = v_1\sqrt{\frac{m}{k}} = (1.50\text{ m/s})\sqrt{\frac{0.100\text{ kg}}{20.0\text{ N/m}}}$$
$$= 0.106\text{ m} = 10.6\text{ cm}$$

The stretched spring subsequently pulls the glider back to the left, so the glider is at rest only instantaneously.

b) If the air is turned off, we must also include the work done by the constant force of kinetic friction. The normal force n is equal in magnitude to the weight of the glider, since the track is horizontal and there are no other vertical forces. The magnitude of the kinetic friction force is then $f_k = \mu_k n = \mu_k mg$. The friction force is directed opposite to the displacement, so the work done by friction is

$$W_{fric} = f_k d\cos 180° = -f_k d = -\mu_k mgd$$

The total work is the sum of W_{fric} and the work done by the spring, $-\frac{1}{2}kd^2$. Hence

$$-\mu_k mgd - \frac{1}{2}kd^2 = 0 - \frac{1}{2}mv_1^2$$

$$-(0.47)(0.100\text{ kg})(9.8\text{ m/s}^2)d - \frac{1}{2}(20.0\text{ N/m})d^2$$

$$= -\frac{1}{2}(0.100\text{ kg})(1.50\text{ m/s})^2$$

$$(10.0\text{ N/m})d^2 + (0.461\text{ N})d - (0.113\text{ N}\cdot\text{m}) = 0$$

This is a quadratic equation for d. The solutions are

$$d = \frac{-(0.461\text{ N}) \pm \sqrt{(0.461\text{ N})^2 - 4(10.0\text{ N/m})(-0.113\text{ N}\cdot\text{m})}}{2(10.0\text{ N/m})}$$

$$= 0.086\text{ m} \quad\text{or}\quad -0.132\text{ m}$$

We have used d as the symbol for a positive displacement, so only the positive value of d makes sense. Thus with friction the glider moves a distance

$$d = 0.086\text{ m} = 8.6\text{ cm}$$

EVALUATE: With friction present, the glider goes a shorter distance and the spring stretches less, as you might expect. Again the glider stops instantaneously, and again the spring force pulls the glider to the left; whether it moves or not depends on how great the *static* friction force is. How large would the coefficient of static friction μ_s have to be to keep the glider from springing back to the left?

Work-Energy Theorem for Motion Along a Curve

We can generalize our definition of work further to include a force that varies in direction as well as magnitude, and a displacement that lies along a curved path. Suppose a particle moves from point P_1 to P_2 along a curve, as shown in Fig. 6.19a. We divide the portion of the curve between these points into many infinitesimal vector displacements, and we call a typical one of these $d\vec{l}$. Each $d\vec{l}$ is tangent to the path at its position. Let $\vec{F}$ be the force at a typical point along the path, and let ϕ be the angle between $\vec{F}$ and $d\vec{l}$ at this point. Then the small element of work dW done on the particle during the displacement $d\vec{l}$ may be written as

$$dW = F\cos\phi\, dl = F_{\parallel}\, dl = \vec{F}\cdot d\vec{l}$$

where $F_{\parallel} = F\cos\phi$ is the component of $\vec{F}$ in the direction parallel to $d\vec{l}$ (Fig. 6.19b). The total work done by $\vec{F}$ on the particle as it moves from P_1 to P_2 is then

$$W = \int_{P_1}^{P_2} F\cos\phi\, dl = \int_{P_1}^{P_2} F_{\parallel}\, dl = \int_{P_1}^{P_2} \vec{F}\cdot d\vec{l} \qquad (6.14)$$

(work done on a curved path)

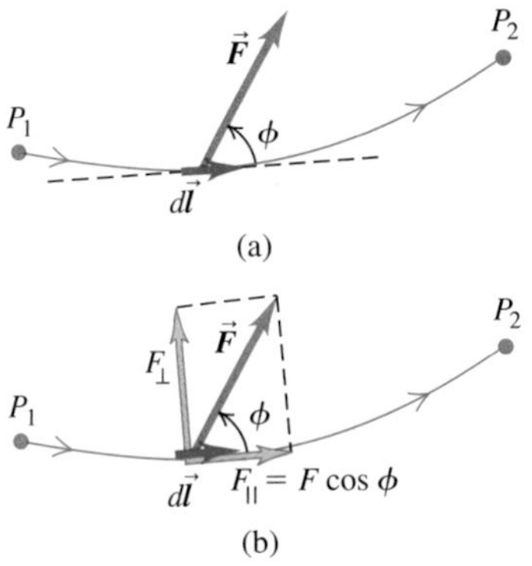

6.19 (a) A particle moves along a curved path from point P_1 to P_2, acted on by a force $\vec{F}$ that varies in magnitude and direction. During an infinitesimal displacement $d\vec{l}$ (a small segment of the path), the work dW done by the force is given by $dW = \vec{F}\cdot d\vec{l} = F\cos\phi\, dl$. (b) The force contributing to the work is the component of force parallel to the displacement, $F_{\parallel} = F\cos\phi$.

We can now show that the work-energy theorem, Eq. (6.6), holds true even with varying forces and a displacement along a curved path. The force $\vec{F}$ is essentially constant over any given infinitesimal segment $d\vec{l}$ of the path, so we can apply the work-energy theorem for straight-line motion to that segment. Thus the change in the particle's kinetic energy K over that segment equals the work $dW = F_{\parallel}\, dl = \vec{F}\cdot d\vec{l}$ done on the particle. Adding up these infinitesimal quantities of work from all the segments along the whole path gives the total work done, Eq. (6.14), which equals the total change in kinetic energy over the whole path. So $W_{tot} = \Delta K = K_2 - K_1$ is true *in general,* no matter what the path and no matter what the character of the forces. This can be proved more rigorously by using steps like those in Eqs. (6.11) through (6.13) (see Challenge Problem 6.104).

Note that only the component of the net force parallel to the path, $F_{\parallel}$, does work on the particle, so only this component can change the speed and kinetic energy of the particle. The component perpendicular to the path, $F_{\perp} = F\sin\phi$, has no effect on the particle's speed; it only acts to change the particle's direction.

The integral in Eq. (6.14) is called a *line integral.* To evaluate this integral in a specific problem, we need some sort of detailed description of the path and of the way in which $\vec{F}$ varies along the path. We usually express the line integral in terms of some scalar variable, as in the following example.

Example 6.9 Motion on a curved path I

At a family picnic you are appointed to push your obnoxious cousin Throckmorton in a swing (Fig. 6.20a). His weight is w, the length of the chains is R, and you push Throcky until the chains make an angle θ_0 with the vertical. To do this, you exert a varying horizontal force $\vec{F}$ that starts at zero and gradually increases just enough so that Throcky and the swing move very slowly and remain very nearly in equilibrium. What is the total work done on Throcky by all forces? What is the work done by the tension T in the chains? What is the work you do by exerting the force $\vec{F}$? (Neglect the weight of the chains and seat.)

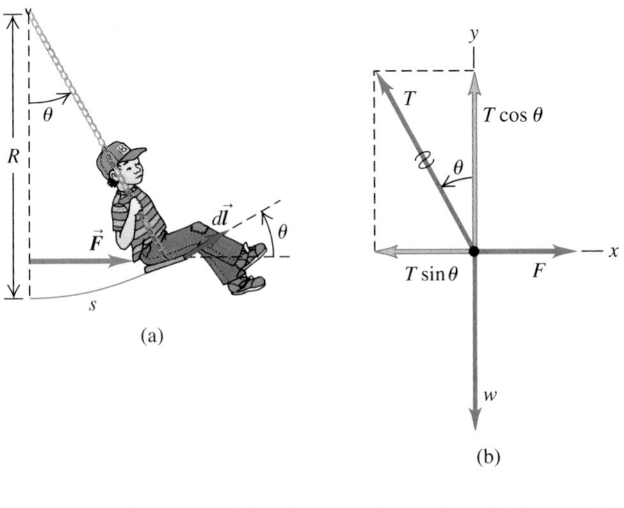

6.20 (a) Pushing cousin Throckmorton in a swing. (b) Free-body diagram for Throcky, considered as a particle and neglecting the small weight of chains and seat.

SOLUTION

IDENTIFY and SET UP: The motion is along a curve, so we will use Eq. (6.14) to calculate the work done by the tension force, by the force $\vec{F}$, and by the net force. Figure 6.20b shows the free-body diagram as well as the coordinate system. We have replaced the tensions in the two chains with a single tension T.

EXECUTE: There are two ways to find the total work done during the motion: (1) by calculating the work done by each force and then adding the quantities of work together, and (2) by calculating the work done by the net force. The second approach is far easier, because in this situation Throcky is in equilibrium at every point. Hence the net force on him is zero, so the integral of the net force in Eq. (6.14) is zero and the total work done on him by all forces is zero.

It's also easy to calculate the work done by the chain tension on Throcky, because the chain tension is perpendicular to the direction of motion at all points along the path. Hence at all points the angle between the chain tension and the displacement vector $d\vec{l}$ is 90° and the scalar product in Eq. (6.14) is zero. Thus the chain tension does zero work.

To compute the work done by $\vec{F}$, we need to know how this force varies with the angle θ. Throcky is in equilibrium at every point, so from $\sum F_x = 0$ we get

$$F + (-T\sin\theta) = 0$$

and from $\sum F_y = 0$ we find

$$T\cos\theta + (-w) = 0$$

By eliminating T from these two equations, we obtain

$$F = w\tan\theta$$

The point where $\vec{F}$ is applied swings through the arc s. The arc length s equals the radius R of the circular path multiplied by the length θ (in radians), so $s = R\theta$. Therefore the displacement $d\vec{l}$ corresponding to a small change of angle $d\theta$ has a magnitude $dl = ds = R\,d\theta$. The work done by $\vec{F}$ is

$$W = \int \vec{F}\cdot d\vec{l} = \int F\cos\theta\,ds$$

Now we express everything in terms of the varying angle θ:

$$W = \int_0^{\theta_0} (w\tan\theta)\cos\theta\,(R\,d\theta) = wR\int_0^{\theta_0}\sin\theta\,d\theta$$
$$= wR\,(1 - \cos\theta_0)$$

EVALUATE: If $\theta_0 = 0$, there is no displacement; in that case, $\cos\theta_0 = 1$ and $W = 0$, as we should expect. If $\theta_0 = 90°$, then $\cos\theta_0 = 0$ and $W = wR$. In that case the work you do is the same as if you had lifted Throcky straight up a distance R with a force equal to his weight w. In fact, the quantity $R\,(1 - \cos\theta_0)$ is the increase in his height above the ground during the displacement, so for any value of θ_0 the work done by force $\vec{F}$ is the change in height multiplied by the weight. This is an example of a more general result that we'll prove in Section 7.1.

Example 6.10 Motion on a curved path II

In Example 6.9 the infinitesimal vector displacement $d\vec{l}$ shown in Fig. 6.20a has a magnitude of ds, an x-component of $ds\cos\theta$, and a y-component of $ds\sin\theta$. Hence we can write $d\vec{l} = \hat{\imath}ds\cos\theta + \hat{\jmath}ds\sin\theta$. Use this expression and Eq. (6.14) to calculate the work done during the motion by the chain tension, by the force of gravity, and by the force $\vec{F}$.

SOLUTION

IDENTIFY and SET UP: We use the same free-body diagram, Fig. 6.20b, as in Example 6.9. The only difference is that here we calculate the scalar product in Eq. (6.14) by using Eq. (1.21) for the scalar product in terms of components.

EXECUTE: From Fig. 6.20b, we can write the three forces in terms of unit vectors:

$$\vec{T} = \hat{\imath}(-T\sin\theta) + \hat{\jmath}T\cos\theta$$
$$\vec{w} = \hat{\jmath}(-w)$$
$$\vec{F} = \hat{\jmath}F$$

To use Eq. (6.14), we must calculate the scalar product of each of these forces with $d\vec{l}$. Using Eq. (1.21),

$$\vec{T}\cdot d\vec{l} = (-T\sin\theta)(ds\cos\theta) + (T\cos\theta)(ds\sin\theta) = 0$$
$$\vec{w}\cdot d\vec{l} = (-w)(ds\sin\theta) = -w\sin\theta\,ds$$
$$\vec{F}\cdot d\vec{l} = F(ds\cos\theta) = F\cos\theta\,ds$$

Since $\vec{T}\cdot d\vec{l} = 0$, the integral of this quantity is zero and the work done by the chain tension is zero (just as we found in Example 6.9). Using $ds = Rd\theta$ as in Example 6.9, the work done by the force of gravity is

$$\int \vec{w}\cdot d\vec{l} = \int(-w\sin\theta)\,Rd\theta) = -wR\int_0^{\theta_0}\sin\theta\,d\theta$$
$$= -wR(1 - \cos\theta_0)$$

The work done by gravity is negative because gravity pulls down while Throcky moves upward. Finally, the work done by the force $\vec{F}$ is the integral $\int\vec{F}\cdot d\vec{l} = \int F\cos\theta\,ds$, which we calculated in Example 6.9; the answer is $+wR(1 - \cos\theta_0)$.

EVALUATE: As a check on our answers, note that the sum of all three quantities of work is zero. This is just what we concluded in Example 6.9 using the work-energy theorem.

The method of components is often the most convenient way to calculate scalar products. Use it when it will make your life easier!

Test Your Understanding

A 1000-kg car being tested for crashworthiness is driven at 20 m/s into a giant spring of force constant 1.6×10^4 N/m. How much work does the car do on the spring as it comes to a halt? How much work does the spring do on the car? How much does the spring compress? Ignore friction.

6.4 | Power

The definition of work makes no reference to the passage of time. If you lift a barbell weighing 400 N through a vertical distance of 0.5 m at constant velocity, you do $(400\ \text{N})(0.5\ \text{m}) = 200$ J of work whether it takes you 1 second, 1 hour, or 1 year to do it. But often we need to know how quickly work is done. We describe this in terms of *power*. In ordinary conversation the word "power" is often synonymous with "energy" or "force." In physics we use a much more precise definition: **power** is the time *rate* at which work is done. Like work and energy, power is a scalar quantity.

When a quantity of work ΔW is done during a time interval Δt, the average work done per unit time or **average power** P_{av} is defined to be

$$P_{\text{av}} = \frac{\Delta W}{\Delta t} \quad \text{(average power)} \tag{6.15}$$

The rate at which work is done might not be constant. Even when it varies, we can define **instantaneous power** P as the limit of the quotient in Eq. (6.15) as Δt approaches zero:

$$P = \lim_{\Delta t \to 0} \frac{\Delta W}{\Delta t} = \frac{dW}{dt} \quad \text{(instantaneous power)} \tag{6.16}$$

The SI unit of power is the **watt** (W), named for the English inventor James Watt. One watt equals one joule per second $(1\ \text{W} = 1\ \text{J/s})$. The kilowatt $(1\ \text{kW} = 10^3\ \text{W})$ and the megawatt $(1\ \text{MW} = 10^6\ \text{W})$ are also commonly used. In the British system, work is expressed in foot-pounds, and the unit of power is the foot-pound per second. A larger unit called the *horsepower* (hp) is also used:

$$1\ \text{hp} = 550\ \text{ft} \cdot \text{lb/s} = 33{,}000\ \text{ft} \cdot \text{lb/min}$$

That is, a 1-hp motor running at full load does 33,000 ft · lb of work every minute. A useful conversion factor is

$$1\ \text{hp} = 746\ \text{W} = 0.746\ \text{kW}$$

or 1 horsepower equals about $\frac{3}{4}$ of a kilowatt.

The watt is a familiar unit of *electrical* power; a 100-W light bulb converts 100 J of electrical energy into light and heat each second. But there's nothing inherently electrical about a watt or a kilowatt. A light bulb could be rated in horsepower, and some automobile manufacturers rate their engines in kilowatts rather than horsepower.

The units of power can be used to define new units of work or energy. The *kilowatt-hour* (kWh) is the usual commercial unit of electrical energy. One kilowatt-hour is the total work done in 1 hour (3600 s) when the power is 1 kilowatt $(10^3\ \text{J/s})$, so

$$1\ \text{kWh} = (10^3\ \text{J/s})(3600\ \text{s}) = 3.6 \times 10^6\ \text{J} = 3.6\ \text{MJ}$$

The kilowatt-hour is a unit of *work* or *energy,* not power.

In mechanics we can also express power in terms of force and velocity. Suppose that a force $\vec{F}$ acts on a body while it undergoes a vector displacement $\Delta\vec{s}$. If $F_{\parallel}$ is the component of $\vec{F}$ tangent to the path (parallel to $\Delta\vec{s}$), then the work done by the force is $\Delta W = F_{\parallel}\,\Delta s$, and the average power is

$$P_{av} = \frac{F_{\parallel}\Delta s}{\Delta t} = F_{\parallel}\frac{\Delta s}{\Delta t} = F_{\parallel}v_{av} \tag{6.17}$$

Instantaneous power P is the limit of this expression as $\Delta t \rightarrow 0$:

$$P = F_{\parallel}v \tag{6.18}$$

where v is the magnitude of the instantaneous velocity. We can also express Eq. (6.18) in terms of the scalar product:

$$P = \vec{F}\cdot\vec{v} \tag{6.19}$$

(instantaneous rate at which force $\vec{F}$ does work on a particle)

Example 6.11 Force and power

Each of the two jet engines on a Boeing 767 airliner develops a thrust (a forward force on the airplane) of 197,000 N (44,300 lb). When the airplane is flying at 250 m/s (900 km/h, or roughly 560 mi/h), what horsepower does each engine develop?

SOLUTION

IDENTIFY and SET UP: The thrust is in the direction of motion, so $F_{\parallel}$ in Eq. (6.18) is just equal to the thrust. Our target variable is the instantaneous power P.

EXECUTE: At $v = 250$ m/s, each engine develops power P given by Eq. (6.18):

$$P = F_{\parallel}v = (1.97 \times 10^5\ \text{N})(250\ \text{m/s}) = 4.93 \times 10^7\ \text{W}$$
$$= (4.93 \times 10^7\ \text{W})\frac{1\ \text{hp}}{746\ \text{W}} = 66{,}000\ \text{hp}$$

EVALUATE: The speed of modern airliners is directly related to the power of their engines (Fig. 6.21). The largest engines on propeller-driven airliners of the 1950s developed about 3400 hp $(2.5 \times 10^6\ \text{W})$, and these airliners had maximum speeds of about 600 km/h (370 mi/h). The power of each of the engines of a Boeing 767 is nearly 20 times greater, enabling it to fly at about 900 km/h (560 mi/h) and to carry a much heavier load.

What if the engines are running at maximum thrust while the airliner is at rest on the ground (with the brakes held)? Because $v = 0$, the engines develop *zero* power. Force and power are not the same thing!

(a)

(b)

6.21 (a) Propeller-driven and (b) jet airliners.

Example 6.12 A "power climb"

As part of a charity fund-raising drive, a Chicago marathon runner with mass 50.0 kg runs up the stairs to the top of the 443-m-tall Sears Tower, the tallest building in the United States (Fig. 6.22). To lift herself to the top in 15.0 minutes, what must be her average power output in watts? In kilowatts? In horsepower?

6.22 How much power is required to run up the stairs of Chicago's Sears Tower in 15 minutes?

SOLUTION

IDENTIFY and SET UP: We'll treat the runner as a particle of mass m. We can calculate her average power output in two ways: (1) by first determining how much work she must do and then dividing by the elapsed time, as in Eq. (6.15), or (2) by calculating the average upward force she must exert (in the direction of the climb) and multiplying it by her upward velocity, as in Eq. (6.17).

EXECUTE: As in Example 6.9, lifting a mass m against gravity requires an amount of work equal to the weight mg multiplied by the height h it is lifted. Hence the work she must do is

$$W = mgh = (50.0\ \text{kg})(9.80\ \text{m/s}^2)(443\ \text{m}) = 2.17 \times 10^5\ \text{J}$$

The time is 15.0 min = 900 s, so from Eq. (6.15) the average power is

$$P_{\text{av}} = \frac{2.17 \times 10^5\ \text{J}}{900\ \text{s}} = 241\ \text{W} = 0.241\ \text{kW} = 0.323\ \text{hp}$$

Let's try the calculation again using the alternative approach with Eq. (6.17). The force exerted is vertical, and the average vertical component of velocity is $(443\ \text{m})/(900\ \text{s}) = 0.492\ \text{m/s}$, so the average power is

$$P_{\text{av}} = F_{\parallel}v_{\text{av}} = (mg)v_{\text{av}} = (50.0\ \text{kg})(9.80\ \text{m/s}^2)(0.492\ \text{m/s}) = 241\ \text{W}$$

which is the same result as before.

EVALUATE: In fact the runner's *total* power output will be several times greater than we have calculated. The reason is that the runner isn't really a particle but a collection of parts that exert forces on each other and do work, such as the work done to inhale and exhale and to make her arms and legs swing. What we've calculated is only the part of her power output that goes into lifting her to the top of the building.

Test Your Understanding

In Example 6.2, what is the power associated with the force F_T when the sled is moving at 2.0 m/s? What is the power associated with the friction force at this speed?

SUMMARY

When a constant force $\vec{F}$ acts on a particle that undergoes a straight-line displacement $\vec{s}$, the work done by the force on the particle is defined to be the scalar product of $\vec{F}$ and $\vec{s}$. The unit of work in SI units is 1 joule = 1 newton-meter $(1\ \text{J} = 1\ \text{N}\cdot\text{m})$. Work is a scalar quantity; it has an algebraic sign (positive or negative) but no direction in space. (See Examples 6.1 through 6.3)

$$W = \vec{F}\cdot\vec{s} = Fs\cos\phi$$
$$\phi = \text{angle between } \vec{F} \text{ and } \vec{s} \qquad (6.2), (6.3)$$

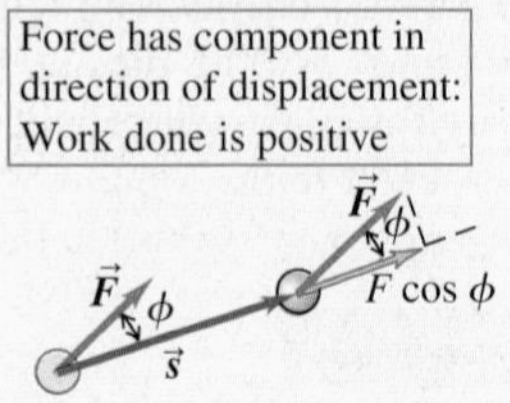

The kinetic energy K of a particle equals the amount of work required to accelerate the particle from rest to speed v. It is also equal to the amount of work the particle can do in the process of being brought to rest. Kinetic energy is a scalar quantity that has no direction in space; it is always positive or zero. Its units are the same as the units of work: $1\ \text{J} = 1\ \text{N}\cdot\text{m} = 1\ \text{kg}\cdot\text{m}^2/\text{s}^2$.

$$K = \frac{1}{2}mv^2 \qquad (6.5)$$

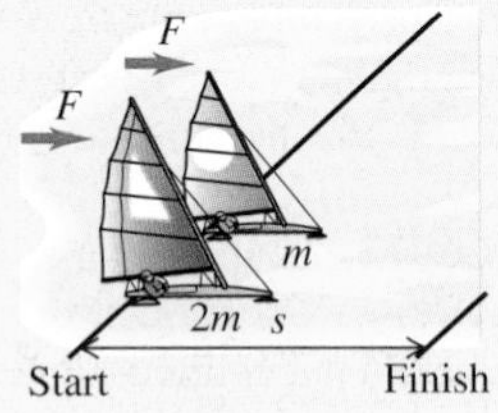

When forces act on a particle while it undergoes a displacement, the particle's kinetic energy changes by an amount equal to the total work done on the particle by all the forces. This relation, called the work-energy theorem, is valid whether the forces are constant or varying and whether the particle moves along a straight or curved path. It is applicable only to bodies that can be treated as a particle. (See Examples 6.4 through 6.6)

$$W_{tot} = K_2 - K_1 = \Delta K \qquad (6.6)$$

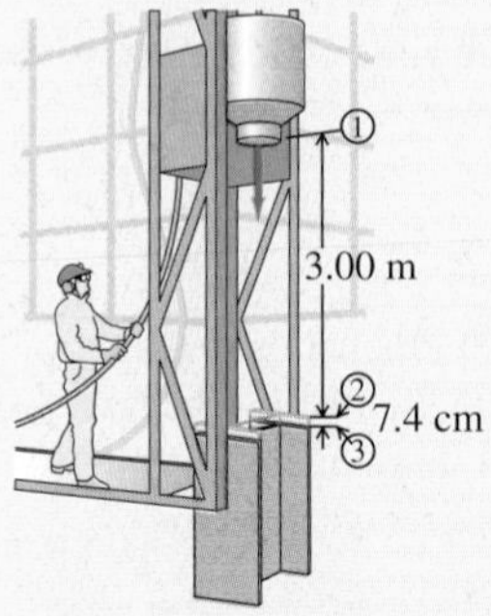

When a force varies during a straight-line displacement, the work done by the force is given by an integral [Eq. (6.7)]. (See Examples 6.7 and 6.8)

$$W = \int_{x_1}^{x_2} F_x\,dx \qquad (6.7)$$

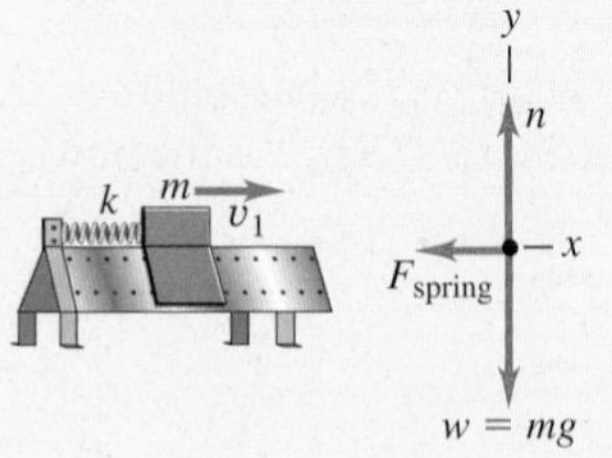

If the particle follows a curved path, then the work done by a force $\vec{F}$ is given by an integral that involves the angle ϕ between the force and the displacement. This expression is valid even if the force magnitude and the angle ϕ vary during the displacement. (See Examples 6.9 and 6.10)

$$W = \int_{P_1}^{P_2} F\cos\phi\,dl = \int_{P_1}^{P_2} F_{\parallel}\,dl$$
$$= \int_{P_1}^{P_2} \vec{F}\cdot d\vec{l} \qquad (6.14)$$

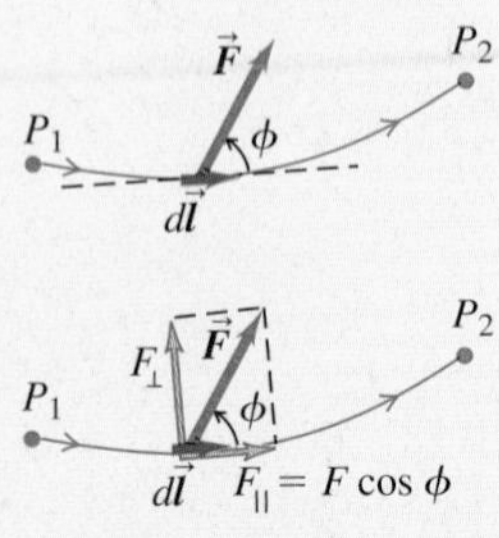

Power is the time rate of doing work. The average power P_{av} is the amount of work ΔW done in time Δt divided by that time. The instantaneous power is the limit of the average power as Δt goes to zero. When a force $\vec{F}$ acts on a particle moving with velocity $\vec{v}$, the instantaneous power (rate at which the force does work) is the scalar product of $\vec{F}$ and $\vec{v}$. Like work and kinetic energy, power is a scalar quantity. The SI unit of power is 1 watt = 1 joule/second (1 W = 1 J/s). (See Examples 6.11 and 6.12)

$$P_{av} = \frac{\Delta W}{\Delta t} \tag{6.15}$$

$$P = \lim_{\Delta t \to 0} \frac{\Delta W}{\Delta t} = \frac{dW}{dt} \tag{6.16}$$

$$P = \vec{F} \cdot \vec{v} \tag{6.19}$$

Key Terms

work, 208
joule, 208
kinetic energy, 214
work-energy theorem, 214
force constant, 221
Hooke's law, 221
power, 227
average power, 227
instantaneous power, 227
watt, 227

Your Notes

Answer to Chapter Opening Question

It is indeed true that the javelin does work on the athlete. However, because the javelin exerts a backward force on the athlete's hand as her hand moves forward, the work done by the javelin is *negative* (see Section 6.1).

Answers to Test Your Understanding Questions

Section 6.1 In each case the barbell has zero acceleration, so the net force on it is zero. Hence the weight lifter must exert an upward force of the same magnitude as the barbell's weight, 500 N. (a) The force and displacement are in the same direction, so $W = Fs = (500\text{ N})(1.5\text{ m}) = 750\text{ J}$. (b) The displacement is zero, so $W = 0$. (c) The displacement is opposite to the force, so $W = -Fs = -(500\text{ N})(1.5\text{ m}) = -750\text{ J}$.
Section 6.2 The work that the pitcher did on the ball is equal to the kinetic energy change that he imparted to it. This work is equal to the product of the average horizontal force F that he applied and the displacement $s = 2.0$ m. The initial kinetic energy K_1 was zero (the ball began at rest), and the final kinetic energy was $K_2 = \frac{1}{2}mv_2^2 = \frac{1}{2}(0.145\text{ kg})(40\text{ m/s})^2 = 116\text{ J}$. Hence, by the work-energy theorem, $F(2.0\text{ m}) = 116\text{ J} - 0$ and $F = (116\text{ J})/(2.0\text{ m}) = 58\text{ N}$.
Section 6.3 As the car comes to a halt, it does an amount of work on the spring equal to its initial kinetic energy: $\frac{1}{2}mv_1^2 = \frac{1}{2}(1000\text{ kg})(20\text{ m/s})^2 = 2.0 \times 10^5\text{ J}$. The work done by the spring on the car is the negative of this value, -2.0×10^5 J. From Eq. (6.10), the work that the car does on the spring is $\frac{1}{2}kx_2^2 - \frac{1}{2}kx_1^2$; the spring is not compressed initially, so $x_1 = 0$ and $2.0 \times 10^5\text{ J} = \frac{1}{2}kx_2^2 = \frac{1}{2}(1.6 \times 10^4\text{ N/m})x_2^2$. Solving, we find $x_2 = 5.0$ m or -5.0 m; here the minus sign is appropriate because the spring is compressed. Hence the spring is compressed by 5.0 m.
Section 6.4 The component of the force F_T in the direction of the sled's motion is $F_\parallel = F_T \cos 36.9°$, with $F_T = 5000$ N. Hence the power is $P = F_\parallel v = (5000\text{ N})(\cos 36.9°)(2.0\text{ m/s}) = 8000\text{ W}$. For the friction force, $F_\parallel = -f = -3500$ N; the minus sign indicates that the force is opposite to the velocity. So the frictional power is $P = F_\parallel v = (-3500\text{ N})(2.0\text{ m/s}) = -7000\text{ W}$. In other words, the friction force does 7000 J of *negative* work on the sled each second.

Discussion Questions

Q6.1 For a constant force in the direction of the displacement, how can twice the work be done by a force of half the magnitude?
Q6.2 An elevator is hoisted by its cables at constant speed. Is the total work done on the elevator positive, negative, or zero? Explain.
Q6.3 A rope tied to a body is pulled, causing the body to accelerate. But according to Newton's third law, the body pulls back on the rope with an equal and opposite force. Is the total work done then zero? If so, how can the body's kinetic energy change? Explain.
Q6.4 When you use an automobile jack, you lift a car by exerting a force much smaller in magnitude than the weight of the car. Does this mean that you do less work on the car than if you had lifted the car directly? Explain.
Q6.5 If there is a net force on a moving object that is nonzero and constant in magnitude, is it possible for the total work done on the object to be zero? Explain, with an example that illustrates your answer.
Q6.6 In Example 5.5 (Section 5.1), how does the work done on the bucket by the tension in the cable compare to the work done on the cart by the tension in the cable?
Q6.7 In Example 5.21 (Section 5.4), does the force $\vec{F}$ do work on the sled? Do any of the forces on the sled do work on it? Is the speed of the sled constant? Is the velocity of the sled constant? Explain your answers.
Q6.8 In Example 5.11 (Section 5.2), gravity does work on the toboggan as it travels a distance d down the hill, where d is measured parallel to the slope. The magnitude of the gravity force depends on the mass of the toboggan and its contents but is independent of the hill's slope angle α. For a constant distance d, does the *work* done by gravity depend on α? Explain.
Q6.9 A force $\vec{F}$ is in the x-direction and has a magnitude that depends on x. Sketch a possible graph of F versus x such that the force does zero work on an object that moves from x_1 to x_2, even though the force magnitude is not zero at all x in this range.
Q6.10 Does the kinetic energy of a car change more when it speeds up from 10 to 15 m/s or from 15 to 20 m/s? Explain.
Q6.11 A falling brick has a mass of 1.5 kg and is moving straight downward with a speed of 5.0 m/s. A 1.5-kg physics book is sliding across the floor with a speed of 5.0 m/s. A 1.5-kg melon is traveling with a horizontal velocity component 3.0 m/s to the right and a vertical component 4.0 m/s upward. Do these objects all have the same velocity? Do these objects all have the same kinetic energy? For each question, give the reasoning behind your answer.
Q6.12 Can the *total* work done on an object during a displacement be negative? Explain. If the total work is negative, can its magnitude be larger than the initial kinetic energy of the object? Explain.
Q6.13 A net force acts on an object and accelerates it from rest to a speed v_1. In doing so, the force does an amount of work W_1. By what factor must the work done on the object be increased to produce three times the final speed, with the object again starting from rest?
Q6.14 A truck speeding down the highway has a lot of kinetic energy relative to a stopped state trooper, but no kinetic energy relative to the truck driver. In these two frames of reference, is the same amount of work required to stop the truck? Explain.
Q6.15 You are holding a briefcase by the handle, with your arm straight down by your side. Does the force your hand exerts do work on the briefcase when i) you walk at constant speed down a horizontal hallway; ii) you ride an escalator from the first to second floor of a building? In each case justify your answers.
Q6.16 When a book slides along a tabletop, the force of friction does negative work on it. Can friction ever do *positive* work? Explain. (*Hint:* Think of a box in the back of an accelerating truck, with friction between the box and the truck floor.)
Q6.17 Time yourself while running up a flight of steps and compute the average rate at which you do work against the force of gravity. Express your answer in watts and in horsepower.
Q6.18 When a constant force is applied to a body moving with constant acceleration, is the power of the force constant? If not,

how would the force have to vary with speed for the power to be constant?

Q6.19 An advertisement for a portable electrical generating unit claims that the unit's diesel engine produces 28,000 hp to drive an electrical generator that produces 30 MW of electrical power. Is this possible? Explain.

Q6.20 A car speeds up while the engine delivers constant power. Is the acceleration greater at the beginning of this process or at the end? Explain.

Q6.21 Consider a graph of instantaneous power versus time, with the vertical P axis starting at $P = 0$. What is the physical significance of the area under the P versus t curve between vertical lines at t_1 and t_2? How could you find the average power from the graph? Draw a P versus t curve that consists of two straight-line sections and for which the peak power is equal to twice the average power.

Q6.22 A nonzero net force acts on an object. Is it possible for any of the following quantities to be constant: i) the particle's speed; ii) the particle's velocity; iii) the particle's kinetic energy?

Exercises

Section 6.1 Work

6.1 You push your physics book 1.50 m along a horizontal tabletop with a horizontal force of 2.40 N. The opposing force of friction is 0.600 N. a) How much work does your 2.40-N force do on the book? b) What is the work done on the book by the friction force? c) What is the total work done on the book?

6.2 An old oaken bucket of mass 6.75 kg hangs in a well at the end of a rope. The rope passes over a frictionless pulley at the top of the well, and you pull horizontally on the end of the rope to raise the bucket slowly a distance of 4.00 m. a) How much work do you do on the bucket in pulling it up? b) How much work is done by the gravitational force acting on the bucket? c) What is the total work done on the bucket?

6.3 A fisherman reels in 12.0 m of line while pulling in a fish that exerts a constant resisting force of 25.0 N. If the fish is pulled in at constant velocity, how much work is done on it by the tension in the line?

6.4 A factory worker pushes a 30.0-kg crate a distance of 4.5 m along a level floor at constant velocity by pushing horizontally on it. The coefficient of kinetic friction between the crate and floor is 0.25. a) What magnitude of force must the worker apply? b) How much work is done on the crate by this force? c) How much work is done on the crate by friction? d) How much work is done by the normal force? By gravity? e) What is the total work done on the crate?

6.5 Suppose the worker in Exercise 6.4 pushes downward at an angle of 30° below the horizontal. a) What magnitude of force must the worker apply to move the crate at constant velocity? b) How much work is done on the crate by this force when the crate is pushed a distance of 4.5 m? c) How much work is done on the crate by friction during this displacement? d) How much work is done by the normal force? By gravity? e) What is the total work done on the crate?

6.6 A boat with a horizontal tow rope pulls a water skier. She skis off to the side, so the rope makes an angle of 15.0° with her direction of motion, and then continues in a straight line. The tension in the rope is 180 N. How much work is done on the skier by the rope during a displacement of 300 m?

6.7 Two tugboats pull a disabled supertanker. Each tug exerts a constant force of 1.80×10^6 N, one 14° west of north and the other 14° east of north, as they pull the tanker 0.75 km toward the north. What is the total work they do on the supertanker?

6.8 A loaded grocery cart is rolling across a parking lot in a strong wind. You apply a constant force $\vec{F} = (30\ \text{N})\hat{\imath} - (40\ \text{N})\hat{\jmath}$ to the cart as it undergoes a displacement $\vec{s} = (-9.0\ \text{m})\hat{\imath} - (3.0\ \text{m})\hat{\jmath}$. How much work does the force you apply do on the grocery cart?

6.9 A 0.800-kg ball is tied to the end of a string 1.60 m long and swung in a vertical circle. a) During one complete circle, starting anywhere, calculate the total work done on the ball by: (i) the tension in the string (ii) gravity. b) Repeat part (a) for motion along the semicircle from the lowest to the highest point on the path.

Section 6.2 Work and Kinetic Energy

6.10 a) Compute the kinetic energy, in joules, of a 1600-kg automobile traveling at 50.0 km/h. b) By what factor does the kinetic energy change if the speed is doubled?

6.11 T. Rex. The dinosaur *Tyrannosaurus rex* is thought to have had a mass of about 7000 kg. a) Treat the dinosaur as a particle and estimate its kinetic energy at a walking speed of 4.0 km/h. b) With what speed would a 70-kg person have to move to have the same kinetic energy as a walking *T. rex?*

6.12 a) How many joules of kinetic energy does a walking person have? What about a running person? b) How many joules of kinetic energy does a large car going at freeway speeds have? c) If you drop a 1-kg weight (about 2 lb) from shoulder height, how many joules of kinetic energy will it have when it reaches the ground?

6.13 You are a member of an Alpine Rescue Team and must project a box of supplies up an incline of constant slope angle α so that it reaches a stranded skier who is a vertical distance h above the bottom of the incline. The incline is slippery, but there is some friction present, with kinetic friction coefficient μ_k. Use the work-energy theorem to calculate the minimum speed you must give the box at the bottom of the incline so that it will reach the skier. Express your answer in terms of g, h, μ_k, and α.

6.14 You throw a 20-N rock vertically into the air from ground level. You observe that when it is 15.0 m above the ground, it is traveling at 25.0 m/s upward. Use the work-energy theorem to find a) its speed just as it left the ground; b) its maximum height.

6.15 A mass m slides down a smooth inclined plane from an initial vertical height h, making an angle α with the horizontal. a) The work done by a force is the sum of the work done by the components of the force. Consider the components of gravity parallel and perpendicular to the surface of the plane. Calculate the work done on the mass by each of the components and use these results to show that the work done by gravity is exactly the same as if the mass had fallen straight down through the air from a height h. b) Use the work-energy theorem to prove that the speed of the mass at the bottom of the incline is the same as if it had been dropped from height h, independent of the angle α of the incline. Explain how this speed can be independent of the slope angle. c) Use the results of part (b) to find the speed of a rock that slides down an icy frictionless hill, starting from rest 15.0 m above the bottom.

6.16 A car is stopped by a constant friction force that is independent of the car's speed. By what factor is the stopping distance changed if the car's initial speed is doubled? (Solve using work-energy methods.)

6.17 A baseball leaves a pitcher's hand at a speed of 32.0 m/s. The mass of the baseball is 0.145 kg. You can ignore air resistance. How much work has the pitcher done on the ball by throwing it?

6.18 Iceboat. In Conceptual Example 6.6 (Section 6.2), call the iceboat with mass m, boat A and the other iceboat, with mass $2m$, boat B. a) At the finish line what is the ratio v_A/v_B of the speeds of the two iceboats? b) Let t_A be the elapsed time it takes boat A to reach the finish line and t_B the elapsed time for boat B. What is the ratio t_B/t_A of these two times?

6.19 A moving electron has kinetic energy K_1. After a net amount of work W has been done on it, the electron is moving one-quarter as fast in the opposite direction. a) Find W in terms of K_1. b) Does your answer depend on the final direction of the electron's motion?

6.20 A sled with mass 8.00 kg moves in a straight line on a frictionless horizontal surface. At one point in its path, its speed is 4.00 m/s; after it has traveled 2.50 m beyond this point, its speed is 6.00 m/s. Use the work-energy theorem to find the force acting on the sled, assuming that this force is constant and that it acts in the direction of the sled's motion.

6.21 A soccer ball with mass 0.420 kg is initially moving with speed 2.00 m/s. A soccer player kicks the ball, exerting a constant force of magnitude 40.0 N in the same direction as the ball's motion. Over what distance must her foot be in contact with the ball to increase the ball's speed to 6.00 m/s?

6.22 A 12-pack of Omni-Cola (mass 4.30 kg) is initially at rest on a horizontal floor. It is then pushed in a straight line for 1.20 m by a trained dog who exerts a horizontal force with magnitude 36.0 N. Use the work-energy theorem to find the final speed of the 12-pack if a) there is no friction between the 12-pack and the floor; b) the coefficient of kinetic friction between the 12-pack and the floor is 0.30.

6.23 A pitcher throws a baseball with mass 0.145 kg straight upward with an initial speed of 25.0 m/s. a) How much work has gravity done on the baseball when it reaches a height of 20.0 m above the pitcher's hand? b) Use the work-energy theorem to calculate the speed of the baseball at a height of 20.0 m above the pitcher's hand. You can ignore air resistance. c) Does the answer to part (b) depend on whether the baseball is moving upward or downward at a height of 20.0 m? Explain.

6.24 A 4.80-kg watermelon is dropped (zero initial speed) from the roof of a 25.0-m-tall building. a) Calculate the work done by gravity on the watermelon during its displacement from the roof to the ground. b) What is the kinetic energy of the watermelon just before it strikes the ground? You can ignore air resistance.

6.25 A little red wagon with mass 7.00 kg moves in a straight line on a frictionless horizontal surface. It has an initial speed of 4.00 m/s and then is pushed 3.0 m in the direction of the initial velocity by a force with a magnitude of 10.0 N. a) Use the work-energy theorem to calculate the wagon's final speed. b) Calculate the acceleration produced by the force. Use this acceleration in the kinematic relations of Chapter 2 to calculate the wagon's final speed. Compare this result to that calculated in part (a).

6.26 A block of ice with mass 2.00 kg slides 0.750 m down an inclined plane that slopes downward at an angle of 36.9° below the horizontal. If the block of ice starts from rest, what is its final speed? You can ignore friction.

6.27 Stopping Distance. A car is traveling on a level road with speed v_0 at the instant when the brakes lock, so that the tires slide rather than roll. a) Use the work-energy theorem to calculate the minimum stopping distance of the car in terms of v_0, g, and the coefficient of kinetic friction μ_k between the tires and the road. b) The car stops in a distance of 91.2 m if $v_0 = 80.0$ km/h. What is the stopping distance if $v_0 = 60.0$ km/h? Assume that the value of μ_k remains the same.

Section 6.3 Work and Energy with Varying Forces

6.28 To stretch a spring 3.00 cm from its unstretched length, 12.0 J of work must be done. How much work must be done to compress this spring 4.00 cm from its unstretched length?

6.29 A force of 160 N stretches a spring 0.050 m beyond its unstretched length. a) What magnitude of force is required to stretch the spring 0.015 m beyond its unstretched length? To compress the spring 0.020 m? b) How much work must be done to stretch the spring 0.015 m beyond its unstretched length? How much work to compress the spring 0.020 m from its unstretched length?

6.30 A child applies a force $\vec{F}$ parallel to the x-axis to a 10.0-kg sled moving on the frozen surface of a small pond. As the child controls the speed of the sled, the x-component of the force she applies varies with the x-coordinate of the sled as shown in Fig. 6.23. Calculate the work done by the force $\vec{F}$ when the sled moves a) from $x = 0$ to $x = 8.0$ m; b) from $x = 8.0$ m to $x = 12.0$ m; c) from $x = 0$ to 12.0 m.

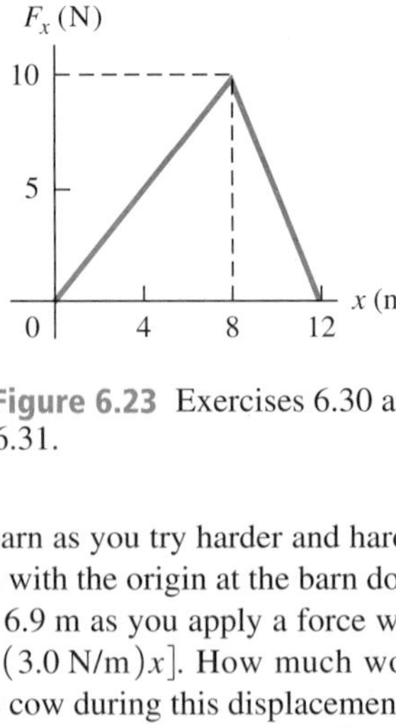

Figure 6.23 Exercises 6.30 and 6.31.

6.31 Suppose the sled in Exercise 6.30 is initially at rest at $x = 0$. Use the work-energy theorem to find the speed of the sled at a) $x = 8.0$ m; b) $x = 12.0$ m. You can ignore friction between the sled and the surface of the pond.

6.32 A balky cow is leaving the barn as you try harder and harder to push her back in. In coordinates with the origin at the barn door, the cow walks from $x = 0$ to $x = 6.9$ m as you apply a force with x-component $F_x = -[20.0\text{ N} + (3.0\text{ N/m})x]$. How much work does the force you apply do on the cow during this displacement?

6.33 A 6.0-kg box moving at 3.0 m/s on a horizontal, frictionless surface runs into a light spring of force constant 75 N/cm. Use the work-energy theorem to find the maximum compression of the spring.

6.34 Leg Presses. As part of your daily workout, you lie on your back and push with your feet against a platform attached to two stiff springs arranged side by side so that they are parallel to each other. When you push the platform, you compress the springs. You do 80.0 J of work when you compress the springs 0.200 m from their uncompressed length. a) What magnitude of force must you apply to hold the platform in this position? b) How much *additional* work must you do to move the platform 0.200 m *farther*, and what maximum force must you apply?

6.35 a) In Example 6.8 (Section 6.3) it is calculated that with the air track turned off, the glider travels 8.6 cm before it stops instantaneously. How large would the coefficient of static friction μ_s have to be to keep the glider from springing back to the left? b) If the coefficient of static friction between the glider and the track is $\mu_s = 0.60$, what is the maximum initial speed v_1 that the glider can be given and still remain at rest after it stops instantaneously? With the air track turned off, the coefficient of kinetic friction is $\mu_k = 0.47$.

6.36 A 4.00-kg block of ice is placed against a horizontal spring that has force constant $k = 200$ N/m and is compressed 0.025 m. The spring is released and accelerates the block along a horizontal surface. You can ignore friction and the mass of the spring. a) Calculate the work done on the block by the spring during the motion of the block from its initial position to where the spring has returned to its uncompressed length. b) What is the speed of the block after it leaves the spring?

6.37 A force $\vec{F}$ is applied to a 2.0-kg radio-controlled model car parallel to the x-axis as it moves along a straight track. The x-component of the force varies with the x-coordinate of the car as shown in Fig. 6.24. Calculate the work done by the force $\vec{F}$ when the car moves from a) $x = 0$ to $x = 3.0$ m; b) $x = 3.0$ m to $x = 4.0$ m; c) $x = 4.0$ m to $x = 7.0$ m; d) $x = 0$ to $x = 7.0$ m; e) $x = 7.0$ m to $x = 2.0$ m.

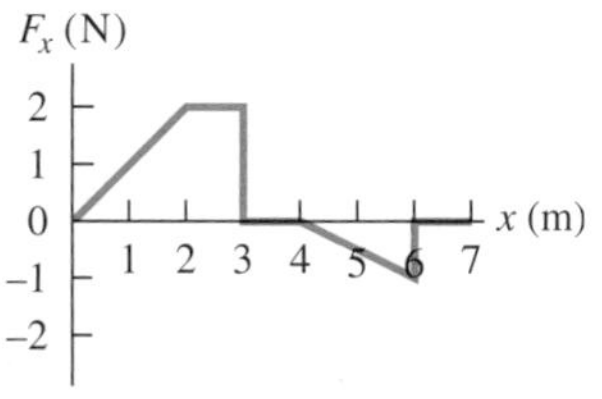

Figure 6.24 Exercises 6.37 and 6.38.

6.38 Suppose the 2.0-kg model car in Exercise 6.37 is initially at rest at $x = 0$ and $\vec{F}$ is the net force acting on the car. Use the work-energy theorem to find the speed of the car at a) $x = 3.0$ m; b) $x = 4.0$ m; c) $x = 7.0$ m.

6.39 At a waterpark, sleds with riders are sent along a slippery, horizontal surface by the release of a large, compressed spring. The spring with a force constant $k = 4000$ N/m and negligible mass rests on the frictionless horizontal surface. One end is in contact with a stationary wall. A sled and rider with total mass 70.0 kg are pushed against the other end, compressing the spring 0.375 m. The sled is then released with zero initial velocity. What is the sled's speed when the spring a) returns to its uncompressed length? b) is still compressed 0.200 m?

6.40 In Example 6.9 (Section 6.3), instead of applying a varying horizontal force $\vec{F}$ that maintains Throcky very nearly in equilibrium, you apply a constant horizontal force with magnitude $F = 2w$, where w is Throcky's weight. As in Example 6.9, consider Throcky to be a particle and ignore the small weight of the chains and seat. You push Throcky until the chains make an angle θ_0 with the vertical. a) Use Eq. (6.14) to calculate the work done on Throcky by the force $\vec{F}$ that you apply. b) At the angle θ_0, how does the magnitude of $\vec{F}$ in this exercise compare to that in Example 6.9? c) How does the work done by $\vec{F}$ in this exercise compare to that in Example 6.7?

6.41 A small glider is placed against a compressed spring at the bottom of an air track that slopes upward at an angle of 40.0° above the horizontal. The glider has mass 0.0900 kg. The spring has $k = 640$ N/m and negligible mass. When the spring is released, the glider travels a maximum distance of 1.80 m along the air track before sliding back down. Before reaching this maximum distance, the glider loses contact with the spring. a) What distance was the spring originally compressed? b) When the glider has traveled along the air track 0.80 m from its initial position against the compressed spring, is it still in contact with the spring? What is the kinetic energy of the glider at this point?

6.42 Bricklaying. An ingenious bricklayer builds a device for shooting bricks up to the top of the wall where he is working. He places a brick on a vertical compressed spring with force constant $k = 450$ N/m and negligible mass. When the spring is released, the brick is propelled upward. If the brick has mass 1.80 kg and is to reach a maximum height of 3.6 m above its initial position on the compressed spring, what distance must the bricklayer compress the spring initially? (The brick loses contact with the spring when the spring returns to its uncompressed length. Why?)

Section 6.5 Power

6.43 How many joules of energy does a 100-watt light bulb use per hour? How fast would a 70-kg person have to run to have that amount of energy?

6.44 A 20.0-kg rock is sliding on a rough, horizontal surface at 8.00 m/s and eventually stops due to friction. The coefficient of kinetic friction between the rock and the surface is 0.200. What average thermal power is produced as the rock stops?

6.45 A tandem (two-person) bicycle team must overcome a force of 165 N to maintain a speed of 9.00 m/s. Find the power required per rider, assuming that each contributes equally. Express your answer in watts and in horsepower.

6.46 The total consumption of electrical energy in the United States is about 1.0×10^{19} joules per year. a) What is the average rate of electrical energy consumption in watts? b) If the population of the United States is 260 million, what is the average rate of electrical energy consumption per person? c) The sun transfers energy to the earth by radiation at a rate of approximately 1.0 kW per square meter of surface. If this energy could be collected and converted to electrical energy with 40% efficiency, how great an area (in square kilometers) would be required to collect the electrical energy used by the United States?

6.47 When its 75-kW (100-hp) engine is generating full power, a small single-engine airplane with mass 700 kg gains altitude at a rate of 2.5 m/s (150 m/min, or 500 ft/min). What fraction of the engine power is being used to make the airplane climb? (The remainder is used to overcome the effects of air resistance and of inefficiencies in the propeller and engine.)

6.48 Working Like a Horse. Your job is to lift 30-kg crates a vertical distance of 0.90 m from the ground onto the bed of a truck. a) How many crates would you have to load onto the truck in one minute for the average power output you use to lift the crates to equal 0.50 hp? b) How many crates for an average power output of 100 W?

6.49 An elevator has a mass 600 kg, not including passengers. The elevator is designed to ascend, at constant speed, a vertical distance of 20.0 m (five floors) in 16.0 s, and it is driven by a motor that can provide up to 40 hp to the elevator. What is the maximum number of passengers that can ride in the elevator? Assume that an average passenger has mass 65.0 kg.

6.50 The hammer of a pile driver weighs 3800 N and must be lifted 2.80 m vertically at constant speed in 4.00 s. What horsepower must the engine provide to the hammer?

6.51 The aircraft carrier *John F. Kennedy* has mass 7.4×10^7 kg. When its engines are developing their full power of 280,000 hp, the *John F. Kennedy* travels at its top speed of 35 knots (65 km/h). If 70% of the power output of the engines is applied to pushing the ship through the water, what is the magnitude of the force of water resistance that opposes the carrier's motion at this speed?

6.52 A ski tow operates on a 15.0° slope of length 300 m. The rope moves at 12.0 km/h and provides power for 50 riders at one time, with an average mass per rider of 70.0 kg. Estimate the power required to operate the tow.

6.53 A particle is accelerated from rest by a constant net force. a) Show that the instantaneous power supplied by the net force is ma^2t. b) To triple the acceleration at any given time, by what factor must the power be increased? c) At $t = 5.0$ s the power supplied by the net force is 36 W. What must the power be at $t = 15.0$ s to maintain constant acceleration?

6.54 Show that the instantaneous power P supplied by the net force acting on a particle is related to the particle's kinetic energy K by $P = dK/dt$.

6.55 A typical flying insect applies an average force equal to twice its weight during each downward stroke while hovering. Take the mass of the insect to be 10 g and assume the wings move an average downward distance of 1.0 cm during each stroke. Assuming 100 downward strokes per second, estimate the average power output of the insect.

Problems

6.56 Rotating Bar. A thin, uniform 12.0-kg bar that is 2.00 m long rotates uniformly about a pivot at one end, making 5.00 complete revolutions every 3.00 seconds. What is the kinetic energy of this bar? (*Hint*: Different points in the bar have different speeds. Break the bar up into infinitesimal segments of mass dm and integrate to add up the kinetic energy of all these segments.)

6.57 A luggage handler pulls a 20.0-kg suitcase up a ramp inclined at 25.0° above the horizontal by a force $\vec{F}$ of magnitude 140 N that acts parallel to the ramp. The coefficient of kinetic friction between the ramp and the incline is $\mu_k = 0.300$. If the suitcase travels 3.80 m along the ramp, calculate a) the work done on the suitcase by the force $\vec{F}$; b) the work done on the suitcase by the gravitational force; c) the work done on the suitcase by the normal force; d) the work done on the suitcase by the friction force; e) the total work done on the suitcase. f) If the speed of the suitcase is zero at the bottom of the ramp, what is its speed after it has traveled 3.80 m along the ramp?

6.58 Chin-Ups. While doing a chin-up, a man lifts his body 0.40 m. a) How much work must the man do per kilogram of body mass? b) The muscles involved in doing a chin-up can generate about 70 J of work per kilogram of muscle mass. If the man can just barely do a 0.40-m chin-up, what percentage of his body's mass do these muscles constitute? (For comparison, the *total* percentage of muscle in a typical 70-kg man with 14% body fat is about 43%.) c) Repeat part (b) for the man's young son, who has arms half as long as his father's but whose muscles can also generate 70 J of work per kilogram of muscle mass. d) Adults and children have about the same percentage of muscle in their bodies. Explain why children can commonly do chin-ups more easily than their fathers.

6.59 Simple Machines. Ramps for the disabled are used because a large weight w can be raised by relatively small force equal to $w \sin \alpha$ plus the small friction force. Such inclined planes are an example of a class of devices called *simple machines*. An input force F_{in} is applied to the system and results in an output force F_{out} applied to the object that is moved. For a simple machine the ratio of these forces, F_{out}/F_{in}, is called the actual mechanical advantage (AMA). The inverse ratio of the distances that the points of application of these forces move through during the motion of the object, s_{in}/s_{out}, is called the ideal mechanical advantage (IMA). a) Find the IMA for an inclined plane. b) What can we say about the relation between the work supplied to the machine, W_{in}, and the work output of the machine, W_{out}, if AMA = IMA? c) Sketch a single pulley arranged to give IMA = 2. d) We define the efficiency e of a simple machine to equal the ratio of the output work to the input work, $e = W_{out}/W_{in}$. Show that e = AMA/IMA.

6.60 A woman stands in an elevator that has a constant upward acceleration while the elevator travels upward a distance of 18.0 m. During the 18.0-m displacement, the normal force exerted by the elevator floor does 8.25 kJ of work on her and gravity does −7.35 kJ of work on her. a) What is the mass of the woman? b) What is the normal force that the elevator floor exerts on her? c) What is the acceleration of the elevator?

6.61 The space shuttle *Endeavour*, with mass 86,400 kg, is in a circular orbit of radius 6.66×10^6 m around the earth. It takes 90.1 min for the shuttle to complete each orbit. On a repair mission, it is cautiously moving 1.00 m closer to a disabled satellite every 3.00 s. Calculate the shuttle's kinetic energy a) relative to the earth; b) relative to the satellite.

6.62 A 5.00-kg package slides 1.50 m down a long ramp that is inclined at 12.0° below the horizontal. The coefficient of kinetic friction between the package and the ramp is $\mu_k = 0.310$. Calculate a) the work done on the package by friction; b) the work done on the package by gravity; c) the work done on the package by the normal force; d) the total work done on the package. e) If the package has a speed of 2.20 m/s at the top of the ramp, what is its speed after sliding 1.50 m down the ramp?

6.63 The package in Problem 6.62 has a speed of 2.20 m/s at the top of the ramp. Use the work-energy theorem to find how far down the ramp the package slides before it stops.

6.64 An object is attracted toward the origin with a force given by $F_x = -k/x^2$. (Gravitational and electrical forces have this distance dependence.) a) Calculate the work done by the force F_x when the object moves in the x-direction from x_1 to x_2. If $x_2 > x_1$, is the work done by F_x positive or negative? b) The only other force acting on the object is a force that you exert with your hand to move the object slowly from x_1 to x_2. How much work do you do? If $x_2 > x_1$, is the work you do positive or negative? c) Explain the similarities and differences between your answers to parts (a) and (b).

6.65 The gravitational pull of the earth on an object is inversely proportional to the square of the distance of the object from the center of the earth. At the earth's surface this force is equal to the object's normal weight mg, where $g = 9.8\ \text{m/s}^2$, and at large distances, the force is zero. If a 20,000-kg asteroid falls to earth from a very great distance away, what will be its minimum speed as it strikes the earth's surface, and how much kinetic energy will it impart to our planet? You can ignore the effects of the earth's atmosphere.

6.66 Varying Coefficient of Friction. A box is sliding with a speed of 4.50 m/s on a horizontal surface when, at point P, it encounters a rough section. On the rough section, the coefficient of friction is not constant, but starts at 0.100 at P and increases linearly with distance past P, reaching a value of 0.600 at 12.5 m past point P. a) Use the work-energy theorem to find how far this box slides before stopping. b) What is the coefficient of friction at the stopping point? c) How far would the box have slid if the friction coefficient didn't increase, but instead had the constant value of 0.100?

6.67 An object that can move along the x-axis is attracted toward the origin with a force of magnitude $F = \alpha x^3$, where $\alpha = 4.00\ \text{N/m}^3$. What is the force F when the object is a) at $x = 1.00$ m? b) at $x = 2.00$ m? c) How much work is done by the force F when the object moves from $x = 1.00$ m to $x = 2.00$ m? Is this work positive or negative?

6.68 Consider a spring that does not obey Hooke's law very faithfully. One end of the spring is fixed. To keep the spring stretched or compressed an amount x, a force along the x-axis with x-component $F_x = kx - bx^2 + cx^3$ must be applied to the free end. Here $k = 100$ N/m, $b = 700\ \text{N/m}^2$, and $c = 12{,}000\ \text{N/m}^3$. Note that $x > 0$ when the spring is stretched and $x < 0$ when it is compressed. a) How much work must be done to stretch this spring by 0.050 m from its unstretched length? b) How much work must be done to *compress* this spring by 0.050 m from its unstretched length? c) Is it easier to stretch or compress this spring? Explain why in terms of the dependence of F_x on x. (Many real springs behave qualitatively in the same way.)

6.69 A small block with a mass of 0.120 kg is attached to a cord passing through a hole in a frictionless, horizontal surface (Fig. 6.25). The block is originally revolving at a distance of 0.40 m from the hole with a speed of 0.70 m/s. The cord is then pulled from below, shortening the radius of the circle in which the block revolves to 0.10 m. At this new distance, the speed of the block is observed to be 2.80 m/s. a) What is the tension in the cord in the original situation when the block has speed $v = 0.70$ m/s? b) What is the tension in the cord in the final situation when the block has speed $v = 2.80$ m/s? c) How much work was done by the person who pulled on the cord?

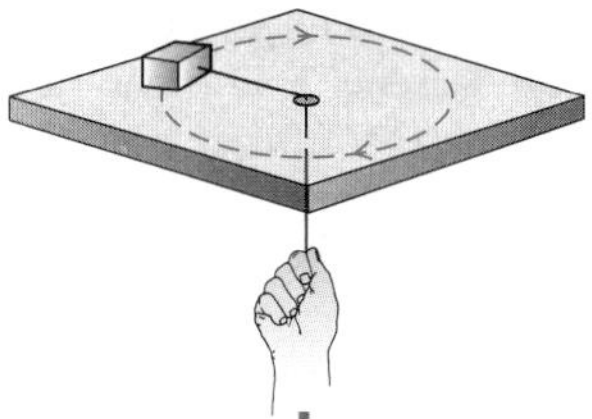

Figure 6.25 Problem 6.69.

6.70 Proton Bombardment. A proton with mass 1.67×10^{-27} kg is propelled at an initial speed of 3.00×10^5 m/s directly toward a uranium nucleus 5.00 m away. The proton is repelled by the uranium nucleus with a force of magnitude $F = \alpha/x^2$, where x is the separation between the two objects and $\alpha = 2.12 \times 10^{-26}\ \text{N} \cdot \text{m}^2$. Assume that the uranium nucleus remains at rest. a) What is the speed of the proton when it is 8.00×10^{-10} m from the uranium nucleus? b) As the proton approaches the uranium nucleus, the repulsive force slows down the proton until it comes momentarily to rest, after which the proton moves away from the uranium nucleus. How close to the uranium nucleus does the proton get? c) What is the speed of the proton when it is again 5.00 m away from the uranium nucleus?

6.71 A block of ice with mass 6.00 kg is initially at rest on a frictionless, horizontal surface. A worker then applies a horizontal force $\vec{F}$ to it. As a result, the block moves along the x-axis such that its position as a function of time is given by $x(t) = \alpha t^2 + \beta t^3$, where $\alpha = 0.200\ \text{m/s}^2$ and $\beta = 0.0200\ \text{m/s}^3$. a) Calculate the velocity of the object when $t = 4.00$ s. b) Calculate the magnitude of $\vec{F}$ when $t = 4.00$ s. c) Calculate the work done by the force $\vec{F}$ during the first 4.00 s of the motion.

6.72 A net force with magnitude $(5.00\ \text{N/m}^2)x^2$ and directed at a constant angle of 31.0° with the $+x$-axis acts on an object of mass 0.250 kg as the object moves parallel to the x-axis. How fast is the object moving at $x = 1.50$ m if it has a speed of 4.00 m/s at $x = 1.00$ m?

6.73 You and your bicycle have combined mass 80.0 kg. When you reach the base of an bridge, you are traveling along the road at 5.00 m/s (Fig. 6.26). At the top of the bridge, you have climbed a vertical distance of 5.20 m and have slowed to 1.50 m/s. You can ignore work done by friction and any inefficiency in the bike or your legs. a) What is the total work done on you and your bicycle when you go from the base to the top of the bridge? b) How much work have you done with the force you apply to the pedals?

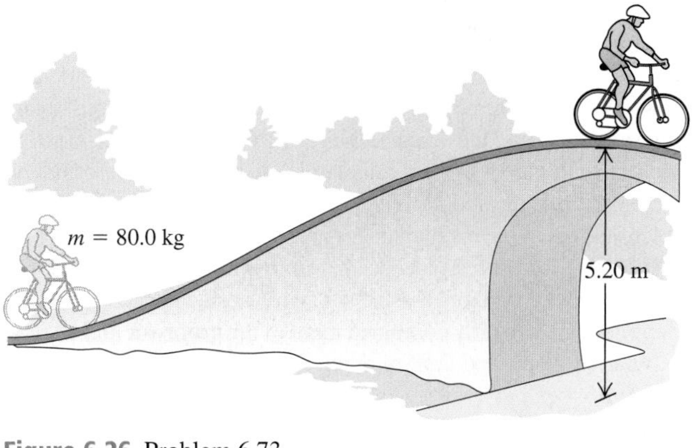

Figure 6.26 Problem 6.73.

6.74 A force in the $+x$-direction has magnitude $F = b/x^n$, where b and n are constants. a) For $n > 1$, calculate the work done on a particle by this force when the particle moves along the x-axis from $x = x_0$ to infinity. b) Show that for $0 < n < 1$, even though F becomes zero as x becomes very large, an infinite amount of work is done by F when the particle moves from $x = x_0$ to infinity.

6.75 You are asked to design spring bumpers for the walls of a parking garage. A freely rolling 1200-kg car moving at 0.65 m/s is to compress the spring no more than 0.070 m before stopping. What should be the force constant of the spring? Assume that the spring has negligible mass.

6.76 The spring of a spring gun has force constant $k = 400$ N/m and negligible mass. The spring is compressed 6.00 cm and a ball with mass 0.0300 kg is placed in the horizontal barrel against the compressed spring. The spring is then released, and the ball is propelled out the barrel of the gun. The barrel is 6.00 cm long, so the ball leaves the barrel at the same point that it loses contact with the spring. The gun is held so the barrel is horizontal. a) Calculate the speed with which the ball leaves the barrel if you can ignore friction. b) Calculate the speed of the ball as it leaves the barrel if a constant resisting force of 6.00 N acts on the ball as it moves along the barrel. c) For the situation in part (b), at what position along the barrel does the ball have the greatest speed, and what is that speed? (In this case, the maximum speed does not occur at the end of the barrel.)

6.77 A 2.50-kg textbook is forced against a horizontal spring of negligible mass and force constant 250 N/m, compressing the spring a distance of 0.250 m. When released, the textbook slides on a horizontal table top with coefficient of kinetic friction $\mu_k = 0.30$. Use the work-energy theorem to find how far the textbook moves from its initial position before coming to rest.

6.78 Pushing a Cat. Your cat "Ms." (mass 7.00 kg) is trying to make it to the top of a frictionless ramp 2.00 m long and inclined upward at 30.0° above the horizontal. Since the poor cat can't get any traction on the ramp, you push her up the entire length of the ramp by exerting a constant 100-N force parallel to the ramp. If Ms. takes a running start so that she is moving at 2.40 m/s at the bottom of the ramp, what is her speed when she reaches the top of the incline? Use the work-energy theorem.

6.79 Crash Barrier. A student proposes a design for an automobile crash barrier in which a 1700-kg sport utility vehicle moving at 20.0 m/s crashes into a spring of negligible mass that slows it to a stop. To avoid injuring the passengers, the acceleration of the vehicle as it slows can be no more than 5.00*g*. a) Find the required spring constant *k*, and find the distance the spring will compress in slowing the vehicle to a stop. In your calculation, disregard any deformation or crumpling of the vehicle and the friction between the vehicle and the ground. b) What disadvantages are there to this design?

6.80 A physics professor is pushed up a ramp inclined upward at 30.0° above the horizontal as he sits in his desk chair that slides on frictionless rollers. The combined mass of the professor and chair is 85.0 kg. He is pushed 2.50 m along the incline by a group of students who together exert a constant horizontal force of 600 N. The professor's speed at the bottom of the ramp is 2.00 m/s. Use the work-energy theorem to find his speed at the top of the ramp.

6.81 A 5.00-kg block is moving at $v_0 = 6.00$ m/s along a frictionless, horizontal surface toward a spring with force constant $k = 500$ N/m that is attached to a wall (Fig. 6.27). The spring has negligible mass. a) Find the maximum distance the spring will be compressed. b) If the spring is to compress by no more than 0.150 m, what should be the maximum value of v_0?

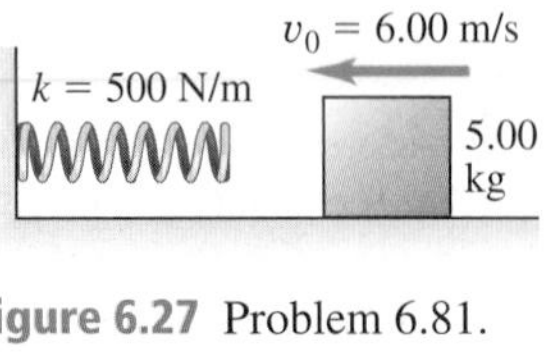

Figure 6.27 Problem 6.81.

6.82 Consider the system shown in Fig. 6.28. The rope and pulley have negligible mass, and the pulley is frictionless. The coefficient of kinetic friction between the 8.00-kg block and the tabletop is $\mu_k = 0.250$. The blocks are released from rest. Use energy methods to calculate the speed of the 6.00-kg block after it has descended 1.50 m.

6.83 Consider the system shown in Fig. 6.28. The rope and pulley have negligible mass, and the pulley is frictionless. Initially the 6.00-kg block is moving downward and the 8.00-kg block is moving to the right, both with a speed of 0.900 m/s. The blocks come to rest after moving 2.00 m. Use the work-energy theorem to calculate the coefficient of kinetic friction between the 8.00-kg block and the tabletop.

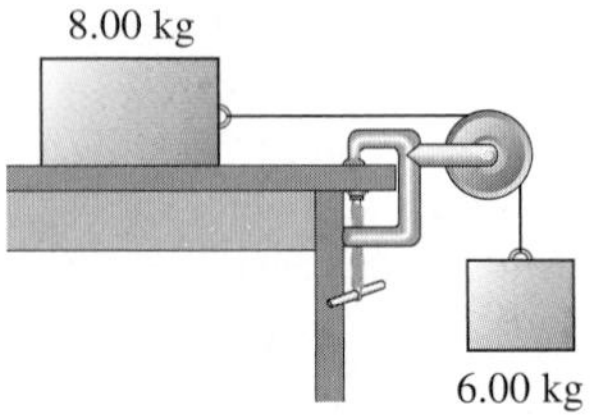

Figure 6.28 Problems 6.82 and 6.83.

6.84 Bow and Arrow. Figure 6.29 shows how the force exerted by the string of a compound bow on an arrow varies as a function of how far back the arrow is pulled (*the draw length*). Assume that the same force is exerted on the arrow as it moves forward after being released. Full draw for this bow is at a draw length of 75.0 cm. If the bow shoots a 0.0250-kg arrow from full draw, what is the speed of the arrow as it leaves the bow?

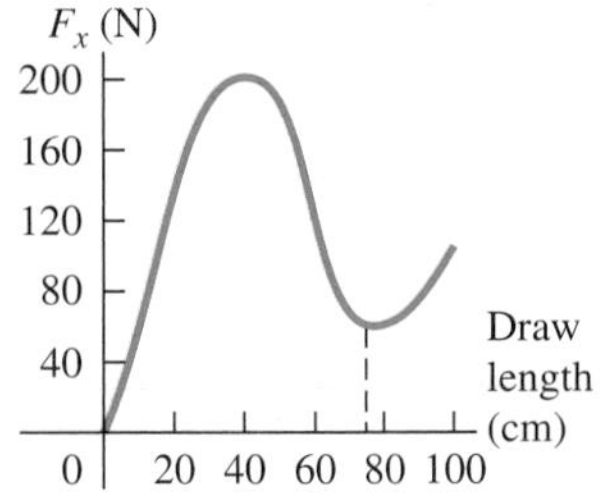

Figure 6.29 Problem 6.84.

6.85 On an essentially frictionless, horizontal ice rink, a skater moving at 3.0 m/s encounters a rough patch that reduces her speed by 45% due to a friction force that is 25% of her weight. Use the work-energy theorem to find the length of this rough patch.

6.86 Rescue. Your friend (mass 65.0 kg) is standing on the ice in the middle of a frozen pond. There is very little friction between her feet and the ice, so she is unable to walk. Fortunately, there is a light rope tied around her waist and you stand on the bank holding the other end. You pull on the rope for 3.00 s and accelerate your friend from rest to a speed of 6.00 m/s while you remain at rest. What is the average power supplied by the force you applied?

6.87 A pump is required to lift 800 kg of water (about 210 gallons) per minute from a well 14.0 m deep and eject it with a speed of 18.0 m/s. a) How much work is done per minute in lifting the water? b) How much in giving the water the kinetic energy it has when ejected? c) What must be the power output of the pump?

6.88 Find the power output of the worker in Problem 6.71 as a function of time. What is the numerical value of the power (in watts) at $t = 4.00$ s?

6.89 A physics student spends part of her day walking between classes or for recreation, during which time she expends energy at an average rate of 280 W. The remainder of the day she is sitting in class, studying, or resting; during these activities, she expends energy at an average rate of 100 W. If she expends a total of 1.1×10^7 J of energy in a 24-hour day, how much of the day did she spend walking?

6.90 All birds, independent of their size, must maintain a power output of 10 to 25 watts per kilogram of body mass in order to fly by flapping their wings. (a) The Andean giant hummingbird *(Patagona gigas)* has mass 70 g and flaps its wings ten times per second while hovering. Estimate the amount of work done by such a hummingbird in each wingbeat. (b) A 70-kg athlete can maintain a power output of 1.4 kW for no more than a few seconds; the *steady* power output of a typical athlete is only 500 W or so. Is it possible for a human-powered aircraft to fly for extended periods by flapping its wings? Explain.

6.91 The Grand Coulee Dam is 1270 m long and 170 m high. The electrical power output from generators at its base is approximately 2000 MW. How many cubic meters of water must flow from the top of the dam per second to produce this amount of power if 92% of the work done on the water by gravity is converted to electrical energy? (Each cubic meter of water has a mass of 1000 kg.)

6.92 The engine of a car with mass m supplies a constant power P to the wheels to accelerate the car. You can ignore rolling friction and air resistance. The car is initially at rest. a) Show that the speed of the car is given as a function of time by $v = (2Pt/m)^{1/2}$. b) Show that the acceleration of the car is not constant, but is given as a function of time by $a = (P/2mt)^{1/2}$. c) Show that the displacement as a function of time is given by $x - x_0 = (8P/9m)^{1/2}t^{3/2}$.

6.93 Power of the Human Heart. The human heart is a powerful and extremely reliable pump. Each day it takes in and discharges about 7500 L of blood. Assume that the work done by the heart is equal to the work required to lift this amount of blood a height equal to that of the average American female (1.63 m). The density (mass per unit volume) of blood is 1.05×10^3 kg/m^3. a) How much work does the heart do in a day? b) What is its power output in watts?

6.94 Six diesel units in series can provide 13.4 MW of power to the lead car of a freight train. The diesel units have total mass 1.10×10^6 kg. The average car in the train has mass 8.2×10^4 kg and requires a horizontal pull of 2.8 kN to move at a constant 27 m/s on level tracks. a) How many cars can be in the train under these conditions? b) However, this would leave no power for accelerating or climbing hills. Show that the extra force needed to accelerate the train is about the same for a 0.10 m/s^2 acceleration or a 1.0% slope (slope angle $\alpha = \arctan 0.010$). c) With the 1.0% slope, show that an extra 2.9 MW of power is needed to maintain the 27 m/s speed of the diesel units. d) With 2.9 MW less power available, how many cars can the six diesel units pull up a 1.0% slope at a constant 27 m/s?

6.95 It takes a force of 53 kN on the lead car of a 16-car passenger train with mass 9.1×10^5 kg to pull it at a constant 45 m/s (101 mi/h) on level tracks. a) What power must the locomotive provide to the lead car? b) How much more power to the lead car than calculated in part (a) would be needed to give the train an acceleration of 1.5 m/s^2, at the instant that the train has a speed of 45 m/s on level tracks? c) How much more power to the lead car than that calculated in part (a) would be needed to move the train up a 1.5% grade (slope angle $\alpha = \arctan 0.015$) at a constant 45 m/s?

6.96 An object has several forces acting on it. One of these forces is $\vec{F} = axy\hat{\imath}$, a force in the x-direction whose magnitude depends on the position of the object, with $\alpha = 2.50$ N/m^2. Calculate the work done on the object by this force for the following displacements of the object: a) The object starts at the point $x = 0$, $y = 3.00$ m and moves parallel to the x-axis to the point $x = 2.00$ m, $y = 3.00$ m. b) The object starts at the point $x = 2.00$ m, $y = 0$ and moves in the y-direction to the point $x = 2.00$ m, $y = 3.00$ m. c) The object starts at the origin and moves on the line $y = 1.5x$ to the point $x = 2.00$ m, $y = 3.00$ m.

6.97 Cycling. For a touring bicyclist the drag coefficient $C(f_{\text{air}} = \frac{1}{2}CA\rho v^2)$ is 1.00, the frontal area A is 0.463 m^2, and the coefficient of rolling friction is 0.0045. The rider has mass 50.0 kg, and her bike has mass 12.0 kg. a) To maintain a speed of 12.0 m/s (about 27 mi/h) on a level road, what must the rider's power output to the rear wheel be? b) For racing, the same rider uses a different bike with a coefficient of rolling friction 0.0030 and mass 9.00 kg. She also crouches down, reducing her drag coefficient to 0.88 and reducing her frontal area to 0.366 m^2. What must her power output to the rear wheel be then to maintain a speed of 12.0 m/s? c) For the situation in part (b), what power output is required to maintain a speed of 6.0 m/s? Note the great drop in power requirement when the speed is only halved. (For more on aerodynamic speed limitations for a wide variety of human-powered vehicles see "The Aerodynamics of Human-Powered Land Vehicles," *Scientific American,* December, 1983.)

6.98 Automotive Power I. A truck engine transmits 28.0 kW (37.5 hp) to the driving wheels when the truck is traveling at a constant velocity of magnitude 60.0 km/h (37.3 mi/h) on a level road. a) What is the resisting force acting on the truck? b) Assume that 65% of the resisting force is due to rolling friction, and that the remainder is due to air resistance. If the force of rolling friction is independent of speed, and the force of air resistance is proportional to the square of the speed, what power will drive the truck at 30.0 km/h? At 120.0 km/h? Give your answers in kilowatts and in horsepower.

6.99 Automotive Power II. a) If 8.00 hp are required to drive a 1800-kg automobile at 60.0 km/h on a level road, what is the total retarding force due to friction, air resistance, and so on? b) What power is necessary to drive the car at 60.0 km/h up a 10.0% grade (a hill rising 10.0 m vertically in 100.0 m horizontally)? c) What power is necessary to drive the car at 60.0 km/h *down* a 1.00% grade? d) Down what percent grade would the car coast at 60.0 km/h?

Challenge Problems

6.100 On a winter's day in Maine, a warehouse worker is shoving boxes up a rough plank inclined at an angle α above the horizontal. The plank is partially covered with ice, with more ice near the bottom of the plank than near the top, so that the coefficient of friction increases with the distance x along the plank: $\mu = Ax$, where A is a positive constant and the bottom of the plank is at $x = 0$. (For this plank the coefficients of kinetic and static friction are equal: $\mu_k = \mu_s = \mu$.) The worker shoves a box up the plank so that it leaves the bottom of the plank moving at speed v_0. Show that when the box first comes to rest, it will remain at rest if

$$v_0^2 \geq \frac{3g \sin^2 \alpha}{A \cos \alpha}$$

6.101 A Spring with Mass. We usually ignore the kinetic energy of the moving coils of a spring, but let's try to get a reasonable approximation to this. Consider a spring of mass M, equilibrium length L_0, and spring constant k. The work done to stretch or compress the spring by a distance L is $\frac{1}{2}kX^2$, where $X = L - L_0$. a) Consider a spring, as described above, that has one end fixed and the other end moving with speed v. Assume that the speed of points along the length of the spring varies linearly with distance l from the fixed end. Assume also that the mass M of the spring is distributed uniformly along the length of the spring. Calculate the kinetic energy of the spring in terms of M and v. (*Hint:* Divide the spring into pieces of length dl; find the speed of each piece in terms of l, v, and L; find the mass of each piece in terms of dl, M, and L; and integrate from 0 to L. The result is *not* $\frac{1}{2}Mv^2$, since not all of the spring moves with the same speed.) In a spring gun, a spring of mass 0.243 kg and force constant 3200 N/m is compressed 2.50 cm from its unstretched length. When the trigger is pulled, the spring pushes horizontally on a 0.053-kg ball. The work done by friction is negligible. Calculate the ball's speed when the spring reaches its uncompressed length b) ignoring the mass of the spring; c) including, using the results of part (a), the mass of the spring. d) In part (c), what is the final kinetic energy of the ball and of the spring?

6.102 An airplane in flight is subject to an air resistance force proportional to the square of its speed v. But there is an additional resistive force because the airplane has wings. Air flowing over the wings is pushed down and slightly forward, so from Newton's third law the air exerts a force on the wings and airplane that is up and slightly backward (Fig. 6.30). The upward force is the lift force that keeps the airplane aloft, and the backward force is called *induced drag*. At flying speeds, induced drag is inversely proportional to v^2, so that the total air resistance force can be expressed by $F_{air} = \alpha v^2 + \beta/v^2$, where α and β are positive constants that depend on the shape and size of the airplane and the density of the air. For a Cessna 150, a small single-engine airplane, $\alpha = 0.30\ \text{N}\cdot\text{s}^2/\text{m}^2$ and $\beta = 3.5 \times 10^5\ \text{N}\cdot\text{m}^2/\text{s}^2$. In steady flight, the engine must provide a forward force that exactly balances the air resistance force. a) Calculate the speed (in km/h) at which this airplane will have the maximum *range* (that is, travel the greatest distance) for a given quantity of fuel. b) Calculate the speed (in km/h) for which the airplane will have the maximum *endurance* (that is, will remain in the air the longest time).

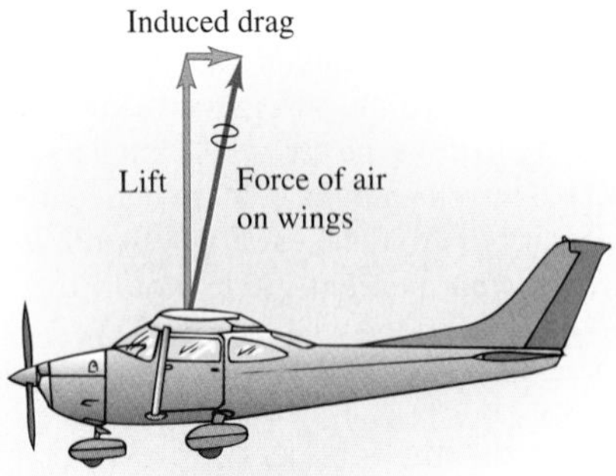

Figure 6.30 Challenge Problem 6.102.

6.103 Figure 6.31 shows the oxygen consumption rate of men walking and running at different speeds. The vertical axis shows the volume of oxygen (in cm^3) that a man consumes per kilogram of body mass per minute. Note the transition from walking to running that occurs naturally at about 9 km/h. The metabolism of 1 cm^3 of oxygen releases about 20 J of energy. Using the data in the graph, calculate the energy required for a 70-kg man to travel 1 km on foot at each of the following speeds a) 5 km/h (walking); b) 10 km/h (running); c) 15 km/h (running). d) Which speed is the most efficient, that is, requires the least energy to travel 1 km?

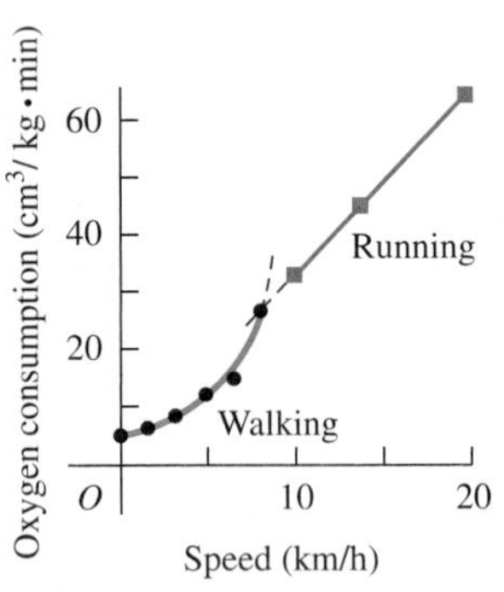

Figure 6.31 Challenge Problem 6.103.

6.104 General Proof of the Work-Energy Theorem. Consider a particle that moves along a curved path in space from (x_1, y_1, z_1) to (x_2, y_2, z_2). At the initial point, the particle has velocity $\vec{v} = v_{1x}\hat{\imath} + v_{1y}\hat{\jmath} + v_{1z}\hat{k}$. The path that the particle follows may be divided into infinitesimal segments $d\vec{l} = dx\hat{\imath} + dy\hat{\jmath} + dz\hat{k}$. As the particle moves, it is acted on by a net force $\vec{F} = F_x\hat{\imath} + F_y\hat{\jmath} + F_z\hat{k}$. The force components F_x, F_y, and F_z are in general functions of position. By the same sequence of steps used in Eqs. (6.11) through (6.13), prove the work-energy theorem for this general case. That is, prove that

$$W_{tot} = K_2 - K_1$$

where

$$W_{tot} = \int_{(x_1, y_1, z_1)}^{(x_2, y_2, z_2)} \vec{F}\cdot d\vec{l} = \int_{(x_1, y_1, z_1)}^{(x_2, y_2, z_2)} (F_x\,dx + F_y\,dy + F_z\,dz)$$

7 POTENTIAL ENERGY AND ENERGY CONSERVATION

As a competitive diver plummets toward the water, gravitational potential energy is converted into the diver's kinetic energy. When the diver enters the water, he loses speed and his kinetic energy is converted into thermal energy of the water, raising its temperature slightly. Careful measurements show that no energy is lost in this process—instead, the energy merely changes form.

? As the diver enters the water, is the water doing positive or negative work on him? Is the force of gravity doing positive or negative work on him?

When a diver jumps off a high board into a swimming pool, he hits the water moving pretty fast, with a lot of kinetic energy. Where does that energy come from? The answer we learned in Chapter 6 was that the gravitational force (his weight) does work on the diver as he falls. The diver's kinetic energy—energy associated with his *motion*—increases by an amount equal to the work done.

However, there is a very useful alternative way to think about work and kinetic energy. This new approach is based on the concept of *potential energy,* which is energy associated with the *position* of a system rather than its motion. In this approach, there is *gravitational potential energy* even while the diver is standing on the high board. Energy is not added to the earth–diver system as the diver falls, but rather a storehouse of energy is *transformed* from one form (potential energy) to another (kinetic energy) as he falls. In this chapter we'll see how the work-energy theorem explains this transformation.

If the diver bounces on the end of the board before he jumps, the bent board stores a second kind of potential energy called *elastic potential energy.* We'll discuss elastic potential energy of simple systems such as a stretched or compressed spring. (An important third kind of potential energy is associated with the positions of electrically charged particles relative to each other. We'll encounter this potential energy in Chapter 23.)

We will prove that in some cases the sum of a system's kinetic and potential energy, called the *total mechanical energy* of the system, is constant during the motion of the system. This will lead us to the general statement of the *law of conservation of energy,* one of the most fundamental and far-reaching principles in all of science.

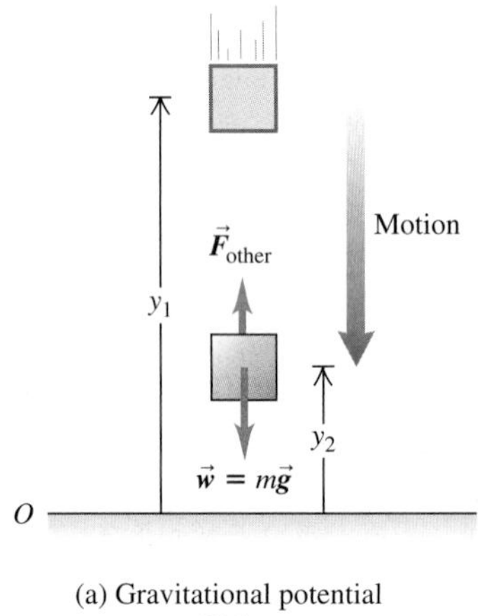

7.1 Work is done by the gravitational force $\vec{w}$ during the vertical motion of a body from an initial height y_1 to a final height y_2. The gravitational potential energy (a) decreases if the body moves downward and (b) increases if it moves upward.

7.1 | Gravitational Potential Energy

A particle gains or loses kinetic energy because it interacts with other objects that exert forces on it. We learned in Chapter 6 that during any interaction the change in a particle's kinetic energy is equal to the total work done on the particle by the forces that act on it.

In many situations it seems as though energy has been stored in a system, to be recovered later. It's like a savings account in a bank; you deposit money, and then you can draw it out later. For example, you must do work on the hammerhead of a pile driver to lift it. It seems reasonable that in hoisting the hammerhead into the air you are storing energy in the system, energy that is later converted into kinetic energy as the hammerhead falls. Or consider your cousin Throckmorton on a swing. Suppose you give him a shove, which gives him some kinetic energy, and then you let him swing freely back and forth. He stops momentarily when he reaches the front and back end points of his arc, so at these points he has no kinetic energy. But he regains his kinetic energy as he passes through the low point in the arc. It seems as though at the high points the energy is stored in some other form, related to his height above the ground, and is converted back to kinetic energy as he swings toward the low point.

Both these examples point to the idea of an energy associated with the *position* of bodies in a system. This kind of energy is a measure of the *potential* or *possibility* for work to be done; when a hammerhead is raised into the air, there is a potential for work to be done on it by the gravitational force, but only if the hammerhead is allowed to fall to the ground. For this reason, energy associated with position is called **potential energy.** Our discussion suggests that there is potential energy associated with a body's weight and its height above the ground. We call this *gravitational potential energy.*

When a body falls without air resistance, gravitational potential energy decreases and the body's kinetic energy increases. But in Chapter 6 we used the work-energy theorem to conclude that the kinetic energy of a falling body increases because the force of the earth's gravity (the body's weight) does work on the body. Let's use the work-energy theorem to show that these two descriptions of a falling body are equivalent and to derive the expression for gravitational potential energy.

Let's consider a body with mass m that moves along the (vertical) y-axis, as in Fig. 7.1. The forces acting on it are its weight, with magnitude $w = mg$, and possibly some other forces; we call the vector sum (resultant) of all the other forces $\vec{F}_{other}$.

We'll assume that the body stays close enough to the earth's surface that the weight is constant. (We'll find in Chapter 12 that weight decreases with altitude.) We want to find the work done by the weight when the body drops from a height y_1 above the origin to a lower height y_2 (Fig. 7.1a). The weight and displacement are in the same direction, so the work W_{grav} done on the body by its weight is positive;

$$W_{grav} = Fs = w(y_1 - y_2) = mgy_1 - mgy_2 \tag{7.1}$$

This expression also gives the correct work when the body moves upward and y_2 is greater than y_1 (Fig. 7.1b). In that case the quantity $(y_1 - y_2)$ is negative, and W_{grav} is negative because the weight and displacement are opposite in direction.

Equation (7.1) shows that we can express W_{grav} in terms of the values of the quantity mgy at the beginning and end of the displacement. This quantity, the product of the weight mg and the height y above the origin of coordinates, is called the **gravitational potential energy,** U:

$$U = mgy \qquad \text{(gravitational potential energy)} \tag{7.2}$$

Its initial value is $U_1 = mgy_1$ and its final value is $U_2 = mgy_2$. The change in U is the final value minus the initial value, or $\Delta U = U_2 - U_1$. We can express the work W_{grav} done by the gravitational force during the displacement from y_1 to y_2 as

$$W_{grav} = U_1 - U_2 = -(U_2 - U_1) = -\Delta U \tag{7.3}$$

The negative sign in front of ΔU is *essential.* When the body moves up, y increases, the work done by the gravitational force is negative, and the gravitational potential energy increases $(\Delta U > 0)$. When the body moves down, y decreases, the gravitational force does positive work, and the gravitational potential energy decreases $(\Delta U < 0)$. It's like drawing money out of the bank (decreasing U) and spending it (doing positive work). As Eq. (7.3) shows, the unit of potential energy is the joule (J), the same unit as is used for work.

CAUTION Although you might be tempted to do so, it is not correct to call $U = mgy$ the "gravitational potential energy of the body." The reason is that gravitational potential energy is a *shared* property of the body and the earth. Gravitational potential energy increases if the earth stays fixed and the height of the body increases; it also increases if the body is fixed in space and the earth is moved away from the body. Notice that the formula $U = mgy$ involves characteristics of both the body (its mass m) and the earth (the value of g).

Conservation of Mechanical Energy (Gravitational Forces Only)

To see what gravitational potential energy is good for, suppose the body's weight is the *only* force acting on it, so $\vec{F}_{other} = \mathbf{0}$. The body is then falling freely with no air resistance, and can be moving either up or down. Let its speed at point y_1 be v_1 and let its speed at y_2 be v_2. The work-energy theorem, Eq. (6.6), says that the total work done on the body equals the change in the body's kinetic energy; $W_{tot} = \Delta K = K_2 - K_1$. If gravity is the only force that acts, then from Eq. (7.3), $W_{tot} = W_{grav} = -\Delta U = U_1 - U_2$. Putting these together, we get

$$\Delta K = -\Delta U \quad \text{or} \quad K_2 - K_1 = U_1 - U_2$$

which we can rewrite as

$$K_1 + U_1 = K_2 + U_2 \qquad \text{(if only gravity does work)} \tag{7.4}$$

or

$$\frac{1}{2}mv_1^2 + mgy_1 = \frac{1}{2}mv_2^2 + mgy_2 \qquad \text{(if only gravity does work)} \tag{7.5}$$

We now define the sum $K + U$ of kinetic and potential energy to be E, the **total mechanical energy of the system.** By "system" we mean the body of mass m and the earth considered together, because gravitational potential energy U is a shared property of both bodies. Then $E_1 = K_1 + U_1$ is the total mechanical energy at y_1 and $E_2 = K_2 + U_2$ is the total mechanical energy at y_2. Equation (7.4) says that when the body's weight is the only force doing work on it, $E_1 = E_2$. That is, E is constant; it has the same value at y_1 and y_2. But since the positions y_1 and y_2 are arbitrary points in the motion of the body, the total mechanical energy E has the same value at *all* points during the motion;

$$E = K + U = \text{constant} \qquad \text{(if only gravity does work)}$$

A quantity that always has the same value is called a *conserved* quantity. *When only the force of gravity does work, the total mechanical energy is constant, that is, is conserved* (Fig. 7.2). This is our first example of the **conservation of mechanical energy.**

When we throw a ball into the air, its speed decreases on the way up as kinetic energy is converted to potential energy; $\Delta K < 0$ and $\Delta U > 0$. On the way back down, potential energy is converted back to kinetic energy and the ball's speed increases; $\Delta K > 0$ and $\Delta U < 0$. But the *total* mechanical energy (kinetic plus potential) is the same at every point in the motion, provided that no force other than gravity does work on the ball (that is, air resistance must be negligible). It's still true that the gravitational force does work on the body as it moves up or down, but we no longer have to calculate work directly; keeping track of changes in the value of U takes care of this completely.

CAUTION An important point about gravitational potential energy is that it doesn't matter what height we choose to be $y = 0$, the origin of coordinates. If we shift the origin for y, the values of y_1 and y_2 both change, but the *difference*

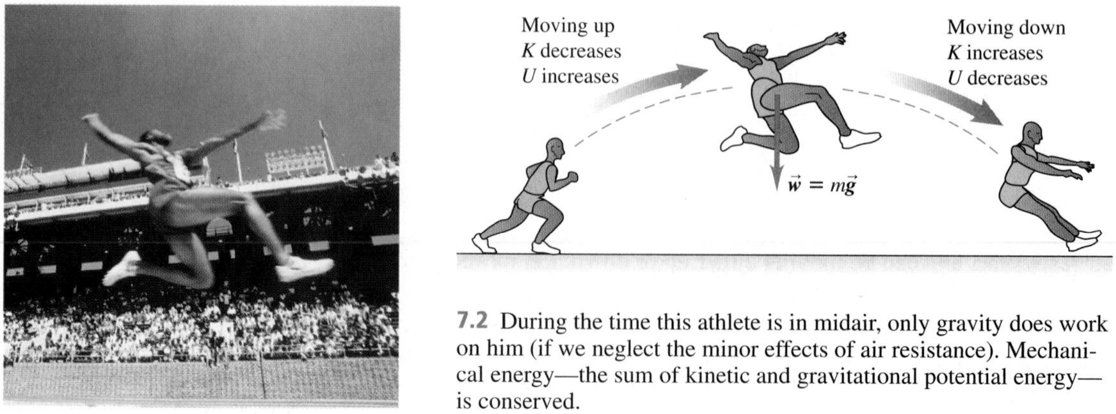

7.2 During the time this athlete is in midair, only gravity does work on him (if we neglect the minor effects of air resistance). Mechanical energy—the sum of kinetic and gravitational potential energy—is conserved.

$y_2 - y_1$ stays the same. It follows that although U_1 and U_2 depend on where we place the origin, the difference $U_2 - U_1 = mg(y_2 - y_1)$ does not. The physically significant quantity is not the value of U at a particular point, but only the *difference* in U between two points. So we can define U to be zero at whatever point we choose without affecting the physics of the situation. This is shown in the following example.

Example 7.1 Height of a baseball from energy conservation

You throw a 0.145-kg baseball straight up in the air, giving it an initial upward velocity of magnitude 20.0 m/s. Use conservation of energy to find how high it goes, ignoring air resistance.

SOLUTION

IDENTIFY: After the ball leaves your hand, the only force doing work on the ball is its weight. Hence we can use conservation of mechanical energy.

SET UP: We'll use Eqs. (7.4) and (7.5), taking point 1 to be where the ball leaves your hand and point 2 to be where the ball reaches its maximum height. As in Fig. 7.1, we take the y-axis to point vertically upward. The ball's speed at point 1 is $v_1 = 20.0$ m/s. The ball is instantaneously at rest at the high point of its motion (point 2), so $v_2 = 0$.

Our target variable is how far the ball moves vertically between these two points, so our target variable is the displacement $y_2 - y_1$. For simplicity, let's take the origin to be at point 1, where the ball leaves your hand. Then $y_1 = 0$ (Fig. 7.3) and the target variable is just y_2.

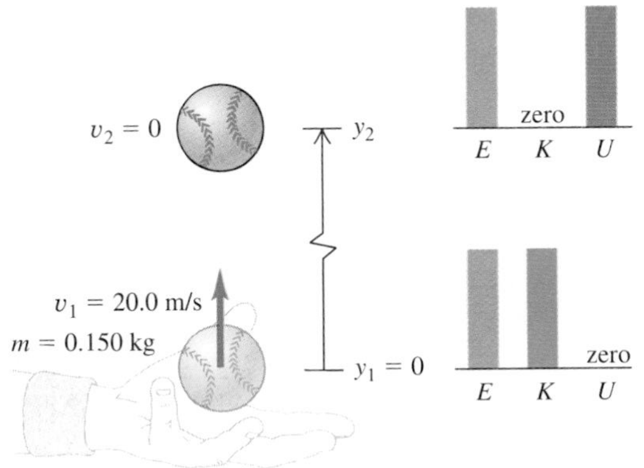

7.3 After a baseball leaves your hand, the only force acting on it is its weight (neglecting air resistance), so mechanical energy $E = K + U$ is conserved. The energy bar graphs show the values of E, K, and U at $y_1 = 0$ and y_2.

EXECUTE: Since $y_1 = 0$, the potential energy at point 1 is $U_1 = mgy_1 = 0$. Furthermore, since the ball is at rest at point 2, the kinetic energy at that point is $K_2 = \frac{1}{2}mv_2^2 = 0$. Hence Eq. (7.4), which says that $K_1 + U_1 = K_2 + U_2$, becomes

$$K_1 = U_2$$

As the energy bar graphs in Fig. 7.3 show, the kinetic energy of the ball at point 1 is completely converted into gravitational potential energy at point 2. At point 1 the kinetic energy is

$$K_1 = \frac{1}{2}mv_1^2 = \frac{1}{2}(0.145\text{ kg})(20.0\text{ m/s})^2 = 29.0\text{ J}$$

This equals the gravitational potential energy $U_2 = mgy_2$ at point 2, so

$$y_2 = \frac{U_2}{mg} = \frac{29.0\text{ J}}{(0.145\text{ kg})(9.80\text{ m/s}^2)} = 20.4\text{ m}$$

We can also solve the equation $K_1 = U_2$ algebraically for y_2:

$$\frac{1}{2}mv_1^2 = mgy_2$$

$$y_2 = \frac{v_1^2}{2g} = \frac{(20.0\text{ m/s})^2}{2(9.80\text{ m/s}^2)} = 20.4\text{ m}$$

EVALUATE: The mass divides out, as we should expect; we learned in Chapter 2 that the motion of a body in free fall doesn't depend on its mass. Indeed, we could have derived the result $y_2 = v_1^2/2g$ using Eq. (2.13).

In carrying out the calculation above, we chose the origin to be at point 1, so $y_1 = 0$ and $U_1 = 0$. What happens if we make a different choice? As an example, suppose we choose the origin to be 5.0 m below point 1, so $y_1 = 5.0$ m. With this choice the total mechanical energy at point 1 is part kinetic and part potential, while at point 2 it's purely potential energy. If you work through the calculation again with this choice of origin, you'll find $y_2 = 25.4$ m; this is 20.4 m above point 1, just as with the first choice of origin. In every problem, it's up to you to choose the height at which $U = 0$; don't agonize over the choice, however, because the physics of the answer doesn't depend on your choice.

ActivPhysics Online

5.2 Upward-Moving Elevator Stops

5.3 Stopping a Downward-Moving Elevator

5.6 Skier Speed

Effect of Other Forces

When other forces act on the body in addition to its weight, then $\vec{F}_{other}$ in Fig. 7.1 is *not* zero. For the pile driver described in Example 6.5 (Section 6.2), the force applied by the hoisting cable and the friction with the vertical guide rails are examples of forces that might be included in $\vec{F}_{other}$. The gravitational work W_{grav} is still given by Eq. (7.3), but the total work W_{tot} is then the sum of W_{grav} and the work done by $\vec{F}_{other}$. We will call this additional work W_{other}, so the total work done by all forces is $W_{tot} = W_{grav} + W_{other}$. Equating this to the change in kinetic energy, we have

$$W_{other} + W_{grav} = K_2 - K_1 \tag{7.6}$$

Also, from Eq. (7.3), $W_{grav} = U_1 - U_2$, so

$$W_{other} + U_1 - U_2 = K_2 - K_1$$

which we can rearrange in the form

$$K_1 + U_1 + W_{other} = K_2 + U_2 \tag{7.7}$$

(if forces other than gravity do work)

Finally, using the appropriate expressions for the various energy terms, we obtain

$$\frac{1}{2}mv_1^2 + mgy_1 + W_{other} = \frac{1}{2}mv_2^2 + mgy_2 \tag{7.8}$$

(if forces other than gravity do work)

The meaning of Eqs. (7.7) and (7.8) is this: *The work done by all forces* other than the gravitational force *equals the change in the total mechanical energy* $E = K + U$ *of the system, where U is the gravitational potential energy.* When W_{other} is positive, E increases, and $K_2 + U_2$ is greater than $K_1 + U_1$. When W_{other} is negative, E decreases. In the special case in which no forces other than the body's weight do work, $W_{other} = 0$. The total mechanical energy is then constant, and we are back to Eq. (7.4) or (7.5).

Problem-Solving Strategy: Problems Using Mechanical Energy

IDENTIFY *the relevant concepts:* First decide whether the problem should be solved by energy methods, by using $\Sigma\vec{F} = m\vec{a}$ directly, or by a combination of these strategies. The energy approach is particularly useful when the problem involves motion with varying forces, along a curved path (discussed later in this section), or both. But if the problem involves elapsed time, the energy approach is usually *not* the best choice, because this approach doesn't involve time directly.

SET UP *the problem using the following steps:*

1. When using the energy approach, first decide what the initial and final states (the positions and velocities) of the system are. Use the subscript 1 for the initial state and the subscript 2 for the final state. It helps to draw sketches showing the initial and final states.
2. Define your coordinate system, particularly the level at which $y = 0$. You will use it to compute gravitational potential energies. Equation (7.2) assumes that the positive direction for y is upward; we suggest that you use this choice consistently.
3. Identify all nongravitational forces that do work. A free-body diagram is always helpful. If some of the quantities you need are unknown, represent them by algebraic symbols.
4. List the unknown and known quantities, including the coordinates and velocities at each point. Decide which unknowns are your target variables.

EXECUTE *the solution:* Write expressions for the initial and final kinetic and potential energies—that is, K_1, K_2, U_1, and U_2. In general some of these quantities will be known and some will be unknown. Then relate the kinetic and potential energies and the nongravitational work W_{other} using Eq. (7.7). (You will have to calculate W_{other} in terms of the nongravitational forces.) If there is no nongravitational work, this expression becomes Eq. (7.5). It's helpful to draw bar graphs showing the initial and final values of K, U, and $E = K + U$. Then solve to find whatever unknown quantity is required.

EVALUATE *your answer:* Check whether your answer makes physical sense. Keep in mind, here and in later sections, that the work done by each force must be represented either in $U_1 - U_2 = -\Delta U$ or as W_{other}, but *never* in both places. The gravitational work is included in ΔU, so make sure you did not include it again in W_{other}.

Example 7.2 Work and energy throwing a baseball

In Example 7.1, suppose your hand moves up 0.50 m while you are throwing the ball, which leaves your hand with an upward velocity of 20.0 m/s. Again ignore air resistance. a) Assuming that your hand exerts a constant upward force on the ball, find the magnitude of that force. b) Find the speed of the ball at a point 15.0 m above the point where it leaves your hand.

SOLUTION

IDENTIFY: In Example 7.1 we used conservation of mechanical energy because only gravity did work. In this example, however, we must also include the nongravitational work done by your hand.

SET UP: Figure 7.4 shows a diagram of the situation, including a free-body diagram for the ball while it is being thrown. The ball's motion occurs in two stages: while it is in contact with your hand and after it leaves your hand. To keep track of these stages, we let point 1 be where your hand first starts to move, point 2 be where the ball leaves your hand, and point 3 be where the ball is 15.0 m above point 2. The nongravitational force $\vec{F}$ of your hand acts only between points 1 and 2. Using the same coordinate system as in Example 7.1, we have $y_1 = -0.50$ m, $y_2 = 0$, and $y_3 = 15.0$ m. The ball starts at rest at point 1, so $v_1 = 0$, and we are given that the ball's speed as it leaves your hand is $v_2 = 20.0$ m/s. Our target variables are (a) the magnitude F of the force of your hand and (b) the speed v_3 at point 3.

EXECUTE:

a) To determine the magnitude of $\vec{F}$, we'll first use Eq. (7.7) to calculate the work W_{other} done by this force. We have

$$K_1 = 0$$
$$U_1 = mgy_1 = (0.145\text{ kg})(9.80\text{ m/s}^2)(-0.50\text{ m}) = -0.71\text{ J}$$
$$K_2 = \frac{1}{2}mv_2^2 = \frac{1}{2}(0.145\text{ kg})(20.0\text{ m/s})^2 = 29.0\text{ J}$$
$$U_2 = mgy_2 = (0.145\text{ kg})(9.80\text{ m/s}^2)(0) = 0$$

The initial potential energy U_1 is negative because the ball was initially below the origin. According to Eq. (7.7), $K_1 + U_1 + W_{other} = K_2 + U_2$, so

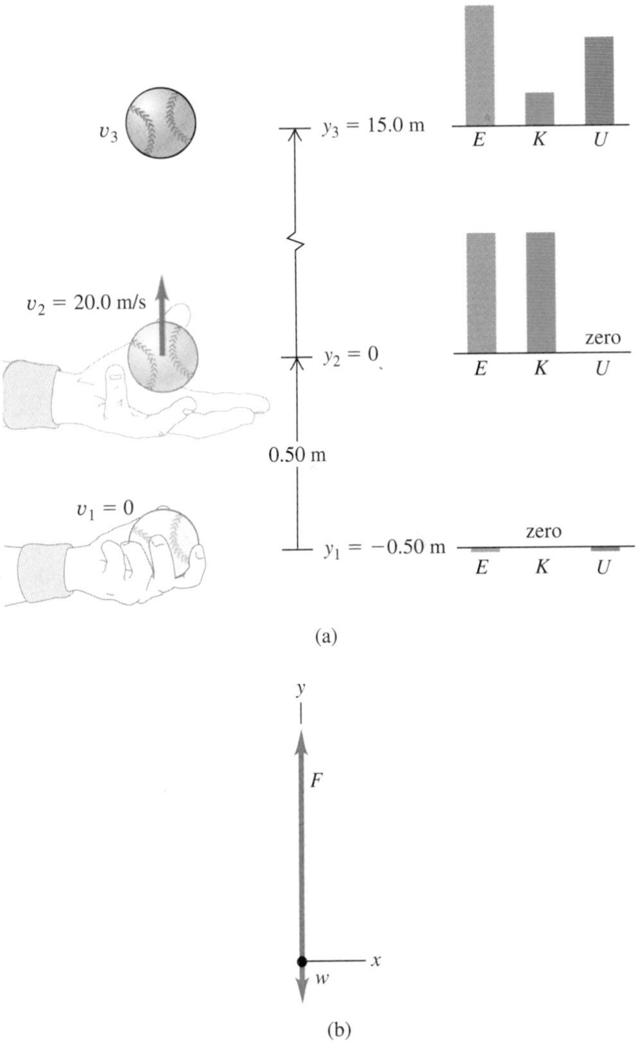

7.4 (a) Throwing a ball vertically upward. (b) The free-body diagram for the ball as the force $\vec{F}$ applied by the hand does the work W_{other} on the ball. The force $\vec{F}$ and gravity both act between y_1 and y_2. From y_2 to y_3, only gravity acts on the ball.

$$W_{other} = (K_2 - K_1) + (U_2 - U_1)$$
$$= (29.0\text{ J} - 0) + (0 - (-0.71\text{ J})) = 29.7\text{ J}$$

The kinetic energy of the ball increases by $K_2 - K_1 = 29.0$ J, and the potential energy increases by $U_2 - U_1 = 0.71$ J; the sum is $E_2 - E_1$, the change in total mechanical energy, which is equal to W_{other}.

Assuming that the upward force $\vec{F}$ that your hand applies is constant, the work W_{other} done by this force is equal to the magnitude F of the force multiplied by the upward displacement $y_2 - y_1$ over which it acts:

$$W_{other} = F(y_2 - y_1)$$
$$F = \frac{W_{other}}{y_2 - y_1} = \frac{29.7\text{ J}}{0.50\text{ m}} = 59\text{ N}$$

This is about 40 times greater than the weight of the ball.

b) To find the speed at point 3, note that between points 2 and 3, total mechanical energy is conserved; the force of your hand no longer acts, so $W_{other} = 0$. We can then find the kinetic energy at point 3 using Eq. (7.4):

$$K_2 + U_2 = K_3 + U_3$$
$$U_3 = mgy_3 = (0.145\text{ kg})(9.80\text{ m/s}^2)(15.0\text{ m}) = 21.3\text{ J}$$
$$K_3 = (K_2 + U_2) - U_3 = (29.0\text{ J} + 0\text{ J}) - 21.3\text{ J} = 7.7\text{ J}$$

Since $K_3 = \frac{1}{2}mv_{3y}^2$, where v_{3y} is the y-component of the ball's velocity at point 3, we have

$$v_{3y} = \pm\sqrt{\frac{2K_3}{m}} = \pm\sqrt{\frac{2(7.7\text{ J})}{0.145\text{ kg}}} = \pm 10\text{ m/s}$$

The significance of the plus-or-minus sign is that the ball passes point 3 *twice*, once on the way up and again on the way down. The total mechanical energy E is constant and equal to 29.0 J while the ball is in free fall, and the potential energy at point 3 is $U_3 = 21.3$ J whether the ball is moving up or down. So at point 3, the ball's kinetic energy K_3 and *speed* don't depend on the direction the ball is moving. The velocity v_{3y} is positive $(+10\text{ m/s})$ when the ball is moving up and negative (-10 m/s) when it is moving down; the speed v_3 is 10 m/s in either case.

EVALUATE: As a check on our result, recall from Example 7.1 that the ball reaches a maximum height $y = 20.4$ m. At that point all of the kinetic energy that the ball had when it left your hand at $y = 0$ has been converted to gravitational potential energy. At $y = 15.0$ m, the ball is about three-fourths of the way to its maximum height, so about three-fourths of its mechanical energy should be in the form of potential energy. Can you show that this is true from our results for K_3 and U_3?

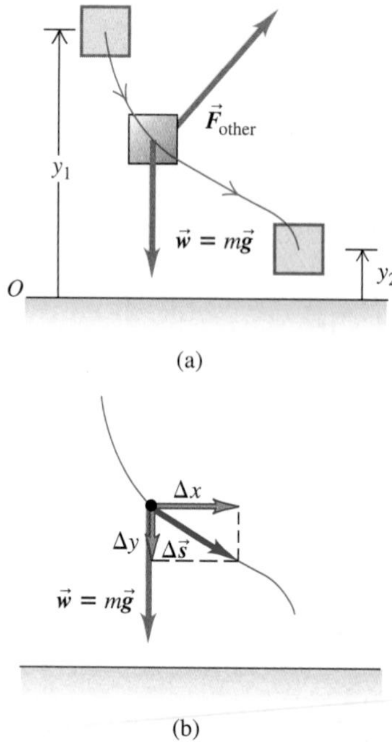

7.5 (a) A displacement along a curved path. (b) The work done by the gravitational force $\vec{w} = m\vec{g}$ depends only on the vertical component of displacement Δy (in this figure, Δy is negative).

Gravitational Potential Energy for Motion Along a Curved Path

In our first two examples the body moved along a straight vertical line. What happens when the path is slanted or curved (Fig. 7.5a)? The body is acted on by the gravitational force $\vec{w} = m\vec{g}$ and possibly by other forces whose resultant we call $\vec{F}_{other}$. To find the work done by the gravitational force during this displacement, we divide the path up into small segments $\Delta\vec{s}$; a typical segment is shown in Fig. 7.5b. The work done by the gravitational force over this segment is the scalar product of the force and the displacement. In terms of unit vectors, the force is $\vec{w} = m\vec{g} = -mg\hat{j}$ and the displacement is $\Delta\vec{s} = \Delta x\hat{i} + \Delta y\hat{j}$, so the work done by the gravitational force is

$$\vec{w}\cdot\Delta\vec{s} = -mg\hat{j}\cdot(\Delta x\hat{i} + \Delta y\hat{j}) = -mg\Delta y$$

The work done by gravity is the same as though the body had been displaced vertically a distance Δy, with no horizontal displacement. This is true for every segment, so the *total* work done by the gravitational force is $-mg$ multiplied by the *total* vertical displacement $(y_2 - y_1)$:

$$W_{grav} = -mg(y_2 - y_1) = mgy_1 - mgy_2 = U_1 - U_2$$

This is the same as Eq. (7.1) or (7.3), in which we assumed a purely vertical path. So even if the path a body follows between two points is curved, the total work done by the gravitational force depends only on the difference in height between the two points of the path. This work is unaffected by any horizontal motion that may occur. So *we can use the same expression for gravitational potential energy whether the body's path is curved or straight.*

Conceptual Example 7.3 Energy in projectile motion

A batter hits two identical baseballs with the same initial speed and height but different initial angles. Prove that at a given height h, both balls have the same speed if air resistance can be neglected.

SOLUTION

If there is no air resistance, the only force acting on each ball after it is hit is its weight. Hence the total mechanical energy for each ball is constant. Figure 7.6 shows the trajectories of two balls batted at the same height with the same initial speed, and thus the same total mechanical energy, but with different initial angles. At all points at the same height the potential energy is the same. Thus the kinetic energy at this height must be the same for both balls, and the speeds are the same.

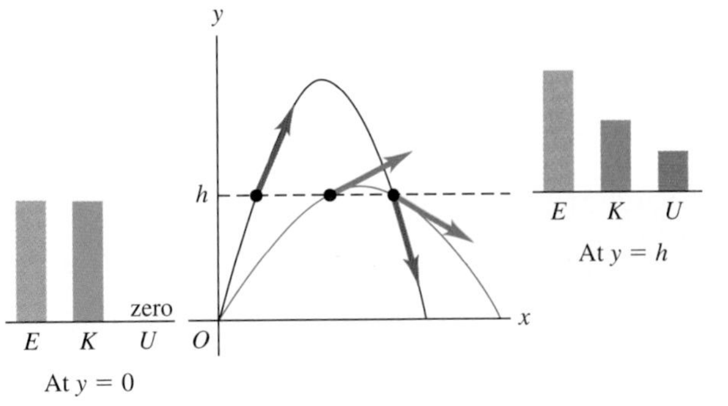

7.6 For the same initial speed and initial height, the speed of a projectile at a given elevation h is always the same, neglecting air resistance.

Example 7.4 Maximum height of a projectile, using energy methods

In Example 3.10 (Section 3.3) we derived an expression for the maximum height h of a projectile launched with initial speed v_0 at initial angle α_0:

$$h = \frac{v_0^2 \sin^2 \alpha_0}{2g}$$

Derive this expression using energy considerations.

SOLUTION

IDENTIFY and SET UP: We neglect air resistance, so just as in Conceptual Example 7.3, total mechanical energy is conserved. Let point 1 at $y = 0$ be the launch point, where the speed is $v_1 = v_0$, and let point 2 at $y = h$ be the highest point on the trajectory (Fig. 7.7). Our target variable is the maximum height h, at which the kinetic energy is minimum and the gravitational potential energy is maximum.

This appears to be an easy problem: the potential energy at point 2 is $U_2 = mgh$, so it may seem that all we need to do is solve the energy-conservation equation $K_1 + U_1 = K_2 + U_2$ for U_2. However, while we know the initial kinetic and potential energies $(K_1 = \frac{1}{2}mv_1^2 = \frac{1}{2}mv_0^2$ and $U_1 = 0)$, we don't know the speed or kinetic energy at point 2. To get around this issue, we'll use two results from our study of projectile motion in Chapter 3: (1) the x-component of acceleration is zero, so the x-component of velocity is constant, and (2) the y-component of velocity is zero at point 2 (the highest point of the trajectory).

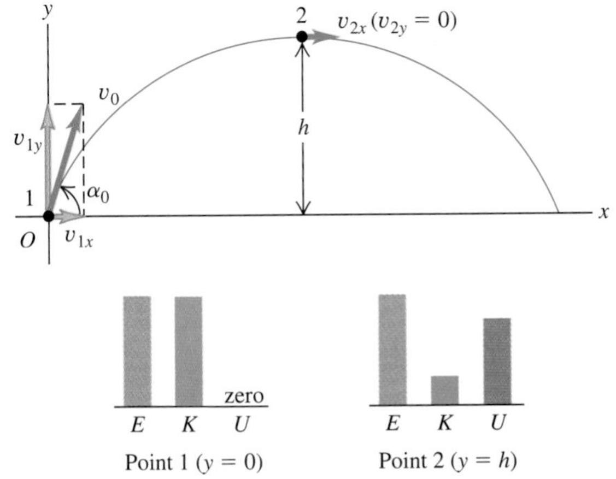

7.7 Trajectory of a projectile.

EXECUTE: We can express the kinetic energy at each point in terms of the components of velocity, using $v^2 = v_x^2 + v_y^2$:

$$K_1 = \frac{1}{2}m(v_{1x}^2 + v_{1y}^2)$$

$$K_2 = \frac{1}{2}m(v_{2x}^2 + v_{2y}^2)$$

Conservation of energy then gives $K_1 + U_1 = K_2 + U_2$, so

$$\frac{1}{2}m(v_{1x}^2 + v_{1y}^2) + 0 = \frac{1}{2}m(v_{2x}^2 + v_{2y}^2) + mgh$$

To simplify this, we multiply through by $2/m$ to obtain

$$v_{1x}^2 + v_{1y}^2 = v_{2x}^2 + v_{2y}^2 + 2gh$$

Now we use our results from projectile motion. Since the x-component of velocity doesn't change, $v_{1x} = v_{2x}$ and we can cancel the v_x^2 terms from both sides of the equation given above. Furthermore, because the projectile has zero vertical velocity at the highest point of its motion, $v_{2y} = 0$. Hence we get

$$v_{1y}^2 = 2gh$$

But v_{1y} is just the y-component of initial velocity, which is equal to $v_0 \sin \alpha_0$. Making this substitution and solving for h, we find

$$h = \frac{v_0^2 \sin^2 \alpha_0}{2g}$$

EVALUATE: This agrees with the result of Example 3.10, as it must.

Example 7.5 Calculating speed along a vertical circle

Your cousin Throckmorton skateboards down a curved playground ramp. Treating Throcky and his skateboard as a particle, he moves through a quarter-circle with radius R (Fig. 7.8). The total mass of Throcky and his skateboard is 25.0 kg. He starts from rest and there is no friction. (a) Find his speed at the bottom of the ramp. (b) Find the normal force that acts on him at the bottom of the curve.

SOLUTION

IDENTIFY: We can't use the equations of motion with constant acceleration; the acceleration isn't constant because the slope decreases as Throcky descends. Instead, we'll use the energy approach. Since Throcky moves along a circular arc, we'll also use what we learned about circular motion in Section 5.4.

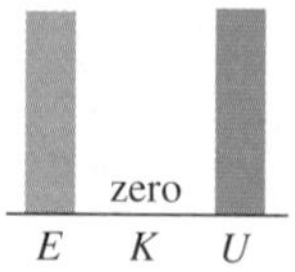

SET UP: Since there is no friction, the only force other than Throcky's weight is the normal force $\vec{n}$ exerted by the ramp (Fig. 7.8b). Although this force acts all along the path, it does *zero* work because $\vec{n}$ is perpendicular to Throcky's velocity at every point. Hence $W_{\text{other}} = 0$ and mechanical energy is conserved.

Take point 1 at the starting point and point 2 at the bottom of the curved ramp, and let $y = 0$ be at the bottom of the ramp (Fig. 7.8a). Then $y_1 = R$ and $y_2 = 0$. (We are treating Throcky as if his entire mass were concentrated at his center.) Throcky starts at rest at the top, so $v_1 = 0$. Our target variable in part (a) is his speed at the bottom, v_2. In part (b) we want to find the magnitude n of the normal force at point 2. As this force does no work, it doesn't appear in the energy equation, so we'll use Newton's second law instead.

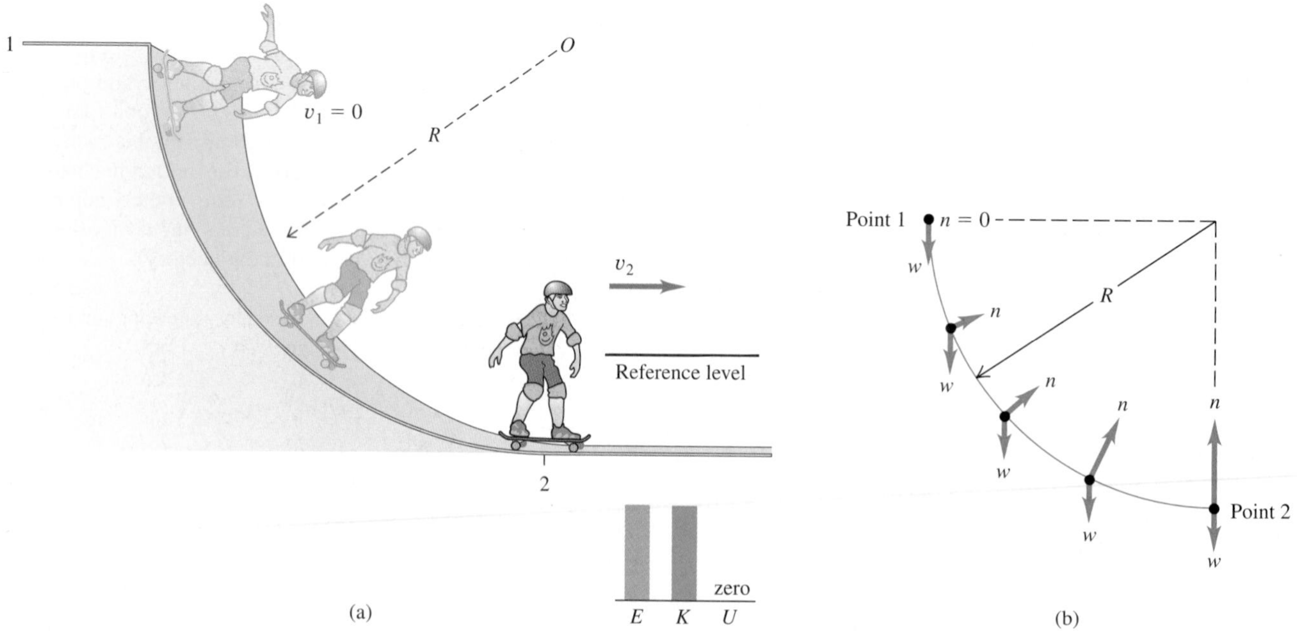

7.8 (a) Throcky skateboarding down a frictionless circular ramp. The total mechanical energy is constant. (b) Free-body diagrams for Throcky and his skateboard at various points on the ramp.

EXECUTE:
a) The various energy quantities are

$$K_1 = 0 \qquad U_1 = mgR$$

$$K_2 = \frac{1}{2}mv_2^2 \qquad U_2 = 0$$

From conservation of energy,

$$K_1 + U_1 = K_2 + U_2$$

$$0 + mgR = \frac{1}{2}mv_2^2 + 0$$

$$v_2 = \sqrt{2gR}$$

The speed is the same as if Throcky had fallen vertically through a height R, and it is independent of his mass.

As a numerical example, let $R = 3.00$ m. Then

$$v_2 = \sqrt{2(9.80\ \text{m/s}^2)(3.00\ \text{m})} = 7.67\ \text{m/s}$$

Notice that this answer doesn't depend on the ramp being circular; no matter what the shape of the ramp, Throcky will have the same speed $v_2 = \sqrt{2gR}$ at the bottom. This would be true even if the wheels of his skateboard lost contact with the ramp during the ride, because only the gravitational force would still do work.

b) To find n at point 2 using Newton's second law, we need the free-body diagram at that point (Fig. 7.8b). At point 2, Throcky is moving at speed $v_2 = \sqrt{2gR}$ in a circle of radius R; his acceleration is toward the center of the circle and has magnitude

$$a_{\text{rad}} = \frac{v_2^2}{R} = \frac{2gR}{R} = 2g$$

If we take the positive y-direction to be upward, the y-component of Newton's second law is

$$\sum F_y = n + (-w) = ma_{\text{rad}} = 2mg$$
$$n = w + 2mg = 3mg$$

At point 2 the normal force is three times Throcky's weight. This result is independent of the radius of the circular ramp. We learned in Example 5.10 (Section 5.2) and Example 5.25 (Section 5.4) that the magnitude of n is the *apparent weight,* so Throcky feels as though he weighs three times his true weight mg. But as soon as he reaches the horizontal part of the ramp to the right of point 2, the normal force decreases to $w = mg$ and Throcky feels normal again. Can you see why?

EVALUATE: This example shows a general rule about the role of forces in problems in which we use energy techniques: What matters is not simply whether a force acts, but whether that force does work. If the force does no work, like the normal force $\vec{n}$ in this example, then it does not appear at all in Eq. (7.7), $K_1 + U_1 + W_{\text{other}} = K_2 + U_2$.

Notice we had to use *both* the energy approach and Newton's second law to solve this problem; energy conservation gave us the speed and $\sum\vec{F} = m\vec{a}$ gave us the normal force. For each part of the problem we used the technique that most easily led us to the answer.

Example 7.6 A vertical circle with friction

In Example 7.5, suppose that the ramp is not frictionless and that Throcky's speed at the bottom is only 6.00 m/s. What work was done by the friction force acting on him? Use $R = 3.00$ m.

SOLUTION

IDENTIFY and SET UP: We use the same coordinate system and the same initial and final points as in Example 7.5 (Fig. 7.9). Again the normal force does no work, but now there is a friction force $\vec{f}$ that *does* do work. Hence the nongravitational work done on Throcky between points 1 and 2, W_{other}, is just equal to the work done by friction, W_f. This is our target variable, which we'll find using Eq. (7.7).

EXECUTE: The energy quantities are

$$K_1 = 0$$
$$U_1 = mgR = (25.0\ \text{kg})(9.80\ \text{m/s}^2)(3.00\ \text{m}) = 735\ \text{J}$$
$$K_2 = \frac{1}{2}mv_2^2 = \frac{1}{2}(25.0\ \text{kg})(6.00\ \text{m/s})^2 = 450\ \text{J}$$
$$U_2 = 0$$

From Eq. (7.7),

$$W_f = K_2 + U_2 - K_1 - U_1$$
$$= 450\ \text{J} + 0 - 0 - 735\ \text{J} = -285\ \text{J}$$

The work done by the friction force is -285 J, and the total mechanical energy *decreases* by 285 J. Do you see why W_f has to be negative?

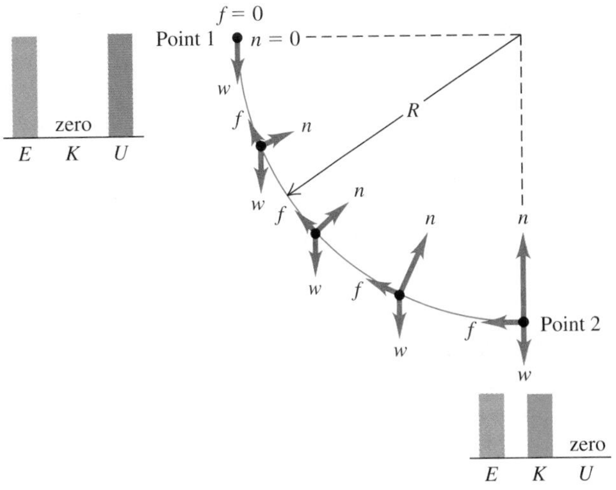

7.9 Free-body diagram for Throcky skateboarding down a ramp with friction. The total mechanical energy decreases as Throcky moves down the ramp.

EVALUATE: Throcky's motion is determined by Newton's second law, $\sum \vec{F} = m\vec{a}$. But it would be very difficult to apply the second law directly to this problem because the normal and friction forces and the acceleration are continuously changing in both magnitude and direction as Throcky skates down. The energy approach, by contrast, relates the motions at the top and bottom of the ramp without involving the details of what happens in between. Many problems are easy if energy considerations are used but very complex if we try to use Newton's laws directly.

Example 7.7 An inclined plane with friction

We want to load a 12-kg crate into a truck by sliding it up a ramp 2.5 m long, inclined at 30°. A worker, giving no thought to friction, calculates that he can get the crate up the ramp by giving it an initial speed of 5.0 m/s at the bottom and letting it go. But friction is *not* negligible; the crate slides 1.6 m up the ramp, stops, and slides back down (Fig. 7.10). a) Assuming that the friction force acting on the crate is constant, find its magnitude. b) How fast is the crate moving when it reaches the bottom of the ramp?

SOLUTION

IDENTIFY: The friction force does work on the crate as it slides. As in Example 7.2, we'll use the energy approach in part (a) to find the magnitude of the nongravitational force that does work (in this case, friction). Once we know the magnitude of the friction force, we can calculate how much nongravitational work this force does as the crate slides back down. We can then use the energy approach again to find the crate's speed at the bottom of the ramp.

SET UP: The first part of the motion is from point 1, at the bottom of the ramp, to point 2, where the crate stops instantaneously (Fig. 7.10a). In the second part of the motion, the crate returns to the bottom of the ramp, which we'll call point 3 (Fig. 7.10b). We'll take $y = 0$ (and hence $U = 0$) to be at ground level, so $y_1 = 0$, $y_2 = (1.6\text{ m})\sin 30° = 0.80\text{ m}$, and $y_3 = 0$. We are given that $v_1 = 5.0\text{ m/s}$ and $v_2 = 0$ (the crate is instantaneously at rest at point 2). Our target variable in part (a) is f, the magnitude of the friction force, which we'll find using Eq. (7.7). In part (b) our target variable is v_3, the speed at the bottom of the ramp.

EXECUTE:

a) The energy quantities are

$$K_1 = \frac{1}{2}(12\text{ kg})(5.0\text{ m/s})^2 = 150\text{ J} \qquad U_1 = 0$$

$$K_2 = 0 \qquad U_2 = (12\text{ kg})(9.8\text{ m/s}^2)(0.80\text{ m}) = 94\text{ J}$$

$$W_{\text{other}} = -fs$$

Here f is the unknown magnitude of the friction force and $s = 1.6$ m. Using Eq. (7.7), we find

$$K_1 + U_1 + W_{\text{other}} = K_2 + U_2$$

$$W_{\text{other}} = -fs = (K_2 + U_2) - (K_1 + U_1)$$

$$f = -\frac{(K_2 + U_2) - (K_1 + U_1)}{s}$$

$$= -\frac{[(0 + 94\text{ J}) - (150\text{ J} + 0)]}{1.6\text{ m}} = 35\text{ N}$$

The friction force of 35 N, acting over 1.6 m, causes the mechanical energy of the crate to decrease from 150 J to 94 J (Fig. 7.10c).

b) The crate returns to point 3 at the bottom of the ramp; $y_3 = 0$ and

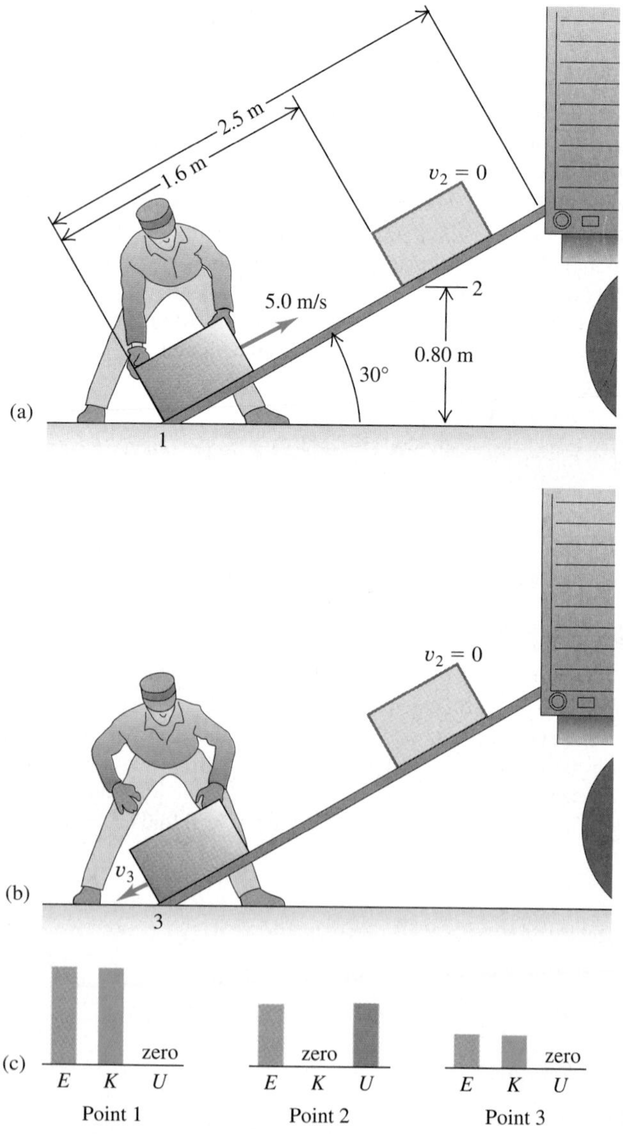

7.10 (a) A crate slides part way up the ramp, stops, and (b) slides back down. (c) Energy bar graphs for points 1, 2, and 3.

$U_3 = 0$ (Fig. 7.10b). On the way down, the friction force and the displacement both reverse direction but have the same magnitudes, so the frictional work has the same negative value for each half of the trip. The total work done by friction between points 1 and 3 is

$$W_{other} = W_{fric} = -2fs = -2(35\text{ N})(1.6\text{ m}) = -112\text{ J}$$

From part (a), $K_1 = 150$ J and $U_1 = 0$. Equation (7.7) then gives

$$K_1 + U_1 + W_{other} = K_3 + U_3$$

$$K_3 = K_1 + U_1 - U_3 + W_{other}$$
$$= 150\text{ J} + 0 - 0 + (-112\text{ J}) = 38\text{ J}$$

The crate returns to the bottom of the ramp with only 38 J of the original 150 J of mechanical energy (Fig. 7.10c). Using $K_3 = \frac{1}{2}mv_3^2$, we get

$$v_3 = \sqrt{\frac{2K_3}{m}} = \sqrt{\frac{2(38\text{ J})}{12\text{ kg}}} = 2.5\text{ m/s}$$

EVALUATE: Notice that the crate's speed when it returns to the bottom of the ramp, $v_3 = 2.5$ m/s, is less than the speed $v_1 = 5.0$ m/s at which it left that point. That's good—energy was lost due to friction.

Notice also that in part (b) we applied Eq. (7.7) to points 1 and 3, considering the entire round trip as a whole. Alternatively, we could have considered the second part of the motion by itself and applied Eq. (7.7) to points 2 and 3. Try it and see whether you get the same result for v_3.

Test Your Understanding

Figure 7.11 shows two different frictionless ramps. The heights y_1 and y_2 are the same for each ramp. If a block of mass m is released from rest at the left-hand end of each ramp, which block arrives at the right-hand end with the greater speed?

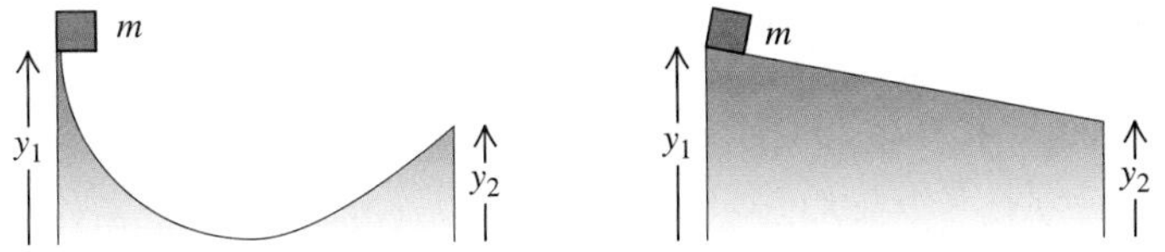

7.11 Two ramps with the same y_1 and y_2.

7.2 | Elastic Potential Energy

When a railroad car runs into a spring bumper at the end of the track, the spring is compressed as the car is brought to a stop. If there is no friction, the bumper springs back and the car moves away in the opposite direction with its original speed. During the interaction with the spring, the car's kinetic energy has been "stored" in the elastic deformation of the spring. Something very similar happens in a rubber-band slingshot. Work is done on the rubber band by the force that stretches it, and that work is stored in the rubber band until you let it go. Then the rubber band gives kinetic energy to the projectile.

This is the same pattern we saw with the pile-driver in Section 7.1: do work on the system to store energy, which can later be converted to kinetic energy. We'll describe the process of storing energy in a deformable body such as a spring or rubber band in terms of *elastic potential energy* (Fig. 7.12). A body is called *elastic* if it returns to its original shape and size after being deformed. To be specific, we'll consider storing energy in an ideal spring, like the ones we discussed in Section 6.3. To keep such an ideal spring stretched by a distance x, we must exert a force $F = kx$, where k is the force constant of the spring. The ideal

7.12 The Achilles tendon, which runs along the back of the ankle to the heel bone, acts like a natural spring. When it stretches and then relaxes, this tendon stores and then releases elastic potential energy. This spring action reduces the amount of work your leg muscles must do as you run.

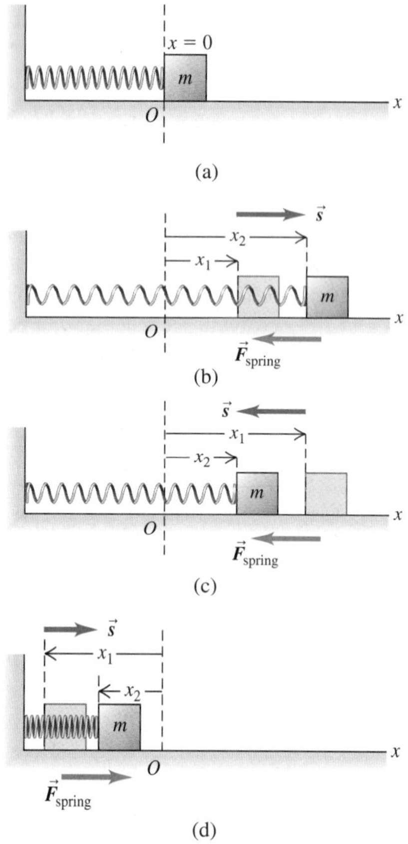

7.13 (a) A block attached to a spring in equilibrium $(x = 0)$ on a horizontal surface. (b) As the spring is stretched, it does negative work on the block. (c) As the spring relaxes, it does positive work on the block. (d) A compressed spring also does positive work on the block as it relaxes.

spring is a useful idealization because many other elastic bodies show this same direct proportionality between force $\vec{F}$ and displacement x, provided that x is sufficiently small.

We proceed just as we did for gravitational potential energy. We begin with the work done by the elastic (spring) force and then combine this with the work-energy theorem. The difference is that gravitational potential energy is a shared property of a body and the earth, but elastic potential energy is stored just in the spring (or other deformable body).

Figure 7.13 shows the ideal spring from Fig. 6.15, with its left end held stationary and its right end attached to a block with mass m that can move along the x-axis. In Fig. 7.13a the body is at $x = 0$ when the spring is neither stretched nor compressed. We move the block to one side, thereby stretching or compressing the spring, and then let it go. As the block moves from one position x_1 to another position x_2, how much work does the elastic (spring) force do on the block?

We found in Section 6.3 that the work we must do *on* the spring to move one end from an elongation x_1 to a different elongation x_2 is

$$W = \frac{1}{2}kx_2^2 - \frac{1}{2}kx_1^2 \qquad \text{(work done } on \text{ a spring)}$$

where k is the force constant of the spring. If we stretch the spring farther, we do positive work on the spring; if we let the spring relax while holding one end, we do negative work on it. We also saw that this expression for work is still correct if the spring is compressed, not stretched, so that x_1 or x_2 or both are negative. Now we need to find the work done *by* the spring. From Newton's third law the two quantities of work are just negatives of each other. Changing the signs in this equation, we find that in a displacement from x_1 to x_2 the spring does an amount of work W_{el} given by

$$W_{el} = \frac{1}{2}kx_1^2 - \frac{1}{2}kx_2^2 \qquad \text{(work done } by \text{ a spring)}$$

The subscript "el" stands for *elastic*. When x_1 and x_2 are both positive and $x_2 > x_1$ (Fig. 7.13b), the spring does negative work on the block, which moves in the $+x$-direction while the spring pulls on it in the $-x$-direction. The spring stretches farther, and the block slows down. When x_1 and x_2 are both positive and $x_2 < x_1$ (Fig. 7.13c), the spring does positive work as it relaxes and the block speeds up. If the spring can be compressed as well as stretched, x_1 or x_2 or both may be negative, but the expression for W_{el} is still valid. In Fig. 7.13d, both x_1 and x_2 are negative, but x_2 is less negative than x_1; the compressed spring does positive work as it relaxes, speeding the block up.

Just as for gravitational work, we can express the work done by the spring in terms of a given quantity at the beginning and end of the displacement. This quantity is $\frac{1}{2}kx^2$, and we define it to be the **elastic potential energy:**

$$U = \frac{1}{2}kx^2 \qquad \text{(elastic potential energy)} \qquad (7.9)$$

Figure 7.14 is a graph of Eq. (7.9). The unit of U is the joule (J), the unit used for *all* energy and work quantities; to see this from Eq. (7.9), recall that the units of k are N/m and that $1\ \text{N} \cdot \text{m} = 1\ \text{J}$.

We can use Eq. (7.9) to express the work W_{el} done on the block by the elastic force in terms of the change in potential energy:

$$W_{el} = \frac{1}{2}kx_1^2 - \frac{1}{2}kx_2^2 = U_1 - U_2 = -\Delta U \qquad (7.10)$$

When a stretched spring is stretched farther, as in Fig. 7.13b, W_{el} is negative and *U increases;* a greater amount of elastic potential energy is stored in the spring. When a stretched spring relaxes, as in Fig. 7.13c, x decreases, W_{el} is positive, and *U decreases;* the spring loses elastic potential energy. Negative values of x refer to a compressed spring. But, as Fig. 7.14 shows, U is positive for both positive and negative x, and Eqs. (7.9) and (7.10) are valid for both cases. Thus, when a compressed spring is compressed farther, $W_{el} < 0$ and U increases; when a compressed spring relaxes (Fig. 7.13d), $W_{el} > 0$ and U decreases. The more a spring is compressed *or* stretched, the greater its elastic potential energy.

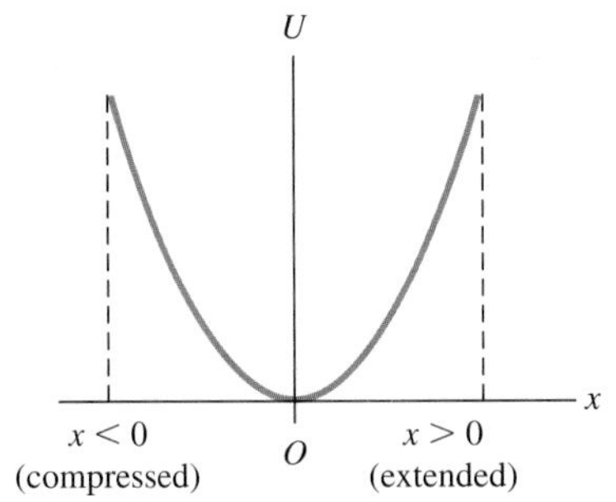

7.14 The graph of elastic potential energy for an ideal spring is a parabola: $U = \frac{1}{2}kx^2$, where x is the extension or compression of the spring. For extension (stretching), x is positive. For compression (when that is possible), x is negative. Elastic potential energy U is never negative.

CAUTION An important difference between gravitational potential energy $U = mgy$ and elastic potential energy $U = \frac{1}{2}kx^2$ is that we do *not* have the freedom to choose $x = 0$ to be wherever we wish. To be consistent with Eq. (7.9), $x = 0$ *must* be the position at which the spring is neither stretched nor compressed. At that position, its elastic potential energy and the force that it exerts are both zero.

The work-energy theorem says that $W_{tot} = K_2 - K_1$, no matter what kind of forces are acting on a body. If the elastic force is the *only* force that does work on the body, then

$$W_{tot} = W_{el} = U_1 - U_2$$

The work-energy theorem $W_{tot} = K_2 - K_1$ then gives us

$$K_1 + U_1 = K_2 + U_2 \qquad \text{(if only the elastic force does work)} \qquad (7.11)$$

Here U is given by Eq. (7.9), so

$$\frac{1}{2}mv_1^2 + \frac{1}{2}kx_1^2 = \frac{1}{2}mv_2^2 + \frac{1}{2}kx_2^2 \qquad (7.12)$$
$$\text{(if only the elastic force does work)}$$

In this case the total mechanical energy $E = K + U$ (the sum of kinetic and elastic potential energy) is *conserved.* An example of this is the motion of the block in Fig. 7.13, provided that the horizontal surface is frictionless so that no force does work other than that exerted by the spring.

For Eq. (7.12) to be strictly correct, the ideal spring that we've been discussing must also be *massless.* If the spring has a mass, it also has kinetic energy as the coils of the spring move back and forth. We can neglect the kinetic energy of the spring if its mass is much less than the mass m of the body attached to the spring. For instance, a typical automobile has a mass of 1200 kg or more. The springs in its suspension have masses of only a few kilograms, so their mass can be neglected if we want to study how a car bounces on its suspension.

If forces other than the elastic force also do work on the body, we call their work W_{other}, as before. Then the total work is $W_{tot} = W_{el} + W_{other}$, and the work-energy theorem gives

$$W_{el} + W_{other} = K_2 - K_1$$

The work done by the spring is still $W_{el} = U_1 - U_2$, so again

$$K_1 + U_1 + W_{other} = K_2 + U_2 \quad (7.13)$$
(if forces other than the elastic force do work)

and

$$\frac{1}{2}mv_1^2 + \frac{1}{2}kx_1^2 + W_{other} = \frac{1}{2}mv_2^2 + \frac{1}{2}kx_2^2 \quad (7.14)$$
(if forces other than the elastic force do work)

This equation shows that *the work done by all forces other than the elastic force equals the change in the total mechanical energy $E = K + U$ of the system, where U is the elastic potential energy.* The "system" is made up of the body of mass m and the spring of force constant k. When W_{other} is positive, E increases; when W_{other} is negative, E decreases. You should compare Eq. (7.14) to Eq. (7.8), which describes situations in which there is gravitational potential energy but no elastic potential energy.

7.15 The fall of a bungee jumper involves an interplay among kinetic energy, gravitational potential energy, and elastic potential energy. Mechanical energy is not conserved, however, because work is also done by frictional forces within the bungee cord and by air resistance. This is a good thing—if mechanical energy were conserved, the bungee jumper would keep bouncing up and down forever!

Situations with Both Gravitational and Elastic Potential Energy

Equations (7.11), (7.12), (7.13), and (7.14) are valid when the only potential energy in the system is elastic potential energy. What happens when we have *both* gravitational and elastic forces, such as a block attached to the lower end of a vertically hanging spring? We can still use Eq. (7.13), but now U_1 and U_2 are the initial and final values of the *total* potential energy, including both gravitational and elastic potential energies. That is, $U = U_{grav} + U_{el}$. Thus *the most general statement* of the relationship between kinetic energy, potential energy, and work done by other forces is

$$K_1 + U_{grav,1} + U_{el,1} + W_{other} = K_2 + U_{grav,2} + U_{el,2} \quad (7.15)$$
(valid in general)

That is, **the work done by all forces other than the gravitational force or elastic force equals the change in the total mechanical energy $E = K + U$ of the system, where U is the sum of the gravitational potential energy and the elastic potential energy.** If the gravitational and elastic forces are the *only* forces that do work on the body, then $W_{other} = 0$ and the total mechanical energy (including both gravitational and elastic potential energy) is conserved.

Bungee jumping (Fig. 7.15) is an example of transformations among kinetic energy, elastic potential energy, and gravitational potential energy. As the jumper falls, gravitational potential energy decreases and is converted into the kinetic energy of the jumper and the elastic potential energy of the bungee cord. Beyond a certain point in the fall, the jumper's speed decreases so that both gravitational potential energy and kinetic energy are converted into elastic potential energy.

The strategy outlined in Section 7.1 is equally useful in solving problems that involve elastic forces as well as gravitational forces. The only new wrinkle is that the potential energy U now includes the elastic potential energy $U_{el} = \frac{1}{2}kx^2$, where x is the displacement of the spring *from its unstretched length.* The work done by the gravitational and elastic forces is accounted for by their potential energies; the work of the other forces, W_{other}, has to be included separately.

ActivPhysics ONLINE

5.4 Inverse Bungee Jumper

5.5 Spring-Launched Bowler

Example 7.8 Motion with elastic potential energy

In Fig. 7.16a a glider with mass $m = 0.200$ kg sits on a frictionless horizontal air track, connected to a spring with force constant $k = 5.00$ N/m. You pull on the glider, stretching the spring 0.100 m, and then release it with no initial velocity (Fig. 7.16b). The glider begins to move back toward its equilibrium position $(x = 0)$. What is its x-velocity when $x = 0.080$ m?

SOLUTION

IDENTIFY: Because the spring force varies with position, this problem can't be solved with the equations for motion with constant acceleration. Instead, we'll use the energy method to find a simple solution. In particular, we'll use the idea that as the glider starts to move, elastic potential energy is converted into kinetic energy. (The glider remains at the same height throughout the motion, so gravitational potential energy is not a factor.)

SET UP: The spring force is the only force doing work on the glider, so $W_{other} = 0$ and we may use Eq. (7.11). Let point 1 be where the glider is released (Fig. 7.16b), and let point 2 be at $x = 0.080$ m (Fig. 7.16c). We know the velocity at point 1 $(v_{1x} = 0)$; our target variable is the x-velocity at point 2, v_{2x}.

EXECUTE: The energy quantities are

$$K_1 = \frac{1}{2}(0.200\ \text{kg})(0)^2 = 0$$

$$U_1 = \frac{1}{2}(5.00\ \text{n/m})(0.100\ \text{m})^2 = 0.0250\ \text{J}$$

$$K_2 = \frac{1}{2}mv_{2x}^2$$

$$U_2 = \frac{1}{2}(5.00\ \text{N/m})(0.080\ \text{m})^2 = 0.0160\ \text{J}$$

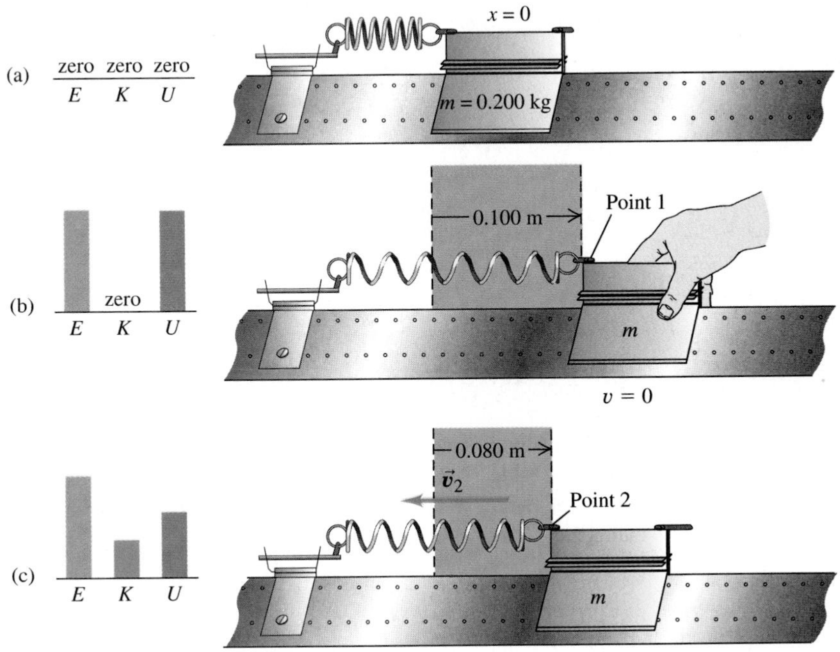

7.16 (a) An air-track glider attached to a spring. (b) Elastic potential energy is added to the system by stretching the spring. (c) Elastic potential energy is transformed to kinetic energy as the glider moves back toward its equilibrium position.

Then from Eq. (7.11),

$$K_2 = K_1 + U_1 - U_2 = 0 + 0.0250\text{ J} - 0.0160\text{ J} = 0.0090\text{ J}$$

$$v_{2x} = \pm\sqrt{\frac{2K_2}{m}} = \pm\sqrt{\frac{2(0.0090\text{ J})}{0.200\text{ kg}}} = \pm 0.30\text{ m/s}$$

We choose the negative root because the glider is moving in the $-x$-direction; the answer we want is $v_{2x} = -0.30$ m/s.

EVALUATE: What is the meaning of the second solution, $v_{2x} = +0.30$ m/s? Eventually the spring will compress and push the glider back to the right in the positive x-direction (see Fig. 7.13d). The second solution tells us that when the glider passes through $x = 0.080$ m while moving to the right, its speed will be 0.30 m/s—the same speed as when it passed through this point while moving to the left.

When the glider passes through the point $x = 0$, the spring is relaxed and all of the mechanical energy is in the form of kinetic energy. Can you show that the speed of the glider at this point is 0.50 m/s?

Example 7.9 Motion with elastic potential energy and work done by other forces

For the system of Example 7.8, suppose the glider is initially at rest at $x = 0$, with the spring unstretched. You then apply a constant force $\vec{F}$ in the $+x$-direction with magnitude 0.610 N to the glider. What is the glider's velocity when it has moved to $x = 0.100$ m?

SOLUTION

IDENTIFY: Although the force $\vec{F}$ you apply is constant, the spring force isn't, so the acceleration of the glider won't be constant. Total mechanical energy is not conserved because of the work done by the force $\vec{F}$, but we can still use the energy relation as given by Eq. (7.13).

SET UP: Let point 1 be at $x = 0$, where the velocity is $v_{1x} = 0$, and let point 2 be at $x = 0.100$ m. (These points are different from the ones labeled in Fig. 7.16.) Our target variable is v_{2x}, the velocity at point 2.

EXECUTE: The energy quantities are

$$K_1 = 0 \qquad U_1 = 0$$

$$K_2 = \frac{1}{2}mv_{2x}^2 \qquad U_2 = \frac{1}{2}(5.00\text{ N/m})(0.100\text{ m})^2 = 0.0250\text{ J}$$

$$W_{\text{other}} = (0.610\text{ N})(0.100\text{ m}) = 0.0610\text{ J}$$

(To calculate W_{other} we multiplied the magnitude of the force by the displacement, since both are in the $+x$-direction.) Initially, the total mechanical energy is zero; the work done by the force $\vec{F}$ increases the total mechanical energy to 0.0610 J, of which 0.0250 J is elastic potential energy. The remainder is kinetic energy. From Eq. (7.13),

$$K_1 + U_1 + W_{\text{other}} = K_2 + U_2$$

$$K_2 = K_1 + U_1 + W_{\text{other}} - U_2 = 0 + 0 + 0.0610\text{ J} - 0.0250\text{ J} = 0.0360\text{ J}$$

$$v_{2x} = \sqrt{\frac{2K_2}{m}} = \sqrt{\frac{2(0.0360\text{ J})}{0.200\text{ kg}}} = 0.60\text{ m/s}$$

We choose the positive square root because the glider is moving in the $+x$-direction.

EVALUATE: To test our answer, think what would be different if we disconnected the glider from the spring. Then $\vec{F}$ would be the only force doing work, there would be zero potential energy at all times, and Eq. (7.13) would give us

$$K_2 = K_1 + W_{\text{other}} = 0 + 0.0610\text{ J}$$

$$v_{2x} = \sqrt{\frac{2K_2}{m}} = \sqrt{\frac{2(0.0610\text{ J})}{0.200\text{ kg}}} = 0.78\text{ m/s}$$

We found a lower velocity than this value because the spring does negative work on the glider as it stretches (see Fig. 7.13b).

Example 7.10 Motion with elastic potential energy after other forces have ceased

In Example 7.9, suppose the force $\vec{F}$ is removed when the glider reaches the point $x = 0.100$ m. How much farther does the glider move before coming to rest?

SOLUTION

IDENTIFY: After $\vec{F}$ is removed, the spring force is the only force doing work. Hence for this part of the motion the mechanical energy $E = K + U$ is conserved.

SET UP: We'll let point 2 be at $x = 0.100$ m, as in Example 7.9, and let point 3 be where the glider comes instantaneously to rest. Our target variable is the coordinate x_3 of this point. We'll find its value using the conservation of energy expressions, Eq. (7.11), along with the relation $U = \frac{1}{2}kx^2$ for elastic potential energy.

EXECUTE: We saw in Example 7.9 that the kinetic and potential energies at point 2 are $K_2 = 0.0360$ J and $U_2 = 0.0250$ J. The total mechanical energy at and beyond this point is therefore

$K_2 + U_2 = 0.0610$ J. When the glider comes to rest at $x = x_3$, the kinetic energy K_3 is zero and the elastic potential energy U_3 is equal to the total mechanical energy 0.0610 J. We can also see this from $K_2 + U_2 = K_3 + U_3$:

$$U_3 = K_2 + U_2 - K_3 = 0.0360 \text{ J} + 0.0250 \text{ J} - 0 = 0.0610 \text{ J}$$

But $U_3 = \frac{1}{2}kx_3^2$, so

$$x_3 = \sqrt{\frac{2U_3}{k}} = \sqrt{\frac{2(0.0610 \text{ J})}{5.00 \text{ N/m}}} = 0.156 \text{ m}$$

The body moves an additional 0.056 m after the force is removed at $x = 0.100$ m.

EVALUATE: The total mechanical energy for the motion from point 2 to point 3 is 0.0610 J, the same as the work W_{other} done by the force $\vec{F}$ in Example 7.8. Is this just coincidence? Not at all; the system of glider and spring had zero mechanical energy initially (at point 1 in Example 7.9), so all the mechanical energy it has came from the work done by $\vec{F}$.

Example 7.11 Motion with gravitational, elastic, and friction forces

In a "worst-case" design scenario, a 2000-kg elevator with broken cables is falling at 25 m/s when it first contacts a cushioning spring at the bottom of the shaft. The spring is supposed to stop the elevator, compressing 3.00 m as it does so (Fig. 7.17). During the motion a safety clamp applies a constant 17,000-N frictional force to the elevator. As a design consultant, you are asked to determine what the force constant of the spring should be.

SOLUTION

Identify: We'll use the energy approach to determine the force constant, which appears in the expression for elastic potential energy. Note that this problem involves *both* gravitational and elastic potential energy. Furthermore, total mechanical energy is not conserved because the friction force does negative work W_{other} on the elevator.

SET UP: Since mechanical energy isn't conserved and more than one kind of potential energy is involved, we'll use the most general form of the energy relation, Eq. (7.15). We take point 1 as the position of the bottom of the elevator when it initially contacts the spring, and take point 2 as its position when it is at rest. We choose the origin to be at point 1, so $y_1 = 0$ and $y_2 = -3.00$ m. With this choice the coordinate of the upper end of the spring is the same as the coordinate of the elevator, so the elastic potential energy at any point between point 1 and point 2 is $U_{el} = \frac{1}{2}ky^2$. (The gravitational potential energy is $U_{grav} = mgy$ as usual.) We know the initial and final speeds of the elevator and the magnitude of the friction force, so the only unknown is the force constant k (our target variable).

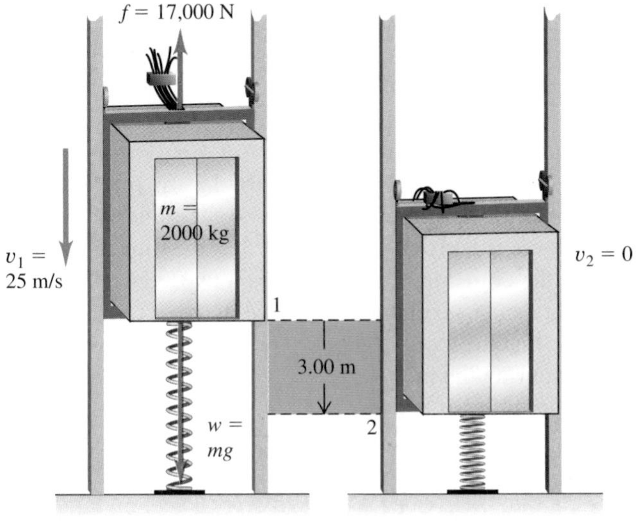

7.17 The fall of an elevator is stopped by a spring and by a constant friction force.

EXECUTE: The elevator's initial speed is $v_1 = 25$ m/s, so the initial kinetic energy is

$$K_1 = \frac{1}{2}mv_1^2 = \frac{1}{2}(2000 \text{ kg})(25 \text{ m/s})^2 = 625{,}000 \text{ J}$$

The elevator stops at point 2, so $K_2 = 0$. The potential energy at point 1, U_1, is zero; U_{grav} is zero because $y_1 = 0$, and $U_{el} = 0$ because the spring is not yet compressed. At point 2 there is both gravitational and elastic potential energy, so

$$U_2 = mgy_2 + \frac{1}{2}ky_2^2$$

The gravitational potential energy at point 2 is

$$mgy_2 = (2000 \text{ kg})(9.80 \text{ m/s}^2)(-3.00 \text{ m}) = -58{,}800 \text{ J}$$

The other force is the 17,000-N friction force, acting opposite to the 3.00-m displacement, so

$$W_{other} = -(17{,}000 \text{ N})(3.00 \text{ m}) = -51{,}000 \text{ J}$$

Putting these terms into $K_1 + U_1 + W_{other} = K_2 + U_2$, we have

$$K_1 + 0 + W_{other} = 0 + \left(mgy_2 + \frac{1}{2}ky_2^2\right)$$

so the force constant of the spring is

$$k = \frac{2(K_1 + W_{other} - mgy_2)}{y_2^2}$$
$$= \frac{2[625{,}000 \text{ J} + (-51{,}000 \text{ J}) - (-58{,}800 \text{ J})]}{(-3.00 \text{ m})^2}$$
$$= 1.41 \times 10^5 \text{ N/m}$$

This is comparable to the springs in an automobile suspension.

EVALUATE: Let's note what might seem to be a paradox in this problem. The elastic potential energy in the spring at point 2 is

$$\frac{1}{2}ky_2^2 = \frac{1}{2}(1.41 \times 10^5 \text{ N/m})(-3.00 \text{ m})^2 = 632{,}800 \text{ J}$$

This is *more* than the total mechanical energy at point 1,

$$E_1 = K_1 + U_1 = 625{,}000 \text{ J} + 0 = 625{,}000 \text{ J}$$

But the friction force caused the mechanical energy of the system to *decrease* by 51,000 J between point 1 and point 2. Does this mean that energy appeared from nowhere? Don't panic; there is no paradox. At point 2 there is also *negative* gravitational potential energy $mgy_2 = -58{,}800$ J because point 2 is below the origin. The total mechanical energy at point 2 is

$$E_2 = K_2 + U_2 = 0 + \frac{1}{2}ky_2^2 + mgy_2$$

$$= 632{,}800 \text{ J} + (-58{,}800 \text{ J}) = 574{,}000 \text{ J}$$

This is just the initial mechanical energy of 625,000 J, minus 51,000 J lost to friction.

Your next job as design consultant would be to tell your client that the elevator won't stay at the bottom of the shaft. Instead, it will bounce back up. The reason is that at point 2 the compressed spring exerts an upward force of magnitude $F_{\text{spring}} = (1.41 \times 10^5 \text{ N/m})(3.00 \text{ m}) = 422{,}000$ N. The weight of the elevator is only $w = mg = (2000 \text{ kg})(9.80 \text{ m/s}^2) = 19{,}600$ N, so the net force will be upward. The elevator bounces back upward even if the safety clamp now exerts a *downward* friction force of magnitude $f = 17{,}000$ N; the spring force is greater than the sum of f and mg. The elevator will bounce again and again until enough mechanical energy has been removed by friction for it to stop.

Can you also show that the acceleration of the falling elevator when it hits the spring is unacceptably high?

Test Your Understanding

First the value of y when the elevator in Example 7.11 finally comes to a stop, and find the values of K, U_{grav}, U_{el}, and E at this point. Assume that the safety clamp exerts an upward force of 17,000 N. (*Hint:* What is the relationship between the compression of the spring and the force it exerts on the elevator?) How does the mechanical energy at this point compare to the value at point 1?

7.3 | Conservative and Nonconservative Forces

In our discussions of potential energy we have talked about "storing" kinetic energy by converting it to potential energy. We always have in mind that later we may retrieve it again as kinetic energy. When you throw a ball up in the air, it slows down as kinetic energy is converted into potential energy. But on the way down, the conversion is reversed, and the ball speeds up as potential energy is converted back to kinetic energy. If there is no air resistance, the ball is moving just as fast when you catch it as when you threw it.

If a glider moving on a frictionless horizontal air track runs into a spring bumper at the end of the track, the spring compresses and the glider stops. But then it bounces back, and if there is no friction, the glider has the same speed and kinetic energy it had before the collision. Again, there is a two-way conversion from kinetic to potential energy and back. In both cases we find that we can define a potential-energy function so that the total mechanical energy, kinetic plus potential, is constant or *conserved* during the motion.

A force that offers this opportunity of two-way conversion between kinetic and potential energies is called a **conservative force.** We have seen two examples of conservative forces: the gravitational force and the spring force. An essential feature of conservative forces is that their work is always *reversible.* Anything that we deposit in the energy "bank" can later be withdrawn without loss. Another important aspect of conservative forces is that a body may move from point 1 to point 2 by various paths, but the work done by a conservative force is the same for

7.18 For any conservative force, the work done by that force depends only on the end points, not on the path taken. Thus the gravitational force, which is conservative, does the same work on the runner no matter what path he takes from point 1 to point 2.

all of these paths (Fig. 7.18). Thus, if a body stays close to the surface of the earth, the gravitational force $m\vec{g}$ is independent of height, and the work done by this force depends only on the change in height. If the body moves around a closed path, ending at the same point where it started, the *total* work done by the gravitational force is always zero.

The work done by a conservative force *always* has these properties:

1. It can always be expressed as the difference between the initial and final values of a *potential energy* function.
2. It is reversible.
3. It is independent of the path of the body and depends only on the starting and ending points.
4. When the starting and ending points are the same, the total work is zero.

When the only forces that do work are conservative forces, the total mechanical energy $E = K + U$ is constant.

Not all forces are conservative. Consider the friction force acting on the crate sliding on a ramp in Example 7.7 (Section 7.1). When the body slides up and then back down to the starting point, the total work done on it by the friction force is *not* zero. When the direction of motion reverses, so does the friction force, and friction does *negative* work in *both* directions. When a car with its brakes locked skids across the pavement with decreasing speed (and decreasing kinetic energy), the lost kinetic energy cannot be recovered by reversing the motion or in any other way, and mechanical energy is *not* conserved. There is no potential-energy function for the friction force.

In the same way, the force of fluid resistance (Section 5.3) is not conservative. If you throw a ball up in the air, air resistance does negative work on the ball while it's rising *and* while it's descending. The ball returns to your hand with less speed and less kinetic energy than when it left, and there is no way to get back the lost mechanical energy.

A force that is not conservative is called a **nonconservative force.** The work done by a nonconservative force *cannot* be represented by a potential-energy function. Some nonconservative forces, like kinetic friction or fluid resistance, cause mechanical energy to be lost or dissipated; a force of this kind is called a **dissipative force.** There are also nonconservative forces that *increase* mechanical energy. The fragments of an exploding firecracker fly off with very large kinetic energy, thanks to a chemical reaction of gunpowder with oxygen. The forces unleashed by this reaction are nonconservative because the process is not reversible. Imagine the fragments spontaneously reassembling themselves into a complete firecracker!

Example 7.12 Frictional work depends on the path

You are rearranging your furniture and wish to move a 40.0-kg futon 2.50 m across the room (Fig. 7.19). However, the straight-line path is blocked by a heavy coffee table that you don't want to move. Instead, you slide the futon in a dogleg path over the floor; the doglegs are 2.00 m and 1.50 m long. Compared to the straight-line path, how much more work must you do to push the futon in the dogleg path? The coefficient of kinetic friction is 0.200.

SOLUTION

IDENTIFY: Here work is done both by you and by the force of friction, so we must use the energy relation that includes forces other than elastic or gravitational forces. We'll use this relation to find a connection between the work that *you* do and the work done by *friction.*

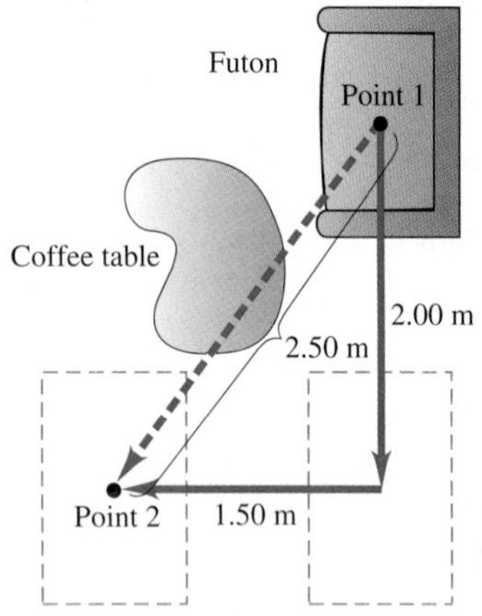

7.19 A view from above of furniture being moved. How much more work must you do to move the futon on a dogleg path?

SET UP: Figure 7.19 shows the initial and final points. The futon is at rest at both point 1 and point 2, so $K_1 = K_2 = 0$. The gravitational potential energy does not change because the futon moves only horizontally; to be specific, we'll say $U_1 = U_2 = 0$. From Eq. (7.7) it follows that $W_{other} = 0$. The other work done on the futon is the sum of the positive work you do, W_{you}, and the negative work W_{fric} done by the kinetic friction force. Since the sum of these is zero, we have

$$W_{you} = -W_{fric}$$

Thus to determine W_{you}, we'll calculate the work done by friction.

EXECUTE: Because the floor is horizontal, the normal force on the futon equals its weight mg, and the magnitude of the friction force is $f_k = \mu_k n = \mu_k mg$. The work you must do over each path is then

$$\begin{aligned} W_{you} &= -W_{fric} = -(-f_k s) = +\mu_k mgs \\ &= (0.200)(40.0\text{ kg})(9.80\text{ m/s}^2)(2.50\text{ m}) \\ &= 196\text{ J} \qquad \text{(straight-line path)} \\ W_{you} &= -W_{fric} \\ &= (0.200)(40.0\text{ kg})(9.80\text{ m/s}^2)(2.00\text{ m} + 1.50\text{ m}) \\ &= 274\text{ J} \qquad \text{(dogleg path)} \end{aligned}$$

The extra work you must do is 274 J − 196 J = 78 J.

EVALUATE: The work done by friction is $W_{fric} = -W_{you} = -196$ J on the straight-line path, and −274 J on the dogleg. The work done by friction depends on the path taken, which illustrates that friction is a nonconservative force.

Example 7.13 Conservative or nonconservative?

In a certain region of space the force on an electron is $\vec{F} = Cx\hat{j}$, where C is a positive constant. The electron moves in a counterclockwise direction around a square loop in the xy-plane (Fig. 7.20). The corners of the square are at $(x, y) = (0, 0)$, $(L, 0)$, (L, L) and $(0, L)$. Calculate the work done on the electron by the force $\vec{F}$ during one complete trip around the square. Is this force conservative or nonconservative?

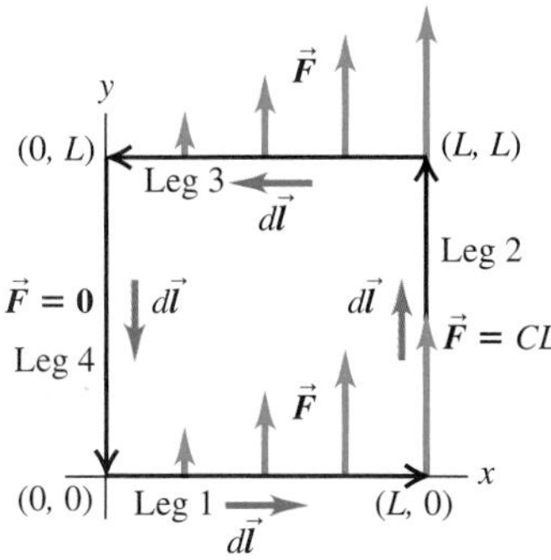

7.20 An electron moving around a square loop while being acted on by the force $\vec{F} = Cx\hat{j}$.

SOLUTION

IDENTIFY AND SET UP: In Example 7.12 the force of friction was constant in magnitude and always opposite to the displacement, so it was easy to calculate the work done. Here, however, the force $\vec{F}$ is not constant and in general is not in the same direction as the displacement. So we'll use the more general expression for work, Eq. (6.14):

$$W = \int_{P_1}^{P_2} \vec{F} \cdot d\vec{l}$$

where $d\vec{l}$ is an infinitesimal displacement. Let's calculate the work done by the force $\vec{F}$ on each leg of the square and then add the results to find the work done on the round trip.

EXECUTE: On the first leg, from $(0, 0)$ to $(L, 0)$, the force varies but is everywhere perpendicular to the displacement. So $\vec{F} \cdot d\vec{l} = 0$, and the work done on the first leg is $W_1 = 0$. The force has the same value $\vec{F} = CL\hat{j}$ everywhere on the second leg from $(L, 0)$ to (L, L). The displacement on this leg is in the $+y$-direction, so $d\vec{l} = dy\hat{j}$ and

$$\vec{F} \cdot d\vec{l} = CL\hat{j} \cdot dy\hat{j} = CL\,dy$$

The work done on the second leg is then

$$W_2 = \int_{(L,0)}^{(L,L)} \vec{F} \cdot d\vec{l} = \int_{y=0}^{y=L} CL\,dy = CL\int_0^L dy = CL^2$$

On the third leg, from (L, L) to $(0, L)$, $\vec{F}$ is again perpendicular to the displacement so $W_3 = 0$. The force is zero on the final leg, from $(0, L)$ to $(0, 0)$, so no work is done and $W_4 = 0$. The work done by the force $\vec{F}$ on the round trip is

$$W = W_1 + W_2 + W_3 + W_4 = 0 + CL^2 + 0 + 0 = CL^2$$

The starting and ending points are the same, but the total work done by $\vec{F}$ is not zero. This is a nonconservative force; it cannot be represented by a potential energy function.

EVALUATE: Because W is positive, the mechanical energy of the electron *increases* as it goes around the loop. This is not a mathematical curiosity; it's a description of what happens in an electrical generating plant. A loop of wire is moved through a magnetic field, which gives rise to a nonconservative force similar to the one in this example. Electrons in the wire gain energy as they move around the loop, and this energy is carried via transmission lines to the consumer. (We'll discuss how this works in detail in Chapter 29.) All the electrical energy used in the home and in industry comes from work done by nonconservative forces!

How would the value of W change if the electron went around the loop clockwise instead of counterclockwise? The force $\vec{F}$ would be unaffected, but the direction of each infinitesimal displacement $d\vec{l}$ would reverse. Thus the sign of work would also reverse, and the work for a clockwise round trip would be $W = -CL^2$. This is a different behavior than the nonconservative friction force. When a body slides over a stationary surface with friction, the work done by friction is always negative, no matter what the direction of motion (see Example 7.7 in Section 7.1).

The Law of Conservation of Energy

Nonconservative forces cannot be represented in terms of potential energy. But we can describe the effects of these forces in terms of kinds of energy other than kinetic and potential energy. When a car with locked brakes skids to a stop, the tires and the road surface both become hotter. The energy associated with this change in the state of the materials is called **internal energy.** Raising the temperature of a body increases its internal energy; lowering the body's temperature decreases its internal energy.

To see the significance of internal energy, let's consider a block sliding on a rough surface. Friction does *negative* work on the block as it slides, and the change in internal energy of the block and surface (both of which get hotter) is *positive*. Careful experiments show that the increase in the internal energy is *exactly* equal to the absolute value of the work done by friction. In other words,

$$\Delta U_{int} = -W_{other}$$

7.21 When a liter of gasoline is burned in an automotive engine, it releases 3.3×10^7 J of internal energy. Hence $\Delta U_{\text{int}} = -3.3 \times 10^7$ J, where the minus sign means that the amount of energy stored in the gasoline has decreased. This energy can be converted into kinetic energy (making the car go faster) or into potential energy (enabling the car to climb uphill).

where ΔU_{int} is the change in internal energy. If we substitute this into Eq. (7.7), (7.13), or (7.15), we find

$$K_1 + U_1 - \Delta U_{\text{int}} = K_2 + U_2$$

Writing $\Delta K = K_2 - K_1$ and $\Delta U = U_2 - U_1$, we can finally express this as

$$\Delta K + \Delta U + \Delta U_{\text{int}} = 0 \qquad \text{(law of conservation of energy)} \qquad (7.16)$$

This remarkable statement is the general form of the **law of conservation of energy.** In a given process, the kinetic energy, potential energy, and internal energy of a system may all change. But the *sum* of those changes is always zero. If there is a decrease in one form of energy, it is made up for by an increase in the other forms (Fig. 7.21). When we expand our definition of energy to include internal energy, Eq. (7.16) says that *energy is never created or destroyed; it only changes form.* No exception to this rule has ever been found.

Notice that the concept of work has been banished from Eq. (7.16). Instead, this equation invites us to think purely in terms of the conversion of energy from one form to another. When you throw a baseball straight up, you convert a portion of the internal energy of your molecules into kinetic energy of the baseball. This is converted into gravitational potential energy as the ball climbs and back to kinetic energy as the ball falls. If there is air resistance, part of the energy is used to heat up the air and the ball and increase their internal energy. Energy is converted back into the kinetic form as the ball falls. If you catch the ball in your hand, whatever energy was not lost to the air once again becomes internal energy; the ball and your hand are now warmer than they were at the beginning.

ActivPhysics ONLINE

5.7 Modified Atwood Machine

In a hydroelectric generating station, falling water is used to drive turbines ("water wheels"), which in turn run electric generators. Fundamentally, gravitational potential energy is released as the water falls, and the generating station converts this into electrical energy. Even if we know nothing about the details of how this is done, we can still use the law of conservation of energy, Eq. (7.16), to draw an important conclusion: the amount of electrical energy produced can be no greater than the amount of gravitational potential energy lost. (Because of friction, some of the potential energy goes into raising the temperatures of the water and the mechanism.)

In later chapters we will study the relation of internal energy to temperature changes, heat, and work. This is the heart of the area of physics called *thermodynamics.*

Example 7.14 Work done by friction

Let's look again at Example 7.6 in Section 7.1, in which your cousin Throcky skateboards down a curved ramp. He starts with zero kinetic energy and 735 J of potential energy, and at the bottom he has 450 J of kinetic energy and zero potential energy. So $\Delta K = +450$ J and $\Delta U = -735$ J. The work $W_{\text{other}} = W_{\text{fric}}$ done by the nonconservative friction forces is -285 J, so the change in internal energy is $\Delta U_{\text{int}} = -W_{\text{other}} = +285$ J. The wheels, the bearings, and the ramp all get a little warmer as Throcky rolls down. In accordance with Eq. (7.16), the sum of the energy changes equals zero:

$$\Delta K + \Delta U + \Delta U_{\text{int}} = +450\text{ J} + (-735\text{ J}) + 285\text{ J} = 0$$

The total energy of the system (including nonmechanical forms of energy) is conserved.

Test Your Understanding

For the situation in Example 7.12, calculate the total amount of work you must do to move the futon from point 1 to point 2 along the 2.00-m and 1.50-m doglegs, then back to point 1 via the straight-line path shown as a dashed line in Fig. 7.19. (Assume that the coffee table was first moved out of the way.) Since the futon ended up back where it started, why isn't the total work done on the futon equal to zero?

7.4 | Force and Potential Energy

For the two kinds of conservative forces (gravitational and elastic) we have studied, we started with a description of the behavior of the force and derived from that an expression for the potential energy. For example, for a body with mass m in a uniform gravitational field, the gravitational force is $F_y = -mg$. We found that the corresponding potential energy is $U(y) = mgy$. To stretch an ideal spring by a distance x, we exert a force equal to $+kx$. By Newton's third law the force that an ideal spring exerts on a body is opposite this, or $F_x = -kx$. The corresponding potential energy function is $U(x) = \frac{1}{2}kx^2$.

In studying physics, however, you'll encounter situations in which you are given an expression for the *potential energy* as a function of position and have to find the corresponding *force*. We'll see several examples of this kind when we study electric forces later in this book: it's often far easier to calculate the electric potential energy first, then determine the corresponding electric force afterward.

Here's how we find the force that corresponds to a given potential-energy expression. First let's consider motion along a straight line, with coordinate x. We denote the x-component of force, a function of x, by $F_x(x)$, and the potential energy as $U(x)$. This notation reminds us that both F_x and U are *functions* of x. Now we recall that in any displacement, the work W done by a conservative force equals the negative of the change ΔU in potential energy:

$$W = -\Delta U$$

Let's apply this to a small displacement Δx. The work done by the force $F_x(x)$ during this displacement is approximately equal to $F_x(x)\,\Delta x$. We have to say "approximately" because $F_x(x)$ may vary a little over the interval Δx. But it is at least approximately true that

$$F_x(x)\Delta x = -\Delta U \quad \text{and} \quad F_x(x) = -\frac{\Delta U}{\Delta x}$$

You can probably see what's coming. We take the limit as $\Delta x \to 0$; in this limit, the variation of F_x becomes negligible, and we have the exact relation

$$F_x(x) = -\frac{dU(x)}{dx} \quad \text{(force from potential energy, one dimension)} \quad (7.17)$$

This result makes sense; in regions where $U(x)$ changes most rapidly with x (that is, where $dU(x)/dx$ is large), the greatest amount of work is done during a given displacement, and this corresponds to a large force magnitude. Also, when $F_x(x)$ is in the positive x-direction, $U(x)$ *decreases* with increasing x. So $F_x(x)$ and $dU(x)/dx$

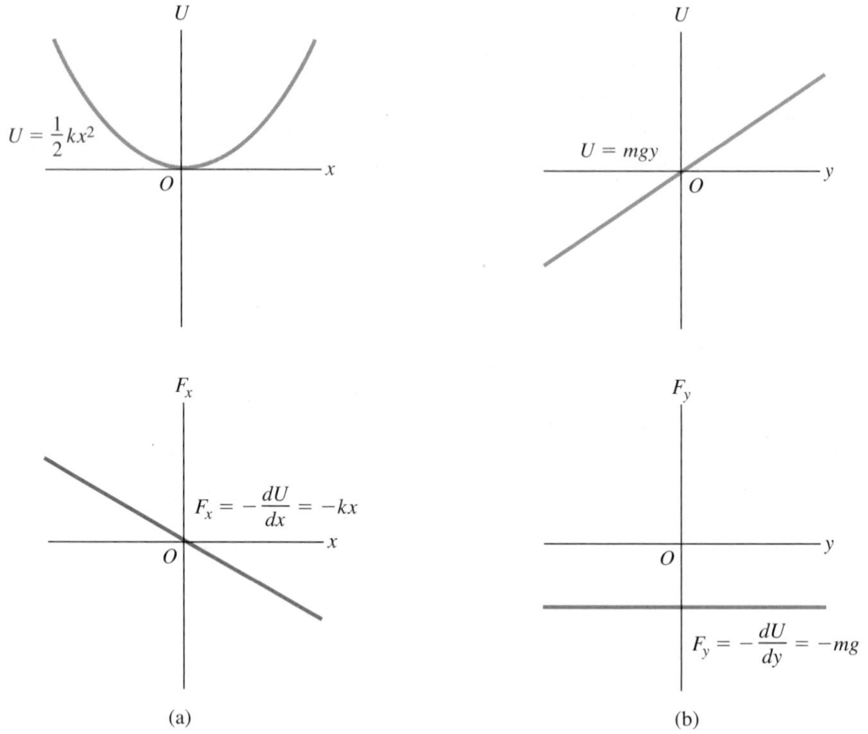

7.22 Graphs of potential energy and force versus position for (a) the spring force and (b) the gravitational force. In each case the force is the negative derivative of the potential energy.

should indeed have opposite signs. The physical meaning of Eq. (7.17) is that *a conservative force always acts to push the system toward lower potential energy.*

As a check, let's consider the function for elastic potential energy, $U(x) = \frac{1}{2}kx^2$. Substituting this into Eq. (7.17) yields

$$F_x(x) = -\frac{d}{dx}\left(\frac{1}{2}kx^2\right) = -kx$$

which is the correct expression for the force exerted by an ideal spring (Fig. 7.22a). Similarly, for gravitational potential energy we have $U(y) = mgy$; taking care to change x to y for the choice of axis, we get $F_y = -dU/dy = -d(mgy)/dy = -mg$, which is the correct expression for gravitational force (Fig. 7.22b).

Example 7.15 An electric force and its potential energy

An electrically charged particle is held at rest at the point $x = 0$, while a second particle with equal charge is free to move along the positive x-axis. The potential energy of the system is

$$U(x) = \frac{C}{x}$$

where C is a positive constant that depends on the magnitude of the charges. Derive an expression for the x-component of force acting on the movable charge, as a function of its position.

SOLUTION

IDENTIFY AND SET UP: We are given the potential-energy function $U(x)$. Hence we can use Eq. (7.17) to find our target, the function $F_x(x)$.

EXECUTE: The derivative with respect to x of the function $1/x$ is $-1/x^2$. So the force on the movable charge for $x > 0$ is

$$F_x(x) = -\frac{dU(x)}{dx} = -C\left(-\frac{1}{x^2}\right) = \frac{C}{x^2}$$

EVALUATE: The x-component of force is positive, corresponding to a repulsive interaction between like electric charges. The potential energy is very large for small x and approaches zero as x becomes large; the force pushes the movable charge toward large positive values of x, for which the potential energy is less. The force varies as $1/x^2$; it is small when the particles are far apart (large x) but becomes large when the particles are close together (small x). This is an example of Coulomb's law for electric interactions, which we will study in greater detail in Chapter 21.

Force and Potential Energy in Three Dimensions

We can extend this analysis to three dimensions, where the particle may move in the x-, y-, or z-direction, or all at once, under the action of a conservative force that has components F_x, F_y, and F_z. Each component of force may be a function of the coordinates x, y, and z. The potential-energy function U is also a function of all three space coordinates. We can now use Eq. (7.17) to find each component of force. The potential-energy change ΔU when the particle moves a small distance Δx in the x-direction is again given by $-F_x \Delta x$; it doesn't depend on F_y and F_z, which represent force components that are perpendicular to the displacement and do no work. So we again have the approximate relation

$$F_x = -\frac{\Delta U}{\Delta x}$$

The y- and z-components of force are determined in exactly the same way:

$$F_y = -\frac{\Delta U}{\Delta y} \qquad F_z = -\frac{\Delta U}{\Delta z}$$

To make these relations exact, we need to take the limits $\Delta x \rightarrow 0$, $\Delta y \rightarrow 0$, and $\Delta z \rightarrow 0$ so that these ratios become derivatives. Because U may be a function of all 3 coordinates, we need to remember that when we calculate each of these derivatives, only one coordinate changes at a time. We compute the derivative of U with respect to x by assuming that y and z are constant and only x varies, and so on. Such a derivative is called a *partial derivative*. The usual notation is $\partial U/\partial x$ and so on; the symbol ∂ is a modified d to remind us of the nature of this operation. So we write

$$F_x = -\frac{\partial U}{\partial x} \qquad F_y = -\frac{\partial U}{\partial y} \qquad F_z = -\frac{\partial U}{\partial z} \qquad (7.18)$$

(force from potential energy)

We can use unit vectors to write a single compact vector expression for the force $\vec{F}$:

$$\vec{F} = -\left(\frac{\partial U}{\partial x}\hat{\imath} + \frac{\partial U}{\partial y}\hat{\jmath} + \frac{\partial U}{\partial z}\hat{k}\right) \quad \text{(force from potential energy)} \quad (7.19)$$

The expression inside the parentheses represents a particular operation on the function U, in which we take the partial derivative of U with respect to each coordinate, multiply by the corresponding unit vector, and take the vector sum. This operation is called the **gradient** of U and is often abbreviated as $\vec{\nabla} U$. Thus the force is the negative of the gradient of the potential-energy function:

$$\vec{F} = -\vec{\nabla} U \qquad (7.20)$$

As a check, let's substitute into Eq. (7.20) the function $U = mgy$ for gravitational potential energy:

$$\vec{F} = -\vec{\nabla}(mgy) = -\left(\frac{\partial(mgy)}{\partial x}\hat{\imath} + \frac{\partial(mgy)}{\partial y}\hat{\jmath} + \frac{\partial(mgy)}{\partial z}\hat{k}\right) = (-mg)\hat{\jmath}$$

This is just the familiar expression for the gravitational force.

Example 7.16 Force and potential energy in two dimensions

A puck slides on a level, frictionless air-hockey table. The coordinates of the puck are x and y. It is acted on by a conservative force described by the potential-energy function

$$U(x, y) = \frac{1}{2}k(x^2 + y^2)$$

Derive an expression for the force acting on the puck, and find an expression for the magnitude of the force as a function of position.

SOLUTION

IDENTIFY AND SET UP: We'll find the components of the force from the function $U(x, y)$ using Eq. (7.18), then determine the magnitude of the force using the formula for the magnitude of a vector: $F = \sqrt{F_x^2 + F_y^2}$.

EXECUTE: The components of the force are

$$F_x = -\frac{\partial U}{\partial x} = -kx \qquad F_y = -\frac{\partial U}{\partial y} = -ky$$

From Eq. (7.19) this corresponds to the vector expression

$$\vec{F} = -k(x\hat{\imath} + y\hat{\jmath})$$

Now $x\hat{\imath} + y\hat{\jmath}$ is just the position vector $\vec{r}$ of the particle, so we can rewrite this expression as $\vec{F} = -k\vec{r}$. This represents a force that at each point is opposite in direction to the position vector of the point—that is, a force that at each point is directed toward the origin. The potential energy is minimum at the origin, so again the force pushes in the direction of decreasing potential energy.

The *magnitude* of the force at any point is

$$F = \sqrt{(-kx)^2 + (-ky)^2} = k\sqrt{x^2 + y^2} = kr$$

where r is the particle's distance from the origin. This is the force exerted by a spring that obeys Hooke's law and has a negligibly small length (compared to the other distances in the problem) when it is not stretched. So the motion of the puck is the same as if it were attached to one end of an ideal spring of negligible unstretched length; the other end is attached to the air-hockey table at the origin.

EVALUATE: We can check our result by noting that the potential-energy function can also be expressed as $U = \frac{1}{2}kr^2$. Written this way, U is a function of a single coordinate r, so we can find the force using Eq. (7.17) with x replaced by r:

$$F_r = -\frac{dU}{dr} = -\frac{d}{dr}\left(\frac{1}{2}kr^2\right) = -kr$$

Just as we calculated above, the force has magnitude kr; the minus sign indicates that the force is radially inward (toward the origin).

Test Your Understanding

A particle moving along the x-axis is acted on by a conservative force F_x. At a certain point, the force is zero. What does this tell you about the value of the potential-energy function $U(x)$ at that point? What about the value of the derivative of $U(x)$ at that point?

7.5 | Energy Diagrams

When a particle moves along a straight line under the action of a conservative force, we can get a lot of insight into the possible motions by looking at the graph of the potential-energy function $U(x)$. Figure 7.23a shows a glider with mass m that moves along the x-axis on an air track. The spring exerts on the glider a force with x-component $F_x = -kx$. Figure 7.23b is a graph of the corresponding potential-energy function $U(x) = \frac{1}{2}kx^2$. If the elastic force of the spring is the *only* horizontal force acting on the glider, the total mechanical energy $E = K + U$ is constant, independent of x. A graph of E as a function of x is thus a straight horizontal line.

The vertical distance between the U and E graphs at each point represents the difference $E - U$, equal to the kinetic energy K at that point. We see that K is greatest at $x = 0$. It is zero at the values of x where the two graphs cross, labeled A and $-A$ in the diagram. Thus the speed v is greatest at $x = 0$, and it is zero at $x = \pm A$, the points of *maximum* possible displacement from $x = 0$ for a given value of the total energy E. The potential energy U can never be greater than the total energy E; if it were, K would be negative, and that's impossible. The motion is a back-and-forth oscillation between the points $x = A$ and $x = -A$.

At each point, the force F_x on the glider is equal to the negative of the slope of the $U(x)$ curve: $F_x = -dU/dx$. When the particle is at $x = 0$, the slope and the force are zero, so this is an *equilibrium* position. When x is positive, the slope of the $U(x)$ curve is positive and the force F_x is negative, directed toward the origin. When x is negative, the slope is negative and F_x is positive, again toward the origin. Such a force is sometimes called a *restoring force;* when the glider is displaced to either side of $x = 0$, the resulting force tends to "restore" it back to $x = 0$. An analogous situation is a marble rolling around in a round-bottomed salad bowl. We say that $x = 0$ is a point of **stable equilibrium.** More generally, *any minimum in a potential-energy curve is a stable equilibrium position.*

Figure 7.24a shows a hypothetical but more general potential-energy function $U(x)$. Figure 7.24b shows the corresponding force $F_x = -dU/dx$. Points x_1 and x_3 are stable equilibrium points. At each of these points, F_x is zero because the slope

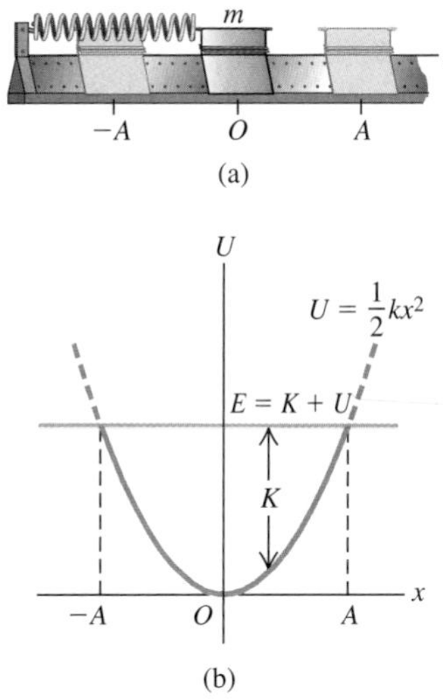

7.23 (a) A glider on an air track. The spring exerts a force $F_x = -kx$. (b) The potential-energy function. The limits of the motion are the points where the U curve intersects the horizontal line representing the total mechanical energy E.

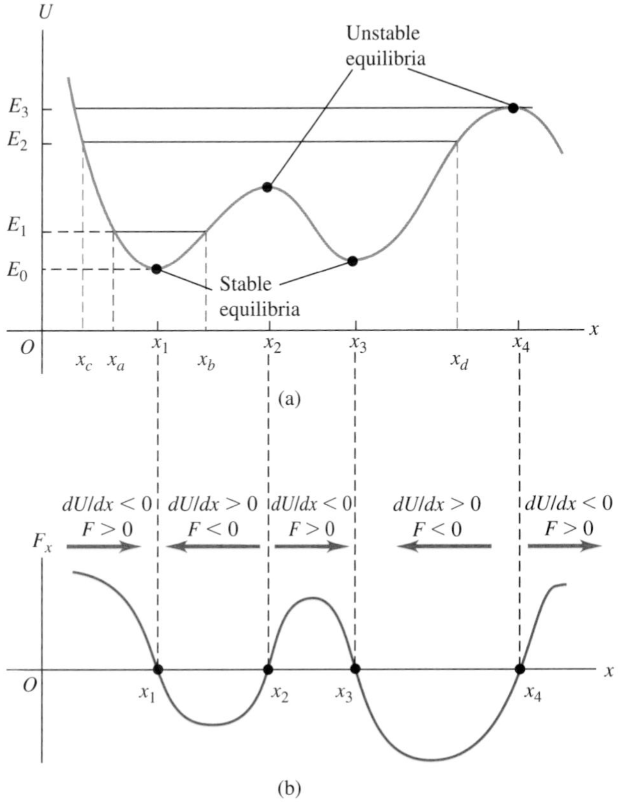

7.24 (a) A hypothetical potential-energy function $U(x)$. (b) The corresponding force $F_x = -dU/dx$. Maxima and minima of $U(x)$ correspond to points where $F_x = 0$.

of the $U(x)$ curve is zero. When the particle is displaced to either side, the force pushes back toward the equilibrium point. The slope of the $U(x)$ curve is also zero at points x_2 and x_4, and these are also equilibrium points. But when the particle is displaced a little to the right of either point, the slope of the $U(x)$ curve becomes negative, corresponding to a positive F_x that tends to push the particle still farther from the point. When the particle is displaced a little to the left, F_x is negative, again pushing away from equilibrium. This is analogous to a marble rolling on the top of a bowling ball. Points x_2 and x_4 are called **unstable equilibrium** points; *any maximum in a potential-energy curve is an unstable equilibrium position.*

CAUTION The direction of the force on a body is *not* determined by the sign of the potential energy U. Rather, it's the sign of $F_x = -dU/dx$ that matters. As we discussed in Section 7.1, the physically significant quantity is the *difference* in the value of U between two points, which is just what the derivative $F_x = -dU/dx$ measures. This means that you can always add a constant to the potential-energy function without changing the physics of the situation.

If the total energy is E_1 and the particle is initially near x_1, it can move only in the region between x_a and x_b determined by the intersection of the E_1 and U graphs (Fig. 7.24a). Again, U cannot be greater than E_1 because K can't be negative. We speak of the particle as moving in a *potential well,* and x_a and x_b are the *turning points* of the particle's motion (since at these points, the particle stops and reverses direction). If we increase the total energy to the level E_2, the particle can move over a wider range, from x_c to x_d. If the total energy is greater than E_3, the particle can "escape" and move to indefinitely large values of x. At the other extreme, E_0 represents the least possible total energy the system can have.

Test Your Understanding

The potential-energy function for a particle moving along the positive x-axis is $U(x) = (0.600\ \mathrm{N})x + (2.40\ \mathrm{N \cdot m^2})/x$. Where is the point of equilibrium? (Just consider positive values of x.) What is the potential energy at that point? What is the force at that point?

SUMMARY

The work done on a particle by a constant gravitational force can be represented as a change in the gravitational potential energy $U = mgy$. This energy is a shared property of the particle and the earth.

$$W_{grav} = mgy_1 - mgy_2 = U_1 - U_2 = -\Delta U \quad (7.1), (7.3)$$

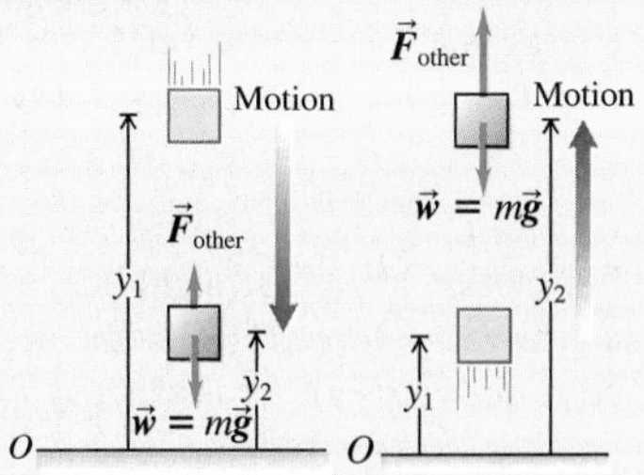

An ideal stretched or compressed spring exerts an elastic force $F_x = -kx$ on a particle, where x is the amount of stretch or compression. The work done by this force can be represented as a change in the elastic potential energy of the spring, $U = \frac{1}{2}kx^2$.

$$W_{el} = \frac{1}{2}kx_1^2 - \frac{1}{2}kx_2^2 = U_1 - U_2 = -\Delta U \quad (7.10)$$

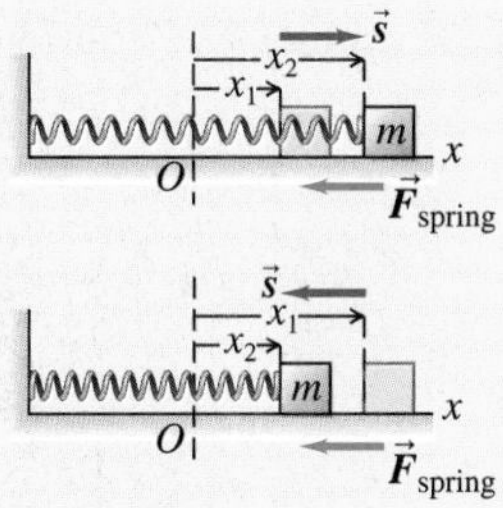

The total potential energy is the sum of the gravitational and elastic potential energy. If no forces other than the gravitational and elastic forces do work on a particle, the sum of kinetic and potential energy is conserved. This sum $E = K + U$ is called the total mechanical energy. (See Examples 7.1, 7.3, 7.4, 7.5, 7.8, and 7.10)

$$K_1 + U_1 = K_2 + U_2 \quad (7.11)$$

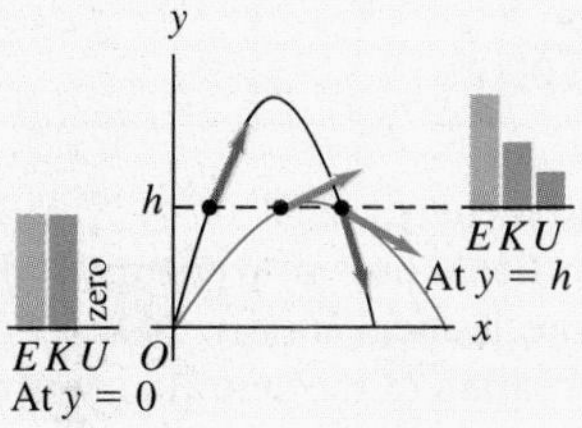

When forces other than the gravitational and elastic forces do work on a particle, the work W_{other} done by these other forces equals the change in total mechanical energy (kinetic energy plus total potential energy)
(See Examples 7.2, 7.6, 7.7, 7.9, and 7.11)

$$K_1 + U_1 + W_{other} = K_2 + U_2 \quad (7.13)$$

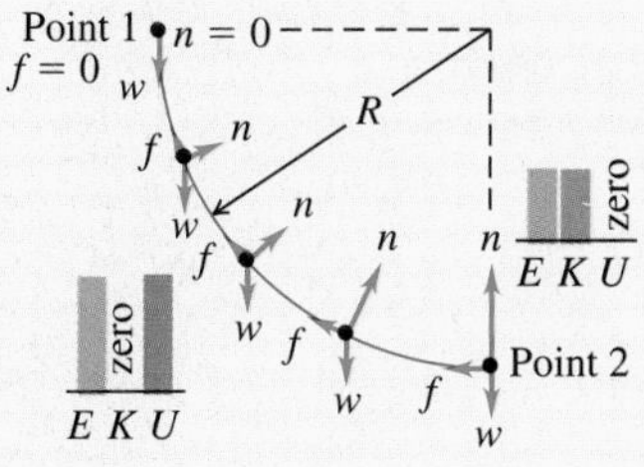

All forces are either conservative or nonconservative. A conservative force is one for which the work-kinetic energy relation is completely reversible. The work of a conservative force can always be represented by a potential-energy function, but the work of a nonconservative force cannot.

The work done by nonconservative forces manifests itself as changes in the internal energy of bodies. The sum of kinetic, potential, and internal energy is always conserved.
(See Examples 7.12 through 7.14)

$$\Delta K + \Delta U + \Delta U_{\text{int}} = 0 \quad (7.16)$$

For motion along a straight line, a conservative force $F_x(x)$ is the negative derivative of its associated potential-energy function U. In three dimensions, the components of a conservative force are negative partial derivatives of U.
(See Examples 7.15 and 7.16)

$$F_x(x) = -\frac{dU(x)}{dx} \quad (7.17)$$

$$F_x = -\frac{\partial U}{\partial x}, \quad F_y = -\frac{\partial U}{\partial y} \quad F_z = -\frac{\partial U}{\partial y} \quad (7.18)$$

$$\vec{F} = -\left(\frac{\partial U}{\partial x}\hat{\imath} + \frac{\partial U}{\partial y}\hat{\jmath} + \frac{\partial U}{\partial z}\hat{k}\right) \quad (7.19)$$

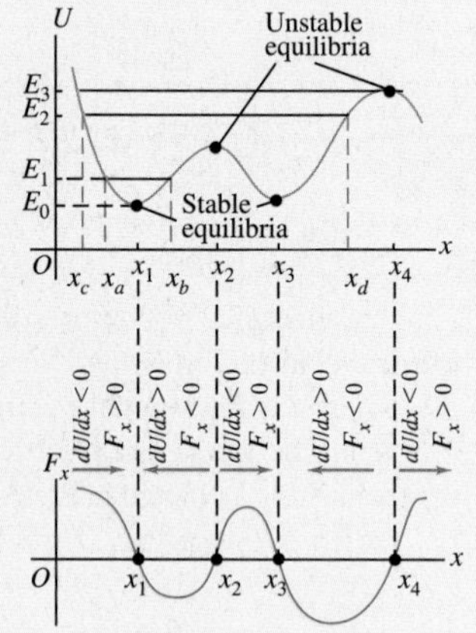

Key Terms

potential energy, 242
gravitational potential energy, 243
total mechanical energy, 244
conservation of mechanical energy, 244
elastic potential energy, 254
conservative force, 260
nonconservative force, 262
dissipative force, 262
internal energy, 263
law of conservation of energy, 264
gradient, 267
stable equilibrium, 269
unstable equilibrium, 270

Your Notes

Answer to Chapter Opening Question

The water is doing negative work on the diver; it exerts an upward force of fluid resistance as he moves downward. Gravity is doing positive work, since this force is in the same downward direction as the diver's displacement.

Answers to Test Your Understanding Questions

Section 7.1 The initial kinetic energy $K_1 = 0$, the initial potential energy $U_1 = mgy_1$, and the final potential energy $U_2 = mgy_2$ are the same for both ramps. Mechanical energy is conserved in both cases, so the final kinetic energy $K = \frac{1}{2}mv_2^2$ is also the same for both ramps. Hence the speed at the right-hand end is the *same* in both cases!

Section 7.2 When the elevator finally comes to a stop, the net force on it is zero. Its weight is $(2000\text{ kg})(9.8\text{ m/s}^2) = 19{,}600\text{ N}$, and friction provides an upward force of 17,000 N. Hence the spring must provide an additional 2600 N of upward force. Using the relationship (force) = (force constant)(compression) for a spring, we find that the spring is compressed by $(2600\text{ N})/(1.41 \times 10^5\text{ N/m}) = 0.018\text{ m}$, or 1.8 cm. Hence $y = -0.018\text{ m}$. At this point $K = 0$, $U_{grav} = mgy = (2000\text{ kg})(9.8\text{ m/s}^2)(-0.018\text{ m}) = -360\text{ J}$, $U_{el} = \frac{1}{2}ky^2 = \frac{1}{2}(1.41 \times 10^5\text{ N/m})(-0.018\text{ m})^2 = 23\text{ J}$, and $E = K + U_{grav} + U_{el} = -337\text{ J}$. The initial mechanical energy of 625,000 J was completely lost to friction, plus another 337 J.

Section 7.3 The total work *you* would do is $274\text{ J} + 196\text{ J} = 470\text{ J}$. The *friction* force would do -470 J of work, so the total work done by *all* forces is, in fact, zero: $470\text{ J} - 470\text{ J} = 0$. The moral is that it's important to include *all* work done by forces other than gravity or spring forces.

Section 7.4 If $F_x = 0$ at a point, then the derivative of $U(x)$ is zero at that point. However, this tells us absolutely nothing about the *value* of $U(x)$ at that point.

Section 7.5 Equilibrium occurs where $F_x = 0$. The force is

$$F_x = -dU/dx = -0.600\text{ N} + (2.40\text{ N}\cdot\text{m}^2)/x^2$$

Solving for the value of x where $F_x = 0$ (the equilibrium position) gives $x^2 = (2.40\text{ N}\cdot\text{m}^2)/(0.600\text{ N}) = 4.00\text{ m}^2$, or $x = 2.00\text{ m}$. At that point $U = (0.600\text{ N})(2.00\text{ m}) + (2.40\text{ N}\cdot\text{m}^2)/(2.00\text{ m}) = 2.40\text{ N}\cdot\text{m} = 2.40\text{ J}$; the force is zero.

Discussion Questions

Q7.1 A baseball is thrown straight up with initial speed v_1, as in Example 7.1 (Section 7.1). If air resistance cannot be ignored when the ball returns to its initial height its speed is less than v_1. Explain, using energy concepts.

Q7.2 In Fig. 7.6 the projectile has the same initial kinetic energy in each case. Why does it not rise to the same maximum height in each case?

Q7.3 In Example 7.5 (Section 7.1), does Throckmorton's speed at the bottom depend on the shape of the ramp or just the difference in height of points 1 and 2? Explain. Answer this same question if the ramp is not frictionless, as in Example 7.6 (Section 7.1).

Q7.4 An egg is released from rest at the roof of a building and falls to the ground. Its fall is observed by a student on the roof of the building, who uses coordinates with origin at the roof and by a student on the ground, who uses coordinates with origin at the ground. Do the two students assign the same or different values to the initial gravitational potential energy, the final gravitational potential energy, the change in gravitational potential energy, and the kinetic energy of the egg just before it strikes the ground? Explain.

Q7.5 A physics teacher had a bowling ball suspended from a very long rope attached to the high ceiling of a large lecture hall. To illustrate his faith in conservation of energy, he would back up to one side of the stage, pull the ball far to one side until the taut rope brought it just to the end of his nose, and then release it. The massive ball would swing in a mighty arc across the stage and then return to stop momentarily just in front of the nose of the stationary, unflinching teacher. However, one day after the demonstration he looked up just in time to see a student at the other side of the stage *push* the ball away from his nose as he tried to duplicate the demonstration. Tell the rest of the story and explain the reason for the potentially tragic outcome.

Q7.6 When a returning space shuttle touches down on the runway, it has lost almost all the kinetic energy it had in orbit. The gravitational potential energy has also decreased considerably. Where did all that energy go?

Q7.7 Is it possible for a frictional force to *increase* the mechanical energy of a system? If so, give examples.

Q7.8 A woman bounces on a trampoline, going a little higher with each bounce. Explain how she increases the total mechanical energy.

Q7.9 A box slides down a ramp and work is done on the box by the forces of gravity and friction. Can the work of each of these forces be expressed in terms of the change in a potential energy function? For each force explain why or why not.

Q7.10 A compressed spring is clamped in its compressed position and then is dissolved in acid. What becomes of its potential energy?

Q7.11 When an object moves away from the earth the gravitational potential energy increases; when it moves closer the potential energy decreases. But the potential energy for a spring increases both when the spring is stretched and when it is compressed. Explain the reason for the difference in behavior of these two potential energies.

Q7.12 Since only changes in potential energy are important in any problem, a student decides to let the elastic potential energy of a spring be zero when the spring is stretched a distance x_1. The student decides, therefore, to let $U = \frac{1}{2}k(x - x_1)^2$. Is this correct? Explain.

Q7.13 Figure 7.22a shows the potential energy function for the force $F_x = -kx$. Sketch the potential energy function for the force $F_x = +kx$. For this force, is $x = 0$ a point of equilibrium? Is this equilibrium stable or unstable? Explain.

Q7.14 Figure 7.22b shows the potential energy function associated with the gravitational force between an object and the earth. Use this graph to explain why objects always fall toward the earth when they are released.

Q7.15 For a system of two particles we often let the potential energy for the force between the particles approach zero as the separation of the particles approaches infinity. If this choice is made, explain why the potential energy at noninfinite separation is positive if the particles repel one another and negative if they attract.

Q7.16 Explain why the points $x = A$ and $x = -A$ in Fig. 7.23b are called *turning points*. How are the values of E and U related at a turning point?

Q7.17 A particle is in *neutral equilibrium* if the net force on it is zero and remains zero if the particle is displaced slightly in any direction. Sketch the potential energy function near a point of neutral equilibrium, for the case of one-dimensional motion. Give an example of an object in neutral equilibrium.

Q7.18 The net force on a particle of mass m has the potential-energy function graphed in Figure 7.24a. If the total energy is E_1, graph the speed v of the particle versus its position x. At what value of x is the speed greatest? Sketch v versus x if the total energy is E_2.

Q7.19 A ball moves from point 1 to point 2. The gravitational potential energy at point 2 is *greater* than at point 1. When the ball moves from point 1 to point 2, does gravity do positive or negative work?

Q7.20 The potential energy function for a force $\vec{F}$ is $U = \alpha x^3$, where α is a positive constant. What is the direction of $\vec{F}$?

Exercises

Section 7.1 Gravitational Potential Energy

7.1 What is the potential energy for an 800-kg elevator at the top of Chicago's Sears Tower, 440 m above street level? Let the potential energy be zero at street level.

7.2 A 5.00-kg sack of flour is lifted vertically at a constant speed of 3.50 m/s through a height of 15.0 m. a) How great a force is required? b) How much work is done on the sack by the lifting force? What becomes of this work?

7.3 Repeat part (a) of Example 6.5 (Section 6.2) using Eq. (7.7).

7.4 A mail bag with a mass of 120 kg is suspended by a vertical rope 6.0 m long. a) What horizontal force is necessary to hold the bag in a position displaced sideways 3.0 m from its initial position? b) How much work is done by the worker in moving the bag to this position?

7.5 A baseball is thrown from the roof of a 22.0-m-tall building with an initial velocity of magnitude 12.0 m/s and directed at an angle of 53.1° above the horizontal. a) What is the speed of the ball just before it strikes the ground? Use energy methods and ignore air resistance. b) What is the answer for part (a) if the initial velocity is at an angle of 53.1° *below* the horizontal? c) If the effects of air resistance are included, will part (a) or (b) give the higher speed?

7.6 a) In Example 7.7 (Section 7.1), calculate the minimum initial speed required for the crate to reach the top of the ramp. b) If the initial speed of the crate in Example 7.7 is 11.0 m/s, what is its speed at the top of the ramp?

7.7 Answer part (b) of Example 7.7 (Section 7.1) by applying Eq. (7.7) to points 2 and 3, rather than to points 1 and 3 as was done in the example.

7.8 An empty crate slides down a ramp, starting with a speed v_0 and reaching the bottom with a speed v and kinetic energy K. Some books are placed in the box, so that the total mass is quadrupled. The coefficient of kinetic friction is constant and air resistance is negligible. Starting again with v_0 at the top of the ramp, what are the speed and kinetic energy at the bottom? Explain the reasoning behind your answers.

7.9 A small rock with mass 0.20 kg is released from rest at point A, which is at the top edge of a large, hemispherical bowl with radius $R = 0.50$ m (Fig. 7.25). Assume that the size of the rock is small compared to R, so that the rock can be treated as a particle, and assume that the rock slides rather than rolling. The work done by friction on the rock when it moves from point A to point B at the bottom of the bowl is -0.22 J. What is the speed of the rock when it reaches point B?

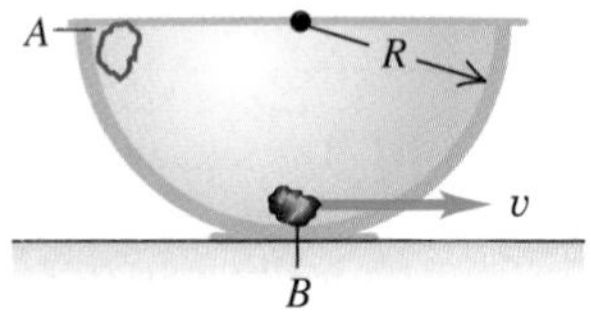

Figure 7.25 Exercise 7.9.

7.10 Jumping Flea. High-speed motion pictures (3500 frames/second) of a jumping 210 μg flea yielded the data to plot the graph shown in Fig. 7.26 (see "The Flying Leap of the Flea," by M. Rothschild, Y. Schlein, K. Parker, C. Neville, and S. Sternberg in the Nov. 1973 *Scientific American*). This flea was about 2 mm long and jumped at a nearly vertical take-off angle. a) Find the maximum height above the ground reached by the flea. (Can you ignore the distance traveled by the flea before its speed becomes constant? Explain.) b) How much work was done by the flea in launching itself into the air?

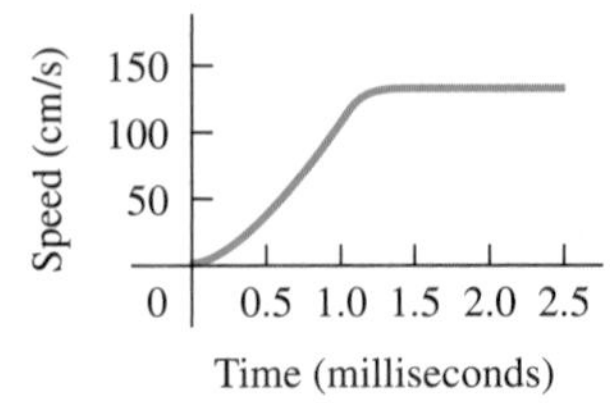

Figure 7.26 Exercise 7.10.

7.11 You are testing a new amusement park roller coaster with an empty car with mass 120 kg. One part of the track is a vertical loop with radius 12.0 m. At the bottom of the loop (point A) the car has speed 25.0 m/s and at the top of the loop (point B) has speed 8.0 m/s. As the car rolls from point A to point B, how much work is done by friction?

7.12 Tarzan and Jane. Tarzan, in one tree, sights Jane in another tree. He grabs the end of a vine with length 20 m that makes an angle of 45° with the vertical, steps off his tree limb, and swings down and then up to Jane's open arms. When he arrives, his vine makes an angle of 30° with the vertical. Determine whether he gives her a tender embrace or knocks her off her limb by calculating Tarzan's speed just before he reaches Jane. You can ignore air resistance and the mass of the vine.

7.13 A 10.0-kg microwave oven is pushed 8.00 m up the sloping surface of a loading ramp inclined at an angle of 36.9° above the horizontal, by a constant force $\vec{F}$ with a magnitude 110 N and acting parallel to the ramp. The coefficient of kinetic friction between the oven and the ramp is 0.250. a) What is the work done on the oven by the force $\vec{F}$? b) What is the work done on the oven by the friction force? c) Compute the increase in potential energy for the oven. d) Use your answers to parts (a), (b), and (c) to calculate the increase in the oven's kinetic energy. e) Use $\sum\vec{F} = m\vec{a}$ to calculate the acceleration of the oven. Assuming that the oven is initially at rest, use the acceleration to calculate the oven's speed after traveling 8.00 m. From this, compute the increase in the oven's kinetic energy, and compare it to the answer you got in part (d).

7.14 Pendulum. A small rock with mass 0.12 kg is fastened to a massless string with length 0.80 m to form a pendulum. The pendulum is swinging so as to make a maximum angle of 45° with the vertical. Air resistance is negligible. a) What is the speed of the rock when the string passes through the vertical position? b) What is the tension in the string when it makes an angle of 45° with the vertical? c) What is the tension in the string as it passes through the vertical?

Section 7.2 Elastic Potential Energy

7.15 A force of 800 N stretches a certain spring a distance of 0.200 m. a) What is the potential energy of the spring when it is stretched 0.200 m? b) What is its potential energy when it is compressed 5.00 cm?

7.16 A force of 720 N stretches a certain spring a distance of 0.150 m. What is the potential energy of the spring when a 60.0-kg mass hangs vertically from it?

7.17 A spring of negligible mass has force constant $k = 1600$ N/m. a) How far must the spring be compressed for 3.20 J of potential energy to be stored in it? b) You place the spring vertically with one end on the floor. You then drop a 1.20-kg book onto it from a height of 0.80 m above the top of the spring. Find the maximum distance the spring will be compressed.

7.18 A slingshot will shoot a 10-g pebble 22.0 m straight up. a) How much potential energy is stored in the slingshot's rubber band? b) With the same potential energy stored in the rubber band, how high can the slingshot shoot a 25-g pebble? c) What physical effects did you ignore in solving this problem?

7.19 A 1.20-kg piece of cheese is placed on a vertical spring of negligible mass and force constant $k = 1800$ N/m that is compressed 15.0 cm. When the spring is released, how high does the cheese rise from this initial position? (The cheese and the spring are *not* attached.)

7.20 Consider the glider of Example 7.8 (Section 7.2) and Fig. 7.16. As in the example, the glider is released from rest with the spring stretched 0.100 m. What is the displacement x of the glider from its equilibrium position when its speed is 0.20 m/s? (You should get more than one answer. Explain why.)

7.21 Consider the glider of Example 7.8 (Section 7.2) and Fig. 7.16. a) As in the example, the glider is released from rest with the spring stretched 0.100 m. What is the speed of the glider when it returns to $x = 0$? b) What must the initial displacement of the glider be if its maximum speed in the subsequent motion is to be 2.50 m/s?

7.22 Consider the glider of Example 7.8 (Section 7.2) and Fig. 7.16. As in the example, the glider is released from rest with the spring stretched 0.100 m. But now the air track is turned off so that there is a kinetic friction force acting on the glider. a) If the coefficient of kinetic friction between the air track and the glider is $\mu_k = 0.050$, what is the speed of the glider when it has traveled 0.020 m, so that $x = 0.080$ m? b) If $\mu_k = 0.050$, what is the speed of the glider when it has traveled 0.100 m, so that $x = 0$? c) What must the value of μ_k be so that the glider reaches the $x = 0$ position with zero speed?

7.23 a) For the elevator of Example 7.11 (Section 7.2), what is the speed of the elevator after it has moved downward 1.00 m from point 1 in Fig. 7.17? b) When the elevator is 1.00 m below point 1 in Fig. 7.17, what is its acceleration?

7.24 You are asked to design a spring that will give a 1160-kg satellite a speed of 2.50 m/s relative to an orbiting space shuttle. Your spring is to give the satellite a maximum acceleration of $5.00g$. The spring's mass, the recoil kinetic energy of the shuttle, and changes in gravitational potential energy will all be negligible. a) What must the force constant of the spring be? b) What distance must the spring be compressed?

Section 7.3 Conservative and Nonconservative Forces

7.25 A 0.75-kg book moves vertically upward a distance of 16 m, and then vertically downward 16 m, returning to its initial position. a) How much work is done by gravity during the upward motion of the book? b) How much work is done by gravity during the downward motion of the book? c) What is the total work done on the book by gravity during the complete round trip? d) On the basis of your answer to part (c), would you say that the gravitational force is conservative or nonconservative? Explain.

7.26 A 0.050-kg rock moves from the origin to the point (3.0 m, 5.0 m) in a coordinate system in which the positive y-direction is upward. a) The rock first moves horizontally from the origin to the point (3.0 m, 0). Then it moves vertically from (3.0 m, 0) to (3.0 m, 5.0 m). Sketch the path of the rock in the xy-plane. How much work is done on the rock by gravity during the displacement? b) Instead of the path in part (a), let the rock first move vertically from the origin to (0, 5.0 m) and then horizontally from (0, 5.0 m) to (3.0 m, 5.0 m). Sketch the path of the rock in the xy-plane. How much work is done on the rock by gravity during the displacement? c) Compare the answers to parts (a) and (b). From your results would you say that the gravitational force is conservative or nonconservative? Explain.

7.27 In an experiment, one of the forces exerted on a proton is $\vec{F} = -\alpha x^2\hat{\imath}$, where $\alpha = 12$ N/m². a) How much work does $\vec{F}$ do when the proton moves along the straight-line path from the point (0.10 m, 0) to the point (0.10 m, 0.40 m)? b) Along the straight-line path from the point (0.10 m, 0) to the point (0.30 m, 0)? c) Along the straight-line path from the point (0.30 m, 0) to the point (0.10 m, 0)? d) Is the force $\vec{F}$ conservative? Explain. If $\vec{F}$ is conservative, what is the potential energy function for it? Let $U = 0$ when $x = 0$.

7.28 Consider the electron and force $\vec{F}$ of Example 7.13 (Section 7.3). a) The electron travels from the point (0, 0) to the point (L, L) along the straight line from (0, 0) to $(0, L)$ and then along the straight line from $(0, L)$ to (L, L). What is the work done by the force $\vec{F}$ during this displacement? b) The electron travels from the point (0, 0) to the point (L, L) along the straight line from (0, 0) to $(L, 0)$ and then along the straight line from $(L, 0)$ to (L, L). What is the work done by the force $\vec{F}$ during this displacement? c) The electron travels from the point (0, 0) to the point (L, L) along the straight line from (0, 0) to (L, L). What is the work done by the force $\vec{F}$ during this displacement? d) Compare the answers in parts (a), (b), and (c) and explain the results of this comparison.

7.29 A 0.60-kg book slides on a horizontal table. The kinetic friction force on the book has magnitude 1.2 N. a) How much work is done on the book by friction during a displacement of 3.0 m to the left? b) The book now slides 3.0 m to the right, returning to its starting point. During this second 3.0-m displacement, how much work is done on the book by friction? c) What is the total work done on the book by friction during the complete round trip? d) On the basis of your answer to part (c), would you say that the friction force is conservative or nonconservative? Explain.

7.30 A 30.0-kg packing crate in a warehouse is pushed to the loading dock by a worker who applies a horizontal force. The coefficient of kinetic friction between the crate and the floor is 0.20. The loading dock is 15.0 m southwest of the initial position of the crate. a) If the crate is pushed 10.6 m south and then 10.6 m west, what is the total work done on the crate by friction? b) If the crate is pushed along a straight-line path to the dock, so that it travels 15.0 m southwest, what is the work done on the crate by friction? c) Sketch the paths along which the crate moves in parts (a) and (b). Based on your answers to parts (a) and (b), would you say that the friction force is conservative or nonconservative? Explain.

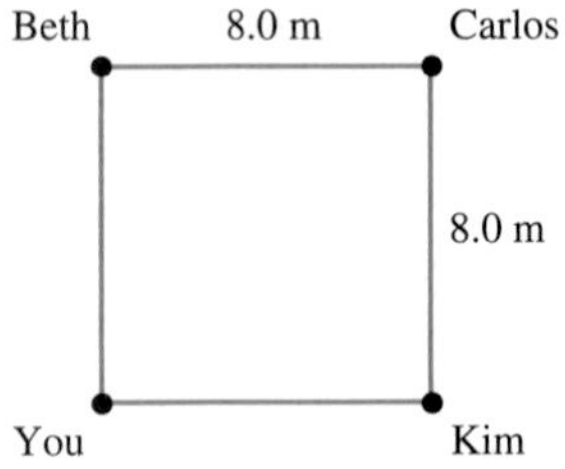

Figure 7.27 Exercise 7.31.

7.31 You and three friends stand at the corners of a square whose sides are 8.0 m long in the middle of the gym floor, as shown in Fig. 7.27. You take your physics book and push it from one person to the other. The book has a mass of 1.5 kg, and the coefficient of kinetic friction between the book and the floor is $\mu_k = 0.25$. a) The book slides from you to Beth and then from Beth to Carlos, along the lines connecting these people. What is the work done by friction during this displacement? b) You slide the book from you to Carlos along the diagonal of the square. What is the work done by friction during this displacement? c) You slide the book to Kim who then slides it back to you. What is the total work done by friction during this motion of the book? d) Is the friction force on the book conservative or nonconservative? Explain.

7.32 A block with mass m is attached to an ideal spring that has force constant k. a) The block moves from x_1 to x_2, where $x_2 > x_1$. How much work does the spring force do during this displacement? b) The block moves from x_1 to x_2 and then from x_2 to x_1. How much work does the spring force do during the displacement from x_2 to x_1? What is the total work done by the spring during the entire $x_1 \to x_2 \to x_1$ displacement? Explain why you obtained the answer you did. c) The block moves from x_1 to x_3, where $x_3 > x_2$. How much work does the spring force do during this displacement? The block then moves from x_3 to x_2. How much work does the spring force do during this displacement? What is the total work done by the spring force during the $x_1 \to x_3 \to x_2$ displacement? Compare your answer to the answer to part (a), where the starting and ending points are the same but the path is different.

Section 7.4 Force and Potential Energy

7.33 The potential energy of a pair of hydrogen atoms separated by a large distance x is given by $U(x) = -C_6/x^6$, where C_6 is a positive constant. What is the force that one atom exerts on the other? Is this force attractive or repulsive?

7.34 A force parallel to the x-axis acts on a particle moving along the x-axis. This force produces a potential energy $U(x)$ given by $U(x) = \alpha x^4$, where $\alpha = 1.20\ \text{J/m}^4$. What is the force (magnitude and direction) when the particle is at $x = -0.800$ m?

7.35 An experimental device moving in the xy-plane is acted on by a conservative force described by the potential-energy function $U(x, y) = k(x^2 + y^2) + k'xy$, where k and k' are positive constants. Derive an expression for the force expressed in terms of the unit vectors $\hat{\imath}$ and $\hat{\jmath}$.

7.36 An object moving in the xy-plane is acted on by a conservative force described by the potential-energy function $U(x, y) = \alpha(1/x^2 + 1/y^2)$, where α is a positive constant. Derive an expression for the force expressed in terms of the unit vectors $\hat{\imath}$ and $\hat{\jmath}$.

Section 7.5 Energy Diagrams

7.37 The potential energy of two atoms in a diatomic molecule is approximated by $U(r) = a/r^{12} - b/r^6$, where r is the spacing between atoms and a and b are positive constants. a) Find the force $F(r)$ on one atom as a function of r. Make two graphs, one of $U(r)$ versus r and one of $F(r)$ versus r. b) Find the equilibrium distance between the two atoms. Is this equilibrium stable? c) Suppose the distance between the two atoms is equal to the equilibrium distance found in part (b). What minimum energy must be added to the molecule to *dissociate* it, that is, to separate the two atoms to an infinite distance apart? This is called the *dissociation energy* of the molecule. d) For the molecule CO, the equilibrium distance between the carbon and oxygen atoms is 1.13×10^{-10} m and the dissociation energy is 1.54×10^{-18} J per molecule. Find the values of the constants a and b.

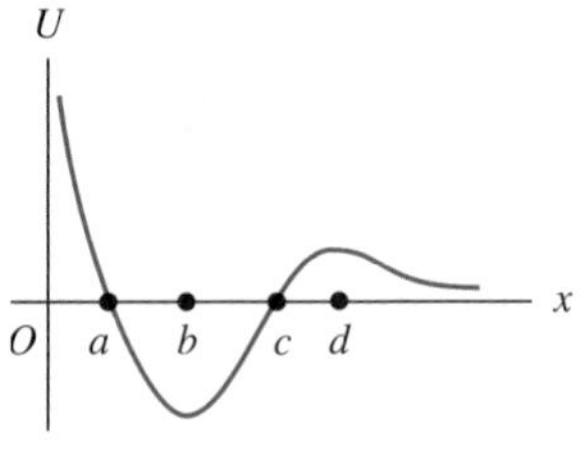

Figure 7.28 Exercise 7.38.

7.38 A marble moves along the x-axis. The potential-energy function is shown in Fig. 7.28. a) At which of the labeled x-coordinates is the force on the marble zero? b) Which of the labeled x-coordinates is a position of stable equilibrium? c) Which of the labeled x-coordinates is a position of unstable equilibrium?

Problems

7.39 A man with mass 70.0 kg sits on a platform suspended from a movable pulley, as shown in Fig. 7.29, and raises himself at constant speed by a rope passing over a fixed pulley. The platform and the pulleys have

Figure 7.29 Problem 7.39.

negligible mass. Assume that there are no friction losses. a) Find the force he must exert. b) Find the increase in the energy of the system when he raises himself 1.20 m. (Answer by calculating the increase in potential energy and also by computing the product of the force on the rope and the length of the rope passing through his hands.)

7.40 Two blocks with different mass are attached to either end of a light rope that passes over a light, frictionless pulley that is suspended from the ceiling. The masses are released from rest, and the more massive one starts to descend. After this block has descended 1.20 m, its speed is 3.00 m/s. If the total mass of the two blocks is 15.0 kg, what is the mass of each block?

7.41 Legal Physics. In an auto accident, a car hit a pedestrian and the driver then slammed on the brakes to stop the car. During the subsequent trial, the driver's lawyer claimed that he was obeying the posted 35 mph speed limit, but that the legal speed was too high to allow him to see and react to the pedestrian in time. You have been called in as the state's expert witness. Your investigation of the accident yields the following measurements: the skid marks made while the brakes were applied were 280 ft long, and the tread on the tires produced a coefficient of kinetic friction of 0.30 with the road. a) In your testimony in court, will you say that the driver was obeying the posted speed? You must be able to back up your conclusion with clear reasoning, because one of the lawyers will surely cross-examine you. b) If the driver's speeding ticket were \$10 for each mph he was driving above the posted speed limit, would he have to pay a ticket, and if so, how much would it be?

7.42 A 2.00-kg block is pushed against a spring with negligible mass and force constant $k = 400$ N/m, compressing it 0.220 m. When the block is released, it moves along a frictionless, horizontal surface and then up a frictionless incline with slope 37.0° (Fig. 7.30). a) What is the speed of the block as it slides along the horizontal surface after having left the spring? b) How far does the block travel up the incline before starting to slide back down?

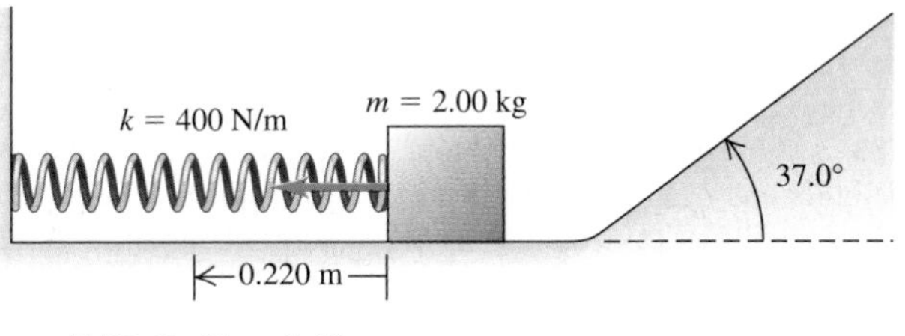

Figure 7.30 Problem 7.42.

7.43 A block with mass 0.50 kg is forced against a horizontal spring of negligible mass, compressing the spring a distance of 0.20 m (Fig. 7.31). When released, the block moves on a horizontal table top for 1.00 m before coming to rest. The spring constant k is 100 N/m. What is the coefficient of kinetic friction, μ_k, between the block and the table?

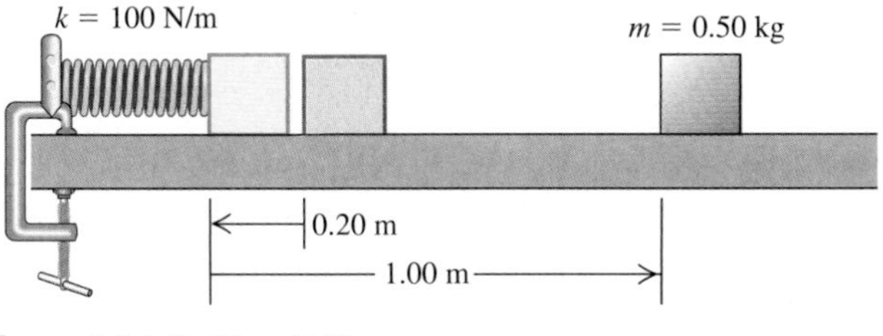

Figure 7.31 Problem 7.43.

7.44 On a horizontal surface, a crate with mass 50.0 kg is placed against a spring that stores 360 J of energy. The spring is released and the crate slides 5.60 m before coming to rest. What is the speed of the crate when it is 2.00 m from its initial position?

7.45 Bouncing Ball. A 650-gram rubber ball is dropped from an initial height of 2.50 m and on each bounce it returns to 75% of its previous height. a) What is the initial mechanical energy of the ball, just after it is released from its initial height? b) How much mechanical energy does the ball lose during its first bounce? What happens to this energy? c) How much mechanical energy is lost during the second bounce?

7.46 Riding a Loop-the-loop. A car in an amusement park ride rolls without friction around the track shown in Fig. 7.32. It starts from rest at point A at a height h above the bottom of the loop. Treat the car as a particle. a) What is the minimum value of h (in terms of R) such that the car moves around the loop without falling off at the top (point B)? b) If $h = 3.50R$ and $R = 20.0$ m, compute the speed, radial acceleration, and tangential acceleration of the passengers when the car is at point C, which is at the end of a horizontal diameter. Show these acceleration components in a diagram, approximately to scale.

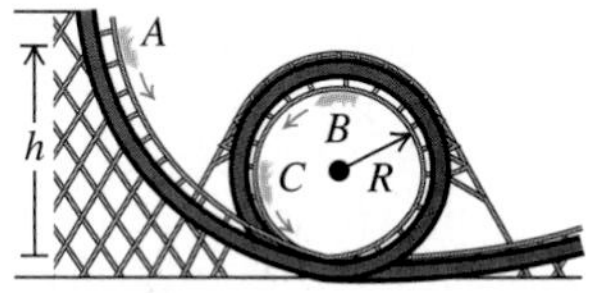

Figure 7.32 Problem 7.46.

7.47 A 2.0-kg piece of wood slides on the surface shown in Fig. 7.33. The curved sides are perfectly smooth, but the rough horizontal bottom is 30 m long and has a kinetic friction coefficient of 0.20 with the wood. The piece of wood starts from rest 4.0 m above the rough bottom. a) Where will this wood eventually come to rest? b) For the motion from the initial release until the piece of wood comes to rest, what is the total amount of work done by friction?

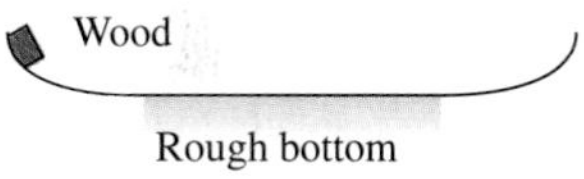

Figure 7.33 Problem 7.47.

7.48 Up and Down the Hill. A 28-kg rock approaches the foot of a hill with a speed of 15 m/s. This hill slopes upward at a constant angle of 40.0° above the horizontal. The coefficients of static and kinetic friction between the hill and the rock are 0.75 and 0.20, respectively. a) Use energy conservation to find the maximum height above the foot of the hill reached by the rock. b) Will the rock remain at rest at its highest point, or will it slide back down the hill? c) If the rock does slide back down, find its speed when it returns to the bottom of the hill.

7.49 A 15.0-kg stone slides down a snow-covered hill (Fig. 7.34), leaving point A with a speed of 10.0 m/s. There is no friction on the hill between points A and B, but there is friction on the level ground at the

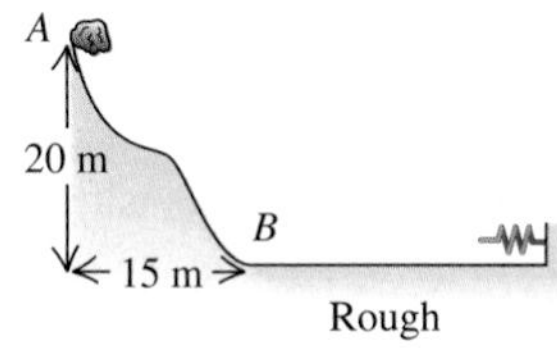

Figure 7.34 Problem 7.49.

bottom of the hill, between B and the wall. After entering the rough horizontal region, the stone travels 100 m and then runs into a very long, light spring with force constant 2.00 N/m. The coefficients of kinetic and static friction between the stone and the horizontal ground are 0.20 and 0.80, respectively. a) What is the speed of the stone when it reaches point B? b) How far will the stone compress the spring? c) Will the stone move again after it has been stopped by the spring?

7.50 A 2.8-kg block slides over the smooth, icy hill shown in Fig. 7.35. The top of the hill is horizontal and 70 m higher than its base. What minimum speed must the block have at the base of the hill so that it will not fall into the pit on the far side of the hill?

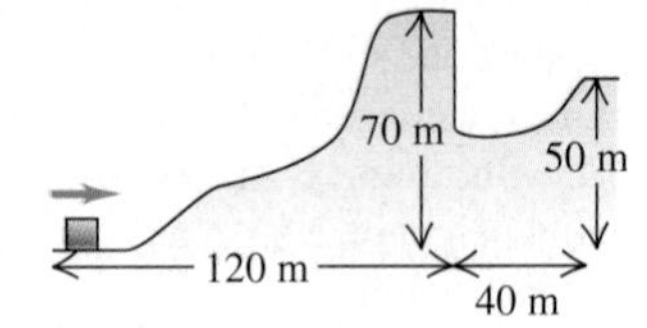

Figure 7.35 Problem 7.50.

7.51 Bungee Jump. A bungee cord is 30.0 m long and when stretched a distance x, it exerts a restoring force of magnitude kx. Your father-in-law (mass 95.0 kg) stands on a platform 45.0 m above the ground, and one end of the cord is tied securely to his ankle and the other end to the platform. You have promised him that when he steps off the platform he will fall a maximum distance of only 41.0 m before the cord stops him. You had several bungee cords to select from and tested them by stretching them out, tying one end to a tree, and pulling on the other end with a force of 380.0 N. When you do this, what distance will the bungee cord that you should select have stretched?

7.52 Ski Jump Ramp. You are designing a ski jump ramp for the next Winter Olympics. You need to calculate the vertical height h from the starting gate to the bottom of the ramp. The skiers push off hard with their ski poles at the start, just above the starting gate, so they typically have a speed of 2.0 m/s as they reach the gate. For safety, the skiers should have a speed of no more than 30.0 m/s when they reach the bottom of the ramp. You determine that for a 85.0-kg skier with good form, friction and air resistance will do a total of -4000 J of work on him during his run down the slope. What is the maximum height h for which the maximum safe speed will not be exceeded?

7.53 The Great Sandini is a 60-kg circus performer who is shot from a cannon (actually a spring gun). You don't find many men of his caliber, so you help him design a new gun. This new gun has a very large spring with a very small mass and a force constant of 1100 N/m that he will compress with a force of 4400 N. The inside of the gun barrel is coated with Teflon, so the average friction force will be only 40 N during the 4.0 m he moves in the barrel. At what speed will he emerge from the end of the barrel, 2.5 m above his initial rest position?

7.54 You are designing a delivery ramp for crates containing exercise equipment. The 1470-N crates will move at 1.8 m/s at the top of a ramp that slopes downward at 22.0°. The ramp exerts a 550-N kinetic friction force on each crate, and the maximum static friction force also has this value. Each crate will compress a spring at the bottom of the ramp and will come to rest after traveling a total distance of 8.0 m along the ramp. Once stopped, a crate must not rebound back up the ramp. Calculate the force constant of the spring that will be needed in order to meet the design criteria.

7.55 A system of two paint buckets connected by a lightweight rope is released from rest with the 12.0-kg bucket 2.00 m above the floor (Fig. 7.36). Use the principle of conservation of energy to find the speed with which this bucket strikes the floor. You can ignore friction and the inertia of the pulley.

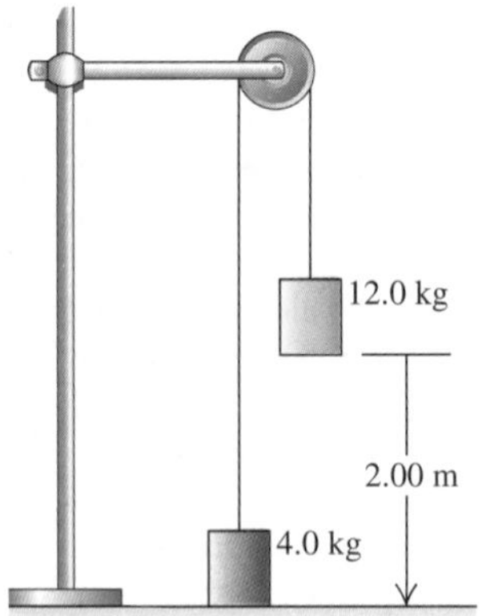

Figure 7.36 Problem 7.55.

7.56 a) For the elevator of Example 7.11 (Section 7.2), how much potential energy is stored in the spring when the elevator is at point 2 in Fig. 7.17? b) What maximum distance does the elevator travel upward from point 2 before it starts to slide down the shaft again? c) As the elevator slides back down, what is its speed just before it reaches the spring? d) When the elevator compresses the spring the second time, what is the maximum energy that is stored in the spring and what is the force that the spring exerts on the elevator?

7.57 Redesign the elevator safety system of Example 7.11 (Section 7.2) so that the elevator does not bounce but stays at rest the first time its speed becomes zero. The mass of the elevator is 2000 kg and its speed when it first touches the spring is 25 m/s. There is a kinetic friction force of 17,000 N, and the maximum static friction force on the elevator is also 17,000 N. You can ignore the mass of the spring. a) What spring constant is required, and what distance is the spring compressed when the elevator is stopped? Do you think this design is practical? Explain. b) What is the maximum magnitude of the acceleration of the elevator?

7.58 A wooden rod of negligible mass and length 80.0 cm is pivoted about a horizontal axis through its center. A white rat with mass 0.500 kg clings to one end of the stick, and a mouse with mass 0.200 kg clings to the other end. The system is released from rest with the rod horizontal. If the animals can manage to hold on, what are their speeds as the rod swings through a vertical position?

7.59 A 0.100-kg potato is tied to a string with length 2.50 m, and the other end of the string is tied to a rigid support. The potato is held straight out horizontally from the point of support, with the string pulled taut, and is then released. a) What is the speed of the potato at the lowest point of its motion? b) What is the tension in the string at this point?

7.60 The following is data from a computer simulation for a batted baseball with mass 0.145 kg, including air resistance:

t	x	y	v_x	v_y
0	0	0	30.0 m/s	40.0 m/s
3.05 s	70.2 m	53.6 m	18.6 m/s	0
6.59 s	124.4 m	0	11.9 m/s	-28.7 m/s

a) How much work was done by the air on the baseball as it moved from its initial position to its maximum height? b) How much work was done by the air on the baseball as it moved from its maximum height back to the starting elevation? c) Explain why the magnitude of the answer to part (b) is smaller than the magnitude of the answer in part (a).

7.61 Down the Pole. A fireman of mass m slides a distance d down a pole. He starts from rest. He moves as fast at the bottom as if he had stepped off a platform a distance $h \leq d$ above the ground and descended with negligible air resistance. a) What average friction force did the fireman exert on the pole? Does your answer make sense in the special cases of $h = d$ and $h = 0$? b) Find a numerical value for the average friction force a 75-kg fireman exerts, for $d = 2.5$ m and $h = 1.0$ m. c) In terms of g, h, and d, what is the speed of the fireman when he is a distance y above the bottom of the pole?

7.62 A 60.0-kg skier starts from rest at the top of a ski slope 65.0 m high. a) If frictional forces do -10.5 kJ of work on her as she descends, how fast is she going at the bottom of the slope? b) Now moving horizontally, the skier crosses a patch of soft snow, where $\mu_k = 0.20$. If the patch is 82.0 m wide and the average force of air resistance on the skier is 160 N, how fast is she going after crossing the patch? c) The skier hits a snowdrift and penetrates 2.5 m into it before coming to a stop. What is the average force exerted on her by the snowdrift as it stops her?

7.63 A skier starts at the top of a very large, frictionless snowball, with a very small, initial speed, and skis straight down the side (Fig. 7.37). At what point does she lose contact with the snowball and fly off at a tangent? That is, at the instant she loses contact with the snowball, what angle α does a radial line from the center of the snowball to the skier make with the vertical?

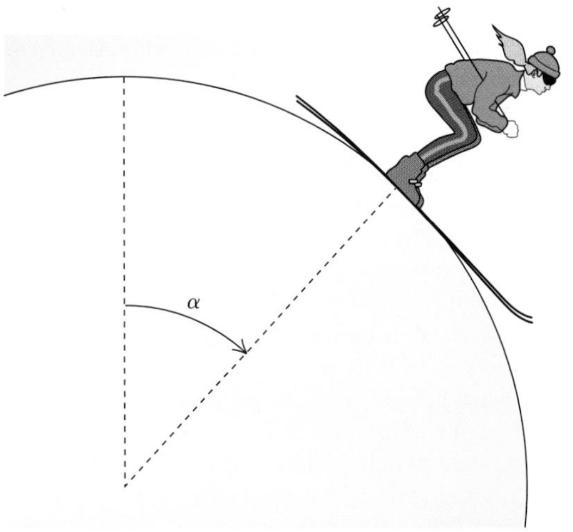

Figure 7.37 Problem 7.63.

7.64 A rock is tied to a cord and the other end of the cord is held fixed. The rock is given an initial tangential velocity that causes it to rotate in a vertical circle. Prove that the tension in the cord at the lowest point exceeds that at the highest point by six times the weight of the rock.

7.65 In a truck-loading station at a post office, a small 0.200-kg package is released from rest at point A on a track that is one-quarter of a circle with radius 1.60 m (Fig. 7.38). The size of the package is much less than 1.60 m, so the package can be treated as a particle. It slides down the track and reaches point B with a speed of 4.80 m/s. From point B, it slides on a level surface a distance of 3.00 m to point C, where it comes to rest. a) What is the coefficient of kinetic friction on the horizontal surface? b) How much work is done on the package by friction as it slides down the circular arc from A to B?

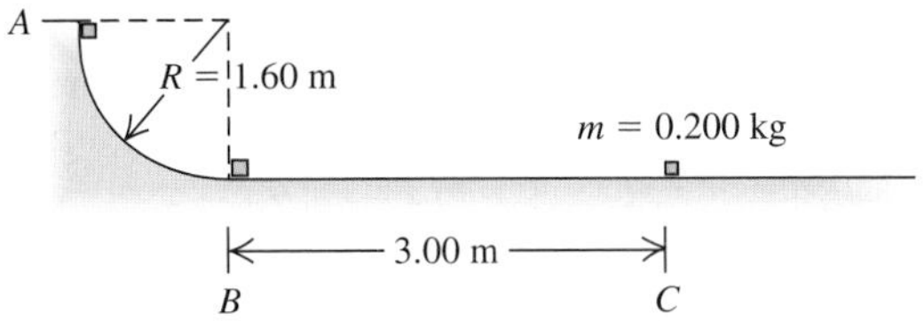

Figure 7.38 Problem 7.65.

7.66 A truck with mass m has a brake failure while going down an icy mountain road of constant downward slope angle α (Fig. 7.39). Initially the truck is moving downhill at speed v_0. After careening downhill a distance L with negligible friction, the truck driver steers the runaway vehicle onto a runaway truck ramp of constant upward slope angle β. The truck ramp has a soft sand surface for which the coefficient of rolling friction is μ_r. What is the distance that the truck moves up the ramp before coming to a halt? Use energy methods.

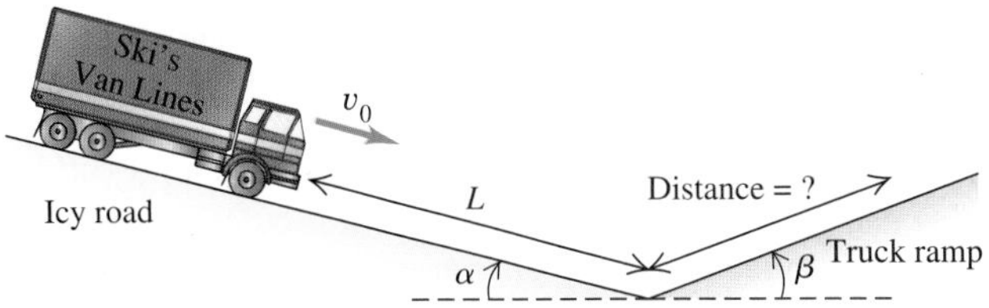

Figure 7.39 Problem 7.66.

7.67 A certain spring is found *not* to obey Hooke's law; it exerts a restoring force $F_x(x) = -\alpha x - \beta x^2$ if it is stretched or compressed, where $\alpha = 60.0$ N/m and $\beta = 18.0$ N/m^2. The mass of the spring is negligible. a) Calculate the potential energy function $U(x)$ for this spring. Let $U = 0$ when $x = 0$. b) An object with mass 0.900 kg on a frictionless, horizontal surface is attached to this spring, pulled a distance 1.00 m to the right (the $+x$-direction) to stretch the spring, and released. What is the speed of the object when it is 0.50 m to the right of the $x = 0$ equilibrium position?

7.68 A variable force $\vec{F}$ is maintained tangent to a frictionless, semicircular surface (Fig. 7.40). By slowly varying the force, a

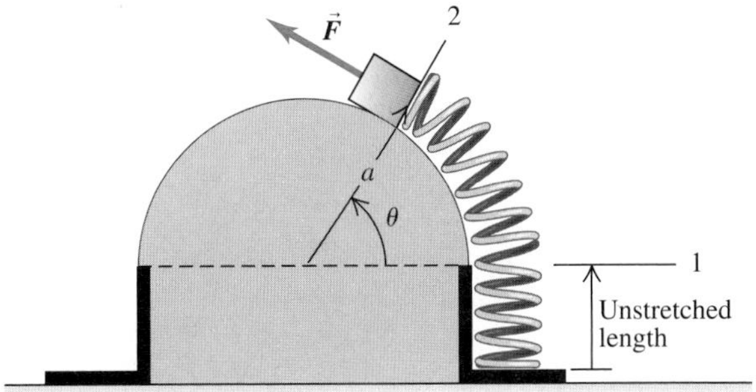

Figure 7.40 Problem 7.68.

block with weight w is moved, and the spring to which it is attached is stretched from position 1 to position 2. The spring has negligible mass and force constant k. The end of the spring moves in an arc of radius a. Calculate the work done by the force $\vec{F}$.

7.69 A 0.150-kg block of ice is placed against a horizontal, compressed spring mounted on a horizontal tabletop that is 1.20 m above the floor. The spring has force constant 1900 N/m and is initially compressed 0.045 m. The mass of the spring is negligible. The spring is released, and the block slides along the table, goes off the edge, and travels to the floor. If there is negligible friction between the block of ice and the table, what is the speed of the block of ice when it reaches the floor?

7.70 An 80.0-kg man jumps from a height of 2.50 m onto a platform mounted on springs. As the springs compress, he pushes the platform down a maximum distance of 0.240 m below its initial position, and then it rebounds. The platform and springs have negligible mass. a) What is the man's speed at the instant he depresses the platform 0.120 m? b) If the man just steps gently onto the platform, what maximum distance would he push it down?

7.71 An experimental apparatus with mass m is placed on a vertical spring of negligible mass and pushed down until the spring is compressed a distance x. The apparatus is then released and reaches its maximum height at a distance h above the point where it is released. The apparatus is not attached to the spring, and at its maximum height it is no longer in contact with the spring. The maximum magnitude of acceleration the apparatus can have without being damaged is a, where $a > g$. a) What should the force constant of the spring be? b) What distance x must the spring be compressed initially?

7.72 If a fish is attached to a vertical spring and slowly lowered to its equilibrium position, it is found to stretch the spring by an amount d. If the same fish is attached to the end of the unstretched spring and then allowed to fall from rest, through what maximum distance does it stretch the spring? (*Hint:* Calculate the force constant of the spring in terms of the distance d and the mass m of the fish.)

7.73 A wooden block with mass 1.50 kg is placed against a compressed spring at the bottom of an incline of slope 30.0° (point A). When the spring is released, it projects the block up the incline. At point B, a distance of 6.00 m up the incline from A, the block is moving up the incline at 7.00 m/s and is no longer in contact with the spring. The coefficient of kinetic friction between the block and incline is $\mu_k = 0.50$. The mass of the spring is negligible. Calculate the amount of potential energy that was initially stored in the spring.

7.74 A 2.00-kg package is released on a 53.1° incline, 4.00 m from a long spring with force constant 120 N/m that is attached at the bottom of the incline (Fig. 7.41). The coefficients of friction between the package and the incline are $\mu_s = 0.40$ and $\mu_k = 0.20$. The mass of the spring is negligible. a) What is the speed of the package just before it reaches the spring? b) What is the maximum compression of the spring? c) The package rebounds back up the incline. How close does it get to its initial position?

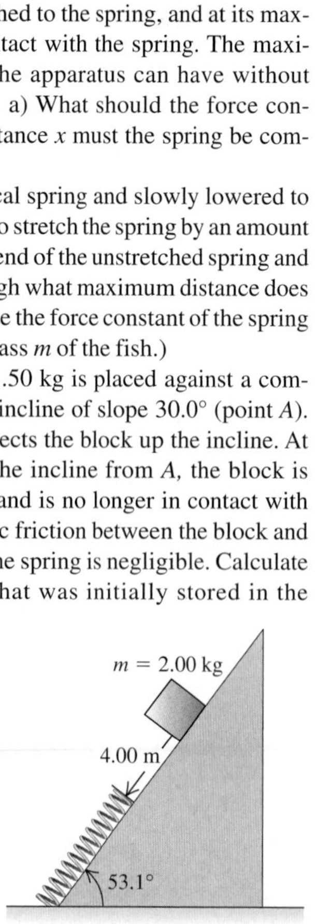

Figure 7.41 Problem 7.74.

7.75 A 0.500-kg block, attached to a spring with length 0.60 m and force constant 40.0 N/m is at rest with the back of the block at point A on a frictionless, horizontal air table (Fig. 7.42). The mass of the spring is negligible. You move the block to the right along the surface by pulling with a constant 20.0-N horizontal force. a) What is the block's speed when the back of the block reaches point B, which is 0.25 m to the right of point A? b) When the back of the block reaches point B, you let go of the block. In the subsequent motion, how close does the block get to the wall where the left end of the spring is attached?

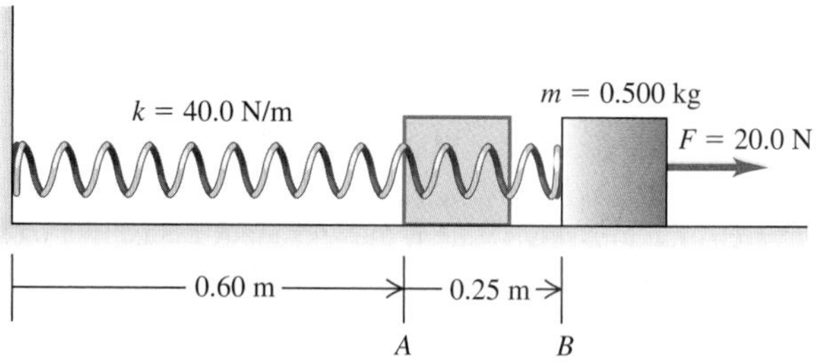

Figure 7.42 Problem 7.75.

7.76 A box of mass m is pushed against a spring of negligible mass and force constant k, compressing it a distance x. The box is then released and travels up a ramp that is at an angle α above the horizontal. The coefficient of kinetic friction between the box and the ramp is μ_k, where $\mu_k < 1$. The box is still moving up the ramp after traveling a distance $s > |x|$ along the ramp. Calculate the angle α for which the speed of the box after traveling distance s is a minimum. Explain why the minimum speed doesn't occur when $\alpha = 90°$, even though for this α there is the maximum increase in gravitational potential energy.

7.77 Fraternity Physics. The brothers of Iota Eta Pi fraternity build a platform, supported at all four corners by vertical springs, in the basement of their frat house. A brave fraternity brother wearing a football helmet stands in the middle of the platform; his weight compresses the springs by 0.18 m. Then four of his fraternity brothers, pushing down at the corners of the platform, compress the springs another 0.53 m until the top of the brave brother's helmet is 0.90 m below the basement ceiling. They then simultaneously release the platform. You can ignore the masses of the springs and platform. a) When the dust clears, the fraternity asks you to calculate their fraternity brother's speed just before his helmet hit the flimsy ceiling. b) Without the ceiling, how high would he have gone? c) In discussing their probation, the dean of students suggests that the next time they try this, they do it outdoors on another planet. Would the answer to part (b) be the same if this stunt were performed on a planet with a different value of g? Assume that the fraternity brothers push the platform down 0.53 m as before. Explain your reasoning.

7.78 A particle with mass m is acted on by a conservative force and moves along a path given by $x = x_0 \cos \omega_0 t$ and $y = y_0 \sin \omega_0 t$, where x_0, y_0 and ω_0 are constants. a) Find the components of the force that acts on the particle. b) Find the potential energy of the particle as a function of x and y. Take $U = 0$ when $x = 0$ and $y = 0$.

c) Find the total energy of the particle when: (i) $x = x_0$, $y = 0$ (ii) $x = 0$, $y = y_0$.

7.79 When it is burned, one gallon of gasoline produces 1.3×10^8 J of energy. A 1500-kg car accelerates from rest to 37 m/s in 10 s. The engine of this car is only 15% efficient (which is typical), meaning that only 15% of the energy from the combustion of the gasoline is used to accelerate the car. The rest goes into things like the internal kinetic energy of the engine parts, as well as heating of the exhaust air and engine. a) How many gallons of gasoline does this car use during the acceleration? b) How many such accelerations will it take to burn up one gallon of gas?

7.80 A hydroelectric dam holds back a lake of surface area 3.0×10^6 m^2 that has vertical sides below the water level. The water level in the lake is 150 m above the base of the dam. When the water passes through turbines at the base of the dam, its mechanical energy is converted into electrical energy with 90% efficiency. a) If gravitational potential energy is taken to be zero at the base of the dam, how much energy is stored in the top meter of the water in the lake? The density of water is 1000 kg/m^3. b) What volume of water must pass through the dam to produce 1000 kilowatt-hours of electrical energy? What distance does the level of water in the lake fall when this much water passes through the dam?

7.81 How much total energy is stored in the lake in Problem 7.80? As in Problem 7.80, take the gravitational potential energy to be zero at the base of the dam. Express your answer in joules and in kilowatt-hours. (*Hint:* Break the lake up into infinitesimal, horizontal layers of thickness dy and integrate to find the total potential energy.)

7.82 a) Is the force $\vec{F} = Cy^2\hat{j}$, where C is a negative constant with units of N/m^2, conservative or nonconservative? Justify your answer. b) Is the force $\vec{F} = Cy^2\hat{i}$, where C is a negative constant with units of N/m^2, conservative or nonconservative? Justify your answer.

7.83 A cutting tool under microprocessor control has several forces acting on it. One force is $\vec{F} = -\alpha xy^2\hat{j}$, a force in the negative y-direction whose magnitude depends on the position of the tool. The constant is $\alpha = 2.50$ N/m^3. Consider the displacement of the tool from the origin to the point $x = 3.00$ m, $y = 3.00$ m. a) Calculate the work done on the tool by $\vec{F}$ if this displacement is along the straight line $y = x$ that connects these two points. b) Calculate the work done on the tool by $\vec{F}$ if the tool is first moved out along the x-axis to the point $x = 3.00$ m, $y = 0$ and then moved parallel to the y-axis to $x = 3.00$ m, $y = 3.00$ m. c) Compare the work done by $\vec{F}$ along these two paths. Is $\vec{F}$ conservative or nonconservative? Explain.

7.84 An object has several forces acting on it. One force is $\vec{F} = \alpha xy\hat{i}$, a force in the x-direction whose magnitude depends on the position of the object. (See Problem 6.96.) The constant is $\alpha = 2.00$ N/m^2. The object moves along the following path: (1) it starts at the origin and moves along the y-axis to the point $x = 0$, $y = 1.50$ m; (2) it moves parallel to the x-axis to the point $x = 1.50$ m, $y = 1.50$ m; (3) it moves parallel to the y-axis to the point $x = 1.50$ m, $y = 0$; (4) it moves parallel to the x-axis back to the origin. a) Sketch this path in the xy-plane. b) Calculate the work done on the object by $\vec{F}$ for each leg of the path and for the complete round trip. c) Is $\vec{F}$ conservative or nonconservative? Explain.

7.85 A Hooke's law force $-kx$ and a constant conservative force F in the $+x$-direction act on an atomic ion. a) Show that a possible potential-energy function for this combination of forces is $U(x) = \frac{1}{2}kx^2 - Fx - F^2/2k$. Is this the *only* possible function? Explain. b) Find the stable equilibrium position. c) Graph $U(x)$ (in units of F^2/k) versus x (in units of F/k) for values of x between $-5F/k$ to $5F/k$. d) Are there any unstable equilibrium positions? e) If the total energy is $E = F^2/k$, what are the maximum and minimum values of x that the ion reaches in its motion? f) If the ion has mass m, find its maximum speed if the total energy is $E = F^2/k$. For what value of x is the speed maximum?

7.86 A particle moves along the x-axis while acted on by a single conservative force parallel to the x-axis. The force corresponds to the potential-energy function graphed in Fig. 7.43. The particle is released from rest at point A. a) What is the direction of the force on the particle when it is at point A? b) At point B? c) At what value of x is the kinetic energy of the particle a maximum? d) What is the force on the particle when it is at point C? e) What is the largest value of x reached by the particle during its motion? f) What value or values of x correspond to points of stable equilibrium? g) Of unstable equilibrium?

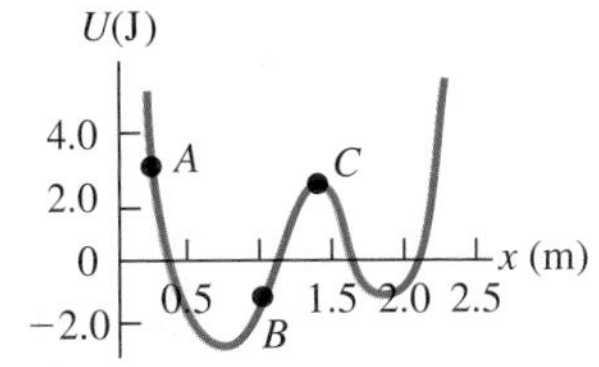

Figure 7.43 Problem 7.86.

Challenge Problem

7.87 A proton with mass m moves in one dimension. The potential-energy function is $U(x) = \alpha/x^2 - \beta/x$, where α and β are positive constants. The proton is released from rest at $x_0 = \alpha/\beta$. a) Show that $U(x)$ can be written as

$$U(x) = \frac{\alpha}{x_0^2}\left[\left(\frac{x_0}{x}\right)^2 - \frac{x_0}{x}\right]$$

Graph $U(x)$. Calculate $U(x_0)$ and thereby locate the point x_0 on the graph. b) Calculate $v(x)$ the speed of the proton as a function of position. Graph $v(x)$ and give a qualitative description of the motion. c) For what value of x is the speed of the proton a maximum? What is the value of that maximum speed? d) What is the force on the proton at the point in part (c)? e) Let the proton be released instead at $x_1 = 3\alpha/\beta$. Locate the point x_1 on the graph of $U(x)$. Calculate $v(x)$ and give a qualitative description of the motion. f) For each release point ($x = x_0$ and $x = x_1$), what are the maximum and minimum values of x reached during the motion?

8 MOMENTUM, IMPULSE, AND COLLISIONS

When one football player tackles another, the dynamics of the collision between their bodies can be quite complex. Yet the collision is governed by a simple physical principle: if you calculate the product of mass and vector velocity for each player, and then add together the results for each player, the sum has the same value just before the collision as it does just after the collision. This is an example of the law of conservation of momentum.

? Which could potentially cause you the greater injury: being tackled by a lightweight, fast-moving football player, or being tackled by a player with double the mass but moving at half the speed?

There are many questions involving forces that cannot be answered by directly applying Newton's second law, $\sum\vec{F} = m\vec{a}$. For example, when an eighteen-wheeler collides head-on with a compact car, what determines which way the wreckage goes after the collision? In playing pool, how do you decide how to aim the cue ball in order to knock the eight ball into the pocket? And when a meteorite collides with the earth, how much of the meteorite's kinetic energy is released in the impact?

A common theme of all of these questions is that they involve forces about which we know very little: the forces between the car and the eighteen-wheeler, between the two pool balls, or between the meteorite and the earth. Remarkably, we will find in this chapter that we don't have to know *anything* about these forces to answer questions of this kind!

Our approach uses two new concepts, *momentum* and *impulse,* and a new conservation law, *conservation of momentum.* This conservation law is every bit as important as that of conservation of energy. The law of conservation of momentum is valid even in situations in which Newton's laws are inadequate, such as bodies moving at very high speeds (near the speed of light) or objects on a very small scale (such as the constituents of atoms). Within the domain of Newtonian mechanics, conservation of momentum enables us to analyze many situations that would be very difficult if we tried to use Newton's laws directly. Among these are *collision* problems, in which two bodies collide and can exert very large forces on each other for a short time.

8.1 | Momentum and Impulse

In Chapter 6 we re-expressed Newton's second law for a particle, $\sum\vec{F} = m\vec{a}$, in terms of the work-energy theorem. This theorem helped us tackle a great number of physics problems and led us to the law of conservation of energy. Let's now return to $\sum\vec{F} = m\vec{a}$ and see yet another useful way in which this fundamental law can be restated.

Consider a particle of constant mass m. (Later in this chapter we'll see how to deal with situations in which the mass of a body changes.) Because $\vec{a} = d\vec{v}/dt$, we can write Newton's second law for this particle as

$$\sum\vec{F} = m\frac{d\vec{v}}{dt} = \frac{d}{dt}(m\vec{v}) \qquad (8.1)$$

We can take the mass m inside the derivative because it is constant. Thus, Newton's second law says that the net force $\sum\vec{F}$ acting on a particle equals the time rate of change of the combination $m\vec{v}$, the product of the particle's mass and velocity. We'll call this combination the **momentum,** or **linear momentum,** of the particle. Using the symbol $\vec{p}$ for momentum, we have

$$\vec{p} = m\vec{v} \quad \text{(definition of momentum)} \qquad (8.2)$$

Momentum is a vector quantity that has a magnitude (mv) and a direction (the same as the velocity vector $\vec{v}$). The momentum of a car driving north at 20 m/s is different from the momentum of the same car driving east at the same speed. A fastball thrown by a major-league pitcher has greater magnitude of momentum than the same ball thrown by a child because the speed is greater. An eighteen-wheeler going 65 mi/h has greater magnitude of momentum than a Saturn automobile with the same speed because the truck's mass is greater.

The units of the magnitude of momentum are units of mass times speed; the SI units of momentum are kg·m/s. The plural of momentum is "momenta."

Substituting Eq. (8.2) into Eq. (8.1), we have

$$\sum\vec{F} = \frac{d\vec{p}}{dt} \quad \text{(Newton's second law in terms of momentum)} \qquad (8.3)$$

The net force (vector sum of all forces) acting on a particle equals the time rate of change of momentum of the particle. This, not $\sum\vec{F} = m\vec{a}$, is the form in which Newton originally stated his second law (though he called momentum the "quantity of motion"). This law is valid only in inertial frames of reference.

According to Eq. (8.3), a rapid change in momentum requires a large net force, while a gradual change in momentum requires less net force. This principle is used in the design of automobile safety devices such as air bags. The driver of a fast-moving automobile has a large momentum (the product of the driver's mass and velocity). If the car stops suddenly in a collision, the driver's momentum becomes zero. An air bag causes the driver to lose momentum more gradually than would an abrupt collision with the steering wheel, reducing the force exerted on the driver (and the possibility for injury). The same principle applies to the padding used to package fragile objects for shipping.

We often express the momentum of a particle in terms of its components. If the particle has velocity components v_x, v_y, and v_z, its momentum components p_x, p_y, and p_z are given by

$$p_x = mv_x \qquad p_y = mv_y \qquad p_z = mv_z \tag{8.4}$$

These three component equations are equivalent to Eq. (8.2).

A particle's momentum $\vec{p} = m\vec{v}$ and its kinetic energy $K = \frac{1}{2}mv^2$ both depend on the mass and velocity of the particle. What is the fundamental difference between these two quantities? A purely mathematical answer is that momentum is a vector whose magnitude is proportional to speed, while kinetic energy is a scalar proportional to the speed squared. But to see the *physical* difference between momentum and kinetic energy, we must first define a quantity closely related to momentum called *impulse*.

Let's first consider a particle acted on by a *constant* net force $\sum\vec{F}$ during a time interval Δt from t_1 to t_2. (We'll look at the case of varying forces shortly.) The **impulse** of the net force, denoted by $\vec{J}$, is defined to be the product of the net force and the time interval:

$$\vec{J} = \sum\vec{F}(t_2 - t_1) = \sum\vec{F}\Delta t \qquad \text{(assuming constant net force)} \tag{8.5}$$

Impulse is a vector quantity; its direction is the same as the net force $\sum\vec{F}$. Its magnitude is the product of the magnitude of the net force and the length of time that the net force acts. The SI unit of impulse is the newton-second $(\text{N}\cdot\text{s})$. Because $1\text{ N} = 1\text{ kg}\cdot\text{m/s}^2$, an alternative set of units for impulse is kg·m/s, the same as the units of momentum.

To see what impulse is good for, let's go back to Newton's second law as restated in terms of momentum, Eq. (8.3). If the net force $\sum\vec{F}$ is constant, then $d\vec{p}/dt$ is also constant. In that case, $d\vec{p}/dt$ is equal to the *total* change in momentum $\vec{p}_2 - \vec{p}_1$ during the time interval $t_2 - t_1$, divided by the interval:

$$\sum\vec{F} = \frac{\vec{p}_2 - \vec{p}_1}{t_2 - t_1}$$

Multiplying this equation by $(t_2 - t_1)$, we have

$$\sum\vec{F}(t_2 - t_1) = \vec{p}_2 - \vec{p}_1$$

Comparing to Eq. (8.5), we end up with a result called the **impulse-momentum theorem:**

$$\vec{J} = \vec{p}_2 - \vec{p}_1 \qquad \text{(impulse-momentum theorem)} \tag{8.6}$$

The change in momentum of a particle during a time interval equals the impulse of the net force that acts on the particle during that interval.

The impulse-momentum theorem also holds when forces are not constant. To see this, we integrate both sides of Newton's second law $\sum\vec{F} = d\vec{p}/dt$ over time between the limits t_1 and t_2:

$$\int_{t_1}^{t_2}\sum\vec{F}\,dt = \int_{t_1}^{t_2}\frac{d\vec{p}}{dt}dt = \int_{\vec{p}_1}^{\vec{p}_2}d\vec{p} = \vec{p}_2 - \vec{p}_1$$

The integral on the left is defined to be the impulse $\vec{J}$ of the net force $\sum\vec{F}$ during this interval:

$$\vec{J} = \int_{t_1}^{t_2} \sum\vec{F}\, dt \qquad \text{(general definition of impulse)} \qquad (8.7)$$

With this definition the impulse-momentum theorem $\vec{J} = \vec{p}_2 - \vec{p}_1$, Eq. (8.6), is valid even when the net force $\sum\vec{F}$ varies with time.

We can define an *average* net force $\vec{F}_{av}$ such that even when $\sum\vec{F}$ is not constant, the impulse $\vec{J}$ is given by

$$\vec{J} = \vec{F}_{av}(t_2 - t_1) \qquad (8.8)$$

When $\sum\vec{F}$ is constant, $\sum\vec{F} = \vec{F}_{av}$ and Eq. (8.8) reduces to Eq. (8.5).

Figure 8.1 shows a graph of the x-component of net force $\sum F_x$ as a function of time during a collision. This might represent the force on a soccer ball that is in contact with a player's foot from time t_1 to t_2. The x-component of impulse during this interval is represented by the blue area under the curve between t_1 and t_2. This area is equal to the rectangular area bounded by t_1, t_2, and $(F_{av})_x$, so $(F_{av})_x(t_2 - t_1)$ is equal to the impulse of the actual time-varying force during the same interval.

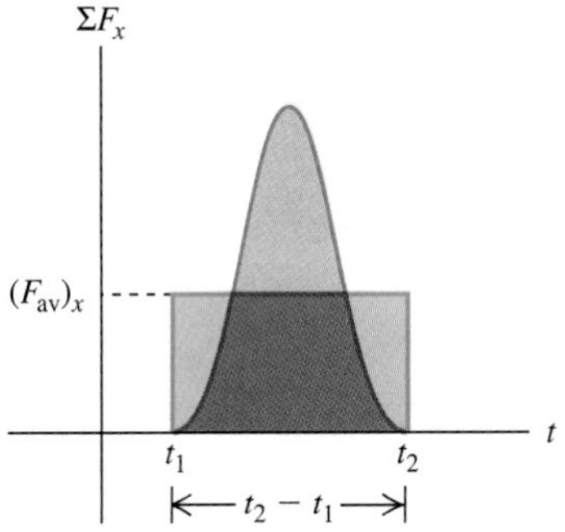

8.1 The x-component of the impulse of $\sum F_x$ between t_1 and t_2 equals the blue area under the $\sum F_x$–t curve, which also equals the area under the rectangle with height $(F_{av})_x$.

Impulse and momentum are both vector quantities, and Eqs. (8.5) through (8.8) are all vector equations. In specific problems it is often easiest to use them in component form:

$$J_x = \int_{t_1}^{t_2} \sum F_x\, dt = (F_{av})_x(t_2 - t_1) = p_{2x} - p_{1x} = mv_{2x} - mv_{1x}$$

$$J_y = \int_{t_1}^{t_2} \sum F_y\, dt = (F_{av})_y(t_2 - t_1) = p_{2y} - p_{1y} = mv_{2y} - mv_{1y} \qquad (8.9)$$

and similarly for the z-component.

Momentum and Kinetic Energy Compared

We can now see the fundamental difference between momentum and kinetic energy. The impulse-momentum theorem $\vec{J} = \vec{p}_2 - \vec{p}_1$ says that changes in a particle's momentum are due to impulse, which depends on the *time* over which the net force acts. By contrast, the work-energy theorem $W_{tot} = K_2 - K_1$ tells us that kinetic energy changes when work is done on a particle; the total work depends on the *distance* over which the net force acts. Consider a particle that starts from rest at t_1 so that $\vec{v} = \mathbf{0}$. Its initial momentum is $\vec{p}_1 = m\vec{v}_1 = \mathbf{0}$, and its initial kinetic energy is $K_1 = \frac{1}{2}mv_1^2 = 0$. Now let a constant net force equal to $\vec{F}$ act on that particle from time t_1 until time t_2. During this interval the particle moves a distance s in the direction of the force. From Eq. (8.6), the particle's momentum at time t_2 is

$$\vec{p}_2 = \vec{p}_1 + \vec{J} = \vec{J}$$

where $\vec{J} = \vec{F}(t_2 - t_1)$ is the impulse that acts on the particle. So *the momentum of a particle equals the impulse that accelerated it from rest to its present speed;* impulse is the product of the net force that accelerated the particle and the *time* required for the acceleration. By comparison, the kinetic energy of the particle at t_2 is $K_2 = W_{tot} = Fs$, the total *work* done on the particle to accelerate it from rest. The total work is the product of the net force and the *distance* required to accelerate the particle (Fig. 8.2).

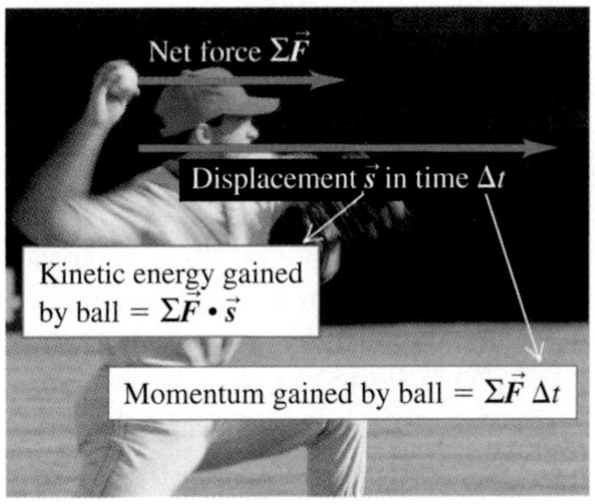

8.2 The kinetic energy of a pitched baseball is equal to the work the pitcher does on it (force multiplied by the distance the ball moves during the throw). The momentum of the ball is equal to the impulse the pitcher imparts to it (force multiplied by the time it took to bring the ball up to speed).

?

Here's an application of the distinction between momentum and kinetic energy. Suppose you have a choice between catching a 0.50-kg ball moving at 4.0 m/s or a 0.10-kg ball moving at 20 m/s. Which will be easier to catch? Both have the same magnitude of momentum, $p = mv = (0.50\ \text{kg})(4.0\ \text{m/s}) = (0.10\ \text{kg})(20\ \text{m/s}) = 2.0\ \text{kg}\cdot\text{m/s}$. However, the two balls have very different values of kinetic energy $K = \frac{1}{2}mv^2$; the large, slow-moving ball has $K = 4.0\ \text{J}$, while the small, fast-moving ball has $K = 20\ \text{J}$. Since the momentum is the same for both balls, both require the same *impulse* to be brought to rest. But stopping the 0.10-kg ball with your hand requires five times more *work* than stopping the 0.50-kg ball, because the smaller ball has five times as much kinetic energy. Hence for a given force that you exert with your hand, it takes the same amount of time (the duration of the catch) to stop either ball, but your hand and arm will be pushed back five times farther if you choose to catch the small, fast-moving ball. To minimize arm strain, you should choose to catch the 0.50-kg ball with its lower kinetic energy. (Kinetic energy may not be the only factor to consider, however. A pitched baseball and a .22 bullet fired from a rifle have roughly the same kinetic energy, and the bullet has less momentum than does the baseball. Nonetheless, you'd probably prefer to catch the baseball rather than the bullet. Why?)

ActivPhysics ONLINE

6.1 Momentum and Energy Change

Both the impulse-momentum and work-energy theorems are relationships between force and motion, and both rest on the foundation of Newton's laws. They are *integral* principles, relating the motion at two different times separated by a finite interval. By contrast, Newton's second law itself (in either of the forms $\sum\vec{F} = m\vec{a}$ or $\sum\vec{F} = d\vec{p}/dt$) is a *differential* principle, relating the forces to the rate of change of velocity or momentum at each instant.

Conceptual Example 8.1

Momentum versus kinetic energy

Consider again the race described in Conceptual Example 6.6 (Section 6.2) between two iceboats on a frictionless frozen lake. The iceboats have masses m and $2m$, respectively, and the wind exerts the same constant horizontal force $\vec{F}$ on each iceboat (Fig. 6.10). The two iceboats start from rest and cross the finish line a distance s away. Which iceboat crosses the finish line with greater momentum?

SOLUTION

In Conceptual Example 6.6 we asked how the kinetic energies of the iceboats compare when they cross the finish line. The way to determine this was not by using the formula $K = \frac{1}{2}mv^2$, but rather by remembering that a body's kinetic energy is equal to the total work done to accelerate it from rest. Both iceboats started from rest, and the total work done between the starting and finish lines was the same for both iceboats (because the net force and displacement were the same for both). Hence both iceboats cross the finish line with the same kinetic energy.

Similarly, the best way to compare the momenta of the iceboats is *not* to use the formula for momentum, $\vec{p} = m\vec{v}$. By itself this formula isn't enough to decide which iceboat has greater momentum at the finish line. The iceboat of mass $2m$ has greater mass, which suggests greater momentum; but this iceboat crosses the finish line going slower than the other one, which suggests less momentum.

Instead, we use the idea that the momentum of each iceboat equals the impulse that accelerated it from rest. For each iceboat the downward force of gravity and the upward normal force add to zero, so the net force equals the constant horizontal wind force $\vec{F}$. Let Δt be the time an iceboat takes to reach the finish line, so that the impulse on the iceboat during that time is $\vec{J} = \vec{F}\,\Delta t$. Since the iceboat starts from rest, this just equals the iceboat's momentum $\vec{p}$ at the finish line:

$$\vec{p} = \vec{F}\,\Delta t$$

Both iceboats are subjected to the same force $\vec{F}$, but they do not take the same amount of time Δt to reach the finish line. The more massive iceboat accelerates more slowly and takes a longer time to travel the distance s; thus there is a greater impulse on this iceboat between the starting and finish lines. So the iceboat of mass $2m$ crosses the finish line with a greater magnitude of momentum than the iceboat of mass m (but with the same kinetic energy). Can you show that the iceboat of mass $2m$ has $\sqrt{2}$ times as much momentum at the finish line as the iceboat of mass m?

Example 8.2 A ball hits a wall

Suppose you throw a ball with a mass of 0.40 kg against a brick wall. It hits the wall moving horizontally to the left at 30 m/s and rebounds horizontally to the right at 20 m/s. a) Find the impulse of the net force on the ball during its collision with the wall. b) If the ball is in contact with the wall for 0.010 s, find the average horizontal force that the wall exerts on the ball during the impact.

SOLUTION

IDENTIFY: We're given enough information to determine the initial and final values of the ball's momentum, so we can use the impulse-momentum theorem to find the impulse. We'll then use the definition of impulse to determine the average force.

SET UP: The motion is purely horizontal, so we just need a single axis; we'll take the x-axis to be horizontal and the positive direction to be to the right (Fig. 8.3). Our target variable in part (a) is the x-component of impulse, J_x, which we'll find from the x-components of momentum before and after the impact, using Eqs. (8.9). In part (b), our target variable is the average x-component of force $(F_{av})_x$; once we know J_x, we can also find this force by using Eqs. (8.9).

EXECUTE: a) With our choice of x-axis, the initial and final x-components of momentum of the ball are

$$p_{1x} = mv_{1x} = (0.40\text{ kg})(-30\text{ m/s}) = -12\text{ kg}\cdot\text{m/s}$$
$$p_{2x} = mv_{2x} = (0.40\text{ kg})(+20\text{ m/s}) = +8.0\text{ kg}\cdot\text{m/s}$$

From the x-equation in Eqs. (8.9), the x-component of impulse equals the *change* in the x-component of momentum:

$$\begin{aligned} J_x &= p_{2x} - p_{1x} \\ &= 8.0\text{ kg}\cdot\text{m/s} - (-12\text{ kg}\cdot\text{m/s}) = 20\text{ kg}\cdot\text{m/s} = 20\text{ N}\cdot\text{s} \end{aligned}$$

b) The collision time is $t_2 - t_1 = \Delta t = 0.010$ s. From the x-equation in Eqs. (8.9), $J_x = (F_{av})_x(t_2 - t_1) = (F_{av})_x\Delta t$, so

$$(F_{av})_x = \frac{J_x}{\Delta t} = \frac{20\text{ N}\cdot\text{s}}{0.010\text{ s}} = 2000\text{ N}$$

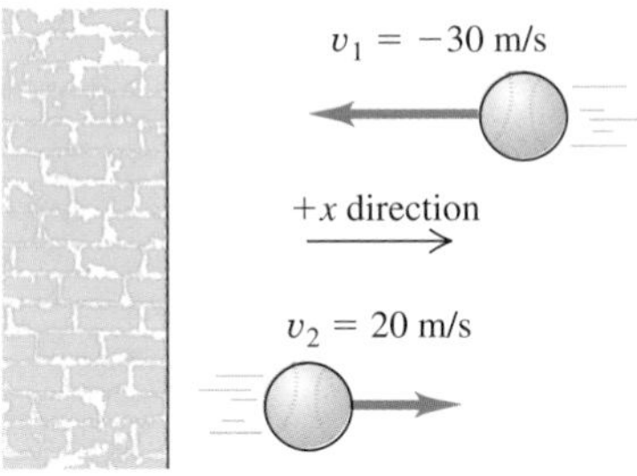

8.3 A ball moving to the left, hitting a wall, and rebounding to the right.

8.4 Typically, a tennis ball is in contact with the racket for approximately 0.01 s. The ball flattens noticeably due to the tremendous force exerted by the racket.

EVALUATE: The x-component of impulse is positive—that is, to the right in Fig. 8.3. This is as it should be: the impulse represents the "kick" that the wall imparts to the ball, and this "kick" is certainly to the right.

CAUTION Because momentum is a vector, we had to include the negative sign in p_{1x}. Had we been careless and omitted it, we would have calculated the impulse to be $8.0\text{ kg}\cdot\text{m/s} - (12\text{ kg}\cdot\text{m/s}) = -4\text{ kg}\cdot\text{m/s}$. This answer is clearly nonsensical, because it says that the wall would somehow have given the ball a kick to the *left!* Make sure that you account for the direction of momentum while carrying out your calculations.

The average force that the wall exerts on the ball is quite substantial, 2000 N (equal to 450 lb, or the weight of a 200-kg object). The magnitude of this force has to be large to change the ball's momentum in such a short time interval. Other forces that act on the ball during the collision are very weak by comparison; for instance, the gravitational force is only 3.9 N. Thus, during the brief time that the collision lasts, we can ignore all other forces on the ball to a very good approximation. Figure 8.4 is a photograph showing the impact of a tennis ball and racket.

Note that the 2000-N value we calculated is just the *average* horizontal force that the wall exerts on the ball during the impact. It corresponds to the horizontal line $(F_{av})_x$ in Fig. 8.1. The horizontal force is zero before impact, rises to a maximum, and then decreases to zero when the ball loses contact with the wall. If the ball is relatively rigid, like a baseball or golf ball, the collision lasts a short time and the maximum force is large, as in curve (a) in Fig. 8.5. If

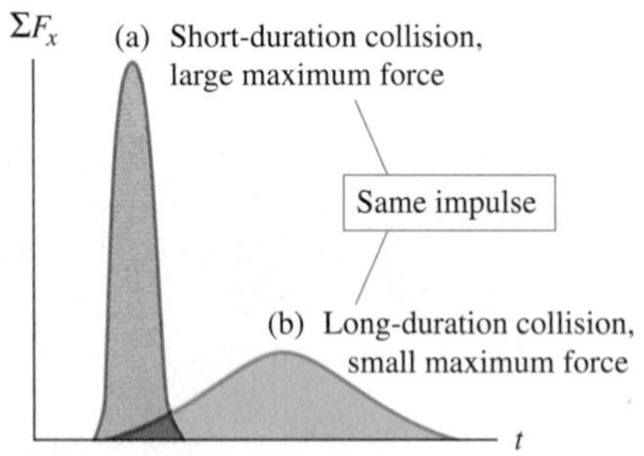

8.5 (a) A "hard" collision, such as that of a golf ball. (b) A "soft" collision, such as that of a tennis ball.

the ball is softer, like a tennis ball, the collision time is longer and the maximum force is less, as in curve (b). In either case, the impulse is the same: it equals the change in the ball's momentum. From Eq. (8.7), we know that impulse is the integral of force with respect to time, which means that the *area* under each curve in Fig. 8.5 is the same.

Example 8.3 Kicking a soccer ball

A soccer ball has a mass of 0.40 kg. Initially, it is moving to the left at 20 m/s, but then it is kicked and given a velocity at 45° upward and to the right, with a magnitude of 30 m/s (Fig. 8.6a). Find the impulse of the net force and the average net force, assuming a collision time $\Delta t = 0.010$ s.

SOLUTION

IDENTIFY: This example uses the same principles as Example 8.2. The key difference is that the initial and final velocities are not along the same line, so we have to be careful to treat momentum and impulse as vector quantities, using their x- and y-components.

SET UP: We take the x-axis to be horizontally to the right and the y-axis to be vertically upward. Our target variables are the components of the net impulse on the ball, J_x and J_y, and the components of the average net force on the ball, $(F_{av})_x$ and $(F_{av})_y$. We'll find them using the x- and y-components of Eqs. (8.9).

EXECUTE: With our choice of axes, we find the following velocity components:

$$v_{1x} = -20 \text{ m/s} \qquad v_{1y} = 0$$
$$v_{2x} = v_{2y} = (0.707)(30 \text{ m/s}) = 21.2 \text{ m/s}$$
$$(\text{since } \cos 45° = \sin 45° = 0.707)$$

The x-component of impulse is equal to the x-component of momentum change, and the same is true for the y-components:

$$J_x = p_{2x} - p_{1x} = m(v_{2x} - v_{1x})$$
$$= (0.40 \text{ kg})[21.2 \text{ m/s} - (-20 \text{ m/s})] = 16.5 \text{ kg}\cdot\text{m/s}$$
$$J_y = p_{2y} - p_{1y} = m(v_{2y} - v_{1y})$$
$$= (0.40 \text{ kg})(21.2 \text{ m/s} - 0) = 8.5 \text{ kg}\cdot\text{m/s}$$

The components of the average net force on the ball are

$$(F_{av})_x = \frac{J_x}{\Delta t} = 1650 \text{ N} \qquad (F_{av})_y = \frac{J_y}{\Delta t} = 850 \text{ N}$$

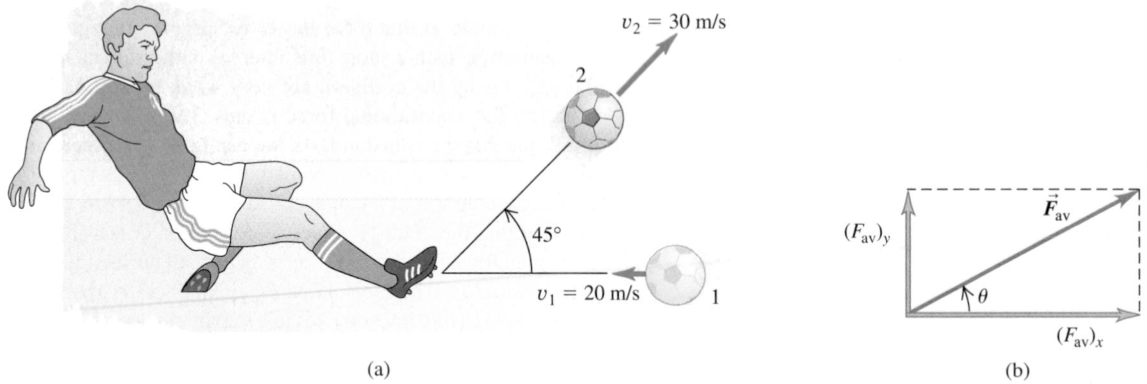

8.6 (a) A soccer ball (1) before and (2) after being kicked. (b) Finding the average force from its components.

The magnitude and direction of the average force are

$$F_{av} = \sqrt{(1650\text{ N})^2 + (850\text{ N})^2} = 1.9 \times 10^3\text{ N}$$

$$\theta = \arctan\frac{850\text{ N}}{1650\text{ N}} = 27°$$

where θ is measured upward from the $+x$-axis (Fig. 8.6b). Note that because the ball was not initially at rest, the ball's final velocity does *not* have the same direction as the average force acting on it.

EVALUATE: The average net force $\vec{F}_{av}$ includes the effects of the force of gravity, though these are small; the weight of the ball is only 3.9 N. As in Example 8.2, the average force acting during the collision is exerted almost entirely by the object that the ball hit (in this case, the soccer player's foot).

Test Your Understanding

Consider the falling hammerhead in Example 6.4 (Section 6.2). (a) Find the hammerhead's y-component of momentum just before it hit the I-beam and the impulse of the net force that acted on the hammerhead to bring it to a stop. (b) Use impulse ideas to find the time it took for this net force to bring the hammerhead to a stop.

8.2 Conservation of Momentum

The concept of momentum is particularly important in situations in which we have two or more interacting bodies. Let's consider first an idealized system consisting of two bodies that interact with each other but not with anything else—for example, two astronauts who touch each other as they float freely in the zero-gravity environment of outer space (Fig. 8.7a). Think of the astronauts as particles. Each particle exerts a force on the other; according to Newton's third law, the

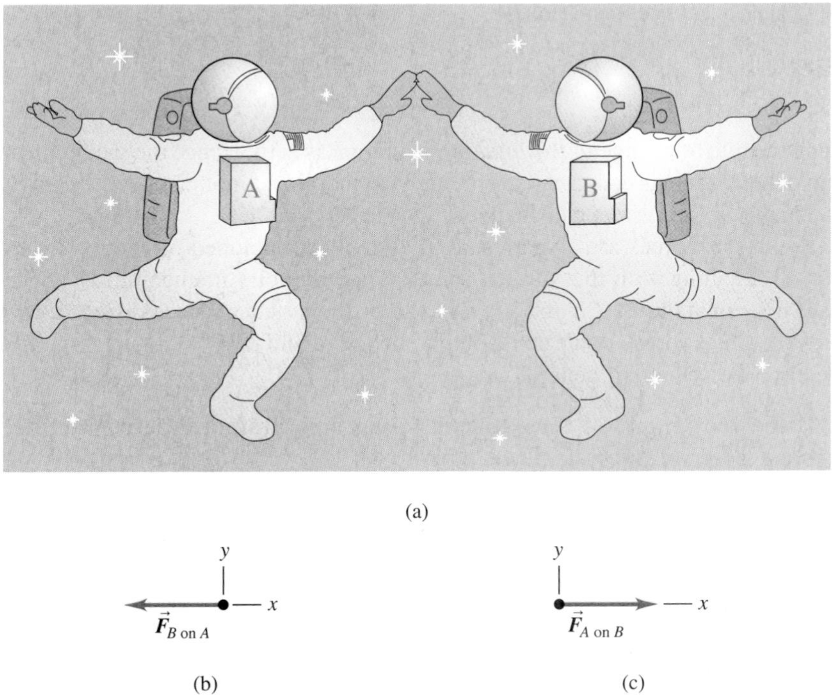

8.7 (a) Two astronauts push each other as they float freely in the zero-gravity environment of space. (b) Free-body diagram for astronaut A. (c) Free-body diagram for astronaut B. The only forces that act are internal forces; they don't change the total momentum of the two astronauts.

two forces are always equal in magnitude and opposite in direction. Hence, the *impulses* that act on the two particles are equal and opposite, and the changes in momentum of the two particles are equal and opposite.

Let's go over that again with some new terminology. For any system, the forces that the particles of the system exert on each other are called **internal forces.** Forces exerted on any part of the system by some object outside it are called **external forces.** For the system we have described, the internal forces are $\vec{F}_{B \text{ on } A}$, exerted by particle B on particle A, and $\vec{F}_{A \text{ on } B}$ exerted by particle A on particle B (Figs. 8.7b, c). There are *no* external forces; when this is the case, we have an **isolated system.**

The net force on particle A is $\vec{F}_{B \text{ on } A}$ and the net force on particle B is $\vec{F}_{A \text{ on } B}$, so from Eq. (8.3) the rates of change of the momenta of the two particles are

$$\vec{F}_{B \text{ on } A} = \frac{d\vec{p}_A}{dt} \qquad \vec{F}_{A \text{ on } B} = \frac{d\vec{p}_B}{dt} \tag{8.10}$$

ActivPhysics ONLINE

6.3 Momentum Conservation and Collisions

6.7 Explosion Problems

6.10 Pendulum Person-Projectile Bowling

The momentum of each particle changes, but these changes are not independent; according to Newton's third law, the two forces $\vec{F}_{B \text{ on } A}$ and $\vec{F}_{A \text{ on } B}$ are always equal in magnitude and opposite in direction. That is, $\vec{F}_{B \text{ on } A} = -\vec{F}_{A \text{ on } B}$, so $\vec{F}_{B \text{ on } A} + \vec{F}_{A \text{ on } B} = \mathbf{0}$. Adding together the two equations in Eq. (8.10), we have

$$\vec{F}_{B \text{ on } A} + \vec{F}_{A \text{ on } B} = \frac{d\vec{p}_A}{dt} + \frac{d\vec{p}_B}{dt} = \frac{d(\vec{p}_A + \vec{p}_B)}{dt} = \mathbf{0} \tag{8.11}$$

The rates of change of the two momenta are equal and opposite, so that the rate of change of the vector sum $\vec{p}_A + \vec{p}_B$ is zero. We now define the **total momentum $\vec{P}$** of the system of two particles as the vector sum of the momenta of the individual particles. That is,

$$\vec{P} = \vec{p}_A + \vec{p}_B \tag{8.12}$$

Then Eq. (8.11) becomes, finally,

$$\vec{F}_{B \text{ on } A} + \vec{F}_{A \text{ on } B} = \frac{d\vec{P}}{dt} = \mathbf{0} \tag{8.13}$$

The time rate of change of the *total* momentum $\vec{P}$ is zero. Hence the total momentum of the system is constant, even though the individual momenta of the particles that make up the system can change.

If external forces are also present, they must be included on the left side of Eq. (8.13) along with the internal forces. Then the total momentum is, in general, not constant. But if the vector sum of the external forces is zero, as in Fig. 8.8, these forces don't contribute to the sum, and $d\vec{P}/dt$ is again zero. Thus we have the following general result:

If the vector sum of the external forces on a system is zero, the total momentum of the system is constant.

This is the simplest form of the **principle of conservation of momentum.** This principle is a direct consequence of Newton's third law. What makes this principle useful is that it doesn't depend on the detailed nature of the internal forces that act between members of the system. This means that we can apply conservation of momentum even if (as is often the case) we know very little about the internal forces. We have used Newton's second law to derive this principle, so we have to be careful to use it only in inertial frames of reference.

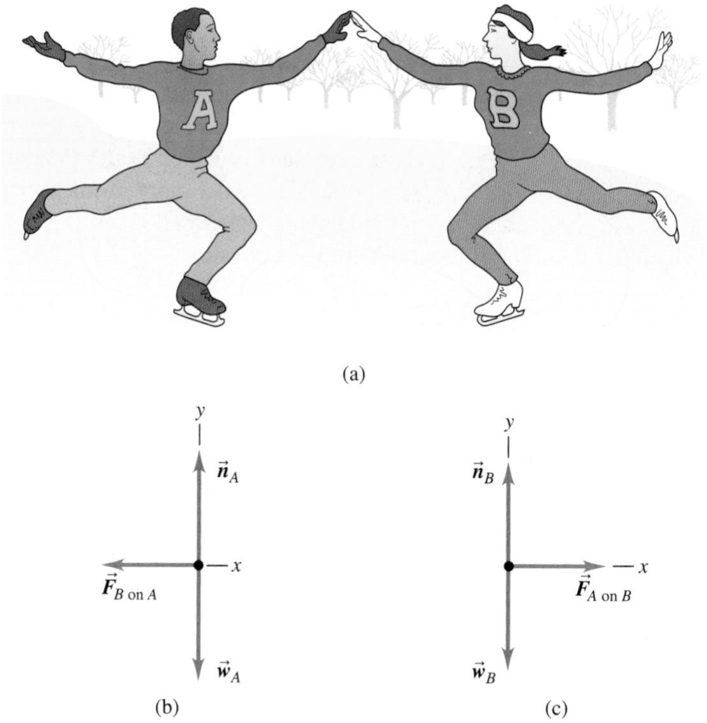

8.8 (a) Two ice skaters push each other as they skate on a frictionless, horizontal surface. (b) Free-body diagram for skater A. (c) Free-body diagram for skater B. The normal forces and gravitational forces are external forces, but the vector sum of the external forces is zero and total momentum is conserved. (Compare to Fig. 8.7.)

We can generalize this principle for a system containing any number of particles A, B, C, . . . interacting only with each other. The total momentum of such a system is

$$\vec{P} = \vec{p}_A + \vec{p}_B + \cdots = m_A\vec{v}_A + m_B\vec{v}_B + \cdots \quad \text{(total momentum of a system of particles)} \quad (8.14)$$

We make the same argument as before; the total rate of change of momentum of the system due to each action-reaction pair of internal forces is zero. Thus the total rate of change of momentum of the entire system is zero whenever the vector sum of the external forces acting on it is zero. The internal forces can change the momenta of individual particles in the system but not the *total* momentum of the system.

CAUTION When you apply the conservation of momentum to a system, it's absolutely essential to remember that momentum is a *vector* quantity. Hence you must use vector addition to compute the total momentum of a system (Fig. 8.9). Using components is usually the simplest method. If p_{Ax}, p_{Ay}, and p_{Az} are the components of momentum of particle A, and similarly for the other particles, then Eq. (8.14) is equivalent to the component equations

$$P_x = p_{Ax} + p_{Bx} + \cdots$$
$$P_y = p_{Ay} + p_{By} + \cdots$$
$$P_z = p_{Az} + p_{Bz} + \cdots \quad (8.15)$$

If the vector sum of the external forces on the system is zero, then P_x, P_y, and P_z are all constant.

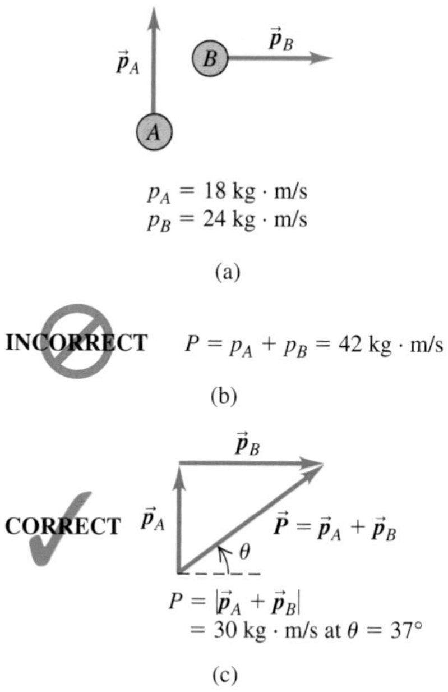

8.9 (a) A system of two particles. Note that their momenta do *not* point in the same direction. (b) The magnitude of the total momentum is *not* the sum of the individual momentum magnitudes. (c) The correct way to find the total momentum is by vector addition.

In some ways the principle of conservation of momentum is more general than the principle of conservation of mechanical energy. For example, mechanical energy is conserved only when the internal forces are *conservative*—that is, when the forces allow two-way conversion between kinetic and potential energy—but conservation of momentum is valid even when the internal forces are *not* conservative. In this chapter we will analyze situations in which both momentum and mechanical energy are conserved, and others in which only momentum is conserved. These two principles play a fundamental role in all areas of physics, and we will encounter them throughout our study of physics.

Problem-Solving Strategy

Conservation of Momentum

IDENTIFY *the relevant concepts:* Before applying conservation of momentum to a problem, you must first decide whether momentum *is* conserved! This will be true *only* if the vector sum of the external forces acting on the system of particles is zero. If this is not the case, you can't use conservation of momentum.

SET UP *the problem using the following steps:*

1. Define a coordinate system. Make a sketch showing the coordinate axes, including the positive direction for each. Often it is easiest to choose the x-axis to have the direction of one of the initial velocities. Make sure you are using an inertial frame of reference. Most of the problems in this chapter deal with two-dimensional situations, in which the vectors have only x- and y-components; all of the following statements can be generalized to include z-components when necessary.
2. Treat each body as a particle. Draw "before" and "after" sketches, and include vectors on each to represent all known velocities. Label the vectors with magnitudes, angles, components, or whatever information is given, and give each unknown magnitude, angle, or component an algebraic symbol. You may find it helpful to use the subscripts 1 and 2 for velocities before and after the interaction, respectively; if you use these subscripts, use letters (not numbers) to label each particle.
3. As always, identify the target variable(s) from among the unknowns.

EXECUTE *the solution as follows:*

1. Write an equation in terms of symbols equating the total *initial* x-component of momentum (that is, before the interaction) to the total *final* x-component of momentum (that is, after the interaction), using $p_x = mv_x$ for each particle. Write another equation for the y-components, using $p_y = mv_y$ for each particle. Remember that the x- and y-components of velocity or momentum are *never* added together in the same equation! Even when all the velocities lie along a line (such as the x-axis), the components of velocity along this line can be positive or negative; be careful with signs!
2. Solve these equations to determine whatever results are required. In some problems you will have to convert from the x- and y-components of a velocity to its magnitude and direction, or the reverse.
3. In some problems, energy considerations give additional relationships among the various velocities, as we will see later in this chapter.

EVALUATE *your answer:* Does your answer make physical sense? If your target variable is a certain body's momentum, check that the direction of the momentum is reasonable.

Example 8.4

Recoil of a rifle

A marksman holds a rifle of mass $m_R = 3.00$ kg loosely in his hands, so as to let it recoil freely when fired. He fires a bullet of mass $m_B = 5.00$ g horizontally with a velocity relative to the ground of $v_{Bx} = 300$ m/s (Fig. 8.10). What is the recoil velocity v_{Rx} of the rifle? What are the final momentum and kinetic energy of the bullet? Of the rifle?

SOLUTION

IDENTIFY: We consider an idealized model in which the horizontal forces the marksman exerts on the rifle are negligible. Then there is no net horizontal force on the system (the bullet and rifle) during the firing of the rifle, and so the total horizontal momentum of the system is the same before and after the rifle is fired (i.e., is conserved).

SET UP: Take the positive x-axis to be the direction the rifle is aimed. Initially, both the rifle and bullet are at rest, so the initial x-component of total momentum is zero. After the shot is fired, the x-component of the bullet's momentum is $p_{Bx} = m_B v_{Bx}$ and the x-component of the rifle's momentum is $p_{Rx} = m_R v_{Rx}$. Our target variables are v_R, p_{Bx}, p_{Rx}, and $K_B = \frac{1}{2} m_B v_{Bx}^2$ and, $K_R = \frac{1}{2} m_R v_{Rx}^2$ (the final kinetic energies of the bullet and rifle, respectively).

EXECUTE: Conservation of the x-component of total momentum gives

$$P_x = 0 = m_B v_{Bx} + m_R v_{Rx}$$

$$v_{Rx} = -\frac{m_B}{m_R} v_{Bx} = -\left(\frac{0.00500 \text{ kg}}{3.00 \text{ kg}}\right)(300 \text{ m/s}) = -0.500 \text{ m/s}$$

The negative sign means that the recoil is in the direction opposite to that of the bullet. If the butt of a rifle were to hit your shoulder traveling at this speed, you'd feel it. That's what the "kick" of a rifle is all about. It's more comfortable to hold the rifle tightly against your shoulder when you fire it; then m_R is replaced by the sum of your mass and the rifle's mass, and the recoil speed is much less.

The final momentum and kinetic energy of the bullet are

$$p_{Bx} = m_B v_{Bx} = (0.00500 \text{ kg})(300 \text{ m/s}) = 1.50 \text{ kg} \cdot \text{m/s}$$

$$K_B = \frac{1}{2} m_B v_{Bx}^2 = \frac{1}{2}(0.00500 \text{ kg})(300 \text{ m/s})^2 = 225 \text{ J}$$

For the rifle, the momentum and final kinetic energy are

$$p_{Rx} = m_R v_{Rx} = (3.00 \text{ kg})(-0.500 \text{ m/s}) = -1.50 \text{ kg} \cdot \text{m/s}$$

$$K_R = \frac{1}{2} m_R v_{Rx}^2 = \frac{1}{2}(3.00 \text{ kg})(-0.500 \text{ m/s})^2 = 0.375 \text{ J}$$

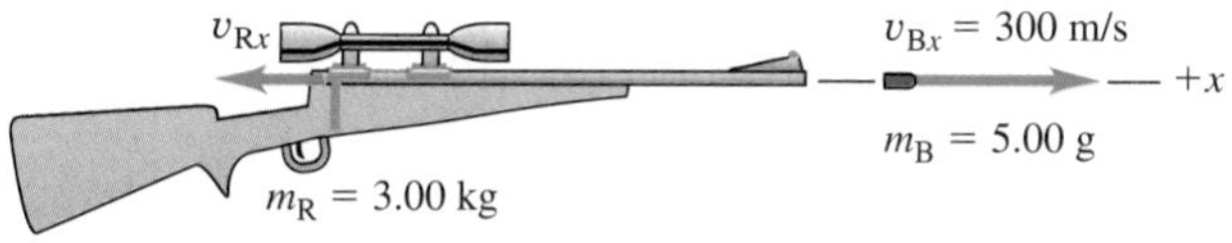

8.10 The ratio of bullet speed to recoil speed is the inverse of the ratio of bullet mass to rifle mass.

EVALUATE: The bullet and the rifle have equal and opposite momenta after the interaction because they were subjected to equal and opposite interaction forces for the same amount of *time* (i.e., equal and opposite impulses). But the bullet acquires much greater kinetic energy than the rifle because the bullet travels a much greater *distance* than the rifle during the interaction. Thus the force on the bullet does more work than the force on the rifle. The ratio of the two kinetic energies, 600 : 1, is equal to the inverse ratio of the masses; in fact, it can be shown that this always happens in recoil situations. We leave the proof as a problem (see Exercise 8.21).

Notice that our calculation doesn't depend at all on the details of how the rifle works. In a real rifle, the bullet is propelled forward by an explosive charge; if instead a very stiff spring were used to give the same bullet velocity, the answers would have been exactly the same.

Example 8.5 Collision along a straight line

Two gliders move toward each other on a frictionless linear air track (Figure 8.11a). After they collide, glider B moves away with a final velocity of +2.0 m/s (Fig. 8.11b). What is the final velocity of glider A? How do the changes in momentum and in velocity compare for the two gliders?

SOLUTION

IDENTIFY: The total vertical force on each glider is zero; the net force on each glider is the horizontal force exerted on it by the other glider. The net *external* force on the two gliders together is zero, so the total momentum is conserved.

SET UP: We take the x-axis to lie along the air track, with the positive direction to the right. We are given the masses and initial velocities of both gliders, as well as the final velocity of glider B (see Fig. 8.11). Our target variables are v_{A2x}, the final x-component of velocity of glider A, and the changes in momentum and in velocity of the two gliders (that is, the value after the collision minus the value before the collision).

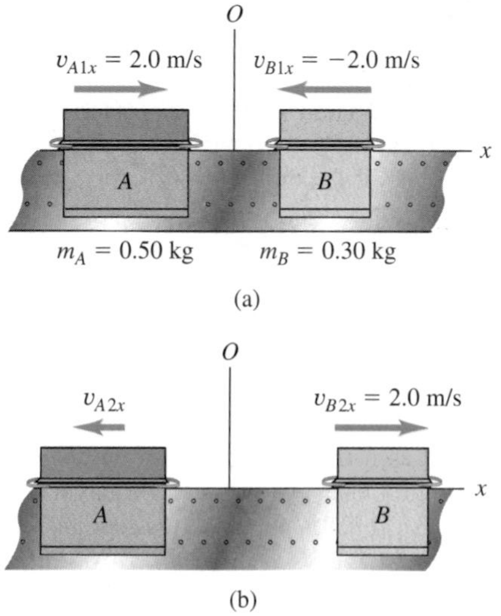

8.11 Two gliders (a) before and (b) after their collision.

EXECUTE: The x-component of total momentum before the collision is

$$\begin{aligned} P_x &= m_A v_{A1x} + m_B v_{B1x} \\ &= (0.50\ \text{kg})(2.0\ \text{m/s}) + (0.30\ \text{kg})(-2.0\ \text{m/s}) \\ &= 0.40\ \text{kg}\cdot\text{m/s} \end{aligned}$$

This is positive (to the right in Fig. 8.11) because glider A has a greater magnitude of momentum before the collision than does glider B. The x-component of total momentum has the same value after the collision, so

$$P_x = m_A v_{A2x} + m_B v_{B2x}$$

Solving this equation for v_{A2x}, the final velocity of A, we find

$$\begin{aligned} v_{A2x} &= \frac{P_x - m_B v_{B2x}}{m_A} = \frac{0.40\ \text{kg}\cdot\text{m/s} - (0.30\ \text{kg})(2.0\ \text{m/s})}{0.50\ \text{kg}} \\ &= -0.40\ \text{m/s} \end{aligned}$$

The change in momentum of glider A is

$$\begin{aligned} m_A v_{A2x} - m_A v_{A1x} &= (0.50\ \text{kg})(-0.40\ \text{m/s}) \\ &\quad - (0.50\ \text{kg})(2.0\ \text{m/s}) = -1.2\ \text{kg}\cdot\text{m/s} \end{aligned}$$

and the change in momentum of glider B is

$$\begin{aligned} m_B v_{B2x} - m_B v_{B1x} &= (0.30\ \text{kg})(2.0\ \text{m/s}) \\ &\quad - (0.30\ \text{kg})(-2.0\ \text{m/s}) = +1.2\ \text{kg}\cdot\text{m/s} \end{aligned}$$

The two interacting gliders undergo changes in momentum that are equal in magnitude and opposite in direction. The same is not true of their changes in velocity, however. For A, $v_{A2x} - v_{A1x} = (-0.40\ \text{m/s}) - 2.0\ \text{m/s} = -2.4\ \text{m/s}$; for B, $v_{B2x} - v_{B1x} = 2.0\ \text{m/s} - (-2.0\ \text{m/s}) = +4.0\ \text{m/s}$.

EVALUATE: Why do the momentum changes have the same magnitude for the two gliders, while the velocity changes do not? By Newton's third law, both gliders were acted on for equal amounts of time by an interaction force of the same magnitude. Hence both gliders experienced impulses of the same magnitude, as well as equal-magnitude changes in momentum. But by Newton's second law, the less massive glider (B) had a greater magnitude of acceleration and hence a greater velocity change.

When a sports utility vehicle (SUV) collides with a car of normal size, both cars undergo equal changes in momentum. The occupants of the normal-size car, however, are subjected to significantly greater acceleration (and significantly greater chance of injury) than the occupants of the SUV. An even more extreme example is what happens when an SUV collides with an insect: the SUV driver won't notice the resulting acceleration at all, but the insect surely will!

Example 8.6 Collision in a horizontal plane

Figure 8.12a shows two battling robots sliding on a frictionless surface. Robot A, with mass 20 kg, initially moves at 2.0 m/s parallel to the x-axis. It collides with robot B, which has mass 12 kg and is initially at rest. After the collision, robot A is moving at 1.0 m/s in a direction that makes an angle $\alpha = 30°$ with its initial direction (Fig. 8.12b). What is the final velocity of robot B?

SOLUTION

IDENTIFY: There are no horizontal (x or y) external forces, so both the x-component and the y-component of the total momentum of the system are conserved in the collision.

SET UP: The coordinate axes are shown in Fig. 8.12. The velocities are not all along a single line, so we have to treat momentum as a vector quantity, using components of each momentum in the x- and y-directions. Then momentum conservation requires that the sum of the x-components *before* the collision (subscript 1) must equal the sum *after* the collision (subscript 2), and similarly for the y-components. Just as with force equilibrium problems, we write a separate equation for each component. Our target variable is $\vec{v}_{B2}$, the final velocity of robot B. We'll express our answer in terms of the x- and y-components as well as in terms of magnitude and direction.

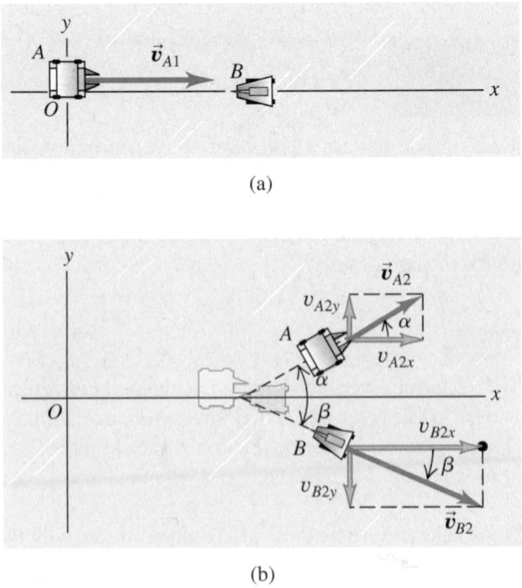

8.12 Views from above of the velocities (a) before and (b) after the collision.

EXECUTE: For the x-components (before and after) we have

$$m_A v_{A1x} + m_B v_{B1x} = m_A v_{A2x} + m_B v_{B2x}$$

$$v_{B2x} = \frac{m_A v_{A1x} + m_B v_{B1x} - m_A v_{A2x}}{m_B}$$

$$= \frac{\begin{bmatrix}(20\text{ kg})(2.0\text{ m/s}) + (12\text{ kg})(0)\\ -(20\text{ kg})(1.0\text{ m/s})(\cos 30°)\end{bmatrix}}{12\text{ kg}}$$

$$= 1.89\text{ m/s}.$$

Conservation of the y-component of total momentum gives

$$m_A v_{A1y} + m_B v_{B1y} = m_A v_{A2y} + m_B v_{B2y}$$

$$v_{B2y} = \frac{m_A v_{A1y} + m_B v_{B1y} - m_A v_{A2y}}{m_B}$$

$$= \frac{\begin{bmatrix}(20\text{ kg})(0) + (12\text{ kg})(0)\\ -(20\text{ kg})(1.0\text{ m/s})(\sin 30°)\end{bmatrix}}{12\text{ kg}}$$

$$= -0.83\text{ m/s}$$

After the collision, robot B moves in the positive x-direction and the negative y-direction (Figure 8.12b). The magnitude of $\vec{v}_{B2}$ is

$$v_{B2} = \sqrt{(1.89\text{ m/s})^2 + (-0.83\text{ m/s})^2} = 2.1\text{ m/s}$$

and the angle of its direction from the positive x-axis is

$$\beta = \arctan \frac{-0.83\text{ m/s}}{1.89\text{ m/s}} = -24°$$

EVALUATE: A useful way to check our answer is to look at the values of the momentum components before and after the collision. Initially all of the momentum is in robot A, which has x-momentum $m_A v_{A1x} = (20\text{ kg})(2.0\text{ m/s}) = 40\text{ kg}\cdot\text{m/s}$ and zero y-momentum. After the collision, robot A has x-momentum $m_A v_{A2x} = (20\text{ kg})(1.0\text{ m/s}) \times (\cos 30°) = 17\text{ kg}\cdot\text{m/s}$, while robot B has x-momentum $m_B v_{B2x} = (12\text{ kg})(1.89\text{ m/s}) = 23\text{ kg}\cdot\text{m/s}$; the total x-momentum is $40\text{ kg}\cdot\text{m/s}$, the same as before the collision (as it should be). In the y-direction, the momentum of robot A after the collision is $m_A v_{A2y} = (20\text{ kg})(1.0\text{ m/s})(\sin 30°) = 10\text{ kg}\cdot\text{m/s}$, while robot B has y-momentum $m_B v_{B2y} = (12\text{ kg})(-0.83\text{ m/s}) = -10\text{ kg}\cdot\text{m/s}$. Note that while robot A acquires a positive y-momentum in the collision, robot B acquires a negative y-momentum of the same magnitude. Thus the *total* y-component of momentum after the collision has the same value (zero) as before the collision.

Test Your Understanding

A spring-loaded toy of mass 0.60 kg sits at rest on a horizontal frictionless surface. When the spring releases, the toy breaks into three equal-mass pieces, A, B, and C, which slide along the surface. Piece A moves off in the negative x-direction at 3.0 m/s, while piece B moves off in the negative y-direction at 4.0 m/s. What is the velocity (magnitude and direction) of piece C?

8.3 | Inelastic Collisions

ActivPhysics ONLINE

6.4 Collision Problems
6.8 Skier and Cart

To the average person on the street the term *collision* is likely to mean some sort of automotive disaster. We'll use it in that sense, but we'll also broaden the meaning to include any strong interaction between bodies that lasts a relatively short time. So we include not only car accidents but also balls colliding on a billiard table, neutrons hitting atomic nuclei in a nuclear reactor, a bowling ball striking the pins, the impact of a meteor on the Arizona desert, and a close encounter of a spacecraft with the planet Saturn.

If the forces between the bodies are much larger than any external forces, as is the case in most collisions, we can neglect the external forces entirely and treat the bodies as an *isolated* system. Then momentum is conserved and the total momentum of the system has the same value before and after the collision. Two cars colliding at an icy intersection provide a good example. Even two cars colliding on dry pavement can be treated as an isolated system during the collision if the forces between the cars are much larger than the friction forces of pavement against tires.

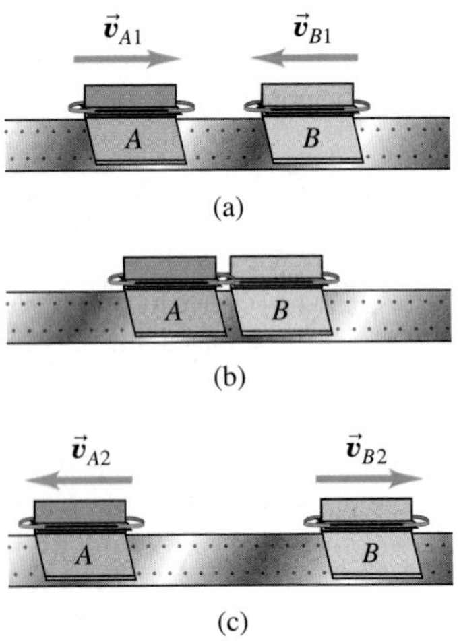

8.13 Model of an elastic collision: Glider *A* and glider *B* approach each other on a frictionless surface. Each glider has a steel spring bumper on the end to ensure an elastic collision.

If the forces between the bodies are also *conservative,* so that no mechanical energy is lost or gained in the collision, the total *kinetic* energy of the system is the same after the collision as before. Such a collision is called an **elastic collision.** A collision between two marbles or two billiard balls is almost completely elastic. Figure 8.13 shows a model for an elastic collision. When the bodies collide, the springs are momentarily compressed and some of the original kinetic energy is momentarily converted to elastic potential energy. Then the bodies bounce apart, the springs expand, and this potential energy is converted back to kinetic energy.

A collision in which the total kinetic energy after the collision is *less* than that before the collision is called an **inelastic collision.** A meatball landing on a plate of spaghetti and a bullet embedding itself in a block of wood are examples of inelastic collisions. An inelastic collision in which the colliding bodies stick together and move as one body after the collision is often called a **completely inelastic collision.** An example is shown in Fig. 8.14; we have replaced the spring bumper in Fig. 8.13 with a ball of putty that squashes and sticks the two bodies together.

CAUTION It's a common misconception that the *only* inelastic collisions are those in which the colliding bodies stick together. In fact, inelastic collisions include many situations in which the bodies do *not* stick. If two cars bounce off each other in a "fender-bender," the work done to deform the fenders cannot be recovered as kinetic energy of the cars, and so the collision is inelastic (Fig. 8.15).

Remember the following rule: **In any collision in which external forces can be neglected, momentum is conserved and the total momentum before equals the total momentum after; in elastic collisions *only*, the total kinetic energy before equals the total kinetic energy after.**

Completely Inelastic Collisions

Let's look at what happens to momentum and kinetic energy in a *completely* inelastic collision of two bodies (*A* and *B*), as in Fig. 8.14. Because the two bodies stick together after the collision, their final velocities must be equal:

$$\vec{v}_{A2} = \vec{v}_{B2} = \vec{v}_2$$

Conservation of momentum gives the relation

$$m_A\vec{v}_{A1} + m_B\vec{v}_{B1} = (m_A + m_B)\vec{v}_2 \qquad \text{(completely inelastic collision)} \quad (8.16)$$

If we know the masses and initial velocities, we can compute the common final velocity $\vec{v}_2$.

Suppose, for example, that a body with mass m_A and initial x-component of velocity v_1 collides inelastically with a body with mass m_B that is initially at rest ($v_{B1} = 0$). From Eq. (8.16) the common x-component of velocity v_2 of both bodies after the collision is

$$v_2 = \frac{m_A}{m_A + m_B}v_1 \qquad \text{(completely inelastic collision, } B \text{ initially at rest)} \quad (8.17)$$

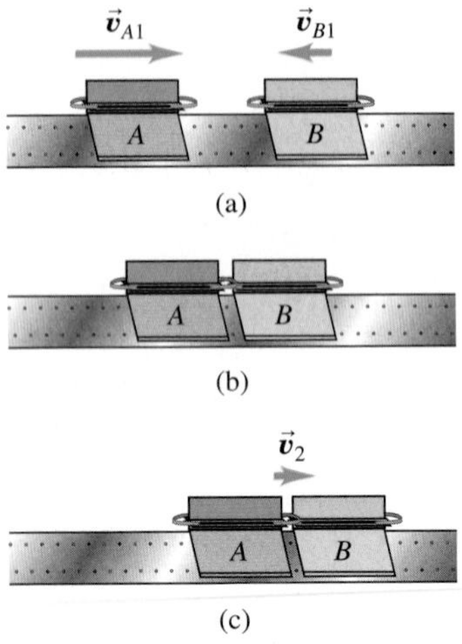

8.14 Model of a completely inelastic collision. The spring bumpers on the gliders are replaced by putty, so the gliders stick together after collision.

Let's verify that the total kinetic energy after this completely inelastic collision is less than before the collision. The kinetic energies K_1 and K_2 before and after the collision, respectively, are

$$K_1 = \frac{1}{2}m_A v_1^2$$

$$K_2 = \frac{1}{2}(m_A + m_B)v_2^2 = \frac{1}{2}(m_A + m_B)\left(\frac{m_A}{m_A + m_B}\right)^2 v_1^2$$

The ratio of final to initial kinetic energy is

$$\frac{K_2}{K_1} = \frac{m_A}{m_A + m_B} \quad \text{(completely inelastic collision, } B \text{ initially at rest)} \quad (8.18)$$

The right side is always less than unity because the denominator is always greater than the numerator. Even when the initial velocity of m_B is not zero, it is not hard to verify that the kinetic energy after a completely inelastic collision is always less than before.

Please note: We don't recommend memorizing Eqs. (8.17) or (8.18). We derived them only to prove that kinetic energy is always lost in a completely inelastic collision.

8.15 Automobile collisions are intended to be inelastic, so that the structure of the car absorbs as much of the energy of the collision as possible. This absorbed energy cannot be recovered, since it goes into a permanent deformation of the car.

Example 8.7 A completely inelastic collision

Suppose that in the collision described in Example 8.5 (Section 8.2), the two gliders do not bounce but stick together after the collision. The masses and initial velocities are as shown in Figure 8.11a. Find the common final velocity v_{2x}, and compare the initial and final kinetic energies.

SOLUTION

IDENTIFY and SET UP: As in Example 8.5, we take the positive x-axis to point to the right in Fig. 8.11a. There are no external forces in the x-direction, so the x-component of momentum is conserved. Our target variables are the final velocity v_{2x} and the initial and final kinetic energies of the system.

EXECUTE: From conservation of the x-component of momentum,

$$m_A v_{A1x} + m_B v_{B1x} = (m_A + m_B)v_{2x}$$

$$v_{2x} = \frac{m_A v_{A1x} + m_B v_{B1x}}{m_A + m_B}$$

$$= \frac{(0.50 \text{ kg})(2.0 \text{ m/s}) + (0.30 \text{ kg})(-2.0 \text{ m/s})}{(0.50 \text{ kg} + 0.30 \text{ kg})}$$

$$= 0.50 \text{ m/s}$$

Because v_{2x} is positive, the gliders move together to the right (the $+x$-direction) after the collision.

Before the collision, the kinetic energies of gliders A and B are

$$K_A = \frac{1}{2}m_A v_{A1x}^2 = \frac{1}{2}(0.50 \text{ kg})(2.0 \text{ m/s})^2 = 1.0 \text{ J}$$

$$K_B = \frac{1}{2}m_B v_{B1x}^2 = \frac{1}{2}(0.30 \text{ kg})(-2.0 \text{ m/s})^2 = 0.60 \text{ J}$$

(Note that the kinetic energy of glider B is positive, even though the x-components of its velocity v_{B1x} and momentum $m_B v_{B1x}$ are both negative.) The *total* kinetic energy before the collision is 1.6 J. The kinetic energy after the collision is

$$\frac{1}{2}(m_A + m_B)v_{2x}^2 = \frac{1}{2}(0.50 \text{ kg} + 0.30 \text{ kg})(0.50 \text{ m/s})^2 = 0.10 \text{ J}$$

EVALUATE: The final kinetic energy is only $\frac{1}{16}$ of the original; $\frac{15}{16}$ is converted from mechanical energy to various other forms. If there is a ball of chewing gum between the gliders, it squashes and becomes warmer. If the gliders couple together like two freight cars, the energy goes into elastic vibrations that eventually dissipate. If there is a spring between the gliders that is compressed as they lock together, then the energy is stored as potential energy of the spring. In all of these cases the *total* energy of the system is conserved, although *kinetic* energy is not. However, in an isolated system, momentum is *always* conserved, whether the collision is elastic or not.

Example 8.8 The ballistic pendulum

Figure 8.16 shows a ballistic pendulum, a system for measuring the speed of a bullet. The bullet, with mass m, is fired into a block of wood with mass M, suspended like a pendulum, and makes a completely inelastic collision with it. After the impact of the bullet, the block swings up to a maximum height y. Given the values of y, m, and M, what is the initial velocity v_x of the bullet?

SOLUTION

IDENTIFY: We'll analyze this event in two stages: (1) the embedding of the bullet in the block and (2) the subsequent swinging of the block on its strings. Let's first find out which physical quantities are conserved in each stage.

During the first stage, the bullet embeds itself in the block so quickly that the block has no time to swing appreciably away from its straight-down position. During this short-duration impact the supporting strings remain very nearly vertical, and negligible external horizontal force acts on the system of bullet plus block. Thus, in this first stage, the *horizontal component of momentum* is conserved. Note that mechanical energy is *not* conserved in this stage because a nonconservative force acts on the system (the force of friction between the bullet and the block).

In the second stage, after the collision, the block and bullet move as a unit. The only forces acting on this unit are gravity (a conservative force) and the string tensions (which do no work). Thus, as the block swings upward and to the right, *mechanical energy* is conserved. Momentum is *not* conserved during this stage because there is a net external force (the sum of gravity and the string tensions, which don't cancel when the strings are inclined).

SET UP: We take the positive x-axis to be to the right and the positive y-axis to be upward in Fig. 8.16. Our target variable is v_y.

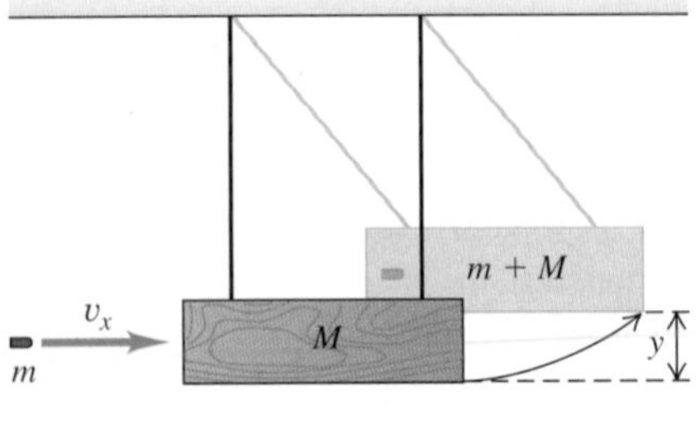

8.16 A ballistic pendulum.

Another unknown quantity is the x-component of velocity of the block and bullet as a unit just after the collision (that is, just at the end of the first stage); we'll call it V_x. We'll use momentum conservation in the first stage to relate v_x to V_x, and we'll use energy conservation in the second stage to relate V_x to the (given) maximum height y.

EXECUTE: In the first stage, momentum conservation gives

$$mv_x = (m + M)V_x \qquad v_x = \frac{m + M}{m}V_x$$

At the beginning of the second stage, the block–bullet unit has kinetic energy $K = \frac{1}{2}(m + M)v_x^2$. (As in Eq. (8.18), this is less than the kinetic energy before the collision; the collision is inelastic!) The block–bullet unit swings up and comes to rest for an instant at a height y, where its kinetic energy is zero and its potential energy is $(m + M)gy$; it then swings back down. Energy conservation gives

$$\frac{1}{2}(m + M)V_x^2 = (m + M)gy \qquad V_x = \sqrt{2gy}$$

Now we substitute this expression into the momentum equation to find an expression for our target variable v_x:

$$v_x = \frac{m + M}{m}\sqrt{2gy}$$

By measuring m, M, and y, we can compute the original speed of the bullet.

EVALUATE: Let's check our answers by plugging in some realistic numbers. If $m = 5.00\text{ g} = 0.00500\text{ kg}$, $M = 2.00\text{ kg}$, and $y = 3.00\text{ cm} = 0.0300\text{ m}$, the initial velocity of the bullet is

$$v_x = \frac{2.00\text{ kg} + 0.00500\text{ kg}}{0.00500\text{ kg}}\sqrt{2(9.80\text{ m/s}^2)(0.0300\text{ m})}$$
$$= 307\text{ m/s}$$

The velocity V_x of the block just after impact is

$$V_x = \sqrt{2gy} = \sqrt{2(9.80\text{ m/s}^2)(0.0300\text{ m})}$$
$$= 0.767\text{ m/s}$$

The kinetic energy of the bullet just before impact is $\frac{1}{2}(0.00500\text{ kg}) \times (307\text{ m/s})^2 = 236\text{ J}$. Just after impact the kinetic energy of the bullet and block is $\frac{1}{2}(2.005\text{ kg})(0.767\text{ m/s})^2 = 0.589\text{ J}$. Nearly all the kinetic energy disappears as the wood splinters and the bullet and block become hotter.

Example 8.9 Analysis of an automobile collision

A small compact car with a mass of 1000 kg is traveling north on Morewood Avenue at 15 m/s. At the intersection of Morewood and Fifth Avenue it collides with an enormous SUV with a mass of 2000 kg, traveling east on Fifth Avenue at 10 m/s (Fig. 8.17a). Fortunately, all occupants are wearing seat belts and there are no injuries, but the two cars are thoroughly tangled and move away from the impact point as one mass. a) Treating each car as a particle, find the total momentum just before the collision. b) The insurance adjustor has asked you to find the velocity of the wreckage just after impact. What do you tell her?

SOLUTION

IDENTIFY: We are given the masses and initial velocities of the cars, so it's a straightforward procedure to find the total momentum before the collision. To find out what happens *after* the collision, we'll assume that we can treat the cars as an isolated system during the collision. We can do so because the horizontal forces that the cars exert on each other during the collision have very large magnitudes, great enough to crumple the cars' metal skins. Compared to these forces, we can neglect any external forces such as friction. (We'll justify this assumption later.) Hence the momentum of the system of two cars has the same value just before and just after the collision.

SET UP: In Fig. 8.17, we have drawn coordinate axes and have labeled the small car A and the sports utility vehicle B. Our first target variable is the total momentum before the collision, $\vec{P}$, which we'll find using Eqs. (8.15). The momentum has the same value just after the collision; hence, once we've found $\vec{P}$, we'll be able to find the velocity $\vec{V}$ just after the collision (our second target variable) using the relationship $\vec{P} = M\vec{V}$, where M is the combined mass of the wreckage.

EXECUTE:
a) From Eqs. (8.15) the components of the total momentum $\vec{P}$ are

$$\begin{aligned} P_x &= p_{Ax} + p_{Bx} = m_A v_{Ax} + m_B v_{Bx} \\ &= (1000\text{ kg})(0) + (2000\text{ kg})(10\text{ m/s}) \\ &= 2.0 \times 10^4\text{ kg}\cdot\text{m/s} \\ P_y &= p_{Ay} + p_{By} = m_A v_{Ay} + m_B v_{By} \\ &= (1000\text{ kg})(15\text{ m/s}) + (2000\text{ kg})(0) \\ &= 1.5 \times 10^4\text{ kg}\cdot\text{m/s} \end{aligned}$$

The magnitude of $\vec{P}$ is

$$\begin{aligned} P &= \sqrt{(2.0 \times 10^4\text{ kg}\cdot\text{m/s})^2 + (1.5 \times 10^4\text{ kg}\cdot\text{m/s})^2} \\ &= 2.5 \times 10^4\text{ kg}\cdot\text{m/s} \end{aligned}$$

and its direction is given by the angle θ shown in Fig. 8.17b, where

$$\tan\theta = \frac{P_y}{P_x} = \frac{1.5 \times 10^4\text{ kg}\cdot\text{m/s}}{2.0 \times 10^4\text{ kg}\cdot\text{m/s}} = 0.75 \qquad \theta = 37^\circ$$

b) The total momentum just after the collision is the same as just before, which we found to be 2.5×10^4 kg·m/s in a direction 37° north of due east. Assuming that no parts fall off, the total mass of wreckage is $M = m_A + m_B = 3000$ kg. Using $\vec{P} = M\vec{V}$, the direction of the velocity $\vec{V}$ just after the collision is the same as that of the momentum, and its magnitude is

$$V = \frac{P}{M} = \frac{2.5 \times 10^4\text{ kg}\cdot\text{m/s}}{3000\text{ kg}} = 8.3\text{ m/s}$$

EVALUATE: This is an inelastic collision, so we expect the total kinetic energy to be less after the collision than before. We invite you to carry out the calculations; you will find that the initial

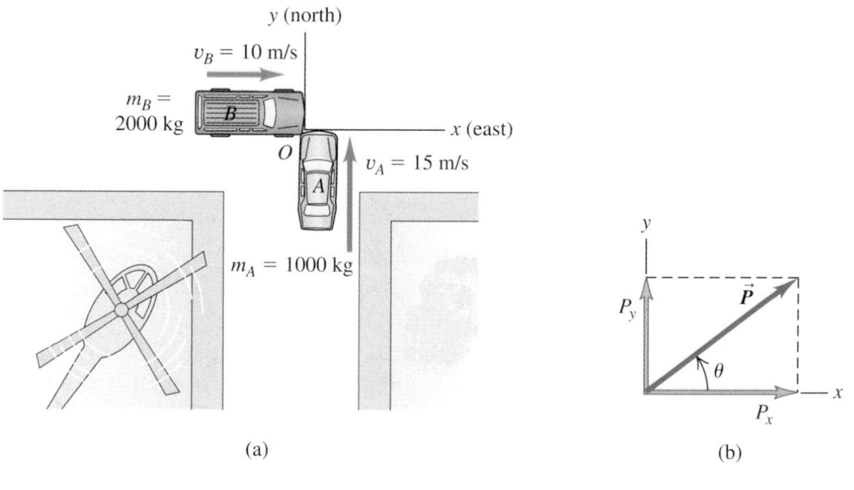

8.17 (a) View from a traffic helicopter overhead of the start of the collision at Morewood and Fifth. (b) Finding the total momentum just before the collision.

kinetic energy is 2.1×10^5 J and the final value is 1.0×10^5 J. Over half of the initial kinetic energy is converted to other forms.

We still need to justify our assumption that we can neglect the external forces on the vehicles during the collision. To do so, let's look at some numbers. The mass of the SUV is 2000 kg, its weight is about 20,000 N, and, if the coefficient of friction is about 0.5, the friction force when it slides across the pavement is about 10,000 N. The SUV's kinetic energy just before the impact is $\frac{1}{2}(2000\text{ kg}) \times (10\text{ m/s})^2 = 1.0 \times 10^5$ J. The car may crumple 0.2 m or so; to do -1.0×10^5 J of work on the car (required to stop it) in a distance of 0.2 m would require a force of 5.0×10^5 N, which is 50 times greater than the friction force. So it's quite reasonable to ignore the external force of friction as compared to the internal forces that the vehicles exert on each other.

CAUTION You may have been tempted to try to find the final velocity by taking the vector sum of the initial velocities. But there is *no* "law of conservation of velocities." The conserved quantity is the total momentum of the system, and the above analysis is the only correct way to do the problem.

Test Your Understanding

Find the kinetic energies of the robots in Example 8.6 (Section 8.2) before and after the collision. Was the collision inelastic? Was it totally inelastic?

8.4 | Elastic Collisions

ActivPhysics ONLINE

6.2 Collisions and Elasticity

6.5 Car Collisions: Two Dimensions

6.9 Pendulum Bashes Box

As discussed in Section 8.3, an *elastic collision* in an isolated system is a collision in which kinetic energy (as well as momentum) is conserved. Elastic collisions occur when the forces between the colliding bodies are *conservative*. When two steel balls collide, they squash a little near the surface of contact, but then they spring back. Some of the kinetic energy is stored temporarily as elastic potential energy, but at the end it is reconverted to kinetic energy (Fig. 8.18).

Let's look at an elastic collision between two bodies A and B. We start with a one-dimensional collision, in which all the velocities lie along the same line; we choose this line to be the x-axis. Each momentum and velocity then has only an x-component. We call the x-components of velocity before the collision v_{A1x} and v_{B1x}, and those after the collision v_{A2x} and v_{B2x}. From conservation of kinetic energy we have

$$\frac{1}{2}m_A v_{A1x}^2 + \frac{1}{2}m_B v_{B1x}^2 = \frac{1}{2}m_A v_{A2x}^2 + \frac{1}{2}m_B v_{B2x}^2$$

and conservation of momentum gives

$$m_A v_{A1x} + m_B v_{B1x} = m_A v_{A2x} + m_B v_{B2x}$$

If the masses m_A and m_B and the initial velocities v_{A1x} and v_{B1x} are known, these two equations can be solved simultaneously to find the two final velocities v_{A2x} and v_{B2x}.

8.18 Billiard balls deform very little when they collide, and quickly spring back from any deformation they do undergo. Hence the force of interaction between the balls is almost perfectly conservative, and the collision is elastic.

The general solution is a little complicated, so we will concentrate on the particular case in which body B is at rest before the collision. Think of it as a target for body A to hit. We can then simplify the velocity notation; we let v_x be the initial x-component of velocity of A, and v_{Ax} and v_{Bx} be the final x-components for A and B. Then the kinetic energy and momentum conservation equations are, respectively,

$$\frac{1}{2}m_A v_x^2 = \frac{1}{2}m_A v_{Ax}^2 + \frac{1}{2}m_B v_{Bx}^2 \tag{8.19}$$

$$m_A v_x = m_A v_{Ax} + m_B v_{Bx} \tag{8.20}$$

We may solve for v_{Ax} and v_{Bx} in terms of the masses and the initial velocity v_x. This involves some fairly strenuous algebra, but it's worth it. No pain, no gain! The simplest approach is somewhat indirect, but along the way it uncovers an additional interesting feature of elastic collisions.

First rearrange Eqs. (8.19) and (8.20) as follows:

$$m_B v_{Bx}^2 = m_A(v_x^2 - v_{Ax}^2) = m_A(v_x - v_{Ax})(v_x + v_{Ax}) \tag{8.21}$$

$$m_B v_{Bx} = m_A(v_x - v_{Ax}) \tag{8.22}$$

Now divide Eq. (8.21) by Eq. (8.22) to obtain

$$v_{Bx} = v_x + v_{Ax} \tag{8.23}$$

Substitute this expression back into Eq. (8.22) to eliminate v_{Bx} and then solve for v_{Ax}:

$$m_B(v_x + v_{Ax}) = m_A(v_x - v_{Ax})$$

$$v_{Ax} = \frac{m_A - m_B}{m_A + m_B}v_x \tag{8.24}$$

Finally, substitute this result back into Eq. (8.23) to obtain

$$v_{Bx} = \frac{2m_A}{m_A + m_B}v_x \tag{8.25}$$

Now we can interpret the results. Suppose body A is a ping-pong ball and body B is a bowling ball. Then we expect A to bounce off after the collision with a velocity nearly equal to its original value but in the opposite direction (Fig. 8.19a), and we expect B's velocity to be much smaller. That's just what the equations predict. When m_A is much smaller than m_B, the fraction in Eq. (8.24) is approximately equal to (-1), so v_{Ax} is approximately equal to $-v_x$. The fraction in Eq. (8.25) is much smaller than unity, so v_{Bx} is much less than v_x. Figure 8.19b shows the opposite case, in which A is the bowling ball and B the ping-pong ball and m_A is much larger than m_B. What do you expect to happen then? Check your predictions against Eqs. (8.24) and (8.25).

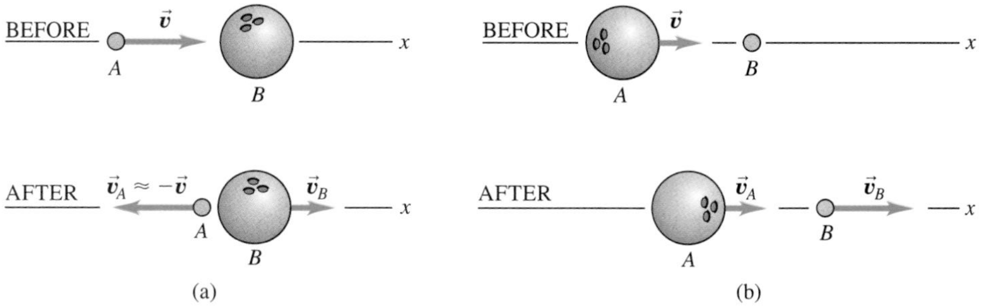

8.19 Collisions between a ping-pong ball and a bowling ball. (a) Initially stationary bowling ball. (b) Initially stationary ping-pong ball.

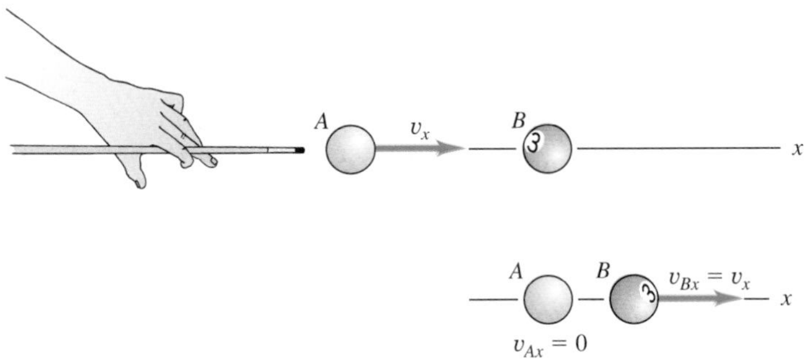

8.20 In a one-dimensional elastic collision between bodies of equal mass, if one body is at rest, it receives all the momentum and kinetic energy of the moving body.

Another interesting case occurs when the masses are equal (Fig. 8.20). If $m_A = m_B$, then Eqs. (8.24) and (8.25) give $v_{Ax} = 0$ and $v_{Bx} = v_x$. That is, the body that was moving stops dead; it gives all its momentum and kinetic energy to the body that was at rest. This behavior is familiar to all pool players and marbles shooters.

Let's return to the general case in which A and B have different masses. Equation (8.23) can be rewritten as

$$v_x = v_{Bx} - v_{Ax} \tag{8.26}$$

Here $v_{Bx} - v_{Ax}$ is just the velocity of B relative to A *after* the collision; from Eq. (8.26), this equals v_x, which is the *negative* of the velocity of B relative to A *before* the collision. (We discussed relative velocity in Section 3.5.) The relative velocity has the same magnitude, but opposite sign, before and after the collision. The sign changes because A and B are approaching each other before the collision but receding from each other after the collision. If we view this collision from a second coordinate system moving with constant velocity relative to the first, the velocities of the bodies are different but the *relative* velocities are the same. Hence our statement about relative velocities holds for *any* straight-line elastic collision, even when neither body is at rest initially. *In a straight-line elastic collision of two bodies, the relative velocities before and after the collision have the same magnitude but opposite sign.* This can be expressed using our original notation as

$$v_{B2x} - v_{A2x} = -(v_{B1x} - v_{A1x}) \tag{8.27}$$

It turns out that a *vector* relationship similar to Eq. (8.27) is a general property of *all* elastic collisions, even when both bodies are moving initially and the velocities do not all lie along the same line. This result provides an alternative and equivalent definition of an elastic collision: *In an elastic collision the relative velocity of the two bodies has the same magnitude before and after the collision.* Whenever this condition is satisfied, the total kinetic energy is also conserved.

When an elastic two-body collision isn't head-on, the velocities don't all lie along a single line. If they all lie in a plane, each final velocity has two unknown components, and there are four unknowns in all. Conservation of energy and conservation of the x- and y-components of momentum give only three equations. To determine the final velocities uniquely, we need additional information, such as the direction or magnitude of one of the final velocities.

Example 8.10 An elastic straight-line collision

The situation in Fig. 8.21 is the same as for Example 8.5 in Section 8.2, but now we've added ideal spring bumpers to the gliders so that the collision is elastic. What are the velocities of A and B after the collision?

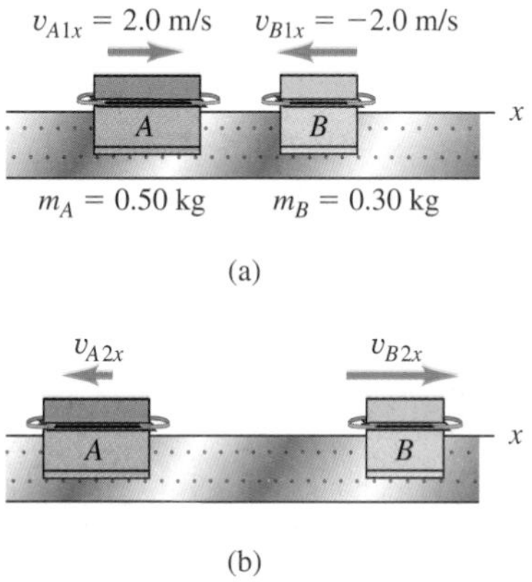

8.21 Air-track gliders (a) before and (b) after an elastic collision.

SOLUTION

IDENTIFY and SET UP: As in Example 8.5, the net external force on the system of two gliders is zero and the momentum of the system is conserved. We again choose the positive x-axis to point to the right. We'll find our target variables, v_{A2x} and v_{B2x}, using Eq. (8.27) and the equation of momentum conservation.

EXECUTE: From conservation of momentum,

$$m_A v_{A1} + m_B v_{B1} = m_A v_{A2} + m_B v_{B12}$$
$$(0.50\text{ kg})(2.0\text{ m/s}) + (0.30\text{ kg})(-2.0\text{ m/s})$$
$$= (0.50\text{ kg})v_{A2x} + (0.30\text{ kg})v_{B2x}$$
$$0.50v_{A2x} + 0.30v_{B2x} = 0.40\text{ m/s}$$

(In the last equation we have divided through by the unit "kg.") From Eq. (8.27), the relative velocity relation for an elastic collision,

$$v_{B2x} - v_{A2x} = -(v_{B1x} - v_{A1x})$$
$$= -(-2.0\text{ m/s} - 2.0\text{ m/s}) = 4.0\text{ m/s}$$

Before the collision, the velocity of B relative to A is to the left at 4.0 m/s; after the collision, the velocity of B relative to A is to the right at 4.0 m/s. Solving these equations simultaneously, we find

$$v_{A2x} = -1.0\text{ m/s} \qquad v_{B2x} = 3.0\text{ m/s}$$

EVALUATE: Both bodies reverse their directions of motion; A moves to the left at 1.0 m/s and B moves to the right at 3.0 m/s. This is different from the result of Example 8.5 because that collision was *not* elastic. The total kinetic energy after the elastic collision is

$$\frac{1}{2}(0.50\text{ kg})(-1.0\text{ m/s})^2 + \frac{1}{2}(0.30\text{ kg})(3.0\text{ m/s})^2 = 1.6\text{ J}$$

As expected, this equals the total kinetic energy before the collision (which we calculated in Example 8.7, Section 8.3).

CAUTION You might have been tempted to solve this problem using Eqs. (8.24) and (8.25). These equations apply *only* to situations in which body B is initially at rest, which isn't the case here. When in doubt, always solve the problem at hand using equations that are applicable to a broad variety of cases.

Example 8.11 Moderator in a nuclear reactor

High-speed neutrons are produced in a nuclear reactor during nuclear fission processes. Before a neutron can trigger additional fissions, it has to be slowed down by collisions with nuclei in the *moderator* of the reactor. The first nuclear reactor (built in 1942 at the University of Chicago) and the reactor involved in the 1986 Chernobyl accident both used carbon (graphite) as the moderator material. A neutron (mass 1.0 u) traveling at 2.6×10^7 m/s undergoes a head-on elastic collision with a carbon nucleus (mass 12 u) initially at rest. The external forces during the collision are negligible. What are the velocities after the collision? (1 u is the *atomic mass unit,* equal to 1.66×10^{-27} kg.)

SOLUTION

IDENTIFY and SET UP: We are given that the external forces can be neglected (so momentum is conserved in the collision) and that the collision is elastic. We take the x-axis to be in the direction that the neutron is moving initially. Because the collision is head-on, both the neutron and the carbon nucleus move along this same axis after the collision. Furthermore, because one body is initially at rest, we can use Eqs. (8.24) and (8.25) with

$$m_A = 1.0\text{ u} \qquad m_B = 12\text{ u} \quad \text{and} \quad v_x = 2.6 \times 10^7\text{ m/s}$$

to solve for the target variables v_{Ax} and v_{Bx} (the final velocities of the neutron and the carbon nucleus, respectively).

EXECUTE: We'll let you do the arithmetic; the results are

$$v_{Ax} = -2.2 \times 10^7 \text{ m/s} \qquad v_{Bx} = 0.4 \times 10^7 \text{ m/s}$$

EVALUATE: The neutron ends up with $\frac{11}{13}$ of its initial speed, and the speed of the recoiling carbon nucleus is $\frac{2}{13}$ of the neutron's initial speed. [These ratios are just the factors $(m_A - m_B)/(m_A + m_B)$ and $2m_A/(m_A + m_B)$ that appear in Eqs. (8.24) and (8.25).] Kinetic energy is proportional to speed squared, so the neutron's final kinetic energy is $(\frac{11}{13})^2$, or about 0.72 of its original value. If it makes a second such collision, its kinetic energy is $(0.72)^2$, or about half its original value, and so on. After several collisions the neutron will be moving quite slowly and will be able to trigger a fission reaction in a uranium nucleus.

Example 8.12 The gravitational slingshot effect

Figure 8.22 shows the planet Saturn moving in the negative x-direction at its orbital speed (with respect to the sun) of 9.6 km/s. The mass of Saturn is 5.69×10^{26} kg. A spacecraft with mass 2150 kg approaches Saturn, moving initially in the $+x$-direction at 10.4 km/s. The gravitational attraction of Saturn (a conservative force) causes the spacecraft to swing around it and head off in the opposite direction. Find the final speed of the spacecraft after it is far enough away to be nearly free of Saturn's gravitational pull.

SOLUTION

IDENTIFY: Here the "collision" is not an impact but rather a gravitational interaction. Nonetheless, we can treat it as a one-dimensional elastic collision because the interaction force is conservative and the spacecraft departs in the direction opposite to that in which it arrives.

SET UP: Let the spacecraft be body A and let Saturn be body B. We can assume that Saturn's speed is essentially constant during the interaction because its mass is so much greater than that of the spacecraft. If we take the positive x-direction to be the direction in which the spacecraft is moving initially, we have $v_{A1x} = 10.4$ km/s and $v_{B1x} = v_{B2x} = -9.6$ km/s. To find our target variable, the magnitude of v_{A2x}, we'll use the result given by Eq. (8.27): that the relative velocity of the spacecraft and Saturn has the same magnitude before and after the interaction.

EXECUTE: Being very careful with signs, we find from Eq. (8.27) that

$$\begin{aligned} v_{A2x} &= v_{B2x} + v_{B1x} - v_{A1x} \\ &= (-9.6 \text{ km/s}) + (-9.6 \text{ km/s}) - 10.4 \text{ km/s} \\ &= -29.6 \text{ km/s} \end{aligned}$$

The spacecraft's final speed (relative to the sun) is 29.6 km/s, nearly three times as great as before the encounter with Saturn. Not only does the spacecraft get a close look at Saturn, but its kinetic energy increases by a factor of $(29.6/10.4)^2 = 8.1$ in the process.

EVALUATE: How can the spacecraft acquire so much additional speed and kinetic energy? The reason is that Saturn is *moving* in its orbit around the sun. If Saturn were initially at rest, the spacecraft would reverse direction but would undergo hardly any change in

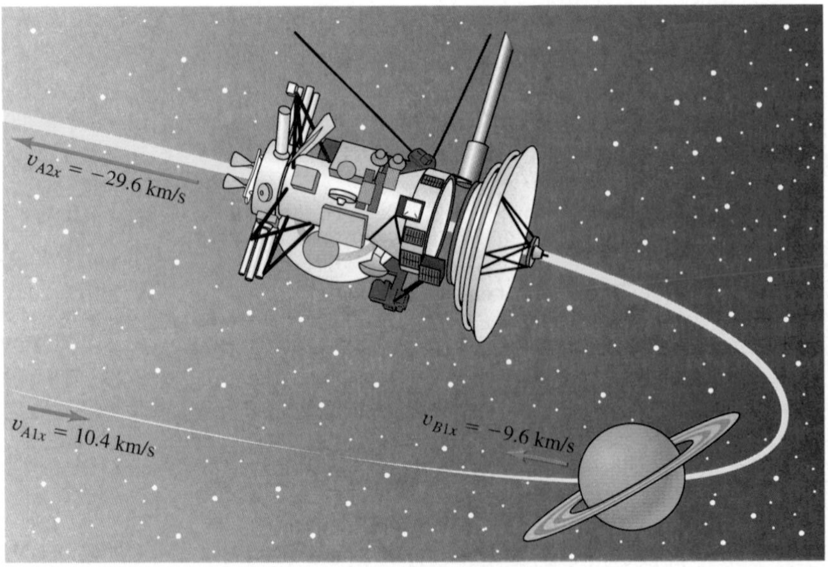

8.22 The gravitational slingshot effect: a simplified model of a spacecraft making a non-impact "collision" with the planet Saturn.

speed; a baseball behaves similarly when it bounces off a bat that's held stationary (a "bunt"). But when a baseball collides with a bat that's swinging toward it, the baseball leaves the bat with a much greater speed, perhaps enough to make a home run. In our example, Saturn takes the place of the swinging bat, and the spacecraft gains speed much like a batted baseball.

This is a simplified version of the *gravitational slingshot effect,* used to give an extra boost to spacecraft. (The actual motion of a spacecraft is never purely one-dimensional as we have assumed here. We'll discuss the trajectories of spacecraft and planets in detail in Chapter 12.) The *Voyager 2* spacecraft, launched in 1977, used the slingshot effect in flying past Jupiter, Saturn, and Uranus. Thanks to the kinetic energy gained in this way, *Voyager 2* reached the planet Neptune in 1989; without the slingshot effect, *Voyager 2* would not have arrived at Neptune until 2008. The *Cassini* spacecraft, launched in 1997, used an even more elaborate trajectory that involved slingshot flybys of Venus, Earth, and Jupiter to help propel it toward a rendezvous with Saturn in 2004.

In this bit of celestial gymnastics, does the spacecraft get something for nothing? Where does its extra energy come from?

Example 8.13 A two-dimensional elastic collision

Figure 8.23 shows an elastic collision of two pucks on a frictionless air-hockey table. Puck *A* has mass $m_A = 0.500$ kg and puck *B* has mass $m_B = 0.300$ kg. Puck *A* has an initial velocity of 4.00 m/s in the positive *x*-direction and a final velocity of 2.00 m/s in an unknown direction. Puck *B* is initially at rest. Find the final speed v_{B2} of puck *B* and the angles α and β in the figure.

SOLUTION

IDENTIFY and SET UP: The target variables are given in the statement of the problem. Although the collision is elastic, it is *not* one-dimensional, so we can't use any of the one-dimensional formulas derived in this section. Instead, we'll use the equations for conservation of energy, conservation of *x*-momentum, and conservation of *y*-momentum. These three equations should be enough to solve for our three target variables.

EXECUTE: Because the collision is elastic, the initial and final kinetic energies are equal:

$$\frac{1}{2}m_A v_{A1}^2 = \frac{1}{2}m_A v_{A2}^2 + \frac{1}{2}m_B v_{B2}^2$$

$$v_{B2}^2 = \frac{m_A v_{A1}^2 - m_A v_{A2}^2}{m_B}$$

$$= \frac{(0.500\text{ kg})(4.00\text{ m/s})^2 - (0.500\text{ kg})(2.00\text{ m/s})^2}{0.300\text{ kg}}$$

$$v_{B2} = 4.47\text{ m/s}$$

Conservation of the *x*-component of total momentum gives

$$m_A v_{A1x} = m_A v_{A2x} + m_B v_{B2x}$$

$$(0.500\text{ kg})(4.00\text{ m/s}) = (0.500\text{ kg})(2.00\text{ m/s})(\cos\alpha) + (0.300\text{ kg})(4.47\text{ m/s})(\cos\beta)$$

and conservation of the *y*-component gives

$$0 = m_A v_{A2y} + m_B v_{B2y}$$

$$0 = (0.500\text{ kg})(2.00\text{ m/s})(\sin\alpha) - (0.300\text{ kg})(4.47\text{ m/s})(\sin\beta)$$

These are two simultaneous equations for α and β. The simplest solution is to eliminate β as follows: We solve the first equation for $\cos\beta$ and the second for $\sin\beta$; we then square each equation and add. Since $\sin^2\beta + \cos^2\beta = 1$, this eliminates β and leaves an equation that we can solve for $\cos\alpha$ and hence for α. We can then substitute this value back into either of the two equations and solve the result for β. We leave the details for you to work out as a problem; the results are

$$\alpha = 36.9^\circ \qquad \beta = 26.6^\circ$$

EVALUATE: A quick way to check the answers is to make sure that the *y*-momentum, which was zero before the collision, is still zero after the collision. The *y*-momenta of the pucks are

$$p_{A2y} = (0.500\text{ kg})(2.00\text{ m/s})(\sin 36.9^\circ) = +0.600\text{ kg}\cdot\text{m/s}$$

$$p_{B2y} = -(0.300\text{ kg})(4.47\text{ m/s})(\sin 26.6^\circ) = -0.600\text{ kg}\cdot\text{m/s}$$

The sum of these values is zero, as it should be.

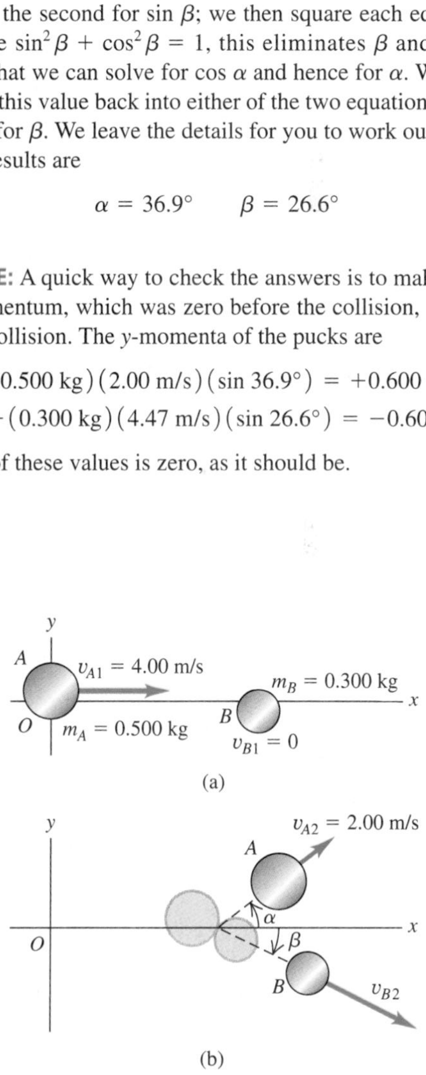

8.23 (a) Before and (b) after an elastic collision that isn't head-on.

The examples in this and the preceding sections show that we can classify collisions according to energy considerations. A collision in which kinetic energy is conserved is called *elastic.* A collision in which the total kinetic energy decreases is called *inelastic.* When the two bodies have a common final velocity, we say that the collision is *completely inelastic.* There are also cases in which the final kinetic energy is *greater* than the initial value. Rifle recoil, discussed in Example 8.4 (Section 8.2), is an example.

Finally, we emphasize again that we can sometimes use momentum conservation even when there are external forces acting on the system, if the net external force acting on the colliding bodies is small in comparison to the internal forces during the collision.

Test Your Understanding

Most present-day nuclear reactors use water as a moderator (see Example 8.11). After a neutron collides with a water molecule (mass 18 u), what fraction of its initial speed does the neutron have? What fraction of its initial kinetic energy? Are water molecules a better or worse moderator than carbon atoms? (One advantage of water is that it also acts as a coolant for the reactor's radioactive core.)

8.5 | Center of Mass

We can restate the principle of conservation of momentum in a useful way by using the concept of **center of mass.** Suppose we have several particles, with masses m_1, m_2, and so on. Let the coordinates of m_1 be (x_1, y_1), those of m_2 be (x_2, y_2), and so on. We define the center of mass of the system as the point having coordinates (x_{cm}, y_{cm}) given by

$$x_{cm} = \frac{m_1 x_1 + m_2 x_2 + m_3 x_3 + \cdots}{m_1 + m_2 + m_3 + \cdots} = \frac{\sum_i m_i x_i}{\sum_i m_i}$$

$$y_{cm} = \frac{m_1 y_1 + m_2 y_2 + m_3 y_3 + \cdots}{m_1 + m_2 + m_3 + \cdots} = \frac{\sum_i m_i y_i}{\sum_i m_i} \qquad \text{(center of mass)} \qquad (8.28)$$

The position vector $\vec{r}_{cm}$ of the center of mass can be expressed in terms of the position vectors $\vec{r}_1, \vec{r}_2, \ldots$ of the particles as

$$\vec{r}_{cm} = \frac{m_1\vec{r}_1 + m_2\vec{r}_2 + m_3\vec{r}_3 + \cdots}{m_1 + m_2 + m_3 + \cdots} = \frac{\sum_i m_i \vec{r}_i}{\sum_i m_i} \qquad \text{(center of mass)} \qquad (8.29)$$

In statistical language, the center of mass is a *mass-weighted average* position of the particles.

Example 8.14 Center of mass of a water molecule

Figure 8.24 shows a simple model of the structure of a water molecule. The separation between atoms is $d = 9.57 \times 10^{-11}$ m. Each hydrogen atom has mass 1.0 u and the oxygen atom has mass 16.0 u. Find the position of the center of mass.

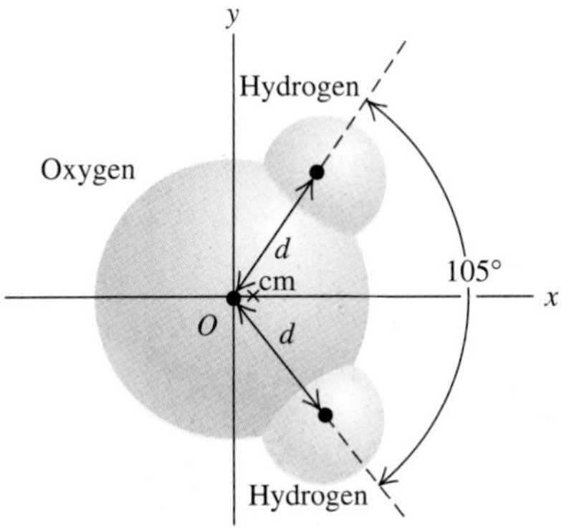

8.24 Where is the center of mass of a water molecule?

SOLUTION

IDENTIFY and SET UP: The coordinate system is shown in Fig. 8.24. Nearly all the mass of each atom is concentrated in its nucleus, which is only about 10^{-5} times the overall radius of the atom, so we can safely represent each atom as a point. We'll use Eqs. (8.28) to determine the coordinates x_{cm} and y_{cm}.

EXECUTE: The x-coordinate of each hydrogen atom is $d \cos(105°/2)$; the y-coordinates of the upper and lower hydrogen atoms are $+d \sin(105°/2)$ and $-d \sin(105°/2)$, respectively. The coordinates of the oxygen atom are $x = 0$, $y = 0$. From Eqs. (8.28) the x-coordinate of the center of mass is

$$x_{cm} = \frac{\begin{bmatrix} (1.0\,\text{u})(d\cos 52.5°) + (1.0\,\text{u}) \\ \times\ (d\cos 52.5°) + (16.0\,\text{u})(0) \end{bmatrix}}{1.0\,\text{u} + 1.0\,\text{u} + 16.0\,\text{u}} = 0.068\,d$$

and the y-coordinate is

$$y_{cm} = \frac{\begin{bmatrix} (1.0\,\text{u})(d\sin 52.5°) + (1.0\,\text{u}) \\ \times\ (-d\sin 52.5°) + (16.0\,\text{u})(0) \end{bmatrix}}{1.0\,\text{u} + 1.0\,\text{u} + 16.0\,\text{u}} = 0$$

The actual value of d for the water molecule is 9.57×10^{-11} m, so we find

$$x_{cm} = (0.068)(9.57 \times 10^{-11}\,\text{m}) = 6.5 \times 10^{-12}\,\text{m}$$

EVALUATE: The center of mass is much closer to the oxygen atom than to either hydrogen atom because the oxygen atom is much more massive. Notice that the center of mass lies along the x-axis, which is the *axis of symmetry* of this molecule. If the molecule is rotated by 180° around this axis, it looks exactly the same as before. The position of the center of mass can't be affected by this rotation, so it must lie on the axis of symmetry.

For solid bodies, in which we have (at least on a macroscopic level) a continuous distribution of matter, the sums in Eqs. (8.28) have to be replaced by integrals. The calculations can get quite involved, but we can say some general things about such problems. First, whenever a homogeneous body has a geometric center, such as a billiard ball, a sugar cube, or a can of frozen orange juice, the center of mass is at the geometric center. Second, whenever a body has an axis of symmetry, such as a wheel or a pulley, the center of mass always lies on that axis. Third, there is no law that says the center of mass has to be within the body. For example, the center of mass of a doughnut is right in the middle of the hole. You can probably think of other examples.

We'll talk a little more about locating the center of mass in Chapter 11 in connection with the related concept *center of gravity.*

Motion of the Center of Mass

To see the significance of the center of mass of a collection of particles, we must ask what happens to the center of mass when the particles move. The x- and y-components of velocity of the center of mass, $v_{cm\text{-}x}$ and $v_{cm\text{-}y}$, are the time deriva-

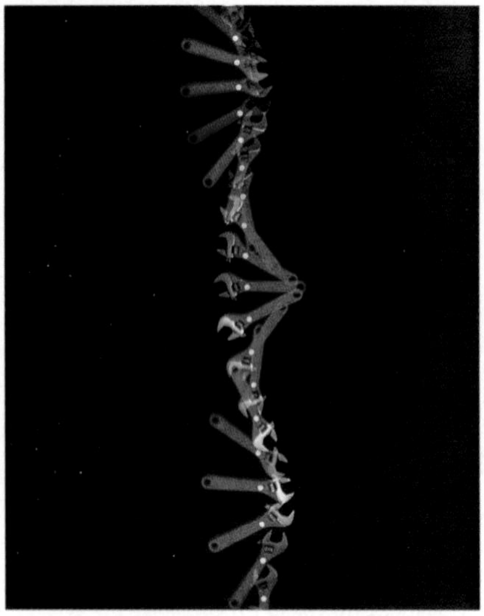

8.25 The center of mass of this wrench is marked with a white dot. The total external force acting on the wrench is almost zero. As it spins on a smooth horizontal surface, the center of mass moves in a straight line with nearly constant velocity.

tives of x_{cm} and y_{cm}. Also, dx_1/dt is the x-component of velocity of particle 1, and so on, so $dx_1/dt = v_{1x}$, and so on. Taking time derivatives of Eqs. (8.28), we get

$$v_{cm\text{-}x} = \frac{m_1 v_{1x} + m_2 v_{2x} + m_3 v_{3x} + \cdots}{m_1 + m_2 + m_3 + \cdots}$$

$$v_{cm\text{-}y} = \frac{m_1 v_{1y} + m_2 v_{2y} + m_3 v_{3y} + \cdots}{m_1 + m_2 + m_3 + \cdots} \tag{8.30}$$

These equations are equivalent to the single vector equation obtained by taking the time derivative of Eq. (8.29):

$$\vec{v}_{cm} = \frac{m_1 \vec{v}_1 + m_2 \vec{v}_2 + m_3 \vec{v}_3 + \cdots}{m_1 + m_2 + m_3 + \cdots} \tag{8.31}$$

We denote the *total* mass $m_1 + m_2 + \cdots$ by M. We can then rewrite Eq. (8.31) as

$$M\vec{v}_{cm} = m_1 \vec{v}_1 + m_2 \vec{v}_2 + m_3 \vec{v}_3 + \cdots = \vec{P} \tag{8.32}$$

The right side is simply the total momentum $\vec{P}$ of the system. Thus we have proved that *the total momentum is equal to the total mass times the velocity of the center of mass.* When you catch a baseball, you are really catching a collection of a very large number of molecules of masses $m_1, m_2, m_3, \ldots$. The impulse you feel is due to the total momentum of this entire collection. But this impulse is the same as if you were catching a single particle of mass $M = m_1 + m_2 + m_3 + \cdots$ moving with velocity $\vec{v}_{cm}$, the velocity of the collection's center of mass. So Eq. (8.32) helps to justify representing an extended body as a particle.

For a system of particles on which the net external force is zero, so that the total momentum $\vec{P}$ is constant, the velocity of the center of mass $\vec{v}_{cm} = \vec{P}/M$ is also constant. Suppose we mark the center of mass of an adjustable wrench, which lies at a point partway down the handle. We then slide the wrench with a spinning motion across a smooth, horizontal table top (Fig. 8.25). The overall motion appears complicated, but the center of mass follows a straight line, as though all the mass were concentrated at that point.

Example 8.15 A tug-of-war on the ice

James and Ramon are standing 20.0 m apart on the slippery surface of a frozen pond. Ramon has mass 60.0 kg and James has mass 90.0 kg. Midway between the two men a mug of their favorite beverage sits on the ice (Fig. 8.26). They pull on the ends of a light rope that is stretched between them. When James has moved 6.0 m toward the mug, how far and in what direction has Ramon moved?

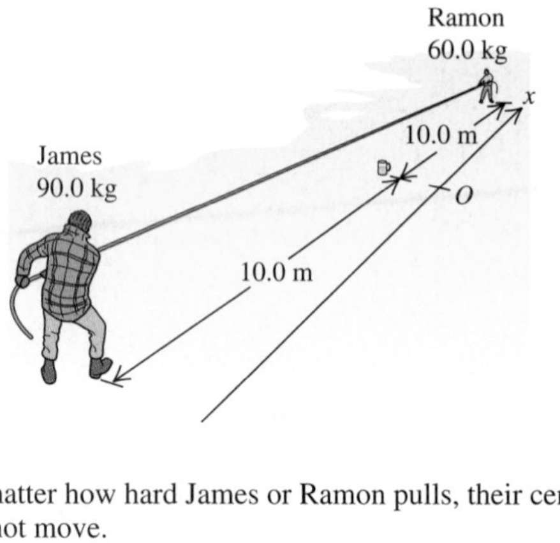

8.26 No matter how hard James or Ramon pulls, their center of mass will not move.

SOLUTION

IDENTIFY: The frozen surface is horizontal and essentially frictionless, so the net external force on the system of James, Ramon, and the rope is zero. Hence their total momentum is conserved. Initially, there is no motion, so the total momentum is zero; thus the velocity of the center of mass is zero, and the center of mass remains at rest. We can use this to relate the positions of James and Ramon.

SET UP: Let's take the origin at the position of the mug, and let the $+x$-axis extend from the mug toward Ramon. Since the rope is light, we can neglect its mass in calculating the position of the center of mass with Eq (8.28).

EXECUTE: The initial x-coordinates of James and Ramon are -10.0 m and $+10.0$ m, respectively, so the x-coordinate of the center of mass is

$$x_{cm} = \frac{(90.0\text{ kg})(-10.0\text{ m}) + (60.0\text{ kg})(10.0\text{ m})}{90.0\text{ kg} + 60.0\text{ kg}} = -2.0\text{ m}$$

When James moves 6.0 m toward the mug, his new x-coordinate is -4.0 m; we'll call Ramon's new x-coordinate x_2. The center of mass doesn't move, so

$$x_{cm} = \frac{(90.0\text{ kg})(-4.0\text{ m}) + (60.0\text{ kg})x_2}{90.0\text{ kg} + 60.0\text{ kg}} = -2.0\text{ m}$$

$$x_2 = 1.0\text{ m}$$

James has moved 6.0 m in the positive x-direction and is still 4.0 m from the mug, but Ramon has moved 9.0 m in the negative x-direction and is only 1.0 m from it.

EVALUATE: The ratio of how far each man moved, $(6.0\text{ m})/(9.0\text{ m}) = \frac{2}{3}$, equals the inverse ratio of their masses. Can you see why? If the two men keep moving (and if the surface is frictionless, they will!), Ramon will reach the mug first. This result is completely independent of how hard either person pulls; pulling harder just helps Ramon quench his thirst sooner.

External Forces and Center-of-Mass Motion

If the net external force on a system of particles is not zero, then total momentum is not conserved and the velocity of the center of mass changes. Let's look at the relation between the motion of the center of mass and the forces acting on the system.

Equations (8.31) and (8.32) give the *velocity* of the center of mass in terms of the velocities of the individual particles. Proceeding one additional step, we take the time derivatives of these equations to show that the accelerations are related in the same way. Let $\vec{a}_{cm} = d\vec{v}_{cm}/dt$ be the acceleration of the center of mass; then we find

$$M\vec{a}_{cm} = m_1\vec{a}_1 + m_2\vec{a}_2 + m_3\vec{a}_3 + \cdots \qquad (8.33)$$

Now $m_1\vec{a}_1$ is equal to the vector sum of forces on the first particle, and so on, so the right side of Eq. (8.33) is equal to the vector sum $\sum\vec{F}$ of *all* the forces on *all* the particles. Just as we did in Section 8.2, we may classify each force as *internal* or *external.* The sum of all forces on all the particles is then

$$\sum\vec{F} = \sum\vec{F}_{ext} + \sum\vec{F}_{int} = M\vec{a}_{cm}$$

Because of Newton's third law, the internal forces all cancel in pairs, and $\sum\vec{F}_{int} = \mathbf{0}$. What survives on the left side is only the sum of the *external* forces, and we have

$$\sum\vec{F}_{ext} = M\vec{a}_{cm} \quad \text{(body or collection of particles)} \qquad (8.34)$$

When a body or a collection of particles is acted on by external forces, the center of mass moves just as though all the mass were concentrated at that point and it were acted on by a net force equal to the sum of the external forces on the system.

This result may not sound very impressive, but in fact it is central to the whole subject of mechanics. In fact, we've been using this result all along; without it, we would not be able to represent an extended body as a point particle when we apply Newton's laws. It explains why only *external* forces can affect the motion of an extended body. If you pull upward on your belt, your belt exerts an equal downward force on your hands; these are *internal* forces that cancel and have no effect on the overall motion of your body.

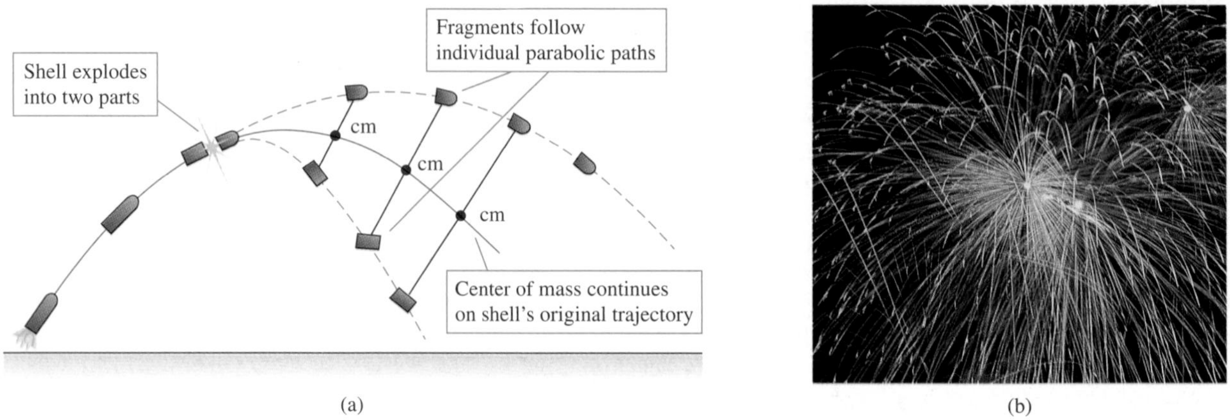

8.27 (a) A shell explodes into two parts in flight. If air resistance is ignored, the center of mass continues on the same trajectory as the shell's path before exploding—until one of the fragments hits the ground. (b) The same effect occurs with exploding fireworks.

Suppose a cannon shell traveling in a parabolic trajectory (neglecting air resistance) explodes in flight, splitting into two fragments with equal mass (Fig. 8.27a). The fragments follow new parabolic paths, but the center of mass continues on the original parabolic trajectory, just as though all the mass were still concentrated at that point. A skyrocket exploding in air (Fig. 8.27b) is a spectacular example of this effect.

This property of the center of mass is important when we analyze the motion of rigid bodies. We describe the motion of an extended body as a combination of translational motion of the center of mass and rotational motion about an axis through the center of mass. We will return to this topic in Chapter 10. This property also plays an important role in the motion of astronomical objects. It's not correct to say that the moon orbits the earth; rather, the earth and moon both move in orbits around their center of mass.

There's one more useful way to describe the motion of a system of particles. Using $\vec{a}_{\text{cm}} = d\vec{v}_{\text{cm}}/dt$, we can rewrite Eq. (8.33) as

$$M\vec{a}_{\text{cm}} = M\frac{d\vec{v}_{\text{cm}}}{dt} = \frac{d(M\vec{v}_{\text{cm}})}{dt} = \frac{d\vec{P}}{dt} \tag{8.35}$$

The total system mass M is constant, so we're allowed to take it inside the derivative. Substituting Eq. (8.35) into Eq. (8.34), we find

$$\sum \vec{F}_{\text{ext}} = \frac{d\vec{P}}{dt} \quad \text{(extended body or system of particles)} \tag{8.36}$$

This equation looks like Eq. (8.3). The difference is that Eq. (8.36) describes a *system* of particles, such as an extended body, while Eq. (8.3) describes a single particle. The interactions between the particles that make up the system can change the individual momenta of the particles, but the *total* momentum $\vec{P}$ of the system can be changed only by external forces acting from outside the system.

Finally, we note that if the net external force is zero, Eq. (8.34) shows that the acceleration $\vec{a}_{\text{cm}}$ of the center of mass is zero. So the center-of-mass velocity $\vec{v}_{\text{cm}}$ is constant, as for the wrench in Figure 8.25. From Eq. (8.36) the total momentum $\vec{P}$ is also constant. This reaffirms our statement in Section 8.3 of the principle of conservation of momentum.

Test Your Understanding

Find the location of the center of mass of the earth–moon system using information from Appendix F. Where is the center of mass relative to the earth's surface?

*8.6 | Rocket Propulsion

Momentum considerations are particularly useful for analyzing a system in which the masses of parts of the system change with time. In such cases we can't use Newton's second law $\sum\vec{F} = m\vec{a}$ directly because m changes. Rocket propulsion offers a typical and interesting example of this kind of analysis. A rocket is propelled forward by rearward ejection of burned fuel that initially was in the rocket. The forward force on the rocket is the reaction to the backward force on the ejected material. The total mass of the system is constant, but the mass of the rocket itself decreases as material is ejected.

ActivPhysics ONLINE

6.6 Saving an Astronaut

As a simple example, we consider a rocket fired in outer space, where there is no gravitational force and no air resistance. We let m denote the mass of the rocket, which will change as it expends fuel. We choose our x-axis to be along the rocket's direction of motion. Figure 8.28a shows the rocket at a time t, when its mass is m and the x-component of its velocity relative to our coordinate system is v. (For simplicity, we will drop the subscript x in this discussion.) The x-component of total momentum at this instant is $P_1 = mv$. In a short time interval dt the mass of the rocket changes by an amount dm. This is an inherently negative quantity because the rocket's mass m *decreases* with time. During dt a *positive* mass $-dm$ of burned fuel is ejected from the rocket. Let v_{ex} be the exhaust *speed* of this material *relative to the rocket;* the burned fuel is ejected opposite the direction of motion, so its x-component of *velocity* relative to the rocket is $-v_{ex}$. The x-component of velocity v_{fuel} of the burned fuel relative to our coordinate system is then

$$v_{fuel} = v + (-v_{ex}) = v - v_{ex}$$

and the x-component of momentum of the ejected mass $(-dm)$ is

$$(-dm)v_{fuel} = (-dm)(v - v_{ex})$$

As shown in Fig. 8.28b, at the end of the time interval dt, the x-component of velocity of the rocket and unburned fuel has increased to $v + dv$, and its mass has decreased to $m + dm$ (remember that dm is negative). The rocket's momentum at this time is

$$(m + dm)(v + dv)$$

Thus the *total* x-component of momentum P_2 of rocket plus ejected fuel at time $t + dt$ is

$$P_2 = (m + dm)(v + dv) + (-dm)(v - v_{ex})$$

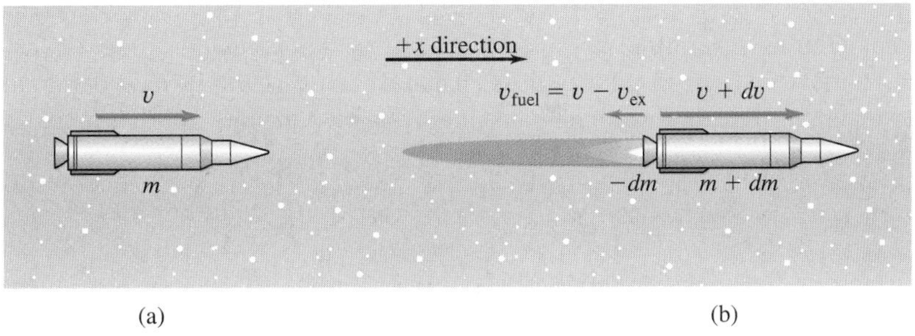

8.28 (a) A rocket moving in gravity-free outer space with mass m and x-component of velocity v at time t. (b) At time $t + dt$ the rocket's mass has decreased, so dm is negative. The mass of the rocket (including the unburned fuel) is $m + dm$, and its velocity component is $v + dv$. The mass of burned fuel in the exhaust is $-dm$, and its velocity relative to the coordinate system shown is $v_{fuel} = v - v_{ex}$.

According to our initial assumption, the rocket and fuel are an isolated system. Thus momentum is conserved, and the total x-component of momentum of the system must be the same at time t and at time $t + dt$: $P_1 = P_2$. Hence

$$mv = (m + dm)(v + dv) + (-dm)(v - v_{ex})$$

This can be simplified to

$$m\,dv = -dm\,v_{ex} - dm\,dv$$

We can neglect the term $(-dm\,dv)$ because it is a product of two small quantities and thus is much smaller than the other terms. Dropping this term, dividing by dt, and rearranging, we find

$$m\frac{dv}{dt} = -v_{ex}\frac{dm}{dt} \tag{8.37}$$

Now dv/dt is the acceleration of the rocket, so the left side of this equation (mass times acceleration) equals the net force F, or *thrust,* on the rocket,

$$F = -v_{ex}\frac{dm}{dt} \tag{8.38}$$

The thrust is proportional both to the relative speed v_{ex} of the ejected fuel and to the mass of fuel ejected per unit time, $-dm/dt$. (Remember that dm/dt is negative because it is the rate of change of the rocket's mass.)

The x-component of acceleration of the rocket is

$$a = \frac{dv}{dt} = -\frac{v_{ex}}{m}\frac{dm}{dt} \tag{8.39}$$

This is positive because v_{ex} is positive (remember, it's the exhaust *speed*) and dm/dt is negative. The rocket's mass m decreases continuously while the fuel is being consumed. If v_{ex} and dm/dt are constant, the acceleration increases until all the fuel is gone.

Equation (8.38) tells us that an effective rocket burns fuel at a rapid rate (large $-dm/dt$) and ejects the burned fuel at a high relative speed (large v_{ex}), as in Fig. 8.29. In the early days of rocket propulsion, people who didn't understand conservation of momentum thought that a rocket couldn't function in outer space because "it doesn't have anything to push against." On the contrary, rockets work *best* in outer space, where there is no air resistance! The rocket in Fig. 8.29 is *not* "pushing against the ground" to get into the air.

8.29 Launch of a space shuttle, a dramatic example of rocket propulsion. To provide enough upward thrust to overcome gravity, space launch vehicles can consume over 10,000 kg/s of fuel and eject the burned fuel at speeds of over 4000 m/s.

If the exhaust speed v_{ex} is constant, we can integrate Eq. (8.39) to find a relation between the velocity v at any time and the remaining mass m. At time $t = 0$, let the mass be m_0 and the velocity v_0. Then we rewrite Eq. (8.39) as

$$dv = -v_{ex}\frac{dm}{m}$$

We change the integration variables to v' and m', so we can use v and m as the upper limits (the final speed and mass). Then we integrate both sides, using limits v_0 to v and m_0 to m, and take the constant v_{ex} outside the integral:

$$\int_{v_0}^{v} dv' = -\int_{m_0}^{m} v_{ex}\frac{dm'}{m'} = -v_{ex}\int_{m_0}^{m}\frac{dm'}{m'}$$

$$v - v_0 = -v_{ex}\ln\frac{m}{m_0} = v_{ex}\ln\frac{m_0}{m} \tag{8.40}$$

The ratio m_0/m is the original mass divided by the mass after the fuel has been exhausted. In practical spacecraft this ratio is made as large as possible to maximize the speed gain, which means that the initial mass of the rocket is almost all fuel. The final velocity of the rocket will be greater in magnitude (and is often *much* greater) than the relative speed v_{ex} if $\ln(m_0/m) > 1$, that is, if $m_0/m > e = 2.71828\ldots$.

We've assumed throughout this analysis that the rocket is in gravity-free outer space. However, gravity must be taken into account when a rocket is launched from the surface of a planet, as in Fig. 8.29 (see Problem 8.104).

Example 8.16 Acceleration of a rocket

A rocket is in outer space, far from any planet, when the rocket engine is turned on. In the first second of firing, the rocket ejects $\frac{1}{120}$ of its mass with a relative speed of 2400 m/s. What is the rocket's initial acceleration?

SOLUTION

IDENTIFY and SETUP: Equation (8.39) gives the acceleration of the rocket—our target variable—in terms of the exhaust speed v_{ex}, the current value m of the rocket's mass, and the rate of change of the mass dm/dt. While we are not given the mass or the rate of change, we are told what fraction of the initial mass is lost during a given time interval; this information should be all we need.

EXECUTE: The initial rate of change of mass is

$$\frac{dm}{dt} = -\frac{m_0/120}{1\text{ s}} = -\frac{m_0}{120\text{ s}}$$

where m_0 is the initial $(t = 0)$ mass of the rocket. From Eq. (8.39) the initial acceleration is

$$a = -\frac{v_{ex}}{m_0}\frac{dm}{dt} = -\frac{2400\text{ m/s}}{m_0}\left(-\frac{m_0}{120\text{ s}}\right) = 20\text{ m/s}^2$$

EVALUATE: Note that the answer didn't depend on the value of m_0. If v_{ex} is the same, the initial acceleration is the same for a 120,000-kg spacecraft that ejects 1000 kg/s as for a 60-kg astronaut equipped with a small rocket that ejects 0.5 kg/s.

Example 8.17 Speed of a rocket

Suppose that $\frac{3}{4}$ of the initial mass m_0 of the rocket in Example 8.16 is fuel, so the final mass is $m = m_0/4$, and that the fuel is completely consumed at a constant rate in a total time $t = 90$ s. If the rocket starts from rest in our coordinate system, find its speed at the end of this time.

SOLUTION

IDENTIFY and SET UP: We are given the initial velocity v_0 (equal to zero), the exhaust speed v_{ex}, and the final mass m in terms of the initial mass m_0. Hence we can use Eq. (8.40) directly to find the final speed v.

EXECUTE: We have $m_0/m = 4$, so from Eq. (8.40),

$$v = v_0 + v_{ex}\ln\frac{m_0}{m} = 0 + (2400\text{ m/s})(\ln 4) = 3327\text{ m/s}$$

EVALUATE: Let's examine what happens as the rocket gains speed. At the start of the flight, when the velocity of the rocket is zero, the ejected fuel is moving to the left, relative to our coordinate system, at 2400 m/s. At the end of the first second ($t = 1$ s) the rocket is moving at 20 m/s, and the fuel's speed relative to our system is 2380 m/s. During the next second the acceleration, given by Eq. (8.39), is a little greater. At $t = 2$ s the rocket is moving at a little over 40 m/s, and the fuel's speed is a little less than 2360 m/s. Detailed calculation shows that at about $t = 75.6$ s, the rocket's velocity v in our coordinate system equals 2400 m/s. The burned fuel ejected after this time moves *forward,* not backward, in our system. Since the final velocity of the rocket is 3327 m/s and the relative velocity is 2400 m/s, the last portion of the ejected fuel has a forward velocity (relative to our frame of reference) of $(3327 - 2400)$ m/s $= 927$ m/s. (To illustrate our point, we are using more figures than are significant.)

Test Your Understanding

If a rocket in gravity-free outer space has the same thrust at all times, is its acceleration constant, increasing, or decreasing? If the rocket has the same acceleration at all times, is the thrust constant, increasing, or decreasing?

SUMMARY

The momentum $\vec{p}$ of a particle is a vector quantity equal to the product of the particle's mass m and velocity $\vec{v}$. Newton's second law says that the net force on a particle is equal to the rate of change of the particle's momentum.

$$\vec{p} = m\vec{v} \qquad (8.2)$$

$$\sum \vec{F} = \frac{d\vec{p}}{dt} \qquad (8.3)$$

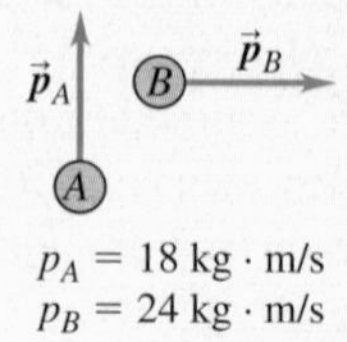

If a constant net force $\sum \vec{F}$ acts on a particle for a time interval Δt from t_1 to t_2, the impulse $\vec{J}$ of the net force is the product of the net force and the time interval. If $\sum \vec{F}$ varies with time, $\vec{J}$ is the integral of the net force over the time interval. In any case, the change in a particle's momentum during a time interval equals the impulse of the net force that acted on it during that interval. The momentum of a particle equals the impulse that accelerated it from rest to its present speed. (See Examples 8.1 through 8.3)

$$\vec{J} = \sum \vec{F}(t_2 - t_1) = \sum \vec{F}\Delta t \qquad (8.5)$$

$$\vec{J} = \int_{t_1}^{t_2} \sum \vec{F}\, dt \qquad (8.7)$$

$$\vec{J} = \vec{p}_2 - \vec{p}_1 \qquad (8.6)$$

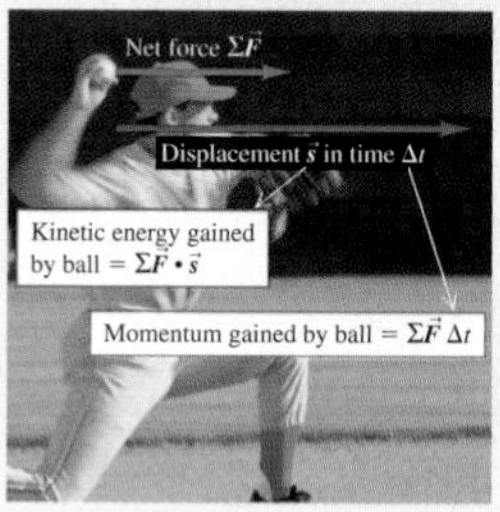

The total momentum of a system of particles A, B, C, . . . is the vector sum of the momenta of the individual particles.

$$\vec{P} = \vec{p}_A + \vec{p}_B = \cdots$$
$$= m_A\vec{v}_A + m_B\vec{v}_B + \cdots \qquad (8.14)$$

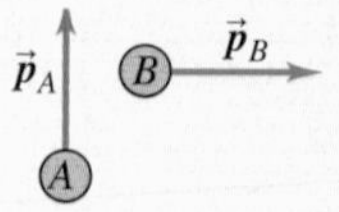

An internal force is a force exerted by one part of a system on another. An external force is a force exerted on any part of a system by something outside the system. If the net external force on a system is zero, the total momentum of the system is constant (conserved); each component of total momentum is separately conserved. (See Examples 8.4 through 8.6)

$$\text{If } \sum \vec{F} = \mathbf{0} \qquad \vec{P} = \text{constant}$$

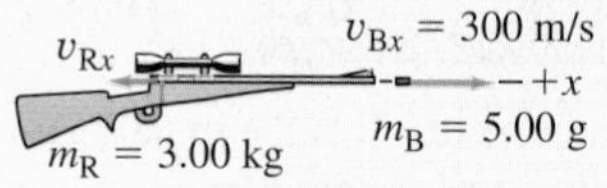

In collisions of all kinds, the initial and final total momenta are equal. In an elastic collision between two bodies, the initial and final total kinetic energies are also equal, and the initial and final relative velocities have the same magnitude. In an inelastic two-body collision, the total kinetic energy is less after the collision than before. If the two bodies have the same final velocity, the collision is completely inelastic. (See Examples 8.7 through 8.13)

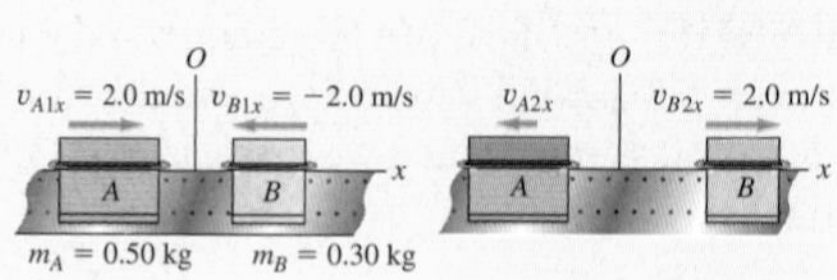

The position vector of the center of mass of a system of particles, $\vec{r}_{\text{cm}}$, is a weighted average of the positions $\vec{r}_1, \vec{r}_2, \ldots$ of the individual particles. (See Example 8.14)

$$\vec{r}_{\text{cm}} = \frac{m_1\vec{r}_1 + m_2\vec{r}_2 + m_3\vec{r}_3 + \cdots}{m_1 + m_2 + m_3 + \cdots}$$

$$= \frac{\sum_i m_i\vec{r}_i}{\sum_i m_i} \qquad (8.29)$$

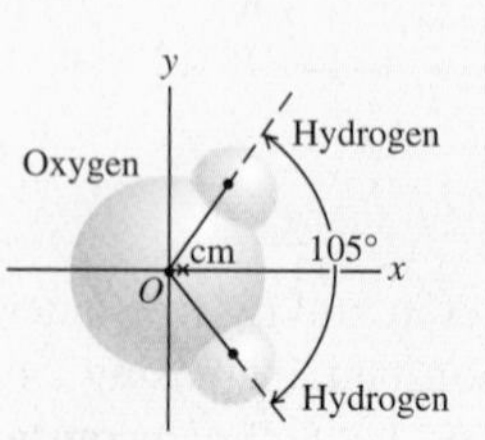

The total momentum $\vec{P}$ of a system equals its total mass M multiplied by the velocity of its center of mass, $\vec{v}_{cm}$. The center of mass moves as through all the mass M was concentrated at that point. If the net external force on the system is zero, the center-of-mass velocity $\vec{v}_{cm}$ is constant. If the net external force is not zero, the center of mass accelerates as though it was a particle of mass M being acted on by the same net external force. (See Example 8.15)

$$\vec{P} = m_1\vec{v}_1 + m_2\vec{v}_2 + m_3\vec{v}_3 + \cdots$$
$$= M\vec{v}_{cm} \qquad (8.32)$$

$$\sum \vec{F}_{ext} = M\vec{a}_{cm} \qquad (8.34)$$

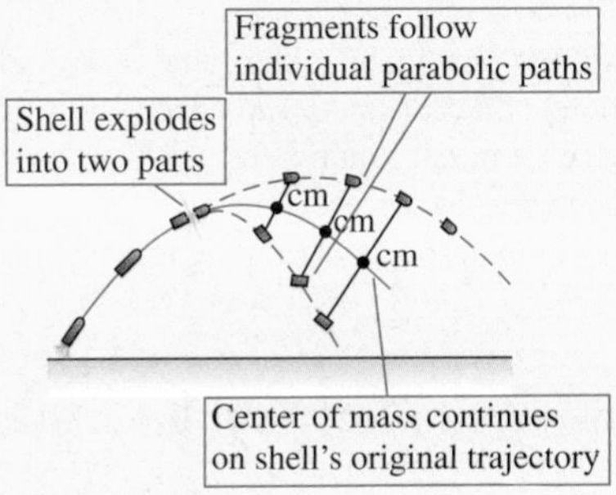

In rocket propulsion, the mass of a rocket changes as the fuel is used up and ejected from the rocket. Analysis of the motion of the rocket must include the momentum carried away by the spent fuel as well as the momentum of the rocket itself. (See Examples 8.16 and 8.17)

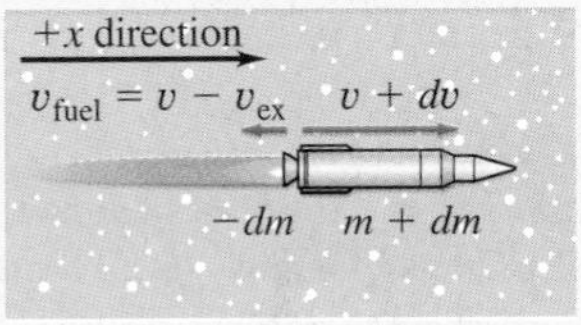

Key Terms

momentum (linear momentum), 283
impulse, 284
impulse-momentum theorem, 284
internal force, 290
external force, 290
isolated system, 290
total momentum, 290
principle of conservation of momentum, 290
elastic collision, 296
inelastic collision, 296
completely inelastic collision, 296
center of mass, 306

Your Notes

Answer to Chapter Opening Question

The two players have the same magnitude of momentum $p = mv$ (the product of mass and speed), but the lightweight player has twice as much kinetic energy $K = \frac{1}{2}mv^2$. Hence, the lightweight player can do twice as much work on you (and twice as much damage) in the process of coming to a halt.

Answers to Test Your Understanding Questions

Section 8.1 (a) The velocity was $v_y = -7.55$ m/s (downward), so $p_y = mv_y = (200\text{ kg})(-7.55\text{ m/s}) = -1510\text{ kg}\cdot\text{m/s}$. The final momentum was zero, so $J_y = 0 - (-1510\text{ kg}\cdot\text{m/s}) = +1510\text{ kg}\cdot\text{m/s} = +1510\text{ N}\cdot\text{s}$. (b) The net upward force was $F_y = n + f - w = 79{,}000\text{ N} + 60\text{ N} - 1960\text{ N} = 77{,}100\text{ N}$. From the definition of impulse, $J_y = F_y\Delta t$, so the elapsed time was $\Delta t = J_y/F_y = (1510\text{ N}\cdot\text{s})/(77{,}100\text{ N}) = 0.0196\text{ s}$.

Section 8.2 There are no external horizontal forces, so the x- and y-components of the total momentum of the system are both conserved. Both components of the total momentum are zero before the spring releases, so they must be zero after the spring releases. Hence

$$P_x = 0 = m_A v_{A2x} + m_B v_{B2x} + m_C v_{C2x}$$

$$P_y = 0 = m_A v_{A2y} + m_B v_{B2y} + m_C v_{C2y}$$

We are given that $m_A = m_B = m_C = 0.20$ kg, $v_{A2x} = -3.0$ m/s, $v_{A2y} = 0$, $v_{B2x} = 0$, and $v_{B2y} = -4.0$ m/s. You can solve the above equations to show that $v_{C2x} = +3.0$ m/s and $v_{C2y} = +4.0$ m/s. Hence the velocity of piece C has magnitude $\sqrt{(3.0\text{ m/s})^2 + (4.0\text{ m/s})^2} = 5.0$ m/s and is at an angle from the positive x-axis of $\arctan[(4.0\text{ m/s})/(3.0\text{ m/s})] = 53°$.

Section 8.3 The kinetic energy before the collision is $K_1 = \frac{1}{2}m_A v_{A1}^2 = \frac{1}{2}(20\text{ kg})(2.0\text{ m/s})^2 = 40$ J. After the collision, the kinetic energy is $K_2 = \frac{1}{2}m_A v_{A2}^2 + \frac{1}{2}m_B v_{B2}^2 = \frac{1}{2}(20\text{ kg})(1.0\text{ m/s})^2 + \frac{1}{2}(12\text{ kg}) \times (2.1\text{ m/s})^2 = 36$ J, which is less than K_1. Because kinetic energy was lost, the collision was inelastic. It was not, however, *totally* inelastic, because the two robots did not stick together.

Section 8.4 After the collision, the speed of the neutron is $(m_A - m_B)/(m_A + m_B) = (18\text{ u} - 1.0\text{ u})/(18\text{ u} + 1.0\text{ u}) = (\frac{17}{19})$ of its initial speed, and its kinetic energy is $(\frac{17}{19})^2 = 0.80$ of the initial value. Hence a water molecule is a poorer moderator than a carbon atom.

Section 8.5 Let the subscripts 1 and 2 represent the earth and the moon, respectively, and take the origin to be at the center of the earth. Then, using data from Appendix F,

$$\begin{aligned} x_{cm} &= (m_1x_1 + m_2x_2)/(m_1 + m_2) \\ &= \frac{\left[\begin{array}{c}(5.97\times10^{24}\text{ kg})(0) + (7.35\times10^{22}\text{ kg})\\ \times(3.84\times10^{8}\text{ m})\end{array}\right]}{(5.97\times10^{24}\text{ kg} + 7.35\times10^{22}\text{ kg})} \\ &= 4.67\times10^{6}\text{ m} = 4670\text{ km}\end{aligned}$$

The radius of the earth is 6.38×10^6 m $= 6380$ km. Hence the center of mass of the earth–moon system is actually within the earth's interior, 1710 km below the surface.

Section 8.6 From Eqs. (8.37) and (8.38), the thrust F is equal to $m\,dv/dt$, where m is the rocket's mass and dv/dt is its acceleration. Because m decreases with time, if the thrust F is constant, then the acceleration must increase with time (the same force acts on a smaller mass); if the acceleration dv/dt is constant, then the thrust must decrease with time (a smaller force is all that's needed to accelerate a smaller mass).

Discussion Questions

Q8.1 In splitting logs with a hammer and wedge, is a heavy hammer more effective than a lighter hammer? Why?

Q8.2 Suppose you catch a baseball and then someone invites you to catch a bowling ball with either the same momentum or with the same kinetic energy as the baseball. Which would you choose? Explain.

Q8.3 When rain falls from the sky, what happens to its momentum as it hits the ground? Is your answer also valid for Newton's famous apple?

Q8.4 A car has the same kinetic energy when it is traveling south at 30 m/s as when it is traveling northwest at 30 m/s. Is the momentum of the car the same in both cases? Explain.

Q8.5 A truck is accelerating as it speeds down the highway. One inertial frame of reference is attached to the ground with its origin at a fence post. A second frame of reference is attached to a police car that is traveling down the highway at constant velocity. Is the momentum of the truck the same in these two reference frames? Explain. Is the rate of change of the truck's momentum the same in these two frames? Explain.

Q8.6 When a large, heavy truck collides with a passenger car, the occupants of the car are more likely to be hurt than the truck driver. Why?

Q8.7 A woman holding a large rock stands on a frictionless, horizontal sheet of ice. She throws the rock with speed v_0 at an angle α above the horizontal. Consider the system consisting of the woman plus the rock. Is the momentum of the system conserved? Why or why not? Is any component of the momentum of the system conserved? Again, why or why not?

Q8.8 In Example 8.7 (Section 8.3), where the two gliders in Fig. 8.11a stick together after the collision, the collision is inelastic since $K_2 < K_1$. In Example 8.5 (Section 8.2), is the collision inelastic? Explain.

Q8.9 In a completely inelastic collision between two objects, where the objects stick together after the collision, is it possible for the final kinetic energy of the system to be zero? If so, give an example in which this would occur. If the final kinetic energy is zero, what must the initial momentum of the system be? Is the initial kinetic energy of the system zero? Explain.

Q8.10 Since for a particle the kinetic energy is given by $K = \frac{1}{2}mv^2$ and the momentum by $\vec{p} = m\vec{v}$, it is easy to show that $K = p^2/2m$. How, then, is it possible to have an event during which the total momentum of the system is constant, but the total kinetic energy changes?

Q8.11 In each of Examples 8.10, 8.11, 8.12, and 8.13 (Section 8.4), verify that the relative velocity vector of the two bodies has the same magnitude before and after the collision. In each case what happens to the *direction* of the relative velocity vector?
Q8.12 A glass dropped on the floor is more likely to break if the floor is concrete than if it is wood. Why? (Refer to Fig. 8.5.)
Q8.13 In Fig. 8.22 the kinetic energy of the spacecraft is larger after its interaction with Saturn than before. From where does the extra energy come? Describe the event in terms of conservation of energy.
Q8.14 A machine gun is fired at a steel plate. Is the average force on the plate from the bullet impact greater if the bullets bounce off or if they are squashed and stick to the plate? Explain.
Q8.15 A net force of 4 N acts on an object initially at rest for 0.25 s and gives it a final speed of 5 m/s. How could a net force of 2 N produce the same final speed?
Q8.16 A net force with x-component $\sum F_x$ acts on an object from time t_1 to time t_2. The x-component of the momentum of the object is the same at t_1 as it is at t_2, but $\sum F_x$ is not zero at all times between t_1 and t_2. What can you say about the graph of $\sum F_x$ versus t?
Q8.17 A tennis player hits a tennis ball with a racket. Consider the system made up of the ball and the racket. Is the total momentum of the system the same just before and just after the hit? Is the total momentum just after the hit the same as two seconds later, when the ball is in midair at the high point of its trajectory? Explain any differences between the two cases.
Q8.18 In Example 8.4 (Section 8.2), consider the system consisting of the rifle plus the bullet. What is the speed of the system's center of mass after the rifle is fired? Explain.
Q8.19 An egg is released from rest from the roof of a building and falls to the ground. As the egg falls, what happens to the momentum of the system of the egg plus the earth?
Q8.20 A woman stands in the middle of a perfectly smooth, frictionless, frozen lake. She can set herself in motion by throwing things, but suppose she has nothing to throw. Can she propel herself to shore *without* throwing anything?
Q8.21 In a zero-gravity environment, can a rocket-propelled spaceship ever attain a speed greater than the relative speed with which the burnt fuel is exhausted?
Q8.22 When an object breaks into two pieces (explosion, radioactive decay, recoil, etc.), the lighter fragment gets more kinetic energy than the heavier one. This is a consequence of momentum conservation, but can you also explain it using Newton's laws of motion?

Exercises

Section 8.1 Momentum and Impulse

8.1 a) What is the magnitude of the momentum of a 10,000-kg truck whose speed is 12.0 m/s? b) What speed would a 2,000-kg sport utility vehicle have to attain in order to have i) the same momentum? ii) the same kinetic energy?
8.2 In Conceptual Example 8.1 (Section 8.1), show that the iceboat with mass $2m$ has $\sqrt{2}$ times as much momentum at the finish line as does the iceboat with mass m.
8.3 a) Show that the kinetic energy K and the momentum magnitude p of a particle with mass m are related by $K = p^2/2m$. b) A 0.040-kg cardinal (*Richmondena cardinalis*) and a 0.145-kg baseball have the same kinetic energy. Which has the greater magnitude of momentum? What is the ratio of the cardinal's magnitude of momentum to the baseball's? c) A 700-N man and a 450-N woman have the same momentum. Who has the greater kinetic energy? What is the ratio of the man's kinetic energy to that of the woman?
8.4 A soccer ball with mass 0.420 kg is traveling with a speed of 4.50 m/s at an angle of 20.0° counterclockwise from the $+x$-axis (Fig. 8.30). What are the ball's x- and y-components of momentum?

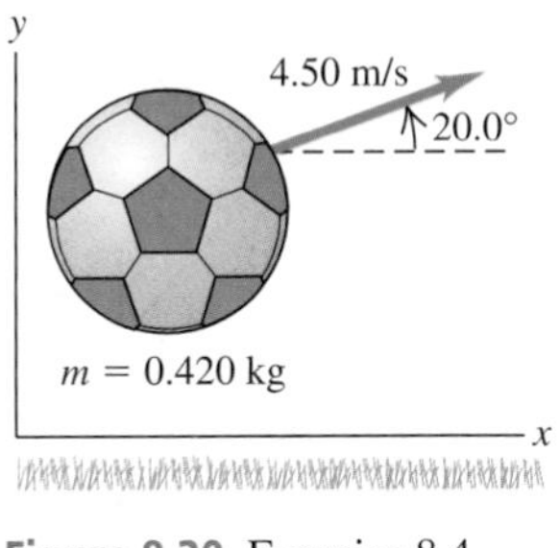

Figure 8.30 Exercise 8.4.

8.5 A baseball with mass 0.145 kg is moving in the $+y$-direction with a speed of 1.30 m/s, and a tennis ball with mass 0.0570 kg is moving in the $-y$-direction with a speed of 7.80 m/s. What are the magnitude and direction of the total momentum of the system consisting of the two balls?
8.6 A golf ball with mass 0.045 kg is moving in the $+x$-direction with a speed of 9.00 m/s, and a baseball with mass 0.145 kg is moving in the $-y$-direction with a speed of 7.00 m/s. What are the magnitude and direction of the total momentum of the system that consists of the two balls?
8.7 Force of a Golf Swing. A 0.0450-kg golf ball initially at rest is given a speed of 25.0 m/s when a club strikes. If the club and ball are in contact for 2.00 ms, what average force acts on the ball? Is the effect of the ball's weight during the time of contact significant? Why or why not?
8.8 Force of a Baseball Swing. A baseball has mass 0.145 kg. a) If the velocity of a pitched ball has a magnitude of 45.0 m/s and the batted ball's velocity is 55.0 m/s in the opposite direction, find the magnitude of the change in momentum of the ball and of the impulse applied to it by the bat. b) If the ball remains in contact with the bat for 2.00 ms, find the magnitude of the average force applied by the bat.
8.9 A 0.160-kg hockey puck is moving on an icy, frictionless, horizontal surface. At $t = 0$ the puck is moving to the right at 3.00 m/s. a) Calculate the velocity of the puck (magnitude and direction) after a force of 25.0 N directed to the right has been applied for 0.050 s. b) If instead, a force of 12.0 N directed to the left is applied from $t = 0$ to $t = 0.050$ s, what is the final velocity of the puck?
8.10 An engine of the orbital maneuvering system (OMS) on a space shuttle exerts a force of $(26{,}700\ \text{N})\hat{\jmath}$ for 3.90 s, exhausting a negligible mass of fuel relative to the 95,000-kg mass of the shuttle. a) What is the impulse of the force for this 3.90 s? b) What is the shuttle's change in momentum from this impulse? c) What is the shuttle's change in velocity from this impulse? d) Why can't we find the resulting change in kinetic energy of the shuttle?
8.11 Coach Johnson's bat exerts a horizontal force on a 0.145-kg baseball of $\vec{F} = [(1.60 \times 10^7\ \text{N/s})t - (6.00 \times 10^9\ \text{N/s}^2)t^2]\hat{\imath}$

between $t = 0$ and $t = 2.50$ ms. At $t = 0$ the baseball's velocity is $-(40.0\hat{\imath} + 5.0\hat{\jmath})$ m/s. a) Calculate the impulse exerted by the bat on the ball during the 2.50 ms that they are in contact. b) Calculate the impulse exerted by gravity on the ball during this time interval. c) Calculate the average force exerted by the bat on the ball during this time interval. d) Calculate the momentum and the velocity of the baseball at $t = 2.50$ ms.

8.12 A bat strikes a 0.145-kg baseball. Just before impact, the ball is traveling horizontally to the right at 50.0 m/s, and it leaves the bat traveling to the left at an angle of 30° above horizontal with a speed of 65.0 m/s. If the ball and bat are in contact for 1.75 ms, find the horizontal and vertical components of the average force on the ball.

8.13 A net force $\Sigma F_x(t) = A + Bt^2$ in the $+x$-direction is applied to a girl on roller skates. The girl has mass m. The force starts at $t_1 = 0$ and continues until $t = t_2$. a) What is the impulse J_x of the force? b) If the girl is initially at rest, what is her speed at time t_2?

Section 8.2 Conservation of Momentum

8.14 Frustrated by having the goalie block his shots, a 75.0-kg hockey player standing on ice throws a 0.160-kg puck horizontally at the net with a speed of 20.0 m/s. With what speed and in what direction will the hockey player begin to move if there is no friction between his feet and the ice?

8.15 You are standing on a sheet of ice that covers the football stadium parking lot in Buffalo; there is negligible friction between your feet and the ice. A friend throws you a 0.400-kg ball that is traveling horizontally at 10.0 m/s. Your mass is 70.0 kg. a) If you catch the ball, with what speed do you and the ball move afterwards? b) If the ball hits you and bounces off your chest, so afterwards it is moving horizontally at 8.0 m/s in the opposite direction, what is your speed after the collision?

8.16 On a frictionless, horizontal air table, puck A (with mass 0.250 kg) is moving toward puck B (with mass 0.350 kg), that is initially at rest. After the collision, puck A has a velocity of 0.120 m/s to the left, and puck B has velocity 0.650 m/s to the right. a) What was the speed of puck A before the collision? b) Calculate the change in the total kinetic energy of the system that occurs during the collision.

8.17 Energy Change During a Hip Check. Ice-hockey star Wayne Gretzky is skating at 13.0 m/s toward a defender, who in turn is skating at 5.00 m/s toward Gretzky (Fig. 8.31). Gretzky's weight is 756 N; that of the defender is 900 N. Immediately after the collision, Gretzky is moving at 1.50 m/s in his original direction. You can ignore external horizontal forces applied by the ice to the skaters during the collision. a) What is the velocity of the defender immediately after the collision? b) Calculate the change in the combined kinetic energy of the two players.

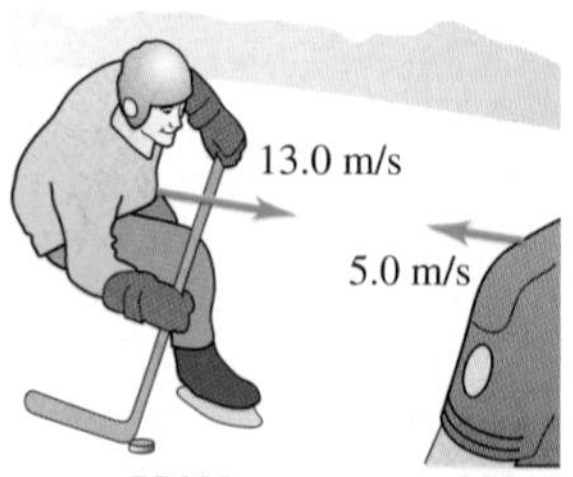

Figure 8.31 Exercise 8.17.

8.18 The expanding gases that leave the muzzle of a rifle also contribute to the recoil. A .30 caliber bullet has mass 0.00720 kg and a speed of 601 m/s relative to the muzzle when fired from a rifle that has mass 2.80 kg. The loosely held rifle recoils at a speed of 1.85 m/s relative to the earth. Find the momentum of the propellant gases in a coordinate system attached to the earth as they leave the muzzle of the rifle.

8.19 Block A in Fig. 8.32 has mass 1.00 kg, and block B has mass 3.00 kg. The blocks are forced together, compressing a spring S between them; then the system is released from rest on a level, frictionless surface. The spring, that has negligible mass, is not fastened to either block and drops to the surface after it has expanded. Block B acquires a speed of 1.20 m/s. a) What is the final speed of block A? b) How much potential energy was stored in the compressed spring?

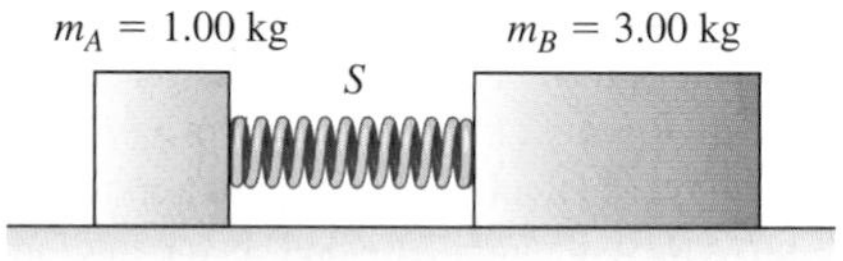

Figure 8.32 Exercise 8.19.

8.20 One of James Bond's adversaries is standing on a frozen lake; there is no friction between his feet and the ice. He throws his steel-lined hat with a velocity of 22.0 m/s at 36.9° above the horizontal, hoping to hit James. If his mass is 120 kg and that of his hat is 4.50 kg, what is the magnitude of his horizontal recoil velocity?

8.21 A ceramic penguin on top of your television set suddenly bursts apart into two pieces. One piece, mass m_A, travels off to the left with speed v_A. The other piece, mass m_B, travels off to the right with speed v_B. a) Use conservation of momentum to solve for v_B in terms of m_A, m_B, and v_A. b) Use the results of part (a) to show that $K_A/K_B = m_B/m_A$, where K_A and K_B are the kinetic energies of the two pieces.

8.22 The nucleus of ^{214}Po decays radioactively by emitting an alpha particle (mass 6.65×10^{-27} kg) with kinetic energy 1.23×10^{-12} J, as measured in the laboratory reference frame. Assuming that the Po was initially at rest in this frame, find the recoil velocity of the nucleus that remains after the decay.

8.23 You are standing on a large sheet of frictionless ice and are holding a large rock. In order to get off the ice, you throw the rock so it has velocity relative to the earth of 12.0 m/s at an angle of 35.0° above the horizontal. If your mass is 70.0 kg and the rock's mass is 15.0 kg, what is your speed after you throw the rock (see Discussion Question Q8.7)?

8.24 Two ice skaters, Daniel (mass 65.0 kg) and Rebecca (mass 45.0 kg), are practicing. Daniel stops to tie his shoelace and, while at rest, is struck by Rebecca, who is moving at 13.0 m/s before she collides with him. After the collision, Rebecca has a velocity of magnitude 8.00 m/s at an angle of 53.1° from her initial direction. Both skaters move on the frictionless, horizontal surface of the rink. a) What are the magnitude and direction of Daniel's velocity after the collision? b) What is the change in total kinetic energy of the two skaters as a result of the collision?

8.25 Ken and Kim are skating together on a rink at 3.00 m/s. Ken keeps asking Kim how much she weighs. Annoyed, Kim pushes away from Ken so that she speeds up to 4.00 m/s and he slows down to 2.25 m/s in the same direction. Friction, in the physics sense, is negligible in this drama. If Ken weighs 700 N, what does Kim weigh?

8.26 Changing Mass. An open-topped freight car with mass 24,000 kg is coasting without friction along a level track. It is raining very hard, and the rain is falling vertically downward. Originally, the car is empty and moving with a speed of 4.00 m/s. What is the speed of the car after it has collected 3000 kg of rainwater?

8.27 A hockey puck B rests on a smooth ice surface and is struck by a second puck A, which was originally traveling at 40.0 m/s and which is deflected 30.0° from its original direction (Fig. 8.33). Puck B acquires a velocity at a 45.0° angle to the original direction of A. The pucks have the same mass. a) Compute the speed of each puck after the collision. b) What fraction of the original kinetic energy of puck A dissipates during the collision?

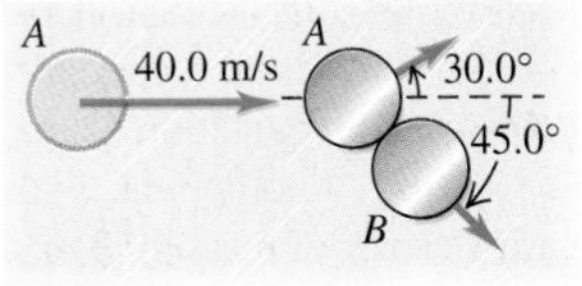

Figure 8.33 Exercise 8.27.

Section 8.3 Inelastic Collisions

8.28 On a greasy, essentially frictionless lunch counter, a 0.500-kg submarine sandwich, moving 3.00 m/s to the left, collides with an 0.250-kg grilled cheese sandwich moving 1.20 m/s to the right. a) If the two sandwiches stick together, what is the final velocity? b) How much mechanical energy dissipates in the collision?

8.29 Your 1050-kg sports car, parked on a hill without the parking brake set, has rolled to the bottom of the hill and is moving westbound on a level road at a speed of 15.0 m/s. A truck driver driving east on the same road sees the sports car coming, and decides to stop the car by running his 6320-kg truck into it. The two vehicles remain locked together after the collision. a) If the truck is moving at 10.0 m/s when it hits your sports car head-on, what is the velocity (magnitude and direction) of the two vehicles just after the collision? b) At what speed should the truck be moving so that the truck and car are both stopped in the collision? c) Find the change in kinetic energy of the system of two vehicles for the situation of part (a) and the situation of part (b). For which situation is the change in kinetic energy greater in magnitude?

8.30 On a very muddy football field, a 110-kg linebacker tackles an 85-kg halfback. Immediately before the collision, the linebacker is slipping with a velocity of 8.8 m/s north and the halfback is sliding with a velocity of 7.2 m/s east. What is the velocity (magnitude and direction) at which the two players move together immediately after the collision?

8.31 Two frisky grasshoppers collide in midair at the top of their respective trajectories and grab onto each other, holding tight thereafter. One is a robust 250-g beast initially moving south at 20.0 cm/s, while the other is a svelte 150-g creature initially moving north at 60.0 cm/s. Calculate the decrease in kinetic energy that results from the collision. What happens to the "lost" kinetic energy?

8.32 Two cars, one a compact with mass 1200 kg and the other a large gas-guzzler with mass 3000 kg, collide head-on at typical freeway speeds. a) Which car has a greater magnitude of momentum change? Which car has a greater velocity change? Calculate the change in the velocity of the small car relative to that of the large car. b) Which car's occupants would you expect to sustain greater injuries? Explain.

8.33 In Dallas, the morning after a winter ice storm, a 1400-kg automobile going west on Chestnut Street at 35.0 km/h collides with a 2800-kg truck going south *across* Chestnut Street at 50.0 km/h. If they become coupled on collision, what are the magnitude and direction of their velocity after colliding? You can ignore forces between the vehicles and the icy road.

8.34 At the intersection of Texas Avenue and University Drive, a blue, subcompact car with mass 950 kg traveling east on University collides with a maroon pickup truck with mass 1900 kg that is traveling north on Texas and ran a red light (Fig. 8.34). The two vehicles stick together as a result of the collision and, after the collision, the wreckage is sliding at 16.0 m/s in the direction 24.0° east of north. Calculate the speed of each vehicle before the collision. The collision occurs during a heavy rainstorm; you can ignore friction forces between the vehicles and the wet road.

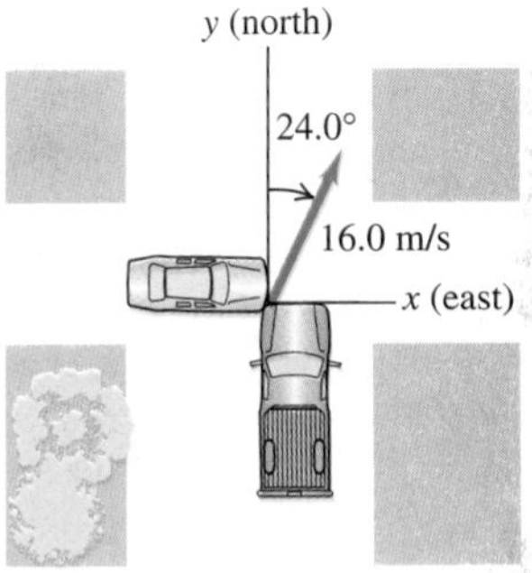

Figure 8.34 Exercise 8.34.

8.35 A 5.00-g bullet is fired horizontally into a 1.20-kg wooden block resting on a horizontal surface. The coefficient of kinetic friction between block and surface is 0.20. The bullet remains embedded in the block, which is observed to slide 0.230 m along the surface before stopping. What was the initial speed of the bullet?

8.36 A Ballistic Pendulum. A 12.0-g rifle bullet is fired with a speed of 380 m/s into a ballistic pendulum with with mass 6.00 kg, suspended from a cord 70.0 cm long (see Example 8.8 in Section 8.3). Compute a) the vertical height through which the pendulum rises; b) the initial kinetic energy of the bullet; c) the kinetic energy of the bullet and pendulum immediately after the bullet becomes embedded in the pendulum.

8.37 You and your friends are doing physics experiments on a frozen pond that serves as a frictionless, horizontal surface. Sam, with mass 80.0 kg, is given a push and slides eastward. Abigail, with mass 50.0 kg, is sent sliding northward. They collide and, after the collision, Sam is moving at 37.0° north of east with a speed of 6.00 m/s and Abigail is moving at 23.0° south of east with a speed of 9.00 m/s. a) What was the speed of each person before the collision? b) By how much did the total kinetic energy of the two people decrease during the collision?

Section 8.4 Elastic Collisions

8.38 Blocks A (mass 2.00 kg) and B (mass 10.00 kg) move on a frictionless, horizontal surface. Initially, block B is at rest and block A is moving toward it at 2.00 m/s. The blocks are equipped with ideal spring bumpers, as in Example 8.10. The collision is head-on, so all motion before and after the collision is along a straight line. a) Find the maximum energy stored in the spring bumpers and the velocity of each block at that time. b) Find the velocity of each block after they have moved apart.

8.39 A 0.150-kg glider is moving to the right on a frictionless, horizontal air track with a speed of 0.80 m/s. It has a head-on collision with a 0.300-kg glider that is moving to the left with a speed of 2.20 m/s. Find the final velocity (magnitude and direction) of each glider if the collision is elastic.

8.40 A 10.0-g marble slides to the left with a velocity of magnitude 0.400 m/s on the frictionless, horizontal surface of an icy, New York sidewalk and has a head-on elastic collision with a larger 30.0-g marble sliding to the right with a velocity of magnitude 0.200 m/s (Fig. 8.35). a) Find the velocity of each marble (magnitude and direction) after the collision. (Since the collision is head-on, all the motion is along a line.) b) Calculate the *change in momentum* (that is, the momentum after the collision minus the momentum before the collision) for each marble. Compare the values you get for each marble. c) Calculate the *change in kinetic energy* (that is, the kinetic energy after the collision minus the kinetic energy before the collision) for each marble. Compare the values you get for each marble.

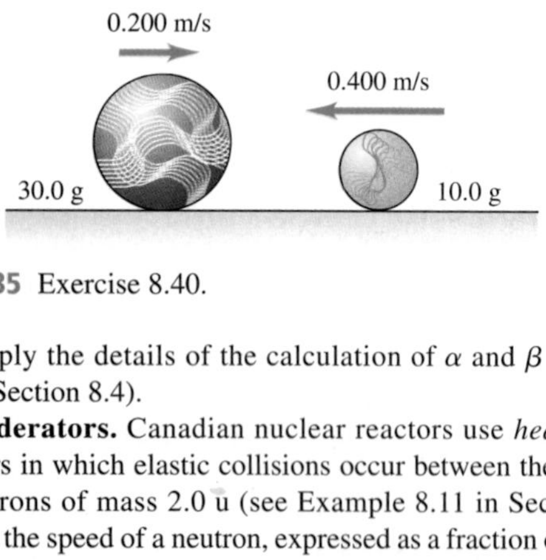

Figure 8.35 Exercise 8.40.

8.41 Supply the details of the calculation of α and β in Example 8.13 (Section 8.4).

8.42 Moderators. Canadian nuclear reactors use *heavy water* moderators in which elastic collisions occur between the neutrons and deuterons of mass 2.0 u (see Example 8.11 in Section 8.4). a) What is the speed of a neutron, expressed as a fraction of its original speed, after a head-on, elastic collision with a deuteron which is initially at rest? b) What is its kinetic energy, expressed as a fraction of its original kinetic energy? c) How many such successive collisions will reduce the speed of a neutron to 1/59,000 of its original value?

8.43 You are at the controls of a particle accelerator, sending a beam of 1.50×10^7 m/s protons (mass m) at a gas target of an unknown element. Your detector tells you that some protons bounce straight back after a collision with one of the nuclei of the unknown element. All such protons rebound with a speed of 1.20×10^7 m/s. Assume that the initial speed of the target nucleus is negligible and that the collision is elastic. a) Find the mass of one nucleus of the unknown element. Express your answer in terms of the proton mass m. b) What is the speed of the unknown nucleus immediately after such a collision?

Section 8.5 Center of Mass

8.44 Three odd-shaped blocks of chocolate have the following masses and center-of-mass coordinates: (1) 0.300 kg, (0.200 m, 0.300 m); (2) 0.400 kg, (0.100 m, −0.400 m); (3) 0.200 kg, (−0.300 m, 0.600 m). Find the coordinates of the center of mass of the system of three chocolate blocks.

8.45 Find the position of the center of mass of the system of the sun and Jupiter. (Since Jupiter is more massive than the rest of the planets put together, this is essentially the position of the center of mass of the solar system.) Does the center of mass lie inside or outside the sun? Use the data in Appendix F.

8.46 A 1200-kg station wagon is moving along a straight highway at 12.0 m/s. Another car, with mass 1800 kg and speed 20.0 m/s, has its center of mass 40.0 m ahead of the center of mass of the station wagon (Fig. 8.36). a) Find the position of the center of mass of the system consisting of the two automobiles. b) Find the magnitude of the total momentum of the system from the above data. c) Find the speed of the center of mass of the system. d) Find the total momentum of the system, using the speed of the center of mass. Compare your result with that of part (b).

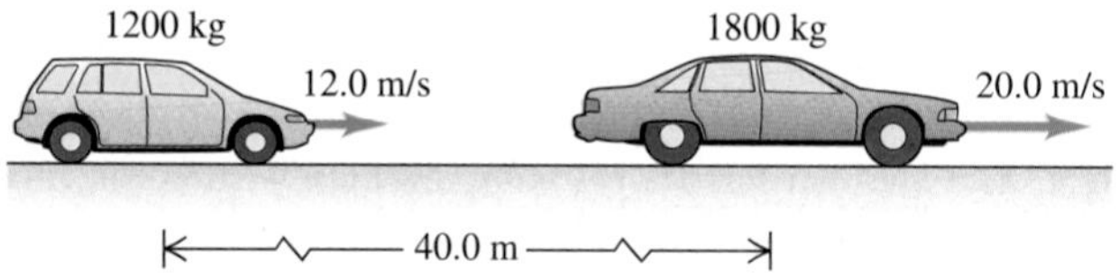

Figure 8.36 Exercise 8.46.

8.47 At one instant, the center of mass of a system of two particles is located on the x-axis at $x = 2.0$ m and has a velocity of $(5.0 \text{ m/s})\hat{\imath}$. One of the particles is at the origin. The other particle has a mass of 0.10 kg and is at rest on the x-axis at $x = 8.0$ m. a) What is the mass of the particle at the origin? b) Calculate the total momentum of this system. c) What is the velocity of the particle at the origin?

8.48 In Example 8.15 (Section 8.5) Ramon pulls on the rope to give himself a speed of 0.70 m/s. What is the speed of James?

8.49 A system consists of two particles. At $t = 0$ one particle is at the origin; the other, which has a mass of 0.50 kg, is on the y-axis at $y = 6.0$ m. At $t = 0$ the center of mass of the system is on the y-axis at $y = 2.4$ m. The velocity of the center of mass is given by $(0.75 \text{ m/s}^3)t^2\hat{\imath}$. a) Find the total mass of the system. b) Find the acceleration of the center of mass at any time t. c) Find the net external force acting on the system at $t = 3.0$ s.

8.50 A radio-controlled model airplane has a momentum given by $[(-0.75 \text{ kg}\cdot\text{m/s}^3)t^2 + (3.0 \text{ kg}\cdot\text{m/s})]\hat{\imath} + (0.25 \text{ kg}\cdot\text{m/s}^2)t\hat{\jmath}$. What are the x-, y-, and z-components of the net force on the airplane?

*Section 8.6 Rocket Propulsion

***8.51** A small rocket burns 0.0500 kg of fuel per second, ejecting it as a gas with a velocity relative to the rocket of magnitude 1600 m/s. a) What is the thrust of the rocket? b) Would the rocket operate in outer space where there is no atmosphere? If so, how would you steer it? Could you brake it?

***8.52** A 70-kg astronaut floating in space in a 110-kg MMU (manned maneuvering unit) experiences an acceleration of 0.029 m/s^2 when he fires one of the MMU's thrusters. a) If the speed of the escaping N_2 gas relative to the astronaut is 490 m/s, how much gas is used by the thruster in 5.0 s? b) What is the thrust of the thruster?

***8.53** A rocket is fired in deep space, where gravity is negligible. If the rocket has an initial mass of 6000 kg and ejects gas at a relative velocity of magnitude 2000 m/s, how much gas must it eject in the first second to have an initial acceleration of 25.0 m/s^2?

***8.54** A rocket is fired in deep space, where gravity is negligible. In the first second it ejects $\frac{1}{160}$ of its mass as exhaust gas and has an acceleration of 15.0 m/s^2. What is the speed of the exhaust gas relative to the rocket?

***8.55** A C 6-5 model rocket engine has an impulse of 10.0 N · s for 1.70 s, while burning 0.0125 kg of propellant. It has a maximum thrust of 13.3 N. The initial mass of the engine plus propellant is 0.0258 kg. a) What fraction of the maximum thrust is the average thrust? b) Calculate the relative speed of the exhaust gases, assuming it is constant. c) Assuming that the relative speed of the exhaust gases is constant, find the final speed of the engine if it was attached to a very light frame and fired from rest in gravity-free outer space.

***8.56** A single-stage rocket is fired from rest from a deep-space platform, where gravity is negligible. If the rocket burns its fuel in a time of 50.0 s and the relative speed of the exhaust gas is $v_{ex} = 2100$ m/s, what must the mass ratio m_0/m be for a final speed v of 8.00 km/s (about equal to the orbital speed of an earth satellite)?

***8.57** Obviously, we can make rockets to go very fast, but what is a reasonable top speed? Assume that a rocket is fired from rest at a space station in deep space, where gravity is negligible. a) If the rocket ejects gas at a relative speed of 2000 m/s and you want the rocket's speed eventually to be $1.00 \times 10^{-3}\,c$, where c is the speed of light, what fraction of the initial mass of the rocket and fuel is *not* fuel? b) What is this fraction if the final speed is to be 3000 m/s?

Problems

8.58 A steel ball with mass 40.0 g is dropped from a height of 2.00 m onto a horizontal steel slab. The ball rebounds to a height of 1.60 m. a) Calculate the impulse delivered to the ball during impact. b) If the ball is in contact with the slab for 2.00 ms, find the average force on the ball during impact.

8.59 The net force acting on a 2.00-kg discus while it is being thrown is $(\alpha t^2)\hat{\imath} + (\beta + \gamma t)\hat{\jmath}$, where $\alpha = 25.0$ N/s^2, $\beta = 30.0$ N, and $\gamma = 5.0$ N/s. If the discus was originally at rest, what is its velocity after the net force has acted for 0.500 s? Express your answer in terms of the unit vectors $\hat{\imath}$ and $\hat{\jmath}$.

8.60 Just before it is struck by a racket, a tennis ball weighing 0.560 N has a velocity of $(20.0\text{ m/s})\hat{\imath} - (4.0\text{ m/s})\hat{\jmath}$. During the 3.00 ms that the racket and ball are in contact, the net force on the ball is constant and equal to $-(380\text{ N})\hat{\imath} + (110\text{ N})\hat{\jmath}$. a) What are the x- and y-components of the impulse of the net force applied to the ball? b) What are the x- and y-components of the final velocity of the ball?

8.61 Three coupled railroad cars roll along and couple with a fourth car, which is initially at rest. These four roll along and couple with a fifth car initially at rest. This process continues until the speed of the final collection of railroad cars is one-fifth the speed of the initial three railroad cars. All the cars are identical. Ignoring friction, how many cars are in the final collection?

8.62 A 1500-kg, blue convertible is traveling south, and a 2000-kg red sport utility vehicle is traveling west. If the total momentum of the system consisting of the two cars is 8000 kg · m/s directed at 60.0° west of south, what is the speed of each car?

8.63 Three identical pucks on a horizontal air table have repelling magnets. They are held together and then released simultaneously. Each has the same speed at any instant. One puck moves due west. What is the direction of the velocity of each of the other two pucks?

8.64 Spheres A (mass 0.020 kg), B (mass 0.030 kg), and C (mass 0.050 kg), are each approaching the origin as they slide on a frictionless air table (Fig. 8.37). The initial velocities of A and B are given in the figure. All three spheres arrive at the origin at the same time and stick together. a) What must the x- and y-components of the initial velocity of C be if all three objects are to end up moving at 0.50 m/s in the $+x$-direction after the collision? b) If C has the velocity found in part (a), what is the change in the kinetic energy of the system of three spheres as a result of the collision?

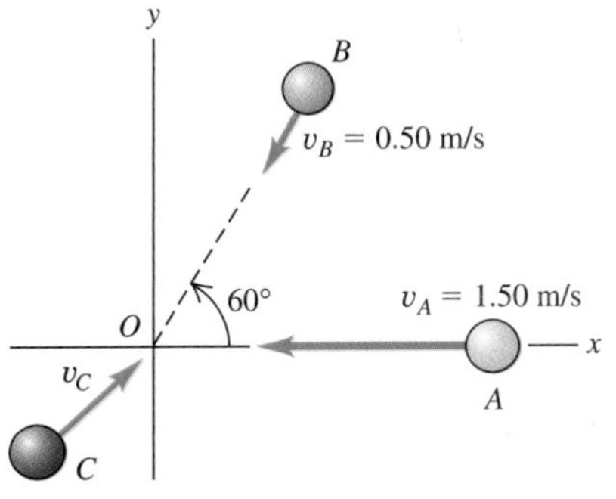

Figure 8.37 Problem 8.64.

8.65 A railroad handcar is moving along straight, frictionless tracks with negligible air resistance. In the following cases, the car initially has a total mass (car and contents) 200 kg and is traveling east with a velocity of magnitude 5.00 m/s. Find the *final velocity* of the car in each case, assuming that the handcar does not leave the tracks. a) A 25.0-kg mass is thrown sideways out of the car with a velocity of magnitude 2.00 m/s relative to the car's initial velocity.

b) A 25.0-kg mass is thrown backward out of the car with a velocity of 5.00 m/s relative to the initial motion of the car. c) A 25.0-kg mass is thrown into the car with a velocity of 6.00 m/s relative to the ground and opposite in direction to the initial velocity of the car.

8.66 Changing Mass. A railroad hopper car filled with sand is rolling with an initial speed of 15.0 m/s on straight, horizontal tracks. You can ignore frictional forces on the railroad car. The total mass of the car plus sand is 85,000 kg. The hopper door is not fully closed so sand leaks out the bottom. After 20 minutes, 13,000 kg of sand has leaked out. Then what is the speed of the railroad car? (Compare your analysis to that used to solve Exercise 8.26.)

8.67 At a classic auto show, a 840-kg 1955 Nash Metropolitan motors by at 9.0 m/s, followed by a 1620-kg 1957 Packard Clipper purring past at 5.0 m/s. a) Which car has the greater kinetic energy? What is the ratio of the kinetic energy of the Nash to that of the Packard? b) Which car has the greater magnitude of momentum? What is the ratio of the magnitude of momentum of the Nash to that of the Packard? c) Let F_N be the net force required to stop the Nash in time t, and let F_P be the net force required to stop the Packard in the same time. Which is larger, F_N or F_P? What is the ratio F_N/F_P of these two forces? d) Now let F_N be the net force required to stop the Nash in a distance d, and let F_P be the net force required to stop the Packard in the same distance. Which is larger, F_N or F_P? What is the ratio F_N/F_P?

8.68 A soldier on a firing range fires an 8-shot burst from an assault weapon at a full automatic rate of 1000 rounds per minute. Each bullet has a mass of 7.45 g and a speed of 293 m/s relative to the ground as it leaves the barrel of the weapon. Calculate the average recoil force exerted on the weapon during that burst.

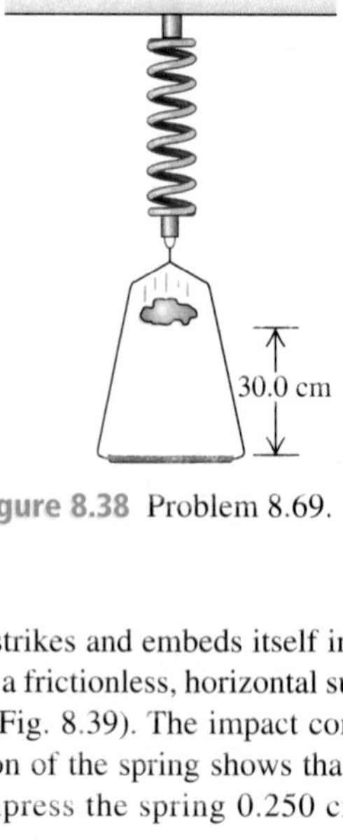

Figure 8.38 Problem 8.69.

8.69 A 0.150-kg frame, when suspended from a coil spring, stretches the spring 0.050 m. A 0.200-kg lump of putty is dropped from rest onto the frame from a height of 30.0 cm (Fig. 8.38). Find the maximum distance the frame moves downward from its initial position.

8.70 A rifle bullet with mass 8.00 g strikes and embeds itself in a block with mass 0.992 kg that rests on a frictionless, horizontal surface and is attached to a coil spring (Fig. 8.39). The impact compresses the spring 15.0 cm. Calibration of the spring shows that a force of 0.750 N is required to compress the spring 0.250 cm.

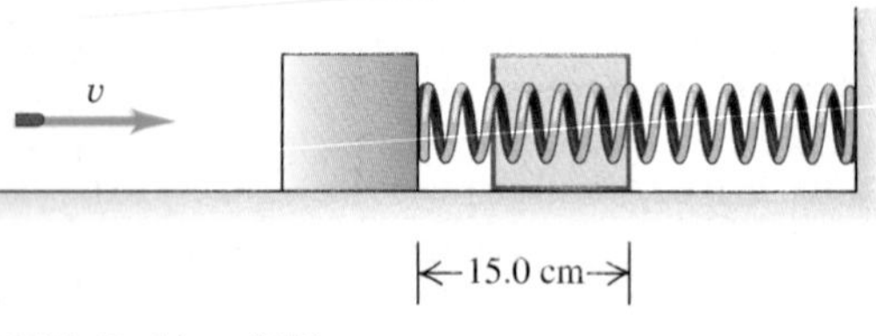

Figure 8.39 Problem 8.70.

a) Find the magnitude of the block's velocity just after impact. b) What was the initial speed of the bullet?

8.71 A Ricocheting Bullet. A 0.100-kg stone rests on a frictionless, horizontal surface. A bullet of mass 6.00 g, traveling horizontally at 350 m/s, strikes the stone and rebounds horizontally at right angles to its original direction with a speed of 250 m/s. a) Compute the magnitude and direction of the velocity of the stone after it is struck. b) Is the collision perfectly elastic?

8.72 A movie stuntman (mass 80.0 kg) stands on a window ledge 5.0 m above the floor (Fig. 8.40). Grabbing a rope attached to a chandelier, he swings down to grapple with the movie's villain (mass 70.0 kg), who is standing directly under the chandelier. (Assume that the stuntman's center of mass moves downward 5.0 m. He releases the rope just as he reaches the villain.) a) With what speed do the entwined foes start to slide across the floor? b) If the coefficient of kinetic friction of their bodies with the floor is $\mu_k = 0.250$, how far do they slide?

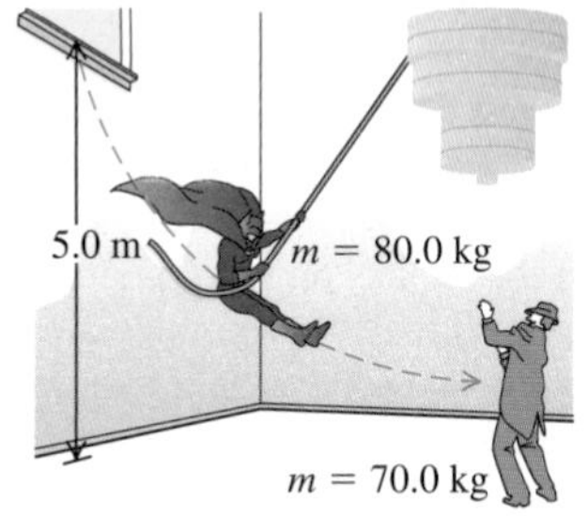

Figure 8.40 Problem 8.72.

8.73 Two identical masses are released from rest in a smooth hemispherical bowl of radius R, from the positions shown in Fig. 8.41. You can ignore friction between the masses and the surface of the bowl. If they stick together when they collide, how high above the bottom of the bowl will the masses go after colliding?

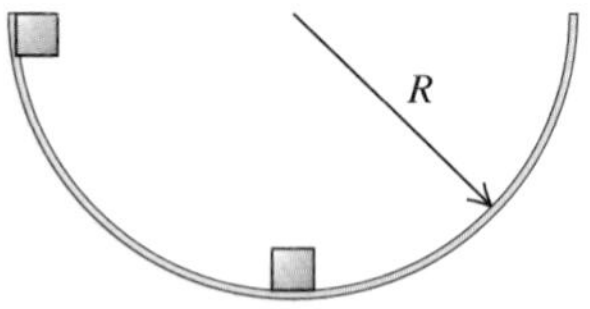

Figure 8.41 Problem 8.73.

8.74 A ball with mass M, moving horizontally at 5.00 m/s, collides elastically with a block with mass $3M$ that is initially hanging at rest from the ceiling on the end of a 50.0-cm wire. Find the maximum angle through which the block swings after it is hit.

8.75 A 20.00-kg lead sphere is hanging from a hook by a thin wire 3.50 m long, and is free to swing in a complete circle. Suddenly it is struck horizontally by a 5.00-kg steel dart that embeds itself in the lead sphere. What must be the minimum initial speed of the dart, so that the combination makes a complete circular loop after the collision?

8.76 A 8.00-kg ball, hanging from the ceiling by a wire 135 cm long, is struck in an elastic collision by a 2.00-kg ball moving with speed v_0 just before the collision. If the maximum tension that the wire can have without breaking is 1600 N, what is the largest that v_0 can be without the wire breaking?

8.77 A 4.00-g bullet, traveling horizontally with a velocity of magnitude 400 m/s, is fired into a wooden block with mass 0.800 kg, initially at rest on a level surface. The bullet passes through the block and emerges with its speed reduced to 120 m/s. The block slides a distance of 45.0 cm along the surface from its initial position. a) What is the coefficient of kinetic friction between block and surface? b) What is the decrease in kinetic energy of the bullet? c) What is the kinetic energy of the block at the instant after the bullet passes through it?

8.78 A 5.00-g bullet is shot *through* a 1.00-kg wood block suspended on a string 2.000 m long. The center of mass of the block rises a distance of 0.45 cm. Find the speed of the bullet as it emerges from the block if its initial speed is 450 m/s.

8.79 A neutron with mass m makes a head-on, elastic collision with a nucleus of mass M, which is initially at rest. a) Show that if the neutron's initial kinetic energy is K_0, the kinetic energy that it loses during the collision is $4mMK_0/(M+m)^2$. b) For what value of M does the incident neutron lose the most energy? c) When M has the value calculated in part (b), what is the speed of the neutron after the collision?

8.80 Energy Sharing in Elastic Collisions. A stationary object with mass m_B is struck head-on by an object with mass m_A that is moving initially at speed v_0. a) If the collision is elastic, what percentage of the original energy does each object have after the collision? b) What does your answer in part (a) give for the following special cases: i) $m_A = m_B$; ii) $m_A = 5m_B$. c) For what values, if any, of the mass ratio m_A/m_B is the original kinetic energy shared equally by the two objects after the collision?

8.81 In a shipping company distribution center, an open cart of mass 50.0 kg is rolling to the left at a speed of 5.00 m/s (Fig. 8.42). You can ignore friction between the cart and the floor. A 15.0-kg package slides down a chute that is inclined at 37.0° from the horizontal and leaves the end of the chute with a speed of 3.00 m/s. The package lands in the cart and they roll off together. If the lower end of the chute is a vertical distance of 4.00 m above the bottom of the cart, what is a) the speed of the package just before it lands in the cart; b) the final speed of the cart?

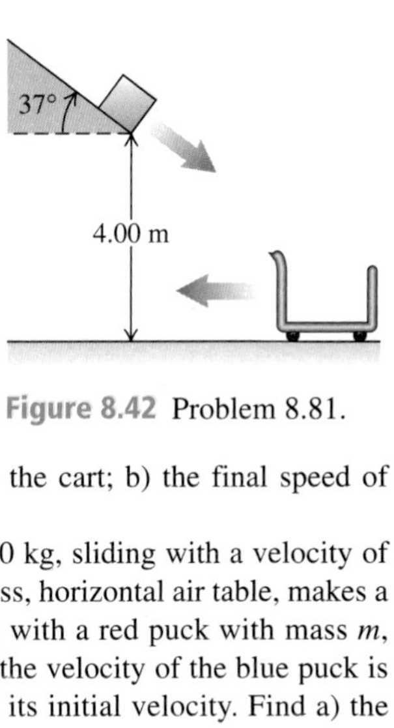

Figure 8.42 Problem 8.81.

8.82 A blue puck with mass 0.0400 kg, sliding with a velocity of magnitude 0.200 m/s on a frictionless, horizontal air table, makes a perfectly elastic, head-on collision with a red puck with mass m, initially at rest. After the collision, the velocity of the blue puck is 0.050 m/s in the same direction as its initial velocity. Find a) the velocity (magnitude and direction) of the red puck after the collision; b) the mass m of the red puck.

8.83 Two asteroids with masses m_A and m_B are moving with velocities $\vec{v}_A$ and $\vec{v}_B$ with respect to an astronomer in a space vehicle. a) Show that the total kinetic energy as measured by the astronomer is

$$K = \frac{1}{2}Mv_{\text{cm}}^2 + \frac{1}{2}(m_A v_A'^2 + m_B v_B'^2)$$

with $\vec{v}_{\text{cm}}$ and M defined as in Section 8.5, $\vec{v}_A' = \vec{v}_A - \vec{v}_{\text{cm}}$ and $\vec{v}_B' = \vec{v}_B - \vec{v}_{\text{cm}}$. In this expression the total kinetic energy of the two asteroids is the energy associated with their center of mass plus that associated with the internal motion relative to the center of mass. b) If the asteroids collide with each other, what is the *minimum* possible kinetic energy they can have after the collision, as measured by the astronomer? Explain.

8.84 Suppose you hold a small ball in contact with, and directly over, the center of a large ball. If you then drop the small ball a short time after dropping the large ball, the small ball rebounds with surprising speed. To show the extreme case, ignore air resistance and suppose the large ball makes an elastic collision with the floor, and then rebounds to make an elastic collision with the still-descending small ball. Just before the collision between the two balls, the large ball is moving upward with velocity $\vec{v}$, and the small ball has velocity $-\vec{v}$. (Do you see why?) Assume the large ball has a much greater mass than the small ball. a) What is the velocity of the small ball immediately after its collision with the large ball? b) From the answer to part (a), what is the ratio of the small ball's rebound distance to the distance it fell before the collision?

8.85 Jack and Jill are standing on a crate at rest on the frictionless, horizontal surface of a frozen pond. Jack has mass 75.0 kg, Jill has mass 45.0 kg, and the crate has mass 15.0 kg. They remember that they must fetch a pail of water, so each jumps horizontally from the top of the crate. Just after each jumps, that person is moving away from the crate with a speed of 4.00 m/s relative to the crate. a) What is the final speed of the crate if both Jack and Jill jump simultaneously and in the same direction? (*Hint:* Use an inertial coordinate system attached to the ground.) b) What is the final speed of the crate if Jack jumps first and then a few seconds later Jill jumps in the same direction? c) What is the final speed of the crate if Jill jumps first and then Jack, again in the same direction?

8.86 Energy Sharing. An object with mass m, initially at rest, explodes into two fragments, one with mass m_A and the other with mass m_B, where $m_A + m_B = m$. a) If energy Q is released in the explosion, how much kinetic energy does each fragment have immediately after the collision? b) What percentage of the total energy released does each fragment get when one fragment has four times the mass of the other?

8.87 Neutron Decay. A neutron at rest decays (breaks up) into a proton and an electron. Energy is released in the decay, and appears as kinetic energy of the proton and electron. The mass of a proton is 1836 times the mass of an electron. What fraction of the total energy released goes into the kinetic energy of the proton?

8.88 A ^{232}Th (thorium) nucleus at rest decays to a ^{228}Ra (radon) nucleus with the emission of an alpha particle. The total kinetic energy of the decay fragments is 6.54×10^{-13} J. An alpha particle has 1.76% of the mass of a ^{228}Ra nucleus. Calculate the kinetic energy of a) the recoiling ^{228}Ra nucleus; b) the alpha particle.

8.89 Antineutrino. In beta decay, a nucleus emits an electron. A ^{210}Bi (bismuth) nucleus at rest undergoes beta decay to ^{210}Po (polonium). Suppose that the emitted electron moves to the right with a momentum of 5.60×10^{-22} kg·m/s. The ^{210}Po nucleus, with mass 3.50×10^{-25} kg, recoils to the left at a speed of 1.14×10^{-3} m/s. Momentum conservation requires that a second particle, called an antineutrino, must also be emitted. Calculate the magnitude and direction of the momentum of the antineutrino that is emitted in this decay.

8.90 A proton moving with speed v_{A1} in the $+x$-direction makes an elastic, off-center collision with an identical proton originally at rest. After impact, the first proton moves with speed v_{A2} in the first quadrant at an angle α with the x-axis, and the second moves with speed v_{B2} in the fourth quadrant at an angle β with the x-axis (Fig. 8.12). a) Write the equations expressing conservation of linear momentum in the x- and y-directions. b) Square the equations from part (a) and add them. c) At this point, introduce the fact that the collision is elastic. d) Prove that $\alpha + \beta = \pi/2$. (You have shown that this equation is obeyed in any elastic, off-center collision between objects of equal mass when one object is initially at rest.)

8.91 Hockey puck B rests on a smooth ice surface and is struck by a second puck A, which has the same mass. Puck A is initially traveling at 15.0 m/s and is deflected 25.0° from its initial direction. Assume that the collision is perfectly elastic. Find the final speed of each puck and the direction of B's velocity after the collision. (*Hint:* Use the relation derived in part (d) of Problem 8.90.)

8.92 Jonathan and Jane are sitting in a sleigh that is at rest on frictionless ice. Jonathan's weight is 800 N, Jane's weight is 600 N, and that of the sleigh is 1000 N. They see a poisonous spider on the floor of the sleigh and immediately jump off. Jonathan jumps to the left with a velocity (relative to the ice) of 5.00 m/s at 30.0° above the horizontal, and Jane jumps to the right at 7.00 m/s at 36.9° above the horizontal (relative to the ice). Calculate the sleigh's horizontal velocity (magnitude and direction) after they jump out.

8.93 The objects in Fig. 8.43 are constructed of uniform wire bent into the shapes shown. Find the position of the center of mass of each.

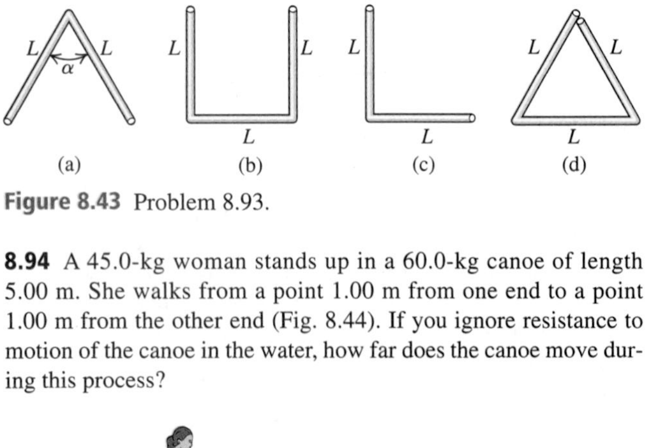

Figure 8.43 Problem 8.93.

8.94 A 45.0-kg woman stands up in a 60.0-kg canoe of length 5.00 m. She walks from a point 1.00 m from one end to a point 1.00 m from the other end (Fig. 8.44). If you ignore resistance to motion of the canoe in the water, how far does the canoe move during this process?

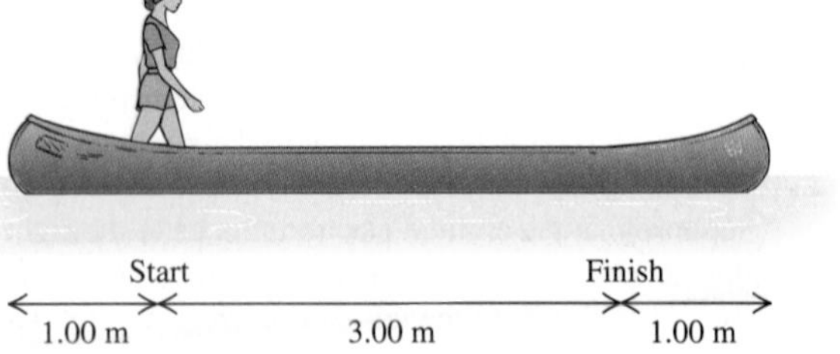

Figure 8.44 Problem 8.94.

8.95 You are standing on a concrete slab, that in turn is resting on a frozen lake. Assume that there is no friction between the slab and the ice. The slab has a weight five times your weight. If you begin walking forward at 2.00 m/s, relative to the ice, with what speed, relative to the ice, does the slab move?

8.96 A 20.0-kg projectile is fired at an angle of 60.0° above the horizontal and with a speed of 80.0 m/s. At the highest point of its trajectory the projectile explodes into two fragments with equal mass, one of which falls vertically with zero initial speed. You can ignore air resistance. a) How far from the point of firing does the other fragment strike if the terrain is level? b) How much energy is released during the explosion?

8.97 A fireworks rocket is fired vertically upward. At its maximum height of 80.0 m it explodes and breaks into two pieces, one with mass 1.40 kg and the other with mass 0.28 kg. In the explosion, 860 J of chemical energy is converted into kinetic energy of the two fragments. a) What is the speed of each fragment just after the explosion? b) It is observed that the two fragments hit the ground at the same time. What is the distance between the points on the ground where they land? Assume that the ground is level and that air resistance can be ignored.

8.98 A 12.0-kg shell is launched at an angle of 55.0° above the horizontal with an initial speed of 150 m/s. When it is at its highest point, the shell exploded into two fragments, one three times as heavy as the other. The two fragments reach the ground at the same time. Assume that air resistance can be ignored. If the heavier fragment lands back at the same point from which the shell was launched, where will the lighter fragment land and how much energy was released in the explosion?

8.99 A Nuclear Reaction. Fission, the process that supplies energy in nuclear power plants, occurs when a heavy nucleus is split into two medium-sized nuclei. One such reaction occurs when a neutron colliding with a ^{235}U (uranium) nucleus splits that nucleus into a ^{141}Ba (barium) nucleus and a ^{92}Kr (krypton) nucleus. In this reaction, two neutrons also are split off from the original ^{235}U. Before the collision, we have the arrangement in Fig. 8.45a. After the collision, we have the ^{141}Ba nucleus moving in the $+z$-direction and the ^{92}Kr nucleus in the $-z$-direction. The three neutrons are moving in the xy-plane, as shown in Fig. 8.45b. If the incoming neutron has an initial velocity of magnitude 3.0×10^3 m/s and a final velocity of magnitude 2.0×10^3 m/s in the directions shown, what are the speeds of the other two neutrons, and what can you say about the speeds of the ^{141}Ba and ^{92}Kr nuclei?

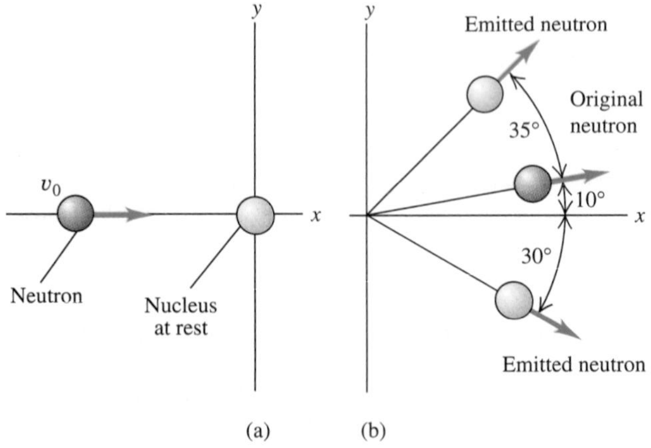

Figure 8.45 Problem 8.99.

(The mass of the ^{141}Ba nucleus is approximately 2.3×10^{-25} kg, and that of ^{92}Kr is about 1.5×10^{-25} kg.)

8.100 Center-of-Mass Coordinate System. Puck A (mass m_A) is moving on a frictionless, horizontal air table in the $+x$-direction with velocity $\vec{v}_{A1}$ and makes an elastic, head-on collision with puck B (mass m_B) that is initially at rest. After the collision, both pucks are moving along the x-axis. a) Calculate the velocity of the center of mass of the two-puck system before the collision. b) Consider a coordinate system whose origin is at the center of mass and moves with it. Is this an inertial reference frame? c) What are the initial velocities $\vec{u}_{A1}$ and $\vec{u}_{B1}$ of the two pucks in this center-of-mass reference frame? What is the total momentum in this frame? d) Use conservation of momentum and energy, applied in the center-of-mass reference frame, to relate the final momentum of each puck to its initial momentum and thus the final velocity of each puck to its initial velocity. Your results should show that a one-dimensional elastic collision has a very simple description in center-of-mass coordinates. e) Let $m_A = 0.400$ kg, $m_B = 0.200$ kg, and $v_{A1} = 6.00$ m/s. Find the center-of-mass velocities $\vec{u}_{A1}$ and $\vec{u}_{B1}$, apply the simple result found in part (d), and transform back to velocities in a stationary frame to find the final velocities of the pucks. Does your result agree with Eqs. (8.24) and (8.25)?

8.101 The *coefficient of restitution* ϵ for a head-on collision is defined as the ratio of the relative speed after the collision to the relative speed before. a) What is ϵ for a completely inelastic collision? b) What is ϵ for an elastic collision? c) A ball is dropped from a height h onto a stationary surface and rebounds back to a height H_1. Show that $\epsilon = \sqrt{H_1/h}$. d) A properly inflated basketball should have a coefficient of restitution 0.85. When dropped from a height of 1.2 m above a solid wood floor, to what height should a properly inflated basketball bounce? e) The height of the first bounce is H_1. If ϵ is constant, show that the height of the nth bounce is $H_n = \epsilon^{2n}h$. f) If ϵ is constant, what is the height of the eighth bounce of a properly inflated basketball dropped from 1.2 m?

8.102 Binding Energy of the Hydrogen Molecule. When two hydrogen atoms of mass m combine to form a diatomic hydrogen molecule (H_2), the potential energy of the system after they combine is $-\Delta$, where Δ is a positive quantity called the *binding energy* of the molecule. a) Show that in a collision that only involves two hydrogen atoms, it is *impossible* to form an H_2 molecule because momentum and energy cannot simultaneously be conserved. (*Hint:* If you can show this to be true in one frame of reference, then it is true in all frames of reference. Can you see why?) b) An H_2 molecule can be formed in a collision that involves *three* hydrogen atoms. Suppose that before such a collision, each of the three atoms has speed 1.00×10^3 m/s, and they are approaching at 120° angles so that at any instant, the atoms lie at the corners of an equilateral triangle. Find the speeds of the H_2 molecule and of the single hydrogen atom that remains after the collision. The binding energy of H_2 is $\Delta = 7.23 \times 10^{-19}$ J, and the mass of the hydrogen atom is 1.67×10^{-27} kg.

8.103 A wagon with two boxes of gold, having total mass 300 kg, is cut loose from the horses by an outlaw when the wagon is at rest 50 m up a 6.0° slope (Fig. 8.46). The outlaw plans to have the wagon roll down the slope and across the level ground, and then fall into a canyon where his confederates wait. But in a tree 40 m

Figure 8.46 Problem 8.103.

from the canyon edge wait the Lone Ranger (mass 75.0 kg) and Tonto (mass 60.0 kg). They drop vertically into the wagon as it passes beneath them. a) If they require 5.0 s to grab the gold and jump out, will they make it before the wagon goes over the edge? The wagon rolls with negligible friction. b) When the two heroes drop into the wagon, is the kinetic energy of the system of the heroes plus the wagon conserved? If not, does it increase or decrease, and by how much?

***8.104** In Section 8.6 we considered a rocket fired in outer space where there is no air resistance and where gravity is negligible. Suppose instead that the rocket is accelerating vertically upward from rest on the earth's surface. Continue to ignore air resistance and consider only that part of the motion where the altitude of the rocket is small so that g may be assumed to be constant. a) How is Eq. (8.37) modified by the presence of the gravity force? b) Derive an expression for the acceleration a of the rocket, analogous to Eq. (8.39). c) What is the acceleration of the rocket in Example 8.16 (Section 8.6) if it is near the earth's surface rather than in outer space? You can ignore air resistance. d) Find the speed of the rocket in Example 8.17 (Section 8.6) after 90 s if the rocket is fired from the earth's surface rather than in outer space. You can ignore air resistance. How does your answer compare to the rocket speed calculated in Example 8.17?

***8.105 A Multistage Rocket.** Suppose the first stage of a two-stage rocket has a total mass 12,000 kg, of which 9000 kg is fuel. The total mass of the second stage is 1000 kg, of which 700 kg is fuel. Assume that the relative speed v_{ex} of ejected material is constant, and ignore any effect of gravity. (The latter effect is small during the firing period if the rate of fuel consumption is large.) a) Suppose the entire fuel supply carried by the two-stage rocket is utilized in a single-stage rocket with the same total mass of 13,000 kg. In terms of v_{ex}, what is the speed of the rocket, starting from rest, when its fuel is exhausted? b) For the two-stage rocket, what is the speed when the fuel of the first stage is exhausted if the first stage carries the second stage with it to this point? This speed then becomes the initial speed of the second stage. At this point, the second stage separates from the first stage. c) What is the final speed of the second stage? d) What value of v_{ex} is required to give the second stage of the rocket a speed of 7.00 km/s?

***8.106** The equation $F = -v_{ex}(dm/dt)$ for the thrust on a rocket, Eq. (8.38), can also be applied to an airplane propeller. In fact, there are *two* contributions to the thrust: one positive and one negative. The positive contribution comes from air pushed backward, away from the propeller (so $dm/dt < 0$), at a speed v_{ex} relative to the propeller. The negative contribution comes from this same quantity of air flowing into the front of the propeller (so $dm/dt > 0$) at speed v, equal to the speed of the airplane through the air. a) Write an equation for the *net* thrust developed by an aircraft propeller in terms of v, v_{ex}, and the absolute value $|dm/dt|$. b) For a Cessna 182 (a single-engine airplane) flying at 130 km/h, 150 kg of air flows through the propeller each second and the propeller develops a net thrust of 1300 N. Determine the speed increase (in km/h) that the propeller imparts to the air.

***8.107** For the rocket described in Examples 8.16 and 8.17 (Section 8.6), the mass of the rocket as a function of time is

$$m(t) = \begin{cases} m_0 & \text{for } t < 0 \\ m_0\left(1 - \dfrac{t}{120\text{ s}}\right) & \text{for } 0 \le t \le 90\text{ s} \\ m_0/4 & \text{for } t \ge 90\text{ s} \end{cases}$$

a) Calculate and graph the velocity of the rocket as a function of time from $t = 0$ s to $t = 100$ s. b) Calculate and graph the acceleration of the rocket as a function of time from $t = 0$ s to $t = 100$ s. c) A 75-kg astronaut lies on a reclined chair during the firing of the rocket. What is the maximum net force exerted by the chair on the astronaut during the firing? How does your answer compare to her weight on earth?

Challenge Problems

8.108 You are going to entertain your Aunt Emma by pulling a tablecloth out from under the dishes at her birthday party. The birthday cake is resting on a tablecloth at the center of a round table of radius $r = 0.90$ m. The tablecloth is the same size as the table top. You grab the edge of the tablecloth and pull sharply. The tablecloth and cake are in contact for time t after you start pulling. Then the sliding cake is stopped (you hope) by the friction between the cake and table top. The coefficient of kinetic friction between the cake and tablecloth is $\mu_{k1} = 0.30$, and that between the cake and table top is $\mu_{k2} = 0.40$. Apply the impulse-momentum relation (Eq. 8.9) and the work-energy theorem (Eq. 6.6) to calculate the maximum value of t if the cake is not to end up on the floor. (*Hint:* Assume that the cake moves a distance d while still on the tablecloth and therefore a distance $r - d$ while sliding on the table top. Assume that the friction forces are independent of the relative speed of the sliding surfaces. You can easily try this trick yourself by pulling a sheet of paper out from under a glass of water, but have a mop handy just in case!)

8.109 In Section 8.5 we calculated the center of mass by considering objects composed of a *finite* number of point masses or objects that, by symmetry, could be represented by a finite number of point masses. For a solid object whose mass distribution does not allow for a simple determination of the center of mass by symmetry, the sums of Eqs. (8.28) must be generalized to integrals

$$x_{cm} = \frac{1}{M}\int x\,dm \qquad y_{cm} = \frac{1}{M}\int y\,dm$$

where x and y are the coordinates of the small piece of the object that has mass dm. The integration is over the whole of the object. Consider a thin rod of length L, mass M, and cross-sectional area A. Let the origin of coordinates be at the left end of the rod and the positive x-axis lie along the rod. a) If the density $\rho = M/V$ of the object is uniform, perform the integration described above to show that the x-coordinate of the center of mass of the rod is at its geometrical center. b) If the density of the object varies linearly with x—that is, $\rho = \alpha x$, where α is a positive constant—calculate the x-coordinate of the rod's center of mass.

8.110 Use the methods of Challenge Problem 8.109 to calculate the x- and y-coordinates of the center of mass of a semicircular metal plate with uniform density ρ and thickness t. Let the radius of the plate be a. The mass of the plate is thus $M = \frac{1}{2}\rho\pi a^2 t$. Use the coordinate system indicated in Fig. 8.47.

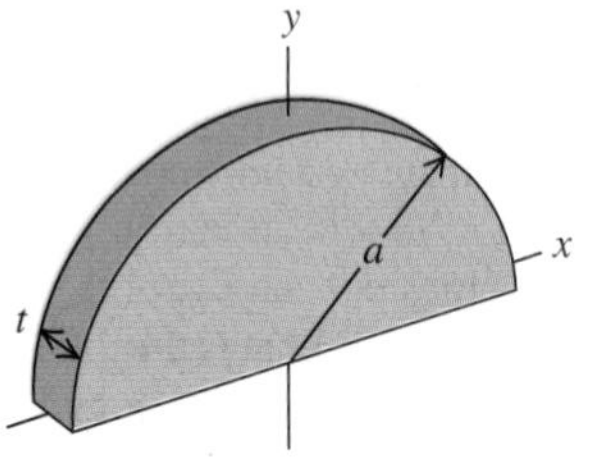

Figure 8.47 Challenge Problem 8.110.

8.111 One-fourth of a rope of length l is hanging down over the edge of a frictionless table. The rope has a uniform, linear density (mass per unit length) λ ("lambda"), and the end already on the table is held by a person. How much work does the person do when she pulls on the rope to raise the rest of the rope slowly onto the table? Do the problem in two ways as follows. a) Find the force that the person must exert to raise the rope and from this the work done. Note that this is a variable force because at different times different amounts of rope are hanging over the edge. b) Suppose the segment of the rope initially hanging over the edge of the table has all of its mass concentrated at its center of mass. Find the work necessary to raise this to table height. You will probably find this approach simpler than that of part (a). How do the answers compare, and why is this so?

***8.112 A Variable-Mass Raindrop.** In a rocket-propulsion problem the mass is variable. Another such problem is a raindrop falling through a cloud of small water droplets. Some of these small droplets adhere to the raindrop, thereby *increasing* its mass as it falls. The force on the raindrop is

$$F_{ext} = \frac{dp}{dt} = m\frac{dv}{dt} + v\frac{dm}{dt}$$

Suppose the mass of the raindrop depends on the distance x that it has fallen. Then $m = kx$, where k is a constant, and $dm/dt = kv$. This gives, since $F_{ext} = mg$

$$mg = m\frac{dv}{dt} + v(kv)$$

Or, dividing by k

$$xg = x\frac{dv}{dt} + v^2$$

This is a differential equation that has a solution of the form $v = at$, where a is the acceleration and is constant. Take the initial velocity of the raindrop to be zero. a) Using the proposed solution for v, find the acceleration a. b) Find the distance the raindrop has fallen in $t = 3.00$ s. c) Given that $k = 2.00$ g/m, find the mass of the raindrop at $t = 3.00$ s. For many more intriguing aspects of this problem see K. S. Krane, *Amer. Jour. Phys.*, Vol. 49 (1981), pp. 113–117.

9 ROTATION OF RIGID BODIES

No matter where you sit on a rotating amusement park ride, you have the same angular velocity and angular acceleration. But your *linear* velocity and *linear* acceleration depend on how far you sit from the axis of rotation. In general, the farther you are from the axis, the greater the linear velocity and acceleration and the wilder the ride.

? **If you double your distance from the ride's axis of rotation, how many times greater is your speed? How many times greater is your acceleration toward the axis of rotation?**

What do the motions of a compact disc, a Ferris wheel, a circular saw blade, and a ceiling fan have in common? None of these can be represented adequately as a moving *point;* each involves a body that *rotates* about an axis that is stationary in some inertial frame of reference.

Rotation occurs at all scales, from the motion of electrons in atoms to the motions of entire galaxies. We need to develop some general methods for analyzing the motion of a rotating body. In this chapter and the next we consider bodies that have definite size and definite shape, and that in general can have rotational as well as translational motion.

Real-world bodies can be very complicated; the forces that act on them can deform them—stretching, twisting, and squeezing them. We'll neglect these deformations for now and assume that the body has a perfectly definite and unchanging shape and size. We call this idealized model a **rigid body.** This chapter and the next are mostly about rotational motion of a rigid body.

We begin with kinematic language for *describing* rotational motion. Next we look at the kinetic energy of rotation, the key to using energy methods for rotational motion. Then in Chapter 10 we'll develop dynamic principles that relate the forces on a body to its rotational motion.

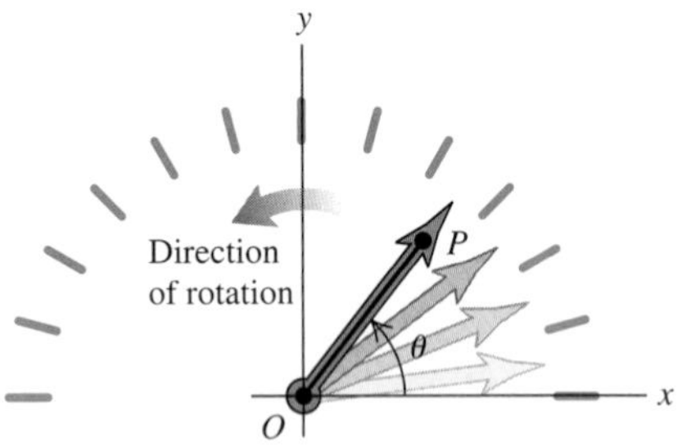

9.1 A speedometer needle (an example of a rigid body) rotating counterclockwise about a fixed axis. The axis passes through O and is perpendicular to the plane of the diagram.

9.1 | Angular Velocity and Acceleration

In analyzing rotational motion, let's think first about a rigid body that rotates about a fixed axis. By *fixed axis* we mean an axis that is at rest in some inertial frame of reference and does not change direction relative to that frame. The rotating rigid body might be a motor shaft, a chunk of roast beef on a barbecue skewer, or a merry-go-round.

Figure 9.1 shows a rigid body (in this case, the indicator needle of a speedometer) rotating about a fixed axis. The axis passes through point O and is perpendicular to the plane of the diagram, which we choose to call the xy-plane. One way to describe the rotation of this body would be to choose a particular point P on the body and to keep track of the x- and y-coordinates of this point. This isn't a terribly convenient method, since it takes two numbers (the two coordinates x and y) to specify the rotational position of the body. Instead, we notice that the line OP is fixed in the body and rotates with it. The angle θ that this line makes with the $+x$-axis describes the rotational position of the body; we will use this single quantity θ as a *coordinate* for rotation.

The angular coordinate θ of a rigid body rotating around a fixed axis can be positive or negative. If we choose positive angles to be measured counterclockwise from the positive x-axis, then the angle θ in Fig. 9.1 is positive. If we instead choose the positive rotation direction to be clockwise, then θ in Fig. 9.1 is negative. When we considered the motion of a particle along a straight line, it was essential to specify the direction of positive displacement along that line; in discussing rotation around a fixed axis, it's just as essential to specify the direction of positive rotation.

In describing rotational motion, the most natural way to measure the angle θ is not in degrees, but in **radians.** As shown in Fig. 9.2a, one radian (1 rad) is the angle subtended at the center of a circle by an arc with a length equal to the radius of the circle. In Fig. 9.2b an angle θ is subtended by an arc of length s on a circle of radius r. The value of θ (in radians) is equal to s divided by r:

$$\theta = \frac{s}{r} \quad \text{or} \quad s = r\theta \tag{9.1}$$

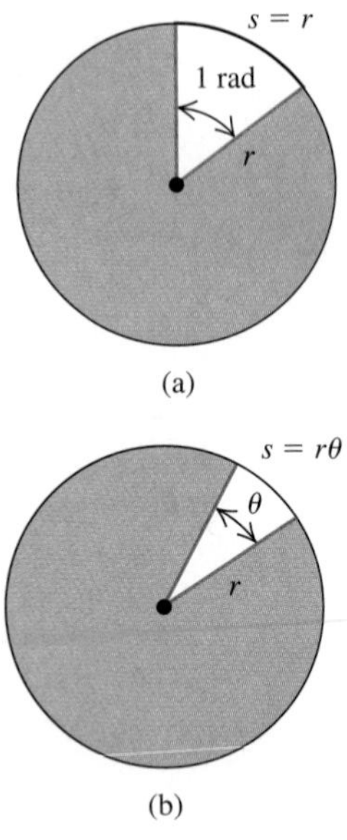

9.2 An angle θ in radians is defined as the ratio of the arc length s to the radius r.

An angle in radians is the ratio of two lengths, so it is a pure number, without dimensions. If $s = 3.0$ m and $r = 2.0$ m, then $\theta = 1.5$, but we will often write this as 1.5 rad to distinguish it from an angle measured in degrees or revolutions.

The circumference of a circle (that is, the arc length all the way around the circle) is 2π times the radius, so there are 2π (about 6.283) radians in one complete revolution $(360°)$. Therefore

$$1 \text{ rad} = \frac{360°}{2\pi} = 57.3°$$

Similarly, $180° = \pi$ rad, $90° = \pi/2$ rad, and so on. If we had insisted on measuring the angle θ in degrees, we would have needed to include an extra factor of $(2\pi/360)$ on the right-hand side of $s = r\theta$ in Eq. (9.1). By measuring angles in radians, we keep the relationship between angle and distance along an arc as simple as possible.

Angular Velocity

The coordinate θ shown in Fig. 9.1 specifies the rotational position of a rigid body at a given instant. We can describe the rotational *motion* of such a rigid body in terms of the rate of change of θ. We'll do this in an analogous way to our descrip-

9.3 (a) Angular displacement $\Delta\theta$ of a rotating body. (b) Every part of a rigid body has the same angular velocity.

tion of straight-line motion in Chapter 2. In Fig. 9.3a a reference line OP in a rotating body makes an angle θ_1 with the $+x$-axis at time t_1. At a later time t_2 the angle has changed to θ_2. We define the **average angular velocity** $\omega_{\text{av-}z}$ (the Greek letter "omega") of the body in the time interval $\Delta t = t_2 - t_1$ as the ratio of the **angular displacement** $\Delta\theta = \theta_2 - \theta_1$ to Δt:

$$\omega_{\text{av-}z} = \frac{\theta_2 - \theta_1}{t_2 - t_1} = \frac{\Delta\theta}{\Delta t} \tag{9.2}$$

The subscript z indicates that the body in Fig. 9.3a is rotating about the z-axis, which is perpendicular to the plane of the diagram. The **instantaneous angular velocity** ω_z is the limit of $\omega_{\text{av-}z}$ as Δt approaches zero, that is, the derivative of θ with respect to t:

$$\omega_z = \lim_{\Delta t \to 0} \frac{\Delta\theta}{\Delta t} = \frac{d\theta}{dt} \quad \text{(definition of angular velocity)} \tag{9.3}$$

When we refer simply to "angular velocity," we mean the instantaneous angular velocity, not the average angular velocity.

The angular velocity ω_z can be positive or negative, depending on the direction in which the rigid body is rotating. The angular *speed* ω, which we will use extensively in Sections 9.3 and 9.4, is the magnitude of angular velocity. Like ordinary (linear) speed v, the angular speed is never negative.

CAUTION Keep in mind the distinction between angular velocity ω_z and ordinary velocity v_x (Section 2.2). If an object has a velocity v_x, the object as a whole is *moving* along the x-axis. By contrast, if an object has an angular velocity ω_z, then it is *rotating* around the z-axis. We do *not* mean that the object is moving along the z-axis.

Different points on a rotating rigid body move different distances in a given time interval, depending on how far the point lies from the rotation axis. But because the body is rigid, *all* points rotate through the same angle in the same time (Fig. 9.3b). Hence *at any instant, every part of a rotating rigid body has the same angular velocity.* The angular velocity is positive if the body is rotating in the direction of increasing θ and negative if it is rotating in the direction of decreasing θ.

If the angle θ is in radians, the unit of angular velocity is the radian per second (rad/s). Other units, such as the revolution per minute (rev/min or rpm), are often used. Since 1 rev $= 2\pi$ rad, two useful conversions are

$$1\text{ rev/s} = 2\pi\text{ rad/s} \quad\text{and}\quad 1\text{ rev/min} = 1\text{ rpm} = \frac{2\pi}{60}\text{ rad/s}$$

That is, 1 rad/s is about 10 rpm.

Example 9.1 Calculating angular velocity

Figure 9.4 shows the flywheel in a car engine under test. The angular position θ of the flywheel is given by

$$\theta = (2.0\text{ rad/s}^3)t^3$$

The diameter of the flywheel is 0.36 m. a) Find the angle θ, in radians and in degrees, at times $t_1 = 2.0$ s and $t_2 = 5.0$ s. b) Find the distance that a particle on the rim moves during that time interval. c) Find the average angular velocity, in rad/s and in rev/min (rpm), between $t_1 = 2.0$ s and $t_2 = 5.0$ s. d) Find the instantaneous angular velocity at time $t = t_2 = 5.0$ s.

SOLUTION

IDENTIFY and SET UP: We're given the angular position θ as a function of time. Hence we can easily find our first two target variables, the values θ_1 and θ_2 of the angular position at times t_1 and t_2. Once we have these values, we can calculate the angular displacement $\Delta\theta$ between t_1 and t_2, and from this result find the distance and the average angular velocity using Eqs. (9.1) and (9.2), respectively. To find the instantaneous angular velocity, we'll take the derivative of θ with respect to time, as in Eq. (9.3).

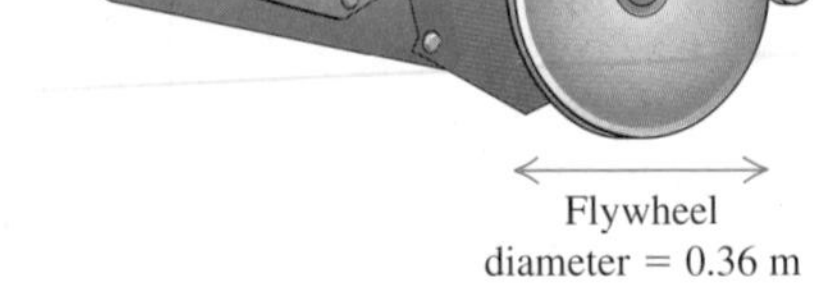

9.4 Rotating flywheel in a car engine.

EXECUTE: a) We substitute the values of t into the given equation:

$$\theta_1 = (2.0\text{ rad/s}^3)(2.0\text{ s})^3 = 16\text{ rad}$$
$$= (16\text{ rad})\frac{360°}{2\pi\text{ rad}} = 920°$$
$$\theta_2 = (2.0\text{ rad/s}^3)(5.0\text{ s})^3 = 250\text{ rad}$$
$$= (250\text{ rad})\frac{360°}{2\pi\text{ rad}} = 14{,}000°$$

b) The flywheel turns through an angular displacement of $\Delta\theta = \theta_2 - \theta_1 = 250\text{ rad} - 16\text{ rad} = 234\text{ rad}$. The radius r is half the diameter, or 0.18 m. Equation (9.1) gives

$$s = r\theta = (0.18\text{ m})(234\text{ rad}) = 42\text{ m}$$

To use Eq. (9.1), the angle *must* be expressed in radians. We drop "radians" from the unit for s because θ is really a dimensionless pure number; s is a distance and is measured in meters, the same unit as r.

c) In Eq. (9.2) we have

$$\omega_{\text{av-}z} = \frac{\theta_2 - \theta_1}{t_2 - t_1} = \frac{250\text{ rad} - 16\text{ rad}}{5.0\text{ s} - 2.0\text{ s}} = 78\text{ rad/s}$$
$$= \left(78\frac{\text{rad}}{\text{s}}\right)\left(\frac{1\text{ rev}}{2\pi\text{ rad}}\right)\left(\frac{60\text{ s}}{1\text{ min}}\right) = 740\text{ rev/min}$$

d) We use Eq. (9.3):

$$\omega_z = \frac{d\theta}{dt} = \frac{d}{dt}[(2.0\text{ rad/s}^3)t^3] = (2.0\text{ rad/s}^3)(3t^2)$$
$$= (6.0\text{ rad/s}^3)t^2$$

At time $t = 5.0$ s,

$$\omega_z = (6.0\text{ rad/s}^3)(5.0\text{ s})^2 = 150\text{ rad/s}$$

EVALUATE: Our result in part (d) shows that ω_z is proportional to t^2 and hence increases with time. Our numerical results are consistent with this result: the 150-rad/s instantaneous angular velocity at $t = 5.0$ s is greater than the 78-rad/s average angular velocity for the 2.0-s interval leading up to that time (from $t_1 = 3.0$ s to $t_2 = 5.0$ s).

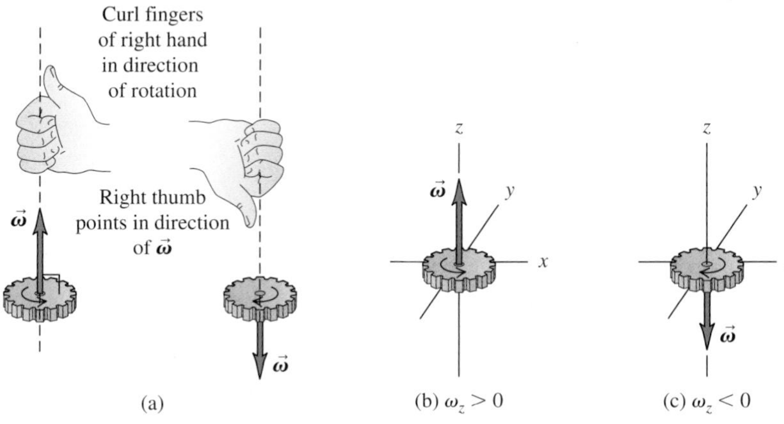

9.5 (a) The right-hand rule for the direction of the angular velocity vector $\vec{\omega}$. Reversing the direction of rotation reverses the direction of $\vec{\omega}$. If the rotation is along the z-axis, the sign of ω_z depends on whether $\vec{\omega}$ (b) points along or (c) points opposite to the positive z-direction.

As we have seen, our notation for the angular velocity ω_z about the z-axis is reminiscent of the notation v_x for the ordinary velocity along the x-axis (see Section 2.2). Just as v_x is the x-component of the velocity vector $\vec{v}$, ω_z is the z-component of an angular velocity *vector* $\vec{\omega}$ directed along the axis of rotation. As Fig. 9.5a shows, the direction of $\vec{\omega}$ is given by the right-hand rule that we used to define the vector product in Section 1.10. If the rotation is about the z-axis, then $\vec{\omega}$ has only a z-component; this component is positive if $\vec{\omega}$ is along the positive z-axis (Fig. 9.5b) and negative if $\vec{\omega}$ is along the negative z-axis (Fig. 9.5c).

The vector formulation is especially useful in situations in which the direction of the rotation axis *changes.* We'll examine such situations briefly at the end of Chapter 10. In this chapter, however, we'll consider only situations in which the rotation axis is fixed. Hence throughout this chapter we'll use "angular velocity" to refer to ω_z, the component of the angular velocity vector $\vec{\omega}$ along the axis.

Angular Acceleration

When the angular velocity of a rigid body changes, it has an *angular acceleration.* When you pedal your bicycle harder to make the wheels turn faster or apply the brakes to bring the wheels to a stop, you're giving the wheels an angular acceleration. You also impart an angular acceleration whenever you change the rotation speed of a piece of spinning machinery, such as an automobile engine's crankshaft or the rotor of a helicopter (Fig. 9.6).

9.6 The pilot of a helicopter can change the angular velocity of the rotor—that is, give it an angular acceleration—by adjusting a control in the cockpit.

If ω_{1z} and ω_{2z} are the instantaneous angular velocities at times t_1 and t_2, we define the **average angular acceleration** $\alpha_{\text{av-}z}$ over the interval $\Delta t = t_2 - t_1$ as the change in angular velocity divided by Δt:

$$\alpha_{\text{av-}z} = \frac{\omega_{2z} - \omega_{1z}}{t_2 - t_1} = \frac{\Delta\omega_z}{\Delta t} \tag{9.4}$$

The **instantaneous angular acceleration** α_z is the limit of $\alpha_{\text{av-}z}$ as $\Delta t \rightarrow 0$:

$$\alpha_z = \lim_{\Delta t \to 0} \frac{\Delta\omega_z}{\Delta t} = \frac{d\omega_z}{dt} \quad \text{(definition of angular acceleration)} \tag{9.5}$$

The usual unit of angular acceleration is the radian per second per second, or rad/s^2. Henceforth we will use the term "angular acceleration" to mean the instantaneous angular acceleration rather than the average angular acceleration.

Because $\omega_z = d\theta/dt$, we can also express angular acceleration as the second derivative of the angular coordinate:

$$\alpha_z = \frac{d}{dt}\frac{d\theta}{dt} = \frac{d^2\theta}{dt^2} \qquad (9.6)$$

You have probably noticed that we are using Greek letters for angular kinematic quantities: θ for angular position, ω_z for angular velocity, and α_z for angular acceleration. These are analogous to x for position, v_x for velocity, and a_x for acceleration, respectively, in straight-line motion. In each case, velocity is the rate of change of position with respect to time and acceleration is the rate of change of velocity with respect to time. We will sometimes use the terms *linear* velocity and *linear* acceleration for the familiar quantities we defined in Chapters 2 and 3 to make sure we distinguish clearly between these and the *angular* quantities introduced in this chapter.

In rotational motion, if the angular acceleration α_z is positive, then the angular velocity ω_z is increasing; if α_z is negative, then ω_z is decreasing. The rotation is speeding up if α_z and ω_z have the same sign and slowing down if α_z and ω_z have opposite signs. (These are exactly the same relationships as those between *linear* acceleration a_x and *linear* velocity v_x for straight-line motion; see Section 2.3.)

Example 9.2 Calculating angular acceleration

In Example 9.1 we found that the instantaneous angular velocity ω_z of the flywheel at any time t is given by

$$\omega_z = (6.0\ \text{rad/s}^3)t^2$$

a) Find the average angular acceleration between $t_1 = 2.0$ s and $t_2 = 5.0$ s. b) Find the instantaneous angular acceleration at time $t_2 = 5.0$ s.

SOLUTION

IDENTIFY and SET UP: We'll use Eqs. (9.4) and (9.5) to find our two target variables, the average angular acceleration $\alpha_{\text{av-}z}$ between t_1 and t_2 and the instantaneous angular acceleration α_z at $t = t_2$.

EXECUTE: a) The values of ω_z at the two times are

$$\omega_{1z} = (6.0\ \text{rad/s}^3)(2.0\ \text{s})^2 = 24\ \text{rad/s}$$
$$\omega_{2z} = (6.0\ \text{rad/s}^3)(5.0\ \text{s})^2 = 150\ \text{rad/s}$$

From Eq. (9.4) the average angular acceleration is

$$\alpha_{\text{av-}z} = \frac{150\ \text{rad/s} - 24\ \text{rad/s}}{5.0\ \text{s} - 2.0\ \text{s}} = 42\ \text{rad/s}^2$$

b) From Eq. (9.5) the instantaneous angular acceleration at any time t is

$$\alpha_z = \frac{d\omega_z}{dt} = \frac{d}{dt}[(6.0\ \text{rad/s}^3)(t^2)] = (6.0\ \text{rad/s}^3)(2t)$$
$$= (12\ \text{rad/s}^3)t$$

At time $t = 5.0$ s,

$$\alpha_z = (12\ \text{rad/s}^3)(5.0\ \text{s}) = 60\ \text{rad/s}^2$$

EVALUATE: Note that the angular acceleration is *not* constant in this situation. The angular velocity ω_z is always increasing because α_z is always positive. Furthermore, the rate at which the angular velocity increases is itself increasing, since α_z increases with time.

Just as we did for angular velocity, it's useful to define an angular acceleration *vector* $\vec{\alpha}$. Mathematically, $\vec{\alpha}$ is the time derivative of the angular velocity vector $\vec{\omega}$. If the object rotates around the fixed z-axis, then $\vec{\alpha}$ has only a z-component; the quantity α_z is just that component. In this case, $\vec{\alpha}$ points in the same direction as

$\vec{\omega}$ if the rotation is speeding up and opposite to $\vec{\omega}$ if the rotation is slowing down (Fig. 9.7).

The angular acceleration vector will be particularly useful in Chapter 10 when we discuss what happens when the rotation axis can change direction. In this chapter, however, the rotation axis will always be fixed and we need use only the z-component α_z.

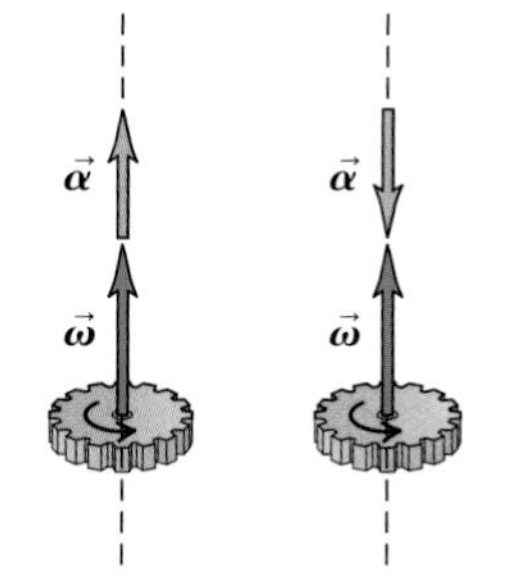

9.7 When the rotation axis is fixed, the angular acceleration and angular velocity vectors both lie along that axis.

Test Your Understanding

Figure 9.8 is a graph of ω_z and α_z versus time for a particular rotating body. During which time intervals is the rotation speeding up? During which time intervals is the rotation slowing down?

9.2 | Rotation with Constant Angular Acceleration

In Chapter 2 we found that straight-line motion is particularly simple when the acceleration is constant. This is also true of rotational motion about a fixed axis. When the angular acceleration is constant, we can derive equations for angular velocity and angular position using exactly the same procedure that we used for straight-line motion in Section 2.4. In fact, the equations we are about to derive are identical to Eqs. (2.8), (2.12), (2.13), and (2.14) if we replace x with θ, v_x with ω_z, and a_x with α_z. We suggest that you review Section 2.4 before continuing.

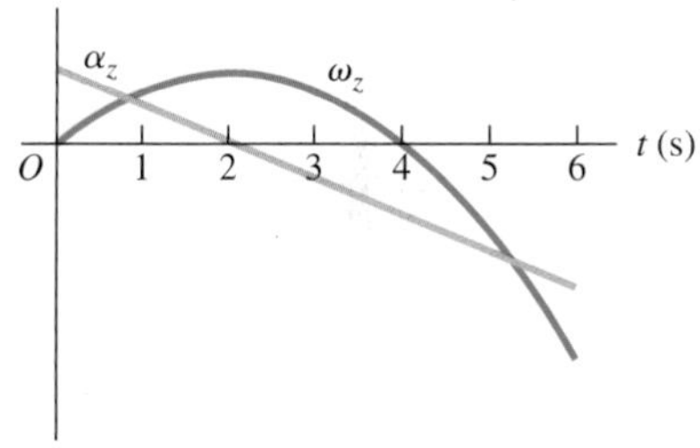

9.8 Graph of angular velocity ω_z and angular acceleration α_z versus time for a rotating body.

Let ω_{0z} be the angular velocity of a rigid body at time $t = 0$ and let ω_z be its angular velocity at any later time t. The angular acceleration α_z is constant and equal to the average value for any interval. Using Eq. (9.4) with the interval from 0 to t, we find

$$\alpha_z = \frac{\omega_z - \omega_{0z}}{t - 0} \quad \text{or}$$

$$\omega_z = \omega_{0z} + \alpha_z t \qquad \text{(constant angular acceleration only)} \qquad (9.7)$$

The product $\alpha_z t$ is the total change in ω_z between $t = 0$ and the later time t; the angular velocity ω_z at time t is the sum of the initial value ω_{0z} and this total change.

With constant angular acceleration, the angular velocity changes at a uniform rate, so its average value between 0 and t is the average of the initial and final values:

$$\omega_{\text{av-}z} = \frac{\omega_{0z} + \omega_z}{2} \qquad (9.8)$$

ActivPhysics Online

7.7 Rotational Kinematics

We also know that $\omega_{\text{av-}z}$ is the total angular displacement $(\theta - \theta_0)$ divided by the time interval $(t - 0)$;

$$\omega_{\text{av-}z} = \frac{\theta - \theta_0}{t - 0} \qquad (9.9)$$

When we equate Eqs. (9.8) and (9.9) and multiply the result by t, we get

$$\theta - \theta_0 = \frac{1}{2}(\omega_{0z} + \omega_z)t \qquad \text{(constant angular acceleration only)} \qquad (9.10)$$

To obtain a relation between θ and t that doesn't contain ω_z, we substitute Eq. (9.7) into Eq. (9.10):

$$\theta - \theta_0 = \frac{1}{2}[\omega_{0z} + (\omega_{0z} + \alpha_z t)]t$$

$$\theta = \theta_0 + \omega_{0z}t + \frac{1}{2}\alpha_z t^2 \qquad \text{(constant angular acceleration only)} \qquad (9.11)$$

That is, if at the initial time $t = 0$ the body is at angular position θ_0 and has angular velocity ω_{0z}, then its angular position θ at any later time t is the sum of three terms: its initial angular position θ_0, plus the rotation $\omega_{0z}t$ it would have if the angular velocity were constant, plus an additional rotation $\frac{1}{2}\alpha_z t^2$ caused by the changing angular velocity.

Following the same procedure as for straight-line motion in Section 2.4, we can combine Eqs. (9.7) and (9.11) to obtain a relation between θ and ω_z that does not contain t. We invite you to work out the details, following the same procedure we used to get Eq. (2.13). (See Exercise 9.12.) In fact, because of the perfect analogy between straight-line and rotational quantities, we can simply take Eq. (2.13) and replace each straight-line quantity by its rotational analog. We get

$$\omega_z^2 = \omega_{0z}^2 + 2\alpha_z(\theta - \theta_0) \qquad \text{(constant angular acceleration only)} \qquad (9.12)$$

CAUTION Keep in mind that all of these results are valid *only* when the angular acceleration α_z is *constant;* be careful not to try to apply them to problems in which α_z is *not* constant. Table 9.1 shows the analogy between Eqs. (9.7), (9.10), (9.11), and (9.12) for fixed-axis rotation with constant angular acceleration and the corresponding equations for straight-line motion with constant linear acceleration.

Table 9.1 Comparison of Linear and Angular Motion with Constant Acceleration

Straight-Line Motion with Constant Linear Acceleration	Fixed-Axis Rotation with Constant Angular Acceleration
$a_x = \text{constant}$	$\alpha_z = \text{constant}$
$v_x = v_{0x} + a_x t$	$\omega_z = \omega_{0z} + \alpha_z t$
$x = x_0 + v_{0x}t + \frac{1}{2}a_x t^2$	$\theta = \theta_0 + \omega_{0z}t + \frac{1}{2}\alpha_z t^2$
$v_x^2 = v_{0x}^2 + 2a_x(x - x_0)$	$\omega_z^2 = \omega_{0z}^2 + 2\alpha_z(\theta - \theta_0)$
$x - x_0 = \frac{1}{2}(v_x + v_{0x})t$	$\theta - \theta_0 = \frac{1}{2}(\omega_z + \omega_{0z})t$

Example 9.3 Rotation with constant angular acceleration

You have just finished watching a movie on DVD and the disc is slowing to a stop. The angular velocity of the disc at $t = 0$ is 27.5 rad/s and its angular acceleration is a constant $-10.0\ \text{rad/s}^2$. A line PQ on the surface of the disc lies along the $+x$-axis at $t = 0$ (Fig. 9.9). a) What is the disc's angular velocity at $t = 0.300$ s? b) What angle does the line PQ make with the $+x$-axis at this time?

SOLUTION

IDENTIFY: The angular acceleration of the disc is constant, so we can use any of the equations derived in this section. Our target variables are the angular velocity and the angular displacement at $t = 0.300$ s.

SET UP: We are given the initial angular velocity $\omega_{0z} = 27.5$ rad/s, the initial angle $\theta_0 = 0$ between the line PQ and the $+x$-axis, the angular acceleration $\alpha_z = -10.0$ rad/s^2, and the time $t = 0.300$ s of interest. With this information it's easiest to use Eqs. (9.7) and (9.11) to find the target variables ω_z and θ, respectively.

EXECUTE: a) From Eq. (9.7), at $t = 0.300$ s we have

$$\omega_z = \omega_{0z} + \alpha_z t = 27.5\ \text{rad/s} + (-10.0\ \text{rad/s}^2)(0.300\ \text{s})$$
$$= 24.5\ \text{rad/s}$$

b) From Eq. (9.11),

$$\theta = \theta_0 + \omega_{0z}t + \frac{1}{2}\alpha_z t^2$$
$$= 0 + (27.5\ \text{rad/s})(0.300\ \text{s}) + \frac{1}{2}(-10.0\ \text{rad/s}^2)(0.300\ \text{s})^2$$
$$= 7.80\ \text{rad} = 7.80\ \text{rad}\left(\frac{1\ \text{rev}}{2\pi\ \text{rad}}\right) = 1.24\ \text{rev}$$

The DVD has turned through one complete revolution plus an additional 0.24 rev—that is, through an additional angle of (0.24 rev)(360°/rev) = 87°. Hence the line PQ is at an angle of 87° with the $+x$-axis.

EVALUATE: Our answer to part (a) tells us that the angular velocity has decreased. This is as it should be, since α_z is negative. We can also use our answer for ω_z in part (a) to check our result for θ in part (b). To do so, we solve Eq. (9.12), $\omega_z^2 = \omega_{0z}^2 + 2\alpha_z(\theta - \theta_0)$, for the angle θ:

$$\theta = \theta_0 + \left(\frac{\omega_z^2 - \omega_{0z}^2}{2\alpha_z}\right)$$
$$= 0 + \frac{(24.5\ \text{rad/s})^2 - (27.5\ \text{rad/s})^2}{2(-10.0\ \text{rad/s}^2)} = 7.80\ \text{rad}$$

which agrees with the result we found earlier.

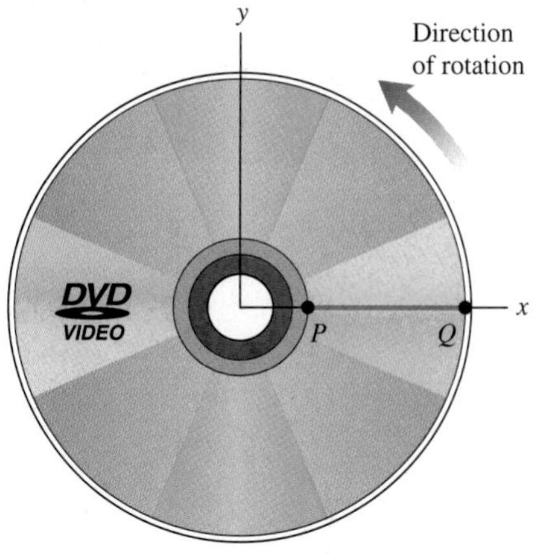

9.9 A line PQ on a rotating DVD at $t = 0$. What angle does this line make with the $+x$-axis at $t = 0.300$ s?

Test Your Understanding

At what time t does the DVD in Example 9.3 come to rest? Through how many revolutions does it rotate between $t = 0$ and coming to rest?

9.3 | Relating Linear and Angular Kinematics

How do we find the linear velocity and acceleration of a particular point in a rotating rigid body? We need to answer this question to proceed with our study of rotation. For example, to find the kinetic energy of a rotating body, we have to start from $K = \frac{1}{2}mv^2$ for a particle, and this requires knowing v for each particle in the body. So it's worthwhile to develop general relations between the *angular* velocity and acceleration of a rigid body rotating about a fixed axis and the *linear* velocity and acceleration of a specific point or particle in the body.

When a rigid body rotates about a fixed axis, every particle in the body moves in a circular path. The circle lies in a plane perpendicular to the axis and is centered on the axis. The speed of a particle is directly proportional to the body's angular velocity; the faster the body rotates, the greater the speed of each particle. In Fig. 9.10, point P is a constant distance r from the axis of rotation, so it moves in a circle of radius r. At any time the angle θ (in radians) and the arc length s are related by

$$s = r\theta$$

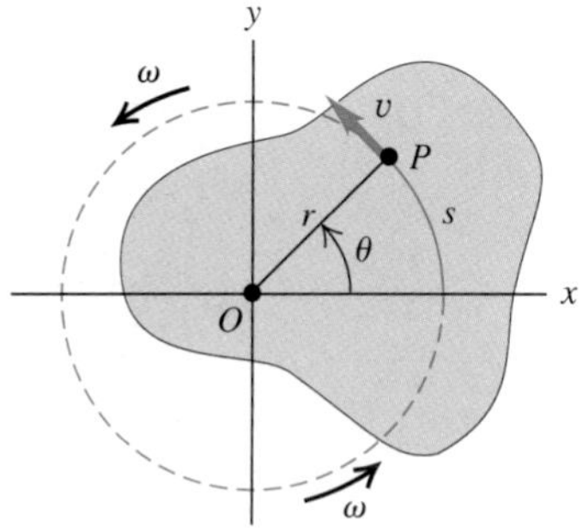

9.10 A rigid body rotating about a fixed axis through point O. The distance s that point P on the body moves through equals $r\theta$ if θ is measured in radians. The linear speed v of point P equals $r\omega$ if the angular speed ω is measured in rad/s.

We take the time derivative of this, noting that r is constant for any specific particle, and take the absolute value of both sides:

$$\left|\frac{ds}{dt}\right| = r\left|\frac{d\theta}{dt}\right|$$

Now $|ds/dt|$ is the absolute value of the rate of change of arc length, which is equal to the instantaneous *linear* speed v of the particle. Analogously, $|d\theta/dt|$, the absolute value of the rate of change of the angle, is the instantaneous **angular speed** ω—that is, the magnitude of the instantaneous angular velocity in rad/s. Thus

$$v = r\omega \quad \text{(relation between linear and angular speed)} \tag{9.13}$$

The farther a point is from the axis, the greater its linear speed. The *direction* of the linear velocity *vector* is tangent to its circular path at each point (Fig. 9.10).

CAUTION Keep in mind the distinction between the linear and angular *speeds* v and ω (which appear in Eq. (9.13)) and the linear and angular *velocities* v_x and ω_z. The quantities without subscripts, v and ω, are never negative; they are the magnitudes of the vectors $\vec{v}$ and $\vec{\omega}$, respectively, and their values tell you only how fast a particle is moving (v) or how fast a body is rotating (ω). The corresponding quantities with subscripts, v_x and ω_z, can be either positive or negative; their signs tell you the direction of the motion.

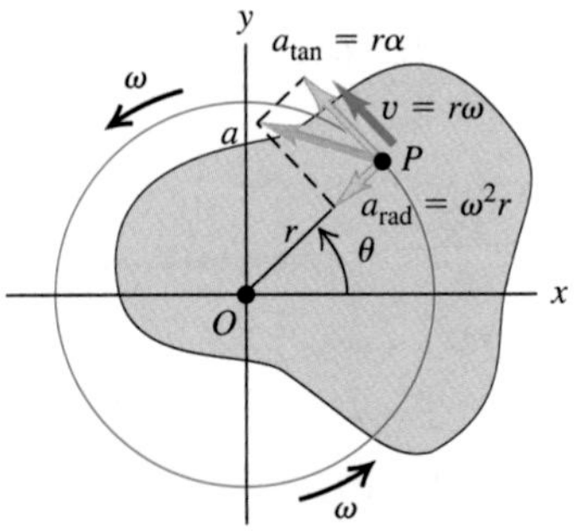

9.11 The component of acceleration of point P parallel to $\vec{v}$ is $a_{tan} = r\alpha$; the component perpendicular to $\vec{v}$, radially inward toward the rotation axis, is $a_{rad} = \omega^2 r$. The rigid body shown here is speeding up.

We can represent the acceleration of a particle moving in a circle in terms of its centripetal and tangential components, a_{rad} and a_{tan} (Fig. 9.11), as we did in Section 3.4. It would be a good idea to review that section now. We found that the **tangential component of acceleration** a_{tan}, the component parallel to the instantaneous velocity, acts to change the *magnitude* of the particle's velocity (i.e., the speed) and is equal to the rate of change of speed. Taking the derivative of Eq. (9.13), we find

$$a_{tan} = \frac{dv}{dt} = r\frac{d\omega}{dt} = r\alpha \tag{9.14}$$

(tangential acceleration of a point on a rotating body)

This component of a particle's acceleration is always tangent to the circular path of the particle.

The quantity $\alpha = d\omega/dt$ in Eq. (9.14) is the rate of change of the angular *speed.* It is not quite the same as $\alpha_z = d\omega_z/dt$, which is the rate of change of the angular *velocity.* For example, consider a body rotating so that its angular velocity vector points in the $-z$-direction (Fig. 9.5c). If the body is gaining angular speed at a rate of 10 rad/s per second, then $\alpha = 10\ \text{rad/s}^2$. But ω_z is negative and becoming more negative as the rotation gains speed, so $\alpha_z = -10\ \text{rad/s}^2$. The rule for rotation about a fixed axis is that α is equal to α_z if ω_z is positive but equal to $-\alpha_z$ if ω_z is negative.

?

The component of the particle's acceleration directed toward the rotation axis, the **centripetal component of acceleration** a_{rad}, is associated with the change of *direction* of the particle's velocity. In Section 3.4 we worked out the relation $a_{rad} = v^2/r$. We can express this in terms of ω by using Eq. (9.13):

$$a_{rad} = \frac{v^2}{r} = \omega^2 r \qquad (9.15)$$

(centripetal acceleration of a point on a rotating body)

This is true at each instant *even when ω and v are not constant.* The centripetal component always points toward the axis of rotation.

The vector sum of the centripetal and tangential components of acceleration of a particle in a rotating body is the linear acceleration $\vec{a}$ (Fig. 9.11).

CAUTION It's important to remember that Eq. (9.1), $s = r\theta$, is valid *only* when θ is measured in radians. The same is true of any equation derived from this, including Eqs. (9.13), (9.14), and (9.15). When you use these equations, you *must* express the angular quantities in radians, not revolutions or degrees (Fig. 9.12).

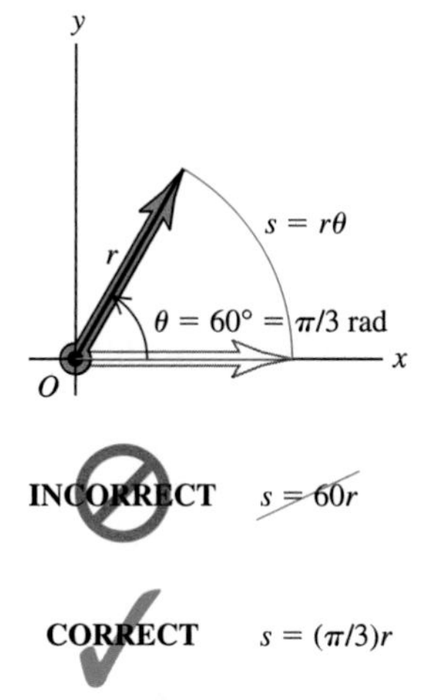

9.12 In any equation that relates linear quantities to angular quantities, the angles must be expressed in radians—never in degrees or revolutions.

Equations (9.1), (9.13), and (9.14) also apply to any particle that has the same tangential velocity as a point in a rotating rigid body. For example, when a rope wound around a circular cylinder unwraps without stretching or slipping, its speed and acceleration at any instant are equal to the speed and tangential acceleration of the point at which it is tangent to the cylinder. The same principle holds for situations such as bicycle chains and sprockets, belts and pulleys that turn without slipping, and so on. We will have several opportunities to use these relations later in this chapter and in Chapter 10. Note that Eq. (9.15) for the centripetal component a_{rad} is applicable to the rope or chain *only* at points that are in contact with the cylinder or sprocket. Other points do not have the same acceleration toward the center of the circle that points on the cylinder or sprocket have.

Example 9.4 Throwing a discus

A discus thrower moves the discus in a circle of radius 80.0 cm. At a certain instant, the thrower is spinning at an angular speed of 10.0 rad/s and the angular speed is increasing at 50.0 rad/s^2. At this instant, find the tangential and centripetal components of the acceleration of the discus and the magnitude of the acceleration.

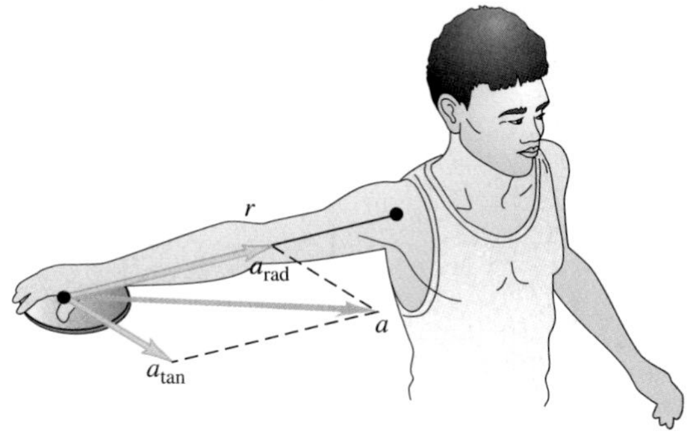

9.13 Whirling a discus in a circle.

SOLUTION

IDENTIFY and SET UP: We model the discus as a particle traveling on a circular path (Fig. 9.13), so we can use the ideas developed in this section. We are given the radius $r = 0.800$ m, the angular speed $\omega = 10.0$ rad/s, and the rate of change of angular speed $\alpha = 50.0$ rad/s^2. The first two target variables are the acceleration components a_{tan} and a_{rad}, which we'll find with Eqs. (9.14) and (9.15), respectively. Given these components of the acceleration vector, we'll find its magnitude (the third target variable) using the Pythagorean theorem (see Fig. 9.11).

EXECUTE: From Eqs. (9.14) and (9.15),

$$a_{tan} = r\alpha = (0.800\text{ m})(50.0\text{ rad/s}^2) = 40.0\text{ m/s}^2$$
$$a_{rad} = \omega^2 r = (10.0\text{ rad/s})^2(0.800\text{ m}) = 80.0\text{ m/s}^2$$

The magnitude of the acceleration vector is

$$a = \sqrt{a_{tan}{}^2 + a_{rad}{}^2} = 89.4\text{ m/s}^2$$

EVALUATE: Note that we have dropped the unit "radian" from our results for a_{tan}, a_{rad}, and a. We can do this because "radian" is a dimensionless quantity.

The magnitude a is about nine times g, the acceleration due to gravity. Can you show that if the angular speed doubles to 20.0 rad/s while α remains the same, the acceleration magnitude a increases to 322 m/s^2, or almost $33g$?

Example 9.5 Designing a propeller

You are asked to design an airplane propeller to turn at 2400 rpm. The forward air speed of the plane is to be 75.0 m/s (270 km/h, or about 168 mi/h), and the speed of the tips of the propeller blades through the air must not exceed 270 m/s (Fig. 9.14a). (This is about 0.80 times the speed of sound in air. If the propeller tips were to move at the speed of sound, they would produce a tremendous amount of noise. Keeping the speed comfortably below this keeps the noise level acceptably low.) a) What is the maximum radius the propeller can have? b) With this radius, what is the acceleration of the propeller tip?

SOLUTION

IDENTIFY: The object of interest in this example is a particle at the tip of the propeller; our target variables are the particle's distance from the axis and its acceleration. Note that the speed of this particle through the air (which cannot exceed 270 m/s) is due to both the propeller's rotation *and* the forward motion of the airplane.

SET UP: As Fig. 9.14b shows, the velocity $\vec{v}_{total}$ of a particle at the propeller tip is the vector sum of its tangential velocity due to the propeller's rotation (magnitude v_P, given by Eq. (9.13)) and the forward velocity of the airplane (magnitude $v_A = 75.0$ m/s). The rotation plane of the propeller is perpendicular to the direction of flight, so these two vectors are perpendicular and we can use the Pythagorean theorem to relate v_P and v_A to v_{total}. We will then set $v_{total} = 270$ m/s and solve for the radius r. Note that the angular speed of the propeller is constant, so the acceleration of the propeller tip has only a radial component; we'll find it using Eq. (9.15).

EXECUTE: We first convert ω to rad/s (see Fig. 9.12):

$$\omega = 2400 \text{ rpm} = \left(2400 \frac{\text{rev}}{\text{min}}\right)\left(\frac{2\pi \text{ rad}}{1 \text{ rev}}\right)\left(\frac{1 \text{ min}}{60 \text{ s}}\right)$$
$$= 251 \text{ rad/s}$$

a) From Fig. 9.14b and Eq. (9.13), the velocity magnitude v_{total} is given by

$$v_{total}^2 = v_A^2 + v_P^2 = v_A^2 + r^2\omega^2$$

so

$$r^2 = \frac{v_{total}^2 - v_A^2}{\omega^2} \quad \text{and} \quad r = \frac{\sqrt{v_{total}^2 - v_A^2}}{\omega}$$

If $v_{total} = 270$ m/s, the propeller radius is

$$r = \frac{\sqrt{(270 \text{ m/s})^2 - (75.0 \text{ m/s})^2}}{251 \text{ rad/s}} = 1.03 \text{ m}$$

b) The purely centripetal acceleration is

$$a_{rad} = \omega^2 r$$
$$= (251 \text{ rad/s})^2(1.03 \text{ m}) = 6.5 \times 10^4 \text{ m/s}^2$$

EVALUATE: From $\Sigma\vec{F} = m\vec{a}$, the propeller must exert a force of 6.5×10^4 N on each kilogram of material at its tip! This is why propellers are made out of tough material (usually aluminum alloy).

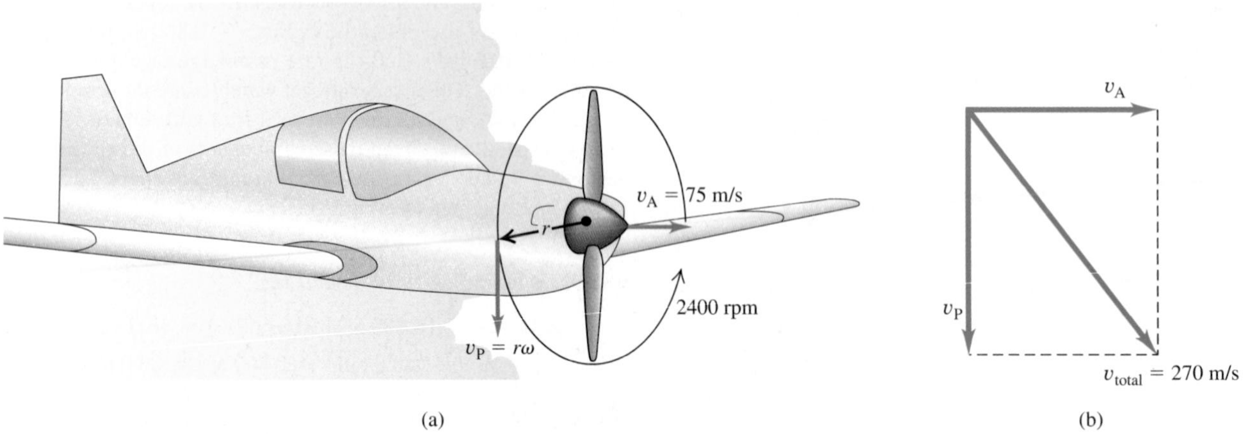

9.14 Designing an airplane propeller.

Conceptual Example 9.6 Bicycle gears

How are the angular speeds of the two bicycle sprockets in Fig. 9.15 related to the number of teeth on each sprocket?

SOLUTION

The chain does not slip or stretch, so it moves at the same tangential speed v on both sprockets. From Eq. (9.13),

$$v = r_1\omega_1 = r_2\omega_2 \quad \text{so} \quad \frac{\omega_2}{\omega_1} = \frac{r_1}{r_2}$$

The angular speed is inversely proportional to the radius. This relationship also holds for pulleys connected by a belt, provided that the belt doesn't slip. For chain sprockets the teeth must be equally spaced on the circumferences of both sprockets for the chain to mesh properly with both. Let N_1 and N_2 be the numbers of teeth; the condition that the tooth spacing is the same on both sprockets is

$$\frac{2\pi r_1}{N_1} = \frac{2\pi r_2}{N_2} \quad \text{or} \quad \frac{r_1}{r_2} = \frac{N_1}{N_2}$$

Combining this with the other equation, we get

$$\frac{\omega_2}{\omega_1} = \frac{N_1}{N_2}$$

The angular speed of each sprocket is inversely proportional to the number of teeth. On a multi-speed bike, you get the highest angular speed ω_2 of the rear wheel for a given pedaling rate ω_1 when the ratio N_1/N_2 is maximum; this means using the largest-radius front sprocket (largest N_1) and the smallest-radius rear sprocket (smallest N_2).

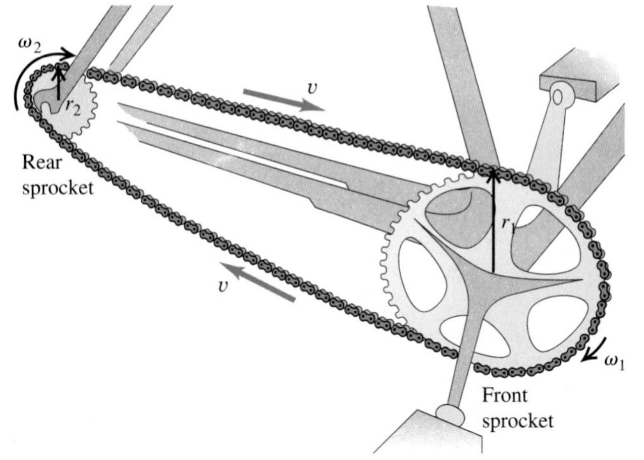

9.15 How are the angular velocities of the two sprockets related to their number of teeth?

Test Your Understanding

Information is stored on a CD or DVD (Fig. 9.9) in a coded pattern of tiny pits. The pits are arranged in a track that spirals outward toward the rim of the disc. As the disc spins inside a player, the track is scanned at a constant *linear* speed. How must the rotation speed of the disc change as the player's scanning head moves over the track?

9.4 | Energy in Rotational Motion

A rotating rigid body consists of mass in motion, so it has kinetic energy. We can express this kinetic energy in terms of the body's angular velocity and a new quantity called *moment of inertia* that we define in this section. To develop this relationship, we think of the body as being made up of a large number of particles, with masses $m_1, m_2, \ldots,$ at distances $r_1, r_2, \ldots$ from the axis of rotation. We label the particles with the index i: The mass of the ith particle is m_i and its distance from the axis of rotation is r_i. The particles don't necessarily all lie in the same plane, so we specify that r_i is the *perpendicular* distance from the axis to the ith particle.

When a rigid body rotates about a fixed axis, the speed v_i of the ith particle is given by Eq. (9.13), $v_i = r_i\omega$, where ω is the body's angular speed. Different particles have different values of r, but ω is the same for all (otherwise, the body wouldn't be rigid). The kinetic energy of the ith particle can be expressed as

$$\frac{1}{2}m_i v_i^2 = \frac{1}{2}m_i r_i^2 \omega^2$$

The *total* kinetic energy of the body is the sum of the kinetic energies of all its particles:

$$K = \frac{1}{2}m_1 r_1^2 \omega^2 + \frac{1}{2}m_2 r_2^2 \omega^2 + \cdots = \sum_i \frac{1}{2}m_i r_i^2 \omega^2$$

Taking the common factor $\omega^2/2$ out of this expression, we get

$$K = \frac{1}{2}(m_1 r_1^2 + m_2 r_2^2 + \cdots)\omega^2 = \frac{1}{2}\left(\sum_i m_i r_i^2\right)\omega^2$$

The quantity in parentheses, obtained by multiplying the mass of each particle by the square of its distance from the axis of rotation and adding these products, is denoted by I and is called the **moment of inertia** of the body for this rotation axis:

$$I = m_1 r_1^2 + m_2 r_2^2 + \cdots = \sum_i m_i r_i^2 \qquad (9.16)$$

(definition of moment of inertia)

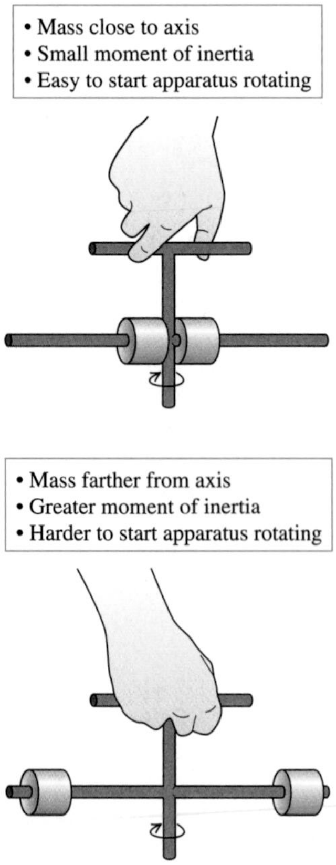

9.16 An apparatus free to rotate around a vertical axis. The moment of inertia can be varied by locking the two equal-mass cylinders into different positions on the horizontal shaft.

The word "moment" means that I depends on how the body's mass is distributed in space; it has nothing to do with a "moment" of time. For a body with a given rotation axis and a given total mass, the greater the distance from the axis to the particles that make up the body, the greater the moment of inertia. In a rigid body, the distances r_i are all constant and I is independent of how the body is rotating around the given axis. The SI unit of moment of inertia is the kilogram-meter^2 $(\text{kg}\cdot\text{m}^2)$.

In terms of moment of inertia I, the **rotational kinetic energy** K of a rigid body is

$$K = \frac{1}{2}I\omega^2 \quad \text{(rotational kinetic energy of a rigid body)} \qquad (9.17)$$

The kinetic energy given by Eq. (9.17) is not a new form of energy; it's the sum of the kinetic energies of the individual particles that make up the rigid body, written in a compact and convenient form in terms of the moment of inertia. When using Eq. (9.17), ω *must* be measured in radians per second, not revolutions or degrees per second, to give K in joules; this is because we used $v_i = r_i\omega$ in our derivation.

Equation (9.17) gives a simple physical interpretation of moment of inertia: the greater the moment of inertia, the greater the kinetic energy of a rigid body rotating with a given angular speed ω. We learned in Chapter 6 that the kinetic energy of a body equals the amount of work done to accelerate that body from rest. So the greater a body's moment of inertia, the harder it is to start the body rotating if it's at rest and the harder it is to stop its rotation if it's already rotating (Fig. 9.16). For this reason, I is also called the *rotational inertia*.

Example 9.7 shows how *changing* the rotation axis can affect the value of I.

Example 9.7 Moments of inertia for different rotation axes

An engineer is designing a one-piece machine part consisting of three heavy connectors linked by light molded struts (Fig. 9.17). a) What is the moment of inertia of this body about an axis through point A, perpendicular to the plane of the diagram? b) What is the moment of inertia about an axis coinciding with rod BC? c) If the body rotates about an axis through A perpendicular to the plane of the diagram, with angular speed $\omega = 4.0$ rad/s, what is its kinetic energy?

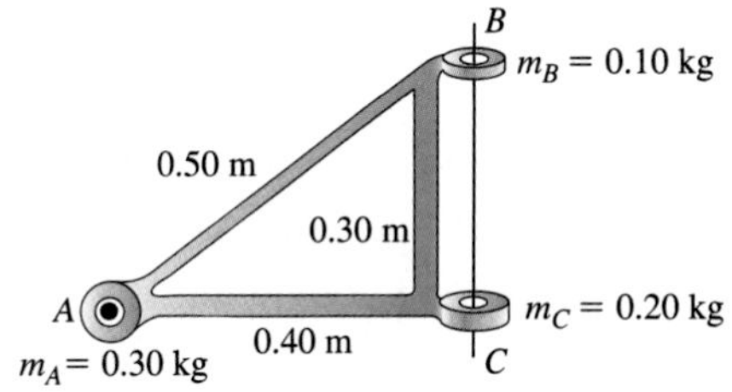

9.17 A strangely shaped machine part.

SOLUTION

IDENTIFY: We'll consider the connectors as massive particles and the lightweight struts as massless rods. Then we can use the ideas of this section to calculate the moment of inertia of this collection of three particles.

SET UP: In parts (a) and (b), we'll use Eq. (9.16) to find the moments of inertia for each of the two axes. Given the moment of inertia for axis A, we'll use Eq. (9.17) in part (c) to find the rotational kinetic energy.

EXECUTE: a) The particle at point A lies *on* the axis. Its distance r from the axis is zero, so it contributes nothing to the moment of inertia. Equation (9.16) gives

$$I = \sum m_i r_i^2 = (0.10\text{ kg})(0.50\text{ m})^2 + (0.20\text{ kg})(0.40\text{ m})^2$$
$$= 0.057\text{ kg}\cdot\text{m}^2$$

b) The particles at B and C both lie *on* the axis, so for them $r = 0$ and neither contributes to the moment of inertia. Only A contributes, and we have

$$I = \sum m_i r_i^2 = (0.30\text{ kg})(0.40\text{ m})^2 = 0.048\text{ kg}\cdot\text{m}^2$$

c) From Eq. (9.17),

$$K = \frac{1}{2}I\omega^2 = \frac{1}{2}(0.057\text{ kg}\cdot\text{m}^2)(4.0\text{ rad/s})^2 = 0.46\text{ J}$$

EVALUATE: Our results show that the moment of inertia for the axis through A is greater than that for the axis through B and C. Hence, of the two axes, it's easier to make the machine part rotate about axis BC.

CAUTION The results of parts (a) and (b) of Example 9.7 show that the moment of inertia of a body depends on the location and orientation of the axis. It's not enough to just say, "The moment of inertia of this body is $0.048\text{ kg}\cdot\text{m}^2$." We have to be specific and say, "The moment of inertia of this body *about axis BC* is $0.048\text{ kg}\cdot\text{m}^2$."

ActivPhysics Online
7.6 Rotational Inertia

In Example 9.7 we represented the body as several point masses, and we evaluated the sum in Eq. (9.16) directly. When the body is a *continuous* distribution of matter, such as a solid cylinder or plate, the sum becomes an integral, and we need to use calculus to calculate the moment of inertia. We will give several examples of such calculations in Section 9.6; meanwhile, Table 9.2 gives moments of inertia for several familiar shapes in terms of the masses and dimensions. Each body shown in Table 9.2 is *uniform;* that is, the density has the same value at all points within the solid parts of the body.

CAUTION You may be tempted to try to compute the moment of inertia of a body by assuming that all the mass is concentrated at the center of mass and then multiplying the total mass by the square of the distance from the center of mass to the axis. Resist that temptation; it doesn't work! For example, when a

Table 9.2 Moments of Inertia of Various Bodies

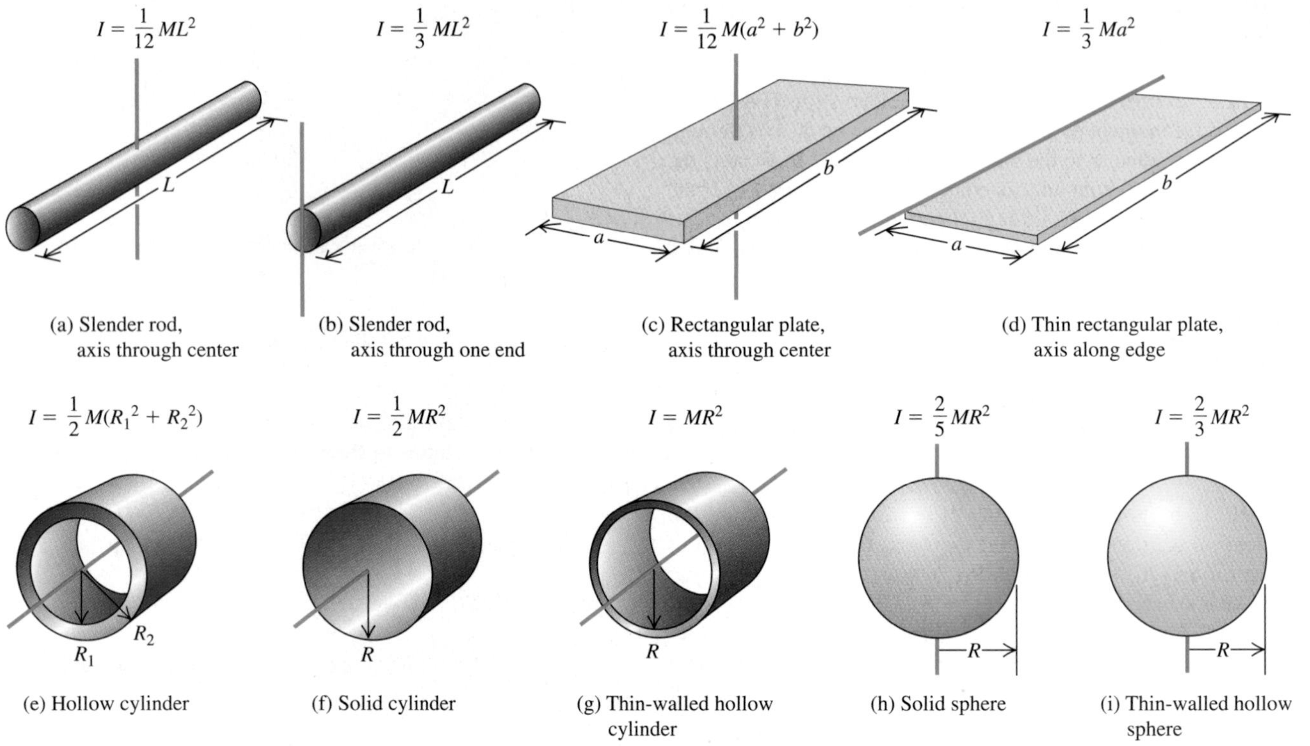

uniform thin rod of length L and mass M is pivoted about an axis through one end, perpendicular to the rod, the moment of inertia is $I = ML^2/3$ (case (b) in Table 9.2). If we took the mass as concentrated at the center, a distance $L/2$ from the axis, we would obtain the *incorrect* result $I = M(L/2)^2 = ML^2/4$.

Now that we know how to calculate the kinetic energy of a rotating rigid body, we can apply the energy principles of Chapter 7 to rotational motion. Here are some points of strategy and some examples.

Problem-Solving Strategy: Rotational Energy

IDENTIFY *the relevant concepts:* You can use work–energy relations and conservation of energy to find relations involving position and motion of a rigid body rotating around a fixed axis. As we saw in Chapter 7, the energy method is usually not helpful for problems that involve elapsed time. In Chapter 10 we'll see how to approach rotational problems of this kind.

SET UP *the problem* using the same steps as in the Problem-Solving Strategy in Section 7.1, with the following addition:

5. Many problems involve a rope or cable wrapped around a rotating rigid body, which functions as a pulley. In these situations, remember that the point on the pulley that contacts the rope has the same linear speed as the rope, provided the rope doesn't slip on the pulley. You can then take advantage of Eqs. (9.13) and (9.14), which relate the linear speed and tangential acceleration of a point on a rigid body to the angular velocity and angular acceleration of the body. Examples 9.8 and 9.9 illustrate this point.

EXECUTE *the solution:* As in Chapter 7, write expressions for the initial and final kinetic and potential energies $(K_1, K_2, U_1,$ and $U_2)$ and the nonconservative work W_{other} (if any). The new feature is rotational kinetic energy, which is expressed in terms of the body's moment of inertia I for the given axis and its angular speed ω $\left(K = \frac{1}{2}I\omega^2\right)$ instead of its mass m and speed v. Substitute these expressions into $K_1 + U_1 + W_{\text{other}} = K_2 + U_2$ (if nonconservative work is done) or $K_1 + U_1 = K_2 + U_2$ (if only conservative work is done) and solve for the target variable(s). As in Chapter 7, it's helpful to draw bar graphs showing the initial and final values of K, U, and $E = K + U$.

EVALUATE *your answer:* As always, check whether your answer makes physical sense.

Example 9.8 An unwinding cable I

A light, flexible, nonstretching cable is wrapped several times around a winch drum, a solid cylinder of mass 50 kg and diameter 0.120 m, which rotates about a stationary horizontal axis held by frictionless bearings (Fig. 9.18). The free end of the cable is pulled with a constant force of magnitude 9.0 N for a distance of 2.0 m. It unwinds without slipping, turning the cylinder as it does so. If the cylinder is initially at rest, find its final angular speed and the final speed of the cable.

SOLUTION

IDENTIFY: We will solve this problem using energy methods. Point 1 is when the cylinder first begins to move and point 2 is when the cable has moved 2.0 m. Because the cable is light, we'll assume that it's massless, so that only the cylinder has kinetic energy. The vertical position of the cylinder doesn't change, so there are no changes in gravitational potential energy. There is friction between the cable and the cylinder, which is what makes the cylinder rotate when the cable is pulled. But because the cable doesn't slip, there is no sliding of the cable relative to the cylinder and no mechanical energy is lost in friction. Because the cable is massless, the force that the cable exerts on the cylinder rim is equal to the applied force F.

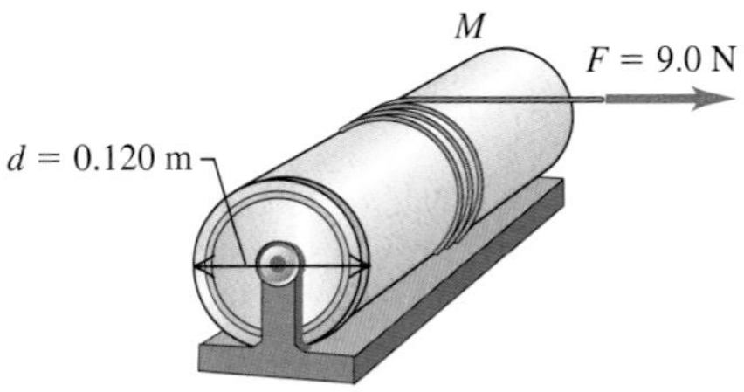

9.18 A cable unwinds from a cylinder. The force F does work and increases the kinetic energy of the cylinder.

SET UP: The cylinder is initially at rest, so the initial kinetic energy is $K_1 = 0$. Between points 1 and 2, the force F does work on the cylinder over a distance $s = 2.0$ m. As a result, the kinetic energy at point 2 is $K_2 = \frac{1}{2}I\omega^2$. One of our target variables is ω; the other is the speed of the cable at point 2, which is equal to the tangential speed v of the cylinder at that point. We'll find v from ω by using Eq. (9.13).

EXECUTE: The work done on the cylinder is $W_{\text{other}} = Fs = (9.0\text{ N})(2.0\text{ m}) = 18$ J. From Table 9.2 the moment of inertia is

$$I = \frac{1}{2}MR^2 = \frac{1}{2}(50\text{ kg})(0.060\text{ m})^2 = 0.090\text{ kg}\cdot\text{m}^2$$

(The radius R is just half the diameter of the cylinder.) The relationship $K_1 + U_1 + W_{\text{other}} = K_2 + U_2$ then gives

$$0 + 0 + W_{\text{other}} = \frac{1}{2}I\omega^2 + 0$$

$$\omega = \sqrt{\frac{2W_{\text{other}}}{I}} = \sqrt{\frac{2(18\text{ J})}{0.090\text{ kg}\cdot\text{m}^2}}$$

$$= 20\text{ rad/s}$$

The final tangential speed of the cylinder, and hence the final speed of the cable, is

$$v = R\omega = (0.060\text{ m})(20\text{ rad/s}) = 1.2\text{ m/s}$$

EVALUATE: We assumed that the cable has negligible mass. If the mass of the cable can't be neglected, then some of the work done would go into the kinetic energy of the cable. Hence the cylinder would end up with less kinetic energy and a smaller angular speed than we calculated here.

Example 9.9 An unwinding cable II

In a lab experiment to test conservation of energy in rotational motion, we wrap a light, flexible cable around a solid cylinder with mass M and radius R. The cylinder rotates with negligible friction about a stationary horizontal axis (Fig. 9.19). We tie the free end of the cable to an object of mass m and release the object with no initial velocity at a distance h above the floor. As the object falls, the

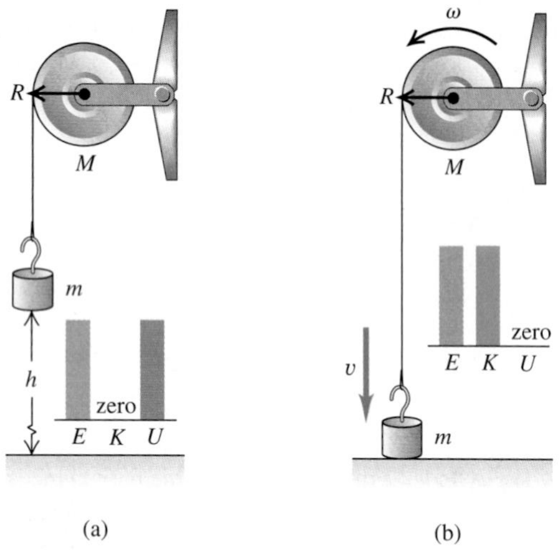

9.19 (a) A cable wrapped around a cylinder is attached to a hanging object. (b) As the cable unwinds, gravitational potential energy is converted into kinetic energy of the falling object and the rotating cylinder.

cable unwinds without stretching or slipping, turning the cylinder. Find the speed of the falling object and the angular speed of the cylinder just as the object strikes the floor.

SOLUTION

IDENTIFY: As in Example 9.8, the cable doesn't slip and friction does no work. (The cable does no net work; at its upper end the force and displacement are in the same direction, and at its lower end they are in opposite directions. Thus the total work done by both ends of the cable is zero.) The only work done is by gravity, so mechanical energy is conserved.

SET UP: Initially the system has no kinetic energy (Fig. 9.19a), so $K_1 = 0$. We take the potential energy to be zero when the object is at floor level; then $U_1 = mgh$ and $U_2 = 0$. (We can ignore the gravitational potential energy for the rotating cylinder, since its height doesn't change.) Just before the falling object hits the floor, both this object and the cylinder have kinetic energy (Fig. 9.19b). The total kinetic energy K_2 at that instant is

$$K_2 = \frac{1}{2}mv^2 + \frac{1}{2}I\omega^2$$

From Table 9.2 the moment of inertia of the cylinder is $I = \frac{1}{2}MR^2$. Also, v and ω are related by $v = R\omega$, since the speed of the falling mass must be equal to the tangential speed at the outer surface of the cylinder. We'll use these relationships to solve for the target variables v and ω at point 2 (Fig. 9.19b).

EXECUTE: Using our expressions for K_1, U_1, K_2, and U_2 and the relation $\omega = v/R$ in the energy-conservation equation $K_1 + U_1 = K_2 + U_2$, we find

$$0 + mgh = \frac{1}{2}mv^2 + \frac{1}{2}\left(\frac{1}{2}MR^2\right)\left(\frac{v}{R}\right)^2 + 0 = \frac{1}{2}\left(m + \frac{1}{2}M\right)v^2$$

$$v = \sqrt{\frac{2gh}{1 + M/2m}}$$

The final angular speed ω is obtained from $\omega = v/R$.

EVALUATE: Let's check some particular cases. When M is much larger than m, v is very small, as we would expect. When M is much smaller than m, v is nearly equal to $\sqrt{2gh}$, which is the speed of a body that falls freely from an initial height h. Does it surprise you that v doesn't depend on the radius of the cylinder?

ActivPhysics ONLINE

7.12 Woman and Flywheel Elevator—Energy Approach

7.13 Rotoride—Energy Approach

Gravitational Potential Energy for an Extended Body

In Example 9.9 the cable was of negligible mass, so we could ignore its kinetic energy as well as the gravitational potential energy associated with it. If the mass is *not* negligible, we need to know how to calculate the *gravitational potential energy* associated with such an extended body. If the acceleration of gravity g is the same at all points on the body, the gravitational potential energy is the same as though all the mass were concentrated at the center of mass of the body. Suppose we take the y-axis vertically upward. Then for a body with total mass M, the gravitational potential energy U is simply

$$U = Mgy_{cm} \tag{9.18}$$

(gravitational potential energy for an extended body)

where y_{cm} is the y-coordinate of the center of mass. This expression applies to any extended body, whether it is rigid or not (Fig. 9.20).

To prove Eq. (9.18), we again represent the body as a collection of mass elements m_i. The potential energy for element m_i is $m_i g y_i$, so the total potential energy is

$$U = m_1 g y_1 + m_2 g y_2 + \cdots = (m_1 y_1 + m_2 y_2 + \cdots) g$$

But from Eq. (8.28), which defines the coordinates of the center of mass,

$$m_1 y_1 + m_2 y_2 + \cdots = (m_1 + m_2 + \cdots) y_{cm} = M y_{cm}$$

where $M = m_1 + m_2 + \cdots$ is the total mass. Combining this with the above expression for U, we find $U = Mgy_{cm}$, in agreement with Eq. (9.18).

We leave the application of Eq. (9.18) to the problems. We'll make use of this relationship in Chapter 10 in the analysis of rigid-body problems where the axis of rotation moves.

9.20 In a technique called the "Fosbury flop" after its innovator, this athlete arches his body as he passes over the bar in the high jump. As a result, his center of mass actually passes *under* the bar. This technique requires a smaller increase in gravitational potential energy [Eq. (9.18)] than the older method of straddling the bar.

Test Your Understanding

Suppose the cylinder and hanging object in Example 9.9 have the same mass, so $m = M$. Just before the object strikes the floor, what fraction of the total kinetic energy of the system is in the motion of the object and what fraction is in the rotation of the cylinder?

9.5 | Parallel-Axis Theorem

We pointed out in Section 9.4 that a body doesn't have just one moment of inertia. In fact, it has infinitely many, because there are infinitely many axes about which it might rotate. But there is a simple relationship between the moment of inertia I_{cm} of a body of mass M about an axis through its center of mass and the moment of inertia I_P about any other axis parallel to the original one but displaced from it by a distance d. This relationship, called the **parallel-axis theorem,** states that

$$I_P = I_{cm} + Md^2 \qquad \text{(parallel-axis theorem)} \qquad (9.19)$$

To prove this theorem, we consider two axes, both parallel to the z-axis, one through the center of mass and the other through a point P (Fig. 9.21). First we take a very thin slice of the body, parallel to the xy-plane and perpendicular to the z-axis. We take the origin of our coordinate system to be at the center of mass of the body; the coordinates of the center of mass are then $x_{cm} = y_{cm} = z_{cm} = 0$. The axis through the center of mass passes through this thin slice at point O, and the parallel axis passes through point P, whose x- and y-coordinates are (a, b). The distance of this axis from the axis through the center of mass is d, where $d^2 = a^2 + b^2$.

We can write an expression for the moment of inertia I_P about the axis through point P. Let m_i be a mass element in our slice, with coordinates (x_i, y_i, z_i). Then the moment of inertia I_{cm} of the slice about the axis through the center of mass (at O) is

$$I_{cm} = \sum_i m_i (x_i^2 + y_i^2)$$

The moment of inertia of the slice about the axis through P is

$$I_P = \sum_i m_i [(x_i - a)^2 + (y_i - b)^2]$$

These expressions don't involve the coordinates z_i measured perpendicular to the slices, so we can extend the sums to include *all* particles in *all* slices. Then I_P

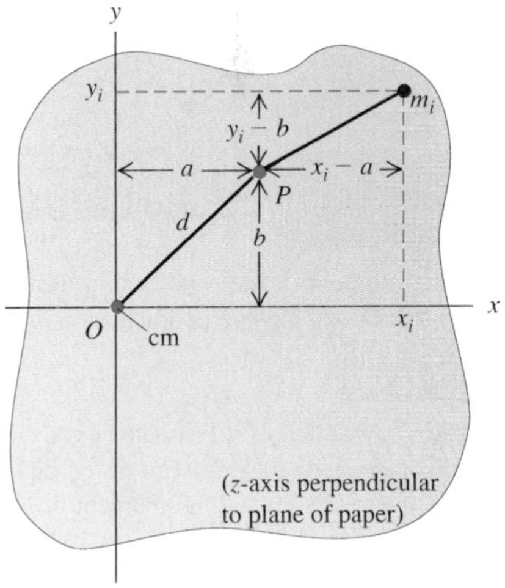

9.21 The mass element m_i has coordinates (x_i, y_i) with respect to an axis of rotation through the center of mass and perpendicular to the plane of the figure. The mass element has coordinates $(x_i - a, y_i - b)$ with respect to the parallel axis through point P.

becomes the moment of inertia of the *entire* body for an axis through P. We then expand the squared terms and regroup, and obtain

$$I_P = \sum_i m_i(x_i^2 + y_i^2) - 2a\sum_i m_i x_i - 2b\sum_i m_i y_i + (a^2 + b^2)\sum_i m_i$$

The first sum is I_{cm}. From Eq. (8.28), the definition of the center of mass, the second and third sums are proportional to x_{cm} and y_{cm}; these are zero because we have taken our origin to be the center of mass. The final term is d^2 multiplied by the total mass, or Md^2. This completes our proof that $I_P = I_{cm} + Md^2$.

As Eq. (9.19) shows, a rigid body has a lower moment of inertia about an axis through its center of mass than about any other parallel axis. Thus it's easier to start a body rotating if the rotation axis passes through the center of mass. This suggests that it's somehow most natural for a rotating body to rotate about an axis through its center of mass; we'll make this idea more quantitative in Chapter 10.

Example 9.10 Using the parallel-axis theorem I

A part of a mechanical linkage (Fig. 9.22) has a mass of 3.6 kg. We measure its moment of inertia about an axis 0.15 m from its center of mass to be $I_P = 0.132\ \text{kg}\cdot\text{m}^2$. What is the moment of inertia I_{cm} about a parallel axis through the center of mass?

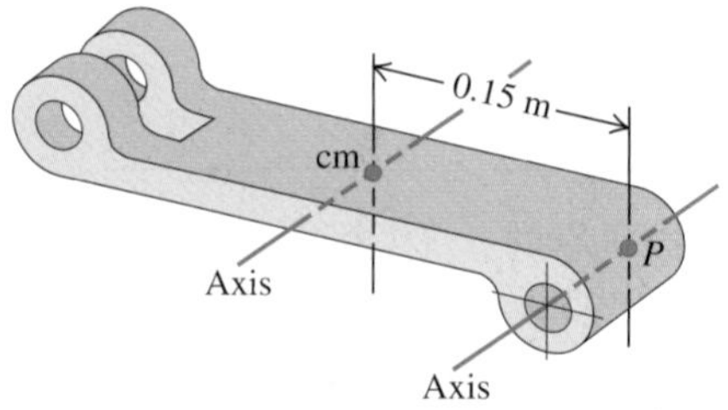

9.22 Calculating I_{cm} from a measurement of I_P.

SOLUTION

IDENTIFY AND SET UP: We'll use Eq. (9.19) to determine the target variable I_{cm}.

EXECUTE: Rearranging the equation and substituting the values,

$$I_{cm} = I_P - Md^2 = 0.132\ \text{kg}\cdot\text{m}^2 - (3.6\ \text{kg})(0.15\ \text{m})^2 = 0.051\ \text{kg}\cdot\text{m}^2$$

EVALUATE: Our result shows that I_{cm} is less than I_P. This is as it should be: as we saw earlier, the moment of inertia for an axis through the center of mass is lower than for any other parallel axis.

Example 9.11 Using the parallel-axis theorem II

Find the moment of inertia of a thin uniform disk of mass M and radius R about an axis perpendicular to its plane at the edge.

SOLUTION

IDENTIFY AND SET UP: Here the target variable in Eq. (9.19) is I_P. The distance from P (at the rim of the disk) to the disk's center of mass is the radius R, and the moment of inertia through the center of mass is given in Table 9.2: $I_{cm} = MR^2/2$.

EXECUTE: The parallel-axis theorem gives

$$I_P = I_{cm} + Md^2 = \frac{MR^2}{2} + MR^2 = \frac{3MR^2}{2}$$

EVALUATE: As in the previous example, $I_{cm} < I_P$. This is a very simple solution of a problem that would be fairly complicated if we tried to solve it using the integration techniques to be discussed in the next section.

Test Your Understanding

The discus in Example 9.4 (Section 9.3) is a uniform disk of mass 2.00 kg and radius 11.0 cm. Find its moment of inertia about the thrower's shoulder, 80.0 cm from the center of the disk (Fig. 9.13).

*9.6 | Moment of Inertia Calculations

When a rigid body cannot be represented by a few point masses but is a continuous distribution of mass, the sum of masses and distances that defines the moment of inertia [Eq. (9.16)] becomes an integral. Imagine dividing the body into small mass elements dm so that all points in a particular element are at essentially the same perpendicular distance from the axis of rotation. We call this distance r, as before. Then the moment of inertia is

$$I = \int r^2\, dm \qquad (9.20)$$

To evaluate the integral, we have to represent r and dm in terms of the same integration variable. When we have an effectively one-dimensional object, such as the slender rods (a) and (b) in Table 9.2, we can use a coordinate x along the length and relate dm to an increment dx. For a three-dimensional object it is usually easiest to express dm in terms of an element of volume dV and the *density* ρ of the body. Density is mass per unit volume, $\rho = dm/dV$, so we may also write Eq. (9.20) as

$$I = \int r^2 \rho\, dV$$

This expression tells us that a body's moment of inertia depends on how its density varies within its volume (Fig. 9.23). If the body is uniform in density, then we may take ρ outside the integral:

$$I = \rho \int r^2\, dV \qquad (9.21)$$

To use this equation, we have to express the volume element dV in terms of the differentials of the integration variables, such as $dV = dx\, dy\, dz$. The element dV must always be chosen so that all points within it are at very nearly the same distance from the axis of rotation. The limits on the integral are determined by the shape and dimensions of the body. For regularly shaped bodies, this integration can often be carried out quite easily.

9.23 By measuring small variations in the orbits of satellites, geophysicists can measure the earth's moment of inertia. This tells us how our planet's mass is distributed within its interior. The data show that the earth is far denser at the core than in its outer layers.

Example 9.12 Uniform thin rod, axis perpendicular to length

Figure 9.24 shows a slender uniform rod with mass M and length L. It might be a baton held by a twirler in a marching band (less the rubber end caps). Compute its moment of inertia about an axis through O, at an arbitrary distance h from one end.

SOLUTION

IDENTIFY AND SET UP: We choose as an element of mass a short section of rod with length dx at a distance x from point O. The ratio of the mass dm of this element to the total mass M is equal to the ratio of its length dx to the total length L:

$$\frac{dm}{M} = \frac{dx}{L} \quad \text{so} \quad dm = \frac{M}{L}dx$$

We'll determine I from Eq. (9.20) with r replaced by x (see Fig. 9.24).

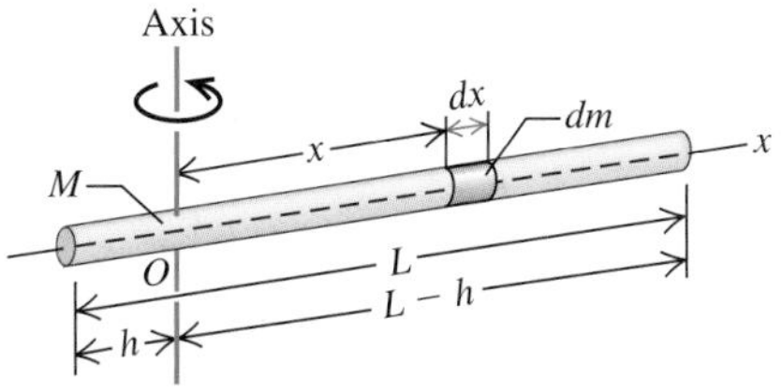

9.24 Finding the moment of inertia of a thin rod about an axis through O. The mass element is a segment with length dx.

EXECUTE: Figure 9.24 shows that the integration limits on x are from $-h$ to $(L - h)$. Hence we obtain

$$I = \int x^2\, dm = \frac{m}{L}\int_{-h}^{L-h} x^2\, dx$$
$$= \left[\frac{M}{L}\left(\frac{x^3}{3}\right)\right]_{-h}^{L-h} = \frac{1}{3}M(L^2 - 3Lh + 3h^2)$$

EVALUATE: From this general expression we can find the moment of inertia about an axis through any point on the rod. For example, if the axis is at the left end, $h = 0$ and

$$I = \frac{1}{3}ML^2$$

If the axis is at the right end, we should get the same result. Putting $h = L$, we again get

$$I = \frac{1}{3}ML^2$$

If the axis passes through the center, the usual place for a twirled baton, then $h = L/2$ and

$$I = \frac{1}{12}ML^2$$

These results agree with the expressions in Table 9.2.

Example 9.13 Hollow or solid cylinder, rotating about axis of symmetry

Figure 9.25 shows a hollow, uniform cylinder with length L, inner radius R_1, and outer radius R_2. It might be a steel cylinder in a printing press or a sheet-steel rolling mill. Find the moment of inertia about the axis of symmetry of the cylinder.

SOLUTION

IDENTIFY AND SET UP: We choose as a volume element a thin cylindrical shell of radius r, thickness dr, and length L. All parts of this element are at very nearly the same distance from the axis. The volume of the element is very nearly equal to that of a flat sheet with thickness dr, length L, and width $2\pi r$ (the circumference of the shell). Then

$$dm = \rho\, dV = \rho(2\pi rL\, dr)$$

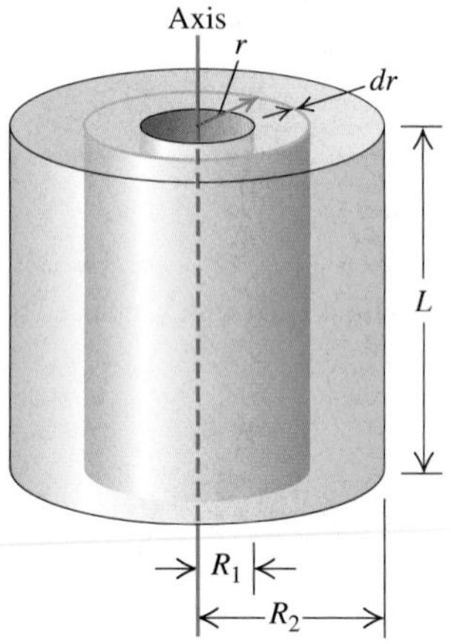

9.25 Finding the moment of inertia of a hollow cylinder about its symmetry axis. The mass element is a cylindrical shell with radius r and thickness dr.

We will use this expression in Eq. (9.20) and integrate from $r = R_1$ to $r = R_2$.

EXECUTE: The moment of inertia is given by

$$I = \int r^2\, dm = \int_{R_1}^{R_2} r^2 \rho(2\pi rL\, dr)$$
$$= 2\pi\rho L\int_{R_1}^{R_2} r^3\, dr$$
$$= \frac{2\pi\rho L}{4}(R_2^4 - R_1^4)$$
$$= \frac{\pi\rho L}{2}(R_2^2 - R_1^2)(R_2^2 + R_1^2)$$

It is usually more convenient to express the moment of inertia in terms of the total mass M of the body, which is its density ρ multiplied by the total volume V. The volume is

$$V = \pi L(R_2^2 - R_1^2)$$

so the total mass M is

$$M = \rho V = \pi L\rho(R_2^2 - R_1^2)$$

and the moment of inertia is

$$I = \frac{1}{2}M(R_1^2 + R_2^2)$$

EVALUATE: This agrees with Table 9.2, case (e). If the cylinder is solid, such as a lawn roller, $R_1 = 0$. Calling the outer radius R_2 simply R, we find that the moment of inertia of a solid cylinder of radius R is

$$I = \frac{1}{2}MR^2$$

If the cylinder has a very thin wall (like a stovepipe), R_1 and R_2 are very nearly equal; if R represents this common radius,

$$I = MR^2$$

We could have predicted this last result; in a thin-walled cylinder, all the mass is the same distance $r = R$ from the axis, so $I = \int r^2\,dm = R^2 \int dm = MR^2$.

Note that the moment of inertia of a cylinder about an axis coinciding with its axis of symmetry depends on its mass and radii but not on its length L. Two hollow cylinders with the same inner and outer radii, one of wood and one of brass, but having the same mass M, have equal moments of inertia even though the length of the wood cylinder is much greater. The moment of inertia depends only on the *radial* distribution of mass, not on its distribution along the axis. Thus the above results hold also for a thin disk, such as a washer or a compact disc (CD).

Example 9.14 Uniform sphere with radius R, axis through center

The object in Fig. 9.26 might be a billiard ball, a steel ball bearing, or the world's largest ball of string. Find the moment of inertia about an axis through the center of the sphere.

SOLUTION

IDENTIFY AND SET UP: We divide the sphere into thin disks of thickness dx, whose moment of inertia we know from Example 9.13. The radius r of the disk shown is

$$r = \sqrt{R^2 - x^2}$$

Its volume is

$$dV = \pi r^2\,dx = \pi(R^2 - x^2)\,dx$$

and its mass is

$$dm = \rho\,dV = \pi\rho(R^2 - x^2)\,dx$$

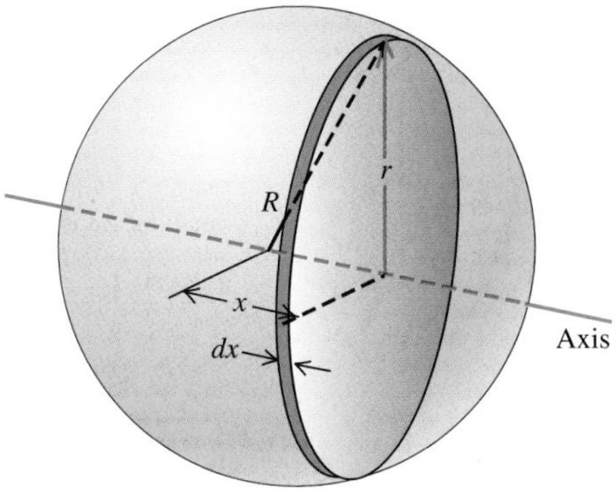

9.26 Finding the moment of inertia of a sphere about an axis through its center. The mass element is a disk with thickness dx.

EXECUTE: From Example 9.13, the moment of inertia of a disk of radius r and mass dm is

$$dI = \frac{1}{2}r^2\,dm = \frac{1}{2}(\sqrt{R^2 - x^2})^2[\pi\rho(R^2 - x^2)\,dx]$$
$$= \frac{\pi\rho}{2}(R^2 - x^2)^2\,dx$$

Integrating this expression from $x = 0$ to $x = R$ gives the moment of inertia of the right hemisphere. From symmetry, the total I for the entire sphere is just twice this:

$$I = (2)\frac{\pi\rho}{2}\int_0^R (R^2 - x^2)^2\,dx$$

Carrying out the integration, we obtain

$$I = \frac{8\pi\rho}{15}R^5$$

The mass M of the sphere of volume $V = 4\pi R^3/3$ is

$$M = \rho V = \frac{4\pi\rho R^3}{3}$$

By comparing the expressions for I and M, we find

$$I = \frac{2}{5}MR^2$$

EVALUATE: This result agrees with the expression in Table 9.2, case (h). Note that the moment of inertia of a solid sphere of mass M and R is less than the moment of inertia of a solid *cylinder* of the same mass and radius, $I = \frac{1}{2}MR^2$. The reason is that more of the sphere's mass is located close to the axis.

Test Your Understanding

Consider the rod in Example 9.12. Use calculus to show that the value of h that gives the smallest moment of inertia is $h = L/2$.

SUMMARY

When a rigid body rotates about a stationary axis (usually called the z-axis), its position is described by an angular coordinate θ. The angular velocity ω_z is the time derivative of the angular coordinate θ: The angular acceleration α_z is the time derivative of the angular velocity ω_z or the second derivative of the angular coordinate θ. (See Examples 9.1 and 9.2)

$$\omega_z = \lim_{\Delta t \to 0} \frac{\Delta\theta}{\Delta t} = \frac{d\theta}{dt} \quad (9.3)$$

$$\alpha_z = \lim_{\Delta t \to 0} \frac{\Delta\omega_z}{\Delta t} = \frac{d\omega_z}{dt} = \frac{d^2\theta}{dt^2} \quad (9.5, 9.6)$$

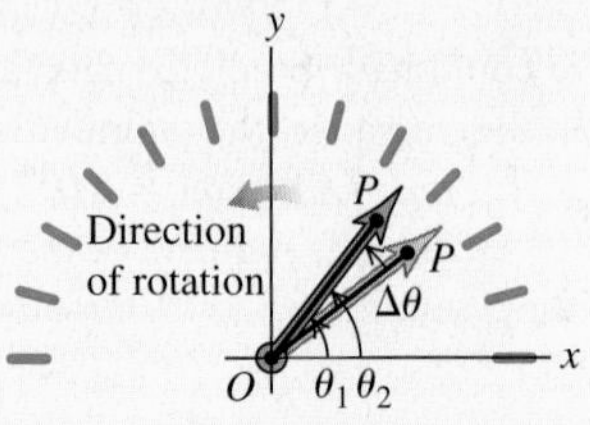

When a rigid body rotates about a fixed axis with constant angular acceleration, the angular coordinate, velocity, and acceleration are related by simple kinematic equations. These involve θ_0 and ω_{0z}, the initial values of the angular coordinate and angular velocity, respectively. (See Example 9.3)

$$\theta = \theta_0 + \omega_{0z}t + \frac{1}{2}\alpha_z t^2 \quad (9.11)$$

$$\theta - \theta_0 = \frac{1}{2}(\omega_z + \omega_{0z})t \quad (9.10)$$

$$\omega_z = \omega_{0z} + \alpha_z t \quad (9.7)$$

$$\omega_z^2 = \omega_{0z}^2 + 2\alpha_z(\theta - \theta_0) \quad (9.12)$$

The angular speed ω of a rigid body is the magnitude of its angular velocity. The rate of change of ω is $\alpha = d\omega/dt$. For a particle in the body a distance r from the rotation axis, the speed v and the components of the acceleration $\vec{a}$ are related to ω and α. (See Examples 9.4 through 9.6)

$$v = r\omega \quad (9.13)$$

$$a_{tan} = \frac{dv}{dt} = r\frac{d\omega}{dt} = r\alpha \quad (9.14)$$

$$a_{rad} = \frac{v^2}{r} = \omega^2 r \quad (9.15)$$

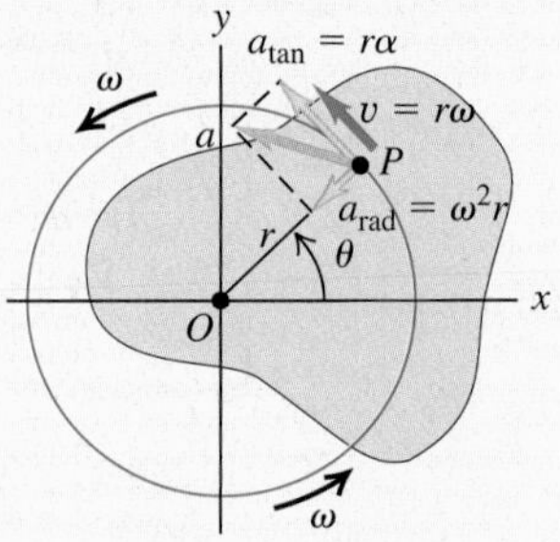

The moment of inertia I of a body about a given axis is a measure of its rotational inertia: the greater the value of I, the more difficult it is to change the state of the body's rotation. The moment of inertia can be expressed as a sum over the particles m_i that make up the body, each of which is at its own perpendicular distance r_i from the axis. (See Examples 9.7 and 9.12 through 9.14)

$$I = m_1 r_1^2 + m_2 r_2^2 + \cdots = \sum_i m_i r_i^2 \quad (9.16)$$

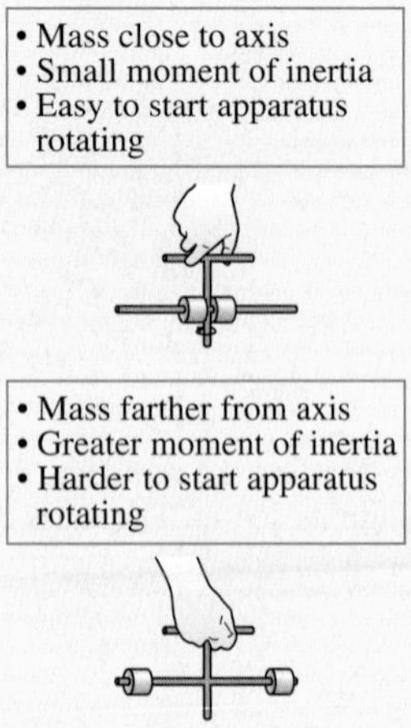

The rotational kinetic energy of a rigid body rotating about a fixed axis depends on the angular speed ω and the moment of inertia I for that rotation axis. (See Examples 9.8 and 9.9)	$K = \frac{1}{2}I\omega^2$ (9.17)	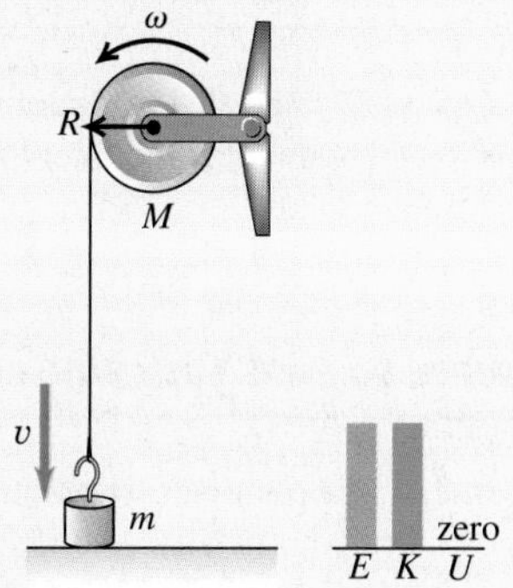
The moment of inertia I_{cm} of a body of mass M about an axis through the center of mass is related to the moment of inertia I_P about a parallel axis at a distance d from the first axis. (See Examples 9.10 and 9.11)	$I_P = I_{cm} + Md^2$ (9.19)	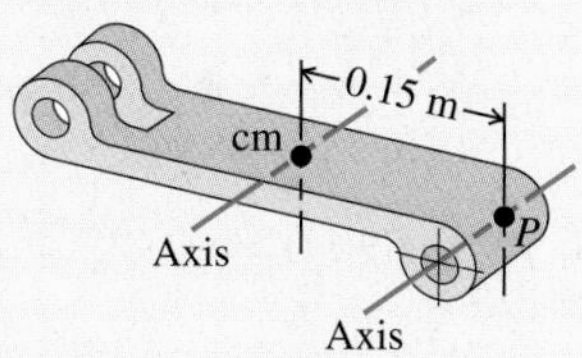

Key Terms

rigid body, 327
radian, 328
average angular velocity, 329
angular displacement, 329
instantaneous angular velocity, 329
average angular acceleration, 331
instantaneous angular acceleration, 331
angular speed, 336
tangential component of acceleration, 336
centripetal component of acceleration, 336
moment of inertia, 340
rotational kinetic energy, 340
parallel-axis theorem, 345

Your Notes

Answer to Chapter Opening Question

From Eqs. (9.13) and (9.15), doubling the distance r for the same angular speed ω doubles the linear speed $v = r\omega$ and doubles the radial acceleration $a_{rad} = \omega^2 r$.

Answers to Test Your Understanding Questions

Section 9.1 The rotation is speeding up between $t = 0$ and $t = 2$ s (ω_z and α_z are both positive) and between $t = 4$ s and $t = 6$ s (ω_z and α_z are both negative). The rotation is slowing down between $t = 2$ s and $t = 4$ s (ω_z is positive and α_z is negative).
Section 9.2 When the DVD comes to rest, $\omega_z = 0$. From Eq. (9.7), the time when this occurs is

$$t = \frac{\omega_z - \omega_{0z}}{\alpha_z} = \frac{0 - 27.5\text{ rad/s}}{-10.0\text{ rad/s}^2} = 2.75\text{ s}$$

Either Eq. (9.10), (9.11), or (9.12) can be used to find the angle θ. Using Eq. (9.10),

$$\theta = \theta_0 + \frac{1}{2}(\omega_z + \omega_{0z})t = 0 + \frac{1}{2}(0 + 27.5\text{ rad/s})(2.75\text{ s})$$

$$= 37.8\text{ rad} = 37.8\text{ rad}\left(\frac{1\text{ rev}}{2\pi\text{ rad}}\right) = 6.02\text{ rev}$$

Section 9.3 From Eq. (9.13), $v = r\omega$. To maintain a constant linear speed v, the angular speed ω must decrease as the scanning head moves outward (greater r).
Section 9.4 The kinetic energy in the falling object is $\frac{1}{2}mv^2$ and the kinetic energy in the rotating cylinder is $\frac{1}{2}I\omega^2 = \frac{1}{2}(\frac{1}{2}mR^2)(\frac{v}{R})^2 = \frac{1}{4}mv^2$. Hence the total kinetic energy of the system is $\frac{3}{4}mv^2$, of which two-thirds is in the object and one-third in the cylinder.
Section 9.5 From Table 9.2, the moment of inertia of the discus around an axis through its center of mass and perpendicular to the plane of the discus is $I_{cm} = MR^2/2 = (2.00\text{ kg})(0.110\text{ m})^2/2 = 0.0121\text{ kg}\cdot\text{m}^2$. The discus is rotating about an axis parallel to this one, so we use the parallel-axis theorem: the moment of inertia around the thrower's shoulder is $I_P = I_{cm} + Md^2 = 0.0121\text{ kg}\cdot\text{m}^2 + (2.00\text{ kg})(0.800\text{ m})^2 = 1.29\text{ kg}\cdot\text{m}^2$.
Section 9.6 To find the minimum of I, take its derivative with respect to h and set it equal to zero:

$$\frac{dI}{dh} = \frac{d}{dh}\frac{1}{3}M(L^2 - 3Lh + 3h^2) = M(-L + 2h) = 0$$

This has solution $h = L/2$. The second derivative $d^2I/dh^2 = 2M$ is positive, as it must be at a minimum.

Discussion Questions

Q9.1 When rewinding an audio or video tape, why does the tape wind up fastest at the end?
Q9.2 Does a body rotating about a fixed axis have to be *perfectly* rigid for all points on the body to have the same angular velocity and the same angular acceleration? Explain.
Q9.3 What is the difference between tangential and radial acceleration for a point on a rotating body?
Q9.4 In Fig. 9.15, all points on the chain have the same linear speed v. Is the magnitude a of the linear acceleration also the same for all points on the chain? How are the angular accelerations of the two sprockets related? Explain.
Q9.5 In Fig. 9.15, how are the radial accelerations of points at the teeth of the two sprockets related? Explain the reasoning behind your answer.
Q9.6 A flywheel rotates with constant angular velocity. Does a point on its rim have a tangential acceleration? A radial acceleration? Are these accelerations constant in magnitude? In direction? In each case give the reasoning behind your answer.
Q9.7 What is the purpose of the spin cycle of a washing machine? Explain in terms of acceleration components.
Q9.8 Although angular velocity and angular acceleration can be treated as vectors, the angular displacement θ, despite having a magnitude and a direction, cannot. This is because θ does not follow the commutative law of vector addition (Eq. 1.4). Prove this to yourself in the following way. Lay your physics textbook flat on the desk in front of you with the cover side up so you can read the writing on it. Rotate the farthest edge toward you by 90° about a horizontal axis. Call this angular displacement θ_1. Then rotate the left edge toward you by 90° about a vertical axis. Call this angular displacement θ_2. The spine of the book should now face you, with the writing on it oriented so that you can read. Now start over again but carry out the two rotations in the reverse order. Do you get a different result? That is, is $\theta_1 + \theta_2$ equal to $\theta_2 + \theta_1$? Now repeat this experiment but this time with an angle of 1° rather than 90°. Do you think that the infinitesimal displacement $d\vec{\boldsymbol{\theta}}$ obeys the commutative law of addition and hence qualifies as a vector? If so, how is the direction of $d\vec{\boldsymbol{\theta}}$ related to the direction of $\vec{\boldsymbol{\omega}}$?
Q9.9 Can you think of a body that has the same moment of inertia for all possible axes? If so, give an example, and if not, explain why this is not possible. Can you think of a body that has the same moment of inertia for all axes passing through a certain point? If so, give an example and indicate where the point is located.
Q9.10 To maximize the moment of inertia of a flywheel while minimizing its weight, what shape and distribution of mass should it have? Explain.
Q9.11 How might you determine experimentally the moment of inertia of an irregularly shaped body about a given axis?
Q9.12 A cylindrical body has mass M and radius R. Can the mass be distributed within the body in such a way that its moment of inertia about its axis of symmetry is greater than MR^2? Explain.
Q9.13 Describe how you could use part (b) of Table 9.2 to derive the result in part (d).
Q9.14 The moment of inertia I of a rigid body for an axis through its center of mass is I_{cm}. Is there any other axis parallel to this axis for which I is less than I_{cm}? Explain.
Q9.15 For the equations for I given in parts (a) and (b) of Table 9.2 to be valid, must the rod have a circular cross section? Is there any

restriction on the size of the cross section in order for these equations to apply? Explain.

Q9.16 In part (d) of Table 9.2, the thickness of the plate must be much less than a for the expression given for I to apply. But in part (c), the expression given for I applies no matter how thick the plate is. Explain.

Q9.17 In Fig. 5.32a, use the expressions $K = \frac{1}{2}I\omega^2$ and $K = \frac{1}{2}mv^2$ to calculate the kinetic energy of the sled (treated as a single particle). How do these two results compare? Explain.

Q9.18 Equation (9.18) says to use y_{cm} to calculate U for an extended rigid body. But in Example 9.9 (Section 9.4), y is measured to the bottom of the hanging mass rather than to the center of mass. Is this a mistake? Explain.

Q9.19 You can use any angular measure—radians, degrees, or revolutions—in some of the equations in Chapter 9, but you can use only radian measure in others. Identify those for which using radians is necessary and those for which it is not, and in each case give the reasoning behind your answer.

Q9.20 When calculating the moment of inertia of an object, can we treat all its mass as if it were concentrated at the center of mass of the object? Justify your answer.

Q9.21 A wheel is rotating about an axis perpendicular to the plane of the wheel and passing through the center of the wheel. The angular speed of the wheel is increasing at a constant rate. Point A is on the rim of the wheel and point B is midway between the rim and center of the wheel. For each of the following quantities, is its magnitude larger at point A, at point B, or is it the same at both points: angular speed, tangential speed, angular acceleration, tangential acceleration, and radial acceleration. Justify each of your answers.

Exercises

Section 9.1 Angular Velocity and Acceleration

9.1 a) What angle in radians is subtended by an arc 1.50 m in length on the circumference of a circle of radius 2.50 m? What is this angle in degrees? b) An arc 14.0 cm in length on the circumference of a circle subtends an angle of 128°. What is the radius of the circle? c) The angle between two radii of a circle with radius 1.50 m is 0.700 rad. What length of arc is intercepted on the circumference of the circle by the two radii?

9.2 An airplane propeller is rotating at 1900 rev/min. a) Compute the propeller's angular velocity in rad/s. b) How long in seconds does it take for the propeller to turn through 35°?

9.3 Consider the flywheel in Examples 9.1 and 9.2 (Section 9.1). a) Calculate the instantaneous angular acceleration at $t = 3.5$ s. Explain why your result is equal to the average angular acceleration for the time interval from 2.0 s to 5.0 s. b) Calculate the instantaneous angular velocity at $t = 3.5$ s. Explain why your result is *not* equal to the average angular velocity for the 2.0 s to 5.0 s time interval, even though 3.5 s is at the middle of this time interval.

9.4 A fan blade rotates with angular velocity given by $\omega_z(t) = \gamma - \beta t^2$, where $\gamma = 5.00$ rad/s and $\beta = 0.800$ rad/s^3. a) Calculate the angular acceleration as a function of time. b) Calculate the instantaneous angular acceleration α_z at $t = 3.00$ s and the average angular acceleration $\alpha_{av\text{-}z}$ for the time interval $t = 0$ to $t = 3.00$ s. How do these two quantities compare? If they are different, why are they different?

9.5 A child is pushing a merry-go-round. The angle through which the merry-go-round has turned varies with time according to $\theta(t) = \gamma t + \beta t^3$, where $\gamma = 0.400$ rad/s and $\beta = 0.0120$ rad/s^3. a) Calculate the angular velocity of the merry-go-round as a function of time. b) What is the initial value of the angular velocity? c) Calculate the instantaneous value of the angular velocity ω_z at $t = 5.00$ s and the average angular velocity $\omega_{av\text{-}z}$ for the time interval $t = 0$ to $t = 5.00$ s. Show that $\omega_{av\text{-}z}$ is *not* equal to the average of the instantaneous angular velocities at $t = 0$ and $t = 5.00$ s, and explain why it is not.

9.6 At $t = 0$ the current to a dc electric motor is reversed, resulting in an angular displacement of the motor shaft given by $\theta(t) = (250 \text{ rad/s})t - (20.0 \text{ rad/s}^2)t^2 - (1.50 \text{ rad/s}^3)t^3$. a) At what time is the angular velocity of the motor shaft zero? b) Calculate the angular acceleration at the instant that the motor shaft has zero angular velocity. c) How many revolutions does the motor shaft turn through between the time when the current is reversed and the instant when the angular velocity is zero? d) How fast was the motor shaft rotating at $t = 0$, when the current was reversed? e) Calculate the average angular velocity for the time period from $t = 0$ to the time calculated in part (a).

9.7 The angle θ through which a bicycle wheel turns is given by $\theta(t) = a + bt^2 - ct^3$, where a, b, and c are positive constants such that for t in seconds, θ will be in radians. a) Calculate the angular acceleration of the wheel as a function of time. b) At what time is the angular velocity of the wheel instantaneously not changing?

9.8 A wheel is rotating about an axis that is in the z-direction. The angular velocity ω_z is -6.00 rad/s at $t = 0$, increases linearly with time and is $+8.00$ m/s at $t = 7.00$ s. We have taken counterclockwise rotation to be positive. a) Is the angular acceleration during this time interval positive or negative? b) During what time interval is the speed of the wheel increasing? Decreasing? c) What is the angular displacement of the wheel at $t = 7.00$ s?

Section 9.2 Rotation with Constant Angular Acceleration

9.9 A high-speed flywheel in a motor is spinning at 500 rpm when a power failure suddenly occurs. The flywheel has mass 40.0 kg and diameter 75.0 cm. The power is off for 30.0 s and during this time the flywheel slows due to friction in its axle bearings. During the time the power is off, the flywheel makes 200 complete revolutions. a) At what rate is the flywheel spinning when the power comes back on? b) How long after the beginning of the power failure would it have taken the flywheel to stop if the power had not come back on, and how many revolutions would the wheel have made during this time?

9.10 A bicycle wheel has an initial angular velocity of 1.50 rad/s. a) If its angular acceleration is constant and equal to 0.300 rad/s^2, what is its angular velocity at $t = 2.50$ s? b) Through what angle has the wheel turned between $t = 0$ and $t = 2.50$ s?

9.11 An electric fan is turned off, and its angular velocity decreases uniformly from 500 rev/min to 200 rev/min in 4.00 s.

a) Find the angular acceleration in rev/s^2 and the number of revolutions made by the motor in the 4.00-s interval. b) How many more seconds are required for the fan to come to rest if the angular acceleration remains constant at the value calculated in part (a)?

9.12 a) Derive Eq. (9.12) by combining Eqs. (9.7) and (9.11) to eliminate t. b) The angular velocity of an airplane propeller increases from 12.0 rad/s to 16.0 rad/s while turning through 7.00 rad. What is the angular acceleration in rad/s^2?

9.13 The rotating blade of a blender turns with constant angular acceleration 1.50 rad/s^2. a) How much time does it take to reach an angular velocity of 36.0 rad/s, starting from rest? b) Through how many revolutions does the blade turn in this time interval?

9.14 A flywheel requires 4.00 s to rotate through 162 rad. Its angular velocity at the end of this time is 108 rad/s. Find a) the angular velocity at the beginning of the 4.00-s interval; b) the constant angular acceleration.

9.15 Emilie's potter's wheel rotates with a constant 2.25 rad/s^2 angular acceleration. After 4.00 s the wheel has rotated through an angle of 60.0 rad. What was the angular velocity of the wheel at the beginning of the 4.00 s interval?

9.16 A circular saw blade 0.200 m in diameter starts from rest. In 6.00 s it accelerates with constant angular acceleration to an angular velocity of 140 rad/s. Find the angular acceleration and the angle through which the blade has turned.

9.17 A safety device brings the blade of a power mower from an initial angular speed of ω_1 to rest in 1.00 revolution. At the same constant acceleration, how many revolutions would it take the blade to come to rest from an initial angular speed ω_3 that was three times as great, $\omega_3 = 3\omega_1$?

9.18 A straight piece of reflecting tape extends from the center of a wheel to its rim. You darken the room and use a camera and strobe unit that flashes once every 0.050 s to take pictures of the wheel as it rotates counterclockwise. You trigger the strobe so that the first flash ($t = 0$) occurs when the tape is horizontal to the right at an angular displacement of zero. For the following situations draw a sketch of the photo you will get for the time exposure over five flashes (at $t = 0$, 0.050 s, 0.100 s, 0.150 s, and 0.200 s) and graph θ versus t and ω versus t for $t = 0$ to $t = 0.200$ s. a) The angular velocity is constant at 10.0 rev/s. b) The wheel starts from rest with a constant angular acceleration of 25.0 rev/s^2. c) The wheel is rotating at 10.0 rev/s at $t = 0$ and changes angular velocity at a constant rate of -50.0 rev/s^2.

9.19 At $t = 0$ a grinding wheel has an angular velocity of 24.0 rad/s. It has a constant angular acceleration of 30.0 rad/s^2 until a circuit breaker trips at $t = 2.00$ s. From then on, it turns through 432 rad as it coasts to a stop at constant angular acceleration. a) Through what total angle did the wheel turn between $t = 0$ and the time it stopped? b) At what time did it stop? c) What was its acceleration as it slowed down?

9.20 a) Derive a constant-angular-acceleration equation that gives $\theta - \theta_0$ in terms of ω_z, α_z, and t (no ω_{0z} in the equation). b) At $t = 8.0$ s a gear is rotating about a fixed axis at 4.50 rad/s. During the preceding 8.0 s, it has turned through 40.0 rad. Use the result of part (a) to calculate the gear's constant angular acceleration. c) What was the angular velocity of the gear at $t = 0$?

Section 9.3 Relating Linear and Angular Kinematics

9.21 The main rotor of a helicopter is turning in a horizontal plane at 90.0 rev/min. The distance from the center of the rotor shaft to each blade tip is 5.00 m. Calculate the speed through the air of the blade tip if a) the helicopter is sitting on the ground; b) the helicopter is rising vertically at 4.00 m/s.

9.22 Compact Disc. A compact disc (CD) stores music in a coded pattern of tiny pits 10^{-7} m deep. The pits are arranged in a track that spirals outward toward the rim of the disc; the inner and outer radii of this spiral are 25.0 mm and 58.0 mm, respectively. As the disc spins inside a CD player, the track is scanned at a constant *linear* speed of 1.25 m/s. a) What is the angular speed of the CD when scanning the innermost part of the track? The outermost part of the track? b) The maximum playing time of a CD is 74.0 min. What would be the length of the track on such a maximum-duration CD if it were stretched out in a straight line? c) What is the average angular acceleration of a maximum-duration CD during its 74.0-min playing time? Take the direction of rotation of the disc to be positive.

9.23 A wheel rotates with a constant angular velocity of 6.00 rad/s. a) Compute the radial acceleration of a point 0.500 m from the axis, using the relation $a_{\text{rad}} = \omega^2 r$. b) Find the tangential speed of the point and compute its radial acceleration from the relation $a_{\text{rad}} = v^2/r$.

9.24 Find the required angular speed (in rev/min) of an ultracentrifuge for the radial acceleration of a point 2.50 cm from the axis to equal 400,000g (that is, 400,000 times the acceleration due to gravity).

9.25 A flywheel with a radius of 0.300 m starts from rest and accelerates with a constant angular acceleration of 0.600 rad/s^2. Compute the magnitude of the tangential acceleration, the radial acceleration and the resultant acceleration of a point on its rim a) at the start; b) after it has turned through 60.0°; c) after it has turned through 120.0°.

9.26 An electric ceiling fan with blades 0.750 m in diameter is rotating about a fixed axis with an initial angular velocity of 0.250 rev/s. The angular acceleration is 0.900 rev/s^2. a) Compute the angular velocity after 0.200 s. b) Through how many revolutions has the blade turned in this time interval? c) What is the tangential speed of a point on the tip of the blade at $t = 0.200$ s? d) What is the magnitude of the resultant acceleration of a point on the tip of the blade at $t = 0.200$ s?

9.27 Centrifuge. An advertisement claims that a centrifuge takes up only 0.127 m of bench space but can produce a radial acceleration of 3000g at 5000 rev/min. Calculate the required radius of the centrifuge. Is the claim a realistic one?

9.28 a) Derive an equation for the radial acceleration that includes v and ω, but not r. b) You are designing a merry-go-round for which a point on the rim will have a radial acceleration of 0.500 m/s^2 when the tangential velocity of that point has magnitude 2.00 m/s. What angular velocity is required to achieve these values?

9.29 A Drill Problem. When drilling a 12.7-mm-diameter hole in wood, plastic, or aluminum, a shop manual recommends a drill speed of 1250 rev/min. For a 12.7-mm-diameter drill bit turning at a constant 1250 rev/min, find a) the maximum linear speed of any part of the bit; b) the maximum radial acceleration of any part of the bit.

9.30 At $t = 3.00$ s a point on the rim of a 0.200-m-radius wheel has a tangential speed of 50.0 m/s as the wheel slows down with a tangential acceleration of constant magnitude 10.0 m/s^2. a) Calculate the wheel's constant angular acceleration. b) Calculate the angular velocities at $t = 3.00$ s and $t = 0$. c) Through what angle did the wheel turn between $t = 0$ and $t = 3.00$ s? d) At what time will the radial acceleration equal g?

9.31 The spin cycles of a washing machine have two angular speeds, 423 rev/min and 640 rev/min. The internal diameter of the drum is 0.470 m. a) What is the ratio of the maximum radial force on the laundry for the higher angular speed to that for the lower speed? b) What is the ratio of the maximum tangential speed of the laundry for the higher angular speed to that for the lower speed? c) Find the laundry's maximum tangential speed and the maximum radial acceleration, in terms of g.

9.32 You are to design a rotating cylindrical axle to lift 800-N buckets of cement from the ground to a rooftop 78.0 m above the ground. The buckets will be attached to a hook on the free end of a cable that wraps around the rim of the axle; as the axle turns, the buckets will rise. a) What should the diameter of the axle be in order to raise the buckets at a steady 2.00 cm/s when it is turning at 7.5 rpm? b) If instead the axle must give the buckets an upward acceleration of 0.400 m/s^2, what should the angular acceleration of the axle be?

9.33 While riding a multi-speed bicycle, the rider can select the radius of the rear sprocket, that is fixed to the rear axle. The front sprocket of a bicycle has radius 12.0 cm. If the angular speed of the front sprocket is 0.600 rev/s, what is the radius of the rear sprocket for which the tangential speed of a point on the rim of the rear wheel will be 5.00 m/s? The rear wheel has radius 0.330 m.

Section 9.4 Energy in Rotational Motion

9.34 Small blocks, each with mass m, are clamped at the ends and at the center of a light rod of length L. Compute the moment of inertia of the system about an axis perpendicular to the rod and passing through a point $\frac{1}{4}$ of the length from one end. You can ignore the moment of inertia of the light rod.

9.35 A twirler's baton is made of a slender metal cylinder of mass M and length L. Each end has a rubber cap of mass m and you can accurately treat each cap as a particle in this problem. Find the total moment of inertia of the baton about the usual twirling axis (perpendicular to the baton through its center).

9.36 Find the moment of inertia about each of the following axes for a rod that is 0.300 cm in diameter and 1.50 m long, and with mass 0.0420 kg. Use the formulas of Table 9.2. a) About an axis perpendicular to the rod and passing through its center. b) About an axis perpendicular to the rod and passing through one end. c) About a longitudinal axis passing through the center of the rod.

9.37 Four small spheres, each of which you can regard as a point of mass 0.200 kg, are arranged in a square 0.400 m on a side and connected by light rods (Fig. 9.27). Find the moment of inertia of the system about an axis a) through the center of the square, perpendicular to its plane (an axis through point O in the figure); b) bisecting two opposite sides of the square (an axis along the line AB in the figure); c) that passes through the centers of the upper left and lower right spheres and through point O.

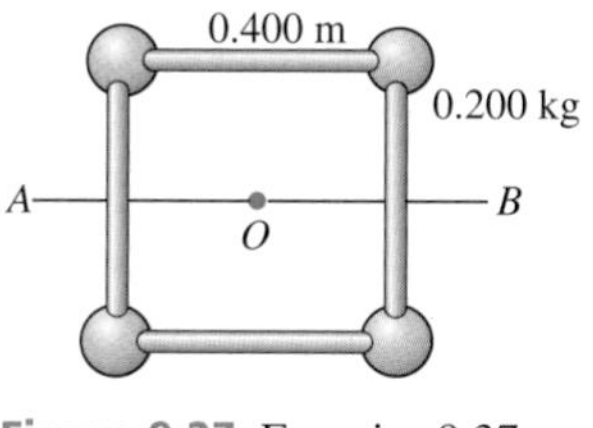

Figure 9.27 Exercise 9.37.

9.38 A uniform bar has two small balls glued to its ends. The bar is 2.00 m long and with mass 4.00 kg, while the balls each have mass 0.500 kg and can be treated as point masses. Find the moment of inertia of this combination about each of the following axes: a) an axis perpendicular to the bar through its center; b) an axis perpendicular to the bar through one of the balls; c) an axis parallel to the bar through both balls; d) an axis parallel to the bar and 0.500 m from it.

9.39 A compound disk of outside diameter 140.0 cm is made up of a uniform solid disk of radius 50.0 cm and area density 3.00 g/cm^2 surrounded by a concentric ring of inner radius 50.0 cm, outer radius 70.0 cm and area density 2.00 g/cm^2. Find the moment of inertia of this object about an axis perpendicular to the plane of the object and passing through its center.

9.40 How I Scales. If we multiply all design dimensions of an object by a scaling factor f, its volume and mass will be multiplied by f^3. a) By what factor will its moment of inertia be multiplied? b) If a $\frac{1}{48}$-scale model has a rotational kinetic energy of 2.5 J, what will be the kinetic energy for the full-scale object of the same material rotating at the same angular velocity?

9.41 A wagon wheel is constructed as shown in Fig. 9.28. The radius of the wheel is 0.300 m, and the rim has mass 1.40 kg. Each of the eight spokes, that lie along a diameter and are 0.300 m long, has mass 0.280 kg. What is the moment of inertia of the wheel about an axis through its center and perpendicular to the plane of the wheel? (Use the formulas given in Table 9.2.)

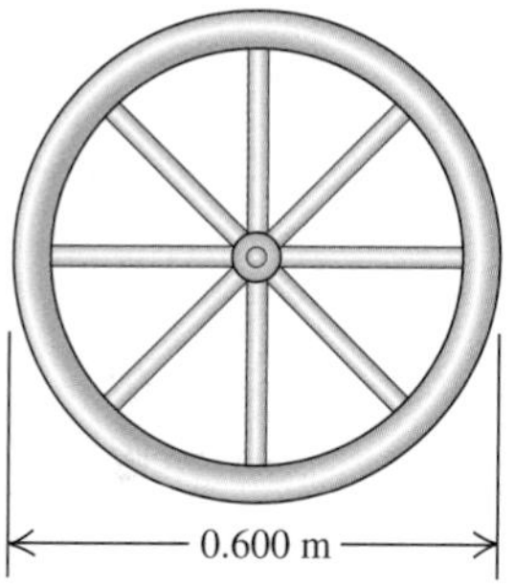

Figure 9.28 Exercise 9.41.

9.42 An airplane propeller is 2.08 m in length (from tip to tip) with mass 117 kg. The propeller is rotating at 2400 rev/min about an axis through its center. a) What is its rotational kinetic energy? Treat the propeller as a slender rod. b) If it were not rotating, how far would it have to drop in free fall to acquire the same kinetic energy?

9.43 a) Show that $\frac{1}{2}I\omega^2$ has units equivalent to joules. Explain why you need not include "rads" in the units. b) Often ω is expressed in rev/min rather than rad/s. Write an expression for rotational kinetic energy such that if I is in kg $\cdot$ m^2 and ω is in rev/min, the energy will be in joules.

9.44 An antique phonograph turntable has a kinetic energy of 0.0250 J when turning at 45.0 rev/min. What is the moment of inertia of the turntable about the rotation axis?

9.45 The flywheel of a gasoline engine is required to give up 500 J of kinetic energy while its angular velocity decreases from 650 rev/min to 520 rev/min. What moment of inertia is required?

9.46 A light, flexible rope is wrapped several times around a *hollow* cylinder with a weight of 40.0 N and a radius of 0.25 m, that rotates without friction about a fixed horizontal axis. The cylinder is attached to the axle by spokes of a negligible moment of inertia. The cylinder is initially at rest. The free end of the rope is pulled with a constant force P for a distance of 5.00 m, at which point the end of the rope is moving at 6.00 m/s. If the rope does not slip on the cylinder, what is the value of P?

9.47 Energy is to be stored in a 70.0-kg flywheel in the shape of a uniform solid disk with radius $R = 1.20$ m. To prevent structural failure of the flywheel, the maximum allowed radial acceleration of a point on its rim is 3500 m/s^2. What is the maximum kinetic energy that can be stored in the flywheel?

9.48 Suppose the solid cylinder in the apparatus described in Example 9.9 (Section 9.4) is replaced by a thin-walled, hollow cylinder with the same mass M and radius R. The cylinder is attached to the axle by spokes of a negligible moment of inertia. a) Find the speed of the hanging mass m just as it strikes the floor. b) Is the answer to part (a) greater than, equal to, or less than the speed found in Example 9.9? Explain why, using energy concepts.

9.49 Rate of Kinetic Energy Loss. A rigid body with a moment of inertia I spins once every T seconds. The rotation is slowing down, so that $dT/dt > 0$. a) Express the kinetic energy of rotation of the body in terms of I and T. b) Express the rate of change of the kinetic energy of rotation in terms of I, T, and dT/dt. c) A large flywheel has $I = 8.0\ \text{kg} \cdot \text{m}^2$. What is the kinetic energy of the flywheel when the period of rotation is 1.5 s? d) What is the rate of change of the kinetic energy of the flywheel in part (c) at the instant when the period is 1.5 s and is changing at the rate of $dT/dt = 0.0060$?

9.50 A uniform rope 10.0 m long and with mass 3.00 kg is hanging with one end attached to a gymnasium ceiling and the other end just touching the floor. The upper end of the rope is released, and the rope falls to the floor. What is the change in the gravitational potential energy if the rope ends up flat on the floor (not coiled up)?

9.51 Center of Mass of an Extended Object. How much work must a wrestler do to raise the center of mass of his 120-kg opponent a vertical distance of 0.700 m?

Section 9.5 Parallel-Axis Theorem

9.52 Find the moment of inertia of a hoop (a thin-walled, hollow ring) with mass M and radius R about an axis perpendicular to the hoop's plane at an edge.

9.53 About what axis will a uniform, balsa-wood sphere have the same moment of inertia as for a thin-walled, hollow, lead sphere of the same mass and radius, with the axis along a diameter?

9.54 Use the parallel-axis theorem to show that the moments of inertia given in parts (a) and (b) of Table 9.2 are consistent.

9.55 A thin, rectangular sheet of metal has mass M and sides of length a and b. Use the parallel-axis theorem to calculate the moment of inertia of the sheet for an axis that is perpendicular to the plane of the sheet and that passes through one corner of the sheet.

9.56 a) For the thin rectangular plate shown in part (d) of Table 9.2, find the moment of inertia about an axis that lies in the plane of the plate, passes through the center of the plate, and is parallel to the axis shown in the figure. b) Find the moment of inertia of the plate for an axis that lies in the plane of the plate, passes through the center of the plate, and is perpendicular to the axis in part (a).

*Section 9.6 Moment of Inertia Calculations

***9.57** Using the information in Table 9.2 and the parallel-axis theorem, find the moment of inertia of the slender rod with mass M and length L shown in Fig. 9.24 about an axis through O, at an arbitrary distance h from one end. Compare your result to that found by integration in Example 9.12 (Section 9.6).

***9.58** Use Eq. (9.20) to calculate the moment of inertia of a uniform, solid disk with mass M and radius R for an axis perpendicular to the plane of the disk and passing through its center.

***9.59** Use Eq. (9.20) to calculate the moment of inertia of a slender, uniform rod with mass M and length L about an axis at one end, perpendicular to the rod.

***9.60** A slender rod with length L has a mass per unit length that varies with distance from the left-hand end, where $x = 0$, according to $dm/dx = \gamma x$, where γ has units of kg/m^2. a) Calculate the total mass of the rod in terms of γ and L. b) Use Eq. (9.20) to calculate the moment of inertia of the rod for an axis at the left-hand end, perpendicular to the rod. Use the expression you derived in part (a) to express I in terms of M and L. How does your result compare to that for a uniform rod? Explain this comparison. c) Repeat part (b) for an axis at the right-hand end of the rod. How do the results for parts (b) and (c) compare? Explain this result.

Problems

9.61 Sketch a wheel lying in the plane of your paper and rotating counterclockwise. Choose a point on the rim and draw a vector $\vec{r}$ from the center of the wheel to that point. a) What is the direction of $\vec{\omega}$? b) Show that the velocity of the point is $\vec{v} = \vec{\omega} \times \vec{r}$. c) Show that the radial acceleration of the point is $\vec{a}_{\text{rad}} = \vec{\omega} \times \vec{v} = \vec{\omega} \times (\vec{\omega} \times \vec{r})$ (see Exercise 9.28).

9.62 Trip to Mars. You are working on a project with NASA to launch a rocket to Mars, with the rocket blasting off from earth when earth and Mars are just aligned along a straight line from the sun. As a first step in doing the calculation, assume circular orbits for both planets. If Mars is now 60° ahead of earth in its orbit around the sun, when should you launch the rocket? (*Note:* All the planets orbit the sun in the same direction, and the year on Mars is 1.9 earth-years.)

9.63 Speedometer. Your car's speedometer converts the angular speed of the wheels to the linear speed of the car, assuming standard-size tires and no slipping on the pavement. a) If your car's standard tires are 24 inches in diameter, at what rate (in rpm) are your wheels rotating when you are driving at a freeway speed of 60 mph? b) Suppose you put oversize, 30-inch diameter tires on your car. How fast are you really going when your speedometer reads 60 mph? c) If you now put on undersize, 20-inch diameter tires, what will the speedometer read when you are actually traveling at 50 mph?

9.64 a) Prove that when an object starts from rest and rotates about a fixed axis with constant angular acceleration, the radial acceleration of a point in the object is directly proportional to its angular displacement. b) Through what angle has the object turned at the instant when the resultant acceleration of a point makes an angle of 36.9° with the radial direction?

9.65 A roller in a printing press turns through an angle $\theta(t)$ given by $\theta(t) = \gamma t^2 - \beta t^3$, where $\gamma = 3.20\ \text{rad/s}^2$ and $\beta = 0.500\ \text{rad/s}^3$. a) Calculate the angular velocity of the roller as a function of time. b) Calculate the angular acceleration of the roller as a function of time. c) What is the maximum positive angular velocity, and at what value of t does it occur?

***9.66** A bicycle wheel with a radius of 0.33 m turns with angular acceleration $\alpha(t) = \gamma - \beta t$, where $\gamma = 1.80\ \text{rad/s}^2$ and $\beta = 0.25\ \text{rad/s}^3$. It is at rest at $t = 0$. a) Calculate the angular velocity and angular displacement as functions of time. b) Calculate the maximum positive angular velocity and maximum positive angular displacement of the wheel. (*Hint:* See Section 2.6.)

9.67 When a toy car is rapidly scooted across the floor, it stores energy in a flywheel. The car has mass 0.180 kg, and its flywheel has a moment of inertia of $4.00 \times 10^{-5}\ \text{kg} \cdot \text{m}^2$. The car is 15.0 cm long. An advertisement claims that the car can travel at a scale speed of up to 700 km/h (440 mi/h). The scale speed is the speed of the toy car multiplied by the ratio of the length of an actual car to the length of the toy. Assume a length of 3.0 m for a real car. a) For a scale speed of 700 km/h, what is the actual translational speed of the car? b) If all the kinetic energy that is initially in the flywheel is converted to the translational kinetic energy of the toy, how much energy is originally stored in the flywheel? c) What initial angular velocity of the flywheel was needed to store the amount of energy calculated in part (b)?

9.68 A classic 1957 Chevrolet Corvette mass 1240 kg starts from rest and speeds up with a constant tangential acceleration of $3.00\ \text{m/s}^2$ on a circular test track of radius 60.0 m. Treat the car as a particle. a) What is its angular acceleration? b) What is its angular speed 6.00 s after it starts? c) What is its radial acceleration at this time? d) Sketch a view from above showing the circular track, the car, the velocity vector, and the acceleration component vectors 6.00 s after the car starts. e) What are the magnitudes of the total acceleration and net force for the car at this time? f) What angle do the total acceleration and net force make with the car's velocity at this time?

9.69 The flywheel of a punch press has a moment of inertia of $16.0\ \text{kg} \cdot \text{m}^2$ and runs at 300 rev/min. The flywheel supplies all the energy needed in a quick punching operation. a) Find the speed in rev/min to which the flywheel will be reduced by a sudden punching operation requiring 4000 J of work. b) What must the constant power supply to the flywheel (in watts) be to bring it back to its initial speed in a time of 5.00 s?

9.70 A totally inedible cafeteria meatball with mass 40.0 g is attached to the free end of a 2.50-m piece of string that is attached to the ceiling. The meatball is pulled to the left so that the string makes a 36.9° angle with the vertical and is then released. a) What is the magnitude and direction of the angular velocity of the meatball the first time the angular acceleration is zero? b) What is it the second time that $\alpha_z = 0$? c) At the times described in parts (a) and (b), what is the magnitude and direction of the meatball's radial acceleration? d) Show that the answer to part (c) is independent of the length of the string.

9.71 A vacuum cleaner belt is looped over a shaft of radius 0.45 cm and a wheel of radius 2.00 cm. The arrangement of the belt, shaft, and wheel is similar to that of the chain and sprockets in Fig. 9.15. The motor turns the shaft at 60.0 rev/s and the moving belt turns the wheel, that in turn is connected by another shaft to the roller that beats the dirt out of the rug being vacuumed. Assume that the belt doesn't slip on either the shaft or the wheel. a) What is the speed of a point on the belt? b) What is the angular velocity of the wheel in rad/s?

9.72 The motor of a table saw is rotating at 3450 rev/min. A pulley attached to the motor shaft drives a second pulley of half the diameter by means of a V-belt. A circular saw blade of diameter 0.208 m is mounted on the same rotating shaft as the second pulley. a) The operator is careless and the blade catches and throws back a small piece of wood. This piece of wood moves with linear speed equal to the tangential speed of the rim of the blade. What is this speed? b) Calculate the radial acceleration of points on the outer edge of the blade to see why sawdust doesn't stick to its teeth.

9.73 A wheel changes its angular velocity with a constant angular acceleration while rotating about a fixed axis through its center. a) Show that the change in the magnitude of the radial acceleration during any time interval of a point on the wheel is twice the product of the angular acceleration, the angular displacement, and the perpendicular distance of the point from the axis. b) The radial acceleration of a point on the wheel that is 0.250 m from the axis changes from $25.0\ \text{m/s}^2$ to $85.0\ \text{m/s}^2$ as the wheel rotates through 15.0 rad. Calculate the tangential acceleration of this point. c) Show that the change in the wheel's kinetic energy during any time interval is the product of the moment of inertia about the axis, the angular acceleration, and the angular displacement. d) During the 15.0-rad angular displacement of part (b), the kinetic energy of the wheel increases from 20.0 J to 45.0 J. What is the moment of inertia of the wheel about the rotation axis?

9.74 A sphere consists of a solid wooden ball of uniform density $800\ \text{kg/m}^3$ and radius 0.20 m and is covered with a thin coating of lead foil with area density $20\ \text{kg/m}^2$. Calculate the moment of inertia of this sphere about an axis passing through its center.

9.75 Estimate your own moment of inertia about a vertical axis through the center of the top of your head when you are standing up straight with your arms outstretched. Make reasonable approximations and measure or estimate necessary quantities.

9.76 A thin uniform rod 50.0 cm long and with mass 0.320 kg is bent at its center into a V shape, with a 70.0° angle at its vertex. Find the moment of inertia of this V-shaped object about an axis perpendicular to the plane of the V at its vertex.

9.77 It has been argued that power plants should make use of off-peak hours (such as late at night) to generate mechanical energy and store it until it is needed during peak load times, such as the middle of the day. One suggestion has been to store the energy in large flywheels spinning on nearly frictionless ball-bearings. Consider a flywheel made of iron (density $7800\ \text{kg/m}^3$) and in the shape of a 10.0 cm thick uniform disk. a) What would the diameter of such a disk need to be if it is to store 10.0 megajoules of kinetic energy when spinning at 90.0 rpm about an axis perpendicular to the disk at its center? b) What would be the centripetal acceleration of a point on its rim when spinning at this rate?

9.78 The three uniform objects shown in Fig. 9.29 have the same mass m. Object A is a solid cylinder with radius R. Object B is a hollow, thin cylinder with radius R. Object C is a solid cube whose sides are of length $2R$. The axis of rotation of each object is perpendicular to the page and through the object's center of mass. a) Which object has the smallest moment of inertia? Explain. b) Which object has the largest moment of inertia? Explain. c) Where would the moment of inertia of a uniform solid sphere rank if its radius is R, its mass is m, and the axis of rotation is along a diameter of the sphere? Explain.

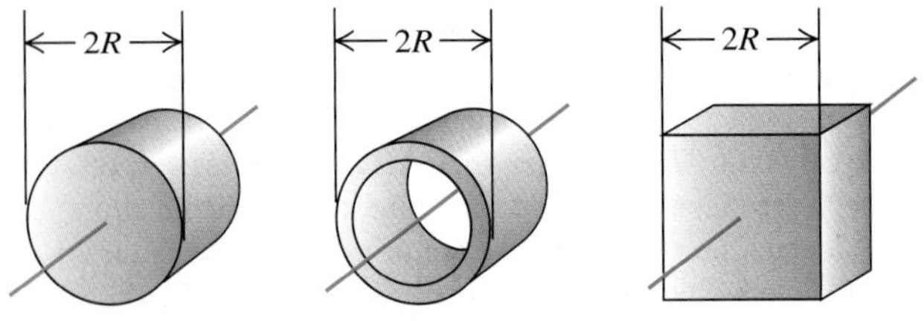

Figure 9.29 Problem 9.78.

9.79 The earth, which is not a uniform sphere, has a moment of inertia of $0.3308MR^2$ about an axis through its north and south poles. It takes the earth 86,164 s to spin once about this axis. Use Appendix F to calculate a) the earth's kinetic energy due to its rotation about this axis and b) the earth's kinetic energy due to its orbital motion around the sun. c) Explain how the value of the earth's moment of inertia tells us that the mass of the earth is concentrated toward the planet's center.

9.80 A uniform, solid disk with mass m and radius R is pivoted about a horizontal axis through its center. A small object of the same mass m is glued to the rim of the disk. If the disk is released from rest with the small object at the end of a horizontal radius, find the angular speed when the small object is directly below the axis.

9.81 A metal sign for a car dealership is a thin, uniform, right triangle with base length b and height h. The sign has mass M. a) What is the moment of inertia of the sign for rotation about the side of length h? b) If $M = 5.40$ kg, $b = 1.60$ m, and $h = 1.20$ m, what is the kinetic energy of the sign when it is rotating about an axis along the 1.20-m side at 2.00 rev/s?

9.82 Measuring *I*. As an intern with an engineering firm, you are asked to measure the moment of inertia of a large wheel, for rotation about an axis through its center. Since you were a good physics student, you know what to do. You measure the diameter of the wheel to be 0.740 m and find that it weighs 280 N. You mount the wheel, using frictionless bearings, on a horizontal axis through the wheel's center. You wrap a light rope around the wheel and hang a 8.00-kg mass from the free end of the rope, as shown in Fig. 9.19. You release the mass from rest; the mass descends and the wheel turns as the rope unwinds. You find that the mass has speed 5.00 m/s after it has descended 2.00 m. a) What is the moment of inertia of the wheel for an axis perpendicular to the wheel at its center? b) Your boss tells you that a larger I is needed. He asks you to design a wheel of the same mass and radius that has $I = 19.0\ \text{kg}\cdot\text{m}^2$. How do you reply?

9.83 A meter stick with a mass of 0.160 kg is pivoted about one end so it can rotate without friction about a horizontal axis. The meter stick is held in a horizontal position and released. As it swings through the vertical, calculate a) the change in gravitational potential energy that has occurred; b) the angular speed of the stick; c) the linear speed of the end of the stick opposite the axis. d) Compare the answer in part (c) to the speed of a particle that has fallen 1.00 m, starting from rest.

9.84 Exactly one turn of a flexible rope with mass m is wrapped around a uniform cylinder with mass M and radius R. The cylinder rotates without friction about a horizontal axle along the cylinder axis. One end of the rope is attached to the cylinder. The cylinder starts with an angular speed ω_0. After one revolution of the cylinder the rope has unwrapped and, at this instant, hangs vertically down, tangent to the cylinder. Find the angular speed of the cylinder and the linear speed of the lower end of the rope at this time. You can ignore the thickness of the rope. (*Hint:* Use Eq. (9.18).)

9.85 The pulley in Fig. 9.30 has radius R and a moment of inertia I. The rope does not slip over the pulley, and the pulley spins on a frictionless axle. The coefficient of kinetic friction between block A and the tabletop is μ_k. The system is released from rest, and block B descends. Block A has mass m_A and block B has mass m_B. Use energy methods to calculate the speed of block B as a function of the distance d that it has descended.

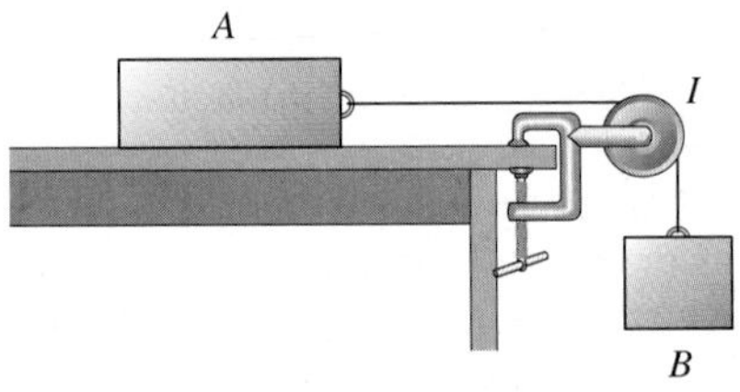

Figure 9.30 Problem 9.85.

9.86 The pulley in Fig. 9.31 has radius 0.160 m and a moment of inertia $0.480\ \text{kg}\cdot\text{m}^2$. The rope does not slip on the pulley rim. Use energy methods to calculate the speed of the 4.00-kg block just before it strikes the floor.

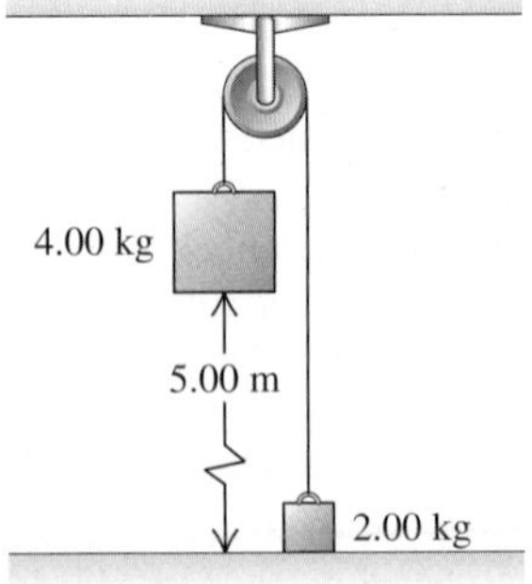

Figure 9.31 Problem 9.86.

9.87 You hang a thin hoop with radius R over a nail at the rim of the hoop. You displace it to the side through an angle β from its equilibrium position and let it go. What is its angular speed when it returns to its equilibrium position? (*Hint:* Use Eq. (9.18).)

9.88 A passenger bus in Zurich, Switzerland derived its motive power from the energy stored in a large flywheel. The wheel was brought up to speed periodically, when the bus stopped at a station, by an electric motor, which could then be attached to the electric power lines. The flywheel was a solid cylinder with mass 1000 kg and diameter 1.80 m; its top angular speed was 3000 rev/min. a) At this angular speed, what is the kinetic energy of the flywheel? b) If the average power required to operate the bus is 1.86×10^4 W, how long could it operate between stops?

9.89 Two metal disks, one with radius $R_1 = 2.50$ cm and mass $M_1 = 0.80$ kg and the other with radius $R_2 = 5.00$ cm and mass $M_2 = 1.60$ kg, are welded together and mounted on a frictionless axis through their common center (Fig. 9.32). a) What is the total moment of inertia of the two disks? b) A light string is wrapped around the edge of the smaller disk, and a 1.50-kg block, suspended from the free end of the string. If the block is released from rest at a distance of 2.00 m above the floor, what is its speed just before it strikes the floor? c) Repeat the calculation of part (b), this time with the string wrapped around the edge of the larger disk. In which case is the final speed of the block the greatest? Explain why this is so.

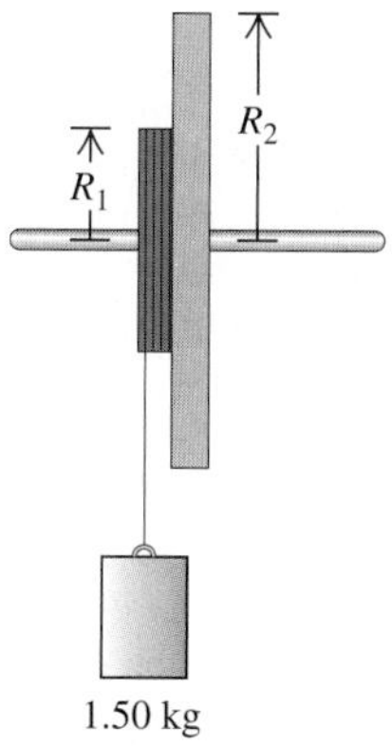

Figure 9.32 Problem 9.89.

9.90 In the cylinder and mass combination described in Example 9.9 (Section 9.4), suppose the falling mass m is made of ideal rubber, so that no mechanical energy is lost when the mass hits the ground. a) If the cylinder is originally not rotating and the mass m is released from rest at a height h above the ground, to what height will this mass rebound if it bounces straight back up from the floor? b) Explain, in terms of energy, why the answer to (a) is *less* than h.

9.91 In the system shown in Fig. 9.19, a 12.0-kg mass is released from rest and falls, causing the uniform 10.0-kg cylinder of diameter 30.0 cm to turn about a frictionless axle through its center. How far will the mass have to descend to give the cylinder 250 J of kinetic energy?

9.92 In Fig. 9.33, the cylinder and pulley turn without friction about stationary horizontal axles that pass through their centers. A light rope is wrapped around the cylinder, passes over the pulley, and has a 3.00-kg box suspended from its free end. There is no slipping between the rope and the pulley surface. The uniform cylinder has mass 5.00 kg and radius 40.0 cm. The pulley is a uniform disk with mass 2.00 kg and radius 20.0 cm. The box is released from rest and descends as the rope unwraps from the cylinder. Find the speed of the box when it has fallen 1.50 m.

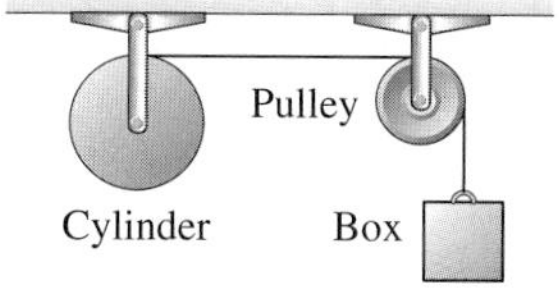

Figure 9.33 Problem 9.92.

9.93 A thin, flat, uniform disk has mass M and radius R. A circular hole of radius $R/4$, centered at a point $R/2$ from the disk's center, is then punched in the disk. a) Find the moment of inertia of the disk with the hole about an axis through the original center of the disk, perpendicular to the plane of the disk. (*Hint:* Find the moment of inertia of the piece punched from the disk.) b) Find the moment of inertia of the disk with the hole about an axis through the center of the hole, perpendicular to the plane of the disk.

9.94 A pendulum is made of a uniform solid sphere with mass M and radius R suspended from the end of a light rod. The distance from the pivot at the upper end of the rod to the center of the sphere is L. The pendulum's moment of inertia I_P for rotation about the pivot is usually approximated as ML^2. a) Use the parallel-axis theorem to show that if R is 5% of L and the mass of the rod is ignored, I_P is only 0.1% larger than ML^2. b) If the mass of the rod is 1% of M and R is much less than L, what is the ratio of I_{rod} for an axis at the pivot to ML^2?

9.95 Perpendicular-Axis Theorem. Consider a rigid body that is a thin, plane sheet of arbitrary shape. Take the body to lie in the xy-plane and let the origin O of coordinates be located at any point within or outside the body. Let I_x and I_y be the moments of inertia about the x- and y-axes and let I_O be the moment of inertia about an axis through O perpendicular to the plane. a) By considering mass elements m_i with coordinates (x_i, y_i), show that $I_x + I_y = I_O$. This is called the perpendicular-axis theorem. Note that point O does not have to be the center of mass. b) For a thin washer with mass M and with inner and outer radii R_1 and R_2, use the perpendicular-axis theorem to find the moment of inertia about an axis that is in the plane of the washer and that passes through its center. You may use the information in Table 9.2. c) Use the perpendicular-axis theorem to show that for a thin, square sheet with mass M and side L, the moment of inertia about *any* axis in the plane of the sheet that passes through the center of the sheet is $\frac{1}{12}ML^2$. You may use the information in Table 9.2.

9.96 A thin, uniform rod is bent into a square of side length a. If the total mass is M, find the moment of inertia about an axis through the center and perpendicular to the plane of the square. (*Hint:* Use the parallel-axis theorem.)

***9.97** A cylinder with radius R and mass M has density that increases linearly with distance r from the cylinder axis, $\rho = \alpha r$, where α is a positive constant. a) Calculate the moment of inertia of the cylinder about a longitudinal axis through its center in terms of M and R. b) Is your answer greater or smaller than the moment of inertia of a cylinder of the same mass and radius but of uniform density? Explain why this result makes qualitative sense.

9.98 Neutron Stars and Supernova Remnants. The Crab Nebula is a cloud of glowing gas about 10 light years across, located about 6500 light years from the earth (Fig. 9.34). It is the remnant of a star that underwent a *supernova explosion,* seen on earth in 1054 A.D. Energy is released by the Crab

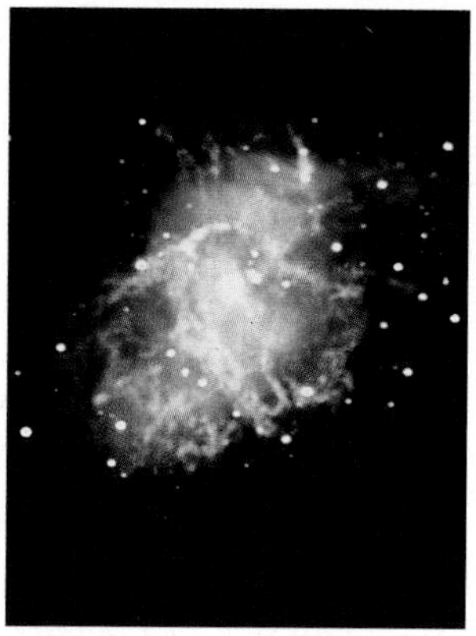
Figure 9.34 Problem 9.98.

Nebula at a rate of about 5×10^{31} W, about 10^5 times the rate at which the sun radiates energy. The Crab Nebula obtains its energy from the rotational kinetic energy of a rapidly spinning *neutron star* at its center. This object rotates once every 0.0331 s, and this period is increasing by 4.22×10^{13} s for each second of time that elapses. a) If the rate at which energy is lost by the neutron star is equal to the rate at which energy is released by the nebula, find the moment of inertia of the neutron star. (Use the expression derived in Exercise 9.49.) b) Theories of supernovae predict that the neutron star in the Crab Nebula has a mass of about 1.4 times that of the sun. Modeling the neutron star as a solid uniform sphere, calculate its radius in kilometers. c) What is the linear speed of a point on the equator of the neutron star? Compare to the speed of light. d) Assume that the neutron star is uniform and calculate its density. Compare to the density of ordinary rock (3000 kg/m^3) and to the density of an atomic nucleus (about 10^{17} kg/m^3). Justify the statement that a neutron star is essentially a large atomic nucleus.

Challenge Problems

9.99 The moment of inertia of a sphere with uniform density about an axis through its center is $\frac{2}{5}MR^2 = 0.400MR^2$. Satellite observations show that the earth's moment of inertia is $0.3308MR^2$. Geophysical data suggests the earth consists of five main regions: the inner core ($r = 0$ to $r = 1220$ km) of average density 12,900 kg/m^3, the outer core ($r = 1220$ km to $r = 3480$ km) of average density 10,900 kg/m^3, the lower mantle ($r = 3480$ km to $r = 5700$ km) of average density 4900 kg/m^3, the upper mantle ($r = 5700$ km to $r = 6350$ km) of average density 3600 kg/m^3, and the outer crust and oceans ($r = 6350$ km to $r = 6370$ km) of average density 2400 kg/m^3. a) Show that the moment of inertia about a diameter of a uniform spherical shell of inner radius R_1, outer radius R_2, and density ρ is $I = \rho(8\pi/15)(R_2^5 - R_1^5)$. (*Hint:* Form the shell by superposition of a sphere of density ρ and a smaller sphere of density $-\rho$.) b) Check the given data by using them to calculate the mass of the earth. c) Use the given data to calculate the earth's moment of inertia in terms of MR^2.

***9.100** Calculate the moment of inertia of a uniform solid cone about an axis through its center (Fig. 9.35). The cone has mass M and altitude h. The radius of its circular base is R.

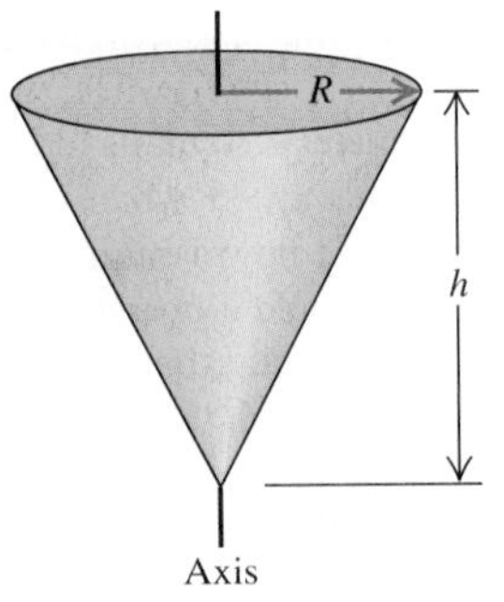

Figure 9.35 Challenge Problem 9.100.

9.101 On a compact disc (CD), music is coded in a pattern of tiny pits arranged in a track that spirals outward toward the rim of the disc. As the disc spins inside a CD player, the track is scanned at a constant *linear* speed of $v = 1.25$ m/s. Because the radius of the track varies as it spirals outward, the *angular* speed of the disc must change as the CD is played. (See Exercise 9.22.) Let's see what angular acceleration is required to keep v constant. The equation of a spiral is $r(\theta) = r_0 + \beta\theta$, where r_0 is the radius of the spiral at $\theta = 0$ and β is a constant. On a CD, r_0 is the inner radius of the spiral track. If we take the rotation direction of the CD to be positive, β must be positive so that r increases as the disc turns and θ increases. a) When the disc rotates through a small angle $d\theta$, the distance scanned along the track is $ds = r\,d\theta$. Using the above expression for $r(\theta)$, integrate ds to find the total distance s scanned along the track as a function of the total angle θ through which the disc has rotated. b) Since the track is scanned at a constant linear speed v, the distance s found in part (a) is equal to vt. Use this to find θ as a function of time. There will be two solutions for θ; choose the positive one, and explain why this is the solution to choose. c) Use your expression for $\theta(t)$ to find the angular velocity ω_z and angular acceleration α_z as functions of time. Is α_z constant? d) On a CD, the inner radius of the track is 25.0 mm, the track radius increases by 1.55 μm per revolution, and the playing time is 74.0 min. Find the values of r_0 and β, and find the total number of revolutions made during the playing time. e) Using your results from parts (c) and (d), make graphs of ω_z (in rad/s) versus t and α_z (in rad/s^2) versus t between $t = 0$ and $t = 74.0$ min.

DYNAMICS OF ROTATIONAL MOTION 10

This tumbling acrobat is not a rigid body, which enables him to vary his rotational speed in mid-air. Moving his arms and legs outward makes his rotation slow down, while pulling his arms and legs inward makes him rotate more rapidly.

? **If the acrobat isn't touching the ground, how can he change his rotation speed? What physical principle is at work here?**

We learned in Chapters 4 and 5 that a net force applied to a body gives that body an acceleration. But what does it take to give a body an *angular* acceleration? That is, what does it take to start a stationary body rotating or to bring a spinning body to a halt? A force is required, but it must be applied in a way that gives a twisting or turning action.

In this chapter we will define a new physical quantity, *torque,* that describes the twisting or turning effort of a force. We'll find that the net torque acting on a rigid body determines its angular acceleration, in the same way that the net force on a body determines its linear acceleration. We'll also look at work and power in rotational motion so as to understand such problems as how energy is transmitted by the rotating drive shaft in a car. Finally, we will develop a new conservation principle, *conservation of angular momentum,* that is tremendously useful for understanding the rotational motion of both rigid and nonrigid bodies.

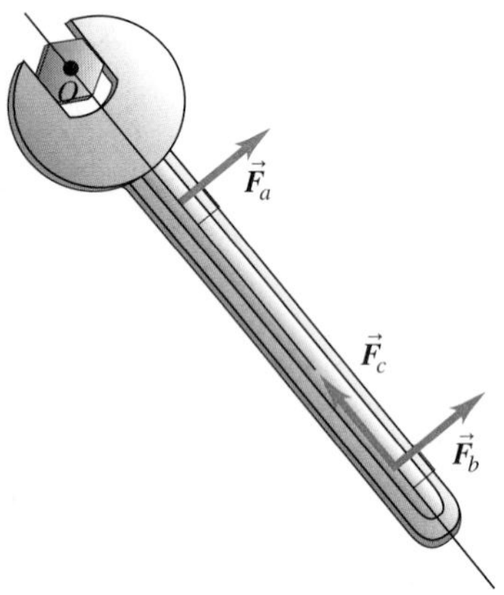

10.1 Which of these three equal-magnitude forces is most likely to loosen the tight bolt?

ActivPhysics ONLINE

7.1 Calculating Torques

We'll finish this chapter by studying *gyroscopes,* rotating devices that seemingly defy common sense and don't fall over when you might think they should—but that actually behave in perfect accordance with the dynamics of rotational motion.

10.1 | Torque

What is it about a force that determines how effective it is in causing or changing rotational motion? The magnitude and direction of the force are important, but so is the position of the point where the force is applied. When you try to swing a heavy door open, it's a lot more effective to push farther away from the axis of rotation (near the knob) than close to the axis (near the hinges). In Fig. 10.1 a wrench is being used to loosen a tight bolt. Force $\vec{F}_b$, applied near the end of the handle, is more effective than an equal force $\vec{F}_a$ applied near the bolt. Force $\vec{F}_c$ doesn't do any good at all; it's applied at the same point and has the same magnitude as $\vec{F}_b$, but it's directed along the length of the handle.

The quantitative measure of the tendency of a force to cause or change the rotational motion of a body is called *torque.* Figure 10.2 shows a body that can rotate about an axis that passes through point O and is perpendicular to the plane of the figure. The body is acted on by three forces, $\vec{F}_1$, $\vec{F}_2$, and $\vec{F}_3$, in the plane of the figure. The tendency of $\vec{F}_1$ to cause a rotation about point O depends on its magnitude F_1. It also depends on the *perpendicular* distance l_1 between the **line of action** of the force (that is, the line along which the force vector lies) and point O. We call the distance l_1 the **lever arm** (or **moment arm**) of force $\vec{F}_1$ about O. The twisting effort is directly proportional to both F_1 and l_1. We define the **torque** (or *moment*) of the force $\vec{F}_1$ with respect to point O as the product $F_1 l_1$. We use the Greek letter τ ("tau") for torque. For a force of magnitude F whose line of action is a perpendicular distance l from the point O, the torque is

$$\tau = Fl \tag{10.1}$$

Physicists usually use the term "torque," while engineers usually use "moment" (unless they are talking about a rotating shaft). Both groups use the term "lever arm" or "moment arm" for the distance l.

The lever arm of $\vec{F}_1$ in Fig. 10.2 is the perpendicular distance OA or l_1, and the lever arm of $\vec{F}_2$ is the perpendicular distance OB or l_2. The line of action of $\vec{F}_3$ passes through the reference point O, so the lever arm for $\vec{F}_3$ is zero and its torque with respect to point O is zero. In the same way, force $\vec{F}_c$ in Fig. 10.1 has zero torque with respect to point O. On the other hand, $\vec{F}_b$ has a greater torque than $\vec{F}_a$ because its lever arm is greater.

CAUTION Note that torque is always defined with reference to a specific point, often (but not always) the origin of a coordinate system. If we shift the position of this point, the torque of each force may also change. For example, the torque of force $\vec{F}_3$ in Fig. 10.2 is zero with respect to point O, but is *not* zero about point A or B. When describing the torque of a certain force, it's not enough to call it "the torque of $\vec{F}$"; you must say "the torque of $\vec{F}$ with respect to point X" or "the torque of $\vec{F}$ about point X."

Torque =
(Force magnitude) × (lever arm)

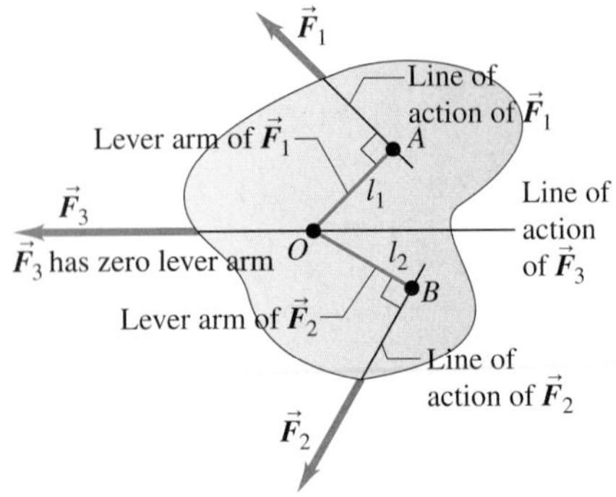

10.2 The torque or moment of a force about a point is the product of the force magnitude and the lever arm of the force.

Force $\vec{F}_1$ in Fig. 10.2 tends to cause *counterclockwise* rotation about O, while $\vec{F}_2$ tends to cause *clockwise* rotation. To distinguish between these two possibilities, we need to choose a positive sense of rotation. With the choice that *counterclockwise torques are positive and clockwise torques are negative,* the torques of $\vec{F}_1$ and $\vec{F}_2$ about O are

$$\tau_1 = +F_1 l_1 \qquad \tau_2 = -F_2 l_2$$

We will often use the symbol

to indicate our choice of the positive sense of rotation.

The SI unit of torque is the newton-meter. In our discussion of work and energy we called this combination the joule. But torque is *not* work or energy, and torque should be expressed in newton-meters, *not* joules.

Figure 10.3 shows a force $\vec{F}$ applied at a point P described by a position vector $\vec{r}$ with respect to the chosen point O. There are several ways to calculate the torque of this force. One is to find the lever arm l and use $\tau = Fl$. Or we can determine the angle ϕ between the vectors $\vec{r}$ and $\vec{F}$; the lever arm is $r \sin \phi$, so $\tau = rF \sin \phi$. A third method is to represent $\vec{F}$ in terms of a radial component F_{rad} along the direction of $\vec{r}$ and a tangential component F_{tan} at right angles, perpendicular to $\vec{r}$. (We call this a tangential component because if the body rotates, the point where the force acts moves in a circle, and this component is tangent to that circle.) Then $F_{tan} = F \sin \phi$ and $\tau = r(F \sin \phi) = F_{tan} r$. The component F_{rad} has no torque with respect to O because its lever arm with respect to that point is zero (compare to forces $\vec{F}_c$ in Fig. 10.1 and $\vec{F}_3$ in Fig. 10.2). Summarizing these three expressions for torque, we have

$$\tau = Fl = rF \sin \phi = F_{tan} r \qquad \text{(magnitude of torque)} \qquad (10.2)$$

We saw in Section 9.1 that angular velocity and angular acceleration can be represented as vectors; the same is true for torque. To see how to do this, note that the quantity $rF \sin \phi$ in Eq. (10.2) is the magnitude of the *vector product* $\vec{r} \times \vec{F}$ that we defined in Section 1.10. You should review that definition. We now generalize the definition of torque as follows: When a force $\vec{F}$ acts at a point having a position vector $\vec{r}$ with respect to an origin O, as in Fig. 10.3, the torque $\vec{\tau}$ of the force with respect to O is the *vector* quantity

$$\vec{\tau} = \vec{r} \times \vec{F} \qquad \text{(definition of torque vector)} \qquad (10.3)$$

The torque as defined in Eq. (10.2) is just the magnitude of the torque vector $\vec{r} \times \vec{F}$. The direction of $\vec{\tau}$ is perpendicular to both $\vec{r}$ and $\vec{F}$. In particular, if both $\vec{r}$ and $\vec{F}$ lie in a plane perpendicular to the axis of rotation, as in Fig. 10.3, then the torque vector $\vec{\tau} = \vec{r} \times \vec{F}$ is directed along the axis of rotation, with a sense given by the right-hand rule (Fig. 1.20). The direction relationships are shown in Fig. 10.4.

In diagrams that involve $\vec{r}$, $\vec{F}$, and $\vec{\tau}$, it's common to have one of the vectors oriented perpendicular to the page. (Indeed, by the very nature of the cross product, $\vec{\tau} = \vec{r} \times \vec{F}$ *must* be perpendicular to the plane of the vectors $\vec{r}$ and $\vec{F}$.) We use a dot (•) to represent a vector that points out of the page (see Fig. 10.4) and a cross (×) to represent a vector that points into the page.

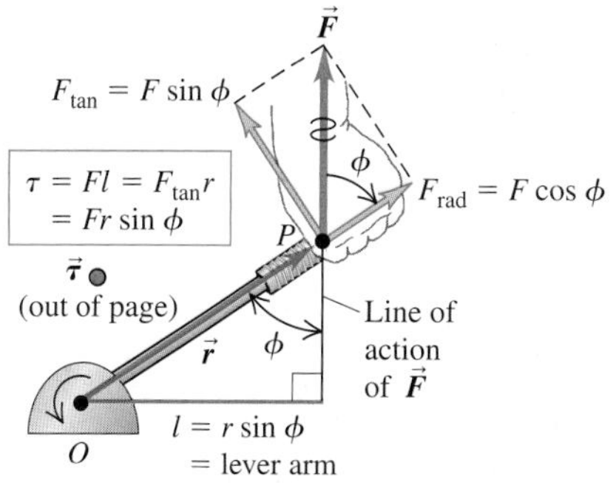

10.3 The torque of the force $\vec{F}$ about the point O is defined by $\vec{\tau} = \vec{r} \times \vec{F}$. The magnitude of $\vec{\tau}$ is $Fr \sin \phi$. In this figure, $\vec{r}$ and $\vec{F}$ are in the plane of the page and $\vec{\tau}$ points out of the page toward you.

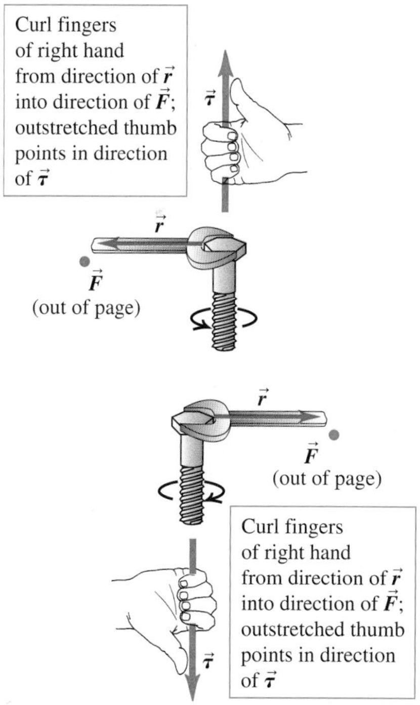

10.4 The torque vector $\vec{\tau} = \vec{r} \times \vec{F}$ is directed along the axis of the bolt, perpendicular to both $\vec{r}$ and $\vec{F}$. The direction of $\vec{\tau}$ is given by the right-hand rule. Note that the fingers of the right hand curl in the direction of the rotation that the torque tends to cause.

In the following sections we will usually be concerned with rotation of a body about an axis oriented in a specified constant direction. In that case, only the component of torque along that axis is of interest, and we often call that component the torque with respect to the specified axis.

Example 10.1 Applying a torque

A weekend plumber, unable to loosen a pipe fitting, slips a piece of scrap pipe (a "cheater") over his wrench handle. He then applies his full weight of 900 N to the end of the cheater by standing on it. The distance from the center of the fitting to the point where the weight acts is 0.80 m, and the wrench handle and cheater make an angle of 19° with the horizontal (Fig. 10.5a). Find the magnitude and direction of the torque he applies about the center of the pipe fitting.

SOLUTION

IDENTIFY and SET UP: We'll use Eq. (10.1) or (10.2) to find the magnitude of the torque and the right-hand rule with Eq. (10.3) to find the torque direction. Figure 10.5b shows the vectors $\vec{r}$ and $\vec{F}$ and the angle between them ($\phi = 109°$).

EXECUTE: To use Eq. (10.1), we first calculate the lever arm l. As Fig. 10.5b shows, it is the perpendicular distance from O to the line of action of the force:

$$l = (0.80\text{ m})\sin 109° = (0.80\text{ m})\sin 71° = 0.76\text{ m}$$

Then Eq. (10.1) tells us that the magnitude of the torque is

$$\tau = Fl = (900\text{ N})(0.76\text{ m}) = 680\text{ N}\cdot\text{m}$$

Or, from Eq. (10.2),

$$\tau = rF\sin\phi = (0.80\text{ m})(900\text{ N})(\sin 109°) = 680\text{ N}\cdot\text{m}$$

Alternatively, we can find $F_{\tan}$, the tangential component of $\vec{F}$. This is the component that acts perpendicular to $\vec{r}$ (that is, perpendicular to the "cheater"). The vector $\vec{r}$ is oriented 19° from the horizontal, so the perpendicular to $\vec{r}$ is oriented 19° from the vertical. Since $\vec{F}$ is vertical, this means $F_{\tan} = F(\cos 19°) = (900\text{ N})(\cos 19°) = 851\text{ N}$. Then the torque is

$$\tau = F_{\tan}r = (851\text{ N})(0.80\text{ m}) = 680\text{ N}\cdot\text{m}$$

If you curl the fingers of your right hand from the direction of $\vec{r}$ (in the plane of Fig. 10.5b, to the right and up) into the direction of $\vec{F}$ (straight down), your right thumb points into the plane of the figure. This is the direction of the torque $\vec{\tau}$.

EVALUATE: We've already checked our answer for the magnitude τ by calculating it in three different ways. To check our result for the direction of the torque, note that the force in Fig. 10.5 tends to produce a clockwise rotation about O. If you curl the fingers of your right hand in a clockwise direction, the thumb points into the plane of Fig. 10.5, which is indeed the direction of the torque.

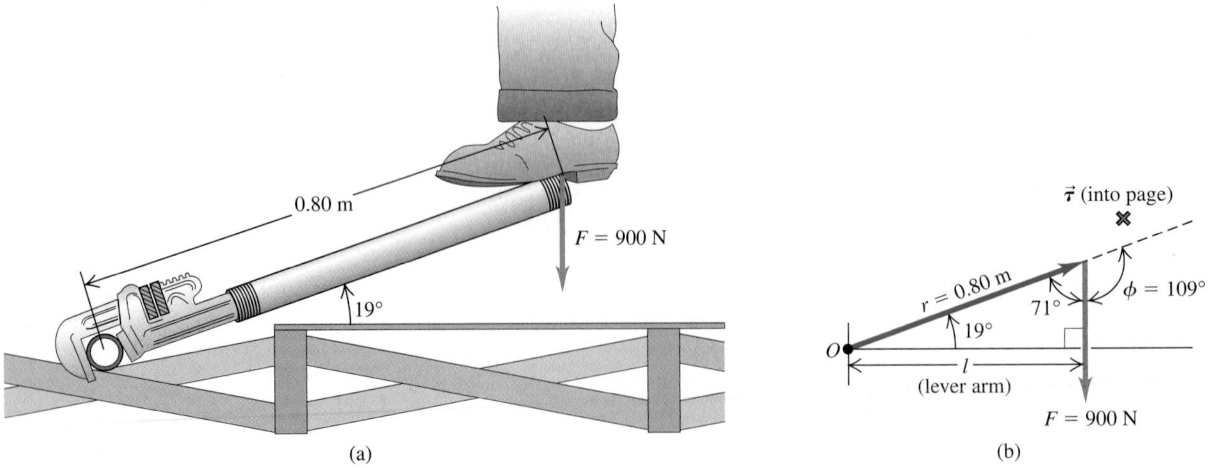

10.5 (a) A weekend plumber tries to loosen a pipe fitting by standing on a "cheater." (b) Vector diagram to find the torque about O.

Test Your Understanding

How great a downward force would the weekend plumber in Example 10.1 have to exert to produce the same torque without the "cheater"? The wrench by itself is 25 cm long.

10.2 | Torque and Angular Acceleration for a Rigid Body

We are now ready to develop the fundamental relation for the rotational dynamics of a rigid body. We will show that the angular acceleration of a rotating rigid body is directly proportional to the sum of the torque components along the axis of rotation. The proportionality factor is the moment of inertia.

To develop this relation, we again imagine the body as being made up of a large number of particles. We choose the axis of rotation to be the z-axis; the first particle has mass m_1 and distance r_1 from this axis (Fig. 10.6). The *net force* acting on this particle has a component $F_{1,\text{rad}}$ along the radial direction, a component $F_{1,\text{tan}}$ that is tangent to the circle of radius r_1 in which the particle moves as the body rotates, and a component F_{1y} along the axis of rotation. Newton's second law for the tangential component is

$$F_{1,\text{tan}} = m_1 a_{1,\text{tan}} \tag{10.4}$$

We can express the tangential acceleration of the first particle in terms of the angular acceleration α_z, using Eq. (9.14): $a_{1,\text{tan}} = r_1 \alpha_z$. Using this relation and multiplying both sides of Eq. (10.4) by r_1, we obtain

$$F_{1,\text{tan}} r_1 = m_1 r_1^2 \alpha_z \tag{10.5}$$

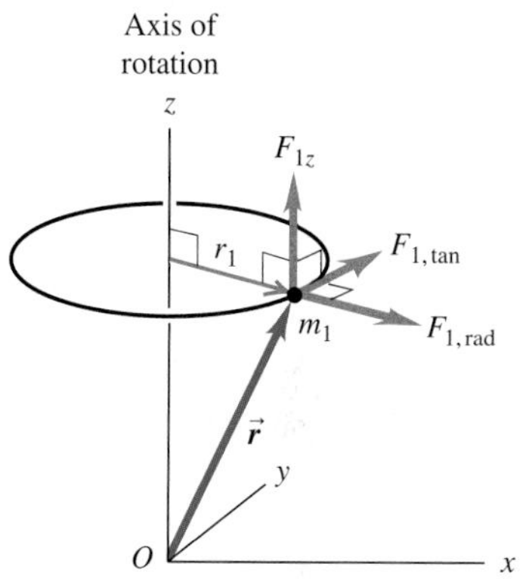

10.6 Three components of the net force acting on one of the particles of a rigid body. Only $F_{1,\text{tan}}$ has a z-component of torque about O.

From Eq. (10.2), $F_{1,\text{tan}} r_1$ is just the *torque* of the net force with respect to the rotation axis, equal to the component τ_{1z} of the torque vector along the rotation axis. The subscript z is a reminder that the torque affects rotation around the z-axis, in the same way that the subscript on F_{1z} is a reminder that this force affects the motion of particle 1 along the z-axis.

Neither of the components $F_{1,\text{rad}}$ or F_{1z} contributes to the torque about the z-axis, since neither tends to change the particle's rotation about that axis. So $\tau_{1z} = F_{1,\text{tan}} r_1$ is the total torque acting on the particle with respect to the rotation axis. Also, $m_1 r_1^2$ is I_1, the moment of inertia of the particle about the rotation axis. With this in mind, we rewrite Eq. (10.5) as

$$\tau_{1z} = I_1 \alpha_z = m_1 r_1^2 \alpha_z$$

We write an equation like this for every particle in the body and then add all these equations:

$$\tau_{1z} + \tau_{2z} + \cdots = I_1 \alpha_z + I_2 \alpha_z + \cdots = m_1 r_1^2 \alpha_z + m_2 r_2^2 \alpha_z + \cdots$$

or

$$\sum \tau_{iz} = \left(\sum m_i r_i^2\right) \alpha_z$$

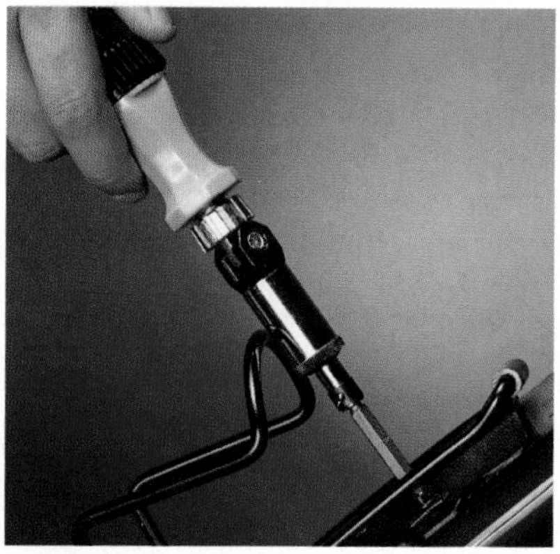

10.7 Loosening or tightening a screw requires giving it an angular acceleration and hence applying a torque. This is made easier by using a screwdriver with a large-radius handle, which provides a large lever arm for the force you apply with your hand.

The left side of this equation is the sum of all the torques about the rotation axis that act on all the particles. The right side is $I = \sum m_i r_i^2$, the total moment of inertia about the rotation axis, multiplied by the angular acceleration α_z. Note that α_z

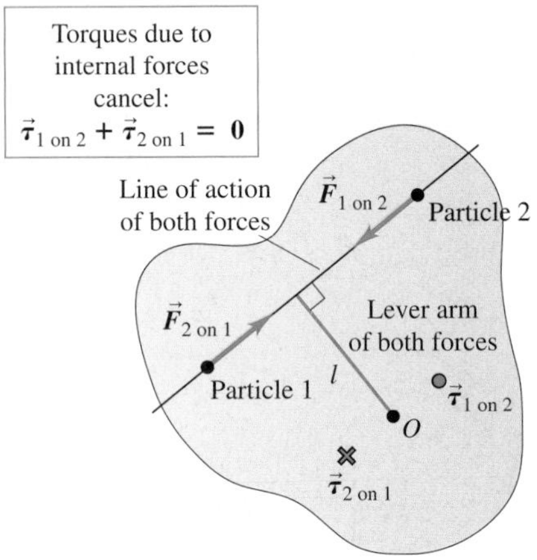

10.8 Two particles in a rigid body exert equal and opposite forces on each other. If the forces act along the line joining the particles, the lever arms of the forces are the same and the torques due to the two forces are equal and opposite. Only *external* torques affect the body's rotation.

is the same for every particle because this is a *rigid* body. Thus for the entire rigid body we have the *rotational analog of Newton's second law:*

$$\sum \tau_z = I\alpha_z \qquad (10.6)$$

(rotational analog of Newton's second law for a rigid body)

Just as Newton's second law says that the net force on a particle equals the particle's mass times its acceleration, Eq. (10.6) says that the net torque on a rigid body equals the body's moment of inertia about the rotation axis times its angular acceleration (Fig. 10.7).

We emphasize that Eq. (10.6) is valid *only* for *rigid* bodies. If the body is not rigid, like a rotating tank of water or a swirling tornado of air, the angular acceleration α_z is different for different particles in the body, and the derivation of Eq. (10.6) isn't valid. Also note that since our derivation used Eq. (9.14), $a_{tan} = r\alpha_z$, α_z must be measured in rad/s^2.

The torque on each particle is due to the net force on that particle, which is the vector sum of external and internal forces (defined in Section 8.2). According to Newton's third law, the *internal* forces that any pair of particles in the rigid body exert on each other are equal and opposite (Fig. 10.8). If these forces act along the line joining the two particles, their lever arms with respect to any axis are also equal. So the torques for each such pair are equal and opposite, and add to zero. Indeed, *all* the internal torques add to zero, so the sum $\sum \tau_z$ in Eq. (10.6) includes only the torques of the *external* forces.

Often, one of the important external forces acting on a body is its *weight.* This force is not concentrated at a single point; it acts on every particle in the entire body. Nevertheless, it turns out that if the value of $\vec{g}$ is the same at all points, we always get the correct torque (about any specified axis) if we assume that all the weight is concentrated at the *center of mass* of the body. We will prove this statement in Chapter 11, but meanwhile we will use it for some of the problems in this chapter.

ActivPhysics ONLINE

7.8 Rotoride—Dynamics Approach
7.9 Falling Ladder
7.10 Woman and Flywheel Elevator—Dynamics Approach

Problem-Solving Strategy: Rotational Dynamics for Rigid Bodies

Our strategy for solving problems in rotational dynamics is very similar to the strategy presented in Section 5.1 for solving problems that involve Newton's second law.

IDENTIFY *the relevant concepts:* The equation $\sum \tau_z = I\alpha_z$ is useful whenever torques act on a rigid body—that is, whenever forces act on a rigid body in such a way as to change the state of the body's rotation.

In some cases you may be able to use an energy approach instead, as we did in Section 9.4. However, if the target variable is a force, a torque, an acceleration, an angular acceleration, or an elapsed time, the approach using $\sum \tau_z = I\alpha_z$ is almost always the most efficient one.

SET UP *the problem using the following steps:*

1. Draw a sketch of the situation and select the body or bodies to be analyzed.
2. For each body, draw a free-body diagram isolating the body and including all the forces (and *only* those forces) that act on the body, including its weight. Label unknown quantities with algebraic symbols. A new consideration is that you must show the *shape* of the body accurately, including all dimensions and angles you will need for torque calculations.
3. Choose coordinate axes for each body and indicate a positive sense of rotation for each rotating body. If there is a linear acceleration, it's usually simplest to pick a positive axis in its direction. If you know the sense of α_z in advance, picking it as the positive sense of rotation simplifies the calculations. When you represent a force in

terms of its components, cross out the original force to avoid including it twice.

EXECUTE *the solution as follows:*

1. For each body in the problem, decide whether it undergoes translational motion, rotational motion, or both. Depending on the behavior of the body in question, apply $\sum\vec{F} = m\vec{a}$ (as in Section 5.1), $\sum\tau_z = I\alpha_z$, or both to the body. Be careful to write separate equations of motion for each body.
2. There may be *geometrical* relations between the motions of two or more bodies, as with a string that unwinds from a pulley while turning it or a wheel that rolls without slipping (to be discussed in Section 10.3). Express these relations in algebraic form, usually as relations between two linear accelerations or between a linear acceleration and an angular acceleration.
3. Check that the number of equations matches the number of unknown quantities. Then solve the equations to find the target variable(s).

EVALUATE *your answer:* Check that the algebraic signs of your results make sense. As an example, suppose the problem is about a spool of thread. If you are pulling thread off the spool, your answers should *not* tell you that the spool is turning in the direction that rolls the thread back on the spool! Whenever possible, check the results for special cases or extreme values of quantities and compare them with your intuitive expectations. Ask yourself: "Does this result make sense?"

Example 10.2 An unwinding cable

Figure 10.9a shows the same situation that we analyzed in Example 9.8 (Section 9.4) using energy methods. A cable is wrapped several times around a uniform solid cylinder that can rotate about its axis. The cylinder has diameter 0.120 m and mass 50 kg. The cable is pulled with a force of 9.0 N. Assuming that the cable unwinds without stretching or slipping, what is its acceleration?

SOLUTION

IDENTIFY: Our target variable is the acceleration of the cable, which we cannot find directly using the energy method of Section 9.4 (which does not involve acceleration). Instead, we'll apply rotational dynamics to the cylinder. To obtain the acceleration of the cable, we'll find a relationship between the motion of the cable and the motion of the rim of the cylinder.

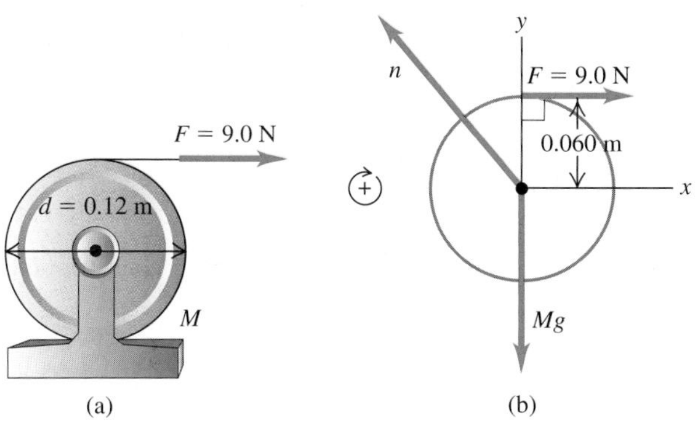

10.9 (a) Cylinder and cable. (b) Free-body diagram for the cylinder.

SET UP: Figure 10.9b shows the free-body diagram for the cylinder of mass $M = 50$ kg. The cylinder rotates clockwise when the cable is pulled, so we take the clockwise rotation to be positive. The net force on the cylinder must be zero because its center of mass remains at rest. Note that the weight (magnitude Mg) and the normal force (magnitude n) exerted by the cylinder bearings act along lines through the rotation axis. Hence these forces produce no torque with respect to that axis.

EXECUTE: The only torque about the rotation axis is due to the force F, which has a lever arm equal to the radius R of the cylinder: $l = R = 0.060$ m, so $\tau_z = FR$. (This torque is positive as it tends to cause a clockwise rotation.) From Example 9.8, the moment of inertia of the cylinder about the rotation axis is $I = \frac{1}{2}MR^2$. Hence Eq. (10.6) gives us the *angular* acceleration of the cylinder:

$$\alpha_z = \frac{\tau_z}{I} = \frac{FR}{MR^2/2} = \frac{2F}{MR} = \frac{2(9.0\ \text{N})}{(50\ \text{kg})(0.060\ \text{m})} = 6.0\ \text{rad/s}^2$$

(Be certain to check that these units are correct. We can add the "rad" to our result because a radian is a dimensionless quantity.)

To get the *linear* acceleration of the cable, we need a kinematic relation. We remarked in Section 9.3 that the acceleration of a cable unwinding from a cylinder is the same as the tangential component of acceleration of a point on the surface of the cylinder where the cable is tangent to it. This tangential acceleration is given by Eq. (9.14):

$$a_x = R\alpha = (0.060\ \text{m})(6.0\ \text{rad/s}^2) = 0.36\ \text{m/s}^2$$

EVALUATE: Can you use this result, together with an equation from Chapter 2, to determine the speed of the cable after it has been pulled 2.0 m? Try it, and compare your result with Example 9.8, in which we found this speed using work and energy considerations.

Example 10.3 An unwinding cable II

Figure 10.10a shows the same situation that we analyzed in Example 9.9 (Section 9.4) using energy methods. Find the acceleration of the object of mass m.

SOLUTION

IDENTIFY: We'll apply translational dynamics to the hanging object and rotational dynamics to the cylinder. Because the cable doesn't slip on the cylinder, there is a relationship between the linear acceleration of the hanging object (our target variable) and the angular acceleration of the cylinder.

SET UP: We have to treat the two bodies separately. Figure 10.10b shows a free-body diagram for each body. We take the positive sense of rotation for the cylinder to be counterclockwise and the positive direction of the y-coordinate for the object to be downward.

EXECUTE: For the object, Newton's second law gives

$$\sum F_y = mg + (-T) = ma_y$$

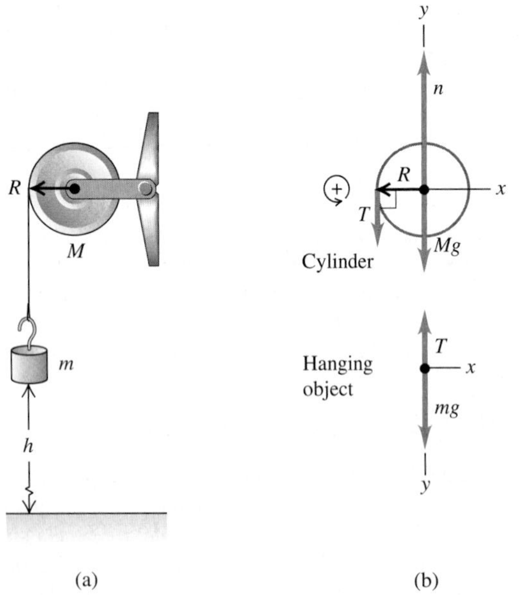

10.10 (a) Cylinder, object, and cable. (b) Free-body diagrams for the cylinder and the hanging object. The cable is assumed to have negligible mass.

For the cylinder, the weight Mg and the normal force n (exerted by the bearing) have no torques with respect to the rotation axis because they act along lines through that axis, just as in Example 10.2. The only torque is that due to the cable tension T. Applying Eq. (10.6) to the cylinder gives

$$\sum \tau_z = RT = I\alpha_z = \frac{1}{2}MR^2\alpha_z$$

As in Example 10.2, the acceleration of the cable is the same as the tangential acceleration of a point on the cylinder rim. According to Eq. (9.14), this acceleration is given by $a_y = a_{\text{tan}} = R\alpha_z$. We use this to replace $(R\alpha_z)$ with a_y in the cylinder equation above, and then divide by R; the result is

$$T = \frac{1}{2}Ma_y$$

Now we substitute this expression for T into Newton's second law for the object and solve for the acceleration a_y:

$$mg - \frac{1}{2}Ma_y = ma_y$$

$$a_y = \frac{g}{1 + M/2m}$$

EVALUATE: The acceleration is positive (in the downward direction) and less than g, as it should be since the cable is holding the object back. To see how much force the cable exerts, substitute our expression for a_y back into Newton's second law for the object to find T:

$$T = mg - ma_y = mg - m\left(\frac{g}{1 + M/2m}\right) = \frac{mg}{1 + 2m/M}$$

The tension in the cable is *not* equal to the weight mg of the object; if it were, the object could not accelerate.

Let's check some particular cases. When M is much larger than m, the tension is nearly equal to mg, and the acceleration is correspondingly much less than g. When M is zero, $T = 0$ and $a_y = g$; the object then falls freely. If the object starts from a height h above the floor with an initial speed v_0, its speed v when it strikes the ground is given by $v^2 = v_0^2 + 2a_y h$. If it starts from rest, $v_0 = 0$ and

$$v = \sqrt{2a_y h} = \sqrt{\frac{2gh}{1 + M/2m}}$$

This is the same result we obtained from energy considerations in Example 9.9.

Example 10.4 Two masses and a rotating pulley

In Fig. 10.11a a glider of mass m_1 slides without friction on a horizontal air track. It is attached to an object of mass m_2 by a massless string. The pulley is a thin cylindrical shell (with massless spokes) with mass M and radius R, and the string turns the pulley without slipping or stretching. Find the acceleration of each body, the angular acceleration of the pulley, and the tension in each part of the string.

SOLUTION

IDENTIFY: We will use translational dynamics to describe the motion of the glider and hanging object, and rotational dynamics to describe the motion of the pulley. Because the string doesn't stretch, both the glider and the object have the same magnitude of acceleration; the rim of the pulley has the same magnitude of *tangential* acceleration because the string doesn't slip.

SET UP: Figure 10.11b shows the free-body diagrams and coordinate systems for the three bodies. With our choices of coordinates the glider and object accelerate in their positive x- and y-directions, respectively. We likewise choose the positive sense of rotation to be clockwise (the same sense as the pulley's actual angular acceleration). We have five target variables: the acceleration of the glider (a_{1x}), the acceleration of the object (a_{2y}), the pulley's angular acceleration α_z, and the two tensions (T_1 and T_2). This may sound like an awesome problem, but we will have as many equations as we have unknowns, and solving them will be a lot easier than you might imagine.

CAUTION We considered a similar situation in Example 5.13 (Section 5.2). In that example the string slid without friction over a fixed pulley, and the tension was the same throughout the massless string. But with a rotating pulley, and with friction between the pulley and string to prevent slipping, the two tensions T_1 and T_2 *cannot* be equal. If they were, the pulley could not have an angular acceleration. Labeling the tension in both parts of the string as simply T would be a serious error. Be on your guard against this error in any problem that involves a rotating pulley.

EXECUTE: The equations of motion for the glider and the object are

$$\text{Glider: } \sum F_x = T_1 = m_1 a_{1x} \tag{10.7}$$

$$\text{Object: } \sum F_y = m_2 g + (-T_2) = m_2 a_{2y} \tag{10.8}$$

The unknown normal force n_2 acts on a line through the pulley's axis of rotation, so it has no lever arm or torque with respect to that axis. From Table 9.2 the moment of inertia of the pulley about this axis is $I = MR^2$. Then the pulley's equation of motion is

$$\text{Pulley: } \sum \tau_z = T_2 R + (-T_1 R) = I\alpha_z = (MR^2)\alpha_z \tag{10.9}$$

Because the string does not stretch or slip, we have the additional *kinematic* relations

$$a_{1x} = a_{2y} = R\alpha_z \tag{10.10}$$

(The accelerations of the glider and the object have different directions but the same magnitude.)

Equations (10.7) through (10.10) are five simultaneous equations for the five unknowns a_{1x}, a_{2y}, α, T_1, and T_2. (Equation (10.10) is actually two equations.) First we use Eqs. (10.10) to eliminate a_{2y} and α_z from Eqs. (10.7) through (10.9). We then have three equations for the three unknowns T_1, T_2, and a_{1x}:

$$T_1 = m_1 a_{1x}$$
$$m_2 g - T_2 = m_2 a_{1x}$$
$$T_2 - T_1 = M a_{1x}$$

The easiest way to solve these is simply to *add* the three equations, which eliminates T_1 and T_2, and then solve for a_{1x}:

$$a_{1x} = \frac{m_2 g}{m_1 + m_2 + M}$$

From Eq. (10.10), the acceleration a_{2y} of the hanging object is equal to a_{1x} and the angular acceleration α_z of the pulley is equal to a_{1x} divided by R. We can then substitute a_{1x} back into Eqs. (10.7) and (10.8) to find the tensions. The results are

$$T_1 = \frac{m_1 m_2 g}{m_1 + m_2 + M} \qquad T_2 = \frac{(m_1 + M) m_2 g}{m_1 + m_2 + M}$$

EVALUATE: Let's check some special cases to see whether these results make sense. First, if either m_1 or M is much larger than m_2, the accelerations are very small and T_2 is approximately $m_2 g$. But if m_2 is much *larger* than either m_1 or M, the acceleration is approximately g. Both of these results are what we should expect. When $M = 0$, do we get the same result as Example 5.13 (Section 5.2)? Why or why not? Can you think of other special cases to check?

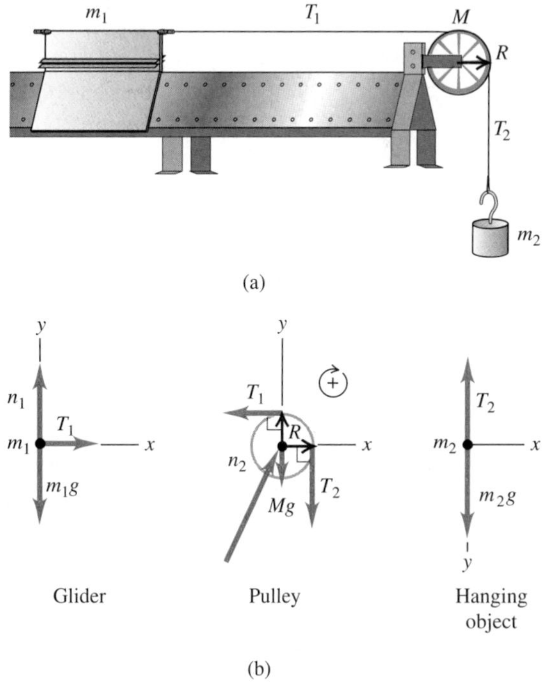

10.11 (a) An air-track glider pulled by an object hanging over a pulley. (b) Free-body diagrams for all three bodies.

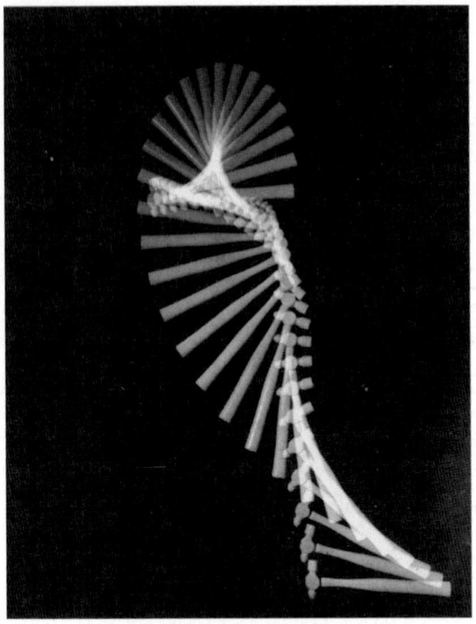

10.12 The motion of a rigid body like this thrown hammer is a combination of translational motion of the center of mass and rotation around the center of mass.

Test Your Understanding

Suppose the system in Example 10.4 is initially in motion, with the glider moving to the left, the hanging object rising, and the pulley rotating counterclockwise. In this situation, what are the linear accelerations of the glider and the object and the angular acceleration of the pulley?

10.3 | Rigid-Body Rotation About a Moving Axis

We can extend our analysis of the dynamics of rotational motion to some cases in which the axis of rotation moves. When that happens, the motion of the body is **combined translation and rotation.** The key to understanding such situations is this: every possible motion of a rigid body can be represented as a combination of *translational motion of the center of mass* and *rotation about an axis through the center of mass.* This is true even when the center of mass accelerates, so that it is not at rest in any inertial frame. A graphic example is the motion of a hammer thrown upward (Fig. 10.12). The center of mass of the hammer follows a parabolic curve, as though the hammer were a particle located at the center of mass. At the same time, the hammer rotates with constant angular velocity about the center of mass (compare the motion of the wrench in Fig. 8.25). The translation of the center of mass and the rotation about the center of mass can be treated as separate but related problems. Other examples of such motion are a ball rolling down a hill and a yo-yo unwinding at the end of a string.

Combined Translation and Rotation: Energy Relations

It's beyond the scope of this book to prove that the motion of a rigid body can always be divided into translation of the center of mass and rotation about the center of mass. But we can show that this is true for the *kinetic energy* of a rigid body that has both translational and rotational motion. In this case, the body's kinetic energy is the sum of a part $\frac{1}{2}Mv_{\text{cm}}{}^2$ associated with motion of the center of mass and a part $\frac{1}{2}I_{\text{cm}}\omega^2$ associated with rotation about an axis through the center of mass:

$$K = \frac{1}{2}Mv_{\text{cm}}{}^2 + \frac{1}{2}I_{\text{cm}}\omega^2$$

(rigid body with both translation and rotation) (10.11)

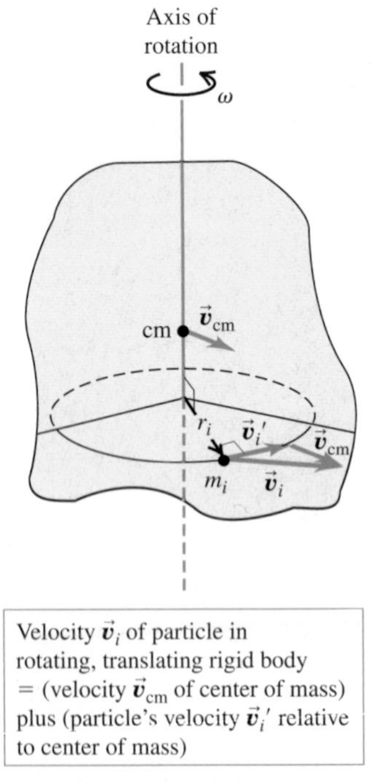

10.13 A rigid body with both translation and rotation.

To prove this relation, we again imagine the rigid body to be made up of particles. Consider a typical particle with mass m_i as shown in Fig. 10.13. The velocity $\vec{v}_i$ of this particle relative to an inertial frame is the vector sum of the velocity $\vec{v}_{\text{cm}}$ of the center of mass and the velocity $\vec{v}_i'$ of the particle *relative to* the center of mass:

$$\vec{v}_i = \vec{v}_{\text{cm}} + \vec{v}_i' \qquad (10.12)$$

The kinetic energy K_i of this particle in the inertial frame is $\frac{1}{2}m_iv_i^2$, which we can also express as $\frac{1}{2}m_i(\vec{v}_i \cdot \vec{v}_i)$. Substituting Eq. (10.12) into this, we get

$$\begin{aligned} K_i &= \frac{1}{2}m_i(\vec{v}_{\text{cm}} + \vec{v}_i') \cdot (\vec{v}_{\text{cm}} + \vec{v}_i') \\ &= \frac{1}{2}m_i(\vec{v}_{\text{cm}} \cdot \vec{v}_{\text{cm}} + 2\vec{v}_{\text{cm}} \cdot \vec{v}_i' + \vec{v}_i' \cdot \vec{v}_i') \\ &= \frac{1}{2}m_i(v_{\text{cm}}{}^2 + 2\vec{v}_{\text{cm}} \cdot \vec{v}_i' + v_i'^2) \end{aligned}$$

The total kinetic energy is the sum $\sum K_i$ for all the particles making up the body. Expressing the three terms in this equation as separate sums, we get

$$K = \sum K_i = \sum\left(\frac{1}{2}m_i v_{cm}^2\right) + \sum\left(m_i \vec{v}_{cm} \cdot \vec{v}_i'\right) + \sum\left(\frac{1}{2}m_i v_i'^2\right)$$

The first and second terms have common factors that can be taken outside the sum:

$$K = \frac{1}{2}\left(\sum m_i\right)v_{cm}^2 + \vec{v}_{cm} \cdot \left(\sum m_i \vec{v}_i'\right) + \sum\left(\frac{1}{2}m_i v_i'^2\right) \quad (10.13)$$

Now comes the reward for our effort. In the first term, $\sum m_i$ is the total mass M. The second term is zero because $\sum m_i \vec{v}_i'$ is M times the velocity of the center of mass *relative to the center of mass,* and this is zero by definition. The last term is the sum of kinetic energies of the particles computed by using their speeds with respect to the center of mass; this is just the kinetic energy of rotation around the center of mass. Using the same steps that led to Eq. (9.17) for the rotational kinetic energy of a rigid body, we can write this last term as $\frac{1}{2}I_{cm}\omega^2$, where I_{cm} is the moment of inertia with respect to the axis through the center of mass and ω is the angular speed. So Eq. (10.13) becomes Eq. (10.11):

$$K = \frac{1}{2}Mv_{cm}^2 + \frac{1}{2}I_{cm}\omega^2$$

An important case of combined translation and rotation is **rolling without slipping,** such as the motion of the wheel shown in Fig. 10.14. The wheel is symmetrical, so its center of mass is at its geometric center. We view the motion in an inertial frame of reference in which the surface on which the wheel rolls is at rest. In this frame, the point on the wheel that contacts the surface must be instantaneously *at rest* so that it does not slip. Hence the velocity $\vec{v}_1'$ of the point of contact relative to the center of mass must have the same magnitude but opposite direction as the center-of-mass velocity $\vec{v}_{cm}$. If the radius of the wheel is R and its angular speed about the center of mass is ω, then the magnitude of $\vec{v}_1'$ is $R\omega$; hence we must have

$$v_{cm} = R\omega \quad \text{(condition for rolling without slipping)} \quad (10.14)$$

As Fig. 10.14 shows, the velocity of a point on the wheel is the vector sum of the velocity of the center of mass and the velocity of the point relative to the center of mass. Thus while point 1, the point of contact, is instantaneously at rest,

ActivPhysics ONLINE

7.11 Race Between a Block and a Disk

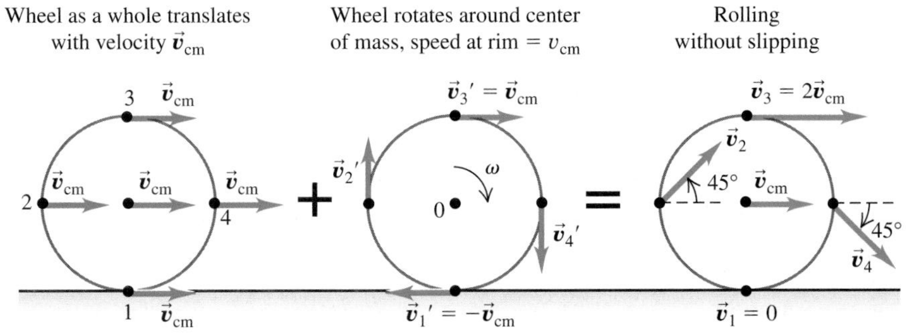

10.14 The motion of a rolling wheel is the sum of the translational motion of the center of mass plus the rotational motion of the wheel around the center of mass. For rolling without slipping, the speed of the rim relative to the center of mass must be equal to the magnitude of $\vec{v}_{cm}$. Note that the wheel is instantaneously at rest at the point where it contacts the ground.

10.15 The smoke rising from this drag racer's rear tires shows that the tires are slipping on the road, so v_{cm} is *not* equal to $R\omega$.

point 3 at the top of the wheel is moving forward *twice as fast* as the center of mass, and points 2 and 4 at the sides have velocities at 45° to the horizontal.

At any instant we can think of the wheel as rotating about an "instantaneous axis" of rotation that passes through the point of contact with the ground. The angular velocity ω is the same for this axis as for an axis through the center of mass; an observer at the center of mass sees the rim make the same number of revolutions per second as does an observer at the rim watching the center of mass spin around him. If we think of the motion of the rolling wheel in Fig. 10.14 in this way, the kinetic energy of the wheel is $K = \frac{1}{2}I_1\omega^2$, where I_1 is the moment of inertia of the wheel about an axis through point 1. But by the parallel-axis theorem, Eq. (9.19), $I_1 = I_{cm} + MR^2$, where M is the total mass of the wheel and I_{cm} is the moment of inertia with respect to an axis through the center of mass. Using Eq. (10.14), the kinetic energy of the wheel is

$$K = \frac{1}{2}I_1\omega^2 = \frac{1}{2}I_{cm}\omega^2 + \frac{1}{2}MR^2\omega^2 = \frac{1}{2}I_{cm}\omega^2 + \frac{1}{2}Mv_{cm}{}^2$$

which is the same as Eq. (10.11).

CAUTION It's important to note that the relationship $v_{cm} = R\omega$ holds only if there is rolling without slipping. When a drag racer first starts to move, the rear tires are spinning very fast while the racer is hardly moving, so $R\omega$ is greater than v_{cm} (Fig. 10.15). If a driver applies the brakes too heavily so that the car skids, the tires will spin hardly at all and $R\omega$ is less than v_{cm}.

If a rigid body changes height as it moves, we must also consider gravitational potential energy. As we discussed in Section 9.4, the gravitational potential energy associated with any extended body of mass M, rigid or not, is the same as if we replace the body by a particle of mass M located at the body's center of mass. That is,

$$U = Mgy_{cm}$$

Example 10.5 A rolling cylindrical shell

A hollow cylindrical shell with mass M and radius R rolls without slipping with speed v_{cm} on a flat surface. What is its kinetic energy?

SOLUTION

IDENTIFY and SET UP: We'll use Eq. (10.11) to find the kinetic energy. The moment of inertia is $I = MR^2$ from Table 9.2, and the angular speed is $\omega = v_{cm}/R$ since there is rolling without slipping.

EXECUTE: Substituting these expressions into Eq. (10.11), we get

$$K = \frac{1}{2}Mv_{cm}{}^2 + \frac{1}{2}(MR^2)\left(\frac{v_{cm}}{R}\right)^2$$
$$= Mv_{cm}{}^2$$

EVALUATE: The kinetic energy is twice as great as it would be if the cylindrical shell were sliding at speed v_{cm} without rolling. Half of the total kinetic energy is translational and half is rotational.

Example 10.6 Speed of a primitive yo-yo

A primitive yo-yo is made by wrapping a string several times around a solid cylinder with mass M and radius R (Fig. 10.16). You hold the end of the string stationary while releasing the cylinder with no initial motion. The string unwinds but does not slip or stretch as the cylinder drops and rotates. Use energy considerations to find the speed v_{cm} of the center of mass of the solid cylinder after it has dropped a distance h.

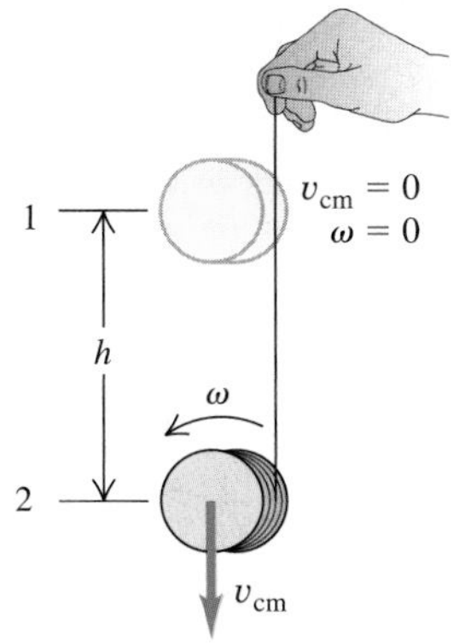

10.16 Calculating the speed of a primitive yo-yo.

SOLUTION

IDENTIFY: The upper end of the string is held fixed, not pulled upward, so the hand in Fig. 10.16 does no work on the system of string and cylinder. As in Example 9.8 (Section 9.4), there is friction between the string and the cylinder, but because the string never slips on the surface of the cylinder, no mechanical energy is lost. Thus we can use conservation of mechanical energy.

SET UP: The potential energies are $U_1 = Mgh$ and $U_2 = 0$. The string has no kinetic energy because it's massless. The initial kinetic energy of the cylinder is $K_1 = 0$, and its final kinetic energy K_2 is given by Eq. (10.11). The moment of inertia is $I = \frac{1}{2}MR^2$, and $\omega = v_{cm}/R$ because the cylinder does not slip on the string.

EXECUTE: Using Eq. (10.11), the kinetic energy at point 2 is

$$K_2 = \frac{1}{2}Mv_{cm}^2 + \frac{1}{2}\left(\frac{1}{2}MR^2\right)\left(\frac{v_{cm}}{R}\right)^2$$
$$= \frac{3}{4}Mv_{cm}^2$$

Then, conservation of energy gives

$$K_1 + U_1 = K_2 + U_2$$
$$0 + Mgh = \frac{3}{4}Mv_{cm}^2 + 0$$

and

$$v_{cm} = \sqrt{\frac{4}{3}gh}$$

EVALUATE: This is less than the speed $\sqrt{2gh}$ that a dropped object would have, because one-third of the potential energy released as the cylinder falls appears as rotational kinetic energy.

Example 10.7 Race of the rolling bodies

In a physics lecture demonstration, an instructor "races" various round rigid bodies by releasing them from rest at the top of an inclined plane (Fig. 10.17). What shape should a body have to reach the bottom of the incline first?

SOLUTION

IDENTIFY: We can again use conservation of energy because there is no sliding of the rigid bodies over the inclined plane. No work is done by kinetic friction if the bodies roll without slipping. We can also ignore the effects of *rolling friction,* introduced in Section 5.3, provided that the bodies and the surface on which they roll are perfectly rigid. (Later in this section we'll explain why this is so.)

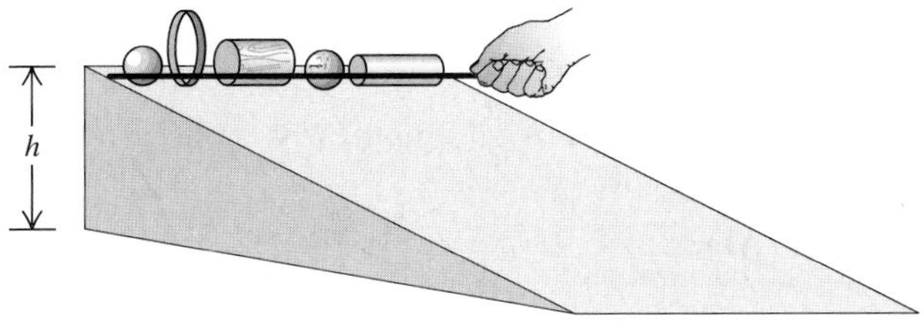

10.17 Which body rolls down the incline fastest, and why?

SET UP: Each body starts from rest at the top of an incline with height *h,* so $K_1 = 0$, $U_1 = Mgh$, and $U_2 = 0$. The kinetic energy at the bottom of the incline is given by Eq. (10.11). If the bodies roll without slipping, $\omega = v_{cm}/R$. The moments of inertia of all the round bodies in Table 9.2 (about axes through their centers of mass) can be expressed as $I_{cm} = cMR^2$, where *c* is a pure number less than or equal to 1 that depends on the shape of the body. Our goal is to find the value of *c* that gives the body the greatest speed at the bottom of the incline.

EXECUTE: From conservation of energy,

$$K_1 + U_1 = K_2 + U_2$$
$$0 + Mgh = \frac{1}{2}Mv_{cm}^2 + \frac{1}{2}cMR^2\left(\frac{v_{cm}}{R}\right)^2$$
$$= \frac{1}{2}(1 + c)Mv_{cm}^2$$

so the speed at the bottom of the incline is

$$v_{cm} = \sqrt{\frac{2gh}{1 + c}}$$

EVALUATE: This is a fairly amazing result; the speed doesn't depend on either the mass M of the body or its radius R. All uniform solid cylinders have the same speed at the bottom, even if their masses and radii are different, because they have the same c. All solid spheres have the same speed, and so on. The smaller the value of c, the faster the body is moving at the bottom (and at any point on the way down). Small-c bodies always beat large-c bodies, because they have less of their kinetic energy tied up in rotation and have more available for translation. Reading the values of c from Table 9.2, we see that the order of finish is as follows: any solid sphere, any solid cylinder, any thin-walled hollow sphere, and any thin-walled hollow cylinder.

10.18 The axle of a bicycle wheel passes through the wheel's center of mass and is an axis of symmetry. Hence the rotation of the wheel is described by Eq. (10.16), provided the bicycle doesn't turn or tilt to one side (which would change the orientation of the axle).

Combined Translation and Rotation: Dynamics

We can also analyze the combined translational and rotational motion of a rigid body from the standpoint of dynamics. We showed in Section 8.5 that for a body with total mass M, the acceleration $\vec{a}_{cm}$ of the center of mass is the same as that of a point mass M acted on by all the external forces on the actual body:

$$\sum \vec{F}_{ext} = M\vec{a}_{cm} \tag{10.15}$$

The rotational motion about the center of mass is described by the rotational analog of Newton's second law, Eq. (10.6):

$$\sum \tau_z = I_{cm}\alpha_z \tag{10.16}$$

where I_{cm} is the moment of inertia with respect to an axis through the center of mass and the sum $\sum \tau_z$ includes all external torques with respect to this axis. It's not immediately obvious that Eq. (10.16) should apply to the motion of a translating rigid body; after all, our derivation of $\sum \tau_z = I\alpha_z$ in Section 10.2 assumed that the axis of rotation was stationary. But in fact, Eq. (10.16) is valid *even when the axis of rotation moves,* provided the following two conditions are met:

1. The axis through the center of mass must be an axis of symmetry.
2. The axis must not change direction.

These conditions are satisfied for many types of rotation (Fig. 10.18). Note that in general this moving axis of rotation is *not* at rest in an inertial frame of reference.

We can now solve dynamics problems involving a rigid body that undergoes translational and rotational motion at the same time, provided that the rotation axis satisfies the two conditions just mentioned. The problem-solving strategy outlined in Section 10.2 is equally useful here, and you should review it now. Keep in mind that when a body undergoes translational and rotational motion at the same time, we need two separate equations of motion *for the same body.* One of these, Eq. (10.15), describes the translational motion of the center of mass. The other equation of motion, Eq. (10.16), describes the rotational motion about the axis through the center of mass.

Example 10.8 Acceleration of a primitive yo-yo

For the primitive yo-yo in Example 10.6, find the downward acceleration of the cylinder and the tension in the string.

SOLUTION

IDENTIFY and SET UP: We'll use Eqs. (10.15) and (10.16), along with the condition that the string does not slip on the cylinder. Figure 10.19 shows a free-body diagram for the yo-yo, including the choice of positive coordinate directions. With these coordinates, our target variable is $a_{cm\text{-}y}$.

EXECUTE: The equation for the translational motion of the center of mass is

$$\sum F_y = Mg + (-T) = Ma_{cm\text{-}y} \tag{10.17}$$

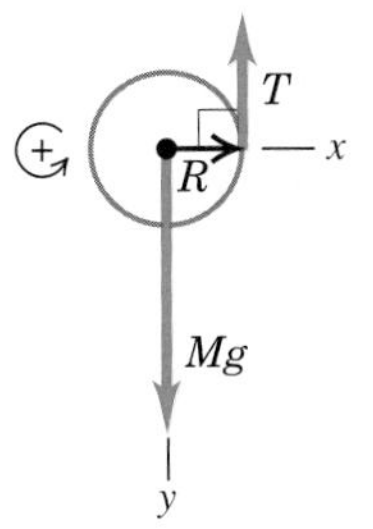

10.19 Free-body diagram for a primitive yo-yo (see Fig. 10.16).

The moment of inertia for an axis through the center of mass is $I_{cm} = \frac{1}{2}MR^2$. Only the tension force has a torque with respect to the axis through the center of mass, so the equation for rotational motion about this axis is

$$\sum \tau_z = TR = I_{cm}\alpha_z = \frac{1}{2}MR^2\alpha_z \qquad (10.18)$$

The string unwinds without slipping, so $v_{cm} = R\omega$ from Eq. (10.14); the derivative of this relation with respect to time is

$$a_{cm\text{-}y} = R\alpha_z \qquad (10.19)$$

We now use Eq. (10.19) to eliminate α_z from Eq. (10.18) and then solve Eqs. (10.17) and (10.18) simultaneously for T and $a_{cm\text{-}y}$. The results are amazingly simple:

$$a_{cm\text{-}y} = \frac{2}{3}g \qquad T = \frac{1}{3}Mg$$

Using the constant-acceleration formula $v_{cm\text{-}y}{}^2 = v_{cm\text{-}0y}{}^2 + 2a_{cm\text{-}y}h$, you can show that the speed of the yo-yo after it has fallen a distance h is $v_{cm} = \sqrt{\frac{4}{3}gh}$, just as we found in Example 10.6.

EVALUATE: From the standpoint of dynamics, the tension force is essential; it causes the yo-yo's acceleration to be less than g, and its torque is what causes the yo-yo to turn. Yet when we analyzed this situation using energy methods in Example 10.6, we didn't have to consider the tension force at all! Because no mechanical energy was lost or gained, from the energy standpoint the string is significant only as a way of helping to convert some of the gravitational potential energy into rotational kinetic energy.

Example 10.9 Acceleration of a rolling sphere

A solid bowling ball rolls without slipping down the return ramp at the side of the alley (Fig. 10.20a). The ramp is inclined at an angle β to the horizontal. What is the ball's acceleration? Treat the ball as a uniform solid sphere, ignoring the finger holes.

SOLUTION:

IDENTIFY: As in Example 10.8, we use Eq. (10.15) to describe the translational motion and Eq. (10.16) to describe the rotational motion. Our target variable is the acceleration of the ball's center of mass.

SET UP: Figure 10.20b shows a free-body diagram, with positive coordinate directions indicated. From Table 9.2 the moment of inertia of a solid sphere is $I_{cm} = \frac{2}{5}MR^2$. The equations of motion for translation and for rotation about the axis through the center of mass, respectively, are

$$\sum F_x = Mg\sin\beta + (-f) = Ma_{cm\text{-}x} \qquad (10.20)$$

$$\sum \tau_z = fR = I_{cm}\alpha_z = \left(\frac{2}{5}MR^2\right)\alpha_z \qquad (10.21)$$

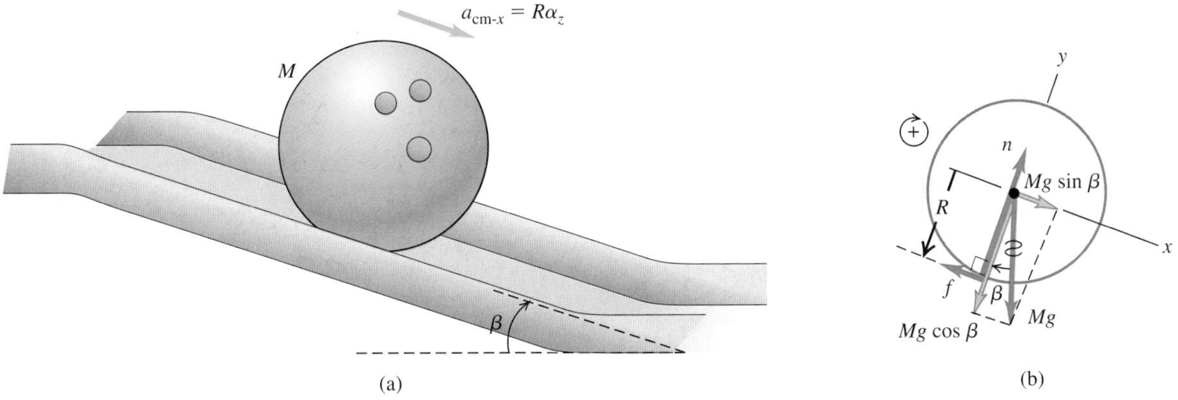

10.20 (a) A bowling ball rolling down a ramp. (b) Free-body diagram for the bowling ball.

Note that only the friction force f has a torque with respect to the axis through the center of mass. If the ball rolls without slipping, we have the same kinematic relation $a_{\text{cm-}x} = R\alpha_z$ as in Example 10.8. We use this to eliminate α_z from Eq. (10.21):

$$fR = \frac{2}{5}MRa_{\text{cm-}x}$$

This equation and Eq. (10.20) are two equations for two unknowns, $a_{\text{cm-}x}$ and f. We solve Eq. (10.20) for f, substitute the expression into the above equation to eliminate f, and then solve for $a_{\text{cm-}x}$ to obtain

$$a_{\text{cm-}x} = \frac{5}{7}g \sin\beta$$

The acceleration is just $\frac{5}{7}$ as large as it would be if the ball could *slide* without friction down the slope, like the toboggan in Example 5.9 (Section 5.2). Finally, we substitute this back into Eq. (10.20) and solve for f:

$$f = \frac{2}{7}Mg \sin\beta$$

EVALUATE: Because the ball does not slip at the instantaneous point of contact with the ramp, the friction force f is a *static* friction force; it prevents slipping and gives the ball its angular acceleration. We can derive an expression for the minimum coefficient of static friction μ_s needed to prevent slipping. The normal force is $n = Mg\cos\beta$. The maximum force of static friction equals $\mu_s n$, so the coefficient of friction must be at least as great as

$$\mu_s = \frac{f}{n} = \frac{\frac{2}{7}Mg\sin\beta}{Mg\cos\beta} = \frac{2}{7}\tan\beta$$

If the plane is tilted only a little, β is small, and only a small value of μ_s is needed to prevent slipping. But as the angle increases, the required value of μ_s increases, as we might expect intuitively. If the ball begins to slip, Eqs. (10.20) and (10.21) are both still valid, but it's no longer true that $v_{\text{cm-}x} = R\omega_z$ or $a_{\text{cm-}x} = R\alpha_z$; we have only two equations for three unknowns ($a_{\text{cm-}x}$, α_z, and f). To solve the problem of rolling *with* slipping requires taking *kinetic* friction into account (see Challenge Problem 10.101).

If the bowling ball descends a vertical distance h as it moves down the ramp, the displacement along the ramp is $h/\sin\beta$. You should be able to show that the speed of the ball at the bottom of the ramp would be $v_{\text{cm}} = \sqrt{\frac{10}{7}gh}$, which is just the result you found in Example 10.7 with $c = \frac{2}{5}$.

If the ball were rolling *uphill,* the force of friction would still be directed uphill as in Fig. 10.20b. Can you see why?

Rolling Friction

In Example 10.7 we said that we can ignore rolling friction if both the rolling body and the surface over which it rolls are perfectly rigid. In Fig. 10.21a a perfectly rigid sphere is rolling down a perfectly rigid incline. The line of action of the normal force passes through the center of the sphere, so its torque is zero; there is no sliding at the point of contact, so the friction force does no work. Fig. 10.21b shows a more realistic situation, in which the surface "piles up" in front of the sphere and the sphere rides in a shallow trench. Because of these deformations, the contact forces on the sphere no longer act along a single point, but over an area; the forces are concentrated on the front of the sphere as shown. As a

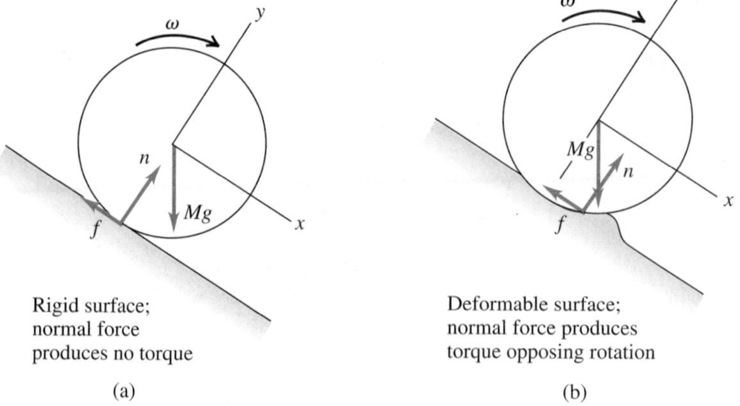

10.21 (a) Forces on a perfectly rigid sphere rolling down a perfectly rigid incline. (b) If the sphere or incline is deformable, the contact forces act at different positions. The normal force produces a counterclockwise torque that opposes the clockwise rotation. The deformation is shown greatly exaggerated.

result, the normal force now exerts a torque that opposes the rotation. In addition, there is some sliding of the sphere over the surface due to the deformation, causing mechanical energy to be lost. The combination of these two effects is the phenomenon of *rolling friction.* Rolling friction also occurs if the rolling body is deformable, such as an automobile tire. Often the rolling body and surface are rigid enough that rolling friction can be ignored, as we have assumed in all the examples in this section.

Test Your Understanding

In Example 10.9, what would be the acceleration and the static friction force if the ball were a hollow sphere?

10.4 | Work and Power in Rotational Motion

When you pedal a bicycle, you apply forces to a rotating body and do work on it. Similar things happen in many other real-life situations, such as a rotating motor shaft driving a power tool or a car engine propelling the vehicle. We can express this work in terms of torque and angular displacement.

Suppose a tangential force $\vec{F}_{tan}$ acts at the rim of a pivoted disk—for example, a child running while pushing on a playground merry-go-round (Fig. 10.22a). The disk rotates through an infinitesimal angle $d\theta$ about a fixed axis during an infinitesimal time interval dt (Fig. 10.22b). The work dW done by the force $\vec{F}_{tan}$ while a point on the rim moves a distance ds is $dW = F_{tan}\,ds$. If $d\theta$ is measured in radians, then $ds = R\,d\theta$ and

$$dW = F_{tan} R\,d\theta$$

Now $F_{tan}R$ is the *torque* τ_z due to the force $\vec{F}_{tan}$, so

$$dW = \tau_z\,d\theta \qquad (10.22)$$

The total work W done by the torque during an angular displacement from θ_1 to θ_2 is

$$W = \int_{\theta_1}^{\theta_2} \tau_z\,d\theta \qquad \text{(work done by a torque)} \qquad (10.23)$$

If the torque is *constant* while the angle changes by a finite amount $\Delta\theta = \theta_2 - \theta_1$, then

$$W = \tau_z(\theta_2 - \theta_1) = \tau_z\Delta\theta \qquad \text{(work done by a constant torque)} \qquad (10.24)$$

The work done by a *constant* torque is the product of torque and the angular displacement. If torque is expressed in newton-meters $(\text{N}\cdot\text{m})$ and angular displacement in radians, the work is in joules. Equation (10.24) is the rotational analog of Eq. (6.1), $W = Fs$, and Eq. (10.23) is the analog of Eq. (6.7), $W = \int F_x\,dx$, for the work done by a force in a straight-line displacement.

If the force in Fig. 10.22 had an axial or radial component, that component would do no work because the displacement of the point of application has only a tangential component. An axial or radial component of force would also make no contribution to the torque about the axis of rotation, so Eqs. (10.23) and (10.24) are correct for any force, no matter what its components.

(a)

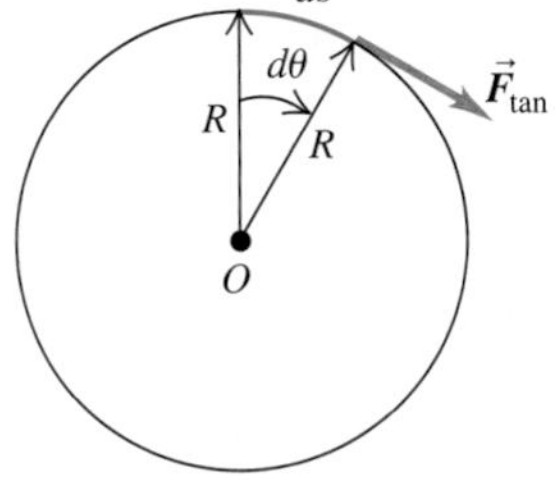

(b)

10.22 A tangential force applied to a rotating body does work.

When a torque does work on a rotating rigid body, the kinetic energy changes by an amount equal to the work done. We can prove this by using exactly the same procedure that we used in Eqs. (6.11) through (6.13) for a particle. We first let τ_z represent the *net* torque on the body so that from Eq. (10.6), $\tau_z = I\alpha_z$. By using this equation, we are assuming that the body is rigid so that the moment of inertia I is constant. We then transform the integrand in Eq. (10.23) into an integral on ω_z as follows:

$$\tau_z d\theta = (I\alpha_z)\,d\theta = I\frac{d\omega_z}{dt}\,d\theta = I\frac{d\theta}{dt}\,d\omega_z = I\omega_z d\omega_z$$

Since τ_z is the net torque, the integral in Eq. (10.23) is the *total* work done on the rotating rigid body. This equation then becomes

$$W_{\text{tot}} = \int_{\omega_1}^{\omega_2} I\omega_z\,d\omega_z = \frac{1}{2}I\omega_2^2 - \frac{1}{2}I\omega_1^2 \qquad (10.25)$$

The change in the rotational kinetic energy of a *rigid* body equals the work done by forces exerted from outside the body. This equation is analogous to Eq. (6.13), the work-energy theorem for a particle.

What about the *power* associated with work done by a torque acting on a rotating body? When we divide both sides of Eq. (10.22) by the time interval dt during which the angular displacement occurs, we find

$$\frac{dW}{dt} = \tau_z\frac{d\theta}{dt}$$

But dW/dt is the rate of doing work, or *power P*, and $d\theta/dt$ is angular velocity ω_z, so

$$P = \tau_z\omega_z \qquad (10.26)$$

When a torque τ_z (with respect to the axis of rotation) acts on a body that rotates with angular velocity ω_z, its power (rate of doing work) is the product of τ_z and ω_z. This is the analog of the relation $P = \vec{F}\cdot\vec{v}$ that we developed in Section 6.4 for particle motion.

Example 10.10

Engine power and torque

The power output of an automobile engine is advertised to be 200 hp at 6000 rpm. What is the corresponding torque?

SOLUTION

IDENTIFY and SET UP: We are given the power output P and the angular velocity ω_z, so we can find the torque using Eq. (10.26).

EXECUTE: First we have to convert the power to watts and the angular velocity to rad/s:

$$P = 200\text{ hp} = 200\text{ hp}\left(\frac{746\text{ W}}{1\text{ hp}}\right) = 1.49 \times 10^5\text{ W}$$

$$\omega_z = 6000\text{ rev/min} = \left(\frac{6000\text{ rev}}{1\text{ min}}\right)\left(\frac{2\pi\text{ rad}}{1\text{ rev}}\right)\left(\frac{1\text{ min}}{60\text{ s}}\right)$$
$$= 628\text{ rad/s}$$

From Eq. (10.26),

$$\tau_z = \frac{P}{\omega_z} = \frac{1.49 \times 10^5\text{ N}\cdot\text{m/s}}{628\text{ rad/s}} = 237\text{ N}\cdot\text{m}$$

EVALUATE: You could apply this much torque by using a wrench 0.25 m long and applying a force of 948 N (213 lb) to the end of its handle. Could you do it?

Example 10.11 Calculating power from torque

An electric motor exerts a constant torque of 10 N·m on a grindstone mounted on its shaft. The moment of inertia of the grindstone about the shaft is 2.0 kg·m^2. If the system starts from rest, find the work done by the motor in 8.0 seconds and the kinetic energy at the end of this time. What was the average power delivered by the motor?

SOLUTION

IDENTIFY and SET UP: We use the rotational version of Newton's second law, $\sum\tau_z = I\alpha_z$, to find the angular acceleration of the grindstone. Given this we use the kinematics equations from Section 9.2 to find the angle through which the grindstone turns in 8.0 s (which, through Eq. (10.24), tells us the work done) and the angular velocity at that time (which tells us the kinetic energy). We find the average power by dividing the work done by the time interval.

EXECUTE: We have $\sum\tau_z = 10\ \text{N}\cdot\text{m}$ (the only torque acting is that due to the motor) and $I = 2.0\ \text{kg}\cdot\text{m}^2$, so from $\sum\tau_z = I\alpha_z$ the angular acceleration is 5.0 rad/s^2. From Eq. (9.11) the total angle through which the system turns in 8.0 s is

$$\Delta\theta = \frac{1}{2}\alpha_z t^2 = \frac{1}{2}(5.0\ \text{rad/s}^2)(8.0\ \text{s})^2 = 160\ \text{rad}$$

and the total work done by the torque is

$$W = \tau_z\Delta\theta = (10\ \text{N}\cdot\text{m})(160\ \text{rad}) = 1600\ \text{J}$$

From Eqs. (9.7) and (9.17), the angular velocity and kinetic energy at $t = 8.0$ s are

$$\omega_z = \alpha_z t = (5.0\ \text{rad/s}^2)(8.0\ \text{s}) = 40\ \text{rad/s}$$

$$K = \frac{1}{2}I\omega_z^2 = \frac{1}{2}(2.0\ \text{kg}\cdot\text{m}^2)(40\ \text{rad/s})^2 = 1600\ \text{J}$$

The initial kinetic energy was zero, so the work done equals the increase in kinetic energy [see Eq. (10.25)].

The average power is

$$P_{\text{av}} = \frac{1600\ \text{J}}{8.0\ \text{s}} = 200\ \text{J/s} = 200\ \text{W}$$

EVALUATE: We can check our answer for *average* power by considering the *instantaneous* power $P = \tau_z\omega_z$. Note that because ω_z increases continuously, P increases continuously as well; its value is zero at $t = 0$ and increases to $(10\ \text{N}\cdot\text{m})(40\ \text{rad/s}) = 400$ W at $t = 8.0$ s. The angular velocity and the power increase uniformly with time, so the *average* power is just half this maximum value, or 200 W.

Test Your Understanding

You apply equal torques to two different cylinders, one of which has a moment of inertia twice as large as the other cylinder. Each cylinder is initially at rest. After one complete rotation, which cylinder has the greater kinetic energy?

10.5 | Angular Momentum

Every rotational quantity that we have encountered in Chapters 9 and 10 is the analog of some quantity in the translational motion of a particle. The analog of *momentum* of a particle is **angular momentum,** a vector quantity denoted as $\vec{L}$. Its relation to momentum $\vec{p}$ (which we will sometimes call *linear momentum* for clarity) is exactly the same as the relation of torque to force, $\vec{\tau} = \vec{r} \times \vec{F}$. For a particle with constant mass m, velocity $\vec{v}$, momentum $\vec{p} = m\vec{v}$, and position vector $\vec{r}$ relative to the origin O of an inertial frame, we define angular momentum $\vec{L}$ as

$$\vec{L} = \vec{r} \times \vec{p} = \vec{r} \times m\vec{v} \quad \text{(angular momentum of a particle)} \quad (10.27)$$

The value of $\vec{L}$ depends on the choice of origin O, since it involves the position vector of the particle relative to the origin. The units of angular momentum are kg·m^2/s.

In Fig. 10.23 a particle moves in the xy-plane; its position vector $\vec{r}$ and momentum $\vec{p} = m\vec{v}$ are shown. The angular momentum vector $\vec{L}$ is perpendicular to the

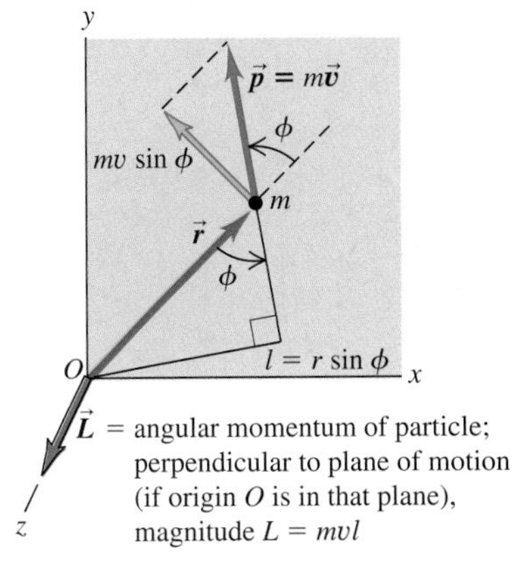

10.23 Calculating the angular momentum $\vec{L} = \vec{r} \times m\vec{v} = \vec{r} \times \vec{p}$ of a particle with mass m moving in the xy-plane.

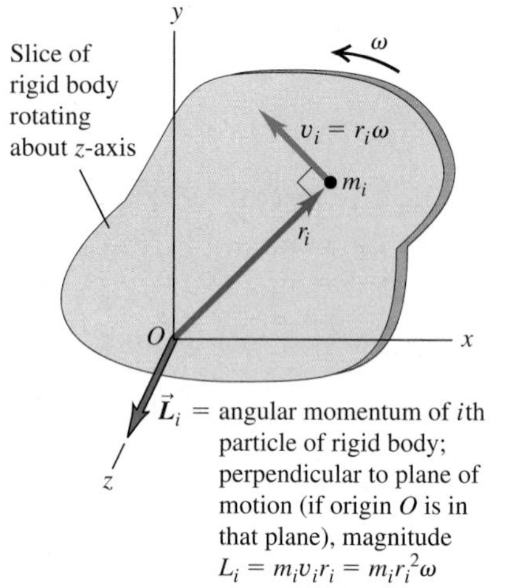

10.24 Calculating the angular momentum of a particle of mass m_i in a rotating rigid body. (Compare Fig. 10.23.) Every particle in the body moves in a circle about the rotation axis with the same angular speed ω.

xy-plane. The right-hand rule for vector products shows that its direction is along the $+z$-axis, and its magnitude is

$$L = mvr \sin\phi = mvl \tag{10.28}$$

where l is the perpendicular distance from the line of $\vec{v}$ to O. This distance plays the role of "lever arm" for the momentum vector.

When a net force $\vec{F}$ acts on a particle, its velocity and momentum change, so its angular momentum may also change. We can show that the *rate of change* of angular momentum is equal to the torque of the net force. We take the time derivative of Eq. (10.27), using the rule for the derivative of a product:

$$\frac{d\vec{L}}{dt} = \left(\frac{d\vec{r}}{dt} \times m\vec{v}\right) + \left(\vec{r} \times m\frac{d\vec{v}}{dt}\right) = (\vec{v} \times m\vec{v}) + (\vec{r} \times m\vec{a})$$

The first term is zero because it contains the vector product of the vector $\vec{v} = d\vec{r}/dt$ with itself. In the second term we replace $m\vec{a}$ with the net force $\vec{F}$, obtaining

$$\frac{d\vec{L}}{dt} = \vec{r} \times \vec{F} = \vec{\tau} \quad \text{(for a particle acted on by net force } \vec{F}\text{)} \tag{10.29}$$

The rate of change of angular momentum of a particle equals the torque of the net force acting on it. Compare this result to Eq. (8.3), which states that the rate of change $d\vec{p}/dt$ of the *linear* momentum of a particle equals the net force that acts on it.

We can use Eq. (10.28) to find the total angular momentum of a rigid body rotating about the z-axis with angular speed ω. First consider a thin slice of the body lying in the xy-plane (Fig. 10.24). Each particle in the slice moves in a circle centered at the origin, and at each instant its velocity $\vec{v}_i$ is perpendicular to its position vector $\vec{r}_i$, as shown. Hence in Eq. (10.28), $\phi = 90°$ for every particle. A particle with mass m_i at a distance r_i from O has a speed v_i equal to $r_i\omega$. From Eq. (10.28) the magnitude L_i of its angular momentum is

$$L_i = m_i(r_i\omega)r_i = m_i r_i^2\omega \tag{10.30}$$

The direction of each particle's angular momentum, as given by the right-hand rule for the vector product, is along the $+z$-axis.

The *total* angular momentum of the slice of the body lying in the xy-plane is the sum $\sum L_i$ of the angular momenta L_i of the particles. Summing Eq. (10.30), we have

$$L = \sum L_i = \left(\sum m_i r_i^2\right)\omega = I\omega$$

where I is the moment of inertia of the slice about the z-axis.

We can do this same calculation for the other slices of the body, all parallel to the xy-plane. For points that do not lie in the xy-plane, a complication arises because the $\vec{r}$ vectors have components in the z-direction as well as the x- and y-directions; this gives the angular momentum of each particle a component perpendicular to the z-axis. But *if the z-axis is an axis of symmetry,* the perpendicular components for particles on opposite sides of this axis add up to zero (Fig. 10.25). So when a body rotates about an axis of symmetry, its angular momentum vector $\vec{L}$ lies along the symmetry axis, and its magnitude is $L = I\omega$.

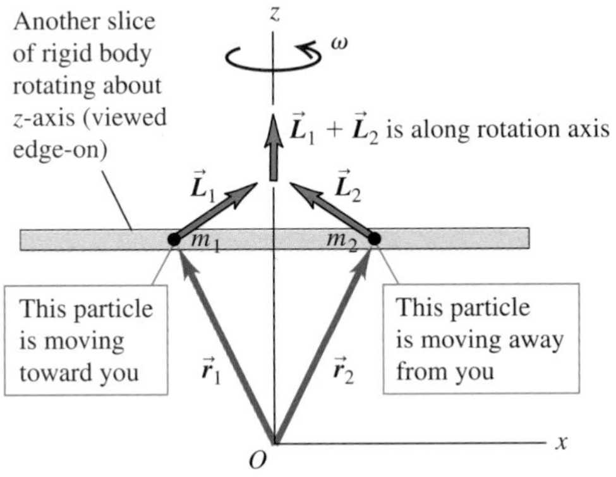

10.25 Catching the angular momentum of two particles in a rotating rigid body. The two particles have the same mass and are located symmetrically on either side of the rotation axis. While the angular momentum vectors $\vec{L}_1$ and $\vec{L}_2$ of the two particles do not lie along the rotation axis, their vector sum $\vec{L}_1 + \vec{L}_2$ does.

The angular velocity vector $\vec{\omega}$ also lies along the rotation axis, as we discussed at the end of Section 9.1. Hence for a rigid body rotating around an axis of symmetry, $\vec{L}$ and $\vec{\omega}$ are in the same direction (Fig. 10.26). So we have the *vector* relationship

$$\vec{L} = I\vec{\omega} \quad \text{(for a rigid body rotating around a symmetry axis)} \quad (10.31)$$

From Eq. (10.29) the rate of change of angular momentum of a particle equals the torque of the net force acting on the particle. For any system of particles (including both rigid and nonrigid bodies), the rate of change of the *total* angular momentum equals the sum of the torques of all forces acting on all the particles. The torques of the *internal* forces add to zero if these forces act along the line from one particle to another, as in Fig. 10.8, and so the sum of torques includes only the torques of the *external* forces. (A similar cancellation occurred in our discussion of center-of-mass motion in Section 8.5.) If the total angular momentum of the system of particles is $\vec{L}$ and the sum of external torques is $\sum\vec{\tau}$, then

$$\sum\vec{\tau} = \frac{d\vec{L}}{dt} \quad \text{(for any system of particles)} \quad (10.32)$$

Finally, if the system of particles is a rigid body rotating about a symmetry axis (the z-axis), then $L_z = I\omega_z$ and I is constant. If this axis has a fixed direction in space, then the vectors $\vec{L}$ and $\vec{\omega}$ change only in magnitude, not in direction. In that case, $dL_z/dt = I\,d\omega_z/dt = I\alpha_z$, or

$$\sum\tau_z = I\alpha_z$$

which is again our basic relation for the dynamics of rigid-body rotation. If the body is *not* rigid, I may change, and in that case, L changes even when ω is constant. For a non-rigid body, Eq. (10.32) is still valid, even though Eq. (10.6) is not.

When the axis of rotation is *not* a symmetry axis, the angular momentum is in general *not* parallel to the axis (Fig. 10.27). As the body turns, the angular momentum vector $\vec{L}$ traces out a cone around the rotation axis. Because $\vec{L}$ changes, there must be a net external torque acting on the body even though the angular velocity magnitude ω may be constant. If the body is an unbalanced wheel on a car, this torque is provided by friction in the bearings, which causes the bearings to wear out. "Balancing" a wheel means distributing the mass so that the rotation axis is an axis of symmetry; then $\vec{L}$ points along the rotation axis, and no net torque is required to keep the wheel turning.

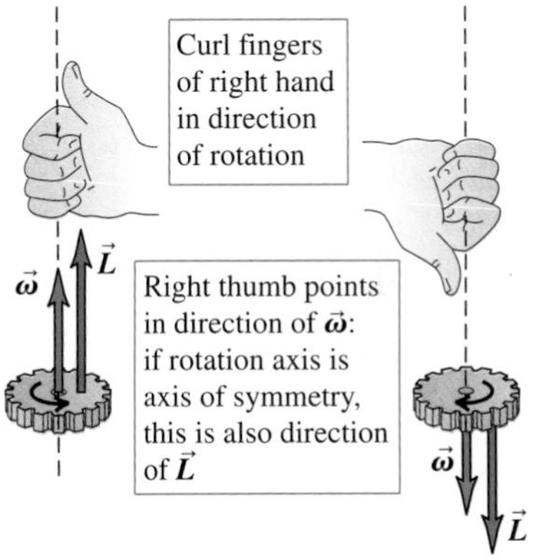

10.26 For rotation about an axis of symmetry, $\vec{\omega}$ and $\vec{L}$ are parallel and along the axis. The directions of both vectors are given by the right-hand rule (compare Fig. 9.5).

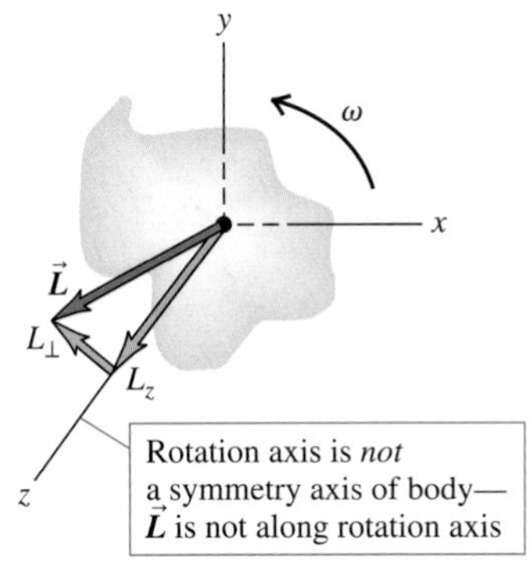

10.27 If the rotation axis of a rigid body is not a symmetry axis, $\vec{L}$ does not in general lie along the rotation axis. Even if $\vec{\omega}$ is constant, the direction of $\vec{L}$ changes and a net torque is required to maintain rotation.

In fixed-axis rotation we often use the term "angular momentum of the body" to refer just to the *component* of $\vec{L}$ along the rotation axis of the body (the z-axis in Fig. 10.27), with a sign to indicate the sense of rotation just as with angular velocity.

Example 10.12 Angular momentum and torque

A turbine fan in a jet engine (Fig. 10.28) has a moment of inertia of $2.5\ \text{kg}\cdot\text{m}^2$ about its axis of rotation. As the turbine is starting up, its angular velocity as a function of time is

$$\omega_z = (40\ \text{rad/s}^3)t^2$$

a) Find the fan's angular momentum as a function of time, and find its value at time $t = 3.0$ s. b) Find the net torque acting on the fan as a function of time, and find the torque at time $t = 3.0$ s.

SOLUTION

IDENTIFY and SET UP: Just like an electric fan, the turbine fan rotates about an axis of symmetry. Hence we can use Eq. (10.31) to find L_z from ω_z and use Eq. (10.32) to relate the net torque to the time derivative of L_z.

EXECUTE:
a) The only component of angular momentum is along the rotation (z) axis:

$$L_z = I\omega_z = (2.5\ \text{kg}\cdot\text{m}^2)(40\ \text{rad/s}^3)t^2 = (100\ \text{kg}\cdot\text{m}^2/\text{s}^3)t^2$$

(We dropped "rad" from the answer because a radian is a dimensionless quantity.) At time $t = 3.0$ s, $L_z = 900\ \text{kg}\cdot\text{m}^2/\text{s}$.
b) The direction of the angular momentum does not change, so the torque is also along the rotation axis. From Eq. (10.32), its component along that axis is

$$\tau_z = \frac{dL_z}{dt} = (100\ \text{kg}\cdot\text{m}^2/\text{s}^3)(2t) = (200\ \text{kg}\cdot\text{m}^2/\text{s}^3)t$$

At time $t = 3.0$ s,

$$\tau_z = (200\ \text{kg}\cdot\text{m}^2/\text{s}^3)(3.0\ \text{s}) = 600\ \text{kg}\cdot\text{m}^2/\text{s}^2 = 600\ \text{N}\cdot\text{m}$$

EVALUATE: As a check on our result, note that the angular acceleration of the turbine fan is $\alpha_z = d\omega_z/dt = (40\ \text{rad/s}^2)(2t) = (80\ \text{rad/s}^2)t$. From the rotational equivalent of Newton's second law, the torque on the fan is $\tau_z = I\alpha_z = (2.5\ \text{kg}\cdot\text{m}^2)(80\ \text{rad/s}^2)t = (200\ \text{kg}\cdot\text{m}^2/\text{s}^3)t$, just as we calculated above.

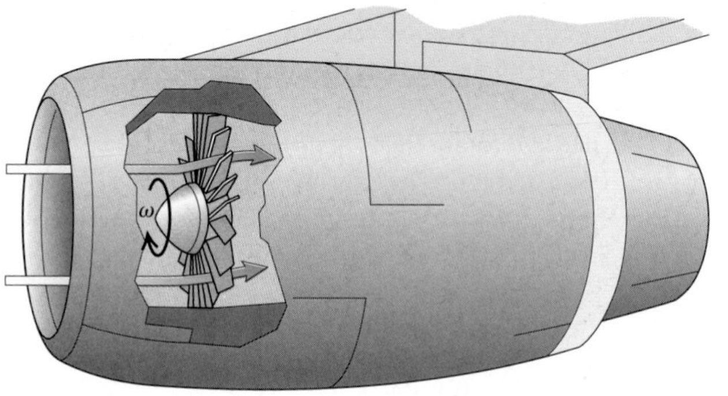

10.28 A turbine fan is used to force air into a turbojet engine.

Test Your Understanding

A ball is attached to one end of a piece of string. You hold the other end of the string and whirl the ball in a circle around your hand. If the ball moves at a constant speed, is its linear momentum $\vec{p}$ constant? Is its angular momentum $\vec{L}$ constant? Why is there a difference?

10.6 | Conservation of Angular Momentum

We have just seen that angular momentum can be used for an alternative statement of the basic dynamic principle for rotational motion. It also forms the basis for the **principle of conservation of angular momentum.** Like conservation of energy and of linear momentum, this principle is a universal conservation law, valid at all scales from atomic and nuclear systems to the motions of galaxies. This principle follows directly from Eq. (10.32): $\sum \vec{\tau} = d\vec{L}/dt$. If $\sum \vec{\tau} = \mathbf{0}$, then $d\vec{L}/dt = \mathbf{0}$, and $\vec{L}$ is constant. **When the net external torque acting on a system is zero, the total angular momentum of the system is constant (conserved).**

ActivPhysics ONLINE

7.14 Ball Hits Bat

A circus acrobat, a diver, and an ice skater pirouetting on the toe of one skate all take advantage of this principle. Suppose an acrobat has just left a swing with arms and legs extended and rotating counterclockwise about her center of mass. When she pulls her arms and legs in, her moment of inertia I_{cm} with respect to her center of mass changes from a large value I_1 to a much smaller value I_2. The only external force acting on her is her weight, which has no torque with respect to an axis through her center of mass. So her angular momentum $L_z = I_{cm}\omega_z$ remains constant, and her angular velocity ω_z increases as I_{cm} decreases. That is,

$$I_1\omega_{1z} = I_2\omega_{2z} \quad \text{(zero net external torque)} \tag{10.33}$$

When a skater or ballerina spins with arms outstretched and then pulls her arms in, her angular velocity increases as her moment of inertia decreases. In each case there is conservation of angular momentum in a system in which the net external torque is zero.

When a system has several parts, the internal forces that the parts exert on each other cause changes in the angular momenta of the parts, but the *total* angular momentum doesn't change. Here's an example. Consider two bodies A and B that interact with each other but not with anything else, such as the astronauts we discussed in Section 8.2 (Fig. 8.7). Suppose body A exerts a force $\vec{F}_{A \text{ on } B}$ on body B; the corresponding torque (with respect to whatever point we choose) is $\vec{\tau}_{A \text{ on } B}$. According to Eq. (10.32), this torque is equal to the rate of change of angular momentum of B:

$$\vec{\tau}_{A \text{ on } B} = \frac{d\vec{L}_B}{dt}$$

At the same time, body B exerts a force $\vec{F}_{B \text{ on } A}$ on body A, with a corresponding torque $\vec{\tau}_{B \text{ on } A}$, and

$$\vec{\tau}_{B \text{ on } A} = \frac{d\vec{L}_A}{dt}$$

From Newton's third law, $\vec{F}_{B \text{ on } A} = -\vec{F}_{A \text{ on } B}$. Furthermore, if the forces act along the same line, as in Fig. 10.8, their lever arms with respect to the chosen axis are equal. Thus the *torques* of these two forces are equal and opposite, and $\vec{\tau}_{B \text{ on } A} = -\vec{\tau}_{A \text{ on } B}$. So if we add the two previous equations, we find

$$\frac{d\vec{L}_A}{dt} + \frac{d\vec{L}_B}{dt} = \mathbf{0}$$

or, because $\vec{L}_A + \vec{L}_B$ is the *total* angular momentum $\vec{L}$ of the system,

$$\frac{d\vec{L}}{dt} = \mathbf{0} \quad \text{(zero net external torque)} \tag{10.34}$$

That is, the total angular momentum of the system is constant. The torques of the internal forces can transfer angular momentum from one body to the other, but they can't change the *total* angular momentum of the system (Fig. 10.29).

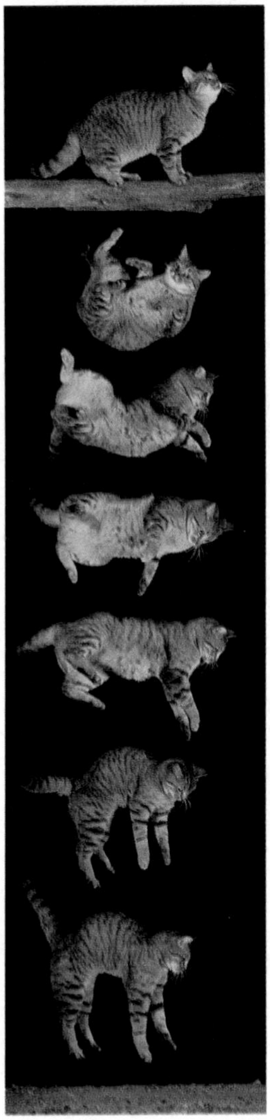

10.29 A falling cat twists different parts of its body in different directions so that it lands feet first. At all times during this process the angular momentum of the cat as a whole remains zero.

Example 10.13 Anyone can be a ballerina

An acrobatic physics professor stands at the center of a turntable, holding his arms extended horizontally with a 5.0-kg dumbbell in each hand (Fig. 10.30). He is set rotating about a vertical axis, making one revolution in 2.0 s. Find the prof's new angular velocity if he pulls the dumbbells in to his stomach, and discuss how this affects the kinetic energy. His moment of inertia (without the dumbbells) is $3.0\ \text{kg}\cdot\text{m}^2$ when his arms are outstretched, dropping to $2.2\ \text{kg}\cdot\text{m}^2$ when his hands are at his stomach. The dumbbells are 1.0 m from the axis initially and 0.20 m from it at the end. Treat the dumbbells as particles.

SOLUTION

IDENTIFY and SET UP: If we neglect friction in the turntable, no external torques act about the vertical (z) axis. Hence the angular momentum about this axis is constant and we can use Eq. (10.33) to find our target variable, the final angular velocity ω_{2z}.

10.30 Fun with conservation of angular momentum—if you don't get dizzy.

EXECUTE: The moment of inertia of the system is $I = I_{\text{prof}} + I_{\text{dumbbells}}$. Each dumbbell of mass m contributes mr^2 to $I_{\text{dumbbells}}$, where r is the perpendicular distance from the rotation axis to the dumbbell. Initially we have

$$I_1 = 3.0\ \text{kg}\cdot\text{m}^2 + 2(5.0\ \text{kg})(1.0\ \text{m})^2 = 13\ \text{kg}\cdot\text{m}^2$$

$$\omega_{1z} = \frac{1\ \text{rev}}{2.0\ \text{s}} = 0.50\ \text{rev/s}$$

The final moment of inertia is

$$I_2 = 2.2\ \text{kg}\cdot\text{m}^2 + 2(5.0\ \text{kg})(0.20\ \text{m})^2 = 2.6\ \text{kg}\cdot\text{m}^2$$

From Eq. (10.33), the final angular velocity is

$$\omega_{2z} = \frac{I_1}{I_2}\omega_{1z} = \frac{13\ \text{kg}\cdot\text{m}^2}{2.6\ \text{kg}\cdot\text{m}^2}(0.50\ \text{rev/s}) = 2.5\ \text{rev/s}$$

That is, the angular velocity increases by a factor of 5 while the angular momentum remains constant. Note that we didn't have to change "revolutions" to "radians" in this calculation. Why not?

EVALUATE: To calculate the kinetic energy, we must express ω_1 and ω_2 in rad/s. (Why?) We have $\omega_{1z} = (0.50\ \text{rev/s})(2\pi\ \text{rad/rev}) = 3.14\ \text{rad/s}$ and $\omega_{2z} = (2.5\ \text{rev/s})(2\pi\ \text{rad/rev}) = 15.7\ \text{rad/s}$. The initial kinetic energy is

$$K_1 = \frac{1}{2}I_1\omega_{1z}^2 = \frac{1}{2}(13\ \text{kg}\cdot\text{m}^2)(3.14\ \text{rad/s})^2 = 64\ \text{J}$$

and the final kinetic energy is

$$K_2 = \frac{1}{2}I_2\omega_{2z}^2 = \frac{1}{2}(2.6\ \text{kg}\cdot\text{m}^2)(15.7\ \text{rad/s})^2 = 320\ \text{J}$$

Where did the extra energy come from?

Example 10.14 A rotational "collision" I

Figure 10.31 shows two disks: one an engine flywheel, and the other a clutch plate attached to a transmission shaft. Their moments of inertia are I_A and I_B; initially, they are rotating with constant angular speeds ω_A and ω_B, respectively. We then push the disks together with forces acting along the axis, so as not to apply any torque on either disk. The disks rub against each other and eventually reach a common final angular speed ω. Derive an expression for ω.

SOLUTION

IDENTIFY: The only torque acting on either disk is the torque applied by the other disk; there are no external torques. Thus the total angular momentum of the system of two disks is the same before and after they are pushed together. At the end they rotate together as one body with total moment of inertia $I = I_A + I_B$ and angular speed ω, which is our target variable.

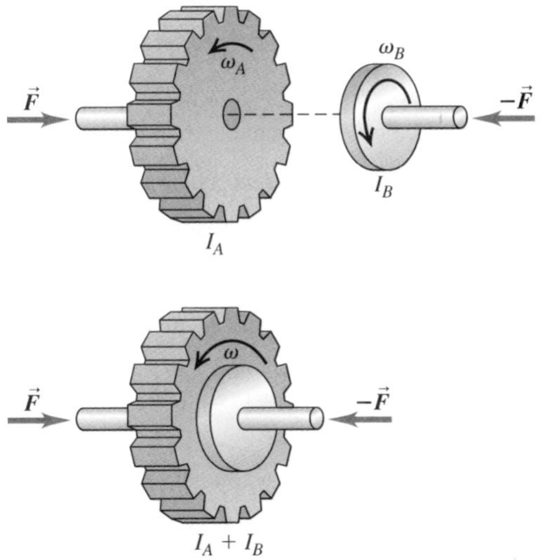

10.31 When the net external torque is zero, angular momentum is conserved. The forces shown are along the axis of rotation and thus exert no torque about any point on it.

EXECUTE: Figure 10.31 shows that all of the angular velocities are in the same direction, so we can regard ω_A, ω_B, and ω as the components of angular velocity along the rotation axis. Conservation of angular momentum then gives

$$I_A\omega_A + I_B\omega_B = (I_A + I_B)\omega$$

$$\omega = \frac{I_A\omega_A + I_B\omega_B}{I_A + I_B}$$

EVALUATE: This "collision" between two disks is analogous to a completely inelastic collision (Section 8.3). When two objects in translational motion along the same axis come together and stick, the linear momentum of the system is conserved. In the situation shown in Fig. 10.31, two objects in *rotational* motion along the same axis come together and stick, and the *angular* momentum is conserved. The kinetic energy of the system decreases in a completely inelastic collision; in Example 10.15 we'll see what becomes of the kinetic energy in the "collision" of two rotating disks.

Example 10.15 A rotational "collision" II

In Example 10.14, suppose flywheel A has a mass of 2.0 kg, a radius of 0.20 m, and an initial angular speed of 50 rad/s (about 500 rpm) and that clutch plate B has a mass of 4.0 kg, a radius of 0.10 m, and an initial angular speed of 200 rad/s. Find the common final angular speed ω after the disks are pushed into contact. What happens to the kinetic energy during this process?

SOLUTION

IDENTIFY and SET UP: We'll use the result of Example 10.14 and the expression $K = \frac{1}{2}I\omega^2$ for rotational kinetic energy.

EXECUTE: The moments of inertia of the two disks are

$$I_A = \frac{1}{2}m_A r_A^2 = \frac{1}{2}(2.0\text{ kg})(0.20\text{ m})^2 = 0.040\text{ kg}\cdot\text{m}^2$$

$$I_B = \frac{1}{2}m_B r_B^2 = \frac{1}{2}(4.0\text{ kg})(0.10\text{ m})^2 = 0.020\text{ kg}\cdot\text{m}^2$$

From Example 10.14 we have

$$\begin{aligned}\omega &= \frac{I_A\omega_A + I_B\omega_B}{I_A + I_B}\\ &= \frac{(0.040\text{ kg}\cdot\text{m}^2)(50\text{ rad/s}) + (0.020\text{ kg}\cdot\text{m}^2)(200\text{ rad/s})}{0.040\text{ kg}\cdot\text{m}^2 + 0.020\text{ kg}\cdot\text{m}^2}\\ &= 100\text{ rad/s}\end{aligned}$$

The initial kinetic energy is

$$\begin{aligned}K_1 &= \frac{1}{2}I_A\omega_A^2 + \frac{1}{2}I_B\omega_B^2\\ &= \frac{1}{2}(0.040\text{ kg}\cdot\text{m}^2)(50\text{ rad/s})^2\\ &\quad + \frac{1}{2}(0.020\text{ kg}\cdot\text{m}^2)(200\text{ rad/s})^2\\ &= 450\text{ J}\end{aligned}$$

The final kinetic energy is

$$\begin{aligned}K_2 &= \frac{1}{2}(I_A + I_B)\omega^2\\ &= \frac{1}{2}(0.040\text{ kg}\cdot\text{m}^2 + 0.020\text{ kg}\cdot\text{m}^2)(100\text{ rad/s})^2 = 300\text{ J}\end{aligned}$$

EVALUATE: One-third of the initial kinetic energy was lost during this "angular collision," the rotational analog of a completely inelastic collision. We shouldn't expect kinetic energy to be conserved, even though the resultant external force and torque are zero, because nonconservative (frictional) internal forces act while the two disks rub together and gradually approach a common angular velocity.

Example 10.16 Angular momentum in a crime bust

A door 1.00 m wide, of mass 15 kg, is hinged at one side so that it can rotate without friction about a vertical axis. It is unlatched. A police officer fires a bullet with a mass of 10 g and a speed of 400 m/s into the exact center of the door, in a direction perpendicular to the plane of the door (Fig. 10.32). Find the angular speed of the door just after the bullet embeds itself in the door. Is kinetic energy conserved?

SOLUTION

IDENTIFY: We consider the door and bullet together as a system. There is no external torque about the axis defined by the hinges, so angular momentum about this axis is conserved.

SET UP: The initial angular momentum is wholly in the bullet and is given by Eq. (10.28). The final angular momentum is that of a rigid body composed of the door and embedded bullet. We'll set these two equal to each other and solve for the angular speed ω of the door and bullet just after the collision.

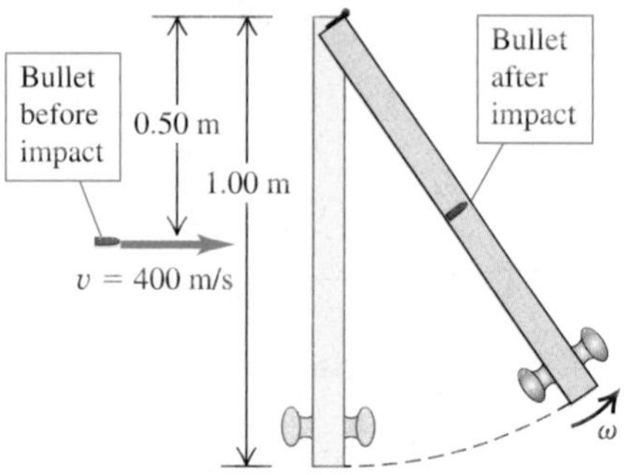

10.32 Shooting open a door (top view). The bullet embeds itself in the center of the door.

EXECUTE: The initial angular momentum of the bullet is:

$$L = mvl = (0.010\ \text{kg})(400\ \text{m/s})(0.50\ \text{m}) = 2.0\ \text{kg}\cdot\text{m}^2/\text{s}$$

The final angular momentum is $I\omega$, where $I = I_{\text{door}} + I_{\text{bullet}}$. From Table 9.2, for a door of width d,

$$I_{\text{door}} = \frac{Md^2}{3} = \frac{(15\ \text{kg})(1.0\ \text{m})^2}{3} = 5.0\ \text{kg}\cdot\text{m}^2$$

The moment of inertia of the bullet (with respect to the axis along the hinges) is

$$I_{\text{bullet}} = ml^2 = (0.010\ \text{kg})(0.50\ \text{m})^2 = 0.0025\ \text{kg}\cdot\text{m}^2$$

Conservation of angular momentum requires that $mvl = I\omega$, or

$$\omega = \frac{mvl}{I} = \frac{2.0\ \text{kg}\cdot\text{m}^2/\text{s}}{5.0\ \text{kg}\cdot\text{m}^2 + 0.0025\ \text{kg}\cdot\text{m}^2} = 0.40\ \text{rad/s}$$

The collision of bullet and door is inelastic because nonconservative forces act during the impact. Thus we do not expect kinetic energy to be conserved. To check, we calculate initial and final kinetic energies:

$$K_1 = \frac{1}{2}mv^2 = \frac{1}{2}(0.010\ \text{kg})(400\ \text{m/s})^2 = 800\ \text{J}$$

$$K_2 = \frac{1}{2}I\omega^2 = \frac{1}{2}(5.0025\ \text{kg}\cdot\text{m}^2)(0.40\ \text{rad/s})^2$$

$$= 0.40\ \text{J}$$

The final kinetic energy is only 1/2000 of the initial value!

EVALUATE: The final angular speed of the door is quite slow: at 0.40 rad/s, the door takes 3.9 s to swing through 90° ($\pi/2$ radians). Can you see that the speed would double if the bullet were shot into the edge of the door near the doorknob?

Test Your Understanding

If the polar ice caps were to completely melt due to global warming, the melted ice would redistribute itself over the earth. Use angular momentum ideas to explain how this change would affect the length of the day (the time needed for the earth to rotate once on its axis). Assume that the sun, moon, and planets exert negligibly small torques on the earth.

10.7 | Gyroscopes and Precession

In all of the situations we've looked at so far in this chapter, the axis of rotation either has stayed fixed or has moved and kept the same direction (such as rolling without slipping). But a variety of new physical phenomena, some quite unexpected, can occur when the axis of rotation can change direction. For example, consider a toy gyroscope that's supported at one end (Fig. 10.33). If we hold it with the flywheel axis horizontal and let go, the free end of the axis simply drops owing to gravity—*if* the flywheel isn't spinning. But if the flywheel *is* spinning, what happens is quite different. One possible motion is a steady circular motion of the axis in a horizontal plane, combined with the spin motion of the flywheel about the axis. This surprising, non-intuitive motion of the axis is called **precession.** Precession is found in nature, as well as in rotating machines such as gyroscopes. As you read these words, the earth itself is precessing; its spin axis (through the north and south poles) slowly changes direction, going through a complete cycle of precession every 26,000 years.

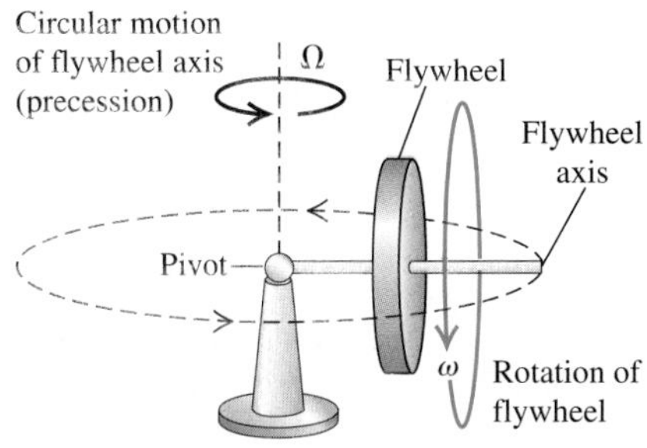

10.33 A gyroscope supported at one end. Because the flywheel is spinning at angular speed ω, the flywheel and axis do not drop but instead move in a horizontal circular motion called precession. The angular speed of precession is Ω.

To study this strange phenomenon of precession, we must remember that angular velocity, angular momentum, and torque are all *vector* quantities. In particular, we need the general relation between the net torque $\sum\vec{\tau}$ that acts on a body and the rate of change of the body's angular momentum $\vec{L}$, given by Eq. (10.32), $\sum\vec{\tau} = d\vec{L}/dt$. Let's first apply this equation to the case in which the flywheel is *not* spinning (Fig. 10.34a). We take the origin O at the pivot and assume that the flywheel is symmetrical, with mass M and moment of inertia I about the flywheel axis. The flywheel axis is initially along the x-axis. The only external forces on the gyroscope are the normal force $\vec{n}$ acting at the pivot (assumed to be frictionless) and the weight $\vec{w}$ of the flywheel that acts at its center of mass, a distance r from the pivot. The normal force has zero torque with respect to the pivot and the weight has a torque $\vec{\tau}$ in the y-direction, as shown in Fig. 10.34a. Initially, there is no rotation, and the initial angular momentum $\vec{L}_i$ is zero. From Eq. (10.32) the *change* $d\vec{L}$ in angular momentum in a short time interval dt following this is

$$d\vec{L} = \vec{\tau}\,dt \tag{10.35}$$

This change is in the y-direction because $\vec{\tau}$ is. As each additional time interval dt elapses, the angular momentum changes by additional increments $d\vec{L}$ in the y-direction because the direction of the torque is constant (Fig. 10.34b). The steadily increasing horizontal angular momentum means that the gyroscope rotates downward faster and faster around the y-axis until it hits either the stand or the table on which it sits.

Now let's see what happens if the flywheel *is* spinning initially, so the initial angular momentum $\vec{L}_i$ is not zero (Fig. 10.35a). Since the flywheel rotates around its symmetry axis, $\vec{L}_i$ lies along the axis. But each change in angular momentum $d\vec{L}$ is perpendicular to the axis, because the torque $\vec{\tau} = \vec{r} \times \vec{w}$ is perpendicular to the axis (Fig. 10.35b). This causes the *direction* of $\vec{L}$ to change, but not its magni-

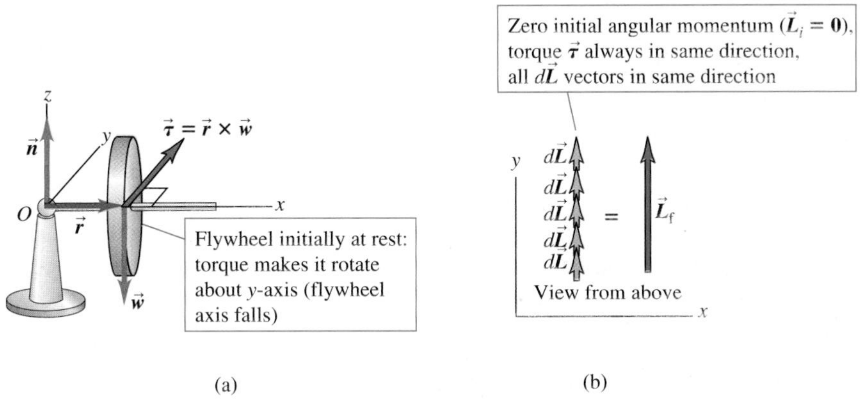

10.34 (a) The flywheel is initially not spinning. The torque $\vec{\tau}$ is due to the weight $\vec{w}$. (b) Looking down from directly over the gyroscope. In each successive time interval dt, the torque produces a change $d\vec{L} = \vec{\tau}dt$ in the angular momentum. In this case the final angular momentum $\vec{L}_f$ is in the same direction as $\vec{\tau}$, and the flywheel axis falls.

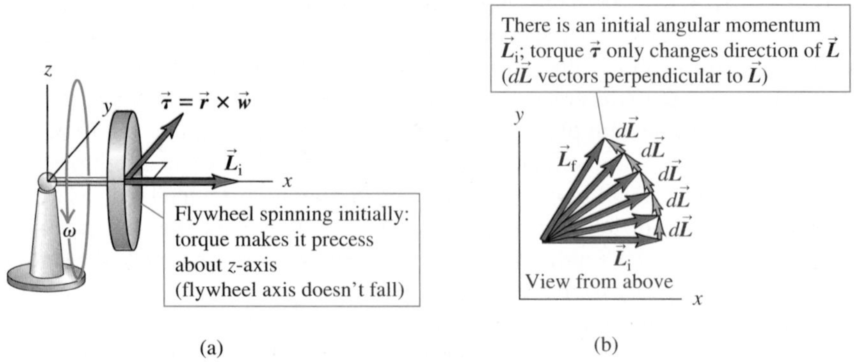

10.35 (a) The flywheel is spinning initially with angular momentum $\vec{L}_i$. The forces (not shown) are the same as those in Fig. 10.34a. (b) Because the initial angular momentum is not zero, each change $d\vec{L} = \vec{\tau}\,dt$ in angular momentum is perpendicular to $\vec{L}$. As a result, the magnitude of $\vec{L}$ remains the same but its direction changes continuously.

tude. The changes $d\vec{L}$ are always in the horizontal xy-plane, so the angular momentum vector and the flywheel axis with which it moves are always horizontal. In other words, the axis doesn't fall—it just precesses.

If this still seems mystifying to you, think about a ball attached to a string. If the ball is initially at rest and you pull the string toward you, the ball moves toward you also. But if the ball is initially moving and you continuously pull the string in a direction perpendicular to the ball's motion, the ball moves in a circle around your hand; it does not approach your hand at all. In the first case the ball has zero linear momentum $\vec{p}$ to start with; when you apply a force $\vec{F}$ toward you for a time dt, the ball acquires a momentum $d\vec{p} = \vec{F}\,dt$, which is also toward you. But if the ball already has linear momentum $\vec{p}$, a change in momentum $d\vec{p}$ that's perpendicular to $\vec{p}$ changes the direction of motion, not the speed. Replace $\vec{p}$ with $\vec{L}$ and $\vec{F}$ with $\vec{\tau}$ in this argument, and you'll see that precession is simply the rotational analog of uniform circular motion.

At the instant shown in Fig. 10.35a, the gyroscope has angular momentum $\vec{L}$. A short time interval dt later, the angular momentum is $\vec{L} + d\vec{L}$; the infinitesimal change in angular momentum is $d\vec{L} = \vec{\tau}\,dt$, which is perpendicular to $\vec{L}$. As the vector diagram in Fig. 10.36 shows, this means that the flywheel axis of the gyroscope has turned through a small angle $d\phi$ given by $d\phi = |d\vec{L}|/|\vec{L}|$. The rate at which the axis moves, $d\phi/dt$, is called the **precession angular speed;** denoting this quantity by Ω, we find

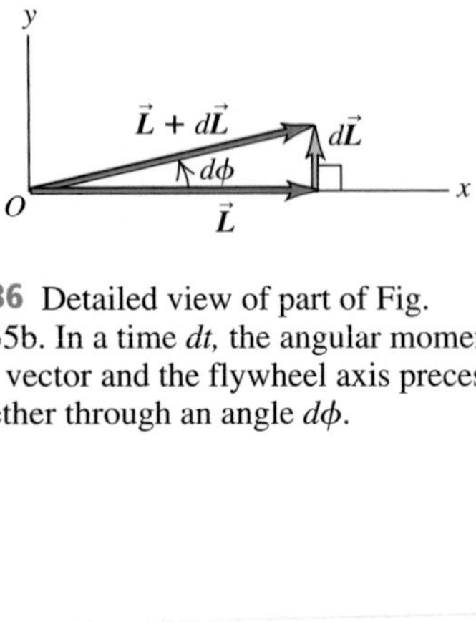

10.36 Detailed view of part of Fig. 10.35b. In a time dt, the angular momentum vector and the flywheel axis precess together through an angle $d\phi$.

$$\Omega = \frac{d\phi}{dt} = \frac{|d\vec{L}|/|\vec{L}|}{dt} = \frac{\tau_z}{L_z} = \frac{wr}{I\omega} \tag{10.36}$$

Thus the precession angular speed is *inversely* proportional to the angular speed of spin about the axis. A rapidly spinning gyroscope precesses slowly; if friction in its bearings causes the flywheel to slow down, the precession angular speed *increases!* The precession angular speed of the earth is very slow (1 rev/26,000 yr) because its spin angular momentum L_z is large and the torque τ_z, due to the gravitational influences of the moon and sun, is relatively small.

As a gyroscope precesses, its center of mass moves in a circle with radius r in a horizontal plane. Its vertical component of acceleration is zero, so the upward normal force $\vec{n}$ exerted by the pivot must be just equal in magnitude to the weight. The circular motion of the center of mass with angular speed Ω requires a force $\vec{F}$ directed toward the center of the circle, with magnitude $F = M\Omega^2 r$. This force must also be supplied by the pivot.

One key assumption that we made in our analysis of the gyroscope was that the angular momentum vector $\vec{L}$ is associated only with the spin of the flywheel and is purely horizontal. But there will also be a vertical component of angular momentum associated with the precessional motion of the gyroscope. By ignoring this, we've tacitly assumed that the precession is *slow*—that is, that the precession angular speed Ω is very much less than the spin angular speed ω. As Eq. (10.36) shows, a large value of ω automatically gives a small value of Ω, so this approximation is reasonable. When the precession is not slow, additional effects show up, including an up-and-down wobble or *nutation* of the flywheel axis that's superimposed on the precessional motion. You can see nutation occurring in a gyroscope as its spin slows down, so that Ω increases and the vertical component of $\vec{L}$ can no longer be ignored.

Example 10.17 A precessing gyroscope

Figure 10.37a shows a top view of a cylindrical gyroscope wheel that has been set spinning by an electric motor. The pivot is at *O*, and the mass of the axle is negligible. a) As seen from above, is the precession clockwise or counterclockwise? b) If the gyro takes 4.0 s for one revolution of precession, at what angular speed does the wheel spin?

SOLUTION

IDENTIFY and SET UP: We'll determine the direction of precession using the right-hand rule as in Fig. 10.35, which shows the same kind of gyroscope as Fig. 10.37. We'll use the relationship between precession angular speed Ω and the spin angular speed ω, Eq. (10.36), to find the value of ω.

EXECUTE: a) The right-hand rule shows that $\vec{\omega}$ and $\vec{L}$ are to the left (Fig. 10.37b). The weight $\vec{w}$ points into the page in this top view and acts at the center of mass (denoted by an $\times$); the torque $\vec{\tau} = \vec{r} \times \vec{w}$ is toward the top of the page; and $d\vec{L}/dt$ is also toward the top of the page. Adding a small $d\vec{L}$ to the $\vec{L}$ that we have initially changes the direction of $\vec{L}$ as shown, so the precession is clockwise as seen from above.

b) Be careful not to confuse ω and Ω! We are given $\Omega = (1\ \text{rev})/(4.0\ \text{s}) = (2\pi\ \text{rad})/(4.0\ \text{s}) = 1.57\ \text{rad/s}$. The weight is equal to mg, and the moment of inertia about its symmetry axis of a solid cylinder with radius R is $I = \frac{1}{2}mR^2$. Solving Eq. (10.36) for ω, we find

$$\omega = \frac{wr}{I\Omega} = \frac{mgr}{(mR^2/2)\Omega} = \frac{2gr}{R^2\Omega}$$

$$= \frac{2(9.8\ \text{m/s}^2)(2.0 \times 10^{-2}\ \text{m})}{(3.0 \times 10^{-2}\ \text{m})^2(1.57\ \text{rad/s})} = 280\ \text{rad/s} = 2600\ \text{rev/min}$$

EVALUATE: The precession angular speed Ω is very much less than the spin angular speed ω, so this is an example of slow precession.

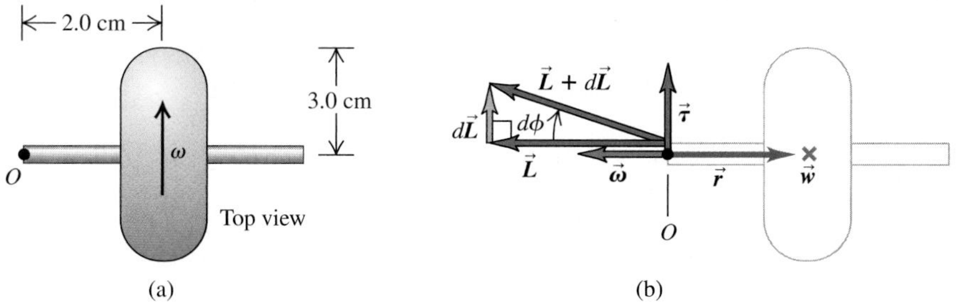

10.37 In which direction does this gyroscope precess?

Test Your Understanding

Suppose the mass of the flywheel in Fig. 10.35 were doubled but all other dimensions and the spin angular speed remained the same. What effect would this change have on the precession angular speed?

SUMMARY

When a force $\vec{F}$ acts on a body, the torque of that force with respect to a point O has a magnitude given by the product of the force magnitude F and the lever arm l. More generally, torque is a vector $\vec{\tau}$ equal to the vector product of $\vec{r}$ (the position vector of the point at which the force acts) and $\vec{F}$.
(See Example 10.1)

$$\tau = Fl \quad (10.2)$$

$$\vec{\tau} = \vec{r} \times \vec{F} \quad (10.3)$$

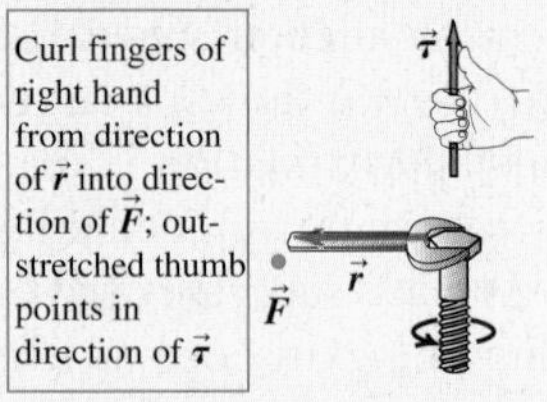

The rotational analog of Newton's second law says that the net torque acting on a body equals the product of the body's moment of inertia and its angular acceleration.
(See Examples 10.2 through 10.4)

$$\sum \tau_z = I\alpha_z \quad (10.6)$$

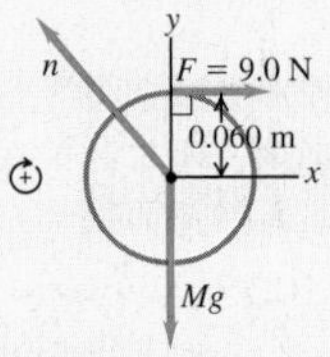

If a rigid body is both moving through space and rotating, its motion may be regarded as a superposition of translational motion of the center of mass and rotational motion about an axis through the center of mass. This approach can be used for the kinetic energy, which is a sum of a translational and rotational kinetic energies. It can also be applied to dynamics: Newton's second law describes the motion of the center of mass, and the rotational equivalent of Newton's second law describes rotation about the center of mass. In the case of rolling without slipping, there is a special relationship between the motion of the center of mass and the rotational motion. (See Example 10.5 through 10.9)

$$K = \frac{1}{2}Mv_{cm}^2 + \frac{1}{2}I_{cm}\omega^2 \quad (10.11)$$

$$\sum \vec{F}_{ext} = M\vec{a}_{cm} \quad (10.15)$$

$$\sum \tau_z = I_{cm}\alpha_z \quad (10.16)$$

$$v_{cm} = R\omega \quad (10.14)$$

(rolling without slipping)

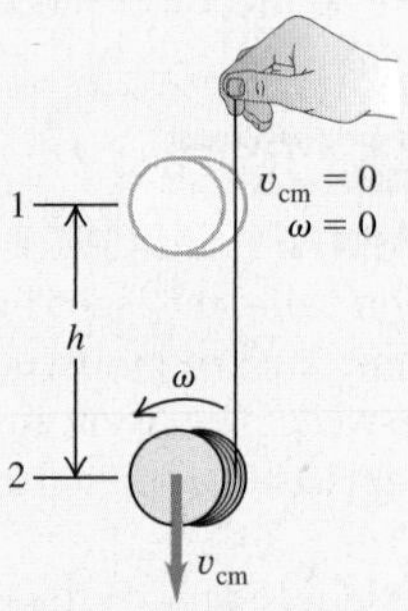

When a torque acts on a rigid body that undergoes an angular displacement, the torque does work on the body. The work can be expressed as an integral of the torque or, if the torque is constant, the product of the torque and the angular displacement. The work—energy theorem for rotational motion of a rigid body—says that the total rotational work done on a body is equal to the change in rotational kinetic energy. The power, or rate at which the torque does work, is the product of the torque and the angular velocity.
(See Examples 10.10 and 10.11)

$$W = \int_{\theta_1}^{\theta_2} \tau_z d\theta \quad (10.23)$$

$$W = \tau_z(\theta_2 - \theta_1) = \tau_z \Delta\theta \quad (10.24)$$

$$W_{tot} = \frac{1}{2}I\omega_2^2 - \frac{1}{2}I\omega_1^2 \quad (10.25)$$

$$P = \tau_z\omega_z \quad (10.26)$$

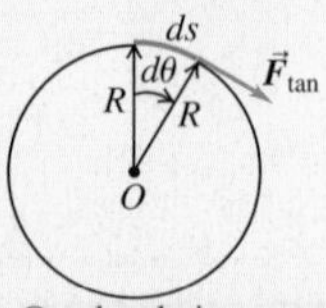

Overhead view of merry-go-round

The angular momentum of a particle with respect to point O is the vector product of the particle's position vector $\vec{r}$ relative to O and its momentum $\vec{p} = m\vec{v}$. When a symmetric body rotates about a stationary axis of symmetry, its angular momentum is the product of its moment of inertia and its angular velocity vector $\vec{\omega}$. If the body is not symmetric or the rotation (z) axis is not an axis of symmetry, the component of angular momentum along the rotation axis is $I\omega_z$. (See Example 10.12)

$$\vec{L} = \vec{r} \times \vec{p} = \vec{r} \times m\vec{v} \qquad (10.27)$$
(particle)

$$\vec{L} = I\vec{\omega} \qquad (10.31)$$
(rigid body rotating about axis of symmetry)

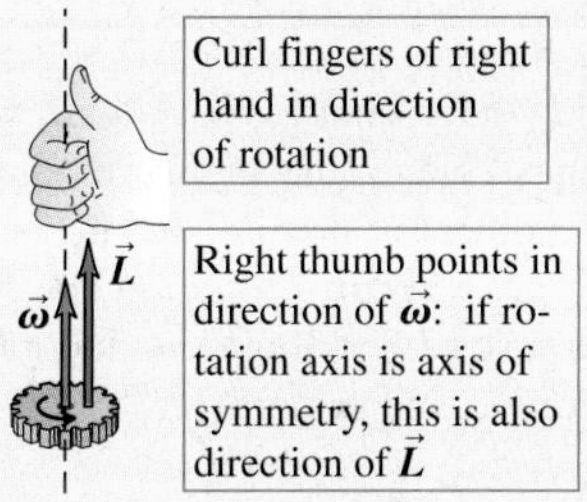

The basic dynamic relation for the rotational motion of any system is that the net external torque is equal to the rate of change of angular momentum. If there is zero net external torque on a system, the total angular momentum of the system is constant (conserved). (See Examples 10.13 through 10.16)

$$\sum \vec{\tau} = \frac{d\vec{L}}{dt} \qquad (10.32)$$

Key Terms

line of action, 362
lever arm (moment arm), 362
torque, 362
combined translation and rotation, 370
rolling without slipping, 371
angular momentum, 379
principle of conservation of angular momentum, 382
precession, 387
precession angular speed, 388

Your Notes

Answer to Chapter Opening Question

When the gymnast is in mid-air, no net torque acts about his center of mass. Hence the angular momentum of his body (the product of the moment of inertia I and the angular speed ω) around the center of mass remains constant. By moving his limbs outward, he increases I and hence ω decreases; if he pulls his limbs in, I decreases and ω increases.

Answers to Test Your Understanding Questions

Section 10.1 The torque is proportional to the product of the distance r and the force magnitude F. Without the "cheater," the distance r is smaller by a factor of $(0.40)/(0.80) = 0.50$. To produce the same torque, the force must therefore be increased by a factor of $(0.80)/(0.40) = 2.0$. Hence the required force is $(2.0)(900\text{ N}) = 1800\text{ N}$.

Section 10.2 The answers would be unchanged. The reason is that none of the forces acting on the glider, pulley, or hanging object depends on the direction of motion, or on whether there is any motion at all.

Section 10.3 With a hollow sphere rather than a solid one, the moment of inertia increases from $\frac{2}{5}MR^2$ to $\frac{2}{3}MR^2$ (see Table 9.2). Repeating the calculation of Example 10.9, we find $a_{\text{cm-}x} = \frac{3}{5}g \sin \beta$. Since $3/5 = 0.600$ is less than $5/7 = 0.714$, the acceleration must be less for the hollow ball than for the solid ball. We would expect this result, since the hollow ball has a greater moment of inertia. Following the steps in Example 10.9 also tells us that the friction force on a hollow ball is $f = \frac{2}{5}Mg \sin \beta$. This is greater than for a solid ball $(2/5 = 0.400$ is greater than $2/7 = 0.286)$, which gives us another way to understand why the hollow ball accelerates more slowly down the incline.

Section 10.4 You apply the same torque over the same angular displacement to both cylinders. Hence, by Eq. (10.24), you do the same amount of work to both cylinders and impart the same kinetic energy to both. (The one with the smaller moment of inertia ends up with a greater angular speed, but that isn't what was asked for. Compare Conceptual Example 6.2 in Section 6.2.)

Section 10.5 As the ball goes around the circle, the magnitude of $\vec{p} = m\vec{v}$ remains the same (the speed is constant) but its direction changes, so the linear momentum vector isn't constant. But $\vec{L} = \vec{r} \times \vec{p}$ *is* constant: it maintains a constant magnitude (the speed and the perpendicular distance from your hand to the ball are both constant) and a constant direction (along the rotation axis, perpendicular to the plane of the ball's motion). The linear momentum changes because there is a net force $\vec{F}$ on the ball (toward the center of the circle). The angular momentum remains constant because there is no net torque; the vector $\vec{r}$ points from your hand to the ball and the force $\vec{F}$ on the ball is directed toward your hand, so the vector product $\vec{\tau} = \vec{r} \times \vec{F}$ is zero.

Section 10.6 In the absence of any external torques, the earth's angular momentum $L_z = I\omega_z$ would remain constant. The melted ice would move from the poles toward the equator—that is, away from our planet's rotation axis—and the earth's moment of inertia I would increase slightly. Hence the angular velocity ω_z would decrease slightly and the length of the day would be slightly longer.

Section 10.7 Doubling the flywheel mass would double both its moment of inertia I and its weight w, so the ratio I/w would be unchanged. Equation (10.36) shows that the precession angular speed depends on this ratio, so there would be *no* effect on the value of Ω.

Discussion Questions

Q10.1 When tightening cylinder-head bolts in an automobile engine, the critical quantity is the *torque* applied to the bolts. Why is this more important than the actual *force* applied to the wrench handle?

Q10.2 Can a single force applied to a body change both its translational and rotational motion? Explain.

Q10.3 Suppose you could use wheels of any type in the design of a soapbox-derby racer (an unpowered, four-wheel vehicle that coasts from rest down a hill). Conforming to the rules on the total weight of the vehicle and rider, should you design with large massive wheels or small light wheels? Should you use solid wheels or wheels with most of the mass at the rim? Explain.

Q10.4 Unless $\vec{r}$ and $\vec{F}$ are perpendicular, there are always two angles between their directions that give the same torque for given magnitudes of $\vec{r}$ and $\vec{F}$. Explain why. Illustrate your answer with a sketch.

Q10.5 The crankshaft of an automobile engine has a flywheel to increase the moment of inertia about the rotation axis. Why is this desirable?

Q10.6 The harder you hit the brakes while driving forward, the more the front end of your car will move down (and the rear end move up). Why? What happens when accelerating? Why do drag racers not use front-wheel drive only?

Q10.7 When an acrobat walks on a tightrope, she extends her arms straight out from her sides. She does this to make it easier for her to catch herself if she should tip to one side or the other. Explain how this works. (*Hint:* Think about Eq. (10.6).)

Q10.8 When you turn on an electric motor, it takes longer to come up to final speed if there is a grinding wheel attached to the shaft. Why?

Q10.9 Experienced cooks can tell whether an egg is raw or hard-boiled by rolling it down a slope (taking care to catch it at the bottom). How is this possible? What are they looking for?

Q10.10 The work done by a force is the product of force and distance. The torque due to a force is a product of force and distance. Does this mean that torque and work are equivalent? Explain.

Q10.11 A valued client brings a treasured ball to your engineering firm, wanting to know if the ball is solid or hollow. He has tried tapping on it, but that has given insufficient information. Design a simple, inexpensive experiment that you could perform quickly, without injuring the precious ball, to find out if it is solid or hollow.

Q10.12 Force A has a magnitude twice that of force B. Is it possible for force B to exert a greater torque on an object than force A?

Q10.13 An electric grinding wheel coasts for a minute or more after the power is turned off, but an electric drill coasts for only a few seconds. Why is there a difference?

Q10.14 The force of gravity acts on the hammer in Fig. 10.12 and forces produce torques that cause a body's angular velocity to change. Why, then, is the angular velocity of the hammer in the figure constant?

Q10.15 In Example 10.6 (Section 10.3) suppose you pulled your hand and the end of the string upward. Would you conserve mechanical energy in this situation? Why or why not?

Q10.16 A wheel is rolling without slipping on a horizontal surface. In an inertial frame of reference in which the surface is at rest, is there any point on the wheel that has a velocity that is purely vertical? Is there any point that has a horizontal velocity component opposite to the velocity of the center of mass? Explain. Do your answers change if the wheel is slipping as it rolls? Why or why not?

Q10.17 Part of the kinetic energy of a moving automobile is in the rotational motion of its wheels. When the brakes are applied hard on an icy street, the wheels "lock," and the car starts to slide. What becomes of the rotational kinetic energy?

Q10.18 A hoop, a uniform solid cylinder, a spherical shell, and a uniform solid sphere are released from rest at the top of an incline. What is the order in which they arrive at the bottom of the incline? Does it matter whether or not the masses and radii of the objects are all the same? Explain.

Q10.19 A ball is rolling along at speed v without slipping on a horizontal surface when it comes to a hill that rises at a constant angle above the horizontal. In which case will it go higher up the hill: if the hill has enough friction to prevent slipping or if the hill is perfectly smooth? Justify your answers in both cases in terms of energy conservation and in terms of Newton's second law.

Q10.20 You are standing at the center of a large, horizontal turntable in a carnival funhouse. The turntable is set rotating on frictionless bearings, and rotates freely (that is, there is no motor driving the turntable). As you walk toward the edge of the turntable, what happens to the combined angular momentum of you and the turntable? What happens to the rotation speed of the turntable? Explain your answer.

Q10.21 A point particle travels in a circle at constant speed. With respect to the origin at the center of the circle, is there a net torque acting on the particle? A net force? What if the speed of the particle is changing? In each case, explain your answers.

Q10.22 A point particle travels in a straight line at constant speed and the closest distance it comes to the origin of coordinates is a distance l. With respect to this origin, does the particle have nonzero angular momentum? As the particle moves along its straight-line path, does its angular momentum with respect to origin change?

Q10.23 In Example 10.13 (Section 10.6) the angular speed ω changes and this must mean that there is nonzero angular acceleration. But there is no torque about the rotation axis if the forces the professor applies to the weights are directly, radially inward. Then by Eq. (10.6), α_z must be zero. Explain what is wrong with this reasoning that leads to this apparent contradiction.

Q10.24 In Example 10.13 (Section 10.6) the rotational kinetic energy of the professor and dumbbells increases. But since there are no external torques, no work is being done to change the rotational kinetic energy. Then, by Eq. (10.25) the kinetic energy must remain the same! Explain what is wrong with this reasoning that leads to this apparent contradiction. Where *does* the extra kinetic energy come from?

Q10.25 As discussed in Section 10.6, the angular momentum of a circus acrobat is conserved as she tumbles through the air. Is her *linear* momentum conserved? Why or why not?

Q10.26 If you stop a spinning raw egg for the shortest instant you can and then release it, the egg will start spinning again. If you do the same to a hard-boiled egg, it will remain stopped. Try it. Explain it.

Q10.27 A helicopter has a large main rotor that rotates in a horizontal plane and provides lift. There is also a small rotor on the tail that rotates in a vertical plane. What is the purpose of the tail rotor? (*Hint:* if there were no tail rotor, what would happen when the pilot changed the angular speed of the main rotor?) Some helicopters have no tail rotor, but instead have two large main rotors that rotate in a horizontal plane. Why is it important that the two main rotors rotate in opposite directions?

Q10.28 In a common design for a gyroscope, the flywheel and flywheel axis are enclosed in a light, spherical frame with the flywheel at the center of the frame. The gyroscope is then balanced on top of a pivot so that the flywheel is directly above the pivot. Does the gyroscope precess if it is released while the flywheel is spinning? Explain.

Q10.29 A gyroscope takes 3.8 s to precess 1.0 revolution about a vertical axis. Two minutes later, it takes only 1.9 s to precess 1.0 revolution. No one has touched the gyroscope. Explain.

Q10.30 A gyroscope is precessing as in Fig. 10.33. What happens if you gently add some weight to the end of the flywheel axis farthest from the pivot?

Q10.31 A bullet emerges from a rifle spinning on its axis. Explain how this prevents the bullet from tumbling and keeps the streamlined end pointed forward.

Exercises

Section 10.1 Torque

10.1 Calculate the torque (magnitude and direction) about point O due to the force $\vec{F}$ in each of the situations sketched in Fig. 10.38. In each case, the force $\vec{F}$ and the rod both lie in the plane of the

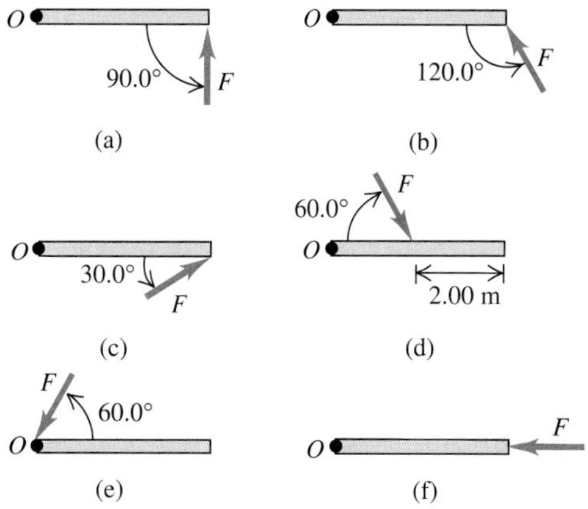

Figure 10.38 Exercise 10.1.

page, the rod has length 4.00 m, and the force has magnitude $F = 10.0$ N.

10.2 Calculate the net torque about point O for the two forces applied as in Fig. 10.39. The rod and both forces are in the plane of the page.

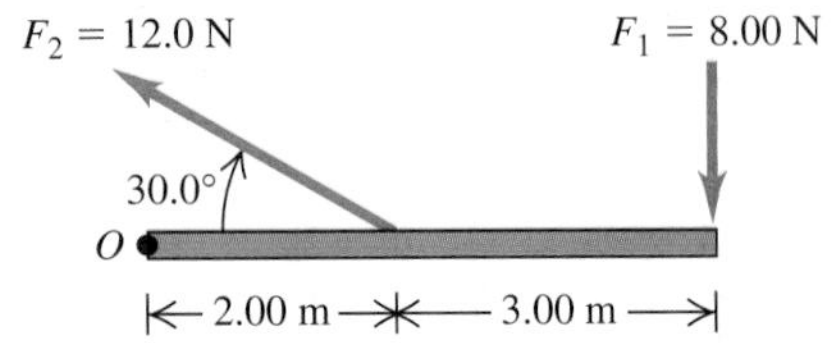

Figure 10.39 Exercise 10.2.

10.3 A square metal plate 0.180 m on each side is pivoted about an axis through point O at its center and perpendicular to the plate (Fig. 10.40). Calculate the net torque about this axis due to the three forces shown in the figure if the magnitudes of the forces are $F_1 = 18.0$ N, $F_2 = 26.0$ N, and $F_3 = 14.0$ N. The plate and all forces are in the plane of the page.

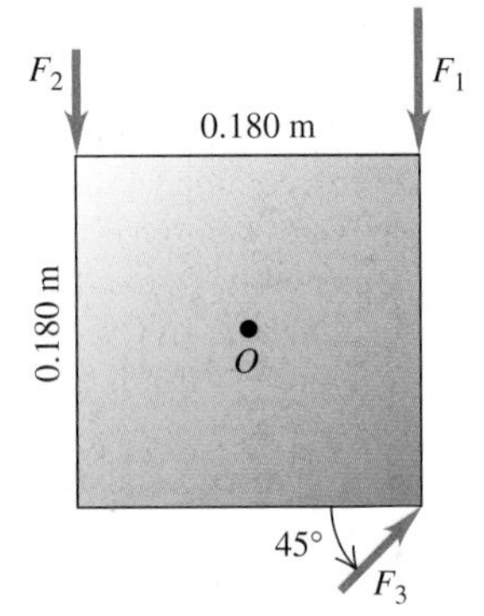

Figure 10.40 Exercise 10.3.

10.4 Forces $F_1 = 7.50$ N and $F_2 = 5.30$ N are applied tangentially to a wheel with radius 0.330 m, as shown in Fig. 10.41. What is the net torque on the wheel due to these two forces for an axis perpendicular to the wheel and passing through its center?

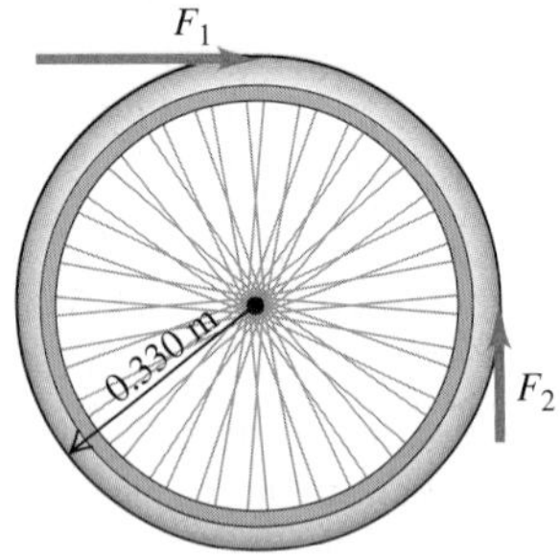

Figure 10.41 Exercise 10.4.

10.5 One force acting on a machine part is $\vec{F} = (-5.00\text{ N})\hat{\imath} + (4.00\text{ N})\hat{\jmath}$. The vector from the origin to the point where the force is applied is $\vec{r} = (-0.450\text{ m})\hat{\imath} + (0.150\text{ m})\hat{\jmath}$. a) In a sketch, show $\vec{r}$, $\vec{F}$, and the origin. b) Use the right-hand rule to determine the direction of the torque. c) Calculate the vector torque produced by this force. Verify that the direction of the torque is the same as you obtained in (b).

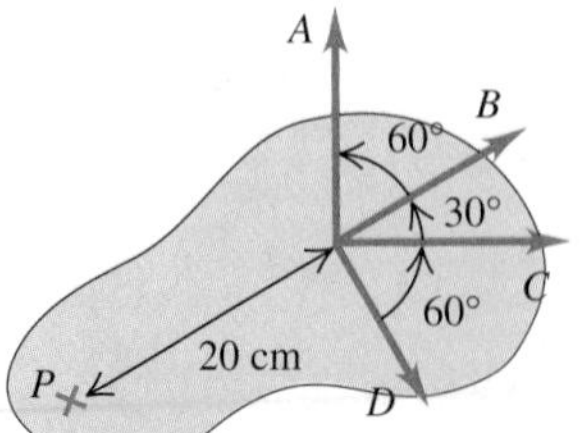

Figure 10.42 Exercise 10.6.

10.6 In Fig. 10.42, the forces $\vec{A}$, $\vec{B}$, $\vec{C}$, and $\vec{D}$ each have magnitude 50 N and act at the same point on the object. a) What torque (magnitude and direction) does each of these forces exert on the object about point P? b) What is the total torque about point P?

Section 10.2 Torque and Angular Acceleration for a Rigid Body

10.7 A uniform, 8.40-kg, spherical shell 50.0 cm in diameter has four small 2.00 kg masses attached to its outer surface and equally spaced around it. This combination is spinning about an axis running through the center of the sphere and two of the small masses (Fig. 10.43). What friction torque is needed to reduce its angular speed from 75.0 rpm to 50.0 rpm in 30.0 s?

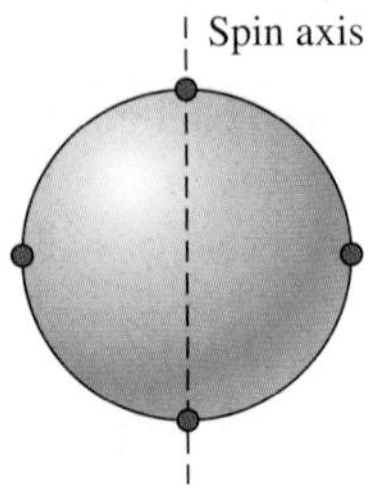

Figure 10.43 Exercise 10.7.

10.8 The flywheel of an engine has a moment of inertia $2.50\text{ kg}\cdot\text{m}^2$ about its rotation axis. a) What constant torque is required to bring it up to an angular speed of 400 rev/min in 8.00 s, starting from rest? b) What is its final kinetic energy?

10.9 Using the value of a_x calculated in Example 10.2 (Section 10.2), what is the speed of the cable after it has been pulled 2.0 m? Compare your result with Example 9.8 (Section 9.4).

10.10 A cord is wrapped around the rim of a wheel 0.250 m in radius, and a steady pull of 40.0 N is exerted on the cord. The wheel is mounted on frictionless bearings on a horizontal shaft through its center. The moment of inertia of the wheel about this shaft is $5.00\text{ kg}\cdot\text{m}^2$. Compute the angular acceleration of the wheel.

10.11 a) Find the magnitude n of the normal force for the situation described in Example 10.3 (Section 10.2). b) Is your answer to part (a) less than, equal to, or greater than the total weight $(M + m)g$ of the cylinder and mass? Explain why this should be so. c) Suppose the cylinder is initially rotating clockwise so that the hanging mass m is initially moving upward at speed v_0 (the cable remains taut). What effect does this have on the tension T and normal force n? Explain.

10.12 a) In the situation described in Example 10.2 of Section 10.2 (Fig. 10.9), the normal force $\vec{n}$ exerted on the cylinder by the bearing is directed up and to the left. Why must the normal force be in this direction? b) Calculate the magnitude and direction of $\vec{n}$.

10.13 A grindstone in the shape of a solid disk with diameter 0.520 m and a mass of 50.0 kg is rotating at 850 rev/min. You press an ax against the rim with a normal force of 160 N (Fig. 10.44), and the grindstone comes to rest in 7.50 s. Find the coefficient of friction

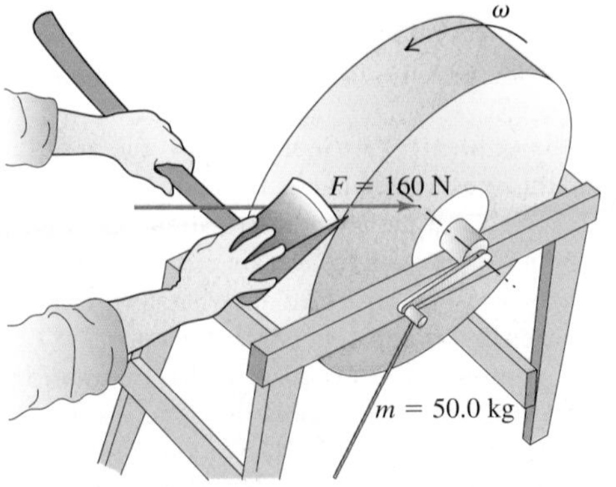

Figure 10.44 Exercise 10.13 and Problem 10.52.

between the ax and the grindstone. You can ignore friction in the bearings.

10.14 A stone is suspended from the free end of a wire that is wrapped around the outer rim of a pulley, as shown in Fig. 10.10. The pulley is a uniform disk with mass 10.0 kg and radius 50.0 cm and turns on frictionless bearings. You measure that the stone travels 12.6 m in the first 3.00 s starting from rest. Find a) the mass of the stone; b) the tension in the wire.

10.15 A solid, uniform cylinder with mass 8.25 kg and diameter 15.0 cm is spinning at 220 rpm on a thin, frictionless axle that passes along the cylinder axis. You design a simple friction-brake to stop the cylinder by pressing the brake against the outer rim with a normal force. The coefficient of kinetic friction between the brake and rim is 0.333. What must the applied normal force be to bring the cylinder to rest after it has turned through 5.25 revolutions?

10.16 A 15.0-kg bucket of water is suspended by a rope wrapped around a windlass, that is a solid cylinder 0.300 m in diameter with mass 12.0 kg. The cylinder pivots on a frictionless axle through its center. The bucket is released from rest at the top of a well and falls 10.0 m to the water. You can ignore the weight of the rope. a) What is the tension in the rope while the bucket is falling? b) With what speed does the bucket strike the water? c) What is the time of fall? d) While the bucket is falling, what is the force exerted on the cylinder by the axle?

10.17 A 2.00-kg textbook rests on a frictionless, horizontal surface. A cord attached to the book passes over a pulley whose diameter is 0.150 m, to a hanging book with mass 3.00 kg. The system is released from rest, and the books are observed to move 1.20 m in 0.800 s. a) What is the tension in each part of the cord? b) What is the moment of inertia of the pulley about its rotation axis?

10.18 A thin, horizontal rod with length l and mass M pivots about a vertical axis at one end. A force with constant magnitude F is applied to the other end, causing the rod to rotate in a horizontal plane. The force is maintained perpendicular to the rod and to the axis of rotation. Calculate the magnitude of the angular acceleration of the rod.

Section 10.3 Rigid-Body Rotation about a Moving Axis

10.19 A string is wrapped several times around the rim of a small hoop with radius 0.0800 m and mass 0.180 kg. If the free end of the string is held in place and the hoop is released from rest (Fig. 10.45), calculate a) the tension in the string while the hoop descends as the string unwinds; b) the time it takes the hoop to descend 0.750 m; c) the angular speed of the rotating hoop after it has descended 0.750 m.

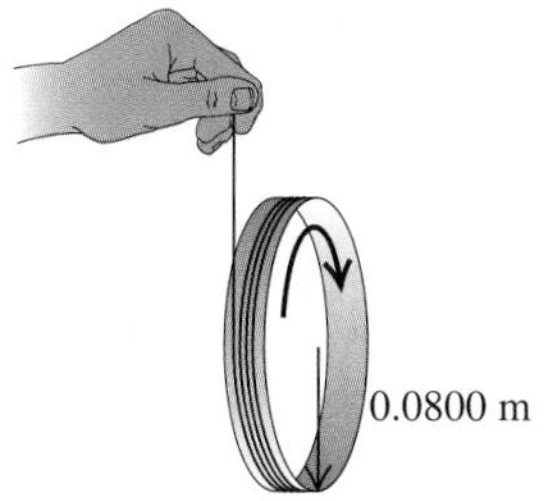

Figure 10.45 Exercise 10.19 and Problem 10.71.

10.20 Repeat part (c) of Exercise 10.19, this time using energy considerations.

10.21 In Example 10.5 (Section 10.3) we found that for a hollow, cylindrical shell rolling without slipping on a horizontal surface, half of the total kinetic energy is translational and half is rotational. What fraction of the total kinetic energy is rotational for the following objects rolling without slipping on a horizontal surface? a) A uniform solid cylinder. b) A uniform sphere. c) A thin-walled hollow sphere. d) A hollow, cylinder with outer radius R and inner radius $R/2$.

10.22 A hollow, spherical shell with mass 2.00 kg rolls without slipping down a 38.0° slope. a) Find the acceleration, the friction force, and the minimum coefficient of friction needed to prevent slipping. b) How would your answers to part (a) change if the mass were doubled to 4.00 kg?

10.23 A solid ball is released from rest and slides down a hillside that slopes downward at 65.0° from the horizontal. a) What minimum value must the coefficient of static friction between the hill and ball surfaces have for no slipping to occur? b) Would the coefficient of friction calculated in part (a) be sufficient to prevent a hollow ball (such as a soccer ball) from slipping? Justify your answer. c) In part (a), why did we use the coefficient of static friction and not the coefficient of kinetic friction?

10.24 A uniform marble rolls down a symmetric bowl, starting from rest at the top of the left side. The top of each side is a distance h above the bottom of the bowl. The left half of the bowl is rough enough to cause the marble to roll without slipping, but the right half has no friction because it is coated with oil. a) How far up the smooth side will the marble go, measured vertically from the bottom? b) How high would the marble go if both sides were as rough as the left side? c) How do you account for the fact that the marble goes *higher* with friction on the right side than without friction?

10.25 A 392-N wheel comes off a moving truck and rolls without slipping along a highway. At the bottom of a hill it is rotating at 25.0 rad/s. The radius of the wheel is 0.600 m and its moment of inertia about its rotation axis is $0.800MR^2$. Friction does work on the wheel as it rolls up the hill to a stop, a height h above the bottom of the hill; this work has absolute value 3500 J. Calculate h.

10.26 A Ball Rolling Uphill. A bowling ball rolls without slipping up a ramp that slopes upward at an angle β to the horizontal (see Example 10.9 in Section 10.3). Treat the ball as a uniform, solid sphere, ignoring the finger holes. a) Draw the free-body diagram for the ball. Explain why the friction force must be directed *uphill*. b) What is the acceleration of the center of mass of the ball? c) What minimum coefficient of static friction is needed to prevent slipping?

Section 10.4 Work and Power in Rotational Motion

10.27 A playground merry-go-round has radius 2.40 m and a moment of inertia 2100 $kg \cdot m^2$ about a vertical axle through its center, and turns with negligible friction. a) A child applies an 18.0-N force tangentially to the edge of the merry-go-round for 15.0 s. If the merry-go-round is initially at rest, what is its angular speed after this 15.0-s interval? b) How much work did the child do on the merry-go-round? c) What is the average power supplied by the child?

10.28 Example 9.5 (Section 9.3) described the design of an aircraft propeller. The engine delivers 175 hp to the propeller at 2400 rev/min. a) How much torque does the aircraft engine provide? b) How much work does the engine do in one revolution of the propeller?

10.29 A 1.50-kg grinding wheel is in the form of a solid cylinder of radius 0.100 m. a) What constant torque will bring it from rest to an angular speed of 1200 rev/min in 2.5 s? b) Through what angle has it turned during that time? c) Use Eq. (10.24) to calculate the work done by the torque. d) What is the grinding wheel's kinetic energy when it is rotating at 1200 rev/min? Compare your answer to the result in part (c).

10.30 What is the power output in horsepower of an electric motor turning at 4800 rev/min and developing a torque of 4.30 N·m?

10.31 The carbide tips of the cutting teeth of a circular saw are 8.6 cm from the axis of rotation. a) The no-load speed of the saw, when it is not cutting anything, is 4800 rev/min. Why is its no-load power output negligible? b) While cutting lumber, the angular speed of the saw slows to 2400 rev/min and the power output is 1.9 hp. What is the tangential force that the wood exerts on the carbide tips?

10.32 An airplane propeller is 2.08 m in length (from tip to tip) and has a mass of 117 kg. When the airplane's engine is first started, it applies a constant torque of 1950 N·m to the propeller, which starts from rest. a) What is the angular acceleration of the propeller? Treat the propeller as a slender rod. (*Hint:* See Table 9.2.) b) What is the propeller's angular speed after making 5.00 rev? c) How much work is done by the engine during the first 5.00 rev? d) What is the average power output of the engine during the first 5.00 rev? e) What is the instantaneous power output of the motor at the instant that the propeller has turned through 5.00 rev?

10.33 a) Compute the torque developed by an industrial motor whose output is 150 kW at an angular speed of 4000 rev/min. b) A drum with negligible mass, 0.400 m in diameter, is attached to the motor shaft and the power output of the motor is used to raise a weight hanging from a rope wrapped around the drum. How heavy a weight can the motor lift at constant speed? c) At what constant speed will the weight rise?

Section 10.5 Angular Momentum

10.34 A woman with mass 50 kg is standing on the rim of a large disk that is rotating at 0.50 rev/s about an axis through its center. The disk has mass 110 kg and a radius 4.0 m. Calculate the magnitude of the total angular momentum of the woman-plus-disk system. (Assume that you can treat the woman as a point.)

10.35 A 2.00-kg rock has a horizontal velocity of magnitude 12.0 m/s when it is at point *P* in Fig. 10.46. a) At this instant, what is the magnitude and direction of its angular momentum relative to point *O*? b) If the only force acting on the rock is its weight, what is the rate of change (magnitude and direction) of its angular momentum at this instant?

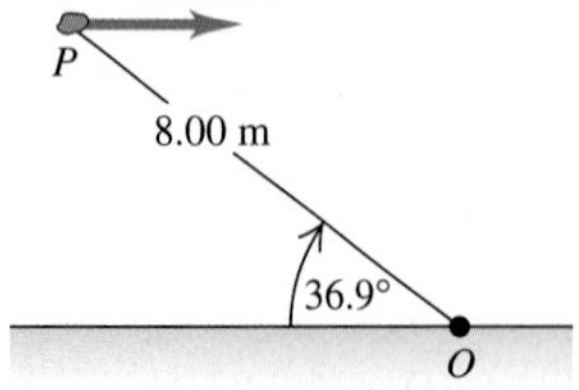

Figure 10.46 Exercise 10.35.

10.36 a) Calculate the magnitude of the angular momentum of the earth considered as a particle orbiting the sun. The mass of the earth is 5.97×10^{24} kg. Treat it as moving in a circular orbit of radius 1.50×10^{11} m at a speed of 2.98×10^{4} m/s. b) Calculate the magnitude of the angular momentum of the earth due to its rotation around an axis through the north and south poles. Treat the earth as a uniform sphere of radius 6.38×10^{6} m that makes one revolution in 24.0 hours.

10.37 Find the magnitude of the angular momentum of the second hand on a clock about an axis through the center of the clock face. The clock hand has a length of 15.0 cm and a mass of 6.00 g. Take the second hand to be a slender rod rotating with constant angular velocity about one end.

Section 10.6 Conservation of Angular Momentum

10.38 Under some circumstances, a star can collapse into an extremely dense object made mostly of neutrons and called a *neutron star.* The density of a neutron star is roughly 10^{14} times as great as that of ordinary solid matter. Suppose we represent the star as a uniform, solid, rigid sphere, both before and after the collapse. The star's initial radius was 7.0×10^{5} km (comparable to our sun); its final radius is 16 km. If the original star rotated once in 30 days, find the angular speed of the neutron star.

10.39 A small block on a frictionless, horizontal surface has a mass of 0.0250 kg. It is attached to a massless cord passing through a hole in the surface (Fig. 10.47). The block is originally revolving at a distance of 0.300 m from the hole with an angular speed of 1.75 rad/s. The cord is then pulled from below, shortening the radius of the circle in which the block revolves to 0.150 m. You may treat the block as a particle. a) Is angular momentum conserved? Why or why not? b) What is the new angular speed? c) Find the change in kinetic energy of the block. d) How much work was done in pulling the cord?

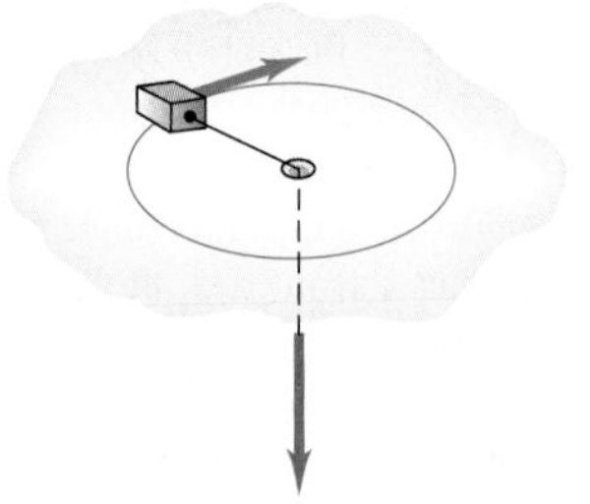
Figure 10.47 Exercise 10.39, Problem 10.92, and Challenge Problem 10.103.

10.40 The Spinning Figure Skater. The outstretched hands and arms of a figure skater preparing for a spin can be considered a slender rod pivoting about an axis through its center (Fig. 10.48). When his hands and arms are brought in and wrapped around his

Figure 10.48 Exercise 10.40.

body to execute the spin, the hands and arms can be considered a thin-walled hollow cylinder. His hands and arms have a combined mass 8.0 kg. When outstretched, they span 1.8 m; when wrapped, they form a cylinder of radius 25 cm. The moment of inertia about the rotation axis of the remainder of his body is constant and equal to 0.40 kg · m^2. If his original angular speed is 0.40 rev/s, what is his final angular speed?

10.41 A diver comes off a board with arms straight up and legs straight down, giving her a moment of inertia about her rotation axis of 18 kg · m^2. She then tucks into a small ball, decreasing this moment of inertia to 3.6 kg · m^2. While tucked, she makes two complete revolutions in 1.0 s. If she hadn't tucked at all, how many revolutions would she have made in the 1.5 s from board to water?

10.42 A large turntable rotates about a fixed vertical axis, making one revolution in 6.00 s. The moment of inertia of the turntable about this axis is 1200 kg · m^2. A child of mass 40.0 kg, initially standing at the center of the turntable, runs out along a radius. What is the angular speed of the turntable when the child is 2.00 m from the center? (Assume that you can treat the child as a particle.)

10.43 A large, wooden turntable in the shape of a flat disk has a radius of 2.00 m and a total mass of 120 kg. The turntable is initially rotating at 3.00 rad/s about a vertical axis through its center. Suddenly, a 70.0-kg parachutist makes a soft landing on the turntable at a point near the outer edge. a) Find the angular speed of the turntable after the parachutist lands. (Assume that you can treat the parachutist as a particle.) b) Compute the kinetic energy of the system before and after the parachutist lands. Why are these kinetic energies not equal?

10.44 A solid wood door 1.00 m wide and 2.00 m high is hinged along one side and has a total mass 40.0 kg. Initially open and at rest, the door is struck at its center by a handful of sticky mud with mass 0.500 kg, traveling perpendicular to the door at 12.0 m/s just before impact. Find the final angular speed of the door. Does the mud make a significant contribution to the moment of inertia?

10.45 A small 10.0-g bug stands at one end of a thin, uniform bar which is initially at rest on a smooth, horizontal table. The other end of the bar pivots about a nail driven into the table and can rotate freely, without friction. The bar has mass 50.0 g and is 100 cm in length. The bug jumps off in the horizontal direction, perpendicular to the bar, with a speed of 20.0 cm/s relative to the table. a) What is the angular speed of the bar just after the frisky insect leaps? b) What is the total kinetic energy of the system just after the bug leaps? c) Where does this energy come from?

10.46 A thin, uniform, metal bar, 2.00 m long and weighing 90.0 N, is hanging vertically from the ceiling by a frictionless pivot. Suddenly it is struck 1.50 m below the ceiling by a small 3.00-kg ball, initially traveling horizontally at 10.0 m/s. The ball rebounds in the opposite direction with a speed of 6.00 m/s. a) Find the angular speed of the bar just after the collision. b) During the collision, why is the angular momentum conserved, but not the linear momentum?

Section 10.7 Gyroscopes and Precession

10.47 Draw a top view of the gyroscope shown in Fig. 10.33. a) Draw labeled arrows on your sketch for $\vec{\omega}$, $\vec{L}$, and $\vec{\tau}$. Draw $d\vec{L}$ produced by $\vec{\tau}$. Draw $\vec{L} + d\vec{L}$. Determine the sense of precession by examining the directions of $\vec{L}$ and $\vec{L} + d\vec{L}$. b) Reverse the direction of the spin angular velocity of the rotor and repeat all steps in part (a). c) Move the pivot to the other end of the shaft, with the same direction of spin angular velocity as in part (b), and repeat all the steps. d) Keeping the pivot as in part (c), reverse the spin angular velocity of the rotor and repeat all steps.

10.48 The rotor (flywheel) of a toy gyroscope has mass 0.140 kg. Its moment of inertia about its axis is 1.20×10^{-4} kg · m^2. The mass of the frame is 0.0250 kg. The gyroscope is supported on a single pivot (Fig. 10.49) with its center of mass a horizontal distance of 4.00 cm from the pivot. The gyroscope is precessing in a horizontal plane at the rate of one revolution in 2.20 s. a) Find the upward force exerted by the pivot. b) Find the angular speed with which the rotor is spinning about its axis, expressed in rev/min. c) Copy the diagram and draw vectors to show the angular momentum of the rotor and the torque acting on it.

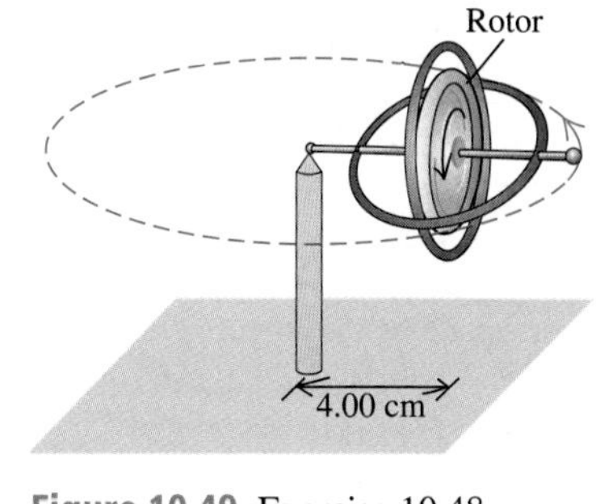

Figure 10.49 Exercise 10.48.

10.49 A Gyro Stabilizer. The stabilizing gyroscope of a ship is a solid disk with mass 60,000 kg; its radius is 2.00 m, and it rotates about a vertical axis with an angular speed of 500 rev/min. a) How much time is required to bring it up to speed, starting from rest, with a constant power input of 7.46×10^4 W? b) Find the torque needed to cause the axis to precess in a vertical fore-and-aft plane at the rate of 1.00°/s.

10.50 A gyroscope is precessing about a vertical axis. Describe what happens to the precession angular speed if the following changes in the variables are made, with all other variables remaining the same: a) the angular speed of the spinning flywheel is doubled; b) the total weight is doubled; c) the moment of inertia about the axis of the spinning flywheel is doubled; d) the distance from the pivot to the center of gravity is doubled. e) What happens if all four of the variables in parts (a) through (d) are doubled?

10.51 The earth precesses once every 26,000 years, and spins on its axis once a day. Estimate the magnitude of the torque that causes the precession of the earth. You may need some data from Appendix F. Make the estimate by assuming i) the earth is a uniform sphere and ii) the precession of the earth is like that of the gyroscope shown in Fig. 10.34. In this model, the precession axis and rotation axis are perpendicular. Actually, the angle between these two axes for the earth is only $23\frac{1}{2}°$; this affects the calculated torque by about a factor of 2.

Problems

10.52 A 55.0-kg grindstone is a solid disk 0.520 m in diameter. You press an ax down on the rim with a normal force of 160 N (Fig. 10.44). The coefficient of kinetic friction between the blade and the stone is 0.60, and there is a constant friction torque of

6.50 N·m between the axle of the stone and its bearings. a) How much force must be applied tangentially at the end of a crank handle 0.500 m long to bring the stone from rest to 120 rev/min in 9.00 s? b) After the grindstone attains an angular speed of 120 rev/min, what tangential force at the end of the handle is needed to maintain a constant angular speed of 120 rev/min? c) How much time does it take the grindstone to come from 120 rev/min to rest if it is acted on by the axle friction alone?

10.53 An experimental bicycle wheel is placed on a test stand so that it is free to turn on its axle. If a constant net torque of 5.00 N·m is applied to the tire for 2.00 s, the angular speed of the tire increases from zero to 100 rev/min. The external torque is then removed, and the wheel is brought to rest by friction in its bearings in 125 s. Compute a) the moment of inertia of the wheel about the rotation axis; b) the friction torque; c) the total number of revolutions made by the wheel in the 125-s time interval.

10.54 A flywheel 0.600 m in diameter pivots on a horizontal axis. A rope is wrapped around the outside of the flywheel, and a steady pull of 40.0 N is exerted on the rope. The flywheel starts from rest, and 5.00 m of rope are unwound in 2.00 s. a) What is the angular acceleration of the flywheel? b) What is its final angular speed? c) What is its final kinetic energy? d) What is its moment of inertia about the rotation axis?

10.55 A wheel starts from rest and rotates with constant angular acceleration about a fixed axis. a) Prove that the power at any given time is proportional to the square of the net torque about the axis. b) If the power at $t = 3.00$ s is 500 W with a constant net torque of 20.0 N·m, what would the power have been at $t = 3.00$ s with a constant net torque of 60.0 N·m? c) Prove that the power at any given angular displacement is proportional to the $\frac{3}{2}$ power of the total torque about the axis at that angular displacement. d) If the power after rotating 37.5 rad with a torque of 20.0 N·m is 500 W, what would the power have been after rotating 37.5 rad with a torque of 60.0 N·m? e) Do the answers to parts (a) and (b) contradict the answers to parts (c) and (d)? Why or why not?

10.56 A beam of length l lies on the $+x$-axis with its left end at the origin. A cable pulls on the beam in the $+y$-direction with a force $\vec{F}$ with a magnitude that depends on where along the beam it is exerted, $F = F_0(1 - x/l)$, where F_0 is a constant equal to the force magnitude when it is applied at the left end of the beam. a) What is the direction of the torque due to $\vec{F}$? The rotation axis is perpendicular to the beam and passes through the origin. b) Graph F versus x from $x = 0$ to $x = l$. Express F in terms of F_0 and x in terms of l. c) Graph the torque versus x from $x = 0$ to $x = l$. Express the torque in terms of $F_0 l$, and express x in terms of l. d) Where along the beam should the force be applied to produce maximum torque and what is that maximum torque?

10.57 While exploring a castle, Exena the Exterminator is spotted by a dragon who chases her down a hallway. Exena runs into a room and attempts to swing the heavy door shut before the dragon gets her. The door is initially perpendicular to the wall, so it must be turned through 90° to close. The door is 3.00 m tall and 1.25 m wide, and it weighs 750 N. You can ignore the friction at the hinges. If Exena applies a force of 220 N at the edge of the door and perpendicular to it, how much time does it take her to close the door?

10.58 A thin rod of length l lies on the $+x$-axis with its left end at the origin. A string pulls on the rod with a force $\vec{F}$ directed toward a point P a distance h above the rod. Where along the rod should you attach the string to get the greatest torque about the origin if point P is a) above the right end of the rod? b) above the left end of the rod? c) above the center of the rod?

10.59 Balancing Act. Attached to one end of a long, thin, uniform rod of length L and mass M is a small blob of clay of the same mass M. a) Locate the position of the center of mass of the system of rod and clay. Note this position on a drawing of the rod. b) You carefully balance the rod on a frictionless tabletop so that it is standing vertically, with the end without the clay touching the table. If the rod is now tipped so that it is a small angle θ away from the vertical, determine its angular acceleration at this instant. Assume that the end without the clay remains in contact with the tabletop. (*Hint:* See Table 9.2.) c) You again balance the rod on the frictionless tabletop so that it is standing vertically, but now the end of the rod *with* the clay is touching the table. If the rod is again tipped so that it is a small angle θ away from the vertical, determine its angular acceleration at this instant. Assume that the end with the clay remains in contact with the tabletop. How does this compare to the angular acceleration in part (b)? d) A pool cue is a tapered wooden rod that is thick at one end and thin at the other. You can easily balance a pool cue vertically on one finger if the thin end is in contact with your finger; this is quite a bit harder to do if the thick end is in contact with your finger. Explain why there is a difference.

10.60 You connect a light string to a point on the edge of a uniform vertical disk with radius R and mass M. The disk is free to rotate without friction about a stationary horizontal axis through its center. Initially, the disk is at rest with the string connection at the highest point on the disk. You pull the string with a constant horizontal force $\vec{F}$ until the wheel has made exactly one-quarter revolution about a horizontal axis through its center, and then let go. a) Use Eq. (10.23) to find the work done by the string. b) Use Eq. (6.14) to find the work done by the string. Do you obtain the same result as in part (a)? c) Find the final angular speed of the disk. d) Find the maximum tangential acceleration of a point on the disk. e) Find the maximum radial (centripetal) acceleration of a point on the disk.

10.61 The mechanism shown in Fig. 10.50 is used to raise a crate of supplies from a ship's hold. The crate has total mass 50 kg. A rope is wrapped around a wooden cylinder that turns on a metal axle. The cylinder has radius 0.25 m and a moment of inertia $I = 2.9\ \text{kg}\cdot\text{m}^2$ about the axle. The crate is suspended from the free end of the rope. One end of the axle pivots on frictionless bearings; a crank handle is attached to the other end. When the crank is turned, the end of the handle rotates about the axle in a vertical circle of radius 0.12 m, the cylinder turns, and the crate is raised. What magnitude of the force $\vec{F}$ applied tangentially to the rotating crank is required to raise the crate with an acceleration of 0.80 m/s^2? (You can ignore the mass of the rope as well as the moments of inertia of the axle and the crank.)

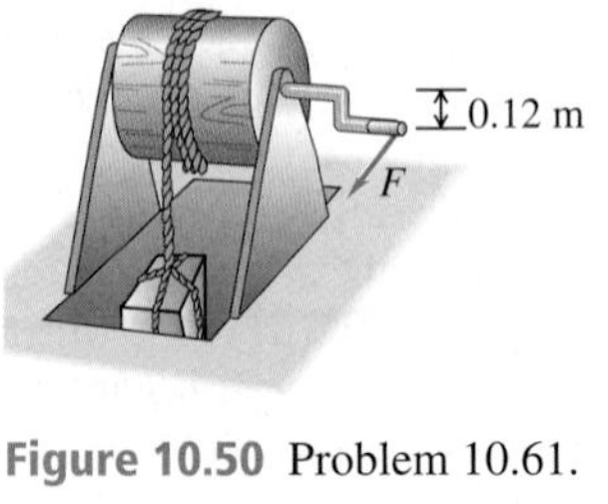

Figure 10.50 Problem 10.61.

10.62 A large 16.0-kg roll of paper with radius $R = 18.0$ cm rests against the wall and is held in place by a bracket attached to a rod through the center of the roll (Fig. 10.51). The rod turns without friction in the bracket, and the moment of inertia of the paper and rod about the axis is $0.260\ \text{kg}\cdot\text{m}^2$. The other end of the bracket is attached by a frictionless hinge to the wall such that the bracket makes an angle of 30.0° with the wall. The weight of the bracket is negligible. The coefficient of kinetic friction between the paper and the wall is $\mu_k = 0.25$. A constant vertical force $F = 40.0$ N is applied to the paper, and the paper unrolls. a) What is the magnitude of the force that the rod exerts on the paper as it unrolls? b) What is the angular acceleration of the roll?

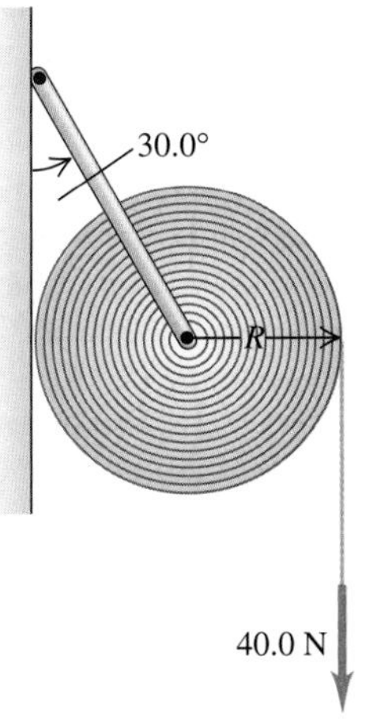

Figure 10.51 Problem 10.62.

10.63 A block with mass $m = 5.00$ kg slides down a surface inclined 36.9° to the horizontal (Fig. 10.52). The coefficient of kinetic friction is 0.25. A string attached to the block is wrapped around a flywheel on a fixed axis at O. The flywheel has mass 25.0 kg and moment of inertia $0.500\ \text{kg}\cdot\text{m}^2$ with respect to the axis of rotation. The string pulls without slipping at a perpendicular distance of 0.200 m from that axis. a) What is the acceleration of the block down the plane? b) What is the tension in the string?

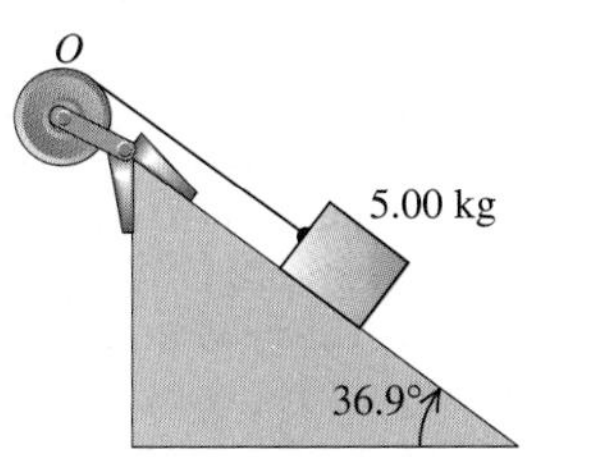

Figure 10.52 Problem 10.63.

10.64 Two metal disks, one with radius $R_1 = 2.50$ cm and mass $M_1 = 0.80$ kg and the other with radius $R_2 = 5.00$ cm and mass $M_2 = 1.60$ kg, are welded together and mounted on a frictionless axis through their common center, as in Problem 9.89. a) A light string is wrapped around the edge of the smaller disk and a 1.50-kg block is suspended from the free end of the string. What is the magnitude of the downward acceleration of the block after it is released? b) Repeat the calculation of part (a), this time with the string wrapped around the edge of the larger disk. In which case is the acceleration of the block the greatest? Does your answer make sense?

10.65 A lawn roller in the form of a thin-walled, hollow cylinder with mass M is pulled horizontally with a constant horizontal force F applied by a handle attached to the axle. If it rolls without slipping, find the acceleration and the friction force.

10.66 Atwood's Machine. Fig. 10.53 shows an Atwood's machine. Find the linear accelerations of blocks A and B, the angular acceleration of the wheel C, and the tension in each side of the cord if there is no slipping between the cord and the surface of the wheel. The masses of blocks A and B are m_A and m_B, respectively, the moment of inertia of the wheel about its axis is I, and the radius of the semicircle in which the string moves is R.

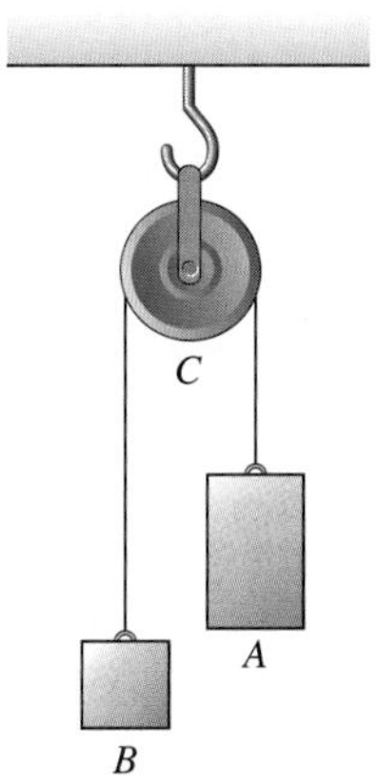

Figure 10.53 Problem 10.66.

10.67 A solid disk is rolling without slipping on a level surface at a constant speed of 2.50 m/s. a) If the disk rolls up a 30.0° ramp, how far along the ramp will it move before it stops? b) Explain why your answer to (a) does not depend on either the mass or the radius of the disk.

10.68 The Yo-yo. A yo-yo is made from two uniform disks, each with mass m and radius R, connected by a light axle of radius b. A light, thin string is wound several times around the axle and then held stationary while the yo-yo is released from rest, dropping as the string unwinds. Find the linear acceleration and angular acceleration of the yo-yo and the tension in the string.

10.69 A uniform marble of radius r starts from rest with its center of mass a height h above the lowest point of a loop-the-loop track of radius R, a track shaped like the one shown in Fig. 7.32. It rolls without slipping. Rolling friction and air resistance are negligible. a) What is the minimum value of h such that the marble will not leave the track at the top of the loop? (*Note:* The radius r is *not* negligible compared to the radius R.) b) What is the answer to part (a) if the track is well lubricated, making friction negligible?

10.70 Fig. 10.54 shows three identical yo-yos initially at rest on a horizontal surface. For each yo-yo, the string is pulled in the direction shown. In each case, there is sufficient friction for the yo-yo to roll without slipping. Draw the free-body diagram for each yo-yo. In what direction will each yo-yo rotate? (Try it!) Explain your answers.

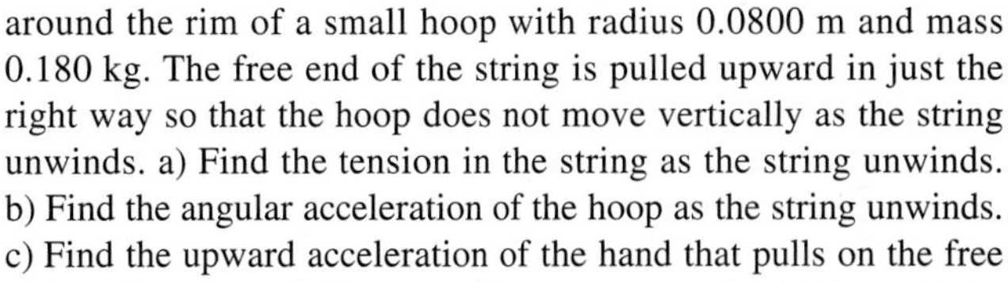

Figure 10.54 Problem 10.70.

10.71 As shown in Fig. 10.45, a string is wrapped several times around the rim of a small hoop with radius 0.0800 m and mass 0.180 kg. The free end of the string is pulled upward in just the right way so that the hoop does not move vertically as the string unwinds. a) Find the tension in the string as the string unwinds. b) Find the angular acceleration of the hoop as the string unwinds. c) Find the upward acceleration of the hand that pulls on the free end of the string. d) How would your answers be different if the hoop were replaced by a solid disk of the same mass and radius?

10.72 Starting from rest, a constant force $F = 100$ N is applied to the free end of a 50-m cable wrapped around the outer rim of a uniform solid cylinder, similar to the situation shown in Fig. 10.9(a).

The cylinder has mass 4.00 kg and diameter 30.0 cm and is free to turn about a fixed, frictionless axle through its center. a) How long does it take to unwrap all the cable, and how fast is the cable moving just as the last bit comes off? b) Now suppose that the cylinder is replaced by a uniform hoop, with all other quantities remaining unchanged. In this case, would the answers to the questions in part (a) be larger or smaller? Explain.

10.73 A uniform marble rolls without slipping down the path shown in Fig. 10.55, starting from rest. a) Find the minimum height h required for the marble not to fall into the pit. b) The moment of inertia of the marble depends on its radius. Explain why the answer to part (a) does not depend on the radius of the marble. c) Solve part (a) for a block that slides without friction instead of the rolling marble. How does the minimum h in this case compare to the answer in part (a)?

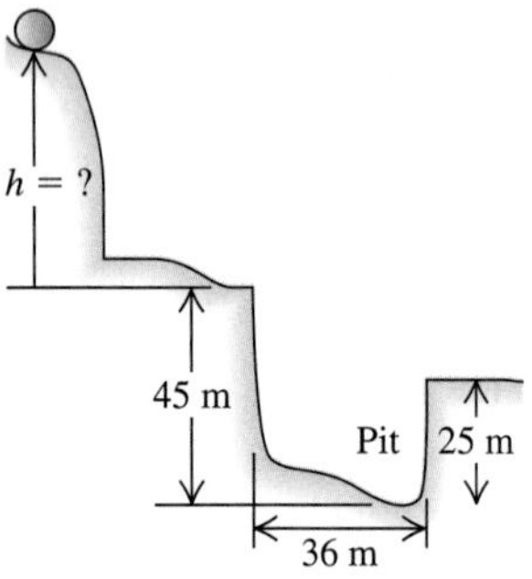

Figure 10.55 Exercise 10.73.

10.74 Rolling Stones. A solid, uniform spherical boulder starts from rest and rolls down a 50.0-m high hill, as shown in Fig. 10.56. The top half of the hill is rough enough to cause the boulder to roll without slipping, but the lower half is covered with ice and there is no friction. What is the translational speed of the boulder when it reaches the bottom of the hill?

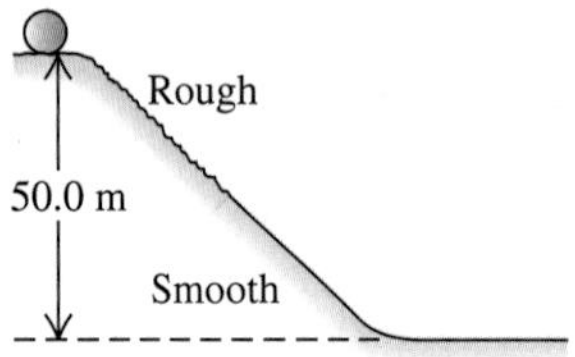

Figure 10.56 Problem 10.74.

10.75 A solid, uniform ball rolls without slipping up a hill, as shown in Fig. 10.57. At the top of the hill, it is moving horizontally, and then it goes over the vertical cliff. a) How far from the foot of the cliff does the ball land and how fast is it moving just before it lands? b) Notice that when the balls lands, it has a larger translational speed than when it was at the bottom of the hill. Does this mean that the ball somehow gained energy? Explain!

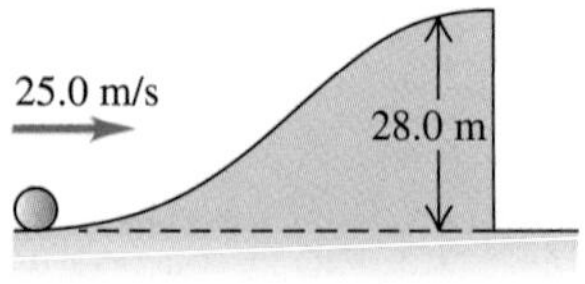

Figure 10.57 Problem 10.75.

10.76 A 42.0-cm diameter wheel, consisting of a rim and six spokes, is constructed from a thin rigid plastic material having a linear mass density of 25.0 g/cm. This wheel is released from rest at the top of a hill 58.0 m high. a) How fast is it rolling when it reaches the bottom of the hill? b) How would your answer change if the linear mass density and the diameter of the wheel were each doubled?

10.77 A high-wheel antique bicycle has a large front wheel with the foot-powered crank mounted on its axle and a small rear wheel turning independently of the front wheel; there is no chain connecting the wheels. The radius of the front wheel is 65.5 cm and the radius of the rear wheel is 22.0 cm. Your modern bike has a wheel diameter of 66.0 cm (26 inches) and front and rear sprockets with radii of 11.0 cm and 5.5 cm, respectively. The rear sprocket is rigidly attached to the axle of the rear wheel. You ride your modern bike and turn the front sprocket at 1.00 rev/s. The wheels of both bikes roll along the ground without slipping. a) What is your linear speed when you ride your modern bike? b) At what rate must you turn the crank of the antique bike in order to travel at the same speed as in part (a)? c) What then is the angular speed (in rev/s) of the small rear wheel of the antique bike?

10.78 In a lab experiment you let a uniform ball roll down a curved track. The ball starts from rest and rolls without slipping. While on the track, the ball descends a vertical distance h. The lower end of the track is horizontal and extends over the edge of the lab table; the ball leaves the track traveling horizontally. While free falling after leaving the track, the ball moves a horizontal distance x and a vertical distance y. a) Calculate x in terms of h and y, ignoring the work done by friction. b) Would the answer to part (a) be any different on the moon? c) In doing the experiment very carefully, your measured value of x is consistently a bit smaller than that calculated in part (a). Why? d) What would x be for the same h and y as in part (a) if you let a silver dollar roll down the track? You can ignore the work done by friction.

10.79 In a spring gun, a spring of force constant 400 N/m is compressed 0.15 m. When fired, 80.0% of the elastic potential energy stored in the spring is eventually converted into kinetic energy of a 0.0590-kg uniform ball that is rolling without slipping at the base of a ramp. The ball continues to roll without slipping up the ramp with 90.0% of the kinetic energy at the bottom converted into an increase in gravitational potential energy at the instant it stops. a) What is the speed of the ball's center of mass at the base of the ramp? b) At this position, what is the speed of a point at the top of the ball? c) At this position, what is the speed of a point at the bottom of the ball? d) What maximum vertical height up the ramp does the ball move?

10.80 If a wheel rolls along a horizontal surface at constant speed, the coordinates of a certain point on the rim of the wheel are $x(t) = R[(2\pi t/T) - \sin(2\pi t/T)]$ and $y(t) = R[1 - \cos(2\pi t/T)]$, where R and T are constants. a) Sketch the trajectory of the point from $t = 0$ to $t = 2T$. A curve with this shape is called a *cycloid.* b) What are the meanings of the constants R and T? c) Find the x- and y-components of the velocity and of the acceleration of the point at any time t. d) Find the times at which the point is instantaneously at rest. What are the x- and y-components of the acceleration at these times? e) Find the magnitude of the acceleration of the point. Does it depend on time? Compare to the magnitude of the acceleration of a particle in uniform circular motion, $a_{rad} = 4\pi^2 R/T^2$. Explain your result for the magnitude of the acceleration of the point on the rolling wheel, using the idea that rolling is a combination of rotational and translational motion.

10.81 A child rolls a 0.600-kg basketball up a long ramp. The basketball can be considered a thin-walled hollow sphere. When the child releases the basketball at the bottom of the ramp, it has a speed of 8.0 m/s. When the ball returns to her after rolling up the ramp and then rolling back down, it has a speed of 4.0 m/s. Assume the work done by friction on the basketball is the same when the ball moves up or down the ramp and that the basketball rolls without slipping. Find the maximum vertical height increase of the ball as it rolls up the ramp.

10.82 A wheel rotates from rest about a fixed axis through its center of mass such that $\theta = bt^3$, where b is a positive constant with units rad/s^3. a) Use Eq. (10.23) to show that the work done by the net torque on the wheel when it has turned through an angle θ is $\frac{9}{2}I_{cm}b^{2/3}\theta^{4/3}$. b) Use Eq. (9.3) to calculate the angular speed of the wheel when it has turned through an angle θ. c) Use the result of part (b) to calculate the kinetic energy of the wheel when it has turned through an angle θ. Is the work-energy theorem, Eq. (10.25), obeyed? Explain.

10.83 A uniform, solid cylinder with mass M and radius $2R$ rests on a horizontal tabletop. A string is attached by a yoke to a frictionless axle through the center of the cylinder so that the cylinder can rotate about the axle. The string runs over a disk-shaped pulley mass M and radius R that is mounted on a frictionless axle through its center. A block of mass M is suspended from the free end of the string (Fig. 10.58). The string doesn't slip over the pulley surface, and the cylinder rolls without slipping on the table top. Find the magnitude of the acceleration of the block after the system is released from rest.

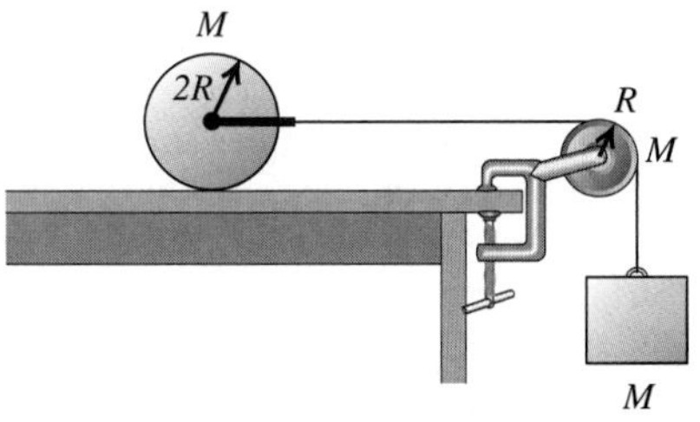

Figure 10.58 Problem 10.83.

10.84 A uniform drawbridge 8.00 m long is attached to the roadway by a frictionless hinge at one end and can be raised by a cable attached to the other end. The bridge is at rest, suspended at 60.0° above the horizontal, when the cable suddenly breaks. a) Find the angular acceleration of the drawbridge just after the cable breaks. b) Could you use Eq.(9.7) to calculate the angular speed of the drawbridge at a later time? Explain. c) What is the angular speed of the drawbridge as it becomes horizontal?

10.85 A 5.00-kg ball is dropped from a height of 12.0 m above one end of a uniform bar that pivots at its center. The bar has mass 8.00 kg and is 4.00 m in length. At the other end of the bar sits another 5.00 kg ball, unattached to the bar. The dropped ball sticks to the bar after the collision. How high will the other ball go after the collision?

10.86 A uniform, 0.0300-kg rod of length 0.400 m rotates in a horizontal plane about a fixed axis through its center and perpendicular to the rod. Two small rings, each with mass 0.0200 kg, are mounted so that they can slide along the rod. They are initially held by catches at positions 0.0500 m on each side of the center of the rod, and the system is rotating at 30.0 rev/min. Without otherwise changing the system, the catches are released, and the rings slide outward along the rod and fly off at the ends. a) What is the angular speed of the system at the instant when the rings reach the ends of the rod? b) What is the angular speed of the rod after the rings leave it?

10.87 A uniform rod of length L rests on a frictionless horizontal surface. The rod pivots about a fixed, frictionless axis at one end. The rod is initially at rest. A bullet traveling parallel to the horizontal surface and perpendicular to the rod with speed v strikes the rod at its center and becomes embedded in it. The mass of the bullet is one-fourth the mass of the rod. a) What is the final angular speed of the rod? b) What is the ratio of the kinetic energy of the system after the collision to the kinetic energy of the bullet before the collision?

10.88 The solid wood door of a gymnasium is 1.00 m wide and 2.00 m high, has total mass 35.0 kg, and is hinged along one side. The door is open and at rest when a stray basketball hits the center of the door head-on, applying an average force of 1500 N to the door for 8.00 ms. Find the angular speed of the door after the impact. (*Hint:* Integrating Eq. (10.32) yields $\Delta L_z = \int_{t_1}^{t_2}(\sum \tau_z)\,dt = (\sum \tau_z)_{av}\Delta t$. The quantity $\int_{t_1}^{t_2}(\sum \tau_z)\,dt$ is called the *angular impulse.*)

10.89 A target in a shooting gallery consists of a vertical square wooden board, 0.250 m on a side and with mass 0.750 kg, that pivots on a horizontal axis along its top edge. The board is struck face-on at its center by a bullet with mass 1.90 g that is traveling at 360 m/s and that remains embedded in the board. a) What is the angular speed of the board just after the bullet's impact? b) What maximum height above the equilibrium position does the center of the board reach before starting to swing down again? c) What minimum bullet speed would be required for the board to swing all the way over after impact?

10.90 Neutron Star Glitches. Occasionally, a rotating neutron star (Exercise 10.38) undergoes a sudden and unexpected speedup called a *glitch.* One explanation is that a glitch occurs when the crust of the neutron star settles slightly, decreasing the moment of inertia about the rotation axis. A neutron star with angular speed $\omega_0 = 70.4$ rad/s underwent such a glitch in October 1975 that increased its angular speed to $\omega = \omega_0 + \Delta\omega$, where $\Delta\omega/\omega_0 = 2.01 \times 10^{-6}$. If the radius of the neutron star before the glitch was 11 km, by how much did its radius decrease in the starquake? Assume that the neutron star is a uniform sphere.

10.91 Disks A and B are mounted on a shaft SS and may be connected or disconnected by a clutch C (Fig. 10.59). Disk A is made of a lighter material than disk B so that the moment of inertia of disk A about the shaft is one-third that of disk B. The moments of inertia of the shaft and clutch are negligible. With the clutch disconnected, A is brought up to an angular speed ω_0. The accelerating torque is then removed from A, and A is coupled to disk B by the clutch. (You can ignore bearing friction.) It is found that 2400 J of

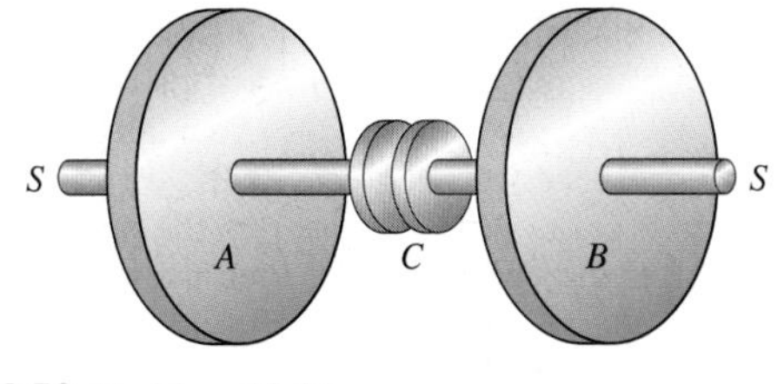

Figure 10.59 Problem 10.91.

thermal energy is developed in the clutch when the connection is made. What was the original kinetic energy of disk A?

10.92 A small block with mass 0.250 kg is attached to a string passing through a hole in a frictionless, horizontal surface (see Fig. 10.47). The block is originally revolving in a circle with a radius of 0.800 m about the hole with a tangential speed of 4.00 m/s. The string is then pulled slowly from below, shortening the radius of the circle in which the block revolves. The breaking strength of the string is 30.0 N. What is the radius of the circle when the string breaks?

10.93 A horizontal, plywood disk with mass 7.00 kg and diameter 1.00 m pivots on frictionless bearings about a vertical axis through its center. You attach a circular model-railroad track of negligible mass and average diameter 0.95 m to the disk. A 1.20-kg, battery-driven model train rests on the tracks. To demonstrate conservation of angular momentum, you switch on the train's engine. The train moves counterclockwise, soon attaining a constant speed of 0.600 m/s relative to the tracks. Find the magnitude and direction of the angular velocity of the disk relative to the earth.

10.94 One particle of mass m moves at a constant speed v in a circle of radius R that is a distance R above the xz-plane. Another particle of mass m moves the same way at the same speed in another circle of radius R that is a distance R below the xz-plane. The two particles are one-half a revolution apart, so that when one is at (x, R, z), the other is at $(-x, -R, -z)$. Thus their center of mass is at the origin but their rotation axis (the y-axis) is not an axis of symmetry. a) Sketch these two particles at the instant they are at $(R, R, 0)$ and $(-R, -R, 0)$, and show their position, velocity, and angular momentum vectors with respect to the origin. b) Show that at any time the two particles have equal angular momentum. c) What angle do $\vec{\omega}$ (the angular velocity vector of the system of two particles) and the total angular momentum of the system make with each other? d) Show that the y-component of the system's total angular momentum is constant and equal to $L_y = 2mvR$. e) What is the y-component of the total torque acting on the system? f) Find the magnitude of the net force acting on each particle and the magnitude of the total torque acting on the system. g) Show, using your sketch from part (a), the direction of the net torque on the system and that this torque rotates parallel to the xz-plane.

10.95 In a physics laboratory you do the following ballistic pendulum experiment. You shoot a ball of mass m horizontally from a spring gun with a speed v. The ball is immediately caught a distance r below a frictionless pivot by a pivoted catcher assembly of mass M. The moment of inertia of this assembly about its rotation axis through the pivot is I. The distance r is much greater than the radius of the ball. a) Use conservation of angular momentum to show that the angular speed of the ball and catcher just after the ball is caught is $\omega = mvr/(mr^2 + I)$. b) After the ball is caught, the center of mass of the ball-catcher assembly system swings up with a maximum height increase h. Use conservation of energy to show that $\omega = \sqrt{2(M + m)gh/(mr^2 + I)}$. c) Your lab partner says that linear momentum is conserved in the collision and derives the expression $mv = (m + M)V$, where V is the speed of the ball immediately after the collision. She then uses conservation of energy to derive that $V = \sqrt{2gh}$, so that $mv = (m + M)\sqrt{2gh}$. Use the results of parts (a) and (b) to show that this equation is satisfied only for the special case when r is given by $I = Mr^2$.

10.96 A 55-kg runner runs around the edge of a horizontal turntable mounted on a vertical, frictionless axis through its center. The runner's velocity relative to the earth has magnitude 2.8 m/s. The turntable is rotating in the opposite direction with an angular velocity of magnitude 0.20 rad/s relative to the earth. The radius of the turntable is 3.0 m, and its moment of inertia about the axis of rotation is 80 kg·m^2. Find the final angular velocity of the system if the runner comes to rest relative to the turntable. (You can treat the runner as a particle.)

10.97 A Falling Bicycle. The moment of inertia of the front wheel of a bicycle about its axle is 0.085 kg·m^2, its radius is 0.33 m, and the forward speed of the bicycle is 6.00 m/s. With what angular velocity must the front wheel be turned about a vertical axis to counteract the capsizing torque due to a mass 50.0 kg located 0.040 m horizontally to the right or left of the line of contact of wheels and ground? (Bicycle riders: Compare your own experience and decide whether your answer seems reasonable.)

10.98 Center of Percussion. A baseball bat rests on a frictionless, horizontal surface. The bat has a length of 0.900 m, a mass of 0.800 kg, and its center of mass is 0.600 m from the handle end of the bat (Fig. 10.60). The moment of inertia of the bat about its center of mass is 0.0530 kg·m^2. The bat is struck by a baseball traveling perpendicular to the bat. The impact applies an impulse $J = \int_{t_1}^{t_2} F\,dt$ at a point a distance x from the handle end of the bat. What must x be so that the handle end of the bat remains at rest as the bat begins to move? (*Hint:* Consider the motion of the center of mass and the rotation about the center of mass. Find x so that these two motions combine to give $v = 0$ for the end of the bat just after the collision. Also, note that integration of Eq. (10.32) gives $\Delta L = \int_{t_1}^{t_2} (\sum \tau)\,dt$ (Problem 10.88).) The point on the bat you have located is called the *center of percussion*. Hitting a pitched ball at the center of percussion of the bat minimizes the "sting" the batter experiences on the hands.

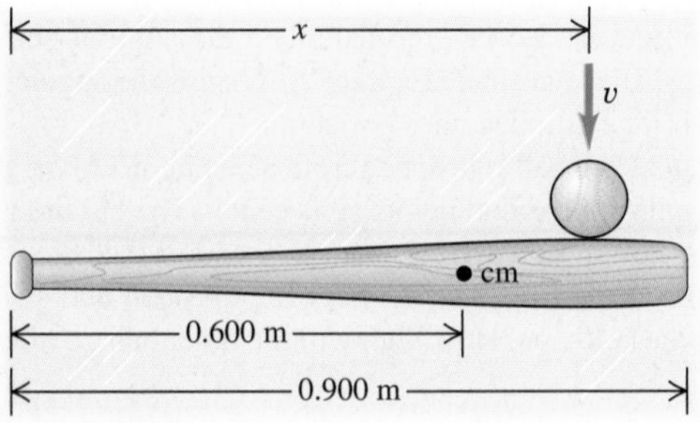

Figure 10.60 Problem 10.98.

10.99 Consider a gyroscope with an axis that is not horizontal, but is inclined from the horizontal by an angle β. Show that the precession angular frequency does not depend on the value of β, but is given by Eq. (10.36).

Challenge Problems

10.100 A uniform ball of radius R rolls without slipping between two rails such that the horizontal distance is d between the two contact points of the rails to the ball. a) In a sketch, show that at any instant $v_{cm} = \omega\sqrt{R^2 - d^2/4}$. Discuss this expression in the limits $d = 0$ and $d = 2R$. b) For a uniform ball starting from rest and descending a vertical distance h while rolling without slipping down a ramp, $v_{cm} = \sqrt{10gh/7}$. Replacing the ramp with the two rails, show that

$$v_{cm} = \sqrt{\frac{10gh}{5 + 2/(1 - d^2/4R^2)}}$$

In each case, the work done by friction has been ignored. c) Which speed in part (b) is smaller? Why? Answer in terms of how the loss of potential energy is shared between the gain in translational and rotational kinetic energies. d) For which value of the ratio d/R do the two expressions for the speed in part (b) differ by 5.0%? By 0.50%?

10.101 When an object is rolling without slipping, the rolling friction force is much less than the friction force when the object is sliding; a silver dollar will roll on its edge much farther than it will slide on its flat side (see Section 5.3). When an object is rolling without slipping on a horizontal surface, we can approximate the friction force to be zero, so that a_x and α_z are approximately zero and v_x and ω_z are approximately constant. Rolling without slipping means $v_x = r\omega_z$ and $a_x = r\alpha_z$. If an object is set into motion on a surface *without* these equalities, sliding (kinetic) friction will act on the object as it slips until rolling without slipping is established. A solid cylinder with mass M and radius R, rotating with angular speed ω_0 about an axis through its center, is set on a horizontal surface for which the kinetic friction coefficient is μ_k. a) Draw a free-body diagram for the cylinder on the surface. Think carefully about the direction of the kinetic friction force on the cylinder. Calculate the accelerations a_x of the center of mass and α_z of rotation about the center of mass. b) The cylinder is initially slipping completely, so initially $\omega_z = \omega_0$ but $v_x = 0$. Rolling without slipping sets in when $v_x = R\omega_z$. Calculate the *distance* the cylinder rolls before slipping stops. c) Calculate the work done by the friction force on the cylinder as it moves from where it was set down to where it begins to roll without slipping.

10.102 A demonstration gyroscope wheel is constructed by removing the tire from a bicycle wheel 0.650 m in diameter, wrapping lead wire around the rim, and taping it in place. The shaft projects 0.200 m at each side of the wheel, and a woman holds the ends of the shaft in her hands. The mass of the system is 8.00 kg; its entire mass may be assumed to be located at its rim. The shaft is horizontal and the wheel is spinning about the shaft at 5.00 rev/s. Find the magnitude and direction of the force each hand exerts on the shaft a) when the shaft is at rest; b) when the shaft is rotating in a horizontal plane about its center at 0.050 rev/s; c) when the shaft is rotating in a horizontal plane about its center at 0.300 rev/s. d) At what rate must the shaft rotate in order that it may be supported at one end only?

10.103 A block with mass m is revolving with linear speed v_1 in a circle of radius r_1 on a frictionless horizontal surface (see Fig. 10.47). The string is slowly pulled from below until the radius of the circle in which the block is revolving is reduced to r_2. a) Calculate the tension T in the string as a function of r, the distance of the block from the hole. Your answer will be in terms of the initial velocity v_1 and the radius r_1. b) Use $W = \int_{r_1}^{r_2} \vec{T}(r) \cdot d\vec{r}$ to calculate the work done by $\vec{T}$ when r changes from r_1 to r_2. c) Compare the results of part (b) to the change in the kinetic energy of the block.

11 EQUILIBRIUM AND ELASTICITY

Flamingos often stand with one foot tucked up against the body, which keeps that foot warm and conserves body heat. To remain in equilibrium, a standing flamingo must keep its center of gravity—the point where the flamingo's weight can be considered to act—over the foot that touches the ground.

? Flamingos have broad, webbed feet. How does this enhance the bird's stability when standing on one leg?

We've devoted a good deal of effort to understanding why and how bodies accelerate in response to the forces that act on them. But very often we're interested in making sure that bodies *don't* accelerate. Any building, from the Petronas Towers in Malaysia (the world's tallest) to the humblest shed, must be designed so that it won't topple over. Similar concerns occur with a suspension bridge, a ladder leaning against a wall, or a crane hoisting a bucket full of concrete.

A body that can be modeled as a *particle* is in equilibrium whenever the vector sum of the forces acting on it is zero. But for the situations we've just described, that condition isn't enough. If forces act at different points on an extended body, an additional requirement must be satisfied to ensure that the body has no tendency to *rotate:* the sum of the *torques* about any point must be zero. This requirement is based on the principles of rotational dynamics developed in Chapter 10. We can compute the torque due to the weight of a body using the concept of center of mass from Section 8.5 and the related concept of center of gravity, which we introduce in this chapter.

Rigid bodies don't bend, stretch, or squash when forces act on them. But the rigid body is an idealization; all real materials are *elastic* and do deform to some extent. Elastic properties of materials are tremendously important. You want the wings of an airplane to be able to bend a little, but you'd rather not have them break off. The steel frame of an earthquake-resistant building has to be able to flex, but not too much. Many of the necessities of everyday life, from rubber bands to suspension bridges, depend on the elastic properties of materials. In this chapter we'll introduce the concepts of *stress, strain,* and *elastic modulus* and a simple principle called *Hooke's law* that helps us predict what deformations will occur when forces are applied to a real (not perfectly rigid) body.

11.1 | Conditions For Equilibrium

We learned in Sections 4.2 and 5.1 that a particle is in equilibrium—that is, the particle does not accelerate—in an inertial frame of reference if the vector sum of all the forces acting on the particle is zero, $\sum \vec{\boldsymbol{F}} = \mathbf{0}$. The equivalent statement for an *extended* body is that the center of mass of the body has zero acceleration if the vector sum of all external forces acting on the body is zero, as discussed in Section 8.5. This is often called the **first condition for equilibrium.** In vector and component forms,

$$\sum \vec{F} = \mathbf{0}$$
$$\sum F_x = 0 \qquad \sum F_y = 0 \qquad \sum F_z = 0 \qquad (11.1)$$
$$\text{(first condition for equilibrium)}$$

where the sum includes *external* forces only.

A second condition for an extended body to be in equilibrium is that the body must have no tendency to *rotate.* This condition is based on the dynamics of rotational motion in exactly the same way that the first condition is based on Newton's first law. A rigid body that, in an inertial frame, is not rotating about a certain point has zero angular momentum about that point, $\vec{\boldsymbol{L}} = \mathbf{0}$. If it is not to start rotating about that point, the rate of change of angular momentum $d\vec{\boldsymbol{L}}/dt$ must *also* be zero. From the discussion in Section 10.5, particularly Eq. (10.32), this means that the sum of torques $\sum \vec{\tau}$ due to all the external forces acting on the body must be zero. A rigid body in equilibrium can't have any tendency to start rotating about *any* point, so the sum of external torques must be zero about any point. This is the **second condition for equilibrium:**

$$\sum \vec{\tau} = \mathbf{0} \quad \text{about any point} \qquad \text{(second condition for equilibrium)} \qquad (11.2)$$

The sum of the torques due to all external forces acting on the body, with respect to any specified point, must be zero.

In this chapter we will apply the first and second conditions for equilibrium to situations in which a rigid body is at rest (no translation or rotation). Such a body is said to be in **static equilibrium.** But the same conditions apply to a rigid body in uniform *translational* motion (without rotation), such as an airplane in flight with constant speed, direction, and altitude. Such a body is in equilibrium but is not static.

Test Your Understanding

Which of the following situations satisfy both the first and second conditions for equilibrium: (a) a seagull gliding at a constant angle below the horizontal and at a constant speed; (b) an automobile crankshaft turning at an increasing angular speed in the engine of a parked car; (c) a thrown baseball that does not rotate as it sails through the air?

11.2 | Center of Gravity

In most equilibrium problems, one of the forces acting on the body is its weight. We need to be able to calculate the *torque* of this force. The weight doesn't act at a single point; it is distributed over the entire body. But we can always calculate the torque due to the body's weight by assuming that the entire force of gravity (weight) is concentrated at a point called the **center of gravity** (abbreviated "cg"). The acceleration due to gravity $\vec{g}$ decreases with altitude; but if we can ignore this variation over the vertical dimension of the body, then the body's center of gravity is identical with its *center of mass,* which we defined in Section 8.5. We stated this result without proof in Section 10.2, and now we'll prove it.

First let's review the definition of the center of mass. For a collection of particles with masses $m_1, m_2, \ldots$ and coordinates $(x_1, y_1, z_1), (x_2, y_2, z_2), \ldots$, the coordinates x_{cm}, y_{cm}, and z_{cm} of the center of mass are given by

$$x_{cm} = \frac{m_1x_1 + m_2x_2 + m_3x_3 + \cdots}{m_1 + m_2 + m_3 + \cdots} = \frac{\sum_i m_ix_i}{\sum_i m_i}$$

$$y_{cm} = \frac{m_1y_1 + m_2y_2 + m_3y_3 + \cdots}{m_1 + m_2 + m_3 + \cdots} = \frac{\sum_i m_iy_i}{\sum_i m_i}$$

$$z_{cm} = \frac{m_1z_1 + m_2z_2 + m_3z_3 + \cdots}{m_1 + m_2 + m_3 + \cdots} = \frac{\sum_i m_iz_i}{\sum_i m_i} \quad \text{(center of mass)} \quad (11.3)$$

Also, x_{cm}, y_{cm}, and z_{cm} are the components of the position vector $\vec{r}_{cm}$ of the center of mass, so Eqs. (11.3) are equivalent to the vector equation

$$\vec{r}_{cm} = \frac{m_1\vec{r}_1 + m_2\vec{r}_2 + \cdots}{m_1 + m_2 + \cdots} = \frac{\sum_i m_i\vec{r}_i}{\sum_i m_i} \quad (11.4)$$

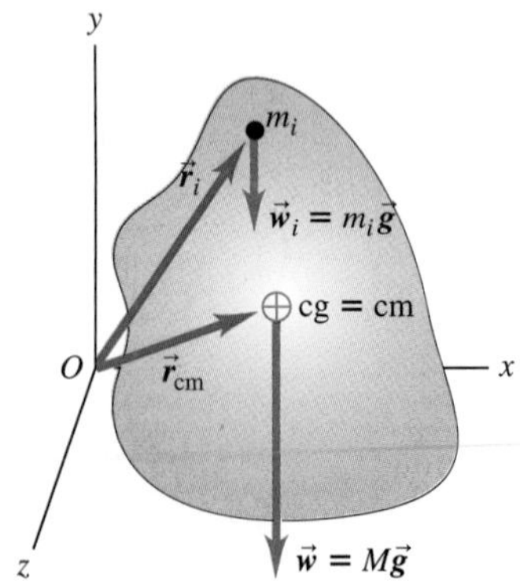

11.1 The gravitational torque about any point can be found by assuming that all the weight of the body acts at its center of gravity (cg). The cg is identical to the center of mass (cm) if $\vec{g}$ is the same at all points on the body.

Now let's consider the gravitational torque on a body of arbitrary shape (Fig. 11.1). We assume that the acceleration due to gravity $\vec{g}$ has the same magnitude and direction at every point in the body. Every particle in the body experiences a gravitational force, and the total weight of the body is the vector sum of a large number of parallel forces. A typical particle has mass m_i and weight $\vec{w}_i = m_i\vec{g}$. If $\vec{r}_i$ is the position vector of this particle with respect to an arbitrary origin O, the torque vector $\vec{\tau}_i$ of the weight $\vec{w}_i$ with respect to O is, from Eq. (10.3),

$$\vec{\tau}_i = \vec{r}_i \times \vec{w}_i = \vec{r}_i \times m_i\vec{g}$$

The *total* torque due to the gravitational forces on all the particles is

$$\begin{aligned}\vec{\tau} &= \sum_i \vec{\tau} = \vec{r}_1 \times m_1\vec{g} + \vec{r}_2 \times m_2\vec{g} + \cdots \\ &= (m_1\vec{r}_1 + m_2\vec{r}_2 + \cdots) \times \vec{g} \\ &= \left(\sum_i m_i\vec{r}_i\right) \times \vec{g}\end{aligned}$$

When we multiply and divide this by the total mass of the body,

$$M = m_1 + m_2 + \cdots = \sum_i m_i$$

we get

$$\vec{\tau} = \frac{m_1\vec{r}_1 + m_2\vec{r}_2 + \cdots}{m_1 + m_2 + \cdots} \times M\vec{g} = \frac{\sum_i m_i\vec{r}_i}{\sum_i m_i} \times M\vec{g}$$

The fraction in this equation is just the position vector $\vec{r}_{cm}$ of the center of mass, with components x_{cm}, y_{cm}, and z_{cm}, as given by Eq. (11.4), and $M\vec{g}$ is equal to the total weight $\vec{w}$ of the body. Thus

$$\vec{\tau} = \vec{r}_{cm} \times M\vec{g} = \vec{r}_{cm} \times \vec{w} \tag{11.5}$$

The total gravitational torque, given by Eq. (11.5), is the same as though the total weight $\vec{w}$ were acting on the position $\vec{r}_{cm}$ of the center of mass, which we also call the *center of gravity.* **If $\vec{g}$ has the same value at all points on a body, its center of gravity is identical to its center of mass.** Note, however, that the center of mass is defined independently of any gravitational effect.

While the value of $\vec{g}$ does vary somewhat with elevation, the variation is extremely slight (Fig. 11.2). Hence we will assume throughout this chapter that the center of gravity and center of mass are identical unless explicitly stated otherwise.

11.2 The acceleration due to gravity at the bottom of the 452-m-tall Petronas Towers in Malaysia is only 0.014% greater than at the top. The center of gravity of the towers is only about 2 cm below the center of mass.

Finding and Using the Center of Gravity

We can often use symmetry considerations to locate the center of gravity of a body, just as we did for center of mass. The center of gravity of a homogeneous sphere, cube, circular sheet, or rectangular plate is at its geometric center. The center of gravity of a right circular cylinder or cone is on its axis of symmetry.

For a body of more complex shape, we can sometimes locate the center of gravity by thinking of the body as being made of symmetrical pieces. For example, we could approximate the human body as a collection of solid cylinders, with a sphere for the head. Then we can compute the coordinates of the center of gravity of the combination from Eqs. (11.3), letting $m_1, m_2, \ldots$ be the masses of the individual pieces and $(x_1, y_1, z_1), (x_2, y_2, z_2), \ldots$ be the coordinates of their centers of gravity.

When a body acted on by gravity is supported or suspended at a single point, the center of gravity is always at or directly above or below the point of suspension. If it were anywhere else, the weight would have a torque with respect to the point of suspension, and the body could not be in rotational equilibrium. This fact can be used to determine experimentally the location of the center of gravity of an irregular body. To see how this works, hold one corner of a piece of paper gently between your thumb and forefinger so that the paper hangs below your hand. Draw a line on the paper that extends vertically downward from the point of suspension; the center of gravity must lie somewhere along this line. Then hold the paper by a different corner and draw a line vertically downward from the new point of suspension. The point where the lines intersect is the paper's center of gravity. (Try this with a coffee cup suspended first from the upper part of its handle, then from the lower part of its handle.)

Using the same reasoning, we can see that a body supported at several points must have its center of gravity somewhere within the area bounded by the sup-

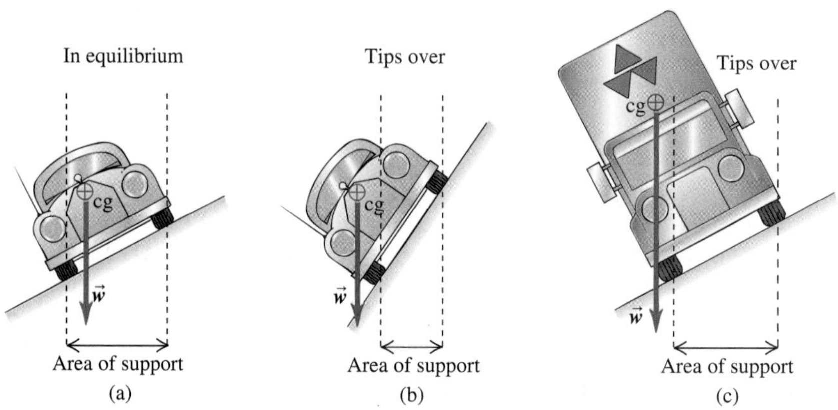

11.3 In (a) the center of gravity is within the area bounded by the supports, and the car is in equilibrium. The car in (b) and the truck in (c) will tip over because their centers of gravity lie outside the area of support.

ActivPhysics ONLINE

7.2 A Tilted Beam: Torques and Equilibrium

7.3 Arm Levers

ports. This explains why a car can drive on a straight but slanted road if the slant angle is relatively small (Fig. 11.3a) but will tip over if the angle is too great (Fig. 11.3b). The truck in Fig. 11.3c has a higher center of gravity than the car and will tip over on a shallower incline; when a truck overturns on a highway and blocks traffic for hours, it's the high center of gravity that's to blame.

The lower the center of gravity and the larger the area of support, the more difficult it is to overturn a body. Four-legged animals such as deer and horses have a large area of support bounded by their legs; hence they are naturally stable and need only small feet or hooves. Animals that walk erect on two legs, such as humans and birds, need relatively large feet to give them a reasonable area of support. If a two-legged animal holds its body approximately horizontal, like a chicken or the dinosaur *Tyrannosaurus rex,* it must perform a delicate balancing act as it walks to keep its center of gravity over the foot that is on the ground. A chicken does this by moving its head; *T. rex* probably did it by moving its massive tail.

Example 11.1 Walking the plank

A uniform wooden plank of length $L = 6.0$ m and mass $M = 90$ kg rests on top of two sawhorses separated by $D = 1.5$ m, located equal distances from the center of the plank (Fig. 11.4). Your cousin Throckmorton tries to stand on the right-hand end of the plank. If the plank is to remain at rest, how massive can Throckmorton be?

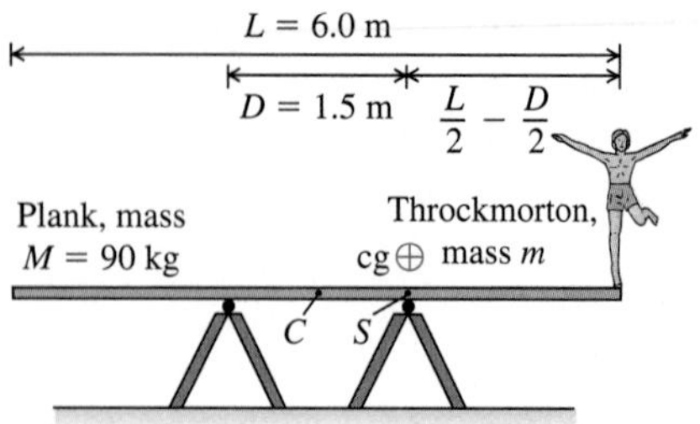

11.4 Throckmorton and the plank, just barely in equilibrium.

SOLUTION

IDENTIFY: If the system of plank and Throckmorton is just in balance, the center of gravity of this system will be directly over the right-hand sawhorse (just barely within the area bounded by the two supports). The target variable is Throcky's mass.

SET UP: We take the origin at C, the geometrical center and center of gravity of the uniform plank, and take the positive x-axis to point horizontally to the right. Then the x-coordinates of the centers of gravity of the plank (mass M) and Throcky (unknown mass m) are $x_P = 0$ and $x_T = l/2 = 3.0$ m, respectively. We will use Eqs. (11.3) to locate the center of gravity of the system of plank and Throcky.

EXECUTE: From the first of Eqs. (11.3),

$$x_{cg} = \frac{M(0) + m(L/2)}{M + m} = \frac{m}{M + m}\frac{L}{2}$$

Setting this equal to $D/2$, the x-coordinate of the right-hand sawhorse, we have

$$\frac{m}{M+m}\frac{L}{2} = \frac{D}{2}$$

$$mL = (M+m)D$$

$$m = M\frac{D}{L-D} = (90\text{ kg})\frac{1.5\text{ m}}{6.0\text{ m} - 1.5\text{ m}}$$

$$= 30\text{ kg}$$

EVALUATE: To check our result, let's repeat the calculation with a different choice of origin. Now we take the origin to be at S, the position of the right-hand sawhorse, so that $x_{cg} = 0$. The centers of gravity of the plank and Throcky are now at $x_P = -D/2$ and $x_T = (L/2) - (D/2)$ respectively, so

$$x_{cg} = \frac{M(-D/2) + m[(L/2) - (D/2)]}{M+m} = 0$$

$$m = \frac{MD/2}{(L/2) - (D/2)} = M\frac{D}{L-D} = 30\text{ kg}$$

The mass doesn't depend on our arbitrary choice of origin.

A 60-kg child could only stand halfway between the right-hand sawhorse and the end of the plank. Can you see why?

Test Your Understanding

You are constructing a mobile with two wooden fish on one side of a lightweight rod and a counterweight on the other side (Fig. 11.5). What must be the mass m of the counterweight if the rod is to remain horizontal?

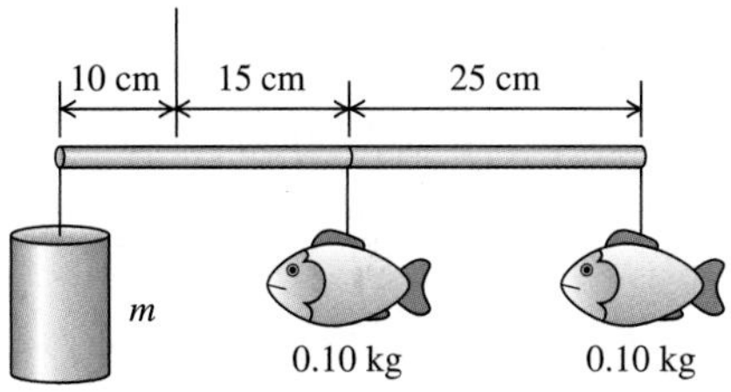

11.5 What is the unknown mass m?

11.3 | Solving Rigid-Body Equilibrium Problems

There are just two key principles of rigid-body equilibrium: the vector sum of the forces on the body must be zero, and the sum of torques about any point must be zero. To keep things simple, we'll restrict our attention to situations in which we can treat all forces as acting in a single plane, which we'll call the xy-plane. Then we can ignore the condition $\sum F_z = 0$ in Eq. (11.1), and in Eq. (11.2) we need consider only the z-components of torque (perpendicular to the plane). The first and second conditions for equilibrium are then

$$\sum F_x = 0 \quad \text{and} \quad \sum F_y = 0$$
(first condition for equilibrium, forces in xy-plane)

$$\sum \tau_z = 0 \qquad (11.6)$$
(second condition for equilibrium, forces in xy-plane)

ActivPhysics Online

7.4 Two Painters on a Beam

7.5 Lecturing from a Beam

CAUTION Although the choice of reference point for calculating torques in $\sum \tau_z$ is arbitrary, once you choose a point you must use the *same* point to calculate *all* of the torques on a body. It helps to pick the point so as to simplify the calculations as much as possible.

The challenge is to apply these simple principles to specific problems. The following strategy is very similar to the suggestions in Section 5.2 for equilibrium of a particle. You should compare it with the strategy for rotational dynamics problems given in Section 10.2.

Problem-Solving Strategy

Equilibrium of a Rigid Body

IDENTIFY *the relevant concepts:* The first and second conditions for equilibrium are useful whenever there is a rigid body that is not rotating and not accelerating in space.

SET UP *the problem using the following steps:*

1. Draw a sketch of the physical situation, including dimensions, and select the body in equilibrium to be analyzed.

2. Draw a free-body diagram showing the forces acting *on* the selected body and no others. *Do not* include forces exerted *by* this body on other bodies. Be careful to show correctly the point at which each force acts; this is crucial for correct torque calculations. You can't represent a rigid body as a point.
3. Choose coordinate axes and specify a positive sense of rotation for torques. Represent forces in terms of their components with respect to the axes you have chosen; when you do this, cross out the original force so that you don't include it twice.
4. In choosing a point to compute torques, note that if a force has a line of action that goes *through* a particular point, the torque of the force with respect to that point is zero. You can often eliminate unknown forces or components from the torque equation by a clever choice of point for your calculation. The body doesn't actually have to be pivoted about an axis through the chosen point.

EXECUTE *the solution as follows:*

1. Write equations expressing the equilibrium conditions. Remember that $\Sigma F_x = 0$, $\Sigma F_y = 0$, and $\Sigma \tau_z = 0$ are always separate equations; *never* add x- and y-components in a single equation. Also remember that when a force is represented in terms of its components, you can compute the torque of that force by finding the torque of each component separately, each with its appropriate lever arm and sign, and adding the results. This is often easier than determining the lever arm of the original force.
2. You always need as many equations as you have unknowns. Depending on the number of unknowns, you may need to compute torques with respect to two or more axes to obtain enough equations. Often, there are several equally good sets of force and torque equations for a particular problem; there is usually no single "right" combination of equations. When you have as many independent equations as unknowns, you can solve the equations simultaneously.

EVALUATE *your answer:* A useful way to check your results is to rewrite the second condition for equilibrium, $\Sigma \tau_z = 0$, using a different choice of origin. If you've done everything correctly, you'll get the same answers using this new choice of origin as you did with your original choice.

Example 11.2 Weight distribution for a car

An auto magazine reports that a certain sports car (Fig. 11.6a) has 53% of its weight on the front wheels and 47% on its rear wheels, with a 2.46-m wheelbase. This means that the total normal force on the front wheels is $0.53w$ and that on the rear wheels is $0.47w$, where w is the total weight. The wheelbase is the distance between front and rear axles. How far in front of the rear axle is the car's center of gravity?

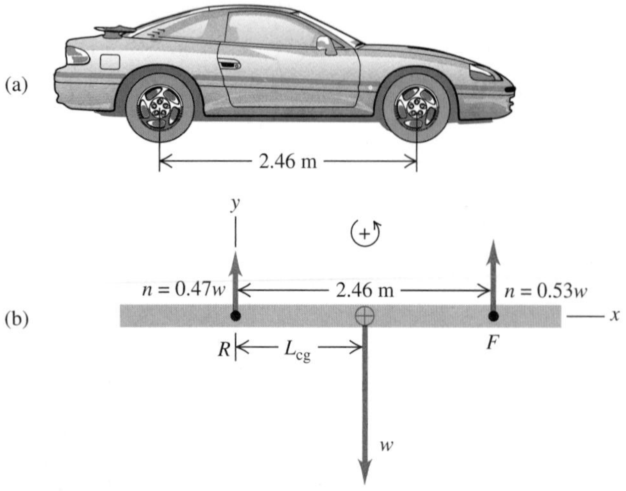

11.6 (a) The car at rest on a horizontal surface. (b) Free-body diagram for the car.

SOLUTION

IDENTIFY: We can use the two conditions for equilibrium since the car is assumed to be at rest. The conditions also apply when the car is driving in a straight line at constant speed, since the net force and net torque on the car are also zero in that situation. The target variable is the coordinate of the car's center of gravity.

SET UP: Figure 11.6 shows a free-body diagram for the car, along with the x- and y-axes and our convention that counterclockwise torques are positive. We have drawn the weight w as acting at the center of gravity, and the distance we want is L_{cg}. It is the lever arm of the weight with respect to the rear axle R, so it is reasonable to take torques with respect to R. Note that the torque due to the weight is negative because it tends to cause a clockwise rotation about R. The torque due to the upward normal force at the front wheels is positive because it tends to cause a counterclockwise rotation about R.

EXECUTE: You can see from Fig. 11.6b that the first condition for equilibrium is satisfied: $\Sigma F_x = 0$ because there aren't any x-components of force and $\Sigma F_y = 0$ because $0.47w + 0.53w + (-w) = 0$. The force equation doesn't involve the target variable L_{cg}, so we must solve for it using the torque equation for point R:

$$\sum \tau_R = 0.47w(0) - wL_{cg} + 0.53w(2.46\text{ m}) = 0$$
$$L_{cg} = 1.30\text{ m}$$

EVALUATE: Note that the cg is between the two supports, as it must be (see Section 11.2). You can check the numerical result for the cg position by writing the torque equation about the front axle F. You'll find that the cg is 1.16 m behind the front axle, or $(2.46\text{ m}) - (1.16\text{ m}) = 1.30\text{ m}$ in front of the rear axle.

Can you show that if f is the fraction of the weight on the front wheels and d is the wheelbase, then the center of gravity is a distance fd in front of the rear wheels? In terms of the center of gravity, why do owners of rear-wheel-drive vehicles put bags of sand in their trunks to improve traction on snow and ice? Would this strategy help with a front-wheel-drive car? With a four-wheel-drive car?

Example 11.3 A heroic rescue

Sir Lancelot is trying to rescue the Lady Elayne from Castle Von Doom by climbing a uniform ladder that is 5.0 m long and weighs 180 N. Lancelot, who weighs 800 N, stops a third of the way up the ladder (Fig. 11.7a). The bottom of the ladder rests on a horizontal stone ledge and leans across the moat in equilibrium against a vertical wall that is frictionless because of a thick layer of moss. The ladder makes an angle of 53.1° with the horizontal, conveniently forming a 3–4–5 right triangle. a) Find the normal and friction forces on the ladder at its base. b) Find the minimum coefficient of static friction needed to prevent slipping at the base. c) Find the magnitude and direction of the contact force on the ladder at the base.

SOLUTION

IDENTIFY: The system of ladder and Lancelot is stationary, so we can use the two conditions for equilibrium. These conditions by themselves are enough to solve part (a). In part (b), we also need the relationship given in Section 5.3 between the static friction force, the coefficient of static friction, and the normal force. The contact force asked for in part (c) is the vector sum of the normal force and the friction force acting at the base of the ladder, which we find in part (a).

SET UP: Figure 11.7b shows the free-body diagram for the system of the ladder and Lancelot. We choose the x- and y-directions as shown and take counterclockwise torques to be positive. The ladder is described as "uniform," so its center of gravity is at its geometrical center. Lancelot's weight acts at a point on the ladder one-third of the way from the base toward the wall.

The frictionless wall exerts only a normal force n_1 at the top of the ladder. The forces at the base are the upward normal force n_2 and the static friction force f_s, which must point to the right to prevent slipping; the magnitudes n_2 and f_s are the target variables in part (a). From Eq. (5.6), these magnitudes are related by $f_s \le \mu_s n_2$, where μ_s is the coefficient of static friction (the target variable in part (b)).

EXECUTE: a) From Eqs. (11.6), the first condition for equilibrium gives

$$\sum F_x = f_s + (-n_1) = 0$$
$$\sum F_y = n_2 + (-800\text{ N}) + (-180\text{ N}) = 0$$

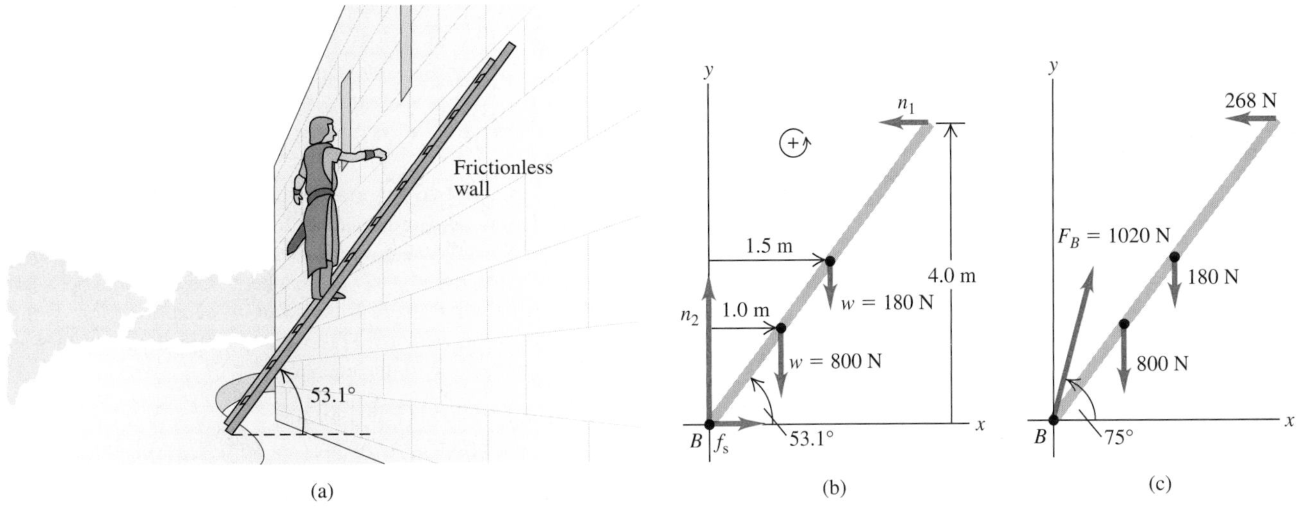

11.7 (a) Sir Lancelot pauses a third of the way up the ladder, fearing it will slip. (b) Free-body diagram for the system of Sir Lancelot and the ladder. (c) The contact force at B is the superposition of the normal force and the static friction force.

These are two equations for the three unknowns n_1, n_2, and f_s. The first equation tells us that the two horizontal forces must be equal and opposite, and the second equation gives

$$n_2 = 980\ \text{N}$$

The ground pushes up with a force of 980 N to balance the total (downward) weight (800 N + 180 N).

We don't yet have enough equations, but now we can use the second condition for equilibrium. We can take torques about any point we choose. The smart choice is the point that will give us the fewest terms and fewest unknowns in the torque equation. With this in mind, we choose point B, at the base of the ladder. The two forces n_2 and f_s have no torque about that point. From Fig. 11.7b we see that the lever arm for the ladder's weight is 1.5 m, the lever arm for Lancelot's weight is 1.0 m, and the lever arm for n_1 is 4.0 m. The torque equation for point B is

$$\sum \tau_B = n_1(4.0\ \text{m}) - (180\ \text{N})(1.5\ \text{m}) - (800\ \text{N})(1.0\ \text{m}) + n_2(0) + f_s(0) = 0$$

Solving for n_1, we get $n_1 = 268$ N. We now substitute this back into the $\sum F_x = 0$ equation to get

$$f_s = 268\ \text{N}$$

b) The static friction force f_s cannot exceed $\mu_s n_2$, so the *minimum* coefficient of static friction to prevent slipping is

$$(\mu_s)_{\text{min}} = \frac{f_s}{n_2} = \frac{268\ \text{N}}{980\ \text{N}} = 0.27$$

c) The components of the contact force $\vec{F}_B$ at the base are the static friction force f_s and the normal force n_2, so

$$\vec{F}_B = f_s\hat{\imath} + n_2\hat{\jmath} = (268\ \text{N})\hat{\imath} + (980\ \text{N})\hat{\jmath}$$

The magnitude and direction of $\vec{F}_B$ (Fig. 11.7c) are then

$$F_B = \sqrt{(268\ \text{N})^2 + (980\ \text{N})^2} = 1020\ \text{N}$$

$$\theta = \arctan\frac{980\ \text{N}}{268\ \text{N}} = 75°$$

EVALUATE: As Fig. 11.7c shows, the contact force $\vec{F}_B$ is *not* directed along the length of the ladder. You may be surprised by this, but there's really no good reason why the two directions should be the same. Can you show that if $\vec{F}_B$ were directed along the ladder, there would be a net counterclockwise torque with respect to the top of the ladder, and equilibrium would be impossible?

Here are a few final comments. First, as Lancelot climbs higher on the ladder, the lever arm and torque of his weight about B increase; this increases the values of n_1, f_s, and $(\mu_s)_{\text{min}}$. At the top, his lever arm would be nearly 3 m, giving a minimum coefficient of static friction of nearly 0.7. The value of μ_s would not be this large for a present-day aluminum ladder on a wood floor, which is why such ladders are usually equipped with nonslip rubber pads. Lancelot's medieval ladder is unlikely to have this feature, and the ladder is likely to slip as he climbs.

Second, a larger ladder angle would decrease the lever arms with respect to B of the weights of the ladder and Lancelot and increase the lever arm of n_1, all of which would decrease the required friction force. The R. D. Werner Ladder Co. recommends that its ladders be used at an angle of 75°. (Why not 90°?)

Finally, if we had assumed friction on the wall as well as on the floor, the problem would be impossible to solve by using the equilibrium conditions alone. (Try it!) Such a problem is said to be *statically indeterminate.* The difficulty is that it's no longer adequate to treat the body as being perfectly rigid. Another simple example of such a problem is a four-legged table; there is no way to use the equilibrium conditions alone to find the force on each separate leg. In a sense, four legs are one too many; three, properly located, are sufficient for stability.

Example 11.4 Another rescue attempt

After Lancelot falls into the moat, Sir Gawain tosses a grappling hook through an open upstairs window. In Fig. 11.8a he is resting partway up the rope. Gawain weighs 700 N, his body makes an angle of 60° with the wall, and his center of gravity is 0.85 m from his feet. The rope force acts 1.30 m from his feet, and the rope makes an angle of 20° with the vertical. Find the tension in the rope and the forces exerted on his feet by the moss-free region of the wall.

SOLUTION

IDENTIFY and SET UP: We use the conditions for equilibrium as in the previous example. Figure 11.8b shows the free-body diagram for Gawain. The forces on his feet include a (horizontal) normal component n and a (vertical) frictional component f. The rope tension is T. The angles needed to find the torques are shown.

EXECUTE: Because two unknown forces (n and f) act at point B at Gawain's feet, we take torques about B, obtaining a torque equation in which T is the only unknown. The easiest way to find the torque of each force is to use $\tau = rF \sin\phi$, where ϕ is the angle between the position vector $\vec{r}$ (from B to the point of application of the force) and the force vector $\vec{F}$. Taking counterclockwise torques as positive, we obtain the torque equation

$$\sum \tau_B = (1.30\ \text{m})T(\sin 140°) - (0.85\ \text{m})(700\ \text{N})(\sin 60°) + (0)n + (0)f = 0$$

We solve this equation for T; the result is

$$T = 617\ \text{N}$$

Note that the rope tension is *less than* Gawain's weight.

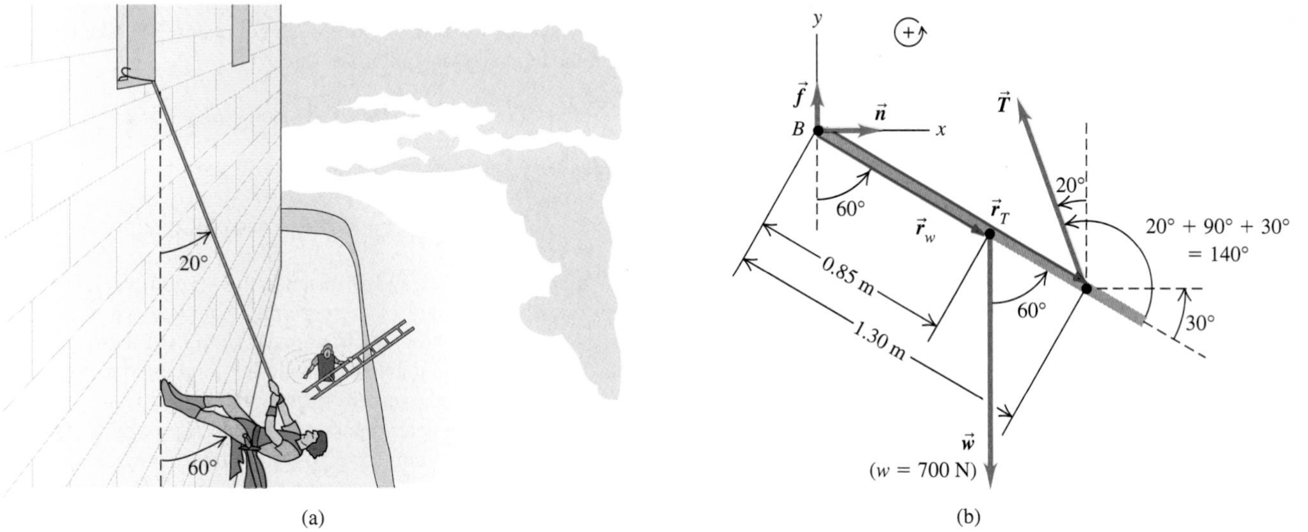

11.8 (a) Gawain pausing to catch his breath. (b) Free-body diagram for Gawain, considered as a rod.

To find the components of force n and f at Gawain's feet, we use the first equilibrium condition. From Fig. 11.8b, T_x is negative and T_y is positive, so $T_x = -T \sin 20°$ and $T_y = T \cos 20°$. Then

$$\sum F_x = n + (-T \sin 20°) = 0$$
$$\sum F_y = f + T \cos 20° + (-700 \text{ N}) = 0$$

When we substitute the value of T into these equations, we get

$$n = 211 \text{ N} \quad \text{and} \quad f = 120 \text{ N}$$

EVALUATE: As a check, note that the sum of the vertical friction force $(f = 120 \text{ N})$ and the vertical component of the tension $(T \cos 20° = 580 \text{ N})$ equals Gawain's 700-N weight.

Example 11.5 Equilibrium and pumping iron

Figure 11.9a shows a human arm lifting a dumbbell, and Fig. 11.9b is a free-body diagram for the forearm. The forearm is in equilibrium under the action of the weight w of the dumbbell, the tension T in the tendon connected to the biceps muscle, and the force E exerted on the forearm by the upper arm at the elbow joint. For clarity the point A where the tendon is attached is drawn farther away from the elbow than its actual position. The weight w and the angle θ are given; we want to find the tendon tension and the two components of force at the elbow (three unknown scalar quantities in all). We neglect the weight of the forearm itself.

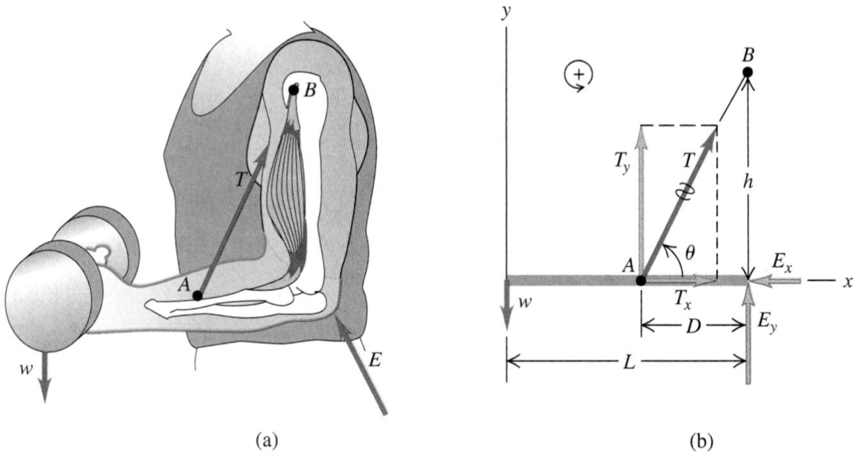

11.9 (a) Building up the biceps. (b) Free-body diagram for the forearm. The weight of the forearm is neglected, and the distance D is greatly exaggerated for clarity.

SOLUTION

IDENTIFY and SET UP: The system is at rest, so once again we use the conditions for equilibrium. As shown in Fig. 11.9b, we represent the tendon force in terms of its components T_x and T_y, using the given angle θ and the unknown magnitude T:

$$T_x = T\cos\theta \qquad T_y = T\sin\theta$$

We also represent the force at the elbow in terms of its components E_x and E_y. We'll guess that the directions of these components are as shown in Fig. 11.9b; there's no need to agonize over this guess, since the results for E_x and E_y will tell us the actual directions. Our target variables are the magnitude T of the tendon tension and the components E_x and E_y of the force at the elbow.

EXECUTE: The simplest way to find the tension T is to take torques about the elbow joint. The resulting torque equation does not contain E_x, E_y, or T_x because the lines of action of all these forces pass through this point. The torque equation is then simply

$$\sum\tau_E = Lw - DT_y = 0$$

From this we find

$$T_y = \frac{Lw}{D} \quad \text{and} \quad T = \frac{Lw}{D\sin\theta}$$

To find E_x and E_y, we use the first conditions for equilibrium, $\sum F_x = 0$ and $\sum F_y = 0$:

$$\sum F_x = T_x + (-E_x) = 0$$

$$E_x = T_x = T\cos\theta = \frac{Lw}{D\sin\theta}\cos\theta = \frac{Lw}{D}\cot\theta$$

$$= \frac{Lw}{D}\frac{D}{h} = \frac{Lw}{h}$$

$$\sum F_y = T_y + E_y + (-w) = 0$$

$$E_y = w - \frac{Lw}{D} = -\frac{(L-D)w}{D}$$

The negative sign shows that our guess for the direction of E_y, shown in Fig. 11.9b, was wrong; it is actually vertically *downward.*

EVALUATE: We can check our results by finding E_x and E_y in a different way that uses two more torque equations. We take torques about the tendon attach point, A:

$$\sum\tau_A = (L-D)w + DE_y = 0 \quad \text{and} \quad E_y = -\frac{(L-D)w}{D}$$

Finally, we take torques about point B in the figure:

$$\sum\tau_B = Lw - hE_x = 0 \quad \text{and} \quad E_x = \frac{Lw}{h}$$

We chose points A and B because the tendon tension T has zero torque about either of these points. (Can you see why from Fig. 11.9b?) Notice how much we have simplified these calculations by choosing the point for calculating torques so as to eliminate one or more of the unknown quantities.

In our alternative determination of E_x and E_y, we didn't explicitly use the first condition for equilibrium (that the vector sum of the forces is zero). As a check, compute $\sum F_x$ and $\sum F_y$ to verify that they really *are* zero. Consistency checks are always a good idea!

As a specific example, suppose $w = 200\ \text{N}$, $D = 0.050\ \text{m}$, $L = 0.30\ \text{m}$, and $\theta = 80°$. Then from $\tan\theta = h/D$, we find

$$h = D\tan\theta = (0.050\ \text{m})(5.67) = 0.28\ \text{m}$$

From the previous general results we find

$$T = \frac{Lw}{D\sin\theta} = \frac{(0.30\ \text{m})(200\ \text{N})}{(0.050\ \text{m})(0.98)} = 1220\ \text{N}$$

$$E_y = -\frac{(L-D)w}{D} = -\frac{(0.30\ \text{m} - 0.050\ \text{m})(200\ \text{N})}{0.050\ \text{m}}$$

$$= -1000\ \text{N}$$

$$E_x = \frac{Lw}{h} = \frac{(0.30\ \text{m})(200\ \text{N})}{0.28\ \text{m}} = 210\ \text{N}$$

The magnitude of the force at the elbow is

$$E = \sqrt{E_x^{\,2} + E_y^{\,2}} = 1020\ \text{N}$$

In view of the magnitudes of our results, neglecting the weight of the forearm itself, which may be 20 N or so, will cause only relatively small errors in our results.

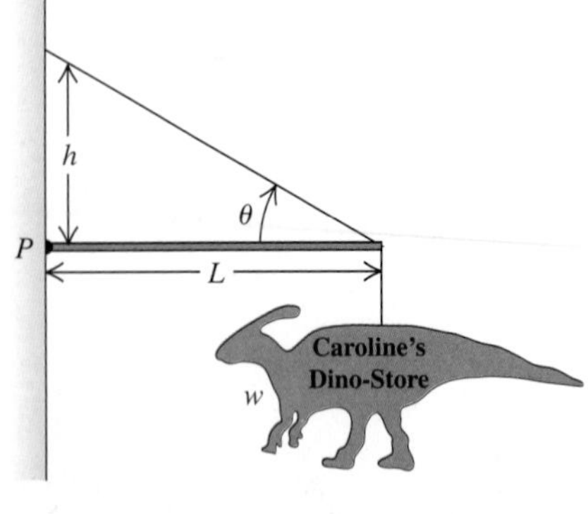

11.10 What is the tension in the diagonal wire and the force exerted by the hinge at P?

Test Your Understanding

A metal advertising sign (weight w) for a specialty shop is suspended from the end of a horizontal rod of length L and negligible mass (Fig. 11.10). The rod is supported by a cable at an angle θ from the horizontal and by a hinge at point P. Find the tension in the cable and the components of force exerted on the rod by the hinge at P.

11.4 | Stress, Strain, and Elastic Moduli

The rigid body is a useful idealized model, but the stretching, squeezing, and twisting of real bodies when forces are applied are often too important to ignore. Figure 11.11 shows three examples. We want to study the relation between the forces and deformations for each case.

(a) Tensile stress

(b) Bulk stress

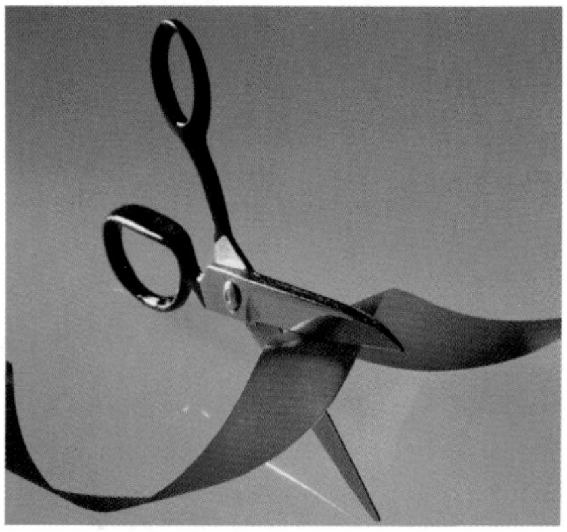

(c) Shear stress

11.11 Three types of stress. (a) A cable under *tensile stress,* being stretched by forces acting at its ends. (b) A dolphin under *bulk stress,* being squeezed from all sides by forces due to water pressure. (c) A ribbon under *shear stress,* being deformed and eventually cut by forces exerted by the scissors.

For each kind of deformation we will introduce a quantity called **stress** that characterizes the strength of the forces causing the deformation, on a "force per unit area" basis. Another quantity, **strain,** describes the resulting deformation. When the stress and strain are small enough, we often find that the two are directly proportional, and we call the proportionality constant an **elastic modulus.** The harder you pull on something, the more it stretches; and the more you squeeze it, the more it compresses. The general pattern that emerges can be formulated as

$$\frac{\text{Stress}}{\text{Strain}} = \text{Elastic modulus} \qquad \text{(Hooke's law)} \tag{11.7}$$

The proportionality of stress and strain (under certain conditions) is called **Hooke's law,** after Robert Hooke (1635–1703), a contemporary of Newton. We used one form of Hooke's law in Sections 6.3 and 7.2: the elongation of an ideal spring is proportional to the stretching force. Remember that Hooke's law is not really a general law but an experimental finding that is valid only over a limited range. The last section of this chapter discusses what this limited range is.

Tensile and Compressive Stress and Strain

The simplest elastic behavior to understand is the stretching of a bar, rod, or wire when its ends are pulled (Fig. 11.11a). Figure 11.12 shows an object that initially has uniform cross-sectional area A and length l_0. We then apply forces of equal magnitude $F_\perp$ but opposite directions at the ends (this ensures that the object has no tendency to move left or right). We say that the object is in **tension.** We've already talked a lot about tension in ropes and strings; it's the same concept here. The subscript $\perp$ is a reminder that the forces act perpendicular to the cross section.

We define the **tensile stress** at the cross section as the ratio of the force $F_\perp$ to the cross-sectional area A:

$$\text{Tensile stress} = \frac{F_\perp}{A} \tag{11.8}$$

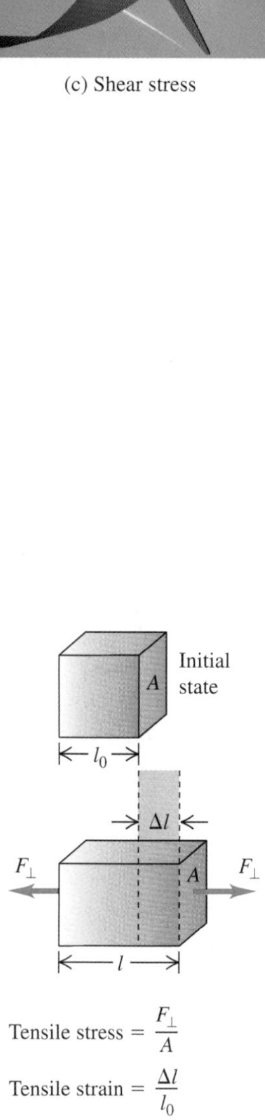

11.12 An object in tension. The net force on the object is zero, but the object deforms. The tensile stress (the ratio of the force to the cross-sectional area) produces a tensile strain (the elongation divided by the initial length). The elongation Δl is exaggerated for clarity.

This is a *scalar* quantity because $F_\perp$ is the *magnitude* of the force. The SI unit of stress is the **pascal** (abbreviated Pa, and named for the 17th-century French scientist and philosopher Blaise Pascal). Equation (11.8) shows that one pascal equals one newton per square meter (N/m^2):

$$1 \text{ pascal} = 1 \text{ Pa} = 1 \text{ N/m}^2$$

In the British system the logical unit of stress would be the pound per square foot, but the pound per square inch (lb/in.^{2} or psi) is more commonly used. The conversion factors are

$$1 \text{ psi} = 6895 \text{ Pa} \quad \text{and} \quad 1 \text{ Pa} = 1.450 \times 10^{-4} \text{ psi}$$

The units of stress are the same as those of *pressure,* which we will encounter often in later chapters. Air pressure in automobile tires is typically around 3×10^5 Pa $= 300$ kPa, and steel cables are commonly required to withstand tensile stresses of the order of 10^8 Pa.

The object shown in Fig. 11.12 stretches to a length $l = l_0 + \Delta l$ when under tension. The elongation Δl does not occur only at the ends; every part of the bar stretches in the same proportion. The **tensile strain** of the object is equal to the fractional change in length, which is the ratio of the elongation Δl to the original length l_0:

$$\text{Tensile strain} = \frac{l - l_0}{l_0} = \frac{\Delta l}{l_0} \tag{11.9}$$

Tensile strain is stretch per unit length. It is a ratio of two lengths, always measured in the same units, and so is a pure (dimensionless) number with no units.

Experiment shows that for a sufficiently small tensile stress, stress and strain are proportional, as in Eq. (11.7). The corresponding elastic modulus is called **Young's modulus,** denoted by Y:

$$Y = \frac{\text{Tensile stress}}{\text{Tensile strain}} = \frac{F_\perp / A}{\Delta l / l_0} = \frac{F_\perp}{A}\frac{l_0}{\Delta l} \qquad \text{(Young's modulus)} \tag{11.10}$$

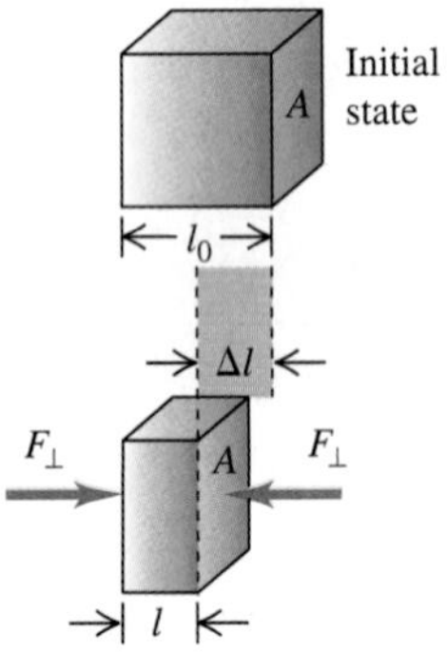

$$\text{Compressive stress} = \frac{F_\perp}{A}$$

$$\text{Compressive strain} = \frac{\Delta l}{l_0}$$

11.13 An object in compression. The compressive stress and compressive strain are defined in the same way as tensile stress and strain (see Fig. 11.12), except that Δl now denotes the distance that the object contracts.

Since strain is a pure number, the units of Young's modulus are the same as those of stress: force per unit area. Some typical values are listed in Table 11.1. (This table also gives values of two other elastic moduli that we will discuss later in this chapter.) A material with a large value of Y is relatively unstretchable; a large stress is required for a given strain. For example, the value of Y for cast steel $(2 \times 10^{11} \text{ Pa})$ is much larger than that for rubber $(5 \times 10^8 \text{ Pa})$.

When the forces on the ends of a bar are pushes rather than pulls (Fig. 11.13), the bar is in **compression** and the stress is a **compressive stress.** The **compres-**

Table 11.1 Approximate Elastic Moduli

Material	Young's Modulus, Y (Pa)	Bulk Modulus, B (Pa)	Shear Modulus, S (Pa)
Aluminum	7.0×10^{10}	7.5×10^{10}	2.5×10^{10}
Brass	9.0×10^{10}	6.0×10^{10}	3.5×10^{10}
Copper	11×10^{10}	14×10^{10}	4.4×10^{10}
Crown glass	6.0×10^{10}	5.0×10^{10}	2.5×10^{10}
Iron	21×10^{10}	16×10^{10}	7.7×10^{10}
Lead	1.6×10^{10}	4.1×10^{10}	0.6×10^{10}
Nickel	21×10^{10}	17×10^{10}	7.8×10^{10}
Steel	20×10^{10}	16×10^{10}	7.5×10^{10}

sive strain of an object in compression is defined in the same way as tensile strain, but Δl has the opposite direction. Hooke's law and Eq. (11.10) are valid for compression as well as tension if the compressive stress is not too great. For many materials, Young's modulus has the same value for both tensile and compressive stresses; composite materials such as concrete are an exception.

In many situations, bodies can experience both tensile and compressive stresses at the same time. As an example, a horizontal beam supported at each end sags under its own weight. As a result, the top of the beam is under compression, while the bottom of the beam is under tension (Fig. 11.14a). To minimize the stress and hence the bending strain, the top and bottom of the beam are given a large cross-sectional area. There is neither compression nor tension along the centerline of the beam, so this part can have a small cross section; this helps to keep the weight of the bar to a minimum and further helps to reduce the stress. The result is an I-beam of the familiar shape used in building construction (Fig. 11.14b).

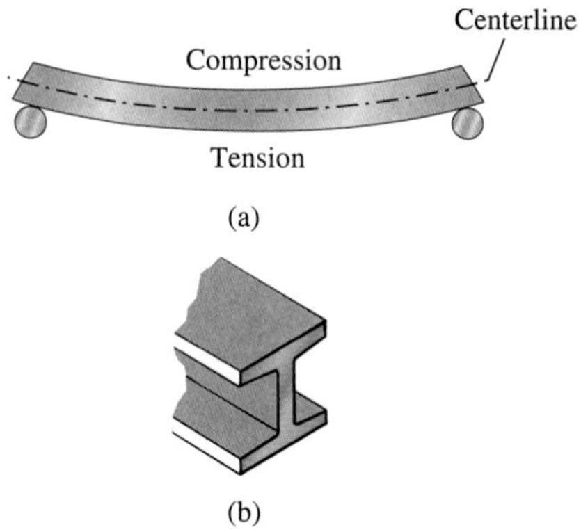

11.14 (a) A beam supported at both ends is under both compression and tension. (b) The cross-sectional shape of an I-beam minimizes both stress and weight.

Example 11.6 Tensile stress and strain

A steel rod 2.0 m long has a cross-sectional area of 0.30 cm^2. The rod is now hung by one end from a support structure and a 550-kg milling machine is hung from the rod's lower end. Determine the stress, the strain, and the elongation of the rod.

SOLUTION

IDENTIFY and SET UP: We use the definitions of stress, strain, and Young's modulus given by Eqs. (11.8), (11.9), and (11.10), respectively, and the value of Y for steel from Table 11.1.

EXECUTE: We find

$$\text{Stress} = \frac{F_\perp}{A} = \frac{(550\text{ kg})(9.8\text{ m/s}^2)}{3.0 \times 10^{-5}\text{ m}^2} = 1.8 \times 10^8\text{ Pa}$$

$$\text{Strain} = \frac{\Delta l}{l_0} = \frac{\text{Stress}}{Y} = \frac{1.8 \times 10^8\text{ Pa}}{20 \times 10^{10}\text{ Pa}} = 9.0 \times 10^{-4}$$

$$\text{Elongation} = \Delta l = (\text{Strain}) \times l_0 = (9.0 \times 10^{-4})(2.0\text{ m}) = 0.0018\text{ m} = 1.8\text{ mm}$$

EVALUATE: The small size of this elongation, which results from a load of over half a ton, is a testament to the stiffness of steel.

Bulk Stress and Strain

When a dolphin plunges deep into the ocean, the water exerts nearly uniform pressure everywhere on its surface and squeezes it to a slightly smaller volume (Fig. 11.11b). This is a different situation from the tensile and compressive stresses and strains we have discussed. The stress is now a uniform pressure on all sides, and the resulting deformation is a volume change. We use the terms **bulk stress** (or **volume stress**) and **bulk strain** (or **volume strain**) to describe these quantities.

If an object is immersed in a fluid (liquid or gas) at rest, the fluid exerts a force on any part of the object's surface; this force is *perpendicular* to the surface. (If we tried to make the fluid exert a force parallel to the surface, the fluid would slip sideways to counteract the effort.) The force $F_\perp$ per unit area that the fluid exerts on the surface of an immersed object is called the **pressure** p in the fluid:

$$p = \frac{F_\perp}{A} \quad \text{(pressure in a fluid)} \tag{11.11}$$

The pressure in a fluid increases with depth. For example, the pressure of the air is about 21% greater at sea level than in Denver (at an elevation of 1.6 km, or 1.0 mi). If an immersed object is relatively small, however, we can ignore pressure differences due to depth for the purpose of calculating bulk stress. Hence we will treat the pressure as having the same value at all points on an immersed object's surface.

Pressure has the same units as stress; commonly used units include 1 Pa $(=1\ \text{N/m}^2)$ and 1 lb/in.^{2} (1 psi). Also in common use is the **atmosphere,** abbreviated atm. One atmosphere is the approximate average pressure of the earth's atmosphere at sea level:

$$1\ \text{atmosphere} = 1\ \text{atm} = 1.013 \times 10^5\ \text{Pa} = 14.7\ \text{lb/in.}^2$$

CAUTION Unlike force, pressure has no intrinsic direction: the pressure on the surface of an immersed object is the same no matter how the surface is oriented. Hence pressure is a *scalar* quantity, not a vector quantity.

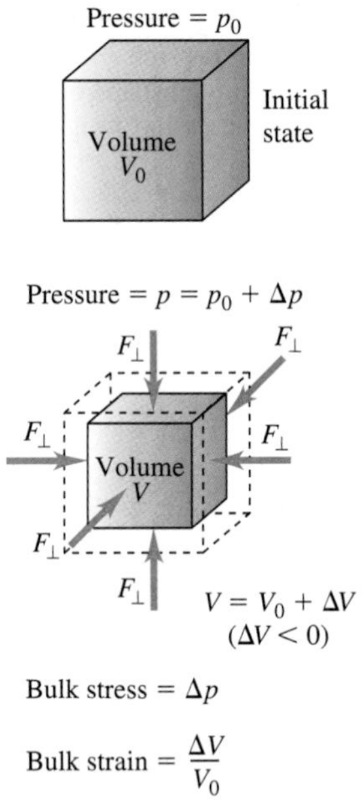

11.15 An object under bulk stress. Without the stress, the cube has volume V_0; when the stress is applied, the cube has a smaller volume V. The volume change ΔV is exaggerated for clarity.

Pressure plays the role of stress in a volume deformation. The corresponding strain is the fractional change in volume (Fig. 11.15), that is, the ratio of the volume change ΔV to the original volume V_0:

$$\text{Bulk (volume) strain} = \frac{\Delta V}{V_0} \tag{11.12}$$

Volume strain is the change in volume per unit volume. Like tensile or compressive strain, it is a pure number, without units.

When Hooke's law is obeyed, an increase in pressure (bulk stress) produces a *proportional* bulk strain (fractional change in volume). The corresponding elastic modulus (ratio of stress to strain) is called the **bulk modulus,** denoted by B. When the pressure on a body changes by a small amount Δp, from p_0 to $p_0 + \Delta p$, and the resulting bulk strain is $\Delta V/V_0$, Hooke's law takes the form

$$B = \frac{\text{Bulk stress}}{\text{Bulk strain}} = -\frac{\Delta p}{\Delta V/V_0} \qquad \text{(bulk modulus)} \tag{11.13}$$

We include a minus sign in this equation because an *increase* of pressure always causes a *decrease* in volume. In other words, if Δp is positive, ΔV is negative. The bulk modulus B itself is a positive quantity.

For small pressure changes in a solid or a liquid, we consider B to be constant. The bulk modulus of a *gas,* however, depends on the initial pressure p_0. Table 11.1 includes values of the bulk modulus for several solid materials. Its units, force per unit area, are the same as those of pressure (and of tensile or compressive stress).

The reciprocal of the bulk modulus is called the **compressibility** and is denoted by k. From Eq. (11.13),

$$k = \frac{1}{B} = -\frac{\Delta V/V_0}{\Delta p} = -\frac{1}{V_0}\frac{\Delta V}{\Delta p} \qquad \text{(compressibility)} \tag{11.14}$$

Compressibility is the fractional decrease in volume, $-\Delta V/V_0$, per unit increase Δp in pressure. The units of compressibility are those of *reciprocal pressure,* Pa^{-1} or atm^{-1}.

Values of compressibility k for several liquids are listed in Table 11.2. From this table the compressibility of water is $46.4 \times 10^{-6}\ \text{atm}^{-1}$. This means that for

Table 11.2 Compressibilities of Liquids

	Compressibility, k	
Liquid	Pa^{-1}	atm^{-1}
Carbon disulfide	93×10^{-11}	94×10^{-6}
Ethyl alcohol	110×10^{-11}	111×10^{-6}
Glycerine	21×10^{-11}	21×10^{-6}
Mercury	3.7×10^{-11}	3.8×10^{-6}
Water	45.8×10^{-11}	46.4×10^{-6}

each atmosphere increase in pressure, the volume of water decreases by 46.4 parts per million. Materials with small bulk modulus and large compressibility are easier to compress.

Example 11.7 Bulk stress and strain

A hydraulic press contains $0.25\ \text{m}^3$ (250 L) of oil. Find the decrease in volume of the oil when it is subjected to a pressure increase $\Delta p = 1.6 \times 10^7$ Pa (about 160 atm or 2300 psi). The bulk modulus of the oil is $B = 5.0 \times 10^9$ Pa (about 5.0×10^4 atm), and its compressibility is $k = 1/B = 20 \times 10^{-6}\ \text{atm}^{-1}$.

SOLUTION

IDENTIFY and SET UP: We are given both the bulk modulus and the compressibility, so we can use either Eq. (11.13) or Eq. (11.14) to find the volume change ΔV.

EXECUTE: Solving Eq. (11.13) for ΔV, we find

$$\Delta V = -\frac{V_0\,\Delta p}{B} = -\frac{(0.25\ \text{m}^3)(1.6 \times 10^7\ \text{Pa})}{5.0 \times 10^9\ \text{Pa}}$$
$$= -8.0 \times 10^{-4}\ \text{m}^3 = -0.80\ \text{L}$$

Alternatively, we can use Eq. (11.14). Solving for ΔV and using the approximate unit conversions given above, we get

$$\Delta V = -kV_0\,\Delta p = -(20 \times 10^{-6}\ \text{atm}^{-1})(0.25\ \text{m}^3)(160\ \text{atm})$$
$$= -8.0 \times 10^{-4}\ \text{m}^3$$

EVALUATE: We get the same result for ΔV with either approach, as we should. Note that ΔV is negative, indicating that the volume decreases when the pressure increases. Note that even though the pressure increase is very large, the *fractional* change in volume is very small:

$$\Delta V/V_0 = (-8.0 \times 10^{-4}\ \text{m}^3)/(0.25\ \text{m}^3) = -0.0032,\ \text{or}\ -0.32\%$$

Shear Stress and Strain

The third kind of stress-strain situation is called *shear.* The ribbon in Fig. 11.11c is under **shear stress:** one part of the ribbon is being pushed up while an adjacent part is being pushed down, producing a deformation of the ribbon. Figure 11.16 shows a body being deformed by a shear stress. In the figure, forces of equal magnitude but opposite direction act *tangent* to the surfaces of opposite ends of the object. We define the shear stress as the force $F_{\parallel}$ acting tangent to the surface, divided by the area A on which it acts:

$$\text{Shear stress} = \frac{F_{\parallel}}{A} \qquad (11.15)$$

Shear stress, like the other two types of stress, is a force per unit area.

Figure 11.16 shows that one face of the object under shear stress is displaced by a distance x relative to the opposite face. We define **shear strain** as the ratio of the displacement x to the transverse dimension h:

$$\text{Shear strain} = \frac{x}{h} \qquad (11.16)$$

In real-life situations, x is nearly always much smaller than h. Like all strains, shear strain is a dimensionless number; it is a ratio of two lengths.

If the forces are small enough that Hooke's law is obeyed, the shear strain is *proportional* to the shear stress. The corresponding elastic modulus (ratio of shear stress to shear strain) is called the **shear modulus,** denoted by S:

$$S = \frac{\text{Shear stress}}{\text{Shear strain}} = \frac{F_{\parallel}/A}{x/h} = \frac{F_{\parallel}}{A}\frac{h}{x} \qquad \text{(shear modulus)} \qquad (11.17)$$

with x and h defined as in Fig. 11.16.

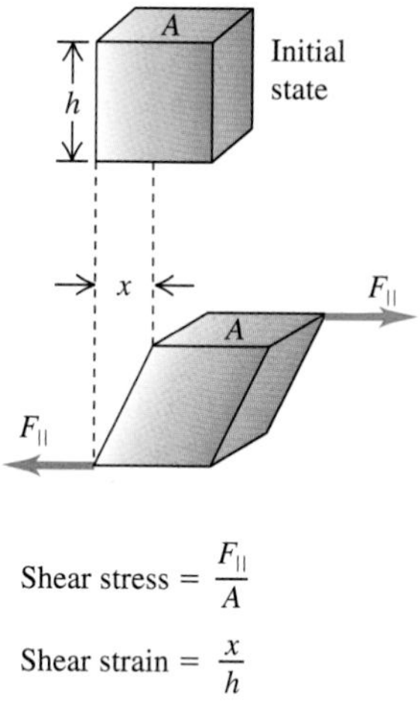

11.16 An object under shear stress. Forces are applied tangent to opposite surfaces of the object (in contrast to the situation in Fig. 11.12, in which the forces act perpendicular to the surfaces). The deformation x is exaggerated for clarity.

Several values of shear modulus are given in Table 11.1. For a given material, S is usually one-third to one-half as large as Young's modulus Y for tensile stress. Keep in mind that the concepts of shear stress, shear strain, and shear modulus apply to *solid* materials only. The reason is that the shear forces in Fig. 11.16 are required to deform the solid block, and the block tends to return to its original shape if the shear forces are removed. By contrast, gases and liquids do not have a definite shape.

Example 11.8

Shear stress and strain

Suppose the object in Fig. 11.16 is the brass base plate of an outdoor sculpture; it experiences shear forces as a result of an earthquake. The frame is 0.80 m square and 0.50 cm thick. How large a force must be exerted on each of its edges if the displacement x (see Fig. 11.16b) is 0.16 mm?

SOLUTION

IDENTIFY and SET UP: We first find the shear strain using Eq. (11.16), then determine the shear stress using Eq. (11.17). We can then solve for the target variable $F_{\parallel}$ using Eq. (11.15). The values of all the other quantities are given, including the shear modulus of brass (from Table 11.1, $S = 3.5 \times 10^{10}$ Pa). Note that h in Fig. 11.16 represents the 0.80-m length of each side of the square plate, while the area A is the product of the 0.80-m length and the 0.50-cm thickness.

EXECUTE: The shear strain is

$$\text{Shear strain} = \frac{x}{h} = \frac{1.6 \times 10^{-4}\ \text{m}}{0.80\ \text{m}} = 2.0 \times 10^{-4}$$

From Eq. (11.17) the shear stress equals the shear strain multiplied by the shear modulus S,

$$\begin{aligned}\text{Stress} &= (\text{Shear strain}) \times S \\ &= (2.0 \times 10^{-4})(3.5 \times 10^{10}\ \text{Pa}) = 7.0 \times 10^{6}\ \text{Pa}\end{aligned}$$

From Eq. (11.15), the force at each edge is the shear stress multiplied by the area of the edge:

$$\begin{aligned}F_{\parallel} &= (\text{Shear stress}) \times A \\ &= (7.0 \times 10^{6}\ \text{Pa})(0.80\ \text{m})(0.0050\ \text{m}) = 2.8 \times 10^{4}\ \text{N}\end{aligned}$$

EVALUATE: The required force is more than three tons! Brass has a large shear modulus, which means that it's intrinsically difficult to deform. Furthermore, the plate is relatively thick (0.50 cm), so the area A is relatively large and a large force $F_{\parallel}$ is needed to provide the necessary stress $F_{\parallel}/A$.

Test Your Understanding

Which has greater tensile strain: a copper rod of cross-sectional area 0.500 cm^2 and length 1.00 m that is elongated by 2.00×10^{-2} mm, or a steel rod of the same cross-sectional area but 0.100 m in length and elongated by 2.00×10^{-3} mm? Which rod is under greater tensile stress?

11.5 | Elasticity and Plasticity

Hooke's law—the proportionality of stress and strain in elastic deformations—has a limited range of validity. In the preceding section we used phrases such as "provided that the forces are small enough that Hooke's law is obeyed." Just what *are* the limitations of Hooke's law? We know that if you pull, squeeze, or twist *anything* hard enough, it will bend or break. Can we be more precise than that?

Let's look at tensile stress and strain again. Suppose we plot a graph of stress as a function of strain. If Hooke's law is obeyed, the graph is a straight line with a slope equal to Young's modulus. Figure 11.17 shows a typical stress-strain graph for a metal such as copper or soft iron. The strain is shown as the *percent* elongation; the horizontal scale is not uniform beyond the first portion of the curve, up to a strain of less than 1%. The first portion is a straight line, indicating Hooke's law behavior with stress directly proportional to strain. This straight-line portion ends at point a; the stress at this point is called the *proportional limit.*

From a to b, stress and strain are no longer proportional, and Hooke's law is *not* obeyed. If the load is gradually removed, starting at any point between O and b, the curve is retraced until the material returns to its original length. The deformation is *reversible,* and the forces are conservative; the energy put into the material to cause the deformation is recovered when the stress is removed. In region Ob we say that the material shows *elastic behavior.* Point b, the end of this region, is called the *yield point;* the stress at the yield point is called the *elastic limit.*

When we increase the stress beyond point b, the strain continues to increase. But now when we remove the load at some point beyond b, say c, the material does not come back to its original length. Instead, it follows the red line in Fig. 11.17. The length at zero stress is now greater than the original length; the material has undergone an irreversible deformation and has acquired what we call a *permanent set.* Further increase of load beyond c produces a large increase in strain for a relatively small increase in stress, until a point d is reached at which *fracture* takes place. The behavior of the material from b to d is called *plastic flow* or *plastic deformation.* A plastic deformation is irreversible; when the stress is removed, the material does not return to its original state.

For some materials a large amount of plastic deformation takes place between the elastic limit and the fracture point. Such a material is said to be *ductile.* But if fracture occurs soon after the elastic limit is passed, the material is said to be *brittle.* A soft iron wire that can have considerable permanent stretch without breaking is ductile, while a steel piano string that breaks soon after its elastic limit is reached is brittle.

Figure 11.18 shows a stress-strain curve for vulcanized rubber that has been stretched by over seven times its original length. The stress is not proportional to the strain, but the behavior is elastic because when the load is removed, the material returns to its original length. However, the material follows different curves for increasing and decreasing stress. This is called *elastic hysteresis.* The work done by the material when it returns to its original shape is less than the work required to deform it; there are nonconservative forces associated with internal friction. Rubber with large elastic hysteresis is very useful for absorbing vibrations, such as in engine mounts and shock-absorber bushings for cars.

The stress required to cause actual fracture of a material is called the *breaking stress,* the *ultimate strength,* or (for tensile stress) the *tensile strength.* Two materials, such as two types of steel, may have very similar elastic constants but vastly different breaking stresses. Table 11.3 gives a few typical values of breaking stress for several materials in tension.

The conversion factor 6.9×10^8 Pa $= 100{,}000$ psi will help put these numbers in perspective. For example, if the breaking stress of a particular steel is 6.9×10^8 Pa, then a bar with 1-in.^{2} cross section has a breaking strength of 100,000 lb.

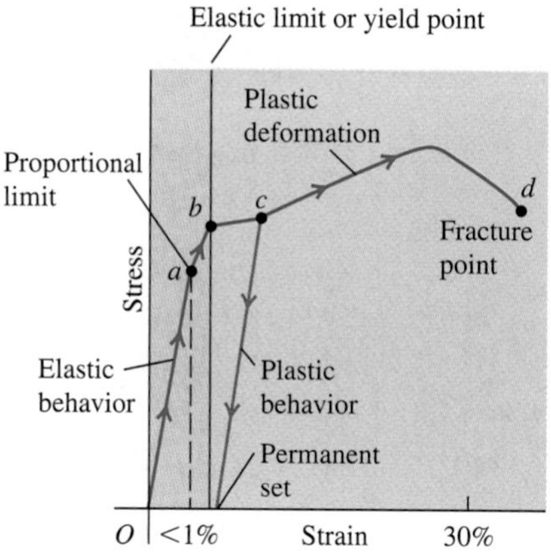

11.17 Typical stress-strain diagram for a ductile metal under tension.

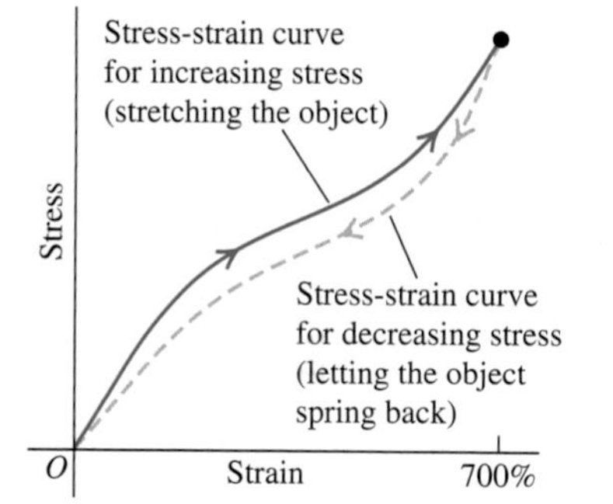

11.18 Typical stress-strain diagram for vulcanized rubber. The curves are different for increasing and decreasing stress, a phenomenon called elastic hysteresis.

Table 11.3 Approximate Breaking Stresses of Materials

Material	Breaking Stress (Pa or N/m^2)
Aluminum	2.2×10^8
Brass	4.7×10^8
Glass	10×10^8
Iron	3.0×10^8
Phosphor bronze	5.6×10^8
Steel	$5–20 \times 10^8$

Test Your Understanding

While parking your car on a crowded street, you accidentally back into a steel post. You pull forward until the car no longer touches the post and then get out to inspect the damage. What does your rear bumper look like if the strain in the impact was (a) less than at the proportional limit; (b) greater than at the proportional limit, but less than at the yield point; (c) greater than at the yield point, but less than at the fracture point; or (d) greater than at the fracture point?

SUMMARY

For a rigid body to be in equilibrium, two conditions must be satisfied. First, the vector sum of forces must be zero. Second, the sum of torques about any point must be zero. To compute the torque due to a force, find the torque of each force component by using its appropriate lever arm and sign, and then add these values. (See Examples 11.2 through 11.5)

$$\sum F_x = 0 \quad \sum F_y = 0 \quad \sum F_z = 0 \qquad (11.1)$$

$$\sum \vec{\tau} = \mathbf{0} \text{ about } \textit{any} \text{ point} \qquad (11.2)$$

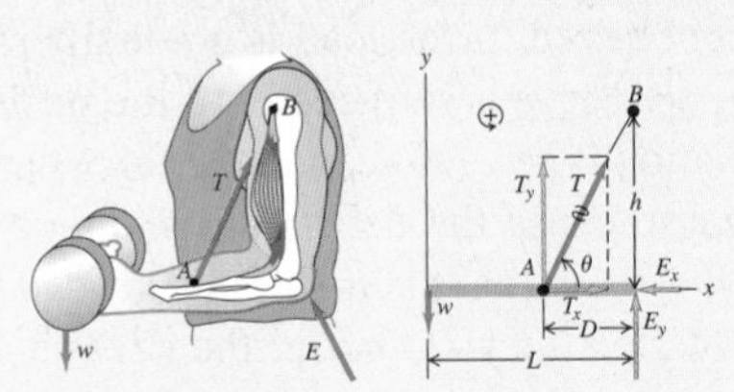

The torque due to the weight of a body can be obtained by assuming the entire weight to be concentrated at the center of gravity. If $\vec{g}$ has the same value at all points, the coordinates of the center of gravity are the same as those of the center of mass. (See Example 11.1)

$$\vec{r}_{cm} = \frac{m_1\vec{r}_1 + m_2\vec{r}_2 + m_3\vec{r}_3 + \cdots}{m_1 + m_2 + m_3 + \cdots} = \frac{\sum_i m_i\vec{r}_i}{\sum_i m_i} \qquad (11.4)$$

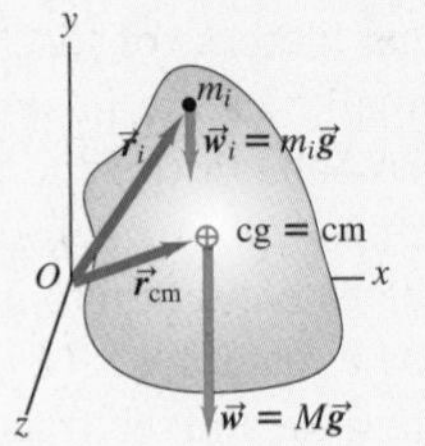

Hooke's law states that in elastic deformations, stress (force per unit area) is proportional to strain (fractional deformation). The proportionality constant is called the elastic modulus.

$$\frac{\text{Stress}}{\text{Strain}} = \text{Elastic modulus} \qquad (11.7)$$

Tensile stress is tensile force per unit area, $F_\perp/A$. Tensile strain is fractional change in length, $\Delta l/l_0$. The elastic modulus is called Young's modulus Y. Compressive stress and strain are defined in the same way as tensile stress and strain. (See Example 11.6)

$$Y = \frac{\text{Tensile stress}}{\text{Tensile strain}} = \frac{F_\perp/A}{\Delta l/l_0} = \frac{F_\perp}{A}\frac{l_0}{\Delta l} \qquad (11.10)$$

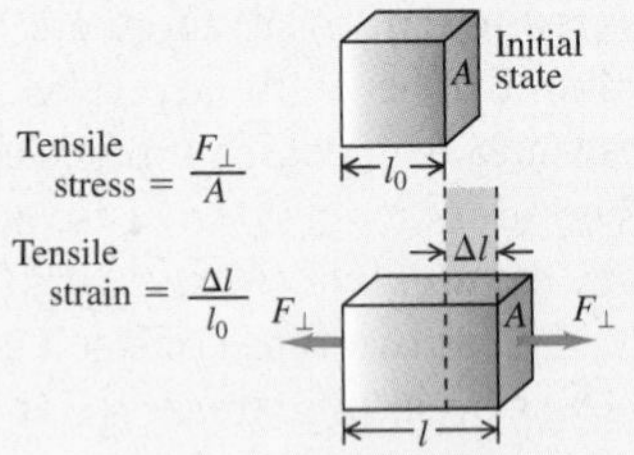

Pressure in a fluid is force per unit area. Bulk stress is pressure change, Δp, and bulk strain is fractional volume change, $\Delta V/V_0$. The elastic modulus is called the bulk modulus, B. Compressibility, k, is the reciprocal of bulk modulus: $k = 1/B$. (See Example 11.7)

$$p = \frac{F_\perp}{A} \qquad (11.11)$$

$$B = \frac{\text{Bulk stress}}{\text{Bulk strain}} = -\frac{\Delta p}{\Delta V/V_0} \qquad (11.13)$$

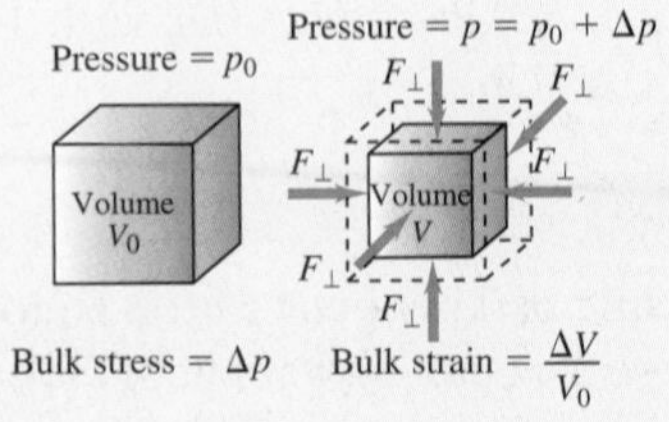

Shear stress is force per unit area, $F_{\parallel}/A$, for a force applied tangent to a surface. Shear strain is the displacement x of one side divided by the transverse dimension h. The elastic modulus is called the shear modulus, S. (See Example 11.8)

$$S = \frac{\text{Shear stress}}{\text{Shear strain}} = \frac{F_{\parallel}/A}{x/h} = \frac{F_{\parallel}}{A}\frac{h}{x} \tag{11.17}$$

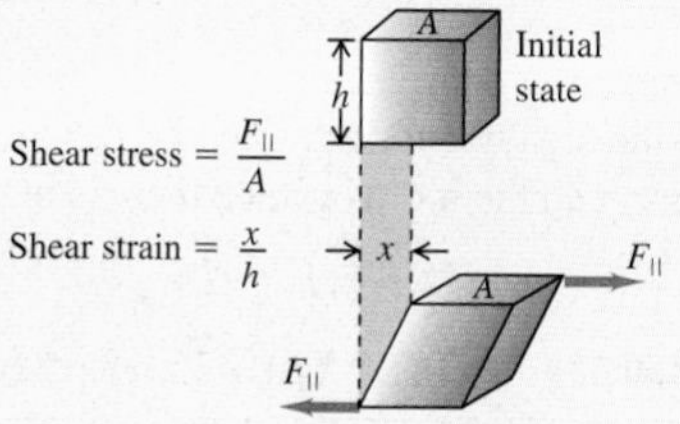

The proportional limit is the maximum stress for which stress and strain are proportional. Beyond the proportional limit, Hooke's law is not valid. The elastic limit is the stress beyond which irreversible deformation occurs. The breaking stress, or ultimate strength, is the stress at which the material breaks.

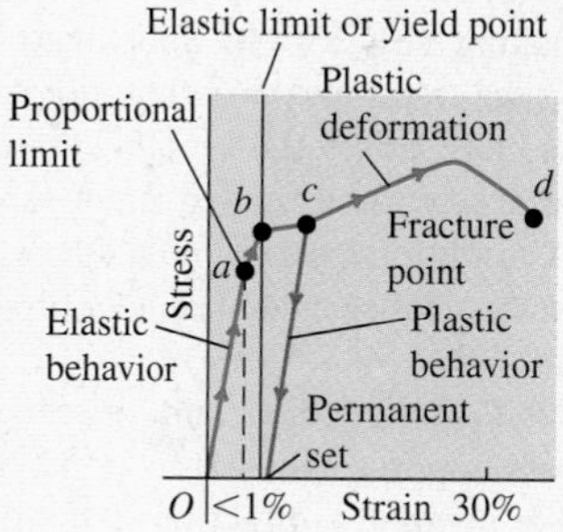

Key Terms

first condition for equilibrium, 405
second condition for equilibrium, 405
static equilibrium, 405
center of gravity, 406
stress, 415
strain, 415
elastic modulus, 415
Hooke's law, 415
tension, 415
tensile stress, 415
pascal, 416
tensile strain, 416
Young's modulus, 416
compression, 416
compressive stress, 416
compressive strain, 416
bulk stress (volume stress), 417
bulk strain (volume strain), 417
pressure, 417
atmosphere, 418
bulk modulus, 418
compressibility, 418
shear stress, 419
shear strain, 419
shear modulus, 419

Your Notes

Answer to Chapter Opening Question

To remain in equilibrium, the flamingo's center of gravity must be over the area of support. The broader the foot, the larger the area of support and the more latitude there is in the location of the center of gravity.

Answers to Test Your Understanding Questions

Section 11.1 Situation (a) satisfies both equilibrium conditions; the seagull has zero acceleration (so $\sum \vec{F} = \mathbf{0}$) and no tendency to start rotating (so $\sum \vec{\tau} = \mathbf{0}$). Situation (b) satisfies the first condition, because the crankshaft as a whole does not accelerate through space, but does not satisfy the second condition; the crankshaft has an angular acceleration, so $\sum \vec{\tau}$ is not zero. Situation (c) satisfies the second condition (there is no tendency to rotate), but not the first one; the baseball accelerates in its flight (due to gravity), so $\sum \vec{F}$ is not zero.

Section 11.2 In equilibrium, the center of gravity must be at the point of support. Choose this point to be the origin, so $x_{cg} = 0$. With the positive x-direction to the right, the coordinates of the three objects are $x = -10$ cm for the counterweight, $x = 15$ cm for the first fish, and $x = 40$ cm for the second fish. (We treat the rod as being massless.) From the x-equation in Eqs. (11.3),

$$x_{cg} = 0 = \frac{m(-10\text{ cm}) + (0.10\text{ kg})(15\text{ cm}) + (0.10\text{ kg})(40\text{ cm})}{m + (0.10\text{ kg}) + (0.10\text{ kg})}$$

so

$$m = \frac{(0.10\text{ kg})(15\text{ cm}) + (0.10\text{ kg})(40\text{ cm})}{10\text{ cm}} = 0.55\text{ kg}$$

Section 11.3 This is the same situation as described in Example 11.5, with the rod replacing the forearm, the hinge replacing the elbow, and the cable replacing the tendon. The cable attachment point is at the end of the rod, so the distance D in Fig. 11.9b is equal to L. From Example 11.5, the tension is

$$T = \frac{Lw}{L\sin\theta} = \frac{w}{\sin\theta}$$

and the horizontal and vertical forces exerted by the hinge are

$$E_x = \frac{Lw}{h} = \frac{w}{\tan\theta} \quad\text{and}\quad E_y = -\frac{(L-L)}{L}w = 0$$

In this situation, the hinge exerts no vertical force.

Section 11.4 The tensile strain $\Delta l/l_0$ is the same for both rods. The copper rod has ten times the elongation Δl of the steel rod, but it also has ten times the original length l_0. The stress is equal to the product of Young's modulus Y and the strain. From Table 11.1, steel has a larger value of Y, so a greater stress is required to produce the same strain.

Section 11.5 In both cases (a) and (b), the bumper will have sprung back to its original shape (although the paint may be scratched). In case (c), the bumper will have a permanent dent or deformation. In case (d), the bumper will be torn or broken.

Discussion Questions

Q11.1 Does a rigid body in uniform rotation about a fixed axis satisfy the first and second conditions for equilibrium? Why or why not? An alternative definition of equilibrium states: "For a rigid body to be in equilibrium, every part of the body must be in equilibrium." On the basis of this definition, is a rigid body in uniform rotation about a fixed axis in equilibrium? Explain.

Q11.2 Is it possible for the net force on an object to be zero but the object not to be in equilibrium? Is it possible for an object to be in equilibrium if the net force is not zero? Explain.

Q11.3 Car tires are sometimes "balanced" on a machine that pivots the tire and wheel about the center. Weights are placed around the wheel rim until it does not tip from the horizontal plane. Discuss this procedure in terms of the center of gravity.

Q11.4 Does the center of gravity of a solid body always lie within the material of the body? If not, give a counterexample.

Q11.5 In Section 11.2 we always assumed that the value of g was the same at all points on the body. This is *not* a good approximation if the dimensions of the body are great enough, because the value of g decreases with altitude. If this is taken into account, will the center of gravity of a long, vertical rod be above, below, or at its center of mass? Explain how this can be used to keep the long axis of an orbiting spacecraft pointed toward the earth. (This would be useful for a weather satellite that must always keep its camera lens trained on the earth.) The moon is not exactly spherical, but is somewhat elongated. Explain why this same effect is responsible for keeping the same face of the moon pointed toward the earth at all times.

Q11.6 You are balancing a wrench by suspending it at a single point. Is the equilibrium stable, unstable, or neutral if the point is above, at, or below the wrench's center of gravity? In each case give the reasoning behind your answer. (For rotation, a rigid body is in *stable* equilibrium if a small rotation of the body produces a torque that tends to return the body to equilibrium; it is in *unstable* equilibrium if a small rotation produces a torque that tends to take the body farther from equilibrium; and it is in *neutral* equilibrium if a small rotation produces no torque.)

Q11.7 You can probably stand flatfooted on the floor, and then rise up and balance on your tiptoes. Why are you unable do it if your toes are touching the wall of your room? (Try it!)

Q11.8 You freely pivot a horseshoe from a horizontal nail through one of its nail holes. You then hang a long string with a weight at its bottom from the same nail, so that the string hangs vertically in front of the horseshoe without touching it. How do you know that the horseshoe's center of gravity is along the line behind the string? How can you locate the center of gravity by repeating the process at another nail hole? Will the center of gravity be within the solid material of the horseshoe?

Q11.9 A stunt driver drives a car in a straight line at constant speed while balanced on the two right-hand wheels. Where is the car's center of gravity?

Q11.10 When four-legged animals walk, they always have three of their legs on the ground at any instant. Explain why, using the concept of center of gravity.

Q11.11 Why must a water skier moving with constant velocity lean backwards? What determines how far back she must lean? Draw a free-body diagram for the water skier to justify your answers.

Q11.12 In pioneer days, when a Conestoga wagon was stuck in the mud, people would grasp the wheel spokes and try to turn the wheels, rather than simply pushing the wagon. Why?

Q11.13 A physical therapist says a physicist-patient has weak stomach muscles because he can't do a partial sit-up without his feet being held down. The physicist, who has short, slender legs and a long, heavy, upper torso, correctly disagrees. Explain, using the concept of center of gravity.

Q11.14 Why is it easier to hold a 10-kg dumbbell in your hand at your side than it is to hold it with your arm extended horizontally?

Q11.15 Figure 11.3 shows that an object at rest on an incline will tip over if its center of gravity is not within the area of support. Explain why this result does *not* depend on the value of the coefficient of static friction between the object and the surface of the incline. Draw a free-body diagram as part of your explanation.

Q11.16 In Example 11.3 (Section 11.3), explain why n_1 is negative in the $\Sigma F_x = 0$ equation, but gives a positive torque in the $\Sigma \tau_B = 0$ equation.

Q11.17 Why is a tapered water glass with a narrow base easier to tip over than one with straight sides? Does it matter whether the glass is full or empty?

Q11.18 When a tall, heavy refrigerator is pushed across a rough floor, what determines whether it slides or tips?

Q11.19 Climbing ropes used by mountaineers are usually made of nylon. Would a steel cable of equal strength be just as good? What advantages and disadvantages would it have in comparison to nylon?

Q11.20 Why is concrete with steel reinforcing rods embedded in it stronger than plain concrete?

Q11.21 Electric power lines are sometimes made by using wires with a steel core and a copper jacket or with strands of copper and steel twisted together. Why?

Q11.22 Compare the mechanical properties of a steel cable, made by twisting many thin wires together, with those of a solid steel rod of the same diameter. What advantages does each have?

Q11.23 The material in human bones and elephant bones is essentially the same, but an elephant has much thicker legs. Explain why, in terms of breaking stress.

Q11.24 There is a small but appreciable amount of elastic hysteresis in the large tendon at the back of a horse's leg. Explain how this can cause damage to the tendon if a horse runs too hard for too long a time.

Q11.25 When rubber mounting blocks are used to absorb machine vibrations through elastic hysteresis, as mentioned in Section 11.5, what becomes of the energy associated with the vibrations?

Exercises

Section 11.2 Center of Gravity

11.1 A ball with radius $r_1 = 0.080$ m and mass 1.00 kg is attached by a light rod 0.400 m in length to a second ball with radius $r_2 = 0.100$ m and mass 2.00 kg (Fig. 11.19). Where is the center of gravity of this system?

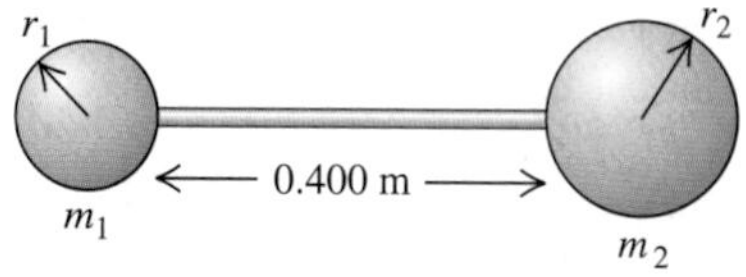

Figure 11.19 Exercise 11.1.

11.2 Suppose the rod in Exercise 11.1 is uniform with mass 1.50 kg. Where is the system's center of gravity? How does your result compare to that for Exercise 11.1?

11.3 For the plank of Example 11.1 (Section 11.2), show that a 60-kg child could stand no farther to the right than halfway between the right-hand sawhorse and the end of the plank without the plank tipping over.

Section 11.3 Solving Rigid-Body Equilibrium Problems

11.4 A uniform 300-N trapdoor in a ceiling is hinged at one side. Find the net upward force needed to begin to open it and the total force exerted on the door by the hinges, a) if the upward force is applied at the center; b) if the upward force is applied at the center of the edge opposite the hinges.

11.5 Raising a Ladder. A ladder carried by a fire truck is 20.0 m long. The ladder weighs 2800 N and its center of gravity is at its center. The ladder is pivoted at one end (A) about a pin (Fig. 11.20); you can ignore the friction torque at the pin. The ladder is raised into position by a force applied by a hydraulic piston at C. Point C is 8.0 m from A, and the force $\vec{F}$ exerted by the piston makes an angle of 40° with the ladder. What magnitude must $\vec{F}$ have to just lift the ladder off the support bracket at B?

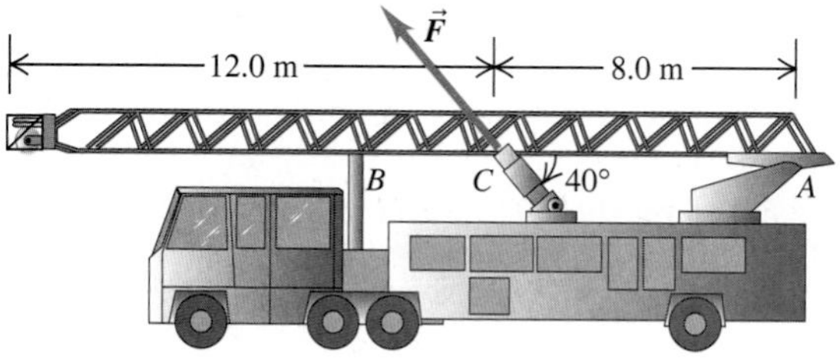

Figure 11.20 Exercise 11.5.

11.6 Two people are carrying a uniform wooden board that is 3.00 m long and weighs 160 N. If one person applies an upward force equal to 60 N at one end, at what point does the other person lift?

11.7 Two people carry a heavy electric motor by placing it on a light board 2.00 m long. One person lifts at one end with a force of

400 N, and the other lifts the opposite end with a force of 600 N. What is the weight of the motor, and where along the board is its center of gravity located?

11.8 Suppose the board in Exercise 11.7 is not light but weighs 200 N, with its center of gravity at its center. The two people each exert the same forces as before. What is the weight of the motor in this case, and where is its center of gravity located?

11.9 Show that the torques about point B due to T_x and T_y in Example 11.5 (Section 11.3) sum to zero. Do this for the general case, rather than for the specific numerical values given at the end of the example.

11.10 A uniform ladder 5.0 m long rests against a frictionless, vertical wall with its lower end 3.0 m from the wall. The ladder weighs 160 N. The coefficient of static friction between the foot of the ladder and the ground is 0.40. A man weighing 740 N climbs slowly up the ladder. a) What is the maximum frictional force that the ground can exert on the ladder at its lower end? b) What is the actual frictional force when the man has climbed 1.0 m along the ladder? c) How far along the ladder can the man climb before the ladder starts to slip?

11.11 A diving board 3.00 m long is supported at a point 1.00 m from the end, and a diver weighing 500 N stands at the free end (Fig. 11.21). The diving board is of uniform cross section and weighs 280 N. Find a) the force at the support point and b) the force at the end that is held down.

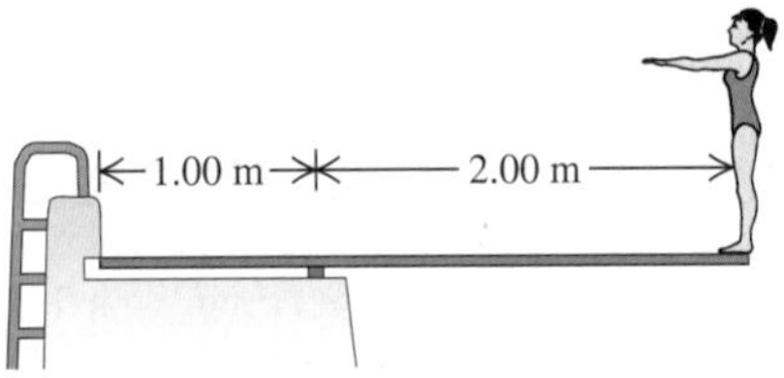

Figure 11.21 Exercise 11.11.

11.12 A uniform, aluminum beam 9.00 m long, weighing 300 N, rests symmetrically on two supports 5.00 m apart (Fig. 11.22). A boy weighing 600 N starts at point A and walks toward the right. a) In the same diagram construct two graphs showing the upward forces F_A and F_B exerted on the beam at points A and B, as functions of the coordinate x of the boy. Let 1 cm = 100 N vertically, and 1 cm = 1.00 m horizontally. b) From your diagram, how far beyond point B can the boy walk before the beam tips? c) How far from the right end of the beam should support B be placed so that the boy can walk just to the end of the beam without causing it to tip?

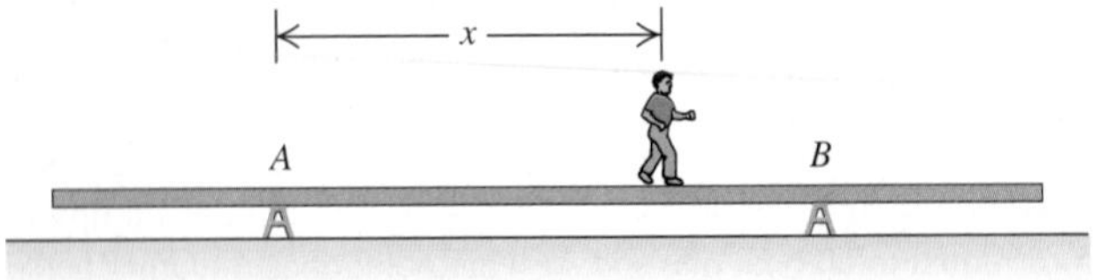

Figure 11.22 Exercise 11.12.

11.13 Find the tension T in each cable and the magnitude and direction of the force exerted on the strut by the pivot in each of the arrangements in Fig. 11.23. In each case let w be the weight of the suspended crate full of priceless art objects. The strut is uniform and also has weight w.

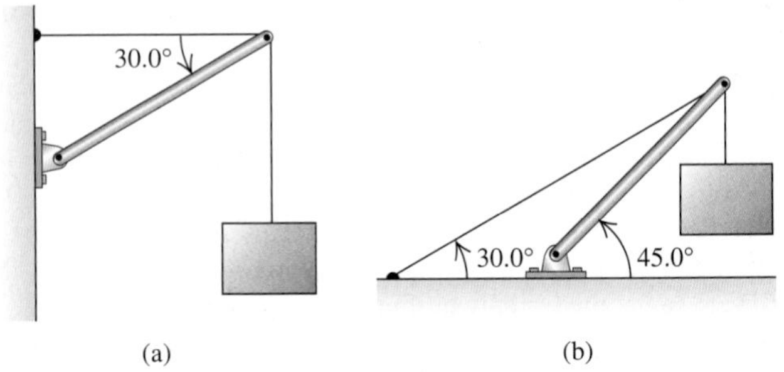

Figure 11.23 Exercise 11.13.

11.14 The horizontal beam in Fig. 11.24 weighs 150 N, and its center of gravity is at its center. Find a) the tension in the cable and b) the horizontal and vertical components of the force exerted on the beam at the wall.

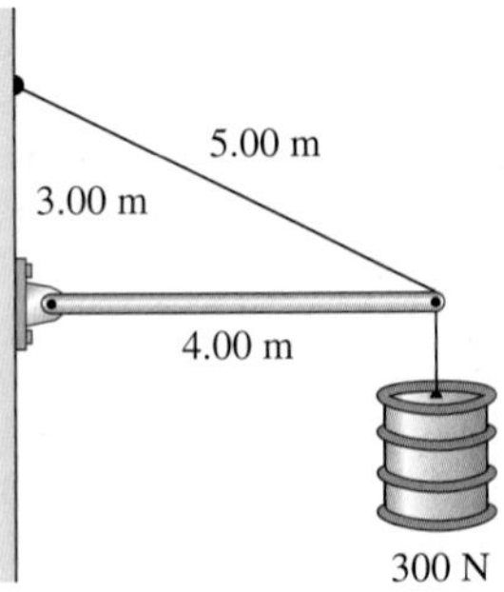

Figure 11.24 Exercise 11.14.

11.15 A door 1.00 m wide and 2.00 m high weighs 280 N and is supported by two hinges, one 0.50 m from the top and the other 0.50 m from the bottom. Each hinge supports half the total weight of the door. Assuming that the door's center of gravity is at its center, find the horizontal components of force exerted on the door by each hinge.

11.16 Suppose that you can lift no more than 450 N (about 100 lb) unaided. a) How much can you lift using a 2.0-m long wheelbarrow? Assume that the wheelbarrow weighs 80 N and that its center of gravity is 0.70 m from the wheel. Assume that the load carried in the wheelbarrow is also 0.70 m from the wheel. b) Where does the force come from to enable you to lift more than 450 N when you use the wheelbarrow?

11.17 You take your dog Clea to the vet, and the doctor decides he must locate the little beast's center of gravity. It would be awkward to hang the pooch from the ceiling, so the vet must devise another method. He places Clea's front feet on one scale and her hind feet on another. The front scale reads 157 N while the rear scale reads 89 N. The vet next measures Clea and finds that her rear feet are 0.95 m behind her hind feet. How much does Clea weigh and where is her center of gravity?

11.18 The boom in Fig. 11.25 weighs 2600 N. The boom is

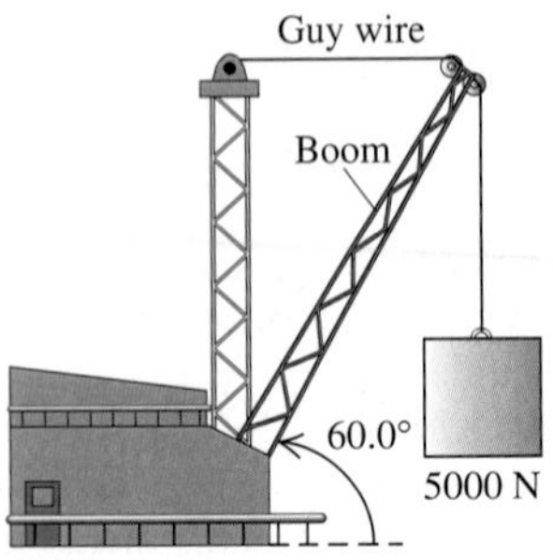

Figure 11.25 Exercise 11.18.

attached with a frictionless pivot at its lower end. It is not uniform; the distance of its center of gravity from the pivot is 35% of its length. a) Find the tension in the guy wire and the horizontal and vertical components of the force exerted on the boom at its lower end. b) Does the line of action of the force exerted on the boom at its lower end lie along the boom?

11.19 A 3.00-m-long, 240-N uniform rod at the zoo is held in a horizontal position by two ropes at its ends (Fig. 11.26). The left rope makes an angle of 150° with the rod and the right rope makes an angle θ with the horizontal. A 90-N howler monkey (*Alouatta seniculus*) hangs motionless 0.50 m from the right end of the rod as he carefully studies you. Calculate the tensions in the two ropes and the angle θ.

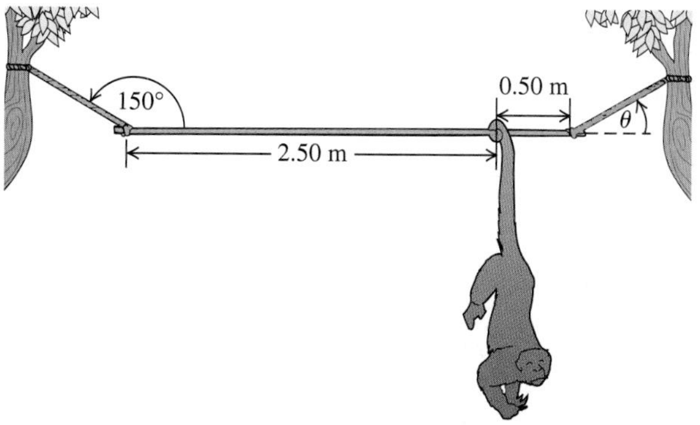

Figure 11.26 Exercise 11.19.

11.20 A non-uniform beam 4.50 m long and weighing 1.00 kN makes an angle of 25.0° with the horizontal. It is held in position by a frictionless pivot at its upper-right end and by a cable 3.00 m farther down the beam and perpendicular to it. The center of gravity of the beam is 2.00 m down the beam from the pivot. Lighting equipment exerts a 5.00-kN downward force on the lower-left end of the beam. Find the tension T in the cable and the horizontal and vertical components of the force exerted on the beam by the pivot.

11.21 A Couple. Two forces equal in magnitude and opposite in direction, acting on an object at two different points, form what is called a *couple*. Two antiparallel forces with equal magnitudes $F_1 = F_2 = 8.00\,N$ are applied to a rod as shown in Fig. 11.27. a) What should the distance l between the forces be if they are to provide a net torque of 6.40 N·m about the left-hand end of the rod? b) Is the sense of this torque clockwise or counterclockwise? c) Repeat parts (a) and (b) for a pivot at the point on the rod where $\vec{F}_2$ is applied.

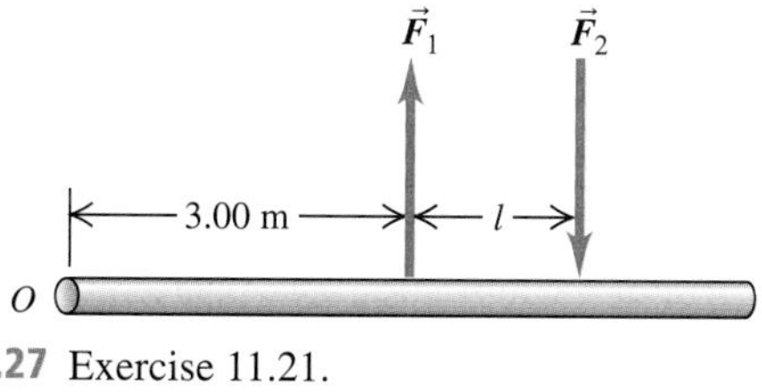

Figure 11.27 Exercise 11.21.

Section 11.4 Stress, Strain, and Elastic Moduli

11.22 A relaxed biceps muscle requires a force of 25.0 N for an elongation of 3.0 cm; the same muscle under maximum tension requires a force of 500 N for the same elongation. Find Young's modulus for the muscle tissue under each of these conditions if the muscle is assumed to be a uniform cylinder with length 0.200 m and cross-sectional area 50.0 cm^2.

11.23 A circular steel wire 2.00 m long must stretch no more than 0.25 cm when a tensile force of 400 N is applied to each end of the wire. What minimum diameter is required for the wire?

11.24 Two round rods, one steel and the other copper, are joined end to end. Each rod is 0.750 m long and 1.50 cm in diameter. The combination is subjected to a tensile force with magnitude 4000 N. For each rod, what is a) the strain? b) the elongation?

11.25 A metal rod that is 4.00 m long and 0.50 cm^2 in cross-sectional area is found to stretch 0.20 cm under a tension of 5000 N. What is Young's modulus for this metal?

11.26 Stress on a Mountaineer's Rope. A nylon rope used by mountaineers elongates 1.10 m under the weight of an 65.0-kg climber. If the rope is 45.0 m in length and 7.0 mm in diameter, what is Young's modulus for this material?

11.27 In constructing a large mobile, an artist hangs an aluminum sphere of mass 6.0 kg from a vertical steel wire 0.50 m long and 2.5×10^{-3} cm^2 in cross-sectional area. On the bottom of the sphere he attaches a similar steel wire, from which he hangs a brass cube of mass 10.0 kg. For each wire, compute a) the tensile strain and b) the elongation.

11.28 A vertical solid steel post 25 cm in diameter and 2.50 m long is required to support a load of 8000 kg. You can ignore the weight of the post. What is a) the stress in the post? b) the strain in the post? c) the change in the post's length when the load is applied?

11.29 Outside a house 1.0 km from ground zero of a 100-kiloton nuclear bomb explosion, the pressure will rapidly rise to as high as 2.8 atm while the pressure inside the house remains 1.0 atm. If the area of the front of the house is 50 m^2, what is the resulting net force exerted by the air on the front of the house?

11.30 A solid gold bar is pulled up from the hold of the sunken RMS *Titanic.* a) What happens to its volume as it goes from the pressure at the ship to the lower pressure at the ocean's surface? b) The pressure difference is proportional to the depth. How many times greater would the volume change have been had the ship been twice as deep? c) The bulk modulus of lead is one-fourth that of gold. Find the ratio of the volume change of a solid lead bar to that of a gold bar of equal volume for the same pressure change.

11.31 A petite young woman distributes her 500 N weight equally over the heels of her high-heeled shoes. Each heel has an area of 0.750 cm^2. a) What is the pressure exerted on the floor by each heel? b) With the same pressure, how much weight could be supported by two flat-bottomed sandals, each of area 200 cm^2?

11.32 In the Challenger Deep of the Marianas Trench, the depth of seawater is 10.9 km and the pressure is 1.16×10^8 Pa (about 1.15×10^3 atm). a) If a cubic meter of water is taken from the surface to this depth, what is the change in its volume? (Normal atmospheric pressure is about 1.0×10^5 Pa. Assume that k for seawater is the same as the freshwater value given in Table 11.2.) b) What is the density of seawater at this depth? (At the surface, seawater has a density of 1.03×10^3 kg/m^3.)

11.33 A specimen of oil having an initial volume of 600 cm^3 is subjected to a pressure increase of 3.6×10^6 Pa, and the volume is found to decrease by 0.45 cm^3. What is the bulk modulus of the material? The compressibility?

11.34 A square steel plate is 10.0 cm on a side and 0.500 cm thick. a) Find the shear strain that results if a force of magnitude 9.0×10^5 N is applied to each of the four sides, parallel to the side. b) Find the displacement x in centimeters.

11.35 Shear forces are applied to a rectangular solid. The same forces are applied to another rectangular solid of the same material, but with three times each edge length. In each case the forces are small enough that Hooke's law is obeyed. What is the ratio of the shear strain for the larger object to that of the smaller object?

11.36 Two strips of brass, each 3.00 cm wide and 45.0 cm long, are placed with their ends overlapping by 1.00 cm. The overlapping ends are then riveted together by four rivets, each with diameter 0.250 cm. Tests show that when a tension of at least 1.20×10^4 N is applied to the ends of the riveted strip, the rivets fail and shear apart. What is the shearing stress on each rivet when it fails? Assume that each rivet carries one quarter of the load.

Section 11.5 Elasticity and Plasticity

11.37 In a materials testing laboratory, a metal wire made from a new alloy is found to break when a tensile force of 90.8 N is applied perpendicular to each end. If the diameter of the wire is 1.84 mm, what is the breaking stress of the alloy?

11.38 A 4.0-m-long steel wire has a cross-sectional area of 0.050 cm^2. Its proportional limit has a value of 0.0016 times its Young's modulus (Table 11.1). Its breaking stress has a value of 0.0065 times its Young's modulus. The wire is fastened at its upper end and hangs vertically. a) How great a weight can be hung from the wire without exceeding the proportional limit? b) How much will the wire stretch under this load? c) What is the maximum weight that the wire can support?

11.39 A steel cable with cross-sectional area 3.00 cm^2 has an elastic limit of 2.40×10^8 Pa. Find the maximum upward acceleration that can be given a 1200-kg elevator supported by the cable if the stress is not to exceed one-third of the elastic limit.

11.40 A brass wire is to withstand a tensile force of 350 N without breaking. What minimum diameter must the wire have?

Problems

11.41 Center of Gravity of a Car. For an automobile with weight w, the front wheels support a fraction f of the weight, so the normal force exerted at the front wheels is fw, and at the rear wheels the normal force is $(1 - f)w$. The distance between the front and rear axles is d. a) Show that the center of gravity of the car is a distance fd in front of the rear wheels. b) Show that the general result derived in part (a) reproduces the numerical answer of Example 11.2 (Section 11.3).

11.42 Sir Lancelot rides slowly out of the castle at Camelot and onto the 12.0-m-long drawbridge that passes over the moat (Fig. 11.28). Unbeknownst to him, his enemies have partially severed the vertical cable holding up the front end of the bridge so that it will break under a tension of 5.80×10^3 N. The bridge has mass 200 kg and its center of gravity is at its center. Lancelot, his lance, his armor, and his horse together have a combined mass of 600 kg. Will the cable break before Lancelot reaches the end of the drawbridge? If so, how far from the castle end of the bridge will the center of gravity of the horse plus rider be when the cable breaks?

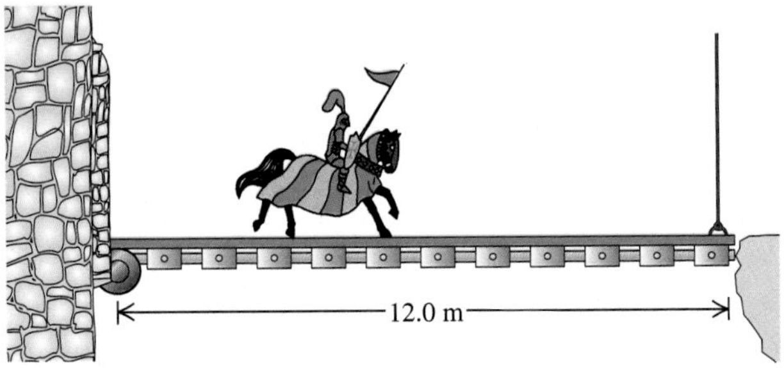

Figure 11.28 Problem 11.42.

11.43 Three vertical forces act on an airplane when it is flying at a constant altitude and with a constant velocity. These are the weight of the airplane, an aerodynamic force on the wing of the airplane, and an aerodynamic force on the airplane's horizontal tail. (The aerodynamic forces are exerted by the surrounding air, and are reactions to the forces that the wing and tail exert on the air as the airplane flies through it.) For a particular light airplane with a weight of 6700 N, the center of gravity is 0.30 m in front of the point where the wing's vertical aerodynamic force acts and 3.66 m in front of the point where the tail's vertical aerodynamic force acts. Determine the magnitude and direction (upward or downward) of each of the two vertical aerodynamic forces.

11.44 A pickup truck has a wheelbase of 3.00 m. Ordinarily, 10,780 N rests on the front wheels and 8820 N on the rear wheels when the truck is parked on a level road. a) A box weighing 3600 N is now placed on the tailgate, 1.00 m behind the rear axle. How much total weight now rests on the front wheels? On the rear wheels? b) How much weight would need to be placed on the tailgate to make the front wheels come off the ground?

11.45 A uniform seesaw is 4.00 m long and weighs 240 N. The seesaw is pivoted about a point on its bottom surface and the location of this pivot along the length of the seesaw can be adjusted as far as 0.20 m from the center of the seesaw. Little Karen, who weighs 420 N, has her center of gravity over the right end, with the seesaw set so she exerts the most torque about the pivot. The seesaw balances horizontally when Elwood has his center of gravity 0.15 m from the left end. Both have their feet off the ground. a) What does Elwood weigh? b) Is this equilibrium stable, unstable, or neutral? (*Hint:* See Question Q11.6.)

11.46 A piece of steel will not balance with the tip of one end just barely resting on the edge of a table (point O in Fig. 11.29). But suppose we fasten on two other pieces as shown in Fig. 11.29. All pieces have the same square cross section and density. You can ignore the weight of the fasteners. The system can now balance if we find the right length l of the third piece. a) What is the length l for which the system will balance with the first and third pieces horizontal, as shown in the figure? b) How would your answer change if the pieces were made of copper rather than steel? Explain.

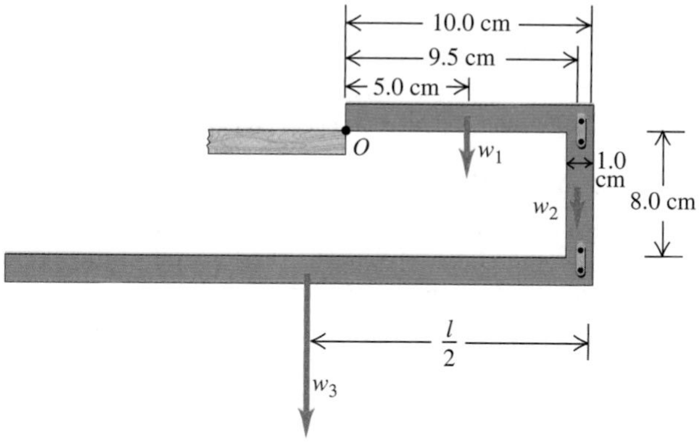

Figure 11.29 Problem 11.46.

11.47 Several external forces $\vec{F}_i$ are applied to a rigid body. With respect to an origin O, force $\vec{F}_1$ is applied at the point located at $\vec{r}_1$, force $\vec{F}_2$ at $\vec{r}_2$, and so on. Point P is located at $\vec{r}_P$. If the body is in translational equilibrium, so that $\Sigma \vec{F}_i = \mathbf{0}$, prove that the sum of the torques about point P equals the sum of torques about point O. (A special case of this result is when $\Sigma \vec{\tau}_O = \mathbf{0}$; for a body in translational equilibrium, if the resultant torque about one point is zero, the resultant torque about any point is zero.)

11.48 A claw hammer is used to pull a nail out of a board (Fig. 11.30). The nail is at an angle of 60° to the board, and a force $\vec{F}_1$ of magnitude 500 N applied to the nail is required to pull it from the board. The hammer head contacts the board at point A, which is 0.080 m from where the nail enters the board. A horizontal force $\vec{F}_2$ is applied to the hammer handle at a distance of 0.300 m above the board. What magnitude of force $\vec{F}_2$ is required to apply the required 500-N force (F_1) to the nail? (You can ignore the weight of the hammer.)

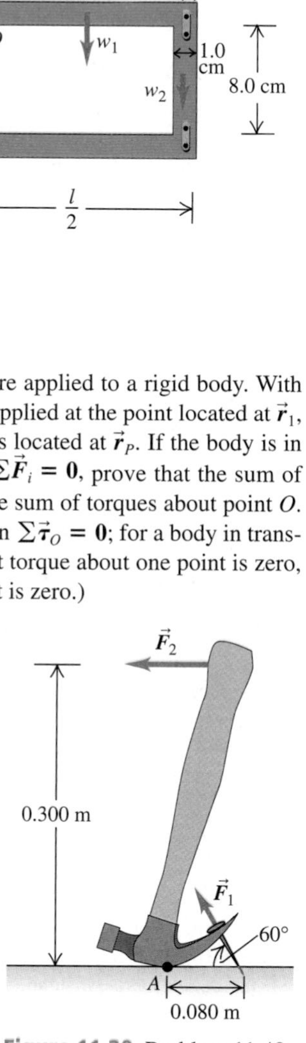

Figure 11.30 Problem 11.48.

11.49 End A of the bar AB in Fig. 11.31 rests on a frictionless horizontal surface, and end B is hinged. A horizontal force $\vec{F}$ of magnitude 120 N is exerted on end A. You can ignore the weight of the bar. What are the horizontal and vertical components of the force exerted by the bar on the hinge at B?

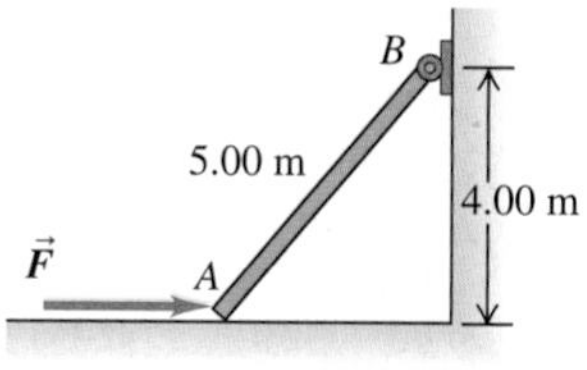

Figure 11.31 Problem 11.49.

11.50 A single additional force is to be applied to the bar in Fig. 11.32 to maintain it in equilibrium in the position shown. You can ignore the weight of the bar. a) What are the horizontal and vertical components of the required force? b) What is the angle the force must make with the bar? c) What is the magnitude of the required force? d) Where should the force be applied?

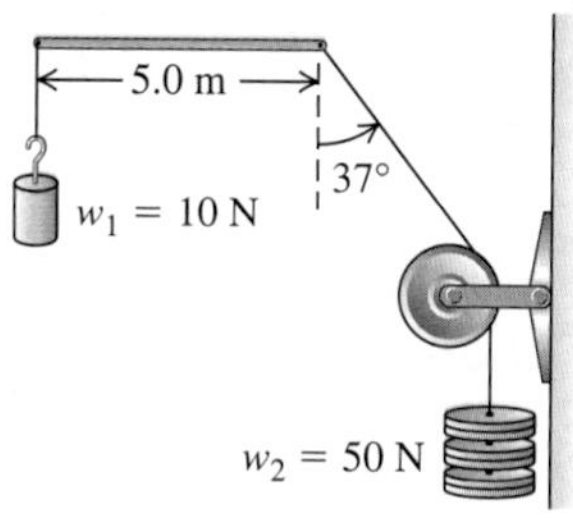

Figure 11.32 Problem 11.50.

11.51 Your dog Nikita is 0.90 m long (nose to hindlegs). Her forelegs are 0.18 m horizontally from her nose, her center of gravity is 0.28 m horizontally from her hindlegs, and she weighs 190 N. a) How much force does a level floor exert on each of Nikita's front feet, and on each of her hind feet? b) If Nikita picks up a 25-N bone and holds it in her mouth (directly under her nose), what is the force exerted by the floor on each of her front feet and each of her hind feet?

11.52 A Truck on a Drawbridge. A loaded cement mixer drives onto an old drawbridge, where it stalls with its center of gravity three-quarters of the way across the span. The truck driver radios for help, sets the hand-brake, and waits. Meanwhile, a boat approaches, so the drawbridge is raised by means of a cable attached to the end opposite the hinge (Fig. 11.33). The drawbridge span is 40.0 m long and has a mass of 12,000 kg; its center of gravity is at its midpoint. The cement mixer, with driver, has mass 30,000 kg. When the drawbridge has been raised to an angle of 30° above the horizontal, the cable makes an angle of 70° with the surface of the bridge. a) What is the tension T in the cable when the drawbridge is held in this position? b) What are the horizontal and vertical components of the force the hinge exerts on the span?

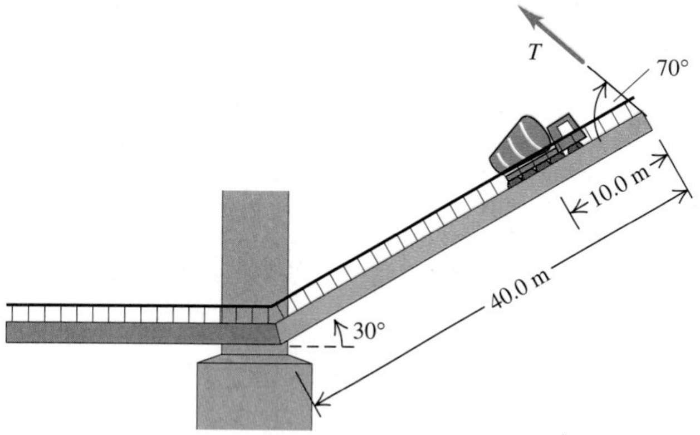

Figure 11.33 Problem 11.52.

11.53 Couples. a) A force $\vec{F}_1$ is applied to a rod at a distance x from some point P on the rod. A second force $\vec{F}_2$ is applied at a distance $x + l$ from point P. The forces are equal in magnitude and opposite in direction, and hence, form a couple (Exercise 11.21). The angle between the rod and the direction of each force is ϕ.

Derive an expression for the net torque about P produced by these two forces and show that it is independent of x. This shows that a couple produces the same torque about any pivot point. b) Two forces with magnitude $F_1 = F_2 = 14.0$ N are applied to a rod as shown in Fig. 11.34, with $\phi = 37°$. Calculate the net torque about point O due to these two forces by calculating the torque due to each separate force. Calculate the net torque about point P due to these two forces by calculating the torque due to each separate force. Compare your results to the general result derived in part (a).

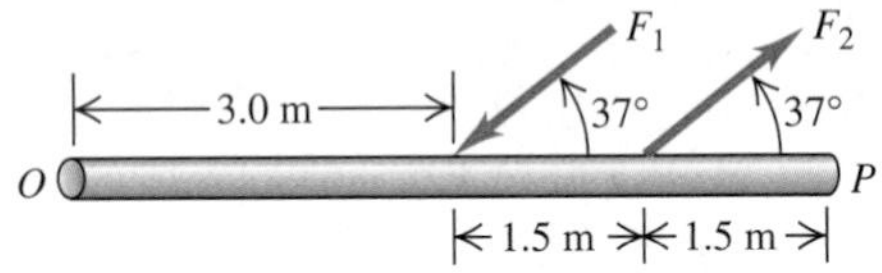

Figure 11.34 Problem 11.53.

11.54 A non-uniform fire escape ladder is 6.0 m long when extended to the icy alley below. It is held at the top by a frictionless pivot, and there is negligible frictional force from the icy surface at the bottom. The ladder weighs 250 N, and its center of gravity is 2.0 m along the ladder from its bottom. A mother and child of total weight 750 N are on the ladder 1.5 m from the pivot. The ladder makes an angle θ with the horizontal. Find the magnitude and direction of a) the force exerted by the icy alley on the ladder and b) the force exerted by the ladder on the pivot. c) Do your answers in parts (a) and (b) depend on the angle θ?

11.55 A uniform strut of mass m makes an angle θ with the horizontal. It is supported by a frictionless pivot located at $\frac{1}{3}$ its length from its lower-left end and a horizontal rope at its upper-right end. A cable and package of total weight w hang from its upper-right end. a) Find the vertical and horizontal components V and H of the pivot's force on the strut as well as the tension T in the rope. b) If the maximum safe tension in the rope is 700 N and the mass of the strut is 20.0 kg, find the maximum safe weight of the cable and package when the strut makes an angle of 55.0° with the horizontal. c) For what angle θ can no weight be safely suspended from the right end of the strut?

11.56 You are asked to design the decorative mobile shown in Fig. 11.35. The strings and rods have negligible weight, and the rods are to hang horizontally. a) Draw a free-body diagram for each rod. b) Find the weights of the balls A, B, and C. Find the tensions in the strings S_1, S_2, and S_3. c) What can you say about the horizontal location of the mobile's center of gravity? Explain.

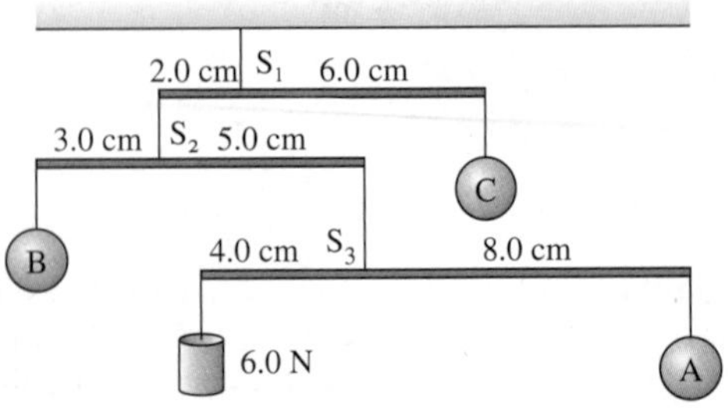

Figure 11.35 Problem 11.56.

11.57 A uniform 7.5-m-long beam weighing 9000 N is hinged to a wall and supported by a thin cable attached 1.5 m from the free end of the beam. The cable runs between the beam and the wall and makes a 40° angle with the beam. What is the tension in the cable when the beam is at an angle of 30° above the horizontal?

11.58 A uniform drawbridge must be held at a 37° angle above the horizontal to allow ships to pass underneath. The drawbridge weighs 45,000 N and is 14.0 m long. A cable is connected 3.5 m from the hinge where the bridge pivots and pulls horizontally on the bridge to hold it in place. a) What is the tension in the cable? b) Find the magnitude and direction of the force the hinge exerts on the bridge.

11.59 A uniform, 250-kg beam is supported using a cable connected to the ceiling, as shown in Fig. 11.36. The lower end of the beam rests on the floor. a) What is the tension in the cable? b) What is the minimum coefficient of static friction between the beam and the floor required for the beam to remain in this position?

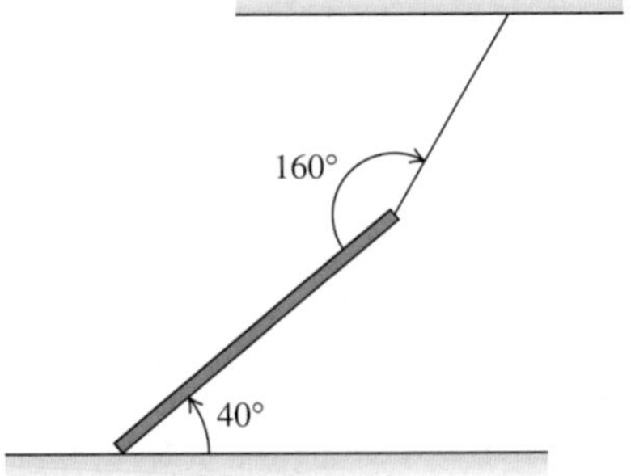

Figure 11.36 Problem 11.59.

11.60 a) In Fig. 11.37 a 6.00-m-long, uniform beam is hanging from a point 1.00 m to the right of its center. The beam weighs 140 N and makes an angle of 30.0° with the vertical. At the right-hand end of the beam a 100-N weight is hung; an unknown weight w hangs at the other end. If the system is in equilibrium, what is w? You can ignore the thickness of the beam. b) If the beam makes, instead, an angle of 45.0° with the vertical, what is w?

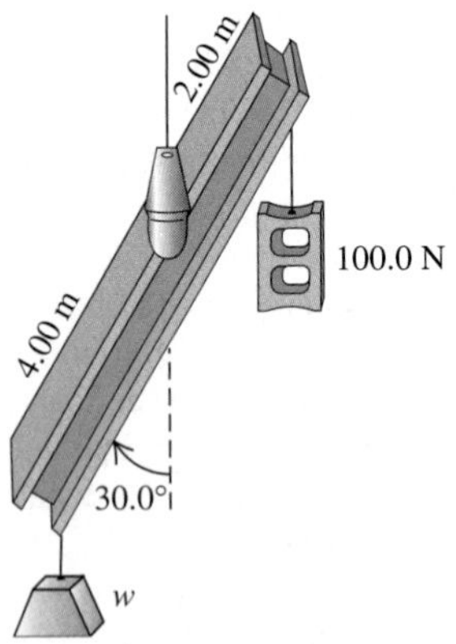

Figure 11.37 Problem 11.60.

11.61 A uniform, horizontal flagpole 5.00 m in length and with a weight of 200 N is hinged to a vertical wall at one end. A 600-N stuntwoman hangs from its other end. The flagpole is supported by a guy wire running from its outer end to a point on the wall directly above the pole. a) If the tension in this wire is not to exceed 1000 N, what is the minimum height above the pole at which it may be fastened to the wall? b) If the flagpole remains horizontal, by how many newtons would the tension be increased if the wire were fastened 0.50 m below this point?

11.62 A holiday decoration consists of two shiny glass spheres with masses 0.0240 kg and 0.0360 kg suspended, as shown in Fig. 11.38, from a uniform rod with mass 0.120 kg and length 1.00 m. The rod is suspended from the ceiling by a vertical cord at each end, so that it is horizontal. Calculate the tension in each of the cords A through F.

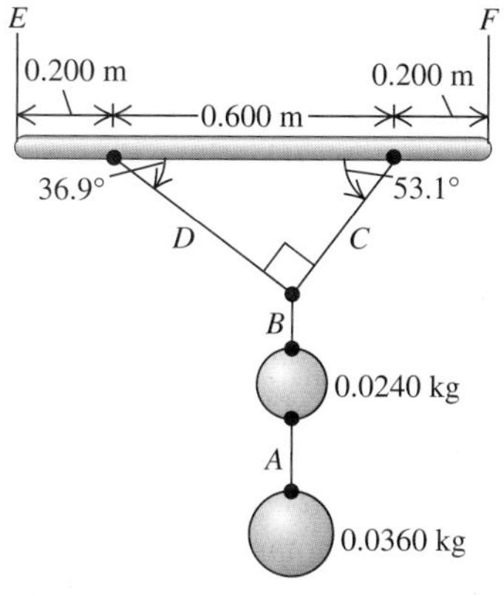

Figure 11.38 Problem 11.62.

11.63 The sign for Lew's Bait Shop and Sushi Restaurant is a uniform rectangle, 1.50 m wide and 0.800 m tall, weighing 300 N. It is supported by a wire at its upper-right corner and a pivot at its lower-left corner. a) Late one night, Lew's competitor from across the street cuts the wire. The sign stays in place because the pivot has rusted tight. After the wire has been cut, what are the magnitude and direction of the force and of the torque (about an axis at the pivot) exerted by the pivot on the sign? b) When the sign was brand new, the pivot was frictionless and the wire was at the angle that minimized the tension in it. What was the tension in the wire and the magnitude and direction of the force exerted on the sign by the pivot?

11.64 When you stretch a wire, rope, or rubber band, it gets thinner as well as longer. When Hooke's law holds, the fractional decrease in width is proportional to the tensile strain. If w_0 is the original width and Δw is the change in width, then $\Delta w/w_0 = -\sigma \Delta l/l_0$, where the minus sign reminds us that width decreases when length increases. The dimensionless constant, σ, different for different materials, is called *Poisson's ratio.* a) If the steel rod of Example 11.6 (Section 11.4) has a circular cross section and a Poisson's ratio of 0.23, what is its change in diameter when the milling machine is hung from it? b) A cylinder made of nickel (Poisson's ratio = 0.42) has radius 2.0 cm. What tensile force $F_\perp$ must be applied perpendicular to each end of the cylinder to cause its radius to decrease by 0.10 mm? Assume that the breaking stress and proportional limit for the metal are extremely large and are not exceeded.

11.65 A circular disk 0.500 m in diameter, pivoted about a horizontal axis through its center, has a cord wrapped around its rim. The cord passes over a frictionless pulley P and is attached to an object that weighs 240 N. A uniform rod 2.00 m long is fastened to the disk, with one end at the center of the disk. The apparatus is in equilibrium, with the rod horizontal (Fig. 11.39). a) What is the weight of the rod? b) What is the new equilibrium direction of the rod when a second object weighing 20.0 N is suspended from the other end of the rod, as shown by the broken line? That is, what angle does the rod then make with the horizontal?

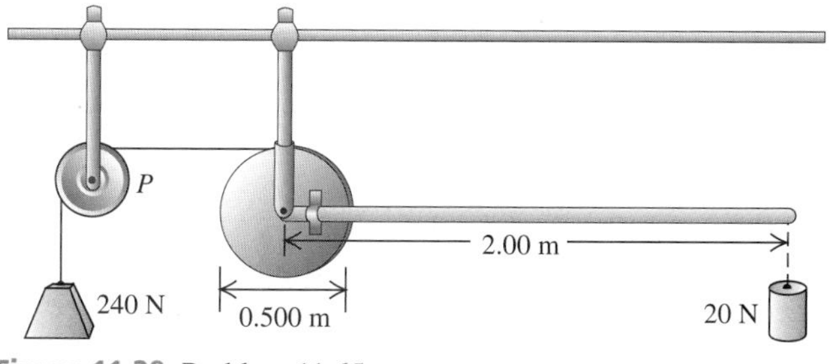

Figure 11.39 Problem 11.65.

11.66 One end of a uniform meter stick is placed against a vertical wall (Fig. 11.40). The other end is held by a lightweight cord that makes an angle θ with the stick. The coefficient of static friction between the end of the meter stick and the wall is 0.40. a) What is the maximum value the angle θ can have if the stick is to remain in equilibrium? b) Let the angle θ be 15°. A block of the same weight as the meter stick is suspended from the stick, as shown, at a distance x from the wall. What is the minimum value of x for which the stick will remain in equilibrium? c) When $\theta = 15°$, how large must the coefficient of static friction be so that the block can be attached 10 cm from the left end of the stick without causing it to slip?

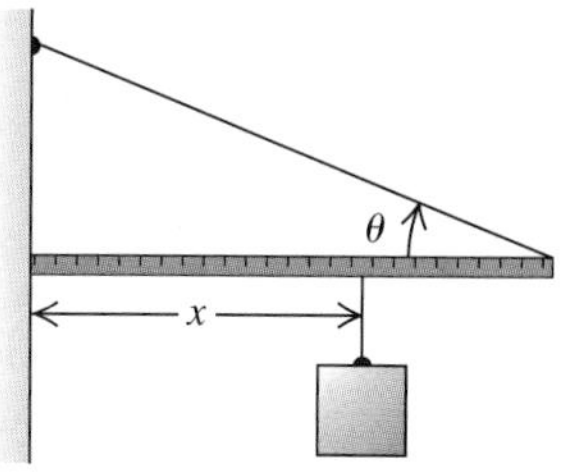

Figure 11.40 Problem 11.66.

11.67 Two friends are carrying a 200-kg crate up a flight of stairs. The crate is 1.25 m long and 0.50 m high, and its center of gravity is at its center. The stairs make a 45.0° angle with respect to the floor. The crate also is carried at a 45.0° angle, so that its bottom side is parallel to the slope of the stairs (Fig. 11.41). If the force each person applies is vertical, what is the magnitude of each of these forces? Is it best to be the person above or below on the stairs?

Figure 11.41 Problem 11.67.

11.68 Forearm. In the human arm, the forearm and hand pivot about the elbow joint. Consider a simplified model in which the biceps muscle is attached to the forearm 3.80 cm from the elbow joint. Assume that the person's hand and forearm together weigh 15.0 N and that their center of gravity is 15.0 cm from the elbow (not quite half-way to the hand). The forearm is held horizontally at a right angle to the upper arm, with the biceps muscle exerting its force perpendicular to the forearm. a) Draw a free-body diagram for the forearm and find the force exerted by the biceps when the hand is empty. b) Now the person holds a 80.0-N weight in his hand, with the forearm still horizontal. Assume that the center of gravity of this weight is 33.0 cm from the elbow. Construct a free-body diagram for the forearm and find the force now exerted by the biceps. Explain why the biceps muscle needs to be very strong. c) Under the conditions of part (b), find the magnitude and direction of the force which the elbow joint exerts on the forearm. d) While holding the 80.0-N weight, the person raises their forearm until it is at an angle of 53.0° above the horizontal. If the biceps muscle continues to exert its force perpendicular to the forearm, what is this force when the forearm is in this position? Has the force increased or decreased from its value in part (b)? Explain why this is so and test your answer by actually doing this with your own arm.

11.69 Refer to the discussion of holding a dumbbell in Example 11.5. The maximum weight that can be held in this way is limited by the maximum allowable tendon tension T (determined by the strength of the tendons) and by the distance D from the elbow to where the tendon attaches to the forearm. a) Let T_{max} represent the maximum value of the tendon tension. Use the results of Example 11.5 to express w_{max} (the maximum weight that can be held) in terms of T_{max}, L, D, and h. Your expression should *not* include the angle θ. b) The tendons of different primates are attached to the forearm at different values of D. Calculate the derivative of w_{max} with respect to D, and determine whether the derivative is positive or negative. c) A chimpanzee tendon is attached to the forearm at a point farther from the elbow than for humans. Use this to explain why chimpanzees have stronger arms than humans. (The disadvantage is that chimpanzees have less flexible arms than do humans.)

11.70 A uniform 90.0-N table is 3.6 m long, 1.0 m high, and 1.2 m wide. A 1500-N weight is placed 0.50 m from one end of the table, a distance of 0.60 m from each of the two legs at that end. Draw a free-body diagram for the table and find the force that each of the four legs exerts on the floor.

11.71 Flying Buttress. a) A symmetric building has a roof sloping upward at 35.0° above the horizontal on each side. If each side of the uniform roof weighs 10,000 N, find the horizontal force that this roof exerts at the top of the wall, which tends to push out the walls. Which type of building would be more in danger of collapsing, one with tall walls or one with short walls? Explain. b) As you saw in part (a), tall walls are in danger of collapsing from the weight of the roof. This problem plagued the ancient builders of large structures. A solution used in the great Gothic cathedrals during the 1200s was the flying buttress, a stone support running between the walls and the ground, that helped to hold in the walls. A Gothic church has a uniform roof weighing a total of 20,000 N and rising at 40° above the horizontal at each wall. The walls are 40 m tall, and a flying buttress meets each wall 10 m below the base of the roof. What horizontal force must this flying buttress apply to the wall?

11.72 You are trying to raise a bicycle wheel of mass m and radius R up over a curb of height h. To do this, you apply a horizontal force $\vec{F}$. What is the least magnitude of the force $\vec{F}$ that will succeed in raising the wheel onto the curb when the force is applied a) at the center of the wheel (Fig. 11.42)? b) at the top of the wheel? c) In which case is less force required?

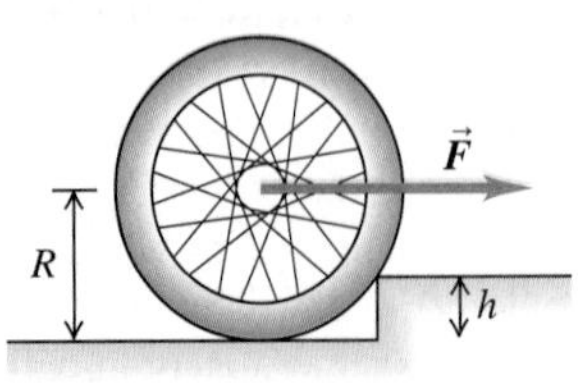

Figure 11.42 Problem 11.72.

11.73 The Farmyard Gate. A gate 4.00 m wide and 2.00 m high weighs 500 N. Its center of gravity is at its center, and it is hinged at A and B. To relieve the strain on the top hinge, a wire CD is connected as shown in Fig. 11.43. The tension in CD is increased until the horizontal force at hinge A is zero. a) What is the tension in the wire CD? b) What is the magnitude of the horizontal component of the force at hinge B? c) What is the combined vertical force exerted by hinges A and B?

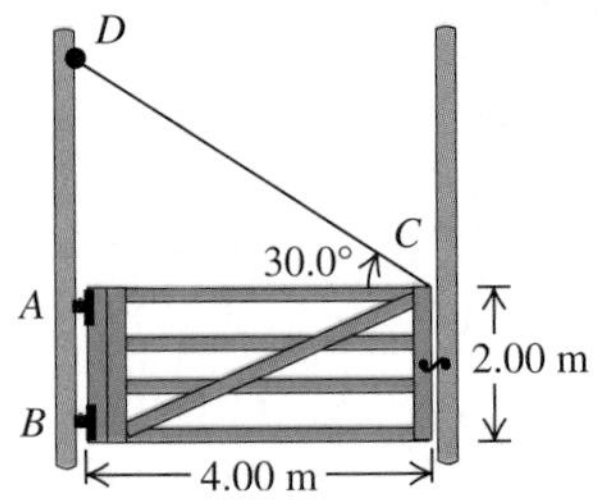

Figure 11.43 Problem 11.73.

11.74 If you put a uniform block at the edge of a table, the center of the block must be over the table for the block not to fall off. a) If you stack two identical blocks at the table edge, the center of the top block must be over the bottom block, and the center of gravity of the two blocks together must be over the table. In terms of the length L of each block, what is the maximum overhang possible (Fig. 11.44)? b) Repeat part (a) for three identical blocks and for four identical blocks. c) Is it possible to make a stack of blocks such that the uppermost block is not directly over the table at all? How many blocks would it take to do this? (Try this with your friends using copies of this book.)

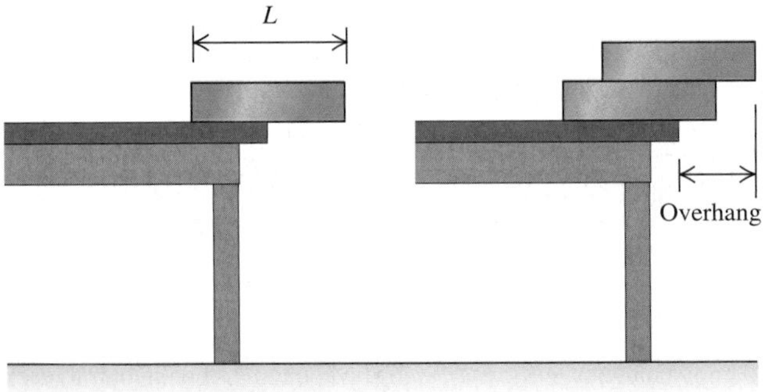

Figure 11.44 Problem 11.74.

11.75 Two uniform 75.0-g marbles 2.00 cm in diameter are stacked as shown in Fig. 11.45 in a container that is 3.00 cm wide. a) Find the force that the container exerts on the marbles at the points of contact A, B, and C. b) What force does each marble exert on the other?

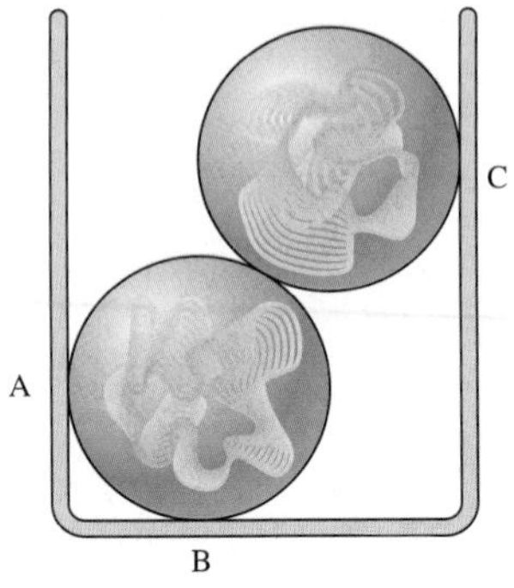

Figure 11.45 Problem 11.75.

11.76 Two identical, uniform beams weighing 260 N each are connected at one end by a frictionless hinge. A light horizontal crossbar attached at the midpoints of the beams maintains an angle of 53.0° between the beams. The beams are suspended from the ceiling by vertical wires such that they form a "V," as shown in Fig. 11.46. a) What force does the crossbar exert on each beam? b) Is the crossbar under tension or compression? c) What force (magnitude and direction) does the hinge at point A exert on each beam?

Figure 11.46 Problem 11.76.

11.77 An engineer is designing a conveyor system for loading hay bales into a wagon (Fig. 11.47). Each bale is 0.25 m wide, 0.50 m high, and 0.80 m long (the dimension perpendicular to the plane of the figure), with mass 30.0 kg. The center of gravity of each bale is at its geometrical center. The coefficient of static friction between a bale and the conveyor belt is 0.60, and the belt moves with constant speed. a) The angle β of the conveyor is slowly increased. At some critical angle a bale will tip (if it doesn't slip first), and at some different critical angle it will slip (if it doesn't tip first). Find the two critical angles and determine which happens at the smaller angle. b) Would the outcome of part (a) be different if the coefficient of friction were 0.40?

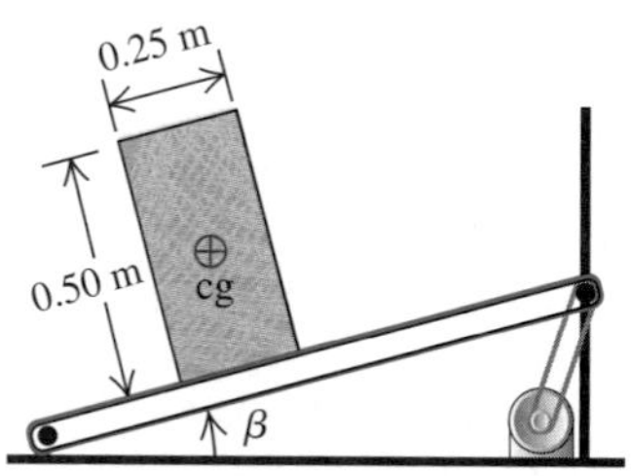

Figure 11.47 Problem 11.77.

11.78 The hay bale of Problem 11.77 is dragged along a horizontal surface with constant speed by a force $\vec{F}$ (Fig. 11.48). The coefficient of kinetic friction is 0.35. a) Find the magnitude of the force $\vec{F}$. b) Find the value of h at which the bale just begins to tip.

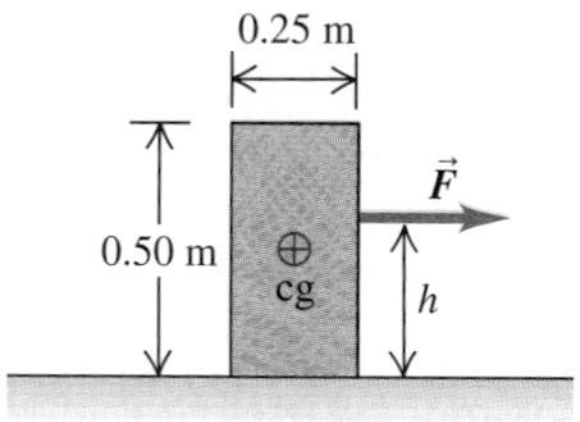

Figure 11.48 Problem 11.78.

11.79 A garage door is mounted on an overhead rail (Fig. 11.49). The wheels at A and B have rusted so that they do not roll, but rather slide along the track. The coefficient of kinetic friction is 0.52. The distance between the wheels is 2.00 m, and each is 0.50 m from the vertical sides of the door. The door is uniform and weighs 950 N. It is pushed to the left at constant speed by a horizontal force $\vec{F}$. a) If the distance h is 1.60 m, what is the vertical component of the force exerted on each wheel by the track? b) Find the maximum value h can have without causing one wheel to leave the track.

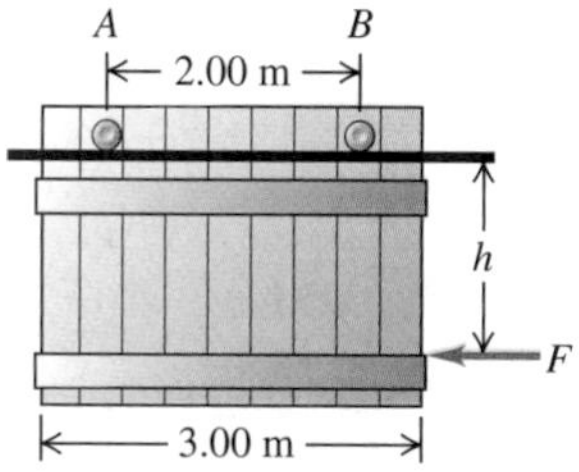

Figure 11.49 Problem 11.79.

11.80 A horizontal boom is supported at its left end by a frictionless pivot. It is held in place by a cable attached to the right-hand end of the boom. A chain and crate of total weight w hang from somewhere along the boom. The boom's weight w_b cannot be ignored and the boom may or may not be uniform. a) Show that the tension in the cable is the same whether the cable makes an angle θ or an angle $180° - \theta$ with the horizontal, and that the horizontal force component exerted on the boom by the pivot has equal magnitude but opposite direction for the two angles. b) Show that the cable cannot be horizontal. c) Show that the tension in the cable is a minimum when the cable is vertical, pulling upward on the right end of the boom. d) Show that when the cable is vertical, the force exerted by the pivot on the boom is vertical.

11.81 Prior to being placed in its hole, a 5700-N, 9.0-m-long, uniform utility pole makes some nonzero angle with the vertical. A vertical cable attached 2.0 m below its upper end holds it in place while its lower end rests on the ground. a) Find the tension in the cable and the magnitude and direction of the force exerted by the ground on the pole. b) Why don't we need the angle the pole makes with the vertical, as long as it is not zero?

11.82 Hooke's law for a tensile stress can be written as $F_x = kx$, where x is the object's change in length and k is the force constant. a) What is the force constant of a rod of length l_0, cross-sectional area A, and Young's modulus Y? b) In terms of l_0, A, and Y, how much work is required to stretch the object a distance x?

11.83 A 12.0-kg mass, fastened to the end of an aluminum wire with an unstretched length of 0.50 m, is whirled in a vertical circle with a constant angular speed of 120 rev/min. The cross-sectional area of the wire is 0.014 cm^2. Calculate the elongation of the wire when the mass is a) at the lowest point of the path and b) at the highest point of its path.

11.84 A metal wire 3.50 m long and 0.70 mm in diameter was given the following test. A load weighing 20 N was originally hung from the wire to keep it taut. The position of the lower end of the wire was read on a scale as load was added.

Added Load (N)	Scale Reading (cm)
0	3.02
10	3.07
20	3.12
30	3.17
40	3.22
50	3.27
60	3.32
70	4.27

a) Graph these values, plotting the increase in length horizontally and the added load vertically. b) Calculate the value of Young's modulus. c) The proportional limit occurred at a scale reading of 3.34 cm. What was the stress at this point?

11.85 A 1.05-m-long rod of negligible weight is supported at its ends by wires A and B of equal length (Fig. 11.50). The cross-sectional area of A is 2.00 mm^2 and that of B is 4.00 mm^2. Young's modulus for wire A is 1.80×10^{11} Pa; that for B is 1.20×10^{11} Pa. At what point along the rod should a weight w be suspended to produce a) equal stresses in A and B? b) equal strains in A and B?

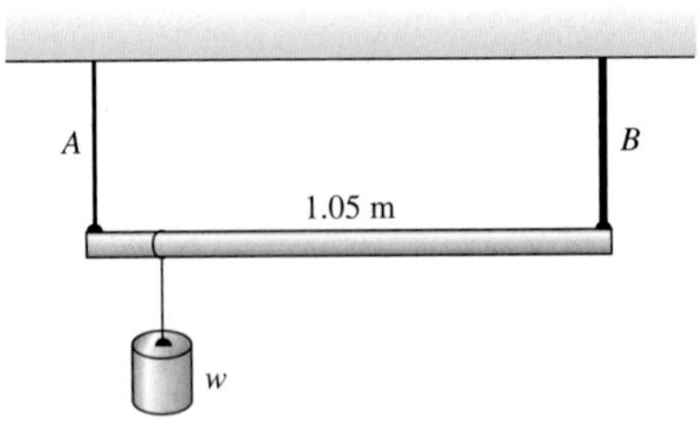

Figure 11.50 Problem 11.85.

11.86 An amusement park ride consists of airplane-shaped cars attached to steel rods (Fig. 11.51). Each rod has a length of 15.0 m and a cross-sectional area of 8.00 cm^2. a) How much is the rod stretched when the ride is at rest? (Assume that each car plus two people seated in it has a total weight of 1900 N.) b) When operating, the ride has a maximum angular speed of 8.0 rev/min. How much is the rod stretched then?

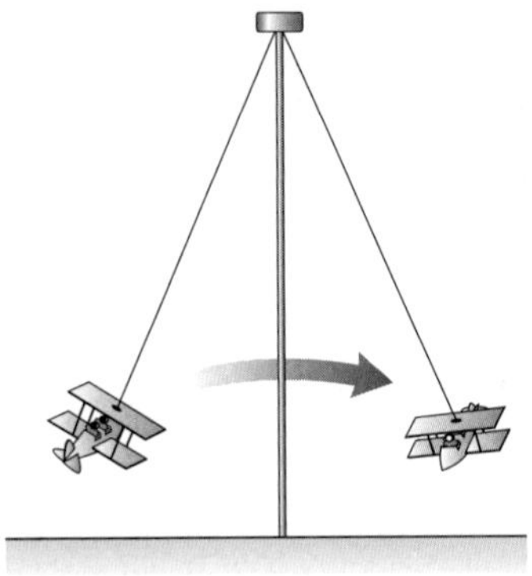

Figure 11.51 Problem 11.86.

11.87 A brass rod with a length of 1.40 m and a cross-sectional area of 2.00 cm^2 is fastened end-to-end to a nickel rod with length L and cross-sectional area 1.00 cm^2. The compound rod is subjected to equal and opposite pulls of magnitude 4.00×10^4 N at its ends. a) Find the length L of the nickel rod if the elongations of the two rods are equal. b) What is the stress in each rod? c) What is the strain in each rod?

11.88 Stress on the Shin Bone. Compressive strength of our bones is important in everyday life. Young's modulus for bone is about 1.4×10^{10} Pa. Bone can take only about a 1.0% change in its length before fracturing. a) What is the maximum force that can be applied to a bone whose minimum cross-sectional area is 3.0 cm^2? (This is approximately the cross-sectional area of a tibia, or shin bone, at its narrowest point.) b) Estimate the maximum height from which a 70-kg man could jump and not fracture the tibia. Take the time between when he first touches the floor and when he has stopped to be 0.030 s, and assume that the stress is distributed equally between his legs.

11.89 You hang a floodlamp from the end of a vertical steel wire. The floodlamp stretches the wire 0.18 mm and the stress is proportional to the strain. How much would it have stretched a) if the wire were twice as long? b) if the wire had the same length but twice the diameter? c) for a copper wire of the original length and diameter?

11.90 A moonshiner produces pure ethanol (ethyl alcohol) late at night and stores it in a stainless steel tank in the form of a cylinder 0.300 m in diameter with a tight-fitting piston at the top. The total volume of the tank is 250 L $(0.250\ m^3)$. In an attempt to squeeze a little more into the tank, the moonshiner piles 1420 kg of lead bricks on top of the piston. What additional volume of ethanol can the moonshiner squeeze into the tank? (Assume that the wall of the tank is perfectly rigid.)

11.91 A bar with cross-sectional area A is subjected to equal and opposite tensile forces $\vec{F}$ at its ends. Consider a plane through the bar making an angle θ with a plane at right angles to the bar (Fig. 11.52). a) What is the tensile (normal) stress at this plane in terms of F, A, and θ? b) What is the shear (tangential) stress at the plane in terms of F, A, and θ? c) For what value of θ is the tensile stress a maximum? d) For what value of θ is the shear stress a maximum?

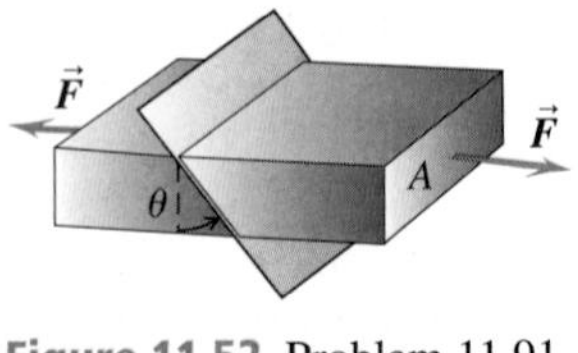

Figure 11.52 Problem 11.91.

11.92 A horizontal, uniform, copper rod has an original length l_0, a cross-sectional area A, a Young's modulus Y, and a mass m. It is supported by a frictionless pivot at its right end and by a cable at its left end. Both pivot and cable are attached so that they exert their forces uniformly over the rod's cross section. The cable makes an angle θ with the rod and compresses it. a) Find the stress exerted by the cable and pivot on the rod. b) Find the change in length of the rod due to this stress. c) The mass of the rod equals $\rho A l_0$, where ρ is the density. Show that the answers to parts (a) and (b) are independent of the cross-sectional area of the rod. d) The density of copper is 8900 kg/m^3. Take Y for compression as given for copper in Table 11.1. Find the stress and change in length for an original length of 1.8 m and an angle of 30°. e) By how much would you multiply the answers of part (d) if the rod were twice as long?

Challenge Problems

11.93 A bookcase weighing 1500 N rests on a horizontal surface for which the coefficient of static friction is $\mu_s = 0.40$. The bookcase is 1.80 m tall and 2.00 m wide; its center of gravity is at its geometrical center. The bookcase rests on four short legs that are each 0.10 m from the edge of the bookcase. A person pulls on a rope attached to an upper corner of the bookcase with a force $\vec{F}$ that makes an angle θ with the bookcase (Fig. 11.53). a) If $\theta = 90°$, so $\vec{F}$ is horizontal, show that as F is increased from zero, the bookcase will start to slide before it tips, and calculate the magnitude of $\vec{F}$ that will start the bookcase sliding. b) If $\theta = 0°$, so $\vec{F}$ is vertical, show that the bookcase will tip over rather than slide, and calculate the magnitude of $\vec{F}$ that will cause the bookcase to start to tip. c) Calculate as a function of θ the magnitude of $\vec{F}$ that will cause the bookcase to start to slide and the magnitude that will cause it to start to tip. What is the smallest value that θ can have so that the bookcase will still start to slide before it starts to tip?

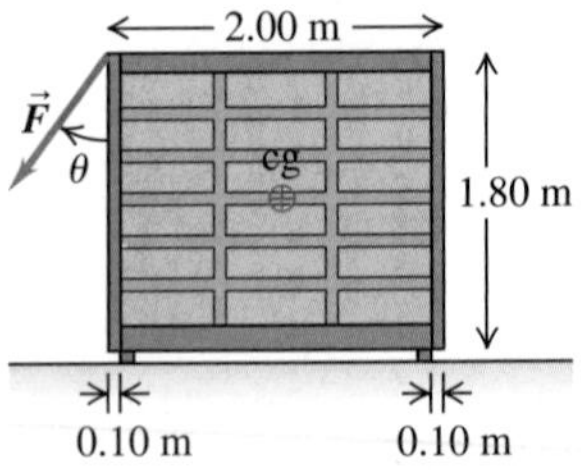

Figure 11.53 Challenge Problem 11.93.

11.94 Knocking Over a Post. One end of a post weighing 400 N and with height h rests on a rough horizontal surface with $\mu_s = 0.30$. The upper end is held by a rope fastened to the surface and making an angle of 36.9° with the post (Fig. 11.54). A horizontal force $\vec{F}$ is exerted on the post as shown. a) If the force $\vec{F}$ is applied at the midpoint of the post, what is the largest value it can have without causing the post to slip? b) How large can the force be without causing the post to slip if its point of application is $\frac{6}{10}$ of the way from the ground to the top of the post? c) Show that if the point of application of the force is too high, the post cannot be made to slip, no matter how great the force. Find the critical height for the point of application.

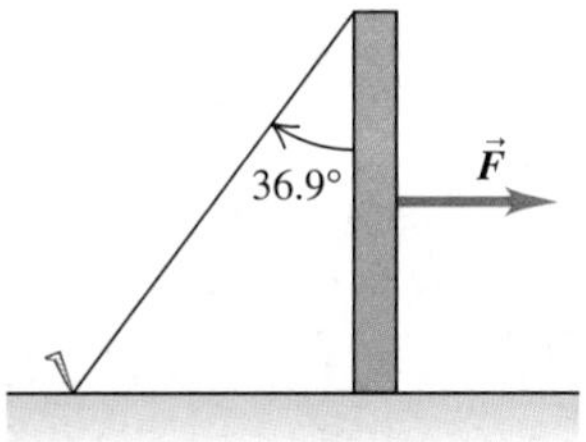

Figure 11.54 Challenge Problem 11.94.

11.95 Minimizing the Tension. A heavy horizontal girder of length L has several objects suspended from it. It is supported by a frictionless pivot at its left end and a cable of negligible weight that is attached to an I-beam at a point a distance h directly above the girder's center. Where should the other end of the cable be attached to the girder so that the cable's tension is a minimum? (*Hint:* In evaluating and presenting your answer, don't forget that the maximum distance of the point of attachment from the pivot is the length L of the beam.)

11.96 Two ladders, 4.00 m and 3.00 m long, are hinged at point A and tied together by a horizontal rope 0.90 m above the floor (Fig. 11.55). The ladders weigh 480 N and 360 N, respectively, and the center of gravity of each is at its center. Assume that the floor is freshly waxed and frictionless. a) Find the upward force at the bottom of each ladder. b) Find the tension in the rope. c) Find the magnitude of the force one ladder exerts on the other at point A. d) If an 800-N painter stands at point A, find the tension in the horizontal rope.

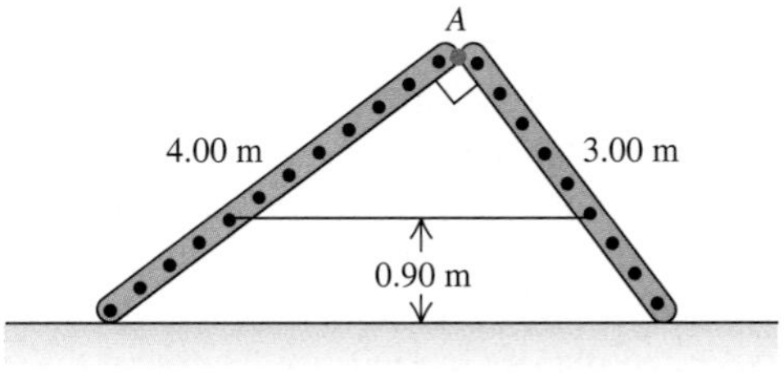

Figure 11.55 Challenge Problem 11.96.

11.97 A device for measuring compressibility consists of a cylinder filled with oil and fitted with a piston at one end. A block of sodium is immersed in the oil and a force is applied to the piston. Assume that the piston and walls of the cylinder are perfectly rigid and that there is no friction and no oil leak. Compute the compressibility of the sodium in terms of the applied force F, the piston displacement x, the piston area A, the initial volume of the oil V_O, the initial volume of the sodium V_S, and the compressibility of the oil k_O.

11.98 Bulk Modulus of an Ideal Gas. The equation of state (the equation relating pressure, volume, and temperature) for an ideal gas is $pV = nRT$, where n and R are constants. a) Show that if the gas is compressed while the temperature T is held constant, the bulk modulus is equal to the pressure. b) When an ideal gas is compressed without the transfer of any heat into or out of it, the pressure and volume are related by $pV^\gamma = \text{constant}$, where γ is a constant having different values for different gases. Show that, in this case, the bulk modulus is given by $B = \gamma p$.

11.99 An angler hangs a 4.50-kg fish from a vertical steel wire 1.50 m long and $5.00 \times 10^{-3}\ \text{cm}^2$ in cross-sectional area. The upper end of the wire is securely fastened to a support. a) Calculate the amount the wire is stretched by the hanging fish. The angler now applies a force $\vec{F}$ to the fish, pulling it very slowly downward by 0.500 mm from its equilibrium position. For this downward motion, calculate b) the work done by gravity; c) the work done by the force $\vec{F}$; d) the work done by the force the wire exerts on the fish; and e) the change in the elastic potential energy (the potential energy associated with the tensile stress in the wire). Compare the answers in parts (d) and (e). (*Hint:* See Problem 11.82.)

12 GRAVITATION

These two immense galaxies—each containing billions of stars and located 114 million light years from earth—have been undergoing a slow-motion collision for tens of millions of years. This "collision" is actually a gravitational interaction between the two galaxies. (The individual stars of the two galaxies are so widely separated that they are unlikely to actually collide with each other.)

? These two galaxies together have a mass more than 10^{11} times larger than the sun's mass. Why, then, isn't the earth pulled away from the sun and toward these galaxies?

Some of the earliest investigations in physical science started with questions that people asked about the night sky. Why doesn't the moon fall to earth? Why do the planets move across the sky? And why doesn't the earth fly off into space rather than remaining in orbit around the sun? The study of gravitation provides the answers for these and many related questions.

As we remarked in Chapter 5, gravitation is one of the four classes of interactions found in nature, and it was the earliest of the four to be studied extensively. Newton discovered in the 17th century that the same interaction that makes an apple fall out of a tree also keeps the planets in their orbits around the sun. This was the beginning of *celestial mechanics,* the study of the dynamics of objects in space. Today, our knowledge of celestial mechanics allows us to determine how to put a satellite into any desired orbit around the earth or to choose just the right trajectory to send a spacecraft to another planet.

In this chapter you will learn the basic law that governs gravitational interactions. This law is *universal:* gravity acts in the same fundamental way between the earth and your body, between the sun and a planet, and between a planet and one of its moons. We'll apply the law of gravitation to phenomena such as the variation of weight with altitude, the orbits of satellites around the earth, and the orbits of planets around the sun.

12.1 | Newton's Law of Gravitation

The example of gravitational attraction that's probably most familiar to you is your *weight,* the force that attracts you toward the earth. During his study of the motions of the planets and of the moon, Newton discovered the fundamental character of the gravitational attraction between *any* two bodies. Along with his three laws of motion, Newton published the **law of gravitation** in 1687. It may be stated as follows:

Every particle of matter in the universe attracts every other particle with a force that is directly proportional to the product of the masses of the particles and inversely proportional to the square of the distance between them.

Translating this into an equation, we have

$$F_g = \frac{Gm_1m_2}{r^2} \quad \text{(law of gravitation)} \tag{12.1}$$

where F_g is the magnitude of the gravitational force on either particle, m_1 and m_2 are their masses, r is the distance between them (Fig. 12.1), and G is a fundamental physical constant called the **gravitational constant.** The numerical value of G depends on the system of units used.

Equation (12.1) tells us that the gravitational force between two particles decreases with increasing distance r: if the distance is doubled, the force is only one-fourth as great, and so on. Although many of the stars in the night sky are far more massive than the sun, they are so far away that their gravitational force on the earth is negligibly small.

CAUTION Because the symbols g and G are almost the same, it's common to confuse the two gravitational quantities that these symbols represent. Lowercase g is the acceleration due to gravity, which relates the weight w of a body to its mass m: $w = mg$. The value of g is different at different locations on the earth's surface and on the surfaces of different planets. By contrast, capital G relates the gravitational force between any two bodies to their masses and the distance between them. We call G a *universal* constant because it has the same value for any two bodies, no matter where in space they are located. In the next section we'll see how the values of g and G are related, but remember that they're two very different quantities!

Gravitational forces always act along the line joining the two particles, and they form an action-reaction pair. Even when the masses of the particles are different, the two interaction forces have equal magnitude (Fig. 12.1). The attractive force that your body exerts on the earth has the same magnitude as the force that it exerts on you. When you fall from a diving board into a swimming pool, the entire earth rises up to meet you! (Why don't you notice this? The earth's mass is greater than yours by a factor of about 10^{23}, so its acceleration is only 10^{-23} as great as yours.)

We have stated the law of gravitation in terms of the interaction between two *particles.* It turns out that the gravitational interaction of any two bodies having *spherically symmetric* mass distributions (such as solid spheres or spherical shells) is the same as though we concentrated all the mass of each at its center, as

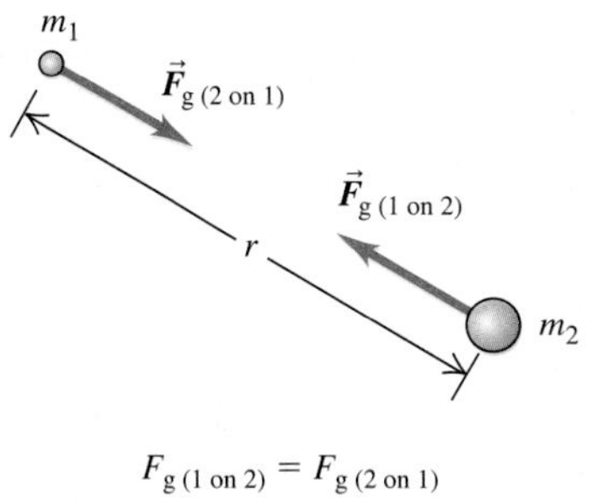

12.1 Even if two particles have very different masses, they exert equally strong forces of gravitational attraction on each other.

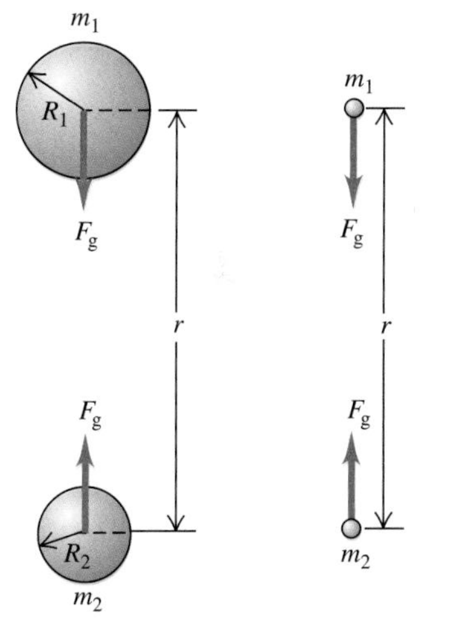

(a) Two spherically symmetric masses m_1 and m_2

(b) Two particles with masses m_1 and m_2

12.2 The gravitational effect *outside* any spherically symmetric mass distribution is the same as though all of the mass were concentrated at its center.

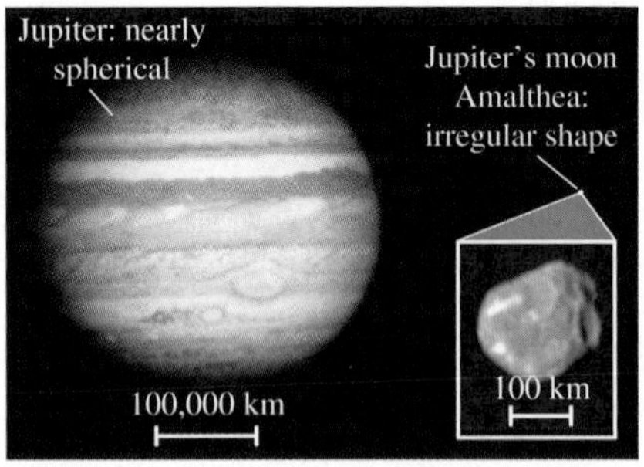

12.3 Thanks to the planet Jupiter's immense mass of 1.90×10^{27} kg, the mutual gravitational attraction of its atoms has pulled the planet into a nearly spherical shape. This has not happened for Amalthea (one of Jupiter's moons, shown in close-up). With a relatively tiny mass of 7.17×10^{18} kg, Amalthea has only weak mutual gravitation and so has an irregular shape.

in Fig. 12.2. Thus, if we model the earth as a spherically symmetric body with mass m_E, the force it exerts on a particle or a spherically symmetric body with mass m, at a distance r between centers, is

$$F_g = \frac{Gm_E m}{r^2} \tag{12.2}$$

provided that the body lies outside the earth. A force of the same magnitude is exerted *on* the earth by the body. (We will prove these statements in Section 12.6.)

At points *inside* the earth the situation is different. If we could drill a hole to the center of the earth and measure the gravitational force on a body at various depths, we would find that toward the center of the earth the force *decreases,* rather than increasing as $1/r^2$. As the body enters the interior of the earth (or other spherical body), some of the earth's mass is on the side of the body opposite from the center and pulls in the opposite direction. Exactly at the center, the earth's gravitational force on the body is zero.

Spherically symmetric bodies are an important case because moons, planets, and stars all tend to be spherical. Since all particles in a body gravitationally attract each other, the particles tend to move to minimize the distance between them. As a result, the body naturally tends to assume a spherical shape, just as a lump of clay forms into a sphere if you squeeze it with equal forces on all sides. This effect is greatly reduced in celestial bodies of low mass since the gravitational attraction is less, and these bodies tend *not* to be spherical (Fig. 12.3).

Determining the Value of G

To determine the value of the gravitational constant G, we have to *measure* the gravitational force between two bodies of known masses m_1 and m_2 at a known distance r. The force is extremely small for bodies that are small enough to be brought into the laboratory, but it can be measured with an instrument called a *torsion balance,* which Sir Henry Cavendish used in 1798 to determine G.

A modern version of the Cavendish torsion balance is shown in Fig. 12.4. A light, rigid rod shaped like an inverted T is supported by a very thin, vertical quartz fiber. Two small spheres, each of mass m_1, are mounted at the ends of the horizontal arms of the T. When we bring two large spheres, each of mass m_2, to the positions shown, the attractive gravitational forces twist the T through a small angle. To

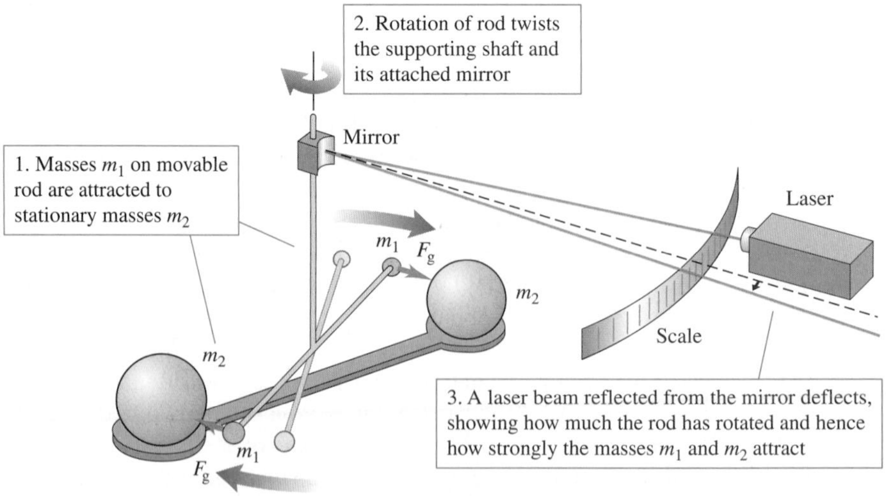

12.4 The principle of the Cavendish balance, used for determining the value of G. The angle of deflection has been exaggerated here for clarity.

measure this angle, we shine a beam of light on a mirror fastened to the T. The reflected beam strikes a scale, and as the T twists, the reflected beam moves along the scale.

After calibrating the Cavendish balance, we can measure gravitational forces and thus determine G. The presently accepted value (in SI units) is

$$G = 6.673(10) \times 10^{-11}\ \mathrm{N \cdot m^2/kg^2}$$

To three significant figures, $G = 6.67 \times 10^{-11}\ \mathrm{N \cdot m^2/kg^2}$. Because $1\ \mathrm{N} = 1\ \mathrm{kg \cdot m/s^2}$, the units of G can also be expressed (in fundamental SI units) as $\mathrm{m^3/(kg \cdot s^2)}$.

Gravitational forces combine vectorially. If each of two masses exerts a force on a third, the *total* force on the third mass is the vector sum of the individual forces of the first two. Example 12.3 makes use of this property, which is often called *superposition of forces.*

Example 12.1 Calculating gravitational force

The mass m_1 of one of the small spheres of a Cavendish balance is 0.0100 kg, the mass m_2 of one of the large spheres is 0.500 kg, and the center-to-center distance between each large sphere and the nearer small one is 0.0500 m. Find the gravitational force F_g on each sphere due to the nearest other sphere.

SOLUTION

IDENTIFY and SET UP: We will use the law of gravitation, Eq. (12.1), to determine F_g.

EXECUTE: Each sphere experiences the *same* magnitude of force from the other sphere, even though the masses may be very different (as they are in this case). The magnitude of each force is

$$F_g = \frac{(6.67 \times 10^{-11}\ \mathrm{N \cdot m^2/kg^2})(0.0100\ \mathrm{kg})(0.500\ \mathrm{kg})}{(0.0500\ \mathrm{m})^2}$$

$$= 1.33 \times 10^{-10}\ \mathrm{N}$$

EVALUATE: This is a very small force, which is what we expect. After all, we don't experience noticeable gravitational pulls from ordinary low-mass objects in our environment. It takes a truly massive object such as the earth to exert a substantial gravitational force.

Example 12.2 Acceleration due to gravitational attraction

Suppose one large sphere and one small sphere are detached from the apparatus in Example 12.1 and placed 0.0500 m (between centers) from each other at a point in space far removed from all other bodies. What is the magnitude of the acceleration of each, relative to an inertial system?

SOLUTION

IDENTIFY and SET UP: We found the magnitude of the force on each sphere in Example 12.1. (The system of two spheres is so distant from other bodies that we can neglect any other forces.) To determine the magnitude of each sphere's acceleration, we'll use Newton's second law.

EXECUTE: The acceleration a_1 of the smaller sphere has magnitude

$$a_1 = \frac{F_g}{m_1} = \frac{1.33 \times 10^{-10}\ \mathrm{N}}{0.0100\ \mathrm{kg}} = 1.33 \times 10^{-8}\ \mathrm{m/s^2}$$

The acceleration a_2 of the larger sphere has magnitude

$$a_2 = \frac{F_g}{m_2} = \frac{1.33 \times 10^{-10}\ \mathrm{N}}{0.500\ \mathrm{kg}} = 2.66 \times 10^{-10}\ \mathrm{m/s^2}$$

EVALUATE: Although the forces on the bodies are equal in magnitude, the magnitudes of the two accelerations are *not* equal. Also, the accelerations are not constant; the gravitational forces increase as the spheres start to move toward each other.

Example 12.3 Superposition of gravitational forces

Many stars in the sky are actually systems of two or more stars that remain together due to their mutual gravitational attraction. Figure 12.5 shows a three-star system at an instant when the stars are at the vertices of a 45° right triangle. (For comparison, the mass of the sun—a rather ordinary star—is 1.99×10^{30} kg and the earth–sun distance is 1.50×10^{11} m.) We assume that the stars are spherical so that we can replace each star by a point mass at its center, as in Fig. 12.2. Find the magnitude and direction of the total gravitational force exerted on the small star by the two large ones.

SOLUTION

IDENTIFY: We use the principle of superposition: the total force on the small star is the vector sum of the forces due to each large star.

SET UP: We assume that the stars are spheres so that we can use the law of gravitation for each force. We first calculate the magnitude of each force using Eq. (12.1), then compute the vector sum using components along the axes shown in Fig. 12.5.

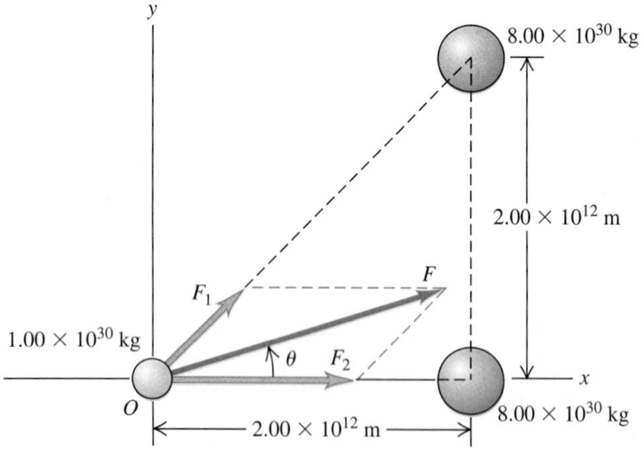

12.5 The total gravitational force on the smaller star is the vector sum of the forces exerted on it by the two larger stars.

EXECUTE: The magnitude F_1 of the force on the small star due to the upper large one is

$$F_1 = \frac{\left[\begin{matrix}(6.67 \times 10^{-11}\ \text{N}\cdot\text{m}^2/\text{kg}^2)\\ \times (8.00 \times 10^{30}\ \text{kg})(1.00 \times 10^{30}\ \text{kg})\end{matrix}\right]}{(2.00 \times 10^{12}\ \text{m})^2 + (2.00 \times 10^{12}\ \text{m})^2}$$
$$= 6.67 \times 10^{25}\ \text{N}$$

The magnitude F_2 of the force due to the lower large star is

$$F_2 = \frac{\left[\begin{matrix}(6.67 \times 10^{-11}\ \text{N}\cdot\text{m}^2/\text{kg}^2)\\ \times (8.00 \times 10^{30}\ \text{kg})(1.00 \times 10^{30}\ \text{kg})\end{matrix}\right]}{(2.00 \times 10^{12}\ \text{m})^2}$$
$$= 1.33 \times 10^{26}\ \text{N}$$

The x- and y-components of these forces are

$$F_{1x} = (6.67 \times 10^{25}\ \text{N})(\cos 45°) = 4.72 \times 10^{25}\ \text{N}$$
$$F_{1y} = (6.67 \times 10^{25}\ \text{N})(\sin 45°) = 4.72 \times 10^{25}\ \text{N}$$
$$F_{2x} = 1.33 \times 10^{26}\ \text{N}$$
$$F_{2y} = 0$$

The components of the total force on the small star are

$$F_x = F_{1x} + F_{2x} = 1.81 \times 10^{26}\ \text{N}$$
$$F_y = F_{1y} + F_{2y} = 4.72 \times 10^{25}\ \text{N}$$

The magnitude of this force is

$$F = \sqrt{F_x^2 + F_y^2} = \sqrt{(1.81 \times 10^{26}\ \text{N})^2 + (4.72 \times 10^{25}\ \text{N})^2}$$
$$= 1.87 \times 10^{26}\ \text{N}$$

and its direction relative to the x-axis is

$$\theta = \arctan\frac{F_y}{F_x} = \arctan\frac{4.72 \times 10^{25}\ \text{N}}{1.81 \times 10^{26}\ \text{N}} = 14.6°$$

EVALUATE: While the total force on the small star is tremendous, the magnitude of the resulting acceleration is not: $a = F/m = (1.87 \times 10^{26}\ \text{N})/(1.00 \times 10^{30}\ \text{kg}) = 1.87 \times 10^{-4}\ \text{m/s}^2$.

Can you show that the total force on the small star is *not* directed toward the center of mass of the two large stars? (See Problem 12.48.)

Comparing Examples 12.1 and 12.3 shows that gravitational forces are negligible between ordinary household-sized objects, but very substantial between objects that are the size of stars. Indeed, gravitation is *the* most important force on the scale of planets, stars, and galaxies (like those shown in the photograph that opens this chapter). It is responsible for holding our earth together and for keeping the planets in orbit about the sun. The mutual gravitational attraction between different parts of the sun compresses material at the sun's core to very high densities and temperatures, making it possible for nuclear reactions to take place there. These reactions generate the sun's energy output, which makes it possible for life to exist on earth and for you to read these words.

The gravitational force is so important on the cosmic scale because it acts *at a distance,* without any direct contact between bodies. Electric and magnetic forces have this same remarkable property, but they are less important on astronomical scales because large accumulations of matter are electrically neutral—that is, they contain equal amounts of positive and negative charge. As a result, the electric and magnetic forces between stars or planets are very small or zero. The strong and weak interactions that we discussed in Section 5.5 also act at a distance, but their influence is negligible at distances much greater than the diameter of an atomic nucleus (about 10^{-14} m).

A useful way to describe forces that act at a distance is in terms of a *field.* One body sets up a disturbance or field at all points in space, and the force that acts on a second body at a particular point is its response to the first body's field at that point. There is a field associated with each force that acts at a distance, and so we refer to gravitational fields, electric fields, magnetic fields, and so on. We won't need the field concept for our study of gravitation in this chapter, so we won't discuss it further here. But in later chapters we'll find that the field description is an extraordinarily powerful tool for describing electric and magnetic interactions.

Test Your Understanding

Calculate the magnitude of the gravitational force between two 900-kg automobiles when they are 3.0 m apart. Do gravitational forces play an appreciable role in automobile collisions?

12.2 | Weight

We defined the *weight* of a body in Section 4.4 as the attractive gravitational force exerted on it by the earth. We can now broaden our definition. **The weight of a body is the total gravitational force exerted on the body by all other bodies in the universe.** When the body is near the surface of the earth, we can neglect all other gravitational forces and consider the weight as just the earth's gravitational attraction. At the surface of the *moon* we consider a body's weight to be the gravitational attraction of the moon, and so on.

If we again model the earth as a spherically symmetric body with radius R_E and mass m_E, the weight w of a small body of mass m at the earth's surface (distance R_E from its center) is

$$w = F_g = \frac{Gm_E m}{R_E^2} \quad \text{(weight of a body of mass } m \text{ at the earth's surface)} \quad (12.3)$$

But we also know from Section 4.4 that the weight w of a body is the force that causes the acceleration g of free fall, so by Newton's second law, $w = mg$. Equating this with Eq. (12.3) and dividing by m, we find

$$g = \frac{Gm_E}{R_E^2} \quad \text{(acceleration due to gravity at the earth's surface)} \quad (12.4)$$

12.6 In an airliner at high altitude, you are farther from the center of the earth than when on the ground and you actually weigh less. The effect is quite small, though measurable. Can you show that at an altitude of 10 km above the surface, you weigh just 0.3% less than you do on the ground?

The acceleration due to gravity g is independent of the mass m of the body because m doesn't appear in this equation. We already knew that, but we can now see how it follows from the law of gravitation.

We can *measure* all the quantities in Eq. (12.4) except for m_E, so this relation allows us to compute the mass of the earth. Solving Eq. (12.4) for m_E and using $R_E = 6380\text{ km} = 6.38 \times 10^6\text{ m}$ and $g = 9.80\text{ m/s}^2$, we find

$$m_E = \frac{gR_E^2}{G} = 5.98 \times 10^{24}\text{ kg}$$

very close to the currently accepted value of 5.974×10^{24} kg. Once Cavendish had measured G, he computed the mass of the earth in just this way.

At a point above the earth's surface a distance r from the center of the earth (a distance $r - R_E$ above the surface), the weight of a body is given by Eq. (12.3) with R_E replaced by r:

$$w = F_g = \frac{Gm_E m}{r^2} \tag{12.5}$$

The weight of a body decreases inversely with the square of its distance from the earth's center (Fig. 12.6). Figure 12.7 shows how the weight varies with height above the earth for an astronaut who weighs 700 N at the earth's surface.

The *apparent* weight of a body on earth differs slightly from the earth's gravitational force because the earth rotates and is therefore not precisely an inertial frame of reference. We have ignored this effect in the above discussion and have assumed that the earth *is* an inertial system. We will return to the effect of the earth's rotation in Section 12.7.

In our discussion of weight, we've used the fact that the earth is an approximately spherically symmetric distribution of mass. But this does *not* mean that the earth is uniform. To demonstrate that it cannot be uniform, let's first calculate the average *density,* or mass per unit volume, of the earth. If we assume a spherical earth, the volume is

$$V_E = \frac{4}{3}\pi R_E^3 = \frac{4}{3}\pi(6.38 \times 10^6\text{ m})^3 = 1.09 \times 10^{21}\text{ m}^3$$

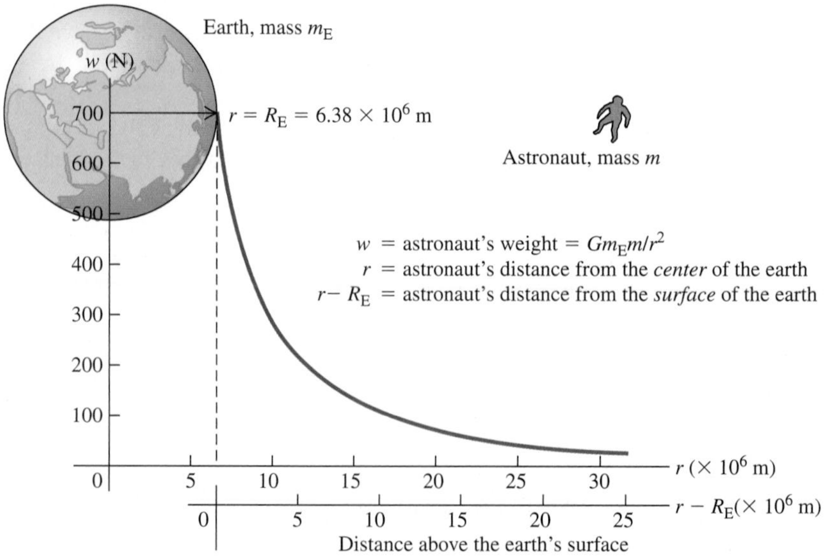

12.7 An astronaut weighing 700 N at the earth's surface experiences less gravitational attraction when above the surface. The relevant distance r is from the astronaut to the *center* of the earth (*not* from the astronaut to the earth's surface).

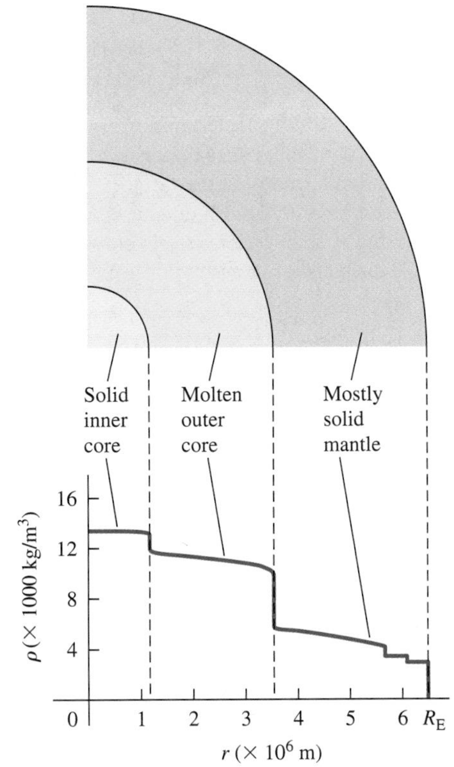

12.8 The density of the earth decreases with increasing distance from its center.

The average density ρ (the Greek letter "rho") of the earth is the total mass divided by the total volume:

$$\rho = \frac{m_E}{V_E} = \frac{5.97 \times 10^{24}\ \text{kg}}{1.09 \times 10^{21}\ \text{m}^3}$$

$$= 5500\ \text{kg/m}^3 = 5.5\ \text{g/cm}^3$$

(For comparison, the density of water is $1000\ \text{kg/m}^3 = 1.00\ \text{g/cm}^3$.) If the earth were uniform, we would expect the density of individual rocks near the earth's surface to have this same value. In fact, the density of igneous surface rocks such as granite or gneiss is about $3000\ \text{kg/m}^3 = 3\ \text{g/cm}^3$, although some basaltic rock has a density of about $5000\ \text{kg/m}^3 = 5\ \text{g/cm}^3$. So the earth *cannot* be uniform, and the interior of the earth must be much more dense than the surface in order that the *average* density be $5500\ \text{kg/m}^3 = 5.5\ \text{g/cm}^3$. According to geophysical models of the earth's interior, the maximum density at the center is about $13{,}000\ \text{kg/m}^3 = 13\ \text{g/cm}^3$. Figure 12.8 is a graph of density as a function of distance from the center.

Example 12.4 Gravity on Mars

You're involved in the design of a mission carrying humans to the surface of the planet Mars, which has a radius $r_M = 3.40 \times 10^6$ m and a mass $m_M = 6.42 \times 10^{23}$ kg. The earth weight of the Mars lander is 39,200 N. Calculate its weight F_g and the acceleration g_M due to the gravity of Mars: a) 6.0×10^6 m above the surface of Mars (the distance at which the moon Phobos orbits Mars); b) at the surface of Mars. Neglect the gravitational effects of the (very small) moons of Mars.

SOLUTION

IDENTIFY and SET UP: We find the weight F_g using Eq. (12.5) with m_E (the mass of the earth) replaced with m_M (the mass of Mars). Note that the value of G is the same everywhere in the universe; it is a fundamental physical constant. We then find the acceleration g_M using $F_g = mg_M$, where m is the mass of the lander. We're not given the value of this mass, but we can determine it from the lander's weight on earth.

EXECUTE: a) The distance r from the *center* of Mars is

$$r = (6.0 \times 10^6 \text{ m}) + (3.40 \times 10^6 \text{ m}) = 9.4 \times 10^6 \text{ m}$$

The mass m of the lander is its earth weight w divided by the acceleration of gravity g on earth:

$$m = \frac{w}{g} = \frac{39{,}200 \text{ N}}{9.8 \text{ m/s}^2} = 4000 \text{ kg}$$

The mass is the same whether the lander is on the earth, on Mars, or in between. From Eq. (12.5),

$$\begin{aligned} F_g &= \frac{Gm_M m}{r^2} \\ &= \frac{(6.67 \times 10^{-11} \text{ N}\cdot\text{m}^2/\text{kg}^2)(6.42 \times 10^{23} \text{ kg})(4000 \text{ kg})}{(9.4 \times 10^6 \text{ m})^2} \\ &= 1940 \text{ N} \end{aligned}$$

The acceleration due to the gravity of Mars at this point is

$$g_M = \frac{F_g}{m} = \frac{1940 \text{ N}}{4000 \text{ kg}} = 0.48 \text{ m/s}^2$$

This is also the acceleration experienced by Phobos in its orbit, 6.0×10^6 m above the surface of Mars.

b) To find F_g and g_M at the surface, we repeat the above calculations, replacing $r = 9.4 \times 10^6$ m with $R_M = 3.40 \times 10^6$ m. Alternatively, because F_g and g_M are inversely proportional to $1/r^2$ (at any point outside the planet), we can multiply the results of part (a) by the factor

$$\left(\frac{9.4 \times 10^6 \text{ m}}{3.40 \times 10^6 \text{ m}}\right)^2$$

You should use both methods to show that at the surface $F_g = 15{,}000$ N and $g_M = 3.7 \text{ m/s}^2$.

EVALUATE: The results for part (b) show that an object's weight and the acceleration due to gravity are roughly 40% as large on the surface of Mars as they are on the earth's surface. Science-fiction films and stories set on Mars commonly describe the planet's lower temperatures and thinner atmosphere, but they seldom focus on the experience of being in a low-gravity environment.

Test Your Understanding

The mass of Mars is only 11% as large as the earth's mass. Why, then, isn't the surface gravity on Mars 11% as great as on earth?

12.3 | Gravitational Potential Energy

When we first developed the concept of gravitational potential energy in Section 7.1, we assumed that the gravitational force on a body is constant in magnitude and direction. This led to the expression $U = mgy$. But we now know that the earth's gravitational force on a body of mass m at any point outside the earth is given more generally by Eq. (12.2), $F_g = Gm_E m/r^2$, where m_E is the mass of the earth and r is the distance of the body from the earth's center. For problems in which r changes enough that the gravitational force can't be considered constant, we need a more general expression for gravitational potential energy.

To find this expression, we follow the same basic sequence of steps as in Section 7.1. We consider a body of mass m outside the earth, and first compute the work W_{grav} done by the gravitational force when the body moves directly away from or toward the center of the earth from $r = r_1$ to $r = r_2$, as in Fig. 12.9. This work is given by

$$W_{grav} = \int_{r_1}^{r_2} F_r \, dr \tag{12.6}$$

where F_r is the radial component of the gravitational force $\vec{F}$, that is, the component in the direction *outward* from the center of the earth. Because $\vec{F}$ points directly *inward* toward the center of the earth, F_r is negative. It differs from Eq. (12.2), the magnitude of the gravitational force, by a minus sign:

$$F_r = -\frac{Gm_E m}{r^2} \tag{12.7}$$

Substituting Eq. (12.7) into Eq. (12.6), we see that W_{grav} is given by

$$W_{grav} = -Gm_E m \int_{r_1}^{r_2} \frac{dr}{r^2} = \frac{Gm_E m}{r_2} - \frac{Gm_E m}{r_1} \tag{12.8}$$

The path doesn't have to be a straight line; it could also be a curve like the one in Fig. 12.9. By an argument similar to that in Section 7.1, this work depends only on the initial and final values of r, not on the path taken. This also proves that the gravitational force is always *conservative.*

We now define the corresponding potential energy U so that $W_{grav} = U_1 - U_2$, as in Eq. (7.3). Comparing this with Eq. (12.8), we see that the appropriate definition for **gravitational potential energy** is

$$U = -\frac{Gm_E m}{r} \quad \text{(gravitational potential energy)} \tag{12.9}$$

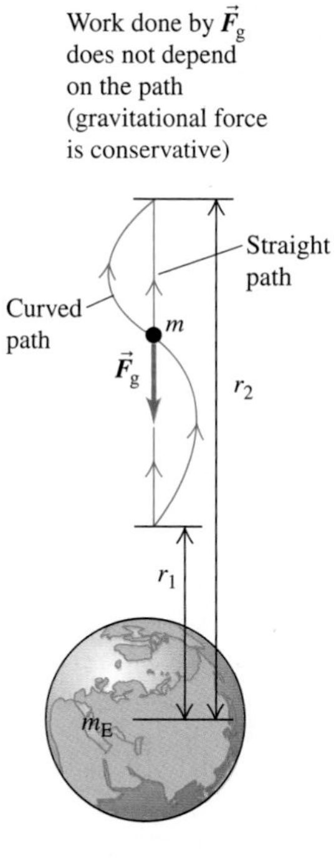

12.9 When a body moves from radial coordinate r_1 to r_2, the work done on it by the gravitational force is the same whether the body follows a straight path or any curved path.

Figure 12.10 shows how the gravitational potential energy depends on the distance r between the body of mass m and the center of the earth. When the body moves away from the earth, r increases, the gravitational force does negative work, and U increases (i.e., becomes less negative). When the body "falls" toward earth, r decreases, the gravitational work is positive, and the potential energy decreases (i.e., becomes more negative).

You may be troubled by Eq. (12.9) because it states that gravitational potential energy is always negative. But in fact you've seen negative values of U before. In using the formula $U = mgy$ in Section 7.1, we found that U was negative whenever the body of mass m was at a value of y below the arbitrary height we chose to be $y = 0$—that is, whenever the body and the earth were closer together than some certain arbitrary distance. (See, for instance, Example 7.2 in Section 7.1.) In defining U by Eq. (12.9), we have chosen U to be zero when the body of mass m is infinitely far away from the earth $(r = \infty)$. As the body moves toward the earth, gravitational potential energy decreases and so becomes negative. If we wanted, we could make $U = 0$ at the surface of the earth, where $r = R_E$, by simply adding the quantity $Gm_E m/R_E$ to Eq. (12.9). This would make U positive when $r > R_E$, but at the price of making the expression for U more complicated. This added term would not affect the *difference* in U between any two points, which is the only physically significant quantity; that is why we omit this term and use Eq. (12.9) for potential energy.

CAUTION Be careful not to confuse the expressions for gravitational force, Eq. (12.7), and gravitational potential energy, Eq. (12.9). The force F_r is proportional to $1/r^2$, while potential energy U is proportional to $1/r$.

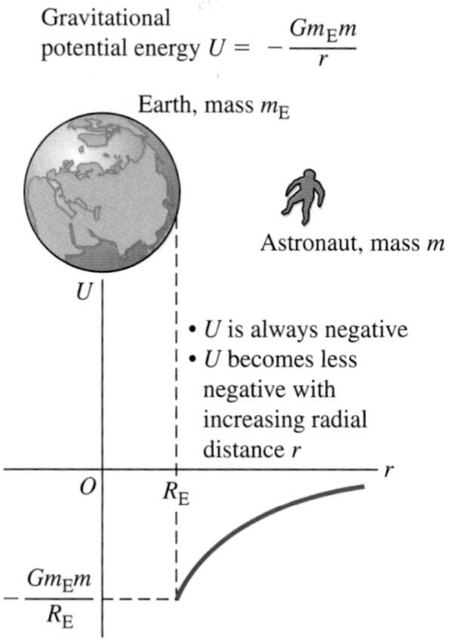

12.10 A graph of the gravitational potential energy U for the system of the earth (mass m_E) and an astronaut (mass m) versus the astronaut's distance r from the center of the earth.

Armed with Eq. (12.9), we can now use general energy relations for problems in which the $1/r^2$ behavior of the earth's gravitational force has to be included. If the gravitational force on the body is the only force that does work, the total mechanical energy of the system is constant, or *conserved.* In the following example we'll use this principle to calculate **escape speed,** the speed required for a body to escape completely from a planet.

Example 12.5 "From the earth to the moon"

In Jules Verne's 1865 story with this title, three men were shot to the moon in a shell fired from a giant cannon sunk in the earth in Florida. a) Find the muzzle speed needed to shoot the shell straight up to a height above the earth equal to the earth's radius. (b) Find the *escape speed*—that is, the muzzle speed that would allow the shell to escape from the earth completely. Neglect air resistance, the earth's rotation, and the gravitational pull of the moon. The earth's radius is $R_E = 6380 \text{ km} = 6.38 \times 10^6 \text{ m}$, and its mass is $m_E = 5.97 \times 10^{24} \text{ kg}$ (see Appendix F).

SOLUTION

IDENTIFY: Once the shell leaves the muzzle of the cannon, only the (conservative) gravitational force does work and mechanical energy is conserved. We use this fact to find the shell's speed as it leaves the muzzle in part (a), for which the shell comes to a halt at a distance of two earth radii from the planet's center, and in part (b), for which the shell comes to a halt at an infinite distance from earth.

SET UP: The equation to use in both parts is the statement of energy conservation, $K_1 + U_1 = K_2 + U_2$, where the potential energy U is given by Eq. (12.9). Point 1 is where the shell leaves the cannon with speed v_1 (the target variable). At this point the distance from the center of the earth is $r_1 = R_E$, the radius of the earth. Point 2 is where the shell reaches its maximum height; in part (a) it is at $r_2 = 2R_E$ (Fig. 12.11), and in part (b) it is infinitely far from the earth at $r_2 = \infty$. In either case the shell is at rest at point 2, so $v_2 = 0$. Let m be the mass of the shell (with passengers).

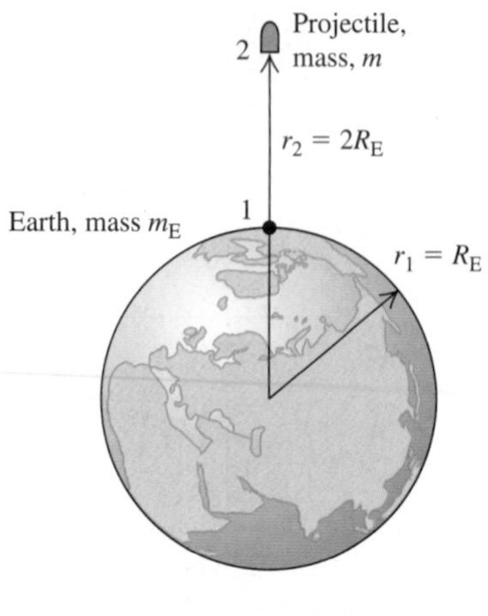

12.11 A projectile fired from the surface of the earth to a height equal to the earth's radius.

EXECUTE: a) We can determine v_1 from the energy-conservation equation

$$K_1 + U_1 = K_2 + U_2$$

$$\frac{1}{2}mv_1^2 + \left(-\frac{Gm_E m}{R_E}\right) = 0 + \left(-\frac{Gm_E m}{2R_E}\right)$$

Rearranging this, we find that

$$v_1 = \sqrt{\frac{Gm_E}{R_E}}$$

$$= \sqrt{\frac{(6.67 \times 10^{-11} \text{ N}\cdot\text{m}^2/\text{kg}^2)(5.97 \times 10^{24} \text{ kg})}{6.38 \times 10^6 \text{ m}}}$$

$$= 7900 \text{ m/s } (=28{,}400 \text{ km/h} = 17{,}700 \text{ mi/h})$$

b) We want the shell barely to be able to "reach" $r_2 = \infty$, with no kinetic energy left over. Hence $K_2 = 0$ and $U_2 = 0$ (the potential energy goes to zero at infinity; see Fig. 12.10). The total energy is therefore zero, and when the shell is fired its positive kinetic energy K_1 and negative potential energy U_1 must also add to zero:

$$\frac{1}{2}mv_1^2 + \left(-\frac{Gm_E m}{R_E}\right) = 0$$

$$v_1 = \sqrt{\frac{2Gm_E}{R_E}}$$

$$= \sqrt{\frac{2(6.67 \times 10^{-11} \text{ N}\cdot\text{m}^2/\text{kg}^2)(5.97 \times 10^{24} \text{ kg})}{6.38 \times 10^6 \text{ m}}}$$

$$= 1.12 \times 10^4 \text{ m/s } (=40{,}200 \text{ km/h} = 25{,}000 \text{ mi/h})$$

EVALUATE: This result does not depend on the mass of the shell, nor does it depend on the direction in which the shell is launched. Modern spacecraft launched from Florida must attain essentially this same speed to escape the earth. A spacecraft on the ground at Cape Canaveral is already moving at 410 m/s to the east because of the earth's rotation; by launching to the east, the spacecraft takes advantage of this "free" contribution toward escape speed.

Generalizing our result, the initial speed v_1 needed for a body to escape from the surface of a spherical mass M with radius R (ignoring air resistance) is

$$v_1 = \sqrt{\frac{2GM}{R}} \quad \text{(escape speed)}$$

You can use this result to compute the escape speed for other bodies. You will find 5.02×10^3 m/s for Mars, 5.95×10^4 m/s for Jupiter, and 6.18×10^5 m/s for the sun.

More on Gravitational Force and Potential Energy

The relationship between the gravitational force on a body, given by Eq. (12.7), and its gravitational potential energy, given by Eq. (12.9), can be expressed in a different way by using the methods of Section 7.4. Equations (7.17) and (7.18) show that the component of force in a given direction equals the negative of the derivative of U with respect to the corresponding coordinate. For motion along the x-axis,

$$F_x = -\frac{dU}{dx} \tag{12.10}$$

The gravitational force has a component only in the radial direction, so we replace x by r in Eq. (12.10). We find

$$F_r = -\frac{dU}{dr} = -\frac{d}{dr}\left(-\frac{Gm_{\text{E}}m}{r}\right) = -\frac{Gm_{\text{E}}m}{r^2} \tag{12.11}$$

This agrees with the F_r we started with in Eq. (12.7). As we remarked before, F_r is negative, showing that the force points opposite to the direction of increasing r.

As a final note, let's show that when we are close to the earth's surface, Eq. (12.9) reduces to the familiar $U = mgy$ from Chapter 7. We first rewrite Eq. (12.8) as

$$W_{\text{grav}} = Gm_{\text{E}}m\frac{r_1 - r_2}{r_1 r_2}$$

If the body stays close to the earth, then in the denominator we may replace r_1 and r_2 by R_{E}, the earth's radius, so

$$W_{\text{grav}} = Gm_{\text{E}}m\frac{r_1 - r_2}{R_{\text{E}}^2}$$

According to Eq. (12.4), $g = Gm_{\text{E}}/R_{\text{E}}^2$, so

$$W_{\text{grav}} = mg(r_1 - r_2)$$

If we replace the r's by y's, this is just Eq. (7.1) for the work done by a constant gravitational force. In Section 7.1 we used this equation to derive Eq. (7.2), $U = mgy$, so we may consider this expression for gravitational potential energy to be a special case of the more general Eq. (12.9).

Test Your Understanding

Is it possible for a planet to have the same surface gravity (that is, the same value of g at the surface) as the earth and yet have a greater escape speed?

12.4 | The Motion of Satellites

Artificial satellites orbiting the earth are a familiar fact of contemporary life (Fig. 12.12). But how do they stay in orbit, and what determines the properties of their orbits? We can use Newton's laws and the law of gravitation to provide the answers. We'll see in the next section that the motion of planets can be analyzed in the same way.

To begin, think back to the discussion of projectile motion in Section 3.3. In Example 3.7 a motorcycle rider rides horizontally off the edge of a cliff,

12.12 With a length of 13.2 m and a mass of 11,000 kg, the Hubble Space Telescope is among the largest satellites placed in orbit.

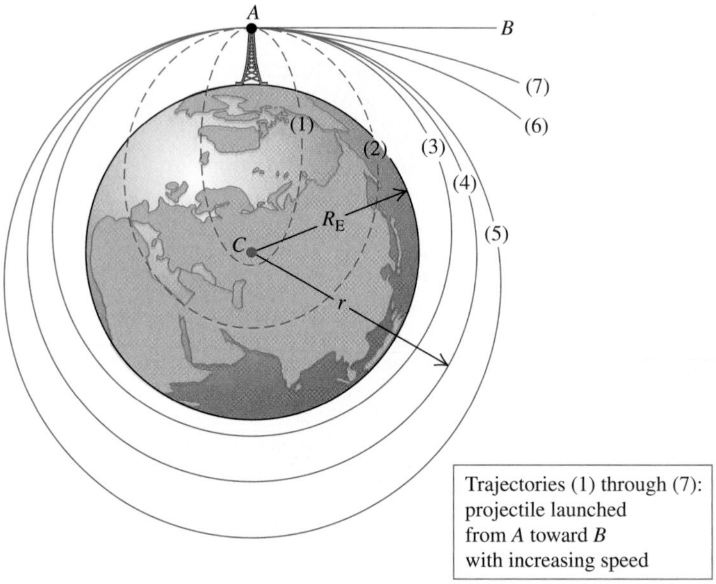

12.13 Trajectories of a projectile launched from point A in the direction AB with different initial speeds. Orbits (1) and (2) would be completed as shown if the earth were a point mass at C. (This illustration is based on one in Isaac Newton's *Principia.*)

launching himself into a parabolic path that ends on the flat ground at the base of the cliff. If he survives and repeats the experiment with increased launch speed, he will land farther from the starting point. We can imagine him launching himself with great enough speed that the earth's curvature becomes significant. As he falls, the earth curves away beneath him. If he is going fast enough, and if his launch point is high enough that he clears the mountain tops, he may be able to go right on around the earth without ever landing.

Figure 12.13 shows a variation on this theme. We launch a projectile from point A in the direction AB, tangent to the earth's surface. Trajectories (1) to (7) show the effect of increasing the initial speed. In trajectories (3) through (5) the projectile misses the earth and becomes a satellite. If there is no retarding force, the projectile's speed when it returns to point A is the same as its initial speed and it repeats its motion indefinitely.

Trajectories (1) through (5) close on themselves and are called **closed orbits.** All closed orbits are ellipses or segments of ellipses; trajectory (4) is a circle, a special case of an ellipse. Trajectories (6) and (7) are **open orbits.** For these paths the projectile never returns to its starting point but travels ever farther away from the earth.

A circular orbit like trajectory (4) in Fig. 12.13 is the simplest case. It is also an important case, since many artificial satellites have nearly circular orbits and the orbits of the planets around the sun are also fairly circular. The only force acting on a satellite in circular orbit around the earth is the earth's gravitational attraction, which is directed toward the center of the earth and hence toward the center of the orbit (Fig. 12.14). As we discussed in Section 5.4, this means that the satellite is in *uniform* circular motion and its speed is constant. The satellite isn't falling *toward* the earth; rather, it's constantly falling *around* the earth. In a circu-

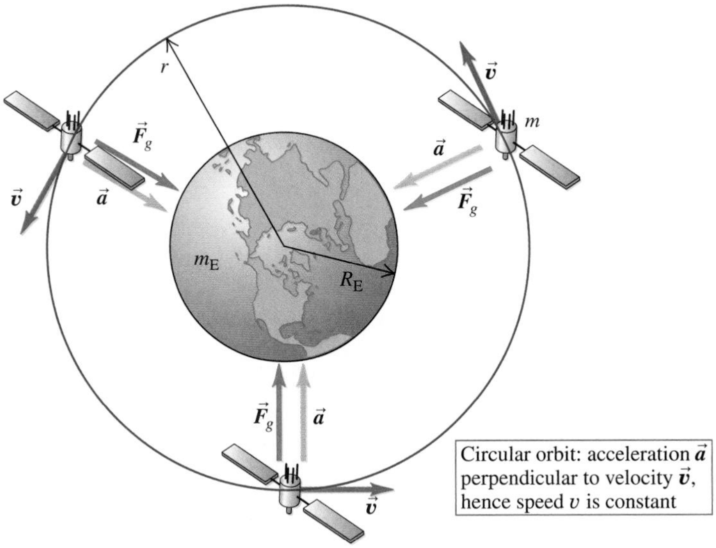

12.14 The force $\vec{F}_g$ due to the earth's gravitational attraction provides the centripetal acceleration that keeps a satellite in orbit. Compare to Fig. 5.29.

lar orbit the speed is just right to keep the distance from the satellite to the center of the earth constant.

How can we find the constant speed v of a satellite in a circular orbit? The radius of the orbit is r, measured from the *center* of the earth; the acceleration of the satellite has magnitude $a_{rad} = v^2/r$ and is always directed toward the center of the circle. By the law of gravitation, the net force (gravitational force) on the satellite of mass m has magnitude $F_g = Gm_Em/r^2$ and is in the same direction as the acceleration. Newton's second law $(\Sigma\vec{F} = m\vec{a})$ then tells us that

$$\frac{Gm_Em}{r^2} = \frac{mv^2}{r}$$

ActivPhysics Online

4.6 Satellites Orbit

Solving this for v, we find

$$v = \sqrt{\frac{Gm_E}{r}} \quad \text{(circular orbit)} \qquad (12.12)$$

This relation shows that we can't choose the orbit radius r and the speed v independently; for a given radius r, the speed v for a circular orbit is determined.

Equation (12.12) also shows that the satellite's motion does not depend on its mass m because m doesn't appear in the equation. If we could cut a satellite in half without changing its speed, each half would continue on with the original motion. An astronaut on board a space shuttle is herself a satellite of the earth, held by the earth's gravitational attraction in the same orbit as the shuttle. The astronaut has the same velocity and acceleration as the shuttle, so nothing is pushing her against the floor or walls of the shuttle. She is in a state of *apparent weightlessness,* as in a freely falling elevator; see the discussion following Example 5.10 in Section 5.2. (*True* weightlessness would occur only if the astronaut were infinitely far from any other masses, so that the gravitational force on her would be zero.) Indeed, every part of her body is apparently weightless; she feels nothing pushing her stomach against her intestines or her head against her shoulders (Fig. 12.15).

12.15 These space shuttle astronauts are in a state of apparent weightlessness. Which are right side up and which are upside down?

(a)

(b)

12.16 (a) The International Space Station orbits at an altitude of about 400 km, or about 6800 km from the center of the earth. Its orbital speed is about 7.7 km/s and its orbital period is about 93 minutes. (b) The moon orbits the earth at a much larger distance of about 384,000 km. As a result, it has a slow orbital speed (about 1.0 km/s) and a long orbital period (about 27.3 days).

Apparent weightlessness is not just a feature of circular orbits; it occurs whenever gravity is the only force acting on a spacecraft. Hence it occurs for orbits of any shape, including open orbits such as (6) and (7) in Fig. 12.13.

We can derive a relation between the radius r of a circular orbit and the period T, the time for one revolution. The speed v is the distance $2\pi r$ traveled in one revolution, divided by the period:

$$v = \frac{2\pi r}{T} \tag{12.13}$$

To get an expression for T, we solve Eq. (12.13) for T and substitute v from Eq. (12.12):

$$T = \frac{2\pi r}{v} = 2\pi r\sqrt{\frac{r}{Gm_E}} = \frac{2\pi r^{3/2}}{\sqrt{Gm_E}} \quad \text{(circular orbit)} \tag{12.14}$$

Equations (12.12) and (12.14) show that larger orbits correspond to slower speeds and longer periods (Fig. 12.16).

It's interesting to compare Eq. (12.12) to the calculation of escape speed in Example 12.5. We see that the escape speed from a spherical body with radius R is just $\sqrt{2}$ times as large as the speed of a satellite in a circular orbit at that radius. If our spacecraft is in circular orbit around *any* planet, we have to multiply our speed by a factor of $\sqrt{2}$ to escape to infinity, regardless of the planet's mass.

Since the speed v in a circular orbit is determined by Eq. (12.12) for a given orbit radius r, the total mechanical energy $E = K + U$ is determined as well. Using Eqs. (12.9) and (12.12), we have

$$E = K + U = \frac{1}{2}mv^2 + \left(-\frac{Gm_E m}{r}\right) = \frac{1}{2}m\left(\frac{Gm_E}{r}\right) - \frac{Gm_E m}{r}$$

$$E = -\frac{Gm_E m}{2r} \quad \text{(circular orbit)} \tag{12.15}$$

The total mechanical energy in a circular orbit is negative and equal to one-half the potential energy. Increasing the orbit radius r means increasing the mechanical energy (that is, making E less negative). If the satellite is in a relatively low orbit that encounters the outer fringes of earth's atmosphere, mechanical energy decreases due to negative work done by the force of air resistance; as a result, the orbit radius decreases until the satellite hits the ground or burns up in the atmosphere.

We have talked mostly about earth satellites, but we can apply the same analysis to the circular motion of *any* body under its gravitational attraction to a stationary body. Other examples include the earth's moon, the moons of other planets, and the rings of Saturn. If Saturn's rings were a rotating rigid body, all parts of the rings would have the same orbital period T, and by Eq. (12.13) the outer regions of the rings (large r) would have a greater speed v than the inner regions (small r). But since the rings are actually composed of many individual orbiting particles, Eqs. (12.12) and (12.14) tell us that particles in the outer regions of the rings move at a slower speed and have a longer period than do particles in the inner regions (Fig. 12.17).

12.17 The rings of Saturn are made up of countless numbers of icy particles, each of which is an independent satellite of Saturn.

Example 12.6 A satellite orbit

Suppose you want to place a 1000-kg weather satellite into a circular orbit 300 km above the earth's surface. a) What speed, period, and radial acceleration must it have? b) How much work has to be done to place this satellite in orbit? c) How much additional work would have to be done to make this satellite escape the earth? The earth's radius is $R_E = 6380$ km and its mass is $m_E = 5.97 \times 10^{24}$ kg.

SOLUTION

IDENTIFY: The satellite is in a circular orbit, so we can use the equations derived in this section.

SET UP: In part (a), given the satellite's altitude we first find the radius of its orbit. We then calculate the speed v and period T using Eqs. (12.12) and (12.14). The acceleration in a circular orbit is given by the familiar formula from Chapter 3, $a_{rad} = v^2/r$. For the processes described in parts (b) and (c), the work required is the difference between the initial and final mechanical energy, which for a circular orbit is given by Eq. (12.15).

EXECUTE: a) The radius of the satellite's orbit is

$$r = 6380 \text{ km} + 300 \text{ km} = 6680 \text{ km} = 6.68 \times 10^6 \text{ m}$$

From Eq. (12.12),

$$v = \sqrt{\frac{Gm_E}{r}} = \sqrt{\frac{(6.67 \times 10^{-11} \text{ N}\cdot\text{m}^2/\text{kg}^2)(5.97 \times 10^{24} \text{ kg})}{6.68 \times 10^6 \text{ m}}}$$
$$= 7720 \text{ m/s}$$

From Eq. (12.14),

$$T = \frac{2\pi r}{v} = \frac{2\pi(6.68 \times 10^6 \text{ m})}{7720 \text{ m/s}}$$
$$= 5440 \text{ s} = 90.6 \text{ min}$$

The radial acceleration is

$$a_{rad} = \frac{v^2}{r} = \frac{(7720 \text{ m/s})^2}{6.68 \times 10^6 \text{ m}}$$
$$= 8.92 \text{ m/s}^2$$

This is the value of g at a height of 300 km above the earth's surface; it is somewhat less than the value of g at the surface.

b) The work required is the difference between E_2, the total mechanical energy when the satellite is in orbit, and E_1, the original mechanical energy when the satellite was at rest on the launch pad back on earth. In orbit, if we use Eq. (12.15), the energy is

$$E_2 = -\frac{Gm_E m}{2r}$$
$$= -\frac{(6.67 \times 10^{-11} \text{ N}\cdot\text{m}^2/\text{kg}^2)(5.97 \times 10^{24} \text{ kg})(1000 \text{ kg})}{2(6.68 \times 10^6 \text{ m})}$$
$$= -2.99 \times 10^{10} \text{ J}$$

At rest on the earth's surface the energy is purely potential:

$$E_1 = K_1 + U_1 = 0 + \left(-\frac{Gm_E m}{R_E}\right)$$
$$= -\frac{(6.67 \times 10^{-11} \text{ N}\cdot\text{m}^2/\text{kg}^2)(5.97 \times 10^{24} \text{ kg})(1000 \text{ kg})}{6.68 \times 10^6 \text{ m}}$$
$$= -6.25 \times 10^{10} \text{ J}$$

and so

$$W_{required} = E_2 - E_1 = -2.99 \times 10^{10} \text{ J} - (-6.25 \times 10^{10} \text{ J})$$
$$= 3.26 \times 10^{10} \text{ J}$$

c) We saw in part (b) of Example 12.5 that for a satellite to escape to infinity, the total mechanical energy must be zero. The total mechanical energy in the circular orbit is $E_2 = -2.99 \times 10^{10}$ J; to increase this to zero, an amount of work equal to 2.99×10^{10} J would have to be done. This extra energy could be supplied by rocket engines attached to the satellite.

EVALUATE: In part (b) we ignored the satellite's initial kinetic energy while it was still on the launch pad due to the rotation of the earth. You should check to see how much difference this makes (see Example 12.5 for useful data).

Test Your Understanding

A space shuttle and a satellite are in the same circular orbit but on opposite sides of the earth. Explain how the shuttle could rendezvous with the satellite by maneuvering into a circular orbit with a slightly smaller radius. Could this rendezvous also be achieved by maneuvering into a circular orbit of slightly *larger* radius?

12.5 | Kepler's Laws and the Motion of Planets

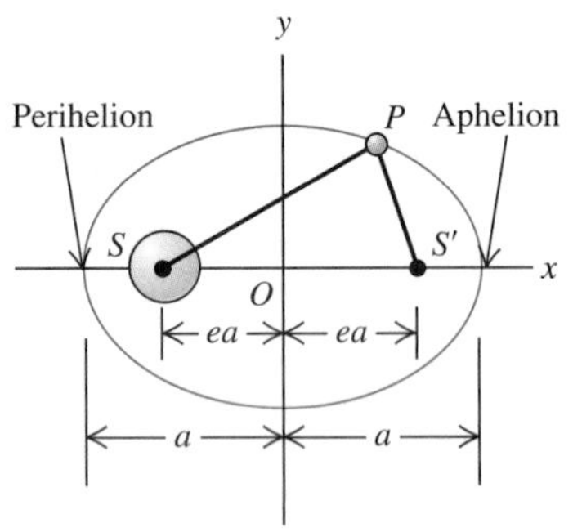

12.18 Geometry of an ellipse. The sum of the distances SP and $S'P$ is the same for every point on the curve. The sizes of the sun (S) and planet (P) are exaggerated for clarity.

The name *planet* comes from a Greek word meaning "wanderer," and indeed the planets continually change their positions in the sky relative to the background of stars. One of the great intellectual accomplishments of the 16th and 17th centuries was the threefold realization that the earth is also a planet, that all planets orbit the sun, and that the apparent motions of the planets as seen from the earth can be used to precisely determine the orbits of the planets.

The first and second of these ideas were published by Nicolaus Copernicus in Poland in 1543. The determination of planetary orbits was carried out between 1601 and 1619 by the German astronomer and mathematician Johannes Kepler, using a voluminous set of precise data on apparent planetary motions compiled by his mentor, the Danish astronomer Tycho Brahe. By trial and error, Kepler discovered three empirical laws that accurately described the motions of the planets:

1. **Each planet moves in an elliptical orbit, with the sun at one focus of the ellipse.**
2. **A line from the sun to a given planet sweeps out equal areas in equal times.**
3. **The periods of the planets are proportional to the $\frac{3}{2}$ powers of the major axis lengths of their orbits.**

Kepler did not know *why* the planets moved in this way. Three generations later, when Newton turned his attention to the motion of the planets, he discovered that each of Kepler's laws can be *derived;* they are consequences of Newton's laws of motion and the law of gravitation. Let's see how each of Kepler's laws arises.

First consider the elliptical orbits described in Kepler's first law. Figure 12.18 shows the geometry of the ellipse. The longest dimension is the *major axis,* with half-length a; this half-length is called the **semi-major axis.** The sum of the distances from S to P and from S' to P is the same for all points on the curve. S and S' are the *foci* (plural of focus). The sun is at S, and the planet is at P; we think of them both as points because the size of each is very small in comparison to the distance between them. There is nothing at the other focus S'.

The distance of each focus from the center of the ellipse is ea, where e is a dimensionless number between 0 and 1 called the **eccentricity.** If $e = 0$, the ellipse is a circle. The actual orbits of the planets are fairly circular; their eccentricities range from 0.007 for Venus to 0.248 for Pluto. (The earth's orbit has $e = 0.017$.) The point in the planet's orbit closest to the sun is the *perihelion,* and the point most distant from the sun is the *aphelion.*

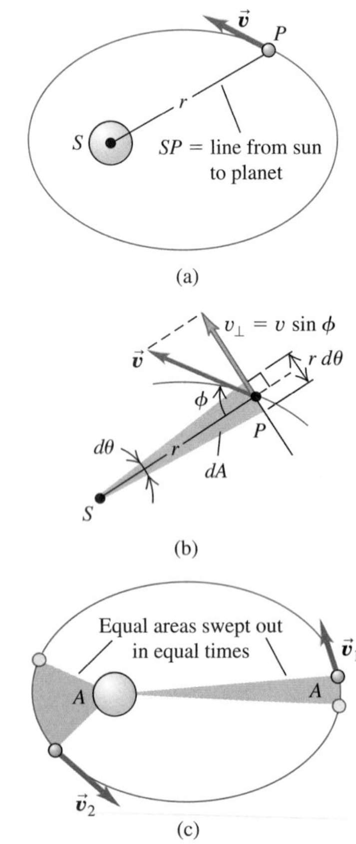

12.19 (a) The planet (P) moves about the sun (S) in an elliptical orbit. (b) In a time dt the line SP sweeps out an area $dA = \frac{1}{2}(r\,d\theta)r = \frac{1}{2}r^2\,d\theta$. (c) The planet's speed varies so that the line SP sweeps out the same area A in a given time t regardless of the planet's position in its orbit.

Newton was able to show that for a body acted on by an attractive force proportional to $1/r^2$, the only possible closed orbits are a circle or an ellipse; he also showed that open orbits (trajectories (6) and (7) in Fig. 12.13) must be parabolas or hyperbolas. These results can be derived by a straightforward application of Newton's laws and the law of gravitation, together with a lot more differential equations than we're ready for.

Kepler's second law is shown in Fig. 12.19. In a small time interval dt, the line from the sun S to the planet P turns through an angle $d\theta$. The area swept out is the colored triangle with height r, base length $r\,d\theta$, and area $dA = \frac{1}{2}r^2\,d\theta$ (Fig. 12.19b). The rate at which area is swept out, dA/dt, is called the *sector velocity:*

$$\frac{dA}{dt} = \frac{1}{2}r^2\frac{d\theta}{dt} \tag{12.16}$$

The essence of Kepler's second law is that the sector velocity has the same value at all points in the orbit. When the planet is close to the sun, r is small and $d\theta/dt$ is large; when the planet is far from the sun, r is large and $d\theta/dt$ is small.

To see how Kepler's second law follows from Newton's laws, we express dA/dt in terms of the velocity vector $\vec{v}$ of the planet P. The component of $\vec{v}$ perpendicular to the radial line is $v_\perp = v \sin\phi$. From Fig. 12.19b the displacement along the direction of $v_\perp$ during time dt is $r\,d\theta$, so we also have $v_\perp = r\,d\theta/dt$. Using this relation in Eq. (12.16), we find

$$\frac{dA}{dt} = \frac{1}{2}rv\sin\phi \qquad \text{(sector velocity)} \tag{12.17}$$

Now $rv\sin\phi$ is the magnitude of the vector product $\vec{r} \times \vec{v}$, which in turn is $1/m$ times the angular momentum $\vec{L} = \vec{r} \times m\vec{v}$ of the planet with respect to the sun. So we have

$$\frac{dA}{dt} = \frac{1}{2m}\left|\vec{r} \times m\vec{v}\right| = \frac{L}{2m} \tag{12.18}$$

Thus Kepler's second law, that sector velocity is constant, means that angular momentum is constant!

It is easy to see why the angular momentum of the planet *must* be constant. According to Eq. (10.29), the rate of change of $\vec{L}$ equals the torque of the gravitational force $\vec{F}$ acting on the planet:

$$\frac{d\vec{L}}{dt} = \vec{\tau} = \vec{r} \times \vec{F}$$

In our situation, $\vec{r}$ is the vector from the sun to the planet, and the force $\vec{F}$ is directed from the planet to the sun. So these vectors always lie along the same line, and their vector product $\vec{r} \times \vec{F}$ is zero. Hence $d\vec{L}/dt = 0$. This conclusion does not depend on the $1/r^2$ behavior of the force; angular momentum is conserved for *any* force that acts always along the line joining the particle to a fixed point. Such a force is called a *central force.* (Kepler's first and third laws are valid *only* for a $1/r^2$ force.)

Conservation of angular momentum also explains why the orbit lies in a plane. The vector $\vec{L} = \vec{r} \times m\vec{v}$ is always perpendicular to the plane of the vectors $\vec{r}$ and $\vec{v}$; since $\vec{L}$ is constant in magnitude *and* direction, $\vec{r}$ and $\vec{v}$ always lie in the same plane, which is just the plane of the planet's orbit.

We have already derived Kepler's third law for the particular case of circular orbits. Equation (12.14) shows that the period of a satellite or planet in a circular orbit is proportional to the $\frac{3}{2}$ power of the orbit radius. Newton was able to show that this same relationship holds for an elliptical orbit, with the orbit radius r replaced by the semi-major axis a:

$$T = \frac{2\pi a^{3/2}}{\sqrt{Gm_S}} \qquad \text{(elliptical orbit around the sun)} \tag{12.19}$$

Since the planet orbits the sun, not the earth, we have replaced the earth's mass m_E in Eq. (12.14) with the sun's mass m_S. Note that the period does not depend on the eccentricity e. An asteroid in an elongated elliptical orbit with semi-major axis a will have the same orbital period as a planet in a circular orbit of radius a. The key difference is that the asteroid moves at different speeds at different points in its elliptical orbit (Fig. 12.19c), while the planet's speed is constant around its circular orbit.

Conceptual Example 12.7 Orbital speeds

At what point in an elliptical orbit (Fig. 12.18) does a planet have the greatest speed?

SOLUTION

Mechanical energy is conserved as the planet moves around its orbit. The planet's kinetic energy $K = \frac{1}{2}mv^2$ is maximum when the potential energy $U = -Gm_{S}m/r$ is minimum (that is, most negative; see Fig. 12.10), which occurs when r is a minimum. Hence the speed v is maximum at perihelion.

Your intuition about falling bodies is helpful here. As the planet falls inward toward the sun, it picks up speed, and its speed is maximum when closest to the sun. By the same reasoning, the planet slows down as it moves away from the sun, and its speed is minimum at aphelion.

Example 12.8 Kepler's third law

The asteroid Pallas has an orbital period of 4.62 years and an orbital eccentricity of 0.233. Find the semi-major axis of its orbit.

SOLUTION

IDENTIFY and SET UP: Asteroids orbit the sun, so we can use Kepler's third law in the form of Eq. (12.19) to determine the semi-major axis a from the given period T. Note that we don't need the value of the eccentricity.

EXECUTE: From Eq. (12.19), $a^{3/2} = (\sqrt{Gm_S}\,T)/2\pi$. To solve for a, raise this expression to the $\frac{2}{3}$ power:

$$a = \left(\frac{Gm_S T^2}{4\pi^2}\right)^{1/3}$$

Since $G = 6.67 \times 10^{-11}\ \text{N}\cdot\text{m}^2/\text{kg}^2$ and $m_S = 1.99 \times 10^{30}$ kg (the mass of the sun from Appendix F) are given in SI units, we must express the period T in seconds rather than years using a conversion factor from Appendix E: $T = (4.62\ \text{yr})(3.156 \times 10^7\ \text{s/yr}) = 1.46 \times 10^8$ s. Using this value, we find $a = 4.15 \times 10^{11}$ m. (Plug in the numbers yourself to check.)

EVALUATE: Our result is intermediate between the semi-major axes of Mars and Jupiter (see Appendix F). Indeed, most known asteroids orbit in an "asteroid belt" between the orbits of these two planets.

As a historical note, Pallas wasn't discovered until 1802, almost two centuries after Kepler's publication of his third law. While Kepler deduced his three laws from the motions of the five planets (other than the earth) known in his time, these laws have proven to apply equally well to all of the planets, asteroids, and comets subsequently discovered to be orbiting the sun.

Example 12.9 Comet Halley

Comet Halley moves in an elongated elliptical orbit around the sun (Fig. 12.20). At perihelion, the comet is 8.75×10^7 km from the sun; at aphelion it is 5.26×10^9 km from the sun. Find the semi-major axis, eccentricity, and period of the orbit.

SOLUTION

IDENTIFY and SET UP: Given the perihelion and aphelion distances, we can find the semi-major axis and eccentricity from Fig. 12.18. Once we know the semi-major axis, we can find the orbital period from Eq. (12.19).

EXECUTE: From Fig. 12.18 the length of the major axis equals the sum of the comet–sun distance at perihelion plus the comet–sun distance at aphelion. The length of the major axis is $2a$, so

$$a = \frac{8.75 \times 10^7\ \text{km} + 5.26 \times 10^9\ \text{km}}{2} = 2.67 \times 10^9\ \text{km}$$

Further inspection of Fig. 12.18 shows that the comet–sun distance at perihelion is given by

$$a - ea = a(1 - e)$$

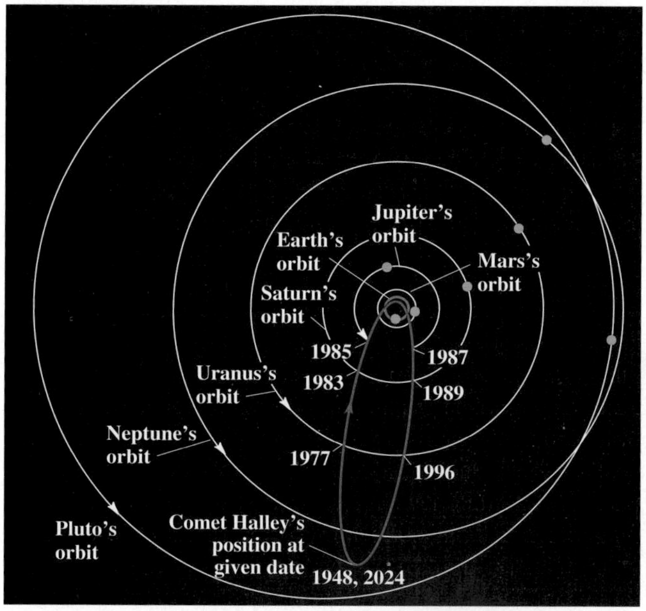

(a)

(b)

12.20 (a) The orbit of Comet Halley. (b) Comet Halley as it appeared in 1986. At the heart of the comet is an icy body, called the nucleus, that is about 10 km across. When the comet's orbit carries it close to the sun, the heat of sunlight causes the nucleus to partially evaporate. the evaporated material forms the tail, which can be tens of millions of kilometers long.

Since we are given that this distance is 8.75×10^7 km, the eccentricity is

$$e = 1 - \frac{8.75 \times 10^7\ \text{km}}{a} = 1 - \frac{8.75 \times 10^7\ \text{km}}{2.67 \times 10^9\ \text{km}} = 0.967$$

The period is given by Eq. (12.19):

$$T = \frac{2\pi a^{3/2}}{\sqrt{Gm_S}} = \frac{2\pi(2.67 \times 10^{12}\ \text{m})^{3/2}}{\sqrt{(6.67 \times 10^{-11}\ \text{N}\cdot\text{m}^2/\text{kg}^2)(1.99 \times 10^{30}\ \text{kg})}}$$
$$= 2.38 \times 10^9\ \text{s} = 75.5\ \text{years}$$

EVALUATE: The eccentricity is very close to 1, so the comet has a very elongated orbit (see Fig. 12.20a). Comet Halley was at perihelion in early 1986; it will next reach perihelion one period later, in 2061.

We have assumed that as a planet or comet orbits the sun, the sun remains absolutely stationary. Of course, this can't be correct; because the sun exerts a gravitational force on the planet, the planet exerts a gravitational force on the sun of the same magnitude but opposite direction. In fact, *both* the sun and the planet orbit around their common center of mass (Fig. 12.21). We've made only a small error by ignoring this effect, however; the sun's mass is about 750 times the total mass of all the planets combined, so the center of mass of the solar system is not far from the center of the sun. Remarkably, astronomers have used this effect to detect the presence of planets orbiting other stars. Sensitive telescopes are able to detect the apparent "wobble" of a star as it orbits the common center of mass of the star and an unseen companion planet. (The planets are too faint to observe directly.) By analyzing the details of these "wobbles," astronomers have discovered planets in orbit around dozens of other stars.

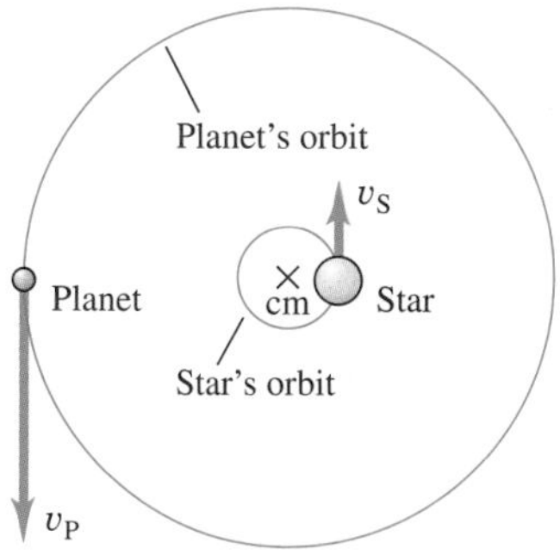

12.21 A star and its planet both orbit about their common center of mass. The star is much more massive than the planet and is much closer to the center of mass (far closer than shown here). The planet and star are on opposite sides of the center of mass at all times.

Newton's analysis of planetary motions is used on a daily basis by modern-day astronomers. But the most remarkable result of Newton's work is that the motions of bodies in the heavens obey the *same* laws of motion as do bodies on the earth. This *Newtonian synthesis,* as it has come to be called, is one of the great unifying principles of science. It has had profound effects on the way that humanity looks at the universe—not as a realm of impenetrable mystery, but as a direct extension of our everyday world, subject to scientific study and calculation.

Test Your Understanding

The orbit of Comet X has a semi-major axis that is four times larger than the semi-major axis of Comet Y. What is the ratio of the two comets' orbital periods?

*12.6 | Spherical Mass Distributions

We have used without proof the statement that the gravitational interaction between two spherically symmetric mass distributions is the same as though all the mass of each were concentrated at its center. Now we're ready to prove this statement. Newton searched for a proof for several years, and he delayed publication of the law of gravitation until he found one.

Here's our program. Rather than starting with two spherically symmetric masses, we'll tackle the simpler problem of a point mass m interacting with a thin spherical shell with total mass M. We will show that when m is outside the sphere, the *potential energy* associated with this gravitational interaction is the same as though M were all concentrated at the center of the sphere. According to Eq. (12.10), the force is the negative derivative of the potential energy, so the *force* on m is also the same as for a point mass M. Any spherically symmetric mass distribution can be thought of as being made up of many concentric spherical shells, so our result will also hold for *any* spherically symmetric M.

We start by considering a ring on the surface of the shell (Fig. 12.22a), centered on the line from the center of the shell to m. The reason we do this is that all of the particles that make up the ring are the same distance s from the point mass m. From Eq. (12.9) the potential energy of interaction between the earth (mass m_E) and a point mass m, separated by a distance r, is $U = -Gm_E m/r$. By changing notation in this expression, we see that in the situation shown in Fig. 12.22a, the potential energy of interaction between the point mass m and a particle of mass m_i within the ring is given by

$$U_i = -\frac{Gmm_i}{s}$$

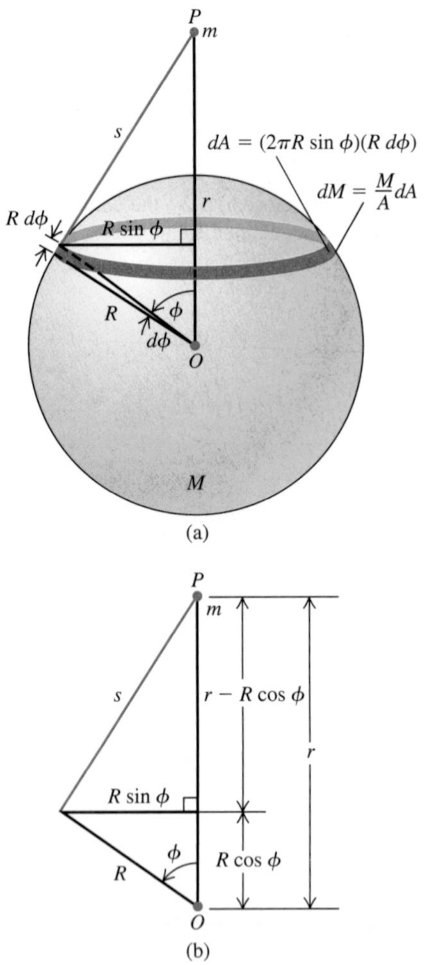

12.22 (a) To calculate the gravitational effects *outside* a spherical shell, all the mass M can be considered concentrated at the center. (b) The distance s is the hypotenuse of a right triangle with sides $(r - R\cos\phi)$ and $R\sin\phi$.

To find the potential energy of interaction between m and the entire ring of mass $dM = \sum_i m_i$, we sum this expression for U_i over all particles in the ring. Calling this potential energy dU, we find

$$dU = \sum_i U_i = \sum_i \left(-\frac{Gmm_i}{s}\right) = -\frac{Gm}{s}\sum_i m_i = -\frac{Gm\,dM}{s} \qquad (12.20)$$

To proceed, we need to know the mass dM of the ring. We can find this with the aid of a little geometry. The radius of the shell is R, so in terms of the angle ϕ shown in the figure, the radius of the ring is $R\sin\phi$, and its circumference is $2\pi R\sin\phi$. The width of the ring is $R\,d\phi$, and its area dA is approximately equal to its width times its circumference:

$$dA = 2\pi R^2 \sin\phi\,d\phi$$

The ratio of the ring mass dM to the total mass M of the shell is equal to the ratio of the area dA of the ring to the total area $A = 4\pi R^2$ of the shell:

$$\frac{dM}{M} = \frac{2\pi R^2 \sin\phi\, d\phi}{4\pi R^2} = \frac{1}{2}\sin\phi\, d\phi \qquad (12.21)$$

Now we solve Eq. (12.21) for dM and substitute the result into Eq. (12.20) to find the potential energy of interaction between the point mass m and the ring:

$$dU = -\frac{GMm \sin\phi\, d\phi}{2s} \qquad (12.22)$$

The total potential energy of interaction between the point mass and the *shell* is the integral of Eq. (12.22) over the whole sphere as ϕ varies from 0 to π (*not* 2π!) and s varies from $r - R$ to $r + R$. To carry out the integration, we have to express the integrand in terms of a single variable; we choose s. To express ϕ and $d\phi$ in terms of s, we have to do a little more geometry. Figure 12.22b shows that s is the hypotenuse of a right triangle with sides $(r - R\cos\phi)$ and $R\sin\phi$, so the Pythagorean theorem gives

$$\begin{aligned} s^2 &= (r - R\cos\phi)^2 + (R\sin\phi)^2 \\ &= r^2 - 2rR\cos\phi + R^2 \end{aligned} \qquad (12.23)$$

We take differentials of both sides:

$$2s\, ds = 2rR \sin\phi\, d\phi$$

Next we divide this by $2rR$ and substitute the result into Eq. (12.22):

$$dU = -\frac{GMm}{2s}\frac{s\, ds}{rR} = -\frac{GMm}{2rR}ds \qquad (12.24)$$

We can now integrate Eq. (12.24), recalling that s varies from $r - R$ to $r + R$:

$$U = -\frac{GMm}{2rR}\int_{r-R}^{r+R} ds = -\frac{GMm}{2rR}[(r + R) - (r - R)] \qquad (12.25)$$

Finally, we have

$$U = -\frac{GMm}{r} \qquad \text{(point mass } m \text{ outside spherical shell } M\text{)} \qquad (12.26)$$

This is equal to the potential energy of two point masses m and M at a distance r. So we have proved that the gravitational potential energy of the spherical shell M and the point mass m at any distance r is the same as though they were point masses. Because the force is given by $F_r = -dU/dr$, the force is also the same.

Any spherically symmetric mass distribution can be thought of as a combination of concentric spherical shells. Because of the principle of superposition of forces, what is true of one shell is also true of the combination. So we have proved half of what we set out to prove, that the gravitational interaction between any spherically symmetric mass distribution and a point mass is the same as though all the mass of the spherically symmetric distribution were concentrated at its center.

The other half is to prove that *two* spherically symmetric mass distributions interact as though they were both points. That's easier. In Fig. 12.22 the forces the two bodies exert on each other are an action-reaction pair, and they obey Newton's third law. So we have also proved that the force that m exerts *on* the sphere

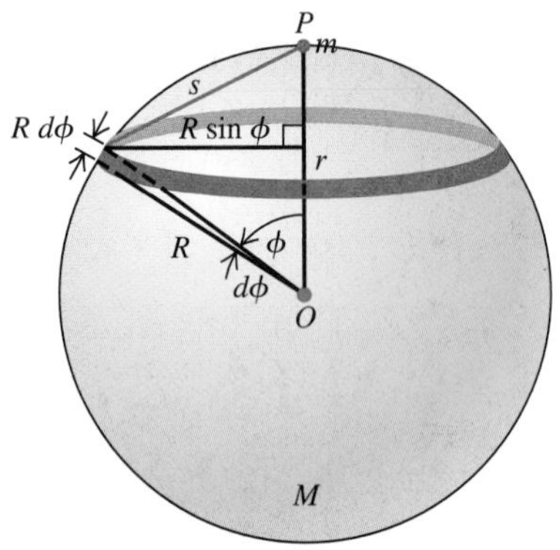

12.23 When a point mass m is *inside* a uniform spherical shell of mass M, the potential energy is the same no matter where inside the shell the point mass is located. The force from the masses' mutual gravitational interaction is zero.

M is the same as though M were a point. But now if we replace m with a spherically symmetric mass distribution centered at m's location, the resulting gravitational force on any part of M is the same as before, and so is the total force. This completes our proof.

A Point Mass Inside a Spherical Shell

We assumed at the beginning that the point mass m was outside the spherical shell, so our proof is valid only when m is outside a spherically symmetric mass distribution. When m is *inside* a spherical shell, the geometry is as shown in Fig. 12.23. The entire analysis goes just as before; Eqs. (12.20) through (12.24) are still valid. But when we get to Eq. (12.25), the limits of integration have to be changed to $R - r$ and $R + r$. We then have

$$U = -\frac{GMm}{2rR}\int_{R-r}^{R+r} ds = -\frac{GMm}{2rR}[(R + r) - (R - r)] \qquad (12.27)$$

and the final result is

$$U = -\frac{GMm}{R} \qquad \text{(point mass } m \text{ inside spherical shell } M) \qquad (12.28)$$

Compare this result to Eq. (12.26): instead of having r, the distance between m and the center of M, in the denominator, we have R, the radius of the shell. This means that U in Eq. (12.28) doesn't depend on r and thus has the same value everywhere inside the shell. When m moves around inside the shell, no work is done on it, so the force on m at any point inside the shell must be zero.

More generally, at any point in the interior of any spherically symmetric mass distribution (not necessarily a shell), at a distance r from its center, the gravitational force on a point mass m is the same as though we removed all the mass at points farther than r from the center and concentrated all the remaining mass at the center.

Example 12.10 "Journey to the center of the earth"

Suppose we drill a hole through the earth (radius R_E, mass m_E) along a diameter and drop a mail pouch (mass m) down the hole. Derive an expression for the gravitational force on the pouch as a function of its distance r from the center. Assume that the density of the earth is uniform (not a very realistic model; see Fig. 12.8).

SOLUTION

IDENTIFY and SET UP: According to the statements above, the gravitational force at distance r from the center is determined only by the mass M within a sphere of radius r (Fig. 12.24).

EXECUTE: With uniform density the mass is proportional to the volume of the sphere, which is $\frac{4}{3}\pi r^3$ for this sphere and $\frac{4}{3}\pi R_E^3$ for the entire earth. So we have

$$\frac{M}{m_E} = \frac{\frac{4}{3}\pi r^3}{\frac{4}{3}\pi R_E^3} = \frac{r^3}{R_E^3}$$

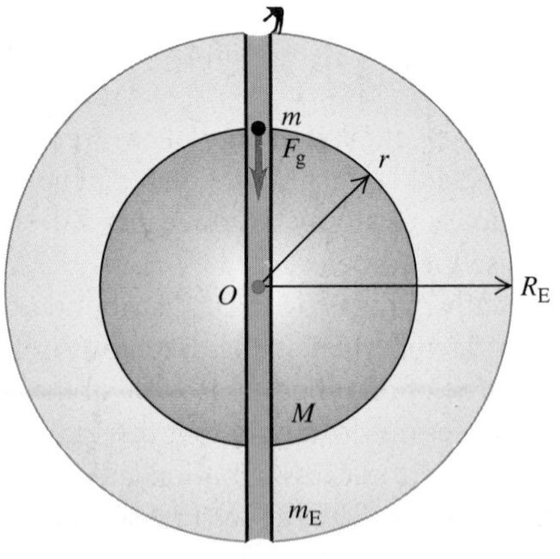

12.24 A hole through the center of the earth (assumed to be uniform). When an object is a distance r from the center, only the mass inside a sphere of radius r exerts a net gravitational force on it.

The magnitude of the gravitational force on m is given by

$$F_g = \frac{GMm}{r^2} = \frac{Gm_E m}{r^2}\frac{r^3}{R_E^3} = \frac{Gm_E m}{R_E^3}r$$

EVALUATE: At points inside the earth, F_g is *directly proportional* to the distance r from the center, rather than proportional to $1/r^2$ as it is outside the sphere. Right at the surface, where $r = R_E$, the above expression gives $F_g = Gm_E m/R_E^2$, as we should expect. In the next chapter we'll learn how to compute the time it would take for the mail pouch to emerge on the other side of the earth. Meanwhile, keep in mind that the assumption of uniform density of the earth is, like the hole, pure fantasy. We discussed the variation of density in the earth in Section 12.2.

Test Your Understanding

In the classic 1913 science-fiction novel *At the Earth's Core* by Edgar Rice Burroughs, explorers discover that the earth is a hollow sphere and that an entire civilization lives on the inside of the sphere. Would it be possible to stand and walk around on the inner surface of a hollow planet?

*12.7 | Apparent Weight and the Earth's Rotation

Because the earth rotates on its axis, it is not precisely an inertial frame of reference. For this reason the apparent weight of a body on earth is not precisely equal to the earth's gravitational attraction, which we will call the **true weight** $\vec{w}_0$ of the body. Figure 12.25 is a cutaway view of the earth, showing three observers. Each one holds a spring scale with a body of mass m hanging from it. Each scale applies a tension force $\vec{F}$ to the body hanging from it, and the reading on each scale is the magnitude F of this force. If the observers are unaware of the earth's rotation, each one *thinks* that the scale reading equals the weight of the body because he

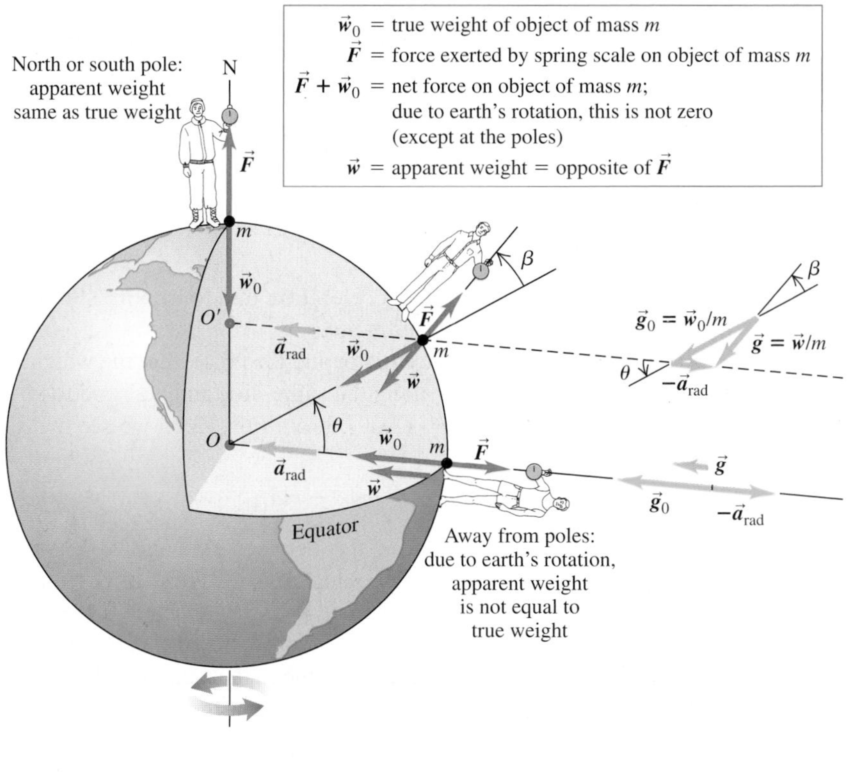

12.25 Except at the poles, the reading for an object being weighed on a scale (the *apparent weight*) is less than the gravitational force of attraction on the object (the *true weight*). The reason is that a net force is needed to provide a centripetal acceleration as the object rotates with the earth. For clarity, the illustration greatly exaggerates the angle β between the true and apparent weight vectors.

thinks the body on his spring scale is in equilibrium. So each observer thinks that the tension $\vec{F}$ must be opposed by an equal and opposite force $\vec{w}$, which we call the **apparent weight.** But if the bodies are rotating with the earth, they are *not* precisely in equilibrium. Our problem is to find the relation between the apparent weight $\vec{w}$ and the true weight $\vec{w}_0$.

If we assume that the earth is spherically symmetric, then the true weight $\vec{w}_0$ has magnitude Gm_Em/R_E^2, where m_E and R_E are the mass and radius of the earth. This value is the same for all points on the earth's surface. If the center of the earth can be taken as the origin of an inertial coordinate system, then the body at the north pole really *is* in equilibrium in an inertial system, and the reading on that observer's spring scale is equal to w_0. But the body at the equator is moving in a circle of radius R_E with speed v, and there must be a net inward force equal to the mass times the centripetal acceleration:

$$w_0 - F = \frac{mv^2}{R_E}$$

So the magnitude of the apparent weight (equal to the magnitude of F) is

$$w = w_0 - \frac{mv^2}{R_E} \qquad \text{(at the equator)} \qquad (12.29)$$

If the earth were not rotating, the body when released would have a free-fall acceleration $g_0 = w_0/m$. Since the earth *is* rotating, the falling body's actual acceleration relative to the observer at the equator is $g = w/m$. Dividing Eq. (12.29) by m and using these relations, we find

$$g = g_0 - \frac{v^2}{R_E} \qquad \text{(at the equator)}$$

To evaluate v^2/R_E, we note that in 86,164 s a point on the equator moves a distance equal to the earth's circumference, $2\pi R_E = 2\pi(6.38 \times 10^6\ \text{m})$. (The solar day, 86,400 s, is $\frac{1}{365}$ longer than this because in one day the earth also completes $\frac{1}{365}$ of its orbit around the sun.) Thus we find

$$v = \frac{2\pi(6.38 \times 10^6\ \text{m})}{86{,}164\ \text{s}} = 465\ \text{m/s}$$

$$\frac{v^2}{R_E} = \frac{(465\ \text{m/s})^2}{6.38 \times 10^6\ \text{m}} = 0.0339\ \text{m/s}^2$$

So for a spherically symmetric earth the acceleration due to gravity should be about 0.03 m/s^2 less at the equator than at the poles.

At locations intermediate between the equator and the poles, the true weight $\vec{w}_0$ and the centripetal acceleration are not along the same line, and we need to write a vector equation corresponding to Eq. (12.29). From Fig. 12.25 we see that the appropriate equation is

$$\vec{w} = \vec{w}_0 - m\vec{a}_{\text{rad}} = m\vec{g}_0 - m\vec{a}_{\text{rad}} \qquad (12.30)$$

The difference in the magnitudes of g and g_0 lies between zero and 0.0339 m/s^2. As shown in Fig. 12.25, the *direction* of the apparent weight differs from the direction toward the center of the earth by a small angle β, which is 0.1° or less.

Table 12.1 gives the values of g at several locations, showing variations with latitude. There are also small additional variations due to the lack of perfect spherical symmetry of the earth, local variations in density, and differences in elevation.

Table 12.1 Variations of g with Latitude and Elevation

Station	North Latitude	Elevation (m)	g (m/s^2)
Canal Zone	9°	0	9.78243
Jamaica	18°	0	9.78591
Bermuda	32°	0	9.79806
Denver	40°	1638	9.79609
Pittsburgh, Pa.	40.5°	235	9.80118
Cambridge, Mass.	42°	0	9.80398
Greenland	70°	0	9.82534

12.26 This orbiting astronaut is acted on by the earth's gravity, but *feels* weightless because his acceleration is equal to $\vec{g}$.

Our discussion of apparent weight can also be applied to the phenomenon of apparent weightlessness in orbiting spacecraft, which we described in Section 12.4. Bodies in an orbiting spacecraft are *not* weightless; the earth's gravitational attraction continues to act on them just as though they were at rest relative to the earth. The apparent weight of a body in a spacecraft is again given by Eq. (12.30):

$$\vec{w} = \vec{w}_0 - m\vec{a}_{\text{rad}} = m\vec{g}_0 - m\vec{a}_{\text{rad}}$$

But for a spacecraft in orbit, as well as any body inside the spacecraft, the acceleration $\vec{a}_{\text{rad}}$ toward the earth's center is equal to the value of the acceleration of gravity $\vec{g}_0$ at the position of the spacecraft. Hence

$$\vec{g}_0 = \vec{a}_{\text{rad}}$$

and the apparent weight is

$$\vec{w} = \mathbf{0}$$

This is what we mean when we say that an astronaut or other body in the spacecraft is apparently weightless. Note that we didn't make any assumptions about the shape of the orbit; as we mentioned in Section 12.4, an astronaut will be apparently weightless no matter what the orbit (Fig. 12.26).

Test Your Understanding

What would the speed v of a point on the equator have to be for an object there to have an apparent weight of *zero?*

12.8 | Black Holes

The concept of a black hole is one of the most interesting and startling products of modern gravitational theory, yet the basic idea can be understood on the basis of Newtonian principles. Think first about the properties of our own sun. Its mass $M = 1.99 \times 10^{30}$ kg and radius $R = 6.96 \times 10^8$ m are much larger than those of any planet, but compared to other stars, our sun is not exceptionally massive.

What's the average *density* ρ of the sun? You can find it the same way we found the average density of the earth in Section 12.2:

$$\rho = \frac{M}{V} = \frac{M}{\frac{4}{3}\pi R^3} = \frac{1.99 \times 10^{30}\ \text{kg}}{\frac{4}{3}\pi(6.96 \times 10^8\ \text{m})^3}$$
$$= 1410\ \text{kg/m}^3$$

The sun's temperatures range from 5800 K (about 5500°C, or 10,000°F) at the surface up to 1.5×10^7 K (about 2.7×10^7 °F) in the interior, so it surely contains no solids or liquids. Yet gravitational attraction pulls the sun's gas atoms together until the sun is, on average, 41% denser than water and about 1200 times as dense as the air we breathe.

Now think about the escape speed for a body at the surface of the sun. In Example 12.5 (Section 12.3) we found that the escape speed from the surface of a spherical mass M with radius R is $v = \sqrt{2GM/R}$. We can relate this to the average density. Substituting $M = \rho V = \rho(\frac{4}{3}\pi R^3)$ into the expression for escape speed gives

$$v = \sqrt{\frac{2GM}{R}} = \sqrt{\frac{8\pi G\rho}{3}}R \qquad (12.31)$$

Using either form of this equation, you can show that the escape speed for a body at the surface of our sun is $v = 6.18 \times 10^5$ m/s (about 2.2 million km/h, or 1.4 million mi/h). This value, roughly 1/500 the speed of light, is independent of the mass of the escaping body; it depends only on the mass and radius (or average density and radius) of the sun.

Now consider various stars with the same average density ρ and different radii R. Equation (12.31) shows that for a given value of density ρ, the escape speed v is directly proportional to R. In 1783 the Rev. John Mitchell, an amateur astronomer, noted that if a body with the same average density as the sun had about 500 times the radius of the sun, the magnitude of its escape velocity would be greater than the speed of light c. With his statement that "all light emitted from such a body would be made to return toward it," Mitchell became the first person to suggest the existence of what we now call a **black hole.**

The first expression for escape speed in Eq. (12.31) also suggests that a body of mass M will act as a black hole if its radius R is less than or equal to a certain critical radius. How can we determine this critical radius? You might think that you can find the answer by simply setting $v = c$ in Eq. (12.31). As a matter of fact, this does give the correct result, but only because of two compensating errors. The kinetic energy of light is *not* $mc^2/2$, and the gravitational potential energy near a black hole is *not* given by Eq. (12.9). In 1916, Karl Schwarzschild used Einstein's general theory of relativity (in part a generalization and extension of Newtonian gravitation theory) to derive an expression for the critical radius R_S, now called the **Schwarzschild radius.** The result turns out to be the same as though we had set $v = c$ in Eq. (12.31), so

$$c = \sqrt{\frac{2GM}{R_S}}$$

Solving for the Schwarzschild radius R_S, we find

$$R_S = \frac{2GM}{c^2} \quad \text{(Schwarzschild radius)} \qquad (12.32)$$

If a spherical, nonrotating body with mass M has a radius less than R_S, then *nothing* (not even light) can escape from the surface of the body, and the body is a black hole (Fig. 12.27). In this case, any other body within a distance R_S of the center of the black hole is trapped by the gravitational attraction of the black hole and cannot escape from it.

(a)

(b)

12.27 (a) When the radius R of a body is greater than the Schwarzschild radius R_S, light can escape from the surface of the body. As it travels away, it is "red-shifted" to longer wavelengths. (b) When the body is within the event horizon (radius R_S), it is a black hole with an escape speed greater than the speed of light. In that case we can obtain very little information about it.

The surface of the sphere with radius R_S surrounding a black hole is called the **event horizon:** since light can't escape from within that sphere, we can't see events occurring inside. All that an observer outside the event horizon can know about a black hole is its mass (from its gravitational effects on other bodies), its electric charge (from the electric forces it exerts on other charged bodies), and its angular momentum (because a rotating black hole tends to drag space—and everything in that space—around with it). All other information about the body is irretrievably lost when it collapses inside its event horizon.

Example 12.11 Black hole calculations

Astrophysical theory suggests that a burned-out star will collapse under its own gravity to form a black hole when its mass is at least three solar masses. If it does, what is the radius of its event horizon?

SOLUTION

IDENTIFY and SET UP: The radius in question is R_S as given by Eq. (12.32). "Three solar masses" means $M = 3(1.99 \times 10^{30}\ \text{kg}) = 6.0 \times 10^{30}\ \text{kg}$.

EXECUTE: From Eq. (12.32),

$$R_S = \frac{2GM}{c^2} = \frac{2(6.67 \times 10^{-11}\ \text{N}\cdot\text{m}^2/\text{kg}^2)(6.0 \times 10^{30}\ \text{kg})}{(3.00 \times 10^8\ \text{m/s})^2}$$
$$= 8.9 \times 10^3\ \text{m} = 8.9\ \text{km}$$

or less than 6 miles.

EVALUATE: If the radius of such an object is just equal to the Schwarzschild radius, the average density has the incredibly large value

$$\rho = \frac{M}{\frac{4}{3}\pi R^3} = \frac{6.0 \times 10^{30}\ \text{kg}}{\frac{4}{3}\pi(8.9 \times 10^3\ \text{m})^3}$$
$$= 2.0 \times 10^{18}\ \text{kg/m}^3$$

This is about 10^{15} times as great as the density of familiar matter on earth and is comparable to the densities of atomic nuclei. In fact, once the body collapses to a radius of R_S, nothing can prevent it from collapsing further. All of the mass ends up being crushed down to a single point called a *singularity* at the center of the event horizon. This point has zero volume and so has *infinite* density.

At points far from a black hole, its gravitational effects are the same as those of any normal body with the same mass. If the sun collapsed to form a black hole, the orbits of the planets would be unaffected. But things get dramatically different close to the black hole. If you decided to become a martyr for science and jump into a black hole, those you left behind would notice several odd effects as you moved toward the event horizon, most of them associated with effects of general relativity. If you carried a radio transmitter to send back your comments on what was happening, they would have to retune their receiver continuously to lower and lower frequencies, an effect called the *gravitational red shift.* Consistent with this shift, they would observe that your clocks (electronic or biological) would appear to run more and more slowly, an effect called *time dilation.* In fact, during their lifetimes they would never see you make it to the event horizon. In your frame of reference, you would make it to the event horizon in a rather short time but in a rather disquieting way. As you fell feet first into the black hole, the gravitational pull on your feet would be greater than that on your head, which would be slightly farther away from the black hole. The *differences* in gravitational force on different parts of your body would be great enough to stretch you along the direction toward the black hole and compress you perpendicular to it. These effects (called *tidal forces*) would rip you to atoms, and then rip your atoms apart, before you reached the event horizon.

If light cannot escape from a black hole and if black holes are as small as Example 12.11 suggests, how can we know that such things exist? The answer is that any gas or dust near the black hole tends to be pulled into an *accretion disk* that swirls around and into the black hole, rather like a whirlpool (Fig. 12.28). Friction within the accretion disk's material causes it to lose mechanical energy and spiral into the black hole; as it moves inward, it is compressed together. This causes heating of the material, just as air compressed in a bicycle pump gets hotter. Temperatures in excess of 10^6 K can occur in the accretion disk, so hot that the disk emits not just vis-

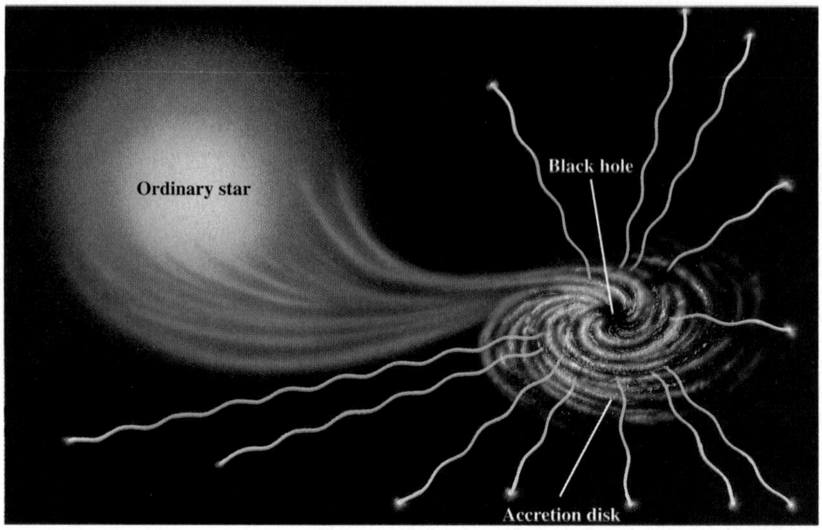

12.28 A binary star system in which an ordinary star and a black hole orbit each other. Matter is pulled from the ordinary star to form an accretion disk around the black hole. The gas in the accretion disk is compressed and heated to such high temperatures that it becomes an intense source of x rays.

ible light (as do bodies that are "red-hot" or "white-hot") but x rays. Astronomers look for these x rays (emitted *before* the material crosses the event horizon) to signal the presence of a black hole. Several promising candidates have been found, and astronomers now express considerable confidence in the existence of black holes.

Black holes in binary star systems like the one depicted in Fig. 12.28 have masses a few times greater than the sun's mass. There is also mounting evidence for the existence of much larger *supermassive black holes.* One example is thought to lie at the center of our Milky Way galaxy, some 26,000 light years from earth in the direction of the constellation Sagittarius. High-resolution images of the galactic center reveal stars moving at speeds greater than 1500 km/s about an unseen object that lies at the position of a source of radio waves called Sgr A* (Fig. 12.29). By analyzing these motions, astronomers can infer the period T and semi-major axis a of each star's orbit. The mass m_X of the unseen object can then be calculated using Kepler's third law in the form given in Eq. (12.19), with the mass of the sun m_S replaced by m_X:

$$T = \frac{2\pi a^{3/2}}{\sqrt{Gm_X}} \quad \text{so} \quad m_X = \frac{4\pi^2 a^3}{GT^2}$$

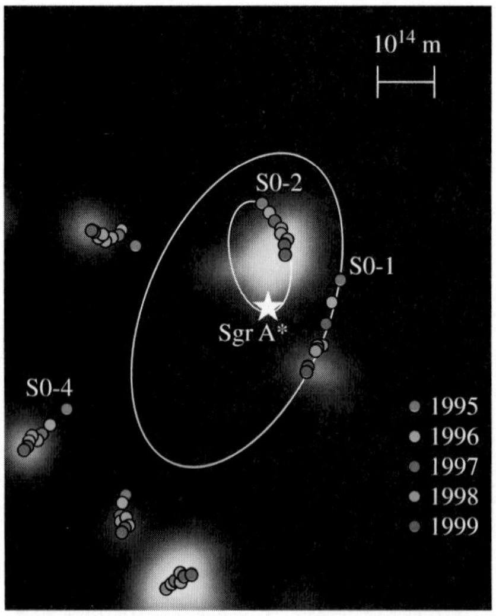

12.29 This false-color image shows the motions of stars at the center of our galaxy over a five-year period. Analyzing these orbits using Kepler's third law indicates that the stars are moving about an unseen object that is some 2.6×10^6 times the mass of the sun. The scale bar indicates a length of 10^{14} m (670 times the distance from the earth to the sun) at the distance of the galactic center.

The conclusion is that the mysterious dark object at the galactic center has a mass of 5.2×10^{36} kg, or 2.6 *million* times the mass of the sun. Yet observations with radio telescopes show that it has a radius of no more than about 10^{11} m, comparable to the distance from the earth to the sun. These observations suggest that this massive, compact object is a black hole with a Schwarzschild radius of 7.8×10^9 m. In the next several years astronomers hope to improve the resolution of their observations and actually be able to see the event horizon of this black hole.

Other lines of research suggest that even larger black holes, in excess of 10^9 solar masses, lie at the centers of other galaxies. Observational and theoretical studies of black holes of all sizes continue to be an exciting area of research in both physics and astronomy.

Test Your Understanding

If the sun somehow collapsed to form a black hole, what effect would this event have on the orbit of the earth?

SUMMARY

Newton's law of gravitation states that any two bodies with masses m_1 and m_2, a distance r apart, attract each other with forces inversely proportional to r^2. These forces form an action-reaction pair and obey Newton's third law. When two or more bodies exert gravitational forces on a particular body, the total gravitational force on that individual body is the vector sum of the forces exerted by the other bodies. (See Examples 12.1 through 12.3)

$$F_g = \frac{Gm_1m_2}{r^2} \qquad (12.1)$$

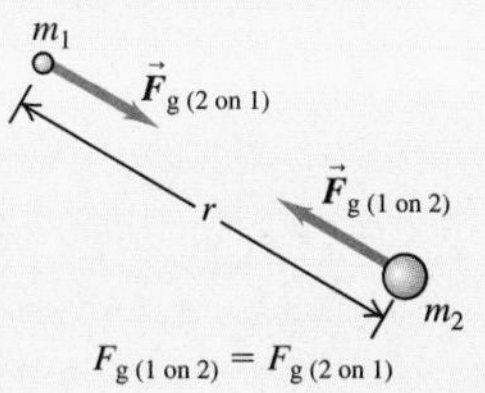

The weight w of a body is the total gravitational force exerted on it by all other bodies in the universe. Near the surface of the earth (mass m_E and radius R_E), it is essentially equal to the gravitational force of the earth alone. (See Example 12.4)

$$w = F_g = \frac{Gm_Em}{R_E^2} \qquad (12.3)$$

(weight at earth's surface)

$$g = \frac{Gm_E}{R_E^2} \qquad (12.4)$$

(acceleration due to gravity at earth's surface)

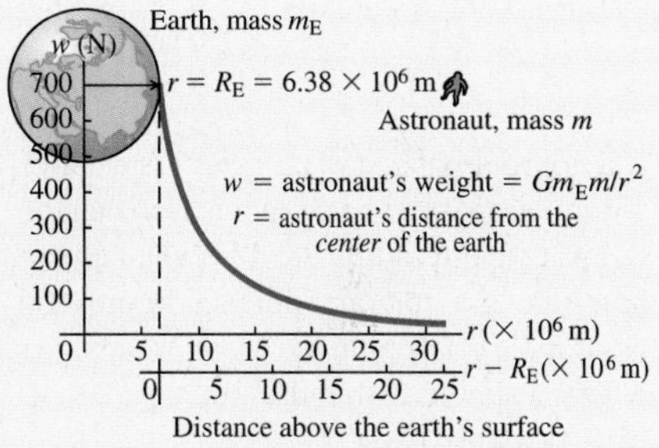

The gravitational potential energy U of two masses m and m_E separated by a distance r is inversely proportional to r. The potential energy is never positive; it is zero only when the two bodies are infinitely far apart. (See Example 12.5)

$$U = -\frac{Gm_Em}{r} \qquad (12.9)$$

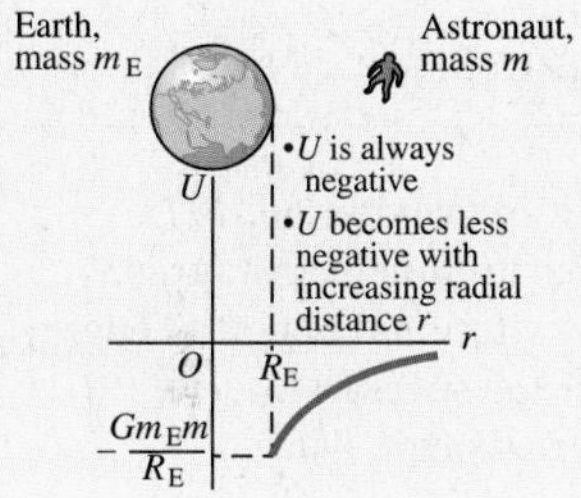

When a satellite moves in a circular orbit, the centripetal acceleration is provided by the gravitational attraction of the earth. (See Example 12.6)

$$v = \sqrt{\frac{Gm_E}{r}} \text{ (speed in circular orbit)} \qquad (12.12)$$

$$T = \frac{2\pi r}{v} = 2\pi r\sqrt{\frac{r}{Gm_E}} = \frac{2\pi r^{3/2}}{\sqrt{Gm_E}} \qquad (12.14)$$

(period in circular orbit)

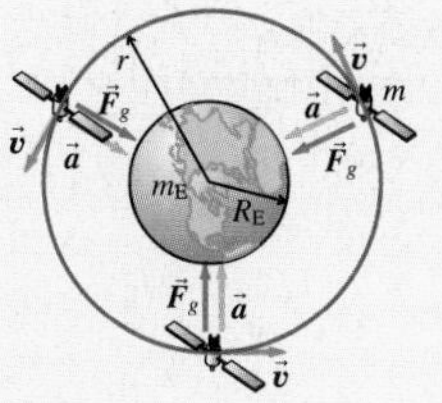

Kepler's three laws describe characteristics of the elliptical orbits of planets around the sun or satellites around a planet. (See Examples 12.7 through 12.9)

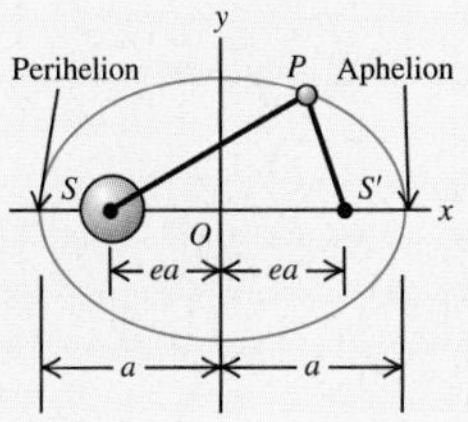

The gravitational interaction of any spherically symmetric mass distribution, at points outside the distribution, is the same as if all the mass were concentrated at the center. (See Example 12.10)

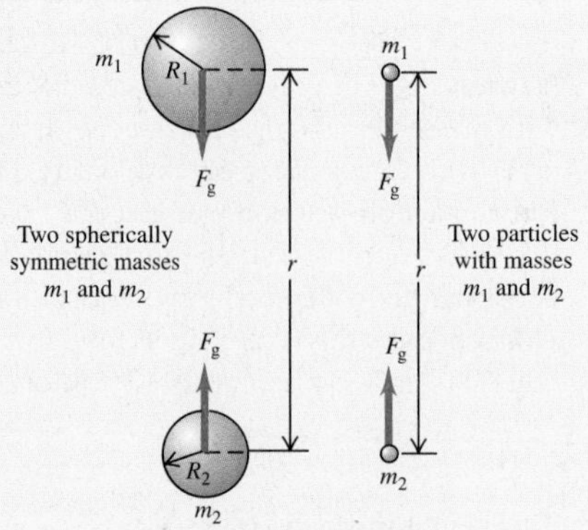

If a nonrotating spherical mass distribution with total mass M has a radius less than its Schwarzschild radius R_S, it is called a black hole. The gravitational interaction prevents anything, including light, from escaping from within a sphere with radius R_S. (See Example 12.11)

$$R_S = \frac{2GM}{c^2} \qquad (12.32)$$

(Schwarzschild radius)

Key Terms

law of gravitation, 437
gravitational constant, 437
gravitational potential energy, 445
escape speed, 446
closed orbit, 448
open orbit, 448
semi-major axis, 452
eccentricity, 452
true weight, 459
apparent weight, 460
black hole, 462
Schwarzschild radius, 462
event horizon, 462

Your Notes

Answer to Chapter Opening Question

The gravitational force that an object of mass m_1 exerts on the earth (mass m_2) is inversely proportional to the distance r to that object: $F = Gm_1m_2/r^2$. While the galaxies are 10^{11} times more massive than the sun, they are 7.5×10^{12} times more distant from the earth and their gravitational force on the earth is only $(10^{11})/(7.5 \times 10^{12})^2 = 1.8 \times 10^{-15}$ times as great as the sun's gravitational force. Hence the galaxies' gravitational effect on the earth is negligible.

Answers to Test Your Understanding Questions

Section 12.1 From Eq. (12.1), the magnitude of the force on either car is

$$F_g = (6.67 \times 10^{-11}\ \text{N}\cdot\text{m}^2/\text{kg}^2)(900\ \text{kg})(900\ \text{kg})/(3.0\ \text{m})^2$$
$$= 6.0 \times 10^{-6}\ \text{N}$$

You can exert a force more than 10,000 times greater than this with your little finger. Gravitational forces are negligible in automobile collisions.

Section 12.2 The acceleration due to gravity g_M at the surface of Mars depends on both the planet's mass m_M and its radius R_M: $g_M = Gm_M/R_M^2$. The radius of Mars is 53% as great as that of the earth, and its smaller radius makes the value of g_M larger than what you would expect based on the mass alone.

Section 12.3 This is indeed possible, because the two quantities depend in different ways on the planet's mass m_P and radius R_P: the value of g at the surface is Gm_P/R_P^2, while the escape speed is $\sqrt{2Gm_P/R_P}$. For the planet Saturn, for example, m_P is about 100 times the earth's mass and R_P is about 10 times the earth's radius. The value of g is different than on earth by a factor $(100)/(10)^2 = 1$ (i.e., it is the same as on earth), while the escape speed is greater by a factor $\sqrt{100/10} = 3.2$.

It may help to remember that the surface gravity tells you about conditions right next to the planet's surface, while the escape speed (which tells you how fast you must travel to escape to infinity) depends on conditions at *all* points between the planet's surface and infinity. Because Saturn has so much more mass than the earth, its gravitational effects are appreciable at much greater distances and its escape speed is higher.

Section 12.4 In a smaller-radius orbit, the shuttle has a faster speed and will eventually catch up with the satellite. The shuttle could then use its thrusters to maneuver into the same orbit as the satellite to complete the rendezvous. The shuttle could also enter a larger-radius orbit; then the satellite would have the greater speed and would eventually catch up with the shuttle.

Section 12.5 Equation (12.9) shows that the orbital period is proportional to the $\frac{3}{2}$ power of the semi-major axis. Hence the orbital period of Comet X is longer than that of Comet Y by a factor of $4^{3/2} = 8$.

Section 12.6 Our analysis shows that there is *zero* gravitational force inside a hollow spherical shell. Hence visitors to the interior of a hollow planet would find themselves quite weightless, and they could not stand or walk on the planet's inner surface.

Section 12.7 We want $g = g_0 - v^2/R_E = 0$. Hence v^2 must equal g_0R_E, or

$$v = \sqrt{g_0R_E} = \sqrt{(9.80\ \text{m/s}^2)(6.38 \times 10^6\ \text{m})} = 7910\ \text{m/s}$$

For this to happen, the earth would have to spin $(7910\ \text{m/s})/(465\ \text{m/s}) = 17.0$ times faster than it does now!

Section 12.8 There would be no effect. If the sun collapsed into a black hole (which, according to our understanding of stars, it cannot do), it would have the same mass but a much smaller radius. Because the gravitational attraction of the sun on the earth does not depend on the sun's radius, the earth's orbit would be unaffected.

Discussion Questions

Q12.1 A student wrote: "The only reason an apple falls downward to meet the earth instead of the earth rising upward to meet the apple is that the earth is much more massive and so exerts a much greater pull." Please comment.

Q12.2 To use the Cavendish balance (Fig. 12.4) to determine G, you must first know the masses m_1 and m_2. How could you determine these masses *without* weighing them?

Q12.3 If all planets had the same average density, how would the acceleration due to gravity at the surface of a planet depend on its radius?

Q12.4 Is a pound of butter on the earth the same amount as a pound of butter on Mars? What about a kilogram of butter? Explain.

Q12.5 Example 12.2 (Section 12.1) shows that the acceleration of each sphere caused by the gravitational force is inversely proportional to the mass of that sphere. So why does the force of gravity give all masses the same acceleration when they are dropped near the surface of the earth?

Q12.6 When will you attract the sun more: today at noon, or tonight at midnight? Explain.

Q12.7 Since the moon is constantly attracted toward the earth by the gravitational interaction, why doesn't it crash into the earth?

Q12.8 Imagine that the sun somehow became twice as massive as it is now. What effect would this have on your weight on the earth (as measured by standing on a scale)? Explain your answer.

Q12.9 The sun pulls on the moon with a force that is more than twice the magnitude of the force with which the earth attracts the moon. Why, then, doesn't the sun take the moon away from the earth?

Q12.10 As defined in Chapter 7, gravitational potential energy is $U = mgy$ and is positive for a body of mass m above the earth's surface (which is at $y = 0$). But in this chapter, gravitational potential energy is $U = -Gm_Em/r^2$, which is *negative* for a body of mass m above the earth's surface (which is at $r = R_E$). How can you reconcile these seemingly incompatible descriptions of gravitational potential energy?

Q12.11 A planet is moving at constant speed in a circular orbit around a star. In one complete orbit, what is the net amount of work done on the planet by the star's gravitational force: positive, negative, or zero? What if the planet's orbit is an ellipse, so that the speed is not constant? Explain your answers.

Q12.12 Does the escape speed for an object at the earth's surface depend on the direction in which it is launched? Explain. Does your answer depend on whether or not you include the effects of air resistance?

Q12.13 If a projectile is fired straight up from the earth's surface, what would happen if the total mechanical energy (kinetic plus potential) is a) less than zero? b) greater than zero? In each case, ignore air resistance and the gravitational effects of the sun, the moon, and the other planets.

Q12.14 Discuss whether the following statement is correct: "In the absence of air resistance, the trajectory of a projectile thrown near the earth's surface is an *ellipse,* not a parabola."

Q12.15 The earth is closer to the sun in November than in May. In which month does it move faster in its orbit? Explain.

Q12.16 A communications firm wants to place a satellite in orbit so that it is always directly above the earth's 45th parallel (latitude 45° north). This means that the plane of the orbit will not pass through the center of the earth. Is such an orbit possible? Why or why not?

Q12.17 At what point in an elliptical orbit is the acceleration maximum? At what point is it minimum? Justify your answers.

Q12.18 Which takes more fuel, a voyage from the earth to the moon or from the moon to the earth? Explain.

Q12.19 What would Kepler's third law be for circular orbits if an amendment to Newton's law of gravitation made the gravitational force inversely proportional to r^3? Would this change affect Kepler's other two laws? Explain.

Q12.20 In the elliptical orbit of Comet Halley shown in Fig. 12.20a, the sun's gravity is responsible for making the comet fall inward from aphelion to perihelion. But what is responsible for making the comet move from perihelion back outward to aphelion?

Q12.21 Many people believe that orbiting astronauts feel weightless because they are "beyond the pull of the earth's gravity." How far from the earth would a spacecraft have to travel to be truly beyond the earth's gravitational influence? If a spacecraft were really unaffected by the earth's gravity, would it remain in orbit? Explain. What is the real reason why astronauts in orbit feel weightless?

Q12.22 As part of their training before going into orbit, astronauts ride in an airliner that is flown along the same parabolic trajectory as a freely falling projectile. Explain why this gives the same experience of apparent weightlessness as being in orbit.

Exercises

Section 12.1 Newton's Law of Gravitation

12.1 What is the ratio of the gravitational pull of the sun on the moon to that of the earth on the moon? (Assume the distance of the moon from the sun can be approximated by the distance of the earth from the sun.) Use the data in Appendix F. Is it more accurate to say that the moon orbits the earth, or that the moon orbits the sun?

12.2 A 2150-kg satellite used in a cellular telephone network is in a circular orbit at a height of 780 km above the surface of the earth. What is the gravitational force on the satellite? What fraction is this of its weight at the surface of the earth?

12.3 A set of two uniform spheres with masses m_1 and m_2 have their centers separated by a distance r_{12}. They exert a gravitational force of magnitude F_{12} on each other. What magnitude of gravitational force does a second set of uniform spheres exert on one another if they have masses nm_1 and nm_2 and a center-to-center separation of nr_{12}, where n is any positive number?

12.4 Two uniform spheres, each with mass M and radius R, touch one another. What is the magnitude of their gravitational force of attraction?

12.5 An interplanetary spaceship passes through the point in space where the gravitational forces from the sun and the earth on the ship exactly cancel. a) How far from the center of the earth is it? Use the data in Appendix F. b) What, if anything, happens to the spaceship when it passes through the point described in part (a)? Explain.

12.6 a) In Fig. 12.30 what are the magnitude and direction of the net gravitational force exerted on the 0.100-kg uniform sphere by the other two uniform spheres? The centers of all three spheres are on the same line. b) According to Newton's third law, does the 0.100-kg sphere exert forces of the same magnitude as your answer to part (a), but opposite direction, on *each* of the other two spheres?

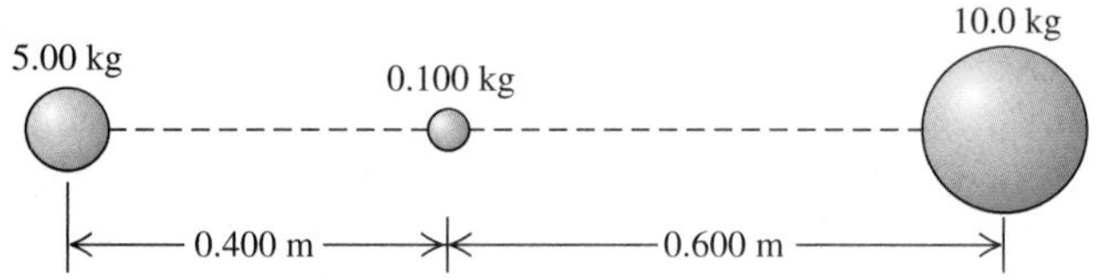

Figure 12.30 Exercise 12.6.

12.7 The typical adult human has a mass of about 70 kg. What force does a full moon exert on such a human when it is directly overhead with its center 378,000 km away?

12.8 The sun has a mass 333,000 times that of the earth. For a person on earth, the average distance to the center of the sun is 23,500 times the distance to the center of the earth. In magnitude, what is the ratio of the sun's gravitational force on you to the earth's gravitational force on you?

12.9 Calculate the magnitude and direction of the net gravitational force on the moon due to the earth and the sun when the moon is in each of the positions shown in Fig. 12.31. (Note that the figure is *not* drawn to scale. Assume that the sun is in the plane of the earth-

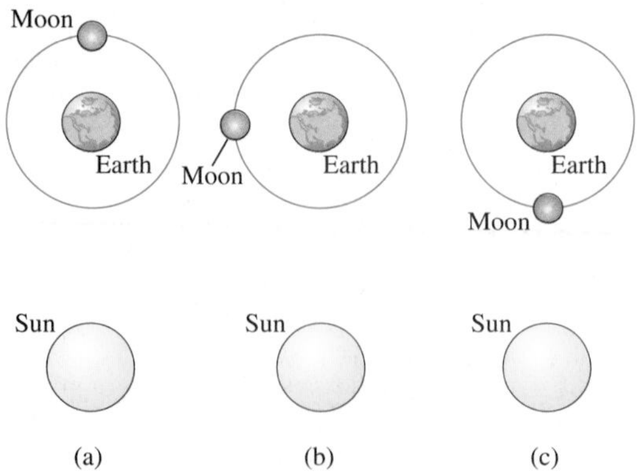

Figure 12.31 Exercise 12.9

moon orbit, even though this is not actually the case.) Use the data in Appendix F.

12.10 Four identical masses of 800 kg each are placed at the corners of a square whose side length is 10.0 cm. What is the net gravitational force (magnitude and direction) on one of the masses, due to the other three?

12.11 Three identical masses of 500 kg each are placed on the x-axis. One mass is at $x = -10$ cm, one is at the origin, and one is at $x = 40$ cm. What is the net gravitational force (magnitude and direction) on the mass at the origin, due to the other two masses?

12.12 Determine the magnitude and direction of the gravitational force on a particle of mass m that is midway between two spherically symmetric objects, one of mass m_1 and the other of mass m_2, where $m_2 > m_1$. The centers of the two spherically symmetric objects are a distance d apart.

12.13 Two uniform spheres, each of mass 0.260 kg, are fixed at points A and B (Fig. 12.32). Find the magnitude and direction of the initial acceleration of a uniform sphere with mass 0.010 kg if released from rest at point P and acted on only by forces of gravitational attraction of the spheres at A and B.

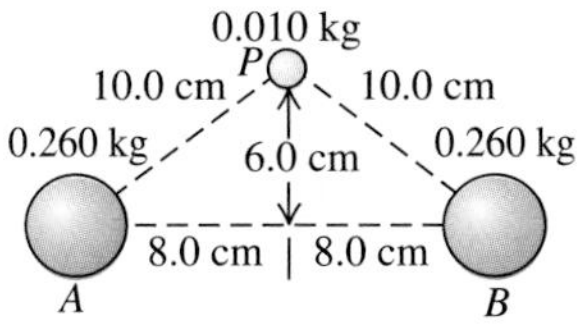

Figure 12.32 Exercise 12.13.

Section 12.2 Weight

12.14 Use the mass and radius of the planet Pluto given in Appendix F to calculate the acceleration due to gravity at the surface of Pluto.

12.15 At what distance above the surface of the earth is the acceleration due to the earth's gravity 0.980 m/s^2 if the acceleration due to gravity at the surface has magnitude 9.80 m/s^2?

12.16 The mass of Venus is 81.5% that of the earth, and its radius is 94.9% that of the earth. a) Compute the acceleration due to gravity on the surface of Venus from these data. b) What is the weight of a 5.00-kg rock on the surface of Venus?

12.17 Titania, the largest moon of the planet Uranus, has $\frac{1}{8}$ the radius of the earth and $\frac{1}{1700}$ the mass of the earth. a) What is the acceleration due to gravity at the surface of Titania? b) What is the average density of Titania? (This is less than the density of rock, which is one piece of evidence that Titania is made primarily of ice.)

12.18 Rhea, one of Saturn's moons, has a radius of 765 km and an acceleration due to gravity of 0.278 m/s^2 at its surface. Calculate its mass and average density.

12.19 Calculate the earth's gravity force on a 75-kg astronaut who is repairing the Hubble Space Telescope 600 km above the earth's surface, and then compare this value with his weight at the earth's surface. In view of your result, explain why we say astronauts are weightless when they orbit the earth in a satellite such as a space shuttle.

12.20 Neutron stars, such as the one at the center of the Crab Nebula, have about the same mass as our sun but a *much* smaller diameter. If you weigh 675 N on the earth, what would you weigh when at the surface of a neutron star that has the same mass as our sun and a diameter of 20 km?

12.21 An experiment using the Cavendish balance to measure the gravitational constant G found that a uniform 0.400-kg sphere attracts another uniform 0.00300-kg sphere with a force of 8.00×10^{-10} N, when the distance between the centers of the spheres is 0.0100 m. The acceleration due to gravity at the earth's surface is 9.80 m/s^2, and the radius of the earth is 6380 km. Compute the mass of the earth from these data.

12.22 In Example 12.4 (Section 12.2) we ignored the gravitational effects of the moons of Mars. The larger of the two moons, Phobos, has a radius of approximately 12 km and an average density of 2000 kg/m^3. a) Calculate the gravitational force of Phobos on the Mars lander of Example 12.4 if the Mars lander were on the surface of Phobos. b) Discuss whether the approximation made in Example 12.4 was appropriate.

Section 12.3 Gravitational Potential Energy

12.23 The asteroid Dactyl, discovered in 1993, has a radius of only about 700 m and a mass of about 3.6×10^{12} kg. Use the results of Example 12.5 (Section 12.3) to calculate the escape speed for an object at the surface of Dactyl. Could a person reach this speed just by walking?

12.24 When in orbit, a communication satellite attracts the earth with a force of 19.0 kN and the earth-satellite gravitational potential energy (relative to zero at infinite separation) is -1.39×10^{11} J. a) Find the satellite's altitude above the earth's surface. b) Find the mass of the satellite.

12.25 Use the results of Example 12.5 (Section 12.3) to calculate the escape speed for a spacecraft a) from the surface of Mars; b) from the surface of Jupiter. Use the data in Appendix F. c) Why is the escape speed for a spacecraft independent of the spacecraft's mass?

12.26 Ten days after it was launched toward Mars in December 1998, the *Mars Climate Orbiter* spacecraft (mass 629 kg) was 2.87×10^6 km from the earth and traveling at 1.20×10^4 km/h relative to the earth. At this time, what were a) the spacecraft's kinetic energy relative to the earth and b) the potential energy of the earth-spacecraft system?

Section 12.4 The Motion of Satellites

12.27 An earth satellite moves in a circular orbit with an orbital speed of 6200 m/s. a) Find the time of one revolution. b) Find the radial acceleration of the satellite in its orbit.

12.28 What is the period of revolution of a satellite with mass m that orbits the earth in a circular path of radius 7880 km (about 1500 km above the surface of the earth)?

12.29 For a satellite to be in a circular orbit 780 km above the surface of the earth, what must its orbital speed be?

12.30 Assume that the earth's orbit around the sun is circular. Use the earth's orbital radius and orbital period given in Appendix F to calculate the mass of the sun.

12.31 Deimos, a moon of Mars, is about 12 km in diameter with mass 2.0×10^{15} kg. Suppose you are stranded alone on Deimos and want to play a one-person game of baseball. You would be the pitcher, and you would be the batter! With what speed would you have to throw a baseball so that it would go into orbit and return to you so you could hit it? Do you think you could actually throw it at this speed?

Section 12.5 Kepler's Laws and the Motion of Planets

12.32 Planet Vulcan. Suppose that a planet were discovered between the sun and Mercury, with a circular orbit of radius equal to $\frac{2}{3}$ of the average orbit radius of Mercury. (Such a planet was once postulated, in part to explain the precession of Mercury's orbit. It was even given the name Vulcan, although we now have no evidence that it actually exists. Mercury's precession has been explained by general relativity.) What would be the orbital period of such a planet?

12.33 The star Rho[1] Cancri is 57 light years from the earth and has a mass 0.85 times that of our sun. A planet has been detected in a circular orbit around Rho[1] Cancri with an orbital radius equal to 0.11 times the radius of the earth's orbit around the sun. What are a) the orbital speed and b) the orbital period of the planet of Rho[1] Cancri?

12.34 Venus orbits the sun in a nearly circular orbit. Use the data in Appendix F for the orbital radius and period of Venus to calculate the mass of the sun.

12.35 a) Use Fig. 12.18 to show that the sun-planet distance at perihelion is $(1 - e)a$, the sun-planet distance at aphelion is $(1 + e)a$, and therefore that the sum of these two distances is $2a$. b) Pluto is usually called the outermost planet, but at perihelion in 1989, it was almost 100 million km closer to the sun than Neptune. The semi-major axes of the orbits of Pluto and Neptune are 5.92×10^{12} m and 4.50×10^{12} m, respectively, and the eccentricities are 0.248 and 0.010. Find Pluto's closest distance and Neptune's farthest distance from the sun. c) How many years after being at perihelion in 1989 will Pluto again be at perihelion?

12.36 The star 70 Virginis is 59 light years from the earth and has a mass of 1.9×10^{30} kg. a) A large planet of mass 1.3×10^{28} kg is known to orbit this star. The planet is attracted to the star by a force of 3.3×10^{26} N, when their centers are separated by a distance equal to the semi-major axis of the planet's orbit. Calculate this distance (in km). b) What is the period (in days) of this planet's orbit? c) Other as yet undiscovered planets may orbit 70 Virginis. If a second planet is in a circular orbit around 70 Virginis with an orbital period equal to eight times the value found in part (b), what is the radius (in km) of the second planet's orbit?

12.37 The *Helios B* spacecraft had a speed of 71 km/s when it was 4.3×10^7 km from the sun. a) Prove that it was not in a circular orbit about the sun. b) Prove that its orbit about the sun was closed and therefore elliptical.

*Section 12.6 Spherical Mass Distributions

***12.38** A thin, uniform rod has length L and mass M. A small uniform sphere of mass m is placed a distance x from one end of the rod, along the axis of the rod (Fig. 12.33). a) Calculate the gravitational potential energy of the rod-sphere system. Take the potential energy to be zero when the rod and sphere are infinitely far apart. Show that your answer reduces to the expected result when x is much larger than L. (*Hint:* Use the power series expansion for $\ln(1 + x)$ given in Appendix B.) b) Use $F_x = -dU/dx$ to find the magnitude and direction of the gravitational force exerted on the sphere by the rod. Show that your answer reduces to the expected result when x is much larger than L.

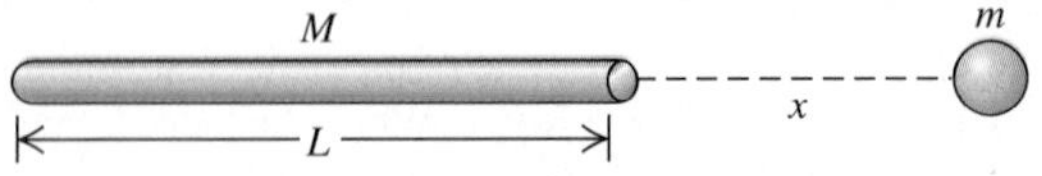

Figure 12.33 Exercise 12.38, Problem 12.83.

***12.39** Consider the ring-shaped body of Fig. 12.34. A particle with mass m is placed a distance x from the center of the ring, along the line through the center of the ring and perpendicular to its plane. a) Calculate the gravitational potential energy U of this system. Take the potential energy to be zero when the two objects are far apart. b) Show that your answer to part (a) reduces to the expected result when x is much larger than the radius a of the ring. c) Use $F_x = -dU/dx$ to find the magnitude and direction of the force on the particle. d) Show that your answer to part (c) reduces to the expected result when x is much larger than a. e) What are the values of U and F_x when $x = 0$? Explain why these results make sense.

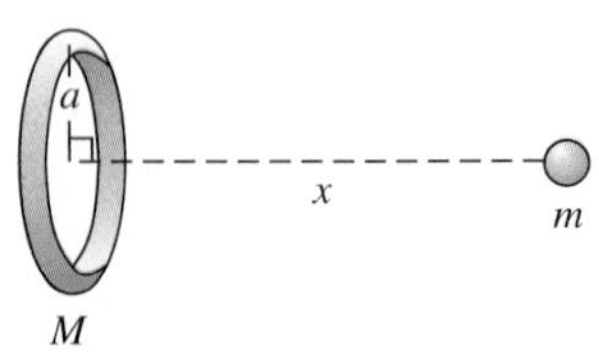

Figure 12.34 Exercise 12.39, Problem 12.82.

Section 12.7 Apparent Weight and the Earth's Rotation

***12.40** The weight of a person at the equator, as determined by a spring balance, is 690 N. By how much does this differ from the true force of gravitational attraction at the same point? Assume that the earth is spherically symmetric.

***12.41** The acceleration due to gravity at the north pole of Neptune is approximately 10.7 m/s^2. Neptune has mass 1.0×10^{26} kg and radius 2.5×10^4 km and rotates once around its axis in about 16 h. a) What is the gravitational force on a 5.0-kg object at the north pole of Neptune? b) What is the apparent weight of this same object at Neptune's equator? (Note that Neptune's "surface" is gaseous, not solid, so it is impossible to stand on it.)

Section 12.8 Black Holes

12.42 a) Show that a black hole attracts an object with mass m with a force of $mc^2R_S/(2r^2)$, where r is the distance between the object and the center of the black hole. b) Calculate the magnitude of the gravitational force exerted by a black hole of Schwarzschild radius 14.0 mm on a 5.00-kg mass 3000 km from it. c) What is the mass of this black hole?

12.43 At the Galaxy's Core. Astronomers have observed a small, massive object at the center of our Milky Way galaxy (Section 12.8). A ring of material orbits this massive object; the ring has a diameter of about 15 light years and an orbital speed of about 200 km/s. a) Determine the mass of the massive object at the center of the Milky Way galaxy. Give your answer both in kilograms and in solar masses (one solar mass is the mass of the sun). b) Observations of stars, as well as theories of the structure of stars, suggest that it is impossible for a single star to have a mass of more than about 50 solar masses. Can this massive object be a single, ordinary star? c) Many astronomers believe that the massive object at the center of the Milky Way galaxy is a black hole. If so, what must the

Schwarzschild radius of this black hole be? Would a black hole of this size fit inside the earth's orbit around the sun?

12.44 Derive the relation $R_S = (3.0 \text{ km/solar mass})M$, where M is in solar masses. (One solar mass is the mass of the sun.) How accurate is this relation?

12.45 To what fraction of its current radius would the earth have to be compressed to become a black hole?

Problems

12.46 Three uniform spheres are fixed at positions shown in Fig. 12.35. a) What is the magnitude and direction of the force on a 0.0150-kg particle placed at P? b) If the spheres are in deep outer space and a 0.0150-kg particle is released from rest 300 m from the origin along a line 45° below the $-x$-axis, what will the particle's speed be when it reaches the origin?

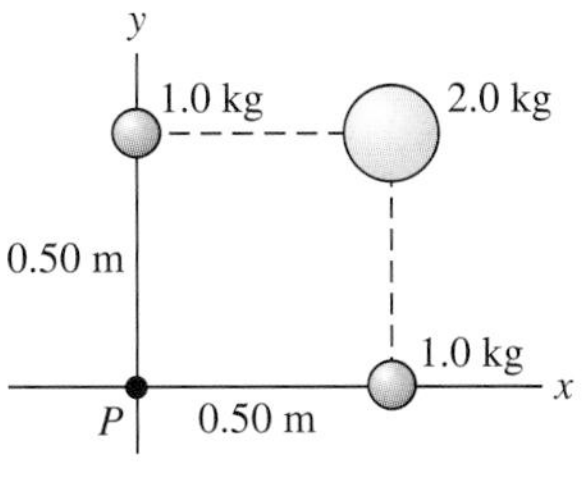

Figure 12.35 Problem 12.46.

12.47 A uniform sphere with mass 60.0 kg is held with its center at the origin and a second uniform sphere with mass 80.0 kg is held with its center at the point $x = 0$, $y = 3.00$ m. a) What are the magnitude and direction of the net gravitational force due to these objects on a third uniform sphere with mass 0.500 kg placed at the point $x = 4.00$ m, $y = 0$? b) Where, other than infinitely far away, could the third sphere be placed such that the net gravitational force acting on it from the other two spheres is equal to zero?

12.48 a) Show that the gravitational force on the small star due to the two large stars in Example 12.3 (Section 12.1) is *not* directed toward the point midway between the two large masses. b) Consider the two large stars as comprising a single, rigid body, as if they were joined by a rod of negligible mass. Calculate the torque exerted by the small star on the rigid body for a pivot at its center of mass. c) Explain how the result in part (b) shows that the center of mass does not coincide with the center of gravity. Why is this the case in this situation?

12.49 At a certain instant, the earth, the moon, and a stationary 1250-kg spacecraft lie at the vertices of an equilateral triangle whose sides are 3.84×10^5 km in length. a) Find the magnitude and direction of the net gravitational force exerted on the spacecraft by the earth and moon. State the direction as an angle measured from a line connecting the earth and the spacecraft. In a sketch, show the earth, the moon, the spacecraft, and the force vector. b) What is the minimum amount of work that you would have to do to move the spacecraft to a point far from the earth and moon? You can ignore any gravitational effects due to the other planets or the sun.

12.50 An experiment is performed in deep space with two uniform spheres, one with mass 25.0 kg and the other with mass 100.0 kg. They have equal radii, $r = 0.20$ m. The spheres are released from rest with their centers 40.0 m apart. They accelerate toward each other because of their mutual gravitational attraction. You can ignore all gravitational forces other than that between the two spheres. a) Explain why linear momentum is conserved. b) When their centers are 20.0 m apart, find i) the speed of each sphere and ii) the magnitude of the relative velocity with which one sphere is approaching the other. c) How far from the initial position of the center of the 25.0-kg sphere do the surfaces of the two spheres collide?

12.51 Assume that the moon orbits the earth in a circular orbit. From the observed orbital period of 27.3 days, calculate the distance of the moon from the center of the earth. Assume that the moon's motion is determined solely by the gravitational force exerted on it by the earth, and use the mass of the earth given in Appendix F.

12.52 Gravitational Field. The *gravitational field* $\vec{g}$ produced by an object is defined as the gravitational force $\vec{F}_g$ that the object exerts on a small test particle with mass m divided by m. That is, $\vec{g} = \vec{F}_g/m$. What are the magnitude and direction of the gravitational field produced by a 20.0-kg spherically symmetric object at a point 1.50 m directly above the center of the object?

12.53 Geosynchronous Satellites. Many satellites are moving in a circle in the earth's equatorial plane. They are at such a height above the earth's surface that they always remain above the same point. a) Find the altitude of these satellites above the earth's surface. (Such an orbit is said to be *geosynchronous*.) b) Explain, with a sketch, why the radio signals from these satellites cannot directly reach receivers on earth that are north of 81.3° N latitude.

12.54 A landing craft with mass 12,500 kg is in a circular orbit 5.75×10^5 m above the surface of a planet. The period of the orbit is 5800 s. The astronauts in the lander measure the diameter of the planet to be 9.60×10^6 m. The lander sets down at the north pole of the planet. What is the weight of a 85.6-kg astronaut as he steps out onto the planet's surface?

12.55 What is the escape speed from a 300-km-diameter asteroid with a density of 2500 kg/m^3?

12.56 a) Asteroids have average densities of about 2500 kg/m^3 and radii from 470 km down to less than a kilometer. Assuming that the asteroid has a spherically symmetric mass distribution, estimate the radius of the largest asteroid from which you could escape simply by jumping off. (*Hint*: You can estimate your jump speed by relating it to the maximum height that you can jump on earth.) b) Europa, one of Jupiter's four large moons, has a radius of 1570 km. The acceleration due to gravity at its surface is 1.33 m/s^2. Calculate its average density.

12.57 a) Suppose you are at the earth's equator and observe a satellite passing directly overhead and moving from west to east in the sky. Exactly 12.0 hours later, you again observe this satellite to be directly overhead. How far above the earth's surface is the satellite's orbit? b) You observe another satellite directly overhead and traveling east to west. This satellite is again overhead in 12.0 hours. How far is this satellite's orbit above the surface of the earth?

12.58 Planet X rotates in the same manner as the earth, around an axis through its north and south poles, and is perfectly spherical. An astronaut who weighs 943.0 N on the earth weighs 915.0 N at the north pole of Planet X and only 850.0 N at its equator. The distance from the north pole to the equator is 18,850 km, measured along the surface of Planet X. a) How long is the day on Planet X? b) If a

45,000 kg satellite is placed in a circular orbit 2000 km above the surface of Planet X, what will be its orbital period?

12.59 There are two equations from which a change in the gravitational potential energy U of the system of a mass m and the earth can be calculated. One is $U = mgy$ (Eq. 7.2). The other is $U = -Gm_E m/r$ (Eq. 12.9). As shown in Section 12.3, the first equation is correct only if the gravitational force is a constant over the change in height Δy. The second is always correct. Actually, the gravitational force is never exactly constant over any change in height, but if the variation is small, we can ignore it. Consider the difference in U between a mass at the earth's surface and a distance h above it using both equations, and find the value of h for which Eq. (7.2) is in error by 1%. Express this value of h as a fraction of the earth's radius, and also obtain a numerical value for it.

12.60 Your starship, the *Aimless Wanderer,* lands on the mysterious planet Mongo. As chief scientist-engineer, you make the following measurements: a 2.50-kg stone thrown upward from the ground at 12.0 m/s returns to the ground in 8.00 s; the circumference of Mongo at the equator is 2.00×10^5 km; and there is no appreciable atmosphere on Mongo. The starship commander, Captain Confusion, asks for the following information: a) What is the mass of Mongo? b) If the *Aimless Wanderer* goes into a circular orbit 30,000 km above the surface of Mongo, how many hours will it take the ship to complete one orbit?

12.61 Calculate the percent difference between your weight in Sacramento, near sea level, and at the top of Mount Everest, which is 8800 m above sea level.

12.62 In Example 12.5 (Section 12.3) we ignored the gravitational effects of the moon on a spacecraft en route from the earth to the moon. In fact, we must include the gravitational potential energy due to the moon as well. For this problem,you can ignore the motion of the earth and moon. a) If the moon has radius R_M and the distance between the centers of the earth and the moon is R_{EM}, find the total gravitational potential energy of the particle-earth and particle-moon systems when a particle with mass m is between the earth and the moon, and a distance r from the center of the earth. Take the gravitational potential energy to be zero when the objects are far from each other. b) There is a point along a line between the earth and the moon where the net gravitational force is zero. Use the expression derived in part (a) and numerical values from Appendix F to find the distance of this point from the center of the earth. With what speed must a spacecraft be launched from the surface of the earth just barely to reach this point? c) If a spacecraft were launched from the earth's surface toward the moon with an initial speed of 11.2 km/s, with what speed would it impact the moon?

12.63 An unmanned spacecraft is in a circular orbit around the moon, observing the lunar surface from an altitude of 50.0 km (see Appendix F). To the dismay of scientists on earth, an electrical fault causes an on-board thruster to fire, decreasing the speed of the spacecraft by 20.0 m/s. If nothing is done to correct its orbit, with what speed (in km/h) will the spacecraft crash into the lunar surface?

***12.64** What would be the length of a day (that is, the time required for one rotation of the earth on its axis) if the rate of rotation of the earth were such that $g = 0$ at the equator?

12.65 Falling Hammer. A hammer with mass m is dropped from rest from a height h above the earth's surface. This height is not necessarily small compared with the radius R_E of the earth. If you ignore air resistance, derive an expression for the speed v of the hammer when it reaches the surface of the earth. Your expression should involve h, R_E, and m_E, the mass of the earth.

12.66 a) Calculate how much work is required to launch a spacecraft of mass m from the surface of the earth (mass m_E, radius R_E) and place it in a circular *low earth orbit*, that is, an orbit whose altitude above the earth's surface is much less than R_E. (As an example, the International Space Station is in low earth orbit at an altitude of about 400 km, much less than $R_E = 6380$ km.) You can ignore the kinetic energy that the spacecraft has on the ground due to the earth's rotation. b) Calculate the minimum amount of additional work required to move the spacecraft from low earth orbit to a very great distance from the earth. You can ignore the gravitational effects of the sun, the moon, and the other planets. c) Justify the statement that "in terms of energy, low earth orbit is halfway to the edge of the universe."

12.67 A spacecraft is to be launched from the surface of the earth so that it will escape from the solar system altogether. a) Find the speed relative to the center of the earth with which the spacecraft must be launched. Take into consideration the gravitational effects of both the earth and the sun, and include the effects of the earth's orbital speed, but ignore air resistance. b) The rotation of the earth can help this spacecraft achieve escape speed. Find the speed that the spacecraft must have relative to the earth's *surface* if the spacecraft is launched from Florida at the point shown in Fig. 12.36. The rotation and orbital motions of the earth are in the same direction. The launch facilities in Florida are 28.5° north of the equator. c) The European Space Agency (ESA) uses launch facilities in French Guiana (immediately north of Brazil), 5.15° north of the equator. What speed relative to the earth's surface would a spacecraft need to escape the solar system if launched from French Guiana?

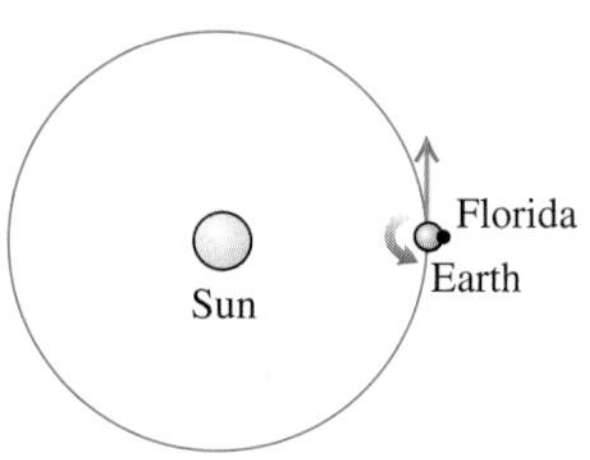

Figure 12.36 Problem 12.67

12.68 Gravitational Potential. We can define a useful quantity called the *gravitational potential* of the earth (denoted by ϕ) by requiring that the product of its value at any point, and the mass of a particle placed at that point, gives us the gravitational potential energy of that earth-particle system. (We will use an analogous quantity, the electric potential V, in the discussion of electrical interactions in Chapter 23.) a) Find the SI units for the gravitational potential and show they are equivalent to m^2/s^2. b) What is the equation for the gravitational potential at any point outside the earth, if it is taken to be zero infinitely far from the earth? c) What is the increase in gravitational potential between the earth's surface and the International Space Station's altitude of 400 km? d) Use the result of part (c) to calculate the amount of work that must be done against the earth's gravitational force to move a 15,000-kg payload from the surface of the earth to an altitude of 400 km.

12.69 Kirkwood Gaps. Hundreds of thousands of asteroids orbit the sun within the *asteroid belt*, which extends from about 3×10^8 km to about 5×10^8 km from the sun. a) Find the orbital period (in years) of i) an asteroid at the inside of the belt and ii) an asteroid at the outside of the belt. Assume circular orbits. b) In 1867 the American astronomer Daniel Kirkwood pointed out that several gaps exist in the asteroid belt where relatively few asteroids are found. It is now understood that these *Kirkwood gaps* are caused by the gravitational attraction of Jupiter, the largest planet, which orbits the sun once every 11.86 years. As an example, if an asteroid has an orbital period half that of Jupiter, or 5.93 years, on every other orbit this asteroid would be at its closest to Jupiter and feel a strong attraction toward the planet. This attraction, acting over and over on successive orbits, could sweep asteroids out of the Kirkwood gap. Use this hypothesis to determine the orbital radius for this Kirkwood gap. c) One of several other Kirkwood gaps appears at a distance from the sun where the orbital period is 0.400 that of Jupiter. Explain why this happens, and find the orbital radius for this Kirkwood gap.

12.70 If a satellite is in a sufficiently low orbit, it will encounter air drag from the earth's atmosphere. Since air drag does negative work (the force of air drag is directed opposite the motion), the mechanical energy will decrease. According to Eq. (12.15), if E decreases (becomes more negative), the radius r of the orbit will decrease. If air drag is relatively small, the satellite can be considered to be in a circular orbit of continually decreasing radius. a) According to Eq. (12.12), if the radius of a satellite's circular orbit decreases, the satellite's orbital speed v *increases*. How can you reconcile this with the statement that the mechanical energy *decreases?* (*Hint:* Is air drag the only force that does work on the satellite as the orbital radius decreases?) b) Due to air drag, the radius of a satellite's circular orbit decreases from r to $r - \Delta r$, where the positive quantity Δr is much less than r. The mass of the satellite is m. Show that the increase in orbital speed is $\Delta v = +(\Delta r/2)\sqrt{Gm_E/r^3}$; that the change in kinetic energy is $\Delta K = +(Gm_E m/2r^2)\Delta r$; that the change in gravitational potential energy is $\Delta U = -2\Delta K = -(Gm_E m/r^2)\Delta r$; and that the amount of work done by the force of air drag is $W = -(Gm_E m/2r^2)\Delta r$. Interpret these results in light of your comments in part (a). c) A satellite with mass 3000 kg is initially in a circular orbit 300 km above the earth's surface. Due to air drag, the satellite's altitude decreases to 250 km. Calculate the initial orbital speed; the increase in orbital speed; the initial mechanical energy; the change in kinetic energy; the change in gravitational potential energy; the change in mechanical energy; and the work done by the force of air drag. d) Eventually a satellite will descend to a low enough altitude in the atmosphere that the satellite burns up and the debris falls to the earth. What becomes of the initial mechanical energy?

12.71 Binary Star—Equal Masses. Two identical stars with mass M orbit around their center of mass. Each orbit is circular and has radius R, so that the two stars are always on opposite sides of the circle. a) Find the gravitational force of one star on the other. b) Find the orbital speed of each star and the period of the orbit. c) How much energy would be required to separate the two stars to infinity?

12.72 Binary Star—Different Masses. Two stars, with mass M_1 and M_2, are in circular orbits around their center of mass. The star with mass M_1 has an orbit of radius R_1; the star with mass M_2 has an orbit of radius R_2. a) Show that the ratio of the orbital radii of the two stars equals the reciprocal of the ratio of their masses, that is, $R_1/R_2 = M_2/M_1$. b) Explain why the two stars have the same orbital period, and show that the period T is given by $T = 2\pi(R_1 + R_2)^{3/2}/\sqrt{G(M_1 + M_2)}$. c) The two stars in a certain binary star system move in circular orbits. The first star, Alpha, has an orbital speed of 36.0 km/s. The second star, Beta, has an orbital speed of 12.0 km/s. The orbital period is 137 d. What are the masses of each of the two stars? d) One of the best candidates for a black hole is found in the binary system called A0620-0090. The two objects in the binary system are an orange star, V616 Monocerotis, and a compact object believed to be a black hole (Fig. 12.21). The orbital period of A0620-0090 is 7.75 hours, the mass of V616 Monocerotis is estimated to be 0.67 times the mass of the sun, and the mass of the black hole is estimated to be 3.8 times the mass of the sun. Assuming that the orbits are circular, find the radius of each object's orbit and the orbital speed of each object. Compare these answers to the orbital radius and orbital speed of the earth in its orbit around the sun.

12.73 Comets travel around the sun in elliptical orbits with large eccentricities. If a comet has speed 2.0×10^4 m/s when at a distance of 2.5×10^{11} m from the center of the sun, what is its speed when at a distance of 5.0×10^{10} m?

12.74 As Mars orbits the sun in its elliptical orbit, its distance of closest approach to the center of the sun (at perihelion) is 2.067×10^{11} m, and its maximum distance from the center of the sun (at aphelion) is 2.492×10^{11} m. If the orbital speed of Mars at aphelion is 2.198×10^4 m/s, what is its orbital speed at perihelion? (You can ignore the influence of the other planets.)

12.75 Consider a spacecraft in an elliptical orbit around the earth. At the low point, or perigee of its orbit, it is 400 km above the earth's surface; at the high point, or apogee, it is 4000 km above the earth's surface. a) What is the period of the spacecraft's orbit? b) Using conservation of angular momentum, find the ratio of the spacecraft's speed at perigee to its speed at apogee. c) Using conservation of energy, find the speed at perigee and the speed at apogee. d) It is necessary to have the spacecraft escape from the earth completely. If the spacecraft's rockets are fired at perigee, by how much would the speed have to be increased to achieve this? What if the rockets were fired at apogee? Which point in the orbit is more efficient to use?

12.76 The planet Uranus has a radius of 25,560 km and a surface acceleration due to gravity of 11.1 m/s^2 at its poles. Its moon Miranda (discovered by Kuiper in 1948) is in a circular orbit about Uranus at an altitude of 104,000 km above the planet's surface. Miranda has a mass of 6.6×10^{19} kg and a radius of 235 km. a) Calculate the mass of Uranus from the above data. b) Calculate the magnitude of Miranda's acceleration due to its orbital motion about Uranus. c) Calculate the acceleration due to Miranda's gravity at the surface of Miranda. d) Do the answers to parts (b) and (c) mean that an object released 1 m above Miranda's surface on the side toward Uranus will fall *up* relative to Miranda? Explain.

12.77 A 3000-kg spacecraft is in a circular orbit 2000 km above the surface of Mars. How much work must the spacecraft engines perform to move the spacecraft to a circular orbit that is 4000 km above the surface?

12.78 In Example 12.9 (Section 12.5) the orbital period of Comet Halley was calculated to be 75.5 years. Many other comets have orbits that extend much farther from the sun, and hence have much longer periods. a) At aphelion, a typical long-period comet is about 8×10^{12} km from the sun; at perihelion, it passes inside the earth's orbit. Estimate the orbital period of such a comet. Give your answer in years. b) Estimate the speed of a typical long-period comet at perihelion. (*Hint:* At aphelion, the comet is moving very slowly.) c) The massive nucleus of a comet contains about 10^{15} kg of material. If the earth were struck by a long-period comet, estimate the kinetic energy the comet would have just before impact. (*Hint:* The motion of the comet is due primarily to the gravitational attraction of the sun, not the earth. Why?) Compare this to the energy released in a very large volcanic eruption, about 6×10^{18} J, and to the energy that would be released by burning all the fossil fuel on earth, about 2×10^{23} J. An impact of this kind apparently occurred 65 million years ago in the Yucatan, and is implicated in the demise of the dinosaurs as well as many other ancient species.

12.79 One of the brightest comets of the twentieth century was Comet Hyakutake, which passed close to the sun in early 1996. The orbital period of this comet is estimated to be about 30,000 years. Find the semi-major axis of this comet's orbit. Compare to the average sun-Pluto distance and to the distance to Alpha Centauri, the nearest star to the sun, which is 4.3 light years distant.

12.80 Planets are not uniform inside. Normally, they are densest at the center and have decreasing density outward toward the surface. Model a spherically symmetric planet, with the same radius as the earth, as having a density that decreases linearly with distance from the center. Let the density be $15.0 \times 10^3\ \text{kg/m}^3$ at the center and $2.0 \times 10^3\ \text{kg/m}^3$ at the surface. What is the acceleration due to gravity at the surface of this planet?

12.81 A uniform wire with mass M and length L is bent into a semicircle. Find the magnitude and direction of the gravitational force this wire exerts on a point with mass m placed at the center of curvature of the semicircle.

***12.82** An object in the shape of a thin ring has radius a and mass M. A uniform sphere with mass m and radius R is placed with its center at a distance x to the right of the center of the ring, along a line through the center of the ring, and perpendicular to its plane (Fig. 12.34). What is the gravitational force that the sphere exerts on the ring-shaped object? Show that your result reduces to the expected result when x is much larger than a.

***12.83** A thin, uniform rod has length L and mass M. Calculate the magnitude of the gravitational force the rod exerts on a particle with mass m that is at a point along the axis of the rod a distance x from one end (Fig. 12.33). Show that your result reduces to the expected result when x is much larger than L.

***12.84** A shaft is drilled from the surface to the center of the earth (Fig. 12.24). As in Example 12.10 (Section 12.6), make the rather unrealistic assumption that the density of the earth is uniform. With this approximation, the gravitational force on an object with mass m, that is inside the earth at a distance r from the center, has magnitude $F_g = Gm_E mr/R_E^3$ (as shown in Example 12.10) and points toward the center of the earth. a) Derive an expression for the gravitational potential energy $U(r)$ of the object-and-earth system as a function of the object's distance from the center of the earth. Take the potential energy to be zero when the object is at the center of the earth. b) If an object is released in the shaft at the earth's surface, what speed will it have when it reaches the center of the earth?

Challenge Problems

12.85 a) When an object is in a circular orbit of radius r around the earth (mass m_E), the period of the orbit is T [given by Eq. (12.14)] and the orbital speed is v, given by Eq. (12.12). Show that when the object is moved into a circular orbit of slightly larger radius $r + \Delta r$, where $\Delta r \ll r$, its new period is $T + \Delta T$ and its new orbital speed is $v - \Delta v$, where Δr, ΔT, and Δv are all positive quantities and

$$\Delta T = \frac{3\pi \Delta r}{v} \quad \text{and} \quad \Delta v = \frac{\pi \Delta r}{T}$$

(*Hint:* Use the expression $(1 + x)^n \approx 1 + nx$, valid for $|x| \ll 1$.) b) The International Space Station (ISS) is in a nearly circular orbit at an altitude of 398.00 km above the surface of the earth. A maintenance crew is about to arrive on the space shuttle that is also in a circular orbit in the same orbital plane as the ISS, but with an altitude of 398.10 km. The crew has come to remove a faulty 125-m electrical cable, one end of which is attached to the ISS and the other end of which is floating free in space. The plan is for the shuttle to snag the free end just at the moment that the shuttle, the ISS, and the center of the earth all lie along the same line. The cable will then break free from the ISS when it becomes taut. How long after the free end is caught by the space shuttle will it detach from the ISS? Give your answer in minutes. c) If the shuttle misses catching the cable, show that the crew must wait a time $t \approx T^2/\Delta T$ before they have a second chance. Find the numerical value of t and explain whether it would be worth the wait.

12.86 Interplanetary Navigation. The most efficient way to send a spacecraft from the earth to another planet is by using a *Hohmann transfer orbit* (Fig. 12.37). If the orbits of the departure and desti-

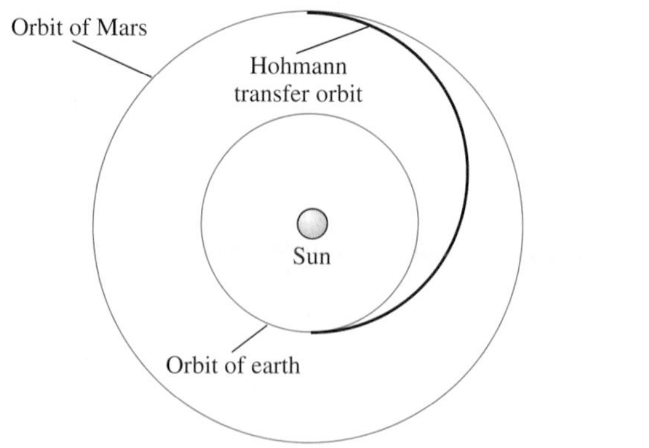

Figure 12.37 Challenge Problem 12.86.

nation planets are circular, the Hohmann transfer orbit is an elliptical orbit whose perihelion and aphelion are tangent to the orbits of the two planets. The rockets are fired briefly at the departure planet to put the spacecraft into the transfer orbit; the spacecraft then coasts until it reaches the destination planet. The rockets are then fired again to put the spacecraft into the same orbit about the sun as the destination planet. a) For a flight from earth to Mars? in what direction must the rockets be fired at the earth and at Mars: in the direction of motion, or opposite the direction of motion? What about from a flight from Mars to the earth? b) How long does a one-way trip from the the earth to Mars take, between the firings of the rockets? c) To reach Mars from the earth, the launch must be timed so that Mars will be at the right spot when the spacecraft reaches Mars's orbit around the sun. At launch, what must the angle between a sun-Mars line and a sun-earth line be? Use data from Appendix F.

12.87 Tidal Forces Near a Black Hole. An astronaut inside a spacecraft, which protects her from harmful radiation, is orbiting a black hole at a distance of 120 km from its center. The black hole is 5.00 times the mass of the sun and has a Schwarzschild radius of 15.0 km. The astronaut is positioned inside the spaceship such that one of her 0.030-kg ears is 6.0 cm farther from the black hole than the center of mass of the spacecraft and the other ear is 6.0 cm closer. a) What is the tension between her ears? Would the astronaut find it difficult to keep from being torn apart by the gravitational forces? (Since her whole body orbits with the same angular velocity, one ear is moving too slowly for the radius of its orbit and the other is moving too fast. Hence her head must exert forces on her ears to keep them in their orbits.) b) Is the center of gravity of her head at the same point as the center of mass? Explain.

***12.88** Mass M is distributed uniformly over a disk of radius a. Find the gravitational force (magnitude and direction) between this disk-shaped mass and a particle with mass m located a distance x above the center of the disk (Fig. 12.38). Does your result reduce to the correct expression as x becomes very large? (*Hint:* Divide the disk into infinitesimally thin concentric rings, use the expression derived in Problem 12.70 for the gravitational force due to each ring, and integrate to find the total force.)

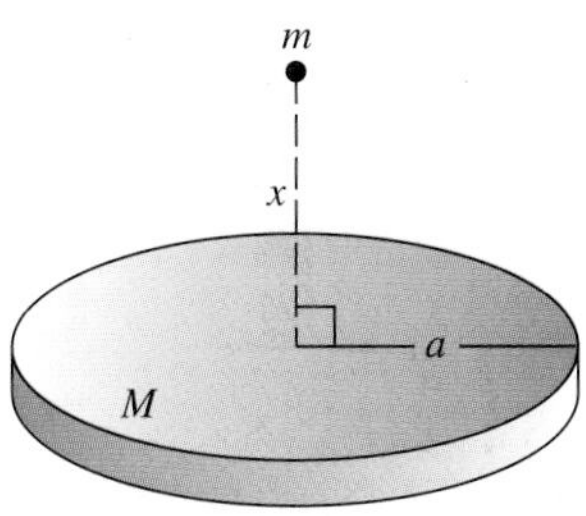

Figure 12.38 Challenge Problem 12.88.

***12.89** Mass M is distributed uniformly along a line of length $2L$. A particle with mass m is at a point that is a distance a above the center of the line on its perpendicular bisector (point P in Fig. 12.39). For the gravitational force that the line exerts on the particle, calculate the components perpendicular and parallel to the line. Does your result reduce to the correct expression as a becomes very large?

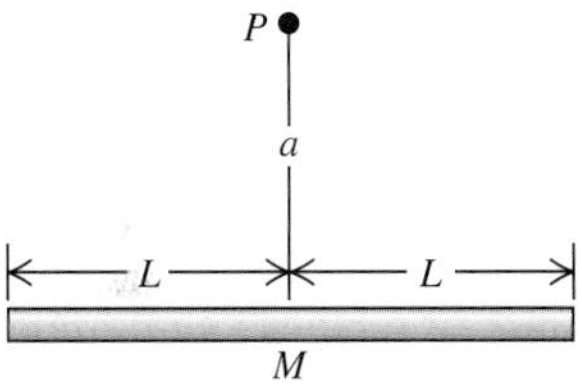

Figure 12.39 Challenge Problem 12.89

13 PERIODIC MOTION

Pendulum clocks have been keeping track of time since the mid-1600s. The underlying principle is that the time for a complete back-and-forth swing depends hardly at all on the size, or amplitude, of the swing. Hence a pendulum clock keeps proper track of time even as the driving mechanism winds down and the pendulum's oscillations die away due to friction.

? **Suppose you doubled the mass of a clock's pendulum (including the rod and the weight at the end) while keeping its dimensions the same. Would the clock run fast or slow?**

Many kinds of motion repeat themselves over and over: the vibration of a quartz crystal in a watch, the swinging pendulum of a grandfather clock, the sound vibrations produced by a clarinet or an organ pipe, and the back-and-forth motion of the pistons in a car engine. We call this **periodic motion** or **oscillation,** and it is the subject of this chapter. Understanding periodic motion will be essential for our later study of waves, sound, alternating electric currents, and light.

A body that undergoes periodic motion always has a stable equilibrium position. When it is moved away from this position and released, a force or torque comes into play to pull it back toward equilibrium. But by the time it gets there, it has picked up some kinetic energy, so it overshoots, stopping somewhere on the other side, and is again pulled back toward equilibrium. Picture a ball rolling back and forth in a round bowl or a pendulum that swings back and forth past its straight-down position.

In this chapter we will concentrate on two simple examples of systems that can undergo periodic motions: spring-mass systems and pendulums. We will also study why oscillations often tend to die out with time and why some oscillations can build up to greater and greater displacements from equilibrium when periodically varying forces act.

13.1 | Describing Oscillation

One of the simplest systems that can have periodic motion is shown in Fig. 13.1a. A body with mass m moves on a frictionless horizontal guide system, such as a linear air track, so it moves only along the x-axis. The body is attached to a spring of negligible mass that can be either stretched or compressed. The left end of the spring is held fixed and the right end is attached to the body. The spring force is the only horizontal force acting on the body; the vertical normal and gravitational forces always add to zero. The quantities x, v_x, a_x, and F_x refer to the x-components of the position, velocity, acceleration, and force vectors, respectively. Like all components, they can be positive, negative, or zero.

It's simplest to define our coordinate system so that the origin O is at the equilibrium position, where the spring is neither stretched nor compressed. Then x is the x-component of the **displacement** of the body from equilibrium and is also the change in length of the spring. The x-component of acceleration a_x is given by $a_x = F_x/m$.

Figure 13.1b shows free-body diagrams for three different positions of the body. Whenever the body is displaced from its equilibrium position, the spring force tends to restore it to the equilibrium position. We call a force with this character a **restoring force.** Oscillation can occur only when there is a restoring force tending to return the system to equilibrium.

Let's analyze how oscillation occurs in this system. If we displace the body to the right to $x = A$ and then let go, the net force and the acceleration are to the left. The speed increases as the body approaches the equilibrium position O. When the body is at O, the net force acting on it is zero, but because of its motion it *overshoots* the equilibrium position. On the other side of the equilibrium position the velocity is to the left but the acceleration is to the right; the speed decreases until the body comes to a stop. We will show later that with an ideal spring, the stopping point is at $x = -A$. The body then accelerates to the right, overshoots equilibrium again, and stops at the starting point $x = A$, ready to repeat the whole process. The body is oscillating! If there is no friction or other force to remove mechanical energy from the system, this motion repeats itself forever; the restoring force perpetually draws the body back toward the equilibrium position, only to have the body overshoot time after time.

In different situations the force may depend on the displacement x from equilibrium in different ways. But oscillation *always* occurs if the force is a *restoring* force that tends to return the system to equilibrium.

Here are some terms that we'll use in discussing periodic motions of all kinds:

The **amplitude** of the motion, denoted by A, is the maximum magnitude of displacement from equilibrium—that is, the maximum value of $|x|$. It is always positive. If the spring in Fig. 13.1 is an ideal one, the total overall range of the motion is $2A$. The SI unit of A is the meter. A complete vibration, or **cycle,** is one complete round trip, say from A to $-A$ and back to A, or from O to A, back through O to $-A$, and back to O. Note that motion from one side to the other (say, $-A$ to A) is a half-cycle, not a whole cycle.

The **period,** T, is the time for one cycle. It is always positive. The SI unit is the second, but it is sometimes expressed as "seconds per cycle."

The **frequency,** f, is the number of cycles in a unit of time. It is always positive. The SI unit of frequency is the hertz:

$$1 \text{ hertz} = 1 \text{ Hz} = 1 \text{ cycle/s} = 1 \text{ s}^{-1}$$

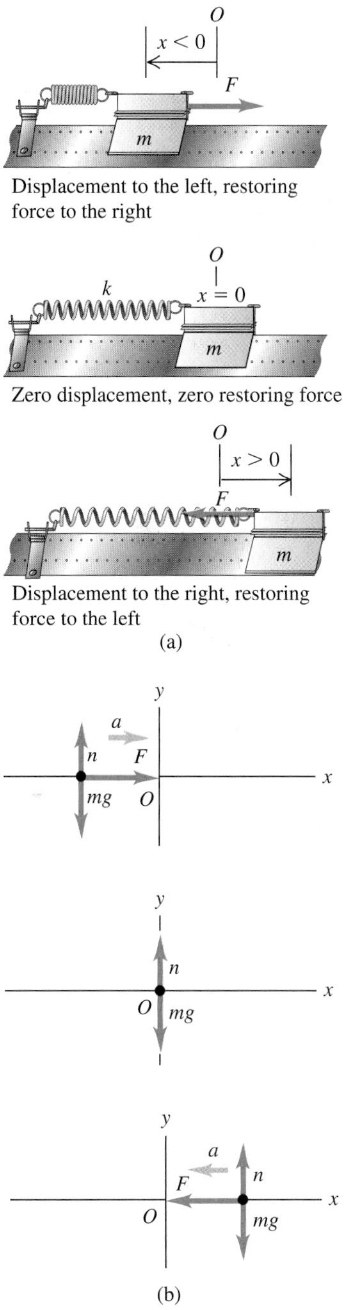

13.1 Model for periodic motion. (a) At the equilibrium position the spring exerts zero force. When the body is displaced from equilibrium, the spring exerts a restoring force back toward the equilibrium position. (b) Free-body diagrams for the three positions.

This unit is named in honor of the German physicist Heinrich Hertz (1857–1894), a pioneer in investigating electromagnetic waves.

The **angular frequency,** ω, is 2π times the frequency:

$$\omega = 2\pi f$$

We'll learn shortly why ω is a useful quantity. It represents the rate of change of an angular quantity (not necessarily related to a rotational motion) that is always measured in radians, so its units are rad/s. Since f is in cycle/s, we may regard the number 2π as having units rad/cycle.

From the definitions of period T and frequency f we see that each is the reciprocal of the other:

$$f = \frac{1}{T} \qquad T = \frac{1}{f} \qquad \text{(relationships between frequency and period)} \quad (13.1)$$

Also, from the definition of ω,

$$\omega = 2\pi f = \frac{2\pi}{T} \qquad \text{(angular frequency)} \quad (13.2)$$

Example 13.1 Period, frequency, and angular frequency

An ultrasonic transducer (a kind of loudspeaker) used for medical diagnosis oscillates at a frequency of 6.7 MHz $= 6.7 \times 10^6$ Hz. How much time does each oscillation take, and what is the angular frequency?

SOLUTION

IDENTIFY and SET UP: Our target variables are the period T and the angular frequency ω. We are given the frequency f, so we can find these variables using Eqs. (13.1) and (13.2), respectively.

EXECUTE: From Eqs. (13.1) and (13.2),

$$T = \frac{1}{f} = \frac{1}{6.7 \times 10^6 \text{ Hz}} = 1.5 \times 10^{-7} \text{ s} = 0.15\ \mu\text{s}$$

$$\begin{aligned}\omega &= 2\pi f = 2\pi(6.7 \times 10^6 \text{ Hz}) \\ &= (2\pi \text{ rad/cycle})(6.7 \times 10^6 \text{ cycle/s}) \\ &= 4.2 \times 10^7 \text{ rad/s}\end{aligned}$$

EVALUATE: This is a very rapid vibration, with large f and ω and small T. A slow vibration has small f and ω and large T.

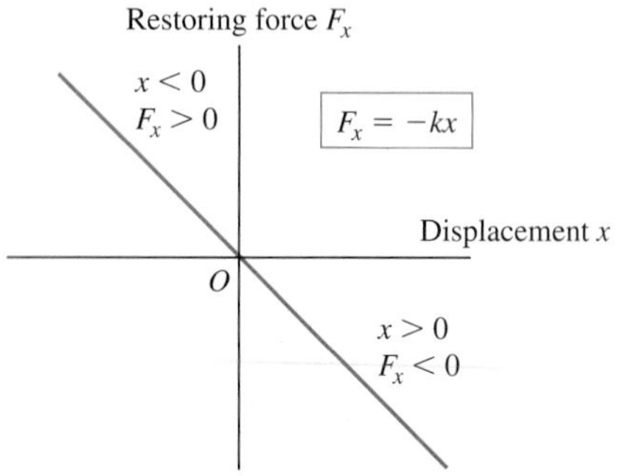

13.2 The restoring force of an idealized spring is directly proportional to the displacement. This is Hooke's law, $F_x = -kx$. Oscillation with a Hooke's law restoring force is called simple harmonic motion.

Test Your Understanding

A boat at anchor bobs up and down with the waves. The boat moves 6.0 cm above and 6.0 cm below its equilibrium position, and makes one complete up-and-down cycle every 5.00 s. What are the amplitude, period, frequency, and angular frequency of the motion?

13.2 | Simple Harmonic Motion

The very simplest kind of oscillation occurs when the restoring force F_x is *directly proportional* to the displacement from equilibrium x. This happens if the spring in Fig. 13.1 is an ideal one that obeys Hooke's law. The constant of proportionality between F_x and x is the force constant k. (You may want to review Hooke's law and the definition of the force constant in Section 6.3.) On either side of the equilibrium position, F_x and x always have opposite signs. In Section 6.3 we represented the force acting *on* a stretched ideal spring as $F_x = kx$. The x-component of force the spring exerts *on the body* is the negative of this, so the x-component of force F_x on the body is

$$F_x = -kx \qquad \text{(restoring force exerted by an ideal spring)} \qquad (13.3)$$

This equation gives the correct magnitude and sign of the force, whether x is positive, negative, or zero (Fig. 13.2). The force constant k is always positive and has units of N/m (a useful alternative set of units is kg/s^2). We are assuming that there is no friction, so Eq. (13.3) gives the *net* force on the body.

When the restoring force is directly proportional to the displacement from equilibrium, as given by Eq. (13.3), the oscillation is called **simple harmonic motion,** *abbreviated* **SHM.** The acceleration $a_x = d^2x/dt^2 = F_x/m$ of a body in SHM is given by

$$a_x = \frac{d^2x}{dt^2} = -\frac{k}{m}x \qquad \text{(simple harmonic motion)} \qquad (13.4)$$

The minus sign means the acceleration and displacement always have opposite signs. This acceleration is *not* constant, so don't even think of using the constant-acceleration equations from Chapter 2. We'll see shortly how to solve this equation to find the displacement x as a function of time. A body that undergoes simple harmonic motion is called a **harmonic oscillator.**

Why is simple harmonic motion important? Keep in mind that not all periodic motions are simple harmonic; in periodic motion in general, the restoring force depends on displacement in a more complicated way than in Eq. (13.3). But in many systems the restoring force is *approximately* proportional to displacement if the displacement is sufficiently small (Fig. 13.3). That is, if the amplitude is small enough, the oscillations of such systems are approximately simple harmonic and therefore approximately described by Eq. (13.4). Thus we can use SHM as an approximate model for many different periodic motions, such as the vibration of the quartz crystal in a watch, the motion of a tuning fork, the electric current in an alternating-current circuit, and the vibrations of atoms in molecules and solids.

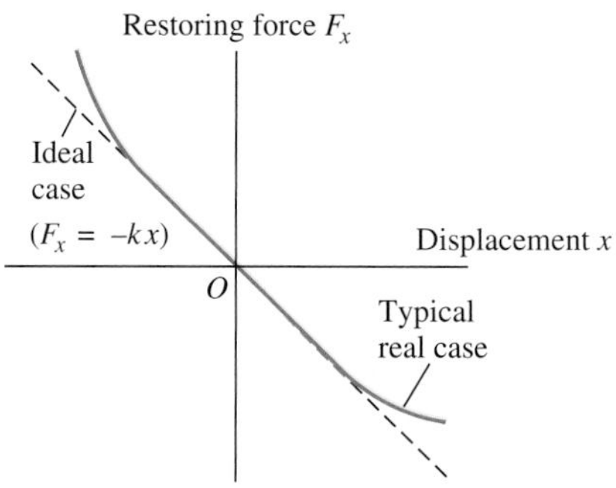

13.3 In most real oscillations the restoring force is not directly proportional to the displacement. However, $F_x = -kx$ is often a good approximation to the force if the displacement x is sufficiently small.

Equations of Simple Harmonic Motion

To explore the properties of simple harmonic motion, we must express the displacement x of the oscillating body as a function of time, $x(t)$. The second derivative of this function, d^2x/dt^2, must be equal to $(-k/m)$ times the function itself, as required by Eq. (13.4). As we mentioned above, the formulas for constant acceleration from Section 2.4 are no help because the acceleration changes constantly as the displacement x changes. Instead, we'll find $x(t)$ by noticing a striking similarity between SHM and another form of motion that we've already studied in detail.

Figure 13.4 shows a top view of a horizontal disk of radius A with a ball attached to its rim at point Q. The disk rotates with constant angular speed ω (measured in rad/s), so the ball moves in uniform circular motion. A horizontal light beam shines on the rotating disk and casts a shadow of the ball on a screen. The shadow at point P oscillates back and forth as the ball moves in a circle. We then arrange a body attached to an ideal spring, like the combination shown in Fig. 13.1, so that the body oscillates parallel to the shadow. We will prove that the motion of the body

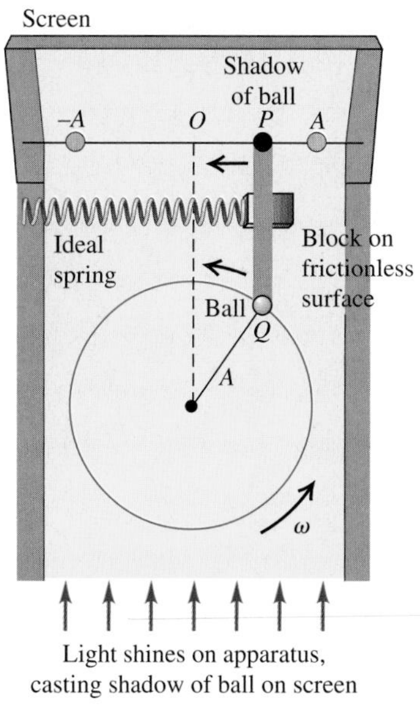

13.4 The ball at point Q rotates counterclockwise in uniform circular motion. Its shadow at point P moves in simple harmonic motion, exactly like a body oscillating on an ideal spring.

and the motion of the ball's shadow are *identical* if the amplitude of the body's oscillation is equal to the disk radius A, and if the angular frequency $2\pi f$ of the oscillating body is equal to the angular speed ω of the rotating disk. That is, *simple harmonic motion is the projection of uniform circular motion onto a diameter.*

We can verify this remarkable statement by finding the acceleration of the shadow at P and comparing it to the acceleration of a body undergoing SHM, given by Eq. (13.4). The circle in which the ball moves so that its projection matches the motion of the oscillating body is called the **circle of reference;** we will call the point Q the *reference point.* We take the circle of reference to lie in the xy-plane, with the origin O at the center of the circle (Fig. 13.5a). At time t the vector OQ from the origin to the reference point Q makes an angle θ with the positive x-axis. As the point Q moves around the reference circle with constant angular speed ω, the vector OQ rotates with the same angular speed. Such a rotating vector is called a **phasor.** (This term was in use long before the invention of the similarly named Star Trek stun gun. The phasor method for analyzing oscillations is useful in many areas of physics. We'll use phasors when we study alternating-current circuits in Chapter 31 and the interference of light in Chapters 35 and 36.)

The x-component of the phasor at time t is just the x-coordinate of the point Q:

$$x = A\cos\theta \tag{13.5}$$

This is also the x-coordinate of the shadow P, which is the *projection* of Q onto the x-axis. Hence the acceleration of the shadow P along the x-axis is equal to the x-component of the acceleration vector of the reference point Q (Fig. 13.5c). Since point Q is in uniform circular motion, its acceleration vector $\vec{a}_Q$ is always directed toward O. Furthermore, the magnitude of $\vec{a}_Q$ is constant and given by the angular velocity squared times the radius of the circle (see Section 3.4):

$$a_Q = \omega^2 A \tag{13.6}$$

Figure 13.5c shows that the x-component of $\vec{a}_Q$ is $a_x = -a_Q\cos\theta$. Combining this with Eqs. (13.5) and (13.6), we get that the acceleration of point P is

$$a_x = -a_Q\cos\theta = -\omega^2 A\cos\theta \tag{13.7}$$

or

$$a_x = -\omega^2 x \tag{13.8}$$

The acceleration of the point P is directly proportional to the displacement x and always has the opposite sign. These are precisely the hallmarks of simple harmonic motion.

Equation (13.8) is *exactly* the same as Eq. (13.4) for the acceleration of a harmonic oscillator, provided that the angular speed ω of the reference point Q is related to the force constant k and mass m of the oscillating body by

$$\omega^2 = \frac{k}{m} \quad \text{or} \quad \omega = \sqrt{\frac{k}{m}} \tag{13.9}$$

We have been using the same symbol ω for the angular *speed* of the reference point Q and the angular *frequency* of the oscillating point P. The reason is that these quantities are equal! If point Q makes one complete revolution in time T, then point P goes through one complete cycle of oscillation in the same time; hence T is the period of the oscillation. During time T the point Q moves through 2π radians, so its angular speed is $\omega = 2\pi/T$. But this is just the same as Eq. (13.2) for the angular frequency of the point P, which verifies our statement about

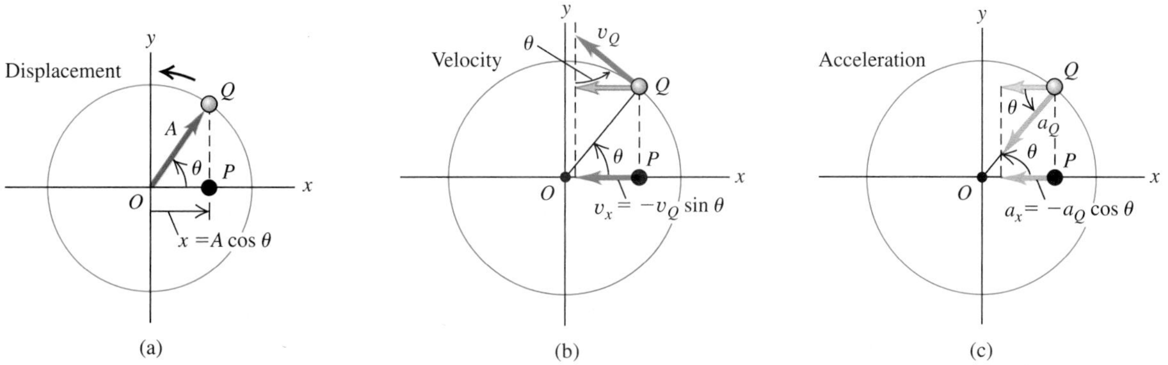

13.5 (a) The x-coordinate of the ball's shadow P (Fig. 13.4) changes as the ball Q moves in uniform circular motion. (b), (c) The x-velocity and x-acceleration of P are the x-components of the velocity and acceleration vectors, respectively, of Q.

the two interpretations of ω. This is why we introduced angular frequency in Section 13.1; this quantity makes the connection between oscillation and circular motion. So we re-interpret Eq. (13.9) as an expression for the angular frequency of simple harmonic motion for a body of mass m, acted on by a restoring force with force constant k:

$$\omega = \sqrt{\frac{k}{m}} \quad \text{(simple harmonic motion)} \tag{13.10}$$

When you start a body oscillating in SHM, the value of ω is not yours to choose; it is predetermined by the values of k and m. The units of k are N/m or kg/s^2, so k/m is in $(\text{kg/s}^2)/\text{kg} = \text{s}^{-2}$. When we take the square root in Eq. (13.10), we get s^{-1}, or more properly rad/s because this is an *angular* frequency (recall that a radian is not a true unit).

According to Eqs. (13.1) and (13.2), the frequency f and period T are

$$f = \frac{\omega}{2\pi} = \frac{1}{2\pi}\sqrt{\frac{k}{m}} \quad \text{(simple harmonic motion)} \tag{13.11}$$

$$T = \frac{1}{f} = \frac{2\pi}{\omega} = 2\pi\sqrt{\frac{m}{k}} \quad \text{(simple harmonic motion)} \tag{13.12}$$

We see from Eq. (13.12) that a larger mass m, with its greater inertia, will have less acceleration, move more slowly, and take a longer time for a complete cycle (Fig. 13.6). In contrast, a stiffer spring (one with a larger force constant k) exerts a greater force at a given deformation x, causing greater acceleration, higher speeds, and a shorter time T per cycle.

CAUTION You can run into trouble if you don't make the distinction between frequency f and angular frequency $\omega = 2\pi f$. Frequency tells you how many cycles of oscillation occur per second, while angular frequency tells you how many radians per second this corresponds to on the reference circle. In solving problems, pay careful attention to whether the goal is to find f or ω.

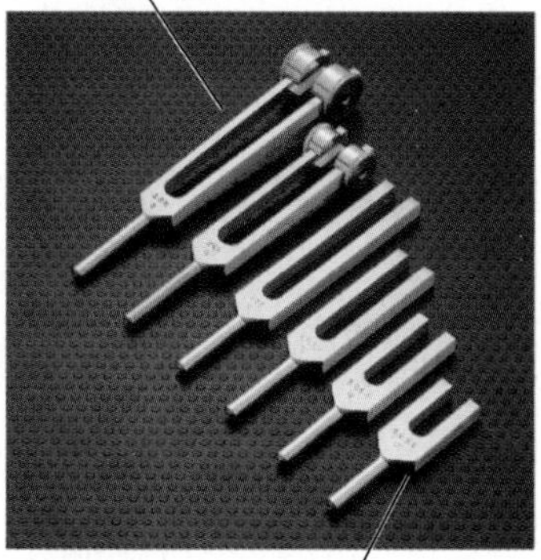

13.6 The greater the mass m in a tuning fork's tines, the lower the frequency of oscillation $f = (\frac{1}{2}\pi)\sqrt{k/m}$ and the lower the pitch of the sound that the tuning fork produces.

Equations (13.11) and (13.12) show that the period and frequency of simple harmonic motion are completely determined by the mass m and the force constant k. *In simple harmonic motion the period and frequency do not depend on the amplitude A.* For given values of m and k the time of one complete oscillation is the same whether the amplitude is large or small. Equation (13.3) shows why we should expect this. Larger A means that the body reaches larger values of $|x|$ and is subjected to larger restoring forces. This increases the average speed of the body over a complete cycle; this exactly compensates for having to travel a larger distance, so the same total time is involved.

The oscillations of a tuning fork are essentially simple harmonic motion, which means that it always vibrates with the same frequency, independent of amplitude. This is why a tuning fork can be used as a standard for musical pitch. If it were not for this characteristic of simple harmonic motion, it would be impossible to make familiar types of mechanical and electronic clocks run accurately or to play most musical instruments in tune. If you encounter an oscillating body with a period that *does* depend on the amplitude, the oscillation is *not* simple harmonic motion.

Example 13.2 Angular frequency, frequency, and period in SHM

A spring is mounted horizontally, with its left end held stationary. By attaching a spring balance to the free end and pulling toward the right (Fig. 13.7), we determine that the stretching force is proportional to the displacement and that a force of 6.0 N causes a displacement of 0.030 m. We remove the spring balance and attach a 0.50-kg body to the end, pull it a distance of 0.020 m, release it, and watch it oscillate. a) Find the force constant of the spring. b) Find the angular frequency, frequency, and period of the oscillation.

SOLUTION

IDENTIFY: Because the spring force (equal in magnitude to the stretching force) is proportional to the displacement, the motion is simple harmonic.

SET UP: We find the value of the force constant k using Hooke's law, Eq. (13.3), and the values of ω, f, and T using Eqs. (13.10), (13.11), and (13.12), respectively.

EXECUTE: a) When $x = 0.030$ m, the force the spring exerts on the spring balance is $F_x = -6.0$ N. From Eq. (13.3),

$$k = -\frac{F_x}{x} = -\frac{-6.0\text{ N}}{0.030\text{ m}} = 200\text{ N/m} = 200\text{ kg/s}^2$$

b) Using $m = 0.50$ kg in Eq. (13.10), we find

$$\omega = \sqrt{\frac{k}{m}} = \sqrt{\frac{200\text{ kg/s}^2}{0.50\text{ kg}}} = 20\text{ rad/s}$$

The frequency f is

$$f = \frac{\omega}{2\pi} = \frac{20\text{ rad/s}}{2\pi\text{ rad/cycle}} = 3.2\text{ cycle/s} = 3.2\text{ Hz}$$

The period T is the reciprocal of the frequency f:

$$T = \frac{1}{f} = \frac{1}{3.2\text{ cycle/s}} = 0.31\text{ s}$$

A period is usually stated in "seconds" rather than "seconds per cycle."

EVALUATE: The amplitude of the oscillation is 0.020 m, the distance to the right that we pulled the body attached to the spring before releasing it. We didn't need to use this information to find the angular frequency, frequency, or period because in SHM, none of these quantities depend on the amplitude.

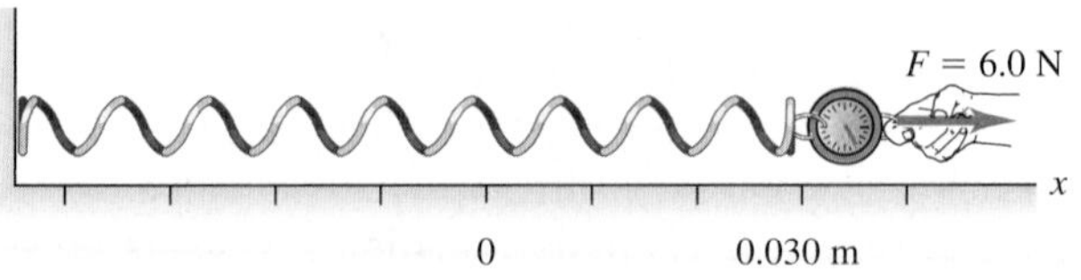

13.7 The force exerted *on* the spring is (shown by the vector F) has x-component $F_x = +6.0$ N. The force exerted *by* the spring has x-component $F_x = -6.0$ N.

Displacement, Velocity, and Acceleration in SHM

We still need to find the displacement x as a function of time for a harmonic oscillator. Equation (13.4) for a body in simple harmonic motion along the x-axis is identical to Eq. (13.8) for the x-coordinate of the reference point in uniform circular motion with constant angular speed $\omega = \sqrt{k/m}$. It follows that Eq. (13.5), $x = A\cos\theta$, describes the coordinate x for both of these situations. If at $t = 0$ the phasor OQ makes an angle ϕ ("phi") with the positive x-axis, then at any later time t this angle is $\theta = \omega t + \phi$. We substitute this into Eq. (13.5) to obtain

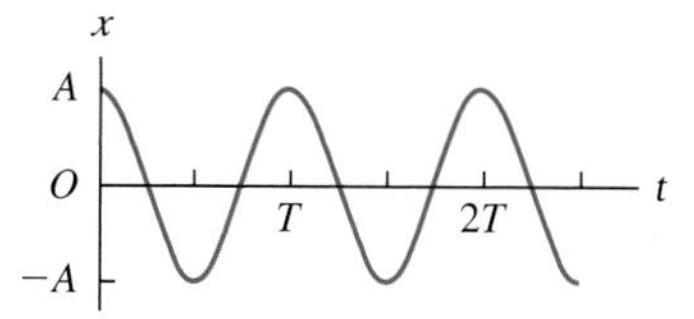

13.8 Graph of x versus t [Eq. (13.13)] for simple harmonic motion. The case shown has $\phi = 0$.

$$x = A\cos(\omega t + \phi) \quad \text{(displacement in SHM)} \tag{13.13}$$

where $\omega = \sqrt{k/m}$. Figure. 13.8 shows a graph of Eq. (13.13) for the particular case $\phi = 0$. The displacement x is a periodic function of time, as expected for SHM. We could also have written Eq. (13.13) in terms of a sine function rather than a cosine by using the identity $\cos\alpha = \sin(\alpha + \pi/2)$. *In simple harmonic motion the position is a periodic, sinusoidal function of time.* There are many other periodic functions, but none so smooth and simple as a sine or cosine function.

The value of the cosine function is always between -1 and 1, so in Eq. (13.13), x is always between $-A$ and A. This confirms that A is the amplitude of the motion. Figure 13.9a shows the graph of x versus t for various values of A.

The period T is the time for one complete cycle of oscillation. The cosine function repeats itself whenever the quantity in parentheses in Eq. (13.13) increases by 2π radians. Thus, if we start at time $t = 0$, the time T to complete one cycle is given by

$$\omega T = \sqrt{\frac{k}{m}}\,T = 2\pi \quad \text{or} \quad T = 2\pi\sqrt{\frac{m}{k}}$$

which is just Eq. (13.12). Changing either m or k changes the period of oscillation, as shown in Figs. 13.9b and 13.9c.

The constant ϕ in Eq. (13.13) is called the **phase angle.** It tells us at what point in the cycle the motion was at $t = 0$ (equivalent to where around the circle the point Q was at $t = 0$). We denote the position at $t = 0$ by x_0. Putting $t = 0$ and $x = x_0$ in Eq. (13.13), we get

$$x_0 = A\cos\phi \tag{13.14}$$

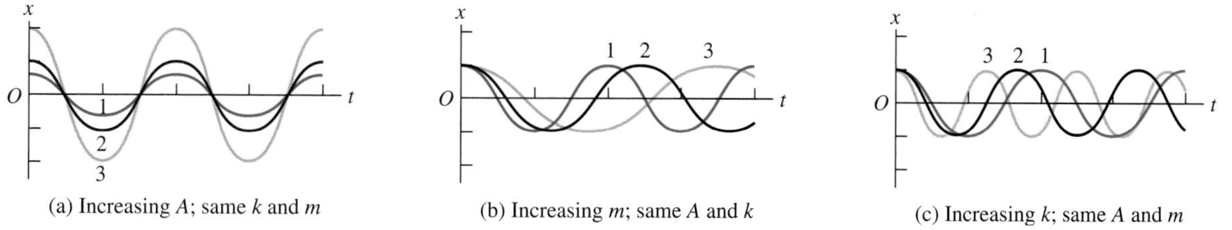

13.9 Variations of simple harmonic motion. All cases shown have $\phi = 0$. (a) Amplitude A increases from curve 1 to 2 to 3. Changing A alone has no effect on the period. (b) Mass m increases from 1 to 2 to 3. Increasing m alone increases the period. (c) Force constant k increases from 1 to 2 to 3. Increasing k alone decreases the period.

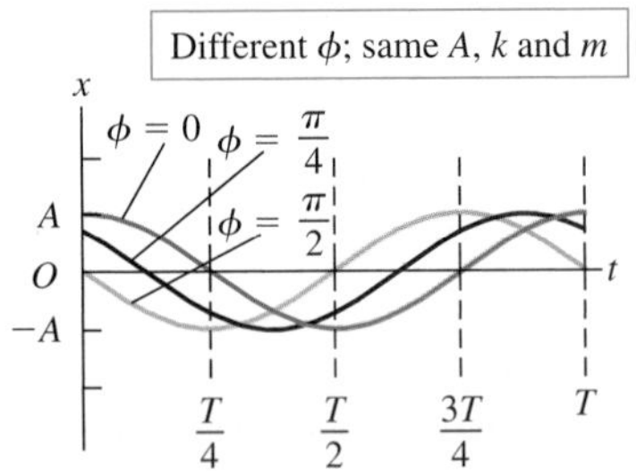

13.10 These three curves show SHM with the same period and amplitude but with different phase angles ϕ.

If $\phi = 0$, then $x_0 = A \cos 0 = A$, and the body starts at its maximum positive displacement. If $\phi = \pi$, then $x_0 = A \cos \pi = -A$, and the particle starts at its maximum *negative* displacement. If $\phi = \pi/2$, then $x_0 = A \cos(\pi/2) = 0$, and the particle is initially at the origin. Figure 13.10 shows the displacement x versus time for different phase angles.

We find the velocity v_x and acceleration a_x as functions of time for a harmonic oscillator by taking derivatives of Eq. (13.13) with respect to time:

$$v_x = \frac{dx}{dt} = -\omega A \sin(\omega t + \phi) \qquad \text{(velocity in SHM)} \qquad (13.15)$$

$$a_x = \frac{dv_x}{dt} = \frac{d^2x}{dt^2} = -\omega^2 A \cos(\omega t + \phi) \qquad \text{(acceleration in SHM)} \qquad (13.16)$$

The velocity v_x oscillates between $v_{max} = +\omega A$ and $-v_{max} = -\omega A$, and the acceleration a_x oscillates between $a_{max} = +\omega^2 A$ and $-a_{max} = -\omega^2 A$ (Fig. 13.11). Comparing Eq. (13.16) with Eq. (13.13) and recalling that $\omega^2 = k/m$ from Eq. (13.9), we see that

$$a_x = -\omega^2 x = -\frac{k}{m}x$$

which is just Eq. (13.4) for simple harmonic motion. This confirms that Eq. (13.13) for x as a function of time is correct.

We actually derived Eq. (13.16) earlier in a geometrical way by taking the x-component of the acceleration vector of the reference point Q. This was done in Fig. 13.5c and Eq. (13.7) (recall that $\theta = \omega t + \phi$). In the same way, we could have derived Eq. (13.15) by taking the x-component of the velocity vector of Q, as shown in Fig. 13.5b. We'll leave the details for you to work out (see Problem 13.79).

Note that the sinusoidal graph of displacement versus time (Fig. 13.11a) is shifted by one-quarter period from the graph of velocity versus time (Fig. 13.11b) and by one-half period from the graph of acceleration versus time (Fig. 13.11c). When the body is passing through the equilibrium position so that the displacement is zero, the velocity equals either v_{max} or $-v_{max}$ (depending on which way the body is moving) and the acceleration is zero. When the body is at either its maximum positive displacement, $x = +A$, or its maximum negative displacement, $x = -A$, the velocity is zero and the body is instantaneously at rest. At these points, the restoring force $F_x = -kx$ and the acceleration of the body have their maximum magnitudes. At $x = +A$ the acceleration is negative and equal to $-a_{max}$. At $x = -A$ the acceleration is positive: $a_x = +a_{max}$.

If we are given the initial position x_0 and initial velocity v_{0x} for the oscillating body, we can determine the amplitude A and the phase angle ϕ. Here's how to do it. The initial velocity v_{0x} is the velocity at time $t = 0$; putting $v_x = v_{0x}$ and $t = 0$ in Eq. (13.15), we find

$$v_{0x} = -\omega A \sin \phi \qquad (13.17)$$

To find ϕ, divide Eq. (13.17) by Eq. (13.14). This eliminates A and gives an equation that we can solve for ϕ:

$$\frac{v_{0x}}{x_0} = \frac{-\omega A \sin \phi}{A \cos \phi} = -\omega \tan \phi$$

$$\phi = \arctan\left(-\frac{v_{0x}}{\omega x_0}\right) \qquad \text{(phase angle in SHM)} \qquad (13.18)$$

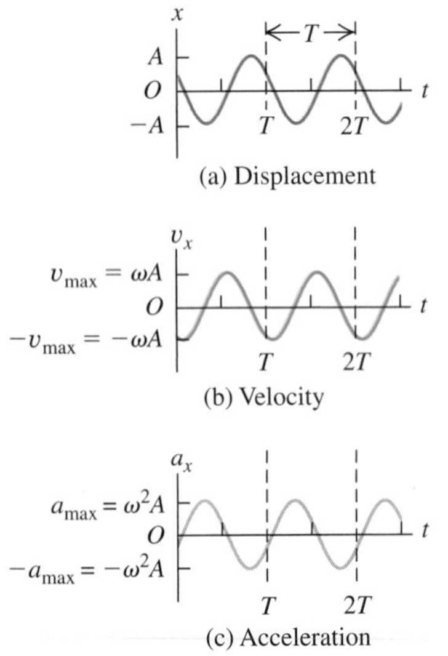

13.11 (a) Graph of x versus t for SHM. In this graph, $\phi = \pi/3$. (b) Graph of v_x versus t for the same motion. The v_x-t graph is shifted by $\frac{1}{4}$ cycle from the x-t graph. (c) Graph of a_x versus t for the same motion. The x-t graph is shifted by $\frac{1}{4}$ cycle from the v_x-t graph and by $\frac{1}{2}$ cycle from the a_x-t graph.

It is also easy to find the amplitude A if we are given x_0 and v_{0x}. We'll sketch the derivation, and you can fill in the details. Square Eq. (13.14); divide Eq. (13.17) by ω, square it, and add to the square of Eq. (13.14). The right side will be $A^2(\sin^2\phi + \cos^2\phi)$, which is equal to A^2. The final result is

$$A = \sqrt{x_0^2 + \frac{v_{0x}^2}{\omega^2}} \quad \text{(amplitude in SHM)} \qquad (13.19)$$

Note that when the body has both an initial displacement x_0 and a nonzero initial velocity v_{0x}, the amplitude A is *not* equal to the initial displacement. That's reasonable; if you start the body at a positive x_0 but give it a positive velocity v_{0x}, it will go *farther* than x_0 before it turns and comes back.

ActivPhysics ONLINE

9.1 Position Graphs and Equations
9.2 Describing Vibrational Motion
9.5 Ape Drops Tarzan

Problem-Solving Strategy: Simple Harmonic Motion I

IDENTIFY *the relevant concepts:* An oscillating system undergoes simple harmonic motion (SHM) *only* if the restoring force is directly proportional to the displacement. Be certain that this is the case for the problem at hand before attempting to use any of the results of this section. As always, identify the target variables.

SET UP *the problem using the following steps:*

1. Identify the known and unknown quantities, and determine which are the target variables.
2. It's useful to distinguish between two kinds of quantities. *Basic properties* of the system include the mass m and force constant k. (In some problems, m, k, or both can be determined from other information.) They also include quantities derived from m and k, such as the period T, frequency f, and angular frequency ω. *Properties of the motion* describe how the system behaves when it is set into motion in a particular way. They include the amplitude A, maximum velocity v_{max}, and phase angle ϕ, as well as the values of x, v_x, and a_x at a particular time.
3. If necessary, define an x-axis as in Fig. 13.6.

EXECUTE *the solution as follows:*

1. Use the equations given in Sections 13.1 and 13.2 to solve for the target variables.
2. If you need to calculate the phase angle, be certain to express it in radians. The quantity ωt in Eq. (13.13) is naturally in radians, so ϕ must be as well.
3. If you need to find the values of x, v_x, and a_x at various times, use Eqs. (13.11), (13.15), and (13.16). If the initial position x_0 and initial velocity v_{0x} are both given, you can determine the phase angle and amplitude from Eqs. (13.18) and (13.19). If the body is given an initial positive displacement x_0 but zero initial velocity $(v_{0x} = 0)$, then the amplitude is $A = x_0$ and the phase angle is $\phi = 0$. If it has an initial positive velocity v_{0x} but no initial displacement $(x_0 = 0)$, the amplitude is $A = v_{0x}/\omega$ and the phase angle is $\phi = -\pi/2$.

EVALUATE *your answer:* Check your results to make sure that they're consistent. As an example, suppose you've used the initial position and velocity to find general expressions for x and v_x at time t. If you substitute $t = 0$ into these expressions, you should get back the correct values of x_0 and v_{0x}.

Example 13.3 Describing SHM

Let's return to the system of mass and horizontal spring we considered in Example 13.2, with $k = 200$ N/m and $m = 0.50$ kg. This time we give the body an initial displacement of $+0.015$ m and an initial velocity of $+0.40$ m/s. a) Find the period, amplitude, and phase angle of the motion. b) Write equations for the displacement, velocity, and acceleration as functions of time.

SOLUTION

IDENTIFY: As in Example 13.2, the oscillations are SHM and we may use the expressions developed in this section.

SET UP: We are given the values of k, m, x_0, and v_{0x}. From them, we calculate the target variables T, A, and ϕ and the expressions for x, v_x, and a_x as functions of time.

EXECUTE: a) The period is exactly the same as in Example 13.2, $T = 0.31$ s. In simple harmonic motion the period does not depend on the amplitude, only on the values of k and m.

In Example 13.2 we found that $\omega = 20$ rad/s. So from Eq. (13.19),

$$A = \sqrt{x_0^2 + \frac{v_{0x}^2}{\omega^2}}$$

$$= \sqrt{(0.015\text{ m})^2 + \frac{(0.40\text{ m/s})^2}{(20\text{ rad/s})^2}}$$

$$= 0.025\text{ m}$$

To find the phase angle ϕ, we use Eq. (13.18):

$$\phi = \arctan\left(\frac{-v_{0x}}{\omega x_0}\right)$$

$$= \arctan\left(\frac{-0.40\text{ m/s}}{(20\text{ rad/s})(0.015\text{ m})}\right) = -53° = -0.93\text{ rad}$$

b) The displacement, velocity, and acceleration at any time are given by Eqs. (13.13), (13.15), and (13.16), respectively. Substituting the values, we get

$$x = (0.025\text{ m})\cos[(20\text{ rad/s})t - 0.93\text{ rad}]$$
$$v_x = -(0.50\text{ m/s})\sin[(20\text{ rad/s})t - 0.93\text{ rad}]$$
$$a_x = -(10\text{ m/s}^2)\cos[(20\text{ rad/s})t - 0.93\text{ rad}]$$

The velocity varies sinusoidally between -0.50 m/s and $+0.50$ m/s, and the acceleration varies sinusoidally between -10 m/s^2 and $+10\text{ m/s}^2$.

EVALUATE: You can check the results for x and v_x as functions of time by substituting $t = 0$ and evaluating the result. You should get $x = x_0 = 0.015$ m and $v_x = v_{0x} = 0.40$ m/s. Do you?

Test Your Understanding

The position of a certain object in SHM is given as a function of time by $x = (0.050\text{ m})\cos[(290\text{ rad/s})t + (2.5\text{ rad})]$. What are the amplitude, period, phase angle, and initial position for this motion?

13.3 | Energy in Simple Harmonic Motion

ActivPhysics ONLINE

9.3 Vibrational Energy
9.4 Two Ways to Weigh Young Tarzan
9.6 Releasing a Vibrating Skier I
9.7 Releasing a Vibrating Skier II

We can learn even more about simple harmonic motion by using energy considerations. Take another look at the body oscillating on the end of a spring in Fig. 13.1. We've already noted that the spring force is the only horizontal force on the body. The force exerted by an ideal spring is a conservative force, and the vertical forces do no work, so the total mechanical energy of the system is *conserved*. We also assume that the mass of the spring itself is negligible.

The kinetic energy of the body is $K = \frac{1}{2}mv^2$ and the potential energy of the spring is $U = \frac{1}{2}kx^2$, just as in Section 7.2. (It would be helpful to review that section.) There are no nonconservative forces that do work, so the total mechanical energy $E = K + U$ is conserved:

$$E = \frac{1}{2}mv_x^2 + \frac{1}{2}kx^2 = \text{constant} \tag{13.20}$$

(Since the motion is one-dimensional, $v^2 = v_x^2$.)

The total mechanical energy E is also directly related to the amplitude A of the motion. When the body reaches the point $x = A$, its maximum displacement from equilibrium, it momentarily stops as it turns back toward the equilibrium position. That is, when $x = A$ (or $-A$), $v_x = 0$. At this point the energy is entirely potential, and $E = \frac{1}{2}kA^2$. Because E is constant, this quantity equals E at any other point. Combining this expression with Eq. (13.20), we get

$$E = \frac{1}{2}mv_x^2 + \frac{1}{2}kx^2 = \frac{1}{2}kA^2 = \text{constant} \tag{13.21}$$

(total mechanical energy in SHM)

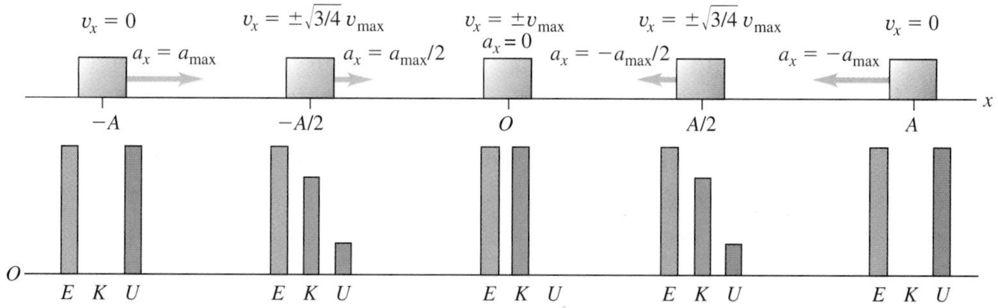

13.12 In SHM the total mechanical energy E is constant. As the body oscillates, the energy continually transforms back and forth between potential energy U and kinetic energy K.

We can verify this equation by substituting x and v_x from Eqs. (13.13) and (13.15) and using $\omega^2 = k/m$ from Eq. (13.9):

$$E = \frac{1}{2}mv_x^2 + \frac{1}{2}kx^2 = \frac{1}{2}m[-\omega A \sin(\omega t + \phi)]^2 + \frac{1}{2}k[A\cos(\omega t + \phi)]^2$$
$$= \frac{1}{2}kA^2\sin^2(\omega t + \phi) + \frac{1}{2}kA^2\cos^2(\omega t + \phi)$$
$$= \frac{1}{2}kA^2$$

(Recall that $\sin^2\alpha + \cos^2\alpha = 1$.) Hence our expressions for displacement and velocity in SHM are consistent with energy conservation, as they must be.

We can use Eq. (13.21) to solve for the velocity v_x of the body at a given displacement x:

$$v_x = \pm\sqrt{\frac{k}{m}}\sqrt{A^2 - x^2} \tag{13.22}$$

The $\pm$ sign means that at a given value of x the body can be moving in either direction. For example, when $x = \pm A/2$,

$$v_x = \pm\sqrt{\frac{k}{m}}\sqrt{A^2 - \left(\pm\frac{A}{2}\right)^2} = \pm\sqrt{\frac{3}{4}}\sqrt{\frac{k}{m}}A$$

Equation (13.22) also shows that the *maximum* speed v_{max} occurs at $x = 0$. Using Eq. (13.10), $\omega = \sqrt{k/m}$, we find that

$$v_{max} = \sqrt{\frac{k}{m}}A = \omega A \tag{13.23}$$

This agrees with Eq. (13.15), which showed that v_x oscillates between $-\omega A$ and $+\omega A$.

Figure 13.12 shows the energy quantities E, K, and U at $x = 0$, $x = \pm A/2$, and $x = \pm A$. Fig. 13.13 is a graphical display of Eq. (13.21); energy (kinetic, potential, and total) is plotted vertically and the coordinate x is plotted horizontally. The parabolic curve in Fig. 13.13a represents the potential energy $U = \frac{1}{2}kx^2$. The horizontal line represents the total mechanical energy E, which is constant and does not vary with x. This line intersects the potential-energy curve at $x = -A$ and $x = A$, where the energy is entirely potential. At any value of x between $-A$ and A, the vertical distance from the x-axis to the parabola is U; since $E = K + U$, the remaining vertical distance up to the horizontal line is K. Figure 13.13b shows

Activ Physics Online

9.8 One- and Two-Spring Vibrating Systems

9.9 Vibro-Ride

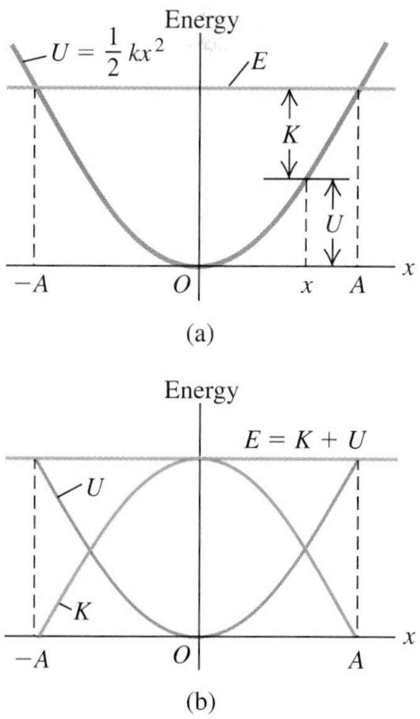

13.13 Kinetic energy K, potential energy U, and total mechanical energy E as functions of position for SHM. At each value of x the sum of the values of K and U equals the constant value of E.

both K and U as functions of x. As the body oscillates between $-A$ and A, the energy is continuously transformed from potential to kinetic and back again.

Figure 13.13a shows the connection between the amplitude A and the corresponding total mechanical energy $E = \frac{1}{2}kA^2$. If we tried to make x greater than A (or less than $-A$), U would be greater than E, and K would have to be negative. But K can never be negative, so x can't be greater than A or less than $-A$.

Problem-Solving Strategy: Simple Harmonic Motion II

The energy equation, Eq. (13.21), is a useful alternative relation between velocity and position, especially when energy quantities are also required. If the problem involves a relation among position, velocity, and acceleration without reference to time, it is usually easier to use Eq. (13.4) (from Newton's second law) or (13.21) (from energy conservation) than to use the general expressions for x, v_x, and a_x as functions of time [Eqs. (13.13), (13.15), and (13.16), respectively]. Because the energy equation involves x^2 and v_x^2, it cannot tell you the sign of x or of v_x; you have to infer the sign from the situation. For instance, if the body is moving from the equilibrium position toward the point of greatest positive displacement, then x is positive and v_x is positive.

Example 13.4: Velocity, acceleration, and energy in SHM

In the oscillation described in Example 13.2, $k = 200$ N/m, $m = 0.50$ kg, and the oscillating mass is released from rest at $x = 0.020$ m. a) Find the maximum and minimum velocities attained by the oscillating body. b) Compute the maximum acceleration. c) Determine the velocity and acceleration when the body has moved halfway to the center from its original position. d) Find the total energy, potential energy, and kinetic energy at this position.

IDENTIFY: Note that the problem refers to the motion at various *positions* in the motion, not at specified *times*. This is a hint that we can use the energy relations found in this section to solve for the target variables.

SET UP: Figure 13.14 shows the choice of x-axis. For any position x we use Eqs. (13.22) and (13.4) to find the velocity v_x and acceleration a_x, respectively. Given the velocity and position, we use Eq. (13.21) to find the energy quantities K, U, and E.

CAUTION The velocity of the body is *not* constant, so these images of the body at equally spaced positions are *not* equally spaced in time.

EXECUTE: a) The velocity v_x at any displacement x is given by Eq. (13.22):

$$v_x = \pm\sqrt{\frac{k}{m}}\sqrt{A^2 - x^2}$$

The maximum velocity occurs when the body is moving to the right through the equilibrium position, where $x = 0$:

$$v_x = v_{\max} = \sqrt{\frac{k}{m}}A = \sqrt{\frac{200\ \text{N/m}}{0.50\ \text{kg}}}(0.020\ \text{m}) = 0.40\ \text{m/s}$$

The minimum (i.e., most negative) velocity occurs when the body is moving to the left through $x = 0$; its value is $-v_{\max} = -0.40$ m/s.

b) From Eq. (13.4),

$$a_x = -\frac{k}{m}x$$

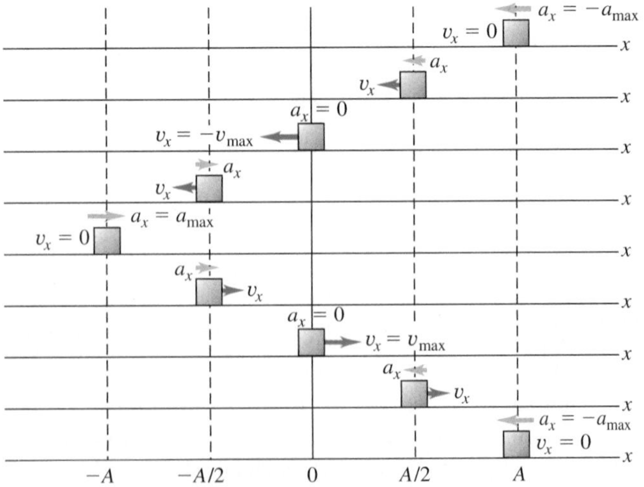

13.14 A body is attached to a spring, pulled a distance A from the equilibrium position and released. This figure shows the velocity and acceleration at nine points in the motion.

The maximum (most positive) acceleration occurs at the most negative value of x, $x = -A$; therefore

$$a_{max} = -\frac{k}{m}(-A) = -\frac{200\ \text{N/m}}{0.50\ \text{kg}}(-0.020\ \text{m}) = 8.0\ \text{m/s}^2$$

The minimum (most negative) acceleration is $-8.0\ \text{m/s}^2$, occurring at $x = +A = +0.020$ m.

c) At a point halfway to the center from the initial position, $x = A/2 = 0.010$ m. From Eq. (13.22),

$$v_x = -\sqrt{\frac{200\ \text{N/m}}{0.50\ \text{kg}}}\sqrt{(0.020\ \text{m})^2 - (0.010\ \text{m})^2} = -0.35\ \text{m/s}$$

We choose the negative square root because the body is moving from $x = A$ toward $x = 0$. From Eq. (13.4),

$$a_x = -\frac{200\ \text{N/m}}{0.50\ \text{kg}}(0.010\ \text{m}) = -4.0\ \text{m/s}^2$$

At this point the velocity and the acceleration have the same sign, so the speed is increasing. The conditions at $x = 0$, $\pm A/2$, and $\pm A$ are shown in Fig. 13.14.

d) The total energy has the same value at all points during the motion:

$$E = \frac{1}{2}kA^2 = \frac{1}{2}(200\ \text{N/m})(0.020\ \text{m})^2 = 0.040\ \text{J}$$

The potential energy is

$$U = \frac{1}{2}kx^2 = \frac{1}{2}(200\ \text{N/m})(0.010\ \text{m})^2 = 0.010\ \text{J}$$

and the kinetic energy is

$$K = \frac{1}{2}mv_x^2 = \frac{1}{2}(0.50\ \text{kg})(-0.35\ \text{m/s})^2 = 0.030\ \text{J}$$

EVALUATE: At the point $x = A/2$, the energy is one-fourth potential energy and three-fourths kinetic energy. You can check this result by inspecting Fig. 13.13b.

Example 13.5 Energy and momentum in SHM

A block with mass M attached to a horizontal spring with force constant k is moving with simple harmonic motion having amplitude A_1. At the instant when the block passes through its equilibrium position, a lump of putty with mass m is dropped vertically onto the block from a very small height and sticks to it (Fig. 13.15a). a) Find the new amplitude and period. b) Repeat part (a) for the case in which the putty is dropped on the block when it is at one end of its path (Fig. 13.15b).

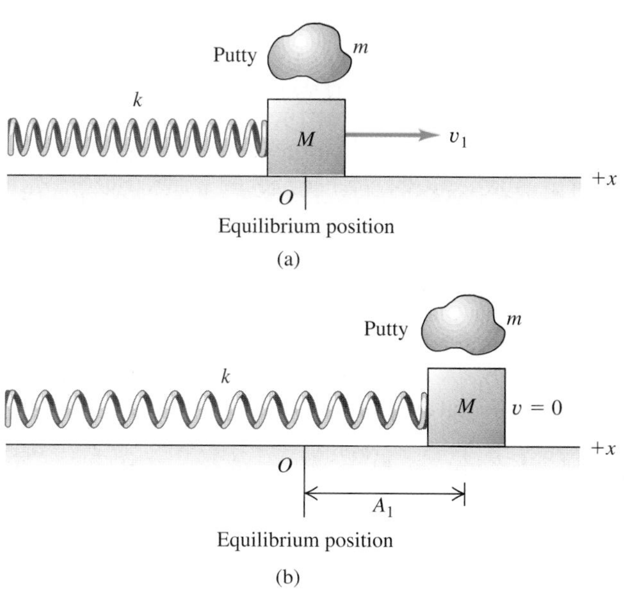

13.15 (a) Dropping a lump of putty on an oscillating block as it passes through equilibrium. (b) Dropping a lump of putty on the block at $x = A_1$.

SOLUTION

IDENTIFY: The problem involves the motion at a given position, not a given time, so we can use energy methods to solve it. Before the putty lands on the block, the mechanical energy of the oscillating block and spring is constant. When the putty lands on the block, it's a completely inelastic collision (Section 8.3); the horizontal component of momentum is conserved, but kinetic energy decreases. Once the collision ends, the mechanical energy remains constant at its new value. We use this principle to find the new amplitude (which is related to the total energy of the system). We find the new period using the relationship between period and mass.

SET UP: Figure 13.15 shows the choice of coordinates. The target variables in each part are the amplitude A_2 and the period T_2 after the collision. In each part we consider what happens before, during, and after the collision.

EXECUTE: a) Before the collision the total mechanical energy of the block and spring is $E_1 = \frac{1}{2}kA_1^2$. Since the block is at the equilibrium position, $U = 0$, and the energy is purely kinetic. If we let v_1 be the speed of the block at the equilibrium position, we have

$$E_1 = \frac{1}{2}Mv_1^2 = \frac{1}{2}kA_1^2 \quad \text{so} \quad v_1 = \sqrt{\frac{k}{m}}A_1$$

During the collision the x-component of momentum of the system of block and putty is conserved. (Why?) Just before the collision this component is the sum of Mv_1 (for the block) and zero (for the putty). Just after the collision the block and putty move together with speed v_2, and their combined x-component of momentum is $(M + m)v_2$. From conservation of momentum,

$$Mv_1 + 0 = (M + m)v_2 \quad \text{so} \quad v_2 = \frac{M}{M + m}v_1$$

The collision lasts a very short time, so just after the collision the block and putty are still at the equilibrium position. The energy is still purely kinetic, but is *less* than before the collision:

$$E_2 = \frac{1}{2}(M + m)v_2^2 = \frac{1}{2}\frac{M^2}{M + m}v_1^2 = \frac{M}{M + m}\left(\frac{1}{2}Mv_1^2\right)$$

$$= \left(\frac{M}{M + m}\right)E_1$$

Since E_2 equals $\frac{1}{2}kA_2^2$, where A_2 is the amplitude after the collision, we have

$$\frac{1}{2}kA_2^2 = \left(\frac{M}{M + m}\right)\frac{1}{2}kA_1^2$$

$$A_2 = A_1\sqrt{\frac{M}{M + m}}$$

The larger the putty mass m, the smaller the final amplitude.

Finding the period of oscillation after the collision is the easy part. Using Eq. (13.12), we have

$$T_2 = 2\pi\sqrt{\frac{M + m}{k}}$$

b) When the putty drops on the block, the block is instantaneously at rest; all the mechanical energy is stored in the spring as potential energy. The x-component of momentum is zero both before and after the collision: the block has zero kinetic energy just before the collision, and the block and putty have zero kinetic energy just after the collision. So in this case, adding the extra mass of the putty has *no effect* on the mechanical energy. That is,

$$E_2 = E_1 = \frac{1}{2}kA_1^2$$

and the amplitude after the collision is unchanged $(A_2 = A_1)$. The period still changes when the putty is added, though; its value doesn't depend on how the mass is added, only on what the total mass is. So T_2 is the same as we found in part (a), $T_2 = 2\pi\sqrt{(M + m)/k}$.

EVALUATE: Why is energy lost in part (a) but not in part (b)? The difference is that in part (a) the putty slides against the moving block during the collision, and energy is dissipated by kinetic friction.

Test Your Understanding

To double the total energy for an oscillating mass-spring system, by what factor must the amplitude increase? What effect does this change have on the frequency? Assume SHM.

13.4 | Applications of Simple Harmonic Motion

So far, we've looked at a grand total of *one* situation in which simple harmonic motion (SHM) occurs: a body attached to an ideal horizontal spring. But SHM can occur in any system in which there is a restoring force that is directly proportional to the displacement from equilibrium, as given by Eq. (13.3), $F_x = -kx$. The restoring force will originate in different ways in different situations, so the force constant k has to be found for each case by examining the net force on the system. Once this is done, it's straightforward to find the angular frequency ω, frequency f, and period T; we just substitute the value of k into Eqs. (13.10), (13.11), and (13.12), respectively. Let's use these ideas to examine several examples of simple harmonic motion.

Vertical SHM

Suppose we hang a spring with force constant k (Fig. 13.16a) and suspend from it a body with mass m. Oscillations will now be vertical; will they still be SHM? In Fig. 13.16b the body hangs at rest, in equilibrium. In this position the spring is stretched an amount Δl just great enough that the spring's upward vertical force $k\Delta l$ on the body balances its weight mg:

$$k\Delta l = mg$$

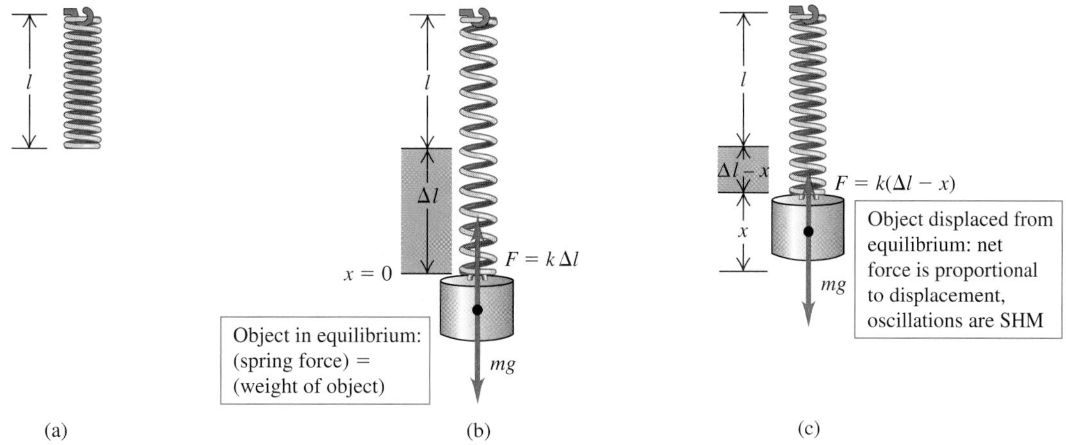

13.16 (a) A hanging spring. (b) A body suspended from the spring. When the spring is stretched just far enough that the upward spring force equals the object's weight, the object is in equilibrium. (c) If the spring obeys Hooke's law, the object's motion will be simple harmonic.

Take $x = 0$ to be this equilibrium position and take the positive x-direction to be upward. When the body is a distance x *above* its equilibrium position (Fig. 13.16c), the extension of the spring is $\Delta l - x$. The upward force it exerts on the body is then $k(\Delta l - x)$, and the net x-component of force on the body is

$$F_{\text{net}} = k(\Delta l - x) + (-mg) = -kx$$

that is, a net downward force of magnitude kx. Similarly, when the body is *below* the equilibrium position, there is a net upward force with magnitude kx. In either case there is a restoring force with magnitude kx. If the body is set in vertical motion, it oscillates in SHM with the same angular frequency as though it were horizontal, $\omega = \sqrt{k/m}$. So vertical SHM doesn't differ in any essential way from horizontal SHM. The only real change is that the equilibrium position $x = 0$ no longer corresponds to the point at which the spring is unstretched. The same ideas hold if a body with weight mg is placed atop a compressible spring (Fig. 13.17) and compresses it a distance Δl.

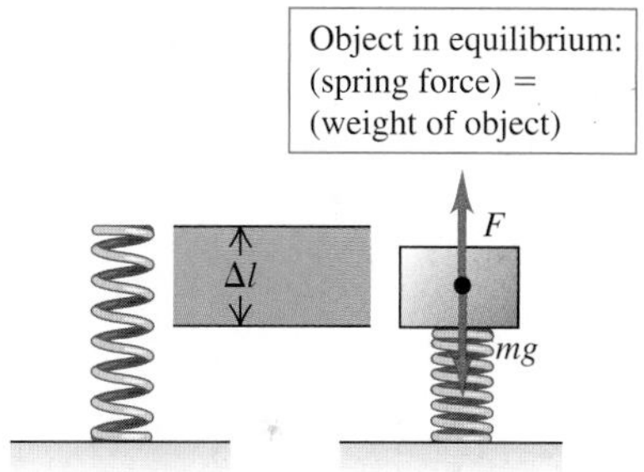

13.17 If the weight mg compresses the spring a distance Δl, the force constant is $k = mg/\Delta l$ and the angular frequency for vertical SHM is $\omega = \sqrt{k/m}$—the same as if the body were suspended from the spring (Fig. 13.16).

Example 13.6 Vertical SHM in an old car

The shock absorbers in an old car with mass 1000 kg are completely worn out. When a 980-N person climbs slowly into the car to its center of gravity, the car sinks 2.8 cm. When the car, with the person aboard, hits a bump, the car starts oscillating up and down in SHM. Model the car and person as a single body on a single spring, and find the period and frequency of the oscillation.

SOLUTION

IDENTIFY and SET UP: The situation is like that shown in Fig. 13.17. The compression of the spring when the extra weight is added tells us the force constant, which we can use to find the period and frequency (the target variables).

EXECUTE: When the force increases by 980 N, the spring compresses an additional 0.028 m, and the coordinate x of the car changes by -0.028 m. Hence the effective force constant (including the effect of the entire suspension) is

$$k = -\frac{F_x}{x} = -\frac{980\ \text{N}}{-0.028\ \text{m}} = 3.5 \times 10^4\ \text{kg/s}^2$$

The person's mass is $w/g = (980\ \text{N})/(9.8\ \text{m/s}^2) = 100\ \text{kg}$. The *total* oscillating mass is $m = 1000\ \text{kg} + 100\ \text{kg} = 1100\ \text{kg}$. The period T is

$$T = 2\pi\sqrt{\frac{m}{k}} = 2\pi\sqrt{\frac{1100\ \text{kg}}{3.5 \times 10^4\ \text{kg/s}^2}} = 1.11\ \text{s}$$

and the frequency is

$$f = \frac{1}{T} = \frac{1}{1.11\ \text{s}} = 0.90\ \text{Hz}$$

EVALUATE: A persistent oscillation with a period of about one second makes for a very unpleasant ride. The purpose of shock absorbers is to make such oscillations die out (see Section 13.7).

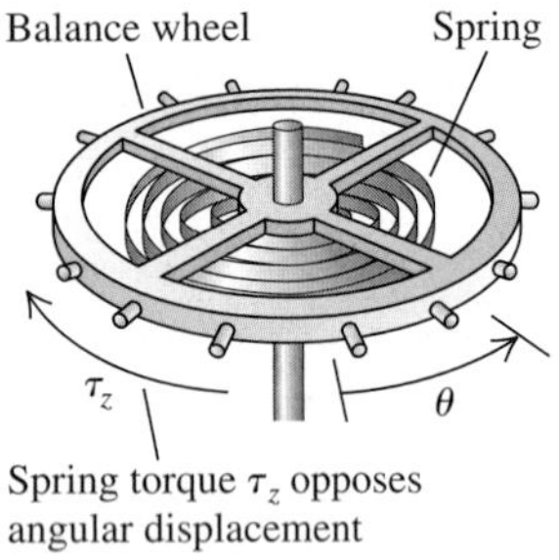

13.18 The balance wheel of a mechanical watch. The restoring torque exerted by the spring is proportional to the angular displacement θ, so the motion is angular SHM.

Angular SHM

Figure 13.18 shows the balance wheel of a mechanical watch. The wheel has a moment of inertia I about its axis. A coil spring exerts a restoring torque τ_z that is proportional to the angular displacement θ from the equilibrium position. We write $\tau_z = -\kappa\theta$, where κ (the Greek letter "kappa") is a constant called the *torsion constant.* Using the rotational analog of Newton's second law for a rigid body, $\sum\tau_z = I\alpha_z = I\,d^2\theta/dt^2$, the equation of motion is

$$-\kappa\theta = I\alpha \quad \text{or} \quad \frac{d^2\theta}{dt^2} = -\frac{\kappa}{I}\theta$$

The form of this equation is exactly the same as Eq. (13.4) for the acceleration in simple harmonic motion, with x replaced by θ and k/m replaced by κ/I. So we are dealing with a form of *angular* simple harmonic motion. The angular frequency ω and frequency f are given by Eqs. (13.10) and (13.11), respectively, with the same replacement:

$$\omega = \sqrt{\frac{\kappa}{I}} \quad \text{and} \quad f = \frac{1}{2\pi}\sqrt{\frac{\kappa}{I}} \qquad \text{(angular SHM)} \qquad (13.24)$$

The motion is described by the function

$$\theta = \Theta\cos(\omega t + \phi)$$

where Θ (capital "theta") plays the role of an angular amplitude.

It's a good thing that the motion of a balance wheel *is* simple harmonic. If it weren't, the frequency might depend on the amplitude, and the watch would run too fast or too slow as the spring ran down.

*Vibrations of Molecules

The following discussion of the vibrations of molecules uses the binomial theorem. If you aren't familiar with this theorem, you should read about it in the appropriate section of your math textbook.

When two atoms are separated from each other by a few atomic diameters, they can exert attractive forces on each other. But if the atoms are so close to each other that their electron shells overlap, the forces between the atoms are repulsive. Between these limits, there can be an equilibrium separation distance at which two atoms form a *molecule*. If these atoms are displaced slightly from equilibrium, they will oscillate. Let's see whether these oscillations can be simple harmonic.

As an example, we'll consider one type of interaction between atoms called the *van der Waals interaction*. Our immediate task here is to study oscillations, so we

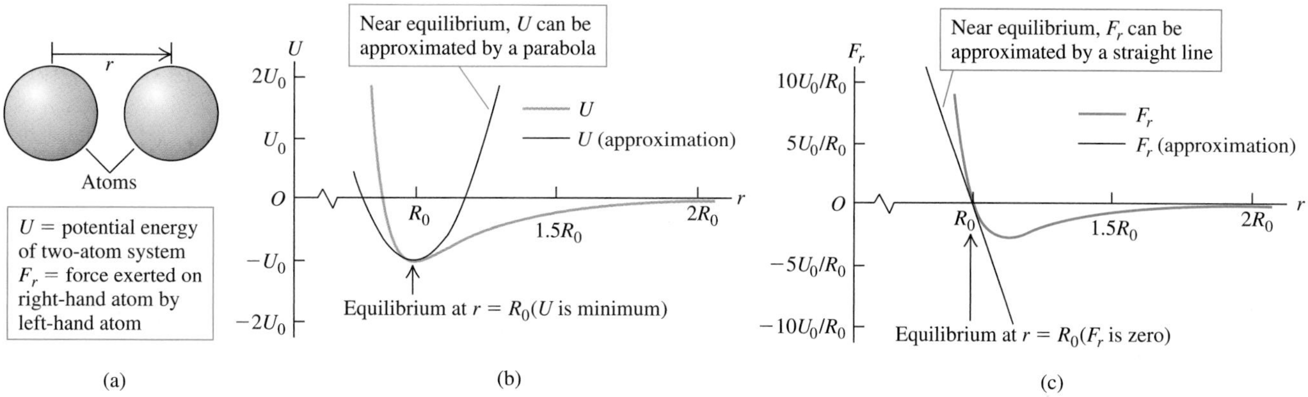

13.19 (a) Two atoms with centers separated by r. (b) Potential energy U in the van der Waals interaction as a function of r. (c) Force F_r on the right-hand atom as function of r.

won't go into the details of how this interaction arises. Let the center of one atom be at the origin and let the center of the other atom be a distance r away (Fig. 13.19a); the equilibrium distance between centers is $r = R_0$. Experiment shows that the van der Waals interaction can be described by the potential energy function

$$U = U_0\left[\left(\frac{R_0}{r}\right)^{12} - 2\left(\frac{R_0}{r}\right)^{6}\right] \tag{13.25}$$

where U_0 is a positive constant with units of joules. When the two atoms are very far apart, $U = 0$; when they are separated by the equilibrium distance $r = R_0$, $U = -U_0$. The force on the second atom is the negative derivative of Eq. (13.25),

$$F_r = -\frac{dU}{dr} = U_0\left[\frac{12R_0^{12}}{r^{13}} - 2\frac{6R_0^{6}}{r^{7}}\right] = 12\frac{U_0}{R_0}\left[\left(\frac{R_0}{r}\right)^{13} - \left(\frac{R_0}{r}\right)^{7}\right] \tag{13.26}$$

The potential energy and force are plotted in Figs. 13.19b and 13.19c, respectively. The force is positive for $r < R_0$ and negative for $r > R_0$, so it is a *restoring* force.

To study small-amplitude oscillations around the equilibrium separation $r = R_0$, we introduce the quantity x to represent the displacement from equilibrium:

$$x = r - R_0 \quad \text{so} \quad r = R_0 + x$$

In terms of x, the force F_r in Eq. (13.26) becomes

$$F_r = 12\frac{U_0}{R_0}\left[\left(\frac{R_0}{R_0 + x}\right)^{13} - \left(\frac{R_0}{R_0 + x}\right)^{7}\right] = 12\frac{U_0}{R_0}\left[\frac{1}{(1 + x/R_0)^{13}} - \frac{1}{(1 + x/R_0)^{7}}\right] \tag{13.27}$$

This looks nothing like Hooke's law, $F_x = -kx$, so we might be tempted to conclude that molecular oscillations cannot be SHM. But let us restrict ourselves to *small-amplitude* oscillations so that the absolute value of the displacement x is small in comparison to R_0 and the absolute value of the ratio x/R_0 is much less than 1. We can then simplify Eq. (13.27) by using the *binomial theorem:*

$$(1 + u)^n = 1 + nu + \frac{n(n-1)}{2!}u^2 + \frac{n(n-1)(n-2)}{3!}u^3 + \cdots \tag{13.28}$$

If $|u|$ is much less than 1, each successive term in Eq. (13.28) is much smaller than the one it follows, and we can safely approximate $(1 + u)^n$ by just the first two terms. In Eq. (13.27), u is replaced by x/R_0 and n equals -13 or -7, so

$$\frac{1}{(1 + x/R_0)^{13}} = (1 + x/R_0)^{-13} \approx 1 + (-13)\frac{x}{R_0}$$

$$\frac{1}{(1 + x/R_0)^{7}} = (1 + x/R_0)^{-7} \approx 1 + (-7)\frac{x}{R_0}$$

$$F_r \approx 12\frac{U_0}{R_0}\left[\left(1 + (-13)\frac{x}{R_0}\right) - \left(1 + (-7)\frac{x}{R_0}\right)\right] = -\left(\frac{72U_0}{R_0^2}\right)x \quad (13.29)$$

This is just Hooke's law, with force constant $k = 72U_0/R_0^2$. (Note that k has the correct units, J/m^2 or N/m.) So oscillations of molecules bound by the van der Waals interaction can be simple harmonic motion, provided that the amplitude is small in comparison to R_0 so that the approximation $|x/R_0| \ll 1$ used in the derivation of Eq. (13.29) is valid.

You can also show that the potential energy U, Eq. (13.25), can be written as $U \approx \frac{1}{2}kx^2 + C$, where $C = -U_0$ and k is again equal to $72U_0/R_0^2$. Adding a constant to the potential energy has no effect on the physics, so the system of two atoms is fundamentally no different from a mass attached to a horizontal spring for which $U = \frac{1}{2}kx^2$. The proof is left as an exercise.

Example 13.7 Molecular vibration

Two argon atoms can form a weakly bound molecule, Ar_2, held together by a van der Waals interaction with $U_0 = 1.68 \times 10^{-21}$ J and $R_0 = 3.82 \times 10^{-10}$ m. Find the frequency for small oscillations of one of the atoms about its equilibrium position.

SOLUTION

IDENTIFY and SET UP: Because the oscillations are small, we can use Eq. (13.11) to obtain the frequency of simple harmonic motion. The force constant is given by Eq. (13.29).

EXECUTE: The force constant is

$$k = \frac{72U_0}{R_0^2} = \frac{72(1.68 \times 10^{-21}\ \text{J})}{(3.82 \times 10^{-10}\ \text{m})^2} = 0.829\ \text{J/m}^2 = 0.829\ \text{N/m}$$

This is comparable to the force constant of a loose, floppy toy spring like a Slinky™.

From the periodic table of the elements (Appendix D), the average atomic mass of argon is $(39.948\ \text{u})(1.66 \times 10^{-27}\ \text{kg/1 u}) = 6.63 \times 10^{-26}$ kg. If one of the argon atoms is fixed and the other atom oscillates, the frequency of oscillation is

$$f = \frac{1}{2\pi}\sqrt{\frac{k}{m}} = \frac{1}{2\pi}\sqrt{\frac{0.829\ \text{N/m}}{6.63 \times 10^{-26}\ \text{kg}}} = 5.63 \times 10^{11}\ \text{Hz}$$

The oscillating mass is very small, so even a loose spring causes very rapid oscillations.

EVALUATE: Our answer for f isn't quite right, however. If there is no net external force acting on the molecule, the center of mass of the molecule (located halfway between the two atoms) doesn't accelerate. To ensure this, *both* atoms must oscillate with the same amplitude in opposite directions. It turns out that we can account for this by replacing m with $m/2$ in the expression for f. (See Problem 13.81.) This makes f larger by a factor of $\sqrt{2}$, so $f = \sqrt{2}(5.63 \times 10^{11}\ \text{Hz}) = 7.96 \times 10^{11}$ Hz. An additional complication is that on the atomic scale we must use *quantum mechanics,* not Newtonian mechanics, to describe oscillation and other motions; happily, the frequency has the same value in quantum mechanics.

Test Your Understanding

Suppose one of the argon atoms in an Ar_2 molecule (Example 13.7) is displaced from equilibrium by 1.00×10^{-11} m and released. What is the magnitude of the atom's initial acceleration?

13.5 | The Simple Pendulum

A **simple pendulum** is an idealized model consisting of a point mass suspended by a massless, unstretchable string. When the point mass is pulled to one side of its straight-down equilibrium position and released, it oscillates about the equilibrium position. Familiar situations such as a wrecking ball on a crane's cable, the plumb bob on a surveyor's transit, and a child on a swing can be modeled as simple pendulums.

The path of the point mass (sometimes called a pendulum bob) is not a straight line but the arc of a circle with radius L equal to the length of the string (Fig. 13.20). We use as our coordinate the distance x measured along the arc. If the motion is simple harmonic, the restoring force must be directly proportional to x or (because $x = L\theta$) to θ. Is it?

In Fig. 13.20 we represent the forces on the mass in terms of tangential and radial components. The restoring force F_θ is the tangential component of the net force:

$$F_\theta = -mg \sin\theta \qquad (13.30)$$

The restoring force is provided by gravity; the tension T merely acts to make the point mass move in an arc. The restoring force is proportional *not* to θ but to $\sin\theta$, so the motion is *not* simple harmonic. However, *if the angle* θ *is small,* $\sin\theta$ is very nearly equal to θ in radians (Fig. 13.21). For example, when $\theta = 0.1$ rad (about 6°), $\sin\theta = 0.0998$, a difference of only 0.2%. With this approximation, Eq. (13.30) becomes

$$F_\theta = -mg\theta = -mg\frac{x}{L} \quad \text{or}$$

$$F_\theta = -\frac{mg}{L}x \qquad (13.31)$$

The restoring force is then proportional to the coordinate *for small displacements,* and the force constant is $k = mg/L$. From Eq. (13.10) the angular frequency ω of a simple pendulum with small amplitude is

$$\omega = \sqrt{\frac{k}{m}} = \sqrt{\frac{mg/L}{m}} = \sqrt{\frac{g}{L}} \qquad (13.32)$$
(simple pendulum, small amplitude)

The corresponding frequency and period relations are

$$f = \frac{\omega}{2\pi} = \frac{1}{2\pi}\sqrt{\frac{g}{L}} \quad \text{(simple pendulum, small amplitude)} \qquad (13.33)$$

$$T = \frac{2\pi}{\omega} = \frac{1}{f} = 2\pi\sqrt{\frac{L}{g}} \quad \text{(simple pendulum, small amplitude)} \qquad (13.34)$$

Note that these expressions do not involve the *mass* of the particle. This is because the restoring force, a component of the particle's weight, is proportional to m. Thus the mass appears on *both* sides of $\Sigma\vec{F} = m\vec{a}$ and cancels out. (This is the same physics that explains why bodies of different masses fall with the same acceleration in a vacuum.) For small oscillations the period of a pendulum for a given value of g is determined entirely by its length.

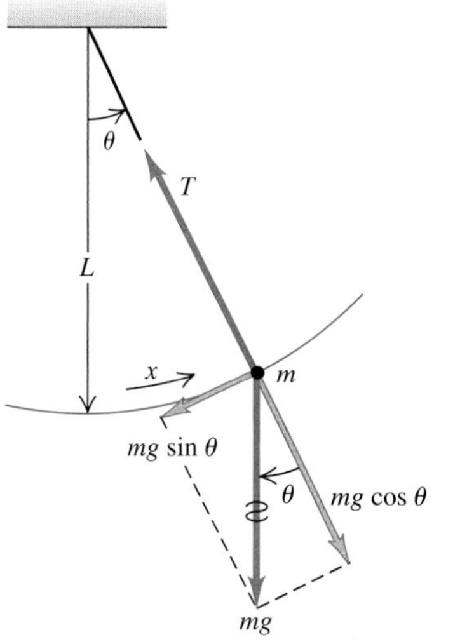

13.20 The dynamics of a simple pendulum.

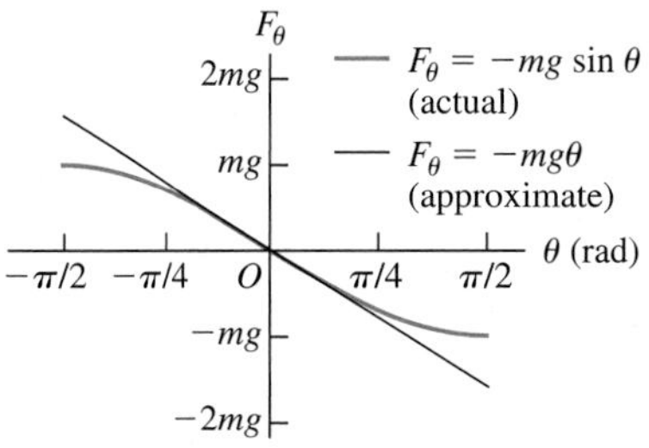

13.21 For small angular displacements θ, the restoring force $F_\theta = -mg\sin\theta$ on a simple pendulum is approximately equal to $-mg\theta$; that is, it is approximately proportional to the displacement θ. Hence for small angles the oscillations are simple harmonic.

ActivPhysics ONLINE

9.10 Pendulum Frequency

9.11 Risky Pendulum Walk

The dependence on L and g in Eqs. (13.32) through (13.34) is just what we should expect. A long pendulum has a longer period than a shorter one. Increasing g increases the restoring force, causing the frequency to increase and the period to decrease.

We emphasize again that the motion of a pendulum is only *approximately* simple harmonic. When the amplitude is not small, the departures from simple harmonic motion can be substantial. But how small is "small"? The period can be expressed by an infinite series; when the maximum angular displacement is Θ, the period T is given by

$$T = 2\pi\sqrt{\frac{L}{g}}\left(1 + \frac{1^2}{2^2}\sin^2\frac{\Theta}{2} + \frac{1^2 \cdot 3^2}{2^2 \cdot 4^2}\sin^4\frac{\Theta}{2} + \cdots\right) \tag{13.35}$$

We can compute the period to any desired degree of precision by taking enough terms in the series. We invite you to check that when $\Theta = 15°$ (on either side of the central position), the true period is longer than that given by the approximate Eq. (13.34) by less than 0.5%.

The usefulness of the pendulum as a timekeeper depends on the period being *very nearly* independent of amplitude, provided that the amplitude is small. Thus, as a pendulum clock runs down and the amplitude of the swings decreases a little, the clock still keeps very nearly correct time.

Example 13.8 A simple pendulum

Find the period and frequency of a simple pendulum 1.000 m long at a location where $g = 9.800\ \text{m/s}^2$.

SOLUTION

IDENTIFY and SET UP: We use Eq. (13.34) to determine the period T of the pendulum from its length, and Eq. (13.1) to find the frequency f from T.

EXECUTE: From Eqs. (13.34) and (13.1),

$$T = 2\pi\sqrt{\frac{L}{g}} = 2\pi\sqrt{\frac{1.000\ \text{m}}{9.800\ \text{m/s}^2}} = 2.007\ \text{s}$$

$$f = \frac{1}{T} = 0.4983\ \text{Hz}$$

EVALUATE: The period is almost exactly 2 s. In fact, when the metric system was first established, the second was defined as half the period of a one-meter pendulum. This wasn't a very good standard for time, however, because the value of g varies from place to place. We discussed more modern time standards in Section 1.3.

Test Your Understanding

Deposits of ore or oil affect the local value of g because their density differs from that of their surroundings. Suppose a 1.000-m simple pendulum has a period of exactly 2.000 s at a certain location. What is the local value of g?

13.6 | The Physical Pendulum

A **physical pendulum** is any *real* pendulum, using a body of finite size, as contrasted to the idealized model of the *simple* pendulum with all the mass concentrated at a single point. For small oscillations, analyzing the motion of a real, physical pendulum is almost as easy as for a simple pendulum. Figure 13.22 shows a body of irregular shape pivoted so that it can turn without friction about an axis through point O. In the equilibrium position the center of gravity is directly below the pivot; in the position shown in the figure, the body is displaced from equilibrium by an angle θ, which we use as a coordinate for the system. The distance from O to the center of gravity is d, the moment of inertia of the body about the axis of rotation through O is I, and the total mass is m. When the body is displaced as shown, the weight mg causes a restoring torque

$$\tau_z = -(mg)(d \sin \theta) \qquad (13.36)$$

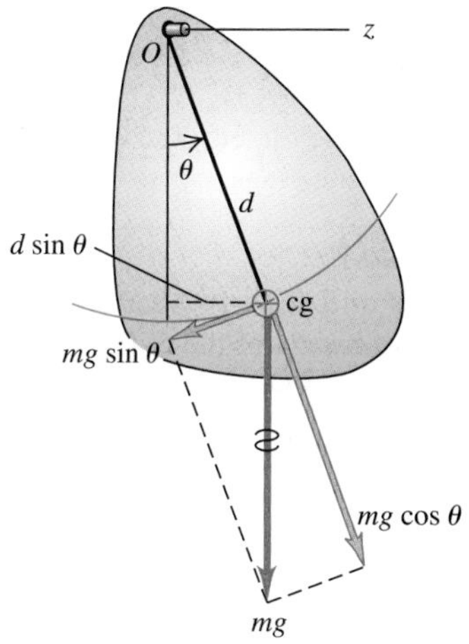

13.22 Dynamics of a physical pendulum.

The negative sign shows that the restoring torque is clockwise when the displacement is counterclockwise, and vice versa.

When the body is released, it oscillates about its equilibrium position. The motion is not simple harmonic because the torque τ_z is proportional to $\sin \theta$ rather than to θ itself. However, if θ is small, we can again approximate $\sin \theta$ by θ in radians, and the motion is *approximately* simple harmonic. With this approximation,

$$\tau_z = -(mgd)\theta$$

The equation of motion is $\sum \tau_z = I\alpha_z$, so

$$-(mgd)\theta = I\alpha_z = I\frac{d^2\theta}{dt^2}$$

$$\frac{d^2\theta}{dt^2} = -\frac{mgd}{I}\theta \qquad (13.37)$$

ActivPhysics ONLINE 9.12 Physical Pendulum

Comparing this with Eq. (13.4), we see that the role of (k/m) for the spring-mass system is played here by the quantity (mgd/I). Thus the angular frequency is given by

$$\omega = \sqrt{\frac{mgd}{I}} \quad \text{(physical pendulum, small amplitude)} \qquad (13.38)$$

The frequency f is $1/2\pi$ times this, and the period T is

$$T = 2\pi\sqrt{\frac{I}{mgd}} \quad \text{(physical pendulum, small amplitude)} \qquad (13.39)$$

Equation (13.39) is the basis of a common method for experimentally determining the moment of inertia of a body with a complicated shape. The center of gravity of the body is first located by balancing. The body is then suspended so that it is free to oscillate about an axis, and the period T of small-amplitude oscillations is measured. By using Eq. (13.39), the moment of inertia I of the body about this axis can then be calculated from T, the body's mass m, and the distance d from the axis to the center of gravity (see Exercise 13.50). Biomechanics researchers use this method to find the moments of inertia of an animal's limbs. This information is important for analyzing how an animal walks, as we'll see in the second of the two following examples.

Example 13.9 Physical pendulum versus simple pendulum

Suppose the body in Fig. 13.22 is a uniform rod with length L, pivoted at one end. Find the period of its motion.

SOLUTION

IDENTIFY and SET UP: We use Table 9.2 (Section 9.4) to find the moment of inertia of the rod, then substitute this value into Eq. (13.39) to determine the period of oscillation.

EXECUTE: From Table 9.2, the moment of inertia of a uniform rod about an axis through one end is $I = \frac{1}{3}ML^2$. The distance from the pivot to the center of gravity is $d = L/2$. From Eq. (13.39),

$$T = 2\pi\sqrt{\frac{\frac{1}{3}ML^2}{MgL/2}} = 2\pi\sqrt{\frac{2L}{3g}}$$

EVALUATE: If the rod is a meter stick $(L = 1.00\text{ m})$ and $g = 9.80\text{ m/s}^2$, then

$$T = 2\pi\sqrt{\frac{2(1.00\text{ m})}{3(9.80\text{ m/s}^2)}} = 1.64\text{ s}$$

The period is smaller by a factor of $\sqrt{2/3} = 0.816$ than the period of a simple pendulum with the same length, calculated in Example 13.8 (Section 13.5).

Example 13.10 *Tyrannosaurus rex* and the physical pendulum

All walking animals, including humans, have a natural walking pace, a number of steps per minute that is more comfortable than a faster or slower pace. Suppose this natural pace is equal to the period of the leg, viewed as a uniform rod pivoted at the hip joint. a) How does the natural walking pace depend on the length L of the leg, measured from hip to foot? b) Fossil evidence shows that *Tyrannosaurus rex*, a two-legged dinosaur that lived about 65 million years ago at the end of the Cretaceous period, had a leg length $L = 3.1$ m and a stride length (the distance from one footprint to the next print of the same foot) $S = 4.0$ m (Fig. 13.23). Estimate the walking speed of *Tyrannosaurus rex*.

SOLUTION

IDENTIFY and SET UP: We treat the leg as a physical pendulum, with a period of oscillation as found in Example 13.9. Our target variables are a) the relationship between the walking pace and the leg length and b) the walking speed of *T. rex*.

EXECUTE: a) From Example 13.9 the period of oscillation of the leg is $T = 2\pi\sqrt{2L/3g}$, which is proportional to $\sqrt{L}$. Each period (a complete back-and-forth swing of the leg) corresponds to *two* steps, so the walking pace in steps per unit time is just twice the oscilla-

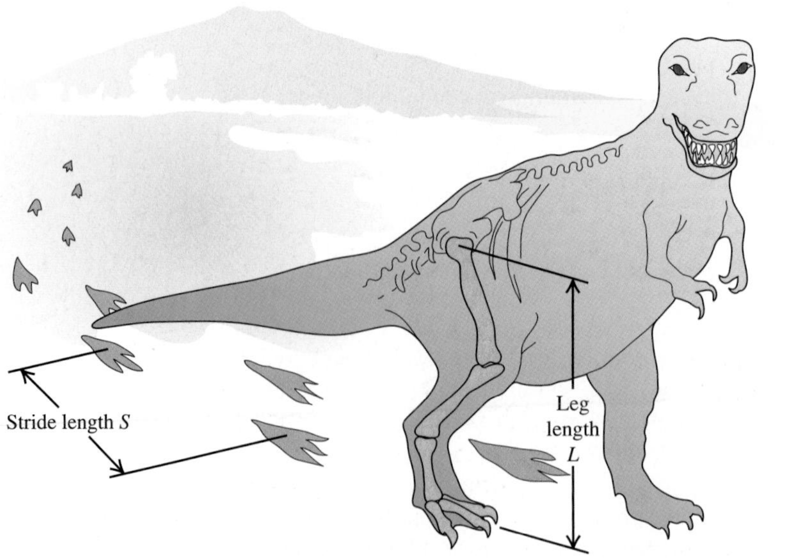

13.23 The walking speed of *Tyrannosaurus rex* can be estimated from its leg length L and its stride length S.

tion frequency $f = 1/T$. Hence the walking pace is proportional to $1/\sqrt{L}$. Animals with short legs (small values of L), such as mice or Chihuahuas, have rapid walking paces; humans, giraffes, and other animals with long legs (large values of L) walk at slower paces.
b) According to our model for the natural walking pace, the elapsed time for one stride of a walking *Tyrannosaurus rex* is

$$T = 2\pi\sqrt{\frac{2L}{3g}} = 2\pi\sqrt{\frac{2(3.1\text{ m})}{3(9.8\text{ m/s}^2)}} = 2.9\text{ s}$$

The distance moved during this time is the stride length S, so the walking speed is

$$v = \frac{S}{T} = \frac{4.0\text{ m}}{2.9\text{ s}} = 1.4\text{ m/s} = 5.0\text{ km/h} = 3.1\text{ mi/h}$$

This is about the same as a typical human walking speed!

EVALUATE: Our estimate must be somewhat in error because a uniform rod isn't a very good model for a leg. The legs of many animals, including *T. rex* as well as humans, are tapered; there is a lot more mass between the knee and the hip than between the knee and the foot. Thus the center of mass is less than $L/2$ from the hip; a reasonable guess would be about $L/4$. The moment of inertia is *considerably* less than $ML^2/3$, probably somewhere around $ML^2/15$. Try these numbers out with the analysis of Example 13.9; you'll get a shorter oscillation period and an even faster walking speed for *T. rex*.

Test Your Understanding

Suppose the body in Fig. 13.22 is a uniform rod of mass m and length L pivoted around its midpoint. The moment of inertia for this pivot point is $I = \frac{1}{12}mL^2$. What is the oscillation period? Interpret your result.

13.7 | Damped Oscillations

The idealized oscillating systems we have discussed so far are frictionless. There are no nonconservative forces, the total mechanical energy is constant, and a system set into motion continues oscillating forever with no decrease in amplitude.

Real-world systems always have some dissipative forces, however, and oscillations do die out with time unless we provide some means for replacing the dissipated mechanical energy (Fig 13.24). A mechanical pendulum clock continues to run because potential energy stored in the spring or a hanging weight system replaces the mechanical energy lost due to friction in the pivot and the gears. But eventually the spring runs down or the weights reach the bottom of their travel. Then no more energy is available, and the pendulum swings decrease in amplitude and stop.

The decrease in amplitude caused by dissipative forces is called **damping,** and the corresponding motion is called **damped oscillation.** The simplest case to analyze in detail is a simple harmonic oscillator with a frictional damping force that is directly proportional to the *velocity* of the oscillating body. This behavior occurs in friction involving viscous fluid flow, such as in shock absorbers or sliding between oil-lubricated surfaces. We then have an additional force on the body due to friction, $F_x = -bv_x$, where $v_x = dx/dt$ is the velocity and b is a constant that describes the strength of the damping force. The negative sign shows that the force is always opposite in direction to the velocity. The *net* force on the body is then

$$\sum F_x = -kx - bv_x \tag{13.40}$$

and Newton's second law for the system is

$$-kx - bv_x = ma_x \quad \text{or} \quad -kx - b\frac{dx}{dt} = m\frac{d^2x}{dt^2} \tag{13.41}$$

13.24 A swinging bell left to itself will eventually stop oscillating due to damping forces (air resistance and friction at the point of suspension).

Equation (13.41) is a differential equation for x; it would be the same as Eq. (13.4), the equation for the acceleration in SHM, except for the added term $-b\,dx/dt$. Solving this equation is a straightforward problem in differential equations, but we won't go into the details here. If the damping force is relatively small, the motion is described by

$$x = Ae^{-(b/2m)t}\cos(\omega' t + \phi) \quad \text{(oscillator with little damping)} \quad (13.42)$$

The angular frequency of oscillation ω' is given by

$$\omega' = \sqrt{\frac{k}{m} - \frac{b^2}{4m^2}} \quad \text{(oscillator with little damping)} \quad (13.43)$$

You can verify that Eq. (13.42) is a solution of Eq. (13.41) by calculating the first and second derivatives of x, substituting them into Eq. (13.41), and checking whether the left and right sides are equal. This is a straightforward but slightly tedious procedure.

The motion described by Eq. (13.42) differs from the undamped case in two ways. First, the amplitude $Ae^{-(b/2m)t}$ is not constant but decreases with time because of the decreasing exponential factor $e^{-(b/2m)t}$. Figure 13.25 is a graph of Eq. (13.42) for the case $\phi = 0$; it shows that the larger the value of b, the more quickly the amplitude decreases.

Second, the angular frequency ω', given by Eq. (13.43), is no longer equal to $\omega = \sqrt{k/m}$ but is somewhat smaller. It becomes zero when b becomes so large that

$$\frac{k}{m} - \frac{b^2}{4m^2} = 0 \quad \text{or} \quad b = 2\sqrt{km} \quad (13.44)$$

When Eq. (13.44) is satisfied, the condition is called **critical damping.** The system no longer oscillates but returns to its equilibrium position without oscillation when it is displaced and released.

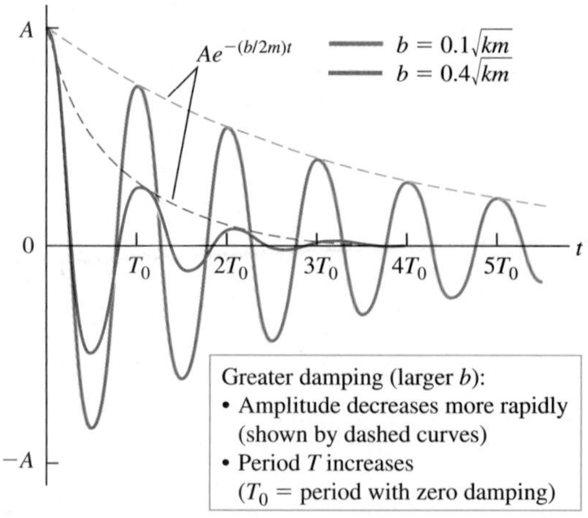

13.25 Graph of displacement versus time for an oscillator with little damping [Eq. (13.42)] with phase angle $\phi = 0$. Graphs are shown for two values of the damping constant b.

If b is greater than $2\sqrt{km}$, the condition is called **overdamping.** Again there is no oscillation, but the system returns to equilibrium more slowly than with critical damping. For the overdamped case the solutions of Eq. (13.41) have the form

$$x = C_1 e^{-a_1 t} + C_2 e^{-a_2 t}$$

where C_1 and C_2 are constants that depend on the initial conditions and a_1 and a_2 are constants determined by m, k, and b.

When b is less than the critical value, as in Eq. (13.42), the condition is called **underdamping.** The system oscillates with steadily decreasing amplitude.

In a vibrating tuning fork or guitar string, it is usually desirable to have as little damping as possible. By contrast, damping plays a beneficial role in the oscillations of an automobile's suspension system. The shock absorbers provide a velocity-dependent damping force so that when the car goes over a bump, it doesn't continue bouncing forever (Fig. 13.26). For optimal passenger comfort, the system should be critically damped or slightly underdamped. As the shocks get old and worn, the value of b decreases and the bouncing persists longer. Not only is this nauseating, but it is also bad for steering because the front wheels have less positive contact with the ground. Thus damping is an advantage in this system. Too much damping would be counterproductive, however; if b is excessively large, the system is overdamped and the suspension returns to equilibrium more slowly. If the suspension is overdamped and the car hits a second bump just after the first one, the springs in the suspension will still be compressed somewhat from the first bump and will be unable to fully absorb the impact.

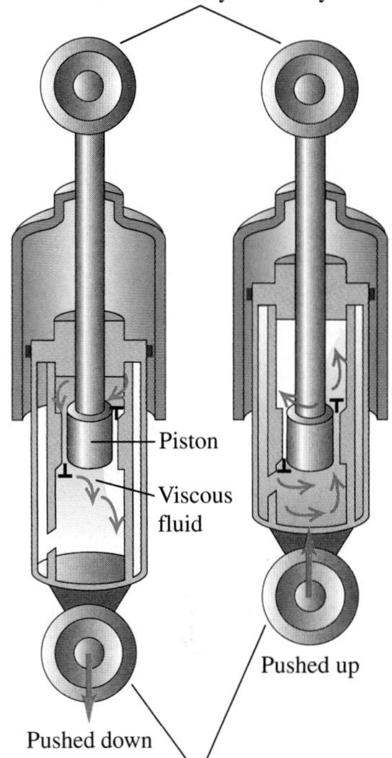

13.26 An automobile shock absorber. The viscous fluid causes a damping force that depends on the relative velocity of the two ends of the unit. This helps control wheel bounce and jounce.

In damped oscillations the damping force is nonconservative; the mechanical energy of the system is not constant but decreases continuously, approaching zero after a long time. To derive an expression for the rate of change of energy, we first write an expression for the total mechanical energy E at any instant:

$$E = \frac{1}{2}mv_x^2 + \frac{1}{2}kx^2$$

To find the rate of change of this quantity, we take its time derivative:

$$\frac{dE}{dt} = mv_x\frac{dv_x}{dt} + kx\frac{dx}{dt}$$

But $dv_x/dt = a_x$ and $dx/dt = v_x$, so

$$\frac{dE}{dt} = v_x(ma_x + kx)$$

From Eq. (13.41), $ma_x + kx = -b\,dx/dt = -bv_x$, so

$$\frac{dE}{dt} = v_x(-bv_x) = -bv_x^2 \qquad \text{(damped oscillations)} \qquad (13.45)$$

The right side of Eq. (13.45) is always negative, whether v is positive or negative. This shows that E continuously decreases, though not at a uniform rate. The term $-bv_x^2 = (-bv_x)v_x$ (force times velocity) is the rate at which the damping force does (negative) work on the system (that is, the damping *power*). This equals the rate of change of the total mechanical energy of the system.

Similar behavior occurs in electric circuits containing inductance, capacitance, and resistance. There is a natural frequency of oscillation, and the resistance plays the role of the damping constant b. In such circuits it is usually desirable to minimize damping, but damping can never be prevented completely. We will study these circuits in detail in Chapters 30 and 31.

Test Your Understanding

Consider again the mass-spring combination in Example 13.2 (Section 13.2), with $k = 200$ N/m and $m = 0.50$ kg. With no damping, this system has an angular frequency $\omega = 20$ rad/s. What is the angular frequency if there is damping with $b = 10$ kg/s? For what value of b is there critical damping?

13.8 | Forced Oscillations and Resonance

A damped oscillator left to itself will eventually stop moving altogether. But we can maintain a constant-amplitude oscillation by applying a force that varies with time in a periodic or cyclic way, with a definite period and frequency. As an example, consider your cousin Throckmorton on a playground swing. You can keep him swinging with constant amplitude by giving him a little push once each cycle. We call this additional force a **driving force.**

If we apply a periodically varying driving force with angular frequency ω_d to a damped harmonic oscillator, the motion that results is called a **forced oscillation** or a *driven oscillation.* It is different from the motion that occurs when the system is simply displaced from equilibrium and then left alone, in which case the system oscillates with a **natural angular frequency** ω' determined by m, k, and b, as in Eq. (13.43). In a forced oscillation, however, the angular frequency with which the mass oscillates is equal to the driving angular frequency ω_d. This does *not* have to be equal to the angular frequency ω' with which the system would oscillate without a driving force. If you grab the ropes of Throckmorton's swing, you can force the swing to oscillate with any frequency you like.

Suppose we force the oscillator to vibrate with an angular frequency ω_d that is nearly *equal* to the angular frequency ω' it would have with no driving force. What happens? The oscillator is naturally disposed to oscillate at $\omega = \omega'$, so we expect the amplitude of the resulting oscillation to be larger than when the two frequencies are very different. Detailed analysis and experiment shows that this is just what happens. The easiest case to analyze is a *sinusoidally* varying force, say, $F(t) = F_{max}\cos\omega_d t$. If we vary the frequency ω_d of the driving force, the amplitude of the resulting forced oscillation varies in an interesting way (Fig. 13.27). When there is very little damping (small b), the amplitude goes through a sharp peak as the driving angular frequency ω_d nears the natural oscillation angular frequency ω'. When the damping is increased (larger b), the peak becomes broader and smaller in height and shifts toward lower frequencies.

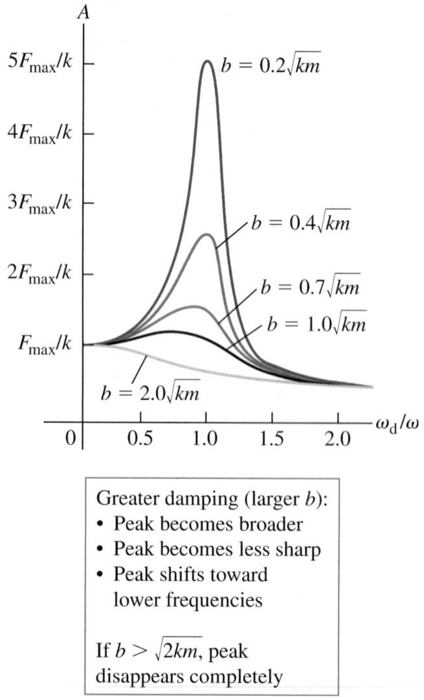

13.27 Graph of the amplitude A of forced oscillation as a function of the angular frequency ω_d of the driving force. The horizontal axis shows the ratio of ω_d to the angular frequency $\omega = \sqrt{k/m}$ of an undamped oscillator. Each curve has a different value of the damping constant b.

We could work out an expression that shows how the amplitude A of the forced oscillation depends on the frequency of a sinusoidal driving force, with maximum value F_{max}. That would involve more differential equations than we're ready for, but here is the result:

$$A = \frac{F_{max}}{\sqrt{(k - m\omega_d^2)^2 + b^2\omega_d^2}} \quad \text{(amplitude of a driven oscillator)} \qquad (13.46)$$

When $k - m\omega_d^2 = 0$, the first term under the radical is zero, so A has a maximum near $\omega_d = \sqrt{k/m}$. The height of the curve at this point is proportional to $1/b$; the less damping, the higher the peak. At the low-frequency extreme, when $\omega_d = 0$, we get $A = F_{max}/k$. This corresponds to a *constant* force F_{max} and a constant displacement $A = F_{max}/k$ from equilibrium, as we might expect.

The fact that there is an amplitude peak at driving frequencies close to the natural frequency of the system is called **resonance.** Physics is full of examples of resonance; building up the oscillations of a child on a swing by pushing with a frequency equal to the swing's natural frequency is one. A vibrating rattle in a car that occurs only at a certain engine speed or wheel-rotation speed is an all-too-familiar example. Inexpensive loudspeakers often have an annoying boom or buzz when a musical note happens to coincide with the resonant frequency of the speaker cone or the speaker housing. A tuned circuit in a radio or television receiver responds strongly to waves having frequencies near its resonant frequency, and this fact is used to select a particular station and reject the others. We will study resonance in electric circuits in detail in Chapter 31.

Resonance in mechanical systems can be destructive. A company of soldiers once destroyed a bridge by marching across it in step; the frequency of their steps was close to a natural vibration frequency of the bridge, and the resulting oscillation had large enough amplitude to tear the bridge apart. Ever since, marching soldiers have been ordered to break step before crossing a bridge. Some years ago, vibrations of the engines of a particular airplane had just the right frequency to resonate with the natural frequencies of its wings. Large oscillations built up, and occasionally the wings fell off.

Nearly everyone has seen the film of the collapse of the Tacoma Narrows suspension bridge in 1940 (Fig. 13.28). This is usually cited as an example of resonance driven by the wind, but there's some doubt as to whether it should be called that. The wind didn't have to vary *periodically* with a frequency close to a natural frequency of the bridge. The air flow past the bridge was turbulent, and vortices were formed in the air with a regular frequency that depended on the flow speed. It is conceivable that this frequency may have coincided with a natural frequency of the bridge. But the cause may well have been something more subtle called a *self-excited oscillation,* in which the aerodynamic forces caused by a *steady* wind blowing on the bridge tended to displace it farther from equilibrium at times when it was already moving away from equilibrium. It is as though we had a damping force such as the $-bv_x$ term in Eq. (13.40) but with the sign reversed. Instead of draining mechanical energy away from the system, this anti-damping force pumps energy into the system, building up the oscillations to destructive amplitudes. The approximate differential equation is Eq. (13.41) with the sign of the b term reversed, and the oscillating solution is Eq. (13.42) with a *positive* sign in the exponent. You can see that we're headed for trouble. Engineers have learned how to stabilize suspension bridges, both structurally and aerodynamically, to prevent such disasters.

13.28 The Tacoma Narrows Bridge collapsed four months and six days after it was opened for traffic. The main span was 2800 ft long and 39 ft wide, with 8-ft-high steel stiffening girders on both sides. The maximum amplitude of the torsional vibrations was 35°; the frequency was about 0.2 Hz.

Test Your Understanding

Earlier we described how a certain airplane had an unwanted resonance between vibrations of the engines and vibrations of the wings. This problem was fixed by stiffening the wing structure. Use the ideas of resonance to explain how this solution worked.

SUMMARY

Periodic motion is motion that repeats itself in a definite cycle. It occurs whenever a body has a stable equilibrium position and a restoring force that acts when it is displaced from equilibrium. Period T is the time for one cycle. Frequency f is the number of cycles per unit time. Angular frequency ω is 2π times the frequency. (See Example 13.1)

$$f = \frac{1}{T} \qquad T = \frac{1}{f} \tag{13.1}$$

$$\omega = 2\pi f = \frac{2\pi}{T} \tag{13.2}$$

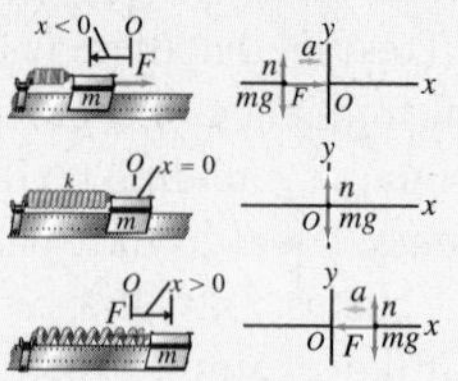

If the net force is a restoring force F_x that is directly proportional to the displacement x, the motion is called simple harmonic motion (SHM). In many cases this condition is satisfied if the displacement from equilibrium is small.

$$F_x = -kx \tag{13.3}$$

$$a_x = \frac{F_x}{m} = -\frac{k}{m}x \tag{13.4}$$

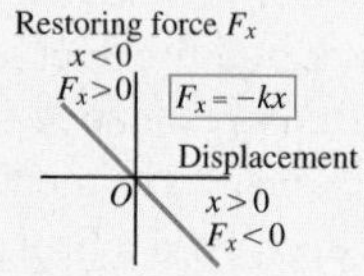

The circle of reference construction uses a rotating vector called a phasor, having a length equal to the amplitude of the motion. Its projection on the horizontal axis represents the actual motion of a body in simple harmonic motion.

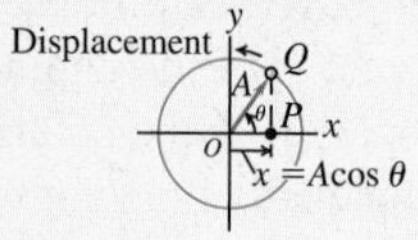

The angular frequency, frequency, and period in SHM do not depend on the amplitude, but only on the mass m and force constant k. (See Examples 13.2, 13.6, and 13.7)

$$\omega = \sqrt{\frac{k}{m}} \tag{13.10}$$

$$f = \frac{\omega}{2\pi} = \frac{1}{2\pi}\sqrt{\frac{k}{m}} \tag{13.11}$$

$$T = \frac{1}{f} = 2\pi\sqrt{\frac{m}{k}} \tag{13.12}$$

In SHM, the displacement, velocity, and acceleration are sinusoidal functions of time. The angular frequency is $\omega = \sqrt{k/m}$; the amplitude A and phase angle ϕ are determined by the initial position and velocity of the body. (See Example 13.3)

$$x = A\cos(\omega t + \phi) \tag{13.13}$$

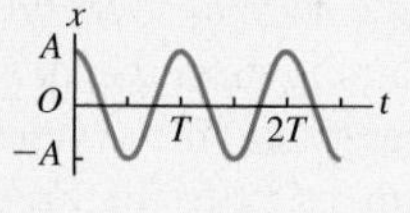

Energy is conserved in SHM. The total energy can be expressed in terms of the force constant k and amplitude A. (See Examples 13.4 and 13.5)

$$E = \frac{1}{2}mv_x^2 + \frac{1}{2}kx^2 = \frac{1}{2}kA^2 = \text{constant} \tag{13.21}$$

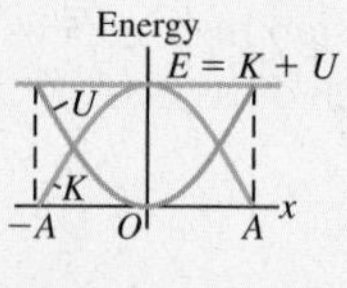

In angular simple harmonic motion, the frequency and angular frequency are related to the moment of inertia I and the torsion constant κ.

$$\omega = \sqrt{\frac{\kappa}{I}} \quad \text{and}$$

$$f = \frac{1}{2\pi}\sqrt{\frac{\kappa}{I}} \tag{13.24}$$

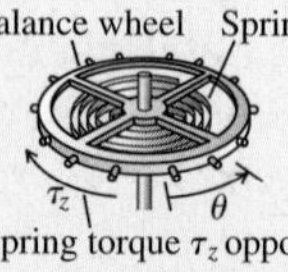

A simple pendulum consists of a point mass m at the end of a massless string of length L. Its motion is approximately simple harmonic for sufficiently small amplitude; the angular frequency, frequency, and period then depend only on g and L, not on the mass or amplitude. (See Examples 13.8)

$$\omega = \sqrt{\frac{g}{L}} \qquad (13.32)$$

$$f = \frac{\omega}{2\pi} = \frac{1}{2\pi}\sqrt{\frac{g}{L}} \qquad (13.33)$$

$$T = \frac{2\pi}{\omega} = \frac{1}{f} = 2\pi\sqrt{\frac{L}{g}} \qquad (13.34)$$

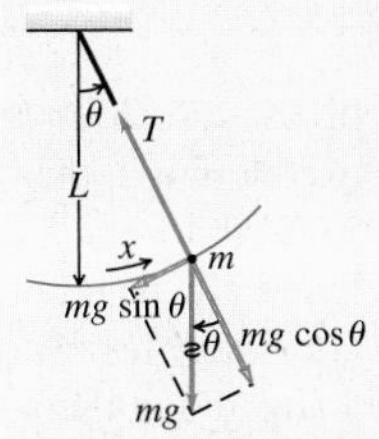

A physical pendulum is a body suspended from an axis of rotation a distance d from its center of gravity. If the moment of inertia about the axis of rotation is I, the angular frequency and period for small-amplitude oscillations are independent of amplitude. (See Examples 13.9 and 13.10)

$$\omega = \sqrt{\frac{mgd}{I}} \qquad (13.38)$$

$$T = 2\pi\sqrt{\frac{I}{mgd}} \qquad (13.39)$$

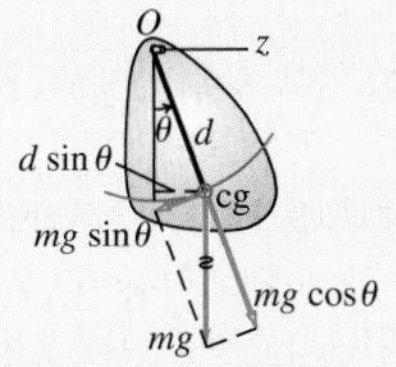

When a damping force $F_x = -bv_x$ proportional to velocity is added to a simple harmonic oscillator, the motion is called a damped oscillation. For a relatively small damping force, the motion is sinusoidal with a decaying amplitude and an angular frequency ω' that is lower than it would be without damping. This situation occurs when $b < 2\sqrt{km}$, a condition called underdamping. When $b = 2\sqrt{km}$, the system is critically damped and no longer oscillates. When $b > 2\sqrt{km}$, the system is overdamped.

$$x = Ae^{-(b/2m)t}\cos\omega' t \qquad (13.42)$$

$$\omega' = \sqrt{\frac{k}{m} - \frac{b^2}{4m^2}} \qquad (13.43)$$

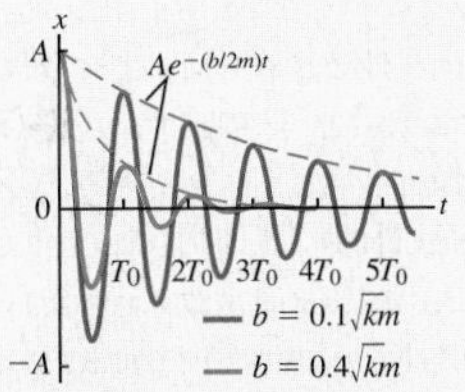

When a sinusoidally varying driving force is added to a damped harmonic oscillator, the resulting motion is called a forced oscillation. The amplitude is a function of driving frequency ω_d and reaches a peak at a driving frequency close to the natural frequency of the system. This behavior is called resonance.

$$A = \frac{F_{max}}{\sqrt{(k - m\omega_d^2)^2 + b^2\omega_d^2}} \qquad (13.46)$$

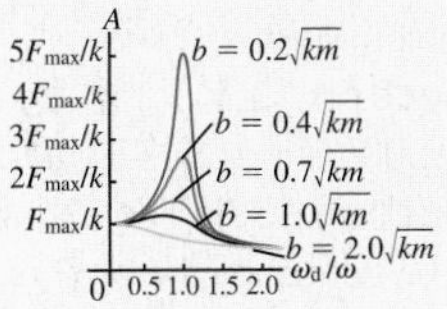

Key Terms

periodic motion (oscillation), 476
displacement, 477
restoring force, 477
amplitude, 477
cycle, 477
period, 477
frequency, 477
angular frequency, 478
simple harmonic motion (SHM), 479
harmonic oscillator, 479
circle of reference, 480
phasor, 480
phase angle, 483
simple pendulum, 495
physical pendulum, 496
damping, 499
damped oscillation, 499
critical damping, 500
overdamping, 501
underdamping, 501
driving force, 502
forced oscillation, 502
natural angular frequency, 502
resonance, 503

Your Notes

Answer to Chapter Opening Question

Neither—the clock would still keep time correctly. If its rod has negligible mass, then the pendulum is a simple pendulum and its period is independent of the mass [Eq. (13.34)]. If the rod's mass is included, the pendulum is a physical pendulum. Doubling its mass m also doubles its moment of inertia I, so the ratio I/m is unchanged and the period $T = 2\pi\sqrt{I/mgd}$ [Eq. (13.39)] remains the same.

Answers to Test Your Understanding Questions

Section 13.1 The amplitude is $A = 6.0$ cm (the maximum displacement from equilibrium). The period is $T = 5.00$ s, the frequency is $f = 1/T = 1/(5.00\text{ s}) = 0.200$ Hz, and the angular frequency is $\omega = 2\pi f = 2\pi(0.200\text{ Hz}) = 1.26$ rad/s.
Section 13.2 Comparing with Eq. (13.13), $x = A\cos(\omega t + \phi)$, we see that the amplitude $A = 0.050$ m, the phase angle $\phi = 2.5$ rad, and the angular frequency $\omega = 290$ rad/s. From Eq. (13.2), the period is $T = 2\pi/\omega = (2\pi\text{ rad})/(290\text{ rad/s}) = 0.022$ s. At $t = 0$ the position is $x_0 = (0.050\text{ m})\cos(2.5\text{ rad}) = -0.040$ m.
Section 13.3 To increase the energy $E = \frac{1}{2}kA^2$ by a factor of 2, the amplitude A must increase by a factor of $\sqrt{2}$. Because the motion is SHM, changing the amplitude has no effect on the frequency.
Section 13.4 From Eq. (13.4), $a_x = -(k/m)x$. With $k = 0.829$ N/m, $m = 6.63 \times 10^{-26}$ kg, and an initial displacement of 1.00×10^{-11} m, the magnitude of the acceleration is $[(0.829\text{ N/m})/(6.63 \times 10^{-26}\text{ kg})] \times (1.00 \times 10^{-11}\text{ m}) = 1.25 \times 10^{14}\text{ m/s}^2$. Microscopic objects can undergo truly immense accelerations!
Section 13.5 From Eq. (13.34), $g = 4\pi^2 L/T^2 = 4\pi^2(1.000\text{ m})/(2.000\text{ s})^2 = 9.870\text{ m/s}^2$.
Section 13.6 The distance from the pivot to the center of gravity (the center of the rod) is $d = 0$. Hence the period $T = 2\pi\sqrt{I/mgd}$ is infinite. In other words, when it is displaced, the rod takes an infinite amount of time to move back. The rod is in equilibrium for *any* value of the angle θ in Fig. 13.22, so there is no restoring torque and no tendency for the rod to move.
Section 13.7 From Eq. (13.43), the angular frequency with $b = 10$ kg/s is

$$\omega' = \sqrt{\frac{k}{m} - \frac{b^2}{4m^2}}$$

$$= \sqrt{\frac{200\text{ N/m}}{0.50\text{ kg}} - \frac{(10\text{ kg/s})^2}{4(0.50\text{ kg})^2}} = 17\text{ rad/s}$$

which is less than with zero damping. Critical damping requires

$$b = 2\sqrt{km} = 2\sqrt{(200\text{ N/m})(0.50\text{ kg})} = 20\text{ kg/s}$$

Section 13.8 The stiffened wing had a higher force constant k and hence a higher natural frequency. As a result, its natural frequency no longer matched the driving frequency of the engines, and resonance no longer occurred.

Discussion Questions

Q13.1 An object is moving with SHM of amplitude A on the end of a spring. If the amplitude is doubled, what happens to the total distance the object travels in one period? What happens to the period? What happens to the maximum speed of the object? Discuss how these answers are related.
Q13.2 Think of several examples in everyday life of motion that are at least, approximately simple harmonic. In what respects does each differ from SHM?
Q13.3 Does a tuning fork or similar tuning instrument undergo simple harmonic motion? Why is this a crucial question for musicians?
Q13.4 A highly elastic "superball" bouncing on a hard floor has a motion that is approximately periodic. In what ways is the motion similar to SHM? In what ways is it different?
Q13.5 If a uniform spring is cut in half, what is the force constant of each half? Justify your answer. How would the frequency of SHM using a half-spring differ from that using the same mass and the entire spring?
Q13.6 The analysis of SHM in this chapter ignored the mass of the spring. How does the spring's mass change the characteristics of the motion?
Q13.7 Two identical gliders on an air track are connected by an ideal spring. Could such a system undergo SHM? Explain. How would the period compare with that of a single glider attached to a spring whose other end is rigidly attached to a stationary object? Explain.
Q13.8 You are captured by aliens, taken into their ship, and drugged. You awake some time later and find yourself locked in a small room with no windows. All they have left you with is your digital watch, your school ring, and your long silver-chain necklace. Explain how you can determine whether or not you are still on earth or if you have been transported to Mars.
Q13.9 The system shown in Fig. 13.16 is mounted in an elevator, which accelerates upward with constant acceleration. Does the period increase, decrease, or remain the same? Justify your answer.
Q13.10 If a pendulum has a period of 2.5 s on earth, what would be its period in a space station orbiting the earth? If a mass hung from a vertical spring has a period of 5.0 s on earth, what would be its period in the space station? Justify each of your answers.
Q13.11 A pendulum is mounted in an elevator that accelerates upward with constant acceleration. Does the period increase, decrease, or remain the same? Explain your answer.
Q13.12 What should you do to the length of the string of a simple pendulum to a) double its frequency? b) double its period? c) double its angular frequency?
Q13.13 If a pendulum clock is taken to a mountaintop, does it gain or lose time, assuming it is correct at a lower elevation? Explain your answer.
Q13.14 When the amplitude of a simple pendulum increases, should its period increase or decrease? Give a qualitative argument; do not rely on Eq. (13.35). Is your argument also valid for a physical pendulum?
Q13.15 Why do short dogs (like Chihuahuas) walk with quicker strides than do tall dogs (like Great Danes)?

Q13.16 At what point in the motion of a simple pendulum is the string tension greatest? Least? In each case give the reasoning behind your answer.

Q13.17 Could a standard of time be based on the period of a certain standard pendulum? What advantages and disadvantages would such a standard have compared to the actual present-day standard discussed in Section 1.3?

Q13.18 The frequency with which dogs pant is the natural frequency of their respiratory system. Why do they choose this frequency?

Q13.19 In designing structures in an earthquake-prone region, how should the natural frequencies of oscillation of a structure relate to typical earthquake frequencies? Why? Should the structure have a large or small amount of damping?

Exercises

Section 13.1 Describing Oscillation

13.1 A piano string sounds a middle A by vibrating primarily at 220 Hz. a) Calculate its period and angular frequency. b) Calculate the period and angular frequency for a soprano singing a "high A," two octaves up, which is four times the frequency of the piano string.

13.2 If an object on a horizontal frictionless surface is attached to a spring, displaced, and then released, it will oscillate. If it is displaced 0.120 m from its equilibrium position and released with zero initial speed, then after 0.800 s its displacement is found to be 0.120 m on the opposite side, and it has passed the equilibrium position once during this interval. Find a) the amplitude; b) the period; c) the frequency.

13.3 The tip of a tuning fork goes through 440 complete vibrations in 0.500 s. Find the angular frequency and the period of the motion.

13.4 The displacement of an oscillating object as a function of time is shown in Fig. 13.29. What is a) the frequency; b) the amplitude; c) the period?

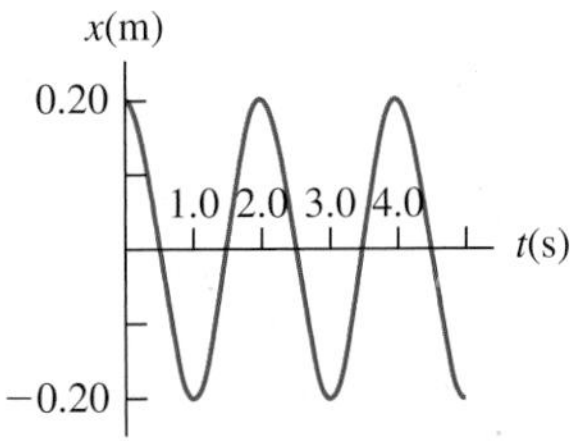

Figure 13.29 Exercise 13.4.

Section 13.2 Simple Harmonic Motion

13.5 A machine part is undergoing SHM with a frequency of 5.00 Hz and amplitude 1.80 cm. How long does it take the part to go from $x = 0$ to $x = -1.80$ cm?

13.6 In a physics lab, you attach a 0.200-kg air-track glider to the end of an ideal spring of negligible mass and start it oscillating. The elapsed time from when the glider first moves through the equilibrium point to the second time it moves through that point is 2.60 s. Find the spring's force constant.

13.7 A body of unknown mass is attached to an ideal spring with force constant 120 N/m. It is found to vibrate with a frequency of 6.00 Hz. Find a) the period; b) the angular frequency; c) the mass of the body.

13.8 A harmonic oscillator is made by using a 0.600-kg frictionless block and an ideal spring of unknown force constant. The oscillator is found to have a period of 0.150 s. Find the force constant of the spring.

13.9 A harmonic oscillator has mass 0.500 kg and an ideal spring with force constant 140 N/m. Find a) the period; b) the frequency; c) the angular frequency.

13.10 Substitute the following equations, in which A, ω, and β are constant, into Eq. (13.4) to see if they describe SHM. If so, what must ω equal? a) $x = A \sin(\omega t + \beta)$. b) $x = A\omega t^2 + \beta$. c) $x = Ae^{i(\omega t + \beta)}$, where $i = \sqrt{-1}$.

13.11 **Jerk.** A guitar string vibrates at a frequency of 440 Hz. A point at its center moves in SHM with an amplitude of 3.0 mm and a phase angle of zero. a) Write an equation for the position of the center of the string as a function of time. b) What are the maximum values of the magnitudes of the velocity and acceleration of the center of the string? c) The derivative of the acceleration with respect to time is a quantity called the *jerk*. Write an equation for the jerk of the center of the string as a function of time, and find the maximum value of the magnitude of the jerk.

13.12 A 2.00-kg frictionless block is attached to an ideal spring with force constant 300 N/m. At $t = 0$ the spring is neither stretched nor compressed and the block is moving in the negative direction at 12.0 m/s. Find a) the amplitude and b) the phase angle. c) Write an equation for the position as a function of time.

13.13 Repeat Exercise 13.12, but assume that at $t = 0$ the block has velocity -4.00 m/s and displacement $+0.200$ m.

13.14 The point of the needle of a sewing machine moves in SHM along the x-axis with a frequency of 2.5 Hz. At $t = 0$ its position and velocity components are $+1.1$ cm and -15 cm/s. a) Find the acceleration component of the needle at $t = 0$. b) Write equations giving the position, velocity, and acceleration components of the point as a function of time.

13.15 An object is undergoing simple harmonic motion with period 1.200 s and amplitude 0.600 m. At $t = 0$ the object is at $x = 0$. How far is the object from the equilibrium position when $t = 0.480$ s?

13.16 A 42.5-kg chair is attached to a spring and allowed to oscillate. When the chair is empty the chair takes 1.30 s to make one complete vibration. But with a person sitting in it, with her feet off the floor, the chair now takes 2.54 s for one cycle. What is the mass of the person?

13.17 A 0.400-kg object undergoing SHM has $a_x = -2.70$ m/s^2, when $x = 0.300$ m. What is the time for one oscillation?

13.18 A 0.500-kg mass on a spring has velocity as a function of time given by $v_x(t) = (3.60\ \text{cm/s}) \sin[(4.71\ \text{s}^{-1})t - \pi/2]$. What is a) the period? b) the amplitude? c) the maximum acceleration of the mass?

13.19 A 1.50-kg mass on a spring has displacement as a function of time given by the equation $x(t) = (7.40\ \text{cm}) \cos[(4.16\ \text{s}^{-1})t - 2.42]$. Find a) the time for one complete vibration; b) the force constant of the spring; c) the maximum speed of the mass; d) the maximum force on the mass; e) the position, speed, and acceleration of the mass at $t = 1.00$ s, and the force on the mass at that time.

13.20 An object is undergoing simple harmonic motion with period 0.300 s and amplitude 6.00 cm. At $t = 0$ the object is instantaneously at rest at $x = 6.00$ cm. Calculate the time it takes the object to go from $x = 6.00$ cm to $x = -1.50$ cm.

Section 13.3 Energy in Simple Harmonic Motion

13.21 Derive Eq. (13.19) a) from Eqs. (13.14) and (13.17); b) from conservation of energy, Eq. (13.21).

13.22 A tuning fork labeled 392 Hz has the tip of each of its two prongs vibrating with an amplitude of 0.600 mm. a) What is the maximum speed of the tip of a prong? b) A housefly (*Musca domestica*) with mass 0.0270 g is holding on to the tip of one of the prongs. As the prong vibrates, what is the fly's maximum kinetic energy? Assume that the fly's mass has a negligible effect on the frequency of oscillation.

13.23 A harmonic oscillator has angular frequency ω and amplitude A. a) What are the magnitudes of the displacement and velocity when the elastic potential energy is equal to the kinetic energy? (Assume that $U = 0$ at equilibrium.) b) How often does this occur in each cycle? What is the time between occurrences? c) At an instant when the displacement is equal to $A/2$, what fraction of the total energy of the system is kinetic and what fraction is potential?

13.24 A 0.500-kg glider, attached to the end of an ideal spring with force constant $k = 450$ N/m, undergoes simple harmonic motion with an amplitude 0.040 m. Compute a) the maximum speed of the glider; b) the speed of the glider when it is at $x = -0.015$ m; c) the magnitude of the maximum acceleration of the glider; d) the acceleration of the glider at $x = -0.015$ m; e) the total mechanical energy of the glider at any point in its motion.

13.25 A cheerleader waves her pom-pom in SHM with an amplitude of 18.0 cm and a frequency of 0.850 Hz. Find a) the maximum magnitude of the acceleration and of the velocity; b) the acceleration and speed when the pom-pom's coordinate is $x = +9.0$ cm; c) the time required to move from the equilibrium position directly to a point 12.0 cm away. d) Which of the quantities asked for in parts (a), (b), and (c) can be found using the energy approach used in Section 13.3, and which cannot? Explain.

13.26 For the situation described in part (a) of Example 13.5, what should be the value of the putty mass m so that the amplitude after the collision is one-half the original amplitude? For this value of m, what fraction of the original mechanical energy is converted into heat?

13.27 A 0.150-kg toy is undergoing SHM on the end of a horizontal spring with force constant $k = 300$ N/m. When the object is 0.0120 m from its equilibrium position, it is observed to have a speed of 0.300 m/s. What is a) the total energy of the object at any point of its motion? b) the amplitude of the motion? c) the maximum speed attained by the object during its motion?

13.28 You are watching an object that is moving in SHM. When the object is displaced 0.600 m to the right of its equilibrium position, it has a velocity of 2.20 m/s to the right and an acceleration of 8.40 m/s^2 to the left. How much farther from this point will the object move before it stops momentarily and then starts to move back to the left?

13.29 An object moves in SHM with a period of 0.500 s. The object's maximum acceleration is 6.40 m/s^2. What is its maximum speed?

Section 13.4 Applications of Simple Harmonic Motion

13.30 The scale of a spring balance reading from zero to 200 N is 12.5 cm long. A fish hanging from the bottom of the spring oscillates vertically at 2.60 Hz. What is the mass of the fish? You can ignore the mass of the spring.

13.31 A proud deep-sea fisherman hangs a 65.0-kg fish from an ideal spring having negligible mass. The fish stretches the spring 0.120 m. a) What is the force constant of the spring? b) What is the period of oscillation of the fish if it is pulled down and released?

13.32 A thrill-seeking cat with mass 4.00 kg is attached by a harness to an ideal spring of negligible mass and oscillates vertically in SHM. The amplitude is 0.050 m, and at the highest point of the motion the spring has its natural unstretched length. Calculate the elastic potential energy of the spring (take it to be zero for the unstretched spring), the kinetic energy of the cat, the gravitational potential energy of the system relative to the lowest point of the motion, and the sum of these three energies when the cat is a) at its highest point; b) at its lowest point; c) at its equilibrium position.

13.33 A 2.00-kg block of Cheddar cheese hangs from an ideal spring of negligible mass. When displaced from equilibrium and released, the cheese oscillates with a period of 0.400 s. How much is the spring stretched when the cheese hangs in equilibrium (at rest)?

13.34 Your boss at the Cut-Rate Cuckoo Clock Company asks you what would happen to the frequency of the angular SHM of the balance wheel if it had the same density and the same coil spring (thus the same torsion constant), but all the balance wheel dimensions were made one-third as great to save material. a) What do you tell her? b) By what factor would the torsion constant need to be changed to make the smaller balance wheel oscillate at the original frequency?

13.35 A certain alarm clock ticks four times each second, each tick representing half a period. The balance wheel consists of a thin rim with radius 0.55 cm, connected to the balance staff by thin spokes of negligible mass. The total mass of the balance wheel is 0.90 g. a) What is the moment of inertia of the balance wheel about its shaft? b) What is the torsion constant of the hairspring?

13.36 A thin metal disk with mass 2.00×10^{-3} kg and radius 2.20 cm is attached at its center to a long fiber (Fig. 13.30). The disk, when twisted and released, oscillates with a period of 1.00 s. Find the torsion constant of the fiber.

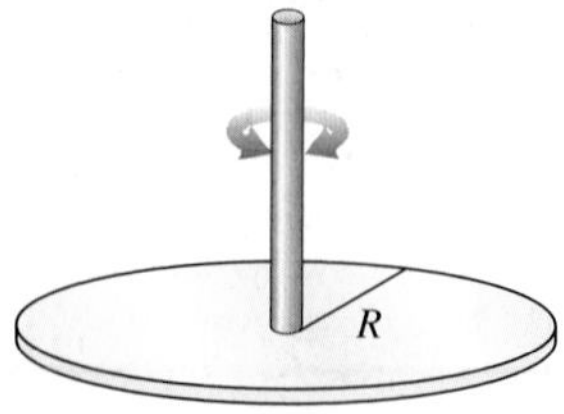

Figure 13.30 Exercise 13.36.

13.37 You want to find the moment of inertia of a complicated machine part about an axis through its center of mass. You suspend it from a wire along this axis. The wire has a torsion constant of 0.450 N·m/rad. You twist the part a small amount about this axis and let it go, timing 125 oscillations in 265 s. What is the moment of inertia you want to find?

13.38 The balance wheel of a watch vibrates with an angular amplitude Θ, angular frequency ω, and phase angle $\phi = 0$. a) Find expressions for the angular velocity $d\theta/dt$ and angular acceleration $d^2\theta/dt^2$ as functions of time. b) Find the balance wheel's angular velocity and angular acceleration when its angular displacement is Θ, and when its angular displacement is $\Theta/2$ and θ is decreasing. (*Hint:* Sketch a graph of θ versus t.)

***13.39** For the van der Waals interaction with potential energy function given by Eq. (13.25), show that when the magnitude of the displacement x from equilibrium $(r = R_0)$ is small, the potential energy can be written approximately as $U \approx \frac{1}{2}kx^2 - U_0$. [*Hint:* In Eq. (13.25), let $r = R_0 + x$ and $u = x/R_0$. Then approximate $(1 + u)^n$ by the first *three* terms in Eq. (13.28).] How does k in this equation compare with the force constant in Eq. (13.29) for the force?

***13.40** When displaced from equilibrium, the two hydrogen atoms in an H_2 molecule are acted on by a restoring force $F_x = -kx$ with $k = 580$ N/m. Calculate the oscillation frequency of the H_2 molecule. (*Hint:* The mass of a hydrogen atom is 1.008 atomic mass units, or 1 u; see Appendix E. As in Example 13.7 of Section 13.4, use $m/2$ instead of m in the expression for f.)

Section 13.5 The Simple Pendulum

13.41 A Pendulum on Mars. A certain simple pendulum has a period on the earth of 1.60 s. What is its period on the surface of Mars, where $g = 3.71\ \text{m/s}^2$?

13.42 You pull a simple pendulum of length 0.240 m to the side through an angle of 3.50° and release it. a) How much time does it take the pendulum bob to reach its highest speed? b) How much time does it take if the pendulum is relased at an angle of 1.75° instead of 3.50°?

13.43 An apple weighs 1.00 N. When you hang it from the end of a long spring of force constant 1.50 N/m and negligible mass, it bounces up and down in SHM. If you stop the bouncing and let the apple swing from side to side through a small angle, the frequency of this simple pendulum is half the bounce frequency. (Because the angle is small, the back and forth swings do not cause any appreciable change in the length of the spring.) What is the unstretched length of the spring?

13.44 A small sphere with mass m is attached to a massless rod of length L that is pivoted at the top, forming a simple pendulum. The pendulum is pulled to one side so that the rod is at an angle Θ from the vertical, and released from rest. a) In a diagram, show the pendulum just after it is released. On the diagram, draw vectors representing the *forces* acting on the small sphere and the *acceleration* of the sphere. Accuracy counts! At this point, what is the linear acceleration of the sphere? b) Repeat part (a) for the instant when the pendulum rod is at an angle $\Theta/2$ from the vertical. c) Repeat part (a) for the instant when the pendulum rod is vertical. At this point, what is the linear speed of the sphere?

13.45 After landing on an unfamiliar planet, a space explorer constructs a simple pendulum of length 50.0 cm. She finds that the pendulum makes 100 complete swings in 136 s. What is the value of g on this planet?

Section 13.6 The Physical Pendulum

13.46 We want to support a thin hoop by a horizontal nail and have the hoop make one complete small-angle oscillation each 2.0 s. What must the hoop's radius be?

13.47 Show that the expression for the period of a physical pendulum reduces to that of a simple pendulum if the physical pendulum consists of a particle with mass m on the end of a massless string of length L.

13.48 A 1.80-kg monkey wrench is pivoted 0.250 m from its center of mass and allowed to swing as a physical pendulum. The period for small-angle oscillations is 0.940 s. a) What is the moment of inertia of the wrench about an axis through the pivot? b) If the wrench is initially displaced 0.400 rad from its equilibrium position, what is the angular speed of the wrench as it passes through the equilibrium position?

13.49 A holiday ornament in the shape of a hollow sphere with mass $M = 0.015$ kg and radius $R = 0.050$ m is hung from a tree limb by a small loop of wire attached to the surface of the sphere. If the ornament is displaced a small distance and released, it swings back and forth as a physical pendulum. Calculate its period. (You can ignore friction at the pivot. The moment of inertia of the sphere about the pivot at the tree limb is $5MR^2/3$.)

13.50 A 1.80-kg connecting rod from a car engine is pivoted about a horizontal knife edge as shown in Fig. 13.31. The center of gravity of the rod was located by balancing and is 0.200 m from the pivot. When it is set into small amplitude oscillation, the rod makes 100 complete swings in 120 s. Calculate the moment of inertia of the rod about the rotation axis through the pivot.

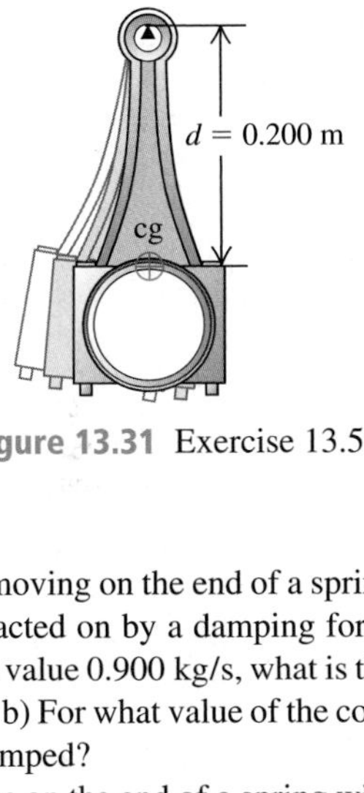

Figure 13.31 Exercise 13.50.

Section 13.7 Damped Oscillations

13.51 An unhappy 0.300-kg mouse, moving on the end of a spring with force constant $k = 2.50$ N/m is acted on by a damping force $F_x = -bv_x$. a) If the constant b has the value 0.900 kg/s, what is the frequency of oscillation of the mouse? b) For what value of the constant b will the motion be critically damped?

13.52 A 50.0-g hard-boiled egg moves on the end of a spring with force constant $k = 25.0$ N/m. Its initial displacement is 0.300 m. A damping force $F_x = -bv_x$ acts on the egg, and the amplitude of the motion decreases to 0.100 m in 5.00 s. Calculate the magnitude of the damping constant b.

13.53 The motion of an underdamped oscillator is described by Eq. (13.42). Let the phase angle ϕ be zero. a) According to this equation, what is the value of x at $t = 0$? b) What are the magnitude and direction of the velocity at $t = 0$? What does the result tell you about the slope of the graph of x versus t near $t = 0$? c) Obtain an expression for the acceleration a_x at $t = 0$. For what value or range of values of the damping constant b (in terms of k and m) is the acceleration at $t = 0$ negative, zero, and positive? Discuss each case in terms of the shape of the graph of x versus t near $t = 0$.

Section 13.8 Forced Oscillations and Resonance

13.54 A sinusoidally varying driving force is applied to a damped harmonic oscillator of force constant k and mass m. If the damping constant has a value b_1, the amplitude is A_1 when the driving angular frequency equals $\sqrt{k/m}$. In terms of A_1, what is the amplitude for the same driving frequency and the same driving force amplitude F_{max}, if the damping constant is a) $3b_1$? b) $b_1/2$?

13.55 A sinusoidally varying driving force is applied to a damped harmonic oscillator. a) What are the units of the damping constant b? b) Show that the quantity $\sqrt{km}$ has the same units as b. c) In terms of F_{max} and k, what is the amplitude for $\omega_d = \sqrt{k/m}$ when i) $b = 0.2\sqrt{km}$? ii) $b = 0.4\sqrt{km}$? Compare your results to Fig. 13.27.

13.56 An experimental package and its support structure which are to be placed on board the International Space Station act as an underdamped spring-mass system with a force constant of 2.1×10^6 N/m and mass 108 kg. A NASA requirement is that resonance for forced oscillations not occur for any frequency below 35 Hz. Does this package meet the requirement?

Problems

13.57 SHM in a Car Engine. The motion of the piston of an automobile engine (Fig. 13.32) is approximately simple harmonic. a) If the stroke of an engine (twice the amplitude) is 0.100 m and the engine runs at 3500 rev/min, compute the acceleration of the piston at the endpoint of its stroke. b) If the piston has mass 0.450 kg, what net force must be exerted on it at this point? c) What are the speed and kinetic energy of the piston at the midpoint of its stroke? d) What average power is required to accelerate the piston from rest to the speed found in part (c)? e) If the engine runs at 7000 rev/min, what are the answers to parts (b), (c), and (d)?

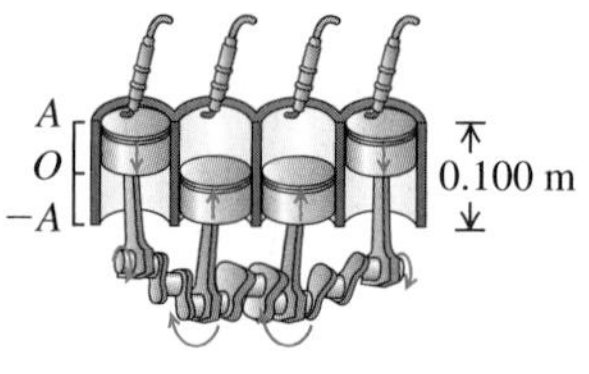

Figure 13.32 Problem 13.57.

13.58 Four passengers with combined mass 250 kg compress the springs of a car with worn-out shock absorbers by 4.00 cm when they enter it. Model the car and passengers as a single body on a single ideal spring. If the loaded car has a period of vibration of 1.08 s, what is the period of vibration of the empty car?

13.59 A glider is oscillating in SHM on an air track with an amplitude A_1. You slow it so that its amplitude is halved. What happens to its a) period, frequency, and angular frequency? b) total mechanical energy? c) maximum speed? d) speed at $x = \pm A_1/4$? e) potential and kinetic energies at $x = \pm A_1/4$?

13.60 You hang an unknown weight from the end of a spring, moving your hand down slowly so that when the weight comes to equilibrium, it has stretched the spring a distance L. If the spring has negligible mass, prove that the weight can oscillate in SHM with the same period as a simple pendulum of length L.

13.61 A child with poor table manners is sliding his 250-g dinner plate back and forth in SHM with an amplitude of 0.100 m on a horizontal surface. At a point 0.060 m away from equilibrium, the speed of the plate is 0.300 m/s. a) What is the period? b) What is the displacement when the speed is 0.160 m/s? c) In the center of the dinner plate is a 10.0-g carrot slice. If the carrot slice is just on the verge of slipping at the end point of the path, what is the coefficient of static friction between the carrot slice and the plate?

13.62 No Bull! While visiting friends at Cal State Chico, you pay a visit to the Crazy Horse Saloon. This fine establishment features a 200-kg mechanical bucking bull, that has a mechanism that makes it move vertically in simple harmonic motion. Whether the "bull" has a rider or not, it moves with the same amplitude (0.250 m) and frequency (1.50 Hz). After watching other saloon patrons hold on to the "bull" while riding, you (mass 75.0 kg) decide to ride it the macho way by *not* holding on. No one is terribly surprised when you come out of the saddle. Later, while waiting for your bruises and pride to heal, you pass the time by calculating the following: a) You leave the saddle when it is moving upward. What is the magnitude of the downward acceleration of the saddle when you lose contact with it? b) How high is the top surface of the saddle above its equilibrium position when you first become airborne? c) How fast are you moving upward when you leave the saddle? d) You are in free fall until you return to the saddle. Show that this happens 0.538 s later. e) What is your speed relative to the saddle at the instant you return?

13.63 A block with mass M rests on a frictionless surface and is connected to a horizontal spring of force constant k. The other end of the spring is attached to a wall (Fig. 13.33). A second block with mass m rests on top of the first block. The coefficient of static friction between the blocks is μ_s. Find the *maximum* amplitude of oscillation such that the top block will not slip on the bottom block.

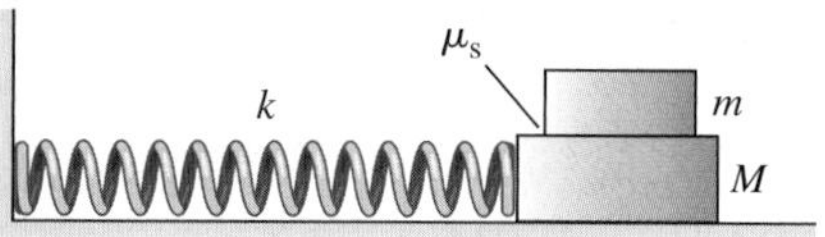

Figure 13.33 Problem 13.63.

13.64 A 10.0-kg mass is traveling to the right with a speed of 2.00 m/s on a smooth, horizontal surface when it collides with and sticks to a second 10.0-kg mass that is initially at rest but is attached to a light spring with force constant 80.0 N/m. a) Find the frequency, amplitude, and period of the subsequent oscillations. b) How long does it take the system to return the first time to the position it had immediately after the collision?

13.65 A block with mass m_1 attached to a horizontal spring that has force constant k, is moving with SHM that has amplitude A_1 and period T_1. a) At the instant that the block passes through its equilibrium position, it suddenly splits into two identical pieces. One piece remains attached to the spring; the other is quickly pushed out of the way. In terms of A_1 and T_1, what are the new amplitude and period of the SHM after the block has split? b) Repeat part (a), for the situation where the block splits when it is at $x = A_1$. c) Compare your results to those of Example 13.5 (Section 13.3) and comment on the similarities and differences.

13.66 An object with mass 0.200 kg is acted on by an elastic restoring force with force constant 10.0 N/m. a) Graph elastic potential energy U as a function of displacement x over a range of x from -0.300 m to $+0.300$ m. On your graph, let 1 cm = 0.05 J vertically and 1 cm = 0.05 m horizontally. The object is set into oscillation with an initial potential energy of 0.140 J and an initial kinetic energy of 0.060 J. Answer the following questions by referring to the graph. b) What is the amplitude of oscillation? c) What is the potential energy when the displacement is one-half the amplitude? d) At what displacement are the kinetic and potential energies equal? e) What is the value of the phase angle ϕ if the initial velocity is positive and the initial displacement is negative?

13.67 Figure 13.16 shows a body with mass m hanging from a spring of force constant k. The positive x-axis is upward and $x = 0$ is the equilibrium position of the body. a) Show that when the body is at coordinate x, the elastic potential energy in the spring is $U_{el} = \frac{1}{2}k(\Delta l - x)^2$. b) Let $x = x_0$ be the coordinate at which the gravitational potential energy is zero. Show that the total potential energy is $U = \frac{1}{2}kx^2 + \frac{1}{2}k(\Delta l)^2 - mgx_0$. c) The expression for the total potential energy in part (b) is of the form $U = \frac{1}{2}kx^2 + C$, where the constant C equals $\frac{1}{2}k(\Delta l)^2 - mgx_0$. Explain why the behavior of the system does not depend on the value of this constant, so that vertical SHM is fundamentally no different than horizontal SHM, for which $U = \frac{1}{2}kx^2$.

13.68 A hanging wire is 1.80 m in length. When a 60.0-kg steel ball is suspended from the wire, the wire stretches by 2.00 mm. If the ball is pulled down a small additional distance and released, at what frequency will it vibrate? Assume that the stress on the wire is less than the proportional limit (see Section 11.5).

13.69 A 5.00-kg partridge is suspended from a pear tree by an ideal spring of negligible mass. When the partridge is pulled down 0.100 m below its equilibrium position and released, it vibrates with a period of 4.20 s. a) What is its speed as it passes through the equilibrium position? b) What is its acceleration when it is 0.050 m above the equilibrium position? c) When it is moving upward, how much time is required for it to move from a point 0.050 m below its equilibrium position to a point 0.050 m above it? d) The motion of the partridge is stopped, and then it is removed from the spring. How much does the spring shorten?

13.70 A 0.0200-kg bolt moves with SHM that has an amplitude of 0.240 m and a period of 1.500 s. The displacement of the bolt is $+0.240$ m when $t = 0$. Compute a) the displacement of the bolt when $t = 0.500$ s; b) the magnitude and direction of the force acting on the bolt when $t = 0.500$ s; c) the minimum time required for the bolt to move from its initial position to the point where $x = -0.180$ m; d) the speed of the bolt when $x = -0.180$ m.

13.71 SHM of a Butcher's Scale. A spring of negligible mass and force constant $k = 400$ N/m is hung vertically, and a 0.200-kg pan is suspended from its lower end. A butcher drops a 2.2-kg steak onto the pan from a height of 0.40 m. The steak makes a totally inelastic collision with the pan and sets the system into vertical simple harmonic motion. What is a) the speed of the pan and steak immediately after the collision? b) the amplitude of the subsequent motion? c) the period of that motion?

13.72 A uniform beam is suspended horizontally by two identical, vertical springs that are attached between the ceiling and each end of the beam. The beam has mass 225 kg and a 175-kg sack of gravel sits on the middle of it. The beam is oscillating in SHM, with an amplitude of 40.0 cm and a frequency of 0.600 cycles/s. a) The sack of gravel falls off the beam when the beam has its maximum upward displacement. What are the frequency and amplitude of the subsequent SHM of the beam? b) If the gravel instead falls off when the beam has its maximum speed, what are the frequency and amplitude of the subsequent SHM of the beam?

13.73 On the planet Newtonia, a simple pendulum having a bob with mass 1.25 kg and a length of 185.0 cm takes 1.42 s, when released from rest, to swing through an angle of 12.5°, where it again has zero speed. The circumference of Newtonia is measured to be 51,400 km. What is the mass of the planet Newtonia?

13.74 A 40.0-N force stretches a vertical spring 0.250 m. a) What mass must be suspended from the spring so that the system will oscillate with a period of 1.00 s? b) If the amplitude of the motion is 0.050 m and the period is that specified in part (a), where is the object and in what direction is it moving 0.35 s after it has passed the equilibrium position, moving downward? c) What force (magnitude and direction) does the spring exert on the object when it is 0.030 m below the equilibrium position, moving upward?

13.75 Don't Miss the Boat. While on a visit to Minnesota ("Land of 10,000 Lakes"), you sign up to take an excursion around one of the larger lakes. When you go to the dock where the 1500-kg boat is tied, you find that the boat is bobbing up and down in the waves, executing simple harmonic motion with amplitude 20 cm. The boat takes 3.5 s to make one complete up-and-down cycle. When the boat is at its highest point, its deck is at the same height as the stationary dock. As you watch the boat bob up and down, you (mass 60 kg) begin to feel a bit woozy, due in part to the previous night's dinner of lutefisk. As a result, you refuse to board the boat unless the level of the boat's deck is within 10 cm of the dock level. How much time do you have to board the boat comfortably during each cycle of up-and-down motion?

13.76 An interesting though highly impractical example of oscillation is the motion of an object dropped down a hole that extends from one side of the earth, through its center, to the other side. With the assumption (not very realistic) that the earth is a sphere of uniform density, prove that the motion is simple harmonic and find the period. *Note:* The gravitational force on the object as a function of the object's distance r from the center of the earth was derived in Example 12.10 (Section 12.6). The motion is simple harmonic if the acceleration a_x and the displacement from equilibrium x are related by Eq. (13.8), and the period is then $T = 2\pi/\omega$.]

13.77 Let t_1 be the time it takes a body undergoing SHM to move from $x = 0$ (when $t = 0$) to $x = A$. Obtain an equation for t_1 in the following way. In Eq. (13.22) replace v_x with dx/dt. Separate variables by writing all factors containing x on one side and all those containing t on the other side so that each side can be integrated. Integrate the equation with limits on t of 0 to t_1 and limits on x of 0 to A and thereby obtain an equation for t_1. How does t_1 compare to the period T?

13.78 For a certain oscillator the net force on the body with mass m is given by $F_x = -cx^3$. a) What is the potential energy function for this oscillator, if we take $U = 0$ at $x = 0$? b) One-quarter of a period is the time for the body to move from $x = 0$ to $x = A$. Calculate this time and hence the period. [*Hint:* Begin with

Eq. (13.20), modified to include the potential energy function you found in part (a), and solve for the velocity v_x as a function of x. Then replace v_x with dx/dt and follow the steps outlined in Problem 13.77. In the x-integral make the change of variable $u = x/A$. The resulting integral can be evaluated by numerical methods on a computer and has the value $\int_0^1 du/\sqrt{1-u^4} = 1.31$.] c) According to the result you obtained in part (b), does the period depend on the amplitude A of the motion? Are the oscillations simple harmonic?

13.79 Consider the circle of reference shown in Fig. 13.5. The x-component of the velocity of Q is the velocity of P. Compute this component and show that the velocity of P is as given by Eq. (13.15).

13.80 To measure g in an unorthodox manner, a student places a ball bearing on the concave side of a lens (Fig. 13.34). She attaches the lens to a simple harmonic oscillator (actually a small stereo speaker) whose amplitude is A and whose frequency f can be varied. She can measure both A and f with a strobe light. a) If the ball bearing has mass m, find the normal force exerted by the lens on the ball bearing as a function of time. Your result should be in terms of A, f, m, g, and a phase angle ϕ. b) The frequency is slowly increased. When it reaches a value f_b, the ball is heard to bounce. What is g in terms of A and f_b?

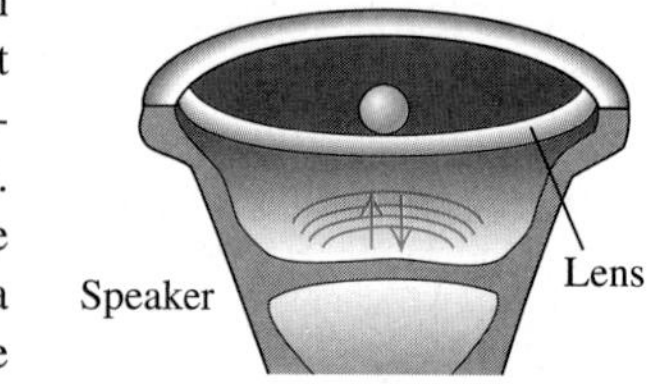

Figure 13.34 Problem 13.80.

***13.81 Diatomic Molecule.** Two identical atoms in a diatomic molecule vibrate as harmonic oscillators. However, their center of mass, midway between them, remains at rest. a) Show that at any instant, the momenta of the atoms relative to the center of mass are $\vec{p}$ and $-\vec{p}$. b) Show that the total kinetic energy K of the two atoms at any instant is the same as that of a single object with mass $m/2$ with a momentum of magnitude p. (Use $K = p^2/2m$.) This result shows why $m/2$ should be used in the expression for f in Example 13.7 (Section 13.4). c) If the atoms are not identical, but have masses m_1 and m_2, show that the result of part (a) still holds and the single object's mass in part (b) is $m_1m_2/(m_1 + m_2)$. The quantity $m_1m_2/(m_1 + m_2)$ is called the *reduced mass* of the system.

***13.82** An approximation for the potential energy of a KCl molecule is $U = A[(R_0^7/8r^8) - 1/r]$, where $R_0 = 2.67 \times 10^{-10}$ m and $A = 2.31 \times 10^{-28}$ J·m. Using this approximation: a) Show that the radial component of the force on each atom is $F_r = A[(R_0^7/r^9) - 1/r^2]$. b) Show that R_0 is the equilibrium separation. c) Find the minimum potential energy. d) Use $r = R_0 + x$ and the first two terms of the binomial theorem (Eq. 13.28) to show that $F_r \approx -(7A/R_0^3)x$, so that the molecule's force constant is $k = 7A/R_0^3$. e) With both the K and Cl atoms vibrating in opposite directions on opposite sides of the molecule's center of mass, $m_1m_2/(m_1 + m_2) = 3.06 \times 10^{-26}$ kg is the mass to use in calculating the frequency (see Problem 13.81). Calculate the frequency of small-amplitude vibrations.

***13.83** Three positively charged particles are maintained in a straight line. The end particles are identical and are held fixed at a distance $2R_0$ apart. The potential energy for the force acting on the center particle can be written as $U = A[1/r - 1/(r - 2R_0)]$, where A is a constant and r is the distance from the left-hand particle to the center particle. a) Show that the force on the center particle is $F_r = A[1/r^2 - 1/(r - 2R_0)^2]$. b) Show that $r = R_0$ at equilibrium. c) Use $r = R_0 + x$ and the first two terms of the binomial theorem (Eq. 13.28) to show that $F_r \approx -(4A/R_0^3)x$, so that the force constant is $k = 4A/R_0^3$. d) What is the frequency of small-amplitude oscillation of the center particle if it has mass m?

13.84 Many real springs are more easily stretched than compressed. We can represent this by using different spring constants for $x > 0$ and for $x < 0$. As an example, consider a spring that exerts the following restoring force:

$$F_x = \begin{cases} -kx & \text{for } x > 0 \\ -2kx & \text{for } x < 0 \end{cases}$$

A mass m on a frictionless, horizontal surface is attached to this spring, displaced to $x = A$ by stretching the spring and released. a) Find the period of the motion. Does the period depend on A? Are the oscillations simple harmonic? b) What is the most negative value of x that the mass m reaches? Is the oscillation symmetric about the point $x = 0$?

13.85 Two solid cylinders connected along their common axis by a short, light rod have radius R and total mass M, and rest on a horizontal tabletop. A spring with force constant k has one end attached to a clamp and the other end attached to a frictionless ring at the center of mass of the cylinders (Fig. 13.35). The cylinders are pulled to the left a distance x, which stretches the spring, and released. There is sufficient friction between the tabletop and the cylinders for the cylinders to roll without slipping as they move back and forth on the end of the spring. Show that the motion of the center of mass of the cylinders is simple harmonic, and calculate its period in terms of M and k. [*Hint:* The motion is simple harmonic if a_x and x are related by Eq. (13.8), and the period then is $T = 2\pi/\omega$. Apply $\sum\tau_z = I_{cm}\alpha_z$ and $\sum F_x = Ma_{cm-x}$ to the cylinders in order to relate a_{cm-x} and the displacement x of the cylinders from their equilibrium position.]

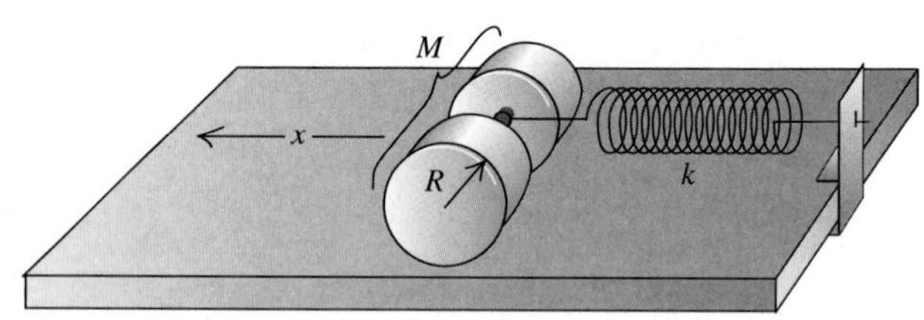

Figure 13.35 Problem 13.85.

13.86 In Fig. 13.36 the upper ball is released from rest, collides with the stationary lower ball, and sticks to it. The strings are both 50.0 cm long. The upper ball has mass 2.00 kg and it is initially 10.0 cm higher than the lower ball, which has mass 3.00 kg. Find the frequency and maximum angular displacement of the motion after the collision.

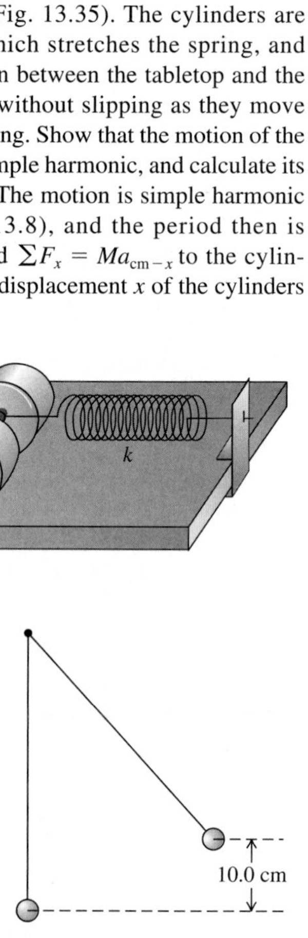

Figure 13.36 Problem 13.86.

13.87 T-rex. Model the leg of the T-rex in Example 13.10 (Section 13.6) as two uniform rods joined end-to-end. Let each rod have length 1.55 m. Let the lower rod have mass M and the upper rod mass $2M$. The composite object is pivoted about the top of the upper rod. Compute the oscillation period of this object for small amplitude oscillations. Compare your result to that of Example 13.10.

13.88 A slender, uniform, metal rod with mass M is pivoted without friction about an axis through its midpoint and perpendicular to the rod. A horizontal spring with force constant k is attached to the lower end of the rod, with the other end of the spring attached to a rigid support. If the rod is displaced by a small angle Θ from the vertical (Fig. 13.37) and released, show that it moves in angular SHM and calculate the period. (*Hint:* Assume that the angle Θ is small enough for the approximations $\sin\Theta \approx \Theta$ and $\cos\Theta \approx 1$ to be valid. The motion is simple harmonic if $d^2\theta/dt^2 = -\omega^2\theta$, and the period is then $T = 2\pi/\omega$.)

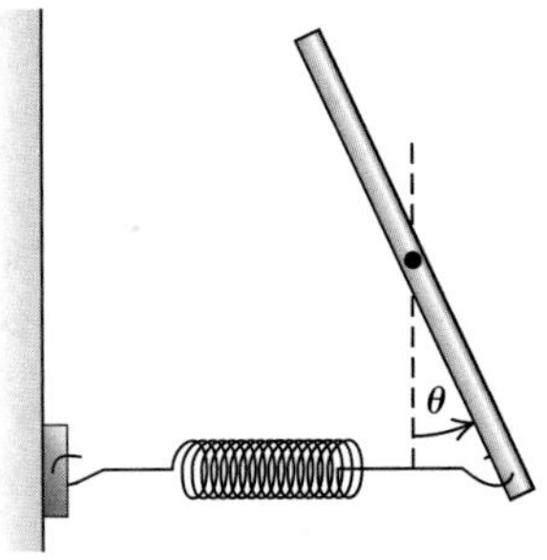

Figure 13.37 Problem 13.88.

13.89 The Silently Ringing Bell Problem. A large bell is hung from a wooden beam so it can swing back and forth with negligible friction. The center of mass of the bell is 0.60 m below the pivot, the bell has mass 34.0 kg, and the moment of inertia of the bell about an axis at the pivot is 18.0 kg·m^2. The clapper is a small, 1.8-kg mass attached to one end of a slender rod that has length L and negligible mass. The other end of the rod is attached to the inside of the bell so it can swing freely about the same axis as the bell. What should be the length L of the clapper rod for the bell to ring silently, that is, for the period of oscillation for the bell to equal that for the clapper?

13.90 Two identical, thin rods, each with mass m and length L, are joined at right angles to form an L-shaped object. This object is balanced on top of a sharp edge (Fig. 13.38). If the L-shaped object is deflected slightly, it oscillates. Find the frequency of oscillation.

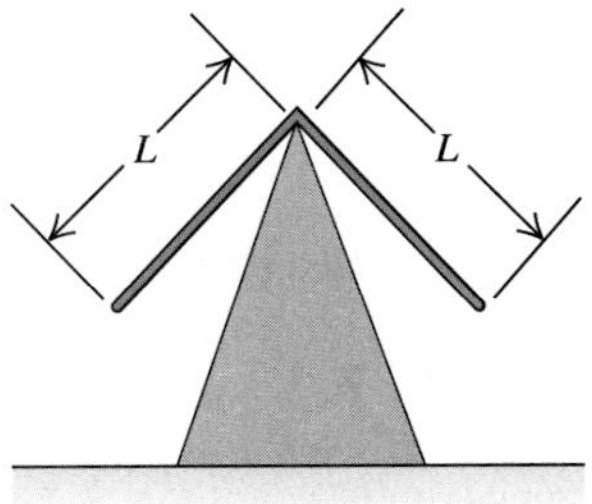

Figure 13.38 Problem 13.90.

13.91 You want to construct a pendulum with a period of 4.00 s at a location where $g = 9.80$ m/s^2. a) What is the length of a *simple* pendulum having this period? b) Suppose the pendulum must be mounted in a case that is not over 0.50 m high. Can you devise a pendulum having a period of 4.00 s that will satisfy this requirement?

13.92 Show that $x(t)$ as given in Eq. (13.42) is a solution of Newton's second law for damped oscillations, Eq. (13.41), if ω' is as defined in Eq. (13.43).

13.93 A uniform rod of length L oscillates through small angles about a point a distance x from its center. a) Prove that its angular frequency is $\sqrt{gx/[(L^2/12) + x^2]}$. b) Show that its maximum angular frequency occurs when $x = L/\sqrt{12}$. c) What is the length of the rod if the maximum angular frequency is 2π rad/s?

13.94 When Is It Simple? To make a pendulum, you hang a uniform solid sphere of radius R on the end of a light string. The distance from the pivot to the center of the sphere is L. a) Prove that the period of your pendulum is $T = T_{sp}\sqrt{1 + 2R^2/5L^2}$, where T_{sp} is the period of a simple pendulum of length L. b) At what value of L is T only 0.10% larger than T_{sp}? c) For a sphere of diameter 2.540 cm, what is the value of L in part (b)?

Challenge Problems

13.95 Two springs, each with unstretched length 0.200 m, but with different force constants k_1 and k_2, are attached to opposite ends of a block with mass m on a level, frictionless surface. The outer ends of the springs are now attached to two pins P_1 and P_2, 0.100 m from the original positions of the ends of the springs (Fig. 13.39). Let $k_1 = 2.00$ N/m, $k_2 = 6.00$ N/m, and $m = 0.100$ kg. a) Find the length of each spring when the block is in its new equilibrium position after the springs have been attached to the pins. b) Find the period of vibration of the block if it is slightly displaced from its new equilibrium position and released.

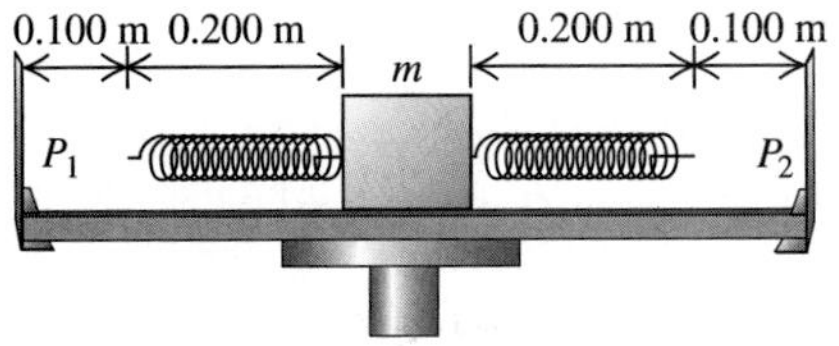

Figure 13.39 Challenge Problem 13.95.

13.96 The Effective Force Constant of Two Springs. Two springs with the same unstretched length, but different force constants k_1 and k_2 are attached to a block mass m on a level, frictionless surface. Calculate the effective force constant k_{eff} in each of the three cases (a), (b), and (c) depicted in Fig. 13.40. (The effective force constant is defined by $\sum F_x = -k_{eff}x$.) d) An object with mass m, suspended from a uniform spring with a force constant k, vibrates with a frequency f_1. When the spring is cut in half and the same object is suspended from one of the halves, the frequency is f_2. What is the ratio f_2/f_1?

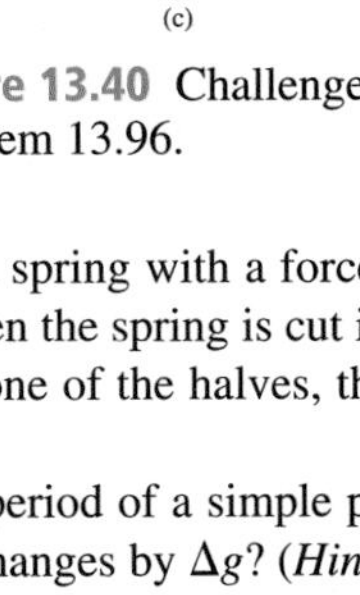

Figure 13.40 Challenge Problem 13.96.

13.97 a) What is the change ΔT in the period of a simple pendulum when the acceleration of gravity g changes by Δg? (*Hint:* The new period $T + \Delta T$ is obtained by substituting $g + \Delta g$ for g:

$$T + \Delta T = 2\pi\sqrt{\frac{L}{g + \Delta g}}$$

To obtain an approximate expression, expand the factor $(g + \Delta g)^{-1/2}$ using the binomial theorem (Appendix B) and keeping only the first two terms:

$$(g + \Delta g)^{-1/2} = g^{-1/2} - \frac{1}{2}g^{-3/2}\Delta g + \cdots$$

The other terms contain higher powers of Δg and are very small if Δg is small.) Express your result as the *fractional* change in period $\Delta T/T$ in terms of the fractional change $\Delta g/g$. b) A pendulum clock keeps correct time at a point where $g = 9.8000\ \text{m/s}^2$, but is found to lose 4.0 s each day at a higher elevation. Use the result of part (a) to find the approximate value of g at this new location.

13.98 A Spring with Mass. The preceding problems in this chapter have assumed that the springs had negligible mass. But of course no spring is completely massless. To find the effect of the spring's mass, consider a spring with mass M, equilibrium length L_0, and spring constant k. When stretched or compressed to a length L, the potential energy is $\frac{1}{2}kx^2$, where $x = L - L_0$. a) Consider a spring, as described above, that has one end fixed and the other end moving with speed v. Assume that the speed of points along the length of the spring varies linearly with distance l from the fixed end. Assume also that the mass M of the spring is distributed uniformly along the length of the spring. Calculate the kinetic energy of the spring in terms of M and v. (*Hint:* Divide the spring into pieces of length dl; find the speed of each piece in terms of l, v, and L; find the mass of each piece in terms of dl, M, and L; and integrate from 0 to L. The result is *not* $\frac{1}{2}Mv^2$, since not all of the spring moves with the same speed.) b) Take the time derivative of the conservation of energy equation, Eq. (13.21), for a mass m moving on the end of a *massless* spring. By comparing your results to Eq. (13.8), which defines ω, show that the angular frequency of oscillation is $\omega = \sqrt{k/m}$. c) Apply the procedure of part (b) to obtain the angular frequency of oscillation ω of the spring considered in part (a). If the *effective mass* M' of the spring is defined by $\omega = \sqrt{k/M'}$, what is M' in terms of M?

13.99 A uniform 1.00-m stick hangs from a horizontal axis at one end and oscillates as a physical pendulum. An object of small dimensions and with mass equal to that of the meter stick can be clamped to the stick at a distance y below the axis. Let T represent the period of the system with the body attached and T_0 the period of the meter stick alone. a) Find the ratio T/T_0. Evaluate your expression for y ranging from 0 to 1.0 m in steps of 0.1 m, and graph T/T_0 versus y. b) Is there any value of y, other than $y = 0$, for which $T = T_0$? If so, find it and explain why the period is unchanged when y has this value.

13.100 You measure the period of a physical pendulum about one pivot point to be T. Then you find another pivot point on the opposite side of the center of mass that gives the same period. The two points are separated by a distance L. Use the parallel-axis theorem to show that $g = L(2\pi/T)^2$. (This result shows a way that you can measure g without knowing the mass or any moments of inertia of the physical pendulum.)

13.101 Resonance in a Mechanical System. A mass m is attached to one end of a massless spring with a force constant k and an unstretched length l_0. The other end of the spring is free to turn about a nail driven into a frictionless, horizontal surface (Fig. 13.41). The mass is made to revolve in a circle with an angular frequency of revolution ω'. a) Calculate the length l of the spring as a function of ω'. b) What happens to the result in part (a) when ω' approaches the natural frequency $\omega = \sqrt{k/m}$ of the mass-spring system? (If your result bothers you, remember that massless springs and frictionless surfaces don't exist as such, but are only approximate descriptions of real springs and surfaces. Also, Hooke's law is itself only an approximation of the way real springs behave; the greater the elongation of the spring, the greater the deviation from Hooke's law.)

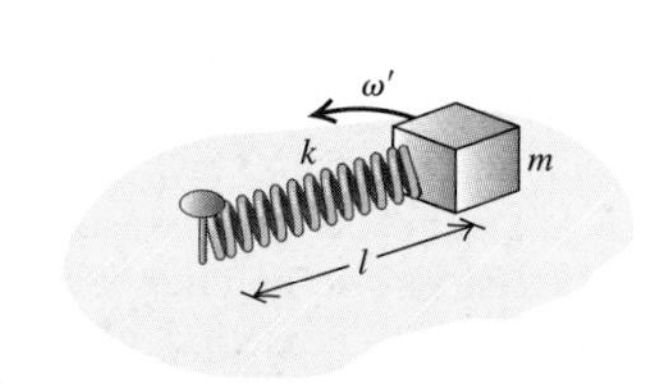

Figure 13.41 Challenge Problem 13.101.

***13.102 Vibration of a Covalently Bonded Molecule.** Many diatomic (two-atom) molecules are bound together by *covalent bonds* that are much stronger than the van der Waals interaction. Examples include H_2, O_2, and N_2. Experiment shows that for many such molecules, the interaction can be described by a force of the form

$$F_r = A\left[e^{-2b(r-R_0)} - e^{-b(r-R_0)}\right]$$

where A and b are positive constants, r is the center-to-center separation of the atoms, and R_0 is the equilibrium separation. For the hydrogen molecule (H_2), $A = 2.97 \times 10^{-8}$ N, $b = 1.95 \times 10^{10}\ \text{m}^{-1}$, and $R_0 = 7.4 \times 10^{-11}$ m. Find the force constant for small oscillations around equilibrium. (*Hint:* Use the expansion for e^x given in Appendix B.) Compare your result to the value given in Exercise 13.40.

FLUID MECHANICS 14

This Olympic cyclist is riding a stationary bike in a wind tunnel. By observing the airflow around his body (aided by trails of smoke), scientists are able to determine which bicycle designs and cycling techniques will minimize air resistance and maximize performance.

? At various places around the cyclist, air is forced through a narrow constriction (for example, between his upper arm and his body). Does this make the air slow down, speed up, or neither?

Fluids play a vital role in many aspects of everyday life. We drink them, breathe them, swim in them. They circulate through our bodies and control our weather. Airplanes fly through them; ships float in them. A fluid is any substance that can flow; we use the term for both liquids and gases. We usually think of a gas as easily compressed and a liquid as nearly incompressible, although there are exceptional cases.

We begin our study with **fluid statics,** the study of fluids at rest in equilibrium situations. Like other equilibrium situations, it is based on Newton's first and third laws. We will explore the key concepts of density, pressure, and buoyancy. **Fluid dynamics,** the study of fluids in motion, is much more complex; indeed, it is one of the most complex branches of mechanics. Fortunately, we can analyze many important situations using simple idealized models and familiar principles such as Newton's laws and conservation of energy. Even so, we will barely scratch the surface of this broad and interesting topic.

14.1 | Density

An important property of any material is its **density,** defined as its mass per unit volume. A homogeneous material such as ice or iron has the same density throughout. We use the Greek letter ρ ("rho") for density. If a mass m of homogeneous material has volume V, the density ρ is

14.1 The price of gold is quoted by weight (say, in dollars per ounce). Because gold is one of the densest of the metals, a fortune in gold can be stored in a small volume.

$$\rho = \frac{m}{V} \quad \text{(definition of density)} \tag{14.1}$$

The density of some materials varies from point to point within the material; some examples are the earth's atmosphere (which is less dense at high altitude) and oceans (which are denser at greater depths). For these materials, Eq. (14.1) describes the *average* density. In general, the density of a material depends on environmental factors such as temperature and pressure.

The SI unit of density is the kilogram per cubic meter ($1\ \text{kg/m}^3$). The cgs unit, the gram per cubic centimeter ($1\ \text{g/cm}^3$), is also widely used. The conversion factor

$$1\ \text{g/cm}^3 = 1000\ \text{kg/m}^3$$

is useful. Densities of several common substances at ordinary temperatures are given in Table 14.1 (see also Fig. 14.1). Note the wide range of magnitudes. The densest material found on earth is the metal osmium ($\rho = 22{,}500\ \text{kg/m}^3$), but its density pales by comparison to the densities of exotic astronomical objects such as white dwarf stars and neutron stars.

Table 14.1 Densities of Some Common Substances

Material	Density (kg/m^3)*	Material	Density (kg/m^3)*
Air (1 atm, 20° C)	1.20	Iron, steel	7.8×10^3
Ethanol	0.81×10^3	Brass	8.6×10^3
Benzene	0.90×10^3	Copper	8.9×10^3
Ice	0.92×10^3	Silver	10.5×10^3
Water	1.00×10^3	Lead	11.3×10^3
Seawater	1.03×10^3	Mercury	13.6×10^3
Blood	1.06×10^3	Gold	19.3×10^3
Glycerin	1.26×10^3	Platinum	21.4×10^3
Concrete	2×10^3	White dwarf star	10^{10}
Aluminum	2.7×10^3	Neutron star	10^{18}

*To obtain the densities in grams per cubic centimeter, simply divide by 10^3.

The **specific gravity** of a material is the ratio of its density to the density of water at 4.0°C, $1000\ \text{kg/m}^3$; it is a pure number without units. For example, the specific gravity of aluminum is 2.7. "Specific gravity" is a poor term, since it has nothing to do with gravity; "relative density" would have been better.

Measuring density is an important analytical technique. For example, we can determine the charge condition of a storage battery by measuring the density of its electrolyte, a sulfuric acid solution. As the battery discharges, the sulfuric acid (H_2SO_4) combines with lead in the battery plates to form insoluble lead sulfate ($PbSO_4$), decreasing the concentration of the solution. The density decreases from about $1.30 \times 10^3\ \text{kg/m}^3$ for a fully charged battery to $1.15 \times 10^3\ \text{kg/m}^3$ for a discharged battery.

Another automotive example is permanent-type antifreeze, which is usually a solution of ethylene glycol ($\rho = 1.12 \times 10^3\ \text{kg/m}^3$) and water. The freezing point of the solution depends on the glycol concentration, which can be determined by measuring the specific gravity. Such measurements are performed routinely by mechanics using a device called a hydrometer, which we'll discuss in Section 14.3.

Example 14.1 The weight of a roomful of air

Find the mass and weight of the air in a living room with a 4.0 m × 5.0 m floor and a ceiling 3.0 m high. What is the mass and weight of an equal volume of water?

SOLUTION

IDENTIFY: We assume that air is homogeneous, so that the density is the same throughout the room. (It is true that air is less dense at high elevations than near sea level. The variation in density over the 3.0-m height of the room, however, is negligibly small; see Section 14.2.)

SET UP: We use Eq. (14.1) to relate the mass (the target variable) to the volume (which we calculate from the dimensions of the room) and the density (from Table 14.1).

EXECUTE: The volume of the room is $V = (3.0\text{ m})(4.0\text{ m}) \times (5.0\text{ m}) = 60\text{ m}^3$. The mass m_{air} of air is given by Eq. (14.1):

$$m_{\text{air}} = \rho_{\text{air}} V = (1.2\text{ kg/m}^3)(60\text{ m}^3) = 72\text{ kg}$$

The weight of the air is

$$w_{\text{air}} = m_{\text{air}} g = (72\text{ kg})(9.8\text{ m/s}^2) = 700\text{ N} = 160\text{ lb}$$

The mass of an equal volume of water is

$$m_{\text{water}} = \rho_{\text{water}} V = (1000\text{ kg/m}^3)(60\text{ m}^3) = 6.0 \times 10^4\text{ kg}$$

The weight is

$$w_{\text{water}} = m_{\text{water}}\, g = (6.0 \times 10^4\text{ kg})(9.8\text{ m/s}^2) = 5.9 \times 10^5\text{ N} = 1.3 \times 10^5\text{ lb} = 66\text{ tons}$$

EVALUATE: A roomful of air weighs about the same as an average adult! Water is nearly a thousand times denser than air, and its mass and weight are larger by the same factor. Indeed, the weight of a roomful of water would certainly collapse the floor of an ordinary house.

Test Your Understanding

What volume of water would have the same mass as one cubic meter of platinum? If this water occupied a cube, what would be the length of each side?

14.2 | Pressure in a Fluid

When a fluid (either liquid or gas) is at rest, it exerts a force perpendicular to any surface in contact with it, such as a container wall or a body immersed in the fluid. This is the force that you feel pressing on your legs when you dangle them in a swimming pool. While the fluid as a whole is at rest, the molecules that make up the fluid are in motion; the force exerted by the fluid is due to molecules colliding with their surroundings.

If we think of an imaginary surface *within* the fluid, the fluid on the two sides of the surface exerts equal and opposite forces on the surface. (Otherwise, the surface would accelerate and the fluid would not remain at rest.) Consider a small surface of area dA centered on a point in the fluid; the normal force exerted by the fluid on each side is $dF_\perp$ (Fig. 14.2). We define the **pressure** p at that point as the normal force per unit area, that is, the ratio of $dF_\perp$ to dA:

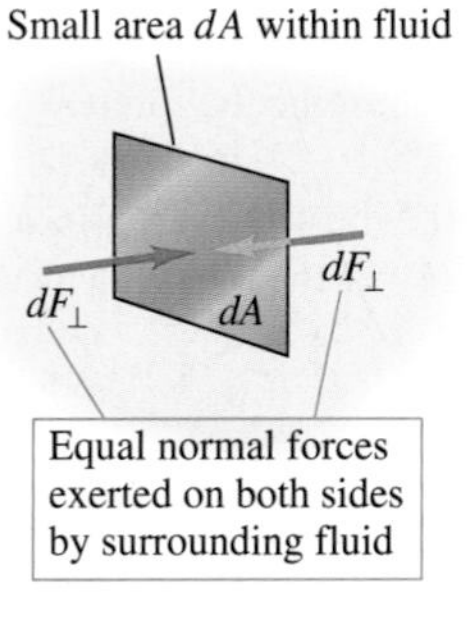

14.2 The pressure acting on either side of a small area within a fluid is $p = dF_\perp/dA$.

$$p = \frac{dF_\perp}{dA} \quad \text{(definition of pressure)} \tag{14.2}$$

If the pressure is the same at all points of a finite plane surface with area A, then

$$p = \frac{F_\perp}{A} \tag{14.3}$$

where $F_\perp$ is the net normal force on one side of the surface. The SI unit of pressure is the **pascal,** where

$$1 \text{ pascal} = 1 \text{ Pa} = \text{N/m}^2$$

We introduced the pascal in Chapter 11. Two related units, used principally in meteorology, are the *bar,* equal to 10^5 Pa, and the *millibar,* equal to 100 Pa.

Atmospheric pressure p_a is the pressure of the earth's atmosphere, the pressure at the bottom of this sea of air in which we live. This pressure varies with weather changes and with elevation. Normal atmospheric pressure at sea level (an average value) is 1 *atmosphere* (atm), defined to be exactly 101,325 Pa. To four significant figures,

$$(p_a)_{av} = 1 \text{ atm} = 1.013 \times 10^5 \text{ Pa}$$
$$= 1.013 \text{ bar} = 1013 \text{ millibar} = 14.70 \text{ lb/in.}^2$$

CAUTION In everyday language the words "pressure" and "force" mean pretty much the same thing. In fluid mechanics, however, these words describe distinct quantities with different characteristics (Fig. 14.3). Fluid pressure acts perpendicular to any surface in the fluid, no matter how that surface is oriented. Hence pressure has no intrinsic direction of its own; it's a scalar. By contrast, force is a vector with a definite direction. Remember, too, that pressure is force per unit area.

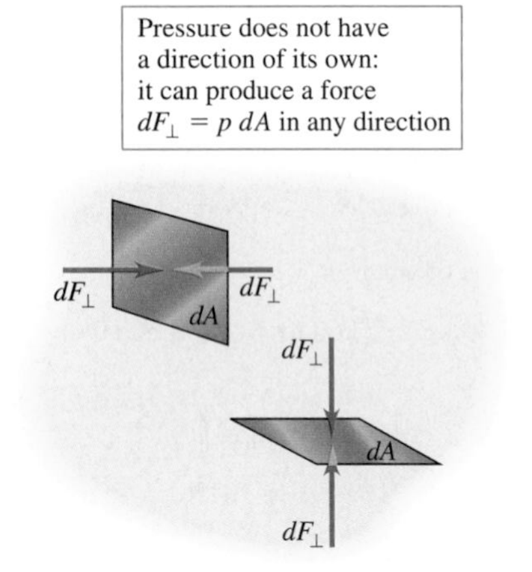

14.3 Pressure is a scalar (it has no intrinsic direction) and has units of newtons per square meter. By contrast, force is a vector with units of newtons.

Example 14.2 The force of air

In the room described in Example 14.1, what is the total downward force on the surface of the floor due to air pressure of 1.00 atm?

SOLUTION

IDENTIFY AND SET UP: The pressure is uniform, so we use Eq. (14.3) to determine the force $F_\perp$ from the pressure and area.

EXECUTE: The floor area is $A = (4.0\text{ m})(5.0\text{ m}) = 20\text{ m}^2$, so from Eq. (14.3) the total downward force is

$$F_\perp = pA = (1.013 \times 10^5 \text{ N/m}^2)(20 \text{ m}^2)$$
$$= 2.0 \times 10^6 \text{ N} = 4.5 \times 10^5 \text{ lb} = 225 \text{ tons}$$

EVALUATE: As in Example 14.1, this is more than enough force to collapse the floor. So why doesn't it collapse? Because there is an upward force on the underside of the floor. If the house has a basement, this upward force is exerted by the air underneath the floor. In this case, if we neglect the thickness of the floor, the *net* force due to air pressure is zero.

Pressure, Depth, and Pascal's Law

If the weight of the fluid can be neglected, the pressure in a fluid is the same throughout its volume. We used that approximation in our discussion of bulk stress and strain in Section 11.4. But often the fluid's weight is *not* negligible. Atmospheric pressure is less at high altitude than at sea level, which is why an airplane cabin has to be pressurized when flying at 35,000 feet. When you dive into deep water, your ears tell you that the pressure increases rapidly with increasing depth below the surface.

We can derive a general relation between the pressure p at any point in a fluid at rest and the elevation y of the point. We'll assume that the density ρ and the

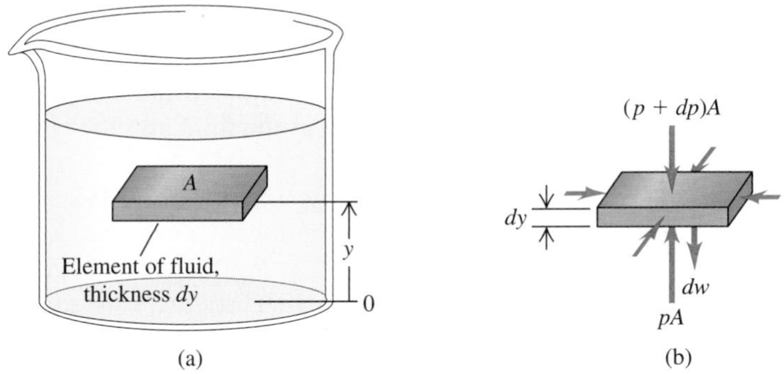

14.4 The forces on an element of fluid in equilibrium.

acceleration due to gravity g are the same throughout the fluid. If the fluid is in equilibrium, every volume element is in equilibrium. Consider a thin element of fluid, with height dy (Fig. 14.4). The bottom and top surfaces each have area A, and they are at elevations y and $y + dy$ above some reference level where $y = 0$. The volume of the fluid element is $dV = A\,dy$, its mass is $dm = \rho\,dV = \rho A\,dy$, and its weight is $dw = dm\,g = \rho g A\,dy$.

What are the other forces on this fluid element? Call the pressure at the bottom surface p; the total y-component of upward force on this surface is pA. The pressure at the top surface is $p + dp$, and the total y-component of (downward) force on the top surface is $-(p + dp)A$. The fluid element is in equilibrium, so the total y-component of force, including the weight and the forces at the bottom and top surfaces, must be zero:

$$\sum F_y = 0 \quad \text{so} \quad pA - (p + dp)A - \rho g A\,dy = 0$$

When we divide out the area A and rearrange, we get

$$\frac{dp}{dy} = -\rho g \tag{14.4}$$

This equation shows that when y increases, p decreases; that is, as we move upward in the fluid, pressure decreases, as we expect. If p_1 and p_2 are the pressures at elevations y_1 and y_2, respectively, and if ρ and g are constant, then

$$p_2 - p_1 = -\rho g(y_2 - y_1) \quad \text{(pressure in a fluid of uniform density)} \tag{14.5}$$

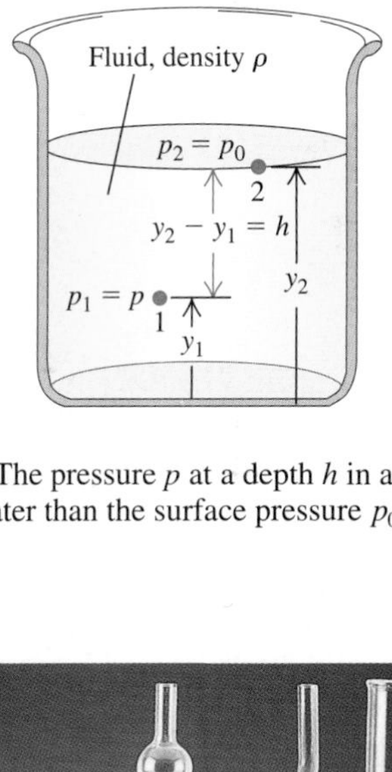

14.5 The pressure p at a depth h in a fluid is greater than the surface pressure p_0 by $\rho g h$.

It's often convenient to express Eq. (14.5) in terms of the *depth* below the surface of a fluid (Fig. 14.5). Take point 1 at any level in the fluid and let p represent the pressure at this point. Take point 2 at the *surface* of the fluid, where the pressure is p_0 (subscript zero for zero depth). The depth of point 1 below the surface is $h = y_2 - y_1$, and Eq. (14.5) becomes

$$p_0 - p = -\rho g(y_2 - y_1) = -\rho g h$$

$$p = p_0 + \rho g h \quad \text{(pressure in a fluid of uniform density)} \tag{14.6}$$

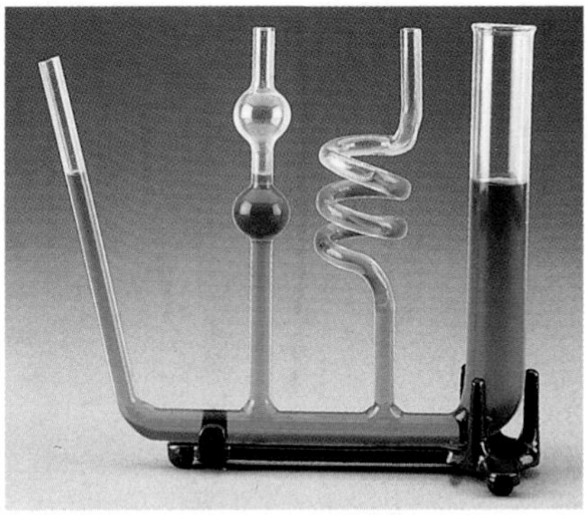

14.6 The pressure at the top of each fluid column is equal to p_0, the pressure of the atmosphere. Pressure depends only on height, not the shape of the container, so each fluid column has the same height.

The pressure p at a depth h is greater than the pressure p_0 at the surface by an amount $\rho g h$. Note that the pressure is the same at any two points at the same level in the fluid. The *shape* of the container does not matter (Fig. 14.6).

Equation (14.6) shows that if we increase the pressure p_0 at the top surface, possibly by using a piston that fits tightly inside the container to push down on

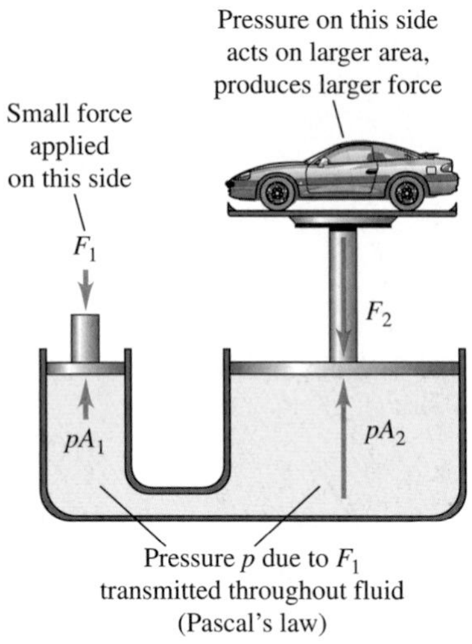

14.7 Principle of the hydraulic lift, an application of Pascal's law. The size of the fluid-filled container is exaggerated for clarity.

the fluid surface, the pressure p at any depth increases by exactly the same amount. This fact was recognized in 1653 by the French scientist Blaise Pascal (1623–1662) and is called **Pascal's law: Pressure applied to an enclosed fluid is transmitted undiminished to every portion of the fluid and the walls of the containing vessel.**

The hydraulic lift shown schematically in Fig. 14.7 illustrates Pascal's law. A piston with small cross-sectional area A_1 exerts a force F_1 on the surface of a liquid such as oil. The applied pressure $p = F_1/A_1$ is transmitted through the connecting pipe to a larger piston of area A_2. The applied pressure is the same in both cylinders, so

$$p = \frac{F_1}{A_1} = \frac{F_2}{A_2} \quad \text{and} \quad F_2 = \frac{A_2}{A_1}F_1 \tag{14.7}$$

The hydraulic lift is a force-multiplying device with a multiplication factor equal to the ratio of the areas of the two pistons. Dentist's chairs, car lifts and jacks, many elevators, and hydraulic brakes all use this principle.

For gases the assumption that the density ρ is uniform is realistic only over short vertical distances. In a room with a ceiling height of 3.0 m filled with air of uniform density 1.2 kg/m^3, the difference in pressure between floor and ceiling, given by Eq. (14.6), is

$$\rho gh = (1.2\ \text{kg/m}^3)(9.8\ \text{m/s}^2)(3.0\ \text{m}) = 35\ \text{Pa}$$

or about 0.00035 atm, a very small difference. But between sea level and the summit of Mount Everest (8882 m) the density of air changes by nearly a factor of 3, and in this case we cannot use Eq. (14.6). Liquids, by contrast, are nearly incompressible, and it is usually a very good approximation to regard their density as independent of pressure. A pressure of several hundred atmospheres will cause only a few percent increase in the density of most liquids.

Absolute Pressure, Gauge Pressure, and Pressure Gauges

If the pressure inside a car tire is equal to atmospheric pressure, the tire is flat. The pressure has to be *greater* than atmospheric to support the car, so the significant quantity is the *difference* between the inside and outside pressures. When we say that the pressure in a car tire is "32 pounds" (actually 32 lb/in.^{2}, equal to 220 kPa or 2.2×10^5 Pa), we mean that it is *greater* than atmospheric pressure (14.7 lb/in.^{2} or 1.01×10^5 Pa) by this amount. The *total* pressure in the tire is then 47 lb/in.^{2} or 320 kPa. The excess pressure above atmospheric pressure is usually called **gauge pressure,** and the total pressure is called **absolute pressure.** Engineers use the abbreviations psig and psia for "pounds per square inch gauge" and "pounds per square inch absolute," respectively. If the pressure is *less* than atmospheric, as in a partial vacuum, the gauge pressure is negative.

Example 14.3 Finding absolute and gauge pressures

A solar water-heating system uses solar panels on the roof, 12.0 m above the storage tank. The water pressure at the level of the panels is one atmosphere. What is the absolute pressure in the tank? The gauge pressure?

SOLUTION

IDENTIFY: Water is nearly incompressible. (Imagine trying to compress a cylinder full of water with a piston—you couldn't do it!) Hence we can treat it as a fluid of uniform density.

SET UP: The level of the panels corresponds to point 2 in Fig. 14.5, and the level of the tank corresponds to point 1. Hence our target variable is p in Eq. (14.6); we are given $p_0 = 1\text{ atm} = 1.01 \times 10^{-5}\text{ Pa}$ and $h = 12.0\text{ m}$.

EXECUTE: From Eq. (14.6), the absolute pressure is

$$\begin{aligned} p &= p_0 + \rho g h \\ &= (1.01 \times 10^5\text{ Pa}) + (1000\text{ kg/m}^3)(9.80\text{ m/s}^2)(12.0\text{ m}) \\ &= 2.19 \times 10^5\text{ Pa} = 2.16\text{ atm} = 31.8\text{ lb/in.}^2 \end{aligned}$$

The gauge pressure is

$$\begin{aligned} p - p_0 &= (2.19 - 1.01) \times 10^5\text{ Pa} \\ &= 1.18 \times 10^5\text{ Pa} = 1.16\text{ atm} = 17.1\text{ lb/in.}^2 \end{aligned}$$

EVALUATE: If such a tank has a pressure gauge, it is usually calibrated to read gauge pressure rather than absolute pressure. As we have mentioned, the variation in *atmospheric* pressure over this height is negligibly small.

The simplest pressure gauge is the open-tube *manometer* (Fig. 14.8a). The U-shaped tube contains a liquid of density ρ, often mercury or water. One end of the tube is connected to the container where the pressure p is to be measured, and the other end is open to the atmosphere at pressure $p_0 = p_a$. The pressure at the bottom of the tube due to the fluid in the left column is $p + \rho g y_1$, and the pressure at the bottom due to the fluid in the right column is $p_a + \rho g y_2$. These pressures are measured at the same point, so they must be equal:

$$\begin{aligned} p + \rho g y_1 &= p_a + \rho g y_2 \\ p - p_a &= \rho g (y_2 - y_1) = \rho g h \end{aligned} \tag{14.8}$$

In Eq. (14.8), p is the *absolute pressure,* and the difference $p - p_a$ between absolute and atmospheric pressure is the gauge pressure. Thus the gauge pressure is proportional to the difference in height $(y_2 - y_1)$ of the liquid columns.

Another common pressure gauge is the **mercury barometer.** It consists of a long glass tube, closed at one end, that has been filled with mercury and then inverted in a dish of mercury (Fig. 14.8b). The space above the mercury column contains only mercury vapor; its pressure is negligibly small, so the pressure p_0 at the top of the mercury column is practically zero. From Eq. (14.6),

$$p_a = p = 0 + \rho g (y_2 - y_1) = \rho g h \tag{14.9}$$

Thus the mercury barometer reads the atmospheric pressure p_a directly from the height of the mercury column.

In many applications, pressures are still commonly described in terms of the height of the corresponding mercury column, as so many "inches of mercury" or "millimeters of mercury" (abbreviated mm Hg). A pressure of 1 mm Hg is called *one torr,* after Evangelista Torricelli, inventor of the mercury barometer. But these units depend on the density of mercury, which varies with temperature, and on the value of g, which varies with location, so the pascal is the preferred unit of pressure.

One common type of blood-pressure gauge, called a *sphygmomanometer*, uses a mercury-filled manometer. Blood-pressure readings, such as 130/80, refer to the maximum and minimum gauge pressures in the arteries, measured in mm Hg or torr. Blood pressure varies with height; the standard reference point is the upper arm, level with the heart.

Many types of pressure gauges use a flexible sealed vessel (Fig. 14.9). A change in the pressure either inside or outside the vessel causes a change in its dimensions. This change is detected optically, electrically, or mechanically.

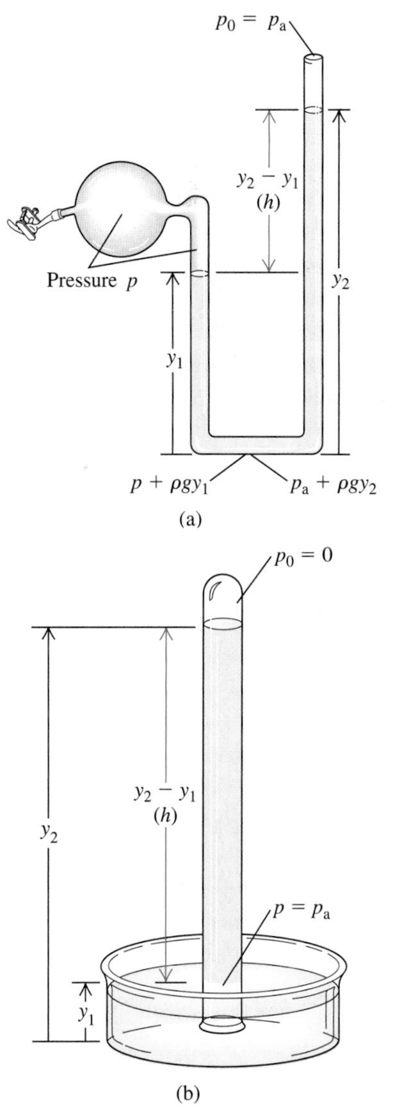

14.8 Two types of pressure gauge. (a) The open-tube manometer. (b) The mercury barometer.

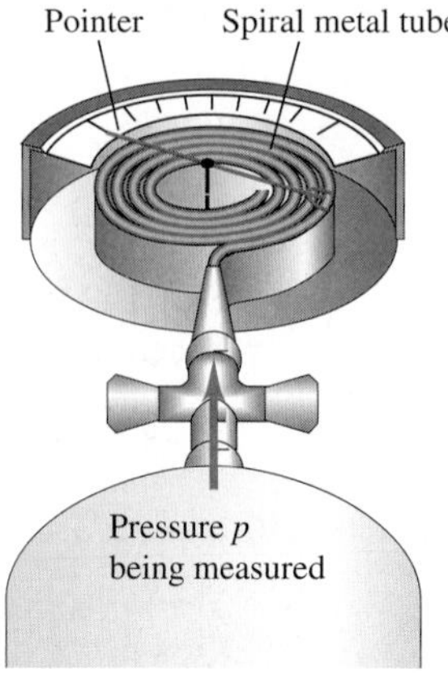

(a)

(b)

14.9 (a) A Bourdon pressure gauge. When the pressure inside the spiral metal tube increases, the tube straightens out a little, deflecting the attached pointer. (b) A Bourdon-type pressure gauge used on a compressed gas tank.

Example 14.4 A tale of two fluids

A manometer tube is partially filled with water. Oil (which does not mix with water, and has a lower density than water) is then poured into the left arm of the tube until the oil–water interface is at the midpoint of the tube (Fig. 14.10). Both arms of the tube are open to the air. Find a relationship between the heights h_{oil} and h_{water}.

SOLUTION

IDENTIFY: The relationship between pressure and depth in a fluid applies only to fluids of uniform density. Hence we can't write a single equation for the oil and the water together. We can, however, write a pressure–depth relation for each fluid separately, noting that both fluid columns have the same pressure at the bottom (where they are in contact and in equilibrium, so the pressures must be the same) and at the top (where both are in contact with and in equilibrium with the air).

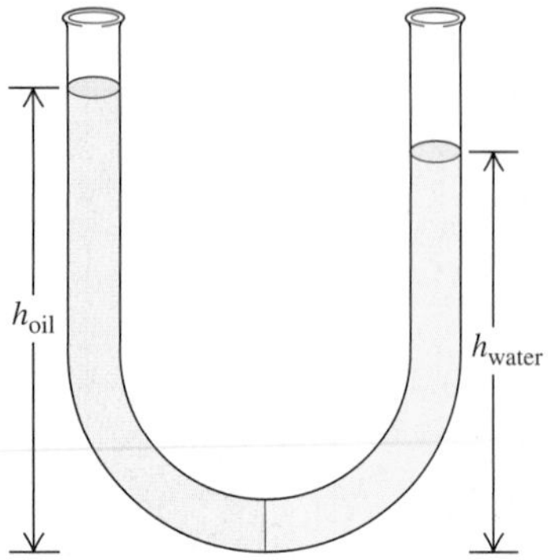

14.10 A U-shaped tube containing oil (on the left) and water (on the right). What is the relationship between the heights of the two fluid columns?

SET UP: Let p_0 be atmospheric pressure and let p be the pressure at the bottom of the tube. The densities of the two fluids are ρ_{water} and ρ_{oil} (which is less than ρ_{water}). We use Eq. (14.6) for each fluid.

EXECUTE: For the two fluids, Eq. (14.6) becomes

$$p = p_0 + \rho_{water} g h_{water}$$
$$p = p_0 + \rho_{oil} g h_{oil}$$

Since the pressure p at the bottom of the tube is the same for both fluids, we set these two expressions equal to each other and solve for h_{oil} in terms of h_{water}. You can show that the result is

$$h_{oil} = \frac{\rho_{water}}{\rho_{oil}} h_{water}$$

EVALUATE: Since oil is less dense than water, the ratio ρ_{water}/ρ_{oil} is greater than unity and h_{oil} is greater than h_{water} (as shown in Fig. 14.10). This result makes sense: a greater height of low-density oil is needed to produce the same pressure p at the bottom of the tube.

Test Your Understanding

Mercury is less dense at high temperatures than at low temperatures. Suppose you move a mercury barometer from the inside of a tightly sealed refrigerator to outdoors on a hot summer day. You find that the column of mercury remains at the same height in the tube. How does the air pressure outdoors compare to the air pressure inside the refrigerator? (Ignore the even smaller change in the dimensions of the glass tube due to the temperature change.)

14.3 | Buoyancy

Buoyancy is a familiar phenomenon: a body immersed in water seems to weigh less than when it is in air. When the body is less dense than the fluid, it floats. The human body usually floats in water, and a helium-filled balloon floats in air.

Archimedes's principle states: **When a body is completely or partially immersed in a fluid, the fluid exerts an upward force on the body equal to the weight of the fluid displaced by the body.** To prove this principle, we consider an arbitrary portion of fluid at rest. In Fig. 14.11a the irregular outline is the surface bounding this portion of fluid. The arrows represent the forces exerted on the boundary surface by the surrounding fluid.

The entire fluid is in equilibrium, so the sum of all the y-components of force on this portion of fluid is zero. Hence the sum of the y-components of the *surface* forces must be an upward force equal in magnitude to the weight mg of the fluid inside the surface. Also, the sum of the torques on the portion of fluid must be zero, so the line of action of the resultant y-component of surface force must pass through the center of gravity of this portion of fluid.

Now we remove the fluid inside the surface and replace it with a solid body having exactly the same shape (Fig. 14.11b). The pressure at every point is exactly the same as before. So the total upward force exerted on the body by the fluid is also the same, again equal in magnitude to the weight mg of the fluid displaced to make way for the body. We call this upward force the **buoyant force** on the solid body. The line of action of the buoyant force again passes through the center of gravity of the displaced fluid (which doesn't necessarily coincide with the center of gravity of the body).

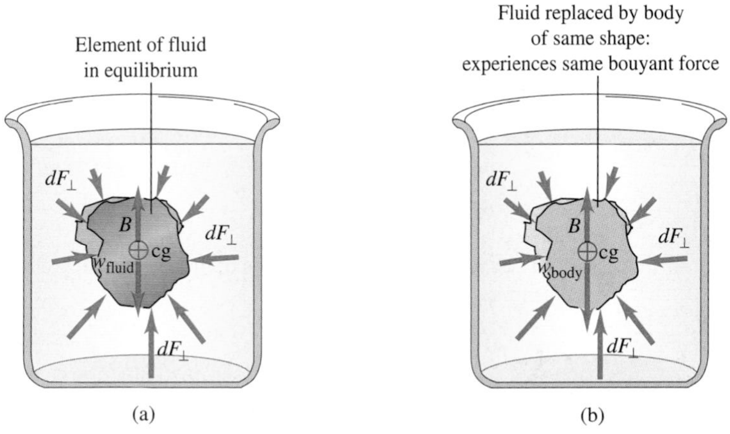

14.11 Archimedes's principle. (a) An element of a fluid in equilibrium. The buoyant force of the surrounding fluid equals the weight of the element. (b) If the fluid element is replaced by a body of identical shape, the body experiences the same buoyant force as in (a). This force is equal to the weight of the displaced fluid.

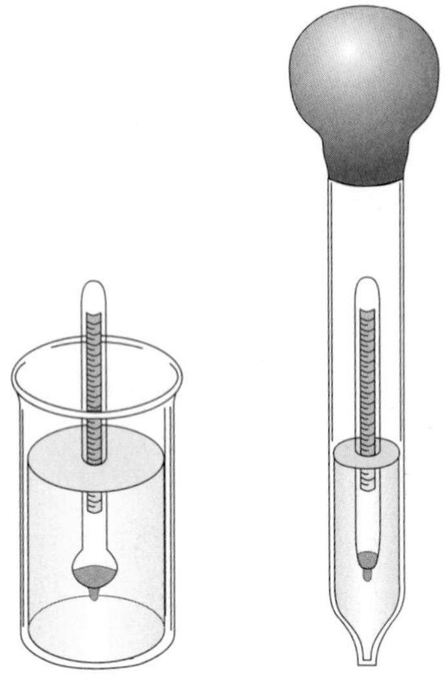

(a) (b)

14.12 (a) A simple hydrometer. (b) Use of a hydrometer as a tester for battery acid or antifreeze.

When a balloon floats in equilibrium in air, its weight (including the gas inside it) must be the same as the weight of the air displaced by the balloon. When a submerged submarine is in equilibrium, its weight must equal the weight of the water it displaces. A body whose average density is less than that of a liquid can float partially submerged at the free upper surface of the liquid. The greater the density of the liquid, the less of the body is submerged. When you swim in seawater (density 1030 kg/m^3), your body floats higher than in fresh water (1000 kg/m^3). Unlikely as it may seem, lead floats in mercury. Very flat-surfaced "float glass" is made by floating molten glass on molten tin and then letting it cool.

Another familiar example is the hydrometer, used to measure the density of liquids (Fig. 14.12a). The calibrated float sinks into the fluid until the weight of the fluid it displaces is exactly equal to its own weight. The hydrometer floats *higher* in denser liquids than in less dense liquids. It is weighted at its bottom end so that the upright position is stable, and a scale in the top stem permits direct density readings. Figure 14.12b shows a type of hydrometer that is commonly used to measure the density of battery acid or antifreeze. The bottom of the large tube is immersed in the liquid; the bulb is squeezed to expel air and is then released, like a giant medicine dropper. The liquid rises into the outer tube, and the hydrometer floats in this sample of the liquid.

Example 14.5 Buoyancy

A 15.0-kg solid gold statue is being raised from a sunken ship (Fig. 14.13a). What is the tension in the hoisting cable when the statue is at rest and a) completely immersed; b) out of the water?

SOLUTION

IDENTIFY: When the statue is immersed, it experiences an upward buoyant force equal in magnitude to the weight of fluid displaced.

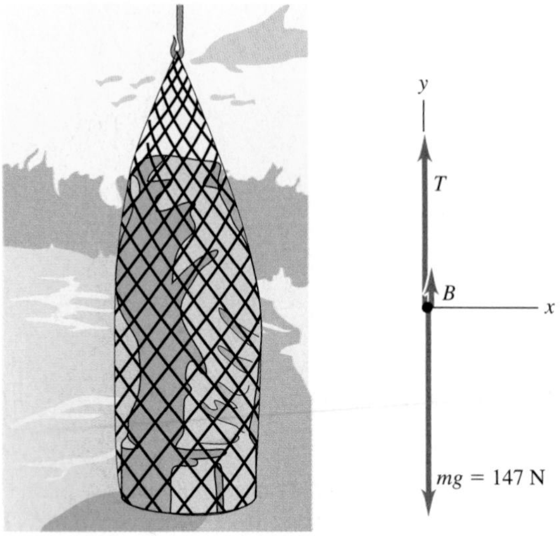

(a) (b)

14.13 (a) The completely immersed statue at rest. (b) Free-body diagram for the immersed statue.

To find the tension, we note that the statue is in equilibrium (it is at rest) and consider the three forces acting on it: weight, the buoyant force, and the tension in the cable.

SET UP: Figure 14.13b shows the free-body diagram for the statue in equilibrium. Our target variable is the tension T. We are given the weight mg, and we can calculate the buoyant force B by using Archimedes's principle. We do this for two cases: (a) when the statue is immersed in water and (b) when it is out of the water and immersed in air.

EXECUTE: a) To find the buoyant force, we first find the volume of the statue, using the density of gold from Table 14.1:

$$V = \frac{m}{\rho_{\text{gold}}} = \frac{15.0\ \text{kg}}{19.3 \times 10^3\ \text{kg/m}^3} = 7.77 \times 10^{-4}\ \text{m}^3$$

Using Table 14.1 again, we find the weight of this volume of seawater:

$$\begin{aligned} w_{\text{sw}} &= m_{\text{sw}}g = \rho_{\text{sw}}Vg \\ &= (1.03 \times 10^3\ \text{kg/m}^3)(7.77 \times 10^{-4}\ \text{m}^3)(9.80\ \text{m/s}^2) \\ &= 7.84\ \text{N} \end{aligned}$$

This equals the buoyant force B.

The statue is at rest, so the net external force acting on it is zero. From Fig. 14.13b,

$$\sum F_y = B + T + (-mg) = 0$$

$$\begin{aligned} T &= mg - B = (15.0\ \text{kg})(9.80\ \text{m/s}^2) - 7.84\ \text{N} \\ &= 147\ \text{N} - 7.84\ \text{N} = 139\ \text{N} \end{aligned}$$

If a spring scale is attached to the upper end of the cable, it will indicate 7.84 N less than if the statue were not immersed in seawater. Hence the submerged statue seems to weigh 139 N, about 5% less than its actual weight of 147 N.

b) The density of air is about 1.2 kg/m^3, so the buoyant force of air on the statue is

$$B = \rho_{air}Vg = (1.2\ \text{kg/m}^3)(7.77 \times 10^{-4}\ \text{m}^3)(9.80\ \text{m/s}^2)$$
$$= 9.1 \times 10^{-3}\ \text{N}$$

This is only 62 parts per million of the statue's actual weight. This effect is not within the precision of our data, and we ignore it. Thus the tension in the cable with the statue in air is equal to the statue's weight, 147 N.

EVALUATE: Note that the buoyant force is proportional to the density of the *fluid,* not the density of the statue. The denser the fluid, the greater the buoyant force and the smaller the cable tension. If the fluid had the same density as the statue, the buoyant force would be equal to the statue's weight and the tension would be zero (the cable would go slack). If the fluid were denser than the statue, the tension would be *negative:* the buoyant force would be greater than the statue's weight, and a downward force would be required to keep the statue from rising upward.

Conceptual Example 14.6 A weighty question

You place a container of seawater on a scale and note the reading on the scale. You now suspend the statue of Example 14.5 in the water (Fig. 14.14). How does the scale reading change?

SOLUTION

Consider the water, the statue, and the container together as a system; the total weight of the system does not depend on whether the statue is immersed. The total supporting force, including the tension T and the upward force F of the scale on the container (equal to the scale reading), is the same in both cases. But we saw in Example 14.5 that T decreases by 7.84 N when the statue is immersed, so the scale reading F must *increase* by 7.84 N. An alternative viewpoint is that the water exerts an upward buoyant force of 7.84 N on the statue, so the statue must exert an equal downward force on the water, making the scale reading 7.84 N greater than the weight of water and container.

14.14 How does the scale reading change when the statue is immersed in water?

Surface Tension

An object less dense than water, such as an air-filled beach ball, floats with part of its volume below the surface. Conversely, a paper clip can rest *atop* a water surface even though its density is several times that of water. This is an example of **surface tension:** the surface of the liquid behaves like a membrane under tension (Fig. 14.15). Surface tension arises because the molecules of the liquid exert attractive forces on each other. There is zero net force on a molecule inside the volume of the liquid, but a surface molecule is drawn into the volume (Fig. 14.16). Thus the liquid tends to minimize its surface area, just as a stretched membrane does.

14.15 The surface of the water acts like a membrane under tension, allowing this water strider to literally "walk on water."

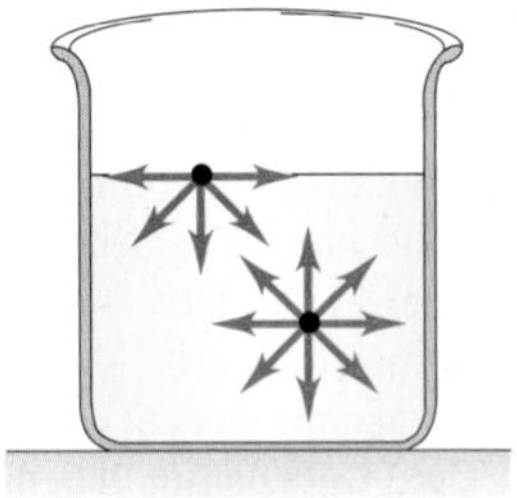

14.16 Each molecule in a liquid is attracted to its fellow molecules. A molecule of the surface is attracted into the bulk liquid, which tends to reduce the liquid's surface area.

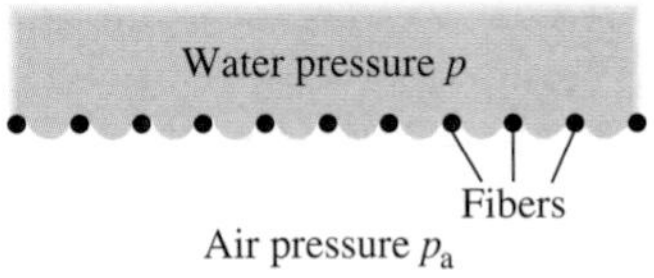

14.17 Surface tension makes it difficult to force water through small crevices. The required water pressure p can be reduced by using hot, soapy water, which has less surface tension.

Surface tension explains why freely falling raindrops are spherical (*not* teardrop-shaped): a sphere has a smaller surface area for its volume than any other shape. It also explains why hot, soapy water is used for washing. To wash clothing thoroughly, water must be forced through the tiny spaces between the fibers (Fig. 14.17). To do so requires increasing the surface area of the water, which is difficult to achieve because of surface tension. The job is made easier by increasing the temperature of the water and adding soap, both of which decrease the surface tension.

Surface tension is important for a millimeter-sized water drop, which has a relatively large surface area for its volume. (A sphere of radius r has surface area $4\pi r^2$ and volume $(4\pi/3)r^3$. The ratio of surface area to volume is $3/r$, which increases with decreasing radius.) For large quantities of liquid, however, the ratio of surface area to volume is relatively small, and surface tension is negligible compared to pressure forces. For the remainder of this chapter, we will consider only fluids in bulk and hence will ignore the effects of surface tension.

Test Your Understanding

An object with uniform density floats in water with one-third of its volume above the surface. How does the density of the object compare to the density of water?

14.4 | Fluid Flow

We are now ready to consider *motion* of a fluid. Fluid flow can be extremely complex, as shown by the currents in river rapids or the swirling flames of a campfire. But some situations can be represented by relatively simple idealized models. An **ideal fluid** is a fluid that is *incompressible* (that is, its density cannot change) and has no internal friction (called **viscosity**). Liquids are approximately incompressible in most situations, and we may also treat a gas as incompressible if the pressure differences from one region to another are not too great. Internal friction in a fluid causes shear stresses when two adjacent layers of fluid move relative to each other, as when fluid flows inside a tube or around an obstacle. In some cases we can neglect these shear forces in comparison with forces arising from gravitation and pressure differences.

The path of an individual particle in a moving fluid is called a **flow line.** If the overall flow pattern does not change with time, the flow is called **steady flow.** In steady flow, every element passing through a given point follows the same flow line. In this case the "map" of the fluid velocities at various points in space remains constant, although the velocity of a particular particle may change in both magnitude and direction during its motion. A **streamline** is a curve whose tangent at any point is in the direction of the fluid velocity at that point. When the flow pattern changes with time, the streamlines do not coincide with the flow lines. We will consider only steady-flow situations, for which flow lines and streamlines are identical.

The flow lines passing through the edge of an imaginary element of area, such as the area A in Fig. 14.18, form a tube called a **flow tube.** From the definition of a flow line, in steady flow no fluid can cross the side walls of a flow tube; the fluids in different flow tubes cannot mix.

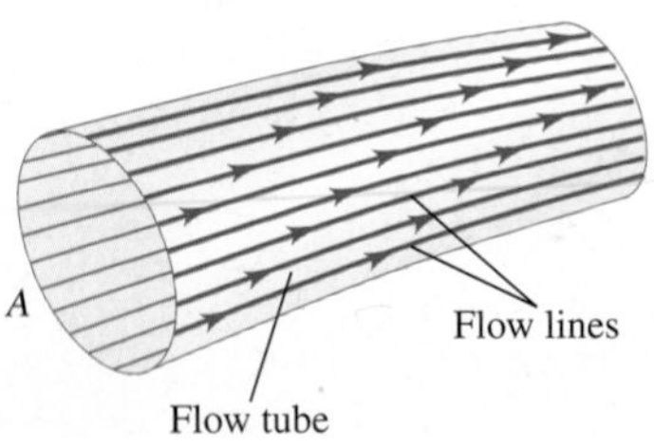

14.18 A flow tube bounded by flow lines. In steady flow, fluid cannot cross the walls of a flow tube.

14.19 Laminar flow around obstacles of different shapes.

Figure 14.19 shows patterns of fluid flow from left to right around a number of obstacles. The photographs were made by injecting dye into water flowing between two closely spaced glass plates. These patterns are typical of **laminar flow,** in which adjacent layers of fluid slide smoothly past each other and the flow is steady. (A *lamina* is a thin sheet.) At sufficiently high flow rates, or when boundary surfaces cause abrupt changes in velocity, the flow can become irregular and chaotic. This is called **turbulent flow** (Fig. 14.20). In turbulent flow there is no steady-state pattern; the flow pattern continuously changes.

14.20 The flow of smoke rising from these incense sticks is laminar up until a certain point, then becomes turbulent.

The Continuity Equation

The mass of a moving fluid doesn't change as it flows. This leads to an important quantitative relationship called the **continuity equation.** Consider a portion of a flow tube between two stationary cross sections with areas A_1 and A_2 (Fig. 14.21). The fluid speeds at these sections are v_1 and v_2, respectively. No fluid flows in or out across the sides of the tube because the fluid velocity is tangent to the wall at every point on the wall. During a small time interval dt, the fluid at A_1 moves a distance $v_1\,dt$, so a cylinder of fluid with height $v_1\,dt$ and volume $dV_1 = A_1 v_1\,dt$ flows into the tube across A_1. During this same interval, a cylinder of volume $dV_2 = A_2 v_2\,dt$ flows out of the tube across A_2.

Let's first consider the case of an incompressible fluid so that the density ρ has the same value at all points. The mass dm_1 flowing into the tube across A_1 in time dt is $dm_1 = \rho A_1 v_1\,dt$. Similarly, the mass dm_2 that flows out across A_2 in the same time is $dm_2 = \rho A_2 v_2\,dt$. In steady flow the total mass in the tube is constant, so $dm_1 = dm_2$ and

$$\rho A_1 v_1\,dt = \rho A_2 v_2\,dt \qquad \text{or}$$

$$A_1 v_1 = A_2 v_2 \qquad \text{(continuity equation, incompressible fluid)} \quad (14.10)$$

The product Av is the *volume flow rate dV/dt*, the rate at which volume crosses a section of the tube:

$$\frac{dV}{dt} = Av \qquad \text{(volume flow rate)} \qquad (14.11)$$

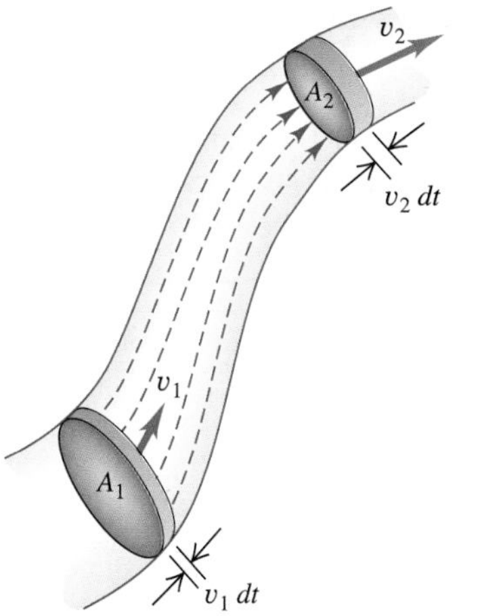

14.21 A flow tube with changing cross-sectional area. If the fluid is incompressible, the product Av has the same value at all points along the tube.

The *mass* flow rate is the mass flow per unit time through a cross section. This is equal to the density ρ times the volume flow rate dV/dt.

Equation (14.10) shows that the volume flow rate has the same value at all points along any flow tube. When the cross section of a flow tube decreases, the speed increases, and vice versa. The deep part of a river has larger cross section and slower current than the shallow part, but the volume flow rates are the same in both. This is the essence of the familiar maxim, "Still waters run deep." The stream of water from a faucet narrows as it gains speed during its fall, but dV/dt is the same everywhere along the stream. If a water pipe with 2-cm diameter is connected to a pipe with 1-cm diameter, the flow speed is four times as great in the 1-cm part as in the 2-cm part.

We can generalize Eq. (14.10) for the case in which the fluid is *not* incompressible. If ρ_1 and ρ_2 are the densities at sections 1 and 2, then

$$\rho_1 A_1 v_1 = \rho_2 A_2 v_2 \qquad \text{(continuity equation, compressible fluid)} \qquad (14.12)$$

We leave the details for an exercise. If the fluid is incompressible so that ρ_1 and ρ_2 are always equal, Eq. (14.12) reduces to Eq. (14.10).

Example 14.7 Incompressible fluid flow

As part of a lubricating system for heavy machinery, oil of density 850 kg/m^3 is pumped through a cylindrical pipe of diameter 8.0 cm at a rate of 9.5 liters per second. a) What is the speed of the oil? What is the mass flow rate? b) If the pipe diameter is reduced to 4.0 cm, what are the new values of the speed and volume flow rate? Assume that the oil is incompressible.

SOLUTION

IDENTIFY AND SET UP: We use the definition of volume flow rate, Eq. (14.11), to determine the speed v_1 in the 8.0-cm-diameter section. The mass flow rate is the product of the density and the volume flow rate. The continuity equation for incompressible flow, Eq. (14.10), allows us to find the speed v_2 in the 4.0-cm-diameter section.

EXECUTE: a) The volume flow rate dV/dt equals the product $A_1 v_1$, where A_1 is the cross-sectional area of the pipe of diameter 8.0 cm and radius 4.0 cm. Hence

$$v_1 = \frac{dV/dt}{A_1} = \frac{(9.5\ \text{L/s})(10^{-3}\ \text{m}^3/\text{L})}{\pi(4.0 \times 10^{-2}\ \text{m})^2} = 1.9\ \text{m/s}$$

The mass flow rate is $\rho\, dV/dt = (850\ \text{kg/m}^3)(9.5 \times 10^{-3}\ \text{m}^3/\text{s}) = 8.1\ \text{kg/s}$.

b) Since the oil is incompressible, the volume flow rate has the same value (9.5 L/s) in both sections of pipe. From Eq. (14.10),

$$v_2 = \frac{A_1}{A_2} v_1 = \frac{\pi(4.0 \times 10^{-2}\ \text{m})^2}{\pi(2.0 \times 10^{-2}\ \text{m})^2}(1.9\ \text{m/s}) = 7.6\ \text{m/s}$$

EVALUATE: The second section of pipe has one-half the diameter and one-fourth the cross-sectional area of the first section. Hence the speed must be four times greater in the second section, which is just what our result shows $(v_2 = 4v_1)$.

Test Your Understanding

Air in the atmosphere is nearly incompressible. Use this fact to explain why particularly fast-moving winds are found in mountain passes.

14.5 | Bernoulli's Equation

According to the continuity equation, the speed of fluid flow can vary along the paths of the fluid. The pressure can also vary; it depends on height as in the static situation (Section 14.2), and it also depends on the speed of flow. We can derive an important relationship called *Bernoulli's equation* that relates the pressure, flow speed, and height for flow of an ideal, incompressible fluid. Bernoulli's equation is an essential tool in analyzing plumbing systems, hydroelectric generating stations, and the flight of airplanes.

The dependence of pressure on speed follows from the continuity equation, Eq. (14.10). When an incompressible fluid flows along a flow tube with varying cross section, its speed *must* change, and so an element of fluid must have an acceleration. If the tube is horizontal, the force that causes this acceleration has to be applied by the surrounding fluid. This means that the pressure *must* be different in regions of different cross section; if it were the same everywhere, the net force on every fluid element would be zero. When a horizontal flow tube narrows and a fluid element speeds up, it must be moving toward a region of lower pressure in order to have a net forward force to accelerate it. If the elevation also changes, this causes an additional pressure difference.

To derive Bernoulli's equation, we apply the work-energy theorem to the fluid in a section of a flow tube. In Fig. 14.22 we consider the element of fluid that at some initial time lies between the two cross sections a and c. The speeds at the lower and upper ends are v_1 and v_2. In a small time interval dt, the fluid that is initially at a moves to b, a distance $ds_1 = v_1\,dt$, and the fluid that is initially at c moves to d, a distance $ds_2 = v_2\,dt$. The cross-sectional areas at the two ends are A_1 and A_2, as shown. The fluid is incompressible; hence by the continuity equation, Eq. (14.10), the volume of fluid dV passing *any* cross section during time dt is the same. That is, $dV = A_1\,ds_1 = A_2\,ds_2$.

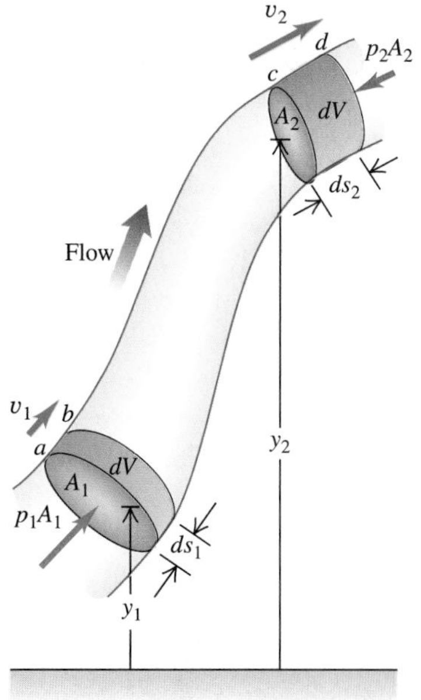

14.22 The net work done on a fluid element by the pressure of the surrounding fluid equals the change in the kinetic energy plus the change in the gravitational potential energy.

Let's compute the *work* done on this fluid element during dt. We assume that there is negligible internal friction in the fluid (i.e., no viscosity), so the only nongravitational forces that do work on the fluid element are due to the pressure of the surrounding fluid. The pressures at the two ends are p_1 and p_2; the force on the cross section at a is p_1A_1, and the force at c is p_2A_2. The net work dW done on the element by the surrounding fluid during this displacement is therefore

$$dW = p_1A_1\,ds_1 - p_2A_2\,ds_2 = (p_1 - p_2)dV \qquad (14.13)$$

The second term has a negative sign because the force at c opposes the displacement of the fluid.

The work dW is due to forces other than the conservative force of gravity, so it equals the change in the total mechanical energy (kinetic energy plus gravitational potential energy) associated with the fluid element. The mechanical energy for the fluid between sections b and c does not change. At the beginning of dt the fluid between a and b has volume $A_1\,ds_1$, mass $\rho A_1\,ds_1$, and kinetic energy $\frac{1}{2}\rho(A_1\,ds_1)v_1^2$. At the end of dt the fluid between c and d has kinetic energy $\frac{1}{2}\rho(A_2\,ds_2)v_2^2$. The net change in kinetic energy dK during time dt is

$$dK = \frac{1}{2}\rho\,dV(v_2^2 - v_1^2) \qquad (14.14)$$

What about the change in gravitational potential energy? At the beginning of dt the potential energy for the mass between a and b is $dm\,gy_1 = \rho\,dV\,gy_1$. At the end of dt the potential energy for the mass between c and d is $dm\,gy_2 = \rho\,dV\,gy_2$. The net change in potential energy dU during dt is

$$dU = \rho\,dV\,g(y_2 - y_1) \qquad (14.15)$$

Combining Eqs. (14.13), (14.14), and (14.15) in the energy equation $dW = dK + dU$, we obtain

$$(p_1 - p_2)dV = \frac{1}{2}\rho\, dV(v_2^2 - v_1^2) + \rho\, dV\, g(y_2 - y_1)$$

$$p_1 - p_2 = \frac{1}{2}\rho(v_2^2 - v_1^2) + \rho g(y_2 - y_1) \tag{14.16}$$

This is **Bernoulli's equation.** It states that the work done on a unit volume of fluid by the surrounding fluid is equal to the sum of the changes in kinetic and potential energies per unit volume that occur during the flow. We may also interpret Eq. (14.16) in terms of pressures. The first term on the right is the pressure difference associated with the change of speed of the fluid. The second term on the right is the additional pressure difference caused by the weight of the fluid and the difference in elevation of the two ends.

We can also express Eq. (14.16) in a more convenient form as

$$p_1 + \rho g y_1 + \frac{1}{2}\rho v_1^2 = p_2 + \rho g y_2 + \frac{1}{2}\rho v_2^2 \quad \text{(Bernoulli's equation)} \tag{14.17}$$

The subscripts 1 and 2 refer to *any* two points along the flow tube, so we can also write

$$p + \rho g y + \frac{1}{2}\rho v^2 = \text{constant} \tag{14.18}$$

Note that when the fluid is *not* moving (so $v_1 = v_2 = 0$), Eq. (14.17) reduces to the pressure relation we derived for a fluid at rest, Eq. (14.5).

CAUTION We stress again that Bernoulli's equation is valid only for incompressible, steady flow of a fluid with no internal friction (no viscosity). It's a simple equation that's easy to use; don't let this tempt you into using it in situations in which it doesn't apply!

Problem-Solving Strategy: Bernoulli's Equation

Bernoulli's equation is derived from the work-energy theorem, so it isn't surprising that much of the problem-solving strategy suggested in Section 7.1 is applicable here.

IDENTIFY *the relevant concepts:* First ensure that the fluid flow is steady and that the fluid is incompressible and has no internal friction. This case is an idealization, but it holds up surprisingly well for fluids flowing through sufficiently large pipes and for flows within bulk fluids (e.g., air flowing around an airplane or water flowing around a fish).

SET UP *the problem using the following steps:*

1. Always begin by identifying clearly the points 1 and 2 referred to in Bernoulli's equation.
2. Define your coordinate system, particularly the level at which $y = 0$.
3. Make lists of the unknown and known quantities in Eq. (14.17). The variables are p_1, p_2, v_1, v_2, y_1, and y_2, and the constants are ρ and g. Decide which unknowns are your target variables.

EXECUTE *the solution as follows:* Write Bernoulli's equation and solve for the unknowns. In some problems you will need to use the continuity equation, Eq. (14.10), to get a relation between the two speeds in terms of cross-sectional areas of pipes or containers. Or perhaps you will know both speeds and need to determine one of the areas. You may also need to use Eq. (14.11) to find the volume flow rate.

EVALUATE *your answer:* As always, verify that the results make physical sense. Double-check that you have used consistent units. In SI units, pressure is in pascals, density in kilograms per cubic meter, and speed in meters per second. Also note that the pressures must be either all absolute pressures or all gauge pressures.

Example 14.8 Water pressure in the home

Water enters a house through a pipe with an inside diameter of 2.0 cm at an absolute pressure of 4.0×10^5 Pa (about 4 atm). A 1.0-cm-diameter pipe leads to the second-floor bathroom 5.0 m above (Fig. 14.23). When the flow speed at the inlet pipe is 1.5 m/s, find the flow speed, pressure, and volume flow rate in the bathroom.

SOLUTION

IDENTIFY AND SET UP: Let points 1 and 2 be at the inlet pipe and at the bathroom, respectively. We are given the speed v_1 and pressure p_1 at the inlet pipe, and the pipe diameters at points 1 and 2 (from which we calculate the areas A_1 and A_2). We take $y_1 = 0$ (at the inlet) and $y_2 = 5.0$ m (at the bathroom). Our first two target variables are the speed v_2 and pressure p_2. Since we have more than one unknown, we use both Bernoulli's equation and the continuity equation. Once we find v_2, we calculate the volume flow rate v_2A_2 at point 2.

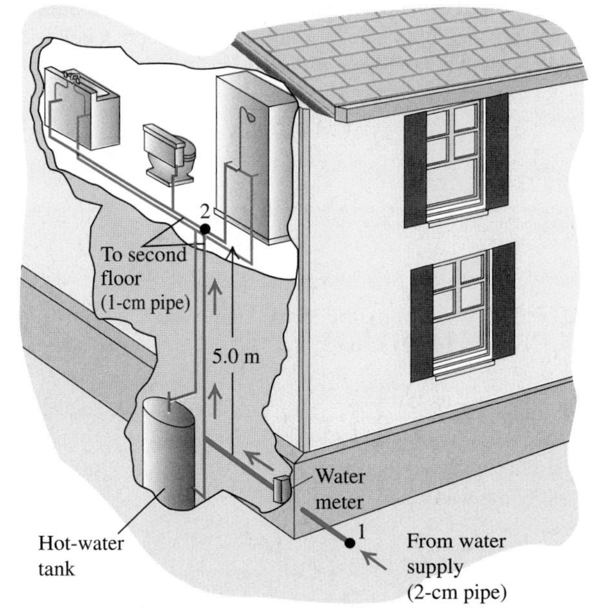

14.23 What is the water pressure in the second-story bathroom of this house?

EXECUTE: The speed v_2 at the bathroom is obtained from the continuity equation, Eq. (14.10):

$$v_2 = \frac{A_1}{A_2}v_1 = \frac{\pi(1.0\text{ cm})^2}{\pi(0.50\text{ cm})^2}(1.5\text{ m/s}) = 6.0\text{ m/s}$$

We are given p_1 and v_1, and can find p_2 from Bernoulli's equation:

$$\begin{aligned}p_2 &= p_1 - \frac{1}{2}\rho(v_2^2 - v_1^2) - \rho g(y_2 - y_1) = 4.0 \times 10^5\text{ Pa}\\&\quad - \frac{1}{2}(1.0 \times 10^3\text{ kg/m}^3)(36\text{ m}^2/\text{s}^2 - 2.25\text{ m}^2/\text{s}^2)\\&\quad -(1.0 \times 10^3\text{ kg/m}^3)(9.8\text{ m/s}^2)(5.0\text{ m})\\&= 4.0 \times 10^5\text{ Pa} - 0.17 \times 10^5\text{ Pa} - 0.49 \times 10^5\text{ Pa}\\&= 3.3 \times 10^5\text{ Pa} = 3.3\text{ atm} = 48\text{ lb/in.}^2\end{aligned}$$

The volume flow rate is

$$\begin{aligned}\frac{dV}{dt} &= A_2v_2 = \pi(0.50 \times 10^{-2}\text{ m})^2(6.0\text{ m/s})\\&= 4.7 \times 10^{-4}\text{ m}^3/\text{s} = 0.47\text{ L/s}\end{aligned}$$

EVALUATE: This is a reasonable flow rate for a bathroom faucet or shower. Note that after the water is turned off, the term $\frac{1}{2}\rho(v_2^2 - v_1^2)$ in the equation for pressure vanishes, and the pressure rises to 3.5×10^5 Pa.

Example 14.9 Speed of efflux

Figure 14.24 shows a gasoline storage tank with cross-sectional area A_1, filled to a depth h. The space above the gasoline contains air at pressure p_0, and the gasoline flows out through a short pipe with area A_2. Derive expressions for the flow speed in the pipe and the volume flow rate.

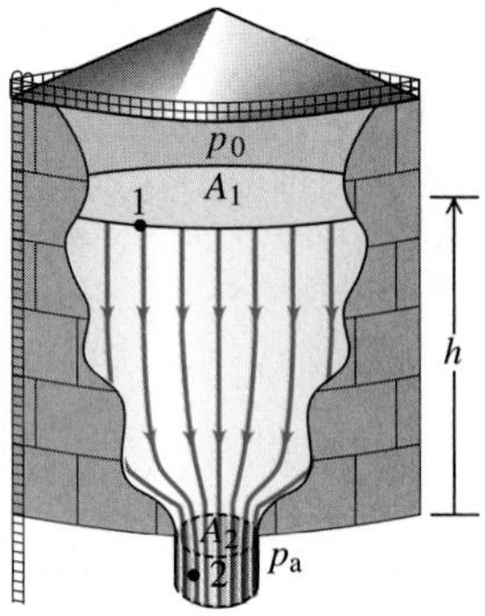

14.24 Calculating the speed of efflux for gasoline flowing out the bottom of a storage tank.

SOLUTION

IDENTIFY: We consider the entire volume of moving liquid as a single flow tube, so we can use Bernoulli's principle.

SET UP: Points 1 and 2 in Fig. 14.24 are at the surface of the gasoline and at the short exit pipe, respectively. At point 1 the pressure is p_0 and at point 2 it is atmospheric pressure p_a. We take $y = 0$ at the exit pipe, so $y_1 = h$ and $y_2 = 0$. Because A_1 is very much larger than A_2, the upper surface of the gasoline will drop very slowly and we can regard v_1 as essentially equal to zero. We find the target variable v_2 from Eq. (14.17) and the volume flow rate from Eq. (14.11).

EXECUTE: Applying Bernoulli's equation to points 1 and 2 and taking $y = 0$ at the bottom of the tank, we find

$$p_0 + \frac{1}{2}\rho v_1^2 + \rho g h = p_a + \frac{1}{2}\rho v_2^2$$

$$v_2^2 = v_1^2 + 2\frac{p_0 - p_a}{\rho} + 2gh$$

Using $v_1 = 0$, we find

$$v_2^2 = 2\frac{p_0 - p_a}{\rho} + 2gh$$

From Eq. (14.11), the volume flow rate is $dV/dt = v_2 A_2$.

EVALUATE: The speed v_2, sometimes called the *speed of efflux,* depends on both the pressure difference $(p_0 - p_a)$ and the height h of the liquid level in the tank. If the top of the tank is vented to the atmosphere, there is no excess pressure: $p_0 = p_a$ and $p_0 - p_a = 0$. In that case,

$$v_2 = \sqrt{2gh}$$

That is, the speed of efflux from an opening at a distance h below the top surface of the liquid is the *same* as the speed a body would acquire in falling freely through a height h. This result is called *Torricelli's theorem.* It is valid not only for an opening in the bottom of a container, but also for a hole in a side wall at a depth h below the surface. In this case the volume flow rate is

$$\frac{dV}{dt} = A_2\sqrt{2gh}$$

Example 14.10 The Venturi meter

Figure 14.25 shows a *Venturi meter,* used to measure flow speed in a pipe. The narrow part of the pipe is called the *throat.* Derive an expression for the flow speed v_1 in terms of the cross-sectional areas A_1 and A_2 and the difference in height h of the liquid levels in the two vertical tubes.

SOLUTION

IDENTIFY AND SET UP: We apply Bernoulli's equation to the wide (point 1) and narrow (point 2) parts of the pipe. The difference in height between the two vertical tubes tells us the pressure difference between points 1 and 2.

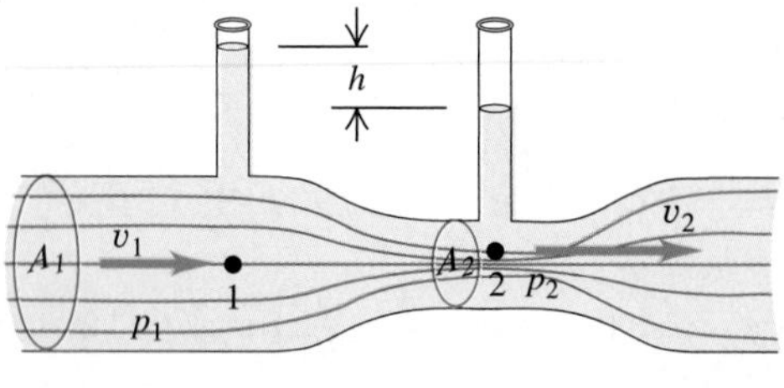

14.25 The Venturi meter.

EXECUTE: The two points are at the same vertical coordinate $(y_1 = y_2)$, so Eq. (14.17) says

$$p_1 + \frac{1}{2}\rho v_1^2 = p_2 + \frac{1}{2}\rho v_2^2$$

From the continuity equation, $v_2 = (A_1/A_2)v_1$. Substituting this and rearranging, we get

$$p_1 - p_2 = \frac{1}{2}\rho v_1^2\left(\frac{A_1^2}{A_2^2} - 1\right)$$

The pressure difference $p_1 - p_2$ is also equal to $\rho g h$, where h is the difference in liquid level in the two tubes. Combining this with the above result and solving for v_1, we get

$$v_1 = \sqrt{\frac{2gh}{(A_1/A_2)^2 - 1}}$$

EVALUATE: Because A_1 is greater than A_2, v_2 is greater than v_1 and the pressure p_2 in the throat is *less* than p_1. A net force to the right accelerates the fluid as it enters the throat, and a net force to the left slows it as it leaves.

Conceptual Example 14.11 Lift on an airplane wing

Figure 14.26 shows flow lines around a cross section of an airplane wing. The flow lines crowd together above the wing, corresponding to increased flow speed and reduced pressure in this region, just as in the Venturi throat. The upward force on the underside of the wing is greater than the downward force on the top side; there is a net upward force, or *lift.* Lift is not simply due to the impulse of air striking the underside of the wing; in fact, it turns out that the reduced pressure on the upper wing surface makes the greatest contribution to the lift. (This highly simplified discussion ignores the formation of vortices; a more complete discussion would take these into account.)

We can also understand the lift force on the basis of momentum changes. Figure 14.26 shows that there is a net *downward* change in the vertical component of momentum of the air flowing past the wing, corresponding to the downward force the wing exerts on the air. The reaction force *on* the wing is *upward,* as we concluded above.

A similar flow pattern and lift force are found in the vicinity of any humped object in a wind (see the airflow over the cyclist's back in the photo that opens this chapter). In a sufficiently strong wind, the lift force on the top of an open umbrella can collapse the umbrella upward. A lift force also acts on a car driving at high speed due to air moving over the car's curved upper surface. Such lift can reduce traction on the car's tires, which is why many cars are equipped with an aerodynamic "spoiler" at the car's tail. The spoiler is shaped like an upside-down wing and provides a downward force on the rear wheels.

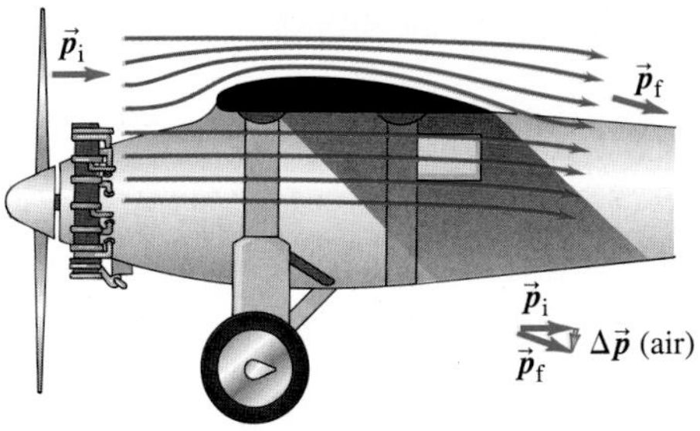

14.26 Flow lines around an airplane wing. The momentum of a parcel of air (relative to the wing) is $\vec{p}_i$ before encountering the wing and $\vec{p}_f$ afterward.

Test Your Understanding

It's not surprising that a wind blowing directly on an open door can make the door slam shut. Use Bernoulli's principle to explain how a wind blowing *across* a doorway (that is, parallel to the opening) can make the door close. (Assume that the door opens inward.)

*14.6 | Viscosity and Turbulence

In our discussion of fluid flow we assumed that the fluid had no internal friction and that the flow was laminar. While these assumptions are often quite valid, in many important physical situations the effects of viscosity (internal friction) and turbulence (nonlaminar flow) are extremely important. Let's take a brief look at some of these situations.

Viscosity

Viscosity is internal friction in a fluid. Viscous forces oppose the motion of one portion of a fluid relative to another. Viscosity is the reason why it takes effort to paddle a canoe through calm water, but it is also the reason why the paddle works. Viscous effects are important in the flow of fluids in pipes, the flow of blood, the lubrication of engine parts, and many other situations.

Fluids that flow readily, such as water or gasoline, have smaller viscosities than do "thick" liquids such as honey or motor oil. Viscosities of all fluids are strongly temperature dependent, increasing for gases and decreasing for liquids as the temperature increases (Fig. 14.27). An important goal in the design of oils for engine lubrication is to *reduce* the temperature variation of viscosity as much as possible.

14.27 Lava is an example of a viscous fluid. The viscosity decreases with increasing temperature: the hotter the lava, the more easily it can flow.

A viscous fluid always tends to cling to a solid surface in contact with it. There is always a thin *boundary layer* of fluid near the surface, in which the fluid is nearly at rest with respect to the surface. That's why dust particles can cling to a fan blade even when it is rotating rapidly, and why you can't get all the dirt off your car by just squirting a hose at it.

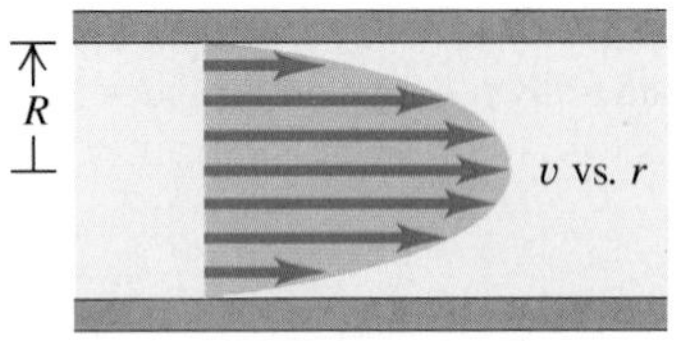

14.28 Velocity profile for a viscous fluid in a cylindrical pipe.

Viscosity has important effects on the flow of liquids through pipes, including the flow of blood in the circulatory system. First think about a fluid with zero viscosity so that we can apply Bernoulli's equation, Eq. (14.17). If the two ends of a long cylindrical pipe are at the same height $(y_1 = y_2)$ and the flow speed is the same at both ends (so $v_1 = v_2$), Bernoulli's equation tells us that the pressure is the same at both ends of the pipe. But this result simply isn't true if we take viscosity into account. To see why, consider Fig. 14.28, which shows the flow-speed profile for laminar flow of a viscous fluid in a long cylindrical pipe. Due to viscosity, the speed is *zero* at the pipe walls (to which the fluid clings) and is greatest at the center of the pipe. The motion is like a lot of concentric tubes sliding relative to one another, with the central tube moving fastest and the outermost tube at rest. Viscous forces between the tubes oppose this sliding, so to keep the flow going we must apply a greater pressure at the back of the flow than at the front. That's why you have to keep squeezing a tube of toothpaste or a packet of ketchup (both viscous fluids) to keep the fluid coming out of its container. Your fingers provide a pressure at the back of the flow that is far greater than the atmospheric pressure at the front of the flow.

The pressure difference required to sustain a given volume flow rate through a cylindrical pipe of length L and radius R turns out to be proportional to L/R^4. If we decrease R by one-half, the required pressure increases by $2^4 = 16$; decreasing R by a factor of 0.90 (a 10% reduction) increases the required pressure difference by a factor of $(1/0.90)^4 = 1.52$ (a 52% increase). This simple relationship explains the connection between a high-cholesterol diet (which tends to narrow the arteries) and high blood pressure. Due to the R^4 dependence, even a small narrowing of the arteries can result in substantially elevated blood pressure and added strain on the heart muscle.

Turbulence

When the speed of a flowing fluid exceeds a certain critical value, the flow is no longer laminar. Instead, the flow pattern becomes extremely irregular and complex, and it changes continuously with time; there is no steady-state pattern. This irregular, chaotic flow is called **turbulence.** Figure 14.20 shows the contrast between laminar and turbulent flow for smoke rising in air. Bernoulli's equation is *not* applicable to regions where there is turbulence, because the flow is not steady.

Whether a flow is laminar or turbulent depends in part on the fluid's viscosity. The greater the viscosity, the greater the tendency for the fluid to flow in sheets or lamina and the more likely the flow is to be laminar. (When we discussed Bernoulli's equation in Section 14.5, we assumed that the flow was laminar and that the fluid had zero viscosity. In fact, a *little* viscosity is needed to ensure that the flow is laminar.)

For a fluid of a given viscosity, flow speed is a determining factor. A flow pattern that is stable at low speeds suddenly becomes unstable when a critical speed is reached. Irregularities in the flow pattern can be caused by roughness in the pipe wall, variations in the density of the fluid, and many other factors. At low flow speeds, these disturbances damp out; the flow pattern is *stable* and tends to maintain its laminar nature. When the critical speed is reached, however the flow pattern becomes unstable. The disturbances no longer damp out, but grow until they destroy the entire laminar-flow pattern.

Normal blood flow in the human aorta is laminar, but a small disturbance such as a heart pathology can cause the flow to become turbulent. Turbulence makes noise, which is why listening to blood flow with a stethoscope is a useful diagnostic technique.

Conceptual Example 14.12 The curve ball

Does a curve ball *really* curve? Yes, it certainly does, and the reason is turbulence. Figure 14.29a shows a ball moving through the air from left to right. To an observer moving with the center of the ball, the air stream appears to move from right to left, as shown by the flow lines in the figure. Because of the large speeds that are ordinarily involved (near 160 km/h, or 100 mi/h), there is a region of *turbulent* flow behind the ball.

Figure 14.29b shows a *spinning* ball with "top spin." Layers of air near the ball's surface are pulled around in the direction of spin by friction between the ball and air and by the air's internal friction (viscosity). The speed of air relative to the ball's surface becomes smaller at the top of the ball than at the bottom, and turbulence occurs farther forward on the top side than on the bottom. This asymmetry causes a pressure difference; the average pressure at the top of the ball is now greater than that at the bottom. The net force deflects the ball downward, as shown in Fig. 14.29c. This is why "top spin" is used in tennis to keep a very fast serve in the court (Fig. 14.29d). In a baseball curve pitch, the ball spins about a nearly vertical axis, and the actual deflection is sideways. In that case, Fig. 14.29c is a *top* view of the situation. A curve ball thrown by a left-handed pitcher curves *toward* a right-handed batter, making it harder to hit (Fig. 14.29e).

A similar effect occurs with golf balls, which always have "back spin" from impact with the slanted face of the golf club. The resulting pressure difference between the top and bottom of the ball causes a lift force that keeps the ball in the air considerably longer than would be possible without spin. A well-hit drive appears from the tee to "float" or even curve *upward* during the initial portion of its flight. This is a real effect, not an illusion. The dimples on the ball play an essential role; the viscosity of air gives an undimpled ball a much shorter trajectory than a dimpled one with the same initial velocity and spin. Figure 14.30 shows the backspin of a golf ball just after it is struck by a club.

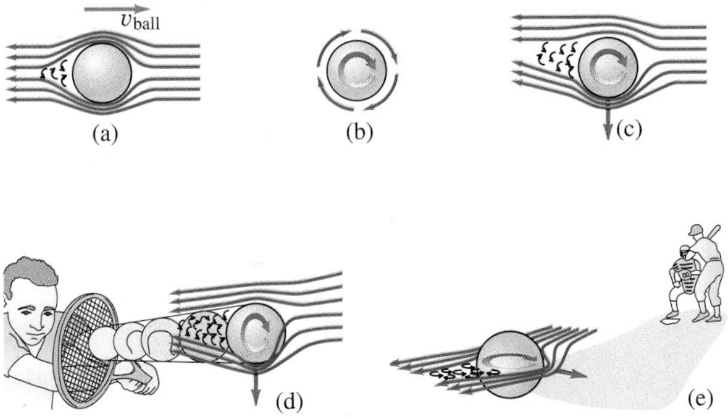

14.29 Motion of the air from right to left relative to the ball corresponds to motion of a ball through still air from left to right. (a) A nonspinning ball has a symmetric region of turbulence behind it. (b) A spinning ball drags along layers of air near its surface. (c) The resulting asymmetric region of turbulence and the deflection of the air stream by the spinning ball. The force shown is exerted on the ball by the air stream; it pushes the ball in the direction of the tangential velocity of the front of the ball. The force can (d) push a tennis ball downward or (e) make a baseball curve.

14.30 Stroboscopic photograph of a golf ball being struck by a club. The picture was taken at 1000 flashes per second. The ball rotates about once in eight pictures, corresponding to an angular speed of 125 rev/s, or 7500 rpm.

Test Your Understanding

How much more thumb pressure must a nurse use to administer an injection with a hypodermic needle of inside diameter 0.40 mm compared to one with inside diameter 0.55 mm? Assume that the two needles have the same length and that the volume flow rate is the same in both cases.

SUMMARY

Density is mass per unit volume. If a mass m of homogeneous material has volume V, its density ρ is the ratio m/V. Specific gravity is the ratio of the density of a material to the density of water. (See Example 14.1)

$$\rho = \frac{m}{V} \qquad (14.1)$$

Pressure is normal force per unit area. Pascal's law states that pressure applied to the surface of an enclosed fluid is transmitted undiminished to every portion of the fluid. Absolute pressure is the total pressure in a fluid; gauge pressure is the difference between absolute pressure and atmospheric pressure. The SI unit of pressure is the pascal (Pa): 1 Pa = 1 N/m^2. (See Example 14.2)

$$p = \frac{dF_\perp}{dA} \qquad (14.2)$$

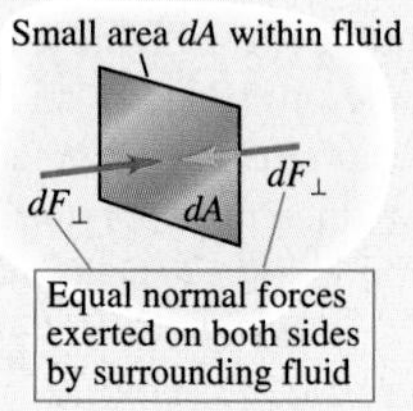

The pressure difference between points 1 and 2 in a static fluid of uniform density ρ (an incompressible fluid) is proportional to the difference between the elevations y_1 and y_2. If the pressure at the surface of an incompressible liquid at rest is p_0, the pressure at a depth h is greater by an amount $\rho g h$. (See Examples 14.3 and 14.4)

$$p_2 - p_1 = -\rho g (y_2 - y_1)$$
(pressure in a fluid of uniform density) (14.5)

$$p = p_0 + \rho g h$$
(pressure in a fluid of uniform density) (14.6)

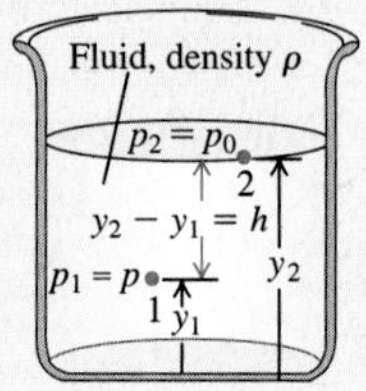

Archimedes's principle states that when a body is immersed in a fluid, the fluid exerts an upward buoyant force on the body equal to the weight of the fluid that the body displaces. (See Examples 14.5 and 14.6)

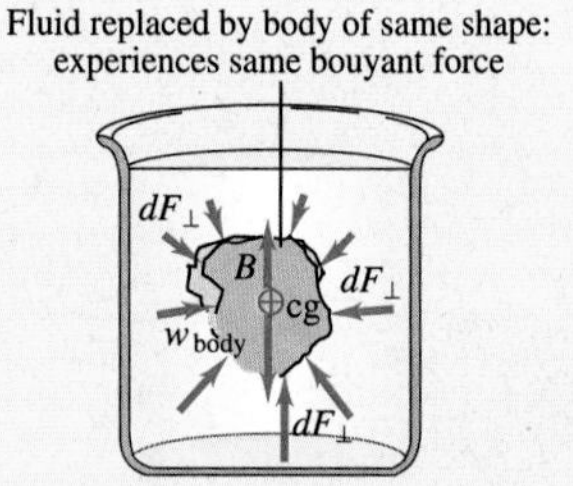

An ideal fluid is incompressible and has no viscosity (no internal friction). A flow line is the path of a fluid particle; a streamline is a curve tangent at each point to the velocity vector at that point. A flow tube is a tube bounded at its sides by flow lines. In laminar flow, layers of fluid slide smoothly past each other. In turbulent flow, there is great disorder and a constantly changing flow pattern.

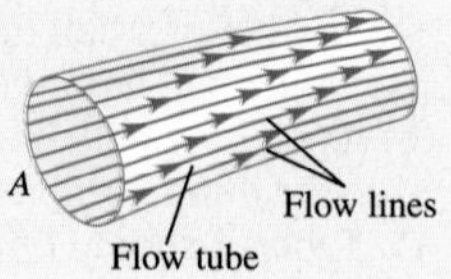

Conservation of mass in an incompressible fluid is expressed by the continuity equation, which relates the flow speeds v_1 and v_2 for two cross sections A_1 and A_2 in a flow tube. The product Av equals the volume flow rate, dV/dt, the rate at which volume crosses a section of the tube. (See Example 14.7)

$$A_1 v_1 = A_2 v_2$$
(continuity equation, incompressible fluid) (14.10)

$$\frac{dV}{dt} = Av$$
(volume flow rate) (14.11)

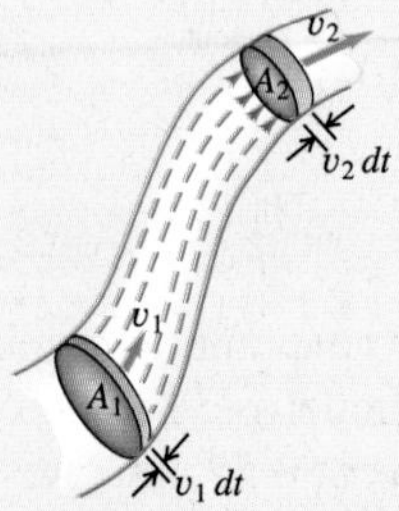

Bernoulli's equation relates the pressure p, flow speed v, and elevation y for any two points, assuming steady flow in an ideal fluid. (See Examples 14.8 through 14.11)	$p_1 + \rho g y_1 + \frac{1}{2}\rho v_1^2 = p_2 + \rho g y_2 + \frac{1}{2}\rho v_2^2$ (Bernoulli's equation) (14.17)	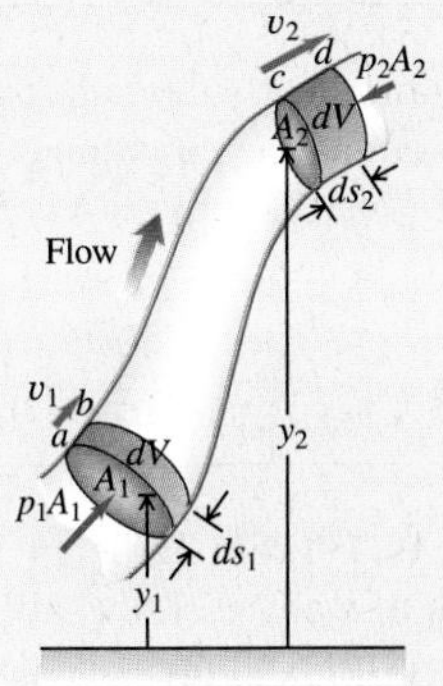

Key Terms

fluid statics, 515
fluid dynamics, 515
density, 515
specific gravity, 516
pressure, 517
pascal, 518
atmospheric pressure, 518
Pascal's law, 520
gauge pressure, 520
absolute pressure, 520
mercury barometer, 521
buoyancy, 523
Archimedes's principle, 523
buoyant force, 523
surface tension, 525
ideal fluid, 526
viscosity, 526
flow line, 526
steady flow, 526
streamline, 526
flow tube, 526
laminar flow, 527
turbulent flow, 527
continuity equation, 527
Bernoulli's equation, 530
viscosity, 533
turbulence, 534

Your Notes

Answer to Chapter Opening Question

Air maintains nearly the same density when passing through such a constriction, so the equation of continuity for an incompressible fluid, Eq. (14.10), applies. A constriction corresponds to a reduced cross-sectional area A, so the speed v must increase.

Answers to Test Your Understanding Questions

Section 14.1 From Table 14.1, the density of platinum is 21.4 times that of water $(21.4 \times 10^3\ \text{kg/m}^3$ versus $1.00 \times 10^3\ \text{kg/m}^3)$. From Eq. (14.1), volume and density are inversely proportional to each other, so the same mass of water has 21.4 times the volume of the platinum, or $21.4\ \text{m}^3$. The length of each side of the cube would be $\sqrt[3]{21.4\ \text{m}^3} = 2.78\ \text{m}$.
Section 14.2 From Eq. (14.9), the outside pressure is equal to the product $\rho g h$. The density ρ decreases while the height h of the mercury column remains the same, so the pressure must be lower outdoors than inside the refrigerator.
Section 14.3 The object displaces two-thirds of its volume V, so the upward buoyant force is $B = \frac{2}{3}\rho_{\text{water}} V g$. The object is in equilibrium, so B is equal to the weight of the object $\rho_{\text{object}} V g$. Hence $\rho_{\text{object}} = \frac{2}{3}\rho_{\text{water}}$. This is an example of a general rule: if an object floats in a liquid with a fraction x of its volume submerged, the object's average density is x times the density of the liquid.
Section 14.4 Because air is nearly incompressible, the volume flow rate of air is essentially constant. When air blows through a constriction such as a mountain pass, the speed increases to keep the volume flow rate the same.
Section 14.5 From Bernoulli's equation, an increase in the flow speed v corresponds to a decrease in the air pressure p. The reduced air pressure on the "outdoor" side of the door makes the door swing toward that side, closing it.
Section 14.6 The required pressure is proportional to $1/R^4$. With the smaller-diameter needle, the pressure is greater by a factor of $[(0.55\ \text{mm})/(0.40\ \text{mm})]^4 = 3.6$.

Discussion Questions

Q14.1 If the weight of a roomful of water is so great (see Example 14.1 in Section 14.1), why don't the first floors of houses built over basements or crawl spaces all collapse when flooded to the ceiling?
Q14.2 A rubber hose is attached to a funnel, and the free end is bent around to point upward. When water is poured into the funnel, it rises in the hose to the same level as in the funnel, even though the funnel has a lot more water in it than the hose. Why?
Q14.3 Comparing Examples 14.1 and 14.2 in Sections 14.1 and 14.2, it seems that 700 N of air is exerting a downward force of 2.0×10^6 N on the floor. How is this possible?
Q14.4 Equation (14.7) shows that an area ratio of 100 to 1 can give 100 times more output force than input force. Doesn't this violate conservation of energy? Explain.
Q14.5 You have probably noticed that the lower the tire pressure, the greater the contact area between the tire and the pavement. Why?
Q14.6 In hot-air ballooning, a large balloon is filled with air heated by a gas burner at the bottom. Why must the air be heated? How does the balloonist control ascent and descent?
Q14.7 In describing the size of a large ship, one uses such expressions as "it displaces 20,000 tons." What does this mean? Can the weight of the ship be obtained from this information?
Q14.8 You drop a solid sphere of aluminum in a bucket of water that sits on the ground. The buoyant force equals the weight of water displaced; this is less than the weight of the sphere, so the sphere sinks to the bottom. If you take the bucket with you on an elevator that accelerates upward, the apparent weight of the water increases and the buoyant force on the sphere increases. Could the acceleration of the elevator be great enough to make the sphere pop up out of the water? Explain.
Q14.9 A rigid, lighter-than-air dirigible filled with helium cannot continue to rise indefinitely. Why not? What determines the maximum height it can attain?
Q14.10 Air pressure decreases with increasing altitude. So why is air near the surface not continuously drawn upward toward the lower-pressure regions above?
Q14.11 The purity of gold can be tested by weighing it in air and in water. How? Do you think you could get away with making a fake gold brick by gold-plating some cheaper material?
Q14.12 During the Great Mississippi Flood of 1993, the levees in St. Louis tended to rupture first at the bottom. Why?
Q14.13 A cargo ship travels from the Atlantic Ocean (salt water) to Lake Ontario (fresh water) via the St. Lawrence River. The ship rides several centimeters lower in the water in Lake Ontario than it did in the ocean. Explain why.
Q14.14 A submarine is more compressible than water. Why, then, can a submarine completely surrounded by water only be in unstable equilibrium?
Q14.15 An old question is "Which weighs more, a pound of feathers or a pound of lead?" If the weight in pounds is the gravitational force, will a pound of feathers balance a pound of lead on opposite pans of an equal-arm balance? Explain, taking into account buoyant forces.
Q14.16 Suppose the door of a room makes an airtight but frictionless fit in its frame. Do you think you could open the door if the air pressure on one side were standard atmospheric pressure and that on the other side differed from standard by 1%? Explain.
Q14.17 A balloon is less compressible than air. Why, then, is there a height at which a helium balloon is in stable equilibrium?
Q14.18 A piece of iron is glued to the top of a block of wood. When the block is placed in a bucket of water with the iron on top, the block floats. The block is now turned over so that the iron is submerged beneath the wood. Does the block float or sink? Does the water level in the bucket rise, drop, or stay the same? Explain your answers.
Q14.19 You take an empty glass jar and push it into a tank of water with the open mouth of the jar downward, so that the air inside the jar is trapped and cannot get out. If you push the jar deeper into the water, does the buoyant force on the jar stay the same? If not, does it increase or decrease? Explain your answer.

Q14.20 You are floating in a canoe in the middle of a swimming pool. Your friend is at the edge of the pool, carefully noting the level of the water on the side of the pool. You have a bowling ball with you in the canoe. If you carefully drop the bowling ball over the side of the canoe and it sinks to the bottom of the pool, does the water level in the pool rise or fall?

Q14.21 You are floating in a canoe in the middle of a swimming pool. A large bird flies up and lights on your shoulder. Does the water level in the pool rise or fall?

Q14.22 You are swimming in a swimming pool and then climb onto an inflatable plastic raft that is floating in the pool. If you are totally out of the water when you are floating on the raft, when you climb onto the raft does the water level in the pool rise or fall?

Q14.23 An ice cube floats in a glass of water. When the ice melts, will the water level in the glass rise, fall, or remain unchanged? Explain.

Q14.24 You are told, "Bernoulli's equation tells us where there is higher fluid speed, there is lower fluid pressure, and vice versa." Is this statement always true, even for an idealized fluid? Explain.

Q14.25 If the velocity at each point in space in steady-state fluid flow is constant, how can a fluid particle accelerate?

Q14.26 In a store-window vacuum cleaner display, a table-tennis ball is suspended in midair in a jet of air blown from the outlet hose of a tank-type vacuum cleaner. The ball bounces around a little but always moves back toward the center of the jet, even if the jet is tilted from the vertical. How does this behavior illustrate Bernoulli's equation?

Q14.27 A tornado consists of a rapidly whirling air vortex. Why is the pressure always much lower in the center than at the outside? How does this condition account for the destructive power of a tornado?

Q14.28 Airports at high elevations have longer runways for takeoffs and landings than do airports at sea level. One reason is that aircraft engines develop less power in the thin air well above sea level. What is another reason?

Q14.29 When a smooth-flowing stream of water comes out of a faucet, it narrows as it falls. Explain why this happens.

Exercises

Section 14.1 Density

14.1 On a part-time job, you are asked to bring a cylindrical iron rod of length 85.8 cm and diameter 2.85 cm from a storage room to a machinist. Will you need a cart? (To answer, calculate the weight of the rod.)

14.2 The moon has radius 1740 km and mass 7.35×10^{22} kg. What is its average density?

14.3 You purchase a rectangular piece of metal that has dimensions $5.0 \times 15.0 \times 30.0$ mm and mass 0.0158 kg. The seller tells you that the metal is gold. To check this, you compute the average density of the piece. What value do you get? Were you cheated?

14.4 A kidnapper demands a 40.0-kg cube of platinum as a ransom. What is the length of a side?

14.5 A uniform, lead sphere and a uniform aluminum sphere have the same mass. What is the ratio of the radius of the aluminum sphere to the radius of the lead sphere?

14.6 a) What is the average density of the sun? b) What is the average density of a neutron star that has the same mass as the sun but a radius of only 20.0 km?

Section 14.2 Pressure in a Fluid

14.7 At what depth in seawater is the gauge pressure 1.00×10^5 Pa?

14.8 In intravenous feeding, a needle is inserted in a vein in the patient's arm and a tube leads from the needle to a reservoir of fluid (density 1050 kg/m^3) located at height h above the arm. The top of the reservoir is open to the air. If the gauge pressure inside the vein is 5980 Pa, what is the minimum value of h that allows fluid to enter the vein? Assume the needle diameter is large enough that you can ignore the viscosity (Section 14.6) of the fluid.

14.9 A barrel contains a 0.120-m layer of oil floating on water that is 0.250 m deep. The density of the oil is 600 kg/m^3 a) What is the gauge pressure at the oil-water interface? b) What is the gauge pressure at the bottom of the barrel?

14.10 An empty sport utility vehicle weighs 16.5 kN. Each of its tires has a gauge pressure of 205 kPa (29.7 lb/in.^2). a) What is the total contact area of the four tires with the pavement? (Assume that the tire walls are flexible so that the pressure exerted by the tire on the pavement equals the air pressure inside the tire.) b) For the same tire pressure, what is that area when the vehicle is loaded with 9.1 kN of passengers and cargo?

14.11 You are designing a diving bell to withstand the pressure of seawater at a depth of 250 m. a) What is the gauge pressure at this depth? (You can ignore changes in the density of the water with depth.) b) At this depth, what is the net force due to the water outside and the air inside the bell on a circular glass window 30.0 cm in diameter if the pressure inside the diving bell equals the pressure at the surface of the water? (You can ignore the small variation of pressure over the surface of the window.)

14.12 What gauge pressure must a pump produce to pump water from the bottom of the Grand Canyon (elevation 730 m) to Indian Gardens (elevation 1370 m)? Express your results in pascals and in atmospheres.

14.13 The liquid in the open-tube manometer in Fig. 14.8a is mercury, $y_1 = 3.00$ cm, and $y_2 = 7.00$ cm. Atmospheric pressure is 980 millibars. a) What is the absolute pressure at the bottom of the U-shaped tube? b) What is the absolute pressure in the open tube at a depth of 4.00 cm below the free surface? c) What is the absolute pressure of the gas in the tank? d) What is the gauge pressure of the gas in pascals?

14.14 There is a maximum depth at which a diver can breathe through a snorkel tube (Fig. 14.31) because as the depth increases, so does the pressure difference, tending to collapse the diver's lungs. Since the snorkel connects the air in the lungs to the atmosphere at the surface, the pressure inside the lungs is atmospheric pressure. What is the external-internal pressure difference when the diver's lungs are at a depth of 6.1 m (about 20 ft)? Assume that the diver is in fresh water. (A scuba diver breathing from compressed air tanks can operate at greater depths than can a snorkeler, since the pressure of the air inside the scuba diver's lungs increases to match the external pressure of the water.)

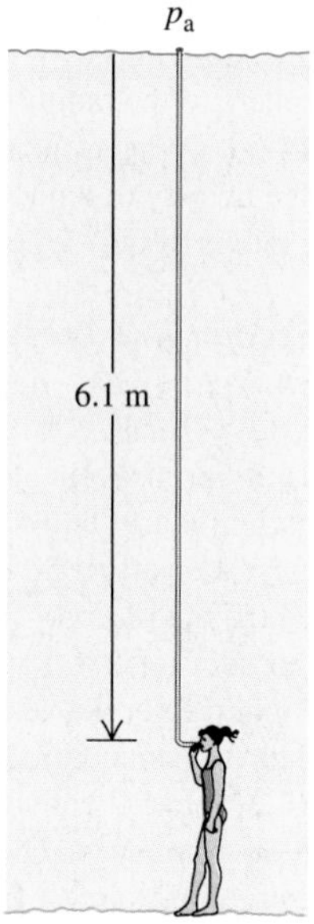

Figure 14.31 Exercise 14.14.

14.15 A tall cylinder with a cross-sectional area 12.0 cm^2 is partially filled with mercury; the surface of the mercury is 5.00 cm above the bottom of the cylinder. Water is slowly poured in on top of the mercury and the two fluids don't mix. What volume of water must be added to double the gauge pressure at the bottom of the cylinder?

14.16 A closed container is partially filled with water. Initially, the air above the water is at atmospheric pressure (1.01×10^5 Pa) and the gauge pressure at the bottom of the water is 2500 Pa. Then additional air is pumped in, increasing the pressure of the air above the water by 1500 Pa. a) What then is the gauge pressure at the bottom of the water? b) By how much must the water level in the container be reduced, by drawing some water out through a valve at the bottom of the container, to return the gauge pressure at the bottom of the water to its original value of 2500 Pa? The pressure of the air above the water is maintained at 1500 Pa above atmospheric pressure.

14.17 An electrical short cuts off all power to a submersible when it is 30 m below the surface of the ocean. The crew must push out a hatch of area 0.75 m^2 and weight 300 N on the bottom to escape. If the pressure inside is 1.0 atm, what downward force must they exert on the hatch to open it?

14.18 You are assigned the design of a cylindrical, pressurized water tank for a future colony on Mars, where the acceleration due to gravity is 3.71 m/s^2. The pressure at the surface of the water will be 130 kPa and the depth of the water will be 14.2 m. The pressure of the air in the building outside the tank will be 93 kPa. Find the net downward force on the tank's flat bottom, of area 2.00 m^2, exerted by the water and air inside the tank and the air outside the tank.

14.19 A tapered, pressurized tank for a rocket contains 0.250 m^3 of kerosene, with mass 205 kg. The pressure at the top of the kerosene is 2.01×10^5 Pa. The kerosene exerts a force of 16.4 kN on the tank's bottom, which has an area of 0.0700 m^2. Find the depth of the kerosene.

14.20 The piston of a hydraulic automobile lift is 0.30 m in diameter. What gauge pressure, in pascals, is required to lift a car with a mass of 1200 kg? Also express this pressure in atmospheres.

Section 14.3 Buoyancy

14.21 A slab of ice floats on a freshwater lake. What minimum volume must the slab have for a 45.0-kg woman to be able to stand on it without getting her feet wet?

14.22 An ore sample weighs 17.50 N in air. When the sample is suspended by a light cord and totally immersed in water, the tension in the cord is 11.20 N. Find the total volume and the density of the sample.

14.23 An object of average density ρ floats at the surface of a fluid of density ρ_{fluid}. a) How must the two densities be related? b) In view of the answer to part (a), how can steel ships float in water? c) In terms of ρ and ρ_{fluid}, what fraction of the object is submerged and what fraction is above the fluid? Check that your answers give the correct limiting behavior as $\rho \to \rho_{fluid}$ and as $\rho \to 0$. d) While on board your yacht, your cousin Throckmorton cuts a rectangular piece (dimensions $5.0 \times 4.0 \times 3.0$ cm) out of a life preserver and throws it into the ocean. The piece has a mass of 42 g. As it floats in the ocean, what percentage of its volume is above the surface?

14.24 A hollow, plastic sphere is held below the surface of a freshwater lake by a cord anchored to the bottom of the lake. The sphere has a volume of 0.650 m^3 and the tension in the cord is 900 N. a) Calculate the buoyant force exerted by the water on the sphere. b) What is the mass of the sphere? c) The cord breaks and the sphere rises to the surface. When the sphere comes to rest, what fraction of its volume will be submerged?

14.25 A cubical block of wood, 10.0 cm on a side, floats at the interface between oil and water with its lower surface 1.50 cm below the interface (Fig. 14.32). The density of the oil is 790 kg/m^3. a) What is the gauge pressure at the upper face of the block? b) What is the gauge pressure at the lower face of the block? c) What is the mass and density of the block?

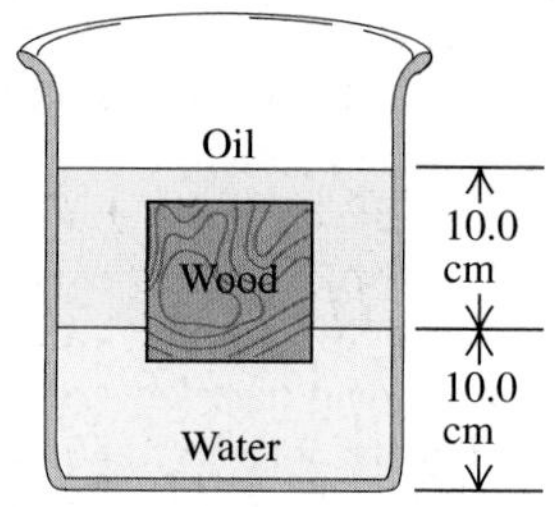

Figure 14.32 Exercise 14.25.

14.26 A solid, aluminum ingot weighs 89 N in air. a) What is its volume? b) The ingot is suspended from a rope and totally immersed in water. What is the tension in the rope (the *apparent* weight of the ingot in water)?

14.27 Two cubical blocks of identical size and shape are suspended by strings and completely immersed in a swimming pool. Block A is aluminum; the top of block A is 0.5 m below the surface of the water. Block B is brass; the top of block B is 1.5 m below the surface of the water. For each of the following, is the quantity larger for block A, larger for B, or the same for both: a) the water pressure at the top of the block; b) the buoyant force exerted on the block by the water; c) the tension in the string attached to the block.

14.28 A rock is suspended by a light string. When the rock is in air, the tension in the string is 39.2 N. When the rock is totally immersed in water, the tension is 28.4 N. When the rock is totally immersed in an unknown liquid, the tension is 18.6 N. What is the density of the unknown liquid?

Section 14.4 Fluid Flow

14.29 A shower head has 20 circular openings, each with radius 1.0 mm. The shower head is connected to a pipe with radius 0.80 cm. If the speed of water in the pipe is 3.0 m/s, what is its speed as it exits the shower-head openings?

14.30 Water is flowing in a pipe with a varying cross-sectional area, and at all points the water completely fills the pipe. At point 1 the cross-sectional area of the pipe is 0.070 m^2, and the magnitude of the fluid velocity is 3.50 m/s. a) What is the fluid speed at points in the pipe where the cross-sectional area is i) 0.105 m^2? ii) 0.047 m^2? b) Calculate the volume of water discharged from the open end of the pipe in 1.00 hour.

14.31 Water is flowing in a circular pipe with varying cross-sectional area, and at all points the water completely fills the pipe. a) At one point in the pipe the radius is 0.150 m. What is the speed of the water at this point if the volume flow rate in the pipe is 1.20 m^3/s? b) At a second point in the pipe the water speed is 3.80 m/s. What is the radius of the pipe at this point?

14.32 a) Derive Eq. (14.12). b) If the density increases by 1.50% from point 1 to point 2, what happens to the volume flow rate?

Section 14.5 Bernoulli's Equation

14.33 A sealed tank containing seawater to a height of 11.0 m also contains air above the water at a gauge pressure of 3.00 atm. Water flows out from the bottom through a small hole. Calculate the efflux speed of the water.

14.34 A small circular hole 6.00 mm in diameter is cut in the side of a large water tank, 14.0 m below the water level in the tank. The top of the tank is open to the air. Find a) the speed of efflux; b) the volume discharged per unit time.

14.35 What gauge pressure is required in the city water mains for a stream from a fire hose connected to the mains to reach a vertical height of 15.0 m? (Assume that the mains have a much larger diameter than the fire hose.)

14.36 At one point in a pipeline the water's speed is 3.00 m/s and the gauge pressure is 5.00×10^4 Pa. Find the gauge pressure at a second point in the line, 11.0 m lower than the first, if the pipe diameter at the second point is twice that at the first.

14.37 Lift on an Airplane. Air streams horizontally past a small airplane's wings such that the speed is 70.0 m/s over the top surface and 60.0 m/s past the bottom surface. If the plane has mass 1340 kg and a wing area of 16.2 m^2, what is the net vertical force (including the effects of gravity) on the airplane? The density of the air is 1.20 kg/m^3.

14.38 A soft drink (mostly water) flows in a pipe at a beverage plant with a mass flow rate that would fill 220 0.355-L cans per minute. At point 2 in the pipe, the gauge pressure is 152 kPa and the cross-sectional area is 8.00 cm^2. At point 1, 1.35 m above point 2, the cross-sectional area is 2.00 cm^2. Find the a) mass flow rate; b) volume flow rate; c) flow speeds at points 1 and 2; d) gauge pressure at point 1.

14.39 Water discharges from a horizontal, cylindrical pipe at the rate of 465 cm^3/s. At a point in the pipe where the radius is 2.05 cm, the absolute pressure is 1.60×10^5 Pa. What is the pipe radius at a constriction if the pressure there is reduced to 1.20×10^5 Pa?

14.40 At a certain point in a horizontal pipeline, the water's speed is 2.50 m/s, and the gauge pressure is 1.80×10^4 Pa. Find the gauge pressure at a second point in the line if the cross-sectional area at the second point is twice that at the first.

14.41 A golf course sprinkler system discharges water from a horizontal pipe at the rate of 7200 cm^3/s. At one point in the pipe, where the radius is 4.00 cm, the water's absolute pressure is 2.40×10^5 Pa. At a second point in the pipe, the water passes through a constriction where the radius is 2.00 cm. What is the water's absolute pressure as it flows through this constriction?

Problems

14.42 In a lecture demonstration, a professor pulls apart two hemispherical, steel shells (diameter D) with ease using their attached handles. She then places them together, pumps out the air to an absolute pressure of p, and hands them to a bodybuilder in the back row to pull apart. a) If atmospheric pressure is p_0, how much force must the bodybuilder exert on each shell? b) Evaluate your answer for the case $p = 0.025$ atm, $D = 10.0$ cm.

14.43 The deepest point known in any of the earth's oceans is in the Marianas Trench, 10.92 km deep. a) Assuming water to be incompressible, what is the pressure at this depth? Use the density of seawater. b) The actual pressure is 1.16×10^8 Pa; your calculated value will be less because the density actually varies with depth. Using the compressibility of water and the actual pressure, find the density of the water at the bottom of the Marianas Trench. What is the percent change in the density of the water?

14.44 A swimming pool measures 5.0 m long × 4.0 m wide × 3.0 m deep. Compute the force exerted by the water against a) the bottom; b) either end. (*Hint:* Calculate the force on a thin, horizontal strip at a depth h, and integrate this over the end of the pool.) Do not include the force due to air pressure.

14.45 The upper edge of a gate in a dam runs along the water surface. The gate is 2.00 m high and 4.00 m wide and is hinged along a horizontal line through its center (Fig. 14.33). Calculate the torque about the hinge arising from the force due to the water. (*Hint:* Use a procedure similar to that used in Problem 14.44; calculate the torque of a thin, horizontal strip at a depth h and integrate this over the gate.)

Figure 14.33 Problem 14.45.

14.46 Force and Torque on a Dam. A dam has the shape of a rectangular solid. The side facing the lake has area A and height H. The surface of the freshwater lake behind the dam is at the top of the dam. a) Show that the net horizontal force exerted by the water on the dam equals $\frac{1}{2}\rho gHA$, that is, the average gauge pressure across the face of the dam times the area (see Problem 14.44). b) Show that the torque exerted by the water about an axis along the bottom of the dam is $\rho gH^2A/6$. c) How do the force and torque depend on the size of the lake?

14.47 An astronaut is standing at the north pole of a newly discovered, spherically symmetric planet of radius R. In his hands he holds a container full of a liquid with mass m and volume V. At the surface of the liquid, the pressure is p_0; at a depth d below the surface, the pressure has a greater value p. From this information, determine the mass of the planet.

14.48 To find the density at a given point within a material, consider a small volume dV centered on that point. If the mass within that volume is dm, the density at the point is $\rho = dm/dV$. Consider a cylindrical rod with mass M, radius R, and length L with a density that is proportional to the square of the distance from one end, $\rho = Cx^2$. a) Show that $C = 3M/\pi R^2L^3$. b) Show that the average density, given by Eq. (14.1), is $\frac{1}{3}$ of the density at the $x = L$ end.

14.49 The earth does not have a uniform density; it is most dense at its center and least dense at its surface. An approximation of its density is $\rho(r) = A - Br$, where $A = 12{,}700\ \text{kg/m}^3$ and $B = 1.50 \times 10^{-3}\ \text{kg/m}^4$. Use $R = 6.37 \times 10^6$ m for the radius of the earth approximated as a sphere. a) Geological evidence indicates that the densities are 13,100 kg/m^3 and 2,400 kg/m^3 at the earth's center and surface. What values does the linear approximation model give for the densities at these two locations? b) Imagine dividing the earth into concentric, spherical shells. Each shell has radius r, thickness dr, volume $dV = 4\pi r^2 dr$, and mass $dm = \rho(r)\ dV$. By integrating from $r = 0$ to $r = R$, show that the mass of the earth in this model is $M = \frac{4}{3}\pi R^3(A - \frac{3}{4}BR)$. c) Show that the given values of A and B give the correct mass of the earth to within 0.4%. d) We saw in Section 12.6 that a uniform spherical shell gives no contribution to g inside it. Show that $g(r) = \frac{4}{3}\pi Gr(A - \frac{3}{4}Br)$ inside the earth in this model. e) Verify that the expression of part (d) gives $g = 0$ at the center of the earth and $g = 9.85\ \text{m/s}^2$ at the surface. f) Show that in this model g does *not* decrease uniformly with depth but rather has a maximum of $4\pi GA^2/9B = 10.01\ \text{m/s}^2$ at $r = 2A/3B = 5640$ km.

14.50 In Example 12.10 (Section 12.6) we saw that inside a planet of uniform density (not a very realistic assumption for the earth) the acceleration due to gravity increases uniformly with distance from the center of the planet. That is, $g(r) = g_s r/R$, where g_s is the acceleration due to gravity at the surface, r is the distance from the center of the planet, and R is the radius of the planet. The interior of the planet can be treated approximately as an incompressible fluid of density ρ. a) Replace the height y in Eq. (14.4) with the radial coordinate r and integrate to find the pressure inside a uniform planet as a function of r. Let the pressure at the surface be zero. (This means ignoring the pressure of the planet's atmosphere.) b) Using this model, calculate the pressure at the center of the earth. (Use a value of ρ equal to the average density of the earth, calculated from the mass and radius given in Appendix F.) c) Geologists estimate the pressure at the center of the earth to be approximately 4×10^{11} Pa. Does this agree with your calculation for the pressure at $r = 0$? What might account for any differences?

14.51 A U-shaped tube open to the air at both ends contains some mercury. A quantity of water is carefully poured into the left arm of the U-shaped tube until the vertical height of the water column is 15.0 cm (Fig. 14.34). a) What is the gauge pressure at the water-mercury interface? b) Calculate the vertical distance h from the top of the mercury in the right-hand arm of the tube to the top of the water in the left-hand arm.

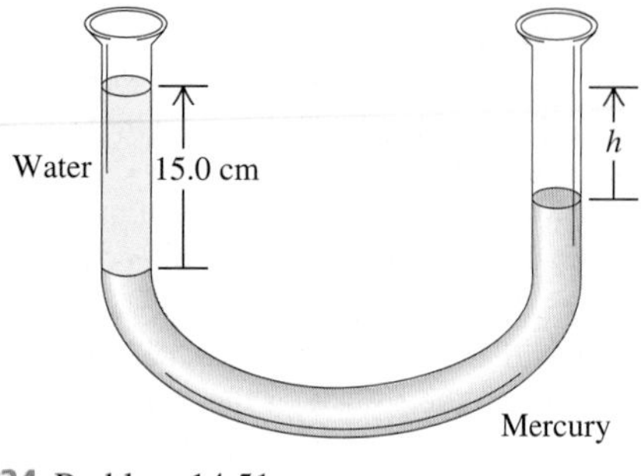

Figure 14.34 Problem 14.51.

14.52 The Great Molasses Flood. On the afternoon of January 15, 1919, an unusually warm day in Boston, a 27.4-m-high, 27.4-m-diameter cylindrical metal tank used for storing molasses ruptured. Molasses flooded into the streets in a 9-m-deep stream, killing pedestrians and horses, and knocking down buildings. The molasses had a density of 1600 kg/m^3. If the tank was full before the accident, what was the total outward force the molasses exerted on its sides? (*Hint:* Consider the outward force on a circular ring of the tank wall of width dy and at a depth y below the surface. Integrate to find the total outward force. Assume that before the tank ruptured, the pressure at the surface of the molasses was equal to the air pressure outside the tank.)

14.53 An open barge has the dimensions shown in Fig. 14.35. If the barge is made out of 4.0-cm-thick steel plate on each of its four sides and its bottom, what mass of coal can the barge carry in fresh water without sinking? Is there enough room in the barge to hold this amount of coal? (The density of coal is about 1500 kg/m^3.)

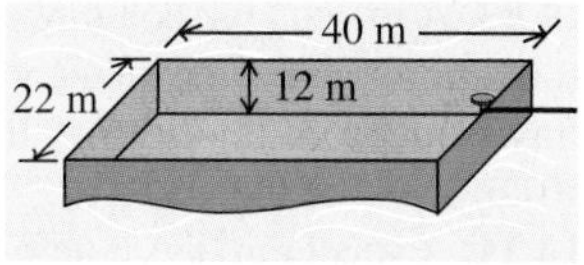

Figure 14.35 Problem 14.53.

14.54 A hot-air balloon has a volume of 2200 m^3. The balloon fabric (the envelope) weighs 900 N. The basket with gear and full propane tanks weighs 1700 N. If the balloon can barely lift an additional 3200 N of passengers, breakfast, and champagne when the outside air density is 1.23 kg/m^3, what is the average density of the heated gases in the envelope?

14.55 Advertisements for a certain small car claim that it floats in water. a) If the car's mass is 900 kg and its interior volume 3.0 m^3, what fraction of the car is immersed when it floats? You can ignore the volume of steel and other materials. b) Water gradually leaks in and displaces the air in the car. What fraction of the interior volume is filled with water when the car sinks?

14.56 A single ice cube with mass 9.70 g floats in a glass completely full of 420 cm^3 of water. You can ignore the water's surface tension and its variation in density with temperature (as long as it remains a liquid). a) What volume of water does the ice cube displace? b) When the ice cube has completely melted, has any water overflowed? If so, how much? If not, explain why this is so. c) Suppose the water in the glass had been very salty water of density 1050 kg/m^3. What volume of salt water would the 9.70-g ice cube displace? d) Redo part (b) for the freshwater ice cube in the salty water.

14.57 A piece of wood is 0.600 m long, 0.250 m wide, and 0.080 m thick. Its density is 600 kg/m^3. What volume of lead must be fastened underneath it to sink the wood in calm water so that its top is just even with the water level? What is the mass of this volume of lead?

14.58 A hydrometer consists of a spherical bulb and a cylindrical stem with a cross-sectional area of 0.400 cm^2 (Fig. 14.12a). The total volume of bulb and stem is 13.2 cm^3. When immersed in water, the hydrometer floats with 8.00 cm of the stem above the water surface. When immersed in an organic fluid, 3.20 cm of the stem is above the surface. Find the density of the organic fluid. (*Note:* This illustrates the precision of such a hydrometer. Relatively small density differences give rise to relatively large differences in hydrometer readings.)

14.59 The densities of air, helium, and hydrogen (at $p = 1.0$ atm and $T = 20°C$) are $1.20\ kg/m^3$, $0.166\ kg/m^3$, and $0.0899\ kg/m^3$, respectively. a) What is the volume in cubic meters displaced by a hydrogen-filled airship that has a total "lift" of 120 kN? (The "lift" is the amount by which the buoyant force exceeds the weight of the gas that fills the airship.) b) What would be the "lift" if helium were used instead of hydrogen? In view of your answer, why is helium used in modern airships like advertising blimps?

14.60 SHM of a Floating Object. An object with height h, mass M, and a uniform cross-sectional area A floats upright in a liquid with density ρ. a) Calculate the vertical distance from the surface of the liquid to the bottom of the floating object at equilibrium. b) A downward force with magnitude F is applied to the top of the object. At the new equilibrium position, how much farther below the surface of the liquid is the bottom of the object than it was in part (a)? (Assume that some of the object remains above the surface of the liquid.) c) Your result in part (b) shows that if the force is suddenly removed, the object will oscillate up and down in SHM. Calculate the period of this motion in terms of the density ρ of the liquid, the mass M, and cross-sectional area A of the object. You can ignore the damping due to fluid friction (Section 13.7).

14.61 A 950-kg cylindrical can buoy floats vertically in salt water. The diameter of the buoy is 0.900 m. a) Calculate the additional distance the buoy will sink when a 70.0-kg man stands on top. (Use the expression derived in part (b) of Problem 14.60.) b) Calculate the period of the resulting vertical SHM when the man dives off. (Use the expression derived in part (c) of Problem 14.60, and as in that problem, you can ignore the damping due to fluid friction.)

14.62 In seawater, a life preserver with a volume of $0.0400\ m^3$ will support a 75.0-kg person (average density $980\ kg/m^3$) with 20% of the person's volume above water when the life preserver is fully submerged. What is the density of the material composing the life preserver?

14.63 A block of balsa wood placed in one scale pan of an equal-arm balance is exactly balanced by a 0.0950-kg brass mass in the other scale pan. Find the true mass of the balsa wood if its density is $150\ kg/m^3$. Explain why it is accurate to ignore the buoyancy in air of the brass but *not* the buoyancy in air of the balsa wood.

14.64 Block A in Fig. 14.36 hangs by a cord from spring balance D and is submerged in a liquid C contained in beaker B. The mass of the beaker is 1.00 kg; the mass of the liquid is 1.80 kg. Balance D reads 3.50 kg and balance E reads 7.50 kg. The volume of block A is $3.80 \times 10^{-3}\ m^3$. a) What is the density of the liquid? b) What will each balance read if block A is pulled up out of the liquid?

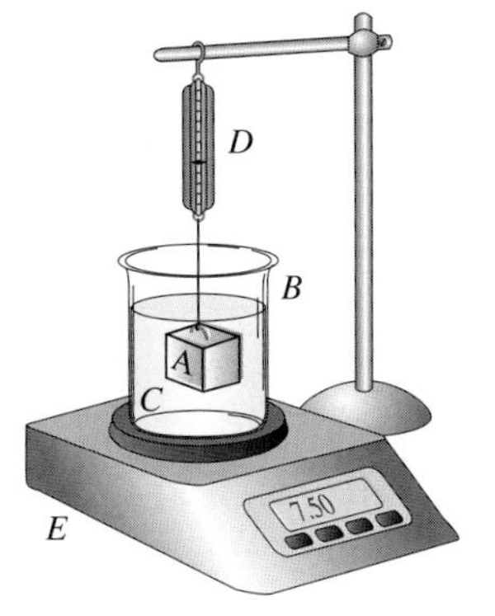

Figure 14.36 Problem 14.64.

14.65 A hunk of aluminum is completely covered with a gold shell to form an ingot of weight 45.0 N. When you suspend the ingot from a spring balance and submerge the ingot in water, the balance reads 39.0 N. What is the weight of the gold in the shell?

14.66 A plastic ball has radius 12.0 cm and floats in water with 16.0% of its volume submerged. a) What force must you apply to the ball to hold it at rest totally below the surface of the water? b) If you let go of the ball, what is its acceleration the instant you release it?

14.67 The weight of a king's solid crown is w. When suspended by a light rope and completely immersed in water, the tension in the rope (the crown's apparent weight) is fw. a) Prove that the crown's relative density (specific gravity) is $1/(1 - f)$. Discuss the meaning of the limits as f approaches 0 and 1. b) If the crown is solid gold and weighs 12.9 N in air, what is its apparent weight when completely immersed in water? c) Repeat part (b) if the crown is solid lead with a very thin gold plating, but still has a weight in air of 12.9 N.

14.68 A piece of steel has a weight w, an apparent weight (see Problem 14.67) w_{water} when completely immersed in water, and an apparent weight w_{fluid} when completely immersed in an unknown fluid. a) Prove that the fluid's density relative to water (specific gravity) is $(w - w_{fluid})/(w - w_{water})$. b) Is this result reasonable for the three cases of w_{fluid} greater than, equal to, or less than w_{water}? c) The apparent weight of the piece of steel in water of density $1000\ kg/m^3$ is 87.2% of its weight. What percentage of its weight will its apparent weight be in formic acid (density $1220\ kg/m^3$)?

14.69 You cast some metal of density ρ_m in a mold, but you are worried that there might be cavities within the casting. You measure the weight of the casting to be w, and the buoyant force when it is completely surrounded by water to be B. a) Show that $V_0 = B/(\rho_{water}g) - w/(\rho_m g)$ is the total volume of any enclosed cavities. b) If your metal is copper, the casting's weight is 156 N, and the buoyant force is 20 N, what is the total volume of any enclosed cavities in your casting? What fraction is this of the total volume of the casting?

14.70 A cubical block of wood 0.100 m on a side and with a density of $550\ kg/m^3$ floats in a jar of water. Oil with a density of $750\ kg/m^3$ is poured on the water until the top of the oil layer is 0.035 m below the top of the block. a) How deep is the oil layer? b) What is the gauge pressure at the block's lower face?

14.71 Dropping Anchor. An iron anchor with mass 35.0 kg and density $7860\ kg/m^3$ lies on the deck of a small barge that has vertical sides and floats in a freshwater river. The area of the bottom of the barge is $8.00\ m^2$. The anchor is thrown overboard but is suspended above the bottom of the river by a rope; the mass and volume of the rope are small enough to ignore. After the anchor is overboard and the barge has finally stopped bobbing up and down, has the barge risen or sunk down in the water? By what vertical distance?

14.72 Assume that crude oil from a supertanker has density $750\ kg/m^3$. The tanker runs aground on a sandbar. To refloat the tanker, its oil cargo is pumped out into steel barrels, each of which has a mass of 15.0 kg when empty and holds $0.120\ m^3$ of oil. You can ignore the volume occupied by the steel from which the barrel is made. a) If a salvage worker accidentally drops a filled, sealed barrel overboard, will it float or sink in the seawater? b) If the barrel floats, what fraction of its volume will be above the water surface? If it sinks, what minimum tension would have to be exerted by a rope to haul the barrel up from the ocean floor? c) Repeat parts (a) and (b) if the density of the oil is 910 kg/m^3 and the mass of each empty barrel is 32.0 kg.

14.73 A cubical block of density ρ_B and with sides of length L floats in a liquid of greater density ρ_L. a) What fraction of the block's volume is above the surface of the liquid? b) The liquid is denser than water (density ρ_W) and does not mix with it. If water is poured on the surface of the liquid, how deep must the water layer be so that the water surface just rises to the top of the block? Express your answer in terms of L, ρ_B, ρ_L, and ρ_W. c) Find the depth of the water layer in part (b) if the liquid is mercury, the block is made of iron, and the side length is 10.0 cm.

14.74 A barge is in a rectangular lock on a freshwater river. The lock is 60.0 m long and 20.0 m wide, and the steel doors on each end are closed. With the barge floating in the lock, a 2.50×10^6 N load of scrap metal is put onto the barge. The metal has density 9000 kg/m^3. a) When the load of scrap metal, initially on the bank, is placed onto the barge, what vertical distance does the water in the lock rise? b) The scrap metal is now pushed overboard into the water. Does the water level in the lock rise, fall, or remain the same? If it rises or falls, by what vertical distance does it change?

14.75 A U-shaped tube with a horizontal portion of length l (Fig. 14.37) contains a liquid. What is the difference in height between the liquid columns in the vertical arms a) if the tube has an acceleration a toward the right? b) if the tube is mounted on a horizontal turntable rotating with an angular speed ω with one of the vertical arms on the axis of rotation? c) Explain why the difference in height does not depend on the density of the liquid or on the cross-sectional area of the tube. Would it be the same if the vertical tubes did not have equal cross-sectional areas? Would it be the same if the horizontal portion were tapered from one end to the other? Explain.

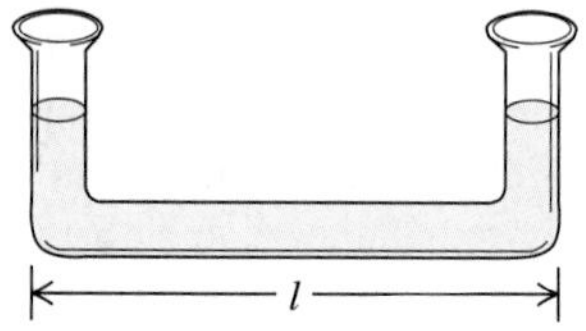

Figure 14.37 Problem 14.75.

14.76 A cylindrical container of an incompressible liquid with density ρ rotates with constant angular speed ω about its axis of symmetry, which we take to be the y-axis (Fig. 14.38). a) Show that the pressure at a given height within the fluid increases in the radial direction (outward from the axis of rotation) according to $\partial p/\partial r = \rho\omega^2 r$. b) Integrate this partial differential equation to find the pressure as a function of distance from the axis of rotation along a horizontal line at $y = 0$. c) Combine the result of part (b) with Eq. (14.5) to show that the surface of the rotating liquid has a *parabolic* shape, that is, the height of the liquid is given by $h(r) = \omega^2 r^2/2g$. (This technique is used for making parabolic telescope mirrors; liquid glass is rotated and allowed to solidify while rotating.)

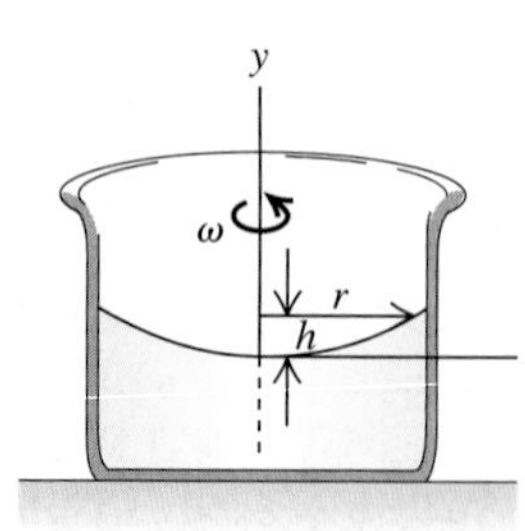

Figure 14.38 Problem 14.76.

14.77 An incompressible fluid with density ρ is in a horizontal test tube of inner cross-sectional area A. The test tube spins in a horizontal circle in an ultracentrifuge at an angular speed ω. Gravitational forces are negligible. Consider a volume element of the fluid of area A and thickness dr' a distance r' from the rotation axis. The pressure on its inner surface is p and on its outer surface is $p + dp$. a) Apply Newton's second law to the volume element to show $dp = \rho\omega^2 r'\,dr'$. b) If the surface of the fluid is at a radius r_0 where the pressure is p_0, show that the pressure p at a distance $r \geq r_0$ is $p = p_0 + \rho\omega^2(r^2 - r_0^2)/2$. c) An object of volume V and density ρ_{ob} has its center of mass at a distance R_{cmob} from the axis. Show that the net horizontal force on the object is $\rho V\omega^2 R_{cm}$, where R_{cm} is the distance from the axis to the center of mass of the displaced fluid. d) Explain why the object will move inward if $\rho R_{cm} > \rho_{ob} R_{cmob}$ and outward if $\rho R_{cm} < \rho_{ob} R_{cmob}$. e) For small objects of uniform density, $R_{cm} = R_{cmob}$. What happens to a mixture of small objects of this kind with different densities in an ultracentrifuge?

14.78 Untethered helium balloons, floating in a car that has all the windows rolled up and outside air vents closed, move in the direction of the car's acceleration, but loose balloons filled with air move in the opposite direction. To show why, consider only the horizontal forces acting on the balloons. Let a be the magnitude of the car's forward acceleration. Consider a horizontal tube of air with a cross-sectional area A that extends from the windshield, where $x = 0$ and $p = p_0$, back along the x-axis. Now consider a volume element of thickness dx in this tube. The pressure on its front surface is p and the pressure on its rear surface is $p + dp$. Assume the air has a constant density ρ. a) Apply Newton's second law to the volume element to show $dp = \rho a\,dx$. b) Integrate the result of part (a) to find the pressure at the front surface in terms of a and x. c) To show that considering ρ constant is reasonable, calculate the pressure difference in atm for a distance as long as 2.5 m and a large acceleration of 5.0 m/s^2. d) Show that the net horizontal force on a balloon of volume V is ρVa. e) For negligible friction forces, show that the acceleration of the balloon (average density ρ_{bal}) is $(\rho/\rho_{bal})a$, so that the acceleration relative to the car is $a_{rel} = [(\rho/\rho_{bal}) - 1]a$. f) Use the expression for a_{rel} in part (e) to explain the movement of the balloons.

14.79 A cubical block of wood 0.30 m on a side is weighted so that its center of gravity is at the point shown in Fig. 14.39a and it floats in water with $\frac{1}{2}$ its volume submerged. If the block is "heeled" at an angle of 45.0° as in Fig. 14.39b, compute the net torque about a horizontal axis perpendicular to the block and passing through its geometrical center.

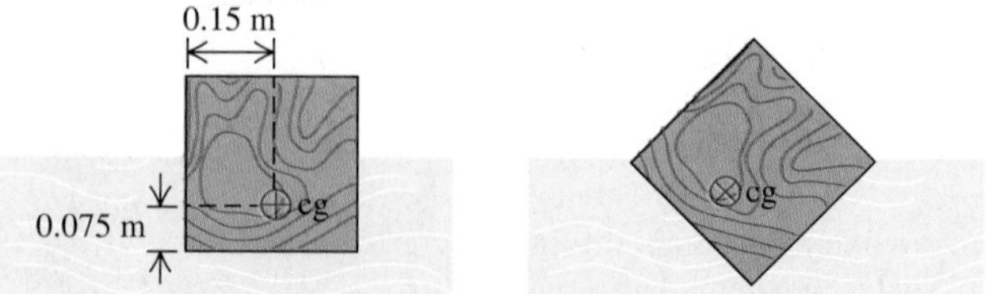

(a) (b)

Figure 14.39 Problem 14.79.

14.80 Water stands at a depth H in a large, open tank whose side walls are vertical (Fig. 14.40). A hole is made in one of the walls at a depth h below the water surface. a) At what distance R from the foot of the wall does the emerging stream strike the floor? b) How far above the bottom of the tank could a second hole be cut so that the stream emerging from it could have the same range as for the first hole?

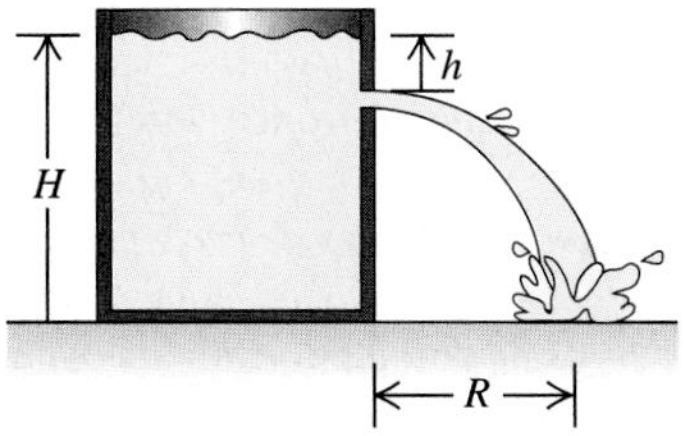

Figure 14.40 Problem 14.80.

14.81 A cylindrical bucket, open at the top, is 25.0 cm high and 10.0 cm in diameter. A circular hole with a cross-sectional area 1.50 cm^2 is cut in the center of the bottom of the bucket. Water flows into the bucket from a tube above it at the rate of 2.40×10^{-4} m^3/s. How high will the water in the bucket rise?

14.82 Water flows steadily from an open tank as in Fig. 14.41. The elevation of point 1 is 10.0 m, and the elevation of points 2 and 3 is 2.00 m. The cross-sectional area at point 2 is 0.0480 m^2; at point 3 it is 0.0160 m^2. The area of the tank is very large compared with the cross-sectional area of the pipe. Assuming that Bernoulli's equation applies, compute a) the discharge rate in cubic meters per second; b) the gauge pressure at point 2.

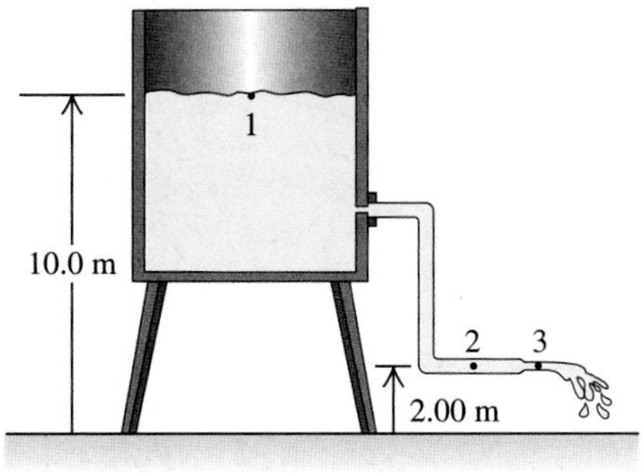

Figure 14.41 Problem 14.82.

14.83 Modern airplane design calls for a lift, due to the net force of the moving air on the wing, of about 2000 N per square meter of wing area. Assume that air flows past the wing of an aircraft with streamline flow. If the speed of flow past the lower wing surface is 120 m/s, what is the required speed over the upper surface to give a lift of 2000 N/m^2? The density of the air is 1.20 kg/m^3.

14.84 In 1993 the radius of Hurricane Emily was about 350 km. The wind speed near the center ("eye") of the hurricane, whose radius was about 30 km, reached about 200 km/h. As air swirled in from the rim of the hurricane toward the eye, its angular momentum remained roughly constant. a) Estimate the wind speed at the rim of the hurricane. b) Estimate the pressure difference at the earth's surface between the eye and the rim. (*Hint:* See Table 14.1.). Where is the pressure greater? c) If the kinetic energy of the swirling air in the eye could be converted completely to gravitational potential energy, how high would the air go? d) In fact the air in the eye is lifted to heights of several kilometers. How can you reconcile this with your answer to part (c)?

14.85 Two very large open tanks A and F (Fig. 14.42) contain the same liquid. A horizontal pipe BCD, having a constriction at C and open to the air at D, leads out of the bottom of tank A, and a vertical pipe E opens into the constriction at C and dips into the liquid in tank F. Assume streamline flow and no viscosity. If the cross-sectional area at C is one half that at D and if D is a distance h_1 below the level of the liquid in A, to what height h_2 will liquid rise in pipe E? Express your answer in terms of h_1.

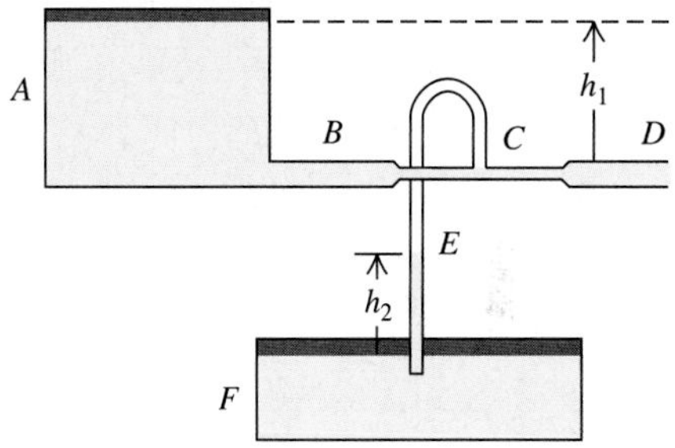

Figure 14.42 Problem 14.85.

14.86 The horizontal pipe, shown in Fig. 14.43, has a cross-sectional area of 40.0 cm^2 at the wider portions and 10.0 cm^2 at the constriction. Water is flowing in the pipe, and the discharge from the pipe is 6.00×10^{-3} m^3/s (6.00 L/s). Find a) the flow speeds at the wide and the narrow portions; b) the pressure difference between these portions; c) the difference in height between the mercury columns in the U-shaped tube.

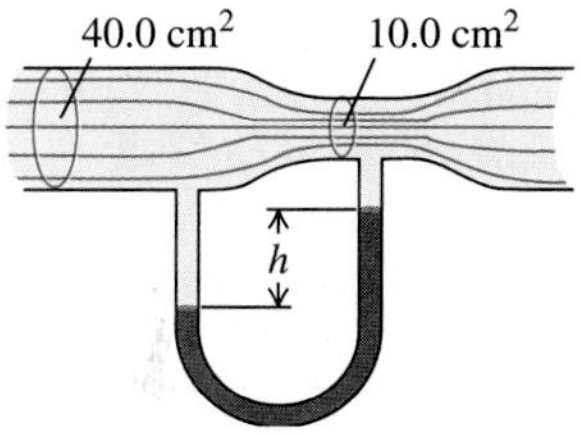

Figure 14.43 Problem 14.86.

14.87 A liquid flowing from a vertical pipe has a very definite shape as it flows from the pipe. To get the equation for this shape, assume that the liquid is in free fall once it leaves the pipe. Just as it leaves the pipe, the liquid has speed v_0 and the radius of the stream of liquid is r_0. a) Find an equation for the speed of the liquid as a function of the distance y it has fallen. Combining this with the equation of continuity, find an expression for the radius of the stream as a function of y. b) If water flows out of a vertical pipe at a speed of 1.20 m/s, how far below the outlet will the radius be $\frac{1}{2}$ the original radius of the stream?

Challenge Problems

14.88 A rock with mass $m = 3.00$ kg is suspended from the roof of an elevator by a light cord. The rock is totally immersed in a bucket of water that sits on the floor of the elevator, but the rock doesn't touch the bottom or sides of the bucket. a) When the elevator is at rest, the tension in the cord is 21.0 N. Calculate the volume of the rock. b) Derive an expression for the tension in the cord when the elevator is accelerating *upward* with an acceleration of magnitude a. Calculate the tension when $a = 2.50\ \text{m/s}^2$ upward. c) Derive an expression for the tension in the cord when the elevator is accelerating *downward* with an acceleration of magnitude a. Calculate the tension when $a = 2.50\ \text{m/s}^2$ downward. d) What is the tension when the elevator is in free fall with a downward acceleration equal to g?

14.89 Suppose a piece of styrofoam, $\rho = 180\ \text{kg/m}^3$, is held completely submerged in water (Fig. 14.44). a) What is the tension in the cord? Find this using Archimedes's principle. b) Use $p = p_0 + \rho gh$ to calculate directly the force exerted by the water on the two sloped sides and the bottom of the styrofoam; then show that the vector sum of these forces is the buoyant force.

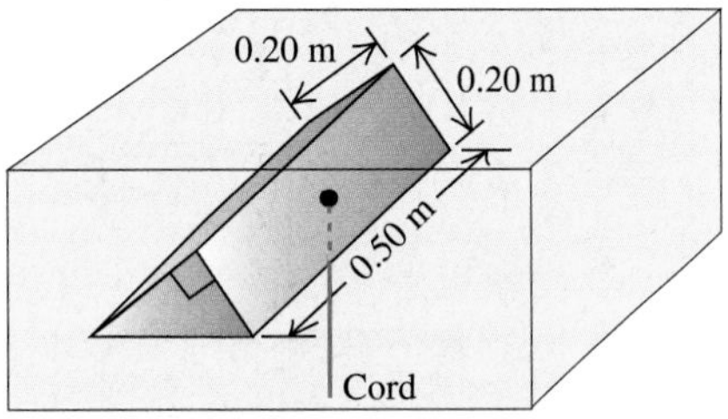

Figure 14.44 Challenge Problem 14.89.

14.90 A large tank with diameter D, open to the air, contains water to a height H. A small hole with diameter $d (d \ll D)$ is made at the base of the tank. Ignoring any effects of viscosity, calculate the time it takes for the tank to drain completely.

14.91 A *siphon*, as shown in Fig. 14.45, is a convenient device for removing liquids from containers. To establish the flow, the tube must be initially filled with fluid. Let the fluid have density ρ and let the atmospheric pressure be p_a. Assume that the cross-sectional area of the tube is the same at all points along it. a) If the lower end of the siphon is at a distance h below the surface of the liquid in the container, what is the speed of the fluid as it flows out the lower end of the siphon? (Assume that the container has a very large diameter, and ignore any effects of viscosity.) b) A curious feature of a siphon is that the fluid initially flows "uphill." What is the greatest height H that the high point of the tube can have if flow is still to occur?

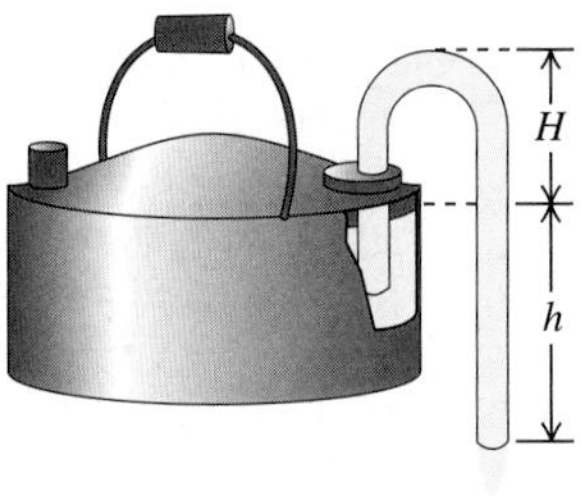

Figure 14.45 Challenge Problem 14.91.

14.92 The following is quoted from a letter. *It is the practice of carpenters hereabouts, when laying out and leveling up the foundations of relatively long buildings, to use a garden hose filled with water, with glass tubes 10 to 12 inches long thrust into the ends of the hose. The theory is that water, seeking a common level, will be the same height in both the tubes and thus effect a level. Now the question rises as to what happens if a bubble of air is left in the hose. Our greybeards contend the air will not affect the reading from one end to the other. Others say that it will cause important inaccuracies.* Can you give a relatively simple answer to this question, together with an explanation? Fig. 14.46 gives a rough sketch of the situation that caused the dispute.

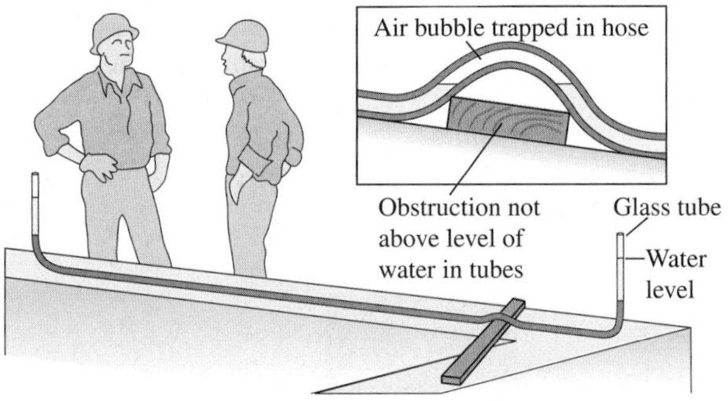

Figure 14.46 Challenge Problem 14.92.

MECHANICAL WAVES

15

When an earthquake strikes, the news of the event travels through the body of the earth in the form of seismic waves. By studying such waves, geophysicists learn about the internal structure of the earth and where future earthquake activity is likely to occur.

? Some seismic waves are more energetic and destructive than others. Which aspects of a seismic wave determine how much power is carried by the wave?

Ripples on a pond, musical sounds, seismic tremors triggered by an earthquake—all these are *wave* phenomena. Waves can occur whenever a system is disturbed from equilibrium and when the disturbance can travel, or *propagate,* from one region of the system to another. As a wave propagates, it carries energy. The energy in light waves from the sun warms the surface of our planet; the energy in seismic waves can crack our planet's crust.

This chapter and the next are about mechanical waves—waves that travel within some material called a *medium.* (Chapter 16 is concerned with sound, an important type of mechanical wave.) We'll begin this chapter by deriving the basic equations for describing waves, including the important special case of *sinusoidal* waves in which the wave pattern is a repeating sine or cosine function. To help us understand waves in general, we'll look at the simple case of waves that travel on a stretched string or rope.

Waves on a string play an important role in music. When a musician strums a guitar or bows a violin, she makes waves that travel in opposite directions along the instrument's strings. What happens when these oppositely directed waves overlap is called *interference*. We'll discover that sinusoidal waves can occur on a guitar or violin string only for certain special frequencies, called *normal-mode frequencies,* determined by the properties of the string. The normal-mode frequencies of a stringed instrument determine the pitch of the musical sounds that the instrument produces. (In the next chapter we'll find that interference also helps explain the pitches of *wind* instruments such as flutes and pipe organs.)

Not all waves are mechanical in nature. *Electromagnetic* waves—including light, radio waves, infrared and ultraviolet radiation, and x-rays—can propagate even in empty space, where there is *no* medium. We'll explore these and other nonmechanical waves in later chapters.

15.1 | Types of Mechanical Waves

A **mechanical wave** is a disturbance that travels through some material or substance called the **medium** for the wave. As the wave travels through the medium, the particles that make up the medium undergo displacements of various kinds, depending on the nature of the wave.

Figure 15.1 shows three varieties of mechanical waves. In Fig. 15.1a the medium is a string or rope under tension. If we give the left end a small upward shake or wiggle, the wiggle travels along the length of the string. Successive sections of string go through the same motion that we gave to the end, but at successively later times. Because the displacements of the medium are perpendicular or *transverse* to the direction of travel of the wave along the medium, this is called a **transverse wave.**

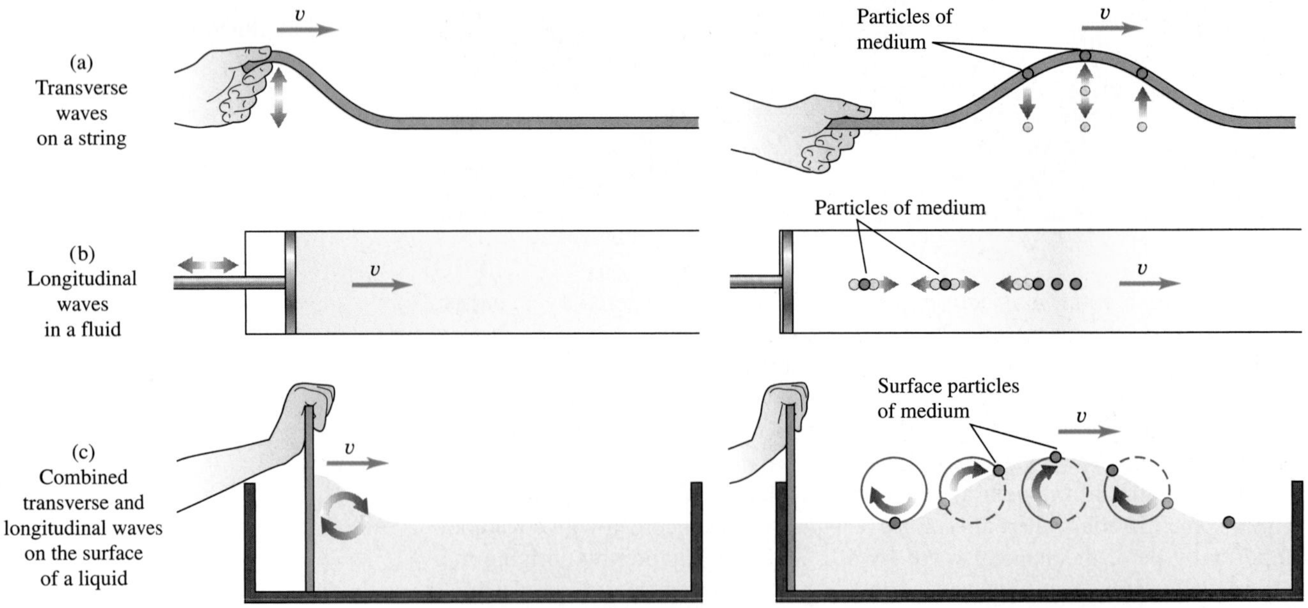

15.1 Three ways to make a wave that moves to the right. (a) The hand moves the string up and then returns, producing a transverse wave. (b) The piston moves to the right, compressing the gas or liquid, then returns, producing a longitudinal wave. (c) The board moves to the right then returns, producing a combination of longitudinal and transverse waves.

In Fig. 15.1b the medium is a liquid or gas in a tube with a rigid wall at the right end and a movable piston at the left end. If we give the piston a single back-and-forth motion, displacement and pressure fluctuations travel down the length of the medium. This time the motions of the particles of the medium are back and forth along the *same* direction that the wave travels. We call this a **longitudinal wave.**

In Fig. 15.1c the medium is water in a channel, such as an irrigation ditch or canal. When we move the flat board at the left end forward and back once, a wave disturbance travels down the length of the channel. In this case the displacements of the water have *both* longitudinal and transverse components.

Each of these systems has an equilibrium state. For the stretched string it is the state in which the system is at rest, stretched out along a straight line. For the fluid in a tube it is a state in which the fluid is at rest with uniform pressure, and for the water in a trough it is a smooth, level water surface. In each case the wave motion is a disturbance from the equilibrium state that travels from one region of the medium to another. And in each case there are forces that tend to restore the system to its equilibrium position when it is displaced, just as the force of gravity tends to pull a pendulum toward its straight-down equilibrium position when it is displaced.

These examples have three things in common. First, in each case the disturbance travels or *propagates* with a definite speed through the medium. This speed is called the speed of propagation, or simply the **wave speed.** It is determined in each case by the mechanical properties of the medium. We will use the symbol v for wave speed. (The wave speed is *not* the same as the speed with which particles move when they are disturbed by the wave. We'll return to this point in Section 15.3.) Second, the medium itself does not travel through space; its individual particles undergo back-and-forth or up-and-down motions around their equilibrium positions. The overall pattern of the wave disturbance is what travels. Third, to set any of these systems into motion, we have to put in energy by doing mechanical work on the system. The wave motion transports this energy from one region of the medium to another. *Waves transport energy, but not matter, from one region to another.*

ActivPhysics ONLINE

10.1 Properties of Mechanical Waves

15.2 | Periodic Waves

The transverse wave on a stretched string in Fig. 15.1a is an example of a *wave pulse.* The hand shakes the string up and down just once, exerting a transverse force on it as it does so. The result is a single "wiggle," or pulse, that travels along the length of the string. The tension in the string restores its straight-line shape once the pulse has passed.

A more interesting situation develops when we give the free end of the string a repetitive, or *periodic,* motion. (You may want to review the discussion of periodic motion in Chapter 13 before going ahead.) Then each particle in the string also undergoes periodic motion as the wave propagates, and we have a **periodic wave.**

In particular, suppose we move the string up and down with *simple harmonic motion* (SHM) with amplitude A, frequency f, angular frequency $\omega = 2\pi f$, and period $T = 1/f = 2\pi/\omega$. Figure 15.2 shows one way to do this. The wave that results is a symmetrical sequence of *crests* and *troughs.* As we will see, periodic waves with simple harmonic motion are particularly easy to analyze; we call them **sinusoidal waves.** It also turns out that *any* periodic wave can be represented as a combination of sinusoidal waves. So this particular kind of wave motion is worth special attention.

In Fig. 15.2 the wave that advances along the string is a *continuous succession* of transverse sinusoidal disturbances. Figure 15.3 shows the shape of a part of the string near the left end at time intervals of $\frac{1}{8}$ of a period, for a total time of one period. The wave shape advances steadily toward the right, as indicated by the

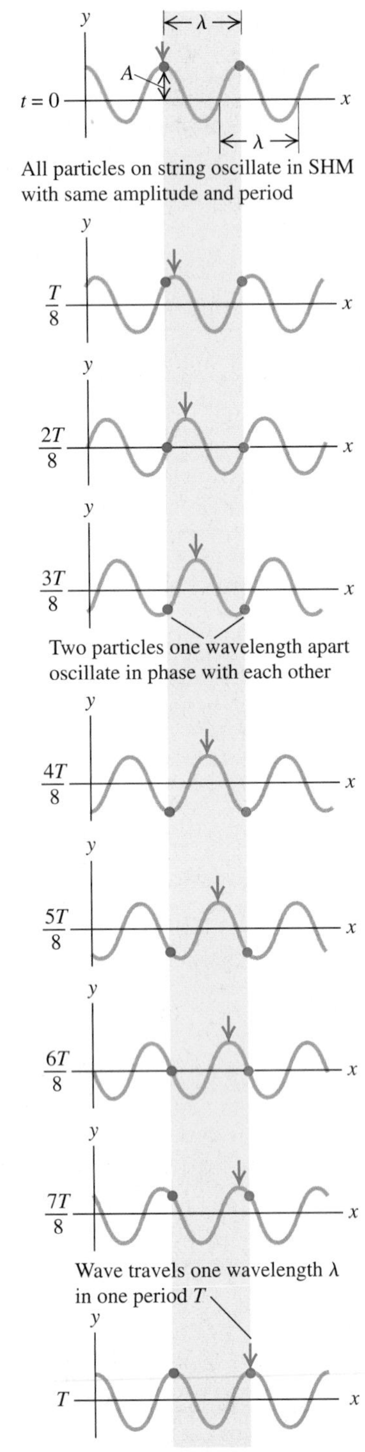

15.3 A sinusoidal transverse wave traveling to the right along a string. The shape of the string is shown at intervals of $\frac{1}{8}$ of a period; the vertical scale is exaggerated. The blue and red dots represent particles in the string whose x-coordinates differ by the wavelength λ.

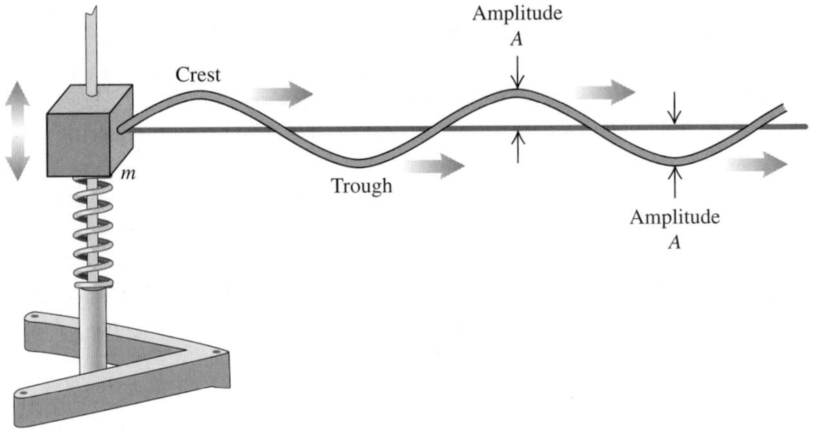

15.2 A block of mass m attached to a spring undergoes simple harmonic motion, producing a sinusoidal wave that travels to the right on the string. The amplitude of the wave is the same as the amplitude of the spring's oscillation. (In a real-life system a driving force would have to be applied to the block to replace the energy carried away by the wave.)

short red arrow pointing to a particular wave crest. As the wave moves, any point on the string (the blue dot, for example) oscillates up and down about its equilibrium position with simple harmonic motion. *When a sinusoidal wave passes through a medium, every particle in the medium undergoes simple harmonic motion with the same frequency.*

CAUTION Be very careful to distinguish between the motion of the *transverse wave* along the string and the motion of a *particle* of the string. The wave moves with constant speed v *along* the length of the string, while the motion of the particle is simple harmonic and *transverse* (perpendicular) to the length of the string.

For a periodic wave, the shape of the string at any instant is a repeating pattern. The length of one complete wave pattern is the distance from one crest to the next, or from one trough to the next, or from any point to the corresponding point on the next repetition of the wave shape. We call this distance the **wavelength** of the wave, denoted by λ (the Greek letter "lambda"). The wave pattern travels with constant speed v and advances a distance of one wavelength λ in a time interval of one period T. So the wave speed v is given by $v = \lambda/T$, or, because $f = 1/T$,

$$v = \lambda f \qquad \text{(periodic wave)} \qquad (15.1)$$

The speed of propagation equals the product of wavelength and frequency. The frequency is a property of the *entire* periodic wave because all points on the string oscillate with the same frequency f.

Waves on a string propagate in just one dimension (in Fig. 15.3, along the x-axis). But the ideas of frequency, wavelength, and amplitude apply equally well to waves that propagate in two or three dimensions. Figure 15.4 shows a wave propagating in two dimensions on the surface of a tank of water. As with waves on a string, the wavelength is the distance from one crest to the next and the amplitude is the height of a crest.

15.4 A series of drops falling into water produces a periodic wave that spreads radially outward. The wave crests and troughs are concentric circles. The wavelength λ is the radial distance between adjacent crests or adjacent troughs.

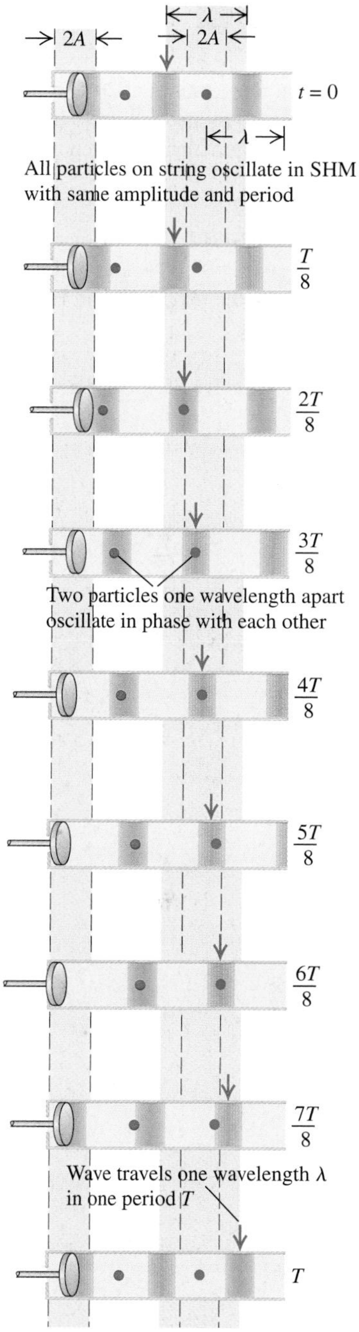

15.5 A sinusoidal longitudinal wave traveling to the right in a fluid. The wave has the same amplitude and period as the oscillation of the piston. The blue and red dots represent particles in the fluid whose x-coordinates differ by the wavelength λ. In equilibrium the particle shown by the red dot is $\frac{1}{2}$ wavelength from the piston; its oscillations are one half-cycle out of phase with the piston.

In many important situations including waves on a string, the wave speed v is determined entirely by the mechanical properties of the medium. In this case, increasing f causes λ to decrease so that the product $v = \lambda f$ remains the same, and waves of *all* frequencies propagate with the same wave speed. In this chapter we will consider *only* waves of this kind. (In later chapters we will study the propagation of light waves in matter for which the wave speed depends on frequency; this turns out to be the reason why prisms break white light into a spectrum and why raindrops create a rainbow.)

To understand the mechanics of a periodic *longitudinal* wave, we consider a long tube filled with a fluid, with a piston at the left end as in Fig. 15.1b. If we push the piston in, we compress the fluid near the piston, increasing the pressure in this region. This region then pushes against the neighboring region of fluid, and so on, and a wave pulse moves along the tube.

Now suppose we move the piston back and forth with simple harmonic motion, along a line parallel to the axis of the tube (Fig. 15.5). This motion forms regions in the fluid where the pressure and density are greater or less than the equilibrium values. We call a region of increased density a *compression;* a region of reduced density is a *rarefaction*. Figure 15.5 shows compressions as darkly shaded areas and rarefactions as lightly shaded areas. The short red arrow in Fig. 15.5 shows the position of one particular compression; the compressions and rarefactions move to the right with constant speed v.

The motion of a single particle of the medium, such as the one shown by a blue dot in Fig. 15.5, is simple harmonic motion parallel to the direction of wave propagation. The wavelength is the distance from one compression to the next or from one rarefaction to the next. The fundamental equation $v = \lambda f$ holds for longitudinal waves as well as for transverse waves, and indeed for *all* types of periodic waves. Just as for transverse waves, in this chapter and the next we will consider only situations in which the speed of longitudinal waves does not depend on the frequency.

Example 15.1 Wavelength of a musical sound

Sound waves are longitudinal waves in air. The speed of sound depends on temperature; at 20°C it is 344 m/s (1130 ft/s). What is the wavelength of a sound wave in air at 20°C if the frequency is 262 Hz (the approximate frequency of middle C on a piano)?

SOLUTION

IDENTIFY AND SET UP: The target variable is the wavelength λ. The wave speed $v = 344$ m/s and frequency $f = 262$ Hz are given, so we can use the relationship in Eq. (15.1) between λ, v, and f for periodic waves.

EXECUTE: We solve Eq. (15.1) for the target variable λ:

$$\lambda = \frac{v}{f} = \frac{344 \text{ m/s}}{262 \text{ Hz}} = \frac{344 \text{ m/s}}{262 \text{ s}^{-1}} = 1.31 \text{ m}$$

Note that the units of frequency are either hertz (Hz) or inverse seconds (s^{-1}).

EVALUATE: What happens to the wavelength if the frequency changes? The speed of sound waves is unaffected by changes in frequency, so the relationship $\lambda = v/f$ tells us that wavelength will change in inverse proportion to the frequency. As an example, the "high C" sung by coloratura sopranos is two octaves above middle C. Each octave corresponds to a factor of two in frequency, so the frequency of high C is four times that of middle C: $f = 4(262 \text{ Hz}) = 1048$ Hz. Hence the *wavelength* of high C is *one-fourth* as large: $\lambda = (1.31 \text{ m})/4 = 0.328$ m.

Test Your Understanding

If you double the wavelength of a wave on a string, what happens to the speed of the wave? What happens to the frequency of the wave?

15.3 | Mathematical Description of a Wave

Many characteristics of periodic waves can be described by using the concepts of wave speed, amplitude period, frequency, and wavelength. Often, though, we need a more detailed description of the positions and motions of individual particles of the medium at particular times during wave propagation. For this description we need the concept of a *wave function,* a function that describes the position of any particle in the medium at any time. We will concentrate on *sinusoidal* waves, in which each particle undergoes simple harmonic motion about its equilibrium position.

As a specific example, let's look at waves on a stretched string. If we ignore the sag of the string due to gravity, the equilibrium position of the string is along a straight line. We take this to be the x-axis of a coordinate system. Waves on a string are *transverse;* during wave motion a particle with equilibrium position x is displaced some distance y in the direction perpendicular to the x-axis. The value of y depends on which particle we are talking about (that is, y depends on x) and also on the time t when we look at it. Thus y is a *function* of both x and t; $y = y(x, t)$. We call $y(x, t)$ the **wave function** that describes the wave. If we know this function for a particular wave motion, we can use it to find the displacement (from equilibrium) of any particle at any time. From this we can find the velocity and acceleration of any particle, the shape of the string, and anything else we want to know about the behavior of the string at any time.

Wave Function for a Sinusoidal Wave

Let's see how to determine the form of the wave function for a sinusoidal wave. Suppose a sinusoidal wave travels from left to right (the direction of increasing x) along the string, as in Fig. 15.3. Every particle of the string oscillates with simple harmonic motion with the same amplitude and frequency. But the oscillations of particles at different points on the string are *not* all in step with each other. The particle marked by the blue dot in Fig. 15.3 is at its maximum positive value of y at $t = 0$ and returns to $y = 0$ at $t = 2T/8$; these same events occur for a particle at the *center* of the colored band at $t = 4T/8$ and $t = 6T/8$, exactly one half-period later. For any two particles of the string, the motion of the particle on the right (in terms of the wave, the "downstream" particle) lags behind the motion of the particle on the left by an amount proportional to the distance between the particles.

Hence the cyclic motions of various points on the string are out of step with each other by various fractions of a cycle. We call these differences *phase differences,* and we say that the *phase* of the motion is different for different points. For example, if one point has its maximum positive displacement at the same time that another has its maximum negative displacement, the two are a half-cycle out of phase. (This is the case for the blue dot in Fig. 15.3 and a point at the center of the colored band.)

Suppose that the displacement of a particle at the left end of the string $(x = 0)$, where the wave originates, is given by

$$y(x = 0, t) = A \cos \omega t = A \cos 2\pi f t \tag{15.2}$$

That is, the particle oscillates in simple harmonic motion with amplitude A, frequency f, and angular frequency $\omega = 2\pi f$. The notation $y(x = 0, t)$ reminds us that the motion of this particle is a special case of the wave function $y(x, t)$ that describes the entire wave. At $t = 0$ the particle at $x = 0$ is at its maximum positive displacement $(y = A)$ and is instantaneously at rest (because the value of y is a maximum).

The wave disturbance travels from $x = 0$ to some point x to the right of the origin in an amount of time given by x/v, where v is the wave speed. So the motion of point x at time t is the same as the motion of point $x = 0$ at the earlier time $t - x/v$. Hence we can find the displacement of point x at time t by simply replacing t in Eq. (15.2) by $(t - x/v)$. When we do that, we find the following expression for the wave function:

$$y(x, t) = A \cos\left[\omega\left(t - \frac{x}{v}\right)\right]$$

Because $\cos(-\theta) = \cos\theta$, we can rewrite the wave function as

$$y(x, t) = A \cos\left[\omega\left(\frac{x}{v} - t\right)\right] = A \cos 2\pi f\left(\frac{x}{v} - t\right) \tag{15.3}$$

(sinusoidal wave moving in $+x$-direction)

The displacement $y(x, t)$ is a function of both the location x of the point and the time t. We could make Eq. (15.3) more general by allowing for different values of the phase angle, as we did for simple harmonic motion in Section 13.2, but for now we omit this.

We can rewrite the wave function given by Eq. (15.3) in several different but useful forms. We can express it in terms of the period $T = 1/f$ and the wavelength $\lambda = v/f$:

$$y(x, t) = A \cos 2\pi\left(\frac{x}{\lambda} - \frac{t}{T}\right) \tag{15.4}$$

(sinusoidal wave moving in $+x$-direction)

We get another convenient form of the wave function if we define a quantity k, called the **wave number:**

$$k = \frac{2\pi}{\lambda} \qquad \text{(wave number)} \tag{15.5}$$

Substituting $\lambda = 2\pi/k$ and $f = \omega/2\pi$ into the wavelength-frequency relation $v = \lambda f$ gives

$$\omega = vk \qquad \text{(periodic wave)} \tag{15.6}$$

We can then rewrite Eq. (15.4) as

$$y(x, t) = A\cos(kx - \omega t) \tag{15.7}$$

(sinusoidal wave moving in $+x$-direction)

Which of these various forms for the wave function $y(x, t)$ we use in any specific problem is a matter of convenience. Note that ω has units rad/s, so for unit consistency in Eqs. (15.6) and (15.7) the wave number k must have the units rad/m. (Some physicists define the wave number as $1/\lambda$ rather than $2\pi/\lambda$. When reading other texts, be sure to determine how this term is defined.)

The wave function $y(x, t)$ is graphed as a function of x for a specific time t in Fig. 15.6a. This graph gives the displacement y of a particle from its equilibrium position as a function of the coordinate x of the particle. If the wave is a transverse wave on a string, the graph in Fig. 15.6a represents the shape of the string at that instant, like a flash photograph of the string. In particular, at time $t = 0$,

$$y(x, t = 0) = A\cos kx = A\cos 2\pi\frac{x}{\lambda}$$

Figure 15.6b is a graph of the wave function versus time t for a specific coordinate x. This graph gives the displacement y of the particle at that coordinate as a function of time—that is, it describes the motion of that particle. In particular, at the position $x = 0$,

$$y(x = 0, t) = A\cos(-\omega t) = A\cos\omega t = A\cos 2\pi\frac{t}{T}$$

This is consistent with our original statement about the motion at $x = 0$, Eq. (15.2).

CAUTION Although they may look the same at first glance, Figs. 15.6a and 15.6b are *not* identical. Figure 15.6a is a picture of the shape of the string at $t = 0$, while Fig. 15.6b is a graph of the displacement y of a particle at $x = 0$ as a function of time.

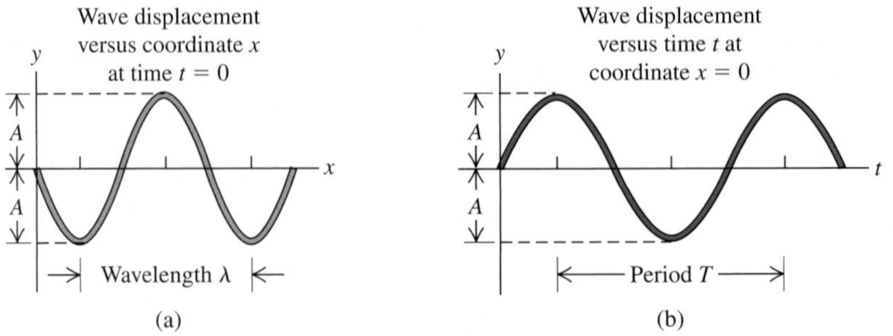

15.6 (a) The graph of $y(x, t)$ versus coordinate x for a specific time t shows the shape of the wave at that time. (b) The graph of $y(x, t)$ versus time t at a specific coordinate x depicts the motion of a particle at that coordinate. The vertical scale is exaggerated in both (a) and (b).

We can modify Eqs. (15.3) through (15.7) to represent a wave traveling in the *negative* x-direction. In this case the displacement of point x at time t is the same as the motion of point $x = 0$ at the *later* time $(t + x/v)$, so in Eq. (15.2) we replace t by $(t + x/v)$. For a wave traveling in the negative x-direction,

$$y(x,t) = A\cos 2\pi f\left(\frac{x}{v} + t\right) = A\cos 2\pi\left(\frac{x}{\lambda} + \frac{t}{T}\right) = A\cos(kx + \omega t)$$

(sinusoidal wave moving in $-x$-direction) (15.8)

In the expression $y(x,t) = A\cos(kx \pm \omega t)$ for a wave traveling in the $-x$ or $+x$-direction, the quantity $(kx \pm \omega t)$ is called the **phase.** It plays the role of an angular quantity (always measured in radians) in Eq. (15.7) or (15.8), and its value for any values of x and t determines what part of the sinusoidal cycle is occurring at a particular point and time. For a crest (where $y = A$ and the cosine function has the value 1), the phase could be 0, 2π, 4π, and so on; for a trough (where $y = -A$ and the cosine has the value -1) it could be π, 3π, 5π, and so on.

The wave speed is the speed with which we have to move along with the wave to keep alongside a point of a given phase, such as a particular crest of a wave on a string. For a wave traveling in the $+x$-direction, that means $kx - \omega t =$ constant. Taking the derivative with respect to t, we find $k\,dx/dt = \omega$, or

$$\frac{dx}{dt} = \frac{\omega}{k}$$

Comparing this with Eq. (15.6), we see that dx/dt is equal to the speed v of the wave. Because of this relationship, v is sometimes called the *phase velocity* of the wave. (*Phase speed* would be a better term.)

Problem-Solving Strategy: Mechanical Waves

IDENTIFY *the relevant concepts:* Wave problems fall into two broad categories. *Kinematics* problems are concerned with describing wave motion; they involve wave speed v, wavelength λ (or wave number k), frequency f (or angular frequency ω), and amplitude A. They may also involve the position, velocity, and acceleration of individual particles in the medium. *Dynamics* problems also use concepts from Newton's laws such as force and mass. As an example, later in this chapter we'll encounter problems that involve the relation of wave speed to the mechanical properties of the wave medium. We'll get into these relations in the next section and in Chapter 16.

As always, make sure that you identify the target variable(s) for the problem. In some cases it will be the wavelength, frequency, or wave speed; in other cases you'll be asked to find an expression for the wave function.

SET UP *the problem using the following steps:*

1. Make a list of the quantities whose values are given. To help you visualize the situation, you'll find it useful to sketch graphs of y versus x (like Fig. 15.6a) and of y versus t (like Fig. 15.6b). Label your graphs with the values of the known quantities.
2. Decide which equations you'll need to use. If any two of v, f, and λ are given, you'll need to use Eq. (15.1) $(v = \lambda f)$ to find the third quantity (see Example 15.1). If the problem involves the angular frequency ω and/or the wave number k, you'll need to use the definitions of those quantities and Eq. (15.6) $(\omega = vk)$. You may also need the various forms of the wave function given in Eqs. (15.3), (15.4), and (15.7).
3. If the wave speed isn't given, and you don't have enough information to determine it using $v = \lambda f$, you may be able to find v using the relationship between v and the mechanical properties of the system. (In the next section we'll develop this relation for waves on a string.)

EXECUTE *the solution as follows:*
Solve for the unknown quantities using the equations you've selected. In some problems all you need to do is find the value of one of the wave variables.

If you're asked to determine the wave function, you need to know A and any two of v, λ, and f (or v, k, and ω). Once you have this information, you can use it in Eq. (15.3), (15.4), or (15.7) to get the specific wave function for the problem at hand. Once you have that, you can find the value of y at any point (value of x) and at any time by substituting into the wave function.

EVALUATE *your answer:* Look at your results with a critical eye. Check to see whether the values of v, f, and λ (or v, ω, and k) agree with the relationships given in Eq. (15.1) or (15.6). If you've calculated the wave function, check one or more special cases for which you can guess what the results ought to be.

Example 15.2 Wave on a clothesline

Your cousin Throckmorton is playing with the clothesline. He unties one end, holds it taut, and wiggles the end up and down sinusoidally with frequency 2.00 Hz and amplitude 0.075 m. The wave speed is $v = 12.0$ m/s. At time $t = 0$ the end has maximum positive displacement and is instantaneously at rest. Assume no wave bounces back from the far end to muddle up the pattern. a) Find the amplitude, angular frequency, period, wavelength, and wave number of the wave. b) Write a wave function describing the wave. c) Write equations for the displacement as a function of time of Throckmorton's end of the clothesline and of a point 3.00 m from his end.

SOLUTION

IDENTIFY: This is a kinematics problem about the motion of the clothesline. Since Throcky moves his hand in a sinusoidal way, he produces a sinusoidal wave that propagates down the clothesline. Hence we can use all of the expressions we've developed in this section. Our target variables in part (a) are amplitude A, angular frequency ω, period T, wavelength λ, and wave number k, so we need to use the equations that relate these quantities. In parts (b) and (c) our target "variables" are actually expressions for displacement; to find these, we use the general equations for the wave function of a sinusoidal wave.

SET UP: A photograph of the clothesline at time $t = 0$ would look just like Fig. 15.6a, with the maximum displacement at $x = 0$ (the end that Throcky has in his hand). We take the positive x-direction to be the direction in which the wave propagates, so we can use Eqs. (15.4) and (15.7) to describe the displacement of the clothesline as a function of position x and time t. We also use the relationships $f = 1/T$, $\omega = 2\pi f$, $k = 2\pi/\lambda$, $v = \lambda f$, and $\omega = vk$.

EXECUTE: a) The amplitude A of the wave is just the amplitude of the motion of the end of the clothesline, $A = 0.075$ m. Similarly, the wave frequency is $f = 2.00$ Hz, the same as the frequency of the end of the clothesline. The angular frequency is

$$\omega = 2\pi f = (2\pi\ \text{rad/cycle})(2.00\ \text{cycles/s}) = 4.00\pi\ \text{rad/s} = 12.6\ \text{rad/s}$$

The period is $T = 1/f = 0.500$ s. We get the wavelength from Eq. (15.1):

$$\lambda = \frac{v}{f} = \frac{12.0\ \text{m/s}}{2.00\ \text{s}^{-1}} = 6.00\ \text{m}$$

We find the wave number from Eq. (15.5) or (15.6):

$$k = \frac{2\pi}{\lambda} = \frac{2\pi\ \text{rad}}{6.00\ \text{m}} = 1.05\ \text{rad/m} \quad \text{or}$$

$$k = \frac{\omega}{v} = \frac{4.00\pi\ \text{rad/s}}{12.0\ \text{m/s}} = 1.05\ \text{rad/m}$$

b) Since we found the values of A, T, and λ in part (a), we can write down the wave function using Eq. (15.4):

$$\begin{aligned} y(x,t) &= A\cos 2\pi\left(\frac{x}{\lambda} - \frac{t}{T}\right) \\ &= (0.075\ \text{m})\cos 2\pi\left(\frac{x}{6.00\ \text{m}} - \frac{t}{0.500\ \text{s}}\right) \\ &= (0.075\ \text{m})\cos\left[(1.05\ \text{rad/m})x - (12.6\ \text{rad/s})t\right] \end{aligned}$$

We can also get this same equation from Eq. (15.7) by using the values of ω and k we obtained in part (a).

c) With our choice of the positive x-direction, the two points in question are at $x = 0$ and $x = +3.00$ m. For each point, we can find the displacement as a function of time by substituting these values of x into the wave function we found in part (b):

$$\begin{aligned} y(x=0,t) &= (0.075\ \text{m})\cos 2\pi\left(\frac{0}{6.00\ \text{m}} - \frac{t}{0.500\ \text{s}}\right) \\ &= (0.075\ \text{m})\cos(12.6\ \text{rad/s})t \\ y(x=+3.00\ \text{m},t) &= (0.075\ \text{m})\cos 2\pi\left(\frac{3.00\ \text{m}}{6.00\ \text{m}} - \frac{t}{0.500\ \text{s}}\right) \\ &= (0.075\ \text{m})\cos\left[\pi - (12.6\ \text{rad/s})t\right] \\ &= -(0.075\ \text{m})\cos(12.6\ \text{rad/s})t \end{aligned}$$

EVALUATE: In part (b), the quantity $(1.05\ \text{rad/m})x - (12.6\ \text{rad/s})t$ is the *phase* of a point x on the string at time t. The phases of the two points in part (c) differ by π because these points are separated by one half-wavelength $(\lambda/2 = (6.00\ \text{m})/2 = 3.00\ \text{m})$. Both points oscillate in SHM with the same frequency and amplitude, but their oscillations are one half-cycle out of phase. Thus, while a graph of y versus t for the point at $x = 0$ is a cosine curve (like Fig. 15.6b), a graph of y versus t for the point $x = 3.00$ m is a *negative* cosine (the same as a cosine curve shifted by one half-cycle).

Using the expression for $y(x = 0, t)$ in part (c), can you show that the end of the string at $x = 0$ is instantaneously at rest at $t = 0$, just as we stated at the beginning of this example? (*Hint:* calculate the velocity at this point by taking the derivative of y with respect to t.)

Particle Velocity and Acceleration in a Sinusoidal Wave

From the wave function we can get an expression for the transverse velocity of any *particle* in a transverse wave. We call this v_y to distinguish it from the wave propagation speed v. To find the transverse velocity v_y at a particular point x, we take the derivative of the wave function $y(x, t)$ with respect to t, keeping x constant. If the wave function is

$$y(x, t) = A \cos(kx - \omega t)$$

then

$$v_y(x, t) = \frac{\partial y(x, t)}{\partial t} = \omega A \sin(kx - \omega t) \qquad (15.9)$$

The ∂ in this expression is a modified d, used to remind us that $y(x, t)$ is a function of *two* variables and that we are allowing only one (t) to vary. The other (x) is constant because we are looking at a particular point on the string. This derivative is called a *partial derivative.* If you haven't reached this point yet in your study of calculus, don't fret; it's a simple idea.

Equation (15.9) shows that the transverse velocity of a particle varies with time, as we expect for simple harmonic motion. The maximum particle speed is ωA; this can be greater than, less than, or equal to the wave speed v, depending on the amplitude and frequency of the wave.

The *acceleration* of any particle is the *second* partial derivative of $y(x, t)$ with respect to t:

$$a_y(x, t) = \frac{\partial^2 y(x, t)}{\partial t^2} = -\omega^2 A \cos(kx - \omega t) = -\omega^2 y(x, t) \qquad (15.10)$$

The acceleration of a particle equals $-\omega^2$ times its displacement, which is the result we obtained in Section 13.2 for simple harmonic motion.

We can also compute partial derivatives of $y(x, t)$ with respect to x, holding t constant. This corresponds to studying the shape of the string at one instant of time, like a flash photo. The first derivative $\partial y(x, t)/\partial x$ is the *slope* of the string at any point. The second partial derivative with respect to x is the *curvature* of the string:

$$\frac{\partial^2 y(x, t)}{\partial x^2} = -k^2 A \cos(kx - \omega t) = -k^2 y(x, t) \qquad (15.11)$$

From Eqs. (15.10) and (15.11) and the relation $\omega = vk$ we see that

$$\frac{\partial^2 y(x, t)/\partial t^2}{\partial^2 y(x, t)/\partial x^2} = \frac{\omega^2}{k^2} = v^2 \quad \text{and}$$

$$\frac{\partial^2 y(x, t)}{\partial x^2} = \frac{1}{v^2}\frac{\partial^2 y(x, t)}{\partial t^2} \qquad \text{(wave equation)} \qquad (15.12)$$

We've derived Eq. (15.12) for a wave traveling in the positive x-direction. You can use the same steps to show that the wave function for a sinusoidal wave propagating in the *negative* x-direction, $y(x, t) = A\cos(kx + \omega t)$, also satisfies this equation.

Equation (15.12), called the **wave equation,** is one of the most important equations in all of physics. Whenever it occurs, we know that a disturbance can propagate as a wave along the x-axis with wave speed v. The disturbance need not be a sinusoidal wave; we'll see in the next section that *any* wave on a string obeys Eq. (15.12), whether the wave is periodic or not (see also Problems 15.54 and 15.59). In Chapter 32 we will find that electric and magnetic fields satisfy the wave equation; the wave speed will turn out to be the speed of light, which will lead us to the conclusion that light is an electromagnetic wave.

Figure 15.7a shows the velocity v_y and acceleration a_y, given by Eqs. (15.9) and (15.10), for several points on a string as a sinusoidal wave passes along it. Note that at points where the string has an upward curvature $(\partial^2 y/\partial x^2 > 0)$, the acceleration of that point is positive $(a_y = \partial^2 y/\partial t^2 > 0)$; this follows from the wave equation, Eq. (15.12). For the same reason the acceleration is negative $(a_y = \partial^2 y/\partial t^2 < 0)$ at points where the string has a downward curvature $(\partial^2 y/\partial x^2 < 0)$, and the acceleration is zero $(a_y = \partial^2 y/\partial t^2 = 0)$ at points of inflection where the curvature is zero $(\partial^2 y/\partial x^2 = 0)$. We emphasize again that v_y and a_y are the *transverse* velocity and acceleration of points on the string; these points move along the y-direction, not along the propagation direction of the wave. Figure 15.7b shows the transverse motions of several points on the string.

The concept of wave function is equally useful with *longitudinal* waves, and everything we have said about wave functions can be adapted to this case. The quantity y still measures the displacement of a particle of the medium from its equilibrium position; the difference is that for a longitudinal wave, this displacement is *parallel* to the x-axis instead of perpendicular to it. We'll discuss longitudinal waves in detail in Chapter 16.

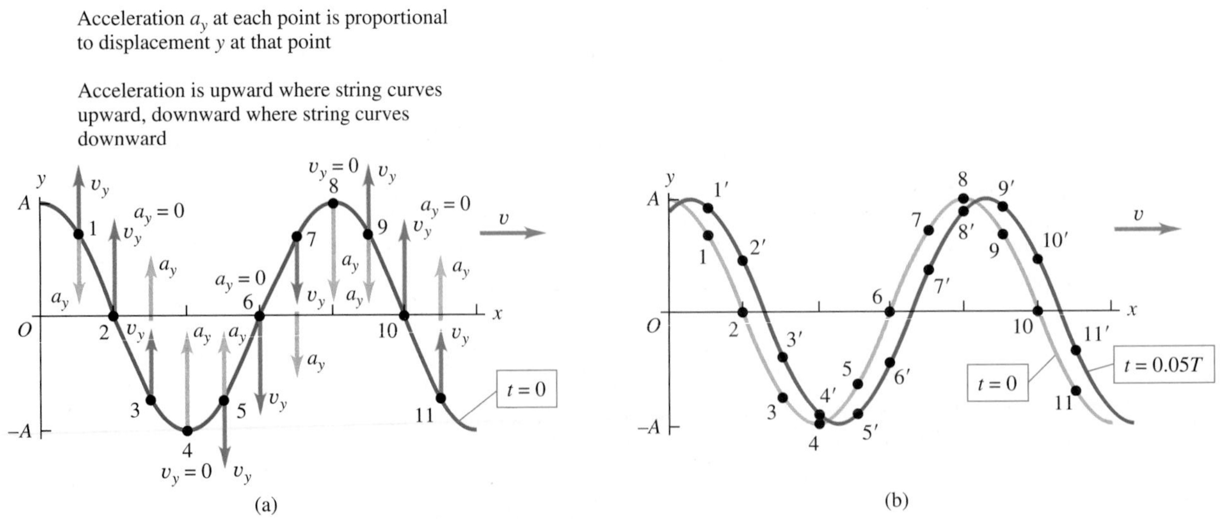

15.7 (a) Another view of the wave at $t = 0$ in Fig. 15.6a. The vectors show the transverse velocity v_y and transverse acceleration a_y at several points on the string. (b) The two curves show the wave at $t = 0$ and $t = 0.05T$. During this interval, a particle at point 1 is displaced to point 1′, a particle at point 2 is displaced to point 2′, and so on.

Test Your Understanding

The wave function for a sinusoidal wave is

$$y(x, t) = (2.50\ \text{mm}) \cos\left[(2.00\ \text{m}^{-1})x + (3.20\ \text{rad/s})t\right]$$

In which direction is this wave propagating? What is its wave number? Its angular frequency? Its amplitude?

15.4 | Speed of a Transverse Wave

10.2 Speed of Waves on a String

One of the key properties of any wave is the wave speed. Light waves in air have a much greater speed of propagation than do sound waves in air (3.00×10^8 m/s versus 344 m/s); that's why you see the flash from a bolt of lightning before you hear the clap of thunder. In this section we'll see what determines the speed of propagation of one particular kind of wave: transverse waves on a string. The speed of these waves is important to understand in its own right because it is an essential part of analyzing stringed musical instruments, as we'll discuss later in this chapter. Furthermore, the speeds of many kinds of mechanical waves turn out to have the same basic mathematical expression as does the speed of waves on a string.

The physical quantities that determine the speed of transverse waves on a string are the *tension* in the string and its *mass per unit length* (also called *linear mass density*). We might guess that increasing the tension should increase the restoring forces that tend to straighten the string when it is disturbed, thus increasing the wave speed. We might also guess that increasing the mass should make the motion more sluggish and decrease the speed. Both these guesses turn out to be right. We'll develop the exact relationship between wave speed, tension, and mass per unit length by two different methods. The first is simple in concept and considers a specific wave shape; the second is more general but also more formal. Choose whichever you like better.

Wave Speed on a String: First Method

We consider a perfectly flexible string (Fig. 15.8). In the equilibrium position the tension is F, and the linear mass density (mass per unit length) is μ. (When portions of the string are displaced from equilibrium, the mass per unit length decreases a little, and the tension increases a little.) We ignore the weight of the string so that when the string is at rest in the equilibrium position, the string forms a perfectly straight line as in Fig. 15.8a.

Starting at time $t = 0$, we apply a constant upward force F_y at the left end of the string. We might expect that the end would move with constant acceleration; that would happen if the force were applied to a *point* mass. But here the effect of the force F_y is to set successively more and more mass in motion. The wave travels with constant speed v, so the division point P between moving and nonmoving portions moves with the same constant speed v (Fig. 15.8b).

Figure 15.8b shows that all particles in the moving portion of the string move upward with constant *velocity* v_y, not constant acceleration. To see why this is so, we note that the *impulse* of the force F_y up to time t is $F_y t$. According to the impulse-momentum theorem (Section 8.1), the impulse is equal to the change in the total transverse component of momentum $(mv_y - 0)$ of the moving part of

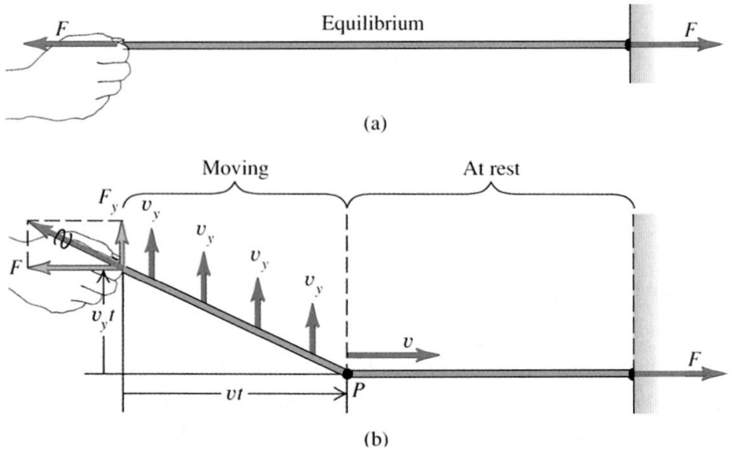

15.8 Propagation of a transverse wave on a string. (a) String in equilibrium; (b) part of the string in motion.

the string. Because the system started with *no* transverse momentum, this is equal to the total momentum at time t:

$$F_y t = mv_y$$

The total momentum thus must increase proportionately with time. But since the division point P moves with constant speed, the length of string that is in motion and hence the total mass m in motion are also proportional to the time t that the force has been acting. So the *change* of momentum must be associated entirely with the increasing amount of mass in motion, not with an increasing velocity of an individual mass element. That is, mv_y changes because m, not v_y, changes.

At time t, the left end of the string has moved up a distance $v_y t$, and the boundary point P has advanced a distance vt. The total force at the left end of the string has components F and F_y. Why F? There is no motion in the direction along the length of the string, so there is no unbalanced horizontal force. Therefore F, the magnitude of the horizontal component, does not change when the string is displaced. In the displaced position the tension is $(F^2 + F_y^2)^{1/2}$ (greater than F), and the string stretches somewhat.

To derive an expression for the wave speed v, we again apply the impulse-momentum theorem to the portion of the string in motion at time t, that is, the portion to the left of P in Fig. 15.8b. The transverse *impulse* (transverse force times time) is equal to the change of transverse *momentum* of the moving portion (mass times transverse component of velocity). The impulse of the transverse force F_y in time t is $F_y t$. In Fig. 15.8b the right triangle whose vertex is at P, with sides $v_y t$ and vt, is similar to the right triangle whose vertex is at the position of the hand, with sides F_y and F. Hence

$$\frac{F_y}{F} = \frac{v_y t}{vt} \qquad F_y = F\frac{v_y}{v}$$

and

$$\text{Transverse impulse} = F_y t = F\frac{v_y}{v}t$$

The mass of the moving portion of the string is the product of the mass per unit length μ and the length vt, or μvt. The transverse momentum is the product of this mass and the transverse velocity v_y:

$$\text{Transverse momentum} = (\mu vt)v_y$$

We note again that the momentum increases with time *not* because mass is moving faster, as was usually the case in Chapter 8, but because *more mass* is brought into motion. But the impulse of the force F_y is still equal to the total change in momentum of the system. Applying this relation, we obtain

$$F\frac{v_y}{v}t = \mu vtv_y$$

Solving this for v, we find

$$v = \sqrt{\frac{F}{\mu}} \quad \text{(speed of a transverse wave on a string)} \quad (15.13)$$

15.9 These cables have a relatively large amount of mass per unit length (μ) and a low tension (F). If the cables are disturbed—say, by a bird landing on them—transverse waves will travel along them at a slow speed $v = \sqrt{F/\mu}$.

Equation (15.13) confirms our prediction that the wave speed v should increase when the tension F increases but decrease when the mass per unit length μ increases (Fig. 15.9).

Note that v_y does not appear in Eq. (15.13); thus the wave speed doesn't depend on v_y. Our calculation considered only a very special kind of pulse, but we can consider *any* shape of wave disturbance as a series of pulses with different values of v_y. So even though we derived Eq. (15.13) for a special case, it is valid for *any* transverse wave motion on a string, including the sinusoidal and other periodic waves we discussed in Section 15.3. Note also that the wave speed doesn't depend on the amplitude or frequency of the wave, in accordance with our assumptions in Section 15.3.

Wave Speed on a String: Second Method

Here is an alternative derivation of Eq. (15.13). If you aren't comfortable with partial derivatives, it can be omitted. We apply Newton's second law, $\sum\vec{F} = m\vec{a}$, to a small segment of string whose length in the equilibrium position is Δx (Fig. 15.10). The mass of the segment is $m = \mu\,\Delta x$; the forces at the ends are represented in terms of their x- and y-components. The x-components have equal magnitude F and add to zero because the motion is transverse and there is no component of acceleration in the x-direction. To obtain F_{1y} and F_{2y}, we note that the ratio F_{1y}/F is

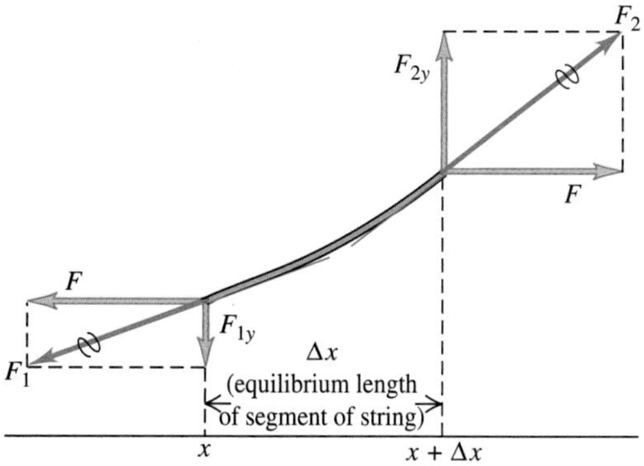

15.10 Free-body diagram for a segment of string. The force at each end of the string is tangent to the string at the point of application.

equal in magnitude to the *slope* of the string at point x and that F_{2y}/F is equal to the slope at point $x + \Delta x$. Taking proper account of signs, we find

$$\frac{F_{1y}}{F} = -\left(\frac{\partial y}{\partial x}\right)_x \qquad \frac{F_{2y}}{F} = \left(\frac{\partial y}{\partial x}\right)_{x+\Delta x} \tag{15.14}$$

The notation reminds us that the derivatives are evaluated at points x and $x + \Delta x$, respectively. From Eq. (15.14) we find that the net y-component of force is

$$F_y = F_{1y} + F_{2y} = F\left[\left(\frac{\partial y}{\partial x}\right)_{x+\Delta x} - \left(\frac{\partial y}{\partial x}\right)_x\right] \tag{15.15}$$

We now equate F_y from Eq. (15.15) to the mass $\mu\,\Delta x$ times the y-component of acceleration $\partial^2 y/\partial t^2$. We obtain

$$F\left[\left(\frac{\partial y}{\partial x}\right)_{x+\Delta x} - \left(\frac{\partial y}{\partial x}\right)_x\right] = \mu\,\Delta x \frac{\partial^2 y}{\partial t^2} \tag{15.16}$$

or, dividing by $F\,\Delta x$,

$$\frac{\left(\frac{\partial y}{\partial x}\right)_{x+\Delta x} - \left(\frac{\partial y}{\partial x}\right)_x}{\Delta x} = \frac{\mu}{F}\frac{\partial^2 y}{\partial t^2} \tag{15.17}$$

We now take the limit as $\Delta x \to 0$. In this limit, the left side of Eq. (15.17) becomes the derivative of $\partial y/\partial x$ with respect to x (at constant t), that is, the *second* (partial) derivative of y with respect to x:

$$\frac{\partial^2 y}{\partial x^2} = \frac{\mu}{F}\frac{\partial^2 y}{\partial t^2} \tag{15.18}$$

Now, finally, comes the punch line of our story. Equation (15.18) has exactly the same form as the *wave equation,* Eq. (15.12), that we derived at the end of Section 15.3. That equation and Eq. (15.18) describe the very same wave motion, so they must be identical. Comparing the two equations, we see that for this to be so, we must have

$$v = \sqrt{\frac{F}{\mu}} \tag{15.19}$$

which is the same expression as Eq. (15.13).

In going through this derivation, we didn't make any special assumptions about the shape of the wave. Since our derivation led us to rediscover Eq. (15.12), the wave equation, we conclude that the wave equation is valid for waves on a string that have *any* shape.

Equation (15.13) or (15.19) gives the wave speed only for the special case of mechanical waves on a stretched string or rope. Remarkably, it turns out that for many types of mechanical waves, including waves on a string, the expression for wave speed has the same general form:

$$v = \sqrt{\frac{(\text{restoring force returning the system to equilibrium})}{(\text{inertia resisting the return to equilibrium})}}$$

To interpret this expression, let's look at the now-familiar case of waves on a string. The tension F in the string plays the role of the restoring force; it tends to bring the string back to its undisturbed, equilibrium configuration. The mass of the string—or, more properly, the linear mass density μ—provides the inertia that prevents the string from returning instantaneously to equilibrium. Hence we have $v = \sqrt{F/\mu}$ for the speed of waves on a string.

In Chapter 16 we'll see a similar expression for the speed of sound waves in a gas. Roughly speaking, the gas pressure provides the force that tends to return the gas to its undisturbed state when a sound wave passes through. The inertia is provided by the density, or mass per unit volume, of the gas.

Example 15.3 Varying the speed of a wave

In Example 15.2 the linear mass density of the clothesline is 0.250 kg/m. a) How much tension does Throcky have to apply to produce the observed wave speed of 12.0 m/s? b) If the tension is increased to four times the value in (a) but the frequency is still 2.00 Hz, what will be the wavelength of the wave on the clothesline?

SOLUTION

IDENTIFY: The target variable is the tension in part (a) and the wavelength in part (b). Both parts of the problem involve *dynamics*, that is, the relationship between the wave speed and the properties of the clothesline itself (tension and linear mass density). Part (b) also involves *kinematics*, since we'll need the relationship between the wave speed, wavelength, and frequency.

SET UP: In part (a) we're given the linear mass density μ and the wave speed v. Hence we can find the tension F using Eq. (15.13), $v = \sqrt{F/\mu}$. In part (b) we use this same relationship to find the new wave speed with the new value of tension, then find the new wavelength λ using Eq. (15.1), $v = \lambda f$.

EXECUTE: (a) Solving Eq. (15.13) for F, we find

$$F = \mu v^2 = (0.250\ \text{kg/m})(12.0\ \text{m/s})^2 = 36.0\ \text{kg-m/s}^2$$
$$= 36.0\ \text{N} = 8.09\ \text{lb}$$

which is well within Throcky's capabilities.

b) Equation (15.13) says that the wave speed is proportional to the square root of the tension F. Furthermore, Eq. (15.1) tells us that $\lambda = v/f$, so that the wavelength is linearly proportional to the wave speed (for constant frequency f, which is the case in this problem). So if we increase the value of F by a factor of 4, the wave speed and the wavelength both increase by a factor of $\sqrt{4} = 2$. Hence, using the old wavelength value from Example 15.2 will give the new wavelength

$$\lambda = 2(6.00\ \text{m}) = 12.0\ \text{m}$$

EVALUATE: We can check our answer to part (b) by using the new tension $F = 4(36.0\ \text{N})$ in the expression $v = \sqrt{F/\mu}$ to find the new wave speed, then calculating the new wavelength using $\lambda = v/f$:

$$v = \sqrt{\frac{144\ \text{N}}{0.250\ \text{kg/m}}} = 24.0\ \text{m/s} \quad \text{so}$$
$$\lambda = \frac{24.0\ \text{m/s}}{2.00\ \text{s}^{-1}} = 12.0\ \text{m}$$

Example 15.4 Calculating wave speed

One end of a nylon rope is tied to a stationary support at the top of a vertical mine shaft 80.0 m deep (Fig. 15.11). The rope is stretched taut by a box of mineral samples with mass 20.0 kg attached at the lower end. The mass of the rope is 2.00 kg. The geologist at the bottom of the mine signals to his colleague at the top by jerking the rope sideways. a) What is the speed of a transverse wave on the rope? b) If a point on the rope is given a transverse simple harmonic motion with a frequency of 2.00 Hz, how many cycles of the wave are there in the rope's length?

SOLUTION

IDENTIFY: This example is similar to the previous one; it involves both dynamics [in part (a)] and kinematics [in part (b)].

In part (a) the target variable is the wave speed. The new feature is that the tension is provided by the weight of the box of samples. In fact, the weight of the rope itself contributes to the tension, which means that the tension is different at the top and bottom of the rope. We'll ignore this effect here since the weight of the rope is small compared to the weight of the samples.

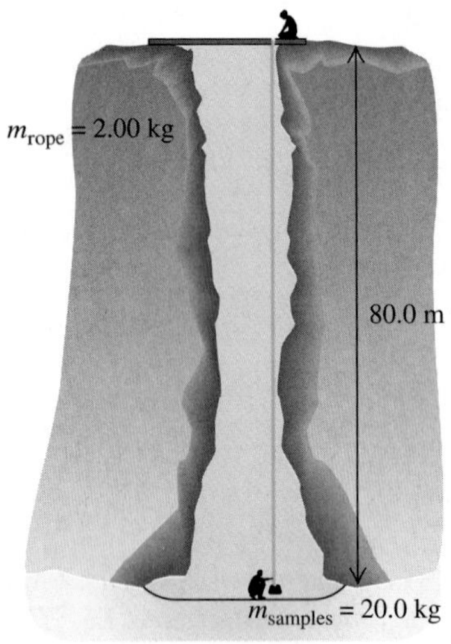

15.11 Sending signals along a vertical rope using transverse waves.

Note that the target variable in part (b) is actually the number of wavelengths that fit into the length of the rope.

SET UP: As in Example 15.3, we use the relationship $v = \sqrt{F/\mu}$ in part (a). If we neglect the weight of the rope itself, the tension F is just equal to the weight of the box. In part (b) we use the equation $v = f\lambda$ to find the wavelength, which we then compare to the 80.0-m length of the rope.

EXECUTE: a) The tension in the rope (due to the sample box) is

$$F = m_{\text{samples}}g = (20.0\text{ kg})(9.80\text{ m/s}^2) = 196\text{ N}$$

and the mass per unit length of the rope is

$$\mu = \frac{m_{\text{rope}}}{L} = \frac{2.00\text{ kg}}{80.0\text{ m}} = 0.0250\text{ kg/m}$$

Hence, from Eq. (15.13), the wave speed is

$$v = \sqrt{\frac{F}{\mu}} = \sqrt{\frac{196\text{ N}}{0.0250\text{ kg/m}}} = 88.5\text{ m/s}$$

b) From Eq. (15.1),

$$\lambda = \frac{v}{f} = \frac{88.5\text{ m/s}}{2.00\text{ s}^{-1}} = 44.3\text{ m}$$

The length of the rope is 80.0 m, so the number of wave cycles in the rope is

$$\frac{80.0\text{ m/s}}{44.3\text{ m/cycle}} = 1.81\text{ cycles}$$

EVALUATE: If we account for the weight of the rope, the tension is greater at the top of the rope than at the bottom. Hence the wave speed increases and the wavelength decreases as the wave travels up the rope. Can you verify that the wave speed at the top is 92.9 m/s?

Test Your Understanding

The six strings of a guitar are of the same length and are under nearly the same tension, but have different thicknesses. On which string do waves travel the fastest?

15.5 | Energy in Wave Motion

Every wave motion has *energy* associated with it. The energy we receive from sunlight and the destructive effects of ocean surf and earthquakes bear this out. To produce any of the wave motions we have discussed in this chapter, we have to apply a force to a portion of the wave medium; the point where the force is applied moves, so we do *work* on the system. As the wave propagates, each portion of the medium exerts a force and does work on the adjoining portion. In this way a wave can transport energy from one region of space to another.

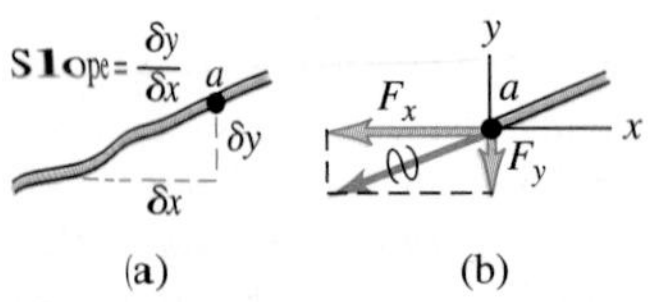

15.12 (a) Point a on a string carrying a wave from left to right. (b) The components of the force exerted on the right part of the string by the left part at point a.

As an example of energy considerations in wave motion, let's look again at transverse waves on a string. How is energy transferred from one portion of string to another? Picture a wave traveling from left to right (the positive x-direction) on the string, and consider a particular point a on the string (Fig. 15.12a). The string to the left of a exerts a force on the string to the right of it, and vice versa. In Fig. 15.12b the string to the left of a has been removed, and the force it exerts at a is represented by the components F and F_y, as we did in Figs. 15.8 and 15.10. We note again that F_y/F is equal to the negative of the *slope* of the string at a, which is also given by $\partial y/\partial x$. Putting these together, we have

$$F_y(x,t) = -F\frac{\partial y(x,t)}{\partial x} \tag{15.20}$$

We need the negative sign because F_y is negative when the slope is positive. We write the vertical force as $F_y(x,t)$ as a reminder that its value may be different at different points along the string and at different times.

When point a moves in the y-direction, the force F_y does *work* on this point and therefore transfers energy into the part of the string to the right of a. The corresponding power P (rate of doing work) at the point a is the transverse force $F_y(x,t)$ at a times the transverse velocity $v_y(x,t) = \partial y(x,t)/\partial t$ of that point:

$$P(x,t) = F_y(x,t)v_y(x,t) = -F\frac{\partial y(x,t)}{\partial x}\frac{\partial y(x,t)}{\partial t} \tag{15.21}$$

This power is the *instantaneous* rate at which energy is transferred along the string. Its value depends on the position x on the string and on the time t. Note that energy is being transferred only at points where the string has a nonzero slope ($\partial y/\partial x$ is nonzero), so that there is a transverse component of the tension force, and where the string has a nonzero transverse velocity ($\partial y/\partial t$ is nonzero) so that the transverse force can do work.

Equation (15.21) is valid for *any* wave on a string, sinusoidal or not. For a sinusoidal wave with wave function given by Eq. (15.7), we have

$$y(x,t) = A\cos(kx - \omega t)$$

$$\frac{\partial y(x,t)}{\partial x} = -kA\sin(kx - \omega t)$$

$$\frac{\partial y(x,t)}{\partial t} = \omega A\sin(kx - \omega t)$$

$$P(x,t) = Fk\omega A^2\sin^2(kx - \omega t) \tag{15.22}$$

By using the relations $\omega = vk$ and $v^2 = F/\mu$, we can also express Eq. (15.22) in the alternative form

$$P(x,t) = \sqrt{\mu F}\,\omega^2A^2\sin^2(kx - \omega t) \tag{15.23}$$

The $\sin^2$ function is never negative, so the instantaneous power in a sinusoidal wave is either positive (so that energy flows in the positive x-direction) or zero (at points where there is no energy transfer). Energy is never transferred in the direction opposite to the direction of wave propagation (Fig. 15.13).

The maximum value of the instantaneous power $P(x,t)$ occurs when the $\sin^2$ function has the value unity:

$$P_{max} = \sqrt{\mu F}\,\omega^2A^2 \tag{15.24}$$

To obtain the *average* power from Eq. (15.23), we note that the *average* value of the $\sin^2$ function, averaged over any whole number of cycles, is $\frac{1}{2}$. Hence the average power is

$$P_{av} = \frac{1}{2}\sqrt{\mu F}\,\omega^2A^2 \quad \text{(average power, sinusoidal wave on a string)} \tag{15.25}$$

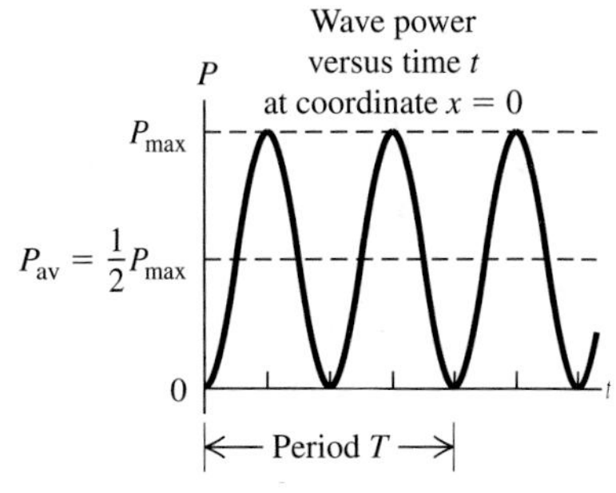

15.13 The instantaneous power $P(x,t)$ in a sinusoidal wave as given by Eq. (15.23), shown as a function of time at coordinate $x = 0$. The power is never negative, which means that energy never flows opposite to the direction of wave propagation.

The average power is just one-half of the maximum instantaneous power (see Fig. 15.13).

?

The average rate of energy transfer is proportional to the square of the amplitude and to the square of the frequency. This proportionality is a general result for mechanical waves of all types, including seismic waves (see the photo that opens this chapter). For a mechanical wave, the rate of energy transfer quadruples if the frequency is doubled (for the same amplitude) or if the amplitude is doubled (for the same frequency).

Electromagnetic waves turn out to be a bit different. While the average rate of energy transfer in an electromagnetic wave is proportional to the square of the amplitude, just as for mechanical waves, it is independent of the value of ω.

Example 15.5 Power in a wave

a) In Examples 15.2 and 15.3, at what maximum rate does Throcky put energy into the rope? That is, what is his maximum instantaneous power? b) What is his average power? c) As Throcky tires, the amplitude decreases. What is the average power when the amplitude has dropped to 7.50 mm?

SOLUTION

IDENTIFY: Our target variable in (a) is the *maximum instantaneous* power, while the target variable in (b) and (c) is the *average* power. As we've seen, these two quantities have different values for a sinusoidal wave. We'll be able to calculate the values of both quantities because we know all the other properties of the wave from Example 15.2.

SET UP: For part (a) we use Eq. (15.24), while for parts (b) and (c) we use Eq. (15.25).

EXECUTE: a) The maximum instantaneous power is

$$\begin{aligned} P_{max} &= \sqrt{\mu F}\omega^2 A^2 \\ &= \sqrt{(0.250 \text{ kg/m})(36.0 \text{ N})}(4.00\pi \text{ rad/s})^2(0.075 \text{ m})^2 \\ &= 2.66 \text{ W} \end{aligned}$$

b) From Eqs. (15.24) and (15.25), the average power is one-half of the maximum instantaneous power, so

$$P_{av} = \tfrac{1}{2}(2.66 \text{ W}) = 1.33 \text{ W}$$

c) The new amplitude is $\frac{1}{10}$ of the value we used in parts (a) and (b). The average power is proportional to the *square* of the amplitude, so now the average power is

$$P_{av} = \left(\frac{1}{10}\right)^2 (1.33 \text{ W}) = 0.0133 \text{ W} = 13.3 \text{ mW}$$

EVALUATE: The *maximum* instantaneous power in part (a) occurs when the quantity $\sin^2(kx - \omega t)$ in Eq. (15.23) is equal to 1. At any given value of x, this happens twice per period of the wave—once when the sine function is equal to $+1$, once when it's equal to -1. The *minimum* instantaneous power is zero; this occurs when $\sin(kx - \omega t) = 0$, which also happens twice per period.

Wave Intensity

Waves on a string carry energy in just one dimension of space (along the direction of the string). But other types of waves, including sound waves in air and seismic waves in the body of the earth, carry energy across all three dimensions of space. For waves that travel in three dimensions, we define the **intensity** (denoted by I) to be *the time average rate at which energy is transported by the wave, per unit area,* across a surface perpendicular to the direction of propagation. That is, intensity I is average power per unit area. It is usually measured in watts per square meter (W/m^2).

If waves spread out equally in all directions from a source, the intensity at a distance r from the source is inversely proportional to r^2. (Fig. 15.14). This follows directly from energy conservation. If the power output of the source is

P, then the average intensity I_1 through a sphere with radius r_1 and surface area $4\pi r_1^2$ is

$$I_1 = \frac{P}{4\pi r_1^2}$$

The average intensity I_2 through a sphere with a different radius r_2 is given by a similar expression. If no energy is absorbed between the two spheres, the power P must be the same for both, and

$$4\pi r_1^2 I_1 = 4\pi r_2^2 I_2$$

$$\frac{I_1}{I_2} = \frac{r_2^2}{r_1^2} \quad \text{(inverse-square law for intensity)} \qquad (15.26)$$

The intensity I at any distance r is therefore inversely proportional to r^2. This relationship is called the *inverse-square law* for intensity.

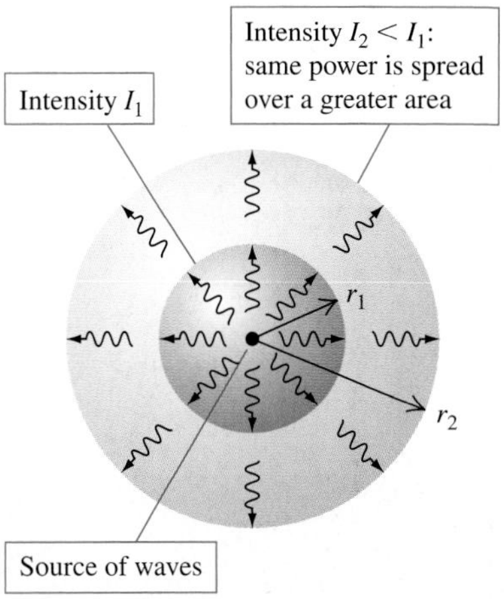

15.14 The greater the distance from a wave source, the greater the area over which the wave power is distributed and the smaller the wave intensity.

Example 15.6 The inverse-square law

A tornado warning siren on top of a tall pole radiates sound waves uniformly in all directions. At a distance of 15.0 m the intensity of the sound is 0.250 W/m². At what distance from the siren is the intensity 0.010 W/m²?

SOLUTION

IDENTIFY: Because waves spread out equally in all directions, we can use the inverse-square law. Our target variable is a distance from the wave source.

SET UP: The relation to use is Eq. (15.26). We are given the distance $r_1 = 15.0$ m at which the intensity is $I_1 = 0.250\ \text{W/m}^2$, and we want to find the distance r_2 at which the intensity is $I_2 = 0.010\ \text{W/m}^2$.

EXECUTE: We solve Eq. (15.26) for r_2:

$$r_2 = r_1\sqrt{\frac{I_1}{I_2}} = (15.0\ \text{m})\sqrt{\frac{0.250\ \text{W/m}^2}{0.010\ \text{W/m}^2}} = 75.0\ \text{m}$$

EVALUATE: As a check on our answer, note that r_2 is five times greater than r_1. By the inverse-square law, the intensity I_2 should be $1/5^2 = 1/25$ as great as I_1, and indeed it is.

By using the inverse-square law we've assumed that the sound waves travel in straight lines away from the siren. A more realistic solution of this problem would account for the reflection of sound waves from the ground. Such a solution is beyond our scope, however.

Test Your Understanding

In Example 15.5, suppose the wave amplitude remained equal to 0.075 m but the frequency increased from 2.00 Hz to 4.00 Hz. What effect would this have on the average power?

15.6 | Wave Interference, Boundary Conditions, and Superposition

Up to this point we've been discussing waves that propagate continuously in the same direction. But when a wave strikes the boundaries of its medium, all or part of the wave is *reflected*. When you yell at a building wall or a cliff face some distance away, the sound wave is reflected from the rigid surface and you hear an echo. When you flip the end of a rope whose far end is tied to a rigid support, a pulse travels the length of the rope and is reflected back to you. In both cases, the initial and reflected waves overlap in the same region of the medium. This overlapping of waves is called **interference.** (In general, the term *interference* refers to what happens when two or more waves pass through the same region at the same time.)

As a simple example of wave reflections and the role of the boundary of a wave medium, let's look again at transverse waves on a stretched string. What happens when a wave pulse or a sinusoidal wave arrives at the *end* of the string?

If the end is fastened to a rigid support, it is a *fixed* end that cannot move. The arriving wave exerts a force on the support; the reaction to this force, exerted *by* the support *on* the string, "kicks back" on the string and sets up a *reflected* pulse or wave traveling in the reverse direction. Figure 15.15 is a series of photographs showing the reflection of a pulse at the fixed end of a long coiled spring. The reflected pulse moves in the opposite direction from the initial, or *incident,* pulse, and its displacement is also opposite. This situation is illustrated for a wave pulse on a string in Fig. 15.16a.

15.15 A series of images of a wave pulse, equally spaced in time from top to bottom. The pulse starts at the right in the top image, travels to the left, and is reflected from the fixed end at the left.

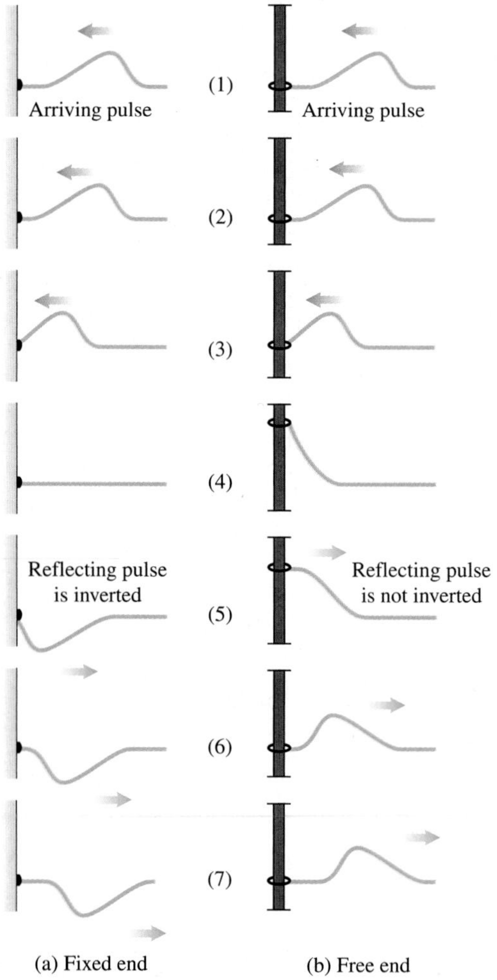

15.16 Reflection of a wave pulse (a) at a fixed end of a string and (b) at a free end. Time increases from top to bottom in each figure.

The opposite situation from an end that is held stationary is a *free* end, one that is perfectly free to move in the direction perpendicular to the length of the string. For example, the string might be tied to a light ring that slides on a frictionless rod perpendicular to the rope, as in Fig. 15.16b. The ring and rod maintain the tension but exert no transverse force. When a wave arrives at this free end, the ring slides along the rod. The ring reaches a maximum displacement, and both it and the string come momentarily to rest, as in drawing (4) in Fig. 15.16b. But the string is now stretched, giving increased tension, so the free end of the string is pulled back down, and again a reflected pulse is produced (drawing (7)). As for a fixed end, the reflected pulse moves in the opposite direction from the initial pulse, but now the direction of the displacement is the same as for the initial pulse. The conditions at the end of the string, such as a rigid support or the complete absence of transverse force, are called **boundary conditions.**

The formation of the reflected pulse is similar to the overlap of two pulses traveling in opposite directions. Figure 15.17 shows two pulses with the same shape, one inverted with respect to the other, traveling in opposite directions. As the pulses overlap and pass each other, the total displacement of the string is the *algebraic sum* of the displacements at that point in the individual pulses. Because these two pulses have the same shape, the total displacement at point O in the middle of the figure is zero at all times. Thus the motion of the right half of the string would be the same if we cut the string at point O, threw away the left side, and held the end at O fixed. The two pulses on the right side then correspond to the incident and reflected pulses, combining so that the total displacement at O is *always* zero. For this to occur, the reflected pulse must be inverted relative to the incident pulse.

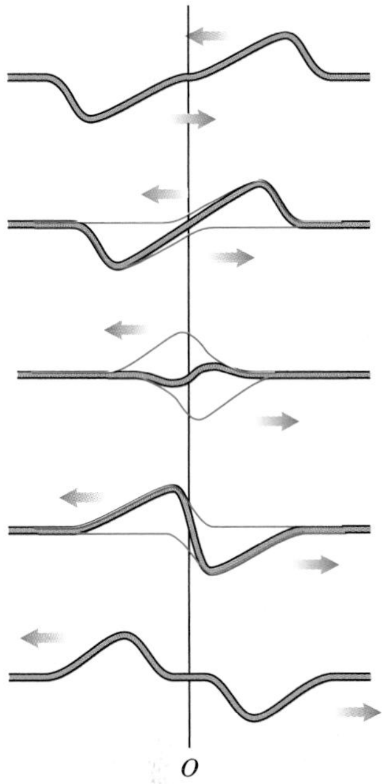

15.17 Overlap of two wave pulses—one right side up, one inverted—traveling in opposite directions. Time increases from top to bottom.

Figure 15.18 shows two pulses with the same shape, traveling in opposite directions but *not* inverted relative to each other. The displacement at point O in the middle of the figure is not zero, but the slope of the string at this point is always zero. According to Eq. (15.20), this corresponds to the absence of any transverse force at this point. In this case the motion of the right half of the string would be the same as if we cut the string at point O and anchored the end with a frictionless sliding ring (Fig. 15.16b) that maintains tension without exerting any transverse force. In other words, this situation corresponds to reflection of a pulse at a free end of a string at point O. In this case the reflected pulse is *not* inverted.

The Principle of Superposition

Combining the displacements of the separate pulses at each point to obtain the actual displacement is an example of the **principle of superposition:** when two waves overlap, the actual displacement of any point on the string at any time is obtained by adding the displacement the point would have if only the first wave were present and the displacement it would have if only the second wave were present. In other words, the wave function $y(x, t)$ that describes the resulting motion in this situation is obtained by *adding* the two wave functions for the two separate waves:

$$y(x, t) = y_1(x, t) + y_2(x, t) \quad \text{(principle of superposition)} \quad (15.27)$$

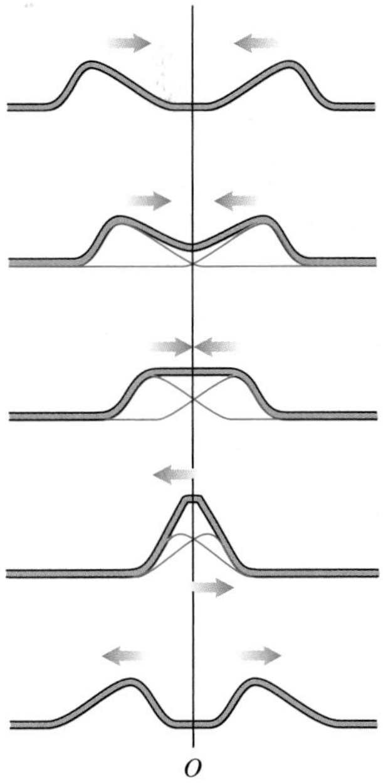

15.18 Overlap of two wave pulses—both right side up—traveling in opposite directions. Time increases from top to bottom. Compare to Fig. 15.17.

Mathematically, this additive property of wave functions follows from the form of the wave equation, Eq. (15.12) or (15.18), which every physically possible wave function must satisfy. Specifically, the wave equation is *linear;* that is, it contains the function $y(x, t)$ only to the first power (there are no terms involving

$y(x, t)^2$, $y(x, t)^{1/2}$, etc.). As a result, if any two functions $y_1(x, t)$ and $y_2(x, t)$ satisfy the wave equation separately, their sum $y_1(x, t) + y_2(x, t)$ also satisfies it and is therefore a physically possible motion. Because this principle depends on the linearity of the wave equation and the corresponding linear-combination property of its solutions, it is also called the *principle of linear superposition.* For some physical systems, such as a medium that does not obey Hooke's law, the wave equation is *not* linear; this principle does not hold for such systems.

The principle of superposition is of central importance in all types of waves. When a friend talks to you while you are listening to music, you can distinguish the sound of speech and the sound of music from each other. This is precisely because the total sound wave reaching your ears is the algebraic sum of the wave produced by your friend's voice and the wave produced by the speakers of your stereo. If two sound waves did *not* combine in this simple linear way, the sound you would hear in this situation would be a hopeless jumble. Superposition also applies to electromagnetic waves (such as light) and many other types of waves.

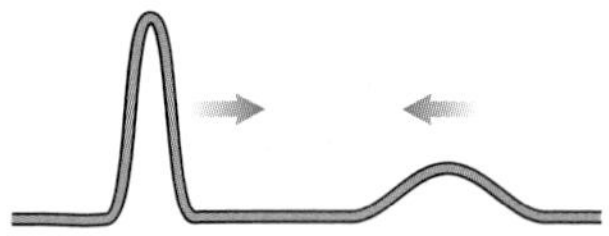

15.19 Two wave pulses with different shapes.

Test Your Understanding

Two wave pulses with different shapes are traveling in opposite directions along a string (Fig. 15.19). Make a series of sketches like Fig. 15.18 showing the shape of the string as the two pulses approach, overlap, and then pass each other.

15.7 | Standing Waves on a String

We have talked about the reflection of a wave *pulse* on a string when it arrives at a boundary point (either a fixed end or a free end). Now let's look at what happens when a *sinusoidal* wave is reflected by a fixed end of a string. We'll again approach the problem by considering the superposition of two waves propagating through the string, one representing the original or incident wave and the other representing the wave reflected at the fixed end.

Figure 15.20 shows a string that is fixed at its left-hand end. Its right-hand end is moved up and down in simple harmonic motion to produce a wave that travels to the left; the wave reflected from the fixed end travels to the right. The resulting motion when the two waves combine no longer looks like two waves traveling in opposite directions. The string appears to be subdivided into a number of segments, as in the time-exposure photographs of Figs. 15.20a, 15.20b, 15.20c, and 15.20d. Figure 15.20e shows two instantaneous shapes of the string in Fig. 15.20b. Let's compare this behavior with the waves we studied in Sections 15.1 through 15.5. In a wave that travels along the string, the amplitude is constant and the wave pattern moves with a speed equal to the wave speed. Here, instead, the wave pattern remains in the same position along the string and its amplitude fluctuates. There are particular points called **nodes** (labeled N in Fig. 15.20e) that never move at all. Midway between the nodes are points called **antinodes** (labeled A in Fig. 15.20e) where the amplitude of motion is greatest. Because the wave pattern doesn't appear to be moving in either direction along the string, it is called a **standing wave.** (To emphasize the difference, a wave that *does* move along the string is called a **traveling wave.**)

The principle of superposition explains how the incident and reflected wave combine to form a standing wave. In Fig. 15.21 the red curves show a wave traveling to the left. The blue curves show a wave traveling to the right with the same propagation speed, wavelength, and amplitude. The waves are shown at nine

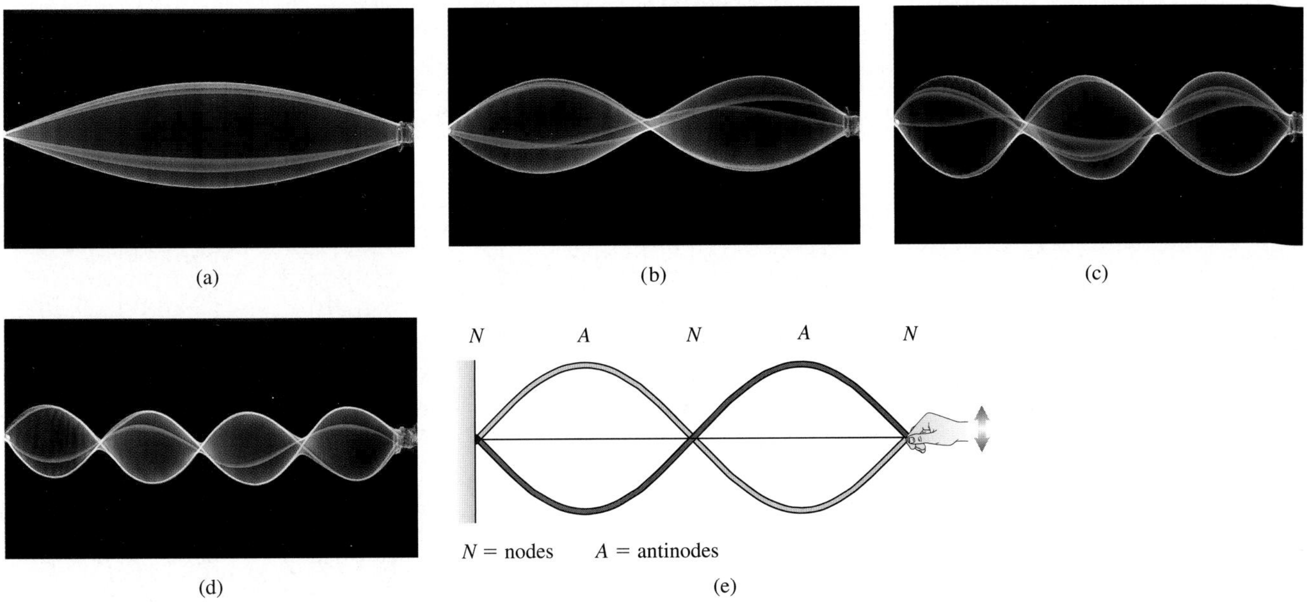

15.20 (a)–(d) Time exposures of standing waves in a stretched string. From (a) to (d), the frequency of oscillation of the right-hand end increases and the wavelength of the standing wave decreases. (e) The extremes of the motion of the standing wave in (b), with nodes at the center and at the ends. The right-hand end of the string moves very little compared to the antinodes and so is essentially a node.

instants, one sixteenth of a period apart. At each point along the string, we add the displacements (the values of y) for the two separate waves; the result is the total wave on the string, shown in black.

At certain instants, such as $t = 4T/16$, the two wave patterns are exactly in phase with each other, and the shape of the string is a sine curve with twice the amplitude of either individual wave. At other instants, such as $t = 8T/16$, the two waves are exactly out of phase with each other, and the total wave at that instant is zero. The resultant displacement is *always* zero at those places marked N at the bottom of Fig. 15.21. These are the *nodes.* At a node the displacements of the two waves in red and blue are always equal and opposite and cancel each other out. This cancellation is called **destructive interference.** Midway between the nodes are the points of *greatest* amplitude or *antinodes,* marked A. At the antinodes the displacements of the two waves in red and blue are always identical, giving a large resultant displacement; this phenomenon is called **constructive interference.** We can see from the figure that the distance between successive nodes or between successive antinodes is one half-wavelength, or $\lambda/2$.

We can derive a wave function for the standing wave of Fig. 15.21 by adding the wave functions $y_1(x, t)$ and $y_2(x, t)$ for two waves with equal amplitude, period, and wavelength traveling in opposite directions. Here $y_1(x, t)$ (the red curves in Fig. 15.21) represents an incoming, or *incident,* wave traveling to the left along the $+x$-axis, arriving at the point $x = 0$ and being reflected; $y_2(x, t)$ (the blue curves in Fig. 15.21) represents the *reflected* wave traveling to the right from $x = 0$. We noted in Section 15.6 that the wave reflected from a fixed end of a string is inverted, so we give a negative sign to one of the waves:

$$y_1(x, t) = -A\cos(kx + \omega t) \qquad \text{(incident wave traveling to the left)}$$

$$y_2(x, t) = A\cos(kx - \omega t) \qquad \text{(reflected wave traveling to the right)}$$

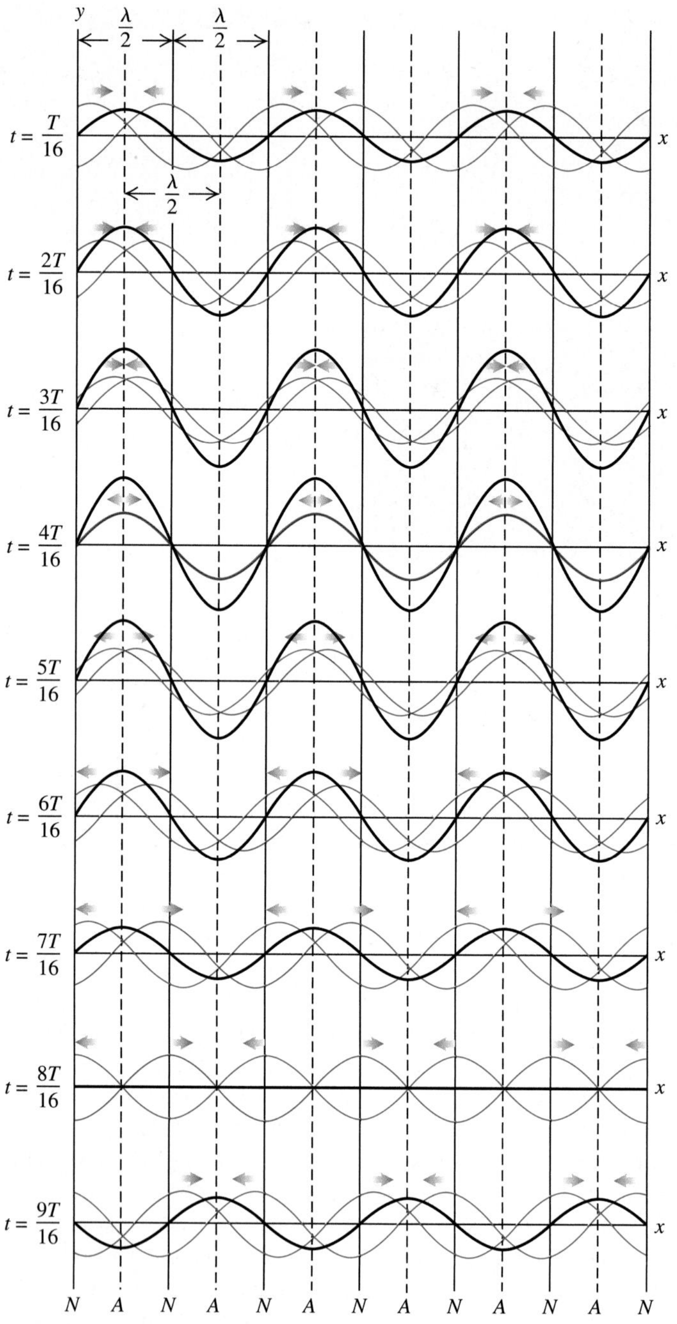

15.21 Formation of a standing wave. A wave traveling to the left (red curves) combines with a wave traveling to the right (blue curves) to form a standing wave (black curves). The horizontal x-axis in each part shows the equilibrium position of the string.

Note also that the change in sign corresponds to a shift in *phase* of 180° or π radians. At $x = 0$ the motion from the reflected wave is $A \cos \omega t$; and the motion from the incident wave is $-A \cos \omega t$, which we can also write as $A \cos(\omega t + \pi)$. From Eq. (15.27), the wave function for the standing wave is the sum of the individual wave functions:

$$y(x, t) = y_1(x, t) + y_2(x, t) = A\left[-\cos(kx + \omega t) + \cos(kx - \omega t)\right]$$

We can rewrite each of the cosine terms by using the identities for the cosine of the sum and difference of two angles: $\cos(a \pm b) = \cos a \cos b \mp \sin a \sin b$. Applying these and combining terms, we obtain the wave function for the standing wave:

$$y(x,t) = y_1(x,t) + y_2(x,t) = (2A \sin kx)\sin \omega t$$

or

$$y(x,t) = (A_{SW} \sin kx)\sin \omega t \tag{15.28}$$

(standing wave on a string, fixed end at $x = 0$)

The standing wave amplitude A_{SW} is twice the amplitude A of either of the original traveling waves:

$$A_{SW} = 2A$$

Equation (15.28) has two factors: a function of x and a function of t. The factor A_{SW} shows that at each instant the shape of the string is a sine curve. But unlike a wave traveling along a string, the wave shape stays in the same position, oscillating up and down as described by the sin ωt factor. This behavior is shown graphically by the black curves in Fig. 15.21. Each point in the string still undergoes simple harmonic motion, but all the points between any successive pair of nodes oscillate *in phase*. This is in contrast to the phase differences between oscillations of adjacent points that we see with a wave traveling in one direction.

We can use Eq. (15.28) to find the positions of the nodes; these are the points for which $\sin kx = 0$, so the displacement is *always* zero. This occurs when $kx = 0, \pi, 2\pi, 3\pi, \ldots$, or, using $k = 2\pi/\lambda$,

$$\begin{aligned} x &= 0, \frac{\pi}{k}, \frac{2\pi}{k}, \frac{3\pi}{k}, \ldots \\ &= 0, \frac{\lambda}{2}, \frac{2\lambda}{2}, \frac{3\lambda}{2}, \ldots \end{aligned} \tag{15.29}$$

(nodes of a standing wave on a string, fixed end at $x = 0$)

In particular, there is a node at $x = 0$, as there should be, since this point is a fixed end of the string.

A standing wave, unlike a traveling wave, *does not* transfer energy from one end to the other. The two waves that form it would individually carry equal amounts of power in opposite directions. There is a local flow of energy from each node to the adjacent antinodes and back, but the *average* rate of energy transfer is zero at every point. If you evaluate the wave power given by Eq. (15.21) using the wave function of Eq. (15.28), you will find that the average power is zero (see Challenge Problem 15.82).

Problem-Solving Strategy — Standing Waves

IDENTIFY *the relevant concepts:* As with traveling waves, it's useful to distinguish between the purely kinematic quantities, such as wave speed v, wavelength λ, and frequency f, and the dynamic quantities involving the properties of the medium, such as F and μ for transverse waves on a string. Once you decide what the target variable is, try to determine whether the problem is only kinematic in nature or whether the properties of the medium are also involved.

SET UP *the problem using the following steps:*

1. In visualizing nodes and antinodes in standing waves, it is always helpful to draw diagrams. For a string you can draw the shape at one instant and label the nodes N and antinodes A. The distance between two adjacent nodes or two adjacent antinodes is always $\lambda/2$, and the distance between a node and the adjacent antinode is always $\lambda/4$.
2. Decide which equations you'll need to use. The wave function for the standing wave is almost always useful (like Eq. (15.28)).
3. You can compute the wave speed if you know either λ and f (or, equivalently, $k = 2\pi/\lambda$ and $\omega = 2\pi f$) or the properties of the medium (for a string, F and μ).

EXECUTE *the solution as follows:*
Solve for the unknown quantities using the equations you've selected. Once you have the wave function, you can find the value of the displacement y at any point in the wave medium (value of x) and at any time. You can find the velocity of a particle in the wave medium by taking the partial derivative of y with respect to time. To find the acceleration of such a particle, take the second partial derivative of y with respect to time.

EVALUATE *your answer:* Compare your numerical answers with your diagram. Check that the wave function is compatible with the boundary conditions (for example, the displacement should be zero at a fixed end).

Example 15.7 Standing waves on a guitar string

One of the strings of a guitar lies along the x-axis when in equilibrium. The end of the string at $x = 0$ (the bridge of the guitar) is tied down. An incident sinusoidal wave, corresponding to the red curves in Fig. 15.21, travels along the string in the $-x$-direction at 143 m/s with an amplitude of 0.750 mm and a frequency of 440 Hz. This wave is reflected from the fixed end at $x = 0$, and the superposition of the incident traveling wave and the reflected traveling wave forms a standing wave. a) Find the equation giving the displacement of a point on the string as a function of position and time. b) Locate the points on the string that don't move at all. c) Find the amplitude, maximum transverse velocity, and maximum transverse acceleration at points of maximum oscillation.

SOLUTION

IDENTIFY: This is a *kinematics* problem in which we are asked to describe the motion of the string (see the Problem-Solving Strategy in Section 15.3). The target variables are the wave function of the standing wave (part (a)), the locations of the points that don't move, or *nodes* (part (b)), and the maximum values of displacement y, transverse velocity v_y, and transverse acceleration a_y. (Waves on a string are transverse waves, so *transverse* means "in the direction of the displacement," that is, in the y-direction.) To find these quantities we use the expression that we derived in this section for a standing wave on a string with a fixed end, as well as other relations from Sections 15.2 and 15.3.

SET UP: Since there is a fixed end at $x = 0$, we may use Eqs. (15.28) and (15.29) to describe this standing wave. We also use the relationships among ω, k, f, λ, and the wave speed v.

EXECUTE: a) To use Eq. (15.28) we need the values of A_{SW}, ω, and k. The amplitude of the incident wave is $A = 0.750 \text{ mm} = 7.50 \times 10^{-4} \text{ m}$; the reflected wave has the same amplitude, and the standing wave amplitude is $A_{SW} = 2A = 1.50 \times 10^{-3}$ m. The angular frequency ω and wave number k are

$$\omega = 2\pi f = (2\pi \text{ rad})(440 \text{ s}^{-1}) = 2760 \text{ rad/s}$$

$$k = \frac{\omega}{v} = \frac{2760 \text{ rad/s}}{143 \text{ m/s}} = 19.3 \text{ rad/m}$$

Then Eq. (15.28) gives

$$\begin{aligned} y(x,t) &= (A_{SW} \sin kx) \sin \omega t \\ &= [(1.50 \times 10^{-3} \text{ m}) \sin (19.3 \text{ rad/m})x] \cos (2760 \text{ rad/s})t \end{aligned}$$

b) The positions of the nodes are given by Eq. (15.29): $x = 0, \lambda/2, \lambda, 3\lambda/2, \ldots$ The wavelength is

$$\lambda = \frac{v}{f} = \frac{143 \text{ m/s}}{440 \text{ Hz}} = 0.325 \text{ m}$$

so the nodes are at the following distances from the fixed end:

$$x = 0,\ 0.163 \text{ m},\ 0.325 \text{ m},\ 0.488 \text{ m},\ \ldots$$

c) From the expression in part (a) for $y(x, t)$, we see that the maximum displacement from equilibrium is $1.50 \times 10^{-3} \text{ m} = 1.50$ mm, which is just twice the amplitude of the incident wave. This maximum occurs at the *antinodes*, which are midway between adjacent nodes (that is, at $x = 0.081$ m, 0.244 m, 0.406 m, . . .).

For a particle on the string at any point x, the transverse (y-) velocity is

$$\begin{aligned} v_y(x,t) &= \frac{\partial y(x,t)}{\partial t} \\ &= [(1.50 \times 10^{-3} \text{ m}) \sin (19.3 \text{ rad/m})x] \\ &\quad \times [(2760 \text{ rad/s}) \cos (2760 \text{ rad/s})t] \\ &= [(4.15 \text{ m/s}) \sin (19.3 \text{ rad/m})x] \cos (2760 \text{ rad/s})t \end{aligned}$$

At an antinode, $\sin(19.3 \text{ rad/m})x = \pm 1$ and the transverse velocity varies in value between 4.15 m/s and -4.15 m/s. As is always the case in simple harmonic motion, the maximum velocity occurs when the particle is passing through the equilibrium position ($y = 0$).

The transverse acceleration $a_y(x, t)$ is the first partial derivative of $v_y(x, t)$ with respect to time (that is, the *second* partial derivative of $y(x, t)$ with respect to time). We leave the calculation to you; the result is

$$a_y(x, t) = \frac{\partial v_y(x, t)}{\partial t} = \frac{\partial^2 y(x, t)}{\partial t^2}$$
$$= [(-1.15 \times 10^4 \text{ m/s}^2) \sin (19.3 \text{ rad/m})x] \times \sin (2760 \text{ rad/s})t$$

At the antinodes, the transverse acceleration varies in value between $+1.15 \times 10^4$ m/s^2 and -1.15×10^4 m/s^2.

EVALUATE: The maximum transverse velocity at an antinode is quite respectable (about 15 km/h, or 9.3 mi/h). But the maximum transverse acceleration is tremendous, 1170 times the acceleration due to gravity! Guitar strings are made of sturdy stuff to be able to withstand such acceleration.

Guitar strings are actually tied down at *both* ends. We'll see the consequences of this in the next section.

Test Your Understanding

Suppose the frequency of the standing wave in Example 15.7 were doubled from 440 Hz to 880 Hz. Would all of the nodes for $f = 440$ Hz also be nodes for $f = 880$ Hz? If so, would there be additional nodes for $f = 880$ Hz? If not, which nodes are absent for $f = 880$ Hz?

15.8 | Normal Modes of a String

When we described standing waves on a string rigidly held at one end, as in Fig. 15.20, we made no assumptions about the length of the string or about what was happening at the other end. Let's now consider a string of a definite length L, rigidly held at *both* ends. Such strings are found in many musical instruments, including pianos, violins, and guitars. When a guitar string is plucked, a wave is produced in the string; this wave is reflected and re-reflected from the ends of the string, making a standing wave. This standing wave on the string in turn produces a sound wave in the air, with a frequency determined by the properties of the string. This is what makes stringed instruments so useful in making music.

ActivPhysics Online

10.4 Standing Waves on Strings

10.5 Tuning a Stringed Instrument: Standing Waves

10.6 String Mass and Standing Waves

To understand these properties of standing waves on a string fixed at both ends, let's first examine what happens when we set up a sinusoidal wave on such a string. The standing wave that results must have a node at *both* ends of the string. We saw in the preceding section that adjacent nodes are one half-wavelength $(\lambda/2)$ apart, so the length of the string must be $\lambda/2$, or $2(\lambda/2)$, or $3(\lambda/2)$, or in general some integer number of half-wavelengths:

$$L = n\frac{\lambda}{2} \qquad (n = 1, 2, 3, \ldots) \qquad \text{(string fixed at both ends)} \qquad (15.30)$$

That is, if a string with length L is fixed at both ends, a standing wave can exist only if its wavelength satisfies Eq. (15.30).

Solving this equation for λ and labeling the possible values of λ as λ_n, we find

$$\lambda_n = \frac{2L}{n} \qquad (n = 1, 2, 3, \ldots) \qquad \text{(string fixed at both ends)} \qquad (15.31)$$

Waves can exist on the string if the wavelength is *not* equal to one of these values, but there cannot be a steady wave pattern with nodes and antinodes, and the total wave cannot be a standing wave. Equation (15.31) is illustrated by the standing waves shown in Figs. 15.20a, 15.20b, 15.20c, and 15.20d; these represent $n = 1, 2, 3$, and 4, respectively.

15.22 Each string of a violin naturally oscillates at one or more of its harmonic frequencies, producing sound waves in the air with the same frequencies.

Corresponding to the series of possible standing-wave wavelengths λ_n is a series of possible standing-wave frequencies f_n, each related to its corresponding wavelength by $f_n = v/\lambda_n$. The smallest frequency f_1 corresponds to the largest wavelength (the $n = 1$ case), $\lambda_1 = 2L$:

$$f_1 = \frac{v}{2L} \qquad \text{(string fixed at both ends)} \qquad (15.32)$$

This is called the **fundamental frequency.** The other standing-wave frequencies are $f_2 = 2v/2L$, $f_3 = 3v/2L$, and so on. These are all integer multiples of the fundamental frequency f_1, such as $2f_1$, $3f_1$, $4f_1$, and so on, and we can express *all* the frequencies as

$$f_n = n\frac{v}{2L} = nf_1 \qquad (n = 1, 2, 3, \dots) \qquad \text{(string fixed at both ends)} \qquad (15.33)$$

These frequencies are called **harmonics,** and the series is called a **harmonic series.** Musicians sometimes call f_2, f_3, and so on **overtones;** f_2 is the second harmonic or the first overtone, f_3 is the third harmonic or the second overtone, and so on. The first harmonic is the same as the fundamental frequency (Fig. 15.22).

For a string with fixed ends at $x = 0$ and $x = L$, the wave function $y(x, t)$ of the nth standing wave is given by Eq. (15.28) (which satisfies the condition that there is a node at $x = 0$), with $\omega = \omega_n = 2\pi f_n$ and $k = k_n = 2\pi/\lambda_n$:

$$y_n(x, t) = A_{SW} \sin k_n x \sin \omega_n t \qquad (15.34)$$

You can easily show that this wave function has nodes at both $x = 0$ and $x = L$, as it must.

A **normal mode** of an oscillating system is a motion in which all particles of the system move sinusoidally with the same frequency. For a system made up of a string of length L fixed at both ends, each of the wavelengths given by Eq. (15.31) corresponds to a possible normal-mode pattern and frequency. There are infinitely many normal modes, each with its characteristic frequency and vibration pattern. Figure 15.23 shows the first four normal-mode patterns and their associated frequencies and wave-

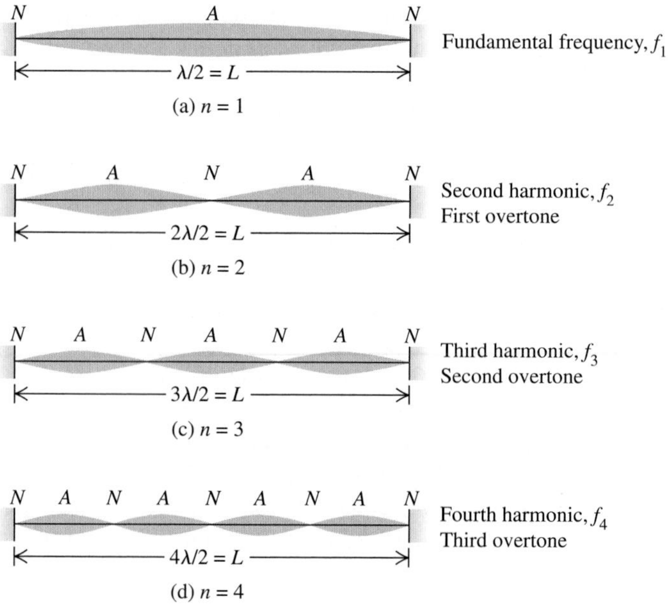

15.23 The first four normal modes of a string fixed at both ends. (Compare these to the photographs in Fig. 15.20.)

lengths; these correspond to Eq. (15.34) with $n = 1, 2, 3,$ and 4. By contrast, a harmonic oscillator, which has only one oscillating particle, has only one normal mode and one characteristic frequency. The string fixed at both ends has infinitely many normal modes because it is made up of a very large (effectively infinite) number of particles. More complicated oscillating systems also have infinite numbers of normal modes, though with more complex normal-mode patterns than a string (Fig. 15.24).

15.24 Astronomers have discovered that the sun oscillates in several different normal modes. In this computer simulation of one such mode, blue denotes places where solar material is moving outward and red denotes where it is moving inward.

If we could displace a string so that its shape is the same as one of the normal-mode patterns and then release it, it would vibrate with the frequency of that mode. Such a vibrating string would displace the surrounding air with the same frequency, producing a traveling sinusoidal sound wave that your ears would perceive as a pure tone. But when a string is struck (as in a piano) or plucked (as is done to guitar strings), the shape of the displaced string is *not* as simple as one of the patterns in Fig. 15.23. The fundamental as well as many overtones are present in the resulting vibration. This motion is therefore a combination or *superposition* of many normal modes. Several simple-harmonic motions of different frequencies are present simultaneously, and the displacement of any point on the string is the sum (or superposition) of displacements associated with the individual modes. The sound produced by the vibrating string is likewise a superposition of traveling sinusoidal sound waves, which you perceive as a rich, complex tone with the fundamental frequency f_1. The standing wave on the string and the traveling sound wave in the air have similar **harmonic content** (the extent to which frequencies higher than the fundamental are present). The harmonic content depends on how the string is initially set into motion. If you pluck the strings of an acoustic guitar in the normal location over the sound hole, the sound that you hear has a different harmonic content than if you pluck the strings next to the fixed end on the guitar body.

It is possible to represent every possible motion of the string as some superposition of normal-mode motions. Finding this representation for a given vibration pattern is called *harmonic analysis.* The sum of sinusoidal functions that represents a complex wave is called a *Fourier series.* Figure 15.25 shows how a standing wave that is produced by plucking a guitar string of length L at a point $L/4$ from one end can be represented as a combination of sinusoidal functions.

As we have seen, the fundamental frequency of a vibrating string is $f_1 = v/2L$. The speed v of waves on the string is determined by Eq. (15.13), $v = \sqrt{F/\mu}$. Combining these, we find

$$f_1 = \frac{1}{2L}\sqrt{\frac{F}{\mu}} \quad \text{(string fixed at both ends)} \tag{15.35}$$

ActivPhysics Online

10.10 Complex Waves: Fourier Analysis

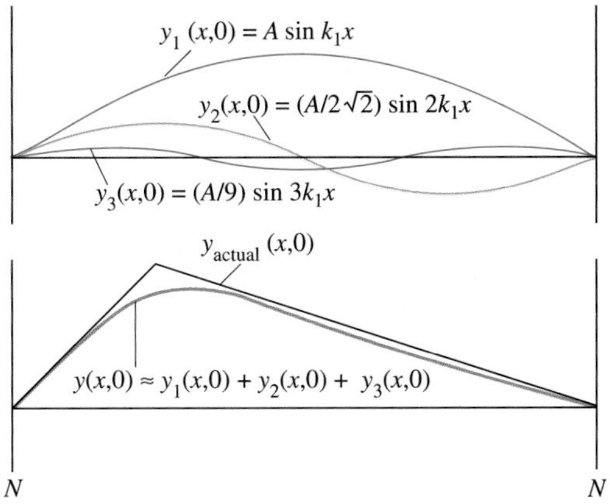

15.25 When a guitar string is plucked (pulled into a triangular shape) and released, a standing wave results. The standing wave is well represented (except at the sharp maximum point) by the sum of just three sinusoidal functions. Including additional sinusoidal functions further improves the representation.

15.26 Comparing the range of a concert grand piano to those of a violin (red bar), a viola (yellow bar), a cello (green bar), and a string bass (blue bar). In all cases, longer strings provide bass notes and shorter strings produce treble notes.

This is also the fundamental frequency of the sound wave created in the surrounding air by the vibrating string. Familiar musical instruments show how f_1 depends on the properties of the string. The inverse dependence of frequency on length L is illustrated by the long strings of the bass (low-frequency) section of the piano or the bass viol compared with the shorter strings on the treble section of the piano or the violin (Fig. 15.26). The pitch of a violin or guitar is usually varied by pressing a string against the fingerboard with the fingers to change the length L of the vibrating portion of the string. Increasing the tension F increases the wave speed v and thus increases the frequency (and the pitch). All string instruments are "tuned" to the correct frequencies by varying the tension; you tighten the string to raise the pitch. Finally, increasing the mass per unit length μ decreases the wave speed and thus the frequency. The lower notes on a steel guitar are produced by thicker strings, and one reason for winding the bass strings of a piano with wire is to obtain the desired low frequency without resorting to a string that is inconveniently long.

Wind instruments such as saxophones and trombones also have normal modes. As for stringed instruments, the frequencies of these normal modes determine the pitch of the musical tones that these instruments produce. We'll discuss these instruments and many other aspects of sound in Chapter 16.

Example 15.8 A giant bass viol

In an effort to get your name in the *Guinness Book of World Records,* you set out to build a bass viol with strings that have a length of 5.00 m between fixed points. One string has a linear mass density of 40.0 g/m and a fundamental frequency of 20.0 Hz (the lowest frequency that the human ear can hear). Calculate (a) the tension of this string, (b) the frequency and wavelength on the string of the second harmonic, and (c) the frequency and wavelength on the string of the second overtone.

SOLUTION

IDENTIFY: The target variable in (a) is the string tension; we find this from the expression for the fundamental frequency of the string, which involves the tension. In parts (b) and (c) the target variables are the frequency and wavelength of different harmonics. We determine these from the given length of the string and fundamental frequency.

SET UP: For part (a), the equation to use is Eq. (15.35); it involves the known values of f_1, L, and μ as well as the target variable F. We solve parts (b) and (c) using Eqs. (15.31) and (15.33).

EXECUTE: a) We solve Eq. (15.35) for the string tension F:

$$F = 4\mu L^2 f_1^2 = 4(40.0 \times 10^{-3}\ \text{kg/m})(5.00\ \text{m})^2(20.0\ \text{s}^{-1})^2$$
$$= 1600\ \text{N} = 360\ \text{lb}$$

b) The second harmonic is denoted by $n = 2$. From Eq. (15.33), the second harmonic frequency is

$$f_2 = 2f_1 = 2(20.0\ \text{Hz}) = 40.0\ \text{Hz}$$

From Eq. (15.31), the wavelength on the string of the second harmonic is

$$\lambda_2 = \frac{2L}{2} = 5.00\ \text{m}$$

c) The second overtone is the "second tone over" (above) the fundamental, that is, $n = 3$. Its frequency and wavelength are

$$f_3 = 3f_1 = 3(20.0\ \text{Hz}) = 60.0\ \text{Hz}$$
$$\lambda_3 = \frac{2L}{3} = 3.33\ \text{m}$$

EVALUATE: The tension in part (a) is a bit larger than in a real bass viol, for which the string tension is typically a few hundred newtons. The wavelengths in parts (b) and (c) are equal to the length of the string and two-thirds the length of the string respectively; these results agree with the drawings of standing waves in Fig. 15.23.

Example 15.9 From waves on a string to sound waves in air

What are the frequency and wavelength of the sound waves produced in the air when the string in the previous example is vibrating at its fundamental frequency? The speed of sound in air at 20°C is 344 m/s.

SOLUTION

IDENTIFY: Our target variables are f and λ for the sound wave produced by the bass viol, *not* for the standing wave on the string. However, when the string vibrates at a particular frequency, the surrounding air is forced to vibrate at the same frequency. So the frequency of the sound wave is the same as that of the standing wave on the string. However, the relationship $\lambda = v/f$ shows that the *wavelength* of the sound wave is typically different from the wavelength of the standing wave on the string, because the two waves have different speeds.

SET UP: The only equation we need is $v = \lambda f$. We apply this to both the standing wave on the string (speed v_{string}) and the traveling sound wave (speed v_{sound}).

EXECUTE: The sound wave frequency is the same as the standing wave fundamental frequency: $f = f_1 = 20.0$ Hz. The wavelength of the sound wave is

$$\lambda_{1\,(\text{sound})} = \frac{v_{\text{sound}}}{f_1} = \frac{344\ \text{m/s}}{20.0\ \text{Hz}} = 17.2\ \text{m}$$

EVALUATE: Note that $\lambda_{1\,(\text{sound})}$ is greater than the wavelength of the standing wave on the string, $\lambda_{1\,(\text{string})} = 2L = 2(5.00\ \text{m}) = 10.0$ m. Fundamentally this is because the speed of sound is greater than the speed of waves on the string, $v_{\text{string}} = \lambda_{1\,(\text{sound})} f_1 = (10.0\ \text{m})(20.0\ \text{Hz}) = 200$ m/s. Hence, for *any* normal mode on this string, the sound wave that is produced has the same frequency as the wave on the string but a wavelength that is greater by a factor of $v_{\text{sound}}/v_{\text{string}} = (344\ \text{m/s})/(200\ \text{m/s}) = 1.72$.

Test Your Understanding

While a guitar string is vibrating, you gently touch the midpoint of the string to ensure that the string does not vibrate at that point. Which normal modes *cannot* be present on the string while you are touching it in this way?

SUMMARY

A wave is any disturbance from an equilibrium condition that propagates from one region to another. A mechanical wave always travels within some material called the medium. The wave disturbance propagates at the wave speed v, which depends on the type of wave and the properties of the medium.

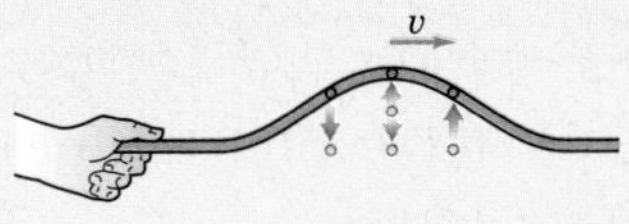

In a periodic wave, the motion of each point of the medium is periodic. A sinusoidal wave is a special periodic wave in which each point moves in simple harmonic motion. For any periodic wave, the frequency f is the number of cycles per unit time, the period T is the time for one cycle, the wavelength λ is the distance over which the wave pattern repeats, and the amplitude A is the maximum displacement of a particle in the medium. The product of λ and f equals the wave speed. (See Example 15.1)

$$v = \lambda f \qquad (15.1)$$

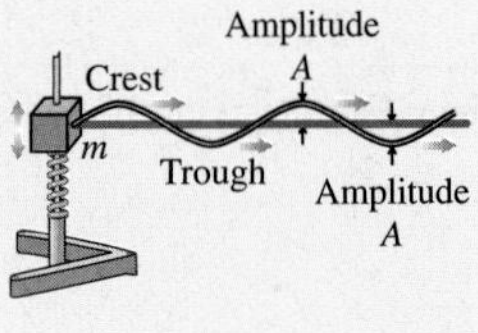

The wave function $y(x, t)$ describes the displacements of individual particles in the medium. Equations (15.3), (15.4), and (15.7) give the wave equation for a sinusoidal wave traveling in the $+x$-direction. If the wave is moving in the $-x$ -direction, the minus signs in the cosine functions are replaced by plus signs. (See Example 15.2)

$$y(x, t) = A\cos\left[\omega\left(\frac{x}{v} - t\right)\right] = A\cos 2\pi f\left(\frac{x}{v} - t\right) \qquad (15.3)$$

$$y(x, t) = A\cos 2\pi\left(\frac{x}{\lambda} - \frac{t}{T}\right) \qquad (15.4)$$

$$y(x, t) = A\cos(kx - \omega t) \qquad (15.7)$$

where $k = 2\pi/\lambda$ and $\omega = 2\pi f = vk$

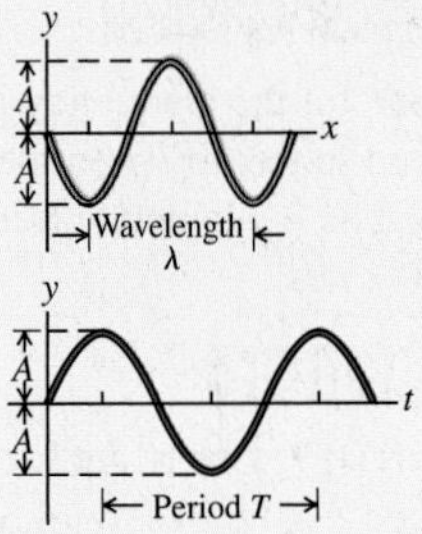

The wave function obeys a partial differential equation called the wave equation.

$$\frac{\partial^2 y(x, t)}{\partial x^2} = \frac{1}{v^2}\frac{\partial^2 y(x, t)}{\partial t^2} \qquad (15.12)$$

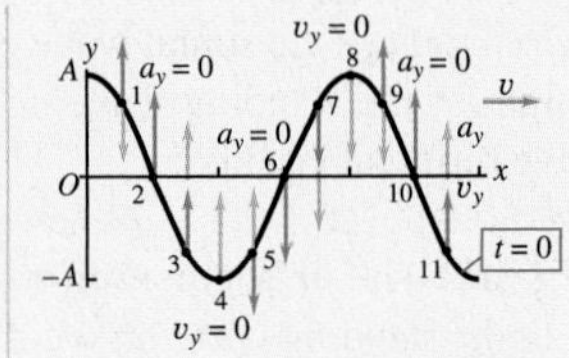

The speed of transverse waves on a string depends on the tension F and mass per unit length μ. (See Examples 15.3 and 15.4)

$$v = \sqrt{\frac{F}{\mu}} \quad \text{(waves on a string)} \qquad (15.13)$$

Wave motion conveys energy from one region to another. For a sinusoidal mechanical wave, the average power P_{av} is proportional to the square of the wave amplitude and the square of the frequency. For waves that spread out in three dimensions, the wave intensity I is inversely proportional to the distance from the source. (See Examples 15.5 and 15.6)

$$P_{av} = \frac{1}{2}\sqrt{\mu F}\,\omega^2 A^2 \qquad (15.25)$$

(average power, sinusoidal wave)

$$\frac{I_1}{I_2} = \frac{r_2^2}{r_1^2} \qquad (15.26)$$

(inverse-square law for intensity)

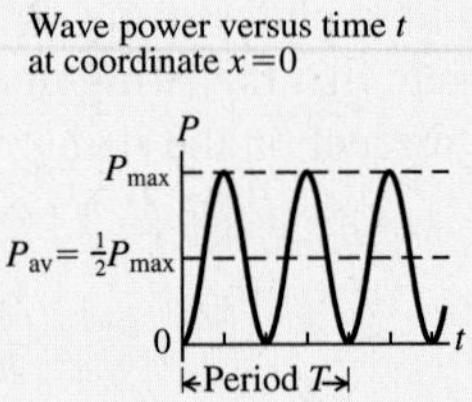

A wave that reaches a boundary of the medium in which it propagates is reflected. The principle of superposition states that the total wave displacement at any point where two or more waves overlap is the sum of the displacements of the individual waves.

$$y(x, t) = y_1(x, t) + y_2(x, t)$$
(principle of superposition) (15.27)

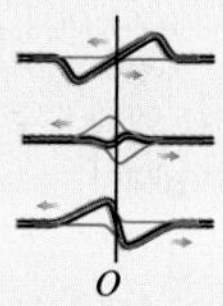

When a sinusoidal wave is reflected from a fixed or free end of a stretched string, the incident and reflected waves combine to form a standing sinusoidal wave with nodes and antinodes. Adjacent nodes are spaced a distance $\lambda/2$ apart, as are adjacent antinodes. (See Example 15.7)

$$y(x, t) = (A_{SW} \sin kx) \sin \omega t$$
(standing wave on a string, fixed end at $x = 0$) (15.28)

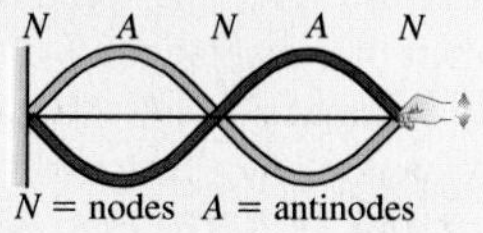

When both ends of a string with length L are held fixed, standing waves can occur only when L is an integer multiple of $\lambda/2$. Each frequency and its associated vibration pattern is called a normal mode. The lowest frequency f_1 is called the fundamental frequency. (See Examples 15.8 and 15.9)

$$f_n = n\frac{v}{2L} = nf_1 \quad (n = 1, 2, 3, \dots) \qquad (15.33)$$

$$f_1 = \frac{1}{2L}\sqrt{\frac{F}{\mu}} \qquad (15.35)$$

(string fixed at both ends)

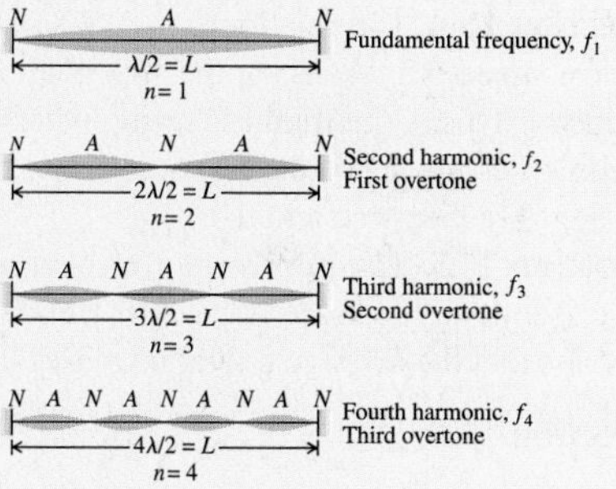

Key Terms

mechanical wave, 548
medium, 548
transverse wave, 548
longitudinal wave, 548
wave speed, 549
periodic wave, 549
sinusoidal wave, 549
wavelength, 550
wave function, 552
wave number, 554
phase, 555
wave equation, 558
intensity, 566
interference, 568
boundry condition, 569
principle of superposition, 569
node, 570
antinode, 570
standing wave, 570
traveling wave, 570
destructive interference, 571
constructive interference, 571
fundamental frequency, 576
harmonics, 576
harmonic series, 576
overtone, 576
normal mode, 576
harmonic content, 577

Your Notes

Answer to Chapter Opening Question

The power of a mechanical wave depends on its frequency and amplitude [see Eq. (15.25)].

Answers to Test Your Understanding Questions

Section 15.2 The speed of waves on a string doesn't change at all when the wavelength is changed. The relation $v = \lambda f$ [Eq. (15.1)] tells us that $f = v/\lambda$, so doubling the wavelength λ will reduce the frequency f by a factor of 1/2.
Section 15.3 The plus sign in the cosine function tells us that the wave is propagating in the negative x-direction. The wave has wave number $k = 2.00\ \text{m}^{-1}$, angular frequency $\omega = 3.20$ rad/s, and amplitude $A = 2.50$ mm.
Section 15.4 The relation $v = \sqrt{F/\mu}$ [Eq. (15.13)] says that the wave speed is greatest on the string with the smallest linear mass density. This is the thinnest string, which has the smallest amount of mass m and hence the smallest linear mass density $\mu = m/L$ (all strings are the same length).
Section 15.5 The average power is proportional to the square of the frequency [see Eq. (15.25)]. Hence doubling the frequency means that the average power increases by a factor of $2^2 = 4$.

Section 15.6

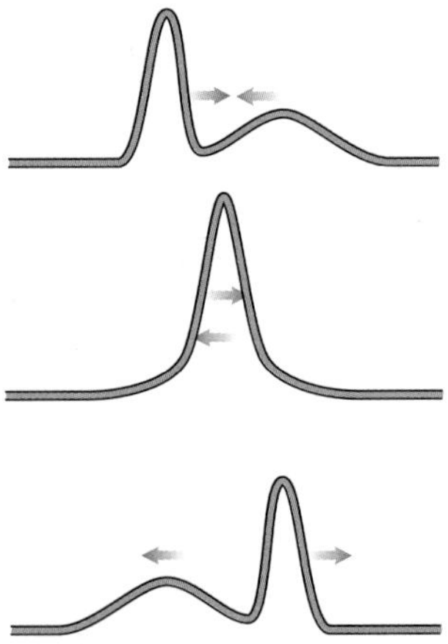

Section 15.7 Doubling the frequency makes the wavelength half as large. Hence the spacing between nodes (equal to $\lambda/2$) is also half as large. There are nodes at all of the previous positions, but there is also a new node between every pair of old nodes.
Section 15.8 When you touch the string at its center, you are demanding that there be a node at the center. Hence only standing waves with a node at $x = L/2$ are allowed. From Figure 15.23 you can see that the normal modes $n = 1, 3, 5, \ldots$ cannot be present.

Discussion Questions

Q15.1 A common pastime at sporting events is "doing the wave," in which the spectators in one section raise and lower their arms, signaling the spectators in the next section to do the same, and so on. In the language of this chapter, is this truly a wave? Why or why not?
Q15.2 Explain why it is necessary to include the factor 2π in Equation (15.4). What unit does 2π have? Could you correctly replace this factor with 360°?
Q15.3 What kinds of energy are associated with waves on a stretched string? How could you detect such energy experimentally?
Q15.4 The amplitude of a wave decreases gradually as the wave travels down a long, stretched string. What happens to the energy of the wave when this happens?
Q15.5 For the wave motions discussed in this chapter, does the speed of propagation depend on the amplitude? How can you tell?
Q15.6 The speed of ocean waves depends on the depth of the water; the deeper the water, the faster the wave travels. Use this to explain why ocean waves crest and "break" as the wave nears the shore.
Q15.7 Is it possible to have a longitudinal wave on a stretched string? Why or why not? Is it possible to have a transverse wave on a steel rod? Again, why or why not? If your answer is yes in either case, explain how you would create such a wave.
Q15.8 An echo is sound reflected from a distant object, such as a wall or a cliff. Explain how you can determine how far away the object is by timing the echo.
Q15.9 Why do you see lightning before you hear the thunder? A familiar rule of thumb is to start counting slowly, once per second, when you see the lightning; when you hear the thunder, divide the number you have reached by 3 to obtain your distance from the lightning in kilometers (or by 5 to obtain your distance in miles). Why does this work, or does it?
Q15.10 For transverse waves on a string, is the wave speed the same as the maximum speed of any part of the string, or is there a different relation between these two speeds? Explain your reasoning.
Q15.11 Children make toy telephones by sticking each end of a long string through a hole in the bottom of a paper cup and knotting it so it will not pull out. When the spring is pulled taut, sound can be transmitted from one cup to the other. How does this work? Why is the transmitted sound louder than the sound traveling through air for the same distance?
Q15.12 The four strings on a violin have different thicknesses, but are all under approximately the same tension. Do waves travel faster on the thick strings or the thin strings? Why? How does the fundamental vibration frequency compare for the thick versus the thin strings?
Q15.13 When a rock is thrown into a pond and the resulting ripples spread in ever-widening circles, the amplitude decreases with increasing distance from the center. Why?
Q15.14 Two strings of different mass per unit length μ_1 and μ_2 are tied together and stretched with a tension F. A wave travels along the string and passes the discontinuity in μ. Which of the following wave properties will be the same on both sides of the discontinuity and which ones will change: Speed of the wave? Frequency? Wavelength? Explain the physical reasoning behind each of your answers.
Q15.15 A long rope with mass m is suspended from the ceiling and hangs vertically. A wave pulse is produced at the lower end of the

rope and the pulse travels up the rope. Does the speed of the wave pulse change as it moves up the rope, and if so, does it increase or decrease?

Q15.16 In a transverse wave on a string, the motion of the string is perpendicular to the length of the string. How, then, is it possible for energy to move along the length of the string?

Q15.17 Justify the following statement: "A wave transfers energy fastest where the particles of the medium are moving fastest." In particular, explain how can this can be true for a transverse wave, in which the particles of the medium move in a direction perpendicular to that of wave propagation.

Q15.18 Energy can be transferred along a string by wave motion. However, in a standing wave on a string, no energy can ever be transferred past a node. Why not?

Q15.19 Can a standing wave be produced on a string by superposing two waves traveling in opposite directions with the same frequency but different amplitudes? Why or why not? Can a standing wave be produced by superposing two waves traveling in opposite directions with different frequencies but the same amplitude? Why or why not?

Q15.20 If you stretch a rubber band and pluck it, you hear a (somewhat) musical tone. How does the frequency of this tone change as you stretch the rubber band further? (Try it!) Does this agree with Eq. (15.35) for a string fixed at both ends? Explain.

Q15.21 A musical interval of an *octave* corresponds to a factor of two in frequency. By what factor must the tension in a guitar or violin string be increased to raise its pitch one octave? To raise it two octaves? Explain your reasoning. Is there any danger in attempting these changes in pitch?

Q15.22 By touching a string lightly at its center while bowing, a violinist can produce a note exactly one octave above the note to which the string is tuned, that is, a note with exactly twice the frequency. Why is this possible?

Q15.23 As we discussed in Section 15.1, water waves are a combination of longitudinal and transverse waves. Defend the following statement: "When water waves hit a vertical wall, the wall is a node of the longitudinal displacement but an antinode of the transverse displacement."

Q15.24 Violins are short instruments, while cellos and basses are long. In terms of the frequency of the waves they produce, explain why this is so.

Q15.25 What is the purpose of the frets on a guitar? In terms of the frequency of the vibration of the strings, explain their use.

Exercises

Section 15.2 Periodic Waves

15.1 A fisherman notices that his boat is moving up and down periodically, owing to waves on the surface of the water. It takes 2.5 s for the boat to travel from its highest point to its lowest, a total distance of 0.62 m. The fisherman sees that the wave crests are spaced 6.0 m apart. a) How fast are the waves traveling? b) What is the amplitude of each wave? c) If the total vertical distance traveled by the boat were 0.30 m, but the other data remained the same, how would the answers to parts (a) and (b) be affected? d) Would you expect that the motion of the boat would be purely vertical? Why or why not?

15.2 Ultrasound Imaging. Ultrasound is the name given to frequencies above the human range of hearing, which is about 20,000 Hz. Waves above this frequency can be used to penetrate the body and to produce images by reflecting from surfaces. In a typical ultrasound scan, the waves travel with a speed of 1500 m/s. For a good detailed image, the wavelength should be no more than 1.0 mm. What frequency is required?

15.3 The speed of sound in air at 20°C is 344 m/s. a) What is the wavelength of a sound wave with a frequency of 784 Hz, corresponding to the note G_5 on a piano? b) What is the frequency of a sound wave with a wavelength of 0.0655 mm? (This frequency is too high for you to hear.)

15.4 The speed of radio waves in vacuum (equal to the speed of light) is 3.00×10^8 m/s. Find the wavelength for a) an AM radio station with frequency 540 kHz; b) an FM radio station with frequency 104.5 MHz.

15.5 Audible Wavelengths. Provided that the amplitude is sufficiently great, the human ear can respond to longitudinal waves over a range of frequencies from about 20.0 Hz to about 20,000 Hz. Compute the wavelengths corresponding to these frequencies a) for waves in air ($v = 344$ m/s); b) for waves in water ($v = 1480$ m/s).

Section 15.3 Mathematical Description of a Wave

15.6 A certain transverse wave is described by

$$y(x,t) = (6.50\text{ mm})\cos 2\pi\left(\frac{x}{28.0\text{ cm}} - \frac{t}{0.0360\text{ s}}\right)$$

Determine the wave's a) amplitude; b) wavelength; c) frequency; d) speed of propagation; e) direction of propagation.

15.7 Transverse waves on a string have wave speed 8.00 m/s, amplitude 0.0700 m, and wavelength 0.320 m. The waves travel in the $-x$-direction, and at $t = 0$ the $x = 0$ end of the string has its maximum upward displacement. a) Find the frequency, period, and wave number of these waves. b) Write a wave function describing the wave. c) Find the transverse displacement of a particle at $x = 0.360$ m at time $t = 0.150$ s. d) How much time must elapse from the instant in part (c) until the particle at $x = 0.360$ m next has maximum upward displacement?

15.8 Redraw Figs. 15.7a and 15.7b for the case in which the wave is propagating to the *left*.

15.9 Show that the following wave functions satisfy the wave equation, Eq. (15.12): a) $y(x,t) = A\cos(kx + \omega t)$; b) $y(x,t) = A\sin(kx + \omega t)$. c) In what directions are these waves traveling? How can you tell? d) For the wave of part (b), write the equations for the transverse velocity and transverse acceleration of a particle at point x.

15.10 a) For a wave described by $y(x,t) = A\cos(kx - \omega t)$, graph y, v_y, and a_y as functions of x for time $t = 0$. b) Consider the following points on the string: i) $x = 0$; ii) $x = \pi/4k$; iii) $x = \pi/2k$; iv) $x = 3\pi/4k$; v) $x = \pi/k$; vi) $x = 5\pi/4k$; vii) $x = 3\pi/2k$; and viii) $x = 7\pi/4k$. For a particle at each of these points at $t = 0$, describe in words whether the particle is moving and in what direction, and whether the particle is speeding up, slowing down, or instantaneously not accelerating.

15.11 A sinusoidal wave is propagating along a stretched string that lies along the x-axis. The displacement of the string as a function of time is graphed in Fig. 15.27 for particles at $x = 0$ and at $x = 0.0900$ m. a) What is the amplitude of the wave? b) What is the period of the wave? c) You are told that the two points $x = 0$ and $x = 0.0900$ m are within one wavelength of each other. If the wave is moving in the $+x$-direction, determine the wavelength and the wave speed. d) If instead the wave is moving in the $-x$-direction, determine the wavelength and the wave speed. e) Would it be possible to determine definitively the wavelength in parts (c) and (d) if you were not told that the two points were within one wavelength of each other? Why or why not?

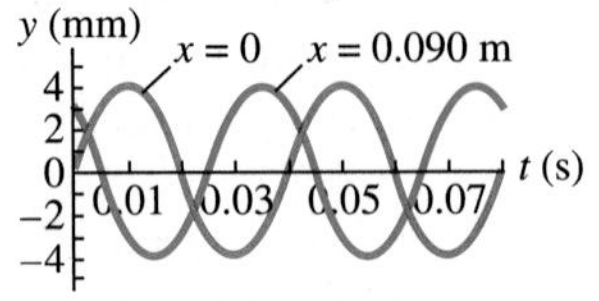

Figure 15.27 Exercise 15.11.

15.12 Speed of Propagation vs. Particle Speed. a) Show that Eq. (15.3) may be written as

$$y(x, t) = A \cos\left[\frac{2\pi}{\lambda}(x - vt)\right]$$

b) Use $y(x, t)$ to find an expression for the transverse velocity v_y of a particle in the string on which the wave travels. c) Find the maximum speed of a particle of the string. Under what circumstances is this equal to the propagation speed v? Less than v? Greater than v?

15.13 A transverse wave on a string has amplitude 0.300 cm, wavelength 12.0 cm, and speed 6.00 cm/s. It is represented by $y(x, t)$ as given in Exercise 15.12. a) At time $t = 0$, compute y at 1.5-cm intervals of x (that is, at $x = 0$, $x = 1.5$ cm, $x = 3.0$ cm, and so on) from $x = 0$ to $x = 12.0$ cm. Graph the results. This is the shape of the string at time $t = 0$. b) Repeat the calculations for the same values of x at times $t = 0.400$ s and $t = 0.800$ s. Graph the shape of the string at these instants. In what direction is the wave traveling?

Section 15.4 Speed of a Transverse Wave

15.14 With what tension must a rope with length 2.50 m and mass 0.120 kg be stretched for transverse waves of frequency 40.0 Hz to have a wavelength of 0.750 m?

15.15 One end of a horizontal rope is attached to a prong of an electrically driven tuning fork that vibrates at 120 Hz. The other end passes over a pulley and supports a 1.50-kg mass. The linear mass density of the rope is 0.0550 kg/m. a) What is the speed of a transverse wave on the rope? b) What is the wavelength? c) How would your answers to parts (a) and (b) change if the mass were increased to 3.00 kg?

15.16 A cowgirl ties one end of a 10.0-m-long rope to a fence post and pulls on the other end so the rope is stretched horizontally with a tension of 140 N. The mass of the rope is 0.800 kg. a) What is the speed of transverse waves on the rope? b) If the cowgirl moves the free end up and down with a frequency of 1.20 Hz, what is the wavelength of the transverse waves on the rope? c) The cowgirl pulls harder on the rope so that the tension is doubled to 280 N. With what frequency must she move the free end of the rope up and down to produce transverse waves of the same wavelength as in part (a)?

15.17 One end of a 14.0-m-long rubber tube, with total mass 0.800 kg, is fastened to a fixed support. A cord attached to the other end passes over a pulley and supports an object with a mass of 7.50 kg. The tube is struck a transverse blow at one end. Find the time required for the pulse to reach the other end.

15.18 Weighty Rope. If in Example 15.4 (Section 15.4) we do *not* neglect the weight of the rope, what is the wave speed a) at the bottom of the rope? b) at the middle of the rope? c) at the top of the rope?

15.19 A simple, harmonic oscillator at the point $x = 0$ generates a wave on a rope. The oscillator operates at a frequency of 40.0 Hz and with an amplitude of 3.00 cm. The rope has a linear mass density of 50.0 g/m and is stretched with a tension of 5.00 N. a) Determine the speed of the wave. b) Find the wavelength. c) Write the wave function $y(x, t)$ for the wave. Assume that the oscillator has its maximum upward displacement at time $t = 0$. d) Find the maximum transverse acceleration of points on the rope. e) In the discussion of transverse waves in this chapter, the force of gravity was ignored. Is that a reasonable approximation for this wave? Explain.

Section 15.5 Energy in Wave Motion

15.20 A piano wire with mass 3.00 g and length 80.0 cm is stretched with a tension of 25.0 N. A wave with frequency 120.0 Hz and amplitude 1.6 mm travels along the wire. a) Calculate the average power carried by the wave. b) What happens to the average power if the wave amplitude is halved?

15.21 Redraw Fig. 15.13 for a position on the string at $x = \lambda/4$. Explain any differences from Fig. 15.13.

15.22 Threshold of Pain. You are investigating the report of a UFO landing in an isolated portion of New Mexico, and encounter a strange object that is radiating sound waves uniformly in all directions. Assume that the sound comes from a point source and that you can ignore reflections. You are slowly walking toward the source. When you are 7.5 m from it, you measure its intensity to be 0.11 W/m^2. An intensity of 1.0 W/m^2 is often used as the "threshold of pain." How much closer to the source can you move before the sound intensity reaches this threshold?

15.23 Energy Output. By measurement you determine that sound waves are spreading out equally in all directions from a point source and that the intensity is 0.026 W/m^2 at a distance of 4.3 m from the source. a) What is the intensity at a distance of 3.1 m from the source? b) How much sound energy does the source emit in one hour if its power output remains constant?

15.24 A fellow student of mathematical bent tells you that the wave function of a traveling wave on a thin rope is $y(x, t) = 2.30 \text{ mm} \cos[(6.98 \text{ rad/m})x + (742 \text{ rad/s})t]$.Being more practical-minded, you measure the rope to have a length of 1.35 m and a mass of 3.38 grams. You are then asked to determine the following: a) amplitude; b) frequency; c) wavelength; d) wave speed; e) direction the wave is traveling; f) tension in the rope; g) average power transmitted by the wave.

15.25 What is the total power output of the siren in Example 15.6?

Section 15.6 Wave Interference, Boundary Conditions, and Superposition

15.26 Reflection. A wave pulse on a string has the dimensions shown in Fig. 15.28 at $t = 0$. The wave speed is 40 cm/s. a) If point O is a fixed end, draw the total wave on the string at $t = 15$ ms, 20 ms, 25 ms, 30 ms, 35 ms, 40 ms, and 45 ms. b) Repeat part (a) for the case in which point O is a free end.

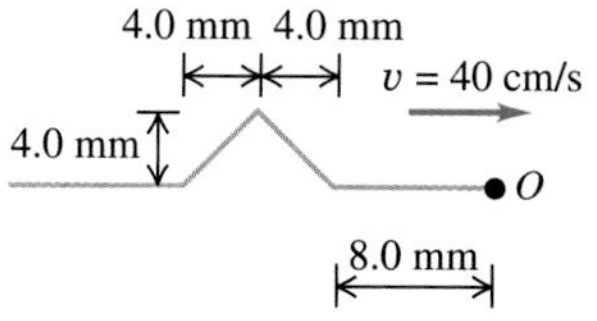

Figure 15.28 Exercise 15.26.

15.27 Reflection. A wave pulse on a string has the dimensions shown in Fig. 15.29 at $t = 0$. The wave speed is 5.0 m/s. a) If point O is a fixed end, draw the total wave on the string at $t = 1.0$ ms, 2.0 ms, 3.0 ms, 4.0 ms, 5.0 ms, 6.0 ms, and 7.0 ms. b) Repeat part (a) for the case in which point O is a free end.

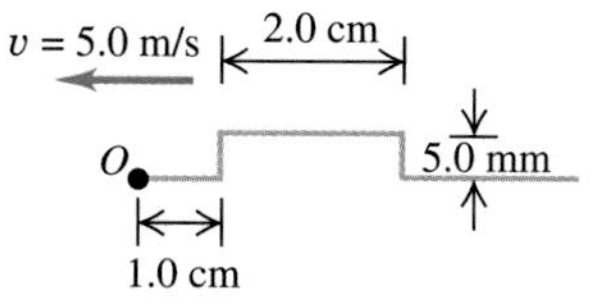

Figure 15.29 Exercise 15.27.

15.28 Interference of Triangular Pulses. Two triangular wave pulses are traveling toward each other on a stretched string as shown in Fig. 15.30. Each pulse is identical to the other and travels at 2.00 cm/s. The leading edges of the pulses are 1.00 cm apart at $t = 0$. Sketch the shape of the string at $t = 0.250$ s, $t = 0.500$ s, $t = 0.750$ s, $t = 1.000$ s, and $t = 1.250$ s.

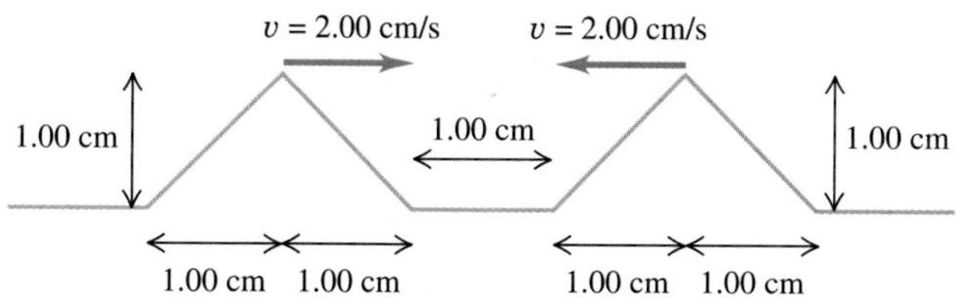

Figure 15.30 Exercise 15.28.

15.29 Interference of Rectangular Pulses. Figure 15.31 shows two rectangular wave pulses on a stretched string traveling toward each other. Each pulse is traveling with a speed of 1.00 mm/s and has the height and width shown in the figure. If the leading edges of the pulses are 8.00 mm apart at $t = 0$, sketch the shape of the string at $t = 4.00$ s, $t = 6.00$ s, and $t = 10.0$ s.

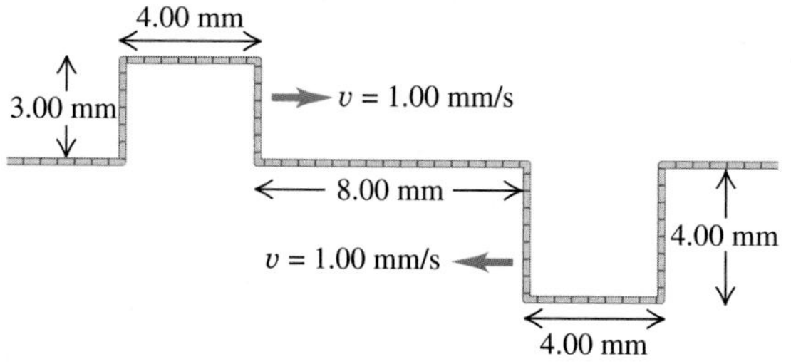

Figure 15.31 Exercise 15.29.

15.30 Two traveling waves move on a string that has a fixed end at $x = 0$. They are identical, expect for opposite velocities. Each has an amplitude of 2.46 mm, a period of 3.65 ms, and a speed of 111 m/s. Write the wave function of the resulting standing wave.

Section 15.7 Standing Waves on a String

15.31 Standing waves on a wire are described by Eq. (15.28), with $A_{SW} = 2.50$ mm, $\omega = 942$ rad/s, and $k = 0.750\pi$ rad/m. The left-hand end of the wire is at $x = 0$. At what distances from the left-hand end are a) the nodes of the standing wave? b) the antinodes of the standing wave?

15.32 Wave Equation and Standing Waves. a) Prove by direct substitution that $y(x, t) = (A_{SW} \sin kx) \sin \omega t$ is a solution of the wave equation Eq. (15.12), for $v = \omega/k$. b) Explain why the relation $v = \omega/k$ for *traveling* waves also applies to *standing* waves.

15.33 Adjacent antinodes of a standing wave on a string are 15.0 cm apart. A particle at an antinode oscillates in simple harmonic motion with amplitude 0.850 cm and period 0.0750 s. The string lies along the $+x$-axis and is fixed at $x = 0$. a) Find the displacement of a point on the string as a function of position and time. b) Find the speed of propagation of a transverse wave in the string. c) Find the amplitude at a point 3.0 cm to the right of an antinode.

15.34 Give the details of the derivation of Eq. (15.28) from $y_1(x, t) + y_2(x, t) = A[-\cos(kx + \omega t) + \cos(kx - \omega t)]$.

15.35 Let $y_1(x, t) = A\cos(k_1 x - \omega_1 t)$ and $y_2(x, t) = A\cos(k_2 x - \omega_2 t)$ be two solutions to the wave equation, Eq. (15.12), for the same v. Show that $y(x, t) = y_1(x, t) + y_2(x, t)$ is also a solution to the wave equation.

Section 15.8 Normal Modes of a String

15.36 A piano tuner stretches a steel piano wire with a tension of 800 N. The steel wire is 0.400 m long and has a mass of 3.00 g. a) What is the frequency of its fundamental mode of vibration? b) What is the number of the highest harmonic that could be heard by a person who is capable of hearing frequencies up to 10,000 Hz?

15.37 A wire with mass 40.0 g is stretched so that its ends are tied down at points 80.0 cm apart. The wire vibrates in its fundamental mode with frequency 60.0 Hz and with an amplitude at the antinodes of 0.300 cm. a) What is the speed of propagation of transverse waves in the wire? b) Compute the tension in the wire. c) Find the maximum transverse velocity and acceleration of particles in the wire.

15.38 Waves on a Stick. A flexible stick 2.0 m long is not fixed in any way and is free to vibrate. Make clear drawings of this stick vibrating in its first three harmonics and then use your drawings to find the wavelengths of each of these harmonics.

15.39 A thin, taut string tied at both ends and oscillating in its third harmonic has its shape described by the equation $y(x, t) = (5.60\text{ cm})\sin[(0.0340\text{ rad/cm})x]\sin[(50.0\text{ rad/s})t]$, where the origin is at the left end of the string, the x-axis is along the string and the y-axis is perpendicular to the string. a) Draw a sketch that shows the standing wave pattern. b) Find the amplitude of the two traveling waves that make up this standing wave. c) What is the length of the string? d) Find the wavelength, frequency, period, and speed of the traveling waves. e) Find the maximum transverse speed of a point on the string. f) What would be the equation $y(x, t)$ for this string if it were vibrating in its eighth harmonic?

15.40 The wave function of a standing wave is $y(x, t) =$ 4.44 mm $\sin[(32.5 \text{ rad/m})x] \sin[(754 \text{ rad/s})t]$. For the two traveling waves that make up this standing wave, find the a) amplitude; b) wavelength; c) frequency; d) wave speed; e) wave functions. f) From the information given, can you determine which harmonic this is? Explain.

15.41 Consider again the rope and traveling wave of Exercise 15.24. Assume that the ends of the rope are held fixed and that there is both this traveling wave and the reflected wave traveling in the opposite direction. a) What is the wavefunction $y(x, t)$ for the standing wave that is produced? b) In which harmonic is the standing wave oscillating? c) What is the frequency of the fundamental oscillation?

15.42 In deriving Eq. (15.29) for a string with a fixed end at $x = 0$, we found that nodes occur at positions x that satisfy $kx = n\pi$, where $n = 0, 1, 2, 3, \ldots$ Apply this condition to a string with fixed ends at $x = 0$ and $x = L$ to rederive Eq. (15.31) for the possible standing-wave wavelengths.

15.43 Playing a Cello. The portion of a cello string between the bridge and upper end of the finger board (that part of the string that is free to vibrate) is 60.0 cm long, and this length of the string has mass 2.00 g. The string sounds an A_4 note (440 Hz) when played. a) Where must the cellist put a finger (what distance x from bridge) to play a D_5 note (587 Hz)? (See Fig. 15.32.) For both the A_4 and D_5 notes, the string vibrates in its fundamental mode. b) Without retuning, is it possible to play a G_4 note (392 Hz) on this string? Why or why not?

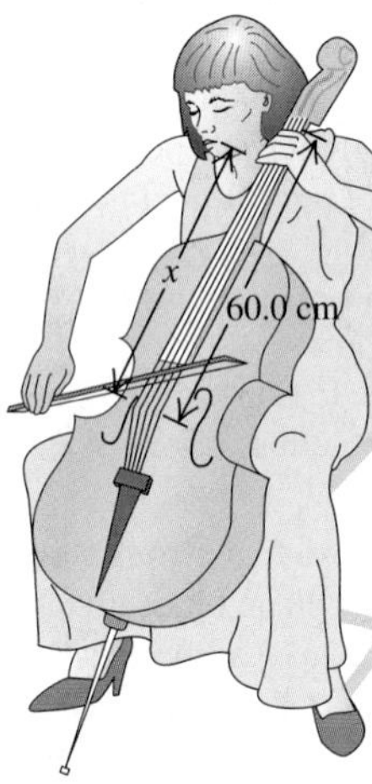

Figure 15.32 Exercise 15.43.

15.44 a) A horizontal string tied at both ends is vibrating in its fundamental mode. The traveling waves have speed v, frequency f, amplitude A, and wavelength λ. Calculate the maximum transverse velocity and maximum transverse acceleration of points located at i) $x = \lambda/2$, ii) $x = \lambda/4$, and iii) $x = \lambda/8$ from the left-hand end of the string. b) At each of the points in part (a), what is the amplitude of the motion? c) At each of the points in part (a), how much time does it take the string to go from its largest upward displacement to its largest downward displacement?

15.45 A 1.50-m-long rope is stretched between two supports with a tension that makes the speed of transverse waves 48.0 m/s. What are the wavelength and frequency of a) the fundamental? b) the second overtone? c) the fourth harmonic?

15.46 Guitar String. One of the 63.5-cm-long strings of an ordinary guitar is tuned to produce the note B_3 (frequency 245 Hz) when vibrating in its fundamental mode. a) Find the speed of transverse waves on this string. b) If the tension in this string is increased by 1.0%, what will be the new fundamental frequency of the string? c) If the speed of sound in the surrounding air is 344 m/s, find the frequency and wavelength of the sound wave produced in the air by the vibration of the B_3 string. How do these compare to the frequency and wavelength of the standing wave on the string?

Problems

15.47 A transverse sine wave with an amplitude of 2.50 mm and a wavelength of 1.80 m travels from left to right along a long, horizontal stretched string with a speed of 36.0 m/s. Take the origin at the left end of the undisturbed string. At time $t = 0$ the left end of the string has its maximum upward displacement. a) What are the frequency, angular frequency, and wave number of the wave? b) What is the function $y(x, t)$ that describes the wave? c) What is $y(t)$ for a particle at the left end of the string? d) What is $y(t)$ for a particle 1.35 m to the right of the origin? e) What is the maximum magnitude of transverse velocity of any particle of the string? f) Find the transverse displacement and the transverse velocity of a particle 1.35 m to the right of the origin at time $t = 0.0625$ s.

15.48 A transverse wave on a rope is given by

$$y(x, t) = (0.750 \text{ cm}) \cos \pi[(0.400 \text{ cm}^{-1})x + (250 \text{ s}^{-1})t]$$

a) Find the amplitude, period, frequency, wavelength, and speed of propagation. b) Sketch the shape of the rope at the following values of t: 0, 0.0005 s, and 0.0010 s. c) Is the wave traveling in the $+x$- or $-x$-direction? d) The mass per unit length of the rope is 0.0500 kg/m. Find the tension. e) Find the average power of this wave.

15.49 Three pieces of string, each of length L, are joined together end-to-end, to make a combined string of length $3L$. The first piece of string has mass per unit length μ_1, the second piece has mass per unit length $\mu_2 = 4\mu_1$, and the third piece has mass per unit length $\mu_3 = \mu_1/4$. a) If the combined string is under tension F, how much time does it take a transverse wave to travel the entire length $3L$? Give your answer in terms of L, F, and μ_1. b) Does your answer to part (a) depend on the order in which the three pieces are joined together? Explain.

15.50 Low-Cost Remote Control. Your roommates have lost the TV remote control, and no amount of searching can find it. Rather than buy a new one, you build a low-cost replacement. You attach one end of a small lever mechanism to the TV's channel-changing button. You plan to attach the other end of the lever to a 3.0-m long string that will run from the TV to the couch. When you pull the string tight and pluck your end of the string, a wave will travel down the string and trigger the lever, changing the channel. Your design assumes that you will disturb the string vertically by only 5.0 mm when you pluck it (so that using your "remote" will not require undue exertion) and that your wave will take only 0.20 s to travel horizontally along the string from your hand to the lever (to give quick channel changes). Unfortunately, your roommates have also taken most of your supply of string. All you can find around the apartment are two different pieces of 1.5-m long string; one piece has a mass of 90 g, while the other piece has a mass of just 10 g. You tie the two pieces together to make a combined 3.0-m long string and attach one end of the combined string to the lever mechanism. You then take the other end in hand and head for the couch. When you stretch the combined string to make it taut, how hard do you have to pull?

15.51 Ant Joy Ride. You place your pet ant Klyde (mass m) on top of a horizontal, stretched rope, where he holds on tightly. The rope has mass M and length L and is under tension F. You start a sinusoidal transverse wave of wavelength λ and amplitude A propagating along the rope. The motion of the rope is in a vertical plane. Klyde's mass is so small that his presence has no effect on the propagation of the wave. a) What is Klyde's top speed as he oscillates up and down? b) Klyde enjoys the ride and begs for more. You decide to double his top speed by changing the tension while keeping the wavelength and amplitude the same. Should the tension be increased or decreased, and by what factor?

15.52 Weightless Ant. An ant with mass m is standing peacefully on top of a horizontal, stretched rope. The rope has mass per unit length μ and is under tension F. Without warning, Throckmorton starts a sinusoidal transverse wave of wavelength λ propagating along the rope. The motion of the rope is in a vertical plane. What minimum wave amplitude will make the ant become momentarily weightless? Assume that m is so small that the presence of the ant has no effect on the propagation of the wave.

15.53 When a transverse sinusoidal wave is present on a string, the particles of the string undergo SHM. This is the same motion as that of a mass m attached to an ideal spring of force constant k', for which the angular frequency of oscillation was found in Chapter 13 to be $\omega = \sqrt{k'/m}$. Consider a string with tension F and mass per unit length μ, along which is propagating a sinusoidal wave with amplitude A and wavelength λ. a) Find the "force constant" k' of the restoring force that acts on a short segment of the string of length Δx (where $\Delta x \ll \lambda$). b) How does the "force constant" calculated in part (b) depend on F, μ, A, and λ? Explain the physical reasons why this should be so.

15.54 Motion of a Wave Pulse. The sinusoidal waves described in Section 15.2 can be produced on a string by continually oscillating one end of the string. If instead the end of the string is given a single shake, a *wave pulse* propagates down the string. A particular wave pulse is described by the function

$$y(x, t) = \frac{A^3}{A^2 + (x - vt)^2}$$

where $A = 1.00$ cm and $v = 20.0$ m/s. a) Sketch the pulse as a function of x at $t = 0$. How far along the string does the pulse extend? b) Sketch the pulse as a function of x at $t = 0.001$ s. c) At the point $x = 4.50$ cm, at what time t is the displacement maximum? At which two times is the displacement at $x = 4.50$ cm equal to half its maximum value? d) Show that $y(x, t)$ satisfies the wave equation Eq. (15.12).

15.55 A Non-Sinusoidal Wave. The shape of a wave on a string at a specific instant is shown in Fig. 15.33. The wave is propagating to the right, in the $+x$-direction. a) Determine the direction of the transverse *velocity* of each of the six labeled points on the string. If the velocity is zero, state it as such. Explain your reasoning. b) Determine the direction of the transverse *acceleration* of each of the six labeled points on the string. Explain your reasoning. c) How would your answers be affected if the wave was propagating to the left, in the $-x$-direction?

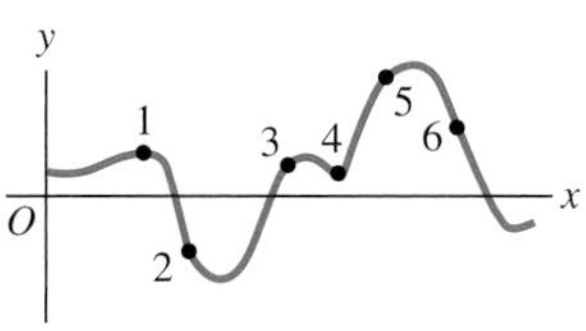

Fig. 15.33 Exercise 15.55.

15.56 A continuous succession of sinusoidal wave pulses are produced at one end of a very long string and travel along the length of the string. The wave has frequency 40.0 Hz, amplitude 5.00 mm, and wavelength 0.600 m. a) How long does it take the wave to travel a distance of 8.00 m along the length of the string? b) How long does it take a point on the string to travel a distance of 8.00 m, once the wave train has reached the point and set it into motion? c) In parts (a) and (b), how does the time change if the amplitude is doubled?

15.57 Two-Dimensional Waves. A stretched string lies along the x-axis. The string is displaced along both the y- and z-direction, so that the transverse displacement of the string is given by

$$y(x, t) = A\cos(kx - \omega t) \qquad z(x, t) = A\sin(kx - \omega t)$$

a) Draw a graph of z versus y for a particle on the string at $x = 0$. This shows the trajectory of the particle as seen by an observer on the $+x$-axis looking back toward $x = 0$. Indicate the position of the particle at $t = 0$, $t = \pi/2\omega$, $t = \pi/\omega$, and $t = 3\pi/2\omega$. b) Find the velocity vector of a particle at an *arbitrary* position x on the string. Show that this represents the tangential velocity of a particle moving in a circle of radius A with angular velocity ω, and show that the speed of the particle is constant (i.e. the particle is in uniform circular motion). (See Problem 3.75.) c) Find the acceleration vector of the particle in part (b). Show that the acceleration is always directed toward the center of the circle and that its magnitude is $a = \omega^2 A$. Explain these results in terms of uniform circular motion. Suppose that the displacement of the string was instead given by

$$y(x, t) = A\cos(kx - \omega t) \qquad z(x, t) = -A\sin(kx - \omega t)$$

Describe how the motion of a particle at x would be different from the motion described in part (a).

15.58 Locating Lightning by Radio. While doing her physics homework, a student in the Boston area is in her dorm room listening to the radio broadcast of a Red Sox–Yankees baseball game at Boston's Fenway Park. In the bottom of the fourth inning a thunderstorm moving from west to east makes its presence known in three ways: (1) The student sees a lightning flash (and hears the electromagnetic pulse on her radio receiver); (2) 3.00 s later, she hears the thunder over the radio. (3) 4.43 s after the lightning flash, the thunder rattles her window. By a previous careful measurement she knows that she is 1.12 km due north of the broadcast booth at the ballpark. The speed of sound is 344 m/s. Where did the lightning flash occur in relation to the ballpark?

15.59 Waves of Arbitrary Shape. a) Explain why *any* wave described by a function of the form $y(x, t) = f(x - vt)$ moves in the $+x$-direction with speed v. b) Show that $y(x, t) = f(x - vt)$ satisfies the wave equation, no matter what the functional form of f. To do this, write $y(x, t) = f(u)$, where $u = x - vt$. Then, to take partial derivatives of $y(x, t)$, use the chain rule:

$$\frac{\partial y(x, t)}{\partial t} = \frac{df(u)}{du}\frac{\partial u}{\partial t} = \frac{df(u)}{du}(-v)$$

$$\frac{\partial y(x,t)}{\partial x} = \frac{df(u)}{du}\frac{\partial u}{\partial x} = \frac{df(u)}{du}$$

c) A wave pulse is described by the function $y(x, t) = De^{-(Bx-Ct)^2}$, where B, C, and D are all positive constants. What is the speed of this wave?

15.60 Equation (15.7) for a sinusoidal wave can be made more general by including a phase angle ϕ, where $0 \le \phi \le 2\pi$ (in radians) Then the wave function $y(x, t)$ becomes

$$y(x,t) = A\cos(kx - \omega t + \phi)$$

a) Sketch the wave as a function of x at $t = 0$ for $\phi = 0$, $\phi = \pi/4$, $\phi = \pi/2$, $\phi = 3\pi/4$, and $\phi = 3\pi/2$. b) Calculate the transverse velocity $v_y = \partial y/\partial t$. c) At $t = 0$, a particle on the string at $x = 0$ has displacement $y = A/\sqrt{2}$. Is this enough information to determine the value of ϕ? In addition, if you are told that a particle at $x = 0$ is moving toward $y = 0$ at $t = 0$, what is the value of ϕ? d) Explain in general what you must know about the wave's behavior at a given instant to determine the value of ϕ.

15.61 a) Show that Eq. (15.25) can also be written as $P_{av} = \frac{1}{2}Fk\omega A^2$, where k is the wave number of the wave. b) If the tension F in the string is quadrupled while the amplitude A is kept the same, how must k and ω each change to keep the average power constant? (*Hint:* Recall Eq. (15.6).)

15.62 Energy in a Triangular Pulse. A triangular wave pulse on a taut string travels in the positive x-direction with speed v. The tension in the string is F and the linear mass density of the string is μ. At $t = 0$, the shape of the pulse is given by

$$y(x,0) = \begin{cases} 0 & \text{if } x < -L \\ h(L + x)/L & \text{for } -L < x < 0 \\ h(L - x)/L & \text{for } 0 < x < L \\ 0 & \text{for } x > L \end{cases}$$

a) Draw the pulse at $t = 0$. b) Determine the wave function $y(x, t)$ at all times t. c) Find the instantaneous power in the wave. Show that the power is zero except for $-L < (x - vt) < L$ and that in this interval the power is constant. Find the value of this constant power.

15.63 A sinusoidal transverse wave travels on a string. The string has length 8.00 m and mass 6.00 grams. The wave speed is 30.0 m/s and the wavelength is 0.200 m. a) If the wave is to have an average power of 50.0 W, what must be the amplitude of the wave? b) For this same string, if the amplitude and wavelength are the same as in part (a), what is the average power for the wave if the tension is increased such that the wave speed is doubled?

15.64 Instantaneous Power in a Wave. a) Graph $y(x, t)$ as given by Eq. (15.7) as a function of x for a given time t (say, $t = 0$). On the same axes, make a graph of the instantaneous power $P(x, t)$ as given by Eq. (15.23). b) Explain the connection between the slope of the graph of $y(x, t)$ versus x and the value of $P(x, t)$. In particular, explain what is happening at points where $P = 0$, where there is no instantaneous energy transfer. c) The quantity $P(x, t)$ always has the same sign. What does this imply about the direction of energy flow? d) Consider a wave moving in the $-x$-direction, for which $y(x, t) = A\cos(kx + \omega t)$. Calculate $P(x, t)$ for this wave, and make a graph of $y(x, t)$ and $P(x, t)$ as functions of x for a given time t (say, $t = 0$). What differences arise from reversing the direction of the wave?

15.65 A metal wire, with density ρ and Young's modulus Y, is stretched between rigid supports. At temperature T, the speed of a transverse wave is found to be v_1. When the temperature is increased to $T + \Delta T$, the speed decreases to $v_2 < v_1$. Determine the coefficient of linear expansion of the wire.

15.66 A vibrating string 50.0 cm long is under a tension of 1.00 N. The results from five successive stroboscopic pictures are shown in Fig. 15.34. The strobe rate is set at 5000 flashes per minute and observations reveal that the maximum displacement occurred at flashes 1 and 5 with no other maxima in between. a) Find the period, frequency, and wavelength for the traveling waves on this string. b) In what normal mode (harmonic) is the string vibrating? c) What is the speed of the traveling waves on the string? d) How fast is point P moving when the string is in: (i) position 1? (ii) position 3? e) What is the mass of this string?

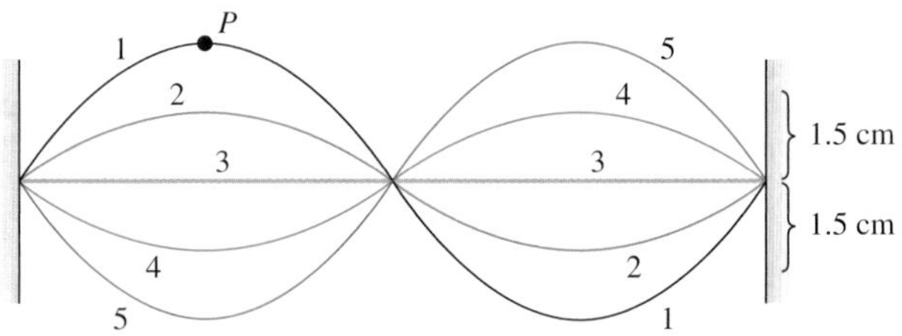

Figure 15.34 Problem 15.66.

15.67 Clothesline Nodes. Cousin Throckmorton is once again playing with the clothesline in Example 15.2. One end of the clothesline is attached to a vertical post. Throcky holds the other end loosely in his hand, so that the speed of waves on the clothesline is a relatively slow 0.720 m/s. He finds several frequencies at which he can oscillate his end of the clothesline so that a light clothespin 45.0 cm from the post doesn't move. What are these frequencies?

15.68 A guitar string is vibrating in its fundamental mode, with nodes at each end. The length of the segment of the string that is free to vibrate is 0.386 m. The maximum transverse acceleration of a point at the middle of the segment is 8.40×10^3 m/s^2 and the maximum transverse velocity is 3.80 m/s. a) What is the amplitude of this standing wave? b) What is the wave speed for the transverse traveling waves on this string?

15.69 As shown in Exercise 15.32, a standing wave given by Eq. (15.28) satisfies the wave equation Eq. (15.12). a) Show that a standing wave given by Eq. (15.28) also satisfies the equation

$$\frac{\partial^2 y(x,t)}{\partial t^2} = -\omega^2 y(x,t)$$

Interpret this equation in terms of what you know about simple harmonic motion. b) Does a traveling wave given by $y(x, t) = A\cos(kx - \omega t)$ also satisfy the equation in part (a)? Interpret this result.

15.70 a) The red and blue waves in Fig. 15.17 combine so that the displacement of the string at O is always zero. To show this mathematically for a wave of arbitrary shape, consider a wave moving to the right along the string in Fig. 15.17 (shown in blue) that, at time T, is given by $y_1(x, T) = f(x)$, where f is some function of x. (The form of $f(x)$ determines the shape of the wave.) If the point O corresponds to $x = 0$, explain why, at time T, the wave moving to the left in Fig. 15.17 (shown in red) is given by the function $y_2(x, T) = -f(-x)$. b) Show that the total wave function $y(x, T) = y_1(x, T) + y_2(x, T)$ is zero at O, independent of the form of the function $f(x)$. c) The red and blue waves in Fig. 15.18 combine so that the slope of the string at O is always zero. To show this mathematically for a wave of arbitrary shape, again let the wave moving to the right in Fig. 15.18 (shown in blue) be given by $y_1(x, T) = f(x)$ at time T. Explain why the wave moving to the left (shown in red) at this same time T is given by $y_2(x, T) = f(-x)$. d) Show that the total wave function $y(x, T) = y_1(x, T) + y_2(x, T)$ has zero slope at O, independent of the form of the function $f(x)$, as long as $f(x)$ has a finite first derivative.

15.71 A string that lies along the $+x$-axis has a free end at $x = 0$. a) By using steps similar to those used to derive Eq. (15.28), show that an incident traveling wave $y_1(x, t) = A \cos(kx + \omega t)$ gives rise to a standing wave $y(x, t) = 2A \cos \omega t \cos kx$. b) Show that the standing wave has an antinode at its free end $(x = 0)$. c) Find the maximum displacement, maximum speed, and maximum acceleration of the free end of the string.

15.72 A string with both ends held fixed is vibrating in its third harmonic. The waves have a speed of 192 m/s and a frequency of 240 Hz. The amplitude of the standing wave at an antinode is 0.400 cm. a) Calculate the amplitude at points on the string a distance of i) 40.0 cm; ii) 20.0 cm; and iii) 10.0 cm from the left-hand end of the string. b) At each point in part (a), how much time does it take the string to go from its largest upward displacement to its largest downward displacement? c) Calculate the maximum transverse velocity and the maximum transverse acceleration of the string at each of the points in part (a).

15.73 Pit and Plank. A wooden plank is placed over a pit that is 5.00 m wide. A physics student stands in the middle of the plank and begins to jump up and down such that she jumps upward from the plank two times each second. The plank oscillates with a large amplitude, with maximum amplitude at its center. a) What is the speed of transverse waves on the plank? b) At what rate does the student have to jump to produce large-amplitude oscillations if she is standing 1.25 m from the edge of the pit? (*Note:* The transverse standing waves of the plank have nodes at the two ends that rest on the ground on either side of the pit.)

15.74 Holding Up under Stress. A string or rope will break apart if it is placed under too much tensile stress [Eq. (11.8)]. Thicker ropes can withstand more tension without breaking because the thicker the rope, the greater the cross-sectional area and the smaller the stress. One type of steel has density 7800 kg/m^3 and will break if the tensile stress exceeds 7.0×10^8 N/m^2. You want to make a guitar string from 4.0 g of this type of steel. In use, the guitar string must be able to withstand a tension of 900 N without breaking. Your job is the following: a) Determine the maximum length and minimum radius the string can have. b) Determine the highest possible fundamental frequency of standing waves on this string, if the entire length of the string is free to vibrate.

15.75 Combining Standing Waves. A guitar string of length L is plucked in such a way that the total wave produced is the sum of the fundamental and the second harmonic. That is, the standing wave is given by

$$y(x, t) = y_1(x, t) + y_2(x, t)$$

where

$$y_1(x, t) = C \sin \omega_1 t \sin k_1 x$$
$$y_2(x, t) = C \sin \omega_2 t \sin k_2 x$$

with $\omega_1 = vk_1$ and $\omega_2 = vk_2$. a) At what values of x are the nodes of y_1? b) At what values of x are the nodes of y_2? c) Graph the total wave at $t = 0$, $t = \frac{1}{8}f_1$, $t = \frac{1}{4}f_1$, $t = \frac{3}{8}f_1$, and $t = \frac{1}{2}f_1$. d) Does the sum of the two standing waves y_1 and y_2 produce a standing wave? Explain.

15.76 A massive aluminum sculpture is hung from a steel wire. The fundamental frequency for transverse standing waves on the wire is 200 Hz. The sculpture is then immersed in water so that $\frac{1}{3}$ of its volume is submerged. a) What is the new fundamental frequency? b) Why is it a good approximation to treat the wire as being fixed at both ends?

15.77 Tuning a Cello. A cellist tunes the C-string of her instrument to a fundamental frequency of 65.4 Hz. The vibrating portion of the string is 0.600 m long and has a mass of 14.4 g. a) With what tension must she stretch it? b) What percent increase in tension is needed to increase the frequency from 65.4 Hz to 73.4 Hz, corresponding to a rise in pitch from C to D?

Challenge Problems

15.78 Longitudinal Waves on a Spring. A long spring such as a Slinky™ is often used to demonstrate longitudinal waves. a) Show that if a spring that obeys Hooke's law has mass m, length L, and force constant k', the speed of longitudinal waves on the spring is $v = L\sqrt{k'/m}$. b) Evaluate v for a spring with $m = 0.250$ kg, $L = 2.00$ m, and $k' = 1.50$ N/m.

15.79 a) Show that for a wave on a string, the kinetic energy *per unit length of string* is

$$u_k(x, t) = \frac{1}{2}\mu v_y^{\,2}(x, t) = \frac{1}{2}\mu\left(\frac{\partial y(x, t)}{\partial t}\right)^2$$

where μ is the mass per unit length. b) Calculate $u_k(x, t)$ for a sinusoidal wave given by Eq. (15.7). c) There is also elastic potential energy in the string, associated with the work required to deform and stretch the string. Consider a short segment of string at position x that has unstretched length Δx, as in Fig. 15.10. Ignoring the (small) curvature of the segment, its slope is $\partial y(x, t)/\partial x$. Assume that the displacement of the string from equilibrium is small, so that $\partial y/\partial x$ has a magnitude much less than unity. Show that the stretched length of the segment is approximately

$$\Delta x\left[1 + \frac{1}{2}\left(\frac{\partial y(x, t)}{\partial x}\right)^2\right]$$

(*Hint:* Use the relationship $\sqrt{1 + u} \approx 1 + \frac{1}{2}u$, valid for $|u| \ll 1$.) d) The potential energy stored in the segment equals the work done by the string tension F (which acts along the string) to stretch the segment from its unstretched length Δx to the length calculated in part (c). Calculate this work and show that the potential energy *per unit length of string* is

$$u_p(x, t) = \frac{1}{2}F\left(\frac{\partial y(x, t)}{\partial x}\right)^2$$

e) Calculate $u_p(x, t)$ for a sinusoidal wave given by Eq. (15.7). f) Show that $u_k(x, t) = u_p(x, t)$ for all x and t. g) Show $y(x, t)$, $u_k(x, t)$, and $u_p(x, t)$ as functions of x for $t = 0$ in one graph with all three functions on the same axes. Explain why u_k and u_p are maximum where y is zero, and vice versa. h) Show that the instantaneous power in the wave, given by Eq. (15.22), is equal to the total energy per unit length multiplied by the wave speed v. Explain why this result is reasonable.

15.80 A deep-sea diver is suspended beneath the surface of Loch Ness by a 100-m long cable that is attached to a boat on the surface (Fig. 15.35). The diver and his suit have a total mass of 120 kg and a volume of 0.0800 m^3. The cable has a diameter of 2.00 cm and a linear mass density of $\mu = 1.10$ kg/m. The diver thinks he sees something moving in the murky depths and jerks the end of the cable back and forth to send transverse waves up the cable as a signal to his companions in the boat. a) What is the tension in the cable at its lower end, where it is attached to the diver? Do not forget to include the buoyant force that the water (density 1000 kg/m^3) exerts on him. b) Calculate the tension in the cable a distance x above the diver. The buoyant force on the cable must be included in your calculation. c) The speed of transverse waves on the cable is given by $v = \sqrt{F/\mu}$ (Eq. 15.13). The speed therefore varies along the cable, since the tension is not constant. (This expression neglects the damping force that the water exerts on the moving cable.) Integrate to find the time required for the first signal to reach the surface.

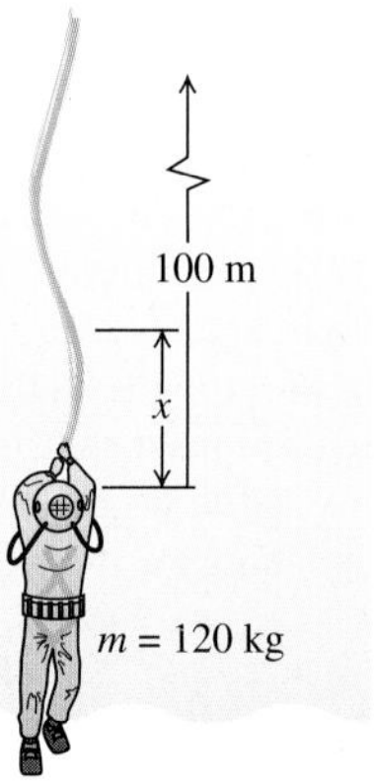

Figure 15.35 Challenge Problem 15.80.

15.81 A uniform rope with length L and mass m is held at one end and whirled in a horizontal circle with angular velocity ω. You can ignore the force of gravity on the rope. Find the time required for a transverse wave to travel from one end of the rope to the other.

15.82 Instantaneous Power in a Standing Wave. From Eq. (15.21), the instantaneous rate at which a wave transmits energy along a string (instantaneous power) is

$$P(x, t) = -F\frac{\partial y(x, t)}{\partial x}\frac{\partial y(x, t)}{\partial t}$$

where F is the tension. a) Evaluate $P(x, t)$ for a standing wave of the form given by Eq. (15.28). b) Show that for all values of x, the *average* power P_{av} carried by the standing wave is zero. (Equation (15.25) does not apply here. Can you see why?) c) For a standing wave given by Eq. (15.28), graph $P(x, t)$ and the displacement $y(x, t)$ as functions of x for $t = 0$, $t = \pi/4\omega$, $t = \pi/2\omega$, and $t = 3\pi/4\omega$. (Positive $P(x, t)$ means energy is flowing in the $+x$-direction; negative $P(x, t)$ means the flow is in the $-x$-direction.) d) The *kinetic* energy per unit length of string is greatest where the string has the greatest transverse speed, and the *potential* energy per unit length of string is greatest where the string has the steepest slope (because there the string is stretched the most). (See Challenge Problem 15.79.) Using these ideas, discuss the flow of energy along the string.

15.83 Out of Tune. The B-string of a guitar is made of steel (density 7800 kg/m^3), is 63.5 cm long, and has diameter 0.406 mm. The fundamental frequency is $f = 247.0$ Hz. a) Find the string tension. b) If the tension F is changed by a small amount ΔF, the frequency f changes by a small amount Δf. Show that

$$\frac{\Delta f}{f} = \frac{1}{2}\frac{\Delta f}{f}$$

c) The string is tuned as in part (a) when its temperature is 18.5°C. Strenuous playing can make the temperature of the string rise, changing its vibration frequency. Find Δf if the temperature of the string rises to 29.5°C. The steel string has a Young's modulus of 2.00×10^{11} Pa and a coefficient of linear expansion of $1.20 \times 10^{-5}\ (\text{C}°)^{-1}$. Assume that the temperature of the body of the guitar remains constant. Will the vibration frequency rise or fall?

SOUND AND HEARING 16

A human voice and a trumpet are both examples of wind instruments. In both cases, a standing sound wave is set up in the air within a long pipe (the singer's vocal tract or the metal tube of the trumpet) in response to a vibration at one end (the singer's vocal folds or the trumpeter's lips). The character of the sound is determined by the properties of the standing wave.

? What is the physical difference between musical sound and noise?

Of all the mechanical waves that occur in nature, the most important in our everyday lives are longitudinal waves in a medium, usually air, called *sound* waves. The reason is that the human ear is tremendously sensitive and can detect sound waves even of very low intensity. Besides their use in spoken communication, our ears allow us to pick up a myriad of cues about our environment, from the welcome sound of a meal being prepared to the warning sound of an approaching car. The ability to hear an unseen nocturnal predator was essential to the survival of our ancestors, so it is no exaggeration to say that we humans owe our existence to our highly evolved sense of hearing.

Up to this point we have described mechanical waves primarily in terms of displacement; however, a description of sound waves in terms of *pressure* fluctuations is often more appropriate, largely because the ear is primarily sensitive to changes in pressure. We'll study the relations among displacement, pressure fluctuation, and intensity and the connections between these quantities and human sound perception.

When a source of sound or a listener moves through the air, the listener may hear a different frequency than the one emitted by the source. This is the Doppler effect, which has important applications in medicine and technology.

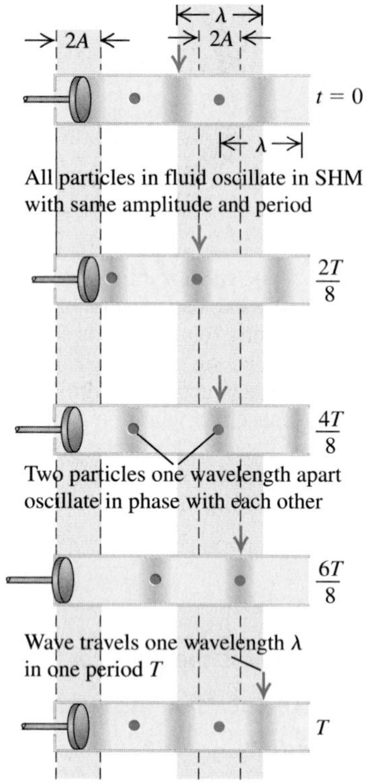

16.1 A sinusoidal longitudinal wave traveling to the right in a fluid. The blue and red dots represent two particles in the fluid that are one-half wavelength apart, and the magenta arrow tracks the motion of a wave compression. (Compare to Fig. 15.5.)

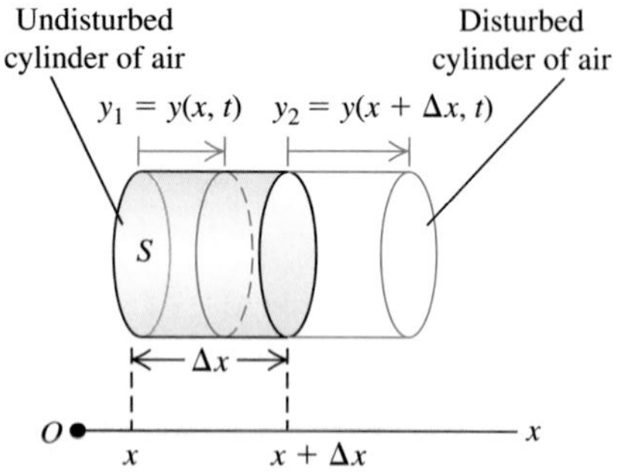

16.2 A cylindrical volume of air with cross-sectional area S and undisturbed length Δx. As a sound wave propagates along the x-axis, the left and right ends undergo different displacements y_1 and y_2. The resulting change in volume is $S(y_2 - y_1)$.

16.1 | Sound Waves

The most general definition of **sound** is that it is a longitudinal wave in a medium. Our main concern in this chapter is with sound waves in air, but sound can travel through any gas, liquid, or solid. You may be all too familiar with the propagation of sound through a solid if your neighbor's stereo speakers are right next to your wall.

The simplest sound waves are sinusoidal waves, which have definite frequency, amplitude, and wavelength. The human ear is sensitive to waves in the frequency range from about 20 to 20,000 Hz, called the **audible range,** but we also use the term *sound* for similar waves with frequencies above **(ultrasonic)** and below **(infrasonic)** the range of human hearing.

Sound waves usually travel out in all directions from the source of sound, with an amplitude that depends on the direction and distance from the source. We'll return to this point in the next section. For now, we concentrate on the idealized case of a sound wave that propagates in the positive x-direction only. As we discussed in Section 15.3, such a wave is described by a wave function $y(x, t)$, which gives the instantaneous displacement y of a particle in the medium at position x at time t. If the wave is sinusoidal, we can express it using Eq. (15.7):

$$y(x, t) = A\cos(kx - \omega t) \qquad (16.1)$$

(sound wave propagating in the $+x$-direction)

Remember that in a longitudinal wave the displacements are *parallel* to the direction of travel of the wave, so distances x and y are measured parallel to each other, not perpendicular as in a transverse wave. The amplitude A is the maximum displacement of a particle in the medium from its equilibrium position (Fig. 16.1). It is also called the **displacement amplitude.**

Sound waves may also be described in terms of variations of *pressure* at various points. In a sinusoidal sound wave in air, the pressure fluctuates above and below atmospheric pressure p_a in a sinusoidal variation with the same frequency as the motions of the air particles. The human ear operates by sensing such pressure variations. A sound wave entering the ear canal exerts a fluctuating pressure on one side of the eardrum; the air on the other side of the eardrum, vented to the outside by the Eustachian tube, is at atmospheric pressure. The pressure difference on the two sides of the eardrum sets it into motion. Microphones and similar devices also usually sense pressure differences, not displacements, so it is very useful to develop a relation between these two descriptions.

Let $p(x, t)$ be the instantaneous pressure fluctuation in a sound wave at any point x at time t. That is, $p(x, t)$ is the amount by which the pressure *differs* from normal atmospheric pressure p_a. Think of $p(x, t)$ as the *gauge pressure* defined in Section 14.2; it can be either positive or negative. The *absolute* pressure at a point is then $p_a + p(x, t)$.

To see the connection between the pressure fluctuation $p(x, t)$ and the displacement $y(x, t)$ in a sound wave propagating in the $+x$-direction, consider an imaginary cylinder of air with cross-sectional area S and axis along the direction of propagation (Fig. 16.2). When no sound wave is present, the cylinder has length Δx and volume $V = S\,\Delta x$, as shown by the shaded volume in Fig. 16.2. When a wave is present, at time t the end of the cylinder that is initially at x is displaced by $y_1 = y(x, t)$, and the end that is initially at $x + \Delta x$ is displaced by $y_2 = y(x + \Delta x, t)$; this is shown by the red lines. If $y_2 > y_1$, as shown in Fig. 16.2, the cylinder's volume has increased, which causes a decrease in pressure. If $y_2 < y_1$, the cylinder's volume has decreased and the pressure has increased. If

$y_2 = y_1$, the cylinder is simply shifted to the left or right; there is no volume change and no pressure fluctuation. The pressure fluctuation depends on the *difference* between the displacement at neighboring points in the medium.

Quantitatively, the change in volume ΔV of the cylinder is

$$\Delta V = S(y_2 - y_1) = S[y(x + \Delta x, t) - y(x, t)]$$

In the limit as $\Delta x \to 0$, the fractional change in volume dV/V is

$$\frac{dV}{V} = \lim_{\Delta x \to 0} \frac{S[y(x + \Delta x, t) - y(x, t)]}{S\Delta x} = \frac{\partial y(x, t)}{\partial x} \tag{16.2}$$

The fractional volume change is related to the pressure fluctuation by the bulk modulus B, which by definition [Eq. (11.13)] is $B = -p(x, t)/(dV/V)$ (see Section 11.4). Solving for $p(x, t)$, we have

$$p(x, t) = -B\frac{\partial y(x, t)}{\partial x} \tag{16.3}$$

The negative sign arises because when $\partial y(x, t)/\partial x$ is positive, the displacement is greater at $x + \Delta x$ than at x, corresponding to an increase in volume and a *decrease* in pressure.

When we evaluate $\partial y(x, t)/\partial x$ for the sinusoidal wave of Eq. (16.1), we find

$$p(x, t) = BkA \sin(kx - \omega t) \tag{16.4}$$

Figure 16.3 shows $y(x, t)$ and $p(x, t)$ for a sinusoidal sound wave at $t = 0$. It also shows how individual particles of the wave are displaced at this time. While $y(x, t)$ and $p(x, t)$ describe the same wave, these two functions are one-quarter cycle out of phase: at any time, the displacement is greatest where the pressure fluctuation is zero, and vice versa. In particular, note that the compressions (points of greatest pressure and density) and rarefactions (points of lowest pressure and density) are points of *zero* displacement.

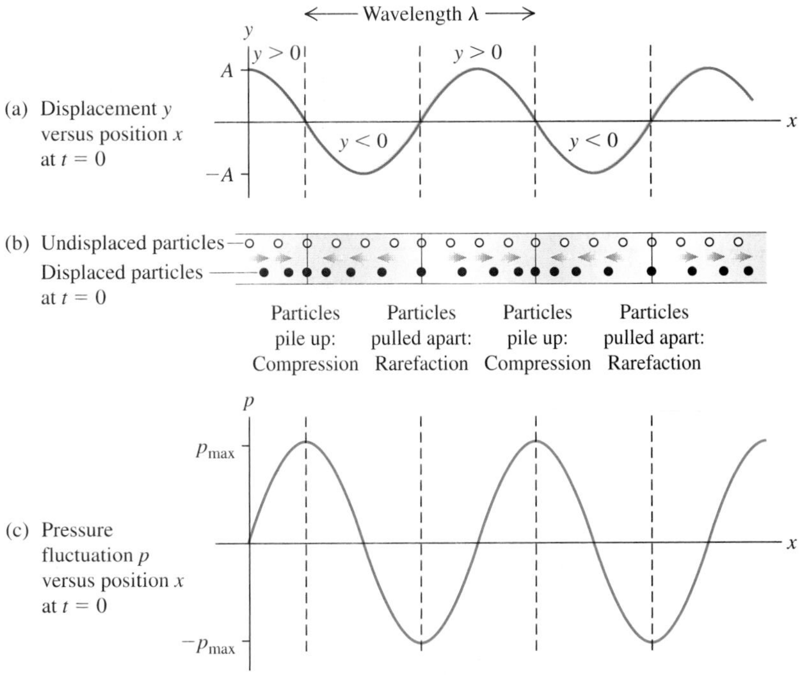

16.3 Three ways to describe a sound wave. (a) A graph of displacement y versus position x [Eq. (16.1)] at $t = 0$. (b) A cartoon showing the displacement of individual particles in the gas at $t = 0$. Particles are displaced to the right where $y > 0$ and to the left where $y < 0$. Points where particles pile up are compressions (dark shading), and points where they are pulled apart are rarefactions (light shading). (c) A graph of pressure fluctuation p versus position x [Eq. (16.4)] at $t = 0$. Note that p is most positive at a compression and most negative at a rarefaction.

CAUTION Keep in mind that the graphs in Fig. 16.3 show the wave at only *one* instant of time. Because the wave is propagating in the $+x$-direction, as time goes by the wave patterns in the functions $y(x, t)$ and $p(x, t)$ move to the right at the wave speed $v = \omega/k$. Hence the positions of the compressions and rarefactions also move to the right at this same speed. The particles, by contrast, simply oscillate back and forth in simple harmonic motion as shown in Fig. 16.1.

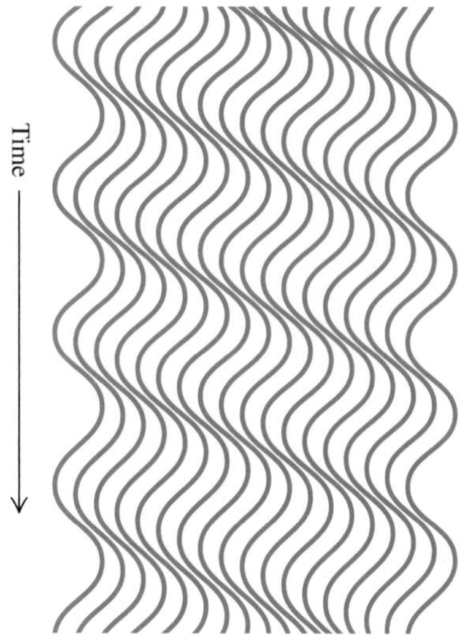

16.4 Diagram for illustrating longitudinal traveling waves.

Visualizing the relationship between particle motion and wave motion is not as easy for longitudinal waves as it is for transverse waves on a string. Figure 16.4 will help you understand these motions. To use this figure, tape two index cards edge-to-edge with a gap of 1 mm or so between them, forming a thin slit. Place the cards over the top of the figure with the slit horizontal, and move them downward with constant speed. The portions of the sine curves that are visible through the slit correspond to a row of particles in a medium in which a longitudinal sinusoidal wave is traveling. Each particle undergoes simple harmonic motion about its equilibrium position, with delays or phase shifts that increase continuously along the slit. The regions of maximum compression and expansion move from left to right with constant speed. Moving the card upward simulates a wave traveling from right to left.

Equation (16.4) shows that the quantity BkA represents the maximum pressure fluctuation. We call this the **pressure amplitude,** denoted by p_{max}:

$$p_{max} = BkA \qquad \text{(sinusoidal sound wave)} \qquad (16.5)$$

The pressure amplitude is directly proportional to the displacement amplitude A, as we might expect, and it also depends on wavelength. Waves of shorter wavelength λ (larger wave number $k = 2\pi/\lambda$) have greater pressure variations for a given amplitude because the maxima and minima are squeezed closer together. A medium with a large value of bulk modulus B requires a relatively large pressure amplitude for a given displacement amplitude because large B means a less compressible medium; that is, greater pressure change is required for a given volume change.

Example 16.1 Amplitude of a sound wave in air

In a sinusoidal sound wave of moderate loudness the maximum pressure variations are of the order of 3.0×10^{-2} Pa above and below atmospheric pressure p_a (nominally 1.013×10^5 Pa at sea level). Find the corresponding maximum displacement if the frequency is 1000 Hz. In air at normal atmospheric pressure and density, the speed of sound is 344 m/s and the bulk modulus is 1.42×10^5 Pa.

SOLUTION

IDENTIFY and SET UP: We are given the pressure amplitude p_{max}, wave speed v, frequency f, and bulk modulus B. Our target variable is the displacement A, which is related to p_{max} by Eq. (16.5). We also use the relationship $\omega = vk$ [Eq. (15.6)] to determine the wave number k from v and the angular frequency $\omega = 2\pi f$.

EXECUTE: From Eq. (16.5), the maximum displacement is $A = p_{max}/Bk$. From Eq. (15.6), the wave number is

$$k = \frac{\omega}{v} = \frac{2\pi f}{v} = \frac{(2\pi\ \text{rad})(1000\ \text{Hz})}{344\ \text{m/s}} = 18.3\ \text{rad/m}$$

Then

$$A = \frac{p_{max}}{Bk} = \frac{3.0 \times 10^{-2}\ \text{Pa}}{(1.42 \times 10^5\ \text{Pa})(18.3\ \text{rad/m})} = 1.2 \times 10^{-8}\ \text{m}$$

EVALUATE: This displacement amplitude is only about $\frac{1}{100}$ the size of a human cell. Remember that the ear actually senses pressure fluctuations; it detects these minuscule displacements only indirectly.

Example 16.2 Amplitude of a sound wave in the inner ear

When a sound wave enters the ear, it sets the eardrum into oscillation, which in turn causes oscillation of the three tiny bones in the middle ear called the *ossicles* (Fig. 16.5). This oscillation is finally transmitted to the fluid-filled inner ear; the motion of the fluid disturbs hair cells within the inner ear, which transmit nerve impulses to the brain with the information that a sound is present. The moving part of the eardrum has an area of about 43 mm^2, and the area of the stirrup (the smallest of the ossicles) where it connects to the inner ear is about 3.2 mm^2. For the sound in the previous example, determine a) the pressure amplitude and b) the displacement amplitude of the wave in the fluid of the inner ear. The speed of sound in this fluid is about 1500 m/s.

SOLUTION

IDENTIFY: Although the sound wave is now traveling in fluid (mostly water) rather than gas, the same principles and relationships among the properties of the wave apply.

SET UP: We can safely neglect the mass of the ossicles (about 58 mg $= 5.8 \times 10^{-5}$ kg), so the force exerted by the ossicles on the fluid in the inner ear is the same as the force exerted on the eardrum and ossicles by the sound wave in air. (We used this same idea in Chapters 4 and 5 when we said that the tension is the same at either end of a massless rope.) Hence the pressure amplitude p_{max} is greater in the inner ear than in the outside air because the same force is exerted on a smaller area (the area of the stirrup versus the area of the eardrum).

Given the pressure amplitude in the inner ear, we find the displacement amplitude using Eq. (16.5). The values of B and k are different than in the air. To determine k, note that the wave in the inner ear has the same angular frequency ω as the wave in the air because the air, eardrum, ossicles, and inner-ear fluid all oscillate together. But because the wave speed v is greater in the inner ear than in the air (1500 m/s versus 344 m/s), the wave number $k = \omega/v$ is greater as well.

EXECUTE: a) Using the area of the eardrum and pressure amplitude found in Example 16.1, the maximum force exerted by the sound wave in air on the eardrum is $F_{max} = p_{max\,(air)}S_{eardrum}$. Hence the pressure amplitude in the inner ear fluid is

$$p_{max\,(inner\ ear)} = \frac{F_{max}}{S_{stirrup}} = \frac{p_{max\,(air)}S_{eardrum}}{S_{stirrup}}$$
$$= \frac{(3.0 \times 10^{-2}\ \text{Pa})(43\ \text{mm}^2)}{3.2\ \text{mm}^2} = 0.40\ \text{Pa}$$

b) To find the maximum displacement, we again use the relationship $A = p_{max}/Bk$ as in Example 16.1. The fluid in the inner ear is mostly water, which has a much greater bulk modulus than air because water is much more difficult to compress. From Table 11.2 the compressibility of water (unfortunately also called k) equals $45.8 \times 10^{-11}\ \text{Pa}^{-1}$, so $B_{fluid} = 1/(45.8 \times 10^{-11}\ \text{Pa}^{-1}) = 2.18 \times 10^9$ Pa.

We determine the value of the wave number k using the value of ω from Example 16.1 and $v = 1500$ m/s for the inner-ear fluid. Hence

$$k_{inner\ ear} = \frac{\omega}{v_{inner\ ear}} = \frac{6283\ \text{rad/s}}{1500\ \text{m/s}} = 4.2\ \text{rad/m}$$

When we put everything together, the maximum displacement of the fluid in the inner ear is

$$A_{inner\ ear} = \frac{p_{max\,(inner\ ear)}}{B_{fluid}\,k_{inner\ ear}} = \frac{0.40\ \text{Pa}}{(2.18 \times 10^9\ \text{Pa})(4.2\ \text{rad/m})}$$
$$= 4.4 \times 10^{-11}\ \text{m}$$

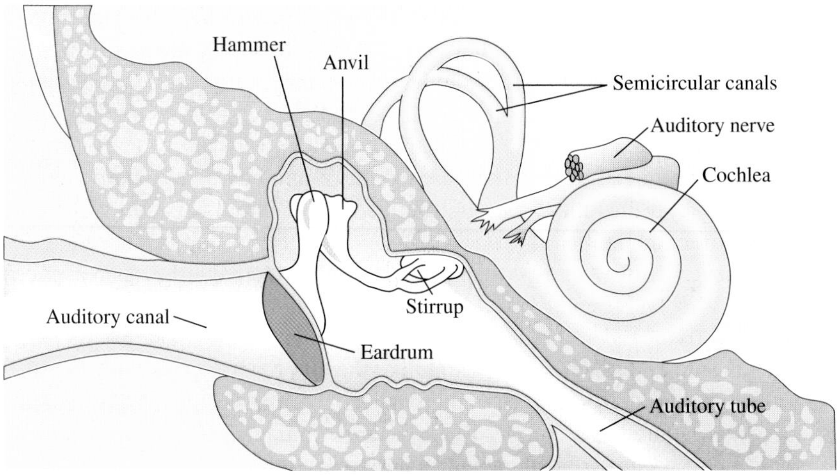

16.5 The anatomy of the human ear. The middle ear is the size of a small marble; the ossicles (hammer, anvil, and stirrup) are the smallest bones in the human body.

EVALUATE: The result in part (a) shows that the effect of the ossicles is to increase the pressure amplitude in the inner ear by a factor of $(43\ \text{mm}^2)/(3.2\ \text{mm}^2) = 13$. This amplification factor helps give the human ear its great sensitivity.

The displacement amplitude in the inner ear is even smaller than in the air. What really matters in the inner ear, however, is the *pressure* amplitude, since the pressure variations within the fluid cause the forces that set the hair cells into motion.

Perception of Sound Waves

The physical characteristics of a sound wave are directly related to the perception of that sound by a listener. For a given frequency, the greater the pressure amplitude of a sinusoidal sound wave, the greater the perceived **loudness.** The relationship between pressure amplitude and loudness is not a simple one, and it varies from one person to another. One important factor is that the ear is not equally sensitive to all frequencies in the audible range. A sound at one frequency may seem louder than one of equal pressure amplitude at a different frequency. At 1000 Hz the minimum pressure amplitude that can be perceived with normal hearing is about 3×10^{-5} Pa; to produce the same loudness at 200 Hz or 15,000 Hz requires about 3×10^{-4} Pa. Perceived loudness also depends on the health of the ear. A loss of sensitivity at the high-frequency end usually happens naturally with age but can be further aggravated by excessive noise levels. Studies have shown that many young rock musicians have suffered permanent ear damage and have hearing that is typical of persons 65 years of age. Portable stereo headsets used at high volume pose similar threats to hearing. Be careful!

The frequency of a sound wave is the primary factor in determining the **pitch** of a sound, the quality that lets us classify the sound as "high" or "low." The higher the frequency of a sound (within the audible range), the higher the pitch that a listener will perceive. Pressure amplitude also plays a role in determining pitch. When a listener compares two sinusoidal sound waves with the same frequency but different pressure amplitudes, the one with the greater pressure amplitude is usually perceived as louder but also as slightly lower in pitch.

Musical sounds have wave functions that are more complicated than a simple sine function. The pressure fluctuation in the sound wave produced by a clarinet is shown in Fig. 16.6a. The pattern is so complex because the column of air in a wind instrument like a clarinet vibrates at a fundamental frequency and at many harmonics at the same time. (In Section 15.8, we described this same behavior for a string that has been plucked, bowed, or struck. We'll examine the physics of wind instruments in Section 16.5.) The sound wave produced in the surrounding air has a similar amount of each harmonic, that is, a similar *harmonic content.* Figure 16.6b shows the harmonic content of the sound of a clarinet. The mathematical process of translating a pressure-time graph like Fig. 16.6a into a graph of harmonic content like Fig. 16.6b is called *Fourier analysis.*

Two tones produced by different instruments might have the same fundamental frequency (and thus the same pitch) but sound different because of the presence of different amounts of the various harmonics. The difference is called *tone color, quality,* or **timbre** and is often described in subjective terms such as reedy, golden, round, mellow, and tinny. A tone that is rich in harmonics, like the clarinet tone in Figs. 16.6a and 16.6b usually sounds thin and "stringy" or "reedy," while a tone containing mostly a fundamental, like the alto recorder tone in Figs. 16.6c and 16.6d is more mellow and flutelike. The same principle applies to the human voice, which is another example of a wind instrument; the vowels "a" and "e" sound different because of differences in harmonic content.

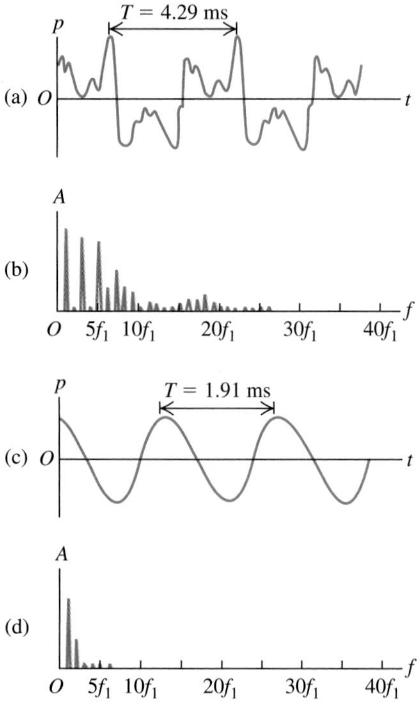

16.6 (a) Graph of pressure fluctuation versus time from a clarinet with fundamental frequency $f_1 = 233$ Hz. (b) Harmonic content of the sound of the clarinet. (c) Graph of pressure fluctuation versus time from an alto recorder with fundamental frequency $f_1 = 523$ Hz. (d) Harmonic content of the sound of the alto recorder. (From R.E. Berg and D.G. Stork, *The Physics of Sound,* Prentice-Hall, 1982.)

Another factor in determining tone quality is the behavior at the beginning (*attack*) and end (*decay*) of a tone. A piano tone begins with a thump and then dies away gradually. A harpsichord tone, in addition to having different harmonic content, begins much more quickly with a click, and the higher harmonics begin before the lower ones. When the key is released, the sound also dies away much more rapidly with a harpsichord than with a piano. Similar effects are present in other musical instruments. With wind and string instruments the player has considerable control over the attack and decay of the tone, and these characteristics help to define the unique characteristics of each instrument.

ActivPhysics Online

10.10 Complex Waves: Fourier Analysis

Unlike the tones made by musical instruments or the vowels in human speech, **noise** is a combination of *all* frequencies, not just frequencies that are integer multiples of a fundamental frequency. (An extreme case is "white noise," which contains equal amounts of all frequencies across the audible range.) Examples include the sound of the wind and the hissing sound you make in saying the consonant "s."

Test Your Understanding

You use an electronic signal generator to produce a sinusoidal sound wave in air. You then increase the frequency of the wave while keeping the pressure amplitude constant. What effect does this have on the displacement amplitude of the sound wave?

16.2 | Speed of Sound Waves

We found in Section 15.4 that the speed v of a transverse wave on a string depends on the string tension F and the linear mass density μ: $v = \sqrt{F/\mu}$. What, we may ask, is the corresponding expression for the speed of sound waves in a gas or liquid? On what properties of the medium does the speed depend?

We can make an educated guess about these questions by remembering a claim that we made in Section 15.4: For mechanical waves in general, the expression for the wave speed is of the form

$$v = \sqrt{\frac{(\text{restoring force returning the system to equilibrium})}{(\text{inertia resisting the return to equilibrium})}}$$

A sound wave in a bulk fluid causes compressions and rarefactions of the fluid, so the restoring-force term in the above expression must be related to how easy or difficult it is to compress the fluid. This is precisely what the bulk modulus B of the medium tells us. According to Newton's second law, inertia is related to mass. The "massiveness" of a bulk fluid is described by its density, or mass per unit volume, ρ. (The corresponding quantity for a string is the mass per unit length, μ.) Hence we expect that the speed of sound waves should be of the form $v = \sqrt{B/\rho}$.

To check our guess, we'll derive the speed of sound (longitudinal) waves in a fluid in a pipe. This is a situation of some importance, since all musical wind instruments are fundamentally pipes in which a longitudinal wave (sound) propagates in a fluid (air). Human speech works on the same principle; sound waves propagate in your vocal tract, which is basically an air-filled pipe connected to the lungs at one end (your larynx) and to the outside air at the other end (your mouth). The steps in our derivation are completely parallel to those we used in Section 15.4 to find the speed of transverse waves, so you'll find it useful to review that section.

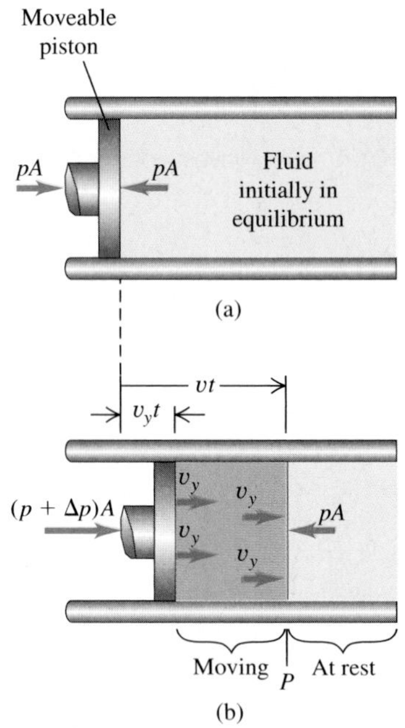

16.7 A sound wave propagating in a fluid confined to a tube. (a) Fluid in equilibrium. (b) A time t after the piston begins moving to the right at speed v_y, the fluid between the piston and point P is in motion. The speed of sound waves is v.

Figure 16.7 shows a fluid (either liquid or gas) with density ρ in a pipe with cross-sectional area A. In the equilibrium state, the fluid is under a uniform pressure p. In Fig. 16.7a the fluid is at rest. We take the x-axis along the length of the pipe. This is also the direction in which we make a longitudinal wave propagate, so the displacement y is also measured along the pipe, just as in Section 16.1 (see Fig. 16.2).

At time $t = 0$ we start the piston at the left end moving toward the right with constant speed v_y. This initiates a wave motion that travels to the right along the length of the pipe, in which successive sections of fluid begin to move and become compressed at successively later times.

Figure 16.7b shows the fluid at time t. All portions of fluid to the left of point P are moving to the right with speed v_y, and all portions to the right of P are still at rest. The boundary between the moving and stationary portions travels to the right with a speed equal to the speed of propagation or wave speed v. At time t the piston has moved a distance $v_y t$, and the boundary has advanced a distance vt. As with a transverse disturbance in a string, we can compute the speed of propagation from the impulse-momentum theorem.

The quantity of fluid set in motion in time t is the amount that originally occupied a section of the cylinder with length vt, cross-sectional area A, and volume vtA. The mass of this fluid is ρvtA, and its longitudinal momentum (that is, momentum along the length of the pipe) is

$$\text{Longitudinal momentum} = (\rho vtA)v_y$$

Next we compute the increase of pressure, Δp, in the moving fluid. The original volume of the moving fluid, Avt, has decreased by an amount $Av_y t$. From the definition of the bulk modulus B, Eq. (11.13) in Section 11.5,

$$B = \frac{-\text{Pressure change}}{\text{Fractional volume change}} = \frac{-\Delta p}{-Av_y t/Avt}$$

$$\Delta p = B\frac{v_y}{v}$$

The pressure in the moving fluid is $p + \Delta p$ and the force exerted on it by the piston is $(p + \Delta p)A$. The net force on the moving fluid (see Fig. 16.7b) is ΔpA, and the longitudinal impulse is

$$\text{Longitudinal impulse} = \Delta pAt = B\frac{v_y}{v}At$$

Because the fluid was at rest at time $t = 0$, the change in momentum up to time t is equal to the momentum at that time. Applying the impulse-momentum theorem, we find

$$B\frac{v_y}{v}At = \rho vtAv_y \tag{16.6}$$

When we solve this expression for v, we get

$$v = \sqrt{\frac{B}{\rho}} \quad \text{(speed of a longitudinal wave in a fluid)} \tag{16.7}$$

which agrees with our educated guess. Thus the speed of propagation of a longitudinal pulse in a fluid depends only on the bulk modulus B and the density ρ of the medium.

While we derived Eq. (16.7) for waves in a pipe, it also applies to longitudinal waves in a bulk fluid. Thus the speed of sound waves traveling in air or water is determined by this equation.

When a longitudinal wave propagates in a *solid* rod or bar, the situation is somewhat different. The rod expands sideways slightly when it is compressed longitudinally, while a fluid in a pipe with constant cross section cannot move sideways. Using the same kind of reasoning that led us to Eq. (16.7), we can show that the speed of a longitudinal pulse in the rod is given by

$$v = \sqrt{\frac{Y}{\rho}} \quad \text{(speed of a longitudinal wave in a solid rod)} \quad (16.8)$$

where Y is Young's modulus, defined in Section 11.4.

CAUTION Equation (16.8) applies only to a rod or bar whose sides are free to bulge and shrink a little as the wave travels. It does not apply to longitudinal waves in a *bulk* solid, since in these materials, sideways motion in any element of material is prevented by the surrounding material. The speed of longitudinal waves in a bulk solid depends on the density, the bulk modulus, and the *shear* modulus; a full discussion is beyond the scope of this book.

As with the derivation for a transverse wave on a string, Eqs. (16.7) and (16.8) are valid for sinusoidal and other periodic waves, not just for the special case discussed here.

Table 16.1 lists the speed of sound in several bulk materials. Sound waves travel more slowly in lead than in aluminum or steel because lead has a lower bulk modulus and shear modulus and a higher density.

Table 16.1 Speed of Sound in Various Bulk Materials

Material	Speed of Sound (m/s)
Gases	
Air (20°C)	344
Helium (20°C)	999
Hydrogen (20°C)	1330
Liquids	
Liquid helium (4 K)	211
Mercury (20°C)	1451
Water (0°C)	1402
Water (20°C)	1482
Water (20°C)	1543
Solids	
Aluminum	6420
Lead	1960
Steel	5941

Example 16.3 Wavelength of sonar waves

A ship uses a sonar system to detect underwater objects (Fig. 16.8). The system emits underwater sound waves and measures the time interval for the reflected wave (echo) to return to the detector. Determine the speed of sound waves in water using Eq. (16.7) and find the wavelength of a 262-Hz wave.

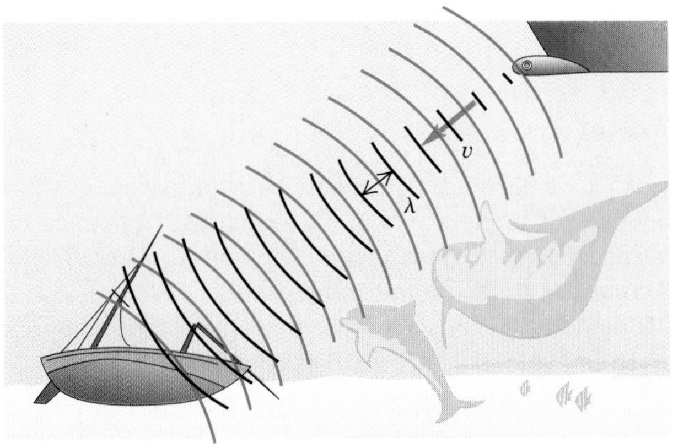

16.8 A sonar system uses underwater sound waves to detect and locate submerged objects.

SOLUTION

IDENTIFY and SET UP: To use Eq. (16.7), we find the bulk modulus of water from the compressibility (Table 11.2) and the density ($\rho = 1.00 \times 10^3\ \text{kg/m}^3$). Given the speed and the frequency $f = 262$ Hz, we find the wavelength from the relationship $v = f\lambda$.

EXECUTE: From Table 11.2 we find that the compressibility of water, which is the reciprocal of the bulk modulus, is $k = 45.8 \times 10^{-11}\ \text{Pa}^{-1}$. Thus $B = (1/45.8) \times 10^{11}$ Pa. We obtain

$$v = \sqrt{\frac{B}{\rho}} = \sqrt{\frac{(1/45.8) \times 10^{11}\ \text{Pa}}{1.00 \times 10^3\ \text{kg/m}^3}}$$
$$= 1480\ \text{m/s}$$

The wavelength is given by

$$\lambda = \frac{v}{f} = \frac{1480\ \text{m/s}}{262\ \text{s}^{-1}} = 5.65\ \text{m}$$

EVALUATE: The calculated value of v agrees well with the experimental value in Table 16.1. Though water is far denser than air (ρ is larger), it is also far more incompressible (B is larger) and the speed $v = \sqrt{B/\rho}$ turns out to be more than four times the speed of sound in air at ordinary temperatures.

We found in Example 15.1 (Section 15.2) that a wave of this frequency in air has a wavelength of 1.31 m. The speed of sound in water is greater than in air, so the wavelength $\lambda = v/f$ must be greater as well; this is just as we calculated.

Dolphins emit high-frequency sound waves (typically 100,000 Hz) and use the echoes for guidance and for hunting. The corresponding wavelength in water is 1.48 cm. With this high-frequency "sonar" system they can sense objects that are roughly as small as the wavelength (but not much smaller). *Ultrasonic imaging* is a medical technique that uses exactly the same physical principle; sound waves of very high frequency and very short wavelength, called *ultrasound,* are scanned over the human body, and the "echoes" from interior organs are used to create an image. With ultrasound of frequency 5 MHz $= 5 \times 10^6$ Hz, the wavelength in water (the primary constituent of the body) is 0.3 mm, and features as small as this can be discerned in the image. Ultrasound is used for the study of heart-valve action, detection of tumors, and prenatal examinations (Fig. 16.9). Ultrasound is more sensitive than x rays in distinguishing various kinds of tissues and does not have the radiation hazards associated with x rays.

Example 16.4 Speed of a longitudinal wave

What is the speed of longitudinal waves in a lead rod?

SOLUTION

IDENTIFY: Waves of this kind are made by clamping the rod in place and striking one end face-on with a hammer. The amplitude of the resulting waves is so small as to be nearly invisible to the naked eye. But the question is about the *speed* of the waves, which does not depend on the amplitude. Note that we cannot simply use the value for lead from Table 16.1, as that value refers to the speed of sound in a *bulk* material, not a rod.

SET UP: Equation (16.8) applies to this situation. We find the values of Young's modulus Y and the density ρ from Tables 11.1 and 14.1, respectively.

EXECUTE: We find $Y = 1.6 \times 10^{10}$ Pa and $\rho = 11.3 \times 10^3$ kg/m^3 (that is, 11.4 times the density of water), so

$$v = \sqrt{\frac{Y}{\rho}} = \sqrt{\frac{1.6 \times 10^{10}\ \text{Pa}}{11.3 \times 10^3\ \text{kg/m}^3}} = 1.2 \times 10^3\ \text{m/s}$$

EVALUATE: This is more than three times the speed of sound in air, but slower than the speed of sound in *bulk* lead (see Table 16.1). The reason is that for lead the bulk modulus is greater than Young's modulus.

Speed of Sound in Gases

ActivPhysics Online

10.3 Speed of Sound in a Gas

Most of the sound waves that we encounter on a daily basis propagate in air. To use Eq. (16.7) to find the speed of sound waves in air, we must keep in mind that the bulk modulus of a gas depends on the pressure of the gas: the greater the pressure applied to a gas to compress it, the more it resists further compression and hence the greater the bulk modulus. (That's why specific values of the bulk modulus for gases are not given in Table 11.1.) The expression for the bulk modulus of a gas for use in Eq. (16.7) is

$$B = \gamma p_0 \tag{16.9}$$

where p_0 is the equilibrium pressure of the gas. The quantity γ (the Greek letter "gamma") is called the *ratio of heat capacities.* It is a dimensionless number that characterizes the thermal properties of the gas. (We'll learn more about this quantity in Chapter 19.) As an example, the ratio of heat capacities for air is $\gamma = 1.40$. At normal atmospheric pressure $p_0 = 1.01 \times 10^5$ Pa, so $B = (1.40)(1.013 \times 10^5 \text{ Pa}) = 1.42 \times 10^5$ Pa. This value is miniscule compared to the bulk modulus of a typical solid (Table 11.1), which is approximately 10^{10} to 10^{11} Pa. This shouldn't be surprising: it's simply a statement that air is far easier to compress than steel.

The density ρ of a gas also depends on the pressure, which in turn depends on the temperature. It turns out that the ratio B/ρ for a given type of ideal gas does *not* depend on the pressure at all, only the temperature. From Eq. (16.7), this means that the speed of sound in a gas is fundamentally a function of temperature T:

$$v = \sqrt{\frac{\gamma RT}{M}} \quad \text{(speed of sound in an ideal gas)} \qquad (16.10)$$

This expression incorporates several quantities that you may recognize from your study of ideal gases in chemistry and which we will study in Chapters 17, 18, and 19. The temperature T is the *absolute* temperature in kelvins (K), equal to the Celsius temperature plus 273.15; thus 20.00°C corresponds to $T = 293.15$ K. The quantity M is the *molar mass*, or mass per mole of the substance of which the gas is composed. The *gas constant* R has the same value for all gases. The current best numerical value of R is

$$R = 8.314472(15) \text{ J/mol}\cdot\text{K}$$

which for practical calculations we can write as 8.314 J/mol · K.

For any particular gas, γ, R, and M are constants, and the wave speed is proportional to the square root of the absolute temperature. We will see in Chapter 18 that Eq. (16.10) is almost identical to the expression for the average speed of molecules in an ideal gas. This shows that sound speeds and molecular speeds are closely related; exploring that relationship in detail would be beyond our scope.

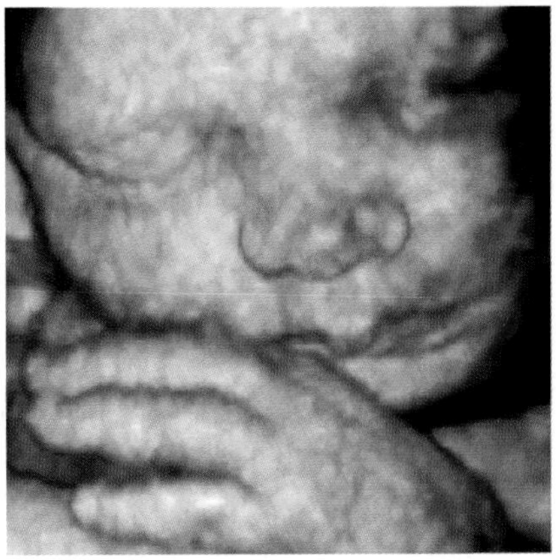

16.9 This three-dimensional image of a fetus in the womb was made using a sequence of ultrasound scans. Each individual scan reveals a two-dimensional "slice" through the fetus; a large number of these slices were then combined digitally to yield this remarkable view.

Example 16.5 Speed of sound in air

Compute the speed of sound waves in air at room temperature ($T = 20°\text{C}$) and find the range of wavelengths in air to which the human ear (which can hear frequencies in the range of 20–20,000 Hz) is sensitive. The mean molar mass for air (a mixture of principally nitrogen and oxygen) is 28.8×10^{-3} kg/mol and the ratio of heat capacities is $\gamma = 1.40$.

SOLUTION

IDENTIFY and SET UP: We use Eq. (16.10) to find the sound speed and the relation $v = f\lambda$ to determine the wavelength that corresponds to each frequency.

EXECUTE: At $T = 20°\text{C} = 293$ K we find

$$v = \sqrt{\frac{\gamma RT}{M}} = \sqrt{\frac{(1.40)(8.314 \text{ J/mol}\cdot\text{K})(293 \text{ K})}{28.8 \times 10^{-3} \text{ kg/mol}}} = 344 \text{ m/s}$$

Using this value of v and the expression $\lambda = v/f$, we find that at 20°C a 20-Hz note corresponds to a wavelength of 17 m and a 20,000-Hz note corresponds to a wavelength of 1.7 cm.

EVALUATE: Our calculated value of v agrees with the measured speed of sound at this temperature to within 0.3%.

It's interesting to note that bats can hear much higher frequencies. Like dolphins, bats use high-frequency sound waves for navigation. A typical frequency is 100 kHz; the corresponding wavelength in air at 20°C is about 3.4 mm, small enough for them to detect the flying insects they eat.

In this discussion we have ignored the *molecular* nature of a gas and have treated it as a continuous medium. A gas is actually composed of molecules in random motion, separated by distances that are large in comparison with their diameters. The vibrations that constitute a wave in a gas are superposed on the random thermal motion. At atmospheric pressure, a molecule travels an average distance of about 10^{-7} m between collisions, while the displacement amplitude of a faint sound may be only 10^{-9} m. We can think of a gas with a sound wave passing through as being comparable to a swarm of bees; the swarm as a whole oscillates slightly while individual insects move about through the swarm, apparently at random.

Test Your Understanding

Calculate the speed of longitudinal waves along a steel railroad track. Explain why you can hear the sound of an approaching train from listening to the rails before you can hear it in the air.

16.3 | Sound Intensity

Traveling sound waves, like all other traveling waves, transfer energy from one region of space to another. We saw in Section 15.5 that a useful way to describe the energy carried by a sound wave is through the *wave intensity I,* equal to the time average rate at which energy is transported per unit area across a surface perpendicular to the direction of propagation. In particular, we will express the intensity of a sound wave in terms of the displacement amplitude A or pressure amplitude p_{max}.

For simplicity, let us consider a sound wave propagating in the $+x$-direction so that we can use our expressions for Section 16.1 for the displacement $y(x, t)$ and pressure fluctuation $p(x, t)$—Eqs. (16.1) and (16.4), respectively. In Section 6.4 we saw that power equals the product of force and velocity [see Eq. (6.18)]. So the power per unit area in this sound wave equals the product of $p(x, t)$ (force per unit area) and the *particle* velocity $v_y(x, t)$. The particle velocity $v_y(x, t)$ is the velocity at time t of that portion of the wave medium at coordinate x.) Using Eqs. (16.1) and (16.4), we find

$$v_y(x, t) = \frac{\partial y(x, t)}{\partial t} = \omega A \sin(kx - \omega t)$$

$$p(x, t)v_y(x, t) = [BkA \sin(kx - \omega t)][\omega A \sin(kx - \omega t)]$$

$$= B\omega k A^2 \sin^2(kx - \omega t)$$

CAUTION Remember that the velocity of the wave as a whole is *not* the same as the particle velocity. While the wave continues to move in the direction of propagation, individual particles in the wave medium merely slosh back and forth, as shown in Fig. (16.1). Furthermore, the maximum speed of a particle of the medium can be very different from the wave speed.

The intensity is, by definition, the time average value of $p(x, t)v_y(x, t)$. For any value of x the average value of the function $\sin^2(kx - \omega t)$ over one period $T = 2\pi/\omega$ is 1/2, so

$$I = \frac{1}{2}B\omega k A^2 \tag{16.11}$$

By using the relations $\omega = vk$ and $v^2 = B/\rho$, we can transform Eq. (16.11) into the form

$$I = \frac{1}{2}\sqrt{\rho B}\,\omega^2 A^2 \qquad \text{(intensity of a sinusoidal sound wave)} \qquad (16.12)$$

This equation shows why in a stereo system, a low-frequency woofer has to vibrate with much larger amplitude than a high-frequency tweeter to produce the same sound intensity.

It is usually more useful to express I in terms of the pressure amplitude p_{max}. Using Eq. (16.5) and the relation $\omega = vk$, we find

$$I = \frac{\omega p_{max}^2}{2Bk} = \frac{v p_{max}^2}{2B} \qquad (16.13)$$

By using the wave speed relation $v^2 = B/\rho$, we can also write Eq. (16.13) in the alternative forms

$$I = \frac{p_{max}^2}{2\rho v} = \frac{p_{max}^2}{2\sqrt{\rho B}} \qquad \text{(intensity of a sinusoidal sound wave)} \qquad (16.14)$$

You should verify these expressions (see Exercise 16.16). Comparison of Eqs. (16.12) and (16.14) shows that sinusoidal sound waves of the same intensity but different frequency have different displacement amplitudes A but the *same* pressure amplitude p_{max}. This is another reason why it is usually more convenient to describe a sound wave in terms of pressure fluctuations, not displacement.

The *total* average power carried across a surface by a sound wave equals the product of the intensity at the surface and the surface area if the intensity over the surface is uniform. The average total sound power emitted by a person speaking in an ordinary conversational tone is about 10^{-5} W, while a loud shout corresponds to about 3×10^{-2} W. If all the residents of New York City were to talk at the same time, the total sound power would be about 100 W, equivalent to the electric power requirement of a medium-sized light bulb. On the other hand, the power required to fill a large auditorium or stadium with loud sound is considerable (see Example 16.9.)

If the sound source emits waves in all directions equally, the intensity decreases with increasing distance r from the source according to the inverse-square law: the intensity is proportional to $1/r^2$. We discussed this law and its consequences in Section 15.5.

The inverse-square relationship does not apply indoors because sound energy can also reach a listener by reflection from the walls and ceiling. Indeed, part of the architect's job in designing an auditorium is to tailor these reflections so that the intensity is as nearly constant as possible over the entire auditorium.

Problem-Solving Strategy

Sound Intensity

IDENTIFY *the relevant concepts:* The relationships between intensity and amplitude of a sound wave are rather straightforward. Quite a few other quantities are involved in these relationships, however, so it's particularly important to decide which is your target variable.

SET UP *the problem using the following steps:*

1. Sort the various physical quantities into categories. The amplitude is described by A or p_{max}, and the frequency f can be determined from ω, k, or λ. These quantities are related through the wave speed v, which in turn is determined by the properties of the medium: B and ρ for a liquid; γ, T, and M for a gas.
2. Determine which quantities are given and which are the unknown target variables. Then start looking for relationships that take you where you want to go.

EXECUTE *the solution as follows:* Use the equations you've selected to solve for the target variables. Be certain that all of the quantities are expressed in the correct units. In particular, if temperature is used to calculate the speed of sound in a gas, make sure that it is expressed in kelvins (Celsius temperature plus 273.15).

EVALUATE *your answer:* There are multiple relationships among the quantities that describe a wave. Try using an alternative one to check your results.

Example 16.6 Intensity of a sound wave in air

Find the intensity of the sound wave in Example 16.1, with $p_{max} = 3.0 \times 10^{-2}$ Pa. Assume the temperature is 20°C so that the density of air is $\rho = 1.20\ \text{kg/m}^3$ and the speed of sound is $v = 344$ m/s.

SOLUTION

IDENTIFY and SET UP: We are given the pressure amplitude p_{max}, density ρ, and wave speed v, and want to find the intensity I. This is done most easily with Eq. (16.14).

EXECUTE: From Eq. (16.14),

$$I = \frac{p_{max}^2}{2\rho v} = \frac{(3.0 \times 10^{-2}\ \text{Pa})^2}{2(1.20\ \text{kg/m}^3)(344\ \text{m/s})}$$
$$= 1.1 \times 10^{-6}\ \text{J/(s}\cdot\text{m}^2) = 1.1 \times 10^{-6}\ \text{W/m}^2$$

EVALUATE: This seems like a very low intensity, but in fact it is well within the range of sound intensities encountered on a daily basis. A very loud sound wave at the threshold of pain has a pressure amplitude of about 30 Pa and an intensity of about 1 W/m^2. The pressure amplitude of the faintest sound wave that can be heard is about 3×10^{-5} Pa, and the corresponding intensity is about 10^{-12} W/m^2. (Try these values of p_{max} in Eq. (16.14) to check that the corresponding intensities are as we have stated.)

Example 16.7 Same intensity, different frequencies

A 20-Hz sound wave has the same intensity as the 1000-Hz sound wave in Examples 16.1 and 16.6. What are the displacement amplitude and pressure amplitude of the 20-Hz sound wave?

SOLUTION

IDENTIFY and SET UP: Given the intensity, we can use Eq. (16.12) to determine the displacement amplitude A. Note that we are not given the value of B. However, ρ and B depend only on the properties of the medium, not the amplitude or frequency, so their values will cancel out if we equate the intensities at 20 Hz and at 1000 Hz. We can also use Eq. (16.14) to find the pressure amplitude at 20 Hz.

EXECUTE:

Inspection of Eq. (16.12) shows that if a wave in a given medium (same ρ and B) has the same intensity I at two different frequencies, then the product ωA must have the same value for both frequencies. From Example 16.1, $A = 1.2 \times 10^{-8}$ m at 1000 Hz, so

$$(20\ \text{Hz})A_{20} = (1000\ \text{Hz})(1.2 \times 10^{-8}\ \text{m})$$
$$A_{20} = 6.0 \times 10^{-7}\ \text{m} = 0.60\ \mu\text{m}$$

Do you understand why we didn't have to convert the frequencies to angular frequencies?

Since the intensity is the same for both frequencies, Eq. (16.14) shows that the *pressure* amplitude p_{max} must also be the same for both. Hence $p_{max} = 3.0 \times 10^{-2}$ Pa for $f = 20$ Hz.

EVALUATE: Our result reinforces the idea that pressure amplitude offers a more convenient description of a sound wave than displacement amplitude. Note also that using Eq. (16.5) and $k = \omega/v$, we get $p_{max} = BkA = (B/v)\omega A$; the bulk modulus B and wave speed v depend only on the medium, so we again conclude that the product ωA must have the same value for both frequencies.

Example 16.8 A high-altitude sound wave

At an altitude of 11,000 m, near the cruising altitude of jetliners (Fig. 16.10), the atmosphere is cold and thin: the temperature is $-57°C$, the pressure is 2.26×10^4 Pa, and the density is 0.364 kg/m^3. What would be the intensity of a 1000-Hz sound wave with the same displacement amplitude as the wave at sea level in Examples 16.1 and 16.6?

16.10 How do sound waves differ in the upper atmosphere?

SOLUTION

IDENTIFY and SET UP: We can use Eq. (16.12) to determine the intensity from the density, bulk modulus, frequency, and amplitude. We are not given the value of B, but we can calculate it using Eq. (16.9).

EXECUTE: Using Eq. (16.9) and the value $\gamma = 1.40$ for air, we find $B = \gamma p_0 = (1.40)(2.26 \times 10^4 \text{ Pa}) = 3.16 \times 10^4$ Pa. Then, from Eq. (16.12),

$$I = \frac{1}{2}\sqrt{(0.364 \text{ kg/m}^3)(2.26 \times 10^4 \text{ Pa})}$$
$$\times (2\pi)^2(1000 \text{ Hz})^2(1.2 \times 10^{-8} \text{ m})^2$$
$$= 3.1 \times 10^{-7} \text{ W/m}^2$$

EVALUATE: This intensity is only 28% as great as that of a wave with the same amplitude at sea level ($3.1 \times 10^{-7} \text{ W/m}^2$ versus $1.1 \times 10^{-6} \text{ W/m}^2$). Hence a loudspeaker oscillating with a certain amplitude would produce a substantially less intense sound at high altitude. At even higher altitude, the atmosphere fades into the near-vacuum of interplanetary space and the density and pressure (and hence bulk modulus) decrease to zero. The intensity likewise decreases, and outer space is nearly silent.

Example 16.9 "Play it loud!"

For an outdoor concert we want the sound intensity at a distance of 20 m from the speaker array to be 1 W/m^2. Assuming that the sound waves have the same intensity in all directions, what acoustic power output is needed from the speaker array?

SOLUTION

IDENTIFY: This example uses the definition of intensity as power per unit area. Here the total power is the target variable, and the area in question is a hemisphere centered on the speaker array.

SET UP: We make the assumptions that the speakers are near ground level and that the acoustic power is spread uniformly over a hemisphere 20 m in radius (that is, we assume that none of the acoustic power is directed into the ground). The surface area of this hemisphere is equal to $(1/2)(4\pi)(20 \text{ m})^2$, or about 2500 m^2. The required power is the product of this area and the intensity.

EXECUTE: The speaker array power is

$$(1 \text{ W/m}^2)(2500 \text{ m}^2) = 2500 \text{ W} = 2.5 \text{ kW}$$

EVALUATE: The electrical power input to the speaker would need to be considerably larger because the efficiency of such devices is not very high (typically a few percent for ordinary speakers, and up to 25% for horn-type speakers).

The Decibel Scale

Because the ear is sensitive over a broad range of intensities, a *logarithmic* intensity scale is usually used. The **sound intensity level** β of a sound wave is defined by the equation

$$\beta = (10 \text{ dB}) \log \frac{I}{I_0} \quad \text{(definition of sound intensity level)} \quad (16.15)$$

In this equation, I_0 is a reference intensity, chosen to be $10^{-12}\ \text{W/m}^2$, approximately the threshold of human hearing at 1000 Hz. Recall that "log" means the logarithm to base 10. Sound intensity levels are expressed in **decibels,** abbreviated dB. A decibel is $\frac{1}{10}$ of a *bel,* a unit named for Alexander Graham Bell (the inventor of the telephone). The bel is inconveniently large for most purposes, and the decibel is the usual unit of sound intensity level.

If the intensity of a sound wave equals I_0 or $10^{-12}\ \text{W/m}^2$, its sound intensity level is 0 dB. An intensity of $1\ \text{W/m}^2$ corresponds to 120 dB. Table 16.2 gives the sound intensity levels in decibels of several familiar sounds. You can use Eq. (16.15) to check the value of sound intensity level β given for each intensity in the table.

Table 16.2 Sound Intensity Levels from Various Sources (Representative Values)

Source or Description of Sound	Sound Intensity Level, β (dB)	Intensity, $I\ (\text{W/m}^2)$
Military jet aircraft 30 m away	140	10^2
Threshold of pain	120	1
Riveter	95	3.2×10^{-3}
Elevated train	90	10^{-3}
Busy street traffic	70	10^{-5}
Ordinary conversation	65	3.2×10^{-6}
Quiet automobile	50	10^{-7}
Quiet radio in home	40	10^{-8}
Average whisper	20	10^{-10}
Rustle of leaves	10	10^{-11}
Threshold of hearing at 1000 Hz	0	10^{-12}

Because the ear is not equally sensitive to all frequencies in the audible range, some sound-level meters weight the various frequencies unequally. One such scheme leads to the so-called dBA scale; this scale deemphasizes the low and very high frequencies, where the ear is less sensitive than at midrange frequencies.

Example 16.10 Temporary deafness

A ten-minute exposure to 120-dB sound will typically shift your threshold of hearing at 1000 Hz from 0 dB up to 28 dB for a while. Ten years of exposure to 92-dB sound will cause a *permanent* shift up to 28 dB. What intensities correspond to 28 dB and 92 dB?

SOLUTION

IDENTIFY and SET UP: We solve Eq. (16.15) to find the intensity I (the target variable) for each value of the sound intensity level β.

EXECUTE: We rearrange Eq. (16.15) by dividing both sides by 10 dB and then using the relationship $10^{\log x} = x$:

$$I = I_0 10^{\beta/(10\ \text{dB})}$$

When $\beta = 28$ dB,

$$I = (10^{-12}\ \text{W/m}^2)10^{(28\ \text{dB}/10\ \text{dB})}$$
$$= (10^{-12}\ \text{W/m}^2)10^{2.8} = 6.3 \times 10^{-10}\ \text{W/m}^2$$

Similarly, for $\beta = 92$ dB,

$$I = (10^{-12}\ \text{W/m}^2)10^{(92\ \text{dB}/10\ \text{dB})} = 1.6 \times 10^{-3}\ \text{W/m}^2$$

EVALUATE: If your answers are a factor of 10 too large, you may have entered 10×10^{-12} in your calculator instead of 1×10^{-12}. Be careful!

Example 16.11 A bird sings in a meadow

Consider an idealized model with a bird (treated as a point source) emitting constant sound power, with intensity inversely proportional to the square of the distance from the bird. By how many decibels does the sound intensity level drop when you move twice as far away from the bird?

SOLUTION

IDENTIFY: Because the decibel scale is logarithmic, the *difference* between two sound intensity levels (the target variable) corresponds to the *ratio* of the corresponding intensities. The ratio of the intensities comes from the inverse-square law.

SET UP: We label the two points 1 and 2 (see Fig. 16.11). We use Eq. (16.15), the definition of sound intensity level, twice (once at each point). We use Eq. (15.26), the statement of the inverse-square law, to relate the intensities at the two points.

EXECUTE: The difference in sound intensity level, $\beta_2 - \beta_1$, is given by

$$\begin{aligned}\beta_2 - \beta_1 &= (10\text{ dB})\left(\log\frac{I_2}{I_0} - \log\frac{I_1}{I_0}\right)\\ &= (10\text{ dB})[(\log I_2 - \log I_0) - (\log I_1 - \log I_0)]\\ &= (10\text{ dB})\log\frac{I_2}{I_1}\end{aligned}$$

Now we use the reciprocal of Eq. (15.26); $I_2/I_1 = r_1^2/r_2^2$, so

$$\begin{aligned}\beta_2 - \beta_1 &= (10\text{ dB})\log\frac{r_1^2}{r_2^2} = (10\text{ dB})\log\frac{r_1^2}{(2r_1)^2}\\ &= (10\text{ dB})\log\frac{1}{4} = -6.0\text{ dB}\end{aligned}$$

A decrease in intensity of a factor of 4 corresponds to a 6-dB decrease in sound intensity level.

EVALUATE: Our result is negative, which tells us (correctly) that the sound intensity level is less at point 2 than at point 1. The 6-dB difference doesn't depend on the value of the sound intensity level at point 1. If point 1 is relatively close to the bird so that $\beta_1 = 56$ dB, then at a point twice as far away, $\beta_2 = 50$ dB; if point 1 is more distant from the bird so that $\beta_1 = 28$ dB, then at a point twice as far away, $\beta_2 = 22$ dB.

It's interesting to note that the perceived *loudness* of a sound is not directly proportional to its intensity. As an example, most people usually interpret an increase of 8 to 10 dB in sound intensity level (corresponding to an intensity that increases by a factor of 6 to 10) as a doubling of loudness.

16.11 When you double your distance from a point source of sound, by how much does the sound intensity level decrease?

Test Your Understanding

You double the intensity of a sound wave in air while leaving the frequency unchanged. (The pressure, density, and temperature of the air remain unchanged as well.) What effect does this have on the displacement amplitude, pressure amplitude, bulk modulus, sound speed, and sound intensity level?

16.4 | Standing Sound Waves and Normal Modes

When longitudinal (sound) waves propagate in a fluid in a pipe with finite length, the waves are reflected from the ends in the same way that transverse waves on a string are reflected at its ends. The superposition of the waves traveling in opposite directions again forms a standing wave. Just as for transverse standing waves on a string (Section 15.7), standing sound waves (normal modes) in a pipe can be used to create sound waves in the surrounding air. This is the operating principle of the human voice as well as many musical instruments, including woodwinds, brasses, and pipe organs.

Transverse waves on a string, including standing waves, are usually described only in terms of the displacement of the string. But, as we have seen, sound waves in a fluid may be described either in terms of the displacement of the fluid or in terms of the pressure variation in the fluid. To avoid confusion, we'll use the terms **displacement node** and **displacement antinode** to refer to points where particles of the fluid have zero displacement and maximum displacement, respectively.

We can demonstrate standing sound waves in a column of gas using an apparatus called Kundt's tube (Fig. 16.12). A horizontal glass tube a meter or so long is closed at one end and has a flexible diaphragm at the other end that can transmit vibrations. A nearby loudspeaker is driven by an audio oscillator and amplifier; this produces sound waves that force the diaphragm to vibrate sinusoidally with a frequency that we can vary. The sound waves within the tube are reflected at the other, closed end of the tube. We spread a small amount of light powder uniformly along the bottom of the tube. As we vary the frequency of the sound, we pass through frequencies at which the amplitude of the standing waves becomes large enough for the powder to be swept along the tube at those points where the gas is in motion. The powder therefore collects at the displacement nodes (where the gas is not moving). Adjacent nodes are separated by a distance equal to $\lambda/2$, and we can measure this distance. Given the wavelength, we can use this experiment to determine the wave speed: We read the frequency f from the

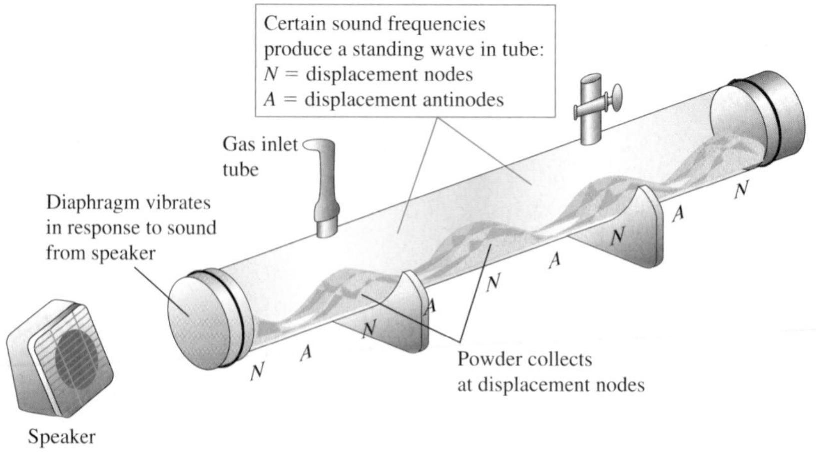

16.12 Demonstrating standing sound waves using a Kundt's tube. The blue shading represents the density of the gas at an instant when the gas pressure at the displacement nodes is a maximum or a minimum.

oscillator dial, and we can then calculate the speed v of the waves from the relation $v = \lambda f$.

Figure 16.13 will help you to visualize standing sound waves; it is analogous to Fig. 16.4 for longitudinal traveling waves. Again tape two index cards together edge-to-edge, with a gap of a millimeter or two forming a thin slit. Place the card over the diagram with the slit horizontal and move it vertically with constant velocity. The portions of the sine curves that appear in the slit correspond to the oscillations of the particles in a longitudinal standing wave. Each particle moves with longitudinal simple harmonic motion about its equilibrium position.

Figure 16.14 is an enlarged version of a portion of Fig. 16.13 centered on a displacement node. Note that particles on opposite sides of a displacement node vibrate in opposite phase. When these particles approach each other, the gas between them is compressed and the pressure rises; when they recede from each other, there is an expansion and the pressure drops. Hence at a displacement *node* the gas undergoes the maximum amount of compression and expansion, and the variations in pressure and density above and below the average have their maximum value. By contrast, particles on opposite sides of a displacement *antinode* vibrate *in phase;* the distance between the particles is nearly constant, and there is *no* variation in pressure or density at a displacement antinode.

We use the term **pressure node** to describe a point in a standing sound wave at which the pressure and density do not vary and the term **pressure antinode** to describe a point at which the variations in pressure and density are greatest. Using these terms, we can summarize our observations about standing sound waves as follows: **a pressure node is always a displacement antinode, and a pressure antinode is always a displacement node.** Figure 16.12 depicts a standing sound wave at an instant at which the pressure variations are greatest; the blue shading shows that the density and pressure of the gas have their maximum and minimum values at the displacement nodes (labeled N).

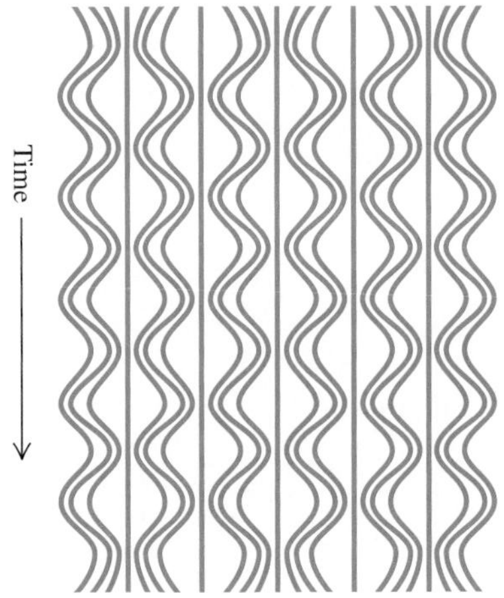

16.13 Diagram for illustrating longitudinal standing waves. The straight vertical lines mark the locations of displacement nodes and pressure antinodes.

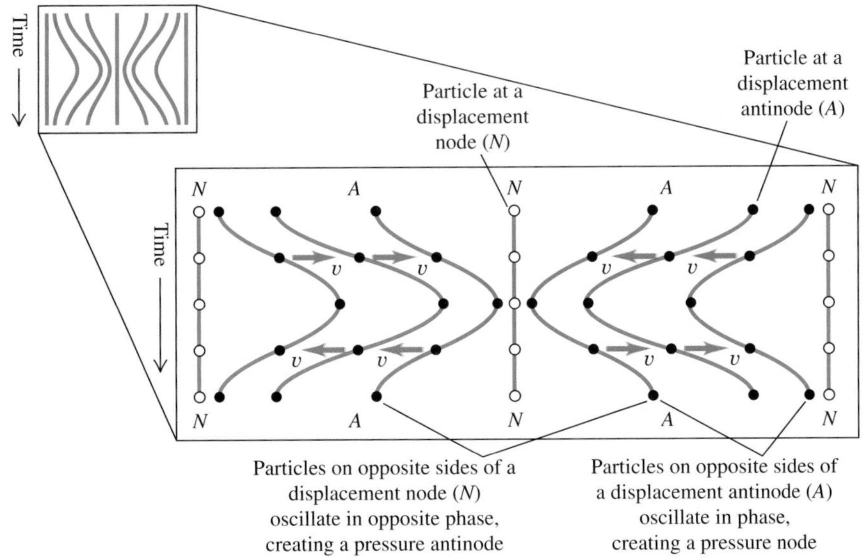

16.14 In a standing longitudinal wave, a displacement node N is a pressure antinode (a point where the pressure fluctuates the most) and a displacement antinode A is a pressure node (a point where the pressure does not fluctuate at all).

When reflection takes place at a *closed* end of a pipe (an end with a rigid barrier or plug), the displacement of the particles at this end must always be zero, analogous to a fixed end of a string. Thus a closed end of a pipe is a displacement node and a pressure antinode; the particles do not move, but the pressure variations are maximum. An *open* end of a pipe is a pressure node because it is open to the atmosphere, where the pressure is constant. Because of this, an open end is always a displacement *antinode,* in analogy to a free end of a string; the particles oscillate with maximum amplitude, but the pressure does not vary. (Strictly speaking, the pressure node actually occurs somewhat beyond an open end of a pipe. But if the diameter of the pipe is small in comparison to the wavelength, which is true for most musical instruments, this effect can safely be neglected.) Thus longitudinal waves in a column of fluid are reflected at the closed and open ends of a pipe in the same way that transverse waves in a string are reflected at fixed and free ends, respectively.

Conceptual Example 16.12 The sound of silence

A directional loudspeaker aims a sound wave of wavelength λ at a wall (Fig. 16.15). At what distances from the wall could you stand and hear no sound at all?

SOLUTION

Your ear detects *pressure* variations in the air; increases or decreases in the pressure outside your eardrum cause it to move slightly in or out, a motion that generates an electrical signal that is sent to the brain. (If you've ever had trouble getting your ears to "pop" on a drive up into the mountains or on an airline flight, you're familiar with just how sensitive your ears are to pressure changes.) Hence you will hear no sound if your ear is at a pressure node, which is a displacement antinode. The wall is a displacement node; the distance from a node to an adjacent antinode is $\lambda/4$, and the distance from one antinode to the next is $\lambda/2$ (Fig. 16.15). Hence the distances d from the wall at which no sound will be heard are

$d = \lambda/4$
(first displacement antinode and pressure node.)

$d = \lambda/4 + \lambda/2 = 3\lambda/4$
(second displacement antiode and pressure node)

$d = 3\lambda/4 + \lambda/2 = 5\lambda/4$
(third displacement antinode and pressure node)

and so on. If the loudspeaker is not highly directional, this effect is hard to notice because of multiple reflections of sound waves from the floor, ceiling, and other points on the walls.

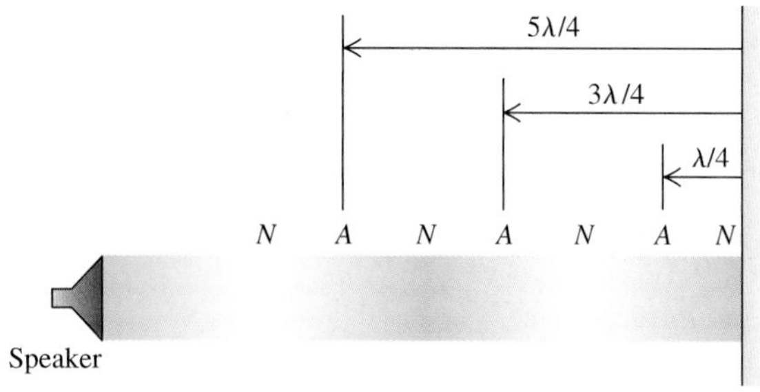

16.15 When a sound wave is directed at a wall, it interferes with the reflected wave to create a standing wave. The N's and A's are *displacement* nodes and antinodes.

Organ Pipes and Wind Instruments

The most important application of standing sound waves is the production of musical tones by wind instruments. Organ pipes are one of the simplest examples (Fig. 16.16). Air is supplied by a blower, at a gauge pressure typically of the order of 10^3 Pa (10^{-2} atm), to the bottom end of the pipe (Fig. 16.17). A stream of air emerges from the narrow opening at the edge of the horizontal surface and is directed against the top edge of the opening, which is called the *mouth* of the pipe. The column of air in the pipe is set into vibration, and there is a series of possible normal modes, just as with the stretched string. The mouth always acts as an open end; thus it is a pressure node and a displacement antinode. The other end of the pipe (at the top in Fig. 16.17) may be either open or closed.

In Fig. 16.18, both ends of the pipe are open, so both ends are pressure nodes and displacement antinodes. An organ pipe that is open at both ends is called an *open pipe*. The fundamental frequency f_1 corresponds to a standing-wave pattern with a displacement antinode at each end and a displacement node in the middle (Fig. 16.18a). The distance between adjacent antinodes is always equal to one half-wavelength, and in this case that is equal to the length L of the pipe; $\lambda/2 = L$. The corresponding frequency, obtained from the relation $f = v/\lambda$, is

$$f_1 = \frac{v}{2L} \quad \text{(open pipe)} \tag{16.16}$$

Figures 16.18b and 16.18c show the second and third harmonics (first and second overtones); their vibration patterns have two and three displacement nodes, respectively. For these, a half-wavelength is equal to $L/2$ and $L/3$, respectively, and the frequencies are twice and three times the fundamental, respectively. That is, $f_2 = 2f_1$ and $f_3 = 3f_1$. For *every* normal mode of an open pipe the length L

16.16 Organ pipes of different sizes produce tones with different frequencies.

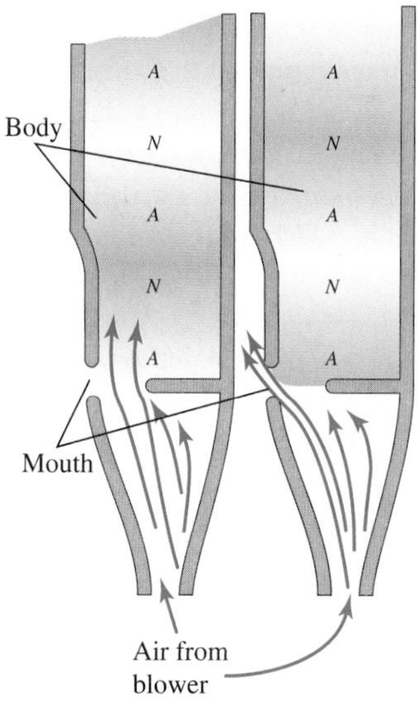

16.17 Cross sections of an organ pipe at two instants one half-period apart. Vibrations from the turbulent air flow set up standing waves in the pipe. The N's and A's are *displacement* nodes and antinodes; as the blue shading shows, these are points of maximum pressure variation and zero pressure variation, respectively.

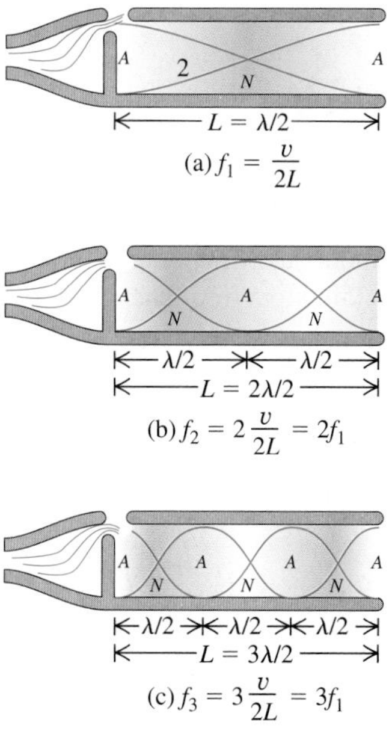

16.18 A cross section of an open pipe showing the first three normal modes. The shading indicates the pressure variations. The red curves are graphs of the displacement along the pipe axis at two instants separated in time by one half-period. The N's and A's are the *displacement* nodes and antinodes; interchange these to show the *pressure* nodes and antinodes.

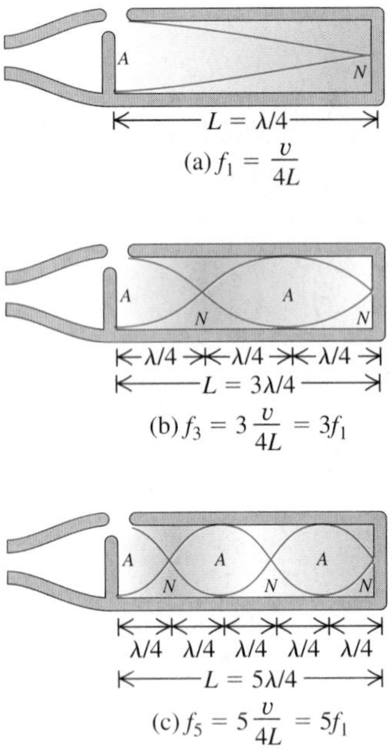

16.19 A cross section of a stopped pipe showing the first three normal modes as well as the *displacement* nodes and antinodes. Only odd harmonics are possible.

must be an integer number of half-wavelengths, and the possible wavelengths λ_n are given by

$$L = n\frac{\lambda_n}{2} \quad \text{or} \quad \lambda_n = \frac{2L}{n} \qquad (n = 1, 2, 3, \dots) \qquad \text{(open pipe)} \qquad (16.17)$$

The corresponding frequencies f_n are given by $f_n = v/\lambda_n$, so all the normal-mode frequencies for a pipe that is open at both ends are given by

$$f_n = \frac{nv}{2L} \qquad (n = 1, 2, 3, \dots) \qquad \text{(open pipe)} \qquad (16.18)$$

The value $n = 1$ corresponds to the fundamental frequency, $n = 2$ to the second harmonic (or first overtone), and so on. Alternatively, we can say

$$f_n = nf_1 \qquad (n = 1, 2, 3, \dots) \qquad \text{(open pipe)} \qquad (16.19)$$

with f_1 given by Eq. (16.16).

Figure 16.19 shows a pipe that is open at the left end but closed at the right end. This is called a *stopped pipe.* The left (open) end is a displacement antinode (pressure node), but the right (closed) end is a displacement node (pressure antinode). The distance between a node and the adjacent antinode is always one quarter-wavelength. Figure 16.19a shows the lowest-frequency mode; the length of the pipe is a quarter-wavelength $(L = \lambda_1/4)$. The fundamental frequency is $f_1 = v/\lambda_1$, or

$$f_1 = \frac{v}{4L} \qquad \text{(stopped pipe)} \qquad (16.20)$$

This is one-half the fundamental frequency for an *open* pipe of the same length. In musical language, the *pitch* of a closed pipe is one octave lower (a factor of 2 in frequency) than that of an open pipe of the same length. Figure 16.19b shows the next mode, for which the length of the pipe is *three-quarters* of a wavelength, corresponding to a frequency $3f_1$. For Fig. 16.19c, $L = 5\lambda/4$ and the frequency is $5f_1$. The possible wavelengths are given by

$$L = n\frac{\lambda_n}{4} \quad \text{or} \quad \lambda_n = \frac{4L}{n} \qquad (n = 1, 3, 5, \dots) \qquad \text{(stopped pipe)} \quad (16.21)$$

The normal-mode frequencies are given by $f_n = v/\lambda_n$, or

$$f_n = \frac{nv}{4L} \qquad (n = 1, 3, 5, \dots) \qquad \text{(stopped pipe)} \qquad (16.22)$$

or

$$f_n = nf_1 \qquad (n = 1, 3, 5, \dots) \qquad \text{(stopped pipe)} \qquad (16.23)$$

with f_1 given by Eq. (16.20). We see that the second, fourth, and all *even* harmonics are missing. In a pipe that is closed at one end, the fundamental frequency is $f_1 = v/4L$, and only the odd harmonics in the series $(3f_1, 5f_1, \dots)$ are possible.

A final possibility is a pipe that is closed at *both* ends, with displacement nodes and pressure antinodes at both ends. This wouldn't be of much use as a musical instrument because there would be no way for the vibrations to get out of the pipe.

Example 16.13 A tale of two pipes

On a day when the speed of sound is 345 m/s, the fundamental frequency of a stopped organ pipe is 220 Hz. a) How long is this stopped pipe? b) The second *overtone* of this pipe has the same wavelength as the third *harmonic* of an open pipe. How long is the open pipe?

SOLUTION

IDENTIFY and SET UP: Since this is a stopped pipe (open at one end and closed at the other), the normal-mode frequencies are given by Eq. (16.22). We use this to determine the length L from the frequency in part (a). In part (b), we must make a comparison to an open pipe, for which the frequencies are given by Eq. (16.18).

EXECUTE: a) For a stopped pipe, $f_1 = v/4L$, so the length of the stopped pipe is

$$L_{\text{stopped}} = \frac{v}{4f_1} = \frac{345\ \text{m/s}}{4(220\ \text{s}^{-1})} = 0.392\ \text{m}$$

b) The frequency of the first overtone of a stopped pipe is $f_3 = 3f_1$, and the frequency of the second overtone is $f_5 = 5f_1$:

$$f_5 = 5f_1 = 5(220\ \text{Hz}) = 1100\ \text{Hz}$$

If the wavelengths are the same, the frequencies are the same, so the frequency of the third harmonic of the open pipe is also 1100 Hz. The third harmonic of an open pipe is at $3f_1 = 3(v/2L)$. If this equals 1100 Hz, then

$$1100\ \text{Hz} = 3\left(\frac{345\ \text{m/s}}{2L_{\text{open}}}\right) \quad \text{and} \quad L_{\text{open}} = 0.470\ \text{m}$$

EVALUATE: The stopped pipe is 0.392 m long and has a fundamental frequency of 220 Hz; the open pipe is longer, 0.470 m, but has a higher fundamental frequency of $(1100\ \text{Hz})/3 = 367\ \text{Hz}$. If this seems like a contradiction, you should again compare Figs. 16.18a and 16.19a.

In an organ pipe in actual use, several modes are always present at once; the motion of the air is a superposition of these modes. This situation is analogous to a string that is struck or plucked, as in Fig. 15.25. Just as for a vibrating string, a complex standing wave in the pipe produces a traveling sound wave in the surrounding air with a harmonic content similar to that of the standing wave. A very narrow pipe produces a sound wave rich in higher harmonics, which we hear as a thin and "stringy" tone; a fatter pipe produces mostly the fundamental mode, heard as a softer, more flute-like tone. The harmonic content also depends on the shape of the pipe's mouth.

We have talked about organ pipes, but this discussion is also applicable to other wind instruments. The flute and the recorder are directly analogous. The most significant difference is that those instruments have holes along the pipe. Opening and closing the holes with the fingers changes the effective length L of the air column and thus changes the pitch. Any individual organ pipe, by comparison, can play only a single note. The flute and recorder behave as *open* pipes, while the clarinet acts as a *stopped* pipe (closed at the reed end, open at the bell).

Equations (16.18) and (16.22) show that the frequencies of any wind instrument are proportional to the speed of sound v in the air column inside the instrument. As Eq. (16.10) shows, v depends on temperature; it increases when temperature increases. Thus the pitch of all wind instruments rises with increasing temperature. An organ that has some of its pipes at one temperature and others at a different temperature is bound to sound out of tune.

Test Your Understanding

If you connect a hose to one end of a metal pipe and blow compressed air into it, the pipe produces a musical tone. If instead you blow compressed helium into the pipe at the same pressure and temperature, the pipe produces a higher tone. Why?

16.5 | Resonance

Many mechanical systems have normal modes of oscillation. As we have seen, these include columns of air (as in an organ pipe) and stretched strings (as in a guitar; see Section 15.8). In each mode, every particle of the system oscillates with simple harmonic motion at the same frequency as the mode. Air columns and stretched strings have an infinite series of normal modes, but the basic concept is closely related to the simple harmonic oscillator, discussed in Chapter 13, which has only a single normal mode (that is, only one frequency at which it oscillates after being disturbed).

Suppose we apply a periodically varying force to a system that can oscillate. The system is then forced to oscillate with a frequency equal to the frequency of the applied force (called the *driving frequency*). This motion is called a *forced oscillation.* We talked about forced oscillations of the harmonic oscillator in Section 13.8, and we suggest that you review that discussion. In particular, we described the phenomenon of mechanical **resonance.** A simple example of resonance is pushing Cousin Throckmorton on a swing. The swing is a pendulum; it has only a single normal mode, with a frequency determined by its length. If we push the swing periodically with this frequency, we can build up the amplitude of the motion. But if we push with a very different frequency, the swing hardly moves at all.

Resonance also occurs when a periodically varying force is applied to a system with many normal modes. An example is shown in Fig. 16.20a. An open organ pipe is placed next to a loudspeaker that is driven by an amplifier and emits pure sinusoidal sound waves of frequency f, which can be varied by adjusting the amplifier. The air in the pipe is forced to vibrate with the same frequency f as the *driving force* provided by the loudspeaker. In general the amplitude of this motion is relatively small, and the air inside the pipe will not move in any of the normal-mode patterns shown in Fig. 16.18. But if the frequency f of the force is close to one of the normal-mode frequencies, the air in the pipe moves in the normal-mode

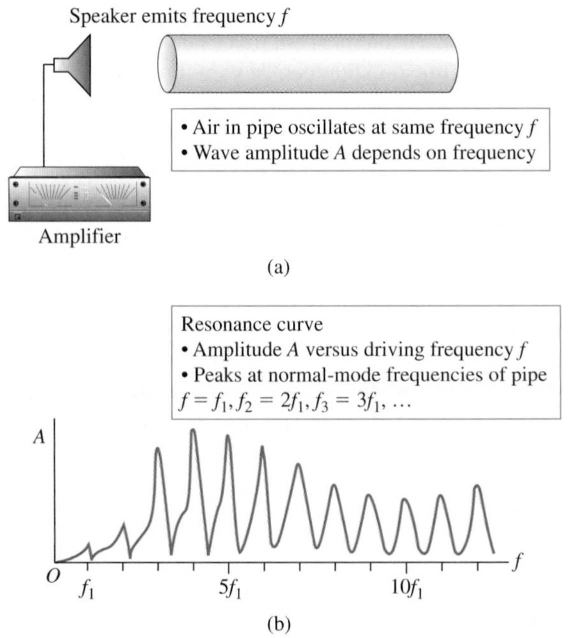

16.20 (a) The air in an open pipe is forced to oscillate at the same frequency as the sinusoidal sound waves coming from the loudspeaker. (b) The resonance curve of the open pipe graphs the amplitude of the standing sound wave in the pipe as a function of the driving frequency.

pattern for that frequency, and the amplitude can become quite large. Figure 16.20b shows the amplitude of oscillation of the air in the pipe as a function of the driving frequency f. The shape of this graph is called the **resonance curve** of the pipe; it has peaks where f equals the normal-mode frequencies of the pipe. The detailed shape of the resonance curve depends on the geometry of the pipe.

If the frequency of the force is precisely *equal* to a normal-mode frequency, the system is in resonance, and the amplitude of the forced oscillation is maximum. If there were no friction or other energy-dissipating mechanism, a driving force at a normal-mode frequency would continue to add energy to the system, and the amplitude would increase indefinitely. In such an idealized case the peaks in the resonance curve of Fig. 16.20b would be infinitely high. But in any real system there is always some dissipation of energy, or damping, as we discussed in Section 13.8; the amplitude of oscillation in resonance may be large, but it cannot be infinite.

The "sound of the ocean" you hear when you put your ear next to a large seashell is due to resonance. The noise of the outside air moving past the seashell is a mixture of sound waves of almost all audible frequencies, which forces the air inside the seashell to oscillate. The seashell behaves like an organ pipe, with a set of normal-mode frequencies; hence the inside air oscillates most strongly at those frequencies, producing the seashell's characteristic sound. To hear a similar phenomenon, uncap a full bottle of your favorite beverage and blow across the open top. The noise is provided by your breath blowing across the top, and the "organ pipe" is the column of air inside the bottle above the surface of the liquid. If you take a drink and repeat the experiment, you will hear a lower tone because the "pipe" is longer and the normal-mode frequencies are lower.

Resonance also occurs when a stretched string is forced to oscillate (Section 15.8). Suppose that one end of a stretched string is held fixed while the other is given a transverse sinusoidal motion with small amplitude, setting up standing waves. If the frequency of the driving mechanism is *not* equal to one of the normal-mode frequencies of the string, the amplitude at the antinodes is fairly small. However, if the frequency is equal to any one of the normal-mode frequencies, the string is in resonance, and the amplitude at the antinodes is very much larger than that at the driven end. The driven end is not precisely a node, but it lies much closer to a node than to an antinode when the string is in resonance. The photographs in Fig. 15.20 were made this way, with the left end of the string fixed and the right end oscillating vertically with small amplitude; large-amplitude standing waves resulted when the frequency of oscillation of the right end was equal to the fundamental frequency or to one of the first three overtones.

It is easy to demonstrate resonance with a piano. Push down the damper pedal (the right-hand pedal) so that the dampers are lifted and the strings are free to vibrate, and then sing a steady tone into the piano. When you stop singing, the piano seems to continue to sing the same note. The sound waves from your voice excite vibrations in the strings that have natural frequencies close to the frequencies (fundamental and harmonics) present in the note you sang.

A more spectacular example is a singer breaking a wine glass with her amplified voice. A good-quality wine glass has normal-mode frequencies that you can hear by tapping it. If the singer emits a loud note with a frequency corresponding exactly to one of these normal-mode frequencies, large-amplitude oscillations can build up and break the glass (Fig. 16.21).

Resonance is a very important concept, not only in mechanical systems but in all areas of physics. In Chapter 31 we will see examples of resonance in electric circuits.

16.21 The frequency of the sound from this trumpet exactly matches one of the normal-mode frequencies of the goblet. The resonant vibrations of the goblet have such large amplitude that the goblet tears itself apart.

Example 16.14 An organ-guitar duet

A stopped organ pipe is sounded near a guitar, causing one of the strings to vibrate with large amplitude. We vary the tension of the string until we find the maximum amplitude. The string is 80% as long as the stopped pipe. If both the pipe and the string vibrate at their fundamental frequency, calculate the ratio of the wave speed on the string to the speed of sound in air.

SOLUTION

IDENTIFY: The large response of the string is an example of resonance. It occurs because the organ pipe and the guitar string have the same fundamental frequency.

SET UP: Letting the subscripts a and s stand for the air in the pipe and the string, respectively, the condition for resonance is $f_{1a} = f_{1s}$. Equation (16.20) gives the fundamental frequency for a stopped pipe, while the fundamental frequency for a guitar string held at both ends is given by Eq. (15.32). These expressions involve the wave speed in air (v_a) and on the string (v_s) and the lengths of the pipe and string; we are given that $L_s = 0.80L_a$, and our target variable is the ratio v_s/v_a.

EXECUTE: From Eqs. (16.20) and (15.32), $f_{1a} = v_a/4L_a$ and $f_{1s} = v_s/2L_s$. Setting these equal to each other, we find

$$\frac{v_a}{4L_a} = \frac{v_s}{2L_s}$$

Substituting $L_s = 0.80\,L_a$ and rearranging, we get

$$\frac{v_s}{v_a} = 0.40$$

EVALUATE: As an example, if the speed of sound in air is 345 m/s, the wave speed on the string is $(0.40)(345\text{ m/s}) = 138\text{ m/s}$. Note that while the standing waves in the pipe and on the string have the same frequency, they have different *wavelengths* $\lambda = v/f$ because the two media have different wave speeds v. Which standing wave has the greater wavelength?

Test Your Understanding

Suppose the open pipe in Fig. 16.21 is 60.0 cm long. If the speed of sound in air is 345 m/s, what frequency emitted by the speaker produces the greatest response in the pipe?

16.6 | Interference of Waves

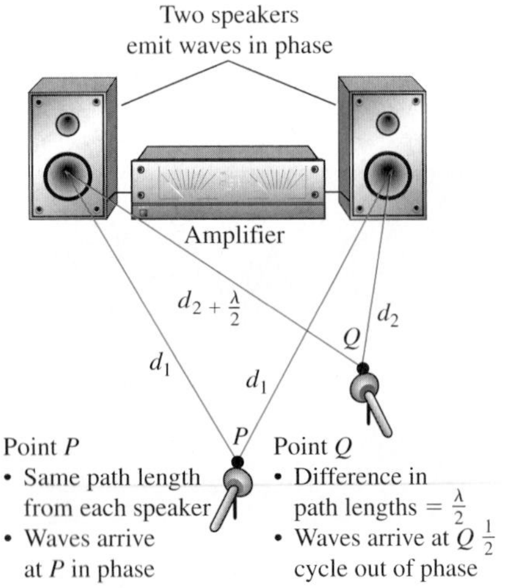

16.22 Two speakers driven by the same amplifier. Constructive interference occurs at point P and destructive interference occurs at point Q.

Wave phenomena that occur when two or more waves overlap in the same region of space are grouped under the heading *interference.* As we have seen, standing waves are a simple example of an interference effect: two waves traveling in opposite directions in a medium combine to produce a standing wave pattern with nodes and antinodes that do not move.

Figure 16.22 shows an example of another type of interference that involves waves that spread out in space. Two speakers, driven in phase by the same amplifier, emit identical sinusoidal sound waves with the same constant frequency. We place a microphone at point P in the figure, equidistant from the speakers. Wave crests emitted from the two speakers at the same time travel equal distances and arrive at point P at the same time; hence the waves arrive in phase, and there is constructive interference. The total wave amplitude at P is twice the amplitude from each individual wave, and we can measure this combined amplitude with the microphone.

Now let's move the microphone to point Q, where the distances from the two speakers to the microphone differ by a half-wavelength. Then the two waves arrive a half-cycle out of step, or *out of phase;* a positive crest from one speaker arrives at the same time as a negative crest from the other. Destructive interference takes place, and the amplitude measured by the microphone is much *smaller* than when only one speaker is present. If the amplitudes from the two speakers are equal, the two waves cancel each other out completely at point Q, and the total amplitude there is zero.

CAUTION Although this situation bears some resemblance to standing waves in a pipe, the total wave in Fig. 16.22 is a *traveling* wave, not a standing wave. To see why, recall that in a standing wave there is no net flow of energy in any direction. By contrast, in Fig. 16.22 there is an overall flow of energy from the speakers into the surrounding air; this is characteristic of a traveling wave. The interference between the waves from the two speakers simply causes the energy flow to be *channeled* into certain directions (for example, toward P) and away from other directions (for example, away from Q). You can see another difference between Fig. 16.22 and a standing wave by considering a point, such as Q, where destructive interference occurs. Such a point is *both* a displacement node *and* a pressure node because there is no wave at all at this point. Compare this to a standing wave, in which a pressure node is a displacement antinode and vice versa.

Constructive interference occurs wherever the distances traveled by the two waves differ by a whole number of wavelengths, $0, \lambda, 2\lambda, 3\lambda, \ldots$; in all these cases the waves arrive at the microphone in phase (Fig. 16.23a). If the distances from the two speakers to the microphone differ by any half-integer number of wavelengths, $\lambda/2, 3\lambda/2, 5\lambda/2, \ldots$, the waves arrive at the microphone out of phase and there will be destructive interference (Fig. 16.23b). In this case, little or no sound energy flows toward the microphone directly in front of the speakers. The energy is instead directed to the sides, where constructive interference occurs.

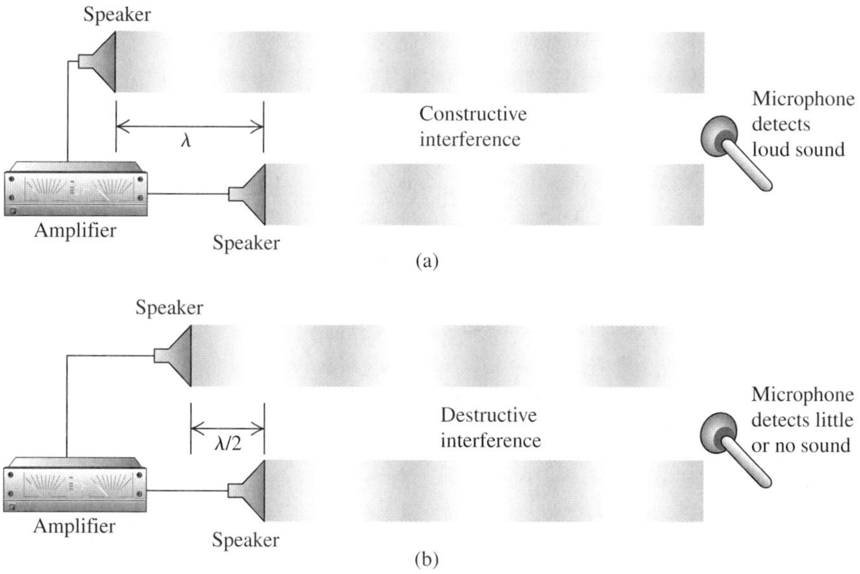

16.23 Two speakers driven by the same amplifier, emitting waves in phase. Only the waves directed toward the microphone are shown, and they are separated for clarity. (a) Constructive interference occurs when the path difference is $0, \lambda, 2\lambda, 3\lambda, \ldots$. (b) Destructive interference occurs when the path difference is $\lambda/2, 3\lambda/2, 5\lambda/2, \ldots$

Example 16.15 Loudspeaker interference

Two small loudspeakers, A and B (Fig. 16.24), are driven by the same amplifier and emit pure sinusoidal waves in phase. If the speed of sound is 350 m/s, a) for what frequencies does constructive interference occur at point P? b) for what frequencies does destructive interference occur at point P?

SOLUTION

IDENTIFY: The nature of the interference at P depends on the difference in path lengths from points A and B to P and how this difference compares to the wavelength.

SET UP: We calculate the path lengths from A to P and from B to P using the Pythagorean theorem. Constructive interference occurs when the difference in path lengths equals a whole number of wavelengths, while destructive interference occurs when the path length difference is a half-integer number of wavelengths. To find the corresponding frequencies, we use the relationship $v = f\lambda$.

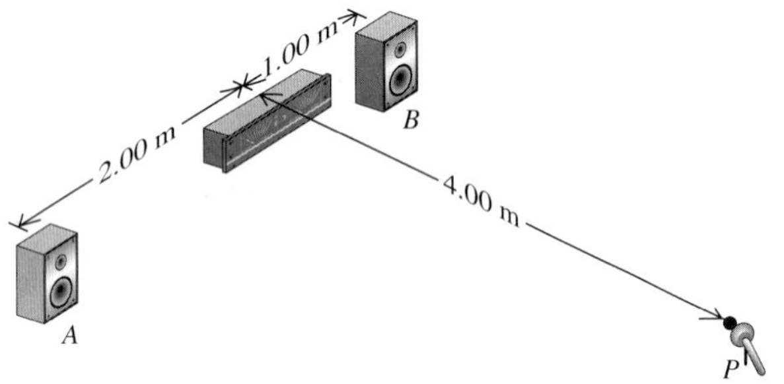

16.24 What sort of interference occurs at P?

EXECUTE: The distance from speaker A to point P is $[(2.00\text{ m})^2 + (4.00\text{ m})^2]^{1/2} = 4.47$ m, and the distance from speaker B to point P is $[(1.00\text{ m})^2 + (4.00\text{ m})^2]^{1/2} = 4.12$ m. The path difference is $d = 4.47\text{ m} - 4.12\text{ m} = 0.35$ m. a) Constructive interference occurs when the path difference is $d = 0, \lambda, 2\lambda, \ldots$ or $d = 0, v/f, 2v/f, \ldots = nv/f$. So the possible frequencies are

$$f_n = \frac{nv}{d} = n\frac{350\text{ m/s}}{0.35\text{ m}} \qquad (n = 1, 2, 3, \ldots)$$
$$= 1000\text{ Hz}, 2000\text{ Hz}, 3000\text{ Hz}, \ldots$$

b) Destructive interference occurs when the path difference is $d = \lambda/2, 3\lambda/2, 5\lambda/2, \ldots$ or $d = v/2f, 3v/2f, 5v/2f, \ldots$. The possible frequencies are

$$f_n = \frac{nv}{2d} = n\frac{350\text{ m/s}}{2(0.35\text{ m})} \qquad (n = 1, 2, 3, \ldots)$$
$$= 500\text{ Hz}, 1500\text{ Hz}, 2500\text{ Hz}, \ldots$$

EVALUATE: As we increase the frequency, the sound at point P alternates between large and small amplitudes; the maxima and minima occur at the frequencies we have found. It can be hard to notice this effect in an ordinary room because of multiple reflections from the walls, floor, and ceiling. Such an experiment is best done either outdoors or in an anechoic chamber, which has walls that absorb almost all sound and thereby eliminate reflections.

Experiments closely analogous to the one in Example 16.15, but using light, have provided both strong evidence for the wave nature of light and a means of measuring its wavelengths. We will discuss these experiments in detail in Chapter 35.

Interference effects are used to control noise from very loud sound sources such as gas-turbine power plants or jet engine test cells. The idea is to use additional sound sources that in some regions of space interfere destructively with the unwanted sound and cancel it out. Microphones in the controlled area feed signals back to the sound sources, which are continuously adjusted for optimum cancellation of noise in the controlled area.

Test Your Understanding

Suppose the frequency from the two loudspeakers in Fig. 16.22 is 250 Hz, so that there is neither constructive nor destructive interference at P. What will the microphone detect if it is moved toward the amplifier along the line in Fig. 16.24 labeled 4.00 m?

16.7 | Beats

In Section 16.6 we talked about *interference* effects that occur when two different waves with the same frequency overlap in the same region of space. Now let's look at what happens when we have two waves with equal amplitude but slightly different frequencies. This occurs, for example, when two tuning forks with slightly different frequencies are sounded together, or when two organ pipes that are supposed to have exactly the same frequency are slightly "out of tune."

Consider a particular point in space where the two waves overlap. The displacements of the individual waves at this point are plotted as functions of time in Fig. 16.25a. The total length of the time axis represents one second, and the frequencies are 16 Hz (blue graph) and 18 Hz (red graph). Applying the principle of superposition, we add the two displacements at each instant of time to find the total displacement at that time. The result is the graph of Fig. 16.25b. At certain times the two waves are in phase; their maxima coincide and their amplitudes add. But because of their slightly different frequencies, the two waves cannot be in phase at all times. Indeed, at certain times (like $t = 0.50$ s in Fig. 16.25) the two waves are exactly *out* of phase. The two waves then cancel each other, and the total amplitude is zero.

The resultant wave in Fig. 16.25b looks like a single sinusoidal wave with a varying amplitude that goes from a maximum to zero and back. In this example the amplitude goes through two maxima and two minima in one second, so the frequency of this amplitude variation is 2 Hz. The amplitude variation causes variations of loudness called **beats,** and the frequency with which the loudness varies is called the **beat frequency.** In this example the beat frequency is the *difference* of the two frequencies. If the beat frequency is a few hertz, we hear it as a waver or pulsation in the tone.

ActivPhysics Online

10.7 Beats and Beat Frequency

We can prove that the beat frequency is *always* the difference of the two frequencies f_a and f_b. Suppose f_a is larger than f_b; the corresponding periods are T_a and T_b, with $T_a < T_b$. If the two waves start out in phase at time $t = 0$, they are again in phase when the first wave has gone through exactly one more cycle than the second. This happens at a value of t equal to T_{beat}, the *period* of the beat. Let n be the number of cycles of the first wave in time T_{beat}; then the number of cycles of the second wave in the same time is $(n - 1)$, and we have the relations

$$T_{\text{beat}} = nT_a \quad \text{and} \quad T_{\text{beat}} = (n-1)T_b$$

Eliminating n between these two equations, we find

$$T_{\text{beat}} = \frac{T_a T_b}{T_b - T_a}$$

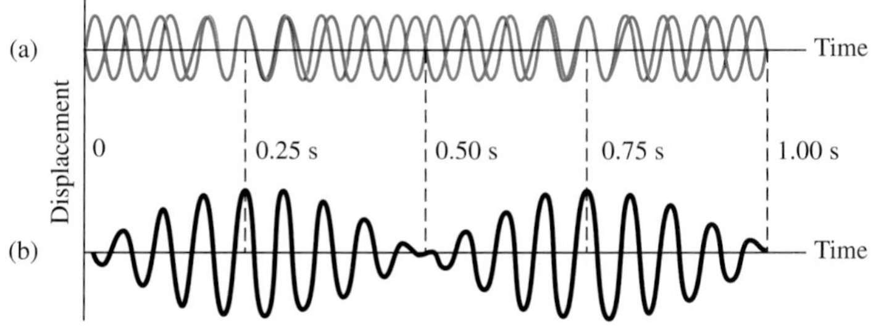

16.25 Beats are fluctuations in amplitude produced by two sound waves of slightly different frequency, 16 Hz and 18 Hz. (a) Individual waves. (b) Resultant wave formed by superposition of the two waves. The beat frequency is 18 Hz − 16 Hz = 2 Hz.

The reciprocal of the beat period is the beat *frequency*, $f_{beat} = 1/T_{beat}$, so

$$f_{beat} = \frac{T_b - T_a}{T_a T_b} = \frac{1}{T_a} - \frac{1}{T_b}$$

and finally

$$f_{beat} = f_a - f_b \quad \text{(beat frequency)} \tag{16.24}$$

As claimed, the beat frequency is the difference of the two frequencies. In using Eq. (16.24), remember that f_a is the higher frequency.

An alternative way to derive Eq. (16.24) is to write functions to describe the curves in Fig. 16.25a and then add them. Suppose that at a certain position the two waves are given by $y_a(t) = A \sin 2\pi f_a t$ and $y_b(t) = -A \sin 2\pi f_b t$. We use the trigonometric identity

$$\sin a - \sin b = 2 \sin \frac{1}{2}(a - b) \cos \frac{1}{2}(a + b)$$

We can then express the total wave $y(t) = y_a(t) + y_b(t)$ as

$$y_a(t) + y_b(t) = \left[2A \sin \frac{1}{2}(2\pi)(f_a - f_b)t \right] \cos \frac{1}{2}(2\pi)(f_a + f_b)t$$

The amplitude factor (the quantity in brackets) varies slowly with frequency $\frac{1}{2}(f_a - f_b)$. The cosine factor varies with a frequency equal to the *average* frequency $\frac{1}{2}(f_a + f_b)$. The *square* of the amplitude factor, which is proportional to the intensity that the ear hears, goes through two maxima and two minima per cycle. So the beat frequency f_{beat} that is heard is twice the quantity $\frac{1}{2}(f_a - f_b)$, or just $f_a - f_b$, in agreement with Eq. (16.24).

Beats between two tones can be heard up to a beat frequency of about 6 or 7 Hz. Two piano strings or two organ pipes differing in frequency by 2 or 3 Hz sound wavery and "out of tune," although some organ stops contain two sets of pipes deliberately tuned to beat frequencies of about 1 to 2 Hz for a gently undulating effect. Listening for beats is an important technique in tuning all musical instruments.

At frequency differences greater than about 6 or 7 Hz, we no longer hear individual beats, and the sensation merges into one of *consonance* or *dissonance*, depending on the frequency ratio of the two tones. In some cases the ear perceives a tone called a *difference tone*, with a pitch equal to the beat frequency of the two tones. For example, if you listen to a whistle that produces sounds at 1800 Hz and 1900 Hz when blown, you will hear not only these tones but also a much lower 100-Hz tone.

The engines on multiengine propeller aircraft have to be synchronized so that the sounds don't cause annoying beats, which are heard as loud throbbing sounds (Fig. 16.26). On some planes this is done electronically; on others the pilot does it by ear, just like tuning a piano.

16.26 If the two propellers on this airplane are not precisely synchronized, the pilots, passengers, and listeners on the ground will hear beats.

Test Your Understanding

One tuning fork vibrates at 440 Hz, while a second tuning fork vibrates at an unknown frequency. When both tuning forks are sounded simultaneously, you hear a tone that rises and falls in intensity three times per second. What do you conclude about the frequency of the second tuning fork?

16.8 | The Doppler Effect

You've probably noticed that when a car approaches you with its horn sounding, the pitch seems to drop as the car passes. This phenomenon, first described by the 19th-century Austrian scientist Christian Doppler, is called the **Doppler effect.** When a source of sound and a listener are in motion relative to each other, the frequency of the sound heard by the listener is not the same as the source frequency. A similar effect occurs for light and radio waves; we'll return to this later in this section.

To analyze the Doppler effect for sound, we'll work out a relation between the frequency shift and the velocities of source and listener relative to the medium (usually air) through which the sound waves propagate. To keep things simple, we consider only the special case in which the velocities of both source and listener lie along the line joining them. Let v_S and v_L be the velocity components along this line for the source and the listener, respectively, relative to the medium. We choose the positive direction for both v_S and v_L to be the direction from the listener L to the source S. The speed of sound relative to the medium, v, is always considered positive.

Activ ONLINE Physics

10.8 Doppler Effect: Conceptual Introduction

10.9 Doppler Effect: Problems

Moving Listener

Let's think first about a listener L moving with velocity v_L toward a stationary source S (Fig. 16.27). The source emits a sound wave with frequency f_S and wavelength $\lambda = v/f_S$. The figure shows several wave crests, separated by equal distances λ. The wave crests approaching the moving listener have a speed of propagation *relative to the listener* of $(v + v_L)$. So the frequency f_L with which the crests arrive at the listener's position (that is, the frequency the listener hears) is

$$f_L = \frac{v + v_L}{\lambda} = \frac{v + v_L}{v/f_S} \tag{16.25}$$

or

$$f_L = \left(\frac{v + v_L}{v}\right) f_S = \left(1 + \frac{v_L}{v}\right) f_S \tag{16.26}$$

(moving listener, stationary source)

So a listener moving toward a source ($v_L > 0$), as in Fig. 16.27, hears a higher frequency (higher pitch) than does a stationary listener. A listener moving away from the source ($v_L < 0$) hears a lower frequency (lower pitch).

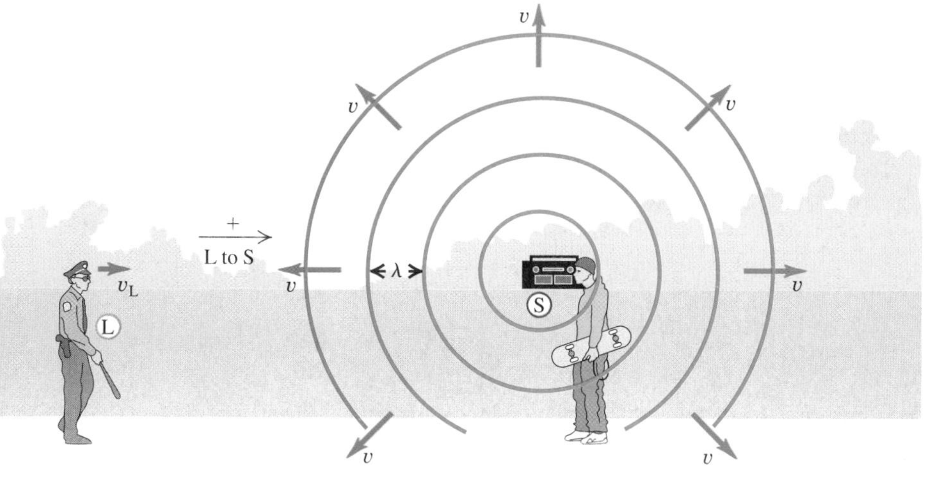

16.27 A listener moving toward a stationary source hears a frequency that is higher than the source frequency because the relative speed of listener and wave is greater than the wave speed v.

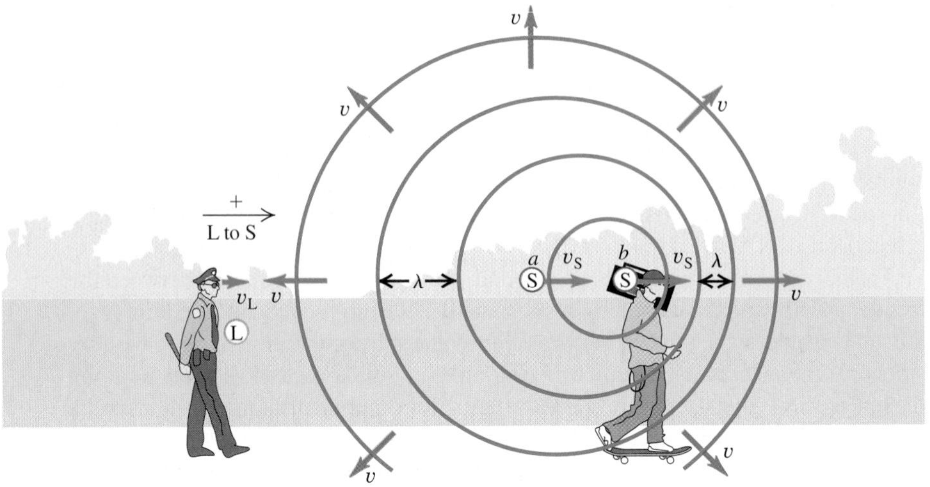

16.28 Wave crests emitted by a moving source are crowded together in front of the source (to the right of this source) and stretched out behind it (to the left of this source).

Moving Source and Moving Listener

Now suppose the source is also moving, with velocity v_S (Fig. 16.28). The wave speed relative to the wave medium (air) is still v; it is determined by the properties of the medium and is not changed by the motion of the source. But the wavelength is no longer equal to v/f_S. Here's why. The time for emission of one cycle of the wave is the period $T = 1/f_S$. During this time, the wave travels a distance $vT = v/f_S$ and the source moves a distance $v_S T = v_S/f_S$. The wavelength is the distance between successive wave crests, and this is determined by the *relative* displacement of source and wave. As Fig. 16.28 shows, this is different in front of and behind the source. In the region to the right of the source in Fig. 16.28 (that is, in front of the source), the wavelength is

$$\lambda = \frac{v}{f_S} - \frac{v_S}{f_S} = \frac{v - v_S}{f_S} \tag{16.27}$$

(wavelength in front of a moving source)

In the region to the left of the source (that is, behind the source), it is

$$\lambda = \frac{v + v_S}{f_S} \quad \text{(wavelength behind a moving source)} \tag{16.28}$$

The waves in front of and behind the source are compressed and stretched out, respectively, by the motion of the source (Fig. 16.29).

To find the frequency heard by the listener behind the source, we substitute Eq. (16.28) into the first form of Eq. (16.25):

$$f_L = \frac{v + v_L}{\lambda} = \frac{v + v_L}{(v + v_S)/f_S}$$

$$f_L = \frac{v + v_L}{v + v_S} f_S \tag{16.29}$$

(Doppler effect, moving source and moving listener)

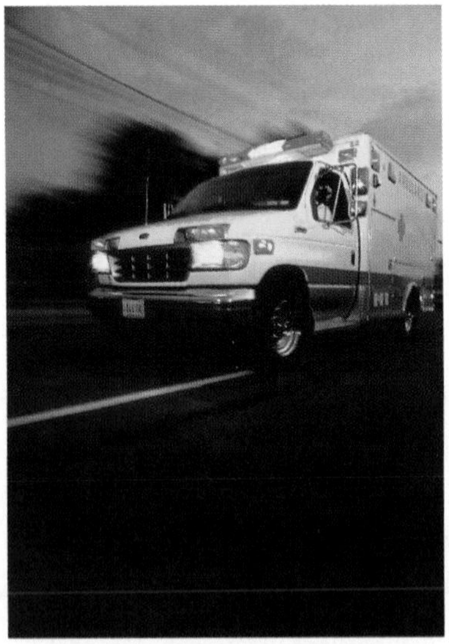

16.29 The Doppler effect explains why the siren on a fire engine or ambulance has a high pitch ($f_L > f_S$) when it is approaching you ($v_S < 0$) and a low pitch ($f_L < f_S$) when it is moving away ($v_S > 0$).

This expresses the frequency f_L heard by the listener in terms of the frequency f_S of the source.

Equation (16.29) includes all possibilities for motion of source and listener (relative to the medium) along the line joining them. If the listener happens to be at rest in the medium, v_L is zero. When both source and listener are at rest or have the same velocity relative to the medium, then $v_L = v_S$ and $f_L = f_S$. Whenever the direction of the source or listener velocity is opposite to the direction from the listener toward the source (which we have defined as positive), the corresponding velocity to be used in Eq. (16.29) is negative.

Problem-Solving Strategy: Doppler Effect

IDENTIFY *the relevant concepts:* The Doppler effect is relevant whenever the source of waves, the wave detector (listener), or both are in motion.

SET UP *the problem using the following steps:*

1. Establish a coordinate system. Define the positive direction to be the direction from the listener to the source, and make sure you know the signs of all relevant velocities. A velocity in the direction from the listener toward the source is positive; a velocity in the opposite direction is negative. Also, the velocities must all be measured relative to the air in which the sound is traveling.
2. Use consistent notation to identify the various quantities: subscript S for source, L for listener.
3. Determine which unknown quantities are your target variables.

EXECUTE *the solution:*

1. Use Eq. (16.29) to relate the frequencies at the source and the listener, the sound speed, and the velocities of the source and the listener. If the source is moving, you can find the wavelength measured by the listener using Eq. (16.27) or (16.28).
2. When a wave is reflected from a surface, either stationary or moving, the analysis can be carried out in two steps. In the first, the surface plays the role of listener; the frequency with which the wave crests arrive at the surface is f_L. Then think of the surface as a new source, emitting waves with this same frequency f_L. Finally, determine what frequency is heard by a listener detecting this new wave.

EVALUATE *your answer:* Ask whether your final result makes sense. If the source and the listener are moving toward each other, $f_L > f_S$; if they are moving apart, $f_L < f_S$. If the source and the listener have no relative motion, $f_L = f_S$.

Example 16.16 Doppler effect I: Wavelengths

A police siren emits a sinusoidal wave with frequency $f_S = 300$ Hz. The speed of sound is 340 m/s. a) Find the wavelength of the waves if the siren is at rest in the air. b) If the siren is moving at 30 m/s (108 km/h, or 67 mi/h), find the wavelengths of the waves ahead of and behind the source.

SOLUTION

IDENTIFY: The Doppler effect is not involved in part (a), since neither the source nor the listener is moving. In part (b), the source is in motion and we must invoke the Doppler effect.

SET UP: We use the relationship $v = \lambda f$ to determine the wavelength when the police siren is at rest. When it is in motion, we find the wavelength on either side of the siren using Eqs. (16.27) and (16.28).

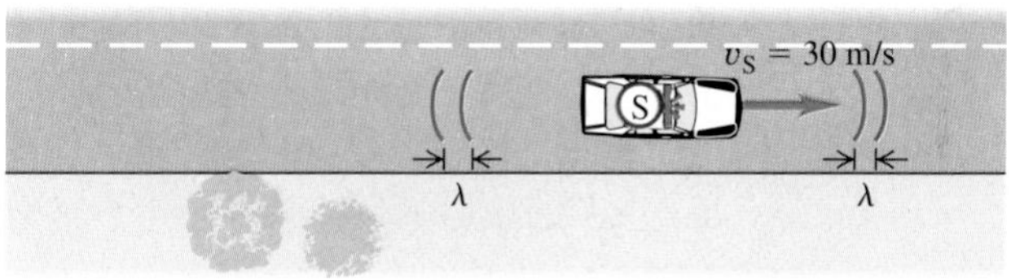

16.30 Wavelengths ahead of and behind the police siren when the siren is moving through the air at 30 m/s.

EXECUTE: a) When the source is at rest,

$$\lambda = \frac{v}{f_S} = \frac{340\text{ m/s}}{300\text{ Hz}} = 1.13\text{ m}$$

b) The situation is shown in Fig. 16.30. From Eq. (16.27), in front of the siren,

$$\lambda = \frac{v - v_S}{f_S} = \frac{340\text{ m/s} - 30\text{ m/s}}{300\text{ Hz}} = 1.03\text{ m}$$

From Eq. (16.28), behind the siren,

$$\lambda = \frac{v + v_S}{f_S} = \frac{340\text{ m/s} + 30\text{ m/s}}{300\text{ Hz}} = 1.23\text{ m}$$

EVALUATE: The wavelength is less in front of the siren and greater behind the siren, as it should be.

Example 16.17 Doppler effect II: Frequencies

If a listener L is at rest and the siren in Example 16.16 is moving away from L at 30 m/s (Fig. 16.31), what frequency does the listener hear?

SOLUTION

IDENTIFY and SET UP: Our target variable is the listener's frequency f_L. We know $f_S = 300$ Hz from Example 16.16, and we have $v_L = 0$ and $v_S = 30$ m/s. (The source velocity v_S is positive because the siren is moving in the same direction as the direction from listener to source.)

EXECUTE: From Eq. (16.29),

$$f_L = \frac{v}{v + v_S} f_S = \frac{340\text{ m/s}}{340\text{ m/s} + 30\text{ m/s}}(300\text{ Hz}) = 276\text{ Hz}$$

EVALUATE: The source and listener are moving apart, so the frequency f_L heard by the listener is less than the frequency f_S emitted by the source.

Here's an alternative approach we can use to check our result. From Example 16.16, the wavelength behind the source (which is where the listener in Fig. 16.31 is located) is 1.23 m, so

$$f_L = \frac{v}{\lambda} = \frac{340\text{ m/s}}{1.23\text{ m}} = 276\text{ Hz}$$

Even though the source is moving, the wave speed v relative to the stationary listener is unchanged.

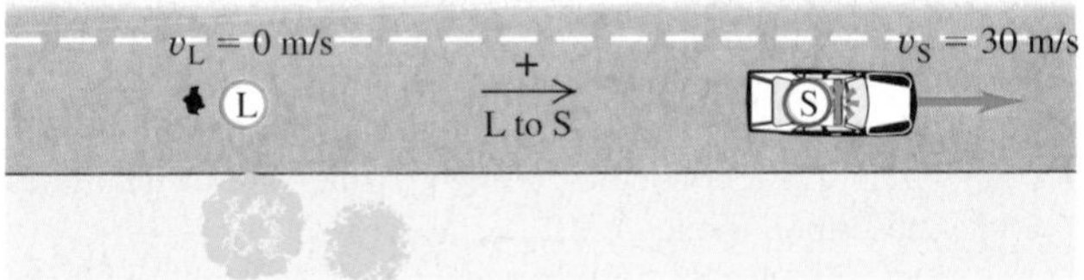

16.31 The listener is at rest, and the siren moves away from the listener at 30 m/s.

Example 16.18 Doppler effect III: A moving listener

If the siren is at rest and the listener is moving toward the left at 30 m/s (Fig. 16.32), what frequency does the listener hear?

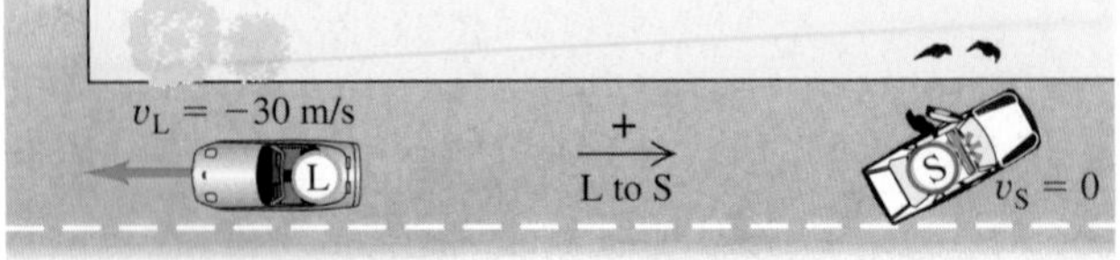

16.32 The siren is at rest, and the listener moves away from it at 30 m/s.

SOLUTION

IDENTIFY and SET UP: The key difference between this example and Example 16.17 is that the source is at rest (so $v_S = 0$) and the listener is in motion. The positive direction (from listener to source) is still from left to right, so $v_L = -30$ m/s.

EXECUTE: From Eq. (16.27),

$$f_L = \frac{v + v_L}{v} f_S = \frac{340\text{ m/s} + (-30\text{ m/s})}{340\text{ m/s}}(300\text{ Hz}) = 274\text{ Hz}$$

EVALUATE: Again the frequency heard by the listener is less than the source frequency. Note that the *relative velocity* of source and listener is the same as in the previous example, but the Doppler shift is different because the velocities relative to the *air* are different.

Example 16.19 Doppler effect IV: Moving source, moving listener

If the siren is moving away from the listener with a speed of 45 m/s relative to the air and the listener is moving toward the siren with a speed of 15 m/s relative to the air (Fig. 16.33), what frequency does the listener hear?

SOLUTION

IDENTIFY and SET UP: Now *both* the listener and the source are in motion, with $v_L = 15$ m/s and $v_S = 45$ m/s. (Both velocities are positive because both velocity vectors point in the direction from listener to source.)

EXECUTE: Once again using Eq. (16.27), we find

$$f_L = \frac{v + v_L}{v + v_S} f_S = \frac{340\text{ m/s} + 15\text{ m/s}}{340\text{ m/s} + 45\text{ m/s}}(300\text{ Hz})$$
$$= 277\text{ Hz}$$

EVALUATE: The frequency heard by the listener is again less than the source frequency, but the value is different than in the previous two examples, even though the source and listener move away from each other at 30 m/s in all three cases. The *sign* of the Doppler shift of frequency (that is, whether f_L is less than or greater than f_S) depends on how the source and the listener are moving relative to each other; to determine the *value* of the Doppler shift of frequency, you must know the velocities of source and listener relative to the air.

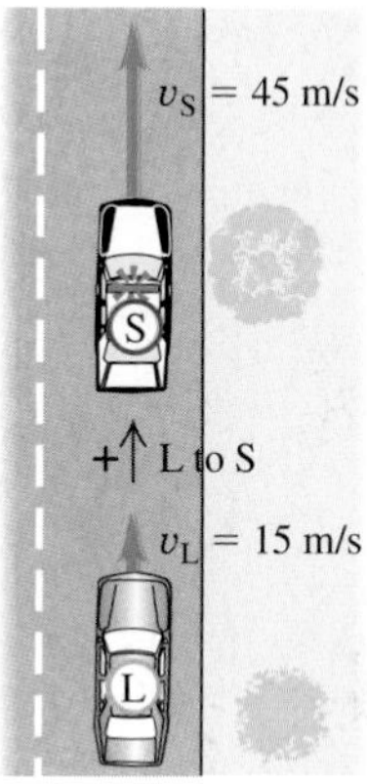

16.33 The velocities relative to the air when the siren is moving away from the listener with a relative velocity of 30 m/s.

Example 16.20 Doppler effect V: A double Doppler shift

The police car with its 300-Hz siren is moving toward a warehouse at 30 m/s, intending to crash through the door. What frequency does the driver of the police car hear reflected from the warehouse?

SOLUTION

IDENTIFY: In this situation there are *two* Doppler shifts, as shown in Fig. 16.34. In the first shift, the warehouse is the stationary "listener." The frequency of sound reaching the warehouse, which we call f_W, is greater than 300 Hz because the source is approaching. In the second shift, the warehouse acts as a source of sound with frequency f_W, and the listener is the driver of the police car; she hears a frequency greater than f_W because she is approaching the source.

SET UP: To determine f_W, we use Eq. (16.27) with f_L replaced by f_W. For this part of the problem, $v_L = v_W = 0$ (the warehouse is at rest) and $v_S = -30$ m/s (the siren is moving in the negative direction from source to listener).

To determine the frequency heard by the driver, which is our target variable, we again use Eq. (16.27) but now with f_S replaced by f_W. For this second part of the problem, $v_S = 0$ because the stationary warehouse is the source and the velocity of the listener (the driver) is $v_L = +30$ m/s. (The listener's velocity is positive because it is in the direction from listener to source.)

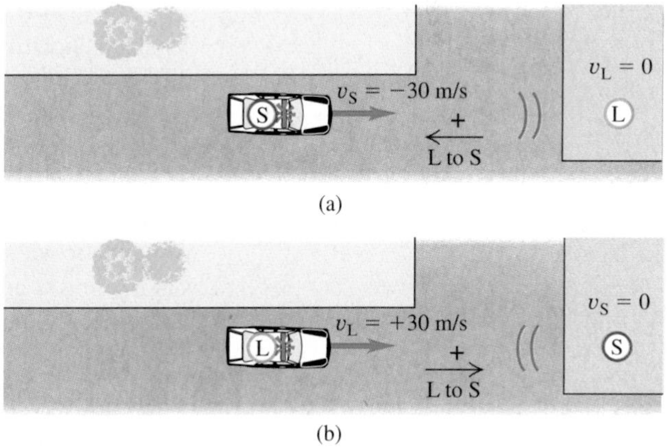

16.34 (a) The police car with siren (source) moves at 30 m/s toward the stationary warehouse (listener). (b) The sound wave is reflected from the stationary warehouse (source) toward the police car (listener).

EXECUTE: The frequency reaching the warehouse is

$$f_W = \frac{v}{v + v_S} f_S = \frac{340 \text{ m/s}}{340 \text{ m/s} + (-30 \text{ m/s})}(300 \text{ Hz}) = 329 \text{ Hz}$$

Then the frequency heard by the driver is

$$f_L = \frac{v + v_L}{v} f_W = \frac{340 \text{ m/s} + 30 \text{ m/s}}{340 \text{ m/s}}(329 \text{ Hz}) = 358 \text{ Hz}$$

EVALUATE: Because there are two Doppler shifts, the reflected sound heard by the driver has an even higher frequency than the sound heard by a stationary listener in the warehouse.

Doppler Effect for Electromagnetic Waves

In the Doppler effect for sound, the velocities v_L and v_S are always measured relative to the *air* or whatever medium we are considering. There is also a Doppler effect for *electromagnetic* waves in empty space, such as light waves or radio waves. In this case there is no medium that we can use as a reference to measure velocities, and all that matters is the *relative* velocity of source and receiver. (By contrast, the Doppler effect for sound does not depend simply on this relative velocity, as discussed in Example 16.19.)

To derive the expression for the Doppler frequency shift for light, we have to use the special theory of relativity. We will discuss this in Chapter 37, but for now we quote the result without derivation. The wave speed is the speed of light, usually denoted by c, and it is the same for both source and receiver. In the frame of reference in which the receiver is at rest, the source is moving away from the receiver with velocity v. (If the source is *approaching* the receiver, v is negative.) The source frequency is again f_S. The frequency f_R measured by the receiver R (the frequency of arrival of the waves at the receiver) is then given by

$$f_R = \sqrt{\frac{c - v}{c + v}} f_S \qquad \text{(Doppler effect for light)} \qquad (16.30)$$

When v is positive, the source is moving directly *away* from the receiver and f_R is always less than f_S; when v is negative, the source is moving directly *toward* the receiver and f_R is *greater* than f_S. The qualitative effect is the same as for sound, but the quantitative relationship is different.

A familiar application of the Doppler effect for radio waves is the radar device mounted on the side window of a police car to check other cars' speeds. The electromagnetic wave emitted by the device is reflected from a moving car, which acts as a moving source, and the wave reflected back to the device is Doppler-shifted in frequency. The transmitted and reflected signals are combined to produce beats, and the speed can be computed from the frequency of the beats. Similar techniques ("Doppler radar") are used to measure wind velocities in the atmosphere.

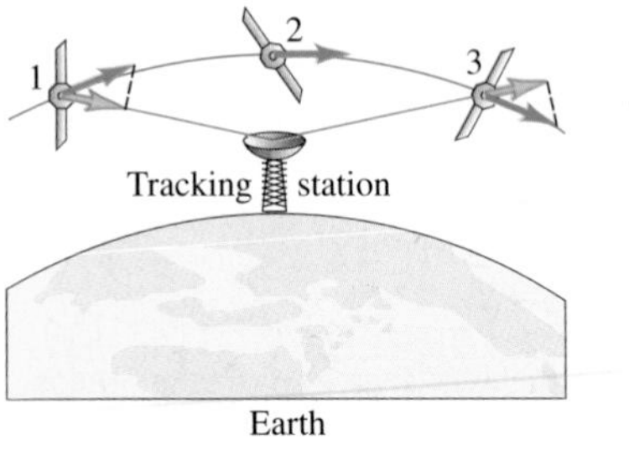

16.35 Change of velocity component along the line of sight of a satellite passing a tracking station. The frequency received at the tracking station changes from high to low as the satellite passes overhead.

The Doppler effect is also used to track satellites and other space vehicles. In Fig. 16.35 a satellite emits a radio signal with constant frequency f_S. As the satellite orbits past, it first approaches and then moves away from the receiver; the frequency f_R of the signal received on earth changes from a value greater than f_S to a value less than f_S as the satellite passes overhead.

The Doppler effect for electromagnetic waves, including visible light, is important in astronomy. Astronomers compare wavelengths of light from distant stars to those emitted by the same elements on earth. For example, in a binary star system, in which two stars orbit about their common center of mass, the light is

Doppler-shifted to higher frequencies when a star is moving toward an observer on earth and to lower frequencies when it's moving away. Measurements of the frequency shifts reveal information about the orbits and masses of the stars that comprise the binary system.

Light from most galaxies is shifted toward the longer-wavelength or red end of the visible spectrum, an effect called the *red shift.* This is often described as a Doppler shift resulting from motion of these galaxies away from us. However, from the point of view of the general theory of relativity, it is something much more fundamental: it is associated with the expansion of space itself. Distant galaxies have large red shifts because their light has been in transit for a long time and has shared in the expansion of all the space through which it moved. Extrapolating this expansion backward to over 10^{10} years ago leads to the "big bang" picture. From this point of view, the big bang was not an explosion in space but the initial rapid expansion of space itself.

Test Your Understanding

You are at an outdoor concert with a wind blowing at 10 m/s from the performers toward you. What is the value of v_S? Of v_L? Is the sound you hear Doppler-shifted? If so, is it shifted to lower or higher frequencies?

*16.9 | Shock Waves

You may have experienced "sonic booms" caused by an airplane flying overhead faster than the speed of sound. We can see qualitatively why this happens from Fig. 16.36. Let v_S denote the *speed* of the airplane relative to the air, so that it is always positive. The motion of the airplane through the air produces sound; if v_S is less than the speed of sound v, the waves in front of the airplane are crowded together with a wavelength given by Eq. (16.27):

$$\lambda = \frac{v - v_S}{f_S}$$

As the speed v_S of the airplane approaches the speed of sound v, the wavelength approaches zero and the wave crests pile up on each other (Fig. 16.36a). The airplane must exert a large force to compress the air in front of it; by Newton's third law, the air exerts an equally large force back on the airplane. Hence there is a large increase in aerodynamic drag (air resistance) as the airplane approaches the speed of sound, a phenomenon known as the "sound barrier."

When v_S is greater in magnitude than v, the source of sound is **supersonic,** and Eqs. (16.27) and (16.29) for the Doppler effect no longer describe the sound wave in front of the source. Figure 16.36b shows a cross section of what happens. As the airplane moves, it displaces the surrounding air and produces sound. A series of wave crests is emitted from the nose of the airplane; each spreads out in a circle centered at the position of the airplane when it emitted the crest. After a time t the crest emitted from point S_1 has spread to a circle with radius vt, and the airplane has moved a greater distance $v_S t$ to position S_2. You can see that the circular crests interfere constructively at points along the green line that makes an angle α with the direction of the airplane velocity, leading to a very-large-amplitude wave crest along this line. This large-amplitude crest is called a **shock wave** (Fig. 16.36c).

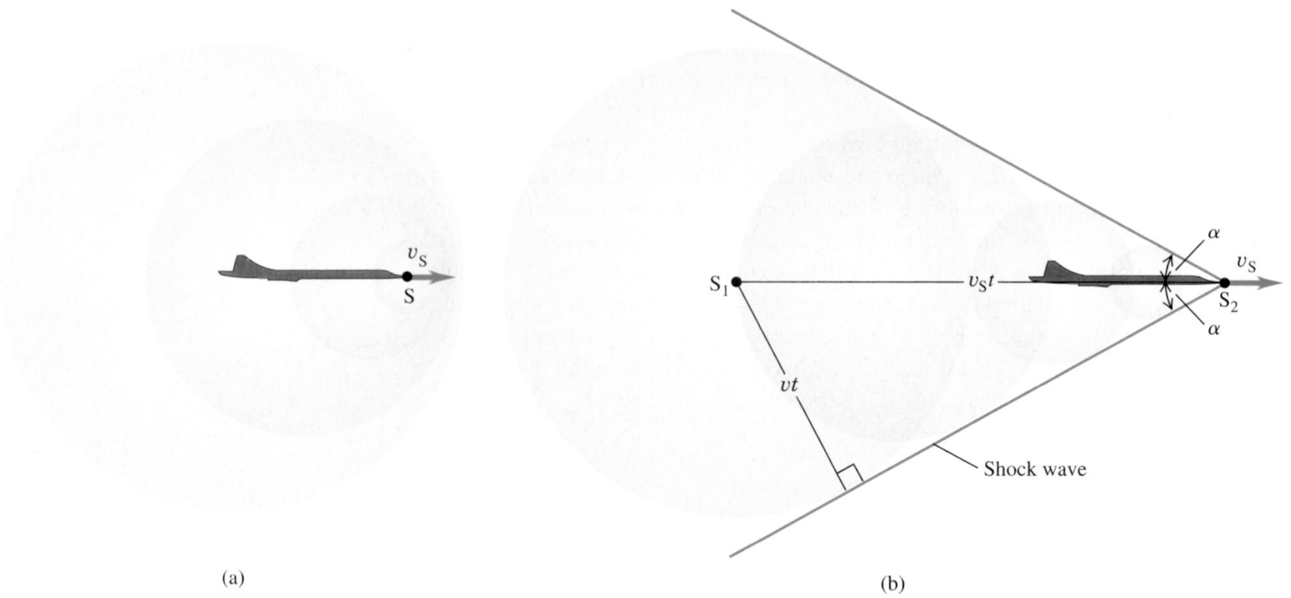

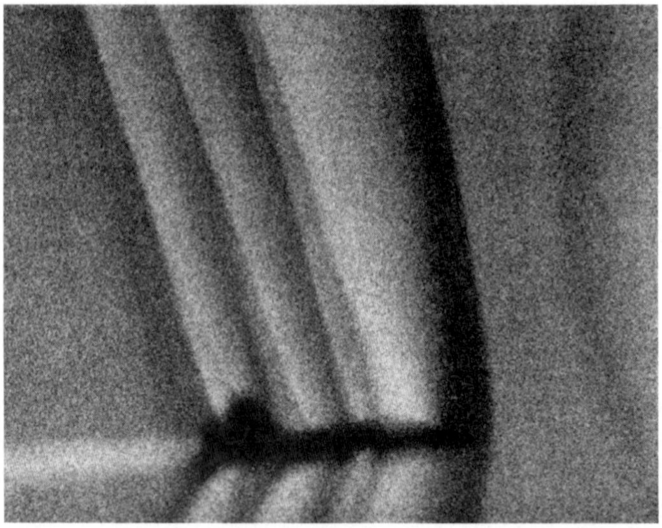

(c)

16.36 (a) As the speed of the source of sound S approaches the speed of sound, the wave crests begin to pile up in front of S. (b) A shock wave forms when the speed of the source is greater than the speed of sound. (c) Photograph of shock waves produced by a T-38 jet aircraft moving at 1.1 times the speed of sound. Separate shock waves are generated by the nose, wings, and tail. The angles of these waves vary because the air is accelerated and decelerated as it moves relative to the airplane, so the relative speed of the airplane and air is different at different points on the airplane.

16.37 The first supersonic airplane, the Bell X-1, was shaped much like a .50-caliber bullet—which was known to be able to travel faster than sound.

From the right triangle in Fig. 16.36b we can see that the angle α is given by

$$\sin\alpha = \frac{vt}{v_S t} = \frac{v}{v_S} \qquad \text{(shock wave)} \qquad (16.31)$$

In this relation, v_S is the *speed* of the source (the magnitude of its velocity) relative to the air and is always positive. The ratio v_S/v is called the **Mach number.** It is greater than unity for all supersonic speeds, and $\sin\alpha$ in Eq. (16.31) is the reciprocal of the Mach number. The first person to break the sound barrier was Capt. Chuck Yeager of the U.S. Air Force, flying the Bell X-1 at Mach 1.06 on October 14, 1947 (Fig. 16.37).

The actual situation is three-dimensional; the shock wave forms a *cone* around the direction of motion of the source. If the source (possibly a supersonic jet airplane or a rifle bullet) moves with constant velocity, the angle α is constant, and the shock-wave cone moves along with the source. It's the arrival of this shock wave that causes the sonic boom you hear after a supersonic airplane has passed by. The larger the airplane, the stronger the sonic boom; the shock wave produced at ground level by the (now retired) Concorde supersonic airliner flying at 12,000 m (40,000 ft) causes a sudden jump in air pressure of about 20 Pa. In front of the shock-wave cone, there is no sound. Inside the cone a stationary listener hears the Doppler-shifted sound of the airplane moving away.

CAUTION We emphasize that a shock wave is produced *continuously* by any object that moves through the air at supersonic speed, not only at the instant that it "breaks the sound barrier." The sound waves that combine to form the shock wave, as in Fig. 16.36b, are created by the motion of the object itself, not by any sound source that the object may carry. The cracking noises of a bullet and of the tip of a circus whip are due to their supersonic motion. A supersonic jet airplane may have very loud engines, but these do not cause the shock wave. Indeed, a space shuttle makes a very loud sonic boom when coming in for a landing; its engines are out of fuel at this point, so it is a supersonic glider.

Shock waves have applications outside of aviation. They are used to break up kidney stones and gallstones without invasive surgery, using a technique with the impressive name *extracorporeal shock-wave lithotripsy.* A shock wave produced outside the body is focused by a reflector or acoustic lens so that as much of it as possible converges on the stone. When the resulting stresses in the stone exceed its tensile strength, it breaks into small pieces and can be eliminated. This technique requires accurate determination of the location of the stone, which may be done using ultrasonic imaging techniques (see Example 16.3 in Section 16.2).

Example 16.21 Sonic boom of the Concorde

The Concorde is flying at Mach 1.75 at an altitude of 8000 m, where the speed of sound is 320 m/s. How long after the plane passes directly overhead will you hear the sonic boom?

SOLUTION

IDENTIFY: The shock wave forms a cone trailing backward from the airplane, so the problem is really asking for how much time elapses from when the Concorde flies overhead to when the shock wave reaches you.

SET UP: Figure 16.38 shows the situation just as the shock wave reaches you at point L. A time t (our target variable) has elapsed since the Concorde passed overhead, during which the airliner flying at speed v_S has traveled a distance $v_S t$. We use trigonometry to solve for t.

EXECUTE: From Eq. (16.31) the angle α of the shock cone is

$$\alpha = \arcsin\frac{1}{1.75} = 34.8°$$

The speed of the plane is the speed of sound multiplied by the Mach number:

$$v_S = (1.75)(320\ \text{m/s}) = 560\ \text{m/s}$$

From Fig. 16.38 we have

$$\tan\alpha = \frac{8000\ \text{m}}{v_S t}$$

$$t = \frac{8000\ \text{m}}{(560\ \text{m/s})(\tan 34.8°)} = 20.5\ \text{s}$$

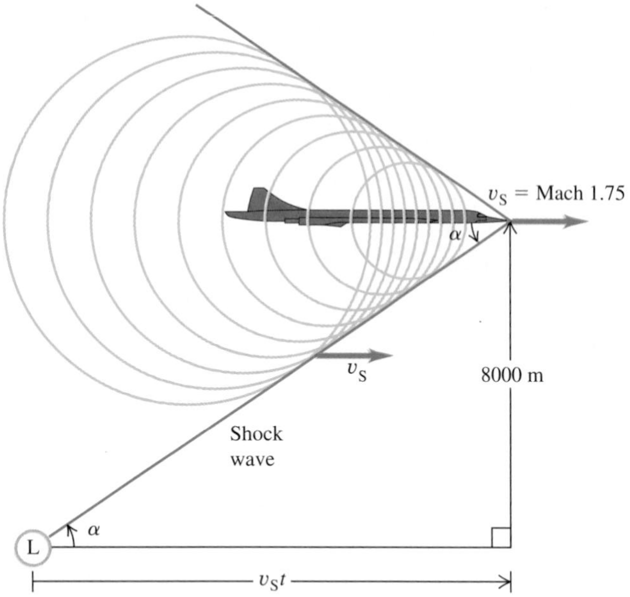

EVALUATE: You hear the boom 20.5 s after the Concorde passes overhead, and at that time it has traveled $(560\ \text{m/s})(20.5\ \text{s}) = 11.5\ \text{km}$ past the straight-overhead point.

In this calculation we assumed that the speed of sound is the same at all altitudes, so $\alpha = \arcsin v/v_S$ is a constant and the shock wave forms a perfect cone. In fact, the speed of sound decreases with increasing altitude. How would this affect the result?

16.38 You hear a sonic boom when the shock wave reaches you at L (*not* just when the plane breaks the sound barrier). A listener to the right of L has not yet heard the sonic boom but will shortly; a listener to the left of L has already heard the sonic boom and now hears the Doppler-shifted sound of the airplane.

Test Your Understanding

Large meteors have been known to produce sonic booms as they descend at supersonic speeds through the earth's atmosphere. If the shock wave from such a meteor has angle $\alpha = 4.0°$, what is the Mach number of the meteor?

SUMMARY

Sound consists of longitudinal waves in a medium. A sinusoidal sound wave is characterized by its frequency f and wavelength λ (or angular frequency ω and wave number k) and by its displacement amplitude A. The pressure amplitude p_{max} is directly proportional to the displacement amplitude; the proportionality constant is the product of the wave number and the bulk modulus B of the wave medium. (See Examples 16.1 and 16.2)

$$p_{max} = BkA \qquad (16.5)$$

(sinusoidal sound wave)

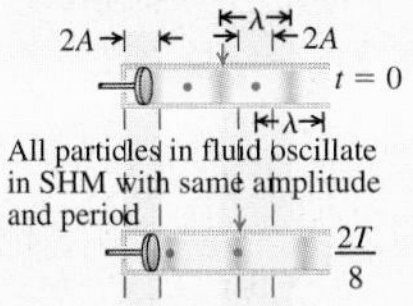

The loudness of a sound depends on its amplitude and frequency, while the pitch depends primarily on its frequency. The tone quality or timbre depends on the harmonic content and the attack and decay characteristics.

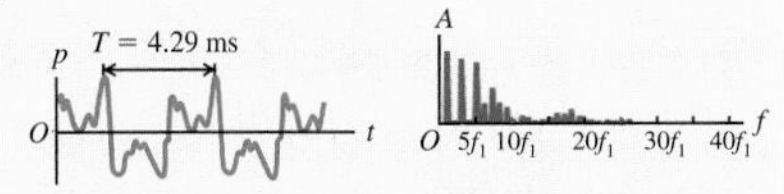

The speed of a longitudinal (sound) wave in a fluid depends on the bulk modulus B and density ρ. If the fluid is an ideal gas, the speed can be expressed in terms of the temperature T, molar mass M, and ratio of heat capacities γ of the gas. The speed of longitudinal waves in a solid rod depends on the density and Young's modulus Y. (See Examples 16.3 through 16.5)

$$v = \sqrt{\frac{B}{\rho}} \quad \text{(longitudinal wave in a fluid)} \qquad (16.7)$$

$$v = \sqrt{\frac{\gamma RT}{M}} \quad \text{(sound wave in an ideal gas)} \qquad (16.7)$$

$$v = \sqrt{\frac{Y}{\rho}} \quad \text{(longitudinal wave in a solid rod)} \qquad (16.8)$$

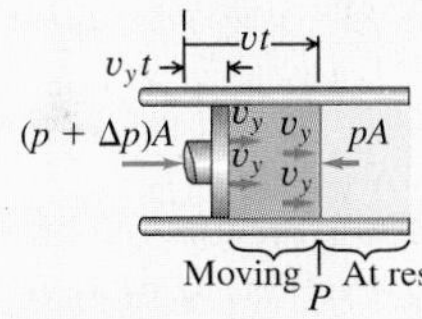

The intensity I of a sound wave is the time average rate at which energy is transported by the wave, per unit area. For a sinusoidal wave, the intensity can be expressed in terms of the displacement amplitude A or the pressure amplitude p_{max}. (See Examples 16.6 through 16.9)

$$I = \frac{1}{2}\sqrt{\rho B}\,\omega^2 A^2 = \frac{p_{max}{}^2}{2\rho v}$$

$$= \frac{p_{max}{}^2}{2\sqrt{\rho B}} \qquad (16.12), (16.14)$$

(intensity of a sinusoidal sound wave)

The sound intensity level β of a sound wave is a logarithmic measure of its intensity. It is measured relative to I_0, an arbitrary intensity defined to be $10^{-12}\ \text{W/m}^2$. Sound intensity levels are expressed in decibels (dB). (See Examples 16.10 and 16.11)

$$\beta = (10\ \text{dB}) \log \frac{I}{I_0}$$

(definition of sound intensity level) (16.15)

Standing sound waves can be set up in a pipe or tube. A closed end is a displacement node and a pressure antinode; an open end is a displacement antinode and a pressure node. For a pipe of length L open at both ends, the normal-mode frequencies are integer multimples of the sound speed divided by $2L$. For a stopped pipe (one that is open at only one end), the normal-mode frequencies are the odd multiples of the sound speed divided by $4L$. (See Examples 16.12 and 16.13)

$$f_n = \frac{nv}{2L} \quad (n = 1, 2, 3, \ldots) \qquad (16.18)$$

(open pipe)

$$f_n = \frac{nv}{4L} \quad (n = 1, 3, 5, \ldots) \qquad (16.22)$$

(stopped pipe)

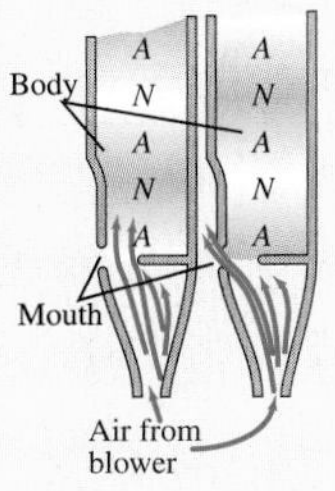

A system with normal modes of oscillation can be driven to oscillate at any frequency. A maximum response, or resonance, occurs if the driving frequency is close to one of the normal-mode frequencies of the system. (See Example 16.14)

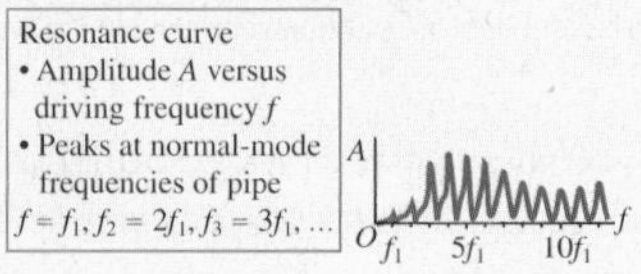

When two or more waves overlap in the same region of space, the resulting effects are called interference. The resulting amplitude can be either larger or smaller than the amplitude of each individual wave, depending on whether the waves are in phase (constructive interference) or out of phase (destructive interference). (See Example 16.15)

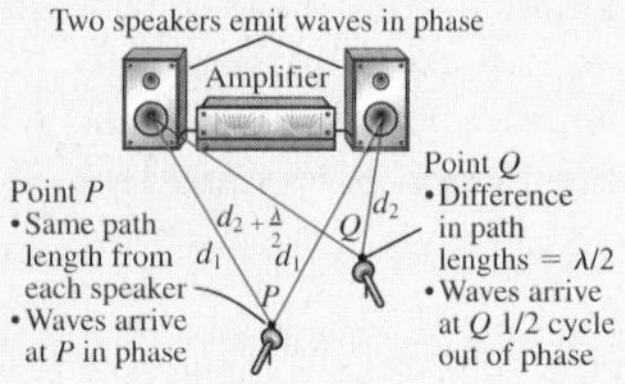

Beats are heard when two tones with slightly different frequencies f_a and f_b are sounded together. The beat frequency f_{beat} is the difference between f_a and f_b.

$$f_{beat} = f_a - f_b \qquad (16.24)$$

(beat frequency)

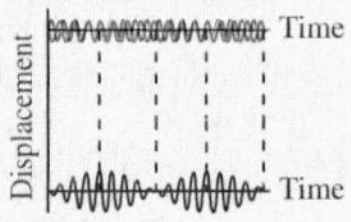

The Doppler effect for sound is the frequency shift that occurs when there is motion of a source of sound, a listener, or both, relative to the medium. The source and listener frequencies f_S and f_L are related by the source and listener velocities v_S and v_L relative to the medium and to the speed of sound v. (See Examples 16.16 through 16.20)

$$f_L = \frac{v + v_L}{v + v_S} f_S \qquad (16.29)$$

(Doppler effect, moving source and moving listener)

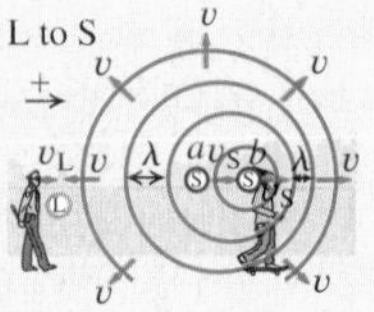

A sound source moving with a speed v_S greater than the speed of sound v creates a shock wave. The wave front is a cone with angle α. (See Example 16.21)

$$\sin \alpha = \frac{v}{v_S} \qquad (16.31)$$

(shock wave)

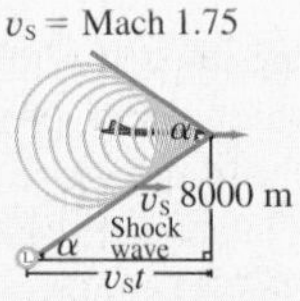

Key Terms

sound, 592
audible range, 592
ultrasonic, 592
infrasonic, 592
displacement amplitude, 592
pressure amplitude, 594
loudness, 596
pitch, 596
timbre, 596
noise, 597
sound intensity level, 605
decibel, 606
displacement node, 608
displacement antinode, 608
pressure node, 609
pressure antinode, 609
resonance, 614
resonance curve, 615
beats, 619
beat frequency, 619
Doppler effect, 621
*supersonic, 627
*shock wave, 627
*Mach number, 628

Your Notes

Answer to Chapter Opening Question

Both musical sounds and noise are made up of a combination of sinusoidal sound waves. The difference is that the frequencies of the sine waves in musical sound are all integer multiples of a fundamental frequency, while *all* frequencies are present in noise.

Answers to Test Your Understanding Questions

Section 16.1 From Eq. (16.5), the displacement amplitude is $A = p_{max}/Bk$. The pressure amplitude p_{max} and bulk modulus B remain the same, but as the frequency f increases the wave number $k = \omega/v = 2\pi f/v$ increases as well. Since A is inversely proportional to k, the displacement amplitude decreases. In other words, at higher frequency a smaller maximum displacement is required to produce the same maximum pressure fluctuation.
Section 16.2 The steel rails are long rods, so the expression to use is Eq. (16.8). Young's modulus for steel (from Table 11.1) is $Y = 20 \times 10^{10}$ Pa and the density (from Table 14.1) is $\rho = 7.8 \times 10^3\ \text{kg/m}^3$. Hence

$$v = \sqrt{\frac{Y}{\rho}} = \sqrt{\frac{20 \times 10^{10}\ \text{Pa}}{7.8 \times 10^3\ \text{kg/m}^3}} = 5.1 \times 10^3\ \text{m/s}$$

This is almost 15 times faster than the speed of sound in air, which is why the train can be heard first through the rails.
Section 16.3 Equations (16.9) and (16.10) show that the bulk modulus B and sound speed v remain the same because the physical properties of the air are unchanged. From Eqs. (16.12) and (16.14), the intensity is proportional to the square of the displacement amplitude or the square of the pressure amplitude. Hence doubling the intensity means that A and p_{max} both increase by a factor of $\sqrt{2}$. Example 16.11 shows that *multiplying* the intensity by a factor of $2(I_2/I_1 = 2)$ corresponds to *adding* to the sound intensity level by $(10\ \text{dB})\log(I_2/I_1) = (10\ \text{dB})\log 2 = 3.0\ \text{dB}$.
Section 16.4 Helium is less dense and has a lower molar mass than air, so sound travels faster in helium than in air. The normal-mode frequencies for a pipe are proportional to the sound speed v, so the frequency and hence the pitch increase when the air in the pipe is replaced with helium.
Section 16.5 The highest peak in the resonance curve of Fig. 16.23b is at $f = 4f_1$, where $f_1 = v/2L$ is the fundamental frequency of the open pipe. Hence the frequency that produces the greatest response is $4v/2L = 2v/L = 2(345\ \text{m/s})/(0.600\ \text{m}) = 1150\ \text{Hz}$.
Section 16.6 If $f = 400$ Hz, the wavelength is $\lambda = v/f = (350\ \text{m/s})/(250\ \text{Hz}) = 1.4\ \text{m}$. As the microphone is moved toward the amplifier, the path difference increases from 0.35 m at P to 1.00 m at the amplifier. At a certain point along the way, the path difference will be equal to $0.70\ \text{m} = \lambda/2$ and destructive interference will occur. So as the microphone is moved inward, the sound intensity decreases to zero at the destructive-interference point, and then increases again as the microphone moves past that point.
Section 16.7 The beat frequency is 3 Hz, so the difference between the two tuning fork frequencies is also 3 Hz. Hence the second tuning fork vibrates at a frequency of either 443 Hz or 437 Hz. You can distinguish between the two possibilities by comparing the pitches of the two tuning forks sounded one at a time: the frequency is 437 Hz if the second tuning fork has a lower pitch and 443 Hz if it has a higher pitch.
Section 16.8 The air (the medium for sound waves) is moving from the source toward the listener. Hence, relative to the air, both the source and the listener are moving in the direction from listener to source. So both velocities are positive and $v_S = v_L = +10\ \text{m/s}$. The equality of these two velocities means that the numerator and the denominator in Eq. (16.29) are the same, so $f_L = f_S$ and there is *no* Doppler shift.
Section 16.9 From Eq. (16.31), $\sin\alpha$ is equal to the ratio of the speed of sound to the speed of the source, which in turn is the reciprocal of the Mach number: $\sin\alpha = v/v_S = 1/M$. Hence $M = 1/(\sin 4.0°) = 14$; that is, the meteor is traveling at 14 times the speed of sound.

Discussion Questions

Q16.1 When sound travels from air into water, does the frequency of the wave change? The speed? The wavelength? Explain your reasoning.
Q16.2 The hero of a western movie listens for an oncoming train by putting his ear to the track. Why does this method give early warning of the approach of a train?
Q16.3 Does the pitch (or frequency) of an organ pipe increase or decrease with increasing temperature? Explain.
Q16.4 In most modern wind instruments the pitch is changed by using keys or valves to change the length of the vibrating air column. The bugle, however, has no valves or keys, yet it can play many notes. How is this possible? Are there restrictions on what notes a bugle can play?
Q16.5 Symphonic musicians always "warm up" their wind instruments by blowing them before a performance. What purpose does this serve?
Q16.6 When you inhale helium, your voice becomes high and squeaky. Why? (*Warning:* Inhaling too much helium can cause unconsciousness or death.)
Q16.7 Lane dividers on highways sometimes have regularly spaced ridges or ripples. When the tires of a moving car roll along such a divider, a musical note is produced. Why? Explain how this phenomenon could be used to measure the car's speed.
Q16.8 The tone quality of an acoustic guitar is different when the strings are plucked near the bridge (the lower end of the strings) than when they are plucked near the sound hole (close to the center of the strings). Why?
Q16.9 Which has a more direct influence on the loudness of a sound wave: the *displacement* amplitude or the *pressure* amplitude? Explain your reasoning.
Q16.10 If the pressure amplitude of a sound wave is halved, by what factor does the intensity of the wave decrease? By what factor must the pressure amplitude of a sound wave be increased in order to increase the intensity by a factor of 16? Explain.

Q16.11 Defend the following statement: "*Adding* a certain number of decibels to the sound intensity level corresponds to *multiplying* the intensity by a corresponding factor."

Q16.12 A small fraction of the energy in a sound wave is absorbed by the air through which the sound passes. How does this modify the inverse-square relationship between intensity and distance from the source? Explain your reasoning.

Q16.13 An organist in a cathedral plays a loud chord and then releases the keys. The sound persists for a few seconds and gradually dies away. Why does it persist? What happens to the sound energy when the sound dies away?

Q16.14 Two vibrating tuning forks have identical frequencies, but one is stationary and the other is mounted at the rim of a rotating platform. What does a listener hear? Explain.

Q16.15 A large church has part of the organ in the front of the church and part in the back. A person walking rapidly down the aisle while both segments are playing at once reports that the two segments sound out of tune. Why?

Q16.16 A sound source and a listener are both at rest on the earth, but a strong wind is blowing from the source toward the listener. Is there a Doppler effect? Why or why not?

Q16.17 Can you think of circumstances in which a Doppler effect would be observed for surface waves in water? For elastic waves propagating in a body of water deep below the surface? If so, describe the circumstances and explain your reasoning. If not, explain why not.

***Q16.18** Does an aircraft make a sonic boom only at the instant its speed exceeds Mach 1? Explain your reasoning.

***Q16.19** If you are riding in a supersonic aircraft, what do you hear? Explain your reasoning. In particular, do you hear a continuous sonic boom? Why or why not?

***Q16.20** The speed of sound is not the same at all altitudes, but decreases with increasing altitude. How does this affect the calculation in Example 16.21 (Section 16.9)? Is the shape of the shock wave still a cone? Explain your reasoning.

***Q16.21** A jet airplane is flying at a constant altitude at a steady speed v_S greater than the speed of sound. Describe what is being heard by observers at points A, B, and C at the instant shown in Fig. 16.39, when the shock wave has just reached point B. Explain your reasoning.

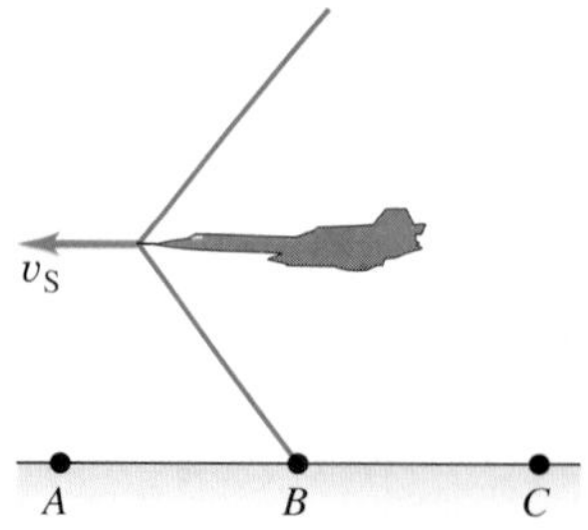

Figure 16.39 Question Q16.21.

Exercises

Unless indicated otherwise, assume the speed of sound in air to be $v = 344$ m/s.

Section 16.1 Sound Waves

16.1 Example 16.1 (Section 16.1) showed that for sound waves in air with frequency 1000 Hz, a displacement amplitude of 1.2×10^{-8} m produces a pressure amplitude of 3.0×10^{-2} Pa. a) What is the wavelength of these waves? b) For 1000-Hz waves in air, what displacement amplitude would be needed for the pressure amplitude to be at the pain threshold, which is 30 Pa? c) For what wavelength and frequency will waves with a displacement amplitude of 1.2×10^{-8} m produce a pressure amplitude of 1.5×10^{-3} Pa?

16.2 Example 16.1 (Section 16.1) showed that for sound waves in air with frequency 1000 Hz, a displacement amplitude of 1.2×10^{-8} m produces a pressure amplitude of 3.0×10^{-2} Pa. Water at 20°C has a bulk modulus of 2.2×10^{9} Pa, and the speed of sound in water at this temperature is 1480 m/s. For 1000-Hz sound waves in 20°C water, what displacement amplitude is produced if the pressure amplitude is 3.0×10^{-2} Pa? Explain why your answer is much less than 1.2×10^{-8} Pa.

16.3 Consider a sound wave in air that has displacement amplitude 0.0200 mm. Calculate the pressure amplitude for frequencies of a) 150 Hz; b) 1500 Hz; c) 15,000 Hz. In each case compare the results to the pain threshold, which is 30 Pa.

16.4 Find the pressure amplitude of the wave traveling in the air at an altitude of 11,000 m that is described in Example 16.8 (Section 16.3). How does your answer compare to the pressure amplitude of the wave at sea level that has this same displacement amplitude (Example 16.1 in Section 16.2)? What does this tell you about the variation of pressure amplitude with altitude, if the frequency and displacement amplitude of the wave is kept fixed?

Section 16.2 Speed of Sound Waves

16.5 a) In a liquid with density 1300 kg/m^3, longitudinal waves with frequency 400 Hz are found to have wavelength 8.00 m. Calculate the bulk modulus of the liquid. b) A metal bar with a length of 1.50 m has density 6400 kg/m^3. Longitudinal sound waves take 3.90×10^{-4} s to travel from one end of the bar to the other. What is Young's modulus for this metal?

16.6 A major earthquake centered on Loma Prieta, California, near San Francisco, occurred at 5:04 P.M. local time on October 17, 1989 (in UTC, Coordinated Universal Time, 0h 4m 15s on October 18, 1989). The primary seismic waves (P waves) from such an earthquake are longitudinal waves that travel through the earth's crust. P waves were detected at Caracas, Venezuela, at 0h 13m 54s UTC; at Kevo, Finland, at 0h 15m 35s UTC; and at Vienna, Austria, at 0h 17m 02s UTC. The distances the P waves traveled from Loma Prieta were 6280 km to Caracas, 8690 km to Kevo, and 9650 km to Vienna. a) Use the arrival times to calculate the average speed of the P waves that traveled to these three cities. How can you account for any differences between the average speeds? b) The average density of the earth's crust is about 3.3 g/cm^3. Use this value to calculate the bulk modulus of the earth's crust along the path traveled by the P waves to each of the three cities. How do your answers compare to the bulk moduli in Table 11.1?

16.7 A submerged scuba diver hears the sound of a boat horn directly above her on the surface of the lake. At the same time, a friend on dry land 22.0 m from the boat also hears the horn (Fig. 16.40). The horn is 1.2 m above the surface of the water. What is the distance (labeled by "?" in Fig. 16.40) from the horn to the diver? Both air and water are at 20°C.

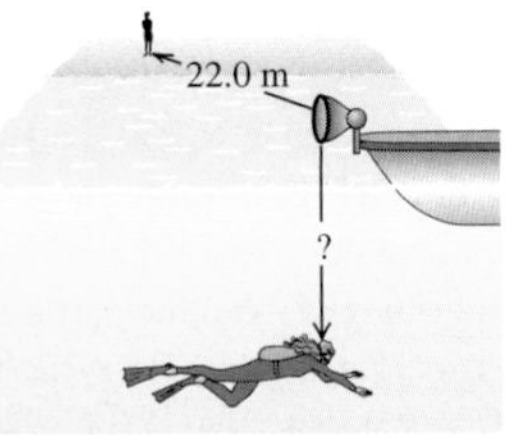

Figure 16.40 Exercise 16.7

16.8 At a temperature of 27.0°C, what is the speed of longitudinal waves in a) hydrogen (molar mass 2.02 g/mol)? b) helium (molar mass 4.00 g/mol)? c) argon (molar mass 39.9 g/mol)? See Table 19.1 for values of γ. d) Compare your answers for parts (a), (b) and (c) with the speed in air at the same temperature.

16.9 A jet airliner is cruising at high altitude at a speed of 850 km/h (about 530 mi/h). This is equal to 0.85 times the speed of sound at that altitude (also called "Mach 0.85"). a) What is the air temperature at this altitude? b) With this data, can you determine the air pressure at this altitude? Explain.

16.10 The speed of sound in air at 20°C was found to be 344 m/s. What is the change in speed for a 1.0 C° change in air temperature?

16.11 An 80.0-m-long brass rod is struck at one end. A person at the other end hears two sounds as a result of two longitudinal waves, one traveling in the metal rod and the other traveling in the air. What is the time interval between the two sounds? The speed of sound in air is 344 m/s; relevant information about brass can be found in Table 11.1 and Table 14.1.

16.12 What is the difference between the speed of longitudinal waves in air at 27.0°C and their speed at −13.0°C?

16.13 What must be the stress (F/A) in a stretched wire of a material whose Young's modulus is Y for the speed of longitudinal waves to equal 30 times the speed of transverse waves?

Section 16.3 Sound Intensity

16.14 A longitudinal wave of frequency 220 Hz travels down a copper rod of radius 8.00 mm. The average power in the wave is 6.50 μW. (See Tables 11.1 and 14.1 for the necessary data about copper.) a) Find the wavelength of the wave. b) Find the amplitude A of the wave. c) Find the maximum longitudinal velocity of a particle in the rod.

16.15 Longitudinal Waves in Different Fluids. a) A longitudinal wave propagating in a water-filled pipe has intensity 3.00×10^{-6} W/m^2 and frequency 3400 Hz. Find the amplitude A and wavelength λ of the wave. Water has density 1000 kg/m^3 and bulk modulus 2.18×10^9 Pa. b) If the pipe is filled with air at pressure 1.00×10^5 Pa and density 1.20 kg/m^3, what will be the amplitude A and wavelength λ of a longitudinal wave with the same intensity and frequency as in part (a)? c) In which fluid is the amplitude larger, water or air? What is the ratio of the two amplitudes? Why is this ratio so different from 1.00?

16.16 Derive Eq. (16.14) from the equations that precede it.

16.17 A sound wave in air at 20°C has a frequency of 150 Hz and a displacement amplitude of 5.00×10^{-3} mm. For this sound wave calculate the a) pressure amplitude (in Pa); b) intensity (in W/m^2); c) sound intensity level (in decibels).

16.18 a) What is the sound intensity level in a car when the sound intensity is 0.500 μW/m^2? b) What is the sound intensity level in the air near a jackhammer when the pressure amplitude of the sound is 0.150 Pa and the temperature is 20.0°C?

16.19 For a person with normal hearing, the faintest sound that can be heard at a frequency of 400 Hz has a pressure amplitude of about 6.0×10^{-5} Pa. Calculate the corresponding intensity and sound intensity level at 20°C.

16.20 The intensity due to a number of independent sound sources is the sum of the individual intensities. a) When four quadruplets cry simultaneously, how many decibels greater is the sound intensity level than when a single one cries? b) To increase the sound intensity level again by the same number of decibels as in part (a), how many more crying babies are required?

16.21 A baby's mouth is 30 cm from her father's ear and 1.50 m from her mother's ear. What is the difference between the sound intensity levels heard by the father and by the mother?

16.22 In October 1991, the Sacramento City Council adopted a law to reduce the allowed sound intensity level of the much despised leaf blowers from their current level of about 95 dB to 70 dB by May 1992. With the new law, what is the ratio of the new allowed intensity to the previously allowed intensity?

16.23 a) By what factor must the sound intensity be increased to increase the sound intensity level by 13.0 dB? b) Explain why you don't need to know the original sound intensity.

Section 16.4 Standing Sound Waves and Normal Modes

16.24 The fundamental frequency of an open pipe is 594 Hz. What is the fundamental frequency if one end is plugged?

16.25 Standing sound waves are produced in a pipe that is 1.20 m long. For the fundamental and first two overtones, determine the locations along the pipe (measured from the left end) of the displacement nodes and the pressure nodes if a) the pipe is open at both ends; b) the pipe is closed at the left end and open at the right end.

16.26 Find the fundamental frequency and the frequency of the first three overtones of a pipe 45.0 cm long a) if the pipe is open at both ends; b) if the pipe is closed at one end. Use $v = 344$ m/s. c) For each of the above cases, what is the number of the highest harmonic that may be heard by a person who can hear frequencies from 20 Hz to 20,000 Hz?

16.27 The human vocal tract is a pipe that extends about 17 cm from the lips to the vocal folds (also called "vocal cords") near the middle of your throat. The vocal folds behave rather like the reed of a clarinet, and the vocal tract acts as a stopped pipe. Estimate the first three standing-wave frequencies of the vocal tract. Use $v = 344$ m/s. (The answers are only an estimate, since the position of lips and tongue affect the motion of air in the vocal tract.)

16.28 A certain pipe produces a fundamental frequency of 262 Hz in air. a) If the pipe is filled with helium at the same temperature, what fundamental frequency does it produce? (The molar mass of air is 28.8 g/mol, and the molar mass of helium is 4.00 g/mol.) b) Does your answer to part (a) depend on whether the pipe is open or stopped? Why or why not?

Section 16.5 Resonance

16.29 You blow across the open mouth of an empty test tube and produce the fundamental standing wave of the air column inside the test tube. The speed of sound in air is 344 m/s and the test tube acts as a stopped pipe. a) If the length of the air column in the test tube is 14.0 cm, what is the frequency of this standing wave? b) What is the frequency of the fundamental standing wave in the air column if the test tube is half-filled with water?

16.30 The D note (588 Hz) from a trumpet causes a guitar string to vibrate in its second overtone with large amplitude. The vibrating portion of the guitar string has length 63.5 cm. What is the speed of transverse waves on the guitar string?

Section 16.6 Interference of Waves

16.31 Two loudspeakers, A and B (Fig. 16.41), are driven by the same amplifier and emit sinusoidal waves in phase. Speaker B is 2.00 m to the right of speaker A. Consider point Q along the extension of the line connecting the speakers, 1.00 m to the right of speaker B. Both speakers emit sound waves that travel directly from the speaker to point Q. a) What is the lowest frequency for which *constructive* interference occurs at point Q? b) What is the lowest frequency for which *destructive* interference occurs at point Q?

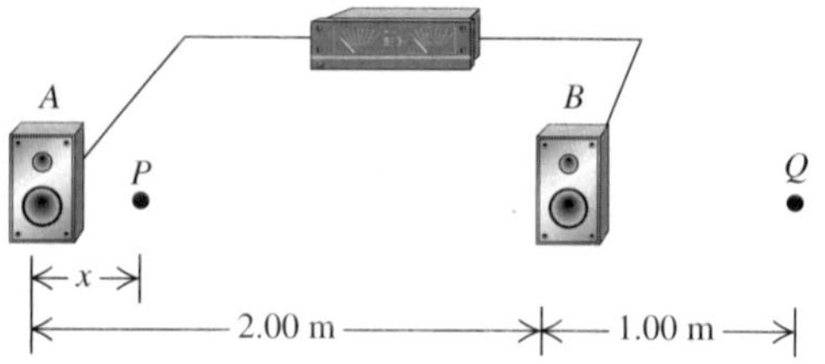

Figure 16.41 Exercises 16.31 and 16.32.

16.32 Two loudspeakers, A and B (Fig. 16.41), are driven by the same amplifier and emit sinusoidal waves in phase. Speaker B is 2.00 m to the right of speaker A. The frequency of the sound waves produced by the loudspeakers is 206 Hz. Consider point P between the speakers and along the line connecting them, a distance x to the right of speaker A. Both speakers emit sound waves that travel directly from the speaker to point P. a) For what values of x will *destructive* interference occur at point P? b) For what values of x will *constructive* interference occur at point P? c) Interference effects like those in parts (a) and (b) are almost never a factor in listening to home stereo equipment. Why not?

16.33 Two loudspeakers, A and B, are driven by the same amplifier and emit sinusoidal waves in phase. Speaker B is 12.0 m to the right of speaker A. The frequency of the waves emitted by each speaker is 688 Hz. You are standing between the speakers, along the line connecting them and are at a point of constructive interference. How far must you walk toward speaker B to move to a point of destructive interference?

16.34 Two loudspeakers, A and B, are driven by the same amplifier and emit sinusoidal waves in phase. The frequency of the waves emitted by each speaker is 172 Hz. You are 8.00 m from A. What is the closest you can be to B and be at a point of destructive interference?

16.35 Two loudspeakers, A and B, are driven by the same amplifier and emit sinusoidal waves in phase. The frequency of the waves emitted by each speaker is 860 Hz. Point P is 12.0 m from A and 13.4 m from B. Is the interference at P constructive or destructive? Give the reasoning behind your answer.

Section 16.7 Beats

16.36 Two identical violin strings, when in tune and stretched with the same tension, have a fundamental frequency of 440.0 Hz. One of the strings is retuned by adjusting its tension. When this is done, 1.5 beats per second are heard when both strings are plucked simultaneously. a) What are the possible fundamental frequencies of the retuned string? b) By what fractional amount was the string tension changed if it was i) increased; ii) decreased?

16.37 Two sinusoidal sound waves with frequencies 108 Hz and 112 Hz arrive at your ear simultaneously. Each wave has an amplitude of 1.5×10^{-8} m when it reaches your ear. a) Describe in detail what you hear. b) What is the maximum amplitude of the total sound wave? What is the minimum amplitude?

Section 16.8 The Doppler Effect

16.38 On the planet Arrakis a male ornithoid is flying toward his mate at 25.0 m/s while singing at a frequency of 1200 Hz. If the stationary female hears a tone of 1240 Hz, what is the speed of sound in the atmosphere of Arrakis?

16.39 In Example 16.20 (Section 16.8), suppose the police car is moving away from the warehouse at 20 m/s. What frequency does the driver of the police car hear reflected from the warehouse?

16.40 Two train whistles, A and B, each have a frequency of 392 Hz. A is stationary and B is moving toward the right (away from A) at a speed of 35.0 m/s. A listener is between the two whistles and is moving toward the right with a speed of 15.0 m/s (Fig. 16.42). No wind is blowing. a) What is the frequency from A as heard by the listener? b) What is the frequency from B as heard by the listener? c) What is the beat frequency detected by the listener?

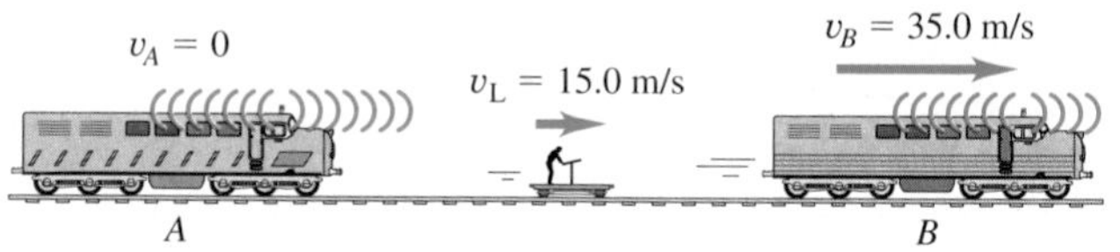

Figure 16.42 Exercise 16.40.

16.41 A railroad train is traveling at 25.0 m/s in still air. The frequency of the note emitted by the locomotive whistle is 400 Hz. What is the wavelength of the sound waves a) in front of the locomotive? b) behind the locomotive? What is the frequency of the sound heard by a stationary listener c) in front of the locomotive? d) behind the locomotive?

16.42 A swimming duck paddles the water with its feet once every 1.6 s, producing surface waves with this period. The duck is moving at constant speed in a pond where the speed of surface waves is 0.32 m/s, and the crests of the waves ahead of the duck are spaced 0.12 m apart. a) What is the duck's speed? b) How far apart are the crests behind the duck?

16.43 Moving Source vs. Moving Listener. a) A sound source producing 1.00 kHz waves moves toward a stationary listener at $\frac{1}{2}$ the speed of sound. What frequency will the listener hear? b) Suppose instead that the source is stationary and the listener moves toward the source at $\frac{1}{2}$ the speed of sound. What frequency does the listener hear? How does your answer compare to that in part (a)? Explain on physical grounds why the two answers differ.

16.44 A car alarm is emitting sound waves of frequency 520 Hz. You are on a motorcycle, traveling directly away from the car. How fast must you be traveling if you detect a frequency of 490 Hz?

16.45 A railroad train is traveling at 30.0 m/s in still air. The frequency of the note emitted by the train whistle is 262 Hz. What frequency is heard by a passenger on a train moving in the opposite direction to the first at 18.0 m/s and a) approaching the first? b) receding from the first?

*Section 16.9 Shock Waves

***16.46** A jet plane flies overhead at Mach 1.70 and at a constant altitude of 950 m. a) What is the angle α of the shock-wave cone? b) How much time after the plane passes directly overhead do you hear the sonic boom? Neglect the variation of the speed of sound with altitude.

Problems

16.47 a) Defend the following statement: "In a sinusoidal sound wave, the pressure variation given by Eq. (16.4) is greatest where the displacement given by Eq. (16.1) is zero." b) For a sinusoidal sound wave given by Eq. (16.1) with amplitude $A = 10.0\ \mu\text{m}$ and wavelength $\lambda = 0.250$ m, graph the displacement y and pressure fluctuation p as functions of x at time $t = 0$. Show at least two wavelengths of the wave on your graphs. c) The displacement y in a *nonsinusoidal* sound wave is shown in Fig. 16.43 as a function of x for $t = 0$. Draw a graph showing the pressure fluctuation p in this wave as a function of x at $t = 0$. This sound wave has the same 10.0-μm amplitude as the wave in part (b). Does it have the same pressure amplitude? Why or why not? d) Is the statement in part (a) necessarily true if the sound wave is *not* sinusoidal? Explain your reasoning.

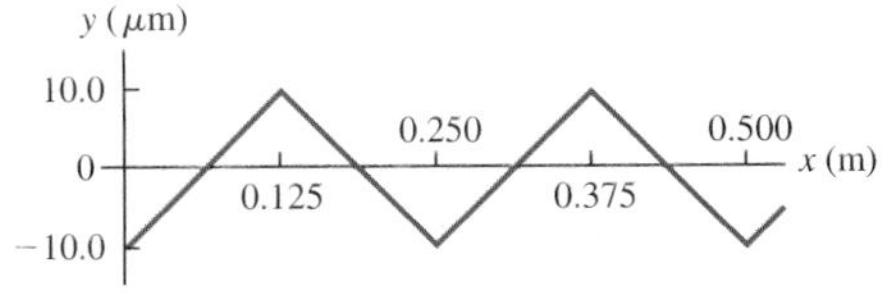

Figure 16.43 Problem 16.47.

16.48 Many airports have noise ordinances restricting the maximum sound intensity that an aircraft may produce when it takes off. At one California airport the maximum allowable sound intensity level is 98.5 dB as measured by a microphone at the end of the 1740-m-long runway. A certain airliner produces a sound intensity level of 100.0 dB on the ground when it flies over at an altitude of 100 m. On takeoff, this airliner rolls for 1200 m along the runway before leaving the ground, at which point it climbs at a 15° angle. Does this airliner violate the noise ordinance? You can ignore any effects due to reflection of the sound waves from the ground.

16.49 The sound from a trumpet radiates uniformly in all directions in 20°C air. At a distance of 5.00 m from the trumpet the sound intensity level is 52.0 dB. The frequency is 587 Hz. a) What is the pressure amplitude at this distance? b) What is the displacement amplitude? c) At what distance is the sound intensity level 30.0 dB?

16.50 A sound wave entering a window of area A has a sound intensity level of β. a) How much acoustic power enters the window? b) Find the value of the acoustic power for the case of a 1.20-m^2 window and 55.0-dB sound.

16.51 A person is playing a small flute 10.75 cm long, open at one end and closed at the other, near a taut string having a fundamental frequency of 600.0 Hz. If the speed of sound is 344.0 m/s, for which harmonics of the flute will the string resonate? In each case, which harmonic of the string is in resonance?

16.52 A New Musical Instrument. You have designed a new musical instrument of very simple construction. Your design consists of a metal tube with length L and diameter $L/10$. You have stretched a string of mass per unit length μ across the open end of the tube. The other end of the tube is closed. To produce the musical effect you're looking for, you want the frequency of the third-harmonic standing wave on the string to be the same as the fundamental frequency for sound waves in the air column in the tube. The speed of sound waves in this air column is v_s. a) What must be the tension of the string to produce the desired effect? b) What happens to the sound produced by the instrument if the tension is changed to twice the value calculated in part (a)? c) For the tension calculated in part (a), what other harmonics of the string, if any, are in resonance with standing waves in the air column?

16.53 An organ pipe has two successive harmonics with frequencies 1372 and 1764 Hz. a) Is this an open or a stopped pipe? Explain. b) What two harmonics are these? c) What is the length of the pipe?

16.54 Longitudinal Standing Waves in a Solid. Longitudinal standing waves can be produced in a solid rod by holding it at some point between the fingers of one hand and stroking it with the other hand. The rod oscillates with antinodes at both ends. a) Why are the ends antinodes and not nodes? b) The fundamental frequency can be obtained by stroking the rod while it is held at its center. Explain why this is the *only* place to hold the rod to obtain the fundamental. c) Calculate the fundamental frequency of a steel rod of length 1.50 m (see Table 16.1). d) What is the next possible standing-wave frequency of this rod? Where should the rod be held to excite a standing wave of this frequency?

16.55 a) Determine the first three normal-mode frequencies for a pipe of length L that is closed at *both* ends. Explain your reasoning. b) Use the results of part (a) to estimate the normal-mode frequencies of a shower stall. Explain the connection between these frequencies and the observation that your singing voice probably sounds better in the shower, especially when you sing at certain frequencies.

16.56 One type of steel has a density of $7.8 \times 10^3\ \text{kg/m}^3$ and a breaking stress of $7.0 \times 10^8\ \text{N/m}^2$. A cylindrical guitar string is to be made of 4.00 g of this steel. a) What are the length and radius of the longest and thinnest string that can be placed under a tension of 900 N without breaking? b) What is the highest fundamental frequency that this string could have?

16.57 A long tube contains air at a pressure of 1.00 atm and a temperature of 77.0°C. The tube is open at one end and closed at the other by a movable piston. A tuning fork near the open end is vibrating with a frequency of 500 Hz. Resonance is produced when the piston is at distances 18.0, 55.5, and 93.0 cm from the open end. a) From these measurements, what is the speed of sound in air at 77.0°C? b) From the result of part (a), what is the value of γ? c) These data show that a displacement antinode is slightly outside of the open end of the tube. How far outside is it?

16.58 The auditory canal of the human ear (Fig. 16.5) extends about 2.5 cm from the outside ear to the eardrum. a) Explain why the human ear is especially sensitive to sounds at frequencies around 3500 Hz. Use $v = 344$ m/s. b) Would you expect the ear to be especially sensitive to frequencies around 7000 Hz? Around 10,500 Hz? Why or why not?

16.59 A platinum wire (density 21.4 g/cm^3) is 225 μm in diameter and 0.450 m long. One end of the wire is attached to the ceiling, while a 420-g mass is attached to the other end so that the wire hangs vertically under tension. If a vibrating tuning fork of just the right frequency is held next to the wire, the wire begins to vibrate as well. a) What tuning-fork frequencies will cause this to happen? You may assume that the bottom end of the wire (to which the mass is attached) is essentially stationary, and that the tension in the wire is essentially constant along its length. b) Justify the assumptions made in part (a).

16.60 The frequency of the note F_4 is 349 Hz. a) If an organ pipe is open at one end and closed at the other, what length must it have for its fundamental mode to produce this note at 20.0°C? b) At what air temperature will the frequency be 370 Hz, corresponding to a rise in pitch from F to F-sharp? (Ignore the change in length of the pipe due to the temperature change.)

16.61 A standing wave with a frequency of 1100 Hz in a column of methane (CH_4) at 20.0°C produces nodes that are 0.200 m apart. What is the value of γ for methane? (The molar mass of methane is 16.0 g/mol.)

16.62 Two identical loudspeakers are located at points A and B, 2.00 m apart. The loudspeakers are driven by the same amplifier and produce sound waves with a frequency of 784 Hz. Take the speed of sound in air to be 344 m/s. A small microphone is moved out from point B along a line perpendicular to the line connecting A and B (line BC in Fig. 16.44). a) At what distances from B will there be *destructive* interference? b) At what distances from B will there be *constructive* interference? c) If the frequency is made low enough, there will be no positions along the line BC at which destructive interference occurs. How low must the frequency be for this to be the case?

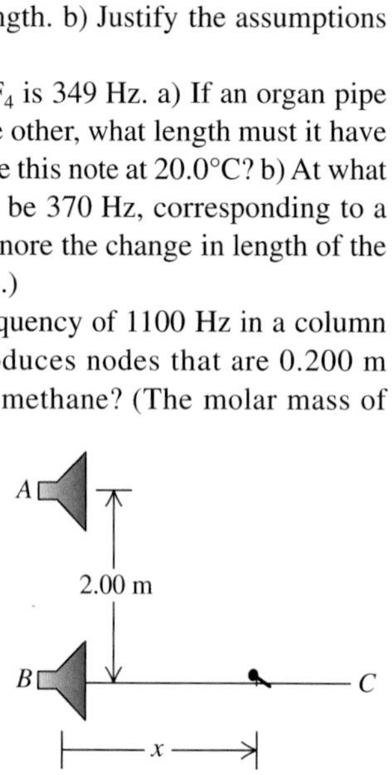

Figure 16.44 Problem 16.62.

16.63 Soap Opera. A man marries a great Wagnerian soprano but alas, he discovers he cannot stand Wagnerian opera. In order to save his eardrums, the unhappy man decides he must silence his lark-like wife for good. His plan is to tie her to the front of his car and send car and soprano speeding towards a brick wall. This soprano, however is quite shrewd, having studied physics in her student days at the music conservatory. She realizes that this wall has a resonant frequency of 600 Hz, which means that if a continuous sound wave of this frequency hits the wall, it will fall down, and she will be saved to sing again. The car is heading toward the wall at a high speed of 30 m/s. a) At what frequency must the soprano sing so that the wall will crumble? b) What frequency will the soprano hear reflected from the wall just before it crumbles?

16.64 A bat flies toward a wall, emitting a steady sound of frequency 2.00 kHz. This bat hears its own sound plus the sound reflected by the wall. How fast should the bat fly in order to hear a beat frequency of 10.0 Hz?

16.65 A small sphere of radius R is arranged to pulsate so that its radius varies in simple harmonic motion between a minimum of $R - \Delta R$ and a maximum of $R + \Delta R$ with frequency f. This produces sound waves in the surrounding air of density ρ and bulk modulus B. a) Find the intensity of sound waves at the surface of the sphere. (The amplitude of oscillation of the sphere is the same as that of the air at the surface of the sphere.) b) Find the total acoustic power radiated by the sphere. c) At a distance $d \gg R$ from the center of the sphere, find the amplitude, pressure amplitude, and intensity of the sound wave.

16.66 A 2.00-MHz sound wave travels through a pregnant woman's abdomen and is reflected from the fetal heart wall of her unborn baby. The heart wall is moving toward the sound receiver as the heart beats. The reflected sound is then mixed with the transmitted sound, and 85 beats per second are detected. The speed of sound in body tissue is 1500 m/s. Calculate the speed of the fetal heart wall at the instant this measurement is made.

16.67 The sound source of a ship's sonar system operates at a frequency of 22.0 kHz. The speed of sound in water (assumed to be at a uniform 20°C) is 1482 m/s. a) What is the wavelength of the waves emitted by the source? b) What is the difference in frequency between the directly radiated waves and the waves reflected from a whale traveling directly toward the ship at 4.95 m/s? The ship is at rest in the water.

16.68 A police siren of frequency f_{siren} is attached to a vibrating platform. The platform and siren oscillate up and down in simple harmonic motion with amplitude A_p and frequency f_p. a) Find the maximum and minimum sound frequencies that you would hear at a position directly above the siren. b) At what point in the motion of the platform is the maximum frequency heard? The minimum frequency? Explain.

16.69 Horseshoe bats (genus *Rhinolophus*) emit sounds from their nostrils, then listen to the frequency of the sound reflected from their prey to determine the prey's speed. (The "horseshoe" that gives the bat its name is a depression around the nostrils that acts like a focusing mirror, so that the bat emits sound in a narrow beam like a flashlight.) A *Rhinolophus* flying at speed v_{bat} emits sound of frequency f_{bat}; the sound it hears reflected from an insect flying toward it has a higher frequency f_{refl}. a) Show that the speed of the insect is

$$v_{insect} = v\left[\frac{f_{refl}(v - v_{bat}) - f_{bat}(v + v_{bat})}{f_{refl}(v - v_{bat}) + f_{bat}(v + v_{bat})}\right]$$

where v is the speed of sound. b) If $f_{bat} = 80.7$ kHz, $f_{refl} = 83.5$ kHz, and $v_{bat} = 3.9$ m/s, calculate the speed of the insect.

16.70 A sound wave with frequency f_0 and wavelength λ_0 travels horizontally toward the right. It strikes and is reflected from a large, rigid, vertical plane surface, perpendicular to the direction of propagation of the wave and moving toward the left with a speed v_1. a) How many positive wave crests strike the surface in a time interval t? b) At the end of this time interval, how far to the left of the surface is the wave that was reflected at the beginning of the time interval? c) What is the wavelength of the reflected waves in terms of λ_0? d) What is the frequency of the reflected waves in terms of f_0? Is your result consistent with the assertion made in the

Problem-Solving Strategy in Section 16.8? e) A listener is at rest at the left of the moving surface. How many beats per second does she detect as a result of the combined effect of the incident and reflected waves?

16.71 a) Show that Eq. (16.30) can be written as

$$f_R = f_S\left(1 - \frac{v}{c}\right)^{1/2}\left(1 + \frac{v}{c}\right)^{-1/2}$$

b) Use the binomial theorem to show that if $v \ll c$, this is approximately equal to

$$f_R = f_S\left(1 - \frac{v}{c}\right)$$

c) A pilotless reconnaissance aircraft emits a radio signal with a frequency of 243 MHz. It is flying directly toward a test engineer on the ground. The engineer detects beats between the received signal and a local signal also of frequency 243 MHz. The beat frequency is 46.0 Hz. What is the speed of the aircraft? (Radio waves travel at the speed of light, $c = 3.00 \times 10^8$ m/s.)

16.72 The gas cloud known as the Crab Nebula (Fig. 9.34) can be seen with even a small telescope. It is the remnant of a *supernova*, a cataclysmic explosion of a star. The explosion was seen on the earth on July 4, 1054 A.D. The streamers in Fig. 9.34 glow with the characteristic red color of heated hydrogen gas. In a laboratory on the earth, heated hydrogen produces red light with frequency 4.568×10^{14} Hz; the red light received from streamers in the Crab Nebula pointed toward the earth has frequency 4.586×10^{14} Hz. a) Estimate the speed with which the outer edges of the Crab Nebula are expanding. Assume that the speed of the center of the nebula relative to the earth is negligible. (You may use the formulas derived in Problem 16.71. The speed of light is 3.00×10^8 m/s.) b) Assuming that the expansion speed has been constant since the supernova explosion, estimate the diameter of the Crab Nebula. Give your answer in meters and in light years. c) The angular diameter of the Crab Nebula as seen from earth is about 5 arc minutes $\left(1 \text{ arc minute} = \frac{1}{60} \text{ of a degree}\right)$. Estimate the distance (in light years) to the Crab Nebula, and estimate the year in which the supernova explosion actually took place.

16.73 A radar installation used for monitoring thunderstorms emits radio waves at a frequency of 2800 MHz. A line of thunderstorms is approaching the installation at 42.0 km/h. a) In the frame of reference of the thunderstorms, is the frequency of the radio waves greater than or less than 2800 MHz? Why? By what amount does the frequency differ from 2800 MHz? (You may use the formulas derived in Problem 16.71. Radio waves travel at the speed of light, $c = 3.00 \times 10^8$ m/s.) b) Radio waves are reflected from the water drops in the thunderstorms, and the reflected waves are detected back at the installation. As measured by the receiver at the installation, is the frequency of these reflected waves greater than or less than 2800 MHz? Why? By what amount does the frequency of the reflected waves differ from 2800 MHz?

16.74 A woman stands at rest in front of a large, smooth wall. She holds a vibrating tuning fork of frequency f_0 directly in front of her (between her and the wall). a) She now runs toward the wall with a speed v_W. She detects beats due to the interference between the sound waves reaching her directly from the fork and those reaching her after being reflected from the wall. How many beats per second will she detect? (*Note:* If the beat frequency is too large, the woman may have to use some instrumentation other than her ears to detect and count the beats.) b) If she instead runs away from the wall, holding the tuning fork at her back so it is between her and the wall, how many beats per second will she detect?

***16.75** On a clear day you see a jet plane flying overhead. From the apparent size of the plane, you determine that it is flying at a constant altitude h. You hear the sonic boom at time T after the plane passes directly overhead. Show that if the speed of sound v is the same at all altitudes, the speed of the plane is

$$v_S = \frac{hv}{\sqrt{h^2 - v^2T^2}}$$

(*Hint:* Trigonometric identities will be useful.)

Challenge Problems

16.76 Figure 16.45 shows the pressure fluctuation p of a nonsinusoidal sound wave as a function of x for $t = 0$. The wave is traveling in the $+x$-direction. a) Graph the pressure fluctuation p as a function of t for $x = 0$. Show at least two cycles of oscillation. b) Graph the displacement y in this sound wave as a function of x at $t = 0$. At $x = 0$, the displacement at $t = 0$ is zero. Show at least two wavelengths of the wave. c) Graph the displacement y as a function of t for $x = 0$. Show at least two cycles of oscillation. d) Calculate the maximum velocity and the maximum acceleration of an element of the air through which this sound wave is traveling. e) Describe how the cone of a loudspeaker must move as a function of time to produce the sound wave in this problem.

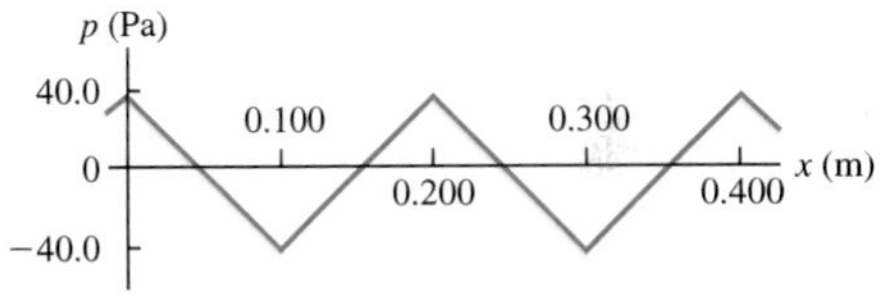

Figure 16.45 Challenge Problem 16.76.

16.77 Two loudspeakers, A and B, radiate sound uniformly in all directions in air at 20°C. The acoustic power output from A is 8.00×10^{-4} W, and from B it is 6.00×10^{-5} W. Both loudspeakers are vibrating in phase at a frequency of 172 Hz. a) Determine the difference in phase of the two signals at a point C along the line joining A and B, 3.00 m from B and 4.00 m from A (Fig. 16.46). b) Determine the intensity and sound intensity level at C from speaker A if speaker B is turned off and the intensity and sound intensity level at point C from speaker B if speaker A is turned off. c) With both speakers on, what are the intensity and sound intensity level at C?

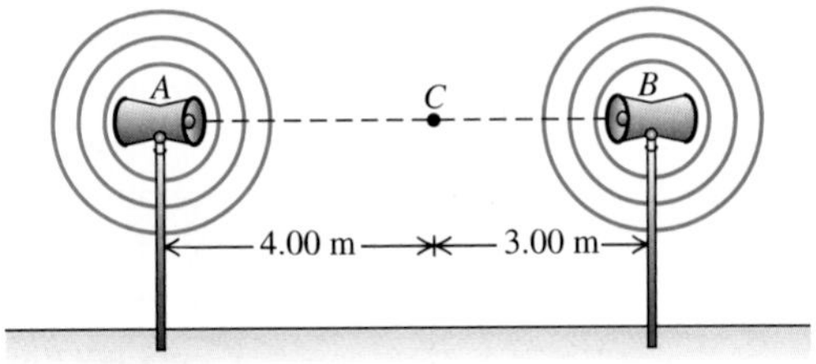

Fig. 16.46 Challenge Problem 16.77.

17 TEMPERATURE AND HEAT

Perhaps the highest-temperature material you will ever see is the sun's outer atmosphere, or corona. At a temperature of about 2,000,000°C (3,600,000°F), the corona glows with a light that is literally unearthly. But because the corona is also very thin, its light is rather faint. You can only see the corona during a total solar eclipse when the sun's disk is covered by the moon, as shown here.

? Is it accurate to say that the corona contains heat?

Whether it's a sweltering summer day or a frozen midwinter night, your body needs to be kept at a nearly constant temperature. It has effective temperature-control mechanisms, but sometimes it needs help. On a hot day you wear less clothing to improve heat transfer from your body to the air and for better cooling by evaporation of perspiration. You probably drink cold beverages possibly with ice in them, and sit near a fan or in an air-conditioned room. On a cold day you wear more clothes or stay indoors where it's warm. When you're outside, you keep active and drink hot liquids to stay warm. The concepts in this chapter will help you understand the basic physics of keeping warm or cool.

The terms *temperature* and *heat* are often used interchangeably in everyday language. In physics, however, these two terms have very different meanings. In this chapter we'll define temperature in terms of how it's measured and see how temperature changes affect the dimensions of objects. We'll see that heat refers to energy transfer caused by temperature differences and learn how to calculate and control such energy transfers.

Our emphasis in this chapter is on the concepts of temperature and heat as they relate to *macroscopic* objects such as cylinders of gas, ice cubes, and the human body. In Chapter 18 we'll look at these same concepts from a *microscopic* viewpoint in terms of the behavior of individual atoms and molecules. These two chapters lay the groundwork for the subject of **thermodynamics,** the study of energy transformations involving heat, mechanical work, and other aspects of energy and how these transformations relate to the properties of matter. Thermodynamics forms an indispensable part of the foundation of physics, chemistry, and the life sciences, and its applications turn up in such places as car engines, refrigerators, biochemical processes, and the structure of stars. We'll explore the key ideas of thermodynamics in Chapters 19 and 20.

17.1 | Temperature and Thermal Equilibrium

The concept of **temperature** is rooted in qualitative ideas of "hot" and "cold" based on our sense of touch. A body that feels hot usually has a higher temperature than a similar body that feels cold. That's pretty vague, and the senses can be deceived. But many properties of matter that we can *measure* depend on temperature. The length of a metal rod, steam pressure in a boiler, the ability of a wire to conduct an electric current, and the color of a very hot glowing object—all these depend on temperature.

Temperature is also related to the kinetic energies of the molecules of a material. In general this relationship is fairly complex, so it's not a good place to start in *defining* temperature. In Chapter 18 we will look at the relationship between temperature and the energy of molecular motion for an ideal gas. It is important to understand, however, that temperature and heat can be defined independently of any detailed molecular picture. In this section we'll develop a *macroscopic* definition of temperature.

To use temperature as a measure of hotness or coldness, we need to construct a temperature scale. To do this, we can use any measurable property of a system that varies with its "hotness" or "coldness." Figure 17.1a shows a familiar system that is used to measure temperature. When the system becomes hotter, the colored liquid (usually mercury or ethanol) expands and rises in the tube, and the value of L increases. Another simple system is a quantity of gas in a constant-volume container (Fig. 17.1b). The pressure p, measured by the gauge, increases or decreases as the gas becomes hotter or colder. A third example is the electrical resistance R of a conducting wire, which also varies when the wire becomes hotter or colder. Each of these properties gives us a number (L, p, or R) that varies with hotness and coldness, so each property can be used to make a **thermometer.**

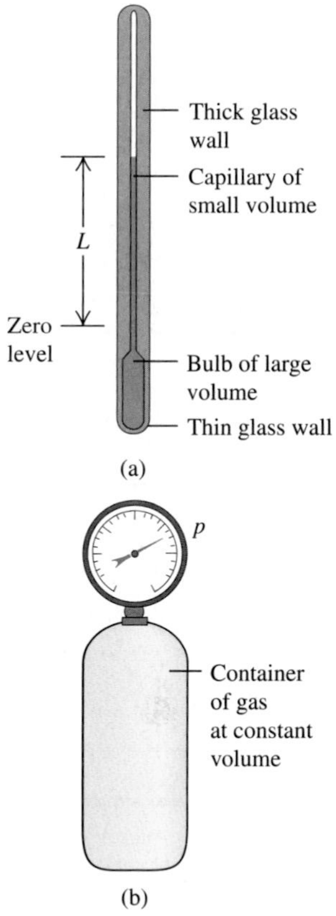

17.1 (a) A system whose temperature is specified by the value of length L. (b) A system whose temperature is given by the value of the pressure p.

To measure the temperature of a body, you place the thermometer in contact with the body. If you want to know the temperature of a cup of hot coffee, you stick the thermometer in the coffee; as the two interact, the thermometer becomes hotter and the coffee cools off a little. After the thermometer settles down to a steady value, you read the temperature. The system has reached an *equilibrium* condition, in which the interaction between the thermometer and the coffee causes no further change in the system. We call this a state of **thermal equilibrium.**

If two systems are separated by an insulating material or **insulator** such as wood, plastic foam, or fiberglass, they influence each other more slowly. Camping coolers are made with insulating materials to delay the ice and cold food inside from warming up and attaining thermal equilibrium with the hot summer air outside. An *ideal insulator* is a material that permits no interaction at all between the two systems. It prevents the systems from attaining thermal equilibrium if they aren't in thermal equilibrium at the start. An ideal insulator is just that, an idealization; real insulators, like those in camping coolers, aren't ideal, so the contents of the cooler will warm up eventually.

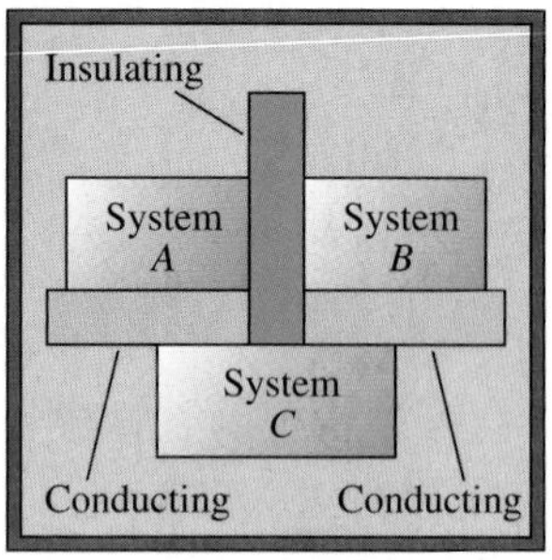

(a) If systems A and B are each in thermal equilibrium with system C ...

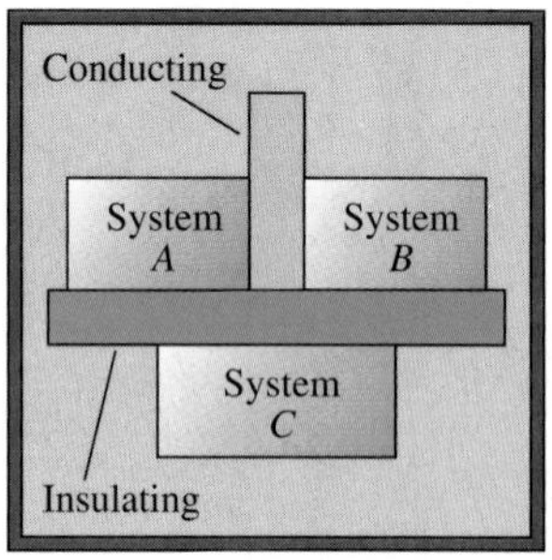

(b) ... then systems A and B are in thermal equilibrium with each other

17.2 The zeroth law of thermodynamics. Blue slabs represent thermal insulators, and orange slabs represent thermal conductors.

We can discover an important property of thermal equilibrium by considering three systems, A, B, and C, that initially are not in thermal equilibrium (Fig. 17.2). We surround them with an ideal insulating box so that they cannot interact with anything except each other. We separate systems A and B with an ideal insulating wall (the blue slab in Fig. 17.2a), but we let system C interact with both systems A and B. This interaction is shown in the figure by an orange slab representing a thermal **conductor,** a material that *permits* thermal interactions through it. We wait until thermal equilibrium is attained; then A and B are each in thermal equilibrium with C. But are they in thermal equilibrium *with each other?*

To find out, we separate system C from systems A and B with an ideal insulating wall (Fig. 17.2b), and then we replace the insulating wall between A and B with a *conducting* wall that lets A and B interact. What happens? Experiment shows that *nothing* happens; there are no additional changes to A or B. We conclude that **if C is initially in thermal equilibrium with both A and B, then A and B are also in thermal equilibrium with each other.** This result is called the **zeroth law of thermodynamics.** (The importance of this law was recognized only after the first, second, and third laws of thermodynamics had been named. Since it is fundamental to all of them, the name "zeroth" seemed appropriate.)

Now suppose system C is a thermometer, such as the tube-and-liquid system of Fig. 17.1a. In Fig. 17.2a the thermometer C is in contact with both A and B. In thermal equilibrium, when the thermometer reading reaches a stable value, the thermometer measures the temperature of both A and B; hence A and B both have the *same* temperature. Experiment shows that thermal equilibrium isn't affected by adding or removing insulators, so the reading of thermometer C wouldn't change if it were in contact only with A or only with B. We conclude that **two systems are in thermal equilibrium if and only if they have the same temperature.** This is what makes a thermometer useful; a thermometer actually measures *its own* temperature, but when a thermometer is in thermal equilibrium with another body, the temperatures must be equal. When the temperatures of two systems are different, they *cannot* be in thermal equilibrium.

Test Your Understanding

Why does a nurse taking your temperature wait for the thermometer reading to stop changing? What object is it whose temperature the nurse is reading?

17.2 | Thermometers and Temperature Scales

To make the liquid-in-tube device shown in Fig. 17.1a into a useful thermometer, we need to mark a scale on the tube wall with numbers on it. These numbers are arbitrary, and historically many different schemes have been used. Suppose we label the thermometer's liquid level at the freezing temperature of pure water "zero" and the level at the boiling temperature "100," and divide the distance between these two points into 100 equal intervals called *degrees.* The result is the **Celsius temperature scale** (formerly called the *centigrade* scale in English-speaking countries). The Celsius temperature for a state colder than freezing water is a negative number. The Celsius scale is used, both in everyday life and in science and industry, almost everywhere in the world.

Another common type of thermometer uses a *bimetallic strip,* made by bonding strips of two different metals together (Fig. 17.3a). When the temperature of the composite strip increases, one metal expands more than the other and the strip bends. This strip is usually formed into a spiral, with the outer end anchored to the thermometer case and the inner end attached to a pointer (Fig. 17.3c). The pointer rotates in response to temperature changes.

In a *resistance thermometer* the changing electrical resistance of a coil of fine wire, a carbon cylinder, or a germanium crystal is measured. Because resistance can be measured very precisely, resistance thermometers are usually more precise than most other types.

Some thermometers don't require physical contact with the object they're measuring. An example is an ear thermometer (Fig. 17.4). Such a thermometer uses a device called a *thermopile* to measure the amount of infrared radiation emitted by the eardrum, which indicates the eardrum's temperature. (We'll see in Section 17.7 that *all* objects emit electromagnetic radiation as a consequence of their temperature.) The advantage of this technique is that it doesn't require touching the fragile and easily damaged eardrum.

In the **Fahrenheit temperature scale,** still used in everyday life in the United States, the freezing temperature of water is 32°F (thirty-two degrees Fahrenheit) and the boiling temperature is 212°F, both at standard atmospheric pressure. There are 180 degrees between freezing and boiling, compared to 100 on the Celsius scale, so one Fahrenheit degree represents only $\frac{100}{180}$, or $\frac{5}{9}$, as great a temperature change as one Celsius degree.

To convert temperatures from Celsius to Fahrenheit, note that a Celsius temperature T_C is the number of Celsius degrees above freezing; the number of Fahrenheit degrees above freezing is $\frac{9}{5}$ of this. But freezing on the Fahrenheit scale is at 32°F, so to obtain the actual Fahrenheit temperature T_F, multiply the Celsius value by $\frac{9}{5}$ and then add 32°. Symbolically,

$$T_F = \frac{9}{5}T_C + 32° \qquad (17.1)$$

To convert Fahrenheit to Celsius, solve this equation for T_C:

$$T_C = \frac{5}{9}(T_F - 32°) \qquad (17.2)$$

In words, subtract 32° to get the number of Fahrenheit degrees above freezing, and then multiply by $\frac{5}{9}$ to obtain the number of Celsius degrees above freezing, that is, the Celsius temperature.

We don't recommend memorizing Eqs. (17.1) and (17.2). Instead, try to understand the reasoning that led to them so that you can derive them on the spot when you need them, checking your reasoning with the relation 100°C = 212°F.

It is useful to distinguish between an actual temperature and a temperature *interval* (a difference or change in temperature). An actual temperature of 20° is stated as 20°C (twenty degrees Celsius), and a temperature *interval* of 10° is 10 C° (ten Celsius degrees). A beaker of water heated from 20°C to 30°C has a temperature change of 10 C°.

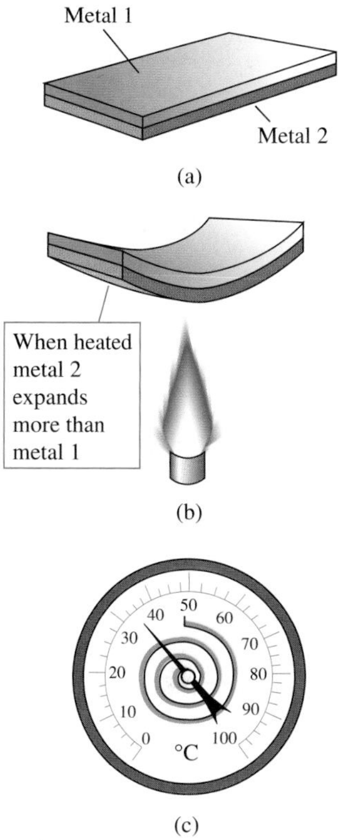

17.3 (a) A bimetallic strip. (b) The strip bends when its temperature is raised. (c) A bimetallic strip used in a thermometer.

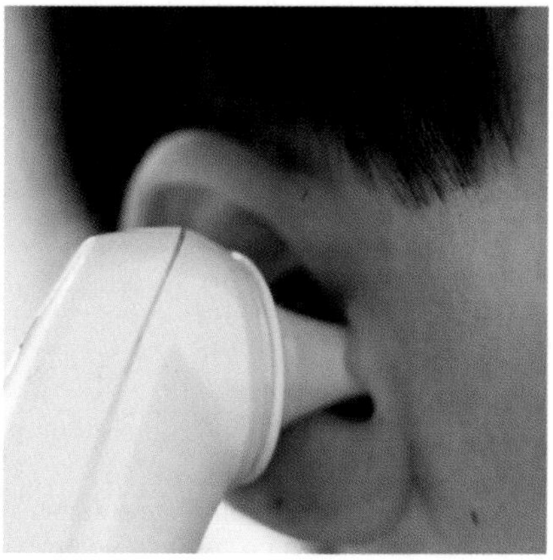

17.4 An ear thermometer measures infrared radiation from the eardrum, which is located far enough inside the head that it gives an excellent indication of the body's internal temperature.

Test Your Understanding

Find the average Fahrenheit temperature on the planet Venus (average Celsius temperature 460°C), and find the temperature at which the Fahrenheit and Celsius scales agree.

17.3 | Gas Thermometers and the Kelvin Scale

When we calibrate two thermometers, such as a liquid-in-tube system and a resistance thermometer, so that they agree at 0°C and 100°C, they may not agree exactly at intermediate temperatures. Any temperature scale defined in this way always depends somewhat on the specific properties of the material used. Ideally, we would like to define a temperature scale that *doesn't* depend on the properties of a particular material. To establish a truly material-independent scale, we first need to develop some principles of thermodynamics. We'll return to this fundamental problem in Chapter 20. Here we'll discuss a thermometer that comes close to the ideal, the *gas thermometer.*

The principle of a gas thermometer is that the pressure of a gas at constant volume increases with temperature. A quantity of gas is placed in a constant-volume container (Fig. 17.5a), and its pressure is measured by one of the devices described in Section 14.2. To calibrate a constant-volume gas thermometer, we measure the pressure at two temperatures, say 0°C and 100°C, plot these points on a graph, and draw a straight line between them. Then we can read from the graph the temperature corresponding to any other pressure. Figure 17.5b shows the results of three such experiments, each using a different type and quantity of gas.

By extrapolating this graph, we see that there is a hypothetical temperature, −273.15°C, at which the absolute pressure of the gas would become zero. We might expect that this temperature would be different for different gases, but it turns out to be the *same* for many different gases (at least in the limit of very low gas density). We can't actually observe this zero-pressure condition. Gases liquefy and solidify at very low temperatures, and the proportionality of pressure to temperature no longer holds.

We use this extrapolated zero-pressure temperature as the basis for a temperature scale with its zero at this temperature. This is the **Kelvin temperature scale,**

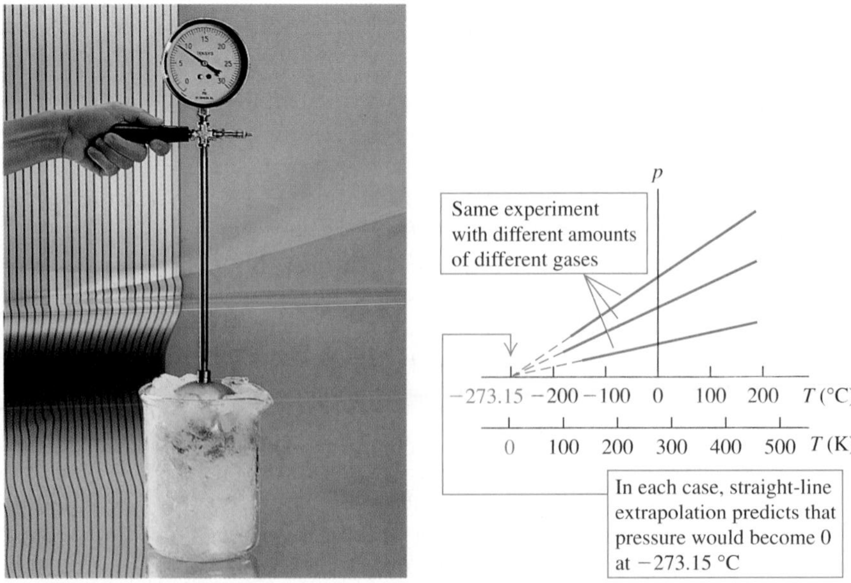

(a) (b)

17.5 (a) A constant-volume gas thermometer. (b) Graphs of absolute pressure versus temperature for a constant-volume low-density gas thermometer. The three graphs are for experiments with different types and quantities of gas: the greater the amount of gas, the higher the graph. The dashed lines are extrapolations of the data to low temperature.

named for the British physicist Lord Kelvin (1824–1907). The units are the same size as those on the Celsius scale, but the zero is shifted so that 0 K = −273.15°C and 273.15 K = 0°C; that is,

$$T_K = T_C + 273.15 \qquad (17.3)$$

This scale is shown in Fig. 17.5b. A common room temperature, 20°C (= 68°F), is 20 + 273.15, or about 293 K.

CAUTION In SI nomenclature, "degree" is not used with the Kelvin scale; the temperature mentioned above is read "293 kelvins," not "degrees Kelvin" (Fig. 17.6). We capitalize Kelvin when it refers to the temperature scale; however, the *unit* of temperature is the *kelvin,* which is not capitalized (but is nonetheless abbreviated as a capital K).

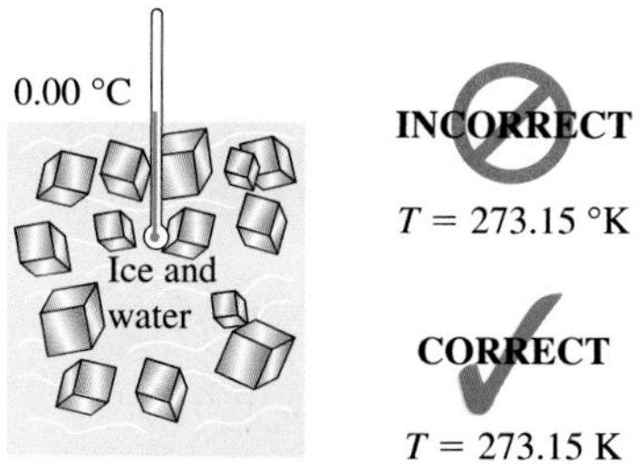

17.6 Kelvin temperatures are measured in kelvins (K), *not* "degrees Kelvin."

Example 17.1 Body temperature

You place a small piece of melting ice in your mouth. Eventually, the water all converts from ice at $T_1 = 32.00°F$ to body temperature, $T_2 = 98.60°F$. Express these temperatures as °C and K, and find $\Delta T = T_2 - T_1$ in both cases.

SOLUTION

IDENTIFY and SET UP: We convert Fahrenheit to Celsius temperatures using Eq. (17.2), and Celsius to Kelvin temperatures using Eq. (17.3).

EXECUTE: First we find the Celsius temperatures. We know that $T_1 = 32.00°F = 0.00°C$, and 98.60°F is 98.60 − 32.00 = 66.60 F° above freezing; multiply this by (5 C°/9 F°) to find 37.00 C° above freezing, or $T_2 = 37.00°C$.

To get the Kelvin temperatures, we just add 273.15 to each Celsius temperature: $T_1 = 273.15$ K and $T_2 = 310.15$ K. "Normal" body temperature is 37.0°C, but if your doctor says that your temperature is 310 K, don't be alarmed.

The temperature *difference* $\Delta T = T_2 - T_1$ is 37.00 C° = 37.00 K.

EVALUATE: The Celsius and Kelvin scales have different zero points but the same size degrees. Therefore any temperature difference is the *same* on the Celsius and Kelvin scales but not the same on the Fahrenheit scale.

The Celsius scale has two fixed points, the normal freezing and boiling temperatures of water. But we can define the Kelvin scale using a gas thermometer with only a single reference temperature. We define the ratio of any two temperatures T_1 and T_2 on the Kelvin scale as the ratio of the corresponding gas-thermometer pressures p_1 and p_2:

$$\frac{T_2}{T_1} = \frac{p_2}{p_1} \quad \text{(constant-volume gas thermometer, } T \text{ in kelvins)} \qquad (17.4)$$

The pressure p is directly proportional to the Kelvin temperature, as shown in Fig. 17.5b. To complete the definition of T, we need only specify the Kelvin temperature of a single specific state. For reasons of precision and reproducibility the state

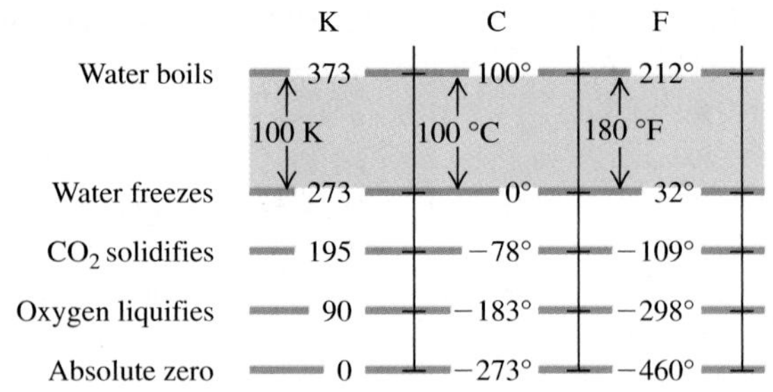

17.7 Relations among Kelvin, Celsius, and Fahrenheit temperature scales. Temperatures have been rounded off to the nearest degree.

chosen is the *triple point* of water. This is the unique combination of temperature and pressure at which solid water (ice), liquid water, and water vapor can all coexist. It occurs at a temperature of 0.01°C and a water-vapor pressure of 610 Pa (about 0.006 atm). (This is the pressure of the *water;* it has nothing to do directly with the gas pressure in the *thermometer.*) The triple-point temperature T_{triple} of water is *defined* to have the value $T_{triple} = 273.16$ K, corresponding to 0.01°C. From Eq. (17.4), if p_{triple} is the pressure in a gas thermometer at temperature T_{triple} and p is the pressure at some other temperature T, then T is given on the Kelvin scale by

$$T = T_{triple}\frac{p}{p_{triple}} = (273.16\ \text{K})\frac{p}{p_{triple}} \tag{17.5}$$

Low-pressure gas thermometers using various gases are found to agree very closely, but they are large, bulky, and very slow to come to thermal equilibrium. They are used principally to establish high-precision standards and to calibrate other thermometers.

The relationships among the three temperature scales we have discussed are shown graphically in Fig. 17.7. The Kelvin scale is called an **absolute temperature scale,** and its zero point ($T = 0\ \text{K} = -273.15°\text{C}$, the temperature at which $p = 0$ in Eq. (17.5)) is called **absolute zero.** At absolute zero a system of molecules (such as a quantity of a gas, a liquid, or a solid) has its *minimum* possible total energy (kinetic plus potential); because of quantum effects, however, it is *not* correct to say that all molecular motion ceases at absolute zero. To define more completely what we mean by absolute zero, we need to use the thermodynamic principles developed in the next several chapters. We will return to this concept in Chapter 20.

Test Your Understanding

The temperature of the solar corona (see the photograph that opens this chapter) is 2.0×10^7 °C, and the temperature at which helium becomes a liquid at standard pressure is −268.93°C. Express these temperatures in kelvins, and explain why the Kelvin scale is usually employed for very high and very low temperatures.

17.4 | Thermal Expansion

Most materials expand when their temperatures increase. Rising temperatures make the liquid expand in a liquid–in–tube thermometer (Fig. 17.1a) and bend bimetallic strips (Fig. 17.3). The decks of bridges need special joints and supports to allow for expansion. A completely filled and tightly capped bottle of water cracks when it is heated, but you can loosen a metal jar lid by running hot water over it. These are all examples of *thermal expansion.*

Linear Expansion

Suppose a rod of material has a length L_0 at some initial temperature T_0. When the temperature changes by ΔT, the length changes by ΔL. Experiments show that if ΔT is not too large (say, less than 100 C° or so), ΔL is *directly proportional* to ΔT. If two rods made of the same material have the same temperature change, but one is twice as long as the other, then the *change* in its length is also twice as great. Therefore ΔL must also be proportional to L_0. Introducing a proportionality constant α (which is different for different materials), we may express these relations in an equation:

$$\Delta L = \alpha L_0 \,\Delta T \qquad \text{(linear thermal expansion)} \qquad (17.6)$$

If a body has length L_0 at temperature T_0, then its length L at a temperature $T = T_0 + \Delta T$ is

$$L = L_0 + \Delta L = L_0 + \alpha L_0 \,\Delta T = L_0(1 + \alpha \,\Delta T) \qquad (17.7)$$

The constant α, which describes the thermal expansion properties of a particular material, is called the **coefficient of linear expansion.** The units of α are K^{-1} or $(C°)^{-1}$. (Remember that a temperature *interval* is the same in the Kelvin and Celsius scales.) For many materials, every linear dimension changes according to Eq. (17.6) or (17.7). Thus L could be the thickness of a rod, the side length of a square sheet, or the diameter of a hole. Some materials, such as wood or single crystals, expand differently in different directions. We won't consider this complication.

We can understand thermal expansion qualitatively on a molecular basis. Picture the interatomic forces in a solid as springs, as in Fig. 17.8. (We explored the analogy between spring forces and interatomic forces in Section 13.4.) Each atom vibrates about its equilibrium position. When the temperature increases, the energy and amplitude of the vibration also increase. The interatomic spring forces are not symmetrical about the equilibrium position; they usually behave like a spring that is easier to stretch than to compress. As a result, when the amplitude of vibration increases, the *average* distance between atoms also increases. As the atoms get farther apart, every dimension increases.

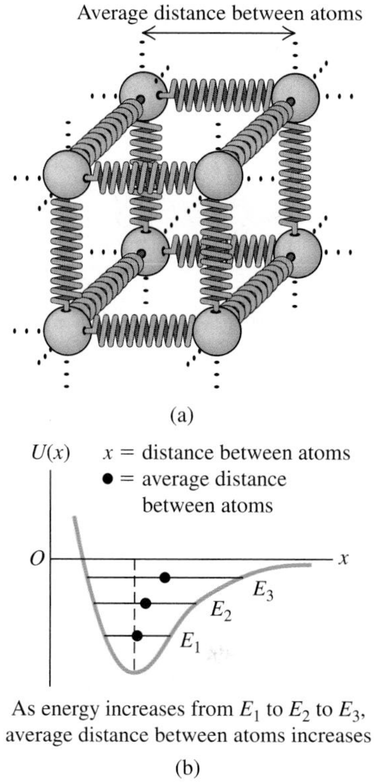

17.8 (a) A model of the forces between neighboring atoms in a solid. The "springs" are easier to stretch than to compress. (b) A graph of the "spring" potential energy U versus distance x between neighboring atoms (compare Fig. 13.19a). The graph is not symmetrical: as the energy increases and the atoms oscillate with greater amplitude, the *average* distance increases.

CAUTION If a solid object has a hole in it, what happens to the size of the hole when the temperature of the object increases? A common misconception is that if the object expands, the hole will shrink because material expands into the hole. But the truth of the matter is that if the object expands, the hole will expand too (Fig. 17.9); as we stated above, *every* linear dimension of an object changes in the same way when the temperature changes. If you're not convinced, think of the atoms in Fig. 17.8a as outlining a cubical hole. When the object expands, the atoms move apart and the hole

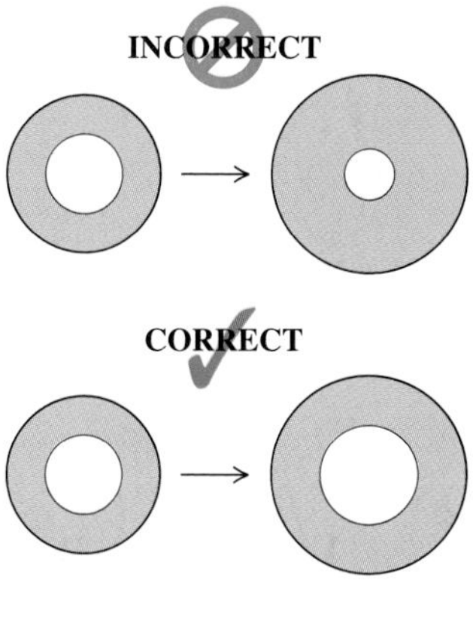

17.9 When an object undergoes thermal expansion, any holes in the object expand as well.

17.10 When this SR-71 aircraft is sitting on the ground, its wing panels fit together so loosely that fuel leaks out of the wings onto the ground. But once it is in flight at over three times the speed of sound, air friction heats the panels so much that they expand to make a perfect fit. (In-flight refueling makes up for the lost fuel.)

increases in size. The only situation in which a "hole" will fill in due to thermal expansion is when two separate objects expand and close the gap between them (Fig. 17.10).

The direct proportionality expressed by Eq. (17.6) is not exact; it is *approximately* correct only for sufficiently small temperature changes. For a given material, α varies somewhat with the initial temperature T_0 and the size of the temperature interval. We'll ignore this complication here, however. Average values of α for several materials are listed in Table 17.1. Within the precision of these values we don't need to worry whether T_0 is 0°C or 20°C or some other temperature. Note that typical values of α are very small; even for a temperature change of 100 C°, the fractional length change $\Delta L/L_0$ is only of the order of 1/1000 for the metals in the table.

Volume Expansion

Increasing temperature usually causes increases in *volume* for both solid and liquid materials. Just as with linear expansion, experiments show that if the temperature change ΔT is not too great (less than 100 C° or so), the increase in volume ΔV is approximately proportional to both the temperature change ΔT and the initial volume V_0:

$$\Delta V = \beta V_0 \Delta T \qquad \text{(volume thermal expansion)} \tag{17.8}$$

The constant β characterizes the volume expansion properties of a particular material; it is called the **coefficient of volume expansion.** The units of β are K^{-1} or $(\text{C}°)^{-1}$. As with linear expansion, β varies somewhat with temperature, and Eq. (17.8) is an approximate relation that is valid only for small temperature changes. For many substances, β decreases at low temperatures. Several values of β in the neighborhood of room temperature are listed in Table 17.2. Note that the values for liquids are generally much larger than those for solids.

For solid materials there is a simple relation between the volume expansion coefficient β and the linear expansion coefficient α. To derive this relation, we consider a cube of material with side length L and volume $V = L^3$. At the initial temperature the values are L_0 and V_0. When the temperature increases by dT, the side length increases by dL and the volume increases by an amount dV given by

$$dV = \frac{dV}{dL}dL = 3L^2\,dL$$

Now we replace L and V by the initial values L_0 and V_0. From Eq. (17.6), dL is

$$dL = \alpha L_0\,dT$$

since $V_0 = L_0^3$, this means that dV can also be expressed as

$$dV = 3L_0^2 \alpha L_0\,dT = 3\alpha V_0\,dT$$

This is consistent with the infinitesimal form of Eq. (17.8), $dV = \beta V_0\,dT$, only if

$$\beta = 3\alpha \tag{17.9}$$

You should check this relation for some of the materials listed in Tables 17.1 and 17.2.

Table 17.1 Coefficients of Linear Expansion

Material	$\alpha[K^{-1} \text{ or } (C°)^{-1}]$
Aluminum	2.4×10^{-5}
Brass	2.0×10^{-5}
Copper	1.7×10^{-5}
Glass	$0.4\text{–}0.9 \times 10^{-5}$
Invar (nickel–iron alloy)	0.09×10^{-5}
Quartz (fused)	0.04×10^{-5}
Steel	1.2×10^{-5}

Table 17.2 Coefficients of Volume Expansion

Solids	$\beta[K^{-1} \text{ or } (C°)^{-1}]$	Liquids	$\beta[K^{-1} \text{ or } (C°)^{-1}]$
Aluminum	7.2×10^{-5}	Ethanol	75×10^{-5}
Brass	6.0×10^{-5}	Carbon disulfide	115×10^{-5}
Copper	5.1×10^{-5}	Glycerin	49×10^{-5}
Glass	$1.2\text{–}2.7 \times 10^{-5}$	Mercury	18×10^{-5}
Invar	0.27×10^{-5}		
Quartz (fused)	0.12×10^{-5}		
Steel	3.6×10^{-5}		

Problem-Solving Strategy

Thermal Expansion

IDENTIFY *the relevant concepts:* Decide whether the problem involves changes in length (linear thermal expansion) or in volume (volume thermal expansion).

SET UP *the problem using the following steps:*

1. Choose Eq. (17.6) for linear expansion and Eq. (17.8) for volume expansion.
2. Identify which quantities in Eq. (17.6) or (17.8) are known and which are the unknown target variables.

EXECUTE *the solution as follows:*

1. Solve for the target variables. Often you will be given two temperatures and asked to compute ΔT. Or you may be given an initial temperature T_0 and asked to find a final temperature corresponding to a given length or volume change. In this case, plan to find ΔT first; then the final temperature is $T_0 + \Delta T$.
2. Unit consistency is crucial, as always. L_0 and ΔL (or V_0 and ΔV) must have the same units, and if you use a value of α or β in K^{-1} or $(C°)^{-1}$, then ΔT must be in kelvins or Celsius degrees (C°). But you can use K and C° interchangeably.

EVALUATE *your answer:* Check whether your results make sense. Remember that the sizes of holes in a material expand with temperature just the same way as any other linear dimension, and the volume of a hole (such as the volume of a container) expands the same way as the corresponding solid shape.

Example 17.2

Length change due to temperature change I

A surveyor uses a steel measuring tape that is exactly 50.000 m long at a temperature of 20°C. What is its length on a hot summer day when the temperature is 35°C?

SOLUTION

IDENTIFY and SET UP: This problem concerns linear expansion, so we use Eq. (17.6). We are given $L = 50.000$ m, $T_0 = 20°C$, and $T = 35°C$, and the value of α is found from Table 17.1. The target variable is the new length L.

EXECUTE: The temperature change is $\Delta T = T - T_0 = 15\ C°$, so from Eq. (17.6) the change in length ΔL and the final length $L = L_0 + \Delta L$ are

$$\Delta L = \alpha L_0 \Delta T = (1.2 \times 10^{-5}\ K^{-1})(50\ m)(15\ K)$$
$$= 9.0 \times 10^{-3}\ m = 9.0\ mm$$
$$L = L_0 + \Delta L = 50.000\ m + 0.009\ m = 50.009\ m$$

Thus the length at 35°C is 50.009 m.

EVALUATE: Note that L_0 is given to five significant figures but that we need only two of them to compute ΔL. Note also that ΔL is proportional to the initial length L_0: a 50-m tape would expand by 0.90 mm, and a 0.50-m (50-cm) tape would expand by a mere 0.090 mm.

This example shows that metals expand very little under moderate temperature changes. Even a metal baking pan in a 200°C (392°F) oven is only slightly larger than it is at room temperature.

Example 17.3 Length change due to temperature change II

In Example 17.2 the surveyor uses the measuring tape to measure a distance when the temperature is 35°C; the value that she reads off the tape is 35.794 m. What is the actual distance? Assume that the tape is calibrated for use at 20°C.

SOLUTION

IDENTIFY and SET UP: As we saw in Example 17.2, at 35°C the tape has expanded slightly. The distance between two successive meter marks is slightly more than one meter, so the scale underestimates the actual distance. Hence the actual distance is *larger* than the distance read off the tape by a factor equal to the ratio of the tape's length L at 35°C to its length L_0 at 20°C.

EXECUTE: The ratio L/L_0 is $(50.009\text{ m})/(50.000\text{ m})$, so the true distance is

$$\frac{50.009\text{ m}}{50.000\text{ m}}(35.794\text{ m}) = 35.800\text{ m}$$

EVALUATE: Although the difference of 0.008 m = 8 mm between the scale reading and the actual distance seems small, it can be important in precision work.

Example 17.4 Volume change due to temperature change

A glass flask with volume 200 cm^3 is filled to the brim with mercury at 20°C. How much mercury overflows when the temperature of the system is raised to 100°C? The coefficient of linear expansion of the glass is $0.40 \times 10^{-5}\text{ K}^{-1}$.

SOLUTION

IDENTIFY: This problem involves the volume expansion of the glass and of the mercury. The amount of overflow depends on the *difference* between the volume changes for these two materials.

SET UP: The amount of overflow is equal to the difference between the values of ΔV for mercury and for glass, both given by Eq. (17.8). For the mercury to overflow, its coefficient of volume expansion β must be larger than that for glass. The value for mercury is $\beta_{\text{mercury}} = 18 \times 10^{-5}\text{ K}^{-1}$ from Table 17.2, and we find the value of β for this type of glass from Eq. (17.9), $\beta = 3\alpha$.

EXECUTE: The coefficient of volume expansion for the glass is

$$\beta_{\text{glass}} = 3\alpha_{\text{glass}} = 3(0.40 \times 10^{-5}\text{ K}^{-1}) = 1.2 \times 10^{-5}\text{ K}^{-1}$$

The increase in volume of the glass flask is

$$\begin{aligned}\Delta V_{\text{glass}} &= \beta_{\text{glass}} V_0 \Delta T \\ &= (1.2 \times 10^{-5}\text{ K}^{-1})(200\text{ cm}^3)(100°\text{C} - 20°\text{C}) \\ &= 0.19\text{ cm}^3\end{aligned}$$

The increase in volume of the mercury is

$$\begin{aligned}\Delta V_{\text{mercury}} &= \beta_{\text{mercury}} V_0 \Delta T \\ &= (18 \times 10^{-5}\text{ K}^{-1})(200\text{ cm}^3)(100°\text{C} - 20°\text{C}) \\ &= 2.9\text{ cm}^3\end{aligned}$$

The volume of mercury that overflows is

$$\Delta V_{\text{mercury}} - \Delta V_{\text{glass}} = 2.9\text{ cm}^3 - 0.19\text{ cm}^3 = 2.7\text{ cm}^3$$

EVALUATE: This is basically how a mercury-in-glass thermometer works, except that instead of letting the mercury overflow and run all over the place, the thermometer has it rise inside a sealed tube as T increases.

As Tables 17.1 and 17.2 show, glass has smaller coefficients of expansion α and β than do most metals. This is why you can use hot water to loosen a metal lid on a glass jar; the metal expands more than the glass does.

Thermal Expansion of Water

Water, in the temperature range from 0°C to 4°C, *decreases* in volume with increasing temperature. In this range its coefficient of volume expansion is *negative.* Above 4°C, water expands when heated (Fig. 17.11). Hence water has its greatest density at 4°C. Water also expands when it freezes, which is why ice humps up in the middle of the compartments in an ice cube tray. By contrast, most materials contract when they freeze.

This anomalous behavior of water has an important effect on plant and animal life in lakes. A lake cools from the surface down; above 4°C, the cooled water at the surface flows to the bottom because of its greater density. But when the surface temperature drops below 4°C, the water near the surface is less dense than the warmer water below. Hence the downward flow ceases, and the water near the surface remains colder than that at the bottom. As the surface freezes, the ice floats because it is less dense than water. The water at the bottom remains at 4°C until nearly the entire lake is frozen. If water behaved like most substances, contracting continuously on cooling and freezing, lakes would freeze from the bottom up. Circulation due to density differences would continuously carry warmer water to the surface for efficient cooling, and lakes would freeze solid much more easily. This would destroy all plant and animal life that cannot withstand freezing. If water did not have this special property, the evolution of life would have taken a very different course.

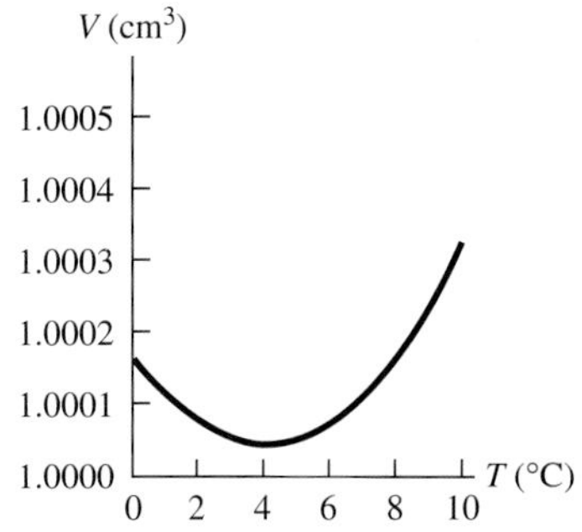

17.11 The volume of one gram of water in the temperature range from 0°C to 10°C. By 100°C the volume has increased to 1.034 cm^3. If the coefficient of volume expansion were constant, the curve would be a straight line.

Thermal Stress

If we clamp the ends of a rod rigidly to prevent expansion or contraction and then change the temperature, tensile or compressive stresses called **thermal stresses** develop. The rod would like to expand or contract, but the clamps won't let it. The resulting stresses may become large enough to strain the rod irreversibly or even break it. (You may want to review the discussion of stress and strain in Section 11.4).

Engineers must account for thermal stress when designing structures. Concrete highways and bridge decks usually have gaps between sections, filled with a flexible material or bridged by interlocking teeth (Fig. 17.12), to permit expansion and contraction of the concrete. Long steam pipes have expansion joints or U-shaped sections to prevent buckling or stretching with temperature changes. If one end of a steel bridge is rigidly fastened to its abutment, the other end usually rests on rollers.

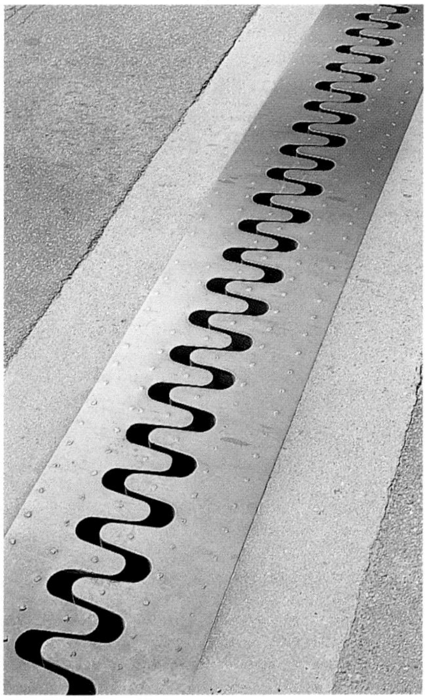

17.12 The interlocking teeth of an expansion joint on a bridge. These joints are needed to accommodate changes in length that result from thermal expansion.

To calculate the thermal stress in a clamped rod, we compute the amount the rod *would* expand (or contract) if not held and then find the stress needed to compress (or stretch) it back to its original length. Suppose that a rod with length L_0 and cross-sectional area A is held at constant length while the temperature is reduced (negative ΔT), causing a tensile stress. The fractional change in length if the rod were free to contract would be

$$\left(\frac{\Delta L}{L_0}\right)_{\text{thermal}} = \alpha\,\Delta T \tag{17.10}$$

Both ΔL and ΔT are negative. The tension must increase by an amount F that is just enough to produce an equal and opposite fractional change in length $(\Delta L/L_0)_{\text{tension}}$. From the definition of Young's modulus, Eq. (11.10),

$$Y = \frac{F/A}{\Delta L/L_0} \quad \text{so} \quad \left(\frac{\Delta L}{L_0}\right)_{\text{tension}} = \frac{F}{AY} \tag{17.11}$$

If the length is to be constant, the *total* fractional change in length must be zero. From Eqs. (17.10) and (17.11), this means that

$$\left(\frac{\Delta L}{L_0}\right)_{\text{thermal}} + \left(\frac{\Delta L}{L_0}\right)_{\text{tension}} = \alpha\,\Delta T + \frac{F}{AY} = 0$$

Solving for the tensile stress F/A required to keep the rod's length constant, we find

$$\frac{F}{A} = -Y\alpha\,\Delta T \qquad \text{(thermal stress)} \tag{17.12}$$

For a decrease in temperature, ΔT is negative, so F and F/A are positive; this means that a *tensile* force and stress are needed to maintain the length. If ΔT is positive, F and F/A are negative, and the required force and stress are *compressive.*

If there are temperature differences within a body, non-uniform expansion or contraction will result and thermal stresses can be induced. You can break a glass bowl by pouring very hot water into it; the thermal stress between the hot and cold parts of the bowl exceeds the breaking stress of the glass, causing cracks. The same phenomenon makes ice cubes crack when dropped into warm water. Heat-resistant glasses such as Pyrex™ have exceptionally low expansion coefficients and high strength.

Example 17.5 Thermal stress

An aluminum cylinder 10 cm long, with a cross-sectional area of 20 cm^2, is to be used as a spacer between two steel walls. At 17.2°C it just slips in between the walls. When it warms to 22.3°C, calculate the stress in the cylinder and the total force it exerts on each wall, assuming that the walls are perfectly rigid and a constant distance apart.

SOLUTION

IDENTIFY and SET UP: We use Eq. (17.12) to relate the stress (our target variable) to the temperature change. The relevant values of Young's modulus Y and the coefficient of linear expansion α are those for aluminum, the material of which the cylinder is made; we find these values from Tables 11.1 and 17.1, respectively.

EXECUTE: For aluminum, $Y = 7.0 \times 10^{10}$ Pa and $\alpha = 2.4 \times 10^{-5}\ \text{K}^{-1}$. The temperature change is $\Delta T = 22.3°\text{C} - 17.2°\text{C} = 5.1\ \text{C}° = 5.1$ K. The stress is F/A; from Eq. (17.12),

$$\frac{F}{A} = -Y\alpha\,\Delta T = -(0.70 \times 10^{11}\ \text{Pa})(2.4 \times 10^{-5}\ \text{K}^{-1})(5.1\ \text{K})$$
$$= -8.6 \times 10^{6}\ \text{Pa}\ (\text{or } -1200\ \text{lb/in.}^2)$$

The negative sign indicates that compressive rather than tensile stress is needed to keep the cylinder's length constant. This stress is independent of the length and cross-sectional area of the cylinder. The total force F is the cross-sectional area times the stress:

$$F = A\left(\frac{F}{A}\right) = (20 \times 10^{-4}\ \text{m}^2)(-8.6 \times 10^{6}\ \text{Pa})$$
$$= -1.7 \times 10^{4}\ \text{N}$$

or nearly two tons. The negative sign indicates compression.

EVALUATE: The stress on the cylinder and the force it exerts on each wall are immense. This points out the importance of accounting for such thermal stresses in engineering.

Test Your Understanding

In the bimetallic strip shown in Fig. 17.3, metal 1 is copper. What are two different materials that could be used for metal 2?

17.5 | Quantity of Heat

When you put a cold spoon into a cup of hot coffee, the spoon warms up and the coffee cools down as they approach thermal equilibrium. The interaction that causes these temperature changes is fundamentally a transfer of *energy* from one substance to another. Energy transfer that takes place solely because of a temperature difference is called *heat flow* or *heat transfer,* and energy transferred in this way is called **heat.**

An understanding of the relation between heat and other forms of energy emerged gradually during the 18th and 19th centuries. Sir James Joule (1818–1889) studied how water can be warmed by vigorous stirring with a paddle wheel (Fig. 17.13a). The paddle wheel adds energy to the water by doing *work* on it, and Joule found that *the temperature rise is directly proportional to the amount of work done.* The same temperature change can also be caused by putting the water in contact with some hotter body (Fig. 17.13b); hence this interaction must also involve an energy exchange. We will explore the relation between heat and mechanical energy in greater detail in Chapters 19 and 20.

CAUTION It is absolutely essential for you to keep clearly in mind the distinction between *temperature* and *heat.* Temperature depends on the physical state of a material and is a quantitative description of its hotness or coldness. In physics the term "heat" always refers to energy in transit from one body or system to another because of a temperature difference, never to the amount of energy contained within a particular system. We can change the temperature of a body by adding heat to it or taking heat away, or by adding or subtracting energy in other ways, such as mechanical work (Fig. 17.13a). If we cut a body in half, each half has the same temperature as the whole; but to raise the temperature of each half by a given interval, we add *half* as much heat as for the whole.

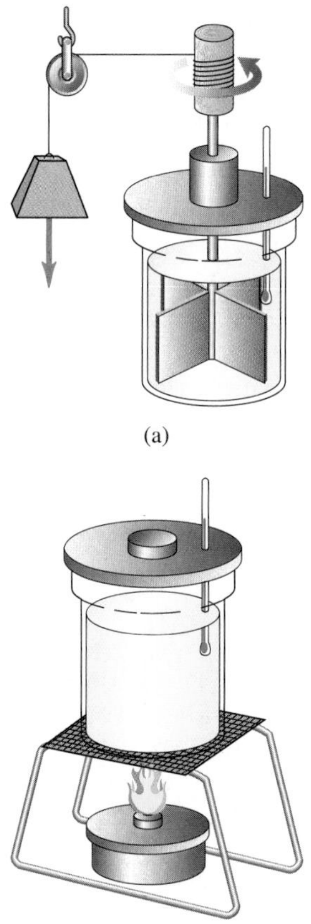

(a)

(b)

17.13 The same temperature change of the same system may be accomplished by (a) doing work on it or (b) adding heat to it.

We can define a *unit* of quantity of heat based on temperature changes of some specific material. The **calorie** (abbreviated cal) is defined as *the amount of heat required to raise the temperature of one gram of water from* 14.5°C *to* 15.5°C. The kilocalorie (kcal), equal to 1000 cal, is also used; a food-value calorie is actually a kilocalorie. A corresponding unit of heat using Fahrenheit degrees and British units is the **British thermal unit,** or Btu. One Btu is the quantity of heat required to raise the temperature of one pound (weight) of water 1 F° from 63°F to 64°F.

Because heat is energy in transit, there must be a definite relation between these units and the familiar mechanical energy units such as the joule. Experiments similar in concept to Joule's have shown that

$$1 \text{ cal} = 4.186 \text{ J}$$

$$1 \text{ kcal} = 1000 \text{ cal} = 4186 \text{ J}$$

$$1 \text{ Btu} = 778 \text{ ft} \cdot \text{lb} = 252 \text{ cal} = 1055 \text{ J}$$

The calorie is not a fundamental SI unit. The International Committee on Weights and Measures recommends using the joule as the basic unit of energy in all forms, including heat. We will follow that recommendation in this book.

Specific Heat

We use the symbol Q for quantity of heat. When it is associated with an infinitesimal temperature change dT, we call it dQ. The quantity of heat Q required to increase the temperature of a mass m of a certain material from T_1 to T_2 is found to be approximately proportional to the temperature change $\Delta T = T_2 - T_1$. It is also proportional to the mass m of material. When you're heating water to make

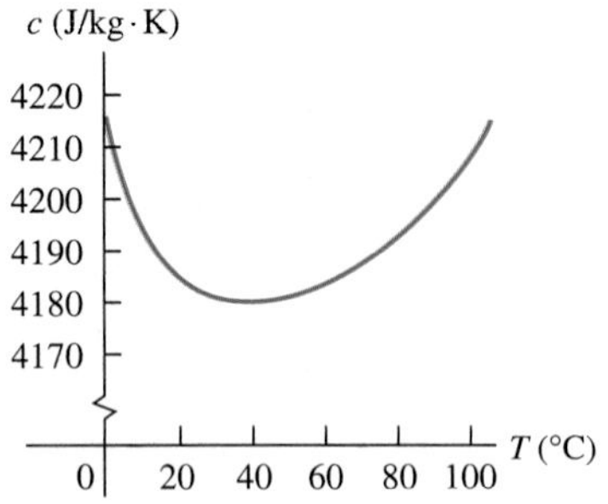

17.14 Specific heat of water as a function of temperature. The value of c varies by less than 1% between 0°C and 100°C.

tea, you need twice as much heat for two cups as for one if the temperature interval is the same. The quantity of heat needed also depends on the nature of the material; raising the temperature of one kilogram of water by 1 C° requires 4190 J of heat, but only 910 J is needed to raise the temperature of one kilogram of aluminum by 1 C°.

Putting all these relationships together, we have

$$Q = mc\,\Delta T \qquad \text{(heat required for temperature change of mass } m) \quad (17.13)$$

where c is a quantity, different for different materials, called the **specific heat** of the material. For an infinitesimal temperature change dT and corresponding quantity of heat dQ,

$$dQ = mc\,dT \qquad (17.14)$$

$$c = \frac{1}{m}\frac{dQ}{dT} \qquad \text{(specific heat)} \qquad (17.15)$$

In Eqs. (17.13), (17.14), and (17.15), Q (or dQ) and ΔT (or dT) can be either positive or negative. When they are positive, heat enters the body and its temperature increases; when they are negative, heat leaves the body and its temperature decreases.

CAUTION Remember that dQ does not represent a change in the amount of heat *contained* in a body; this is a meaningless concept. Heat is always energy *in transit* as a result of a temperature difference. There is no such thing as "the amount of heat in a body."

The specific heat of water is approximately

4190 J/kg·K   1 cal/g·C°   or 1 Btu/lb·F°

The specific heat of a material always depends somewhat on the initial temperature and the temperature interval. Figure 17.14 shows this variation for water. In the problems and examples in this chapter we will usually ignore this small variation.

Example 17.6 Feed a cold, starve a fever

During a bout with the flu an 80-kg man ran a fever of 39.0°C (102.2°F) instead of the normal body temperature of 37.0°C (98.6°F). Assuming that the human body is mostly water, how much heat is required to raise his temperature by that amount?

SOLUTION

IDENTIFY: This problem uses the relationship between heat (the target variable), mass, specific heat, and temperature change.

SET UP: We are given the values of $m = 80\text{ kg}$, $c = 4190\text{ J/kg}\cdot\text{K}$ (for water), and $\Delta T = 39.0°\text{C} - 37.0°\text{C} = 2.0\text{ C}° = 2.0\text{ K}$. We use Eq. (17.13) to determine the required heat.

EXECUTE: From Eq. (17.13),

$$Q = mc\,\Delta T = (80\text{ kg})(4190\text{ J/kg}\cdot\text{K})(2.0\text{ K}) = 6.7 \times 10^5\text{ J}$$

EVALUATE: This corresponds to 160 kcal, or 160 food-value calories. In fact, the specific heat of the human body is more nearly equal to 3480 J/kg·K, about 83% that of water. The difference is due to the presence of protein, fat, and minerals, which have lower specific heats. With this value of c, the required heat is $5.6 \times 10^5\text{ J} =$ 133 kcal. Either result shows us that were it not for the body's temperature-regulating systems, taking in energy in the form of food would produce measurable changes in body temperature. (In the case of a person with the flu, the elevated temperature results from the body's extra activity in fighting infection.)

Example 17.7 Overheating electronics

You are designing an electronic circuit element made of 23 mg of silicon. The electric current through it adds energy at the rate of 7.4 mW $= 7.4 \times 10^{-3}$ J/s. If your design doesn't allow any heat transfer out of the element, at what rate does its temperature increase? The specific heat of silicon is 705 J/kg · K.

SOLUTION

IDENTIFY and SET UP: The target variable is the *rate* of temperature change. From Eq. (17.14), the temperature change ΔT in kelvins is proportional to the heat transferred in joules, so the rate of temperature change in K/s is proportional to the rate of heat transfer in J/s.

EXECUTE: In one second, $Q = (7.4 \times 10^{-3}\ \text{J/s})(1\ \text{s}) = 7.4 \times 10^{-3}$ J. From Eq. (17.13), $Q = mc\,\Delta T$, the temperature change in one second is

$$\Delta T = \frac{Q}{mc} = \frac{7.4 \times 10^{-3}\ \text{J}}{(23 \times 10^{-6}\ \text{kg})(705\ \text{J/kg}\cdot\text{K})} = 0.46\ \text{K}$$

Alternatively, we can divide both sides of Eq. (17.14) by dt and rearrange:

$$\frac{dT}{dt} = \frac{dQ/dt}{mc}$$
$$= \frac{7.4 \times 10^{-3}\ \text{J/s}}{(23 \times 10^{-6}\ \text{kg})(705\ \text{J/kg}\cdot\text{K})} = 0.46\ \text{K/s}$$

EVALUATE: At this rate of temperature rise (27 K every minute), the circuit element would soon self-destruct. Heat transfer is an important design consideration in electronic circuit elements.

Molar Heat Capacity

Sometimes it's more convenient to describe a quantity of substance in terms of the number of *moles* n rather than the *mass* m of material. Recall from your study of chemistry that a mole of any pure substance always contains the same number of molecules. (We will discuss this point in more detail in Chapter 18.) The *molar mass* of any substance, denoted by M, is the mass per mole. (The quantity M is sometimes called *molecular weight,* but *molar mass* is preferable; the quantity depends on the mass of a molecule, not its weight.) For example, the molar mass of water is 18.0 g/mol $= 18.0 \times 10^{-3}$ kg/mol; one mole of water has a mass of 18.0 g $=$ 0.0180 kg. The total mass m of material is equal to the mass per mole M times the number of moles n:

$$m = nM \qquad (17.16)$$

Replacing the mass m in Eq. (17.13) by the product nM, we find

$$Q = nMc\,\Delta T \qquad (17.17)$$

The product Mc is called the **molar heat capacity** (or *molar specific heat*) and is denoted by C (capitalized). With this notation we rewrite Eq. (17.17) as

$$Q = nC\,\Delta T \qquad \text{(heat required for temperature change of } n \text{ moles)} \qquad (17.18)$$

Comparing to Eq. (17.15), we can express the molar heat capacity C (heat per mole per temperature change) in terms of the specific heat c (heat per mass per temperature change) and the molar mass M (mass per mole):

$$C = \frac{1}{n}\frac{dQ}{dT} = Mc \qquad \text{(molar heat capacity)} \qquad (17.19)$$

For example, the molar heat capacity of water is

$$C = Mc = (0.0180\ \text{kg/mol})(4190\ \text{J/kg}\cdot\text{K}) = 75.4\ \text{J/mol}\cdot\text{K}$$

Table 17.3 Approximate Specific Heats and Molar Heat Capacities (Constant Pressure)

Substance	Specific Heat, c (J/kg · K)	M (kg/mol)	Molar Heat Capacity, C (J/mol · K)
Aluminum	910	0.0270	24.6
Beryllium	1970	0.00901	17.7
Copper	390	0.0635	24.8
Ethanol	2428	0.0461	111.9
Ethylene glycol	2386	0.0620	148.0
Ice (near 0°C)	2100	0.0180	37.8
Iron	470	0.0559	26.3
Lead	130	0.207	26.9
Marble ($CaCO_3$)	879	0.100	87.9
Mercury	138	0.201	27.7
Salt (NaCl)	879	0.0585	51.4
Silver	234	0.108	25.3
Water (liquid)	4190	0.0180	75.4

17.15 Water has a much higher specific heat than the glass or metals used to make cookware. This helps explain why it takes several minutes to boil water on a stove, even though the pot or kettle reaches a high temperature very quickly.

Values of specific heat and molar heat capacity for several substances are given in Table 17.3. Note the remarkably large specific heat for water (Fig. 17.15).

CAUTION The term *heat capacity* is unfortunate because it gives the erroneous impression that a body *contains* a certain amount of heat. Remember, heat is energy in transit to or from a body, not the energy residing in the body.

Precise measurements of specific heats and molar heat capacities require great experimental skill. Usually, a measured quantity of energy is supplied by an electric current in a heater wire wound around the specimen. The temperature change ΔT is measured with a resistance thermometer or thermocouple embedded in the specimen. This sounds simple, but great care is needed to avoid or compensate for unwanted heat transfer between the sample and its surroundings. Measurements for solid materials are usually made at constant atmospheric pressure; the corresponding values are called the *specific heat* and *molar heat capacity at constant pressure,* denoted by c_p and C_p. For a gas it is usually easier to keep the substance in a container with constant *volume;* the corresponding values are called the *specific heat* and *molar heat capacity at constant volume,* denoted by c_V and C_V. For a given substance, C_V and C_p are different. If the system can expand while heat is added, there is additional energy exchange through the performance of *work* by the system on its surroundings. If the volume is constant, the system does no work. For gases the difference between C_p and C_V is substantial. We will study heat capacities of gases in detail in Section 19.7.

The last column of Table 17.3 shows something interesting. The molar heat capacities for most elemental solids are about the same, about 25 J/mol · K. This correlation, named the *rule of Dulong and Petit* (for its discoverers), forms the basis for a very important idea. The number of atoms in one mole is the same for all elemental substances. This means that on a *per atom* basis, about the same amount of heat is required to raise the temperature of each of these elements by a given amount, even though the *masses* of the atoms are very different. The heat required for a given temperature increase depends only on *how many* atoms the sample contains, not on the mass of an individual atom. We will see the reason why the rule of Dulong and Petit works so well when we study the molecular basis of heat capacities in greater detail in Chapter 18.

Test Your Understanding

Suppose you want to use the apparatus shown in Fig. 17.13a to heat water to make a cup of tea. Through what distance would a 1.00-kg block have to fall to raise the temperature of 0.250 kg of water from 20.0°C to 90.0°C? Assume that all of the potential energy that the block loses as it falls goes into raising the temperature of the water.

17.6 | Calorimetry and Phase Changes

Calorimetry means "measuring heat." We have discussed the energy transfer (heat) involved in temperature changes. Heat is also involved in *phase changes,* such as the melting of ice or boiling of water. Once we understand these additional heat relationships, we can analyze a variety of problems involving quantity of heat.

Phase Changes

We use the term **phase** to describe a specific state of matter, such as a solid, liquid, or gas. The compound H_2O exists in the *solid phase* as ice, in the *liquid phase* as water, and in the *gaseous phase* as steam. (These are also referred to as **states of matter:** the solid state, the liquid state, and the gaseous state.) A transition from one phase to another is called a **phase change** or *phase transition.* For any given pressure a phase change takes place at a definite temperature, usually accompanied by absorption or emission of heat and a change of volume and density.

A familiar example of a phase change is the melting of ice. When we add heat to ice at 0°C and normal atmospheric pressure, the temperature of the ice *does not* increase. Instead, some of it melts to form liquid water. If we add the heat slowly, to maintain the system very close to thermal equilibrium, the temperature remains at 0°C until all the ice is melted (Fig. 17.16). The effect of adding heat to this system is not to raise its temperature but to change its *phase* from solid to liquid.

17.16 The surrounding air is at room temperature, but this ice–water mixture remains at 0°C until all of the ice has melted and the phase change is complete.

To change 1 kg of ice at 0°C to 1 kg of liquid water at 0°C and normal atmospheric pressure requires 3.34×10^5 J of heat. The heat required per unit mass is called the **heat of fusion** (or sometimes *latent heat of fusion*), denoted by L_f. For water at normal atmospheric pressure the heat of fusion is

$$L_f = 3.34 \times 10^5\ \text{J/kg} = 79.6\ \text{cal/g} = 143\ \text{Btu/lb}$$

More generally, to melt a mass m of material that has a heat of fusion L_f requires a quantity of heat Q given by

$$Q = mL_f$$

This process is *reversible.* To freeze liquid water to ice at 0°C, we have to *remove* heat; the magnitude is the same, but in this case, Q is negative because heat is removed rather than added. To cover both possibilities and to include other kinds of phase changes, we write

$$Q = \pm mL \qquad \text{(heat transfer in a phase change)} \qquad (17.20)$$

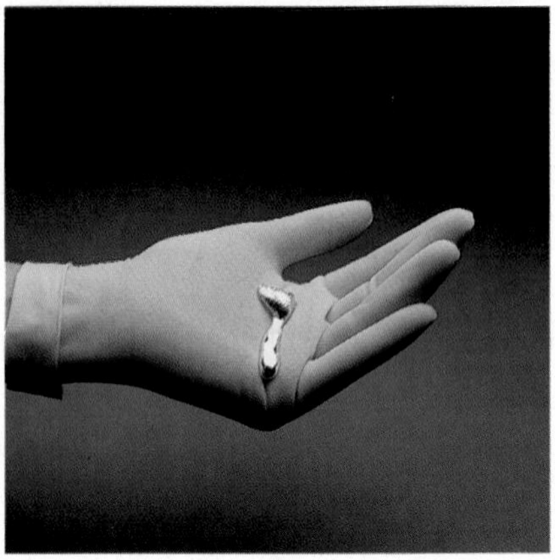

17.17 The metal gallium, shown here melting in a person's hand, is one of the few elements that melt in the vicinity of room temperature. Its melting temperature is 29.8°C, and its heat of fusion is 8.04×10^4 J/kg.

The plus sign (heat entering) is used when the material melts; the minus sign (heat leaving) is used when it freezes. The heat of fusion is different for different materials, and it also varies somewhat with pressure.

For any given material at any given pressure, the freezing temperature is the same as the melting temperature. At this unique temperature the liquid and solid phases (liquid water and ice, for example) can coexist in a condition called **phase equilibrium.**

We can go through this whole story again for *boiling* or *evaporation,* a phase transition between liquid and gaseous phases. The corresponding heat (per unit mass) is called the **heat of vaporization** L_v. At normal atmospheric pressure the heat of vaporization L_v for water is

$$L_v = 2.256 \times 10^6 \text{ J/kg} = 539 \text{ cal/g} = 970 \text{ Btu/lb}$$

That is, it takes 2.256×10^6 J to change 1 kg of liquid water at 100°C to 1 kg of water vapor at 100°C. By comparison, to raise the temperature of a kilogram of water from 0°C to 100°C requires $Q = mc\,\Delta T = (1.00\text{ kg})(4190\text{ J/kg}\cdot\text{C}°) \times (100\text{ C}°) = 4.19 \times 10^5$ J, less than one-fifth as much heat as is required for vaporization at 100°C. This agrees with everyday kitchen experience; a pot of water may reach boiling temperature in a few minutes, but it takes a much longer time to completely evaporate all the water away.

Like melting, boiling is a reversible transition. When heat is removed from a gas at the boiling temperature, the gas returns to the liquid phase, or *condenses,* giving up to its surroundings the same quantity of heat (heat of vaporization) that was needed to vaporize it. At a given pressure the boiling and condensation temperatures are always the same; at this temperature the liquid and gaseous phases can coexist in phase equilibrium.

Both L_v and the boiling temperature of a material depend on pressure. Water boils at a lower temperature (about 95°C) in Denver than in Pittsburgh because Denver is at higher elevation and the average atmospheric pressure is less. The heat of vaporization is somewhat greater at this lower pressure, about 2.27×10^6 J/kg.

Table 17.4 lists heats of fusion and vaporization for several materials and their melting and boiling temperatures at normal atmospheric pressure. Very few *ele-*

Table 17.4 Heats of Fusion and Vaporization

	Normal Melting Point		Heat of Fusion, L_f	Normal Boiling Point		Heat of Vaporization, L_v
Substance	**K**	**°C**	**(J/kg)**	**K**	**°C**	**(J/kg)**
Helium	*	*	*	4.216	−268.93	20.9×10^3
Hydrogen	13.84	−259.31	58.6×10^3	20.26	−252.89	452×10^3
Nitrogen	63.18	−209.97	25.5×10^3	77.34	−195.8	201×10^3
Oxygen	54.36	−218.79	13.8×10^3	90.18	−183.0	213×10^3
Ethanol	159	−114	104.2×10^3	351	78	854×10^3
Mercury	234	−39	11.8×10^3	630	357	272×10^3
Water	273.15	0.00	334×10^3	373.15	100.00	2256×10^3
Sulfur	392	119	38.1×10^3	717.75	444.60	326×10^3
Lead	600.5	327.3	24.5×10^3	2023	1750	871×10^3
Antimony	903.65	630.50	165×10^3	1713	1440	561×10^3
Silver	1233.95	960.80	88.3×10^3	2466	2193	2336×10^3
Gold	1336.15	1063.00	64.5×10^3	2933	2660	1578×10^3
Copper	1356	1083	134×10^3	1460	1187	5069×10^3

*A pressure in excess of 25 atmospheres is required to make helium solidify. At 1 atmosphere pressure, helium remains a liquid down to absolute zero.

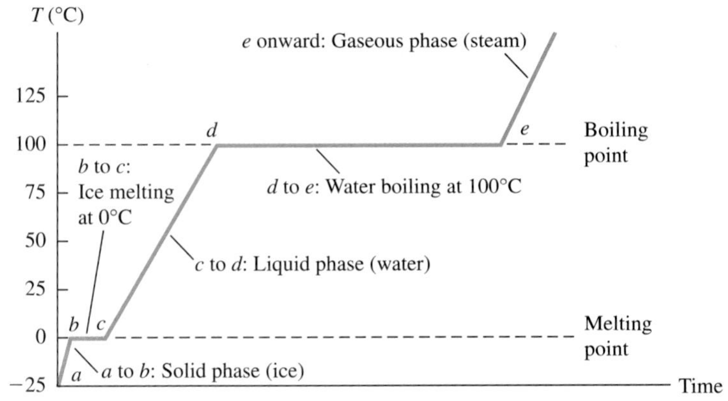

17.18 Graph of temperature versus time for a specimen of water initially in the solid phase (ice). Heat is added to the specimen at a constant rate. The temperature remains constant during each change of phase, provided that the pressure remains constant.

ments have melting temperatures in the vicinity of ordinary room temperatures; one of the few is the metal gallium, shown in Fig. 17.17.

Figure 17.18 shows how the temperature varies when we add heat continuously to a specimen of ice with an initial temperature below 0°C (point *a*). The temperature rises until we reach the melting point (point *b*). As more heat is added, the temperature remains constant until all the ice has melted (point *c*). Then the temperature starts to rise again until the boiling temperature is reached (point *d*). At that point the temperature again is constant until all the water is transformed into the vapor phase (point *e*). If the rate of heat input is constant, the slope of the line for the solid phase (ice) has a steeper slope than does the line for the liquid phase (water). Do you see why? (See Table 17.3.)

A substance can sometimes change directly from the solid to the gaseous phase. This process is called *sublimation,* and the solid is said to *sublime.* The corresponding heat is called the *heat of sublimation,* L_s. Liquid carbon dioxide cannot exist at a pressure lower than about 5×10^5 Pa (about 5 atm), and "dry ice" (solid carbon dioxide) sublimes at atmospheric pressure. Sublimation of water from frozen food causes freezer burn. The reverse process, a phase change from gas to solid, occurs when frost forms on cold bodies such as refrigerator cooling coils.

Very pure water can be cooled several degrees below the freezing temperature without freezing; the resulting unstable state is described as *supercooled.* When a small ice crystal is dropped in or the water is agitated, it crystallizes within a second or less. Supercooled water *vapor* condenses quickly into fog droplets when a disturbance, such as dust particles or ionizing radiation, is introduced. This principle is used in "seeding" clouds, which often contain supercooled water vapor, to cause condensation and rain.

A liquid can sometimes be *superheated* above its normal boiling temperature. Any small disturbance such as agitation causes local boiling with bubble formation.

Steam heating systems for buildings use a boiling–condensing process to transfer heat from the furnace to the radiators. Each kilogram of water that is turned to steam in the boiler absorbs over 2×10^6 J (the heat of vaporization L_v of water) from the boiler and gives it up when it condenses in the radiators. Boiling–condensing processes are also used in refrigerators, air conditioners, and heat pumps. We will discuss these systems in Chapter 20.

The temperature-control mechanisms of many warm-blooded animals make use of heat of vaporization, removing heat from the body by using it to evaporate

17.19 The water may be warm and it may be a hot day, but these children will feel cold when they first step out of the swimming pool. That's because as water evaporates from their skin, it removes the heat of vaporization from their bodies. To stay warm, they will need to dry off immediately.

water from the tongue (panting) or from the skin (sweating). Evaporative cooling enables humans to maintain normal body temperature in hot, dry desert climates where the air temperature may reach 55°C (about 130°F). The skin temperature may be as much as 30°C cooler than the surrounding air. Under these conditions a normal person may perspire several liters per day, and this lost water must be replaced. Old-time desert rats (such as one of the authors) state that in the desert, any canteen that holds less than a gallon should be viewed as a toy! Evaporative cooling also explains why you feel cold when you first step out of a swimming pool (Fig. 17.19).

Evaporative cooling is also used to cool buildings in hot, dry climates and to condense and recirculate "used" steam in coal-fired or nuclear-powered electric-generating plants. That's what goes on in the large, tapered concrete towers that you see at such plants.

Chemical reactions such as combustion are analogous to phase changes in that they involve definite quantities of heat. Complete combustion of one gram of gasoline produces about 46,000 J or about 11,000 cal, so the **heat of combustion** L_c of gasoline is

$$L_c = 46{,}000 \text{ J/g} = 4.6 \times 10^7 \text{ J/kg}$$

Energy values of foods are defined similarly; the U.S. unit of food energy, although called a calorie, is a *kilocalorie,* equal to 1000 cal or 4186 J. When we say that a gram of peanut butter "contains 6 calories," we mean that 6 kcal of heat (6,000 cal or 25,000 J) is released when the carbon and hydrogen atoms in the peanut butter react with oxygen (with the help of enzymes) and are completely converted to CO_2 and H_2O. Not all of this energy is directly useful for mechanical work. We will study the *efficiency* of energy utilization in Chapter 20.

Heat Calculations

Let's look at some examples of calorimetry calculations (calculations with heat). The basic principle is very simple: when heat flow occurs between two bodies that are isolated from their surroundings, the amount of heat lost by one body must equal the amount gained by the other. Heat is energy in transit, so this principle is really just conservation of energy. Calorimetry, dealing entirely with one conserved quantity, is in many ways the simplest of all physical theories!

Problem-Solving Strategy

Calorimetry Problems

IDENTIFY *the relevant concepts:* The basic principle underlying calorimetry calculations (calculations with heat) is very simple: when heat flow occurs between two bodies that are isolated from their surroundings, the amount of heat lost by one body must equal the amount gained by the other body. Heat is energy in transit, so this principle is really just conservation of energy. Calorimetry, dealing entirely with one conserved quantity, is in many ways the simplest of all physical theories!

SET UP *the problem using the following steps:*

1. Identify which objects exchange heat. To avoid confusion with algebraic signs, take each quantity of heat *added* to a body as *positive* and each quantity *leaving* a body as *negative.* When several bodies interact, the *algebraic sum* of the quantities of heat transferred to all the bodies must be zero.
2. Each object will undergo a temperature change with no phase change, a phase change at constant temperature, or both. Use Eq. (17.13) to describe temperature changes and Eq. (17.20) to describe phase changes.
3. Consult Table 17.3 for values of the specific heat or molar heat capacity and Table 17.4 for heats of fusion or vaporization.
4. Be certain to identify which quantities are known and which are the unknown target variables.

EXECUTE *the solution as follows:*

1. Solve Eq. (17.13) and/or Eq. (17.20) for the target variables. Often you will need to find an unknown temperature. Represent it by an algebraic symbol such as T. Then if a body has an initial temperature of 20°C and an unknown final temperature T, the temperature change for the body is $\Delta T = T_{\text{final}} - T_{\text{initial}} = T - 20°\text{C}$ (*not* $20°\text{C} - T$).
2. In problems where a phase change takes place, as when ice melts, you may not know in advance whether *all* the material undergoes a phase change or only part of it. You can always assume one or the other, and if the resulting calculation gives an absurd result (such as a final temperature higher or lower than *any* of the initial temperatures), you know the initial assumption was wrong. Back up and try again!

EVALUATE *your answer:* A common error is to use the wrong algebraic sign for either a Q or ΔT term. Double-check your calculations, and make sure that the final results are physically sensible.

Example 17.8 A temperature change with no phase change

A geologist working in the field drinks her morning coffee out of an aluminum cup. The cup has a mass of 0.120 kg and is initially at 20.0°C when she pours in 0.300 kg of coffee initially at 70.0°C. What is the final temperature after the coffee and the cup attain thermal equilibrium? (Assume that coffee has the same specific heat as water and that there is no heat exchange with the surroundings.)

SOLUTION

IDENTIFY and SET UP: No phase changes occur in this situation, so we need use only Eq. (17.13). The two objects we must consider are the cup and the coffee, and the target variable is their common final temperature.

EXECUTE: By using Table 17.3, the (negative) heat gained by the coffee is

$$Q_{\text{coffee}} = m_{\text{coffee}} c_{\text{water}} \Delta T_{\text{coffee}}$$
$$= (0.300\ \text{kg})(4190\ \text{J/kg}\cdot\text{K})(T - 70.0°\text{C})$$

The (positive) heat gained by the aluminum cup is

$$Q_{\text{aluminum}} = m_{\text{aluminum}} c_{\text{aluminum}} \Delta T_{\text{aluminum}}$$
$$= (0.120\ \text{kg})(910\ \text{J/kg}\cdot\text{K})(T - 20.0°\text{C})$$

We equate the sum of these two quantities of heat to zero, obtaining an algebraic equation for T:

$$Q_{\text{coffee}} + Q_{\text{aluminum}} = 0 \quad \text{or}$$
$$(0.300\ \text{kg})(4190\ \text{J/kg}\cdot\text{K})(T - 70.0°\text{C}) + (0.120\ \text{kg})(910\ \text{J/kg}\cdot\text{K})(T - 20.0°\text{C}) = 0$$

Solution of this equation gives $T = 66.0°\text{C}$.

EVALUATE: The final temperature is much closer to the initial temperature of the coffee than to that of the cup; water has a much larger specific heat than aluminum, and we have more than twice as much mass of water. We can also find the quantities of heat by substituting this value for T back into the original equations. We find that $Q_{\text{coffee}} = -5.0 \times 10^3$ J and $Q_{\text{aluminum}} = +5.0 \times 10^3$ J; Q_{coffee} is negative, which means that the coffee loses heat.

Example 17.9 Changes in both temperature and phase

A physics student wants to cool 0.25 kg of Diet Omni-Cola (mostly water), initially at 25°C, by adding ice initially at −20°C. How much ice should she add so that the final temperature will be 0°C with all the ice melted if the heat capacity of the container may be neglected?

SOLUTION

IDENTIFY and SET UP: The ice and the Omni-Cola are the objects that exchange heat. The Omni-Cola undergoes a temperature change only, while the ice undergoes both a temperature change and a phase change from solid to liquid. The target variable is the mass of ice, m_{ice}.

EXECUTE: The Omni-Cola loses heat, so the heat added to it is negative:

$$Q_{\text{Omni}} = m_{\text{Omni}} c_{\text{water}} \Delta T_{\text{Omni}}$$
$$= (0.25\ \text{kg})(4190\ \text{J/kg}\cdot\text{K})(0°\text{C} - 25°\text{C})$$
$$= -26{,}000\ \text{J}$$

From Table 17.3, the specific heat of ice (not the same as for liquid water) is 2.1×10^3 J/kg·K. Let the mass of ice be m_{ice}; then the heat Q_1 needed to warm it from −20°C to 0°C is

$$Q_1 = m_{ice}c_{ice}\,\Delta T_{ice}$$
$$= m_{ice}(2.1 \times 10^3\ \text{J/kg}\cdot\text{K})[0°\text{C} - (-20°\text{C})]$$
$$= m_{ice}(4.2 \times 10^4\ \text{J/kg})$$

From Eq. (17.20) the additional heat Q_2 needed to melt this mass of ice is the mass times the heat of fusion. Using Table 17.4, we find

$$Q_2 = m_{ice}L_f$$
$$= m_{ice}(3.34 \times 10^5\ \text{J/kg})$$

The sum of these three quantities must equal zero:

$$Q_{Omni} + Q_1 + Q_2 = -26{,}000\ \text{J} + m_{ice}(42{,}000\ \text{J/kg}) + m_{ice}(334{,}000\ \text{J/kg}) = 0$$

Solving this for m_{ice}, we get $m_{ice} = 0.069\ \text{kg} = 69\ \text{g}$.

EVALUATE: This mass of ice corresponds to three or four medium-size ice cubes, which seems reasonable for the quantity of Omni-Cola in this problem.

Example 17.10 What's cooking?

A heavy copper pot of mass 2.0 kg (including the copper lid) is at a temperature of 150°C. You pour 0.10 kg of water at 25°C into the pot, then quickly close the lid of the pot so that no steam can escape. Find the final temperature of the pot and its contents, and determine the phase (liquid or gas) of the water. Assume that no heat is lost to the surroundings.

SOLUTION

IDENTIFY and SET UP: The two objects that exchange heat are the water and the pot. Note that there are three conceivable outcomes in this situation. One, none of the water boils, and the final temperature is less than 100°C; two, a portion of the water boils, giving a mixture of water and steam at 100°C; or three, all the water boils, giving 0.10 kg of steam at a temperature of 100°C or greater.

EXECUTE: The simplest case to calculate is the first possibility, so let's try that first. Let the common final temperature of the liquid water and the copper pot be T. Since we are assuming that no phase changes take place, the sum of the quantities of heat added to the two materials is

$$Q_{water} + Q_{copper} = m_{water}c_{water}(T - 25°\text{C}) + m_{copper}c_{copper}(T - 150°\text{C})$$
$$= (0.10\ \text{kg})(4190\ \text{J/kg}\cdot\text{K})(T - 25°\text{C}) + (2.0\ \text{kg})(390\ \text{J/kg}\cdot\text{K})(T - 150°\text{C})$$
$$= 0$$

Solving this for T, we find $T = 106°\text{C}$. But this is above the boiling point of water, which contradicts our assumption that none of the water boils! So this assumption can't be correct; at least some of the water undergoes a phase change.

If we try the second possibility, in which the final temperature is 100°C, we have to find the fraction of water that changes to the gaseous phase. Let this fraction be x; the (positive) amount of heat needed to vaporize this water is $(xm_{water})L_v$. Setting the final temperature T equal to 100°C, we have

$$Q_{water} = m_{water}c_{water}(100°\text{C} - 25°\text{C}) + xm_{water}L_v$$
$$= (0.10\ \text{kg})(4190\ \text{J/kg}\cdot\text{K})(75\ \text{K}) + x(0.10\ \text{kg})(2.256 \times 10^6\ \text{J/kg})$$
$$= 3.14 \times 10^4\ \text{J} + x(2.256 \times 10^5\ \text{J})$$
$$Q_{copper} = m_{copper}c_{copper}(100°\text{C} - 150°\text{C})$$
$$= (2.0\ \text{kg})(390\ \text{J/kg}\cdot\text{K})(-50\ \text{K}) = -3.90 \times 10^4\ \text{J}$$

The requirement that the sum of all the quantities of heat be zero then gives

$$Q_{water} + Q_{copper} = 3.14 \times 10^4\ \text{J} + x(2.256 \times 10^5\ \text{J}) - 3.90 \times 10^4\ \text{J} = 0$$
$$x = \frac{3.90 \times 10^4\ \text{J} - 3.14 \times 10^4\ \text{J}}{2.256 \times 10^5\ \text{J}} = 0.034$$

This makes sense, and we conclude that the final temperature of the water and copper is 100°C. Of the original 0.10 kg of water, $0.034(0.10\ \text{kg}) = 0.0034\ \text{kg} = 3.4\ \text{g}$ has been converted to steam at 100°C.

EVALUATE: Had x turned out to be greater than 1, we would have again had a contradiction (the fraction of water that vaporized can't be greater than 1). In this case the third possibility would have been the correct description, all the water would have vaporized, and the final temperature would have been greater than 100°C. Can you show that this would have been the case if we had originally poured less than 15 g of 25°C water into the pot?

Example 17.11

Combustion, temperature change, and phase change

In a particular gasoline camp stove, 30% of the energy released in burning the fuel actually goes to heating the water in the pot on the stove. If we heat 1.00 L (1.00 kg) of water from 20°C to 100°C and boil 0.25 kg of it away, how much gasoline do we burn in the process?

SOLUTION

IDENTIFY and SET UP: In this problem, Eqs. (17.13) and (17.20) apply to the water, all of which undergoes a temperature change and part of which also undergoes a phase change from liquid to gas. This requires a certain amount of heat, which we use to determine the amount of gasoline that must be burned (the target variable).

EXECUTE: The heat required to raise the temperature of the water from 20°C to 100°C is

$$Q_1 = mc\,\Delta T = (1.00\text{ kg})(4190\text{ J/kg}\cdot\text{K})(80\text{ K}) = 3.35 \times 10^5\text{ J}$$

To boil 0.25 kg of water at 100°C requires

$$Q_2 = mL_v = (0.25\text{ kg})(2.256 \times 10^6\text{ J/kg}) = 5.64 \times 10^5\text{ J}$$

The total energy needed is the sum of these, or 8.99×10^5 J. This is only 0.30 of the total heat of combustion, so that energy is $(8.99 \times 10^5\text{ J})/0.30 = 3.00 \times 10^6$ J. As we mentioned earlier, one gram of gasoline releases 46,000 J, so the mass of gasoline required is

$$\frac{3.00 \times 10^6\text{ J}}{46{,}000\text{ J/g}} = 65\text{ g}$$

or a volume of about 0.09 L of gasoline.

EVALUATE: This result is a testament to the tremendous amount of energy that can be released by burning even a small quantity of gasoline. Note that most of the heat delivered was used to boil away 0.25 L of water. Can you show that another 123 g of gasoline would be required to boil away the remaining water?

Test Your Understanding

You take a block of ice at 0°C and add heat to it at a steady rate. After a time t, the block of ice has been converted completely to steam at 100°C. What is the temperature of the (ice? water? steam?) at time $t/2$, and what is its phase?

17.7 | Mechanisms of Heat Transfer

We have talked about *conductors* and *insulators,* materials that permit or prevent heat transfer between bodies. Now let's look in more detail at *rates* of energy transfer. In the kitchen you use an aluminum pot for good heat transfer from the stove to whatever you're cooking, but your refrigerator is insulated with a material that *prevents* heat from flowing into the food inside the refrigerator. How do we describe the difference between these two materials?

The three mechanisms of heat transfer are conduction, convection, and radiation. *Conduction* occurs within a body or between two bodies in contact. *Convection* depends on motion of mass from one region of space to another. *Radiation* is heat transfer by electromagnetic radiation, such as sunshine, with no need for matter to be present in the space between bodies.

Conduction

If you hold one end of a copper rod and place the other end in a flame, the end you are holding gets hotter and hotter, even though it is not in direct contact with the flame. Heat reaches the cooler end by **conduction** through the material. On the atomic level, the atoms in the hotter regions have more kinetic energy, on the average, than

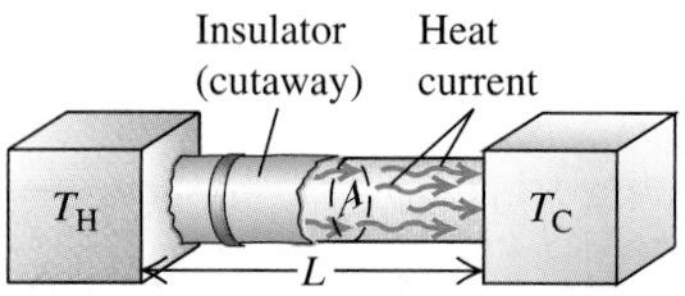

17.20 Steady-state heat flow due to conduction in a uniform rod.

their cooler neighbors. They jostle their neighbors, giving them some of their energy. The neighbors jostle *their* neighbors, and so on through the material. The atoms themselves do not move from one region of material to another, but their energy does.

Most metals also use another, more effective mechanism to conduct heat. Within the metal, some electrons can leave their parent atoms and wander through the crystal lattice. These "free" electrons can rapidly carry energy from the hotter to the cooler regions of the metal, so metals are generally good conductors of heat. A metal rod at 20°C feels colder than a piece of wood at 20°C because heat can flow more easily from your hand into the metal. The presence of "free" electrons also causes most metals to be good electrical conductors.

Heat transfer occurs only between regions that are at different temperatures, and the direction of heat flow is always from higher to lower temperature. Figure 17.20 shows a rod of conducting material with cross-sectional area A and length L. The left end of the rod is kept at a temperature T_H and the right end at a lower temperature T_C, so heat flows from left to right. The sides of the rod are covered by an ideal insulator, so no heat transfer occurs at the sides.

When a quantity of heat dQ is transferred through the rod in a time dt, the rate of heat flow is dQ/dt. We call this rate the **heat current,** denoted by H. That is, $H = dQ/dt$. Experiments show that the heat current is proportional to the cross-sectional area A of the rod and to the temperature difference $(T_H - T_C)$ and is inversely proportional to the rod length L. Introducing a proportionality constant k called the **thermal conductivity** of the material, we have

$$H = \frac{dQ}{dt} = kA\frac{T_H - T_C}{L} \quad \text{(heat current in conduction)} \quad (17.21)$$

The quantity $(T_H - T_C)/L$ is the temperature difference *per unit length;* it is called the magnitude of the **temperature gradient.** The numerical value of k depends on the material of the rod. Materials with large k are good conductors of heat; materials with small k are poor conductors or insulators. Equation (17.21) also gives the heat current through a slab or through *any* homogeneous body with uniform cross section A perpendicular to the direction of flow; L is the length of the heat-flow path.

The units of heat current H are units of energy per time, or power; the SI unit of heat current is the watt $(1\ \text{W} = 1\ \text{J/s})$. We can find the units of k by solving Eq. (17.21) for k; you can show that the SI units are W/m·K. Some numerical values of k are given in Table 17.5.

Table 17.5 Thermal Conductivities

Substance	k (W/m·K)
Metals	
Aluminum	205.0
Brass	109.0
Copper	385.0
Lead	34.7
Mercury	8.3
Silver	406.0
Steel	50.2
Various solids (representative values)	
Brick, insulating	0.15
Brick, red	0.6
Concrete	0.8
Cork	0.04
Felt	0.04
Fiberglass	0.04
Glass	0.8
Ice	1.6
Rock wool	0.04
Styrofoam	0.01
Wood	0.12–0.04
Gases	
Air	0.024
Argon	0.016
Helium	0.14
Hydrogen	0.14
Oxygen	0.023

The thermal conductivity of "dead" (that is, nonmoving) air is very small. A wool sweater keeps you warm because it traps air between the fibers. In fact, many insulating materials such as Styrofoam and fiberglass are mostly dead air. Figure 17.21 shows a ceramic material with very unusual thermal properties, including very small conductivity.

If the temperature varies in a non-uniform way along the length of the conducting rod, we introduce a coordinate x along the length and generalize the temperature gradient to be dT/dx. The corresponding generalization of Eq. (17.21) is

$$H = \frac{dQ}{dt} = -kA\frac{dT}{dx} \quad (17.22)$$

The negative sign shows that heat always flows in the direction of *decreasing* temperature.

For thermal insulation in buildings, engineers use the concept of **thermal resistance,** denoted by R. The thermal resistance R of a slab of material with area A is defined so that the heat current H through the slab is

$$H = \frac{A(T_H - T_C)}{R} \qquad (17.23)$$

where T_H and T_C are the temperatures on the two sides of the slab. Comparing this with Eq. (17.21), we see that R is given by

$$R = \frac{L}{k} \qquad (17.24)$$

where L is the thickness of the slab. The SI unit of R is $1\ m^2 \cdot K/W$. In the units used for commercial insulating materials in the United States, H is expressed in Btu/h, A is in ft^2, and $T_H - T_C$ in F°. $(1\ Btu/h = 0.293\ W.)$ The units of R are then $ft^2 \cdot F° \cdot h/Btu$, though values of R are usually quoted without units; a 6-inch-thick layer of fiberglass has an R value of 19 (that is, $R = 19\ ft^2 \cdot F° \cdot h/Btu$), a two-inch-thick slab of polyurethane foam has a value of 12, and so on. Doubling the thickness doubles the R-value. Common practice in new construction in severe northern climates is to specify R values of around 30 for exterior walls and ceilings. When the insulating material is in layers, such as a plastered wall, fiberglass insulation, and wood exterior siding, the R values are additive. Do you see why? (See Problem 17.110.)

17.21 This protective tile, developed for use in the space shuttle, has extraordinary thermal properties. The extremely small thermal conductivity and small heat capacity of the material make it possible to hold the tile by its edges, even though its temperature is high enough to emit the light for this photograph.

Problem-Solving Strategy: Heat Conduction

IDENTIFY *the relevant concepts:* The concept of heat conduction comes into play whenever two objects at different temperature are placed in contact.

SET UP *the problem using the following steps:*

1. Identify the direction of heat flow in the problem (from hot to cold). In Eq. (17.21), L is always measured along this direction, and A is always an area perpendicular to this direction. Often when a box or other container has an irregular shape but uniform wall thickness, you can approximate it as a flat slab with the same thickness and total wall area.
2. Identify the target variable.

EXECUTE *the solution as follows:*

1. If heat flows through a single object, use Eq. (17.21) to solve for the target variable.
2. In some problems the heat flows through two different materials in succession. The temperature at the interface between the two materials is then intermediate between T_H and T_C; represent it by a symbol such as T. The temperature differences for the two materials are then $(T_H - T)$ and $(T - T_C)$. In steady-state heat flow, the same heat has to pass through both materials in succession, so the heat current H must be *the same* in both materials.
3. If there are two *parallel* heat-flow paths, so that some heat flows through each, then the total H is the sum of the quantities H_1 and H_2 for the separate paths. An example is heat flow from inside to outside a house, both through the glass in a window and through the surrounding frame. In this case the temperature difference is the same for the two paths, but L, A, and k may be different for the two paths.
4. As always, it is essential to use a consistent set of units. If you use a value of k expressed in $W/m \cdot K$, don't use distances in centimeters, heat in calories, or T in degrees Fahrenheit!

EVALUATE *your answer:* As always, ask yourself whether the results are physically reasonable.

Example 17.12 Conduction through a picnic cooler

A Styrofoam box used to keep drinks cold at a picnic has total wall area (including the lid) of 0.80 m^2 and wall thickness 2.0 cm. It is filled with ice, water, and cans of Omni-Cola at 0°C. What is the rate of heat flow into the box if the temperature of the outside wall is 30°C? How much ice melts in one day?

SOLUTION

IDENTIFY and SET UP: The first target variable is the heat current H. The second is the amount of ice melted, which depends on the heat current (heat per unit time), the elapsed time, and the heat of fusion.

EXECUTE: We assume that the total heat flow is approximately the same as it would be through a flat slab of area 0.80 m^2 and thickness 2.0 cm = 0.020 m (Fig. 17.22). We find k from Table 17.5. From Eq. (17.21) the heat current (rate of heat flow) is

$$H = kA\frac{T_H - T_C}{L} = (0.010\ \text{W/m}\cdot\text{K})(0.80\ \text{m}^2)\frac{30°\text{C} - 0°\text{C}}{0.020\ \text{m}}$$
$$= 12\ \text{W} = 12\ \text{J/s}$$

The total heat flow Q in one day (86,400 s) is

$$Q = Ht = (12\ \text{J/s})(86{,}400\ \text{s}) = 1.04 \times 10^6\ \text{J}$$

The heat of fusion of ice is 3.34×10^5 J/kg, so the quantity of ice melted by this quantity of heat is

$$m = \frac{Q}{L_f} = \frac{1.04 \times 10^6\ \text{J}}{3.34 \times 10^5\ \text{J/kg}} = 3.1\ \text{kg}$$

EVALUATE: The low heat current is a result of the low thermal conductivity of Styrofoam. A substantial amount of heat flows in 24 hours, but a relatively small amount of ice melts because the heat of fusion is high.

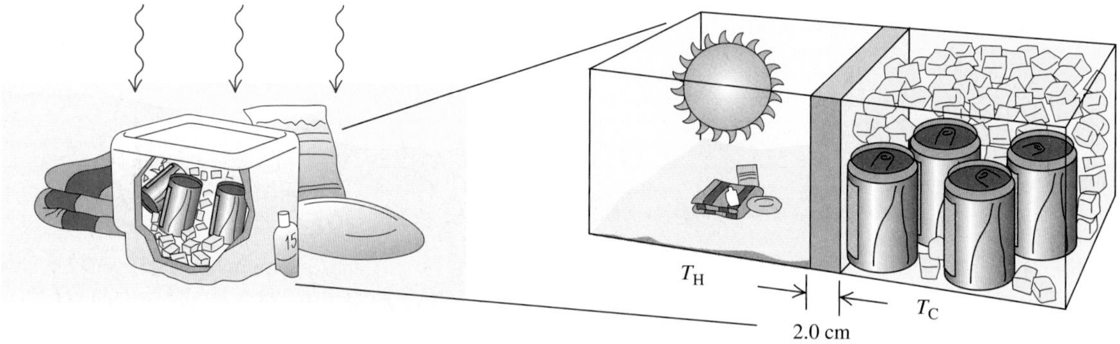

17.22 Conduction of heat. We can approximate heat flow through the walls of a picnic cooler by heat flow through a single flat slab of Styrofoam.

Example 17.13 Conduction through two bars I

A steel bar 10.0 cm long is welded end-to-end to a copper bar 20.0 cm long (Fig. 17.23). Both bars are insulated perfectly on their sides. Each bar has a square cross section, 2.00 cm on a side. The free end of the steel bar is maintained at 100°C by placing it in contact with steam, and the free end of the copper bar is maintained at 0°C by placing it in contact with ice. Find the temperature at the junction of the two bars and the total rate of heat flow.

SOLUTION

IDENTIFY and SET UP: As we discussed in the Problem-Solving Strategy, the heat currents in the two bars must be equal; this idea is the key to the solution. We write Eq. (17.21) twice, once for each bar, and set the heat currents H_{steel} and H_{copper} equal to each other. Both expressions for the heat current involve the temperature T at the junction, which is one of our target variables.

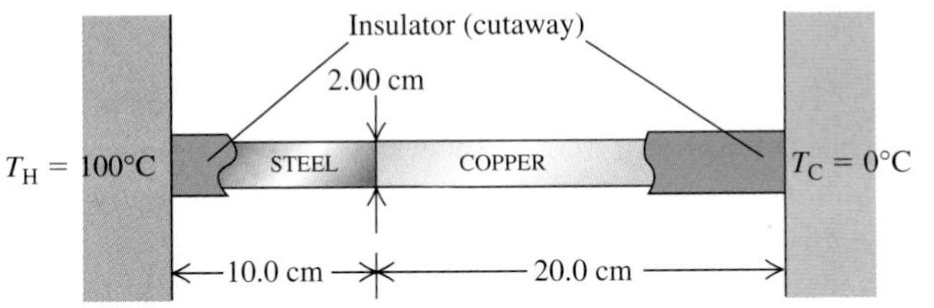

17.23 Heat flow along two metal bars, one of steel and one of copper, connected end-to-end.

EXECUTE: Setting the two heat currents equal,

$$H_{steel} = \frac{k_{steel}A(100°C - T)}{L_{steel}} = H_{copper} = \frac{k_{copper}A(T - 0°C)}{L_{copper}}$$

The areas A are equal and may be divided out. Substituting $L_{steel} = 0.100$ m, $L_{copper} = 0.200$ m, and numerical values of k from Table 17.5, we find

$$\frac{(50.2\ \text{W/m}\cdot\text{K})(100°C - T)}{0.100\ \text{m}} = \frac{(385\ \text{W/m}\cdot\text{K})(T - 0°C)}{0.200\ \text{m}}$$

Rearranging and solving for T, we find

$$T = 20.7°C$$

We can find the total heat current by substituting this value for T back into either of the above expressions:

$$H_{steel} = \frac{(50.2\ \text{W/m}\cdot\text{K})(0.0200\ \text{m})^2(100°C - 20.7°C)}{0.100\ \text{m}}$$

$$= 15.9\ \text{W}$$

or

$$H_{copper} = \frac{(385\ \text{W/m}\cdot\text{K})(0.0200\ \text{m})^2(20.7°C)}{0.200\ \text{m}} = 15.9\ \text{W}$$

EVALUATE: Even though the steel bar is shorter, the temperature drop across it is much greater than across the copper bar (from 100°C to 20.7°C in the steel versus from 20.7°C to 0°C in the copper). This difference arises because steel is a much poorer conductor than copper.

Example 17.14 Conduction through two bars II

In Example 17.13, suppose the two bars are separated. One end of each bar is maintained at 100°C and the other end of each bar is maintained at 0°C (Fig. 17.24). What is the *total* rate of heat flow in the two bars?

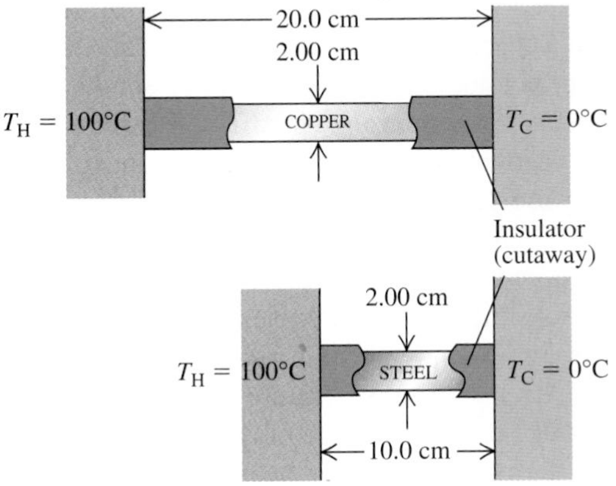

17.24 Heat flow along two metal bars, one of steel and one of copper, parallel to and separated from each other.

SOLUTION

IDENTIFY and SET UP: In this case the bars are in parallel rather than in series. The total heat current is now the *sum* of the currents in the two bars, and for each bar, $T_H - T_C = 100°C - 0°C = 100$ K.

EXECUTE: We write the heat currents for the two rods individually, then add them to get the total heat current:

$$H = H_{steel} + H_{copper} = \frac{k_{steel}A(T_H - T_C)}{L_{steel}} + \frac{k_{copper}A(T_H - T_C)}{L_{copper}}$$

$$= \frac{(50.2\ \text{W/m}\cdot\text{K})(0.0200\ \text{m})^2(100\ \text{K})}{0.100\ \text{m}}$$

$$+ \frac{(385\ \text{W/m}\cdot\text{K})(0.0200\ \text{m})^2(100\ \text{K})}{0.200\ \text{m}}$$

$$= 20.1\ \text{W} + 77.0\ \text{W} = 97.1\ \text{W}$$

EVALUATE: The heat flow in the copper bar is much greater than that in the steel bar, even though it is longer, because the thermal conductivity of copper is much larger. The total heat flow is much greater than in Example 17.13, partly because the total cross section for heat flow is greater and partly because the full 100-K temperature difference appears across each bar.

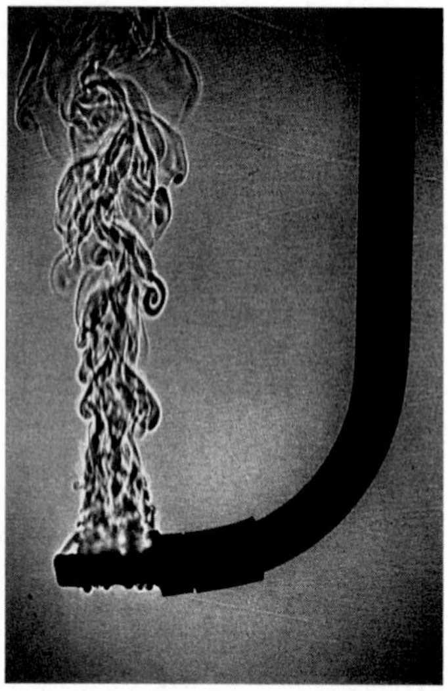

17.25 A heating element in the tip of this submerged tube warms the surrounding water, producing a complex pattern of free convection.

Convection

Convection is the transfer of heat by mass motion of a fluid from one region of space to another. Familiar examples include hot-air and hot-water home heating systems, the cooling system of an automobile engine, and the flow of blood in the body. If the fluid is circulated by a blower or pump, the process is called *forced convection;* if the flow is caused by differences in density due to thermal expansion, such as hot air rising, the process is called *natural convection* or *free convection* (Fig. 17.25).

Free convection in the atmosphere plays a dominant role in determining the daily weather, and convection in the oceans is an important global heat-transfer mechanism. On a smaller scale, soaring hawks and glider pilots make use of thermal updrafts from the warm earth. The most important mechanism for heat transfer within the human body (needed to maintain nearly constant temperature in various environments) is *forced* convection of blood, with the heart serving as the pump.

Convective heat transfer is a very complex process, and there is no simple equation to describe it. Here are a few experimental facts:

1. The heat current due to convection is directly proportional to the surface area. This is the reason for the large surface areas of radiators and cooling fins.
2. The viscosity of fluids slows natural convection near a stationary surface, giving a surface film that on a vertical surface typically has about the same insulating value as 1.3 cm of plywood $(R \text{ value} = 0.7)$. Forced convection decreases the thickness of this film, increasing the rate of heat transfer. This is the reason for the "wind-chill factor"; you get cold faster in a cold wind than in still air with the same temperature.
3. The heat current due to convection is found to be approximately proportional to the $\frac{5}{4}$ power of the temperature difference between the surface and the main body of fluid.

Radiation

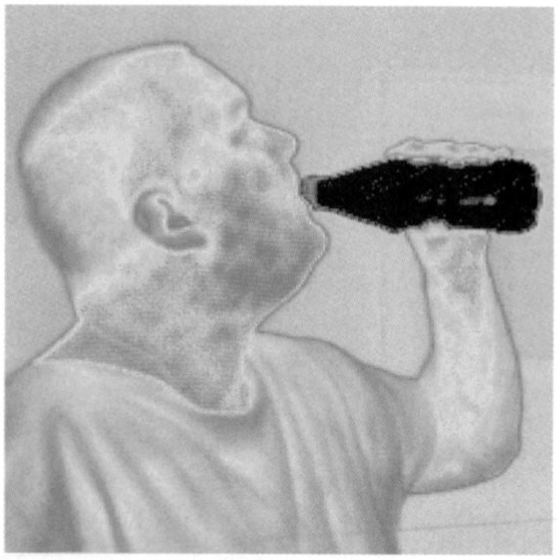

17.26 This false-color infrared photograph reveals radiation emitted by various parts of the man's body. The strongest emission (colored red) comes from the warmest areas, while there is almost no emission from the bottle of cold beverage.

Radiation is the transfer of heat by electromagnetic waves such as visible light, infrared, and ultraviolet radiation. Everyone has felt the warmth of the sun's radiation and the intense heat from a charcoal grill or the glowing coals in a fireplace. Most of the heat from these very hot bodies reaches you not by conduction or convection in the intervening air but by *radiation.* This heat transfer would occur even if there were nothing but vacuum between you and the source of heat.

Every body, even at ordinary temperatures, emits energy in the form of electromagnetic radiation. At ordinary temperatures, say 20°C, nearly all the energy is carried by infrared waves with wavelengths much longer than those of visible light (see Figs. 17.4 and 17.26). As the temperature rises, the wavelengths shift to shorter values. At 800°C a body emits enough visible radiation to be self-luminous and appears "red-hot," although even at this temperature most of the energy is carried by infrared waves. At 3000°C, the temperature of an incandescent lamp filament, the radiation contains enough visible light so the body appears "white-hot."

The rate of energy radiation from a surface is proportional to the surface area A. The rate increases very rapidly with temperature, depending on the fourth power of the absolute (Kelvin) temperature. The rate also depends on the nature of the surface; this dependence is described by a quantity e called the **emissivity.** A dimensionless number between 0 and 1, it represents txhe ratio of the rate of radiation from a particular surface to the rate of radiation from an equal area of an ideal radiating surface at the same temperature. Emissivity also depends somewhat on temperature. Thus the heat current $H = dQ/dt$ due to radiation from a surface area A with emissivity e at absolute temperature T can be expressed as

$$H = Ae\sigma T^4 \qquad \text{(heat current in radiation)} \qquad (17.25)$$

where σ is a fundamental physical constant called the **Stefan-Boltzmann constant.** This relation is called the **Stefan-Boltzmann law** in honor of its late-19th-century discoverers. The current best numerical value of σ is

$$\sigma = 5.670400(40) \times 10^{-8}\ \text{W/m}^2 \cdot \text{K}^4$$

We invite you to check unit consistency in Eq. (17.25). Emissivity (e) is often larger for dark surfaces than for light ones. The emissivity of a smooth copper surface is about 0.3, but e for a dull black surface can be close to unity.

Example 17.15 Heat transfer by radiation

A thin square steel plate, 10 cm on a side, is heated in a blacksmith's forge to a temperature of 800°C. If the emissivity is 0.60, what is the total rate of radiation of energy?

SOLUTION

IDENTIFY and SET UP: We use Eq. (17.25). The target variable is H, the rate of emission of energy. All of the other quantities are given.

EXECUTE: The total surface area, including both sides, is $2(0.10\ \text{m})^2 = 0.020\ \text{m}^2$. We must convert the temperature to the Kelvin scale; 800°C = 1073 K. Then Eq. (17.25) gives

$$\begin{aligned} H &= Ae\sigma T^4 \\ &= (0.020\ \text{m}^2)(0.60)(5.67 \times 10^{-8}\ \text{W/m}^2 \cdot \text{K}^4)(1073\ \text{K})^4 \\ &= 900\ \text{W} \end{aligned}$$

EVALUATE: A blacksmith standing nearby will easily feel heat being radiated from this plate.

While a body at absolute temperature T is radiating, its surroundings at temperature T_s are also radiating, and the body *absorbs* some of this radiation. If it is in thermal equilibrium with its surroundings, $T = T_s$ and the rates of radiation and absorption must be equal. For this to be true, the rate of absorption must be given in general by $H = Ae\sigma T_s^4$. Then the *net* rate of radiation from a body at temperature T with surroundings at temperature T_s is

$$H_{net} = Ae\sigma T^4 - Ae\sigma T_s^4 = Ae\sigma(T^4 - T_s^4) \qquad (17.26)$$

In this equation a positive value of H means a net heat flow *out of* the body. Equation (17.26) shows that for radiation, as for conduction and convection, the heat current depends on the temperature *difference* between two bodies.

Example 17.16 Radiation from the human body

If the total surface area of the human body is 1.20 m² and the surface temperature is 30°C = 303 K, find the total rate of radiation of energy from the body. If the surroundings are at a temperature of 20°C, what is the *net* rate of heat loss from the body by radiation? The emissivity of the body is very close to unity, irrespective of skin pigmentation.

SOLUTION

IDENTIFY and SET UP: The rate of radiation of energy from the body is given by Eq. (17.25), and the net rate of heat loss is given by Eq. (17.26).

EXECUTE: Taking $e = 1$ in Eq. (17.25), we find that the body radiates at a rate

$$H = Ae\sigma T^4$$
$$= (1.20\ \text{m}^2)(1)(5.67 \times 10^{-8}\ \text{W/m}^2\cdot\text{K}^4)(303\ \text{K})^4$$
$$= 574\ \text{W}$$

This loss is partly offset by *absorption* of radiation, which depends on the temperature of the surroundings. The *net* rate of radiative energy transfer is given by Eq. (17.26):

$$H_{\text{net}} = Ae\sigma(T^4 - T_s^4)$$
$$= (1.20\ \text{m}^2)(1)(5.67 \times 10^{-8}\ \text{W/m}^2\cdot\text{K}^4)$$
$$\times [(303\ \text{K})^4 - (293\ \text{K})^4] = 72\ \text{W}$$

EVALUATE: The value of H_{net} is positive because the body is losing heat to its colder surroundings.

Heat transfer by radiation is important in some surprising places. A premature baby in an incubator can be cooled dangerously by radiation if the walls of the incubator happen to be cold, even when the *air* in the incubator is warm. Some incubators regulate the air temperature by measuring the baby's skin temperature.

A body that is a good absorber must also be a good emitter. An ideal radiator, with an emissivity of unity, is also an ideal absorber, absorbing *all* of the radiation that strikes it. Such an ideal surface is called an ideal black body or simply a **blackbody.** Conversely, an ideal *reflector,* which absorbs *no* radiation at all, is also a very ineffective radiator.

This is the reason for the silver coatings on vacuum ("Thermos") bottles, invented by Sir James Dewar (1842–1923). A vacuum bottle has double glass walls. The air is pumped out of the spaces between the walls; this eliminates nearly all heat transfer by conduction and convection. The silver coating on the walls reflects most of the radiation from the contents back into the container, and the wall itself is a very poor emitter. Thus a vacuum bottle can keep coffee or soup hot for several hours. The Dewar flask, used to store very cold liquefied gases, is exactly the same in principle.

Test Your Understanding

The ear thermometer in Fig. 17.4 measures the radiation emitted by the eardrum. By what percentage does the radiation rate increase if the eardrum's temperature increases from 37.00°C to 37.10°C?

SUMMARY

A thermometer measures temperature. Two bodies in thermal equilibrium must have the same temperature. A conducting material between two bodies permits them to interact and come to thermal equilibrium; an insulating material prevents or impedes this interaction.

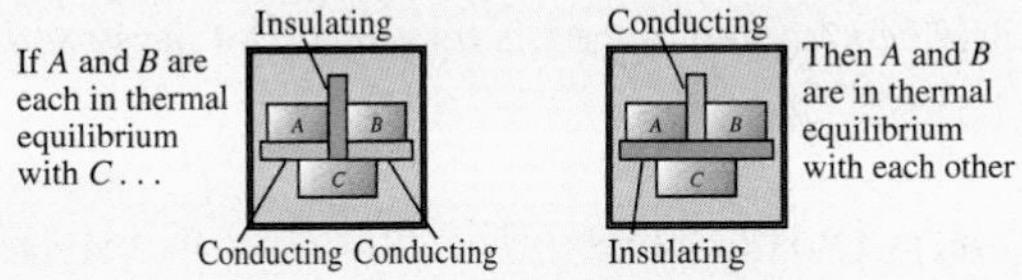

The Celsius and Fahrenheit temperature scales are based on the freezing temperature $(0°\text{C} = 32°\text{F})$ and boiling temperature $(100°\text{C} = 212°\text{F})$. One Celsius degree equals $\frac{9}{5}$ Fahrenheit degrees. (See Example 17.1)

$$T_F = \frac{9}{5}T_C + 32° \quad (17.1)$$

$$T_C = \frac{5}{9}(T_F - 32°) \quad (17.2)$$

The Kelvin scale has its zero at the extrapolated zero-pressure temperature for a gas thermometer, $-273.15°\text{C} = 0\ \text{K}$. In the gas-thermometer scale, the ratio of two temperatures T_1 and T_2 is defined to be equal to the ratio of the two corresponding gas-thermometer pressures p_1 and p_2. The triple-point temperature of water (0.01°C) is defined to be 273.16 K.

$$T_K = T_C + 273.15 \quad (17.3)$$

$$\frac{T_2}{T_1} = \frac{p_2}{p_1} \quad (17.4)$$

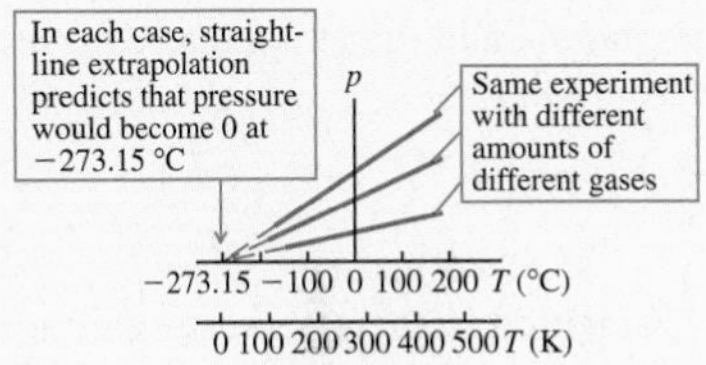

A temperature change ΔT causes a change in any linear dimension L_0 of a solid body. The change ΔL is approximately proportional to L_0 and ΔT. Similarly, a temperature change causes a change ΔV in the volume V_0 of any solid or liquid material that is approximately proportional to V_0 and ΔT. The quantities α and β are the coefficients of linear expansion and volume expansion. For solids, $\beta = 3\alpha$. (See Examples 17.2 through 17.4)

$$\Delta L = \alpha L_0 \,\Delta T \quad (17.6)$$

$$\Delta V = \beta V_0 \,\Delta T \quad (17.8)$$

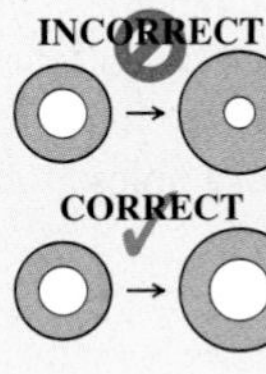

When a material is cooled or heated and held so it cannot contract or expand, it is under a tensile stress F/A. (See Example 17.5)

$$\frac{F}{A} = -Y\alpha\,\Delta T \quad (17.12)$$

Heat is energy in transit from one body to another as a result of a temperature difference. The quantity of heat Q required to raise the temperature of a quantity of material by a small amount ΔT is proportional to ΔT. This proportionality can be expressed either in terms of the mass m and specific heat capacity c or in terms of the number of moles n and the molar heat capacity $C = Mc$. Here M is the molar mass and $m = nM$. (See Examples 17.6 and 17.7)

$$Q = mc\,\Delta T \quad (17.13)$$

$$Q = nC\,\Delta T \quad (17.18)$$

To change a mass m of a material to a different phase at the same temperature (such as liquid to solid or liquid to vapor) requires the addition or subtraction of a quantity of heat. The amount of heat is equal to the product of m and L, the heat of fusion, vaporization, or sublimation.

$$Q = \pm mL \quad (17.20)$$

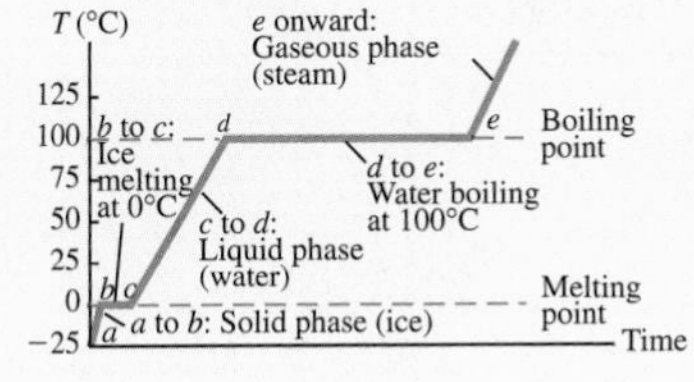

When heat is added to a body, the corresponding Q is positive; when it is removed, Q is negative. The basic principle of calorimetry comes from conservation of energy. In an isolated system whose parts interact by heat exchange, the algebraic sum of the Q's for all parts of the system must be zero. (See Examples 17.8 through 17.11)

Conduction is the transfer of energy of molecular motion within bulk materials without bulk motion of the materials. The heat current H for conduction depends on the area A through which the heat flows, the length L of the heat-flow path, the temperature difference $(T_H - T_C)$, and the thermal conductivity k of the material.
(See Examples 17.2 through 17.4)

$$H = \frac{dQ}{dt} = kA\frac{T_H - T_C}{L} \qquad (17.21)$$

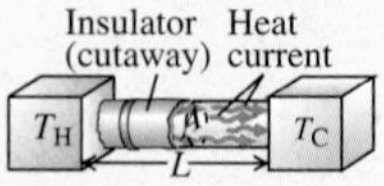

Convection is a complex heat transfer process that involves mass motion from one region to another. It depends on surface area, orientation, and the temperature difference between a body and its surroundings.

Radiation is energy transfer through electromagnetic radiation. The radiation heat current H depends on the surface area A, the emissivity e of the surface (a pure number between 0 and 1), and the Kelvin temperature T. It involves a fundamental constant σ called the Stefan-Boltzmann constant. When a body at temperature T is surrounded by material at temperature T_s, the *net* heat current H_{net} from the body to its surroundings depends on both T and T_s.
(See Examples 17.5 and 17.16)

$$H = Ae\sigma T^4 \qquad (17.25)$$

$$H_{net} = Ae\sigma(T^4 - T_s^4) \qquad (17.26)$$

Key Terms

thermodynamics, 640
temperature, 641
thermometer, 641
thermal equilibrium, 641
insulator, 641
conductor, 642
zeroth law of thermodynamics, 642
Celsius temperature scale, 642
Fahrenheit temperature scale, 643
Kelvin temperature scale, 644
absolute temperature scale, 646
absolute zero, 646
coefficient of linear expansion, 647
coefficient of volume expansion, 648
thermal stress, 651
heat, 653
calorie, 653
British thermal unit, 653
specific heat, 654
molar heat capacity, 655
phase, 657
states of matter, 657
phase change, 657
heat of fusion, 657
phase equilibrium, 658
heat of vaporization, 658
heat of combustion, 660
conduction, 663
heat current, 664
thermal conductivity, 664
temperature gradient, 664
thermal resistance, 665
convection, 668
radiation, 668
emissivity, 668
Stefan–Boltzmann constant, 669
Stefan–Boltzmann law, 669
blackbody, 670

Your Notes

Answer to Chapter Opening Question ?

No. By "heat" we mean energy that is in transit from one body to another as a result of temperature difference between the bodies. Bodies do not *contain* heat.

Answers to Test Your Understanding Questions

Section 17.1 The nurse is waiting for the thermometer and your body to come into thermal equilibrium. The reading actually shows the reading of the thermometer; if the thermometer is in thermal equilibrium with your body, it also indicates body temperature.

Section 17.2 From Eq. (17.1), on Venus $T_F = \frac{9}{5}T_C + 32° = \frac{9}{5}(460°) + 32° = 860°F$. The two scales agree when $T_F = T_C$; again using Eq. (17.1), this means $T_C = \frac{9}{5}T_C + 32°$. Solving, $\frac{4}{5}T_C = -32°$ and $T_C = \frac{5}{4}(-32°C) = -40°C$. Hence $-40°F$ is the same temperature as $-40°C$.

Section 17.3 The Kelvin temperature of the corona is $2.0 \times 10^7 + 273.15 = 2.0 \times 10^7$ K. (Note that the Celsius temperature is given to only two significant figures.) The Kelvin temperature of the boiling point of helium is $-268.93 + 273.15 = 4.22$ K. The Kelvin scale is convenient at high temperatures where the difference between 0 K and 0°C is negligible. At low temperatures, using the Kelvin scale gives smaller, more easily interpreted values.

Section 17.4 Metal 2 must expand more than metal 1 when heated and so must have a larger coefficient of linear expansion α. From Table 17.1, two metals with larger values of α than copper are aluminum and brass.

Section 17.5 The temperature change is $\Delta T = 70$ K, so the required amount of heat is $Q = m_{water}c\,\Delta T = (1.00\text{ kg})(4190\text{ J/kg}\cdot\text{K})(70\text{ K}) = 2.93 \times 10^5$ J. If the block falls a distance h, the amount of potential energy lost is $m_{block}gh$, so $h = Q/m_{block}g = (2.93 \times 10^5\text{ J})/(1.00\text{ kg})(9.80\text{ m/s}^2) = 2.99 \times 10^4$ m, or 29.9 kilometers! Because the specific heat of water is so high, this is a *very* impractical way to make tea.

Section 17.6 In time t the system goes from point b to point e in Fig. 17.18. According to this figure, at time $t/2$ (halfway along the horizontal axis from b to e), the system is at 100°C and is still boiling; that is, it is a mixture of liquid and gas. This says that most of the heat added goes into boiling the water.

Section 17.7 From Eq. (17.25), the radiation rate (heat current) is proportional to the fourth power of the Kelvin temperature. The temperature increases from $(37.00 + 273.15)$ K $= 310.15$ K to $(37.10 + 273.15)$ K $= 310.25$ K, so the radiation rate increases by a factor $[(310.25\text{ K})/(310.15\text{ K})]^4 = 1.0013$. Hence the percentage increase is 0.13%, which is well within the sensitivity of an ear thermometer.

Discussion Questions

Q17.1 A thermometer is laid out in direct sunlight. Does it measure the temperature of the air, or of the sun, or of something else? Explain.

Q17.2 Does it make sense to say that one body is twice as hot as another? Why or why not?

Q17.3 Many automobile engines have cast-iron cylinders and aluminum pistons. What kinds of problems could occur if the engine gets too hot? (The coefficient of volume expansion of cast iron is approximately the same as that of steel.)

Q17.4 Why do frozen water pipes burst? Would a mercury thermometer break if the temperature went below the freezing temperature of mercury? Why or why not?

Q17.5 Two bodies made of the same material have the same external dimensions and appearance, but one is solid and the other is hollow. When their temperature is increased, is the overall volume expansion the same or different? Why?

Q17.6 The inside of an oven is at a temperature of 200°C (392°F). You can put your hand in the oven without injury as long as you don't touch anything. But since the air inside the oven is also at 200°C, why isn't your hand burned just the same?

Q17.7 A newspaper article about the weather states that "the temperature of a body measures how much heat the body contains." Is this description correct? Why or why not?

Q17.8 To raise the temperature of an object, must you add heat to it? If you add heat to an object, must you raise its temperature? Explain.

Q17.9 A student asserts that a suitable unit for specific heat capacity is $1\text{ m}^2/\text{s}^2\cdot\text{C}°$. Is she correct? Why or why not?

Q17.10 In some household air conditioners used in dry climates, air is cooled by blowing it through a water-soaked filter, evaporating some of the water. How does this cool the air? Would such a system work well in a high-humidity climate? Why or why not?

Q17.11 The units of specific heat capacity c are J/kg·K, but the units of heat of fusion L_f or heat of vaporization L_v are simply J/kg. Why do the units of L_f and L_v not include a factor of $(\text{K})^{-1}$ to account for a temperature change?

Q17.12 If you set an ice cube on your kitchen table, it will stay frozen longer if it is wrapped in a wet paper towel. Why?

Q17.13 Why does food cook faster in a pressure cooker than in an open pot of boiling water?

Q17.14 Desert travelers sometimes keep water in a canvas bag. Some water seeps through the bag and evaporates. How does this cool the water inside?

Q17.15 When you first step out of the shower, you feel cold. But as soon as you are dry you feel warmer, even though the room temperature does not change. Why?

Q17.16 The climate of regions adjacent to large bodies of water (like the Pacific and Atlantic coasts) is usually more moderate than that of regions far from large bodies of water (like the prairies). Why?

Q17.17 When water is placed in ice-cube trays in a freezer, why does the water not freeze all at once when the temperature has reached 0°C? In fact the water freezes first in a layer adjacent to the sides of the tray. Why?

Q17.18 Before giving you an injection, a physician swabs your arm with isopropyl alcohol at room temperature. Why does this make your arm feel cold? (*Hint:* The reason is *not* the fear of the injection! The boiling point of isopropyl alcohol is 82.4°C.)

Q17.19 A cold block of metal feels colder than a block of wood at the same temperature. Why? A *hot* block of metal feels hotter than a block of wood at the same temperature. Again, why? Is there any temperature at which the two blocks feel equally hot or cold? What temperature is this?

Q17.20 A person pours a cup of hot coffee, intending to drink it five minutes later. To keep the coffee as hot as possible, should she put cream in it now, or wait until just before she drinks it? Explain.

Q17.21 When a freshly baked apple pie has just been removed from the oven, the crust and filling are both at the same temperature. Yet if you sample the pie, the filling will burn your tongue but the crust will not. Why is there a difference? (*Hint:* The filling is moist while the crust is dry.)

Q17.22 Old-time kitchen lore suggests that things cook better (evenly and without burning) in heavy cast-iron pots. What desirable characteristics do such pots have?

Q17.23 Coastal land has a higher temperature than the ocean during the day, but a lower temperature at night. Explain why. (*Hint:* The specific heat capacity of soil is only 0.2–0.8 times as great as that of water.)

Q17.24 It is well known that a potato bakes faster if a large nail is stuck through it. Why? Is an aluminum nail better than a steel one? Why or why not? (*Note:* Don't try this in a microwave oven!) There is also a gadget on the market to hasten the roasting of meat, consisting of a hollow metal tube containing a wick and some water; this is claimed to be much better than a solid metal rod. How does this work?

Q17.25 Glider pilots in the Midwest know that thermal updrafts are likely to occur in the vicinity of freshly plowed fields. Why?

Q17.26 Some folks claim that ice cubes freeze faster if the trays are filled with hot water, because hot water cools off faster than cold water. What do you think?

Q17.27 We're lucky that the earth isn't in thermal equilibrium with the sun (which has a surface temperature of 5800 K). But why are they not in thermal equilibrium?

Q17.28 When energy shortages occur, magazines sometimes urge us to keep our homes at a constant temperature day and night to conserve fuel. They argue that by turning down the heat at night, the walls, ceilings, etc. cool off and must be reheated in the morning. So, by keeping the temperature constant, these parts of the house would not cool off and would not have to be reheated. Does this argument make sense? Would we really save energy by following this advice?

Exercises

Section 17.2 Thermometers and Temperature Scales

17.1 Convert the following Celsius temperatures to Fahrenheit: a) -62.8°C, the lowest temperature ever recorded in North America (February 3, 1947, Snag, Yukon); b) 56.7°C, the highest temperature ever recorded in the United States (July 10, 1913, Death Valley, California); c) 31.1°C, the world's highest average annual temperature (Lugh Ferrandi, Somalia).

17.2 Find the Celsius temperatures corresponding to a) a winter night in Seattle (41.0°F); b) a hot summer day in Palm Springs (107.0°F); c) a cold winter day in northern Manitoba $(-18.0°\text{F})$.

17.3 You work in a materials testing lab and your boss tells you to increase the temperature of a sample by 40.0C°. The only thermometer you can find at your workbench reads in °F. If the initial temperature of the sample is 68.2°F, what is its temperature in °F when the desired temperature increase has been achieved?

17.4 a) On January 22, 1943, the temperature in Spearfish, South Dakota, rose from -4.0°F to 45.0°F in just two minutes. What was the temperature change in Celsius degrees? b) The temperature in Browning, Montana, was 44.0°F on January 23, 1916. The next day the temperature plummeted to -56.0°F. What was the temperature change in Celsius degrees?

17.5 a) You feel sick and are told that you have a temperature of 40.2°C. What is your temperature in °F? Should you be concerned? b) The morning weather report in Sydney gives a current temperature of 12°C. What is this temperature in °F?

17.6 A "blue norther" passes through Lubbock, Texas, one September afternoon and the temperature drops 11.8 C° in one hour. What is this temperature change in F°?

17.7 Two beakers of water, A and B, initially are at the same temperature. The temperature of the water in beaker A is increased 10 F° and the temperature of the water in beaker B is increased 10 K. After these temperature changes, which beaker of water has the higher temperature? Explain.

17.8 You put a bottle of soft drink in a refrigerator and leave it until its temperature has dropped 10.0 K. What is its temperature change in a) F°; b) C°?

Section 17.3 Gas Thermometers and the Kelvin Scale

17.9 Convert the following record-setting temperatures to the Kelvin scale: a) the lowest temperature recorded in the 48 contiguous states (-70.0°F at Rogers Pass, Montana, on January 20, 1954); b) Australia's highest temperature (127.0°F at Cloncurry, Queensland, on January 16, 1889); c) the lowest temperature recorded in the northern hemisphere (-90.0°F at Verkhoyansk, Siberia, in 1892).

17.10 Convert the following Kelvin temperatures to the Celsius and Fahrenheit scales: a) the midday temperature at the surface of the moon (400 K); b) the temperature at the tops of the clouds in the atmosphere of Saturn (95 K); c) the temperature at the center of the sun $(1.55 \times 10^7\ \text{K})$.

17.11 The normal boiling point of liquid neon is -245.92°C. What is this temperature on the Kelvin scale?

17.12 The ratio of the pressure of a gas at the melting point of platinum and its pressure at the triple point of water, when the gas is kept at constant volume, is found to be 7.476. What is the Celsius temperature of the melting point of platinum?

17.13 A gas thermometer registers an absolute pressure corresponding to 325 mm of mercury when in contact with water at the triple point. What pressure does it read when in contact with water at the normal boiling point?

17.14 Like the Kelvin scale, the *Rankine scale* is an absolute temperature scale: absolute zero is zero degrees Rankine (0°R). However, the units of this scale are the same size as those on the Fahrenheit scale rather than the Celsius scale. What is the numerical value of the triple-point temperature of water on the Rankine scale?

17.15 A Constant-Volume Gas Thermometer. An experimenter using a gas thermometer found the pressure at the triple point of water (0.01°C) to be 4.80×10^4 Pa and the pressure at the normal boiling point (100°C) to be 6.50×10^4 Pa. a) Assuming that the pressure varies linearly with temperature, use these two data points to find the Celsius temperature at which the gas pressure would be zero (that is, find the Celsius temperature of absolute zero). b) Does the gas in this thermometer obey Eq. (17.4) precisely? If that equation were precisely obeyed, and the pressure at 100°C were 6.50×10^4 Pa, what pressure would the experimenter have measured at 0.01°C? (As we will learn in Section 18.1, Eq. (17.4) is accurate only for gases at very low density.)

Section 17.4 Thermal Expansion

17.16 Air Friction and Thermal Expansion. The supersonic airliner Concorde is 62.1 m long when sitting on the ground on a typical (15°C) day. It is primarily made of aluminum. In flight at twice the speed of sound, friction with the air warms the Concorde's skin and causes the aircraft to lengthen by 25 cm. (The passenger cabin is on rollers, and the airplane expands around the passengers.) What is the temperature of the Concorde's skin in flight?

17.17 The Humber Bridge in England has the world's longest single span, 1410 m in length. Calculate the change in length of the steel deck of the span when the temperature increases from −5.0°C to 18.0°C.

17.18 Ensuring a Tight Fit. Aluminum rivets used in airplane construction are made slightly larger than the rivet holes and cooled by "dry ice" (solid CO_2) before being driven. If the diameter of a hole is 4.500 mm, what should be the diameter of a rivet at 23.0°C, if its diameter is to equal that of the hole when the rivet is cooled to −78.0°C, the temperature of dry ice? Assume that the expansion coefficient remains constant at the value given in Table 17.1.

17.19 A U.S. one-cent coin has a diameter of 1.9000 cm at 20.0°C. The coin is made of a metal alloy (mostly zinc) for which the coefficient of linear expansion is $2.6 \times 10^{-5}\ K^{-1}$. What would its diameter be a) on a hot day in Death Valley (48.0°C)? b) on a cold night in the mountains of Greenland (−53.0°C)?

17.20 The pendulum shaft of a clock is made of brass. What is the fractional change in length of the shaft when it is cooled from 19.50°C to 5.00°C?

17.21 A metal rod is 40.125 cm long at 20.0°C and 40.148 cm long at 45.0°C. Calculate the average coefficient of linear expansion of the rod for this temperature range.

17.22 A copper cylinder is initially at 20.0°C. At what temperature will its volume be 0.150% larger than it is at 20.0°C?

17.23 An underground tank with a capacity of 1700 L ($1.70\ m^3$) is filled with ethanol that has an initial temperature of 19.0°C. After the ethanol has cooled off to the temperature of the tank and ground, which is 10.0°C, how much air space will there be above the ethanol in the tank? (Assume that the volume of the tank doesn't change.)

17.24 A steel tank is completely filled with $2.80\ m^3$ of ethanol when both the tank and the ethanol are at a temperature of 32.0°C. When the tank and its contents have cooled to 18.0°C, what additional volume of ethanol can be put into the tank?

17.25 A glass flask whose volume is $1000.00\ cm^3$ at 0.0°C is completely filled with mercury at this temperature. When flask and mercury are warmed to 55.0°C, $8.95\ cm^3$ of mercury overflow. If the coefficient of volume expansion of mercury is $18.0 \times 10^{-5}\ K^{-1}$, compute the coefficient of volume expansion of the glass.

17.26 a) If an area measured on the surface of a solid body is A_0 at some initial temperature and then changes by ΔA when the temperature changes by ΔT, show that

$$\Delta A = (2\alpha)A_0\Delta T$$

where α is the coefficient of linear expansion. b) A circular sheet of aluminum is 55.0 cm in diameter at 15.0°C. By how much does the area of one side of the sheet change when the temperature increases to 27.5°C?

17.27 A machinist bores a hole of diameter 1.350 cm in a steel plate at a temperature of 25°C. What is the cross-sectional area of the hole a) at 25°C; b) when the temperature of the plate is increased to 175°C? Assume that the coefficient of linear expansion remains constant over this temperature range. (*Hint:* See Exercise 17.26.)

17.28 As a new mechanical engineer for Engines Inc., you have been assigned to design brass pistons to slide inside steel cylinders. The engines in which these pistons will be used will operate between 20°C and 150°C. Assume that the coefficients of expansion are constant over this temperature range. a) If the piston just fits inside the chamber at 20°C, will the engines be able to run at higher temperatures? Explain. b) If the cylindrical pistons are 25.000 cm in diameter at 20°C, what should be the minimum diameter of the cylinders at that temperature so the pistons will operate at 150°C?

17.29 The markings on an aluminum ruler and a brass ruler are perfectly aligned at 0°C. How far apart will the 20.0 cm marks be on the two rulers at 100°C, if precise alignment of the left-hand ends of the rulers is maintained?

17.30 A brass rod is 185 cm long and 1.60 cm in diameter. What force must be applied to each end of the rod to prevent it from contracting when it is cooled from 120°C to 10°C?

17.31 a) A wire that is 1.50 m long at 20°C is found to increase in length by 1.9 cm when warmed to 420°C. Compute its average coefficient of linear expansion for this temperature range. b) The wire is stretched just taut (zero tension) at 420°C. Find the stress in the wire if it is cooled to 20°C without being allowed to contract. Young's modulus for the wire is 2.0×10^{11} Pa.

17.32 Steel train rails are laid in 12.0-m-long segments placed end-to-end. The rails are laid on a winter day when their temperature is −2.0°C. a) How much space must be left between adjacent rails if they are just to touch on a summer day when their temperature is 33.0°C? b) If the rails are originally laid in contact, what is the stress in them on a summer day when their temperature is 33.0°C?

Section 17.5 Quantity of Heat

17.33 Heat Loss During Breathing. In very cold weather a significant mechanism for heat loss by the human body is energy expended in warming the air taken into the lungs with each breath. a) On a cold winter day when the temperature is −20°C, what is the amount of heat needed to warm to body temperature (37°C) the

0.50 L of air exchanged with each breath? Assume that the specific heat capacity of air is 1020 J/kg·K and that 1.0 L of air has mass 1.3×10^{-3} kg. b) How much heat is lost per hour if the respiration rate is 20 breaths per minute?

17.34 While running, a 70-kg student generates thermal energy at a rate of 1200 W. To maintain a constant body temperature of 37°C, this energy must be removed by perspiration or other mechanisms. If these mechanisms failed and the heat could not flow out of the student's body, for what amount of time could a student run before irreversible body damage occurs? (Protein structures in the body are irreversibly damaged if body temperature rises to 44°C or above. The specific heat capacity of a typical human body is 3480 J/kg·K, slightly less than that of water. The difference is due to the presence of protein, fat, and minerals, which have lower specific heat capacities.)

17.35 While painting the top of an antenna 225 m in height, a worker accidentally lets a 1.00-L water bottle fall from his lunchbox. The bottle lands in some bushes at ground level and does not break. If a quantity of heat equal to the magnitude of the change in mechanical energy of the water goes into the water, what is its increase in temperature?

17.36 A crate of fruit with mass 35.0 kg and specific heat capacity 3650 J/kg·K slides down a ramp inclined at 36.9° below the horizontal. The ramp is 8.00 m long. a) If the crate was at rest at the top of the incline and has a speed of 2.50 m/s at the bottom, how much work was done on the crate by friction? b) If an amount of heat equal to the magnitude of the work done by friction goes into the crate of fruit and the fruit reaches a uniform final temperature, what is its temperature change?

17.37 An engineer is working on a new engine design. One of the moving parts contains 1.60 kg of aluminum and 0.30 kg of iron and is designed to operate at 210°C. How much heat is required to raise its temperature from 20° to 210°C?

17.38 A nail driven into a board increases in temperature. If we assume that 60% of the kinetic energy delivered by a 1.80-kg hammer with a speed of 7.80 m/s is transformed into heat that flows into the nail and does not flow out, what is the temperature increase of an 8.00-g aluminum nail after it is struck ten times?

17.39 An aluminum tea kettle with mass 1.50 kg and containing 1.80 kg of water is placed on a stove. If no heat is lost to the surroundings, how much heat must be added to raise the temperature from 20.0°C to 85.0°C?

17.40 In an effort to stay awake for an all-night study session, a student makes a cup of coffee by first placing a 200-W electric immersion heater in 0.320 kg of water. a) How much heat must be added to the water to raise its temperature from 20.0°C to 80.0°C? b) How much time is required? Assume that all of the heater's power goes into heating the water.

17.41 A technician measures the specific heat capacity of an unidentified liquid by immersing an electrical resistor in it. Electrical energy is converted to heat transferred to the liquid for 120 s at a constant rate of 65.0 W. The mass of the liquid is 0.780 kg, and its temperature increases from 18.55°C to 22.54°C. a) Find the average specific heat capacity of the liquid in this temperature range. Assume that negligible heat is transferred to the container that holds the liquid and that no heat is lost to the surroundings. b) Suppose that in this experiment heat transfer from the liquid to the container or surroundings cannot be ignored. Is the result calculated in part (a) an *overestimate* or an *underestimate* of the average specific heat capacity? Explain.

17.42 You are given a sample of metal and are asked to determine its specific heat capacity. You weigh the sample and find that its weight is 28.4 N. You carefully add 1.25×10^4 J of heat energy to the sample and find that its temperature rises 18.0 C°. What is the sample's specific heat capacity?

17.43 You add 8950 J of heat to 3.00 moles of iron. a) What is the temperature increase of the iron? b) If this same amount of heat is added to 3.00 kg of iron, what is the iron's temperature increase? c) How do the results of parts (a) and (b) compare? Explain.

Section 17.6 Calorimetry and Phase Changes

17.44 As a physicist, you put heat into a 500-g solid sample at the rate of 10.0 kJ/min, while recording its temperature as a function of time. You plot your data and obtain the graph shown in Fig. 17.27. a) What is the latent heat of fusion for this solid? b) What are the specific heats of the liquid and solid states of the material?

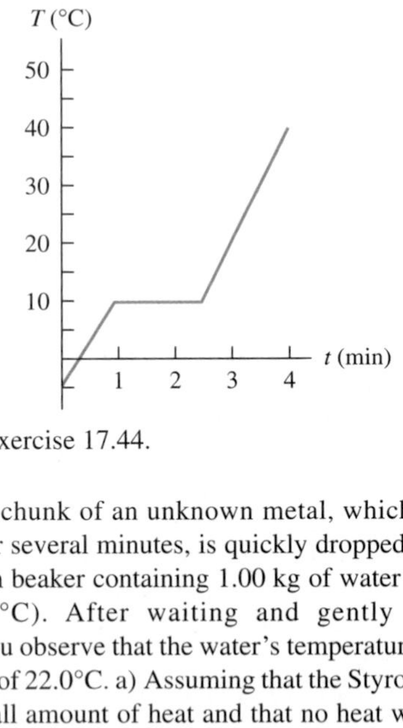

Figure 17.27 Exercise 17.44.

17.45 A 500-g chunk of an unknown metal, which has been in boiling water for several minutes, is quickly dropped into an insulating Styrofoam beaker containing 1.00 kg of water at room temperature (20.0°C). After waiting and gently stirring for 5.00 minutes, you observe that the water's temperature has reached a constant value of 22.0°C. a) Assuming that the Styrofoam absorbs a negligibly small amount of heat and that no heat was lost to the surroundings, what is the specific heat of the metal? b) Which is more useful for storing heat, this metal or an equal weight of water? Explain. c) What if the heat absorbed by the Styrofoam actually is not negligible. How would the specific heat you calculated in part (a) be in error? Would it be too large, too small, or still correct? Explain.

17.46 Before going in for his annual physical, a 70.0-kg man whose body temperature is 37.0°C consumes an entire 0.355-L can of a soft drink (mostly water) at 12.0°C. a) What will be his body temperature after equilibrium is attained? Ignore any heating by the man's metabolism. The specific heat capacity of the man's body is 3480 J/kg·K. b) Is the change in his body temperature great enough to be measured by a medical thermometer?

17.47 In the situation described in Exercise 17.46, the man's metabolism will eventually return the temperature of his body (and of the soft drink that he consumed) to 37.0°C. If his body releases energy at a rate of 7.00×10^3 kJ/day (the *basal metabolic rate,* or BMR), how long does this take? Assume that all of the released energy goes into raising the temperature.

17.48 An ice cube tray of negligible mass contains 0.350 kg of water at 18.0°C. How much heat must be removed to cool the water to 0.0°C and freeze it? Express your answer in joules, calories, and Btu.

17.49 How much heat is required to convert 12.0 g of ice at −10.0°C to steam at 100.0°C? Express your answer in joules, calories, and Btu.

17.50 An open container holds 0.550 kg of ice at −15.0°C. The mass of the container can be ignored. Heat is supplied to the container at the constant rate of 800 J/min for 500 min. a) After how many minutes does the ice *start* to melt? b) After how many minutes, from the time when the heating is first started, does the temperature begin to rise above 0°C? c) Plot a curve showing the temperature as a function of the elapsed time.

17.51 The capacity of commercial air conditioners is sometimes expressed in "tons," the number of tons of ice (1 ton = 2000 lb) that can be frozen from water at 0°C in 24 h by the unit. Express the capacity of a 2-ton air conditioner in Btu/h and in watts.

17.52 Steam Burns vs. Water Burns. What is the amount of heat input to your skin when it receives the heat released a) by 25.0 g of steam initially at 100.0°C, when it is cooled to skin temperature (34.0°C)? b) by 25.0 g of water initially at 100.0°C, when it is cooled to 34.0°C? c) What does this tell you about the relative severity of steam and hot water burns?

17.53 What must the initial speed of a lead bullet be at a temperature of 25°C so that the heat developed when it is brought to rest will be just sufficient to melt it? Assume that all the initial mechanical energy of the bullet is converted to heat and that no heat flows from the bullet to its surroundings. (Typical rifles have muzzle speeds that exceed the speed of sound in air, which is 347 m/s at 25°C.)

17.54 Evaporation of sweat is an important mechanism for temperature control in some warm-blooded animals. a) What mass of water must evaporate from the skin of a 70.0-kg man to cool his body 1.00 C°? The heat of vaporization of water at body temperature (37°C) is 2.42×10^6 J/kg · K. The specific heat capacity of a typical human body is 3480 J/kg · K (see Exercise 17.34). b) What volume of water must the man drink to replenish the evaporated water? Compare to the volume of a soft-drink can ($355\ \text{cm}^3$).

17.55 "The Ship of the Desert." Camels require very little water because they are able to tolerate relatively large changes in their body temperature. While humans keep their body temperatures constant to within one or two Celsius degrees, a dehydrated camel permits its body temperature to drop to 34.0°C overnight and rise to 40.0°C during the day. To see how effective this mechanism is for saving water, calculate how many liters of water a 400-kg camel would have to drink if it attempted to keep its body temperature at a constant 34.0°C by evaporation of sweat during the day (12 hours) instead of rising to 40.0°C. (The specific heat capacity of a camel or other mammal is about the same as that of a typical human, 3480 J/kg · K. The heat of vaporization of water at 34°C is 2.42×10^6 J/kg.)

17.56 In a physics lab experiment a student immersed 200 one-cent coins (each having a mass of 3.00 g) in boiling water. After they reached thermal equilibrium, she fished them out and dropped them into 0.240 kg of water at 20.0°C in an insulated container of negligible mass. What was the final temperature of the coins? (One-cent coins are made of a metal alloy—mostly zinc—with a specific heat capacity of 390 J/kg · K.)

17.57 A copper pot with mass 0.500 kg contains 0.170 kg of water at a temperature of 20.0°C. A 0.250-kg block of iron at 85.0°C is dropped into the pot. Find the final temperature, assuming no heat loss to the surroundings.

17.58 A laboratory technician drops a 0.0850-kg sample of unknown material, at a temperature of 100.0°C, into a calorimeter. The calorimeter can, initially at 19.0°C, is made of 0.150 kg of copper and contains 0.200 kg of water. The final temperature of the calorimeter can is 26.1°C. Compute the specific heat capacity of the sample.

17.59 An insulated beaker with negligible mass contains 0.250 kg of water at a temperature of 75.0°C. How many kilograms of ice at a temperature of −20.0°C must be dropped in the water so that the final temperature of the system will be 30.0°C?

17.60 A glass vial containing a 16.0-g sample of an enzyme is cooled in an ice bath. The bath contains water and 0.120 kg of ice. The sample has specific heat capacity 2250 J/kg · K; the glass vial has mass 6.0 g and specific heat capacity 2800 J/kg · K. How much ice melts in cooling the enzyme sample from room temperature (19.5°C) to the temperature of the ice bath?

17.61 A 4.00-kg silver ingot is taken from a furnace, where its temperature is 750°C, and placed on a large block of ice at 0°C. Assuming that all the heat given up by the silver is used to melt the ice, how much ice is melted?

17.62 A copper calorimeter can with mass 0.100 kg contains 0.160 kg of water and 0.018 kg of ice in thermal equilibrium at atmospheric pressure. If 0.750 kg of lead at a temperature of 255°C is dropped into the calorimeter can, what is the final temperature? Assume that no heat is lost to the surroundings.

17.63 A vessel whose walls are thermally insulated contains 2.40 kg of water and 0.450 kg of ice, all at a temperature of 0.0°C. The outlet of a tube leading from a boiler in which water is boiling at atmospheric pressure is inserted into the water. How many grams of steam must condense inside the vessel (also at atmospheric pressure) to raise the temperature of the system to 28.0°C? You can ignore the heat transferred to the container.

Section 17.7 Mechanisms of Heat Transfer

17.64 Use Eq. (17.21) to show that the SI units of thermal conductivity are W/m · K.

17.65 Suppose that the rod in Fig. 17.20 is made of copper, is 45.0 cm long, and has a cross-sectional area of $1.25\ \text{cm}^2$. Let $T_H = 100.0°\text{C}$ and $T_C = 0.0°\text{C}$. a) What is the final steady-state temperature gradient along the rod? b) What is the heat current in the rod in the final steady state? c) What is the final steady-state temperature at a point in the rod 12.0 cm from its left end?

17.66 One end of an insulated metal rod is maintained at 100°C, and the other end is maintained at 0°C by an ice-water mixture. The rod is 60.0 cm long and has a cross-sectional area of 1.25 cm^2. The heat conducted by the rod melts 8.50 g of ice in 10.0 min. Find the thermal conductivity k of the metal.

17.67 A carpenter builds an exterior house wall with a layer of wood 3.0 cm thick on the outside and a layer of Styrofoam insulation 2.2 cm thick on the inside wall surface. The wood has $k = 0.080$ W/m·K and the Styrofoam has $k = 0.010$ W/m·K. The interior surface temperature is 19.0°C, and the exterior surface temperature is −10.0°C. a) What is the temperature at the plane where the wood meets the Styrofoam? b) What is the rate of heat flow per square meter through this wall?

17.68 An electric kitchen range has a total wall area of 1.40 m^2 and is insulated with a layer of fiberglass 4.0 cm thick. The inside surface of the fiberglass has a temperature of 175°C and its outside surface is at 35°C. The fiberglass has a thermal conductivity of 0.040 W/m·K. a) What is the heat current through the insulation, assuming it may be treated as a flat slab with an area of 1.40 m^2? b) What electric-power input to the heating element is required to maintain this temperature?

17.69 The ceiling of a room has an area of 125 ft^2. The ceiling is insulated to an R value of 30 (in units of ft^2·F°·h/Btu). The surface in the room is maintained at 69°F, and the surface in the attic has a temperature of 35°F. What is the heat flow through the ceiling into the attic in 5.0 h? Express your answer in Btu and in joules.

17.70 A long rod, insulated to prevent heat loss along its sides, is in perfect thermal contact with boiling water (at atmospheric pressure) at one end and with an ice-water mixture at the other (Fig. 17.28). The rod consists of 1.00-m section of copper (one end in steam) joined end-to-end to a length L_2 of steel (one end in ice). Both sections of the rod have cross-sectional areas of 4.00 cm^2. The temperature of the copper-steel junction is 65.0°C after a steady state has been set up. a) How much heat per second flows from the steam bath to the ice-water mixture? b) What is the length L_2 of the steel section?

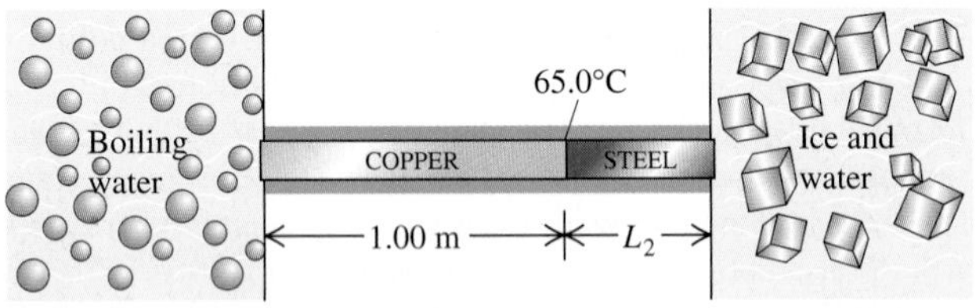

Figure 17.28 Exercise 17.70.

17.71 A pot with a steel bottom 8.50 mm thick rests on a hot stove. The area of the bottom of the pot is 0.150 m^2. The water inside the pot is at 100.0°C, and 0.390 kg are evaporated every 3.00 min. Find the temperature of the lower surface of the pot, which is in contact with the stove.

17.72 You are asked to design a cylindrical steel rod 50.0 cm long, with a circular cross-section, that will conduct 150 J/s from a furnace at 400°C to a container of boiling water under one atmosphere. What must the rod's diameter be?

17.73 A rod 1.300 m long consists of a 0.800-m length of aluminum joined end-to-end to a 0.500-m length of brass. The free end of the aluminum section is maintained at 150.0°C and the free end of the brass piece is maintained at 20.0°C. No heat is lost through the sides of the rod. At a steady state, what is the temperature T of the point where the two metals are joined?

17.74 What is the rate of energy radiation per unit area of a blackbody at a temperature of a) 273 K? b) 2730 K?

17.75 What is the net rate of heat loss by radiation in Example 17.16 (Section 17.7) if the temperature of the surroundings is 5.0°C?

17.76 The emissivity of tungsten is 0.35. A tungsten sphere with radius 1.50 cm is suspended within a large evacuated enclosure whose walls are at 290 K. What power input is required to maintain the sphere at a temperature of 3000 K if heat conduction along the supports is neglected?

17.77 Size of a Light-Bulb Filament. The operating temperature of a tungsten filament in an incandescent light bulb is 2450 K, and its emissivity is 0.35. Find the surface area of the filament of a 150-W bulb if all the electrical energy consumed by the bulb is radiated by the filament as electromagnetic waves. (Only a fraction of the radiation appears as visible light.)

17.78 The Sizes of Stars. The hot glowing surfaces of stars emit energy in the form of electromagnetic radiation. It is a good approximation to assume $e = 1$ for these surfaces. Find the radii of the following stars (assumed to be spherical): a) Rigel, the bright blue star in the constellation Orion, that radiates energy at a rate of 2.7×10^{32} W and has surface temperature 11,000 K; b) Procyon B (only visible using a telescope), that radiates energy at a rate of 2.1×10^{23} W and has surface temperature 10,000 K. c) Compare your answers to the radius of the earth, the radius of the sun, and the distance between the earth and the sun. (Rigel is an example of a *supergiant* star, and Procyon B is an example of a *white dwarf* star.)

Problems

17.79 You propose a new temperature scale, with temperatures given in °M. You define 0.0°M to be the normal melting point of mercury and 100.0° to be the normal boiling point of mercury. a) What is the normal boiling point of water in °M? b) A temperature change of 10.0 M° corresponds to how many C°?

17.80 Suppose that a steel hoop could be constructed to fit just around the earth's equator at a temperature of 20.0°C. What would be the thickness of space between the hoop and the earth if the temperature of the hoop were increased by 0.50 C°?

17.81 At an absolute temperature T_0, a cube has sides of length L_0 and has density ρ_0. The cube is made of a material with coefficient of volume expansion β. a) Show that if the temperature increases to $T_0 + \Delta T$, the density of the cube becomes, approximately,

$$\rho \approx \rho_0(1 - \beta\Delta T)$$

(*Hint:* Use the expression $(1 + x)^n \approx 1 + nx$, valid for $|x| \ll 1$.) Explain why this approximate result is only valid if ΔT is much less than $1/\beta$, and explain why you would expect this to be the case in most situations. b) A copper cube has sides of length 1.25 cm at 20.0°C. Find the change in its volume and density when its temperature is increased to 70.0°C.

17.82 A 250-kg weight is hanging from the ceiling by a thin copper wire. In its fundamental mode, this wire vibrates at the frequency of concert A (440 Hz). You then increase the temperature of the wire by 40 C°. a) By how much will the fundamental frequency change? Will it increase or decrease? b) By what percent will the speed of a wave on the wire change? c) By what percent will the wavelength of the fundamental standing wave change? Will it increase or decrease?

17.83 You are making pesto for your pasta and have a cylindrical measuring cup 10.0 cm high made of ordinary glass $(\beta = 2.7 \times 10^{-5}(\text{C}°)^{-1})$ that is filled with olive oil $(\beta = 6.8 \times 10^{-4}(\text{C}°)^{-1})$ to a height of 1.00 mm below the top of the cup. Initially, the cup and oil are at room temperature (22.0°C). You get a phone call and forget about the olive oil, which you inadvertently leave on the hot stove. The cup and oil heat up slowly, and have a common temperature. At what temperature will the olive oil start to spill out of the cup?

17.84 Use Fig. 17.11 to find the approximate coefficient of volume expansion of water at 2.0°C and at 8°C.

17.85 As in Problem 17.90, a steel rod and an aluminum rod are placed end-to-end between rigid supports. But now 0.20 cm has been cut from each rod, so the steel rod is 0.348 m long, the aluminum rod is 0.248 m long and there is a 0.40 cm gap between them. Initially, each rod has a temperature of 20.0°C. If both rods are either heated or cooled to the same final temperature, at what temperature will the gap just become closed?

17.86 At a temperature of 20.0°C the volume of a certain glass flask, up to a reference mark on the long stem of the flask, is exactly 100 cm^3. The flask is filled to this point with a liquid whose coefficient of volume expansion is $8.00 \times 10^{-4}\ \text{K}^{-1}$, with both flask and liquid at 20.0°C. The coefficient of volume expansion of the glass is $2.00 \times 10^{-5}\ \text{K}^{-1}$. The cross-sectional area of the stem is 50.0 mm^2 and can be considered constant. a) Explain why it is a good approximation to ignore the change in the cross-sectional area of the stem. b) How far will the liquid rise or fall in the stem when the temperature is raised to 50.0°C?

17.87 A metal rod that is 30.0 cm long expands by 0.0650 cm when its temperature is raised from 0°C to 100°C. A rod of a different metal and of the same length expands by 0.0350 cm for the same rise in temperature. A third rod, also 30.0 cm long, is made up of pieces of each of the above metals placed end-to-end and expands 0.0580 cm between 0°C and 100°C. Find the length of each portion of the composite bar.

17.88 On a cool (4°C) Saturday morning, a pilot fills the fuel tanks of her Pitts S-2C (a two-seat aerobatic airplane) to their full capacity of 106.0 L. Before flying on Sunday morning, when the temperature is again 4°C, she checks the fuel level and finds only 103.4 L of gasoline in the tanks. She realizes that it was hot on Saturday afternoon, and that thermal expansion of the gasoline caused the missing fuel to empty out of the tank's vent. a) What was the maximum temperature (in °C) reached by the fuel and the tank on Saturday afternoon? The coefficient of volume expansion of gasoline is $9.5 \times 10^{-4}\ \text{K}^{-1}$, and the tank is made of aluminum. b) In order to have the maximum amount of fuel available for flight, when should the pilot have filled the fuel tanks?

17.89 a) Equation (17.12) gives the stress required to keep the length of a rod constant as its temperature changes. Show that if the length is permitted to change by an amount ΔL when its temperature changes by ΔT, the stress is equal to

$$\frac{F}{A} = Y\left(\frac{\Delta L}{L_0} - \alpha \Delta T\right)$$

where F is the tension on the rod, L_0 is the original length of the rod, A its cross-sectional area, α its coefficient of linear expansion, and Y its Young's modulus. b) A heavy brass bar has projections at its ends, as in Fig. 17.29. Two fine steel wires, fastened between the projections, are just taut (zero tension) when the whole system is at 20°C. What is the tensile stress in the steel wires when the temperature of the system is raised to 140°C? Make any simplifying assumptions you think are justified, but state what they are.

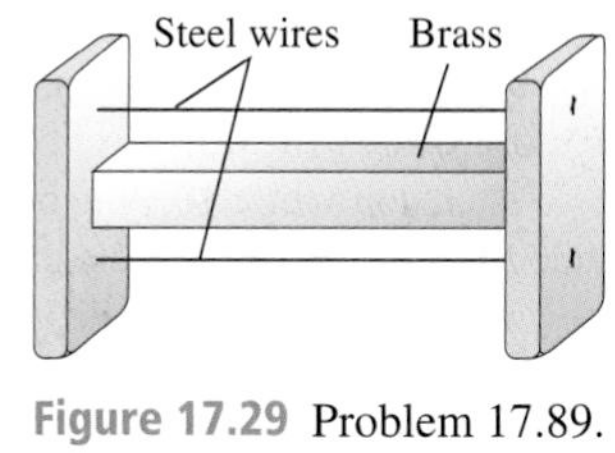

Figure 17.29 Problem 17.89.

17.90 A steel rod with a length of 0.350 m and an aluminum rod with a length of 0.250 m, both with the same diameter, are placed end-to-end between rigid supports with no initial stress in the rods. The temperature of the rods is now raised by 60.0 C°. What is the stress in each rod? (*Hint:* The length of the combined rod remains the same, but the lengths of the individual rods do not. See Problem 17.89.)

17.91 A steel ring with a 2.5000-in. inside diameter at 20.0°C is to be warmed and slipped over a brass shaft with a 2.5020-in. outside diameter at 20.0°C. a) To what temperature should the ring be warmed? b) If the ring and the shaft together are cooled by some means such as liquid air, at what temperature will the ring just slip off the shaft?

17.92 Bulk Stress Due to a Temperature Increase. a) Prove that, if an object under pressure is raised in temperature but not allowed to expand, the increase in pressure is

$$\Delta p = B\beta\Delta T$$

where the bulk modulus B and the average coefficient of volume expansion β are both assumed positive and constant. b) What pressure is necessary to prevent a steel block from expanding when its temperature is increased from 20.0°C to 35.0°C?

17.93 A liquid is enclosed in a metal cylinder that is provided with a piston of the same metal. The system is originally at a pressure of 1.00 atm $(1.013 \times 10^5\ \text{Pa})$ and at a temperature of 30.0°C. The piston is forced down until the pressure on the liquid is increased by 50.0 atm, and then clamped in this position. Find the new temperature at which the pressure of the liquid is again 1.00 atm. Assume that the cylinder is sufficiently strong so that its volume is not altered by changes in pressure, but only by changes in temperature. Use the result derived in Problem 17.92. (*Hint:* See Section 11.4.)

Compressibility of liquid: $k = 8.50 \times 10^{-10}\ \text{Pa}^{-1}$

Coefficient of volume expansion of liquid: $\beta = 4.80 \times 10^{-4}\ \text{K}^{-1}$

Coefficient of volume expansion of metal: $\beta = 3.90 \times 10^{-5}\ \text{K}^{-1}$

17.94 A thirsty mechanic cools a 2.00-L bottle of a soft drink (mostly water) by pouring it into a large aluminum mug with mass

0.257 kg and adding 0.120 kg of ice initially at $-15.0°C$. If the soft drink and mug are initially at 20.0°C, what is the final temperature of the system, assuming no heat losses?

17.95 Spacecraft Reentry. A spacecraft made of aluminum circles the earth at a speed of 7700 m/s. a) Find the ratio of its kinetic energy to the energy required to raise its temperature from 0°C to 600°C. (The melting point of aluminum is 660°C. Assume a constant specific heat capacity of 910 J/kg·K.) b) Discuss the bearing of your answer on the problem of the reentry of a manned space vehicle into the earth's atmosphere.

17.96 A capstan is a rotating drum or cylinder over which a rope or cord slides in order to provide a great amplification of the rope's tension while keeping both ends free (Fig. 17.30). Since the added tension in the rope is due to friction, the capstan generates thermal energy. a) If the difference in tension between the two ends of the rope is 520 N and the capstan has a diameter of 10.0 cm and turns once in 0.900 s, find the rate at which thermal energy is generated. Why does the number of turns not matter? b) If the capstan is made of iron and has mass 6.00 kg, at what rate does its temperature rise? Assume that the temperature in the capstan is uniform and that all the thermal energy generated flows into it.

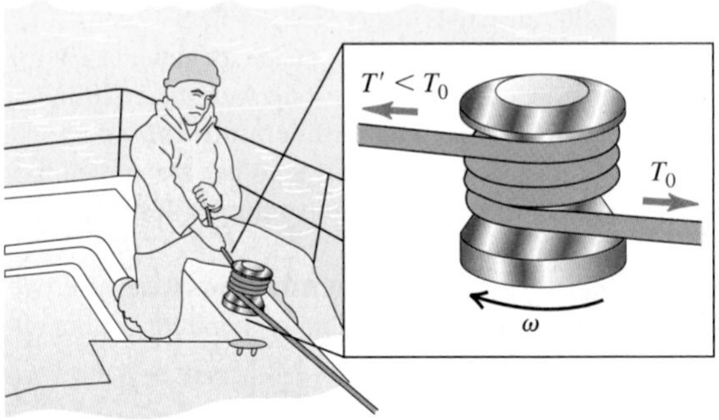

Figure 17.30 Problem 17.96.

17.97 Debye's T^3 Law. At very low temperatures the molar heat capacity of rock salt varies with temperature according to Debye's T^3 law:

$$C = k\frac{T^3}{\Theta^3}$$

where $k = 1940$ J/mol·K and $\Theta = 281$ K. a) How much heat is required to raise the temperature of 1.50 mol of rock salt from 10.0 K to 40.0 K? (*Hint:* Use Eq. (17.18) in the form $dQ = nC\,dT$ and integrate.) b) What is the average molar heat capacity in this range? c) What is the true molar heat capacity at 40.0 K?

17.98 A person of mass 70.0 kg is sitting in the bathtub. The bathtub is 190 cm by 80 cm; before the person got in, the water was 10 cm deep. The water is at a temperature of 37.0°C. Suppose that the water were to cool down spontaneously to form ice at 0.0°C, and that all the energy released was used to launch the hapless bather vertically into the air. How high would the bather go? (As you will see in Chapter 20, this event is allowed by energy conservation but is prohibited by the second law of thermodynamics.)

17.99 Hot Air in a Physics Lecture. a) A typical student listening attentively to a physics lecture has a heat output of 100 W. How much heat energy does a class of 90 physics students release into a lecture hall over the course of a 50-min lecture? b) Assume that all the heat energy in part (a) is transferred to the 3200 m^3 of air in the room. The air has specific heat capacity 1020 J/kg·K and density 1.20 kg/m^3. If none of the heat escapes and the air conditioning system is off, how much will the temperature of the air in the room rise during the 50-min lecture? c) If the class is taking an exam, the heat output per student rises to 280 W. What is the temperature rise during 50 min in this case?

17.100 The molar heat capacity of a certain substance varies with temperature according to the empirical equation

$$C = 29.5\ \text{J/mol}\cdot\text{K} + (8.20 \times 10^{-3}\ \text{J/mol}\cdot\text{K}^2)T$$

How much heat is necessary to change the temperature of 3.00 mol of this substance from 27°C to 227°C? (*Hint:* Use Eq. (17.18) in the form $dQ = nC\,dT$ and integrate.)

17.101 a) An ice cube with mass 0.075 kg is taken from a freezer, where the cube's temperature was $-10.0°C$, and dropped into a glass of water at 0.0°C. If no heat is gained or lost from outside, how much water will freeze onto the cube? b) Is it possible to add so much ice to the glass that all of the water freezes? Explain.

17.102 Hot Water versus Steam Heating. In a household hot-water heating system, water is delivered to the radiators at 70.0°C (158.0°F) and leaves at 28.0°C (82.4°F). The system is to be replaced by a steam system in which steam at atmospheric pressure condenses in the radiators and the condensed steam leaves the radiators at 35.0°C (95.0°F). How many kilograms of steam will supply the same heat as was supplied by 1.00 kg of hot water in the first system?

17.103 A copper calorimeter can with mass 0.446 kg contains 0.0950 kg of ice. The system is initially at 0.0°C. a) If 0.0350 kg of steam at 100.0°C and 1.00 atm pressure is added to the can, what is the final temperature of the calorimeter can and its contents? b) At the final temperature, how many kilograms are there of ice, how many of liquid water, and how many of steam?

17.104 In a container of negligible mass, 0.140 kg of ice initially at $-15.0°C$ is added to 0.190 kg of water that has a temperature of 35.0°C. a) If no heat is lost to the surroundings, what is the final temperature of the system? b) At the final temperature, how many kilograms are there of ice and how many of liquid water?

17.105 In a container of negligible mass, 0.0400 kg of steam at 100°C and atmospheric pressure is added to 0.200 kg of water at 50.0°C. a) If no heat is lost to the surroundings, what is the final temperature of the system? b) At the final temperature, how many kilograms are there of steam and how many of liquid water?

17.106 A tube leads from a 0.150-kg calorimeter to a flask in which water is boiling under atmospheric pressure. The calorimeter has specific heat capacity 420 J/kg·K, and it originally contains 0.340 kg of water at 15.0°C. Steam is allowed to condense in the calorimeter at atmospheric pressure until the temperature of the calorimeter and contents reaches 71.0°C, at which point the total mass of the calorimeter and its contents is found to be 0.525 kg. Compute the heat of vaporization of water from these data.

17.107 In a container of negligible mass, 0.150 kg of ice at 0°C and 0.0950 kg of steam at 100°C are added to 0.200 kg of water that has a temperature of 50.0°C. a) If no heat is lost to the surroundings and the pressure in the container is a constant 1.00 atm, what is the final temperature of the system? b) At the final temperature, how many kilograms are there of ice, how many of liquid water, and how many of steam? c) Repeat parts (a) and (b) if 0.350 kg of ice at 0°C and 0.012 kg of steam at 100°C are added to 0.200 kg of water that has a temperature of 40.0°C.

17.108 One experimental method of measuring an insulating material's thermal conductivity is to construct a box of the material and measure the power input to an electric heater inside the box that maintains the interior at a measured temperature above the outside surface. Suppose that in such an apparatus a power input of 180 W is required to keep the interior surface of the box 65.0 C° (about 120 F°) above the temperature of the outer surface. The total area of the box is 2.18 m^2, and the wall thickness is 3.9 cm. Find the thermal conductivity of the material in SI units.

17.109 Effect of a Window in a Door. A carpenter builds a solid wood door with dimensions 2.00 m $\times$ 0.95 m $\times$ 5.0 cm. Its thermal conductivity is $k = 0.120$ W/m·K. The air films on the inner and outer surfaces of the door have the same combined thermal resistance as an additional 1.8 cm thickness of solid wood. The inside air temperature is 20.0°C and the outside air temperature is −8.0°C. a) What is the rate of heat flow through the door? b) By what factor is the heat flow increased if a window 0.50 m on a side is inserted in the door? The glass is 0.45 cm thick and the glass has a thermal conductivity of 0.80 W/m·K. The air films on the two sides of the glass have a total thermal resistance that is the same as an additional 12.0 cm of glass.

17.110 A wood ceiling with thermal resistance R_1 is covered with a layer of insulation with thermal resistance R_2. Prove that the effective thermal resistance of the combination is $R = R_1 + R_2$.

17.111 Compute the ratio of the rate of heat loss through a single-pane window with area 0.15 m^2 to that for a double-pane window with the same area. The glass of a single pane is 4.2 mm thick, and the air space between the two panes of the double-pane window is 7.0 mm thick. The glass has thermal conductivity 0.80 W/m·K. The air films on the room and outdoor surfaces of either window have a combined thermal resistance of 0.15 m^2·K/W.

17.112 Rods of copper, brass, and steel are welded together to form a Y-shaped figure. The cross-sectional area of each rod is 2.00 cm^2. The free end of the copper rod is maintained at 100.0°C and the free ends of the brass and steel rods at 0.0°C. Assume that there is no heat loss from the surfaces of the rods. The lengths of the rods are as follows: copper, 13.0 cm; brass, 18.0 cm; steel, 24.0 cm. a) What is the temperature of the junction point? b) What is the heat current in each of the three rods?

17.113 Time Needed for a Lake to Freeze Over. a) When the air temperature is below 0°C, the water at the surface of a lake freezes to form an ice sheet. Why doesn't freezing occur throughout the entire volume of the lake? b) Show that the thickness of the ice sheet formed on the surface of a lake is proportional to the square root of the time if the heat of fusion of the water freezing on the underside of the ice sheet is conducted through the sheet. c) Assuming that the upper surface of the ice sheet is at −10°C and that the bottom surface is at 0°C, calculate the time it will take to form an ice sheet 25 cm thick. d) If the lake in part (c) is uniformly 40 m deep, how long would it take to freeze all the water in the lake? Is this likely to occur?

17.114 A rod is initially at a uniform temperature of 0°C throughout. One end is kept at 0°C, and the other is brought into contact with a steam bath at 100°C. The surface of the rod is insulated so that heat can flow only lengthwise along the rod. The cross-sectional area of the rod is 2.50 cm^2, its length is 120 cm, its thermal conductivity is 380 W/m·K, its density is 1.00×10^4 kg/m^3, and its specific heat capacity is 520 J/kg·K. Consider a short cylindrical element of the rod 1.00 cm in length. a) If the temperature gradient at the cooler end of this element is 140 C°/m, how many joules of heat energy flow across this end per second? b) If the average temperature of the element is increasing at the rate of 0.250 C°/s, what is the temperature gradient at the other end of the element?

17.115 If the solar radiation energy incident per second on the frozen surface of a lake is 600 W/m^2 and 70% of this energy is absorbed by the ice, how much time will it take for a 2.50-cm-thick layer of ice to melt? The ice and the water beneath it are at a temperature of 0°C.

17.116 The rate at which radiant energy from the sun reaches the earth's upper atmosphere is about 1.50 kW/m^2. The distance from the earth to the sun is 1.50×10^{11} m, and the radius of the sun is 6.96×10^8 m. a) What is the rate of radiation of energy per unit area from the sun's surface? b) If the sun radiates as an ideal blackbody, what is the temperature of its surface?

17.117 A Thermos for Liquid Helium. A physicist uses a cylindrical metal can 0.250 m high and 0.090 m in diameter to store liquid helium at 4.22 K; at that temperature the heat of vaporization of helium is 2.09×10^4 J/kg. Completely surrounding the metal can are walls maintained at the temperature of liquid nitrogen, 77.3 K, with vacuum between the can and the surrounding walls. How much helium is lost per hour? The emissivity of the metal can is 0.200. The only heat transfer between the metal can and the surrounding walls is by radiation.

17.118 Thermal Expansion of an Ideal Gas. a) The pressure p, volume V, number of moles n, and Kelvin temperature T of an ideal gas are related by the equation $pV = nRT$, where R is a constant. Prove that the coefficient of volume expansion for an ideal gas is equal to the reciprocal of the Kelvin temperature if the expansion occurs at constant pressure. b) Compare the coefficients of volume expansion of copper and air at a temperature of 20°C. Assume that air may be treated as an ideal gas and that the pressure remains constant.

17.119 An engineer is developing an electric water heater to provide a continuous supply of hot water. One trial design is shown in

Fig. 17.31. Water is flowing at the rate of 0.500 kg/min, the inlet thermometer registers 18.0°C, the voltmeter reads 120 V, and the ammeter reads 15.0 A (corresponding to a power input of $(120\ \text{V}) \times (15.0\ \text{A}) = 1800\ \text{W}$). a) When a steady state is finally reached, what is the reading of the outlet thermometer? b) Why is it unnecessary to take into account the heat capacity mc of the apparatus itself?

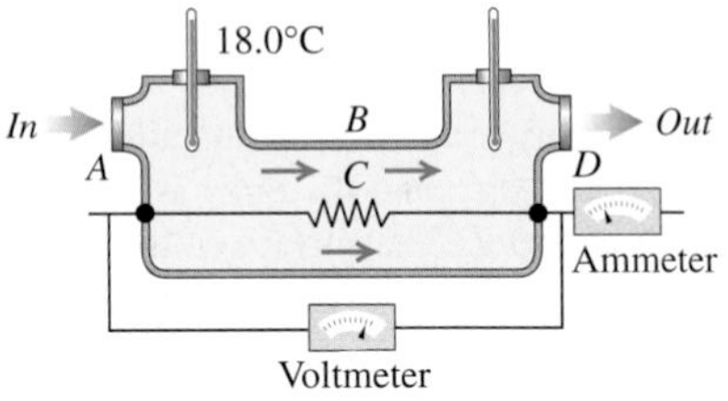

Figure 17.31 Problem 17.119.

17.120 Food Intake of a Hamster. The energy output of an animal engaged in an activity is called the basal metabolic rate (BMR) and is a measure of the conversion of food energy into other forms of energy. A simple calorimeter to measure the BMR consists of an insulated box with a thermometer to measure the temperature of the air. The air has density $1.20\ \text{kg/m}^3$ and specific heat capacity $1020\ \text{J/kg}\cdot\text{K}$. A 50.0-g hamster is placed in a calorimeter that contains $0.0500\ \text{m}^3$ of air at room temperature. a) When the hamster is running in a wheel, the temperature of the air in the calorimeter rises 1.60 C° per hour. How much heat does the running hamster generate in an hour? Assume that all this heat goes into the air in the calorimeter. You can ignore the heat that goes into the walls of the box and into the thermometer, and assume that no heat is lost to the surroundings. b) Assuming that the hamster converts seed into heat with an efficiency of 10% and that hamster seed has a food energy value of 24 J/g, how many grams of seed must the hamster eat per hour to supply this energy?

Challenge Problems

17.121 a) A spherical shell has inner and outer radii a and b, respectively, and the temperatures at the inner and outer surfaces are T_2 and T_1. The thermal conductivity of the material of which the shell is made is k. Derive an equation for the total heat current through the shell. b) Derive an equation for the temperature variation within the shell in part (a); that is, calculate T as a function of r, the distance from the center of the shell. c) A hollow cylinder has length L, inner radius a, and outer radius b, and the temperatures at the inner and outer surfaces are T_2 and T_1. (The cylinder could represent an insulated hot-water pipe, for example.) The thermal conductivity of the material of which the cylinder is made is k. Derive an equation for the total heat current through the walls of the cylinder. d) For the cylinder of part (c), derive an equation for the temperature variation inside the cylinder walls. e) For the spherical shell of part (a) and the hollow cylinder of part (c), show that the equation for the total heat current in each case reduces to Eq. (17.21) for linear heat flow when the shell or cylinder is very thin.

17.122 A steam pipe with a radius of 2.00 cm, carrying steam at 140°C, is surrounded by a cylindrical jacket with inner and outer radii 2.00 cm and 4.00 cm and made of a type of cork with thermal conductivity $4.00 \times 10^{-2}\ \text{W/m}\cdot\text{K}$. This in turn is surrounded by a cylindrical jacket made of a brand of Styrofoam with thermal conductivity $1.00 \times 10^{-2}\ \text{W/m}\cdot\text{K}$ and having inner and outer radii 4.00 cm and 6.00 cm (Fig. 17.32). The outer surface of the Styrofoam is in contact with air at 15°C. Assume that this outer surface has a temperature of 15°C. a) What is the temperature at a radius of 4.00 cm, where the two insulating layers meet? b) What is the total rate of transfer of heat out of a 2.00-m length of pipe? (*Hint:* Use the expression derived in part (c) of Challenge Problem 17.121.)

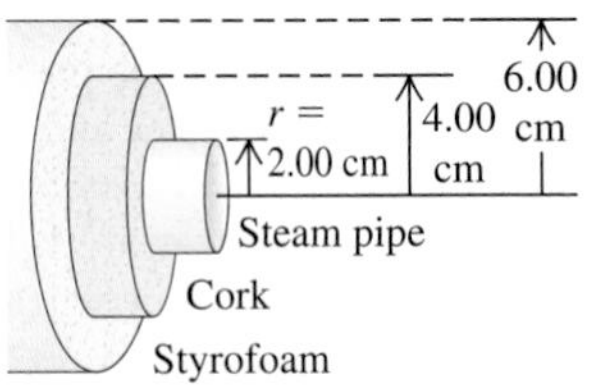

Figure 17.32 Challenge Problem 17.122.

17.123 Suppose that both ends of the rod in Fig. 17.20 are kept at a temperature of 0°C, and that the initial temperature distribution along the rod is given by $T = (100°\text{C}) \sin \pi x/L$, where x is measured from the left end of the rod. Let the rod be copper, with length $L = 0.100$ m and cross-sectional area $1.00\ \text{cm}^2$. a) Show the initial temperature distribution in a diagram. b) What is the final temperature distribution after a very long time has elapsed? c) Sketch curves that you think would represent the temperature distribution at intermediate times. d) What is the initial temperature gradient at the ends of the rod? e) What is the initial heat current from the ends of the rod into the bodies making contact with its ends? f) What is the initial heat current at the center of the rod? Explain. What is the heat current at this point at any later time? g) What is the value of the *thermal diffusivity* $k/\rho c$ for copper, and in what unit is it expressed? (Here k is the thermal conductivity, $\rho = 8.9 \times 10^3\ \text{kg/m}^3$ is the density, and c is the specific heat capacity.) h) What is the initial time rate of change of temperature at the center of the rod? i) How much time would be required for the center of the rod to reach its final temperature if the temperature continued to decrease at this rate? (This time is called the *relaxation time* of the rod.) j) From the graphs in part (c), would you expect the magnitude of the rate of temperature change at the midpoint to remain constant, increase, or decrease as a function of time? k) What is the initial rate of change of temperature at a point in the rod 2.5 cm from its left end?

17.124 Temperature Change in a Clock. A pendulum clock is designed to tick off one second on each side-to-side swing of the pendulum (two ticks per complete period). a) Will a pendulum clock gain time in hot weather and lose it in cold, or the reverse? Explain your reasoning. b) A particular pendulum clock keeps correct time at 20.0°C. The pendulum shaft is of steel, and its mass can be ignored compared with that of the bob. What is the fractional change in length of the shaft when it is cooled to 10.0°C? c) How many seconds per day will the clock gain or lose at 10.0°C? d) How closely must the temperature be controlled if the clock is not to gain or lose more than 1.00 s a day? Does the answer depend on the period of the pendulum?

17.125 A solid cylindrical copper rod 0.200 m long has one end maintained at a temperature of 20.00 K. The other end is blackened and exposed to thermal radiation from surrounding walls at 500 K. The sides of the rod are insulated, so no energy is lost or gained except at the ends of the rod. When equilibrium is reached, what is the temperature of the blackened end? (*Hint:* Since copper is a very good conductor of heat at low temperature, with $k = 1670\ \mathrm{W/m \cdot K}$ at 20 K, the temperature of the blackened end is only slightly greater than 20.00 K.)

17.126 A Walk in the Sun. Consider a poor lost soul walking at 5 km/h on a hot day in the desert, wearing only a bathing suit. This person's skin temperature tends to rise due to four mechanisms: i) energy is generated by metabolic reactions in the body at a rate of 280 W, and almost all of this energy is converted to heat that flows to the skin; ii) heat is delivered to the skin by convection from the outside air at a rate equal to $k'A_{\mathrm{skin}}(T_{\mathrm{air}} - T_{\mathrm{skin}})$, where k' is $54\ \mathrm{J/h \cdot C^\circ \cdot m^2}$, the exposed skin area A_{skin} is $1.5\ \mathrm{m^2}$, the air temperature T_{air} is 47°C, and the skin temperature T_{skin} is 36°C; iii) the skin absorbs radiant energy from the sun at a rate of $1400\ \mathrm{W/m^2}$; iv) the skin absorbs radiant energy from the environment, which has temperature 47°C. a) Calculate the net rate (in watts) at that the person's skin is heated by all four of these mechanisms. Assume that the emissivity of the skin is $e = 1$ and that the skin temperature is initially 36°C. Which mechanism is the most important? b) At what rate (in L/h) must perspiration evaporate from this person's skin to maintain a constant skin temperature? (The heat of vaporization of water at 36°C is 2.42×10^6 J/kg.) c) Suppose instead the person is protected by light-colored clothing $(e \approx 0)$ so that the exposed skin area is only $0.45\ \mathrm{m^2}$. What rate of perspiration is required now? Discuss the usefulness of the traditional clothing worn by desert peoples.

18 THERMAL PROPERTIES OF MATTER

What is the fundamental difference between the warm air of summer and the cold air of winter? The answer is kinetic energy: the higher the temperature of a gas, the greater the average kinetic energy of the gas molecules. At a given temperature, gas molecules with different masses (such as oxygen and nitrogen) have different average speeds but the same average kinetic energy.

? **How much faster do air molecules move on a hot day in the desert $(+40°C)$ than on a frigid arctic night $(-40°C)$?**

The kitchen is a great place to learn about how the properties of matter depend on temperature. When you boil water in a tea kettle, the increase in temperature produces steam that whistles out of the spout at high pressure. If you forget to poke holes in a potato before baking it, the high-pressure steam produced inside the potato can cause it to explode messily. Water vapor in the air can condense into droplets of liquid on the sides of a glass of ice water; if the glass is just out of the freezer, frost will form on the sides as water vapor changes to a solid.

All of these examples show the interrelations between the large-scale or *macroscopic* properties of a substance, such as pressure, volume, temperature, and mass of substance. But we can also describe a substance using a *microscopic* perspective. This means investigating small-scale quantities such as the masses, speeds, kinetic energies, and momenta of the individual molecules that make up a substance.

The macroscopic and microscopic descriptions are intimately related. For example, the (microscopic) collision forces that occur when air molecules strike a solid surface (such as your skin) cause (macroscopic) atmospheric pressure. Standard atmospheric pressure is 1.01×10^5 Pa; to produce this pressure, 10^{32} molecules strike your skin every day with an average speed of over 1700 km/h (1000 mi/h)!

In this chapter we'll begin our study of the thermal properties of matter by looking at some macroscopic aspects of matter in general. We'll pay special attention to the *ideal gas,* one of the simplest types of matter to understand. Using our knowledge of momentum and kinetic energy, we'll relate the macroscopic properties of an ideal gas to the microscopic behavior of its individual molecules. We'll also use microscopic ideas to understand the heat capacities of both gases and solids. Finally, we'll take a look at the various phases of matter—gas, liquid, and solid—and the conditions under which each occurs.

18.1 | Equations of State

The conditions in which a particular material exists are described by physical quantities such as pressure, volume, temperature, and amount of substance. For example, a tank of oxygen in a welding outfit has a pressure gauge and a label stating its volume. We could add a thermometer and place the tank on a scale to determine its mass. These variables describe the *state* of the material and are called **state variables.**

The volume V of a substance is usually determined by its pressure p, temperature T, and amount of substance, described by the mass m or number of moles n. Ordinarily, we can't change one of these variables without causing a change in another. When the tank of oxygen gets hotter, the pressure increases. If the tank gets too hot, it explodes; this happens occasionally with overheated steam boilers.

In a few cases the relationship among p, V, T, and m (or n) is simple enough that we can express it as an equation called the **equation of state.** When it's too complicated for that, we can use graphs or numerical tables. Even then, the relation among the variables still exists; we call it an equation of state even when we don't know the actual equation.

Here's a simple (though approximate) equation of state for a solid material. The temperature coefficient of volume expansion β (Section 17.4) is the fractional volume change $\Delta V/V_0$ per unit temperature change, and the compressibility k (Section 11.4) is the negative of the fractional volume change $\Delta V/V_0$ per unit pressure change. If a certain amount of material has volume V_0 when the pressure is p_0 and the temperature is T_0, the volume V at slightly differing pressure p and temperature T is approximately

$$V = V_0[1 + \beta(T - T_0) - k(p - p_0)] \tag{18.1}$$

(There is a negative sign in front of the term $k(p - p_0)$ because an *increase* in pressure causes a *decrease* in the volume.) Equation (18.1) is called an *equation of state* for the material.

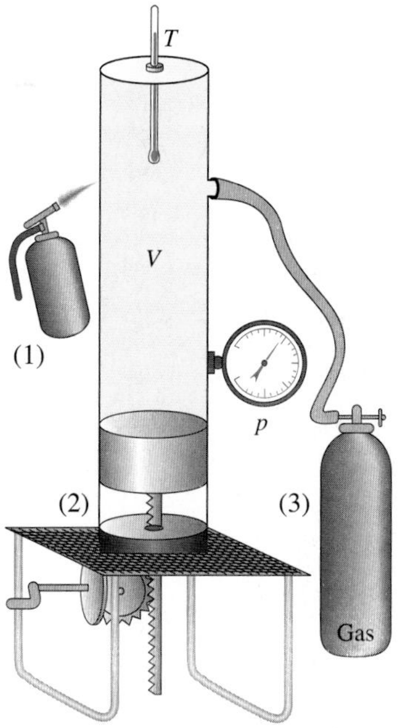

18.1 A hypothetical setup for studying the behavior of gases. By (1) heating the gas, (2) varying the volume with a movable piston, and (3) adding more gas, we can control the gas pressure p, volume V, temperature T, and number of moles n.

The Ideal Gas Equation

Another simple equation of state is the one for an *ideal gas.* Figure 18.1 shows an experimental setup to study the behavior of a gas. The cylinder has a movable piston to vary the volume, the temperature can be varied by heating, and we can pump any desired amount of any gas into the cylinder. We then measure the pressure, volume, temperature, and amount of gas. Note that *pressure* refers both to the force per unit area exerted by the cylinder on the gas and to the force per unit area exerted by the gas on the cylinder; by Newton's third law, these must be equal.

It is usually easiest to describe the amount of gas in terms of the number of moles n, rather than the mass. We did this when we defined molar heat capacity in Section 17.5; you may want to review that section. The **molar mass** M of a compound (sometimes called *molecular weight*) is the mass per mole, and the total mass m_{tot} of a given quantity of that compound is the number of moles n times the mass per mole M:

$$m_{tot} = nM \qquad \text{(total mass, number of moles, and molar mass)} \qquad (18.2)$$

We are calling the total mass m_{tot} because later in the chapter we will use m for the mass of one molecule.

Measurements of the behavior of various gases lead to several conclusions. First, the volume V is proportional to the number of moles n. If we double the number of moles, keeping pressure and temperature constant, the volume doubles.

Second, the volume varies *inversely* with the absolute pressure p. If we double the pressure while holding the temperature T and number of moles n constant, the gas compresses to one-half of its initial volume. In other words, $pV = \text{constant}$ when n and T are constant.

Third, the pressure is proportional to the *absolute* temperature. If we double the absolute temperature, keeping the volume and number of moles constant, the pressure doubles. In other words, $p = (\text{constant})T$ when n and V are constant.

These three relationships can be combined neatly into a single equation, called the **ideal-gas equation:**

ActivPhysics ONLINE

8.4 State Variables and Ideal Gas Law

$$pV = nRT \qquad \text{(ideal-gas equation)} \qquad (18.3)$$

where R is a proportionality constant. An **ideal gas** is one for which Eq. (18.3) holds precisely for *all* pressures and temperatures. This is an idealized model; it works best at very low pressures and high temperatures, when the gas molecules are far apart and in rapid motion. It is reasonably good (within a few percent) at moderate pressures (such as a few atmospheres) and at temperatures well above those at which the gas liquefies.

We might expect that the constant R in the ideal-gas equation would have different values for different gases, but it turns out to have the same value for *all* gases, at least at sufficiently high temperature and low pressure. It is called the **gas constant** (or *ideal-gas constant*). The numerical value of R depends on the units of p, V, and T. In SI units, in which the unit of p is Pa $(1\text{ Pa} = 1\text{ N/m}^2)$ and the unit of V is m^3, the current best numerical value of R is

$$R = 8.314472(15)\text{ J/mol}\cdot\text{K}$$

or $R = 8.314\text{ J/mol}\cdot\text{K}$ to four significant digits. Note that the units of pressure times volume are the same as units of work or energy (for example, N/m^2 times m^3); that's why R has units of energy per mole per unit of absolute temperature. In chemical calculations, volumes are often expressed in liters (L) and pressures in atmospheres (atm). In this system, to four significant digits,

$$R = 0.08206\,\frac{\text{L}\cdot\text{atm}}{\text{mol}\cdot\text{K}}$$

We can express the ideal-gas equation, Eq. (18.3), in terms of the mass m_{tot} of gas, using $m_{tot} = nM$ from Eq. (18.2):

$$pV = \frac{m_{tot}}{M}RT \qquad (18.4)$$

From this we can get an expression for the density $\rho = m_{\text{tot}}/V$ of the gas:

$$\rho = \frac{pM}{RT} \qquad (18.5)$$

CAUTION When using Eq. (18.5), be certain that you distinguish between the Greek letter ρ ("rho") for density and the letter p for pressure.

For a *constant mass* (or constant number of moles) of an ideal gas the product nR is constant, so the quantity pV/T is also constant. If the subscripts 1 and 2 refer to any two states of the same mass of a gas, then

$$\frac{p_1 V_1}{T_1} = \frac{p_2 V_2}{T_2} = \text{constant} \qquad \text{(ideal gas, constant mass)} \qquad (18.6)$$

Notice that you don't need the value of R to use this equation.

The proportionality of pressure to absolute temperature is familiar; in fact, in Chapter 17 we *defined* a temperature scale in terms of pressure in a constant-volume gas thermometer. That may make it seem that the pressure–temperature relation in the ideal-gas equation, Eq. (18.3), is just a result of the way we define temperature. But the equation also tells us what happens when we change the volume or the amount of substance. Also, the gas-thermometer scale turns out to correspond closely to a temperature scale that we will define in Chapter 20 that doesn't depend on the properties of any particular material. For now, consider this equation as being based on this genuinely material-independent temperature scale, even though we haven't defined it yet.

Problem-Solving Strategy — Ideal Gases

IDENTIFY *the relevant concepts:* Unless the problem explicitly states otherwise, you can use the ideal-gas equation for any situation in which you need to find the state (pressure, volume, temperature, and/or number of moles) of a gas.

SET UP *the problem using the following steps:*

1. Identify the target variables.
2. In some problems you will be concerned with only one state of the system, in which case Eq. (18.3) is the relationship to use. Some of the quantities in this equation will be known; others will be unknown. Make a list of what you know and what you have to find. For example, $p = 1.0 \times 10^6$ Pa, $V = 4\ \text{m}^3$, $T = ?$, $n = 2$ mol, or something comparable.
3. In other problems you will compare two different states of the same amount of gas. Decide which is state 1 and which is state 2, and make a list of the quantities for each: p_1, p_2, V_1, V_2, T_1, T_2. If all but one of these quantities are known, you can use Eq. (18.6). Otherwise, use Eq. (18.3). For example, if p_1, V_1, and n are given, you can't use Eq. (18.6) because you don't know T_1.
4. Some problems involve the density ρ (mass per volume) rather than the number of moles n and the volume V. In this case it's most convenient to use Eq. (18.5), $\rho = pM/RT$.

EXECUTE *the solution as follows:*

1. As always, use a consistent set of units. If you need the value of the gas constant R, choose its units and then convert the units of the other quantities accordingly. You may have to convert atmospheres to pascals or liters to cubic meters ($1\ \text{m}^3 = 10^3\ \text{L} = 10^6\ \text{cm}^3$). Sometimes the problem statement will make one system of units clearly more convenient than others. Decide on your system and stick to it.
2. Don't forget that T must always be an *absolute* temperature. If you are given temperatures in °C, be sure to convert to Kelvin temperature by adding 273.15 (to three

significant figures, 273). Likewise, p is always the absolute pressure, never the gauge pressure.

3. You may sometimes have to convert between mass m_{tot} and number of moles n. The relationship is $m_{tot} = Mn$, where M is the molecular mass. Here's a tricky point: If you use Eq. (18.4), you *must* use the same mass units for m_{tot} and M. So if M is in grams per mole (the usual units for molecular mass), then m_{tot} must also be in grams. If you want to use m_{tot} in kilograms, then you must convert M to kg/mol. For example, the molecular mass of oxygen is 32 g/mol or 32×10^{-3} kg/mol. Be careful!
4. Once you have taken care of steps 1–3, solve for the target variables.

EVALUATE *your answer:* Look carefully at your results and see whether they make physical sense. For example, we'll find in Example 18.1 that a mole of gas at one atmosphere pressure and 0°C occupies a volume of 22.4 liters. If you do a calculation of the amount of air inside a soccer ball (radius 11 cm, volume 5.6 L) and get a fantastically large answer like $n = 5000$ moles, you probably converted units incorrectly or made an algebraic error.

Example 18.1 Volume of a gas at STP

The condition called **standard temperature and pressure** (STP) for a gas is defined to be a temperature of 0°C = 273.15 K and a pressure of 1 atm = 1.013×10^5 Pa. If you want to keep a mole of an ideal gas in your room at STP, how big a container do you need?

SOLUTION

IDENTIFY and SET UP: We are asked about the properties of a single state of the system, so we use Eq. (18.3).

EXECUTE: From Eq. (18.3), using R in J/mol·K,

$$V = \frac{nRT}{p} = \frac{(1\text{ mol})(8.314\text{ J/mol}\cdot\text{K})(273.15\text{ K})}{1.013 \times 10^5\text{ Pa}}$$
$$= 0.0224\text{ m}^3 = 22.4\text{ L}$$

EVALUATE: You may be familiar with this result from your study of chemistry. Note that 22.4 L is almost exactly the volume of three basketballs. A cube 0.282 m on a side would also do the job.

Example 18.2 Compressing gas in an automobile engine

In an automobile engine, a mixture of air and gasoline is compressed in the cylinders before being ignited. A typical engine has a compression ratio of 9.00 to 1; this means that the gas in the cylinders is compressed to $1/(9.00)$ of its original volume (Fig. 18.2). The initial pressure is 1.00 atm and the initial temperature is 27°C. If the pressure after compression is 21.7 atm, find the temperature of the compressed gas.

SOLUTION

IDENTIFY and SET UP: Let state 1 be the uncompressed gas and let state 2 be the fully compressed gas. Then $p_1 = 1.00$ atm, $p_2 = 21.7$ atm, and $V_1 = 9.00\, V_2$. Converting temperature to the Kelvin scale by adding 273, $T_1 = 300$ K; the final temperature T_2 is the target variable. The intake and exhaust valves at the top of the cylinder in Fig. 18.2 stay closed during the compression, so the number of moles of gas n is constant. Hence we can use Eq. (18.6):

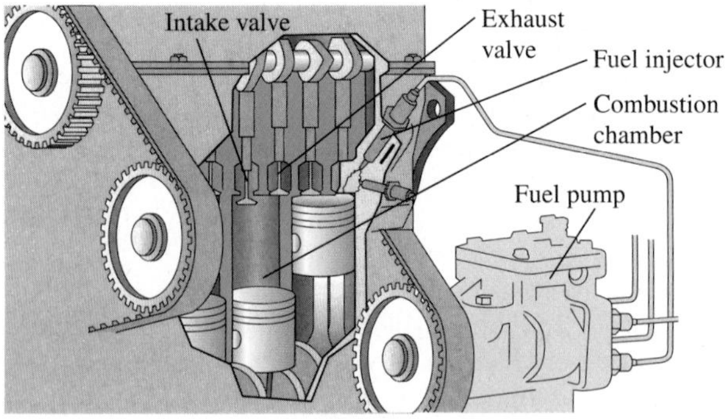

18.2 Cutaway of an automobile engine. While the air–gasoline mixture is being compressed prior to ignition, the intake and exhaust valves are both in the closed (up) position.

EXECUTE: Solving Eq. (18.6) for the temperature T_2 of the compressed gas,

$$T_2 = T_1\frac{p_2 V_2}{p_1 V_1} = (300\text{ K})\frac{(21.7\text{ atm})\, V_2}{(1.00\text{ atm})(9.00\, V_2)}$$
$$= 723\text{ K} = 450°\text{C}$$

We didn't need to know the values of V_1 or V_2, only their ratio.

EVALUATE: Note that T_2 is the temperature of the air–gasoline mixture *before* the mixture is ignited; when burning starts, the temperature becomes higher still.

Example 18.3 Mass of air in a scuba tank

A typical tank used for scuba diving has a volume of 11.0 L (about 0.4 ft^3) and a gauge pressure, when full, of 2.10×10^7 Pa (about 3000 psig). The "empty" tank contains 11.0 L of air at 21°C and 1 atm (1.013×10^5 Pa). When the tank is filled with hot air from a compressor, the temperature is 42°C and the gauge pressure is 2.10×10^7 Pa. What mass of air was added? (Air is a mixture of gases, about 78% nitrogen, 21% oxygen, and 1% miscellaneous; its average molar mass is 28.8 g/mol $= 28.8 \times 10^{-3}$ kg/mol.)

SOLUTION

IDENTIFY and SET UP: Our target variable is the *difference* between the mass present at the beginning (state 1) and at the end (state 2). We are given the molar mass of air, so we can use Eq. (18.2) to find the target variable if we know the number of moles present in states 1 and 2. We determine n_1 and n_2 by applying Eq. (18.3) to each state individually.

EXECUTE: We must remember to convert the temperatures to the Kelvin scale by adding 273 and to convert the pressure to absolute by adding 1.013×10^5 Pa. From Eq. (18.3), the number of moles n_1 in the "empty" tank is

$$n_1 = \frac{p_1V_1}{RT_1} = \frac{(1.013 \times 10^5\ \text{Pa})(11.0 \times 10^{-3}\ \text{m}^3)}{(8.315\ \text{J/mol}\cdot\text{K})(294\ \text{K})} = 0.46\ \text{mol}$$

The volume of the metal tank is hardly affected by the increased pressure, so $V_1 = V_2$. The number of moles in the full tank is

$$n_2 = \frac{p_2V_2}{RT_2} = \frac{(21.1 \times 10^6\ \text{Pa})(11.0 \times 10^{-3}\ \text{m}^3)}{(8.315\ \text{J/mol}\cdot\text{K})(315\ \text{K})} = 88.6\ \text{mol}$$

We added $n_2 - n_1 = 88.6$ mol $- 0.46$ mol $= 88.1$ mol to the tank. From Eq. (18.2), the added mass is $M(n_2 - n_1) = (88.1\ \text{mol})(28.8 \times 10^{-3}\ \text{kg/mol}) = 2.54$ kg.

EVALUATE: The added mass is not insubstantial: you could certainly use a scale to determine whether the tank was empty or full.

Could this problem have been solved in the same way as Example 18.2? The volume is constant, so $p/nT = R/V$ is constant and $p_1/n_1T_1 = p_2/n_2T_2$; this can be solved for n_2/n_1, the ratio of the final and initial number of moles. But we need the *difference* of these two numbers, not the ratio, so this equation by itself isn't enough to solve the problem.

Example 18.4 Variation of atmospheric pressure with elevation

Find the variation of atmospheric pressure with elevation in the earth's atmosphere, assuming that the temperature is 0°C at all elevations. Ignore the variation of g with elevation.

SOLUTION

IDENTIFY: As the elevation increases, both the atmospheric pressure and the density decrease. Hence we have *two* unknown functions of elevation; to solve for them, we need two separate relationships. One of these is the ideal-gas equation, which we can write in terms of pressure and density; the other is the relationship between pressure and density in a fluid in equilibrium, discussed in Section 14-2.

SET UP: In Section 14-2, we found the general equation $dp/dy = -\rho g$, Eq. (14.4), for the variation of pressure p with elevation y as a function of density ρ. Equation (18.5), $\rho = pM/RT$, states the ideal-gas equation in terms of density. We are told to assume that g and T are the same at all elevations; we also assume that the atmosphere has the same chemical composition, and hence the same molar mass M, at all heights. We then combine the two expressions and solve for $p(y)$.

EXECUTE: We substitute $\rho = pM/RT$ into $dp/dy = -\rho g$, separate variables, and integrate, letting p_1 be the pressure at elevation y_1 and p_2 be the pressure at y_2:

$$\frac{dp}{dy} = -\frac{pM}{RT}g$$

$$\int_{p_1}^{p_2} \frac{dp}{p} = -\frac{Mg}{RT}\int_{y_1}^{y_2} dy$$

$$\ln\frac{p_2}{p_1} = -\frac{Mg}{RT}(y_2 - y_1)$$

$$\frac{p_2}{p_1} = e^{-Mg(y_2-y_1)/RT}$$

Now let $y_1 = 0$ be at sea level and let the pressure at that point be $p_0 = 1.013 \times 10^5$ Pa. Then our final expression for the pressure p at any height y is

$$p = p_0 e^{-Mgy/RT}$$

EVALUATE: According to our calculation, the pressure decreases exponentially with elevation. The graph in Fig. 18.3 shows that the slope dp/dy becomes less negative with greater elevation. That result makes sense, since $dp/dy = -\rho g$ and the density also decreases with elevation. At the summit of Mount Everest, where $y = 8863$ m,

$$\frac{Mgy}{RT} = \frac{(28.8 \times 10^{-3}\ \text{kg/mol})(9.80\ \text{m/s}^2)(8863\ \text{m})}{(8.315\ \text{J/mol}\cdot\text{K})(273\ \text{K})} = 1.10$$

$$p = (1.013 \times 10^5\ \text{Pa})e^{-1.10} = 0.337 \times 10^5\ \text{Pa} = 0.33\ \text{atm}$$

The assumption of constant temperature isn't realistic, and g decreases a little with increasing elevation (see Challenge Problem 18.88). Even so, this example shows why mountaineers need to carry oxygen on Mount Everest.

The ability of the human body to absorb oxygen from the atmosphere depends critically on atmospheric pressure. Absorption drops sharply when the pressure is less than about 0.65×10^5 Pa, corresponding to an elevation above sea level of about 4700 m (15,000 ft). There is no permanent human habitation on earth above 6000 m (about 20,000 ft), although survival for short periods of time is possible at higher elevations. Jet airplanes, which typically fly at altitudes of 8000 to 12,000 m, *must* have pressurized cabins for passenger comfort and health.

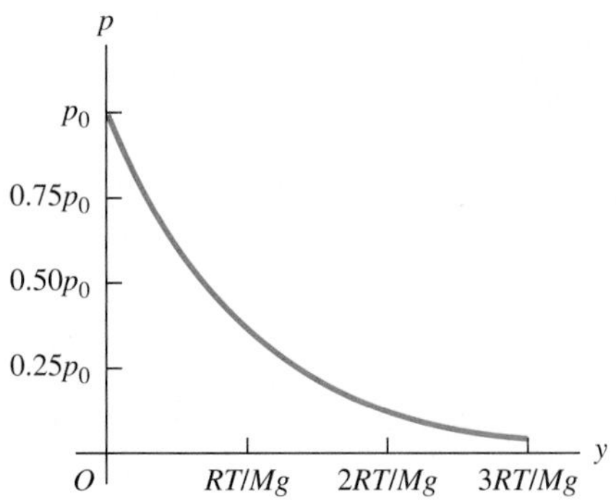

18.3 The variation of atmospheric pressure p with elevation y, assuming a constant temperature T.

The Van Der Waals Equation

The ideal-gas equation, Eq. (18.3), can be obtained from a simple molecular model that ignores the volumes of the molecules themselves and the attractive forces between them. We'll examine that model in Section 18.3. Meanwhile, we mention another equation of state, the **van der Waals equation,** that makes approximate corrections for these two omissions. This equation was developed by the 19th-century Dutch physicist J. D. van der Waals; the interaction between atoms that we discussed in Section 13.4 was named the *van der Waals interaction* after him. The van der Waals equation is

$$\left(p + \frac{an^2}{V^2}\right)(V - nb) = nRT \tag{18.7}$$

The constants a and b are empirical constants, different for different gases. Roughly speaking, b represents the volume of a mole of molecules; the total volume of the molecules is then nb, and the net volume available for the molecules to move around in is $V - nb$. The constant a depends on the attractive intermolecular forces, which reduce the pressure of the gas for given values of n, V, and T by *pulling* the molecules together as they *push* on the walls of the container. The decrease in pressure is proportional to the number of molecules per unit volume in a layer near the wall (which are exerting the pressure on the wall) and is also proportional to the number per unit volume in the next layer beyond the wall (which are doing the attracting). Hence the decrease in pressure due to intermolecular forces is proportional to n^2/V^2.

When n/V is small (that is, when the gas is *dilute*), the average distance between molecules is large, the corrections in the van der Waals equation become insignificant, and Eq. (18.7) reduces to the ideal-gas equation. As an example, for carbon dioxide gas (CO_2) the constants in the van der Waals equation are

$a = 0.364\ \text{J}\cdot\text{m}^3/\text{mol}^2$ and $b = 4.27 \times 10^{-5}\ \text{m}^3/\text{mol}$. We found in Example 18.1 that one mole of an ideal gas at $T = 0°\text{C} = 273.15\ \text{K}$ and $p = 1\ \text{atm} = 1.013 \times 10^5\ \text{Pa}$ occupies a volume $V = 0.0224\ \text{m}^3$; according to Eq. (18.7), one mole of CO_2 occupying this volume at this temperature would be at a pressure 532 Pa less than 1 atm, a difference of only 0.5% from the ideal-gas value.

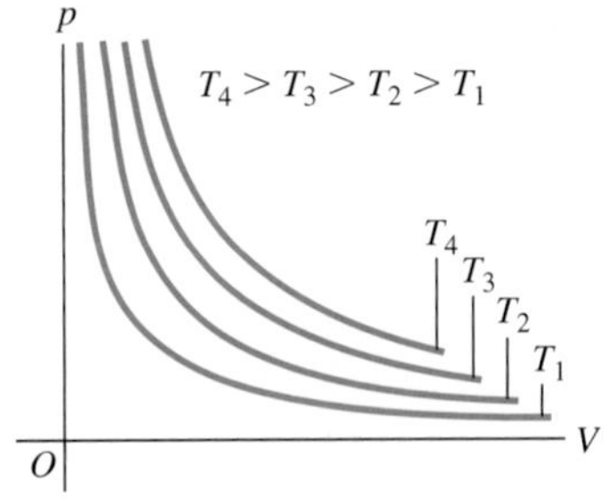

18.4 Isotherms, or constant-temperature curves, for a constant amount of an ideal gas. For each curve, the product $pV = nRT$ is constant, so p is proportional to $1/V$; the proportionality constant increases with increasing T.

pV-Diagrams

We could in principle represent the p-V-T relationship graphically as a *surface* in a three-dimensional space with coordinates p, V, and T. This representation sometimes helps in grasping the overall behavior of the substance, but ordinary two-dimensional graphs are usually more convenient. One of the most useful of these is a set of graphs of pressure as a function of volume, each for a particular constant temperature. Such a diagram is called a ***pV*-diagram.** Each curve, representing behavior at a specific temperature, is called an **isotherm,** or a *pV-isotherm.*

Figure 18.4 shows pV-isotherms for a constant amount of an ideal gas. The highest temperature is T_4; the lowest is T_1. This is a graphical representation of the ideal-gas equation of state. We can read off the volume V corresponding to any given pressure p and temperature T in the range shown.

Figure 18.5 shows a pV-diagram for a material that *does not* obey the ideal-gas equation. At temperatures below T_c the isotherms develop flat regions in which we can compress the material without an increase in pressure. Observation of the gas shows that it is *condensing* from the vapor (gas) to the liquid phase. The flat parts of the isotherms in the shaded area of Fig. 18.5 represent conditions of liquid-vapor *phase equilibrium.* As the volume decreases, more and more material goes from vapor to liquid, but the pressure does not change. (To keep the temperature constant during condensation, we have to remove the heat of vaporization, discussed in Section 17.6.)

When we compress such a gas at a constant temperature T_2 in Fig. 18.5, it is vapor until point a is reached. Then it begins to liquefy; as the volume decreases further, more material liquefies, and *both* the pressure and the temperature remain constant. At point b, all the material is in the liquid state. After this, any further compression results in a very rapid rise of pressure, because liquids are in general much less compressible than gases. At a lower constant temperature T_1, similar behavior occurs, but the condensation begins at lower pressure and greater volume than at the constant temperature T_2. At temperatures greater than T_c, *no* phase transition occurs as the material is compressed; at the highest temperatures, such as T_4, the curves resemble the ideal-gas curves of Fig. 18.4. We call T_c the *critical temperature* for this material. In Section 18.6 we'll discuss what happens to the phase of the gas above the critical temperature.

We will use pV-diagrams often in the next two chapters. We will show that the *area* under a pV-curve (whether or not it is an isotherm) represents the *work* done by the system during a volume change. This work, in turn, is directly related to heat transfer and changes in the *internal energy* of the system, which we'll get to in Chapter 19.

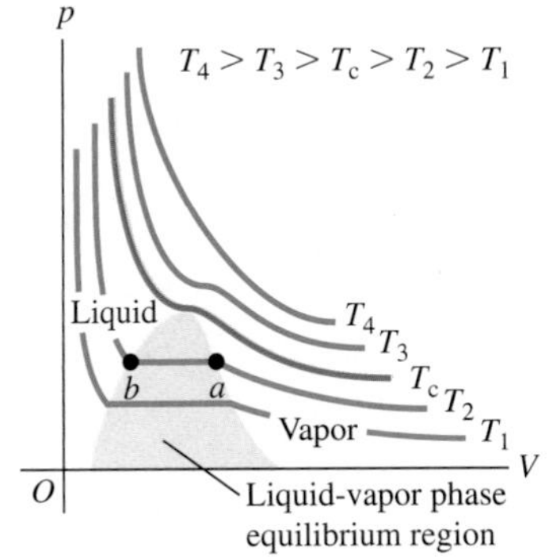

18.5 A pV-diagram for a non-ideal gas, showing isotherms for temperatures above and below the critical temperature T_c. The liquid-vapor equilibrium region is shown as a green shaded area. At still lower temperatures the material might undergo phase transitions from liquid to solid or from gas to solid; these are not shown in this diagram.

Test Your Understanding

Use the ideal-gas equation to explain why it's important to make holes in the crust of an apple pie before baking it.

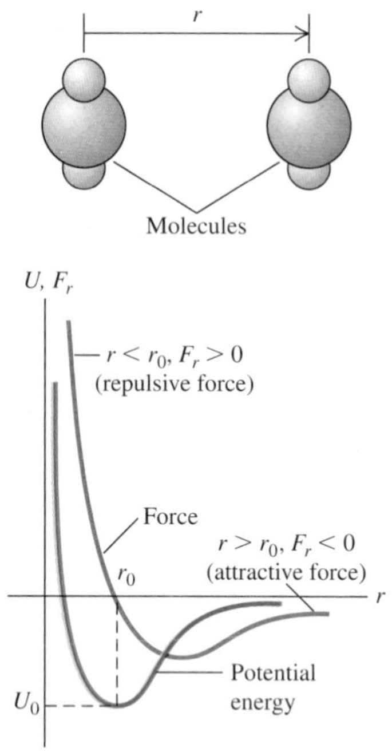

18.6 The force between two molecules (red curve) is zero at a separation $r = r_0$, where the potential energy (blue curve) is a minimum. The force is attractive when the separation is greater than r_0 and repulsive when the separation is less than r_0. (Compare this figure to Fig. 13.19.)

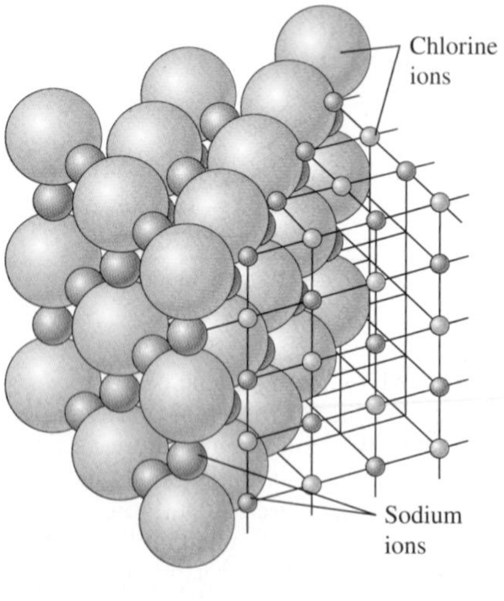

18.7 Schematic representation of the cubic crystal structure of sodium chloride.

18.2 | Molecular Properties of Matter

We have studied several properties of matter in bulk, including elasticity, density, surface tension, heat capacities, and equations of state, with only passing references to molecular structure. Now we want to look in more detail at the relation of bulk behavior to microscopic structure. We begin with a general discussion of the molecular structure of matter. Then in the next two sections we develop the kinetic-molecular model of an ideal gas, obtaining from this molecular model the equation of state and an expression for heat capacity.

All familiar matter is made up of **molecules.** For any specific chemical compound, all the molecules are identical. The smallest molecules contain one atom each and are of the order of 10^{-10} m in size; the largest contain many atoms and are at least 10,000 times that large. In gases the molecules move nearly independently; in liquids and solids they are held together by intermolecular forces that are electrical in nature, arising from interactions of the electrically charged particles that make up the molecules. Gravitational forces between molecules are negligible in comparison with electrical forces.

The interaction of two *point* electric charges is described by a force (repulsive for like charges, attractive for unlike charges) with a magnitude proportional to $1/r^2$, where r is the distance between the points. We will study this relationship, called *Coulomb's law,* in Chapter 21. Molecules are *not* point charges but complex structures containing both positive and negative charge, and their interactions are more complex. The force between molecules in a gas varies with the distance r between molecules somewhat as shown in Fig. 18.6, where a positive F_r corresponds to a repulsive force and a negative F_r to an attractive force. When molecules are far apart, the intermolecular forces are very small and usually attractive. As a gas is compressed and its molecules are brought closer together, the attractive forces increase. The intermolecular force becomes zero at an equilibrium spacing r_0, corresponding roughly to the spacing between molecules in the liquid and solid states. In liquids and solids, relatively large pressures are needed to compress the substance appreciably. This shows that at molecular distances slightly *less* than the equilibrium spacing, the forces become *repulsive* and relatively large.

Figure 18.6 also shows the potential energy as a function of r. This function has a *minimum* at r_0, where the force is zero. The two curves are related by $F_r(r) = -dU/dr$, as we showed in Section 7.4. Such a potential energy function is often called a **potential well.** A molecule at rest at a distance r_0 from a second molecule would need an additional energy $|U_0|$, the "depth" of the potential well, to "escape" to an indefinitely large value of r.

Molecules are always in motion; their kinetic energies usually increase with temperature. At very low temperatures the average kinetic energy of a molecule may be much *less* than the depth of the potential well. The molecules then condense into the liquid or solid phase with average intermolecular spacings of about r_0. But at higher temperatures the average kinetic energy becomes larger than the depth $|U_0|$ of the potential well. Molecules can then escape the intermolecular force and become free to move independently, as in the gaseous phase of matter.

In *solids,* molecules vibrate about more or less fixed points. In a crystalline solid these points are arranged in a recurring *crystal lattice.* Figure 18.7 shows the cubic crystal structure of sodium chloride (ordinary salt). A scanning tunneling microscope image of individual silicon atoms on the surface of a crystal is shown in Fig. 18.8.

The vibration of molecules in a solid about their equilibrium positions may be nearly simple harmonic if the potential well is approximately parabolic in shape at distances close to r_0. (We discussed this kind of simple harmonic motion in Section 13.4.) But if the potential-energy curve rises more gradually for $r > r_0$ than for $r < r_0$, as in Fig. 18.6, the average position shifts to larger r with increasing amplitude. As we pointed out in Section 17.4, this is the basis of thermal expansion.

In a *liquid,* the intermolecular distances are usually only slightly greater than in the solid phase of the same substance, but the molecules have much greater freedom of movement. Liquids show regularity of structure only in the immediate neighborhood of a few molecules. This is called *short-range order,* in contrast with the *long-range order* of a solid crystal.

18.8 A scanning tunneling microscope image of the surface of a silicon crystal. The area shown is only 9.0 nm $(9.0 \times 10^{-9}\ \text{m})$ tall. Each blue "bead" is an individual silicon atom; you can clearly see how these atoms are arranged in a (nearly) perfect array of hexagons.

The molecules of a *gas* are usually widely separated and so have only very small attractive forces. A gas molecule moves in a straight line until it collides with another molecule or with a wall of the container. In molecular terms, an *ideal gas* is a gas whose molecules exert *no* attractive forces on each other and therefore have no *potential* energy.

At low temperatures, most common substances are in the solid phase. As the temperature rises, a substance melts and then vaporizes. From a molecular point of view, these transitions are in the direction of increasing molecular kinetic energy. Thus temperature and molecular kinetic energy are closely related.

We have used the mole as a measure of quantity of substance. One **mole** of any pure chemical element or compound contains a definite number of molecules, the same number for all elements and compounds. The official SI definition is that **one mole is the amount of substance that contains as many elementary entities as there are atoms in 0.012 kilogram of carbon-12.** In our discussion, the "elementary entities" are molecules. (In a monatomic substance such as carbon or helium, each molecule is a single atom, but we'll still call it a molecule here.) Note that atoms of a given element may occur in any of several isotopes, which are chemically identical but have different atomic masses; "carbon-12" refers to a specific isotope of carbon.

The number of molecules in a mole is called **Avogadro's number,** denoted by N_A. The current best numerical value of N_A is

$$N_A = 6.02214199(47) \times 10^{23}\ \text{molecules/mol} \qquad \text{(Avogadro's number)}$$

The *molar mass* M of a compound is the mass of one mole. It is equal to the mass m of a single molecule multiplied by Avogadro's number.

$$M = N_A m \qquad (18.8)$$

(molar mass, Avogadro's number, and mass of a molecule)

When the molecule consists of a single atom, the term *atomic mass* is often used instead of molar mass or molecular weight.

Example 18.5 Atomic and molecular mass

Find the mass of a single hydrogen atom and the mass of an oxygen molecule.

SOLUTION

IDENTIFY and SET UP: We use Eq. (18.8) in the form $m = M/N_A$ and the values of the molar masses from the periodic table of the elements (Appendix D).

EXECUTE: The mass per mole of atomic hydrogen (that is, the atomic mass) is 1.008 g/mol. Therefore the mass m_H of a single hydrogen atom is

$$m_H = \frac{1.008 \text{ g/mol}}{6.002 \times 10^{23} \text{ molecules/mol}} = 1.674 \times 10^{-24} \text{ g/molecule}$$

From Appendix D, the atomic mass of oxygen is 16.0 g/mol, so the molar mass of oxygen, which has diatomic (two-atom) molecules, is 32.0 g/mol. The mass of a single molecule of O_2 is

$$m_{O_2} = \frac{32.0 \text{ g/mol}}{6.022 \times 10^{23} \text{ molecules/mol}} = 53.1 \times 10^{-24} \text{ g/molecule}$$

EVALUATE: We note that the values in Appendix D are for the *average* atomic masses of a natural sample of each element. Such a sample may contain several different isotopes of the element, each with a different atomic mass. Natural samples of hydrogen and oxygen are almost entirely made up of just one isotope; this is not the case for all elements, however.

Test Your Understanding

Suppose you could adjust the value of r_0 for the molecules of a certain chemical compound (Fig. 18.6) by turning a dial. If you doubled the value of r_0, how would this change the density of the solid form of this compound?

18.3 | Kinetic-Molecular Model of an Ideal Gas

The goal of any molecular theory of matter is to understand the *macroscopic* properties of matter in terms of its atomic or molecular structure and behavior. Such theories are of tremendous practical importance; once we have this understanding, we can design materials to have specific desired properties. Such analysis has led to the development of high-strength steels, glasses with special optical properties, semiconductor materials for electronic devices, and countless other materials essential to contemporary technology.

ActivPhysics ONLINE

8.1 Characteristics of a Gas

In this and the following sections we will consider a simple molecular model of an ideal gas. This *kinetic-molecular model* represents the gas as a large number of particles bouncing around in a closed container. In this section we use the kinetic-molecular model to understand how the ideal-gas equation of state, Eq. (18.3), is related to Newton's laws. In the following section we use the kinetic-molecular model to predict the molar heat capacity of an ideal gas. We'll go on to elaborate the model to include "particles" that are not points but have a finite size. We will be able to see why polyatomic gases have larger molar heat capacities than monatomic gases.

The following discussion of the kinetic-molecular model has several steps, and you may need to go over it several times to grasp how it all goes together. Don't get discouraged!

Here are the assumptions of our model:

1. A container with volume V contains a very large number N of identical molecules, each with mass m.
2. The molecules behave as point particles; their size is small in comparison to the average distance between particles and to the dimensions of the container.
3. The molecules are in constant motion; they obey Newton's laws of motion. Each molecule collides occasionally with a wall of the container. These collisions are perfectly elastic.
4. The container walls are perfectly rigid and infinitely massive and do not move.

CAUTION Make sure you don't confuse N, the number of *molecules* in the gas, with n, the number of *moles*. The number of molecules is equal to the number of moles multiplied by Avogadro's number: $N = nN_A$.

During collisions the molecules exert *forces* on the walls of the container; this is the origin of the *pressure* that the gas exerts. In a typical collision (Fig. 18.9) the velocity component parallel to the wall is unchanged, and the component perpendicular to the wall reverses direction but does not change in magnitude.

Our program is first to determine the *number* of collisions that occur per unit time for a certain area A of wall. Then we find the total momentum change associated with these collisions and the force needed to cause this momentum change. From this we can determine the pressure, which is force per unit area, and compare the result to the ideal-gas equation. We'll find a direct connection between the temperature of the gas and the kinetic energy of the gas molecules.

To begin, we will assume that all molecules in the gas have the same *magnitude* of x-velocity, $|v_x|$. This isn't right, but making this temporary assumption helps to clarify the basic ideas. We will show later that this assumption isn't really necessary.

As shown in Fig. 18.9, for each collision the x-component of velocity changes from $-|v_x|$ to $+|v_x|$. So the x-component of momentum changes from $-m|v_x|$ to $+m|v_x|$, and the *change* in the x-component of momentum is $m|v_x| - (-m|v_x|) = 2m|v_x|$.

If a molecule is going to collide with a given wall area A during a small time interval dt, then at the beginning of dt it must be within a distance $|v_x|\,dt$ from the wall (Fig. 18.10) and it must be headed toward the wall. So the number of molecules that collide with A during dt is equal to the number of molecules within a cylinder with base area A and length $|v_x|\,dt$ that have their x-velocity aimed toward the wall. The volume of such a cylinder is $A|v_x|\,dt$. Assuming that the number of molecules per unit volume (N/V) is uniform, the *number* of molecules in this cylinder is $(N/V)(A|v_x|\,dt)$. On the average, half of these molecules are moving toward the wall and half are moving away from it. So the number of collisions with A during dt is

$$\frac{1}{2}\left(\frac{N}{V}\right)(A|v_x|dt)$$

For the system of all molecules in the gas, the total momentum change dP_x during dt is the *number* of collisions multiplied by $2m|v_x|$:

$$dP_x = \frac{1}{2}\left(\frac{N}{V}\right)(A|v_x|dt)(2m|v_x|) = \frac{NAmv_x^2\,dt}{V} \qquad (18.9)$$

(We are using capital P for total momentum and small p for pressure. Be careful!) We wrote v_x^2 rather than $|v_x|^2$ in the final expression because the square of the absolute value of a number is equal to the square of that number. The *rate* of change of momentum component P_x is

$$\frac{dP_x}{dt} = \frac{NAmv_x^2}{V} \qquad (18.10)$$

According to Newton's second law, this rate of change of momentum equals the force exerted by the wall area A on the gas molecules. From Newton's *third* law

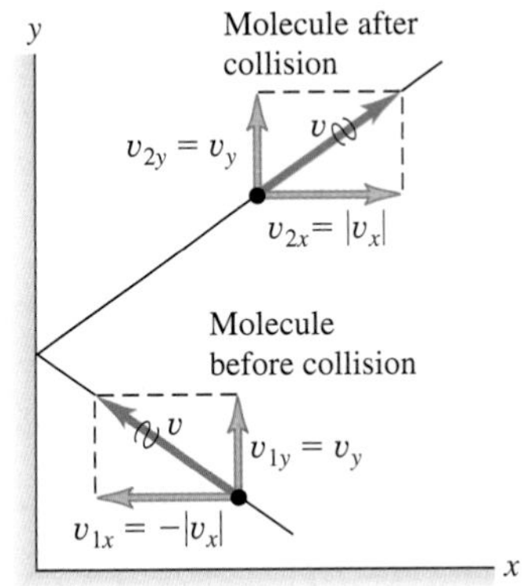

18.9 Elastic collision of a molecule with an idealized container wall. The velocity component parallel to the wall does not change; the velocity component perpendicular to the wall reverses direction. The speed v does not change.

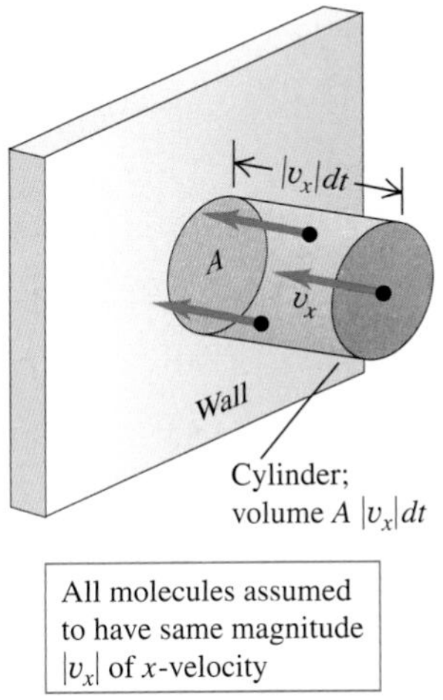

18.10 For a molecule to strike the wall in area A during a time interval dt, the molecule must be headed for the wall and be within the shaded cylinder of height $|v_x|dt$ at the beginning of the interval.

this is equal and opposite to the force exerted *on* the wall *by* the molecules. Pressure p is the magnitude of the force exerted on the wall per unit area, and we obtain

$$p = \frac{F}{A} = \frac{Nmv_x^2}{V} \qquad (18.11)$$

The pressure exerted by the gas depends on the number of molecules per volume (N/V), the mass m per molecule, and the speed of the molecules.

We mentioned that $|v_x|$ is really *not* the same for all the molecules. But we could have sorted the molecules into groups having the same $|v_x|$ within each group, then added up the resulting contributions to the pressure. The net effect of all this is just to replace v_x^2 in Eq. (18.11) by the *average* value of v_x^2, which we denote by $(v_x^2)_{av}$. Furthermore, $(v_x^2)_{av}$ is related simply to the *speeds* of the molecules. The speed v of any molecule is related to the velocity components v_x, v_y, and v_z by

$$v^2 = v_x^2 + v_y^2 + v_z^2$$

We can average this relation over all molecules:

$$(v^2)_{av} = (v_x^2)_{av} + (v_y^2)_{av} + (v_z^2)_{av}$$

But there is no real difference in our model between the x-, y-, and z-directions. (Molecular speeds are very fast in a typical gas, so the effects of gravity are negligibly small.) It follows that $(v_x^2)_{av}$, $(v_y^2)_{av}$, and $(v_z^2)_{av}$ must all be *equal.* Hence $(v^2)_{av}$ is equal to $3(v_x^2)_{av}$, and

$$(v_x^2)_{av} = \frac{1}{3}(v^2)_{av}$$

so Eq. (18.11) becomes

$$pV = \frac{1}{3}Nm(v^2)_{av} = \frac{2}{3}N\left[\frac{1}{2}m(v^2)_{av}\right] \qquad (18.12)$$

We notice that $\frac{1}{2}m(v^2)_{av}$ is the average translational kinetic energy of a single molecule. The product of this and the total number of molecules N equals the total random kinetic energy K_{tr} of translational motion of all the molecules. (The notation K_{tr} reminds us that this energy is associated with *translational* motion. There may be additional energies associated with rotational and vibrational motion of molecules.) The product pV equals two-thirds of the total translational kinetic energy:

$$pV = \frac{2}{3}K_{tr} \qquad (18.13)$$

Now we compare this with the ideal-gas equation,

$$pV = nRT$$

which is based on experimental studies of gas behavior. For the two equations to agree, we must have

$$K_{tr} = \frac{3}{2}nRT \qquad (18.14)$$

(average translational kinetic energy of n moles of ideal gas)

This remarkably simple result shows that K_{tr} is *directly proportional* to the absolute temperature T. We will use this important result several times in the following discussion.

The average translational kinetic energy of a single molecule is the total translational kinetic energy K_{tr} of all molecules divided by the number of molecules, N:

$$\frac{K_{tr}}{N} = \frac{1}{2}m(v^2)_{av} = \frac{3nRT}{2N}$$

Also, the total number of molecules N is the number of moles n multiplied by Avogadro's number N_A, so

$$N = nN_A \qquad \frac{n}{N} = \frac{1}{N_A}$$

and

$$\frac{K_{tr}}{N} = \frac{1}{2}m(v^2)_{av} = \frac{3}{2}\left(\frac{R}{N_A}\right)T \tag{18.15}$$

The ratio R/N_A occurs frequently in molecular theory. It is called the **Boltzmann constant,** k:

$$k = \frac{R}{N_A} = \frac{8.314 \text{ J/mol}\cdot\text{K}}{6.022 \times 10^{23} \text{ molecules/mol}}$$
$$= 1.381 \times 10^{-23} \text{ J/molecule}\cdot\text{K}$$

(The current best numerical value of k is $1.3806503(24) \times 10^{-23}$ J/molecule · K.) In terms of k we can rewrite Eq. (18.15) as

$$\frac{1}{2}m(v^2)_{av} = \frac{3}{2}kT \tag{18.16}$$

(average translational kinetic energy of a gas molecule)

This shows that the average translational kinetic energy *per molecule* depends only on the temperature, not on the pressure, volume, or kind of molecule. We can obtain the average translational kinetic energy *per mole* by multiplying Eq. (18.16) by Avogadro's number and using the relation $M = N_A m$:

$$N_A\frac{1}{2}m(v^2)_{av} = \frac{1}{2}M(v^2)_{av} = \frac{3}{2}RT \tag{18.17}$$

(average translational kinetic energy per mole of gas)

The translational kinetic energy of a mole of ideal-gas molecules depends only on T.

Finally, it is sometimes convenient to rewrite the ideal-gas equation on a molecular basis. We use $N = N_A n$ and $R = N_A k$ to obtain the alternative form of the ideal-gas equation:

$$pV = NkT \tag{18.18}$$

This shows that we can think of the Boltzmann constant k as a gas constant on a "per-molecule" basis instead of the usual "per-mole" basis for R.

Molecular Speeds

From Eqs. (18.16) and (18.17) we can obtain expressions for the square root of $(v^2)_{av}$, called the **root-mean-square speed** (or **rms speed**) v_{rms}:

$$v_{rms} = \sqrt{(v^2)_{av}} = \sqrt{\frac{3kT}{m}} = \sqrt{\frac{3RT}{M}} \qquad (18.19)$$

(root-mean-square speed of a gas molecule)

It might seem more natural to characterize molecular speeds by their *average* value rather than by v_{rms}, but we see that v_{rms} evolves more directly from Eqs. (18.16) and (18.17). To compute the rms speed, we square each molecular speed, add, divide by the number of molecules, and take the square root; v_{rms} is the *root* of the *mean* of the *squares*. Example 18.7 illustrates this procedure.

Equations (18.16) and (18.19) show that at a given temperature T, gas molecules of different mass m have the same average kinetic energy but different root-mean-square speeds. On average, the nitrogen molecules $(M = 28\text{ g/mol})$ in the air around you are moving faster than are the oxygen molecules $(M = 32\text{ g/mol})$. Hydrogen molecules $(M = 2\text{ g/mol})$ are fastest of all; this is why there is hardly any hydrogen in the earth's atmosphere, despite its being the most common element in the universe (Fig. 18.11). A sizable fraction of any H_2 molecules in the atmosphere would have speeds greater than the earth's "escape speed" of 1.12×10^4 m/s (calculated in Example 12.5, Section 12.3) and would escape into space. The heavier, slower-moving gases cannot escape so easily, which is why they dominate our atmosphere.

?

18.11 The relationship between molecular speeds and molecular masses explains why hydrogen—the most common element in the universe—is only a trace constituent of our atmosphere (only 0.00005% hydrogen by volume).

The assumption that individual molecules undergo perfectly elastic collisions with the container wall is actually a little too simple. More detailed investigation has shown that in most cases, molecules actually adhere to the wall for a short time and then leave again with speeds that are characteristic of the temperature *of the wall.* However, the gas and the wall are ordinarily in thermal equilibrium and have the same temperature. So there is no net energy transfer between gas and wall, and this discovery does not alter the validity of our conclusions.

Problem-Solving Strategy

Kinetic-Molecular Theory

IDENTIFY *the relevant concepts:* Use the results of the kinetic-molecular model whenever you are asked to relate macroscopic properties of a gas, such as temperature and pressure, to microscopic properties, such as molecular speeds.

SET UP *the problem using the following steps:*

1. Identify which variables are known and which are the unknown target variables.
2. Choose the equation(s) to be used from among Eqs. (18.14), (18.16), and (18.19).

EXECUTE *the solution as follows:* As you solve for the target variable, be on your guard for inconsistency in units. Special caution is needed in the following places:

1. The usual units for molecular mass M are grams per mole; the molecular mass of oxygen (O_2) is 32 g/mol, for example. These units are often omitted in tables. In equations such as Eq. (18.19), when you use SI units you *must* express M in kilograms per mole by multiplying the table value by $(1\text{ kg}/10^3\text{ g})$. Thus in SI units, M for oxygen is 32×10^{-3} kg/mol.
2. Are you working on a "per molecule" basis or a "per mole" basis? Remember that m is the mass of a single molecule and M is the mass of a mole of molecules; N is the number of molecules and n is the number of moles; k is the gas constant per molecule and R is the gas constant per mole. Although N, the number of molecules, is in one sense a dimensionless number, you can do a complete unit

check if you think of N as having units of "molecules"; then m has units of "mass per molecule," and k has units of "joules per molecule per kelvin."

3. Remember that T is always *absolute* (Kelvin) temperature.

EVALUATE *your answer:* Are your answers reasonable? Keep in mind that typical molecular speeds at room temperature are several hundred meters per second. If your answer seems dramatically different, recheck your calculations.

Example 18.6 Calculating molecular kinetic energy and v_{rms}

a) What is the average translational kinetic energy of a molecule of an ideal gas at a temperature of 27°C? b) What is the total random translational kinetic energy of the molecules in one mole of this gas? c) What is the root-mean-square speed of oxygen molecules at this temperature?

SOLUTION

IDENTIFY and SET UP: We are given temperature $T = 27°C$ and number of moles $n = 1$ mol, and the molecular mass m is that for oxygen. We use Eq. (18.16) to determine the average kinetic energy of a molecule, Eq. (18.14) to find the total molecular kinetic energy, and Eq. (18.19) to find the rms speed of a molecule.

EXECUTE: a) To use Eq. (18.16), we first convert the temperature to the Kelvin scale: 27°C = 300 K. Then

$$\frac{1}{2}m(v^2)_{av} = \frac{3}{2}kT = \frac{3}{2}(1.38 \times 10^{-23}\ \text{J/K})(300\ \text{K})$$
$$= 6.21 \times 10^{-21}\ \text{J}$$

This answer does not depend on the mass of the molecule.

b) From Eq. (18.14), the total translational kinetic energy of a mole of molecules is

$$K_{tr} = \frac{3}{2}nRT = \frac{3}{2}(1\ \text{mol})(8.315\ \text{J/mol}\cdot\text{K})(300\ \text{K})$$
$$= 3740\ \text{J}$$

This is about the same kinetic energy as that of a sprinter in a 100-m dash.

c) From Example 18.5 (Section 18.2), the mass of an oxygen molecule is

$$m_{O_2} = (53.1 \times 10^{-24}\ \text{g})(1\ \text{kg}/10^3\ \text{g}) = 5.31 \times 10^{-26}\ \text{kg}$$

From Eq. (18.19),

$$v_{rms} = \sqrt{\frac{3kT}{m}} = \sqrt{\frac{3(1.38 \times 10^{-23}\ \text{J/K})(300\ \text{K})}{5.31 \times 10^{-26}\ \text{kg}}}$$
$$= 484\ \text{m/s}$$

This is 1740 km/h, or 1080 mi/h! Alternatively,

$$v_{rms} = \sqrt{\frac{3RT}{M}} = \sqrt{\frac{3(8.314\ \text{J/mol}\cdot\text{K})(300\ \text{K})}{32.0 \times 10^{-3}\ \text{kg/mol}}}$$
$$= 484\ \text{m/s}$$

EVALUATE: We can check our result in part (b) by noting that the translational kinetic energy per mole must be equal to the average translational kinetic energy per molecule from part (a) multiplied by Avogadro's number N_A: $K_{tr} = (6.022 \times 10^{23}\ \text{molecules})(6.21 \times 10^{-21}\ \text{J/molecule}) = 3740$ J.

In part (c), note that when we use Eq. (18.19) with R in SI units, we must express M in *kilograms* per mole, not grams per mole. In this example we use $M = 32.0 \times 10^{-3}$ kg/mol, *not* 32.0 g/mol.

Example 18.7 Rms and average speeds

Five gas molecules chosen at random are found to have speeds of 500, 600, 700, 800, and 900 m/s. Find the rms speed. Is it the same as the *average* speed?

SOLUTION

IDENTIFY and SET UP: To solve this problem, we use the definition of the root-mean-square value: square each value, find the average (mean) of the squares, and then take the square root of the result.

EXECUTE: The average value of v^2 for the five molecules is

$$(v^2)_{av} = \frac{\left[\begin{array}{c}(500\ \text{m/s})^2 + (600\ \text{m/s})^2 + (700\ \text{m/s})^2 \\ +(800\ \text{m/s})^2 + (900\ \text{m/s})^2\end{array}\right]}{5}$$
$$= 5.10 \times 10^5\ \text{m}^2/\text{s}^2$$

The square root of this is v_{rms}:

$$v_{rms} = 714\ \text{m/s}$$

The *average* speed v_{av} is given by

$$v_{av} = \frac{500\ \text{m/s} + 600\ \text{m/s} + 700\ \text{m/s} + 800\ \text{m/s} + 900\ \text{m/s}}{5}$$
$$= 700\ \text{m/s}$$

EVALUATE: We see that in general, v_{rms} and v_{av} are *not* the same. Roughly speaking, v_{rms} gives greater weight to the larger speeds than does v_{av}.

Example 18.8 Volume of a gas at STP, revisited

Find the number of molecules and the number of moles in one cubic meter of air at atmospheric pressure and 0°C.

SOLUTION

IDENTIFY and SET UP: We are given the volume V, the pressure p, and the temperature T. We find the first target variable, the number of molecules N, using Eq. (18.18); the number of moles is N divided by Avogadro's number.

EXECUTE: From Eq. (18.18),

$$N = \frac{pV}{kT} = \frac{(1.013 \times 10^5 \text{ Pa})(1 \text{ m}^3)}{(1.38 \times 10^{-23} \text{ J/K})(273 \text{ K})}$$
$$= 2.69 \times 10^{25} \text{ molecules}$$

The number of moles n is

$$n = \frac{N}{N_A} = \frac{2.69 \times 10^{25} \text{ molecules}}{6.022 \times 10^{23} \text{ molecules/mol}} = 44.7 \text{ mol}$$

EVALUATE: To check our result, note that the total volume is 1.00 m^3, so the volume of one mole is

$$\frac{1.00 \text{ m}^3}{44.7 \text{ mol}} = 0.0224 \text{ m}^3\text{/mol} = 22.4 \text{ L/mol}$$

This is the same result that we obtained in Example 18.1 (Section 18.1) using only macroscopic quantities.

Collisions Between Molecules

We have ignored the possibility that two gas molecules might collide. If they are really points, they *never* collide. But consider a more realistic model in which the molecules are rigid spheres with radius r. How often do they collide with other molecules? How far do they travel, on average, between collisions? We can get approximate answers from the following rather primitive model.

Consider N spherical molecules with radius r in a volume V. Suppose only one molecule is moving. When it collides with another molecule, the distance between centers is $2r$. Suppose we draw a cylinder with radius $2r$, with its axis parallel to the velocity of the molecule (Fig. 18.12). The moving molecule collides with any other molecule whose center is inside this cylinder. In a short time dt a molecule with speed v travels a distance $v\,dt$; during this time it collides with any molecule that is in the cylindrical volume of radius $2r$ and length $v\,dt$. The volume of the cylinder is $4\pi r^2 v\,dt$. There are N/V molecules per unit volume, so the number dN with centers in this cylinder is

$$dN = 4\pi r^2 v\,dt\,N/V$$

Thus the number of collisions *per unit time* is

$$\frac{dN}{dt} = \frac{4\pi r^2 vN}{V}$$

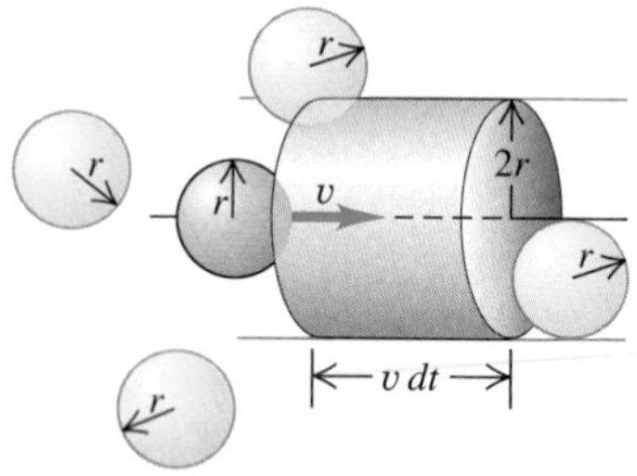

18.12 In a time dt a molecule with radius r will collide with any other molecule within a cylindrical volume of radius $2r$ and length $v\,dt$.

This result assumes that only one molecule is moving. The analysis is quite a bit more involved when all the molecules move at once. It turns out that in this case the collisions are more frequent, and the above equation has to be multiplied by a factor of $\sqrt{2}$:

$$\frac{dN}{dt} = \frac{4\pi\sqrt{2}r^2 vN}{V}$$

The average time t_{mean} between collisions, called the *mean free time*, is the reciprocal of this expression:

$$t_{mean} = \frac{V}{4\pi\sqrt{2}r^2 vN} \tag{18.20}$$

The average distance traveled between collisions is called the **mean free path,** denoted by λ (the Greek letter "lambda"). In our simple model, this is just the molecule's speed v multiplied by t_{mean}:

$$\lambda = vt_{mean} = \frac{V}{4\pi\sqrt{2}r^2N} \quad \text{(mean free path of a gas molecule)} \quad (18.21)$$

The mean free path is inversely proportional to the number of molecules per unit volume (N/V) and inversely proportional to the cross-sectional area πr^2 of a molecule; the more molecules there are and the larger the size of a molecule, the shorter the mean distance between collisions (Fig. 18.13). Note that the mean free path *does not* depend on the speed of the molecule.

We can express Eq. (18.21) in terms of macroscopic properties of the gas, using the ideal-gas equation in the form of Eq. (18.18), $pV = NkT$. We find

$$\lambda = \frac{kT}{4\pi\sqrt{2}r^2p} \quad (18.22)$$

If the temperature is increased at constant pressure, the gas expands, the average distance between molecules increases, and λ increases. If the pressure is increased at constant temperature, the gas compresses and λ decreases.

18.13 If you try to walk through a crowd, your mean free path—the distance you can travel on average without running into another person—depends on how large the people are and how closely they are spaced.

Example 18.9 Calculating mean free path

a) Estimate the mean free path of a molecule of air at 27°C and 1 atm. Model the molecules as spheres with radius $r = 2.0 \times 10^{-10}$ m. b) Estimate the mean free time of an oxygen molecule with $v = v_{rms}$.

SOLUTION

IDENTIFY and SET UP: We use Eq. (18.21) to determine the mean free path λ. To find the mean free time t_{mean} we could use Eq. (18.20), but it's more convenient to use the basic relationship $\lambda = vt_{mean}$ in Eq. (18.21). For the speed v we use the root-mean-square speed for oxygen calculated in Example 18.6.

EXECUTE: a) From Eq. (18.22),

$$\lambda = \frac{kT}{4\pi\sqrt{2}r^2p} = \frac{(1.38 \times 10^{-23}\ \text{J/K})(300\ \text{K})}{4\pi\sqrt{2}(2.0 \times 10^{-10}\ \text{m})^2(1.01 \times 10^5\ \text{Pa})} = 5.8 \times 10^{-8}\ \text{m}$$

The molecule doesn't get very far between collisions, but the distance is still several hundred times the radius of the molecule. To get a mean free path of one meter, the pressure must be about 5.8×10^{-8} atm. Pressures this low are found 100 km or so above the earth's surface, at the outer fringe of our atmosphere.

b) From Example 18.6, for oxygen at 27°C the root-mean-square speed is $v_{rms} = 484$ m/s, so the mean free time for a molecule with this speed is

$$t_{mean} = \frac{\lambda}{v} = \frac{5.8 \times 10^{-8}\ \text{m}}{484\ \text{m/s}} = 1.2 \times 10^{-10}\ \text{s}$$

This molecule undergoes about 10^{10} collisions per second!

EVALUATE: Note that the mean free *path* calculated in part (a) doesn't depend on the molecule's speed, but the mean free *time* does. Slower molecules have a longer average time interval t_{mean} between collisions than do fast ones, but the average *distance* λ between collisions is the same no matter what the molecule's speed.

Test Your Understanding

What is the root-mean-square speed of hydrogen atoms in the sun's corona, shown in the photo that opens Chapter 17? (The corona's temperature is so high and its density so low that hydrogen atoms cannot form into molecules.) Compare your result to the sun's escape speed of 6.18×10^5 m/s.

18.4 | Heat Capacities

When we introduced the concept of heat capacity in Section 17.5, we talked about ways to *measure* the specific heat or molar heat capacity of a particular material. Now we'll see how these numbers can be *predicted* on theoretical grounds. That's a significant step forward.

Heat Capacities of Gases

The basis of our analysis is that heat is *energy* in transit. When we add heat to a substance, we are increasing its molecular energy. In this discussion we will keep the volume of the gas constant so that we don't have to worry about energy transfer through mechanical work. If we were to let the gas expand, it would do work by pushing on moving walls of its container, and this additional energy transfer would have to be included in our calculations. We'll return to this more general case in Chapter 19. For now, with the volume held constant, we are concerned with C_V, the molar heat capacity *at constant volume.*

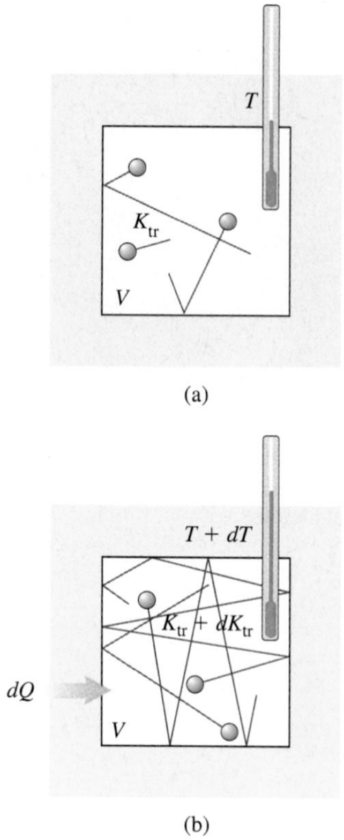

18.14 (a) A fixed volume V of a monatomic ideal gas. (b) When an amount of heat dQ is added to the gas, the total translational kinetic energy increases by $dK_{tr} = dQ$ and the temperature increases by $dT = dQ/nC_V$.

In the simple kinetic-molecular model of Section 18.3 the molecular energy consists only of the translational kinetic energy K_{tr} of the pointlike molecules. This energy is directly proportional to the absolute temperature T, as shown by Eq. (18.14), $K_{tr} = \frac{3}{2}nRT$. When the temperature changes by a small amount dT, the corresponding change in kinetic energy is

$$dK_{tr} = \frac{3}{2}nR\,dT \qquad (18.23)$$

From the definition of molar heat capacity at constant volume, C_V (Section 17.5), we also have

$$dQ = nC_V\,dT \qquad (18.24)$$

where dQ is the heat input needed for a temperature change dT. Now if K_{tr} represents the total molecular energy, as we have assumed, then dQ and dK_{tr} must be *equal* (Fig. 18.14). Equating the expressions given by Eqs. (18.23) and (18.24), we get

$$nC_V\,dT = \frac{3}{2}nR\,dT$$

$$C_V = \frac{3}{2}R \qquad \text{(ideal gas of point particles)} \qquad (18.25)$$

This surprisingly simple result says that the molar heat capacity (at constant volume) of *every* gas whose molecules can be represented as points is equal to $3R/2$.

To see whether this makes sense, let's first check the units. The gas constant *does* have units of energy per mole per kelvin, the correct units for a molar heat capacity. But more important is whether Eq. (18.25) agrees with *measured* values of molar heat capacities. In SI units, Eq. (18.25) gives

$$C_V = \frac{3}{2}(8.314\ \text{J/mol}\cdot\text{K}) = 12.47\ \text{J/mol}\cdot\text{K}$$

For comparison, Table 18.1 gives measured values of C_V for several gases. We see that for *monatomic* gases our prediction is right on the money, but that it is way off for diatomic and polyatomic gases.

Table 18.1 Molar Heat Capacities of Gases

Type of Gas	Gas	C_V (J/mol·K)
Monatomic	He	12.47
	Ar	12.47
Diatomic	H_2	20.42
	N_2	20.76
	O_2	21.10
	CO	20.85
Polyatomic	CO_2	28.46
	SO_2	31.39
	H_2S	25.95

This comparison tells us that our point-molecule model is good enough for monatomic gases but that for diatomic and polyatomic molecules we need something more sophisticated. For example, we can picture a diatomic molecule as *two* point masses, like a little elastic dumbbell, with an interaction force between the atoms of the kind shown in Fig. 18.6. Such a molecule can have additional kinetic energy associated with *rotation* about axes through its center of mass. The atoms may also have a back-and-forth *vibrating* motion along the line joining them, with additional kinetic and potential energies. These possibilities are shown in Fig. 18.15.

When heat flows into a *monatomic* gas at constant volume, *all* of the added energy goes into an increase in random *translational* molecular kinetic energy. Equation (18.23) shows that this gives rise to an increase in temperature. But when the temperature is increased by the same amount in a *diatomic* or *polyatomic* gas, additional heat is needed to supply the increased rotational and vibrational energies. Thus polyatomic gases have *larger* molar heat capacities than monatomic gases, as Table 18.1 shows.

But how do we know how much energy is associated with each additional kind of motion of a complex molecule, compared to the translational kinetic energy? The new principle that we need is called the principle of **equipartition of energy.** It can be derived from sophisticated statistical-mechanics considerations; that derivation is beyond our scope, and we will treat the principle as an axiom.

The principle of equipartition of energy states that each velocity component (either linear or angular) has, on average, an associated kinetic energy per molecule of $\frac{1}{2}kT$, or one-half the product of the Boltzmann constant and the absolute temperature. The number of velocity components needed to describe the motion of a molecule completely is called the number of **degrees of freedom.** For a monatomic gas, there are three degrees of freedom (for the velocity components v_x, v_y, and v_z); this gives a total average kinetic energy per molecule of $3(\frac{1}{2}kT)$, consistent with Eq. (18.16).

For a *diatomic* molecule there are two possible axes of rotation, perpendicular to each other and to the molecule's axis. (We don't include rotation about the molecule's own axis because in ordinary collisions there is no way for this rotational motion to change.) If we assign five degrees of freedom to a diatomic molecule, the average total kinetic energy per molecule is $\frac{5}{2}kT$ instead of $\frac{3}{2}kT$. The total kinetic energy of n moles is $K_{\text{tot}} = nN_A(\frac{5}{2}kT) = \frac{5}{2}n(kN_A)T = \frac{5}{2}nRT$, and the molar heat capacity (at constant volume) is

$$C_V = \frac{5}{2}R \quad \text{(diatomic gas, including rotation)} \qquad (18.26)$$

In SI units,

$$C_V = \frac{5}{2}(8.314 \text{ J/mol}\cdot\text{K}) = 20.79 \text{ J/mol}\cdot\text{K}$$

This agrees within a few percent with the measured values for diatomic gases given in Table 18.1.

Vibrational motion can also contribute to the heat capacities of gases. Molecular bonds are not rigid; they can stretch and bend, and the resulting vibrations lead to additional degrees of freedom and additional energies. For most diatomic gases, however, vibrational motion does *not* contribute appreciably to heat capacity. The reason for this is a little subtle and involves some concepts of quantum mechanics.

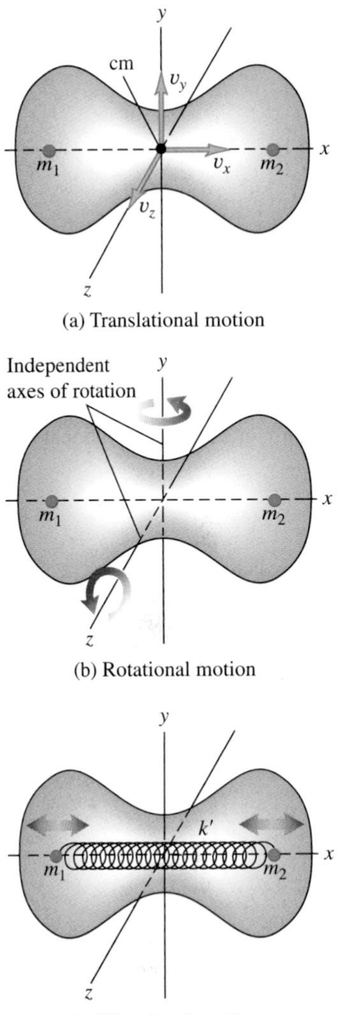

(a) Translational motion

(b) Rotational motion

(c) Vibrational motion

18.15 A diatomic molecule. Almost all the mass of each atom is in its tiny nucleus. (a) The center of mass has three independent velocity components. (b) The molecule has two independent axes of rotation through its center of mass. (c) The atoms and "spring" have additional kinetic and potential energies of vibration.

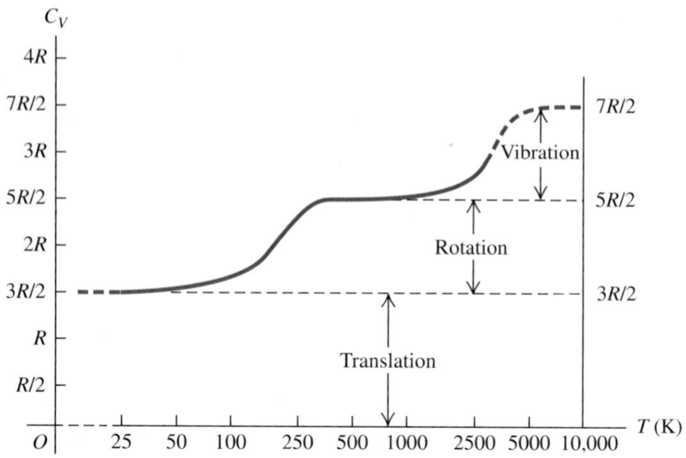

18.16 Experimental values of C_V for hydrogen gas (H_2). The temperature is plotted on a logarithmic scale. Appreciable rotational motion begins to occur above 50 K, and above 600 K the molecule begins to appreciably increase its vibrational motion.

Briefly, vibrational energy can change only in finite steps. If the energy change of the first step is much larger than the energy possessed by most molecules, then nearly all the molecules remain in the minimum-energy state of motion. In that case, changing the temperature does not change their average vibrational energy appreciably, and the vibrational degrees of freedom are said to be "frozen out." In more complex molecules the gaps between permitted energy levels are sometimes much smaller, and then vibration *does* contribute to heat capacity. The rotational energy of a molecule also changes by finite steps, but they are usually much smaller; the "freezing out" of rotational degrees of freedom occurs only in rare instances, such as for the hydrogen molecule below about 100 K.

In Table 18.1 the large values of C_V for some polyatomic molecules show the contributions of vibrational energy. In addition, a molecule with three or more atoms that are not in a straight line has three, not two, rotational degrees of freedom.

From this discussion we expect heat capacities to be temperature-dependent, generally increasing with increasing temperature. Figure 18.16 is a graph of the temperature dependence of C_V for hydrogen gas (H_2), showing the temperatures at which the rotational and vibrational energies begin to contribute to the heat capacity.

Heat Capacities of Solids

We can carry out a similar heat-capacity analysis for a crystalline solid. Consider a crystal consisting of N identical atoms (a *monatomic solid*). Each atom is bound to an equilibrium position by interatomic forces. The elasticity of solid materials shows us that these forces must permit stretching and bending of the bonds. We can think of a crystal as an array of atoms connected by little springs (Fig. 18.17). Each atom can *vibrate* about its equilibrium position.

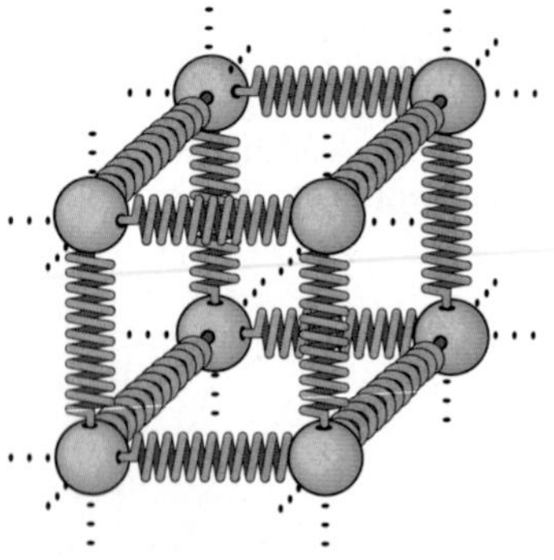

18.17 To visualize the forces between neighboring atoms in a crystal, envision every atom as being attached to its neighbors by springs.

Each atom has three degrees of freedom, corresponding to its three components of velocity. According to the equipartition principle, each atom has an average kinetic energy of $\frac{1}{2}kT$ for each degree of freedom. In addition, each atom has *potential* energy associated with the elastic deformation. For a simple harmonic oscillator (discussed in Chapter 13) it is not hard to show that the average kinetic energy of an atom is *equal* to its average potential energy. In our model of a crystal, each atom is essentially a three-dimensional harmonic oscillator; it can be shown that the equality of average kinetic and potential energies also holds here, provided that the "spring" forces obey Hooke's law.

Thus we expect each atom to have an average kinetic energy $\frac{3}{2}kT$ and an average potential energy $\frac{3}{2}kT$, or an average total energy $3kT$ per atom. If the crystal contains N atoms or n moles, its total energy is

$$E_{\text{tot}} = 3NkT = 3nRT \tag{18.27}$$

From this we conclude that the molar heat capacity of a crystal should be

$$C_V = 3R \quad \text{(ideal monatomic solid)} \tag{18.28}$$

In SI units,

$$C_V = (3)(8.314\ \text{J/mol}\cdot\text{K}) = 24.9\ \text{J/mol}\cdot\text{K}$$

This is the **rule of Dulong and Petit,** which we encountered as an *empirical* finding in Section 17.5: elemental solids all have molar heat capacities of about 25 J/mol · K. Now we have *derived* this rule from kinetic theory. The agreement is only approximate, to be sure, but considering the very simple nature of our model, it is quite significant.

At low temperatures, the heat capacities of most solids *decrease* with decreasing temperature (Fig. 18.18) for the same reason that vibrational degrees of freedom of molecules are frozen out at low temperatures. At very low temperatures the quantity kT is much *smaller* than the smallest energy step the vibrating atoms can take. Hence most of the atoms remain in their lowest energy states because the next higher energy level is out of reach. The average vibrational energy per atom is then *less* than $3kT$, and the heat capacity per molecule is *less* than $3k$. At higher temperatures when kT is *large* in comparison to the minimum energy step, the equipartition principle holds, and the total heat capacity is $3k$ per molecule or $3R$ per mole as the Dulong and Petit relation predicts. Quantitative understanding of the temperature variation of heat capacities was one of the triumphs of quantum mechanics during its initial development in the 1920s.

Test Your Understanding

A cylinder with a fixed volume contains hydrogen gas (H_2) at 25 K. You then add heat to the gas at a constant rate until its temperature reaches 500 K. Does the temperature of the gas increase at a constant rate? Why or why not? If not, does the temperature increase most rapidly near the beginning or near the end of this process?

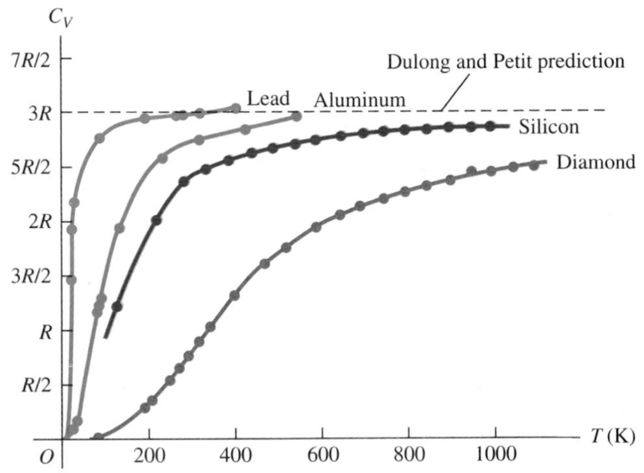

18.18 Experimental values of C_V for lead, aluminum, silicon, and diamond. At high temperatures, C_V for each solid approaches about $3R$, in agreement with the rule of Dulong and Petit. At low temperatures, C_V is much less than $3R$.

*18.5 | Molecular Speeds

As we mentioned in Section 18.3, the molecules in a gas don't all have the same speed. Figure 18.19 shows one experimental scheme for measuring the distribution of molecular speeds. A substance is vaporized in a hot oven; molecules of the vapor escape through an aperture in the oven wall and into a vacuum chamber. A series of slits blocks all molecules except those in a narrow beam, which is aimed at a pair of rotating disks. A molecule passing through the slit in the first disk is blocked by the second disk unless it arrives just as the slit in the second disk is lined up with the beam. The disks function as a speed selector that passes only molecules within a certain narrow speed range. This range can be varied by changing the disk rotation speed, and we can measure how many molecules lie within each of various speed ranges.

To describe the results of such measurements, we define a function $f(v)$ called a *distribution function.* If we observe a total of N molecules, the number dN having speeds in the range between v and $v + dv$ is given by

$$dN = Nf(v)\,dv \qquad (18.29)$$

We can also say that the *probability* that a randomly chosen molecule will have a speed in the interval v to $v + dv$ is $f(v)\,dv$. Hence $f(v)$ is the probability per unit speed *interval;* it is *not* equal to the probability that a molecule has speed exactly equal to v. Since a probability is a pure number, $f(v)$ has units of reciprocal speed (s/m).

Figure 18.20a shows distribution functions for several different temperatures. At each temperature the height of the curve for any value of v is proportional to the number of molecules with speeds near v. The peak of the curve represents the *most probable speed* v_{mp} for the corresponding temperature. As the temperature increases, the average molecular kinetic energy increases, and so the peak of $f(v)$ shifts to higher and higher speeds.

Figure 18.20b shows that the area under a curve between any two values of v represents the fraction of all the molecules having speeds in that range. Every molecule must have *some* value of v, so the integral of $f(v)$ over all v must be unity for any T.

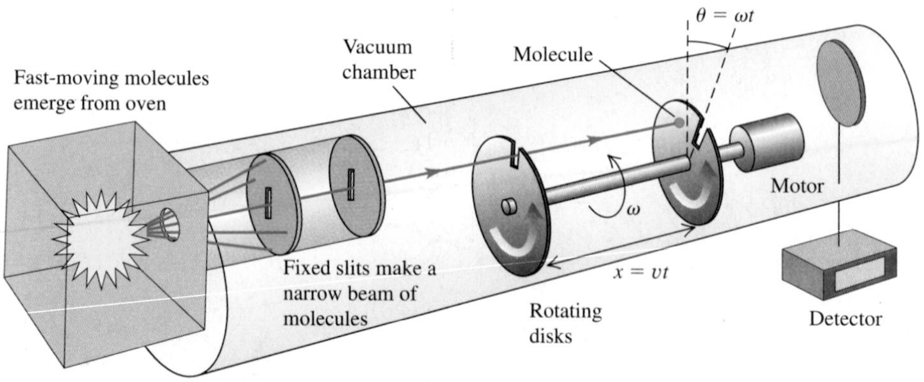

18.19 A molecule with a speed v passes through the slit in the first rotating disk. When the molecule reaches the second rotating disk, the disks have rotated through the offset angle θ. If $v = \omega x/\theta$, the molecule passes through the slit in the second rotating disk and reaches the detector.

If we know $f(v)$, we can calculate the most probable speed v_{mp}, the average speed v_{av}, and the rms speed v_{rms}. To find v_{mp}, we simply find the point where $df/dv = 0$; this gives the value of the speed where the curve has its peak. To find v_{av}, we take the number $Nf(v)\,dv$ having speeds in each interval dv, multiply each number by the corresponding speed v, add all these products (by integrating over all v from zero to infinity), and finally divide by N. That is,

$$v_{av} = \int_0^\infty v f(v)\,dv \tag{18.30}$$

The rms speed is obtained similarly; the average of v^2 is given by

$$(v^2)_{av} = \int_0^\infty v^2 f(v)\,dv \tag{18.31}$$

and v_{rms} is the square root of this.

The function $f(v)$ describing the actual distribution of molecular speeds is called the **Maxwell-Boltzmann distribution.** It can be derived from statistical-mechanics considerations, but that derivation is beyond our scope. Here is the result:

$$f(v) = 4\pi\left(\frac{m}{2\pi kT}\right)^{3/2} v^2 e^{-mv^2/2kT}$$

(Maxwell-Boltzmann distribution) (18.32)

We can also express this function in terms of the translational kinetic energy of a molecule, which we denote by ϵ. That is, $\epsilon = \frac{1}{2}mv^2$. We invite you to verify that when this is substituted into Eq. (18.32), the result is

$$f(v) = \frac{8\pi}{m}\left(\frac{m}{2\pi kT}\right)^{3/2} \epsilon e^{-\epsilon/kT} \tag{18.33}$$

This form shows that the exponent in the Maxwell-Boltzmann distribution function is $-\epsilon/kT$ and that the shape of the curve is determined by the relative magnitude of ϵ and kT at any point. We leave it as an exercise to prove that the *peak* of each curve occurs where $\epsilon = kT$, corresponding to a most probable speed v_{mp} given by

$$v_{mp} = \sqrt{\frac{2kT}{m}} \tag{18.34}$$

To find the average speed, we substitute Eq. (18.32) into Eq. (18.30) and carry out the integration, making a change of variable $v^2 = x$ and then integrating by parts. We leave the details as an exercise; the result is

$$v_{av} = \sqrt{\frac{8kT}{\pi m}} \tag{18.35}$$

Finally, to find the rms speed, we substitute Eq. (18.32) into Eq. (18.31). Evaluating the resulting integral takes some mathematical acrobatics, but we can find it in a table of integrals. The result is

$$v_{rms} = \sqrt{\frac{3kT}{m}} \tag{18.36}$$

This result agrees with Eq. (18.19); it *must* agree if the Maxwell-Boltzmann distribution is to be consistent with the equipartition theorem and our other kinetic-theory calculations.

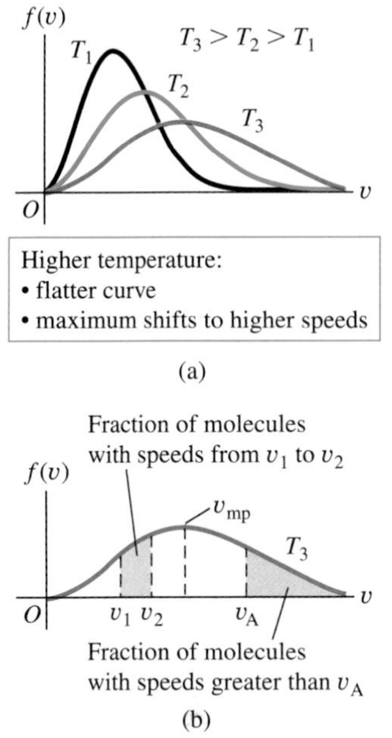

18.20 (a) Curves of the Maxwell-Boltzmann distribution function $f(v)$ for various temperatures. (b)The shaded areas under the curve represent the fractions of molecules within certain speed ranges. The most probable speed v_{mp} for a given temperature is at the peak of the curve.

ActivPhysics Online

8.2 Maxwell-Boltzmann Distribution—Conceptual Analysis

8.3 Maxwell-Boltzmann Distribution—Quantitative Analysis

Table 18.2 Fractions of Molecules in an Ideal Gas with Speeds Less Than Various Multiples of v/v_{rms}

v/v_{rms}	Fraction
0.20	0.011
0.40	0.077
0.60	0.218
0.80	0.411
1.00	0.608
1.20	0.771
1.40	0.882
1.60	0.947
1.80	0.979
2.00	0.993

Table 18.2 shows the fraction of all the molecules in an ideal gas that have speeds *less than* various multiples of v_{rms}. These numbers were obtained by numerical integration; they are the same for all ideal gases.

The distribution of molecular speeds in liquids is similar, although not identical, to that for gases. We can understand the vapor pressure of a liquid and the phenomenon of boiling on this basis. Suppose a molecule must have a speed at least as great as v_A in Fig. 18.20b to escape from the surface of a liquid into the adjacent vapor. The number of such molecules, represented by the area under the "tail" of each curve (to the right of v_A), increases rapidly with temperature. Thus the rate at which molecules can escape is strongly temperature-dependent. This process is balanced by another one in which molecules in the vapor phase collide inelastically with the surface and are trapped back into the liquid phase. The number of molecules suffering this fate per unit time is proportional to the pressure in the vapor phase. Phase equilibrium between liquid and vapor occurs when these two competing processes proceed at exactly the same rate. So if the molecular speed distributions are known for various temperatures, we can make a theoretical prediction of vapor pressure as a function of temperature. When liquid evaporates, it's the high-speed molecules that escape from the surface. The ones that are left have less energy on average; this gives us a molecular view of evaporative cooling.

Rates of chemical reactions are often strongly temperature-dependent, and the reason is contained in the Maxwell-Boltzmann distribution. When two reacting molecules collide, the reaction can occur only when the molecules are close enough for the electric-charge distributions of their electrons to interact strongly. This requires a minimum energy, called the *activation energy,* and thus a certain minimum molecular speed. Figure 18.20a shows that the number of molecules in the high-speed tail of the curve increases rapidly with temperature. Thus we expect the rate of any reaction that depends on an activation energy to increase rapidly with temperature. Similarly, many plant growth processes have strongly temperature-dependent rates, as can be seen by the rapid and diverse growth in tropical rain forests.

Test Your Understanding

The Maxwell-Boltzmann distribution curve has a long "tail" that extends to infinity. Use this fact to explain why v_{av} is greater than v_{mp}.

18.6 | Phases of Matter

We've talked a lot about ideal gases in the last few sections. An ideal gas is the simplest system to analyze from a molecular viewpoint because we ignore the interactions between molecules. But those interactions are the very thing that makes matter condense into the liquid and solid phases under some conditions. So it's not surprising that theoretical analysis of liquid and solid structure and behavior is a lot more complicated than that for gases. We won't try to go far here with a microscopic picture, but we can talk in general about phases of matter, phase equilibrium, and phase transitions.

In Section 17.6 we learned that each phase is stable only in certain ranges of temperature and pressure. A transition from one phase to another ordinarily takes place under conditions of **phase equilibrium** between the two phases, and for a given pressure this occurs at only one specific temperature. We can represent these conditions on a graph with axes p and T, called a **phase diagram;** Fig. 18.21 shows an example. Each point on the diagram represents a pair of values of p and

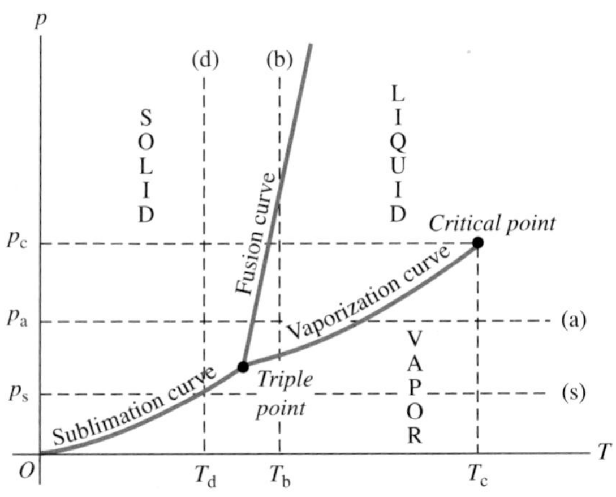

18.21 A typical pT phase diagram, showing regions of temperature and pressure at which the various phases exist and where phase changes occur.

T. Only a single phase can exist at each point, except for points on the solid lines, where two phases can coexist in phase equilibrium.

These lines separate the diagram into solid, liquid, and vapor regions. For example, the fusion curve separates the solid and liquid areas and represents possible conditions of solid-liquid phase equilibrium. Similarly, the vaporization curve separates the liquid and vapor areas, and the sublimation curve separates the solid and vapor areas. The three curves meet at the **triple point,** the only condition under which all three phases can coexist (Fig. 18.22). In Section 17.3 we used the triple-point temperature of water to define the Kelvin temperature scale. Triple-point data for several substances are given in Table 18.3.

Table 18.3 Triple-Point Data

Substance	Temperature (K)	Pressure (Pa)
Hydrogen	13.80	0.0704×10^5
Deuterium	18.63	0.171×10^5
Neon	24.56	0.432×10^5
Nitrogen	63.18	0.125×10^5
Oxygen	54.36	0.00152×10^5
Ammonia	195.40	0.0607×10^5
Carbon dioxide	216.55	5.17×10^5
Sulfur dioxide	197.68	0.00167×10^5
Water	273.16	0.00610×10^5

18.22 Atmospheric pressure on earth is higher than the triple-point pressure of water (see line (a) in Fig. 18.21). Depending on the temperature, water can exist as a vapor (in the atmosphere), as a liquid (in the ocean), or a solid (like the iceberg shown here).

If we add heat to a substance at a constant pressure p_a, it goes through a series of states represented by the horizontal line (a) in Figure 18.21. The melting and boiling temperatures at this pressure are the temperatures at which the line intersects the fusion and vaporization curves, respectively. When the pressure is p_s, constant-pressure heating transforms a substance from solid directly to vapor. This process is called *sublimation;* the intersection of line (s) with the sublimation curve gives the temperature T_s at which it occurs for a pressure p_s. At any pressure less than the triple-point pressure, no liquid phase is possible. The triple-point pressure for carbon dioxide is 5.1 atm. At normal atmospheric pressure, solid carbon dioxide ("dry ice") undergoes sublimation; there is no liquid phase at this pressure.

Line (b) in Fig. 18.21 represents compression at a constant temperature T_b. The material passes from vapor to liquid and then to solid at the points where line (b) crosses the vaporization curve and fusion curve, respectively. Line (d) shows constant-temperature compression at a lower temperature T_d; the material passes from vapor to solid at the point where line (d) crosses the sublimation curve.

We saw in the pV-diagram of Fig. 18.5 that a liquid-vapor phase transition occurs only when the temperature and pressure are less than those at the point lying at the top of the green shaded area labeled "Liquid-vapor phase equilibrium region." This point corresponds to the end point at the top of the vaporization curve in Fig. 18.21. It is called the **critical point,** and the corresponding values of p and T are called the critical pressure and temperature, p_c and T_c. A gas at a pressure *above* the critical pressure does not separate into two phases when it is cooled at constant pressure (along a horizontal line above the critical point in Fig. 18.21). Instead, its properties change gradually and continuously from those we ordinarily associate with a gas (low density, large compressibility) to those of a liquid (high density, small compressibility) *without a phase transition.*

If this stretches credibility, think about liquid-phase transitions at successively higher points on the vaporization curve. As we approach the critical point, the *differences* in physical properties (such as density, bulk modulus, and viscosity) between the liquid and vapor phases become smaller and smaller. Exactly *at* the critical point they all become zero, and at this point the distinction between liquid and vapor disappears. The heat of vaporization also grows smaller and smaller as we approach the critical point, and it too becomes zero at the critical point.

For nearly all familiar materials the critical pressures are much greater than atmospheric pressure, so we don't observe this behavior in everyday life. For example, the critical point for water is at 647.4 K and 221.2×10^5 Pa (about 218 atm or 3210 psi). But high-pressure steam boilers in electric generating plants regularly run at pressures and temperatures well above the critical point.

Many substances can exist in more than one solid phase. A familiar example is carbon, which exists as noncrystalline soot and crystalline graphite and diamond. Water is another example; at least eight types of ice, differing in crystal structure and physical properties, have been observed at very high pressures.

We remarked in Section 18.1 that the equation of state of any material can be represented graphically as a surface in a three-dimensional space with coordinates p, V, and T. Such a surface is seldom useful in representing detailed quantitative information, but it can add to our general understanding of the behavior of materials at various temperatures and pressure. Figure 18.23 shows a typical pVT-surface. The light lines represent pV-isotherms; projecting them onto the pV-plane would give a diagram similar to Fig. 18.5. The pV-isotherms represent contour lines on the pVT-surface, just as contour lines on a topographic map represent the elevation (the third dimension) at each point. The projections of the edges of the surface onto the pT-plane gives the pT-phase diagram of Fig. 18.21.

Line *abcdef* represents constant-pressure heating, with melting along *bc* and vaporization along *de*. Note the volume changes that occur as T increases along this line. Line *ghjklm* corresponds to an isothermal (constant temperature) compression, with liquefaction along *hj* and solidification along *kl*. Between these, segments *gh* and *jk* represent isothermal compression with increase in pressure; the pressure increases are much greater in the liquid region *jk* and the solid region *lm* than in the vapor region *gh*. Finally, *nopq* represents isothermal solidification directly from the vapor phase; this is the process involved in growth of crystals directly from vapor, as in the formation of snowflakes or frost and in the fabrication of some solid-state electronic devices. These three lines on the pVT-surface are worth careful study.

For contrast, Fig. 18.24 shows the much simpler pVT-surface for a substance that obeys the ideal-gas equation of state under all conditions. The projections of

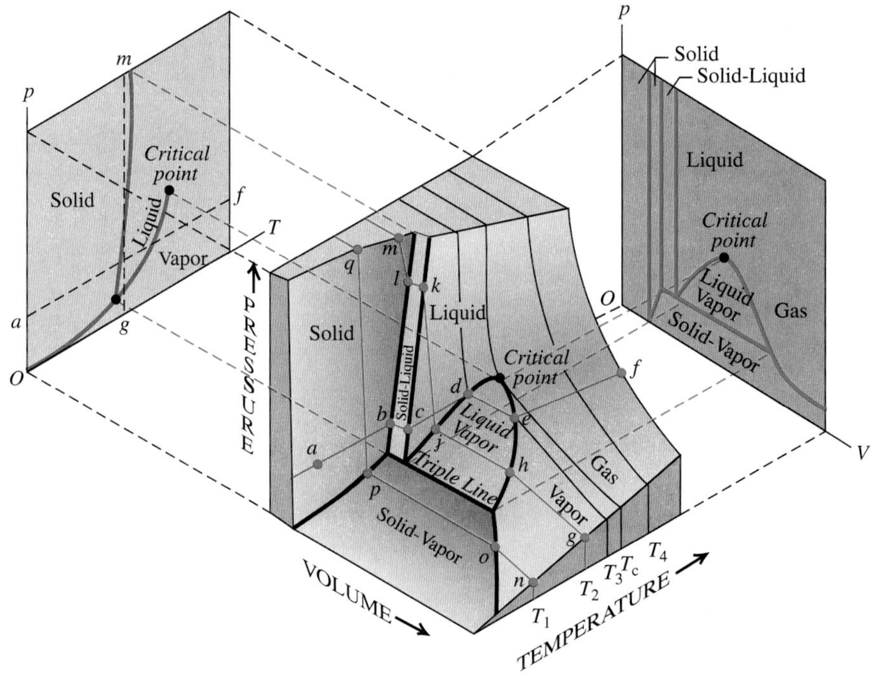

18.23 A pVT-surface for a substance that expands on melting. Projections of the boundaries on the surface on the pT- and pV-planes are also shown.

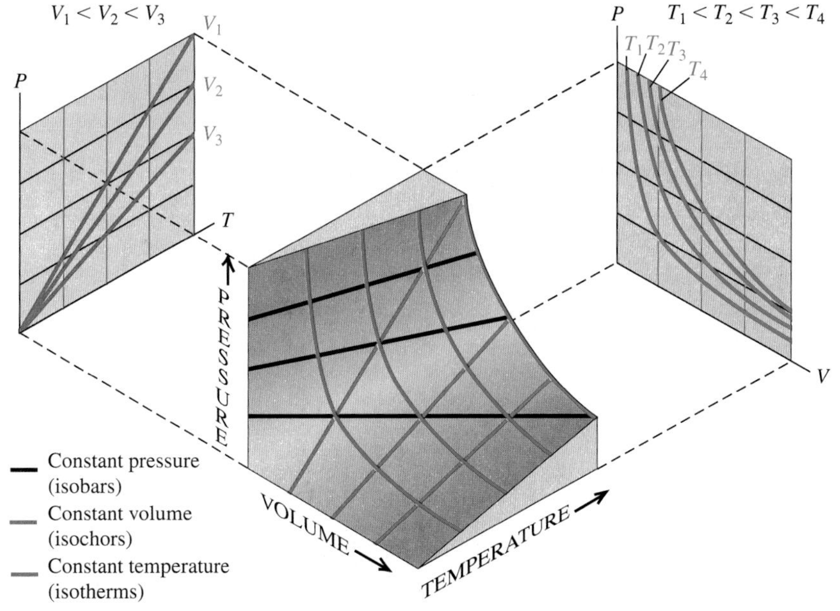

18.24 A pVT-surface for an ideal gas. At the left, each red line corresponds to a certain constant volume; at the right, each blue-green line corresponds to a certain constant temperature.

the constant-temperature curves onto the pV-plane correspond to the curves of Fig. 18.4, and the projections of the constant-volume curves onto the pT-plane show the direct proportionality of pressure to absolute temperature.

Test Your Understanding

The average atmospheric pressure on Mars is 6.0×10^2 Pa. Could there be lakes or rivers on Mars today? What about in the past, when the atmospheric pressure is thought to have been substantially greater than today?

SUMMARY

The pressure p, volume V, and absolute temperature T of a given quantity of a substance are called state variables. They are related by an equation of state. This relation pertains only to equilibrium states, in which p and T are uniform throughout the system. The ideal-gas equation of state relates p, V, T and the number of moles n through a constant R that is the same for all gases. (See Examples 18.1 through 18.4)

$$pV = nRT \qquad (18.3)$$

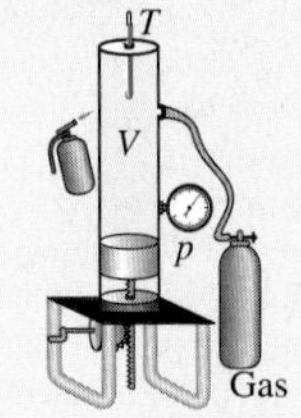

A pV-diagram is a set of graphs, called isotherms, each showing pressure as a function of volume for a constant temperature.

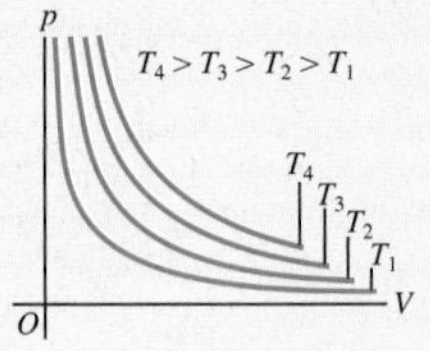

The molar mass M of a pure substance is the mass per mole. The mass m_{tot} of a quantity of substance equals M multiplied by the number of moles n. Avogadro's number N_A is the number of molecules in a mole. The mass m of an individual molecule is M divided by N_A. (See Example 18.5)

$$m_{tot} = nM \qquad (18.2)$$

$$M = N_A m \qquad (18.8)$$

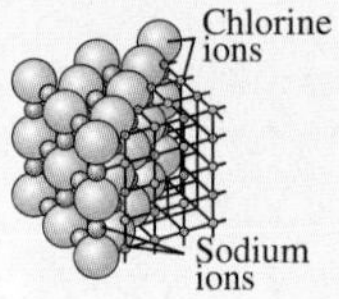

In an ideal gas, the total translational kinetic energy of the gas as a whole (K_{tr}) and the average translational kinetic energy per molecule $[\frac{1}{2}m(v^2)_{av}]$ are proportional to the absolute temperature T. The root-mean-square speed of molecules in an ideal gas is proportional to the square root of T. These expressions involve the Boltzmann constant $k = R/N_A$. (See Examples 18.6 through 18.8)

$$K_{tr} = \frac{3}{2}nRT \qquad (18.14)$$

$$\frac{1}{2}m(v^2)_{av} = \frac{3}{2}kT \qquad (18.16)$$

$$v_{rms} = \sqrt{(v^2)_{av}} = \sqrt{\frac{3kT}{m}} = \sqrt{\frac{3RT}{M}} \qquad (18.19)$$

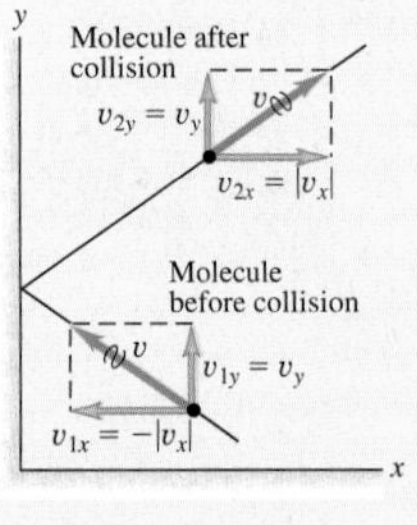

The mean free path λ of molecules in an ideal gas depends on the number of molecules per volume (N/V) and the molecular radius r. (See Example 18.9)

$$\lambda = vt_{mean} = \frac{V}{4\pi\sqrt{2}r^2N} \qquad (18.21)$$

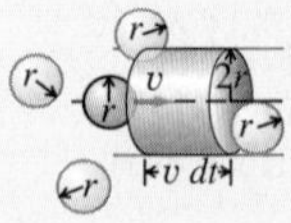

The molar heat capacity C_V at constant volume can be expressed as a simple multiple of the gas constant R for certain idealized cases: an ideal monatomic gas [Eq. (18.25)]; an ideal diatomic gas including rotation energy [Eq. (18.26)]; and an ideal monatomic solid [Eq. (18.28)]. Many real systems are approximated well by these idealizations.

$$C_V = \frac{3}{2}R \qquad (18.25)$$

$$C_V = \frac{5}{2}R \qquad (18.26)$$

$$C_V = 3R \qquad (18.28)$$

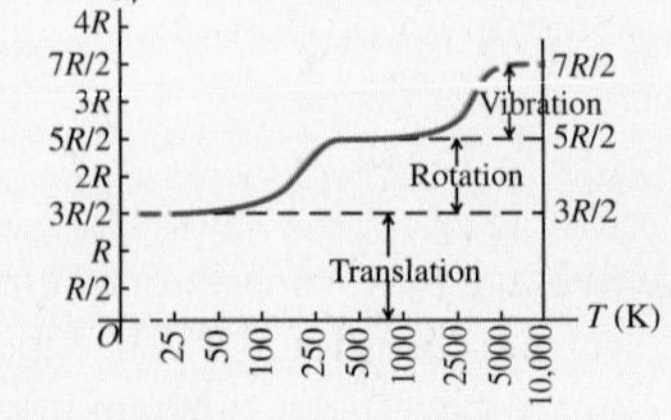

The speeds of molecules in an ideal gas are distributed according to the Maxwell-Boltzmann distribution $f(v)$. The quantity $f(v)\,dv$ describes what fraction of the molecules have speeds between v and $v + dv$.

$$f(v) = 4\pi\left(\frac{m}{2\pi kT}\right)^{3/2} v^2 e^{-mv^2/2kT} \qquad (18.32)$$

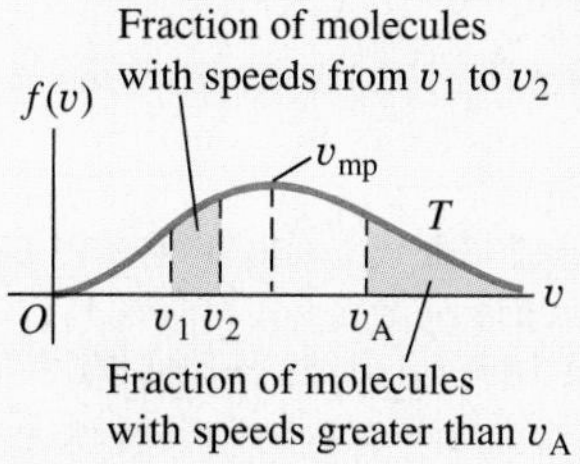

Ordinary matter exists in the solid, liquid, and gas phases. A phase diagram shows conditions under which two phases can coexist in phase equilibrium. All three phases can coexist at the triple point. The vaporization curve ends at the critical point, above which the distinction between the liquid and gas phases disappears.

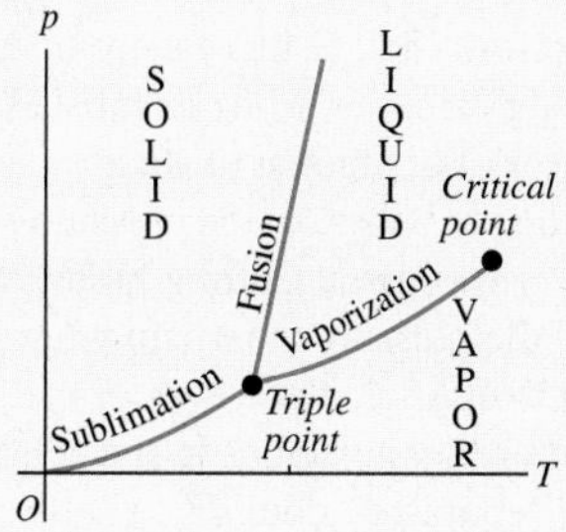

Key Terms

state variables, 685
equation of state, 685
molar mass, 685
ideal-gas equation, 686
ideal gas, 686
gas constant, 686
standard temperature and pressure, 688
van der Waals equation, 690
***pV*-diagram, 691**
isotherm, 691
molecule, 692
potential well, 692
mole, 693
Avogadro's number, 693
Boltzmann constant, 697
root-mean-square speed (rms speed), 698
mean free path, 701
equipartition of energy, 703
degrees of freedom, 703
rule of Dulong and Petit, 705
Maxwell-Boltzmann distribution, 707
phase equilibrium, 708
phase diagram, 708
triple point, 709
critical point, 710

Your Notes

Answer to Chapter Opening Question

From Eq. (18.19), the root-mean-square speed of a gas molecule is proportional to the square root of the absolute temperature T. The temperature range we're considering is from $(-40 + 273.15)$ K = 233 K to $(40 + 273.15)$ K = 313 K. Hence the speeds increase by a factor of $\sqrt{(313\text{ K})/(233\text{ K})} = 1.16$; that is, there is a 16% increase. While 40°C feels far warmer than −40°C, the difference in molecular speeds is relatively minor.

Answers to Test Your Understanding Questions

Section 18.1 If there were no holes in the crust, the steam emanating from the filling as it bakes would be confined to a fixed volume V. The amount of steam n and the temperature T both increase as the pie bakes, so the pressure $p = nRT/V$ increases as well, until the crust bursts. Cutting holes in the crust allows steam to escape and the pressure to remain within limits.
Section 18.2 The value of r_0 determines the equilibrium separation of the molecules in the solid phase, so doubling r_0 means that the separation doubles as well. Hence a solid cube of this compound might grow from 1 cm on a side to 2 cm on a side. The volume would then be $2^3 = 8$ times larger, and the density (mass divided by volume) would be $\frac{1}{8}$ as great.
Section 18.3 The temperature of the corona is 2,000,000°C, or to two significant figures 2.0×10^6 K (the difference between Celsius and Kelvin temperatures is irrelevant). Hydrogen has a molar mass (from Appendix D) of $M = 1.008$ g/mol $= 1.008 \times 10^{-3}$ kg/mol, so

$$v_{\text{rms}} = \sqrt{\frac{3RT}{M}} = \sqrt{\frac{3(8.314\text{ J/mol}\cdot\text{K})(2.0 \times 10^6\text{ K})}{1.008 \times 10^{-3}\text{ kg/mol}}}$$
$$= 1.8 \times 10^5\text{ m/s}$$

This is less than the escape speed, so not all of the hydrogen in the corona can escape into space. But remember that not all the atoms move at the rms speed: some are moving fast enough to escape, and they help make up a continuous flow of particles called the *solar wind*.
Section 18.4 Adding a small amount of heat dQ to the gas changes the temperature by dT, where $dQ = nC_V\,dT$ from Eq. (18.24). Figure 18.16 shows that C_V for H_2 varies with temperature between 25 K and 500 K, so a given amount of heat gives rise to different amounts of temperature change during the process. Hence the temperature will *not* increase at a constant rate. The temperature change $dT = dQ/nC_V$ is inversely proportional to C_V, so the temperature increases most rapidly at the beginning of the process when the temperature is lowest and C_V is smallest (see Fig. 18.16).
Section 18.5 If the curve were symmetrical like a bell curve, the average value would be at the peak of the curve and v_{av} would be equal to v_{mp} (which, by definition, is at the peak of the curve). The curve is not symmetrical, however, and its long tail shifts the average value to the right of the peak. Hence $v_{\text{av}} > v_{\text{mp}}$.
Section 18.6 The triple-point pressure of water from Table 18.3 is 6.10×10^2 Pa. The present-day pressure on Mars is just less than this value, corresponding to the line labeled p_s in Fig. 18.21. Hence liquid water cannot exist on the present-day Martian surface, and there are no rivers or lakes. Planetary scientists conclude that liquid water could have and almost certainly did exist on Mars in the past, when the atmosphere was thicker.

Discussion Questions

Q18.1 Section 18.1 states that ordinarily, pressure, volume, and temperature cannot change individually without one affecting the others. Yet when a liquid evaporates, its volume changes, even though its pressure and temperature are constant. Is this inconsistent? Why or why not?
Q18.2 In the ideal-gas equation, could an equivalent Celsius temperature be used instead of the Kelvin one if an appropriate numerical value of the constant R is used? Why or why not?
Q18.3 On a chilly morning you can "see your breath." Can you really? What are you actually seeing? Does this phenomenon depend on the temperature of the air, the humidity, or both? Explain.
Q18.4 When a car is driven some distance, the air pressure in the tires increases. Why? Should you let out some air to reduce the pressure? Why or why not?
Q18.5 The coolant in an automobile radiator is kept at a pressure higher than atmospheric pressure. Why is this desirable? The radiator cap will release coolant when the gauge pressure of the coolant reaches a certain value, typically 15 lb/in.^{2} or so. Why not just seal the system completely?
Q18.6 Unwrapped food placed in a freezer experiences dehydration, known as "freezer burn." Why?
Q18.7 "Freeze-drying" food involves the same process as "freezer burn," referred to in the previous question. For freeze-drying, the food is usually frozen first, then placed in a vacuum chamber and irradiated with infrared radiation. What is the purpose of the vacuum? The radiation? What advantages might freeze-drying have in comparison to ordinary drying?
Q18.8 A group of students drove from their university (near sea level) up into the mountains for a skiing weekend. Upon arriving at the slopes, they discovered that the bags of potato chips they had brought for snacks had all burst open. What caused this to happen?
Q18.9 How does evaporation of perspiration from your skin cool your body?
Q18.10 Helium is a mixture of two isotopes, one having molar mass 3 g/mol and the other 4 g/mol. In a sample of helium gas which atoms move faster, on the average? Explain why.
Q18.11 Which has more atoms, a kilogram of hydrogen or a kilogram of lead? Which has more mass? Explain.
Q18.12 Use the concepts of the kinetic-molecular model to explain: a) why the pressure of a gas in a rigid container increases as heat is added to the gas; b) why the pressure of a gas increases as we compress it, even if we do not change its temperature.
Q18.13 The proportion of various gases in the earth's atmosphere changes somewhat with altitude. Would you expect the proportion of oxygen at high altitude to be greater or less than at sea level compared to the proportion of nitrogen? Why?

Q18.14 Comment on the following statement: *When two gases are mixed, if they are to be in thermal equilibrium they must have the same average molecular speed.* Is the statement correct? Why or why not?

Q18.15 The kinetic-molecular model contains a hidden assumption about the temperature of the container walls. What is this assumption? What would happen if this assumption were not valid?

Q18.16 The temperature of an ideal gas is directly proportional to the average kinetic energy of its molecules. If a container of ideal gas is moving past you at 2000 m/s, is the temperature of the gas higher than if the container was at rest? Defend your answer.

Q18.17 If the pressure of an ideal monatomic gas is increased while the number of moles is kept constant, what happens to the average translational kinetic energy of one atom of the gas? Is it possible to change *both* the volume and pressure of an ideal gas and keep the average translational kinetic energy of the atoms constant? Explain.

Q18.18 In deriving the ideal-gas equation from the kinetic-molecular model, we ignored potential energy due to the earth's gravity. Is this omission justified? Why or why not?

Q18.19 The derivation of the ideal-gas equation included the assumption that the number of molecules is very large, so that we could compute the average force due to many collisions. However, the ideal-gas equation holds accurately only at low pressures, where the molecules are few and far between. Is this inconsistent? Why or why not?

Q18.20 A gas storage tank has a small leak. The pressure in the tank drops more quickly if the gas is hydrogen or helium than if it is oxygen. Why?

Q18.21 Consider two specimens of ideal gas at the same temperature. Specimen A has the same total mass as specimen B, but the molecules in specimen A have greater molar mass than they do in specimen B. In which specimen is the total kinetic energy of the gas greater? Does your answer depend on the molecular structure of the gases? Why or why not?

Q18.22 The temperature of an ideal monatomic gas is increased from 25°C to 50°C. Does the average translational kinetic energy of each gas atom double? Explain. If your answer is no, what would the final temperature be if the average translational kinetic energy is doubled?

Q18.23 If the root-mean-square speed of the atoms of an ideal gas is to be doubled, by what factor must the Kelvin temperature of the gas be increased? Explain.

Q18.24 Some elements that form solid crystals have molar heat capacities *greater* than $3R$. What could account for this?

***Q18.25** In a gas that contains N molecules, would it be accurate to say that the number of molecules with speed v is equal to $f(v)$? Or would it be accurate to say that this number is given by $Nf(v)$? Explain your answers.

***Q18.26** Fig. 18.19 shows a velocity selector. The two disks are a distance x apart, the slits in the two disks are at an angle θ to each other, and the disks rotate together with angular speed ω. Explain why a molecule that passes through the first slit of the selector will only be able to pass through the second slit if its speed is $v = \omega x/\theta$.

***Q18.27** Imagine a special air filter placed in a window of a house. The tiny holes in the filter only allow air molecules moving faster than a certain speed to exit the house, and only allow air molecules moving slower than that speed to enter the house from outside. What effect would this filter have on the temperature inside the house? (It turns out that the second law of thermodynamics—which we will discuss in Chapter 20—tells us that such a wonderful air filter would be quite impossible to make.)

Q18.28 A beaker of water at room temperature is placed in an enclosure, and the air pressure in the enclosure is slowly reduced. When the air pressure is reduced sufficiently, the water begins to boil. The temperature of the water does not rise when it boils; in fact, the temperature *drops* slightly. Explain these phenomena.

Q18.29 Ice is slippery to walk on, and especially slippery if you wear ice skates. What does this tell you about how the melting temperature of ice depends on pressure? Explain.

Q18.30 Hydrothermal vents are openings in the ocean floor that discharge very hot water. The water emerging from one such vent off the Oregon coast, 2400 m below the surface, has a temperature of 279°C. Despite its high temperature, the water doesn't boil. Why not?

Q18.31 The dark areas on the moon's surface are called *maria,* Latin for "seas," and were once thought to be bodies of water. In fact the maria are not "seas" at all, but plains of solidified lava. Given that there is no atmosphere on the moon, how can you explain the absence of liquid water on the moon's surface?

Q18.32 In addition to the normal cooking directions printed on the back of a box of rice, there are also "high altitude directions." The only difference is that the "high altitude directions" suggest increasing the cooking time and using a greater volume of boiling water in which to cook the rice. Why should the directions depend on the altitude in this way?

Exercises

Section 18.1 Equations of State

18.1 A 20.0-L tank contains 0.225 kg of helium at 18.0°C. The molar mass of helium is 4.00 g/mol. a) How many moles of helium are in the tank? b) What is the pressure in the tank, in pascals and in atmospheres?

18.2 Helium gas with a volume of 2.60 L, under a pressure of 1.30 atm and at a temperature of 41.0°C, is warmed until both pressure and volume are doubled. a) What is the final temperature? b) How many grams of helium are there? The molar mass of helium is 4.00 g/mol.

18.3 A cylindrical tank has a tight-fitting piston that allows the volume of the tank to be changed. The tank originally contains 0.110 m^3 of air at a pressure of 3.40 atm. The piston is slowly pulled out until the volume of the gas is increased to 0.390 m^3. If the temperature remains constant, what is the final value of the pressure?

18.4 A 3.00-L tank contains air at 3.00 atm and 20.0°C. The tank is sealed and cooled until the pressure is 1.00 atm. a) What is the temperature then in degrees Celsius? Assume that the volume of the tank is constant. b) If the temperature is kept at the value found in part (a) and the gas is compressed, what is the volume when the pressure again becomes 3.00 atm?

18.5 a) Use the ideal gas law to estimate the number of air molecules in your physics lab room, assuming all the air is N_2. b) Calculate the particle density in the lab (that is, the number of molecules per cubic centimeter).

18.6 You have several identical balloons. You experimentally determine that a balloon will break if its volume exceeds 0.900 L. The pressure of the gas inside the balloon equals air pressure (1.00 atm). a) If the air inside the balloon is at a constant temperature of 22.0°C and behaves as an ideal gas, what mass of air can you blow into one of the balloons before it bursts? b) Repeat part (a) if the gas is helium rather than air.

18.7 A Jaguar XK8 convertible has an eight-cylinder engine. At the beginning of its compression stroke, one of the cylinders contains 499 cm^3 of air at atmospheric pressure $(1.01 \times 10^5 \text{ Pa})$ and a temperature of 27.0°C. At the end of the stroke the air has been compressed to a volume of 46.2 cm^3 and the gauge pressure has increased to 2.72×10^6 Pa. Compute the final temperature.

18.8 A welder using a tank of volume 0.0750 m^3 fills it with oxygen (molar mass = 32.0 g/mol) at a gauge pressure of 3.00×10^5 Pa and temperature of 37.0°C. The tank has a small leak, and in time some of the oxygen leaks out. On a day when the temperature is 22.0°C, the gauge pressure of the oxygen in the tank is 1.80×10^5 Pa. Find a) the initial mass of oxygen; b) the mass of oxygen that has leaked out.

18.9 A large cylindrical tank contains 0.750 m^3 of nitrogen gas at 27°C and 1.50×10^5 Pa (absolute pressure). The tank has a tight-fitting piston that allows the volume to be changed. What will be the pressure if the volume is decreased to 0.480 m^3 and the temperature is increased to 157°C?

18.10 A room with dimensions 7.00 m × 8.00 m × 2.50 m is filled with pure oxygen at 22.0°C and 1.00 atm. The molar mass of oxygen is 32.0 g/mol. a) How many moles of oxygen are required? b) What is the mass of this oxygen, in kilograms?

18.11 The gas inside a balloon will always have a pressure nearly equal to atmospheric pressure, since that is the pressure applied to the outside of the balloon. You fill a balloon with helium (a nearly ideal gas) to a volume of 0.600 L at a temperature of 19.0°C. What is the volume of the balloon if you cool it to the boiling point of liquid nitrogen (77.3 K)?

18.12 Deviations from the Ideal-Gas Equation. For carbon dioxide gas (CO_2), the constants in the van der Waals equation are $a = 0.364 \text{ J} \cdot \text{m}^3/\text{mol}^2$ and $b = 4.27 \times 10^{-5} \text{ m}^3/\text{mol}$. a) If 1.00 mol of CO_2 gas at 350 K is confined to a volume of 400 cm^3, find the pressure of the gas using the ideal-gas equation and the van der Waals equation. b) Which equation gives a lower pressure? Why? What is the percentage difference of the van der Waals result from the ideal-gas equation result? c) The gas is kept at the same temperature as it expands to a volume of 4000 cm^3. Repeat the calculations of parts (a) and (b). d) Explain how your calculations show that the van der Waals equation is equivalent to the ideal-gas equation if n/V is small.

18.13 The total lung volume for a typical physics student is 6.00 L. A physics student fills her lungs with air at an absolute pressure of 1.00 atm. Then, holding her breath, she compresses her chest cavity, decreasing her lung volume to 5.70 L. What is the pressure of the air in her lungs then? Assume that the temperature of the air remains constant.

18.14 A diver observes a bubble of air rising from the bottom of a lake (where the absolute pressure is 3.50 atm) to the surface (where the pressure is 1.00 atm). The temperature at the bottom is 4.0°C, and the temperature at the surface is 23.0°C. a) What is the ratio of the volume of the bubble as it reaches the surface to its volume at the bottom? b) Would it be safe for the diver to hold his breath while ascending from the bottom of the lake to the surface? Why or why not?

18.15 A metal tank with volume 3.10 L will burst if the absolute pressure of the gas it contains exceeds 100 atm. a) If 11.0 mol of an ideal gas is put into the tank at a temperature of 23.0°C, to what temperature can the gas be warmed before the tank ruptures? You can ignore the thermal expansion of the tank. b) Based on your answer to part (a), is it reasonable to ignore the thermal expansion of the tank? Explain.

18.16 Three moles of ideal gas are in a rigid cubical box with sides of length 0.200 m. a) What is the force that the gas exerts on each of the six sides of the box when the gas temperature is 20.0°C? b) What is the force when the temperature of the gas is increased to 100.0°C?

18.17 With the assumptions of Example 18.4 (Section 18.1), at what altitude above sea level is air pressure 90% of the pressure at sea level?

18.18 Make the same assumptions as in Example 18.4 (Section 18.1). How does the percent decrease in air pressure in going from sea level to an altitude of 100 m compare to that when going from sea level to an altitude of 1000 m? If your second answer is not ten times your first answer, explain why.

18.19 With the assumptions of Example 18.4 (Section 18.1), how does the density of air at sea level compare to that at an altitude of 100 m above sea level?

18.20 With the assumption (not very realistic) that the air temperature is a uniform 0°C, what is the atmospheric pressure at an altitude of 5000 m? This is roughly the maximum altitude usually attained by aircraft without pressurized cabins.

18.21 At an altitude of 11.000 m (a typical cruising altitude for a jet airliner), the air temperature is −56.5°C and the air density is 0.364 kg/m^3. What is the pressure of the atmosphere at that altitude? (Note that the temperature at this altitude is not the same as at the surface of the earth, so that the calculation of Example 18.4 (Section 18.1) doesn't apply.)

Section 18.2 Molecular Properties of Matter

18.22 A large organic molecule has a mass of 1.41×10^{-21} kg. What is the molar mass of this compound?

18.23 What is the volume of 3.00 moles of copper?

18.24 Modern vacuum pumps make it easy to attain pressures of the order of 10^{-13} atm in the laboratory. At a pressure of 9.00×10^{-14} atm and an ordinary temperature (say $T = 300$ K), how many molecules are present in a volume of 1.00 cm^3?

18.25 The Lagoon Nebula (Fig. 18.25) is a cloud of hydrogen gas located 3900 light years from the earth. The cloud is about 45 light years in diameter, and glows because of its high temperature of 7500 K. (The gas is raised to this temperature by the stars that lie within the nebula). The cloud is also very thin; there are only 80 molecules per cubic centimeter. a) Find the gas pressure (in atmospheres) in the Lagoon Nebula. Compare to the laboratory pressure referred to in Exercise 18.24. b) Science fiction films sometimes show starships being buffeted by turbulence as they fly through gas clouds such as the Lagoon Nebula. Does this seem realistic? Why or why not?

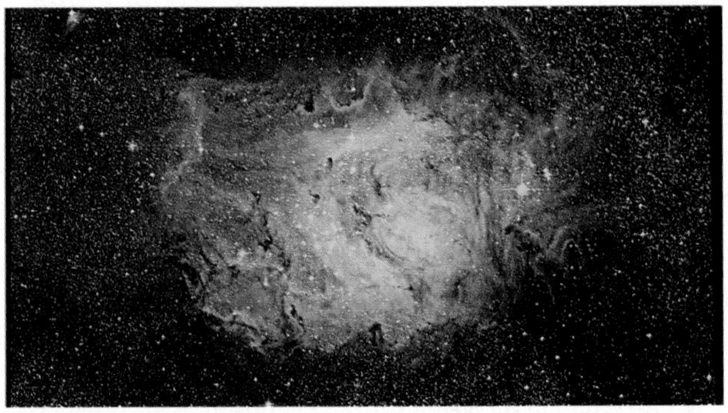

Figure 18.25 Exercise 18.25.

18.26 In a gas at standard conditions, what is the length of the side of a cube that contains a number of molecules equal to the population of the earth (about 6×10^9)?

18.27 How many moles are there in a 1.00-kg bottle of water? How many molecules? The molar mass of water is 18.0 g/mol.

18.28 Consider an ideal gas at 27°C and 1.00 atm pressure. Imagine the molecules to be, on average, uniformly spaced, with each molecule at the center of a small cube. a) What is the length of an edge of each small cube if adjacent cubes touch but don't overlap? b) How does this distance compare with the diameter of a molecule?

18.29 Consider 5.00 mol of liquid water. a) What volume is occupied by this amount of water? The molar mass of water is 18.0 g/mol. b) Imagine the molecules to be, on average, uniformly spaced, with each molecule at the center of a small cube. What is the length of an edge of each small cube if adjacent cubes touch but don't overlap? c) How does this distance compare with the diameter of a molecule?

Section 18.3 Kinetic-Molecular Model of an Ideal Gas

18.30 A flask contains a mixture of neon (Ne), krypton (Kr), and radon (Rn) gases. Compare a) the average kinetic energies of the three types of atoms; b) the root-mean-square speeds. (*Hint:* the periodic table in Appendix D shows the molar mass (in g/mol) of each element under the chemical symbol for that element.)

18.31 Gaseous Diffusion of Uranium. a) A process called *gaseous diffusion* is often used to separate isotopes of uranium, that is, atoms of the elements that have different masses, such as ^{235}U and ^{238}U. The only gaseous compound of uranium at ordinary temperatures is uranium hexafluoride, UF_6. Speculate on how $^{235}UF_6$ and $^{238}UF_6$ molecules might be separated by diffusion. b) The molar masses for $^{235}UF_6$ and $^{238}UF_6$ molecules are 0.349 kg/mol and 0.352 kg/mol, respectively. If uranium hexafluoride acts as an ideal gas, what is the ratio of the root-mean-square speed of $^{235}UF_6$ molecules to that of $^{238}UF_6$ molecules if the temperature is uniform?

18.32 The ideas of average and root-mean-square value can be applied to any distribution. A class of 150 students had the following scores on a 100-point quiz:

Score	Number of Students
10	11
20	12
30	24
40	15
50	19
60	10
70	12
80	20
90	17
100	10

a) Find the average score for the class. b) Find the root-mean-square score for the class.

18.33 We have two equal size boxes, A and B. Each box contains gas that behaves as an ideal gas. We insert a thermometer into each box and find that the gas in box A is at a temperature of 50°C while that in B is at 10°C. This is all we know about the gas in the boxes. Which of the following statements *must* be true? Which *could* be true? a) The pressure in A is higher than in B. b) There are more molecules in A than in B. c) A and B cannot contain the same type of gas. d) The molecules in A have more average kinetic energy per molecule than those in B. e) The molecules in A are moving faster than those in B. Explain the reasoning behind your answers.

18.34 We have two equal size boxes, A and B. Each box contains gas that behaves as an ideal gas. We put a pressure gauge into each box and find that the gauge reads 0.200 atm in box A, but only 0.040 atm in box B. This is all we know about the gas in the boxes. Which of the following statements *must* be true? Which *could* be true? a) There are more molecules in A than in B. b) The molecules in A are moving faster than those in B. c) The temperature in A is higher than in B. d) The molecules in A are heavier than those in B. e) The molecules in A have more average kinetic energy per molecule than those in B. Explain the reasoning behind your answers.

18.35 a) A deuteron, ^2_1H, is the nucleus of a hydrogen isotope and consists of one proton and one neutron. The plasma of deuterons in a nuclear fusion reactor must be heated to about 300 million K. What is the rms speed of the deuterons? Is this a significant fraction of the speed of light $(c = 3.0 \times 10^8 \text{ m/s})$? b) What would the temperature of the plasma be if the deuterons had an rms speed equal to $0.10c$?

18.36 Initially, the translational rms speed of an atom of a monatomic ideal gas is 250 m/s. The pressure and volume of the gas are each doubled while the number of moles of the gas is kept constant. What is the final translational rms speed of the atoms?

18.37 a) Oxygen (O_2) has a molar mass of 32.0 g/mol. What is the average translational kinetic energy of an oxygen molecule at a temperature of 300 K? b) What is the average value of the square of its speed? c) What is the root-mean-square speed? d) What is the momentum of an oxygen molecule traveling at this speed? e) Suppose an oxygen molecule traveling at this speed bounces back and forth between opposite sides of a cubical vessel 0.10 m on a side.

What is the average force it exerts on one of the walls of the container? (Assume that the molecule's velocity is perpendicular to the two sides that it strikes.) f) What is the average force per unit area? g) How many oxygen molecules traveling at this speed are necessary to produce an average pressure of 1 atm? h) Compute the number of oxygen molecules that are actually contained in a vessel of this size at 300 K and atmospheric pressure. i) Your answer for part (h) should be three times as large as the answer for part (g). Where does this discrepancy arise?

18.38 Calculate the mean free path of air molecules at a pressure of 3.50×10^{-13} atm and a temperature of 300 K. (This pressure is readily attainable in the laboratory; see Exercise 18.24.) As in Example 18.9, model the air molecules as spheres of radius 2.0×10^{-10} m.

18.39 At what temperature is the root-mean-square speed of nitrogen molecules equal to the root-mean-square speed of hydrogen molecules at 20.0°C? (*Hint:* The periodic table in Appendix D shows the molar mass (in g/mol) of each element under the chemical symbol for that element. The molar mass of H_2 is twice the molar mass of hydrogen atoms, and similarly for N_2.)

18.40 Smoke particles in the air typically have masses of the order of 10^{-16} kg. The Brownian motion (rapid, irregular movement) of these particles, resulting from collisions with air molecules, can be observed with a microscope. a) Find the root-mean-square speed of Brownian motion for a particle with a mass of 3.00×10^{-16} kg in air at 300 K. b) Would the root-mean-square speed be different if the particle were in hydrogen gas at the same temperature? Explain.

Section 18.4 Heat Capacities

18.41 a) Calculate the specific heat capacity at constant volume of water vapor, assuming the nonlinear triatomic molecule has three translational and three rotational degrees of freedom and that vibrational motion does not contribute. The molar mass of water is 18.0 g/mol. b) The actual specific heat capacity of water vapor at low pressures is about 2000 J/kg·K. Compare this with your calculation and comment on the actual role of vibrational motion.

18.42 a) The specific heat of ice at constant volume is 833 J/kg·K at −180°C, 1640 J/kg·K at −60°C, and 2060 J/kg·K at −5.0°C. Calculate C_V, the molar heat capacity at constant volume for ice, at each of these temperatures. The molar mass of H_2O is 18.0 g/mol. b) Why do you suppose the value of C_V increases with increasing temperature? c) Do the values that you calculate approach the value of $3R$ (given in the rule of Dulong and Petit) as the temperature increases? Speculate about why this should be so.

18.43 a) How much heat does it take to increase the temperature of 2.50 mol of a diatomic ideal gas by 30.0 K near room temperature if the gas is held at constant volume? b) What is the answer to the question in part (a) if the gas is monatomic rather than diatomic?

18.44 a) Compute the specific heat capacity at constant volume of nitrogen (N_2) gas, and compare with the specific heat capacity of liquid water. The molar mass of N_2 is 28.0 g/mol. b) You warm 1.00 kg of water at a constant volume of 1.00 L from 20.0°C to 30.0°C in a kettle. For the same amount of heat, how many kilograms of 20.0°C air would you be able to warm to 30.0°C? What volume (in liters) would this air occupy at 20.0°C and a pressure of 1.00 atm? Make the simplifying assumption that air is 100% N_2.

*Section 18.5 Molecular Speeds

***18.45** For a gas of nitrogen molecules (N_2), what must the temperature be if 94.7% of all the molecules have speeds less than a) 1500 m/s; b) 1000 m/s; c) 500 m/s? Use Table 18.2. The molar mass of N_2 is 28.0 g/mol.

***18.46** Derive Eq. (18.33) from Eq. (18.32).

***18.47** Prove that $f(v)$ as given by Eq. (18.33) is maximum for $\epsilon = kT$. Use this result to obtain Eq. (18.34).

***18.48** For diatomic carbon dioxide gas (CO_2, molar mass = 44.0 g/mol) at $T = 300$ K, calculate a) the most probable speed v_{mp}; b) the average speed v_{av}; c) the root-mean-square speed v_{rms}.

Section 18.6 Phases of Matter

18.49 Puffy cumulus clouds, which are made of water droplets, occur at lower altitudes in the atmosphere. Wispy cirrus clouds, which are made of ice crystals, occur only at higher altitudes. Find the altitude y (measured from sea level) above which only cirrus clouds can occur. On a typical day and at altitudes less than 11 km, the temperature at an altitude y is given by $T = T_0 - \alpha y$, where $T_0 = 15.0$°C and $\alpha = 6.0$ C°/1000 m.

18.50 Solid water (ice) is slowly warmed from a very low temperature. a) What minimum external pressure p_1 must be applied to the solid if a melting phase transition is to be observed? Describe the sequence of phase transitions that occur if the applied pressure p is such that $p < p_1$. b) Above a certain maximum pressure p_2, no boiling transition is observed. What is this pressure? Describe the sequence of phase transitions that occur if $p_1 < p < p_2$.

18.51 A physicist places a piece of ice at 0.00°C and a beaker of water at 0.00°C inside a glass box and closes the lid of the box. All the air is then removed from the box. If the ice, water, and beaker are all maintained at a temperature of 0.00°C, describe the final equilibrium state inside the box.

18.52 The atmosphere of the planet Mars is 95.3% carbon dioxide (CO_2) and about 0.03% water vapor. The atmospheric pressure is only 600 Pa, and the surface temperature varies from −30°C to −100°C. The polar ice caps contain both CO_2 ice and water ice. Could there be *liquid* CO_2 on the surface of Mars? Could there be liquid water? Why or why not?

Problems

18.53 a) Use Eq. (18.1) to estimate the change in the volume of a solid steel sphere of volume 11 L when the temperature and pressure increase from 21°C and 1.013×10^5 Pa to 42°C and 2.10×10^7 Pa. (*Hint:* Consult Chapters 11 and 17 to determine the values of β and k.) b) In Example 18.3 the change in volume of an 11-L steel scuba tank was ignored. Was this a good approximation? Explain.

18.54 At an absolute pressure of 2.00×10^{-13} atm, a partial vacuum easily obtained in laboratories (see Exercise 18.24), calculate the mass of nitrogen present in a volume of 3000 cm^3 if the temperature of the gas is 22.0°C. The molar mass of nitrogen (N_2) is 28.0 g/mol.

18.55 A cylinder 1.00 m tall with inside diameter 0.120 m is used to hold propane gas (molar mass 44.1 g/mol) for use in a barbecue. It is initially filled with gas until the gauge pressure is 1.30×10^6 Pa and the temperature is 22.0°C. The temperature of the gas remains constant as it is partially emptied out of the tank, until the gauge pressure is 2.50×10^5 Pa. Calculate the mass of propane that has been used.

18.56 During a test dive in 1939, prior to being accepted by the U.S. Navy, the submarine *Squalus* sank at a point where the depth of water was 73.0 m. The temperature at the surface was 27.0°C and at the bottom it was 7.0°C. The density of seawater is 1030 kg/m^3. a) A diving bell was used to rescue 33 trapped crewmen from the *Squalus*. The diving bell was in the form of a circular cylinder 2.30 m high, open at the bottom and closed at the top. When the diving bell was lowered to the bottom of the sea, to what height did water rise within the diving bell? (*Hint:* You may ignore the relatively small variation in water pressure between the bottom of the bell and the surface of the water within the bell.) b) At what gauge pressure must compressed air have been supplied to the bell while on the bottom to expel all the water from it?

18.57 A glassblower makes a barometer using a tube 0.900 m long and with a cross-sectional area of 0.620 cm^2. Mercury stands in this tube to a height of 0.750 m. The room temperature is 20.0°C. A small amount of air is accidentally introduced into the evacuated space above the mercury, and the column drops to a height of 0.690 m. How many grams of air were introduced? The molar mass of air is 28.8 g/mol.

18.58 A hot-air balloon stays aloft because hot air at atmospheric pressure is less dense than cooler air at the same pressure; the calculation of the buoyant force is discussed in Chapter 14. If the volume of the balloon is 500 m^3 and the surrounding air is at 15.0°C, what must the temperature of the air in the balloon be for it to lift a total load of 290 kg (in addition to the mass of the hot air)? The density of air at 15.0°C and atmospheric pressure is 1.23 kg/m^3.

18.59 An automobile tire has a volume of 0.0150 m^3 on a cold day when the temperature of the air in the tire is 5.0°C and atmospheric pressure is 1.02 atm. Under these conditions the gauge pressure is measured to be 1.70 atm (about 25 lb/in.^{2}). After the car is driven on the highway for 30 min, the temperature of the air in the tires has risen to 45.0°C and the volume to 0.0159 m^3. What then is the gauge pressure?

18.60 A flask with a volume of 1.50 L, provided with a stopcock, contains ethane gas (C_2H_6) at 300 K and atmospheric pressure (1.013×10^5 Pa). The molar mass of ethane is 30.1 g/mol. The system is warmed to a temperature of 380 K, with the stopcock open to the atmosphere. The stopcock is then closed, and the flask cooled to its original temperature. a) What is the final pressure of the ethane in the flask? b) How many grams of ethane remain in the flask?

18.61 A balloon whose volume is 750 m^3 is to be filled with hydrogen at atmospheric pressure (1.01×10^5 Pa). a) If the hydrogen is stored in cylinders with volumes of 1.90 m^3 at a gauge pressure of 1.20×10^6 Pa, how many cylinders are required? Assume that the temperature of the hydrogen remains constant. b) What is the total weight (in addition to the weight of the gas) that can be supported by the balloon if the gas in the balloon and the surrounding air are both at 15.0°C? The molar mass of hydrogen (H_2) is 2.02 g/mol. The density of air at 15.0°C and atmospheric pressure is 1.23 kg/m^3. See Chapter 14 for a discussion of buoyancy. c) What weight could be supported if the balloon were filled with helium (with a molar mass of 4.00 g/mol) instead of hydrogen, again at 15.0°C?

18.62 A vertical cylindrical tank 0.900 m high has its top end closed by a tightly fitting frictionless piston of negligible weight. The air inside the cylinder is at an absolute pressure of 1.00 atm. The piston is depressed by pouring mercury on it slowly (Fig. 18.26). How far will the piston descend before mercury spills over the top of the cylinder? The temperature of the air is kept constant.

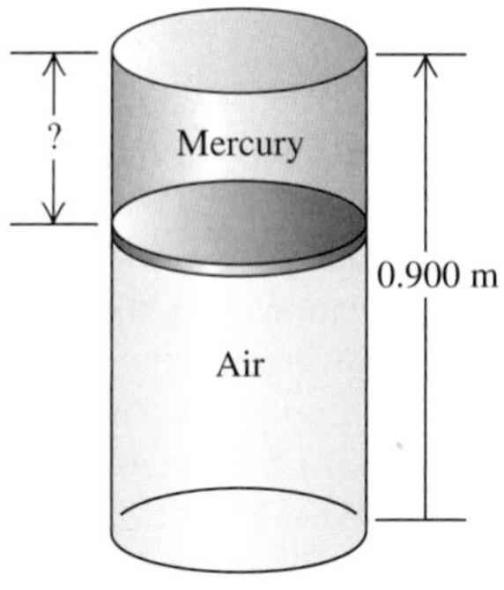

Figure 18.26 Problem 18.62.

18.63 A large tank of water has a hose connected to it, as shown in Fig. 18.27. The tank is sealed at the top and has compressed air between the water surface and the top. When the water height h has the value 3.50 m, the absolute pressure p of the compressed air is 4.20×10^5 Pa. Assume that the air above the water expands at constant temperature, and take the atmospheric pressure to be 1.00×10^5 Pa. a) What is the speed with which water flows out of the hose when $h = 3.50$ m? b) As water flows out of the tank, h decreases. Calculate the speed of flow for $h = 3.00$ m and for $h = 2.00$ m. c) At what value of h does the flow stop?

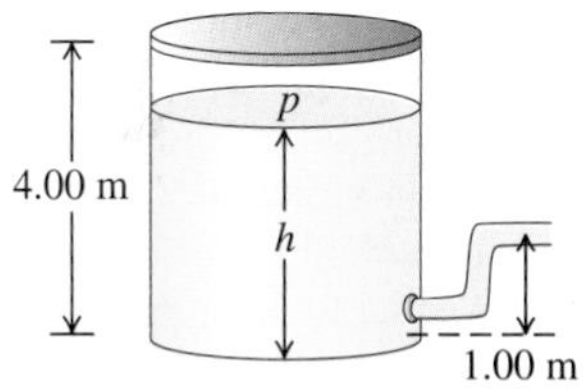

Figure 18.27 Problem 18.63.

18.64 In one hour, an average person at rest consumes 14.5 L of oxygen at a pressure of 1.00 atm and a temperature of 20.0°C. a) Express this rate of oxygen consumption in terms of the number of molecules per second. b) A person at rest inhales and exhales 0.5 L of air with each breath. The inhaled air is 21.0% oxygen, and the exhaled air is 16.3% oxygen. How many breaths per minute will satisfy this person's oxygen requirements? c) Repeat part (b) for a resting person at an elevation of 3000 m, where the pressure is 0.72 atm and the temperature is 0.0°C. Assume that the oxygen percentages and volume per inhalation are the same as stated in part (b). (To maintain its functions, the body still requires the same number of oxygen molecules per second as at sea level.) Explain why some people report "shortness of breath" at such elevations.

18.65 How Many Atoms Are You? Estimate the number of atoms in the body of a 50-kg physics student. Note that the human body is mostly water, which has molar mass 18.0 g/mol, and that each water molecule contains three atoms.

18.66 The size of an oxygen molecule is about 2.0×10^{-10} m. Make a rough estimate of the pressure at which the finite volume of the molecules should cause noticeable deviations from ideal-gas behavior at ordinary temperatures ($T = 300$ K).

18.67 Successive Approximations and the van der Waals Equation. In the ideal-gas equation, the number of moles per volume n/V is simply equal to p/RT. In the van der Waals equation, solving for n/V in terms of the pressure p and temperature T is somewhat more involved. a) Show that the van der Waals equation can be written as

$$\frac{n}{V} = \left(\frac{p + an^2/V^2}{RT}\right)\left(1 - \frac{bn}{V}\right)$$

b) The van der Waals parameters for hydrogen sulfide gas (H_2S) are $a = 0.448\ \text{J} \cdot \text{m}^3/\text{mol}^2$ and $b = 4.29 \times 10^{-5}\ \text{m}^3/\text{mol}$. Determine the number of moles per volume of H_2S gas at 127°C and an absolute pressure of 9.80×10^5 Pa as follows: i) Calculate a first approximation using the ideal-gas equation, $n/V = p/RT$. ii) Substitute this approximation for n/V into the right-hand side of the equation in part (a). The result is a new, improved approximation for n/V. iii) Substitute the new approximation for n/V into the right-hand side of the equation in (a). The result is a further-improved approximation for n/V. iv) Repeat step (iii) until successive approximations agree to the desired level of accuracy (in this case, to three significant figures). c) Compare your final result in part (b) to the result p/RT obtained using the ideal-gas equation. Which result gives a larger value of n/V? Why?

18.68 a) Compute the increase in gravitational potential energy for a nitrogen molecule (molar mass 28.0 g/mol) for an increase in elevation of 400 m near the earth's surface. b) At what temperature is this equal to the average kinetic energy of a nitrogen molecule? c) Is it possible that a nitrogen molecule near sea level where $T = 15.0°\text{C}$ could rise to an altitude of 400 m? Is it likely that it could do so without hitting any other molecules along the way? Explain.

18.69 The Lennard-Jones Potential. A commonly used potential-energy function for the interaction of two molecules (Fig. 18.6) is the Lennard-Jones 6-12 potential

$$U(r) = U_0\left[\left(\frac{R_0}{r}\right)^{12} - 2\left(\frac{R_0}{r}\right)^{6}\right]$$

where r is the distance between the centers of the molecules and U_0 and R_0 are positive constants. The corresponding force $F(r)$ is given in Eq. (13.26). a) Graph $U(r)$ and $F(r)$ versus r. b) Let r_1 be the value of r at which $U(r) = 0$, and let r_2 be the value of r at which $F(r) = 0$. Show the locations of r_1 and r_2 on your graphs of $U(r)$ and $F(r)$. Which of these values represents the equilibrium separation between the molecules? c) Find the values of r_1 and r_2 in terms of R_0, and find the ratio r_1/r_2. d) If the molecules are located a distance r_2 apart (as calculated in part (c)), how much work must be done to pull them apart so that $r \to \infty$?

18.70 a) What is the total random translational kinetic energy of 5.00 L of hydrogen gas, with pressure 1.01×10^5 Pa and temperature 300 K? b) If the tank containing the gas is moved with a speed of 30.0 m/s, by what percentage is the total kinetic energy of the gas increased? The molar mass of hydrogen (H_2) is 2.016 g/mol.

18.71 The speed of propagation of a sound wave in air at 27°C is about 350 m/s. Calculate, for comparison, a) v_{rms} for nitrogen molecules; b) the rms value of v_x at this temperature. The molar mass of nitrogen (N_2) is 28.0 g/mol.

18.72 The surface of the sun has a temperature of about 5800 K and consists largely of hydrogen atoms. a) Find the rms speed of a hydrogen atom at this temperature. (The mass of a single hydrogen atom is 1.67×10^{-27} kg.) b) The escape speed for a particle to leave the gravitational influence of the sun is given by $(2GM/R)^{1/2}$, where M is the sun's mass, R its radius, and G the gravitational constant (Example 12.5 of Section 12.3). Use the data in Appendix F to calculate this escape speed. c) Can appreciable quantities of hydrogen escape from the sun? Can *any* hydrogen escape? Explain.

18.73 a) Show that a projectile with mass m can "escape" from the surface of a planet if it is launched vertically upward with a kinetic energy greater than mgR_p, where g is the acceleration due to gravity at the planet's surface and R_p is the planet's radius. Ignore air resistance. (See Problem 18.72.) b) If the planet in question is the earth, at what temperature does the average translational kinetic energy of a nitrogen molecule (molar mass = 28.0 g/mol) equal that required to escape? What about a hydrogen molecule (molar mass = 2.02 g/mol)? c) Repeat part (b) for the moon, for which $g = 1.63\ \text{m/s}^2$ and $R_p = 1740$ km. d) While the earth and the moon have similar average surface temperatures, the moon has essentially no atmosphere. Use your results from parts (b) and (c) to explain why.

18.74 Planetary Atmospheres. a) The temperature near the top of Jupiter's multicolored cloud layer is about 140 K. The temperature at the top of the earth's troposphere, at an altitude of about 20 km, is about 220 K. Calculate the rms speed of hydrogen molecules in both these environments. Give your answers in m/s and as a fraction of the escape speed from the respective planet (see Problem 18.72). b) Hydrogen gas (H_2) is a rare element in the earth's atmosphere. In the atmosphere of Jupiter, by contrast, 89% of all molecules are H_2. Explain why, using your results from part (a). c) Suppose an astronomer claims to have discovered an oxygen (O_2) atmosphere on the asteroid Ceres. How likely is this? Ceres has a mass equal to 0.014 times the mass of the moon, a density of 2400 kg/m^3, and a surface temperature of about 200 K.

18.75 a) For what mass of molecule or particle is v_{rms} equal to 1.00 mm/s at 300 K? b) If the particle is an ice crystal, how many molecules does it contain? The molar mass of water is 18.0 g/mol. c) Calculate the diameter of the particle if it is a spherical piece of ice. Would it be visible to the naked eye?

18.76 In describing the heat capacities of solids in Section 18.4, we stated that the potential energy $U = \frac{1}{2}kx^2$ of a harmonic oscillator averaged over one period of the motion is equal to the kinetic energy $K = \frac{1}{2}mv^2$ averaged over one period. Prove this result using Eqs. (13.13) and (13.15) for the position and velocity of a simple harmonic oscillator. For simplicity, assume that the initial position and velocity make the phase angle ϕ equal to zero. (*Hint:* Use the trigonometric identities $\cos^2(\theta) = [1 + \cos(2\theta)]/2$ and $\sin^2(\theta) = [1 - \cos(2\theta)]/2$. What is the average value of $\cos(2\omega t)$ averaged over one period?)

18.77 It is possible to make crystalline solids that are only one layer of atoms thick. Such "two-dimensional" crystals can be created by depositing atoms on a very flat surface. a) If the atoms in such a two-dimensional crystal can only move within the plane of the crystal, what will be its molar heat capacity near room temperature? Give your answer as a multiple of R and in J/mol · K. b) At very low temperatures, will the molar heat capacity of a two-dimensional crystal be greater than, less, or equal to the result you found in part (a)? Explain why.

18.78 a) Calculate the total *rotational* kinetic energy of the molecules in 1.00 mol of a diatomic gas at 300 K. b) Calculate the moment of inertia of an oxygen molecule (O_2) for rotation about either the y- or z-axis shown in Fig. 18.15. Treat the molecule as two massive points (representing the oxygen atoms) separated by a distance of 1.21×10^{-10} m. The molar mass of oxygen *atoms* is 16.0 g/mol. c) Find the rms angular velocity of rotation of an oxygen molecule about either the y- or z-axis shown in Fig. 18.15. How does your answer compare to the angular velocity of a typical piece of rapidly rotating machinery (10,000 rev/min)?

18.79 For each polyatomic gas in Table 18.1, compute the value of the molar heat capacity at constant volume, C_V, on the assumption that there is no vibrational energy. Compare with the measured values in the table, and compute the fraction of the total heat capacity that is due to vibration for each of the three gases. (*Note:* CO_2 is linear; SO_2 and H_2S are not. Recall that a linear polyatomic molecule has two rotational degrees of freedom, and a nonlinear molecule has three.)

***18.80** a) Show that $\int_0^\infty f(v)\,dv = 1$, where $f(v)$ is the Maxwell-Boltzmann distribution of Eq. (18.32). b) In terms of the physical definition of $f(v)$, explain why the integral in part (a) *must* have this value.

***18.81** Calculate the integral in Eq. (18.31), $\int_0^\infty v^2 f(v)\,dv$, and compare this result to $(v^2)_{av}$ as given by Eq. (18.16). (*Hint:* You may use the tabulated integral

$$\int_0^\infty x^{2n} e^{-\alpha x^2}\,dx = \frac{1 \cdot 3 \cdot 5 \cdots (2n-1)}{2^{n+1}\alpha^n}\sqrt{\frac{\pi}{\alpha}}$$

where n is a positive integer and α is a positive constant.)

***18.82** Calculate the integral in Eq. (18.30), $\int_0^\infty v f(v)\,dv$, and compare this result to v_{av} as given by Eq. (18.35). (*Hint:* Make the change of variable $v^2 = x$ and use the tabulated integral

$$\int_0^\infty x^n e^{-\alpha x}\,dx = \frac{n!}{\alpha^{n+1}}$$

where n is a positive integer and α is a positive constant.)

***18.83** a) Explain why in a gas of N molecules, the number of molecules having speeds in the *finite* interval v to $v + \Delta v$ is $\Delta N = N\int_v^{v+\Delta v} f(v)\,dv$. b) If Δv is small, then $f(v)$ is approximately constant over the interval and $\Delta N \approx N f(v)\,\Delta v$. For oxygen gas $(O_2$, molar mass $= 32.0$ g/mol$)$ at $T = 300$ K, use this approximation to calculate the number of molecules with speeds within $\Delta v = 20$ m/s of v_{mp}. Express your answer as a multiple of N. c) Repeat part (b) for speeds within $\Delta v = 20$ m/s of $7v_{mp}$. d) Repeat parts (b) and (c) for a temperature of 600 K. e) Repeat parts (b) and (c) for a temperature of 150 K. f) What do your results tell you about the shape of the distribution as a function of temperature? Do your conclusions agree with what is shown in Fig. 18.20?

18.84 The *vapor pressure* is the pressure of the vapor phase of a substance when it is in equilibrium with the solid or liquid phase of the substance. The *relative humidity* is the partial pressure of water vapor in the air divided by the vapor pressure of water at that same temperature, expressed as a percentage. The air is saturated when the humidity is 100%. a) The vapor pressure of water at 20.0°C is 2.34×10^3 Pa. If the air temperature is 20.0°C and the relative humidity is 60%, what is the partial pressure of water vapor in the atmosphere (that is, the pressure due to water vapor alone)? b) Under the conditions of part (a), what is the mass of water in 1.00 m^3 of air? (The molar mass of water is 18.0 g/mol. Assume that water vapor can be treated as an ideal gas.)

18.85 The Dew Point. The vapor pressure of water (Problem 18.84) decreases as the temperature decreases. If the amount of water vapor in the air is kept constant as the air is cooled, a temperature is reached, called the *dew point,* at which the partial pressure and vapor pressure coincide and the vapor is saturated. If the air is cooled further, vapor condenses to liquid until the partial pressure again equals the vapor pressure at that temperature. The temperature in a room is 30.0°C. A meteorologist cools a metal can by gradually adding cold water. When the can temperature reaches 16.0°C, water droplets form on its outside surface. What is the relative humidity of the 30.0°C air in the room? The table below lists the vapor pressure of water at various temperatures:

Temperature (°C)	Vapor Pressure (Pa)
10.0	1.23×10^3
12.0	1.40×10^3
14.0	1.60×10^3
16.0	1.81×10^3
18.0	2.06×10^3
20.0	2.34×10^3
22.0	2.65×10^3
24.0	2.99×10^3
26.0	3.36×10^3
28.0	3.78×10^3
30.0	4.25×10^3

18.86 Altitude at Which Clouds Form. On a spring day in the midwestern United States, the air temperature at the surface is 28.0°C. Puffy cumulus clouds form at an altitude where the air temperature equals the dew point (Problem 18.85). If the air temperature decreases with altitude at a rate of 0.6 C°/100 m, at approximately what height above the ground will clouds form if the relative humidity at the surface is a) 35% b) 80%? (*Hint:* Use the table in Problem 18.85.)

Challenge Problems

18.87 Dark Nebulae and the Interstellar Medium. The dark area in Fig. 18.28 that appears devoid of stars is a *dark nebula,* a cold gas cloud in interstellar space that contains enough material to block out light from the stars behind it. A typical dark nebula is about 20 light years in diameter and contains about 50 hydrogen atoms per cubic centimeter (monatomic hydrogen, *not* H_2) at a temperature of about 20 K. (A light year is the distance light travels in vacuum in one year and is equal to 9.46×10^{15} m.) a) Estimate the mean free path for a hydrogen atom in a dark nebula. The radius of a hydrogen atom is 5.0×10^{-11} m. b) Estimate the rms speed of a hydrogen

Figure 18.28 Challenge Problem 18.87.

atom and the mean free time (the average time between collisions for a given atom). Based on this result, do you think that atomic collisions, such as those leading to H_2 molecule formation, are very important in determining the composition of the nebula? c) Estimate the pressure inside a dark nebula. d) Compare the rms speed of a hydrogen atom to the escape speed at the surface of the nebula (assumed spherical). If the space around the nebula were a vacuum, would such a cloud be stable or would it tend to evaporate? e) The stability of dark nebulae is explained by the presence of the *interstellar medium* (ISM), an even thinner gas that permeates space and into which the dark nebulae are embedded. Show that for dark nebulae to be in equilibrium with the ISM, the numbers of atoms per volume (N/V) and the temperatures (T) of dark nebulae and the ISM must be related by

$$\frac{(N/V)_{\text{nebula}}}{(N/V)_{\text{ISM}}} = \frac{T_{\text{ISM}}}{T_{\text{nebula}}}$$

f) In the vicinity of the sun, the ISM contains about 1 hydrogen atom per 200 cm^3. Estimate the temperature of the ISM in the vicinity of the sun. Compare to the temperature of the sun's surface, about 5800 K. Would a spacecraft coasting through interstellar space burn up? Why or why not?

18.88 In the *troposphere,* the part of the atmosphere that extends from the surface to an altitude of about 11 km, the temperature is not uniform but decreases with increasing elevation. a) Show that if the temperature variation is approximated by the linear relation

$$T = T_0 - \alpha y$$

where T_0 is the temperature at the earth's surface and T is the temperature at height y, the pressure p at height y is given by

$$\ln\left(\frac{p}{p_0}\right) = \frac{Mg}{R\alpha}\ln\left(\frac{T_0 - \alpha y}{T_0}\right)$$

where p_0 is the pressure at the earth's surface and M is the molar mass for air. The coefficient α is called the lapse rate of temperature. It varies with atmospheric conditions, but an average value is about 0.6 C°/100 m. b) Show that the above result reduces to the result of Example 18.4 (Section 18.1) in the limit that $\alpha \to 0$. c) With $\alpha = 0.6$ C°/100 m, calculate p for $y = 8863$ m and compare your answer to the result of Example 18.4. Take $T_0 = 288$ K and $p_0 = 1.00$ atm.

18.89 Van der Waals Equation and Critical Points. a) In pV-diagrams the slope $\partial p/\partial V$ along an isotherm is never positive. Explain why. b) Regions where $\partial p/\partial V = 0$ represent equilibrium between two phases; volume can change with no change in pressure, as when water boils at atmospheric pressure. We can use this to determine the temperature, pressure, and volume per mole at the critical point using the equation of state $p = p(V, T, n)$. If $T > T_c$, then $p(V)$ has no maximum along an isotherm, but if $T < T_c$, then $p(V)$ has a maximum. Explain how this leads to the following condition for determining the critical point:

$$\frac{\partial p}{\partial V} = 0 \quad \text{and} \quad \frac{\partial^2 p}{\partial V^2} = 0 \qquad \text{at the critical point}$$

c) Solve the van der Waals equation (Eq. 18.7) for p. That is, find $p(V, T, n)$. Find $\partial p/\partial V$ and $\partial^2 p/\partial V^2$. Set these equal to zero to obtain two equations for V, T, and n. d) Simultaneous solution of the two equations obtained in part (c) gives the temperature and volume per mole at the critical point, T_c and $(V/n)_c$. Find these constants in terms of a and b. (*Hint:* Divide one equation by the other to eliminate T.) e) Substitute these values into the equation of state to find p_c, the pressure at the critical point. f) Use the results from parts (d) and (e) to find the ratio $RT_c/p_c(V/n)_c$. This should not contain either a or b, and so should have the same value for all gases. g) Compute the ratio $RT_c/p_c(V/n)_c$ for the gases H_2, N_2, and H_2O using the critical point data given below.

Gas	T_c (K)	p_c (Pa)	$(V/n)_c$ (m^3/mol)
H_2	33.3	13.0×10^5	65.0×10^{-6}
N_2	126.2	33.9×10^5	90.1×10^{-6}
H_2O	647.4	221.2×10^5	56.0×10^{-6}

h) Discuss how well the results of part (g) compare to the prediction of part (f) based on the van der Waals equation. What do you conclude about the accuracy of the van der Waals equation as a description of the behavior of gases near the critical point?

***18.90** In Example 18.7 (Section 18.3) we saw that $v_{\text{rms}} > v_{\text{av}}$. It is not difficult to show that this is *always* the case. (The only exception is when the particles have the same speed, in which case $v_{\text{rms}} = v_{\text{av}}$.) a) For two particles with speeds v_1 and v_2, show that $v_{\text{rms}} \geq v_{\text{av}}$, regardless of the numerical values of v_1 and v_2. Then show that $v_{\text{rms}} > v_{\text{av}}$ if $v_1 \neq v_2$. b) Suppose that for a collection of N particles you know that $v_{\text{rms}} > v_{\text{av}}$. Another particle, with speed u, is added to the collection of particles. If the new rms and average speeds are denoted as v'_{rms} and v'_{av}, show that

$$v'_{\text{rms}} = \sqrt{\frac{Nv_{\text{rms}}^2 + u^2}{N + 1}} \quad \text{and} \quad v'_{\text{av}} = \frac{Nv_{\text{av}} + u}{N + 1}$$

c) Use the expressions in part (b) to show that $v'_{\text{rms}} > v'_{\text{av}}$ regardless of the numerical value of u. d) Explain why your results for (a) and (c) together show that $v_{\text{rms}} > v_{\text{av}}$ for any collection of particles if the particles do not all have the same speed.

THE FIRST LAW OF THERMODYNAMICS 19

This steam locomotive operates using the first law of thermodynamics. When heat is generated by burning wood or coal in the locomotive's engine, part of the energy goes into increasing the temperature of the engine's water (which boils and produces steam). The remainder is used by the expanding steam to do work and propel the locomotive forward.

? Would it be possible for the steam to propel the locomotive by doing work as it *condenses?*

Every time you drive a car, turn on an air conditioner, or cook a meal, you reap the practical benefits of *thermodynamics,* the study of relationships involving heat, mechanical work, and other aspects of energy and energy transfer. For example, in a car engine heat is generated by the chemical reaction of oxygen and vaporized gasoline in the engine's cylinders. The heated gas pushes on the pistons within the cylinders, doing mechanical work that is used to power the car. This is an example of a *thermodynamic process.*

The first law of thermodynamics, central to the understanding of such processes, is an extension of the principle of conservation of energy. It broadens this principle to include energy exchange by both heat transfer and mechanical work and introduces the concept of the *internal energy* of a system. Conservation of energy plays a vital role in every area of physical science, and the first law has extremely broad usefulness. To state energy relationships precisely, we need the concept of a *thermodynamic system.* We'll discuss *heat* and *work* as two means of transferring energy into or out of such a system.

19.1 | Thermodynamic Systems

We have studied energy transfer through mechanical work (Chapter 6) and through heat transfer (Chapters 17 and 18). Now we are ready to combine and generalize these principles.

19.1 The popcorn in the pot is a thermodynamic system. In the thermodynamic process shown here, heat is added to the system, and the system does work on its surroundings to lift the lid of the pot.

We always talk about energy transfer to or from some specific *system.* The system might be a mechanical device, a biological organism, or a specified quantity of material, such as the refrigerant in an air conditioner or steam expanding in a turbine. In general, a **thermodynamic system** is any collection of objects that is convenient to regard as a unit, and that may have the potential to exchange energy with its surroundings. A familiar example is a quantity of popcorn kernels in a pot with a lid. When the pot is placed on a stove, energy is added to the popcorn by conduction of heat. As the popcorn pops and expands, it does work as it exerts an upward force on the lid and moves it through a displacement (Fig. 19.1). The *state* of the popcorn changes in this process, since the volume, temperature, and pressure of the popcorn all change as it pops. A process such as this one, in which there are changes in the state of a thermodynamic system, is called a **thermodynamic process.**

In mechanics we used the concept of *system* on a regular basis in connection with free-body diagrams and conservation of energy and momentum. With *thermodynamic* systems, as with all others, it is essential to define clearly at the start exactly what is and is not included in the system. Only then can we describe unambiguously the energy transfers into and out of that system. For instance, in our popcorn example we defined the system to include the popcorn but not the pot, lid, or stove.

Thermodynamics has its roots in many practical problems other than popping popcorn (Fig. 19.2). The engine in an automobile and the jet engines in an airplane use the heat of combustion of their fuel to perform mechanical work in propelling the vehicle. Muscle tissue in living organisms metabolizes chemical energy in food and performs mechanical work on the organism's surroundings. A steam engine or steam turbine uses the heat of combustion of coal or other fuel to perform mechanical work such as driving an electric generator or pulling a train.

Signs For Heat and Work in Thermodynamics

We describe the energy relations in any thermodynamic process in terms of the quantity of heat Q added *to* the system and the work W done *by* the system. Both Q and W may be positive, negative, or zero (Fig. 19.3). A positive value of Q represents heat flow *into* the system, with a corresponding input of energy to it; neg-

(a) (b)

19.2 (a) A rocket engine uses the heat of combustion of its fuel to do work propelling the launch vehicle. (b) Humans and other biological organisms are more complicated systems than we can analyze fully in this book, but the same basic principles of thermodynamics apply to them.

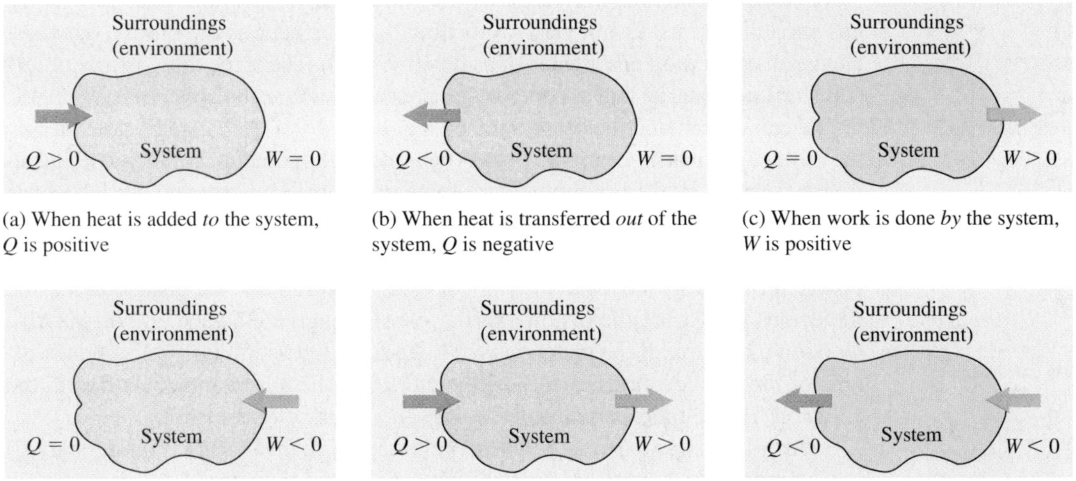

(a) When heat is added *to* the system, Q is positive

(b) When heat is transferred *out* of the system, Q is negative

(c) When work is done *by* the system, W is positive

(d) When work is done *on* the system, W is negative

(e) Heat is added to the system and work is done by the system

(f) Heat is transferred out of the system and work is done on the system

19.3 A thermodynamic system may exchange energy with its surroundings (environment) by means of heat [(a) and (b)], work. [(c) and (d)], or heat and work simultaneously [(e) and (f)]. Note the sign conventions for Q and W.

ative Q represents heat flow *out of* the system. A positive value of W represents work done *by* the system against its surroundings, such as work done by an expanding gas, and hence corresponds to energy *leaving* the system. Negative W, such as work done during compression of a gas in which work is done *on the gas* by its surroundings, represents energy *entering* the system. We will use these conventions consistently in the examples in this chapter and the next.

CAUTION Note that our sign rule for work is *opposite* to the one we used in mechanics, in which we always spoke of the work done by the forces acting *on* a body. In thermodynamics it is usually more convenient to call W the work done *by* the system so that when a system expands, the pressure, volume change, and work are all positive. Take care to use the sign rules for work and heat consistently!

Test Your Understanding

In Example 17.8 (Section 17.6), what is the sign of Q for the coffee? For the aluminum cup? If a block slides along a horizontal surface with friction, what is the sign of W for the block?

19.2 | Work Done During Volume Changes

A simple but common example of a thermodynamic system is a quantity of gas enclosed in a cylinder with a movable piston. Internal-combustion engines, steam engines, and compressors in refrigerators and air conditioners all use some version of such a system. In the next several sections we will use the gas-in-cylinder system to explore several kinds of processes involving energy transformations.

We'll use a microscopic viewpoint, based on the kinetic and potential energies of individual molecules in a material, to develop intuition about thermodynamic quantities. But it is important to understand that the central principles of thermodynamics can be treated in a completely *macroscopic* way, without reference to microscopic models. Indeed, part of the great power and generality of thermodynamics is that it does *not* depend on details of the structure of matter.

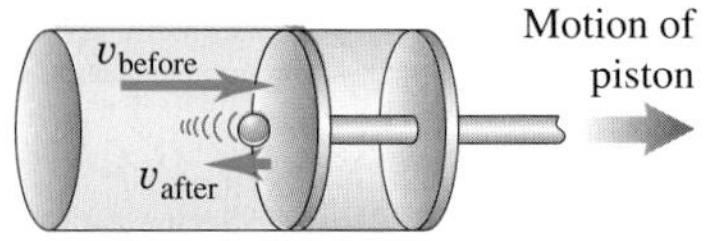

(a) Piston moves away from molecule during collision: molecule loses kinetic energy, does positive work on piston

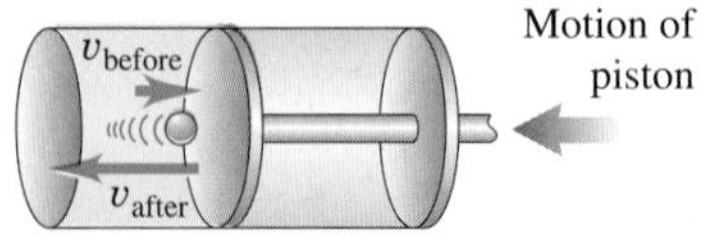

(b) Piston moves toward molecule during collision: molecule gains kinetic energy, does negative work on piston

19.4 (a) A molecule striking a piston (a) does positive work if the piston is moving away from the molecule and (b) does negative work if the piston is moving toward the molecule. Hence a gas does positive work when it expands as in (a) but does negative work when it compresses as in (b).

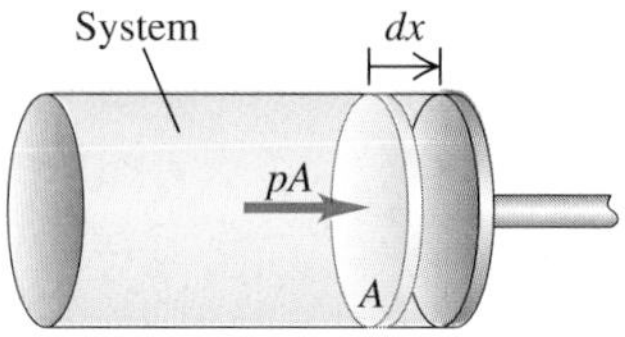

19.5 The infinitesimal work done by the system during the small expansion dx is $dW = pA\,dx$.

19.6 When fuel is burned in one of the cylinders of an automotive engine, the heated fuel–air mixture expands and does an amount of work given by Eq. (19.2).

First we consider the *work* done by the system during a volume change. When a gas expands, it pushes outward on its boundary surfaces as they move outward. Hence an expanding gas always does positive work. The same thing is true of any solid or fluid material that expands under pressure, such as the popcorn in Fig. 19.1.

We can understand the work done by a gas in a volume change by considering the molecules that make up the gas. When one such molecule collides with a stationary surface, it exerts a momentary force on the wall but does no work because the wall does not move. But if the surface is moving, such as a piston in a gasoline engine, the molecule *does* do work on the surface during the collision. If the piston in Fig. 19.4a moves to the right, so that the volume of the gas increases, the molecules that strike the piston exert a force through a distance and do *positive* work on the piston. If the piston moves toward the left as in Fig. 19.4b, so the volume of the gas decreases, then positive work is done *on* the molecule during the collision. Hence the gas molecules do *negative* work on the piston.

Figure 19.5 shows a system whose volume can change (a gas, liquid, or solid) in a cylinder with a movable piston. Suppose that the cylinder has a cross-section area A and that the pressure exerted by the system at the piston face is p. The total force F exerted by the system on the piston is $F = pA$. When the piston moves out an infinitesimal distance dx, the work dW done by this force is

$$dW = F\,dx = pA\,dx$$

But

$$A\,dx = dV$$

where dV is the infinitesimal change of volume of the system. Thus we can express the work done by the system in this infinitesimal volume change as

$$dW = p\,dV \qquad (19.1)$$

In a finite change of volume from V_1 to V_2,

$$W = \int_{V_1}^{V_2} p\,dV \qquad \text{(work done in a volume change)} \qquad (19.2)$$

In general, the pressure of the system may vary during the volume change. For example, this is the case in the cylinders of an automobile engine as the pistons move back and forth (Fig. 19.6). To evaluate the integral in Eq. (19.2), we have to know how the pressure varies as a function of volume. We can represent this relationship as a graph of p as a function of V (a pV-diagram, described at the end of Section 18.1). Figure 19.7a shows a simple example. In this figure, Eq. (19.2) is represented graphically as the *area* under the curve of p versus V between the limits V_1 and V_2. (In Section 6.3 we used a similar interpretation of the work done by a force F as the area under the curve of F versus x between the limits x_1 and x_2.)

According to the rule we stated in Section 19.1, work is *positive* when a system *expands*. In an expansion from state 1 to state 2 in Fig. 19.7a the area under the curve and the work are positive. A *compression* from 1 to 2 in Fig. 19.7b gives a *negative* area; when a system is compressed, its volume decreases and it does *negative* work on its surroundings (see also Fig. 19.4b).

CAUTION When using Eq. (19.2), always remember that V_1 is the *initial* volume and V_2 is the *final* volume. That's why the labels 1 and 2 are reversed in Fig. 19.7b compared to Fig. 19.7a, even though both processes move between the same two thermodynamic states.

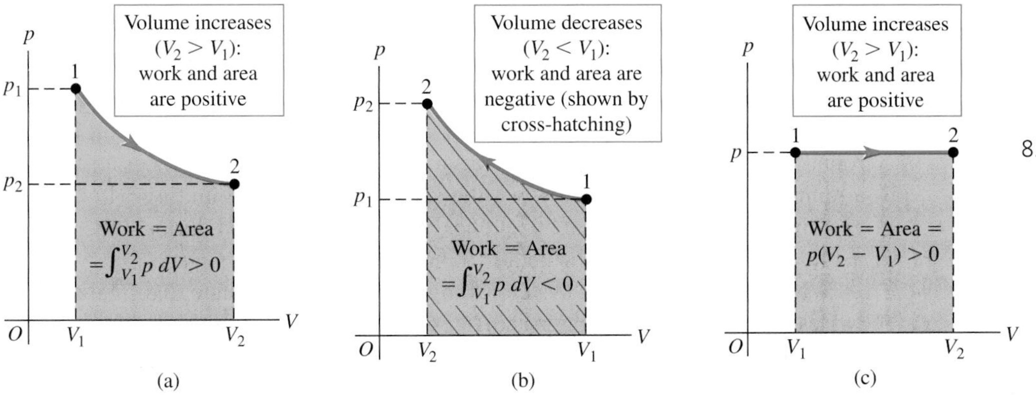

ActivPhysics Online 8.5 Work Done By a Gas

19.7 The work done equals the area under the curve on a pV-diagram. (a) In an expansion, the work done is positive. (b) In a compression, the work done is negative. (c) In a constant-volume process, the work done is easy to calculate because the area (in this case positive) is rectangular.

If the pressure p remains constant while the volume changes from V_1 to V_2 (Fig. 19.7c), the work done by the system is

$$W = p(V_2 - V_1) \quad \text{(work done in a volume change at constant pressure)} \tag{19.3}$$

In any process in which the volume is *constant,* the system does no work because there is no displacement.

Example 19.1 Isothermal expansion of an ideal gas

An ideal gas undergoes an *isothermal* (constant-temperature) *expansion* at temperature T, during which its volume changes from V_1 to V_2. How much work does the gas do?

SOLUTION

IDENTIFY and SET UP: We are asked to calculate the work done due to a volume change, which means that we must use Eq. (19.2). To use this equation we need to know the pressure as a function of volume; for this we use the ideal-gas law, Eq. (18.3). Although it may be tempting to do so, we *cannot* use Eq. (19.3) because the temperature, not the pressure, is held constant.

EXECUTE: From Eq. (19.2),

$$W = \int_{V_1}^{V_2} p\,dV$$

From Eq. (18.3) the pressure p of n moles of ideal gas occupying volume V at absolute temperature T is

$$p = \frac{nRT}{V}$$

where R is the gas constant. We substitute this into the integral, take the constants n, R, and T outside, and evaluate the integral:

$$W = nRT\int_{V_1}^{V_2} \frac{dV}{V} = nRT \ln\frac{V_2}{V_1}$$
$$\text{(ideal gas, isothermal process)}$$

Also, when T is constant,

$$p_1V_1 = p_2V_2 \quad \text{or} \quad \frac{V_2}{V_1} = \frac{p_1}{p_2}$$

so the isothermal work may also be expressed as

$$W = nRT\ln\frac{p_1}{p_2} \quad \text{(ideal gas, isothermal process)}$$

EVALUATE: We check our result by noting that in an expansion $V_2 > V_1$ and the ratio V_2/V_1 is greater than 1. The logarithm of a number greater than 1 is positive, so $W > 0$, as it should be. As an additional check, look at our second expression for W: in an isothermal expansion the volume increases and the pressure drops, so $p_2 < p_1$, the ratio $p_1/p_2 > 1$, and $W = nRT\ln(p_1/p_2)$ is again positive.

These results also apply to an isothermal *compression* of a gas, for which $V_2 < V_1$ and $p_2 > p_1$.

Test Your Understanding

In an isothermal expansion of n moles of an ideal gas at temperature T, how many times larger than the initial volume must the final volume be if the gas does work nRT? If the gas does work $2nRT$?

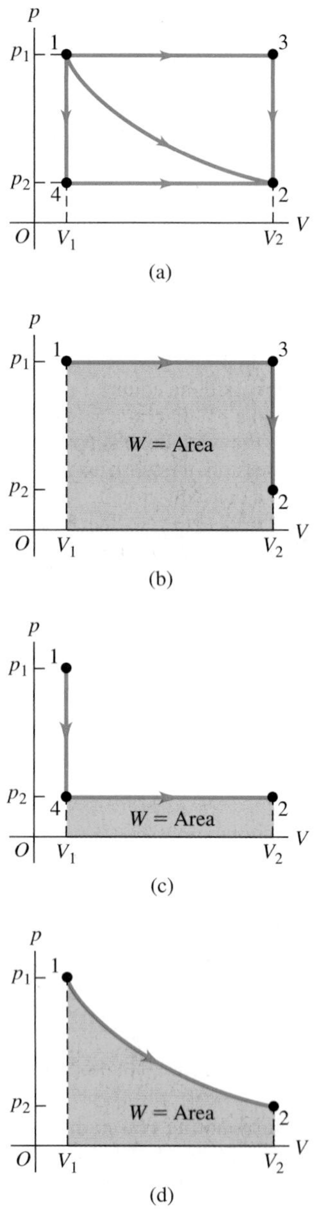

19.8 (a) Three different paths between state 1 and state 2. (b)–(d) The work done by the system during a transition between two states depends on the path chosen.

19.3 | Paths Between Thermodynamic States

We've seen that if a thermodynamic process involves a change in volume, the system undergoing the process does work (either positive or negative) on its surroundings. Heat also flows into or out of the system during the process if there is a temperature difference between the system and its surroundings. Let's now examine how the work done by and heat added to the system during a thermodynamic process depend on the details of how the process takes place.

When a thermodynamic system changes from an initial state to a final state, it passes through a series of intermediate states. We call this series of states a **path.** There are always infinitely many different possibilities for these intermediate states. When they are all equilibrium states, the path can be plotted on a pV-diagram (Fig. 19.8a). Point 1 represents an initial state with pressure p_1 and volume V_1, and point 2 represents a final state with pressure p_2 and volume V_2. To pass from state 1 to state 2, we could keep the pressure constant at p_1 while the system expands to volume V_2 (point 3 in Fig. 19.8b), then reduce the pressure to p_2 (probably by decreasing the temperature) while keeping the volume constant at V_2 (to point 2 on the diagram). The work done by the system during this process is the area under the line $1 \rightarrow 3$; no work is done during the constant-volume process $3 \rightarrow 2$. Or the system might traverse the path $1 \rightarrow 4 \rightarrow 2$ (Fig. 19.8c); in that case the work is the area under the line $4 \rightarrow 2$, since no work is done during the constant-volume process $1 \rightarrow 4$. The smooth curve from 1 to 2 is another possibility (Fig. 19.8d), and the work for this path is different from that for either of the other paths.

We conclude that *the work done by the system depends not only on the initial and final states, but also on the intermediate states, that is, on the path.* Furthermore, we can take the system through a series of states forming a closed loop, such as $1 \rightarrow 3 \rightarrow 2 \rightarrow 4 \rightarrow 1$. In this case the final state is the same as the initial state, but the total work done by the system is *not* zero. (In fact, it is represented on the graph by the area enclosed by the loop; can you prove that? See Exercise 19.8.) It follows that it doesn't make sense to talk about the amount of work *contained in* a system. In a particular state, a system may have definite values of the state coordinates p, V, and T, but it wouldn't make sense to say that it has a definite value of W.

Like work, the *heat* added to a thermodynamic system when it undergoes a change of state depends on the path from the initial state to the final state. Here's an example. Suppose we want to change the volume of a certain quantity of an ideal gas from 2.0 L to 5.0 L while keeping the temperature constant at $T = 300$ K. Figure 19.9 shows two different ways in which we can do this. In Fig. 19.9a the gas is contained in a cylinder with a piston, with an initial volume of 2.0 L. We let the gas expand slowly, supplying heat from the electric heater to keep the temperature at 300 K. After expanding in this slow, controlled, isothermal manner, the gas reaches its final volume of 5.0 L; it absorbs a definite amount of heat in the process.

Figure 19.9b shows a different process leading to the same final state. The container is surrounded by insulating walls and is divided by a thin, breakable partition into two compartments. The lower part has volume 2.0 L and the upper part has volume 3.0 L. In the lower compartment we place the same amount of the same gas as in Fig. 19.9a, again at $T = 300$ K. The initial state is the same as before. Now we break the partition; the gas undergoes a rapid, uncontrolled expansion, with no heat passing through the insulating walls. The final volume is 5.0 L, the same as in Fig. 19.9a. The gas does no work during this expansion because it doesn't push against anything that moves. This uncontrolled expansion

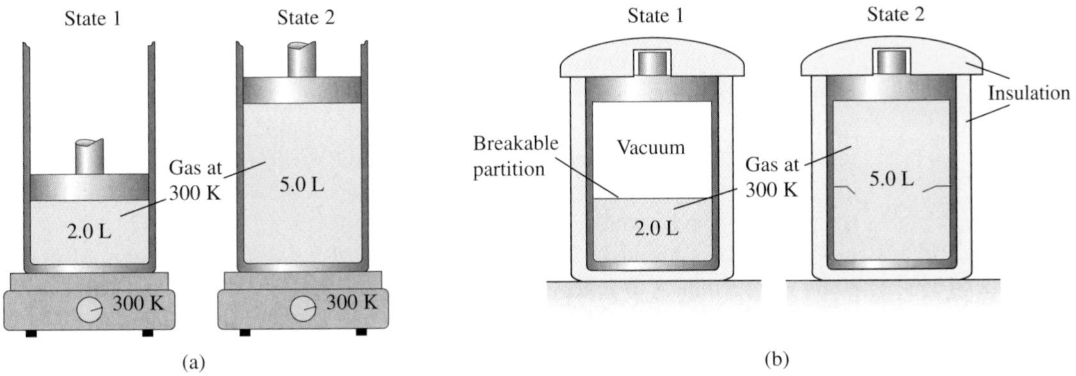

19.9 (a) Slow, controlled isothermal expansion of a gas from an initial state 1 to a final state 2 with the same temperature but lower pressure. (b) Rapid, uncontrolled expansion of the same gas starting at the same state 1 and ending at the same state 2.

of a gas into vacuum is called a **free expansion;** we will discuss it further in Section 19.6.

Experiments have shown that when an ideal gas undergoes a free expansion, there is no temperature change. Therefore the final state of the gas is the same as in Fig. 19.9a. The intermediate states (pressures and volumes) during the transition from state 1 to state 2 are entirely different in the two cases; Figs. 19.9a and 19.9b represent *two different paths* connecting the *same states* 1 and 2. For the path in Fig. 19.9b, *no* heat is transferred into the system, and the system does no work. Like work, *heat depends not only on the initial and final states but also on the path.*

Because of this path dependence, it would not make sense to say that a system "contains" a certain quantity of heat. To see this, suppose we assign an arbitrary value to "the heat in a body" in some reference state. Then presumably the "heat in the body" in some other state would equal the heat in the reference state plus the heat added when the body goes to the second state. But that's ambiguous, as we have just seen; the heat added depends on the *path* we take from the reference state to the second state. We are forced to conclude that there is *no* consistent way to define "heat in a body"; it is not a useful concept.

While it doesn't make sense to talk about "work in a body" or "heat in a body," it *does* make sense to speak of the amount of *internal energy* in a body. This important concept is our next topic.

Test Your Understanding

If there is one mole of gas in the systems shown in Fig. 19.9, how much work is done by the gas in the process shown in (a)? In the process shown in (b)?

19.4 | Internal Energy and the First Law of Thermodynamics

Internal energy is one of the most important concepts in thermodynamics. In Section 7.3, when we discussed energy changes for a body sliding with friction, we stated that warming a body increased its internal energy and that cooling the body decreased its internal energy. But what *is* internal energy? We can look at it in various ways; let's start with one based on the ideas of mechanics. Matter consists of atoms and molecules, and these are made up of particles having kinetic and potential energies. We *tentatively* define the **internal energy** of a system as the sum of the kinetic energies of all of its constituent particles, plus the sum of all the potential energies of interaction among these particles.

Note that internal energy does *not* include potential energy arising from the interaction between the system and its surroundings. If the system is a glass of water, placing it on a high shelf increases the gravitational potential energy arising from the interaction between the glass and the earth. But this has no effect on the interaction between the molecules of the water, and so the internal energy of the water does not change.

We use the symbol U for internal energy. (We used this same symbol in our study of mechanics to represent potential energy. You may have to remind yourself occasionally that U has a different meaning in thermodynamics.) During a change of state of the system the internal energy may change from an initial value U_1 to a final value U_2. We denote the change in internal energy as $\Delta U = U_2 - U_1$.

We know that heat transfer is energy transfer. When we add a quantity of heat Q to a system and it does no work during the process, the internal energy increases by an amount equal to Q; that is, $\Delta U = Q$. When a system does work W by expanding against its surroundings and no heat is added during the process, energy leaves the system and the internal energy decreases. That is, when W is positive, ΔU is negative, and vice versa. So $\Delta U = -W$. When *both* heat transfer and work occur, the *total* change in internal energy is

$$U_2 - U_1 = \Delta U = Q - W \qquad \text{(first law of thermodynamics)} \qquad (19.4)$$

We can rearrange this to the form

$$Q = \Delta U + W \qquad (19.5)$$

The message of Eq. (19.5) is that in general, when heat Q is added to a system, some of this added energy remains within the system, changing its internal energy by an amount ΔU; the remainder leaves the system again as the system does work W against its surroundings. Because W and Q may be positive, negative, or zero, ΔU can be positive, negative, or zero for different processes (Fig. 19.10).

Equation (19.4) or (19.5) is the **first law of thermodynamics.** It is a generalization of the principle of conservation of energy to include energy transfer through heat as well as mechanical work. As you will see in later chapters, this principle can be extended to ever broader classes of phenomena by identifying additional forms of energy and energy transfer. In every situation in which it seems that the total energy in all known forms is not conserved, it has been possible to identify a new form of energy such that the total energy, including the new form, *is* conserved. There is energy associated with electric fields, with magnetic fields, and, according to the theory of relativity, even with mass itself.

At the beginning of this discussion we tentatively defined internal energy in terms of microscopic kinetic and potential energies. This has drawbacks, however. Actually *calculating* internal energy in this way for any real system would be hopelessly complicated. Furthermore, this definition isn't an *operational* one because it doesn't describe how to determine internal energy from physical quantities that we can measure directly.

So let's look at internal energy in another way. Starting over, we define the *change* in internal energy ΔU during any change of a system as the quantity given by Eq. (19.4), $\Delta U = Q - W$. This *is* an operational definition, because we can measure Q and W. It does not define U itself, only ΔU. This is not a shortcoming, because we can *define* the internal energy of a system to have a specified value in

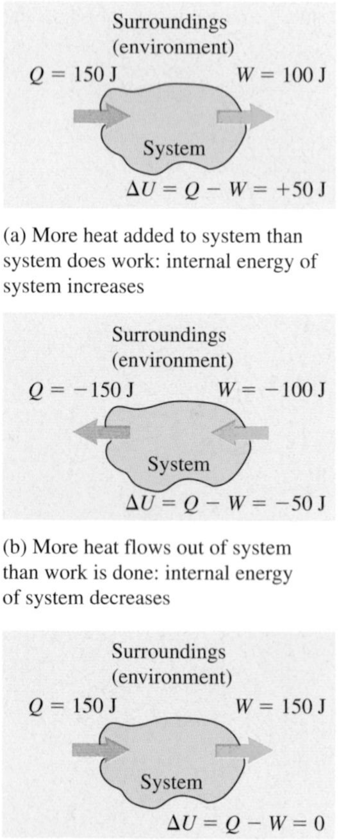

(a) More heat added to system than system does work: internal energy of system increases

(b) More heat flows out of system than work is done: internal energy of system decreases

(c) Heat added to system equals work done by system: internal energy of system unchanged

19.10 In a thermodynamic process, the internal energy U of a system may (a) increase $(\Delta U > 0)$, (b) decrease $(\Delta U < 0)$, or (c) remain the same $(\Delta U = 0)$.

some reference state, and then use Eq. (19.4) to define the internal energy in any other state. This is analogous to our treatment of potential energy in Chapter 7, in which we arbitrarily defined the potential energy of a mechanical system to be zero at a certain position.

This new definition trades one difficulty for another. If we define ΔU by Eq. (19.4), then when the system goes from state 1 to state 2 by two different paths, how do we know that ΔU is the same for the two paths? We have already seen that Q and W are, in general, *not* the same for different paths. If ΔU, which equals $Q - W$, is also path-dependent, then ΔU is ambiguous. If so, the concept of internal energy of a system is subject to the same criticism as the erroneous concept of quantity of heat in a system, as we discussed at the end of Section 19.3.

The only way to answer this question is through *experiment.* For various materials we measure Q and W for various changes of state and various paths to learn whether ΔU is or is not path-dependent. The results of many such investigations are clear and unambiguous: While Q and W depend on the path, $\Delta U = Q - W$ *is independent of path. The change in internal energy of a system during any thermodynamic process depends only on the initial and final states, not on the path leading from one to the other.*

Experiment, then, is the ultimate justification for believing that a thermodynamic system in a specific state has a unique internal energy that depends only on that state. An equivalent statement is that the internal energy U of a system is a function of the state coordinates p, V, and T (actually, any two of these, since the three variables are related by the equation of state).

To say that the first law of thermodynamics, given by Eq. (19.4) or (19.5), represents conservation of energy for thermodynamic processes is correct, as far as it goes. But an important *additional* aspect of the first law is the fact that internal energy depends only on the state of a system (Fig. 19.11). In changes of state, the change in internal energy is path-independent.

19.11 The internal energy of a cup of coffee just depends on its thermodynamic state—how much water and ground coffee it contains, and what its temperature is. It does not depend on the history of how the coffee was prepared—that is, the thermodynamic path that led to its current state.

All this may seem a little abstract if you are satisfied to think of internal energy as microscopic mechanical energy. There's nothing wrong with that view, and we will make use of it at various times during our discussion. But in the interest of precise *operational* definitions, internal energy, like heat, can and must be defined in a way that is independent of the detailed microscopic structure of the material.

Two special cases of the first law of thermodynamics are worth mentioning. A process that eventually returns a system to its initial state is called a *cyclic* process. For such a process, the final state is the same as the initial state, and so the *total* internal energy change must be zero. Then

$$U_2 = U_1 \quad \text{and} \quad Q = W$$

If a net quantity of work W is done by the system during this process, an equal amount of energy must have flowed into the system as heat Q. But there is no reason why either Q or W individually has to be zero (Fig. 19.12).

ActivPhysics ONLINE

8.6 Heat, Internal Energy, and First Law of Thermodynamics

Another special case occurs in an *isolated* system, one that does no work on its surroundings and has no heat flow to or from its surroundings. For any process taking place in an isolated system,

$$W = Q = 0$$

and therefore

$$U_2 = U_1 = \Delta U = 0$$

In other words, *the internal energy of an isolated system is constant.*

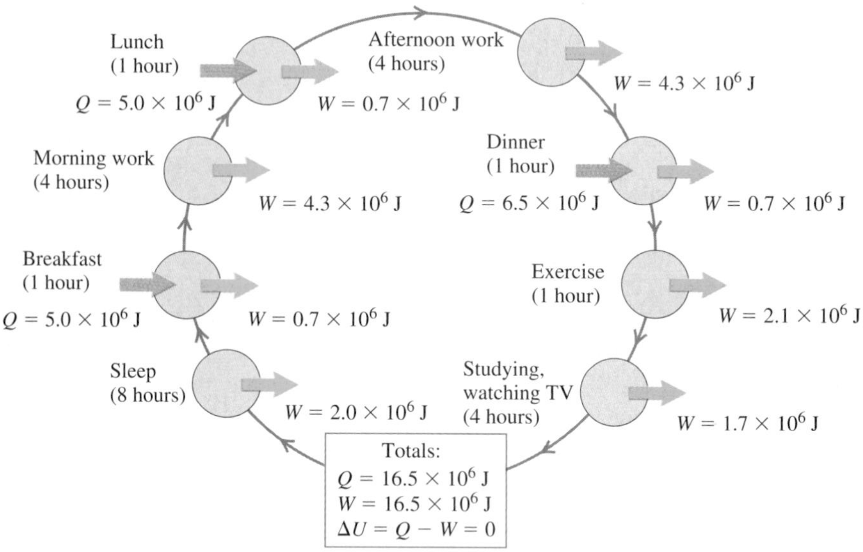

19.12 Every day, your body (a thermodynamic system) goes through a cyclic thermodynamic process like this one. Heat Q is added by metabolizing food, and your body does work W in breathing, walking, and other activities. If you return to the same state at the end of the day, $Q = W$ and the net change in your internal energy is zero.

Problem-Solving Strategy: The First Law of Thermodynamics

IDENTIFY *the relevant concepts:* The first law of thermodynamics is the statement of the law of conservation of energy in its most general form. You can apply it to *any* situation in which you are concerned with changes in the internal energy of a system, with heat flow into or out of a system, and/or with work done by or on a system.

SET UP *the problem using the following steps:*

1. Carefully define what the thermodynamic system is.
2. The first law of thermodynamics focuses on systems that go through thermodynamic processes. Some problems involve processes with more than one step, so make sure that you identify the initial and final state for each step.
3. Identify the known quantities and the target variables.
4. Check whether you have enough equations. The first law, $\Delta U = Q - W$, can be applied just once to each step in a thermodynamic process, so you will often need additional equations. These often include Eq. (19.2) for the work done in a volume change and the equation of state of the material that makes up the thermodynamic system (for an ideal gas, $pV = nRT$).

EXECUTE *the solution as follows:*

1. You shouldn't be surprised to be told that consistent units are essential. If p is in Pa and V in m^3, then W is in joules. Otherwise, you may want to convert the pressure and volume units into units of Pa and m^3. If a heat capacity is given in terms of calories, usually the simplest procedure is to convert it to joules. Be especially careful with moles. When you use $n = m_{tot}/M$ to convert between total mass and number of moles, remember that if m_{tot} is in kilograms, M must be in *kilograms* per mole. The usual units for M are *grams* per mole; be careful!
2. The internal energy change ΔU in any thermodynamic process or series of processes is independent of the path, whether the substance is an ideal gas or not. This point is of the utmost importance in the problems in this chapter and the next. Sometimes you will be given enough information about one path between the given initial and final states to calculate ΔU for that path. Since ΔU is the same for every possible path between the same two states, you can then relate the various energy quantities for other paths.
3. When a process consists of several distinct steps, it often helps to make a chart showing Q, W, and ΔU for each step. Put these quantities for each step on a different line, and arrange them so the Q's, W's, and ΔU's form columns. Then you can apply the first law to each line; in addition, you can add each column and apply the first law to the sums. Do you see why?
4. Using steps (1)–(3), solve for the target variables.

EVALUATE *your answer:* Check your results for reasonableness. In particular, make sure that each of your answers has the correct algebraic sign. Remember that a positive Q means that heat flows *into* the system, and that a negative Q means that heat flows *out of* the system. A positive W means that work is done *by* the system on its environment, while a negative W means that work is done *on* the system by its environment.

Example 19.2 Working off your dessert

You propose to eat a 900-calorie hot fudge sundae (with whipped cream) and then run up several flights of stairs to work off the energy you have taken in. How high do you have to climb? Assume that your mass is 60.0 kg.

SOLUTION

IDENTIFY: Eating the hot fudge sundae corresponds to a heat flow into your body, and running up stairs means that you do work. We can relate these quantities using the first law of thermodynamics.

SET UP: The system consists of your body. We are given that 900 food calories of heat flows into your body. The purpose of running up the stairs is to make sure that the final state of the system is the same as the initial state (no fatter, no leaner), so there is no net change in internal energy: $\Delta U = 0$. The work you must do to raise your mass m a height h is $W = mgh$; our target variable is h.

EXECUTE: Using the first law of thermodynamics, $\Delta U = 0 = Q - W$, so $W = Q$: the work that you do running up the stairs must just equal the heat input from the sundae. Using $W = mgh$, the height that you climb is $h = Q/mg$. Before substituting values into this equation, we first convert units: one food-value calorie is 1 kcal = 1000 cal = 4190 J (to three significant figures), so

$$Q = 900\ \text{kcal}\,(4190\ \text{J}/1\ \text{kcal}) = 3.77 \times 10^6\ \text{J}$$

Then

$$h = \frac{Q}{mg} = \frac{3.77 \times 10^6\ \text{J}}{(60.0\ \text{kg})(9.80\ \text{m/s}^2)}$$
$$= 6410\ \text{m} \qquad (\text{about } 21{,}000\ \text{ft})$$

EVALUATE: Good luck! We have assumed 100% efficiency in the conversion of food energy into mechanical work; this isn't very realistic. We'll talk more about efficiency later.

Example 19.3 A cyclic process

Figure 19.13 shows a pV-diagram for a *cyclic* process, one in which the initial and final states are the same. It starts at point a and proceeds counterclockwise in the pV-diagram to point b, then back to a, and the total work is $W = -500$ J. a) Why is the work negative? b) Find the change in internal energy and the heat added during this process.

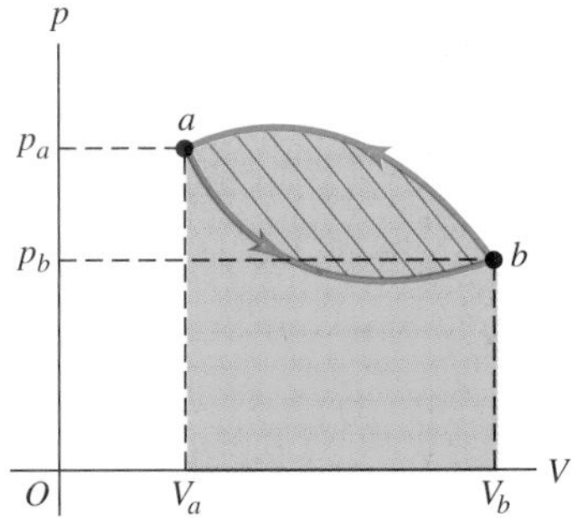

19.13 The net work done by the system in the process aba is -500 J. What would it have been if the process had proceeded clockwise in this pV-diagram?

SOLUTION

IDENTIFY and SET UP: The thermodynamic process here has two steps: $a \rightarrow b$ via the lower curve in Fig. 19.13 and $b \rightarrow a$ via the upper curve. But note that the questions in (a) and (b) are about the *entire* cyclic process $a \rightarrow b \rightarrow a$ (around the loop in Fig. 19.13).

EXECUTE: a) The work done equals the area under the curve, with the area taken as positive for increasing volume and negative for decreasing volume. The area under the lower curve from a to b is positive, but it is smaller than the absolute value of the negative area under the upper curve from b back to a. Therefore the net area (the area enclosed by the path, shown with red stripes) and the work are negative. In other words, 500 more joules of work is done *on* the system than *by* the system.
b) For this and any other cyclic process (in which the beginning and end points are the same), $\Delta U = 0$, so $Q = W = -500$ J. That is, 500 joules of heat must come *out of* the system.

EVALUATE: This example illustrates a general principle about pV-diagrams of cyclic processes: The total work is positive if the process goes around the cycle in a clockwise direction, and the total work is negative if the process goes around the cycle in a counterclockwise direction (as in Fig. 19.13).

Example 19.4 Comparing thermodynamic processes

A series of thermodynamic processes is shown in the pV-diagram of Fig. 19.14. In process ab, 150 J of heat is added to the system, and in process bd, 600 J of heat is added. Find a) the internal energy change in process ab; b) the internal energy change in process abd (shown in light blue); and c) the total heat added in process acd (shown in dark blue).

SOLUTION

IDENTIFY and SET UP: In each process we use $\Delta U = Q - W$. We are given $Q_{ab} = +150$ J and $Q_{bd} = +600$ J (both values are positive because heat is *added* to the system). Our target variables are (a) ΔU_{ab}, (b) ΔU_{abd}, and (c) Q_{acd}.

EXECUTE: a) No volume change occurs during process ab, so $W_{ab} = 0$ and $\Delta U_{ab} = Q_{ab} = 150$ J.

b) Process bd occurs at constant pressure, so the work done by the system during this expansion is

$$\begin{aligned} W_{bd} &= p(V_2 - V_1) \\ &= (8.0 \times 10^4\ \text{Pa})(5.0 \times 10^{-3}\ \text{m}^3 - 2.0 \times 10^{-3}\ \text{m}^3) \\ &= 240\ \text{J} \end{aligned}$$

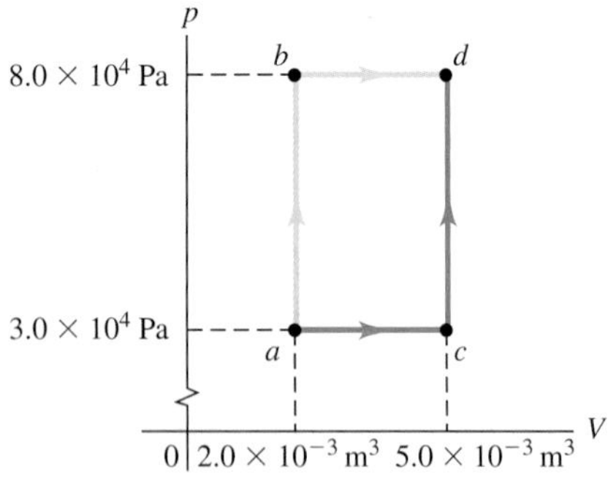

19.14 A pV-diagram showing the various thermodynamic processes.

The total work for process abd is

$$W_{abd} = W_{ab} + W_{bd} = 0 + 240\ \text{J} = 240\ \text{J}$$

and the total heat is

$$Q_{abd} = Q_{ab} + Q_{bd} = 150\ \text{J} + 600\ \text{J} = 750\ \text{J}$$

Applying Eq. (19.4) to abd, we find

$$\Delta U_{abd} = Q_{abd} - W_{abd} = 750\ \text{J} - 240\ \text{J} = 510\ \text{J}$$

c) Because ΔU is independent of path, the internal energy change is the same for path acd as for path abd; that is,

$$\Delta U_{acd} = \Delta U_{abd} = 510\ \text{J}$$

The total work for the path acd is

$$\begin{aligned} W_{acd} &= W_{ac} + W_{cd} = p(V_2 - V_1) + 0 \\ &= (3.0 \times 10^4\ \text{Pa})(5.0 \times 10^{-3}\ \text{m}^3 - 2.0 \times 10^{-3}\ \text{m}^3) \\ &= 90\ \text{J} \end{aligned}$$

Now we apply Eq. (19.5) to process acd:

$$Q_{acd} = \Delta U_{acd} + W_{acd} = 510\ \text{J} + 90\ \text{J} = 600\ \text{J}$$

Here is a tabulation of the various quantities:

STEP	Q	W	$\Delta U = Q - W$	STEP	Q	W	$\Delta U = Q - W$
ab	150 J	0 J	150 J	ac	?	90 J	?
bd	600 J	240 J	360 J	cd	?	0 J	?
abd	750 J	240 J	510 J	acd	600 J	90 J	510 J

EVALUATE: We see that although ΔU is the same (510 J) for abd and acd, W (240 J versus 90 J) and Q (750 J versus 600 J) are quite different for the two processes.

Notice that we don't have enough information to find Q or ΔU for the processes ac and cd. We were nonetheless able to analyze the composite process acd by comparing it to the process abd, which has the same initial and final states and for which we have more complete information.

Example 19.5 Thermodynamics of boiling water

One gram of water ($1\ \text{cm}^3$) becomes $1671\ \text{cm}^3$ of steam when boiled at a constant pressure of 1 atm (1.013×10^5 Pa). The heat of vaporization at this pressure is $L_v = 2.256 \times 10^6$ J/kg. Compute a) the work done by the water when it vaporizes and b) its increase in internal energy.

SOLUTION

IDENTIFY and SET UP: The water is boiled at a constant pressure, so we can calculate the work W done by the water using Eq. (19.3). We can calculate the heat Q added to the water from the mass and the heat of vaporization, and can then find the internal energy change using $\Delta U = Q - W$.

EXECUTE: a) From Eq. (19.2), the work done by the vaporizing water is

$$\begin{aligned} W &= p(V_2 - V_1) \\ &= (1.013 \times 10^5\ \text{Pa})(1671 \times 10^{-6}\ \text{m}^3 - 1 \times 10^{-6}\ \text{m}^3) \\ &= 169\ \text{J} \end{aligned}$$

b) Using Eq. (17.20), the heat added to the water to vaporize it is

$$Q = mL_v = (10^{-3}\ \text{kg})(2.256 \times 10^6\ \text{J/kg}) = 2256\ \text{J}$$

From the first law of thermodynamics, Eq. (19.4), the change in internal energy is

$$\Delta U = Q - W = 2256\ \text{J} - 169\ \text{J} = 2087\ \text{J}$$

EVALUATE: To vaporize one gram of water, we have to add 2256 J of heat. Most (2087 J) of this added energy remains in the system as an increase in internal energy. The remaining 169 J leaves the system again as it does work against the surroundings while expanding from liquid to vapor. The increase in internal energy is associated mostly with the intermolecular forces that hold the molecules together in the liquid state. These forces are attractive, so the associated potential energies are greater after work has been done to pull the molecules apart, forming the vapor state. It's like increasing gravitational potential energy by pulling an elevator farther from the center of the earth.

Infinitesimal Changes of State

In the preceding examples the initial and final states differ by a finite amount. Later we will consider *infinitesimal* changes of state in which a small amount of heat dQ is added to the system, the system does a small amount of work dW, and its internal energy changes by an amount dU. For such a process we state the first law in differential form as

$$dU = dQ - dW \qquad (19.6)$$

(first law of thermodynamics, infinitesimal process)

For the systems we will discuss, the work dW is given by $dW = p\,dV$, so we can also state the first law as

$$dU = dQ - p\,dV \qquad (19.7)$$

Test Your Understanding

Consider a thermodynamic process that goes directly from state a to state d in Example 19.4. In the pV-diagram of Fig. 19.14, this process is a straight line from a to d. What is the internal energy change for this process?

19.5 | Kinds of Thermodynamic Processes

In this section we describe four specific kinds of thermodynamic processes that occur often in practical situations. These can be summarized briefly as "no heat transfer" or *adiabatic,* "constant volume" or *isochoric,* "constant pressure" or *isobaric,* and "constant temperature" or *isothermal.* For some of these processes we can use a simplified form of the first law of thermodynamics.

Adiabatic Process

An **adiabatic process** (pronounced "ay-dee-ah-*bat*-ic") is defined as one with no heat transfer into or out of a system; $Q = 0$. We can prevent heat flow either by surrounding the system with thermally insulating material or by carrying out the process so quickly that there is not enough time for appreciable heat flow. From the first law we find that for every adiabatic process,

$$U_2 - U_1 = \Delta U = -W \qquad \text{(adiabatic process)} \tag{19.8}$$

ActivPhysics ONLINE

8.4 State Variables and Ideal Gas Law

When a system expands adiabatically, W is positive (the system does work on its surroundings), so ΔU is negative and the internal energy decreases. When a system is *compressed* adiabatically, W is negative (work is done on the system by its surroundings) and U increases. In many (but not all) systems an increase of internal energy is accompanied by a rise in temperature.

The compression stroke in an internal-combustion engine is an approximately adiabatic process. The temperature rises as the air–fuel mixture in the cylinder is compressed. The expansion of the burned fuel during the power stroke is also an approximately adiabatic expansion with a drop in temperature. In Section 19.8 we'll consider adiabatic processes in an ideal gas.

Isochoric Process

An **isochoric process** (pronounced "eye-so-*kor*-ic") is a *constant-volume* process. When the volume of a thermodynamic system is constant, it does no work on its surroundings. Then $W = 0$ and

$$U_2 - U_1 = \Delta U = Q \qquad \text{(isochoric process)} \tag{19.9}$$

In an isochoric process, all the energy added as heat remains in the system as an increase in internal energy. Heating a gas in a closed constant-volume container is an example of an isochoric process. The processes *ab* and *cd* in Example 19.4 are also examples of isochoric processes. (Note that there are types of work that do not involve a volume change. For example, we can do work on a fluid by stirring it. In some literature, "isochoric" is used to mean that no work of any kind is done.)

19.15 Most cooking involves isobaric processes. That's because the air pressure above a saucepan or frying pan, or inside a microwave oven, remains essentially constant while the food is being heated. (But see Fig. 19.20 for an example of an *adiabatic* process in the kitchen.)

Isobaric Process

An **isobaric process** (pronounced "eye-so-*bear*-ic") is a *constant-pressure* process. In general, none of the three quantities ΔU, Q, and W is zero in an isobaric process, but calculating W is easy nonetheless. From Eq. (19.3),

$$W = p(V_2 - V_1) \qquad \text{(isobaric process)} \tag{19.10}$$

Example 19.5 concerns an isobaric process, boiling water at constant pressure (Fig. 19.15).

Isothermal Process

An **isothermal process** is a *constant-temperature* process. For a process to be isothermal, any heat flow into or out of the system must occur slowly enough that thermal equilibrium is maintained. In general, none of the quantities ΔU, Q, or W is zero in an isothermal process.

In some special cases the internal energy of a system depends *only* on its temperature, not on its pressure or volume. The most familiar system having this special property is an ideal gas, as we'll discuss in the next section. For such systems, if the temperature is constant, the internal energy is also constant; $\Delta U = 0$ and $Q = W$. That is, any energy entering the system as heat Q must leave it again as work W done by the system. Example 19.1, involving an ideal gas, is an example

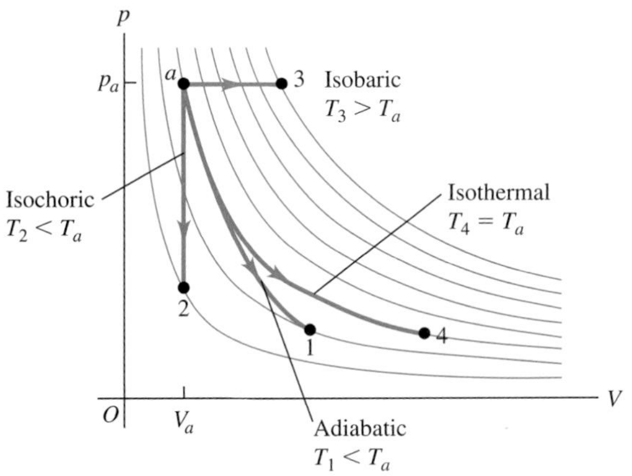

19.16 Four different processes for a constant amount of an ideal gas, all starting at state a. For the adiabatic process, $Q = 0$; for the isochoric process, $W = 0$; and for the isothermal process, $\Delta U = 0$. The temperature increases only during the isobaric expansion.

of an isothermal process in which U is also constant. For most systems other than ideal gases the internal energy depends on pressure as well as temperature, so U may vary even when T is constant.

Figure 19.16 shows a pV-diagram for each of these four processes for a constant amount of an ideal gas. The path followed in an adiabatic process (a to 1) is called an **adiabat.** A vertical line (constant volume) is an **isochor,** a horizontal line (constant pressure) is an **isobar,** and a curve of constant temperature (shown as light blue lines in Fig. 19.16) is an **isotherm.**

Test Your Understanding

Which of the processes in Fig. 19.8 are isochoric? Which are isobaric? Is it possible to tell if any of the processes are isothermal or adiabatic?

19.6 | Internal Energy of an Ideal Gas

We now show that for an ideal gas, the internal energy U depends only on temperature, not on pressure or volume. Let's think again about the free-expansion experiment described in Section 19.3. A thermally insulated container with rigid walls is divided into two compartments by a partition (Fig. 19.17). One compartment has a quantity of an ideal gas and the other is evacuated.

When the partition is removed or broken, the gas expands to fill both parts of the container. The gas does no work on its surroundings because the walls of the container don't move, and there is no heat flow through the insulation. So both Q and W are zero and the internal energy U is constant. This is true of any substance, whether it is an ideal gas or not.

Does the *temperature* change during a free expansion? Suppose it *does* change, while the internal energy stays the same. In that case we have to conclude that the internal energy depends on both the temperature and the volume or on both the temperature and the pressure, but certainly not on the temperature alone. But if T is constant during a free expansion, for which we know that U is constant even though both p and V change, then we have to conclude that U depends only on T, not on p or V.

Many experiments have shown that when a low-density gas undergoes a free expansion, its temperature *does not* change. Such a gas is essentially an ideal gas.

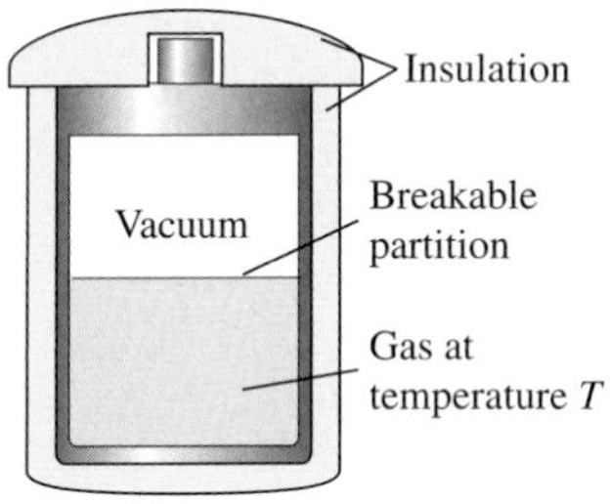

19.17 The partition is broken (or removed) to start the free expansion of gas into the vacuum region.

The conclusion is: **The internal energy of an ideal gas depends only on its temperature, not on its pressure or volume.** This property, in addition to the ideal-gas equation of state, is part of the ideal-gas model. Make sure that you understand that U depends only on T for an ideal gas, for we will make frequent use of this fact.

For non-ideal gases, some temperature change occurs during free expansions, even though the internal energy is constant. This shows that the internal energy cannot depend *only* on temperature; it must depend on pressure as well. From the microscopic viewpoint, in which internal energy U is the sum of the kinetic and potential energies for all the particles that make up the system, this is not surprising. Non-ideal gases usually have attractive intermolecular forces, and when molecules move farther apart, the associated potential energies increase. If the total internal energy is constant, the kinetic energies must decrease. Temperature is directly related to molecular *kinetic* energy, and for such a gas a free expansion is usually accompanied by a *drop* in temperature.

Test Your Understanding

Is the internal energy of a solid likely to be independent of its volume, as is the case for an ideal gas? Explain your reasoning. (*Hint:* See Fig. 18.17.)

19.7 | Heat Capacities of an Ideal Gas

We defined specific heat and molar heat capacity in Section 17.5. We also remarked at the end of that section that the specific heat or molar heat capacity of a substance depends on the conditions under which the heat is added. It is usually easiest to measure the heat capacity of a gas in a closed container under constant-volume conditions. The corresponding heat capacity is the **molar heat capacity at constant volume,** denoted by C_V. Heat capacity measurements for solids and liquids are usually carried out in the atmosphere under constant atmospheric pressure, and we call the corresponding heat capacity the **molar heat capacity at constant pressure,** C_p. If neither p nor V is constant, we have an infinite number of possible heat capacities.

Let's consider C_V and C_p for an ideal gas. To measure C_V, we raise the temperature of an ideal gas in a rigid container with constant volume (neglecting its thermal expansion). To measure C_p, we let the gas expand just enough to keep the pressure constant as the temperature rises.

Why should these two molar heat capacities be different? The answer lies in the first law of thermodynamics. In a constant-volume temperature increase, the system does no work, and the change in internal energy ΔU equals the heat added Q. In a constant-pressure temperature increase, on the other hand, the volume *must* increase; otherwise, the pressure (given by the ideal-gas equation of state $p = nRT/V$) could not remain constant. As the material expands, it does an amount of work W. According to the first law,

$$Q = \Delta U + W \tag{19.11}$$

For a given temperature increase, the internal energy change ΔU of an ideal gas has the same value no matter what the process (remember that the internal energy of an ideal gas depends only on temperature, not pressure or volume). Equation (19.11) then shows that the heat input for a constant-pressure process must be *greater* than that for a constant-volume process because additional energy must be supplied to account for the work W done during the expansion. So C_p is greater than C_V for an ideal gas. The pV-diagram in Fig. 19.18 shows this relationship. For air, C_p is 40% greater than C_V.

For a very few substances (one of which is water between 0°C and 4°C) the volume *decreases* during heating. In this case, W is negative, the heat input is *less* than in the constant-volume case, and C_p is *less* than C_V.

We can derive a simple relation between C_p and C_V for an ideal gas. First consider the constant-*volume* process. We place n moles of an ideal gas at temperature T in a constant-volume container. We place it in thermal contact with a hotter body; an infinitesimal quantity of heat dQ flows into the gas, and its temperature increases by an infinitesimal amount dT. By the definition of C_V, the molar heat capacity at constant volume,

$$dQ = nC_V\,dT \tag{19.12}$$

The pressure increases during this process, but the gas does no work $(dW = 0)$ because the volume is constant. The first law in differential form, Eq. (19.6), is $dQ = dU + dW$. Since $dW = 0$, $dQ = dU$ and Eq. (19.12) can also be written as

$$dU = nC_V\,dT \tag{19.13}$$

Now consider a constant-*pressure* process with the same temperature change dT. We place the same gas in a cylinder with a piston that we can allow to move just enough to maintain constant pressure. Again we bring the system into contact with a hotter body. As heat flows into the gas, it expands at constant pressure and does work. By the definition of C_p, the molar heat capacity at constant pressure, the amount of heat dQ entering the gas is

$$dQ = nC_p\,dT \tag{19.14}$$

The work dW done by the gas in this constant-pressure process is

$$dW = p\,dV$$

We can also express dW in terms of the temperature change dT by using the ideal-gas equation of state, $pV = nRT$. Because p is constant, the change in V is proportional to the change in T:

$$dW = p\,dV = nR\,dT \tag{19.15}$$

Now we substitute Eqs. (19.14) and (19.15) into the first law, $dQ = dU + dW$. We obtain

$$nC_p\,dT = dU + nR\,dT \tag{19.16}$$

Now here comes the crux of the calculation. The internal energy change dU for the constant-pressure process is again given by Eq. (19.13), $dU = nC_V\,dT$, *even though now the volume is not constant.* Why is this so? Recall the discussion of Section 19.6; one of the special properties of an ideal gas is that its internal energy depends *only* on temperature. Thus the *change* in internal energy during any process must be determined only by the temperature change. If Eq. (19.13) is valid for an ideal gas for one particular kind of process, it must be valid for an ideal gas for *every* kind of process with the same dT. So we may replace dU in Eq. (19.16) by $nC_V\,dT$:

$$nC_p\,dT = nC_V\,dT + nR\,dT$$

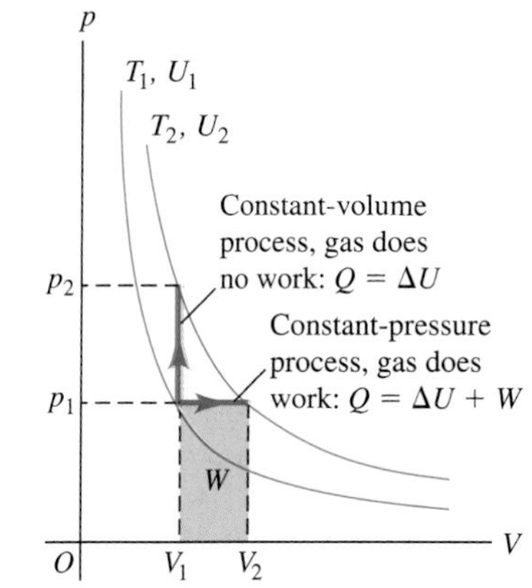

19.18 Raising the temperature of an ideal gas from T_1 to T_2 by a constant-volume or a constant-pressure process. For an ideal gas, U depends only on T, so ΔU is the same for both processes. But for the constant-pressure process, more heat Q must be added to both increase U and do work W. Hence $C_p > C_V$.

ActivPhysics Online

8.7 Heat Capacity
8.8 Isochoric Process
8.9 Isobaric Process
8.10 Isothermal Process

When we divide each term by the common factor $n\,dT$, we get

$$C_p = C_V + R \qquad \text{(molar heat capacities of an ideal gas)} \qquad (19.17)$$

As we predicted, the molar heat capacity of an ideal gas at constant pressure is *greater* than the molar heat capacity at constant volume; the difference is the gas constant R. (Of course, R must be expressed in the same units as C_p and C_V, such as J/mol·K.)

We have used the ideal-gas model to derive Eq. (19.17), but it turns out to be obeyed to within a few percent by many real gases at moderate pressures. Measured values of C_p and C_V are given in Table 19.1 for several real gases at low pressures; the difference in most cases is approximately $R = 8.314$ J/mol·K.

Table 19.1 Molar Heat Capacities of Gases at Low Pressure

Type of Gas	Gas	C_V (J/mol·K)	C_p (J/mol·K)	$C_p - C_V$ (J/mol·K)	$\gamma = C_p/C_V$
Monatomic	He	12.47	20.78	8.31	1.67
	Ar	12.47	20.78	8.31	1.67
Diatomic	H_2	20.42	28.74	8.32	1.41
	N_2	20.76	29.07	8.31	1.40
	O_2	20.85	29.17	8.31	1.40
	CO	20.85	29.16	8.31	1.40
Polyatomic	CO_2	28.46	36.94	8.48	1.30
	SO_2	31.39	40.37	8.98	1.29
	H_2S	25.95	34.60	8.65	1.33

The table also shows that the molar heat capacity of a gas is related to its molecular structure, as we discussed in Section 18.4. In fact, the first two columns of Table 19.1 are the same as Table 18.1.

The last column of Table 19.1 lists the values of the dimensionless **ratio of heat capacities,** C_p/C_V, denoted by the Greek letter γ ("gamma"):

$$\gamma = \frac{C_p}{C_V} \qquad \text{(ratio of heat capacities)} \qquad (19.18)$$

(This is sometimes called the "ratio of specific heats.") For gases, C_p is always greater than C_V and γ is always greater than unity. This quantity plays an important role in *adiabatic* processes for an ideal gas, which we will study in the next section.

We can use our kinetic-theory discussion of the molar heat capacity of an ideal gas (Section 18.4) to predict values of γ. As an example, an ideal monatomic gas has $C_V = \frac{3}{2}R$. From Eq. (19.17),

$$C_p = C_V + R = \frac{3}{2}R + R = \frac{5}{2}R$$

so

$$\gamma = \frac{C_p}{C_V} = \frac{\frac{5}{2}R}{\frac{3}{2}R} = \frac{5}{3} = 1.67$$

As Table 19.1 shows, this agrees well with values of γ computed from measured heat capacities. For most diatomic gases near room temperature, $C_V = \frac{5}{2}R$, $C_p = C_V + R = \frac{7}{2}R$, and

$$\gamma = \frac{C_p}{C_V} = \frac{\frac{7}{2}R}{\frac{5}{2}R} = \frac{7}{5} = 1.40$$

also in good agreement with measured values.

Here's a final reminder: For an ideal gas the internal energy change in *any* process is given by $\Delta U = nC_V\,\Delta T$, *whether the volume is constant or not.* This relation, which comes in handy in the following example, holds for other substances *only* when the volume is constant.

Example 19.6 Cooling your room

A typical dorm room or bedroom contains about 2500 moles of air. Find the change in the internal energy of this much air when it is cooled from 23.9°C to 11.6°C at a constant pressure of 1.00 atm. Treat the air as an ideal gas with $\gamma = 1.400$.

SOLUTION

IDENTIFY and SET UP: This is a constant-pressure process. Your first impulse may be to find C_p and then calculate Q from $Q = nC_p\,\Delta T$; find the volume change and find the work done by the gas from $W = p\,\Delta V$; then finally use the first law to find ΔU. This would be perfectly correct, but there's a much easier way. For an ideal gas the internal-energy change is $\Delta U = nC_V\,\Delta T$ for *every* process, *whether the volume is constant or not.* So all we have to do is find C_V and use this expression for ΔU. We are given the number of moles and the initial and final temperatures.

EXECUTE: We are given the value of γ for air, so we use Eqs. (19.17) and (19.18) to determine C_V:

$$\gamma = \frac{C_p}{C_V} = \frac{C_V + R}{C_V} = 1 + \frac{R}{C_V}$$

$$C_V = \frac{R}{\gamma - 1} = \frac{8.314\ \text{J/mol}\cdot\text{K}}{1.400 - 1} = 20.79\ \text{J/mol}\cdot\text{K}$$

Then

$$\begin{aligned}\Delta U &= nC_V\,\Delta T\\ &= (2500\ \text{mol})(20.79\ \text{J/mol}\cdot\text{K})(11.6^\circ\text{C} - 23.9^\circ\text{C})\\ &= -6.39 \times 10^5\ \text{J}\end{aligned}$$

EVALUATE: A room air conditioner must extract this much internal energy from the air in your room and transfer it to the air outside. We'll discuss how this is done in Chapter 20.

Test Your Understanding

Would it be easier to cool your room (Example 19.6) if air were an ideal monatomic gas? If so, how much easier would it be?

19.8 | Adiabatic Processes for an Ideal Gas

An adiabatic process, defined in Section 19.5, is a process in which no heat transfer takes place between a system and its surroundings. Zero heat transfer is an idealization, but a process is approximately adiabatic if the system is well insulated or if the process takes place so quickly that there is not enough time for appreciable heat flow to occur.

In an adiabatic process, $Q = 0$, so from the first law, $\Delta U = -W$. An adiabatic process for an ideal gas is shown in the pV-diagram of Fig. 19.19. As the gas expands from volume V_a to V_b, it does positive work, so its internal energy decreases and its temperature drops. If point a, representing the initial state, lies on an isotherm at

ActivPhysics Online
8.11 Adiabatic Process

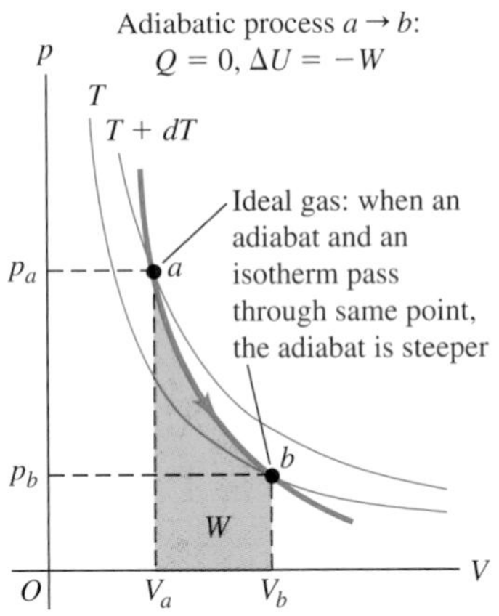

19.19 A pV-diagram of an adiabatic $(Q = 0)$ process for an ideal gas. As the gas expands from V_a to V_b, it does positive work W on its environment, its internal energy decreases $(\Delta U = -W < 0)$, and its temperature drops from $T + dT$ to T. (An adiabatic process is also shown in Fig. 19.16.)

temperature $T + dT$, then point b for the final state is on a different isotherm at a lower temperature T. For an ideal gas an adiabatic curve (adiabat) at any point is always *steeper* than the isotherm passing through the same point. For an adiabatic *compression* from V_b to V_a the situation is reversed and the temperature rises.

The air in the output pipes of air compressors used in gasoline stations, in paint-spraying equipment, and to fill scuba tanks is always warmer than the air entering the compressor; this is because the compression is rapid and hence approximately adiabatic. Adiabatic *cooling* occurs when you open a bottle of your favorite carbonated beverage. The gas just above the beverage surface expands rapidly in a nearly adiabatic process; the temperature of the gas drops so much that water vapor in the gas condenses, forming a miniature cloud.

CAUTION Keep in mind that when we talk about "adiabatic heating" and "adiabatic cooling," we really mean "raising the temperature" and "lowering the temperature," respectively. In an adiabatic process, the temperature change is due to work done by or on the system; there is *no* heat flow at all.

We can derive a relation between volume and temperature changes for an infinitesimal adiabatic process in an ideal gas. Equation (19.13) gives the internal energy change dU for *any* process for an ideal gas, adiabatic or not, so we have $dU = nC_V\,dT$. Also, the work done by the gas during the process is given by $dW = p\,dV$. Then, since $dU = -dW$ for an adiabatic process, we have

$$nC_V\,dT = -p\,dV \tag{19.19}$$

To obtain a relation containing only the volume V and temperature T, we eliminate p using the ideal-gas equation in the form $p = nRT/V$. Substituting this into Eq. (19.19) and rearranging, we get

$$nC_V\,dT = -\frac{nRT}{V}\,dV$$

$$\frac{dT}{T} + \frac{R}{C_V}\frac{dV}{V} = 0$$

The coefficient R/C_V can be expressed in terms of $\gamma = C_p/C_V$. We have

$$\frac{R}{C_V} = \frac{C_p - C_V}{C_V} = \frac{C_p}{C_V} - 1 = \gamma - 1$$

$$\frac{dT}{T} + (\gamma - 1)\frac{dV}{V} = 0 \tag{19.20}$$

19.20 Hot steam escapes from the top of this pressure cooker at high speed. Hence it has no time to exchange heat with its surroundings, and its expansion is nearly adiabatic. As the steam's volume increases, its temperature drops so much [see Eq. (19.22)] that it feels cool on this chef's hand.

Because γ is always greater than unity for a gas, $(\gamma - 1)$ is always positive. This means that in Eq. (19.20), dV and dT always have opposite signs. An adiabatic *expansion* of an ideal gas $(dV > 0)$ always occurs with a *drop* in temperature $(dT < 0)$, and an adiabatic *compression* $(dV < 0)$ always occurs with a *rise* in temperature $(dT > 0)$; this confirms our earlier prediction (Fig. 19.20).

For finite changes in temperature and volume we integrate Eq. (19.20), obtaining

$$\ln T + (\gamma - 1)\ln V = \text{constant}$$

$$\ln T + \ln V^{\gamma-1} = \text{constant}$$

$$\ln(TV^{\gamma-1}) = \text{constant}$$

and finally,

$$TV^{\gamma-1} = \text{constant} \tag{19.21}$$

Thus for an initial state (T_1, V_1) and a final state (T_2, V_2),

$$T_1V_1^{\gamma-1} = T_2V_2^{\gamma-1} \quad \text{(adiabatic process, ideal gas)} \tag{19.22}$$

Because we have used the ideal-gas equation in our derivation of Eqs. (19.21) and (19.22), the T's must always be *absolute* (Kelvin) temperatures.

We can also convert Eq. (19.21) into a relation between pressure and volume by eliminating T, using the ideal-gas equation in the form $T = pV/nR$. Substituting this into Eq. (19.21), we find

$$\frac{pV}{nR}V^{\gamma-1} = \text{constant}$$

or, because n and R are constant,

$$pV^{\gamma} = \text{constant} \tag{19.23}$$

For an initial state (p_1, V_1) and a final state (p_2, V_2), Eq. (19.23) becomes

$$p_1V_1^{\gamma} = p_2V_2^{\gamma} \quad \text{(adiabatic process, ideal gas)} \tag{19.24}$$

We can also calculate the *work* done by an ideal gas during an adiabatic process. We know that $Q = 0$ and $W = -\Delta U$ for *any* adiabatic process. For an ideal gas, $\Delta U = nC_V(T_2 - T_1)$. If the number of moles n and the initial and final temperatures T_1 and T_2 are known, we have simply

$$W = nC_V(T_1 - T_2) \quad \text{(adiabatic process, ideal gas)} \tag{19.25}$$

We may also use $pV = nRT$ in this equation to obtain

$$W = \frac{C_V}{R}(p_1V_1 - p_2V_2) = \frac{1}{\gamma - 1}(p_1V_1 - p_2V_2) \quad \text{(adiabatic process, ideal gas)} \tag{19.26}$$

(We used the result $C_V = R/(\gamma - 1)$ from Example 19.6.) If the process is an expansion, the temperature drops, T_1 is greater than T_2, p_1V_1 is greater than p_2V_2, and the work is *positive*, as we should expect. If the process is a compression, the work is negative.

Throughout this analysis of adiabatic processes we have used the ideal-gas equation of state, which is valid only for *equilibrium* states. Strictly speaking, our results are valid only for a process that is fast enough to prevent appreciable heat exchange with the surroundings (so that $Q = 0$ and the process is adiabatic), yet slow enough that the system does not depart very much from thermal and mechanical equilibrium. Even when these conditions are not strictly satisfied, though, Eqs. (19.22), (19.24), and (19.26) give useful approximate results.

Example 19.7 Adiabatic compression in a diesel engine

The compression ratio of a diesel engine is 15 to 1; this means that air in the cylinders is compressed to $\frac{1}{15}$ of its initial volume (Fig. 19.21). If the initial pressure is 1.01×10^5 Pa and the initial temperature is 27°C (300 K), find the final pressure and the temperature after compression. Air is mostly a mixture of diatomic oxygen and nitrogen; treat it as an ideal gas with $\gamma = 1.40$.

SOLUTION

IDENTIFY and SET UP: We are given the initial pressure $p_1 = 1.01 \times 10^5$ Pa and the initial temperature $T_1 = 300$ K, and we are told that the ratio of initial and final volumes is $V_1/V_2 = 15$. Since the gas is ideal and the process is adiabatic, we can find the final temperature T_2 using Eq. (19.22) and the final pressure p_2 using Eq. (19.24).

EXECUTE: From Eq. (19.22),

$$T_2 = T_1\left(\frac{V_1}{V_2}\right)^{\gamma-1} = (300\text{ K})(15)^{0.40} = 886\text{ K} = 613°\text{C}$$

From Eq. (19.24),

$$p_2 = p_1\left(\frac{V_1}{V_2}\right)^{\gamma} = (1.01 \times 10^5\text{ Pa})(15)^{1.40}$$
$$= 44.8 \times 10^5\text{ Pa} = 44\text{ atm}$$

EVALUATE: If the compression had been isothermal, the final pressure would have been 15 atm, but because the temperature also increases during an adiabatic compression, the final pressure is much greater. When fuel is injected into the cylinders near the end of the compression stroke, the high temperature of the air attained during compression causes the fuel to ignite spontaneously without the need for spark plugs.

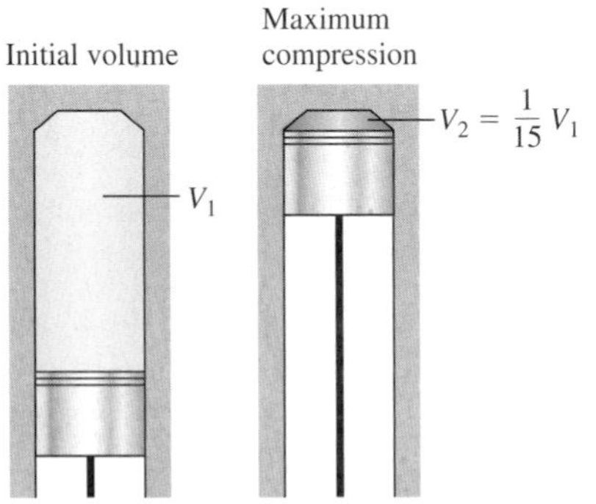

19.21 Adiabatic compression of air in a cylinder of a diesel engine.

Example 19.8 Work done in an adiabatic process

In Example 19.7, how much work does the gas do during the compression if the initial volume of the cylinder is $1.00\text{ L} = 1.00 \times 10^{-3}\text{ m}^3$? Assume that C_V for air is 20.8 J/mol·K and $\gamma = 1.40$.

SOLUTION

IDENTIFY and SET UP: We use Eq. (19.25) to determine the work done. We are given the initial temperature T_1 and we calculated the final temperature T_2 in Example 19.7. We are not given the number of moles n, but we can calculate it from the given information using the ideal-gas law $pV = nRT$.

EXECUTE: The number of moles is

$$n = \frac{p_1V_1}{RT_1} = \frac{(1.01 \times 10^5\text{ Pa})(1.00 \times 10^{-3}\text{ m}^3)}{(8.314\text{ J/mol}\cdot\text{K})(300\text{ K})}$$
$$= 0.0405\text{ mol}$$

and Eq. (19.25) gives

$$W = nC_V(T_1 - T_2)$$
$$= (0.0405\text{ mol})(20.8\text{ J/mol}\cdot\text{K})(300\text{ K} - 886\text{ K})$$
$$= -494\text{ J}$$

EVALUATE: We can check our result using Eq. (19.26), the alternative expression for work done by an ideal gas in an adiabatic process:

$$W = \frac{1}{\gamma - 1}(p_1V_1 - p_2V_2)$$
$$= \frac{1}{1.40 - 1}\left[\begin{array}{l}(1.01 \times 10^5\text{ Pa})(1.00 \times 10^{-3}\text{ m}^3)\\ -(44.8 \times 10^5\text{ Pa})\left(\dfrac{1.00 \times 10^{-3}\text{ m}^3}{15}\right)\end{array}\right]$$
$$= -494\text{ J}$$

The work is negative because the gas is compressed.

Test Your Understanding

If you compress a certain quantity of a monatomic ideal gas to one-half of its initial volume, by what factor does the pressure change if the compression is done isothermally? If it is done adiabatically? Would these answers be different if the gas were diatomic?

SUMMARY

A thermodynamic system has the potential to exchange energy with its surroundings by heat transfer or by mechanical work. When a system at pressure p changes volume from V_1 to V_2, it does an amount of work W given by the integral of p with respect to volume. If the pressure is constant, the work done is equal to p times the change in volume. A negative value of W means that work is done on the system. (See Example 19.1)

$$W = \int_{V_1}^{V_2} p\,dV \qquad (19.2)$$

$$W = p(V_2 - V_1) \qquad (19.3)$$
(constant pressure only)

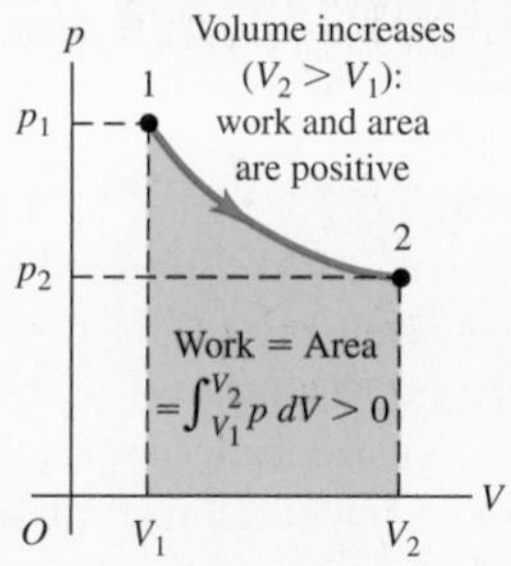

In any thermodynamic process, the heat added to the system and the work done by the system depend not only on the initial and final states, but also on the path (the series of intermediate states through which the system passes).

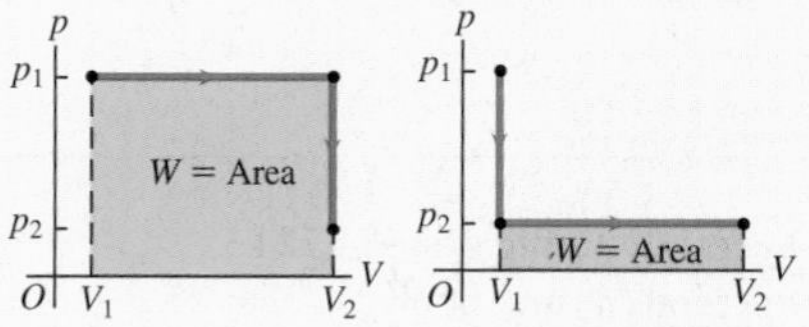

The first law of thermodynamics states that when heat Q is added to a system while it does work W, the internal energy U changes by an amount equal to $Q - W$. This law can also be expressed for an infinitesimal process. (See Examples 19.2, 19.3, and 19.5)

$$\Delta U = Q - W \qquad (19.4)$$

$$dU = dQ - dW \qquad (19.6)$$
(infinitesimal process)

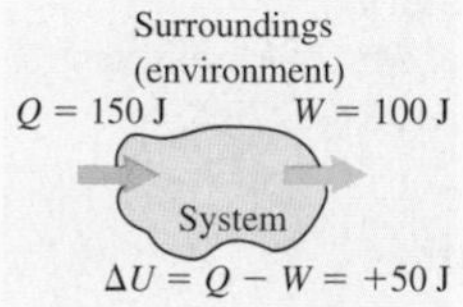

The internal energy of any thermodynamic system depends only on its state. The change in internal energy in any process depends only on the initial and final states, not on the path. The internal energy of an isolated system is constant. (See Example 19.4)

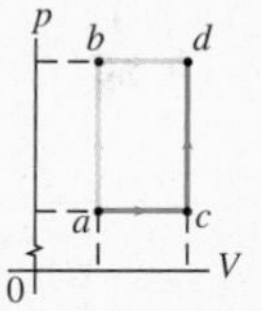

Some important types of thermodynamic processes:

- Adiabatic process: No heat transfer in or out of a system; $Q = 0$.
- Isochoric process: Constant volume; $W = 0$.
- Isobaric process: Constant pressure; $W = p(V_2 - V_1)$.
- Isothermal process: Constant temperature.

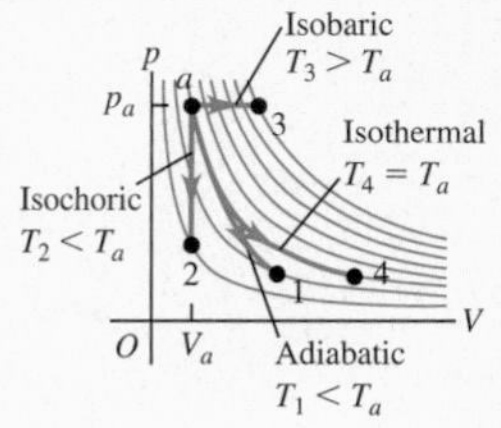

The internal energy of an ideal gas depends only on its temperature, not its pressure or volume. For other substances the internal energy generally depends on both pressure and temperature.

The molar heat capacities C_V and C_p of an ideal gas differ by R, the ideal gas constant. The dimensionless ratio of heat capacities, C_p/C_V, is denoted by γ. (See Example 19.6)

$$C_p = C_V + R \quad (19.17)$$

$$\gamma = \frac{C_p}{C_V} \quad (19.18)$$

For an adiabatic process for an ideal gas, the quantities $TV^{\gamma-1}$ and pV^{γ} are constant. The work done by an ideal gas during an adiabatic expansion can be expressed in terms of the initial and final values of temperature, or in terms of the initial and final values of pressure and volume. (See Examples 19.7 and 19.8)

$$W = nC_V(T_1 - T_2) = \frac{C_V}{R}(p_1V_1 - p_2V_2) = \frac{1}{\gamma - 1}(p_1V_1 - p_2V_2) \quad (19.25), (19.26)$$

Key Terms

thermodynamic system, 724
thermodynamic process, 724
path, 728
free expansion, 729
internal energy, 729
first law of thermodynamics, 730
adiabatic process, 735
isochoric process, 736
isobaric process, 736
isothermal process, 736
adiabat, 737
isochor, 737
isobar, 737
isotherm, 737
molar heat capacity at constant volume, 738
molar heat capacity at constant pressure, 738
ratio of heat capacities, 740

Your Notes

Answer to Chapter Opening Question ?

No. The work done by a gas as its volume changes from V_1 to V_2 is equal to the integral $\int p\,dV$ between those two volume limits. If the volume of the gas contracts, the final volume V_2 is less than the initial volume V_1 and the gas does negative work. Propelling the locomotive requires that the gas do positive work, so the gas doesn't contribute to propulsion while contracting.

Answers to Test Your Understanding Questions

Section 19.1 Heat flows out of the coffee, so $Q_{\text{coffee}} < 0$; heat flows into the aluminum cup, so $Q_{\text{aluminum}} > 0$. In mechanics, we would say that negative work is done *on* the truck, since the car exerts a force on the truck that opposes the truck's motion. But in thermodynamics we use the opposite convention and say that $W > 0$, which means that positive work is done *by* the truck on the car (thus pushing the car forward).

Section 19.2 We use the expression $W = nRT\ln(V_2/V_1)$. If $W = nRT$, then $\ln(V_2/V_1) = 1$ and $V_2/V_1 = e = 2.718$. If $W = 2nRT$, then $\ln(V_2/V_1) = 2$ and $V_2/V_1 = e^2 = 7.389$. Note that the expansion ratio V_2/V_1 must more than double in order to double the amount of work done.

Section 19.3 The situation in Fig. 19.9a is an isothermal expansion, so from Example 19.1 the gas does work $W = nRT\ln(V_2/V_1) = (1.00\text{ mol})\ (8.314\text{ J/mol}\cdot\text{K})\ (300\text{ K})\ \ln[(5.0\text{ L})\,/\,(2.0\text{ L})] = 2.3 \times 10^3\text{ J}$. The gas does zero work in the free expansion shown in Fig. 19.9b.

Section 19.4 The internal energy change in a thermodynamic process depends only on the initial and final states, not on the particular path taken. Hence the internal energy change for the process *ad* is the same as for either of the processes *abd* or *acd* in Example 19.4: $\Delta U_{ad} = 510\text{ J}$.

Section 19.5 In a *pV*-diagram like those shown in Fig. 19.8, isochoric processes are represented by vertical lines (lines of constant volume) and isobaric processes are represented by horizontal lines (lines of constant pressure). Hence the processes $1 \rightarrow 4$ and $3 \rightarrow 2$ are isochoric and the processes $1 \rightarrow 3$ and $4 \rightarrow 2$ are isobaric. The process $1 \rightarrow 2$ in Fig. 19.8 is shown as a curved line, which superficially resembles the adiabatic and isothermal processes for an ideal gas in Fig. 19.16. Without more information we can't tell whether process $1 \rightarrow 2$ is isothermal, adiabatic, or neither.

Section 19.6 Using the model of a solid in Fig. 18.17, we can see that the internal energy of a solid *does* depend on its volume. Compressing the solid means compressing the "springs" between the atoms, thereby increasing their stored potential energy and hence the internal energy of the solid.

Section 19.7 The molar heat capacity of an ideal monatomic gas at constant volume is $C_V = \frac{3}{2}R = 12.47\text{ J/mol}\cdot\text{K}$, which is $\frac{3}{5}$ as great as the actual value of C_V for air calculated in Example 19.6. Hence, for the same number of moles n and temperature change ΔT, the amount of energy that would have to be extracted would be only $\frac{3}{5}$ as much.

Section 19.8 In an isothermal process, $pV = \text{constant}$. If the volume decreases to one-half of its initial value, the pressure must double. By contrast, in an adiabatic process $pV^{\gamma} = \text{constant}$ and the pressure increases by a factor of $2^{\gamma} = 2^{5/3} = 3.17$. (Because the final pressure is greater in the adiabatic case, more work is done in the adiabatic compression.) If the gas were diatomic, the isothermal result would be unaffected (the ideal-gas law doesn't change), but the adiabatic result would be different due to the different value of γ. Can you show that if $\gamma = 1.40$ for the diatomic gas, the pressure would increase by a factor of 2.64?

Discussion Questions

Q19.1 It is not correct to say that a body contains a certain amount of heat, yet a body can transfer heat to another body. How can a body give away something it does not have in the first place?

Q19.2 If you are told the initial and final states of a system and the associated change in internal energy, can you determine whether the internal energy change was due to work or to heat transfer? Explain.

Q19.3 Discuss the application of the first law of thermodynamics to a mountaineer who eats food, gets warm and perspires a lot during a climb, and does a lot of mechanical work in raising herself to the summit. What about the descent? The mountaineer also gets warm during the descent. Is the source of this energy the same as during the ascent?

Q19.4 When ice melts at 0°C, its volume decreases. Is the internal energy change greater than, less than, or equal to the heat added? How can you tell?

Q19.5 Example 19.5 (Section 19.4) describes a process that occurs at a constant temperature of 100°C. Why, then, can't we use in this example the expressions for *W* derived for an isothermal process in Example 19.1 (Section 19.2)?

Q19.6 Translated literally from the Greek, "isothermal" means "same heat." Give as many reasons as you can why such a literal translation is misleading.

Q19.7 Imagine a gas made up entirely of negatively charged electrons. Like charges repel, so the electrons exert repulsive forces on each other. Would you expect that the temperature of such a gas would rise, fall, or stay the same in a free expansion? Why?

Q19.8 There are a few materials that contract when their temperature is increased, such as water between 0°C and 4°C. Would you expect C_p for such materials to be greater or less than C_V? Why or why not?

Q19.9 In a constant-volume process, $dU = nC_V\,dT$. But in a constant-pressure process, it is *not* true that $dU = nC_p\,dT$. Why not?

Q19.10 When you blow on the back of your hand with your mouth wide open, your breath feels warm. But if you partially close your mouth to form an "o" and then blow on your hand, your breath feels cool. Why?

Q19.11 On a warm summer day, a large cylinder of compressed gas (propane or butane) was used to supply several large gas burners at a cookout. After a while, frost formed on the outside of the tank. Why?

Q19.12 In the carburetor of an aircraft or automobile engine, air flows through a relatively small aperture, then expands. In cool, foggy weather, ice sometimes forms in this aperture even though the outside air temperature is above freezing. Why?

Q19.13 For each of the following equations, does the equation apply only to isothermal, or adiabatic, or isobaric, or isochoric processes, or does it apply to all processes: $p_1V_1^\gamma = p_2V_2^\gamma$; $Q = nC_p\Delta T$; $Q = nC_V\Delta T$; $\Delta U = nC_V\Delta T$; $Q = \Delta U + W$; $W = p\Delta V$?

Q19.14 When you use a hand pump to inflate the tires of your bicycle, the pump gets warm after a while. Why? What happens to the temperature of the air in the pump as you compress it? Why does this happen? When you raise the pump handle to draw outside air into the pump, what happens to the temperature of the air taken in? Again, why does this happen?

Q19.15 The prevailing winds on the Hawaiian island of Kauai blow from the northeast. The winds cool as they go up the slope of Mt. Waialeale (elevation 1523 m), causing water vapor to condense and rain to fall. There is much more precipitation at the summit than at the base of the mountain. In fact, Mt. Waialeale is the rainiest spot on earth, averaging 11.7 m of rainfall a year. But what makes the winds cool?

Q19.16 Applying the same considerations as in Question 19.15, explain why the island of Niihau, a few kilometers to the southwest of Kauai, is almost a desert, and farms there need to be irrigated.

Q19.17 When a gas is compressed adiabatically into the surrounding air, its temperature rises even though there is no heat input to the gas. Where does the energy come from to raise the temperature?

Q19.18 When a gas expands adiabatically, it does work on its surroundings. But if there is no heat input to the gas, where does the energy come from to do the work?

Exercises

Section 19.2 Work Done During Volume Changes, and Section 19.3 Paths Between Thermodynamic States

19.1 Two moles of an ideal gas are heated at constant pressure from $T = 27°C$ to 107°C. a) Draw a pV-diagram for this process. b) Calculate the work done by the gas.

19.2 Three moles of an ideal gas have an initial temperature of 127.0°C. While the temperature is kept constant, the volume is increased until the pressure drops to 40.0% of its original value. a) Draw a pV-diagram for this process. b) Calculate the work done by the gas.

19.3 Five moles of an ideal gas are kept at a constant temperature of 53.0°C while the pressure of the gas is increased from 1.00 atm to 3.00 atm. a) Draw a pV-diagram for this process. b) Calculate the work done by the gas.

19.4 Six moles of an ideal gas are in a cylinder fitted at one end with a movable piston. The initial temperature of the gas is 27.0°C and the pressure is constant. As part of a machine design project, calculate the final temperature of the gas after it has done 1.75×10^3 J of work.

19.5 A metal cylinder with rigid walls contains 2.50 mol of oxygen gas. The gas is cooled until the pressure decreases to 30.0% of its original value. You can ignore the thermal contraction of the cylinder. a) Draw a pV-diagram for this process. b) Calculate the work done by the gas.

19.6 A gas under a constant pressure of 1.50×10^5 Pa and with an initial volume of $0.0900\ m^3$ is cooled until its volume becomes $0.0600\ m^3$. a) Draw a pV-diagram for this process. b) Calculate the work done by the gas.

19.7 A gas undergoes two processes. In the first, the volume remains constant at $0.200\ m^3$ and the pressure increases from 2.00×10^5 Pa to 5.00×10^5 Pa. The second process is a compression to a volume of $0.120\ m^3$ at a constant pressure of 5.00×10^5 Pa. a) In a pV-diagram, show both processes. b) Find the total work done by the gas during both processes.

19.8 Work Done in a Cyclic Process. a) In Fig. 19.8a, consider the closed loop $1 \to 3 \to 2 \to 4 \to 1$. This is a *cyclic* process in which the initial and final states are the same. Find the total work done by the system in this cyclic process, and show that it is equal to the area enclosed by the loop. b) How is the work done for the process in part (a) related to the work done if the loop is traversed in the opposite direction, $1 \to 4 \to 2 \to 3 \to 1$? Explain.

Section 19.4 Internal Energy and the First Law of Thermodynamics

19.9 In a certain chemical process, a lab technician supplies 254 J of heat to a system. At the same time, 73 J of work are done on the system by its surroundings. What is the increase in the internal energy of the system?

19.10 A gas in a cylinder expands from a volume of $0.110\ m^3$ to $0.320\ m^3$. Heat flows into the gas just rapidly enough to keep the pressure constant at 1.80×10^5 Pa during the expansion. The total heat added is 1.15×10^5 J. a) Find the work done by the gas. b) Find the change in internal energy of the gas. c) Does it matter whether or not the gas is ideal? Why or why not?

19.11 Five moles of an ideal monatomic gas with an initial temperature of 127°C expand and, in the process, absorb 1200 J of heat and do 2100 J of work. What is the final temperature of the gas?

19.12 The pressure of a system is decreased while the volume is kept constant. If heat flows into the system during this process, does the internal energy of the system increase or decrease? Explain your reasoning.

19.13 A gas in a cylinder is held at a constant pressure of 2.30×10^5 Pa and is cooled and compressed from $1.70\ m^3$ to $1.20\ m^3$. The internal energy of the gas decreases by 1.40×10^5 J. a) Find the work done by the gas. b) Find the absolute value $|Q|$ of the heat flow into or out of the gas, and state the direction of the heat flow. c) Does it matter whether or not the gas is ideal? Why or why not?

19.14 A system is taken from state a to state b along the three paths shown in Fig. 19.22. a) Along which path is the work done by the system the greatest? The least? b) If $U_b > U_a$, along which path is the absolute value $|Q|$ of the heat transfer the greatest? For this path, is heat absorbed or liberated by the system?

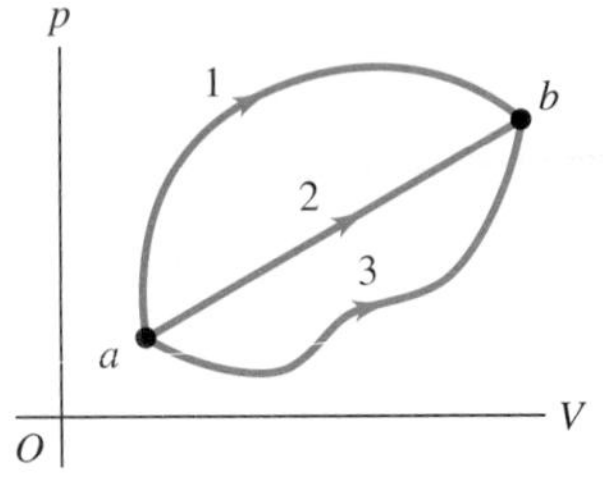

Figure 19.22 Exercise 19.14.

19.15 Doughnuts: Breakfast of Champions! A typical doughnut contains 2.0 g of protein, 17.0 g of carbohydrates, and 7.0 g of fat. The average food-energy values of these substances are

4.0 kcal/g for protein and carbohydrates and 9.0 kcal/g for fat. a) During heavy exercise, an average person uses energy at a rate of 510 kcal/h. How long would you have to exercise to "work off" one doughnut? b) If the energy in the doughnut could somehow be converted into the kinetic energy of your body as a whole, how fast could you move after eating the doughnut? Take your mass to be 60 kg, and express your answer in m/s and km/h.

19.16 A liquid is irregularly stirred in a well-insulated container and thereby undergoes a rise in temperature. Regard the liquid as the system. a) Has heat been transferred? How can you tell? b) Has work been done? How can you tell? Why is it important that the stirring is irregular? c) What is the sign of ΔU? How can you tell?

19.17 A system is taken around the cycle shown in Fig. 19.23, from state a to state b and then back to state a. The absolute value of the heat transfer during one cycle is 7200 J. a) Does the system absorb or liberate heat when it goes around the cycle in the direction shown in the figure? How can you tell? b) What is the work W done by the system in one cycle? c) If the system goes around the cycle in a counterclockwise direction, does it absorb or liberate heat in one cycle? What is the magnitude of the heat absorbed or liberated in one counterclockwise cycle?

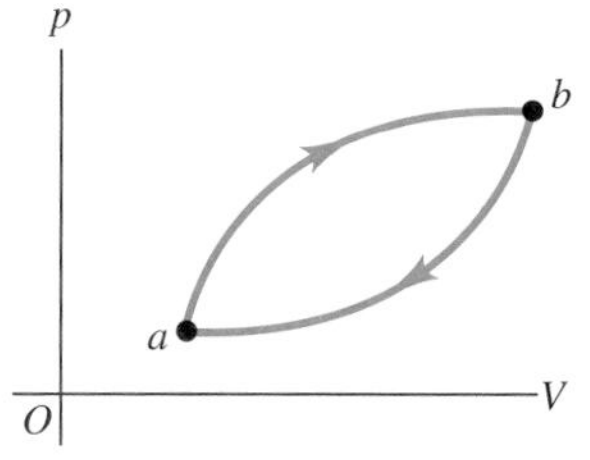

Figure 19.23 Exercise 19.17.

19.18 A thermodynamic system undergoes a cyclic process as shown in Fig. 19.24. The cycle consists of two closed loops, loop I and loop II. a) Over one complete cycle, does the system do positive or negative work? b) In each of loops I and II, is the net work done by the system positive or negative? c) Over one complete cycle, does heat flow into or out of the system? d) In each of loops I and II, does heat flow into or out of the system?

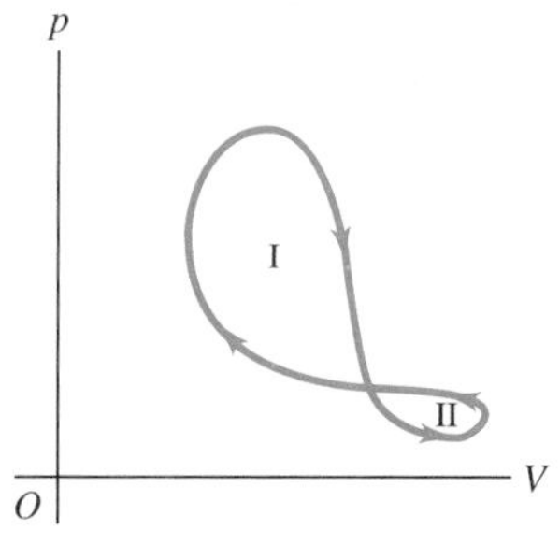

Figure 19.24 Exercise 19.18.

19.19 A student performs a combustion experiment by burning a mixture of fuel and oxygen in a constant-volume metal can surrounded by a water bath. During the experiment the temperature of the water is observed to rise. Regard the mixture of fuel and oxygen as the system. a) Has heat been transferred? How can you tell? b) Has work been done? How can you tell? c) What is the sign of ΔU? How can you tell?

19.20 Boiling Water at High Pressure. When water is boiled under a pressure of 2.00 atm, the heat of vaporization is 2.20×10^6 J/kg and the boiling point is 120°C. At this pressure, 1.00 kg of water has a volume of 1.00×10^{-3} m^3, and 1.00 kg of steam has a volume of 0.824 m^3. a) Compute the work done when 1.00 kg of steam is formed at this temperature. b) Compute the increase in internal energy of the water.

Section 19.5 Kinds of Thermodynamic Processes, Section 19.6 Internal Energy of an Ideal Gas, and Section 19.7 Heat Capacities of an Ideal Gas

19.21 In an experiment to simulate conditions within an automobile engine, 645 J of heat is transferred to 0.185 mol of air contained within a cylinder of volume 40.0 cm^3. Initially the air is at a pressure of 3.00×10^6 Pa and a temperature of 780 K. a) If the volume of the cylinder is held fixed, what is the final temperature of the air? Assume that the air is essentially all nitrogen gas, and use the data in Table 19.1 even though the pressure is not low. Draw a pV-diagram for this process. b) Find the final temperature of the air if the volume of the cylinder is allowed to increase while the pressure remains constant. Draw a pV-diagram for this process.

19.22 A cylinder contains 0.0100 mol of helium at $T = 27.0$°C. a) How much heat is needed to raise the temperature to 67.0°C while keeping the volume constant? Draw a pV-diagram for this process. b) If instead the pressure of the helium is kept constant, how much heat is needed to raise the temperature from 27.0°C to 67.0°C? Draw a pV-diagram for this process. c) What accounts for the difference between your answers to parts (a) and (b)? In which case is more heat required? What becomes of the additional heat? d) If the gas is ideal, what is the change in its internal energy in part (a)? In part (b)? How do the two answers compare? Why?

19.23 Five moles of gas have their temperature increased from -10.0°C to $+20.0$°C. Calculate the heat that must be transferred into the gas if the gas is a) He at a constant pressure of 1.5 atm; b) Ar at a fixed volume of 8.2 m^3; c) CO_2 at a constant pressure of 20,000 Pa.

19.24 When a quantity of monatomic ideal gas expands at a constant pressure of 4.00×10^4 Pa, the volume of the gas increases from 2.00×10^{-3} m^3 to 8.00×10^{-3} m^3. What is the change in the internal energy of the gas?

19.25 An ideal gas expands while the pressure is kept constant. During this process, does heat flow into the gas or out of the gas? Justify your answer.

19.26 Heat Q flows into a monatomic ideal gas, and the volume increases while the pressure is kept constant. What fraction of the heat energy is used to do the expansion work of the gas?

19.27 The temperature of 0.150 mol of an ideal gas is held constant at 77.0°C while its volume is reduced to 25.0% of its initial volume. The initial pressure of the gas is 1.25 atm. a) Determine the work done by the gas. b) What is the change in its internal energy? c) Does the gas exchange heat with its surroundings? If so, how much? Does the gas absorb or liberate heat?

19.28 During an isothermal compression of an ideal gas, 335 J of heat must be removed from the gas to maintain constant temperature. How much work is done by the gas during the process?

19.29 Propane gas (C_3H_8) behaves like an ideal gas with $\gamma = 1.127$. Determine the molar heat capacity at constant volume and the molar heat capacity at constant pressure.

19.30 A cylinder contains 0.250 mol of carbon dioxide (CO_2) gas at a temperature of 27.0°C. The cylinder is provided with a frictionless piston, which maintains a constant pressure of 1.00 atm on the gas. The gas is heated until its temperature increases to 127.0°C. Assume that the CO_2 may be treated as an ideal gas. a) Draw a

pV-diagram for this process. b) How much work is done by the gas in this process? c) On what is this work done? d) What is the change in internal energy of the gas? e) How much heat was supplied to the gas? f) How much work would have been done if the pressure had been 0.50 atm?

19.31 Gaseous ethane (C_2H_6) has $\gamma = 1.220$ and may be treated as an ideal gas. a) If 2.40 mol of ethane is to be heated from 20.0°C to 25.0°C at a constant pressure of 1.00 atm, how much heat is required? b) What will be the change in internal energy of the ethane?

Section 19.8 Adiabatic Processes for an Ideal Gas

19.32 A monatomic ideal gas that is initially at a pressure of 1.50×10^5 Pa and with a volume of 0.0800 m^3 is compressed adiabatically to a volume of 0.0400 m^3. a) What is the final pressure? b) How much work is done by the gas? c) What is the ratio of the final temperature of the gas to its initial temperature? Is the gas heated or cooled by this compression?

19.33 The engine of a Ferrari F355 F1 sports car takes in air at 20.0°C and 1.00 atm and compresses it adiabatically to 0.0900 times the original volume. The air may be treated as an ideal gas with $\gamma = 1.40$. a) Draw a pV-diagram for this process. b) Find the final temperature and pressure.

19.34 Two moles of carbon monoxide (CO) start at a pressure of 1.2 atm and a volume of 30 liters. The gas is then compressed adiabatically to $\frac{1}{3}$ this volume. Assume that the gas may be treated as ideal. What is the change in the internal energy of the gas? Does the internal energy increase or decrease? Does the temperature of the gas increase or decrease during this process? Explain.

19.35 In an adiabatic process for an ideal gas the pressure decreases. In this process does the internal energy of the gas increase or decrease? Explain your reasoning.

19.36 An ideal gas initially at 4.00 atm and 350 K is permitted to expand adiabatically to 1.50 times its initial volume. Find the final pressure and temperature if the gas is a) monatomic; b) diatomic with $C_V = \frac{5}{2}R$.

19.37 During an adiabatic expansion the temperature of 0.450 mol of argon (Ar) drops from 50.0°C to 10.0°C. The argon may be treated as an ideal gas. a) Draw a pV-diagram for this process. b) How much work does the gas do? c) What is the change in internal energy of the gas?

19.38 A cylinder contains 0.100 mol of an ideal monatomic gas. Initially the gas is at a pressure of 1.00×10^5 Pa and occupies a volume of 2.50×10^{-3} m^3. a) Find the initial temperature of the gas in kelvins. b) If the gas is allowed to expand to twice the initial volume, find the final temperature (in kelvins) and pressure of the gas if the expansion is i) isothermal; ii) isobaric; iii) adiabatic.

19.39 A quantity of sulfur dioxide gas (SO_2) occupies a volume of 5.00×10^{-3} m^3 at a pressure of 1.10×10^5 Pa. The gas expands adiabatically to a volume of 1.00×10^{-2} m^3. Assume that the gas may be treated as ideal. a) Find the final pressure of the gas. (*Hint:* See Table 19.1.) b) How much work does the gas do on its surroundings? c) What is the ratio of the final temperature of the gas to its initial temperature?

Problems

19.40 A quantity of air is taken from state a to state b along a path that is a straight line in the pV-diagram (Fig. 19.25). a) In this process, does the temperature of the gas increase, decrease, or stay the same? Explain. b) If $V_a = 0.0700$ m^3, $V_b = 0.1100$ m^3, $p_a = 1.00 \times 10^5$ Pa, and $p_b = 1.40 \times 10^5$ Pa, what is the work W done by the gas in this process? Assume that the gas may be treated as ideal.

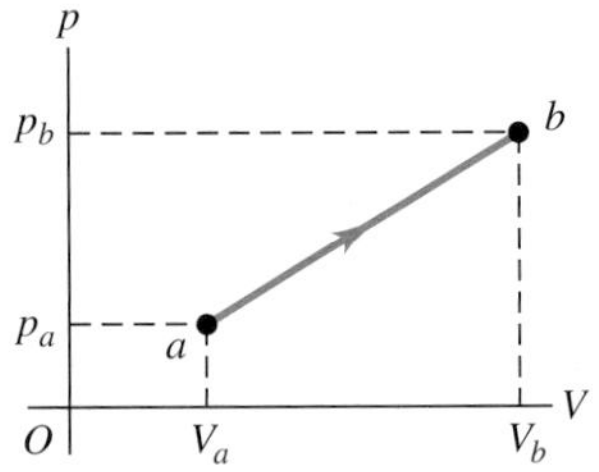

Figure 19.25 Problem 19.40.

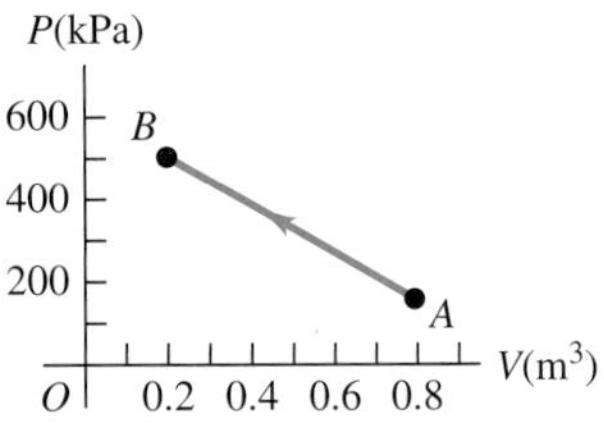

Figure 19.26 Problem 19.41.

19.41 Four moles of O_2 are taken from A to B by the process shown in the pV-diagram in Fig. 19.26. Assume the gas may be treated as ideal. What is the heat flow Q during this process? Does heat enter or leave the gas?

19.42 a) One-third of a mole of He gas is taken along the path abc that is shown as the solid line in Fig. 19.27. Assume the gas may be treated as ideal. How much heat is transferred into or out of the gas? b) If the gas instead went from state a to state c along the horizontal dashed line in Fig. 19.27, how much heat would be transferred into or out of the gas? c) How does Q in part (b) compare to Q in part (a)? Explain.

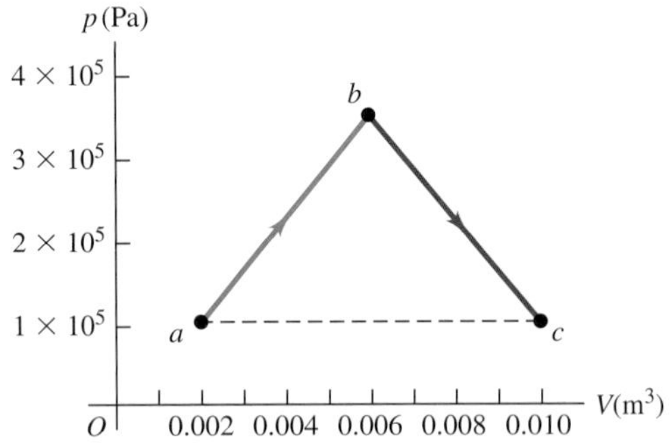

Figure 19.27 Problem 19.42.

19.43 When a system is taken from state a to state b in Fig. 19.28 along the path acb, 90.0 J of heat flows into the system and 60.0 J of work is done by the system. a) How much heat flows into the system along path adb if the work done by the system is 15.0 J? b) When the system is returned from b to a along the curved path,

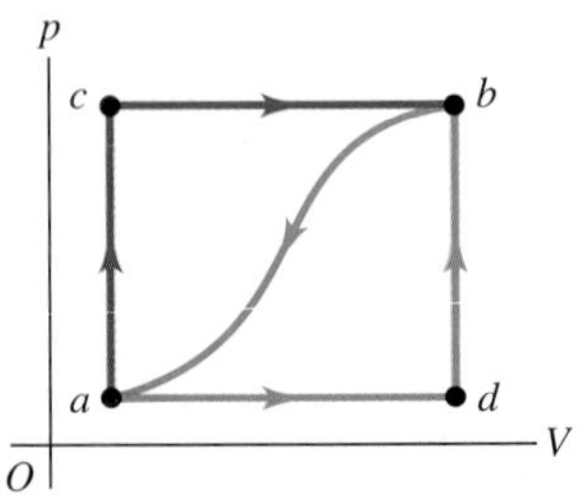

Figure 19.28 Problem 19.43.

the absolute value of the work done by the system is 35.0 J. Does the system absorb or liberate heat? How much heat? c) If $U_a = 0$ and $U_d = 8.0$ J, find the heat absorbed in the processes *ad* and *db*.

19.44 A thermodynamic system is taken from state *a* to state *c* in Fig. 19.29 along either path *abc* or path *adc*. Along path *abc* the work *W* done by the system is 450 J. Along path *adc*, *W* is 120 J. The internal energies of each of the four states shown in the figure are $U_a = 150$ J, $U_b = 240$ J, $U_c = 680$ J, and $U_d = 330$ J. Calculate the heat flow *Q* for each of the four processes *ab*, *bc*, *ad*, and *dc*. In each process, does the system absorb or liberate heat?

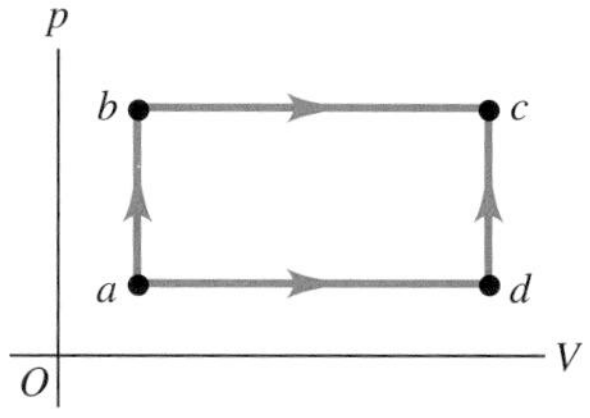

Figure 19.29 Problems 19.44 and 19.45.

19.45 Fig. 19.29 shows four states of a thermodynamic system, *a*, *b*, *c*, and *d*. The volume of the system is V_a for both state *a* and state *b*, and is V_c for both state *c* and state *d*. The pressure of the system is p_a for both state *a* and state *d*, and is p_c for both state *b* and state *c*. The internal energies of the four states are U_a, U_b, U_c, and U_d. For each of the processes *ab*, *bc*, *ad*, and *dc*, find a) the work done by the system during the process and b) the heat flow into the system during the process. c) The system can be taken from state *a* to state *c* along either path *abc* or path *adc*. Find the net heat flow into the system and the net work done by the system for each path. For which path is the net heat flow greater? For which path is the net work greater? d) A friend tells you that the amounts of heat flow should be the same for both path *abc* and path *adc*, since the initial state (*a*) and final state (*c*) of the system are the same for both paths. How would you respond?

19.46 Two moles of N_2 gas undergo the cycle *abcd*. The pressure of the gas in each state is: $p_a = p_d = 6000$ Pa; $p_b = p_c = 2000$ Pa. The volume of the gas in each state is: $V_a = V_b = 2.00\ \text{m}^3$; $V_c = V_d = 9.00\ \text{m}^3$. The gas may be treated as ideal. a) Draw the *pV*-diagram for the cycle. b) Find the total work done on (or by) the gas in the complete cycle. Does the gas do work or is work done on it? c) Find the net heat put into (or taken out of) the gas in one cycle. Does heat go into the gas or out of the gas?

19.47 Two moles of an ideal monatomic gas go through the cycle *abc*. For the complete cycle, 800 J of heat flows out of the gas. Process *ab* is at constant pressure and process *bc* is at constant volume. States *a* and *b* have temperatures $T_a = 200$ K and $T_b = 300$ K. a) Sketch the *pV*-diagram for the cycle. b) What is the work *W* for the process *ca*?

19.48 Three moles of an ideal gas are taken around the cycle *abc* shown in Fig. 19.30. For this gas, $C_p = 29.1$ J/mol·K. Process *ac* is at constant pressure, process *ba* is at constant volume, and process *cb* is adiabatic. The temperature of the gas in states *a*, *c* and *b* is $T_a = 300$ K, $T_c = 492$ K, and $T_b = 600$ K. Calculate the total work *W* for the cycle.

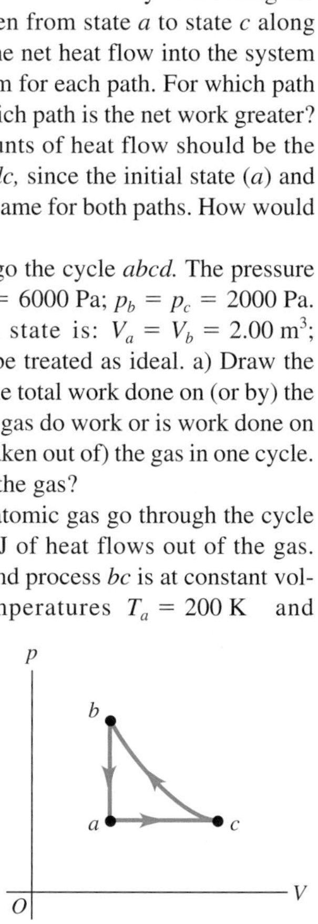

Figure 19.30 Problem 19.48.

19.49 An ideal monatomic gas first has its pressure increased while its volume is held fixed (process *ab*). Its volume is then decreased at constant pressure until its temperature returns to the value it had in state *a* (process *bc*). The pressure for states *b* and *c* is 6.00×10^4 Pa; the volumes for these states are $V_b = 0.700\ \text{m}^3$ and $V_c = 0.300\ \text{m}^3$. a) Sketch the processes *ab* and *bc* on the same *pV*-diagram. b) What is the total heat flow for these two processes?

19.50 Nitrogen gas in an expandable container is cooled from 50.0°C to 10.0°C with the pressure held constant at 3.00×10^5 Pa. The total heat liberated by the gas is 2.50×10^4 J. Assume that the gas may be treated as ideal. a) Find the number of moles of gas. b) Find the change in internal energy of the gas. c) Find the work done by the gas. d) How much heat would be liberated by the gas for the same temperature change if the volume were constant?

19.51 In a certain process, 2.15×10^5 J of heat is liberated by a system, and at the same time the system contracts under a constant external pressure of 9.50×10^5 Pa. The internal energy of the system is the same at the beginning and end of the process. Find the change in volume of the system. (The system is *not* an ideal gas.)

19.52 A cylinder with a frictionless, movable piston like that shown in Fig. 19.5, contains a quantity of helium gas. Initially the gas is at a pressure of 1.00×10^5 Pa, has a temperature of 300 K, and occupies a volume of 1.50 L. The gas then undergoes two processes. In the first, the gas is heated and the piston is allowed to move to keep the temperature equal to 300 K. This continues until the pressure reaches 2.50×10^4 Pa. In the second process, the gas is compressed at constant pressure until it returns to its original volume of 1.50 L. Assume that the gas may be treated as ideal. a) In a *pV*-diagram, show both processes. b) Find the volume of the gas at the end of the first process, and find the pressure and temperature at the end of the second process. c) Find the total work done by the gas during both processes. d) What would you have to do to the gas to return it to its original pressure and temperature?

19.53 A Thermodynamic Process in a Liquid. A chemical engineer is studying the properties of liquid methanol (CH_3OH). She uses a steel cylinder with a cross-sectional area of $0.0200\ \text{m}^2$ and containing $1.20 \times 10^{-2}\ \text{m}^3$ of methanol. The cylinder is equipped with a tightly fitting piston that supports a load of 3.00×10^4 N. The temperature of the system is increased from 20.0°C to 50.0°C. For methanol, the coefficient of volume expansion is $1.20 \times 10^{-3}\ \text{K}^{-1}$, the density is $791\ \text{kg/m}^3$, and the specific heat capacity at constant pressure is $c_p = 2.51 \times 10^3$ J/kg·K. You can ignore the expansion of the steel cylinder. Find a) the increase in volume of the methanol; b) the mechanical work done by the methanol against the 3.00×10^4 N force; c) the amount of heat added to the methanol; d) the change in internal energy of the methanol. e) Based on your results, explain whether there is any substantial difference between the specific heat capacities c_p (at constant pressure) and c_V (at constant volume) for methanol under these conditions.

19.54 A Thermodynamic Process in a Solid. A cube of copper 2.00 cm on a side is suspended by a string. (The physical properties of copper are given in Tables 14.1, 17.2, and 17.3.) The cube is heated with a burner from 20.0°C to 90.0°C. The air surrounding the cube is at atmospheric pressure $(1.01 \times 10^5\ \text{Pa})$. Find a) the increase in volume of the cube; b) the mechanical work done by the cube to expand against the pressure of the surrounding air; c) the amount of heat added to the cube; d) the change in internal energy of the cube. e) Based on your results, explain whether there is any substantial difference between the specific heat capacities c_p (at constant pressure) and c_V (at constant volume) for copper under these conditions.

19.55 A Thermodynamic Process in an Insect. The African bombardier beetle *Stenaptinus insignis* can emit a jet of defensive spray from the moveable tip of its abdomen (Fig. 19.31). The beetle's body has reservoirs of two different chemicals; when the beetle is disturbed, these chemicals are combined in a reaction chamber, producing a compound that is warmed from 20°C to 100°C by the heat of reaction. The high pressure produced allows the compound to be sprayed out at speeds up to 19 m/s (68 km/h), scaring away predators of all kinds. (The beetle shown in the figure is tethered to a wire fastened to its back by wax. It is responding to having its left foreleg pinched by forceps. The beetle is 2 cm long.) Calculate the heat of reaction of the two chemicals (in J/kg). Assume that the specific heat capacity of the two chemicals and the spray is the same as that of water, $4.19 \times 10^3\ \text{J/kg} \cdot \text{K}$, and that the initial temperature of the chemicals is 20°C.

Figure 19.31 Problem 19.55.

19.56 A Compressed-Air Engine. You are designing an engine that runs on compressed air. Air enters the engine at a pressure of 1.60×10^6 Pa and leaves at a pressure of 2.80×10^5 Pa. What must the temperature of the compressed air be for there to be no possibility of frost forming in the exhaust ports of the engine? Assume that the expansion is adiabatic. (*Note:* Frost will form if the moist air is cooled below 0°C in the expansion.)

19.57 Chinook. During certain seasons strong winds called "chinooks" blow from the west across the eastern slopes of the Rockies and downhill into Denver and adjoining areas. Although the mountains are cool, the wind in Denver is very hot; within a few minutes after the chinook wind arrives, the temperature can climb 20 C° ("chinook" is a Native American word meaning "snow eater"). Similar winds occur in the Alps (called "foehns") and in southern California (called "Santa Anas"). a) Explain why the temperature of the chinook wind rises as it descends the slopes. Why is it important that the wind be fast-moving? b) Suppose a strong wind is blowing toward Denver (elevation 1630 m) from Grays Peak (80 km west of Denver, at an elevation of 4350 m), where the air pressure is 5.60×10^4 Pa and the air temperature is -15.0°C. The temperature and pressure in Denver before the wind arrives are 2.0°C and 8.12×10^4 Pa. By how many Celsius degrees will the temperature in Denver rise when the chinook arrives?

19.58 A certain ideal gas has molar heat capacity at constant volume C_V. A sample of this gas initially occupies a volume V_0 at pressure p_0 and absolute temperature T_0. The gas expands isobarically to a volume $2V_0$, then expands further adiabatically to a final volume of $4V_0$. a) Draw a pV-diagram for this sequence of processes. b) Compute the total work done by the gas for this sequence of processes. c) Find the final temperature of the gas. d) Find the absolute value $|Q|$ of the total heat flow into or out of the gas for this sequence of processes, and state the direction of heat flow.

19.59 An air pump has a cylinder 0.250 m long with a movable piston. The pump is used to compress air from the atmosphere (at absolute pressure 1.01×10^5 Pa) into a very large tank at 4.20×10^5 Pa gauge pressure. (For air, $C_V = 20.8\ \text{J/mol} \cdot \text{K}$.) a) The piston begins the compression stroke at the open end of the cylinder. How far down the length of the cylinder has the piston moved when air first begins to flow from the cylinder into the tank? Assume that the compression is adiabatic. b) If the air is taken into the pump at 27.0°C, what is the temperature of the compressed air? c) How much work does the pump do in putting 20.0 mol of air into the tank?

19.60 Engine Turbochargers and Intercoolers. The power output of an automobile engine is directly proportional to the mass of air that can be forced into the volume of the engine's cylinders to react chemically with gasoline. Many cars have a *turbocharger* which compresses the air before it enters the engine, giving a greater mass of air per volume. This rapid, essentially adiabatic compression also heats the air. To compress it further, the air then passes through an *intercooler* in which the air exchanges heat with its surroundings at essentially constant pressure. The air is then drawn into the cylinders. In a typical installation, air is taken into the turbocharger at atmospheric pressure $(1.01 \times 10^5\ \text{Pa})$, density $\rho = 1.23\ \text{kg/m}^3$, and temperature 15.0°C. It is compressed adiabatically to 1.45×10^5 Pa. In the intercooler, the air is cooled to the original temperature of 15.0°C at a constant pressure of 1.45×10^5 Pa. a) Draw a pV-diagram for this sequence of processes. b) If the volume of one of the engine's cylinders is 575 cm^3, what mass of air exiting from the intercooler will fill the cylinder at 1.45×10^5 Pa? Compared to the power output of an engine that takes in air at 1.01×10^5 Pa at 15.0°C, what percentage increase in power is obtained by using the turbocharger and intercooler? c) If the intercooler is not used, what mass of air exiting from the turbocharger will fill the cylinder at 1.45×10^5 Pa? Compared to the power output of an engine that takes in air at 1.01×10^5 Pa at 15.0°C, what percentage increase in power is obtained by using the turbocharger alone?

19.61 A monatomic ideal gas expands slowly to twice its original volume, doing 300 J of work in the process. Find the heat added to the gas and the change in internal energy of the gas if the process is a) isothermal; b) adiabatic; c) isobaric.

19.62 A cylinder with a piston contains 0.250 mol of oxygen at 2.40×10^5 Pa and 355 K. The oxygen may be treated as an ideal gas. The gas first expands isobarically to twice its original volume. It is then compressed isothermally back to its original volume, and finally it is cooled isochorically to its original pressure. a) Show the series of processes on a pV-diagram. b) Compute the temperature during the isothermal compression. c) Compute the maximum pressure. d) Compute the total work done by the piston on the gas during the series of processes.

19.63 Use the conditions and processes of Problem 19.62 to compute a) the work done by the gas, the heat added to it, and its internal-energy change during the initial expansion; b) the work done, the heat added, and the internal-energy change during the final cooling; c) the internal-energy change during the isothermal compression.

19.64 A cylinder with a piston contains 0.150 mol of nitrogen at 1.80×10^5 Pa and 300 K. The nitrogen may be treated as an ideal gas. The gas is first compressed isobarically to $\frac{1}{2}$ its original volume. It then expands adiabatically back to its original volume, and finally it is heated isochorically to its original pressure. a) Show the series of processes in a pV-diagram. b) Compute the temperatures at the beginning and end of the adiabatic expansion. c) Compute the minimum pressure.

19.65 Use the conditions and processes of Problem 19.64 to compute a) the work done by the gas, the heat added to it, and its internal-energy change during the initial compression; b) the work done by the gas, the heat added to it, and its internal-energy change during the adiabatic expansion; c) the work done, the heat added, and the internal-energy change during the final heating.

19.66 Comparing Thermodynamic Processes. In a cylinder, 1.20 mol of an ideal monatomic gas, initially at 3.60×10^5 Pa and 300 K, expands until its volume triples. Compute the work done by the gas if the expansion is a) isothermal; b) adiabatic; c) isobaric. d) Show each process in a pV-diagram. In which case is the absolute value of the work done by the gas greatest? Least? e) In which case is the absolute value of the heat transfer greatest? Least? f) In which case is the absolute value of the change in internal energy of the gas greatest? Least?

19.67 A flexible balloon contains 0.350 mol of hydrogen sulfide gas (H_2S). Initially the H_2S has a volume of 7.00×10^3 cm^3 and a temperature of 27.0°C. The H_2S first expands isobarically until the volume doubles. Then it expands adiabatically until the temperature returns to its initial value. Assume that the H_2S may be treated as an ideal gas. a) Draw a diagram of the process in the pV-plane. b) What is the total heat supplied to the H_2S in the process? c) What is the total change in the internal energy of the H_2S? d) What is the total work done by the H_2S? e) What is the final volume?

Challenge Problems

19.68 Oscillations of a Piston. A vertical cylinder of radius r contains a quantity of ideal gas, and is fitted with a piston with mass m that is free to move (Fig. 19.32). The piston and the walls of the cylinder are frictionless and are made of a perfect thermal insulator. The outside air pressure is p_0. In equilibrium, the piston sits at a height h above the bottom of the cylinder. a) Find the absolute pressure of the gas trapped below the piston when in equilibrium. b) The piston is pulled up by a small distance and released. Find the net force acting on the piston when its base is a distance $h + y$ above the bottom of the cylinder, where y is much less than h. c) After the piston is displaced from equilibrium and released, it oscillates up and down. Find the frequency of these small oscillations. If the displacement is not small, are the oscillations simple harmonic? How can you tell?

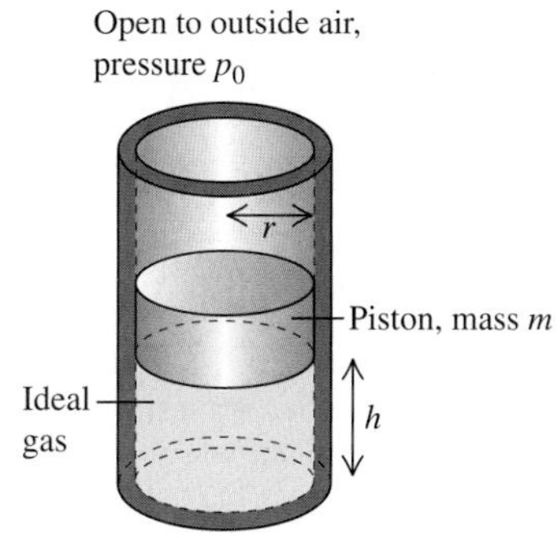

Figure 19.32 Challenge Problem 19.68.

19.69 The van der Waals equation of state, an approximate representation of the behavior of gases at high pressure, is given by Eq. (18.7):

$$\left(p + \frac{an^2}{V^2}\right)(V - nb) = nRT$$

where a and b are constants having different values for different gases. (In the special case of $a = b = 0$, this is the ideal-gas equation.) a) Calculate the work done by a gas with this equation of state in an isothermal expansion from V_1 to V_2. Show that your answer agrees with the ideal-gas result found in Example 19.1 (Section 19.2) when you set $a = b = 0$. b) For ethane gas (C_2H_6), $a = 0.554$ J·m^3/mol^2 and $b = 6.38 \times 10^{-5}$ m^3/mol. Calculate the work W done by 1.80 mol of ethane when it expands from 2.00×10^{-3} m^3 to 4.00×10^{-3} m^3 at a constant temperature of 300 K. Do the calculation using i) the van der Waals equation of state and ii) the ideal-gas equation of state. c) How large is the difference between the two results for W in part (b)? For which equation of state is W larger? Use the interpretation of the terms a and b given in Section 18.1 to explain why this should be so. Are the differences between the two equations of state important in this case?

20 THE SECOND LAW OF THERMODYNAMICS

Could heat spontaneously flow out of these fried eggs, thus "unfrying" them and making them cooler, and flow back into the hot frying pan? Such a process would conserve energy and is allowed by the first law of thermodynamics. But such a process is in fact impossible because it would violate the *second* law of thermodynamics: heat cannot by itself flow from one body into a hotter body.

? **Is it *ever* possible for heat to flow from a cold body into a hot one?**

Many thermodynamic processes proceed naturally in one direction but not the opposite. For example, heat always flows from a hot body to a cooler body, never the reverse. Heat flow from a cool body to a hot body would not violate the first law of thermodynamics; energy would be conserved. But it doesn't happen in nature. Why not? As another example, note that it is easy to convert mechanical energy completely into heat; this happens every time we use a car's brakes to stop it. Going in the reverse direction, there are plenty of devices that convert heat *partially* into mechanical energy. (An automobile engine is an example.) But even the cleverest would-be inventors have never succeeded in building a machine that converts heat *completely* into mechanical energy. Again, why not?

The answer to both of these questions has to do with the *directions* of thermodynamic processes and is called the *second law of thermodynamics.* This law places fundamental limitations on the efficiency of an engine or a power plant. It also places limitations on the minimum energy input needed to operate a refrigerator. So the second law is directly relevant for many important practical problems.

We can also state the second law in terms of the concept of *entropy,* a quantitative measure of the degree of disorder or randomness of a system. The idea of entropy helps explain why ink mixed with water never spontaneously unmixes and why we never observe a host of other seemingly possible processes.

20.1 | Directions of Thermodynamic Processes

Thermodynamic processes that occur in nature are all **irreversible processes.** These are processes that proceed spontaneously in one direction but not the other (Fig. 20.1a). The flow of heat from a hot body to a cooler body is irreversible, as is the free expansion of a gas discussed in Sections 19.3 and 19.6. Sliding a book across a table converts mechanical energy into heat by friction; this process is irreversible, for no one has ever observed the reverse process (in which a book initially at rest on the table would spontaneously start moving and the table and book would cool down). Our main topic for this chapter is the *second law of thermodynamics,* which determines the preferred direction for such processes.

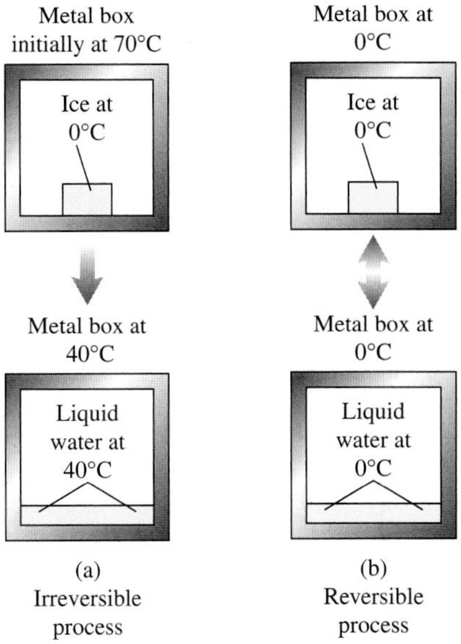

20.1 (a) A block of ice melts *irreversibly* when we place it in a hot meal box. Heat flows from the box into the ice and water, never the reverse. (b) We can melt ice *reversibly* in a metal box at 0°C. By raising or lowering the temperature of the box by an infinitesimal amount, we can make heat flow out of the box into the ice (thus melting it) or into the box from the water (thus refreezing it).

Despite this preferred direction for every natural process, we can think of a class of idealized processes that *would* be reversible. A system that undergoes such an idealized **reversible process** is always very close to being in thermodynamic equilibrium within itself and with its surroundings. Any change of state that takes place can then be reversed (made to go the other way) by making only an infinitesimal change in the conditions of the system. For example, heat flow between two bodies whose temperatures differ only infinitesimally can be reversed by making only a very small change in one temperature or the other (Fig. 20.1b).

Reversible processes are thus **equilibrium processes,** with the system always in thermodynamic equilibrium. Of course, if a system were *truly* in thermodynamic equilibrium, no change of state would take place. Heat would not flow into or out of a system with truly uniform temperature throughout, and a system that is truly in mechanical equilibrium would not expand and do work against its surroundings. A reversible process is an idealization that can never be precisely attained in the real world. But by making the temperature gradients and the pressure differences in the substance very small, we can keep the system very close to equilibrium states and make the process nearly reversible. That's why we call a reversible process a *quasi-equilibrium process.*

By contrast, heat flow with finite temperature difference, free expansion of a gas, and conversion of work to heat by friction are all *irreversible* processes; no small change in conditions could make any of them go the other way. They are also all *nonequilibrium* processes, in that the system is not in thermodynamic equilibrium at any point until the end of the process.

There is a relationship between the direction of a process and the *disorder* or *randomness* of the resulting state. For example, imagine a tedious sorting job, such as alphabetizing a thousand book titles written on file cards. Throw the alphabetized stack of cards into the air. Do they come down in alphabetical order? Alas, no: their tendency is to come down in a random or disordered state. In the free expansion of a gas discussed in Sections 19.3 and 19.6, the air is more disordered after it has expanded into the entire box than when it was confined in one side, just as your clothes are more disordered when scattered all over your floor than when confined to your closet.

Similarly, macroscopic kinetic energy is energy associated with organized, coordinated motions of many molecules, but heat transfer involves changes in energy of random, disordered molecular motion. Therefore conversion of mechanical energy into heat involves an increase of randomness or disorder.

In the following sections we will introduce the second law of thermodynamics by considering two broad classes of devices: *heat engines,* which are partly successful in converting heat into work, and *refrigerators,* which are partly successful in transporting heat from cooler to hotter bodies.

Test Your Understanding

Your left and right hands are normally at the same temperature, just like the metal box and ice in Fig. 20.1b. Is rubbing your hands together to warm them a reversible process? Explain.

20.2 | Heat Engines

The essence of our technological society is the ability to use sources of energy other than muscle power. Sometimes, mechanical energy is directly available; water power is an example. But most of our energy comes from the burning of fossil fuels (coal, oil, and gas) and from nuclear reactions. They supply energy that is transferred as *heat.* This is directly useful for heating buildings, for cooking, and for chemical processing, but to operate a machine or propel a vehicle, we need *mechanical* energy.

Thus it's important to know how to take heat from a source and convert as much of it as possible into mechanical energy or work. This is what happens in gasoline engines in automobiles, jet engines in airplanes, steam turbines in electric power plants, and many other systems. Closely related processes occur in the animal kingdom; food energy is "burned" (that is, carbohydrates combine with oxygen to yield water, carbon dioxide, and energy) and partly converted to mechanical energy as an animal's muscles do work on its surroundings.

ActivPhysics ONLINE

8.12 Cyclic Process—Strategies

8.13 Cyclic Process—Problems

Any device that transforms heat partly into work or mechanical energy is called a **heat engine.** Usually, a quantity of matter inside the engine undergoes inflow and outflow of heat, expansion and compression, and sometimes change of phase. We call this matter the **working substance** of the engine. In internal-combustion engines the working substance is a mixture of air and fuel; in a steam turbine it is water (Fig. 20.2).

The simplest kind of engine to analyze is one in which the working substance undergoes a **cyclic process,** a sequence of processes that eventually leaves the substance in the same state in which it started. In a steam turbine the water is recycled and used over and over. Internal-combustion engines do not use the same air over and over, but we can still analyze them in terms of cyclic processes that approximate their actual operation.

20.2 All motorized vehicles (except electric-powered vehicles) use heat engines for propulsion.

All heat engines *absorb* heat from a source at a relatively high temperature, perform some mechanical work, and *discard* or *reject* some heat at a lower temperature. As far as the engine is concerned, the discarded heat is wasted. In internal-combustion engines the waste heat is that discarded in the hot exhaust gases and the cooling system; in a steam turbine it is the heat that must flow out of the used steam to condense and recycle the water.

When a system is carried through a cyclic process, its initial and final internal energies are equal. For any cyclic process, the first law of thermodynamics requires that

$$U_2 - U_1 = 0 = Q - W \quad \text{so} \quad Q = W$$

That is, the net heat flowing into the engine in a cyclic process equals the net work done by the engine.

When we analyze heat engines, it helps to think of two bodies with which the working substance of the engine can interact. One of these, called the *hot reservoir,* represents the heat source; it can give the working substance large amounts of heat at a constant temperature T_H without appreciably changing its own temperature. The other body, called the *cold reservoir,* can absorb large amounts of discarded heat from the engine at a constant lower temperature T_C. In a steam-turbine system the flames and hot gases in the boiler are the hot reservoir, and the cold water and air used to condense and cool the used steam are the cold reservoir.

We denote the quantities of heat transferred from the hot and cold reservoirs as Q_H and Q_C, respectively. A quantity of heat Q is positive when heat is transferred *into* the working substance and is negative when heat leaves the working substance. Thus in a heat engine, Q_H is positive but Q_C is negative, representing heat *leaving* the working substance. This sign convention is consistent with the rules we stated in Section 19.1; we will continue to use those rules here. Frequently, it clarifies the relationships to state them in terms of the absolute values of the Q's and W's because absolute values are always positive. When we do this, our notation will show it explicitly.

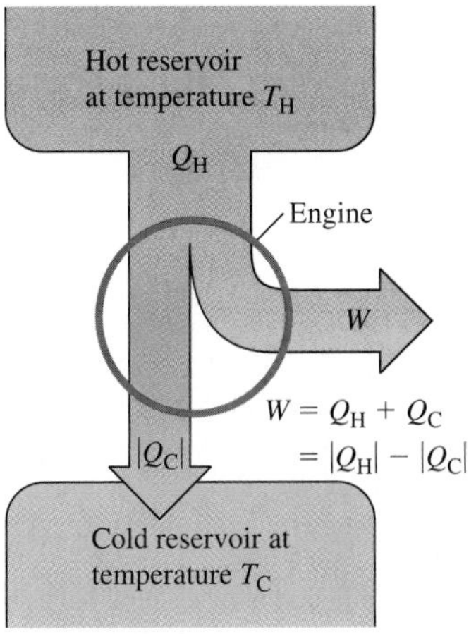

20.3 Schematic energy-flow diagram for a heat engine.

We can represent the energy transformations in a heat engine by the *energy-flow diagram* of Fig. 20.3. The engine itself is represented by the circle. The amount of heat Q_H supplied to the engine by the hot reservoir is proportional to the width of the incoming "pipeline" at the top of the diagram. The width of the outgoing pipeline at the bottom is proportional to the magnitude $|Q_C|$ of the heat rejected in the exhaust. The branch line to the right represents the portion of the heat supplied that the engine converts to mechanical work, W.

When an engine repeats the same cycle over and over, Q_H and Q_C represent the quantities of heat absorbed and rejected by the engine *during one cycle;* Q_H is positive, and Q_C is negative. The *net* heat Q absorbed per cycle is

$$Q = Q_H + Q_C = |Q_H| - |Q_C| \tag{20.1}$$

The useful output of the engine is the net work W done by the working substance. From the first law,

$$W = Q = Q_H + Q_C = |Q_H| - |Q_C| \tag{20.2}$$

Ideally, we would like to convert *all* the heat Q_H into work; in that case we would have $Q_H = W$ and $Q_C = 0$. Experience shows that this is impossible; there is always some heat wasted, and Q_C *is never zero.* We define the **thermal efficiency** of an engine, denoted by e, as the quotient

$$e = \frac{W}{Q_H} \tag{20.3}$$

The thermal efficiency e represents the fraction of Q_H that *is* converted to work. To put it another way, e is what you get divided by what you pay for. This is always less than unity, an all-too-familiar experience! In terms of the flow diagram of Fig. 20.3, the most efficient engine is the one for which the branch pipeline representing the work output is as wide as possible and the exhaust pipeline representing the heat thrown away is as narrow as possible.

When we substitute the two expressions for W given by Eq. (20.2) into Eq. (20.3), we get the following equivalent expressions for e:

$$e = \frac{W}{Q_H} = 1 + \frac{Q_C}{Q_H} = 1 - \left|\frac{Q_C}{Q_H}\right| \quad \text{(thermal efficiency of an engine)} \tag{20.4}$$

Note that e is a quotient of two energy quantities and thus is a pure number, without units. Of course, we must always express W, Q_H, and Q_C in the same units.

Problem-Solving Strategy — Heat Engines

Problems involving heat engines are, first and foremost, problems in the first law of thermodynamics. Hence the Problem-Solving Strategy in Section 19.4 is equally useful throughout the present chapter, and we suggest that you reread it.

IDENTIFY *the relevant concepts:* A heat engine is any device that converts work partially to heat, as shown schematically in Fig. 20.3. We will see in Section 20.4 that a refrigerator is essentially a heat engine running in reverse, so many of the same concepts apply.

SET UP *the problem* as suggested in the Problem-Solving Strategy in Section 19.4. Equation (20.4) is useful in situations for which the thermal efficiency of the engine is relevant.

EXECUTE *the solution as follows:*

1. Be very careful with the sign conventions for W and the various Q's. W is positive when the system expands and does work; W is negative when the system is compressed. Each Q is positive if it represents heat entering the system and is negative if it represents heat leaving the system. When you know that a quantity is negative, such as Q_C in the above discussion, it sometimes helps to write it as $Q_C = -|Q_C|$.
2. Some problems deal with power rather than energy quantities. Power is work per unit time $(P = W/t)$, and rate of heat transfer (heat current) H is heat transfer per unit time $(H = Q/t)$. In such problems it helps to ask, "What is W or Q in one second (or one hour)?"
3. Keeping steps (1) and (2) in mind, solve for the target variables.

EVALUATE *your answer:* Use the first law of thermodynamics to check your results, paying particular attention to algebraic signs.

Example 20.1 — Analyzing a heat engine

A gasoline engine in a large truck takes in 10,000 J of heat and delivers 2000 J of mechanical work per cycle. The heat is obtained by burning gasoline with heat of combustion $L_c = 5.0 \times 10^4$ J/g. a) What is the thermal efficiency of this engine? b) How much heat is discarded in each cycle? c) How much gasoline is burned in each cycle? d) If the engine goes through 25 cycles per second, what is its power output in watts? In horsepower? e) How much gasoline is burned per second? Per hour?

SOLUTION

IDENTIFY and SET UP: The problem is about a heat engine, for which we are told the amount of work done per cycle $(W = 2000\text{ J})$ and the amount of heat taken in per cycle $(Q_H = 10{,}000\text{ J})$. Hence we use the first form of Eq. (20.4) to find the thermal efficiency. The first law of thermodynamics tells us the amount of heat rejected per cycle, and the heat of combustion tells us how much gasoline must be burned per cycle and hence per unit time.

EXECUTE: a) From the first expression in Eq. (20.4), the thermal efficiency is

$$e = \frac{W}{Q_H} = \frac{2000\text{ J}}{10{,}000\text{ J}} = 0.20 = 20\%$$

This is a fairly typical figure for cars and trucks if W includes only the work actually delivered to the wheels. b) From Eq. (20.2), $W = Q_H + Q_C$ so

$$Q_C = W - Q_H = 2000\text{ J} - 10{,}000\text{ J}$$
$$= -8000\text{ J}$$

That is, 8000 J of heat leaves the engine during each cycle.

c) Let m be the mass of gasoline burned during each cycle. Then

$$Q_H = mL_c$$
$$m = \frac{Q_H}{L_c} = \frac{10{,}000\text{ J}}{5.0 \times 10^4\text{ J/g}} = 0.20\text{ g}$$

d) The power P (rate of doing work) is the work per cycle multiplied by the number of cycles per second:

$$P = (2000\text{ J/cycle})(25\text{ cycles/s}) = 50{,}000\text{ W} = 50\text{ kW}$$
$$= (50{,}000\text{ W})\frac{1\text{ hp}}{746\text{ W}} = 67\text{ hp}$$

e) The mass of gasoline burned per second is the mass per cycle multiplied by the number of cycles per second:

$$(0.20\text{ g/cycle})(25\text{ cycles/s}) = 5.0\text{ g/s}$$

The mass burned per hour is

$$(5.0\text{ g/s})\frac{3600\text{ s}}{1\text{ h}} = 18{,}000\text{ g/h} = 18\text{ kg/h}$$

EVALUATE: We can check our result in part (e) by converting it to a more familiar quantity, the amount of fuel consumed per unit distance.The density of gasoline is about 0.70 g/cm^3, so this is about 25,700 cm^3, 25.7 L, or 6.8 gallons of gasoline per hour. If the truck is traveling at 55 mi/h (88 km/h), this represents fuel consumption of 8.1 miles/gallon (3.4 km/L). This is substantially worse fuel consumption than a passenger car, but fairly typical of large trucks.

Test Your Understanding

How much heat is rejected into the environment during a 24-hour period by 1000 trucks like those in Example 20.1 driving around a city?

20.3 | Internal-Combustion Engines

The gasoline engine, used in automobiles and many other types of machinery, is a familiar example of a heat engine. Let's look at its thermal efficiency. Figure 20.4 shows the operation of one type of gasoline engine. First a mixture of air and gasoline vapor flows into a cylinder through an open intake valve while the piston descends, increasing the volume of the cylinder from a minimum of V (when the piston is all the way up) to a maximum of rV (when it is all the way down). The quantity r is called the **compression ratio;** for present-day automobile engines its value is typically 8 to 10. At the end of this *intake stroke,* the intake valve closes and the mixture is compressed, approximately adiabatically, to volume V during the *compression stroke.* The mixture is then ignited by the spark plug, and the heated gas expands, approximately adiabatically, back to volume rV, pushing on the piston and doing work; this is the *power stroke.* Finally, the exhaust valve opens, and the combustion products are pushed out (during the *exhaust stroke*), leaving the cylinder ready for the next intake stroke.

The Otto Cycle

Figure 20.5 is a pV-diagram for an idealized model of the thermodynamic processes in a gasoline engine. This model is called the **Otto cycle.** At point a the gasoline–air mixture has entered the cylinder. The mixture is compressed adiabatically to point b and is then ignited. Heat Q_H is added to the system by the burn-

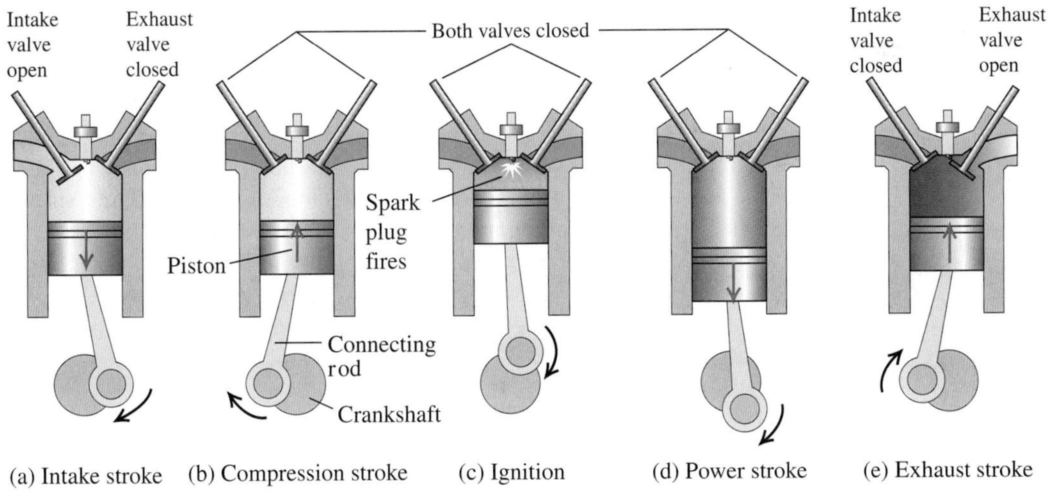

20.4 Cycle of a four-stroke internal-combustion engine. (a) *Intake stroke:* Piston moves down, causing a partial vacuum in cylinder; gasoline–air mixture flows through open intake valve into cylinder. (b) *Compression stroke:* Intake valve closes, and mixture is compressed as piston moves up. (c) *Ignition:* Spark plug ignites mixture. (d) *Power stroke:* Hot burned mixture pushes piston down, doing work. (e) *Exhaust stroke:* Exhaust valve opens and piston moves up, pushing burned mixture out of cylinder. Engine is now ready for next intake stroke, and the cycle repeats.

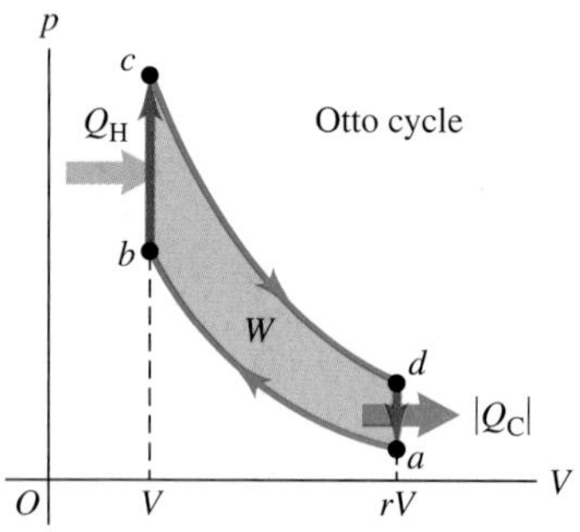

$a \to b$: Compression stroke (adiabatic compression)
$b \to c$: Ignite fuel (heating at constant volume)
$c \to d$: Power stroke (adiabatic expansion)
$d \to a$: Reject heat to environment (cooling at constant volume)

20.5 The pV-diagram for the Otto cycle, an idealized model of the thermodynamic processes in a gasoline engine.

ing gasoline along line bc, and the power stroke is the adiabatic expansion to d. The gas is cooled to the temperature of the outside air along line da; during this process, heat $|Q_C|$ is rejected. In practice, this gas leaves the engine as exhaust and does not enter the engine again. But since an equivalent amount of gasoline and air enters, we may consider the process to be cyclic.

We can calculate the efficiency of this idealized cycle. Processes bc and da are constant-volume, so the heats Q_H and Q_C are related simply to the temperatures:

$$Q_H = nC_V(T_c - T_b) > 0$$

$$Q_C = nC_V(T_a - T_d) < 0$$

The thermal efficiency is given by Eq. (20.4). Inserting the above expressions and cancelling out the common factor nC_V, we find

$$e = \frac{Q_H + Q_C}{Q_H} = \frac{T_c - T_b + T_a - T_d}{T_c - T_b} \tag{20.5}$$

To simplify this further, we use the temperature-volume relation for adiabatic processes for an ideal gas, Eq. (19.22). For the two adiabatic processes ab and cd,

$$T_a(rV)^{\gamma-1} = T_b V^{\gamma-1} \quad \text{and} \quad T_d(rV)^{\gamma-1} = T_c V^{\gamma-1}$$

We divide each of these equations by the common factor $V^{\gamma-1}$ and substitute the resulting expressions for T_b and T_c back into Eq. (20.5). The result is

$$e = \frac{T_d r^{\gamma-1} - T_a r^{\gamma-1} + T_a - T_d}{T_d r^{\gamma-1} - T_a r^{\gamma-1}} = \frac{(T_d - T_a)(r^{\gamma-1} - 1)}{(T_d - T_a)r^{\gamma-1}}$$

Dividing out the common factor $(T_d - T_a)$, we get

$$e = 1 - \frac{1}{r^{\gamma-1}} \quad \text{(thermal efficiency in Otto cycle)} \tag{20.6}$$

The thermal efficiency given by Eq. (20.6) is always less than unity, even for this idealized model. With $r = 8$ and $\gamma = 1.4$ (the value for air) the theoretical efficiency is $e = 0.56$, or 56%. The efficiency can be increased by increasing r. However, this also increases the temperature at the end of the adiabatic compression of the air–fuel mixture. If the temperature is too high, the mixture explodes spontaneously during compression instead of burning evenly after the spark plug ignites it. This is called *pre-ignition* or *detonation;* it causes a knocking sound and can damage the engine. The octane rating of a gasoline is a measure of its antiknock qualities. The maximum practical compression ratio for high-octane, or "premium," gasoline is about 10. Higher ratios can be used with more exotic fuels.

The Otto cycle, which we have just described, is a highly idealized model. It assumes that the mixture behaves as an ideal gas; it neglects friction, turbulence, loss of heat to cylinder walls, and many other effects that combine to reduce the efficiency of a real engine. Another source of inefficiency is incomplete combustion. A mixture of gasoline vapor with just enough air for complete combustion of the hydrocarbons to H_2O and CO_2 does not ignite readily. Reliable ignition requires a mixture that is "richer" in gasoline. The resulting incomplete combustion leads to CO and unburned hydrocarbons in the exhaust. The heat obtained from the gasoline is then less than the total heat of combustion; the difference is wasted, and the exhaust products contribute to air pollution. Efficiencies of real gasoline engines are typically around 35%.

The Diesel Cycle

The Diesel engine is similar in operation to the gasoline engine. The most important difference is that there is no fuel in the cylinder at the beginning of the compression stroke. A little before the beginning of the power stroke, the injectors start to inject fuel directly into the cylinder, just fast enough to keep the pressure approximately constant during the first part of the power stroke. Because of the high temperature developed during the adiabatic compression, the fuel ignites spontaneously as it is injected; no spark plugs are needed.

The idealized **Diesel cycle** is shown in Fig. 20.6. Starting at point a, air is compressed adiabatically to point b, heated at constant pressure to point c, expanded adiabatically to point d, and cooled at constant volume to point a. Because there is no fuel in the cylinder during most of the compression stroke, pre-ignition cannot occur, and the compression ratio r can be much higher than for a gasoline engine. This improves efficiency and ensures reliable ignition when the fuel is injected (because of the high temperature reached during the adiabatic compression). Values of r of 15 to 20 are typical; with these values and $\gamma = 1.4$, the theoretical efficiency of the idealized Diesel cycle is about 0.65 to 0.70. As with the Otto cycle, the efficiency of any actual engine is substantially less than this. While Diesel engines are very efficient, they must be built to much tighter tolerances than gasoline engines and the fuel-injection system requires careful maintenance.

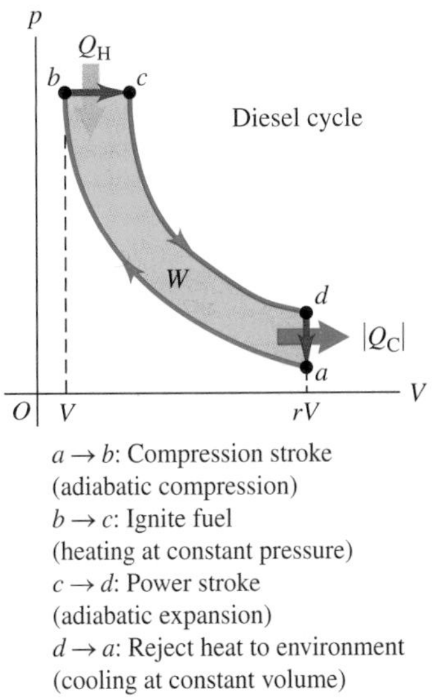

20.6 The pV-diagram for the idealized Diesel cycle.

Test Your Understanding

Use Fig. 20.5, *not* Eq. (20.6), to explain why the efficiency of an Otto-cycle engine increases with increasing compression ratio.

20.4 | Refrigerators

We can think of a **refrigerator** as a heat engine operating in reverse. A heat engine takes heat from a hot place and gives off heat to a colder place. A refrigerator does the opposite; it takes heat from a cold place (the inside of the refrigerator) and gives it off to a warmer place (usually the air in the room where the refrigerator is located). A heat engine has a net *output* of mechanical work; the refrigerator requires a net *input* of mechanical work. Using the sign conventions from Section 20.2, for a refrigerator Q_C is positive but both W and Q_H are negative; hence $|W| = -W$ and $|Q_H| = -Q_H$.

A flow diagram for a refrigerator is shown in Fig. 20.7. From the first law for a cyclic process,

$$Q_H + Q_C - W = 0 \quad \text{or} \quad -Q_H = Q_C - W$$

or, because both Q_H and W are negative,

$$|Q_H| = Q_C + |W| \tag{20.7}$$

Thus, as the diagram shows, the heat $|Q_H|$ leaving the working substance and given to the hot reservoir is always *greater* than the heat Q_C taken from the cold reservoir. Note that the absolute-value relation

$$|Q_H| = |Q_C| + |W| \tag{20.8}$$

is valid for both heat engines and refrigerators.

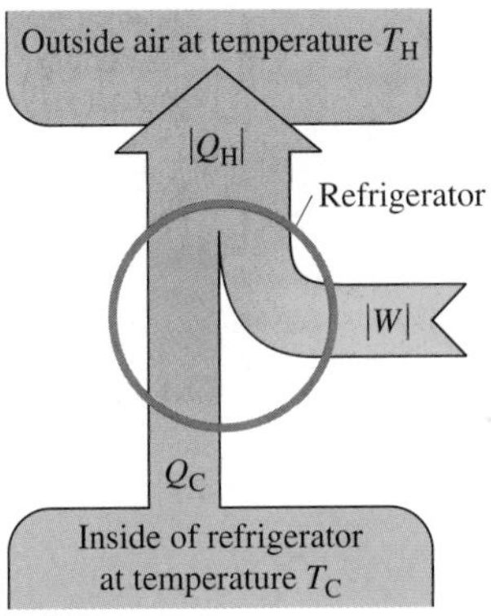

20.7 Schematic energy-flow diagram of a refrigerator.

From an economic point of view, the best refrigeration cycle is one that removes the greatest amount of heat $|Q_C|$ from the inside of the refrigerator for the least expenditure of mechanical work, $|W|$. The relevant ratio is therefore $|Q_C|/|W|$; the larger this ratio, the better the refrigerator. We call this ratio the **coefficient of performance,** denoted by K. From Eq. (20.8), $|W| = |Q_H| - |Q_C|$, so

$$K = \frac{|Q_C|}{|W|} = \frac{|Q_C|}{|Q_H| - |Q_C|} \tag{20.9}$$

(coefficient of performance of a refrigerator)

As always, we measure Q_H, Q_C, and W all in the same energy units; K is then a dimensionless number.

The principles of the common refrigeration cycle are shown schematically in Fig. 20.8a. The fluid "circuit" contains a refrigerant fluid (the working substance). The left side of the circuit (including the cooling coils inside the refrigerator) is at low temperature and low pressure; the right side (including the condenser coils outside the refrigerator) is at high temperature and high pressure. Ordinarily, both sides contain liquid and vapor in phase equilibrium.

The compressor takes in fluid, compresses it adiabatically, and delivers it to the condenser coil at high pressure. The fluid temperature is then higher than that of the air surrounding the condenser, so the refrigerant gives off heat $|Q_H|$ and partially condenses to liquid. The fluid then expands adiabatically into the evaporator at a rate controlled by the expansion valve. As the fluid expands, it cools considerably, enough that the fluid in the evaporator coil is colder than its surroundings. It absorbs heat $|Q_C|$ from its surroundings, cooling them and partially vaporizing. The fluid then enters the compressor to begin another cycle. The compressor, usually driven by an electric motor (Fig. 20.8b), requires energy input and does work $|W|$ *on* the working substance during each cycle.

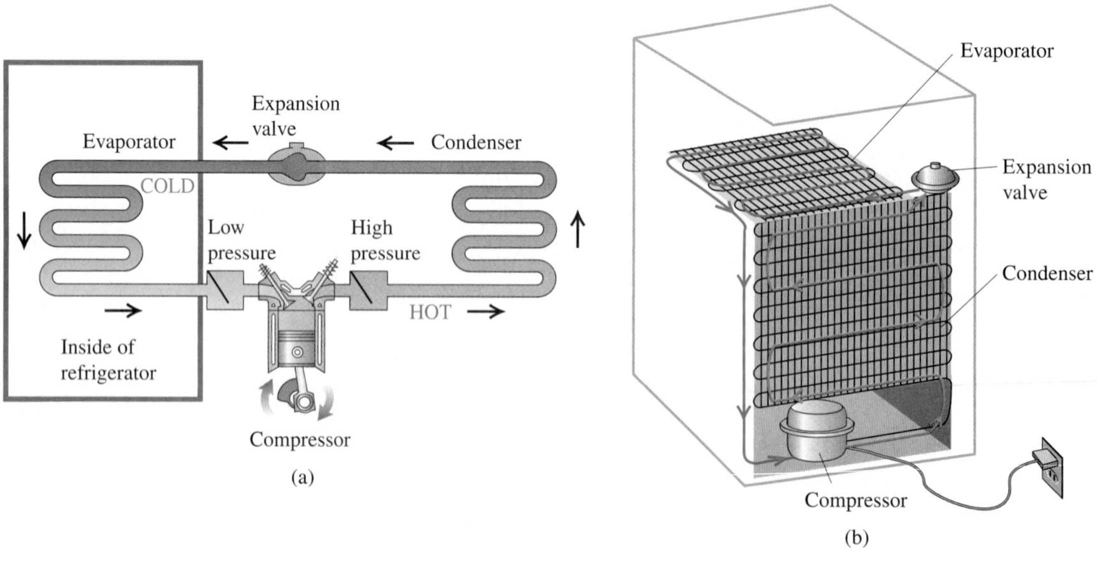

20.8 (a) Principle of the mechanical refrigeration cycle. (b) How the key elements are arranged in a practical refrigerator.

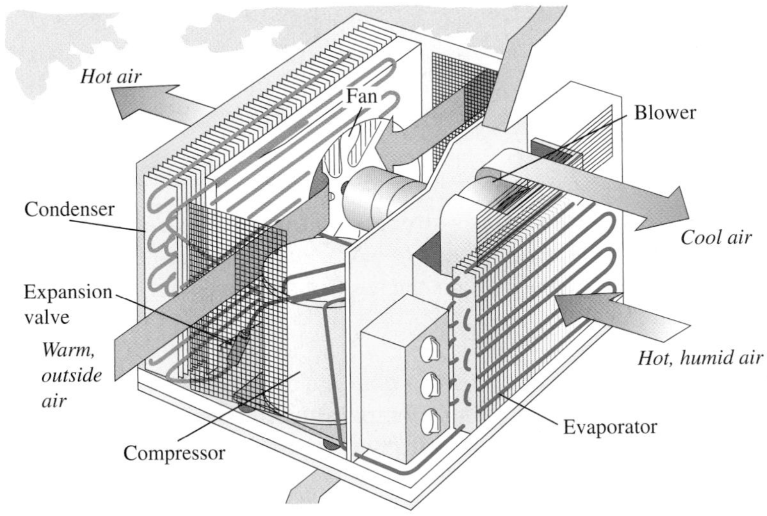

20.9 An air conditioner works on the same principle as a refrigerator.

An air conditioner operates on exactly the same principle. In this case the refrigerator box becomes a room or an entire building. The evaporator coils are inside, the condenser is outside, and fans circulate air through these (Fig. 20.9). In large installations the condenser coils are often cooled by water. For air conditioners the quantities of greatest practical importance are the *rate* of heat removal (the heat current H from the region being cooled) and the *power* input $P = W/t$ to the compressor. If heat $|Q_C|$ is removed in time t, then $H = |Q_C|/t$. Then we can express the coefficient of performance as

$$K = \frac{|Q_C|}{|W|} = \frac{Ht}{Pt} = \frac{H}{P}$$

Typical room air conditioners have heat removal rates H of 5000 to 10,000 Btu/h, or about 1500–3000 W, and require electric power input of about 600 to 1200 W. Typical coefficients of performance are about 2.5; the actual values depend on the inside and outside temperatures.

Unfortunately, K is often expressed commercially in mixed units, with H in Btu per hour and P in watts. In these units, H/P is called the **energy efficiency rating** (EER); the units, customarily omitted, are (Btu/h)/W. Because 1 W = 3.413 Btu/h, the EER is numerically 3.413 times as large as the dimensionless K. Room air conditioners typically have an EER of 7 to 10.

A variation on this theme is the **heat pump,** used to heat buildings by cooling the outside air. It functions like a refrigerator turned inside out. The evaporator coils are outside, where they take heat from cold air, and the condenser coils are inside, where they give off heat to the warmer air. With proper design, the heat $|Q_H|$ delivered to the inside per cycle can be considerably greater than the work $|W|$ required to get it there.

Work is *always* needed to transfer heat from a colder to a hotter body. Heat flows spontaneously from hotter to colder, and to reverse this flow requires the addition of work from the outside. Experience shows that it is impossible to make a refrigerator that transports heat from a colder body to a hotter body without the addition of work. If no work were needed, the coefficient of performance would be infinite. We call such a device a *workless refrigerator;* it is a mythical beast, like the unicorn and the free lunch.

?

Test Your Understanding

Why can't you cool your house by leaving the refrigerator door open?

20.5 | The Second Law of Thermodynamics

Experimental evidence suggests strongly that it is *impossible* to build a heat engine that converts heat completely to work, that is, an engine with 100% thermal efficiency. This impossibility is the basis of one statement of the **second law of thermodynamics,** as follows:

> **It is impossible for any system to undergo a process in which it absorbs heat from a reservoir at a single temperature and converts the heat completely into mechanical work, with the system ending in the same state in which it began.**

We will call this the "engine" statement of the second law. (It is also known to physicists as the *Kelvin-Planck statement* of this law.)

The basis of the second law of thermodynamics lies in the difference between the nature of internal energy and that of macroscopic mechanical energy. In a moving body the molecules have random motion, but superimposed on this is a coordinated motion of every molecule in the direction of the body's velocity. The kinetic energy associated with this *coordinated* macroscopic motion is what we call the kinetic energy of the moving body. The kinetic and potential energies associated with the *random* motion constitute the internal energy.

When a body sliding on a surface comes to rest as a result of friction, the organized motion of the body is converted to random motion of molecules in the body and in the surface. Since we cannot control the motions of individual molecules, we cannot convert this random motion completely back to organized motion. We can convert *part* of it, and this is what a heat engine does.

If the second law were *not* true, we could power an automobile or run a power plant by cooling the surrounding air. Neither of these impossibilities violates the *first* law of thermodynamics. The second law, therefore, is not a deduction from the first but stands by itself as a separate law of nature. The first law denies the possibility of creating or destroying energy; the second law limits the *availability* of energy and the ways in which it can be used and converted.

Our analysis of refrigerators in Section 20.4 forms the basis for an alternative statement of the second law of thermodynamics. Heat flows spontaneously from hotter to colder bodies, never the reverse. A refrigerator does take heat from a colder to a hotter body, but its operation requires an input of mechanical energy or work. Generalizing this observation, we state:

> **It is impossible for any process to have as its sole result the transfer of heat from a cooler to a hotter body.**

We'll call this the "refrigerator" statement of the second law (it is also known as the *Clausius statement*). It may not seem to be very closely related to the "engine" statement. In fact, though, the two statements are completely equivalent. For example, if we could build a workless refrigerator, violating the second or "refrigerator" statement of the second law, we could use it in conjunction with a heat engine, pumping the heat rejected by the engine back to the hot reservoir to be

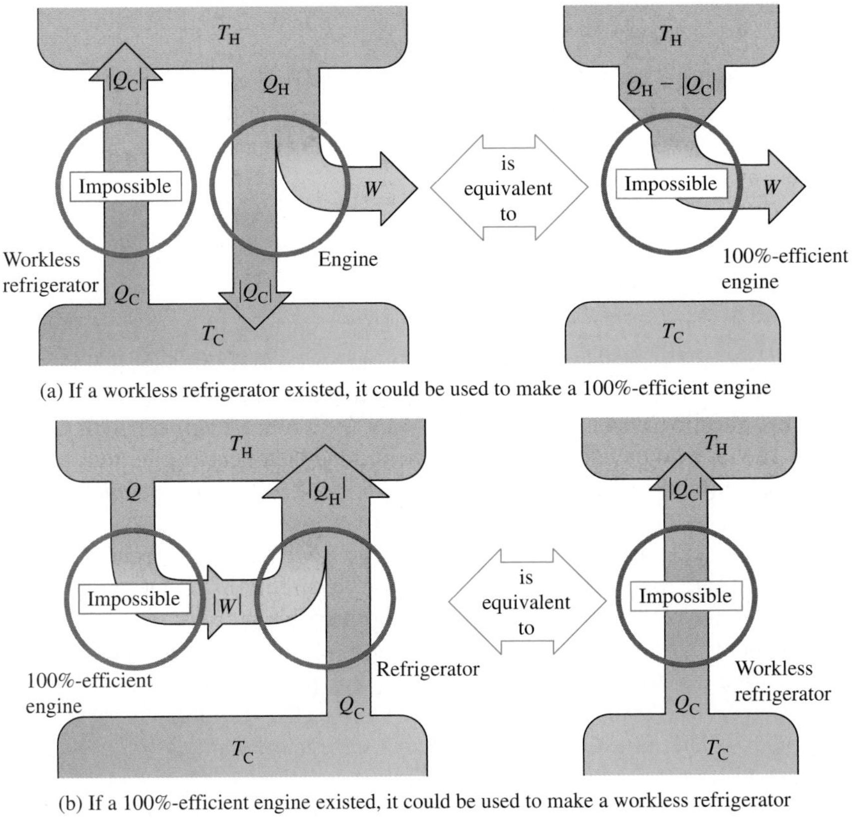

20.10 Energy-flow diagrams showing that the two forms of the second law are equivalent. (a) Using an impossible workless refrigerator and an ordinary heat engine to form a composite device that functions as an equally impossible 100%-efficient engine. (b) Using an impossible 100%-efficient engine and an ordinary refrigerator to form an equally impossible workless refrigerator.

reused. This composite machine (Fig. 20.10a) would violate the "engine" statement of the second law because its net effect would be to take a net quantity of heat $Q_H - |Q_C|$ from the hot reservoir and convert it completely to work W.

Alternatively, if we could make an engine with 100% thermal efficiency, in violation of the first statement, we could run it using heat from the hot reservoir and use the work output to drive a refrigerator that pumps heat from the cold reservoir to the hot (Fig. 20.10b). This composite device would violate the "refrigerator" statement because its net effect would be to take heat Q_C from the cold reservoir and deliver it to the hot reservoir without requiring any input of work. Thus any device that violates one form of the second law can be used to make a device that violates the other form. If violations of the first form are impossible, so are violations of the second!

The conversion of work to heat, as in friction or viscous fluid flow, and heat flow from hot to cold across a finite temperature gradient, are *irreversible* processes. The "engine" and "refrigerator" statements of the second law state that these processes can be only partially reversed. We could cite other examples. Gases always seep spontaneously through an opening from a region of high pressure to a region of low pressure; gases and miscible liquids left by themselves always tend to mix, not to unmix. The second law of thermodynamics is an expression of the inherent one-way aspect of these and many other irreversible processes. Energy conversion is an essential aspect of all plant and animal life and of human technology, so the second law of thermodynamics is of the utmost fundamental importance in the world we live in.

Test Your Understanding

You partially fill a test tube with water and put a cork in the top of the tube. You place a burner under the tube to boil the water, causing the cork to pop off the tube. Explain how this process converts heat into mechanical work, and explain why the process is not 100% efficient.

20.6 | The Carnot Cycle

ActivPhysics ONLINE

8.14 Carnot Cycle

According to the second law, no heat engine can have 100% efficiency. How great an efficiency *can* an engine have, given two heat reservoirs at temperatures T_H and T_C? This question was answered in 1824 by the French engineer Sadi Carnot (1796-1832), who developed a hypothetical, idealized heat engine that has the maximum possible efficiency consistent with the second law. The cycle of this engine is called the **Carnot cycle.**

To understand the rationale of the Carnot cycle, we return to a recurrent theme in this chapter: *reversibility* and its relation to directions of thermodynamic processes. Conversion of work to heat is an irreversible process; the purpose of a heat engine is a *partial* reversal of this process, the conversion of heat to work with as great an efficiency as possible. For maximum heat-engine efficiency, therefore, *we must avoid all irreversible processes.* This requirement turns out to be enough to determine the basic sequence of steps in the Carnot cycle, as we will show next.

Heat flow through a finite temperature drop is an irreversible process. Therefore, during heat transfer in the Carnot cycle there must be *no* finite temperature difference. When the engine takes heat from the hot reservoir at temperature T_H, the working substance of the engine must also be at T_H; otherwise, irreversible heat flow would occur. Similarly, when the engine discards heat to the cold reservoir at T_C, the engine itself must be at T_C. That is, every process that involves heat transfer must be *isothermal* at either T_H or T_C.

Conversely, in any process in which the temperature of the working substance of the engine is intermediate between T_H and T_C, there must be *no* heat transfer between the engine and either reservoir because such heat transfer could not be reversible. Therefore any process in which the temperature T of the working substance changes must be *adiabatic.*

The bottom line is that every process in our idealized cycle must be either isothermal or adiabatic. In addition, thermal and mechanical equilibrium must be maintained at all times so that each process is completely reversible.

The Carnot Cycle

The Carnot cycle consists of two reversible isothermal and two reversible adiabatic processes. Figure 20.11 shows a Carnot cycle using as its working substance an ideal gas in a cylinder with a piston. It consists of the following steps:

1. The gas expands isothermally at temperature T_H, absorbing heat Q_H (*ab*).
2. It expands adiabatically until its temperature drops to T_C (*bc*).
3. It is compressed isothermally at T_C, rejecting heat $|Q_C|$ (*cd*).
4. It is compressed adiabatically back to its initial state at temperature T_H (*da*).

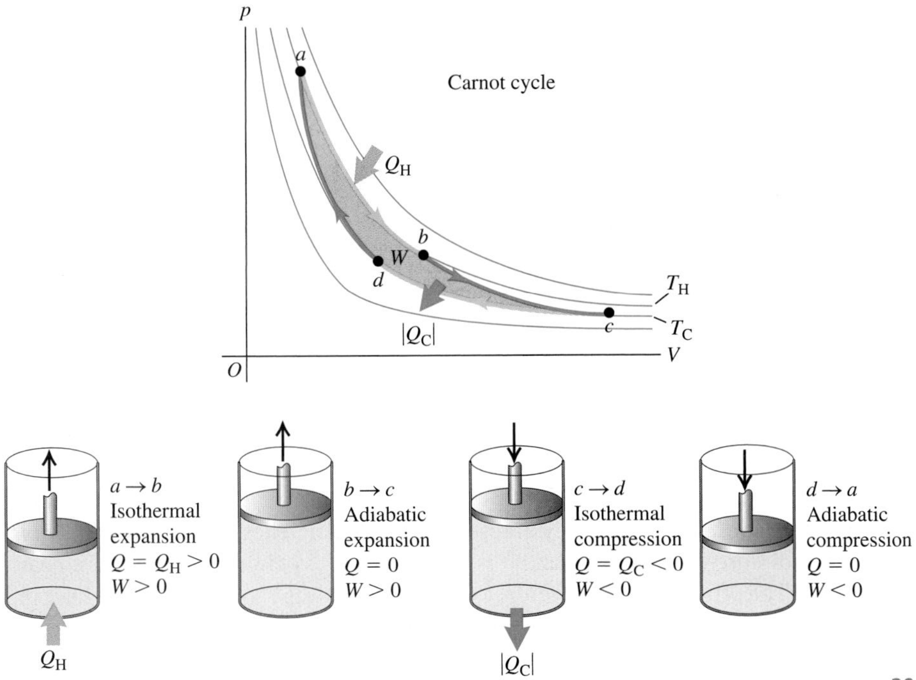

20.11 The Carnot cycle for an ideal gas. The light blue lines in the pV-diagram are isotherms (curves of constant temperature) and the dark blue lines are adiabats (curves of zero heat flow).

We can calculate the thermal efficiency e of a Carnot engine in the special case in which the working substance is an *ideal gas*. To carry out this calculation, we will first find the ratio Q_C/Q_H of the quantities of heat transferred in the two isothermal processes and then use Eq. (20.4) to find e.

For an ideal gas the internal energy U depends only on temperature and is thus constant in any isothermal process. For the isothermal expansion ab, $\Delta U_{ab} = 0$ and Q_H is equal to the work W_{ab} done by the gas during its isothermal expansion at temperature T_H. We calculated this work in Example 19.1 (Section 19.2); using that result, we have

$$Q_H = W_{ab} = nRT_H \ln \frac{V_b}{V_a} \tag{20.10}$$

Similarly,

$$Q_C = W_{cd} = nRT_C \ln \frac{V_d}{V_c} = -nRT_C \ln \frac{V_c}{V_d} \tag{20.11}$$

Because V_d is less than V_c, Q_C is negative $(Q_C = -|Q_C|)$; heat flows out of the gas during the isothermal compression at temperature T_C.

The ratio of the two quantities of heat is thus

$$\frac{Q_C}{Q_H} = -\left(\frac{T_C}{T_H}\right) \frac{\ln (V_c/V_d)}{\ln (V_b/V_a)} \tag{20.12}$$

This can be simplified further by use of the temperature-volume relation for an adiabatic process, Eq. (19.22). We find for the two adiabatic processes:

$$T_H V_b^{\gamma-1} = T_C V_c^{\gamma-1} \quad \text{and} \quad T_H V_a^{\gamma-1} = T_C V_d^{\gamma-1}$$

Dividing the first of these by the second, we find

$$\frac{V_b^{\gamma-1}}{V_a^{\gamma-1}} = \frac{V_c^{\gamma-1}}{V_d^{\gamma-1}} \quad \text{and} \quad \frac{V_b}{V_a} = \frac{V_c}{V_d}$$

Thus the two logarithms in Eq. (20.12) are equal, and that equation reduces to

$$\frac{Q_C}{Q_H} = -\frac{T_C}{T_H} \quad \text{or} \quad \frac{|Q_C|}{|Q_H|} = \frac{T_C}{T_H} \quad \text{(heat transfer in a Carnot engine)} \quad (20.13)$$

The ratio of the heat rejected at T_C to the heat absorbed at T_H is just equal to the ratio T_C/T_H. Then from Eq. (20.4) the efficiency of the Carnot engine is

$$e_{\text{Carnot}} = 1 - \frac{T_C}{T_H} = \frac{T_H - T_C}{T_H} \quad \text{(efficiency of a Carnot engine)} \quad (20.14)$$

This simple result says that the efficiency of a Carnot engine depends only on the temperatures of the two heat reservoirs. The efficiency is large when the temperature *difference* is large, and it is very small when the temperatures are nearly equal. The efficiency can never be exactly unity unless $T_C = 0$; we'll see later that this, too, is impossible.

CAUTION In all calculations involving the Carnot cycle, you must make sure that you use *absolute* (Kelvin) temperatures only. That's because Eqs. (20.10) through (20.14) come from the ideal-gas equation $pV = nRT$, in which T is absolute temperature.

Example 20.2 Analyzing a Carnot engine I

A Carnot engine takes 2000 J of heat from a reservoir at 500 K, does some work, and discards some heat to a reservoir at 350 K. How much work does it do, how much heat is discarded, and what is the efficiency?

SOLUTION

IDENTIFY and SET UP: For this Carnot engine we are given $Q_H = 2000$ J, the amount of heat absorbed, and the temperatures $T_H = 500$ K and $T_C = 350$ K of the hot and cold reservoirs, respectively. We find the amount of heat discarded using Eq. (20.13), then calculate the amount of work done using the first law of thermodynamics: the work done in a complete cycle is the sum of the heat absorbed and the (negative) heat discarded (see Eq. (20.2)). We find the efficiency from the two temperatures using Eq. (20.14).

EXECUTE: From Eq. (20.13) the heat Q_C discarded by the engine is

$$Q_C = -Q_H \frac{T_C}{T_H} = -(2000\text{ J})\frac{350\text{ K}}{500\text{ K}}$$
$$= -1400\text{ J}$$

Then from the first law, the work W done by the engine is

$$W = Q_H + Q_C = 2000\text{ J} + (-1400\text{ J})$$
$$= 600\text{ J}$$

From Eq. (20.14) the thermal efficiency is

$$e = 1 - \frac{T_C}{T_H} = 1 - \frac{350\text{ K}}{500\text{ K}} = 0.30 = 30\%$$

EVALUATE: The negative sign of Q_C is correct: it shows that heat flows *out* of the engine and into the cold reservoir. Note that we can check our result for e by using the basic definition of thermal efficiency,

$$e = \frac{W}{Q_H} = \frac{600\text{ J}}{2000\text{ J}} = 0.30 = 30\%$$

Example 20.3 Analyzing a Carnot engine II

Suppose 0.200 mol of an ideal diatomic gas $(\gamma = 1.40)$ undergoes a Carnot cycle with temperatures 227°C and 27°C. The initial pressure is $p_a = 10.0 \times 10^5$ Pa, and during the isothermal expansion at the higher temperature the volume doubles. a) Find the pressure and volume at each of points a, b, c, and d in the pV-diagram of Fig. 20.11. b) Find Q, W, and ΔU for each step and for the entire cycle. c) Determine the efficiency directly from the results of part (b), and compare it with the result from Eq. (20.14).

SOLUTION

IDENTIFY: This problem involves the properties of the Carnot cycle as well as those of an ideal gas.

SET UP: We are given the number of moles and the pressure and temperature at point a (which is at the higher of the two reservoir temperatures), so we can find the volume at a using the ideal-gas equation. We then find the pressure and volume at the other points from the equations given in this section in combination with the ideal-gas equation. Next, for each step in the cycle, we use Eqs. (20.10) and (20.11) to find the heat flow and work done and we use Eq. (19.13) to calculate the internal energy change. As in the preceding example, we find the efficiency using Eq. (20.14).

EXECUTE: a) We first remember to convert the Celsius temperatures to absolute temperatures: The higher temperature is $T_H = (227 + 273.15)$ K $= 500$ K and the lower temperature is $T_C = (27 + 273.15)$ K $= 300$ K. We then use the ideal-gas equation to find V_a:

$$V_a = \frac{nRT_H}{p_a} = \frac{(0.200\text{ mol})(8.314\text{ J/mol}\cdot\text{K})(500\text{ K})}{10.0 \times 10^5\text{ Pa}}$$
$$= 8.31 \times 10^{-4}\text{ m}^3$$

The volume doubles during the isothermal expansion $a \rightarrow b$, so

$$V_b = 2V_a = 2(8.31 \times 10^{-4}\text{ m}^3)$$
$$= 16.6 \times 10^{-4}\text{ m}^3$$

Also, during the isothermal expansion $a \rightarrow b$, $p_aV_a = p_bV_b$, so

$$p_b = \frac{p_aV_a}{V_b} = 5.00 \times 10^5\text{ Pa}$$

For the adiabatic expansion $b \rightarrow c$, $T_H V_b^{\gamma-1} = T_C V_c^{\gamma-1}$, so

$$V_c = V_b\left(\frac{T_H}{T_C}\right)^{1/(\gamma-1)} = (16.6 \times 10^{-4}\text{ m}^3)\left(\frac{500\text{ K}}{300\text{ K}}\right)^{2.5}$$
$$= 59.6 \times 10^{-4}\text{ m}^3$$

Using the ideal-gas equation again for point c, we find

$$p_c = \frac{nRT_C}{V_c} = \frac{(0.200\text{ mol})(8.314\text{ J/mol}\cdot\text{K})(300\text{ K})}{59.6 \times 10^{-4}\text{ m}^3}$$
$$= 0.837 \times 10^5\text{ Pa}$$

For the adiabatic compression $d \rightarrow a$, $T_C V_d^{\gamma-1} = T_H V_a^{\gamma-1}$, and

$$V_d = V_a\left(\frac{T_H}{T_C}\right)^{1/(\gamma-1)} = (8.31 \times 10^{-4}\text{ m}^3)\left(\frac{500\text{ K}}{300\text{ K}}\right)^{2.5}$$
$$= 29.8 \times 10^{-4}\text{ m}^3$$
$$p_d = \frac{nRT_C}{V_d} = \frac{(0.200\text{ mol})(8.314\text{ J/mol}\cdot\text{K})(300\text{ K})}{29.8 \times 10^{-4}\text{ m}^3}$$
$$= 1.67 \times 10^5\text{ Pa}$$

b) For the isothermal expansion $a \rightarrow b$, $\Delta U_{ab} = 0$. To find W_{ab} $(= Q_H)$, we use Eq. (20.10):

$$W_{ab} = Q_H = nRT_H \ln\frac{V_b}{V_a}$$
$$= (0.200\text{ mol})(8.314\text{ J/mol}\cdot\text{K})(500\text{ K})(\ln 2) = 576\text{ J}$$

For the adiabatic expansion $b \rightarrow c$, $Q_{bc} = 0$. From the first law of thermodynamics, $\Delta U_{bc} = Q_{bc} - W_{bc} = -W_{bc}$; hence the work W_{bc} done by the gas in this process equals the negative of the change in internal energy of the gas. From Eq. (19.13) we have $\Delta U = nC_V\Delta T$, where $\Delta T = T_C - T_H$ (final temperature minus initial temperature). Using $C_V = 20.8$ J/mol·K for an ideal diatomic gas, we find

$$W_{bc} = -\Delta U_{bc} = -nC_V(T_C - T_H) = nC_V(T_H - T_C)$$
$$= (0.200\text{ mol})(20.8\text{ J/mol}\cdot\text{K})(500\text{ K} - 300\text{ K}) = 832\text{ J}$$

For the isothermal compression $c \rightarrow d$, $\Delta U_{cd} = 0$; Eq. (20.11) gives

$$W_{cd} = Q_C = nRT_C \ln\frac{V_d}{V_c}$$
$$= (0.200\text{ mol})(8.314\text{ J/mol}\cdot\text{K})(300\text{ K})\left(\ln\frac{29.8 \times 10^{-4}\text{ m}^3}{59.6 \times 10^{-4}\text{ m}^3}\right)$$
$$= -346\text{ J}$$

For the adiabatic compression $d \rightarrow a$, $Q_{da} = 0$, and

$$W_{da} = -\Delta U_{da} = -nC_V(T_H - T_C) = nC_V(T_C - T_H)$$
$$= (0.200\text{ mol})(20.8\text{ J/mol}\cdot\text{K})(300\text{ K} - 500\text{ K})$$
$$= -832\text{ J}$$

We can tabulate the results as follows:

PROCESS	Q	W	ΔU
$a \rightarrow b$	576 J	576 J	0
$b \rightarrow c$	0	832 J	−832 J
$c \rightarrow d$	−346 J	−346 J	0
$d \rightarrow a$	0	−832 J	832 J
Total	230 J	230 J	0

c) From the table, $Q_H = 576$ J and the total work is 230 J. Thus

$$e = \frac{W}{Q_H} = \frac{230\ \text{J}}{576\ \text{J}} = 0.40 = 40\%$$

We can compare this with the result from Eq. (20.14):

$$e = \frac{T_H - T_C}{T_H} = \frac{500\ \text{K} - 300\ \text{K}}{500\ \text{K}} = 0.40 = 40\%$$

EVALUATE: In the table of results in part (b), note that for the entire cycle $Q = W$ and $\Delta U = 0$. These results are just what we would expect: in a complete cycle, the net heat input is used to do work with zero net change in the internal energy of the system. Note also that the quantities of work in the two adiabatic processes are negatives of each other. Can you show from the analysis leading to Eq. (20.13) that this must *always* be the case in a Carnot cycle?

Note that the efficiency in this example is greater than that obtained in Example 20.2. That's because the ratio of the high and low temperatures is higher, (500 K)/(300 K) as compared to (500 K)/(350 K).

The Carnot Refrigerator

Because each step in the Carnot cycle is reversible, the *entire cycle* may be reversed, converting the engine into a refrigerator. The coefficient of performance of the Carnot refrigerator is obtained by combining the general definition of K, Eq. (20.9), with Eq. (20.13) for the Carnot cycle. We first rewrite Eq. (20.9) as

$$K = \frac{|Q_C|}{|Q_H| - |Q_C|} = \frac{|Q_C|/|Q_H|}{1 - |Q_C|/|Q_H|}$$

Then we substitute Eq. (20.13), $|Q_C|/|Q_H| = T_C/T_H$, into this expression. The result is

$$K_{\text{Carnot}} = \frac{T_C}{T_H - T_C} \qquad (20.15)$$

(coefficient of performance of a Carnot refrigerator)

When the temperature difference $T_H - T_C$ is small, K is much larger than unity; in this case a lot of heat can be "pumped" from the lower to the higher temperature with only a little expenditure of work. But the greater the temperature difference, the smaller the value of K and the more work is required to transfer a given quantity of heat.

Example 20.4 Analyzing a Carnot refrigerator

If the cycle described in Example 20.3 is run backward as a refrigerator, what is its coefficient of performance?

SOLUTION

IDENTIFY and SET UP: Equation (20.9) gives the coefficient of performance of *any* refrigerator in terms of the heat extracted from the cold reservoir per cycle and the work that must be done per cycle.

EXECUTE: In Example 20.3 we found that in one cycle the Carnot engine rejects heat $Q_C = -346$ J to the cold reservoir and does work $W = 230$ J. Hence, when run in reverse as a refrigerator, the system extracts heat $Q_C = +346$ J from the cold reservoir while requiring a work input of $W = -230$ J. From Eq. (20.9),

$$K = \frac{|Q_C|}{|W|} = \frac{346\ \text{J}}{230\ \text{J}} = 1.50$$

Because the cycle is a Carnot cycle, we may also use Eq. (20.15):

$$K = \frac{T_C}{T_H - T_C} = \frac{300\ \text{K}}{500\ \text{K} - 300\ \text{K}} = 1.50$$

EVALUATE: For a Carnot cycle, e and K depend only on the temperatures, as shown by Eqs. (20.14) and (20.15), and we don't need to calculate Q and W. For cycles containing irreversible processes, however, these two equations are not valid, and more detailed calculations are necessary.

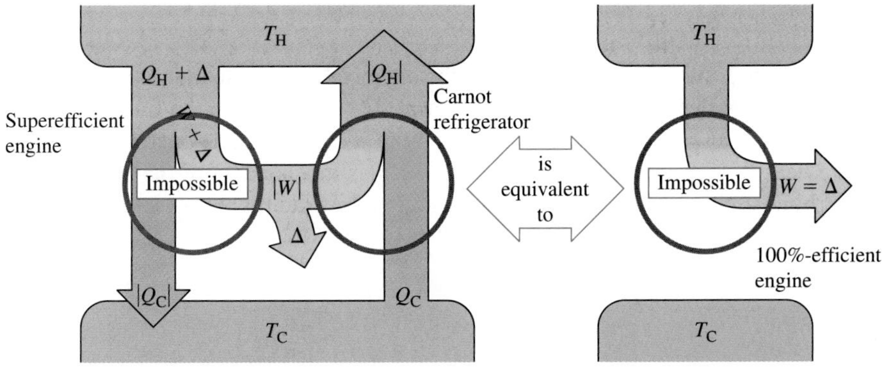

20.12 Proving that the Carnot engine has the highest possible efficiency. A "superefficient" engine (more efficient than a Carnot engine) combined with a Carnot refrigerator could convert heat completely into work with no net heat transfer to the cold reservoir. This would violate the second law of thermodynamics.

The Carnot Cycle and the Second Law

We can prove that **no engine can be more efficient than a Carnot engine operating between the same two temperatures.** The key to the proof is the above observation that since each step in the Carnot cycle is reversible, the *entire cycle* may be reversed. Run backward, the engine becomes a refrigerator. Suppose we have an engine that is more efficient than a Carnot engine (Fig. 20.12). Let the Carnot engine, run backward as a refrigerator by negative work $-|W|$, take in heat Q_C from the cold reservoir and expel heat $|Q_H|$ to the hot reservoir. The superefficient engine expels heat $|Q_C|$, but to do this, it takes in a greater amount of heat $Q_H + \Delta$. Its work output is then $W + \Delta$, and the net effect of the two machines together is to take a quantity of heat Δ and convert it completely into work. This violates the engine statement of the second law. We could construct a similar argument that a superefficient engine could be used to violate the refrigerator statement of the second law. Note that we don't have to assume that the superefficient engine is reversible. In a similar way we can show that *no refrigerator can have a greater coefficient of performance than a Carnot refrigerator operating between the same two temperatures.*

Thus the statement that no engine can be more efficient than a Carnot engine is yet another equivalent statement of the second law of thermodynamics. It also follows directly that **all Carnot engines operating between the same two temperatures have the same efficiency, irrespective of the nature of the working substance.** Although we derived Eq. (20.14) for a Carnot engine using an ideal gas as its working substance, it is in fact valid for *any* Carnot engine, no matter what its working substance.

Equation (20.14), the expression for the efficiency of a Carnot engine, sets an upper limit to the efficiency of a real engine such as a steam turbine. To maximize this upper limit and the actual efficiency of the real engine, the designer must make the intake temperature T_H as high as possible and the exhaust temperature T_C as low as possible (Fig. 20.13).

The exhaust temperature cannot be lower than the lowest temperature available for cooling the exhaust. For a steam turbine at an electric power plant, T_C may be the temperature of river or lake water; then we want the boiler temperature T_H to be as high as possible. The vapor pressures of all liquids increase rapidly with temperature, so we are limited by the mechanical strength of the boiler. At 500°C the vapor pressure of water is about 240×10^5 Pa (235 atm); this is about the maximum practical pressure in large present-day steam boilers.

20.13 To maximize efficiency, the temperatures inside a jet engine are made as high as possible. Exotic ceramic materials are used that can withstand temperatures in excess of 1000°C without melting or becoming soft.

*The Kelvin Temperature Scale

In Chapter 17 we expressed the need for a temperature scale that doesn't depend on the properties of any particular material. We can now use the Carnot cycle to define such a scale. The thermal efficiency of a Carnot engine operating between two heat reservoirs at temperatures T_H and T_C is independent of the nature of the working substance and depends only on the temperatures. From Eq. (20.4), this thermal efficiency is

$$e = \frac{Q_H + Q_C}{Q_H} = 1 + \frac{Q_C}{Q_H}$$

Therefore the ratio Q_C/Q_H is the same for *all* Carnot engines operating between two given temperatures T_H and T_C.

Kelvin proposed that we *define* the ratio of the temperatures, T_C/T_H, to be equal to the magnitude of the ratio Q_C/Q_H of the quantities of heat absorbed and rejected:

$$\frac{T_C}{T_H} = \frac{|Q_C|}{|Q_H|} = -\frac{Q_C}{Q_H} \quad \text{(definition of Kelvin temperature)} \qquad (20.16)$$

Equation (20.16) looks identical to Eq. (20.13), but there is a subtle and crucial difference. The temperatures in Eq. (20.13) are based on an ideal-gas thermometer, as defined in Section 17.3, while Eq. (20.16) *defines* a temperature scale based on the Carnot cycle and the second law of thermodynamics and is independent of the behavior of any particular substance. Thus the **Kelvin temperature scale** is truly *absolute*. To complete the definition of the Kelvin scale, we assign, as in Section 17.3, the arbitrary value of 273.16 K to the temperature of the triple point of water. When a substance is taken around a Carnot cycle, the ratio of the heats absorbed and rejected, $|Q_H|/|Q_C|$, is equal to the ratio of the temperatures of the reservoirs *as expressed on the gas-thermometer scale* defined in Section 17.3. Since the triple point of water is chosen to be 273.16 K in both scales, it follows that *the Kelvin and ideal-gas scales are identical.*

The zero point on the Kelvin scale is called **absolute zero.** Absolute zero can be interpreted on a molecular level; at absolute zero the system has its *minimum* possible total internal energy (kinetic plus potential). Because of quantum effects, however, it is *not* true that at $T = 0$, all molecular motion ceases. There are theoretical reasons for believing that absolute zero cannot be attained experimentally, although temperatures below 10^{-7} K have been achieved. The more closely we approach absolute zero, the more difficult it is to get closer. One statement of the *third law of thermodynamics* is that it is impossible to reach absolute zero in a finite number of thermodynamic steps.

Test Your Understanding

An inventor looking for financial support comes to you with an idea for a gasoline engine that runs on a novel type of thermodynamic cycle. His design is made entirely of copper and is air-cooled. He claims that the engine will be 85% efficient. Should you invest in this marvelous new engine? (*Hint:* See Table 17.4.)

20.7 | Entropy

The second law of thermodynamics, as we have stated it, is rather different in form from many familiar physical laws. It is not an equation or a quantitative relationship but rather a statement of *impossibility.* However, the second law *can* be stated as a quantitative relation with the concept of *entropy,* the subject of this section.

We have talked about several processes that proceed naturally in the direction of increasing disorder. Irreversible heat flow increases disorder because the molecules are initially sorted into hotter and cooler regions; this sorting is lost when the system comes to thermal equilibrium. Adding heat to a body increases its disorder because it increases average molecular speeds and therefore the randomness of molecular motion. Free expansion of a gas increases its disorder because the molecules have greater randomness of position after the expansion than before.

Entropy and Disorder

Entropy provides a *quantitative* measure of disorder. To introduce this concept, let's consider an infinitesimal isothermal expansion of an ideal gas. We add heat dQ and let the gas expand just enough to keep the temperature constant. Because the internal energy of an ideal gas depends only on its temperature, the internal energy is also constant; thus from the first law, the work dW done by the gas is equal to the heat dQ added. That is,

$$dQ = dW = p\,dV = \frac{nRT}{V}\,dV \quad \text{so} \quad \frac{dV}{V} = \frac{dQ}{nRT}$$

The gas is in a more disordered state after the expansion than before because the molecules are moving in a larger volume and have more randomness of position. Thus the fractional volume change dV/V is a measure of the increase in disorder, and the above equation shows that it is proportional to the quantity dQ/T. We introduce the symbol S for the entropy of the system, and we define the infinitesimal entropy change dS during an infinitesimal reversible process at absolute temperature T as

$$dS = \frac{dQ}{T} \quad \text{(infinitesimal reversible process)} \qquad (20.17)$$

If a total amount of heat Q is added during a reversible isothermal process at absolute temperature T, the total entropy change $\Delta S = S_2 - S_1$ is given by

$$\Delta S = S_2 - S_1 = \frac{Q}{T} \quad \text{(reversible isothermal process)} \qquad (20.18)$$

Entropy has units of energy divided by temperature; the SI unit of entropy is 1 J/K.

We can see how the quotient Q/T is related to the increase in disorder. Higher temperature means greater randomness of motion. If the substance is initially cold, with little molecular motion, adding heat Q causes a substantial fractional increase in molecular motion and randomness. But if the substance is already hot, the same quantity of heat adds relatively little to the greater molecular motion already present. So the quotient Q/T is an appropriate characterization of the increase in randomness or disorder when heat flows into a system.

Example 20.5 Entropy change in melting

One kilogram of ice at 0°C is melted and converted to water at 0°C. Compute its change in entropy, assuming that the melting is done reversibly. The heat of fusion of water is $L_f = 3.34 \times 10^5$ J/kg.

SOLUTION

IDENTIFY: The melting occurs at a constant temperature of 0°C, so this is a reversible isothermal process.

SET UP: We are given the amount of heat added (in terms of the heat of fusion) and the temperature $T = 273$ K. (Note that in entropy calculations we must always use absolute, or Kelvin, temperatures.) We can then calculate the entropy change using Eq. (20.18).

EXECUTE: The heat needed to melt the ice is $Q = mL_f = 3.34 \times 10^5$ J. From Eq. (20.18) the increase in entropy of the system is

$$\Delta S = S_2 - S_1 = \frac{Q}{T} = \frac{3.34 \times 10^5\ \text{J}}{273\ \text{K}} = 1.22 \times 10^3\ \text{J/K}$$

EVALUATE: This increase corresponds to the increase in disorder when the water molecules go from the highly ordered state of a crystalline solid to the much more disordered state of a liquid (Fig. 20.14).

In any *isothermal* reversible process, the entropy change equals the heat transferred divided by the absolute temperature. When we refreeze the water, Q has the opposite sign, and the entropy change of the water is $\Delta S = -1.22 \times 10^3$ J/K. The water molecules rearrange themselves into a crystal to form ice, so disorder and entropy both decrease.

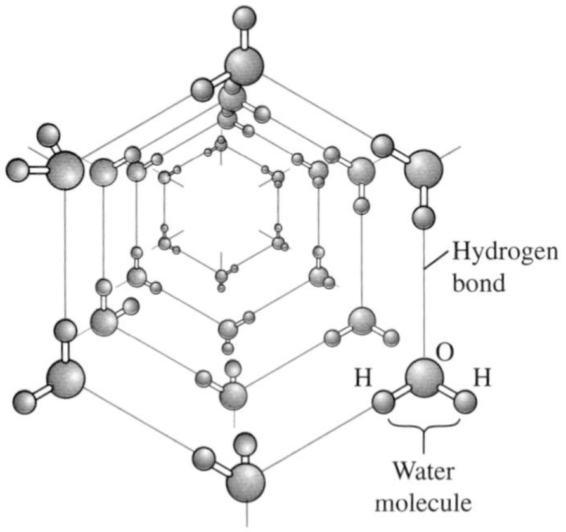

20.14 Water molecules are arranged in a regular, ordered way in an ice crystal. When the ice melts, the hydrogen bonds between molecules are broken, increasing the water's disorder and its entropy.

We can generalize the definition of entropy change to include *any* reversible process leading from one state to another, whether it is isothermal or not. We represent the process as a series of infinitesimal reversible steps. During a typical step, an infinitesimal quantity of heat dQ is added to the system at absolute temperature T. Then we sum (integrate) the quotients dQ/T for the entire process; that is,

$$\Delta S = \int_1^2 \frac{dQ}{T} \quad \text{(entropy change in a reversible process)} \qquad (20.19)$$

The limits 1 and 2 refer to the initial and final states.

Because entropy is a measure of the disorder of a system in any specific state, it must depend only on the current state of the system, not on its past history. We will show later that this is indeed the case. When a system proceeds from an initial state with entropy S_1 to a final state with entropy S_2, the change in entropy $\Delta S = S_2 - S_1$ defined by Eq. (20.19) does not depend on the path leading from the initial to the final state but is the same for *all possible* processes leading from state 1 to state 2. Thus the entropy of a system must also have a definite value for any given state of the system. We recall that *internal energy,* introduced in Chapter 19, also has this property, although entropy and internal energy are very different quantities.

Since entropy is a function only of the state of a system, we can also compute entropy changes in *irreversible* (non-equilibrium) processes for which Eqs. (20.17) and (20.19) are not applicable. We simply invent a path connecting the given initial and final states that *does* consist entirely of reversible, equilibrium processes and compute the total entropy change for that path. It is not the actual path, but the entropy change must be the same as for the actual path.

As with internal energy, the above discussion does not tell us how to calculate entropy itself, but only the change in entropy in any given process. Just as with internal energy, we may arbitrarily assign a value to the entropy of a system in a specified reference state and then calculate the entropy of any other state with reference to this.

Example 20.6 Entropy change in a temperature change

One kilogram of water at 0°C is heated to 100°C. Compute its change in entropy.

SOLUTION

IDENTIFY and SET UP: In practice, the process described would be done irreversibly, perhaps by setting a pan of water on an electric range whose cooking surface is maintained at 100°C. But the entropy change of the water depends only on the initial and final states of the system, and is the same whether the process is reversible or irreversible. Hence we can imagine that the temperature of the water is increased reversibly in a series of infinitesimal steps, in each of which the temperature is raised by an infinitesimal amount dT. We then use Eq. (20.19) to integrate over all these steps and calculate the entropy change for the total process.

EXECUTE: From Eq. (17.14) the heat required to carry out each such infinitesimal step is $dQ = mc\,dT$. Substituting this into Eq. (20.19) and integrating, we find

$$\Delta S = S_2 - S_1 = \int_1^2 \frac{dQ}{T} = \int_{T_1}^{T_2} mc\frac{dT}{T} = mc\ln\frac{T_2}{T_1}$$

$$= (1.00\text{ kg})(4190\text{ J/kg}\cdot\text{K})\left(\ln\frac{373\text{ K}}{273\text{ K}}\right)$$

$$= 1.31 \times 10^3\text{ J/K}$$

EVALUATE: The entropy change is positive, as it must be for a process in which the system absorbs heat.

In this calculation we assumed that the specific heat c doesn't depend on temperature. That's a pretty good approximation, since c for water increases by only 1% between 0°C and 100°C.

CAUTION In solving this problem you might be tempted to avoid doing an integral by using the simpler expression in Eq. (20.18), $\Delta S = Q/T$. This would be incorrect, however, because Eq. (20.18) is applicable only to *isothermal* processes and the initial and final temperatures in our example are *not* the same. If you persist and use $\Delta S = Q/T$ anyway, you would know what to use for the numerator $[Q = (4190\text{ J/kg}\cdot\text{K})(1.00\text{ kg})(100\text{ K}) = 4.19 \times 10^5\text{ J}]$ but you would be confused about whether to use $T = 273$ K (the initial temperature T_1), 373 K (the final temperature T_2), or perhaps 323 K (the average of T_1 and T_2). It doesn't matter, because none of these choices would give you the right answer! (Try it and see.) The *only* correct way to find the entropy change in a process with different initial and final temperatures is to use Eq. (20.19).

Conceptual Example 20.7 A reversible adiabatic process

A gas expands adiabatically and reversibly. What is its change in entropy?

SOLUTION

In an adiabatic process, no heat enters or leaves the system. Hence $dQ = 0$ and there is *no* change in entropy in this reversible process: $\Delta S = 0$. Every *reversible* adiabatic process is a constant-entropy process. (For this reason, reversible adiabatic processes are also called *isentropic* processes.) The increase in disorder resulting from the gas occupying a greater volume is exactly balanced by the decrease in disorder associated with the lowered temperature and reduced molecular speeds.

Example 20.8 Entropy change in a free expansion

A thermally insulated box is divided by a partition into two compartments, each having volume V (Fig. 20.15). Initially, one compartment contains n moles of an ideal gas at temperature T, and the other compartment is evacuated. We then break the partition, and the gas expands to fill both compartments. What is the entropy change in this free-expansion process?

SOLUTION

IDENTIFY: For this process, $Q = 0$, $W = 0$, $\Delta U = 0$, and therefore (because the system is an ideal gas) $\Delta T = 0$. We might think that the entropy change is zero because there is no heat exchange. But Eq. (20.19) can be used to calculate entropy changes for *reversible* processes only; this free expansion is *not* reversible, and there *is* an entropy change. The process is adiabatic because $Q = 0$, but it is not isentropic because $\Delta S \neq 0$. As we mentioned at the beginning of this section, entropy increases in a free expansion because the positions of the molecules are more random than before the expansion.

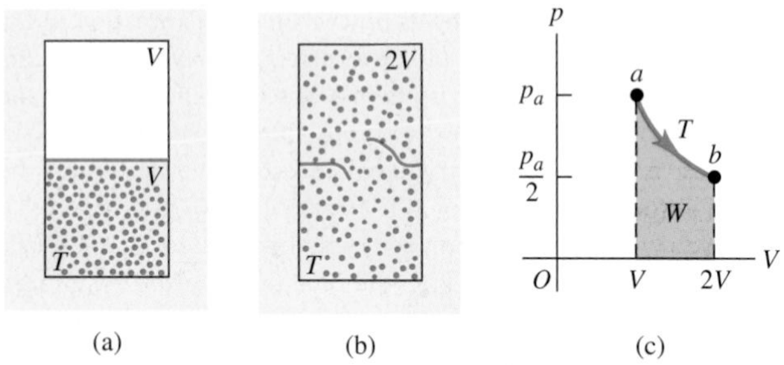

20.15 (a,b) Free expansion of an insulated ideal gas. (c) The free-expansion process doesn't pass through equilibrium states from a to b. However, the entropy change $S_b - S_a$ can be calculated by using the isothermal path shown or *any* reversible path from a to b.

SET UP: To calculate ΔS, we recall that the entropy change depends only on the initial and final states. We can devise a *reversible* process having the same endpoints, use Eq. (20.19) to calculate its entropy change, and thus determine the entropy change in the original process. An appropriate reversible process in this case is an isothermal expansion from V to $2V$ at temperature T. The gas does work W during this substitute expansion, so an equal amount of heat Q must be supplied to keep the internal energy constant. We find the entropy change for this reversible isothermal process using Eq. (20.18); the entropy change for the free expansion will be the same.

EXECUTE: We found in Example 19.1 (Section 19.2) that the work done by n moles of ideal gas in an isothermal expansion from V_1 to V_2 is $W = nRT \ln (V_2/V_1)$. Using $V_1 = V$ and $V_2 = 2V$, we have

$$Q = W = nRT \ln \frac{2V}{V} = nRT \ln 2$$

Thus the entropy change is

$$\Delta S = \frac{Q}{T} = nR \ln 2$$

which is also the entropy change for the free expansion with the same initial and final states. For one mole,

$$\Delta S = (1 \text{ mol})(8.314 \text{ J/mol} \cdot \text{K})(\ln 2) = 5.76 \text{ J/K}$$

EVALUATE: The entropy change is positive, as we predicted. The factor ($\ln 2$) in our answer is a result of the volume having increased by a factor of 2. Can you show that if the volume had increased in the free expansion from V to xV, where x is an arbitrary number, the entropy change would have been $\Delta S = nR \ln x$?

Example 20.9 Entropy and the Carnot cycle

For the Carnot engine in Example 20.2 (Section 20.6), find the total entropy change in the engine during one cycle.

SOLUTION

IDENTIFY and SET UP: The Carnot cycle is made up of four reversible steps, shown in Fig. 20.11. We find the entropy change ΔS for each step, then add the entropy changes to get the total ΔS for the cycle as a whole.

EXECUTE: There is no entropy change during the adiabatic expansion or adiabatic compression. During the isothermal expansion at $T_H = 500$ K the engine takes in 2000 J of heat, and its entropy change, from Eq. (20.18), is

$$\Delta S_H = \frac{Q_H}{T_H} = \frac{2000\text{ J}}{500\text{ K}} = 4.0\text{ J/K}$$

During the isothermal compression at $T_C = 350$ K the engine gives off 1400 J of heat, and its entropy change is

$$\Delta S_C = \frac{Q_C}{T_C} = \frac{-1400\text{ J}}{350\text{ K}} = -4.0\text{ J/K}$$

The total entropy change in the engine during one cycle is $\Delta S_{total} = \Delta S_H + \Delta S_C = 4.0\text{ J/K} + (-4.0\text{ J/K}) = 0$.

EVALUATE: The result $\Delta S_{total} = 0$ tells us that when the Carnot engine completes a cycle, it has the same entropy as it did at the beginning of the cycle. We'll explore this result in the following subsection.

What is the total entropy change of the engine's *environment* during this cycle? The hot (500 K) reservoir gives off 2000 J of heat during the reversible isothermal expansion, so its entropy change is $(-2000\text{ J})/(500\text{ K}) = -4.0$ J/K; the cold (350 K) reservoir absorbs 1400 J of heat during the reversible isothermal compression, so its entropy change is $(+1400\text{ J})/(350\text{ K}) = +4.0$ J/K. Thus each individual reservoir has an entropy change; however, the sum of these changes—that is, the total entropy change of the system's environment—is zero.

These results apply to the special case of the Carnot cycle, for which *all* of the processes are reversible. In this case we find that the total entropy change of the system and the environment together is zero. We will see that if the cycle includes irreversible processes (as is the case for the Otto cycle or Diesel cycle of Section 20.3), the total entropy change of the system and the environment *cannot* be zero, but rather must be positive.

Entropy in Cyclic Processes

Example 20.9 showed that the total entropy change for a cycle of a particular Carnot engine, which uses an ideal gas as its working substance, is zero. This result follows directly from Eq. (20.13), which we can rewrite as

$$\frac{Q_H}{T_H} + \frac{Q_C}{T_C} = 0 \qquad (20.20)$$

The quotient Q_H/T_H equals ΔS_H, the entropy change of the engine that occurs at $T = T_H$. Likewise, Q_C/T_C equals ΔS_C, the (negative) entropy change of the engine that occurs at $T = T_C$. Hence Eq. (20.20) says that $\Delta S_H + \Delta S_C = 0$; that is, there is zero net entropy change in one cycle.

What about Carnot engines that use a different working substance? According to the second law, *any* Carnot engine operating between given temperatures T_H and T_C has the same efficiency $e = 1 - T_C/T_H$ (Eq. (20.14)). Combining this expression for e with Eq. (20.4), $e = 1 + Q_C/Q_H$, just reproduces Eq. (20.20). So Eq. (20.20) is valid for any Carnot engine working between these temperatures, whether its working substance is an ideal gas or not. We conclude that *the total entropy change in one cycle of any Carnot engine is zero.*

This result can be generalized to show that the total entropy change during *any* reversible cyclic process is zero. A reversible cyclic process appears on a pV-diagram as a closed path (Fig. 20.16a). We can approximate such as path as closely as we like by a sequence of isothermal and adiabatic processes forming parts of

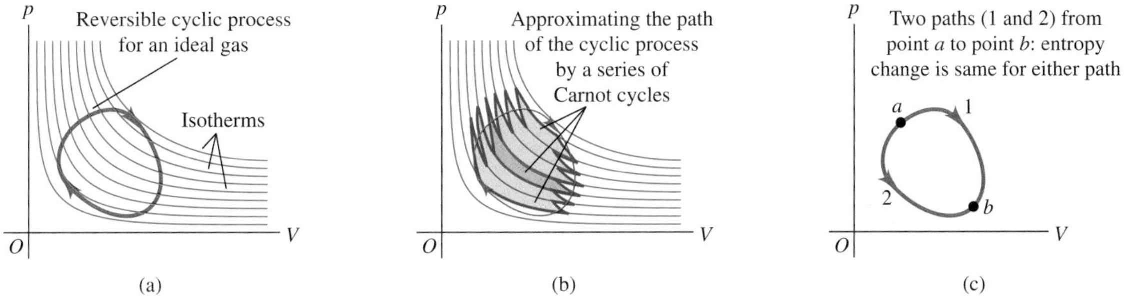

20.16 (a) A reversible cyclic process for an ideal gas is shown as a red closed path on a pV-diagram. Several ideal-gas isotherms are shown in blue. (b) We can approximate the path in (a) by a series of long, thin Carnot cycles; one cycle is highlighted in gold. The total entropy change is zero for each Carnot cycle and for the actual cyclic process. (c) The entropy change between points a and b is independent of the path.

many long, thin Carnot cycles (Fig. 20.16b). The total entropy change for the full cycle is the sum of the entropy changes for each small Carnot cycle, each of which is zero. So **the total entropy change during *any* reversible cycle is zero:**

$$\int \frac{dQ}{T} = 0 \qquad \text{(reversible cyclic process)} \qquad (20.21)$$

It follows that when a system undergoes a reversible process leading from any state a to any other state b, *the entropy change of the system is independent of the path* (Fig. 20.16c). If the entropy change for path 1 were different from the change for path 2, the system could be taken along path 1 and then backward along path 2 to the starting point, with a nonzero net change in entropy. This would violate the conclusion that the total entropy change in such a cyclic process must be zero. Because the entropy change in such processes is independent of path, we conclude that in any given state, the system has a definite value of entropy that depends only on the state, not on the processes that led to that state.

Entropy in Irreversible Processes

In an idealized, reversible process involving only equilibrium states, the total entropy change of the system and its surroundings is zero. But all *irreversible* processes involve an increase in entropy. Unlike energy, *entropy is not a conserved quantity.* The entropy of an isolated system *can* change, but as we shall see, it can never decrease. The free expansion of a gas, described in Example 20.8, is an irreversible process in an isolated system in which there is an entropy increase.

Example 20.10 An irreversible process

Suppose 1.00 kg of water at 100°C is placed in thermal contact with 1.00 kg of water at 0°C. What is the total change in entropy? Assume that the specific heat of water is constant at 4190 J/kg·K over this temperature range.

SOLUTION

IDENTIFY and SET UP: This process involves irreversible heat flow because of the temperature differences. We nonetheless calculate the entropy changes for the (initially) hot water and the (initially) cold water in the same way as in Example 20.6 by assuming that the process occurs reversibly. We calculate ΔS for each substance using Eq. (20.19) because the temperatures change in the process.

EXECUTE: The final temperature is 50°C = 323 K. The entropy change of the hot water is

$$\Delta S_{\text{hot}} = mc\int_{T_1}^{T_2} \frac{dT}{T} = (1.00\ \text{kg})(4190\ \text{J/kg}\cdot\text{K})\int_{373\ \text{K}}^{323\ \text{K}} \frac{dT}{T}$$

$$= (4190\ \text{J/K})\left(\ln \frac{323\ \text{K}}{373\ \text{K}}\right) = -603\ \text{J/K}$$

The entropy change of the cold water is

$$\Delta S_{\text{cold}} = (4190\ \text{J/K})\left(\ln \frac{323\ \text{K}}{273\ \text{K}}\right) = +705\ \text{J/K}$$

The *total* entropy change of the system is

$$\Delta S_{\text{total}} = \Delta S_{\text{hot}} + \Delta S_{\text{cold}} = (-603\ \text{J/K}) + 705\ \text{J/K} = +102\ \text{J/K}$$

EVALUATE: An irreversible heat flow in an isolated system is accompanied by an increase in entropy. We could have reached the same end state by simply mixing the two quantities of water. This, too, is an irreversible process; because the entropy depends only on the state of the system, the total entropy change would be the same, 102 J/K.

It's worth noting that the entropy of the system increases *continuously* as the two quantities of water come to equilibrium. For example, the first 4190 J of heat transferred cools the hot water to 99°C and warms the cold water to 1°C. The net change in entropy for this step is approximately

$$\Delta S = \frac{-4190\ \text{J}}{373\ \text{K}} + \frac{4190\ \text{J}}{273\ \text{K}} = +4.1\ \text{J/K}$$

Can you show in a similar way that the net entropy change is positive for *any* one-degree temperature change leading to the equilibrium condition?

Entropy and the Second Law

The results of Example 20.10 about the flow of heat from a higher to a lower temperature, or the mixing of substances at different temperatures, are characteristic of *all* natural (that is, irreversible) processes. When we include the entropy changes of all the systems taking part in the process, the increases in entropy are always greater than the decreases. In the special case of a *reversible* process, the increases and decreases are equal. Hence we can state the general principle: **When all systems taking part in a process are included, the entropy either remains constant or increases.** In other words, **no process is possible in which the total entropy decreases, when all systems taking part in the process are included.** This is an alternative statement of the second law of thermodynamics in terms of entropy. Thus it is equivalent to the "engine" and "refrigerator" statements discussed earlier. Figure 20.17 shows a specific example of this general principle.

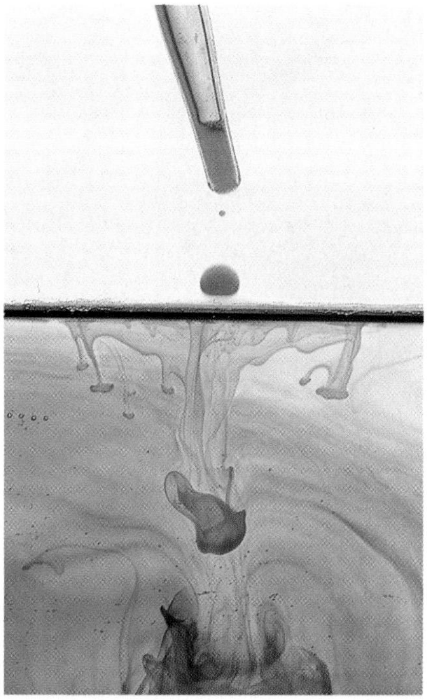

20.17 The mixing of colored ink and water starts from a state of relative order (low entropy) in which each fluid is separate and distinct from the other. The final state after mixing is more disordered (has greater entropy). Spontaneous unmixing of the ink and water, a process in which there would be a net decrease in entropy, is never observed.

The increase of entropy in every natural, irreversible process measures the increase of disorder or randomness in the universe associated with that process. Consider again the example of mixing hot and cold water (Example 20.10). We *might* have used the hot and cold water as the high- and low-temperature reservoirs of a heat engine. While removing heat from the hot water and giving heat to the cold water, we could have obtained some mechanical work. But once the hot and cold water have been mixed and have come to a uniform temperature, this opportunity to convert heat to mechanical work is lost irretrievably. The lukewarm water will never *unmix* itself and separate into hotter and colder portions. No decrease in *energy* occurs when the hot and cold water are mixed. What has been lost is not *energy,* but *opportunity,* the opportunity to convert part of the heat from the hot water into mechanical work. Hence when entropy increases, energy becomes less *available,* and the universe becomes more random or "run down."

Test Your Understanding

Suppose 2.00 kg of water at 50°C spontaneously changes temperature, so that half of the water cools to 0°C while the other half spontaneously warms to 100°C. (All of the water remains liquid, so it doesn't freeze or boil.) What would be the entropy change of the water? Is this process possible? Why or why not?

*20.8 | Microscopic Interpretation of Entropy

We described in Section 19.4 how the internal energy of a system could be calculated, at least in principle, by adding up all the kinetic energies of its constituent particles and all the potential energies of interaction among the particles. This is called a *microscopic* calculation of the internal energy. We can also make a microscopic calculation of the entropy S of a system. Unlike energy, however, entropy is not something that belongs to each individual particle or pair of particles in the system. Rather, entropy is a measure of the disorder of the system as a whole. To see how to calculate entropy microscopically, we first have to introduce the idea of *macroscopic* and *microscopic states.*

Suppose you toss N identical coins on the floor, and half of them show heads and half show tails. This is a description of the large-scale or **macroscopic state** of the system of N coins. A description of the **microscopic state** of the system includes information about each individual coin: Coin 1 was heads, coin 2 was tails, coin 3 was tails, and so on. There can be many microscopic states that cor-

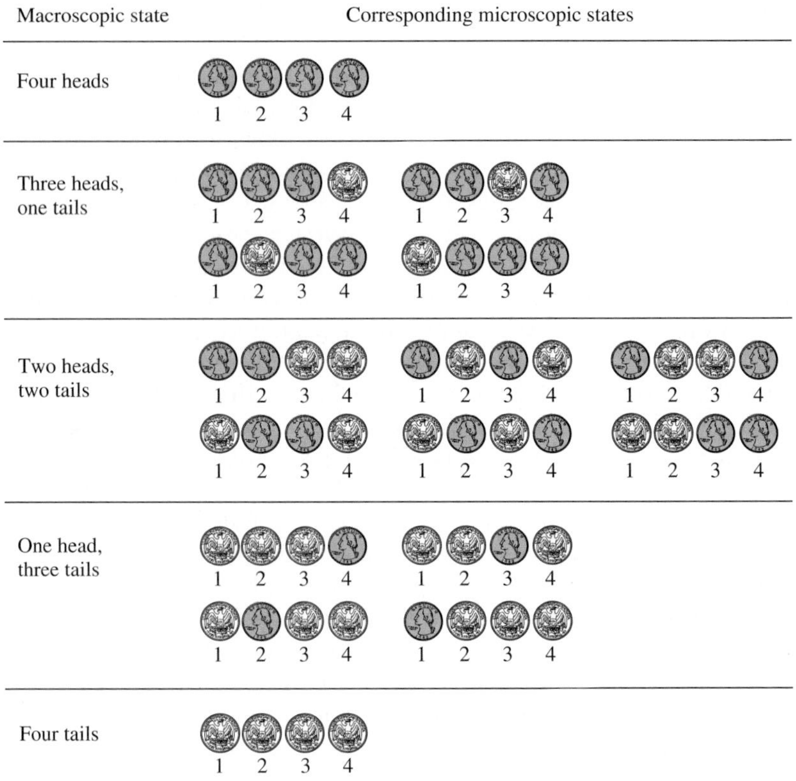

20.18 All possible microscopic states of four coins. There can be several possible microscopic states for each macroscopic state.

respond to the same macroscopic description. For instance, with $N = 4$ coins there are six possible states in which half are heads and half are tails (Fig. 20.18). The number of microscopic states grows rapidly with increasing N; for $N = 100$ there are $2^{100} = 1.27 \times 10^{30}$ microscopic states, of which 1.01×10^{29} are half heads and half tails.

The least probable outcomes of the coin toss are the states that are either all heads or all tails. It is certainly possible that you could throw 100 heads in a row, but don't bet on it; the probability of doing this is only 1 in 1.27×10^{30}. The most probable outcome of tossing N coins is that half are heads and half are tails. The reason is that this *macroscopic* state has the greatest number of corresponding *microscopic* states, as shown in Fig. 20.18.

To make the connection to the concept of entropy, note that N coins that are all heads constitutes a completely ordered macroscopic state; the description "all heads" completely specifies the state of each one of the N coins. The same is true if the coins are all tails. But the macroscopic description "half heads, half tails" by itself tells you very little about the state (heads or tails) of each individual coin. We say that the system is *disordered* because we know so little about its microscopic state. Compared to the state "all heads" or "all tails," the state "half heads, half tails" has a much greater number of possible microscopic states, much greater disorder, and hence much greater entropy (which is a quantitative measure of disorder).

Now instead of N coins, consider a mole of an ideal gas containing Avogadro's number of molecules. The macroscopic state of this gas is given by its pressure p, volume V, and temperature T; a description of the microscopic state involves stating the position and velocity for each molecule in the gas. At a given pressure, volume, and temperature, the gas may be in any one of an astronomically large number of microscopic states, depending on the positions and velocities of its 6.02×10^{23} molecules. If the gas undergoes a free expansion into a greater volume, the range of possible positions increases, as does the number of possible microscopic states. The system becomes more disordered, and the entropy increases as calculated in Example 20.8 (Section 20.7).

We can draw the following general conclusion: *For any system, the most probable macroscopic state is the one with the greatest number of corresponding microscopic states, which is also the macroscopic state with the greatest disorder and the greatest entropy.*

Let w represent the number of possible microscopic states for a given macroscopic state. (For the four coins shown in Fig. 20.18 the state of four heads has $w = 1$, the state of three heads and one tails has $w = 4$, and so on.) Then the entropy S of a macroscopic state can be shown to be given by

$$S = k \ln w \qquad \text{(microscopic expression for entropy)} \qquad (20.22)$$

where $k = R/N_A$ is the Boltzmann constant (gas constant per molecule) introduced in Section 18.3. As Eq. (20.22) shows, increasing the number of possible microscopic states w increases the entropy S.

What matters in a thermodynamic process is not the absolute entropy S but the *difference* in entropy between the initial and final states. Hence an equally valid and useful definition would be $S = k \ln w + C$, where C is a constant, since C cancels in any calculation of an entropy difference between two states. But it's convenient to set this constant equal to zero and use Eq. (20.22). With this choice, since the smallest possible value of w is unity, the smallest possible value of S for any system is $k \ln 1 = 0$. Entropy can *never* be negative.

In practice, calculating w is a difficult task, so Eq. (20.22) is typically used only to calculate the absolute entropy S of certain special systems. But we can use this relation to calculate *differences* in entropy between one state and another. Consider a system that undergoes a thermodynamic process that takes it from macroscopic state 1, for which there are w_1 possible microscopic states, to macroscopic state 2, with w_2 associated microscopic states. The change in entropy in this process is

$$\Delta S = S_2 - S_1 = k \ln w_2 - k \ln w_1 = k \ln \frac{w_2}{w_1} \qquad (20.23)$$

The *difference* in entropy between the two macroscopic states depends on the *ratio* of the numbers of possible microscopic states.

As the following example shows, using Eq. (20.23) to calculate a change in entropy from one macroscopic state to another gives the same results as considering a reversible process connecting those two states and using Eq. (20.19).

Example 20.11 A microscopic calculation of entropy change

Use Eq. (20.23) to calculate the entropy change in the free expansion of n moles of gas at temperature T described in Example 20.8.

SOLUTION

IDENTIFY and SET UP: The situation is shown in Figure 20.19. When the partition is broken, the velocities of the molecules are unaffected, since no work is done. But each molecule now has twice as much volume in which it can move and hence has twice the number of possible positions. This is all we need to calculate the entropy change using Eq. (20.23).

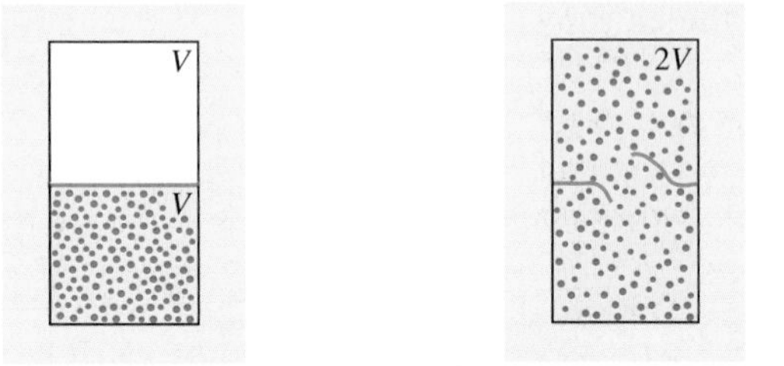

(a) Gas occupies volume V; number of microstates $= w_1$

(b) Gas occupies volume $2V$; number of microstates $= w_2 = 2^N w_1$

20.19 In a free expansion of N molecules in which the volume doubles, the number of possible microscopic states increases by 2^N.

EXECUTE: Let w_1 be the number of microscopic states of the system as a whole when the gas occupies volume V_1 (Fig. 20.19a). The number of molecules is $N = nN_A$, and each molecule has twice as many possible states after the partition is broken. Hence the number w_2 of microscopic states when the volume occupies volume $2V$ (Fig. 20.19b) is greater by a factor of 2^N; that is, $w_2 = 2^N w_1$. The change in entropy in this process is

$$\Delta S = k \ln \frac{w_2}{w_1} = k \ln \frac{2^N w_1}{w_1} = k \ln 2^N$$
$$= Nk \ln 2$$

Since $N = nN_A$ and $k = R/N_A$, this becomes

$$\Delta S = (nN_A)(R/N_A) \ln 2 = nR \ln 2$$

EVALUATE: We have found the same result as in Example 20.8, but without any reference to the thermodynamic path taken.

The relationship between entropy and the number of microscopic states gives us new insight into the entropy statement of the second law of thermodynamics, that the entropy of a closed system can never decrease. From Eq. (20.22) this means that a closed system can never spontaneously undergo a process that decreases the number of possible microscopic states.

An example of such a forbidden process would be if all of the air in your room spontaneously moved to one half of the room, leaving a vacuum in the other half. Such a "free compression" would be the reverse of the free expansion of Examples 20.8 and 20.11. This would decrease the number of possible microscopic states by a factor of 2^N. Strictly speaking, this process is not impossible! The probability of finding a given molecule in one half of the room is 1/2, so the probability of finding all of the molecules in one half of the room at once is $(1/2)^N$. (This is exactly the same as the probability of having a tossed coin come up heads N times in a row.) This probability is *not* zero. But lest you worry about suddenly finding yourself gasping for breath in the evacuated half of your room, consider that a typical room might hold 1000 moles of air, and so $N = 1000N_A = 6.02 \times 10^{26}$ molecules. The probability of all the molecules being in the same half of the room is therefore $(1/2)^{6.02 \times 10^{26}}$. Expressed as a decimal, this number has more than 10^{26} zeros to the right of the decimal point!

Because the probability of such a "free compression" taking place is so vanishingly small, it has almost certainly never occurred anywhere in the universe since the beginning of time. We conclude that for all practical purposes the second law of thermodynamics is never violated.

Test Your Understanding

A quantity of N molecules of an ideal gas initially occupies volume V. By what factor does the number of microscopic states of the gas increase if the gas expands to volume $2V$ in (a) a reversible isothermal process or (b) a reversible adiabatic process? Are the answers the same or different? Why?

SUMMARY

A reversible process is one whose direction can be reversed by an infinitesimal change in the conditions of the process, and in which the system is always in or very close to thermal equilibrium. All other thermodynamic processes are irreversible.

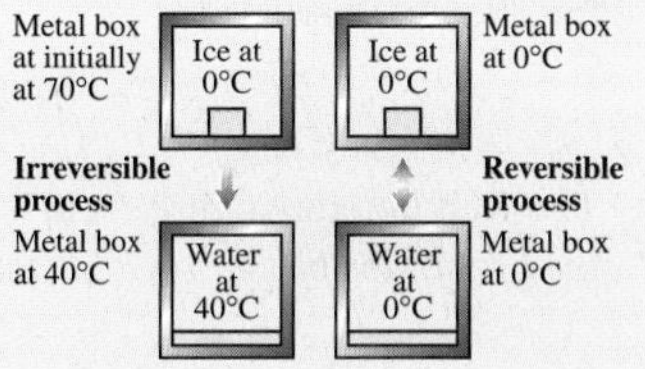

A heat engine takes heat Q_H from a source, converts part of it to work W, and discards the remainder $|Q_C|$ at a lower temperature. The thermal efficiency e of a heat engine measures how much of the absorbed heat is converted to work. (See Example 20.1)

$$e = \frac{W}{Q_H} = 1 + \frac{Q_C}{Q_H} = 1 - \left|\frac{Q_C}{Q_H}\right| \quad (20.4)$$

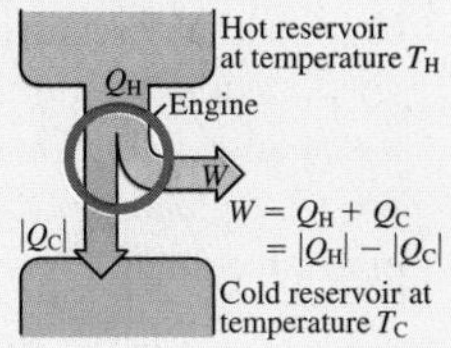

A gasoline engine operating on the Otto cycle has a theoretical maximum thermal efficiency e that depends on the compression ratio r and the ratio of heat capacities γ of the working substance.

$$e = 1 - \frac{1}{r^{\gamma - 1}} \quad (20.6)$$

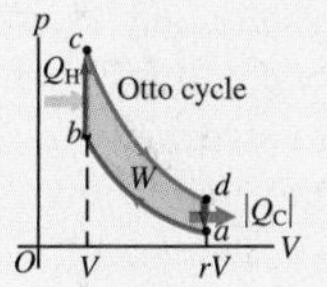

A refrigerator takes heat Q_C from a colder place, has a work input $|W|$, and discards heat $|Q_H|$ at a warmer place. The effectiveness of the refrigerator is given by its coefficient of performance K.

$$K = \frac{|Q_C|}{|W|} = \frac{|Q_C|}{|Q_H| - |Q_C|} \quad (20.9)$$

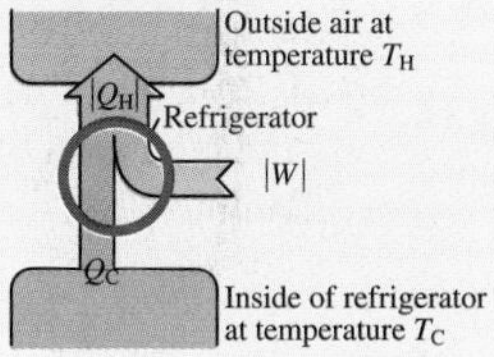

The second law of thermodynamics describes the directionality of natural thermodynamic processes. It can be stated in several equivalent forms. The *engine* statement is that no cyclic process can convert heat completely into work. The *refrigerator* statement is that no cyclic process can transfer heat from a colder place to a hotter place with no input of mechanical work.

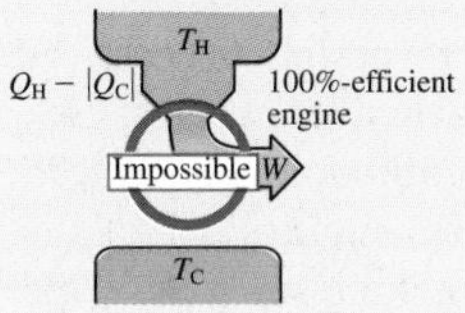

The Carnot cycle operates between two heat reservoirs at temperatures T_H and T_C and uses only reversible processes. Its thermal efficiency depends only on T_H and T_C. An additional equivalent statement of the second law is that no engine operating between the same two temperatures can be more efficient than a Carnot engine. (See Examples 20.2 and 20.3)

$$e_{Carnot} = 1 - \frac{T_C}{T_H} = \frac{T_H - T_C}{T_H} \quad (20.14)$$

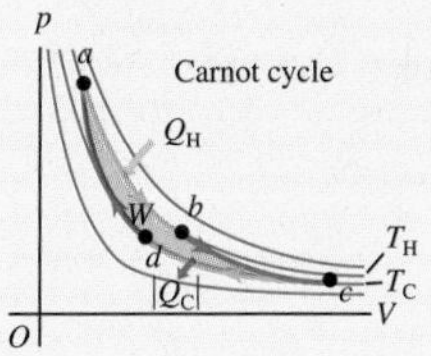

A Carnot engine run backward is a Carnot refrigerator. Its coefficient of performance depends only on T_H and T_C. Another form of the second law states that no refrigerator operating between the same two temperatures can have a larger coefficient of performance than a Carnot refrigerator. (See Example 20.4)

$$K_{Carnot} = \frac{T_C}{T_H - T_C} \quad (20.15)$$

The Kelvin temperature scale is based on the efficiency of the Carnot cycle and is independent of the properties of any specific material. The zero point on the Kelvin scale is called absolute zero.

Entropy is a quantitative measure of the disorder of a system. The entropy change in any reversible process depends on the amount of heat flow and the absolute temperature T. Entropy depends only on the state of the system, and the change in entropy between given initial and final states is the same for all processes leading from one state to the other. This fact can be used to find the entropy change in an irreversible process. (See Examples 20.5 through 20.10)

$$\Delta S = \int_1^2 \frac{dQ}{T} \quad (20.19)$$

(reversible process)

An important statement of the second law of thermodynamics is that the entropy of an isolated system may increase but can never decrease. When a system interacts with its surroundings, the total entropy change of system and surroundings can never decrease. When the interaction involves only reversible processes, the total entropy is constant and $\Delta S = 0$; when there is any irreversible process, the total entropy increases and $\Delta S > 0$.

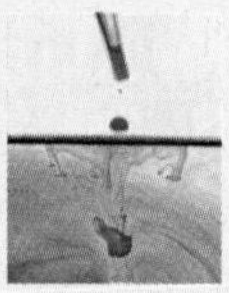

When a system is in a particular macroscopic state, the particles that make up the system may be in any of w possible microscopic states. The greater the number w, the greater the entropy. (See Example 20.11)

$$S = k \ln w \quad (20.22)$$

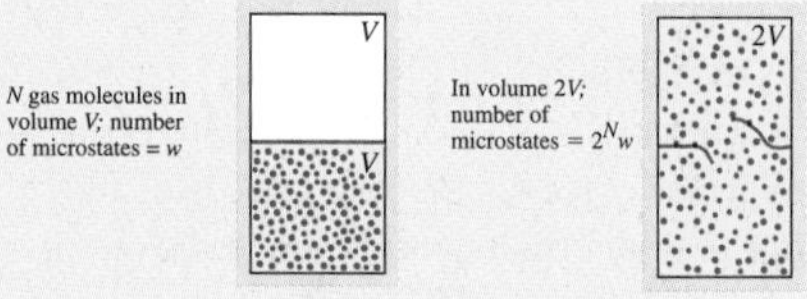

Key Terms

irreversible process, 755
reversible process, 755
equilibrium process, 755
heat engine, 756
working substance, 756
cyclic process, 756
thermal efficiency, 757
compression ratio, 759
Otto cycle, 759
Diesel cycle, 761
refrigerator, 761
coefficient of performance, 762
energy efficiency rating, 763
heat pump, 763
second law of thermodynamics, 764
Carnot cycle, 766
Kelvin temperature scale, 772
absolute zero, 772
entropy, 773
macroscopic state, 779
microscopic state, 779

Your Notes

Answer to Chapter Opening Question

Yes. That's what a refrigerator does: it makes heat flow from the cold interior of the refrigerator to the warm outside. The second law of thermodynamics says that heat cannot *spontaneously* flow from a cold body to a hot one. A refrigerator has a motor that does work on the system to *force* the heat to flow in that way.

Answers to Test Your Understanding Questions

Section 20.1 This process is irreversible. Like sliding a book across a table, rubbing your hands together uses friction to convert mechanical energy into heat. The (impossible) reverse process would involve your hands spontaneously getting colder, with the released energy forcing your hands to move rhythmically back and forth!

Section 20.2 Each truck's engine rejects 8000 J of heat per cycle and goes through 25 cycles per second. The total heat released in 24 hours is therefore

$$|Q_{\mathrm{C-total}}| = (1000 \text{ trucks})\left(\frac{8000 \text{ J}}{\text{truck} \cdot \text{cycle}}\right)\left(\frac{25 \text{ cycles}}{\text{s}}\right) \times \left(\frac{60 \text{ s}}{\text{min}}\right)\left(\frac{60 \text{ min}}{\text{h}}\right)(24 \text{ h}) = 1.7 \times 10^{13} \text{ J}$$

This is equivalent to 4.8×10^6 kilowatt-hours (a kilowatt-hour is a unit of energy used in electric power systems, equal to 10^3 watts of power delivered for one hour). A kilowatt-hour of electric energy costs a few cents, so this discarded heat is equivalent to approximately \$100,000 of lost energy.

Section 20.3 Let Q_H, the heat input at volume V, be constant. Then a greater compression ratio r means a greater maximum volume rV, so the points d and a are farther to the right in the pV-diagram of Fig. 20.5 and the area enclosed by the loop *abcd* is greater. The area within this loop is W, the work done by the engine, so a greater compression ratio means more work for the same input of heat and hence a greater efficiency $e = W/Q_H$.

Section 20.4 A refrigerator uses an input of work to transfer heat from one system (the refrigerator's interior) to another system (its exterior, which includes the house in which the refrigerator is installed). If the door is open, these two systems are really the *same* system and will eventually come to the same temperature. By the first law of thermodynamics, all of the work input to the refrigerator motor will be converted into heat and the temperature in your house will actually *increase*. To cool the house you need a system that will transfer heat from it to the outside world, such as an air conditioner or heat pump.

Section 20.5 The expanding steam does work to push the cork upward and pop it into the air. This process is not 100% efficient because only part of the heat absorbed from the burner goes into doing work; the remainder goes into raising the temperature of the water and the test tube.

Section 20.6 The efficiency can be no better than that of a Carnot engine running between the same two temperature limits, $e_{\mathrm{Carnot}} = 1 - (T_C/T_H)$ [Eq. (20.14)]. The temperature T_C of the cold reservoir for this air-cooled engine is about 300 K (ambient temperature) and the temperature T_H of the hot reservoir cannot exceed the melting point of copper, 1356 K (see Table 17.4). Hence the maximum possible Carnot efficiency is $e = 1 - (300 \text{ K})/(1356 \text{ K}) = 0.78$, or 78%. The temperature of any real engine would be less than this, so it would be impossible for the inventor's engine to attain 85% efficiency. You should invest your money elsewhere.

Section 20.7 The process described is exactly the opposite of the process used in Example 20.10, so the net entropy change would be -102 J/K. This result violates the second law of thermodynamics, which states that the entropy of an isolated system cannot decrease. This remarkable process is quite impossible.

Section 20.8 (a) We saw in Example 20.8 (Section 20.7) that for an ideal gas, the entropy change in a free expansion is the same as in an isothermal expansion. From Eq. (20.23), this implies that the ratio of the number of microscopic states after and before the expansion, w_2/w_1, is also the same for these two cases. From Example 20.11, $w_2/w_1 = 2^N$, so the number of microscopic states increases by a factor 2^N. (b) In a reversible expansion the entropy change is $\Delta S = \int dQ/T = 0$; if the expansion is adiabatic there is no heat flow, so $\Delta S = 0$. From Eq. (20.23), $w_2/w_1 = 1$ and there is *no* change in the number of microscopic states. The difference is that in an adiabatic expansion the temperature drops and the molecules move more slowly, so they have fewer microscopic states available to them than in an isothermal expansion.

Discussion Questions

Q20.1 A pot is half-filled with water, and a lid is placed on the pot, forming a tight seal so that no water vapor can escape. The pot is heated on a stove, forming water vapor inside the pot. The heat is then turned off and the water vapor condenses back to liquid. Is this cycle reversible or irreversible? Why?

Q20.2 Give two examples of reversible processes and two examples of irreversible processes in purely mechanical systems, such as blocks sliding on planes, springs, pulleys, and strings. Explain what makes each process reversible or irreversible.

Q20.3 What irreversible processes occur in a gasoline engine? Why are they irreversible?

Q20.4 Suppose you try to cool the kitchen of your house by leaving the refrigerator door open. What happens? Why? Would the result be the same if you left open a picnic cooler full of ice? Explain the reason for any differences.

Q20.5 A member of the U.S. Congress proposed a scheme to produce energy as follows. Water molecules (H_2O) are to be broken apart to produce hydrogen and oxygen. The hydrogen is then burned (that is, combined with oxygen), releasing energy in the process. The only product of this combustion is water, so there is no pollution. In light of the second law of thermodynamics, what do you think of this energy-producing scheme?

Q20.6 Some critics of biological evolution claim that it violates the second law of thermodynamics, since evolution involves simple life forms developing into more complex and more highly ordered organisms. Explain why this is not a valid argument against evolution.

Q20.7 Is it a violation of the second law of thermodynamics to convert mechanical energy completely into heat? To convert heat completely into work? Explain your answers.

Q20.8 A growing plant creates a highly complex and organized structure out of simple materials such as air, water, and trace minerals. Does this violate the second law of thermodynamics? Why or why not? What is the plant's ultimate source of energy? Explain your reasoning.

Q20.9 Imagine a special air filter placed in a window of a house. The tiny holes in the filter only allow air molecules moving faster than a certain speed to exit the house, and only allows air molecules moving slower than that speed to enter the house from outside. Explain why such an air filter would cool the house, and why the second law of thermodynamics makes building such a filter an impossible task.

Q20.10 An electric motor has its shaft coupled to that of an electric generator. The motor drives the generator, and some current from the generator is used to run the motor. The excess current is used to light a home. What is wrong with this scheme?

Q20.11 When a wet cloth is hung up in a hot wind in the desert, it is cooled by evaporation to a temperature that may be 20 C° or so below that of the air. Discuss this process in light of the second law of thermodynamics.

Q20.12 If no real engine can be as efficient as a Carnot engine operating between the same two temperatures, what is the point of developing and using Eq. (20.14)?

Q20.13 Suppose you want to increase the efficiency of a heat engine. Would it be better to increase T_H or to decrease T_C by an equal amount? Why?

Q20.14 What would be the efficiency of a Carnot engine operating with $T_H = T_C$? What would be the efficiency if $T_C = 0$ K and T_H were any temperature above 0 K? Interpret your answers.

Q20.15 Real heat engines, like the gasoline engine in a car, always have some friction between their moving parts, although lubricants keep the friction to a minimum. Would a heat engine with completely frictionless parts be 100% efficient? Why or why not? Does the answer depend on whether or not the engine runs on the Carnot cycle? Again, why or why not?

Q20.16 Does a refrigerator full of food consume more power if the room temperature is 20°C than if it is 15°C? Or is the power consumption the same? Explain your reasoning.

Q20.17 Explain why each of the following processes are examples of increasing disorder or randomness: mixing hot and cold water; free expansion of a gas; irreversible heat flow; and developing heat by mechanical friction. Are entropy increases involved in all of these? Why or why not?

Q20.18 Are the earth and sun in thermal equilibrium? Are there entropy changes associated with the transmission of energy from the sun to the earth? Does radiation differ from other modes of heat transfer with respect to entropy changes? Explain your reasoning.

Q20.19 Discuss the entropy changes involved in the preparation and consumption of a hot fudge sundae.

Q20.20 If you run a movie film backwards, it is as if the direction of time is reversed. In the time-reversed movie, would you see processes that violate conservation of energy? Conservation of linear momentum? Would you see processes that violate the second law of thermodynamics? In each case, if law-breaking processes could occur, give some examples.

Exercises

Section 20.2 Heat Engines

20.1 A diesel engine performs 2200 J of mechanical work and discards 4300 J of heat each cycle. a) How much heat must be supplied to the engine in each cycle? b) What is the thermal efficiency of the engine?

20.2 An aircraft engine takes in 9000 J of heat and discards 6400 J each cycle. a) What is the mechanical work output of the engine during one cycle? b) What is the thermal efficiency of the engine?

20.3 A Gasoline Engine. A gasoline engine takes in 1.61×10^4 J of heat and delivers 3700 J of work per cycle. The heat is obtained by burning gasoline with a heat of combustion of 4.60×10^4 J/g. a) What is the thermal efficiency? b) How much heat is discarded in each cycle? c) What mass of fuel is burned in each cycle? d) If the engine goes through 60.0 cycles per second, what is its power output in kilowatts? In horsepower?

20.4 A gasoline engine has a power output of 180 kW (about 241 hp). Its thermal efficiency is 28.0%. a) How much heat must be supplied to the engine per second? b) How much heat is discarded by the engine per second?

20.5 A certain nuclear-power plant has a mechanical-power output (used to drive an electric generator) of 330 MW. Its rate of heat input from the nuclear reactor is 1300 MW. a) What is the thermal efficiency of the system? b) At what rate is heat discarded by the system?

Section 20.3 Internal-Combustion Engines

20.6 What compression ratio r must an Otto cycle have to achieve an ideal efficiency of 65.0% if $\gamma = 1.40$?

20.7 For an Otto cycle with $\gamma = 1.40$ and $r = 9.50$, the temperature of the gasoline-air mixture when it enters the cylinder is 22.0°C (point a of Fig. 20.5). a) What is the temperature at the end of the compression stroke (point b)? b) The initial pressure of the gasoline-air mixture (point a) is 8.50×10^4 Pa, slightly below atmospheric pressure. What is the pressure at the end of the compression stroke?

20.8 The Otto-cycle engine in a Mercedes-Benz SLK230 has a compression ratio of 8.8. a) What is the ideal efficiency of the engine? Use $\gamma = 1.40$. b) The engine in a Dodge Viper GT2 has a slightly higher compression ratio of 9.6. How much increase in the ideal efficiency results from this increase in the compression ratio?

Section 20.4 Refrigerators

20.9 A refrigerator has a coefficient of performance of 2.10. Each cycle it absorbs 3.40×10^4 J of heat from the cold reservoir. a) How much mechanical energy is required each cycle to operate the refrigerator? b) During each cycle, how much heat is discarded to the high-temperature reservoir?

20.10 Liquid refrigerant at a pressure of 1.34×10^5 Pa leaves the expansion valve of a refrigerator at $-23.0°C$. It then flows through the evaporation coils inside the refrigerator and leaves as vapor at the same pressure and at $-20.5°C$, the same temperature as the inside of the refrigerator. The boiling point of the refrigerant at this pressure is $-23.0°C$, the heat of vaporization is 1.60×10^5 J/kg, and the specific heat capacity of the vapor at constant pressure is 485 J/kg·K. The coefficient of performance of the refrigerator is $K = 2.8$. If 8.00 kg of refrigerant flows through the refrigerator each hour, find the electric power that must be supplied to the refrigerator.

20.11 A window air-conditioner unit absorbs 9.80×10^4 J of heat per minute from the room being cooled and in the same time period deposits 1.44×10^5 J of heat into the outside air. a) What is the power consumption of the unit in watts? b) What is the energy efficiency rating of the unit?

20.12 A freezer has a coefficient of performance of 2.40. The freezer is to convert 1.80 kg of water at 25.0°C to 1.80 kg of ice at $-5.0°C$ in one hour. a) What amount of heat must be removed from the water at 25.0°C to convert it to ice at $-5.0°C$? b) How much electrical energy is consumed by the freezer during this hour? c) How much wasted heat is rejected to the room in which the freezer sits?

Section 20.6 The Carnot Cycle

20.13 A Carnot engine whose high-temperature reservoir is at 620 K takes in 550 J of heat at this temperature in each cycle and gives up 335 J to the low-temperature reservoir. a) How much mechanical work does the engine perform during each cycle? b) What is the temperature of the low-temperature reservoir? c) What is the thermal efficiency of the cycle?

20.14 A Carnot engine is operated between two heat reservoirs at temperatures of 520 K and 300 K. a) If the engine receives 6.45 kJ of heat energy from the reservoir at 520 K in each cycle, how many joules per cycle does it reject to the reservoir at 300 K? b) How much mechanical work is performed by the engine during each cycle? c) What is the thermal efficiency of the engine?

20.15 An ice-making machine operates in a Carnot cycle. It takes heat from water at 0.0°C and rejects heat to a room at 24.0°C. Suppose that 85.0 kg of water at 0.0°C are converted to ice at 0.0°C. a) How much heat is rejected to the room? b) How much energy must be supplied to the device?

20.16 A Carnot refrigerator is operated between two heat reservoirs at temperatures of 320 K and 270 K. a) If in each cycle the refrigerator receives 415 J of heat energy from the reservoir at 270 K, how many joules of heat energy does it deliver to the reservoir at 320 K? b) If the refrigerator goes through 165 cycles each minute, what power input is required to operate the refrigerator? c) What is the coefficient of performance of the refrigerator?

20.17 A Carnot device extracts 5.00 kJ of heat from a body at $-10.0°C$. How much work is done if the device exhausts heat into the environment at a) 25.0°C? b) 0.0°C?c) $-25.0°C$? In each case, is the device acting as an engine or as a refrigerator?

20.18 The Kwik-Freez Appliance Co. wants you to design a food freezer that will keep the freezing compartment at $-5.0°C$ and will operate in a room at 20.0°C. This freezer is to make 5.00 kg of ice at 0.0°C. starting with water at 20.0°C. Find the least possible amount of electrical energy needed to make this ice and the smallest possible amount of heat expelled into the room.

20.19 An ideal Carnot engine operates between 500°C and 100°C with a heat input of 250 J per cycle. What minimum number of cycles are necessary for the engine to lift a 500-kg rock through a height of 100 m?

20.20 A Carnot heat engine has a thermal efficiency of 0.600 and the temperature of its hot reservoir is 800 K. If 3000 J of heat is rejected to the cold reservoir in one cycle, what is the work output of the engine during one cycle?

20.21 A Carnot heat engine uses a hot reservoir consisting of a large amount of boiling water and a cold reservoir consisting of a large tub of ice and water. In five minutes of operation of the engine, the heat rejected by the engine melts 0.0400 kg of ice. During this time, how much work W is performed by the engine?

20.22 An inventor claims to have developed an engine that in each cycle takes in 2.60×10^8 J of heat at a temperature of 400 K, does 42.0 kWh of mechanical work, and rejects heat at a temperature of 250 K. Would you advise investing money to put this engine on the market? Why or why not?

20.23 a) Show that the efficiency e of a Carnot engine and the coefficient of performance K of a Carnot refrigerator are related by $K = (1 - e)/e$. The engine and refrigerator operate between the same hot and cold reservoirs. b) What is K for the limiting values $e \rightarrow 1$ and $e \rightarrow 0$? Explain.

Section 20.7 Entropy

20.24 A sophomore with nothing better to do adds heat to 0.350 kg of ice at 0.0°C until it is all melted. a) What is the change in entropy of the water? b) The source of heat is a very massive body at a temperature of 25.0°C. What is the change in entropy of this body? c) What is the total change in entropy of the water and the heat source?

20.25 You pour 100 g of water at 80.0°C into the ocean, which is at 20.0°C, and wait about ten minutes. Treat the water you pour and the ocean as an isolated system. a) Is this process reversible or irreversible? Explain your reasoning using simple physical reasoning without resorting to any equations. b) Calculate the net entropy change of the system during this process. Explain whether or not this result is consistent with your answer to part (a).

20.26 A 15.0-kg block of ice at 0.0°C melts inside a large room that has a temperature of 20.0°C. Treat the ice and the room as an isolated system and assume that the room is large enough for its temperature change to be ignored. a) Is this process reversible or irreversible? Explain your reasoning using simple physical reasoning without resorting to any equations. b) Calculate the net entropy change of the system during this process. Explain whether or not this result is consistent with your answer to part (a).

20.27 Calculate the entropy change that occurs when 1.00 kg of water at 20.0°C is mixed with 2.00 kg of water at 80.0°C.

20.28 Three moles of an ideal gas undergo a reversible isothermal compression at 20.0°C. During this compression, 1850 J of work is done on the gas. What is the change of entropy of the gas?

20.29 What is the change in entropy of 0.130 kg of helium gas at the normal boiling point of helium when it all condenses isothermally to 1.00 L of liquid helium? (*Hint:* See Table 17.4 in Section 17.6.)

20.30 a) Calculate the change in entropy when 1.00 kg of water at 100°C is vaporized and converted to steam at 100°C (see Table 17.4). b) Compare your answer to the change in entropy when 1.00 kg of ice is melted at 0°C, calculated in Example 20.5 (Section 20.7). Is the change in entropy greater for melting or for vaporization? Interpret your answer, using the idea that entropy is a measure of the randomness of a system.

20.31 a) Calculate the change in entropy when 1.00 mol of water (molecular mass 18.0 g/mol) at 100°C evaporates to form water vapor at 100°C. b) Repeat the calculation of part (a) for 1.00 mol of liquid nitrogen, 1.00 mol of silver, and 1.00 mol of mercury when each is vaporized at its normal boiling point. (See Table 17.4 for the heats of vaporization, and Appendix D for the molar masses. Note that the nitrogen molecule is N_2.) c) Your results in parts (a) and (b) should be in relatively close agreement. (This is called the *rule of Drepez and Trouton.*) Explain why this should be so, using the idea that entropy is a measure of the randomness of a system.

20.32 A block of copper with a mass of 3.50 kg, initially at 100.0°C, is dropped into 0.800 kg of water initially at 0.0°C. a) What is the final temperature of the system? b) What is the total change in entropy of the system?

20.33 Two moles of an ideal gas undergo a reversible isothermal expansion from 0.0280 m^3 to 0.0420 m^3 at a temperature of 25.0°C. What is the change in entropy of the gas?

*Section 20.8 Microscopic Interpretation of Entropy

***20.34** A box is separated by a partition into two parts of equal volume. The left side of the box contains 500 molecules of nitrogen gas; the right side contains 100 molecules of oxygen gas. The two gases are at the same temperature. The partition is punctured, and equilibrium is eventually attained. Assume that the volume of the box is large enough for each gas to undergo a free expansion and not change temperature. a) On average, how many molecules of each type will there be in either half of the box? b) What is the change in entropy of the system when the partition is punctured? c) What is the probability that the molecules will be found in the same distribution as they were before the partition was punctured, that is, 500 nitrogen molecules in the left half and 100 oxygen molecules in the right half?

***20.35** Two moles of an ideal gas occupy a volume V. The gas expands isothermally and reversibly to a volume $3V$. a) Is the velocity distribution changed by the isothermal expansion? Explain. b) Use Eq. (20.23) to calculate the change in entropy of the gas. c) Use Eq. (20.18) to calculate the change in entropy of the gas. Compare this result to that obtained in part (b).

***20.36** You toss four identical coins on the floor. Each coin has an equal probability of showing heads or tails. a) What is the probability of all four coins showing heads? Of all four showing tails? b) What is the probability of three coins showing heads and one tails? What is the probability of three coins showing tails and one heads? c) What is the probability of two coins showing heads and two showing tails? d) What is the sum of the five probabilities calculated in parts (a) through (c)? Explain.

Problems

20.37 You are designing a Carnot engine that has two moles of CO_2 as its working substance; the gas may be treated as ideal. The gas is to have a maximum temperature of 527°C and a maximum pressure of 5.00 atm. With a heat input of 400 J per cycle, you want 300 J of useful work. a) Find the temperature of the cold reservoir. b) For how many cycles must this engine run to melt completely a 10.0-kg block of ice originally at 0.0°C, using only the heat rejected by the engine?

20.38 A Carnot engine whose low-temperature reservoir is at −90.0°C has an efficiency of 40.0%. An engineer is assigned the problem of increasing this to 45.0%. a) By how many Celsius degrees must the temperature of the high-temperature reservoir be increased if the temperature of the low-temperature reservoir remains constant? b) By how many Celsius degrees must the temperature of the low-temperature reservoir be decreased if that of the high-temperature reservoir remains constant?

20.39 A heat engine takes 0.350 mol of a diatomic ideal gas around the cycle shown in the pV-diagram of Fig. 20.20. Process $1 \rightarrow 2$ is at constant volume, process $2 \rightarrow 3$ is adiabatic, and process $3 \rightarrow 1$ is at a constant pressure of 1.00 atm. The value of γ for this gas is 1.40. a) Find the pressure and volume at points 1, 2, and 3. b) Calculate Q, W, and ΔU for each of the three processes. c) Find the net work done by the gas in the cycle. d) Find the net heat flow into the engine in one cycle. e) What is the thermal efficiency of the engine? How does this compare to the efficiency of a Carnot-cycle engine operating between the same minimum and maximum temperatures T_1 and T_2?

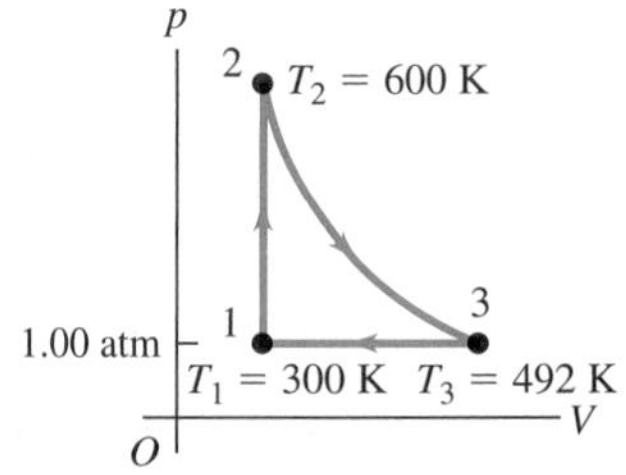

Figure 20.20 Problem 20.39.

20.40 A heat engine operates with cycle *abcd*. The working substance of the engine is CO_2 gas; the gas can be treated as ideal. The maximum temperature of the gas during the cycle is 1000 K. The pressure and volume of the gas in each state is $p_a = p_d = 2.00 \times 10^5$ Pa; $p_b = p_c = 6.00 \times 10^5$ Pa; $V_a = V_b = 0.0100$ m^3; $V_c = V_d = 0.0300$ m^3. Calculate a) the number of moles of CO_2; b) the heat input in each cycle; c) the waste heat per cycle; d) the work performed by the engine in each cycle; e) the thermal efficiency of the engine.

20.41 Heat Pump. A heat pump is a heat engine run in reverse. In winter it pumps heat from the cold air outside into the warmer air inside the building, maintaining the building at a comfortable temperature. In summer, it pumps heat from the cooler air inside the building to the warmer air outside, acting as an air conditioner. a) If the outside temperature in winter is −5.0°C and the inside temperature is 17.0°C, how many joules of heat will the heat pump deliver to the inside for each joule of electrical energy used to run the unit, assuming an ideal Carnot cycle? b) Suppose you have the option of using electrical resistance heating rather than a heat pump. How much electrical energy would you need in order to deliver the same amount of heat to the inside of the house as in part (a)? Consider a Carnot heat pump delivering heat to the inside of a house to maintain it at 68°F. Show that the heat pump delivers less heat for each joule of electrical energy used to operate the unit as the outside temperature decreases. Notice that this behavior is opposite to the dependence of the efficiency of a Carnot heat engine on the difference in the reservoir temperatures. Explain why this is so.

20.42 A heat engine operates using the cycle shown in Fig. 20.21. The working substance is 2.00 moles of helium gas, which reaches a maximum temperature of 327°C. Assume the helium can be treated as an ideal gas. Process bc is isothermal. The pressure in states a and c is 1.00×10^5 Pa and the pressure in state b is 3.00×10^5 Pa. a) How much heat enters the gas and how much leaves the gas each cycle? b) How much work does the engine do each cycle and what is its efficiency? c) Compare this engine's efficiency with the maximum possible efficiency attainable with the hot and cold reservoirs used by this cycle.

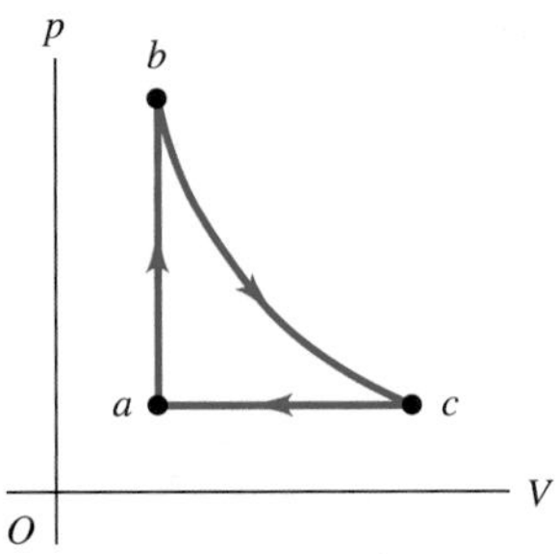

Figure 20.21 Problem 20.42.

20.43 As a budding mechanical engineer, you are called upon to design a Carnot engine that has 2.00 moles of a monatomic ideal gas as its working substance and that operates from a high-temperature reservoir at 500°C. The engine is to lift a 15.0-kg weight 2.00 m per cycle, using 500 J of heat input. The gas in the engine chamber can have a minimum volume of 5.00 L during the cycle. a) Draw a pV-diagram for this cycle. Show in your diagram where heat enters and leaves the gas. b) What must be the temperature of the cold reservoir? c) What is the thermal efficiency of the engine? d) How much heat energy does this engine waste per cycle? e) What is the maximum pressure that the gas chamber will have to withstand?

20.44 An experimental power plant at the Natural Energy Laboratory of Hawaii generates electricity from the temperature gradient of the ocean. The surface and deep-water temperatures are 27°C and 6°C, respectively. a) What is the maximum theoretical efficiency of this power plant? b) If the power plant is to produce 210 kW of power, at what rate must heat be extracted from the warm water? At what rate must heat be absorbed by the cold water? Assume the maximum theoretical efficiency. c) The cold water that enters the plant leaves it at a temperature of 10°C. What must be the flow rate of cold water through the system? Give your answer in kg/h and L/h.

20.45 What is the thermal efficiency of an engine that operates by taking n moles of diatomic ideal gas through the cycle $1 \rightarrow 2 \rightarrow 3 \rightarrow 4 \rightarrow 1$ shown in Fig. 20.22?

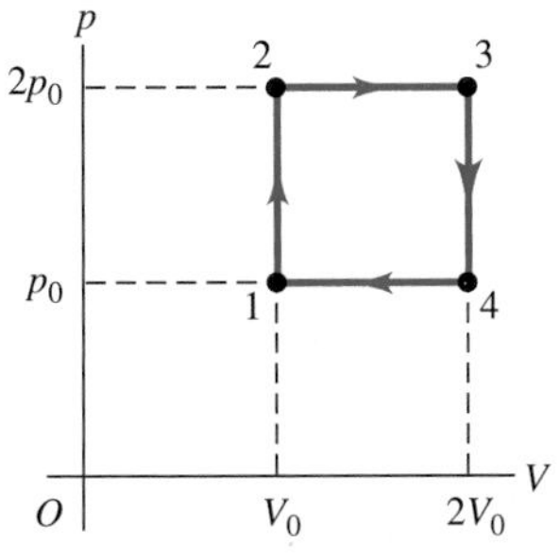

Figure 20.22 Problem 20.45.

20.46 A cylinder contains oxygen at a pressure of 2.00 atm. The volume is 4.00 L, and the temperature is 300 K. Assume that the oxygen may be treated as an ideal gas. The oxygen is carried through the following processes:

(i) Heated at constant pressure from the initial state (state 1) to state 2, which has $T = 450$ K.
(ii) Cooled at constant volume to 250 K (state 3).
(iii) Compressed at constant temperature to a volume of 4.00 L (state 4).
(iv) Heated at constant volume to 300 K, which takes the system back to state 1.

a) Show these four processes in a pV-diagram, giving the numerical values of p and V in each of the four states. b) Calculate Q and W for each of the four processes. c) Calculate the net work done by the oxygen. d) What is the efficiency of this device as a heat engine? How does this compare to the efficiency of a Carnot-cycle engine operating between the same minimum and maximum temperatures of 250 K and 450 K?

20.47 Thermodynamic Processes for a Refrigerator. A refrigerator operates on the cycle shown in Fig. 20.23. The compression ($d \rightarrow a$) and expansion ($b \rightarrow c$) steps are adiabatic. The temperature, pressure, and volume of the coolant in each of the four states a, b, c, and d are given in the table below.

State	T (°C)	P (kPa)	V (m^3)	U (kJ)	Percentage That Is Liquid
a	80	2305	0.0682	1969	0
b	80	2305	0.00946	1171	100
c	5	363	0.2202	1005	54
d	5	363	0.4513	1657	5

a) In each cycle, how much heat is taken from inside the refrigerator into the coolant while the coolant is in the evaporator? b) In each cycle, how much heat is exhausted from the coolant into the air outside the refrigerator while the coolant is in the condenser? c) In each cycle, how much work is done by the motor that operates the compressor? d) Calculate the coefficient of performance of the refrigerator.

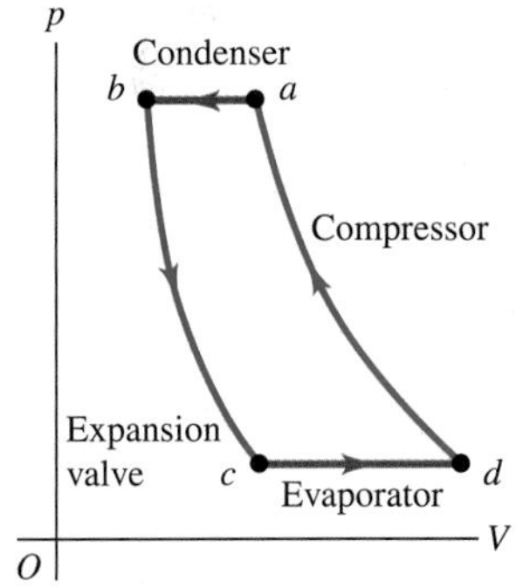

Figure 20.23 Problem 20.47.

20.48 A monatomic ideal gas is taken around the cycle shown in Fig. 20.24 in the direction

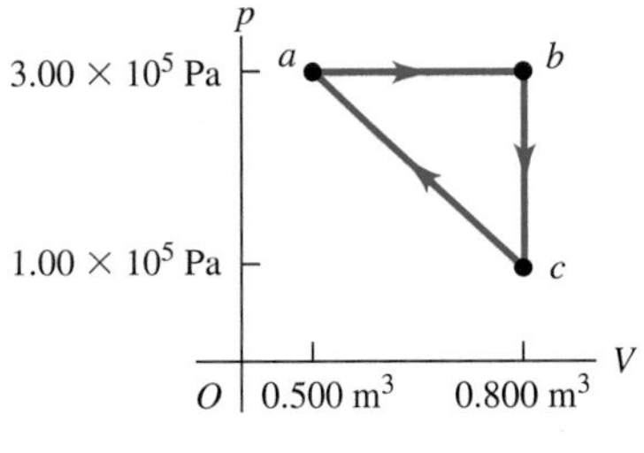

Figure 20.24 Problem 20.48.

shown in the figure. The path for process $c \to a$ is a straight line in the pV-diagram. a) Calculate Q, W, and ΔU for each process $a \to b$, $b \to c$, and $c \to a$. b) What are Q, W, and ΔU for one complete cycle? c) What is the efficiency of the cycle?

20.49 A Stirling-cycle Engine. The *Stirling cycle* is similar to the Otto cycle, except that the compression and expansion of the gas is done at constant temperature, not adiabatically as in the Otto cycle. The Stirling cycle is used in *external* combustion engines, which means that the gas inside the cylinder is not used in the combustion process. Heat is supplied by burning fuel steadily outside the cylinder, instead of explosively inside the cylinder as in the Otto cycle. For this reason Stirling-cycle engines are quieter than Otto-cycle engines, since there are no intake and exhaust valves (a major source of engine noise). While small Stirling engines are used for a variety of purposes, Stirling engines for automobiles have not been successful because they are larger, heavier, and more expensive than conventional automobile engines. In the cycle, the working fluid goes through the following sequence of steps (Fig. 20.25):

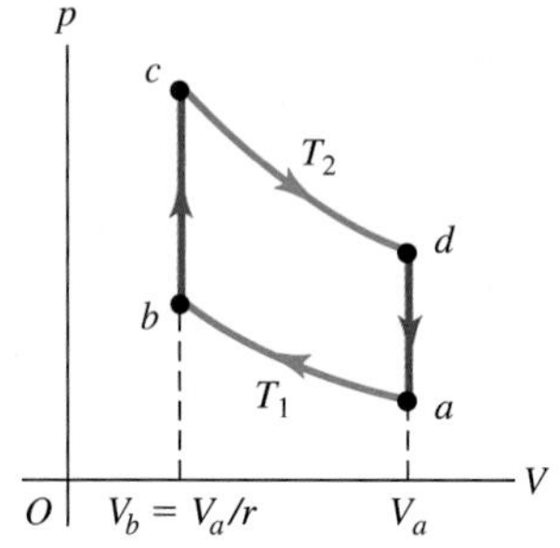

Figure 20.25 Problem 20.49.

(i) Compressed isothermally at temperature T_1 from the initial state a to state b, with a compression ratio r.
(ii) Heated at constant volume to state c at temperature T_2.
(iii) Expanded isothermally at T_2 to state d.
(iv) Cooled at constant volume back to the initial state a.

Assume that the working fluid is n moles of an ideal gas (for which C_V is independent of temperature). a) Calculate Q, W, and ΔU for each of the processes $a \to b$, $b \to c$, $c \to d$, and $d \to a$. b) In the Stirling cycle, the heat transfers in the processes $b \to c$ and $d \to a$ do not involve external heat sources, but rather use *regeneration:* the same substance that transfers heat to the gas inside the cylinder in the process $b \to c$ also absorbs heat back from the gas in the process $d \to a$. Hence the heat transfers $Q_{b\to c}$ and $Q_{d\to a}$ do not play a role in determining the efficiency of the engine. Explain this last statement by comparing the expressions for $Q_{b\to c}$ and $Q_{d\to a}$ calculated in part (a). c) Calculate the efficiency of a Stirling-cycle engine in terms of the temperatures T_1 and T_2. How does this compare to the efficiency of a Carnot-cycle engine operating between these same two temperatures? (Historically, the Stirling cycle was devised before the Carnot cycle.) Does this result violate the second law of thermodynamics? Explain. Unfortunately, actual Stirling-cycle engines cannot achieve this efficiency, due to problems with the heat-transfer processes and pressure losses in the engine.

20.50 A Carnot engine operates between two heat reservoirs at temperatures T_H and T_C. An inventor proposes to increase the efficiency by running one engine between T_H and an intermediate temperature T' and a second engine between T' and T_C using as input the heat expelled by the first engine. Compute the efficiency of this composite system and compare it to that of the original engine.

20.51 The maximum power that can be extracted by a wind turbine from an air stream is approximately

$$P = kd^2v^3$$

where d is the blade diameter, v is the wind speed, and the constant $k = 0.5\ \mathrm{W \cdot s^3/m^5}$. a) Explain the dependence of P on d and on v by considering a cylinder of air that passes over the turbine blades in time t (Fig. 20.26). This cylinder has diameter d, length $L = vt$, and density ρ. b) The Mod-5B wind turbine at Kahaku on the Hawaiian island of Oahu has a blade diameter of 97 m (slightly longer than a football field) and sits atop a 58-m tower. It can produce 3.2 MW of electric power. Assuming 25% efficiency, what wind speed is required to produce this amount of power? Give your answer in m/s and in km/h. c) Commercial wind turbines are commonly located in or downwind of mountain passes. Why?

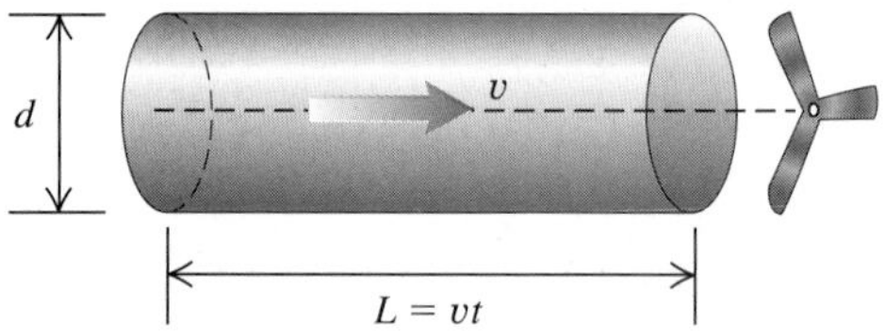

Figure 20.26 Problem 20.51.

20.52 Fuel Economy and Automobile Performance. The Otto-cycle engine of a Volvo V70 has a compression ratio $r = 8.5$. The U.S. Environmental Protection Agency (EPA) fuel economy rating of this car is 25 miles per gallon on the highway (105 km/h, or 65 mi/h). Gasoline has a heat of combustion of 4.60×10^7 J/kg, and its density is 740 kg/m^3. a) At 105 km/h, what is the rate of gasoline consumption in L/h? b) What is the theoretical efficiency of the engine? Use $\gamma = 1.40$. c) How much power is the engine producing at 105 km/h? Assume that the engine is operating at its theoretical efficiency, and give your answer in watts and in horsepower. By comparison, the engine of the Volvo V70 has a maximum power of 236 hp. d) Due to heat losses and friction in the drive train, the actual efficiency is approximately 15%. Repeat part (c) using this information. What fraction of the maximum possible power is used for highway driving?

20.53 Automotive Thermodynamics. A Volkswagen Passat has a six-cylinder Otto-cycle engine with compression ratio $r = 10.6$. The diameter of each cylinder, called the *bore* of the engine, is 82.5 mm. The distance that the piston moves during the compression in Fig. 20.4, called the *stroke* of the engine, is 86.4 mm. The initial pressure of the air-fuel mixture (at point a in Fig. 20.5) is 8.50×10^4 Pa, and the initial temperature is 300 K (the same as the outside air). Assume that 200 J of heat is added to each cylinder in each cycle by the burning gasoline, and that the gas has $C_V = 20.5$ J/mol·K and $\gamma = 1.40$. a) Calculate the total work done in one cycle in each cylinder of the engine, and the heat released when the gas is cooled to the temperature of the outside air. b) Calculate the volume of the air-fuel mixture at point a in the cycle. c) Calculate the pressure, volume, and temperature of the gas at points b, c, and d in the cycle. In a pV-diagram, show the numerical values of p, V, and T for each of the four states. d) Compare the efficiency of this engine with the efficiency of a Carnot-cycle engine operating between the same maximum and minimum temperatures.

20.54 In a factory, an insulated iron bar 65.0 cm long having thermal conductivity 79.5 W/m·K and cross-sectional area 15.0 cm^2 conducts heat from a furnace at 250.0°C to a cool water reservoir at 40.0°C, a) By how much does the entropy of the factory change each second due to this process? b) Interpret your answer in terms of the reversibility or irreversibility of this process.

20.55 Unavailable Energy. The discussion of entropy and the second law that follows Example 20.10 (Section 20.7) says that the increase in entropy in an irreversible process is associated with energy becoming less available. Consider a Carnot cycle that uses a low temperature reservoir with Kelvin temperature T_c. This is a true reservoir, that is large enough not to change temperature when it accepts heat from the engine. Let the engine accept heat from an object of temperature T', where $T' > T_c$. The object is of finite size, so cools as heat is extracted from it. The engine continues to operate until $T' = T_c$. a) Show that the total magnitude of heat rejected to the low temperature reservoir is $T_c|\Delta S_h|$, where ΔS_h is the change in entropy of the high-temperature reservoir. b) Apply the result of part (a) to 1.00 kg of water initially at a temperature of 373 K as the heat source for the engine and $T_c = 273$ K. How much total mechanical work can be performed by the engine until it stops? c) Repeat part (b) for 2.00 kg of water at 323 K. d) Compare the amount of work that can be obtained from the energy in the water of Example 20.10 before and after it is mixed. Discuss whether your result shows that energy has become less available.

20.56 Recalculate the change in entropy between points a and b of Fig. 20.15c if the reversible path is a) an isobaric expansion to $2V$ followed by an isochoric process; b) an isochoric cooling to $p_a/2$ followed by an isobaric expansion.

20.57 A 0.0500-kg cube of ice at an initial temperature of −15.0°C is placed in 0.600 kg of water at $T = 45.0$°C in an insulated container of negligible mass. Calculate the change in entropy of the system.

20.58 a) For the Otto cycle shown in Fig. 20.5, calculate the changes in entropy of the gas in each of the constant-volume processes $b \to c$ and $d \to a$ in terms of the temperatures T_a, T_b, T_c, and T_d and the number of moles n and the heat capacity C_V of the gas. b) What is the total entropy change in the engine during one cycle? (*Hint:* Use the relation between T_a and T_b and between T_d and T_c.) c) The processes $b \to c$ and $d \to a$ occur irreversibly in a real Otto engine. Explain how can this be reconciled with your result in part (b).

20.59 A *TS*-Diagram. a) Graph a Carnot cycle, plotting Kelvin temperature vertically and entropy horizontally. This is called a temperature-entropy diagram, or *TS*-diagram. b) Show that the area under any curve representing a reversible path in a temperature-entropy diagram represents the heat absorbed by the system. c) Derive from your diagram the expression for the thermal efficiency of a Carnot cycle. d) Draw a temperature-entropy diagram for the Stirling cycle, described in Problem 20.49. Use this diagram to relate the efficiency of the Carnot and Stirling cycles.

20.60 A physics student immerses one end of a copper rod in boiling water at 100°C and the other end in an ice-water mixture at 0°C. The sides of the rod are insulated. After steady-state conditions have been achieved in the rod, 0.160 kg of ice melts in a certain time interval. For this time interval find a) the entropy change of the boiling water; b) the entropy change of the ice water mixture; c) the entropy change of the copper rod; d) the total entropy change of the entire system.

20.61 To heat one cup of water (250 cm^3) to make coffee, you place an electric heating element in the cup. As the water temperature increases from 20°C to 65°C, the temperature of the heating element remains at a constant 120°C. Calculate the change in entropy of a) the water; b) the heating element; c) the system of water and heating element. (Make the same assumption about the specific heat of water as in Example 20.10 (Section 20.7), and ignore the heat that flows into the ceramic coffee cup itself.) d) Is this process reversible or irreversible? Explain.

20.62 An object of mass m_1, specific heat capacity c_1, and temperature T_1 is placed in contact with a second object of mass m_2, specific heat capacity c_2, and temperature $T_2 > T_1$. As a result, the temperature of the first object increases to T and the temperature of the second object decreases to T'. a) Show that the entropy increase of the system is

$$\Delta S = m_1 c_1 \ln \frac{T}{T_1} + m_2 c_2 \ln \frac{T'}{T_2}$$

and show that energy conservation requires that

$$m_1 c_1 (T - T_1) = m_2 c_2 (T_2 - T')$$

b) Show that the entropy change ΔS, considered as a function of T, is a *maximum* if $T = T'$, which is just the condition of thermodynamic equilibrium. c) Discuss the result of part (b) in terms of the idea of entropy as a measure of disorder.

Challenge Problem

20.63 Consider a Diesel cycle that starts (at point a in Fig. 20.6) with air at temperature T_a. The air may be treated as an ideal gas. a) If the temperature at point c is T_c, derive an expression for the efficiency of the cycle in terms of the compression ratio r. b) What is the efficiency if $T_a = 300$ K, $T_c = 950$ K, $\gamma = 1.40$, and $r = 21.0$?

21 ELECTRIC CHARGE AND ELECTRIC FIELD

Most of the forces on this water skier are electric. Electric interactions between adjacent molecules give rise to the force of the water on the ski, the tension in the tow rope, and the resistance of the air on the skier's body. Electric interactions also hold the skier's body together! Only one wholly nonelectric force acts on the skier: the force of gravity.

? Water makes life possible: the cells of your body could not function without the molecules dissolved in the water within the cells. What electrical properties of water make it such a good solvent?

Back in Chapter 5, we briefly mentioned the four kinds of fundamental forces. To this point the only one of these forces that we have examined in any detail is gravity. Now that we have learned more of the basic concepts of physics, including the behavior of waves and the rules of energy transfer, we are ready to investigate the properties of other forces. By far the most common of these forces in our daily lives is *electromagnetism,* which encompasses both electricity and magnetism. Our exploration of electromagnetic phenomena will occupy our attention for most of the remainder of this book.

Electromagnetic interactions involve particles that have a property called *electric charge,* an attribute that is as fundamental as mass. Just as objects with mass are accelerated by gravitational forces, so electrically charged objects are accelerated by electric forces. The annoying electric spark you feel when you scuff your shoes across a carpet and then reach for a metal doorknob is due to charged particles leaping between your finger and the doorknob. (A lightning bolt is a similar phenomenon on a far grander scale.) Electric currents, such as those in a flashlight, a portable CD player, or a television, are simply streams of charged particles flowing within wires in response to electric forces. Even the forces that hold atoms together to form solid matter, and that keep the atoms of solid objects from passing through each other, are fundamentally due to electric interactions between the charged particles within atoms.

We begin our study of electromagnetism in this chapter by examining the nature of electric charge. We'll find that electric charge is quantized and that it obeys a conservation principle. We then turn to a discussion of the interactions of electric charges that are at rest in our frame of reference, called *electrostatic* interactions. Such interactions are of tremendous importance in chemistry and biology and have many technological applications. Electrostatic interactions are governed by a simple relationship known as *Coulomb's law* and are most conveniently described by using the concept of *electric field.* We'll explore all these concepts in this chapter and expand on them in the three chapters that follow. In later chapters we'll expand our discussion to include electric charges in motion. This will lead us into an understanding of magnetism and, remarkably, of the nature of light.

While the key ideas of electromagnetism are conceptually simple, applying them to practical problems makes use of many of your mathematical skills, especially your knowledge of geometry and integral calculus. For this reason you may find this chapter and those that follow to be more mathematically demanding than previous chapters. The reward for your extra effort will be a deeper understanding of principles that are at the heart of modern physics and technology.

21.1 | Electric Charge

The ancient Greeks discovered as early as 600 B.C. that when they rubbed amber with wool, the amber could then attract other objects. Today we say that the amber has acquired a net **electric charge,** or has become *charged.* The word "electric" is derived from the Greek word *elektron,* meaning amber. When you scuff your shoes across a nylon carpet, you become electrically charged, and you can charge a comb by passing it through dry hair.

Plastic rods and fur (real or fake) are particularly good for demonstrating **electrostatics,** the interactions between electric charges that are at rest (or nearly so). Figure 21.1a shows two plastic rods and a piece of fur. After we charge each rod by rubbing it with the piece of fur, we find that the rods repel each other (Fig. 21.1b).

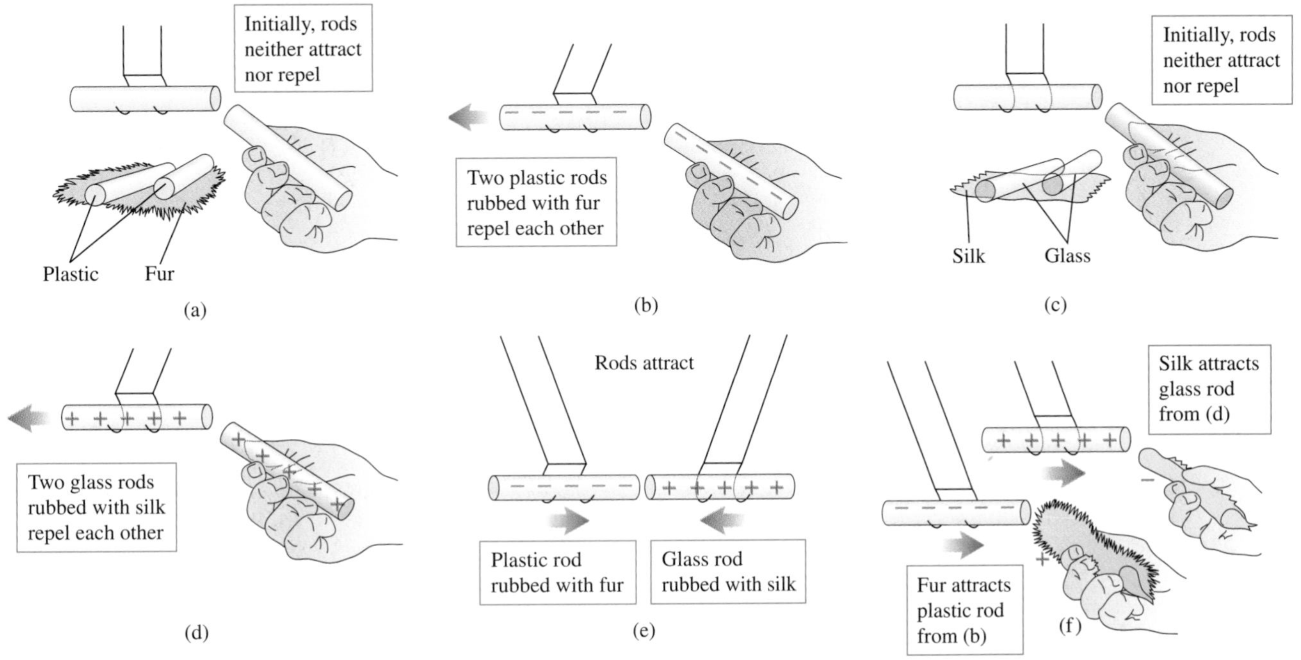

21.1 Experiments in electrostatics. (a, b) Plastic rods rubbed with fur acquire a negative charge and repel each other. (c, d) Glass rods rubbed with silk acquire a negative charge and repel each other. (e, f) Positively charged objects and negatively charged objects attract each other.

When we rub glass rods (Fig. 21.1c) with silk, the glass rods also become charged and repel each other (Fig. 21.1d). But a charged plastic rod *attracts* a charged glass rod (Fig. 21.1e). Furthermore, the plastic rod and the fur attract each other, and the glass rod and the silk attract each other (Fig. 21.1f).

These experiments and many others like them have shown that there are exactly two kinds of electric charge: the kind on the plastic rod rubbed with fur and the kind on the glass rod rubbed with silk. Benjamin Franklin (1706–1790) suggested calling these two kinds of charge *negative* and *positive,* respectively, and these names are still used. The plastic rod and the silk have negative charge; the glass rod and the fur have positive charge. **Two positive charges or two negative charges repel each other. A positive charge and a negative charge attract each other.**

CAUTION The attraction and repulsion of two charged objects is sometimes summarized as "Like charges repel, and opposite charges attract." But keep in mind that the phrase "like charges" does *not* mean that the two charges are exactly identical, only that both charges have the same algebraic *sign* (both positive or both negative). "Opposite charges" means that both objects have an electric charge, and those charges have different signs (one positive and the other negative).

One technological application of forces between charged bodies is in a laser printer (Fig. 21.2). Initially the printer's light-sensitive imaging drum is given a positive charge. As the drum rotates, a laser beam shines on selected areas of the drum, leaving those areas with a *negative* charge. Positively charged particles of toner adhere only to the areas of the drum "written" by the laser. When a piece of paper is placed in contact with the drum, the toner particles stick to the paper and form an image.

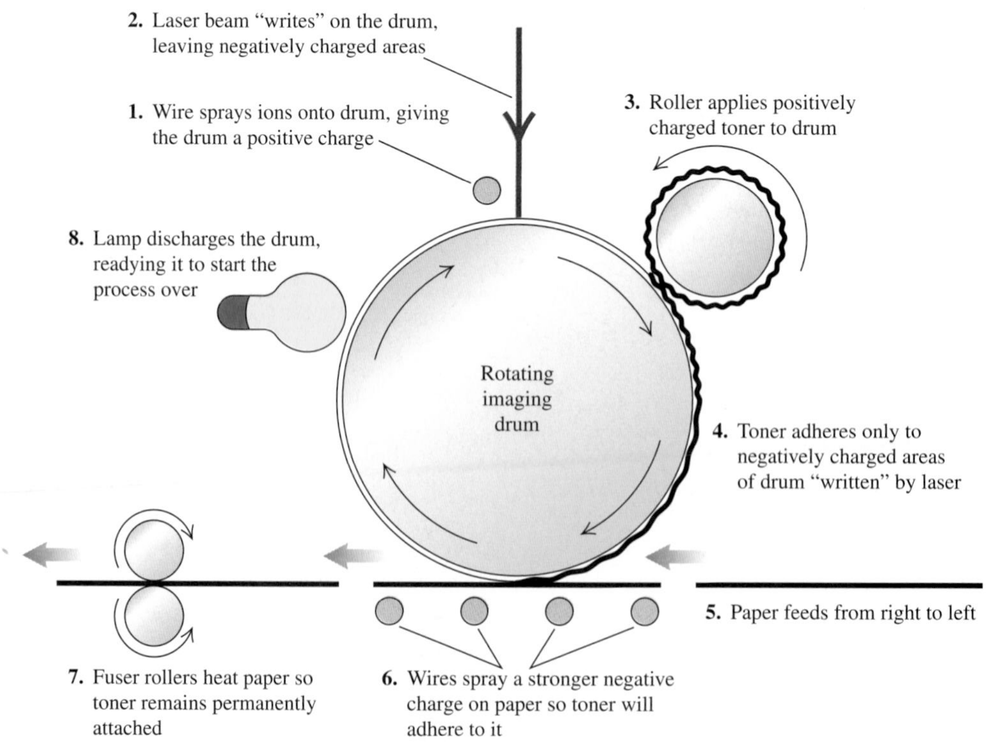

21.2 Schematic diagram of the operation of a laser printer.

Electric Charge and the Structure of Matter

When you charge a rod by rubbing it with fur or silk as in Fig. 21.1, there is no visible change in the appearance of the rod. What, then, actually happens to the rod when you charge it? To answer this question, we must look more closely at the structure and electric properties of atoms, the building blocks of ordinary matter of all kinds.

The structure of atoms can be described in terms of three particles: the negatively charged **electron** (Fig. 21.3), the positively charged **proton,** and the uncharged **neutron.** The proton and neutron are combinations of other entities called *quarks,* which have charges of $\pm\frac{1}{3}$ and $\pm\frac{2}{3}$ times the electron charge. Isolated quarks have not been observed, and there are theoretical reasons to believe that it is impossible in principle to observe a quark in isolation.

The protons and neutrons in an atom make up a small, very dense core called the **nucleus,** with dimensions of the order of 10^{-15} m. Surrounding the nucleus are the electrons, extending out to distances of the order of 10^{-10} m from the nucleus. If an atom were a few kilometers across, its nucleus would be the size of a tennis ball. The negatively charged electrons are held within the atom by the attractive electric forces exerted on them by the positively charged nucleus. (The protons and neutrons are held within the stable atomic nuclei by an attractive interaction, called the *strong nuclear force,* that overcomes the electric repulsion of the protons. The strong nuclear force has a short range, and its effects do not extend far beyond the nucleus.)

The masses of the individual particles, to the precision that they are presently known, are

$$\text{Mass of electron} = m_e = 9.10938188(72) \times 10^{-31}\text{ kg}$$
$$\text{Mass of proton} = m_p = 1.67262158(13) \times 10^{-27}\text{ kg}$$
$$\text{Mass of neutron} = m_n = 1.67492716(13) \times 10^{-27}\text{ kg}$$

The numbers in parentheses are the uncertainties in the last two digits. Note that the masses of the proton and neutron are nearly equal and are roughly 2000 times the mass of the electron. Over 99.9% of the mass of any atom is concentrated in its nucleus.

The negative charge of the electron has (within experimental error) *exactly* the same magnitude as the positive charge of the proton. In a neutral atom the number of electrons equals the number of protons in the nucleus, and the net electric charge (the algebraic sum of all the charges) is exactly zero (Fig. 21.4a). The num-

21.3 The electron—the first constituent of the atom to be isolated—was discovered in 1897 by the English physicist J. J. Thomson. This discovery revolutionized our understanding of the structure of matter, and led to the later discoveries of the proton and neutron. Thomson was awarded the 1906 Nobel Prize in physics and was knighted in 1908.

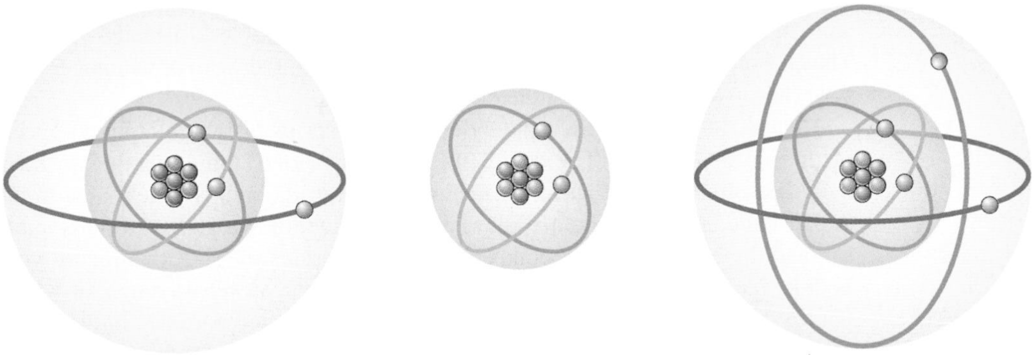

(a) Neutral lithium atom (Li): Nucleus has three protons (red) and four neutrons (purple); three electrons (blue) orbit nucleus

(b) Positive lithium ion (Li^+): Made by removing an electron from neutral lithium atom

(c) Negative lithium ion (Li^-): Made by adding an electron to neutral lithium atom

21.4 (a) A neutral atom has as many electrons as it does protons. (b) A positive ion has a deficit of electrons. (c) A negative ion has an excess of electrons. (The electron "orbits" are a schematic representation of the actual electron distribution, a diffuse cloud many times larger than the nucleus.)

ber of protons or electrons in a neutral atom of an element is called the **atomic number** of the element. If one or more electrons are removed, the remaining positively charged structure is called a **positive ion** (Fig. 21.4b). A **negative ion** is an atom that has *gained* one or more electrons (Fig. 21.4c). This gaining or losing of electrons is called **ionization.**

When the total number of protons in a macroscopic body equals the total number of electrons, the total charge is zero and the body as a whole is electrically neutral. To give a body an excess negative charge, we may either *add negative* charges to a neutral body or *remove positive* charges from that body. Similarly, we can create an excess positive charge by either *adding positive* charge or *removing negative* charge. In most cases, negatively charged (and highly mobile) electrons are added or removed, and a "positively charged body" is one that has lost some of its normal complement of electrons. When we speak of the charge of a body, we always mean its *net* charge. The net charge is always a very small fraction (typically no more than 10^{-12}) of the total positive charge or negative charge in the body.

Implicit in the foregoing discussion are two very important principles. First is the **principle of conservation of charge: The algebraic sum of all the electric charges in any closed system is constant.** If we rub together a plastic rod and a piece of fur, both initially uncharged, the rod acquires a negative charge (since it takes electrons from the fur) and the fur acquires a positive charge of the *same* magnitude (since it has lost as many electrons as the rod has gained). Hence the total electric charge on the two bodies together does not change. In any charging process, charge is not created or destroyed; it is merely *transferred* from one body to another.

Conservation of charge is thought to be a *universal* conservation law. No experimental evidence for any violation of this principle has ever been observed. Even in high-energy interactions in which particles are created and destroyed, such as the creation of electron–positron pairs, the total charge of any closed system is exactly constant.

The second important principle is that **the magnitude of charge of the electron or proton is a natural unit of charge.** Every observable amount of electric charge is always an integer multiple of this basic unit. We say that charge is *quantized.* A familiar example of quantization is money. When you pay cash for an item in a store, you have to do it in one-cent increments. Cash can't be divided into amounts smaller than one cent, and electric charge can't be divided into amounts smaller than the charge of one electron or proton. (The quark charges, $\pm\frac{1}{3}$ and $\pm\frac{2}{3}$ of the electron charge, are probably not observable as isolated charges.) Thus the charge on any macroscopic body is always either zero or an integer multiple (negative or positive) of the electron charge.

Understanding the electric nature of matter gives us insight into many aspects of the physical world. The chemical bonds that hold atoms together to form molecules are due to electric interactions between the atoms. They include the strong ionic bonds that hold sodium and chlorine atoms together to make table salt and the relatively weak bonds between the strands of DNA that record your body's genetic code. The normal force exerted on you by the chair in which you're sitting arises from electric forces between charged particles in the atoms of your seat and in the atoms of your chair. The tension force in a stretched string and the adhesive force of glue are likewise due to the electric interactions of atoms.

Test Your Understanding

Strictly speaking, does the plastic rod in Fig. 21.1 weigh more, less, or the same after rubbing it with fur? What about the glass rod after rubbing it with silk? And what about the fur and the silk?

21.2 | Conductors, Insulators, and Induced Charges

Some materials permit electric charge to move easily from one region of the material to another, while others do not. For example, Fig. 21.5a shows a copper wire supported by a glass rod. Suppose you touch one end of the wire to a charged plastic rod and attach the other end to a metal ball that is initially uncharged; you then remove the charged rod and the wire. When you bring another charged body up close to the ball (Figs. 21.5b and 21.5c), the ball is attracted or repelled, showing that the ball has become electrically charged. Electric charge has been transferred through the copper wire between the ball and the surface of the plastic rod.

The wire is called a **conductor** of electricity. If you repeat the experiment using a rubber band or nylon thread in place of the wire, you find that *no* charge is transferred to the ball. These materials are called **insulators.** Conductors permit the easy movement of charge through them, while insulators do not. As an example, carpet fibers on a dry day are good insulators. As you walk across a carpet, the rubbing of your shoes against the fibers causes charge to build up on you, and this charge remains on you because it can't flow through the insulating fibers. If you then touch a conducting object such as a doorknob, a rapid charge transfer takes place between your finger and the doorknob, and you feel a shock. One way to prevent this is to wind some of the carpet fibers around conducting cores so that any charge that builds up on you can be transferred harmlessly to the carpet. Another solution is to coat the carpet fibers with an antistatic layer that does not easily transfer electrons to or from your shoes; this prevents any charge from building up on you in the first place.

Most metals are good conductors, while most nonmetals are insulators. Within a solid metal such as copper, one or more outer electrons in each atom become detached and can move freely throughout the material, just as the molecules of a gas can move through the spaces between the grains in a bucket of sand. The motion of these negatively charged electrons carries charge through the metal. The other electrons remain bound to the positively charged nuclei, which themselves are bound in nearly fixed positions within the material. In an insulator there are no, or very few, free electrons, and electric charge cannot move freely through the material. Some materials called *semiconductors* are intermediate in their properties between good conductors and good insulators.

We can charge a metal ball by touching it with an electrically charged plastic rod, as in Fig. 21.5a. In this process, some of the excess electrons on the rod are transferred from it to the ball, leaving the rod with a smaller negative charge. There is a different technique in which the plastic rod can give another body a charge of *opposite* sign without losing any of its own charge. This process is called charging by **induction.**

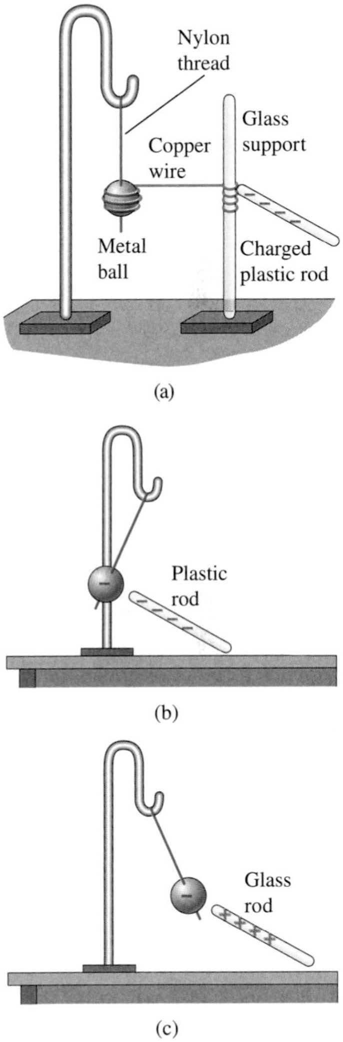

21.5 Copper is a good conductor of electricity; glass and nylon are good insulators. (a) The wire conducts charge between the metal ball and charged plastic rod to charge the ball negatively. (b) Afterward, the metal ball is repelled by a negatively charged plastic rod and (c) attracted to a positively charged glass rod.

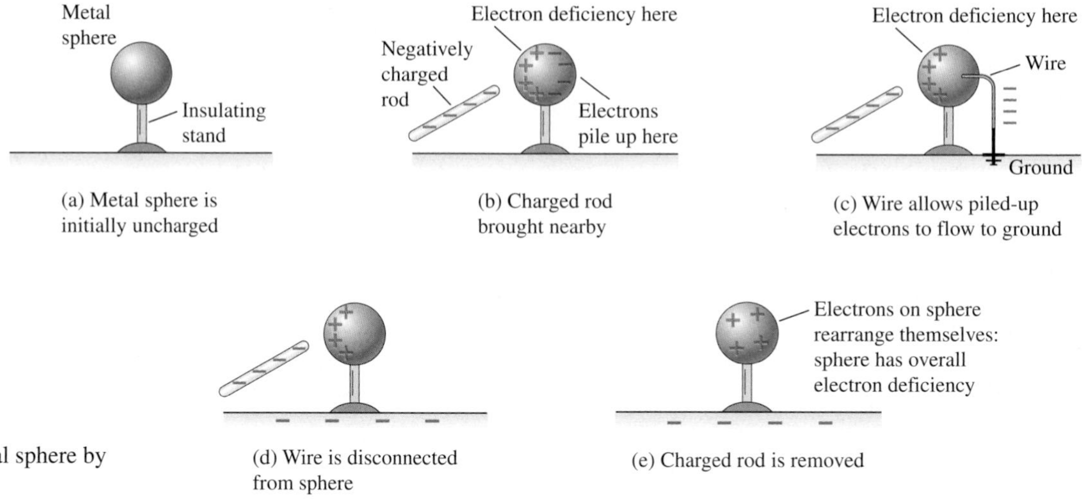

21.6 Charging a metal sphere by induction.

Figure 21.6a shows an example of charging by induction. A metal sphere is supported on an insulating stand. When you bring a negatively charged rod near it, without actually touching it (Fig. 21.6b), the free electrons in the metal sphere are repelled by the excess electrons on the rod, and they shift toward the right, away from the rod. They cannot escape from the sphere because the supporting stand and the surrounding air are insulators. So we get excess negative charge at the right surface of the sphere and a deficiency of negative charge (that is, a net positive charge) at the left surface. These excess charges are called **induced charges.**

Not all of the free electrons move to the right surface of the sphere. As soon as any induced charge develops, it exerts forces toward the *left* on the other free electrons. These electrons are repelled by the negative induced charge on the right and attracted toward the positive induced charge on the left. The system reaches an equilibrium state in which the force toward the right on an electron, due to the charged rod, is just balanced by the force toward the left due to the induced charge. If we remove the charged rod, the free electrons shift back to the left, and the original neutral condition is restored.

What happens if, while the plastic rod is nearby, you touch one end of a conducting wire to the right surface of the sphere and the other end to the earth (Fig. 21.6c)? The earth is a conductor, and it is so large that it can act as a practically infinite source of extra electrons or sink of unwanted electrons. Some of the negative charge flows through the wire to the earth. Now suppose you disconnect the wire (Fig. 21.6d) and then remove the rod (Fig. 21.6e); a net positive charge is left on the sphere. The charge on the negatively charged rod has not changed during this process. The earth acquires a negative charge that is equal in magnitude to the induced positive charge remaining on the sphere.

Charging by induction would work just as well if the mobile charges in the spheres were positive charges instead of negatively charged electrons, or even if both positive and negative mobile charges were present. In a metallic conductor the mobile charges are always negative electrons, but it is often convenient to describe a process *as though* the moving charges were positive. In ionic solutions and ionized gases, both positive and negative charges are mobile.

Finally, we note that a charged body can exert forces even on objects that are *not* charged themselves. If you rub a balloon on the rug and then hold the balloon against the ceiling, it sticks, even though the ceiling has no net electric charge. After you electrify a comb by running it through your hair, you can pick up uncharged bits of paper with the comb. How is this possible?

This interaction is an induced-charge effect. In Fig. 21.6b the plastic rod exerts a net attractive force on the conducting sphere even though the total charge on the sphere is zero, because the positive charges are closer to the rod than the negative charges. Even in an insulator, electric charges can shift back and forth a little when there is charge nearby. This is shown in Fig. 21.7a; the negatively charged plastic comb causes a slight shifting of charge within the molecules of the neutral insulator, an effect called *polarization.* The positive and negative charges in the material are present in equal amounts, but the positive charges are closer to the plastic comb and so feel an attraction that is stronger than the repulsion felt by the negative charges, giving a net attractive force. (In Section 21.3 we will study how electric forces depend on distance.) Note that a neutral insulator is also attracted to a *positively* charged comb (Fig. 21.7b). Now the charges in the insulator shift in the opposite direction; the negative charges in the insulator are closer to the comb and feel an attractive force that is stronger than the repulsion felt by the positive charges in the insulator. Hence a charged object of *either* sign exerts an attractive force on an uncharged insulator.

The attraction between a charged object and an uncharged one has many important practical applications, including the electrostatic painting process used in the automobile industry (Fig. 21.8). A metal object to be painted is connected to the earth, and the paint droplets are given an electric charge as they exit the sprayer nozzle. Induced charges of the opposite sign appear in the object as the droplets approach, just as in Fig. 21.6b, and they attract the droplets to the surface. This process minimizes overspray from clouds of stray paint particles and gives a particularly smooth finish.

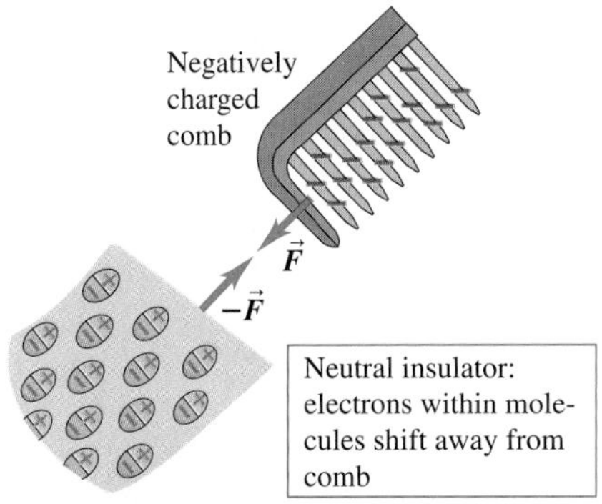

(a) Positive charges in molecules are closer to negative comb than are negative charges: net force is attractive

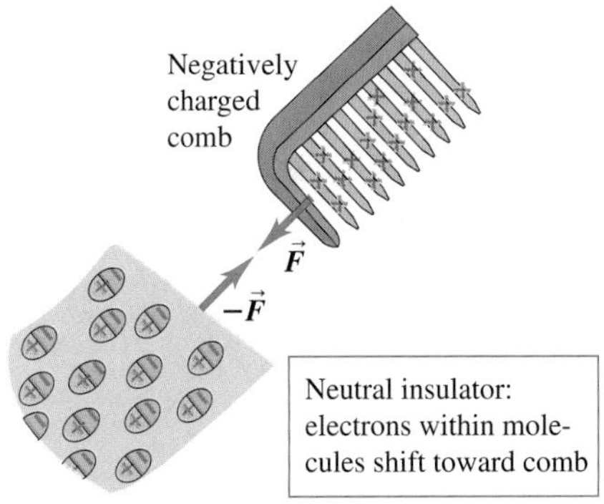

(b) Negative charges in molecules are closer to positive comb than are positive charges: net force is attractive

21.7 The charges within the molecules of an insulating material can shift slightly. As a result, a comb with either sign of charge attracts a neutral insulator. By Newton's third law the neutral insulator exerts an equal-magnitude attractive force on the comb.

Test Your Understanding

Starting with the situation shown in Fig. 21.6a, describe how you could use a charged rod to give the metal sphere a *positive* charge.

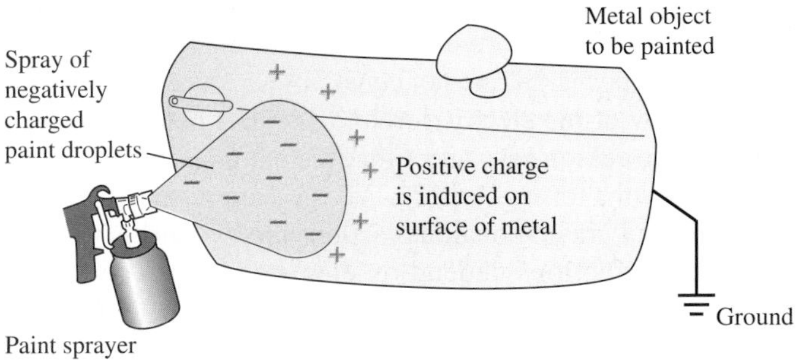

21.8 The electrostatic painting process (compare Figs. 21.6b and 21.6c).

21.3 | Coulomb's Law

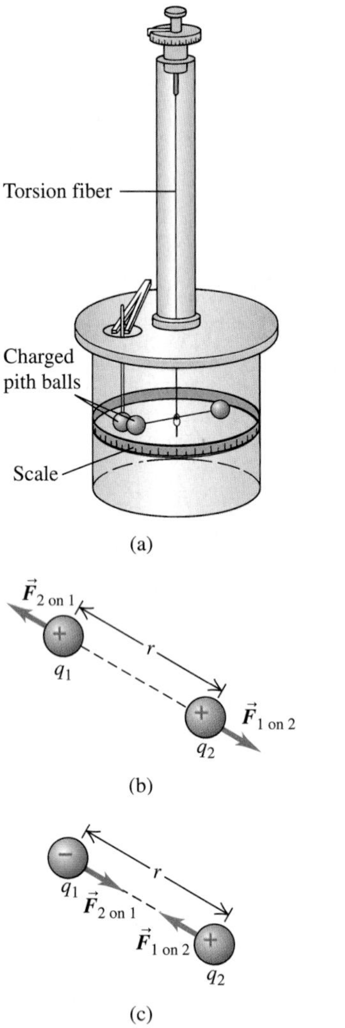

21.9 A torsion balance of the type used by Coulomb to measure the electric force. (b) Electric charges of the same sign repel each other. (c) Electric charges of opposite sign attract each other. In each case the forces obey Newton's third law: $\vec{F}_{1 \text{ on } 2} = -\vec{F}_{2 \text{ on } 1}$.

Charles Augustin de Coulomb (1736–1806) studied the interaction forces of charged particles in detail in 1784. He used a torsion balance (Fig. 21.9a) similar to the one used 13 years later by Cavendish to study the much weaker gravitational interaction, as we discussed in Section 12.1. For **point charges,** charged bodies that are very small in comparison with the distance r between them, Coulomb found that the electric force is proportional to $1/r^2$. That is, when the distance r doubles, the force decreases to $\frac{1}{4}$ of its initial value; when the distance is halved, the force increases to four times its initial value.

The electric force between two point charges also depends on the quantity of charge on each body, which we will denote by q or Q. To explore this dependence, Coulomb divided a charge into two equal parts by placing a small charged spherical conductor into contact with an identical but uncharged sphere; by symmetry, the charge is shared equally between the two spheres. (Note the essential role of the principle of conservation of charge in this procedure.) Thus he could obtain one half, one quarter, and so on, of any initial charge. He found that the forces that two point charges q_1 and q_2 exert on each other are proportional to each charge and therefore are proportional to the *product* q_1q_2 of the two charges.

Thus Coulomb established what we now call **Coulomb's law:**

The magnitude of the electric force between two point charges is directly proportional to the product of the charges and inversely proportional to the square of the distance between them.

In mathematical terms, the magnitude F of the force that each of two point charges q_1 and q_2 a distance r apart exerts on the other can be expressed as

$$F = k\frac{|q_1q_2|}{r^2} \tag{21.1}$$

where k is a proportionality constant whose numerical value depends on the system of units used. The absolute value bars are used in Eq. (21.1) because the charges q_1 and q_2 can be either positive or negative, while the force magnitude F is always positive.

The directions of the forces the two charges exert on each other are always along the line joining them. When the charges q_1 and q_2 have the same sign, either both positive or both negative, the forces are repulsive (Fig. 21.9b); when the charges have opposite signs, the forces are attractive (Fig. 21.9c). The two forces obey Newton's third law; they are always equal in magnitude and opposite in direction, even when the charges are not equal.

The proportionality of the electric force to $1/r^2$ has been verified with great precision. There is no reason to suspect that the exponent is anything different from precisely 2. Thus the form of Eq. (21.1) is the same as that of the law of gravitation. But electric and gravitational interactions are two distinct classes of phenomena. Electric interactions depend on electric charges and can be either attractive or repulsive, while gravitational interactions depend on mass and are always attractive (because there is no such thing as negative mass).

The value of the proportionality constant k in Coulomb's law depends on the system of units used. In our study of electricity and magnetism we will use SI units exclusively. The SI electric units include most of the familiar units such as the volt, the ampere, the ohm, and the watt. (There is *no* British system of electric units.) The SI unit of electric charge is called one **coulomb** (1 C). In SI units the constant k in Eq. (21.1) is

$$k = 8.987551787 \times 10^9\ \text{N}\cdot\text{m}^2/\text{C}^2 \cong 8.988 \times 10^9\ \text{N}\cdot\text{m}^2/\text{C}^2$$

The value of k is known to such a large number of significant digits because this value is closely related to the speed of light in vacuum. (We will show this in Chapter 32 when we study electromagnetic radiation.) As we discussed in Section 1.3, this speed is *defined* to be exactly $c = 2.99792458 \times 10^8$ m/s. The numerical value of k is defined in terms of c to be precisely

$$k = (10^{-7}\ \text{N}\cdot\text{s}^2/\text{C}^2)c^2$$

You should check this expression to confirm that k has the right units.

In principle we can measure the electric force F between two equal charges q at a measured distance r and use Coulomb's law to determine the charge. Thus we could regard the value of k as an operational definition of the coulomb. For reasons of experimental precision it is better to define the coulomb instead in terms of a unit of electric *current* (charge per unit time), the *ampere*, equal to one coulomb per second. We will return to this definition in Chapter 28.

ActivPhysics Online

11.1 Electric Force: Coulomb's Law

11.2 Electric Force: Superposition Principle

11.3 Electric Force: Superposition Principle (Quantitative)

In SI units we usually write the constant k in Eq. (21.1) as $1/4\pi\epsilon_0$, where ϵ_0 ("epsilon-nought" or "epsilon-zero") is another constant. This appears to complicate matters, but it actually simplifies many formulas that we will encounter in later chapters. From now on, we will usually write Coulomb's law as

$$F = \frac{1}{4\pi\epsilon_0}\frac{|q_1q_2|}{r^2} \qquad (21.2)$$

(Coulomb's law: force between two point charges)

The constants in Eq. (21.2) are approximately

$$\epsilon_0 = 8.854 \times 10^{-12}\ \text{C}^2/\text{N}\cdot\text{m}^2 \quad \text{and} \quad \frac{1}{4\pi\epsilon_0} = k = 8.988 \times 10^9\ \text{N}\cdot\text{m}^2/\text{C}^2$$

In examples and problems we will often use the approximate value

$$\frac{1}{4\pi\epsilon_0} = 9.0 \times 10^9\ \text{N}\cdot\text{m}^2/\text{C}^2$$

which is within about 0.1% of the correct value.

As we mentioned in Section 21.1, the most fundamental unit of charge is the magnitude of the charge of an electron or a proton, which is denoted by e. The most precise value available as of the writing of this book is

$$e = 1.602176462(63) \times 10^{-19}\ \text{C}$$

One coulomb represents the negative of the total charge of about 6×10^{18} electrons. For comparison, a copper cube 1 cm on a side contains about 2.4×10^{24} electrons. About 10^{19} electrons pass through the glowing filament of a flashlight bulb every second.

In electrostatics problems (that is, problems that involve charges at rest), it's very unusual to encounter charges as large as one coulomb. Two 1-C charges separated by 1 m would exert forces on each other of magnitude 9×10^9 N (about 1 million tons)! The total charge of all the electrons in a copper penny is even greater, about 1.4×10^5 C, which shows that we can't disturb electric neutrality very much without using enormous forces. More typical values of charge range from about 10^{-9} to about 10^{-6} C. The microcoulomb $(1\ \mu\text{C} = 10^{-6}\ \text{C})$ and the nanocoulomb $(1\ \text{nC} = 10^{-9}\ \text{C})$ are often used as practical units of charge.

Example 21.1 Electric force versus gravitational force

An α particle ("alpha") is the nucleus of a helium atom. It has a mass $m = 6.64 \times 10^{-27}$ kg and a charge $q = +2e = 3.2 \times 10^{-19}$ C. Compare the force of the electric repulsion between two α particles with the force of gravitational attraction between them.

SOLUTION

IDENTIFY and SET UP: The magnitude F_e of the electric force is given by Eq. (21.2),

$$F_e = \frac{1}{4\pi\epsilon_0}\frac{q^2}{r^2}$$

The magnitude F_g of the gravitational force is given by Eq. (12.1),

$$F_g = G\frac{m^2}{r^2}$$

We compare these two magnitudes by calculating their ratio.

EXECUTE: The ratio of the electric force to the gravitational force is

$$\frac{F_e}{F_g} = \frac{1}{4\pi\epsilon_0 G}\frac{q^2}{m^2} = \frac{9.0 \times 10^{9}\ \text{N}\cdot\text{m}^2/\text{C}^2}{6.67 \times 10^{-11}\ \text{N}\cdot\text{m}^2/\text{kg}^2}\frac{(3.2 \times 10^{-19}\ \text{C})^2}{(6.64 \times 10^{-27}\ \text{kg})^2}$$
$$= 3.1 \times 10^{35}$$

EVALUATE: This astonishingly large number shows that the gravitational force in this situation is completely negligible in comparison to the electric force. This is always true for interactions of atomic and subatomic particles. (Notice that this result doesn't depend on the distance r between the two α particles.) But within objects the size of a person or a planet, the positive and negative charges are nearly equal in magnitude, and the net electric force is usually much *smaller* than the gravitational force.

Superposition of Forces

Coulomb's law as we have stated it describes only the interaction of two *point* charges. Experiments show that when two charges exert forces simultaneously on a third charge, the total force acting on that charge is the *vector sum* of the forces that the two charges would exert individually. This important property, called the **principle of superposition of kforces,** holds for any number of charges. By using this principle, we can apply Coulomb's law to *any* collection of charges. Several of the examples at the end of this section show applications of the superposition principle.

Strictly speaking, Coulomb's law as we have stated it should be used only for point charges *in a vacuum.* If matter is present in the space between the charges, the net force acting on each charge is altered because charges are induced in the molecules of the intervening material. We will describe this effect later. As a practical matter, though, we can use Coulomb's law unaltered for point charges in air. At normal atmospheric pressure, the presence of air changes the electric force from its vacuum value by only about one part in 2000.

Problem-Solving Strategy Coulomb's Law

IDENTIFY *the relevant concepts:* Coulomb's law comes into play whenever you need to know the electric force acting between charged particles.

SET UP *the problem using the following steps:*

1. Make a drawing showing the locations of the charged particles and label each particle with its charge. This step is particularly important if more than two charged particles are present.
2. If three or more charges are present and they do not all lie on the same line, set up an xy-coordinate system.
3. Often you will need to find the electric force on just one particle. If so, identify that particle.

EXECUTE *the solution as follows:*

1. For each particle that exerts a force on the particle of interest, calculate the magnitude of that force using Eq. (21.2).
2. Sketch the electric force vectors acting on the particle(s) of interest due to each of the other particles (that is, make a free-body diagram). Remember that the force exerted by particle 1 on particle 2 points from particle 2 toward particle 1 if the two charges have opposite signs, but points from particle 2 directly away from particle 1 if the charges have the same sign.
3. Calculate the total electric force on the particle(s) of interest. Remember that the electric force, like any force, is a *vector.* When the forces acting on a charge are caused by two or more other charges, the total force on the charge is the *vector sum* of the individual forces. You may want to go back and review the vector algebra in Sections 1.7 through 1.9. It's often helpful to use components in an *xy*-coordinate system. Be sure to use correct vector notation; if a symbol represents a vector quantity, put an arrow over it. If you get sloppy with your notation, you will also get sloppy with your thinking.
4. As always, using consistent units is essential. With the value of $k = 1/4\pi\epsilon_0$ given above, distances *must* be in meters, charge in coulombs, and force in newtons. If you are given distances in centimeters, inches, or furlongs, don't forget to convert! When a charge is given in microcoulombs (μC) or nanocoulombs (nC), remember that $1\ \mu\text{C} = 10^{-6}$ C and $1\ \text{nC} = 10^{-9}$ C.
5. Some examples and problems in this and later chapters involve a continuous distribution of charge along a line or over a surface. In these cases the vector sum described in Step 3 becomes a vector integral, usually carried out by use of components. We divide the total charge distribution into infinitesimal pieces, use Coulomb's law for each piece, and then integrate to find the vector sum. Sometimes this process can be done without explicit use of integration.
6. In many situations the charge distribution will be *symmetrical.* For example, you might be asked to find the force on a charge Q in the presence of two other identical charges q, one above and to the left of Q and the other below and to the left of Q. If the distances from Q to each of the other charges are the same, the force on Q from each charge has the same magnitude; if each force vector makes the same angle with the horizontal axis, adding these vectors to find the net force is particularly easy. Whenever possible, exploit any symmetries to simplify the problem-solving process.

EVALUATE *your answer:* Check whether your numerical results are reasonable, and confirm that the direction of the net electric force agrees with the principle that like charges repel and opposite charges attract.

Example 21.2 Force between two point charges

Two point charges, $q_1 = +25$ nC and $q_2 = -75$ nC, are separated by a distance of 3.0 cm (Fig. 21.10a). Find the magnitude and direction of a) the electric force that q_1 exerts on q_2; b) the electric force that q_2 exerts on q_1.

SOLUTION

IDENTIFY and SET UP: We use Coulomb's law, Eq. (21.2), to calculate the magnitude of the force that each particle exerts on the other. The problem asks us for the force on each particle due to the other particle, so we use Newton's third law.

EXECUTE: a) Converting charge to coulombs and distance to meters, the magnitude of the force that q_1 exerts on q_2 is

$$F_{1\,\text{on}\,2} = \frac{1}{4\pi\epsilon_0}\frac{|q_1 q_2|}{r^2}$$

$$= (9.0 \times 10^9\ \text{N}\cdot\text{m}^2/\text{C}^2)\frac{|(+25 \times 10^{-9}\ \text{C})(-75 \times 10^{-9}\ \text{C})|}{(0.030\ \text{m})^2}$$

$$= 0.019\ \text{N}$$

Since the two charges have opposite signs, the force is attractive; that is, the force that acts on q_2 is directed toward q_1 along the line joining the two charges, as shown in Fig. 21.10b.

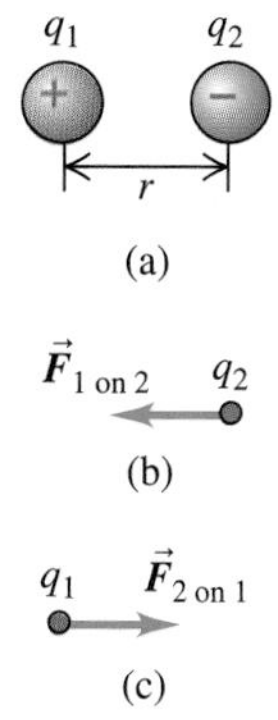

21.10 What force does q_1 exert on q_2, and what force does q_2 exert on q_1? Gravitational forces are negligible. (a) The two charges. (b) Free-body diagram for charge q_2. (c) Free-body diagram for charge q_1.

b) Remember that Newton's third law applies to the electric force. Even though the charges have different magnitudes, the magnitude of the force that q_2 exerts on q_1 is the *same* as the magnitude of the force that q_1 exerts on q_2:

$$F_{2\text{ on }1} = 0.019\text{ N}$$

Newton's third law also states that the direction of the force that q_2 exerts on q_1 is exactly opposite the direction of the force that q_1 exerts on q_2; this is shown in Fig. 21.10c.

EVALUATE: Note that the force on q_1 is directed toward q_2, as it must be, since charges of opposite sign attract each other.

Example 21.3 Vector addition of electric forces on a line

Two point charges are located on the positive x-axis of a coordinate system (Fig. 21.11a). Charge $q_1 = 1.0\text{ nC}$ is 2.0 cm from the origin, and charge $q_2 = -3.0\text{ nC}$ is 4.0 cm from the origin. What is the total force exerted by these two charges on a charge $q_3 = 5.0\text{ nC}$ located at the origin? Gravitational forces are negligible.

SOLUTION

IDENTIFY: Here there are *two* electric forces acting on the charge q_3, and we must add these forces to find the total force.

SET UP: Figure 21.11a shows the coordinate system. Our target variable is the net electric force exerted *on* charge q_3 by the other two charges. This is the vector sum of the forces due to q_1 and q_2 individually.

EXECUTE: Figure 21.11b is a free-body diagram for charge q_3. Note that q_3 is repelled by q_1 (which has the same sign) and attracted to q_2 (which has the opposite sign). Converting charge to coulombs and distance to meters, we use Eq. (21.2) to find the magnitude $F_{1\text{ on }3}$ of the force of q_1 on q_3:

21.11 What is the total force exerted on the point charge q_3 by the other two point charges? (a) The three charges. (b) Free-body diagram for charge q_3.

$$F_{1\text{ on }3} = \frac{1}{4\pi\epsilon_0}\frac{|q_1q_3|}{r^2}$$
$$= (9.0\times10^9\text{ N}\cdot\text{m}^2/\text{C}^2)\frac{(1.0\times10^{-9}\text{ C})(5.0\times10^{-9}\text{ C})}{(0.020\text{ m})^2}$$
$$= 1.12\times10^{-4}\text{ N} = 112\ \mu\text{N}$$

This force has a negative x-component because q_3 is repelled (that is, pushed in the negative x-direction) by q_1.

The magnitude $F_{2\text{ on }3}$ of the force of q_2 on q_3 is

$$F_{2\text{ on }3} = \frac{1}{4\pi\epsilon_0}\frac{|q_2q_3|}{r^2}$$
$$= (9.0\times10^9\text{ N}\cdot\text{m}^2/\text{C}^2)\frac{(3.0\times10^{-9}\text{ C})(5.0\times10^{-9}\text{ C})}{(0.040\text{ m})^2}$$
$$= 8.4\times10^{-5}\text{ N} = 84\ \mu\text{N}$$

This force has a positive x-component because q_3 is attracted (that is, pulled in the positive x-direction) by q_2. The sum of the x-components is

$$F_x = -112\ \mu\text{N} + 84\mu\text{ N} = -28\ \mu\text{N}$$

There are no y- or z-components. Thus the total force on q_3 is directed to the left, with magnitude $28\ \mu\text{N} = 2.8\times10^{-5}\text{ N}$.

EVALUATE: To check the magnitudes of the individual forces, note that q_2 has three times as much charge (in magnitude) as q_1 but is twice as far from q_3. From Eq. (21.2) this means that $F_{2\text{ on }3}$ must be $3/2^2 = \frac{3}{4}$ as large as $F_{1\text{ on }3}$. Indeed, our results show that this ratio is $(84\ \mu\text{N})/(112\ \mu\text{N}) = 0.75$. The direction of the net force also makes sense: $\vec{F}_{1\text{ on }3}$ is opposite to and has a larger magnitude than $\vec{F}_{2\text{ on }3}$, so the net force is in the direction of $\vec{F}_{1\text{ on }3}$.

Example 21.4 Vector addition of electric forces in a plane

In Fig. 21.12, two equal positive point charges $q_1 = q_2 = 2.0\ \mu\text{C}$ interact with a third point charge $Q = 4.0\ \mu\text{C}$. Find the magnitude and direction of the total (net) force on Q.

SOLUTION

IDENTIFY and SET UP: As in Example 21.3, we have to compute the force each charge exerts on Q and then find the vector sum of the forces. The easiest way to do this is to use components.

EXECUTE: Figure 21.12 shows the force on Q due to the upper charge q_1. From Coulomb's law the magnitude F of this force is

$$F_{1\text{ on }Q} = (9.0 \times 10^9\ \text{N}\cdot\text{m}^2/\text{C}^2)\frac{(4.0 \times 10^{-6}\ \text{C})(2.0 \times 10^{-6}\ \text{C})}{(0.50\ \text{m})^2}$$
$$= 0.29\ \text{N}$$

The angle α is below the x-axis, so the components of this force are given by

$$(F_{1\text{ on }Q})_x = (F_{1\text{ on }Q})\cos\alpha = (0.29\ \text{N})\frac{0.40\ \text{m}}{0.50\ \text{m}} = 0.23\ \text{N}$$

$$(F_{1\text{ on }Q})_y = -(F_{1\text{ on }Q})\sin\alpha = -(0.29\ \text{N})\frac{0.30\ \text{m}}{0.50\ \text{m}} = -0.17\ \text{N}$$

The lower charge q_2 exerts a force with the same magnitude but at an angle α *above* the x-axis. From symmetry we see that its $x$component is the same as that due to the upper charge, but its $y$component has the opposite sign. So the components of the total force $\vec{F}$ on Q are

$$F_x = 0.23\ \text{N} + 0.23\ \text{N} = 0.46\ \text{N}$$
$$F_y = -0.17\ \text{N} + 0.17\ \text{N} = 0$$

The total force on Q is in the $+x$-direction, with magnitude 0.46 N.

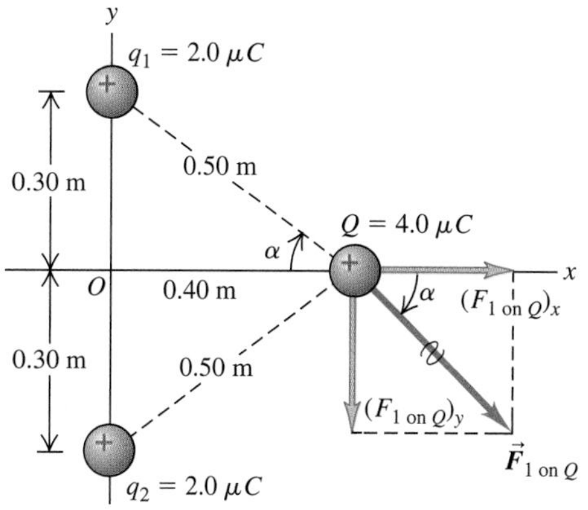

21.12 $\vec{F}_{1\text{ on }Q}$ is the force on Q due to the upper charge q_1.

EVALUATE: The total force on Q is in a direction that points neither directly away from q_1 nor directly away from q_2. Rather, this direction is a compromise that points away from the *system* of charges q_1 and q_2. Can you see that the total force would *not* be in the $+x$-direction if q_1 and q_2 were not equal or if the geometrical arrangement of the changes were not so symmetrical?

Test Your Understanding

Suppose that charge q_2 in Example 21.4 were equal to $-2.0\ \mu\text{C}$. Show that the total electric force on Q would be in the negative y-direction and have magnitude 0.34 N.

21.4 | Electric Field and Electric Forces

When two electrically charged particles in empty space interact, how does each one know the other is there? What goes on in the space between them to communicate the effect of each one to the other? We can begin to answer these questions, and at the same time reformulate Coulomb's law in a very useful way, by using the concept of *electric field.*

To introduce this concept, let's look at the mutual repulsion of two positively charged bodies A and B (Fig. 21.13a). Suppose B has charge q_0, and let $\vec{F}_0$ be the

21.13 A charged body creates an electric field in the space around it.

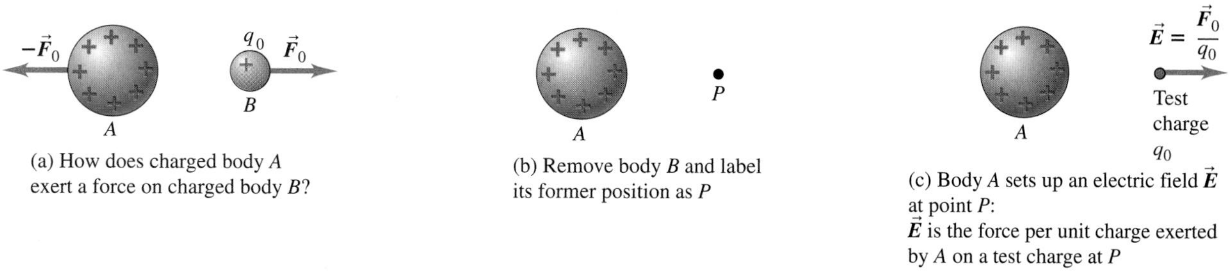

electric force of A on B. One way to think about this force is as an "action-at-a-distance" force—that is, as a force that acts across empty space without needing any matter (such as a push rod or a rope) to transmit it through the intervening space. (Gravity can also be thought of as an "action-at-a-distance" force.) But a more fruitful way to visualize the repulsion between A and B is as a two-stage process. We first envision that body A, as a result of the charge that it carries, somehow *modifies the properties of the space around it.* Then body B, as a result of the charge that *it* carries, senses how space has been modified at its position. The response of body B is to experience the force $\vec{F}_0$.

To elaborate how this two-stage process occurs, we first consider body A by itself: We remove body B and label its former position as point P (Fig. 21.13b). We say that the charged body A produces or causes an **electric field** at point P (and at all other points in the neighborhood). This electric field is present at P even if there is no other charge at P; it is a consequence of the charge on body A only. If a point charge q_0 is then placed at point P, it experiences the force $\vec{F}_0$. We take the point of view that this force is exerted on q_0 *by the field* at P (Fig. 21.13c). Thus the electric field is the intermediary through which A communicates its presence to q_0. Because the point charge q_0 would experience a force at *any* point in the neighborhood of A, the electric field that A produces exists at all points in the region around A.

We can likewise say that the point charge q_0 produces an electric field in the space around it and that this electric field exerts the force $-\vec{F}_0$ on body A. For each force (the force of A on q_0 and the force of q_0 on A), one charge sets up an electric field that exerts a force on the second charge. We emphasize that this is an *interaction* between *two* charged bodies. A single charge produces an electric field in the surrounding space, but this electric field cannot exert a net force on the charge that created it; this is an example of the general principle that a body cannot exert a net force on itself, as discussed in Section 4.3. (If this principle wasn't valid, you would be able to lift yourself to the ceiling by pulling up on your belt!) **The electric force on a charged body is exerted by the electric field created by *other* charged bodies.**

To find out experimentally whether there is an electric field at a particular point, we place a small charged body, which we call a **test charge,** at the point (Fig. 21.13c). If the test charge experiences an electric force, then there is an electric field at that point. This field is produced by charges other than q_0.

Force is a vector quantity, so electric field is also a vector quantity. (Note the use of vector signs, as well as boldface letters and plus, minus, and equals signs in the following discussion.) We define the *electric field* $\vec{E}$ at a point as the electric force $\vec{F}_0$ experienced by a test charge q_0 at the point, divided by the charge q_0. That is, the electric field at a certain point is equal to the *electric force per unit charge* experienced by a charge at that point:

$$\vec{E} = \frac{\vec{F}_0}{q_0} \tag{21.3}$$

(definition of electric field as electric force per unit charge)

In SI units, in which the unit of force is 1 N and the unit of charge is 1 C, the unit of electric field magnitude is 1 newton per coulomb (1 N/C).

If the field $\vec{E}$ at a certain point is known, rearranging Eq. (21.3) gives the force $\vec{F}_0$ experienced by a point charge q_0 placed at that point. This force is just equal to the electric field $\vec{E}$ produced at that point by charges other than q_0, multiplied by the charge q_0:

$$\vec{F}_0 = q_0\vec{E} \quad \text{(force exerted on a point charge } q_0 \text{ by an electric field } \vec{E}) \qquad (21.4)$$

The charge q_0 can be either positive or negative. If q_0 is *positive,* the force $\vec{F}_0$ experienced by the charge is the same direction as $\vec{E}$; if q_0 is *negative,* $\vec{F}_0$ and $\vec{E}$ are in opposite directions (Fig. 21.14).

While the electric field concept may be new to you, the basic idea—that one body sets up a field in the space around it, and a second body responds to that field—is one that you've actually used before. Compare Eq. (21.4) to the familiar expression for the gravitational force $\vec{F}_g$ that the earth exerts on a mass m_0:

$$\vec{F}_g = m_0\vec{g} \qquad (21.5)$$

In this expression, $\vec{g}$ is the acceleration due to gravity. If we divide both sides of Eq. (21.5) by the mass m_0, we obtain

$$\vec{g} = \frac{\vec{F}_g}{m_0}$$

Thus $\vec{g}$ can be regarded as the gravitational force per unit mass. By analogy to Eq. (21.3), we can interpret $\vec{g}$ as the *gravitational field.* Thus we treat the gravitational interaction between the earth and the mass m_0 as a two-stage process: The earth sets up a gravitational field $\vec{g}$ in the space around it, and this gravitational field exerts a force given by Eq. (21.5) on the mass m_0 (which we can regard as a *test mass*). In this sense, you've made use of the field concept every time you've used Eq. (21.5) for the force of gravity. The gravitational field $\vec{g}$, or gravitational force per unit mass, is a useful concept because it does not depend on the mass of the body on which the gravitational force is exerted; likewise, the electric field $\vec{E}$, or electric force per unit charge, is useful because it does not depend on the charge of the body on which the electric force is exerted.

CAUTION The electric force experienced by a test charge q_0 can vary from point to point, so the electric field can also be different at different points. For this reason, Eq. (21.4) can be used only to find the electric force on a *point* charge. If a charged body is large enough in size, the electric field $\vec{E}$ may be noticeably different in magnitude and direction at different points on the body, and calculating the net electric force on the body can become rather complicated.

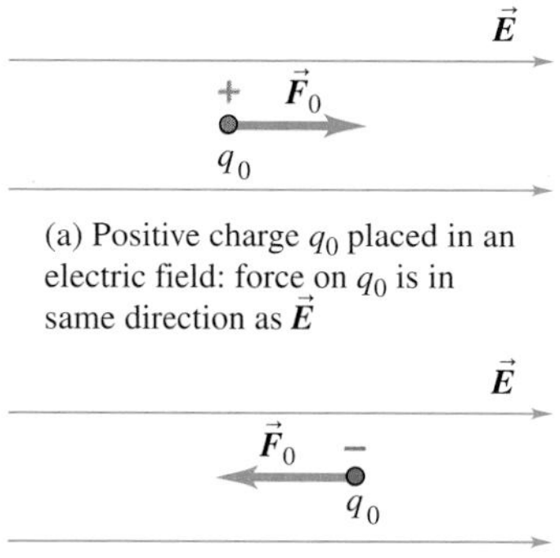

(a) Positive charge q_0 placed in an electric field: force on q_0 is in same direction as $\vec{E}$

(b) Negative charge q_0 placed in an electric field: force on q_0 is in opposite direction from $\vec{E}$

21.14 The force $\vec{F}_0 = q_0\vec{E}$ exerted on a point charge q_0 by an electric field $\vec{E}$.

ActivPhysics Online

11.4 Electric Field: Point Charge

11.9 Motion of a Charge in an Electric Field: Introduction

11.10 Motion in an Electric Field: Problems

We have so far ignored a subtle but important difficulty with our definition of electric field: In Fig. 21.13 the force exerted by the test charge q_0 on the charge distribution on body A may cause this distribution to shift around. This is especially true if body A is a conductor, on which charge is free to move. So the electric field around A when q_0 is present may not be the same as when q_0 is absent. But if q_0 is very small, the redistribution of charge on body A is also very small. So to make a completely correct definition of electric field, we take the *limit* of Eq. (21.3) as the test charge q_0 approaches zero and as the disturbing effect of q_0 on the charge distribution becomes negligible:

$$\vec{E} = \lim_{q_0 \to 0} \frac{\vec{F}_0}{q_0}$$

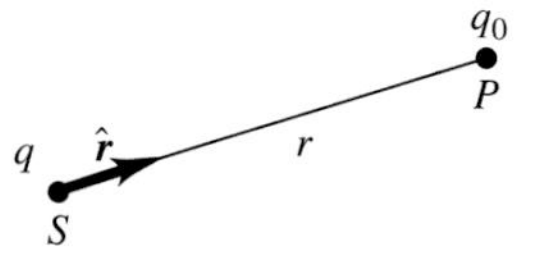

(a) Unit vector $\hat{\boldsymbol{r}}$ points from source point S to field point P

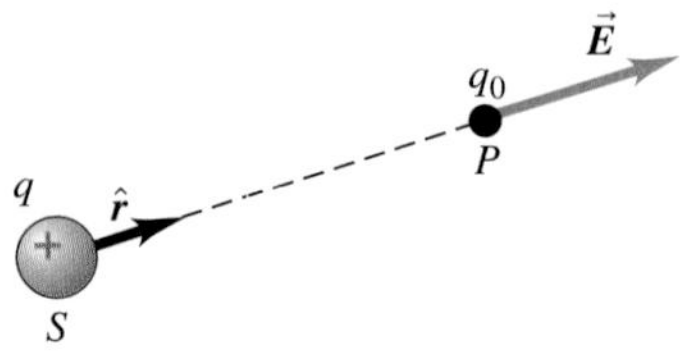

(b) At each point P, the electric field set up by an isolated *positive* point charge q points directly *away* from the charge in the *same* direction as $\hat{\boldsymbol{r}}$

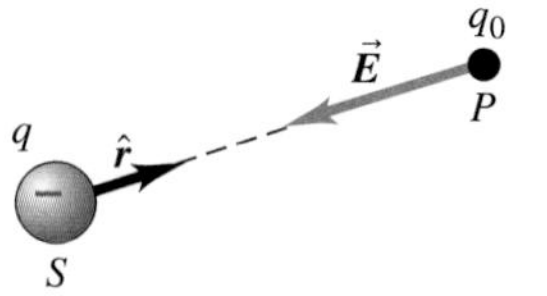

(c) At each point P, the electric field set up by an isolated *negative* point charge q points directly *toward* from the charge in the *opposite* direction from $\hat{\boldsymbol{r}}$

21.15 The electric field $\vec{E}$ produced at point P by an isolated point charge q at S. Note that in both (b) and (c), $\vec{E}$ is *produced* by q [see Eq. (21.7)] but *acts* on the charge q_0 at point P [see Eq. (21.4)].

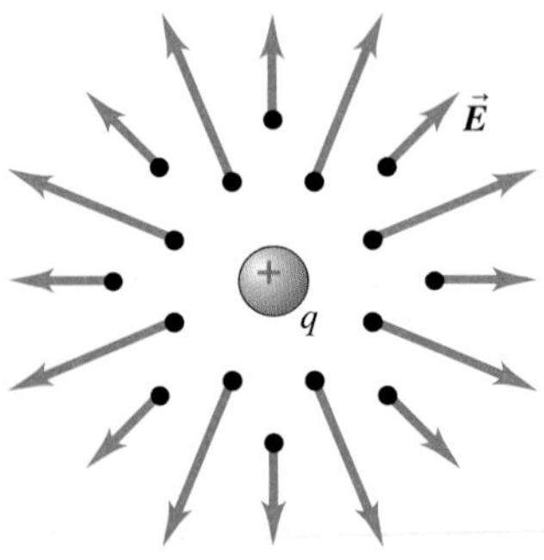

21.16 A point charge q produces an electric field $\vec{E}$ at *all* points in space. The field strength decreases with increasing distance. The field pattern shown here is for a positive charge; in the pattern for a negative charge, the field vectors point toward the charge (see Fig. 21.15c).

In practical calculations of the electric field $\vec{E}$ produced by a charge distribution, we will consider the charge distribution to be fixed, and so we will not need this limiting process.

If the source distribution is a point charge q, it is easy to find the electric field that it produces. We call the location of the charge the **source point,** and we call the point P where we are determining the field the **field point.** It is also useful to introduce a *unit vector* $\hat{\boldsymbol{r}}$ that points along the line from source point to field point (Fig. 21.15a). This unit vector is equal to the displacement vector $\vec{r}$ from the source point to the field point, divided by the distance $r = |\vec{r}|$ between these two points; that is, $\hat{\boldsymbol{r}} = \vec{r}/r$. If we place a small test charge q_0 at the field point P, at a distance r from the source point, the magnitude F_0 of the force is given by Coulomb's law, Eq. (21.2):

$$F_0 = \frac{1}{4\pi\epsilon_0}\frac{|qq_0|}{r^2}$$

From Eq. (21.3) the magnitude E of the electric field at P is

$$E = \frac{1}{4\pi\epsilon_0}\frac{|q|}{r^2} \quad \text{(magnitude of electric field of a point charge)} \qquad (21.6)$$

Using the unit vector $\hat{\boldsymbol{r}}$, we can write a *vector* equation that gives both the magnitude and direction of the electric field $\vec{E}$:

$$\vec{E} = \frac{1}{4\pi\epsilon_0}\frac{q}{r^2}\hat{\boldsymbol{r}} \quad \text{(electric field of a point charge)} \qquad (21.7)$$

By definition, the electric field of a point charge always points *away from* a positive charge (that is, in the same direction as $\hat{\boldsymbol{r}}$; see Fig. 21.15b) but *toward* a negative charge (that is, in the direction opposite $\hat{\boldsymbol{r}}$; see Fig. 21.15c).

We have emphasized calculating the electric field $\vec{E}$ at a certain point. But since $\vec{E}$ can vary from point to point, it is not a single vector quantity but rather an *infinite* set of vector quantities, one associated with each point in space. This is an example of a **vector field.** Figure 21.16 shows a number of the field vectors produced by a point charge. If we use a rectangular (xyz) coordinate system, each component of $\vec{E}$ at any point is in general a function of the coordinates (x, y, z) of the point. We can represent the functions as $E_x(x, y, z)$, $E_y(x, y, z)$, and $E_z(x, y, z)$. Vector fields are an important part of the language of physics, not just in electricity and magnetism. One everyday example of a vector field is the velocity $\vec{v}$ of wind currents; the magnitude and direction of $\vec{v}$, and hence its vector components, vary from point to point in the atmosphere.

In some situations the magnitude and direction of the field (and hence its vector components) have the same values everywhere throughout a certain region; we then say that the field is *uniform* in this region. An important example of this is the electric field inside a *conductor.* If there is an electric field within a conductor, the field exerts a force on every charge in the conductor, giving the free charges a net motion. By definition an electrostatic situation is one in which the charges have *no* net motion. We conclude that *in electrostatics the electric field at every point within the material of a conductor must be zero.* (Note that we are not saying that the field is necessarily zero in a *hole* inside a conductor.)

With the concept of electric field, our description of electric interactions has two parts. First, a given charge distribution acts as a source of electric field. Second, the electric field exerts a force on any charge that is present in the field. Our analysis often has two corresponding steps: first, calculating the field caused by a source charge distribution; second, looking at the effect of the field in terms of force and motion. The second step often involves Newton's laws as well as the principles of electric interactions. In the next section we show how to calculate fields caused by various source distributions, but first here are some examples of calculating the field due to a point charge and of finding the force on a charge due to a given field $\vec{E}$.

Example 21.5 Electric-field magnitude for a point charge

What is the magnitude of the electric field at a field point 2.0 m from a point charge $q = 4.0$ nC? (The point charge could represent any small charged object with this value of q, provided the dimensions of the object are much less than the distance from the object to the field point.)

SOLUTION

IDENTIFY and SET UP: We are given the magnitude of the charge and the distance from the object to the field point, so we use Eq. (21.6) to calculate the field magnitude E.

EXECUTE: From Eq. (21.6),

$$E = \frac{1}{4\pi\epsilon_0}\frac{|q|}{r^2} = (9.0 \times 10^9\ \mathrm{N \cdot m^2/C^2})\frac{4.0 \times 10^{-9}\ \mathrm{C}}{(2.0\ \mathrm{m})^2}$$
$$= 9.0\ \mathrm{N/C}$$

EVALUATE: To check our result, we use the definition of electric field as the electric force per unit charge. We can first use Coulomb's law, Eq. (21.2), to find the magnitude F_0 of the force on a test charge q_0 placed 2.0 m from q:

$$F_0 = \frac{1}{4\pi\epsilon_0}\frac{|qq_0|}{r^2} = (9.0 \times 10^9\ \mathrm{N \cdot m^2/C^2})\frac{4.0 \times 10^{-9}\ \mathrm{C}\,|q_0|}{(2.0\ \mathrm{m})^2}$$
$$= (9.0\ \mathrm{N/C})\,|q_0|$$

Then, from Eq. (21.3), the magnitude of $\vec{E}$ is

$$E = \frac{F_0}{|q_0|} = 9.0\ \mathrm{N/C}$$

Because q is positive, the *direction* of $\vec{E}$ at this point is along the line from q toward q_0, as shown in Fig. 21.15b. However, the magnitude and direction of $\vec{E}$ do not depend on the sign of q_0. Do you see why not?

Example 21.6 Electric-field vector for a point charge

A point charge $q = -8.0$ nC is located at the origin. Find the electric-field vector at the field point $x = 1.2$ m, $y = -1.6$ m (Fig. 21.17).

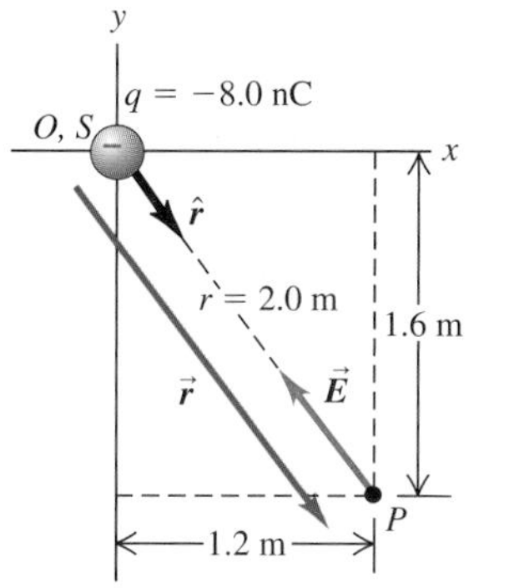

21.17 The vectors $\vec{r}$, $\hat{r}$, and $\vec{E}$ for a point charge.

SOLUTION

IDENTIFY and SET UP: The electric field is given in vector form by Eq. (21.17). To use this equation, we utilize the coordinate system shown in Fig. 21.17 to find the distance from the source point S (the position of the charge q) to the field point P, as well as the unit vector $\hat{r}$ that points in the direction from S to P.

EXECUTE: The distance from the charge at the source point S (which in this example is at the origin O) to the field point P is

$$r = \sqrt{x^2 + y^2} = \sqrt{(1.2\ \mathrm{m})^2 + (-1.6\ \mathrm{m})^2} = 2.0\ \mathrm{m}$$

The unit vector $\hat{r}$ is directed from the source point to the field point. This is equal to the displacement vector $\vec{r}$ from the source point to the field point (shown shifted to one side in Fig. 21.17 so as not to obscure the other vectors), divided by its magnitude r:

$$\hat{r} = \frac{\vec{r}}{r} = \frac{x\hat{\imath} + y\hat{\jmath}}{r}$$

$$= \frac{(1.2\text{ m})\hat{\imath} + (-1.6\text{ m})\hat{\jmath}}{2.0\text{ m}} = 0.60\hat{\imath} - 0.80\hat{\jmath}$$

Hence the electric field vector is

$$\vec{E} = \frac{1}{4\pi\epsilon_0}\frac{q}{r^2}\hat{r}$$

$$= (9.0 \times 10^9\text{ N}\cdot\text{m}^2/\text{C}^2)\frac{(-8.0 \times 10^{-9}\text{ C})}{(2.0\text{ m})^2}(0.60\hat{\imath} - 0.80\hat{\jmath})$$

$$= (-11\text{ N/C})\hat{\imath} + (14\text{ N/C})\hat{\jmath}$$

EVALUATE: Since q is negative, $\vec{E}$ points from the field point to the charge (the source point), in the direction opposite to $\hat{r}$ (compare Fig. 21.15c). We leave the calculation of the magnitude and direction of $\vec{E}$ as an exercise.

Example 21.7 Electron in a uniform field

When the terminals of a battery are connected to two large parallel conducting plates, the resulting charges on the plates cause an electric field $\vec{E}$ in the region between the plates that is very nearly uniform. (We will see the reason for this uniformity in the next section. Charged plates of this kind are used in common electrical devices called *capacitors,* to be discussed in Chapter 24.) If the plates are horizontal and separated by 1.0 cm and the plates are connected to a 100-volt battery, the magnitude of the field is $E = 1.00 \times 10^4$ N/C. Suppose the direction of $\vec{E}$ is vertically upward, as shown by the vectors in Fig. 21.18. a) If an electron is released from rest at the upper plate, what is its acceleration? b) What speed and kinetic energy does it acquire while traveling 1.0 cm to the lower plate? c) How much time is required for it to travel this distance? An electron has a charge $-e = -1.60 \times 10^{-19}$ C and a mass $m = 9.11 \times 10^{-31}$ kg.

SOLUTION

IDENTIFY: This example involves several concepts: the relation between electric field and electric force, the relation between force and acceleration, the definition of kinetic energy, and the kinematic relations between acceleration, distance, velocity, and time.

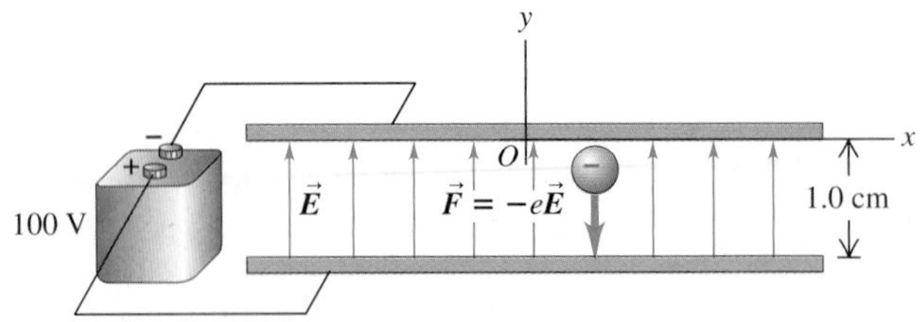

21.18 A uniform electric field between two parallel conducting plates connected to a 100-volt battery. (The separation of the plates is exaggerated in this figure relative to the dimensions of the plates.)

SET UP: Figure 21.18 shows a coordinate system. We are given the electric field, so we use Eq. (21.4) to find the force on the electron and Newton's second law to find its acceleration. Because the field is uniform between the plates, the force and acceleration are constant and we can use the constant-acceleration formulas from Chapter 3 to find the electron's velocity and travel time. We find the kinetic energy using the definition $K = \frac{1}{2}mv^2$.

EXECUTE: a) Note that $\vec{E}$ is upward (in the $+y$-direction) but $\vec{F}$ is downward because the charge of the electron is negative. Thus F_y is negative.

Because F_y is constant, the electron moves with constant acceleration a_y given by

$$a_y = \frac{F_y}{m} = \frac{-eE}{m} = \frac{(-1.60 \times 10^{-19}\text{ C})(1.00 \times 10^4\text{ N/C})}{9.11 \times 10^{-31}\text{ kg}}$$

$$= -1.76 \times 10^{15}\text{ m/s}^2$$

This is an enormous acceleration! To give a 1000-kg car this acceleration, we would need a force of about 2×10^{18} N (about 2×10^{14} tons). The gravitational force on the electron is completely negligible compared to the electric force.

b) The electron starts from rest, so its motion is in the y-direction only (the direction of the acceleration). We can find the electron's speed at any position using the constant-acceleration formula $v_y^2 = v_{0y}^2 + 2a_y(y - y_0)$. We have $v_{0y} = 0$ and $y_0 = 0$, so the speed $|v_y|$ when $y = -1.0\text{ cm} = -1.0 \times 10^{-2}$ m is

$$|v_y| = \sqrt{2a_y y} = \sqrt{2(-1.76 \times 10^{15}\text{ m/s}^2)(-1.0 \times 10^{-2}\text{ m})}$$

$$= 5.9 \times 10^6\text{ m/s}$$

The velocity is downward, so its y-component is $v_y = -5.9 \times 10^6$ m/s. The electron's kinetic energy is

$$K = \frac{1}{2}mv^2 = \frac{1}{2}(9.11 \times 10^{-31}\text{ kg})(5.9 \times 10^6\text{ m/s})^2$$

$$= 1.6 \times 10^{-17}\text{ J}$$

c) From the constant-acceleration formula $v_y = v_{0y} + a_y t$, we find that the time required is very brief:

$$t = \frac{v_y - v_{0y}}{a_y} = \frac{(-5.9 \times 10^6 \text{ m/s}) - (0 \text{ m/s})}{-1.76 \times 10^{15} \text{ m/s}^2}$$
$$= 3.4 \times 10^{-9} \text{ s}$$

(We could also have found the time by solving the equation $y = y_0 + v_{0y}t + \frac{1}{2}a_y t^2$ for t.)

EVALUATE: This example shows that when working problems about subatomic particles such as electrons, many quantities—including acceleration, speed, kinetic energy, and time—will have *very* different values from what we have seen for ordinary objects such as baseballs and automobiles.

Example 21.8 An electron trajectory

If we launch an electron into the electric field of Example 21.7 with an initial horizontal velocity v_0 (Fig. 21.19), what is the equation of its trajectory?

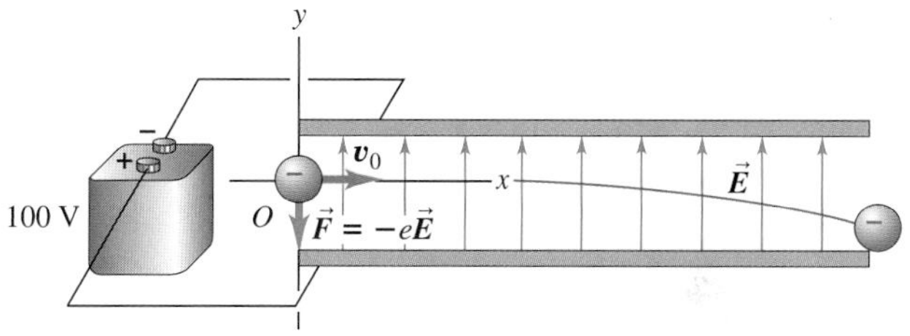

21.19 The parabolic trajectory of an electron in a uniform electric field.

SOLUTION

IDENTIFY and SET UP: The force and acceleration are constant and the same as in Example 21.7, and there is no acceleration in the x-direction. Hence we can use the kinematic equations from Chapter 3 for two-dimensional motion with constant acceleration.

EXECUTE: We have $a_x = 0$ and $a_y = (-e)E/m$. At $t = 0$, $x_0 = y_0 = 0$, $v_{0x} = v_0$, and $v_{0y} = 0$; hence at time t,

$$x = v_0 t \quad \text{and} \quad y = \frac{1}{2}a_y t^2 = -\frac{1}{2}\frac{eE}{m}t^2$$

Eliminating t between these equations, we get

$$y = -\frac{1}{2}\frac{eE}{mv_0^2}x^2$$

EVALUATE: This is the equation of a parabola, just like the trajectory of a projectile launched horizontally in the earth's gravitational field (discussed in Section 3.3). For a given initial velocity of the electron, the curvature of the trajectory depends on the field magnitude E. If we reverse the signs of the charges on the two plates in Fig. 21.19, the direction of $\vec{E}$ reverses, and the electron trajectory will curve up, not down. Hence we can "steer" the electron by varying the charges on the plates. The electric field between charged conducting plates is used in this way to control the trajectory of electron beams in oscilloscopes.

Test Your Understanding

In Example 21.4 (Section 21.3), what is the electric field due to the charges q_1 and q_2 at the point $x = 0.40$ m, $y = 0$?

21.5 | Electric-Field Calculations

Equation (21.7) gives the electric field caused by a single point charge. But in most realistic situations that involve electric fields and forces, we encounter charge that is *distributed* over space. The charged plastic and glass rods in Fig. 21.1 have electric charge distributed over their surfaces, as does the imaging drum of a laser printer (Fig. 21.2). In this section we'll learn to calculate electric fields caused by various distributions of electric charge. Calculations of this kind are of tremendous importance for technological applications of electric forces. To determine the trajectories of electrons in a TV tube, of atomic nuclei in an accelerator for cancer radiotherapy, or of charged particles in a semiconductor electronic device, you have to know the detailed nature of the electric field acting on the charges.

To find the field caused by a charge distribution, we imagine the distribution to be made up of many point charges $q_1, q_2, q_3, \ldots$. (This is actually quite a realistic description, since we have seen that charge is carried by electrons and protons that are so small as to be almost pointlike.) At any given point P, each point charge produces its own electric field $\vec{E}_1, \vec{E}_2, \vec{E}_3, \ldots$, so a test charge q_0 placed at P experiences a force $\vec{F}_1 = q_0\vec{E}_1$ from charge q_1, a force $\vec{F}_2 = q_0\vec{E}_2$ from charge q_2, and so on. From the principle of superposition of forces discussed in Section 21.3, the *total* force $\vec{F}_0$ that the charge distribution exerts on q_0 is the vector sum of these individual forces:

$$\vec{F}_0 = \vec{F}_1 + \vec{F}_2 + \vec{F}_3 + \cdots = q_0\vec{E}_1 + q_0\vec{E}_2 + q_0\vec{E}_3 + \cdots$$

The combined effect of all the charges in the distribution is described by the *total* electric field $\vec{E}$ at point P. From the definition of electric field, Eq. (21.3), this is

ActivPhysics Online

11.5 Electric Field Due to a Dipole

11.6 Electric Field: Problems

$$\vec{E} = \frac{\vec{F}_0}{q_0} = \vec{E}_1 + \vec{E}_2 + \vec{E}_3 + \cdots$$

The total electric field at P is the vector sum of the fields at P due to each point charge in the charge distribution. This is the **principle of superposition of electric fields.**

When charge is distributed along a line, over a surface, or through a volume, a few additional terms are useful. For a line charge distribution (such as a long, thin, charged plastic rod), we use λ ("lambda") to represent the **linear charge density** (charge per unit length, measured in C/m). When charge is distributed over a surface (such as the surface of the imaging drum of a laser printer), we use σ ("sigma") to represent the **surface charge density** (charge per unit area, measured in C/m^2). And when charge is distributed through a volume, we use ρ ("rho") to represent the **volume charge density** (charge per unit volume, C/m^3).

Some of the calculations in the following examples may look fairly intricate; in electric-field calculations a certain amount of mathematical complexity is in the nature of things. After you've worked through the examples one step at a time, the process will seem less formidable. We will use many of the calculational techniques in these examples in Chapter 28 to calculate the *magnetic* fields caused by charges in motion.

Problem-Solving Strategy

Electric-Field Calculations

IDENTIFY *the relevant concepts:* Use the principle of superposition whenever you need to calculate the electric field due to a charge distribution (two or more point charges, a distribution over a line, surface, or volume, or a combination of these).

SET UP *the problem using the following steps:*

1. Make a drawing that clearly shows the locations of the charges and your choice of coordinate axes.
2. On your drawing, indicate the position of the *field point* (the point at which you want to calculate the electric field $\vec{E}$). Sometimes the field point will be at some arbitrary position along a line. For example, you may be asked to find $\vec{E}$ at any point on the x-axis.

EXECUTE *the solution as follows:*

1. Be sure to use a consistent set of units. Distances must be in meters and charge must be in coulombs. If you are given centimeters or nanocoulombs, don't forget to convert.
2. When adding up the electric fields caused by different parts of the charge distribution, remember that electric

field is a vector, so you *must* use vector addition. Don't simply add together the magnitudes of the individual fields; the directions are important, too.

3. Take advantage of any symmetries in the charge distribution. For example, if a positive charge and a negative charge of equal magnitude are placed symmetrically with respect to the field point, they produce electric fields of the same magnitude but with mirror-image directions. Exploiting these symmetries will simplify your calculations.
4. Most often you will use components to compute vector sums. Use the methods you learned in Chapter 1; review them if necessary. Use proper vector notation; distinguish carefully between scalars, vectors, and components of vectors. Be certain the components are consistent with your choice of coordinate axes.
5. In working out the directions of $\vec{E}$ vectors, be careful to distinguish between the *source point* and the *field point.* The field produced by a point charge always points from source point to field point if the charge is positive; it points in the opposite direction if the charge is negative.
6. In some situations you will have a continuous distribution of charge along a line, over a surface, or through a volume. Then you must define a small element of charge that can be considered as a point, find its electric field at point P, and find a way to add the fields of all the charge elements. Usually it is easiest to do this for each component of $\vec{E}$ separately, and often you will need to evaluate one or more integrals. Make certain the limits on your integrals are correct; especially when the situation has symmetry, make sure you don't count the charge twice.

EVALUATE *your answer:* Check that the direction of $\vec{E}$ is reasonable. If your result for the electric-field magnitude E is a function of position (say, the coordinate x), check your result in any limits for which you know what the magnitude should be. When possible, check your answer by calculating it in a different way.

Example 21.9 Field of an electric dipole

Point charges q_1 and q_2 of $+12$ nC and -12 nC, respectively, are placed 0.10 m apart (Fig. 21.20). This combination of two charges with equal magnitude and opposite sign is called an *electric dipole.* (Such combinations occur frequently in nature. For example, in Fig. 21.7, each molecule in the neutral insulator is an electric dipole. We'll study dipoles in more detail in Section 21.7.) Compute the electric field caused by q_1, the field caused by q_2, and the total field a) at point a; b) at point b; and c) at point c.

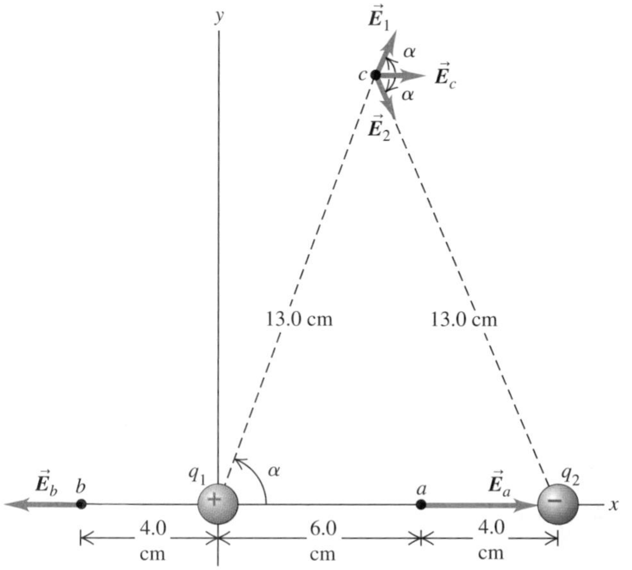

21.20 Electric field at three points, a, b, and c, set up by charges q_1 and q_2, which form an electric dipole.

SOLUTION

IDENTIFY and SET UP: Figure 21.20 shows the coordinate system and the locations of the three field points a, b, and c.

EXECUTE: a) At point a the field $\vec{E}_1$ caused by the positive charge q_1 and the field $\vec{E}_2$ caused by the negative charge q_2 are both directed toward the right. The magnitudes of $\vec{E}_1$ and $\vec{E}_2$ are

$$E_1 = \frac{1}{4\pi\epsilon_0}\frac{|q_1|}{r^2} = (9.0 \times 10^9\ \text{N}\cdot\text{m}^2/\text{C}^2)\frac{12 \times 10^{-9}\ \text{C}}{(0.060\ \text{m})^2}$$
$$= 3.0 \times 10^4\ \text{N/C}$$
$$E_2 = \frac{1}{4\pi\epsilon_0}\frac{|q_2|}{r^2} = (9.0 \times 10^9\ \text{N}\cdot\text{m}^2/\text{C}^2)\frac{12 \times 10^{-9}\ \text{C}}{(0.040\ \text{m})^2}$$
$$= 6.8 \times 10^4\ \text{N/C}$$

The components of $\vec{E}_1$ and $\vec{E}_2$ are

$$E_{1x} = 3.0 \times 10^4\ \text{N/C} \qquad E_{1y} = 0$$
$$E_{2x} = 6.8 \times 10^4\ \text{N/C} \qquad E_{2y} = 0$$

Hence at point a the total electric field $\vec{E}_a = \vec{E}_1 + \vec{E}_2$ has components

$$(E_a)_x = E_{1x} + E_{2x} = (3.0 + 6.8) \times 10^4\ \text{N/C}$$
$$(E_a)_y = E_{1y} + E_{2y} = 0$$

At point a the total field has magnitude 9.8×10^4 N/C and is directed toward the right, so

$$\vec{E}_a = (9.8 \times 10^4 \text{ N/C})\hat{\imath}$$

b) At point b the field $\vec{E}_1$ due to q_1 is directed toward the left, while the field $\vec{E}_2$ due to q_2 is directed toward the right. The magnitudes of $\vec{E}_1$ and $\vec{E}_2$ are

$$E_1 = \frac{1}{4\pi\epsilon_0}\frac{|q_1|}{r^2} = (9.0 \times 10^9 \text{ N}\cdot\text{m}^2/\text{C}^2)\frac{12 \times 10^{-9}\text{ C}}{(0.040 \text{ m})^2}$$
$$= 6.8 \times 10^4 \text{ N/C}$$

$$E_2 = \frac{1}{4\pi\epsilon_0}\frac{|q_2|}{r^2} = (9.0 \times 10^9 \text{ N}\cdot\text{m}^2/\text{C}^2)\frac{12 \times 10^{-9}\text{ C}}{(0.140 \text{ m})^2}$$
$$= 0.55 \times 10^4 \text{ N/C}$$

The components of $\vec{E}_1$, $\vec{E}_2$, and the total field $\vec{E}_b$ at point b are

$$E_{1x} = -6.8 \times 10^4 \text{ N/C} \qquad E_{1y} = 0$$
$$E_{2x} = 0.55 \times 10^4 \text{ N/C} \qquad E_{2y} = 0$$
$$(E_b)_x = E_{1x} + E_{2x} = (-6.8 + 0.55) \times 10^4 \text{ N/C}$$
$$(E_b)_y = E_{1y} + E_{2y} = 0$$

That is, the electric field at b has magnitude 6.2×10^4 N/C and is directed toward the left, so

$$\vec{E}_b = (-6.2 \times 10^4 \text{ N/C})\hat{\imath}$$

c) At point c, both $\vec{E}_1$ and $\vec{E}_2$ have the same magnitude, since this point is equidistant from both charges and the charge magnitudes are the same:

$$E_1 = E_2 = \frac{1}{4\pi\epsilon_0}\frac{|q|}{r^2} = (9.0 \times 10^9 \text{ N}\cdot\text{m}^2/\text{C}^2)\frac{12 \times 10^{-9}\text{ C}}{(0.13 \text{ m})^2}$$
$$= 6.39 \times 10^3 \text{ N/C}$$

The directions of $\vec{E}_1$ and $\vec{E}_2$ are shown in Fig 21.20. The x-components of both vectors are the same:

$$E_{1x} = E_{2x} = E_1 \cos\alpha = (6.39 \times 10^3 \text{ N/C})\left(\frac{5}{13}\right)$$
$$= 2.46 \times 10^3 \text{ N/C}$$

From symmetry the y-components E_{1y} and E_{2y} are equal and opposite and so add to zero. Hence the components of the total field $\vec{E}_c$ are

$$(E_c)_x = E_{1x} + E_{2x} = 2(2.46 \times 10^3 \text{ N/C}) = 4.9 \times 10^3 \text{ N/C}$$
$$(E_c)_y = E_{1y} + E_{2y} = 0$$

So at point c the total electric field has magnitude 4.9×10^3 N/C and is directed toward the right, so

$$\vec{E}_c = (4.9 \times 10^3 \text{ N/C})\hat{\imath}$$

Does it surprise you that the field at point c is parallel to the line between the two charges?

EVALUATE: An alternative way to find the electric field at c is to use the vector expression for the field of a point charge, Eq. (21.7). The displacement vector $\vec{r}_1$ from q_1 to point c, a distance $r = 13.0$ cm away, is

$$\vec{r}_1 = r\cos\alpha\,\hat{\imath} + r\sin\alpha\,\hat{\jmath}$$

Hence the unit vector that points from q_1 to c is

$$\hat{r}_1 = \frac{\vec{r}_1}{r} = \cos\alpha\,\hat{\imath} + \sin\alpha\,\hat{\jmath}$$

and the field due to q_1 at point c is

$$\vec{E}_1 = \frac{1}{4\pi\epsilon_0}\frac{q_1}{r^2}\hat{r}_1 = \frac{1}{4\pi\epsilon_0}\frac{q_1}{r^2}(\cos\alpha\,\hat{\imath} + \sin\alpha\,\hat{\jmath})$$

By symmetry the unit vector $\hat{r}_2$ that points from q_2 to point c has the opposite x-component but the same y-component, so the field at c due to q_2 is

$$\vec{E}_2 = \frac{1}{4\pi\epsilon_0}\frac{q_2}{r^2}\hat{r}_2 = \frac{1}{4\pi\epsilon_0}\frac{q_2}{r^2}(-\cos\alpha\,\hat{\imath} + \sin\alpha\,\hat{\jmath})$$

Since $q_2 = -q_1$, the total field at c is

$$\vec{E}_c = \vec{E}_1 + \vec{E}_2$$
$$= \frac{1}{4\pi\epsilon_0}\frac{q_1}{r^2}(\cos\alpha\,\hat{\imath} + \sin\alpha\,\hat{\jmath})$$
$$+ \frac{1}{4\pi\epsilon_0}\frac{(-q_1)}{r^2}(-\cos\alpha\,\hat{\imath} + \sin\alpha\,\hat{\jmath})$$
$$= \frac{1}{4\pi\epsilon_0}\frac{q_1}{r^2}(2\cos\alpha\,\hat{\imath})$$
$$= (9.0 \times 10^9 \text{ N}\cdot\text{m}^2/\text{C}^2)\frac{12 \times 10^{-9}\text{ C}}{(0.13 \text{ m})^2}\left(2\left(\frac{5}{13}\right)\right)\hat{\imath}$$
$$= (4.9 \times 10^3 \text{ N/C})\hat{\imath}$$

as before.

Example 21.10 Field of a ring of charge

A ring-shaped conductor with radius a carries a total charge Q uniformly distributed around it (Fig. 21.21). Find the electric field at a point P that lies on the axis of the ring at a distance x from its center.

SOLUTION

IDENTIFY and SET UP: The field point is an arbitrary point on the x-axis in Fig. 21.21. Our target variable is the electric field at such a point as a function of the coordinate x.

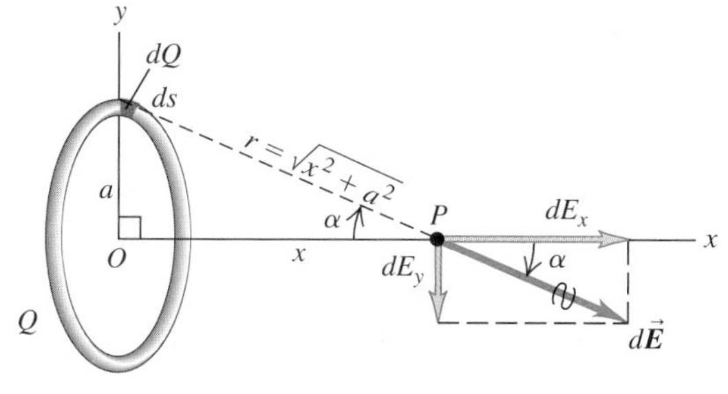

21.21 Calculating the electric field on the axis of a ring of charge. In this figure, the charge is assumed to be positive.

EXECUTE: As shown in Fig. 21.21, we imagine the ring divided into infinitesimal segments of length ds. Each segment has charge dQ and acts as a point-charge source of electric field. Let $d\vec{E}$ be the electric field from one such segment; the net electric field at P is then the sum of all contributions $d\vec{E}$ from all the segments that make up the ring. (This same technique works for any situation in which charge is distributed along a line or a curve.)

The calculation of $\vec{E}$ is greatly simplified because the field point P is on the symmetry axis of the ring. Consider two segments at the top and bottom of the ring: the contributions $d\vec{E}$ to the field at P from these segments have the same x-component but opposite y-components. Hence the total y-component of field due to this pair of segments is zero. When we add up the contributions from all such pairs of segments, the total field $\vec{E}$ will have only a component along the ring's symmetry axis (the x-axis), with no component perpendicular to that axis (that is, no y-component or z-component). So the field at P is described completely by its x-component E_x.

To calculate E_x, note that the square of the distance r from a ring segment to the point P is $r^2 = x^2 + a^2$. Hence the magnitude of this segment's contribution $d\vec{E}$ to the electric field at P is

$$dE = \frac{1}{4\pi\epsilon_0}\frac{dQ}{x^2 + a^2}$$

Using $\cos\alpha = x/r = x/(x^2 + a^2)^{1/2}$, the x-component dE_x of this field is

$$dE_x = dE\cos\alpha = \frac{1}{4\pi\epsilon_0}\frac{dQ}{x^2 + a^2}\frac{x}{\sqrt{x^2 + a^2}} = \frac{1}{4\pi\epsilon_0}\frac{x\,dQ}{(x^2 + a^2)^{3/2}}$$

To find the *total* x-component E_x of the field at P, we integrate this expression over all segments of the ring:

$$E_x = \int \frac{1}{4\pi\epsilon_0}\frac{x\,dQ}{(x^2 + a^2)^{3/2}}$$

Since x does not vary as we move from point to point around the ring, all the factors on the right side except dQ are constant and can be taken outside the integral. The integral of dQ is just the total charge Q, and we finally get

$$\vec{E} = E_x\hat{\imath} = \frac{1}{4\pi\epsilon_0}\frac{Qx}{(x^2 + a^2)^{3/2}}\hat{\imath} \qquad (21.8)$$

EVALUATE: Our result for $\vec{E}$ shows that at the center of the ring $(x = 0)$ the field is zero. We should expect this; charges on opposite sides of the ring would push in opposite directions on a test charge at the center, and the forces would add to zero. When the field point P is much farther from the ring than its size (that is, $x \gg a$), the denominator in Eq. (21.8) becomes approximately equal to x^3, and the expression becomes approximately

$$\vec{E} = \frac{1}{4\pi\epsilon_0}\frac{Q}{x^2}\hat{\imath}$$

In other words, when we are so far from the ring that its size a is negligible in comparison to the distance x, its field is the same as that of a point charge. To an observer far from the ring, the ring would appear like a point, and the electric field reflects this.

In this example we used a *symmetry argument* to conclude that $\vec{E}$ had only an x-component at a point on the ring's axis of symmetry. We'll use symmetry arguments many times in this and subsequent chapters. Keep in mind, however, that such arguments can be used only in special cases. At a point in the xy-plane that is not on the x-axis in Fig. 21.21, the symmetry argument doesn't apply, and the field has in general both x- and y-components.

Example 21.11 Field of a line of charge

Positive electric charge Q is distributed uniformly along a line with length $2a$, lying along the y-axis between $y = -a$ and $y = +a$. (This might represent one of the charged rods in Fig. 21.1.) Find the electric field at point P on the x-axis at a distance x from the origin.

SOLUTION

IDENTIFY and SET UP: Figure 21.22 shows the charge distribution and the coordinate system. As in Example 21.10, our target variable is the electric field at P as a function of the coordinate x.

EXECUTE: We divide the line charge into infinitesimal segments, each of which acts as a point charge; let the length of a typical segment at height y be dy. If the charge is distributed uniformly, the linear charge density λ at any point on the line is equal to $Q/2a$ (the total charge divided by the total length). Hence the charge dQ in a segment of length dy is

$$dQ = \lambda dy = \frac{Q\,dy}{2a}$$

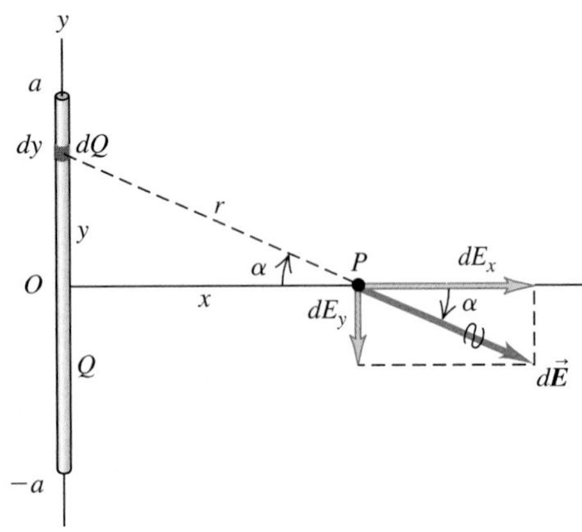

21.22 Finding the electric field at point P on the perpendicular bisector of a line of charge with length $2a$ and total charge Q. In this figure, the charge is assumed to be positive.

The distance r from this segment to P is $(x^2 + y^2)^{1/2}$, so the magnitude of field dE at P due to this segment is

$$dE = \frac{Q}{4\pi\epsilon_0}\frac{dy}{2a(x^2 + y^2)}$$

We represent this field in terms of its x- and y-components:

$$dE_x = dE\cos\alpha \qquad dE_y = -dE\sin\alpha$$

We note that $\sin\alpha = y/(x^2 + y^2)^{1/2}$ and $\cos\alpha = x/(x^2 + y^2)^{1/2}$; combining these with the expression for dE, we find

$$dE_x = \frac{Q}{4\pi\epsilon_0}\frac{x\,dy}{2a(x^2 + y^2)^{3/2}}$$

$$dE_y = -\frac{Q}{4\pi\epsilon_0}\frac{y\,dy}{2a(x^2 + y^2)^{3/2}}$$

To find the total field components E_x and E_y, we integrate these expressions, noting that to include all of Q, we must integrate from $y = -a$ to $y = +a$. We invite you to work out the details of the integration; an integral table is helpful. The final results are

$$E_x = \frac{1}{4\pi\epsilon_0}\frac{Qx}{2a}\int_{-a}^{a}\frac{dy}{(x^2 + y^2)^{3/2}} = \frac{Q}{4\pi\epsilon_0}\frac{1}{x\sqrt{x^2 + a^2}}$$

$$E_y = -\frac{1}{4\pi\epsilon_0}\frac{Q}{2a}\int_{-a}^{a}\frac{y\,dy}{(x^2 + y^2)^{3/2}} = 0$$

or, in vector form,

$$\vec{E} = \frac{1}{4\pi\epsilon_0}\frac{Q}{x\sqrt{x^2 + a^2}}\hat{\imath} \qquad (21.9)$$

EVALUATE: Using a symmetry argument as in Example 21.10, we could have guessed that E_y would be zero; if we place a positive test charge at P, the upper half of the line of charge pushes downward on it, and the lower half pushes up with equal magnitude.

To explore our result, let's first see what happens in the limit that x is much larger than a. Then we can neglect a in the denominator of Eq. (21.9), and our result becomes

$$\vec{E} = \frac{1}{4\pi\epsilon_0}\frac{Q}{x^2}\hat{\imath}$$

This means that if point P is very far from the line charge in comparison to the length of the line, the field at P is the same as that of a point charge. We found a similar result for the charged ring in Example 21.10.

We get an added dividend from our exact result for $\vec{E}$, Eq. (21.9), if we express it in terms of the linear charge density $\lambda = Q/2a$. Substituting $Q = 2a\lambda$ into Eq. (21.9) and simplifying, we get

$$\vec{E} = \frac{1}{2\pi\epsilon_0}\frac{\lambda}{x\sqrt{(x^2/a^2) + 1}}\hat{\imath} \qquad (21.10)$$

Now what happens if we make the line of charge longer and longer, adding charge in proportion to the total length so that λ, the charge per unit length, remains constant? What is $\vec{E}$ at a distance x from a *very* long line of charge? To answer the question, we take the *limit* of Eq. (21.10) as a becomes very large. In this limit, the term x^2/a^2 in the denominator becomes much smaller than unity and can be thrown away. We are left with

$$\vec{E} = \frac{\lambda}{2\pi\epsilon_0 x}\hat{\imath}$$

The field magnitude depends only on the distance of point P from the line of charge. So at any point P at a perpendicular distance r from the line in any direction, $\vec{E}$ has magnitude

$$E = \frac{\lambda}{2\pi\epsilon_0 r} \qquad \text{(infinite line of charge)}$$

Thus the electric field due to an infinitely long line of charge is proportional to $1/r$ rather than to $1/r^2$ as for a point charge. The direction of $\vec{E}$ is radially outward from the line if λ is positive and radially inward if λ is negative.

Of course, there's really no such thing in nature as an infinite line of charge. But when the field point is close enough to the line, there's very little difference between the result for an infinite line and the real-life finite case. For example, if the distance r of the field point from the center of the line is 1% of the length of the line, the value of E differs from the infinite-length value by less than 0.02%.

Example 21.12 Field of a uniformly charged disk

Find the electric field caused by a disk of radius R with a uniform positive surface charge density (charge per unit area) σ, at a point along the axis of the disk a distance x from its center. Assume that x is positive.

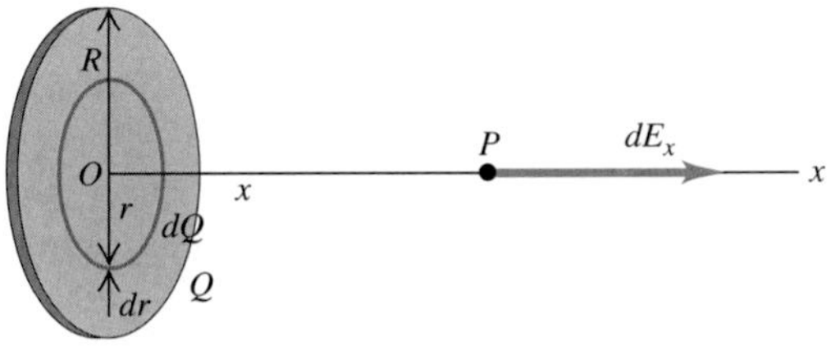

21.23 Finding the electric field on the axis of a uniformly charged disk. In this figure, the charge is assumed to be positive.

SOLUTION

IDENTIFY and SET UP: The situation and coordinate system are shown in Fig. 21.23. This example is similar to Examples 21.10 and 21.11 in that our target variable is the electric field as a function of the coordinate x along the x-axis. We can represent the charge distribution as a collection of concentric rings of charge dQ, as shown in Fig. 21.23. From Example 21.10 we know the field of a single ring on its axis of symmetry, so all we have to do is add the contributions of the rings.

EXECUTE: A typical ring has charge dQ, inner radius r, and outer radius $r + dr$ (Fig. 21.23). Its area dA is approximately equal to its width dr times its circumference $2\pi r$, or $dA = 2\pi r\,dr$. The charge per unit area is $\sigma = dQ/dA$, so the charge of the ring is $dQ = \sigma\,dA = \sigma\,(2\pi r\,dr)$, or

$$dQ = 2\pi\sigma r\,dr$$

We use this in place of Q in the expression for the field due to a ring found in Example 21.10, Eq. (21.8), and also replace the ring radius a with r. The field component dE_x at point P due to charge dQ is

$$dE_x = \frac{1}{4\pi\epsilon_0}\frac{(2\pi\sigma r\,dr)x}{(x^2 + r^2)^{3/2}}$$

To find the total field due to all the rings, we integrate dE_x over r from $r = 0$ to $r = R$ (*not* from $-R$ to R):

$$E_x = \int_0^R \frac{1}{4\pi\epsilon_0}\frac{(2\pi\sigma r\,dr)x}{(x^2 + r^2)^{3/2}} = \frac{\sigma x}{2\epsilon_0}\int_0^R \frac{r\,dr}{(x^2 + r^2)^{3/2}}$$

Remember that x is a constant during the integration and that the integration variable is r. The integral can be evaluated by use of the substitution $z = x^2 + r^2$. We'll let you work out the details; the result is

$$E_x = \frac{\sigma x}{2\epsilon_0}\left[-\frac{1}{\sqrt{x^2 + R^2}} + \frac{1}{x}\right]$$

$$= \frac{\sigma}{2\epsilon_0}\left[1 - \frac{1}{\sqrt{(R^2/x^2) + 1}}\right] \qquad (21.11)$$

EVALUATE: We saw in Example 21.10 that at a point on the symmetry axis of a uniformly charged ring, the electric field due to the ring has no components perpendicular to the axis. Hence at point P in Fig. 21.23, $dE_y = dE_z = 0$ for each ring, and the total field has $E_y = E_z = 0$.

Again we can ask what happens if the charge distribution gets very large. Suppose we keep increasing the radius R of the disk, simultaneously adding charge so that the surface charge density σ (charge per unit area) is constant. In the limit that R is much larger than the distance x of the field point from the disk, the term $1/\sqrt{(R^2/x^2) + 1}$ in Eq. (21.11) becomes negligibly small, and we get

$$E = \frac{\sigma}{2\epsilon_0} \qquad (21.12)$$

Our final result does not contain the distance x from the plane. This means that the electric field produced by an *infinite* plane sheet of charge is *independent of the distance from the sheet.* The field direction is everywhere perpendicular to the sheet, away from it. Again, there is no such thing as an infinite sheet of charge, but if the dimensions of the sheet are much larger than the distance x of the field point P from the sheet, the field is very nearly given by Eq. (21-11).

If P is to the left of the plane $(x < 0)$ instead of to the right, the result is the same except that the direction of $\vec{E}$ is to the left instead of the right. Also, if the surface charge density is negative, the directions of the fields on both sides of the plane are toward it rather than away from it.

Example 21.13 Field of two oppositely charged infinite sheets

Two infinite plane sheets are placed parallel to each other, separated by a distance d (Fig. 21.24). The lower sheet has a uniform positive surface charge density σ, and the upper sheet has a uniform negative surface charge density $-\sigma$ with the same magnitude. Find the electric field between the two sheets, above the upper sheet, and below the lower sheet.

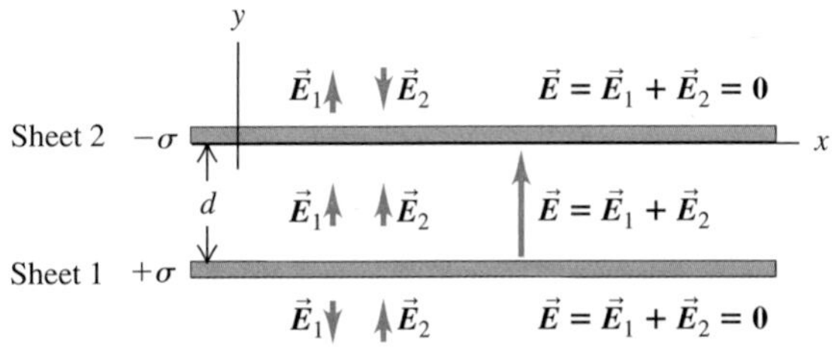

21.24 Finding the electric field due to two oppositely charged infinite sheets. The sheets are seen edge-on; only a portion of the infinite sheets can be shown!

SOLUTION

IDENTIFY and SET UP: In Example 21.12 we found the electric field due to a single infinite plane sheet of charge. We use this result and the principle of superposition to find the total field due to the two infinite planes in Fig. 21.24.

EXECUTE: Let sheet 1 be the lower sheet of positive charge, and let sheet 2 be the upper sheet of negative charge; the fields due to each sheet are $\vec{E}_1$ and $\vec{E}_2$, respectively. From Eq. (21.12) of Example 21.12, both $\vec{E}_1$ and $\vec{E}_2$ have the same magnitude at all points, no matter how far from either sheet:

$$E_1 = E_2 = \frac{\sigma}{2\epsilon_0}$$

At all points, the direction of $\vec{E}_1$ is away from the positive charge of sheet 1, and the direction of $\vec{E}_2$ is toward the negative charge of sheet 2. These fields, as well as the x- and y-axes, are shown in Fig. 21.24.

CAUTION: You may be surprised that $\vec{E}_1$ is unaffected by the presence of sheet 2 and that $\vec{E}_2$ is unaffected by the presence of sheet 1. Indeed, you may have thought that the field of one sheet would be unable to "penetrate" the other sheet. You might conclude this if you think of the electric field as some kind of physical substance that "flows" into or out of charges. But in fact there is no such substance, and the electric fields $\vec{E}_1$ and $\vec{E}_2$ depend only on the individual charge distributions that create them. The *total* field is just the vector sum of $\vec{E}_1$ and $\vec{E}_2$.

At points between the sheets, $\vec{E}_1$ and $\vec{E}_2$ reinforce each other; at points above the upper sheet or below the lower sheet, $\vec{E}_1$ and $\vec{E}_2$ cancel each other. Thus the total field is

$$\vec{E} = \vec{E}_1 + \vec{E}_2 = \begin{cases} \mathbf{0} & \text{above the upper sheet} \\ \dfrac{\sigma}{\epsilon_0}\hat{\jmath} & \text{between the sheets} \\ \mathbf{0} & \text{below the lower sheet} \end{cases}$$

Because we considered the sheets to be infinite, our result does not depend on the separation d.

EVALUATE: Note that the field between the oppositely charged sheets is uniform. We used this in Examples 21.7 and 21.8, in which two large parallel conducting plates were connected to the terminals of a battery. The battery causes the two plates to become oppositely charged, giving a field between the plates that is essentially uniform if the plate separation is much smaller than the dimensions of the plates. In Chapter 23 we will examine how a battery can produce such separation of positive and negative charge. An arrangement of two oppositely charged conducting plates is called a *capacitor;* these devices prove to be of tremendous practical utility and are the principal subject of Chapter 24.

Test Your Understanding

Suppose that the line of charge in Fig. 21.22 (Example 21.11) had charge $+Q$ distributed uniformly between $y = 0$ and $y = +a$ and had charge $-Q$ distributed uniformly between $y = 0$ and $y = -a$. What would be the direction of the electric field at P?

21.6 | Electric Field Lines

The concept of an electric field can be a little elusive because you can't see an electric field directly. Electric field *lines* can be a big help for visualizing electric fields and making them seem more real. An **electric field line** is an imaginary line or curve drawn through a region of space so that its tangent at any point is in the direction of the electric-field vector at that point. The basic idea is shown in Fig. 21.25. (We used a similar concept in our discussion of fluid flow in Section 14.5. A *streamline* is a line or curve whose tangent at any point is in the direction of the velocity of the fluid at that point. However, the similarity between electric field lines and fluid streamlines is a mathematical one only; there is nothing "flowing" in an electric field.) The English scientist Michael Faraday (1791–1867) first introduced the concept of field lines. He called them "lines of force," but the term "field lines" is preferable.

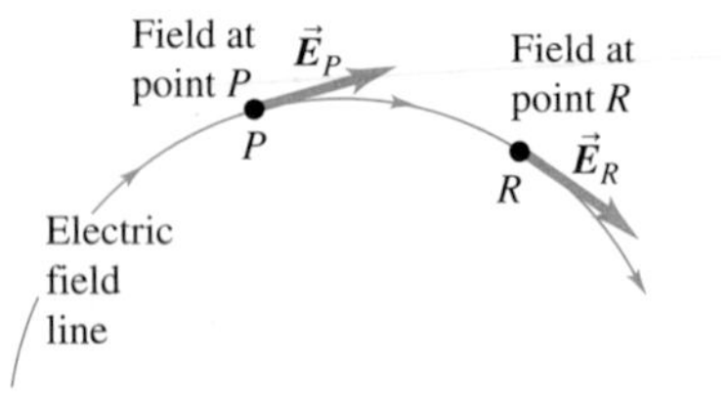

21.25 The direction of the electric field at any point is tangent to the field line through that point.

Electric field lines show the direction of $\vec{E}$ at each point, and their spacing gives a general idea of the *magnitude* of $\vec{E}$ at each point. Where $\vec{E}$ is strong, we draw lines bunched closely together; where $\vec{E}$ is weaker, they are farther apart. At any particular point, the electric field has a unique direction, so only one field line can pass through each point of the field. In other words, *field lines never intersect.*

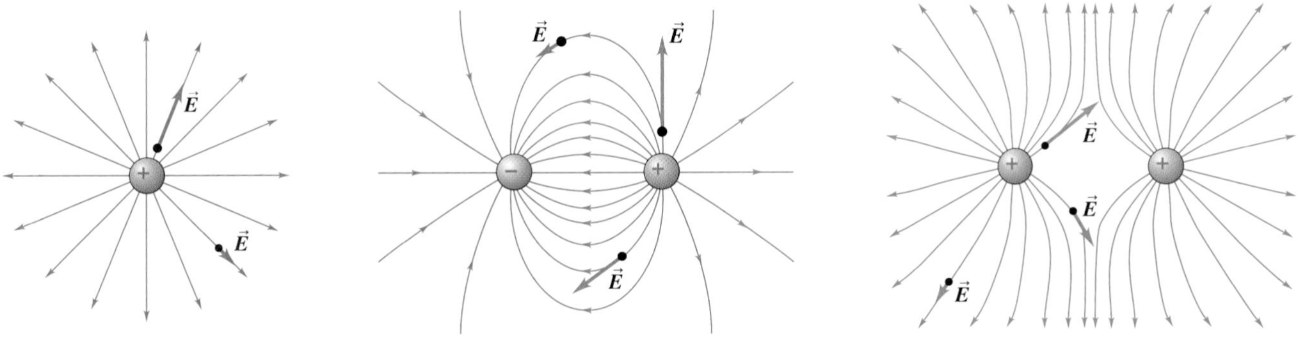

(a) A single positive charge (compare Figure 21.16)

(b) A positive charge and a negative charge of equal magnitude (an electric dipole)

(c) Two equal positive charges

21.26 Electric field lines for three different charge distributions. In general, the magnitude of $\vec{E}$ is different at different points along a given field line.

Figure 21.26 shows some of the electric field lines in a plane containing (a) a single positive charge; (b) two equal-magnitude charges, one positive and one negative (a dipole); and (c) two equal positive charges. Diagrams such as these are sometimes called *field maps;* they are cross sections of the actual three-dimensional patterns. The direction of the total electric field at every point in each diagram is along the tangent to the electric field line passing through the point. Arrowheads indicate the direction of the $\vec{E}$-field vector along each field line. The actual field vectors have been drawn at several points in each pattern. Notice that in general, the magnitude of the electric field is different at different points on a given field line; a field line is *not* a curve of constant electric field magnitude!

Figure 21.26 shows that field lines are directed *away* from positive charges (since close to a positive point charge, $\vec{E}$ points away from the charge) and *toward* negative charges (since close to a negative point charge, $\vec{E}$ points toward the charge). In regions where the field magnitude is large, such as between the positive and negative charges in Fig. 21.26b, the field lines are drawn close together. In regions where the field magnitude is small, such as between the two positive charges in Fig. 21.26c, the lines are widely separated. In a *uniform* field, the field lines are straight, parallel, and uniformly spaced, as in Fig. 21.18.

Figure 21.27a is a view from above of a demonstration setup for visualizing electric field lines. In the arrangement shown here, the tips of two positively charged wires are inserted in a container of insulating liquid, and some grass seeds are floated on the liquid. The grass seeds are electrically neutral insulators, but the electric field of the two charged wires causes *polarization* of the grass seeds; there is a slight shifting of the positive and negative charges within the molecules of each seed, like that shown in Fig. 21.7. The positively charged end of each grass seed is pulled in the direction of $\vec{E}$ and the negatively charged end is pulled opposite $\vec{E}$. Hence the long axis of each grass seed tends to orient parallel to the electric field, in the direction of the field line that passes through the position of the seed (Fig. 21.27b).

CAUTION It's a common misconception that if a charged particle of charge q is in motion where there is an electric field, the particle must move along an electric field line. Because $\vec{E}$ at any point is tangent to the field line that passes through that point, it is indeed true that the *force* $\vec{F} = q\vec{E}$ on the particle, and hence the particle's acceleration, are tangent to the field line. But we learned

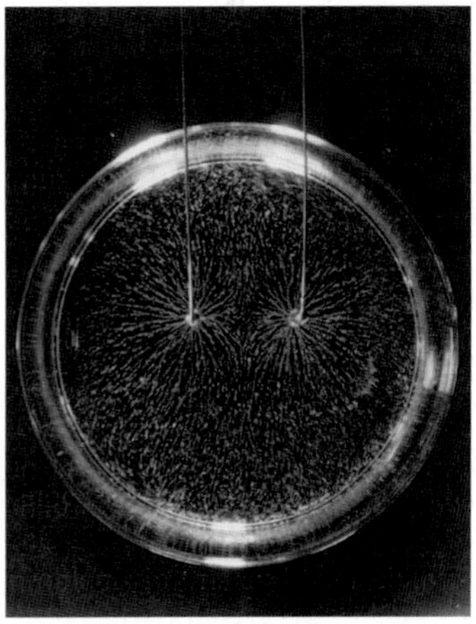

(a)

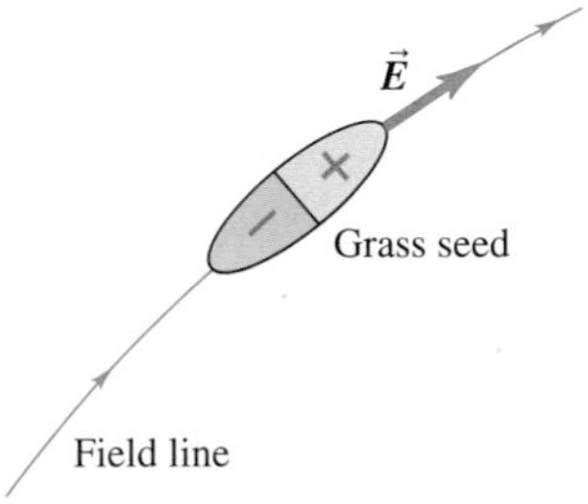

(b)

21.27 (a) Electric field lines produced by two equal point charges. The pattern is formed by grass seeds floating on a liquid above two charged wires. Compare this pattern with Fig. 21.26c. (b) The electric field causes polarization of the grass seeds, which in turn causes the seeds to align with the field.

in Chapter 3 that when a particle moves on a curved path, its acceleration *cannot* be tangent to the path. So in general, the trajectory of a charged particle is *not* the same as a field line.

Test Your Understanding

Suppose the electric field lines in a region of space are straight lines. If a charged particle is released from rest in that region, the trajectory of the particle *will* be a straight line. Explain why.

21.7 | Electric Dipoles

An **electric dipole** is a pair of point charges with equal magnitude and opposite sign (a positive charge q and a negative charge $-q$) separated by a distance d. We introduced electric dipoles in Example 21.9 (Section 21.5); the concept is worth exploring further because many physical systems, from molecules to TV antennas, can be described as electric dipoles. We will also use this concept extensively in our discussion of dielectrics in Chapter 24.

Figure 21.28a shows a molecule of water (H_2O), which in many ways behaves like an electric dipole. The water molecule as a whole is electrically neutral, but the chemical bonds within the molecule cause a displacement of charge; the result is a net negative charge on the oxygen end of the molecule and a net positive charge on the hydrogen end, forming an electric dipole. The effect is equivalent to shifting one electron only about 4×10^{-11} m (about the radius of a hydrogen atom), but the consequences of this shift are profound. Water is an excellent solvent for ionic substances such as table salt (sodium chloride, NaCl) precisely because the water molecule is an electric dipole (Fig. 21.28b). When dissolved in water, salt dissociates into a positive sodium ion (Na^+) and a negative chlorine ion (Cl^-), which tend to be attracted to the negative and positive ends, respectively, of water molecules; this holds the ions in solution. If water molecules were not electric dipoles, water would be a poor solvent, and almost all of the chemistry that occurs in aqueous solutions would be impossible. This includes all of the biochemical reactions that occur in all of the life on earth. In a very real sense, your existence as a living being depends on electric dipoles!

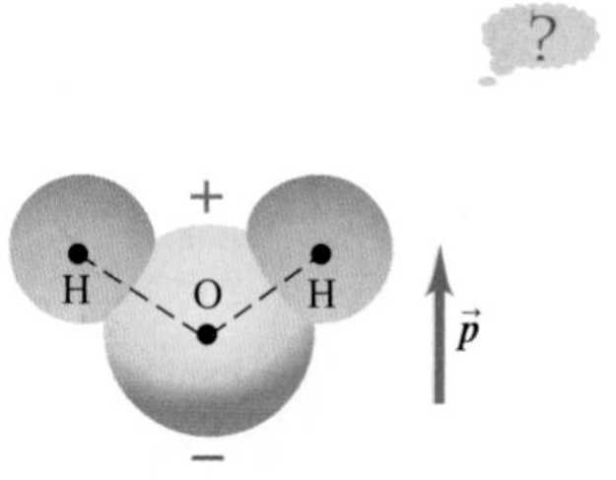

(a)

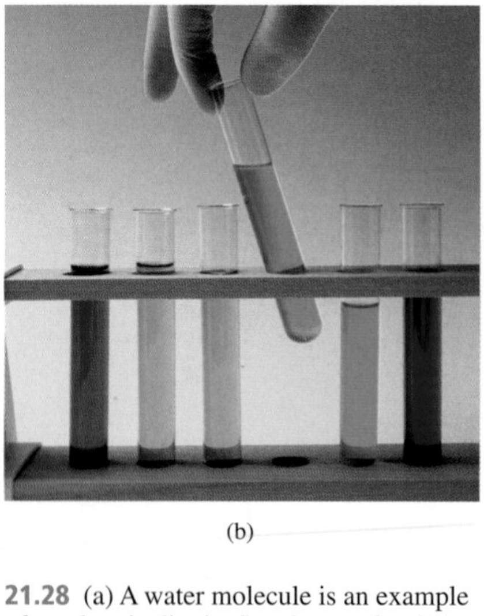

(b)

21.28 (a) A water molecule is an example of an electric dipole. See the text for the definition of the electric dipole moment vector $\vec{p}$. (b) Each test tube contains a solution of a different substance in water. The large electric dipole moment of water makes it an excellent solvent.

We examine two questions about electric dipoles. First, what forces and torques does an electric dipole experience when placed in an external electric field (that is, a field set up by charges outside the dipole)? Second, what electric field does an electric dipole itself produce?

Force and Torque on an Electric Dipole

To start with the first question, let's place an electric dipole in a *uniform* external electric field $\vec{E}$, as shown in Fig. 21.29. The forces $\vec{F}_+$ and $\vec{F}_-$ on the two charges both have magnitude qE, but their directions are opposite, and they add to zero. *The net force on an electric dipole in a uniform external electric field is zero.*

However, the two forces don't act along the same line, so their *torques* don't add to zero. We calculate torques with respect to the center of the dipole. Let the angle between the electric field $\vec{E}$ and the dipole axis be ϕ; then the lever arm for both $\vec{F}_+$ and $\vec{F}_-$ is $(d/2)\sin\phi$. The torque of $\vec{F}_+$ and the torque of $\vec{F}_-$ both have the same magnitude of $(qE)(d/2)\sin\phi$, and both torques tend to rotate the dipole clockwise (that is, $\vec{\tau}$ is directed into the page in Fig. 21.29). Hence the magnitude of the net torque is just twice the magnitude of either individual torque:

$$\tau = (qE)(d \sin \phi) \qquad (21.13)$$

where $d \sin \phi$ is the perpendicular distance between the lines of action of the two forces.

The product of the charge q and the separation d is the magnitude of a quantity called the **electric dipole moment,** denoted by p:

$$p = qd \quad \text{(magnitude of electric dipole moment)} \qquad (21.14)$$

The units of p are charge times distance $(\text{C}\cdot\text{m})$. For example, the magnitude of the electric dipole moment of a water molecule is $p = 6.13 \times 10^{-30}\ \text{C}\cdot\text{m}$.

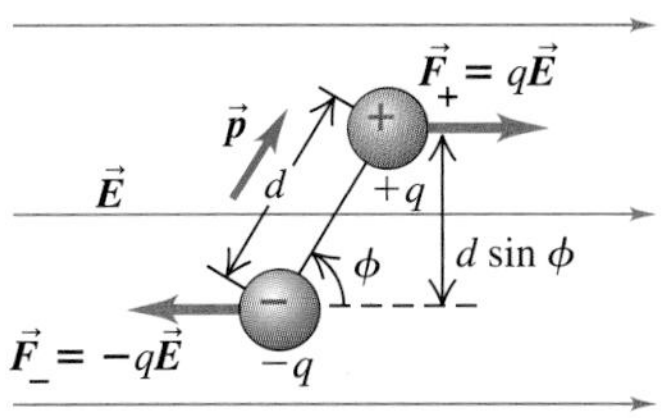

21.29 The net force on this electric dipole is zero, but there is a torque directed into the page that tends to rotate the dipole clockwise.

CAUTION Be careful not to confuse dipole moment with momentum or pressure. There aren't as many letters in the alphabet as there are physical quantities, so some letters are used several times. The context usually makes it clear what we mean, but be careful.

We further define the electric dipole moment to be a *vector* quantity $\vec{p}$. The magnitude of $\vec{p}$ is given by Eq. (21.14), and its direction is along the dipole axis from the negative charge to the positive charge as shown in Fig. 21.29.

In terms of p, Eq. (21.13) for the magnitude τ of the torque exerted by the field becomes

$$\tau = pE \sin \phi \quad \text{(magnitude of the torque on an electric dipole)} \qquad (21.15)$$

Since the angle ϕ in Fig. 21.29 is the angle between the directions of the vectors $\vec{p}$ and $\vec{E}$, this is reminiscent of the expression for the magnitude of the *vector product* discussed in Section 1.10. (You may want to review that discussion.) Hence we can write the torque on the dipole in vector form as

$$\vec{\tau} = \vec{p} \times \vec{E} \quad \text{(torque on an electric dipole, in vector form)} \qquad (21.16)$$

You can use the right-hand rule for the vector product to verify that in the situation shown in Fig. 21.29, $\vec{\tau}$ is directed into the page. The torque is greatest when $\vec{p}$ and $\vec{E}$ are perpendicular and is zero when they are parallel or antiparallel. The torque always tends to turn $\vec{p}$ to line it up with $\vec{E}$. The position $\phi = 0$, with $\vec{p}$ parallel to $\vec{E}$, is a position of stable equilibrium, and the position $\phi = \pi$, with $\vec{p}$ and $\vec{E}$ antiparallel, is a position of unstable equilibrium. The polarization of a grass seed in the apparatus of Fig. 21.27a gives it an electric dipole moment; the torque exerted by $\vec{E}$ then causes the seed to align with $\vec{E}$ and hence with the field lines.

When a dipole changes direction in an electric field, the electric-field torque does *work* on it, with a corresponding change in potential energy. The work dW done by a torque τ during an infinitesimal displacement $d\phi$ is given by Eq. (10.22): $dW = \tau\, d\phi$. Because the torque is in the direction of *decreasing* ϕ, we must write the torque as $\tau = -pE \sin \phi$, and

$$dW = \tau\, d\phi = -pE \sin \phi\, d\phi$$

In a finite displacement from ϕ_1 to ϕ_2 the total work done on the dipole is

$$W = \int_{\phi_1}^{\phi_2} (-pE \sin \phi)\, d\phi$$
$$= pE \cos \phi_2 - pE \cos \phi_1$$

The work is the negative of the change of potential energy, just as in Chapter 7: $W = U_1 - U_2$. So we see that a suitable definition of potential energy U for this system is

$$U(\phi) = -pE \cos \phi \qquad (21.17)$$

In this expression we recognize the *scalar product* $\vec{p} \cdot \vec{E} = pE \cos \phi$, so we can also write

$$U = -\vec{p} \cdot \vec{E} \quad \text{(potential energy for a dipole in an electric field)} \qquad (21.18)$$

The potential energy has its minimum value $U = -pE$ (i.e., its most negative value) at the stable equilibrium position, where $\phi = 0$ and $\vec{p}$ is parallel to $\vec{E}$. The potential energy is maximum when $\phi = \pi$ and $\vec{p}$ is antiparallel to $\vec{E}$; then $U = +pE$. At $\phi = \pi/2$, where $\vec{p}$ is perpendicular to $\vec{E}$, U is zero. We could of course define U differently so that it is zero at some other orientation of $\vec{p}$, but our definition is simplest.

Equation (21.18) gives us another way to look at the effect shown in Fig. 21.27a. The electric field $\vec{E}$ gives each grass seed an electric dipole moment, and the grass seed then aligns itself with $\vec{E}$ to minimize the potential energy.

Example 21.14 Force and torque on an electric dipole

Figure 21.30a shows an electric dipole in a uniform electric field with magnitude 5.0×10^5 N/C directed parallel to the plane of the figure. The charges are $\pm 1.6 \times 10^{-19}$ C; both lie in the plane and are separated by 0.125 nm $= 0.125 \times 10^{-9}$ m. (Both the charge magnitude and the distance are typical of molecular quantities.) Find a) the net force exerted by the field on the dipole; b) the magnitude and direction of the electric dipole moment; c) the magnitude and direction of the torque; d) the potential energy of the system in the position shown.

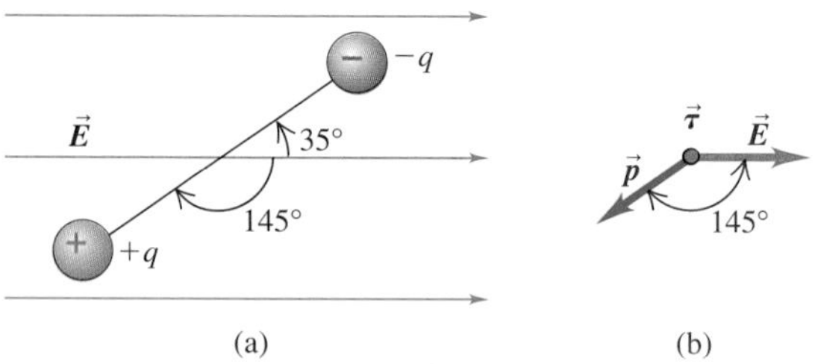

21.30 (a) An electric dipole. (b) Directions of the electric dipole moment, electric field, and torque.

SOLUTION

IDENTIFY and SET UP: We use the relationship $\vec{F} = q\vec{E}$ for each point charge to find the force on the dipole as a whole. Equation (21.14) tells us the dipole moment, Eq. (21.16) tells us the torque on the dipole, and Eq. (21.18) tells us the potential energy of the system.

EXECUTE: a) Since the field is uniform, the forces on the two charges are equal and opposite, and the total force is zero.

b) The magnitude p of the electric dipole moment $\vec{p}$ is

$$p = qd = (1.6 \times 10^{-19}\ \text{C})(0.125 \times 10^{-9}\ \text{m})$$
$$= 2.0 \times 10^{-29}\ \text{C} \cdot \text{m}$$

The direction of $\vec{p}$ is from the negative to the positive charge, 145° clockwise from the electric-field direction (Fig. 21.30b).

c) The magnitude of the torque is

$$\tau = pE \sin \phi = (2.0 \times 10^{-29}\ \text{C})(5.0 \times 10^5\ \text{N/C})(\sin 145°)$$
$$= 5.7 \times 10^{-24}\ \text{N} \cdot \text{m}$$

From the right-hand rule for vector products (Section 1.10), the direction of the torque $\vec{\tau} = \vec{p} \times \vec{E}$ is out of the page. This corresponds to a counterclockwise torque that tends to align $\vec{p}$ with $\vec{E}$.

d) The potential energy is

$$U = -pE \cos \phi$$
$$= -(2.0 \times 10^{-29}\ \text{C} \cdot \text{m})(5.0 \times 10^5\ \text{N/C})(\cos 145°)$$
$$= 8.2 \times 10^{-24}\ \text{J}$$

EVALUATE: The dipole moment, torque, and potential energy are all exceedingly small. Don't be surprised by this result: remember that we are looking at a single molecule, which is a very small object indeed!

In this discussion we have assumed that $\vec{E}$ is uniform, so there is no net force on the dipole. If $\vec{E}$ is not uniform, the forces at the ends may not cancel completely, and the net force may not be zero. Thus a body with zero net charge but an electric dipole moment can experience a net force in a nonuniform electric field. As we mentioned in Section 21.1, an uncharged body can be polarized by an electric field, giving rise to a separation of charge and an electric dipole moment. This is how uncharged bodies can experience electrostatic forces (see Fig. 21.7).

Field of an Electric Dipole

Now let's think of an electric dipole as a *source* of electric field. What does the field look like? The general shape of things is shown by the field map of Fig. 21.26b. At each point in the pattern the total $\vec{E}$ field is the vector sum of the fields from the two individual charges, as in Example 21.9 (Section 21.5). Try drawing diagrams showing this vector sum for several points.

To get quantitative information about the field of an electric dipole, we have to do some calculating, as illustrated in Example 21.15. Notice the use of the principle of superposition of electric fields to add up the contributions to the field of the individual charges. Also notice that we need to use approximation techniques even for the relatively simple case of a field due to two charges. Field calculations often become very complicated, and computer analysis is typically used to determine the field due to an arbitrary charge distribution.

Example 21.15 Field of an electric dipole, revisited

In Fig. 21.31 an electric dipole is centered at the origin, with $\vec{p}$ in the direction of the $+y$-axis. Derive an approximate expression for the electric field at a point on the y-axis for which y is much larger than d. Use the binomial expansion of $(1 + x)^n$, that is, $(1 + x)^n \cong 1 + nx + n(n-1)x^2/2 + \cdots$, for the case $|x| < 1$. (This problem illustrates a useful calculational technique.)

SOLUTION

IDENTIFY: We use the principle of superposition: the total electric field is the vector sum of the field produced by the positive charge and the field produced by the negative charge.

SET UP: At the field point shown in Fig. 21.31, the field of the positive charge has a positive (upward) y-component and the field of the negative charge has a negative (downward) y-component. We add these components to find the total field, then apply the approximation that y is much greater than d.

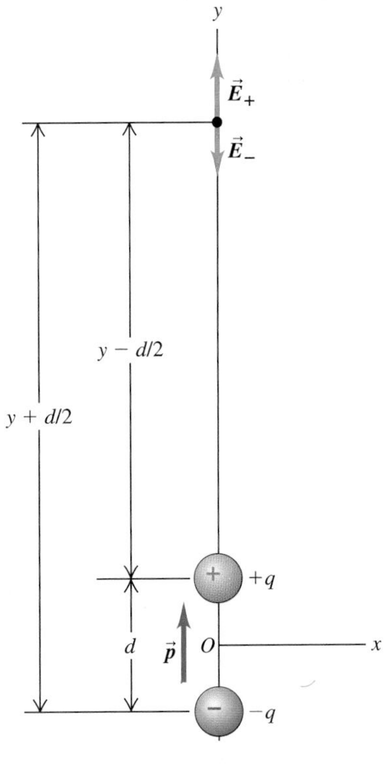

21.31 Finding the electric field of an electric dipole at a point on its axis.

EXECUTE: The total y-component E_y of electric field from the two charges is

$$E_y = \frac{q}{4\pi\epsilon_0}\left[\frac{1}{(y - d/2)^2} - \frac{1}{(y + d/2)^2}\right]$$
$$= \frac{q}{4\pi\epsilon_0 y^2}\left[\left(1 - \frac{d}{2y}\right)^{-2} - \left(1 + \frac{d}{2y}\right)^{-2}\right]$$

We used this exact approach in Example 21.9 (Section 21.5). Now comes the approximation. When y is much greater than d, that is, when we are far away from the dipole compared to its size, the quantity $d/2y$ is much smaller than 1. With $n = -2$ and $d/2y$ playing the role of x in the binomial expansion, we keep only the first two terms. The terms we discard are much smaller than those we keep, and we have

$$\left(1 - \frac{d}{2y}\right)^{-2} \cong 1 + \frac{d}{y} \quad \text{and} \quad \left(1 + \frac{d}{2y}\right)^{-2} \cong 1 - \frac{d}{y}$$

Hence E_y is given approximately by

$$E \cong \frac{q}{4\pi\epsilon_0 y^2}\left[1 + \frac{d}{y} - \left(1 - \frac{d}{y}\right)\right]$$
$$= \frac{qd}{2\pi\epsilon_0 y^3}$$
$$= \frac{p}{2\pi\epsilon_0 y^3}$$

EVALUATE: An alternative route to this expression is to put the fractions in the E_y expression over a common denominator and combine, then approximate the denominator $(y - d/2)^2(y + d/2)^2$ as y^4. We leave the details as a problem.

For points P off the coordinate axes, the expressions are more complicated, but at *all* points far away from the dipole (in any direction) the field drops off as $1/r^3$. We can compare this with the $1/r^2$ behavior of a point charge, the $1/r$ behavior of a long line charge, and the independence of r for a large sheet of charge. There are charge distributions for which the field drops off even more quickly. An *electric quadrupole* consists of two equal dipoles with opposite orientation, separated by a small distance. The field of a quadrupole at large distances drops off as $1/r^4$.

Test Your Understanding

Use the information given in this section to calculate the electric dipole moment of a water molecule. If a water molecule is oriented with its dipole moment at a 90° angle to an electric field of magnitude 2.5×10^4 N/C, what is the torque on the molecule?

SUMMARY

The fundamental quantity in electrostatics is electric charge. There are two kinds of charge, positive and negative. Charges of the same sign repel each other; charges of opposite sign attract. Charge is conserved; the total charge in an isolated system is constant.

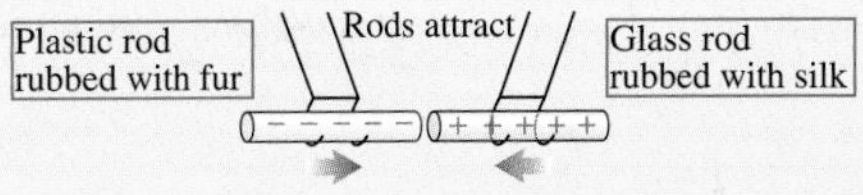

All ordinary matter is made of protons, neutrons, and electrons. The positive protons and electrically neutral neutrons in the nucleus of an atom are bound together by the nuclear force; the negative electrons surround the nucleus at distances much greater than the nuclear size. Electric interactions are chiefly responsible for the structure of atoms, molecules, and solids.

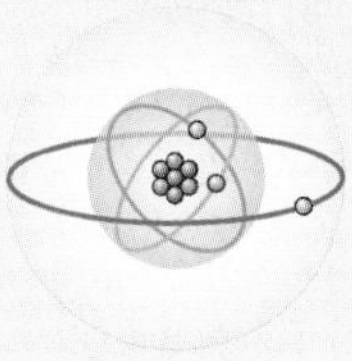

Conductors are materials that permit electric charge to move easily within them. Insulators permit charge to move much less readily. Most metals are good conductors; most nonmetals are insulators.

Coulomb's law is the basic law of interaction for point electric charges. For charges q_1 and q_2 separated by a distance r, the magnitude of the force on either charge is proportional to the product $q_1 q_2$ and inversely proportional to r^2. The force on each charge is along the line joining the two charges—repulsive if q_1 and q_2 have the same sign, attractive if they have opposite signs. The forces form an action-reaction pair and obey Newton's third law. In SI units the unit of electric charge is the coulomb, abbreviated C. (See Examples 21.1 and 21.2)

$$F = \frac{1}{4\pi\epsilon_0}\frac{|q_1 q_2|}{r^2} \tag{21.2}$$

$$\frac{1}{4\pi\epsilon_0} = 8.988 \times 10^9\ \mathrm{N \cdot m^2/C^2}$$

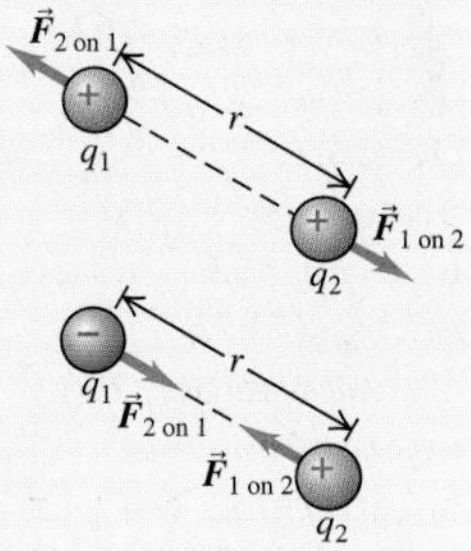

The principle of superposition of forces states that when two or more charges each exert a force on a charge, the total force on that charge is the vector sum of the forces exerted by the individual charges. (See Examples 21.3 and 21.4)

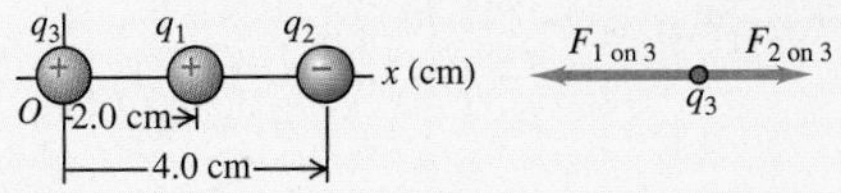

Electric field $\vec{E}$, a vector quantity, is the force per unit charge exerted on a test charge at any point, provided the test charge is small enough that it does not disturb the charges that cause the field. The electric field produced by a point charge is directly radially away from or toward the charge.
(See Examples 21.5 through 21.8)

$$\vec{E} = \frac{\vec{F}_0}{q_0} \tag{21.3}$$

$$\vec{E} = \frac{1}{4\pi\epsilon_0}\frac{q}{r^2}\hat{r} \tag{21.7}$$

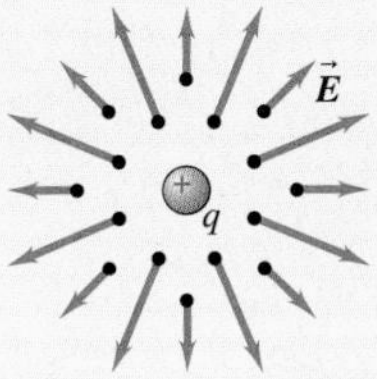

The principle of superposition of electric fields states that the electric field $\vec{E}$ of any combination of charges is the vector sum of the fields caused by the individual charges. To calculate the electric field caused by a continuous distribution of charge, divide the distribution into small elements, calculate the field caused by each element, and then carry out the vector sum or each component sum, usually by integrating. Charge distributions are described by linear charge density λ, surface charge density σ, and volume charge density ρ. (See Examples 21.9 through 21.13)

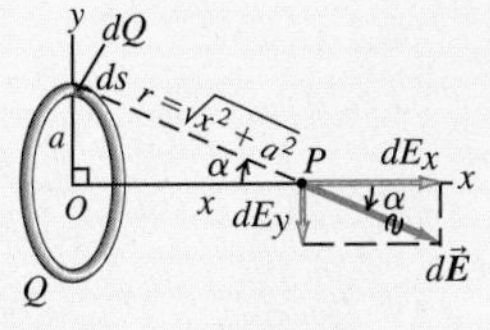

Field lines provide a graphical representation of electric fields. At any point on a field line, the tangent to the line is in the direction of $\vec{E}$ at that point. The number of lines per unit area (perpendicular to their direction) is proportional to the magnitude of $\vec{E}$ at the point.

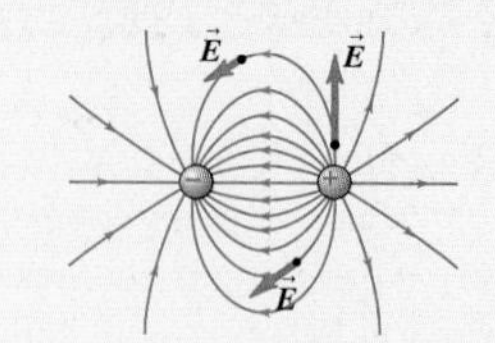

An electric dipole is a pair of electric charges of equal magnitude q but opposite sign, separated by a distance d. The electric dipole moment $\vec{p}$ is defined to have magnitude $p = qd$. The direction of $\vec{p}$ is from negative toward positive charge. An electric dipole in an electric field $\vec{E}$ experiences a torque $\vec{\tau}$ equal to the vector product of $\vec{p}$ and $\vec{E}$. The magnitude of the torque depends on the angle ϕ between $\vec{p}$ and $\vec{E}$. The potential energy U for an electric dipole in an electric field also depends on the relative orientation of $\vec{p}$ and $\vec{E}$. (See Examples 21.14 and 21.15)

$$\tau = pE \sin \phi \qquad (21.15)$$

$$\vec{\tau} = \vec{p} \times \vec{E} \qquad (21.16)$$

$$U = -\vec{p} \cdot \vec{E} \qquad (21.18)$$

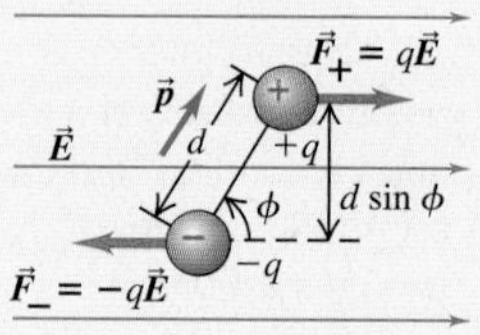

Key Terms

electric charge, 793
electrostatics, 793
electron, 795
proton, 795
neutron, 795
nucleus, 795
atomic number, 796
positive ion, 796
negative ion, 796
ionization, 796
principle of conservation of charge, 796
conductor, 797
insulator, 797
induction, 797
induced charge, 798
point charge, 800
Coulomb's law, 800
coulomb, 801
principle of superposition of forces, 802
electric field, 806
test charge, 806
source point, 808
field point, 808
vector field, 808
principle of superposition of electric fields, 812
linear charge density, 812
surface charge density, 812
volume charge density, 812
electric field line, 818
electric dipole, 820
electric dipole moment, 821

Your Notes

Answer to Chapter Opening Question ?

Water molecules have a permanent electric dipole moment: one end of the molecule has a positive charge and the other end has a negative charge. These ends attract negative and positive ions, respectively, holding the ions apart in solution. Water is less effective as a solvent for materials whose molecules do not ionize (called *nonionic* substances), such as oils.

Answers to Test Your Understanding Questions

Section 21.1 The plastic rod gets a negative charge by taking electrons from the fur, so the rod weighs a little more and the fur weighs a little less after the rubbing. By contrast, the glass rod gets a positive charge by giving electrons to the silk. Hence, after rubbing them together, the glass rod weighs a little less and the silk weighs a little more. The weight change is *very* small: the number of electrons transferred is a small fraction of a mole, and a mole of electrons has a mass of only $(6.02 \times 10^{23}\ \text{electrons})(9.11 \times 10^{-31}\ \text{kg/electron}) = 5.48 \times 10^{-7}\ \text{kg} = 0.548$ milligram!
Section 21.2 You would go through the same series of steps as shown in Fig. 21.6, but would bring a positively charged rod next to the sphere (for example, a glass rod rubbed with silk; see Fig. 21.1).
Section 21.3 The force exerted by q_1 on Q is still as in Example 21.4. The magnitude of the force exerted by q_2 on Q is still equal to $F_{1\ \text{on}\ Q}$, but the direction of the force is now *toward* q_2 at an angle α below the x-axis. Hence both components of this force are negative:

$$(F_{2\ \text{on}\ Q})_x = -(F_{2\ \text{on}\ Q})\cos\alpha = -0.23\ \text{N}$$
$$(F_{2\ \text{on}\ Q})_y = -(F_{2\ \text{on}\ Q})\sin\alpha = -0.17\ \text{N}$$

Now the x-components of the two forces cancel while the y-components add together, so the components of the total force on Q are

$$F_x = 0.23\ \text{N} + (-0.23\ \text{N}) = 0$$
$$F_y = (-0.17\ \text{N}) + (-0.17\ \text{N}) = -0.34\ \text{N}$$

The total force is in the $-y$-direction with magnitude 0.34 N.
Section 21.4 We found in Example 21.3 that the electric force on a third charge $Q = 4.0\ \mu\text{C}$ at this point is $\vec{F} = (0.46\ \text{N})\hat{\imath}$. From Eq. (21.3) the electric field at this point due to the first two charges is the force on Q divided by Q itself, or $\vec{E} = \vec{F}/Q = [(0.46\ \text{N})/(4.0 \times 10^{-6}\ \text{C})]\hat{\imath} = (1.2 \times 10^5\ \text{N/C})\hat{\imath}$.
Section 21.5 In this case $\vec{E}$ would point in the negative y-direction. Think of a pair of segments of length dy, one at coordinate $y > 0$ and the other at coordinate $-y < 0$. The upper segment has a positive charge and produces an electric field $d\vec{E}$ at P that points away from the segment, so this $d\vec{E}$ has a positive x-component and a negative y-component, like the vector $d\vec{E}$ in Fig. 21.22. The lower segment has the same amount of negative charge. It produces a $d\vec{E}$ that has the same magnitude but points *toward* the lower segment, so it has a negative x-component and a negative y-component. By symmetry, the two x-components are equal but opposite, so they cancel. Thus the total electric field has only a negative y-component.
Section 21.6 If the field lines are straight, $\vec{E}$ must point in the same direction throughout the region. Hence the force $\vec{F} = q\vec{E}$ on a particle of charge q is always in the same direction. A particle released from rest accelerates in a straight line the direction of $\vec{F}$, and so its trajectory is a straight line.
Section 21.7 In a water molecule, an amount of charge q equal to $e = 1.60 \times 10^{-19}\ \text{C}$ (the magnitude of the charge of an electron) is displaced by a distance $d = 4 \times 10^{-11}$ m. Hence the magnitude of the electric dipole moment is $p = qd = (4 \times 10^{-11}\ \text{m})(1.60 \times 10^{-19}\ \text{C}) = 6.4 \times 10^{-30}\ \text{C}\cdot\text{m}$. The direction is as shown in Fig. 21.28a. The angle between $\vec{p}$ and $\vec{E}$ is 90°, so the magnitude of the torque is $\tau = pE\sin 90° = pE = (6.4 \times 10^{-30}\ \text{C}\cdot\text{m})(2.5 \times 10^4\ \text{N/C}) = 1.6 \times 10^{-26}\ \text{N}\cdot\text{m}$. The direction of torque is such as to align $\vec{p}$ with $\vec{E}$.

Discussion Questions

Q21.1 Two metal spheres are hanging from nylon threads. When you bring the spheres close to each other, they tend to attract. Based on this information alone, discuss all the possible ways that the spheres could be charged. Is it possible that after the spheres touch, they will cling together? Explain.
Q21.2 The electric force between two charged particles becomes weaker with increasing distance. Suppose instead that the electric force were *independent* of distance. In this case, would a charged comb still cause a neutral insulator to become polarized as in Fig. 21.7? Why or why not? Would the neutral insulator still be attracted to the comb? Again, why or why not?
Q21.3 Your clothing tends to cling together after going through the dryer. Why? Would you expect more or less clinging if all your clothing is made of the same material (say, cotton) than if you dry different kinds of clothing together? Again, why? (You may want to experiment with your next load of laundry.)
Q21.4 An uncharged metal sphere hangs from a nylon thread. When a positively charged glass rod is brought close to the metal sphere, the sphere is drawn toward the rod. But if the sphere touches the rod, it suddenly flies away from the rod. Explain why the sphere is first attracted and then repelled.
Q21.5 The free electrons in a metal are gravitationally attracted toward the earth. Why, then, don't they all settle to the bottom of the conductor, like sediment settling to the bottom of a river?
Q21.6 Some of the free electrons in a good conductor (such as a piece of copper) move at speeds of 10^6 m/s or more. Why do these electrons not fly out of the conductor completely?
Q21.7 Good electrical conductors, such as metals, are typically good conductors of heat; electrical insulators, such as wood, are typically poor conductors of heat. Explain why there should be a relationship between electrical conduction and heat conduction in these materials.
Q21.8 Defend the following statement: "If there were only one electrically charged particle in the entire universe, the concept of electric charge would be meaningless."
Q21.9 Two identical metal objects are mounted on insulating stands. Describe how you could place charges of opposite sign but exactly equal magnitude on the two objects.
Q21.10 You can use plastic food wrap to cover a container by stretching the material across the top and pressing the overhanging

material against the sides. What makes it stick? (*Hint:* The answer involves the electric force.) Does the food wrap stick to itself with equal tenacity? Why or why not? Does it work with metallic containers? Again, why or why not?

Q21.11 If you walk across a nylon rug and then touch a large metal object such as a doorknob, you may get a spark and a shock. Why does this tend to happen more on dry days than on humid days? (*Hint:* See Fig. 21.28.) Why are you less likely to get a shock if you touch a *small* metal object, such as a paper clip?

Q21.12 You have a negatively charged object. How can you use it to place a net negative charge on an insulated metal sphere? To place a net positive charge on the sphere?

Q21.13 You touch a positively charged rod to an initially uncharged, insulated metal sphere. The metal sphere acquires a net positive charge and the rod loses some of its charge. Does this mean that protons were transferred from the rod to the sphere?

Q21.14 Two equal point charges exert equal forces on each other. But if one charge is twice the other, do they still exert equal forces on each other or does one exert twice the force than the other?

Q21.15 A proton is placed in a uniform electric field and then released. Then an electron is placed at this same point and released. Do these two particles experience the same force? The same acceleration? Do they move in the same direction when released?

Q21.16 The electric force between an electron and a proton, between two electrons, or between two protons is much stronger than the gravitational force between any of these pairs of particles. Yet even though the sun and planets contain electrons and protons, it is the gravitational force that holds the planets in their orbits around the sun. Explain this seeming contradiction.

Q21.17 What similarities do electrical forces have to gravitational forces? What are the most significant differences?

Q21.18 Atomic nuclei are made of protons and neutrons. This shows that there must be another kind of interaction in addition to gravitational and electric forces. Explain.

Q21.19 Sufficiently strong electric fields can cause atoms to become positively ionized, that is, to lose one or more electrons. Explain how this can happen. What determines how strong the field must be to make this happen?

Q21.20 When you pull transparent plastic tape off a roll and try to position it precisely on a piece of paper, the tape often jumps over and sticks where it is not wanted. Why?

Q21.21 A particle with positive charge Q is held fixed at the origin. A second particle with positive charge q is fired at the first particle, and follows the trajectory shown in Fig. 21.31. Is the angular momentum of the second particle constant? Why or why not? (*Hint:* How much torque does the first particle exert on the second particle?)

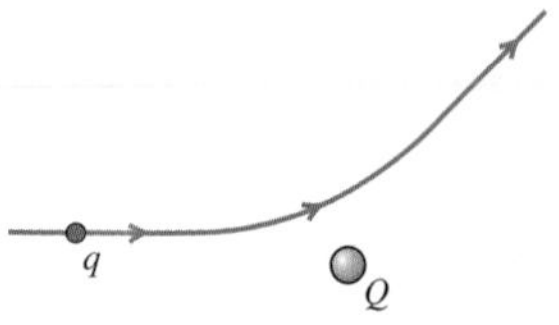

Figure 21.31 Question Q21.21.

Q21.22 The air temperature and the velocity of the air have different values at different places in the earth's atmosphere. Is the air velocity a vector field? Why or why not? Is the air temperature a vector field? Again, why or why not?

Q21.23 Suppose the charge shown in Fig. 21.26a is fixed in position. A small, positively charged particle is then placed at some point in the figure and released. Will the trajectory of the particle follow an electric field line? Why or why not? Suppose instead that the particle is placed at some point in Fig. 21.26b and released (the positive and negative charges shown in the figure are fixed in position). Will its trajectory follow an electric field line? Again, why or why not? Explain any differences between your answers for the two different situations.

Q21.24 The water molecule (H_2O) has a large dipole moment, while the benzene molecule (C_6H_6) has zero dipole moment. Use these facts to explain why ordinary table salt (NaCl, or sodium chloride) dissolves very easily in water, but dissolves poorly in benzene.

Exercises

Section 21.3 Coulomb's Law

21.1 Excess electrons are placed on a small lead sphere with mass 8.00 g so that its net charge is -3.20×10^{-9} C. a) Find the number of excess electrons on the sphere. b) How many excess electrons are there per lead atom? The atomic number of lead is 82, and its atomic mass is 207 g/mol.

21.2 Lightning occurs when there is a flow of electric charge (principally electrons) between the ground and a thundercloud. The maximum rate of charge flow in a lightning bolt is about 20,000 C/s; this lasts for 100 μs or less. How much charge flows between the ground and the cloud in this time? How many electrons flow during this time?

21.3 Estimate how many electrons there are in your body. Make any assumptions you feel are necessary, but clearly state what they are. (*Hint:* Most of the atoms in your body have equal numbers of electrons, protons, and neutrons.) What is the combined charge of all these electrons?

21.4 Particles in a Gold Ring. You have a pure (24-karat) gold ring with mass 17.7 g. Gold has an atomic mass of 197 g/mol and an atomic number of 79. a) How many protons are in the ring, and what is their total positive charge? b) If the ring carries no net charge, how many electrons are in it?

21.5 What is the total charge, in coulombs, of all the electrons in 1.80 mol of hydrogen atoms?

21.6 Two small spheres spaced 20.0 cm apart have equal charge. How many excess electrons must be present on each sphere if the magnitude of the force of repulsion between them is 4.57×10^{-21} N?

21.7 Two small plastic spheres are given positive electrical charges. When they are 15.0 cm apart, the repulsive force between them has magnitude 0.220 N. What is the charge on each sphere a) if the two charges are equal? b) if one sphere has four times the charge of the other?

21.8 Two small aluminum spheres, each having mass 0.0250 kg, are separated by 80.0 cm. a) How many electrons does each sphere contain? (The atomic mass of aluminum is 26.982 g/mol, and its atomic number is 13.) b) How many electrons would have to be removed from one sphere and added to the other to cause an attractive force between the spheres of magnitude 1.00×10^4 N (roughly one ton)? Assume that the spheres may be treated as point charges. c) What fraction of all the electrons in each sphere does this represent?

21.9 How far does the electron of a hydrogen atom have to be removed from the nucleus for the force of attraction to equal the weight of the electron at the surface of the earth?

21.10 a) A glass rod is rubbed with silk and acquires a charge of magnitude 7.50 nC. What is the change in mass of the rod? b) A plastic rod is rubbed with fur and acquires a charge of magnitude 7.50 nC. What is the change in mass of the rod?

21.11 Three point charges are arranged on a line. Charge $q_3 = +5.00$ nC and is at the origin. Charge $q_2 = -3.00$ nC and is at $x = +4.00$ cm. Charge q_1 is at $x = +2.00$ cm. What is q_1 (magnitude and sign) if the net force on q_3 is zero?

21.12 A negative charge $-0.550\ \mu$C exerts an upward 0.200-N force on an unknown charge 0.300 m directly below it. a) What is the unknown charge (magnitude and sign)? b) What are the magnitude and direction of the force that the unknown charge exerts on the $-0.550\ \mu$C charge?

21.13 A point charge of $+3.50\ \mu$C is placed 0.800 m to the left of a second, identical point charge. What are the magnitudes and directions of the forces that each charge exerts on the other?

21.14 In Example 21.4 (Section 21.3), suppose the point charge on the y-axis at $y = -0.30$ m has negative charge $-2.0\ \mu$C, and the other charges remain the same. Find the magnitude and direction of the net force on Q. How does your answer differ from that in Example 21.3? Explain the differences.

21.15 In Example 21.3 (Section 21.3), calculate the net force on charge q_1.

21.16 In Example 21.4 (Section 21.3), what is the net force (magnitude and direction) on charge q_1 exerted by the other two charges?

21.17 Three point charges are arranged along the x-axis. Charge $q_1 = +3.00\ \mu$C is at the origin, and charge $q_2 = -5.00\ \mu$C is at $x = 0.200$ m. Charge $q_3 = -8.00\ \mu$C. Where is q_3 located if the net force on q_1 is 7.00 N in the $-x$-direction?

21.18 Repeat Exercise 21.17, for $q_3 = +8.00\ \mu$C.

21.19 Two point charges are located on the y-axis as follows: charge $q_1 = -1.50$ nC at $y = -0.600$ m, and charge $q_2 = +3.20$ nC at the origin $(y = 0)$. What is the total force (magnitude and direction) exerted by these two charges on a third charge $q_3 = +5.00$ nC located at $y = -0.400$ m?

21.20 Two point charges are placed on the x-axis as follows: Charge $q_1 = +4.00$ nC is located at $x = 0.200$ m, and charge $q_2 = +5.00$ nC is at $x = -0.300$ m. What are the magnitude and direction of the total force exerted by these two charges on a negative point charge $q_3 = -6.00$ nC that is placed at the origin?

21.21 A positive point charge q is placed on the $+y$-axis at $y = a$, and a negative point charge $-q$ is placed on the $-y$-axis at $y = -a$. A negative point charge $-Q$ is located at some point on the $+x$-axis. a) In a free-body diagram, show the forces that act on the charge $-Q$. b) Find the x- and y-components of the net force that the two charges q and $-q$ exert on $-Q$. (Your answer should only involve k, q, Q, a, and the coordinate x of the third charge.) c) What is the net force on the charge $-Q$ when it is at the origin $(x = 0)$? d) Graph the y-component of the net force on the charge $-Q$ as a function of x for values of x between $-4a$ and $+4a$.

21.22 Two positive point charges q are placed on the y-axis at $y = a$ and $y = -a$. A negative point charge $-Q$ is located at some point on the $+x$-axis. a) In a free-body diagram, show the forces that act on the charge $-Q$. b) Find the x- and y-components of the net force that the two positive charges exert on $-Q$. (Your answer should only involve k, q, Q, a, and the coordinate x of the third charge.) c) What is the net force on the charge $-Q$ when it is at the origin $(x = 0)$? d) Graph the x-component of the net force on the charge $-Q$ as a function of x for values of x between $-4a$ and $+4a$.

21.23 Four identical charges q are placed at the corners of a square of side L. a) In a free-body diagram, show all of the forces that act on one of the charges. b) Find the magnitude and direction of the total force exerted on one charge by the other three charges.

Section 21.4 Electric Field and Electric Forces

21.24 A particle has charge -3.00 nC. a) Find the magnitude and direction of the electric field due to this particle at a point 0.250 m directly above it. b) At what distance from this particle does its electric field have a magnitude of 12.0 N/C?

21.25 An alpha particle (charge $+2e$ and mass 6.64×10^{-27} kg) is traveling to the right at 1.50 km/s. What uniform electric field (magnitude and direction) is needed to cause it to travel to the left at the same speed after 2.65 μs?

21.26 An electron is released from rest in a uniform electric field. The electron accelerates vertically upward, traveling 4.50 m in the first 3.00 μs after it is released. a) What are the magnitude and direction of the electric field? b) Are we justified in ignoring the effects of gravity? Justify your answer quantitatively.

21.27 a) What must the charge (sign and magnitude) of a 1.45-g particle be for it to remain stationary when placed in a downward-directed electric field of magnitude 650 N/C? b) What is the magnitude of an electric field in which the electric force on a proton is equal in magnitude to its weight?

21.28 a) What is the electric field of an iron nucleus at a distance of 6.00×10^{-10} m from the nucleus? The atomic number of iron is 26. Assume that the nucleus may be treated as a point charge. b) What is the electric field of a proton at a distance of 5.29×10^{-11} m from the proton? (This is the radius of the electron orbit in the Bohr model for the ground state of the hydrogen atom.)

21.29 A small object carrying a charge of $-55.0\ \mu$C experiences a downward force of 6.20×10^{-9} N when placed at a certain point in an electric field. a) What are the magnitude and direction of the electric field at this point? b) What would be the magnitude and direction of the force acting on a copper nucleus (atomic number = 29, atomic mass = 63.5 g/mol) placed at this same point in the electric field?

21.30 Electric Field of the Earth. The earth has a net electric charge that causes a field at points near its surface equal to 150 N/C and directed in toward the center of the earth. a) What magnitude and sign of charge would a 60-kg human have to acquire to overcome his or her weight by the force exerted by the earth's electric field? b) What would be the force of repulsion between two people each with the charge calculated in part (a) and separated by a distance of 100 m? Is use of the earth's electric field a feasible means of flight? Why or why not?

21.31 An electron is projected with an initial speed $v_0 = 1.60 \times 10^6$ m/s into the uniform field between the parallel plates in Fig. 21.32. Assume that the field between the plates is uniform and directed vertically downward, and that the field outside the plates is zero. The electron enters the field at a point midway between the plates. a) If the electron just misses the upper plate as it emerges from the field, find the magnitude of the electric field. b) Suppose that in Fig. 21.32 the electron is replaced by a proton with the same initial speed v_0. Would the proton hit one of the plates? If the proton would not hit one of the plates, what would be the magnitude and direction of its vertical displacement as it exits the region between the plates? c) Compare the paths traveled by the electron and the proton and explain the differences. d) Discuss whether it is reasonable to ignore the effects of gravity for each particle.

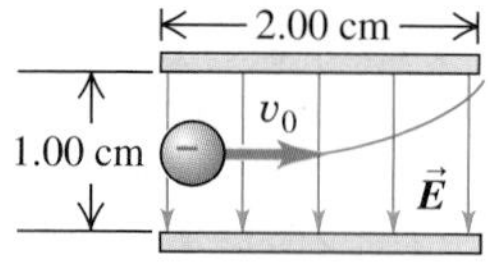

Figure 21.32 Exercise 21.31.

21.32 Point charge $q_1 = -5.00$ nC is at the origin and point charge $q_2 = +3.00$ nC is on the x-axis at $x = 3.00$ cm. Point P is on the y-axis at $y = 4.00$ cm. a) Calculate the electric fields $\vec{E}_1$ and $\vec{E}_2$ at point P due to the charges q_1 and q_2. Express your results in terms of unit vectors (see Example 21.6). b) Use the results of part (a) to obtain the resultant field at P, expressed in unit vector form.

21.33 In Exercise 21.31, what is the speed of the electron as it emerges from the field?

21.34 a) Calculate the magnitude and direction (relative to the $+x$-axis) of the electric field in Example 21.6 (Section 21.4). b) A -2.5-nC point charge is placed at the point P in Fig. 21.17. Find the magnitude and direction of i) the force that the -8.0-nC charge at the origin exerts on this charge and ii) the force that this charge exerts on the -8.0-nC charge at the origin.

21.35 a) For the electron in Examples 21.7 and 21.8 (Section 21.4), compare the weight of the electron to the magnitude of the electric force on the electron. Is it appropriate to ignore the gravitational force on the electron in these examples? Explain. b) A particle with charge $+e$ is placed at rest between the charged plates in Fig. 21.18. What must the mass of this object be if it is to remain at rest? Give your answer in kilograms and in multiples of the electron mass. c) Does the answer to part (b) depend on where between the plates the object is placed? Why or why not?

21.36 A uniform electric field exists in the region between two oppositely charged plane parallel plates. A proton is released from rest at the surface of the positively charged plate and strikes the surface of the opposite plate, 1.60 cm distant from the first, in a time interval of 1.50×10^{-6} s. a) Find the magnitude of the electric field. b) Find the speed of the proton when it strikes the negatively charged plate.

21.37 A point charge is at the origin. With this point charge as the source point, what is the unit vector $\hat{r}$ in the direction of a) the field point at $x = 0$, $y = -1.35$ m; b) the field point at $x = 12.0$ cm, $y = 12.0$ cm; c) the field point at $x = -1.10$ m, $y = 2.60$ m? Express your results in terms of the unit vectors $\hat{\imath}$ and $\hat{\jmath}$.

21.38 According to the safety standards of the Institute of Electrical and Electronics Engineers (IEEE), humans should avoid prolonged exposure to electric fields of magnitudes greater than 614 N/C. a) At a point where $E = 614$ N/C, what is the magnitude of the electric force on a single electron? b) Atomic and molecular dimensions are on the order of 10^{-10} m. What is the magnitude of the electric force on an electron that is 1.0×10^{-10} m from a proton? c) How do the answers in parts (a) and (b) compare? What do you think would happen to a person placed in an electric field that produced a force equal to that calculated in part (b)?

21.39 a) An electron is moving east in a uniform electric field of 1.50 V/m directed to the west. At point A, the velocity of the electron is 4.50×10^5 m/s toward the east. What is the speed of the electron when it reaches point B, 0.375 m east of point A? b) A proton is moving in the uniform electric field of part (a). At point A, the velocity of the proton is 1.90×10^4, east. What is the speed of the proton at point B?

Section 21.5 Electric-Field Calculations

21.40 Two particles having charges $q_1 = 0.500$ nC and $q_2 = 8.00$ nC are separated by a distance of 1.20 m. At what point along the line connecting the two charges is the total electric field due to the two charges equal to zero?

21.41 Two positive point charges q are placed on the x-axis, one at $x = a$ and one at $x = -a$. a) Find the magnitude and direction of the electric field at $x = 0$. b) Derive an expression for the electric field at points on the x-axis. Use your result to graph the x-component of the electric field as a function of x, for values of x between $-4a$ and $+4a$.

21.42 Repeat Exercise 21.40, but now let $q_1 = -4.00$ nC.

21.43 A $+2.00$-nC point charge is at the origin, and a second -5.00-nC point charge is on the x-axis at $x = 0.800$ m. a) Find the electric field (magnitude and direction) at each of the following points on the x-axis: i) $x = 0.200$ m; ii) $x = 1.20$ m; iii) $x = -0.200$ m. b) Find the net electric force that the two charges would exert on an electron placed at each point in part (a).

21.44 A positive point charge q is placed at $x = a$, and a negative point charge $-q$ is placed at $x = -a$. a) Find the magnitude and direction of the electric field at $x = 0$. b) Derive an expression for the electric field at points on the x-axis. Use your result to graph the x-component of the electric field as a function of x, for values of x between $-4a$ and $+4a$.

21.45 In a rectangular coordinate system a positive point charge $q = 6.00 \times 10^{-9}$ C is placed at the point $x = +0.150$ m, $y = 0$, and an identical point charge is placed at $x = -0.150$ m, $y = 0$. Find the x- and y-components, the magnitude, and the direction of the electric field at the following points: a) the origin; b) $x = 0.300$ m, $y = 0$; c) $x = 0.150$ m, $y = -0.400$ m; d) $x = 0$, $y = 0.200$ m.

21.46 A point charge $q_1 = -4.00$ nC is at the point $x = 0.600$ m, $y = 0.800$ m, and a second point charge $q_2 = +6.00$ nC is at the point $x = 0.600$ m, $y = 0$. Calculate the magnitude and direction of the net electric field at the origin due to these two point charges.

21.47 Repeat Exercise 21.45 for the case where the point charge at $x = +0.150$ m, $y = 0$ is positive and the other is negative, each with magnitude 6.00×10^{-9} C.

21.48 A very long, straight wire has charge per unit length 1.50×10^{-10} C/m. At what distance from the wire is the electric field magnitude equal to 2.50 N/C?

21.49 Positive electric charge is distributed along the y-axis with charge per unit length λ. a) Consider the case where charge is distributed only between the points $y = a$ and $y = -a$. For points on the $+x$-axis, graph the x-component of the electric field as a function of x for values of x between $x = a/2$ and $x = 4a$. b) Consider instead the case where charge is distributed along the entire y-axis with the same charge per unit length λ. Using the same graph as in part (a), plot the x-component of the electric field as a function of x for values of x between $x = a/2$ and $x = 4a$. Label which graph refers to which situation.

21.50 A ring-shaped conductor with radius $a = 2.50$ cm has a total positive charge $Q = +0.125$ nC uniformly distributed around it, as shown in Fig. 21.21. The center of the ring is at the origin of coordinates O. a) What is the electric field (magnitude and direction) at point P, which is on the x-axis at $x = 40.0$ cm? b) A point charge $q = -2.50\ \mu$C is placed at the point P described in part (a). What are the magnitude and direction of the force exerted *by* the charge q *on* the ring?

21.51 A uniformly charged disk of radius R carries positive charge per unit area σ, as in Fig. 21.23. For points on the $+x$-axis, graph the x-component of the electric field as a function of x for values of x between $x = 0$ and $x = 4R$.

21.52 Near the earth's surface, the electric field in the open air has magnitude 150 N/C and is directed down toward the ground. If this is regarded as being due to a large sheet of charge lying on the earth's surface, calculate the charge per unit area in the sheet. What is the sign of the charge?

21.53 Each square centimeter of the surface of an infinite plane sheet of paper has 2.50×10^6 excess electrons. Find the magnitude and direction of the electric field at a point 5.00 cm from the surface of the sheet, if the sheet is large enough to be treated as an infinite plane.

21.54 Two horizontal, infinite, plane sheets of charge are separated by a distance d. The lower sheet has negative charge with uniform surface charge density $-\sigma < 0$. The upper sheet has positive charge with uniform surface charge density $\sigma > 0$. What is the electric field (magnitude, and direction if the field is nonzero) a) above the upper sheet? b) below the lower sheet? c) between the sheets?

Section 21.6 Electric Field Lines

21.55 Two large parallel sheets of charge are separated by a distance d. One sheet has positive surface charge density $\sigma > 0$, and the other has negative surface charge density $-\sigma < 0$. Sketch the electric field lines at points near the center of the sheets and therefore well away from the edges.

21.56 Sketch the electric field lines for a disk of radius R with a positive uniform surface charge density σ. Use what you know about the electric field very close to the disk and very far from the disk to make your sketch.

21.57 a) Sketch the electric field lines for an infinite line of charge. You may find it helpful to show the field lines in a plane containing the line of charge in one sketch and the field lines in a plane perpendicular to the line of charge in a second sketch. b) Explain how your sketches show i) that the magnitude E of the electric field depends only on the distance r from the line of charge and ii) that E decreases like $1/r$.

21.58 Fig. 21.33 shows some of the electric field lines due to three point charges arranged along the vertical axis. All three charges have the same magnitude. a) What are the signs of each of the three charges? Explain your reasoning. b) At what point(s) is the magnitude of the electric field the smallest? Explain your reasoning. Explain how the fields produced by each individual point charge combine to give a small net field at this point or points.

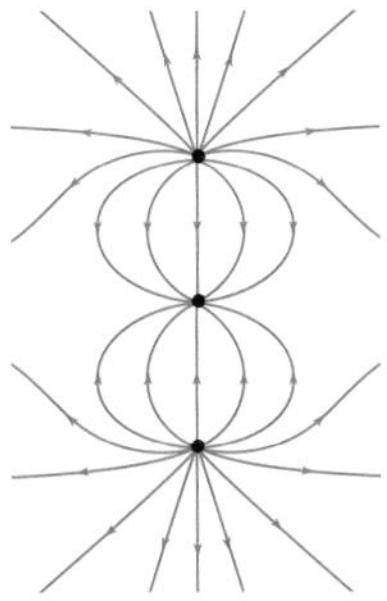

Figure 21.33 Exercise 21.58.

Section 21.7 Electric Dipoles

21.59 Point charges $q_1 = -4.5$ nC and $q_2 = +4.5$ nC are separated by 3.1 mm, forming an electric dipole. a) Find the electric dipole moment (magnitude and direction). b) The charges are in a uniform electric field whose direction makes an angle of 36.9° with the line connecting the charges. What is the magnitude of this field if the torque exerted on the dipole has magnitude 7.2×10^{-9} N·m?

21.60 The potassium chloride molecule (KCl) has a dipole moment of 8.9×10^{-30} C·m. a) Assuming that this dipole moment arises from two charges $\pm 1.6 \times 10^{-19}$ C separated by distance d, calculate d. b) What is the maximum magnitude of the torque that a uniform electric field with magnitude 6.0×10^5 N/C can exert on a KCl molecule? Sketch the relative orientations of the electric dipole moment $\vec{p}$ and the electric field $\vec{E}$ when the torque is a maximum.

21.61 The ammonia molecule (NH_3) has a dipole moment of 5.0×10^{-30} C·m. Ammonia molecules in the gas phase are placed in a uniform electric field $\vec{E}$ with magnitude 1.6×10^6 N/C. a) What is the change in electric potential energy when the dipole moment of a molecule changes its orientation with respect to $\vec{E}$ from parallel to perpendicular? b) At what absolute temperature T is the average translational kinetic energy $\frac{3}{2}kT$ of a molecule equal to the change in potential energy calculated in part (a)? (Above this temperature, thermal agitation prevents the dipoles from aligning with the electric field.)

21.62 The dipole moment of the water molecule (H_2O) is 6.17×10^{-30} C·m. Consider a water molecule located at the origin whose dipole moment $\vec{p}$ points in the $+x$-direction. A chlorine ion (Cl^-), of charge -1.60×10^{-19} C, is located at $x = 3.00 \times 10^{-9}$ m. Find the magnitude and direction of the electric force that the water molecule exerts on the chlorine ion. Is this force attractive or repulsive? Assume that x is much larger than the separation d between the charges in the dipole, so that the approximate expression for the electric field along the dipole axis derived in Example 21.15 (Section 21.7) can be used.

21.63 In Example 21.15 (Section 21.7), the approximate result $E \cong p/2\pi\epsilon_0 y^3$ was derived for the electric field of a dipole at points on the dipole axis. a) Rederive this result by putting the fractions in the expression for E_y over a common denominator, as described in Example 21.15. b) Explain why the approximate result also gives the correct approximate expression for E_y for $y < 0$.

21.64 Consider the electric dipole of Example 21.15 (Section 21.7). a) Derive an expression for the magnitude of the electric field produced by the dipole at a point on the x-axis in Fig. 21.31. What is the direction of this electric field? b) How does the electric field at points on the x-axis depend on x when x is very large?

21.65 Surface Tension. The surface of a polar liquid, such as water, can be viewed as a series of dipoles strung together in the stable arrangement in which the dipole moment vectors are parallel to the surface and all point in the same direction. Suppose now that something presses inward on the surface, distorting the dipoles as shown in Fig. 21.34. a) Show that the two slanted dipoles exert a net upward force on the dipole between them, and hence oppose the downward external force. Show also that the dipoles attract each other and hence resist being separated. The force between dipoles opposes penetration of the liquid's surface, and is a simple model for surface tension (Section 14.3 and Fig. 14.15).

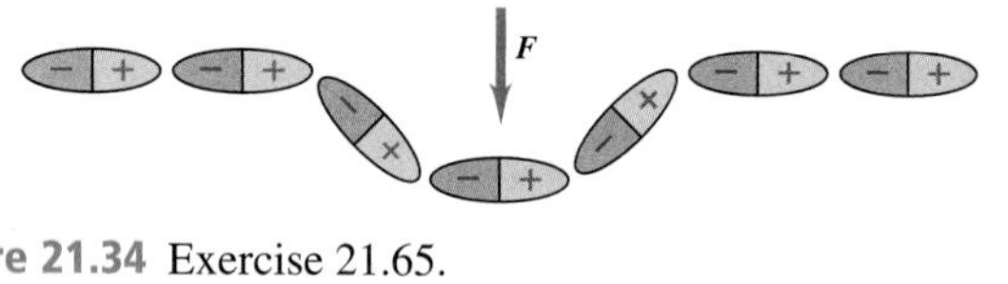

Figure 21.34 Exercise 21.65.

21.66 Torque on a Dipole. An electric dipole with dipole moment $\vec{p}$ is in a uniform electric field $\vec{E}$. a) Find the orientations of the dipole for which the torque on the dipole is zero. b) Which of the orientations in part (a) is stable, and which is unstable? (*Hint:* Consider a small displacement away from the equilibrium position and see what happens.) c) Show that for the stable orientation in (b), the dipole's own electric field tends to oppose the external field.

21.67 Three charges are at the corners of an isosceles triangle as shown in Fig. 21.35. The $\pm 5.00\text{-}\mu\text{C}$ charges form a dipole. a) Find the force (magnitude and direction)which the $-10.00\text{-}\mu\text{C}$ charge exerts on the dipole. b) For an axis perpendicular to the line connecting the $\pm 5.00\text{-}\mu\text{C}$ charges at the midpoint of this line, find the torque (magnitude and direction) exerted on the dipole by the $-10.00\text{-}\mu\text{C}$ charge.

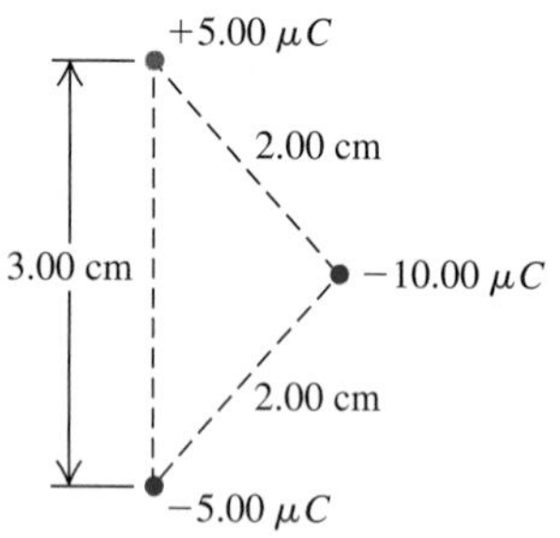

Figure 21.35 Problem 21.67.

Problems

21.68 A charge $q_1 = +5.00$ nC is placed at the origin of an xy-coordinate system, and a charge $q_2 = -2.00$ nC is placed on the positive x-axis at $x = 4.00$ cm. a) If a third charge $q_3 = +6.00$ nC is now placed at the point $x = 4.00$ cm, $y = 3.00$ cm, find the x- and y-components of the total force exerted on this charge by the other two. b) Find the magnitude and direction of this force.

21.69 Two positive point charges Q are held fixed on the x-axis at $x = a$ and $x = -a$. A third positive point charge q, with mass m, is placed on the x-axis away from the origin at a coordinate x such that $|x| \ll a$. The charge q, which is free to move along the x axis, is then released. a) Find the frequency of oscillation of the charge q. (*Hint:* Review the definition of simple harmonic motion in Section 13.2. Use the binomial expansion $(1 + z)^n = 1 + nz + n(n-1)z^2/2 + \cdots$, valid for the case $|z| < 1$.) b) Suppose instead that the charge q were placed on the y-axis at a coordinate y such that $|y| \ll a$, and then released. If this charge is free to move anywhere in the xy-plane, what will happen to it? Explain your answer.

21.70 Two identical spheres with mass m are hung from silk threads of length L, as shown in Fig. 21.36. Each sphere has the same charge, so $q_1 = q_2 = q$. The radius of each sphere is very small compared to the distance between the spheres, so they may be treated as point charges. Show that if the angle θ is small, the equilibrium separation d between the spheres is $d = (q^2L/2\pi\epsilon_0 mg)^{1/3}$. (*Hint:* If θ is small, then $\tan\theta \cong \sin\theta$.)

21.71 Two small spheres with mass $m = 15.0$ g are hung by silk threads of length $L = 1.20$ m from a common point (Fig. 21.36). When the spheres are given equal quantities of negative charge, so that $q_1 = q_2 = q$, each thread hangs at $\theta = 25.0°$ from the vertical. a) Draw a diagram showing the forces on each sphere. Treat the spheres as point charges. b) Find the magnitude of q. c) Both threads are now shortened to length $L = 0.600$ m, while the charges q_1 and q_2 remain unchanged. What new angle will each thread make with the vertical? (*Hint:* This part of the problem can be solved numerically by using trial values for θ and adjusting the values of θ until a self-consistent answer is obtained.)

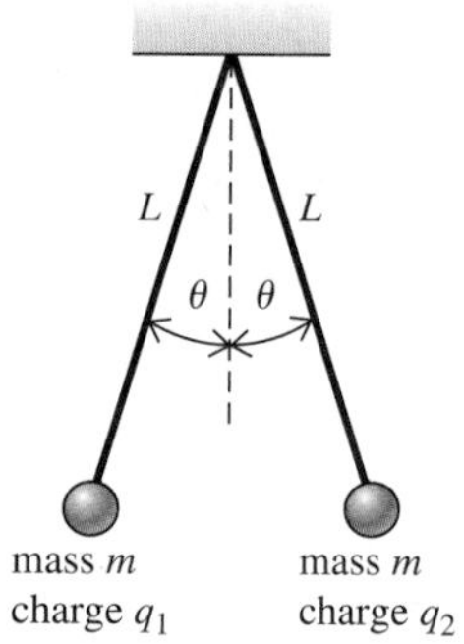

Figure 21.36 Problems 21.70, 21.71, and 21.72.

21.72 Two identical spheres are each attached to silk threads of length $L = 0.500$ m and hung from a common point (Fig. 21.36). Each sphere has mass $m = 8.00$ g. The radius of each sphere is very small compared to the distance between the spheres, so they may be treated as point charges. One sphere is given positive charge q_1, and the other a different positive charge q_2; this causes the spheres to separate so that when the spheres are in equilibrium, each thread makes an angle $\theta = 20.0°$ with the vertical. a) Draw a free-body diagram for each sphere when in equilibrium, and label all the forces that act on each sphere. b) Determine the magnitude of the electrostatic force that acts on each sphere, and determine the tension in each thread. c) Based on the information you have been given, what can you say about the magnitudes of q_1 and q_2? Explain your answers. d) A small wire is now connected between the spheres, allowing charge to be transferred from one sphere to the other until the two spheres have equal charges; the wire is then removed. Each thread now makes an angle of 30.0° with the vertical. Determine the original charges. (*Hint:* The total charge on the pair of spheres is conserved.)

21.73 Sodium chloride (NaCl, ordinary table salt) is made up of positive sodium ions (Na^+) and negative chlorine ions (Cl^-). a) If a point charge with the same charge and mass as all the Na^+

ions in 0.100 mol of NaCl is 2.00 cm from a point charge with the same charge and mass as all the Cl^- ions, what is the magnitude of the attractive force between these two point charges? b) If the positive point charge in part (a) is held in place and the negative point charge is released from rest, what is its initial acceleration? (See Appendix D for atomic masses.) c) Does it seem reasonable that the ions in NaCl could be separated in this way? Why or why not? (In fact, when sodium chloride dissolves in water it breaks up into Na^+ and Cl^- ions. However, in this situation there are additional electric forces exerted by the water molecules on the ions.)

21.74 Three point charges are arranged along the x-axis. Charge $q_1 = -4.50$ nC is located at $x = 0.200$ m, and charge $q_2 = +2.50$ nC is at $x = -0.300$ m. A positive point charge q_3 is located at the origin. a) What must the value of q_3 be for the net force on this point charge to have magnitude 4.00×10^{-6} N? b) What is the direction of the net force on q_3? c) Where along the x-axis can q_3 be placed and the net force on it be zero, other than the trivial answers of $x = \pm\infty$?

21.75 Three identical point charges q are placed at each of three corners of a square of side L. Find the magnitude and direction of the net force on a point charge $-3q$ placed a) at the center of the square; b) at the vacant corner of the square. In each case, draw a free-body diagram showing the forces exerted on the $-3q$ charge by each of the other three charges.

21.76 Three point charges are placed on the y-axis: a charge q at $y = a$, a charge $-2q$ at the origin, and a charge q at $y = -a$. Such an arrangement is called an *electric quadrupole.* a) Find the magnitude and direction of the electric field at points on the y-axis for $y > a$. b) Use the binomial expansion to show that very far from the quadrupole, so that $y \gg a$, the electric field is proportional to y^{-4}. Contrast this behavior to that of the electric field of a point charge and that of the electric field of a dipole.

21.77 a) For the arrangement of charges described in Problem 21.76, find the magnitude and direction of the electric field at points on the positive x-axis. b) Use the binomial expansion to find an approximate expression for the electric field valid for $x \gg a$. Contrast this behavior to that of the electric field of a point charge and that of the electric field of a dipole.

21.78 a) Suppose all the electrons in 20.0 g of carbon atoms were located at the North Pole of the earth and all the protons at the South Pole. What would be the total force of attraction exerted on each group of charges by the other? The atomic number of carbon is 6, and the atomic mass of carbon is 12.0 g/mol. b) What would be the magnitude and direction of the force exerted by the charges in part (a) on a third charge that is equal to the charge at the South Pole, and located at a point on the surface of the earth at the equator? Draw a diagram showing the locations of the charges and the forces on the charge at the equator.

21.79 If Atoms Were Not Neutral... Because the charges on the electron and proton have the same absolute value, atoms are electrically neutral. Suppose this were not precisely true, and the absolute value of the charge of the electron were less than the charge of the proton by 0.00100%. a) Estimate what the net charge of this textbook would be under these circumstances. Make any assumptions you feel are justified, but state clearly what they are. (*Hint:* Most of the atoms in this textbook have equal numbers of electrons, protons, and neutrons.) b) What would be the magnitude of the electric force between two textbooks placed 5.0 m apart? Would this force be attractive or repulsive? Estimate what the acceleration of each book would be if they were 5.0 m apart and there were no non-electrical forces on them. c) Discuss how the fact that ordinary matter is stable shows that the absolute values of the charges on the electron and proton must be identical to a *very* high level of accuracy.

21.80 Vibrations in Crystals. As a simplified model of a crystal, consider three atoms lying along a line, with a distance b between adjacent atoms. The atoms each have a net charge q and mass m. Suppose the middle charge is displaced a very small distance x from its equilibrium position and then released. Show that the net electrical force on the displaced charge is given approximately by $F = (q^2/b^3\pi\epsilon_0)x$, when $x \ll b$. What is the direction of this force? b) Find the frequency of vibration of the displaced charge after it is released in terms of the crystal parameters (q, b, and m). c) If the atoms are singly ionized carbon and are separated by an equilibrium distance of 4.0×10^{-10} m, what is the numerical value of their vibration frequency?

21.81 Two small, copper spheres each have radius 1.00 mm. a) How many atoms does each sphere contain? b) Assume that each copper atom contains 29 protons and 29 electrons. We know that electrons and protons have charges of exactly the same magnitude, but let's explore the effect of small differences (see also Problem 21.79). If the charge of a proton is $+e$ and the magnitude of the charge of an electron is 0.100% smaller, what is the net charge of each sphere and what force would one sphere exert on the other if they were separated by 1.00 m?

21.82 Operation of an Inkjet Printer. In an inkjet printer, letters are built up by squirting drops of ink at the paper from a rapidly moving nozzle. The pattern on the paper is controlled by an electrostatic valve that determines at each nozzle position whether ink is squirted onto the paper or not. The ink drops, 15 μm in radius, leave the nozzle and travel toward the paper at 20 m/s. The drops pass through a charging unit that gives each drop a positive charge q when the drop loses some electrons. The drops then pass between parallel deflecting plates 2.0 cm in length where there is a uniform vertical electric field with magnitude 8.0×10^4 N/C. If a drop is to be deflected 0.30 mm by the time it reaches the end of the deflection plate, what magnitude of charge must be given to the drop? (Assume that the density of the ink drop is the same as that of water, 1000 kg/m^3.)

21.83 A proton is projected into a uniform electric field that points vertically upward and has magnitude E. The initial velocity of the proton has a magnitude v_0 and is directed at an angle α below the horizontal. a) Find the maximum distance h_{max} that the proton descends vertically below its initial elevation. You can ignore gravitational forces. b) After what horizontal distance d does the proton return to its original elevation? c) Sketch the trajectory of the proton. d) Find the numerical values of h_{max} and d if $E = 500$ N/C, $v_0 = 4.00 \times 10^5$ m/s, and $\alpha = 30.0°$.

21.84 A negative point charge $q_1 = -4.00$ nC is on the x-axis at $x = 0.60$ m. A second point charge q_2 is on the x-axis at $x = -1.20$ m. What must the sign and magnitude of q_2 be for the net electric field at the origin to be a) 50.0 N/C in the $+x$-direction? b) 50.0 N/C in the $-x$-direction?

21.85 A charge of 12.0 nC is at the origin; a second, unknown charge is at $x = 3.00$ m, $y = 0$; and a third charge of -16.0 nC is at $x = 5.00$ m, $y = 0$. What are the sign and magnitude of the unknown charge if the net field at $x = 8.00$ m, $y = 0$ has magnitude 12.0 N/C and is in the $+x$-direction?

21.86 Positive charge Q is distributed uniformly along the x-axis from $x = 0$ to $x = a$. A positive point charge q is located on the positive x-axis at $x = a + r$, a distance r to the right of the end of Q (Fig. 21.37). a) Calculate the x- and y-components of the electric field produced by the charge distribution Q at points on the positive x-axis where $x > a$. b) Calculate the force (magnitude and direction) that the charge distribution Q exerts on q. c) Show that if $r \gg a$, the magnitude of the force in part (b) is approximately $Qq/4\pi\epsilon_0 r^2$. Explain why this result is obtained.

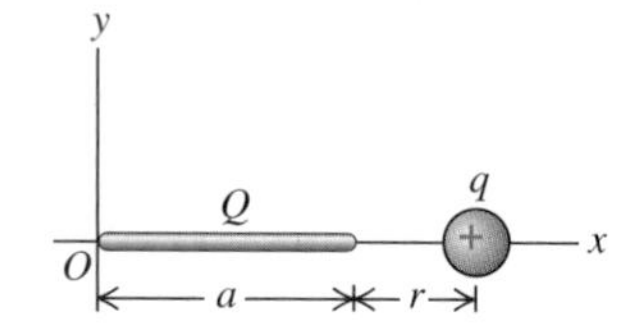

Figure 21.37 Problem 21.86.

21.87 Positive charge Q is distributed uniformly along the positive y-axis between $y = 0$ and $y = a$. A negative point charge $-q$ lies on the positive x-axis, a distance x from the origin (Fig. 21.38). a) Calculate the x- and y-components of the electric field produced by the charge distribution Q at points on the positive x-axis. b) Calculate the x- and y-components of the force that the charge distribution Q exerts on q. c) Show that if $x \gg a$, $F_x \cong -Qq/4\pi\epsilon_0 x^2$ and $F_y \cong +Qqa/8\pi\epsilon_0 x^3$. Explain why this result is obtained.

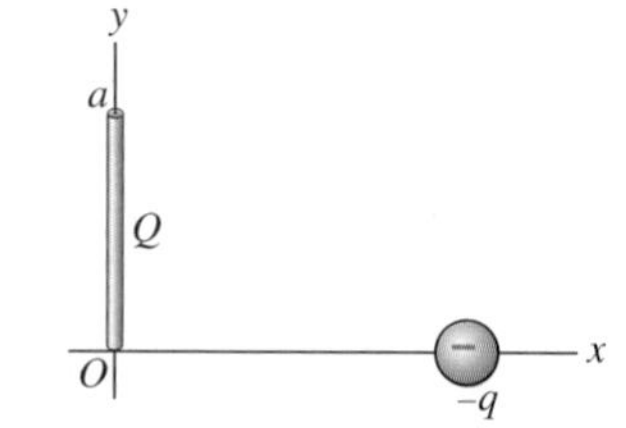

Figure 21.38 Problem 21.87.

21.88 A charged line like that shown in Fig. 21.22 extends from $y = 2.50$ cm to $y = -2.50$ cm. The total charge distributed uniformly along the line is -9.00 nC. a) Find the electric field (magnitude and direction) on the x-axis at $x = 0.25$ cm. b) Is the magnitude of the electric field you calculated in part (a) larger or smaller than the electric field 0.25 cm from an infinite line of charge that has the same charge per unit length as this finite line of charge? In terms of the approximation used to derive $E = \lambda/2\pi\epsilon_0 r$ for an infinite line from Eq. (21.9), explain why this is so. c) At what distance x does the result for the infinite line of charge differ by 1.0% from that for the finite line?

21.89 A charged line like that shown in Fig. 21.22 extends from $y = 2.50$ cm to $y = -2.50$ cm. The total charge distributed uniformly along the line is -9.00 nC. a) Find the electric field (magnitude and direction) on the x-axis at $x = 10.0$ cm. b) Is the magnitude of the electric field you calculated in part (a) larger or smaller than the electric field 10.0 cm from a point charge that has the same total charge as this finite line of charge? In terms of the approximation used to derive $E = Q/4\pi\epsilon_0 x^2$ for a point charge from Eq. (21.9), explain why this is so. c) At what distance x does the result for the finite line of charge differ by 1.0% from that for the point charge?

21.90 A uniformly charged disk like that in Fig. 21.23 has radius 2.50 cm and carries a total charge of 4.0×10^{-12} C. a) Find the electric field (magnitude and direction) on the x-axis at $x = 0.20$ cm. b) Is the magnitude of the electric field you calculated in part (a) larger or smaller than the electric field 0.20 cm from an infinite sheet of charge with the same charge per unit area as the disk? In terms of the approximation used to derive Eq. (21.12) from Eq. (21.11), explain why this is so. c) What is the percent difference between the electric field produced by the finite disk and by an infinite sheet with the same charge per unit area at i) $x = 0.20$ cm? ii) $x = 0.40$ cm?

21.91 A uniformly charged disk like that in Fig. 21.23 has radius 2.50 cm and carries a total charge of 4.0×10^{-12} C. a) Find the electric field (magnitude and direction) on the x-axis at $x = 20.0$ cm. b) Show that for $x \gg R$, Eq. (21.11) becomes $E = Q/4\pi\epsilon_0 x^2$, where Q is the total charge on the disk. c) Is the magnitude of the electric field you calculated in part (a) larger or smaller than the electric field 20.0 cm from a point charge that has the same total charge as this disk? In terms of the approximation used in part (b) to derive $E = Q/4\pi\epsilon_0 x^2$ for a point charge from Eq. (21.11), explain why this is so. d) What is the percent difference between the electric field produced by the finite disk and by a point charge with the same charge at $x = 20.0$ cm and at $x = 10.0$ cm?

21.92 a) Let $f(x)$ be an even function of x so that $f(x) = f(-x)$. Show that $\int_{-a}^{a} f(x)dx = 2\int_0^a f(x)dx$. (*Hint:* Write the integral from $-a$ to a as the sum of the integral from $-a$ to 0 and the integral from 0 to a. In the first integral, make the change of variable $x' = -x$.) b) Let $g(x)$ be an odd function of x so that $g(x) = -g(-x)$. Use the method given in the hint for part (a) to show that $\int_{-a}^{a} g(x)dx = 0$. c) Use the result of part (b) to show why E_y in Example 21.11 (Section 21.5) is zero.

21.93 Positive charge $+Q$ is distributed uniformly along the $+x$-axis from $x = 0$ to $x = a$. Negative charge $-Q$ is distributed uniformly along the $-x$-axis from $x = 0$ to $x = -a$. a) A positive point charge q lies on the positive y-axis, a distance y from the origin. a) Find the force (magnitude and direction) that the positive and negative charge distributions together exert on q. Show that this force is proportional to y^{-3} for $y \gg a$. b) Suppose instead that the positive point charge q lies on the positive x-axis, a distance $x > a$ from the origin. Find the force (magnitude and direction) that the charge distribution exerts on q. Show that this force is proportional to x^{-3} for $x \gg a$.

21.94 Positive charge Q is uniformly distributed around a semicircle of radius a (Fig. 21.39). Find the electric field (magnitude and direction) at the center of curvature P.

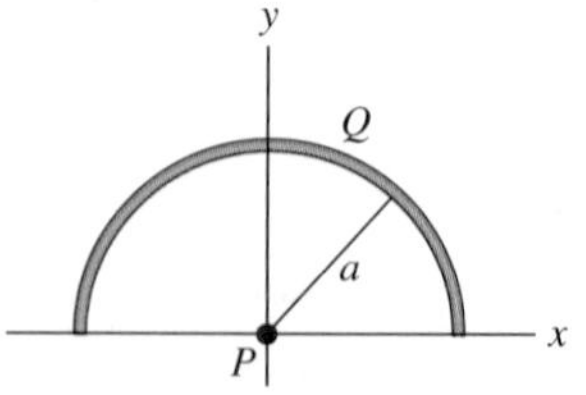

Figure 21.39 Problem 21.94.

21.95 Negative charge $-Q$ is distributed uniformly around a quarter-circle of radius a that lies in the first quadrant, with the center of curvature at the origin. Find the x- and y-components of the net electric field at the origin.

21.96 A small sphere with mass m carries a positive charge q and is attached to one end of a silk fiber of length L. The other end of the fiber is attached to a large vertical insulating sheet that has a positive surface charge density σ. Show that when the sphere is in equilibrium, the fiber makes an angle equal to $\arctan(q\sigma/2mg\epsilon_0)$ with the vertical sheet.

21.97 The imaging drum of a photocopier is positively charged to attract negatively charged toner particles. Near the surface of the drum, its electric field has magnitude 1.40×10^5 N/C. A toner particle is to be attracted to the drum with a force that is ten times the weight of the particle. a) What must be the ratio of the mass of a toner particle to the magnitude of its net charge? b) If the toner particles are made of carbon (atomic number 6, atomic mass 12.0 g/mol), how many carbon atoms are there for each excess electron on a toner particle?

21.98 Electric charge is distributed uniformly along each side of a square. Two adjacent sides have positive charge with total charge $+Q$ on each. a) If the other two sides have negative charge with total charge $-Q$ on each (Fig. 21.40), what are the x- and y-components of the net electric field at the center of the square? Each side of the square has length a. b) Repeat the calculation of part (a) if all four sides have positive charge $+Q$.

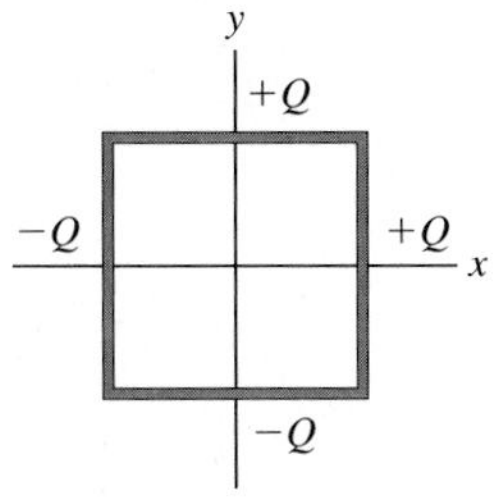

Figure 21.40 Problem 21.98.

21.99 Three large, parallel, insulating sheets have surface charge densities $+0.0200\ \text{C/m}^2$, $+0.0100\ \text{C/m}^2$, and $-0.0200\ \text{C/m}^2$, respectively (Fig. 21.41). Adjacent sheets are a distance of 0.300 m from each other. Calculate the net electric field (magnitude and direction) due to all three sheets at a) point P (0.150 m to the left of sheet I); b) point R (midway between sheets I and II); c) point S (midway between sheets II and III); d) point T (0.150 m to the right of sheet III).

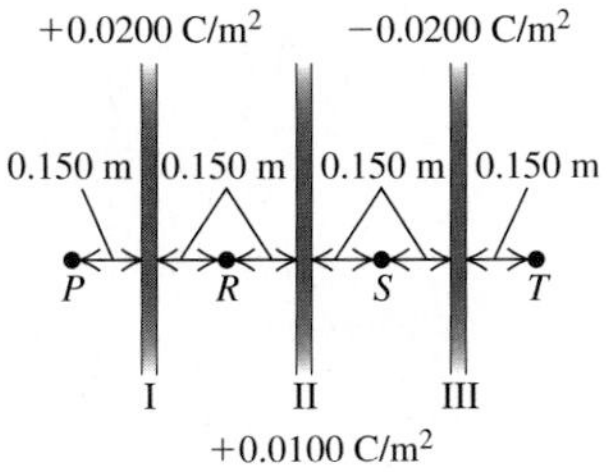

Figure 21.41 Problems 21.99 and 21.100.

21.100 For the situation described in Problem 21.99 (Fig. 21.41), find the force per unit area (magnitude and direction) exerted on each of sheets I, II, and III by the other two sheets.

21.101 An infinite sheet with positive charge per unit area σ lies in the xy-plane. A second infinite sheet with negative charge per unit area $-\sigma$ lies in the yz-plane. Find the net electric field at all points that do not lie in either of these planes. Express your answer in terms of the unit vectors $\hat{\imath}$, $\hat{\jmath}$, and $\hat{k}$.

21.102 A thin disk with a circular hole at its center, called an *annulus,* has inner radius R_1 and outer radius R_2 (Fig. 21.42). The disk has a uniform positive surface charge density σ on its surface. a) Determine the total electric charge on the annulus. b) The annulus lies in the yz-plane, with its center at the origin. For an arbitrary point on the x-axis (the axis of the annulus), find the magnitude and direction of the electric field $\vec{E}$. Consider points both above and below the annulus in Fig. 21.42. c) Show that at points on the x-axis that are sufficiently close to the origin, the magnitude of the electric field is approximately proportional to the distance between the center of the annulus and the point. How close is "sufficiently close"? d) A point particle with mass m and negative charge $-q$ is free to move along the x-axis (but cannot move off the axis). The particle is originally placed at rest at $x = 0.01R_1$ and released. Find the frequency of oscillation of the particle. (*Hint:* Review Section 13.2. The annulus is held stationary.)

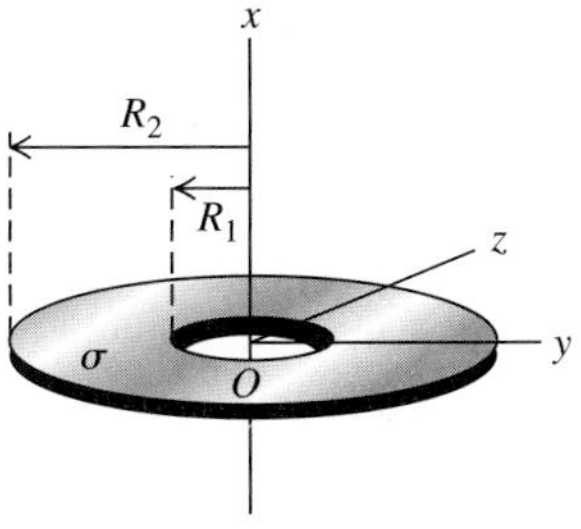

Figure 21.42 Problem 21.102.

Challenge Problems

21.103 Three charges are placed as shown in Fig. 21.43. The magnitude of q_1 is 2.00 μC, but its sign and the value of the charge q_2 are not known. Charge q_3 is $+4.00\ \mu$C, and the net force $\vec{F}$ on q_3 is entirely in the negative x-direction a) Considering the different possible signs of q_1 and q_2, there are four possible force diagrams representing the forces $\vec{F}_1$ and $\vec{F}_2$ that q_1 and q_2 exert on q_3. Sketch these four possible force configurations. b) Using the sketches from part (a) and the direction of $\vec{F}$, deduce the signs of the charges q_1 and q_2. c) Calculate the magnitude of q_2. d) Determine F, the magnitude of the net force on q_3.

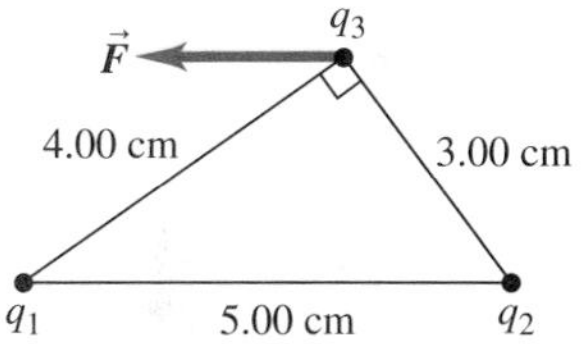

Figure 21.43 Challenge Problem 21.103.

21.104 Two charges are placed as shown in Fig. 21.44. The magnitude of q_1 is 3.00 μC, but its sign and the value of the charge q_2 are not known. The direction of the net electric field $\vec{E}$ at point P is entirely in the negative y-direction a) Considering the different possible signs of q_1 and q_2, there are four possible diagrams that could represent the electric fields $\vec{E}_1$ and $\vec{E}_2$ produced by q_1 and q_2. Sketch the four possible electric field configurations. b) Using the sketches from part (a) and the direction of $\vec{E}$, deduce the signs of q_1 and q_2. c) Determine the magnitude of $\vec{E}$.

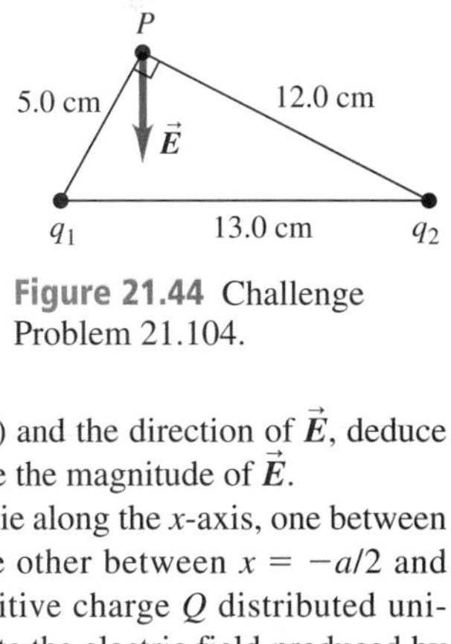

Figure 21.44 Challenge Problem 21.104.

21.105 Two thin rods of length L lie along the x-axis, one between $x = a/2$ and $x = a/2 + L$ and the other between $x = -a/2$ and $x = -a/2 - L$. Each rod has positive charge Q distributed uniformly along its length. a) Calculate the electric field produced by the second rod at points along the positive x-axis. b) Show that the magnitude of the force that one rod exerts on the other is

$$F = \frac{Q^2}{4\pi\epsilon_0 L^2}\ln\left[\frac{(a + L)^2}{a(a + 2L)}\right]$$

c) Show that if $a \gg L$, the magnitude of this force reduces to $F = Q^2/4\pi\epsilon_0 a^2$. (*Hint:* Use the expansion $\ln(1 + z) = z - z^2/2 + z^3/3 - \cdots$, valid for $|z| \ll 1$. Carry *all* expansions to at least order L^2/a^2.) Interpret this result.

22 GAUSS'S LAW

Gauss's law is another way to describe how electric charges and fields behave. One consequence of this law is that static charges on a conductor are found on the conductor's surface, not its interior. That's why this child acquires an electric charge by touching the charged metal sphere. The charged hairs on the child's head repel and stand out.

? If the child stands *inside* a large, charged metal sphere, will her hair stand on end?

Often, there's both an easy way and a hard way to do a job; the easy way may involve nothing more than using the right tools. In physics, an important tool for simplifying problems is the use of *symmetry properties* of systems. Many physical systems have symmetry; for example, a cylindrical body doesn't look any different after you've rotated it around its axis, and a charged metal sphere looks just the same after you've turned it about any axis through its center.

Gauss's law is part of the key to using symmetry considerations to simplify electric-field calculations. For example, the field of a straight-line or plane-sheet charge distribution, which we derived in Section 21.5 using some fairly strenuous integrations, can be obtained in a few lines with the help of Gauss's law. But Gauss's law is more than just a way to make certain calculations easier. Indeed, it is a fundamental statement about the relationship between electric charges and electric fields. Among other things, Gauss's law can help us understand how electric charge distributes itself over conducting bodies.

Here's what Gauss's law is all about. Given any general distribution of charge, we surround it with an imaginary surface that encloses the charge. Then we look at the electric field at various points on this imaginary surface. Gauss's law is a relation between the field at *all* the points on the surface and the total charge enclosed within the surface. This may sound like a rather indirect way of expressing things, but it turns out to be a tremendously useful relationship. Above and

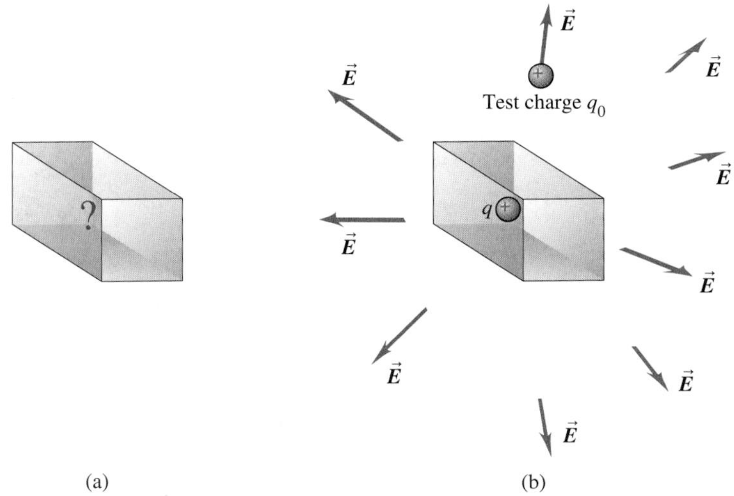

22.1 (a) A box containing an unknown amount of charge. (b) The charge inside the box can be probed by using a test charge q_0 to measure the electric field outside the box.

beyond its use as a calculational tool, Gauss's law can help us gain deeper insights into electric fields. We will make use of these insights repeatedly in the next several chapters as we pursue our study of electromagnetism.

22.1 | Charge and Electric Flux

The discussion of Gauss's law in this section is based on and inspired by the innovative ideas of Ruth W. Chabay and Bruce A. Sherwood (*Electric and Magnetic Interactions,* John Wiley & Sons, 1994).

In Chapter 21 we asked the question, "Given a charge distribution, what is the electric field produced by that distribution at a point P?" We saw that the answer could be found by representing the distribution as an assembly of point charges, each of which produces an electric field $\vec{E}$ given by Eq. (21.7). The total field at P is then the vector sum of the fields due to all the point charges.

But there is an alternative relationship between charge distributions and electric fields. To discover this relationship, let's stand the question of Chapter 21 on its head and ask, "If the electric field pattern is known in a given region, what can we determine about the charge distribution in that region?"

Here's an example. Consider the box shown in Fig. 22.1a, which may or may not contain electric charge. We'll imagine that the box is made of a material that has no effect on any electric fields; it's of the same breed as the massless rope, the frictionless incline, and the free college education. Better still, let the box represent an *imaginary* surface that may or may not enclose some charge. We'll refer to the box as a **closed surface** because it completely encloses a volume. How can you determine how much (if any) electric charge lies within the box?

Knowing that a charge distribution produces an electric field and that an electric field exerts a force on a test charge, you move a test charge q_0 around the vicinity of the box. By measuring the force $\vec{F}$ experienced by the test charge at different positions, you make a three-dimensional map of the electric field $\vec{E} = \vec{F}/q_0$ outside the box. In the case shown in Fig. 22.1b, the map turns out to be the same as that of the electric field produced by a positive point charge (Fig. 21.26a). From the details of the map, you can find the exact value of the point charge inside the box.

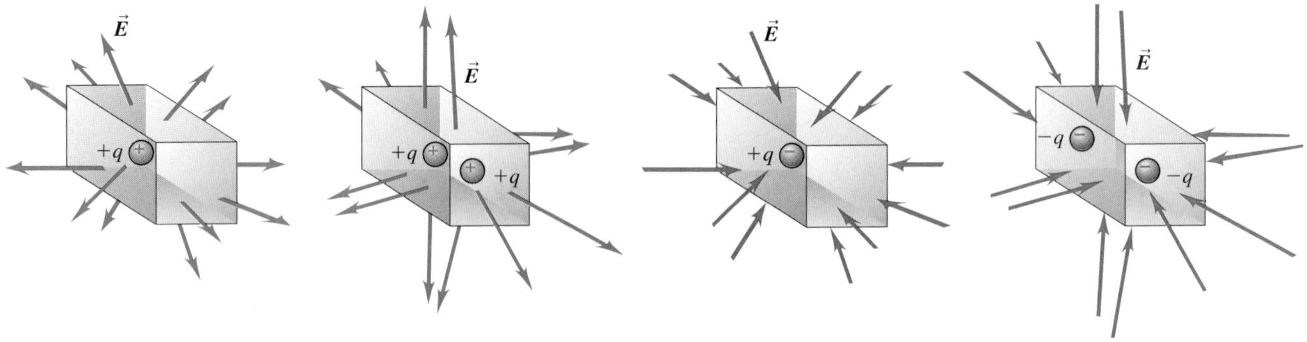

(a) Positive charge inside box, outward flux

(b) Positive charge inside box, outward flux

(c) Negative charge inside box, inward flux

(d) Negative charge inside box, inward flux

22.2 The electric field on the surface of boxes containing (a) a single positive point charge, (b) two positive point charges, (c) a single negative point charge, or (d) two negative point charges.

To determine the contents of the box, we actually need to measure $\vec{E}$ only on the *surface* of the box. In Fig. 22.2a there is a single positive point charge inside the box, and in Fig. 22.2b there are two such charges. The field patterns on the surfaces of the boxes are different in detail, but in both cases the electric field points out of the box. Figures 22.2c and 22.2d show cases with one and two negative point charges, respectively, inside the box. Again, the details of $\vec{E}$ on the surface of the box are different, but in both cases the field points into the box.

In Section 21.4 we mentioned the analogy between electric field vectors and the velocity vectors of a fluid in motion. This analogy can be helpful, even though an electric field does not actually "flow." Using this analogy, in Figs. 22.2a and 22.2b, in which the electric field vectors point out of the surface, we say that there is an outward **electric flux.** (The word *flux* comes from a Latin word meaning "flow.") In Figs. 22.2c and 22.2d the $\vec{E}$ vectors point into the surface, and the electric flux is *inward.*

ActivPhysics ONLINE

11.7 Electric Flux

Figure 22.2 suggests a simple relationship: Positive charge inside the box goes with an outward electric flux through the box's surface, and negative charge inside goes with an inward electric flux. What happens if there is *zero* charge inside the box? In Fig. 22.3a the box is empty and $\vec{E} = \mathbf{0}$ everywhere, so there is no elec-

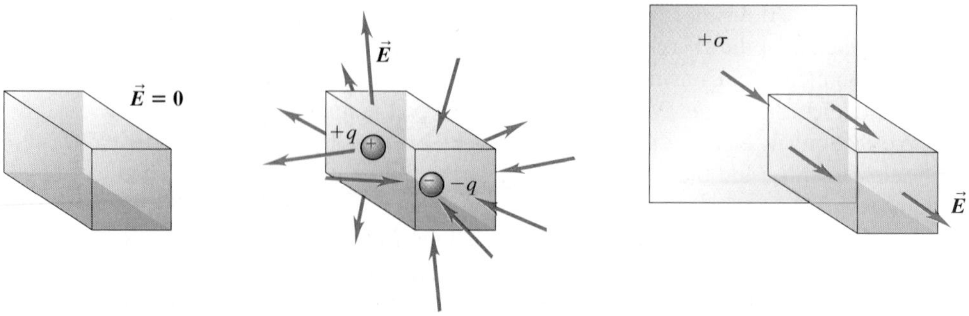

(a) No charge inside box, zero flux

(b) Zero *net* charge inside box, inward flux cancels outward flux

(c) No charge inside box, inward flux cancels outward flux

22.3 Three cases in which there is zero *net* charge inside a box and no net electric flux through the surface of the box. (a) An empty box with $\vec{E} = \mathbf{0}$. (b) A box containing one positive and one equal-magnitude negative point charge. (c) An empty box immersed in a uniform electric field.

tric flux into or out of the box. In Fig. 22.3b, one positive and one negative point charge of equal magnitude are enclosed within the box, so the *net* charge inside the box is zero. There is an electric field, but it "flows into" the box on half of its surface and "flows out of" the box on the other half. Hence there is no *net* electric flux into or out of the box.

The box is again empty in Fig. 22.3c. However, there is charge present *outside* the box; the box has been placed with one end parallel to a uniformly charged infinite sheet, which produces a uniform electric field perpendicular to the sheet (as we learned in Example 21.12 of Section 21.5). On one end of the box, $\vec{E}$ points into the box; on the opposite end, $\vec{E}$ points out of the box; and on the sides, $\vec{E}$ is parallel to the surface and so points neither into nor out of the box. As in Fig. 22.3b, the inward electric flux on one part of the box exactly compensates for the outward electric flux on the other part. So in all of the cases shown in Fig. 22.3, there is no *net* electric flux through the surface of the box, and no *net* charge is enclosed in the box.

Figures 22.2 and 22.3 demonstrate a connection between the *sign* (positive, negative, or zero) of the *net* charge enclosed by a closed surface and the sense (outward, inward, or none) of the net electric flux through the surface. There is also a connection between the *magnitude* of the net charge inside the closed surface and the *strength* of the net "flow" of $\vec{E}$ over the surface. In both Figs. 22.4a and 22.4b there is a single point charge inside the box, but in Fig. 22.4b the magnitude of the charge is twice as great, and so $\vec{E}$ is everywhere twice as great in magnitude as in Fig. 22.4a. Keeping in mind the fluid flow analogy, this means that the net outward electric flux is also twice as great in Fig. 22.4b as in Fig. 22.4a. This suggests that the net electric flux through the surface of the box is *directly proportional* to the magnitude of the net charge enclosed by the box.

This conclusion is independent of the size of the box. In Fig. 22.4c the point charge $+q$ is enclosed by a box with twice the linear dimensions of the box in Fig. 22.4a. The magnitude of the electric field of a point charge decreases with distance according to $1/r^2$, so the average magnitude of $\vec{E}$ on each face of the large box in Fig. 22.4c is just $\frac{1}{4}$ of the average magnitude on the corresponding face in Fig. 22.4a. But each face of the large box has exactly four times the area of the corresponding face of the small box. Hence the outward electric flux is the *same* for the two boxes if we *define* electric flux as follows: For each face of the box, take the product of the average perpendicular component of $\vec{E}$ and the area of that face; then add up the results from all faces of the box. With this definition the net electric flux due to a single point charge inside the box is independent of the size of the box and depends only on the net charge inside the box.

We have seen that there is a relationship between the net amount of charge inside a closed surface and the electric flux through that surface. For the special cases of a closed surface in the shape of a rectangular box and charge distributions made up of point charges or infinite charged sheets, we have found:

1. Whether there is a net outward or inward electric flux through a closed surface depends on the sign of the enclosed charge.
2. Charges *outside* the surface do not give a net electric flux through the surface.
3. The net electric flux is directly proportional to the net amount of charge enclosed within the surface but is otherwise independent of the size of the closed surface.

These observations are a qualitative statement of *Gauss's law*.

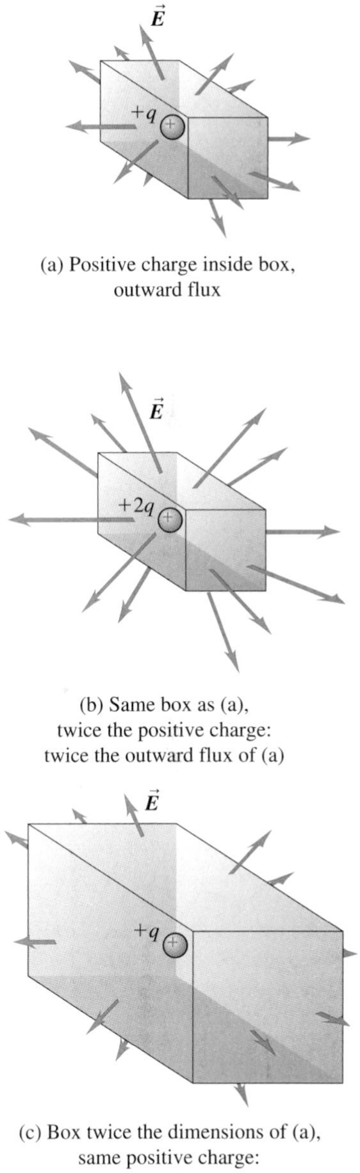

22.4 (a) A box enclosing a positive point charge $+q$. (b) Doubling the charge causes the magnitude of $\vec{E}$ to double, and it doubles the electric flux through the surface. (c) If the charge stays the same but the dimensions of the box are doubled, the flux stays the same. The magnitude of $\vec{E}$ on the surface decreases by a factor of $\frac{1}{4}$, but the area through which $\vec{E}$ "flows" increases by a factor of 4.

Do these observations hold true for other kinds of charge distributions and for closed surfaces of arbitrary shape? The answer to these questions will prove to be "yes." But to explain why this is so, we need a precise mathematical statement of what we mean by electric flux. This is developed in the next section.

Test Your Understanding

If all of the dimensions of the box in Fig. 22.2a are increased by a factor of three, what effect will this change have on the electric flux through the box?

22.2 | Calculating Electric Flux

In the preceding section we introduced the concept of *electric flux.* Qualitatively, the electric flux through a surface is a description of whether the electric field $\vec{E}$ points into or out of the surface. We used this to give a rough qualitative statement of Gauss's law: The net electric flux through a closed surface is directly proportional to the net charge inside that surface. To be able to make full use of this law, we need to know how to *calculate* electric flux. To do this, let's again make use of the analogy between an electric field $\vec{E}$ and the field of velocity vectors $\vec{v}$ in a flowing fluid. (Again, keep in mind that this is only an analogy; an electric field is *not* a flow.)

Figure 22.5 shows a fluid flowing steadily from left to right. Let's examine the volume flow rate dV/dt (in, say, cubic meters per second) through the wire rectangle with area A. When the area is perpendicular to the flow velocity $\vec{v}$ (Fig. 22.5a) and the flow velocity is the same at all points in the fluid, the volume flow rate dV/dt is the area A multiplied by the flow speed v:

$$\frac{dV}{dt} = vA$$

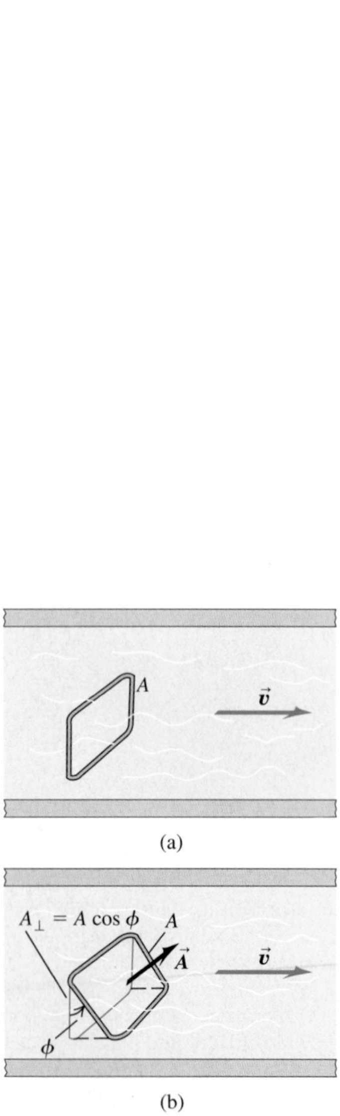

22.5 The volume flow rate of fluid through the wire rectangle (a) is vA when the area of the rectangle is perpendicular to $\vec{v}$ and (b) is $vA\cos\phi$ when the rectangle is tilted at an angle ϕ.

When the rectangle is tilted at an angle ϕ (Fig. 22.5b) so that its face is not perpendicular to $\vec{v}$, the area that counts is the silhouette area that we see when we look in the direction of $\vec{v}$. This area, which is outlined in red and labeled $A_{\perp}$ in Fig. 22.5b, is the *projection* of the area A onto a surface perpendicular to $\vec{v}$. Two sides of the projected rectangle have the same length as the original one, but the other two are foreshortened by a factor of $\cos\phi$, so the projected area $A_{\perp}$ is equal to $A\cos\phi$. Then the volume flow rate through A is

$$\frac{dV}{dt} = vA\cos\phi$$

If $\phi = 90°$, $dV/dt = 0$; the wire rectangle is edge-on to the flow, and no fluid passes through the rectangle.

Also, $v\cos\phi$ is the component of the vector $\vec{v}$ perpendicular to the plane of the area A. Calling this component $v_{\perp}$, we can rewrite the volume flow rate as

$$\frac{dV}{dt} = v_{\perp}A$$

We can express the volume flow rate more compactly by using the concept of *vector area* $\vec{A}$, a vector quantity with magnitude A and a direction perpendicular to the plane of the area we are describing. The vector area $\vec{A}$ describes both the size of an area and its orientation in space. In terms of $\vec{A}$, we can write the volume flow rate of fluid through the rectangle in Fig. 22.5b as a scalar (dot) product:

$$\frac{dV}{dt} = \vec{v} \cdot \vec{A}$$

Using the analogy between electric field and fluid flow, we now define electric flux in the same way as we have just defined the volume flow rate of a fluid; we simply replace the fluid velocity $\vec{v}$ by the electric field $\vec{E}$. The symbol that we use for electric flux is Φ_E (the capital Greek letter "phi"; the subscript E is a reminder that this is *electric* flux). Consider first a flat area A perpendicular to a uniform electric field $\vec{E}$ (Fig. 22.6a). We define the electric flux through this area to be the product of the field magnitude E and the area A:

$$\Phi_E = EA$$

Roughly speaking, we can picture Φ_E in terms of the field lines passing through A. Increasing the area means that more lines of $\vec{E}$ pass through the area, increasing the flux; stronger field means more closely spaced lines of $\vec{E}$ and therefore more lines per unit area, so again the flux increases.

If the area A is flat but not perpendicular to the field $\vec{E}$, then fewer field lines pass through it. In this case the area that counts is the silhouette area that we see when looking in the direction of $\vec{E}$. This is the area $A_\perp$ in Fig. 22.6b and is equal to $A \cos\phi$ (compare to Fig. 22.5b). We generalize our definition of electric flux for a uniform electric field to

$$\Phi_E = EA\cos\phi \qquad \text{(electric flux for uniform } \vec{E}\text{, flat surface)} \qquad (22.1)$$

Since $E\cos\phi$ is the component of $\vec{E}$ perpendicular to the area, we can rewrite Eq. (22.1) as

$$\Phi_E = E_\perp A \qquad \text{(electric flux for uniform } \vec{E}\text{, flat surface)} \qquad (22.2)$$

In terms of the vector area $\vec{A}$ perpendicular to the area, we can write the electric flux as the scalar product of $\vec{E}$ and $\vec{A}$:

$$\Phi_E = \vec{E} \cdot \vec{A} \qquad \text{(electric flux for uniform } \vec{E}\text{, flat surface)} \qquad (22.3)$$

Equations (22.1), (22.2), and (22.3) express the electric flux for a *flat* surface and a *uniform* electric field in different but equivalent ways. The SI unit for electric flux is $1\ \text{N}\cdot\text{m}^2/\text{C}$. Note that if the area is edge-on to the field, $\vec{E}$ and $\vec{A}$ are perpendicular and the flux is zero (Fig. 22.6c).

We can represent the direction of a vector area $\vec{A}$ by using a *unit vector* $\hat{n}$ perpendicular to the area; $\hat{n}$ stands for "normal." Then

$$\vec{A} = A\hat{n} \qquad (22.4)$$

A surface has two sides, so there are two possible directions for $\hat{n}$ and $\vec{A}$. We must always specify which direction we choose. In Section 22.1 we related the charge inside a *closed* surface to the electric flux through the surface. With a closed surface we will always choose the direction of $\hat{n}$ to be *outward,* and we will speak of the flux *out of* a closed surface. Thus what we called "outward electric flux" in Section 22.1 corresponds to a *positive* value of Φ_E, and what we called "inward electric flux" corresponds to a *negative* value of Φ_E.

What happens if the electric field $\vec{E}$ isn't uniform but varies from point to point over the area A? Or what if A is part of a curved surface? Then we divide A into

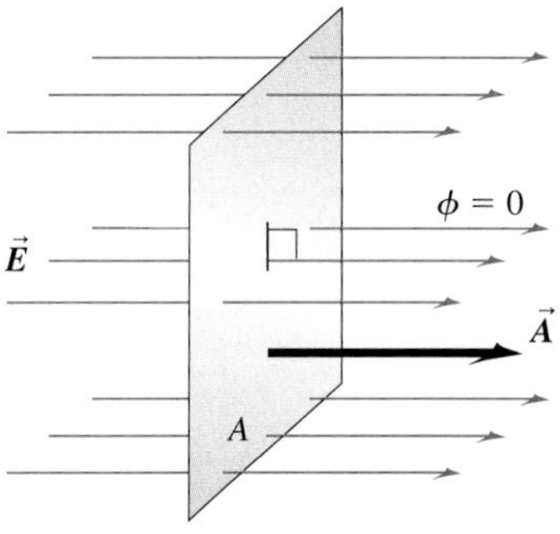

(a) Surface face-on to electric field
$\vec{E}$ and $\vec{A}$ parallel
angle between $\vec{E}$ and $\vec{A}$ is $\phi = 0$
flux $\Phi_E = \vec{E} \cdot \vec{A} = EA$

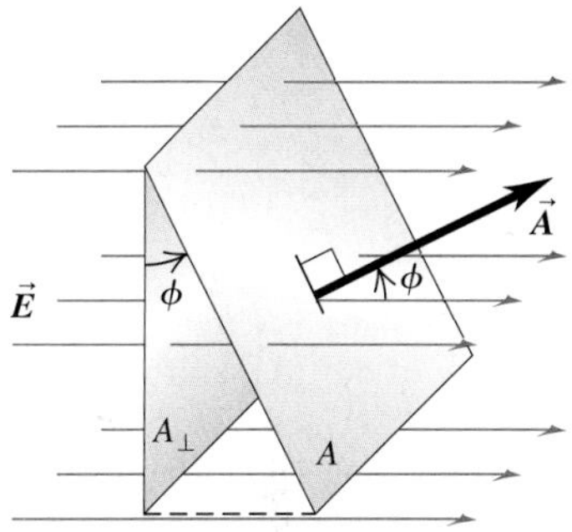

(b) Surface tilted from face-on orientation by an angle ϕ
angle between $\vec{E}$ and $\vec{A}$ is ϕ
flux $\Phi_E = \vec{E} \cdot \vec{A} = EA\cos\phi$

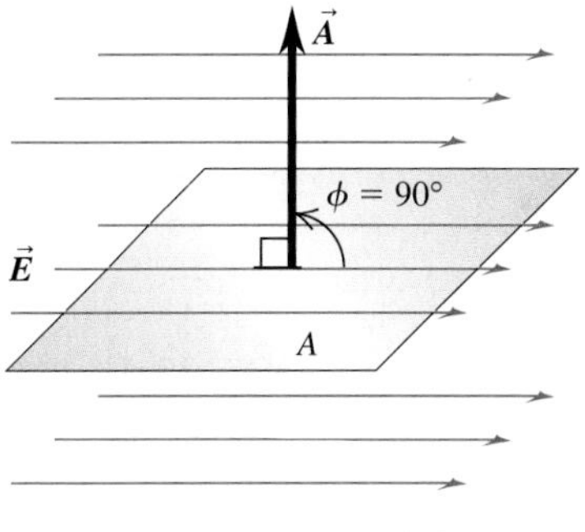

(c) Surface edge-on to electric field
$\vec{E}$ and $\vec{A}$ perpendicular angle between $\vec{E}$ and $\vec{A}$ is $\phi = 90°$
flux $\Phi_E = \vec{E} \cdot \vec{A} = EA\cos 90° = 0$

22.6 A flat surface in a uniform electric field. The electric flux Φ_E through the surface equals the scalar product of the electric field $\vec{E}$ and the area vector $\vec{A}$.

many small elements dA, each of which has a unit vector $\hat{\boldsymbol{n}}$ perpendicular to it and a vector area $d\vec{A} = \hat{\boldsymbol{n}}\, dA$. We calculate the electric flux through each element and integrate the results to obtain the total flux:

$$\Phi_E = \int E \cos\phi \, dA = \int E_\perp \, dA = \int \vec{E} \cdot d\vec{A} \qquad (22.5)$$

(general definition of electric flux)

We call this integral the **surface integral** of the component $E_\perp$ over the area, or the surface integral of $\vec{E} \cdot d\vec{A}$. The various forms of the integral all express the same thing in different terms. In specific problems, one form is sometimes more convenient than another. Example 22.3 at the end of this section illustrates the use of Eq. (22.5).

In Eq. (22.5) the electric flux $\int E_\perp \, dA$ is just equal to the *average* value of the perpendicular component of the electric field, multiplied by the area of the surface. This is the same definition of electric flux that we were led to in Section 22.2, now expressed more mathematically. In the next section we will see the connection between the total electric flux through *any* closed surface, no matter what its shape, and the amount of charge enclosed within that surface.

Example 22.1 Electric flux through a disk

A disk with radius 0.10 m is oriented with its normal unit vector $\hat{\boldsymbol{n}}$ at an angle of 30° to a uniform electric field $\vec{E}$ with magnitude 2.0×10^3 N/C (Fig. 22.7). (Since this isn't a closed surface, it has no "inside" or "outside." That's why we have to specify the direction of $\hat{\boldsymbol{n}}$ in the figure.) a) What is the electric flux through the disk? b) What is the flux through the disk if it is turned so that its normal is perpendicular to $\vec{E}$? c) What is the flux through the disk if its normal is parallel to $\vec{E}$?

SOLUTION

IDENTIFY: This problem is about a flat surface in a uniform electric field, so we can apply the ideas of this section.

SET UP: The orientation of the disk is like that of the rectangle in Fig. 22.6b. We calculate the electric flux using Eq. (22.1).

EXECUTE: a) The area is $A = \pi(0.10 \text{ m})^2 = 0.0314 \text{ m}^2$ and the angle between $\vec{E}$ and $\vec{A} = A\hat{\boldsymbol{n}}$ is $\phi = 30°$, so

$$\Phi_E = EA\cos\phi = (2.0 \times 10^3 \text{ N/C})(0.0314 \text{ m}^2)(\cos 30°)$$
$$= 54 \text{ N}\cdot\text{m}^2/\text{C}$$

b) The normal to the disk is now perpendicular to $\vec{E}$, so $\phi = 90°$, $\cos\phi = 0$, and $\Phi_E = 0$. There is no flux through the disk.

c) The normal to the disk is parallel to $\vec{E}$, so $\phi = 0$, $\cos\phi = 1$, and the flux has its maximum possible value. From Eq. (22.1),

$$\Phi_E = EA\cos\phi = (2.0 \times 10^3 \text{ N/C})(0.0314 \text{ m}^2)(1)$$
$$= 63 \text{ N}\cdot\text{m}^2/\text{C}$$

EVALUATE: As a check on our results, note that the answer to part (a) is smaller than the answer to part (c). Is this as it should be?

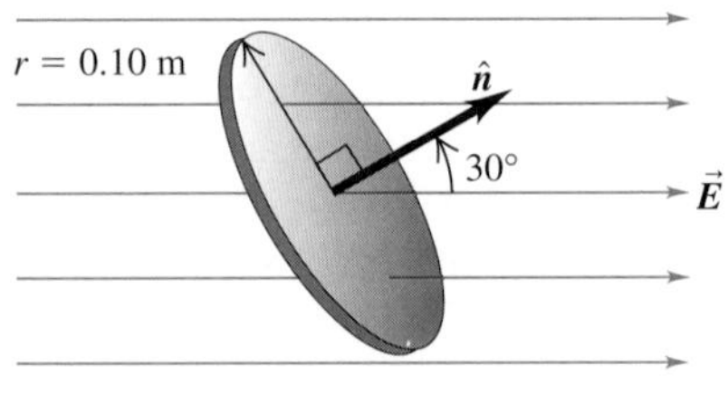

22.7 The electric flux Φ_E through a disk depends on the angle between its normal $\hat{\boldsymbol{n}}$ and the electric field $\vec{E}$.

Example 22.2 Electric flux through a cube

A cube of side L is placed in a region of uniform electric field $\vec{E}$. Find the electric flux through each face of the cube and the total flux through the cube when a) it is oriented with two of its faces perpendicular to the field $\vec{E}$, as in Fig. 22.8a; b) when the cube is turned by an angle θ, as in Fig. 22.8b.

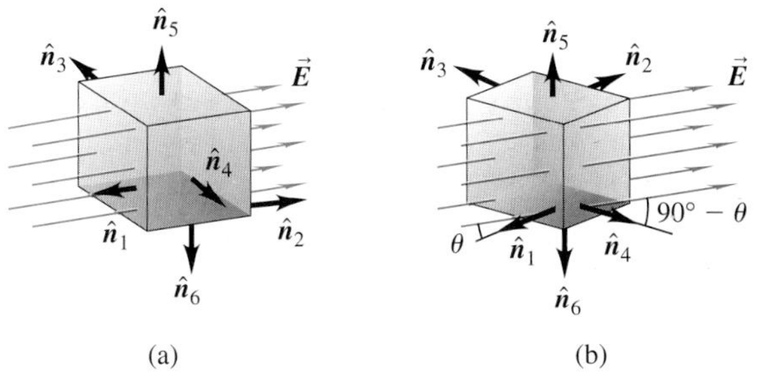

22.8 Electric flux of a uniform field $\vec{E}$ through a cubical box of side L in two orientations.

SOLUTION

IDENTIFY and SET UP: Since $\vec{E}$ is uniform and each of the six faces of the cube is a flat surface, we find the flux through each face using Eqs. (22.3) and (22.4). We then calculate the total flux through the cube by adding the six individual fluxes.

EXECUTE: a) The unit vectors for each face ($\hat{n}_1$ through $\hat{n}_6$) are shown in the figure; the direction of each unit vector is *outward* from the closed surface of the cube. The angle between $\vec{E}$ and $\hat{n}_1$ is 180°; the angle between $\vec{E}$ and $\hat{n}_2$ is 0°; and the angle between $\vec{E}$ and each of the other four unit vectors is 90°. Each face of the cube has area L^2, so the fluxes through each of the faces are

$$\Phi_{E1} = \vec{E}\cdot\hat{n}_1 A = EL^2\cos 180° = -EL^2$$

$$\Phi_{E2} = \vec{E}\cdot\hat{n}_2 A = EL^2\cos 0° = +EL^2$$

$$\Phi_{E3} = \Phi_{E4} = \Phi_{E5} = \Phi_{E6} = EL^2\cos 90° = 0$$

The flux is negative on face 1, where $\vec{E}$ is directed into the cube, and positive on face 2, where $\vec{E}$ is directed out of the cube. The *total* flux through the cube is the sum of the fluxes through the six faces:

$$\Phi_E = \Phi_{E1} + \Phi_{E2} + \Phi_{E3} + \Phi_{E4} + \Phi_{E5} + \Phi_{E6}$$
$$= -EL^2 + EL^2 + 0 + 0 + 0 + 0 = 0$$

b) The fluxes through faces 1 and 3 are negative, since $\vec{E}$ is directed into those faces; the field is directed out of faces 2 and 4, so the fluxes through those faces are positive. We find

$$\Phi_{E1} = \vec{E}\cdot\hat{n}_1 A = EL^2\cos(180° - \theta) = -EL^2\cos\theta$$

$$\Phi_{E2} = \vec{E}\cdot\hat{n}_2 A = +EL^2\cos\theta$$

$$\Phi_{E3} = \vec{E}\cdot\hat{n}_3 A = EL^2\cos(90° + \theta) = -EL^2\sin\theta$$

$$\Phi_{E4} = \vec{E}\cdot\hat{n}_4 A = EL^2\cos(90° - \theta) = +EL^2\sin\theta$$

$$\Phi_{E5} = \Phi_{E6} = EL^2\cos 90° = 0$$

The total flux $\Phi_E = \Phi_{E1} + \Phi_{E2} + \Phi_{E3} + \Phi_{E4} + \Phi_{E5} + \Phi_{E6}$ through the surface of the cube is again zero.

EVALUATE: It's no surprise that the total flux is zero for both orientations. We came to this same conclusion in our discussion of Fig. 22.3c in Section 22.1. There we observed that there was zero net flux of a uniform electric field through a closed surface that contains no electric charge.

Example 22.3 Electric flux through a sphere

A positive point charge $q = 3.0\ \mu\text{C}$ is surrounded by a sphere with radius 0.20 m centered on the charge (Fig. 22.9). Find the electric flux through the sphere due to this charge.

SOLUTION

IDENTIFY: Here the surface is not flat and the electric field is not uniform, so we must use the general definition of electric flux.

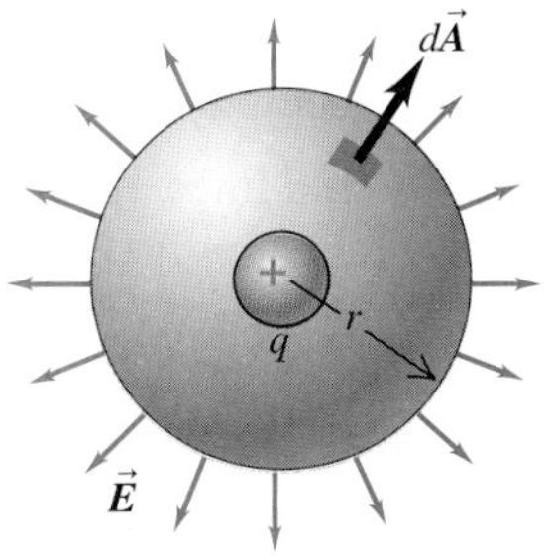

22.9 Electric flux through a sphere centered on a point charge.

SET UP: We use Eq. (22.5) to calculate the electric flux (our target variable). Because the sphere is centered on the point charge, at any point on the spherical surface, $\vec{E}$ is directed out of the sphere perpendicular to the surface. The positive direction for both $\hat{n}$ and $E_\perp$ is outward, so $E_\perp = E$ and the flux through a surface element dA is $\vec{E}\cdot d\vec{A} = E\,dA$. This greatly simplifies the integral in Eq. (22.5).

EXECUTE: At any point on the sphere the magnitude of $\vec{E}$ is

$$E = \frac{q}{4\pi\epsilon_0 r^2} = (9.0\times 10^9\ \text{N}\cdot\text{m}^2/\text{C}^2)\frac{3.0\times 10^{-6}\ \text{C}}{(0.20\ \text{m})^2}$$
$$= 6.75\times 10^5\ \text{N/C}$$

Because E is the same at every point, it can be taken outside the integral $\Phi_E = \int E\,dA$ in Eq. (22.5). What remains is the integral $\int dA$, which is just the total area $A = 4\pi r^2$ of the spherical surface. Thus the total flux out of the sphere is

$$\Phi_E = EA = (6.75\times 10^5\ \text{N/C})(4\pi)(0.20\ \text{m})^2$$
$$= 3.4\times 10^5\ \text{N}\cdot\text{m}^2/\text{C}$$

EVALUATE: Notice that we divided by $r^2 = (0.20\text{ m})^2$ to find E, then multiplied by $r^2 = (0.20\text{ m})^2$ to find Φ_E; hence the radius r of the sphere cancels out of the result for Φ_E. We would have obtained the same flux with a sphere of radius 2.0 m or 200 m. We came to essentially the same conclusion in our discussion of Fig. 22.4 in Section 22.1, where we considered rectangular closed surfaces of two different sizes enclosing a point charge. There we found that the flux of $\vec{E}$ was independent of the size of the surface; the same result holds true for a spherical surface. Indeed, the flux through *any* surface enclosing a single point charge is independent of the shape or size of the surface, as we'll soon see.

22.10 These solar panels are tilted to face the sun. How is the tilt related to the definition of electric flux?

Test Your Understanding

To generate the maximum amount of electric power, solar panels are installed so that they are as nearly face-on to the sun as possible (Fig. 22.10). Explain how this orientation is analogous to maximizing the electric flux through a plane surface.

22.3 | Gauss's Law

Gauss's law is an alternative to Coulomb's law. While completely equivalent to Coulomb's law, Gauss's law provides a different way to express the relationship between electric charge and electric field. It was formulated by Carl Friedrich Gauss (1777–1855), one of the greatest mathematicians of all time. Many areas of mathematics bear the mark of his influence, and he made equally significant contributions to theoretical physics (Fig. 22.11).

Gauss's law states that the total electric flux through any closed surface (a surface enclosing a definite volume) is proportional to the total (net) electric charge inside the surface. In Section 22.1 we observed this relationship qualitatively for certain special cases; now we'll develop it more rigorously. We'll start with the field of a single positive point charge q. The field lines radiate out equally in all directions. We place this charge at the center of an imaginary spherical surface with radius R. The magnitude E of the electric field at every point on the surface is given by

$$E = \frac{1}{4\pi\epsilon_0}\frac{q}{R^2}$$

At each point on the surface, $\vec{E}$ is perpendicular to the surface, and its magnitude is the same at every point, just as in Example 22.3 (Section 22.2). The total electric flux is just the product of the field magnitude E and the total area $A = 4\pi R^2$ of the sphere:

$$\Phi_E = EA = \frac{1}{4\pi\epsilon_0}\frac{q}{R^2}(4\pi R^2) = \frac{q}{\epsilon_0} \tag{22.6}$$

22.11 Carl Friedrich Gauss helped develop several branches of mathematics, including differential geometry, real analysis, and number theory. The "bell curve" of statistics is one of his inventions. Gauss also made state-of-the-art investigations of the earth's magnetism and calculated the orbit of the first asteroid to be discovered.

The flux is independent of the radius R of the sphere. It depends only on the charge q enclosed by the sphere.

We can also interpret this result in terms of field lines. Figure 22.12 shows two spheres with radii R and $2R$, respectively, centered on the point charge q. Every field line that passes through the smaller sphere also passes through the larger sphere, so the total flux through each sphere is the same.

What is true of the entire sphere is also true of any portion of its surface. In Fig. 22.12 an area dA is outlined on a sphere of radius R and then projected onto the sphere of radius $2R$ by drawing lines from the center through points on the boundary of dA. The area projected on the larger sphere is clearly $4\,dA$. But since the electric field due to a point charge is inversely proportional to r^2, the field magnitude is $\frac{1}{4}$ as great on the sphere of radius $2R$ as on the sphere of radius R. Hence the electric flux is the same for both areas and is independent of the radius of the sphere.

This projection technique shows us how to extend this discussion to nonspherical surfaces. Instead of a second sphere, let us surround the sphere of radius R by a surface of irregular shape, as in Fig. 22.13a. Consider a small element of area dA on the irregular surface; we note that this area is *larger* than the corresponding element on a spherical surface at the same distance from q. If a normal to dA makes an angle ϕ with a radial line from q, two sides of the area projected onto the spherical surface are foreshortened by a factor $\cos\phi$ (Fig. 22.13b). The other two sides are unchanged. Thus the electric flux through the spherical surface element is equal to the flux $E\,dA\cos\phi$ through the corresponding irregular surface element.

We can divide the entire irregular surface into elements dA, compute the electric flux $E\,dA\cos\phi$ for each, and sum the results by integrating, as in Eq. (22.5). Each of the area elements projects onto a corresponding spherical surface element. Thus the *total* electric flux through the irregular surface, given by any of the forms of Eq. (22.5), must be the same as the total flux through a sphere, which Eq. (22.6) shows is equal to q/ϵ_0. Thus, for the irregular surface,

$$\Phi_E = \oint \vec{E}\cdot d\vec{A} = \frac{q}{\epsilon_0} \tag{22.7}$$

Equation (22.7) holds for a surface of *any* shape or size, provided only that it is a *closed* surface enclosing the charge q. The circle on the integral sign reminds us that the integral is always taken over a *closed* surface.

The area elements $d\vec{A}$ and the corresponding unit vectors $\hat{n}$ always point *out of* the volume enclosed by the surface. The electric flux is then positive in areas where the electric field points out of the surface and negative where it points inward. Also, $E_\perp$ is positive at points where $\vec{E}$ points out of the surface and negative at points where $\vec{E}$ points into the surface.

If the point charge in Fig. 22.13 is negative, the $\vec{E}$ field is directed radially *inward;* the angle ϕ is then greater than 90°, its cosine is negative, and the integral in Eq. (22.7) is negative. But since q is also negative, Eq. (22.7) still holds.

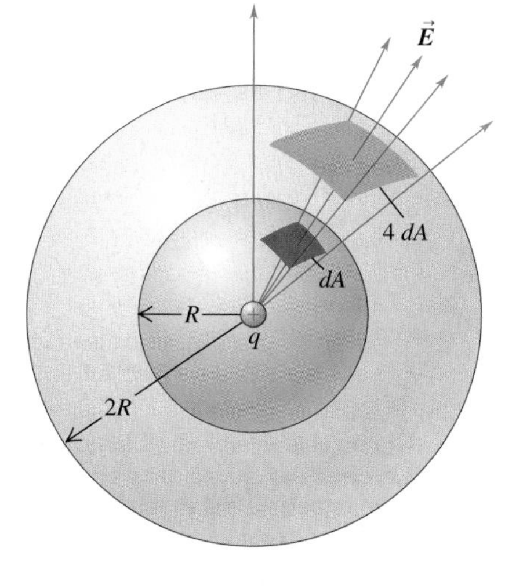

22.12 Projection of an element of area dA of a sphere of radius R onto a concentric sphere of radius $2R$. The projection multiplies each linear dimension by 2, so the area element on the larger sphere is $4\,dA$. The same number of field lines and the same flux pass through each area element.

Activ ONLINE Physics

11.8 Gauss's Law

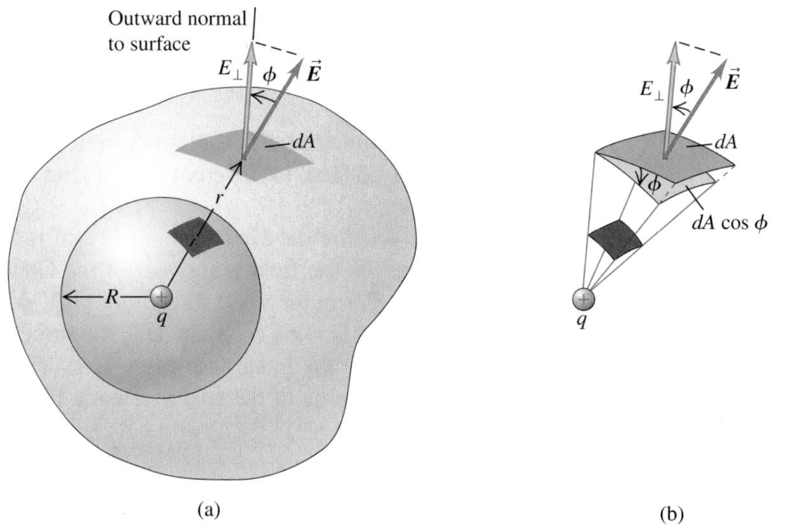

(a) (b)

22.13 (a) The outward normal to the surface makes an angle ϕ with the direction of $\vec{E}$. (b) The projection of the area element dA onto the spherical surface is $dA\cos\phi$.

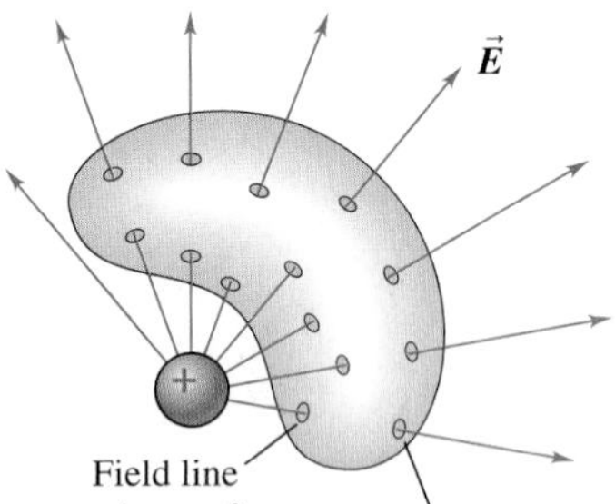

22.14 A point charge *outside* a closed surface that encloses no charge. If an electric field line from the external charge enters the surface at one point, it must leave at another.

For a closed surface enclosing *no* charge,

$$\Phi_E = \oint \vec{E} \cdot d\vec{A} = 0$$

This is the mathematical statement that when a region contains no charge, any field lines caused by charges *outside* the region that enter on one side must leave again on the other side. (In Section 22.1 we came to the same conclusion by considering the special case of a rectangular box in a uniform field.) Figure 22.14 illustrates this point. *Electric field lines can begin or end inside a region of space only when there is charge in that region.*

Now comes the final step in obtaining the general form of Gauss's law. Suppose the surface encloses not just one point charge q but several charges q_1, q_2, q_3, ... The total (resultant) electric field $\vec{E}$ at any point is the vector sum of the $\vec{E}$ fields of the individual charges. Let Q_{encl} be the *total* charge enclosed by the surface: $Q_{encl} = q_1 + q_2 + q_3 + \cdots$. Also let $\vec{E}$ be the *total* field at the position of the surface area element $d\vec{A}$, and let $E_\perp$ be its component perpendicular to the plane of that element (that is, parallel to $d\vec{A}$). Then we can write an equation like Eq. (22.7) for each charge and its corresponding field and add the results. When we do, we obtain the general statement of Gauss's law:

$$\Phi_E = \oint \vec{E} \cdot d\vec{A} = \frac{Q_{encl}}{\epsilon_0} \quad \text{(Gauss's law)} \qquad (22.8)$$

The total electric flux through a closed surface is equal to the total (net) electric charge inside the surface, divided by ϵ_0.

CAUTION Remember that the closed surface in Gauss's law is *imaginary;* there need not be any material object at the position of the surface. We often refer to a closed surface used in Gauss's law as a **Gaussian surface.**

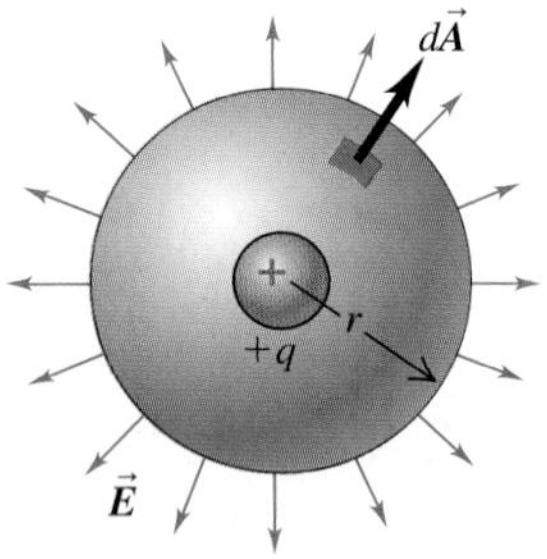

(a) Gaussian surface around positive charge: positive (outward) flux

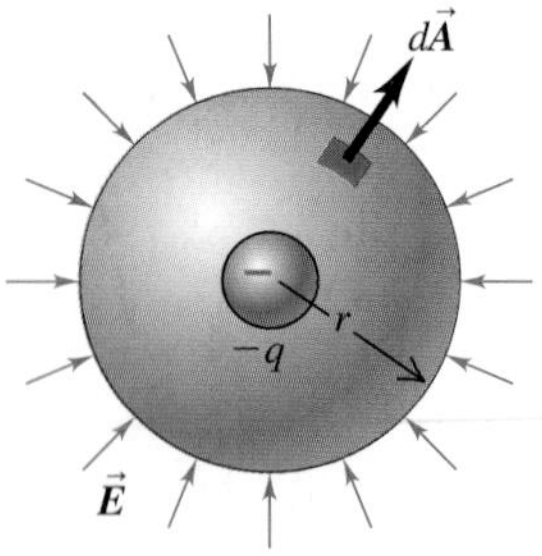

(b) Gaussian surface around negative charge: negative (inward) flux

22.15 Spherical Gaussian surfaces around (a) a positive point charge and (b) a negative point charge.

Using the definition of Q_{encl} and the various ways to express electric flux given in Eq. (22.5), we can express Gauss's law in the following equivalent forms:

$$\Phi_E = \oint E \cos\phi \, dA = \oint E_\perp \, dA = \oint \vec{E} \cdot d\vec{A} = \frac{Q_{encl}}{\epsilon_0} \qquad (22.9)$$

(various forms of Gauss's law)

As in Eq. (22.5), the various forms of the integral all express the same thing, the total electric flux through the Gaussian surface, in different terms. One form is sometimes more convenient than another.

As an example, Fig. 22.15a shows a spherical Gaussian surface of radius r around a positive point charge $+q$. The electric field points out of the Gaussian surface, so at every point on the surface $\vec{E}$ is in the same direction as $d\vec{A}$, $\phi = 0$, and $E_\perp$ is equal to the field magnitude $E = q/4\pi\epsilon_0 r^2$. Since E is the same at all points on the surface, we can take it outside the integral in Eq. (22.9). Then the remaining integral is $\int dA = A = 4\pi r^2$, the area of the sphere. Hence Eq. (22.9) becomes

$$\Phi_E = \oint E_\perp \, dA = \oint \frac{q}{4\pi\epsilon_0 r^2} dA = \frac{q}{4\pi\epsilon_0 r^2} \oint dA = \frac{q}{4\pi\epsilon_0 r^2} 4\pi r^2 = \frac{q}{\epsilon_0}$$

The enclosed charge Q_{encl} is just the charge $+q$, so this agrees with Gauss's law. If the Gaussian surface encloses a *negative* point charge as in Fig. 22.15b, then $\vec{E}$ points *into* the surface at each point in the direction opposite $d\vec{A}$. Then $\phi = 180°$ and $E_\perp$ is equal to the negative of the field magnitude: $E_\perp = -E = -|-q|/4\pi\epsilon_0 r^2 = -q/4\pi\epsilon_0 r^2$. Equation (22.9) then becomes

$$\Phi_E = \oint E_\perp\, dA = \oint\left(\frac{-q}{4\pi\epsilon_0 r^2}\right) dA = \frac{-q}{4\pi\epsilon_0 r^2}\oint dA = \frac{-q}{4\pi\epsilon_0 r^2}4\pi r^2 = \frac{-q}{\epsilon_0}$$

This again agrees with Gauss's law because the enclosed charge in Fig. 22.15b is $Q_{encl} = -q$.

In Eqs. (22.8) and (22.9), Q_{encl} is always the algebraic sum of all the positive and negative charges enclosed by the Gaussian surface, and $\vec{E}$ is the *total* field at each point on the surface. Also note that in general, this field is caused partly by charges inside the surface and partly by charges outside. But as Fig. 22.14 shows, the outside charges do *not* contribute to the total (net) flux through the surface. So Eqs. (22.8) and (22.9) are correct even when there are charges outside the surface that contribute to the electric field at the surface. When $Q_{encl} = 0$, the total flux through the Gaussian surface must be zero, even though some areas may have positive flux and others may have negative flux (see Fig. 22.3b).

Gauss's law is the definitive answer to the question we posed at the beginning of Section 22.1: "If the electric field pattern is known in a given region, what can we determine about the charge distribution in that region?" It provides a relationship between the electric field on a closed surface and the charge distribution within that surface. But in some cases we can use Gauss's law to answer the reverse question: "If the charge distribution is known, what can we determine about the electric field that the charge distribution produces?" Gauss's law may seem like an unappealing way to address this question, since it may look as though evaluating the integral in Eq. (22.8) is a hopeless task. Sometimes it is, but other times it is surprisingly easy. Here's an example in which *no* integration is involved at all; we'll work out several more examples in the next section.

Conceptual Example 22.4 Electric flux and enclosed charge

Figure 22.16 shows the field produced by two point charges $+q$ and $-q$ of equal magnitude but opposite sign (an electric dipole). Find the electric flux through each of the closed surfaces A, B, C, and D.

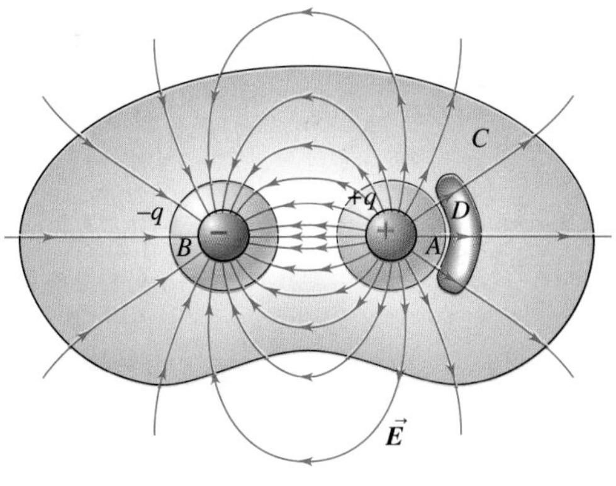

22.16 The net number of field lines leaving a closed surface is proportional to the total charge enclosed by that surface.

SOLUTION

The definition of electric flux given in Eq. (22.5) involves a surface integral, and so it might seem that integration is called for. But Gauss's law says that the total electric flux through a closed surface is equal to the total enclosed charge divided by ϵ_0. By inspection of Fig. 22.16, surface A (shown in red) encloses the positive charge, so $Q_{encl} = +q$; surface B (shown in blue) encloses the negative charge, so $Q_{encl} = -q$; surface C (shown in yellow), which encloses *both* charges, has $Q_{encl} = +q + (-q) = 0$; and surface D (shown in purple), which has no charges enclosed within it, also has $Q_{\text{encl}} = 0$. Hence without having to do any integration, we can conclude that the total fluxes for the various surfaces are $\Phi_E = +q/\epsilon_0$ for surface A, $\Phi_E = -q/\epsilon_0$ for surface B, and $\Phi_E = 0$ for both surface C and surface D.

These results depend only on the charges enclosed within each Gaussian surface, not on the precise shapes of the surfaces. For example, compare surface C to the rectangular surface shown in Fig. 22.3b, which also encloses both charges of an electric dipole. In that case as well, we concluded that the net flux of $\vec{E}$ was zero; the inward flux on one part of the surface exactly compensates for the outward flux on the remainder of the surface.

We can draw similar conclusions by examining the electric field lines. Surface A encloses only the positive charge; in Fig. 22.16, 18 lines are depicted crossing A in an outward direction. Surface B encloses only the negative charge; it is crossed by these same 18 lines, but in an inward direction. Surface C encloses *both* charges. It is intersected by lines at 16 points; at 8 intersections the lines are outward, and at 8 they are inward. The *net* number of lines crossing in an outward direction is zero, and the net charge inside the surface is also zero. Surface D is intersected at 6 points; at 3 points the lines are outward, and at the other 3 they are inward. The net number of lines crossing in an outward direction and the total charge enclosed are both zero. There are points on the surfaces where $\vec{E}$ is not perpendicular to the surface, but this doesn't affect the counting of the field lines.

Test Your Understanding

What is the total electric flux through a surface that completely encloses a negative lithium ion (Fig. 21.4c)? How is your answer affected if the surface is made larger, but still encloses the ion (and no other charge)?

22.4 | Applications of Gauss's Law

Gauss's law is valid for *any* distribution of charges and for *any* closed surface. Gauss's law can be used in two ways. If we know the charge distribution, and if it has enough symmetry to let us evaluate the integral in Gauss's law, we can find the field. Or if we know the field, we can use Gauss's law to find the charge distribution, such as charges on conducting surfaces.

In this section we present examples of both kinds of applications. As you study them, watch for the role played by the symmetry properties of each system. We will use Gauss's law to calculate the electric fields caused by several simple charge distributions; the results are collected in a table in the chapter summary.

In practical problems we often encounter situations in which we want to know the electric field caused by a charge distribution on a conductor. These calculations are aided by the following remarkable fact: *When excess charge is placed on a solid conductor and is at rest, it resides entirely on the surface, not in the interior of the material.* (By *excess* we mean charges other than the ions and free electrons that make up the neutral conductor.) Here's the proof. We know from Section 21.4 that in an electrostatic situation (with all charges at rest) the electric field $\vec{E}$ at every point in the interior of a conducting material is zero. If $\vec{E}$ were *not* zero, the charges would move. Suppose we construct a Gaussian surface inside the conductor, such as surface A in Fig. 22.17. Because $\vec{E} = \mathbf{0}$ everywhere on this surface, Gauss's law requires that the net charge inside the surface is zero. Now imagine shrinking the surface like a collapsing balloon until it encloses a region so small that we may consider it as a point P; then the charge at that point must be zero. We can do this anywhere inside the conductor, so *there can be no excess charge at any point within a solid conductor; any excess charge must reside on the conductor's surface.* (This result is for a *solid* conductor. In the next section we'll discuss what can happen if the conductor has cavities in its interior.) We will make use of this fact frequently in the examples that follow.

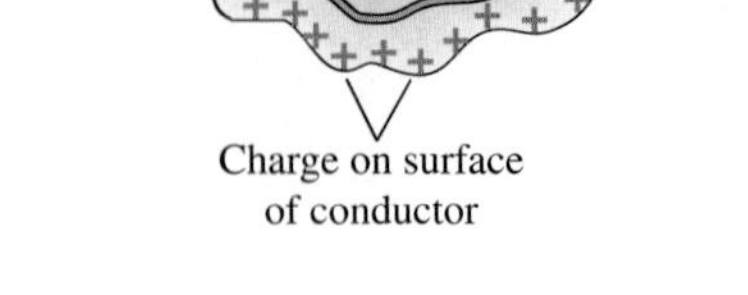

22.17 Under electrostatic conditions (charges not in motion), any excess charge on a solid conductor resides entirely on the conductor's surface.

Problem-Solving Strategy

Gauss's Law

IDENTIFY *the relevant concepts:* Gauss's law is most useful in situations where the charge distribution has spherical or cylindrical symmetry or is distributed uniformly over a plane. In these situations we determine the direction of $\vec{E}$ from the symmetry of the charge distribution. If we are given the charge distribution, we can use Gauss's law to find the magnitude of $\vec{E}$. Alternatively, if we are given the field, we can use Gauss's law to determine the details of the charge distribution. In either case, begin your analysis by asking the question, "What is the symmetry?"

SET UP *the problem using the following steps:*

1. Select the surface that you will use with Gauss's law. We often call it a *Gaussian surface.* If you are trying to find the field at a particular point, then that point must lie on your Gaussian surface.
2. The Gaussian surface does not have to be a real physical surface, such as a surface of a solid body. Often the appropriate surface is an imaginary geometric surface; it may be in empty space, embedded in a solid body, or both.
3. Usually you can evaluate the integral in Gauss's law (without using a computer) only if the Gaussian surface and the charge distribution have some symmetry property. If the charge distribution has cylindrical or spherical symmetry, choose the Gaussian surface to be a coaxial cylinder or a concentric sphere, respectively.

EXECUTE *the solution as follows:*

1. Carry out the integral in Eq. (22.9). This may look like a daunting task, but the symmetry of the charge distribution and your careful choice of a Gaussian surface makes it straightforward.
2. Often you can think of the closed Gaussian surface as being made up of several separate surfaces, such as the sides and ends of a cylinder. The integral $\oint E_\perp \, dA$ over the entire closed surface is always equal to the sum of the integrals over all the separate surfaces. Some of these integrals may be zero, as in points 4 and 5 below.
3. If $\vec{E}$ is *perpendicular* (normal) at every point to a surface with area A, if it points *outward* from the interior of the surface, and if it also has the same *magnitude* at every point on the surface, then $E_\perp = E =$ constant, and $\int E_\perp \, dA$ over that surface is equal to EA. If instead $\vec{E}$ is perpendicular and *inward,* then $E_\perp = -E$ and $\int E_\perp \, dA = -EA$.
4. If $\vec{E}$ is *tangent* to a surface at every point, then $E_\perp = 0$ and the integral over that surface is zero.
5. If $\vec{E} = \mathbf{0}$ at every point on a surface, the integral is zero.
6. In the integral $\oint E_\perp \, dA$, $E_\perp$ is always the perpendicular component of the *total* electric field at each point on the closed Gaussian surface. In general, this field may be caused partly by charges within the surface and partly by charges outside it. Even when there is *no* charge within the surface, the field at points on the Gaussian surface is not necessarily zero. In that case, however, the *integral* over the Gaussian surface—that is, the total electric flux through the Gaussian surface—is always zero.
7. Once you have evaluated the integral, use Eq. (22.9) to solve for your target variable.

EVALUATE *your answer:* Often your result will be a *function* that describes how the magnitude of the electric field varies with position. Examine this function with a critical eye to see whether it makes sense.

Example 22.5

Field of a charged conducting sphere

We place positive charge q on a solid conducting sphere with radius R (Fig. 22.18). Find $\vec{E}$ at any point inside or outside the sphere.

SOLUTION

IDENTIFY: As we discussed earlier in this section, all the charge must be on the surface of the sphere. The system has spherical symmetry.

SET UP: To take advantage of the symmetry, we take as our Gaussian surface an imaginary sphere of radius r centered on the conductor. To calculate the field outside the conductor, we take r to be greater than the conductor's radius R; to calculate the field inside, we take r to be less than R. In either case, the point where we want to calculate $\vec{E}$ lies on the Gaussian surface.

EXECUTE: The role of symmetry deserves careful discussion before doing any calculations. When we say that the system is spherically symmetric, we mean that if we rotate it through any angle about any axis through the center, the system after rotation is indistinguishable from the original unrotated system. There is nothing about the system that distinguishes one direction or orientation in space from another. The charge is free to move on the conductor, and there is nothing about the conductor that would make it tend to concentrate more in some regions than others. If it were nonuniform, then when we rotate the system, the sphere would look the same but the charge distribution would look different, and there is no property of the sphere that can make this happen. So we conclude that the charge is distributed *uniformly* over the surface.

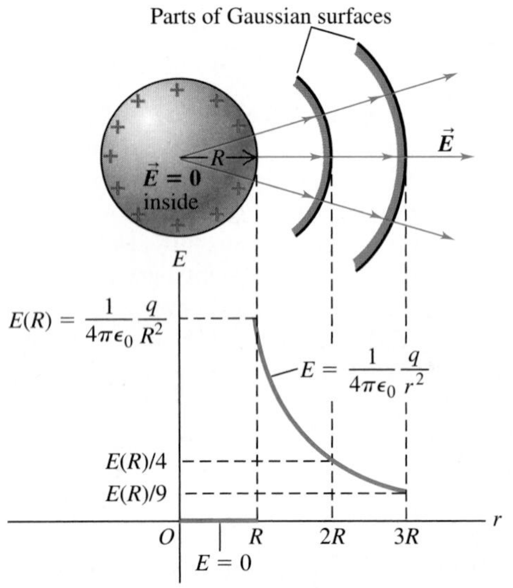

22.18 Under electrostatic conditions the electric field inside a solid conducting sphere is zero. Outside the sphere the electric field drops off as $1/r^2$, as though all of the excess charge on the sphere were concentrated at its center.

A similar argument shows that the direction of the electric field must be *radial*, as shown in Fig. 22.18. If we again rotate the system, the field pattern of the rotated system must be identical to that of the original system. If the field had a component at some point that was perpendicular to the radial direction, that component would have to be different after at least some rotations. Thus there can't be such a component, and the field must be radial. For the same reason the magnitude E of the field can depend only on the distance r from the center and must have the same value at all points on a spherical surface concentric with the conductor.

Our choice of a sphere as a Gaussian surface takes advantage of these symmetry properties. We first consider the field outside the conductor, so we choose $r > R$. The entire conductor is within the Gaussian surface, so the enclosed charge is q. The area of the Gaussian surface is $4\pi r^2$; $\vec{E}$ is uniform over the surface and perpendicular to it at each point. The flux integral $\oint E_\perp \, dA$ in Gauss's law is therefore just $E(4\pi r^2)$, and Eq. (22.8) gives

$$E(4\pi r^2) = \frac{q}{\epsilon_0} \quad \text{and}$$

$$E = \frac{1}{4\pi\epsilon_0}\frac{q}{r^2} \quad \text{(outside a charged conducting sphere)}$$

This expression for the field at any point *outside* the sphere $(r > R)$ is the same as for a point charge; the field due to the charged sphere is the same as though the entire charge were concentrated at its center. Just outside the surface of the sphere, where $r = R$,

$$E = \frac{1}{4\pi\epsilon_0}\frac{q}{R^2}$$

(at the surface of a charged conducting sphere)

CAUTION Remember that we have chosen the charge q to be *positive.* If the charge is negative, the electric field is radially *inward* instead of radially outward, and the electric flux through the Gaussian surface is negative. The electric field magnitudes outside and at the surface of the sphere are given by the same expressions as above, except that q denotes the *magnitude* (absolute value) of the charge.

To find $\vec{E}$ inside the conductor, we use a spherical Gaussian surface with radius $r < R$. The spherical symmetry again tells us that $E(4\pi r^2) = Q_{encl}/\epsilon_0$. But because all of the charge is on the surface of the conductor, our Gaussian surface (which lies entirely within the conductor) encloses *no* charge. So $Q_{encl} = 0$ and, therefore, the electric field inside the conductor is zero.

EVALUATE: We already knew that $\vec{E} = \mathbf{0}$ inside the conductor, as it must be inside any solid conductor when the charges are at rest. Figure 22.18 shows E as a function of the distance r from the center of the sphere. Note that in the limit as $R \to 0$, the sphere becomes a point charge; there is then only an "outside," and the field is everywhere given by $E = q/4\pi\epsilon_0 r^2$. Thus we have deduced Coulomb's law from Gauss's law. (In Section 22.3 we deduced Gauss's law from Coulomb's law, so this completes the demonstration of their logical equivalence.)

We can also use this method for a conducting spherical *shell* (a spherical conductor with a concentric spherical hole in the center) if there is no charge inside the hole. We use a spherical Gaussian surface with radius r less than the radius of the hole. If there *were* a field inside the hole, it would have to be radial and spherically symmetric as before, so $E = Q_{encl}/4\pi\epsilon_0 r^2$. But now there is no enclosed charge, so $Q_{encl} = 0$ and $E = 0$ inside the hole.

Can you use this same technique to find the electric field in the interspace between a charged sphere and a concentric hollow conducting sphere that surrounds it?

Example 22.6 Field of a line charge

Electric charge is distributed uniformly along an infinitely long, thin wire. The charge per unit length is λ (assumed positive). Find the electric field. (This is an approximate representation of the field of a uniformly charged *finite* wire, provided that the distance from the field point to the wire is much less than the length of the wire.)

SOLUTION

IDENTIFY: The system has *cylindrical* symmetry. The field must point away from the positive charges. To determine the direction of $\vec{E}$ more precisely, as well as how its magnitude can depend on position, we use symmetry as in Example 22.5.

SET UP: Cylindrical symmetry means that we can rotate the system through any angle about its axis, and we can shift it by any amount along the axis; in each case the resulting system is indistinguishable from the original. Using the same argument as in Example 22.5, we conclude that $\vec{E}$ at each point doesn't change when either of these operations is carried out. The field can't have any component parallel to the wire; if it did, we would have to explain why the field lines that begin on the wire pointed in one direction parallel to the wire and not the other. Also, the field can't have any component tangent to a circle in a plane perpendicular to the wire with its center on the wire. If it did, we would have to explain why the component pointed in one direction around the wire rather than the other. All that's left is a component radially outward from the wire at each point. So the field lines outside a uniformly charged, infinite wire are *radial* and lie in planes perpendicular to the wire. The field *magnitude* can depend only on the radial distance from the wire.

These symmetry properties suggest that we use as a Gaussian surface a *cylinder* with arbitrary radius r and arbitrary length l, with its ends perpendicular to the wire (Fig. 22.19).

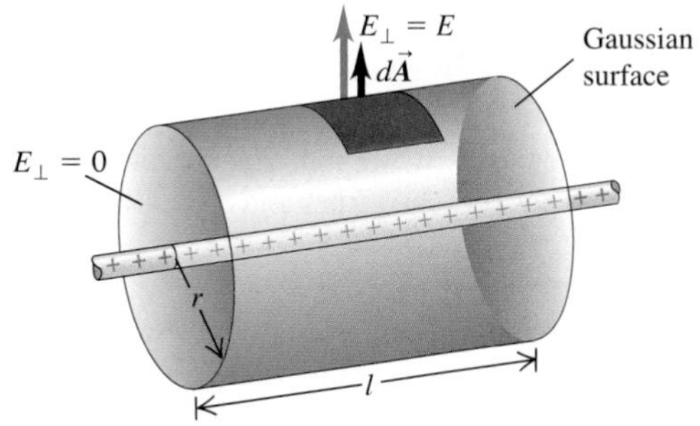

22.19 A coaxial cylindrical Gaussian surface is used to find the electric field outside an infinitely long charged wire.

EXECUTE: We break the surface integral for the flux Φ_E into an integral over each flat end and one over the curved side walls. There is no flux through the ends because $\vec{E}$ lies in the plane of the surface and $E_\perp = 0$. To find the flux through the side walls, note that $\vec{E}$ is perpendicular to the surface at each point, so $E = E_\perp$; by symmetry, E has the same value everywhere on the walls. The area of the side walls is $2\pi rl$. (To make a paper cylinder with radius r and height l, you need a paper rectangle with width $2\pi r$, height l, and area $2\pi rl$.) Hence the total flux Φ_E through the entire cylinder is the sum of the flux through the side walls, which is $(E)(2\pi rl)$, and the zero flux through the two ends. Finally, we need the total enclosed charge, which is the charge per unit length multiplied by the length of wire inside the Gaussian surface, or $Q_{encl} = \lambda l$. From Gauss's law, Eq. (22.8),

$$\Phi_E = (E)(2\pi rl) = \frac{\lambda l}{\epsilon_0} \quad \text{and}$$

$$E = \frac{1}{2\pi\epsilon_0}\frac{\lambda}{r} \quad \text{(field of an infinite line of charge)}$$

This is the same result that we found in Example 22.11 (Section 21.5) by much more laborious means.

We have assumed that λ is *positive*. If it is *negative*, $\vec{E}$ is directed radially inward toward the line of charge, and in the above expression for the field magnitude E we must interpret λ as the *magnitude* (absolute value) of the charge per unit length.

EVALUATE: Note that although the *entire* charge on the wire contributes to the field, only the part of the total charge that is within the Gaussian surface is considered when we apply Gauss's law. This may seem strange; it looks as though we have somehow obtained the right answer by ignoring part of the charge and that the field of a *short* wire of length l would be the same as that of a very long wire. But we *do* include the entire charge on the wire when we make use of the *symmetry* of the problem. If the wire is short, the symmetry with respect to shifts along the axis is not present, and the field is not uniform in magnitude over our Gaussian surface. Gauss's law is then no longer useful and *cannot* be used to find the field; the problem is best handled by the integration technique used in Example 21.11.

We can use a Gaussian surface like that in Fig. 22.19 to show that the field at points outside a long, uniformly charged cylinder is the same as though all the charge were concentrated on a line along its axis. We can also calculate the electric field in the space between a charged cylinder and a coaxial hollow conducting cylinder surrounding it. This is a model of a coaxial cable, like those used to connect your TV to a cable television "feed" (Fig. 22.20). We leave these calculations as problems.

22.20 Inside a coaxial cable is a conducting wire within a hollow conducting cylinder. Gauss's law is used to calculate the static electric field for such an arrangement.

Example 22.7 Field of an infinite plane sheet of charge

Find the electric field caused by a thin, flat, infinite sheet on which there is a uniform positive charge per unit area σ.

SOLUTION

IDENTIFY: The field must point away from the positively charged sheet. As in Examples 22.5 and 22.6, before doing calculations we use the symmetry (in this case, *planar* symmetry) to learn more about the direction and position dependence of $\vec{E}$.

SET UP: Planar symmetry means that the charge distribution doesn't change if we slide it in any direction parallel to the sheet. From this we conclude that at each point, $\vec{E}$ is perpendicular to the sheet. The symmetry also tells us that the field must have the same magnitude E at any given distance on either side of the sheet. To take advantage of these symmetry properties, we use as our Gaussian surface a cylinder with its axis perpendicular to the sheet of charge, with ends of area A (Fig. 22.21).

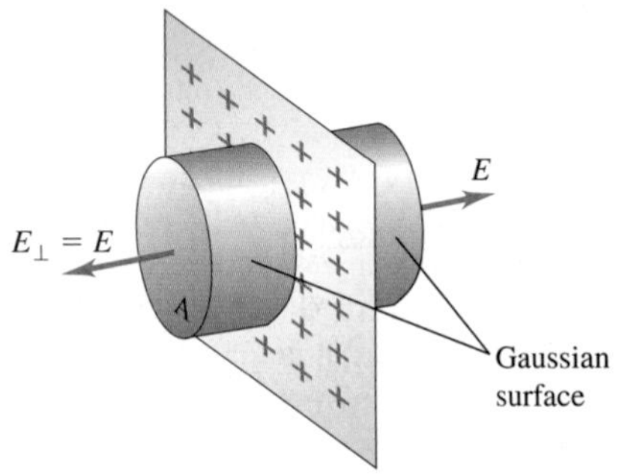

22.21 A cylindrical Gaussian surface is used to find the field of an infinite plane sheet of charge.

EXECUTE: The charged sheet passes through the middle of the cylinder's length, so the cylinder ends are equidistant from the sheet. At each end of the cylinder, $\vec{E}$ is perpendicular to the surface and $E_\perp$ is equal to E; hence the flux through each end is $+EA$.

Because $\vec{E}$ is perpendicular to the charged sheet, it is parallel to the curved *side* walls of the cylinder, so $E_\perp$ at these walls is zero and there is no flux through these walls. The total flux integral in Gauss's law is then $2EA$ (EA from each end and zero from the side walls). The net charge within the Gaussian surface is the charge per unit area multiplied by the sheet area enclosed by the surface, or $Q_{encl} = \sigma A$. Hence Gauss's law, Eq. (22.8), gives

$$2EA = \frac{\sigma A}{\epsilon_0} \quad \text{and}$$

$$E = \frac{\sigma}{2\epsilon_0} \qquad \text{(field of an infinite sheet of charge)}$$

This is the same result that we found in Example 21.12 (Section 21.5) using a much more complex calculation. The field is uniform and directed perpendicular to the plane of the sheet. Its magnitude is *independent* of the distance from the sheet. The field lines are therefore straight, parallel to each other, and perpendicular to the sheet.

If the charge density is negative, $\vec{E}$ is directed *toward* the sheet, the flux through the Gaussian surface in Fig. 22.21 is negative, and σ in the expression $E = \sigma/2\epsilon_0$ denotes the magnitude (absolute value) of the charge density.

EVALUATE: The assumption that the sheet is infinitely large is an idealization; nothing in nature is really infinitely large. But the result $E = \sigma/2\epsilon_0$ is a good approximation for points that are close to the sheet (compared to the sheet's dimensions) and not too near its edges. At such points, the field is very nearly uniform and perpendicular to the plane.

Example 22.8 Field between oppositely charged parallel conducting plates

Two large plane parallel conducting plates are given charges of equal magnitude and opposite sign; the charge per unit area is $+\sigma$ for one and $-\sigma$ for the other. Find the electric field in the region between the plates.

SOLUTION

IDENTIFY: The field between and around the plates is approximately as shown in Fig. 22.22a. Because opposite charges attract, most of the charge accumulates at the opposing faces of the plates. A small amount of charge resides on the *outer* surfaces of the plates, and there is some spreading or "fringing" of the field at the edges. But if the plates are very large in comparison to the distance between them, the amount of charge on the outer surfaces is negligibly small, and the fringing can be neglected except near the edges. In this case we can assume that the field is uniform in the interior region between plates, as in Fig. 22.22b and that the charges are distributed uniformly over the opposing surfaces.

SET UP: To exploit this symmetry, we can use the shaded Gaussian surfaces S_1, S_2, S_3, and S_4. These surfaces are cylinders with ends of area A like the one shown in perspective in Fig. 22.21; they are shown in a side view in Fig. 22.22b. One end of each surface lies within one of the conducting plates.

EXECUTE: For the surface labeled S_1, the left-hand end is within plate 1 (the positive plate). Since the field is zero within the volume

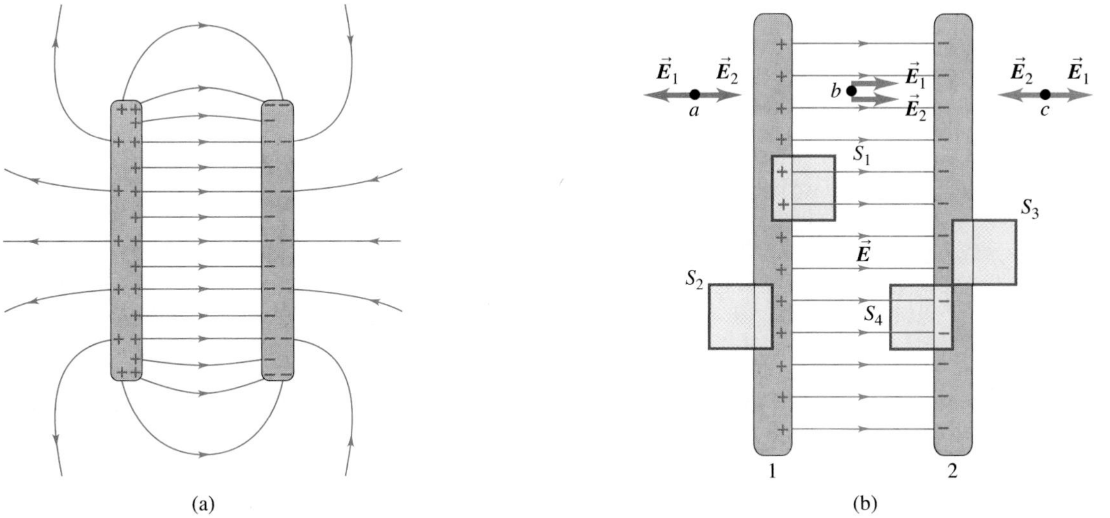

22.22 Electric field between oppositely charged parallel plates. (a) The field between the plates is quite uniform except near the edges. (b) Idealized case in which "fringing" at the edges is ignored.

of any solid conductor under electrostatic conditions, there is no electric flux through this end. The electric field between the plates is perpendicular to the right-hand end, so on that end, $E_\perp$ is equal to E and the flux is EA; this is positive, since $\vec{E}$ is directed out of the Gaussian surface. There is no flux through the side walls of the cylinder, since these walls are parallel to $\vec{E}$. So the total flux integral in Gauss's law is EA. The net charge enclosed by the cylinder is σA, so Eq. (22.8) yields

$$EA = \frac{\sigma A}{\epsilon_0} \quad \text{and}$$

$$E = \frac{\sigma}{\epsilon_0}$$

(field between oppositely charged conducting plates)

The field is uniform and perpendicular to the plates, and its magnitude is independent of the distance from either plate. This same result can be obtained by using the Gaussian surface S_4; furthermore, the surfaces S_2 and S_3 can be used to show that $E = 0$ to the left of plate 1 and to the right of plate 2. We leave these calculations as a problem.

EVALUATE: We obtained the above results in Example 21.13 (Section 21.5) by using the principle of superposition of electric fields. The fields due to the two sheets of charge (one on each plate) are $\vec{E}_1$ and $\vec{E}_2$; from Example 22.7, both of these have magnitude $\sigma/2\epsilon_0$. The total (resultant) electric field at any point is the vector sum $\vec{E} = \vec{E}_1 + \vec{E}_2$. At points a and c in Fig. 22.22b, $\vec{E}_1$ and $\vec{E}_2$ have opposite directions, and their resultant is zero. This is also true at every point within the material of each plate, consistent with the requirement that with charges at rest there can be no field within a solid conductor. At any point b between the plates, $\vec{E}_1$ and $\vec{E}_2$ have the same direction; their resultant has magnitude $E = \sigma/\epsilon_0$, just as we found above using Gauss's law.

Example 22.9 Field of a uniformly charged sphere

Positive electric charge Q is distributed uniformly *throughout the volume* of an *insulating* sphere with radius R. Find the magnitude of the electric field at a point P a distance r from the center of the sphere.

SOLUTION

IDENTIFY and SET UP: As in Example 22.5, the system is spherically symmetric. Hence we can use the conclusions of that example about the direction and magnitude of $\vec{E}$. To make use of the symmetry, we choose as our Gaussian surface a sphere with radius r, concentric with the charge distribution.

EXECUTE: From symmetry the magnitude E of the electric field has the same value at every point on the Gaussian surface, and the direction of $\vec{E}$ is radial at every point on the surface, so $E_\perp = E$. Hence the total electric flux through the Gaussian surface is the product of E and the total area of the surface $A = 4\pi r^2$, that is, $\Phi_E = 4\pi r^2 E$.

The amount of charge enclosed within the Gaussian surface depends on the radius r. Let's first find the field magnitude *inside* the charged sphere of radius R; the magnitude E is evaluated at the radius of the Gaussian surface, so we choose $r < R$. The volume charge density ρ is the charge Q divided by the volume of the entire charged sphere of radius R:

$$\rho = \frac{Q}{4\pi R^3/3}$$

The volume V_{encl} enclosed by the Gaussian surface is $\frac{4}{3}\pi r^3$, so the total charge Q_{encl} enclosed by that surface is

$$Q_{\text{encl}} = \rho V_{\text{encl}} = \left(\frac{Q}{4\pi R^3/3}\right)\left(\frac{4}{3}\pi r^3\right) = Q\frac{r^3}{R^3}$$

Then Gauss's law, Eq. (22.8), becomes

$$4\pi r^2 E = \frac{Q}{\epsilon_0}\frac{r^3}{R^3} \quad \text{or}$$

$$E = \frac{1}{4\pi\epsilon_0}\frac{Qr}{R^3} \qquad \text{(field inside a uniformly charged sphere)}$$

The field magnitude is proportional to the distance r of the field point from the center of the sphere. At the center $(r = 0)$, $E = 0$.

To find the field magnitude *outside* the charged sphere, we use a spherical Gaussian surface of radius $r > R$. This surface encloses the entire charged sphere, so $Q_{\text{encl}} = Q$, and Gauss's law gives

$$4\pi r^2 E = \frac{Q}{\epsilon_0} \quad \text{or}$$

$$E = \frac{1}{4\pi\epsilon_0}\frac{Q}{r^2} \qquad \text{(field outside a uniformly charged sphere)}$$

For *any* spherically symmetric charged body the electric field outside the body is the same as though the entire charge were concentrated at the center. (We made this same observation in Example 22.5.)

Figure 22.23 shows a graph of E as a function of r for this problem. For $r < R$, E is directly proportional to r, and for $r > R$, E varies as $1/r^2$. If the charge is negative instead of positive, $\vec{E}$ is radially *inward* and Q in the expressions for E is interpreted as the magnitude (absolute value) of the charge.

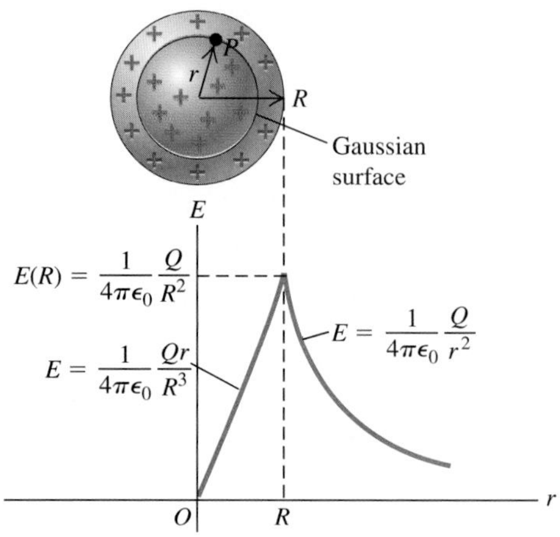

22.23 The magnitude of the electric field of a uniformly charged insulating sphere. Compare this with the field for a conducting sphere (Fig. 22.18).

EVALUATE: Notice that if we set $r = R$ in either of the two expressions for E (inside or outside the sphere), we get the same result $E = Q/4\pi\epsilon_0 R^2$ for the magnitude of the field at the surface of the sphere; this is because the magnitude E is a *continuous* function of r. By contrast, for the charged conducting sphere of Example 22.5 the electric-field magnitude is *discontinuous* at $r = R$ (it jumps from $E = 0$ just inside the sphere to $E = Q/4\pi\epsilon_0 R^2$ just outside the sphere). In general, the electric field $\vec{E}$ is discontinuous in magnitude, direction, or both wherever there is a *sheet* of charge, such as at the surface of a charged conducting sphere (Example 22.5), at the surface of an infinite charged sheet (Example 22.7), or at the surface of a charged conducting plate (Example 22.8).

The general technique used in this example can be applied to *any* spherically symmetric distribution of charge, whether it is uniform or not. Such charge distributions occur within many atoms and atomic nuclei, which is why Gauss's law is a useful tool in atomic and nuclear physics.

Example 22.10 Field of a hollow charged sphere

A thin-walled hollow sphere of radius 0.250 m has an unknown amount of charge distributed uniformly over its surface. At a distance of 0.300 m from the center of the sphere, the electric field points directly toward the center of the sphere and has magnitude 1.80×10^2 N/C. How much charge is on the sphere?

SOLUTION

IDENTIFY and SET UP: The charge distribution is spherically symmetric. As in Examples 22.5 and 22.9, it follows that the electric field is radial everywhere and its magnitude is a function only of the radial distance r from the center of the sphere. We again use a spherical Gaussian surface that is concentric with the charge distribution and that passes through the point of interest at $r = 0.300$ m.

EXECUTE: The charge distribution is the same as if the charge were on the surface of a 0.250-m radius conducting sphere. Hence we can borrow the results of Example 22.5. A key difference from that example is that because the electric field here is directed toward the sphere, the charge must be *negative*. Furthermore, because the electric field is directed into the Gaussian surface, $E_\perp = -E$ and the flux is $\oint E_\perp \, dA = -E(4\pi r^2)$.

By Gauss's law, the flux is equal to the charge q on the sphere (all of which is enclosed by the Gaussian surface) divided by ϵ_0. Solving for q, we find

$$q = -E(4\pi\epsilon_0 r^2) = -(1.80 \times 10^2\ \text{N/C})(4\pi) \times (8.854 \times 10^{-12}\ \text{C}^2/\text{N}\cdot\text{m}^2)(0.300\ \text{m})^2 = -8.01 \times 10^{-10}\ \text{C} = -0.801\ \text{nC}$$

EVALUATE: To determine the charge, we had to know the electric field at *all* points on the Gaussian surface so that we could calculate the flux integral. This was possible here because the charge distribution is highly symmetric. If the charge distribution is irregular or lacks symmetry, however, Gauss's law is not very useful for calculating the charge distribution from the field, or vice versa.

Test Your Understanding

You place a known amount of charge Q on the irregularly shaped conductor shown in Fig. 22.17. If you know the size and shape of the conductor, can you use Gauss's law to calculate the electric field at an arbitrary position outside the conductor?

22.5 | Charges on Conductors

We have learned that in an electrostatic situation (in which there is no net motion of charge) the electric field at every point within a conductor is zero and that any excess charge on a solid conductor is located entirely on its surface (Fig. 22.24a). But what if there is a *cavity* inside the conductor (Fig. 22.24b)? If there is no charge within the cavity, we can use a Gaussian surface such as A (which lies completely within the material of the conductor) to show that the *net* charge on the *surface of the cavity* must be zero, because $\vec{E} = \mathbf{0}$ everywhere on the Gaussian surface. In fact, we can prove in this situation that there can't be any charge *anywhere* on the cavity surface. We will postpone detailed proof of this statement until Chapter 23.

Suppose we place a small body with a charge q inside a cavity within a conductor (Fig. 22.24c). The conductor is uncharged and is insulated from the charge q. Again $\vec{E} = \mathbf{0}$ everywhere on surface A, so according to Gauss's law the *total* charge inside this surface must be zero. Therefore there must be a charge $-q$ distributed on the surface of the cavity, drawn there by the charge q inside the cavity. The *total* charge on the conductor must remain zero, so a charge $+q$ must appear either on its outer surface or inside the material. But we showed in Section 22.4 that in an electrostatic situation there can't be any excess charge within the material of a conductor. So we conclude that the charge $+q$ must appear on the outer surface. By the same reasoning, if the conductor originally had a charge q_C, then the total charge on the outer surface must be $q + q_C$ after the charge q is inserted into the cavity.

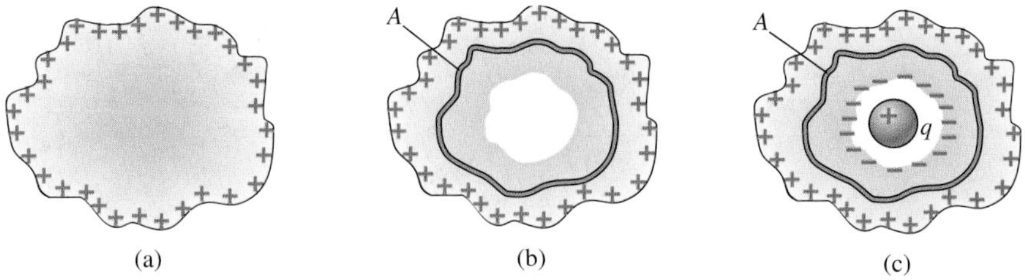

22.24 (a) Charge on a solid conductor resides entirely on its outer surface. (b) If there is no charge inside the conductor's cavity, the net charge on the surface of the cavity is zero. (c) If there is a charge q inside the cavity, the total charge on the cavity surface is $-q$.

Conceptual Example 22.11 A conductor with a cavity

The conductor shown in cross section in Fig. 22.25 carries a total charge of $+3$ nC. The charge within the cavity, insulated from the conductor, is -5 nC. How much charge is on each surface (inner and outer) of the conductor?

SOLUTION

If the charge in the cavity is $q = -5$ nC, the charge on the inner cavity surface must be $-q = -(-5 \text{ nC}) = +5$ nC. The conductor carries a *total* charge of $+3$ nC, none of which is in the interior of the material. If $+5$ nC is on the inner surface of the cavity, then there must be $(+3 \text{ nC}) - (+5 \text{ nC}) = -2$ nC on the outer surface of the conductor.

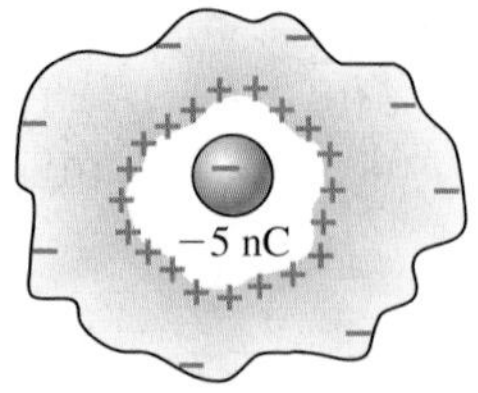

22.25 There is no excess charge in the bulk material of this conductor. Charge is located only on the inner and outer surfaces.

Testing Gauss's Law Experimentally

We can now consider a historic experiment, shown in Fig. 22.26. We mount a conducting container, such as a metal pail with a lid, on an insulating stand. The container is initially uncharged. Then we hang a charged metal ball from an insulating thread (Fig. 22.26a), lower it into the pail, and put the lid on (Fig. 22.26b). Charges are induced on the walls of the container, as shown. But now we let the ball *touch* the inner wall (Fig. 22.26c). The surface of the ball becomes, in effect, part of the cavity surface. The situation is now the same as Fig. 22.24b; if Gauss's law is correct, the net charge on the cavity surface must be zero. Thus the ball must lose all its charge. Finally, we pull the ball out; we find that it has indeed lost all its charge.

This experiment was performed in the nineteenth century by the English scientist Michael Faraday, using a metal icepail with a lid, and it is called **Faraday's icepail experiment.** (Similar experiments were carried out in the eighteenth cen-

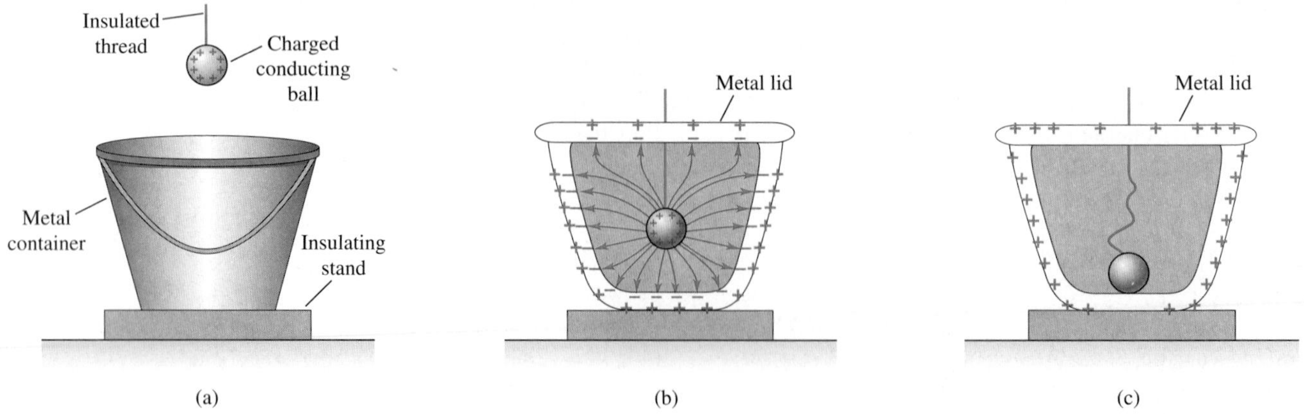

22.26 (a) A charged conducting ball suspended by an insulating thread outside a conducting container on an insulating stand. (b) The ball is lowered into the container, and the lid is put on. Charges are induced on the walls of the container. (c) When the ball is touched to the inner surface of the container, all its charge is transferred to the container and appears on the container's outer surface.

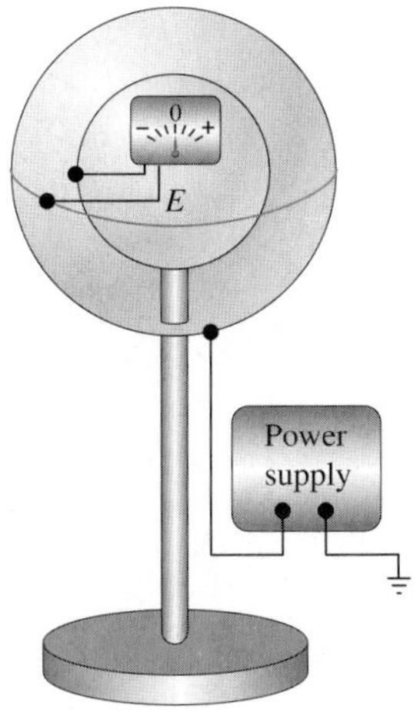

22.27 The outer spherical shell can be alternately charged and discharged by the power supply. If there were any flow of charge between the inner and outer shells, it would be detected by the electrometer inside the inner shell.

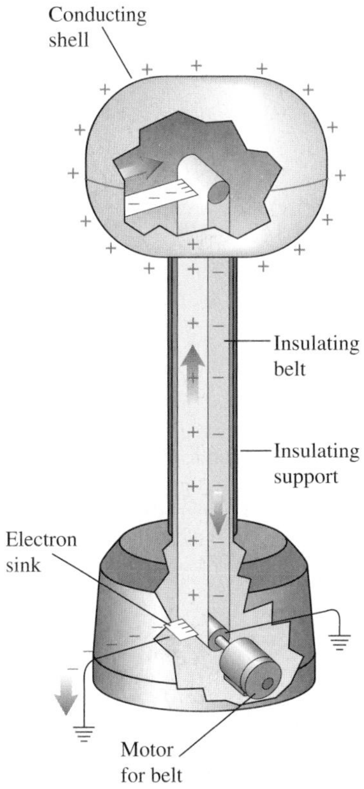

22.28 Cutaway view of the essential parts of a Van de Graaff electrostatic generator. The electron sink at the bottom draws electrons from the belt, giving it a positive charge; at the top the belt attracts electrons away from the conducting shell, giving the shell a positive charge.

tury by Benjamin Franklin in America and Joseph Priestley in England, although with much less precision.) The result confirms the validity of Gauss's law and therefore of Coulomb's law. Faraday's result was significant because Coulomb's experimental method, using a torsion balance and dividing of charges, was not very precise; it is very difficult to confirm the $1/r^2$ dependence of the electrostatic force with great precision by direct force measurements. By contrast, experiments like Faraday's test the validity of Gauss's law, and therefore of Coulomb's law, with much greater precision.

A modern version of Faraday's experiment is shown in Fig. 22.27. The details of the box labeled "power supply" aren't important; its job is to place charge on the outer sphere and remove it, on demand. The inner box with a dial is a sensitive *electrometer,* an instrument that can detect motion of extremely small amounts of charge between the outer and inner spheres. If Gauss's law is correct, there can never be any charge on the inner surface of the outer sphere. If so, there should be no flow of charge between spheres while the outer sphere is being charged and discharged. The fact that no flow is actually observed is a very sensitive confirmation of Gauss's law and therefore of Coulomb's law. The precision of the experiment is limited mainly by the electrometer, which can be astonishingly sensitive. Experiments have shown that the exponent 2 in the $1/r^2$ of Coulomb's law does not differ from precisely 2 by more than 10^{-16}. So there is no reason to suspect that it is anything other than exactly 2.

The same principle behind Faraday's icepail experiment is used in a *Van de Graaff electrostatic generator* (Fig. 22.28). The charged conducting ball of Fig. 22.26 is replaced by a charged belt that continuously carries charge to the inside of a conducting shell, only to have it carried away to the outside surface of the shell. As a result, the charge on the shell and the electric field around it can become very large very rapidly. The Van de Graaff generator is used as an accelerator of charged particles and for physics demonstrations.

This principle also forms the basis for *electrostatic shielding.* Suppose we have a very sensitive electronic instrument that we want to protect from stray electric fields that might cause erroneous measurements. We surround the instrument with a conducting box, or we line the walls, floor, and ceiling of the room with a conducting material such as sheet copper. The external electric field redistributes the free electrons in the conductor, leaving a net positive charge on the outer surface

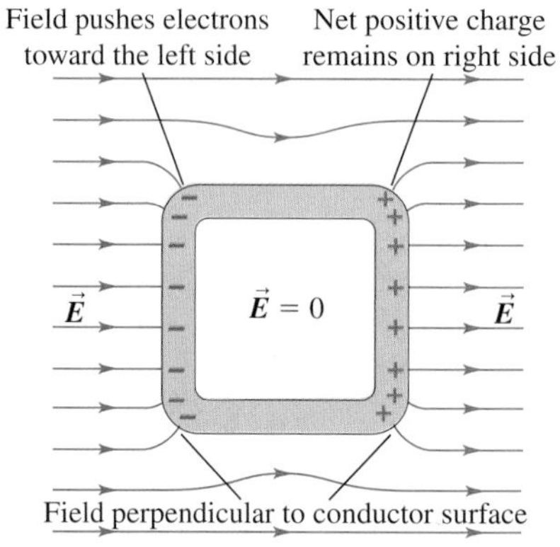

(a)

(b)

22.29 (a) A conducting box (a Faraday cage) immersed in a uniform electric field. The field of the induced charges on the box combines with the uniform field to give zero total field inside the box. (b) Electrostatic shielding can protect you from a dangerous electric discharge.

in some regions and a net negative charge in others (Fig. 22.29). This charge distribution causes an additional electric field such that the *total* field at every point inside the box is zero, as Gauss's law says it must be. The charge distribution on the box also alters the shapes of the field lines near the box, as the figure shows. Such a setup is often called a *Faraday cage.* The same physics tells you that one of the safest places to be in a lightning storm is inside an automobile; if the car is struck by lightning, the charge tends to remain on the metal skin of the vehicle, and little or no electric field is produced inside the passenger compartment.

Field at the Surface of a Conductor

Finally, we note that there is a direct relation between the $\vec{E}$ field at a point just outside any conductor and the surface charge density σ at that point. In general, σ varies from point to point on the surface. We will show in Chapter 23 that at any such point, the direction of $\vec{E}$ is always *perpendicular* to the surface (see Fig. 22.29a).

To find a relation between σ at any point on the surface and the perpendicular component of the electric field at that point, we construct a Gaussian surface in the form of a small cylinder (Fig. 22.30). One end face, with area A, lies within the conductor and the other lies just outside. The electric field is zero at all points within the conductor. Outside the conductor the component of $\vec{E}$ perpendicular to the side walls of the cylinder is zero, and over the end face the perpendicular component is equal to $E_\perp$. (If σ is positive, the electric field points out of the conductor and $E_\perp$ is positive; if σ is negative, the field points inward and $E_\perp$ is negative.) Hence the total flux through the surface is $E_\perp A$. The charge enclosed within the Gaussian surface is σA, so from Gauss's law,

$$E_\perp A = \frac{\sigma A}{\epsilon_0} \quad \text{and} \quad E_\perp = \frac{\sigma}{\epsilon_0} \tag{22.10}$$

(field at the surface of a conductor)

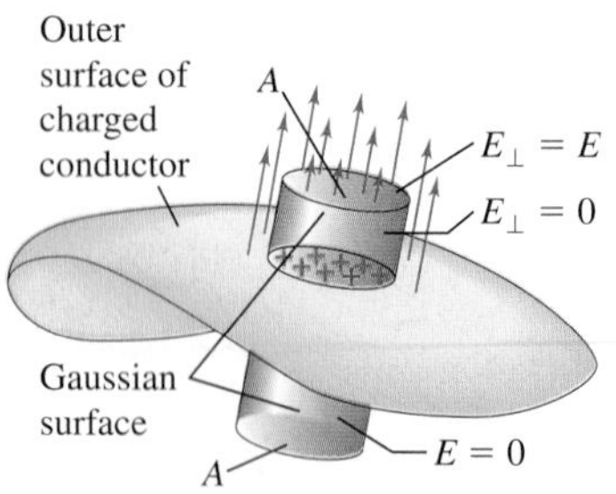

22.30 The field just outside a charged conductor is perpendicular to the surface, and its perpendicular component $E_\perp$ is equal to σ/ϵ_0.

We can check this with the results we have obtained for spherical, cylindrical, and plane surfaces.

We showed in Example 22.8 that the field magnitude between two infinite flat oppositely charged conducting plates also equals σ/ϵ_0. In this case the field magnitude is the same at *all* distances from the plates, but in all other cases it decreases with increasing distance from the surface.

Conceptual Example 22.12 Field at the surface of a conducting sphere

Verify Eq. (22.10) for a conducting sphere with radius R and total charge q.

SOLUTION

In Example 22.5 (Section 22.4) we showed that the electric field just outside the surface is

$$E = \frac{1}{4\pi\epsilon_0}\frac{q}{R^2}$$

The surface charge density is uniform and equal to q divided by the surface area of the sphere:

$$\sigma = \frac{q}{4\pi R^2}$$

Comparing these two expressions, we see that $E = \sigma/\epsilon_0$, as Eq. (22.10) states.

Example 22.13 Electric field of the earth

The earth (a conductor) has a net electric charge. The resulting electric field near the surface can be measured with sensitive electronic instruments; its average value is about 150 N/C, directed toward the center of the earth. a) What is the corresponding surface charge density? b) What is the *total* surface charge of the earth?

SOLUTION

IDENTIFY and SET UP: Given the perpendicular electric field, we determine the surface charge density σ using Eq. (22.10). The total surface charge on the earth is then the product of σ and the earth's surface area.

EXECUTE: a) We know from the direction of the field that σ is negative (corresponding to $\vec{E}$ being directed *into* the surface, so $E_\perp$ is negative). From Eq. (22.10),

$$\begin{aligned}\sigma = \epsilon_0 E_\perp &= (8.85 \times 10^{-12}\ \text{C}^2/\text{N}\cdot\text{m}^2)(-150\ \text{N/C})\\ &= -1.33 \times 10^{-9}\ \text{C/m}^2 = -1.33\ \text{nC/m}^2\end{aligned}$$

b) The earth's surface area is $4\pi R_E^2$, where $R_E = 6.38 \times 10^6$ m is the radius of the earth (see Appendix F). The total charge Q is the product $4\pi R_E^2\sigma$, or

$$\begin{aligned}Q &= 4\pi(6.38 \times 10^6\ \text{m})^2(-1.33 \times 10^{-9}\ \text{C/m}^2)\\ &= -6.8 \times 10^5\ \text{C} = -680\ \text{kC}\end{aligned}$$

Solving for Q, we find

$$\begin{aligned}Q &= 4\pi\epsilon_0 R^2 E_\perp\\ &= \frac{1}{9.0 \times 10^9\ \text{N}\cdot\text{m}^2/\text{C}^2}(6.38 \times 10^6\ \text{m})^2(-150\ \text{N/C})\\ &= -6.8 \times 10^5\ \text{C}\end{aligned}$$

One electron has a charge of -1.60×10^{-19} C. Hence this much excess negative electric charge corresponds to there being $(-6.8 \times 10^5\ \text{C})/(-1.60 \times 10^{-19}\ \text{C}) = 4.2 \times 10^{24}$ excess electrons on the earth, or about 7 moles of excess electrons. This is compensated by an equal *deficiency* of electrons in the earth's upper atmosphere, so the combination of the earth and its atmosphere is electrically neutral.

EVALUATE: You can check our result in part (b), using the result of Example 22.5.

Test Your Understanding

Sensitive computer components are sometimes shipped in electrically conductive containers. How does this practice protect the components?

SUMMARY

Electric flux is a measure of the "flow" of electric field through a surface. It is equal to the product of an area element and the perpendicular component of $\vec{E}$, integrated over a surface.
(See Examples 22.1 through 22.3)

$$\Phi_E = \int E\cos\phi\,dA$$
$$= \int E_\perp\,dA = \int \vec{E}\cdot d\vec{A} \qquad (22.5)$$

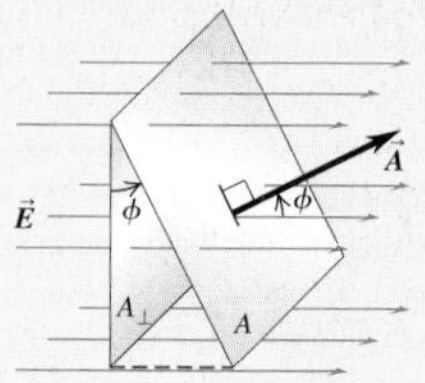

Gauss's law states that the total electric flux through a closed surface, which can be written as the surface integral of the component of $\vec{E}$ normal to the surface, equals a constant times the total charge Q_{encl} enclosed by the surface. Gauss's law is logically equivalent to Coulomb's law, but its use greatly simplifies problems with a high degree of symmetry.
(See Examples 22.4 through 22.10)

$$\Phi_E = \oint E\cos\phi\,dA$$
$$= \oint E_\perp\,dA = \oint \vec{E}\cdot d\vec{A}$$
$$= \frac{Q_{encl}}{\epsilon_0} \qquad (22.8), (22.9)$$

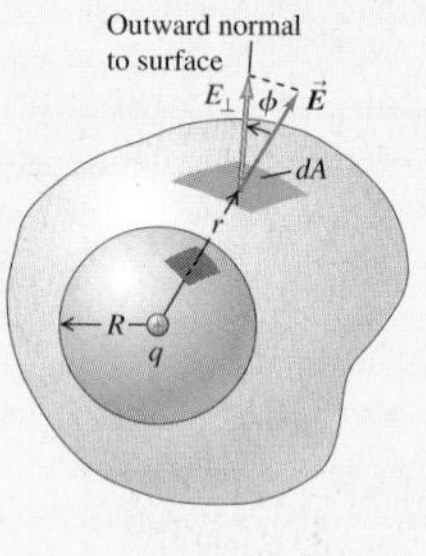

When excess charge is placed on a conductor and is at rest, it resides entirely on the surface, and $\vec{E} = \mathbf{0}$ everywhere in the material of the conductor. (See Examples 22.11 through 22.13)

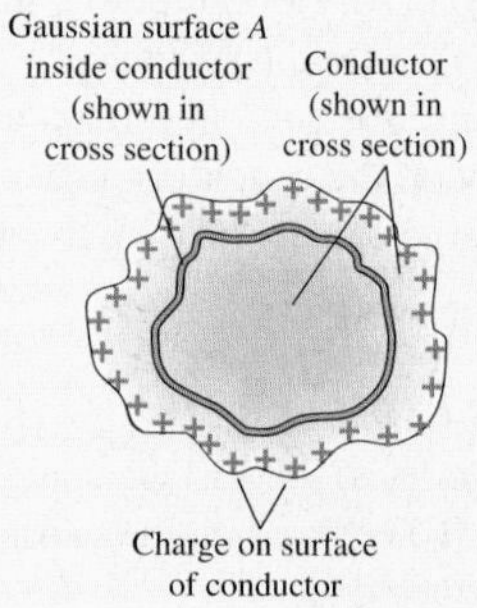

The following table lists electric fields caused by several symmetric charge distributions. In the table, q, Q, λ, and σ refer to the *magnitudes* of the quantities.

Charge Distribution	**Point in Electric Field**	**Electric Field Magnitude**
Single point charge q	Distance r from q	$E = \dfrac{1}{4\pi\epsilon_0}\dfrac{q}{r^2}$
Charge q on surface of conducting sphere with radius R	Outside sphere, $r > R$	$E = \dfrac{1}{4\pi\epsilon_0}\dfrac{q}{r^2}$
	Inside sphere, $r < R$	$E = 0$
Infinite wire, charge per unit length λ	Distance r from wire	$E = \dfrac{1}{2\pi\epsilon_0}\dfrac{\lambda}{r}$

Charge Distribution	Point in Electric Field	Electric Field Magnitude
Infinite conducting cylinder with radius R, charge per unit length λ	Outside cylinder, $r > R$	$E = \dfrac{1}{2\pi\epsilon_0}\dfrac{\lambda}{r}$
	Inside cylinder, $r < R$	$E = 0$
Solid insulating sphere with radius R, charge Q distributed uniformly throughout volume	Outside sphere, $r > R$	$E = \dfrac{1}{4\pi\epsilon_0}\dfrac{Q}{r^2}$
	Inside sphere, $r < R$	$E = \dfrac{1}{4\pi\epsilon_0}\dfrac{Qr}{R^3}$
Infinite sheet of charge with uniform charge per unit area σ	Any point	$E = \dfrac{\sigma}{2\epsilon_0}$
Two oppositely charged conducting plates with surface charge densities $+\sigma$ and $-\sigma$	Any point between plates	$E = \dfrac{\sigma}{\epsilon_0}$

Key Terms

closed surface, 837
electric flux, 838
surface integral, 842
Gauss's law, 844
Gaussian surface, 846
Faraday's icepail experiment, 856

Your Notes

Answer to Chapter Opening Question ?

No. The electric field inside a cavity within a conductor is zero, so there is no electric effect on the child. (See Section 22.5.)

Answers to Test Your Understanding Questions

Section 22.1 Each part of the surface of the box will be three times farther from the charge $+q$, so the electric field will be $1/3^2 = 1/9$ as strong. But the area of the box will increase by a factor $3^2 = 9$. Hence the electric flux will be multiplied by a factor of $(1/9)(9) = 1$. In other words, the flux will be unchanged.
Section 22.2 To maximize the electric flux through a plane surface, the area vector $\vec{A}$ should be in the same direction as the electric field $\vec{E}$ so that the surface is face-on to the field (Fig. 22.6a). In an analogous way, the orientation of a solar panel is chosen to maximize the *light* flux, so that the greatest possible amount of solar energy strikes the panel.
Section 22.3 From Eq. (22.9) the total flux through the surface is equal to Q_{encl}/ϵ_0. A negative lithium ion has a charge of $-e = -1.60 \times 10^{-19}$ C, so $\Phi_E = (-1.60 \times 10^{-19}\ \text{C})/(8.854 \times 10^{-12}\ \text{C}^2/\text{N}\cdot\text{m}^2) = -1.81 \times 10^{-8}\ \text{N}\cdot\text{m}^2/\text{C}$. The flux is negative because the enclosed charge is negative and the electric field is directed toward the ion. Increasing the size of the Gaussian surface has no effect on the flux, because the enclosed charge remains the same.
Section 22.4 No. You might be tempted to draw a Gaussian surface that is an enlarged version of the conductor, with the same shape and placed so that it completely encloses the conductor. While you know the flux through this Gaussian surface (by Gauss's law, it's $\Phi_E = Q/\epsilon_0$), the direction of the electric field need not be perpendicular to the surface and the magnitude of the field need not be the same at all points on the surface. It's not possible to do the flux integral $\oint E_\perp\, dA$, and we can't calculate the electric field. Gauss's law is useful for calculating the electric field only when the charge distribution is *highly* symmetric.
Section 22.5 Computer components are designed to operate with very small amounts of electric current. If they are exposed to strong electric fields, the flow of charge within the components might exceed the design limits and cause damage. Shipping such components in a conducting box shields them from any electric fields they might encounter in transit (see Fig. 22.29).

Discussion Questions

Q22.1 A spherical Gaussian surface encloses a point charge q. If the point charge is moved from the center of the sphere to a point away from the center, does the electric field at a point on the surface change? Does the total flux through the Gaussian surface change? Explain.
Q22.2 A certain region of space bounded by an imaginary closed surface contains no charge. Is the electric field always zero everywhere on the surface? If not, under what circumstances is it zero on the surface?
Q22.3 A rubber balloon has a single point charge in its interior. Does the electric flux through the balloon depend on whether or not it is fully inflated? Explain your reasoning.
Q22.4 Are Coulomb's law and Gauss's law *completely* equivalent? Are there any situations in electrostatics in which one is valid and the other is not? Explain your reasoning.
Q22.5 In Fig. 22.16, suppose a third point charge were placed outside the yellow Gaussian surface C. Would this affect the electric flux through any of the surfaces A, B, C, or D in the figure? Why or why not?
Q22.6 If the electric field of a point charge were proportional to $1/r^3$ instead of $1/r^2$, would Gauss's law still be valid? Explain your reasoning. (*Hint:* Consider a spherical Gaussian surface centered on a single point charge.)
Q22.7 A solid, right circular cylinder with radius R and height $h = R$ has charge distributed uniformly throughout its volume. Can Gauss's law be used to calculate the electric field at all points inside the cylinder? What about at all points outside the cylinder? Explain your reasoning.
Q22.8 It was shown in the text that the electric field inside an empty cavity in a conductor is zero. Is this statement true no matter what the shape of the cavity? Why or why not?
Q22.9 The electric field $\vec{E}$ is uniform throughout a certain region of space. A small conducting sphere that carries a net charge Q is then placed in this region. What is the electric field inside the sphere? Explain your reasoning.
Q22.10 In a conductor, one or more electrons from each atom are free to roam throughout the volume of the conductor. Does this contradict the statement that any excess charge on a solid conductor must reside on its surface? Why or why not?
Q22.11 Explain the following statement: "In a static situation, the electric field at the surface of a conductor can have no component parallel to the surface, because this would violate the condition that the charges on the surface are at rest." Would this same statement be a valid one for the electric field at the surface of an *insulator?* Explain your answer, and the reason for any differences between the cases of a conductor and an insulator.
Q22.12 The magnitude of $\vec{E}$ at the surface of an irregularly shaped solid conductor must be greatest in regions where the surface curves most sharply, such as point A in Fig. 22.31, and must be least in flat regions such as point B in Fig. 22.31. Explain why this must be so by considering how electric field lines must be arranged near a conducting surface. How does the surface charge density compare at points A and B? Explain.

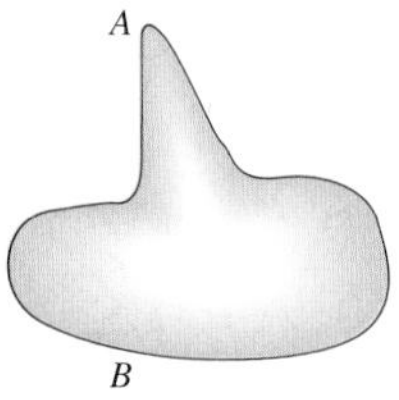

Figure 22.31 Question Q22.12.

Q22.13 A lightning rod is a pointed copper rod mounted on top of a building and welded to a heavy copper cable running down into the ground. Lightning rods are used to protect houses and barns from lightning; the lightning current runs through the copper rather than through the building. Why? Why should the end of the rod be pointed? (*Hint:* The answer to Q22.12 may be helpful.)
Q22.14 A solid conductor has a cavity in its interior. Would the presence of a point charge inside the cavity affect the electric field outside the conductor? Why or why not? Would the presence of a

point charge outside the conductor affect the electric field inside the cavity? Again, why or why not?

Q22.15 Some modern aircraft are made primarily of composite materials that do not conduct electricity. The U.S. Federal Aviation Administration requires that such aircraft have conducting wires embedded in their surfaces to provide protection when flying near thunderstorms. Explain the physics behind this requirement.

Q22.16 You find a sealed box on your doorstep. You suspect that the box contains several charged metal spheres packed in insulating material. How can you determine the total net charge inside the box without opening the box? Or is this not possible?

Q22.17 A solid copper sphere has a net positive charge. The charge is distributed uniformly over the surface of the sphere and the electric field inside the sphere is zero. Then a negative point charge outside the sphere is brought close to the surface of the sphere. Is all the net charge on the sphere still on its surface? If so, is this charge still distributed uniformly over the surface? If it is not uniform, how is it distributed? Is the electric field inside the sphere still zero? In each case justify your answers.

Exercises

Section 22.2 Calculating Electric Flux

22.1 A flat sheet of paper of area 0.250 m^2 is oriented so that the normal to the sheet is at an angle of 60° to a uniform electric field of magnitude 14 N/C. a) Find the magnitude of the electric flux through the sheet. b) Does the answer to part (a) depend on the shape of the sheet? Why or why not? c) For what angle ϕ between the normal to the sheet and the electric field is the magnitude of the flux through the sheet i) largest? ii) smallest? Explain your answers.

22.2 The cube in Fig. 22.32 has sides of length $L = 10.0$ cm. The electric field is uniform, has a magnitude $E = 4.00 \times 10^3$ N/C, and is parallel to the xy-plane at an angle of 36.9° measured from the $+x$-axis toward the $+y$-axis. a) What is the electric flux through each of the six cube faces S_1, S_2, S_3, S_4, S_5, and S_6? b) What is the total electric flux through all faces of the cube?

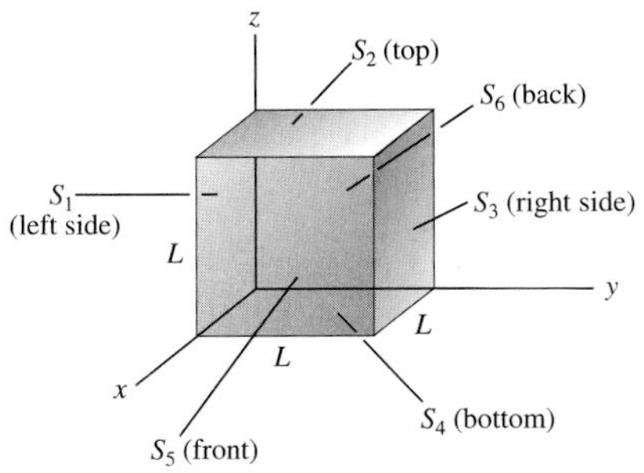

Figure 22.32 Exercises 22.2 and 22.3, and Problem 22.30.

22.3 A cube has sides of length L. It is placed with one corner at the origin as shown in Fig. 22.32. The electric field is uniform and given by $\vec{E} = -B\hat{\imath} + C\hat{\jmath} - D\hat{k}$, where B, C, and D are positive constants. a) Find the electric flux through each of the six cube faces S_1, S_2, S_3, S_4, S_5, and S_6. b) Find the electric flux through the entire cube.

22.4 A flat sheet is in the shape of a rectangle with sides of length 0.400 m and 0.600 m. The sheet is immersed in a uniform electric field of magnitude 75.0 N/C that is directed at 20° from the plane of the sheet (Fig. 22.33). Find the magnitude of the electric flux through the sheet.

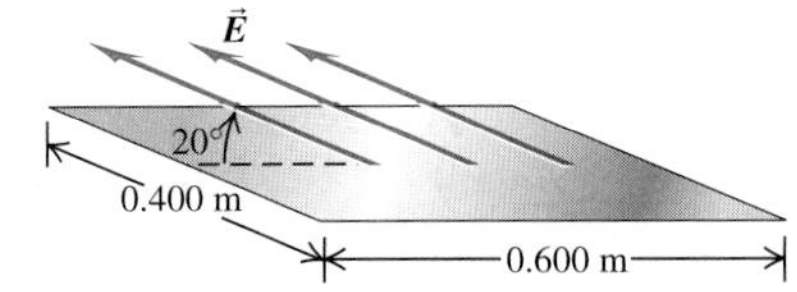

Figure 22.33 Exercise 22.4.

22.5 It was shown in Example 21.11 (Section 21.5) that the electric field due to an infinite line of charge is perpendicular to the line and has magnitude $E = \lambda/2\pi\epsilon_0 r$. Consider an imaginary cylinder with radius $r = 0.250$ m and length $l = 0.400$ m that has an infinite line of positive charge running along its axis. The charge per unit length on the line is $\lambda = 6.00\ \mu$C/m. a) What is the electric flux through the cylinder due to this infinite line of charge? b) What is the flux through the cylinder if its radius is increased to $r = 0.500$ m? c) What is the flux through the cylinder if its length is increased to $l = 0.800$ m?

Section 22.3 Gauss's Law

22.6 The three small spheres shown in Fig. 22.34 carry charges $q_1 = 4.00$ nC, $q_2 = -7.80$ nC, and $q_3 = 2.40$ nC. Find the net electric flux through each of the following closed surfaces shown in cross section in the figure: a) S_1; b) S_2; c) S_3; d) S_4; e) S_5. f) Do your answers to parts (a) through (e) depend on how the charge is distributed over each small sphere? Why or why not?

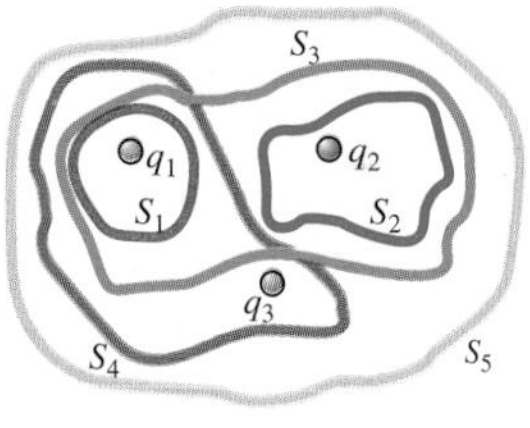

Surface	What it encloses
S_1	q_1
S_2	q_2
S_3	q_1 and q_2
S_4	q_1 and q_3
S_5	q_1 and q_2 and q_3

Figure 22.34 Exercise 22.6.

22.7 a) A closed surface encloses a net charge of $-3.60\ \mu$C. What is the net electric flux through the surface? b) The electric flux through a closed surface is found to be 780 N·m^2/C. What quantity of charge is enclosed by the surface? c) The closed surface in part (b) is a cube with sides of length 2.50 cm. From the information given in part (b), is it possible to tell where within the cube the charge is located? Explain.

22.8 A point charge $q_1 = 4.00$ nC is located on the x-axis at $x = 2.00$ m, and a second point charge $q_2 = -6.00$ nC is on the y-axis at $y = 1.00$ m. What is the total electric flux due to these two point charges through a spherical surface centered at the origin and with radius a) 0.500 m? b) 1.50 m? c) 2.50 m?

22.9 In a certain region of space, the electric field $\vec{E}$ is uniform. a) Use Gauss's law to prove that this region of space must be electrically neutral; that is, the volume charge density ρ must be zero. b) Is the converse true? That is, in a region of space where there is no charge, must $\vec{E}$ be uniform? Explain.

22.10 a) In a certain region of space, the volume charge density ρ has a uniform positive value. Can $\vec{E}$ be uniform in this region? Explain.

b) Suppose that in this region of uniform positive ρ there is a "bubble" within which $\rho = 0$. Can $\vec{E}$ be uniform within this bubble? Explain.

22.11 A 9.60-μC point charge is at the center of a cube with sides of length 0.500 m. a) What is the electric flux through one of the six faces of the cube? b) How would your answer to part (a) change if the sides were of length 0.250 m? Explain.

22.12 A small cube of volume 8.0 cm^3 is 0.30 cm from a metal sphere that has charge 2.00 μC. If the cube is empty, what is the total flux through it?

22.13 A sphere centered at the origin has radius 0.200 m. A -5.00 μC point charge is on the x-axis at $x = 0.300$ m. The net flux through the sphere is 360 N$\cdot$m^2/C. What is the total charge inside the sphere?

Section 22.4 Applications of Gauss's Law and Section 22.5 Charges on Conductors

22.14 A solid metal sphere with radius 0.450 m carries a net charge of 0.250 nC. Find the magnitude of the electric field a) at a point 0.100 m outside the surface of the sphere; b) at a point inside the sphere, 0.100 m below the surface.

22.15 In a physics lecture demonstration a charge of -0.180 μC is placed on the spherical dome of a Van de Graaff generator. a) How far away from the center of the dome should you sit for the electric field at your location to be within the recommended maximum of 614 N/C (Exercise 21.38)? b) The answer to part (a) doesn't depend on the radius of the dome. Why not?

22.16 The cylindrical imaging drum of a photocopier is to have an electric field just outside its surface of 1.40×10^5 N/C. a) If the drum has a surface area of 0.0610 m^2 (the area of a $8\frac{1}{2} \times 11$ in. sheet of paper), what total quantity of charge must reside on the surface of the drum? b) If the surface area of the drum is increased to 0.122 m^2 so that larger sheets of paper can be used, what total quantity of charge is required to produce the same 1.40×10^5 N/C electric field just above the surface?

22.17 How many excess electrons must be added to an isolated spherical conductor 32.0 cm in diameter to produce an electric field of 1150 N/C just outside the surface?

22.18 The electric field 0.400 m from a very long uniform line of charge is 840 N/C. How much charge is contained in a 2.00-cm section of the line?

22.19 A very long uniform line of charge has charge per unit length 4.80 μC/m and lies along the x-axis. A second long uniform line of charge has charge per unit length -2.40 μC/m and is parallel to the x-axis at $y = 0.400$ m. What is the net electric field (magnitude and direction) at the following points on the y-axis: a) $y = 0.200$ m? b) $y = 0.600$ m?

22.20 a) At a distance of 0.200 cm from the center of a charged conducting sphere with radius 0.100 cm, the electric field is 480 N/C. What is the electric field 0.600 cm from the center of the sphere? b) At a distance of 0.200 cm from the axis of a very long charged conducting cylinder with radius 0.100 cm, the electric field is 480 N/C. What is the electric field 0.600 cm from the axis of the cylinder? c) At a distance of 0.200 cm from a large uniform sheet of charge, the electric field is 480 N/C. What is the electric field 1.20 cm from the sheet?

22.21 Two insulating spheres with radius 0.080 m are separated by a center-to-center distance of 0.500 m. One sphere has a net charge -1.80 μC and the other has charge $+3.80$ μC. The charge is uniformly distributed within the volume of each sphere. What is the electric field (magnitude and direction) midway between the spheres?

22.22 An insulating sphere with radius 0.150 m has uniform charge density 7.50 nC/m^3 throughout its volume. What is the electric field a) just outside the surface of the sphere; b) outside the sphere, 0.300 m from its center; c) inside the sphere, 0.075 m from its center?

22.23 An insulating sphere with radius 0.220 m has charge distributed uniformly through its volume. What must the total charge of the sphere be if the electric field 0.110 m from the center of the sphere is 950 N/C?

22.24 A conductor with an inner cavity like that shown in Fig. 22.24c, carries a total charge of $+5.00$ nC. The charge within the cavity, insulated from the conductor, is -6.00 nC. How much charge is on a) the inner surface of the conductor? b) the outer surface of the conductor?

22.25 Apply Gauss's law to the Gaussian surfaces S_2, S_3, and S_4 in Fig. 22.22b to calculate the electric field between and outside the plates.

22.26 A square insulating sheet 80.0 cm on a side is held horizontally. The sheet has 7.50 nC of charge spread uniformly over its area. a) Calculate the electric field at a point 0.100 mm above the center of the sheet. b) Estimate the electric field at a point 100 m above the center of the sheet. c) Would the answers to parts (a) and (b) be different if the sheet were made of a conducting material? Why or why not?

22.27 An infinitely long cylindrical conductor has radius R and uniform surface charge density σ. a) In terms of σ and R, what is the charge per unit length λ for the cylinder? b) In terms of σ, what is the magnitude of the electric field produced by the charged cylinder at a distance $r > R$ from its axis? c) Express the result of part (b) in terms of λ and show that the electric field outside the cylinder is the same as if all the charge were on the axis. Compare your result to the result for a line of charge in Example 22.6 (Section 22.4).

22.28 Two very, large nonconducting plastic sheets, each 10.0 cm thick, carry uniform charge densities σ_1, σ_2, σ_3, and σ_4 on their surfaces, as shown in Fig. 22.35. These surface charge densities have the values $\sigma_1 = -6.00$ μC/m^2, $\sigma_2 = +5.00$ μC/m^2, $\sigma_3 = +2.00$ μC/m^2, and $\sigma_4 = +4.00$ μC/m^2. Use Gauss's law to find the magnitude and direction of the electric field at the following points, far from the edges of these sheets. a) Point A, 5.00 cm from the left face of the left-hand sheet. b) Point B, 1.25 cm from the inner surface of the right-hand sheet. c) Point C, in the middle of the right-hand sheet.

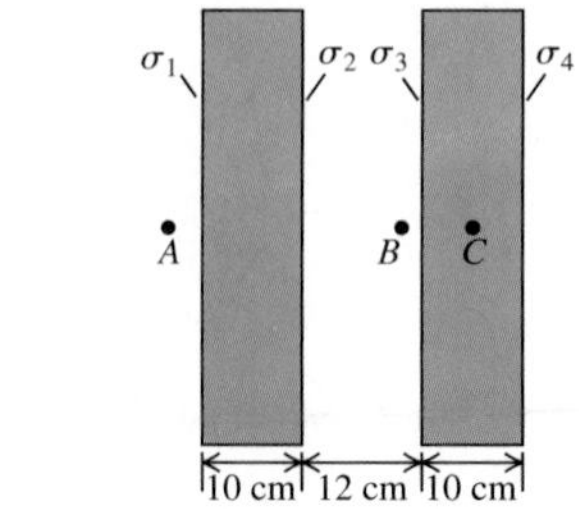

Figure 22.35 Exercise 22.28.

22.29 A negative charge $-Q$ is placed inside the cavity of a hollow metal solid. The outside of the solid is grounded, by connecting a conducting wire between it and the earth. a) Is there any excess charge induced on the inner surface of the piece of metal? If so, find its sign and magnitude.

b) Is there any excess charge on the outside of the piece of metal? Why, or why not? c) Is there an electric field in the cavity? Explain. d) Is there an electric field within the metal? Why, or why not? Is there an electric field outside the piece of metal? Explain why or why not. e) Would someone outside the solid measure an electric field due to the charge $-Q$? Is it reasonable to say that the grounded conductor has *shielded* the region from the effects of the charge $-Q$? In principle, could the same thing be done for gravity. Why or why not?

Problems

22.30 A cube has sides of length $L = 0.300$ m. It is placed with one corner at the origin as shown in Fig. 22.32. The electric field is not uniform, but is given by $\vec{E} = (-5.00 \text{ N/C}\cdot\text{m})x\hat{\imath} + (3.00 \text{ N/C}\cdot\text{m})z\hat{k}$. a) Find the electric flux through each of the six cube faces S_1, S_2, S_3, S_4, S_5, and S_6. b) Find the total electric charge inside the cube.

22.31 A flat, square surface with sides of length L is described by the equations

$$x = L \qquad 0 \le y \le L \qquad 0 \le z \le L$$

a) Draw this square and show the x-, y-, and z-axes. b) Find the electric flux through the square due to a positive point charge q located at the origin ($x = 0$, $y = 0$, $z = 0$). (*Hint:* Think of the square as part of a cube centered on the origin.)

22.32 The electric field $\vec{E}$ in Fig. 22.36 is everywhere parallel to the x-axis, so the components E_y and E_z are zero. The x-component of the field E_x depends on x but not on y and z. At points in the yz-plane (where $x = 0$), $E_x = 125$ N/C. a) What is the electric flux through surface I in Fig. 22.36? b) What is the electric flux through surface II? c) The volume shown in the figure is a small section of a very large insulating slab 1.0 m thick. If there is a total charge -24.0 nC within the volume shown, what are the magnitude and direction of $\vec{E}$ at the face opposite surface I? d) Is the electric field produced only by charges within the slab, or is the field also due to charges out-side the slab? How can you tell?

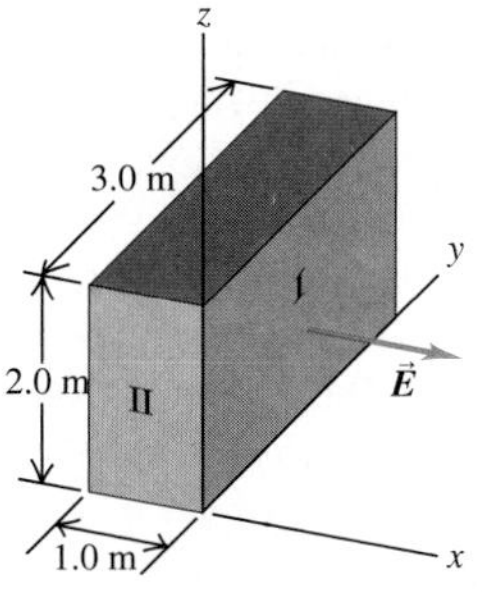

Figure 22.36 Problem 22.32.

22.33 The electric field $\vec{E}_1$ at one face of a parallelepiped is uniform over the entire face and is directed out of the face. At the opposite face, the electric field $\vec{E}_2$ is also uniform over the entire face and is directed into that face (Fig. 22.37). The two faces in question are inclined at 30.0° from the horizontal, while $\vec{E}_1$ and $\vec{E}_2$ are both horizontal; $\vec{E}_1$ has a magnitude of 2.50×10^4 N/C, and $\vec{E}_2$ a magnitude of 7.00×10^4 N/C. a) Assuming that no other electric field lines cross the surfaces of the parallelepiped, determine the net charge contained within. b) Is the electric field produced only by the charges within the parallelepiped, or is the field also due to charges outside the parallelepiped? How can you tell?

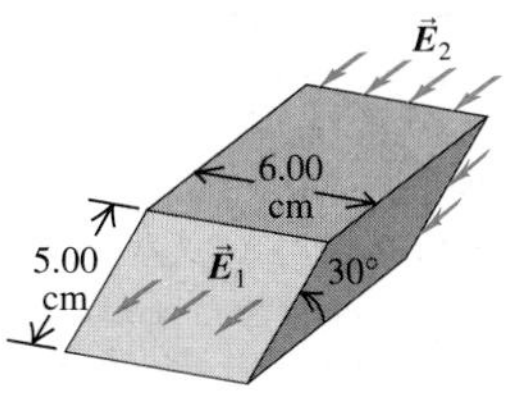

Figure 22.37 Problem 22.33.

22.34 A long line carrying a uniform linear charge density $+50.0$ μC/m runs parallel to and 10.0 cm from the surface of a large, flat plastic sheet that has a uniform surface charge density of -100 $\mu\text{C/m}^2$ on one side. Find the location of all points where an α particle would feel no force due to this arrangement of charged objects.

22.35 An insulating sphere with radius 0.120 m has 0.900 nC of charge uniformly distributed throughout its volume. The center of the sphere is 0.240 m above a large uniform sheet that has charge density -8.00 nC/m^2. Find all points inside the sphere where the electric field is zero. Or, show that there are no such points.

22.36 A Sphere in a Sphere. A solid conducting sphere carrying charge q has radius a. It is inside a concentric hollow conducting sphere with inner radius b and outer radius c. The hollow sphere has no net charge. a) Derive expressions for the electric field magnitude in terms of the distance r from the center for the regions $r < a$, $a < r < b$, $b < r < c$, and $r > c$. b) Graph the magnitude of the electric field as a function of r from $r = 0$ to $r = 2c$. c) What is the charge on the inner surface of the hollow sphere? d) On the outer surface? e) Represent the charge of the small sphere by four plus signs. Sketch the field lines of the system within a spherical volume of radius $2c$.

22.37 A solid conducting sphere with radius R, that carries positive charge Q, is concentric with a very thin insulating shell of radius $2R$ that also carries charge Q. The charge Q is distributed uniformly over the insulating shell. a) Find the electric field (magnitude and direction) in each of the regions $0 < r < R$, $R < r < 2R$, and $r > 2R$. b) Graph the electric field magnitude as a function of r.

22.38 A conducting spherical shell with inner radius a and outer radius b has a positive point charge Q located at its center. The total charge on the shell is $-3Q$, and it is insulated from its surroundings (Fig. 22.38). a) Derive expressions for the electric field magnitude in terms of the distance r from the center for the regions $r < a$, $a < r < b$, and $r > b$. b) What is the surface charge density on the inner surface of the conducting shell? c) What is the surface charge density on the outer surface of the conducting shell? d) Sketch the electric field lines and the location of all charges. e) Graph the electric field magnitude as a function of r.

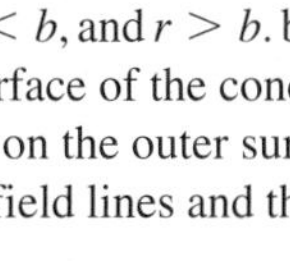

Figure 22.38 Problem 22.38.

22.39 Concentric Spherical Shells. A small conducting spherical shell with inner radius a and outer radius b is concentric with a larger conducting spherical shell with inner radius c and outer radius d (Fig. 22.39). The inner shell has total charge $+2q$,

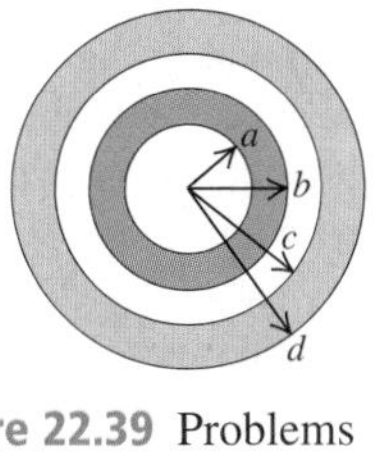

Figure 22.39 Problems 22.39, 22.40, 22.41, and 22.44.

and the outer shell has charge $+4q$. a) Calculate the electric field (magnitude and direction) in terms of q and the distance r from the common center of the two shells for i) $r < a$; ii) $a < r < b$; iii) $b < r < c$; iv) $c < r < d$; v) $r > d$. Show your results in a graph of the radial component of $\vec{E}$ as a function of r. b) What is the total charge on the i) inner surface of the small shell; ii) outer surface of the small shell; iii) inner surface of the large shell; iv) outer surface of the large shell?

22.40 Repeat Problem 22.39, but now let the outer shell have charge $-2q$. As in Problem 22.39, the inner shell has charge $+2q$.

22.41 Repeat Problem 22.39, but now let the outer shell have charge $-4q$. As in Problem 22.39, the inner shell has charge $+2q$.

22.42 A solid conducting sphere with radius R carries a positive total charge Q. The sphere is surrounded by an insulating shell with inner radius R and outer radius $2R$. The insulating shell has a uniform charge density ρ. a) Find the value of ρ so that the net charge of the entire system is zero. b) If ρ has the value found in part (a), find the electric field (magnitude and direction) in each of the regions $0 < r < R$, $R < r < 2R$, and $r > 2R$. Show your results in a graph of the radial component of $\vec{E}$ as a function of r. c) As a general rule, the electric field is only discontinuous at locations where there is a thin sheet of charge. Explain how your results in part (b) agree with this rule.

22.43 Negative charge $-Q$ is distributed uniformly over the surface of a thin spherical insulating shell with radius R. Calculate the force (magnitude and direction) that the shell exerts on a positive point charge q located a) a distance $r > R$ from the center of the shell (outside the shell); b) a distance $r < R$ from the center of the shell (inside the shell).

22.44 A small, *insulating*, spherical shell with inner radius a and outer radius b is concentric with a larger *insulating* spherical shell with inner radius c and outer radius d (Fig. 22.39). The inner shell has total charge $+q$ distributed uniformly over its volume, and the outer shell has charge $-q$ distributed uniformly over its volume. a) Calculate the charge densities in the inner shell and outer shell. b) Calculate the electric field (magnitude and direction) in terms of q and the distance r from the common center of the two shells for i) $r < a$; ii) $a < r < b$; iii) $b < r < c$; iv) $c < r < d$; v) $r > d$. Show your results in a graph of the radial component of $\vec{E}$ as a function of r.

22.45 The Coaxial Cable. A long coaxial cable consists of an inner cylindrical conductor with radius a and an outer coaxial cylinder with inner radius b and outer radius c. The outer cylinder is mounted on insulating supports and has no net charge. The inner cylinder has a uniform positive charge per unit length λ. Calculate the electric field a) at any point between the cylinders, a distance r from the axis; b) at any point outside the outer cylinder. c) Graph the magnitude of the electric field as a function of the distance r from the axis of the cable, from $r = 0$ to $r = 2c$. d) Find the charge per unit length on the inner surface and on the outer surface of the outer cylinder.

22.46 A very long conducting tube (hollow cylinder) has inner radius a and outer radius b. It carries charge per unit length $+\alpha$, where α is a positive constant with units of C/m. A line of charge lies along the axis of the tube. The line of charge has charge per unit length $+\alpha$. a) Calculate the electric field in terms of α and the distance r from the axis of the tube for i) $r < a$; ii) $a < r < b$; iii) $r > b$. Show your results in a graph of E as a function of r. b) What is the charge per unit length on i) the inner surface of the tube; ii) the outer surface of the tube?

22.47 Repeat Problem 22.46, but now let the conducting tube have charge per unit length $-\alpha$. As in Problem 22.46, the line of charge has charge per unit length $+\alpha$.

22.48 A very long, solid cylinder with radius R has positive charge uniformly distributed throughout it, with charge per unit volume ρ. a) Derive the expression for the electric field inside the volume at a distance r from the axis of the cylinder in terms of the charge density ρ. b) What is the electric field at a point outside the volume in terms of the charge per unit length λ in the cylinder? c) Compare the answers to parts (a) and (b) for $r = R$. d) Graph the electric field magnitude as a function of r from $r = 0$ to $r = 3R$.

22.49 A single, isolated, large conducting plate (Fig. 22.40) has a charge per unit area σ on its surface. Because the plate is a conductor, the electric field at its surface is perpendicular to the surface and has magnitude $E = \sigma/\epsilon_0$. a) In Example 22.7 (Section 22.4) it was shown that the field caused by a large, uniformly charged sheet with charge per unit area σ has magnitude $E = \sigma/2\epsilon_0$, exactly *half* as much as for a charged conducting plate. Why is there a difference? b) By regarding the charge distribution on the conducting plate as being two sheets of charge (one on each surface), each with charge per unit area σ, use the result of Example 22.7 and the principle of superposition to show that $E = 0$ inside the plate and $E = \sigma/\epsilon_0$ outside the plate.

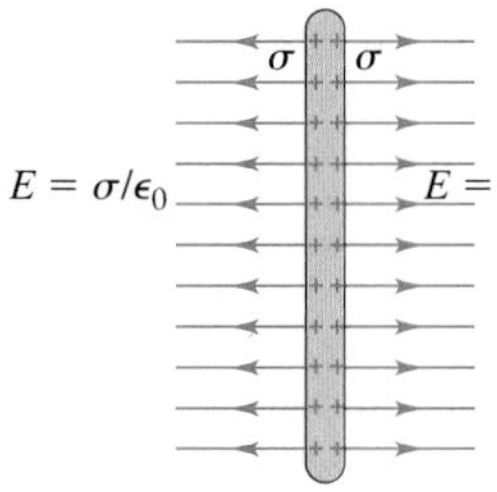

Figure 22.40 Problem 22.49.

22.50 Thomson's Model of the Atom. In the early years of the twentieth century, a leading model of the structure of the atom was that of the English physicist J. J. Thomson (the discoverer of the electron). In Thomson's model, an atom consisted of a sphere of positively charged material into which were embedded negatively charged electrons, like chocolate chips in a ball of cookie dough. Consider such an atom consisting of one electron with mass m and charge $-e$, which may be regarded as a point charge, and a uniformly charged sphere of charge $+e$ and radius R. a) Explain why the equilibrium position of the electron is at the center of the nucleus. b) In Thomson's model, it was assumed that the positive material provided little or no resistance to the motion of the electron. If the electron is displaced from equilibrium by a distance less than R, show that the resulting motion of the electron will be simple harmonic, and calculate the frequency of oscillation. (*Hint:* Review the definition of simple harmonic motion in Section 13.2. If it can be shown that the net force on the electron is of this form, then it follows that the motion is simple harmonic. Conversely, if the net force on the electron does not follow this form, the motion is not simple harmonic.) c) By Thomson's time, it was known that excited atoms emit light waves of only certain frequencies. In his model, the frequency of emitted light is the same as the oscillation frequency of the electron or electrons in the atom. What would the radius of a Thomson-model atom have to be for it to produce red light of frequency 4.57×10^{14} Hz? Compare your answer to the radii of real atoms, which are of order 10^{-10} m (see Appendix F for data about the electron). d) If the electron were displaced from equilibrium by a distance greater than R, would the electron oscillate? Would its motion be simple harmonic? Explain your reasoning. (*Historical note:* In 1910, the atomic nucleus was discovered, proving the Thomson model to be incorrect. An atom's positive charge is not spread over its volume as Thomson supposed, but is concentrated in the tiny nucleus of radius 10^{-14} to 10^{-15} m.)

22.51 Thomson's Model of the Atom, Continued. Using Thomson's (outdated) model of the atom described in Problem 22.50, consider an atom consisting of two electrons, each of charge $-e$, embedded in a sphere of charge $+2e$ and radius R. In equilibrium, each electron is a distance d from the center of the atom (Fig. 22.41). Find the distance d in terms of the other properties of the atom.

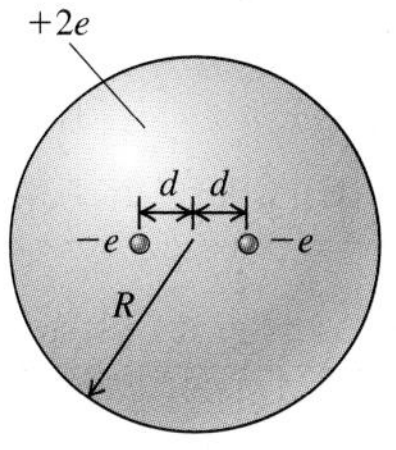

Figure 22.41 Problem 22.51.

22.52 Electric Field Inside a Hydrogen Atom. A hydrogen atom is made up of a proton of charge $+Q = 1.60 \times 10^{-19}$ C and an electron of charge $-Q = -1.60 \times 10^{-19}$ C. The proton may be regarded as a point charge at $r = 0$, the center of the atom. The motion of the electron causes its charge to be "smeared out" into a spherical distribution around the proton, so that the electron is equivalent to a charge per unit volume of

$$\rho(r) = -\frac{Q}{\pi a_0^3} e^{-2r/a_0}$$

where $a_0 = 5.29 \times 10^{-11}$ m is called the *Bohr radius.* a) Find the total amount of the hydrogen atom's charge that is enclosed within a sphere with radius r centered on the proton. Show that as $r \to \infty$, the enclosed charge goes to zero. Explain this result. b) Find the electric field (magnitude and direction) caused by the charge of the hydrogen atom as a function of r. c) Graph the electric field magnitude E as a function of r.

22.53 Probing Atomic Nuclei with Electron Scattering. To study the structure of the lead nucleus, electrons (charge $-e = -1.60 \times 10^{-19}$ C, mass 9.11×10^{-31} kg) are fired at a lead target. Some of the electrons actually enter the nuclei of the target, and the deflection of these electrons is measured. The deflection is caused by the charge of the nucleus, which is distributed approximately uniformly over the spherical volume of the nucleus. A lead nucleus has a charge of $+82e$ and a radius of $R = 7.1 \times 10^{-15}$ m. Find the acceleration of an electron at the following distances from the center of a calcium nucleus: a) $2R$; b) R; c) $R/2$; d) zero (at the center).

22.54 A Uniformly Charged Slab. A slab of insulating material has thickness $2d$ and is oriented so that its faces are parallel to the yz-plane and given by the planes $x = d$ and $x = -d$. The y- and z-dimensions of the slab are very large compared to d and may be treated as essentially infinite. The slab has a uniform positive charge density ρ. a) Explain why the electric field due to the slab is zero at the center of the slab $(x = 0)$. b) Using Gauss's law, find the electric field due to the slab (magnitude and direction) at all points in space.

22.55 A Non–Uniformly Charged Slab. Repeat Problem 22.54, but now let the charge density of the slab be given by $\rho(x) = \rho_0(x/d)^2$, where ρ_0 is a positive constant.

22.56 Can Electric Forces Alone Give Stable Equilibrium? In Chapter 21, several examples were given of calculating the force exerted on a point charge by other point charges in its surroundings. a) Consider a positive point charge $+q$. Give an example of how you would place two other point charges of your choosing so that the net force on charge $+q$ will be zero. b) If the net force on charge $+q$ is zero, then that charge is in equilibrium. The equilibrium will be *stable* if, when the charge $+q$ is displaced slightly in *any* direction from its position of equilibrium, the net force on the charge pushes it back toward the equilibrium position. For this to be the case, what must the direction of the electric field $\vec{E}$ be due to the other charges at points surrounding the equilibrium position of $+q$? c) Imagine that the charge $+q$ is moved very far away, and imagine a small Gaussian surface centered on the position where $+q$ was in equilibrium. By applying Gauss's law to this surface, show that it is *impossible* to satisfy the condition for stability described in part (b). In other words, a charge $+q$ cannot be held in stable equilibrium by electrostatic forces alone. This result is known as *Earnshaw's theorem.* d) Parts (a) through (c) referred to the equilibrium of a positive point charge $+q$. Prove that Earnshaw's theorem also applies to a negative point charge $-q$.

22.57 A non-uniform, but spherically symmetric, distribution of charge has a charge density $\rho(r)$ given as follows:

$$\rho(r) = \rho_0(1 - r/R) \quad \text{for} \quad r \le R$$
$$\rho(r) = 0 \quad \text{for} \quad r \ge R$$

where $\rho_0 = 3Q/\pi R^3$ is a positive constant. a) Show that the total charge contained in the charge distribution is Q. b) Show that the electric field in the region $r \ge R$ is identical to that produced by a point charge Q at $r = 0$. c) Obtain an expression for the electric field in the region $r \le R$. d) Graph the electric field magnitude E as a function of r. e) Find the value of r at which the electric field is maximum, and find the value of that maximum field.

22.58 A non-uniform, but spherically symmetric, distribution of charge has a charge density $\rho(r)$ given as follows:

$$\rho(r) = \rho_0(1 - 4r/3R) \quad \text{for} \quad r \le R$$
$$\rho(r) = 0 \quad \text{for} \quad r \ge R$$

where ρ_0 is a positive constant. a) Find the total charge contained in the charge distribution. b) Obtain an expression for the electric field in the region $r \ge R$. c) Obtain an expression for the electric field in the region $r \le R$. d) Graph the electric field magnitude E as a function of r. e) Find the value of r at which the electric field is maximum, and find the value of that maximum field.

22.59 Gauss's Law for Gravitation. The gravitational force between two point masses separated by a distance r is proportional to $1/r^2$, just like the electric force between two point charges. Because of this similarity between gravitational and electric interactions, there is also a Gauss's law for gravitation. a) Let $\vec{g}$ be the acceleration due to gravity caused by a point mass m at the origin, so that $\vec{g} = -(Gm/r^2)\hat{r}$. Consider a spherical Gaussian surface with radius r centered on this point mass, and show that the flux of $\vec{g}$ through this surface is given by

$$\oint \vec{g} \cdot d\vec{A} = -4\pi Gm$$

b) By following the same logical steps used in Section 22.3 to obtain Gauss's law for the electric field, show that the flux of $\vec{g}$ through *any* closed surface is given by

$$\oint \vec{g} \cdot d\vec{A} = -4\pi G M_{encl}$$

where M_{encl} is the total mass enclosed within the closed surface.

22.60 Applying Gauss's Law for Gravitation. Using Gauss's law for gravitation (derived in part (b) of Problem 22.59), show that the following statements are true: a) For any spherically symmetric mass distribution with total mass M, the acceleration due to gravity outside the distribution is the same as though all the mass were concentrated at the center. (*Hint:* See Example 22.5 in Section 22.4.) b) At any point inside a spherically symmetric shell of mass, the acceleration due to gravity is zero. (*Hint:* See Example 22.5.) c) If we could drill a hole through a spherically symmetric planet to its center, and if the density were uniform, we would find that the magnitude of $\vec{g}$ is directly proportional to the distance r from the center. (*Hint:* See Example 22.9 in Section 22.4.) We proved these results in Section 12.6 using some fairly strenuous analysis; the proofs using Gauss's law for gravitation are *much* easier.

22.61 a) An insulating sphere with radius a has a uniform charge density ρ. The sphere is not centered at the origin, but at $\vec{r} = \vec{b}$. Show that the electric field inside the sphere is given by $\vec{E} = \rho(\vec{r} - \vec{b})/3\epsilon_0$. b) An insulating sphere of radius R has a spherical hole of radius a located within its volume and centered a distance b from the center of the sphere, where $a < b < R$ (a cross section of the sphere is shown in Fig. 22.42). The solid part of the sphere has a uniform volume charge density ρ. Find the magnitude and direction of the electric field $\vec{E}$ inside the hole, and show that $\vec{E}$ is uniform over the entire hole. (*Hint:* Use the principle of superposition and the result of part (a).)

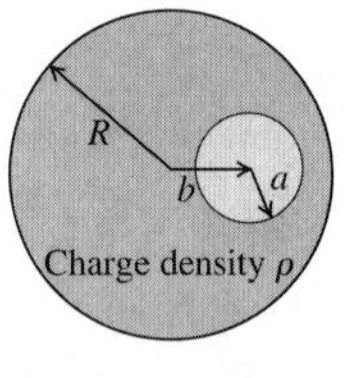

Fig. 22.42 Problem 22.61.

22.62 A very long, solid insulating cylinder with radius R has a cylindrical hole with radius a bored along its entire length. The axis of the hole is a distance b from the axis of the cylinder, where $a < b < R$ (Fig. 22.43). The solid material of the cylinder has a uniform volume charge density ρ. Find the magnitude and direction of the electric field $\vec{E}$ inside the hole, and show that $\vec{E}$ is uniform over the entire hole. (*Hint:* See Problem 22.61.)

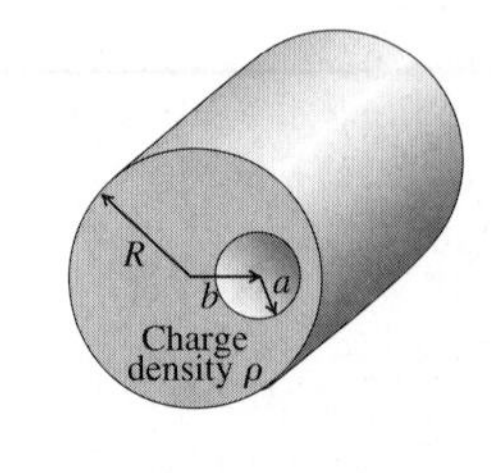

Fig. 22.43 Problem 22.62.

22.63 Positive charge Q is distributed uniformly over each of two spherical volumes with radius R. One sphere of charge is centered at the origin and the other at $x = 2R$ (Fig. 22.44). Find the magnitude and direction of the net electric field due to these two distributions of charge at the following points on the x-axis: a) $x = 0$; b) $x = R/2$; c) $x = R$; d) $x = 3R$.

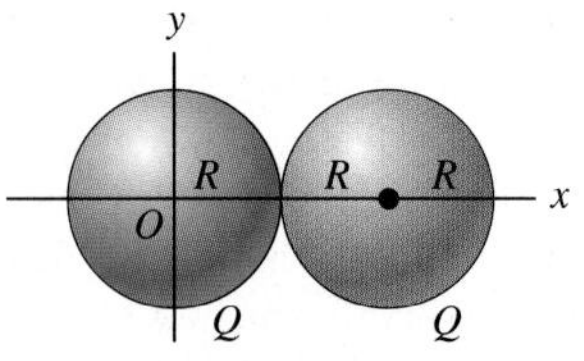

Figure 22.44 Problems 22.63 and 22.64.

22.64 Repeat Problem 22.63, but now let the left-hand sphere have positive charge Q and let the right-hand sphere have negative charge $-Q$.

Challenge Problems

22.65 A region in space contains a total positive charge Q that is distributed spherically such that the volume charge density $\rho(r)$ is given by

$$\rho(r) = \alpha \quad \text{for} \quad r \le R/2$$
$$\rho(r) = 2\alpha(1 - r/R) \quad \text{for} \quad R/2 \le r \le R$$
$$\rho(r) = 0 \quad \text{for} \quad r \ge R$$

Here α is a positive constant having units of C/m^3. a) Determine α in terms of Q and R. b) Using Gauss's law, derive an expression for the magnitude of $\vec{E}$ as a function of r. Do this separately for all three regions. Express your answers in terms of the total charge Q. Be sure to check that your results agree on the boundaries of the regions. c) What fraction of the total charge is contained within the region $r \le R/2$? d) If an electron with charge $q' = -e$ is oscillating back and forth about $r = 0$ (the center of the distribution) with an amplitude less than $R/2$, show that the motion is simple harmonic. (*Hint:* Review the discussion of simple harmonic motion in Section 13.2. If, and only if, the net force on the electron is proportional to its displacement from equilibrium, then the motion is simple harmonic.) e) What is the period of the motion in part (d)? f) If the amplitude of the motion described in part (e) is greater than $R/2$, is the motion still simple harmonic? Why or why not?

22.66 A region in space contains a total positive charge Q that is distributed spherically such that the volume charge density $\rho(r)$ is given by

$$\rho(r) = 3\alpha r/(2R) \quad \text{for} \quad r \le R/2$$
$$\rho(r) = \alpha[1 - (r/R)^2] \quad \text{for} \quad R/2 \le r \le R$$
$$\rho(r) = 0 \quad \text{for} \quad r \ge R$$

Here α is a positive constant having units of C/m^3. a) Determine α in terms of Q and R. b) Using Gauss's law, derive an expression for the magnitude of the electric field as a function of r. Do this separately for all three regions. Express your answers in terms of the total charge Q. c) What fraction of the total charge is contained within the region $R/2 \le r \le R$? d) What is the magnitude of $\vec{E}$ at $r = R/2$? e) If an electron with charge $q' = -e$ is released from rest at any point in any of the three regions, the resulting motion will be oscillatory but not simple harmonic. Why? (See Challenge Problem 22.65.)

ELECTRIC POTENTIAL 23

A lightning bolt releases a staggering amount of electric potential energy. Charge flows between a cloud and the ground at a rate that can exceed 10^5 coulombs per second. The electric potential energy per coulomb—that is, the electric potential—released in this process can be as great as 10^7 joules per coulomb, or 10^7 volts.

? **Many buildings have lightning rods to protect against damage from lightning strikes. How do lightning rods work?**

This chapter is about energy associated with electrical interactions. Every time you turn on a light, a CD player, or an electric appliance, you are making use of electrical energy, an indispensable ingredient of our technological society. In Chapters 6 and 7 we introduced the concepts of *work* and *energy* in the context of mechanics; now we'll combine these concepts with what we've learned about electric charge, electric forces, and electric fields. Just as the energy concept made it possible to solve some kinds of mechanics problems very simply, using energy ideas makes it easier to solve a variety of problems in electricity.

When a charged particle moves in an electric field, the field exerts a force that can do *work* on the particle. This work can always be expressed in terms of electric potential energy. Just as gravitational potential energy depends on the height of a mass above the earth's surface, electric potential energy depends on the position of the charged particle in the electric field. We'll describe electric potential energy using a new concept called *electric potential,* or simply *potential.* In circuits, a difference in potential from one point to another is often called *voltage.* The concepts of potential and voltage are crucial to understanding how electric circuits work and have equally important applications to electron beams in TV picture tubes, high-energy particle accelerators, and many other devices.

23.1 | Electric Potential Energy

The concepts of work, potential energy, and conservation of energy proved to be extremely useful in our study of mechanics. In this section we'll show that these concepts are just as useful for understanding and analyzing electrical interactions.

Let's begin by reviewing several essential points from Chapters 6 and 7. First, when a force $\vec{F}$ acts on a particle that moves from point a to point b, the work $W_{a \to b}$ done by the force is given by a *line integral:*

$$W_{a \to b} = \int_a^b \vec{F} \cdot d\vec{l} = \int_a^b F \cos \phi \, dl \quad \text{(work done by a force)} \tag{23.1}$$

where $d\vec{l}$ is an infinitesimal displacement along the particle's path and ϕ is the angle between $\vec{F}$ and $d\vec{l}$ at each point along the path.

Second, if the force $\vec{F}$ is *conservative,* as we defined the term in Section 7.3, the work done by $\vec{F}$ can always be expressed in terms of a **potential energy** U. When the particle moves from a point where the potential energy is U_a to a point where it is U_b, the change in potential energy is $\Delta U = U_b - U_a$ and the work $W_{a \to b}$ done by the force is

$$W_{a \to b} = U_a - U_b = -(U_b - U_a) = -\Delta U \tag{23.2}$$
$$\text{(work done by a conservative force)}$$

When $W_{a \to b}$ is positive, U_a is greater than U_b, ΔU is negative, and the potential energy *decreases.* That's what happens when a baseball falls from a high point (a) to a lower point (b) under the influence of the earth's gravity; the force of gravity does positive work, and the gravitational potential energy decreases. When a tossed ball is moving upward, the gravitational force does negative work during the ascent, and the potential energy increases.

Third, the work-energy theorem says that the change in kinetic energy $\Delta K = K_b - K_a$ during any displacement is equal to the *total* work done on the particle. If the only work done on the particle is done by conservative forces, then Eq. (23.2) gives the total work, and $K_b - K_a = -(U_b - U_a)$. We usually write this as

$$K_a + U_a = K_b + U_b \tag{23.3}$$

This is, the total mechanical energy (kinetic plus potential) is *conserved* under these circumstances.

Electric Potential Energy in a Uniform Field

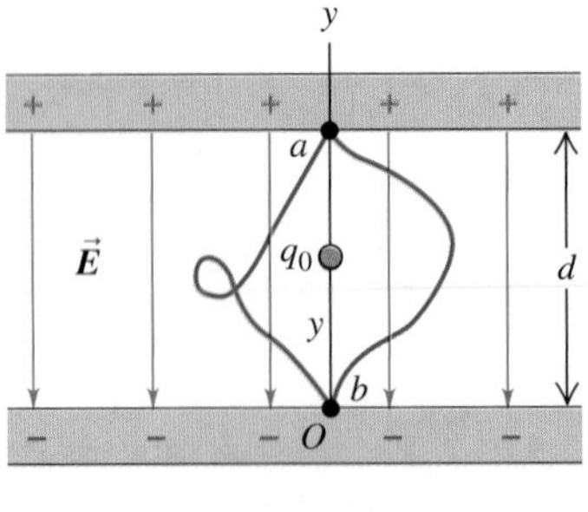

23.1 A test charge q_0 moving from a to b experiences a force of magnitude q_0E. The work done by this force is $W_{a \to b} = q_0Ed$ and is independent of the particle's path.

Let's look at an electrical example of these basic concepts. In Fig. 23.1 a pair of charged parallel metal plates sets up a uniform, downward electric field with magnitude E. The field exerts a downward force with magnitude $F = q_0E$ on a positive test charge q_0. As the charge moves downward a distance d from point a to point b, the force on the test charge is constant and independent of its location. So the work done by the electric field is the product of the force magnitude and the component of displacement in the (downward) direction of the force:

$$W_{a \to b} = Fd = q_0Ed \tag{23.4}$$

This work is positive, since the force is in the same direction as the net displacement of the test charge.

The y-component of the electric force, $F_y = -q_0 E$, is constant, and there is no x- or z-component. This is exactly analogous to the gravitational force on a mass m near the earth's surface; for this force, there is a constant y-component $F_y = -mg$ and the x- and z-components are zero. Because of this analogy, we can conclude that the force exerted on q_0 by the uniform electric field in Fig. 23.1 is *conservative,* just as is the gravitational force. This means that the work $W_{a\to b}$ done by the field is independent of the path the particle takes from a to b. We can represent this work with a *potential energy* function U, just as we did for gravitational potential energy in Section 7.1. The potential energy for the gravitational force $F_y = -mg$ was $U = mgy$; hence the potential energy for the electric force $F_y = -q_0 E$ is

$$U = q_0 E y \tag{23.5}$$

When the test charge moves from height y_a to height y_b, the work done on the charge by the field is given by

$$W_{a\to b} = -\Delta U = -(U_b - U_a) = -(q_0 E y_b - q_0 E y_a) = q_0 E(y_a - y_b) \tag{23.6}$$

When y_a is greater than y_b (Fig. 23.2a), the positive test charge q_0 moves downward, in the same direction as $\vec{E}$; the displacement is in the same direction as the force $\vec{F} = q_0\vec{E}$, so the field does positive work and U decreases. (In particular, if $y_a - y_b = d$ as in Fig. 23.1, Eq. (23.6) gives $W_{a\to b} = q_0 E d$, in agreement with Eq. (23.4).) When y_a is less than y_b (Fig. 23.2b), the positive test charge q_0 moves upward, in the opposite direction to $\vec{E}$; the displacement is opposite the force, the field does negative work, and U increases.

If the test charge q_0 is *negative,* the potential energy increases when it moves with the field and decreases when it moves against the field (Fig. 23.3).

Whether the test charge is positive or negative, the following general rules apply: *U increases* if the test charge q_0 moves in the direction *opposite* to the electric force $\vec{F} = q_0\vec{E}$ (Figs. 23.2b and 23.3a); *U decreases* if q_0 moves in the *same* direction as $\vec{F} = q_0\vec{E}$ (Figs. 23.2a and 23.3b). This is the same behavior as for gravitational potential energy, which increases if a mass m moves upward (oppo-

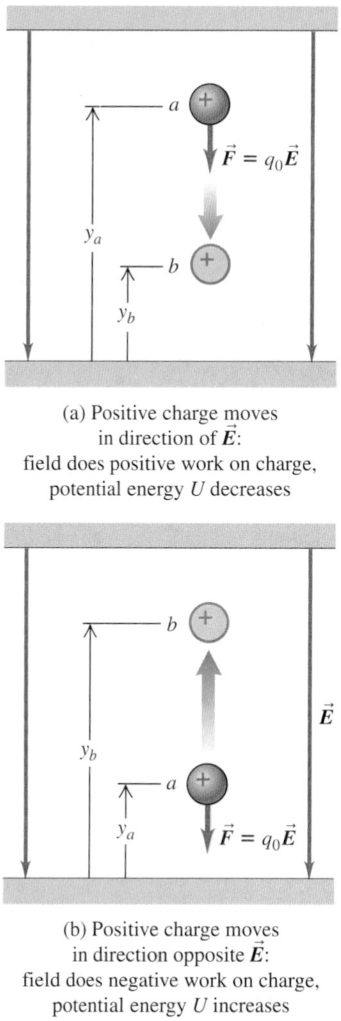

23.2 A positive charge moving (a) in the direction of the electric field $\vec{E}$ and (b) in the direction opposite $\vec{E}$.

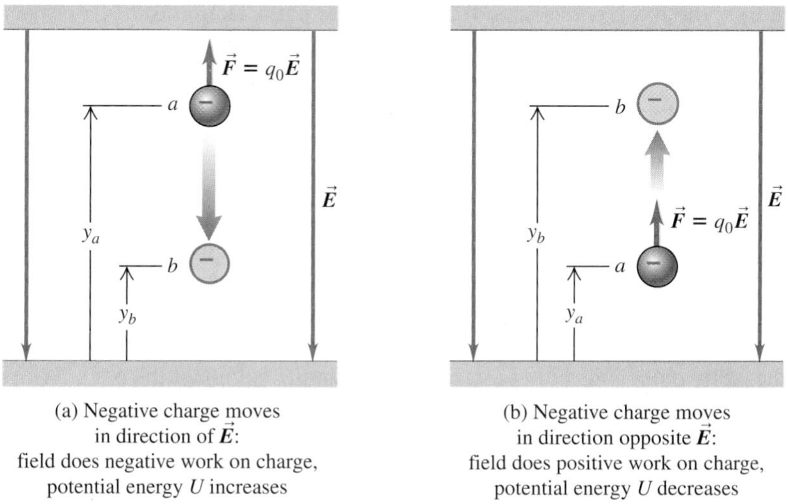

23.3 A negative charge moving (a) in the direction of the electric field $\vec{E}$ and (b) in the direction opposite $\vec{E}$. Compare with Fig. 23.2.

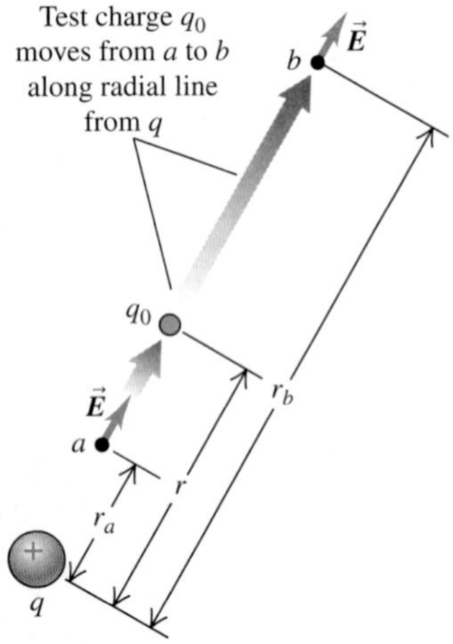

23.4 Test charge q_0 moves along a straight line extending radially from charge q. As it moves from a to b, the distance varies from r_a to r_b.

site the direction of the gravitational force) and decreases if m moves downward (in the same direction as the gravitational force).

CAUTION The relationship between electric potential energy change and motion in an electric field is an important one that we'll use often. It's also a relationship that takes a little effort to truly understand. Take the time to review the preceding paragraph thoroughly and to study Figs. 23.2 and 23.3 carefully. Doing so now will help you tremendously later!

Electric Potential Energy of Two Point Charges

The idea of electric potential energy isn't restricted to the special case of a uniform electric field. Indeed, we can apply this concept to a point charge in *any* electric field caused by a static charge distribution. Recall from Chapter 21 that we can represent any charge distribution as a collection of point charges. Therefore it's useful to calculate the work done on a test charge q_0 moving in the electric field caused by a single, stationary point charge q.

We'll consider first a displacement along the *radial* line in Fig. 23.4, from point a to point b. The force on q_0 is given by Coulomb's law, and its radial component is

$$F_r = \frac{1}{4\pi\epsilon_0}\frac{qq_0}{r^2} \tag{23.7}$$

If q and q_0 have the same sign ($+$ or $-$), the force is repulsive and F_r is positive; if the two charges have opposite signs, the force is attractive and F_r is negative. The force is *not* constant during the displacement, and we have to integrate to calculate the work $W_{a\to b}$ done on q_0 by this force as q_0 moves from a to b. We find

$$W_{a\to b} = \int_{r_a}^{r_b} F_r\,dr = \int_{r_a}^{r_b} \frac{1}{4\pi\epsilon_0}\frac{qq_0}{r^2}\,dr = \frac{qq_0}{4\pi\epsilon_0}\left(\frac{1}{r_a} - \frac{1}{r_b}\right) \tag{23.8}$$

The work done by the electric force for this particular path depends only on the end points.

In fact, the work is the same for *all possible* paths from a to b. To prove this, we consider a more general displacement (Fig. 23.5) in which a and b do not lie on the same radial line. From Eq. (23.1) the work done on q_0 during this displacement is given by

$$W_{a\to b} = \int_{r_a}^{r_b} F\cos\phi\,dl = \int_{r_a}^{r_b} \frac{1}{4\pi\epsilon_0}\frac{qq_0}{r^2}\cos\phi\,dl$$

But the figure shows that $\cos\phi\,dl = dr$. That is, the work done during a small displacement $d\vec{l}$ depends only on the change dr in the distance r between the charges, which is the *radial component* of the displacement. Thus Eq. (23.8) is valid even for this more general displacement; the work done on q_0 by the electric field $\vec{E}$ produced by q depends only on r_a and r_b, not on the details of the path. Also, if q_0 returns to its starting point a by a different path, the total work done in the round-trip displacement is zero (the integral in Eq. (23.8) is from r_a back to r_a). These are the needed characteristics for a conservative force, as we defined it in Section 7.3. Thus the force on q_0 is a *conservative* force.

We see that Eqs. (23.2) and (23.8) are consistent if we define $qq_0/4\pi\epsilon_0 r_a$ to be the potential energy U_a when q_0 is at point a, a distance r_a from q, and we define

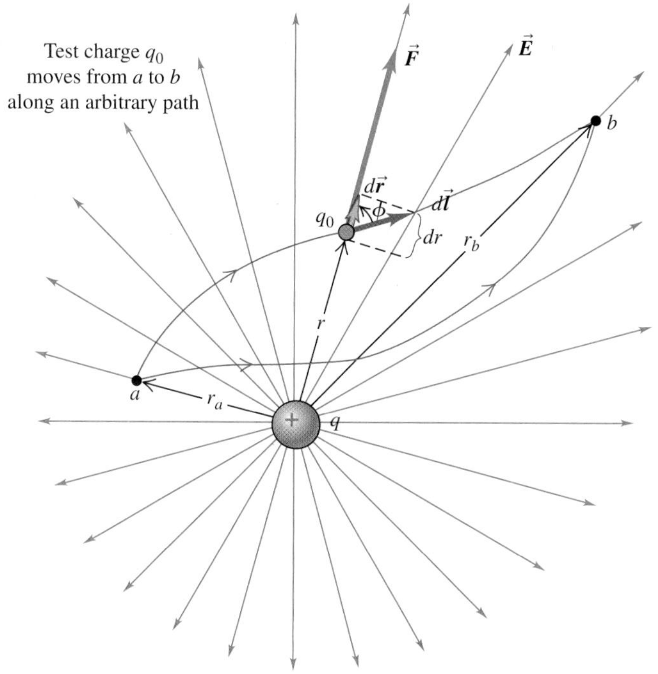

23.5 The work done on charge q_0 by the electric field of charge q does not depend on the path taken, but only on the distances r_a and r_b.

$qq_0/4\pi\epsilon_0 r_b$ to be the potential energy U_b when q_0 is at point b, a distance r_b from q. Thus the potential energy U when the test charge q_0 is at *any* distance r from charge q is

$$U = \frac{1}{4\pi\epsilon_0}\frac{qq_0}{r} \tag{23.9}$$

(electric potential energy of two point charges q and q_0)

Note that we have *not* assumed anything about the signs of q and q_0; Eq. (23.9) is valid for any combination of signs. The potential energy is positive if the charges q and q_0 have the same sign (Fig. 23.6a) and negative if they have opposite signs (Fig. 23.6b).

CAUTION Be careful not to confuse Eq. (23.9) for the potential energy of two point charges with the similar expression in Eq. (23.7) for the radial component of the electric force that one charge exerts on the other. The potential energy U is proportional to $1/r$, while the force component F_r is proportional to $1/r^2$.

Potential energy is always defined relative to some reference point where $U = 0$. In Eq. (23.9), U is zero when q and q_0 are infinitely far apart and $r = \infty$. Therefore U represents the work that would be done on the test charge q_0 by the field of q if q_0 moved from an initial distance r to infinity. If q and q_0 have the same sign, the interaction is repulsive, this work is positive, and U is positive at any finite separation (Fig. 23.6a). If the charges have opposite signs, the interaction is attractive, the work done is negative, and U is negative (Fig. 23.6b).

We emphasize that the potential energy U given by Eq. (23.9) is a *shared* property of the two charges q and q_0; it is a consequence of the *interaction* between these two bodies. If the distance between the two charges is changed from r_a to r_b, the change in potential energy is the same whether q is held fixed and q_0 is moved or q_0 is held fixed and q is moved. For this reason, we never use the phrase "the electric potential energy *of* a point charge." (Likewise, if a mass m is at a height h above the earth's surface, the gravitational potential energy is a shared property of the mass m and the earth. We emphasized this in Sections 7.1 and 12.3.)

Gauss's law tells us that the electric field outside any spherically symmetric charge distribution is the same as though all the charge were concentrated at the center. Therefore Eq. (23.9) also holds if the test charge q_0 is outside any spherically symmetric charge distribution with total charge q at a distance r from the center.

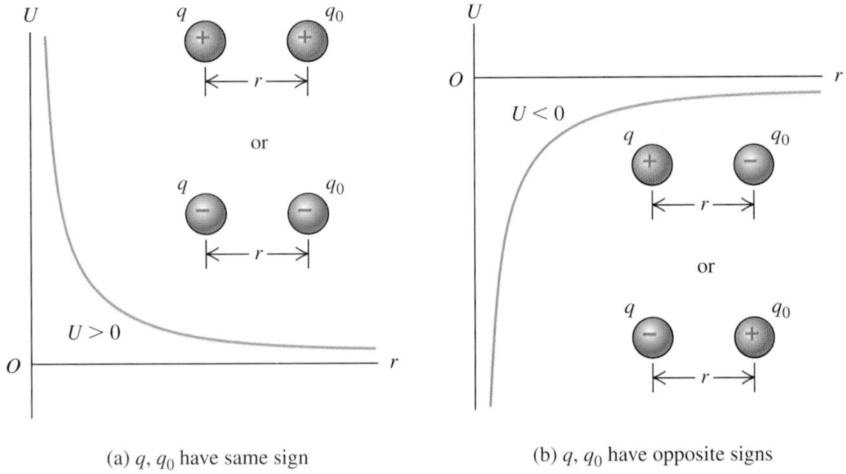

23.6 Graphs of the potential energy U of two point charges q and q_0 versus their separation r.

Example 23.1

Conservation of energy with electric forces

A positron (the antiparticle of the electron) has a mass of 9.11×10^{-31} kg and a charge $+e = +1.60 \times 10^{-19}$ C. Suppose a positron moves in the vicinity of an alpha particle, which has a charge $+2e = 3.20 \times 10^{-19}$ C. The alpha particle is over 7000 times as massive as the positron, so we assume that it is at rest in some inertial frame of reference. When the positron is 1.00×10^{-10} m from the alpha particle, it is moving directly away from the alpha particle at a speed of 3.00×10^{6} m/s. a) What is the positron's speed when the two particles are 2.00×10^{-10} m apart? b) What is the positron's speed when it is very far away from the alpha particle? c) How would the situation change if the moving particle were an electron (same mass as the positron but opposite charge)?

SOLUTION

IDENTIFY: The electric force between the positron and alpha particle is conservative, so mechanical energy (kinetic plus potential) is conserved.

SET UP: The kinetic and potential energies at any two points a and b are related by Eq. (23.3), $K_a + U_a = K_b + U_b$, and the potential energy at any distance r is given by Eq. (23.9). We are given complete information about the system at a point a where the two charges are 1.00×10^{-10} m apart. We use Eqs. (23.3) and (23.9) to find the speed at two different values of r in parts (a) and (b), and for the case where the charge $+e$ is replaced by $-e$ in part (c).

EXECUTE: a) In this part, $r_b = 2.00 \times 10^{-10}$ m and we want to find the final speed v_b of the positron. This appears in the expression for the final kinetic energy, $K_b = \frac{1}{2}mv_b^2$; solving the energy-conservation equation for K_b, we have

$$K_b = K_a + U_a - U_b$$

The values of the energies on the right-hand side of this expression are

$$K_a = \frac{1}{2}mv_a^2 = \frac{1}{2}(9.11 \times 10^{-31}\text{ kg})(3.00 \times 10^{6}\text{ m/s})^2$$
$$= 4.10 \times 10^{-18}\text{ J}$$

$$U_a = \frac{1}{4\pi\epsilon_0}\frac{qq_0}{r_a}$$

$$= (9.0 \times 10^9\ \mathrm{N \cdot m^2/C^2}) \frac{(3.20 \times 10^{-19}\ \mathrm{C})(1.60 \times 10^{-19}\ \mathrm{C})}{1.00 \times 10^{-10}\ \mathrm{m}}$$

$$= 4.61 \times 10^{-18}\ \mathrm{J}$$

$$U_b = (9.0 \times 10^9\ \mathrm{N \cdot m^2/C^2}) \frac{(3.20 \times 10^{-19}\ \mathrm{C})(1.60 \times 10^{-19}\ \mathrm{C})}{2.00 \times 10^{-10}\ \mathrm{m}}$$

$$= 2.30 \times 10^{-18}\ \mathrm{J}$$

Hence the final kinetic energy is

$$K_b = \frac{1}{2}mv_b^2$$
$$= 4.10 \times 10^{-18}\ \mathrm{J} + 4.61 \times 10^{-18}\ \mathrm{J} - 2.30 \times 10^{-18}\ \mathrm{J}$$
$$= 6.41 \times 10^{-18}\ \mathrm{J}$$

and the final speed of the positron is

$$v_b = \sqrt{\frac{2K_b}{m}} = \sqrt{\frac{2(6.41 \times 10^{-18}\ \mathrm{J})}{9.11 \times 10^{-31}\ \mathrm{kg}}} = 3.8 \times 10^6\ \mathrm{m/s}$$

The force is repulsive, so the positron speeds up as it moves away from the stationary alpha particle.

b) When the final positions of the positron and alpha particle are very far apart, the separation r_b approaches infinity and the final potential energy U_b approaches zero. Then the final kinetic energy of the positron is

$$K_b = K_a + U_a - U_b = 4.10 \times 10^{-18}\ \mathrm{J} + 4.61 \times 10^{-18}\ \mathrm{J} - 0$$
$$= 8.71 \times 10^{-18}\ \mathrm{J}$$

and its final speed is

$$v_b = \sqrt{\frac{2K_b}{m}} = \sqrt{\frac{2(8.71 \times 10^{-18}\ \mathrm{J})}{9.11 \times 10^{-31}\ \mathrm{kg}}} = 4.4 \times 10^6\ \mathrm{m/s}$$

Comparing to part (a), we see that as the positron moves from $r = 2.00 \times 10^{-10}$ m to infinity, the additional work done on it by the electric field of the alpha particle increases the speed by only about 16%. This is because the electric force decreases rapidly with distance.

c) If the moving charge is negative, the force on it is attractive rather than repulsive, and we expect it to slow down rather than speed up. The only difference in the above calculations is that both potential-energy quantities are negative. From part (a), at a distance $r_b = 2.00 \times 10^{-10}$ m we have

$$K_b = K_a + U_a - U_b$$
$$= 4.10 \times 10^{-18}\ \mathrm{J} + (-4.61 \times 10^{-18}\ \mathrm{J})$$
$$- (-2.30 \times 10^{-18}\ \mathrm{J})$$
$$= 1.79 \times 10^{-18}\ \mathrm{J}$$
$$v_b = \sqrt{\frac{2K_b}{m}} = 2.0 \times 10^6\ \mathrm{m/s}$$

From part (b), at $r_b = \infty$ the kinetic energy of the electron would seem to be

$$K_b = K_a + U_a - U_b = 4.10 \times 10^{-18}\ \mathrm{J}$$
$$+ (-4.61 \times 10^{-18}\ \mathrm{J}) - 0$$
$$= -5.1 \times 10^{-19}\ \mathrm{J}$$

But kinetic energies can *never* be negative! This result means that the electron can never reach $r_b = \infty$; the attractive force brings the electron to a halt at a finite distance from the alpha particle. The electron will then begin to move back toward the alpha particle. You can solve for the distance r_b at which the electron comes momentarily to rest by setting K_b equal to zero in the equation for conservation of mechanical energy (see Exercise 23.8).

EVALUATE: It's useful to compare our calculations with Fig. 23.6. In parts (a) and (b), the charges have the same sign; since $r_b > r_a$, the potential energy U_b is less than U_a. In part (c), the charges have opposite signs; since $r_b > r_a$, the potential energy U_b is greater (that is, less negative) than U_a.

Electric Potential Energy with Several Point Charges

Suppose the electric field $\vec{E}$ in which charge q_0 moves is caused by *several* point charges $q_1, q_2, q_3, \ldots$ at distances $r_1, r_2, r_3, \ldots$ from q_0, as in Fig. 23.7. For example, q_0 could be a positive ion moving in the presence of other ions

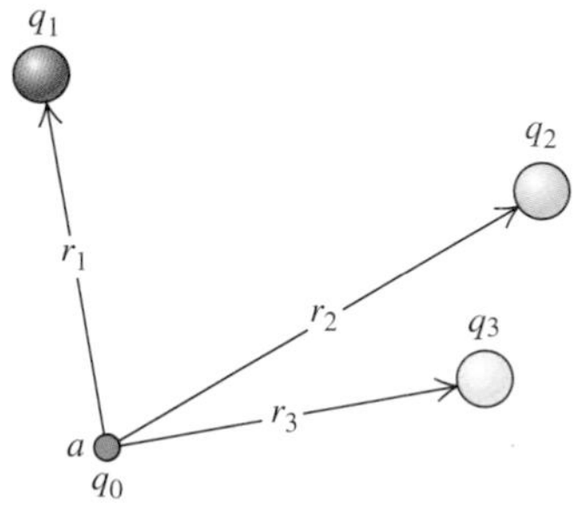

23.7 The potential energy associated with a charge q_0 at point a depends on the other charges q_1, q_2, and q_3 and on their distances r_1, r_2, and r_3 from point a.

23.8 This ion engine for spacecraft ejects a stream of positive xenon ions (Xe^+) at speeds in excess of 30 km/s. The thrust produced is very low (about 0.09 newton) but can be maintained continuously for days, in contrast to chemical rockets, which produce a large thrust for a short time (see Fig. 8.29). Such ion engines have been used for maneuvering interplanetary spacecraft.

(Fig. 23.8). The total electric field at each point is the *vector sum* of the fields due to the individual charges, and the total work done on q_0 during any displacement is the sum of the contributions from the individual charges. From Eq. (23.9) we conclude that the potential energy associated with the test charge q_0 at point a in Fig. 23.7 is the *algebraic* sum (*not* a vector sum)

$$U = \frac{q_0}{4\pi\epsilon_0}\left(\frac{q_1}{r_1} + \frac{q_2}{r_2} + \frac{q_3}{r_3} + \cdots\right) = \frac{q_0}{4\pi\epsilon_0}\sum_i \frac{q_i}{r_i} \qquad (23.10)$$

(point charge q_0 and collection of charges q_i)

When q_0 is at a different point b, the potential energy is given by the same expression, but $r_1, r_2, \ldots$ are the distances from $q_1, q_2, \ldots$ to point b. The work done on charge q_0 when it moves from a to b along any path is equal to the difference $U_a - U_b$ between the potential energies when q_0 is at a and at b.

We can represent *any* charge distribution as a collection of point charges, so Eq. (23.10) shows that we can always find a potential-energy function for *any* static electric field. It follows that **for every electric field due to a static charge distribution, the force exerted by that field is conservative.**

Equations (23.9) and (23.10) define U to be zero when all the distances $r_1, r_2, \ldots$ are infinite—that is, when the test charge q_0 is very far away from all the charges that produce the field. As with any potential-energy function, the point where $U = 0$ is arbitrary; we can always add a constant to make U equal zero at any point we choose. In electrostatics problems it's usually simplest to choose this point to be at infinity. When we analyze electric circuits in Chapters 25 and 26, other choices will be more convenient.

Equation (23.10) gives the potential energy associated with the presence of the test charge q_0 in the $\vec{E}$ field produced by $q_1, q_2, q_3, \ldots$ But there is also potential energy involved in assembling these charges. If we start with charges $q_1, q_2, q_3, \ldots$ all separated from each other by infinite distances and then bring them together so that the distance between q_i and q_j is r_{ij}, the *total* potential energy U is the sum of the potential energies of interaction for each pair of charges. We can write this as

$$U = \frac{1}{4\pi\epsilon_0}\sum_{i<j}\frac{q_i q_j}{r_{ij}} \qquad (23.11)$$

This sum extends over all *pairs* of charges; we don't let $i = j$ (because that would be an interaction of a charge with itself), and we include only terms with $i < j$ to make sure that we count each pair only once. Thus, to account for the interaction between q_3 and q_4, we include a term with $i = 3$ and $j = 4$ but not a term with $i = 4$ and $j = 3$.

Interpreting Electric Potential Energy

As a final comment, here are two viewpoints on electric potential energy. We have defined it in terms of the work done *by the electric field* on a charged particle moving in the field, just as in Chapter 7 we defined potential energy in terms of the work done by gravity or by a spring. When a particle moves from point a to point b, the work done on it by the electric field is $W_{a\to b} = U_a - U_b$. Thus the potential-energy difference $U_a - U_b$ equals *the work that is done by the electric force when the particle moves from a to b.* When U_a is greater than U_b, the field does positive work on the particle as it "falls" from a point of higher potential energy (a) to a point of lower potential energy (b).

An alternative but equivalent viewpoint is to consider how much work we would have to do to "raise" a particle from a point b where the potential energy is U_b to a point a where it has a greater value U_a (pushing two positive charges closer together, for example). To move the particle slowly (so as not to give it any kinetic energy), we need to exert an additional external force $\vec{F}_{ext}$ that is equal and opposite to the electric-field force and does positive work. The potential-energy difference $U_a - U_b$ is then defined as *the work that must be done by an external force to move the particle slowly from b to a against the electric force.* Because $\vec{F}_{ext}$ is the negative of the electric-field force and the displacement is in the opposite direction, this definition of the potential difference $U_a - U_b$ is equivalent to that given above. This alternative viewpoint also works if U_a is less than U_b, corresponding to "lowering" the particle; an example is moving two positive charges away from each other. In this case, $U_a - U_b$ is again equal to the work done by the external force, but now this work is negative.

We will use both of these viewpoints in the next section to interpret what is meant by electric *potential,* or potential energy per unit charge.

Example 23.2 A system of point charges

Two point charges are located on the x-axis, $q_1 = -e$ at $x = 0$ and $q_2 = +e$ at $x = a$. a) Find the work that must be done by an external force to bring a third point charge $q_3 = +e$ from infinity to $x = 2a$. b) Find the total potential energy of the system of three charges.

SOLUTION

IDENTIFY and SET UP: Figure 23.9 shows the final arrangement of the three charges. To find the work required to bring q_3 in from infinity, we use Eq. (23.10) to find the potential energy associated with q_3 in the presence of q_1 and q_2. We then use Eq. (23.11) to find the total potential energy of the system.

EXECUTE: a) The work that must be done on q_3 by an external force $\vec{F}_{ext}$ is equal to the difference between two quantities: the potential energy U associated with q_3 when it is at $x = 2a$ and the potential energy when it is infinitely far away. The second of these is zero, so the work that must be done is equal to U. The distances between the charges are $r_{13} = 2a$ and $r_{23} = a$, so from Eq. (23.10),

$$W = U = \frac{q_3}{4\pi\epsilon_0}\left(\frac{q_1}{r_{13}} + \frac{q_2}{r_{23}}\right) = \frac{+e}{4\pi\epsilon_0}\left(\frac{-e}{2a} + \frac{+e}{a}\right) = \frac{+e^2}{8\pi\epsilon_0 a}$$

If q_3 is brought in from infinity along the $+x$-axis, it is attracted by q_1 but is repelled more strongly by q_2; hence positive work must be done to push q_3 to the position at $x = 2a$.

b) The total potential energy of the assemblage of three charges is given by Eq. (23.11):

$$U = \frac{1}{4\pi\epsilon_0}\sum_{i<j}\frac{q_i q_j}{r_{ij}} = \frac{1}{4\pi\epsilon_0}\left(\frac{q_1 q_2}{r_{12}} + \frac{q_1 q_3}{r_{13}} + \frac{q_2 q_3}{r_{23}}\right)$$
$$= \frac{1}{4\pi\epsilon_0}\left(\frac{(-e)(e)}{a} + \frac{(-e)(e)}{2a} + \frac{(e)(e)}{a}\right) = \frac{-e^2}{8\pi\epsilon_0 a}$$

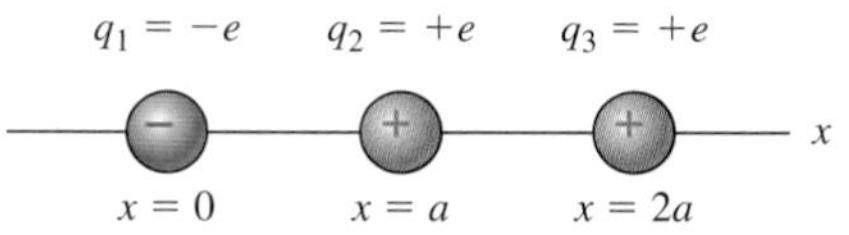

23.9 What is the total potential energy of this system of charges?

EVALUATE: Since our result in part (b) is negative, the system has lower potential energy than it would if the three charges were infinitely far apart. An external force would have to do *negative* work to bring the three charges from infinity to assemble this entire arrangement and would have to do *positive* work to move the three charges back to infinity.

Test Your Understanding

What is the total potential energy of the system of three charges in Example 21.4 (Section 21.3)? Is the result positive or negative? What is the physical interpretation of this sign?

23.2 | Electric Potential

In the preceding section we looked at the potential energy U associated with a test charge q_0 in an electric field. Now we want to describe this potential energy on a "per unit charge" basis, just as electric field describes the force per unit charge on a charged particle in the field. This leads us to the concept of *electric potential,* often called simply *potential.* This concept is very useful in calculations involving energies of charged particles. It also facilitates many electric-field calculations because electric potential is closely related to the electric field $\vec{E}$. When we need to determine an electric field, it is often easier to determine the potential first and then find the field from it.

Potential is *potential energy per unit charge.* We define the potential V at any point in an electric field as the potential energy U *per unit charge* associated with a test charge q_0 at that point:

$$V = \frac{U}{q_0} \quad \text{or} \quad U = q_0 V \tag{23.12}$$

Potential energy and charge are both scalars, so potential is a scalar quantity. From Eq. (23.12) its units are found by dividing the units of energy by those of charge. The SI unit of potential, called one **volt** (1 V) in honor of the Italian scientist and electrical experimenter Alessandro Volta (1745–1827), equals 1 joule per coulomb:

$$1\text{ V} = 1\text{ volt} = 1\text{ J/C} = 1\text{ joule/coulomb}$$

Let's put Eq. (23.2), which equates the work done by the electric force during a displacement from a to b to the quantity $-\Delta U = -(U_b - U_a)$, on a "work per unit charge" basis. We divide this equation by q_0, obtaining

$$\frac{W_{a\rightarrow b}}{q_0} = -\frac{\Delta U}{q_0} = -\left(\frac{U_b}{q_0} - \frac{U_a}{q_0}\right) = -(V_b - V_a) = V_a - V_b \tag{23.13}$$

where $V_a = U_a/q_0$ is the potential energy per unit charge at point a, and similarly for V_b. We call V_a and V_b the *potential at point a* and *potential at point b,* respectively. Thus the work done per unit charge by the electric force when a charged body moves from a to b is equal to the potential at a minus the potential at b.

23.10 The voltage of this battery equals the difference in potential $V_{ab} = V_a - V_b$ between its positive terminal (point a) and its negative terminal (point b).

The difference $V_a - V_b$ is called the *potential of a with respect to b*; we sometimes abbreviate this difference as $V_{ab} = V_a - V_b$ (note the order of the subscripts). This is often called the potential difference between a and $b,$ but that's ambiguous unless we specify which is the reference point. In electric circuits, which we will analyze in later chapters, the potential difference between two points is often called **voltage** (Fig. 23.10). Equation (23.13) then states that $\boldsymbol{V_{ab}}$**, the potential of *a* with respect to *b*, equals the work done by the electric force when a UNIT charge moves from *a* to *b*.**

Another way to interpret the potential difference V_{ab} in Eq. (23.13) is to use the alternative viewpoint mentioned at the end of Section 23.1. In that viewpoint, $U_a - U_b$ is the amount of work that must be done by an *external* force to move a particle of charge q_0 slowly from b to a against the electric force. The work that must be done *per unit charge* by the external force is then $(U_a - U_b)/q_0 = V_a - V_b = V_{ab}$. In other words, $\boldsymbol{V_{ab}}$**, the potential of *a* with respect to *b*, equals the work that must be done to move a UNIT charge slowly from *b* to *a* against the electric force.**

An instrument that measures the difference of potential between two points is called a *voltmeter.* In Chapter 26 we will discuss the principle of the common type of moving-coil voltmeter. There are also much more sensitive potential-measuring devices that use electronic amplification. Instruments that can measure a potential difference of 1 μV are common, and sensitivities down to 10^{-12} V can be attained.

To find the potential V due to a single point charge q, we divide Eq. (23.9) by q_0:

$$V = \frac{U}{q_0} = \frac{1}{4\pi\epsilon_0}\frac{q}{r} \quad \text{(potential due to a point charge)} \qquad (23.14)$$

where r is the distance from the point charge q to the point at which the potential is evaluated. If q is positive, the potential that it produces is positive at all points; if q is negative, it produces a potential that is negative everywhere. In either case, V is equal to zero at $r = \infty$, an infinite distance from the point charge. Note that potential, like electric field, is independent of the test charge q_0 that we use to define it.

Similarly, we divide Eq. (23.10) by q_0 to find the potential due to a collection of point charges:

$$V = \frac{U}{q_0} = \frac{1}{4\pi\epsilon_0}\sum_i \frac{q_i}{r_i} \qquad (23.15)$$

(potential due to a collection of point charges)

In this expression, r_i is the distance from the ith charge, q_i, to the point at which V is evaluated. Just as the electric field due to a collection of point charges is the *vector* sum of the fields produced by each charge, the electric potential due to a collection of point charges is the *scalar* sum of the potentials due to each charge. When we have a continuous distribution of charge along a line, over a surface, or through a volume, we divide the charge into elements dq, and the sum in Eq. (23.15) becomes an integral:

$$V = \frac{1}{4\pi\epsilon_0}\int \frac{dq}{r} \qquad (23.16)$$

(potential due to a continuous distribution of charge)

where r is the distance from the charge element dq to the field point where we are finding V. We'll work out several examples of such cases. The potential defined by Eqs. (23.15) and (23.16) is zero at points that are infinitely far away from *all* the charges. Later we'll encounter cases in which the charge distribution itself extends to infinity. We'll find that in such cases we cannot set $V = 0$ at infinity, and we'll need to exercise care in using and interpreting Eqs. (23.15) and (23.16).

CAUTION Before getting too involved in the details of how to calculate electric potential, you should stop and remind yourself what potential is. The electric *potential* at a certain point is the potential energy that would be associated with a *unit* charge placed at that point. That's why potential is measured in joules per coulomb, or volts. Keep in mind, too, that there doesn't have to be a charge at a given point for a potential V to exist at that point. (In the same way, an electric field can exist at a given point even if there's no charge there to respond to it.)

ActivPhysics Online: 11.13 Electrical Potential Energy and Potential

When we are given a collection of point charges, Eq. (23.15) is usually the easiest way to calculate the potential V. But in some problems in which the electric field is known or can be found easily, it is easier to determine V from $\vec{E}$. The force $\vec{F}$ on a test charge q_0 can be written as $\vec{F} = q_0\vec{E}$, so from Eq. (23.1) the work done by the electric force as the test charge moves from a to b is given by

$$W_{a\rightarrow b} = \int_a^b \vec{F}\cdot d\vec{l} = \int_a^b q_0\vec{E}\cdot d\vec{l}$$

If we divide this by q_0 and compare the result with Eq. (23.13), we find

$$V_a - V_b = \int_a^b \vec{E}\cdot d\vec{l} = \int_a^b E\cos\phi\, dl \tag{23.17}$$

(potential difference as an integral of $\vec{E}$)

The value of $V_a - V_b$ is independent of the path taken from a to b, just as the value of $W_{a\rightarrow b}$ is independent of the path. To interpret Eq. (23.17), remember that $\vec{E}$ is the electric force per unit charge on a test charge. If the line integral $\int_a^b \vec{E}\cdot d\vec{l}$ is positive, the electric field does positive work on a positive test charge as it moves from a to b. In this case the electric potential energy decreases as the test charge moves, so the potential energy per unit charge decreases as well; hence V_b is less than V_a and $V_a - V_b$ is positive.

Figure 23.11a shows a positive point charge. The electric field is directed away from the charge, and $V = q/4\pi\epsilon_0 r$ is positive at any finite distance from the charge. If you move away from the charge, in the direction of $\vec{E}$, you move toward lower values of V; if you move toward the charge, in the direction opposite $\vec{E}$, you move toward greater values of V. For the negative point charge in Fig. 23.11b, $\vec{E}$ is directed toward the charge and $V = q/4\pi\epsilon_0 r$ is negative at any finite distance from the charge. In this case, if you move toward the charge, you are moving in the direction of $\vec{E}$ and in the direction of decreasing (more negative) V. Moving away from the charge, in the direction opposite $\vec{E}$, moves you toward increasing (less negative) values of V. The general rule, valid for *any* electric field, is: moving *with* the direction of $\vec{E}$ means moving in the direction of *decreasing* V, and moving *against* the direction of $\vec{E}$ means moving in the direction of *increasing* V.

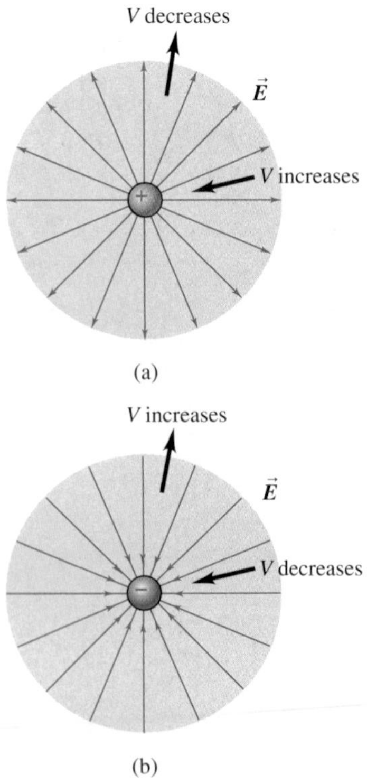

23.11 (a) A positive point charge. (b) A negative point charge. In both cases, if you move in the direction of $\vec{E}$, electric potential V decreases; if you move in the direction opposite $\vec{E}$, V increases.

Also, a positive test charge q_0 experiences an electric force in the direction of $\vec{E}$, toward lower values of V; a negative test charge experiences a force opposite to $\vec{E}$, toward higher values of V. Thus a positive charge tends to "fall" from a high-potential region to a lower-potential one. The opposite is true for a negative charge.

Notice that Eq. (23.17) can be rewritten as follows:

$$V_a - V_b = -\int_b^a \vec{E}\cdot d\vec{l} \tag{23.18}$$

This has a negative sign compared to the integral in Eq. (23.17), and the limits are reversed; hence Eqs. (23.17) and (23.18) are equivalent. But Eq. (23.18) has a slightly different interpretation. To move a unit charge slowly against the electric force, we must apply an *external* force per unit charge equal to $-\vec{E}$, equal and opposite to the electric force per unit charge $\vec{E}$. Equation (23.18) says that $V_a - V_b = V_{ab}$, the potential of a with respect to b, equals the work done per unit charge by this external force to move a unit charge from b to a. This is the same alternative interpretation we discussed under Eq. (23.13).

Equations (23.17) and (23.18) show that the unit of potential difference (1 V) is equal to the unit of electric field (1 N/C) multiplied by the unit of distance (1 m). Hence the unit of electric field can be expressed as 1 *volt per meter* (1 V/m), as well as 1 N/C:

$$1 \text{ V/m} = 1 \text{ volt/meter} = 1 \text{ N/C} = 1 \text{ newton/coulomb}$$

In practice, the volt per meter is the usual unit of electric field magnitude.

The magnitude e of the electron charge can be used to define a unit of energy that is useful in many calculations with atomic and nuclear systems. When a particle with charge q moves from a point where the potential is V_b to a point where it is V_a, the change in the potential energy U is

$$U_a - U_b = q(V_a - V_b) = qV_{ab}$$

If the charge q equals the magnitude e of the electron charge, 1.602×10^{-19} C, and the potential difference is $V_{ab} = 1$ V, the change in energy is

$$U_a - U_b = (1.602 \times 10^{-19}\text{ C})(1\text{ V}) = 1.602 \times 10^{-19}\text{ J}$$

This quantity of energy is defined to be 1 **electron volt** (1 eV):

$$1 \text{ eV} = 1.602 \times 10^{-19}\text{ J}$$

The multiples meV, keV, MeV, GeV, and TeV are often used.

CAUTION Remember that the electron volt is a unit of energy, *not* a unit of potential or potential difference!

When a particle with charge e moves through a potential difference of one volt, the change in potential *energy* is 1 eV. If the charge is some multiple of e, say Ne, the change in potential energy in electron volts is N times the potential difference in volts. For example, when an alpha particle, which has charge $2e$, moves between two points with a potential difference of 1000 V, the change in potential energy is $2(1000\text{ eV}) = 2000$ eV. To confirm this, we write

$$U_a - U_b = qV_{ab} = (2e)(1000\text{ V}) = (2)(1.602 \times 10^{-19}\text{ C})(1000\text{ V})$$
$$= 3.204 \times 10^{-16}\text{ J} = 2000\text{ eV}$$

Although we have defined the electron volt in terms of *potential* energy, we can use it for *any* form of energy, such as the kinetic energy of a moving particle. When we speak of a "one-million-electron-volt proton," we mean a proton with a kinetic energy of one million electron volts (1 MeV), equal to $(10^6)(1.602 \times 10^{-19}\text{ J}) = 1.602 \times 10^{-13}$ J (Fig. 23.12).

23.12 This accelerator at the Fermi National Accelerator Laboratory in Illinois gives protons a kinetic energy of 400 MeV $(4 \times 10^8\text{ eV})$. Additional acceleration stages increase their kinetic energy to 980 GeV, or 0.98 TeV $(9.8 \times 10^{11}\text{ eV})$.

Example 23.3 Electric force and electric potential

A proton (charge $+e = 1.602 \times 10^{-19}$ C) moves in a straight line from point a to point b inside a linear accelerator, a total distance $d = 0.50$ m. The electric field is uniform along this line, with magnitude $E = 1.5 \times 10^7$ V/m $= 1.5 \times 10^7$ N/C in the direction from a to b. Determine a) the force on the proton; b) the work done on it by the field; c) the potential difference $V_a - V_b$.

SOLUTION

IDENTIFY and SET UP: We are given the electric field, so it is straightforward to find the electric force on the proton. Calculating the work done on the proton by this force is also straightforward because $\vec{E}$ is uniform, which means that the force is constant. Once the work is known, we find the potential difference using Eq. (23.13).

EXECUTE: a) The force is in the same direction as the electric field, and its magnitude is

$$F = qE = (1.602 \times 10^{-19}\ \text{C})(1.5 \times 10^{7}\ \text{N/C})$$
$$= 2.4 \times 10^{-12}\ \text{N}$$

b) The force is constant and in the same direction as the displacement, so the work done is

$$W_{a\to b} = Fd = (2.4 \times 10^{-12}\ \text{N})(0.50\ \text{m}) = 1.2 \times 10^{-12}\ \text{J}$$
$$= (1.2 \times 10^{-12}\ \text{J})\frac{1\ \text{eV}}{1.602 \times 10^{-19}\ \text{J}}$$
$$= 7.5 \times 10^{6}\ \text{eV} = 7.5\ \text{MeV}$$

c) From Eq. (23.13) the potential difference is the work per unit charge, which is

$$V_a - V_b = \frac{W_{a\to b}}{q} = \frac{1.2 \times 10^{-12}\ \text{J}}{1.602 \times 10^{-19}\ \text{C}} = 7.5 \times 10^{6}\ \text{J/C}$$
$$= 7.5 \times 10^{6}\ \text{V} = 7.5\ \text{MV}$$

We can get this same result even more easily by remembering that one electron volt equals one volt multiplied by the charge e. Since the work done is 7.5×10^6 eV and the charge is e, the potential difference is $(7.5 \times 10^6\ \text{eV})/e = 7.5 \times 10^6$ V.

EVALUATE: We can check our result in part (c) by using Eq. (23.17) or (23.18) to calculate an integral of the electric field. The angle ϕ between the constant field $\vec{E}$ and the displacement is zero, so Eq. (23.17) becomes

$$V_a - V_b = \int_a^b E \cos\phi\, dl = \int_a^b E\, dl = E\int_a^b dl$$

The integral of dl from a to b is just the distance d, so we again find

$$V_a - V_b = Ed = (1.5 \times 10^{7}\ \text{V/m})(0.50\ \text{m}) = 7.5 \times 10^{6}\ \text{V}$$

Example 23.4 Potential due to two point charges

An electric dipole consists of two point charges, $q_1 = +12$ nC and $q_2 = -12$ nC, placed 10 cm apart (Fig. 23.13). Compute the potentials at points a, b, and c by adding the potentials due to either charge, as in Eq. (23.15).

SOLUTION

IDENTIFY and SET UP: This is the same arrangement of charges as in Example 21.9 (Section 21.5). In that example we calculated electric *field* at each point by doing a *vector* sum. Here we calculate the electric *potential* at each point by doing the *algebraic* sum in Eq. (23.15):

$$V = \frac{1}{4\pi\epsilon_0}\sum_i \frac{q_i}{r_i}$$

EXECUTE: At point a the potential due to the positive charge q_1 is

$$\frac{1}{4\pi\epsilon_0}\frac{q_1}{r_1} = (9.0 \times 10^{9}\ \text{N}\cdot\text{m}^2/\text{C}^2)\frac{12 \times 10^{-9}\ \text{C}}{0.060\ \text{m}}$$
$$= 1800\ \text{N}\cdot\text{m/C}$$
$$= 1800\ \text{J/C} = 1800\ \text{V}$$

and the potential due to the negative charge q_2 is

$$\frac{1}{4\pi\epsilon_0}\frac{q_2}{r_2} = (9.0 \times 10^{9}\ \text{N}\cdot\text{m}^2/\text{C}^2)\frac{(-12 \times 10^{-9}\ \text{C})}{0.040\ \text{m}}$$
$$= -2700\ \text{N}\cdot\text{m/C}$$
$$= -2700\ \text{J/C} = -2700\ \text{V}$$

The potential V_a at point a is the sum of these:

$$V_a = 1800\ \text{V} + (-2700\ \text{V}) = -900\ \text{V}$$

By similar calculations you can show that at point b the potential due to the positive charge is +2700 V, the potential due to the negative charge is −770 V, and

$$V_b = 2700\ \text{V} + (-770\ \text{V}) = 1930\ \text{V}$$

At point c the potential due to the positive charge is

$$\frac{1}{4\pi\epsilon_0}\frac{q_1}{r_1} = (9.0 \times 10^{9}\ \text{N}\cdot\text{m}^2/\text{C}^2)\frac{12 \times 10^{-9}\ \text{C}}{0.13\ \text{m}} = 830\ \text{V}$$

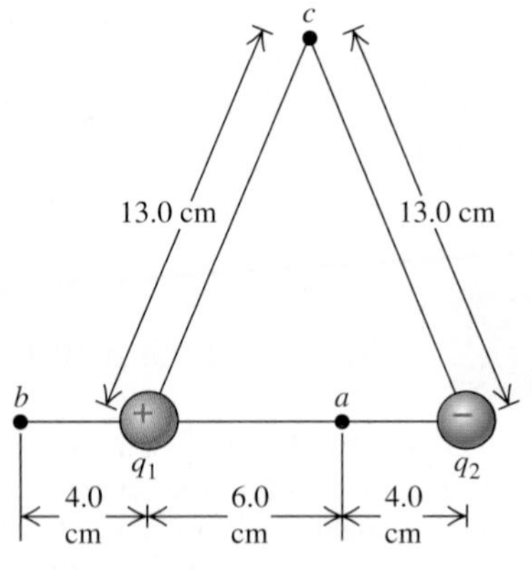

23.13 What are the potentials at points a, b, and c due to this electric dipole?

The potential due to the negative charge is -830 V, and the total potential is zero:

$$V_c = 830\text{ V} + (-830\text{ V}) = 0$$

The potential is also equal to zero at infinity (infinitely far from both charges).

EVALUATE: Comparing this example with Example 21.9 shows that it's much easier to calculate electric potential (a scalar) than electric field (a vector). We'll take advantage of this simplification whenever possible.

Example 23.5 Potential and potential energy

Compute the potential energy associated with a point charge of $+4.0$ nC if it is placed at points a, b, and c in Fig. 23.13.

SOLUTION

IDENTIFY and SET UP: For any point charge q, the associated potential energy is $U = qV$. We use the values of V from Example 23.4.

EXECUTE: At point a,

$$U_a = qV_a = (4.0 \times 10^{-9}\text{ C})(-900\text{ J/C}) = -3.6 \times 10^{-6}\text{ J}$$

At point b,

$$U_b = qV_b = (4.0 \times 10^{-9}\text{ C})(1930\text{ J/C}) = 7.7 \times 10^{-6}\text{ J}$$

At point c,

$$U_c = qV_c = 0$$

All of these values correspond to U and V being zero at infinity.

EVALUATE: Note that *no* net work is done on the 4.0-nC charge if it moves from point c to infinity *by any path.* In particular, let the path be along the perpendicular bisector of the line joining the other two charges q_1 and q_2 in Fig. 23.13. As shown in Example 21.9 (Section 21.5), at points on the bisector the direction of $\vec{E}$ is perpendicular to the bisector. Hence the force on the 4.0-nC charge is perpendicular to the path, and no work is done in any displacement along it.

Example 23.6 Finding potential by integration

By integrating the electric field as in Eq. (23.17), find the potential at a distance r from a point charge q.

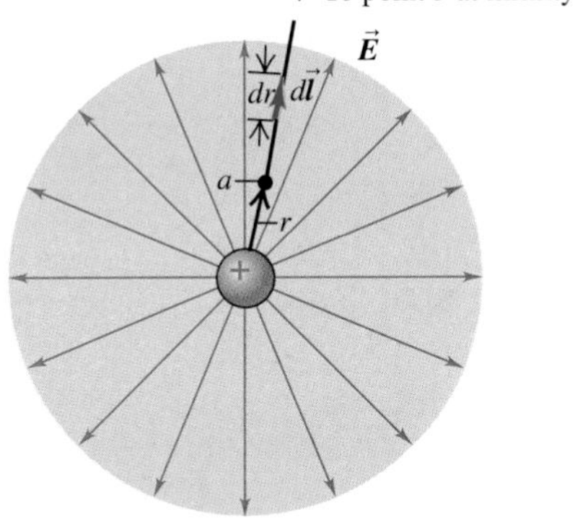

23.14 Calculating the potential by integrating $\vec{E}$ for a single point charge.

SOLUTION

IDENTIFY and SET UP: To find the potential V at a distance r from the point charge, we let point a in Eq. (23.17) be at distance r and let point b be at infinity (Fig. 23.14). As usual, we choose the potential to be zero at an infinite distance from the charge.

EXECUTE: To carry out the integral, we can choose any path we like between points a and b. The most convenient path is a straight radial line as shown in Fig. 23.14, so that $d\vec{l}$ is in the radial direction and has magnitude dr. If q is positive, $\vec{E}$ and $d\vec{l}$ are always parallel, so $\phi = 0$ and Eq. (23.17) becomes

$$V - 0 = \int_r^\infty E\,dr = \int_r^\infty \frac{q}{4\pi\epsilon_0 r^2}\,dr$$
$$= -\frac{q}{4\pi\epsilon_0 r}\bigg|_r^\infty = 0 - \left(-\frac{q}{4\pi\epsilon_0 r}\right)$$
$$V = \frac{q}{4\pi\epsilon_0 r}$$

This agrees with Eq. (23.14). If q is negative, $\vec{E}$ is radially inward while $d\vec{l}$ is still radially outward, so $\phi = 180°$. Since cos 180° = −1, this adds a minus sign to the above result. However, the field magnitude E is always positive, and since q is negative, we must write $E = |q|/4\pi\epsilon_0 r = -q/4\pi\epsilon_0 r$, giving another minus sign. The two minus signs cancel, and the above result for V is valid for point charges of either sign.

EVALUATE: We can get the same result by using Eq. (23.7) for the electric field, which is valid for either sign of q, and writing $d\vec{l} = \hat{r}\,dr$:

$$V - 0 = V = \int_r^\infty \vec{E}\cdot d\vec{l} = \int_r^\infty \frac{1}{4\pi\epsilon_0}\frac{q}{r^2}\hat{r}\cdot\hat{r}\,dr = \int_r^\infty \frac{q}{4\pi\epsilon_0 r^2}\,dr$$

$$V = \frac{q}{4\pi\epsilon_0 r}$$

Example 23.7 Moving through a potential difference

In Fig. 23.15 a dust particle with mass $m = 5.0 \times 10^{-9}$ kg = 5.0 μg and charge $q_0 = 2.0$ nC starts from rest at point a and moves in a straight line to point b. What is its speed v at point b?

SOLUTION

IDENTIFY and SET UP: This problem would be difficult to solve without using energy techniques, since the force that acts on the particle varies in magnitude as the particle moves from a to b. Only the conservative electric force acts on the particle, so mechanical energy is conserved:

$$K_a + U_a = K_b + U_b$$

EXECUTE: For this situation, $K_a = 0$ and $K_b = \frac{1}{2}mv^2$. We get the potential energies (U) from the potentials (V) using Eq. (23.12): $U_a = q_0V_a$ and $U_b = q_0V_b$. Substituting these into the energy-conservation equation and solving for v, we find

$$0 + q_0V_a = \frac{1}{2}mv^2 + q_0V_b$$

$$v = \sqrt{\frac{2q_0(V_a - V_b)}{m}}$$

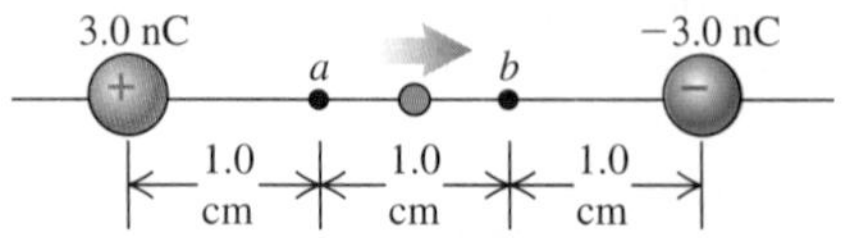

23.15 The particle moves from point a to point b; its acceleration is not constant.

We calculate the potentials using Eq. (23.15), just as we did in Example 23.4:

$$V_a = (9.0 \times 10^9\ \text{N}\cdot\text{m}^2/\text{C}^2) \times \left(\frac{3.0 \times 10^{-9}\ \text{C}}{0.010\ \text{m}} + \frac{(-3.0 \times 10^{-9}\ \text{C})}{0.020\ \text{m}}\right) = 1350\ \text{V}$$

$$V_b = (9.0 \times 10^9\ \text{N}\cdot\text{m}^2/\text{C}^2) \times \left(\frac{3.0 \times 10^{-9}\ \text{C}}{0.020\ \text{m}} + \frac{(-3.0 \times 10^{-9}\ \text{C})}{0.010\ \text{m}}\right) = -1350\ \text{V}$$

$$V_a - V_b = (1350\ \text{V}) - (-1350\ \text{V}) = 2700\ \text{V}$$

Finally,

$$v = \sqrt{\frac{2(2.0 \times 10^{-9}\ \text{C})(2700\ \text{V})}{5.0 \times 10^{-9}\ \text{kg}}} = 46\ \text{m/s}$$

EVALUATE: Our result makes sense: the positive test charge gains speed as it moves away from the positive charge and toward the negative charge. To check unit consistency in the final line of the calculation, note that 1 V = 1 J/C, so the numerator under the radical has units of J or kg·m²/s².

We can use exactly this same method to find the speed of an electron accelerated across a potential difference of 500 V in an oscilloscope tube or 20 kV in a TV picture tube. The end-of-chapter problems include several examples of such calculations.

Test Your Understanding

If the electric *potential* at a certain point is zero, does the electric *field* at that point have to be zero? (*Hint:* Consider point c in Example 23.4 and Example 21.9.)

23.3 | Calculating Electric Potential

When calculating the potential due to a charge distribution, we usually follow one of two routes. If we know the charge distribution, we can use Eq. (23.15) or (23.16). Or if we know how the electric field depends on position, we can use Eq. (23.17), defining the potential to be zero at some convenient place. Some problems require a combination of these approaches.

As you read through these examples, compare them with the related examples of calculating electric *field* in Section 21.5. You'll see how much easier it is to calculate scalar electric potentials than vector electric fields. The moral is clear: whenever possible, solve problems using an energy approach (using electric potential and electric potential energy) rather than a dynamics approach (using electric fields and electric forces).

Problem-Solving Strategy: Calculating Electric Potential

IDENTIFY *the relevant concepts:* Remember that potential is *potential energy per unit charge.* Understanding this statement can get you a long way.

SET UP *the problem using the following steps:*

1. Make a drawing that clearly shows the locations of the charges (which may be point charges or a continuous distribution of charge) and your choice of coordinate axes.
2. Indicate on your drawing the position of the point at which you want to calculate the electric potential V. Sometimes this position will be an arbitrary one (say, a point a distance r from the center of a charged sphere).

EXECUTE *the solution as follows:*

1. To find the potential due to a collection of point charges, use Eq. (23.15). If you are given a continuous charge distribution, devise a way to divide it into infinitesimal elements and then use Eq. (23.16). Carry out the integration, using appropriate limits to include the entire charge distribution. In the integral, be careful about which geometric quantities vary and which are held constant.
2. If you are given the electric field, or if you can find it using any of the methods presented in Chapter 21 or 22, it may be easier to use Eq. (23.17) or (23.18) to calculate the potential difference between points a and b. When appropriate, make use of your freedom to define V to be zero at some convenient place, and choose this place to be point b. (For point charges, this will usually be at infinity. For other distributions of charge—especially those that themselves extend to infinity—it may be convenient or necessary to define V_b to be zero at some finite distance from the charge distribution. This is just like defining U to be zero at ground level in gravitational problems.) Then the potential at any other point, say a, can by found from Eq. (23.17) or (23.18) with $V_b = 0$.
3. Remember that potential is a *scalar* quantity, not a *vector*. It doesn't have components! However, you may have to use components of the vectors $\vec{E}$ and $d\vec{l}$ when you use Eq. (23.17) or (23.18).

EVALUATE *your answer:* Check whether your answer agrees with your intuition. If your result gives V as a function of position, make a graph of this function to see if it makes sense. If you know the electric field, you can make a rough check of your result for V by verifying that V decreases if you move in the direction of $\vec{E}$.

Example 23.8 A charged conducting sphere

A solid conducting sphere of radius R has a total charge q. Find the potential everywhere, both outside and inside the sphere.

SOLUTION

IDENTIFY and SET UP: We used Gauss's law in Example 22.5 (Section 22.4) to find the electric field at all points for this charge distribution. We use that result to determine V at all points.

EXECUTE: From Example 22.5, at all points *outside* the sphere the field is the same as if the sphere were removed and replaced by a point charge q. We take $V = 0$ at infinity, as we did for a point charge. Then the potential at a point outside the sphere at a distance r from its center is the same as the potential due to a point charge q at the center:

$$V = \frac{1}{4\pi\epsilon_0}\frac{q}{r}$$

The potential at the surface of the sphere is $V_{surface} = q/4\pi\epsilon_0 R$.

Inside the sphere, $\vec{E}$ is zero everywhere; otherwise, charge would move within the sphere. Hence if a test charge moves from any point to any other point inside the sphere, no work is done on that charge. This means that the potential is the same at every point inside the sphere and is equal to its value $q/4\pi\epsilon_0 R$ at the surface.

EVALUATE: Figure 23.16 shows the field and potential as a function of r for a positive charge q. In this case the electric field points radially away from the sphere. As you move away from the sphere, in the direction of $\vec{E}$, V decreases (as it should). The electric field at the surface has magnitude $E_{surface} = |q|/4\pi\epsilon_0 R^2$.

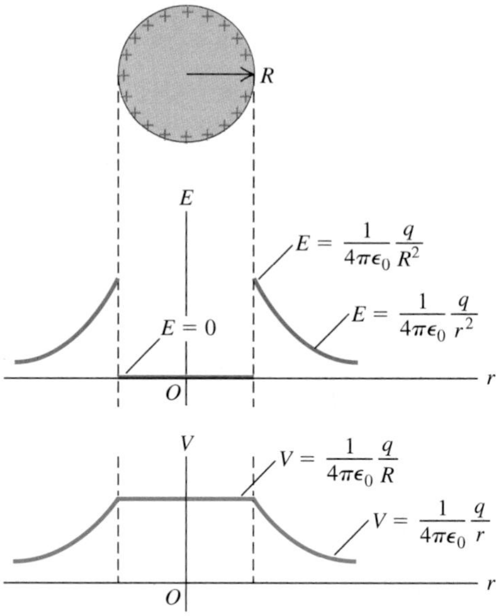

23.16 Electric field magnitude E and potential V at points inside and outside a positively charged spherical conductor.

Ionization and Corona Discharge

The results of Example 23.8 have numerous practical consequences. One consequence relates to the maximum potential to which a conductor in air can be raised. This potential is limited because air molecules become *ionized,* and air becomes a conductor, at an electric-field magnitude of about 3×10^6 V/m. Assume for the moment that q is positive. When we compare the expressions in Example 23.8 for the potential $V_{surface}$ and field magnitude $E_{surface}$ at the surface of a charged conducting sphere, we note that $V_{surface} = E_{surface}R$. Thus, if E_m represents the electric-field magnitude at which air becomes conductive (known as the *dielectric strength* of air), then the maximum potential V_m to which a spherical conductor can be raised is

$$V_m = RE_m$$

For a conducting sphere 1 cm in radius in air, $V_m = (10^{-2}\text{ m})(3 \times 10^6\text{ V/m}) = 30{,}000$ V. No amount of "charging" could raise the potential of a conducting sphere of this size in air higher than about 30,000 V; attempting to raise the potential further by adding extra charge would cause the surrounding air to become ionized and conductive, and the extra added charge would leak into the air.

To attain even higher potentials, high-voltage machines such as Van de Graaff generators use spherical terminals with very large radii (see Fig. 22.28 and the photograph that opens Chapter 22). For example, a terminal of radius $R = 2$ m has a maximum potential $V_m = (2\text{ m})(3 \times 10^6\text{ V/m}) = 6 \times 10^6\text{ V} = 6$ MV. Such machines are sometimes placed in pressurized tanks filled with a gas such as sulfur hexafluoride (SF_6) that has a larger value of E_m than does air and, therefore, can withstand even larger fields without becoming conductive.

Our result in Example 23.8 also explains what happens with a charged conductor with a very *small* radius of curvature, such as a sharp point or thin wire. Because the maximum potential is proportional to the radius, even relatively small potentials applied to sharp points in air produce sufficiently high fields just outside the point to ionize the surrounding air, making it become a conductor. The resulting current and its associated glow (visible in a dark room) are called *corona*.

? Laser printers and photocopying machines use corona from fine wires to spray charge on the imaging drum (see Fig. 21.2). On a larger scale, a lightning rod has a sharp end so that lightning bolts will pass through a conducting path in the air that leads to the rod, and not to other nearby structures that could be damaged (Fig. 23.17). To prevent the charge acquired by the rod from entering the building, a conducting wire leads from the lightning rod to the ground. Car radio antennas have a ball on the end to help *prevent* corona that would cause static.

23.17 The metal mast at the top of the Empire State Building acts as a lightning rod. It is struck by lightning as many as 500 times each year.

Example 23.9 Oppositely charged parallel plates

Find the potential at any height y between the two oppositely charged parallel plates discussed in Section 23.1.

SOLUTION

IDENTIFY and SET UP: The potential energy U for a test charge q_0 at a point a distance y above the bottom plate (Fig. 23.18) is given by Eq. (23.5), $U = q_0Ey$. We use this expression to determine the potential V at such a point.

EXECUTE: The potential $V(y)$ at coordinate y is the potential energy per unit charge:

$$V(y) = \frac{U(y)}{q_0} = \frac{q_0Ey}{q_0} = Ey$$

We have chosen $U(y)$, and therefore $V(y)$, to be zero at point b, where $y = 0$. Even if we choose the potential to be different from zero at b, it is still true that

$$V(y) - V_b = Ey$$

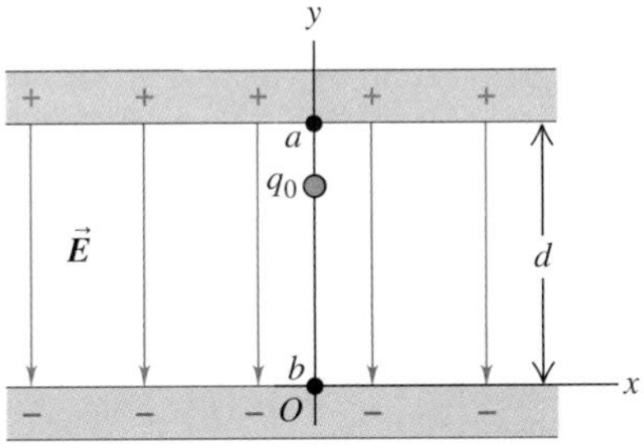

23.18 The charged parallel plates from Fig. 23.1.

The potential decreases as we move from the upper to the lower plate in the direction of $\vec{E}$. At point a, where $y = d$ and $V(y) = V_a$,

$$V_a - V_b = Ed \quad \text{and} \quad E = \frac{V_a - V_b}{d} = \frac{V_{ab}}{d}$$

where V_{ab} is the potential of the positive plate with respect to the negative plate. That is, the electric field equals the potential difference between the plates divided by the distance between them. (This relation between E and V_{ab} holds *only* for the planar geometry we have described. It does *not* work for situations such as concentric cylinders or spheres in which the electric field is not uniform.)

EVALUATE: Our result provides a practical way to measure the charge density on the charges on the two plates in Fig. 23.18. In Example 22.8 (Section 22.4), we derived the expression $E = \sigma/\epsilon_0$ for the electric field E between two conducting plates having surface charge densities $+\sigma$ and $-\sigma$, respectively. Setting this expression equal to $E = V_{ab}/d$ gives

$$\sigma = \frac{\epsilon_0 V_{ab}}{d}$$

The surface charge density on the positive plate is directly proportional to the potential difference between the plates, and its value σ can be determined by measuring V_{ab}. This technique is useful because no instruments are available that read surface charge density directly. On the negative plate the surface charge density is $-\sigma$.

CAUTION You might think that if a conducting body has zero potential, it must necessarily also have zero net charge. But that just isn't so! As an example, the plate at $y = 0$ in Fig. 23.18 has zero potential ($V = 0$) but has a nonzero charge per unit area $-\sigma$. Remember that there's nothing particularly special about the place where potential is zero; we can *define* this place to be wherever we want it to be.

Example 23.10 An infinite line charge or charged conducting cylinder

Find the potential at a distance r from a very long line of charge with linear charge density (charge per unit length) λ.

SOLUTION

IDENTIFY and SET UP: One approach to this problem is to divide the line of charge into infinitesimal elements, as we did in Example 21.11 (Section 21.5) to find the electric field produced by such a line. We could then integrate as in Eq. (23.16) to find the net potential V.

In this case, however, our task is greatly simplified because we already know the electric field. In both Example 21.11 and Example 22.6 (Section 22.4), we found that the electric field at a distance r from a long straight-line charge (Fig. 23.19a) has only a radial component, given by

$$E_r = \frac{1}{2\pi\epsilon_0}\frac{\lambda}{r}$$

We use this expression to find the potential by integrating $\vec{E}$ as in Eq. (23.17).

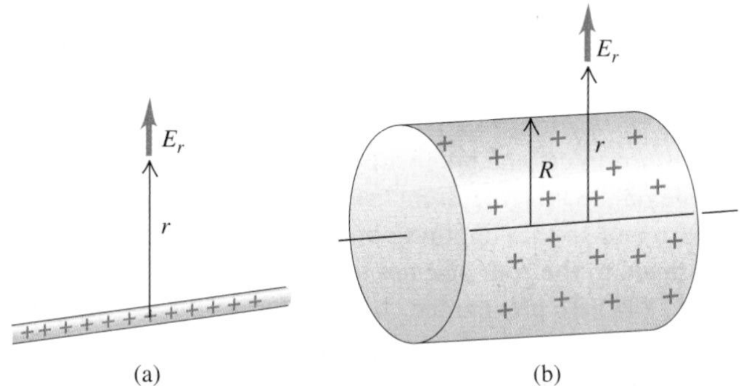

23.19 Electric field outside (a) a long positively charged wire; (b) a long positively charged cylinder.

EXECUTE: Since the field has only a radial component, the scalar product $\vec{E}\cdot d\vec{l}$ is equal to $E_r dr$. Hence the potential of any point a with respect to any other point b, at radial distances r_a and r_b from the line of charge, is

$$V_a - V_b = \int_a^b \vec{E}\cdot d\vec{l} = \int_a^b E_r dr = \frac{\lambda}{2\pi\epsilon_0}\int_{r_a}^{r_b}\frac{dr}{r} = \frac{\lambda}{2\pi\epsilon_0}\ln\frac{r_b}{r_a}$$

If we take point b at infinity and set $V_b = 0$, we find that V_a is *infinite:*

$$V_a = \frac{\lambda}{2\pi\epsilon_0}\ln\frac{\infty}{r_a} = \infty$$

This shows that if we try to define V to be zero at infinity, then V must be infinite at *any* finite distance from the line charge. Hence this is *not* a useful way to define V for this problem! The difficulty, as we mentioned earlier, is that the charge distribution itself extends to infinity.

To get around this difficulty, remember that we can define V to be zero at any point we like. We set $V_b = 0$ at point b at an arbitrary radial distance r_0. Then the potential $V = V_a$ at point a at a radial distance r is given by $V - 0 = (\lambda/2\pi\epsilon_0)\ln(r_0/r)$, or

$$V = \frac{\lambda}{2\pi\epsilon_0}\ln\frac{r_0}{r}$$

EVALUATE: According to our result, if λ is positive then V decreases as r increases. This is as it should be: V decreases as we move in the direction of $\vec{E}$.

From Example 22.6, the expression for E_r with which we started also applies outside a long charged conducting cylinder with charge per unit length λ (Fig. 23.19b). Hence our result also gives the potential in the field for such a cylinder, but only for values of r equal to or greater than the radius R of the cylinder. If we choose r_0 to be the cylinder radius R, so that $V = 0$ when $r = R$, then at any point for which $r > R$,

$$V = \frac{\lambda}{2\pi\epsilon_0}\ln\frac{R}{r}$$

where r is the distance from the axis of the cylinder. Inside the cylinder, $\vec{E} = \mathbf{0}$, and V has the same value (zero) as on the cylinder's surface.

Example 23.11 A ring of charge

Electric charge is distributed uniformly around a thin ring of radius a, with total charge Q (Fig. 23.20). Find the potential at a point P on the ring axis at a distance x from the center of the ring.

SOLUTION

IDENTIFY and SET UP: We could solve this problem using the same approach as in the preceding example: we found the electric field at all points along the x–axis in Example 21.10 (Section 21.5), so we could integrate $\vec{E}$ as in Eq. (23.17) to find V along this axis. But in this case, however, it's far easier to find V by integrating over the charge distribution as in Eq. (23.16), because all parts of the ring (that is, all elements of the charge distribution) are the same distance r from point P.

EXECUTE: We divide the ring into infinitesimal elements of charge dq. Figure 23.20 shows that the distance from each such element to

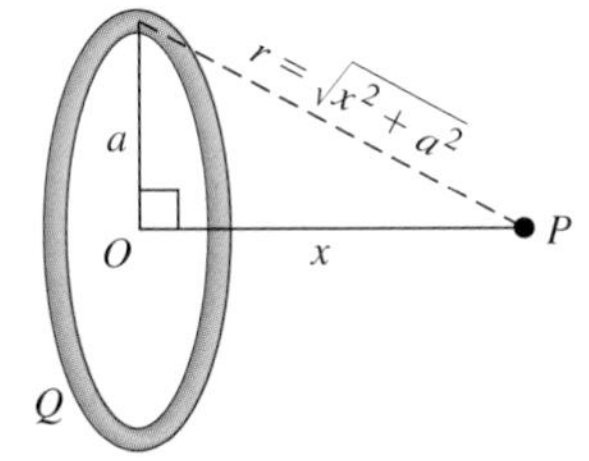

23.20 All the charge in a ring of charge Q is the same distance r from a point P on the ring axis.

the point P is $r = \sqrt{x^2 + a^2}$. Hence we can take the factor $1/r$ outside the integral in Eq. (23.16), and

$$V = \frac{1}{4\pi\epsilon_0}\int \frac{dq}{r} = \frac{1}{4\pi\epsilon_0}\frac{1}{\sqrt{x^2 + a^2}}\int dq = \frac{1}{4\pi\epsilon_0}\frac{Q}{\sqrt{x^2 + a^2}}$$

Potential is a *scalar* quantity; there is no need to consider components of vectors in this calculation, as we had to do when we found the electric field at P. So the potential calculation is a lot simpler than the field calculation.

EVALUATE: Note that when x is much larger than a, the above expression for V becomes approximately equal to $V = Q/4\pi\epsilon_0 x$. This corresponds to the potential of a point charge Q at distance x. When we are very far away from a charged ring, it looks like a point charge. (We drew a similar conclusion about the electric field of a ring in Example 21.10.)

These results for V can also be found by integrating the expression for E_x found in Example 21.10 (see Problem 23.67).

Example 23.12 A line of charge

Electric charge Q is distributed uniformly along a line or thin rod of length $2a$. Find the potential at a point P along the perpendicular bisector of the rod at a distance x from its center.

SOLUTION

IDENTIFY and SET UP: This is the same situation as in Example 21.11 (Section 21.5), where we found an expression for the electric field $\vec{E}$ at an arbitrary point on the x-axis. We could integrate $\vec{E}$ using Eq. (23.17) to find V. Instead, we'll integrate over the charge distribution using Eq. (23.16) to get a bit more experience with this approach.

EXECUTE: As in Example 21.11, the element of charge dQ corresponding to an element of length dy on the rod is given by $dQ = (Q/2a)dy$ (Fig. 23.21). The distance from dQ to P is $\sqrt{x^2 + y^2}$, and the contribution dV that it makes to the potential at P is

$$dV = \frac{1}{4\pi\epsilon_0}\frac{Q}{2a}\frac{dy}{\sqrt{x^2 + y^2}}$$

To get the potential at P due to the entire rod, we integrate dV over the length of the rod from $y = -a$ to $y = a$:

$$V = \frac{1}{4\pi\epsilon_0}\frac{Q}{2a}\int_{-a}^{a}\frac{dy}{\sqrt{x^2 + y^2}}$$

You can look up the integral in a table. The final result is

$$V = \frac{1}{4\pi\epsilon_0}\frac{Q}{2a}\ln\left(\frac{\sqrt{a^2 + x^2} + a}{\sqrt{a^2 + x^2} - a}\right)$$

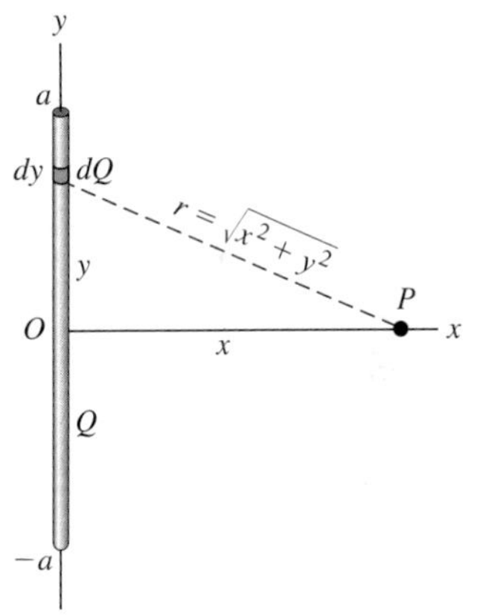

23.21 Finding the electric potential on the perpendicular bisector of a uniformly charged rod of length $2a$.

EVALUATE: We can check our result by letting x approach infinity. In this limit the point P is infinitely far from all of the charge, so we expect V to approach zero; we invite you to verify that it does so.

As in Example 23.11, this problem is simpler than finding $\vec{E}$ at point P because potential is a scalar quantity and no vector calculations are involved.

Test Your Understanding

If the electric *field* at a certain point is zero, does the electric *potential* at that point have to be zero? (*Hint:* Consider the center of the ring in Example 23.11 and Example 21.10.)

23.4 | Equipotential Surfaces

Field lines (Section 21.6) help us visualize electric fields. In a similar way, the potential at various points in an electric field can be represented graphically by *equipotential surfaces*. These use the same fundamental idea as topographic maps like those used by hikers and mountain climbers (Fig. 23.22). On a topographic map, contour lines are drawn through points that are all at the same elevation. Any number of these could be drawn, but typically only a few contour lines are shown at equal spacings of elevation. If a mass m is moved over the terrain along such a contour line, the gravitational potential energy mgy does not change because the elevation y is constant. Thus contour lines on a topographic map are really curves of constant gravitational potential energy. Contour lines are close together in regions where the terrain is steep and there are large changes in elevation over a small horizontal distance; the contour lines are farther apart where the terrain is gently sloping. A ball allowed to roll downhill will experience the greatest downhill gravitational force where contour lines are closest together.

By analogy to contour lines on a topographic map, an **equipotential surface** is a three-dimensional surface on which the *electric potential* V is the same at every point. If a test charge q_0 is moved from point to point on such a surface, the *electric* potential energy q_0V remains constant. In a region where an electric field is present, we can construct an equipotential surface through any point. In diagrams we usually show only a few representative equipotentials, often with equal potential differences between adjacent surfaces. No point can be at two different potentials, so equipotential surfaces for different potentials can never touch or intersect.

Because potential energy does not change as a test charge moves over an equipotential surface, the electric field can do no work on such a charge. It follows that $\vec{E}$ must be perpendicular to the surface at every point so that the electric force $q_0\vec{E}$ is always perpendicular to the displacement of a charge moving on the surface. **Field lines and equipotential surfaces are always mutually perpendicular.** In general, field lines are curves, and equipotentials are curved surfaces. For the special case of a *uniform* field, in which the field lines are straight, parallel, and equally spaced, the equipotentials are parallel *planes* perpendicular to the field lines.

23.22 Contour lines on a topographic map are curves of constant elevation and hence of constant gravitational potential energy.

Figure 23.23 shows several arrangements of charges. The field lines in the plane of the charges are represented by red lines, and the intersections of the equipotential surfaces with this plane (that is, cross sections of these surfaces) are shown as blue lines. The actual equipotential surfaces are three-dimensional. At each crossing of an equipotential and a field line, the two are perpendicular.

In Fig. 23.23 we have drawn equipotentials so that there are equal potential differences between adjacent surfaces. In regions where the magnitude of $\vec{E}$ is large, the equipotential surfaces are close together because the field does a relatively large amount of work on a test charge in a relatively small displacement. This is the case near the point charge in Fig. 23.23a or between the two point charges in Fig. 23.23b; note that in these regions the field lines are also closer together. This

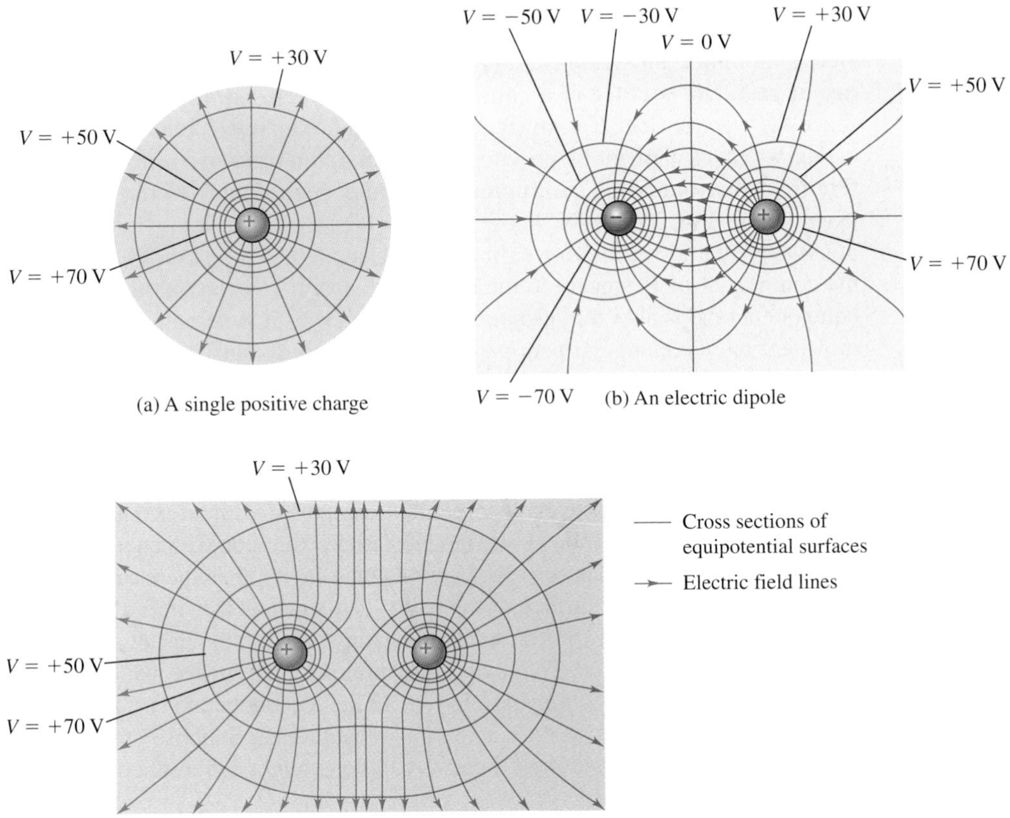

(a) A single positive charge

(b) An electric dipole

(c) Two equal positive charges

23.23 Cross sections of equipotential surfaces (blue lines) and electric field lines (red lines) for assemblies of point charges. There are equal potential differences between adjacent surfaces. Compare these diagrams to those in Fig. 21.26, which showed only the electric field lines.

is directly analogous to the downhill force of gravity being greatest in regions on a topographic map where contour lines are close together. Conversely, in regions where the field is weaker, the equipotential surfaces are farther apart; this happens at larger radii in Fig. 23.23a, to the left of the negative charge or the right of the positive charge in Fig. 23.23b, and at greater distances from both charges in Fig. 23.23c. (It may appear that two equipotential surfaces intersect at the center of Fig. 23.23c, in violation of the rule that this can never happen. In fact this is a single figure-8-shaped equipotential surface.)

CAUTION On a given equipotential surface, the potential V has the same value at every point. In general, however, the electric-field magnitude E is *not* the same at all points on an equipotential surface. For instance, on the equipotential surface labeled "$V = -30\ V$" in Fig. 23.23b, the magnitude E is less to the left of the negative charge than it is between the two charges. On the figure-8-shaped equipotential surface in Fig. 23.23c, $E = 0$ at the middle point halfway between the two charges; at any other point on this surface, E is nonzero.

Equipotentials and Conductors

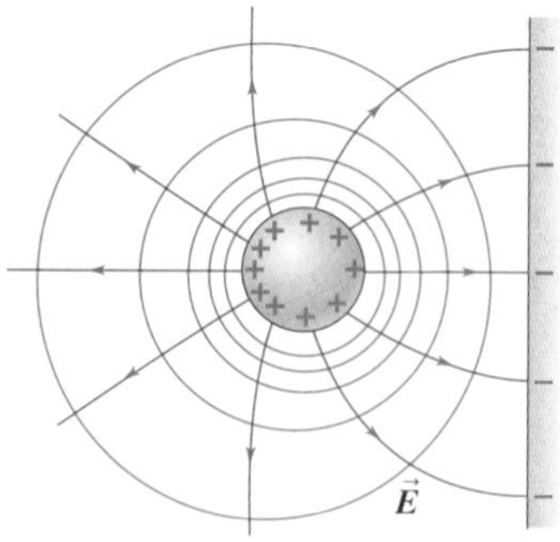

23.24 When charges are at rest, a conducting surface is always an equipotential surface. Field lines (shown in red) are perpendicular to a conducting surface.

Here's an important statement about equipotential surfaces: **When all charges are at rest, the surface of a conductor is always an equipotential surface.** Since the electric field $\vec{E}$ is always perpendicular to an equipotential surface, we can prove this statement by proving that **when all charges are at rest, the electric field just outside a conductor must be perpendicular to the surface at every point** (Fig. 23.24). We know that $\vec{E} = \mathbf{0}$ everywhere inside the conductor; otherwise, charges would move. In particular, at any point just inside the surface the component of $\vec{E}$ tangent to the surface is zero. It follows that the tangential component of $\vec{E}$ is also zero just *outside* the surface. If it were not, a charge could move around a rectangular path partly inside and partly outside (Fig. 23.25) and return to its starting point with a net amount of work having been done on it. This would violate the conservative nature of electrostatic fields, so the tangential component of $\vec{E}$ just outside the surface must be zero at every point on the surface. Thus $\vec{E}$ is perpendicular to the surface at each point, proving our statement.

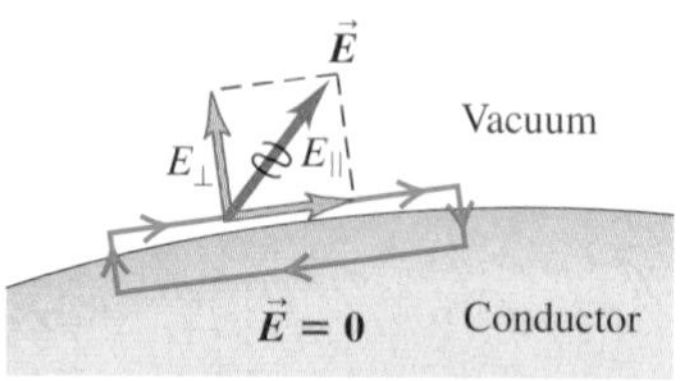

23.25 Inside the conductor, $\vec{E} = \mathbf{0}$. If $\vec{E}$ just outside the conductor had a component $E_\parallel$ parallel to the conductor surface, then a test charge moving around the rectangular loop and returning to the starting point would have nonzero total work done on it by the field. Because the $\vec{E}$ field is conservative, this is impossible. Therefore $E_\parallel$ must be zero, and $\vec{E}$ just outside the surface must be perpendicular to it.

Finally, we can now prove a theorem that we quoted without proof in Section 22.5. The theorem is as follows: In an electrostatic situation, if a conductor contains a cavity and if no charge is present inside the cavity, then there can be no net charge *anywhere* on the surface of the cavity. This means that if you're inside a charged conducting box, you can safely touch any point on the inside walls of the box without being shocked. To prove this theorem, we first prove that *every point in the cavity is at the same potential.* In Fig. 23.26 the conducting surface A of the cavity is an equipotential surface, as we have just proved. Suppose point P in the cavity is at a different potential; then we can construct a different equipotential surface B including point P.

Now consider a Gaussian surface, shown in Fig. 23.26, between the two equipotential surfaces. Because of the relation between $\vec{E}$ and the equipotentials, we know that the field at every point between the equipotentials is from A toward B, or else at every point it is from B toward A, depending on which equipotential surface is at higher potential. In either case the flux through this Gaussian surface is certainly not zero. But then Gauss's law says that the charge enclosed by the Gaussian surface cannot be zero. This contradicts our initial assumption that there is *no* charge in the cavity. So the potential at P *cannot* be different from that at the cavity wall.

The entire region of the cavity must therefore be at the same potential. But for this to be true, *the electric field inside the cavity must be zero everywhere.* Finally, Gauss's law shows that the electric field at any point on the surface of a conductor is proportional to the surface charge density σ at that point. We conclude that *the surface charge density on the wall of the cavity is zero at every point.* This chain of reasoning may seem tortuous, but it is worth careful study.

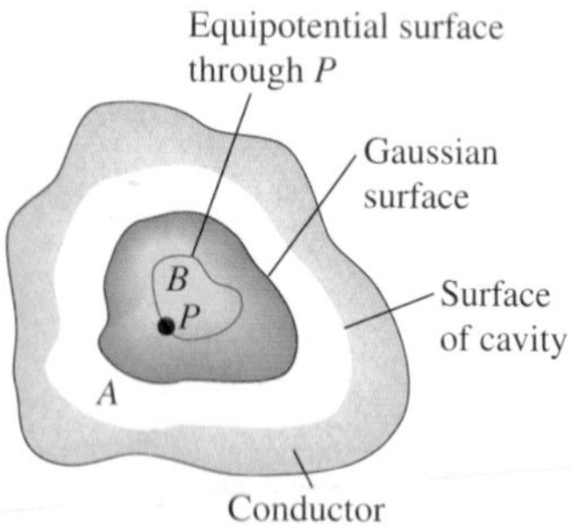

23.26 A cavity in a conductor. If the cavity contains no charge, every point in the cavity is at the same potential, the electric field is zero everywhere in the cavity, and there is no charge anywhere on the surface of the cavity.

CAUTION Don't confuse equipotential surfaces with the Gaussian surfaces we encountered in Chapter 22. Gaussian surfaces have relevance only when we are using Gauss's law, and we can choose *any* Gaussian surface that's convenient. We are *not* free to choose the shape of equipotential surfaces; the shape is determined by the charge distribution.

Test Your Understanding

How would the diagrams in Fig. 23.23 change if the sign of each charge were reversed?

23.5 | Potential Gradient

Electric field and potential are closely related. Equation (23.17), restated below, expresses one aspect of that relationship:

$$V_a - V_b = \int_a^b \vec{E} \cdot d\vec{l}$$

If we know $\vec{E}$ at various points, we can use this equation to calculate potential differences. In this section we show how to turn this around; if we know the potential V at various points, we can use it to determine $\vec{E}$. Regarding V as a function of the coordinates (x, y, z) of a point in space, we will show that the components of $\vec{E}$ are directly related to the *partial derivatives* of V with respect to x, y, and z.

In Eq. (23.17), $V_a - V_b$ is the potential of a with respect to b—that is, the change of potential when a point moves from b to a. We can write this as

$$V_a - V_b = \int_b^a dV = -\int_a^b dV$$

where dV is the infinitesimal change of potential accompanying an infinitesimal element $d\vec{l}$ of the path from b to a. Comparing to Eq. (23.17), we have

$$-\int_a^b dV = \int_a^b \vec{E} \cdot d\vec{l}$$

These two integrals must be equal for *any* pair of limits a and b, and for this to be true the *integrands* must be equal. Thus, for *any* infinitesimal displacement $d\vec{l}$,

$$-dV = \vec{E} \cdot d\vec{l}$$

To interpret this expression, we write $\vec{E}$ and $d\vec{l}$ in terms of their components: $\vec{E} = \hat{\imath}\, E_x + \hat{\jmath}\, E_y + \hat{k}\, E_z$ and $d\vec{l} = \hat{\imath}\, dx + \hat{\jmath}\, dy + \hat{k}\, dz$. Then we have

$$-dV = E_x\, dx + E_y\, dy + E_z\, dz$$

Suppose the displacement is parallel to the x-axis, so $dy = dz = 0$. Then $-dV = E_x\, dx$ or $E_x = -(dV/dx)_{y,\, z \text{ constant}}$, where the subscript reminds us that only x varies in the derivative; recall that V is in general a function of x, y, and z. But this is just what is meant by the partial derivative $\partial V/\partial x$. The y- and z-components of $\vec{E}$ are related to the corresponding derivatives of V in the same way, so we have

ActivPhysics ONLINE

11.12.3 Electric Potential, Field, and Force

$$E_x = -\frac{\partial V}{\partial x} \qquad E_y = -\frac{\partial V}{\partial y} \qquad E_z = -\frac{\partial V}{\partial z} \qquad (23.19)$$

(components of $\vec{E}$ in terms of V)

This is consistent with the units of electric field being V/m. In terms of unit vectors we can write $\vec{E}$ as

$$\vec{E} = -\left(\hat{\imath}\frac{\partial V}{\partial x} + \hat{\jmath}\frac{\partial V}{\partial y} + \hat{k}\frac{\partial V}{\partial z}\right) \qquad (\vec{E} \text{ in terms of } V) \qquad (23.20)$$

In vector notation the following operation is called the **gradient** of the function f:

$$\vec{\nabla} f = \left(\hat{\imath}\frac{\partial}{\partial x} + \hat{\jmath}\frac{\partial}{\partial y} + \hat{k}\frac{\partial}{\partial z}\right) f \qquad (23.21)$$

The operator denoted by the symbol $\vec{\nabla}$ is called "grad" or "del." Thus in vector notation,

$$\vec{E} = -\vec{\nabla}V \tag{23.22}$$

This is read "$\vec{E}$ is the negative of the gradient of V" or "$\vec{E}$ equals negative grad V." The quantity $\vec{\nabla}V$ is called the *potential gradient.*

At each point, the potential gradient points in the direction in which V *increases* most rapidly with a change in position. Hence at each point the direction of $\vec{E}$ is the direction in which V *decreases* most rapidly and is always perpendicular to the equipotential surface through the point. This agrees with our observation in Section 23.2 that moving in the direction of the electric field means moving in the direction of decreasing potential.

Equation (23.22) doesn't depend on the particular choice of the zero point for V. If we were to change the zero point, the effect would be to change V at every point by the same amount; the derivatives of V would be the same.

If $\vec{E}$ is radial with respect to a point or an axis and r is the distance from the point or the axis, the relation corresponding to Eqs. (23.19) is

$$E_r = -\frac{\partial V}{\partial r} \quad \text{(radial electric field)} \tag{23.23}$$

Often we can compute the electric field caused by a charge distribution in either of two ways: directly, by adding the $\vec{E}$ fields of point charges, or by first calculating the potential and then taking its gradient to find the field. The second method is often easier because potential is a *scalar* quantity, requiring at worst the integration of a scalar function. Electric field is a *vector* quantity, requiring computation of components for each element of charge and a separate integration for each component. Thus, quite apart from its fundamental significance, potential offers a very useful computational technique in field calculations. Below, we present several examples in which a knowledge of V is used to find the electric field.

We stress once more that if we know $\vec{E}$ as a function of position, we can calculate V using Eq. (23.17) or (23.18), and if we know V as a function of position, we can calculate $\vec{E}$ using Eq. (23.19), (23.20), or (23.23). Deriving V from $\vec{E}$ requires integration, and deriving $\vec{E}$ from V requires differentiation.

Example 23.13 Potential and field of a point charge

From Eq. (23.14) the potential at a radial distance r from a point charge q is $V = q/4\pi\epsilon_0 r$. Find the vector electric field from this expression for V.

SOLUTION

IDENTIFY and SET UP: By symmetry, the electric field has only a radial component E_r; we use Eq. (23.23) to find this component.

EXECUTE: From Eq. (23.23),

$$E_r = -\frac{\partial V}{\partial r} = -\frac{\partial}{\partial r}\left(\frac{1}{4\pi\epsilon_0}\frac{q}{r}\right) = \frac{1}{4\pi\epsilon_0}\frac{q}{r^2}$$

so the vector electric field is

$$\vec{E} = \hat{r}E_r = \frac{1}{4\pi\epsilon_0}\frac{q}{r^2}\hat{r}$$

EVALUATE: Our result agrees with Eq. (21.7), as it must.

An alternative approach is to ignore the radial symmetry, write the radial distance as $r = \sqrt{x^2 + y^2 + z^2}$, and take the derivatives of V with respect to x, y, and z as in Eq. (23.20). We find

$$\frac{\partial V}{\partial x} = \frac{\partial}{\partial x}\left(\frac{1}{4\pi\epsilon_0}\frac{q}{\sqrt{x^2 + y^2 + z^2}}\right) = -\frac{1}{4\pi\epsilon_0}\frac{qx}{(x^2 + y^2 + z^2)^{3/2}}$$

$$= -\frac{qx}{4\pi\epsilon_0 r^3}$$

and similarly

$$\frac{\partial V}{\partial y} = -\frac{qy}{4\pi\epsilon_0 r^3} \qquad \frac{\partial V}{\partial z} = -\frac{qz}{4\pi\epsilon_0 r^3}$$

From Eq. (23.30), the electric field is

$$\vec{E} = -\left[\hat{i}\left(-\frac{qx}{4\pi\epsilon_0 r^3}\right) + \hat{j}\left(-\frac{qy}{4\pi\epsilon_0 r^3}\right) + \hat{k}\left(-\frac{qz}{4\pi\epsilon_0 r^3}\right)\right]$$

$$= \frac{1}{4\pi\epsilon_0}\frac{q}{r^2}\left(\frac{x\hat{i} + y\hat{j} + z\hat{k}}{r}\right) = \frac{1}{4\pi\epsilon_0}\frac{q}{r^2}\hat{r}$$

This approach gives us the same answer, but with a bit more effort. Clearly it's best to exploit the symmetry of the charge distribution whenever possible.

Example 23.14 Potential and field outside a charged conducting cylinder

In Example 23.10 (Section 23.31) we found that the potential outside a charged conducting cylinder with radius R and charge per unit length λ is

$$V = \frac{\lambda}{2\pi\epsilon_0}\ln\frac{R}{r} = \frac{\lambda}{2\pi\epsilon_0}(\ln R - \ln r)$$

Find the components of the electric field outside the cylinder.

SOLUTION

IDENTIFY and SET UP: As in Example 23.13, the symmetry tells us that $\vec{E}$ has only a radial component E_r, so we again use Eq. (23.23).

EXECUTE: The derivative of $\ln R$ (a constant) is zero, and the derivative of $\ln r$ is $1/r$, so

$$E_r = -\frac{\partial V}{\partial r} = \frac{\lambda}{2\pi\epsilon_0}\frac{\partial(\ln r)}{\partial r} = \frac{1}{2\pi\epsilon_0}\frac{\lambda}{r}$$

EVALUATE: This is the same expression for E_r that we found in Example 22.6 (Section 22.4).

Example 23.15 Potential and field of a ring of charge

In Example 23.11 (Section 23.3) we found that for a ring of charge with radius a and total charge Q, the potential at a point P on the ring axis a distance x from the center is

$$V = \frac{1}{4\pi\epsilon_0}\frac{Q}{\sqrt{x^2 + a^2}}$$

Find the electric field at P.

SOLUTION

IDENTIFY and SET UP: From the symmetry of the charge distribution shown in Fig. 23.20, the electric field along the symmetry axis of the ring can have only an x-component. We find it using the first of Eqs. (23.19).

EXECUTE: The x-component of the electric field is

$$E_x = -\frac{\partial V}{\partial x} = \frac{1}{4\pi\epsilon_0}\frac{Qx}{(x^2 + a^2)^{3/2}}$$

EVALUATE: This agrees with the result that we obtained in Example 21.10 (Section 21.5).

CAUTION In this example, V does not appear to be a function of y or z, but it would *not* be correct to conclude that $\partial V/\partial y = \partial V/\partial z = 0$ and $E_y = E_z = 0$ everywhere. The reason is that our expression for V is valid *only for points on the x-axis,* where $y = z = 0$. If we had the complete expression for V valid at *all* points in space, then we could use it to find the components of $\vec{E}$ at any point using Eq. (23.19).

Test Your Understanding

In a certain region of space the potential is given by $V = A + Bx + Cy^3 + Dz^2$, where A, B, C, and D are constants. What is the electric field in this region?

SUMMARY

The electric force caused by any collection of charges at rest is a conservative force. The work W done by the electric force on a charged particle moving in an electric field can be represented by the change in a potential-energy function U.

$$W_{a\to b} = U_a - U_b \qquad (23.2)$$

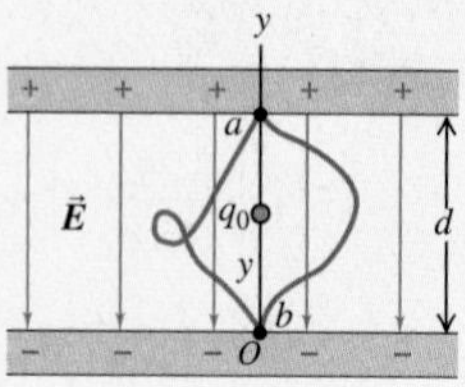

The electric potential energy for two point charges q and q_0 depends on their separation r. The electric potential energy for a charge q_0 in the presence of a collection of charges q_1, q_2, q_3 depends on the distance from q_0 to each of these other charges. (See Examples 23.1 and 23.2)

$$U = \frac{1}{4\pi\epsilon_0}\frac{qq_0}{r} \qquad (23.9)$$

(two point charges)

$$U = \frac{q_0}{4\pi\epsilon_0}\left(\frac{q_1}{r_1} + \frac{q_2}{r_2} + \frac{q_3}{r_3} + \cdots\right) \qquad (23.10)$$

$$= \frac{q_0}{4\pi\epsilon_0}\sum_i \frac{q_i}{r_i}$$

(q_0 in presence of other point charges)

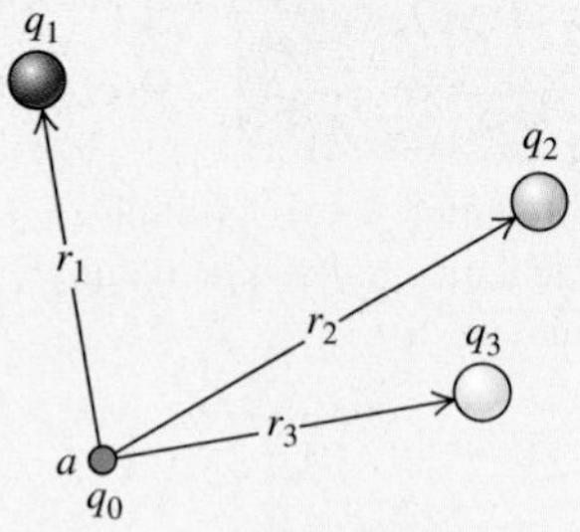

Potential, denoted by V, is potential energy per unit charge. The potential difference between two points equals the amount of work that would be required to move a unit positive test charge between those points. The potential V due to a quantity of charge can be calculated by summing (if the charge is a collection of point charges) or by integrating (if the charge is a distribution). (See Examples 23.3, 23.4, 23.5, 23.7, 23.11, and 23.12)

$$V = \frac{U}{q_0} = \frac{1}{4\pi\epsilon_0}\frac{q}{r} \qquad (23.14)$$

(due to a point charge)

$$V = \frac{U}{q_0} = \frac{1}{4\pi\epsilon_0}\sum_i \frac{q_i}{r_i} \qquad (23.15)$$

(due to a collection of point charges)

$$V = \frac{1}{4\pi\epsilon_0}\int \frac{dq}{r} \qquad (23.16)$$

(due to a charge distribution)

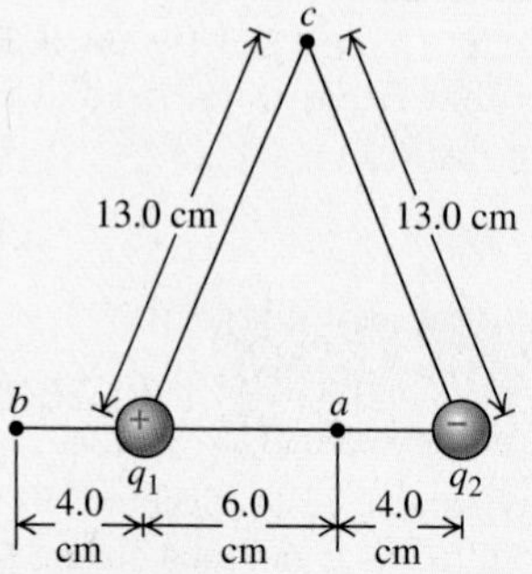

The potential difference between two points a and b, also called the potential of a with respect to b, is given by the line integral of $\vec{E}$. The potential at a given point can be found by first finding $\vec{E}$ and then carrying out this integral. (See Examples 23.6, 23.8, 23.9, and 23.10)

$$V_a - V_b = \int_a^b \vec{E}\cdot d\vec{l} = \int_a^b E\cos\phi\, dl \qquad (23.17)$$

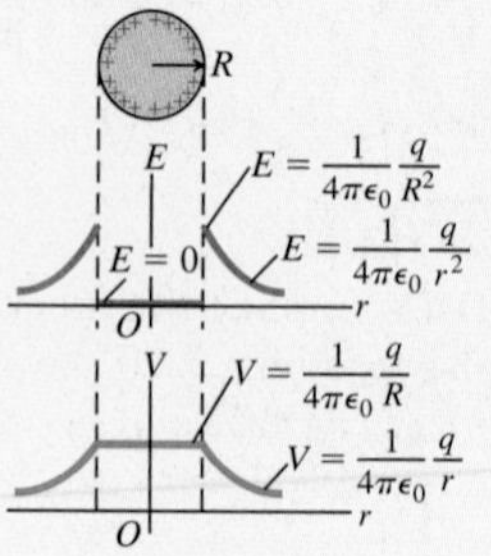

Two equivalent sets of units for electric-field magnitude are volts per meter (V/m) and newtons per coulomb (N/C). One volt is one joule per coulomb $(1\ \text{V} = 1\ \text{J/C})$. A useful unit of energy is the electron volt (eV), which is the energy corresponding to a particle with a charge equal to that of an electron moving through a potential difference of one volt. The conversion factor is $1\ \text{eV} = 1.602 \times 10^{-19}\ \text{J}$.

An equipotential surface is a surface on which the potential has the same value at every point. At a point where a field line crosses an equipotential surface, the two are perpendicular. When all charges are at rest, the surface of a conductor is always an equipotential surface and all points in the interior of a conductor are at the same potential. When a cavity within a conductor contains no charge, the entire cavity is an equipotential region and there is no surface charge anywhere on the surface of the cavity.

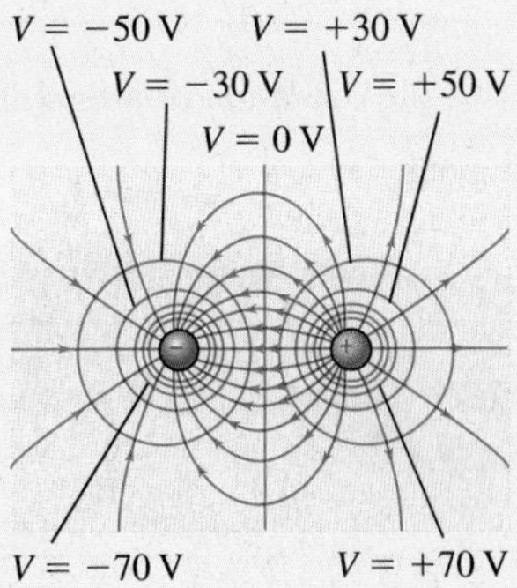

If the potential V is known as a function of the coordinates x, y, and z, the components of electric field $\vec{E}$ at any point are given by partial derivatives of V. (See Examples 23.13 through 23.15)

$$E_x = -\frac{\partial V}{\partial x} \quad E_y = -\frac{\partial V}{\partial y} \quad E_z = -\frac{\partial V}{\partial z} \qquad (23.19)$$

$$\vec{E} = -\left(\hat{\imath}\frac{\partial V}{\partial x} + \hat{\jmath}\frac{\partial V}{\partial y} + \hat{k}\frac{\partial V}{\partial z}\right) \quad \text{(vector form)} \qquad (23.20)$$

Key Terms

(electric) potential energy, 870
(electric) potential, 878
volt, 878
voltage, 878
electron volt, 881
equipotential surface, 890
gradient, 893

Your Notes

Answer to Chapter Opening Question

We describe the operation of a lightning rod after Example 23.8 (Section 23.3).

Answers to Test Your Understanding Questions

Section 23.1 The arrangement of charges is shown in Fig. 21.12. From Eq. (23.11),

$$U = \frac{1}{4\pi\epsilon_0}\left(\frac{q_1 q_2}{0.60\text{ m}} + \frac{q_1 Q}{0.50\text{ m}} + \frac{q_2 Q}{0.50\text{ m}}\right)$$

Using $q_1 = 2.0\ \mu\text{C}$, $q_2 = 2.0\ \mu\text{C}$, and $Q = 4.0\ \mu\text{C}$, we find $U = 0.35$ J. The total energy is positive, indicating that it takes positive work to bring the three charges from infinity to the positions shown in Fig. 21.12.

Section 23.2 If $V = 0$ at a certain point, $\vec{E}$ does *not* have to be zero at that point. An example is point c in Figs. 21.20 and 23.13, for which there is an electric field in the $+x$-direction (see Example 21.9 in Section 21.5) even though $V = 0$ (see Example 23.4). This isn't a surprising result, because V and $\vec{E}$ are quite different quantities: V is the net amount of work required to bring a unit charge from infinity to the point in question, whereas $\vec{E}$ is the electric force that acts on a unit charge when it arrives at that point.

Section 23.3 If $\vec{E} = \mathbf{0}$ at a certain point, V does *not* have to be zero at that point. An example is point O at the center of the charged ring in Figs. 21.21 and 23.20. From Example 21.10 (Section 21.5), the electric field is zero at O, because the electric-field contributions from different parts of the ring completely cancel. From Example 23.11, however, the potential at O is *not* zero: this point corresponds to $x = 0$, so $V = (1/4\pi\epsilon_0)(Q/a)$. This value of V corresponds to the work that would have to be done to move a unit positive test charge along a path from infinity to point O; it is nonzero because the charged ring repels the test charge, so positive work must be done to move the test charge toward the ring.

Section 23.4 If the positive charges in Fig. 23.23 were replaced by negative charges, and vice versa, the electric-field lines would have the same shape but would reverse direction. The equipotential surfaces would be the same as shown in Fig. 23.23, but the sign of the potential would be reversed. For example, the surfaces in Fig. 23.23b with potential $V = +30$ V and $V = -50$ V would have potential $V = -30$ V and $V = +50$ V, respectively.

Section 23.5 From Eq. (23.20), $\vec{E} = -(\hat{\imath}\frac{\partial V}{\partial x} + \hat{\jmath}\frac{\partial V}{\partial x} + \hat{k}\frac{\partial V}{\partial x}) = -(B\hat{\imath} + 3Cy^2\hat{\jmath} + 2Dz\hat{k})$. Note that the constant A has no effect on the electric field. This illustrates that we can add a constant to the potential at all points without changing any of the real physics—that is, without changing $\vec{E}$ or the potential difference between any two points.

Discussion Questions

Q23.1 A student asked, "Since electrical potential is always proportional to potential energy, why bother with the concept of potential at all?" How would you respond?

Q23.2 The potential (relative to a point at infinity) midway between two charges of equal magnitude and opposite sign is zero. Is it possible to bring a test charge from infinity to this midpoint in such a way that no work is done in any part of the displacement? If so, describe how it can be done. If it is not possible, explain why.

Q23.3 Is it possible to have an arrangement of two point charges separated by a finite distance such that the electric potential energy of the arrangement is the same as if the two charges were infinitely far apart? Why or why not? What if there are three charges? Explain your reasoning.

Q23.4 On a physics exam once given at a famous university, students were asked to calculate the potential energy of a particular distribution of point charges. One student did no calculations for this problem, but simply answered "The potential energy can have any value you want." Although this was not the answer that the instructor had in mind when writing the problem, the student nonetheless received full credit for her answer. Why was this answer correct? (*Note:* Don't try this on *your* exams!)

Q23.5 If $\vec{E}$ is zero everywhere along a certain path that leads from point A to point B, what is the potential difference between those two points? Does this mean that $\vec{E}$ is zero everywhere along *any* path from A to B? Explain.

Q23.6 If $\vec{E}$ is zero throughout a certain region of space, is the potential necessarily also zero in this region? Why or why not? If not, what *can* be said about the potential?

Q23.7 If you carry out the integral of the electric field $\int \vec{E} \cdot d\vec{l}$ for a *closed* path like that shown in Fig. 23.27, the integral will *always* be equal to zero, independent of the shape of the path and independent of where charges may be located relative to the path. Explain why.

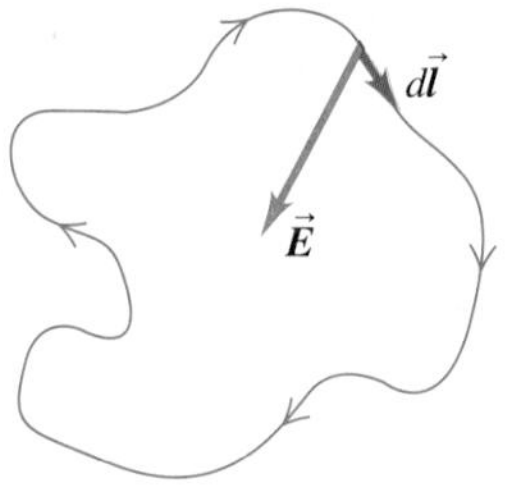

Figure 23.27 Question Q23.7.

Q23.8 The potential difference between the two terminals of an AA battery (used in flashlights and portable stereos) is 1.5 V. If two AA batteries are placed end-to-end with the positive terminal of one battery touching the negative terminal of the other, what is the potential difference between the terminals at the exposed ends of the combination? What if the two positive terminals are touching each other? Explain your reasoning.

Q23.9 It is easy to produce a potential difference of several thousand volts between your body and the floor by scuffing your shoes across a nylon carpet. When you touch a metal doorknob, you get a mild shock. Yet contact with a power line of comparable voltage would probably be fatal. Why is there a difference?

Q23.10 If the electric potential at a single point is known, can $\vec{E}$ at that point be determined? If so, how? If not, why not?

Q23.11 Because electric field lines and equipotential surfaces are always perpendicular, two equipotential surfaces can never cross; if they did, the direction of $\vec{E}$ would be ambiguous at the crossing points. Yet two equipotential surfaces appear to cross at the center of Fig. 23.23c. Explain why there is no ambiguity about the direction of $\vec{E}$ in this particular case.

Q23.12 As you walk through a certain region of space, the potential at your position becomes more and more positive as you move from north to south. The potential doesn't change when you move east, west, up, or down. Is there necessarily an electric field in this region? If so, what is its direction? Explain your reasoning. Suggest a way in which you could cause the potential to vary with position in the way described.

Q23.13 Is potential gradient a scalar quantity or a vector quantity? How can you tell?

Q23.14 A conducting sphere is to be charged by bringing in positive charge a little at a time until the total charge is Q. The total work required for this process is alleged to be proportional to Q^2. Is this correct? Why or why not?

Q23.15 Are there cases in electrostatics in which a conducting surface is *not* an equipotential surface? If so, give an example. If not, explain why not.

Q23.16 A conducting sphere is placed between two charged parallel plates such as those shown in Fig. 23.1. Does the electric field inside the sphere depend on precisely where between the plates the sphere is placed? What about the electric potential inside the sphere? Do the answers to these questions depend on whether or not there is a net charge on the sphere? Explain your reasoning.

Q23.17 A conductor that carries a net charge Q has a hollow, empty cavity in its interior. Does the potential vary from point to point within the material of the conductor? What about within the cavity? How does the potential inside the cavity compare to the potential within the material of the conductor?

Q23.18 A high-voltage dc power line falls on a car, so the entire metal body of the car is at a potential of 10,000 V with respect to the ground. What happens to the occupants a) when they are sitting in the car? b) when they step out of the car? Explain your reasoning.

Q23.19 When a thunderstorm is approaching, sailors at sea sometimes observe a phenomenon called "St. Elmo's fire," a bluish flickering light at the tips of masts. What causes this? Why does it occur at the tips of masts? Why is the effect most pronounced when the masts are wet? (*Hint:* Seawater is a good conductor of electricity.)

Q23.20 A positive point charge is placed near a very large conducting plane. A professor of physics asserted that the field caused by this configuration is the same as would be obtained by removing the plane and placing a negative point charge of equal magnitude in the mirror-image position behind the initial position of the plane. Is this correct? Why or why not? (*Hint:* Inspect Fig. 23.23b.)

Q23.21 In electronics it is customary to define the potential of ground (thinking of the earth as a large conductor) as zero. Is this consistent with the fact that the earth has a net electric charge that is not zero? (Refer to Exercise 21.30.)

Exercises

Section 23.1 Electric Potential Energy

23.1 A point charge $q_1 = +2.40\ \mu\text{C}$ is held stationary at the origin. A second point charge $q_2 = -4.30\ \mu\text{C}$ moves from the point $x = 0.150$ m, $y = 0$ to the point $x = 0.250$ m, $y = 0.250$ m. How much work is done by the electric force on q_2?

23.2 A point charge q_1 is held stationary at the origin. A second charge q_2 is placed at point a, and the electric potential energy of the pair of charges is $+5.4 \times 10^{-8}$ J. When the second charge is moved to point b, the electric force on the charge does -1.9×10^{-8} J of work. What is the electric potential energy of the pair of charges when the second charge is at point b?

23.3 A small metal sphere, carrying a net charge of $q_1 = -2.80\ \mu\text{C}$, is held in a stationary position by insulating supports. A second small metal sphere, with a net charge of $q_2 = -7.80\ \mu\text{C}$ and mass 1.50 g, is projected toward q_1. When the two spheres are 0.800 m apart, q_2 is moving toward q_1 with speed 22.0 m/s (Fig. 23.28). Assume that the two spheres can be treated as point charges. You can ignore the force of gravity. a) What is the speed of q_2 when the spheres are 0.400 m apart? b) How close does q_2 get to q_1?

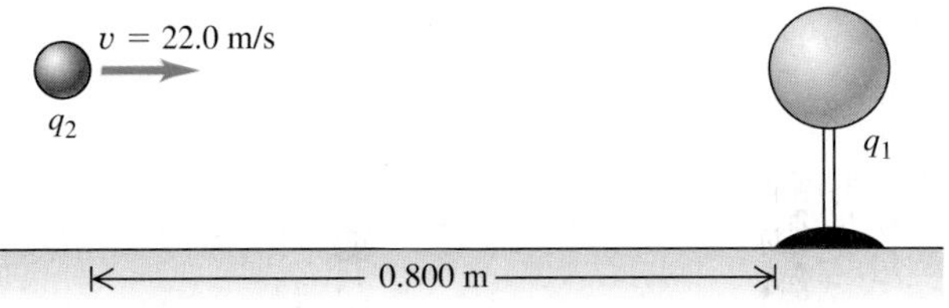

Figure 23.28 Exercise 23.3.

23.4 How far from a -7.20-μC point charge must a $+2.30$-μC point charge be placed for the electric potential energy U of the pair of charges to be -0.400 J? (Take U to be zero when the charges have infinite separation.)

23.5 A point charge $Q = +4.60\ \mu\text{C}$ is held fixed at the origin. A second point charge $q = +1.20\ \mu\text{C}$ with mass of 2.80×10^{-4} kg is placed on the x-axis, 0.250 m from the origin. a) What is the electric potential energy U of the pair of charges? (Take U to be zero when the charges have infinite separation.) b) The second point charge is released from rest. What is its speed when its distance from the origin is i) 0.500 m? ii) 5.00 m? iii) 50.0 m?

23.6 Three equal 1.20-μC point charges are placed at the corners of an equilateral triangle whose sides are 0.500 m. What is the potential energy of the system? (Take as zero the potential energy of the three charges when they are infinitely far apart.)

23.7 A point charge $q_1 = 4.00$ nC is placed at the origin, and a second point charge $q_2 = -3.00$ nC is placed on the x-axis at $x = +20.0$ cm. A third point charge $q_3 = 2.00$ nC is to be placed on the x-axis between q_1 and q_2. (Take as zero the potential energy of the three charges when they are infinitely far apart.) a) What is the potential energy of the system of the three charges if q_3 is placed at $x = +10.0$ cm? b) Where should q_3 be placed to make the potential energy of the system equal to zero?

23.8 In part (c) of Example 23.1 (Section 23.1), calculate the distance between the positron and alpha particle when the positron comes momentarily to rest.

23.9 Three point charges, which initially are infinitely far apart, are placed at the corners of an equilateral triangle with sides d. Two of the point charges are identical and have charge q. If zero net work is required to place the three charges at the corners of the triangle, what must the value of the third charge be?

23.10 A proton, an alpha particle, an electron, and a neutron are at rest at the corners of a square whose side length is 5.00×10^{-10} m with the electron and neutron at opposite corners. How minimum an amount of work must be done to move the particles far from each other?

23.11 Two protons and an alpha particle are held at rest at the corners of an equilateral triangle whose side length is 8.00×10^{-10} m. The particles are released and move apart. What is their total energy when they are far apart?

23.12 Two protons are aimed directly toward each other by a cyclotron accelerator with speeds of 1000 km/s, measured relative to the earth. Find the maximum electrical force that these protons will exert on each other.

Section 23.2 Electric Potential

23.13 A small particle has charge $-5.00\ \mu$C and mass 2.00×10^{-4} kg. It moves from point A, where the electric potential is $V_A = +200$ V, to point B, where the electric potential is $V_B = +800$ V. The electric force is the only force acting on the particle. The particle has speed 5.00 m/s at point A. What is its speed at point B? Is it moving faster or slower at B than at A? Explain.

23.14 Identical point charges $q = +5.00\ \mu$C are placed at opposite corners of a square. The length of each side of the square is 0.200 m. A point charge $q_0 = -2.00\ \mu$C is placed at one of the empty corners. How much work is done on q_0 by the electric force when q_0 is moved to the other empty corner?

23.15 A uniform electric field is directed due east. Point B is 2.00 m west of point A, point C is 2.00 m east of point A and point D is 2.00 m south of A. For each point, B, C, and D, is the potential at that point larger, smaller, or the same as at point A? Give the reasoning behind your answers.

23.16 A particle with a charge of +4.20 nC is in a uniform electric field $\vec{E}$ directed to the left. It is released from rest and moves to the left; after it has moved 6.00 cm, its kinetic energy is found to be $+1.50 \times 10^{-6}$ J. a) What work was done by the electric force? b) What is the potential of the starting point with respect to the endpoint? c) What is the magnitude of $\vec{E}$?

23.17 A charge of 28.0 nC is placed in a uniform electric field that is directed vertically upward and that has a magnitude of 4.00×10^4 V/m. What work is done by the electric force when the charge moves a) 0.450 m to the right; b) 0.670 m upward; c) 2.60 m at an angle of 45.0° downward from the horizontal?

23.18 Two stationary point charges +3.00 nC and +2.00 nC are separated by a distance of 50.0 cm. An electron is released from rest at a point midway between the two charges and moves along the line connecting the two charges. What is the speed of the electron when it is 10.0 cm from the +3.00-nC charge?

23.19 A point charge has a charge of 2.50×10^{-11} C. At what distance from the point charge is the electric potential a) 90.0 V? b) 30.0 V? Take the potential to be zero at an infinite distance from the charge.

23.20 The potential V at a distance of 25.0 cm from a very small charged sphere is 48.0 V, with V taken to be zero at an infinite distance from the sphere. a) If the sphere is treated as a point charge, what is its charge? b) What is the potential at a distance of 75.0 cm from the sphere?

23.21 Two point charges $q_1 = +2.40$ nC and $q_2 = -6.50$ nC are 0.100 m apart. Point A is midway between them; point B is 0.080 m from q_1 and 0.060 m from q_2 (Fig. 23.29). Take the electric potential to be zero at infinity. Find a) the potential at point A; b) the potential at point B; c) the work done by the electric field on a charge of 2.50 nC that travels from point B to point A.

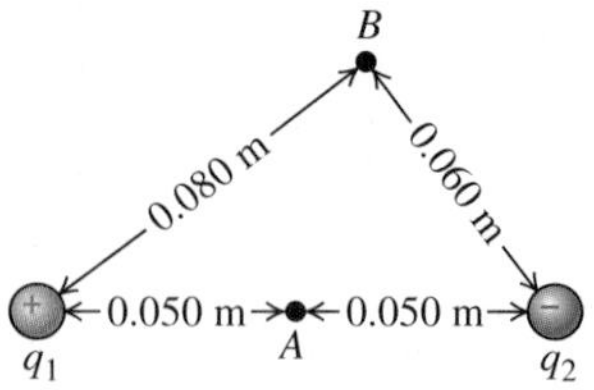

Figure 23.29 Exercise 23.21.

23.22 Two positive point charges, each of magnitude q, are fixed on the y-axis at the points $y = +a$ and $y = -a$. Take the potential to be zero at an infinite distance from the charges. a) Show the positions of the charges in a diagram. b) What is the potential V_0 at the origin? c) Show that the potential at any point on the x-axis is

$$V = \frac{1}{4\pi\epsilon_0}\frac{2q}{\sqrt{a^2 + x^2}}$$

d) Graph the potential on the x-axis as a function of x over the range from $x = -4a$ to $x = +4a$. e) What is the potential when $x \gg a$? Explain why this result is obtained.

23.23 A positive charge $+q$ is located at the point $x = 0$, $y = -a$, and a negative charge $-q$ is located at the point $x = 0$, $y = +a$. a) Show the positions of the charges in a diagram. b) Derive an expression for the potential V at points on the x-axis as a function of the coordinate x. Take V to be zero at an infinite distance from the charges. c) Graph V at points on the x-axis as a function of x over the range from $x = -4a$ to $x = +4a$. d) What is the answer to part (b) if the two charges are interchanged so that $+q$ is at $y = +a$ and $-q$ is at $y = -a$?

23.24 Consider the arrangement of charges described in Exercise 23.23. a) Derive an expression for the potential V at points on the y-axis as a function of the coordinate y. Take V to be zero at an infinite distance from the charges. b) Graph V at points on the y-axis as a function of y over the range from $y = -4a$ to $y = +4a$. c) Show that for $y \gg a$, the potential at a point on the positive y-axis is given by $V = -\left(\frac{1}{4\pi\epsilon_0}\right)2qa/y^2$. d) What are the answers to parts (a) and (c) if the two charges are interchanged so that $+q$ is at $y = +a$ and $-q$ is at $y = -a$?

23.25 A positive charge q is fixed at the point $x = 0$, $y = 0$, and a negative charge $-2q$ is fixed at the point $x = a$, $y = 0$. a) Show the positions of the charges in a diagram. b) Derive an expression for the potential V at points on the x-axis as a function of the coordinate x. Take V to be zero at an infinite distance from the charges. c) At which positions on the x-axis is $V = 0$? d) Graph V at points on the x-axis as a function of x in the range from $x = -2a$ to $x = +2a$. e) What does the answer to part (b) become when $x \gg a$? Explain why this result is obtained.

23.26 Consider the arrangement of point charges described in Exercise 23.25. a) Derive an expression for the potential V at points on the y-axis as a function of the coordinate y. Take V to be zero at an infinite distance from the charges. b) At which positions on the y-axis is $V = 0$? c) Graph V at points on the y-axis as a function of y in the range from $y = -2a$ to $y = +2a$. d) What does the answer to part (a) become when $y \gg a$? Explain why this result is obtained.

23.27 Before the advent of solid-state electronics, vacuum tubes were widely used in radios and other devices. A simple type of vacuum tube known as a *diode* consists essentially of two electrodes within a highly evacuated enclosure. One electrode, the *cathode*, is maintained at a high temperature and emits electrons from its surface. A potential difference of a few hundred volts is maintained between the cathode and the other electrode, known as the *anode*, with the anode at the higher potential. Suppose a diode consists of a cylindrical cathode with radius 0.062 cm, mounted coaxially within a cylindrical anode 0.558 cm in radius. The potential of the anode is 295 V higher than that of the cathode. An electron leaves the surface of the cathode with zero initial speed. Find its speed when it strikes the anode.

23.28 At a certain distance from a point charge, the potential and electric field magnitude due to that charge are 4.98 V and 12.0 V/m, respectively. (Take the potential to be zero at infinity.) a) What is the distance to the point charge? b) What is the magnitude of the charge? c) Is the electric field directed toward or away from the point charge?

23.29 A uniform electric field has magnitude E and is directed in the negative x-direction. The potential difference between point a (at $x = 0.60$ m) and point b (at $x = 0.90$ m) is 240 V. a) Which point, a or b, is at the higher potential? b) Calculate the value of E. c) A negative point charge $q = -0.200\ \mu\text{C}$ is moved from b to a. Calculate the work done on the point charge by the electric field.

23.30 For each of the following arrangements of two point charges, find all the points along the line passing through both charges for which the electric potential V is zero (take $V = 0$ infinitely far from the charges) and for which the electric field E is zero: a) charges $+Q$ and $+2Q$ separated by a distance d; b) charges $-Q$ and $+2Q$ separated by a distance d. c) Are both V and E zero at the same places? Explain.

23.31 a) An electron is to be accelerated from 3.00×10^6 m/s to 8.00×10^6 m/s. Through what potential difference must the electron pass to accomplish this? b) Through what potential difference must the electron pass if it is to be slowed from 8.00×10^6 m/s to a halt?

Section 23.3 Calculating Electric Potential

23.32 A total electric charge of 3.50 nC is distributed uniformly over the surface of a metal sphere with a radius of 24.0 cm. If the potential is zero at a point at infinity, find the value of the potential at the following distances from the center of the sphere: a) 48.0 cm; b) 24.0 cm; c) 12.0 cm.

23.33 A uniformly charged thin ring has radius 15.0 cm and has total charge +24.0 nC. An electron is placed on the ring's axis a distance 30.0 cm from the center of the ring and is constrained to stay on the axis of the ring. The electron is then released from rest. a) Describe the subsequent motion of the electron. b) Find the speed of the electron when it reaches the center of the ring.

23.34 An infinitely long line of charge has linear charge density 5.00×10^{-12} C/m. A proton (mass 1.67×10^{-27} kg, charge $+1.60 \times 10^{-19}$ C) is 18.0 cm from the line and moving directly toward the line at 1.50×10^3 m/s. How close does the proton get to the line of charge?

23.35 Two large, parallel, metal plates carry opposite charges of equal magnitude. They are separated by 45.0 mm, and the potential difference between them is 360 V. a) What is the magnitude of the electric field (assumed to be uniform) in the region between the plates? b) What is the magnitude of the force this field exerts on a particle with charge +2.40 nC? c) Use the results of part (b) to compute the work done by the field on the particle as it moves from the higher-potential plate to the lower. d) Compare the result of part (c) to the change of potential energy of the same charge, computed from the electric potential.

23.36 Two large, parallel, metal sheets carrying opposite electric charges of equal magnitude are separated by a distance of 38.0 mm. The electric field between them is uniform and has magnitude 480 N/C. a) What is the potential difference between the sheets? b) Which sheet is at higher potential, the one with positive charge or the one with negative charge? c) What is the surface charge density σ on the positive sheet?

23.37 A potential difference of 4.75 kV is established between parallel plates in air. a) If the air becomes electrically conducting when the electric field exceeds 3.00×10^6 V/m, what is the minimum separation of the plates? b) When the separation has the minimum value calculated in part (a), what is the surface charge density on each plate?

23.38 Two large parallel conducting plates carrying opposite charges of equal magnitude are separated by 2.20 cm. a) If the surface charge density for each plate has magnitude 47.0 nC/m^2, what is the magnitude of $\vec{E}$ in the region between the plates? b) What is the potential difference between the two plates? c) If the separation between the plates is doubled while the surface charge density is kept constant at the value in part (a), what happens to the magnitude of the electric field and to the potential difference?

23.39 a) Show that V for a spherical shell of radius R, that has charge q distributed uniformly over its surface, is the same as for a solid conductor with radius R and charge q. b) You rub an inflated balloon on the carpet and it acquires a potential that is 1200 V lower than its potential before it became charged. If the charge is uniformly distributed over the surface of the balloon and if the radius of the balloon is 15 cm, what is the net charge on the balloon? c) In light of its 1200 V potential difference relative to you, do you think this balloon is dangerous? Explain.

23.40 The electric field at the surface of a charged, solid, copper sphere with radius 0.200 m is 3800 N/C, directed toward the center of the sphere. What is the potential at the center of the sphere, if we take the potential to be zero infinitely far from the sphere?

Section 23.4 Equipotential Surfaces and Section 23.5 Potential Gradient

23.41 In a certain region of space, the electric potential is $V(x, y, z) = Axy - Bx^2 + Cy$, where A, B, and C are positive constants. a) Calculate the x-, y-, and z-components of the electric field. b) At which points is the electric field equal to zero?

23.42 The potential due to a point charge Q at the origin may be written as

$$V = \frac{Q}{4\pi\epsilon_0 r} = \frac{Q}{4\pi\epsilon_0\sqrt{x^2 + y^2 + z^2}}$$

a) Calculate E_x, E_y, and E_z using Eq. (23.19). b) Show that the result of part (a) agrees with Eq. (21.7) for the electric field of a point charge.

23.43 The potential is given by $V = 0$ for $z < 0$, $V = Cz$ for $0 < z < d$, and $V = Cd$ for $z > d$, where C and d are positive constants. The potential does not depend on x or y. a) Find the electric field (magnitude and direction) at all points. b) What charge distribution could be the source of this field? Explain.

23.44 A metal sphere with radius r_a is supported on an insulating stand at the center of a hollow, metal, spherical shell with radius r_b. There is charge $+q$ on the inner sphere and charge $-q$ on the outer spherical shell. a) Calculate the potential $V(r)$ for i) $r < r_a$; ii) $r_a < r < r_b$; iii) $r > r_b$. (*Hint:* The net potential is the sum of the potentials due to the individual spheres.) Take V to be zero when r is infinite. b) Show that the potential of the inner sphere with respect to the outer is

$$V_{ab} = \frac{q}{4\pi\epsilon_0}\left(\frac{1}{r_a} - \frac{1}{r_b}\right)$$

c) Use Eq. (23.23) and the result from part (a) to show that the electric field at any point between the spheres has magnitude

$$E(r) = \frac{V_{ab}}{(1/r_a - 1/r_b)}\frac{1}{r^2}$$

d) Use Eq. (23.23) and the result from part (a) to find the electric field at a point outside the larger sphere at a distance r from the center, where $r > r_b$. e) Suppose the charge on the outer sphere is not $-q$ but a negative charge of different magnitude, say $-Q$. Show that the answers for parts (b) and (c) are the same as before but the answer for part (d) is different.

23.45 A metal sphere with radius $r_a = 1.20$ cm is supported on an insulating stand at the center of a hollow, metal, spherical shell with radius $r_b = 9.60$ cm. Charge $+q$ is put on the inner sphere and charge $-q$ on the outer spherical shell. The magnitude of q is chosen to make the potential difference between the spheres 500 V, with the inner sphere at higher potential. a) Use the result of Exercise 23.44(b) to calculate q. b) With the help of the result of Exercise 23.44(a), sketch the equipotential surfaces that correspond to 500, 400, 300, 200, 100, and 0 V. c) In your sketch, show the electric field lines. Are the electric field lines and equipotential surfaces mutually perpendicular? Are the equipotential surfaces closer together when the magnitude of $\vec{E}$ is largest?

23.46 A charge Q is uniformly distributed along a rod with length $2a$. In Example 23.12 (Section 23.3), we derived an expression for the potential V at a point on the perpendicular bisector of the rod at a distance x from its center. a) By differentiating this expression, find the x-component E_x of the electric field at such a point. Show that the result is the same expression we found in Example 21.11 (Section 21.5) by direct integration. (*Hint:* Writing the natural logarithm of the fraction as the difference of two logarithms simplifies the calculation.) b) By symmetry, at a point on the perpendicular bisector of the rod, $E_y = 0$ and $E_z = 0$. Explain why it would *not* be correct to draw these conclusions by using Eq. (23.19).

23.47 A potential difference of 480 V is established between large parallel, metal plates. Let the potential of one plate be 480 V and the other be 0 V. The plates are separated by $d = 1.70$ cm. a) Sketch the equipotential surfaces that correspond to 0, 120, 240, 360, and 480 V. b) In your sketch, show the electric field lines. Does your sketch confirm that the field lines and equipotential surfaces are mutually perpendicular?

Problems

23.48 Three small spheres with charge 2.00 μC are arranged in a line, with sphere 2 in the middle. Adjacent spheres are initially 8.00 cm apart. The spheres have masses $m_1 = 20.0$ g, $m_2 = 85.0$ g, and $m_3 = 20.0$ g, and their radii are much smaller than their separation. The three spheres are released from rest. a) What is the acceleration of sphere 1 just after it is released? b) What is the speed of each sphere when they are far apart?

23.49 A particle with charge +7.60 nC is in a uniform electric field directed to the left. Another force, in addition to the electric force, acts on the particle so that when it is released from rest, it moves to the right. After it has moved 8.00 cm, the additional force has done 6.50×10^{-5} J of work and the particle has 4.35×10^{-5} J of kinetic energy. a) What work was done by the electric force? b) What is the potential of the starting point with respect to the endpoint? c) What is the magnitude of the electric field?

23.50 In the *Bohr model* of the hydrogen atom, a single electron revolves around a single proton in a circle of radius r. Assume that the proton remains at rest. a) By equating the electric force to the electron mass times its acceleration, derive an expression for the electron's speed. b) Obtain an expression for the electron's kinetic energy, and show that its magnitude is just half that of the electric potential energy. c) Obtain an expression for the total energy, and evaluate it using $r = 5.29 \times 10^{-11}$ m. Give your numerical result in joules and in electron volts.

23.51 A vacuum tube diode was described in Exercise 23.27. Because of the accumulation of charge near the cathode, the electric potential between the electrodes is not a linear function of the position, even with planar geometry, but is given by

$$V(x) = Cx^{4/3}$$

where x is the distance from the cathode (negative electrode) and C is a constant, characteristic of a particular diode and operating conditions. Assume that the distance between the cathode and anode (positive electrode) is 13.0 mm and the potential difference between electrodes is 240 V. a) Determine the value of C. b) Obtain a formula for the electric field between the electrodes as a function of x. c) Determine the force on an electron when the electron is halfway between the electrodes.

23.52 Refer to Problem 22.51. For the situation shown in Fig. 22.41, calculate the energy required to move the two electrons of charge $-e$ infinitely far from each other and infinitely far from the uniformly charged sphere. Your answer will be in terms of the distances d and R.

23.53 An Ionic Crystal. Figure 23.30 shows eight point charges arranged at the corners of a cube with sides of length d. The values of the charges are $+q$ and $-q$, as shown. This is a model of one cell of a cubic ionic crystal. In sodium chloride (NaCl), for instance, the positive ions are Na^+ and the negative ions are Cl^-. a) Calculate the potential energy U of this arrangement. (Take as zero the potential energy of the eight charges when they are infinitely far apart.) b) In part (a), you should have found that $U < 0$. Explain the relationship between this result and the observation that such ionic crystals exist in nature.

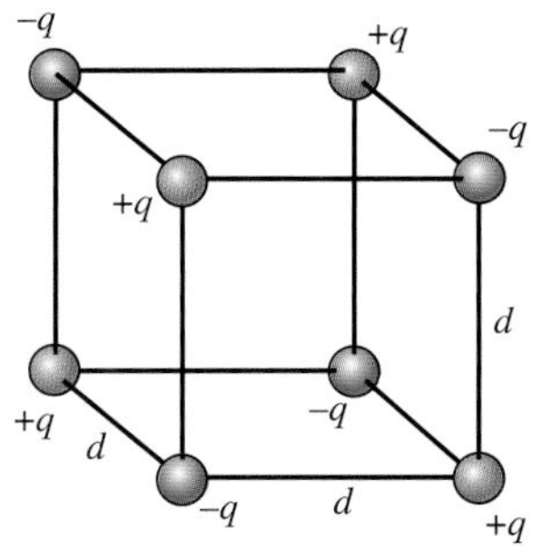

Figure 23.30 Problem 23.53.

23.54 A One-Dimensional "Crystal." While real crystals are three-dimensional, much can be learned from simple one-dimensional models for which calculations can be carried out much more easily. As a one-dimensional model of an ionic crystal such as sodium chloride (NaCl), consider alternating positive and negative ions of charge $+e$ and $-e$, respectively, which are uniformly spaced along the x-axis with spacing d (Fig. 23.31). The charges may be considered to extend to infinity in both directions. a) Consider the potential energy of the interaction between the positive ion at $x = 0$ and all the other ions. This represents the *potential energy per ion* in this one-dimensional "crystal." Write an expression for this potential energy (your expression will be an infinite series). b) Evaluate the infinite series in part (a), using the expansion $\ln(1 + z) = z - z^2/2 + z^3/3 - z^4/4 + \cdots$, valid for the case $|z| \leq 1$. c) Does the potential energy per ion have the same value for negative ions in the "crystal" as for positive ions? Explain your reasoning. d) In the real, three-dimensional NaCl crystal, the spacing between adjacent ions is 2.82×10^{-10} m. Using this spacing for the value of d in Fig. 23.31, calculate the potential energy per ion in the one-dimensional crystal. e) For most real (three-dimensional) ionic crystal substances such as NaCl, the potential energy per ion is about -8×10^{-19} J/ion. How does this compare to your result in part (d)? How good a model is the one-dimensional "crystal" shown in Fig. 23.31?

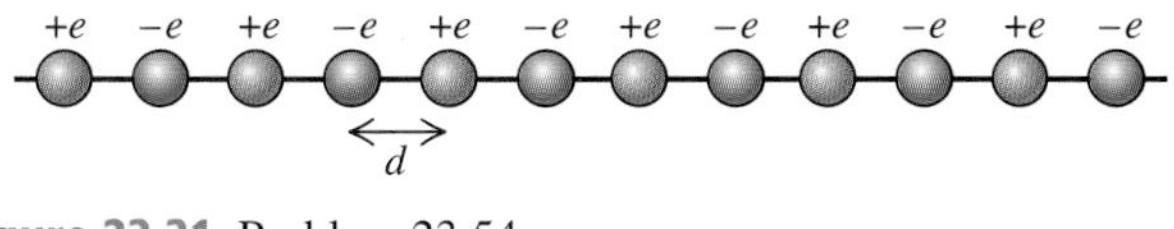

Figure 23.31 Problem 23.54.

23.55 The H_2^+ Ion. The H_2^+ ion is comprised of two protons, each of charge $+e = 1.60 \times 10^{-19}$ C, and an electron of charge $-e$ and mass 9.11×10^{-31} kg. The separation between the protons is 1.07×10^{-10} m. The protons and the electron may be treated as point charges. a) Suppose the electron is located at the point midway between the two protons. What is the potential energy of the interaction between the electron and the two protons? (Do not include the potential energy due to the interaction between the two protons.) b) Suppose the electron in part (a) has a velocity of magnitude 1.50×10^6 m/s in a direction along the perpendicular bisector of the line connecting the two protons. How far from the point midway between the two protons can the electron move? Because the masses of the protons are much greater than the electron mass, the motions of the protons are very slow and can be ignored. (*Note:* A realistic description of the electron motion requires the use of quantum mechanics, not Newtonian mechanics.)

23.56 A small sphere with mass 1.50 g hangs by a thread between two parallel vertical plates 5.00 cm apart. The plates are insulating and have uniform surface charge densities $+\sigma$ and $-\sigma$. The charge on the sphere is $q = 8.90 \times 10^{-6}$ C. What potential difference between the plates will cause the thread to assume an angle of 30.0° with the vertical (Fig. 23.32)?

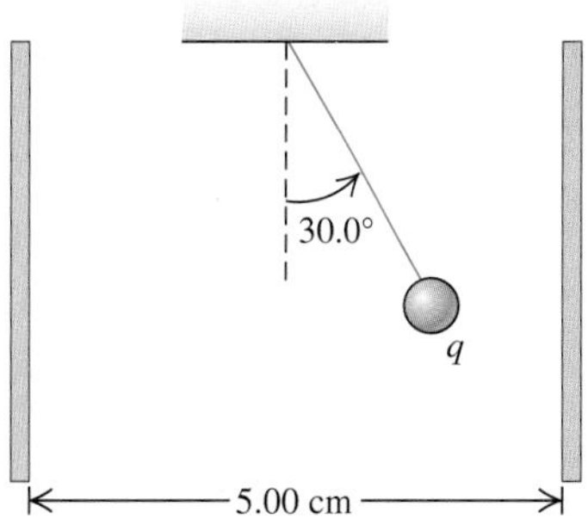

Figure 23.32 Problem 23.56.

23.57 Coaxial Cylinders. A long metal cylinder with radius a is supported on an insulating stand on the axis of a long, hollow, metal tube with radius b. The positive charge per unit length on the inner cylinder is λ, and there is an equal negative charge per unit length on the outer cylinder. a) Calculate the potential $V(r)$ for i) $r < a$; ii) $a < r < b$; iii) $r > b$. (*Hint:* The net potential is the sum of the potentials due to the individual conductors.) Take $V = 0$ at $r = b$. b) Show that the potential of the inner cylinder with respect to the outer is

$$V_{ab} = \frac{\lambda}{2\pi\epsilon_0}\ln\frac{b}{a}$$

c) Use Eq. (23.23) and the result from part (a) to show that the electric field at any point between the cylinders has magnitude

$$E(r) = \frac{V_{ab}}{\ln(b/a)} \cdot \frac{1}{r}$$

d) What is the potential difference between the two cylinders if the outer cylinder has no net charge?

23.58 A *Geiger counter* detects radiation such as alpha particles by using the fact that the radiation ionizes the air along its path. A thin wire lies on the axis of a hollow metal cylinder and is insulated from it (Fig. 23.33). A large potential difference is established

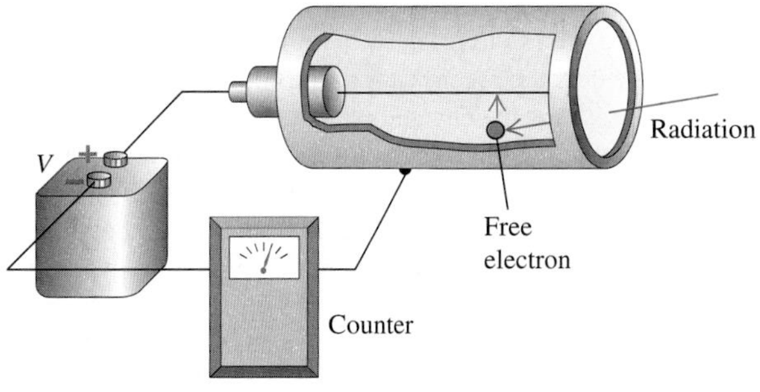

Figure 23.33 Problem 23.58.

between the wire and the outer cylinder, with the wire at higher potential; this sets up a strong electric field directed radially outward. When ionizing radiation enters the device, it ionizes a few air molecules. The free electrons produced are accelerated by the electric field toward the wire and, on the way there, ionize many more air molecules. Thus a current pulse is produced that can be detected by appropriate electronic circuitry and converted to an audible "click." Suppose the radius of the central wire is 145 μm and the radius of the hollow cylinder is 1.80 cm. What potential difference between the wire and the cylinder produces an electric field of 2.00×10^4 V/m at a distance of 1.20 cm from the axis of the wire? (The wire and cylinder are both very long in comparison to their radii, so the results of Problem 23.57 apply.)

23.59 Deflection in a CRT. Cathode-ray tubes (CRTs) are often found in oscilloscopes and computer monitors. In Fig. 23.34 an electron with an initial speed of 6.50×10^6 m/s is projected along the axis midway between the deflection plates of a cathode-ray tube. The uniform electric field between the plates has a magnitude of 1.10×10^3 V/m and is upward. a) What is the force (magnitude and direction) on the electron when it is between the plates? b) What is the acceleration of the electron (magnitude and direction) when acted on by the force in part (a)? c) How far below the axis has the electron moved when it reaches the end of the plates? d) At what angle with the axis is it moving as it leaves the plates? e) How far below the axis will it strike the fluorescent screen S?

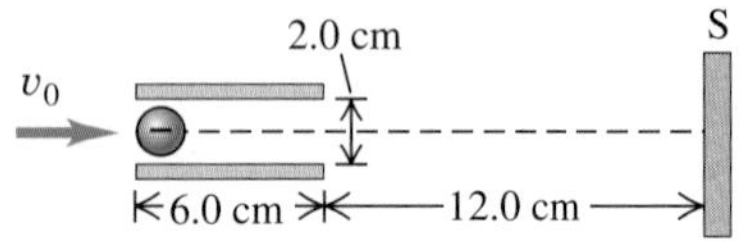

Figure 23.34 Problem 23.59.

23.60 The tube of a Geiger counter (Problem 23.58) has a long, hollow, metal cylinder 2.00 cm in diameter. Along the axis of the tube is a wire 0.127 mm in diameter running its full length. When the tube is operating, a voltage of 850 V is applied between the two conductors. Find the electric field strength at a) the outer surface of the wire; b) the inner surface of the cylinder.

23.61 *Electrostatic precipitators* use electric forces to remove pollutant particles from smoke, in particular in the smokestacks of coal-burning power plants. One form of precipitator consists of a vertical, hollow, metal cylinder with a thin wire, insulated from the cylinder, running along its axis (Fig. 23.35). A large potential difference is established between the wire and the outer cylinder, with the wire at lower potential. This sets up a strong radial electric field directed inward. The field produces a region of ionized air near the wire. Smoke enters the precipitator at the bottom, ash and dust in it pick up electrons, and the charged pollutants are accelerated toward the outer cylinder wall by the electric field. Suppose the radius of the central wire is 90.0 μm, the radius of the cylinder is 14.0 cm, and a potential difference of 50.0 kV is established between the wire and the cylinder. Also assume that the wire and cylinder are both very long in comparison to the cylinder radius, so the results of Problem 23.57 apply. a) What is the magnitude of the electric field midway between the wire and the cylinder wall? b) What magnitude of charge must a 30.0-μg ash particle have if the electric field computed in part (a) is to exert a force ten times the weight of the particle?

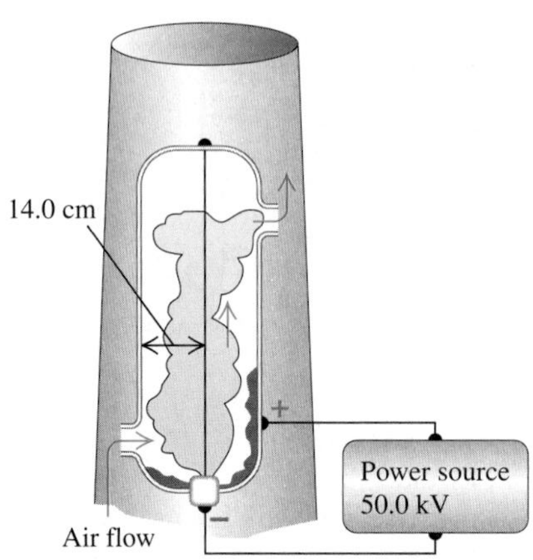

Figure 23.35 Problem 23.61.

23.62 Four line segments of charge are arranged to form a square with sides of length a. The potential is zero at infinity. Calculate the potential at the center of the square if a) two opposite sides are positively charged with charge $+Q$ each and the other two sides are negatively charged with charge $-Q$ each; b) if each side has positive charge $+Q$. (*Hint:* Use the result of Example 23.12 in Section 23.3.)

23.63 A disk with radius R has uniform surface charge density σ. a) By regarding the disk as a series of thin concentric rings, calculate the electric potential V at a point on the disk's axis a distance x from the center of the disk. Assume that the potential is zero at infinity. (*Hint:* Use the result of Example 23.11 in Section 23.3.) b) Calculate $-\partial V/\partial x$. Show that the result agrees with the expression for E_x calculated in Example 21.12 (Section 21.5).

23.64 A line of charge of length $2a$ has total charge Q. The potential at a point along the perpendicular bisector of the rod, at a distance x from its center, was calculated in Example 23.12 (Section 23.3). a) Show that if $x \gg 0$, $V(x)$ approaches $Q/4\pi\epsilon_0 x$. Interpret this result. (*Hint:* For $|z| \ll 1$, $\ln(1 + z) \cong z$.) b) Consider points very close to the rod, such that $x \ll a$. Simplify the expression for $V(x)$ for this special case. Compare to the result $V = (\lambda/2\pi\epsilon_0)\ln(R/r)$ obtained in Example 23.10 (Section 23.3) for an infinitely long charged cylinder. What corresponds to R in this limit? Can you explain this?

23.65 a) From the expression for E obtained in Problem 22.48, find the expressions for the electric potential V as a function of r, both inside and outside the cylinder. Let $V = 0$ at the surface of the cylinder. In each case, express your result in terms of the charge per unit length λ of the charge distribution. b) Graph V and E as functions of r from $r = 0$ to $r = 3R$.

23.66 Refer to Problem 21.68. a) Calculate the potential at the point $x = 3.00$ cm, $y = 0$ and at the point $x = 3.00$ cm, $y = 5.00$ cm due to the first two charges. Let the potential be zero far from the charges. b) If the third charge moves from the point $x = 3.00$ cm, $y = 0$ to the point $x = 3.00$ cm, $y = 5.00$ cm, calculate the work done on it by the field of the first two charges. Comment on the *sign* of this work. Is your result reasonable?

23.67 For the ring of charge described in Example 23.11 (Section 23.3), integrate the expression for E_x found in Example 21.10 (Section 21.5) to find the potential at point P on the ring's axis. Assume that $V = 0$ at infinity. Compare your result to that obtained in Example 23.11 using Eq. (23.16).

23.68 A thin insulating rod is bent into a semicircular arc of radius a, and a total electric charge Q is distributed uniformly along the rod. Calculate the potential at the center of curvature of the arc if the potential is assumed to be zero at infinity.

23.69 For the situation described in Problem 22.30, calculate the magnitude of the potential difference between the following pairs of cube faces (Fig. 22.32): a) S_1 and S_3; b) S_2 and S_4; c) S_5 and S_6. In each case where the potential difference is nonzero, specify which face is at higher potential.

23.70 a) From the expression for $E(r)$ obtained in Example 22.9 (Section 22.4), find the expression for the electric potential $V(r)$ as a function of r both inside and outside the uniformly charged sphere. Assume that $V = 0$ at infinity. b) Graph V and E as functions of r from $r = 0$ to $r = 3R$.

23.71 A solid insulating sphere with radius R has charge Q uniformly distributed throughout its volume. a) Use the results of Problem 23.70 to find the magnitude of the potential difference between the surface of the sphere and its center. b) Which is at higher potential, the surface or the center, if i) Q is positive? ii) Q is negative?

23.72 An insulating spherical shell with inner radius 25.0 cm and outer radius 60.0 cm carries a charge of $+150.0\ \mu\text{C}$ uniformly distributed over its outer surface (see Exercise 23.39). Point a is at the center of the shell, point b is on the inner surface and point c is on the outer surface. a) What will a voltmeter read if it is connected between the following points: i) a and b; ii) b and c; iii) c and infinity; iv) a and c? b) Which is at higher potential i) a or b; ii) b or c; iii) a or c? c) Which, if any, of the answers would change sign if the charges were $-150\ \mu\text{C}$?

23.73 Exercise 23.39 shows that, outside a spherical shell with uniform surface charge, the potential is the same as if all the charge were concentrated into a point charge at the center of the sphere. a) Use this result to show that for two uniformly charged insulating shells, the force they exert on each other and their mutual electrical energy is the same as if all the charge were concentrated at their centers. (*Hint*: See Section 12.6.) b) Does this same result hold for solid insulating spheres, with charge distributed uniformly throughout their volume? c) Does this same result hold for the force between two charged conducting shells? Between two charged solid conductors? Explain.

23.74 Two plastic spheres, each carrying charge uniformly distributed throughout its interior, are initially placed in contact and then released. One sphere is 60.0 cm in diameter, has mass 50.0 g and contains $-10.0\ \mu\text{C}$ of charge. The other sphere is 40.0 cm in diameter, has mass 150.0 g and contains $-30.0\ \mu\text{C}$ of charge. Find the maximum speed and the maximum acceleration achieved by each sphere, assuming that no other forces are acting on them. (Use the results of Problem 23.73.)

23.75 Use the electric field calculated in Problem 22.37 to calculate the potential difference between the solid conducting sphere and the thin insulating shell.

23.76 Consider a solid conducting sphere inside a hollow conducting sphere, with radii and charges specified in Problem 22.36. Take $V = 0$ as $r \to \infty$. Use the electric field calculated in Problem 22.36 to calculate the potential V at the following values of r: a) $r = c$ (at the outer surface of the hollow sphere); b) $r = b$ (at the inner surface of the hollow sphere); c) $r = a$ (at the surface of the solid sphere); d) $r = 0$ (at the center of the solid sphere).

23.77 Use the electric field calculated in Problem 22.54 to calculate the potential difference between the two faces of the uniformly charged slab.

23.78 a) If a spherical raindrop of radius 0.650 mm carries a charge of -1.20 pC uniformly distributed over its volume, what is the potential at its surface? (Take the potential to be zero at an infinite distance from the raindrop.) b) Two identical raindrops, each with radius and charge specified in part (a), collide and merge into one larger raindrop. What is the radius of this larger drop, and what is the potential at its surface, if its charge is uniformly distributed over its volume?

23.79 Electric charge is distributed uniformly along a thin rod of length a, with total charge Q. Take the potential to be zero at infinity. Find the potential at the following points (see Fig. 23.36): a) point P, a distance x to the right of the rod; b) point R, a distance y above the right-hand end of the rod. c) In part (a) and part (b), what does your result reduce to as x or y becomes much larger than a?

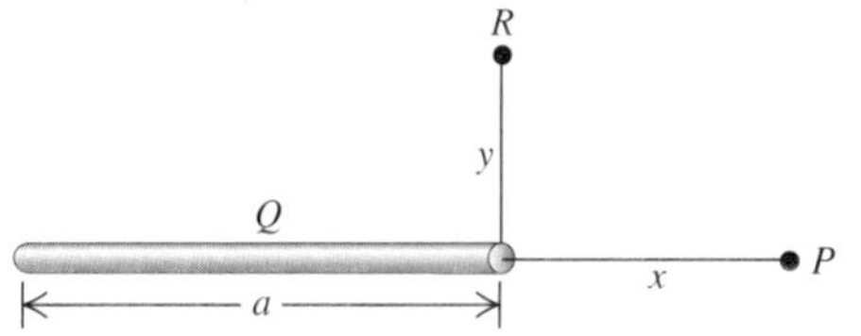

Figure 23.36 Problem 23.79.

23.80 An alpha particle with kinetic energy 11.0 MeV makes a head-on collision with a lead nucleus at rest. What is the distance of closest approach of the two particles? (Assume that the lead nucleus remains stationary and that it may be treated as a point charge. The atomic number of lead is 82. The alpha particle is a helium nucleus, with atomic number 2.)

23.81 Two metal spheres of different sizes are charged such that the electric potential is the same at the surface of each. Sphere A has a radius three times that of sphere B. Let Q_A and Q_B be the charges on each sphere and E_A and E_B be the electric field magnitudes at the surface of each sphere. What is a) the ratio Q_B/Q_A; b) the ratio E_B/E_A?

23.82 Use the charge distribution and electric field calculated in Problem 22.57. a) Show that for $r \geq R$ the potential is identical to that produced by a point charge Q. (Take the potential to be zero at infinity.) b) Obtain an expression for the electric potential valid in the region $r \leq R$.

23.83 A metal sphere with radius R_1 has a charge Q_1. Take the electric potential to be zero at an infinite distance from the sphere. a) What are the electric field and electric potential at the surface of the sphere? This sphere is now connected by a long, thin conducting wire to another sphere of radius R_2 that is several meters from the first sphere. Before the connection is made, this second sphere is uncharged. After electrostatic equilibrium has been reached, what is b) the total charge on each sphere? c) the electric potential at the surface of each sphere? d) the electric field at the surface of each sphere? Assume that the amount of charge on the wire is much less than the charge on each sphere.

23.84 The electric potential V in a region of space is given by

$$V(x, y, z) = A(x^2 - 3y^2 + z^2)$$

where A is a constant. a) Derive an expression for the electric field $\vec{E}$ at any point in this region. b) The work done by the field when a 1.50-μC test charge moves from the point $(x, y, z) = (0, 0, 0.250 \text{ m})$ to the origin is measured to be 6.00×10^{-5} J. Determine A. c) Determine the electric field at the point (0, 0, 0.250 m). d) Show that in every plane parallel to the xz-plane the equipotential contours are circles. e) What is the radius of the equipotential contour corresponding to $V = 1280$ V and $y = 2.00$ m?

23.85 Nuclear Fusion in the Sun. The source of the sun's energy is a sequence of nuclear reactions that occur in its core. The first of these reactions involves the collision of two protons, which fuse together to form a heavier nucleus and release energy. For this process, called *nuclear fusion,* to occur, the two protons must first approach until their surfaces are essentially in contact. a) Assume both protons are moving with the same speed and that they collide head-on. If the radius of the proton is 1.2×10^{-15} m, what is the minimum speed that will allow fusion to occur? The charge distribution within a proton is spherically symmetric, so the electric field and potential outside a proton is the same as if it were a point charge. The mass of the proton is given in Appendix F. b) Another nuclear fusion reaction that occurs in the sun's core involves a collision between two helium nuclei, each of which has 2.99 times the mass of the proton, charge $+2e$, and radius 1.7×10^{-15} m. Assuming the same collision geometry as in part (a), what minimum speed is required for this fusion reaction to take place if the nuclei must approach to a center-to-center distance of about 3.5×10^{-15} m? As for the proton, the charge of the helium nucleus is uniformly distributed throughout its volume. c) In Section 18.3 it was shown that the average translational kinetic energy of a particle with mass m in a gas at absolute temperature T is $\frac{3}{2}kT$, where k is the Boltzmann constant (given in Appendix F). For two protons with kinetic energy equal to this average value to be able to undergo the process described in part (a), what absolute temperature is required? What absolute temperature is required for two average helium nuclei to be able to undergo the process described in part (b)? (At these temperatures, atoms are completely ionized, so nuclei and electrons move separately.) d) The temperature in the sun's core is about 1.5×10^7 K. How does this compare to the temperatures calculated in part (c)? How can the reactions described in parts (a) and (b) occur at all in the interior of the sun? (*Hint:* See the discussion of the distribution of molecular speeds in Section 18.5.)

23.86 Nuclear Fission. The unstable nucleus uranium-236 can be regarded as a uniformly charged sphere of charge $Q = +92e$ and radius $R = 7.4 \times 10^{-15}$ m. In nuclear fission, this can divide into two smaller nuclei, each of $\frac{1}{2}$ the charge and $\frac{1}{2}$ the volume of the original uranium-236 nucleus. This is one of the reactions that occurred in the nuclear weapon that exploded over Hiroshima, Japan in August 1945. a) Find the radii of the two "daughter" nuclei of charge $+46e$. b) In a simple model for the fission process, immediately after the uranium-236 nucleus has undergone fission the "daughter" nuclei are at rest and just touching, as shown in Fig. 23.37. Calculate the kinetic energy that each of the "daughter" nuclei will have when they are very far apart. c) In this model the sum of the kinetic energies of the two "daughter" nuclei, calculated in part (b), is the energy released by the fission of one uranium-236 nucleus. Calculate the energy released by the fission of 10.0 kg of uranium-236. The atomic mass of uranium-236 is 236 u, where 1 u = 1 atomic mass unit = 1.66×10^{-27} kg. Express your answer both in joules and in kilotons of TNT (1 kiloton of TNT releases 4.18×10^{12} J when it explodes). d) In terms of this model, discuss why an atomic bomb could just as well be called an "electric bomb."

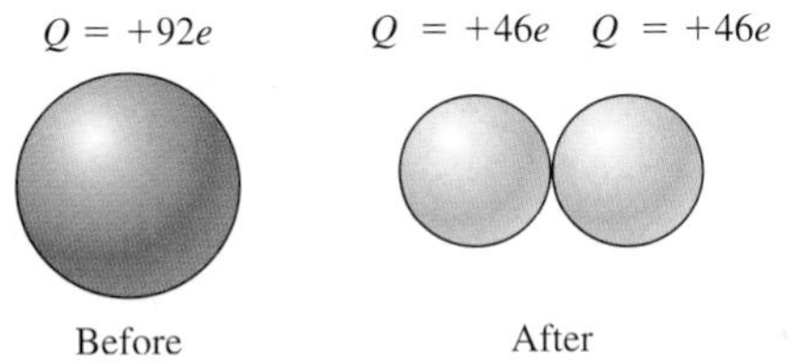

Figure 23.37 Problem 23.86.

Challenge Problems

23.87 In experiments in which atomic nuclei collide, head-on collisions like that described in Problem 23.80 do happen, but "near misses" are more common. Suppose the alpha particle in Problem 23.80 was not "aimed" at the center of the lead nucleus, but had an initial nonzero angular momentum (with respect to the stationary lead nucleus) of magnitude $L = p_0 b$, where p_0 is the magnitude of the initial momentum of the alpha particle and $b = 1.00 \times 10^{-12}$ m. What is the distance of closest approach? Repeat for $b = 1.00 \times 10^{-13}$ m and $b = 1.00 \times 10^{-14}$ m.

23.88 In a certain region, a charge distribution exists that is spherically symmetric but non-uniform. That is, the volume charge density $\rho(r)$ depends on the distance r from the center of the distribution but not on the spherical polar angles θ and ϕ. The electric potential $V(r)$ due to this charge distribution is

$$V(r) = \begin{cases} \dfrac{\rho_0 a^2}{18\epsilon_0}\left[1 - 3\left(\dfrac{r}{a}\right)^2 + 2\left(\dfrac{r}{a}\right)^3\right] & \text{for } r \le a \\ 0 & \text{for } r \ge a \end{cases}$$

where ρ_0 is a constant having units of C/m^3 and a is a constant having units of meters. a) Derive expressions for $\vec{E}$ for the regions $r \le a$ and $r \ge a$. [*Hint:* Use Eq. (23.23).] Explain why $\vec{E}$ has only a radial component. b) Derive an expression for $\rho(r)$ in each of the two regions $r \le a$ and $r \ge a$. [*Hint:* Use Gauss's law for two spherical shells, one of radius r and the other of radius $r + dr$. The charge contained in the infinitesimal spherical shell of radius dr is $dq = 4\pi r^2 \rho(r) dr$.] c) Show that the net charge contained in the volume of a sphere of radius greater than or equal to a is zero. [*Hint:* Integrate the expressions derived in part (b) for $\rho(r)$ over a spherical volume of radius greater than or equal to a.] Is this result consistent with the electric field for $r > a$ that you calculated in part (a)?

23.89 The Millikan Oil-Drop Experiment. The charge of an electron was first measured by the American physicist Robert Millikan during 1909–1913. In his experiment, oil is sprayed in very fine drops (around 10^{-4} mm in diameter) into the space between two parallel horizontal plates separated by a distance d. A potential difference V_{AB} is maintained between the parallel plates, causing a downward electric field between them. Some of the oil drops acquire a negative charge because of frictional effects or because of ionization of the surrounding air by x rays or radioactivity. The drops are observed through a microscope. a) Show that an oil drop of radius r at rest between the plates will remain at rest if the magnitude of its charge is

$$q = \frac{4\pi}{3}\frac{\rho r^3 g d}{V_{AB}}$$

where ρ is the density of the oil. (Ignore the buoyant force of the air.) By adjusting V_{AB} to keep a given drop at rest, the charge on that drop can be determined, provided its radius is known. b) Millikan's oil drops were much too small to measure their radii directly. Instead, Millikan determined r by cutting off the electric field and measuring the *terminal speed* v_t of the drop as it fell. (We discussed the concept of terminal speed in Section 5.3.) The viscous force F on a sphere of radius r moving with speed v through a fluid with viscosity η is given by Stokes's law: $F = 6\pi\eta r v$. When the drop is falling at v_t, the viscous force just balances the weight $w = mg$ of the drop. Show that the magnitude of the charge on the drop is

$$q = 18\pi\frac{d}{V_{AB}}\sqrt{\frac{\eta^3 v_t^3}{2\rho g}}$$

Within the limits of their experimental error, every one of the thousands of drops that Millikan and his co-workers measured had a charge equal to some small integer multiple of a basic charge e. That is, they found drops with charges of $\pm 2e$, $\pm 5e$, and so on, but none with values such as $0.76e$ or $2.49e$. A drop with charge $-e$ has acquired one extra electron; if its charge is $-2e$, it has acquired two extra electrons, and so on. c) A charged oil drop in a Millikan oil-drop apparatus is observed to fall 1.00 mm at constant speed in 39.3 s if $V_{AB} = 0$. The same drop can be held at rest between two plates separated by 1.00 mm if $V_{AB} = 9.16$ V. How many excess electrons has the drop acquired, and what is the radius of the drop? The viscosity of air is $1.81 \times 10^{-5}\ \text{N}\cdot\text{s/m}^2$, and the density of the oil is $824\ \text{kg/m}^3$.

23.90 A hollow, thin-walled insulating cylinder of radius R and length L (like the cardboard tube in a roll of toilet paper) has charge Q uniformly distributed over its surface. a) Calculate the electric potential at all points along the axis of the tube. Take the origin to be at the center of the tube, and take the potential to be zero at infinity. b) Show that if $L \ll R$, the result of part (a) reduces to the potential on the axis of a ring of charge of radius R (Example 23.11 in Section 23.3). c) Use the result of part (a) to find the electric field at all points along the axis of the tube.

23.91 Two point charges are moving to the right along the x-axis. Point charge 1 has charge $q_1 = 2.00\ \mu\text{C}$, mass $m_1 = 6.00 \times 10^{-5}$ kg, and speed v_1. Point charge 2 is to the right of q_1 and has charge $q_2 = -5.00\ \mu\text{C}$, mass $m_2 = 3.00 \times 10^{-5}$ kg, and speed v_2. At a particular instant, the charges are separated by a distance of 9.00 mm and have speeds $v_1 = 400$ m/s and $v_2 = 1300$ m/s. The only forces on the particles are the forces they exert on each other. a) Determine the speed v_{cm} of the center of mass of the system. b) The *relative energy* E_{rel} of the system is defined as the total energy minus the kinetic energy contributed by the motion of the center of mass:

$$E_{rel} = E - \frac{1}{2}(m_1 + m_2)v_{cm}^2$$

where $E = \frac{1}{2}m_1 v_1^2 + \frac{1}{2}m_2 v_2^2 + q_1 q_2/4\pi\epsilon_0 r$ is the total energy of the system and r is the distance between the charges. Show that $E_{rel} = \frac{1}{2}\mu v^2 + q_1 q_2/4\pi\epsilon_0 r$, where $\mu = m_1 m_2/(m_1 + m_2)$ is called the *reduced mass* of the system and $v = v_2 - v_1$ is the relative speed of the moving particles. c) For the numerical values given above, calculate the numerical value of E_{rel}. d) Based on the result of part (c), for the conditions given above, will the particles escape from one another? Explain. e) If the particles do escape, what will be their final relative speed when $r \to \infty$? If the particles do not escape, what will be their distance of maximum separation? That is, what will be the value of r when $v = 0$? f) Repeat parts (c) through (e) for $v_1 = 400$ m/s and $v_2 = 1800$ m/s when the separation is 9.00 mm.

24 CAPACITANCE AND DIELECTRICS

The sensor for automobile air bags is a capacitor: two small, closely spaced metal plates that carry charges $+Q$ and $-Q$. If the car stops suddenly, the rear, low-mass plate moves toward the more massive front plate. This changes the capacitance of the two plates (the ratio of Q to the potential difference between the plates). Circuitry detects this change and deploys the air bags.

? If the charges $+Q$ and $-Q$ on the plates of a capacitor are constant, what happens to the potential difference between the plates when the plates are moved closer together?

When you set an old-fashioned spring mousetrap or pull back the string of an archer's bow, you are storing mechanical energy as elastic potential energy. A capacitor is a device that stores *electric* potential energy and electric charge. To make a capacitor, just insulate two conductors from each other. To store energy in this device, transfer charge from one conductor to the other so that one has a negative charge and the other has an equal amount of positive charge. Work must be done to move the charges through the resulting potential difference between the conductors, and the work done is stored as electric potential energy.

Capacitors have a tremendous number of practical applications in devices such as electronic flash units for photography, pulsed lasers, air bag sensors for cars, and radio and television receivers. We'll encounter many of these applications in later chapters (particularly Chapter 31, in which we'll see the crucial role played by capacitors in the alternating-current circuits that pervade our technological society). In this chapter, however, our emphasis is on the fundamental properties of capacitors. For a particular capacitor, the ratio of the charge on each conductor to the potential difference between the conductors is a constant, called the *capacitance.* The capacitance depends on the sizes and shapes of the conductors and on the insulating material (if any) between them. Compared to the case in which there is only vacuum between the conductors, the capacitance increases when an insulating material (a *dielectric*) is present. This happens because a redistribution of charge, called *polarization,* takes place within the insulating material. Studying polarization will give us added insight into the electrical properties of matter.

Capacitors also give us a new way to think about electric potential energy. The energy stored in a charged capacitor is related to the electric field in the space between the conductors. We will see that electric potential energy can be regarded as being stored *in the field itself.* The idea that the electric field is itself a storehouse of energy is at the heart of the theory of electromagnetic waves and our modern understanding of the nature of light, to be discussed in Chapter 32.

24.1 | Capacitors and Capacitance

Any two conductors separated by an insulator (or a vacuum) form a **capacitor** (Fig. 24.1). In most practical applications, each conductor initially has zero net charge and electrons are transferred from one conductor to the other; this is called *charging* the capacitor. Then the two conductors have charges with equal magnitude and opposite sign, and the *net* charge on the capacitor as a whole remains zero. We will assume throughout this chapter that this is the case. When we say that a capacitor has charge Q, or that a charge Q is *stored* on the capacitor, we mean that the conductor at higher potential has charge $+Q$ and the conductor at lower potential has charge $-Q$ (assuming that Q is positive). Keep this in mind in the following discussion and examples.

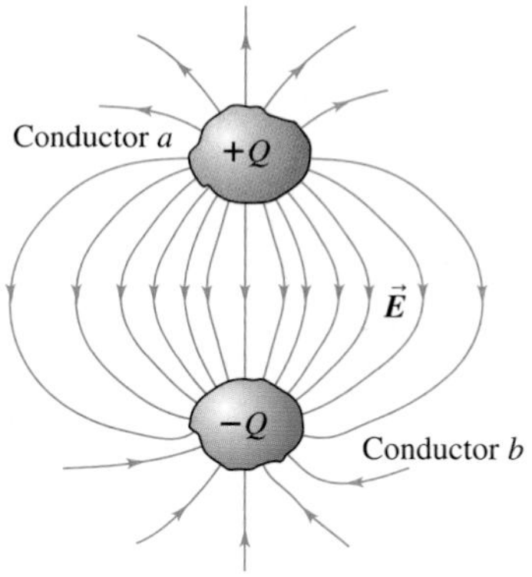

24.1 Any two conductors a and b insulated from one another form a capacitor.

In circuit diagrams a capacitor is represented by either of these symbols:

In either symbol the vertical lines (straight or curved) represent the conductors and the horizontal lines represent wires connected to either conductor. One common way to charge a capacitor is to connect these two wires to opposite terminals of a battery. Once the charges Q and $-Q$ are established on the conductors, the battery is disconnected. This gives a fixed *potential difference* V_{ab} between the conductors (that is, the potential of the positively charged conductor a with respect to the negatively charged conductor b) that is just equal to the voltage of the battery.

The electric field at any point in the region between the conductors is proportional to the magnitude Q of charge on each conductor. It follows that the potential difference V_{ab} between the conductors is also proportional to Q. If we double the magnitude of charge on each conductor, the charge density at each point doubles, the electric field at each point doubles, and the potential difference between conductors doubles; however, the *ratio* of charge to potential difference does not change. This ratio is called the **capacitance** C of the capacitor:

$$C = \frac{Q}{V_{ab}} \quad \text{(definition of capacitance)} \qquad (24.1)$$

The SI unit of capacitance is called one **farad** (1 F), in honor of the nineteenth-century English physicist Michael Faraday. From Eq. (24.1), one farad is equal to one *coulomb per volt* (1 C/V):

$$1\ \text{F} = 1\ \text{farad} = 1\ \text{C/V} = 1\ \text{coulomb/volt}$$

CAUTION Don't confuse the symbol C for capacitance (which is always in italics) with the abbreviation C for coulombs (which is never italicized).

The greater the capacitance C of a capacitor, the greater the magnitude Q of charge on either conductor for a given potential difference V_{ab} and hence the greater the amount of stored energy. (Remember that potential is potential energy per unit charge.) Thus *capacitance is a measure of the ability of a capacitor to store energy.* We will see that the value of the capacitance depends only on the shapes and sizes of the conductors and on the nature of the insulating material between them. (The above remarks about capacitance being independent of Q and V_{ab} do not apply to certain special types of insulating materials. We won't discuss these materials in this book, however.)

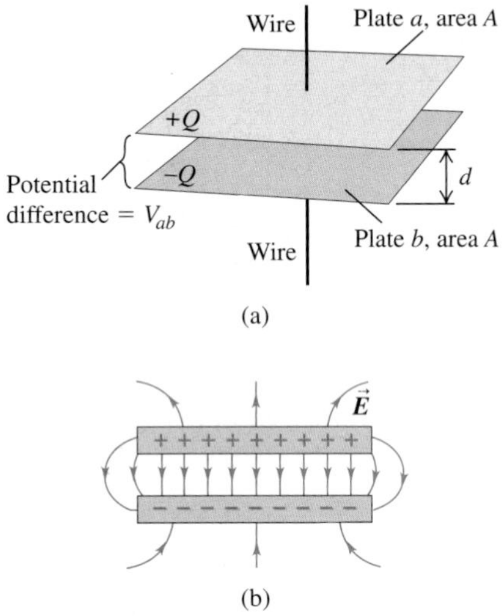

24.2 (a) A charged parallel-plate capacitor. (b) When the separation of the plates is small compared to their size, the fringing of the electric field $\vec{E}$ at the edges is slight.

Calculating Capacitance–Capacitors in Vacuum

We can calculate the capacitance C of a given capacitor by finding the potential difference V_{ab} between the conductors for a given magnitude of charge Q and then using Eq. (24.1). For now we'll consider only *capacitors in vacuum;* that is, we'll assume that the conductors that make up the capacitor are separated by empty space.

The simplest form of capacitor consists of two parallel conducting plates, each with area A, separated by a distance d that is small in comparison with their dimensions (Fig. 24.2a). When the plates are charged, the electric field is almost completely localized in the region between the plates (Fig. 24.2b). As we discussed in Example 22.8 (Section 22.4), the field between such plates is essentially *uniform,* and the charges on the plates are uniformly distributed over their opposing surfaces. We call this arrangement a **parallel-plate capacitor.**

We worked out the electric-field magnitude E for this arrangement in Example 21.13 (Section 21.5) using the principle of superposition of electric fields and again in Example 22.8 (Section 22.4) using Gauss's law. It would be a good idea to review those examples. We found that $E = \sigma/\epsilon_0$, where σ is the magnitude (absolute value) of the surface charge density on each plate. This is equal to the magnitude of the total charge Q on each plate divided by the area A of the plate, or $\sigma = Q/A$, so the field magnitude E can be expressed as

$$E = \frac{\sigma}{\epsilon_0} = \frac{Q}{\epsilon_0 A}$$

The field is uniform, and the distance between the plates is d, so the potential difference (voltage) between the two plates is

$$V_{ab} = Ed = \frac{1}{\epsilon_0}\frac{Qd}{A}$$

From this we see that the capacitance C of a parallel-plate capacitor in vacuum is

$$C = \frac{Q}{V_{ab}} = \epsilon_0 \frac{A}{d} \qquad (24.2)$$

(capacitance of a parallel-plate capacitor in vacuum)

ActivPhysics Online

11.11.6 Electric Potential: Qualitative Introduction

11.12.1 and 11.12.3 Electric Potential, Field, and Force

The capacitance depends only on the geometry of the capacitor; it is directly proportional to the area A of each plate and inversely proportional to their separation d. The quantities A and d are constants for a given capacitor, and ϵ_0 is a universal constant. Thus in a vacuum the capacitance C is a constant independent of the charge on the capacitor or the potential difference between the plates.

When matter is present between the plates, its properties affect the capacitance. We will return to this topic in Section 24.4. Meanwhile, we remark that if the space contains air at atmospheric pressure instead of vacuum, the capacitance differs from the prediction of Eq. (24.2) by less than 0.06%.

In Eq. (24.2), if A is in square meters and d in meters, C is in farads. The units of ϵ_0 are $\text{C}^2/\text{N}\cdot\text{m}^2$, so we see that

$$1\ \text{F} = 1\ \text{C}^2/\text{N}\cdot\text{m} = 1\ \text{C}^2/\text{J}$$

Because $1\ \text{V} = 1\ \text{J/C}$ (energy per unit charge), this is consistent with our definition $1\ \text{F} = 1\ \text{C/V}$. Finally, the units of ϵ_0 can be expressed as $1\ \text{C}^2/\text{N}\cdot\text{m}^2 = 1\ \text{F/m}$, so

$$\epsilon_0 = 8.85 \times 10^{-12}\ \text{F/m}$$

This relation is useful in capacitance calculations, and it also helps us to verify that Eq. (24.2) is dimensionally consistent.

One farad is a very large capacitance, as Example 24.1 shows. In many applications the most convenient units of capacitance are the *microfarad* $(1\ \mu\text{F} = 10^{-6}\ \text{F})$ and the *picofarad* $(1\ \text{pF} = 10^{-12}\ \text{F})$. For example, the power supply of an alternating-current AM radio contains several capacitors with capacitances of the order of 10 or more microfarads, and the capacitances in the tuning circuits are of the order of 10 to 100 picofarads.

For *any* capacitor in vacuum, the capacitance C depends only on the shapes, dimensions, and separation of the conductors that make up the capacitor. If the conductor shapes are more complex than those of the parallel-plate capacitor, the expression for capacitance is more complicated than in Eq. (24.2). In the following examples we show how to calculate C for two other conductor geometries.

Example 24.1 Size of a 1-F capacitor

A parallel-plate capacitor has a capacitance of 1.0 F. If the plates are 1.0 mm apart, what is the area of the plates?

SOLUTION

IDENTIFY and SET UP: We are given the values of C and d for a parallel-plate capacitor, so we use Eq. (24.2) and solve for the target variable A.

EXECUTE: From Eq. (24.2), the area A is

$$A = \frac{Cd}{\epsilon_0} = \frac{(1.0\ \text{F})(1.0 \times 10^{-3}\ \text{m})}{8.85 \times 10^{-12}\ \text{F/m}}$$
$$= 1.1 \times 10^8\ \text{m}^2$$

EVALUATE: This corresponds to a square about 10 km (about 6 miles) on a side! This area is about a third larger than Manhattan Island. Clearly this is not a very practical design for a capacitor.

It used to be considered a good joke to send a newly graduated engineer to the stockroom for a 1-farad capacitor. That's not as funny as it used to be; it's now possible to make 1-F capacitors a few centimeters on a side. One type uses activated carbon granules, 1 gram of which has a surface area of about 1000 m^2.

Example 24.2 Properties of a parallel-plate capacitor

The plates of a parallel-plate capacitor in vacuum are 5.00 mm apart and 2.00 m^2 in area. A potential difference of 10,000 V (10.0 kV) is applied across the capacitor. Compute a) the capacitance; b) the charge on each plate; and c) the magnitude of the electric field in the space between them.

SOLUTION

IDENTIFY and SET UP: We are given the values of the plate area A and the plate spacing d, which we use in Eq. (24.2) to find the capacitance C. We then find the charge Q on each plate using the given potential difference V_{ab} and Eq. (24.1). Once we have Q, we find the electric field between the plates using the relationship $E = Q/\epsilon_0 A$.

EXECUTE: a) From Eq. (24.2),

$$C = \epsilon_0 \frac{A}{d} = \frac{(8.85 \times 10^{-12}\,\text{F/m})(2.00\,\text{m}^2)}{5.00 \times 10^{-3}\,\text{m}}$$
$$= 3.54 \times 10^{-9}\,\text{F} = 0.00354\,\mu\text{F}$$

b) The charge on the capacitor is

$$Q = CV_{ab} = (3.54 \times 10^{-9}\,\text{C/V})(1.00 \times 10^4\,\text{V})$$
$$= 3.54 \times 10^{-5}\,\text{C} = 35.4\,\mu\text{C}$$

The plate at higher potential has charge $+35.4\,\mu\text{C}$ and the other plate has charge $-35.4\,\mu\text{C}$.

c) The electric-field magnitude is

$$E = \frac{\sigma}{\epsilon_0} = \frac{Q}{\epsilon_0 A} = \frac{3.54 \times 10^{-5}\,\text{C}}{(8.85 \times 10^{-12}\,\text{C}^2/\text{N}\cdot\text{m}^2)(2.00\,\text{m}^2)}$$
$$= 2.00 \times 10^6\,\text{N/C}$$

EVALUATE: An alternative way to get the result in part (c) is to recall that the electric field is equal in magnitude to the potential gradient [Eq. (23.22)]. Since the field between the plates is uniform,

$$E = \frac{V_{ab}}{d} = \frac{1.00 \times 10^4\,\text{V}}{5.00 \times 10^{-3}\,\text{m}} = 2.00 \times 10^6\,\text{V/m}$$

(Remember that the newton per coulomb and the volt per meter are equivalent units.)

Example 24.3 A spherical capacitor

Two concentric spherical conducting shells are separated by vacuum. The inner shell has total charge $+Q$ and outer radius r_a, and the outer shell has charge $-Q$ and inner radius r_b (Fig. 24.3). (The inner shell is attached to the outer shell by thin insulating rods that have negligible effect on the capacitance.) Find the capacitance of this spherical capacitor.

SOLUTION

IDENTIFY: This isn't a parallel-plate capacitor, so we can't use the relationships developed for that particular geometry. Instead, we'll go back to the fundamental definition of capacitance: the magnitude of the charge on either conductor divided by the potential difference between the conductors.

SET UP: We use Gauss's law to find the electric field between the spherical conductors. From this value we determine the potential difference V_{ab} between the two conductors; we then use Eq. (24.1) to find the capacitance $C = Q/V_{ab}$.

EXECUTE: Using the same procedure as in Example 22.5 (Section 22.4), we take as our Gaussian surface a sphere with radius r between the two spheres and concentric with them. Gauss's law, Eq. (22.8), states that the electric flux through this surface is equal to the total charge enclosed within the surface, divided by ϵ_0:

$$\oint \vec{E} \cdot d\vec{A} = \frac{Q_{\text{encl}}}{\epsilon_0}$$

By symmetry, $\vec{E}$ is constant in magnitude and parallel to $d\vec{A}$ at every point on this surface, so the integral in Gauss's law is equal to $(E)(4\pi r^2)$. The total charge enclosed is $Q_{\text{encl}} = Q$, so we have

$$(E)(4\pi r^2) = \frac{Q}{\epsilon_0}$$

$$E = \frac{Q}{4\pi\epsilon_0 r^2}$$

The electric field between the spheres is just that due to the charge on the inner sphere; the outer sphere has no effect. We found in Example 22.5 that the charge on a conducting sphere produces zero field *inside* the sphere, which also tells us that the outer conductor makes no contribution to the field between the conductors.

The above expression for E is the same as that for a point charge Q, so the expression for the potential can also be taken to be the same as for a point charge, $V = Q/4\pi\epsilon_0 r$. Hence the potential of the inner (positive) conductor at $r = r_a$ with respect to that of the outer (negative) conductor at $r = r_b$ is

$$V_{ab} = V_a - V_b = \frac{Q}{4\pi\epsilon_0 r_a} - \frac{Q}{4\pi\epsilon_0 r_b}$$
$$= \frac{Q}{4\pi\epsilon_0}\left(\frac{1}{r_a} - \frac{1}{r_b}\right) = \frac{Q}{4\pi\epsilon_0}\frac{r_b - r_a}{r_a r_b}$$

Finally, the capacitance is

$$C = \frac{Q}{V_{ab}} = 4\pi\epsilon_0 \frac{r_a r_b}{r_b - r_a}$$

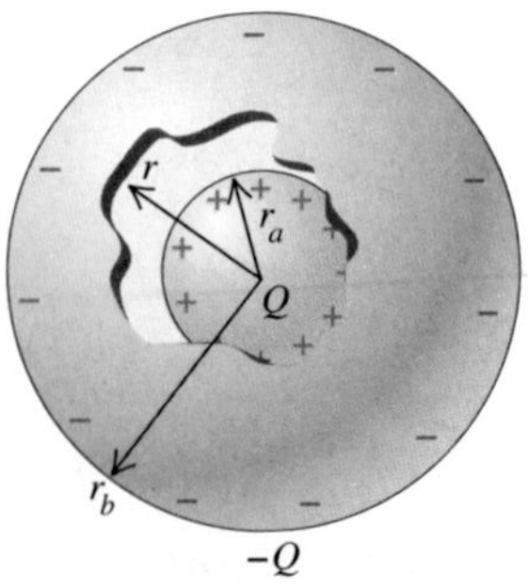

24.3 A spherical capacitor.

As an example, if $r_a = 9.5$ cm and $r_b = 10.5$ cm,

$$C = 4\pi(8.85 \times 10^{-12}\ \text{F/m})\frac{(0.095\ \text{m})(0.105\ \text{m})}{0.010\ \text{m}}$$
$$= 1.1 \times 10^{-10}\ \text{F} = 110\ \text{pF}$$

EVALUATE: We can relate this result to the capacitance of a parallel-plate capacitor. The quantity $4\pi r_a r_b$ is intermediate between the areas $4\pi r_a^2$ and $4\pi r_b^2$ of the two spheres; in fact, it's the *geometric mean* of these two areas, which we can denote by A_{gm}. The distance between spheres is $d = r_b - r_a$, so we can rewrite the above result as $C = \epsilon_0 A_{gm}/d$. This is exactly the same form as for parallel plates: $C = \epsilon_0 A/d$. The point is that if the distance between spheres is very small in comparison to their radii, they behave like parallel plates with the same area and spacing.

Example 24.4 A cylindrical capacitor

A long cylindrical conductor has a radius r_a and a linear charge density $+\lambda$. It is surrounded by a coaxial cylindrical conducting shell with inner radius r_b and linear charge density $-\lambda$ (Fig. 24.4). Calculate the capacitance per unit length for this capacitor, assuming that there is vacuum in the space between cylinders.

SOLUTION

IDENTIFY and SET UP: As in Example 24.3, we use the fundamental definition of capacitance. We first find expressions for the potential difference V_{ab} between the cylinders and the charge Q in a length L of the cylinders; we then find the capacitance of a length L using Eq. (24.1). Our target variable is this capacitance divided by L.

EXECUTE: To find the potential difference between the cylinders, we use a result that we worked out in Example 23.10 (Section 23.3). There we found that at a point outside a charged cylinder a distance r from the axis, the potential due to the cylinder is

$$V = \frac{\lambda}{2\pi\epsilon_0}\ln\frac{r_0}{r}$$

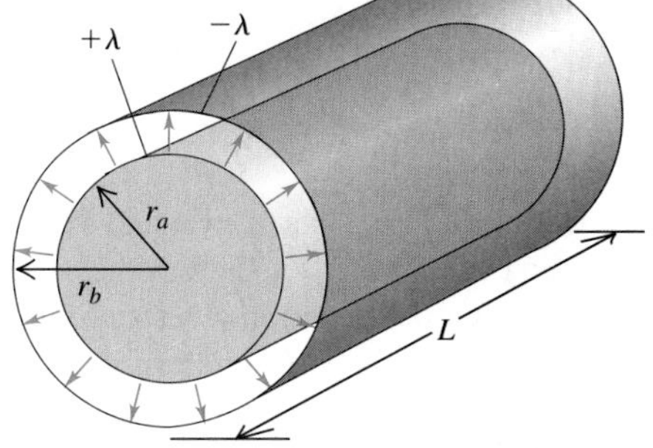

24.4 A long cylindrical capacitor. The linear charge density λ is assumed to be positive in this figure. The magnitude of charge in a length L of either cylinder is λL.

where r_0 is the (arbitrary) radius at which $V = 0$. We can use this same result for the potential *between* the cylinders in the present problem because, according to Gauss's law, the charge on the outer cylinder doesn't contribute to the field between cylinders (see Example 24.3). In our case, we take the radius r_0 to be r_b, the radius of the inner surface of the outer cylinder, so that the outer conducting cylinder is at $V = 0$. Then the potential at the outer surface of the inner cylinder (where $r = r_a$) is just equal to the potential V_{ab} of the inner (positive) cylinder a with respect to the outer (negative) cylinder b, or

$$V_{ab} = \frac{\lambda}{2\pi\epsilon_0}\ln\frac{r_b}{r_a}$$

This potential difference is positive (assuming that λ is positive, as in Fig. 24.4) because the inner cylinder is at higher potential than the outer.

The total charge Q in a length L is $Q = \lambda L$, so from Eq. (24.1) the capacitance C of a length L is

$$C = \frac{Q}{V_{ab}} = \frac{\lambda L}{\dfrac{\lambda}{2\pi\epsilon_0}\ln\dfrac{r_b}{r_a}} = \frac{2\pi\epsilon_0 L}{\ln(r_b/r_a)}$$

The capacitance per unit length is

$$\frac{C}{L} = \frac{2\pi\epsilon_0}{\ln(r_b/r_a)}$$

Substituting $\epsilon_0 = 8.85 \times 10^{-12}$ F/m $= 8.85$ pF/m, we get

$$\frac{C}{L} = \frac{55.6\ \text{pF/m}}{\ln(r_b/r_a)}$$

EVALUATE: We see that the capacitance of the coaxial cylinders is determined entirely by the dimensions, just as for the parallel-plate case. Ordinary coaxial cables are made like this but with an insulating material instead of vacuum between the inner and outer conductors. A typical cable for TV antennas and VCR connections has a capacitance per unit length of 69 pF/m.

Test Your Understanding

If you double the amount of charge on a capacitor, how does the capacitance change? Does your answer depend on the size or shape of the conductors that make up the capacitor? (Assume that there is only vacuum between the conductors.)

24.2 | Capacitors in Series and Parallel

Capacitors are manufactured with certain standard capacitances and working voltages (Fig. 24.5). However, these standard values may not be the ones you actually need in a particular application. You can obtain the values you need by combining capacitors; many combinations are possible, but the simplest combinations are a series connection and a parallel connection.

24.5 An assortment of commercially available capacitors.

Capacitors in Series

Figure 24.6a is a schematic diagram of a **series connection.** Two capacitors are connected in series (one after the other) by conducting wires between points a and b. Both capacitors are initially uncharged. When a constant positive potential difference V_{ab} is applied between points a and b, the capacitors become charged; the figure shows that the charge on *all* conducting plates has the same magnitude. To see why, note first that the top plate of C_1 acquires a positive charge Q. The electric field of this positive charge pulls negative charge up to the bottom plate of C_1 until all of the field lines that begin on the top plate end on the bottom plate. This requires that the bottom plate have charge $-Q$. These negative charges had to come from the top plate of C_2, which becomes positively charged with charge $+Q$. This positive charge then pulls negative charge $-Q$ from the connection at point b onto the bottom plate of C_2. The total charge on the lower plate of C_1 and the upper plate of C_2 together must always be zero because these plates aren't connected to anything except each other. Thus *in a series connection the magnitude of charge on all plates is the same.*

Referring again to Fig. 24.6a, we can write the potential differences between points a and c, c and b, and a and b as

$$V_{ac} = V_1 = \frac{Q}{C_1} \qquad V_{cb} = V_2 = \frac{Q}{C_2}$$

$$V_{ab} = V = V_1 + V_2 = Q\left(\frac{1}{C_1} + \frac{1}{C_2}\right)$$

and so

$$\frac{V}{Q} = \frac{1}{C_1} + \frac{1}{C_2} \tag{24.3}$$

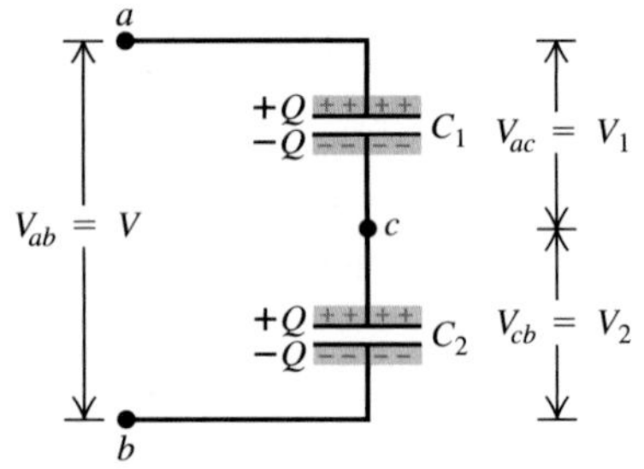

(a)

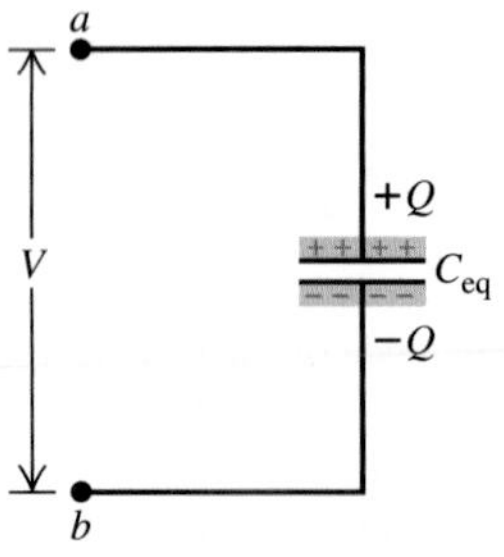

(b)

24.6 (a) Two capacitors in series (b) the equivalent capacitor.

Following a common convention, we use the symbols V_1, V_2, and V to denote the potential *differences* V_{ac} (across the first capacitor), V_{cb} (across the second capacitor), and V_{ab} (across the entire combination of capacitors), respectively.

The **equivalent capacitance** C_{eq} of the series combination is defined as the capacitance of a *single* capacitor for which the charge Q is the same as for the combination, when the potential difference V is the same. In other words, the combination can be replaced by an *equivalent capacitor* of capacitance C_{eq}. For such a capacitor, shown in Fig. 24.6b,

$$C_{eq} = \frac{Q}{V} \quad \text{or} \quad \frac{1}{C_{eq}} = \frac{V}{Q} \tag{24.4}$$

Combining Eqs. (24.3) and (24.4), we find

$$\frac{1}{C_{eq}} = \frac{1}{C_1} + \frac{1}{C_2}$$

We can extend this analysis to any number of capacitors in series. We find the following result for the *reciprocal* of the equivalent capacitance:

$$\frac{1}{C_{eq}} = \frac{1}{C_1} + \frac{1}{C_2} + \frac{1}{C_3} + \cdots \quad \text{(capacitors in series)} \qquad (24.5)$$

The reciprocal of the equivalent capacitance of a series combination equals the sum of the reciprocals of the individual capacitances. In a series connection the equivalent capacitance is always *less than* any individual capacitance.

CAUTION The magnitude of charge is the same on all plates of all the capacitors in a series combination; however, the potential differences of the individual capacitors are *not* the same unless their individual capacitances are the same. The potential differences of the individual capacitors add to give the total potential difference across the series combination: $V_{total} = V_1 + V_2 + V_3 + \cdots$.

Capacitors in Parallel

The arrangement shown in Fig. 24.7a is called a **parallel connection.** Two capacitors are connected in parallel between points *a* and *b*. In this case the upper plates of the two capacitors are connected together by conducting wires to form an equipotential surface, and the lower plates form another. Hence *in a parallel connection the potential difference for all individual capacitors is the same* and is equal to $V_{ab} = V$. The charges Q_1 and Q_2 are not necessarily equal, however, since charges can reach each capacitor independently from the source (such as a battery) of the voltage V_{ab}. The charges are

$$Q_1 = C_1V \quad \text{and} \quad Q_2 = C_2V$$

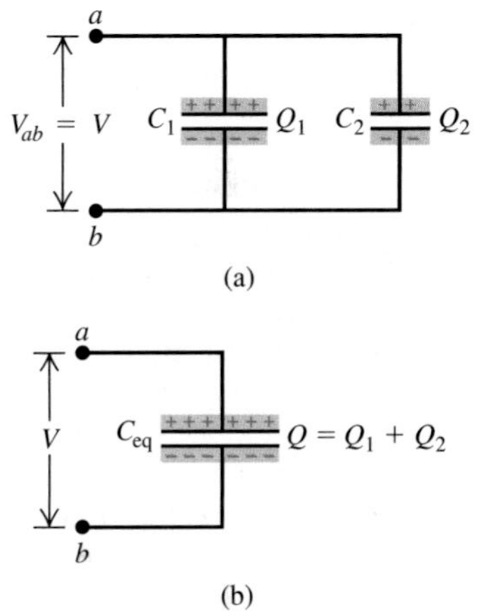

24.7 (a) Two capacitors in parallel and (b) the equivalent capacitor.

The *total* charge Q of the combination, and thus the total charge on the equivalent capacitor, is

$$Q = Q_1 + Q_2 = (C_1 + C_2)V$$

so

$$\frac{Q}{V} = C_1 + C_2 \qquad (24.6)$$

The parallel combination is equivalent to a single capacitor with the same total charge $Q = Q_1 + Q_2$ and potential difference V as the combination (Fig. 24.7b). The equivalent capacitance of the combination, C_{eq}, is the same as the capacitance Q/V of this single equivalent capacitor. So from Eq. (24.6),

$$C_{eq} = C_1 + C_2$$

In the same way we can show that for any number of capacitors in parallel,

$$C_{eq} = C_1 + C_2 + C_3 + \cdots \quad \text{(capacitors in parallel)} \qquad (24.7)$$

The equivalent capacitance of a parallel combination equals the *sum* of the individual capacitances. In a parallel connection the equivalent capacitance is always *greater than* any individual capacitance.

CAUTION The potential differences are the same for all the capacitors in a parallel combination; however, the charges on individual capacitors are *not* the same unless their individual capacitances are the same. The charges on the individual capacitors add to give the total charge on the parallel combination: $Q_{\text{total}} = Q_1 + Q_2 + Q_3 + \cdots$. [Compare these statements to those in the "caution" paragraph following Eq. (24.5).]

Problem-Solving Strategy: Equivalent Capacitance

IDENTIFY *the relevant concepts:* The concept of equivalent capacitance is useful whenever two or more capacitors are connected together.

SET UP *the problem using the following steps:*

1. Make a drawing of the capacitor arrangement.
2. Identify whether the capacitors are connected in series or in parallel. With more complicated combinations, you can sometimes identify parts that are simple series or parallel connections.
3. Keep in mind that when we say a capacitor has charge Q, we always mean that the plate at higher potential has charge $+Q$ and the other plate has charge $-Q$.

EXECUTE *the solution as follows:*

1. When capacitors are connected in series, as in Fig. 24.6a, they always have the same charge, assuming that they were uncharged before they were connected. The potential differences are *not* equal unless the capacitances are equal. The total potential difference across the combination is the sum of the individual potential differences.
2. When capacitors are connected in parallel, as in Fig. 24.7a, the potential difference V is always the same for all of the individual capacitors. The charges on the individual capacitors are *not* equal unless the capacitances are equal. The total charge on the combination is the sum of the individual charges.
3. For more complicated combinations, find the parts that are simple series or parallel connections and replace them with their equivalent capacitances, in a step-by-step reduction. If you then need to find the charge or potential difference for an individual capacitor, you may have to retrace your path to the original capacitors.

EVALUATE *your answer:* Check whether your result makes sense. If the capacitors are connected in series, the equivalent capacitance C_{eq} must be *smaller* than any of the individual capacitances. By contrast, if the capacitors are connected in parallel, C_{eq} must be *greater* than any of the individual capacitances.

Example 24.5 Capacitors in series and in parallel

In Figs. 24.6 and 24.7, let $C_1 = 6.0\ \mu\text{F}$, $C_2 = 3.0\ \mu\text{F}$, and $V_{ab} = 18\ \text{V}$. Find the equivalent capacitance, and find the charge and potential difference for each capacitor when the two capacitors are connected a) in series and b) in parallel.

SOLUTION

IDENTIFY and SET UP: In both parts, one of the target variables is the equivalent capacitance C_{eq}. For the series combination in part (a), it is given by Eq. (24.5); for the parallel combination in part (b), C_{eq} is given by Eq. (24.6). In each part we find the charge and potential difference using the definition of capacitance, Eq. (24.1), and the rules outlined in the Problem-Solving Strategy.

EXECUTE: a) Using Eq. (24.5) for the equivalent capacitance of the series combination (Fig. 24.6a), we find

$$\frac{1}{C_{\text{eq}}} = \frac{1}{C_1} + \frac{1}{C_2} = \frac{1}{6.0\ \mu\text{F}} + \frac{1}{3.0\ \mu\text{F}} \qquad C_{\text{eq}} = 2.0\ \mu\text{F}$$

The charge Q on each capacitor in series is the same as the charge on the equivalent capacitor:

$$Q = C_{\text{eq}}V = (2.0\ \mu\text{F})(18\ \text{V}) = 36\ \mu\text{C}$$

The potential difference across each capacitor is inversely proportional to its capacitance:

$$V_{ac} = V_1 = \frac{Q}{C_1} = \frac{36\ \mu\text{C}}{6.0\ \mu\text{F}} = 6.0\ \text{V}$$

$$V_{cb} = V_2 = \frac{Q}{C_2} = \frac{36\ \mu\text{C}}{3.0\ \mu\text{F}} = 12.0\ \text{V}$$

b) To find the equivalent capacitance of the parallel combination (Fig. 24.7a), we use Eq. (24.6):

$$C_{eq} = C_1 + C_2 = 6.0\ \mu F + 3.0\ \mu F = 9.0\ \mu F$$

The potential difference across each of the two capacitors in parallel is the same as that across the equivalent capacitor, 18 V. The charges Q_1 and Q_2 are directly proportional to the capacitances C_1 and C_2, respectively:

$$Q_1 = C_1V = (6.0\ \mu F)(18\ V) = 108\ \mu C$$
$$Q_2 = C_2V = (3.0\ \mu F)(18\ V) = 54\ \mu C$$

EVALUATE: Note that the equivalent capacitance C_{eq} for the series combination in part (a) is indeed less than either C_1 or C_2, while for the parallel combination in part (b) the equivalent capacitance is indeed greater than either C_1 or C_2.

It's instructive to compare the potential differences and charges in each part of the example. For two capacitors in series, as in part (a), the charge is the same on either capacitor and the *larger* potential difference appears across the capacitor with the *smaller* capacitance. Furthermore, $V_{ac} + V_{cb} = V_{ab} = 18$ V, as it must. By contrast, for two capacitors in parallel, as in part (b), each capacitor has the same potential difference and the *larger* charge appears on the capacitor with the *larger* capacitance. Can you show that the total charge $Q_1 + Q_2$ on the parallel combination is equal to the charge $Q = C_{eq}V$ on the equivalent capacitor?

Example 24.6 A capacitor network

Find the equivalent capacitance of the combination shown in Fig. 24.8a.

SOLUTION

IDENTIFY and SET UP: The five capacitors in Fig. 24.8a are neither all in series nor all in parallel. We can, however, identify portions of the arrangement that *are* either in series or parallel, which we combine to find the net equivalent capacitance.

EXECUTE: We first replace the 12-μF and 6-μF series combination by its equivalent capacitance; calling that C', we use Eq. (24.5):

$$\frac{1}{C'} = \frac{1}{12\ \mu F} + \frac{1}{6\ \mu F} \qquad C' = 4\ \mu F$$

This gives us the equivalent combination shown in Fig. 24.8b. Next we find the equivalent capacitance of the three capacitors in parallel, using Eq. (24.7). Calling their equivalent capacitance C'', we have

$$C'' = 3\ \mu F + 11\ \mu F + 4\ \mu F = 18\ \mu F$$

This gives us the simpler equivalent combination shown in Fig. 24.8c. Finally, we find the equivalent capacitance C_{eq} of these two capacitors in series (Fig. 24.8d):

$$\frac{1}{C_{eq}} = \frac{1}{18\ \mu F} + \frac{1}{9\ \mu F} \qquad C_{eq} = 6\ \mu F$$

EVALUATE: The equivalent capacitance of the network is 6 μF; that is, if a potential difference V_{ab} is applied across the terminals of the network, the net charge on the network is 6 μF times V_{ab}. How is this net charge related to the charges on the individual capacitors in Fig. 24.8a? (See Exercise 24.20.)

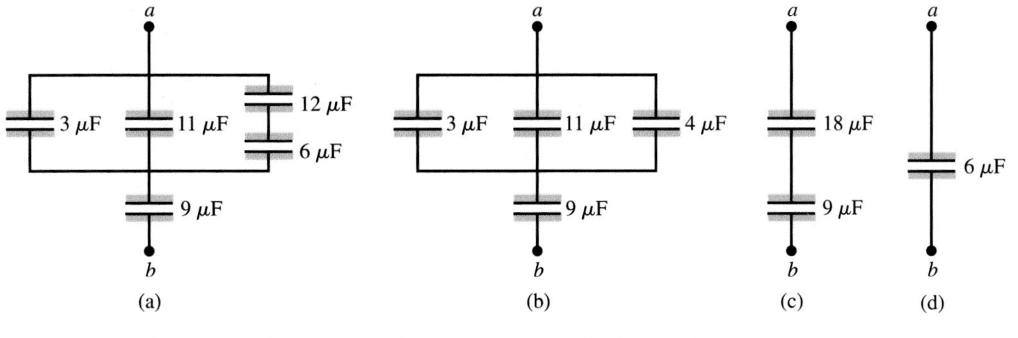

24.8 (a) A capacitor network between points a and b. (b) The 12-μF and 6-μF capacitors in series in (a) are replaced by an equivalent 4-μF capacitor. (c) The 3-μF, 11-μF, and 4-μF capacitors in parallel in (b) are replaced by an equivalent 18-μF capacitor. (d) Finally, the 18-μF and 9-μF capacitors in series in (c) are replaced by an equivalent 6-μF capacitor.

Test Your Understanding

If a 4-μF capacitor and an 8-μF capacitor are connected in parallel, which has the larger potential difference across it? Which has the larger charge?

24.3 | Energy Storage in Capacitors and Electric-Field Energy

Many of the most important applications of capacitors depend on their ability to store energy. The electric potential energy stored in a charged capacitor is just equal to the amount of work required to charge it—that is, to separate opposite charges and place them on different conductors. When the capacitor is discharged, this stored energy is recovered as work done by electrical forces.

We can calculate the potential energy U of a charged capacitor by calculating the work W required to charge it. Suppose that when we are done charging the capacitor, the final charge is Q and the final potential difference is V. From Eq. (24.1) these quantities are related by

$$V = \frac{Q}{C}$$

Let q and v be the charge and potential difference, respectively, at an intermediate stage during the charging process; then $v = q/C$. At this stage the work dW required to transfer an additional element of charge dq is

$$dW = v\,dq = \frac{q\,dq}{C}$$

The total work W needed to increase the capacitor charge q from zero to a final value Q is

$$W = \int_0^W dW = \frac{1}{C}\int_0^Q q\,dq = \frac{Q^2}{2C} \qquad (24.8)$$

(work to charge a capacitor)

This is also equal to the total work done by the electric field on the charge when the capacitor discharges. Then q *decreases* from an initial value Q to zero as the elements of charge dq "fall" through potential differences v that vary from V down to zero.

If we define the potential energy of an *uncharged* capacitor to be zero, then W in Eq. (24.8) is equal to the potential energy U of the charged capacitor. The final stored charge is $Q = CV$, so we can express U (which is equal to W) as

$$U = \frac{Q^2}{2C} = \frac{1}{2}CV^2 = \frac{1}{2}QV \qquad (24.9)$$

(potential energy stored in a capacitor)

When Q is in coulombs, C in farads (coulombs per volt), and V in volts (joules per coulomb), U is in joules.

The last form of Eq. (24.9), $U = \frac{1}{2}QV$, shows that the total work W required to charge the capacitor is equal to the total charge Q multiplied by the *average* potential difference $\frac{1}{2}V$ during the charging process.

The expression $U = \frac{1}{2}(Q^2/C)$ in Eq. (24.9) shows that a charged capacitor is the electrical analog of a stretched spring with elastic potential energy $U = \frac{1}{2}kx^2$. The charge Q is analogous to the elongation x, and the *reciprocal* of the capacitance, $1/C$, is analogous to the force constant k. The energy supplied to a capacitor in the charging process is analogous to the work we do on a spring when we stretch it.

Equations (24.8) and (24.9) tell us that capacitance measures the ability of a capacitor to store both energy and charge. If a capacitor is charged by connecting it to a battery or other source that provides a fixed potential difference V, then increasing the value of C gives a greater charge $Q = CV$ and a greater amount of stored energy $U = \frac{1}{2}CV^2$. If instead the goal is to transfer a given quantity of charge Q from one conductor to another, Eq. (24.8) shows that the work W required is inversely proportional to C; the greater the capacitance, the easier it is to give a capacitor a fixed amount of charge.

Most practical applications of capacitors take advantage of their ability to store and release energy. In electronic flash units used by photographers and in energy-storage units for pulsed lasers, the energy and charge stored in a capacitor are recovered quickly (Fig. 24.9). In other applications, the energy is released more slowly. Springs in the suspension of an automobile, for example, help smooth out the ride by absorbing the energy from sudden jolts and releasing that energy gradually; in an analogous way, a capacitor in an electronic circuit can smooth out unwanted variations in voltage due to power surges. And just as the presence of a spring gives a mechanical system a natural frequency at which it responds most strongly to an applied periodic force, so the presence of a capacitor gives an electric circuit a natural frequency for current oscillations. This idea is used in tuned circuits such as those in radio and television receivers, which respond to broadcast signals at one particular frequency and ignore signals at other frequencies. We discuss these circuits in detail in Chapter 31.

24.9 An electronic flash unit stores potential energy in a capacitor. Depressing the camera's shutter button provides a conducting path from one capacitor plate to the other through the flash tube. Once this path is established and the capacitor discharges, the stored energy is converted into a brief but intense flash of light.

The energy-storage properties of capacitors also have some undesirable practical effects. Adjacent pins on the underside of a computer chip act like a capacitor, and the property that makes capacitors useful for smoothing out voltage variations acts to retard the rate at which the potentials of the chip's pins can be changed. This tendency limits how rapidly the chip can perform computations, an effect that becomes more important as computer chips become smaller and are pushed to operate at faster speeds.

Electric-Field Energy

We can charge a capacitor by moving electrons directly from one plate to another. This requires doing work against the electric field between the plates. Thus we can think of the energy as being stored *in the field* in the region between the plates. To develop this relation, let's find the energy *per unit volume* in the space between the plates of a parallel-plate capacitor with plate area A and separation d. We call this the **energy density,** denoted by u. From Eq. (24.9) the total stored potential energy is $\frac{1}{2}CV^2$ and the volume between the plates is just Ad; hence the energy density is

$$u = \text{Energy density} = \frac{\frac{1}{2}CV^2}{Ad} \tag{24.10}$$

From Eq. (24.2) the capacitance C is given by $C = \epsilon_0 A/d$. The potential difference V is related to the electric field magnitude E by $V = Ed$. If we use these expressions in Eq. (24.10), the geometric factors A and d cancel, and we find

$$u = \frac{1}{2}\epsilon_0 E^2 \quad \text{(electric energy density in a vacuum)} \qquad (24.11)$$

Although we have derived this relation only for a parallel-plate capacitor, it turns out to be valid for any capacitor in vacuum and indeed *for any electric field configuration in vacuum.* This result has an interesting implication. We think of vacuum as space with no matter in it, but vacuum can nevertheless have electric fields and therefore energy. Thus "empty" space need not be truly empty after all. We will use this idea and Eq. (24.11) in Chapter 32 in connection with the energy transported by electromagnetic waves.

CAUTION It's a common misconception that electric-field energy is a new kind of energy, different from the electric potential energy described before. This is *not* the case; it is simply a different way of interpreting electric potential energy. We can regard the energy of a given system of charges as being a shared property of all the charges, or we can think of the energy as being a property of the electric field that the charges create. Either interpretation leads to the same value of the potential energy.

Example 24.7 Transferring charge and energy between capacitors

In Fig. 24.10 we charge a capacitor of capacitance $C_1 = 8.0\ \mu\text{F}$ by connecting it to a source of potential difference $V_0 = 120$ V (not shown in the figure). The switch S is initially open. Once C_1 is charged, the source of potential difference is disconnected. a) What is the charge Q_0 on C_1 if switch S is left open? b) What is the energy stored in C_1 if switch S is left open? c) The capacitor of capacitance $C_2 = 4.0\ \mu\text{F}$ is initially uncharged. After we close switch S, what is the potential difference across each capacitor, and what is the charge on each capacitor? d) What is the total energy of the system after we close switch S?

SOLUTION

IDENTIFY: Initially we have a single capacitor with a given potential difference between its plates. After the switch is closed, one wire connects the upper plates of the two capacitors and another wire connects the lower plates; in other words, the capacitors are connected in parallel.

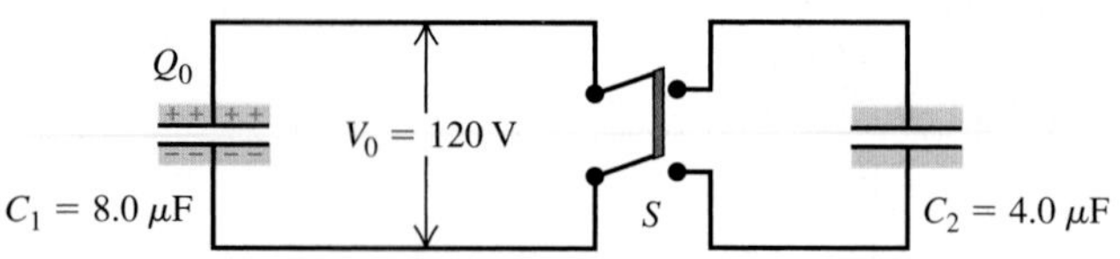

24.10 When the switch S is closed, the charged capacitor C_1 is connected to an uncharged capacitor C_2. The center part of the switch is an insulating handle; charge can only flow between the two upper terminals and between the two lower terminals.

SET UP: In parts (a) and (b) we find the charge and stored energy for capacitor C_1, using Eqs. (24.1) and (24.9), respectively. In part (c) we use the character of the parallel connection to determine how the charge Q_0 is shared between the two capacitors. In part (d) we again use Eq. (24.9) to find the energy stored in capacitors C_1 and C_2; the total energy is the sum of these values.

EXECUTE: a) The charge Q_0 on C_1 is

$$Q_0 = C_1 V_0 = (8.0\ \mu\text{F})(120\ \text{V}) = 960\ \mu\text{C}$$

b) The energy initially stored in the capacitor is

$$U_{\text{initial}} = \frac{1}{2}Q_0 V_0 = \frac{1}{2}(960 \times 10^{-6}\ \text{C})(120\ \text{V}) = 0.058\ \text{J}$$

c) When the switch is closed, the positive charge Q_0 becomes distributed over the upper plates of both capacitors and the negative charge $-Q_0$ is distributed over the lower plates of both capacitors. Let Q_1 and Q_2 be the magnitudes of the final charges on the two capacitors. From conservation of charge,

$$Q_1 + Q_2 = Q_0$$

In the final state, when the charges are no longer moving, both upper plates are at the same potential; they are connected by a conducting wire and so form a single equipotential surface. Both lower plates are also at the same potential, different from that of the upper plates. The final potential difference V between the plates is therefore the same for both capacitors, as we would expect for a parallel connection. The capacitor charges are

$$Q_1 = C_1 V \qquad Q_2 = C_2 V$$

When we combine this with the preceding equation for conservation of charge, we find

$$V = \frac{Q_0}{C_1 + C_2} = \frac{960\ \mu\text{C}}{8.0\ \mu\text{F} + 4.0\ \mu\text{F}} = 80\ \text{V}$$
$$Q_1 = 640\ \mu\text{C} \qquad Q_2 = 320\ \mu\text{C}$$

d) The final energy of the system is the sum of the energies stored in each capacitor:

$$U_{\text{final}} = \frac{1}{2}Q_1 V + \frac{1}{2}Q_2 V = \frac{1}{2}Q_0 V$$
$$= \frac{1}{2}(960 \times 10^{-6}\ \text{C})(80\ \text{V}) = 0.038\ \text{J}$$

EVALUATE: The final energy is less than the original energy $U_{\text{initial}} = 0.058$ J; the difference has been converted to energy of some other form. The conductors become a little warmer because of their resistance, and some energy is radiated as electromagnetic waves. We'll study the circuit behavior of capacitors in detail in Chapters 26 and 31.

Example 24.8 Electric-field energy

Suppose you want to store 1.00 J of electric potential energy in a volume of 1.00 m^3 in vacuum. a) What is the magnitude of the required electric field? b) If the field magnitude is ten times larger, how much energy is stored per cubic meter?

SOLUTION

IDENTIFY and SET UP: We use the given information to find the desired energy density u, then use Eq. (24.11) to find the required value of E. This same equation gives us the relationship between changes in E and the corresponding changes in u.

EXECUTE: a) The desired energy density is $u = (1.00\ \text{J})/(1.00\ \text{m}^3) = 1.00\ \text{J/m}^3$. We solve Eq. (24.11) for E:

$$E = \sqrt{\frac{2u}{\epsilon_0}} = \sqrt{\frac{2(1.00\ \text{J/m}^3)}{8.85 \times 10^{-12}\ \text{C}^2/\text{N}\cdot\text{m}^2}}$$
$$= 4.75 \times 10^5\ \text{N/C} = 4.75 \times 10^5\ \text{V/m}$$

b) Equation (24.11) shows that u is proportional to E^2. If E increases by a factor of 10, u increases by a factor of $10^2 = 100$, and the energy density is 100 J/m^3.

EVALUATE: The electric-field magnitude found in part (a) is sizable, corresponding to a potential difference of nearly a half million volts over a distance of one meter. We will see in Section 24.4 that the field magnitudes in practical conductors can be as great as this or even larger.

Example 24.9 Two ways to calculate energy stored in a capacitor

The spherical capacitor described in Example 24.3 (Section 24.1) has charges $+Q$ and $-Q$ on its inner and outer conductors. Find the electric potential energy stored in the capacitor a) by using the capacitance C found in Example 24.3 and b) by integrating the electric-field energy density.

SOLUTION

IDENTIFY and SET UP: In Example 24.3 we found the capacitance C and the field magnitude E between the conductors. We find the stored energy U in part (a) using the expression for C in Eq. (24.9). In part (b) we use the expression for E in Eq. (24.11) to find the electric-field energy density u between the conductors. The field magnitude depends on the distance r from the center of the capacitor, so u also depends on r. Hence we cannot find U by simply multiplying u by the volume between the conductors; instead, we must integrate u over this volume.

EXECUTE: a) From Example 24.3, the spherical capacitor has capacitance

$$C = 4\pi\epsilon_0 \frac{r_a r_b}{r_b - r_a}$$

where r_a and r_b are the radii of the inner and outer conducting spheres. From Eq. (24.9) the energy stored in this capacitor is

$$U = \frac{Q^2}{2C} = \frac{Q^2}{8\pi\epsilon_0}\frac{r_b - r_a}{r_a r_b}$$

b) The electric field in the volume between the two conducting spheres has magnitude $E = Q/4\pi\epsilon_0 r^2$. The electric field is zero inside the inner sphere and is also zero outside the inner surface of the outer sphere, because a Gaussian surface with radius $r < r_a$ or $r > r_b$ encloses zero net charge. Hence the energy density is nonzero only in the space between the spheres $(r_a < r < r_b)$. In this region,

$$u = \frac{1}{2}\epsilon_0 E^2 = \frac{1}{2}\epsilon_0\left(\frac{Q}{4\pi\epsilon_0 r^2}\right)^2 = \frac{Q^2}{32\pi^2\epsilon_0 r^4}$$

The energy density is *not* uniform; it decreases rapidly with increasing distance from the center of the capacitor. To find the total electric-field energy, we integrate u (the energy per unit volume) over the volume between the inner and outer conducting spheres. Dividing this volume up into spherical shells of radius r, surface area $4\pi r^2$, thickness dr, and volume $dV = 4\pi r^2\,dr$, we have

$$U = \int u\,dV = \int_{r_a}^{r_b}\left(\frac{Q^2}{32\pi^2\epsilon_0 r^4}\right)4\pi r^2\,dr$$

$$= \frac{Q^2}{8\pi\epsilon_0}\int_{r_a}^{r_b}\frac{dr}{r^2} = \frac{Q^2}{8\pi\epsilon_0}\left(-\frac{1}{r_b} + \frac{1}{r_a}\right)$$

$$= \frac{Q^2}{8\pi\epsilon_0}\frac{r_b - r_a}{r_a r_b}$$

EVALUATE: We obtain the same result for U with either approach, as we must. We emphasize that electric potential energy can be regarded as being associated with either the *charges,* as in part (a), or the *field,* as in part (b); regardless of which viewpoint you choose, the amount of stored energy is the same.

Test Your Understanding

If a 4-μF capacitor and an 8-μF capacitor are connected in series, which has the greater amount of stored energy? What if the capacitors are connected in parallel?

24.4 | Dielectrics

Most capacitors have a nonconducting material, or **dielectric,** between their conducting plates. A common type of capacitor uses long strips of metal foil for the plates, separated by strips of plastic sheet such as Mylar. A sandwich of these materials is rolled up, forming a unit that can provide a capacitance of several microfarads in a compact package (Fig. 24.11).

Placing a solid dielectric between the plates of a capacitor serves three functions. First, it solves the mechanical problem of maintaining two large metal sheets at a very small separation without actual contact.

Second, using a dielectric increases the maximum possible potential difference between the capacitor plates. As we described in Example 23.8 (Section 23.3), any insulating material, when subjected to a sufficiently large electric field, experiences **dielectric breakdown,** a partial ionization that permits conduction through it. Many dielectric materials can tolerate stronger electric fields without breakdown than can air. Thus using a dielectric allows a capacitor to sustain a higher potential difference V and so store greater amounts of charge and energy.

Third, the capacitance of a capacitor of given dimensions is *greater* when there is a dielectric material between the plates than when there is a vacuum. We can demonstrate this effect with the aid of a sensitive *electrometer,* a device that measures the potential difference between two conductors without letting any appreciable charge flow from one to the other. Figure 24.12a shows an electrometer connected across a charged capacitor, with magnitude of charge Q on each plate and potential difference V_0. When we insert an uncharged sheet of dielectric, such as glass, paraffin, or polystyrene, between the plates, experiment shows that the

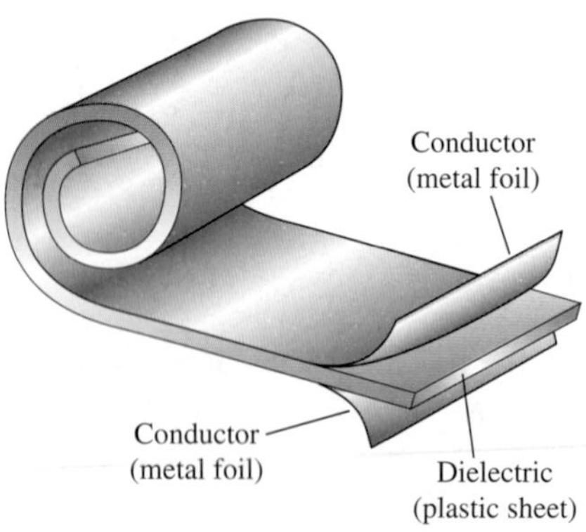

24.11 A common type of capacitor uses a dielectric sheet to separate the two conductors.

potential difference *decreases* to a smaller value V (Fig. 24.12b). When we remove the dielectric, the potential difference returns to its original value V_0, showing that the original charges on the plates have not changed.

The original capacitance C_0 is given by $C_0 = Q/V_0$, and the capacitance C with the dielectric present is $C = Q/V$. The charge Q is the same in both cases, and V is less than V_0, so we conclude that the capacitance C with the dielectric present is *greater* than C_0. When the space between plates is completely filled by the dielectric, the ratio of C to C_0 (equal to the ratio of V_0 to V) is called the **dielectric constant** of the material, K:

$$K = \frac{C}{C_0} \qquad \text{(definition of dielectric constant)} \qquad (24.12)$$

When the charge is constant, $Q = C_0V_0 = CV$ and $C/C_0 = V_0/V$. In this case, Eq. (24.12) can be rewritten as

$$V = \frac{V_0}{K} \qquad \text{(when } Q \text{ is constant)} \qquad (24.13)$$

With the dielectric present, the potential difference for a given charge Q is *reduced* by a factor K.

The dielectric constant K is a pure number. Because C is always greater than C_0, K is always greater than unity. A few representative values of K are given in Table 24.1. For vacuum, $K = 1$ by definition. For air at ordinary temperatures and pressures, K is about 1.0006; this is so nearly equal to 1 that for most purposes an air capacitor is equivalent to one in vacuum. Note that while water has a very large value of K, it is usually not a very practical dielectric for use in capacitors. The reason is that while pure water is a very poor conductor, it is also an excellent ionic solvent. Any ions that are dissolved in the water will cause charge to flow between the capacitor plates, so the capacitor discharges.

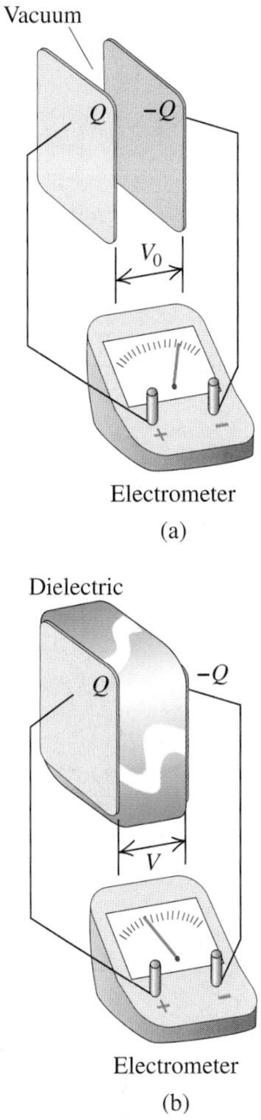

24.12 Effect of a dielectric between the plates of a parallel-plate capacitor. The electrometer measures potential difference. (a) With a given charge, the potential difference is V_0. (b) With the same charge but with a dielectric between the plates, the potential difference V is smaller than V_0.

Table 24.1 Values of Dielectric Constant K at 20°C

Material	K	Material	K
Vacuum	1	Polyvinyl chloride	3.18
Air (1 atm)	1.00059	Plexiglas	3.40
Air (100 atm)	1.0548	Glass	5–10
Teflon	2.1	Neoprene	6.70
Polyethylene	2.25	Germanium	16
Benzene	2.28	Glycerin	42.5
Mica	3–6	Water	80.4
Mylar	3.1	Strontium titanate	310

No real dielectric is a perfect insulator. Hence there is always some *leakage current* between the charged plates of a capacitor with a dielectric. We tacitly ignored this effect in Section 24.2 when we derived expressions for the equivalent capacitances of capacitors in series, Eq. (24.5), and in parallel, Eq. (24.7). But if a leakage current flows for a long enough time to substantially change the charges from the values we used to derive Eqs. (24.5) and (24.7), those equations may no longer be accurate.

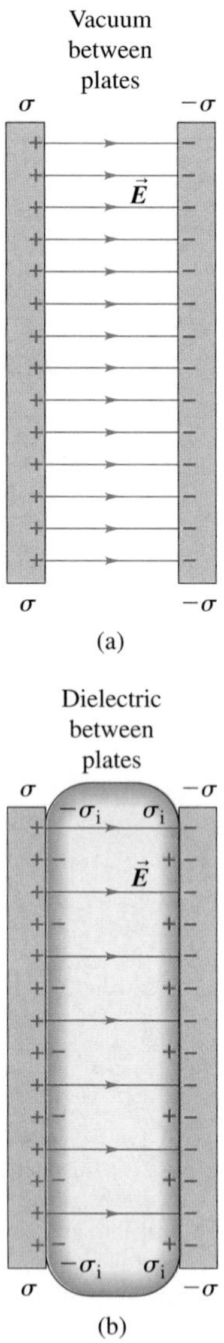

24.13 (a) Electric field lines with vacuum between the plates. (b) The induced charges on the faces of the dielectric decrease the electric field.

Induced Charge and Polarization

When a dielectric material is inserted between the plates while the charge is kept constant, the potential difference between the plates decreases by a factor K. Therefore the electric field between the plates must decrease by the same factor. If E_0 is the vacuum value and E is the value with the dielectric, then

$$E = \frac{E_0}{K} \qquad \text{(when } Q \text{ is constant)} \tag{24.14}$$

Since the electric-field magnitude is smaller when the dielectric is present, the surface charge density (which causes the field) must be smaller as well. The surface charge on the conducting plates does not change, but an *induced* charge of the opposite sign appears on each surface of the dielectric (Fig. 24.13). The dielectric was originally electrically neutral, and is still neutral; the induced surface charges arise as a result of *redistribution* of positive and negative charge within the dielectric material, a phenomenon called **polarization.** We first encountered polarization in Section 21.2, and we suggest that you reread the discussion of Fig. 21.7. We will assume that the induced surface charge is *directly proportional* to the electric-field magnitude E in the material; this is indeed the case for many common dielectrics. (This direct proportionality is analogous to Hooke's law for a spring.) In that case, K is a constant for any particular material. When the electric field is very strong or if the dielectric is made of certain crystalline materials, the relationship between induced charge and the electric field can be more complex; we won't consider such cases here.

We can derive a relation between this induced surface charge and the charge on the plates. Let's denote the magnitude of the charge per unit area induced on the surfaces of the dielectric (the induced surface charge density) by σ_i. The magnitude of the surface charge density on the capacitor plates is σ, as usual. Then the *net* surface charge on each side of the capacitor has magnitude $(\sigma - \sigma_i)$, as shown in Fig. 24.13b. As we found in Example 21.13 (Section 21.5) and in Example 22.8 (Section 22.4), the field between the plates is related to the net surface charge density by $E = \sigma_{net}/\epsilon_0$. Without and with the dielectric, respectively, we have

$$E_0 = \frac{\sigma}{\epsilon_0} \qquad E = \frac{\sigma - \sigma_i}{\epsilon_0} \tag{24.15}$$

Using these expressions in Eq. (24.14) and rearranging the result, we find

$$\sigma_i = \sigma\left(1 - \frac{1}{K}\right) \qquad \text{(induced surface charge density)} \tag{24.16}$$

This equation shows that when K is very large, σ_i is nearly as large as σ. In this case, σ_i nearly cancels σ, and the field and potential difference are much smaller than their values in vacuum.

The product $K\epsilon_0$ is called the **permittivity** of the dielectric, denoted by ϵ:

$$\epsilon = K\epsilon_0 \qquad \text{(definition of permittivity)} \tag{24.17}$$

In terms of ϵ we can express the electric field within the dielectric as

$$E = \frac{\sigma}{\epsilon} \tag{24.18}$$

The capacitance when the dielectric is present is given by

$$C = KC_0 = K\epsilon_0\frac{A}{d} = \epsilon\frac{A}{d} \qquad (24.19)$$

(parallel-plate capacitor, dielectric between plates)

We can repeat the derivation of Eq. (24.11) for the energy density u in an electric field for the case in which a dielectric is present. The result is

$$u = \frac{1}{2}K\epsilon_0 E^2 = \frac{1}{2}\epsilon E^2 \quad \text{(electric energy density in a dielectric)} \quad (24.20)$$

In empty space, where $K = 1$, $\epsilon = \epsilon_0$ and Eqs. (24.19) and (24.20) reduce to Eqs. (24.2) and (24.11), respectively, for a parallel-plate capacitor in a vacuum. For this reason, ϵ_0 is sometimes called the "permittivity of free space" or the "permittivity of vacuum." Because K is a pure number, ϵ and ϵ_0 have the same units, $C^2/N \cdot m^2$ or F/m.

Several practical devices make use of the way in which a capacitor responds to a change in dielectric constant. One example is an electric stud finder, used by home repair workers to locate metal studs hidden behind a wall's surface. It consists of a metal plate with associated circuitry. The plate acts as one half of a capacitor, with the wall acting as the other half. If the stud finder moves over a piece of metal, the effective dielectric constant for the capacitor changes, changing the capacitance and triggering a signal.

Problem-Solving Strategy — Dielectrics

IDENTIFY *the relevant concepts:* The relationships in this section are useful whenever there is an electric field in a dielectric, such as a dielectric between charged capacitor plates. Typically you will be asked to relate the potential difference between the plates, the electric field in the capacitor, the charge density on the capacitor plates, and the induced charge density on the surfaces of the capacitor.

SET UP *the problem using the following steps:*

1. Make a drawing of the situation.
2. Identify the target variables, and choose which of the key equations of this section will help you find those variables.

EXECUTE *the solution as follows:*

1. In problems such as Example 24.10 on the next page, it is easy to get lost in a blizzard of formulas. Ask yourself at each step what kind of quantity each symbol represents. For example, distinguish clearly between charges and charge densities, and between electric fields and electric potential differences.
2. As you calculate, continually check for consistency of units. This effort is a bit more complex with electrical quantities than it was in mechanics. Distances must always be in meters. Remember that a microfarad is 10^{-6} farads, and so on. Don't confuse the numerical value of ϵ_0 with the value of $1/4\pi\epsilon_0$. There are several alternative sets of units for electric field magnitude, including N/C and V/m. The units of ϵ_0 are $C^2/N \cdot m^2$ or F/m.

EVALUATE *your answer:* When you check numerical values, remember that with a dielectric present, (a) the capacitance is always greater than without a dielectric; (b) for a given amount of charge on the capacitor, the electric field and potential difference are less than without a dielectric; and (c) the induced surface charge density σ_i on the dielectric is always less in magnitude than the charge density σ on the capacitor plates.

Example 24.10 A capacitor with and without a dielectric

Suppose the parallel plates in Fig. 24.13 each have an area of 2000 cm^2 $(2.00 \times 10^{-1}\ \text{m}^2)$ and are 1.00 cm $(1.00 \times 10^{-2}\ \text{m})$ apart. The capacitor is connected to a power supply and charged to a potential difference $V_0 = 3000\ \text{V} = 3.00\ \text{kV}$. It is then disconnected from the power supply, and a sheet of insulating plastic material is inserted between the plates, completely filling the space between them. We find that the potential difference decreases to 1000 V while the charge on each capacitor plate remains constant. Compute a) the original capacitance C_0; b) the magnitude of charge Q on each plate; c) the capacitance C after the dielectric is inserted; d) the dielectric constant K of the dielectric; e) the permittivity ϵ of the dielectric; f) the magnitude of the induced charge Q_i on each face of the dielectric; g) the original electric field E_0 between the plates; and h) the electric field E after the dielectric is inserted.

SOLUTION

IDENTIFY and SET UP: Most of the target variables can be obtained in several different ways. The methods used below are a representative sample; we encourage you to think of others and compare your results.

EXECUTE: a) With vacuum between the plates, we use Eq. (24.19) with $K = 1$:

$$C_0 = \epsilon_0 \frac{A}{d} = (8.85 \times 10^{-12}\ \text{F/m}) \frac{2.00 \times 10^{-1}\ \text{m}^2}{1.00 \times 10^{-2}\ \text{m}}$$
$$= 1.77 \times 10^{-10}\ \text{F} = 177\ \text{pF}$$

b) Using the definition of capacitance, Eq. (24.1),

$$Q = C_0 V_0 = (1.77 \times 10^{-10}\ \text{F})(3.00 \times 10^3\ \text{V})$$
$$= 5.31 \times 10^{-7}\ \text{C} = 0.531\ \mu\text{C}$$

c) When the dielectric is inserted, the charge remains the same but the potential decreases to $V = 1000$ V. Hence from Eq. (24.1), the new capacitance is

$$C = \frac{Q}{V} = \frac{5.31 \times 10^{-7}\ \text{C}}{1.00 \times 10^3\ \text{V}} = 5.31 \times 10^{-10}\ \text{F} = 531\ \text{pF}$$

d) From Eq. (24.12), the dielectric constant is

$$K = \frac{C}{C_0} = \frac{5.31 \times 10^{-10}\ \text{F}}{1.77 \times 10^{-10}\ \text{F}} = \frac{531\ \text{pF}}{177\ \text{pF}} = 3.00$$

Alternatively, from Eq. (24.13),

$$K = \frac{V_0}{V} = \frac{3000\ \text{V}}{1000\ \text{V}} = 3.00$$

e) Using K from part (d) in Eq. (24.17), the permittivity is

$$\epsilon = K\epsilon_0 = (3.00)(8.85 \times 10^{-12}\ \text{C}^2/\text{N}\cdot\text{m}^2)$$
$$= 2.66 \times 10^{-11}\ \text{C}^2/\text{N}\cdot\text{m}^2$$

f) Multiplying Eq. (24.15) by the area of each plate gives the induced charge $Q_i = \sigma_i A$ in terms of the charge $Q = \sigma A$ on each plate:

$$Q_i = Q\left(1 - \frac{1}{K}\right) = (5.31 \times 10^{-7}\ \text{C})\left(1 - \frac{1}{3.00}\right)$$
$$= 3.54 \times 10^{-7}\ \text{C}$$

g) Since the electric field between the plates is uniform, its magnitude is just the potential difference divided by the plate separation:

$$E_0 = \frac{V_0}{d} = \frac{3000\ \text{V}}{1.00 \times 10^{-2}\ \text{m}} = 3.00 \times 10^5\ \text{V/m}$$

h) With the new potential difference after the dielectric is inserted,

$$E = \frac{V}{d} = \frac{1000\ \text{V}}{1.00 \times 10^{-2}\ \text{m}} = 1.00 \times 10^5\ \text{V/m}$$

or, from Eq. (24.17),

$$E = \frac{\sigma}{\epsilon} = \frac{Q}{\epsilon A} = \frac{5.31 \times 10^{-7}\ \text{C}}{(2.66 \times 10^{-11}\ \text{C}^2/\text{N}\cdot\text{m}^2)(2.00 \times 10^{-1}\ \text{m}^2)}$$
$$= 1.00 \times 10^5\ \text{V/m}$$

or, from Eq. (24.15),

$$E = \frac{\sigma - \sigma_i}{\epsilon_0} = \frac{Q - Q_i}{\epsilon_0 A}$$
$$= \frac{(5.31 - 3.54) \times 10^{-7}\ \text{C}}{(8.85 \times 10^{-12}\ \text{C}^2/\text{N}\cdot\text{m}^2)(2.00 \times 10^{-1}\ \text{m}^2)}$$
$$= 1.00 \times 10^5\ \text{V/m}$$

or, from Eq. (24.14),

$$E = \frac{E_0}{K} = \frac{3.00 \times 10^5\ \text{V/m}}{3.00} = 1.00 \times 10^5\ \text{V/m}$$

EVALUATE: It's always useful to check the results by finding them in more than one way, as we did in parts (d) and (h). Our results show that inserting the dielectric increased the capacitance by a factor of $K = 3.00$ and reduced the electric field between the plates by a factor of $1/K = 1/3.00$. It did so by developing induced charges on the faces of the dielectric of magnitude $Q(1 - 1/K) = Q(1 - 1/3.00) = 0.667Q$.

Example 24.11

Energy storage with and without a dielectric

Find the total energy stored in the electric field of the capacitor in Example 24.10 and the energy density, both before and after the dielectric sheet is inserted.

SOLUTION

IDENTIFY and SET UP: We use Eq. (24.9) to find the stored energy before and after the dielectric is inserted, and Eq. (24.20) to find the energy density.

EXECUTE: Let the original energy be U_0 and let the energy with the dielectric in place be U. From Eq. (24.9),

$$U_0 = \frac{1}{2}C_0V_0^2 = \frac{1}{2}(1.77 \times 10^{-10}\,\text{F})(3000\,\text{V})^2 = 7.97 \times 10^{-4}\,\text{J}$$

$$U = \frac{1}{2}CV^2 = \frac{1}{2}(5.31 \times 10^{-10}\,\text{F})(1000\,\text{V})^2 = 2.66 \times 10^{-4}\,\text{J}$$

The final energy is one-third of the original energy.

The energy density without the dielectric is given by Eq. (24.20) with $K = 1$:

$$u_0 = \frac{1}{2}\epsilon_0E_0^2 = \frac{1}{2}(8.85 \times 10^{-12}\,\text{C}^2/\text{N}\cdot\text{m}^2)(3.0 \times 10^5\,\text{N/C})^2$$
$$= 0.398\,\text{J/m}^3$$

With the dielectric in place,

$$u = \frac{1}{2}\epsilon E^2 = \frac{1}{2}(2.66 \times 10^{-11}\,\text{C}^2/\text{N}\cdot\text{m}^2)(1.00 \times 10^5\,\text{N/C})^2$$
$$= 0.133\,\text{J/m}^3$$

The energy density with the dielectric is one-third of the original energy density.

EVALUATE: We can check our answer for u_0 by noting that the volume between the plates is $V = (0.200\,\text{m})^2(0.0100\,\text{m}) = 0.00200\,\text{m}^3$. Since the electric field is uniform between the plates, u_0 is uniform as well and the energy density is just the stored energy divided by the volume:

$$u_0 = \frac{U_0}{V} = \frac{7.97 \times 10^{-4}\,\text{J}}{0.00200\,\text{m}^3} = 0.398\,\text{J/m}^3$$

which agrees with our earlier answer. You should use the same approach to check the value for u, the energy density with the dielectric.

We can generalize the results of this example. When a dielectric is inserted into a capacitor while the charge on each plate remains the same, the permittivity ϵ increases by a factor of K (the dielectric constant), the electric field decreases by a factor of $1/K$, and the energy density $u = \frac{1}{2}\epsilon E^2$ decreases by a factor of $1/K$. Where did the energy go? The answer lies in the fringing field at the edges of a real parallel-plate capacitor. As Fig. 24.14 shows, that field tends to pull the dielectric into the space between the plates, doing work on it as it does so. We could attach a spring to the bottom of the dielectric in Fig. 24.14 and use this force to stretch the spring. Because work is done by the field, the field energy density decreases.

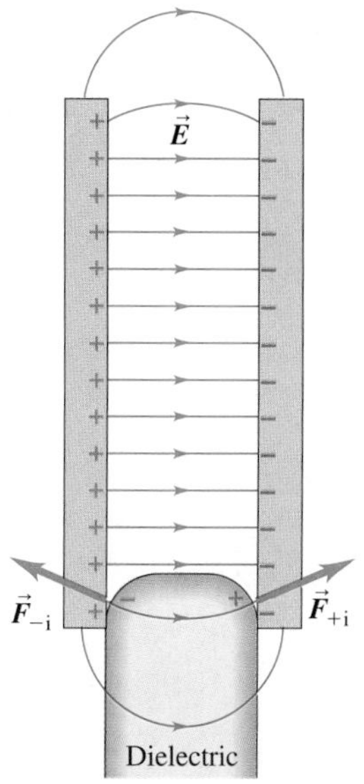

24.14 The fringing field at the edges of the capacitor exerts forces $\vec{F}_{-i}$ and $\vec{F}_{+i}$ on the negative and positive induced surface charges of a dielectric, pulling the dielectric into the capacitor.

Dielectric Breakdown

We mentioned earlier that when any dielectric material is subjected to a sufficiently strong electric field, *dielectric breakdown* takes place and the dielectric becomes a conductor (Fig. 24.15). This occurs when the electric field is so strong that electrons are ripped loose from their molecules and crash into other molecules, liberating even more electrons. This avalanche of moving charge, forming a spark or arc discharge, often starts quite suddenly.

Because of dielectric breakdown, capacitors always have maximum voltage ratings. When a capacitor is subjected to excessive voltage, an arc may form

24.15 A very strong electric field caused dielectric breakdown in a block of Plexiglas. The resulting flow of charge etched this pattern into the block.

through a layer of dielectric, burning or melting a hole in it. This arc creates a conducting path (a short circuit) between the conductors. If a conducting path remains after the arc is extinguished, the device is rendered permanently useless as a capacitor.

The maximum electric-field magnitude that a material can withstand without the occurrence of breakdown is called its **dielectric strength.** This quantity is affected significantly by temperature, trace impurities, small irregularities in the metal electrodes, and other factors that are difficult to control. For this reason we can give only approximate figures for dielectric strengths. The dielectric strength of dry air is about 3×10^6 V/m. A few typical values of dielectric strength for common insulating materials are shown in Table 24.2. Note that the values are all substantially greater than the value for air. For example, a layer of polycarbonate 0.01 mm thick (about the smallest practical thickness) has ten times the dielectric strength of air and can withstand a maximum voltage of about $(3 \times 10^7 \text{ V/m})(1 \times 10^{-5} \text{ m}) = 300 \text{ V}$.

Table 24.2 Dielectric Constant and Dielectric Strength of Some Insulating Materials

Material	Dielectric Constant, K	Dielectric Strength, E_{max} (V/m)
Polycarbonate	2.8	3×10^7
Polyester	3.3	6×10^7
Polypropylene	2.2	7×10^7
Polystyrene	2.6	2×10^7
Pyrex glass	4.7	1×10^7

Test Your Understanding

The space between the plates of an isolated parallel-plate capacitor is filled by a slab of dielectric with dielectric constant K. You then pull out the dielectric slab. The two plates of the capacitor have charges Q and $-Q$, which do not change when you remove the slab. Is the energy stored in the capacitor more, less, or the same after the the dielectric is removed? If it is more, what is the source of the extra energy? If it is less, where did the lost energy go?

*24.5 | Molecular Model of Induced Charge

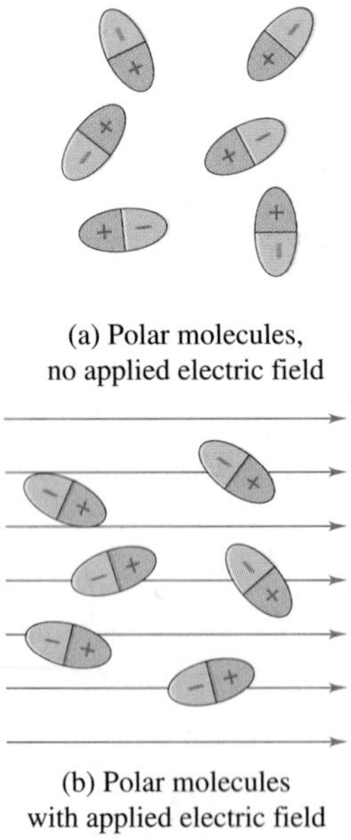

24.16 (a) Polar molecules have random orientations when there is no applied electric field. (b) The molecules tend to line up with an applied electric field $\vec{E}$.

In Section 24.4 we discussed induced surface charges on a dielectric in an electric field. Now let's look at how these surface charges can arise. If the material were a *conductor,* the answer would be simple. Conductors contain charge that is free to move, and when an electric field is present, some of the charge redistributes itself on the surface so that there is no electric field inside the conductor. But an ideal dielectric has *no* charges that are free to move, so how can a surface charge occur?

To understand this, we have to look again at rearrangement of charge at the *molecular* level. Some molecules, such as H_2O and N_2O, have equal amounts of positive and negative charge but a lopsided distribution, with excess positive charge concentrated on one side of the molecule and negative charge on the other. As we described in Section 21.7, such an arrangement is called an *electric dipole,* and the molecule is called a *polar molecule.* When no electric field is present in a gas or liquid with polar molecules, the molecules are oriented randomly (Fig. 24.16a). When they are placed in an electric field, however, they tend to orient themselves as in Fig. 24.16b, as a result of the electric-field torques described in

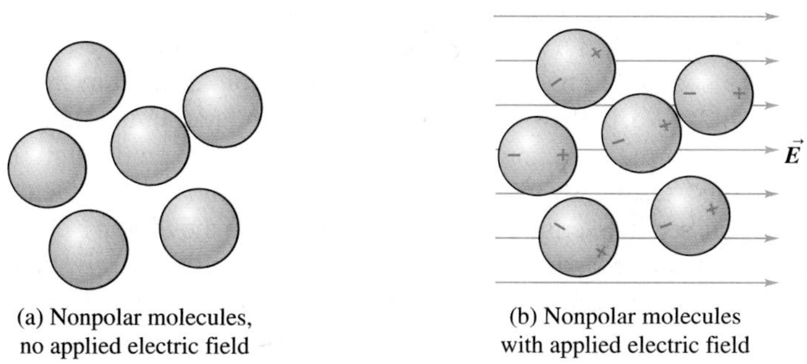

24.17 (a) Nonpolar molecules have their positive and negative charge centers at the same point. (b) These centers become separated slightly by an applied electric field $\vec{E}$.

Section 21.7. Because of thermal agitation, the alignment of the molecules with $\vec{E}$ is not perfect.

Even a molecule that is *not* ordinarily polar *becomes* a dipole when it is placed in an electric field because the field pushes the positive charges in the molecules in the direction of the field and pushes the negative charges in the opposite direction. This causes a redistribution of charge within the molecule (Fig. 24.17). Such dipoles are called *induced* dipoles.

With either polar or nonpolar molecules, the redistribution of charge caused by the field leads to the formation of a layer of charge on each surface of the dielectric material (Fig. 24.18). These layers are the surface charges described in Section 24.4; their surface charge density is denoted by σ_i. The charges are *not* free to move indefinitely, as they would be in a conductor, because each charge is bound to a molecule. They are in fact called **bound charges** to distinguish them from the **free charges** that are added to and removed from the conducting capacitor plates. In the interior of the material the net charge per unit volume remains zero. As we have seen, this redistribution of charge is called *polarization,* and we say that the material is *polarized.*

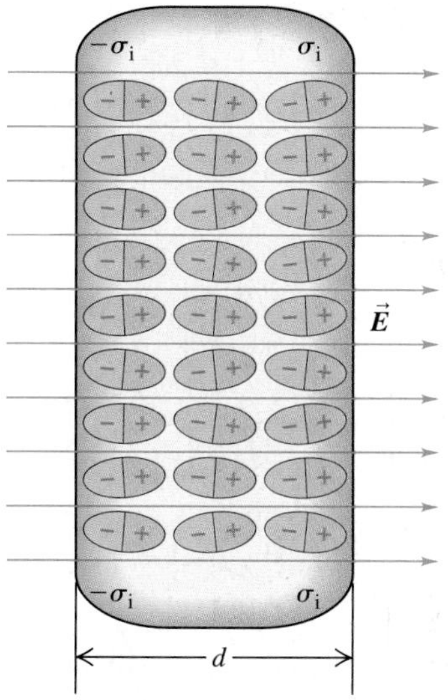

24.18 Polarization of a dielectric in an electric field $\vec{E}$ gives rise to thin layers of bound charges on the surfaces, creating surface charge densities σ_i and $-\sigma_i$. The sizes of the molecules are greatly exaggerated for clarity.

The four parts of Fig. 24.19 show the behavior of a slab of dielectric when it is inserted in the field between a pair of oppositely charged capacitor plates. Figure 24.19a shows the original field. Figure 24.19b is the situation after the dielectric has been inserted but before any rearrangement of charges has occurred. Figure 24.19c shows by thinner arrows the additional field set up in the dielectric by its

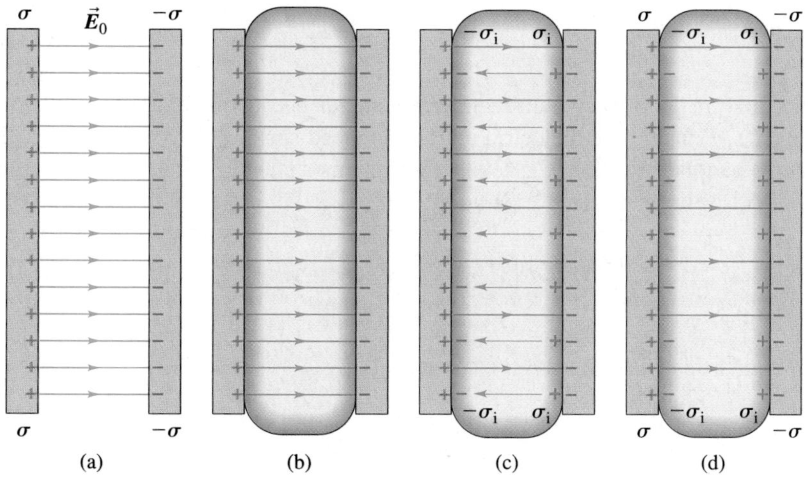

24.19 (a) Electric field of magnitude E_0 between two charged plates. (b) Introduction of a dielectric of dielectric constant K. (c) The induced surface charges and their field (thinner lines). (d) Resultant field of magnitude E_0/K when a dielectric is between charged plates.

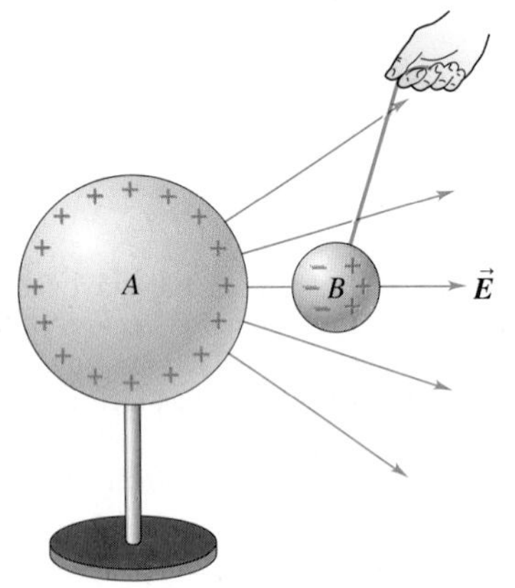

24.20 A neutral sphere B in the radial electric field of a positively charged sphere A is attracted to the charge because of polarization.

induced surface charges. This field is *opposite* to the original field, but it is not great enough to cancel the original field completely because the charges in the dielectric are not free to move indefinitely. The resultant field in the dielectric, shown in Fig. 24.19d, is therefore decreased in magnitude. In the field-line representation, some of the field lines leaving the positive plate go through the dielectric, while others terminate on the induced charges on the faces of the dielectric.

As we discussed in Section 21.2, polarization is also the reason a charged body, such as an electrified plastic rod, can exert a force on an *uncharged* body such as a bit of paper or a pith ball. Figure 24.20 shows an uncharged dielectric sphere B in the radial field of a positively charged body A. The induced positive charges on B experience a force toward the right, while the force on the induced negative charges is toward the left. The negative charges are closer to A, and thus are in a stronger field, than are the positive charges. The force toward the left is stronger than that toward the right, and B is attracted toward A, even though its net charge is zero. The attraction occurs whether the sign of A's charge is positive or negative (see Fig. 21.7). Furthermore, the effect is not limited to dielectrics; an uncharged conducting body would be attracted in the same way.

Test Your Understanding

Why do you suppose water has such a large dielectric constant, as given in Table 24.1? (*Hint:* See Fig. 21.28.)

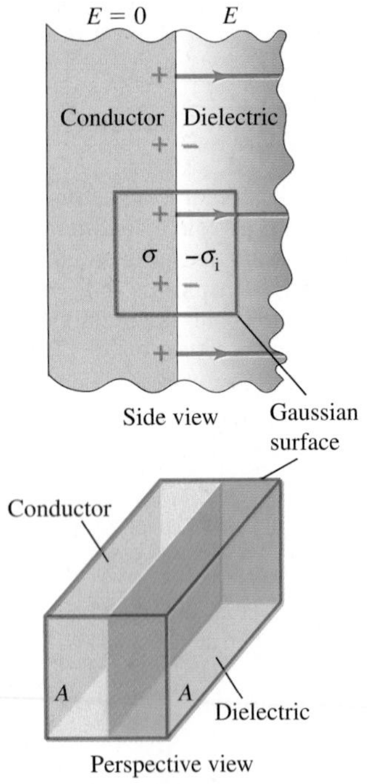

24.21 Gauss's law with a dielectric. This figure shows a close-up of the left-hand capacitor plate in Fig. 24.13b. The Gaussian surface is a rectangular box that lies half in the conductor and half in the dielectric.

*24.6 | Gauss's Law In Dielectrics

We can extend the analysis of Section 24.4 to reformulate Gauss's law in a form that is particularly useful for dielectrics. Figure 24.21 is a close-up view of the left capacitor plate and left surface of the dielectric in Fig. 24.13b. Let's apply Gauss's law to the rectangular box shown in cross section by the purple line; the surface area of the left and right sides is A. The left side is embedded in the conductor that forms the left capacitor plate, and so the electric field everywhere on that surface is zero. The right side is embedded in the dielectric, where the electric field has magnitude E, and $E_\perp = 0$ everywhere on the other four sides. The total charge enclosed, including both the charge on the capacitor plate and the induced charge on the dielectric surface, is $Q_{\text{encl}} = (\sigma - \sigma_{\text{i}})A$, so Gauss's law gives

$$EA = \frac{(\sigma - \sigma_{\text{i}})A}{\epsilon_0} \tag{24.21}$$

This equation is not very illuminating as it stands because it relates two unknown quantities, E inside the dielectric and the induced surface charge density σ_{i}. But now we can use Eq. (24.16), developed for this same situation, to simplify this equation by eliminating σ_{i}. Equation (24.16) is

$$\sigma_{\text{i}} = \sigma\left(1 - \frac{1}{K}\right) \quad \text{or} \quad \sigma - \sigma_{\text{i}} = \frac{\sigma}{K}$$

Combining this with Eq. (24.21), we get

$$EA = \frac{\sigma A}{K\epsilon_0} \quad \text{or} \quad KEA = \frac{\sigma A}{\epsilon_0} \tag{24.22}$$

Equation (24.22) says that the flux of $K\vec{E}$, not $\vec{E}$, through the Gaussian surface in Fig. 24.21 is equal to the enclosed *free* charge σA divided by ϵ_0. It turns out that for *any* Gaussian surface, whenever the induced charge is proportional to the electric field in the material, we can rewrite Gauss's law as

$$\oint K\vec{E}\cdot d\vec{A} = \frac{Q_{\text{encl-free}}}{\epsilon_0} \qquad \text{(Gauss's law in a dielectric)} \qquad (24.23)$$

where $Q_{\text{encl-free}}$ is the total *free* charge (not bound charge) enclosed by the Gaussian surface. The significance of these results is that the right sides contain only the *free* charge on the conductor, not the bound (induced) charge. In fact, although we have not proved it, Eq. (24.23) remains valid even when different parts of the Gaussian surface are embedded in dielectrics having different values of K, provided that the value of K in each dielectric is independent of the electric field (usually the case for electric fields that are not too strong) and that we use the appropriate value of K for each point on the Gaussian surface.

Example 24.12 A spherical capacitor with dielectric

In the spherical capacitor of Example 24.3 (Section 24.1), the volume between the concentric spherical conducting shells is filled with an insulating oil with dielectric constant K. Use Gauss's law to find the capacitance.

SOLUTION

IDENTIFY and SET UP: As we did in Example 24.3, we use a spherical Gaussian surface of radius r between the two spheres. Since a dielectric is present, we use Gauss's law in the form of Eq. (24.23).

EXECUTE: The spherical symmetry of the problem is not changed by the presence of the dielectric, so we have

$$\oint K\vec{E}\cdot d\vec{A} = \oint KE\,dA = KE\oint dA = (KE)\,(4\pi r^2) = \frac{Q}{\epsilon_0}$$

$$E = \frac{Q}{4\pi K\epsilon_0 r^2} = \frac{Q}{4\pi\epsilon r^2}$$

where $\epsilon = K\epsilon_0$ is the permittivity of the dielectric (introduced in Section 24.4). Compared to the case in which there is vacuum between the conducting shells, the electric field is reduced by a factor of $1/K$. The potential difference V_{ab} between the shells is likewise reduced by a factor of $1/K$, and so the capacitance $C = Q/V_{ab}$ is *increased* by a factor of K, just as for a parallel-plate capacitor when a dielectric is inserted. Using the result for the vacuum case in Example 24.3, we find that the capacitance with the dielectric is

$$C = \frac{4\pi K\epsilon_0 r_a r_b}{r_b - r_a} = \frac{4\pi\epsilon r_a r_b}{r_b - r_a}$$

EVALUATE: In this case the dielectric completely fills the volume between the two conductors, so the capacitance is just K times the value with no dielectric. The result is more complicated if the dielectric only partially fills this volume (see Challenge Problem 24.78).

Test Your Understanding

What is the magnitude of the electric field at a distance r from a single point charge q imbedded in a dielectric of dielectric constant K?

SUMMARY

A capacitor is any pair of conductors separated by an insulating material. When the capacitor is charged, there are charges of equal magnitude Q and opposite sign on the two conductors, and the potential V_{ab} of the positively charged conductor with respect to the negatively charged conductor is proportional to Q. The capacitance C is defined as the ratio of Q to V_{ab}. The SI unit of capacitance is the farad (F): 1 F = 1 C/V.

$$C = \frac{Q}{V_{ab}} \quad (24.1)$$

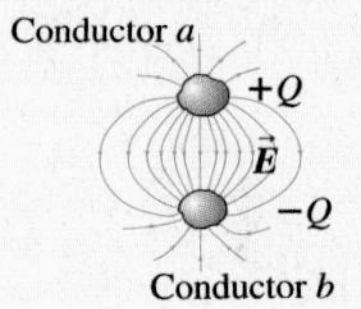

A parallel-plate capacitor consists of two parallel conducting plates, each with area A, separated by a distance d. If they are separated by vacuum, the capacitance depends only on A and d. For other geometries, the capacitance can be found by using the definition $C = Q/V_{ab}$. (See Examples 24.1 through 24.4)

$$C = \frac{Q}{V_{ab}} = \epsilon_0 \frac{A}{d} \quad (24.2)$$

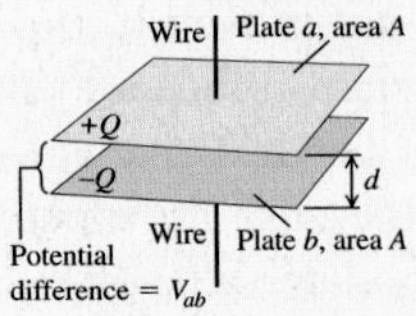

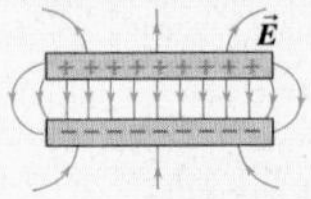

When capacitors with capacitances $C_1, C_2, C_3, \ldots$ are connected in series, the reciprocal of the equivalent capacitance C_{eq} equals the sum of the reciprocals of the individual capacitances. When capacitors are connected in parallel, the equivalent capacitance C_{eq} equals the sum of the individual capacitances. (See Examples 24.5 and 24.6)

$$\frac{1}{C_{eq}} = \frac{1}{C_1} + \frac{1}{C_2} + \frac{1}{C_3} + \cdots \quad (24.5)$$

(capacitors in series)

$$C_{eq} = C_1 + C_2 + C_3 + \cdots \quad (24.5)$$

(capacitors in parallel)

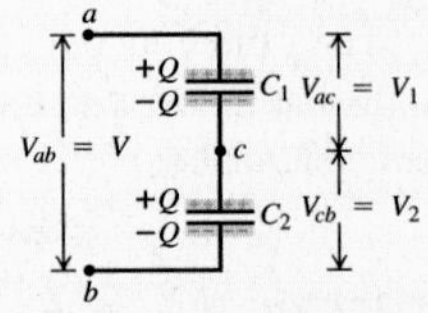

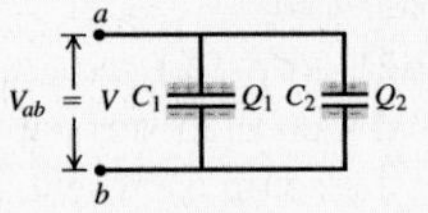

The energy U required to charge a capacitor C to a potential difference V and a charge Q is equal to the energy stored in the capacitor. This energy can be thought of as residing in the electric field between the conductors; the energy density u (energy per unit volume) is proportional to the square of the electric field magnitude. (See Examples 24.7 through 24.9)

$$U = \frac{Q^2}{2C} = \frac{1}{2}CV^2 = \frac{1}{2}QV \quad (24.9)$$

$$u = \frac{1}{2}\epsilon_0 E^2 \quad (24.11)$$

When the space between the conductors is filled with a dielectric material, the capacitance increases by a factor K, called the dielectric constant of the material. The quantity $\epsilon = K\epsilon_0$ is called the permittivity of the dielectric. For a fixed amount of charge on the capacitor plates, induced charges on the surface of the dielectric decrease the electric field and potential difference between the plates by the same factor K. The surface charge results from polarization, a microscopic rearrangement of charge in the dielectric. (See Example 24.10)

$$C = KC_0 = K\epsilon_0 \frac{A}{d} = \epsilon \frac{A}{d} \quad (24.19)$$

(parallel-plate capacitor filled with dielectric)

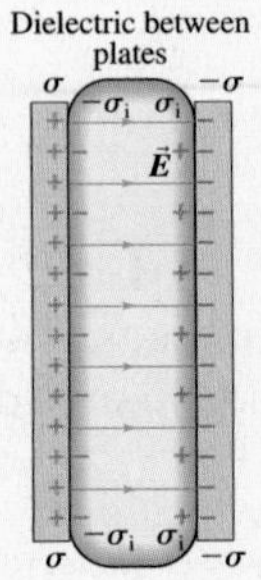

Under sufficiently strong fields, dielectrics become conductors, a situation called dielectric breakdown. The maximum field that a material can withstand without breakdown is called its dielectric strength.

In a dielectric, the expression for the energy density is the same as in vacuum but with ϵ_0 replaced by $\epsilon = K\epsilon$. (See Example 24.11)

$$u = \frac{1}{2}K\epsilon_0 E^2 = \frac{1}{2}\epsilon E^2 \qquad (24.20)$$

Gauss's law in a dielectric has almost the same form as in vacuum, with two key differences: $\vec{E}$ is replaced by $K\vec{E}$ and Q_{encl} is replaced by $Q_{\text{encl-free}}$, which includes only the free charge (not bound charge) enclosed by the Gaussian surface. (See Example 24.12)

$$\oint K\vec{E} \cdot d\vec{A} = \frac{Q_{\text{encl-free}}}{\epsilon_0} \qquad (24.23)$$

Key Terms

capacitor, 909
capacitance, 909
farad, 909
parallel-plate capacitor, 910
series connection, 914
equivalent capacitance, 914
parallel connection, 915
energy density, 919
dielectric, 922
dielectric breakdown, 922
dielectric constant, 923
polarization, 924
permittivity, 924
dielectric strength, 928
bound charge, 929
free charge, 929

Your Notes

Answer to Chapter Opening Question

Decreasing the distance d between the capacitor plates makes the capacitance C increase according to Eq. (24.2). If Q doesn't change, the potential difference $V = Q/C$ decreases.

Answers to Test Your Understanding Questions

Section 24.1 The capacitance does not depend on the value of the charge Q. Doubling the value of Q causes the potential difference V_{ab} to double, so the capacitance $C = Q/V_{ab}$ remains the same. These statements are true no matter what the geometry of the capacitor.

Section 24.2 The potential difference V_{ab} is the *same* across either of the capacitors in a parallel arrangement. The charge on a capacitor is given by $Q = CV_{ab}$, so there is more charge on the 8-μF capacitor (that is, the one with the greater capacitance).

Section 24.3 Capacitors connected in series carry the same charge Q. To compare the amount of energy stored, we use the expression $U = Q^2/2C$ from Eq. (24.9); it shows that the capacitor with the *smaller* capacitance $(C = 4\ \mu\text{F})$ has more stored energy in a series combination. By contrast, capacitors in parallel have the same potential difference V, so to compare them we use $U = \frac{1}{2}CV^2$ from Eq. (24.9). It shows that in a parallel combination, the capacitor with the *larger* capacitance $(C = 8\ \mu\text{F})$ has more stored energy. (If we had instead used $U = \frac{1}{2}CV^2$ to analyze the series combination, we would have to account for the different potential differences across the two capacitors. Likewise, using $U = Q^2/2C$ to study the parallel combination would require us to account for the different charges on the capacitors.)

Section 24.4 Here Q remains the same, so we use $U = Q^2/2C$ from Eq. (24.9) for the stored energy. Removing the dielectric lowers the capacitance by a factor of $1/K$; since U is inversely proportional to C, the stored energy *increases* by a factor of K. It takes work to pull the dielectric slab out of the capacitor, because the fringing field tries to pull the slab back in (Fig. 24.14). The work that you do goes into the energy stored in the capacitor.

Section 24.5 Water molecules are highly polar (see Fig. 21.28). Because water is a liquid, the molecules are free to move and can easily orient themselves with an applied electric field of magnitude E_0. The resulting induced field within the water (like the thin lines in Fig. 24.19c) is so strong that it almost completely cancels the applied field. From Eq. (24.14), the resulting field (the sum of the applied and induced fields) is $E = E_0/K$. Since E is much less than E_0, it follows that the dielectric constant K is large.

Section 24.6 Equation (24.23) shows that this situation is the same as an isolated point charge in vacuum but with $\vec{E}$ replaced by $K\vec{E}$. Hence KE at the point of interest is equal to $q/4\pi\epsilon_0 r^2$, and so $E = q/4\pi K\epsilon_0 r^2 = q/4\pi\epsilon r^2$. As in Example 24.12, filling the space with a dielectric reduces the electric field by a factor of $1/K$.

Discussion Questions

Q24.1 Equation (24.2) shows that the capacitance of a parallel-plate capacitor becomes larger as the plate separation d decreases. However, there is a practical limit to how small d can be made, which places limits on how large C can be. Explain what sets the limit on d. (*Hint:* What happens to the magnitude of the electric field as $d \to 0$?)

Q24.2 A solid slab of metal is placed between the plates of a capacitor without touching either plate. Does the capacitance increase, decrease, or stay the same? Explain your reasoning.

Q24.3 Suppose the two plates of a capacitor have different areas. When the capacitor is charged by connecting it to a battery, do the charges on the two plates have equal magnitude, or may they be different? Explain your reasoning.

Q24.4 At the Fermi National Accelerator Laboratory (Fermilab) in Illinois, protons are accelerated around a ring 2 km in radius to speeds that approach that of light. The energy for this is stored in capacitors the size of a house. When these capacitors are being charged, they make a very loud creaking sound. What is the origin of this sound?

Q24.5 In the parallel-plate capacitor of Fig. 24.2, suppose the plates are pulled apart so that the separation d is much larger than the size of the plates. a) Is it still accurate to say that the electric field between the plates is uniform? Why or why not? b) In the situation shown in Fig. 24.2, the potential difference between the plates is $V_{ab} = Qd/\epsilon_0 A$. If the plates are pulled apart as described above, is V_{ab} more or less than this formula would indicate? Explain your reasoning. c) With the plates pulled apart as described above, is the capacitance more than, less than, or the same as that given by Eq. (24.2)? Explain your reasoning.

Q24.6 A parallel-plate capacitor is charged by being connected to a battery and is kept connected to the battery. The separation between the plates is then doubled. How does the electric field change? The charge on the plates? The total energy? Explain your reasoning.

Q24.7 A parallel-plate capacitor is charged by being connected to a battery and is then disconnected from the battery. The separation between the plates is then doubled. How does the electric field change? The potential difference? The total energy? Explain your reasoning.

Q24.8 According to the text, we can consider the energy in a charged capacitor to be located in the field between the plates. But suppose there is vacuum between the plates; can there be energy in a vacuum? Why or why not? What form could this energy take?

Q24.9 The charged plates of a capacitor attract each other, so to pull the plates farther apart requires work by some external force. What becomes of the energy added by this work? Explain your reasoning.

Q24.10 The two plates of a capacitor are given charges $\pm Q$. The capacitor is then disconnected from the charging device so that the charges on the plates can't change, and the capacitor is immersed in a tank of oil. Does the electric field between the plates increase, decrease, or stay the same? Explain your reasoning. How can this field be measured?

Q24.11 As shown in Table 24.1, water has a very large dielectric constant $K = 80.4$. Why do you think water is not commonly used as a dielectric in capacitors?

Q24.12 Is dielectric strength the same thing as dielectric constant? Explain any differences between the two quantities. Is there a simple relationship between dielectric strength and dielectric constant (see Table 24.2)?

Q24.13 A capacitor made of aluminum foil strips separated by Mylar film was subjected to excessive voltage, and the resulting dielectric breakdown melted holes in the Mylar. After this, the capacitance was found to be about the same as before, but the breakdown voltage was much less. Why?

Q24.14 Two capacitors have equal capacitance, but one has a higher maximum voltage rating than the other. Which capacitor is likely to be bulkier? Why?

Q24.15 The freshness of fish can be measured by placing a fish between the plates of a capacitor and measuring the capacitance. How does this work? (*Hint:* As time passes, the fish dries out. See Table 24.1.)

Q24.16 *Electrolytic* capacitors use as their dielectric an extremely thin layer of nonconducting oxide between a metal plate and a conducting solution. Discuss the advantage of such a capacitor over one constructed using a solid dielectric between the metal plates.

Q24.17 A parallel-plate capacitor is connected to a power supply that maintains a fixed potential difference between the plates. If a sheet of dielectric is then slid between the plates, what happens to i) the electric field between the plates, ii) the magnitude of charge on each plate, and iii) the energy stored in the capacitor? Now suppose that before the dielectric is inserted, the charged capacitor is disconnected from the power supply. In this case, what happens to i) the electric field between the plates, ii) the magnitude of charge on each plate, and iii) the energy stored in the capacitor? Explain any differences between the two situations.

Q24.18 Liquid dielectrics having polar molecules (such as water) always have dielectric constants that decrease with increasing temperature. Why?

Q24.19 A conductor is an extreme case of a dielectric, since if an electric field is applied to a conductor, charges are free to move within the conductor to set up "induced charges." What is the dielectric constant of a perfect conductor? Is it $K = 0$? $K \to \infty$? Or something in between? Explain your reasoning.

Exercises

Section 24.1 Capacitors and Capacitance

24.1 A capacitor has a capacitance of 7.28 μF. What amount of charge must be placed on each of its plates to make the potential difference between its plates equal to 25.0 V?

24.2 The plates of a parallel-plate capacitor are 3.28 mm apart, and each has an area of 12.2 cm^2. Each plate carries a charge of magnitude 4.35×10^{-8} C. The plates are in vacuum. a) What is the capacitance? b) What is the potential difference between the plates? c) What is the magnitude of the electric field between the plates?

24.3 A parallel-plate air capacitor of capacitance 245 pF has a charge of magnitude 0.148 μC on each plate. The plates are 0.328 mm apart. a) What is the potential difference between the plates? b) What is the area of each plate? c) What is the electric field magnitude between the plates? d) What is the surface-charge density on each plate?

24.4 A parallel-plate, air-filled capacitor has circular plates separated by 1.80 mm. The charge per unit area on each plate has magnitude 5.60 pC/m^2. What is the potential difference between the plates of the capacitor?

24.5 A 10.0-μF parallel-plate capacitor with circular plates is connected to a 12.0-V battery. a) What is the charge on each plate? b) How much charge would be on the plates if their separation were doubled while the capacitor remained connected to the battery? c) How much charge would be on the plates if the capacitor were connected to the 12.0-V battery after the radius of each plate was doubled without changing their separation?

24.6 A 10.0-μF parallel-plate capacitor is connected to a 12.0-V battery. After the capacitor is fully charged, the battery is disconnected without loss of any of the charge on the plates. a) A voltmeter is connected across the two plates without discharging them. What does it read? b) What would the voltmeter read if i) the plate separation were doubled; ii) the radius of each plate were doubled, but their separation was unchanged?

24.7 How far apart would parallel pennies have to be to make a 1.00-pF capacitor? Does your answer suggest that you are justified in treating these pennies as infinite sheets? Explain.

24.8 A 5.00-pF parallel-plate air-filled capacitor with circular plates is to be used in a circuit in which it will be subjected to potentials of up to 1.00×10^2 V. The electric field between the plates is to be no greater than 1.00×10^4 N/C. As a budding electrical engineer for Live-Wire Electronics, your tasks are to a) design the capacitor by finding what its physical dimensions and separation must be; b) find the maximum charge these plates can hold.

24.9 A capacitor is made from two hollow, coaxial, iron cylinders, one inside the other. The inner cylinder is negatively charged and the outer is positively charged; the magnitude of the charge on each is 10.0 pC. The inner cylinder has radius 0.50 mm, the outer one has radius 5.00 mm, and the length of each cylinder is 18.0 cm. a) What is the capacitance? b) What applied potential difference is necessary to produce these charges on the cylinders?

24.10 A cylindrical capacitor like that in Example 24.4 (Section 24.1) has a capacitance per length of 31.5 pF/m. a) Find the ratio of the radii of the inner and outer conductors. b) If the potential difference between the inner and outer conductors is 2.60 V, what is the magnitude of the charge per unit length on the conductors?

24.11 A cylindrical capacitor has an inner conductor of radius 1.5 mm and an outer conductor of radius 3.5 mm. The two conductors are separated by vacuum, and the entire capacitor is 2.8 m long. a) What is the capacitance per unit length? b) The potential of the inner conductor is 350 mV higher than that of the outer conductor. Find the charge (magnitude and sign) on both conductors.

24.12 A spherical capacitor is formed from two concentric, spherical, conducting shells separated by vacuum. The inner sphere has radius 15.0 cm and the capacitance is 116 pF. a) What is the radius of the outer sphere? b) If the potential difference between the two spheres is 220 V, what is the magnitude of charge on each sphere?

24.13 A spherical capacitor is formed from two concentric spherical conducting shells separated by vacuum. The inner sphere has radius 12.5 cm and the outer sphere has radius 14.8 cm. A potential difference of 120 V is applied to the capacitor. a) What is the capacitance of the capacitor? b) What is the magnitude of $\vec{E}$ at

$r = 12.6$ cm, just outside the inner sphere? c) What is the magnitude of $\vec{E}$ at $r = 14.7$ cm, just inside the outer sphere? d) For a parallel-plate capacitor $\vec{E}$ is uniform in the region between the plates, except near the edges of the plates. Is this also true for a spherical capacitor?

Section 24.2 Capacitors in Series and Parallel

24.14 In the circuit shown in Fig. 24.22, $C_1 = 3.00\ \mu$F, $C_2 = 5.00\ \mu$F, and $C_3 = 6.00\ \mu$F. The applied potential is $V_{ab} = +24.0$ V. Calculate a) the charge on each capacitor; b) the potential difference across each capacitor; c) the potential difference between points *a* and *d*.

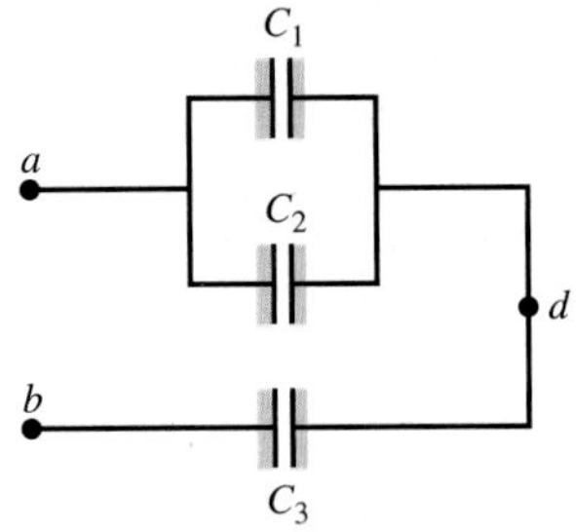

Figure 24.22 Exercises 24.14, 24.22, and 24.23.

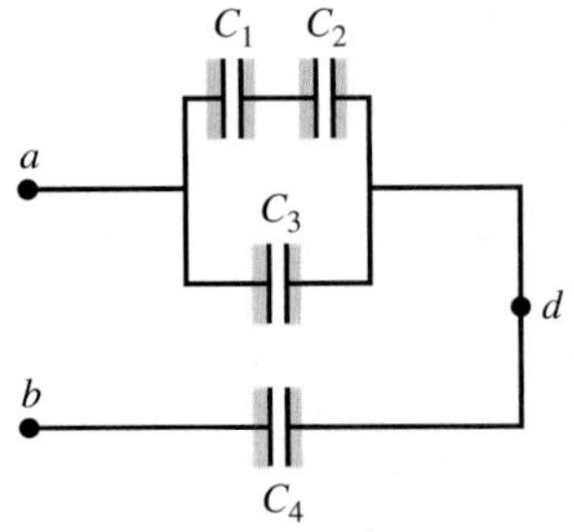

Figure 24.23 Exercise 24.15.

24.15 In Fig. 24.23, each capacitor has $C = 4.00\ \mu$F and $V_{ab} = +28.0$ V. Calculate a) the charge on each capacitor; b) the potential difference across each capacitor; c) the potential difference between points *a* and *d*.

24.16 In Fig. 24.6a, let $C_1 = 3.00\ \mu$F, $C_2 = 5.00\ \mu$F, and $V_{ab} = +52.0$ V. Calculate a) the charge on each capacitor; b) the potential difference across each capacitor.

24.17 In Fig. 24.7a, let $C_1 = 3.00\ \mu$F, $C_2 = 5.00\ \mu$F, and $V_{ab} = +52.0$ V. Calculate a) the charge on each capacitor; b) the potential difference across each capacitor.

24.18 Two parallel-plate vacuum capacitors have plate spacings d_1 and d_2 and equal plate areas *A*. Show that when the capacitors are connected in series, the equivalent capacitance is the same as for a single capacitor with plate area *A* and spacing $d_1 + d_2$.

24.19 Two parallel-plate vacuum capacitors have areas A_1 and A_2 and equal plate spacings *d*. Show that when the capacitors are connected in parallel, the equivalent capacitance is the same as for a single capacitor with plate area $A_1 + A_2$ and spacing *d*.

24.20 For the situation shown in Fig. 24.8a, suppose $V_{ab} = +36$ V. Calculate a) the charge on each capacitor; b) the potential difference across each capacitor. c) The charge on the entire network is equal to $Q = C_{\text{eq}}V_{ab}$, where $C_{\text{eq}} = 6\ \mu$F from Example 24.6 (Section 24.2). Explain why *Q* is equal to the charge on the 9-μF capacitor. Explain why *Q* is also equal to the sum of the charge on the 3-μF capacitor, the 11-μF capacitor, and *either* the 12-μF capacitor or the 6-μF capacitor.

24.21 Suppose the 3-μF capacitor in Fig. 24.8a were removed and replaced by a different one, and that this changes the equivalent capacitance between points *a* and *b* to 8 μF. What would be the capacitance of the replacement capacitor?

24.22 In Fig. 24.22, $C_1 = 6.00\ \mu$F, $C_2 = 3.00\ \mu$F, $C_3 = 5.00\ \mu$F. The capacitor network is connected to an applied potential V_{ab}. After the charges on the capacitors have reached their final values, the charge on C_2 is 40.0 μC. a) What are the charges on capacitors C_1 and C_3? b) What is the applied voltage V_{ab}?

24.23 In Fig. 24.22, $C_1 = 3.00\ \mu$F and $V_{ab} = 120$ V. The charge on capacitor C_1 is 150 μC. Calculate the voltage across the other two capacitors.

Section 24.3 Energy Storage in Capacitors and Electric-Field Energy

24.24 A parallel-plate air capacitor has a capacitance of 920 pF. The charge on each plate is 2.55 μC. a) What is the potential difference between the plates? b) If the charge is kept constant, what will be the potential difference between the plates if the separation is doubled? c) How much work is required to double the separation?

24.25 A 5.80-μF parallel-plate air capacitor has a plate separation of 5.00 mm and is charged to a potential difference of 400 V. Calculate the energy density in the region between the plates, in units of J/m^3.

24.26 An air capacitor is made from two flat parallel plates 1.50 mm apart. The magnitude of charge on each plate is 0.0180 μC when the potential difference is 200 V. a) What is the capacitance? b) What is the area of each plate? c) What maximum voltage can be applied without dielectric breakdown? (Dielectric breakdown for air occurs at an electric field strength of 3.0×10^6 V/m.) d) When the charge is 0.0180 μC, what total energy is stored?

24.27 A 450-μF capacitor is charged to 295 V. Then a wire is connected between the plates. How many joules of thermal energy are produced as the capacitor discharges if all of the energy that was stored goes into heating the wire?

24.28 A capacitor of capacitance *C* is charged to a potential difference V_0. The terminals of the charged capacitor are then connected to those of an uncharged capacitor of capacitance *C*/2. Compute a) the original charge of the system; b) the final potential difference across each capacitor; c) the final energy of the system; d) the decrease in energy when the capacitors are connected. e) Where did the "lost" energy go?

24.29 A parallel-plate vacuum capacitor with plate area *A* and separation *x* has charges $+Q$ and $-Q$ on its plates. The capacitor is disconnected from the source of charge, so the charge on each plate remains fixed. a) What is the total energy stored in the capacitor? b) The plates are pulled apart an additional distance *dx*. What is the change in the stored energy? c) If *F* is the force with which the plates attract each other, then the change in the stored energy must equal the work $dW = F dx$ done in pulling the plates apart. Find an expression for *F*. d) Explain why *F* is *not* equal to *QE*, where *E* is the electric field between the plates.

24.30 A parallel-plate vacuum capacitor has 8.38 J of energy stored in it. The separation between the plates is 2.30 mm. If the separation is decreased to 1.15 mm, what is the energy stored if a) the capacitor is disconnected from the potential source so the charge on the plates remains constant? b) the capacitor remains connected to the potential source so the potential difference between the plates remains constant?

24.31 a) How many excess electrons must be added to one plate and removed from the other to give a 5.00-nF parallel-plate capacitor 25.0 J of stored energy? b) How could you modify the geometry of this capacitor to get it to store 50.0 J of energy without changing the charge on its plates?

24.32 A parallel-plate, air-filled capacitor has plates of area $2.60 \times 10^{-3}\ \text{m}^2$. The magnitude of charge on each plate is 8.20 pC and the potential difference between the plates is 2.40 V. What is the electric-field energy density in the volume between the plates?

24.33 A cylindrical air capacitor of length 15.0 m stores 3.20×10^{-9} J of energy when the potential difference between the two conductors is 4.00 V. a) Calculate the magnitude of the charge on each conductor. b) Calculate the ratio of the radii of the inner and outer conductors.

24.34 A spherical capacitor is formed from two concentric spherical conducting shells separated by vacuum. The inner sphere has radius 10.0 cm, and the separation between the spheres is 1.50 cm. The magnitude of the charge on each sphere is 3.30 nC. a) What is the potential difference between the two spheres? b) What is the electric-field energy stored in the capacitor?

24.35 Consider the spherical capacitor of Exercise 24.13. a) What is the energy density at $r = 12.6$ cm, just outside the inner sphere? b) What is the energy density at $r = 14.7$ cm, just inside the outer sphere? c) For a parallel-plate capacitor the energy density is uniform in the region between the plates, except near the edges of the plates. Is this also true for a spherical capacitor?

24.36 a) Find the electric-field energy density at a point 12.0 cm from an isolated point charge $q = 8.00$ nC. b) If the point charge in part (a) were instead $q = -8.00$ nC, what effect would this have on the electric-field energy density? Explain your reasoning.

24.37 You have two identical capacitors and an external potential source. a) Compare the total energy stored in the capacitors when they are connected to the applied potential in series and in parallel. b) Compare the maximum amount of charge stored in each case. c) Energy storage in a capacitor can be limited by the maximum electric field between the plates. What is the ratio of the electric field for the series and parallel combinations?

Section 24.4 Dielectrics

24.38 A parallel-plate capacitor has capacitance $C_0 = 5.00$ pF when there is air between the plates. The separation between the plates is 1.50 mm. a) What is the maximum magnitude of charge Q that can be placed on each plate if the electric field in the region between the plates is not to exceed 3.00×10^4 V/m? b) A dielectric with $K = 2.70$ is inserted between the plates of the capacitor, completely filling the volume between the plates. Now what is the maximum magnitude of charge on each plate if the electric field between the plates is not to exceed 3.00×10^4 V/m?

24.39 Two parallel plates have equal and opposite charges. When the space between the plates is evacuated, the electric field is $E = 3.20 \times 10^5$ V/m. When the space is filled with dielectric, the electric field is $E = 2.50 \times 10^5$ V/m. a) What is the charge density on each surface of the dielectric? b) What is the dielectric constant?

24.40 Two identical, oppositely-charged, conducting plates of area $2.50\ \text{cm}^2$ are separated by a dielectric 1.80 mm thick, with a dielectric constant of 3.60. The resultant electric field in the dielectric is 1.20×10^6 V/m. Compute a) the charge per unit area on the conducting plate; b) the charge per unit area on the surfaces of the dielectric; c) the total electric-field energy stored in the capacitor.

24.41 The dielectric to be used in a parallel-plate capacitor has a dielectric constant of 3.60 and a dielectric strength of 1.60×10^7 V/m. The capacitor is to have a capacitance of 1.25×10^{-9} F and must be able to withstand a maximum potential difference of 5500 V. What is the minimum area the plates of the capacitor may have?

24.42 Show that Eq. (24.20) holds for a parallel-plate capacitor with a dielectric material between the plates. Use a derivation analogous to that used for Eq. (24.11).

24.43 A capacitor has parallel plates of area 12 cm^2 separated by 2.0 mm. The space between the plates is filled with polystyrene (see Table 24.2). a) Find the permittivity of polystyrene. b) Find the maximum permissible voltage across the capacitor to avoid dielectric breakdown. c) When the voltage equals the value found in part (b), find the surface-charge density on each plate and the induced surface-charge density on the surface of the dielectric.

24.44 A constant potential difference of 12 V is maintained between the terminals of a 0.25-μF parallel-plate air capacitor. a) A sheet of Mylar is inserted between the plates of the capacitor, completely filling the space between the plates. When this is done, how much additional charge flows onto the positive plate of the capacitor (see Table 24.1)? b) What is the total induced charge on either face of the Mylar sheet? c) What effect does the Mylar sheet have on the electric field between the plates? Explain how you can reconcile this with the increase in charge on the plates, which acts to *increase* the electric field.

24.45 When a 360-nF air capacitor ($1\ \text{nF} = 10^{-9}\ \text{F}$) is connected to a power supply, the energy stored in the capacitor is 1.85×10^{-5} J. While the capacitor is kept connected to the power supply, a slab of dielectric is inserted that completely fills the space between the plates. This increases the stored energy by 2.32×10^{-5} J. a) What is the potential difference between the capacitor plates? b) What is the dielectric constant of the slab?

24.46 A parallel-plate capacitor has capacitance $C = 12.5$ pF when the volume between the plates is filled with air. The plates are circular, with radius 3.00 cm. The capacitor is connected to a battery and a charge of magnitude 25.0 pC goes onto each plate. a) With the capacitor still connected to the battery, a slab of dielectric is inserted between the plates, completely filling the space between the plates. After the dielectric has been inserted the charge on each plate has magnitude 45.0 pC. a) What is the dielectric constant K of the dielectric? b) What is the potential difference between the plates before and after the dielectric has been inserted? c) What is the electric field at a point midway between the plates before and after the dielectric has been inserted?

24.47 A parallel-plate capacitor has charge of magnitude 9.00×10^{-6} C on each plate and capacitance 3.00 μF when there is air between the plates. The plates are separated by 2.00 mm. With the charge on the plates kept constant, a dielectric with $K = 5$ is inserted between the plates, completely filling the volume between the plates. a) What is the potential difference between the plates of the capacitor, before and after the dielectric has been inserted? b) What is the electric field at a point midway between the plate before and after the dielectric is inserted?

*Section 24.6 Gauss's Law in Dielectrics

***24.48** A point charge q is imbedded in a solid material of dielectric constant K. a) Use Gauss's law as stated in Eq. (24.23) to find the magnitude of the electric field due to the point charge at a distance d from the charge. b) Use your result from part (a) and Gauss's law in its original form as given in Eq. (22.8) to determine the *total* charge (free *and* bound) within a sphere of radius d centered on the point charge q. c) Find the total bound charge within the sphere described in part (b).

***24.49** A parallel-plate capacitor has the volume between its plates filled with plastic with dielectric constant K. The magnitude of the charge on each plate is Q. Each plate has area A, and the distance between the plates is d. a) Use Gauss's law as stated in Eq. (24.23) to calculate the magnitude of the electric field in the dielectric. b) Use the electric field determined in part (a) to calculate the potential difference between the two plates. c) Use the result of part (b) to determine the capacitance of the capacitor. Compare your result to Eq. (24.12).

Problems

24.50 A parallel-plate air capacitor is made by using two plates 16 cm square, spaced 4.7 mm apart. It is connected to a 12-V battery. a) What is the capacitance? b) What is the charge on each plate? c) What is the electric field between the plates? d) What is the energy stored in the capacitor? e) If the battery is disconnected and then the plates are pulled apart to a separation of 9.4 mm, what are the answers to parts (a), (b), (c), and (d)?

24.51 Suppose the battery in Problem 24.50 remains connected while the plates are pulled apart. What are the answers then to parts (a), (b), (c), and (d) after the plates have been pulled apart?

24.52 Four identical, square, conducting plates have sides of length L. All four plates lie parallel to the xy-plane, with corners at $(x, y) = (L/2, L/2)$, $(-L/2, L/2)$, $(L/2, -L/2)$, and $(-L/2, -L/2)$. Plate 1 is in the plane $z = 0$, plate 2 in the plane $z = d$, plate 3 in the plane $z = 2d$, and plate 4 in the plane $z = 3d$, where $d \ll L$. Plates 1 and 3 each carry a positive charge Q, and plates 2 and 4 each carry a negative charge $-Q$. There is vacuum between the plates. a) Find the total energy stored in this arrangement of plates and charges. b) Plates 2 and 3 are now interchanged without altering the charges that they carry. How much work has to be done to carry out this interchange? [*Hint:* Review the discussion of superposition in Example 21.13 (Section 21.5).]

24.53 Electronic flash units for cameras contain a capacitor for storing the energy used to produce the flash. In one such unit, the flash lasts for $\frac{1}{675}$ s with an average light power output of 2.70×10^5 W. a) If the conversion of electrical energy to light is 95% efficient (the rest of the energy goes to thermal energy), how much energy must be stored in the capacitor for one flash? b) The capacitor has a potential difference between its plates of 125 V when the stored energy equals the value calculated in part (a). What is the capacitance?

24.54 In one type of computer keyboard, each key holds a small metal plate that serves as one plate of a parallel-plate, air-filled capacitor. When the key is depressed, the plate separation decreases and the capacitance increases. Electronic circuitry detects the change in capacitance and thus detects that the key has been pressed. In one particular keyboard, the area of each metal plate is 42.0 mm^2, and the separation between the plates is 0.700 mm before the key is depressed. If the circuitry can detect a change in capacitance of 0.250 pF, how far must the key be depressed before the circuitry detects its depression?

24.55 Consider a cylindrical capacitor like that shown in Fig. 24.4. Let $d = r_b - r_a$ be the spacing between the inner and outer conductors. a) Let the radii of the two conductors be only slightly different, so that $d \ll r_a$. Show that the result derived in Example 24.4 (Section 24.1) for the capacitance of a cylindrical capacitor then reduces to Eq. (24.2), the equation for the capacitance of a parallel-plate capacitor, with A being the surface area of each cylinder. Use the result that $\ln(1 + z) \cong z$ for $|z| \ll 1$. b) Even though the earth is essentially spherical, its surface appears flat to us because its radius is so large. Use this idea to explain why the result of part (a) makes sense from a purely geometrical standpoint.

24.56 In Fig. 24.7a, let $C_1 = 9.0\ \mu$F, $C_2 = 4.0\ \mu$F, and $V_{ab} = 28$ V. Suppose the charged capacitors are disconnected from the source and from each other, then reconnected to each other with plates of *opposite* sign together. By how much does the energy of the system decrease?

24.57 A 4.00-μF capacitor and a 6.00-μF capacitor are connected in parallel across a 660-V supply line. a) Find the charge on each capacitor and the voltage across each. b) The charged capacitors are disconnected from the line and from each other, and then reconnected to each other with terminals of unlike sign together. Find the final charge on each and the voltage across each.

24.58 Several 0.25-μF capacitors are available. The voltage across each is not to exceed 600 V. You need to make a capacitor with capacitance 0.25 μF to be connected across a potential difference of 960 V. a) Show in a diagram how an equivalent capacitor having the desired properties can be obtained. b) No dielectric is a perfect insulator, that would not permit the flow of any charge through its volume. Suppose that the dielectric in one of the capacitors in your diagram is a moderately good conductor. What will happen in this case when your combination of capacitors is connected across the 960-V potential difference?

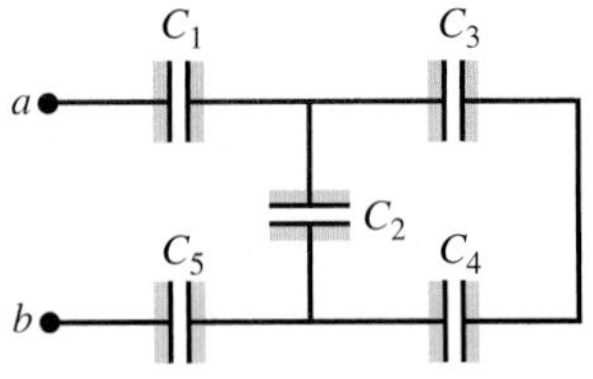

Figure 24.24 Problem 24.59.

24.59 In Fig. 24.24, $C_1 = C_5 = 8.4\ \mu$F and $C_2 = C_3 = C_4 = 4.2\ \mu$F. The applied potential is $V_{ab} = 220$ V. a) What is the equivalent capacitance of the network between points a and b? b) Calculate the charge on each capacitor and the potential difference across each capacitor.

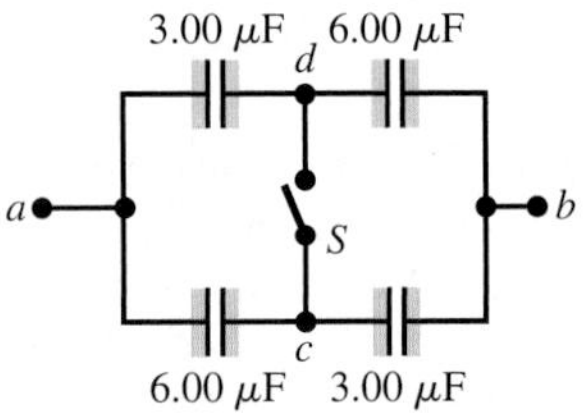

Figure 24.25 Problem 24.60.

24.60 The capacitors in Fig. 24.25 are initially uncharged and are connected, as in the diagram, with switch S open. The applied potential difference is

$V_{ab} = +210$ V. a) What is the potential difference V_{cd}? b) What is the potential difference across each capacitor after switch S is closed? c) How much charge flowed through the switch when it was closed?

24.61 Three capacitors having capacitances of 8.4, 8.4, and 4.2 μF are connected in series across a 36-V potential difference. a) What is the charge on the 4.2-μF capacitor? b) What is the total energy stored in all three capacitors? c) The capacitors are disconnected from the potential difference without allowing them to discharge. They are then reconnected in parallel with each other, with the positively charged plates connected together. What is the voltage across each capacitor in the parallel combination? d) What is the total energy now stored in the capacitors?

24.62 A 4.00-μF capacitor and a 6.00-μF capacitor are connected in series across a 660-V supply line. a) Find the charge on each capacitor and the voltage across each. b) The charged capacitors are disconnected from the line and from each other and then reconnected to each other, with terminals of like sign together. Find the final charge on each and the voltage across each.

24.63 In Fig. 24.26, each capacitance C_1 is 6.9 μF, and each capacitance C_2 is 4.6 μF. a) Compute the equivalent capacitance of the network between points a and b. b) Compute the charge on each of the three capacitors nearest a and b when $V_{ab} = 420$ V. c) With 420 V across a and b, compute V_{cd}.

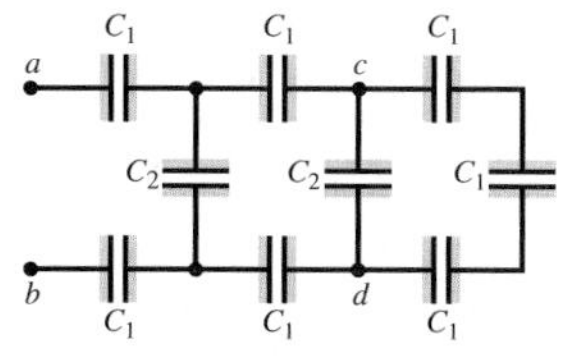

Figure 24.26 Problem 24.63.

24.64 Each combination of capacitors between points a and b in Fig. 24.27 is first connected across a 120-V battery, charging the combination to 120 V. These combinations are then connected to make the circuits shown. When the switch S is thrown, a surge of charge for the discharging capacitors flows to trigger the signal device. How much charge flows through the signal device?

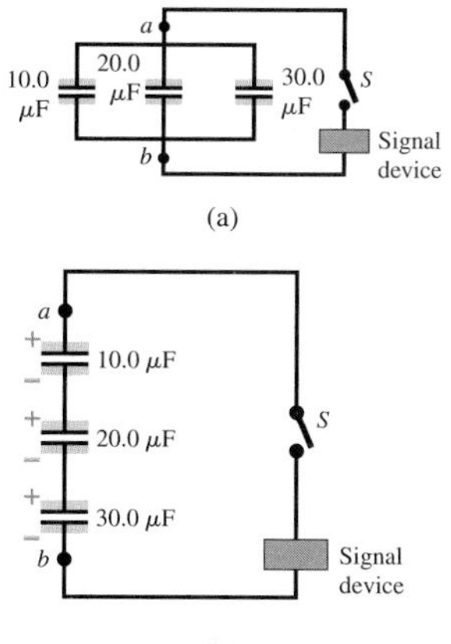

Figure 24.27 Problem 24.64.

24.65 A parallel-plate capacitor with only air between the plates is charged by connecting it to a battery. The capacitor is then disconnected from the battery, without any of the charge leaving the plates. a) A voltmeter reads 45.0 V when placed across the capacitor. When a dielectric is inserted between the plates, completely filling the space, the voltmeter reads 11.5 V. What is the dielectric constant of this material? b) What will the voltmeter read if the dielectric is now pulled partway out so it fills only one-third of the space between the plates?

24.66 An air capacitor is made by using two flat plates, each with area A, separated by a distance d. Then a metal slab having thickness a (less than d) and the same shape and size as the plates is inserted between them, parallel to the plates and not touching either plate (see Fig. 24.28). a) What is the capacitance of this arrangement? b) Express the capacitance as a multiple of the capacitance C_0 when the metal slab is not present. c) Discuss what happens to the capacitance in the limits $a \rightarrow 0$ and $a \rightarrow d$.

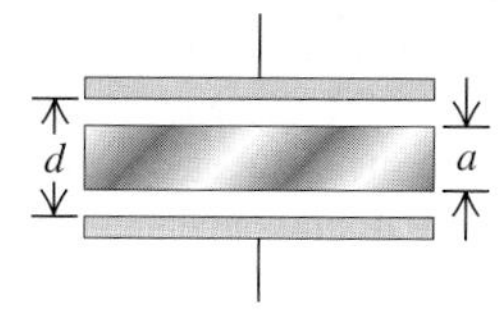

Figure 24.28 Problem 24.66.

24.67 Capacitance of the Earth. a) Discuss how the concept of capacitance can also be applied to a *single* conductor. (*Hint:* In the relationship $C = Q/V_{ab}$, think of the second conductor as being located at infinity.) b) Use Eq. (24.1) to show that $C = 4\pi\epsilon_0 R$ for a solid conducting sphere of radius R. c) Use your result in part (b) to calculate the capacitance of the earth, which is a good conductor of radius 6380 km. Compare to typical capacitors used in electronic circuits, that have capacitances that range from 10 pF to 100 μF.

24.68 A solid conducting sphere of radius R carries a charge Q. Calculate the electric-field energy density at a point a distance r from the center of the sphere for a) $r < R$; b) $r > R$. c) Calculate the total electric-field energy associated with the charged sphere. (*Hint:* Consider a spherical shell of radius r and thickness dr, which has volume $dV = 4\pi r^2\,dr$, and find the energy stored in this volume. Then integrate from $r = 0$ to $r \rightarrow \infty$.) d) Explain why the result of part (c) can be interpreted as the amount of work required to assemble the charge Q on the sphere. e) By using Eq. (24.9) and the result of part (c), show that the capacitance of the sphere is as given in Problem 24.67.

24.69 A uniformly charged sphere of radius R has total charge Q, as described in Example 22.9 (Section 22.4). Calculate the electric-field energy density at a point a distance r from the center of the sphere for a) $r < R$; b) $r > R$. c) Calculate the total electric-field energy. (See the hint in part (c) of Problem 24.68.)

24.70 The inner cylinder of a long, cylindrical capacitor has radius r_a and a linear charge density $+\lambda$. It is surrounded by a coaxial cylindrical conducting shell with inner radius r_b and linear charge density $-\lambda$ (Fig. 24.4). a) What is the energy density in the region between the conductors at a distance r from the axis? b) Integrate the energy density calculated in part (a) over the volume between the conductors in a length L of the capacitor to obtain the total electric-field energy per unit length. c) Use Eq. (24.9) and the capacitance per unit length calculated in Example 24.4 (Section 24.1) to calculate U/L. Does your result agree with that obtained in part (b)?

24.71 A parallel-plate capacitor has the space between the plates filled with two slabs of dielectric, one with constant K_1 and one with constant K_2 (see Fig. 24.29). Each slab has thickness $d/2$, where d, is the plate separation. Show that the capacitance is

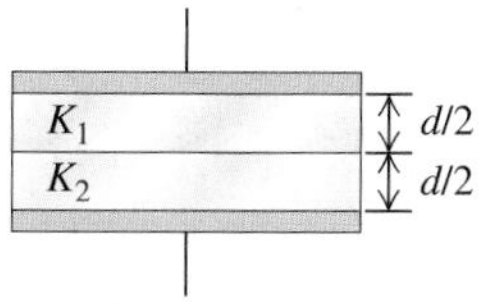

Figure 24.29 Problem 24.71.

$$C = \frac{2\epsilon_0 A}{d}\left(\frac{K_1 K_2}{K_1 + K_2}\right)$$

24.72 A parallel-plate capacitor has the space between the plates filled with two slabs of dielectric, one with constant K_1 and one with constant K_2. (See Fig. 24.30.) The thickness of each slab is the same as the plate separation d, and each slab fills half of the volume between the plates. Show that the capacitance is

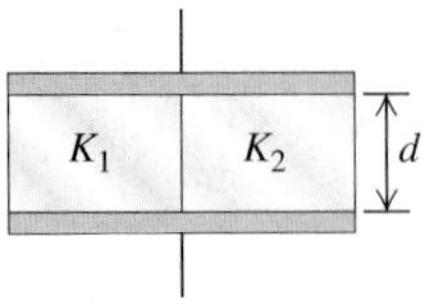

Figure 24.30 Problem 24.72.

$$C = \frac{\epsilon_0 A(K_1 + K_2)}{2d}$$

24.73 Human Cells. Some cell walls in the human body have a layer of negative charge on the inside surface and a layer of positive charge of equal magnitude on the outside surface. Suppose that the surface charge densities are $\pm 0.50 \times 10^{-3}\ \mathrm{C/m^2}$, the cell wall is 5.0×10^{-9} m thick, and the cell-wall material has a dielectric constant of $K = 5.4$. a) Find the magnitude of $\vec{E}$ in the wall between the two charge layers. b) Find the potential difference between the inside and the outside of the cell. Which is at higher potential? c) A typical cell in the human body has volume $10^{-16}\ \mathrm{m^3}$. Estimate the total electric-field energy stored in the wall of a cell of this size. (*Hint:* Assume that the cell is spherical, and calculate the volume of the cell wall.)

24.74 Suppose that the parallel-plate capacitor described in Examples 24.10 and 24.11 (Section 24.4) remains connected to the 3000-V power supply while an insulating plastic sheet with $K = 2.50$ is inserted between the plates, completely filling the space between them. Compute a) the magnitude of charge Q on each plate after the dielectric is inserted; b) the magnitude of induced charge Q_i on each face of the dielectric; c) the electric field E after the dielectric is inserted; d) the total energy stored in the electric field after the dielectric is inserted; e) the energy density after the dielectric is inserted. f) You should have found in part (d) that the stored energy *increased* when the dielectric was inserted, whereas in Example 24.11 the stored energy *decreased* when this was done. Why is there a difference between the two cases? In the present case, where does the extra energy come from?

Challenge Problems

24.75 Capacitors in networks cannot always be grouped into simple series or parallel combinations. As an example, Fig. 24.31a shows three capacitors C_x, C_y, and C_z in a *delta network,* so called because of its triangular shape. This network has *three* terminals a, b, and c and hence cannot be transformed into a single equivalent capacitor. It can be shown that as far as any effect on the external circuit is concerned, a delta network is equivalent to what is called a *Y network.* For example, the delta network of Fig. 24.31a can be replaced by the Y network of Fig. 24.31b. (The name "Y network" also refers to the shape of the network.) a) Show that the transformation equations that give C_1, C_2, and C_3 in terms of C_x, and C_y, and C_z are

$$C_1 = (C_xC_y + C_yC_z + C_zC_x)/C_x$$
$$C_2 = (C_xC_y + C_yC_z + C_zC_x)/C_y$$
$$C_3 = (C_xC_y + C_yC_z + C_zC_x)/C_z$$

(*Hint:* The potential difference V_{ac} must be the same in both circuits, as V_{bc} must be. Also, the charge q_1 that flows from point a along the wire as indicated must be the same in both circuits, as must q_2. Obtain a relationship for V_{ac} as a function of q_1 and q_2 and the capacitances for each network, and obtain a separate relationship for V_{bc} as a function of the charges for each network. The coefficients of corresponding charges in corresponding equations must be the same for both networks.) b) For the network shown in Fig. 24.31c, determine the equivalent capacitance between the terminals at the left end of the network. (*Hint:* Use the delta-Y transformation derived in part (a). Use points a, b, and c to form the delta, and transform the delta into a Y. The capacitors can then be combined using the relationships for series and parallel combinations of capacitors.) c) Determine the charges of, and the potential differences across, each capacitor in Fig. 24.31c.

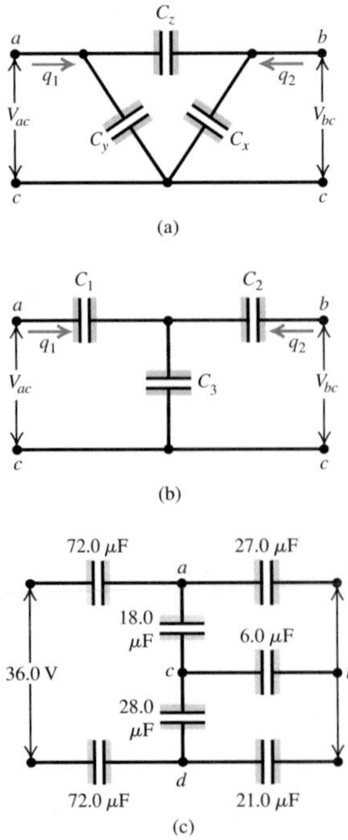

Figure 24.31 Challenge Problem 24.75.

24.76 The parallel-plate air capacitor in Fig. 24.32 consists of two horizontal conducting plates of equal area A. The bottom plate rests on a fixed support, and the top plate is suspended by four springs with spring constant k, positioned at each of the four corners of the top plate as shown in the figure. When uncharged, the plates are separated by a distance z_0. A battery is connected to the plates and produces a potential difference V between them. This causes the plate separation to decrease to z. Neglect any fringing effects. a) Show that the electrostatic force between the charged plates has a magnitude $\epsilon_0 AV^2/2z^2$. (*Hint:* See Exercise 24.29.) b) Obtain an expression that relates the plate separation z to the potential difference V. The resulting equation will be cubic in z. c) Given the values $A = 0.300\ \mathrm{m^2}$, $z_0 = 1.20$ mm, $k = 25.0$ N/m, and $V = 120$ V, find the two values of z for which the top plate will be in equilibrium. (*Hint:* You can solve the cubic equation by plugging a trial value of z into the equation and then adjusting your guess until the equation is satisfied to three significant figures. Locating the roots of the cubic equation graphically can help you in picking starting values of z for this trial-and-error procedure. One root of the cubic equation has a nonphysical negative value.) d) For each of the two

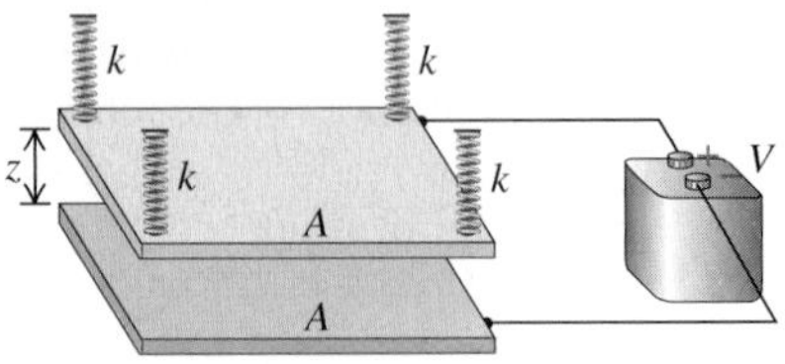

Figure 24.32 Challenge Problem 24.76.

values of z found in part (c), is the equilibrium stable or unstable? For stable equilibrium a small displacement of the object will give rise to a net force tending to return the object to the equilibrium position. For unstable equilibrium a small displacement gives rise to a net force that takes the object farther away from equilibrium.

24.77 Two square conducting plates with sides of length L are separated by a distance D. A dielectric slab with constant K with dimensions $L \times L \times D$ is inserted a distance x into the space between the plates, as shown in Fig. 24.33. a) Find the capacitance C of this system (see Problem 24.72). b) Suppose that the capacitor is connected to a battery that maintains a constant potential difference V between the plates. If the dielectric slab is inserted an additional distance dx into the space between the plates, show that the change in stored energy is

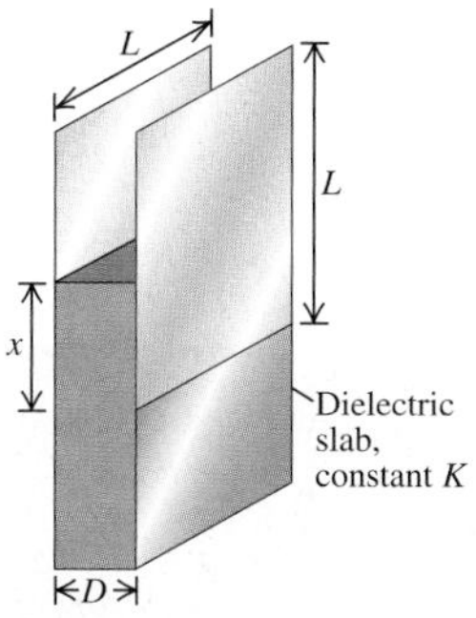

Figure 33 Challenge Problem 24.77.

$$dU = +\frac{(K-1)\epsilon_0 V^2 L}{2D}dx$$

c) Suppose that before the slab is moved by dx, the plates are disconnected from the battery, so that the charges on the plates remain constant. Determine the magnitude of the charge on each plate, and then show that when the slab is moved dx farther into the space between the plates, the stored energy changes by an amount that is the *negative* of the expression for dU given in part (b). d) If F is the force exerted on the slab by the charges on the plates, then dU should equal the work done *against* this force to move the slab a distance dx. Thus $dU = -F\,dx$. Show that applying this expression to the result of part (b) suggests that the electric force on the slab pushes it *out* of the capacitor, while the result of part (c) suggests that the force pulls the slab *into* the capacitor. e) Figure 24.14 shows that the force in fact pulls the slab into the capacitor. Explain why the result of part (b) gives an incorrect answer for the direction of this force, and calculate the magnitude of the force. (This method does not require knowledge of the nature of the fringing field.)

24.78 An isolated spherical capacitor has charge $+Q$ on its inner conductor (radius r_a) and charge $-Q$ on its outer conductor (radius r_b). Half of the volume between the two conductors is then filled with a liquid dielectric of constant K, as shown in cross section in Fig. 24.34. a) Find the capacitance of the half-filled capacitor. b) Find the magnitude of $\vec{E}$ in the volume between the two conductors as a function of the distance r from the center of the capacitor. Give answers for both the upper and lower halves of this volume. c) Find the surface density of free charge on the upper and lower halves of the inner and outer conductors. d) Find the surface density of bound charge on the inner $(r = r_a)$ and outer $(r = r_b)$ surfaces of the dielectric. e) What is the surface density of bound charge on the flat surface of the dielectric? Explain.

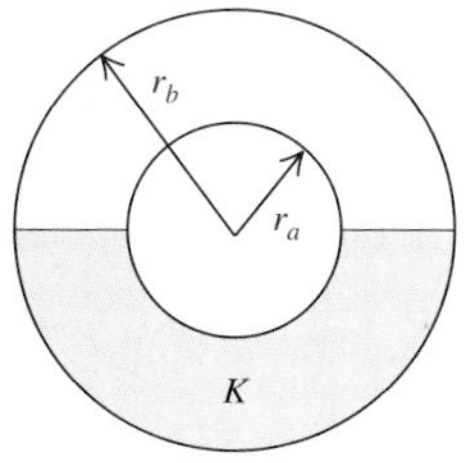

Figure 34 Challenge Problem 24.78.

24.79 Three square metal plates A, B, and C, each 12.0 cm on a side and 1.50 mm thick, are arranged as in Fig. 24.35. The plates are separated by sheets of paper 0.45 mm thick and with dielectric constant 4.2. The outer plates are connected together and connected to point b. The inner plate is connected to point a. a) Copy the diagram and show by plus and minus signs the charge distribution on the plates when point a is maintained at a positive potential relative to point b. b) What is the capacitance between points a and b?

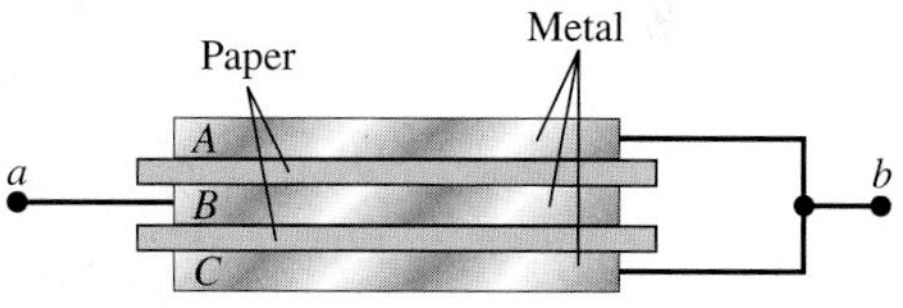

Figure 35 Challenge Problem 24.79.

24.80 A fuel gauge uses a capacitor to determine the height of the fuel in a tank. The effective dielectric constant K_{eff} changes from a value of 1 when the tank is empty to a value of K, the dielectric constant of the fuel, when the tank is full. The appropriate electronic circuitry can determine the effective dielectric constant of the combined air and fuel between the capacitor plates. Each of the two rectangular plates has a width w and a length L (Fig. 24.36). The height of the fuel between the plates is h. You can ignore any fringing effects. a) Derive an expression for K_{eff} as a function of h. b) What is the effective dielectric constant for a tank $\frac{1}{4}$ full, $\frac{1}{2}$ full, and $\frac{3}{4}$ full if the fuel is gasoline $(K = 1.95)$? c) Repeat part (b) for methanol $(K = 33.0)$. d) For which fuel is this fuel gauge more practical?

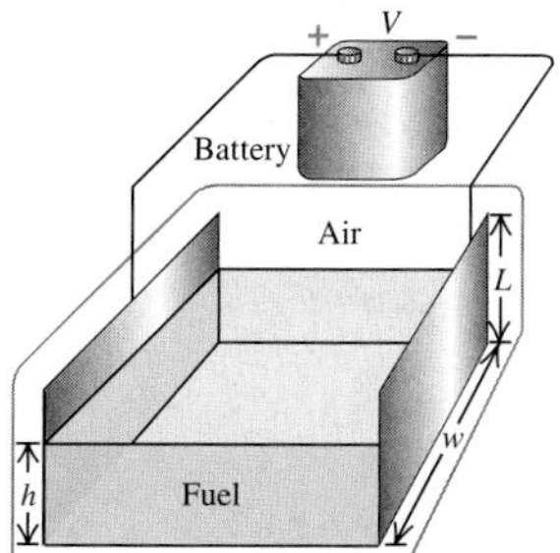

Figure 36 Challenge Problem 24.80.

25 CURRENT, RESISTANCE, AND ELECTROMOTIVE FORCE

A flashlight is a simple example of an electric circuit. The flashlight batteries provide electric potential energy to a current of electrons, which flow through a wire to the flashlight bulb where the potential energy is converted into light and heat. The electrons then return to the batteries to repeat the cycle anew.

? **Is the amount of current that flows out of the flashlight bulb less than, greater than, or equal to the amount of current that flows into the bulb?**

In the past four chapters we studied the interactions of electric charges *at rest;* now we're ready to study charges *in motion.* An *electric current* consists of charges in motion from one region to another. When this motion takes place within a conducting path that forms a closed loop, the path is called an *electric circuit.*

Fundamentally, electric circuits are a means for conveying *energy* from one place to another. As charged particles move within a circuit, electric potential energy is transferred from a source (such as a battery or generator) to a device in which that energy is either stored or converted to another form: into sound in a stereo system or into heat and light in a toaster or light bulb. From a technological standpoint, electric circuits are useful because they allow energy to be transported without any moving parts (other than the moving charged particles themselves). Electric circuits are at the heart of flashlights, CD players, computers, radio and television transmitters and receivers, and household and industrial power distribution systems. The nervous systems of animals and humans are specialized electric circuits that carry vital signals from one part of the body to another.

In Chapter 26 we will see how to analyze electric circuits and will examine some practical applications of circuits. Before we can do so, however, we must understand the basic properties of electric currents. These properties are the subject of this chapter. We'll begin by describing the nature of electric conductors and considering how they are affected by temperature. We'll learn why a short, fat, cold copper wire is a better conductor than a long, skinny, hot steel wire. We'll study the properties of batteries and see how they cause current and energy transfer in a circuit. In this analysis we will use the concepts of current, potential difference (or voltage), resistance, and electromotive force. Finally, we'll look at electric current in a material from a microscopic viewpoint.

25.1 | Current

A **current** is any motion of charge from one region to another. In this section we'll discuss currents in conducting materials. The vast majority of technological applications of charges in motion involve currents of this kind.

In electrostatic situations (discussed in Chapters 21 through 24) the electric field is zero everywhere within the conductor, and there is *no* current. However, this does not mean that all charges within the conductor are at rest. In an ordinary metal such as copper or aluminum, some of the electrons are free to move within the conducting material. These free electrons move randomly in all directions, somewhat like the molecules of a gas but with much greater speeds, of the order of 10^6 m/s. The electrons nonetheless do not escape from the conducting material, because they are attracted to the positive ions of the material. The motion of the electrons is random, so there is no *net* flow of charge in any direction and hence no current.

Now consider what happens if a constant, steady electric field $\vec{E}$ is established inside a conductor. (We'll see later how this can be done.) A charged particle (such as a free electron) inside the conducting material is then subjected to a steady force $\vec{F} = q\vec{E}$. If the charged particle were moving in *vacuum,* this steady force would cause a steady acceleration in the direction of $\vec{F}$, and after a time the charged particle would be moving in that direction at high speed. But a charged particle moving in a *conductor* undergoes frequent collisions with the massive, nearly stationary ions of the material. In each such collision the particle's direction of motion undergoes a random change. The net effect of the electric field $\vec{E}$ is that in addition to the random motion of the charged particles within the conductor, there is also a very slow net motion or *drift* of the moving charged particles as a group in the direction of the electric force $\vec{F} = q\vec{E}$ (Fig. 25.1). This motion is described in terms of the **drift velocity** $\vec{v}_d$ of the particles. As a result, there is a net current in the conductor.

While the random motion of the electrons has a very large average speed of about 10^6 m/s, the drift speed is very slow, often on the order of 10^{-4} m/s. Given that the electrons move so slowly, you may wonder why the light comes on immediately when you turn on the switch of a flashlight. The reason is that the electric field is set up in the wire with a speed approaching the speed of light, and electrons start to move all along the wire at very nearly the same time. The time that it takes any individual electron to get from the switch to the light bulb isn't really relevant. A good analogy is a group of soldiers standing at attention when the sergeant orders them to start marching; the order reaches the soldiers' ears at the speed of sound, which is much faster than their marching speed, so all the soldiers start to march essentially in unison.

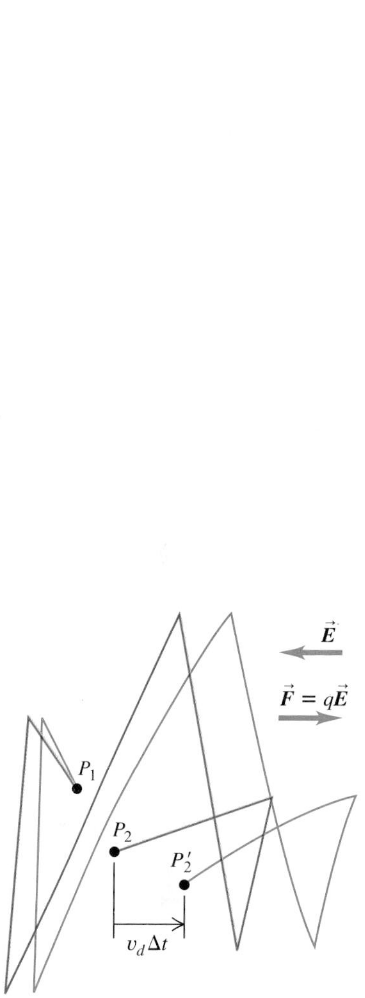

—— Typical electron trajectory in conductor *without* electric field:
- No net electric force on electrons
- Electrons move randomly within conductor
- No net current

—— Typical electron trajectory in conductor *with* electric field:
- Electric force $\vec{F} = q\vec{E}$ imposes a small drift on electron's random motion
- There is a net current

25.1 If there is no electric field inside a conductor, an electron moves randomly from point P_1 to point P_2 in a time Δt. If an electric field $\vec{E}$ is present, the electric force $\vec{F} = q\vec{E}$ imposes a small drift (greatly exaggerated here) that takes the electron to point P_2', a distance $v_d \Delta t$ from P_2 in the direction of the force. The electron has a negative charge q, so the force is in the direction opposite that of $\vec{E}$.

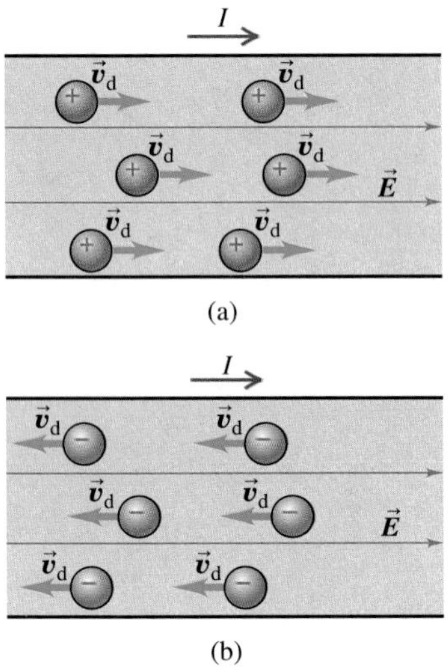

25.2 The same current can be produced by (a) positive charges moving in the direction of the electric field $\vec{E}$ or (b) the same number of negative charges moving at the same speed in the direction opposite to $\vec{E}$.

The drift of moving charges through a conductor can be interpreted in terms of work and energy. The electric field $\vec{E}$ does work on the moving charges. The resulting kinetic energy is transferred to the material of the conductor by means of collisions with the ions, which vibrate about their equilibrium positions in the crystalline structure of the conductor. This energy transfer increases the average vibrational energy of the ions and therefore the temperature of the material. Thus much of the work done by the electric field goes into heating the conductor, *not* into making the moving charges move ever faster and faster. This heating is sometimes useful, as in an electric toaster, but in many situations is simply an unavoidable by-product of current flow.

In different current-carrying materials, the charges of the moving particles may be positive or negative. In metals the moving charges are always (negative) electrons, while in an ionized gas (plasma) or an ionic solution the moving charges may include both electrons and positively charged ions. In a semiconductor material such as germanium or silicon, conduction is partly by electrons and partly by motion of *vacancies,* also known as *holes;* these are sites of missing electrons and act like positive charges.

Figure 25.2 shows segments of two different current-carrying materials. In Fig. 25.2a the moving charges are positive, the electric force is in the same direction as $\vec{E}$, and the drift velocity $\vec{v}_d$ is from left to right. In Fig. 25.2b the charges are negative, the electric force is opposite to $\vec{E}$, and the drift velocity $\vec{v}_d$ is from right to left. In either case there is a net flow of positive charge from left to right, and positive charges end up to the right of negative ones. We *define* the current, denoted by I, to be in the direction in which there is a flow of *positive* charge. Thus we describe currents as though they consisted entirely of positive charge flow, even in cases in which we know that the actual current is due to electrons. Hence the current is to the right in both Fig. 25.2a and Fig. 25.2b. This choice or convention for the direction of current flow is called **conventional current.** While the direction of the conventional current is *not* necessarily the same as the direction in which charged particles are actually moving, we'll find that the sign of the moving charges is of little importance in analyzing electric circuits.

Figure 25.3 shows a segment of a conductor in which a current is flowing. We consider the moving charges to be *positive,* so they are moving in the same direction as the current. We define the current through the cross-sectional area A to be

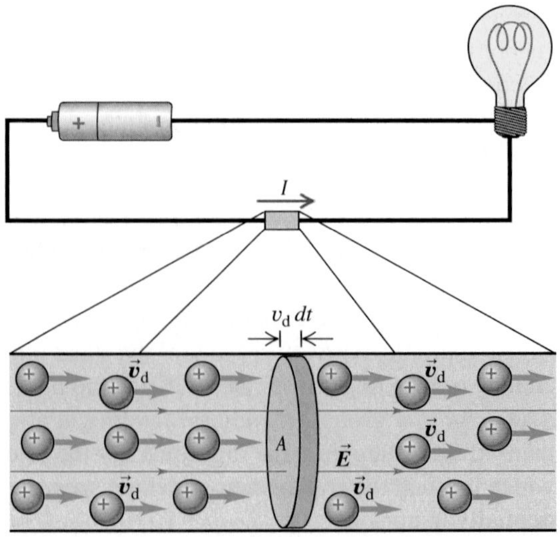

25.3 The current I through the cross-sectional area A is the time rate of charge transfer through A. The random component of each moving charged particle's motion averages to zero, and the current is in the same direction as $\vec{E}$. If the moving charges are positive as shown here, the drift velocity $\vec{v}_d$ is in the same direction as the current and $\vec{E}$; if the moving charges are negative, the drift velocity is in the opposite direction.

the net charge flowing through the area per unit time. Thus, if a net charge dQ flows through an area in a time dt, the current I through the area is

$$I = \frac{dQ}{dt} \qquad \text{(definition of current)} \qquad (25.1)$$

CAUTION Although we refer to the *direction* of a current, current as defined by Eq. (25.1) is *not* a vector quantity. In a current-carrying wire, the current is always along the length of the wire, regardless of whether the wire is straight or curved. No single vector could describe motion along a curved path, which is why current is not a vector. We'll usually describe the direction of current either in words (as in "the current flows clockwise around the circuit") or by choosing a current to be positive if it flows in one direction along a conductor and negative if it flows in the other direction.

The SI unit of current is the **ampere;** one ampere is defined to be *one coulomb per second* $(1\ \text{A} = 1\ \text{C/s})$. This unit is named in honor of the French scientist André Marie Ampère (1775–1836). When an ordinary flashlight (D-cell size) is turned on, the current in the flashlight is about 0.5 to 1 A; the current in the wires of a starter motor used to start a car engine is around 200 A. Currents in radio and television circuits are usually expressed in *milliamperes* $(1\ \text{mA} = 10^{-3}\ \text{A})$ or *microamperes* $(1\ \mu\text{A} = 10^{-6}\ \text{A})$, and currents in computer circuits are expressed in *nanoamperes* $(1\ \text{nA} = 10^{-9}\ \text{A})$ or *picoamperes* $(1\ \text{pA} = 10^{-12}\ \text{A})$.

Current, Drift Velocity, and Current Density

We can express current in terms of the drift velocity of the moving charges. Let's consider again the situation of Fig. 25.3, a conductor with cross-sectional area A and an electric field $\vec{E}$ directed from left to right. To begin with, we'll assume that the free charges in the conductor are positive; then the drift velocity is in the same direction as the field.

Suppose there are n charged particles per unit volume. We call n the **concentration** of particles; its SI unit is m^{-3}. Assume that all the particles move with the same drift velocity with magnitude v_d. In a time interval dt, each particle moves a distance $v_d\,dt$. The particles that flow out of the right end of the shaded cylinder with length $v_d\,dt$ during dt are the particles that were within this cylinder at the beginning of the interval dt. The volume of the cylinder is $Av_d\,dt$, and the number of particles within it is $nAv_d\,dt$. If each particle has a charge q, the charge dQ that flows out of the end of the cylinder during time dt is

$$dQ = q(nAv_d\,dt) = nqv_dA\,dt$$

and the current is

$$I = \frac{dQ}{dt} = nqv_dA$$

The current *per unit cross-sectional area* is called the **current density** J:

$$J = \frac{I}{A} = nqv_d$$

The units of current density are amperes per square meter (A/m^2).

If the moving charges are negative rather than positive, as in Fig. 25.2b, the drift velocity is opposite to $\vec{E}$. But the *current* is still in the same direction as $\vec{E}$ at each point in the conductor. Hence the current I and current density J don't depend on the sign of the charge, and so in the above expressions for I and J we replace the charge q by its absolute value $|q|$:

$$I = \frac{dQ}{dt} = n|q|v_d A \qquad \text{(general expression for current)} \qquad (25.2)$$

$$J = \frac{I}{A} = n|q|v_d \qquad \text{(general expression for current density)} \qquad (25.3)$$

The current in a conductor is the product of the concentration of moving charged particles, the magnitude of charge of each such particle, the magnitude of the drift velocity, and the cross-sectional area of the conductor.

We can also define a *vector* current density $\vec{J}$ that includes the direction of the drift velocity:

$$\vec{J} = nq\vec{v}_d \qquad \text{(vector current density)} \qquad (25.4)$$

There are *no* absolute value signs in Eq. (25.4). If q is positive, $\vec{v}_d$ is in the same direction as $\vec{E}$; if q is negative, $\vec{v}_d$ is opposite to $\vec{E}$. In either case, $\vec{J}$ is in the same direction as $\vec{E}$. Equation (25.3) gives the *magnitude* J of the vector current density $\vec{J}$.

CAUTION Note that current density $\vec{J}$ is a vector, but current I is not. The difference is that the current density $\vec{J}$ describes how charges flow at a certain point, and the vector's direction tells you about the direction of the flow at that point. By contrast, the current I describes how charges flow through an extended object such as a wire. For example, I has the same value at all points in the circuit of Fig. 25.3, but $\vec{J}$ does not: the current density is directed downward in the left-hand side of the loop and upward in the right-hand side. The magnitude of $\vec{J}$ can also vary around a circuit. In Fig. 25.3 the current density magnitude $J = I/A$ is less in the battery (which has a large cross-sectional area A) than in the wires (which have a small cross-sectional area).

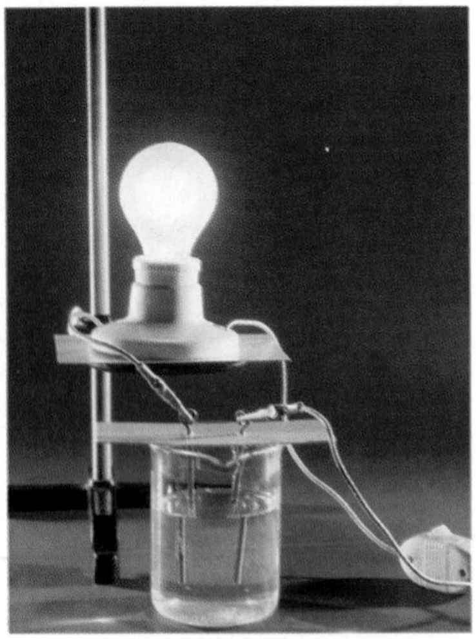

25.4 Part of the electric circuit that includes this light bulb passes through a beaker with a solution of sodium chloride. The current in the solution is carried by both positive charges (Na^+ ions) and negative charges (Cl^- ions).

In general, a conductor may contain several different kinds of moving charged particles having charges $q_1, q_2, \ldots$, concentrations $n_1, n_2, \ldots$, and drift velocities with magnitudes $v_{d1}, v_{d2}, \ldots$. An example is current flow in an ionic solution (Fig. 25.4). In a sodium chloride solution, current can be carried by both positive sodium ions and negative chlorine ions; the total current I is found by adding up the currents due to each kind of charged particle, using Eq. (25.2). Likewise, the total vector current density $\vec{J}$ is found by using Eq. (25.4) for each kind of charged particle and adding up the results.

We will see in Section 25.4 that it is possible to have a current that is *steady* (that is, one that is constant in time) only if the conducting material forms a closed loop, called a *complete circuit*. In such a steady situation, the total charge in every segment of the conductor is constant. Hence the rate of flow of charge *out* at one end of a segment at any instant equals the rate of flow of charge *in* at the other end of the segment, and *the current is the same at all cross sections of the circuit.*

We'll make use of this observation when we analyze electric circuits later in this chapter.

In many simple circuits, such as flashlights or cordless electric drills, the direction of the current is always the same; this is called *direct current.* But home appliances such as toasters, refrigerators, and televisions use *alternating current,* in which the current continuously changes direction. In this chapter we'll consider direct current only. Alternating current has many special features worthy of detailed study, which we'll examine in Chapter 31.

Example 25.1 Current density and drift velocity in a wire

An 18-gauge copper wire (the size usually used for lamp cords) has a nominal diameter of 1.02 mm. This wire carries a constant current of 1.67 A to a 200-watt lamp. The density of free electrons is 8.5×10^{28} electrons per cubic meter. Find the magnitudes of a) the current density and b) the drift velocity.

SOLUTION

IDENTIFY and SET UP: We are given the current and the dimensions of the wire, so we use Eq. (25.3) to find the magnitude J of the current density. We then use Eq. (25.3) again to find the drift speed v_d from J and the concentration of electrons.

EXECUTE: a) The cross-sectional area is

$$A = \frac{\pi d^2}{4} = \frac{\pi(1.02 \times 10^{-3}\ \text{m})^2}{4} = 8.17 \times 10^{-7}\ \text{m}^2$$

The magnitude of the current density is

$$J = \frac{I}{A} = \frac{1.67\ \text{A}}{8.17 \times 10^{-7}\ \text{m}^2} = 2.04 \times 10^6\ \text{A/m}^2$$

b) Solving Eq. (25.3) for the drift velocity magnitude v_d, we find

$$v_d = \frac{J}{n|q|} = \frac{2.04 \times 10^6\ \text{A/m}^2}{(8.5 \times 10^{28}\ \text{m}^{-3})|-1.60 \times 10^{-19}\ \text{C}|}$$
$$= 1.5 \times 10^{-4}\ \text{m/s} = 0.15\ \text{mm/s}$$

EVALUATE: At this speed an electron would require 6700 s, or about 1 hr 50 min, to travel the length of a wire 1 m long. The speeds of random motion of the electrons are of the order of 10^6 m/s. So in this example the drift speed is around 10^{10} times slower than the speed of random motion. Picture the electrons as bouncing around frantically, with a very slow and sluggish drift!

Test Your Understanding

Suppose we replaced the wire in Example 25.1 with 12-gauge copper wire, which has twice the diameter of 18-gauge wire. If the current remains the same, what is the new magnitude of the drift velocity?

25.2 | Resistivity

The current density $\vec{J}$ in a conductor depends on the electric field $\vec{E}$ and on the properties of the material. In general, this dependence can be quite complex. But for some materials, especially metals, at a given temperature, $\vec{J}$ is nearly *directly proportional* to $\vec{E}$, and the ratio of the magnitudes E and J is constant. This relationship, called **Ohm's law,** was discovered in 1826 by the German physicist Georg Simon Ohm (1787–1854). The word "law" should actually be in quotes, since Ohm's law, like the ideal gas equation and Hooke's law, is an *idealized model* that describes the behavior of some materials quite well but is not a general description of *all* matter. In the following discussion we'll assume that Ohm's law is valid, even though there are many situations in which it is not. The situation is comparable to our representation of the behavior of the static and kinetic friction

Table 25.1 Resistivities at Room Temperature (20°C)

	Substance	$\rho\,(\Omega \cdot m)$	Substance	$\rho\,(\Omega \cdot m)$
Conductors			*Semiconductors*	
Metals:	Silver	1.47×10^{-8}	Pure carbon (graphite)	3.5×10^{-5}
	Copper	1.72×10^{-8}	Pure germanium	0.60
	Gold	2.44×10^{-8}	Pure silicon	2300
	Aluminum	2.75×10^{-8}	*Insulators*	
	Tungsten	5.25×10^{-8}	Amber	5×10^{14}
	Steel	20×10^{-8}	Glass	10^{10}–10^{14}
	Lead	22×10^{-8}	Lucite	$>10^{13}$
	Mercury	95×10^{-8}	Mica	10^{11}–10^{15}
Alloys:	Manganin (Cu 84%, Mn 12%, Ni 4%)	44×10^{-8}	Quartz (fused)	75×10^{16}
	Constantan (Cu 60%, Ni 40%)	49×10^{-8}	Sulfur	10^{15}
	Nichrome	100×10^{-8}	Teflon	$>10^{13}$
			Wood	10^{8}–10^{11}

forces; we treated these friction forces as being directly proportional to the normal force, even though we knew that this was at best an approximate description.

We define the **resistivity** ρ of a material as the ratio of the magnitudes of electric field and current density:

$$\rho = \frac{E}{J} \qquad \text{(definition of resistivity)} \qquad (25.5)$$

The greater the resistivity, the greater the field needed to cause a given current density, or the smaller the current density caused by a given field. From Eq. (25.5) the units of ρ are $(\mathrm{V/m})/(\mathrm{A/m^2}) = \mathrm{V \cdot m/A}$. As we will discuss in the next section, 1 V/A is called one *ohm* (1 Ω; we use the Greek letter Ω, or "omega," which is alliterative with "ohm"). So the SI units for ρ are $\Omega \cdot \mathrm{m}$ (ohm-meters). Table 25.1 lists some representative values of resistivity. A perfect conductor would have zero resistivity, and a perfect insulator would have an infinite resistivity. Metals and alloys have the smallest resistivities and are the best conductors. The resistivities of insulators are greater than those of the metals by an enormous factor, on the order of 10^{22}.

25.5 The copper "wires," or traces, on this circuit board are printed directly onto the surface of the dark-colored insulating board. Even though the traces are very close to each other (only about a millimeter apart), the board has such a high resistivity (and low conductivity) compared to the copper that no current can flow between the traces.

The reciprocal of resistivity is **conductivity.** Its units are $(\Omega \cdot m)^{-1}$. Good conductors of electricity have larger conductivity than insulators. Conductivity is the direct electrical analog of thermal conductivity. Comparing Table 25.1 with Table 17.5 (Thermal Conductivities), we note that good electrical conductors, such as metals, are usually also good conductors of heat. Poor electrical conductors, such as ceramic and plastic materials, are also poor thermal conductors. In a metal the free electrons that carry charge in electrical conduction also provide the principal mechanism for heat conduction, so we should expect a correlation between electrical and thermal conductivity. Because of the enormous difference in conductivity between electrical conductors and insulators, it is easy to confine electric currents to well-defined paths or circuits (Fig. 25.5). The variation in *thermal* conductivity is much less, only a factor of 10^3 or so, and it is usually impossible to confine heat currents to that extent.

Semiconductors have resistivities intermediate between those of metals and those of insulators. These materials are important because of the way their resistivities are affected by temperature and by small amounts of impurities.

A material that obeys Ohm's law reasonably well is called an *ohmic* conductor or a *linear* conductor. For such materials, at a given temperature, ρ is a *constant* that does not depend on the value of E. Many materials show substantial departures from Ohm's-law behavior; they are *nonohmic,* or *nonlinear.* In these materials, J depends on E in a more complicated manner.

Analogies with fluid flow can be a big help in developing intuition about electric current and circuits. For example, in making wine or maple syrup, the product is sometimes filtered to remove sediments. A pump forces the fluid through the filter under pressure; if the flow rate (analogous to J) is proportional to the pressure difference between the upstream and downstream sides (analogous to E), the behavior is analogous to Ohm's law.

Resistivity and Temperature

The resistivity of a *metallic* conductor nearly always increases with increasing temperature, as shown in Fig. 25.6a. As temperature increases, the ions of the conductor vibrate with greater amplitude, making it more likely that a moving electron will collide with an ion as in Fig. 25.1; this impedes the drift of electrons through the conductor and hence reduces the current. Over a small temperature range (up to 100°C or so), the resistivity of a metal can be represented approximately by the equation

$$\rho(T) = \rho_0[1 + \alpha(T - T_0)] \qquad (25.6)$$

(temperature dependence of resistivity)

where ρ_0 is the resistivity at a reference temperature T_0 (often taken as 0°C or 20°C) and $\rho(T)$ is the resistivity at temperature T, which may be higher or lower than T_0. The factor α is called the **temperature coefficient of resistivity.** Some representative values are given in Table 25.2. The resistivity of the alloy manganin is practically independent of temperature.

Table 25.2 Temperature Coefficients of Resistivity (Approximate Values Near Room Temperature)

Material	$\alpha\,[(°C)^{-1}]$	Material	$\alpha\,[(°C)^{-1}]$
Aluminum	0.0039	Lead	0.0043
Brass	0.0020	Manganin	0.00000
Carbon (graphite)	−0.0005	Mercury	0.00088
Constantan	0.00001	Nichrome	0.0004
Copper	0.00393	Silver	0.0038
Iron	0.0050	Tungsten	0.0045

The resistivity of graphite (a nonmetal) *decreases* with increasing temperature, since at higher temperatures, more electrons are "shaken loose" from the atoms and become mobile; hence the temperature coefficient of resistivity of graphite is negative. This same behavior occurs for semiconductors (Fig. 25.6b). Measuring the resistivity of a small semiconductor crystal is therefore a sensitive measure of temperature; this is the principle of a type of thermometer called a *thermistor.*

Some materials, including several metallic alloys and oxides, show a phenomenon called *superconductivity.* As the temperature decreases, the resistivity at first decreases smoothly, like that of any metal. But then at a certain critical temperature T_c a phase transition occurs and the resistivity suddenly drops to zero, as shown in Fig. 25.6c. Once a current has been established in a superconducting ring, it continues indefinitely without the presence of any driving field.

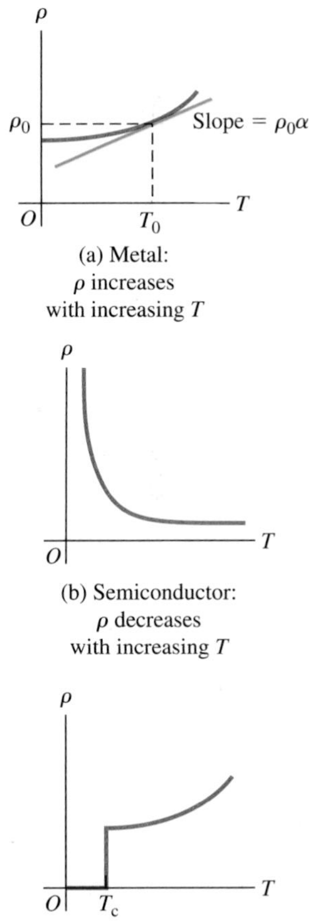

(a) Metal: ρ increases with increasing T

(b) Semiconductor: ρ decreases with increasing T

(c) Superconductor: $\rho = 0$ for $T < T_c$

25.6 Variation of resistivity ρ with absolute temperature T for (a) a normal metal, (b) a semiconductor, and (c) a superconductor. In (a) the linear approximation to ρ as a function of T is shown as a green line; the approximation agrees exactly at $T = T_0$, where $\rho = \rho_0$.

Superconductivity was discovered in 1911 by the Dutch physicist Heike Kamerlingh Onnes (1853–1926). He discovered that at very low temperatures, below 4.2 K, the resistivity of mercury suddenly dropped to zero. For the next 75 years, the highest T_c attained was about 20 K. This meant that superconductivity occurred only when the material was cooled using expensive liquid helium, with a boiling point temperature of 4.2 K, or explosive liquid hydrogen, with a boiling point of 20.3 K. But in 1986 Karl Muller and Johannes Bednorz discovered an oxide of barium, lanthanum, and copper with a T_c of nearly 40 K, and the race was on to develop "high-temperature" superconducting materials.

By 1987 a complex oxide of yttrium, copper, and barium had been found that has a value of T_c well above the 77-K boiling temperature of liquid nitrogen, a refrigerant that is both inexpensive and safe. The current (2003) record for T_c is about 160 K, and materials that are superconductors at room temperature may well become a reality. The implications of these discoveries for power-distribution systems, computer design, and transportation are enormous. Meanwhile, superconducting electromagnets cooled by liquid helium are used in particle accelerators and some experimental magnetic-levitation railroads. Superconductors have other exotic properties that require an understanding of magnetism to explore; we will discuss these further in Chapter 29.

Test Your Understanding

You maintain a constant electric field inside a piece of semiconductor while lowering the semiconductor's temperature. What happens to the current density in the semiconductor?

25.3 | Resistance

For a conductor with resistivity ρ, the current density $\vec{J}$ at a point where the electric field is $\vec{E}$ is given by Eq. (25.5), which we can write as

$$\vec{E} = \rho\vec{J} \tag{25.7}$$

When Ohm's law is obeyed, ρ is constant and independent of the magnitude of the electric field, so $\vec{E}$ is directly proportional to $\vec{J}$. Often, however, we are more interested in the total current in a conductor than in $\vec{J}$ and more interested in the potential difference between the ends of the conductor than in $\vec{E}$. This is so largely because current and potential difference are much easier to measure than are $\vec{J}$ and $\vec{E}$.

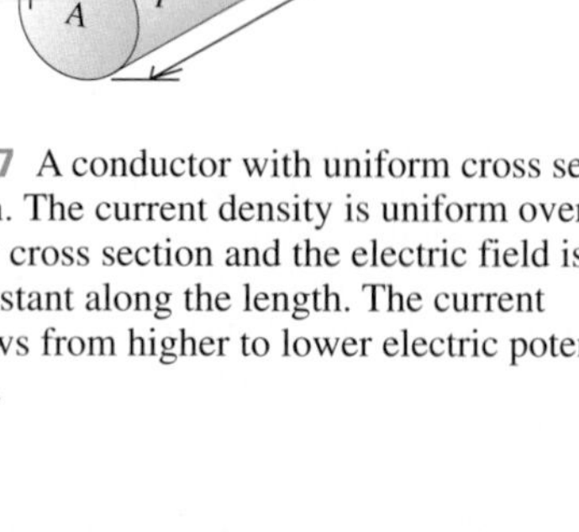

25.7 A conductor with uniform cross section. The current density is uniform over any cross section and the electric field is constant along the length. The current flows from higher to lower electric potential.

Suppose our conductor is a wire with uniform cross-sectional area A and length L, as shown in Fig. 25.7. Let V be the potential difference between the higher-potential and lower-potential ends of the conductor, so that V is positive. The *direction* of the current is always from the higher-potential end to the lower-potential end. That's because current in a conductor flows in the direction of $\vec{E}$, no matter what the sign of the moving charges (Fig. 25.2), and because $\vec{E}$ points in the direction of *decreasing* electric potential (see Section 23.2). As the current flows through the potential difference, electric potential energy is lost; this energy is transferred to the ions of the conducting material during collisions.

We can also relate the *value* of the current I to the potential difference between the ends of the conductor. If the magnitudes of the current density $\vec{J}$ and the electric field $\vec{E}$ are uniform throughout the conductor, the total current I is given by $I = JA$, and the potential difference V between the ends is $V = EL$.

When we solve these equations for J and E, respectively, and substitute the results in Eq. (25.7), we obtain

$$\frac{V}{L} = \frac{\rho I}{A} \quad \text{or} \quad V = \frac{\rho L}{A} I \tag{25.8}$$

This shows that when ρ is constant, the total current I is proportional to the potential difference V.

The ratio of V to I for a particular conductor is called its **resistance** R:

$$R = \frac{V}{I} \tag{25.9}$$

Comparing this definition of R to Eq. (25.8), we see that the resistance R of a particular conductor is related to the resistivity ρ of its material by

$$R = \frac{\rho L}{A} \quad \text{(relationship between resistance and resistivity)} \tag{25.10}$$

If ρ is constant, as is the case for ohmic materials, then so is R.

The equation

$$V = IR \quad \text{(relationship between voltage, current, and resistance)} \tag{25.11}$$

is often called Ohm's law, but it is important to understand that the real content of Ohm's law is the direct proportionality (for some materials) of V to I or of J to E. Equation (25.9) or (25.11) *defines* resistance R for *any* conductor, whether or not it obeys Ohm's law, but only when R is constant can we correctly call this relationship Ohm's law.

Equation (25.10) shows that the resistance of a wire or other conductor of uniform cross section is directly proportional to its length and inversely proportional to its cross-sectional area. It is also proportional to the resistivity of the material of which the conductor is made.

The flowing-fluid analogy is again useful. In analogy to Eq. (25.10), a thin water hose offers more resistance to flow than a fat one, and a long hose has more resistance than a short one. We can increase the resistance to flow by stuffing the hose with cotton or sand; this corresponds to increasing the resistivity. The flow rate is approximately proportional to the pressure difference between the ends. Flow rate is analogous to current, and pressure difference is analogous to potential difference ("voltage"). Let's not stretch this analogy too far, though; the water flow rate in a pipe is usually *not* proportional to its cross-sectional area (see Section 14.6).

The SI unit of resistance is the **ohm,** equal to one volt per ampere $(1\ \Omega = 1\ \text{V/A})$. The *kilohm* $(1\ \text{k}\Omega = 10^3\ \Omega)$ and the *megohm* $(1\ \text{M}\Omega = 10^6\ \Omega)$ are also in common use. A 100-m length of 12-gauge copper wire, the size usually used in household wiring, has a resistance at room temperature of about $0.5\ \Omega$. A 100-W, 120-V light bulb has a resistance (at operating temperature) of $140\ \Omega$. If the same current I flows in both the copper wire and the light bulb, the potential difference $V = IR$ is much greater across the light bulb, and much more potential energy is lost per charge in the light bulb. This lost energy is converted by the light bulb filament into light and heat. You don't want your household wiring to glow white-hot, so its resistance is kept low by using wire of low resistivity and large cross-sectional area.

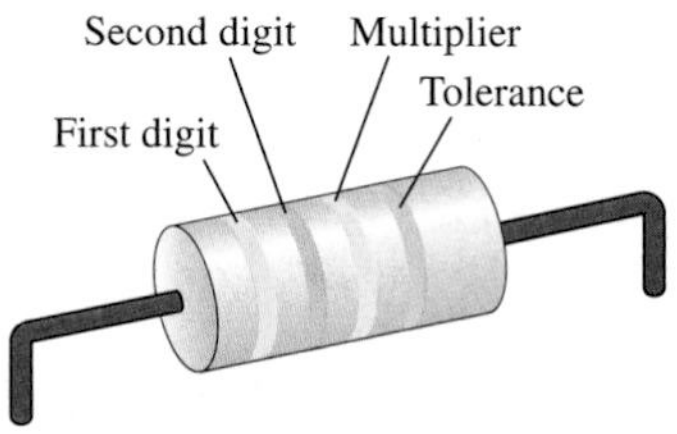

25.8 This resistor has a resistance of 47 kΩ with a precision (tolerance) of ±10%.

Because the resistivity of a material varies with temperature, the resistance of a specific conductor also varies with temperature. For temperature ranges that are not too great, this variation is approximately a linear relation, analogous to Eq. (25.6):

$$R(T) = R_0[1 + \alpha(T - T_0)] \tag{25.12}$$

In this equation, $R(T)$ is the resistance at temperature T and R_0 is the resistance at temperature T_0, often taken to be 0°C or 20°C. The *temperature coefficient of resistance* α is the same constant that appears in Eq. (25.6) if the dimensions L and A in Eq. (25.10) do not change appreciably with temperature; this is indeed the case for most conducting materials (see Problem 25.63). Within the limits of validity of Eq. (25.12), the *change* in resistance resulting from a temperature change $T - T_0$ is given by $R_0\alpha(T - T_0)$.

A circuit device made to have a specific value of resistance between its ends is called a **resistor.** Resistors in the range 0.01 to $10^7\ \Omega$ can be bought off the shelf. Individual resistors used in electronic circuitry are often cylindrical in shape, a few millimeters in diameter and length, with wires coming out of the ends. The resistance may be marked with a standard code using three or four color bands near one end (Fig. 25.8), according to the scheme shown in Table 25.3. The first two bands (starting with the band nearest an end) are digits, and the third is a power-of-ten multiplier, as shown in Fig. 25.8. For example, yellow-violet-orange means $47 \times 10^3\ \Omega$, or 47 kΩ. The fourth band, if present, indicates the precision of the value; no band means ±20%, a silver band ±10%, and a gold band ±5%. Another important characteristic of a resistor is the maximum *power* it can dissipate without damage. We'll return to this point in Section 25.5.

Table 25.3 Color Codes for Resistors

Color	Value as Digit	Value as Multiplier
Black	0	1
Brown	1	10
Red	2	10^2
Orange	3	10^3
Yellow	4	10^4
Green	5	10^5
Blue	6	10^6
Violet	7	10^7
Gray	8	10^8
White	9	10^9

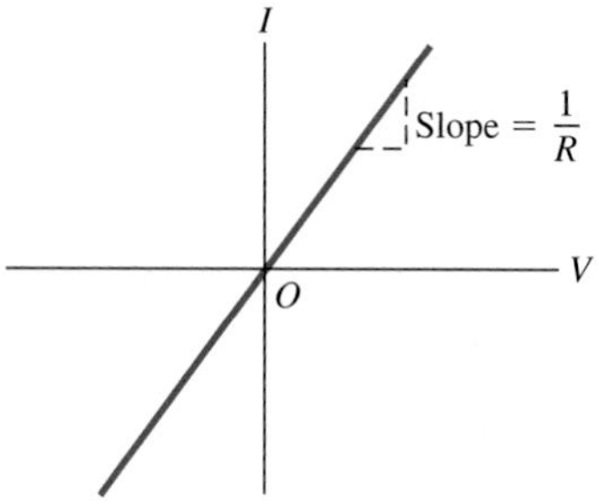

(a) Resistor that obeys Ohm's law

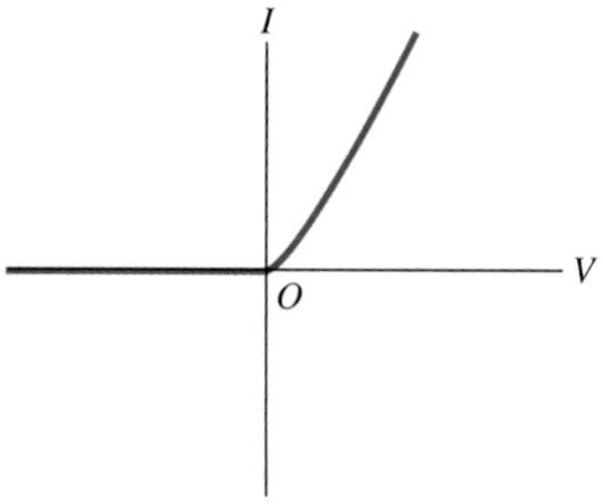

(b) Vacuum diode (does not obey Ohm's law)

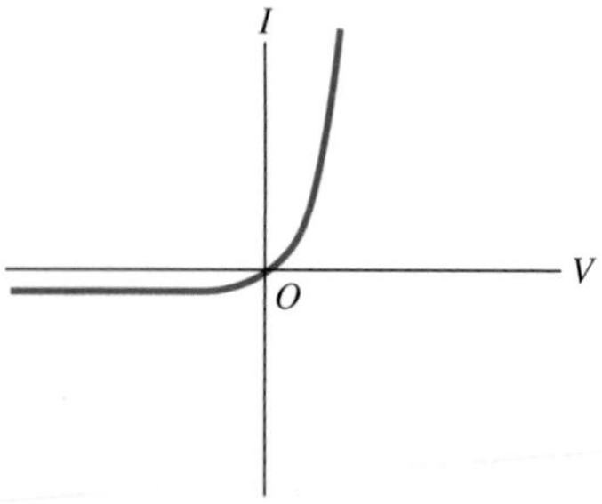

(c) Semiconductor diode (does not obey Ohm's law)

25.9 Current–voltage relations for three devices. Only for a resistor that obeys Ohm's law as in (a) is current I proportional to voltage V.

For a resistor that obeys Ohm's law, a graph of current as a function of potential difference (voltage) is a straight line (Fig. 25.9a). The slope of the line is $1/R$. If the sign of the potential difference changes, so does the sign of the current produced; in Fig. 25.7 this corresponds to interchanging the higher- and lower-potential ends of the conductor, so the electric field, current density, and current all reverse direction. In devices that do not obey Ohm's law, the relation of voltage to current may not be a direct proportion, and it may be different for the two directions of current. Figure 25.9b shows the behavior of a vacuum *diode,* a vacuum tube used to convert high-voltage alternating current to direct current. For positive potentials of the anode (one of two conductors inside the tube) with respect to the cathode (the other conductor), I is approximately proportional to $V^{3/2}$; for negative potentials the current is extremely small. The behavior of semiconductor diodes (Fig. 25.9c) is somewhat different but still strongly asymmetric. For a diode of either type, a positive potential difference V causes a current to flow in the positive direction, but a potential difference of the other sign causes little or no current. Hence a diode acts like a one-way valve in a circuit. Diodes are used to perform a wide variety of logic functions in computer circuitry.

Example 25.2 Electric field, potential difference, and resistance in a wire

The 18-gauge copper wire in Example 25.1 (Section 25.1) has a diameter of 1.02 mm and a cross-sectional area $A = 8.20 \times 10^{-7}\ \text{m}^2$. It carries a current $I = 1.67$ A. Find a) the electric-field magnitude in the wire; b) the potential difference between two points in the wire 50.0 m apart; c) the resistance of a 50.0-m length of this wire.

SOLUTION

IDENTIFY and SET UP: The magnitude of the current density is $J = I/A$ and the resistivity ρ is given in Table 25.1. All of these quantities are given, and we find the electric-field magnitude by using Eq. (25.5), $E = \rho J$. Once we have found E, the potential difference is simply the product of E and the length of the wire. We find the resistance by using Eq. (25.11).

EXECUTE: a) From Table 25.1, the resistivity of copper is $1.72 \times 10^{-8}\ \Omega \cdot \text{m}$. Hence, using Eq. (25.5),

$$E = \rho J = \frac{\rho I}{A} = \frac{(1.72 \times 10^{-8}\ \Omega \cdot \text{m})(1.67\ \text{A})}{8.20 \times 10^{-7}\ \text{m}^2}$$
$$= 0.0350\ \text{V/m}$$

b) The potential difference is given by

$$V = EL = (0.0350\ \text{V/m})(50.0\ \text{m}) = 1.75\ \text{V}$$

c) From Eq. (25.11) the resistance of a 50.0-m length of this wire is

$$R = \frac{V}{I} = \frac{1.75\ \text{V}}{1.67\ \text{A}} = 1.05\ \Omega$$

EVALUATE: To check our result in part (c), we calculate the resistance using Eq. (25.10):

$$R = \frac{\rho L}{A} = \frac{(1.72 \times 10^{-8}\ \Omega \cdot \text{m})(50.0\ \text{m})}{8.20 \times 10^{-7}\ \text{m}^2} = 1.05\ \Omega$$

We emphasize that the resistance of the wire is *defined* to be the ratio of voltage to current. If the wire is made of nonohmic material, then R is different for different values of V but is always given by $R = V/I$. Resistance is also always given by $R = \rho L/A$; if the material is nonohmic, ρ is not constant but depends on E (or, equivalently, on $V = EL$).

Example 25.3 Temperature dependence of resistance

Suppose the resistance of the wire in Example 25.2 is 1.05 Ω at a temperature of 20°C. Find the resistance at 0°C and at 100°C.

SOLUTION

IDENTIFY: This example concerns how resistance (the target variable) depends on temperature. As Table 25.2 shows, this temperature dependence differs for different substances.

SET UP: Our target variables are the values of the wire resistance R at two temperatures, $T = 0°\text{C}$ and $T = 100°\text{C}$. To find these values we use Eq. (25.12). Note that we are given the resistance $R_0 = 1.05\ \Omega$ at a reference temperature $T_0 = 20°\text{C}$, and we know from Example 25.2 that the wire is made of copper.

EXECUTE: From Table 25.2 the temperature coefficient of resistivity of copper is $\alpha = 0.00393\ (\text{C}°)^{-1}$. From Eq. (25.12), the resistance at $T = 0°\text{C}$ is

$$R = R_0[1 + \alpha(T - T_0)]$$
$$= (1.05\ \Omega)(1 + [0.00393\ (\text{C}°)^{-1}][0°\text{C} - 20°\text{C}])$$
$$= 0.97\ \Omega$$

At $T = 100°\text{C}$,

$$R = (1.05\ \Omega)(1 + [0.00393\ (\text{C}°)^{-1}][100°\text{C} - 20°\text{C}])$$
$$= 1.38\ \Omega$$

EVALUATE: The resistance at 100°C is greater than that at 0°C by a factor of $(1.38\ \Omega)/(0.97\ \Omega) = 1.42$. In other words, raising the temperature of ordinary copper wire from 0°C to 100°C increases its resistance by 42%. From Eq. (25.11), $V = IR$, this means that 42% more voltage V is required to produce the same current I at 100°C than at 0°C. This is a substantial effect that must be taken into account in designing electric circuits that are to operate over a wide range of temperatures.

Example 25.4 Calculating resistance

The hollow cylinder shown in Fig. 25.10 has length L and inner and outer radii a and b. It is made of a material with resistivity ρ. A potential difference is set up between the inner and outer surfaces of the cylinder (each of which is an equipotential surface) so that current flows radially through the cylinder. What is the resistance to this radial current flow?

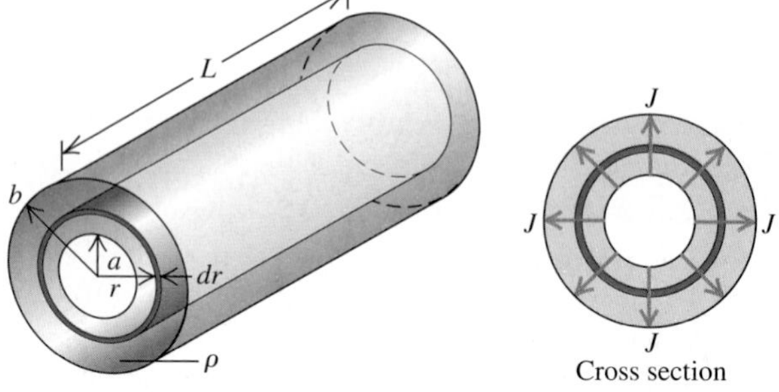

25.10 Finding the resistance for radial current flow.

SOLUTION

IDENTIFY: Figure 25.10 shows that the current flows radially from the inside of the conductor toward the outside, *not* along the length of the conductor as in Fig. 25.7. Hence we must use the ideas of this section to derive a new formula for resistance (our target variable) appropriate for radial current flow.

SET UP: We can't use Eq. (25.10) directly because the cross section through which the charge travels is *not* constant; it varies from $2\pi aL$ at the inner surface to $2\pi bL$ at the outer surface. Instead, we consider a thin cylindrical shell of inner radius r and thickness dr. We calculate the resistance to radial current flow through this shell, then combine the resistances for all such shells between the inner and outer radii of the cylinder.

EXECUTE: The area A for the shell is $2\pi rL$, the surface area that the current encounters as it flows outward. The length of the current path through the shell is dr. The resistance dR of this shell, between inner and outer surfaces, is that of a conductor with length dr and area $2\pi rL$:

$$dR = \frac{\rho\, dr}{2\pi rL}$$

The current has to pass successively through all such shells between the inner and outer radii a and b. From Eq. (25.11) the potential difference across one shell is $dV = I\, dR$, and the total potential difference between the inner and outer surfaces is the sum of the potential differences for all shells. The total current is the same through each shell, so the total resistance is the sum of the resistances of all the shells. If the area $2\pi rL$ were constant, we could just integrate dr from $r = a$ to $r = b$ to get the total length of the current path. But the area increases as the current passes through shells of greater radius, so we have to integrate the above expression for dR. The total resistance is thus given by

$$R = \int dR = \frac{\rho}{2\pi L}\int_a^b \frac{dr}{r} = \frac{\rho}{2\pi L}\ln\frac{b}{a}$$

EVALUATE: The conductor geometry shown in Fig. 25.10 plays an important role in your body's nervous system. Each neuron, or nerve cell, has a long extension called a nerve fiber or *axon*. An axon has a cylindrical membrane shaped much like the resistor in Fig. 25.10, with one conducting fluid inside the membrane and another outside it. Ordinarily all of the inner fluid is at the same potential, so no current tends to flow along the length of the axon. If the axon is stimulated at a certain point along its length, however, charged ions flow radially across the cylindrical membrane at that point, as in Fig. 25.10. This flow causes a potential difference between that point and other points along the length of the axon, which makes a nerve signal flow along that length.

Test Your Understanding

Suppose you increase the voltage across the copper wire in Examples 25.2 and 25.3. The increased voltage causes more current to flow, which makes the temperature of the wire increase. (The same thing happens to the coils of an electric oven or a toaster when a voltage is applied to them. We'll explore this issue in more depth in Section 25.5.) If you double the voltage across the wire, does the current in the wire also double? Why or why not? Is the wire ohmic?

25.4 | Electromotive Force and Circuits

For a conductor to have a steady current, it must be part of a path that forms a closed loop or **complete circuit.** Here's why. If you establish an electric field $\vec{E}_1$ inside an isolated conductor with resistivity ρ that is *not* part of a complete circuit, a current begins to flow with current density $\vec{J} = \vec{E}_1/\rho$ (Fig. 25.11a). As a result a net positive charge quickly accumulates at one end of the conductor and a net negative charge accumulates at the other end (Fig. 25.11b). These charges themselves produce an electric field $\vec{E}_2$ in the direction opposite to $\vec{E}_1$, causing the total electric field and hence the current to decrease. Within a very small fraction of a second, enough charge builds up on the conductor ends that the total electric field $\vec{E} = \vec{E}_1 + \vec{E}_2 = \mathbf{0}$ inside the conductor. Then $\vec{J} = \mathbf{0}$ as well, and the current stops altogether (Fig. 25.11c). So there can be no steady motion of charge in such an *incomplete* circuit.

To see how to maintain a steady current in a *complete* circuit, we recall a basic fact about electric potential energy: If a charge q goes around a complete circuit and returns to its starting point, the potential energy must be the same at the end of the round trip as at the beginning. As described in Section 25.3, there is always a *decrease* in potential energy when charges move through an ordinary conducting material with resistance. So there must be some part of the circuit in which the potential energy *increases.*

The problem is analogous to an ornamental water fountain that recycles its water. The water pours out of openings at the top, cascades down over the terraces and spouts (moving in the direction of decreasing gravitational potential energy), and collects in a basin in the bottom. A pump then lifts it back to the top (increasing the potential energy) for another trip. Without the pump, the water would just fall to the bottom and stay there.

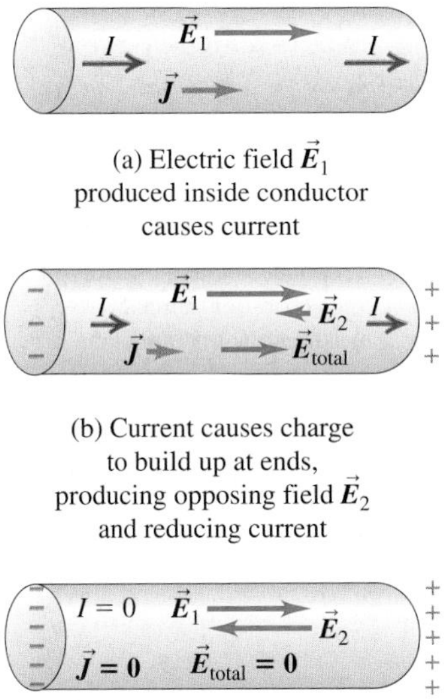

(a) Electric field $\vec{E}_1$ produced inside conductor causes current

(b) Current causes charge to build up at ends, producing opposing field $\vec{E}_2$ and reducing current

(c) After a very short time $\vec{E}_2$ has the same magnitude as $\vec{E}_1$: total field $\vec{E}_{total} = \mathbf{0}$ and current stops completely

25.11 If an electric field is produced inside a conductor that is *not* part of a complete circuit, current flows for only a very short time.

ActivPhysics ONLINE

12.1 DC Series Circuits (Qualitative)

Electromotive Force

In an electric circuit there must be a device somewhere in the loop that acts like the water pump in a water fountain (Fig. 25.12). In this device a charge travels "uphill," from lower to higher potential energy, even though the electrostatic force is trying to push it from higher to lower potential energy. The direction of current in such a device is from lower to higher potential, just the opposite of what happens in an ordinary conductor. The influence that makes current flow from lower to higher potential is called **electromotive force** (abbreviated **emf** and pronounced "ee-em-eff"). This is a poor term because emf is *not* a force but an energy-per-unit-charge quantity, like potential. The SI unit of emf is the same as that for potential, the volt $(1\ \text{V} = 1\ \text{J/C})$. A typical flashlight battery has an emf of 1.5 V; this means that the battery does 1.5 J of work on every coulomb of charge that passes through it. We'll use the symbol $\mathcal{E}$ (a script capital E) for emf.

Every complete circuit with a steady current must include some device that provides emf. Such a device is called a **source of emf.** Batteries, electric generators, solar cells, thermocouples, and fuel cells are all examples of sources of emf. All such devices convert energy of some form (mechanical, chemical, thermal, and so on) into electric potential energy and transfer it into the circuit to which the device is connected. An *ideal* source of emf maintains a constant potential difference between its terminals, independent of the current through it. We define electromotive force quantitatively as the magnitude of this potential difference. As we will see, such an ideal source is a mythical beast, like the frictionless plane and the

25.12 Just as a water fountain requires a pump, an electric circuit requires a source of electromotive force to sustain a steady current.

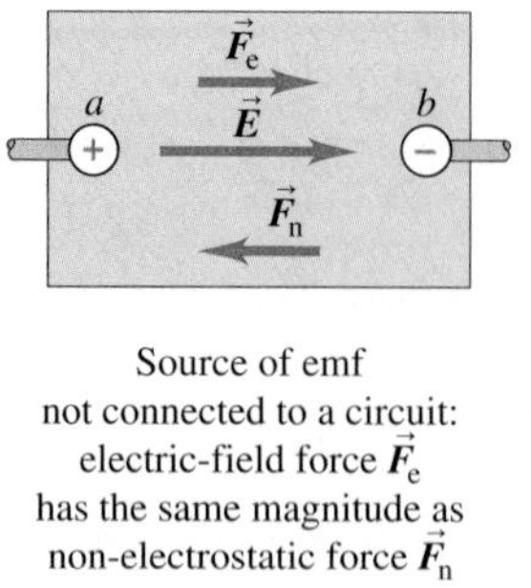

25.13 Schematic diagram of a source of emf in an "open-circuit" situation. The electric-field force $\vec{F}_e = q\vec{E}$ and the non-electrostatic force $\vec{F}_n$ are shown for a positive charge q.

massless rope. We will discuss later how real-life sources of emf differ in their behavior from this idealized model.

Figure 25.13 is a schematic diagram of an ideal source of emf that maintains a potential difference between conductors a and b, called the *terminals* of the device. Terminal a, marked $+$, is maintained at *higher* potential than terminal b, marked $-$. Associated with this potential difference is an electric field $\vec{E}$ in the region around the terminals, both inside and outside the source. The electric field inside the device is directed from a to b, as shown. A charge q within the source experiences an electric force $\vec{F}_e = q\vec{E}$. But the source also provides an additional influence, which we represent as a non-electrostatic force $\vec{F}_n$. This force, operating inside the device, pushes charge from b to a in an "uphill" direction against the electric force $\vec{F}_e$. Thus $\vec{F}_n$ maintains the potential difference between the terminals. If $\vec{F}_n$ were not present, charge would flow between the terminals until the potential difference was zero. The origin of the additional influence $\vec{F}_n$ depends on the kind of source. In a generator it results from magnetic-field forces on moving charges. In a battery or fuel cell it is associated with diffusion processes and varying electrolyte concentrations resulting from chemical reactions. In an electrostatic machine such as a Van de Graaff generator (Fig. 22.28), an actual mechanical force is applied by a moving belt or wheel.

If a positive charge q is moved from b to a inside the source, the non-electrostatic force $\vec{F}_n$ does a positive amount of work $W_n = q\mathcal{E}$ on the charge. This displacement is *opposite* to the electrostatic force $\vec{F}_e$, so the potential energy associated with the charge *increases* by an amount equal to qV_{ab}, where $V_{ab} = V_a - V_b$ is the (positive) potential of point a with respect to point b. For the ideal source of emf that we've described, $\vec{F}_e$ and $\vec{F}_n$ are equal in magnitude but opposite in direction, so the total work done on the charge q is zero; there is an increase in potential energy but *no* change in the kinetic energy of the charge. It's like lifting a book from the floor to a high shelf at constant speed. The increase in potential energy is just equal to the non-electrostatic work W_n, so $q\mathcal{E} = qV_{ab}$, or

$$V_{ab} = \mathcal{E} \qquad \text{(ideal source of emf)} \tag{25.13}$$

Now let's make a complete circuit by connecting a wire with resistance R to the terminals of a source (Fig. 25.14). The potential difference between terminals a and b sets up an electric field within the wire; this causes current to flow around the loop from a toward b, from higher to lower potential. Notice that where the wire bends, equal amounts of positive and negative charge persist on the "inside" and "outside" of the bend. These charges exert the forces that cause the current to follow the bends in the wire.

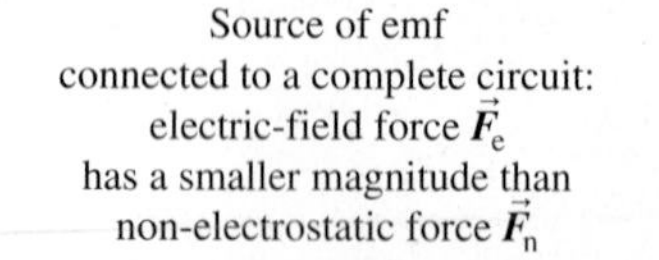

25.14 Schematic diagram of an ideal source of emf in a complete circuit. The electric-field force $\vec{F}_e = q\vec{E}$ and the non-electrostatic force $\vec{F}_n$ are shown for a positive charge q. The current is in the direction from a to b in the external circuit and from b to a within the source.

From Eq. (25.11) the potential difference between the ends of the wire in Fig. 25.14 is given by $V_{ab} = IR$. Combining with Eq. (25.13), we have

$$\mathcal{E} = V_{ab} = IR \qquad \text{(ideal source of emf)} \tag{25.14}$$

That is, when a positive charge q flows around the circuit, the potential *rise* $\mathcal{E}$ as it passes through the ideal source is numerically equal to the potential *drop* $V_{ab} = IR$ as it passes through the remainder of the circuit. Once $\mathcal{E}$ and R are known, this relation determines the current in the circuit.

CAUTION It's a common misconception that in a closed circuit, current is something that squirts out of the positive terminal of a battery and is consumed or "used up" by the time it reaches the negative terminal. In fact the current is the *same* at every point in a simple loop circuit like that in Fig. 25.14, even if the thickness of the wires is different at different points in the circuit. This happens because charge is conserved (that is, it can neither be created nor destroyed) and because charge cannot accumulate in the circuit devices we have described. If charge did accumulate, the potential differences would change with time. It's like the flow of water in an ornamental fountain; water flows out of the top of the fountain at the same rate at which it reaches the bottom, no matter what the dimensions of the fountain. None of the water is "used up" along the way!

Internal Resistance

Real sources in a circuit don't behave in exactly the way we have described; the potential difference across a real source in a circuit is *not* equal to the emf as in Eq. (25.14). The reason is that charge moving through the material of any real source encounters *resistance*. We call this the **internal resistance** of the source, denoted by r. If this resistance behaves according to Ohm's law, r is constant and independent of the current I. As the current moves through r, it experiences an associated drop in potential equal to Ir. Thus, when a current is flowing through a source from the negative terminal b to the positive terminal a, the potential difference V_{ab} between the terminals is

$$V_{ab} = \mathcal{E} - Ir \qquad \text{(terminal voltage, source with internal resistance)} \qquad (25.15)$$

The potential V_{ab}, called the **terminal voltage,** is less than the emf $\mathcal{E}$ because of the term Ir representing the potential drop across the internal resistance r. Expressed another way, the increase in potential energy qV_{ab} as a charge q moves from b to a within the source is now less than the work $q\mathcal{E}$ done by the non-electrostatic force $\vec{F}_n$, since some potential energy is lost in traversing the internal resistance.

A 1.5-V battery has an emf of 1.5 V, but the terminal voltage V_{ab} of the battery is equal to 1.5 V only if no current is flowing through it so that $I = 0$ in Eq. (25.15). If the battery is part of a complete circuit through which current is flowing, the terminal voltage will be less than 1.5 V. *For a real source of emf, the terminal voltage equals the emf only if no current is flowing through the source* (Fig. 25.15). Thus we can describe the behavior of a source in terms of two properties: an emf $\mathcal{E}$, which supplies a constant potential difference independent of current, in series with an internal resistance r.

The current in the external circuit connected to the source terminals a and b is still determined by $V_{ab} = IR$. Combining this with Eq. (25.15), we find

$$\mathcal{E} - Ir = IR$$

or

$$I = \frac{\mathcal{E}}{R + r} \qquad \text{(current, source with internal resistance)} \qquad (25.16)$$

That is, the current equals the source emf divided by the *total* circuit resistance $(R + r)$.

CAUTION You might have thought that a battery or other source of emf always produces the same current, no matter what circuit it's used in. But as Eq. (25.16) shows, the current that a source of emf produces in a given circuit depends on the resistance R of the external circuit (as well as on the internal resistance r of the source). The greater the resistance, the less current the source will produce. It's analogous to pushing an object through a thick, viscous liquid

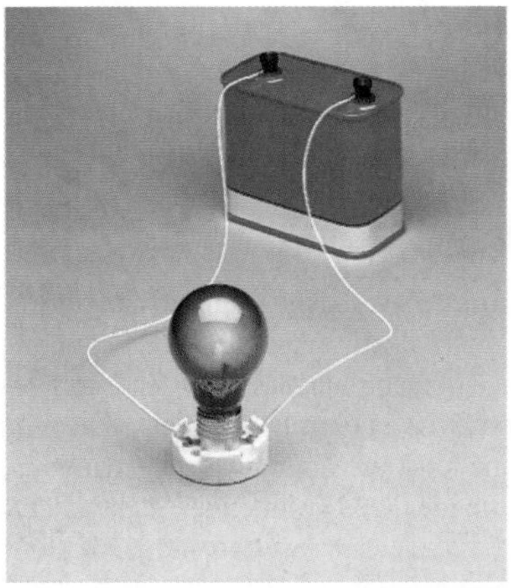

25.15 The emf of this battery—that is, the terminal voltage when it's not connected to anything—is 12 V. But because the battery has internal resistance, the terminal voltage of the battery is less than 12 V when it is supplying current to a light bulb.

such as oil or molasses; if you exert a certain steady push (emf), you can move a small object at high speed (small R, large I) or a large object at low speed (large R, small I).

Symbols for Circuit Diagrams

An important part of analyzing any electric circuit is drawing a schematic *circuit diagram.* Table 25.4 shows the usual symbols used in circuit diagrams. We will use these symbols extensively in this chapter and the next. We usually assume that the wires that connect the various elements of the circuit have negligible resistance; from Eq. (25.11), $V = IR$, the potential difference between the ends of such a wire is zero.

Table 25.4 includes two *meters* that are used to measure the properties of circuits. Idealized meters do not disturb the circuit in which they are connected. A **voltmeter,** introduced in Section 23.2, measures the potential difference between its terminals; an idealized voltmeter has infinitely large resistance and measures potential difference without having any current diverted through it. An **ammeter** measures the current passing through it; an idealized ammeter has zero resistance and has no potential difference between its terminals. Because meters act as part of the circuit in which they are connected, these properties are important to remember.

Table 25.4 Symbols for Circuit Diagrams

	Conductor with negligible resistance
	Resistor
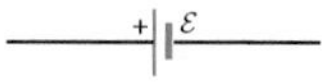	Source of emf (longer vertical line always represent the positive terminal, usually the terminal with higher potential)
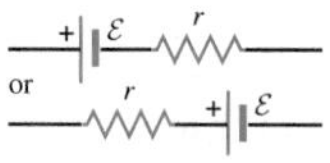	Source of emf with internal resistance r (r can be placed on either side)
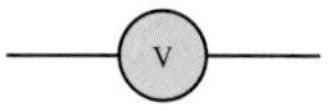	Voltmeter (measures potential difference between its terminals)
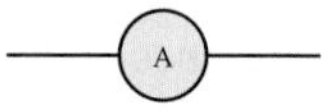	Ammeter (measures current through it)

Conceptual Example 25.5 A source on open circuit

Figure 25.16 shows a source (a battery) with an emf $\mathcal{E}$ of 12 V and an internal resistance r of 2 Ω. (For comparison, the internal resistance of a commercial 12-V lead storage battery is only a few thousandths of an ohm.) The wires to the left of a and to the right of the ammeter A are not connected to anything. Determine the readings of the idealized voltmeter V and the idealized ammeter A.

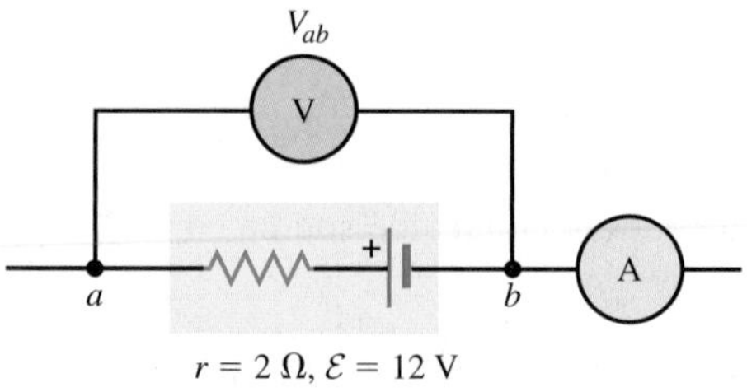

25.16 A source of emf on open circuit.

SOLUTION

There is no current because there is no complete circuit. (There is no current through our idealized voltmeter, with its infinitely large resistance.) Hence the ammeter A reads $I = 0$. Because there is no current through the battery, there is no potential difference across its internal resistance. From Eq. (25.15) with $I = 0$, the potential difference V_{ab} across the battery terminals is equal to the emf. So the voltmeter reads $V_{ab} = \mathcal{E} = 12$ V. The terminal voltage of a real, non-ideal source equals the emf *only* if there is no current flowing through the source, as in this example.

Example 25.6 A source in a complete circuit

Using the battery in Conceptual Example 25.5, we add a 4-Ω resistor to form the complete circuit shown in Fig. 25.17. What are the voltmeter and ammeter readings now?

SOLUTION

IDENTIFY and SET UP: Our first target variable is the current I through the circuit $aa'b'b$ (equal to the ammeter reading), which we find using Eq. (25.16). The second is the potential difference V_{ab} (equal to the voltmeter reading), which we can regard either as the potential difference across the source or as the potential difference around the circuit through the external resistor.

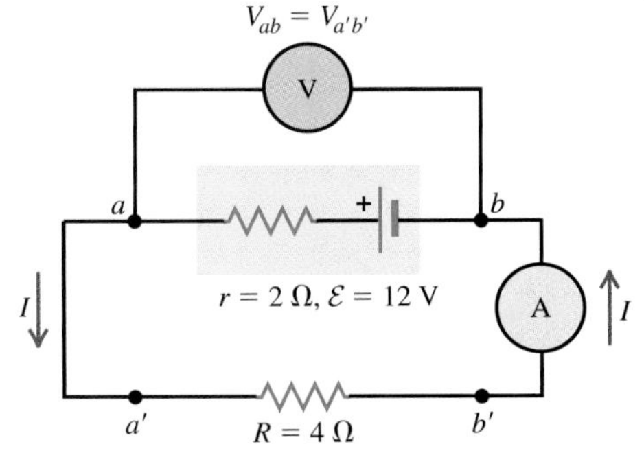

25.17 A source of emf in a complete circuit.

EXECUTE: The ideal ammeter has zero resistance, so the resistance external to the source is $R = 4\ \Omega$. From Eq. (25.16), the current through the circuit $aa'b'b$ is

$$I = \frac{\mathcal{E}}{R + r} = \frac{12\ \text{V}}{4\ \Omega + 2\Omega} = 2\ \text{A}$$

The ammeter A reads $I = 2$ A.

Our idealized conducting wires have zero resistance, and the idealized ammeter A also has zero resistance. So there is no potential difference between points a and a' or between points b and b'. That is, $V_{ab} = V_{a'b'}$. We can find V_{ab} by considering a and b either as the terminals of the resistor or as the terminals of the source. Considering them as terminals of the resistor, we use Ohm's law ($V = IR$):

$$V_{a'b'} = IR = (2\ \text{A})(4\ \Omega) = 8\ \text{V}$$

Considering them as the terminals of the source, we have

$$V_{ab} = \mathcal{E} - Ir = 12\ \text{V} - (2\ \text{A})(2\ \Omega) = 8\ \text{V}$$

Either way, we conclude that the voltmeter reads $V_{ab} = 8$ V.

EVALUATE: With a current flowing through the source, the terminal voltage V_{ab} is less than the emf. The smaller the internal resistance r, the less the difference between V_{ab} and $\mathcal{E}$.

Conceptual Example 25.7 Using voltmeters and ammeters

The voltmeter and ammeter in Example 25.6 are moved to different positions in the circuit. What are the voltmeter and ammeter readings in the situations shown in a) Fig. 25.18a and b) Fig. 25.18b?

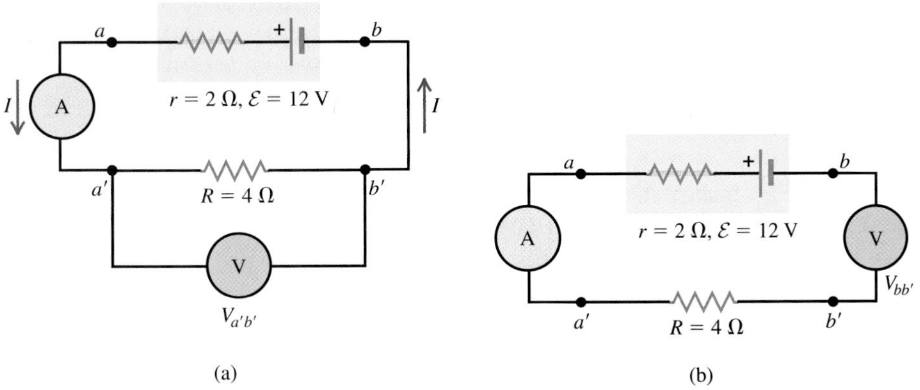

25.18 Different placements of voltmeter and ammeter in a complete circuit.

SOLUTION

a) The voltmeter now measures the potential difference between points a' and b'. But as mentioned in Example 25.6, $V_{ab} = V_{a'b'}$, so the voltmeter reads the same as in Example 25.6: $V_{a'b'} = 8$ V.

CAUTION You might be tempted to conclude that the ammeter in Fig. 25.18a, which is located "upstream" of the resistor, would have a higher reading than the one located "downstream" of the resistor in Fig. 25.17. But this conclusion is based on the misconception that current is somehow "used up" as it moves through a resistor. As charges move through a resistor, there is a decrease in electric potential energy, but there is *no* change in the current. *The current in a simple loop is the same at every point.* An ammeter placed as in Fig. 25.18a reads the same as one placed as in Fig. 25.17: $I = 2\text{ A}$.

b) There is no current through the voltmeter because it has infinitely large resistance. Since the voltmeter is now part of the circuit, there is no current at all in the circuit, and the ammeter reads $I = 0$.

The voltmeter measures the potential difference $V_{bb'}$ between points b and b'. Since $I = 0$, the potential difference across the resistor is $V_{a'b'} = IR = 0$, and the potential difference between the ends a and a' of the idealized ammeter is also zero. So $V_{bb'}$ is equal to V_{ab}, the terminal voltage of the source. As in Conceptual Example 25.5, there is no current flowing, so the terminal voltage equals the emf, and the voltmeter reading is $V_{ab} = \mathcal{E} = 12\text{ V}$.

This example shows that ammeters and voltmeters are circuit elements, too. Moving the voltmeter from the position in Fig. 25.18a to that in Fig. 25.18b changes the current and potential differences in the circuit—in this case rather dramatically. If you want to measure the potential difference between two points in a circuit without disturbing the circuit, use a voltmeter as in Figs. 25.17 or 25.18a, *not* as in Fig. 25.18b.

Example 25.8 A source with a short circuit

Using the same battery as in the preceding three examples, we now replace the 4-Ω resistor with a zero-resistance conductor, as shown in Fig. 25.19. What are the meter readings now?

SOLUTION

IDENTIFY and SET UP: Our target variables are I and V_{ab}, the same as in Example 25.6. The only difference from that example is that the external resistance is now $R = 0$.

EXECUTE: Now we have a zero-resistance path between points a and b, so we must have $V_{ab} = IR = I(0) = 0$, no matter what the current. Knowing this, we can find the current I from Eq. (25.15):

$$V_{ab} = \mathcal{E} - Ir = 0$$

$$I = \frac{\mathcal{E}}{r} = \frac{12\text{ V}}{2\ \Omega} = 6\text{ A}$$

The ammeter reads $I = 6\text{ A}$ and the voltmeter reads $V_{ab} = 0$.

EVALUATE: The current has a different value than in Example 25.6, even though the same battery is used. A source does *not* deliver the same current in all situations; the amount of current depends on the internal resistance r and on the resistance of the external circuit.

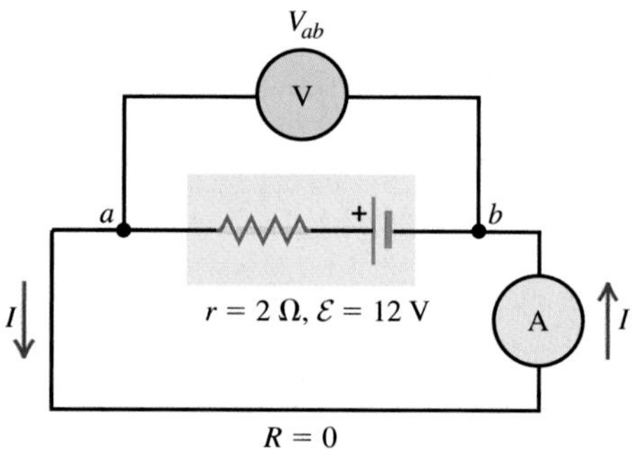

25.19 A source of emf on short circuit. (*Warning:* See the text for the dangers of this arrangement.)

The situation in this example is called a *short circuit.* The terminals of the battery are connected directly to each other, with no external resistance. The short-circuit current is equal to the emf $\mathcal{E}$ divided by the internal resistance r. *Warning:* A short circuit can be an extremely dangerous situation. An automobile battery or a household power line has very small internal resistance (much less than in these examples), and the short-circuit current can be great enough to melt a small wire or cause a storage battery to explode. Don't try it!

The net change in potential energy for a charge q making a round trip around a complete circuit must be zero. Hence the net change in *potential* around the circuit must also be zero; in other words, the algebraic sum of the potential differences and emfs around the loop is zero. We can see this by rewriting Eq. (25.16) in the form

$$\mathcal{E} - Ir - IR = 0$$

A potential gain of $\mathcal{E}$ is associated with the emf and potential drops of Ir and IR are associated with the internal resistance of the source and the external circuit,

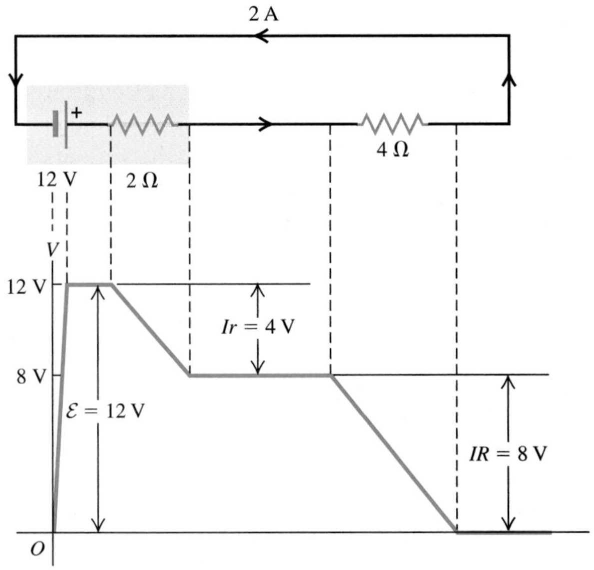

25.20 Potential rises and drops in a circuit.

respectively. Figure 25.20 is a graph showing how the potential varies as we go around the complete circuit of Fig. 25.17. The horizontal axis doesn't necessarily represent actual distances, but rather various points in the loop. If we take the potential to be zero at the negative terminal of the battery, then we have a rise $\mathcal{E}$ and a drop Ir in the battery and an additional drop IR in the external resistor, and as we finish our trip around the loop, the potential is back where it started.

In this section we have considered only situations in which the resistances are ohmic. If the circuit includes a nonlinear device such as a diode (see Figs. 25.9b and 25.9c), Eq. (25.16) is still valid but cannot be solved algebraically because R is not a constant. In such a situation, the current I can be found by using numerical techniques (see Challenge Problem 25.82).

Finally, we remark that Eq. (25.15) is not always an adequate representation of the behavior of a source. The emf may not be constant, and what we have described as an internal resistance may actually be a more complex voltage–current relation that doesn't obey Ohm's law. Nevertheless, the concept of internal resistance frequently provides an adequate description of batteries, generators, and other energy converters. The principal difference between a fresh flashlight battery and an old one is not in the emf, which decreases only slightly with use, but in the internal resistance, which may increase from less than an ohm when the battery is fresh to as much as 1000 Ω or more after long use. Similarly, a car battery can deliver less current to the starter motor on a cold morning than when the battery is warm, not because the emf is appreciably less but because the internal resistance is temperature-dependent, increasing with decreasing temperature. Cold-climate dwellers take a number of measures to avoid this loss, from using special battery warmers to soaking the battery in warm water on very cold mornings.

Test Your Understanding

A battery has an emf of 1.5 V and an internal resistance of 0.10 Ω. When the battery is connected to a resistor, the terminal voltage of the battery is 1.3 V. What is the resistance of the resistor?

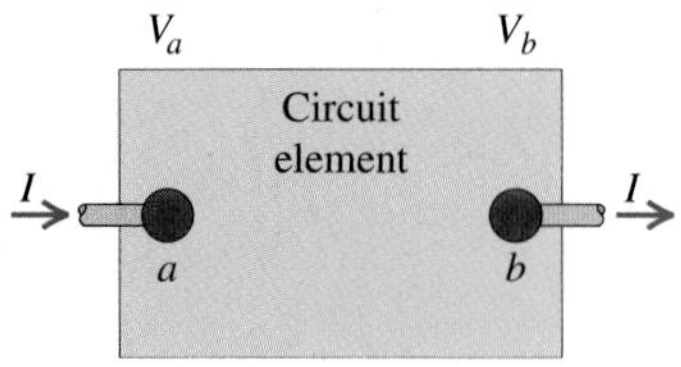

25.21 The power input to the circuit element between a and b is $P = (V_a - V_b)I = V_{ab}I$.

25.5 | Energy and Power in Electric Circuits

Let's now look at some energy and power relations in electric circuits. The box in Fig. 25.21 represents a circuit element with potential difference $V_a - V_b = V_{ab}$ between its terminals and current I passing through it in the direction from a toward b. This element might be a resistor, a battery, or something else; the details don't matter. As charge passes through the circuit element, the electric field does work on the charge. In a source of emf, additional work is done by the force $\vec{F}_n$ that we mentioned in Section 25.4.

As an amount of charge q passes through the circuit element, there is a change in potential energy equal to qV_{ab}. For example, if $q > 0$ and $V_{ab} = V_a - V_b$ is positive, potential energy increases as the charge "falls" from potential V_a to lower potential V_b. The moving charges don't gain *kinetic* energy, because the rate of charge flow (that is, the current) out of the circuit element must be the same as the rate of charge flow into the element. Instead, the quantity qV_{ab} represents electrical energy transferred into the circuit element. This situation occurs in the coils of a toaster or electric oven, in which electrical energy is converted to thermal energy.

It may happen that the potential at b is higher than that at a. In this case V_{ab} is negative, and a net transfer of energy *out* of the circuit element occurs. The element then acts as a source, delivering electrical energy into the circuit to which it is attached. This is the usual situation for a battery, which converts chemical energy into electrical energy and delivers it to the external circuit. Thus qV_{ab} can denote either a quantity of energy delivered to a circuit element or a quantity of energy extracted from that element.

In electric circuits we are most often interested in the *rate* at which energy is either delivered to or extracted from a circuit element. If the current through the element is I, then in a time interval dt an amount of charge $dQ = I\,dt$ passes through the element. The potential energy change for this amount of charge is $V_{ab}\,dQ = V_{ab}I\,dt$. Dividing this expression by dt, we obtain the *rate* at which energy is transferred either into or out of the circuit element. The time rate of energy transfer is *power*, denoted by P, so we write

$$P = V_{ab}I \qquad (25.17)$$

(rate at which energy is delivered to or extracted from a circuit element)

The unit of V_{ab} is one volt, or one joule per coulomb, and the unit of I is one ampere, or one coulomb per second. Hence the unit of $P = V_{ab}I$ is one watt, as it should be:

$$(1\ \text{J/C})(1\ \text{C/s}) = 1\ \text{J/s} = 1\ \text{W}$$

Let's consider a few special cases.

Pure Resistance

If the circuit element in Fig. 25.21 is a resistor, the potential difference is $V_{ab} = IR$. From Eq. (25.17) the electrical power delivered to the resistor by the circuit is

$$P = V_{ab}I = I^2R = \frac{V_{ab}{}^2}{R} \qquad (25.18)$$

(power delivered to a resistor)

In this case the potential at a (where the current enters the resistor) is always higher than that at b (where the current exits). Current enters the higher-potential terminal of the device, and Eq. (25.18) represents the rate of transfer of electric potential energy *into* the circuit element.

What becomes of this energy? The moving charges collide with atoms in the resistor and transfer some of their energy to these atoms, increasing the internal energy of the material. Either the temperature of the resistor increases or there is a flow of heat out of it, or both. In any of these cases we say that energy is *dissipated* in the resistor at a rate I^2R. Every resistor has a *power rating,* the maximum power the device can dissipate without becoming overheated and damaged. In practical applications the power rating of a resistor is often just as important a characteristic as its resistance value. Of course, some devices, such as electric heaters, are designed to get hot and transfer heat to their surroundings. But if the power rating is exceeded, even such a device may melt or even explode.

Power Output of a Source

The upper rectangle in Fig. 25.22a represents a source with emf $\mathcal{E}$ and internal resistance r, connected by ideal (resistanceless) conductors to an external circuit represented by the lower box. This could describe a car battery connected to one of the car's headlights (Fig. 25.22b). Point a is at higher potential than point b, so $V_a > V_b$ and V_{ab} is positive. Note that the current I is *leaving* the source at the higher-potential terminal (rather than entering there). Energy is being delivered to the external circuit, and the rate of its delivery to the circuit is given by Eq. (25.17):

$$P = V_{ab}I$$

For a source that can be described by an emf $\mathcal{E}$ and an internal resistance r, we may use Eq. (25.15):

$$V_{ab} = \mathcal{E} - Ir$$

Multiplying this equation by I, we find

$$P = V_{ab}I = \mathcal{E}I - I^2r \qquad (25.19)$$

What do the terms $\mathcal{E}I$ and I^2r mean? In Section 25.4 we defined the emf $\mathcal{E}$ as the work per unit charge performed on the charges by the non-electrostatic force as the charges are pushed "uphill" from b to a in the source. In a time dt, a charge $dQ = I\,dt$ flows through the source; the work done on it by this non-electrostatic force is $\mathcal{E}\,dQ = \mathcal{E}I\,dt$. Thus $\mathcal{E}I$ is the *rate* at which work is done on the circulating charges by whatever agency causes the non-electrostatic force in the source. This term represents the rate of conversion of nonelectrical energy to electrical energy within the source. The term I^2r is the rate at which electrical energy is *dissipated* in the internal resistance of the source. The difference $\mathcal{E}I - I^2r$ is the *net* electrical power output of the source—that is, the rate at which the source delivers electrical energy to the remainder of the circuit.

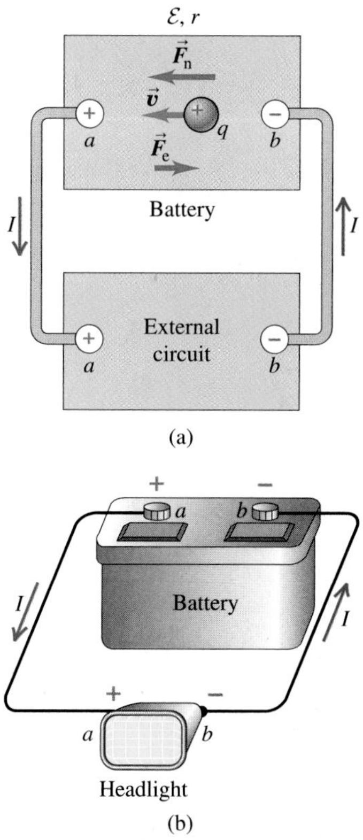

25.22 (a) The rate of conversion of nonelectrical to electrical energy in the source equals $\mathcal{E}I$. The rate of energy dissipation in the source is I^2r. The difference $\mathcal{E}I - I^2r$ is the power output of the source. (b) A car battery connected to a headlight is a real-life example of the generic circuit in part (a).

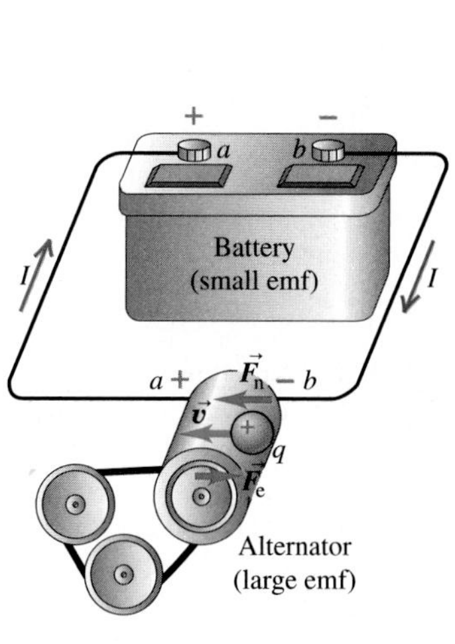

25.23 When two sources are connected in a simple loop circuit, the source with the larger emf delivers energy to the other source.

Power Input to a Source

Suppose that the lower rectangle in Fig. 25.22a is itself a source, with an emf *larger* than that of the upper source and with its emf opposite to that of the upper source. Figure 25.23 shows a practical example, an automobile battery (the upper circuit element) being charged by the car's alternator (the lower element). The current I in the circuit is then *opposite* to that shown in Fig. 25.22; the lower source is pushing current backward through the upper source. Because of this reversal of current, instead of Eq. (25.15) we have for the upper source

$$V_{ab} = \mathcal{E} + Ir$$

and instead of Eq. (25.19), we have

$$P = V_{ab}I = \mathcal{E}I + I^2R \tag{25.20}$$

Work is being done *on,* rather than *by,* the agent that causes the non-electrostatic force in the upper source. There is a conversion of electrical energy into nonelectrical energy in the upper source at a rate $\mathcal{E}I$. The term I^2r in Eq. (25.20) is again the rate of dissipation of energy in the internal resistance of the upper source, and the sum $\mathcal{E}I + I^2r$ is the total electrical power *input* to the upper source. This is what happens when a rechargeable battery (a storage battery) is connected to a charger. The charger supplies electrical energy to the battery; part of it is converted to chemical energy, to be reconverted later, and the remainder is dissipated (wasted) in the battery's internal resistance, warming the battery and causing a heat flow out of it. If you have a power tool or laptop computer with a rechargeable battery, you may have noticed that it gets warm while it is charging.

Problem-Solving Strategy

Power and Energy in Circuits

IDENTIFY *the relevant concepts:* The ideas of electric power input and output can be applied to any electric circuit. In most cases you'll know when these concepts are needed, because the problem will ask you explicitly to consider power or energy.

SET UP *the problem using the following steps:*

1. Make a drawing of the circuit.
2. Identify the circuit elements, including sources of emf and resistors. In later chapters we will include other kinds of circuit elements, including capacitors and inductors (described in Chapter 30).
3. Determine the target variables. Typically they will be the power input or output for each circuit element, or the total amount of energy put into or taken out of a circuit element in a given time.

EXECUTE *the solution as follows:*

1. A source of emf $\mathcal{E}$ delivers power $\mathcal{E}I$ into a circuit when the current I runs through the source from $-$ to $+$. The energy is converted from chemical energy in a battery, from mechanical energy in a generator, or whatever. In this case the source has a *positive* power output to the circuit or, equivalently, a *negative* power input to the source.
2. A source of emf takes power $\mathcal{E}I$ from a circuit—that is, it has a negative power output, or, equivalently, a positive power input—when current passes through the source in the direction from $+$ to $-$. This occurs in charging a storage battery, when electrical energy is converted back to chemical energy. In this case the source has a *negative* power output to the circuit or, equivalently, a *positive* power input to the source.
3. No matter what the direction of the current through a resistor, there is always a *positive* power input to the resistor. It removes energy from a circuit at a rate given by $VI = I^2R = V^2/R$, where V is the potential difference across the resistor.
4. There is also a *positive* power input to the internal resistance r of a source, irrespective of the direction of the current. The internal resistance always removes energy from the circuit, converting it into heat at a rate I^2r.
5. You may need to calculate the total energy delivered to or extracted from a circuit element in a given amount of time. If the power into or out of a circuit element is constant, this integral is just the product of power and elapsed time. (In Chapter 26 we will encounter situations in which the power is not constant. In such cases, calculating the total energy requires an integral.)

EVALUATE *your answer:* Check your results, including a check that energy is conserved. This conservation can be expressed in either of two forms: "net power input = net power output" or "the algebraic sum of the power inputs to the circuit elements is zero."

Example 25.9 Power input and output in a complete circuit

Figure 25.24 shows the same situation that we analyzed in Example 25.6. Find the rate of energy conversion (chemical to electrical) and the rate of dissipation of energy in the battery and the net power output of the battery.

SOLUTION

IDENTIFY and SET UP: Our target variables are the power output $\mathcal{E}I$ of the source of emf, the power input I^2r to the internal resistance, and the net power output of the source given by Eq. (25.19).

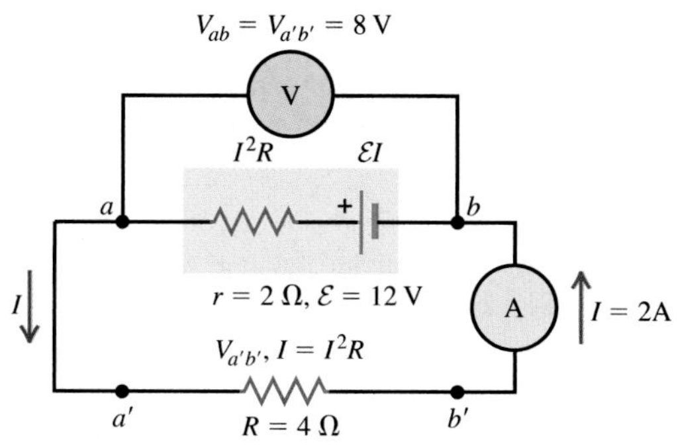

25.24 Power relationships in a simple circuit.

EXECUTE: From Example 25.6 the current in the circuit is $I = 2$ A. The rate of energy conversion in the battery is

$$\mathcal{E}I = (12\text{ V})(2\text{ A}) = 24\text{ W}$$

The rate of dissipation of energy in the battery is

$$I^2r = (2\text{ A})^2(2\ \Omega) = 8\text{ W}$$

The electrical power *output* of the source is the difference between these: $\mathcal{E}I - I^2r = 16$ W.

EVALUATE: The power output is also given by the terminal voltage $V_{ab} = 8$ V (calculated in Example 25.6) multiplied by the current:

$$V_{ab}I = (8\text{ V})(2\text{ A}) = 16\text{ W}$$

The electrical power input to the resistor is

$$V_{a'b'}I = (8\text{ V})(2\text{ A}) = 16\text{ W}$$

This equals the rate of dissipation of electrical energy in the resistor:

$$I^2R = (2\text{ A})^2(4\ \Omega) = 16\text{ W}$$

Note that our results agree with Eq. (25.19), which states that $V_{ab}I = \mathcal{E}I - I^2R$; the left side of this equation equals 16 W, and the right side equals 24 W − 8 W = 16 W. This verifies the consistency of the various power quantities.

Example 25.10 Increasing the resistance

Suppose the 4-Ω resistor in Fig. 25.24 is replaced by an 8-Ω resistor. How does this affect the electrical power dissipated in the resistor?

SOLUTION

IDENTIFY and SET UP: The situation is the same as that in Example 25.9, but with a different value of the external resistance R.

EXECUTE: According to Eq. (25.18), the power dissipated in the resistor is given by $P = I^2R$. If you were in a hurry, you might conclude that since R now has twice the value that it had in Example 25.9, the power should also be twice as great, or $2(16\text{ W}) = 32$ W. Or you might instead try to use the formula $P = V_{ab}^2/R$; this formula would lead you to conclude that the power should be one-half as great as in the preceding example, or $(16\text{ W})/2 = 8$ W. Which answer is correct?

In fact, *both* of these conclusions are *incorrect.* The first is incorrect because changing the resistance R also changes the current in the circuit (remember, a source of emf does *not* generate the same current in all situations). The second conclusion is also incorrect because the potential difference V_{ab} across the resistor changes when the current changes. To get the correct answer, we first use the same technique as in Example 25.6 to find the current:

$$I = \frac{\mathcal{E}}{R + r} = \frac{12\text{ V}}{8\ \Omega + 2\ \Omega} = 1.2\text{ A}$$

The greater resistance causes the current to decrease. The potential difference across the resistor is

$$V_{ab} = IR = (1.2\text{ A})(8\ \Omega) = 9.6\text{ V}$$

which is greater than that with the 4-Ω resistor. We can then find the power dissipated in the resistor in either of two ways:

$$P = I^2R = (1.2\text{ A})^2(8\ \Omega) = 12\text{ W}$$

or

$$P = \frac{V_{ab}^2}{R} = \frac{(9.6\text{ V})^2}{8\ \Omega} = 12\text{ W}$$

EVALUATE: Increasing the resistance R causes a *reduction* in the power input to the resistor. In the expression $P = I^2R$ the decrease in current is more important than the increase in resistance; in the expression $P = V_{ab}^2/R$ the increase in resistance is more important than the increase in V_{ab}. This same principle applies to ordinary light bulbs; a 50-W light bulb has a greater resistance than does a 100-W light bulb.

Can you show that replacing the 4-Ω resistor with an 8-Ω resistor decreases both the rate of energy conversion (chemical to electrical) in the battery and the rate of energy dissipation in the battery?

Example 25.11 Power in a short circuit

Figure 25.25 shows the same short-circuited battery that we analyzed in Example 25.8. Find the rates of energy conversion and energy dissipation in the battery and the net power output of the battery.

SOLUTION

IDENTIFY and SET UP: This is once again the same situation as in Example 25.9, but now the external resistance R is zero.

EXECUTE: We found in Example 25.8 that the current in this situation is $I = 6$ A. The rate of energy conversion (chemical to electrical) in the battery is

$$\mathcal{E}I = (12\text{ V})(6\text{ A}) = 72\text{ W}$$

The rate of dissipation of energy in the battery is

$$I^2r = (6\text{ A})^2(2\ \Omega) = 72\text{ W}$$

The net power output of the source, given by $V_{ab}I$, is zero because the terminal voltage V_{ab} is zero.

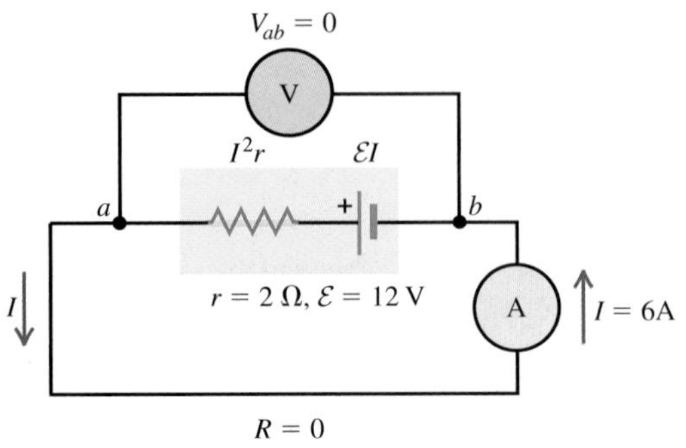

25.25 Power relationships in a short circuit. With ideal wires and ammeter ($R = 0$), no electrical power leaves the battery.

EVALUATE: *All* of the converted energy is dissipated within the source. This is why a short-circuited battery is quickly ruined and in some cases may even explode.

Test Your Understanding

Repeat Example 25.9 for the idealized situation in which the source has no internal resistance. How do your answers compare to those in Example 25.9?

*25.6 | Theory of Metallic Conduction

We can gain additional insight into electrical conduction by looking at the microscopic origin of conductivity. We'll consider a very simple model that treats the electrons as classical particles and ignores their quantum-mechanical, wavelike behavior in solids. Using this model, we'll derive an expression for the resistivity of a metal. Even though this model is not entirely correct conceptually, it will still help you to develop an intuitive idea of the microscopic basis of conduction.

In the simplest microscopic model of conduction in a metal, each atom in the metallic crystal gives up one or more of its outer electrons. These electrons are then free to move through the crystal, colliding at intervals with the stationary positive ions. The motion of the electrons is analogous to the motion of molecules of a gas moving through a porous bed of sand, and they are often referred to as an "electron gas."

If there is no electric field, the electrons move in straight lines between collisions, the directions of their velocities are random, and on average they never get anywhere (Fig. 25.26a). But if an electric field is present, the paths curve slightly because of the acceleration caused by electric-field forces. Figure 25.26b shows a few paths of an electron in an electric field directed from right to left. As we mentioned in Section 25.1, the average speed of random motion is of the order of 10^6 m/s, while the average drift speed is *much* smaller, of the order of 10^{-4} m/s.

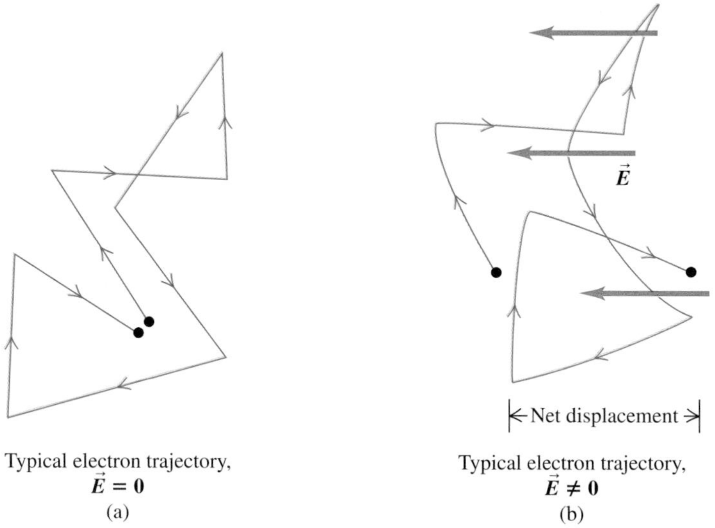

25.26 Random motions of an electron in a metallic crystal (a) with zero electric field; and (b) with an electric field that causes drift. The curvatures of the paths are greatly exaggerated.

The average time between collisions is called the **mean free time,** denoted by τ. Figure 25.27 shows a mechanical analog of this electron motion.

We would like to derive from this model an expression for the resistivity ρ of a material, defined by Eq. (25.5):

$$\rho = \frac{E}{J} \tag{25.21}$$

where E and J are the magnitudes of electric field and current density. The current density $\vec{J}$ is in turn given by Eq. (25.4):

$$\vec{J} = nq\vec{v}_{\text{d}} \tag{25.22}$$

where n is the number of free electrons per unit volume, q is the charge of each, and $\vec{v}_{\text{d}}$ is their average drift velocity. (We also know that $q = -e$ in an ordinary metal; we'll use that fact later.)

We need to relate the drift velocity $\vec{v}_{\text{d}}$ to the electric field $\vec{E}$. The value of $\vec{v}_{\text{d}}$ is determined by a steady-state condition in which, on average, the velocity *gains* of the charges due to the force of the $\vec{E}$ field are just balanced by the velocity *losses* due to collisions.

To clarify this process, let's imagine turning on the two effects one at a time. Suppose that before time $t = 0$ there is no field. The electron motion is then completely random. A typical electron has velocity $\vec{v}_0$ at time $t = 0$, and the value of $\vec{v}_0$ averaged over many electrons (that is, the initial velocity of an average electron) is zero: $(\vec{v}_0)_{\text{av}} = \mathbf{0}$. Then at time $t = 0$ we turn on a constant electric field $\vec{E}$. The field exerts a force $\vec{F} = q\vec{E}$ on each charge, and this causes an acceleration $\vec{a}$ in the direction of the force, given by

$$\vec{a} = \frac{\vec{F}}{m} = \frac{q\vec{E}}{m}$$

where m is the electron mass. *Every* electron has this acceleration.

We wait for a time τ, the average time between collisions, and then "turn on" the collisions. An electron that has velocity $\vec{v}_0$ at time $t = 0$ has a velocity at time $t = \tau$ equal to

$$\vec{v} = \vec{v}_0 + \vec{a}\tau$$

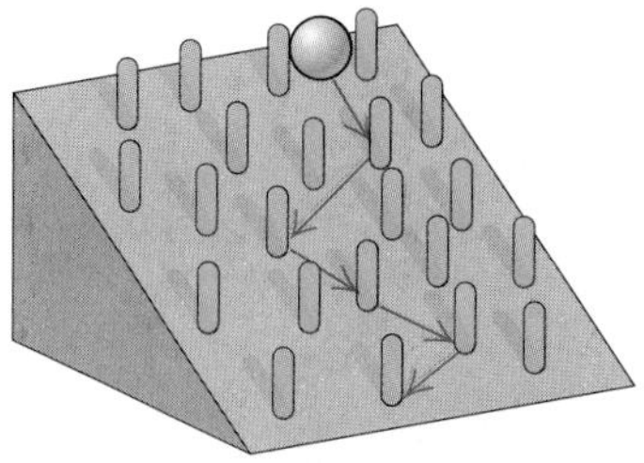

25.27 The motion of a ball rolling down an inclined plane and bouncing off pegs in its path is analogous to the motion of an electron in a metallic conductor with an electric field present.

The velocity $\vec{v}_{av}$ of an *average* electron at this time is the sum of the averages of the two terms on the right. As we have pointed out, the initial velocity $\vec{v}_0$ is zero for an average electron, so

$$\vec{v}_{av} = \vec{a}\tau = \frac{q\tau}{m}\vec{E} \tag{25.23}$$

After time $t = \tau$, the tendency of the collisions to decrease the velocity of an average electron (by means of randomizing collisions) just balances the tendency of the $\vec{E}$ field to increase this velocity. Thus the velocity of an average electron, given by Eq. (25.23), is maintained over time and is equal to the drift velocity $\vec{v}_d$:

$$\vec{v}_d = \frac{q\tau}{m}\vec{E}$$

Now we substitute this equation for the drift velocity $\vec{v}_d$ into Eq. (25.22):

$$\vec{J} = nq\vec{v}_d = \frac{nq^2\tau}{m}\vec{E}$$

Comparing this with Eq. (25.21), which we can rewrite as $\vec{J} = \vec{E}/\rho$, and substituting $q = -e$, we see that the resistivity ρ is given by

$$\rho = \frac{m}{ne^2\tau} \tag{25.24}$$

If n and τ are independent of $\vec{E}$, then the resistivity is independent of $\vec{E}$ and the conducting material obeys Ohm's law.

Turning the interactions on one at a time may seem artificial. But a little thought shows that the derivation would come out the same if each electron had its own clock and the $t = 0$ times were different for different electrons. If τ is the average time between collisions, then $\vec{v}_d$ is still the average electron drift velocity, even though the motions of the various electrons aren't actually correlated in the way we postulated.

What about the temperature dependence of resistivity? In a perfect crystal with no atoms out of place, a correct quantum-mechanical analysis would let the free electrons move through the crystal with no collisions at all. But the atoms vibrate about their equilibrium positions. As the temperature increases, the amplitudes of these vibrations increase, collisions become more frequent, and the mean free time τ decreases. So this theory predicts that the resistivity of a metal increases with temperature. In a superconductor, roughly speaking, there are no inelastic collisions, τ is infinite, and the resistivity ρ is zero.

In a pure semiconductor such as silicon or germanium, the number of charge carriers per unit volume, n, is not constant but increases very rapidly with increasing temperature. This increase in n far outweighs the decrease in the mean free time, and in a semiconductor the resistivity always decreases rapidly with increasing temperature. At low temperatures, n is very small, and the resistivity becomes so large that the material can be considered an insulator.

Electrons gain energy between collisions through the work done on them by the electric field. During collisions they transfer some of this energy to the atoms of the material of the conductor. This leads to an increase in the material's internal energy and temperature; that's why wires carrying current get warm. If the electric field in the material is large enough, an electron can gain enough energy between collisions to knock off electrons that are normally bound to atoms in the material. These can then knock off more electrons, and so on, possibly leading to an avalanche of current. This is the microscopic basis of dielectric breakdown in insulators.

Example 25.12 Mean free time in copper

Calculate the mean free time between collisions in copper at room temperature.

SOLUTION

IDENTIFY and SET UP: We can find an expression for mean free time τ in terms of n, ρ, e, and m by rearranging Eq. (25.24). From Example 25.1 and Table 25.1, for copper $n = 8.5 \times 10^{28}\ \text{m}^{-3}$ and $\rho = 1.72 \times 10^{-8}\ \Omega \cdot \text{m}$. Also, $e = 1.60 \times 10^{-19}\ \text{C}$ and $m = 9.11 \times 10^{-31}\ \text{kg}$ for electrons.

EXECUTE: From Eq. (25.24), we get

$$\tau = \frac{m}{ne^2\rho}$$

$$= \frac{9.11 \times 10^{-31}\ \text{kg}}{(8.5 \times 10^{28}\ \text{m}^{-3})(1.60 \times 10^{-19}\ \text{C})^2(1.72 \times 10^{-8}\ \Omega \cdot \text{m})}$$

$$= 2.4 \times 10^{-14}\ \text{s}$$

EVALUATE: Taking the reciprocal of this time, we find that each electron averages about 4×10^{13} collisions every second!

Test Your Understanding

On average, how far does an electron in the copper wire of Example 25.1 (Section 25.1) drift between collisions? (*Hint:* See Example 25.12.)

SUMMARY

Current is the amount of charge flowing through a specified area, per unit time. The SI unit of current is the ampere, equal to one coulomb per second $(1\ \text{A} = 1\ \text{C/s})$. The current I through an area A depends on the concentration n and charge q of the charge carriers, as well as on the magnitude of their drift velocity $\vec{v}_d$. The current density is current per unit cross-sectional area. Current is conventionally described in terms of a flow of positive charge, even when the actual charge carriers are negative or of both signs. (See Example 25.1)

$$I = \frac{dQ}{dt} = n|q|v_d A \quad (25.2)$$

$$\vec{J} = nq\vec{v}_d \quad (25.4)$$

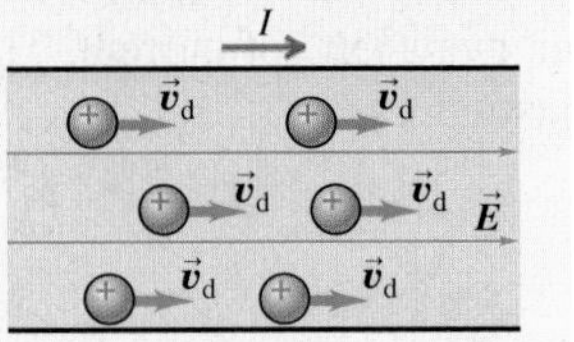

The resistivity ρ of a material is defined by the ratio of the magnitudes of electric field and current density. Good conductors have small resistivity; good insulators have large resistivity. Ohm's law, obeyed approximately by many materials, states that ρ is a constant independent of the value of E. Resistivity usually increases with temperature; for small temperature changes this variation is represented approximately by Eq. (25.6), where α is the temperature coefficient of resistivity.

$$\rho = \frac{E}{J} \quad (25.5)$$

$$\rho(T) = \rho_0[1 + \alpha(T - T_0)] \quad (25.6)$$

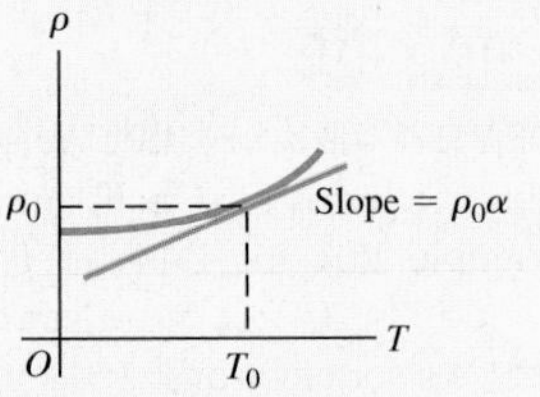

Metal: ρ increases with increasing T

For materials obeying Ohm's law, the potential difference V across a particular sample of material is proportional to the current I through the material. The ratio $V/I = R$ is the resistance of the sample. The SI unit of resistance is the ohm $(1\ \Omega = 1\ V/A)$. The resistance of a cylindrical conductor is related to its resistivity ρ, length L, and cross-sectional area A. (See Examples 25.2 through 25.4)

$$V = IR \quad (25.11)$$

$$R = \frac{\rho L}{A} \quad (25.10)$$

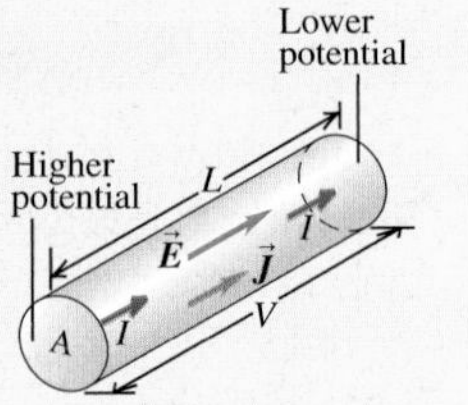

A complete circuit has a continuous current-carrying path. A complete circuit carrying a steady current must contain a source of electromotive force (emf) $\mathcal{E}$. The SI unit of electromotive force is the volt (1 V). An ideal source of emf maintains a constant potential difference, independent of current through the device, but every real source of emf has some internal resistance r. The terminal potential difference V_{ab} then depends on current. (See Examples 25.5 through 25.8)

$$V_{ab} = \mathcal{E} - Ir \quad (25.15)$$

(source with internal resistance)

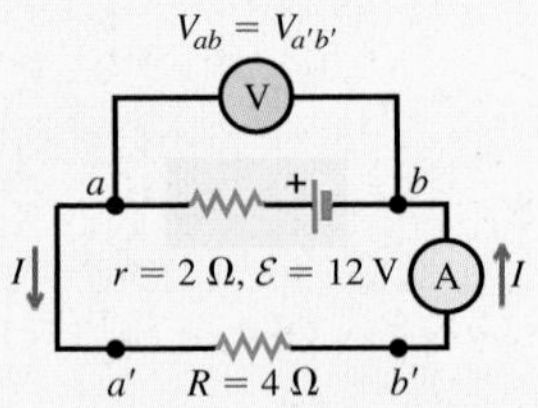

A circuit element with a potential difference $V_a - V_b = V_{ab}$ and a current I puts energy into a circuit if the current direction is from lower to higher potential in the device, and it takes energy out of the circuit if the current is opposite. The power P (rate of energy transfer) is equal to the product of the potential difference and the current. A resistor always takes electrical energy out of a circuit. (See Examples 25.9 through 25.11)

$$P = V_{ab}I \quad (25.17)$$

(general circuit element)

$$P = V_{ab}I = I^2R = \frac{V_{ab}^{\ 2}}{R} \quad (25.18)$$

(power into a resistor)

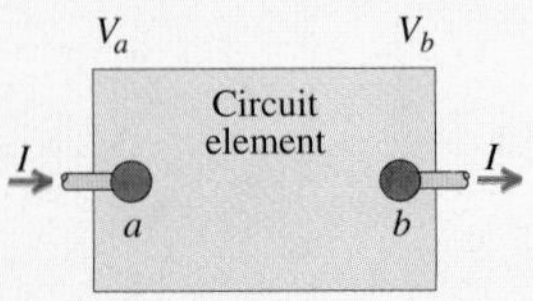

The microscopic basis of conduction in metals is the motion of electrons that move freely through the metallic crystal, bumping into ion cores in the crystal. In a crude classical model of this motion, the resistivity of the material can be related to the electron mass, charge, speed of random motion, density, and mean free time between collisions. (See Example 25.12)

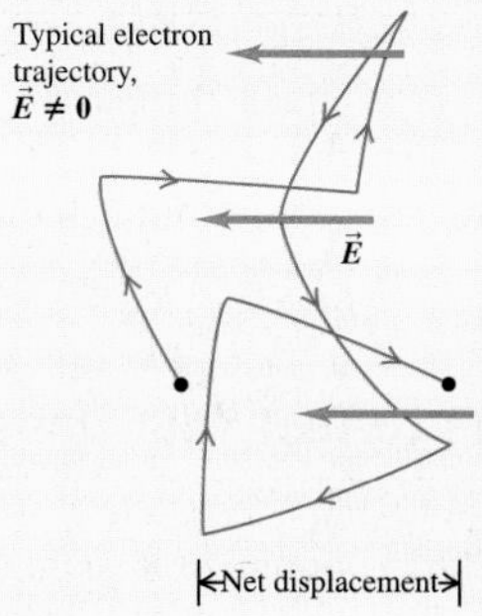

Key Terms

current, 943
drift velocity, 943
conventional current, 944
ampere, 945
concentration, 945
current density, 945
Ohm's law, 945
resistivity, 948
conductivity, 948
temperature coefficient of resistivity, 949
resistance, 951
ohm, 951
resistor, 952
complete circuit, 955
electromotive force (emf), 955
source of emf, 955
internal resistance, 957
terminal voltage, 957
voltmeter, 958
ammeter, 958
mean free time, 967

Your Notes

Answer to Chapter Opening Question

The current out equals the current in. In other words, charge must enter the bulb at the same rate as it exits the bulb. It is not "used up" or consumed as it flows through the bulb.

Answers to Test Your Understanding Questions

Section 25.1 Doubling the diameter increases the cross-sectional area A by a factor of 4. Hence the current density magnitude $J = I/A$ is reduced to $\frac{1}{4}$ of the value in Example 25.1, and the magnitude of the drift velocity $v_d = J/n|q|$ is reduced by the same factor. The new magnitude is $v_d = (0.15\text{ mm/s})/4 = 0.038\text{ mm/s}$. This behavior is the same as that of an incompressible fluid, which slows down when it moves from a narrow pipe to a broader one (see Section 14.4).

Section 25.2 Figure 25.6b shows that the resistivity ρ of a semiconductor increases as the temperature decreases. From Eq. (25.5), the magnitude of the current density is $J = E/\rho$, so the current density decreases as the temperature drops and the resistivity increases.

Section 25.3 Solving Eq. (25.11) for the current shows that $I = V/R$. If the resistance R of the wire remained the same, doubling the voltage V would make the current I double as well. However, we saw in Example 25.3 that the resistance is *not* constant: as the current increases and the temperature increases, R increases as well. Thus doubling the voltage produces a current that is *less* than double the original current. An ohmic conductor is one for which $R = V/I$ has the same value no matter what the voltage, so the wire is *nonohmic.* (In many practical problems the temperature change of the wire is so small that it can be ignored, so we can safely regard the wire as being ohmic. We do so in almost all examples in this book.)

Section 25.4 From Eq. (25.15) the terminal voltage is $V_{ab} = \mathcal{E} - Ir$, so the current in the circuit is $I = (\mathcal{E} - V_{ab})/r = (1.5\text{ V} - 1.3\text{ V})/(0.10\ \Omega) = 2.0\text{ A}$. The terminal voltage of the battery is also the voltage across the resistor, so $V_{ab} = IR$ and $R = V_{ab}/I = (1.3\text{ V})/(2.0\text{ A}) = 0.65\ \Omega$.

Section 25.5 With $r = 0$, the current in the circuit is $I = \mathcal{E}/(R + r) = (12\text{ V})/(4\ \Omega + 0\ \Omega) = 3\text{ A}$. The rate of energy conversion in the battery is $\mathcal{E}I = (12\text{ V})(3\text{ A}) = 36\text{ W}$. Since $r = 0$, this is the same as the electrical power output of the battery. It is also the same as the rate of energy dissipation in the resistor $I^2R = (3\text{ A})^2(4\ \Omega) = 36\text{ W}$. By comparison, with $r = 2\ \Omega$ as in Example 25.9, the battery power output and the rate of energy dissipation in the resistor were both 16 W. This example illustrates how as a battery ages and its internal resistance increases, the power that the battery delivers to the circuit decreases.

Section 25.6 Between collisions an electron drifts at $v_d = 1.5 \times 10^{-4}$ m/s for a time $\tau = 2.4 \times 10^{-14}$ s, covering a distance $v_d\tau = (1.5 \times 10^{-4}\text{ m/s})(2.4 \times 10^{-14}\text{ s}) = 3.6 \times 10^{-18}$ m. By comparison, the radius of a copper atom is a billion times larger, about 10^{-9} m. A single electron drifts by only an infinitesimal distance between collisions!

Discussion Questions

Q25.1 How would you expect the resistivity of a good insulator such as glass or polystyrene to vary with temperature? Explain your reasoning.

Q25.2 Two copper wires of different diameter are joined end-to-end. If a current flows in the wire combination, what happens to electrons when they move from the larger-diameter wire into the smaller-diameter wire? Does their drift speed increase, decrease, or stay the same? If the drift speed changes, what is the force that causes the change? Explain your reasoning.

Q25.3 What is the difference between an emf and a potential difference? Under what circumstances are the potential difference between the terminals of a battery and the emf of the battery equal to each other? Under what circumstances are they not equal?

Q25.4 Can the potential difference between the terminals of a battery ever be opposite in direction to the emf? If it can, give an example. If it cannot, explain why not.

Q25.5 A rule of thumb used to determine the internal resistance of a source is that it is the open-circuit voltage divided by the short-circuit current. Is this correct? Why or why not?

Q25.6 Batteries are always labeled with their emf; for instance, a AA flashlight battery is labeled "1.5 volts." Would it also be appropriate to put a label on batteries stating how much current they provide? Why or why not?

Q25.7 Is there an electric field in the space around an ordinary flashlight battery if it is not connected to anything? Explain.

Q25.8 We have seen that a coulomb is an enormous amount of charge; it is virtually impossible to place a charge of 1 C on an object. Yet, a current of 10 A, 10 C/s, is quite reasonable. Explain this apparent discrepancy.

Q25.9 Electrons in an electric circuit pass through a resistor. The wire on either side of the resistor has the same diameter. a) How does the drift speed of the electrons before entering the resistor compare to the speed after leaving the resistor? Explain your reasoning. b) How does the potential energy for an electron before entering the resistor compare to the potential energy after leaving the resistor? Explain your reasoning.

Q25.10 Why does an electric light bulb nearly always burn out just as you turn on the light, but almost never while the light is shining?

Q25.11 A light bulb glows because it has resistance. The brightness of a light bulb increases with the electrical power dissipated in the bulb. a) In the circuit shown in Fig. 25.28a, the two bulbs A and B are identical. Compared to bulb A, does bulb B glow more brightly, just as brightly, or less brightly? Explain your reasoning.

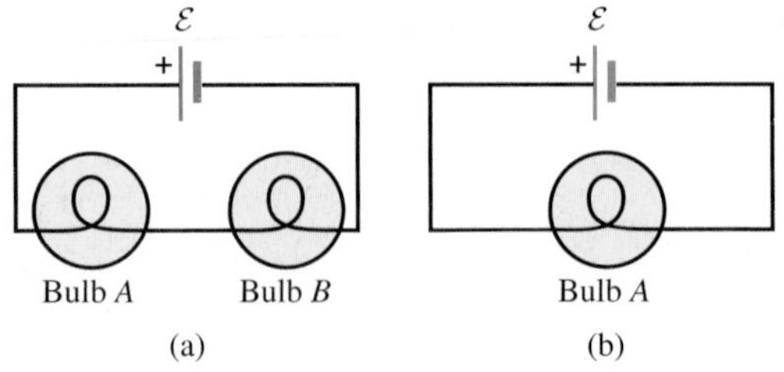

Figure 25.28 Question Q25.11.

b) Bulb B is removed from the circuit and the circuit is completed as shown in Fig. 25.28b. Compared to the brightness of bulb A in Fig. 25.28a, does bulb A now glow more brightly, just as brightly, or less brightly? Explain your reasoning.

Q25.12 (See Question Q25.11.) An ideal ammeter A is placed in a circuit with a battery and a light bulb as shown in Fig. 25.29a, and the ammeter reading is noted. The circuit is then reconnected as in Fig. 25.29b, so that the positions of the ammeter and light bulb are reversed. a) How does the ammeter reading in the situation shown in Fig. 25.29a compare to the reading in the situation shown in Fig. 25.29b? Explain your reasoning. b) In which situation does the light bulb glow more brightly? Explain your reasoning.

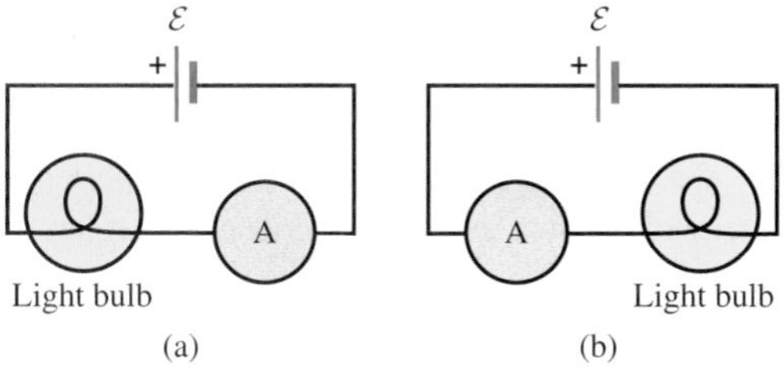

Figure 25.29 Question Q25.12.

Q25.13 (See Question Q25.11.) Will a light bulb glow more brightly if it is connected to a battery as shown in Fig. 25.30a, in which an ideal ammeter A is placed in the circuit, or if it is connected as shown in Fig. 25.30b, in which an ideal voltmeter V is placed in the circuit? Explain your reasoning.

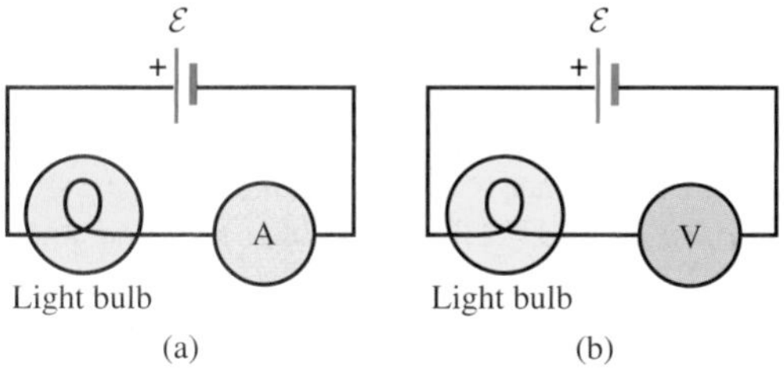

Figure 25.30 Question Q25.13.

Q25.14 The energy that can be extracted from a storage battery is always less than the energy that goes into it while it is being charged. Why?

Q25.15 Eight flashlight batteries in series have an emf of about 12 V, similar to that of a car battery. Could they be used to start a car with a dead battery? Why or why not?

Q25.16 A would-be inventor proposed to increase the power supplied by a battery to a light bulb by using thick wire near the battery and thinner wire near the bulb. In the thin wire the electrons from the thick wire would become more densely packed, more electrons per second would reach the bulb, and the bulb would receive more power than if a wire of constant cross section were used. What do you think of this scheme?

Q25.17 Small aircraft often have 24-V electrical systems rather than the 12-V systems in automobiles, even though the electrical power requirements are roughly the same in both applications. The explanation given by aircraft designers is that a 24-V system weighs less than a 12-V system, because thinner wires can be used. Explain why this is so.

Q25.18 Long-distance, electric-power, transmission lines always operate at very high voltage, sometimes as much as 750 kV. What are the advantages of such high voltages? What are the disadvantages?

Q25.19 Ordinary household electric lines in North America usually operate at 120 V. Why is this a desirable voltage, rather than a value considerably larger or smaller? On the other hand, automobiles usually have 12-V electrical systems. Why is this a desirable voltage?

Q25.20 A fuse is a device designed to break a circuit, usually by melting when the current exceeds a certain value. What characteristics should the material of the fuse have?

Q25.21 High-voltage power supplies are sometimes designed intentionally to have rather large internal resistance as a safety precaution. Why is such a power supply with a large internal resistance safer than one with the same voltage but lower internal resistance?

Q25.22 The text states that good thermal conductors are also good electrical conductors. If so, why don't the cords used to connect toasters, irons, and similar heat-producing appliances get hot by conduction of heat from the heating element?

Exercises

Section 25.1 Current

25.1 A current of 3.6 A flows through an automobile headlight. How many coulombs of charge flow through the headlight in 3.0 h?

25.2 A silver wire 2.6 mm in diameter transfers a charge of 420 C in 80 min. Silver contains 5.8×10^{28} free electrons per cubic meter. a) What is the current in the wire? b) What is the magnitude of the drift velocity of the electrons in the wire?

25.3 Copper has 8.5×10^{28} free electrons per cubic meter. A 71.0-cm length of 12-gauge copper wire, that is 2.05 mm in diameter, carries 4.85 A of current. a) How much time does it take for an electron to travel the length of the wire? b) Repeat part (a) for 6-gauge copper wire (diameter 4.12 mm) of the same length that carries the same current. c) Generally speaking, how does changing the diameter of a wire that carries a given amount of current affect the drift velocity of the electrons in the wire?

25.4 A metallic wire has a diameter of 4.12 mm. When the current in the wire is 8.00 A, the drift velocity is 5.40×10^{-5} m/s. What is the density of free electrons in the metal?

25.5 When a wire carries a current of 1.20 A, the drift velocity is 1.20×10^{-4} m/s. What is the drift velocity when the current is 6.00 A?

25.6 Consider the 18-gauge wire in Example 25.1. How many atoms are there in 1.00 m^3 of copper? With the density of free electrons given in the example, how many free electrons are there per copper atom?

25.7 Assume that in silver metal there is one free electron per silver atom. Compute the free electron density for silver and compare to the value given in Exercise 25.2.

25.8 Current passes through a solution of sodium chloride. In 1.00 s, 2.68×10^{16} Na^+ ions arrive at the negative electrode and

3.92×10^{16} Cl^- ions arrive at the positive electrode. What is the current passing between the electrodes? b) What is the direction of the current?

25.9 The current in a wire varies with time according to the relation $I = 55\text{ A} - (0.65\text{ A/s}^2)t^2$. a) How many coulombs of charge pass a cross section of the wire in the time interval between $t = 0$ and $t = 8.0$ s? b) What constant current would transport the same charge in the same time interval?

Section 25.2 Resistivity and Section 25.3 Resistance

25.10 A copper wire has a square cross section 2.3 mm on a side. The wire is 4.0 m long and carries a current of 3.6 A. The density of free electrons is $8.5 \times 10^{28}/\text{m}^3$. Find the magnitudes of a) the current density in the wire; b) the electric field in the wire. c) How much time is required for an electron to travel the length of the wire?

25.11 In household wiring, copper wire 2.05 mm in diameter is often used. Find the resistance of a 24.0-m length of this wire.

25.12 What length of copper wire, 0.462 mm in diameter, has a resistance of 1.00 Ω?

25.13 In an experiment conducted at room temperature, a current of 0.820 A flows through a wire 3.26 mm in diameter. Find the magnitude of the electric field in the wire if the wire is made of a) tungsten; b) aluminum.

25.14 What diameter must a copper wire have if its resistance is to be the same as that of an equal length of aluminum wire with diameter 3.26 mm?

25.15 You need to produce a set of cylindrical copper wires 3.50 m long that will have a resistance of 0.125 Ω each. What will be the mass of each of these wires?

25.16 A tightly coiled spring having 75 coils, each 3.50 cm in diameter, is made of insulated metal wire 3.25 mm in diameter. An ohmmeter connected across its opposite ends reads 1.74 Ω. What is the resistivity of the metal?

25.17 Critical Current Density in Superconductors. One problem with some of the newer high-temperature superconductors is getting a large enough current density for practical use without causing the resistance to reappear. The maximum current density for which the material will remain a superconductor is called the critical current density of the material. In 1987, IBM research labs had produced thin films with critical current densities of $1.0 \times 10^5\text{ A/cm}^2$. a) How much current could a 18-gauge wire (see Example 25.1 in Section 25.1) of this material carry and still remain superconducting? b) Researchers are trying to develop superconductors with critical current densities of $1.0 \times 10^6\text{ A/cm}^2$. What diameter cylindrical wire of such a material would be needed to carry 1000 A without losing its superconductivity?

25.18 At 20°C, the electric field in a tungsten wire is 0.0560 V/m when the current in the wire is 2.80 A. What electric field in the wire is required for this same current at 320°C?

25.19 An aluminum cube has a side length of 1.80 m. What is the resistance between two opposite faces of the cube?

25.20 A battery-powered light bulb has a tungsten filament. When the switch connecting the bulb to the battery is first turned on and the temperature of the bulb is 20°C, the current in the bulb is 0.860 A. After the bulb has been on for 30 s, the current is 0.220 A. What is then the temperature of the filament?

25.21 As part of a lecture demonstration, a physics professor plans to hold an uninsulated current-carrying wire in her hands. For safety's sake, the potential difference between her hands is to be no more than 1.50 V. She holds her hands 1.20 m apart, with the wire stretched tightly between them. The wire is to carry 6.00 A of current and is to be made of aluminum. What is the minimum wire radius that is consistent with safety?

25.22 You apply a potential difference of 4.50 V between the ends of a wire that is 2.50 m in length and 0.654 mm in radius. The resulting current through the wire is 17.6 A. What is the resistivity of the wire?

25.23 A current-carrying gold wire has diameter 0.84 mm. The electric field in the wire is 0.49 V/m. What is a) the current carried by the wire? b) the potential difference between two points in the wire 6.4 m apart? c) the resistance of a 6.4-m length of this wire?

25.24 A wire of length L and cross-sectional area A has resistance R. What will the resistance of the wire be if it is stretched to twice its original length? Assume that the density and resistivity of the material do not change when the wire is stretched.

25.25 The potential difference between points in a wire 75.0 cm apart is 0.938 V when the current density is $4.40 \times 10^7\text{ A/m}^2$. What is a) the magnitude of $\vec{E}$ in the wire? b) the resistivity of the material of which the wire is made?

25.26 A certain resistor has a resistance of 1.484 Ω at 20.0°C and a resistance of 1.512 Ω at 34.0°C. What is its temperature coefficient of resistivity?

25.27 a) What is the resistance of a Nichrome wire at 0.0°C if its resistance is 100.00 Ω at 11.5°C? b) What is the resistance of a carbon rod at 25.8°C if its resistance is 0.0160 Ω at 0.0°C?

25.28 A carbon resistor is to be used as a thermometer. On a winter day when the temperature is 4.0°C, the resistance of the carbon resistor is 217.3 Ω. What is the temperature on a spring day when the resistance is 215.8 Ω? (Take the reference temperature T_0 to be 4.0°C.)

25.29 A strand of wire has resistance 5.60 μΩ. Find the net resistance of 120 such strands if they are a) placed side by side to form a cable of the same length as a single strand; b) connected end to end to form a wire 120 times as long as a single strand.

25.30 The open-circuit terminal voltage of a battery is 12.6 V. When a resistor $R = 4.00\ \Omega$ is connected between the terminals of the battery, the terminal voltage of the battery is 10.4 V. What is the internal resistance of the battery?

Section 25.4 Electromotive Force and Circuits

25.31 A copper transmission cable 100 km long and 10.0 cm in diameter carries a current of 125 A. a) What is the potential drop across the cable? b) How much electrical energy is dissipated as thermal energy every hour?

25.32 Consider the circuit shown in Fig. 25.31. The terminal voltage of the 24.0-V battery is 21.2 V. What is a) the internal

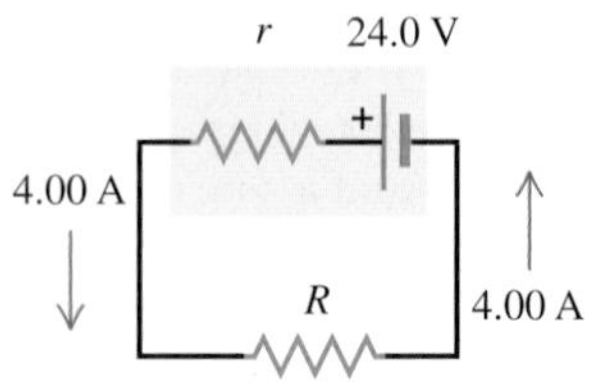

Figure 25.31 Exercise 25.32.

resistance r of the battery; b) the resistance R of the circuit resistor?

25.33 An ideal voltmeter V is connected to a 2.0-Ω resistor and a battery with emf 5.0 V and internal resistance 0.5 Ω as shown in Fig. 25.32. a) What is the current in the 2.0-Ω resistor? b) What is the terminal voltage of the battery? c) What is the reading on the voltmeter? Explain your answers.

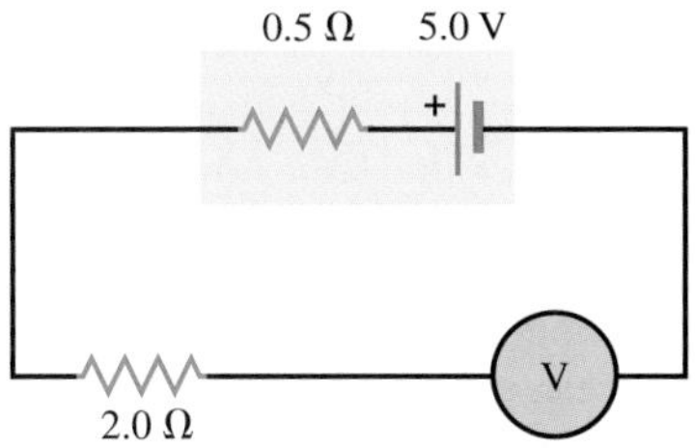

Figure 25.32 Exercise 25.33.

25.34 A complete circuit consists of a 24.0-V battery, a 5.60-Ω resistor, and a switch. The internal resistance of the battery is 0.28 Ω. The switch is opened. What does an ideal voltmeter read when placed a) across the terminals of the battery? b) across the resistor? c) across the switch? d) Repeat parts (a), (b), and (c) for the case when the switch is closed.

25.35 When switch S in Fig. 25.33 is open, the voltmeter V of the battery reads 3.08 V. When the switch is closed, the voltmeter reading drops to 2.97 V, and the ammeter A reads 1.65 A. Find the emf, the internal resistance of the battery, and the circuit resistance R. Assume that the two meters are ideal, so they don't affect the circuit.

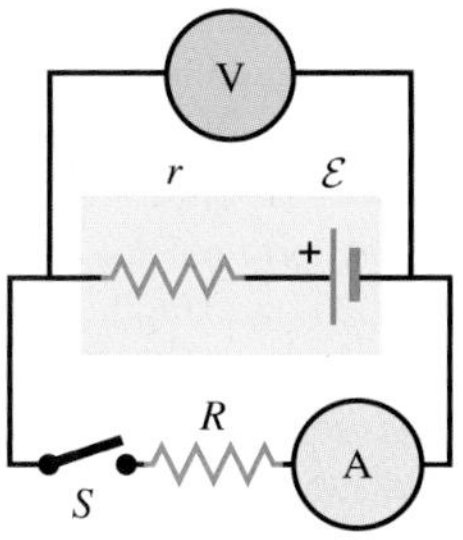

Figure 25.33 Exercise 25.35.

25.36 The circuit shown in Fig. 25.34 contains two batteries, each with an emf and an internal resistance, and two resistors. Find a) the current in the circuit (magnitude *and* direction); b) the terminal voltage V_{ab} of the 16.0-V battery; c) the potential difference V_{ac} of point a with respect to point c. d) Using Fig. 25.20 as a model, graph the potential rises and drops in this circuit.

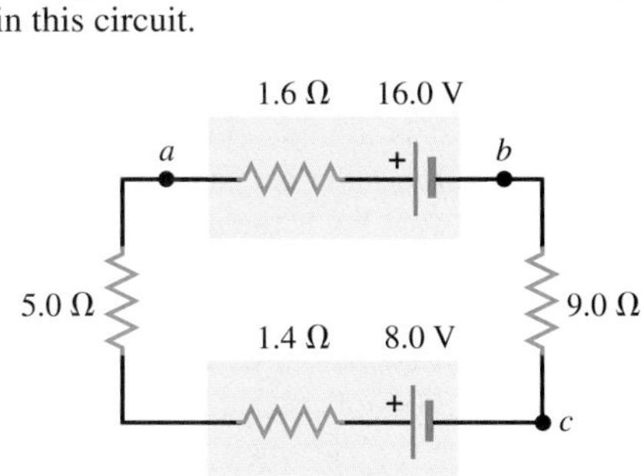

Figure 25.34 Exercises 25.36, 25.37, 25.38, and 25.46.

25.37 In the circuit shown in Fig. 25.34, the 16.0-V battery is removed and reinserted with the opposite polarity, so that its negative terminal is now next to point a. Find a) the current in the circuit (magnitude *and* direction); b) the terminal voltage V_{ba} of the 16.0-V battery; c) the potential difference V_{ac} of point a with respect to point c. d) Graph the potential rises and drops in this circuit (see Fig. 25.20).

25.38 In the circuit of Fig. 25.34, the 5.0-Ω resistor is removed and replaced by a resistor of unknown resistance R. When this is done, an ideal voltmeter connected across the points b and c reads 1.9 V. Find a) the current in the circuit; b) the resistance R. c) Graph the potential rises and drops in this circuit (see Fig. 25.20).

25.39 The following measurements of current and potential difference were made on a resistor constructed of Nichrome wire:

I (A)	0.50	1.00	2.00	4.00
V_{ab} (V)	1.94	3.88	7.76	15.52

a) Graph V_{ab} as a function of I. b) Does Nichrome obey Ohm's law? How can you tell? c) What is the resistance of the resistor in ohms?

25.40 The following measurements were made on a Thyrite resistor:

I (A)	0.50	1.00	2.00	4.00
V_{ab} (V)	2.55	3.11	3.77	4.58

a) Graph V_{ab} as a function of I. b) Does Thyrite obey Ohm's law? How can you tell? c) Graph the resistance $R = V_{ab}/I$ as a function of I.

25.41 The internal resistance of a flashlight battery increases gradually with age, even though the battery is not used. The emf, however, remains fairly constant at about 1.5 V. Batteries may be tested for age at the time of purchase by connecting an ammeter directly across the terminals of the battery and reading the current. The resistance of the ammeter is so small that the battery is practically short-circuited. a) The short-circuit current of a fresh flashlight battery (1.50-V emf) is 14.8 A. What is the internal resistance? b) What is the internal resistance if the short-circuit current is only 6.8 A? c) The short-circuit current of a 12.6-V car battery may be as great as 1000 A. What is its internal resistance?

Section 25.5 Energy and Power in Electric Circuits

25.42 A resistor with a 15.0-V potential difference across its ends develops thermal energy at a rate of 327 W. a) What is its resistance? b) What is the current in the resistor?

25.43 To stun its prey, the electric eel *Electrophorus electricus* generates 0.80-A pulses of current along its skin. This current flows across a 650-V potential difference. At what rate does *Electrophorus* deliver energy to its prey?

25.44 A battery-powered global positioning system (GPS) receiver operating on 9.0 V draws a current of 0.13 A. How much electrical energy does it consume during $1\frac{1}{2}$ h?

25.45 Consider a resistor with length L, uniform cross-sectional area A, and uniform resistivity ρ that is carrying a current with uniform current density J. Use Eq. (25.18) to find the electrical power dissipated per unit volume, p. Express your result in terms of a) E and J; b) J and ρ; c) E and ρ.

25.46 Consider the circuit of Fig. 25.34. a) What is the total rate at which electrical energy is dissipated in the 5.00-Ω and 9.00-Ω resistors? b) What is the power output of the 16.0-V battery? c) At what rate is electrical energy being converted to other forms in the 8.0-V battery? d) Show that the power output of the 16.0-V battery equals the overall rate of dissipation of electrical energy in the rest of the circuit.

25.47 The capacity of a storage battery, such as those used in automobile electrical systems, is rated in ampere-hours (A·h). A 50-A·h battery can supply a current of 50 A for 1.0 h, or 25 A for 2.0 h, and so on. a) What total energy can be supplied by a 12-V, 60-A·h battery if its internal resistance is negligible? b) What volume (in liters) of gasoline has a total heat of combustion equal to the energy obtained in part (a)? (See Section 17.6. The density of gasoline is 900 kg/m^3.) c) If a generator with an average electrical power output of 0.45 kW is connected to the battery, how much time will be required for it to charge the battery fully?

25.48 In the circuit analyzed in Example 25.9 (Section 25.5) the 4.0-Ω resistor is replaced by a 8.0-Ω resistor, as in Example 25.10 (Section 25.5). a) Calculate the rate of conversion of chemical energy to electrical energy in the battery. How does your answer compare to the result calculated in Example 25.9? b) Calculate the rate of electrical energy dissipation in the internal resistance of the battery. How does your answer compare to the result calculated in Example 25.9? c) Use the results of parts (a) and (b) to calculate the net power output of the battery. How does your result compare to the electrical power dissipated in the 8.0-Ω resistor as calculated for this circuit in Example 25.10?

25.49 In the circuit in Fig. 25.35, find a) the rate of conversion of internal (chemical) energy to electrical energy within the battery; b) the rate of dissipation of electrical energy in the battery; c) the rate of dissipation of electrical energy in the external resistor.

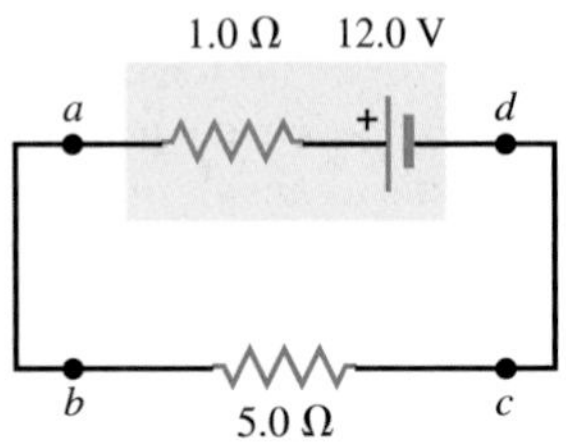

Figure 25.35 Exercise 25.49.

25.50 A typical small flashlight contains two batteries, each having an emf of 1.5 V, connected in series with a bulb having resistance 17 Ω. a) If the internal resistance of the batteries is negligible, what power is delivered to the bulb? b) If the batteries last for 5.0 h, what is the total energy delivered to the bulb? c) The resistance of real batteries increases as they run down. If the initial internal resistance is negligible, what is the combined internal resistance of both batteries when the power to the bulb has decreased to half its initial value? (Assume that the resistance of the bulb is constant. Actually, it will change somewhat when the current through the filament changes, because this changes the temperature of the filament and hence the resistivity of the filament wire.)

25.51 A "540-W" electric heater is designed to operate from 120-V lines. a) What is its resistance? b) What current does it draw? c) If the line voltage drops to 110 V, what power does the heater take? (Assume that the resistance is constant. Actually, it will change because of the change in temperature.) d) The heater coils are metallic, so that the resistance of the heater decreases with decreasing temperature. If the change of resistance with temperature is taken into account, will the electrical power consumed by the heater be larger or smaller than what you calculated in part (c)? Explain.

*Section 25.6 Theory of Metallic Conduction

***25.52** Pure silicon contains approximately 1.0×10^{16} free electrons per cubic meter. a) At room temperature, what mean free time τ gives a value of the resistivity that agrees with the value given in Table 25.1? b) Your answer in part (a) is much greater than the mean free time for copper given in Example 25.12 (Section 25.6). Why, then, does pure silicon have such a high resistivity compared to copper?

Problems

25.53 An electrical conductor designed to carry large currents has a circular cross section 2.50 mm in diameter and is 14.0 m long. The resistance between its ends is 0.104 Ω. a) What is the resistivity of the material? b) If the electric field magnitude in the conductor is 1.28 V/m, what is the total current? c) If the material has 8.5×10^{28} free electrons per cubic meter, find the average drift speed under the conditions of part (b).

25.54 A plastic tube 25.0 m long and 4.00 cm in diameter is dipped into a silver solution, depositing a layer of silver 0.100 mm thick uniformly over the outer surface of the tube. If this coated tube is then connected across a 12.0-V battery, what will be the current?

25.55 On your first day at work as an electrical technician, you are asked to determine the resistance per meter of a long piece of wire. The company you work for is poorly equipped. You find a battery, a voltmeter, and an ammeter, but no meter for directly measuring resistance (an ohmmeter). You put the leads from the voltmeter across the terminals of the battery, and the meter reads 12.6 V. You cut off a 20.0 m length of wire and connect it to the battery, with an ammeter in series with it to measure the current in the wire. The ammeter reads 7.00 A. You then cut off a 40.0 m length of wire and connect it to the battery, again with the ammeter in series to measure the current. The ammeter reads 4.20 A. Even though the equipment you have available to you is limited, your boss assures you of its high quality: the ammeter has very small resistance and the voltmeter has very large resistance. What is the resistance of one meter of wire?

25.56 A 2.0-m length of wire is made by welding the end of a 120-cm-long silver wire to the end of an 80-cm-long copper wire. Each piece of wire is 0.60 mm in diameter. The wire is at room temperature, so the resistivities are as given in Table 25.1. A potential difference of 5.0 V is maintained between the ends of the 2.0-m composite wire. a) What is the current in the copper section? b) What is the current in the silver section? c) What is the magnitude of $\vec{E}$ in the copper? d) What is the magnitude of $\vec{E}$ in the silver? e) What is the potential difference between the ends of the silver section of wire?

25.57 A 3.00-m length of copper wire at 20°C has a 1.20-m-long section with diameter 1.60 mm and a 1.80-m-long section with diameter 0.80 mm. There is a current of 2.5 mA in the 1.60-mm diameter section. a) What is the current in the 0.80-mm-diameter section? b) What is the magnitude of $\vec{E}$ in the 1.60-mm-diameter section? c) What is the magnitude of $\vec{E}$ in the 0.80-mm-diameter section? d) What is the potential difference between the ends of the 3.00-m length of wire?

25.58 a) The available kinetic energy per unit volume, due to the drift velocity of the conduction electrons in a current-carrying conductor, can be defined as $K/\text{volume} = n(\frac{1}{2}mv_d^2)$. Use this expression to evaluate K/volume for the copper wire and current of Example 25.1 (Section 25.1). b) Calculate the total change in electrical potential energy for the conduction electrons in 1.0 cm^3 of copper if they fall through a potential drop of 1.0 V. How does your answer compare to the available kinetic energy in 1.0 cm^3 due to the drift velocity?

25.59 A material of resistivity ρ is formed into a solid, truncated cone of height h and radii r_1 and r_2 at either end (Fig. 25.36). a) Calculate the resistance of the cone between the two flat end faces. (*Hint:* Imagine slicing the cone into very many thin disks, and calculate the resistance of one such disk.) b) Show that your result agrees with Eq. (25.10) when $r_1 = r_2$.

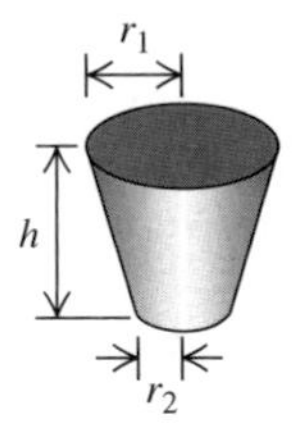

Figure 25.36 Problem 25.59.

25.60 The region between two concentric conducting spheres with radii a and b is filled with a conducting material with resistivity ρ. a) Show that the resistance between the spheres is given by

$$R = \frac{\rho}{4\pi}\left(\frac{1}{a} - \frac{1}{b}\right)$$

b) Derive an expression for the current density as a function of radius, in terms of the potential difference V_{ab} between the spheres. c) Show that the result in part (a) reduces to Eq. (25.10) when the separation $L = b - a$ between the spheres is small.

25.61 Leakage in a Dielectric. Two parallel plates of a capacitor have equal and opposite charges Q. The dielectric has a dielectric constant K and a resistivity ρ. Show that the "leakage" current I carried by the dielectric is given by $I = Q/K\epsilon_0\rho$.

25.62 A rectangular block of metal of resistivity ρ has dimensions $d \times 2d \times 3d$. A potential difference V is to be applied between two opposite faces of the block. a) To which two faces of the block should the potential difference be applied to give the maximum *current density?* What is this maximum current density? b) To which two faces of the block should the potential difference be applied to give the maximum *current?* What is this maximum current?

25.63 The temperature coefficient of resistance α in Eq. (25.12) equals the temperature coefficient of resistivity α in Eq. (25.6) only if the coefficient of thermal expansion is small. A cylindrical column of mercury is in a vertical glass tube. At 20°C, the length of the mercury column is 12.0 cm. The diameter of the mercury column is 1.6 mm and doesn't change with temperature because glass has a small coefficient of thermal expansion. The coefficient of volume expansion of the mercury is given in Table 17.2, its resistivity at 20°C is given in Table 25.1, and its temperature coefficient of resistivity is given in Table 25.2. a) At 20°C, what is the resistance between the ends of the mercury column? b) The mercury column is heated to 60°C. What is the change in its resistivity? c) What is the change in its length? Explain why the coefficient of volume expansion, rather than the coefficient of linear expansion, determines the change in length. d) What is the change in its resistance? (*Hint:* Since the percentage changes in ρ and L are small, you may find it helpful to derive from Eq. (25.10) an equation for ΔR in terms of $\Delta\rho$ and ΔL.) e) What is the temperature coefficient of resistance α for the mercury column, as defined in Eq. (25.12)? How does this value compare with the temperature coefficient of resistivity? Is the effect of the change in length important?

25.64 a) What is the potential difference V_{ad} in the circuit of Fig. 25.37? b) What is the terminal voltage of the 4.00-V battery? c) A battery with emf 10.30 V and internal resistance 0.50 Ω is inserted in the circuit at d, with its negative terminal connected to the negative terminal of the 8.00-V battery. What is the difference of potential V_{bc} between the terminals of the 4.00-V battery now?

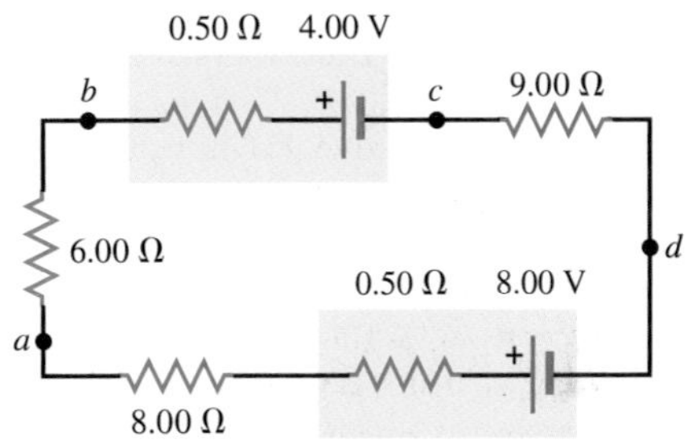

Figure 25.37 Problem 25.64.

25.65 The potential difference across the terminals of a battery is 8.4 V when there is a current of 1.50 A in the battery from the negative to the positive terminal. When the current is 3.50 A in the reverse direction, the potential difference becomes 9.4 V. a) What is the internal resistance of the battery? b) What is the emf of the battery?

25.66 A person with body resistance between his hands of 10 kΩ accidentally grasps the terminals of a 14-kV power supply. a) If the internal resistance of the power supply is 2000 Ω, what is the current through the person's body? b) What is the power dissipated in his body? c) If the power supply is to be made safe by increasing its internal resistance, what should the internal resistance be for the maximum current in the above situation to be 1.00 mA or less?

25.67 The average bulk resistivity of the human body (apart from surface resistance of the skin) is about 5.0 Ω·m. The conducting path between the hands can be represented approximately as a cylinder 1.6 m long and 0.10 m in diameter. The skin resistance can be made negligible by soaking the hands in salt water. a) What is the resistance between the hands if the skin resistance is negligible? b) What potential difference between the hands is needed for a lethal shock current of 100 mA? (Note that your result shows that small potential differences produce dangerous currents when the skin is damp.) c) With the current in part (b), what power is dissipated in the body?

25.68 A semiconductor device that does not obey Ohm's law has the current-voltage relation $V = \alpha I + \beta I^2$, with $\alpha = 2.50\ \Omega$ and $\beta = 0.360\ \Omega/\text{A}$. a) If the device is connected across a potential difference of 4.00 V, what is the current through the device? b) What potential difference is required to produce a current through the device twice that calculated in part (a)?

25.69 A 12.6-V car battery with negligible internal resistance is connected to a series combination of a 3.2-Ω resistor that obeys

Ohm's law and a thermistor that does not obey Ohm's law, but instead has a current-voltage relation $V = \alpha I + \beta I^2$, with $\alpha = 3.8\ \Omega$ and $\beta = 1.3\ \Omega/\text{A}$. What is the current through the 3.2-Ω resistor?

25.70 The open-circuit terminal voltage of a source is 7.86 V, and its short-circuit current is 9.25 A. a) What is the current when a resistor with resistance 2.4 Ω is connected to the terminals of the source? The resistor obeys Ohm's law. b) What is the current in the semiconductor device of Problem 25.68 when connected across the terminals of this source? c) What is the terminal voltage of the source at the current calculated in part (b)?

25.71 A Non-Ideal Ammeter. Unlike the idealized ammeter described in Section 25.4, any real ammeter has a nonzero resistance. a) An ammeter with resistance R_A is connected in series with a resistor R and a battery of emf $\mathcal{E}$ and internal resistance r. The current measured by the ammeter is I_A. Find the current through the circuit if the ammeter is removed so that the battery and the resistor form a complete circuit. Express your answer in terms of I_A, r, R_A, and R. The more "ideal" the ammeter, the smaller the difference between this current and the current I_A. b) If $R = 3.80\ \Omega$, $\mathcal{E} = 7.50$ V, and $r = 0.45\ \Omega$, find the maximum value of the ammeter resistance R_A so that I_A is within 1.0% of the current in the circuit when the ammeter is absent. c) Explain why your answer in part (b) represents a *maximum* value.

25.72 A Non-Ideal Voltmeter. Unlike the idealized voltmeter described in Section 25.4, any real voltmeter has a resistance that is not infinitely large. a) A voltmeter with resistance R_V is connected across the terminals of a battery of emf $\mathcal{E}$ and internal resistance r. Find the potential difference measured by the voltmeter. b) If $\mathcal{E} = 7.50$ V and $r = 0.45\ \Omega$, find the minimum value of the voltmeter resistance R_V so that the voltmeter reading is within 1.0% of the emf of the battery. c) Explain why your answer in part (b) represents a *minimum* value.

25.73 According to the U.S. National Electrical Code, copper wire used for interior wiring of houses, hotels, office buildings, and industrial plants is permitted to carry no more than a specified maximum amount of current. The table below shows the maximum current I_{max} for several common sizes of wire with varnished cambric insulation. The "wire gauge" is a standard method used to describe the diameter of wires. Note that the larger the diameter of the wire, the *smaller* the wire gauge.

Wire gauge	Diameter (cm)	I_{max} (A)
14	0.163	18
12	0.205	25
10	0.259	30
8	0.326	40
6	0.412	60
5	0.462	65
4	0.519	85

a) What considerations determine the maximum current-carrying capacity of household wiring? b) A total of 4200 W of power is to be supplied through the wires of a house to the household electrical appliances. If the potential difference across the group of appliances is 120 V, determine the gauge of the thinnest permissible wire that can be used. c) Suppose the wire used in this house is of the gauge found in part (b) and has total length 42.0 m. At what rate is energy dissipated in the wires? d) The house is built in a community where the consumer cost of electric energy is $0.11 per kilowatt-hour. If the house were built with wire of the next larger diameter than that found in part (b), what would be the savings in electricity costs in one year? Assume that the appliances are kept on for an average of 12 hours a day.

25.74 A toaster using a Nichrome heating element operates on 120 V. When it is switched on at 20°C, the heating element carries an initial current of 1.35 A. A few seconds later the current reaches the steady value of 1.23 A. a) What is the final temperature of the element? The average value of the temperature coefficient of resistivity for Nichrome over the temperature range is $4.5 \times 10^{-4}\ (\text{C}°)^{-1}$. b) What is the power dissipated in the heating element i) initially; ii) when the current reaches a steady value?

25.75 In the circuit of Fig. 25.38, find a) the current through the 8.0-Ω resistor; b) the total rate of dissipation of electrical energy in the 8.0-Ω resistor and in the internal resistance of the batteries. c) In one of the batteries, chemical energy is being converted into electrical energy. In which one is this happening, and at what rate? d) In one of the batteries, electrical energy is being converted into chemical energy. In which one is this happening, and at what rate? e) Show that the overall rate of production of electrical energy equals the overall rate of consumption of electrical energy in the circuit.

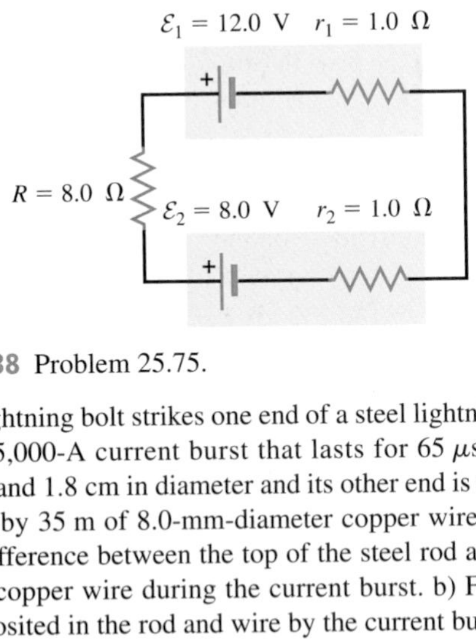

Figure 25.38 Problem 25.75.

25.76 A lightning bolt strikes one end of a steel lightning rod, producing a 15,000-A current burst that lasts for 65 μs. The rod is 2.0 m long and 1.8 cm in diameter and its other end is connected to the ground by 35 m of 8.0-mm-diameter copper wire. a) Find the potential difference between the top of the steel rod and the lower end of the copper wire during the current burst. b) Find the total energy deposited in the rod and wire by the current burst.

25.77 The *Tolman-Stewart experiment* (1916) demonstrated that the free charges in a metal have negative charge and provided a quantitative measurement of their charge to mass ratio, $|q|/m$. The experiment consisted of abruptly stopping a rapidly rotating spool of wire and measuring the potential difference that this produced between the ends of the wire. In a simplified model of this experiment, consider a metal rod of length L that is given a uniform acceleration $\vec{a}$ to the right. Initially the free charges in the metal lag behind the rod's motion, thus setting up an electric field $\vec{E}$ in the rod. In the steady state this field exerts a force on the free charges that makes them accelerate along with the rod. a) Apply $\Sigma\vec{F} = m\vec{a}$ to the free charges to obtain an expression for $|q|/m$ in terms of the magnitudes of the induced electric field $\vec{E}$ and the acceleration $\vec{a}$. b) If all the free charges in the metal rod have the same acceleration, the electric field $\vec{E}$ is the same at all points in the rod. Use this fact to rewrite the expression for $|q|/m$ in terms of the potential V_{bc} between the ends of the rod (Fig. 25.39). c) If the free charges have negative charge, which end of the rod, b or c, is at higher potential? d) If the rod is 0.50 m long and the free charges are electrons (charge $q = -1.60 \times 10^{-19}$ C, mass 9.11×10^{-31} kg), what magnitude of acceleration is required to produce a potential difference of 1.0 mV between the ends of the rod? e) Discuss why the actual experiment used a rotating spool of thin wire rather than a moving bar as in our simplified analysis.

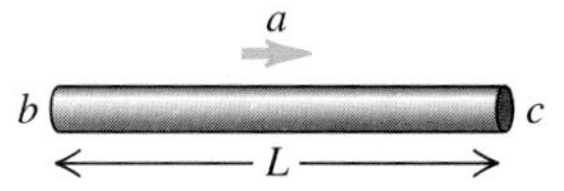

Figure 25.39 Problem 25.77.

25.78 A coil is to be used as an immersion heater for boiling water. The coil is to operate at 120 V and is to heat 250 cm^3 of water from 20°C to 100°C in 6.0 minutes. The specific heat capacity of water is 4190 J/kg·C°. a) What must the resistance of the coil be (assumed temperature-independent)? b) The coil is to be made of 25.0 cm^3 of Nichrome, which will be formed into a wire. What must the total length of the wire used to make the coil be, and what must the radius of the wire be? Assume that the wire has a circular cross section and that the volume of the Nichrome is unaffected by the process that forms it into a wire.

25.79 A 12.0-V battery has an internal resistance of 0.24 Ω and a capacity of 50.0 A·h (see Exercise 25.47). The battery is charged by passing a 10-A current through it for 5.0 h. a) What is the terminal voltage during charging? b) What total electrical energy is supplied to the battery during charging? c) What electrical energy is dissipated in the internal resistance during charging? d) The battery is now completely discharged through a resistor, again with a constant current of 10 A. What is the external circuit resistance? e) What total electrical energy is supplied to the external resistor? f) What total electrical energy is dissipated in the internal resistance? g) Why are the answers to parts (b) and (e) not equal?

25.80 Repeat Problem 25.79 with charge and discharge currents of 30 A. The charging and discharging times will now be 1.7 h rather than 5.0 h. What differences in performance do you see?

Challenge Problems

25.81 The temperature coefficient of resistivity α is given by

$$\alpha = \frac{1}{\rho}\frac{d\rho}{dT}$$

where ρ is the resistivity at the temperature T. Eq. (25.6) then follows if α is assumed constant and much smaller than $(T - T_0)^{-1}$. a) If α is not constant but is given by $\alpha = -n/T$, where T is the Kelvin temperature and n is a constant, show that the resistivity is given by $\rho = a/T^n$ where a is a constant. b) From Fig. 25.9, you can see that such a relation might be used as a rough approximation for a semiconductor. Using the values of ρ and α for carbon from Table 25.1 and Table 25.2, determine a and n. (In Table 25.1, assume that "room temperature" means 293 K.) c) Using your result from part (b), determine the resistivity of carbon at −196°C and 300°C. (Remember to express T in kelvins.)

25.82 The current-voltage relationship of a semiconductor diode is given by

$$I = I_S\left[\exp\left(\frac{eV}{kT}\right) - 1\right]$$

where I and V are the current through, and the voltage across, the diode, respectively. I_S is a constant characteristic of the device, e is the magnitude of the electron charge, k is the Boltzmann constant, and T is the Kelvin temperature. Such a diode is connected in series with a resistor with $R = 1.00\ \Omega$ and a battery with $\mathcal{E} = 2.00$ V. The polarity of the battery is such that the current through the diode is in the forward direction (Fig. 25.40). The battery has negligible internal resistance. a) Obtain an equation for V. Note that you cannot solve for V algebraically. b) The value of V must be obtained by using a numerical method. One approach is to try a value of V, see how the left- and right-hand sides of the equation compare for this V, and use this to refine your guess for V. Using $I_S = 1.50$ mA and $T = 293$ K, obtain a solution (accurate to three significant figures) for the voltage drop V across the diode and the current I through it.

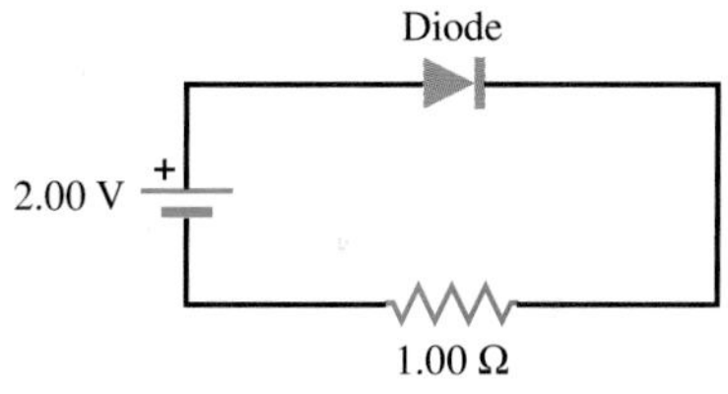

Figure 25.40 Challenge Problem 25.82.

25.83 The resistivity of a semiconductor can be modified by adding different amounts of impurities. A rod of semiconducting material of length L and cross-sectional area A lies along the x-axis between $x = 0$ and $x = L$. The material obeys Ohm's law, and its resistivity varies along the rod according to $\rho(x) = \rho_0\exp(-x/L)$. The end of the rod at $x = 0$ is at a potential V_0 greater than the end at $x = L$. a) Find the total resistance of the rod and the current in the rod. b) Find the electric field magnitude $E(x)$ in the rod as a function of x. c) Find the electric potential $V(x)$ in the rod as a function of x. d) Graph the functions $\rho(x)$, $E(x)$, and $V(x)$ for values of x between $x = 0$ and $x = L$.

25.84 A source with emf $\mathcal{E}$ and internal resistance r is connected to an external circuit. a) Show that the power output of the source is maximum when the current in the circuit is $\frac{1}{2}$ the short-circuit current of the source. b) If the external circuit consists of a resistance R, show that the power output is maximum when $R = r$, and that the maximum power is $\mathcal{E}^2/4r$.

26 DIRECT-CURRENT CIRCUITS

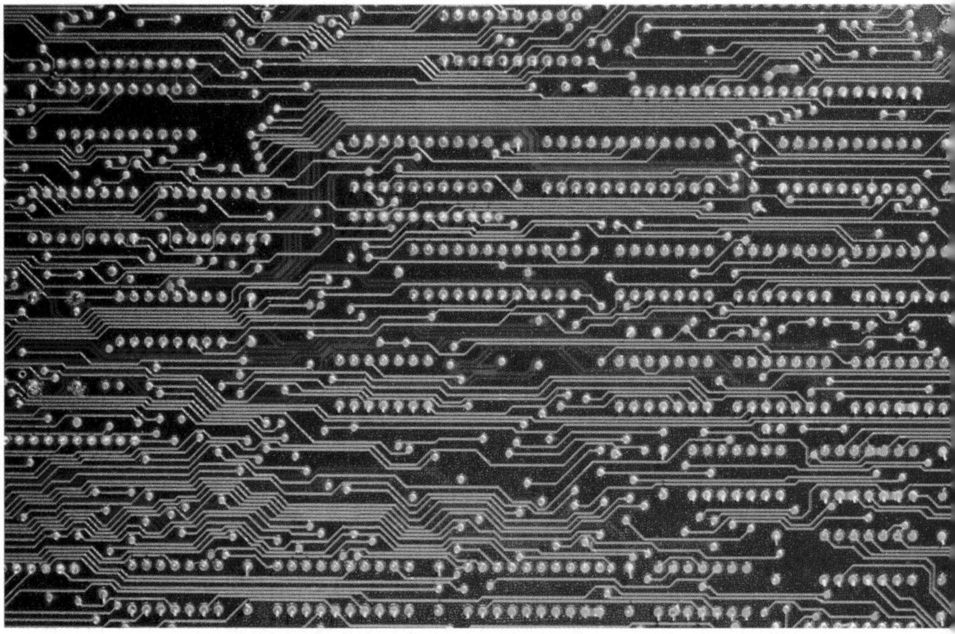

Complex circuits are at the heart of all modern electronic devices. The conducting paths in this printed circuit are thin films (shown in blue-green) deposited on an insulating board. The operation of any such circuit, no matter how complex, can be understood using Kirchhoff's rules—a central topic of this chapter.

? **Is it possible to connect several resistors with different resistances so that they all have the same potential difference? If so, will the current be the same through all of the resistors?**

If you look inside your TV, your computer, or your stereo receiver or under the hood of a car, you will find circuits of much greater complexity than the simple circuits we studied in Chapter 25. Whether connected by wires or integrated in a semiconductor chip, these circuits often include several sources, resistors, and other circuit elements, such as capacitors, transformers, and motors, interconnected in a *network.*

In this chapter we study general methods for analyzing such networks, including how to find unknown voltages, currents, and properties of circuit elements. We'll learn how to determine the equivalent resistance for several resistors connected in series or in parallel. For more general networks we need two rules called *Kirchhoff's rules.* One is based on the principle of conservation of charge applied to a junction; the other is derived from energy conservation for a charge moving around a closed loop. We'll discuss instruments for measuring various electrical quantities. We also look at a circuit containing resistance and capacitance, in which the current varies with time.

Our principal concern in this chapter is with **direct-current** (dc) circuits, in which the direction of the current does not change with time. Flashlights and automobile wiring systems are examples of direct-current circuits. Household electrical power is supplied in the form of **alternating current** (ac), in which the current oscillates back and forth. The same principles for analyzing networks apply to both kinds of circuits, and we conclude this chapter with a look at household wiring systems. We'll discuss alternating-current circuits in detail in Chapter 31.

26.1 | Resistors in Series and Parallel

Resistors turn up in all kinds of circuits, ranging from hair dryers and space heaters to circuits that limit or divide current or reduce or divide a voltage. Such circuits often contain several resistors, so it's appropriate to consider *combinations* of resistors. A simple example is a string of light bulbs used for holiday decorations; each bulb acts as a resistor, and from a circuit-analysis perspective the string of bulbs is simply a combination of resistors.

Suppose we have three resistors with resistances R_1, R_2, and R_3. Figure 26.1 shows four different ways in which they might be connected between points a and b. When several circuit elements such as resistors, batteries, and motors are connected in sequence as in Fig. 26.1a, with only a single current path between the points, we say that they are connected in **series.** We studied *capacitors* in series in Section 24.2; we found that, because of conservation of charge, capacitors in series all have the same charge if they are initially uncharged. In circuits we're often more interested in the *current,* which is charge flow per unit time.

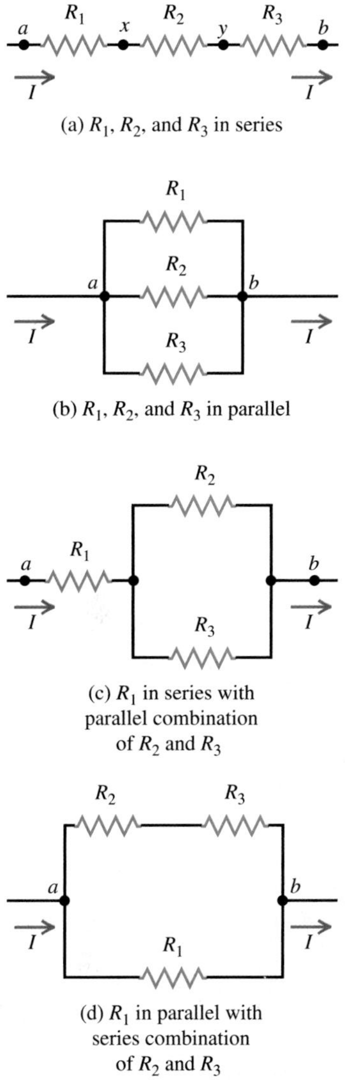

(a) R_1, R_2, and R_3 in series

(b) R_1, R_2, and R_3 in parallel

(c) R_1 in series with parallel combination of R_2 and R_3

(d) R_1 in parallel with series combination of R_2 and R_3

26.1 Four different ways of connecting three resistors.

The resistors in Fig. 26.1b are said to be connected in **parallel** between points a and b. Each resistor provides an alternative path between the points. For circuit elements that are connected in parallel, the *potential difference* is the same across each element. We studied capacitors in parallel in Section 24.2.

In Fig. 26.1c, resistors R_2 and R_3 are in parallel, and this combination is in series with R_1. In Fig. 26.1d, R_2 and R_3 are in series, and this combination is in parallel with R_1.

For any combination of resistors we can always find a *single* resistor that could replace the combination and result in the same total current and potential difference. For example, a string of holiday light bulbs could be replaced by a single, appropriately chosen light bulb that would draw the same current and have the same potential difference between its terminals as the original string of bulbs. The resistance of this single resistor is called the **equivalent resistance** of the combination. If any one of the networks in Fig. 26.1 were replaced by its equivalent resistance R_{eq}, we could write

$$V_{ab} = IR_{eq} \quad \text{or} \quad R_{eq} = \frac{V_{ab}}{I}$$

where V_{ab} is the potential difference between terminals a and b of the network and I is the current at point a or b. To compute an equivalent resistance, we assume a potential difference V_{ab} across the actual network, compute the corresponding current I, and take the ratio V_{ab}/I.

Resistors in Series

We can derive general equations for the equivalent resistance of a series or parallel combination of resistors. If the resistors are in *series,* as in Fig. 26.1a, the current I must be the same in all of them. (As we discussed in Section 25.4, current is *not* "used up" as it passes through a circuit.) Applying $V = IR$ to each resistor, we have

$$V_{ax} = IR_1 \qquad V_{xy} = IR_2 \qquad V_{yb} = IR_3$$

The potential differences across each resistor need not be the same (except for the special case in which all three resistances are equal). The potential difference V_{ab} across the entire combination is the sum of these individual potential differences:

$$V_{ab} = V_{ax} + V_{xy} + V_{yb} = I(R_1 + R_2 + R_3)$$

and so

$$\frac{V_{ab}}{I} = R_1 + R_2 + R_3$$

ActivPhysics ONLINE

12.1 dc Series Circuits (Qualitative)

The ratio V_{ab}/I is, by definition, the equivalent resistance R_{eq}. Therefore

$$R_{eq} = R_1 + R_2 + R_3$$

It is easy to generalize this to any number of resistors:

$$R_{eq} = R_1 + R_2 + R_3 + \cdots \quad \text{(resistors in series)} \tag{26.1}$$

The equivalent resistance of *any number* of resistors in series equals the sum of their individual resistances. The equivalent resistance is *greater than* any individual resistance.

Let's compare this result with Eq. (24.5) for *capacitors* in series. Resistors in series add directly because the voltage across each is directly proportional to its resistance and to the common current. Capacitors in series add reciprocally because the voltage across each is directly proportional to the common charge but *inversely* proportional to the individual capacitance.

Resistors in Parallel

If the resistors are in *parallel,* as in Fig. 26.1b, the current through each resistor need not be the same. But the potential difference between the terminals of each resistor must be the same and equal to V_{ab} (Fig. 26.2). (Remember that the potential difference between any two points does not depend on the path taken between the points.) Let's call the currents in the three resistors I_1, I_2, and I_3. Then from $I = V/R$,

$$I_1 = \frac{V_{ab}}{R_1} \qquad I_2 = \frac{V_{ab}}{R_2} \qquad I_3 = \frac{V_{ab}}{R_3}$$

26.2 A car's headlights are connected in parallel. Hence each headlight is exposed to the full potential difference supplied by the car's electrical system, giving maximum brightness. Another advantage is that if one headlight burns out, the other one keeps shining (see Example 26.2).

In general, the current is different through each resistor. Because charge is not accumulating or draining out of point a, the total current I must equal the sum of the three currents in the resistors:

$$I = I_1 + I_2 + I_3 = V_{ab}\left(\frac{1}{R_1} + \frac{1}{R_2} + \frac{1}{R_3}\right)$$

or

$$\frac{I}{V_{ab}} = \frac{1}{R_1} + \frac{1}{R_2} + \frac{1}{R_3}$$

But by definition of the equivalent resistance R_{eq}, $I/V_{ab} = 1/R_{eq}$, so

$$\frac{1}{R_{eq}} = \frac{1}{R_1} + \frac{1}{R_2} + \frac{1}{R_3}$$

Again it is easy to generalize to *any number* of resistors in parallel:

$$\frac{1}{R_{eq}} = \frac{1}{R_1} + \frac{1}{R_2} + \frac{1}{R_3} + \cdots \quad \text{(resistors in parallel)} \tag{26.2}$$

For *any number* of resistors in parallel, the *reciprocal* of the equivalent resistance equals the *sum of the reciprocals* of their individual resistances. The equivalent resistance is always *less than* any individual resistance.

We can compare this result with Eq. (24.7) for *capacitors* in parallel. Resistors in parallel add reciprocally because the current in each is proportional to the common voltage across them and *inversely* proportional to the resistance of each. Capacitors in parallel add directly because the charge on each is proportional to the common voltage across them and *directly* proportional to the capacitance of each.

For the special case of *two* resistors in parallel,

$$\frac{1}{R_{eq}} = \frac{1}{R_1} + \frac{1}{R_2} = \frac{R_1 + R_2}{R_1 R_2}$$

ActivPhysics ONLINE

12.2 dc Parallel Circuits

and

$$R_{eq} = \frac{R_1 R_2}{R_1 + R_2} \quad \text{(two resistors in parallel)} \qquad (26.3)$$

Because $V_{ab} = I_1 R_1 = I_2 R_2$, it follows that

$$\frac{I_1}{I_2} = \frac{R_2}{R_1} \quad \text{(two resistors in parallel)} \qquad (26.4)$$

This shows that the currents carried by two resistors in parallel are *inversely proportional* to their resistances. More current goes through the path of least resistance.

Problem-Solving Strategy — Resistors in Series and Parallel

IDENTIFY *the relevant concepts:* Many resistor networks are made up of resistors in series, in parallel, or a combination of the two. The key concept is that such a network can be replaced by a single equivalent resistor.

SET UP *the problem using the following steps:*

1. Make a drawing of the resistor network.
2. Determine whether the resistors are connected in series or parallel. Note that you can often consider networks such as those in Figs. 26.1c and 26.1d as combinations of series and parallel arrangements.
3. Determine what the target variables are. They could include the equivalent resistance of the network, the potential difference across each resistor, or the current through each resistor.

EXECUTE *the solution as follows:*

1. Use Eq. (26.1) or (26.2) to find the equivalent resistance for a series or a parallel combination, respectively.
2. If the network is more complex, try reducing it to series and parallel combinations. For example, in Fig. 26.1c we first replace the parallel combination of R_2 and R_3 with its equivalent resistance; this then forms a series combination with R_1. In Fig. 26.1d, the combination of R_2 and R_3 in series forms a parallel combination with R_1.
3. When calculating potential differences, remember that when resistors are connected in series, the total potential difference across the combination equals the sum of the individual potential differences. When they are connected in parallel, the potential difference is the same for every resistor and equals the potential difference across the parallel combination.
4. Keep in mind the analogous statements for current. When resistors are connected in series, the current is the same through every resistor and equals the current through the series combination. When resistors are connected in parallel, the total current through the combination equals the sum of the currents through the individual resistors.

EVALUATE *your answer:* Check whether your results are consistent. If resistors are connected in series, the equivalent resistance should be greater than that of any individual resistor; if they are connected in parallel, the equivalent resistance should be less than that of any individual resistor.

Example 26.1 Equivalent resistance

Compute the equivalent resistance of the network in Fig. 26.3a, and find the current in each resistor. The source of emf has negligible internal resistance.

SOLUTION

IDENTIFY and SET UP: The network of three resistors is a combination of series and parallel connections, just as in Fig. 26.1c. We first determine the equivalent resistance R_{eq} of this network as a whole. Given this value, we find the current in the emf, which is the same as the current in the 4-Ω resistor. This same current is split between the 6-Ω and 3-Ω resistors; we determine how much goes into each resistor by using the principle that the potential difference must be the same across these two resistors (because they are connected in parallel).

EXECUTE: Figures 26.2b and 26.2c show successive stages in reducing the network to a single equivalent resistance. From Eq. (26.2) the 6-Ω and 3-Ω resistors in parallel in Fig. 26.3a are equivalent to the single 2-Ω resistor in Fig. 26.3b:

$$\frac{1}{R_{eq}} = \frac{1}{6\ \Omega} + \frac{1}{3\ \Omega} = \frac{1}{2\ \Omega}$$

(You can find the same result using Eq. (26.3).) From Eq. (26.1) the series combination of this 2-Ω resistor with the 4-Ω resistor is equivalent to the single 6-Ω resistor in Fig. 26.3c.

To find the current in each resistor of the original network, we reverse the steps by which we reduced the network. In the circuit shown in Fig. 26.3d (identical to Fig. 26.3c), the current is $I = V_{ab}/R = (18\ \text{V})/(6\ \Omega) = 3$ A. So the current in the 4-Ω and 2-Ω resistors in Fig. 26.3e (identical to Fig. 26.3b) is also 3 A. The potential difference V_{cb} across the 2-Ω resistor is therefore $V_{cb} = IR = (3\ \text{A})(2\ \Omega) = 6$ V. This potential difference must also be 6 V in Fig. 26.3f (identical to Fig. 26.3a). Using $I = V_{cb}/R$, the currents in the 6-Ω and 3-Ω resistors in Fig. 26.3f are $(6\ \text{V})/(6\ \Omega) = 1$ A and $(6\ \text{V})/(3\ \Omega) = 2$ A, respectively.

EVALUATE: Note that for the two resistors in parallel between points c and b in Fig. 26.3f, there is twice as much current through the 3-Ω resistor as through the 6-Ω resistor; more current goes through the path of least resistance, in accordance with Eq. (26.4). Note also that the total current through these two resistors is 3 A, the same as it is through the 4-Ω resistor between points a and c.

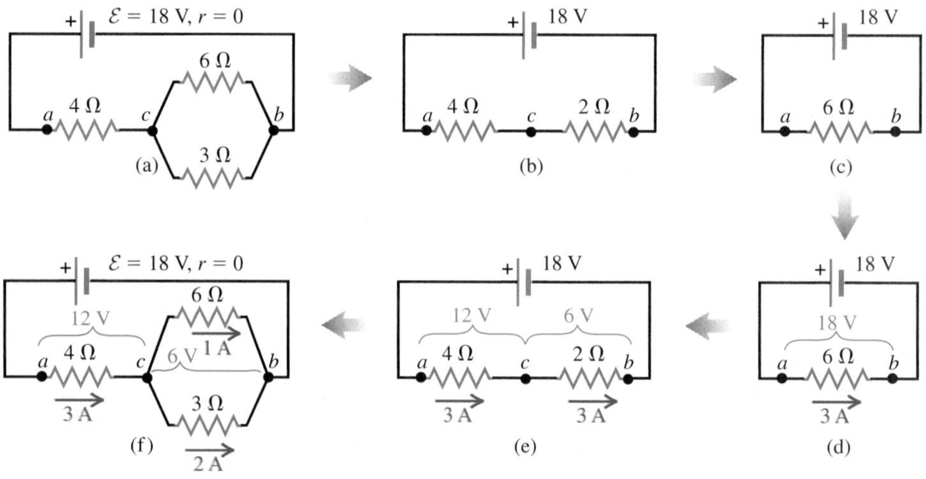

26.3 Steps in reducing a combination of resistors to a single equivalent resistor and finding the current in each resistor.

Example 26.2 Series versus parallel combinations

Two identical light bulbs are to be connected to a source with $\mathcal{E} = 8$ V and negligible internal resistance. Each light bulb has a resistance $R = 2\ \Omega$. Find the current through each bulb, the potential difference across each bulb, and the power delivered to each bulb and to the entire network if the bulbs are connected a) in series, as in Fig. 26.4a; b) in parallel, as in Fig. 26.4b. c) Suppose one of the bulbs burns out; that is, its filament breaks and current can no longer flow through it. What happens to the other bulb in the series case? In the parallel case?

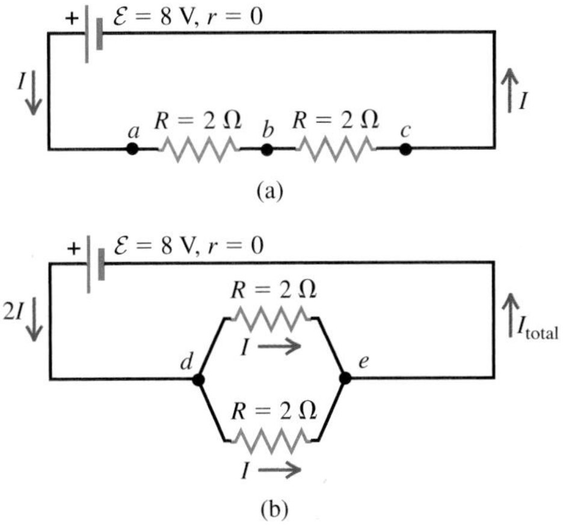

26.4 Circuit diagrams for two light bulbs (a) in series and (b) in parallel.

SOLUTION

IDENTIFY and SET UP: The resistor networks are simple series and parallel connections. We find the power delivered to each resistor using Eq. (25.18), $P = I^2R = V^2/R$.

EXECUTE: a) From Eq. (26.1) the equivalent resistance of the two bulbs between points a and c in Fig. 26.4a is the sum of their individual resistances:

$$R_{eq} = 2R = 2(2\ \Omega) = 4\ \Omega$$

The current is the same through either light bulb in series:

$$I = \frac{V_{ac}}{R_{eq}} = \frac{8\ \text{V}}{4\ \Omega} = 2\ \text{A}$$

Since the bulbs have the same resistance, the potential difference is the same across each bulb:

$$V_{ab} = V_{bc} = IR = (2\ \text{A})(2\ \Omega) = 4\ \text{V}$$

This is one-half of the 8-V terminal voltage of the source. From Eq. (25.18), the power delivered to each light bulb is

$$P = I^2R = (2\ \text{A})^2(2\ \Omega) = 8\ \text{W} \quad \text{or}$$

$$P = \frac{V_{ab}{}^2}{R} = \frac{V_{bc}{}^2}{R} = \frac{(4\ \text{V})^2}{2\ \Omega} = 8\ \text{W}$$

The total power delivered to both bulbs is $P_{total} = 2P = 16$ W. Alternatively, we can find the total power by using the equivalent resistance $R_{eq} = 4\ \Omega$, through which the current is $I = 2$ A and across which the potential difference is $V_{ac} = 8$ V:

$$P_{total} = I^2R_{eq} = (2\ \text{A})^2(4\ \Omega) = 16\ \text{W} \quad \text{or}$$

$$P_{total} = \frac{V_{ac}{}^2}{R_{eq}} = \frac{(8\ \text{V})^2}{4\ \Omega} = 16\ \text{W}$$

b) If the light bulbs are in parallel, as in Fig. 26.4b, the potential difference V_{de} across each bulb is the same and equal to 8 V, the terminal voltage of the source. Hence the current through each light bulb is

$$I = \frac{V_{de}}{R} = \frac{8\ \text{V}}{2\ \Omega} = 4\ \text{A}$$

and the power delivered to each bulb is

$$P = I^2R = (4\ \text{A})^2(2\ \Omega) = 32\ \text{W} \quad \text{or}$$

$$P = \frac{V_{de}{}^2}{R} = \frac{(8\ \text{V})^2}{2\ \Omega} = 32\ \text{W}$$

Both the potential difference across each bulb and the current through each bulb are twice as great as in the series case. Hence the power delivered to each bulb is *four* times greater, and each bulb glows more brightly than in the series case. If the goal is to produce the maximum amount of light from each bulb, a parallel arrangement is superior to a series arrangement.

The total power delivered to the parallel network is $P_{total} = 2P = 64$ W, four times greater than in the series case. The increased power compared to the series case isn't obtained "for free"; energy is extracted from the source four times more rapidly in the parallel case than in the series case. If the source is a battery, it will be used up four times as fast.

We can also find the total power by using the equivalent resistance R_{eq}, given by Eq. (26.2):

$$\frac{1}{R_{eq}} = 2\left(\frac{1}{2\ \Omega}\right) = 1\ \Omega^{-1} \quad \text{or} \quad R_{eq} = 1\ \Omega$$

The total current through the equivalent resistor is $I_{total} = 2I = 2(4\ \text{A}) = 8$ A, and the potential difference across the equivalent resistor is 8 V. Hence the total power is

$$P_{total} = I^2R_{eq} = (8\ \text{A})^2(1\ \Omega) = 64\ \text{W} \quad \text{or}$$

$$P_{total} = \frac{V_{de}{}^2}{R} = \frac{(8\ \text{V})^2}{1\ \Omega} = 64\ \text{W}$$

The potential difference across the equivalent resistance is the same for both the series and parallel cases, but for the parallel case the value of R_{eq} is less, and so $P_{total} = V^2/R_{eq}$ is greater.

c) In the series case the same current flows through both bulbs. If one of the bulbs burns out, there will be no current at all in the circuit, and neither bulb will glow.

In the parallel case the potential difference across either bulb remains equal to 8 V even if one of the bulbs burns out. Hence the current through the functional bulb remains equal to 4 A, and the power delivered to that bulb remains equal to 32 W, the same as before the other bulb burned out. This is another of the merits of a parallel arrangement of light bulbs: If one fails, the other bulbs are unaffected. This principle is used in household wiring systems, which we'll discuss in Section 26.5.

EVALUATE: Our calculation isn't completely accurate, because the resistance $R = V/I$ of real light bulbs is *not* a constant independent of the potential difference V across the bulb. (The resistance of the filament increases with increasing operating temperature and hence with increasing V.) But it is indeed true that light bulbs connected in series across a source glow less brightly than when connected in parallel across the same source (Fig. 26.5).

26.5 When connected to the same source, two light bulbs in series (shown on the left) draw less power and glow less brightly than when they are in parallel (shown on the right).

Test Your Understanding

Suppose all three of the resistors shown in Fig. 26.1 have the same resistance, so $R_1 = R_2 = R_3 = R$. Rank the four arrangements shown in parts (a)–(d) of Fig. 26.1 in order of their equivalent resistance, from lowest to highest.

26.2 | Kirchhoff's Rules

Many practical resistor networks cannot be reduced to simple series-parallel combinations. Figure 26.6a shows a dc power supply with emf $\mathcal{E}_1$ charging a battery with a smaller emf $\mathcal{E}_2$ and feeding current to a light bulb with resistance R. Fig. 26.6b is a "bridge" circuit, used in many different types of measurement and control systems. (One important application of a "bridge" circuit is described in Problem 26.77.) We don't need any new principles to compute the currents in these networks, but there are some techniques that help us handle such problems systematically. We will describe the techniques developed by the German physicist Gustav Robert Kirchhoff (1824–1887).

First, here are two terms that we will use often. A **junction** in a circuit is a point where three or more conductors meet. Junctions are also called *nodes* or *branch points*. A **loop** is any closed conducting path. The circuit in Fig. 26.6a has two junctions, a and b. In Fig. 26.6b, points a, b, c, and d are junctions, but points e and f are not. Some possible loops in Fig. 26.6b are the closed paths *acdba*, *acdefa*, *abdefa*, and *abcdefa*.

Kirchhoff's rules consist of the following two statements:

Kirchhoff's junction rule: *The algebraic sum of the currents into any junction is zero.* That is,

$$\sum I = 0 \qquad \text{(junction rule, valid at any junction)} \qquad (26.5)$$

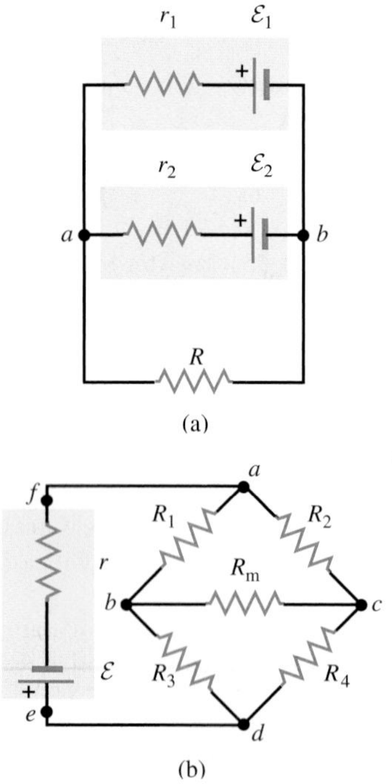

26.6 Two networks that cannot be reduced to simple series-parallel combinations of resistors.

Kirchhoff's loop rule: *The algebraic sum of the potential differences in any loop,* including those associated with emfs and those of resistive elements, *must equal zero.* That is,

$$\sum V = 0 \qquad \text{(loop rule, valid for any closed loop)} \qquad (26.6)$$

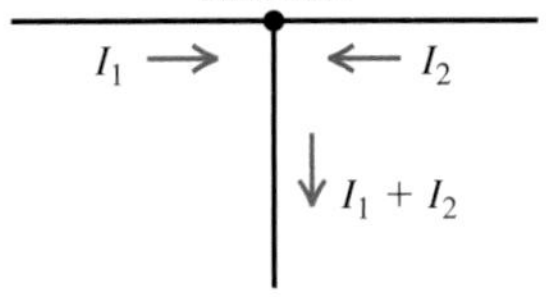

26.7 Kirchhoff's junction rule states that as much current flows into a junction as flows out of it.

The junction rule is based on *conservation of electric charge.* No charge can accumulate at a junction, so the total charge entering the junction per unit time must equal the total charge leaving per unit time (Fig. 26.7). Charge per unit time is current, so if we consider the currents entering a junction to be positive and those leaving to be negative, the algebraic sum of currents into a junction must be zero. It's like a T branch in a water pipe; if you have one liter per minute coming in one pipe, you can't have three liters per minute going out the other two pipes. We may as well confess that we used the junction rule (without saying so) in Section 26.1 in the derivation of Eq. (26.2) for resistors in parallel.

The loop rule is a statement that the electrostatic force is *conservative.* Suppose we go around a loop, measuring potential differences across successive circuit elements as we go. When we return to the starting point, we must find that the *algebraic sum* of these differences is zero; otherwise, we could not say that the potential at this point has a definite value.

ActivPhysics ONLINE

12.3 dc Circuit Puzzles
12.5 Using Kirchhoff's Laws

In applying the loop rule, we need some sign conventions. The following Problem-Solving Strategy describes in detail how to use these, but here's a quick overview. We first assume a direction for the current in each branch of the circuit and mark it on a diagram of the circuit. Then, starting at any point in the circuit, we imagine traveling around a loop, adding emfs and IR terms as we come to them. When we travel through a source in the direction from $-$ to $+$, the emf is considered to be *positive;* when we travel from $+$ to $-$, the emf is considered to be *negative.* When we travel through a resistor in the *same* direction as the assumed current, the IR term is *negative* because the current goes in the direction of decreasing potential. When we travel through a resistor in the direction *opposite* to the assumed current, the IR term is *positive* because this represents a rise of potential.

Kirchhoff's two rules are all we need to solve a wide variety of network problems. Usually, some of the emfs, currents, and resistances are known, and others are unknown. We must always obtain from Kirchhoff's rules a number of independent equations equal to the number of unknowns so that we can solve the equations simultaneously. Often the hardest part of the solution is not in understanding the basic principles but in keeping track of algebraic signs!

Problem-Solving Strategy

Kirchhoff's Rules

IDENTIFY *the relevant concepts:* Kirchhoff's rules are important tools for analyzing any circuit more complicated than a single loop.

SET UP *the problem using the following steps:*

1. Draw a *large* circuit diagram so you have plenty of room for labels. Label all quantities, known and unknown, including an assumed direction for each unknown current and emf. Often you will not know in advance the actual direction of an unknown current or emf, but this doesn't matter. If the actual direction of a particular quantity is opposite to your assumption, the result will come out with a negative sign. If you use Kirchhoff's rules correctly,

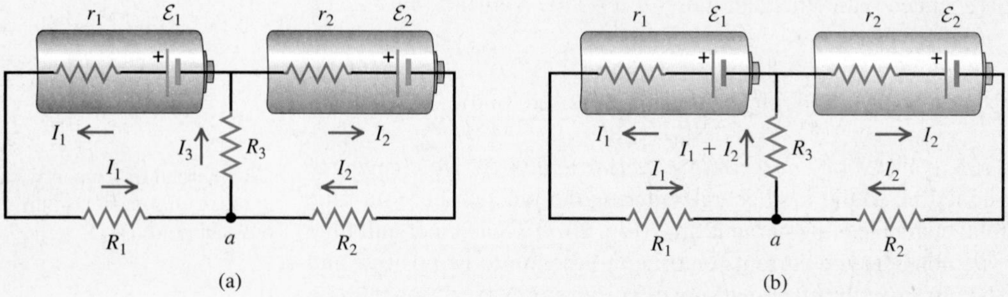

26.8 Application of the junction rule to point a reduces the number of unknown currents from three to two.

they will give you the directions as well as the magnitudes of unknown currents and emfs.

2. When you label currents, it is usually best to use the junction rule immediately to express the currents in terms of as few quantities as possible. For example, Fig. 26.8a shows a circuit correctly labeled; Fig. 26.8b shows the same circuit, relabeled by applying the junction rule to point a to eliminate I_3.
3. Determine which quantities are the target variables.

EXECUTE *the solution as follows:*

1. Choose any closed loop in the network and designate a direction (clockwise or counterclockwise) to travel around the loop when applying the loop rule. The direction doesn't have to be the same as any assumed current direction.
2. Travel around the loop in the designated direction, adding potential differences as you cross them. Remember that a positive potential difference corresponds to an increase in potential and a negative potential difference corresponds to a decrease in potential. An emf is counted as positive when you traverse it from $(-)$ to $(+)$, and negative when you go from $(+)$ to $(-)$. An IR term is negative if you travel through the resistor in the same direction as the assumed current and positive if you pass it in the opposite direction. Figure 26.9 summarizes these sign conventions. In each part of the figure "travel" is the direction that we imagine going around a loop while using Kirchhoff's loop law, not necessarily the direction of current.
3. Equate the sum in Step 2 to zero.
4. If necessary, choose another loop to get a different relation among the unknowns, and continue until you have as many independent equations as unknowns or until every circuit element has been included in at least one of the chosen loops.
5. Solve the equations simultaneously to determine the unknowns. This step involves algebra, not physics, but it can be fairly complex. Be careful with algebraic manipulations; one sign error will prove fatal to the entire solution.
6. You can use this same bookkeeping system to find the potential V_{ab} of any point a with respect to any other point b. Start at b and add the potential changes you encounter in going from b to a, using the same sign rules as in Step 2. The algebraic sum of these changes is $V_{ab} = V_a - V_b$.

EVALUATE *your answer:* Check all the steps in your algebra. A useful strategy is to consider a loop other than the ones you used to solve the problem; if the sum of potential drops around this loop isn't zero, you made an error somewhere in your calculations. As always, ask yourself whether the answers make sense.

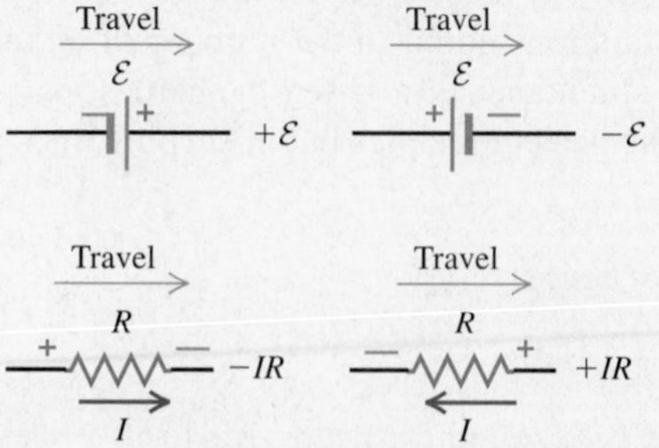

26.9 When using Kirchhoff's rules, follow these sign conventions as you travel around a circuit loop.

Example 26.3 A single-loop circuit

The circuit shown in Fig. 26.10a contains two batteries, each with an emf and an internal resistance, and two resistors. Find a) the current in the circuit, b) the potential difference V_{ab}, and c) the power output of the emf of each battery.

SOLUTION

IDENTIFY and SET UP: This single-loop circuit has no junctions, so we don't need Kirchhoff's junction rule. To apply the loop rule to the single loop, we first assume a direction for the current; let's assume a counterclockwise direction, as shown in Fig. 26.10a.

EXECUTE: a) Starting at a and going counterclockwise, we add potential increases and decreases and equate the sum to zero, as in Eq. (26.6). The resulting equation is

$$-I(4\,\Omega) - 4\text{ V} - I(7\,\Omega) + 12\text{ V} - I(2\,\Omega) - I(3\,\Omega) = 0$$

Collecting terms containing I and solving for I, we find

$$8\text{ V} = I(16\,\Omega) \quad \text{and} \quad I = 0.5\text{ A}$$

The result for I is positive, showing that our assumed current direction is correct. For an exercise, try assuming the opposite direction for I; you should then get $I = -0.5$ A, indicating that the actual current is opposite to this assumption.

b) To find V_{ab}, the potential at a with respect to b, we start at b and add potential changes as we go toward a. There are two possible paths from b to a; taking the lower one first, we find

$$V_{ab} = (0.5\text{ A})(7\,\Omega) + 4\text{ V} + (0.5\text{ A})(4\,\Omega) = 9.5\text{ V}$$

Point a is at 9.5 V higher potential than b. All the terms in this sum, including the IR terms, are positive because each represents an *increase* in potential as we go from b toward a. If we use the upper path instead, the resulting equation is

$$V_{ab} = 12\text{ V} - (0.5\text{ A})(2\,\Omega) - (0.5\text{ A})(3\,\Omega) = 9.5\text{ V}$$

Here the IR terms are negative because our path goes in the direction of the current, with potential decreases through the resistors. The result is the same as for the lower path, as it must be in order for the total potential change around the complete loop to be zero. In each case, potential rises are taken to be positive and drops are taken to be negative.

c) The power output of the emf of the 12-V battery is

$$P = \mathcal{E}I = (12\text{ V})(0.5\text{ A}) = 6\text{ W}$$

and the power output of the emf of the 4-V battery is

$$P = \mathcal{E}I = (-4\text{ V})(0.5\text{ A}) = -2\text{ W}$$

The negative sign in $\mathcal{E}$ for the 4-V battery appears because the current actually runs from the higher-potential side of the battery to the lower-potential side. The negative value of P means that we are *storing* energy in that battery, and it is being *recharged* by the 12-V battery.

EVALUATE: By applying the expression $P = I^2R$ to each of the four resistors in Fig. 26.10a, you should be able to show that the total power dissipated in all four resistors is 4 W. Of the 6 W provided by the emf of the 12-V battery, 2 W goes into storing energy in the 4-V battery and 4 W is dissipated in the resistances.

The circuit shown in Fig. 26.10a is very much like that used when a 12-V automobile battery is used to recharge a run-down battery in another automobile (Fig. 26.10b). The 3-Ω and 7-Ω resistors in Fig. 26.10a represent the resistances of the jumper cables and of the conducting path through the red automobile with the run-down battery. (The values of the resistances in actual automobiles and jumper cables are different than those used in this example.)

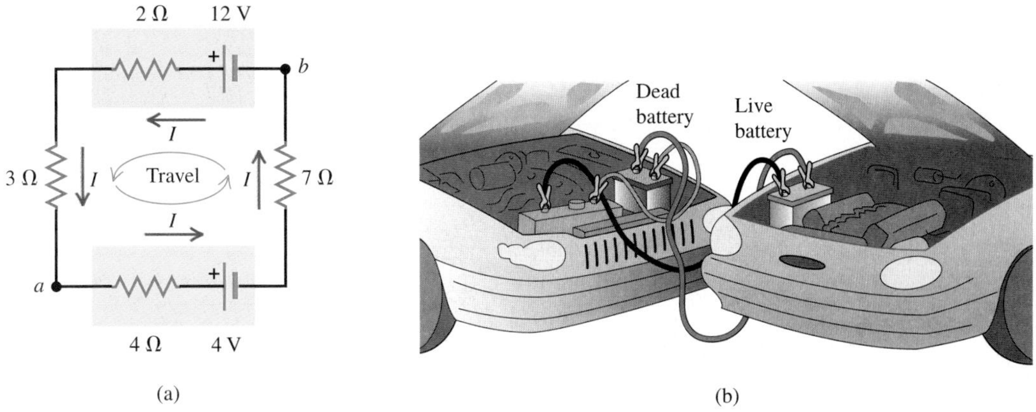

26.10 (a) In this example we travel around the loop in the same direction as the assumed current, so all the IR terms are negative. The potential decreases as we travel from + to − through the bottom emf but increases as we travel from − to + through the top emf.
(b) A real-life example of a circuit of this kind.

Example 26.4 Charging a battery

In the circuit shown in Fig. 26.11, a 12-V power supply with unknown internal resistance r is connected to a run-down rechargeable battery with unknown emf $\mathcal{E}$ and internal resistance 1 Ω and to an indicator light bulb of resistance 3 Ω carrying a current of 2 A. The current through the run-down battery is 1 A in the direction shown. Find the unknown current I, the internal resistance r, and the emf $\mathcal{E}$.

SOLUTION

IDENTIFY and SET UP: We assume the direction of the current through the 12-V power supply to be as shown. This circuit has more than one loop, so we must apply both the junction rule and the loop rule. There are three target variables, so we need three equations.

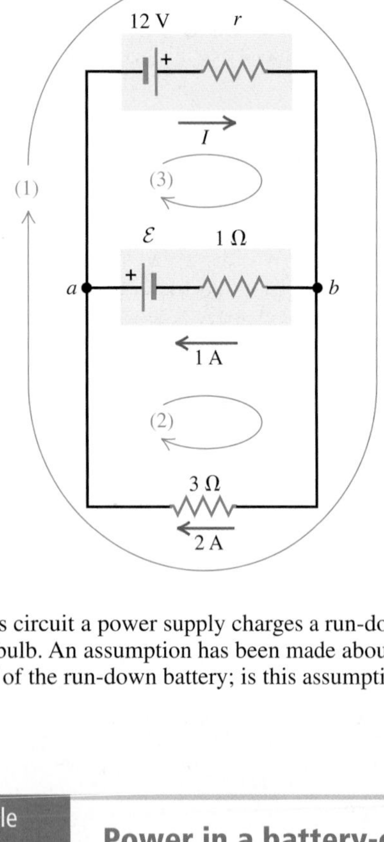

26.11 In this circuit a power supply charges a run-down battery and lights a bulb. An assumption has been made about the polarity of the emf $\mathcal{E}$ of the run-down battery; is this assumption correct?

EXECUTE: First we apply the junction rule, Eq. (26.5), to point a. We find

$$-I + 1\text{ A} + 2\text{ A} = 0 \quad \text{so} \quad I = 3\text{ A}$$

To determine r, we apply the loop rule, Eq. (26.6), to the outer loop labeled (1); we find

$$12\text{ V} - (3\text{ A})r - (2\text{ A})(3\ \Omega) = 0 \quad \text{so} \quad r = 2\ \Omega$$

The terms containing the resistances r and 3 Ω are negative because our loop traverses those elements in the same direction as the current and hence finds potential *drops*. If we had chosen to traverse loop (1) in the opposite direction, every term would have had the opposite sign, and the result for r would have been the same.

To determine $\mathcal{E}$, we apply the loop rule to loop (2):

$$-\mathcal{E} + (1\text{ A})(1\ \Omega) - (2\text{ A})(3\ \Omega) = 0 \quad \text{so} \quad \mathcal{E} = -5\text{ V}$$

The term for the 1-Ω resistor is positive because in traversing it in the direction opposite to the current we find a potential *rise*. The negative value for $\mathcal{E}$ shows that the actual polarity of this emf is opposite to the assumption made in Fig. 26.11; the positive terminal of this source is really on the right side. As in Example 26.3, the battery is being recharged.

EVALUATE: We can check our result for $\mathcal{E}$ by using loop (3), obtaining the equation

$$12\text{ V} - (3\text{ A})(2\ \Omega) - (1\text{ A})(1\ \Omega) + \mathcal{E} = 0$$

from which we again find $\mathcal{E} = -5$ V.

As an additional consistency check, we note that $V_{ba} = V_b - V_a$ equals the voltage across the 3-Ω resistance, which is $(2\text{ A})(3\ \Omega) = 6$ V. Going from a to b by the top branch, we encounter potential differences $+12\text{ V} - (3\text{ A})(2\ \Omega) = +6$ V, and going by the middle branch we find $-(-5\text{ V}) + (1\text{ A})(1\ \Omega) = +6$ V. The three ways of getting V_{ba} give the same results. Make sure that you understand all the signs in these calculations.

Example 26.5 Power in a battery-charging circuit

In the circuit of Example 26.4 (shown in Fig. 26.11), find the power delivered by the 12-V power supply and by the battery being recharged, and find the power dissipated in each resistor.

SOLUTION

IDENTIFY and SET UP: We use the results of Section 25.5, in which we found that the power delivered *from* an emf to a circuit is $\mathcal{E}I$ and the power delivered *to* a resistor from a circuit is $V_{ab}I = I^2R$.

EXECUTE: The power output from the emf of the power supply is

$$P_{\text{supply}} = \mathcal{E}_{\text{supply}}I_{\text{supply}} = (12\text{ V})(3\text{ A}) = 36\text{ W}$$

The power dissipated in the power supply's internal resistance r is

$$P_{r\text{-supply}} = I_{\text{supply}}^2 r_{\text{supply}} = (3\ \text{A})^2(2\ \Omega) = 18\ \text{W}$$

so the power supply's *net* power output is $P_{\text{net}} = 36\ \text{W} - 18\ \text{W} = 18\ \text{W}$. Alternatively, from Example 26.4 the terminal voltage of the battery is $V_{ba} = 6\ \text{V}$, so the net power output is

$$P_{\text{net}} = V_{ba} I_{\text{supply}} = (6\ \text{V})(3\ \text{A}) = 18\ \text{W}$$

The power output of the emf $\mathcal{E}$ of the battery being charged is

$$P_{\text{emf}} = \mathcal{E} I_{\text{battery}} = (-5\ \text{V})(1\ \text{A}) = -5\ \text{W}$$

This is negative because the 1-A current runs through the battery from the higher-potential side to the lower-potential side. (As we mentioned in Example 26.4, the polarity assumed for this battery in Fig. 26.11 was wrong.) We are storing energy in the battery as we charge it. Additional power is dissipated in the battery's internal resistance; this power is

$$P_{r\text{-battery}} = I_{\text{battery}}^2 r_{\text{battery}} = (1\ \text{A})^2(1\ \Omega) = 1\ \text{W}$$

The total power input to the battery is thus $1\ \text{W} + |-5\ \text{W}| = 6\ \text{W}$. Of this, 5 W represents useful energy stored in the battery; the remainder is wasted in its internal resistance.

The power dissipated in the light bulb is

$$P_{\text{bulb}} = I_{\text{bulb}}^2 R_{\text{bulb}} = (2\ \text{A})^2(3\ \Omega) = 12\ \text{W}$$

EVALUATE: As a check, note that all of the power from the supply is accounted for. Of the 18 W of net power from the power supply, 5 W goes to recharge the battery, 1 W is dissipated in the battery's internal resistance, and 12 W is dissipated in the light bulb.

Example 26.6 A complex network

Figure 26.12 shows a "bridge" circuit of the type described at the beginning of this section (see Fig. 26.6b). Find the current in each resistor and the equivalent resistance of the network of five resistors.

SOLUTION

IDENTIFY and SET UP: This network cannot be represented in terms of series and parallel combinations. There are five different currents to determine, but by applying the junction rule to junctions a and b, we can represent them in terms of three unknown currents, as shown in the figure. The current in the battery is $I_1 + I_2$.

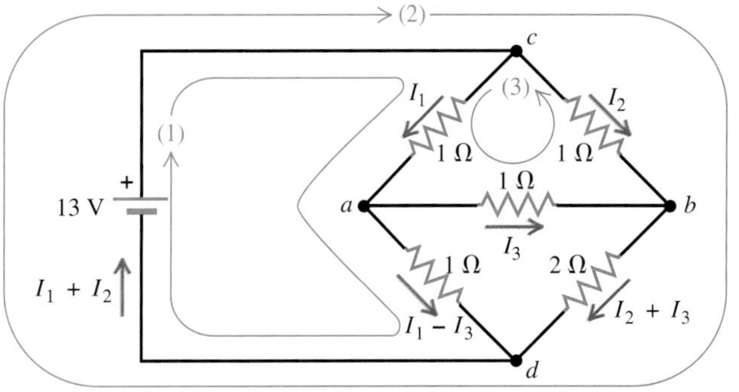

26.12 A network circuit with several resistors.

EXECUTE: We apply the loop rule to the three loops shown, obtaining the following three equations:

$$13\ \text{V} - I_1(1\ \Omega) - (I_1 - I_3)(1\ \Omega) = 0 \quad (1)$$

$$-I_2(1\ \Omega) - (I_2 + I_3)(2\ \Omega) + 13\ \text{V} = 0 \quad (2)$$

$$-I_1(1\ \Omega) - I_3(1\ \Omega) + I_2(1\ \Omega) = 0 \quad (3)$$

This is a set of three simultaneous equations for the three unknown currents. They may be solved by various methods; one straightforward procedure is to solve the third equation for I_2, obtaining $I_2 = I_1 + I_3$, and then substitute this expression into the first two equations to eliminate I_2. When this is done, we are left with the two equations

$$13\ \text{V} = I_1(2\ \Omega) - I_3(1\ \Omega) \quad (1')$$

$$13\ \text{V} = I_1(3\ \Omega) - I_3(5\ \Omega) \quad (2')$$

Now we can eliminate I_3 by multiplying Eq. (1′) by 5 and adding the two equations. We obtain

$$78\ \text{V} = I_1(13\ \Omega) \qquad I_1 = 6\ \text{A}$$

We substitute this result back into Eq. (1′) to obtain $I_3 = -1\ \text{A}$, and finally, from Eq. (3) we find $I_2 = 5\ \text{A}$. The negative value of I_3 tells us that its direction is opposite to our initial assumption.

The total current through the network is $I_1 + I_2 = 11\ \text{A}$, and the potential drop across it is equal to the battery emf—namely, 13 V. The equivalent resistance of the network is

$$R_{\text{eq}} = \frac{13\ \text{V}}{11\ \text{A}} = 1.2\ \Omega$$

EVALUATE: You can check the results $I_1 = 6\ \text{A}$, $I_2 = 5\ \text{A}$, and $I_3 = -1\ \text{A}$ by substituting these values into the three equations (1), (2), and (3). What do you find?

Example 26.7 A potential difference within a complex network

In the circuit of Example 26.6 (Fig. 26.12), find the potential difference V_{ab}.

SOLUTION

IDENTIFY and SET UP: To find $V_{ab} = V_a - V_b$, we start at point b and follow a path to point a, adding potential rises and drops as we go. We can follow any of several paths from b to a; the value of V_{ab} must be independent of which path we choose, which gives us a natural way to check our result.

EXECUTE: The simplest path to follow is through the center 1-Ω resistor. We have found $I_3 = -1\ \text{A}$, showing that the actual current direction in this branch is from right to left. Thus, as we go from b to a, there is a *drop* of potential with magnitude $IR = (1\ \text{A})(1\ \Omega) = 1\ \text{V}$, and $V_{ab} = -1\ \text{V}$. That is, the potential at point a is 1 V less than that at point b.

EVALUATE: To test our result, let's try a path from b to a that goes through the lower two resistors. The currents through these are

$$I_2 + I_3 = 5\ \text{A} + (-1\ \text{A}) = 4\ \text{A} \quad \text{and}$$
$$I_1 - I_3 = 6\ \text{A} - (-1\ \text{A}) = 7\ \text{A}$$

and so

$$V_{ab} = -(4\ \text{A})(2\ \Omega) + (7\ \text{A})(1\ \Omega) = -1\ \text{V}$$

We suggest that you try some other paths from b to a to verify that they also give this result.

Test Your Understanding

Subtract equation (1) from equation (2) in Example 26.6. To which loop in Fig. 26.12 does this equation correspond? Would this equation have simplified the solution of Example 26.6?

26.3 | Electrical Measuring Instruments

We've been talking about potential difference, current, and resistance for two chapters, so it's about time we said something about how to *measure* these quantities. Many common devices, including car instrument panels, battery chargers, and inexpensive electrical instruments, measure potential difference (voltage), current, or resistance using a **d'Arsonval galvanometer** (Fig. 26.13). In the following discussion we'll often just call it a *meter.* A pivoted coil of fine wire is placed in the magnetic field of a permanent magnet (Fig. 26.14). Attached to the coil is a spring, similar to the hairspring on the balance wheel of a watch. In the equilibrium position, with no current in the coil, the pointer is at zero. When there is a current in the coil, the magnetic field exerts a torque on the coil that is proportional to the current. (We'll discuss this magnetic interaction in detail in Chapter 27.) As the coil turns, the spring exerts a restoring torque that is proportional to the angular displacement.

26.13 This ammeter (top) and voltmeter (bottom) are both d'Arsonval galvanometers. The difference has to do with their internal connections (see Fig. 26.15).

Thus the angular deflection of the coil and pointer is directly proportional to the coil current, and the device can be calibrated to measure current. The maximum deflection, typically 90° or so, is called *full-scale deflection.* The essential electrical characteristics of the meter are the current I_{fs} required for full-scale deflection (typically on the order of 10 μA to 10 mA) and the resistance R_c of the coil (typically on the order of 10 to 1000 Ω).

The meter deflection is proportional to the *current* in the coil. If the coil obeys Ohm's law, the current is proportional to the *potential difference* between the terminals of the coil, and the deflection is also proportional to this potential difference. For example, consider a meter whose coil has a resistance $R_c = 20.0\ \Omega$ and

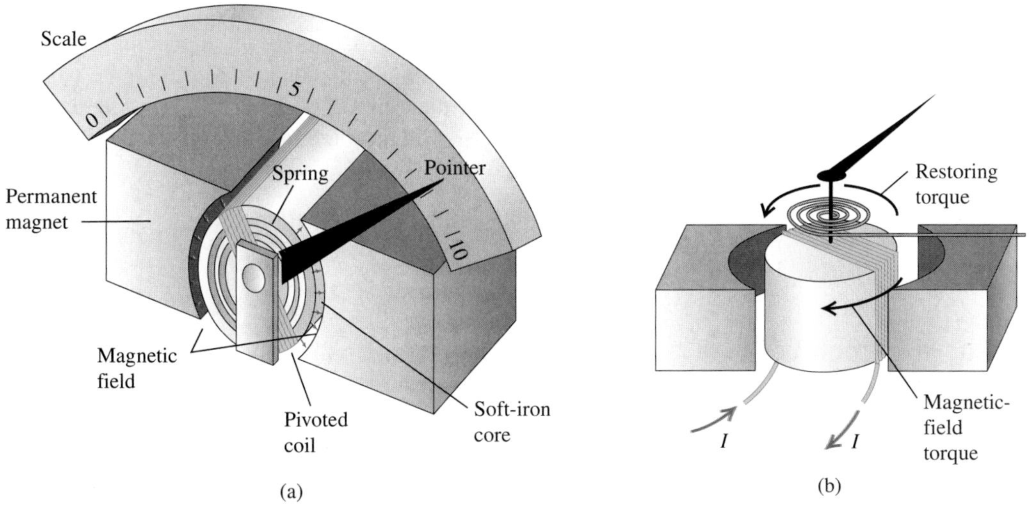

26.14 (a) A d'Arsonval galvanometer, showing a pivoted coil with attached pointer, a permanent magnet supplying a magnetic field that is uniform in magnitude, and a spring to provide restoring torque, which opposes magnetic-field torque. (b) Pivoted coil around a soft-iron core, with supports removed.

that deflects full scale when the current in its coil is $I_{fs} = 1.00$ mA. The corresponding potential difference for full-scale deflection is

$$V = I_{fs}R_c = (1.00 \times 10^{-3}\ \text{A})(20.0\ \Omega) = 0.0200\ \text{V}$$

Ammeters

A current-measuring instrument is usually called an **ammeter** (or milliammeter, microammeter, and so forth, depending on the range). *An ammeter always measures the current passing through it.* An *ideal* ammeter, discussed in Section 25.4, would have *zero* resistance, so including it in a branch of a circuit would not affect the current in that branch. Real ammeters always have some finite resistance, but it is always desirable for an ammeter to have as little resistance as possible.

We can adapt any meter to measure currents that are larger than its full-scale reading by connecting a resistor in parallel with it (Fig. 26.15a) so that some of the current bypasses the meter coil. The parallel resistor is called a **shunt resistor** or simply a *shunt,* denoted as R_{sh}.

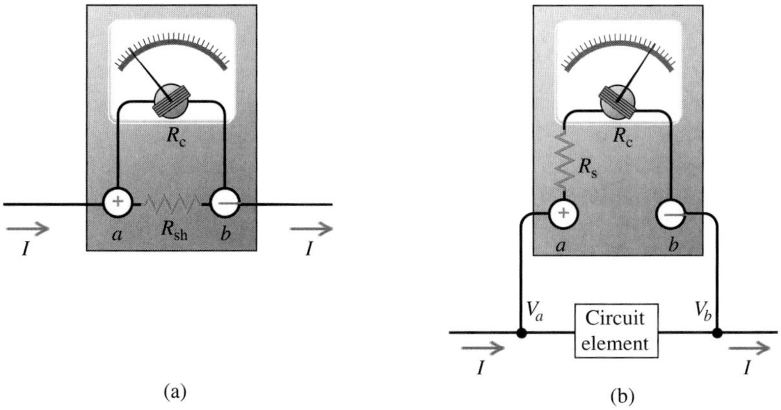

26.15 (a) Internal connections of a moving-coil ammeter. (b) Internal connections of a moving-coil voltmeter.

Suppose we want to make a meter with full-scale current I_{fs} and coil resistance R_c into an ammeter with full-scale reading I_a. To determine the shunt resistance R_{sh} needed, note that at full-scale deflection the total current through the parallel combination is I_a, the current through the coil of the meter is I_{fs}, and the current through the shunt is the difference $I_a - I_{fs}$. The potential difference V_{ab} is the same for both paths, so

$$I_{fs}R_c = (I_a - I_{fs})R_{sh} \qquad \text{(for an ammeter)} \qquad (26.7)$$

Example 26.8 Designing an ammeter

What shunt resistance is required to make the 1.00-mA, 20.0-Ω meter described above into an ammeter with a range of 0 A to 50.0 mA?

SOLUTION

IDENTIFY and SET UP: We want the ammeter to be able to handle a maximum current $I_a = 50.0\ \text{mA} = 50.0 \times 10^{-3}\ \text{A}$. The resistance of the coil is $R_c = 20.0\ \Omega$ and the meter shows full-scale deflection when the current through the coil is $I_{fs} = 1.00 \times 10^{-3}\ \text{A}$. Our target variable is the shunt resistance R_{sh}, which we find using Eq. (26.7).

EXECUTE: Solving Eq. (26.7) for R_{sh}, we find

$$R_{sh} = \frac{I_{fs}R_c}{I_a - I_{fs}} = \frac{(1.00 \times 10^{-3}\ \text{A})(20.0\ \Omega)}{50.0 \times 10^{-3}\ \text{A} - 1.00 \times 10^{-3}\ \text{A}}$$
$$= 0.408\ \Omega$$

EVALUATE: It's useful to consider the equivalent resistance R_{eq} of the ammeter as a whole. From Eq. (26.2),

$$\frac{1}{R_{eq}} = \frac{1}{R_c} + \frac{1}{R_{sh}} = \frac{1}{20.0\ \Omega} + \frac{1}{0.408\ \Omega}$$
$$R_{eq} = 0.400\ \Omega$$

The shunt resistance is so small in comparison to the meter resistance that the equivalent resistance is very nearly equal to the shunt resistance. The result is a low-resistance instrument with the desired range of 0 to 50.0 mA. At full-scale deflection, $I = I_a = 50.0\ \text{mA}$, the current through the galvanometer is 1.00 mA, the current through the shunt resistor is 49.0 mA, and $V_{ab} = 0.0200\ \text{V}$. If the current I is *less* than 50.0 mA, the coil current and the deflection are proportionally less, but the resistance R_{eq} is still 0.400 Ω.

ActivPhysics ONLINE

12.4 Using Ammeters and Voltmeters

Voltmeters

This same basic meter may also be used to measure potential difference or *voltage.* A voltage-measuring device is called a **voltmeter** (or millivoltmeter, and so forth, depending on the range). A voltmeter always measures the potential difference between two points, and its terminals must be connected to these points. (Example 25.7 in Section 25.4 described what can happen if a voltmeter is connected incorrectly.) As we discussed in Section 25.4, an ideal voltmeter would have *infinite* resistance, so connecting it between two points in a circuit would not alter any of the currents. Real voltmeters always have finite resistance, but a voltmeter should have large enough resistance that connecting it in a circuit does not change the other currents appreciably.

For the meter described in Example 26.8 the voltage across the meter coil at full-scale deflection is only $I_{fs}R_c = (1.00 \times 10^{-3}\ \text{A})(20.0\ \Omega) = 0.0200\ \text{V}$. We can extend this range by connecting a resistor R_s in *series* with the coil (Fig. 26.15b). Then only a fraction of the total potential difference appears across the coil itself, and the remainder appears across R_s. For a voltmeter with full-scale reading V_V, we need a series resistor R_s in Fig. 26.15b such that

$$V_V = I_{fs}(R_c + R_s) \qquad \text{(for a voltmeter)} \qquad (26.8)$$

Example 26.9 Designing a voltmeter

How can we make a galvanometer with $R_c = 20.0\ \Omega$ and $I_{fs} = 1.00$ mA into a voltmeter with a maximum range of 10.0 V?

SOLUTION

IDENTIFY and SET UP: The maximum allowable voltage across the voltmeter is $V_V = 10.0$ V. We want this to occur when the current through the coil (of resistance $R_c = 20.0\ \Omega$) is $I_{fs} = 1.00 \times 10^{-3}$ A. Our target variable is the series resistance R_s, which we find with Eq. (26.8).

EXECUTE: From Eq. (26.8),

$$R_s = \frac{V_V}{I_{fs}} - R_c = \frac{10.0\ \text{V}}{0.00100\ \text{A}} - 20.0\ \Omega = 9980\ \Omega$$

EVALUATE: At full-scale deflection, $V_{ab} = 10.0$ V, the voltage across the meter is 0.0200 V, the voltage across R_s is 9.98 V, and the current through the voltmeter is 0.00100 A. In this case most of the voltage appears across the series resistor. The equivalent meter resistance is $R_{eq} = 20.0\ \Omega + 9980\ \Omega = 10{,}000\ \Omega$. Such a meter is described as a "1000 ohms-per-volt meter," referring to the ratio of resistance to full-scale deflection. In normal operation the current through the circuit element being measured (I in Fig. 26.15b) is much greater than 0.00100 A, and the resistance between points a and b in the circuit is much less than 10,000 Ω. So the voltmeter draws off only a small fraction of the current and disturbs only slightly the circuit being measured.

Ammeters and Voltmeters in Combination

A voltmeter and an ammeter can be used together to measure *resistance* and *power*. The resistance R of a resistor equals the potential difference V_{ab} between its terminals divided by the current I; that is, $R = V_{ab}/I$. The power input P to any circuit element is the product of the potential difference across it and the current through it: $P = V_{ab}I$. In principle, the most straightforward way to measure R or P is to measure V_{ab} and I simultaneously.

With practical ammeters and voltmeters this isn't quite as simple as it seems. In Fig. 26.16a, ammeter A reads the current I in the resistor R. Voltmeter V, however, reads the *sum* of the potential difference V_{ab} across the resistor and the potential difference V_{bc} across the ammeter. If we transfer the voltmeter terminal from c to b, as in Fig. 26.16b, then the voltmeter reads the potential difference V_{ab} correctly, but the ammeter now reads the *sum* of the current I in the resistor and the current I_V in the voltmeter. Either way, we have to correct the reading of one instrument or the other unless the corrections are small enough to be negligible.

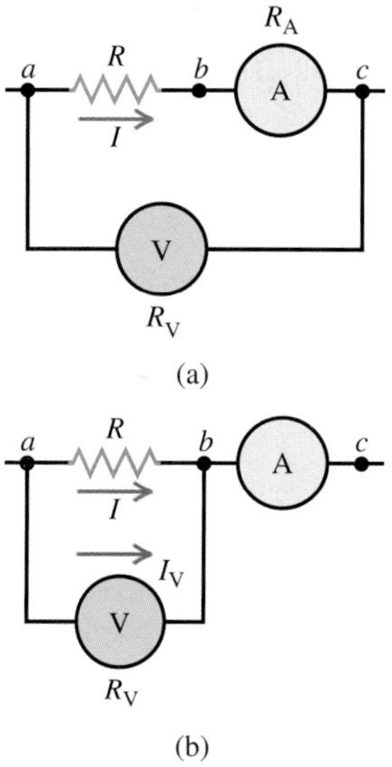

26.16 Ammeter-voltmeter method for measuring resistance.

Example 26.10 Measuring resistance I

Suppose we want to measure an unknown resistance R using the circuit of Fig. 26.16a. The meter resistances are $R_V = 10{,}000\ \Omega$ (for the voltmeter) and $R_A = 2.00\ \Omega$ (for the ammeter). If the voltmeter reads 12.0 V and the ammeter reads 0.100 A, what are the resistance R and the power dissipated in the resistor?

SOLUTION

IDENTIFY and SET UP: The ammeter reads the current $I = 0.100$ A through the resistor and the voltmeter reads the potential difference between a and c. If the ammeter were *ideal* (that is, if $R_A = 0$), there would be zero potential difference between b and c, the voltmeter reading $V = 12.0$ V would be equal to the potential difference V_{ab} across the resistor, and the resistance would simply be equal to $R = V/I = (12.0\ \text{V})/(0.100\ \text{A}) = 120\ \Omega$. The ammeter is *not* ideal, however (its resistance is $R_A = 2.00\ \Omega$), so the voltmeter reading V is actually the sum of the potential differences V_{bc} (across the ammeter) and V_{ab} (across the resistor). We relate these values to the given current by using Ohm's law and solve for V_{ab} and the resistance R. Given these, we are able to calculate the power P into the resistor.

EXECUTE: From Ohm's law, $V_{bc} = IR_A = (0.100\ \text{A})(2.00\ \Omega) = 0.200\ \text{V}$ and $V_{ab} = IR$. The sum of these is $V = 12.0\ \text{V}$, so the potential difference across the resistor is $V_{ab} = V - V_{bc} = (12.0\ \text{V}) - (0.200\ \text{V}) = 11.8\ \text{V}$. Hence the resistance is

$$R = \frac{V_{ab}}{I} = \frac{11.8\ \text{V}}{0.100\ \text{A}} = 118\ \Omega$$

The power dissipated in this resistor is

$$P = V_{ab}I = (11.8\ \text{V})(0.100\ \text{A}) = 1.18\ \text{W}$$

EVALUATE: You can confirm this result for the power by using the alternative formula $P = I^2R$. Do you get the same answer?

Example 26.11 Measuring resistance II

Suppose the meters of Example 26.10 are connected to a different resistor, in the circuit shown in Fig. 26.16b, and the readings obtained on the meters are the same as in Example 26.10. What is the value of this new resistance R, and what is the power dissipated in the resistor?

SOLUTION

IDENTIFY and SET UP: In the preceding example, the ammeter read the actual current through the resistor, but the voltmeter reading was not the same as the potential difference across the resistor. Now the situation is reversed: the voltmeter reading $V = 12.0\ \text{V}$ shows the actual potential difference V_{ab} across the resistor, but the ammeter reading $I_A = 0.100\ \text{A}$ is *not* equal to the current I through the resistor.

Applying the junction rule at b in Fig. 26.16b shows that $I_A = I + I_V$, where I_V is the current through the voltmeter. We find I_V from the given values of V and the voltmeter resistance R_V, and we use this value to find the resistor current I. We then determine the resistance R from I and the voltmeter reading, and calculate the power as in Example 26.10.

EXECUTE: We have $I_V = V/R_V = (12.0\ \text{V})/(10{,}000\ \Omega) = 1.20\ \text{mA}$. The actual current I in the resistor is $I = I_A - I_V = 0.100\ \text{A} - 0.0012\ \text{A} = 0.0988\ \text{A}$, and the resistance is

$$R = \frac{V_{ab}}{I} = \frac{12.0\ \text{V}}{0.0988\ \text{A}} = 121\ \Omega$$

The power dissipated in the resistor is

$$P = V_{ab}I = (12.0\ \text{V})(0.0988\ \text{A}) = 1.19\ \text{W}$$

EVALUATE: Our results for R and P are not too different than the results of Example 26.10, in which the meters are connected in a different way. That's because the ammeter and voltmeter are nearly ideal: compared to the resistance R under test, the ammeter resistance R_A is very small and the voltmeter resistance R_V is very large. Nonetheless, the results of the two examples *are* different, which shows that you must account for how ammeters and voltmeters are used when interpreting their readings.

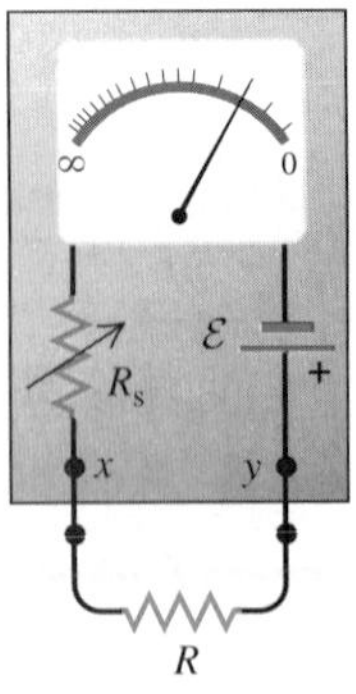

26.17 Ohmmeter circuit. The resistor R_s has a variable resistance, as is indicated by the arrow through the resistor symbol. To use the ohmmeter, first connect x directly to y and adjust R_s until the meter reads zero. Then connect x and y across the resistor R and read the scale.

Ohmmeters

An alternative method for measuring resistance is to use a d'Arsonval meter in an arrangement called an **ohmmeter.** It consists of a meter, a resistor, and a source (often a flashlight battery) connected in series (Fig. 26.17). The resistance R to be measured is connected between terminals x and y.

The series resistance R_s is variable; it is adjusted so that when terminals x and y are short-circuited (that is, when $R = 0$), the meter deflects full-scale. When nothing is connected to terminals x and y, so that the circuit between x and y is *open* (that is, when $R \rightarrow \infty$), there is no current and no deflection. For any intermediate value of R the meter deflection depends on the value of R, and the meter scale can be calibrated to read the resistance R directly. Larger currents correspond to smaller resistances, so this scale reads backward compared to the scale showing the current.

You have probably seen multiple-range meters, or "multimeters," that use d'Arsonval galvanometers. Such a device uses a single-range moving-coil meter; various ranges are provided by switching different resistances in parallel and series with the meter coil. By using appropriate resistances, a multimeter can be used as a voltmeter or as an ammeter. Multimeters also include a battery; placing this in series with the coil makes the meter function as an ohmmeter.

In situations in which high precision is required, instruments containing d'Arsonval meters have been supplanted by electronic instruments with direct digital readouts. These are more precise, stable, and mechanically rugged than d'Arsonval meters. Digital voltmeters can be made with extremely high internal resistance, of the order of 100 MΩ.

The Potentiometer

The *potentiometer* is an instrument that can be used to measure the emf of a source without drawing any current from the source; it also has a number of other useful applications. Essentially, it balances an unknown potential difference against an adjustable, measurable potential difference.

The principle of the potentiometer is shown schematically in Fig. 26.18a. A resistance wire *ab* of total resistance R_{ab} is permanently connected to the terminals of a source of known emf $\mathcal{E}_1$. A sliding contact *c* is connected through the galvanometer G to a second source whose emf $\mathcal{E}_2$ is to be measured. As contact *c* is moved along the resistance wire, the resistance R_{cb} between points *c* and *b* varies; if the resistance wire is uniform, R_{cb} is proportional to the length of wire between *c* and *b*. To determine the value of $\mathcal{E}_2$, contact *c* is moved until a position is found at which the galvanometer shows no deflection; this corresponds to zero current passing through $\mathcal{E}_2$. With $I_2 = 0$, Kirchhoff's loop rule gives

$$\mathcal{E}_2 = IR_{cb}$$

With $I_2 = 0$, the current I produced by the emf $\mathcal{E}_1$ has the same value no matter what the value of the emf $\mathcal{E}_2$. We calibrate the device by replacing $\mathcal{E}_2$ by a source of known emf; then any unknown emf $\mathcal{E}_2$ can be found by measuring the length of wire *cb* for which $I_2 = 0$ (see Exercise 26.31). Note that for this to work, V_{ab} must be greater than $\mathcal{E}_2$.

The term *potentiometer* is also used for any variable resistor, usually having a circular resistance element and a sliding contact controlled by a rotating shaft and knob. The circuit symbol for a potentiometer is shown in Fig. 26.18b.

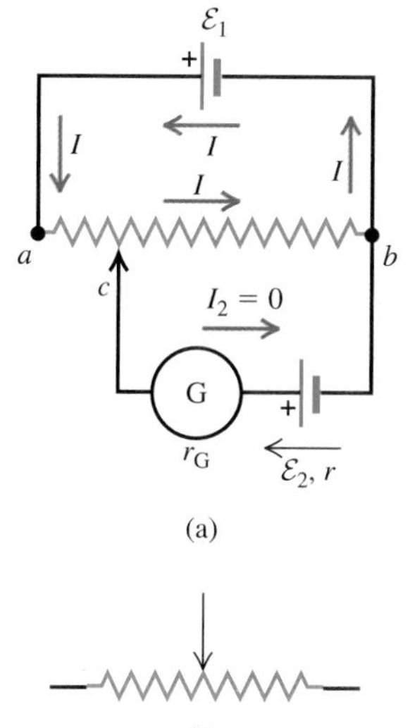

26.18 (a) Potentiometer circuit. (b) Circuit symbol for a potentiometer (variable resistor).

Test Your Understanding

You want to measure the current through and the potential difference across the 2-Ω resistor shown in Fig. 26.12 (Example 26.6 in Section 26.2). How should you connect an ammeter and voltmeter to do this? What resistances should these meters have?

26.4 | *R-C* Circuits

In the circuits we have analyzed up to this point, we have assumed that all the emfs and resistances are *constant* (time-independent) so that all the potentials, currents, and powers are also independent of time. But in the simple act of charging or discharging a capacitor we find a situation in which the currents, voltages, and powers *do* change with time.

Many important devices incorporate circuits in which a capacitor is alternately charged and discharged. These include heart pacemakers (Fig. 26.19), flashing traffic lights, automobile turn signals, and electronic flash units (Fig. 24.9). Understanding what happens in such circuits is thus of great practical importance.

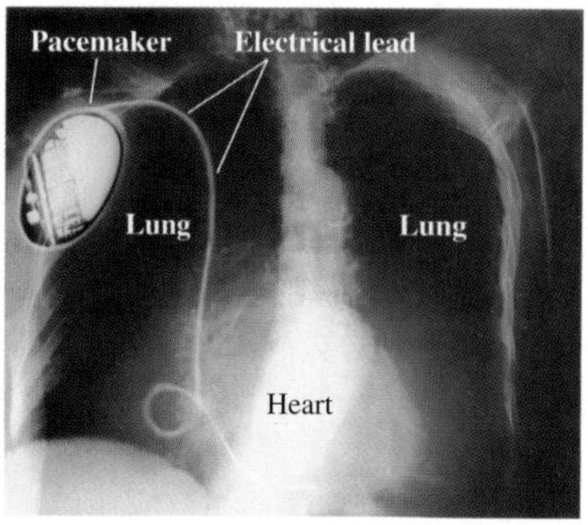

26.19 This colored x-ray image shows a pacemaker surgically implanted in a patient with a malfunctioning sinoatrial node, the part of the heart that generates the electrical signal to trigger heartbeats. To compensate, the pacemaker (located near the collarbone) sends a pulsed electrical signal along the lead to the heart to maintain regular beating.

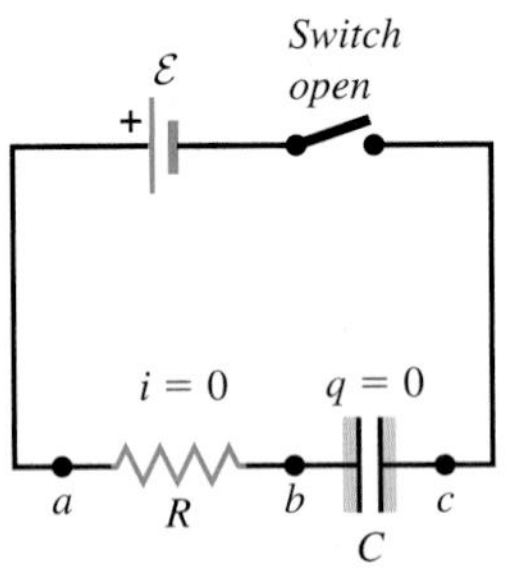

(a) Capacitor initially uncharged

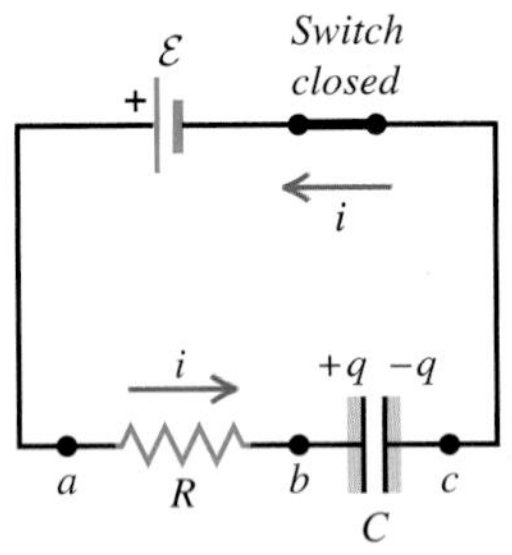

(b) Charging the capacitor

26.20 Charging a capacitor. (a) Just before the switch is closed, the charge q is zero. (b) When the switch closes (at $t = 0$), the current jumps from zero to $\mathcal{E}/R$. As time passes, q approaches Q_f and the current i approaches zero.

ActivPhysics ONLINE

12.6 Capacitance

Charging a Capacitor

Figure 26.20 shows a simple circuit for charging a capacitor. A circuit such as this that has a resistor and a capacitor in series is called an ***R-C* circuit.** We idealize the battery (or power supply) to have a constant emf $\mathcal{E}$ and zero internal resistance $(r = 0)$, and we neglect the resistance of all the connecting conductors.

We begin with the capacitor initially uncharged (Fig. 26.20a); then at some initial time $t = 0$ we close the switch, completing the circuit and permitting current around the loop to begin charging the capacitor (Fig. 26.20b). For all practical purposes, the current begins at the same instant in every conducting part of the circuit, and at each instant the current is the same in every part.

CAUTION Up to this point we have been working with constant potential differences (voltages), currents, and charges and have used *capital* letters V, I, and Q, respectively, to denote these quantities. To distinguish between quantities that vary with time and those that are constant, we will use *lowercase* letters v, i, and q for time-varying voltages, currents, and charges, respectively. We suggest that you follow this same convention in your own work.

Because the capacitor in Fig. 26.20 is initially uncharged, the potential difference v_{bc} across it is zero at $t = 0$. At this time, from Kirchhoff's loop law, the voltage v_{ab} across the resistor R is equal to the battery emf $\mathcal{E}$. The initial $(t = 0)$ current through the resistor, which we will call I_0, is given by Ohm's law: $I_0 = v_{ab}/R = \mathcal{E}/R$.

As the capacitor charges, its voltage v_{bc} increases and the potential difference v_{ab} across the resistor decreases, corresponding to a decrease in current. The sum of these two voltages is constant and equal to $\mathcal{E}$. After a long time the capacitor becomes fully charged, the current decreases to zero, and the potential difference v_{ab} across the resistor becomes zero. Then the entire battery emf $\mathcal{E}$ appears across the capacitor and $v_{bc} = \mathcal{E}$.

Let q represent the charge on the capacitor and i the current in the circuit at some time t after the switch has been closed. We choose the positive direction for the current to correspond to positive charge flowing onto the left-hand capacitor plate, as in Fig. 26.20b. The instantaneous potential differences v_{ab} and v_{bc} are

$$v_{ab} = iR \qquad v_{bc} = \frac{q}{C}$$

Using these in Kirchhoff's loop rule, we find

$$\mathcal{E} - iR - \frac{q}{C} = 0 \tag{26.9}$$

The potential drops by an amount iR as we travel from a to b and by q/C as we travel from b to c. Solving Eq. (26.9) for i, we find

$$i = \frac{\mathcal{E}}{R} - \frac{q}{RC} \tag{26.10}$$

At time $t = 0$, when the switch is first closed, the capacitor is uncharged, and so $q = 0$. Substituting $q = 0$ into Eq. (26.10), we find that the *initial* current I_0 is given by $I_0 = \mathcal{E}/R$, as we have already noted. If the capacitor were not in the circuit, the last term in Eq. (26.10) would not be present; then the current would be *constant* and equal to $\mathcal{E}/R$.

As the charge q increases, the term q/RC becomes larger and the capacitor charge approaches its final value, which we will call Q_f. The current decreases and eventually becomes zero. When $i = 0$, Eq. (26.10) gives

$$\frac{\mathcal{E}}{R} = \frac{Q_f}{RC} \qquad Q_f = C\mathcal{E} \tag{26.11}$$

Note that the final charge Q_f does not depend on R.

The current and the capacitor charge are shown as functions of time in Fig. 26.21. At the instant the switch is closed $(t = 0)$, the current jumps from zero to its initial value $I_0 = \mathcal{E}/R$; after that, it gradually approaches zero. The capacitor charge starts at zero and gradually approaches the final value given by Eq. (26.11), $Q_f = C\mathcal{E}$.

We can derive general expressions for the charge q and current i as functions of time. With our choice of the positive direction for current (Fig. 26.20b), i equals the rate at which positive charge arrives at the left-hand (positive) plate of the capacitor, so $i = dq/dt$. Making this substitution in Eq. (26.10), we have

$$\frac{dq}{dt} = \frac{\mathcal{E}}{R} - \frac{q}{RC} = -\frac{1}{RC}(q - C\mathcal{E})$$

We can rearrange this to

$$\frac{dq}{q - C\mathcal{E}} = -\frac{dt}{RC}$$

and then integrate both sides. We change the integration variables to q' and t' so that we can use q and t for the upper limits. The lower limits are $q' = 0$ and $t' = 0$:

$$\int_0^q \frac{dq'}{q' - C\mathcal{E}} = -\int_0^t \frac{dt'}{RC}$$

When we carry out the integration, we get

$$\ln\left(\frac{q - C\mathcal{E}}{-C\mathcal{E}}\right) = -\frac{t}{RC}$$

Exponentiating both sides (that is, taking the inverse logarithm) and solving for q, we find

$$\frac{q - C\mathcal{E}}{-C\mathcal{E}} = e^{-t/RC}$$

$$q = C\mathcal{E}(1 - e^{-t/RC}) = Q_f(1 - e^{-t/RC}) \tag{26.12}$$

(*R-C* circuit, charging capacitor)

The instantaneous current i is just the time derivative of Eq. (26.12):

$$i = \frac{dq}{dt} = \frac{\mathcal{E}}{R}e^{-t/RC} = I_0 e^{-t/RC} \tag{26.13}$$

(*R-C* circuit, charging capacitor)

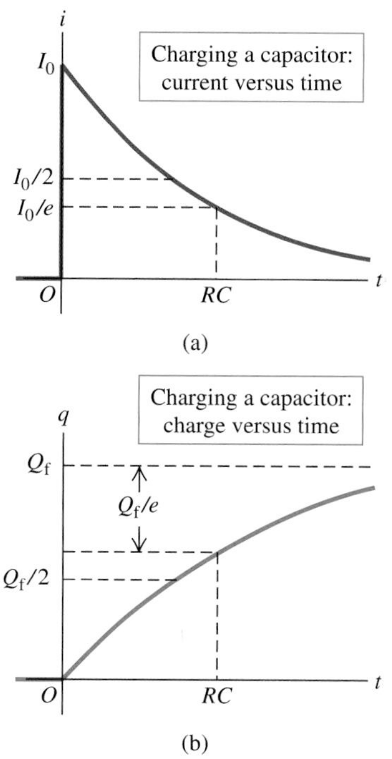

26.21 Current i and capacitor charge q as functions of time for the circuit of Fig. 26.20. The initial current is I_0 and the initial capacitor charge is zero. The current asymptotically approaches zero and the capacitor charge asymptotically approaches a final value of Q_f.

ActivPhysics Online

12.7 Series and Parallel Capacitors

12.8 Circuit Time Constants

The charge and current are both *exponential* functions of time. Figure 26.21a is a graph of Eq. (26.13) and Fig. 26.21b is a graph of Eq. (26.12).

Time Constant

After a time equal to RC, the current in the R-C circuit has decreased to $1/e$ (about 0.368) of its initial value. At this time, the capacitor charge has reached $(1 - 1/e) = 0.632$ of its final value $Q_f = C\mathcal{E}$. The product RC is therefore a measure of how quickly the capacitor charges. We call RC the **time constant,** or the **relaxation time,** of the circuit, denoted by τ:

$$\tau = RC \qquad \text{(time constant for } R\text{-}C \text{ circuit)} \qquad (26.14)$$

When τ is small, the capacitor charges quickly; when it is larger, the charging takes more time. If the resistance is small, it's easier for current to flow, and the capacitor charges more quickly. If R is in ohms and C in farads, τ is in seconds.

In Fig. 26.21a the horizontal axis is an *asymptote* for the curve. Strictly speaking, i never becomes precisely zero. But the longer we wait, the closer it gets. After a time equal to 10 RC, the current has decreased to 0.000045 of its initial value. Similarly, the curve in Fig. 26.21b approaches the horizontal dashed line labeled Q_f as an asymptote. The charge q never attains precisely this value, but after a time equal to $10RC$ the difference between q and Q_f is only 0.000045 of Q_f. We invite you to verify that the product RC has units of time.

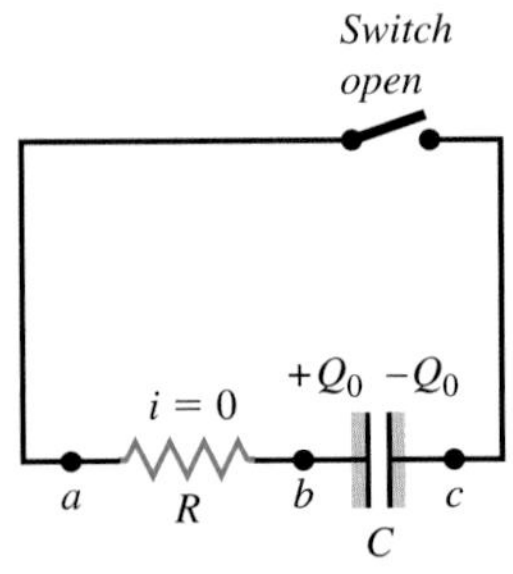

(a) Capacitor initially charged

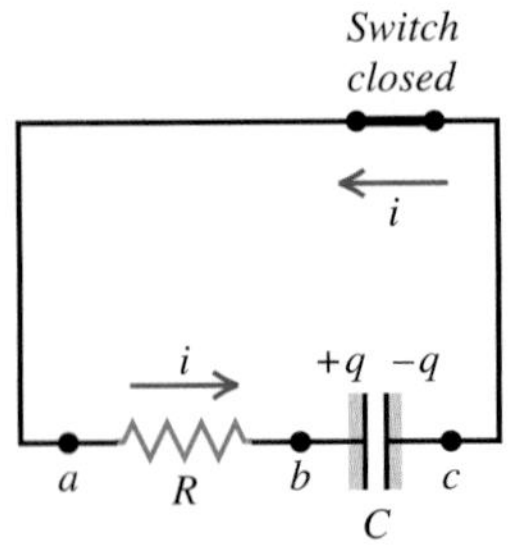

(b) Discharging the capacitor

26.22 Discharging a capacitor. (a) Before the switch is closed at time $t = 0$, the capacitor charge is Q_0 and the current is zero. (b) At time t after the switch is closed, the capacitor charge is q and the current is i. The actual current direction is opposite to the direction shown; i is negative. After a long time, q and i both approach zero.

Discharging a Capacitor

Now suppose that after the capacitor in Fig. 26.20b has acquired a charge Q_0, we remove the battery from our R-C circuit and connect points a and c to an open switch (Fig. 26.22a). We then close the switch and at the same instant reset our stopwatch to $t = 0$; at that time, $q = Q_0$. The capacitor then *discharges* through the resistor, and its charge eventually decreases to zero.

Again let i and q represent the time-varying current and charge at some instant after the connection is made. In Fig. 26.22b we make the same choice of the positive direction for current as in Fig. 26.20b. Then Kirchhoff's loop rule gives Eq. (26.10) but with $\mathcal{E} = 0$; that is,

$$i = \frac{dq}{dt} = -\frac{q}{RC} \qquad (26.15)$$

The current i is now negative; this is because positive charge q is leaving the left-hand capacitor plate in Fig. 26.22b, so the current is in the direction opposite to that shown in the figure. At time $t = 0$, when $q = Q_0$, the initial current is $I_0 = -Q_0/RC$.

To find q as a function of time, we rearrange Eq. (26.15), again change the names of the variables to q' and t', and integrate. This time the limits for q' are Q_0 to q. We get

$$\int_{Q_0}^{q} \frac{dq'}{q'} = -\frac{1}{RC}\int_0^t dt'$$

$$\ln\frac{q}{Q_0} = -\frac{t}{RC}$$

$$q = Q_0 e^{-t/RC} \qquad (R\text{-}C \text{ circuit, discharging capacitor}) \qquad (26.16)$$

The instantaneous current i is the derivative of this with respect to time:

$$i = \frac{dq}{dt} = -\frac{Q_0}{RC}e^{-t/RC} = I_0 e^{-t/RC} \quad (26.17)$$

(*R-C* circuit, discharging capacitor)

The current and the charge are graphed in Fig. 26.23; both quantities approach zero exponentially with time. Comparing these results with Eqs. (26.12) and (26.13), we note that the expressions for the current are identical, apart from the sign of I_0. The capacitor charge approaches zero asymptotically in Eq. (26.16), while the *difference* between q and Q approaches zero asymptotically in Eq. (26.12).

Energy considerations give us additional insight into the behavior of an RC circuit. While the capacitor is charging, the instantaneous rate at which the battery delivers energy to the circuit is $P = \mathcal{E}i$. The instantaneous rate at which electrical energy is dissipated in the resistor is i^2R and the rate at which energy is stored in the capacitor is $iv_{bc} = iq/C$. Multiplying Eq. (26.9) by i, we find

$$\mathcal{E}i = i^2R + iq/C \quad (26.18)$$

This means that of the power $\mathcal{E}i$ supplied by the battery, part (i^2R) is dissipated in the resistor and part (iq/C) is stored in the capacitor.

The *total* energy supplied by the battery during charging of the capacitor equals the battery emf $\mathcal{E}$ multiplied by the total charge Q_f, or $\mathcal{E}Q_f$. The total energy stored in the capacitor, from Eq. (24.9), is $Q_f\mathcal{E}/2$. Thus, of the energy supplied by the battery, *exactly half* is stored in the capacitor, and the other half is dissipated in the resistor. It is a little surprising that this half-and-half division of energy doesn't depend on C, R, or $\mathcal{E}$. This result can also be verified in detail by taking the integral over time of each of the power quantities in Eq. (26.18). We leave this calculation for your amusement (see Problem 26.87).

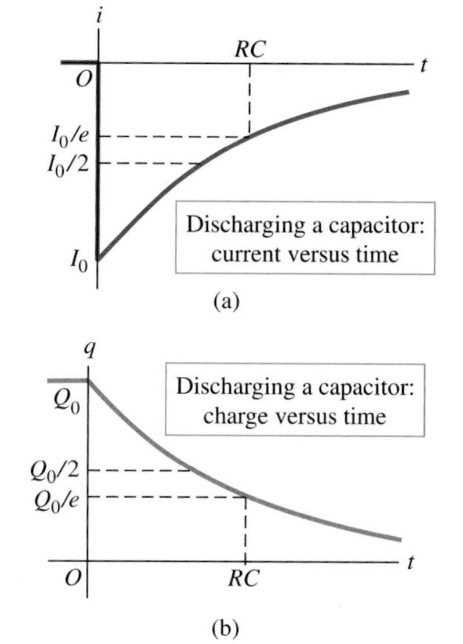

26.23 Current i and capacitor charge q as functions of time for the circuit of Fig. 26.22. The initial current is I_0, and the initial capacitor charge is Q_0; both i and q asymptotically approach zero.

Example 26.12 Charging a capacitor

A resistor with resistance 10 MΩ is connected in series with a capacitor with capacitance 1.0 μF and a battery with emf 12.0 V, as in Fig. 26.20. Before the switch is closed at time $t = 0$, the capacitor is uncharged. a) What is the time constant? b) What fraction of the final charge is on the plates at time $t = 46$ s? c) What fraction of the initial current remains at $t = 46$ s?

SOLUTION

IDENTIFY and SET UP: For a capacitor being charged, the charge is given by Eq. (26.12) and the current by Eq. (26.13). Equation (26.14) gives the time constant.

EXECUTE: a) From Eq. (26.14), the time constant is

$$\tau = RC = (10 \times 10^6\ \Omega)(1.0 \times 10^{-6}\ \text{F}) = 10\ \text{s}$$

b) The fraction of the final capacitor charge is q/Q_f. From Eq. (26.12),

$$\frac{q}{Q_f} = 1 - e^{-t/RC} = 1 - e^{-(46\ \text{s})/(10\ \text{s})} = 0.99$$

The capacitor is 99% charged after a time equal to 4.6 RC, or 4.6 time constants.

c) From Eq. (26.13),

$$\frac{i}{I_0} = e^{-4.6} = 0.010$$

After 4.6 time constants the current has decreased to 1.0% of its initial value.

EVALUATE: The time constant is relatively long because the resistance is very large. The circuit charges more rapidly if a smaller resistance is used.

Example 26.13

Discharging a capacitor

The resistor and capacitor described in Example 26.12 are reconnected as shown in Fig. 26.22. The capacitor is originally given a charge of 5.0 μC, then discharged by closing the switch at $t = 0$. a) At what time will the charge be equal to 0.50 μC? b) What is the current at this time?

SOLUTION

IDENTIFY and SET UP: Now the capacitor is being discharged, so the charge is given by Eq. (26.16) and the current by Eq. (26.17).

EXECUTE: a) Solving Eq. (26.16) for the time t gives

$$t = -RC \ln \frac{q}{Q_0}$$

$$= -(10 \times 10^6\ \Omega)(1.0 \times 10^{-6}\ \text{F}) \ln \frac{0.50\ \mu\text{C}}{5.0\ \mu\text{C}} = 23\ \text{s}$$

This is 2.3 times the time constant $\tau = RC = 10$ s.

b) From Eq. (26.17), with $Q_0 = 5.0\ \mu\text{C} = 5.0 \times 10^{-6}$ C,

$$i = -\frac{Q_0}{RC} e^{-t/RC} = -\frac{5.0 \times 10^{-6}\ \text{C}}{10\ \text{s}} e^{-2.3} = -5.0 \times 10^{-8}\ \text{A}$$

The current has the opposite sign when the capacitor is discharging than when it is charging.

EVALUATE: We could have saved the effort required to calculate $e^{-t/RC}$ by noticing that at the time in question, $q = 0.10\,Q_0$; from Eq. (26.16) this means $e^{-t/RC} = 0.10$.

Test Your Understanding

The energy stored in a capacitor is equal to $q^2/2C$. When a capacitor is discharged, what fraction of the initial energy remains after an elapsed time of one time constant? Does the answer depend on how much energy was stored initially?

26.5 | Power Distribution Systems

We conclude this chapter with a brief discussion of practical household and automotive electric-power distribution systems. Automobiles use direct-current (dc) systems, while nearly all household, commercial, and industrial systems use alternating current (ac) because of the ease of stepping voltage up and down with transformers. Most of the same basic wiring concepts apply to both. We'll talk about alternating-current circuits in greater detail in Chapter 31.

The various lamps, motors, and other appliances to be operated are always connected in *parallel* to the power source (the wires from the power company for houses, or from the battery and alternator for a car). If appliances were connected in series, shutting one appliance off would shut them all off (see Example 26.2 in Section 26.1). The basic idea of house wiring is shown in Fig. 26.24. One side of the "line," as the pair of conductors is called, is called the *neutral* side; it is always connected to "ground" at the entrance panel. For houses, *ground* is an actual electrode driven into the earth (which is usually a good conductor) or sometimes connected to the household water pipes. Electricians speak of the "hot" side and the "neutral" side of the line. Most modern house wiring systems have *two* hot lines with opposite polarity with respect to the neutral. We'll return to this detail later.

Household voltage is nominally 120 V in the United States and Canada, and often 240 V in Europe. (For alternating current, which varies sinusoidally with time, these numbers represent the *root-mean-square* voltage, which is $1/\sqrt{2}$ times the peak voltage. We'll discuss this further in Section 31.1.) The amount of cur-

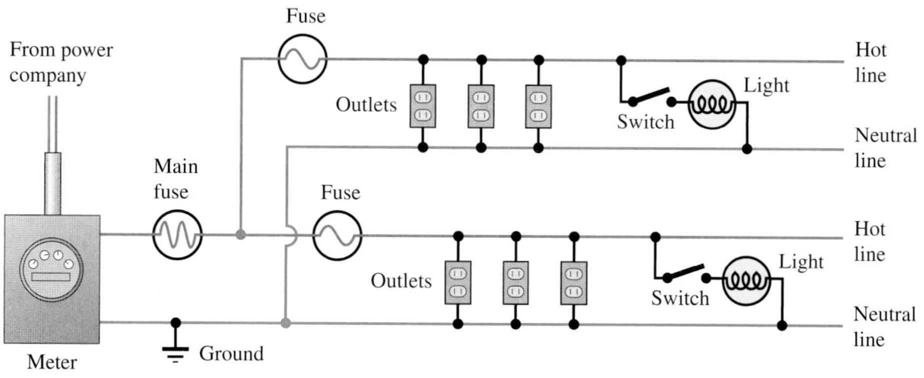

26.24 Schematic diagram of part of a house wiring system. Only two branch circuits are shown; an actual system might have four to thirty branch circuits. Lamps and appliances may be plugged into the outlets. The grounding wires, which normally carry no current, are not shown.

rent I drawn by a given device is determined by its power input P, given by Eq. (25.17): $P = VI$. Hence $I = P/V$. For example, the current in a 100-W light bulb is

$$I = \frac{P}{V} = \frac{100\ \text{W}}{120\ \text{V}} = 0.83\ \text{A}$$

The power input to this bulb is actually determined by its resistance R. Using Eq. (25.18), which states that $P = VI = I^2R = V^2/R$ for a resistor, the resistance of this bulb at operating temperature is

$$R = \frac{V}{I} = \frac{120\ \text{V}}{0.83\ \text{A}} = 144\ \Omega \quad \text{or} \quad R = \frac{V^2}{P} = \frac{(120\ \text{V})^2}{100\ \text{W}} = 144\ \Omega$$

Similarly, a 1500-W waffle iron draws a current of $(1500\ \text{W})/(120\ \text{V}) = 12.5\ \text{A}$ and has a resistance, at operating temperature, of 9.6 Ω. Because of the temperature dependence of resistivity, the resistances of these devices are considerably less when they are cold. If you measure the resistance of a 100-W light bulb with an ohmmeter (whose small current causes very little temperature rise), you will probably get a value of about 10 Ω. When a light bulb is turned on, this low resistance causes an initial surge of current until the filament heats up. That's why a light bulb that's ready to burn out nearly always does so just when you turn it on.

The maximum current available from an individual circuit is limited by the resistance of the wires. As we discussed in Section 25.5, the I^2R power loss in the wires causes them to become hot, and in extreme cases this can cause a fire or melt the wires. Ordinary lighting and outlet wiring in houses usually uses 12-gauge wire. This has a diameter of 2.05 mm and can carry a maximum current of 20 A safely (without overheating). Larger sizes such as 8-gauge (3.26 mm) or 6-gauge (4.11 mm) are used for high-current appliances such as electric ranges and clothes dryers, and 2-gauge (6.54 mm) or larger is used for the main power lines entering a house.

Protection against overloading and overheating of circuits is provided by fuses or circuit breakers. A *fuse* contains a link of lead-tin alloy with a very low melting temperature; the link melts and breaks the circuit when its rated current is exceeded (Fig. 26.25). A *circuit breaker* is an electromechanical device that performs the same function, using an electromagnet or a bimetallic strip to "trip" the

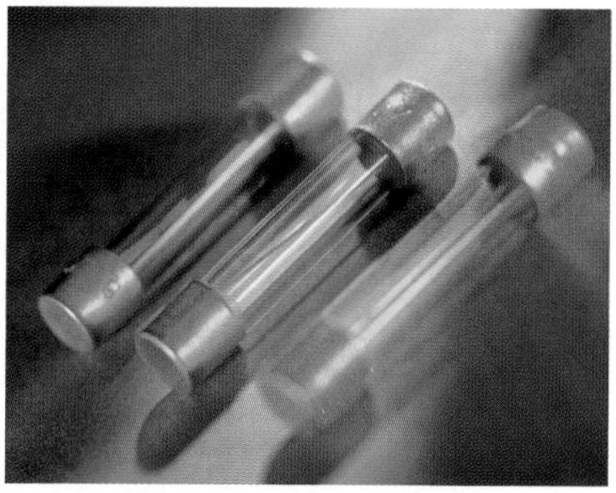

26.25 A set of electrical fuses. A thin wire of lead-tin alloy runs along the length of each fuse, inside the transparent housing.

breaker and interrupt the circuit when the current exceeds a specified value. Circuit breakers have the advantage that they can be reset after they are tripped, while a blown fuse must be replaced. However, fuses are somewhat more reliable in operation than circuit breakers are.

If your system has fuses and you plug too many high-current appliances into the same outlet, the fuse blows. *Do not* replace the fuse with one of larger rating; if you do, you risk overheating the wires and starting a fire. The only safe solution is to distribute the appliances among several circuits. Modern kitchens often have three or four separate 20-A circuits.

Contact between the hot and neutral sides of the line causes a *short circuit.* Such a situation, which can be caused by faulty insulation or by any of a variety of mechanical malfunctions, provides a very low-resistance current path, permitting a very large current that would quickly melt the wires and ignite their insulation if the current were not interrupted by a fuse or circuit breaker (see Example 25.11 in Section 25.5). An equally dangerous situation is a broken wire that interrupts the current path, creating an *open circuit.* This is hazardous because of the sparking that can occur at the point of intermittent contact.

In approved wiring practice, a fuse or breaker is placed *only* in the hot side of the line, never in the neutral side. Otherwise, if a short circuit should develop because of faulty insulation or other malfunction, the ground-side fuse could blow. The hot side would still be live and would pose a shock hazard if you touched the live conductor and a grounded object such as a water pipe. For similar reasons the wall switch for a light fixture is always in the hot side of the line, never the neutral side.

Further protection against shock hazard is provided by a third conductor called the *grounding wire,* included in all present-day wiring. This conductor corresponds to the long round or U-shaped prong of the three-prong connector plug on an appliance or power tool. It is connected to the neutral side of the line at the entrance panel. The grounding wire normally carries no current, but it connects the metal case or frame of the device to ground. If a conductor on the hot side of the line accidentally contacts the frame or case, the grounding conductor provides a current path, and the fuse blows. Without the ground wire, the frame could become "live," that is, at a potential 120 V above ground. Then if you touched it and a water pipe (or even a damp basement floor) at the same time, you could get a dangerous shock (Fig. 26.26). In some situations, especially outlets located outdoors or near a sink or other water pipes, a special kind of circuit breaker called a *ground-fault interrupter* (GFI or GFCI) is used. This device senses the difference in current between the hot and neutral conductors (which is normally zero) and trips when this difference exceeds some very small value, typically 5 mA.

Most modern household wiring systems actually use a slight elaboration of the system described above. The power company provides *three* conductors (Fig. 26.27). One is neutral; the other two are both at 120 V with respect to the neutral but with opposite polarity, giving a voltage between them of 240 V. The power company calls this a *three-wire line,* in contrast to the 120-V two-wire (plus ground wire) line described above. With a three-wire line, 120-V lamps and appliances can be connected between neutral and either hot conductor, and high-power devices requiring 240 V, such as electric ranges and clothes dryers, are connected between the two hot lines.

To help prevent wiring errors, household wiring uses a standardized color code in which the hot side of a line has black insulation (black and red for the two sides of a 240-V line), the neutral side has white insulation, and the grounding conduc-

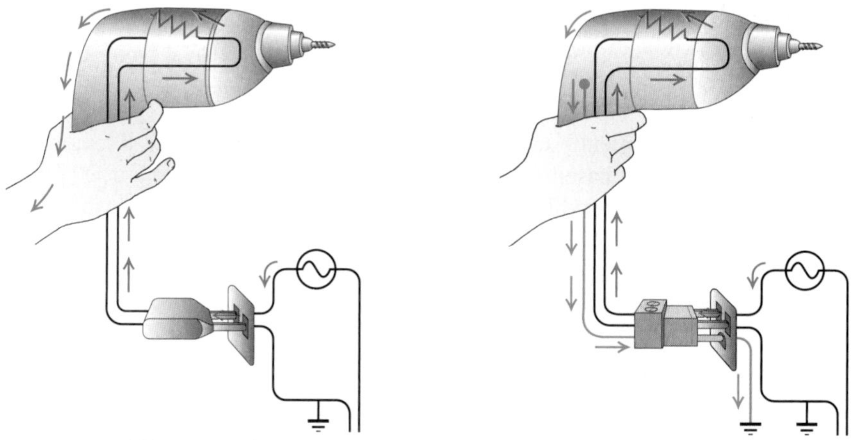

26.26 (a) If a malfunctioning electric drill is connected to a wall socket via a two-prong plug, a person may receive a shock. (b) When the drill malfunctions when connected via a three-prong plug, a person touching it receives no shock, because electric charge flows through the ground wire (shown in green) to the third prong and into the ground rather than into the person's body. If the ground current is appreciable, the fuse blows.

tor is bare or has green insulation. But in electronic devices and equipment the ground or neutral side of the line is usually black. Beware! (Our illustrations do not follow this standard code but use red for the hot line and blue for neutral.)

All of the above discussion can be applied directly to automobile wiring. The voltage is about 13 V (direct current); the power is supplied by the battery and by the alternator, which charges the battery when the engine is running. The neutral side of each circuit is connected to the body and frame of the vehicle. For this low voltage a separate grounding conductor is not required for safety. The fuse or circuit breaker arrangement is the same in principle as in household wiring. Because of the lower voltage (less energy per charge), more current (a greater number of charges per second) is required for the same power; a 100-W headlight bulb requires a current of about $(100\text{ W})/(13\text{ V}) = 8\text{ A}$.

Although we spoke of *power* in the above discussion, what we buy from the power company is *energy*. Power is energy transferred per unit time, so energy is

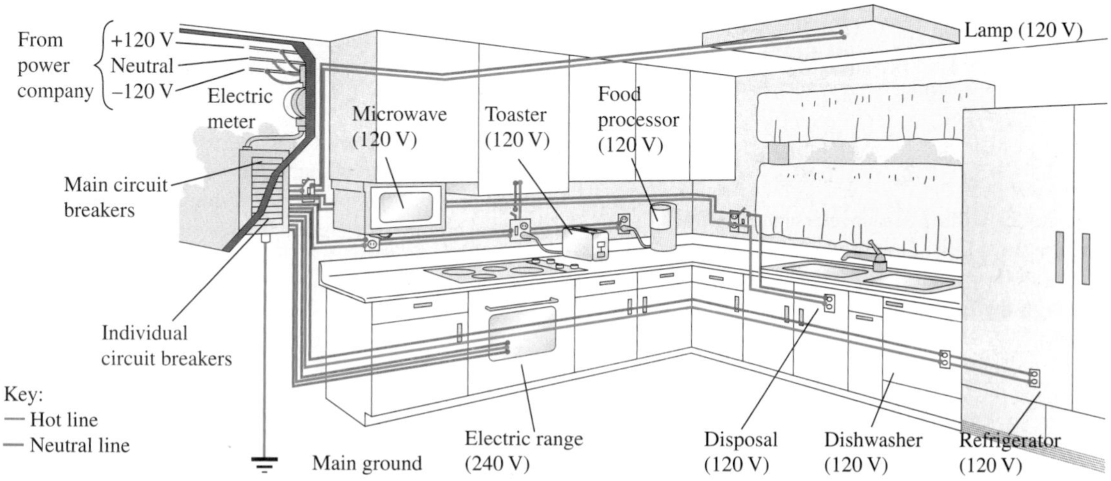

26.27 Diagram of a typical 120–240-V wiring system in a kitchen. Grounding wires are not shown. For each line, the hot side is shown in red, and the neutral line is shown in blue. (Different colors are used in actual household wiring.)

average power multiplied by time. The usual unit of energy sold by the power company is the kilowatt-hour (1 kWh):

$$1\text{ kWh} = (10^3\text{ W})(3600\text{ s}) = 3.6 \times 10^6\text{ W}\cdot\text{s} = 3.6 \times 10^6\text{ J}$$

One kilowatt-hour typically costs 2 to 10 cents, depending on the location and quantity of energy purchased. To operate a 1500-W (1.5-kW) waffle iron continuously for one hour requires 1.5 kWh of energy; at 10 cents per kilowatt-hour, the energy cost is 15 cents. The cost of operating any lamp or appliance for a specified time can be calculated in the same way if the power rating is known. However, many electric cooking utensils (including waffle irons) cycle on and off to maintain a constant temperature, so the average power may be less than the power rating marked on the device.

Example 26.14 A kitchen circuit

An 1800-W toaster, a 1.3-kW electric frying pan, and a 100-W lamp are plugged into the same 20-A, 120-V circuit. a) What current is drawn by each device, and what is the resistance of each device? b) Will this combination blow the fuse?

SOLUTION

IDENTIFY and SET UP: When plugged into the same circuit, the three devices are in parallel. The voltage across each device is $V = 120$ V. We find the current I drawn by each device using the relationship $P = VI$, where P is the power input of the device. To find the resistance R for each device we use the relationship $P = V^2/R$.

EXECUTE: a) To simplify the calculation of current and resistance, we note that $I = P/V$ and $R = V^2/P$. Hence

$$I_{\text{toaster}} = \frac{1800\text{ W}}{120\text{ V}} = 15\text{ A} \qquad R_{\text{toaster}} = \frac{(120\text{ V})^2}{1800\text{ W}} = 8\ \Omega$$

$$I_{\text{frying pan}} = \frac{1300\text{ W}}{120\text{ V}} = 11\text{ A} \qquad R_{\text{frying pan}} = \frac{(120\text{ V})^2}{1300\text{ W}} = 11\ \Omega$$

$$I_{\text{lamp}} = \frac{100\text{ W}}{120\text{ V}} = 0.83\text{ A} \qquad R_{\text{lamp}} = \frac{(120\text{ V})^2}{100\text{ W}} = 144\ \Omega$$

For constant voltage the device with the *least* resistance (in this case the toaster) draws the most current and receives the most power.

b) The total current through the line is the sum of the currents drawn by the three devices:

$$I = I_{\text{toaster}} + I_{\text{frying pan}} + I_{\text{lamp}} = 15\text{ A} + 11\text{ A} + 0.83\text{ A} = 27\text{ A}$$

This exceeds the 20-A rating of the line, and the fuse will indeed blow.

EVALUATE: We could also find the current by first finding the equivalent resistance of the three devices in parallel:

$$\frac{1}{R_{\text{eq}}} = \frac{1}{R_{\text{toaster}}} + \frac{1}{R_{\text{frying pan}}} + \frac{1}{R_{\text{lamp}}} = \frac{1}{8\ \Omega} + \frac{1}{11\ \Omega} + \frac{1}{144\ \Omega} = 0.22\ \Omega^{-1}$$

$$R_{\text{eq}} = 4.5\ \Omega$$

the total current is then $I = V/R_{\text{eq}} = (120\text{ V})/(4.5\ \Omega) = 27$ A, as before. A third way to determine I is to use $I = P/V$ and simply divide the total power delivered to all three devices by the voltage:

$$I = \frac{P_{\text{toaster}} + P_{\text{frying pan}} + P_{\text{lamp}}}{V} = \frac{1800\text{ W} + 1300\text{ W} + 100\text{ W}}{120\text{ V}} = 27\text{ A}$$

Current demands like these are encountered in everyday life in kitchens, which is why modern kitchens have more than one 20-A circuit. In actual practice, the toaster and frying pan should be connected to different circuits; the current in each circuit would then be safely below the 20-A rating.

Test Your Understanding

To prevent the fuse in Example 26.14 from blowing, a home electrician replaces the fuse with one rated at 40 A. Is this a reasonable thing to do?

SUMMARY

When several resistors $R_1, R_2, R_3, \ldots$, are connected in series, the equivalent resistance R_{eq} is the sum of the individual resistances. The same *current* flows through all the resistors in a series connection. When several resistors are connected in parallel, the reciprocal of the equivalent resistance R_{eq} is the sum of the reciprocals of the individual resistances. All resistors in a parallel connection have the same *potential difference* between their terminals. (See Examples 26.1 and 26.2)

$$R_{eq} = R_1 + R_2 + R_3 + \cdots \qquad (26.1)$$

(resistors in series)

$$\frac{1}{R_{eq}} = \frac{1}{R_1} + \frac{1}{R_2} + \frac{1}{R_3} + \cdots \qquad (26.2)$$

(resistors in parallel)

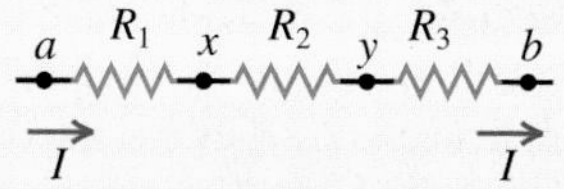

R_1, R_2, and R_3 in series

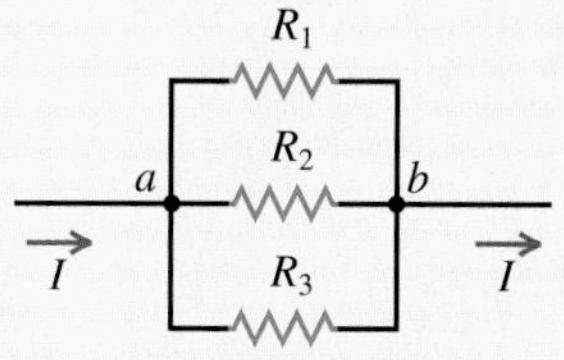

R_1, R_2, and R_3 in parallel

Kirchhoff's junction rule is based on conservation of charge. It states that the algebraic sum of the currents into any junction must be zero. Kirchhoff's loop rule is based on conservation of energy and the conservative nature of electrostatic fields. It states that the algebraic sum of potential differences around any loop must be zero. Careful use of consistent sign rules is essential in applying Kirchhoff's rules.
(See Examples 26.3 through Example 26.7)

$$\sum I = 0 \qquad (26.5)$$

(junction rule)

$$\sum V = 0 \qquad (26.6)$$

(loop rule)

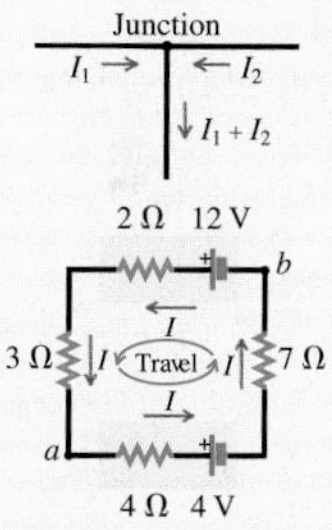

In a d'Arsonval galvanometer, the deflection is proportional to the current in the coil. For a larger current range, a shunt resistor is added, so some of the current bypasses the meter coil. Such an instrument is called an ammeter. If the coil and any additional series resistance included obey Ohm's law, the meter can also be calibrated to read potential difference or voltage. The instrument is then called a voltmeter. A good ammeter has very low resistance; a good voltmeter has very high resistance. (See Examples 26.8 through 26.11)

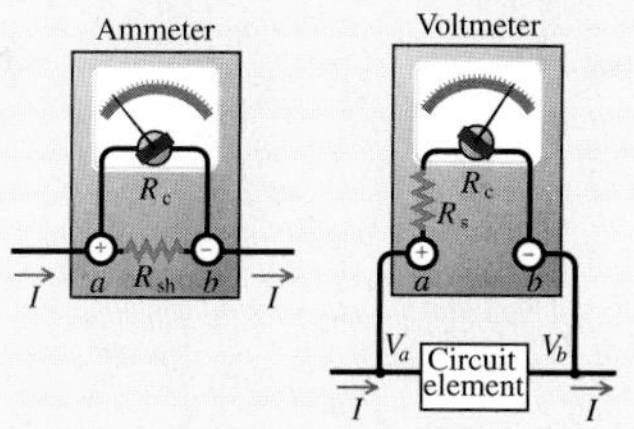

When a capacitor is charged by a battery in series with a resistor, the current and capacitor charge are not constant. The charge approaches its final value asymptotically and the current approaches zero asymptotically. The charge and current in the circuit are given by Eqs. (26.12) and (26.13). After a time $\tau = RC$, the charge has approached within $1/e$ of its final value. This time is called the time constant or relaxation time of the circuit. When the capacitor discharges, the charge and current are given as functions of time by Eqs. (26.16) and (26.17). The time constant is the same for charging and discharging.
(See Examples 26.12 and 26.13)

$$q = C\mathcal{E}(1 - e^{-t/RC}) = Q_f(1 - e^{-t/RC}) \qquad (26.12)$$

(capacitor charging)

$$i = \frac{dq}{dt} = \frac{\mathcal{E}}{R}e^{-t/RC} = I_0 e^{-t/RC} \qquad (26.13)$$

(capacitor charging)

$$q = Q_0 e^{-t/RC} \qquad (26.16)$$

(capacitor discharging)

$$i = \frac{dq}{dt} = -\frac{Q_0}{RC}e^{-t/RC} = I_0 e^{-t/RC} \qquad (26.17)$$

(capacitor discharging)

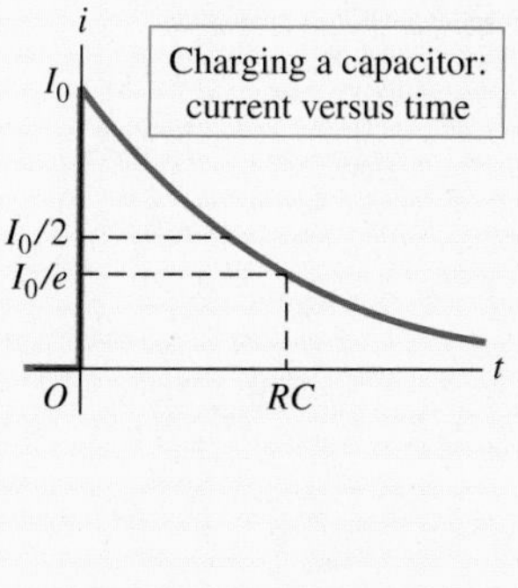

In household wiring systems, the various electrical devices are connected in parallel across the power line, which consists of a pair of conductors, one "hot" and the other "neutral." An additional "ground" wire is included for safety. The maximum permissible current in a circuit is determined by the size of the wires and the maximum temperature they can tolerate. Protection against excessive current and the resulting fire hazard is provided by fuses or circuit breakers.
(See Example 26.14)

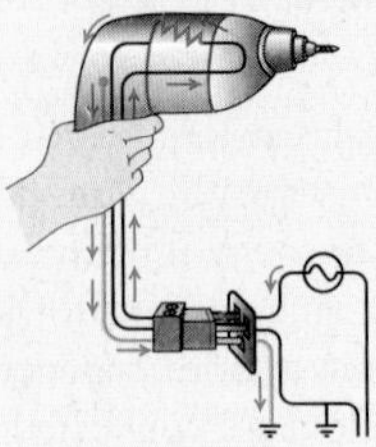

Key Terms

direct current, 980
alternating current, 980
series, 981
parallel, 981
equivalent resistance, 981
junction, 986
loop, 986
Kirchhoff's junction rule, 986
Kirchhoff's loop rule, 987
d'Arsonval galvanometer, 992
ammeter, 993
shunt resistor, 993
voltmeter, 994
ohmmeter, 996
***R-C* circuit, 998**
time constant (relaxation time), 1000

Your Notes

Answer to Chapter Opening Question

The potential difference V is the same across resistors connected in parallel. However, there is a different current I through each resistor if the resistances R are different: $I = V/R$.

Answers to Test Your Understanding Questions

Section 26.1 In order of increasing equivalent resistance R_{eq}, the arrangements are (b), (d), (c), and (a). Here's why: (a) The three resistors in Fig. 26.1a are in series, so $R_{eq} = R + R + R = 3R$. (b) In Fig. 26.1b the three resistors are in parallel, so $1/R_{eq} = 1/R + 1/R + 1/R = 3/R$ and $R_{eq} = R/3$. (c) In Fig. 26.1c the second and third resistors are in parallel, so their equivalent resistance R_{23} is given by $1/R_{23} = 1/R + 1/R = 2/R$; hence $R_{23} = R/2$. This combination is in series with the first resistor, so the three resistors together have equivalent resistance $R_{eq} = R + R/2 = 3R/2$. (d) In Fig. 26.1d the second and third resistors are in series, so their equivalent resistance is $R_{23} = R + R = 2R$. This combination is in parallel with the first resistor, so the equivalent resistance of the three-resistor combination is given by $1/R_{eq} = 1/R + 1/2R = 3/2R$. Hence $R_{eq} = 2R/3$.
Section 26.2 Equation (2) minus equation (1) gives $-I_2(1\,\Omega) - (I_2 + I_3)(2\,\Omega) + (I_1 - I_3)(1\,\Omega) + I_1(1\,\Omega) = 0$. We can obtain this equation by applying the loop rule around the path from c to b to d to a to c in Fig. 26.12. This isn't a new equation, so it would not have helped with the solution of Example 26.6.
Section 26.3 The ammeter should be connected in series with the 2-Ω resistor between points b and d, and the terminals of the voltmeter should be connected to points b and d. Ideally the ammeter would have zero resistance and the voltmeter would have infinite resistance so that their presence would have no effect on either the resistor current or the voltage. Neither of these idealizations is possible, but the ammeter resistance should be much less than 2 Ω and the voltmeter resistance should be much greater than 2 Ω.
Section 26.4 After one time constant, $t = RC$ and the initial charge Q_0 has decreased to $Q_0e^{-t/RC} = Q_0e^{-RC/RC} = Q_0e^{-1} = Q_0/e$. Hence the stored energy has decreased from $Q_0^2/2C$ to $(Q_0/e)^2/2C = Q_0^2/2Ce^2$, a fraction $1/e^2 = 0.135$ of its initial value. This result doesn't depend on the initial value of the energy.
Section 26.5 This is a very dangerous thing to do. The fuse will allow currents up to 40 A, double the rated value of the wiring. The amount of power $P = I^2R$ dissipated in a section of wire can therefore be up to four times the rated value, so the wires could get very warm and start a fire.

Discussion Questions

Q26.1 In which 120-V light bulb does the filament have greater resistance, a 60-W bulb or a 120-W bulb? If the two bulbs are connected to a 120-V line in series, through which bulb will there be the greater voltage drop? What if they are connected in parallel? Explain your reasoning.

Q26.2 Two 120-V light bulbs, one 25-W and one 200-W, were connected in series across a 240-V line. It seemed like a good idea at the time, but one bulb burned out almost instantaneously. Which one burned out, and why?

Q26.3 You connect a number of identical light bulbs to a flashlight battery. What happens to the brightness of each bulb as more and more bulbs are added to the circuit if you connect them in i) series? ii) parallel? Will the battery last longer if the bulbs are in series or in parallel? Explain your reasoning.

Q26.4 In the circuit shown in Fig. 26.28, three identical light bulbs are connected to a flashlight battery. How do the brightnesses of the bulbs compare? Which light bulb has the greatest current passing through it? Which light bulb has the greatest potential difference between its terminals? What happens if bulb A is unscrewed? Bulb B? Bulb C? Explain your reasoning.

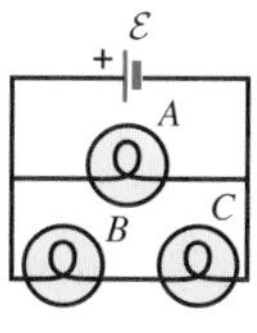

Figure 26.28 Question Q26.4.

Q26.5 Why do the lights on a car become dimmer when the starter is operated?

Q26.6 Resistors R_1 and R_2 are connected in parallel to an emf source that has negligible internal resistance. What happens to the current through R_1 when R_2 is removed from the circuit? Explain your reasoning.

Q26.7 Resistors R_1 and R_2 are connected in series to an emf source that has negligible internal resistance. What happens to the current through R_1 when a third resistor R_3 is connected in parallel with R_2? Explain your reasoning.

Q26.8 Compare the formulas for resistors in series and in parallel with those for capacitors in series and in parallel. What similarities and differences do you see? Sometimes circuit analysis uses the quantity *conductance*, denoted as G and defined as the reciprocal of resistance: $G = 1/R$. Make the corresponding comparison for conductance and capacitance.

Q26.9 Is it possible to connect resistors together in a way that cannot be reduced to some combination of series and parallel combinations? If so, give examples. If not, state why not.

Q26.10 The direction of current in a battery can be reversed by connecting it to a second battery of greater emf with the positive terminals of the two batteries together. When the direction of current is reversed in a battery, does its emf also reverse? Why or why not?

Q26.11 In a two-cell flashlight, the batteries are usually connected in series. Why not connect them in parallel? What possible advantage could there be in connecting several identical batteries in parallel?

Q26.12 Electric rays (genus *Torpedo*) deliver electric shocks to stun their prey and to discourage predators. (In ancient Rome, physicians practiced a primitive form of electroconvulsive therapy by placing electric rays on their patients to cure headaches and gout.) Fig. 26.29a shows *Torpedo* as seen from below. The voltage is produced by thin, wafer-like cells called *electrocytes,* each of

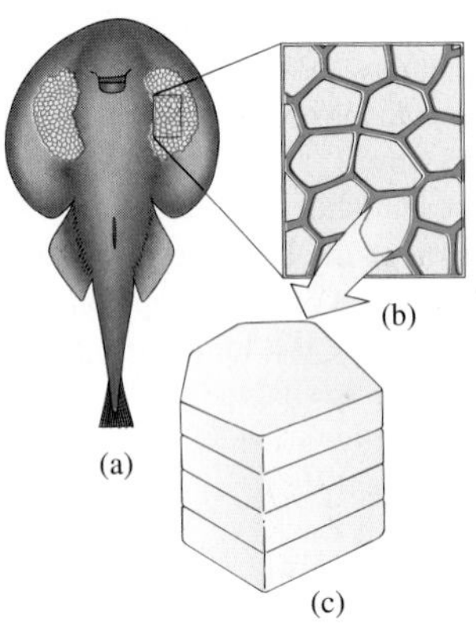

Figure 26.29 Question Q26.12.

which acts as a battery with an emf of about 10^{-4} V. Stacks of electrocytes are arranged side-by-side on the underside of *Torpedo* (Fig. 26.29b); in such a stack, the positive face of each electrocyte touches the negative face of the next electrocyte (Fig. 26.29c). What is the advantage of stacking the electrocytes? Of having the stacks side-by-side?

Q26.13 The emf of a flashlight battery is roughly constant with time, but its internal resistance increases with age and use. What sort of meter should be used to test the freshness of a battery?

Q26.14 Is it possible to have a circuit in which the potential difference across the terminals of a battery in the circuit is zero? If so, give an example. If not, explain why not.

Q26.15 For very large resistances it is easy to construct *R-C* circuits having time constants of several seconds or minutes. How might this fact be used to measure very large resistances, those that are too large to measure by more conventional means?

Q26.16 When a capacitor, battery, and resistor are connected in series, does the resistor affect the maximum charge stored on the capacitor? Why or why not? What purpose does the resistor serve?

Q26.17 The greater the diameter of the wire used in household wiring, the greater the maximum current that can safely be carried by the wire. Why is this? Does the maximum permissible current depend on the length of the wire? Does it depend on what the wire is made of? Explain your reasoning.

Exercises

Section 26.1 Resistors in Series and in Parallel

26.1 A 32-Ω resistor and a 20-Ω resistor are connected in parallel, and the combination is connected across a 240-V dc line. a) What is the resistance of the parallel combination? b) What is the total current through the parallel combination? c) What is the current through each resistor?

26.2 Prove that when two resistors are connected in parallel, the equivalent resistance of the combination is always smaller than that of either resistor.

26.3 If two resistors R_1 and R_2 $(R_2 > R_1)$ are connected in series as shown in Fig. 26.30, which of the following must be true? In each case justify your answer. a) $I_1 = I_2 = I_3$; b) The current is greater in R_1 than in R_2. c) The electrical power consumption is the same for both resistors. d) The electrical power consumption is greater in R_2 than in R_1. e) The potential drop is the same across both resistors. f) The potential at point *a* is the same as at point *c*. g) The potential at point *b* is lower than at point *c*. h) The potential at point *c* is lower than at point *b*.

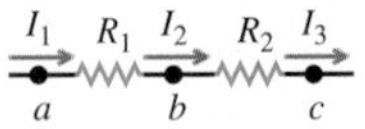

Figure 3.30 Exercise 26.3.

26.4 If two resistors R_1 and R_2 $(R_2 > R_1)$ are connected in parallel as shown in Fig. 26.31, which of the following must be true? In each case justify your answer. a) $I_1 = I_2$. b) $I_3 = I_4$. c) The current is greater in R_1 than in R_2. d) The rate of electrical energy consumption is the same for both resistors. e) The rate of electrical energy consumption is greater in R_2 than in R_1. f) $V_{cd} = V_{ef} = V_{ab}$. g) Point *c* is at higher potential than point *d*. h) Point *f* is at higher potential than point *e*. i) Point *c* is at higher potential than point *e*.

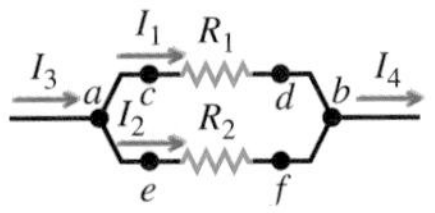

Figure 26.31 Exercise 26.4.

26.5 Three resistors having resistances of 1.60 Ω, 2.40 Ω, and 4.80 Ω are connected in parallel to a 28.0-V battery that has negligible internal resistance. Find a) the equivalent resistance of the combination; b) the current in each resistor; c) the total current through the battery; d) the voltage across each resistor; e) the power dissipated in each resistor. f) Which resistor dissipates the most power: the one with the greatest resistance or the least resistance? Explain why this should be.

26.6 Now the three resistors of Exercise 26.5 are connected in series to the same battery. Answer the same questions for this situation.

26.7 a) The *power rating* of a resistor is the maximum power the resistor can safely dissipate without too great a rise in temperature. The power rating of a 15-kΩ resistor is 5.0 W. What is the maximum allowable potential difference across the terminals of the resistor? b) A 9.0-kΩ resistor is to be connected across a 120-V potential difference. What power rating is required?

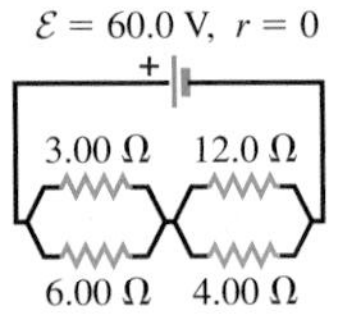

Figure 26.32 Exercise 26.8.

26.8 Compute the equivalent resistance of the network in Fig. 26.32, and find the current in each resistor. The battery has negligible internal resistance.

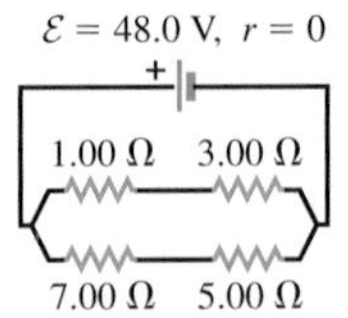

Figure 26.33 Exercise 26.9.

26.9 Compute the equivalent resistance of the network in Fig. 26.33, and find the current in each resistor. The battery has negligible internal resistance.

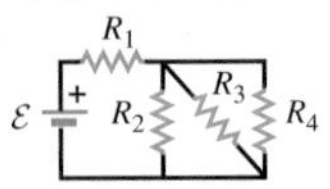

Figure 26.34 Exercises 26.10 and 26.11.

26.10 Four resistors and a battery of negligible internal resistance are assembled to make the circuit in Fig. 26.34. Let $\mathcal{E} = 6.00$ V, $R_1 = 3.50$ Ω, $R_2 = 8.20$ Ω, $R_3 = 1.50$ Ω, and

$R_4 = 4.50\ \Omega$. Find a) the equivalent resistance of the network; b) the current in each resistor.

26.11 In the circuit of Fig. 26.34, each resistor represents a light bulb. Let $R_1 = R_2 = R_3 = R_4 = 4.50\ \Omega$ and $\mathcal{E} = 9.00$ V. a) Find the current in each bulb. b) Find the power dissipated in each bulb. Which bulb or bulbs glows the brightest? c) Bulb R_4 is now removed from the circuit, leaving a break in the wire at its position. Now what is the current in each of the remaining bulbs R_1, R_2, and R_3? d) With bulb R_4 removed, what is the power dissipated in each of the remaining bulbs? e) Which light bulb(s) glow brighter as a result of removing R_4? Which bulb(s) glow less brightly? Discuss why there are different effects on different bulbs.

26.12 Consider the circuit shown in Fig. 26.35. The current through the 6.00 Ω resistor is 4.00 A, in the direction shown. What are the currents through the 25.0 Ω and 20.0 Ω resistors?

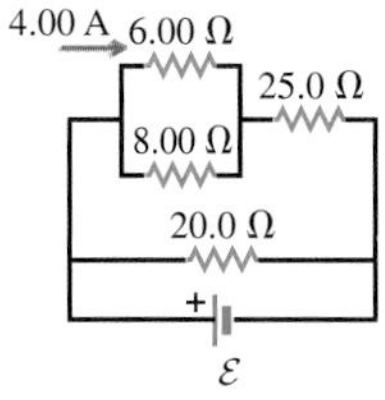

Figure 26.35 Exercise 26.12.

26.13 In the circuit shown in Fig. 26.36, the voltage across the 2.00 Ω resistor is 12.0 V. What are the emf of the battery and the current through the 6.00 Ω resistor?

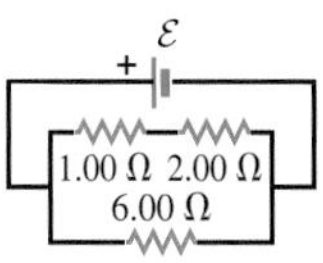

Figure 26.36 Exercise 26.13.

26.14 A Three-Way Light Bulb. A three-way light bulb has three brightness settings (low, medium, and high), but only two filaments. a) A particular three-way light bulb connected across a 120-V line can dissipate 60 W, 120 W, or 180 W. Describe how the two filaments are arranged in the bulb, and calculate the resistance of each filament. b) Suppose the filament with the higher resistance burns out. How much power will the bulb dissipate on each of the three brightness settings? What will be the brightness (low, medium, or high) on each setting? c) Repeat part (b) for the situation where it is the filament with the lower resistance that burns out.

26.15 Light Bulbs in Series and in Parallel. Two light bulbs have resistances of 400 Ω and 800 Ω. If the two light bulbs are connected in series across a 120-V line, find a) the current through each bulb; b) the power dissipated in each bulb, and the total power dissipated in both bulbs. The two light bulbs are now connected in parallel across the 120-V line. Find c) the current through each bulb; d) the power dissipated in each bulb, and the total power dissipated in both bulbs. e) In each situation, which of the two bulbs glows the brightest? In which situation is there a greater total light output from both bulbs combined?

26.16 Light Bulbs in Series. A 60-W, 120-V light bulb and a 200-W, 120-V light bulb are connected in series across a 240-V line. Assume that the resistance of each bulb does not vary with current. (*Note:* This description of a light bulb gives the power it dissipates when connected to the stated potential difference; that is, a 25-W, 120-V light bulb dissipates 25 W when connected to a 120-V line.) a) Find the current through the bulbs. b) Find the power dissipated in each bulb. c) One bulb burns out very quickly. Which one? Why?

26.17 In the circuit in Fig. 26.37, a 20.0-Ω resistor is inside 100 g of pure water that is surrounded by insulating styrofoam. If the water is initially at 10.0°C, how long will it take for its temperature to rise to 58.0°C?

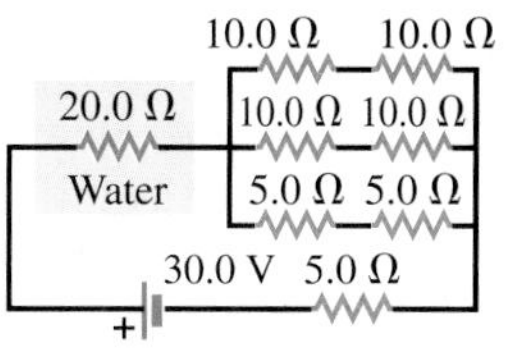

Figure 26.37 Exercise 26.17.

26.18 In the circuit shown in Fig. 26.38, the rate at which R_1 is dissipating electrical energy is 20.0 W. a) Find R_1 and R_2. b) What is the emf of the battery? c) Find the current through both R_2 and the 10.0-Ω resistor. d) Calculate the total electrical power consumption in all the resistors and the electrical power delivered by the battery. Show that your results are consistent with conservation of energy.

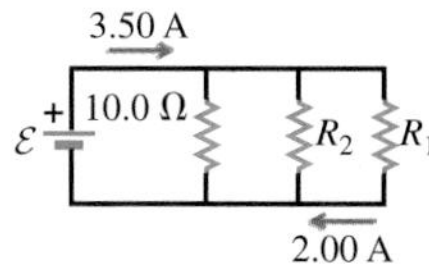

Figure 26.38 Exercise 26.18.

Section 26.2 Kirchhoff's Rules

26.19 In the circuit shown in Fig. 26.39, find a) the current in resistor R; b) the resistance R; c) the unknown emf $\mathcal{E}$. d) If the circuit is broken at point x, what is the current in resistor R?

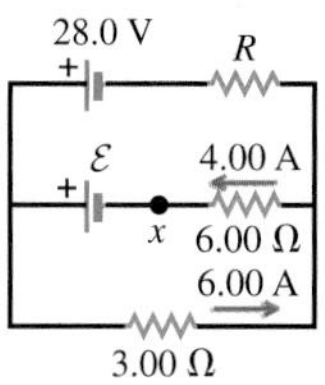

Figure 26.39 Exercise 26.19.

26.20 Find the emf's $\mathcal{E}_1$ and $\mathcal{E}_2$ in the circuit of Fig. 26.40, and find the potential difference of point b relative to point a.

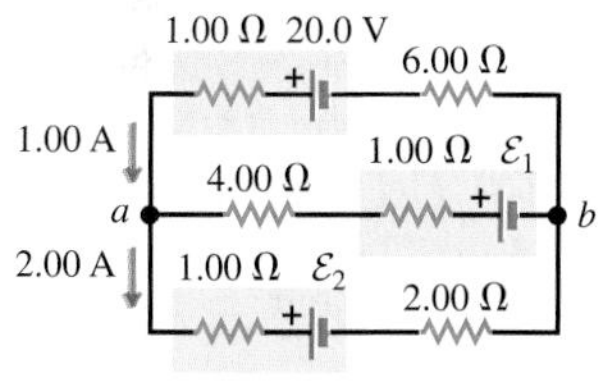

Figure 26.40 Exercise 26.20.

26.21 In the circuit shown in Fig. 26.41, find a) the current in the 3.00-Ω resistor; b) the unknown emf's $\mathcal{E}_1$ and $\mathcal{E}_2$; c) the resistance R. Note that three currents are given.

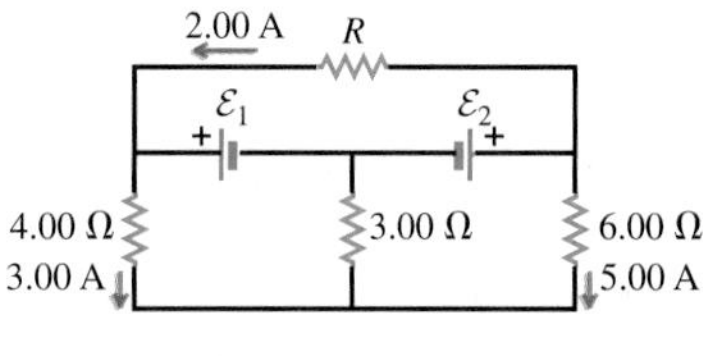

Figure 26.41 Exercise 26.21.

26.22 In the circuit shown in Fig. 26.42, find a) the current in each branch; b) the potential difference V_{ab} of point a relative to point b.

26.23 The 10.00-V battery in Fig. 26.42 is removed from the circuit and reinserted with the opposite polarity, so that its positive terminal is now next to point a. The rest of the circuit is as shown in the figure. Find a) the current in each branch; b) the potential difference V_{ab} of point a relative to point b.

26.24 The 5.00-V battery in Fig. 26.42 is removed from the circuit and replaced by a 20.00-V battery, with its negative terminal next to point b. The rest of the circuit is as shown in the figure. Find a) the current in each branch; b) the potential difference V_{ab} of point a relative to point b.

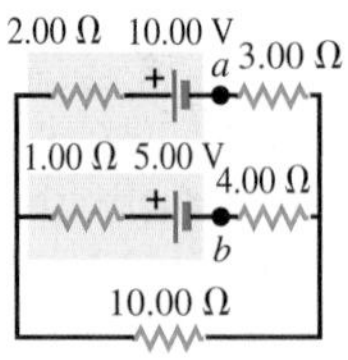

Figure 26.42 Exercises 26.22, 26.23, and 26.24.

26.25 Use the expression $P = I^2R$ to calculate the total power dissipated in all four resistors in Fig. 26.10a.

26.26 In the circuit shown in Fig. 26.10a (Example 26.3 in Section 26.2), the 12-V battery is removed and reinserted with the opposite polarity, so that its positive terminal is now next to point b. The rest of the circuit is as shown in the figure. Find a) the current in the circuit (magnitude *and* direction); b) the potential difference V_{ab}.

26.27 In the circuit shown in Fig. 26.12 (Example 26.6 in Section 26.2), the 2-Ω resistor is replaced by a 1-Ω resistor, and the center 1-Ω resistor (through which the current is I_3) is replaced by a resistor of unknown resistance R. The rest of the circuit is as shown in the figure. a) Calculate the current in each resistor. Draw a diagram of the circuit and label each resistor with the current through it. b) Calculate the equivalent resistance of the network. c) Calculate the potential difference V_{ab}. d) Your answers in parts (a), (b), and (c) do not depend on the value of R. Explain why.

Section 26.3 Electrical Measuring Instruments

26.28 The resistance of a galvanometer coil is 25.0 Ω, and the current required for full-scale deflection is 500 μA. a) Show in a diagram how to convert the galvanometer to an ammeter reading 20.0 mA full scale, and compute the shunt resistance. b) Show how to convert the galvanometer to a voltmeter reading 500 mV full scale, and compute the series resistance.

26.29 The resistance of the coil of a pivoted-coil galvanometer is 9.36 Ω, and a current of 0.0224 A causes it to deflect full scale. We want to convert this galvanometer to an ammeter reading 20.0 A full scale. The only shunt available has a resistance of 0.0250 Ω. What resistance R must be connected in series with the coil (see Fig. 26.43)?

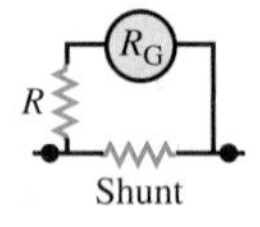

Figure 26.43 Exercise 26.29.

26.30 A 90.0-V battery has an internal resistance of $r = 8.23\ \Omega$. a) What is the reading of a voltmeter having a resistance of $R_V = 425\ \Omega$ when placed across the terminals of the battery? b) What maximum value may the ratio r/R_V have if the error in the reading of the emf of a battery is not to exceed 4.0%?

26.31 Consider the potentiometer circuit of Fig. 26.18. The resistor between a and b is a uniform wire with length l, with a sliding contact c at a distance x from b. An unknown emf $\mathcal{E}_2$ is measured by sliding the contact until the galvanometer G reads zero. a) Show that under this condition the unknown emf is given by $\mathcal{E}_2 = (x/l)\mathcal{E}_1$. b) Why is the internal resistance of the galvanometer not important? c) Suppose $\mathcal{E}_1 = 9.15$ V and $l = 1.000$ m. The galvanometer G reads zero when $x = 0.365$ m. What is the emf $\mathcal{E}_2$?

26.32 Two 150-V voltmeters, one with a resistance of 10.0 kΩ and the other with a resistance of 90.0 kΩ, are connected in series across a 120-V dc line. Find the reading of each voltmeter. (A 150-V voltmeter deflects full scale when the potential difference between its two terminals is 150 V.)

26.33 In the ohmmeter of Fig. 26.17, the coil of the meter has resistance $R_c = 15.0\ \Omega$ and the current required for full-scale deflection is $I_{fs} = 3.60$ mA. The source is a flashlight battery with $\mathcal{E} = 1.50$ V and negligible internal resistance. The ohmmeter is to show a meter deflection of one-half of full scale when connected to a resistor with $R = 600\ \Omega$. What series resistance R_s is required?

26.34 In the ohmmeter in Fig. 26.44, M is a 2.50-mA meter of resistance 65.0 Ω. (A 2.50-mA meter deflects full scale when the current through it is 2.50 mA.) The battery B has an emf of 1.52 V and negligible internal resistance. R is so chosen that when the terminals a and b are shorted $(R_x = 0)$, the meter reads full scale. When a and b are open $(R_x = \infty)$, the meter reads zero. a) What is the resistance of the resistor R? b) What current indicates a resistance R_x of 200 Ω? c) What values of R_x correspond to meter deflections of $\frac{1}{4}$, $\frac{1}{2}$, and $\frac{3}{4}$ of full scale if the deflection is proportional to the current through the galvanometer?

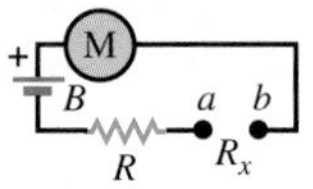

Figure 26.44 Exercise 26.34.

Section 26.4 *R-C* Circuits

26.35 Verify that the product RC has dimensions of time.

26.36 A 4.60-μF capacitor that is initially uncharged is connected in series with a 7.50-kΩ resistor and an emf source with $\mathcal{E} = 125$ V and negligible internal resistance. Just after the circuit is completed, what is a) the voltage drop across the capacitor? b) the voltage drop across the resistor? c) the charge on the capacitor? d) the current through the resistor? e) A long time after the circuit is completed (after many time constants) what are the values of the preceding four quantities?

26.37 A capacitor with capacitance $C = 455$ pF is charged, with charge of magnitude 65.5 nC on each plate. The capacitor is then connected to a voltmeter that has internal resistance 1.28 MΩ. a) What is the current through the voltmeter just after the connection is made? b) What is the time constant of this RC circuit?

26.38 A capacitor is charged to a potential of 12.0 V and is then connected to a voltmeter having an internal resistance of 3.40 MΩ. After a time of 4.00 s the voltmeter reads 3.0 V. What is the capacitance?

26.39 A 12.4-μF capacitor is connected through a 0.895-MΩ resistor to a constant potential difference of 60.0 V. a) Compute the charge on the capacitor at the following times after the connections are made: 0, 5.0 s, 10.0 s, 20.0 s, and 100.0 s. b) Compute the charging currents at the same instants. c) Graph the results of parts (a) and (b) for t between 0 and 20 s.

26.40 A resistor and capacitor are connected in series to an emf source. The time constant for the circuit is 0.870 s. a) A second capacitor, identical to the first, is added in series. What is the time constant for this new circuit? b) In the original circuit a second capacitor, identical to the first, is connected in parallel with the first capacitor. What is the time constant for this new circuit?

26.41 An emf source with $\mathcal{E} = 120$ V, a resistor with $R = 80.0\ \Omega$, and a capacitor with $C = 4.00\ \mu$F are connected in series. As the capacitor charges, when the current in the resistor is 0.900 A, what is the magnitude of the charge on each plate of the capacitor?

26.42 In the circuit shown in Fig. 26.45, $C = 5.90\ \mu$F, $\mathcal{E} = 28.0$ V, and the emf has negligible resistance. Initially the capacitor is uncharged and the switch S is in position 1. The switch is then moved to position 2, so that the capacitor begins to charge. a) What will be the charge on the capacitor a long time after the switch is moved to position 2? b) After the switch has been in position 2 for 3.00 ms, the charge on the capacitor is measured to be 110 μC. What is the value of the resistance R? c) How long after the switch is moved to position 2 will the charge on the capacitor be equal to 99.0% of the final value found in part (a)?

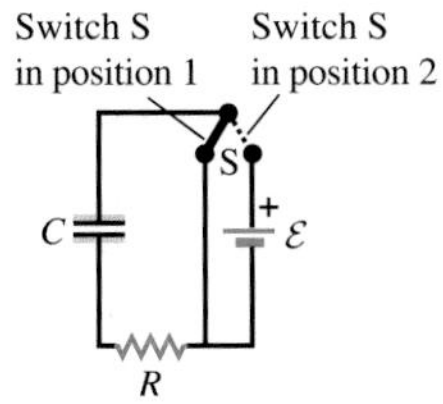

Figure 26.45 Exercises 26.42 and 26.43.

26.43 A capacitor with $C = 1.50 \times 10^{-5}$ F is connected as shown in Fig. 26.45 with a resistor with $R = 980\ \Omega$ and an emf source with $\mathcal{E} = 18.0$ V and negligible internal resistance. Initially the capacitor is uncharged and the switch S is in position 1. The switch is then moved to position 2, so that the capacitor begins to charge. After the switch has been in position 2 for 10.0 ms, the switch is moved back to position 1 so that the capacitor begins to discharge. a) Compute the charge on the capacitor just *before* the switch is thrown from position 2 back to position 1. b) Compute the voltage drops across the resistor and across the capacitor at the instant described in part (a). c) Compute the voltage drops across the resistor and across the capacitor just *after* the switch is thrown from position 2 back to position 1. d) Compute the charge on the capacitor 10.0 ms after the switch is thrown from position 2 back to position 1.

Section 26.5 Power Distribution Systems

26.44 The heating element of an electric dryer is rated at 4.1 kW when connected to a 240-V line. a) What is the current in the heating element? Is 12-gauge wire large enough to supply this current? b) What is the resistance of the dryer's heating element at its operating temperature? c) At 11 cents per kWh, how much does it cost per hour to operate the dryer?

26.45 A 1500-W electric heater is plugged into the outlet of a 120-V circuit that has a 20-A circuit breaker. You plug an electric hair dryer into the same outlet. The hair dryer has power settings of 600 W, 900 W, 1200 W, and 1500 W. You start with the hair dryer on the 600 W setting and increase the power setting until the circuit breaker trips. What power setting caused the breaker to trip?

26.46 How many 90-W, 120-V light bulbs can be connected to a 20-A, 120-V circuit without tripping the circuit breaker? (See the note in Exercise 26.16.)

26.47 The heating element of an electric stove consists of a heater wire embedded within an electrically insulating material, which in turn is inside a metal casing. The heater wire has a resistance of 20 Ω at room temperature $(23.0°\text{C})$ and a temperature coefficient of resistivity $\alpha = 2.8 \times 10^{-3}(\text{C}°)^{-1}$. The heating element operates from a 120-V line. a) When the heating element is first turned on, what current does it draw and what electrical power does it dissipate? b) When the heating element has reached an operating temperature of 280°C $(536°\text{F})$, what current does it draw and what electrical power does it dissipate?

Problems

26.48 a) A resistance R_2 is connected in parallel with a resistance R_1. Derive an expression for the resistance R_3 that must be connected in series with the combination of R_1 and R_2 so that the equivalent resistance is equal to the resistance R_1. Show the arrangement of resistors in a diagram. b) A resistance R_2 is connected in series with a resistance R_1. Derive an expression for the resistance R_3 that must be connected in parallel with the combination of R_1 and R_2 so that the equivalent resistance is equal to R_1. Show the arrangement of resistors in a diagram.

26.49 A 400-Ω, 2.4-W resistor is needed, but only several 400-Ω, 1.2-W resistors are available (see Exercise 26.7). a) What two different combinations of the available units give the required resistance and power rating? b) For each of the resistor networks from part (a), what power is dissipated in each resistor when 2.4 W is dissipated by the combination?

26.50 A 20.0-m-long cable consists of a solid-inner, cylindrical, nickel core 10.0 cm in diameter surrounded by a solid outer cylindrical shell of copper 10.0 cm in inside diameter and 20.0 cm in outside diameter. The resistivity of nickel is $7.8 \times 10^{-8}\ \Omega \cdot \text{m}$. a) What is the resistance of this cable? b) If we think of this cable as a single material, what is its equivalent resistivity?

26.51 Two identical 1.00-Ω wires are laid side-by-side and soldered together so they touch each other for half of their lengths. What is the equivalent resistance of this combination?

26.52 The two identical light bulbs in Example 26.2 (Section 26.1) are connected in parallel to a different source, one with $\mathcal{E} = 8.0$ V and internal resistance 0.8 Ω. Each light bulb has a resistance $R = 2.0\ \Omega$ (assumed independent of the current through the bulb). a) Find the current through each bulb, the potential difference across each bulb, and the power delivered to each bulb. b) Suppose one of the bulbs burns out, so that its filament breaks and current no longer flows through it. Find the power delivered to the remaining bulb. Does the remaining bulb glow more or less brightly after the other bulb burns out than before?

26.53 Each of the three resistors in Fig. 26.46 has a resistance of 2.4 Ω and can dissipate a maximum of 36 W without becoming excessively heated. What is the maximum power the circuit can dissipate?

Figure 26.46 Problem 26.53.

26.54 a) Calculate the equivalent resistance of the circuit of Fig. 26.47 between x and y. b) What is the potential of point a relative to point x if the current in the 8.0-Ω resistor is 2.4 A in the direction from left to right in the figure?

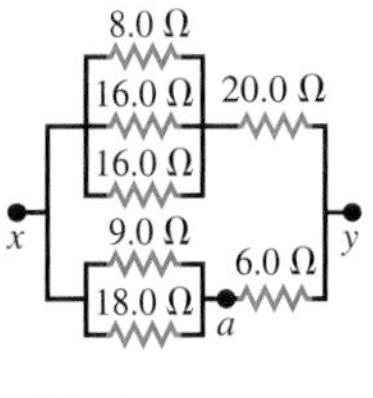

Figure 26.47 Problem 26.54.

26.55 If an ohmmeter is connected between points a and b in each of the circuits shown in Fig. 26.48, what will it read?

Figure 26.48 Problem 26.55.

26.56 For the resistor network shown in Fig. 26.49, the ohmmeter reads 20.2 Ω. What is the resistance of the resistor X?

26.57 Calculate the three currents I_1, I_2, and I_3 indicated in the circuit diagram shown in Fig. 26.50.

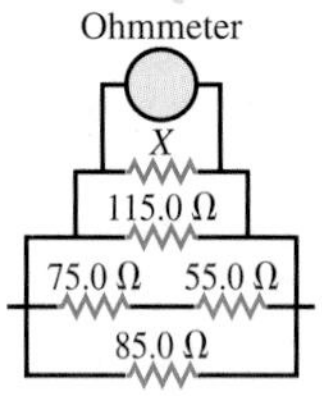

Figure 26.49 Problem 26.56.

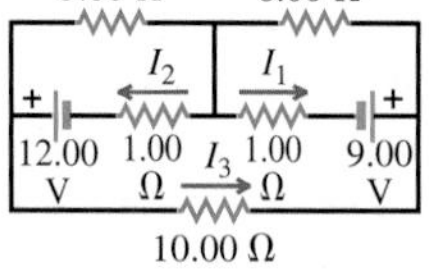

Figure 26.50 Problem 26.57.

26.58 What must the emf $\mathcal{E}$ in Fig. 26.51 be in order for the current through the 7.00-Ω resistor to be 1.80 A? Each emf source has negligible internal resistance.

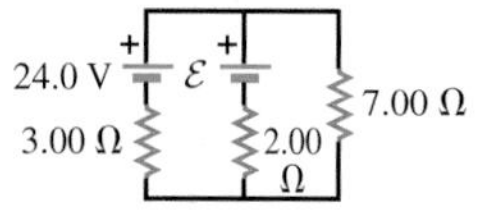

Figure 26.51 Problem 26.58.

26.59 Find the current through each of the three resistors of the circuit shown in Fig. 26.52. The emf sources have negligible internal resistance.

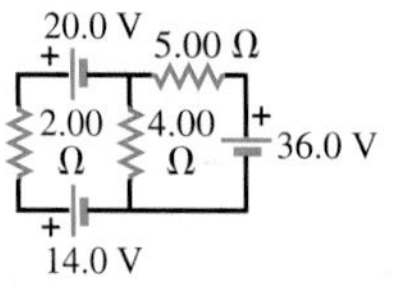

Figure 26.52 Problem 26.59.

26.60 a) Find the current through the battery and each resistor in the circuit shown in Fig. 26.53. b) What is the equivalent resistance of the resistor network?

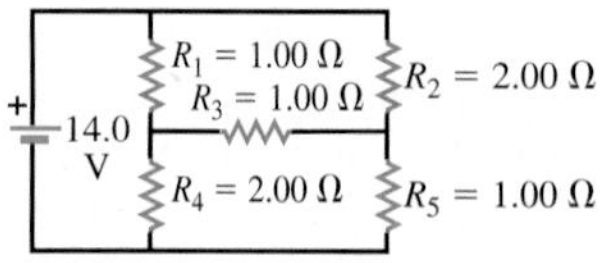

Figure 26.53 Problem 26.60.

26.61 a) Find the potential of point a with respect to point b in Fig. 26.54. b) If points a and b are connected by a wire with negligible resistance, find the current in the 12.0-V battery.

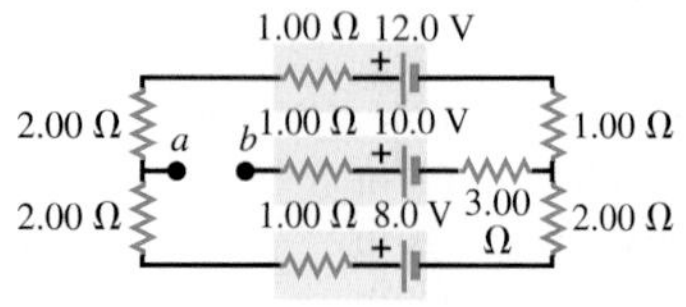

Figure 26.54 Problem 26.61.

26.62 In the circuit shown in Fig. 26.55: a) What must the emf $\mathcal{E}$ of the battery be in order for a current of 2.00 A to flow through the 5.00-V battery, as shown. Is the polarity of the battery correct as shown? b) How long does it take for 60.0 J of thermal energy to be produced in the 10.0-Ω resistor?

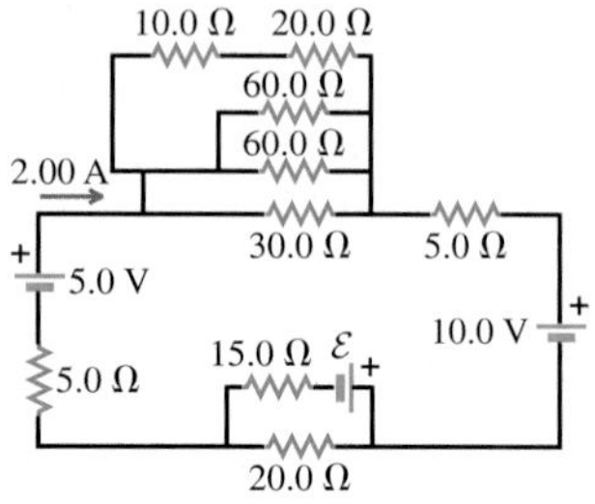

Figure 26.55 Problem 26.62.

26.63 In the circuit shown in Fig. 26.56, the current through the 12.0-V battery is measured to be 70.6 mA, in the direction shown. What is the terminal voltage V_{ab} of the 24.0-V battery?

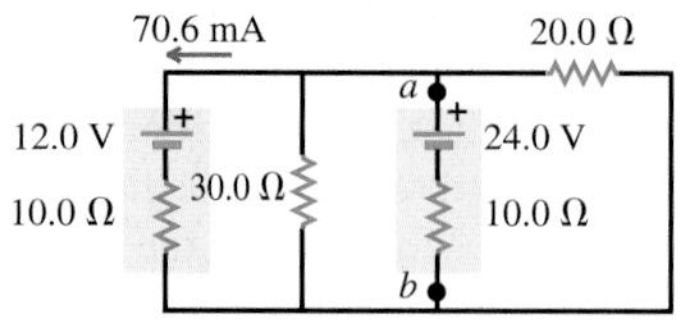

Figure 26.56 Problem 26.63.

26.64 In the circuit shown in Fig. 26.57, all the resistors are rated at a maximum power of 1.00 W. What is the maximum emf $\mathcal{E}$ that the battery can have without burning up any of the resistors?

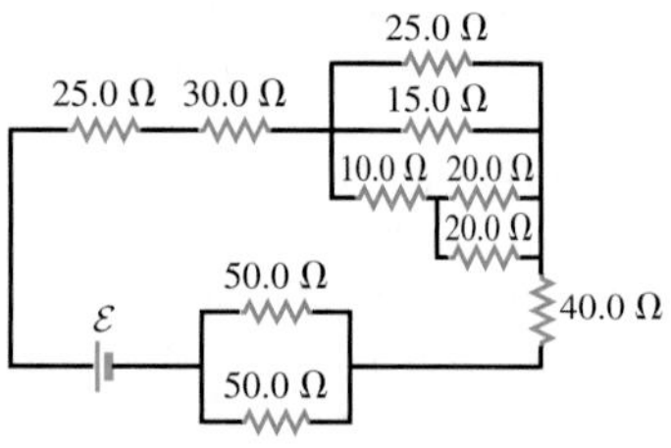

Figure 26.57 Problem 26.64.

26.65 In the circuit shown in Fig. 26.58, the current in the 20.0-V battery is 5.00 A in the direction shown and the voltage across the 8.00-Ω resistor is 16.0 V, with the lower end of the resistor at higher potential. Find a) the emf (including its polarity) of the battery X; b) the current I through the 200.0-V battery (including its direction); c) the resistance R.

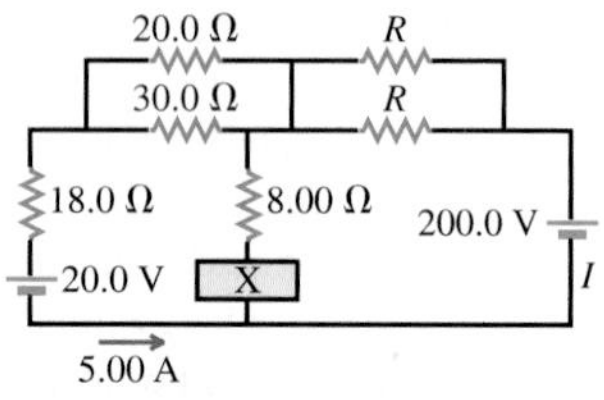

Figure 26.58 Problem 26.65.

26.66 Three identical resistors are connected in series. When a certain potential difference is applied across the combination, the total power dissipated is 27 W. What power would be dissipated if the three resistors were connected in parallel across the same potential difference?

26.67 A resistor R_1 consumes electrical power P_1 when connected to an emf $\mathcal{E}$. When resistor R_2 is connected to the same emf, it consumes electrical power P_2. In terms of P_1 and P_2, what is the total electrical power consumed when they are both connected to this emf source a) in parallel? b) in series?

26.68 The capacitor in Fig. 26.59 is initially uncharged. The switch is closed at $t = 0$. a) Immediately after the switch is closed, what is the current through each resistor? b) What is the final charge on the capacitor?

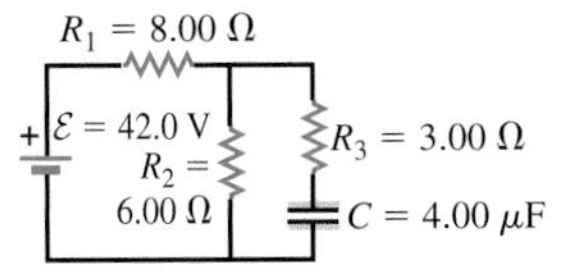

Figure 26.59 Problem 26.68.

26.69 Fig. 26.60 employs a convention often used in circuit diagrams. The battery (or other power supply) is not shown explicitly. It is understood that the point at the top, labeled "36.0 V," is connected to the positive terminal of a 36.0-V battery having negligible internal resistance, and that the "ground" symbol at the bottom is connected to the negative terminal of the battery. The circuit is completed through the battery, even though it is not shown on the diagram. a) What is the potential difference V_{ab}, the potential of point a relative to point b, when the switch S is open? b) What is the current through switch S when it is closed? c) What is the equivalent resistance when switch S is closed?

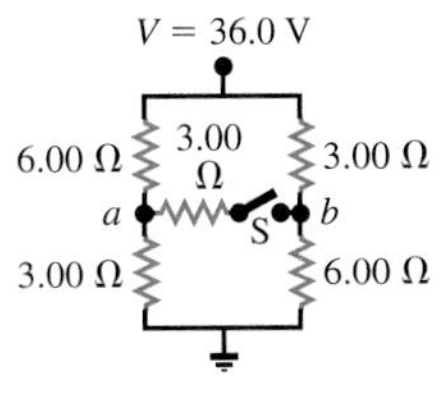

Figure 26.60 Problem 26.69.

26.70 (See Problem 26.69.) a) What is the potential of point a with respect to point b in Fig. 26.61 when switch S is open? b) Which point, a or b, is at the higher potential? c) What is the final potential of point b with respect to ground when switch S is closed? d) How much does the charge on each capacitor change when S is closed?

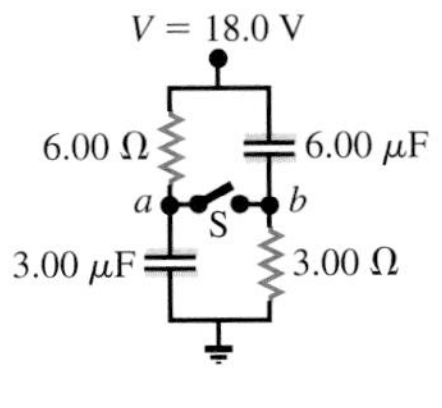

Figure 26.61 Problem 26.70.

26.71 (See Problem 26.69.) a) What is the potential of point a with respect to point b in Fig. 26.62 when switch S is open? b) Which point, a or b, is at the higher potential? c) What is the final potential of point b with respect to ground when switch S is closed? d) How much charge flows through switch S when it is closed?

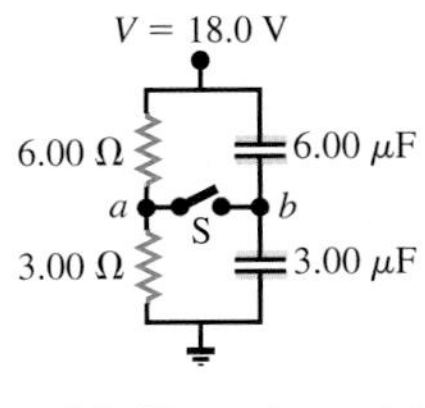

Figure 26.62 Problem 26.71.

26.72 A Multirange Ammeter. The resistance of the moving coil of the galvanometer G in Fig. 26.63 is $48.0\ \Omega$, and the galvanometer deflects full scale with a current of 0.0200 A. When the meter is connected to the circuit being measured, one connection is made to the post marked + and the other to the post marked with the desired current range. Find the magnitudes of the resistances R_1, R_2, and R_3 required to convert the galvanometer to a multirange ammeter deflecting full scale with currents of 10.0 A, 1.00 A, and 0.100 A.

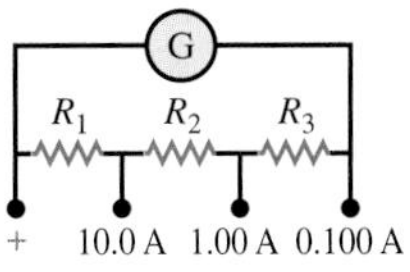

Figure 26.63 Problem 26.72.

26.73 A Multirange Voltmeter. Fig. 26.64 shows the internal wiring of a "three-scale" voltmeter whose binding posts are marked +, 3.00 V, 15.0 V, and 150 V. When the meter is connected to the circuit being measured, one connection is made to the post marked + and the other to the post marked with the desired voltage range. The resistance of the moving coil, R_G, is $40.0\ \Omega$, and a current of 1.00 mA in the coil causes it to deflect full scale. Find the resistances R_1, R_2, and R_3, and the overall resistance of the meter on each of its ranges.

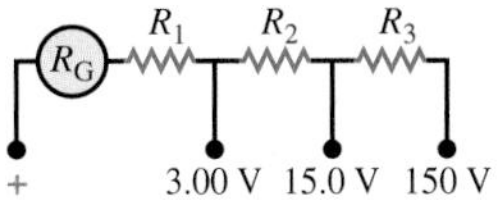

Figure 26.64 Problem 26.73.

26.74 Point a in Fig. 26.65 is maintained at a constant potential of 400 V above ground. (See Problem 26.69.) a) What is the reading of a voltmeter with the proper range and with resistance $5.00 \times 10^4\ \Omega$ when connected between point b and ground? b) What is the reading of a voltmeter with resistance $5.00 \times 10^6\ \Omega$? c) What is the reading of a voltmeter with infinite resistance?

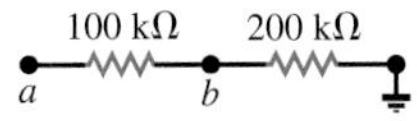

Figure 26.65 Problem 26.74.

26.75 A 150-V voltmeter has a resistance of $30{,}000\ \Omega$. When connected in series with a large resistance R across a 110-V line, the meter reads 68 V. Find the resistance R.

26.76 Let V and I represent, respectively, the readings of the voltmeter and ammeter shown in Fig. 26.16, and let R_V and R_A be their equivalent resistances. Because of the resistances of the meters, the resistance R is not simply equal to V/I. a) When the circuit is connected as in Fig. 26.16a, show that

$$R = \frac{V}{I} - R_A$$

Explain why the true resistance R is always *less* than V/I. b) When the connections are as in Fig. 26.16b, show that

$$R = \frac{V}{I - (V/R_V)}$$

Explain why the true resistance R is always *greater* than V/I. c) Show that the power delivered to the resistor in part (a) is $IV - I^2R_A$ and that in part (b) is $IV - (V^2/R_V)$.

26.77 The Wheatstone Bridge. The circuit shown in Fig. 26.66, called a *Wheatstone bridge,* is

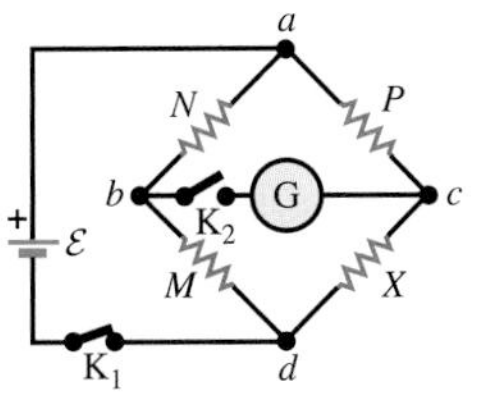

Figure 26.66 Problem 26.77.

used to determine the value of an unknown resistor X by comparison with three resistors M, N, and P whose resistances can be varied. For each setting, the resistance of each resistor is precisely known. With switches K_1 and K_2 closed, these resistors are varied until the current in the galvanometer G is zero; the bridge is then said to be *balanced.* a) Show that under this condition the unknown resistance is given by $X = MP/N$. (This method permits very high precision in comparing resistors.) b) If the galvanometer G shows zero deflection when $M = 850.0\ \Omega$, $N = 15.00\ \Omega$, and $P = 33.48\ \Omega$, what is the unknown resistance X?

26.78 A certain galvanometer has a resistance of 65.0 Ω and deflects full scale with a current of 1.50 mA in its coil. This is to be replaced with a second galvanometer that has a resistance of 38.0 Ω and deflects full scale with a current of 3.60 μA in its coil. Devise a circuit incorporating the second galvanometer such that the equivalent resistance of the circuit equals the resistance of the first galvanometer, and the second galvanometer deflects full scale when the current through the circuit equals the full-scale current of the first galvanometer.

26.79 A 224-Ω resistor and a 589-Ω resistor are connected in series across a 90.0-V line. a) What is the voltage across each resistor? b) A voltmeter connected across the 224-Ω resistor reads 23.8 V. Find the voltmeter resistance. c) Find the reading of the same voltmeter if connected across the 589-Ω resistor. d) The readings on this voltmeter are lower than the "true" voltages (that is, without the voltmeter present). Would it be possible to design a voltmeter that gave readings *higher* than the "true" voltages? Explain.

26.80 A 2.36-μF capacitor that is initially uncharged is connected in series with a 4.26-Ω resistor and an emf source with $\mathcal{E} = 120$ V and negligible internal resistance. a) Just after the connection is made, what is i) the rate at which electrical energy is being dissipated in the resistor? ii) the rate at which the electrical energy stored in the capacitor is increasing? iii) the electrical power output of the source? How do the answers to parts (i), (ii), and (iii) compare? b) Answer the same questions as in part (a) at a long time after the connection is made. c) Answer the same questions as in part (a) at the instant when the charge on the capacitor is one-half its final value.

26.81 A 3.40-μF capacitor is initially charged, then connected in series with a 7.25-kΩ resistor and an emf source with $\mathcal{E} = 180$ V and negligible internal resistance. a) A short time later the charge on the capacitor is 815 μC. At this instant, what is the current and what is its direction: toward the positive plate of the capacitor, or toward the negative plate? b) After a long time has elapsed, what will be the charge on the capacitor?

26.82 A 5.88-kΩ resistor is connected to the plates of a charged capacitor that has capacitance $C = 8.55 \times 10^{-10}$ F. The initial current through the resistor, just after the connection is made, is 0.620 A. What magnitude of charge was initially on each plate of the capacitor?

26.83 A capacitor that is initially uncharged is connected in series with a resistor and an emf source with $\mathcal{E} = 110$ V and negligible internal resistance. Just after the circuit is completed the current through the resistor is 6.5×10^{-5} A. The time constant for the circuit is 6.2 s. What are the resistance of the resistor and the capacitance of the capacitor?

26.84 A resistor with $R = 850\ \Omega$ is connected to the plates of a charged capacitor with capacitance $C = 4.62\ \mu$F. Just before the connection is made, the charge on the capacitor is 8.10 mC. a) What is the energy initially stored in the capacitor? b) What is the electrical power dissipated in the resistor just after the connection is made? c) What is the electrical power dissipated in the resistor at the instant when the energy stored in the capacitor has decreased to half the value calculated in part (a)?

26.85 Strictly speaking, Eq. (26.16) implies that an *infinite* amount of time is required to discharge a capacitor completely. Yet for practical purposes, a capacitor may be considered to be fully discharged after a finite length of time. To be specific, consider a capacitor with capacitance C connected to a resistor R to be fully discharged if its charge q differs from zero by no more than the charge of one electron. a) Calculate the time required to reach this state if $C = 0.920\ \mu$F, $R = 670$ kΩ, and $Q_0 = 7.00\ \mu$C. How many time constants is this? b) For a given Q_0, is the time required to reach this state always the same number of time constants, independent of the values of C and R? Why or why not?

26.86 Two capacitors in series are charged by a 12.0-V battery that has an internal resistance of 1.00 Ω. There is a 5.00-Ω resistor in series between the capacitors (see Fig. 26.67). a) What is the time constant of the charging circuit? b) After the switch has been closed for the time determined in part (a), what is the voltage across the 3.00-μF capacitor?

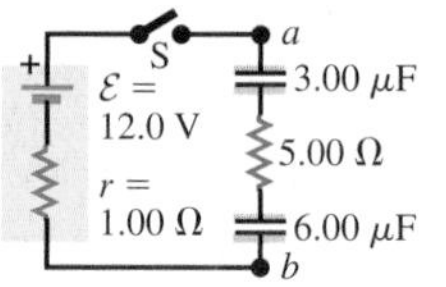

Figure 26.67 Problem 26.86.

26.87 The current in a charging capacitor is given by Eq. (26.13). a) The instantaneous power supplied by the battery is $\mathcal{E}i$. Integrate this to find the total energy supplied by the battery. b) The instantaneous power dissipated in the resistor is i^2R. Integrate this to find the total energy dissipated in the resistor. c) Find the final energy stored in the capacitor, and show that this equals the total energy supplied by the battery less the energy dissipated in the resistor, as obtained in parts (a) and (b). d) What fraction of the energy supplied by the battery is stored in the capacitor? How does this fraction depend on R?

26.88 a) Using Eq. (26.17) for the current in a discharging capacitor, derive an expression for the instantaneous power $P = i^2R$ dissipated in the resistor. b) Integrate the expression for P to find the total energy dissipated in the resistor, and show that this is equal to the total energy initially stored in the capacitor.

Challenge Problems

26.89 According to the theorem of superposition, the response (current) in a circuit is proportional to the stimulus (voltage) that causes it. This is true even if there are multiple sources in a circuit. This theorem can be used to analyze a circuit without resorting to Kirchhoff's rules by considering the currents in the circuit to be the superposition of currents caused by each source independently. In this way the circuit can be analyzed by computing equivalent resistances rather than by using the (sometimes) more cumbersome method of Kirchhoff's rules. Furthermore, by using the superposition theorem it is possible to examine how the modification of a source in one part of the circuit will affect the currents in all parts of the circuit without having to use Kirchhoff's rules to recalculate all of the currents. Consider the circuit shown in Fig. 26.68. If the circuit were redrawn with the 55.0-V and 57.0-V sources replaced by short circuits, the circuit could be analyzed by the method of equivalent resistances without resorting to Kirchhoff's rules, and the current in each branch could be found in a simple manner. Similarly, by redrawing the circuit with the 92.0-V and the 55.0-V sources replaced by short circuits, the circuit could again be analyzed in a simple manner. Finally, by replacing the 92.0-V and the 57.0-V sources with a short circuit the circuit could once again be analyzed simply. By superimposing the respective currents found in each of the branches by using the three simplified circuits, the actual current in each branch can be found. a) Using Kirchhoff's rules, find the branch currents in the 140.0-Ω, 210.0-Ω, and 35.0-Ω resistors. b) Using a circuit similar to the circuit of Fig. 26.68, but with the 55.0-V and 57.0-V sources replaced by a short circuit, determine the currents in each resistance. c) Repeat part (b) by replacing the 92.0-V and 55.0-V sources by short circuits, leaving the 57.0-V source intact. d) Repeat part (b) by replacing the 92.0-V and 57.0-V sources by short circuits, leaving the 55.0-V source intact. e) Verify the superposition theorem by taking the currents calculated in parts (b), (c), and (d) and comparing them with the currents calculated in part (a). f) If the 57.0-V source is replaced by an 80.0-V source, what will be the new currents in all branches of the circuit? [*Hint:* Using the superposition theorem, recalculate the partial currents calculated in part (c) using the fact that those currents are proportional to the source that is being replaced. Then superpose the new partial currents with those found in parts (b) and (d).]

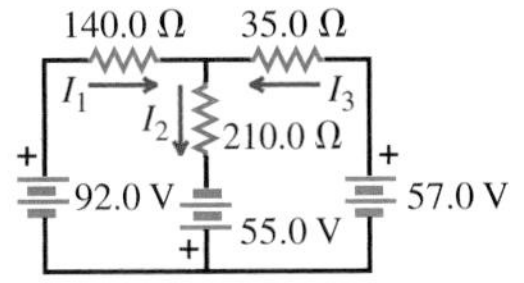

Figure 26.68 Challenge Problem 26.89.

26.90 A Capacitor Burglar Alarm. The capacitance of a capacitor can be affected by dielectric material that, although not inside the capacitor, is near enough to the capacitor to be polarized by the fringing electric field that exists near a charged capacitor. This effect is usually of the order of picofarads (pF), but it can be used with appropriate electronic circuitry to detect a change in the dielectric material surrounding the capacitor. Such a dielectric material might be the human body, and the effect described above might be used in the design of a burglar alarm. Consider the simplified circuit shown in Fig. 26.69. The voltage source has emf $\mathcal{E} = 1000$ V, and the capacitor has capacitance $C = 10.0$ pF. The electronic circuitry for detecting the current, represented as an ammeter in the diagram, has negligible resistance and is capable of detecting a current that persists at a level of at least 1.00 μA for at least 200 μs after the capacitance has changed abruptly from C to C'. The burglar alarm is designed to be activated if the capacitance changes by 10%. a) Determine the charge on the 10.0-pF capacitor when it is fully charged. b) If the capacitor is fully charged before the intruder is detected, assuming that the time taken for the capacitance to change by 10% is small enough to be ignored, derive an equation that expresses the current through the resistor R as a function of the time t since the capacitance has changed. c) Determine the range of values of the resistance R that will meet the design specifications of the burglar alarm. What happens if R is too small? Too large? (*Hint:* You will not be able to solve this part analytically but must use numerical methods. Express R as a logarithmic function of R plus known quantities. Use a trial value of R and calculate from the expression a new value. Continue to do this until the input and output values of R agree to within three significant figures.)

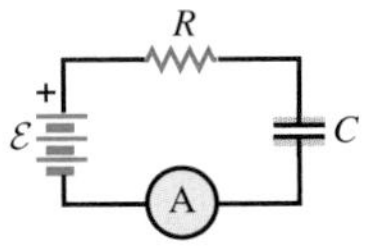

Figure 26.69 Challenge Problem 26.90.

26.91 An Infinite Network. As shown in Fig. 26.70, a network of resistors of resistances R_1 and R_2 extends off to infinity to the right. Prove that the total resistance R_T of the infinite network is equal to

$$R_T = R_1 + \sqrt{R_1^2 + 2R_1R_2}$$

(*Hint:* Since the network is infinite, the resistance of the network to the right of points c and d is also equal to R_T.)

26.92 Suppose a resistor R lies along each edge of a cube (12 resistors in all) with connections at the corners. Find the equivalent resistance between two diagonally opposite corners of the cube (points a and b in Fig. 26.71).

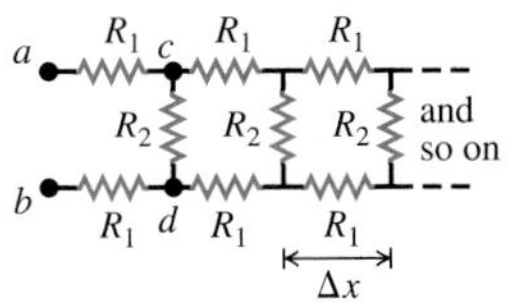

Figure 26.70 Challenge Problems 26.91 and 26.93.

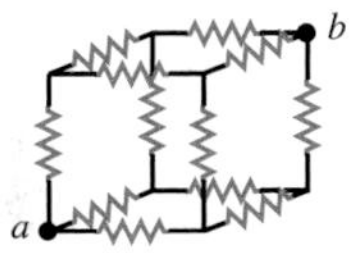

Figure 26.71 Challenge Problem 26.92.

26.93 Attenuator Chains and Axons. The infinite network of resistors shown in Fig. 26.70 is known as an *attenuator chain,* since this chain of resistors causes the potential difference between the upper and lower wires to decrease, or attenuate, along the length of the chain. a) Show that if the potential difference between the points a and b in Fig. 26.70 is V_{ab}, then the potential difference between the points c and d is $V_{cd} = V_{ab}/(1 + \beta)$, where $\beta = 2R_1(R_T + R_2)/R_TR_2$ and R_T, the total resistance of the network, is given in Challenge Problem 26.91. (See the hint given in that problem.) b) If the potential difference between the terminals a and b at the left-hand end of the infinite network is V_0, show that the potential difference between the upper and lower wires n segments from the left-hand end is $V_n = V_0/(1 + \beta)^n$. If $R_1 = R_2$, how many segments are needed to decrease the potential difference V_n to less than 1.0% of V_0? c) An infinite attenuator chain provides a model of the propagation of a voltage pulse along a nerve fiber, or axon. Each segment of the network in Fig. 26.70 represents a short segment of the axon of length Δx. The resistors R_1 represent the resistance of the fluid inside and outside the membrane wall of the axon. The resistance of the membrane to current flowing through the wall is represented by R_2. For an axon segment of length

$\Delta x = 1.0\ \mu\text{m}$, $R_1 = 6.4 \times 10^3\ \Omega$ and $R_2 = 8.0 \times 10^8\ \Omega$ (the membrane wall is a good insulator). Calculate the total resistance R_T and β for an infinitely long axon. (This is a good approximation, since the length of an axon is much greater than its width; the largest axons in the human nervous system are more than 1 m long but only about 10^{-7} m in radius.) d) By what fraction does the potential difference between the inside and outside of the axon decrease over a distance of 2.0 mm? e) The attenuation of the potential difference calculated in part (d) shows that the axon cannot simply be a passive, current-carrying electrical cable; the potential difference must periodically be reinforced along the axon's length. This reinforcement mechanism is slow, so a signal propagates along the axon at only about 30 m/s. In situations where faster response is required, axons are covered with a segmented sheath of fatty myelin. The segments are about 2 mm long, separated by gaps called the *nodes of Ranvier.* The myelin increases the resistance of a 1.0-μm-long segment of the membrane to $R_2 = 3.3 \times 10^{12}\ \Omega$. For such a myelinated axon, by what fraction does the potential difference between the inside and outside of the axon decrease over the distance from one node of Ranvier to the next? This smaller attenuation means the propagation speed is increased.

MAGNETIC FIELD AND MAGNETIC FORCES 27

Charged particles ejected from the sun are steered by the earth's magnetic field toward our planet's north and south poles. When these particles strike gas molecules in the upper atmosphere, the molecules are set aglow. The result is the brilliant light show called the aurora australis (at far southern latitudes) or aurora borealis (at far northern latitudes, as in this photograph from Alaska).

? Why do charged solar particles move predominantly toward the earth's poles rather than the equator?

Everybody uses magnetic forces. They are at the heart of electric motors, TV picture tubes, microwave ovens, loudspeakers, computer printers, and disk drives. The most familiar aspects of magnetism are those associated with permanent magnets, which attract unmagnetized iron objects and can also attract or repel other magnets. A compass needle aligning itself with the earth's magnetism is an example of this interaction. But the *fundamental* nature of magnetism is the interaction of moving electric charges. Unlike electric forces, which act on electric charges whether they are moving or not, magnetic forces act only on *moving* charges.

Although electric and magnetic forces are very different from each other, we use the idea of a *field* to describe both kinds of force. We saw in Chapter 21 that the electric force arises in two stages: (1) a charge produces an electric field in the space around it, and (2) a second charge responds to this field. Magnetic forces also arise in two stages. First, a *moving* charge or a collection of moving charges (that is, an electric current) produces a *magnetic* field. Next, a second current or moving charge responds to this magnetic field, and so experiences a magnetic force.

In this chapter we study the second stage in the magnetic interaction—that is, how moving charges and currents *respond* to magnetic fields. In particular, we will see how to calculate magnetic forces and torques, and we will discover why magnets can pick up iron objects like paper clips. In Chapter 28 we will complete our picture of the magnetic interaction by examining how moving charges and currents *produce* magnetic fields.

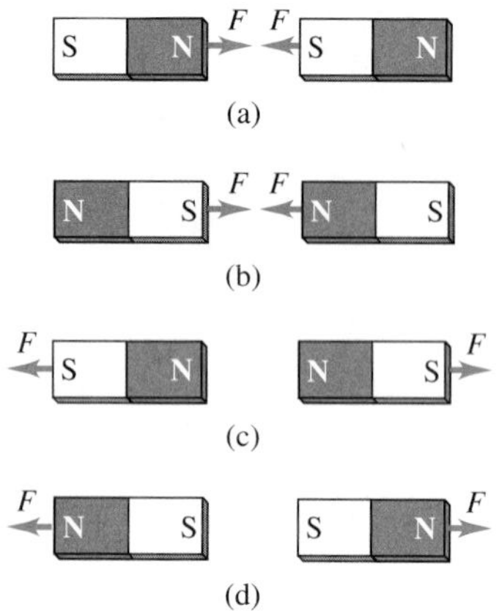

27.1 (a), (b) Two bar magnets attract when opposite poles (N and S, or S and N) are next to each other. (c), (d) The bar magnets repel when like poles (N and N, or S and S) are next to each other.

27.1 | Magnetism

Magnetic phenomena were first observed at least 2500 years ago in fragments of magnetized iron ore found near the ancient city of Magnesia (now Manisa, in western Turkey). These fragments were examples of what are now called **permanent magnets;** you probably have several permanent magnets on your refrigerator door at home. Permanent magnets were found to exert forces on each other as well as on pieces of iron that were not magnetized. It was discovered that when an iron rod is brought in contact with a natural magnet, the rod also becomes magnetized. When such a rod is floated on water or suspended by a string from its center, it tends to line itself up in a north-south direction. The needle of an ordinary compass is just such a piece of magnetized iron.

Before the relation of magnetic interactions to moving charges was understood, the interactions of permanent magnets and compass needles were described in terms of *magnetic poles.* If a bar-shaped permanent magnet, or *bar magnet,* is free to rotate, one end points north. This end is called a *north pole* or *N-pole;* the other end is a *south pole* or *S-pole.* Opposite poles attract each other, and like poles repel each other (Fig. 27.1). An object that contains iron but is not itself magnetized (that is, it shows no tendency to point north or south) is attracted by *either* pole of a permanent magnet (Fig. 27.2). This is the attraction that acts between a magnet and the unmagnetized steel door of a refrigerator. By analogy to electric interactions, we describe the interactions in Figs. 27.1 and 27.2 by saying that a bar magnet sets up a *magnetic field* in the space around it and a second body responds to that field. A compass needle tends to align with the magnetic field at the needle's position.

The earth itself is a magnet. Its north geographical pole is close to a magnetic *south* pole, which is why the north pole of a compass needle points north. The earth's magnetic axis is not quite parallel to its geographic axis (the axis of rotation), so a compass reading deviates somewhat from geographic north. This deviation, which varies with location, is called *magnetic declination* or *magnetic variation.* Also, the magnetic field is not horizontal at most points on the earth's surface; its angle up or down is called *magnetic inclination.* At the magnetic poles the magnetic field is vertical.

Figure 27.3 is a sketch of the earth's magnetic field. The lines, called *magnetic field lines,* show the direction that a compass would point at each location; they are discussed in detail in Section 27.3. The direction of the field at any point can be defined as the direction of the force that the field would exert on a magnetic north pole. In Section 27.2 we'll describe a more fundamental way to define the direction and magnitude of a magnetic field.

The concept of magnetic poles may appear similar to that of electric charge, and north and south poles may seem analogous to positive and negative charge. But the analogy can be misleading. While isolated positive and negative charges exist, there is *no* experimental evidence that a single isolated magnetic pole exists; poles always appear in pairs. If a bar magnet is broken in two, each broken end becomes a pole (Fig. 27.4). The existence of an isolated magnetic pole, or **magnetic monopole,** would have sweeping implications for theoretical physics. Extensive searches for magnetic monopoles have been carried out, but so far without success.

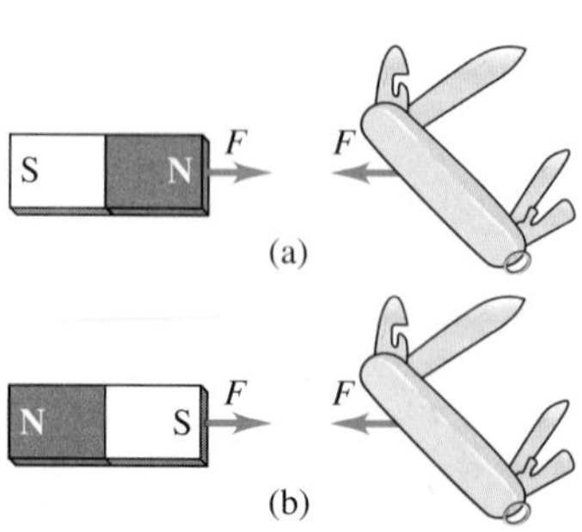

27.2 (a), (b) Either pole of a bar magnet attracts an unmagnetized object that contains iron.

The first evidence of the relationship of magnetism to moving charges was discovered in 1820 by the Danish scientist Hans Christian Oersted. He found that a compass needle was deflected by a current-carrying wire, as shown in Fig. 27.5. Similar investigations were carried out in France by André Ampère. A few years later, Michael Faraday in England and Joseph Henry in the United States discov-

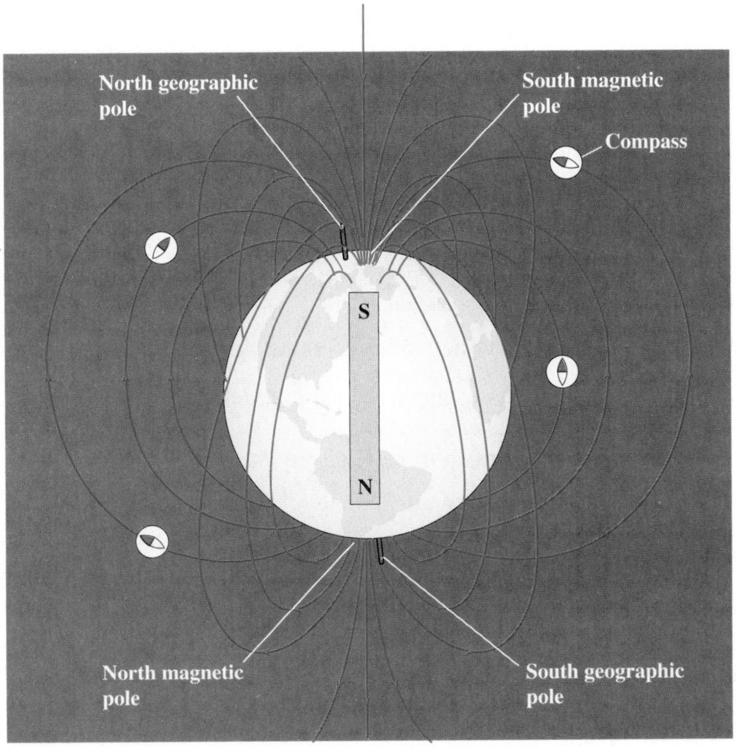

27.3 A compass placed at any location in the earth's magnetic field points in the direction of the field line at that location. Representing the earth's field as that of a tilted bar magnet is only a crude approximation of its fairly complex configuration. The field, which is caused by currents in the earth's molten core, changes with time; geologic evidence shows that it reverses direction entirely at irregular intervals of about a half million years.

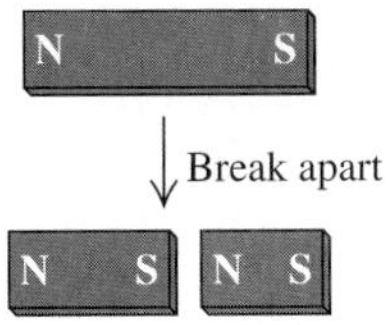

27.4 Breaking a bar magnet. Each piece has a north and south pole, even if the pieces are different sizes. (The smaller the piece, the weaker its magnetism.)

ered that moving a magnet near a conducting loop can cause a current in the loop. We now know that the magnetic forces between two bodies shown in Figs. 27.1 and 27.2 are fundamentally due to interactions between moving electrons in the atoms of the bodies. (There are also *electric* interactions between the two bodies, but these are far weaker than the magnetic interactions because the two bodies are electrically neutral.) Inside a magnetized body such as a permanent magnet, there is a *coordinated* motion of certain of the atomic electrons; in an unmagnetized body these motions are not coordinated. (We'll describe these motions further in Section 27.7, and see how the interactions shown in Figs. 27.1 and 27.2 come about.)

Electric and magnetic interactions prove to be intimately connected. Over the next several chapters we will develop the unifying principles of electromagnetism, culminating in the expression of these principles in *Maxwell's equations.* These equations represent the synthesis of electromagnetism, just as Newton's laws of motion are the synthesis of mechanics, and like Newton's laws they represent a towering achievement of the human intellect.

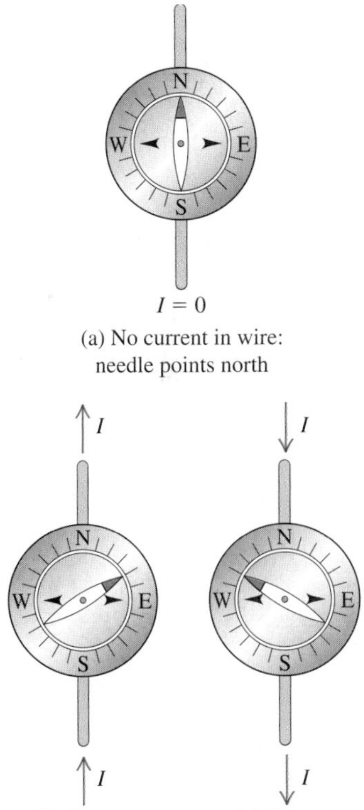

27.5 In Oersted's experiment, a compass is placed directly over a horizontal wire (here viewed from above). When the compass is placed directly under the wire, the compass swings are reversed.

Test Your Understanding

Suppose you cut off the part of the compass needle shown in Fig. 27.5a that is painted white. You discard this part, drill a hole in the remaining red part, and place the red part on the pivot at the center of the compass. Will the red part still swing east and west when a current is applied as in Figs. 27.5b and 27.5c?

27.2 | Magnetic Field

To introduce the concept of magnetic field properly, let's review our formulation of *electric* interactions in Chapter 21, where we introduced the concept of *electric* field. We represented electric interactions in two steps:

1. A distribution of electric charge at rest creates an electric field $\vec{E}$ in the surrounding space.
2. The electric field exerts a force $\vec{F} = q\vec{E}$ on any other charge q that is present in the field.

We can describe magnetic interactions in a similar way:

1. A moving charge or a current creates a **magnetic field** in the surrounding space (in addition to its *electric* field).
2. The magnetic field exerts a force $\vec{F}$ on any other moving charge or current that is present in the field.

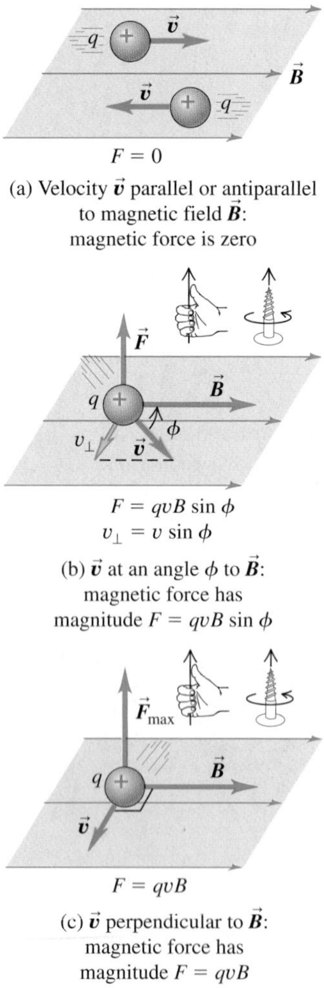

27.6 The magnetic force $\vec{F}$ acting on a positive charge q moving with velocity $\vec{v}$ is perpendicular to both $\vec{v}$ and the magnetic field $\vec{B}$. For given values of the speed v and magnetic field strength B, the force is greatest when $\vec{v}$ and $\vec{B}$ are perpendicular.

In this chapter we'll concentrate on the *second* aspect of the interaction: Given the presence of a magnetic field, what force does it exert on a moving charge or a current? In Chapter 28 we will come back to the problem of how magnetic fields are *created* by moving charges and currents.

Like electric field, magnetic field is a *vector field*—that is, a vector quantity associated with each point in space. We will use the symbol $\vec{B}$ for magnetic field. At any position the direction of $\vec{B}$ is defined as that in which the north pole of a compass needle tends to point. The arrows in Fig. 27.3 suggest the direction of the earth's magnetic field; for any magnet, $\vec{B}$ points out of its north pole and into its south pole.

What are the characteristics of the magnetic force on a moving charge? First, its magnitude is proportional to the magnitude of the charge. If a 1-μC charge and a 2-μC charge move through a given magnetic field with the same velocity, experiments show that the force on the 2-μC charge is twice as great as that on the 1-μC charge. The magnitude of the force is also proportional to the magnitude, or "strength," of the field; if we double the magnitude of the field (for example, by using two identical bar magnets instead of one) without changing the charge or its velocity, the force doubles.

The magnetic force also depends on the particle's velocity. This is quite different from the electric-field force, which is the same whether the charge is moving or not. A charged particle at rest experiences *no* magnetic force. Furthermore, we find by experiment that the magnetic force $\vec{F}$ *does not* have the same direction as the magnetic field $\vec{B}$ but instead is always *perpendicular* to both $\vec{B}$ and the velocity $\vec{v}$. The magnitude F of the force is found to be proportional to the component of $\vec{v}$ perpendicular to the field; when that component is zero (that is, when $\vec{v}$ and $\vec{B}$ are parallel or antiparallel), the force is zero.

Figure 27.6 shows these relationships. The direction of $\vec{F}$ is always perpendicular to the plane containing $\vec{v}$ and $\vec{B}$. Its magnitude is given by

$$F = |q|v_\perp B = |q|vB\sin\phi \qquad (27.1)$$

where $|q|$ is the magnitude of the charge and ϕ is the angle measured from the direction of $\vec{v}$ to the direction of $\vec{B}$, as shown in the figure.

This description does not specify the direction of $\vec{F}$ completely; there are always two directions, opposite to each other, that are both perpendicular to the plane of $\vec{v}$ and $\vec{B}$. To complete the description, we use the same right-hand rule that we used to define the vector product in Section 1.10. (It would be a good idea to review that section before you go on.) Draw the vectors $\vec{v}$ and $\vec{B}$ with their tails together, as in Fig. 27.6b. Imagine turning $\vec{v}$ until it points in the direction of $\vec{B}$ (turning through ϕ, the smaller of the two possible angles). Wrap the fingers of your right hand around the line perpendicular to the plane of $\vec{v}$ and $\vec{B}$ so that they curl around with the sense of rotation from $\vec{v}$ to $\vec{B}$. Your thumb then points in the direction of the force $\vec{F}$ on a *positive* charge. (Alternatively, the direction of the force $\vec{F}$ on a positive charge is the direction in which a right-hand-thread screw would advance if turned the same way.)

This discussion shows that the force on a charge q moving with velocity $\vec{v}$ in a magnetic field $\vec{B}$ is given, both in magnitude and in direction, by

$$\vec{F} = q\vec{v} \times \vec{B} \quad \text{(magnetic force on a moving charged particle)} \quad (27.2)$$

This is the first of several vector products we will encounter in our study of magnetic-field relationships. It's important to note that Eq. (27.2) was *not* deduced theoretically; it is an observation based on *experiment.*

Equation (27.2) is valid for both positive and negative charges. When q is negative, the direction of the force $\vec{F}$ is opposite to that of $\vec{v} \times \vec{B}$. If two charges with equal magnitude and opposite sign move in the same $\vec{B}$ field with the same velocity (Fig. 27.7), the forces have equal magnitude and opposite direction. Figures 27.6 and 27.7 show several examples of the relationships of the directions of $\vec{F}$, $\vec{v}$, and $\vec{B}$ for both positive and negative charges. Be sure you understand the relationships shown in these figures.

ActivPhysics ONLINE

13.4 Magnetic Force on a Particle

Equation (27.1) gives the magnitude of the magnetic force $\vec{F}$ in Eq. (27.2). We can express this magnitude in a different but equivalent way. Since ϕ is the angle between the directions of vectors $\vec{v}$ and $\vec{B}$, we may interpret $B \sin \phi$ as the component of $\vec{B}$ perpendicular to $\vec{v}$, that is, $B_{\perp}$. With this notation the force magnitude is

$$F = |q|vB_{\perp} \quad (27.3)$$

This form is sometimes more convenient, especially in problems involving *currents* rather than individual particles. We will discuss forces on currents later in this chapter.

From Eq. (27.1) the *units* of B must be the same as the units of F/qv. Therefore the SI unit of B is equivalent to $1\ \text{N}\cdot\text{s/C}\cdot\text{m}$, or, since one ampere is one coulomb per second $(1\ \text{A} = 1\ \text{C/s})$, $1\ \text{N/A}\cdot\text{m}$. This unit is called the **tesla** (abbreviated T), in honor of Nikola Tesla (1857–1943), the prominent Serbian-American scientist and inventor:

$$1\ \text{tesla} = 1\ \text{T} = 1\ \text{N/A}\cdot\text{m}$$

Another unit of B, the **gauss** $(1\ \text{G} = 10^{-4}\ \text{T})$, is also in common use. Instruments for measuring magnetic field are sometimes called *gaussmeters.*

The magnetic field of the earth is of the order of 10^{-4} T or 1 G. Magnetic fields of the order of 10 T occur in the interior of atoms and are important in the analysis of atomic spectra. The largest steady magnetic field that can be produced at present in the laboratory is about 45 T. Some pulsed-current electromagnets can produce fields of the order of 120 T for short time intervals of the order of a millisecond. The magnetic field at the surface of a neutron star is believed to be of the order of 10^8 T.

To explore an unknown magnetic field, we can measure the magnitude and direction of the force on a *moving* test charge, then use Eq. (27.2) to determine $\vec{B}$.

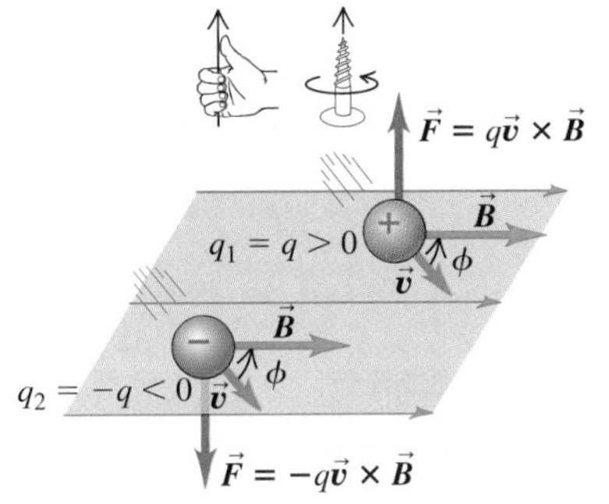

27.7 Two charges of the same magnitude but opposite sign moving with the same velocity in the same magnetic field. The magnetic forces on the charges are equal in magnitude but opposite in direction.

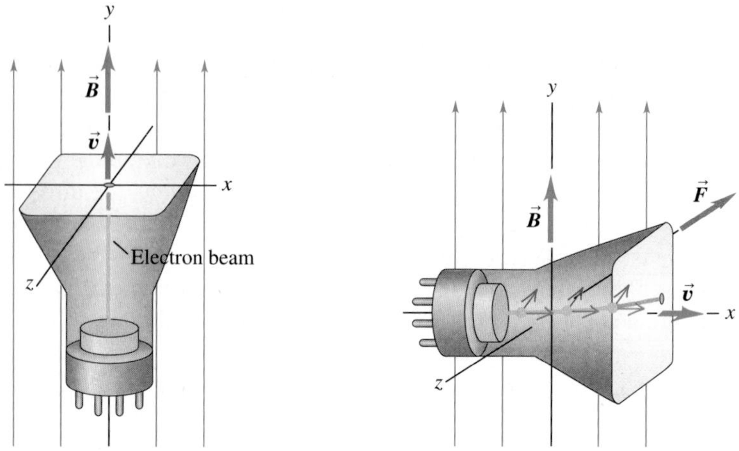

27.8 Determining the direction of a magnetic field using a cathode-ray tube. Because electrons have a negative charge, the magnetic force $\vec{F} = q\vec{v} \times \vec{B}$ in part (b) points in the direction opposite to the rule in Fig. 27.6.

(a) If tube axis is parallel to the y-axis, beam is undeflected: hence $\vec{B}$ is in either $+y$ or $-y$ direction

(b) If tube axis is parallel to the x-axis, beam is deflected in $-z$ direction: hence $\vec{B}$ is in $+y$ direction

The electron beam in a cathode-ray tube, such as that used in a television set, is a convenient device for making such measurements. The electron gun shoots out a narrow beam of electrons at a known speed. If there is no force to deflect the beam, it strikes the center of the screen.

If a magnetic field is present, in general the electron beam is deflected. But if the beam is parallel or antiparallel to the field, then $\phi = 0$ or π in Eq. (27.1) and $F = 0$; there is no force, and hence no deflection. If we find that the electron beam is not deflected when its direction is parallel to a certain axis as in Fig. 27.8a, the $\vec{B}$ vector must point either up or down along that axis.

If we then turn the tube 90° (Fig. 27.8b), $\phi = \pi/2$ in Eq. (27.1) and the magnetic force is maximum; the beam is deflected in a direction perpendicular to the plane of $\vec{B}$ and $\vec{v}$. The direction and magnitude of the deflection determine the direction and magnitude of $\vec{B}$. We can perform additional experiments in which the angle between $\vec{B}$ and $\vec{v}$ is between zero and 90° to confirm Eq. (27.1) or (27.3) and the accompanying discussion. We note that the electron has a negative charge; the force in Fig. 27.8 is opposite in direction to the force on a positive charge.

When a charged particle moves through a region of space where *both* electric and magnetic fields are present, both fields exert forces on the particle. The total force $\vec{F}$ is the vector sum of the electric and magnetic forces:

$$\vec{F} = q(\vec{E} + \vec{v} \times \vec{B}) \qquad (27.4)$$

Problem-Solving Strategy

Magnetic Forces

IDENTIFY *the relevant concepts:* The right-hand rule allows you to determine the magnetic force on a moving charged particle.

SET UP *the problem using the following steps:*

1. Draw the velocity vector $\vec{v}$ and magnetic field $\vec{B}$ with their tails together so that you can visualize the plane in which these two vectors lie.
2. Identify the angle ϕ between the two vectors.

3. Identify the target variables. This may be the magnitude and direction of the force, or it may be the magnitude or direction of $\vec{v}$ or $\vec{B}$.

EXECUTE *the solution as follows:*
1. Express the magnetic force using Eq. (27.2), $\vec{F} = q\vec{v} \times \vec{B}$. The magnitude of the force is given by Eq. (27.1), $F = qvB \sin\phi$.
2. Remember that $\vec{F}$ is perpendicular to the plane of the vectors $\vec{v}$ and $\vec{B}$. The direction of $\vec{v} \times \vec{B}$ is determined by the right-hand rule; keep referring to Figs. 27.6 and 27.7 until you're sure you understand this rule. If q is negative, the force is *opposite* to $\vec{v} \times \vec{B}$.

EVALUATE *your answer:* Whenever you can, solve the problem in two ways. Do it directly from the geometric definition of the vector product. Then find the components of the vectors in some convenient axis system and calculate the vector product algebraically from the components. Verify that the results agree.

Example 27.1 Magnetic force on a proton

A beam of protons ($q = 1.6 \times 10^{-19}$ C) moves at 3.0×10^5 m/s through a uniform magnetic field with magnitude 2.0 T that is directed along the positive z-axis, as in Fig. 27.9. The velocity of each proton lies in the xz-plane at an angle of 30° to the $+z$-axis. Find the force on a proton.

SOLUTION

IDENTIFY and SET UP: Figure 27.9 shows that the vectors $\vec{v}$ and $\vec{B}$ lie in the xz-plane. The angle between these vectors is 30°. The target variables are the magnitude and direction of the force $\vec{F}$.

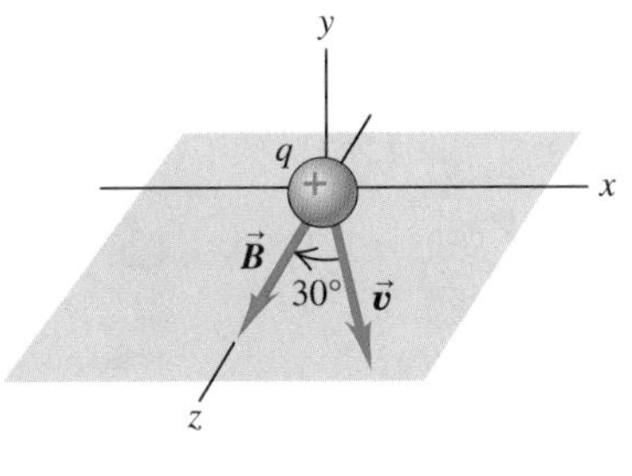

27.9 Directions of $\vec{v}$ and $\vec{B}$ for a proton in a magnetic field.

EXECUTE: The charge is positive, so the force is in the same direction as the vector product $\vec{v} \times \vec{B}$. From the right-hand rule, this direction is along the negative y-axis. The magnitude of the force, from Eq. (27.1), is

$$\begin{aligned} F &= qvB \sin\phi \\ &= (1.6 \times 10^{-19}\,\text{C})(3.0 \times 10^5\,\text{m/s})(2.0\,\text{T})(\sin 30°) \\ &= 4.8 \times 10^{-14}\,\text{N} \end{aligned}$$

EVALUATE: We check our result by evaluating the force using vector language and Eq. (27.2). We have

$$\begin{aligned} \vec{v} &= (3.0 \times 10^5\,\text{m/s})(\sin 30°)\hat{\imath} + (3.0 \times 10^5\,\text{m/s})(\cos 30°)\hat{k} \\ \vec{B} &= (2.0\,\text{T})\hat{k} \\ \vec{F} &= q\vec{v} \times \vec{B} \\ &= (1.6 \times 10^{-19}\,\text{C})(3.0 \times 10^5\,\text{m/s})(2.0\,\text{T}) \\ &\quad \times (\sin 30°\,\hat{\imath} + \cos 30°\hat{k}) \times \hat{k} = (-4.8 \times 10^{-14}\,\text{N})\hat{\jmath} \end{aligned}$$

(Recall that $\hat{\imath} \times \hat{k} = -\hat{\jmath}$ and $\hat{k} \times \hat{k} = \mathbf{0}$.) We again find that the force is in the negative y-direction with magnitude 4.8×10^{-14} N.

If the beam consists of *electrons* rather than protons, the charge is negative ($q = -1.6 \times 10^{-19}$ C), and the direction of the force is reversed. The force is now directed along the *positive* y-axis, but the magnitude is the same as before, $F = 4.8 \times 10^{-14}$ N.

Test Your Understanding

If $\vec{B}$ in Example 27.1 were in the positive x-direction, with all other quantities the same, what would be the force on the proton? What if $\vec{v}$ were in the positive y-direction, with all other quantities the same as in Example 27.1?

27.3 | Magnetic Field Lines and Magnetic Flux

We can represent any magnetic field by **magnetic field lines,** just as we did for the earth's magnetic field in Fig. 27.3. The idea is the same as for the electric field lines we introduced in Section 21.6. We draw the lines so that the line through any point is tangent to the magnetic field vector $\vec{B}$ at that point. Just as with electric field lines, we draw only a few representative lines; otherwise, the lines would fill up all of space. Where adjacent field lines are close together, the field magnitude is large; where these field lines are far apart, the field magnitude is small. Also, because the direction of $\vec{B}$ at each point is unique, field lines never intersect.

CAUTION Magnetic field lines are sometimes called "magnetic lines of force," but that's not a good name for them; unlike electric field lines, they *do not* point in the direction of the force on a charge (Fig. 27.10). Equation (27.2) shows that the force on a moving charged particle is always perpendicular to the magnetic field, and hence to the magnetic field line that passes through the particle's position. The direction of the force depends on the particle's velocity and the sign of its charge, so just looking at magnetic field lines cannot in itself tell you the direction of the force on an arbitrary moving charged particle. Magnetic field lines *do* have the direction that a compass needle would point at each location; this may help you to visualize them.

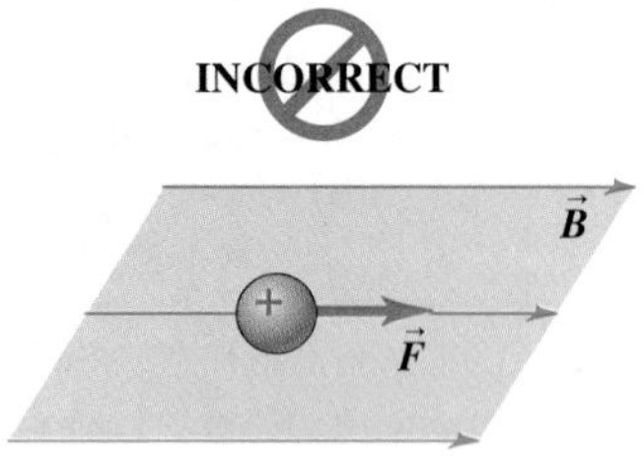

(a)

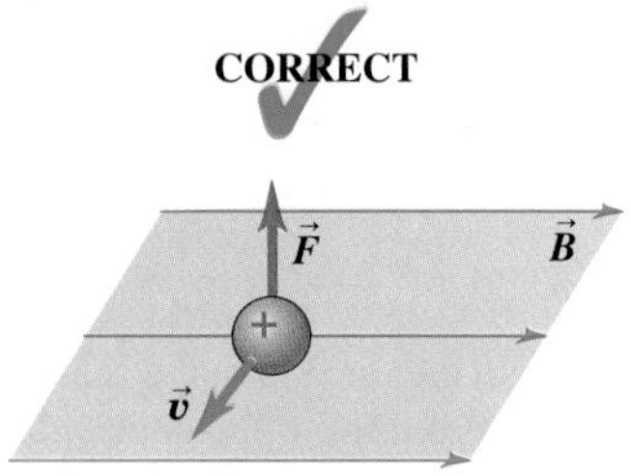

(b)

27.10 (a) Magnetic field lines are *not* "lines of force." The force on a charged particle is not along the direction of a field line. (b) The direction of the magnetic force depends on the velocity $\vec{v}$, as expressed by the magnetic force law $\vec{F} = q\vec{v} \times \vec{B}$.

Figure 27.11 shows magnetic field lines produced by several common sources of magnetic field. In the gap between the poles of the electromagnet shown in Fig. 27.11c, the field lines are approximately straight, parallel, and equally spaced, showing that the magnetic field in this region is approximately *uniform* (that is, constant in magnitude and direction).

Because magnetic field patterns are three-dimensional, it's often necessary to draw magnetic field lines that point into or out of the plane of a drawing. To do this we use a dot (·) to represent a vector directed out of the plane and a cross (×)

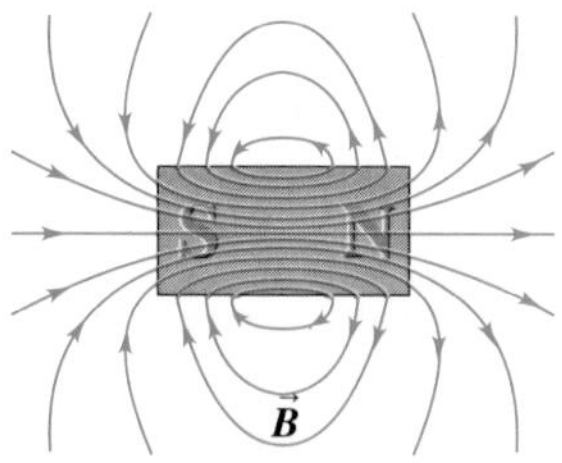

(a) Magnetic field lines through the center of a permanent magnet

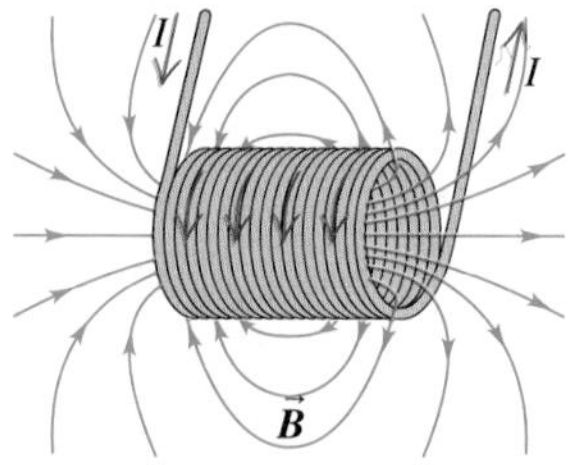

(b) Magnetic field lines through the center of a cylindrical current-carrying coil

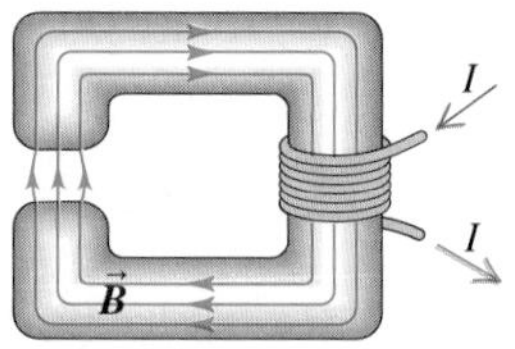

(c) Magnetic field lines through the center of an iron-core electromagnet

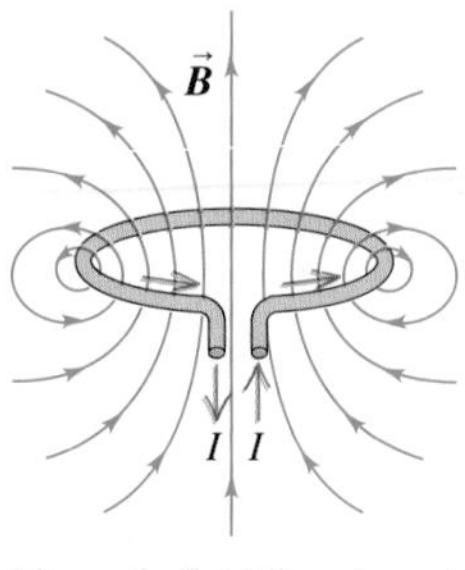

(d) Magnetic field lines in a plan containing the axis of a circular current-carrying loop

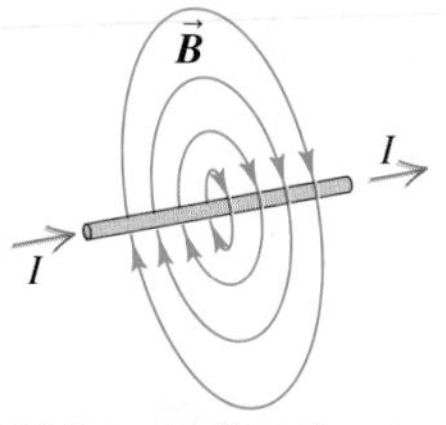

(e) Magnetic field lines in a plane perpendicular to a long, straight, current-carrying wire

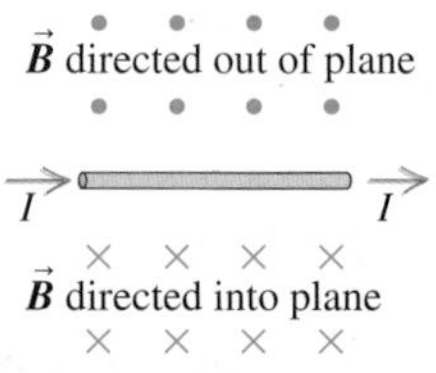

(f) Magnetic field lines in a plane containing a long, straight, current-carrying wire

27.11 (a)-(e) Magnetic field lines produced by several common sources of magnetic field. (f) Dots represent a magnetic field directed out of the plane, and crosses represent a magnetic field directed into the plane.

to represent a vector directed into the plane (Fig. 27.11f). Here's a good way to remember these conventions: think of a dot as the head of an arrow coming directly toward you, and think of a cross as the feathers of an arrow flying directly away from you.

Iron filings, like compass needles, tend to align with magnetic field lines. Hence they provide an easy way to visualize field lines (Fig. 27.12).

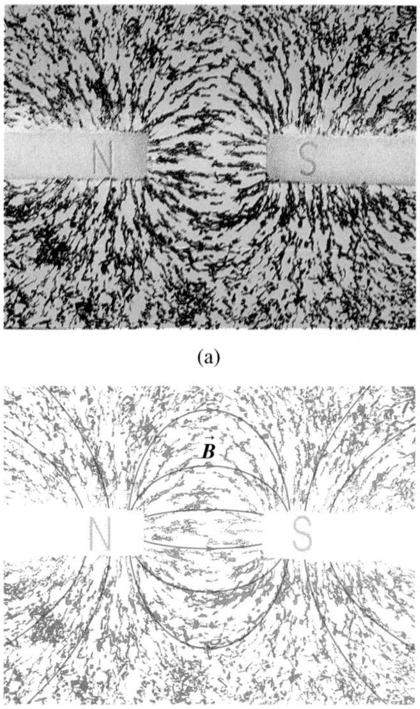

27.12 (a) Like little compass needles, iron filings line up tangent to magnetic field lines. (b) Drawing of the field lines for the situation shown in (a).

Magnetic Flux and Gauss's Law for Magnetism

We define the **magnetic flux** Φ_B through a surface just as we defined electric flux in connection with Gauss's law in Section 22.2. We can divide any surface into elements of area dA (Fig. 27.13). For each element we determine $B_\perp$, the component of $\vec{B}$ normal to the surface at the position of that element, as shown. From the figure, $B_\perp = B\cos\phi$, where ϕ is the angle between the direction of $\vec{B}$ and a line perpendicular to the surface. (Be careful not to confuse ϕ with Φ_B.) In general, this component varies from point to point on the surface. We define the magnetic flux $d\Phi_B$ through this area as

$$d\Phi_B = B_\perp\, dA = B\cos\phi\, dA = \vec{B}\cdot d\vec{A} \tag{27.5}$$

The *total* magnetic flux through the surface is the sum of the contributions from the individual area elements:

$$\Phi_B = \int B_\perp\, dA = \int B\cos\phi\, dA = \int \vec{B}\cdot d\vec{A} \tag{27.6}$$

(magnetic flux through a surface)

(This equation uses the concepts of vector area and surface integral that we introduced in Section 22.2; you may want to review that discussion.)

Magnetic flux is a *scalar* quantity. In the special case in which $\vec{B}$ is uniform over a plane surface with total area A, $B_\perp$ and ϕ are the same at all points on the surface, and

$$\Phi_B = B_\perp A = BA\cos\phi \tag{27.7}$$

If $\vec{B}$ happens to be perpendicular to the surface, then $\cos\phi = 1$ and Eq. (27.7) reduces to $\Phi_B = BA$. We will use the concept of magnetic flux extensively during our study of electromagnetic induction in Chapter 29.

The SI unit of magnetic flux is equal to the unit of magnetic field (1 T) times the unit of area $(1\ \text{m}^2)$. This unit is called the **weber** (1 Wb), in honor of the German physicist Wilhelm Weber (1804–1891):

$$1\ \text{Wb} = 1\ \text{T}\cdot\text{m}^2$$

Also, $1\ \text{T} = 1\ \text{N/A}\cdot\text{m}$, so

$$1\ \text{Wb} = 1\ \text{T}\cdot\text{m}^2 = 1\ \text{N}\cdot\text{m/A}$$

In Gauss's law the total *electric* flux through a closed surface is proportional to the total electric charge enclosed by the surface. For example, if the closed surface encloses an electric dipole, the total electric flux is zero because the total charge is zero. (You may want to review Section 22.3 on Gauss's law.) By analogy, if there were such a thing as a single magnetic charge (magnetic monopole), the total *magnetic* flux through a closed surface would be proportional to the total magnetic charge enclosed. But we have mentioned that no magnetic monopole

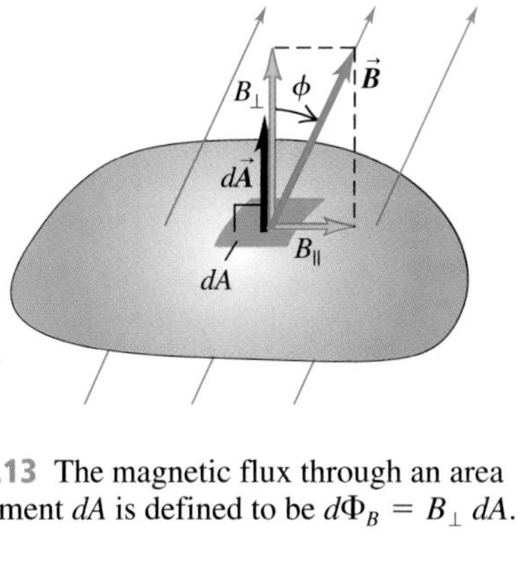

27.13 The magnetic flux through an area element dA is defined to be $d\Phi_B = B_\perp\, dA$.

has ever been observed, despite intensive searches. We conclude that **the total magnetic flux through a closed surface is always zero.** Symbolically,

$$\oint \vec{B} \cdot d\vec{A} = 0 \quad \text{(magnetic flux through any closed surface)} \qquad (27.8)$$

This equation is sometimes called *Gauss's law for magnetism.* You can verify it by examining Fig. 27.11; if you draw a closed surface anywhere in any of the field maps shown in that figure, you will see that every field line that enters the surface also exits from it; the net flux through the surface is zero. It also follows from Eq. (27.8) that magnetic field lines always form closed loops.

CAUTION Unlike electric field lines that begin and end on electric charges, magnetic field lines *never* have end points; such a point would indicate the presence of a monopole. You might be tempted to draw magnetic field lines that begin at the north pole of a magnet and end at a south pole. But as Fig. 27.11a shows, the field lines of a magnet actually continue through the interior of the magnet. Like all other magnetic field lines, they form closed loops.

For Gauss's law, which always deals with *closed* surfaces, the vector area element $d\vec{A}$ in Eq. (27.6) always points *out of* the surface. However, some applications of *magnetic* flux involve an *open* surface with a boundary line; there is then an ambiguity of sign in Eq. (27.6) because of the two possible choices of direction for $d\vec{A}$. In these cases we choose one of the possible sides of the surface to be the "positive" side and use that choice consistently.

If the element of area dA in Eq. (27.5) is at right angles to the field lines, then $B_\perp = B$; calling the area $dA_\perp$, we have

$$B = \frac{d\Phi_B}{dA_\perp} \qquad (27.9)$$

That is, the magnitude of magnetic field is equal to *flux per unit area* across an area at right angles to the magnetic field. For this reason, magnetic field $\vec{B}$ is sometimes called **magnetic flux density.**

Example 27.2 Magnetic flux calculations

Figure 27.14a shows a perspective view of a flat surface with area 3.0 cm^2 in a uniform magnetic field. If the magnetic flux through this area is 0.90 mWb, calculate the magnitude of the magnetic field and find the direction of the area vector.

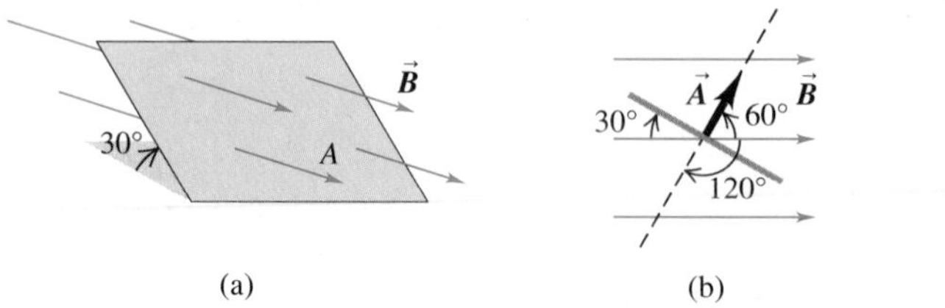

27.14 (a) A flat area A in a uniform magnetic field $\vec{B}$. (b) An edge-on view of the area A. The area vector $\vec{A}$ makes a 60° angle with $\vec{B}$. (If we had chosen $\vec{A}$ to point in the opposite direction, ϕ would have been 120° and the magnetic flux Φ_B would have been negative.)

SOLUTION

IDENTIFY: In many problems we are asked to calculate the flux of a given magnetic field through a given area. In this example, however, we are given the flux, the area, and the direction of the magnetic field. Our target variables are the field magnitude and the direction of the area vector.

SET UP: Because the magnetic field is uniform, B and ϕ are the same at all points on the surface. Hence we can use Eq. (27.7): $\Phi_B = BA \cos \phi$. Our target variable is B.

EXECUTE: The area A is $3.0 \times 10^{-4}\ \text{m}^2$; the direction of $\vec{A}$ is perpendicular to the surface, so ϕ could be either 60° or 120°. But Φ_B, B, and A are all positive, so $\cos \phi$ must also be positive. This rules out 120°, so $\phi = 60°$, and we find

$$B = \frac{\Phi_B}{A \cos \phi} = \frac{0.90 \times 10^{-3}\ \text{Wb}}{(3.0 \times 10^{-4}\ \text{m}^2)(\cos 60°)} = 6.0\ \text{T}$$

The area vector $\vec{A}$ is perpendicular to the area in the direction shown in Fig. 27.14b.

EVALUATE: A good way to check our result is to calculate the product $BA \cos \phi$ to make sure that it is equal to the given value of the magnetic flux Φ_B. Is it?

Test Your Understanding

Imagine moving upward along the axis of the loop in Fig. 27.11d, starting at a point well below the loop and ending at a point well above the loop. How would the magnetic-field strength vary along this path? Where would the field be strongest? How would the field direction change? Explain how you determined your answers.

27.4 | Motion of Charged Particles in a Magnetic Field

When a charged particle moves in a magnetic field, it is acted on by the magnetic force given by Eq. (27.2), and the motion is determined by Newton's laws. Figure 27.15 shows a simple example. A particle with positive charge q is at point O, moving with velocity $\vec{v}$ in a uniform magnetic field $\vec{B}$ directed into the plane of the figure. The vectors $\vec{v}$ and $\vec{B}$ are perpendicular, so the magnetic force $\vec{F} = q\vec{v} \times \vec{B}$ has magnitude $F = qvB$ and a direction as shown in the figure. The force is *always* perpendicular to $\vec{v}$, so it cannot change the *magnitude* of the velocity, only its direction. To put it differently, the magnetic force never has a component parallel to the particle's motion, so the magnetic force can never do *work* on the particle. This is true even if the magnetic field is not uniform. **Motion of a charged particle under the action of a magnetic field alone is always motion with constant speed.**

Using this principle, we see that in the situation shown in Fig. 27.15a the magnitudes of both $\vec{F}$ and $\vec{v}$ are constant. At points such as P and S the directions of force and velocity have changed as shown, but their magnitudes are the same. The particle therefore moves under the influence of a constant-magnitude force that is always at right angles to the velocity of the particle. Comparing these conditions with the discussion of circular motion in Sections 3.4 and 5.4, we see that the particle's path is a *circle,* traced out with constant speed v (Fig. 27.15b). The centripetal acceleration is v^2/R and the only force acting is the magnetic force, so from Newton's second law,

$$F = |q|vB = m\frac{v^2}{R} \qquad (27.10)$$

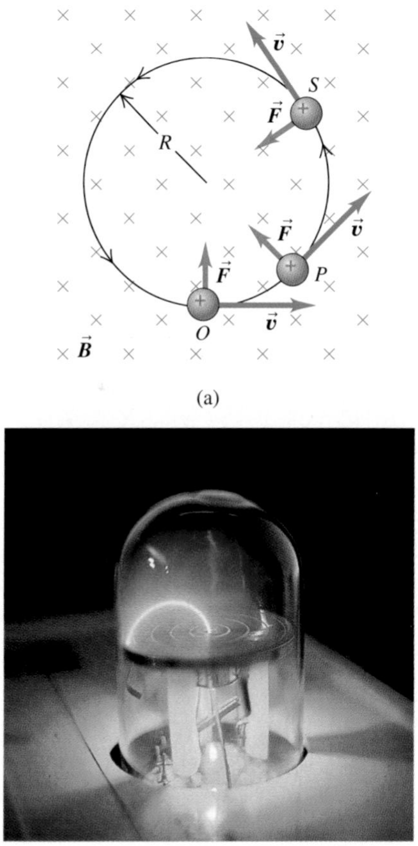

(a)

(b)

27.15 (a) The orbit of a charged particle in a uniform magnetic field $\vec{B}$ is a circle when the initial velocity is perpendicular to the field. The crosses represent a uniform magnetic field directed straight into the page. (b) An electron beam (seen as a blue arc) curving in a magnetic field.

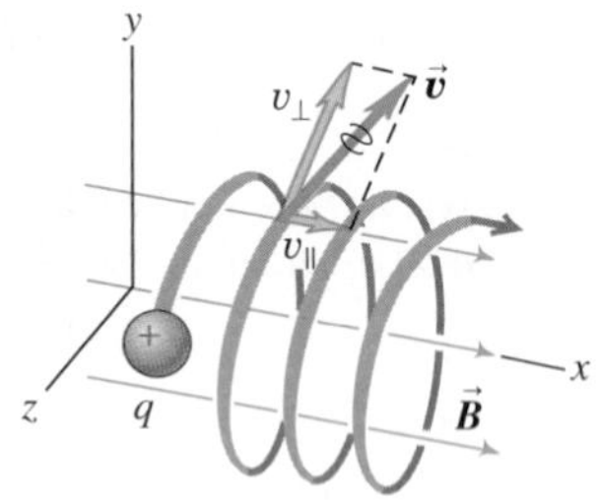

27.16 When a charged particle has velocity components both perpendicular and parallel to a uniform magnetic field, the particle moves in a helical path. The magnetic field does no work on the particle, so its speed and kinetic energy remain constant.

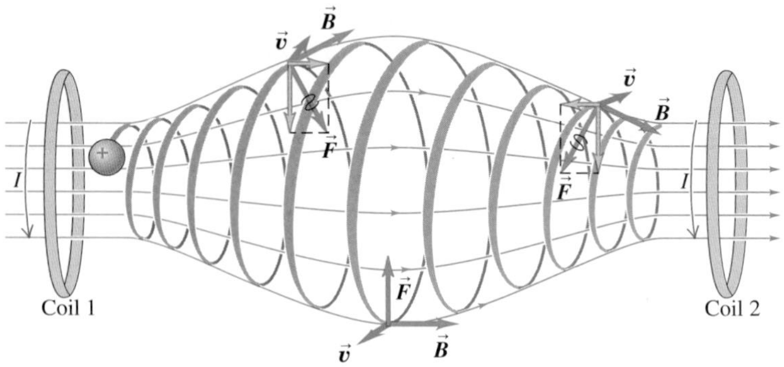

27.17 A magnetic bottle. Particles near either end of the region experience a magnetic force toward the center of the region. This is one way of containing an ionized gas that has a temperature of the order of 10^6 K, which would vaporize any material container.

where m is the mass of the particle. Solving Eq. (27.10) for the radius R of the circular path, we find

$$R = \frac{mv}{|q|B} \quad \text{(radius of a circular orbit in a magnetic field)} \quad (27.11)$$

We can also write this as $R = p/|q|B$, where $p = mv$ is the magnitude of the particle's momentum. If the charge q is negative, the particle moves *clockwise* around the orbit in Fig. 27.15a.

The angular speed ω of the particle can be found from Eq. (9.13), $v = R\omega$. Combining this with Eq. (27.11), we get

$$\omega = \frac{v}{R} = v\frac{|q|B}{mv} = \frac{|q|B}{m} \quad (27.12)$$

The number of revolutions per unit time is $f = \omega/2\pi$. This frequency f is independent of the radius R of the path. It is called the **cyclotron frequency;** in a particle accelerator called a *cyclotron,* particles moving in nearly circular paths are given a boost twice each revolution, increasing their energy and their orbital radii but not their angular speed or frequency. Similarly, one type of *magnetron,* a common source of microwave radiation for microwave ovens and radar systems, emits radiation with a frequency equal to the frequency of circular motion of electrons in a vacuum chamber between the poles of a magnet.

If the direction of the initial velocity is *not* perpendicular to the field, the velocity *component* parallel to the field is constant because there is no force parallel to the field. Then the particle moves in a helix (Fig. 27.16). The radius of the helix is given by Eq. (27.11), where v is now the component of velocity perpendicular to the $\vec{B}$ field.

Motion of a charged particle in a non-uniform magnetic field is more complex. Figure 27.17 shows a field produced by two circular coils separated by some distance. Particles near either coil experience a magnetic force toward the center of the region; particles with appropriate speeds spiral repeatedly from one end of the region to the other and back. Because charged particles can be trapped in such a

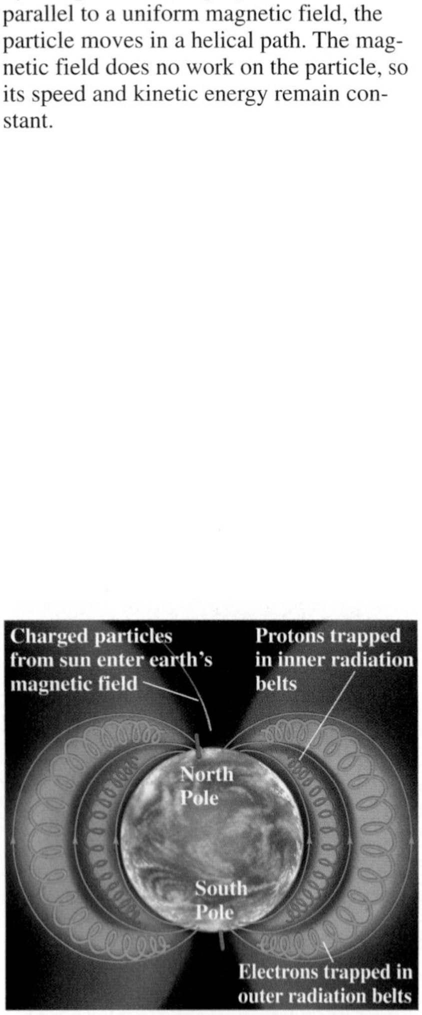

27.18 The Van Allen radiation belts around the earth. Near the poles, charged particles from these belts can enter the atmosphere, producing the aurora borealis ("northern lights") and aurora australis ("southern lights"). (See the photograph that opens this chapter.)

magnetic field, it is called a *magnetic bottle.* This technique is used to confine very hot plasmas with temperatures of the order of 10^6 K. In a similar way the earth's non-uniform magnetic field traps charged particles coming from the sun in doughnut-shaped regions around the earth, as shown in Fig. 27.18. These regions, called the *Van Allen radiation belts,* were discovered in 1958 using data obtained by instruments aboard the Explorer I satellite.

Magnetic forces on charged particles play an important role in studies of elementary particles. Figure 27.19 shows a chamber filled with liquid hydrogen and with a magnetic field directed into the plane of the photograph. A high-energy gamma ray dislodges an electron from a hydrogen atom, sending it off at high speed and creating a visible track in the liquid hydrogen. The track shows the electron curving downward due to the magnetic force. The energy of the collision also produces another electron and a *positron* (a positively charged electron). Because of their opposite charges, the trajectories of the electron and the positron curve in opposite directions. As these particles plow through the liquid hydrogen, they collide with other charged particles, losing energy and speed. As a result, the radius of curvature decreases as suggested by Eq. (27.11). (The electron's speed is comparable to the speed of light, so Eq. (27.11) isn't directly applicable here.) Similar experiments allow physicists to determine the mass and charge of newly discovered particles.

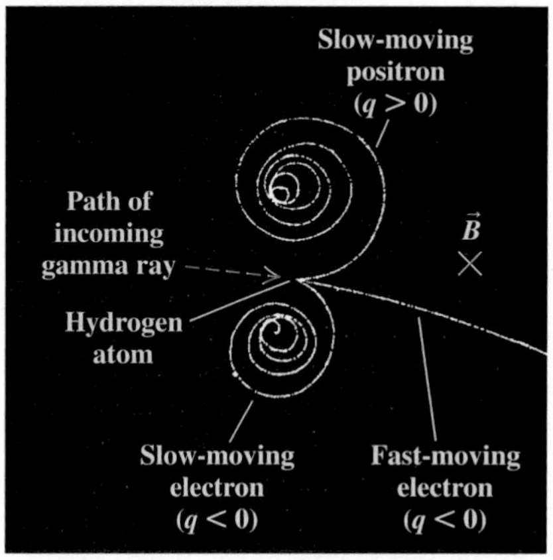

27.19 This bubble chamber image shows the result of a high-energy gamma ray (which does not leave a track) that collides with an electron in a hydrogen atom. This electron flies off to the right at high speed. Some of the energy in the collision is transformed into a second electron and a positron (a positively charged electron). A magnetic field is directed into the plane of the image, which makes the positive and negative charges curve off in different directions.

Problem-Solving Strategy: Motion in Magnetic Fields

IDENTIFY *the relevant concepts:* In analyzing the motion of a charged particle in electric and magnetic fields, you will apply Newton's second law of motion, $\sum\vec{F} = m\vec{a}$, with the net force given by $\sum\vec{F} = q(\vec{E} + \vec{v} \times \vec{B})$. Often, other forces such as gravity can be neglected. Many of the problems are similar to the trajectory and circular-motion problems in Sections 3.3, 3.4, and 5.4; it would be a good idea to review those sections.

SET UP *the problem using the following steps:*
1. Determine the target variable(s).
2. Often the use of components is the most efficient approach. Choose a coordinate system and then express all vector quantities (including $\vec{E}$, $\vec{B}$, $\vec{v}$, $\vec{F}$, and $\vec{a}$) in terms of their components in this system.

EXECUTE *the solution as follows:*
1. If the particle moves perpendicular to a uniform magnetic field, the trajectory is a circle with a radius and angular speed given by Eqs. (27.11) and (27.12).
2. If your calculation involves a more complex trajectory, use $\sum\vec{F} = m\vec{a}$ in component form: $\sum F_x = ma_x$, and so forth. This approach is particularly useful when both electric and magnetic fields are present.

EVALUATE *your answer:* Check whether your results are reasonable.

Example 27.3 Electron motion in a microwave oven

A magnetron in a microwave oven emits electromagnetic waves with frequency $f = 2450$ MHz. What magnetic field strength is required for electrons to move in circular paths with this frequency?

SOLUTION

IDENTIFY and SET UP: Equation (27.12) relates the angular speed in circular motion to the mass and charge of the particle and the magnetic field strength. Our target variable is the magnetic field strength B.

EXECUTE: The angular speed that corresponds to the frequency f is $\omega = 2\pi f = (2\pi)(2450 \times 10^6\ \text{s}^{-1}) = 1.54 \times 10^{10}\ \text{s}^{-1}$. From Eq. (27.12),

$$B = \frac{m\omega}{|q|} = \frac{(9.11 \times 10^{-31}\ \text{kg})(1.54 \times 10^{10}\ \text{s}^{-1})}{1.60 \times 10^{-19}\ \text{C}}$$
$$= 0.0877\ \text{T}$$

EVALUATE: This is a moderate field strength, easily produced with a permanent magnet. Incidentally, 2450-MHz electromagnetic waves are strongly absorbed by water molecules, so they are particularly useful for heating and cooking food.

Example 27.4 Helical particle motion

In a situation like that shown in Fig. 27.16, the charged particle is a proton ($q = 1.60 \times 10^{-19}$ C, $m = 1.67 \times 10^{-27}$ kg) and the uniform magnetic field is directed along the x-axis with magnitude 0.500 T. Only the magnetic force acts on the proton. At $t = 0$ the proton has velocity components $v_x = 1.50 \times 10^5$ m/s, $v_y = 0$, and $v_z = 2.00 \times 10^5$ m/s. a) At $t = 0$, find the force on the proton and its acceleration. b) Find the radius of the helical path, the angular speed of the proton, and the *pitch* of the helix (the distance traveled along the helix axis per revolution).

SOLUTION

IDENTIFY and SET UP: We use the coordinate system depicted in Fig. 27.16. The force is given by $\vec{F} = q\vec{v} \times \vec{B}$ and the acceleration is given by Newton's second law. The force is perpendicular to the velocity, so the speed of the proton does not change. Hence the radius of the helical trajectory is just as given by Eq. (27.11) for circular motion, but with v replaced by the component of velocity perpendicular to $\vec{B}$. The angular speed is given by Eq. (27.12). Given the angular speed, we can determine the time required for one revolution; given the velocity parallel to the magnetic field, we can determine the distance traveled along the helix in this time.

EXECUTE: a) Since $v_y = 0$, the velocity vector is $\vec{v} = v_x\hat{\imath} + v_z\hat{k}$. Using Eq. (27.2) and recalling that $\hat{\imath} \times \hat{\imath} = \mathbf{0}$ and $\hat{k} \times \hat{\imath} = \hat{\jmath}$, we find

$$\vec{F} = q\vec{v} \times \vec{B} = q(v_x\hat{\imath} + v_z\hat{k}) \times B\hat{\imath} = qv_zB\hat{\jmath}$$
$$= (1.60 \times 10^{-19}\ \text{C})(2.00 \times 10^5\ \text{m/s})(0.500\ \text{T})\hat{\jmath}$$
$$= (1.60 \times 10^{-14}\ \text{N})\hat{\jmath}$$

(To check unit consistency, recall from Section 27.2 that $1\ \text{T} = 1\ \text{N/A}\cdot\text{m} = 1\ \text{N}\cdot\text{s/C}\cdot\text{m}$.) This may seem like a very weak force, but the resulting acceleration is tremendous because the proton mass is so small:

$$\vec{a} = \frac{\vec{F}}{m} = \frac{1.60 \times 10^{-14}\ \text{N}}{1.67 \times 10^{-27}\ \text{kg}}\hat{\jmath} = (9.58 \times 10^{12}\ \text{m/s}^2)\hat{\jmath}$$

b) At $t = 0$ the component of velocity perpendicular to $\vec{B}$ is v_z, so

$$R = \frac{mv_z}{|q|B} = \frac{(1.67 \times 10^{-27}\ \text{kg})(2.00 \times 10^5\ \text{m/s})}{(1.60 \times 10^{-19}\ \text{C})(0.500\ \text{T})}$$
$$= 4.18 \times 10^{-3}\ \text{m} = 4.18\ \text{mm}$$

From Eq. (27.12) the angular speed is

$$\omega = \frac{|q|B}{m} = \frac{(1.60 \times 10^{-19}\ \text{C})(0.500\ \text{T})}{1.67 \times 10^{-27}\ \text{kg}} = 4.79 \times 10^7\ \text{rad/s}$$

The time required for one revolution (the period) is $T = 2\pi/\omega = 2\pi/(4.79 \times 10^7\ \text{s}^{-1}) = 1.31 \times 10^{-7}$ s. The pitch is the distance traveled along the x-axis during this time, or

$$v_xT = (1.50 \times 10^5\ \text{m/s})(1.31 \times 10^{-7}\ \text{s})$$
$$= 0.0197\ \text{m} = 19.7\ \text{mm}$$

EVALUATE: The pitch of the helix is almost five times greater than the radius. This helical trajectory is much more "stretched out" than that shown in Fig. 27.16.

Test Your Understanding

If you double the speed of the charged particle in Fig. 27.15a while keeping the magnetic field the same (as well as the charge and the mass), how does this affect the radius of the trajectory? How does it affect the time required for one complete circular orbit?

27.5 | Applications of Motion of Charged Particles

This section describes several applications of the principles introduced in this chapter. Study them carefully, watching for applications of the problem-solving strategy outlined in the preceding section.

Velocity Selector

In a beam of charged particles produced by a heated cathode or a radioactive material, not all particles move with the same speed. Many applications, however, require a beam in which all the particle speeds are the same. Particles of a specific speed can be selected from the beam using an arrangement of electric and magnetic fields called a *velocity selector.* In Fig. 27.20 a charged particle with mass m, charge q, and speed v enters a region of space where the electric and magnetic fields are perpendicular to the particle's velocity and to each other. The electric field $\vec{E}$ is to the left, and the magnetic field $\vec{B}$ is into the plane of the figure. If q is positive, the electric force is to the left, with magnitude qE, and the magnetic force is to the right, with magnitude qvB. For given field magnitudes E and B, for a particular value of v the electric and magnetic forces will be equal in magnitude; the total force is then zero, and the particle travels in a straight line with constant velocity. For zero total force, $\sum F_y = 0$, we need $-qE + qvB = 0$; solving for the speed v for which there is no deflection, we find

$$v = \frac{E}{B} \tag{27.13}$$

Only particles with speeds equal to E/B can pass through without being deflected by the fields. By adjusting E and B appropriately, we can select particles having a particular speed for use in other experiments. Because q divides out in Eq. (27.13), a velocity selector for positively charged particles also works for electrons or other negatively charged particles.

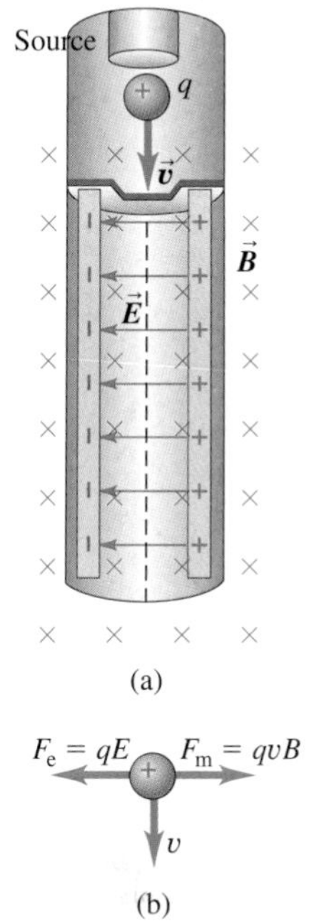

27.20 (a) A velocity selector for charged particles uses perpendicular $\vec{E}$ and $\vec{B}$ fields. Only charged particles with $v = E/B$ move through undeflected. (b) The electric and magnetic forces on a positive charge. The forces are reversed if the charge is negative.

Thomson's *e/m* Experiment

In one of the landmark experiments in physics at the end of the nineteenth century, J. J. Thomson (1856–1940) used the idea just described to measure the ratio of charge to mass for the electron. For this experiment, carried out in 1897 at the Cavendish Laboratory in Cambridge, England, Thomson used the apparatus shown in Fig. 27.21. In a highly evacuated glass container, electrons from the hot

ActivPhysics ONLINE

13.8 Velocity Selector

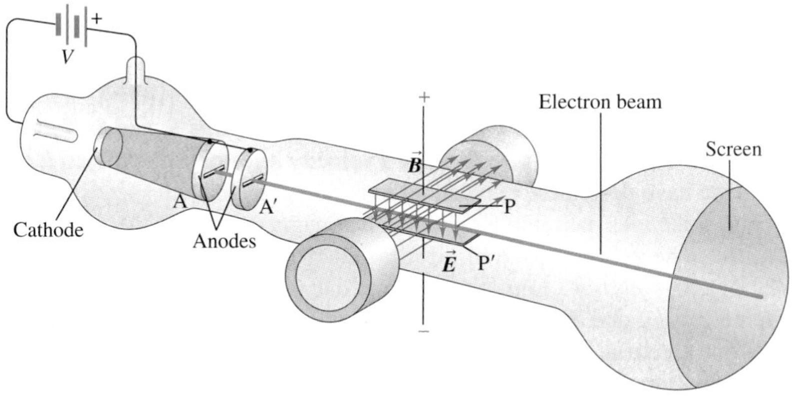

27.21 Thomson's apparatus for measuring the ratio e/m for the electron.

cathode are accelerated and formed into a beam by a potential difference V between the two anodes A and A'. The speed v of the electrons is determined by the accelerating potential V. The kinetic energy $\frac{1}{2}mv^2$ equals the loss of electric potential energy eV, where e is the magnitude of the electron charge:

$$\frac{1}{2}mv^2 = eV \quad \text{or} \quad v = \sqrt{\frac{2eV}{m}} \tag{27.14}$$

The electrons pass between the plates P and P′ and strike the screen at the end of the tube, which is coated with a material that fluoresces (glows) at the point of impact. The electrons pass straight through the plates when Eq. (27.13) is satisfied; combining this with Eq. (27.14), we get

$$\frac{E}{B} = \sqrt{\frac{2eV}{m}} \quad \text{so} \quad \frac{e}{m} = \frac{E^2}{2VB^2} \tag{27.15}$$

All the quantities on the right side can be measured, so the ratio e/m of charge to mass can be determined. It is *not* possible to measure e or m separately by this method, only their ratio.

The most significant aspect of Thomson's e/m measurements was that he found a *single value* for this quantity. It did not depend on the cathode material, the residual gas in the tube, or anything else about the experiment. This independence showed that the particles in the beam, which we now call electrons, are a common constituent of all matter. Thus Thomson is credited with discovery of the first subatomic particle, the electron. He also found that the *speed* of the electrons in the beam was about one-tenth the speed of light, much larger than any previously measured speed of a material particle.

The most precise value of e/m available as of this writing is

$$e/m = 1.758820174(71) \times 10^{11}\ \text{C/kg}$$

In this expression, (71) indicates the likely uncertainty in the last two digits, 74.

Fifteen years after Thomson's experiments, the American physicist Robert Millikan succeeded in measuring the charge of the electron precisely (see Challenge Problem 23.89). This value, together with the value of e/m, enables us to determine the *mass* of the electron. The most precise value available at present is

$$m = 9.10938188(72) \times 10^{-31}\ \text{kg}$$

Mass Spectrometers

Techniques similar to Thomson's e/m experiment can be used to measure masses of ions and thus measure atomic and molecular masses. In 1919, Francis Aston (1877–1945), a student of Thomson's, built the first of a family of instruments called **mass spectrometers.** A variation built by Bainbridge is shown in Fig. 27.22. Positive ions from a source pass through the slits S_1 and S_2, forming a narrow beam. Then the ions pass through a velocity selector with crossed $\vec{E}$ and $\vec{B}$ fields, as we have described, to block all ions except those with speeds v equal to E/B. Finally, the ions pass into a region with a magnetic field $\vec{B}'$ perpendicular to the figure, where they move in circular arcs with radius R determined by Eq. (27.11): $R = mv/qB'$. Ions with different masses strike the photographic plate at different points, and the values of R can be measured. We assume that each ion has lost one electron, so the net charge of each ion is just $+e$. With everything known in this equation except m, we can compute the mass m of the ion.

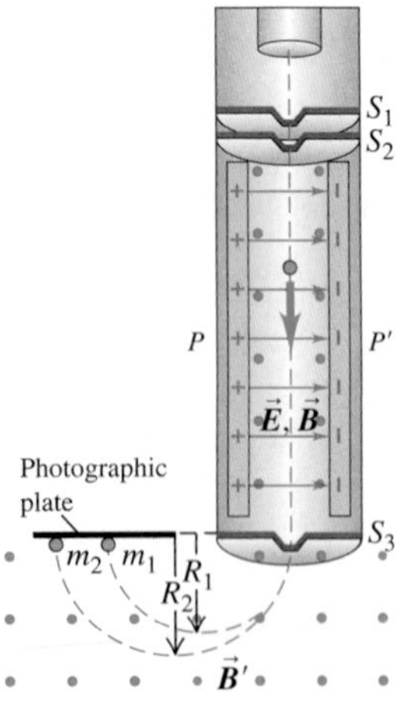

27.22 Bainbridge's mass spectrometer utilizes a velocity selector to produce particles with uniform speed v. In the region of magnetic field B', particles with larger mass ($m_2 > m_1$) travel in paths with larger radius ($R_2 > R_1$).

One of the earliest results from this work was the discovery that neon has two species of atoms, with atomic masses 20 and 22 g/mol. We now call these species **isotopes** of the element. Later experiments have shown that many elements have several isotopes, atoms that are identical in their chemical behavior but different in mass owing to differing numbers of neutrons in their nuclei. This is just one of the many applications of mass spectrometers in chemistry and physics.

Present-day mass spectrometers can measure masses routinely with a precision of better than one part in 10,000. The photographic plate is replaced by a more sophisticated particle detector that scans across the impact region.

ActivPhysics Online

13.7 Mass Spectrometer

Example 27.5 An *e/m* experiment

You set out to reproduce Thomson's *e/m* experiment with an accelerating potential of 150 V and a deflecting electric field of magnitude 6.0×10^6 N/C. a) At what fraction of the speed of light do the electrons move? b) What magnitude of magnetic field will you need? c) With this magnetic field, what will happen to the electron beam if you increase the accelerating potential above 150 V?

SOLUTION

IDENTIFY and SET UP: This is the same situation as depicted in Fig. 27.21. We use Eq. (27.14) to determine the speed of the electrons and Eq. (27.15) to determine the requisite magnetic field.

EXECUTE: a) From Eq. (27.14), the electron speed v is related to the accelerating potential by:

$$v = \sqrt{2(e/m)V} = \sqrt{2(1.76 \times 10^{11}\ \text{C/kg})(150\ \text{V})}$$
$$= 7.27 \times 10^6\ \text{m/s}$$

$$\frac{v}{c} = \frac{7.27 \times 10^6\ \text{m/s}}{3.00 \times 10^8\ \text{m/s}} = 0.024$$

The electrons are traveling at 2.4% of the speed of light.

b) From Eq. (27.13),

$$B = \frac{E}{v} = \frac{6.00 \times 10^6\ \text{N/C}}{7.27 \times 10^6\ \text{m/s}} = 0.83\ \text{T}$$

c) Increasing the accelerating potential V increases the electron speed v. In Fig. 27.21 this doesn't change the upward electric force eE, but it increases the downward magnetic force evB. Therefore the electron beam will be bent *downward* and will hit the end of the tube below the undeflected position.

EVALUATE: The required magnetic field is relatively large. If the maximum available magnetic field B is less than 0.83 T, the electric field strength E would have to be reduced to maintain the desired ratio E/B in Eq. (27.15).

Example 27.6 Finding leaks in a vacuum system

There is almost no helium in ordinary air, so helium sprayed near a leak in a vacuum system will quickly show up in the output of a vacuum pump connected to such a system. You are designing a leak detector that uses a mass spectrometer to detect He^+ ions (charge $+e = +1.60 \times 10^{-19}$ C, mass 6.65×10^{-27} kg). The ions emerge from the velocity selector with a speed of 1.00×10^5 m/s. They are curved in a semicircular path by a magnetic field B' and are detected at a distance of 10.16 cm from the slit S_3 in Fig. 27.22. Calculate the magnitude of the magnetic field B'.

SOLUTION

IDENTIFY and SET UP: The motion of the ion after it passes through slit S_3 in Fig. 27.22 is just motion in a circular path as described in Section 27.4 (see Fig. 27.15). We use Eq. (27.11) to relate the magnetic field strength B' (the target variable) to the radius of curvature of the path and to the mass, charge, and speed of the ion.

EXECUTE: The distance given is the *diameter* of the semicircular path shown in Fig. 27.22, so the radius is $R = \frac{1}{2}(10.16 \times 10^{-2}\ \text{m}) = 5.08 \times 10^{-2}$ m. From Eq. (27.11), $R = mv/qB'$, we get

$$B' = \frac{mv}{qR} = \frac{(6.65 \times 10^{-27}\ \text{kg})(1.00 \times 10^5\ \text{m/s})}{(1.60 \times 10^{-19}\ \text{C})(5.08 \times 10^{-2}\ \text{m})}$$
$$= 0.0817\ \text{T}$$

EVALUATE: Helium leak detectors are actual devices that are widely used for diagnosing problems with high-vacuum systems. Our result shows that only a small magnetic field is required, which makes it possible to build relatively compact leak detectors.

Test Your Understanding

Suppose the He^+ ions in Example 27.5 were replaced by He^{2+} ions, in which both electrons have been removed from the helium atom and the ion charge is $+2e$. Would the velocity selector still work? If so, how would the distance in Fig. 27.22 from slit S_3 to where the ions are detected change?

27.6 | Magnetic Force on a Current-Carrying Conductor

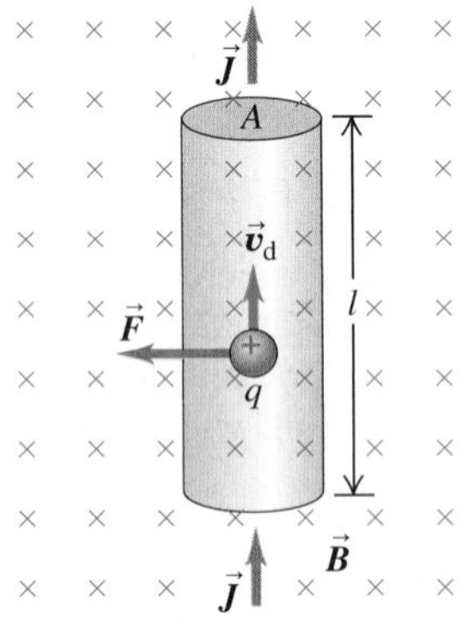

27.23 Forces on a moving positive charge in a current-carrying conductor.

What makes an electric motor work? The forces that make it turn are forces that a magnetic field exerts on a conductor carrying a current. The magnetic forces on the moving charges within the conductor are transmitted to the material of the conductor, and the conductor as a whole experiences a force distributed along its length. The moving-coil galvanometer that we described in Section 26.3 also uses magnetic forces on conductors.

We can compute the force on a current-carrying conductor starting with the magnetic force $\vec{F} = q\vec{v} \times \vec{B}$ on a single moving charge. Figure 27.23 shows a straight segment of a conducting wire, with length l and cross-sectional area A; the current is from bottom to top. The wire is in a uniform magnetic field $\vec{B}$, perpendicular to the plane of the diagram and directed *into* the plane. Let's assume first that the moving charges are positive. Later we'll see what happens when they are negative.

The drift velocity $\vec{v}_d$ is upward, perpendicular to $\vec{B}$. The average force on each charge is $\vec{F} = q\vec{v}_d \times \vec{B}$, directed to the left as shown in the figure; since $\vec{v}_d$ and $\vec{B}$ are perpendicular, the magnitude of the force is $F = qv_dB$.

We can derive an expression for the *total* force on all the moving charges in a length l of conductor with cross-sectional area A, using the same language we used in Eqs. (25.2) and (25.3) of Section 25.1. The number of charges per unit volume is n; a segment of conductor with length l has volume Al and contains a number of charges equal to nAl. The total force $\vec{F}$ on *all* the moving charges in this segment has magnitude

$$F = (nAl)(qv_dB) = (nqv_dA)(lB) \tag{27.16}$$

From Eq. (25.3) the current density is $J = nqv_d$. The product JA is the total current I, so we can rewrite Eq. (27.16) as

$$F = IlB \tag{27.17}$$

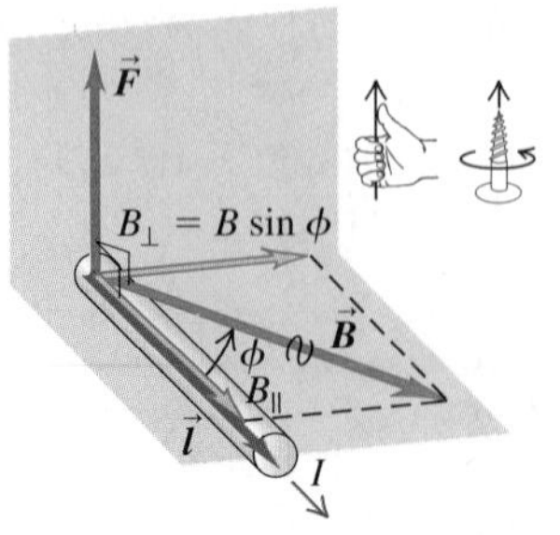

27.24 A straight wire segment of length $\vec{l}$ carries a current I in the direction of $\vec{l}$. The magnetic force on this segment is perpendicular to both $\vec{l}$ and the magnetic field $\vec{B}$.

If the $\vec{B}$ field is not perpendicular to the wire but makes an angle ϕ with it, we handle the situation the same way we did in Section 27.2 for a single charge. Only the component of $\vec{B}$ perpendicular to the wire (and to the drift velocities of the charges) exerts a force; this component is $B_\perp = B \sin \phi$. The magnetic force on the wire segment is then

$$F = IlB_\perp = IlB \sin \phi \tag{27.18}$$

The force is always perpendicular to both the conductor and the field, with the direction determined by the same right-hand rule we used for a moving positive charge (Fig. 27.24). Hence this force can be expressed as a vector product, just like the force on a single moving charge. We represent the segment of wire with a vector $\vec{l}$ along the wire in the direction of the current; then the force $\vec{F}$ on this segment is

$$\vec{F} = I\vec{l} \times \vec{B} \quad \text{(magnetic force on a straight wire segment)} \quad (27.19)$$

Figure 27.25 illustrates the directions of $\vec{B}$, $\vec{l}$, and $\vec{F}$ for several cases.

If the conductor is not straight, we can divide it into infinitesimal segments $d\vec{l}$. The force $d\vec{F}$ on each segment is

$$d\vec{F} = I\,d\vec{l} \times \vec{B} \quad \text{(magnetic force on an infinitesimal wire section)} \quad (27.20)$$

Then we can integrate this expression along the wire to find the total force on a conductor of any shape. The integral is a *line integral,* the same mathematical operation we have used to define work (Section 6.3) and electric potential (Section 23.2).

CAUTION Recall from Section 25.1 that the current I is not a vector. The direction of current flow is described by $d\vec{l}$, not I. If the conductor is curved, the current I is the same at all points along its length, but $d\vec{l}$ changes direction so that it is always tangent to the conductor.

Finally, what happens when the moving charges are negative, such as electrons in a metal? Then in Fig. 27.23 an upward current corresponds to a downward drift velocity. But because q is now negative, the direction of the force $\vec{F}$ is the same as before. Thus Eqs. (27.17) through (27.20) are valid for *both* positive and negative charges and even when *both* signs of charge are present at once. This happens in some semiconductor materials and in ionic solutions.

A common application of the magnetic forces on a current-carrying wire is found in loudspeakers (Fig. 27.26). The radial magnetic field created by the permanent magnet exerts a force on the voice coil that is proportional to the current in the coil; the direction of the force is either to the left or to the right, depending on the direction of the current. The signal from the amplifier causes the current to oscillate in direction and magnitude. The coil and the speaker cone to which it is

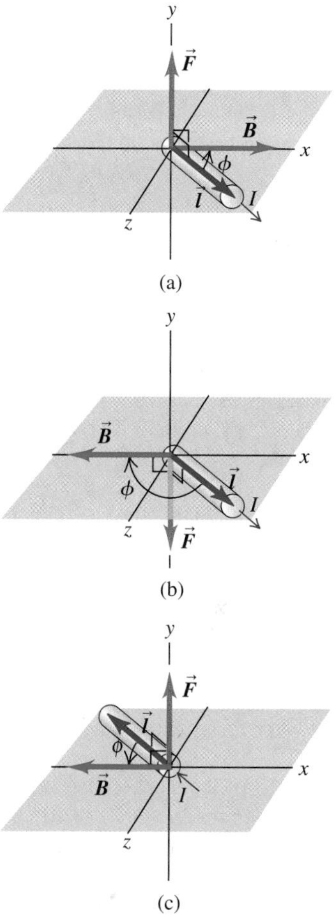

27.25 (a) Magnetic field $\vec{B}$, length $\vec{l}$, and force $\vec{F}$ vectors for a straight wire carrying a current I. (b) Reversing $\vec{B}$ reverses $\vec{F}$. (c) Also reversing the current reverses $\vec{l}$ and returns $\vec{F}$ to the same direction as in (a).

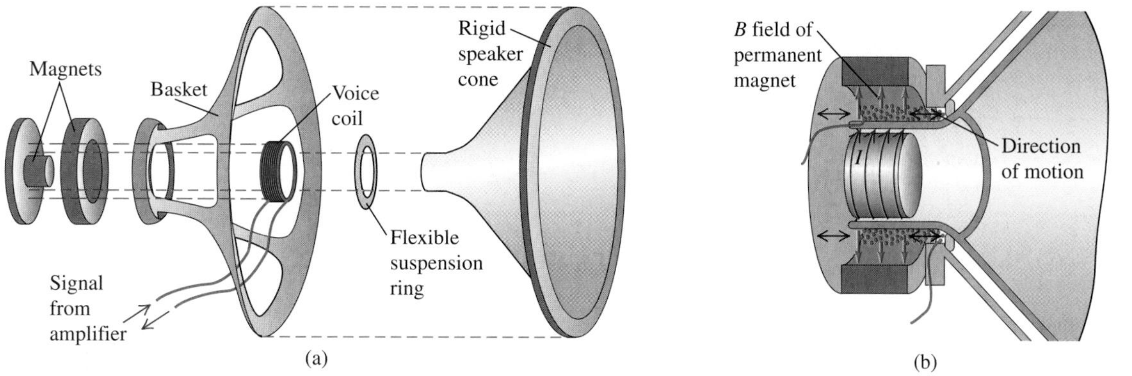

27.26 (a) Components of a loudspeaker. (b) The permanent magnet creates a magnetic field that exerts forces on the current in the voice coil; for a current I in the direction shown, the force is to the right. If the electric current in the voice coil oscillates, the speaker cone attached to the voice coil oscillates at the same frequency.

ActivPhysics ONLINE

13.5 Magnetic Force on a Wire

attached respond by oscillating with an amplitude proportional to the amplitude of the current in the coil. Turning up the volume knob on the amplifier increases the current amplitude and hence the amplitudes of the cone's oscillation and of the sound wave produced by the moving cone.

Example 27.7 Magnetic force on a straight conductor

A straight horizontal copper rod carries a current of 50.0 A from west to east in a region between the poles of a large electromagnet. In this region there is a horizontal magnetic field toward the northeast (that is, 45° north of east) with magnitude 1.20 T, as shown in an overhead view in Fig. 27.27. a) Find the magnitude and direction of the force on a 1.00-m section of rod. b) While keeping the rod horizontal, how should it be oriented to maximize the magnitude of the force? What is the force magnitude in this case?

SOLUTION

IDENTIFY and SET UP: This is a straight wire segment, so we can find the magnitude of the magnetic force using Eq. (27.18) and the direction from the right-hand rule. Alternatively, we can find the force vector (magnitude and direction) using Eq. (27.19).

EXECUTE: a) The angle ϕ between the directions of current and field is 45°. From Eq. (27.18) we obtain

$$F = IlB\sin\phi = (50.0\text{ A})(1.00\text{ m})(1.20\text{ T})(\sin 45°) = 42.4\text{ N}$$

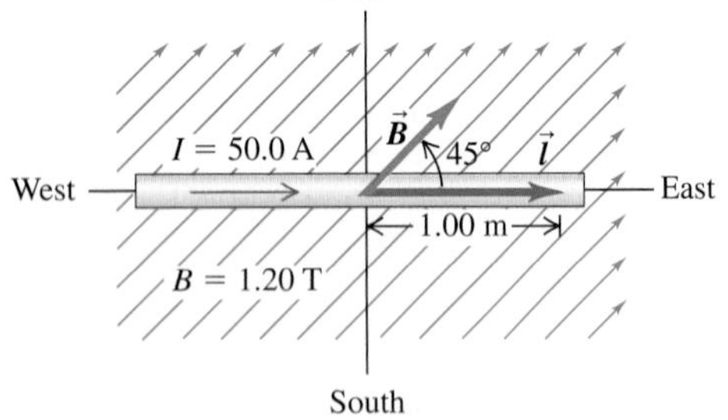

27.27 Overhead view of the copper rod.

The *direction* of the force is perpendicular to the plane of the current and the field, both of which lie in the horizontal plane. Thus the force must be vertical; the right-hand rule shows that it is vertically *upward* (out of the plane of the figure).

Alternatively, we can use a coordinate system with the x-axis pointing east, the y-axis north, and the z-axis up. Then we have

$$\vec{l} = (1.00\text{ m})\hat{\imath} \qquad \vec{B} = (1.20\text{ T})[(\cos 45°)\hat{\imath} + (\sin 45°)\hat{\jmath}]$$

$$\begin{aligned}\vec{F} &= I\vec{l} \times \vec{B}\\ &= (50\text{ A})(1.00\text{ m})\hat{\imath} \times (1.20\text{ T})[(\cos 45°)\hat{\imath} + (\sin 45°)\hat{\jmath}]\\ &= (42.4\text{ N})\hat{k}\end{aligned}$$

If the conductor is in mechanical equilibrium under the action of its weight and the upward magnetic force, its weight is 42.4 N and its mass is

$$m = w/g = (42.4\text{ N})/(9.8\text{ m/s}^2) = 4.33\text{ kg}$$

b) The magnitude of the force is maximum if $\phi = 90°$ so that $\vec{l}$ and $\vec{B}$ are perpendicular. To have the force still be upward, we rotate the rod clockwise by 45° from its orientation in Fig. 27.27 so that the current runs toward the southeast. Then the magnetic force has magnitude

$$F = IlB = (50.0\text{ A})(1.00\text{ m})(1.20\text{ T}) = 60.0\text{ N}$$

and the mass of a rod that can be held up against gravity is $m = w/g = (60.0\text{ N})/(9.8\text{ m/s}^2) = 6.12$ kg.

EVALUATE: This is a simple example of magnetic levitation. Magnetic levitation is also used in special high-speed trains. Conventional electromagnetic technology is used to suspend the train over the tracks; the elimination of rolling friction allows the train to achieve speeds in excess of 400 km/h (250 mi/h).

Example 27.8 Magnetic force on a curved conductor

In Fig. 27.28 the magnetic field $\vec{B}$ is uniform and perpendicular to the plane of the figure, pointing out. The conductor has a straight segment with length L perpendicular to the plane of the figure on the right, with the current opposite to $\vec{B}$; followed by a semicircle with radius R; and finally another straight segment with length L parallel to the x-axis, as shown. The conductor carries a current I. Find the total magnetic force on these three segments of wire.

SOLUTION

IDENTIFY and SET UP: We find the force on the straight segments using Eq. (27.19). For the semicircle, we first use Eq. (27.20) to find the force on an infinitesimal segment, then integrate to find the force on the semicircle as a whole. The total magnetic force on all three segments is the vector sum of the forces on each individual segment.

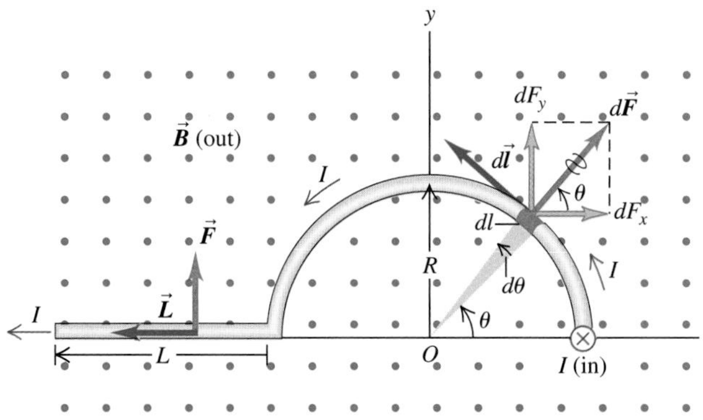

27.28 What is the total magnetic force on the conductor?

EXECUTE: Let's do the easy parts (the straight segments) first. There is *no* force on the segment on the right perpendicular to the plane of the figure because it is antiparallel to $\vec{B}$; $\vec{L} \times \vec{B} = 0$, or $\phi = 180°$ and $\sin\phi = 0$. For the straight segment on the left, $\vec{L}$ points to the left (in the direction of the current), perpendicular to $\vec{B}$. The force has magnitude $F = ILB$, and its direction is up (the $+y$-direction in the figure).

The fun part is the semicircle. The figure shows a segment $d\vec{l}$ with length $dl = R\,d\theta$, at angle θ. The direction of $d\vec{l} \times \vec{B}$ is radially outward from the center; make sure you can verify this direction. Because $d\vec{l}$ and $\vec{B}$ are perpendicular, the magnitude dF of the force on the segment $d\vec{l}$ is just $dF = I\,dl\,B$, so we have

$$dF = I(R\,d\theta)B$$

The components of the force $d\vec{F}$ on segment $d\vec{l}$ are

$$dF_x = IR\,d\theta\,B\cos\theta \qquad dF_y = IR\,d\theta\,B\sin\theta$$

To find the components of the total force, we integrate these expressions, letting θ vary from 0 to π to take in the whole semicircle. We find

$$F_x = IRB\int_0^{\pi}\cos\theta\,d\theta = 0$$

$$F_y = IRB\int_0^{\pi}\sin\theta\,d\theta = 2IRB$$

Finally, adding the forces on the straight and semicircular segments, we find the total force:

$$F_x = 0 \qquad F_y = IB(L + 2R)$$

or

$$\vec{F} = IB(L + 2R)\hat{\jmath}$$

EVALUATE: We could have predicted from symmetry that the x-component of force on the semicircle would be zero. On the right half of the semicircle the x-component of the force is positive (to the right) and on the left half it is negative (to the left); the positive and negative contributions to the integral cancel.

Note that the net force on all three segments together is the same force that would be exerted if we replaced the semicircle with a straight segment along the x-axis. Do you see why?

Test Your Understanding

A straight current-carrying wire is inserted in a region of space where a strong magnetic field is directed vertically downward. How can the wire be oriented so that it is suspended in midair by the magnetic force?

27.7 | Force and Torque on a Current Loop

Current-carrying conductors usually form closed loops, so it is worthwhile to use the results of Section 27.6 to find the *total* magnetic force and torque on a conductor in the form of a loop. Many practical devices make use of the magnetic force or torque on a conducting loop, including loudspeakers (see Fig. 27.26) and galvanometers (see Section 26.3). Hence the results of this section are of substantial practical importance. These results will also help us understand the behavior of bar magnets described in Section 27.1.

As an example, let's look at a rectangular current loop in a uniform magnetic field. We can represent the loop as a series of straight line segments. We will find that the total *force* on the loop is zero but that there can be a net *torque* acting on the loop, with some interesting properties.

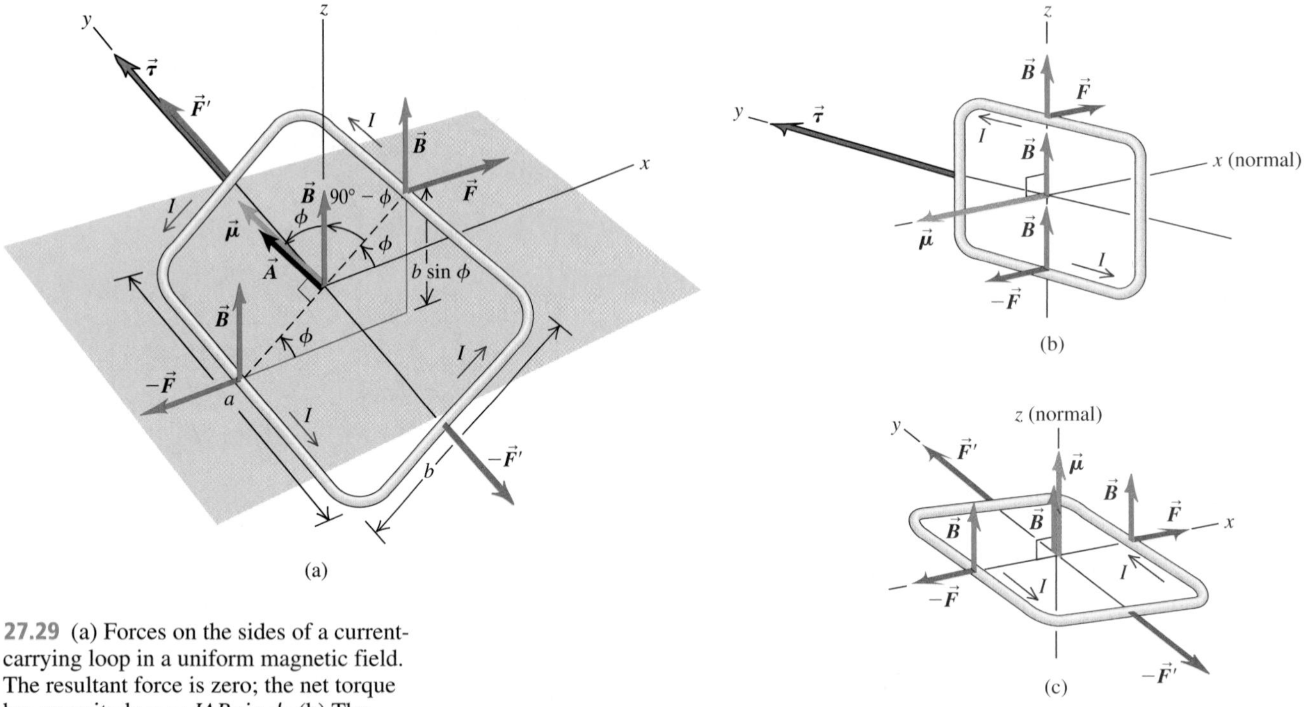

27.29 (a) Forces on the sides of a current-carrying loop in a uniform magnetic field. The resultant force is zero; the net torque has magnitude $\tau = IAB \sin \phi$. (b) The torque is maximum when the normal to the loop is perpendicular to $\vec{B}$. (c) When the normal to the loop is parallel to $\vec{B}$, the torque is zero and the equilibrium is stable. If the normal is antiparallel to $\vec{B}$, the torque is also zero but the equilibrium is unstable.

Figure 27.29 shows a rectangular loop of wire with side lengths a and b. A line perpendicular to the plane of the loop (i.e., a *normal* to the plane) makes an angle ϕ with the direction of the magnetic field $\vec{B}$, and the loop carries a current I. The wires leading the current into and out of the loop and the source of emf are omitted to keep the diagram simple.

The force $\vec{F}$ on the right side of the loop (length a) is to the right, in the $+x$-direction as shown. On this side, $\vec{B}$ is perpendicular to the current direction, and the force on this side has magnitude

$$F = IaB \tag{27.21}$$

A force $-\vec{F}$ with the same magnitude but opposite direction acts on the opposite side of the loop, as shown in the figure.

The sides with length b make an angle $(90° - \phi)$ with the direction of $\vec{B}$. The forces on these sides are the vectors $\vec{F}'$ and $-\vec{F}'$; their magnitude F' is given by

$$F' = IbB \sin(90° - \phi) = IbB \cos \phi$$

The lines of action of both forces lie along the y-axis.

ActivPhysics Online

13.6 Magnetic Torque on a Loop

The *total* force on the loop is zero because the forces on opposite sides cancel out in pairs. **The net force on a current loop in a uniform magnetic field is zero. However, the net *torque* is not in general equal to zero.** (You may find it helpful at this point to review the discussion of torque in Section 10.1.) The two forces $\vec{F}'$ and $-\vec{F}'$ in Fig. 27.29a lie along the same line and so give rise to zero net torque with respect to any point. The two forces $\vec{F}$ and $-\vec{F}$ lie along different lines, and each gives rise to a torque about the y-axis. According to the right-hand rule for determining the direction of torques, the vector torques due to $\vec{F}$ and $-\vec{F}$ are both in the $+y$-direction; hence the net vector torque $\vec{\tau}$ is in the $+y$-direction as well. The moment arm for each of these forces (equal to the perpendicular distance from the rotation axis to the line of action of the force) is $(b/2)\sin\phi$, so the torque due to each force has magnitude $F(b/2)\sin\phi$. If we use Eq. (27.21) for F, the magnitude of the net torque is

$$\tau = 2F(b/2)\sin\phi = (IBa)(b\sin\phi) \qquad (27.22)$$

The torque is greatest when $\phi = 90°$, $\vec{B}$ is in the plane of the loop, and the normal to this plane is perpendicular to $\vec{B}$ (Fig. 27.29b). The torque is zero when ϕ is zero or 180° and the normal to the loop is parallel or antiparallel to the field (Fig. 27.29c). The value $\phi = 0$ is a stable equilibrium position because the torque is zero there, and when the loop is rotated slightly from this position, the resulting torque tends to rotate it back toward $\phi = 0$. The position $\phi = 180°$ is an *unstable* equilibrium position; if displaced slightly from this position, the loop tends to move farther away from $\phi = 180°$. Figure 27.29 shows rotation about the y-axis, but because the net force on the loop is zero, Eq. (27.22) for the torque is valid for *any* choice of axis (see Problem 27.80).

The area A of the loop is equal to ab, so we can rewrite Eq. (27.22) as

$$\tau = IBA\sin\phi \quad \text{(magnitude of torque on a current loop)} \qquad (27.23)$$

The product IA is called the **magnetic dipole moment** or **magnetic moment** of the loop, for which we use the symbol μ (Greek letter "mu"):

$$\mu = IA \qquad (27.24)$$

It is analogous to the electric dipole moment introduced in Section 21.7. In terms of μ, the magnitude of the torque on a current loop is

$$\tau = \mu B\sin\phi \qquad (27.25)$$

where ϕ is the angle between the normal to the loop (the direction of the vector area $\vec{A}$) and $\vec{B}$. The torque tends to rotate the loop in the direction of *decreasing* ϕ, that is, toward its stable equilibrium position in which the loop lies in the xy-plane, perpendicular to the direction of the field $\vec{B}$ (Fig. 27.29c). A current loop, or any other body that experiences a magnetic torque given by Eq. (27.25), is also called a **magnetic dipole.**

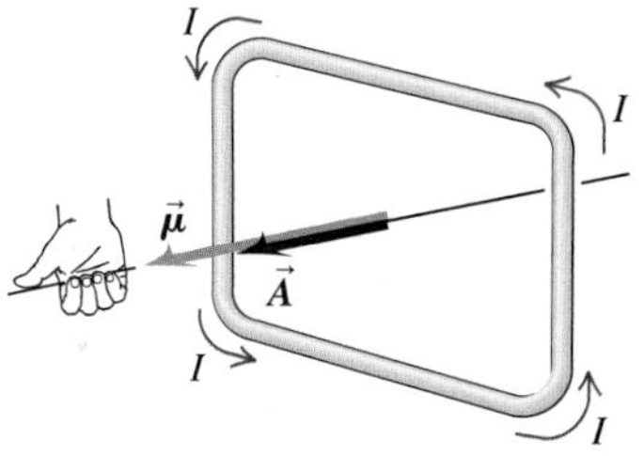

27.30 The right-hand rule determines the direction of the magnetic moment of a current-carrying loop. This is also the direction of the loop's area vector $\vec{A}$; $\vec{\mu} = I\vec{A}$ is a vector equation.

We can also define a vector magnetic moment $\vec{\mu}$ with magnitude IA; this is shown in Fig. 27.29. The direction of $\vec{\mu}$ is defined to be perpendicular to the plane of the loop, with a sense determined by a right-hand rule, as shown in Fig. 27.30. Wrap the fingers of your right hand around the perimeter of the loop in the direction of the current. Then extend your thumb so that it is perpendicular to the plane of the loop; its direction is the direction $\vec{\mu}$ (and of the vector area $\vec{A}$ of the loop). The torque is greatest when $\vec{\mu}$ and $\vec{B}$ are perpendicular and is zero when they are parallel or antiparallel. In the stable equilibrium position, $\vec{\mu}$ and $\vec{B}$ are parallel.

Finally, we can express this interaction in terms of the torque vector $\vec{\tau}$, which we used for *electric*-dipole interactions in Section 21.7. From Eq. (27.25) the magnitude of $\vec{\tau}$ is equal to the magnitude of $\vec{\mu} \times \vec{B}$, and reference to Fig. 27.29 shows that the directions are also the same. So we have

$$\vec{\tau} = \vec{\mu} \times \vec{B} \quad \text{(vector torque on a current loop)} \qquad (27.26)$$

This result is directly analogous to the result we found in Section 21.7 for the torque exerted by an *electric* field $\vec{E}$ on an *electric* dipole with dipole moment $\vec{p}$: $\vec{\tau} = \vec{p} \times \vec{E}$.

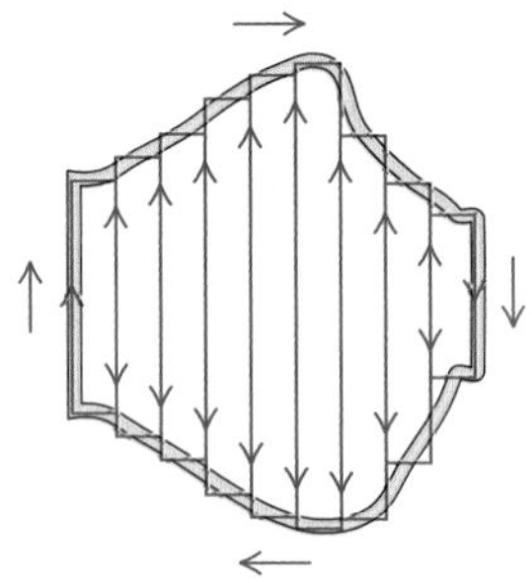

27.31 The collection of rectangles exactly matches the irregular plane loop in the limit as the number of rectangles approaches infinity and the width of each rectangle approaches zero.

When a magnetic dipole changes orientation in a magnetic field, the field does work on it. In an infinitesimal angular displacement $d\phi$ the work dW is given by $\tau d\phi$, and there is a corresponding change in potential energy. As the above discussion suggests, the potential energy is least when $\vec{\mu}$ and $\vec{B}$ are parallel and greatest when they are antiparallel. To find an expression for the potential energy U as a function of orientation, we can make use of the beautiful symmetry between the electric and magnetic dipole interactions. The torque on an *electric* dipole in an *electric* field is $\vec{\tau} = \vec{p} \times \vec{E}$; we found in Section 21.7 that the corresponding potential energy is $U = -\vec{p} \cdot \vec{E}$. The torque on a *magnetic* dipole in a *magnetic* field is $\vec{\tau} = \vec{\mu} \times \vec{B}$, so we can conclude immediately that the corresponding potential energy is

$$U = -\vec{\mu} \cdot \vec{B} = -\mu B \cos\phi \quad \text{(potential energy for a magnetic dipole)} \quad (27.27)$$

With this definition, U is zero when the magnetic dipole moment is perpendicular to the magnetic field.

Although we have derived Eqs. (27.21) through (27.27) for a rectangular current loop, all these relations are valid for a plane loop of any shape at all. Any planar loop may be approximated as closely as we wish by a very large number of rectangular loops, as shown in Fig. 27.31. If these loops all carry equal currents in the same sense, then the forces and torques on the sides of two loops adjacent to each other cancel, and the only forces and torques that do not cancel are due to currents around the boundary. Thus all the above relations are valid for a plane current loop of any shape, with the magnetic moment $\vec{\mu}$ given by $\vec{\mu} = I\vec{A}$.

We can also generalize this whole formulation to a coil consisting of N planar loops close together; the effect is simply to multiply each force, the magnetic moment, the torque, and the potential energy by a factor of N.

An arrangement of particular interest is the **solenoid,** a helical winding of wire, such as a coil wound on a circular cylinder (Fig. 27.32). If the windings are closely spaced, the solenoid can be approximated by a number of circular loops lying in planes at right angles to its long axis. The total torque on a solenoid in a magnetic field is simply the sum of the torques on the individual turns. For a solenoid with N turns in a uniform field B, the magnetic moment is $\mu = NIA$ and

$$\tau = NIAB \sin\phi \quad (27.28)$$

where ϕ is the angle between the axis of the solenoid and the direction of the field. The magnetic moment vector $\vec{\mu}$ is along the solenoid axis. The torque is greatest when the solenoid axis is perpendicular to the magnetic field and zero when they are parallel. The effect of this torque is to tend to rotate the solenoid into a position where its axis is parallel to the field. Solenoids are also useful as *sources* of magnetic field, as we'll discuss in Chapter 28.

The d'Arsonval galvanometer, described in Section 26.3, makes use of a magnetic torque on a coil carrying a current. As Fig. 26.14 shows, the magnetic field is not uniform but is *radial,* so the side thrusts on the coil are always perpendicular to its plane. Thus the angle ϕ in Eq. (27.28) is always 90°, and the magnetic torque is directly proportional to the current, no matter what the orientation of the coil. A restoring torque proportional to the angular displacement of the coil is provided by two hairsprings, which also serve as current leads to the coil. When current is supplied to the coil, it rotates along with its attached pointer until the restoring spring torque just balances the magnetic torque. Thus the pointer deflection is proportional to the current. Such instruments can measure currents as small as 10^{-7} A, and modified versions can measure currents as small as 10^{-10} A.

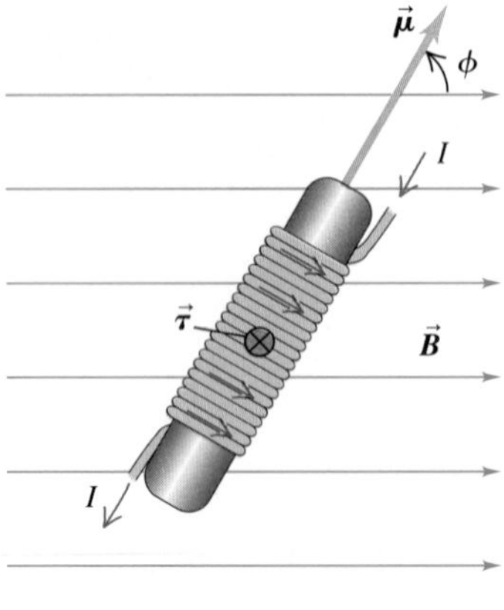

27.32 The torque $\vec{\tau} = \vec{\mu} \times \vec{B}$ on this solenoid in a uniform magnetic field is directed straight into the page. It tends to rotate the solenoid clockwise, aligning the magnetic moment and the axis of the solenoid with the magnetic field. An actual solenoid has many more turns, wrapped closely together.

Example 27.9 Magnetic torque on a circular coil

A circular coil 0.0500 m in radius, with 30 turns of wire, lies in a horizontal plane (Fig. 27.33). It carries a current of 5.00 A in a counterclockwise sense when viewed from above. The coil is in a uniform magnetic field directed toward the right, with magnitude 1.20 T. Find the magnitudes of the magnetic moment and the torque on the coil.

SOLUTION

IDENTIFY and SET UP: The magnitude μ of the magnetic moment of a single turn of wire is given in terms of the current and coil area by Eq. (27.24). For N turns, the magnetic moment is N times greater. The magnitude τ of the torque is found using Eq. (27.25).

EXECUTE: The area of the coil is

$$A = \pi r^2 = \pi(0.0500\ \text{m})^2 = 7.85 \times 10^{-3}\ \text{m}^2$$

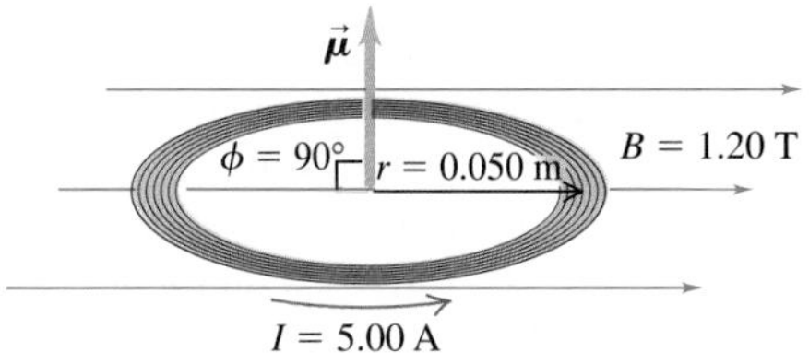

27.33 A circular coil of current-carrying wire in a uniform magnetic field.

The magnetic moment of each turn of the coil is

$$\mu = IA = (5.00\ \text{A})(7.85 \times 10^{-3}\ \text{m}^2) = 3.93 \times 10^{-2}\ \text{A}\cdot\text{m}^2$$

and the total magnetic moment of all 30 turns is

$$\mu_{\text{total}} = (30)(3.93 \times 10^{-2}\ \text{A}\cdot\text{m}^2) = 1.18\ \text{A}\cdot\text{m}^2$$

The angle ϕ between the direction of $\vec{B}$ and the direction of $\vec{\mu}$ (which is along the normal to the plane of the coil) is 90°. From Eq. (27.25),

$$\tau = \mu_{\text{total}} B \sin\phi = (1.18\ \text{A}\cdot\text{m}^2)(1.20\ \text{T})(\sin 90°) = 1.41\ \text{N}\cdot\text{m}$$

Alternatively, from Eq. (27.23), the torque on each turn of the coil is

$$\tau = IBA \sin\phi = (5.00\ \text{A})(1.20\ \text{T})(7.85 \times 10^{-3}\ \text{m}^2)(\sin 90°) = 0.0471\ \text{N}\cdot\text{m}$$

and the total torque on the coil is

$$\tau = (30)(0.0471\ \text{N}\cdot\text{m}) = 1.41\ \text{N}\cdot\text{m}$$

EVALUATE: The torque tends to rotate the right side of the coil down and the left side up, into a position where the normal to its plane is parallel to $\vec{B}$.

Example 27.10 Potential energy for a coil in a magnetic field

If the coil in Example 27.9 rotates from its initial position to a position where its magnetic moment is parallel to $\vec{B}$, what is the change in potential energy?

SOLUTION

IDENTIFY and SET UP: We calculate the potential energy for each orientation using Eq. (27.27). We then take the difference between the final and initial values to find the change in potential energy.

EXECUTE: From Eq. (27.27), the initial potential energy U_1 is

$$U_1 = -\mu_{\text{total}} B \cos\phi_1 = -(1.18\ \text{A}\cdot\text{m}^2)(1.20\ \text{T})(\cos 90°) = 0$$

and the final potential energy U_2 is

$$U_2 = -\mu_{\text{total}} B \cos\phi_2 = -(1.18\ \text{A}\cdot\text{m}^2)(1.20\ \text{T})(\cos 0°) = -1.41\ \text{J}$$

The change in potential energy is $\Delta U = U_2 - U_1 = -1.41$ J.

EVALUATE: The potential energy decreases because the rotation is in the direction of the magnetic torque.

Example 27.11 Equilibrium of a coil in a magnetic field

What vertical forces applied to the left and right edges of the coil of Fig. 27.33 would be required to maintain it in equilibrium in its initial position?

SOLUTION

IDENTIFY and SET UP: From Example 27.9, the right side of the coil tends to rotate down and the left side tends to rotate up. To stabilize the coil we apply an upward force of magnitude F at the right side of the coil and a downward force of the same magnitude at the left side. We choose the magnitude F so that the torque we apply has the same magnitude as the magnetic torque that it opposes.

EXECUTE: An upward force of magnitude F at the right side and a downward force of equal magnitude on the left side would have a total torque of

$$\tau = 2rF = (2)(0.0500\ \text{m})F$$

This applied torque must be equal in magnitude to the magnetic-field torque magnitude, 1.41 N·m. We find that the required forces have magnitude $F = 14.1$ N.

EVALUATE: The orientation of the coil in Fig. 27.33 is the one that gives the largest magnetic torque, and hence requires the largest forces to maintain it in position. In any other orientation, the required forces would be less than 14.1 N.

Magnetic Dipole in a Non-Uniform Magnetic Field

We have seen that a current loop (that is, a magnetic dipole) experiences zero net force in a uniform magnetic field. Figure 27.34 shows two current loops in the *non-uniform* $\vec{B}$ field of a bar magnet; in both cases the net force on the loop is *not* zero. In Fig. 27.34a the magnetic moment $\vec{\mu}$ is in the direction opposite to the field, and the force $d\vec{F} = I\,d\vec{l} \times \vec{B}$ on a segment of the loop has both a radial component and a component to the right. When these forces are summed to find the net force $\vec{F}$ on the loop, the radial components cancel so that the net force is to the right, away from the magnet. Note that in this case the force is toward the region where the field lines are farther apart and the field magnitude B is less. The polarity of the bar magnet is reversed in Fig. 27.34b, so $\vec{\mu}$ and $\vec{B}$ are parallel; now the net force on the loop is to the left, toward the region of greater field magnitude near the magnet. Later in this section we'll use these observations to explain why bar magnets can pick up unmagnetized iron objects.

Magnetic Dipoles and How Magnets Work

The behavior of a solenoid in a magnetic field (Fig. 27.32) resembles that of a bar magnet or compass needle; if free to turn, both the solenoid and the magnet orient themselves with their axes parallel to the $\vec{B}$ field. In both cases this is due to the interaction of moving electric charges with a magnetic field; the difference is that in a bar magnet the motion of charge occurs on the microscopic scale of the atom.

Think of an electron as being like a spinning ball of charge. In this analogy the circulation of charge around the spin axis is like a current loop, and so the electron has a net magnetic moment. (This analogy, while helpful, is inexact; an electron isn't really a spinning sphere. A full explanation of the origin of an electron's magnetic moment involves quantum mechanics, which is beyond our scope here.) In an iron atom a substantial fraction of the electron magnetic moments align with each other, and the atom has a nonzero magnetic moment. (By contrast, the atoms of most elements have little or no net magnetic moment.) In an unmagnetized

27.34 Forces on current loops in a non-uniform $\vec{B}$ field. In each case the axis of the bar magnet is perpendicular to the plane of the loop and passes through the center of the loop.

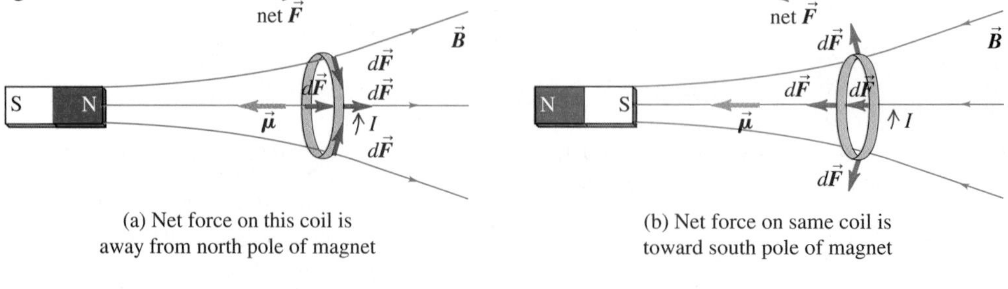

piece of iron there is no overall alignment of the magnetic moments of the atoms; their vector sum is zero, and the net magnetic moment is zero (Fig. 27.35a). But in an iron bar magnet the magnetic moments of many of the atoms are parallel, and there is a substantial net magnetic moment $\vec{\mu}$ (Fig. 27.35b). If the magnet is placed in a magnetic field $\vec{B}$, the field exerts a torque given by Eq. (27.26) that tends to align $\vec{\mu}$ with $\vec{B}$ (Fig. 27.35c). A bar magnet tends to align with a $\vec{B}$ field so that a line from the south pole to the north pole of the magnet is in the direction of $\vec{B}$; hence the real significance of a magnet's north and south poles is that they represent the head and tail, respectively, of the magnet's dipole moment $\vec{\mu}$.

The torque experienced by a current loop in a magnetic field also explains how an unmagnetized iron object like that in Fig. 27.35a becomes magnetized. If an unmagnetized iron paper clip is placed next to a powerful magnet, the magnetic moments of the paper clip's atoms tend to align with the $\vec{B}$ field of the magnet. When the paper clip is removed, its atomic dipoles tend to remain aligned, and the paper clip has a net magnetic moment. The paper clip can be demagnetized by being dropped on the floor or heated; the added internal energy jostles and re-randomizes the atomic dipoles.

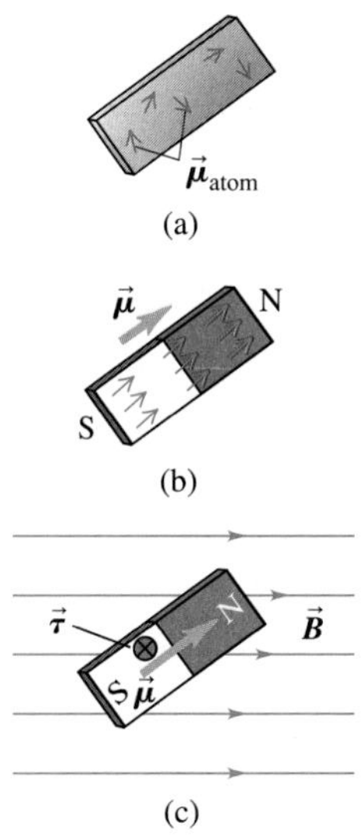

27.35 a) Randomly oriented atomic magnetic moments in an unmagnetized piece of iron. (Only a few representative atomic moments are shown.) (b) Aligned atomic magnetic moments in a magnetized piece of iron (bar magnet). The net magnetic moment of the bar magnet points from its south pole to its north pole. (c) The torque on a bar magnet in a magnetic field tends to align the magnet's dipole moment with the $\vec{B}$ field.

The magnetic-dipole picture of a bar magnet explains the attractive and repulsive forces between bar magnets shown in Fig. 27.1. The magnetic moment $\vec{\mu}$ of a bar magnet points from its south pole to its north pole, so the current loops in Figs. 27.34a and 27.34b are both equivalent to a magnet with its north pole on the left. Hence the situation in Fig. 27.34a is equivalent to two bar magnets with their north poles next to each other; the resultant force is repulsive, just as in Fig. 27.1c. In Fig. 27.34b we again have the equivalent of two bar magnets end to end, but with the south pole of the left-hand magnet next to the north pole of the right-hand magnet. The resultant force is attractive, as in Fig. 27.1b.

Finally, we can explain how a magnet can attract an unmagnetized iron object (Fig. 27.2). It's a two-step process. First, the atomic magnetic moments of the iron tend to align with the $\vec{B}$ field of the magnet, so the iron acquires a net magnetic dipole moment $\vec{\mu}$ parallel to the field. Second, the non-uniform field of the magnet attracts the magnetic dipole. Figure 27.36a shows an example. The north pole of the magnet is closer to the pocket knife (which contains iron), and the magnetic dipole produced in the knife is equivalent to a loop with a current that circulates in a direction opposite to that shown in Fig. 27.34a. Hence the net magnetic force on the knife is opposite to the force on the loop in Fig. 27.34a, and the knife is attracted toward the magnet. Changing the polarity of the magnet, as in Fig. 27.36b, reverses the directions of both $\vec{B}$ and $\vec{\mu}$. The situation is now equivalent to that shown in Fig. 27.34b; like the loop in that figure, the knife is attracted toward the magnet. Hence a previously unmagnetized object containing iron is attracted to *either* pole of a magnet. By contrast, objects made of brass, aluminum, or wood hardly respond at all to a magnet; the atomic magnetic dipoles of these materials, if present at all, have less tendency to align with an external field.

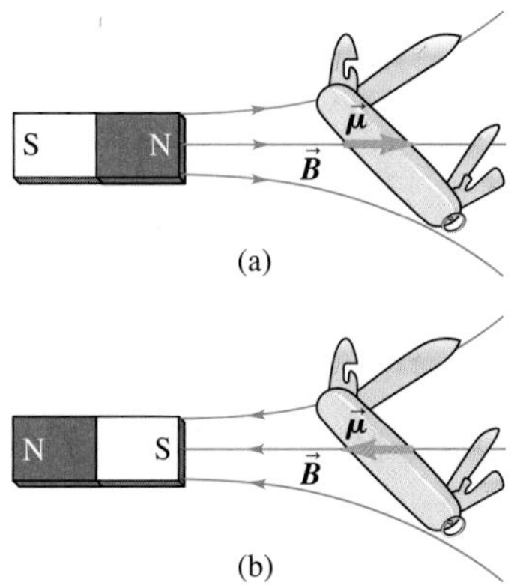

27.36 A bar magnet attracts an unmagnetized iron object in two steps. First, the $\vec{B}$ field of the bar magnet gives rise to a net magnetic moment in the object. Second, because the field of the bar magnet is not uniform, this magnetic dipole is attracted toward the magnet. The attraction is the same whether the object is closer to (a) the magnet's north pole or (b) the magnet's south pole.

Our discussion of how magnets and pieces of iron interact has just scratched the surface of a diverse subject known as *magnetic properties of materials.* We'll discuss these properties in more depth in Section 28.8.

Test Your Understanding

Figure 27.11d depicts the magnetic field lines due to a circular current-carrying loop. What is the direction of the magnetic moment of this loop? Which side of the loop is equivalent to the north pole of a magnet, and which side is equivalent to the south pole?

*27.8 | The Direct-Current Motor

No one needs to be reminded of the importance of electric motors in contemporary society. In a motor a magnetic torque acts on a current-carrying conductor, and electric energy is converted to mechanical energy. As an example, let's look at a simple type of direct-current (dc) motor, shown in Fig. 27.37.

The moving part of the motor is the *rotor,* a length of wire formed into an open-ended loop and free to rotate about an axis. The ends of the rotor wires are attached to circular conducting segments that form a *commutator.* In Fig. 27.37a, each of the two commutator segments makes contact with one of the terminals, or *brushes,* of an external circuit that includes a source of emf. This causes a current to flow into the rotor on one side, shown in red, and out of the rotor on the other side, shown in blue. Hence the rotor is a current loop with a magnetic moment $\vec{\boldsymbol{\mu}}$. The rotor lies between opposing poles of a permanent magnet, so there is a magnetic field $\vec{\boldsymbol{B}}$ that exerts a torque $\vec{\boldsymbol{\tau}} = \vec{\boldsymbol{\mu}} \times \vec{\boldsymbol{B}}$ on the rotor. For the rotor orientation shown in Fig. 27.37a the torque causes the rotor to turn counterclockwise, in the direction that will align $\vec{\boldsymbol{\mu}}$ with $\vec{\boldsymbol{B}}$.

In Fig. 27.37b the rotor has rotated by 90° from its orientation in Fig. 27.37a. If the current through the rotor were constant, the rotor would now be in its equilibrium orientation; it would simply oscillate around this orientation. But here's where the commutator comes into play; each brush is now in contact with *both* segments of the commutator. There is no potential difference between the commutators, so at this instant no current flows through the rotor, and the magnetic moment is zero. The rotor continues to rotate counterclockwise because of its inertia, and current again flows through the rotor as in Fig. 27.37c. But now current enters on the *blue*-colored side of the rotor and exits on the *red*-colored side, just the opposite of the situation in Fig. 27.37a. While the direction of the current has reversed with respect to the rotor, the rotor itself has rotated 180° and the magnetic moment $\vec{\boldsymbol{\mu}}$ is in the same direction with respect to the magnetic field. Hence the magnetic torque $\vec{\boldsymbol{\tau}}$ is in the same direction in Fig. 27.37c as in Fig. 27.37a. Thanks to the commutator, the current reverses after every 180° of rotation, so the torque is always in the direction to rotate the rotor counterclockwise. When the

27.37 Schematic diagram of a simple dc motor. The rotor is a wire loop that is free to rotate about an axis; the rotor ends are attached to the two curved conductors that form the commutator. The commutator segments are insulated from one another.

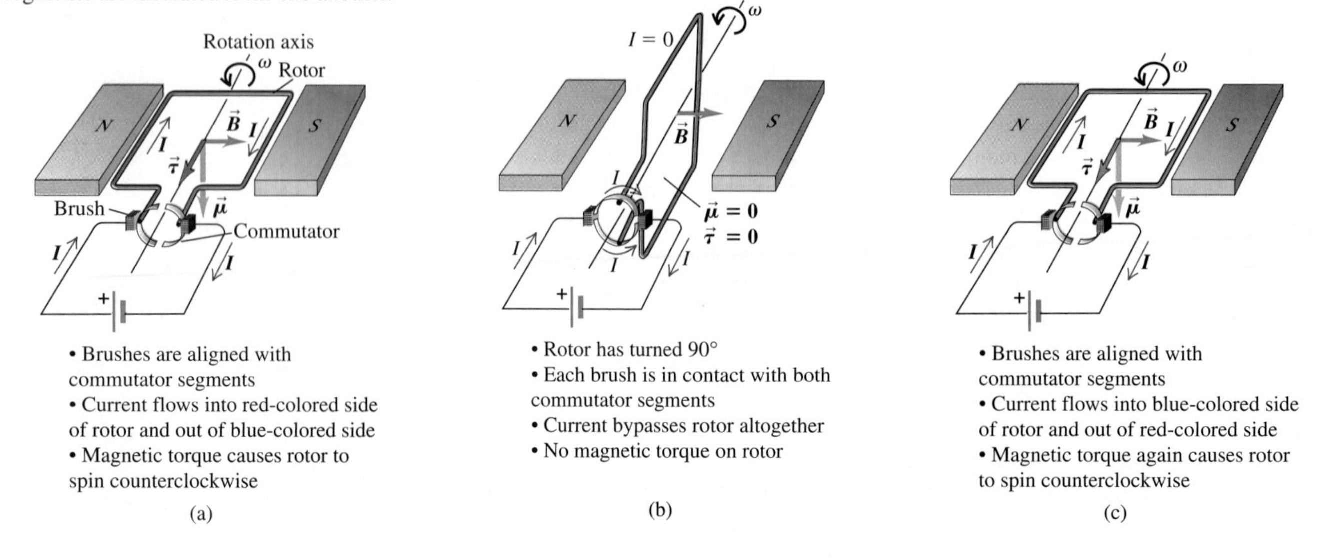

motor has come "up to speed," the average magnetic torque is just balanced by an opposing torque due to air resistance, friction in the rotor bearings, and friction between the commutator and brushes.

The simple motor shown in Fig. 27.37 has only a single turn of wire in its rotor. In practical motors, the rotor has many turns; this increases the magnetic moment and the torque so that the motor can spin larger loads. The torque can also be increased by using a stronger magnetic field, which is why many motor designs use electromagnets instead of a permanent magnet. Another drawback of the simple design in Fig. 27.37 is that the magnitude of the torque rises and falls as the rotor spins. This can be remedied by having the rotor include several independent coils of wire oriented at different angles (Fig. 27.38).

27.38 This motor from a computer disk drive has 12 current-carrying coils. They interact with permanent magnets on the turntable (not shown) to make the turntable rotate. (This design is the reverse of the design in Fig. 27.37, in which the permanent magnets are stationary and the coil rotates.) Because there are multiple coils, the magnetic torque is very nearly constant and the turntable spins at a very constant rate.

Because a motor converts electric energy to mechanical energy or work, it requires electric energy input. If the potential difference between its terminals is V_{ab} and the current is I, then the power input is $P = V_{ab}I$. Even if the motor coils have negligible resistance, there must be a potential difference between the terminals if P is to be different from zero. This potential difference results principally from magnetic forces exerted on the currents in the conductors of the rotor as they rotate through the magnetic field. The associated electromotive force $\mathcal{E}$ is called an *induced* emf; it is also called a *back* emf because its sense is opposite to that of the current. In Chapter 29 we will study induced emfs resulting from motion of conductors in magnetic fields.

In a *series* motor the rotor is connected in series with the electromagnet that produces the magnetic field; in a *shunt* motor they are connected in parallel. In a series motor with internal resistance r, V_{ab} is greater than $\mathcal{E}$, and the difference is the potential drop Ir across the internal resistance. That is,

$$V_{ab} = \mathcal{E} + Ir \tag{27.29}$$

Because the magnetic force is proportional to velocity, $\mathcal{E}$ is *not* constant but is proportional to the speed of rotation of the rotor.

Example 27.12 A series dc motor

A dc motor with its rotor and field coils connected in series has an internal resistance of 2.00 Ω. When running at full load on a 120-V line, it draws a current of 4.00 A. a) What is the emf in the rotor? b) What is the power delivered to the motor? c) What is the rate of dissipation of energy in the resistance of the motor? d) What is the mechanical power developed? e) What is the efficiency of the motor? f) What happens if the machine the motor is driving jams and the rotor suddenly stops turning?

SOLUTION

IDENTIFY and SET UP: We are given the internal resistance $r = 2.00\ \Omega$, the voltage $V_{ab} = 120$ V across the motor, and the current $I = 4.00$ A through the motor. We use Eq. (27.29) to determine the emf $\mathcal{E}$ from these quantities. The power delivered to the motor is $V_{ab}I$, the rate of energy dissipation is I^2r, and the power output by the motor is the difference between the power input and the power dissipated. The efficiency e is the ratio of mechanical power output to electric power input.

EXECUTE: a) From Eq. (27.29), $V_{ab} = \mathcal{E} + Ir$, we have

$$120\text{ V} = \mathcal{E} + (4.0\text{ A})(2.0\ \Omega) \quad \text{and so} \quad \mathcal{E} = 112\text{ V}$$

b) The power delivered to the motor from the source is

$$P_{\text{input}} = V_{ab}I = (120\text{ V})(4.0\text{ A}) = 480\text{ W}$$

c) The power dissipated in the resistance r is

$$P_{\text{dissipated}} = I^2r = (4.0\text{ A})^2(2.0\ \Omega) = 32\text{ W}$$

d) The mechanical power output is the electric power input minus the rate of dissipation of energy in the motor's resistance (assuming that there are no other power losses):

$$P_{\text{output}} = P_{\text{input}} - P_{\text{dissipated}} = 480\text{ W} - 32\text{ W} = 448\text{ W}$$

e) The efficiency e is the ratio of mechanical power output to electric power input:

$$e = \frac{P_{\text{output}}}{P_{\text{input}}} = \frac{448\text{ W}}{480\text{ W}} = 0.93 = 93\%$$

f) With the rotor stalled, the back emf $\mathcal{E}$ (which is proportional to rotor speed) goes to zero. From Eq. (27.29) the current becomes

$$I = \frac{V_{ab}}{r} = \frac{120\ \text{V}}{2.0\ \Omega} = 60\ \text{A}$$

and the power dissipated in the resistance r becomes

$$P_{\text{dissipated}} = I^2 r = (60\ \text{A})^2(2\ \Omega) = 7200\ \text{W}$$

EVALUATE: If this massive overload doesn't blow a fuse or trip a circuit breaker, the coils will quickly melt. When the motor is first turned on, there's a momentary surge of current until the motor picks up speed. This surge causes greater-than-usual voltage drops $(V = IR)$ in the power lines supplying the current. Similar effects are responsible for the momentary dimming of lights in a house when an air conditioner or dishwasher motor starts.

Test Your Understanding

In the circuit shown in Fig. 27.37, you add a switch in series with the source of emf so that the current can be turned on and off. When you close the switch and allow current to flow, will the rotor begin to turn no matter what its original orientation? Why or why not? If not, how could you redesign the motor to remedy this situation?

*27.9 | The Hall Effect

The reality of the forces acting on the moving charges in a conductor in a magnetic field is strikingly demonstrated by the *Hall effect,* an effect analogous to the transverse deflection of an electron beam in a magnetic field in vacuum. (The effect was discovered by the American physicist Edwin Hall in 1879 while he was still a graduate student.) To describe this effect, let's consider a conductor in the form of a flat strip, as shown in Fig. 27.39. The current is in the direction of the $+x$-axis and there is a uniform magnetic field $\vec{B}$ perpendicular to the plane of the strip, in the $+y$-direction. The drift velocity of the moving charges (charge magnitude $|q|$) has magnitude v_d. Figure 27.39a shows the case of negative charges, such as electrons in a metal, and Fig. 27.39b shows positive charges. In both cases the magnetic force is upward, just as the magnetic force on a conductor is the same whether the moving charges are positive or negative. In either case a moving charge is driven toward the *upper* edge of the strip by the magnetic force $F_z = |q|v_d B$.

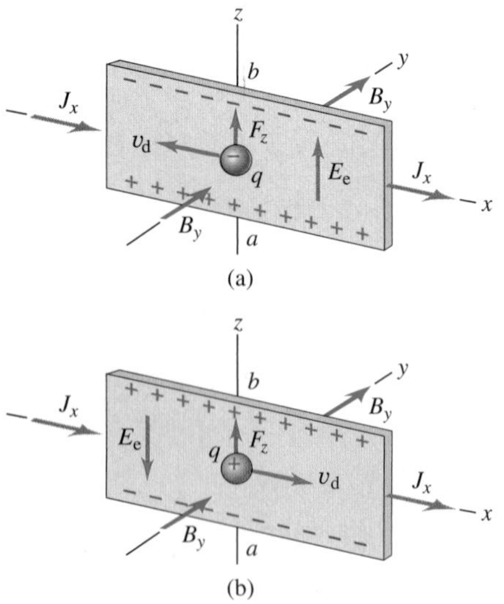

27.39 Forces on charge carriers in a conductor in a magnetic field. (a) Negative current carriers (electrons) are pushed toward the top of the strip, leading to a charge distribution as shown. Point a is at a higher potential than point b. (b) With positive current carriers, the polarity of the potential difference is opposite to that of part (a).

If the charge carriers are electrons, as in Fig. 27.39a, an excess negative charge accumulates at the upper edge of the strip, leaving an excess positive charge at its lower edge. This accumulation continues until the resulting transverse electrostatic field $\vec{E}_e$ becomes large enough to cause a force (magnitude $|q|E_e$) that is equal and opposite to the magnetic force (magnitude $|q|v_d B$). After that, there is no longer any net transverse force to deflect the moving charges. This electric field causes a transverse potential difference between opposite edges of the strip, called the *Hall voltage* or the *Hall emf.* The polarity depends on whether the moving charges are positive or negative. Experiments show that for metals the upper edge of the strip in Fig. 27.39a *does* become negatively charged, showing that the charge carriers in a metal are indeed negative electrons.

However, if the charge carriers are *positive,* as in Fig. 27.39b, then *positive* charge accumulates at the upper edge, and the potential difference is *opposite* to the situation with negative charges. Soon after the discovery of the Hall effect in 1879, it was observed that some materials, particularly some *semiconductors,* show a Hall emf opposite to that of the metals, as if their charge carriers were positively charged. We now know that these materials conduct by a process known as *hole conduction.* Within such a material there are locations, called *holes,* that would normally be occupied by an electron but are actually empty. A missing negative charge is equivalent to a positive charge. When an electron moves in one direction to fill a hole, it leaves another hole behind it. The hole migrates in the direction opposite to that of the electron.

In terms of the coordinate axes in Fig. 27.39b, the electrostatic field $\vec{E}_e$ for the positive-q case is in the $-z$-direction; its z-component E_z is negative. The magnetic field is in the $+y$-direction, and we write it as B_y. The magnetic force (in the $+z$-direction) is qv_dB_y. The current density J_x is in the $+x$-direction. In the steady state, when the forces qE_z and qv_dB_y are equal in magnitude and opposite in direction,

$$qE_z + qv_dB_y = 0 \quad \text{or} \quad E_z = -v_dB_y$$

This confirms that when q is positive, E_z is negative. The current density J_x is

$$J_x = nqv_d$$

Eliminating v_d between these equations, we find

$$nq = \frac{-J_xB_y}{E_z} \quad \text{(Hall effect)} \quad (27.30)$$

Note that this result (as well as the entire derivation) is valid for both positive and negative q. When q is negative, E_z is positive, and conversely.

We can measure J_x, B_y, and E_z, so we can compute the product nq. In both metals and semiconductors, q is equal in magnitude to the electron charge, so the Hall effect permits a direct measurement of n, the concentration of current-carrying charges in the material. The *sign* of the charges is determined by the polarity of the Hall emf, as we have described.

The Hall effect can also be used for a direct measurement of electron drift speed v_d in metals. As we saw in Chapter 25, these speeds are very small, often of the order of 1 mm/s or less. If we move the entire conductor in the opposite direction to the current with a speed equal to the drift speed, then the electrons are at rest with respect to the magnetic field, and the Hall emf disappears. Thus the conductor speed needed to make the Hall emf vanish is equal to the drift speed.

Example 27.13 Using the Hall effect

You place a slab of copper, 2.0 mm thick and 1.50 cm wide, in a uniform magnetic field with magnitude 0.40 T, as shown in Fig. 27.39a. When you run a 75-A current in the $+x$-direction, you find by careful measurement that the potential at the bottom of the slab is 0.81 μV higher than at the top. From this measurement, determine the concentration of mobile electrons in copper.

SOLUTION

IDENTIFY and SET UP: We use Eq. (27.30) to determine the mobile electron concentration n.

EXECUTE: First we find the current density J_x and the electric field E_z:

$$J_x = \frac{I}{A} = \frac{75\text{ A}}{(2.0 \times 10^{-3}\text{ m})(1.50 \times 10^{-2}\text{ m})} = 2.5 \times 10^6\text{ A/m}^2$$

$$E_z = \frac{V}{d} = \frac{0.81 \times 10^{-6}\text{ V}}{1.5 \times 10^{-2}\text{ m}} = 5.4 \times 10^{-5}\text{ V/m}$$

Then, from Eq. (27.30),

$$n = \frac{-J_xB_y}{qE_z} = \frac{-(2.5 \times 10^6\text{ A/m}^2)(0.40\text{ T})}{(-1.60 \times 10^{-19}\text{ C})(5.4 \times 10^{-5}\text{ V/m})}$$
$$= 11.6 \times 10^{28}\text{ m}^{-3}$$

EVALUATE: The actual value of n for copper is $8.5 \times 10^{28}\ \mathrm{m}^{-3}$, which shows that the simple model of the Hall effect in this section, ignoring quantum effects and electron interactions with the ions, must be used with caution. This example also shows that with good conductors, the Hall emf is very small even with large current densities. Hall-effect devices for magnetic-field measurements and other purposes use semiconductor materials, for which moderate current densities give much larger Hall emfs.

Test Your Understanding

You are given a piece of metal with a known concentration of mobile electrons. Explain how you could measure the strength of a magnetic field using only this piece of metal, a source of emf, a voltmeter, and an ammeter.

SUMMARY

Magnetic interactions are fundamentally interactions between moving charged particles. These interactions are described by the vector magnetic field, denoted by $\vec{B}$. A particle with charge q moving with velocity $\vec{v}$ in a magnetic field $\vec{B}$ experiences a force $\vec{F}$ that is perpendicular to both $\vec{v}$ and $\vec{B}$. The SI unit of magnetic field is the tesla $(1\ \mathrm{T} = 1\ \mathrm{N/A \cdot m})$. (See Example 28.1)

$$\vec{F} = q\vec{v} \times \vec{B} \qquad (27.2)$$

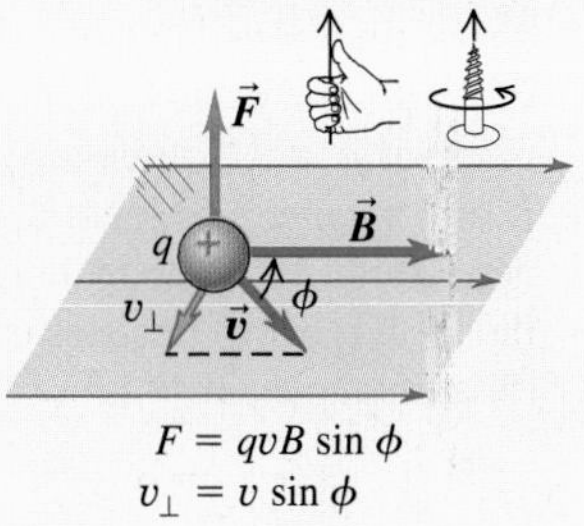

A magnetic field can be represented graphically by magnetic field lines. At each point a magnetic field line is tangent to the direction of $\vec{B}$ at that point. Where field lines are close together the field magnitude is large, and vice versa.

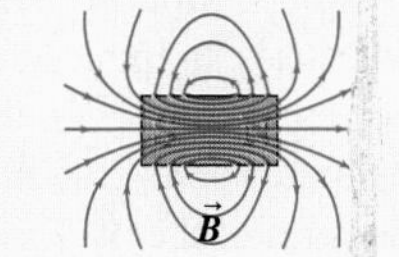

Magnetic flux Φ_B through an area is defined in an analogous way to electric flux. The SI unit of magnetic flux is the weber $(1\ \mathrm{Wb} = 1\ \mathrm{T \cdot m^2})$. The net magnetic flux through any closed surface is zero (Gauss's law for magnetism). As a result, magnetic field lines always close on themselves. (See Example 27.2)

$$\Phi_B = \int B_\perp \, dA = \int B \cos \phi \, dA = \int \vec{B} \cdot d\vec{A} \qquad (27.6)$$

$$\oint \vec{B} \cdot d\vec{A} = 0 \quad \text{(closed surface)} \qquad (27.8)$$

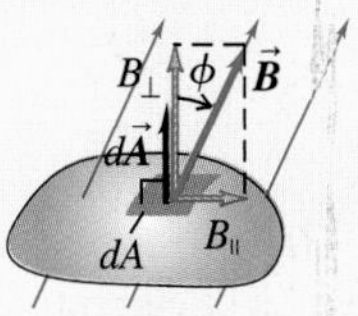

The magnetic force is always perpendicular to $\vec{v}$; a particle moving under the action of a magnetic field alone moves with constant speed. In a uniform field, a particle with initial velocity perpendicular to the field moves in a circle with radius R that depends on the magnetic field strength B and the particle mass m, speed v, and charge q. (See Examples 27.3 and 27.4)

$$R = \frac{mv}{|q|B} \qquad (27.11)$$

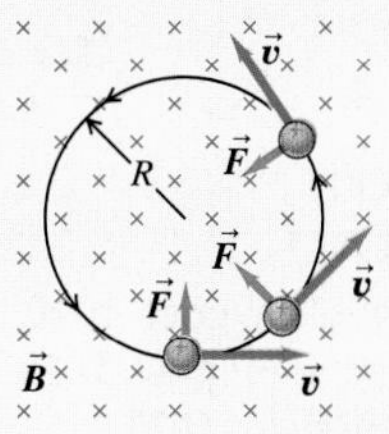

Crossed electric and magnetic fields can be used as a velocity selector. The electric and magnetic forces exactly cancel when $v = E/B$. (See Examples 27.5 and 27.6)

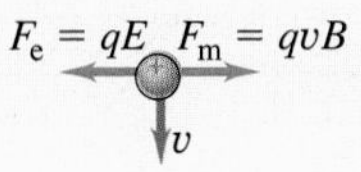

A straight segment of a conductor carrying current I in a magnetic field $\vec{B}$ experiences a force $\vec{F}$ that is perpendicular to both $\vec{B}$ and the vector $\vec{l}$, which points in the direction of the current and has magnitude equal to the length of the segment. A similar relationship gives the force $d\vec{F}$ on an infinitesimal current-carrying segment $d\vec{l}$. (See Examples 27.7 and 27.8)

$$\vec{F} = I\vec{l} \times \vec{B} \qquad (27.19)$$

$$d\vec{F} = I\, d\vec{l} \times \vec{B} \qquad (27.20)$$

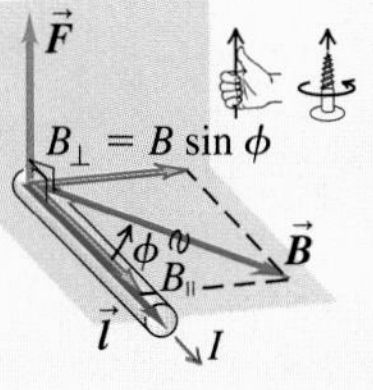

A current loop with area A and current I in a uniform magnetic field $\vec{B}$ experiences no net magnetic force, but does experience a magnetic torque of magnitude τ. The vector torque $\vec{\tau}$ can be expressed in terms of the magnetic moment $\vec{\mu} = I\vec{A}$ of the loop, as can the potential energy U of a magnetic moment in a magnetic field $\vec{B}$. The magnetic moment of a loop depends only on the current and the area; it is independent of the shape of the loop. (See Examples 27.9 through 27.11)

$$\tau = IBA \sin \phi \quad (27.23)$$

$$\vec{\tau} = \vec{\mu} \times \vec{B} \quad (27.26)$$

$$U = -\vec{\mu} \cdot \vec{B} = -\mu B \cos \phi \quad (27.27)$$

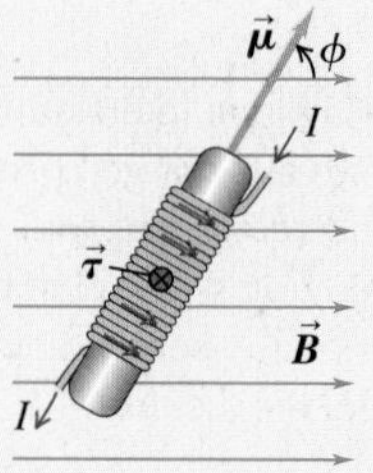

In a dc motor a magnetic field exerts a torque on a current in the rotor. Motion of the rotor through the magnetic field causes an induced emf called a back emf. For a series motor, in which the rotor coil is in parallel with coils that produce the magnetic field, the terminal voltage is the sum of the back emf and the drop Ir across the internal resistance. (See Example 27.12)

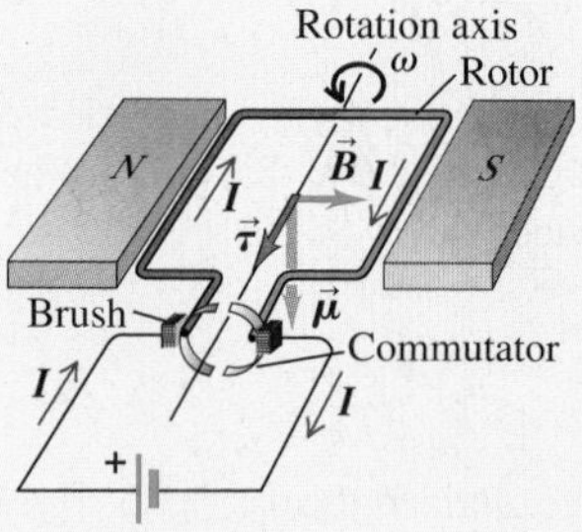

- Brushes are aligned with commutator segments
- Current flows into red-colored side of rotor and out of blue-colored side
- Magnetic torque causes rotor to spin counterclockwise

The Hall effect is a potential difference perpendicular to the direction of current in a conductor, when the conductor is placed in a magnetic field. The Hall potential is determined by the requirement that the associated electric field must just balance the magnetic force on a moving charge. Hall-effect measurements can be used to determine the sign of charge carriers and their concentration n. (See Example 27.13)

$$nq = \frac{-J_x B_y}{E_z} \quad (27.30)$$

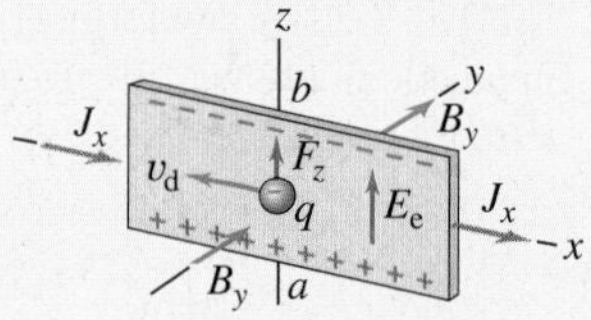

Key Terms

permanent magnet, 1020
magnetic monopole, 1020
magnetic field, 1022
tesla, 1023
gauss, 1023
magnetic field line, 1025
magnetic flux, 1027
weber, 1027
magnetic flux density, 1028
cyclotron frequency, 1030
mass spectrometer, 1034
isotope, 1035
magnetic dipole moment, 1041
magnetic moment, 1041
magnetic dipole, 1041
solenoid, 1042

Your Notes

Answer to Chapter Opening Question

Charged particles are deflected if they move perpendicular to the earth's magnetic field $\vec{B}$, but can easily move either along the direction of $\vec{B}$ or opposite to $\vec{B}$. These "easy" directions naturally lead the particles toward the earth's north and south magnetic poles.

Answers to Test Your Understanding Questions

Section 27.1 Yes. When a magnet is cut apart, each part has a north and south pole (see Fig. 27.4). Hence the small red part behaves much like the original, full-sized compass needle.

Section 27.2 If $\vec{B}$ is in the $+x$-direction, by the right-hand rule $\vec{F}$ is in the positive y-direction. The angle between $\vec{v}$ and $\vec{B}$ is $\phi = 60°$, so the magnitude of the force is $F = qvB \sin\phi = (1.6 \times 10^{-19}\text{ C}) \times (3.0 \times 10^5\text{ m/s})(2.0\text{ T})(\sin 60°) = 8.3 \times 10^{-14}\text{ N}$. If $\vec{B}$ is in the positive y-direction, from the right-hand rule $\vec{F} = q\vec{v} \times \vec{B}$ is in the positive x-direction. The vectors $\vec{v}$ and $\vec{B}$ are mutually perpendicular, so $\phi = 90°$ and the magnitude of the force is $F = qvB \sin 90° = (1.6 \times 10^{-19}\text{ C})(3.0 \times 10^5\text{ m/s})(2.0\text{ T})(1) = 9.6 \times 10^{-14}\,N$.

Section 27.3 The magnitude of $\vec{B}$ would increase as you moved upward, reaching a maximum as you pass through the plane of the loop. As you moved above the plane of the loop, the field magnitude would decrease. You can tell this from the spacing of the field lines: the closer the field lines, the stronger the field. The direction of the field would be directly upward at all points along the path, since the path is along a field line and the direction of $\vec{B}$ at any point is tangent to the field line through that point.

Section 27.4 The radius of the orbit as given by Eq. (27.11) is directly proportional to the speed, so doubling the particle speed causes the radius to double as well. The particle has twice as far to travel to complete one orbit but is traveling at double the speed, so the time for one orbit is unchanged. This result also follows from Eq. (27.12), which states that the angular speed ω is independent of the linear speed v. Hence the time per orbit, $T = 2\pi/\omega$, likewise does not depend on v.

Section 27.5 The velocity selector does not depend on the magnitude or sign of the charge. All it requires is that the particles (in this case, ions) *have* a charge. If the charge q is doubled, it follows from Eq. (27.11) that the radius R of the semicircular path is half as much. Hence the ions will land at a distance $\frac{1}{2}(10.16\text{ cm}) = 5.08\text{ cm}$ from slit S_3.

Section 27.6 There is *no* orientation of the wire that will cause it to be suspended in midair. The force on a current-carrying wire is perpendicular to the direction of the vertical magnetic field, so the magnetic force must be horizontal. As such it cannot oppose the downward force of gravity.

Section 27.7 If you wrap the fingers of your right hand around the coil in the direction of the current, your right thumb points straight upward (perpendicular to the plane of the coil). This is the direction of the magnetic moment $\vec{\mu}$. The magnetic moment points from the south pole to the north pole, so the upper side of the loop is equivalent to a north pole and the lower side is equivalent to a south pole.

***Section 27.8** The rotor will not begin to turn when the switch is closed if the rotor is initially oriented as shown in Fig. 27.37b. In this case there is no current through the rotor and hence no magnetic torque. This situation can be remedied by using multiple rotor coils oriented at different angles around the rotation axis. With this arrangement, there is always a magnetic torque no matter what the orientation.

***Section 27.9** Place the piece of metal so that it is face-on to the magnetic field as in Fig. 27.39a. Attach the source of emf to the piece of metal to make a current flow in the direction indicated by the vectors labeled J_x. Attach the ammeter in series with the source of emf and the piece of metal, and attach the voltmeter across points a and b. The ammeter reading tells you the current I; dividing it by the cross-sectional area of the piece of metal gives you J_x. The voltmeter reading divided by the vertical distance from a to b gives you the electric field E_z. Given the mobile electron concentration n and the electron charge $q = -e$, you can use Eq. (27.30) to solve for the magnetic field B_y.

Discussion Questions

Q27.1 Can a charged particle move through a magnetic field without experiencing any force? If so, how? If not, why not?

Q27.2 At any point in space, the electric field $\vec{E}$ is defined to be in the direction of the electric force on a positively charged particle at that point. Why don't we similarly define the magnetic field $\vec{B}$ to be in the direction of the magnetic force on a moving, positively charged particle?

Q27.3 Section 27.2 describes a procedure for finding the direction of the magnetic force using your right hand. If you use the same procedure, but with your left hand, will you get the correct direction for the force? Explain.

Q27.4 The magnetic force on a moving charged particle is always perpendicular to the magnetic field $\vec{B}$. Is the trajectory of a moving charged particle always perpendicular to the magnetic field lines? Explain your reasoning.

Q27.5 A charged particle is fired into a cubical region of space where there is a uniform magnetic field. Outside this region, there is no magnetic field. Is it possible that the particle will remain inside the cubical region? Why or why not?

Q27.6 If the magnetic force does no work on a charged particle, how can it have any effect on the particle's motion? Are there other examples of forces that do no work but have a significant effect on a particle's motion?

Q27.7 A charged particle moves through a region of space with constant velocity (magnitude and direction). If the external magnetic field is zero in this region, can you conclude that the external electric field in the region is also zero? Explain. (By external we mean fields other than those produced by the charged particle.) If the external electric field is zero in the region, can you conclude that the external magnetic field in the region is also zero?

Q27.8 How might a loop of wire carrying a current be used as a compass? Could such a compass distinguish between north and south? Why or why not?

Q27.9 How could the direction of a magnetic field be determined by making only *qualitative* observations of the magnetic force on a straight wire carrying a current?

Q27.10 A loose, floppy loop of wire is carrying current I. The loop of wire is placed on a horizontal table in a uniform magnetic field $\vec{B}$ perpendicular to the plane of the table. This causes the loop of wire to expand into a circular shape while still lying on the table. In a diagram, show all possible orientations of the current I and magnetic field $\vec{B}$ that could cause this to occur. Explain your reasoning.

Q27.11 Equation (27.19) for the force on a current-carrying wire, $\vec{F} = I\vec{l} \times \vec{B}$, was derived using the assumption that the cross-sectional area of the wire is constant along its length. Is this equation still valid if the cross-sectional area varies along the wire's length? Why or why not?

Q27.12 A student claims that if lightning strikes a metal flagpole, the force exerted by the earth's magnetic field on the current in the pole can be large enough to bend it. Typical lightning currents are of the order of 10^4 to 10^5 A. Is the student's opinion justified? Explain your reasoning.

Q27.13 A student tried to make an electromagnetic compass by suspending a coil or wire from a thread (with the plane of the coil vertical) and passing a current through it. She expected the coil to align itself perpendicular to the horizontal components of the earth's magnetic field; instead, the coil went into what appeared to be angular simple harmonic motion, oscillating back and forth around the expected direction. What was happening? Was the motion truly simple harmonic?

Q27.14 A proton is being accelerated in a cyclotron. If the energy of the proton is doubled, by what factor does the radius of the circular path change? What happens to the orbital period, the time for one revolution of the particle, when its energy is doubled?

Q27.15 An ordinary loudspeaker such as that shown in Fig. 27.26 should not be placed next to a computer monitor or TV screen. Why not?

Q27.16 A permanent magnet can be used to pick up a string of nails, tacks, or paper clips, even though these are not magnets by themselves. How can this be?

Q27.17 If an emf is produced in a dc motor, would it be possible to use the motor somehow as a generator or source, taking power out of it rather than putting power into it? How might this be done?

Q27.18 When the polarity of the voltage applied to a dc motor is reversed, the direction of motion does *not* reverse. Why not? How *could* the direction of motion be reversed?

Q27.19 In a Hall-effect experiment, is it possible that *no* transverse potential difference will be observed? Under what circumstances might this happen?

Q27.20 Hall-effect voltages are much larger for relatively poor conductors (such as germanium) than for good conductors (such as copper), for comparable currents, fields, and dimensions. Why?

Q27.21 Could an accelerator be built in which *all* the forces on the particles, for steering and for increasing speed, are magnetic forces? Why or why not?

Q27.22 The magnetic force acting on a charged particle can never do work, because at every instant the force is perpendicular to the velocity. The torque exerted by a magnetic field can do work on a current loop when the loop rotates. Explain how these seemingly contradictory statements can be reconciled.

Exercises

Section 27.2 Magnetic Field

27.1 A particle with a charge of -1.24×10^{-8} C is moving with instantaneous velocity $\vec{v} = (4.19 \times 10^4 \text{ m/s})\hat{\imath} + (-3.85 \times 10^4 \text{ m/s})\hat{\jmath}$. What is the force exerted on this particle by a magnetic field a) $\vec{B} = (1.40\ T)\hat{\imath}$? b) $\vec{B} = (1.40\ T)\hat{k}$?

27.2 A particle of mass 0.195 g carries a charge of -2.50×10^{-8} C. The particle is given an initial horizontal velocity that is due north and has a magnitude 4.00×10^4 m/s. What are the magnitude and direction of the minimum magnetic field that will keep the particle moving in the earth's gravitational field in the same horizontal, northward direction?

27.3 A particle initially moving south in a vertically downward magnetic field is deflected toward the east. What is the sign of the charge on the particle? Explain your answer with a diagram.

27.4 A particle with mass 1.81×10^{-3} kg and a charge of 1.22×10^{-8} C has, at a given instant, a velocity $\vec{v} = (3.00 \times 10^4 \text{ m/s})\hat{\jmath}$. What are the magnitude and direction of the particle's acceleration produced by a uniform magnetic field $\vec{B} = (1.63 \text{ T})\hat{\imath} + (0.980 \text{ T})\hat{\jmath}$?

27.5 Each of the lettered points at the corners of the cube in Fig. 27.40 represents a positive charge q moving with a velocity of magnitude v in the direction indicated. The region in the figure is in a uniform magnetic field $\vec{B}$, parallel to the x-axis and directed toward the right. Copy the figure, find the magnitude and direction of the force on each charge, and show the force in your diagram.

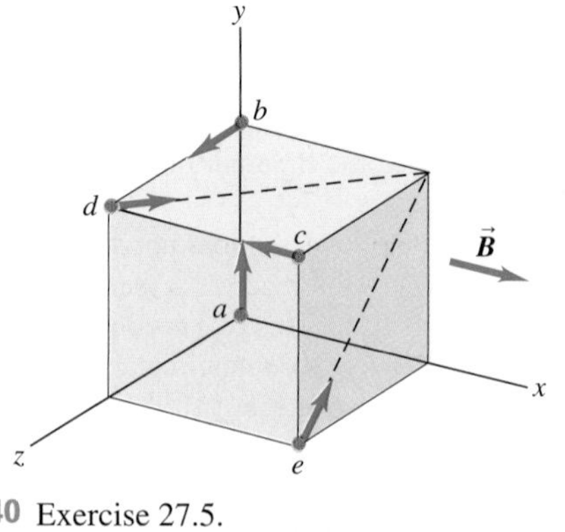

Figure 27.40 Exercise 27.5.

27.6 An electron moves at 2.50×10^6 m/s through a region in which there is a magnetic field of unspecified direction and magnitude 7.40×10^{-2} T. a) What are the largest and smallest possible magnitudes of the acceleration of the electron due to the magnetic field? b) If the actual acceleration of the electron is $\frac{1}{4}$ of the largest magnitude in part (a), what is the angle between the electron velocity and the magnetic field?

27.7 An electron experiences a magnetic force of magnitude 4.60×10^{-15} N when moving at an angle of 60.0° with respect to a magnetic field of magnitude 3.50×10^{-3} T. Find the speed of the electron.

27.8 A particle with charge −5.60 nC is moving in a uniform magnetic field $\vec{B} = -(1.25\ \text{T})\hat{k}$. The magnetic force on the particle is measured to be $\vec{F} = -(3.40 \times 10^{-7}\ \text{N})\hat{\imath} + (7.40 \times 10^{-7}\ \text{N})\hat{\jmath}$. a) Calculate all the components of the velocity of the particle that you can from this information. b) Are there components of the velocity that are not determined by the measurement of the force? Explain. c) Calculate the scalar product $\vec{v} \cdot \vec{F}$. What is the angle between $\vec{v}$ and $\vec{F}$?

27.9 A particle with charge 7.80 μC is moving with velocity $\vec{v} = -(3.80 \times 10^3\ \text{m/s})\hat{\jmath}$. The magnetic force on the particle is measured to be $\vec{F} = +(7.60 \times 10^{-3}\ \text{N})\hat{\imath} - (5.20 \times 10^{-3}\ \text{N})\hat{k}$. a) Calculate all the components of the magnetic field you can from this information. b) Are there components of the magnetic field that are not determined by the measurement of the force? Explain. c) Calculate the scalar product $\vec{B} \cdot \vec{F}$. What is the angle between $\vec{B}$ and $\vec{F}$?

Section 27.3 Magnetic Field Lines and Magnetic Flux

27.10 The magnetic flux through one face of a cube is +0.120 Wb. a) What must the total magnetic flux through the other five faces of the cube be? b) Why did you not need to know the dimensions of the cube in order to answer part (a)? c) Suppose the magnetic flux is due to a permanent magnet like that shown in Fig. 27.11a. In a sketch, show where the cube in part (a) might be located relative to the magnet.

27.11 A circular area with a radius of 6.50 cm lies in the xy-plane. What is the magnitude of the magnetic flux through this circle due to a uniform magnetic field $B = 0.230$ T a) in the $+z$-direction? b) at an angle of 53.1° from the $+z$-direction? c) in the $+y$-direction?

27.12 The magnetic field $\vec{B}$ in a certain region is 0.128 T, and its direction is that of the $+z$-axis in Fig. 27.41. a) What is the magnetic flux across the surface $abcd$ in the figure? b) What is the magnetic flux across the surface $befc$? c) What is the magnetic flux across the surface $aefd$? d) What is the net flux through all five surfaces that enclose the shaded volume?

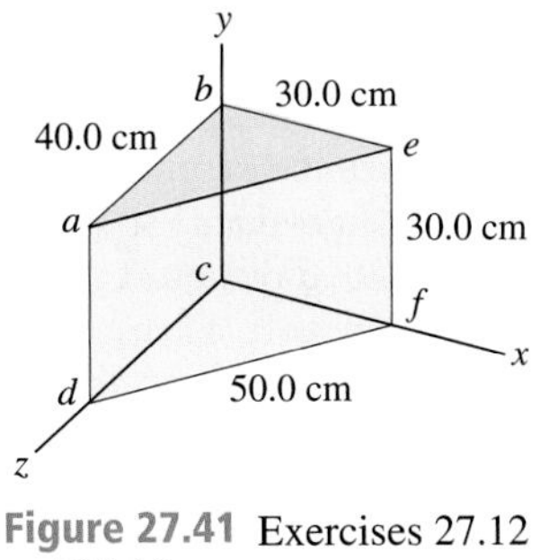

Figure 27.41 Exercises 27.12 and 27.13.

27.13 A physics student claims that she has arranged magnets so that the magnetic field within the shaded volume in Fig. 27.41 is $\vec{B} = (\beta - \gamma y^2)\hat{\jmath}$, where $\beta = 0.300$ T and $\gamma = 2.00\ \text{T/m}^2$. a) Find the net flux of $\vec{B}$ through the five surfaces that enclose the shaded volume in Fig. 27.41. b) Is the student's claim plausible? Why or why not?

Section 27.4 Motion of Charged Particles in a Magnetic Field

27.14 A particle with charge 6.40×10^{-19} C travels in a circular orbit with radius 4.68 mm due to the force exerted on it by a magnetic field with magnitude 1.65 T and perpendicular to the orbit. a) What is the magnitude of the linear momentum $\vec{p}$ of the particle? b) What is the magnitude of the angular momentum $\vec{L}$ of the particle?

27.15 An electron at point A in Fig. 27.42 has a speed v_0 of 1.41×10^6 m/s. Find a) the magnitude and direction of the magnetic field that will cause the electron to follow the semicircular path from A to B; b) the time required for the electron to move from A to B.

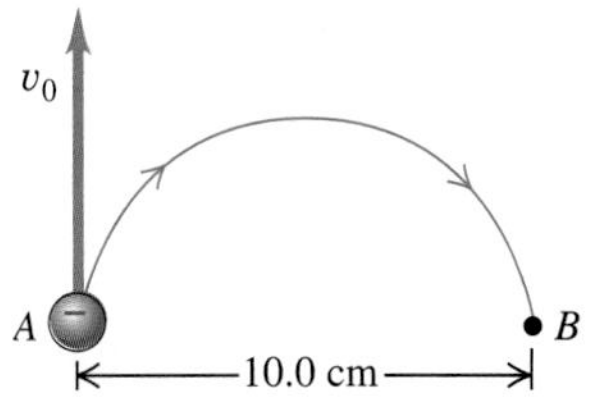

Figure 27.42 Exercises 27.15 and 27.16.

27.16 Repeat Exercise 27.15 for the case in which the particle is a proton rather than an electron.

27.17 Fusion Reactor. If two deuterium nuclei (charge $+e$, mass 3.34×10^{-27} kg) get close enough together, the attraction of the strong nuclear force will fuse them to make an isotope of helium, releasing vast amounts of energy. The range of this force is about 10^{-15} m. This is the principle behind the fusion reactor. The deuterium nuclei are moving much too fast to be contained by physical walls, so they are confined magnetically. a) How fast would two nuclei have to move so that in a head-on collision they would get close enough to fuse? (Treat the nuclei as point charges and assume that a separation of 1.0×10^{-15} is required for fusion.) b) What strength magnetic field is needed to make deuterium nuclei with this speed travel in a circle of diameter 2.50 m?

27.18 a) An ^{16}O nucleus (charge $+8e$) moving horizontally from west to east with a speed of 500 km/s experiences a magnetic force of 0.00320 nN vertically downward. Find the magnitude and direction of the weakest magnetic field required to produce this force. Explain how this same force could be caused by a larger magnetic field. b) An electron moves in a uniform, horizontal 2.10-T magnetic field that is toward the west. What must the magnitude and direction of the minimum velocity of the electron be so that magnetic force on it will be 4.60 pN, vertically upward? Explain how the velocity could be larger than this minimum value and the force still have this same magnitude and direction.

27.19 A 150-g ball containing 4.00×10^8 excess electrons is dropped into a 125-m vertical shaft. At the bottom of the shaft, the ball suddenly enters a uniform horizontal magnetic field that has magnitude 0.250 T and direction from east to west. If air resistance is negligibly small, find the magnitude and direction of the force that this magnetic field exerts on the ball just as it enters the field.

27.20 In an experiment with cosmic rays, a vertical beam of particles that have charge of magnitude $3e$ and mass 12 times the proton mass enters a uniform horizontal magnetic field of 0.250 T and is bent in a semicircle of diameter 95.0 cm, as shown in Fig. 27.43. a) Find the speed of the particles and the sign of their charge. b) Is it reasonable to ignore the gravity force on the particles? c) How does the speed of the particles as they enter the field compare to their speed as they exit the field?

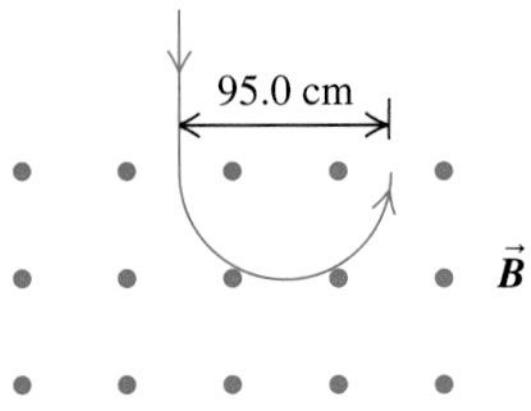

Figure 27.43 Exercise 27.20.

27.21 A deuteron (the nucleus of an isotope of hydrogen) has a mass of 3.34×10^{-27} kg and a charge of $+e$. The deuteron travels in a circular path with a radius of 6.96 mm in a magnetic field with magnitude 2.50 T. a) Find the speed of the deuteron. b) Find the time required for it to make $\frac{1}{2}$ of a revolution. c) Through what potential difference would the deuteron have to be accelerated to acquire this speed?

27.22 Suppose that the electrons in the magnetron of Example 27.3 (Section 27.4) move at 2.80×10^6 m/s in the plane perpendicular to the magnetic field. What are the radii of their orbits?

27.23 A physicist wishes to produce electromagnetic waves of frequency 3.0 THz $(1 \text{ THz} = 1 \text{ terahertz} = 10^{12} \text{ Hz})$ using a magnetron (see Example 27.3 in Section 27.4). a) What magnetic field would be required? Compare this field with the strongest constant magnetic fields yet produced on earth, about 45 T. b) Would there be any advantage to using protons instead of electrons in the magnetron? Why or why not?

27.24 In the situation shown in Fig. 27.16 the charged particle is a proton $(q = 1.60 \times 10^{-19} \text{ C}, m = 1.67 \times 10^{-27} \text{ kg})$ and the uniform magnetic field of magnitude B is directed along the $+x$-axis. The speed of the proton is v, and its velocity $\vec{v}_0$ at $t = 0$ has zero z-component. What must the angle of $\vec{v}_0$ with the xz-plane be for the pitch of the helix to equal its radius (see Example 27.4 in Section 27.4)?

27.25 A proton $(q = 1.60 \times 10^{-19} \text{ C}, m = 1.67 \times 10^{-27} \text{ kg})$ moves in a uniform magnetic field $\vec{B} = (0.500 \text{ T})\hat{\imath}$. At $t = 0$ the proton has velocity components $v_x = 1.50 \times 10^5$ m/s, $v_y = 0$, and $v_z = 2.00 \times 10^5$ m/s (see Example 27.4 in Section 27.4). In addition to the magnetic field there is a uniform electric field in the $+x$-direction, $\vec{E} = (+2.00 \times 10^4 \text{ V/m})\hat{\imath}$. a) Describe the path of the proton. Does the electric field affect the radius of the helix? Explain. b) At $t = T/2$, where T is the period of the circular motion of the proton, what is the x-component of the displacement of the proton from its position at $t = 0$?

27.26 A singly charged ion of ^{7}Li (an isotope of lithium) has a mass of 1.16×10^{-26} kg. It is accelerated through a potential difference of 220 V and then enters a magnetic field with magnitude 0.723 T perpendicular to the path of the ion. What is the radius of the ion's path in the magnetic field?

27.27 An electron in the beam of a TV picture tube is accelerated by a potential difference of 2.00 kV. Then it passes through a region of transverse magnetic field, where it moves in a circular arc with radius 0.180 m. What is the magnitude of the field?

Section 27.5 Applications of Motion of Charged Particles

27.28 a) What is the speed of a beam of electrons when the simultaneous influence of an electric field of 1.56×10^4 V/m and a magnetic field of 4.62×10^{-3} T, with both fields normal to the beam and to each other, produces no deflection of the electrons? b) In a diagram, show the relative orientation of the vectors $\vec{v}$, $\vec{E}$, and $\vec{B}$. c) When the electric field is removed, what is the radius of the electron orbit? What is the period of the orbit?

27.29 Velocity Selector. A velocity selector consists of perpendicular electric and magnetic fields, as shown in Fig. 27.20. a) What strength magnetic field is needed so that only ions of speed 2000 km/s will emerge undeflected, if the electric field is 2.0×10^5 N/C? Will positive ions be separated from negative ions? b) In a certain velocity selector, the electric field is produced by a variable-voltage battery connected across large, parallel, metal plates that are separated by 3.25 cm, and the magnetic field comes from an electromagnet whose field strength can be varied. If the battery voltage can range between 120 V and 560 V and the magnetic field between 0.054 T and 0.180 T, what range of ion velocities can this selector be adjusted to produce?

27.30 Crossed $\vec{E}$ and $\vec{B}$ Fields. A particle with initial velocity $\vec{v}_0 = (5.85 \times 10^3 \text{ m/s})\hat{\jmath}$ enters a region of uniform electric and magnetic fields. The magnetic field in the region is $\vec{B} = -(1.35 \text{ T})\hat{k}$. Calculate the magnitude and direction of the electric field in the region if the particle is to pass through undeflected, for a particle of charge a) $+0.640$ nC; b) -0.320 nC. You can ignore the weight of the particle.

27.31 Determining the Mass of an Isotope. The electric field between the plates of the velocity selector in a Bainbridge mass spectrometer (see Fig. 27.22) is 1.12×10^5 V/m, and the magnetic field in both regions is 0.540 T. A stream of singly charged selenium ions moves in a circular path with a radius of 31.0 cm in the magnetic field. Determine the mass of one selenium ion and the mass number of this selenium isotope. (The mass number is equal to the mass of the isotope in atomic mass units, rounded to the nearest integer. One atomic mass unit $= 1 \text{ u} = 1.66 \times 10^{-27}$ kg.)

27.32 In the Bainbridge mass spectrometer (see Fig. 27.22), the magnetic field magnitude in the velocity selector is 0.650 T, and ions having a speed of 1.82×10^6 m/s pass through undeflected. a) What is the electric field magnitude in the velocity selector? b) If the separation of the plates is 5.20 mm, what is the potential difference between plates P and P'?

Section 27.6 Magnetic Force on a Current-Carrying Conductor

27.33 A horizontal rod with mass m and length L is aligned in a north-south direction. The rod carries a current I in the direction toward the north. When a uniform magnetic field $\vec{B}$ is applied to the entire length of the rod, the magnetic force on the rod keeps it suspended at rest in mid-air. a) Find the minimum magnitude that $\vec{B}$ can have. b) Find the direction of $\vec{B}$ if it has the magnitude found in part (a).

27.34 A straight, vertical wire carries a current of 1.20 A downward in a region between the poles of a large superconducting electromagnet, where the magnetic field has magnitude $B = 0.588$ T and is horizontal. What are the magnitude and direction of the magnetic force on a 1.00-cm section of the wire that is in this uniform magnetic field, if the magnetic field direction is a) east? b) south? c) 30.0° south of west?

27.35 A horizontal rod 0.200 m long is mounted on a balance and carries a current. At the location of the rod a uniform horizontal magnetic field has magnitude 0.067 T and direction perpendicular to the rod. The magnetic force on the rod is measured by the balance and is found to be 0.13 N. What is the current?

27.36 An electromagnet produces a magnetic field of 0.550 T in a cylindrical region of radius 2.50 cm between its poles. A straight wire carrying a current of 10.8 A passes through the center of this region and is perpendicular to both the axis of the cylindrical region and the magnetic field. What magnitude of force is exerted on the wire?

27.37 A wire along the x-axis carries a current of 3.50 A in the negative direction. Calculate the force (expressed in terms of unit vectors) on a 1.00-cm section of the wire exerted by these magnetic fields: a) $\vec{B} = -(0.65\text{ T})\hat{\jmath}$; b) $\vec{B} = +(0.56\text{ T})\hat{k}$; c) $\vec{B} = -(0.31\text{ T})\hat{\imath}$; d) $\vec{B} = +(0.33\text{ T})\hat{\imath} - (0.28\text{ T})\hat{k}$; e) $\vec{B} = +(0.74\text{ T})\hat{\jmath} - (0.36\text{ T})\hat{k}$.

27.38 In Fig. 27.44, a wire carrying current into the plane of the figure is between the north and south poles of two bar magnets. What is the direction of the force exerted by the magnets on the wire?

Figure 27.44 Exercise 27.38.

27.39 A thin, 50.0-cm long metal bar with mass 750 g rests on, but is not attached to, two metallic supports in a uniform 0.450-T magnetic field, as shown in Fig. 27.45. A battery and a 25.0-Ω resistor in series are connected to the supports. a) What is the largest voltage the battery can have without breaking the circuit at the supports? b) The battery voltage has the maximum value calculated in part (a). If the resistor suddenly gets partially short-circuited, decreasing its resistance to 2.0 Ω, find the initial acceleration of the bar.

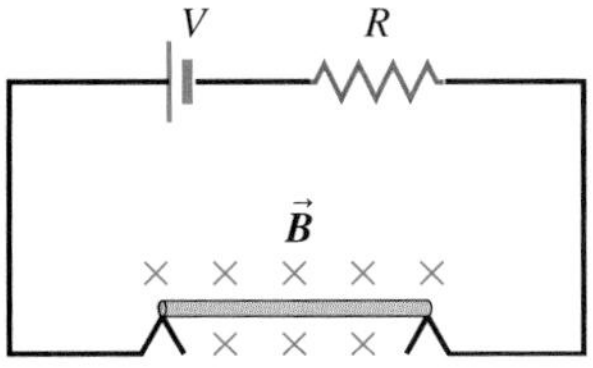

Figure 27.45 Exercise 27.39.

27.40 Magnetic Balance. The circuit shown in Fig. 27.46 is used to make a magnetic balance to weigh objects. The mass m to be measured is hung from the center of the bar, that is in a uniform magnetic field of 1.50 T, directed into the plane of the figure. The battery voltage can be adjusted to vary the current in the circuit. The horizontal bar is 60.0 cm long and is made of extremely light-weight material. It is connected to the battery by thin vertical wires that can support no appreciable tension; all the weight of the suspended mass m is supported by the magnetic force on the bar. A resistor with $R = 5.00\ \Omega$ is in series with the bar; the resistance of the rest of the circuit is much less than this. a) Which point, a or b, should be the positive terminal of the battery? b) If the maximum terminal voltage of the battery is 175 V, what is the greatest mass m that this instrument can measure?

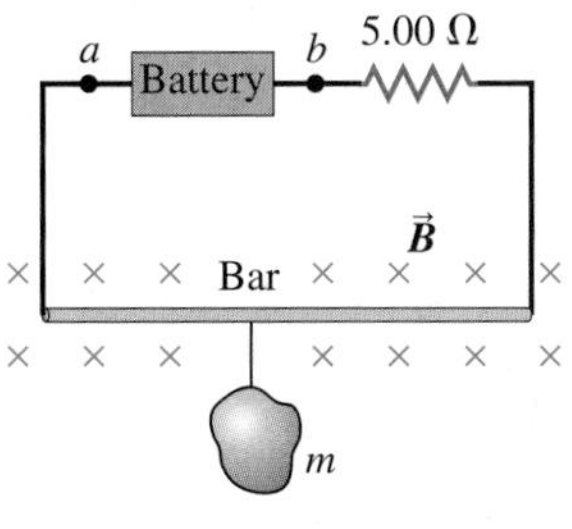

Figure 27.46 Exercise 27.40.

27.41 Consider the conductor and current in Example 27.8. But now let the magnetic field be parallel to the x-axis. a) What are the magnitude and direction of the total magnetic force on the conductor? b) In Example 27.8, the total force is the same as if we replaced the semicircle with a straight segment along the x-axis. Is that still true when the magnetic field is in this different direction? Can you explain why, or why not?

Section 27.7 Force and Torque on a Current Loop

27.42 The plane of a 5.0 cm × 8.0 cm rectangular loop of wire is parallel to a 0.19-T magnetic field. The loop carries a current of 6.2 A. a) What torque acts on the loop? b) What is the magnetic moment of the loop? c) What is the maximum torque that can be obtained with the same total length of wire carrying the same current in this magnetic field?

27.43 A circular coil of wire 8.6 cm in diameter has 15 turns and carries a current of 2.7 A. The coil is in a region where the magnetic field is 0.56 T. a) What orientation of the coil gives the maximum torque on the coil, and what is the maximum torque? b) For what orientation of the coil is the magnitude of the torque 71% of that found in part (a)?

27.44 A rectangular coil of wire, 22.0 cm by 35.0 cm and carrying a current of 1.40 A, is oriented with the plane of its loop perpendicular to a uniform 1.50-T magnetic field, as shown in Fig. 27.47. a) Calculate the net force and torque which the magnetic field exerts on the coil. b) The coil is rotated through a 30.0° angle about the axis shown, the left side coming out of the plane of the figure and the right side going into the plane. Calculate the net force and torque which the magnetic field now exerts on the coil. (*Hint:* In order to help visualize this 3-dimensional problem, make a careful drawing of the coil when viewed along the rotation axis.)

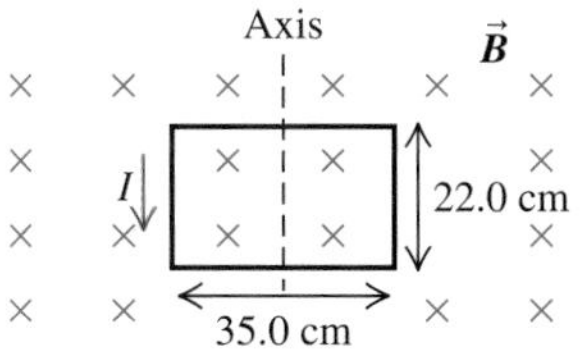

Figure 27.47 Exercise 27.44.

27.45 Magnetic Moment of the Hydrogen Atom. In the Bohr model of the hydrogen atom (Section 39.8), in the lowest energy state the electron orbits the proton at a speed of 2.2×10^6 m/s in a circular orbit of radius 5.3×10^{-11} m. a) What is the orbital period of the electron? b) If the orbiting electron is considered to be a current loop, what is the current I? c) What is the magnetic moment of the atom due to the motion of the electron?

27.46 A circular coil with area A and N turns is free to rotate about a diameter that coincides with the x-axis. Current I is circulating in the coil. There is a uniform magnetic field $\vec{B}$ in the positive y-direction. Calculate the magnitude and direction of the torque $\vec{\tau}$ and the value of the potential energy U, as given in Eq. (27.27), when the coil is oriented as shown in parts (a) through (d) of Fig. 27.48.

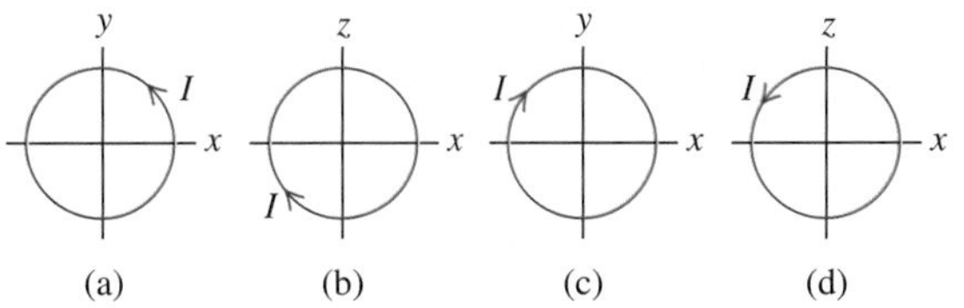

Figure 27.48 Exercise 27.46.

27.47 A coil with magnetic moment $1.45\ \text{A}\cdot\text{m}^2$ is oriented initially with its magnetic moment antiparallel to a uniform 0.835-T magnetic field. What is the change in potential energy of the coil when it is rotated 180° so that its magnetic moment is parallel to the field?

*Section 27.8 The Direct-Current Motor

***27.48** A dc motor with its rotor and field coils connected in series has an internal resistance of 3.2 Ω. When running at full load on a 120-V line, the emf in the rotor is 105 V. a) What is the current drawn by the motor from the line? b) What is the power delivered to the motor? c) What is the mechanical power developed by the motor?

***27.49** In a shunt-wound dc motor with the field coils and rotor connected in parallel (see Fig. 27.49), the resistance R_f of the field coils is 106 Ω, and the resistance R_r of the rotor is 5.9 Ω. When a potential difference of 120 V is applied to the brushes and the motor is running at full speed delivering mechanical power, the current supplied to it is 4.82 A. a) What is the current in the field coils? b) What is the current in the rotor? c) What is the induced emf developed by the motor? d) How much mechanical power is developed by this motor?

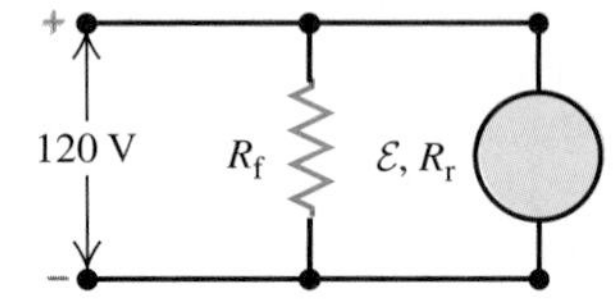

Figure 27.49 Exercises 27.49 and 27.50.

***27.50** A shunt-wound dc motor with the field coils and rotor connected in parallel (see Fig. 27.49) operates from a 120-V dc power line. The resistance of the field windings, R_f, is 218 Ω. The resistance of the rotor, R_r, is 5.9 Ω. When the motor is running, the rotor develops an emf $\mathcal{E}$. The motor draws a current of 4.82 A from the line. Friction losses amount to 45.0 W. Compute a) the field current; b) the rotor current; c) the emf $\mathcal{E}$; d) the rate of development of thermal energy in the field windings; e) the rate of development of thermal energy in the rotor; f) the power input to the motor; g) the efficiency of the motor.

*Section 27.9 The Hall Effect

***27.51** Figure 27.50 shows a portion of a silver ribbon with $z_1 = 11.8$ mm and $y_1 = 0.23$ mm, carrying a current of 120 A in the $+x$-direction. The ribbon lies in a uniform magnetic field, in the y-direction, with magnitude 0.95 T. Apply the simplified model of the Hall effect presented in Section 27.9. If there are 5.85×10^{28} free electrons per cubic meter, find a) the magnitude of the drift velocity of the electrons in the x-direction; b) the magnitude and direction of the electric field in the z-direction due to the Hall effect; c) the Hall emf.

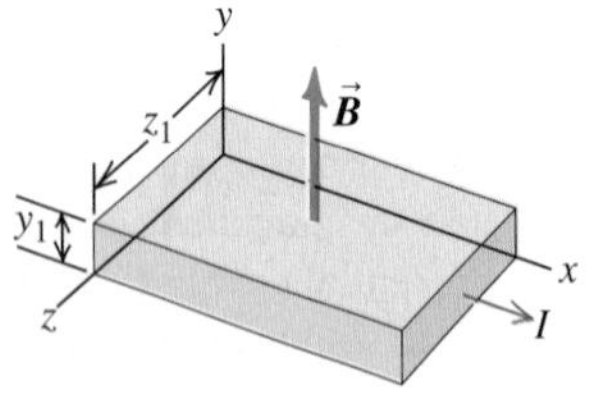

Figure 27.50 Exercises 27.51 and 27.52.

***27.52** Let Fig. 27.50 represent a strip of an unknown metal of the same dimensions as those of the silver ribbon in Exercise 27.51. When the magnetic field is 2.29 T and the current is 78.0 A, the Hall emf is found to be 131 μV. What does the simplified model of the Hall effect presented in Section 27.9 give for the density of free electrons in the unknown metal?

Problems

27.53 When a particle of charge $q > 0$ moves with a velocity of $\vec{v}_1$ that is at 45.0° from the $+x$-axis in the xy-plane, a uniform magnetic field exerts a force $\vec{F}_1$ along the $-z$-axis (see Fig. 27.51). When the same particle moves with a velocity $\vec{v}_2$ with the same magnitude as $\vec{v}_1$ but along the $+z$-axis, a force $\vec{F}_2$ of magnitude F_2 is exerted on it along the $+x$-axis. a) What are the magnitude (in terms of q, v_1, and F_2) and direction of the magnetic field? b) What is the magnitude of $\vec{F}_1$ in terms of F_2?

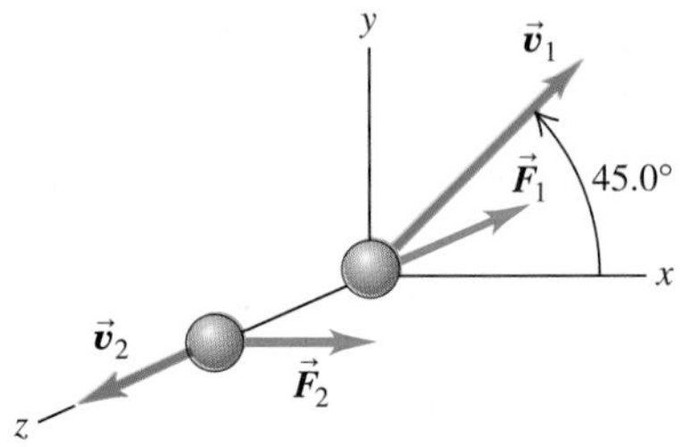

Figure 27.51 Problem 27.53.

27.54 The force on a charged particle moving in a magnetic field can be computed as the vector sum of the forces due to each separate component of the magnetic field. As an example, a particle with charge q is moving with speed v in the $-y$-direction. It is moving in a uniform magnetic field $\vec{B} = B_x\hat{\imath} + B_y\hat{\jmath} + B_z\hat{k}$. a) What are the components of the force $\vec{F}$ exerted on the particle by the magnetic field? b) If $q > 0$, what must the signs of the components of $\vec{B}$ be if the components of $\vec{F}$ are all nonnegative? c) If $q < 0$ and $B_x = B_y = B_z > 0$, find the direction of $\vec{F}$ and find the magnitude of $\vec{F}$ in terms of $|q|$, v, and B_x.

27.55 Magnetic Shooting Gallery. While visiting the Santa Barbara County Fair, you decide to impress your date by winning a stuffed animal at the "magnetic shooting gallery." The object of the game is to hit a target with a lightweight (2.50 g), electrically charged (+0.640 C) particle. The particle is fired horizontally from an air rifle with a muzzle speed of 14.0 m/s, and the center of the target is at the same height as the muzzle of the rifle. What makes the game challenging is that there is a uniform, downward magnetic field of 0.500 T at all points in the shooting gallery. Quickly sizing up the situation, you calculate that if the shooting gallery also had a uniform electric field of a certain strength and direction, you could hit the target by aiming the rifle directly at the target. The gallery operator is so impressed with your correct calculation that he gives you the biggest stuffed animal of all without you having to fire a shot. What was the magnitude and direction of the electric field that you calculated? What effects did you ignore in your calculation? Draw a sketch showing the directions of the particle's motion, the magnetic field, and the electric field.

27.56 In the electron gun of a TV picture tube the electrons (charge $-e$, mass m) are accelerated by a voltage V. After leaving the electron gun, the electron beam travels a distance D to the screen; in this region there is a transverse magnetic field of magnitude B and no electric field. a) Show that the approximate deflection of the beam due to this magnetic field is

$$d = \frac{BD^2}{2}\sqrt{\frac{e}{2mV}}$$

b) Evaluate this expression for $V = 750$ V, $D = 50$ cm, and $B = 5.0 \times 10^{-5}$ T (comparable to the earth's field). Is this deflection significant?

27.57 The magnetic poles of a small cyclotron produce a magnetic field with magnitude 0.85 T. The poles have a radius of 0.40 m, which is the maximum radius of the orbits of the accelerated particles. a) What is the maximum energy to which protons ($q = 1.60 \times 10^{-19}$ C, $m = 1.67 \times 10^{-27}$ kg) can be accelerated by this cyclotron? Give your answer in electron volts and in joules. b) What is the time for one revolution of a proton orbiting at this maximum radius? c) What would the magnetic field magnitude have to be for the maximum energy to which a proton can be accelerated to be twice that calculated in part (a)? d) For $B = 0.85$ T, what is the maximum energy to which alpha particles ($q = 3.20 \times 10^{-19}$ C, $m = 6.65 \times 10^{-27}$ kg) can be accelerated by this cyclotron? How does this compare to the maximum energy for protons?

27.58 A particle of charge $q > 0$ is moving at speed v in the $+z$-direction through a region of uniform magnetic field $\vec{B}$. The magnetic force on the particle is $\vec{F} = F_0(3\hat{\imath} + 4\hat{\jmath})$, where F_0 is a positive constant. a) Determine the components B_x, B_y, and B_z, or at least as many of the three components as is possible from the information given. b) If it is given in addition that the magnetic field has magnitude $6F_0/qv$, determine as much as you can about the remaining components of $\vec{B}$.

27.59 An electron and a nucleus of ^{7}Li (an isotope of lithium that has three protons and four neutrons) both move in a magnetic field in circular paths with the same tangential speed. Compute the ratio of the number of revolutions the electron makes per second to the number made per second by the ^{7}Li nucleus. The mass of the ^{7}Li nucleus is 1.16×10^{-26} kg.

27.60 A cyclotron is to accelerate protons to an energy of 5.4 MeV. The superconducting electromagnet of the cyclotron produces a 3.5-T magnetic field perpendicular to the proton orbits. a) When the protons have achieved a kinetic energy of 2.7 MeV, what is the radius of their circular orbit and what is their angular speed? b) Repeat part (a) when the protons have achieved their final kinetic energy of 5.4 MeV.

27.61 A particle with negative charge q and mass $m = 2.58 \times 10^{-15}$ kg is traveling through a region containing a uniform magnetic field $\vec{B} = -(0.120\text{ T})\hat{k}$. At a particular instant of time the velocity of the particle is $\vec{v} = (1.05 \times 10^6\text{ m/s})(-3\hat{\imath} + 4\hat{\jmath} + 12\hat{k})$ and the force $\vec{F}$ on the particle has a magnitude of 1.25 N. a) Determine the charge q. b) Determine the acceleration $\vec{a}$ of the particle. c) Explain why the path of the particle is a helix, and determine the radius of curvature R of the circular component of the helical path. d) Determine the cyclotron frequency of the particle. e) Although helical motion is not periodic in the full sense of the word, the x- and y-coordinates do vary in a periodic way. If the coordinates of the particle at $t = 0$ are $(x, y, z) = (R, 0, 0)$, determine its coordinates at a time $t = 2T$, where T is the period of the motion in the xy-plane.

27.62 An electron moves in a circular path with radius $r = 4.00$ cm in the space between two concentric cylinders. The inner cylinder is a positively charged wire with radius $a = 1.00$ mm, and the outer cylinder is a negatively charged hollow cylinder with radius $b = 5.00$ cm. The potential difference between the inner and outer cylinders is $V_{ab} = 120$ V, with the wire being at the higher potential. (See Fig. 27.52. Note that the figure is not drawn to scale.) The electric field $\vec{E}$ in the region between the cylinders is radially outward and was shown in Problem 23.57 to have the magnitude $E = V_{ab}/[r \ln(b/a)]$. a) Determine the speed the electron needs to maintain its circular orbit. You can ignore both the gravitational and magnetic fields of the earth. b) Now include the effect of the earth's magnetic field. If the axis of symmetry of the cylinders is positioned parallel to the magnetic field of the earth, at what speed must the electron move to maintain the same circular orbit? Assume that the magnetic field of the earth has magnitude 1.30×10^{-4} T and that its direction is out of the plane of the page in Fig. 27.52. c) Redo the calculation of part (b) for the case in which the magnetic field is in the opposite direction to that of part (b).

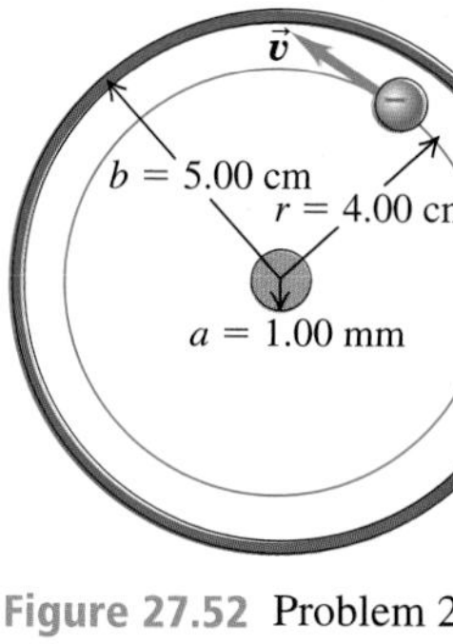

Figure 27.52 Problem 27.62.

27.63 Suppose the electric field between the plates P and P' in Fig. 27.22 is 1.88×10^4 V/m and the magnetic field in both regions is 0.701 T. If the source contains the three isotopes of krypton, ^{82}Kr, ^{84}Kr, and ^{86}Kr, and the ions are singly charged, find the distance between the lines formed by the three isotopes on the photographic plate. Assume the atomic masses of the isotopes (in atomic mass units) are equal to their mass numbers, 82, 84, and 86. (One atomic mass unit = 1 u = 1.66×10^{-27} kg.)

27.64 A particle with charge 9.45×10^{-8} C is moving in a region where there is a uniform magnetic field of 0.450 T in the $+x$-direction. At a particular instant of time the velocity of the particle has components $v_x = -1.68 \times 10^4$ m/s, $v_y = -3.11 \times 10^4$ m/s, and $v_z = 5.85 \times 10^4$ m/s. What are the components of the force on the particle at this time?

27.65 The cube in Fig. 27.53, 75.0 cm on a side, is in a uniform magnetic field of 0.860 T parallel to the x-axis. The wire $abcdef$ carries a current of 6.58 A in the direction indicated. a) Determine the magnitude and direction of the force acting on each of the segments ab, bc, cd, de, and ef. b) What are the magnitude and direction of the total force on the wire?

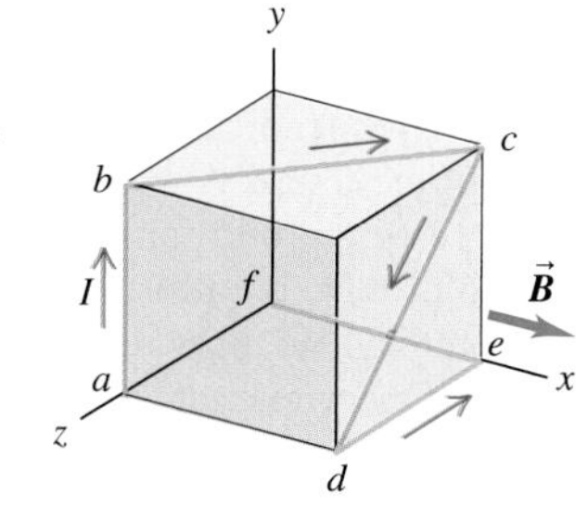

Figure 27.53 Problem 27.65.

27.66 An Electromagnetic Rail Gun. A conducting bar with mass m and length L slides over horizontal rails that are connected to a voltage source. The voltage source maintains a constant current I in the rails and bar, and a constant, uniform, vertical magnetic field $\vec{B}$ fills the region between the rails (see Fig. 27.54). a) Find the magnitude and direction of the net force on the conducting bar. Ignore friction, air resistance, and

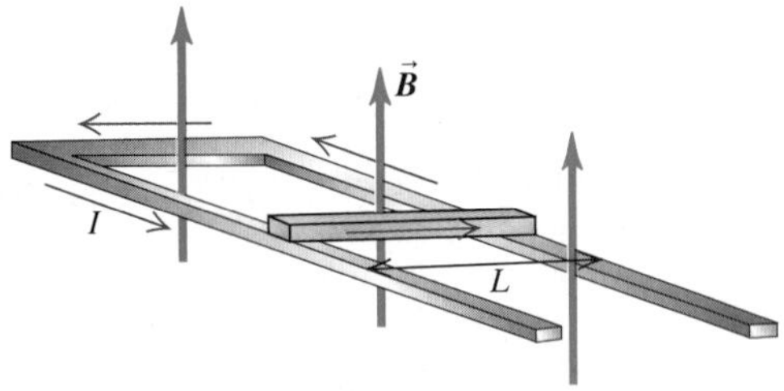

Figure 27.54 Problem 27.66.

electrical resistance. b) If the bar has mass m, find the distance d that the bar must move along the rails from rest to attain speed v. c) It has been suggested that rail guns based on this principle could accelerate payloads into earth orbit or beyond. Find the distance the bar must travel along the rails if it is to reach the escape speed for the earth (11.2 km/s). Let $B = 0.50\ \text{T}$, $I = 2.0 \times 10^3\ \text{A}$, $m = 25\ \text{kg}$, and $L = 50\ \text{cm}$.

27.67 A straight piece of conducting wire with mass M and length L is placed on a frictionless incline tilted at an angle θ from the horizontal (see Fig. 27.55). There is a uniform, vertical magnetic field $\vec{B}$ at all points (produced by an arrangement of magnets not shown in the figure). To keep the wire from sliding down the incline, a voltage source is attached to the ends of the wire. When just the right amount of current flows through the wire, the wire remains at rest. Determine the magnitude and direction of the current in the wire that will cause the wire to remain at rest. Copy the figure and draw the direction of the current on your copy. In addition, show in a free-body diagram all the forces that act on the wire.

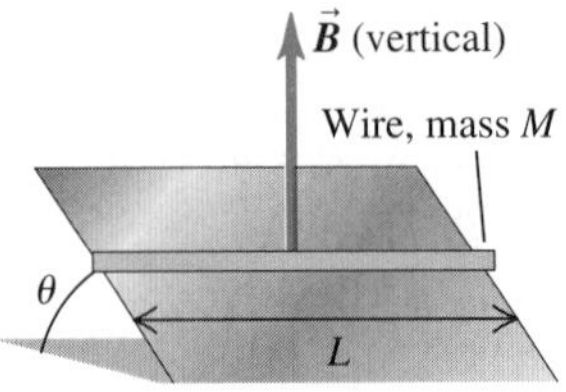

Figure 27.55 Problem 27.67.

27.68 As shown in Fig. 27.56, a thin, flexible wire hangs from point P in a region where there is a uniform, horizontal, magnetic field of magnitude B directed into, and perpendicular to, the plane of the figure. A weight is attached to the bottom of the wire to provide a uniform tension T throughout the wire. (The weight of the wire itself is negligible). When a current I flows from the top to the bottom of the wire, the wire curves into a circular arc of radius R. a) By considering the forces on a small segment of wire that subtends an angle θ, show that the radius of curvature of the wire is $R = T/IB$. (*Hint:* Recall that if θ is small, $\sin\theta \cong \tan\theta \cong \theta$, where θ is in radians.) b) The wire is now removed. A negatively charged particle of charge $-q$ and mass m is launched from the same point P from which the wire was hung, in the same direction as that in which the wire extended from P. Show that the trajectory of the particle will follow the same circular arc as did the wire if the speed of the particle is $v = qT/mI$. (For this reason, weighted wires are utilized in designing magnet systems used for steering beams of charged particles. The bending of the current-carrying wire indicates how particle trajectories will bend.)

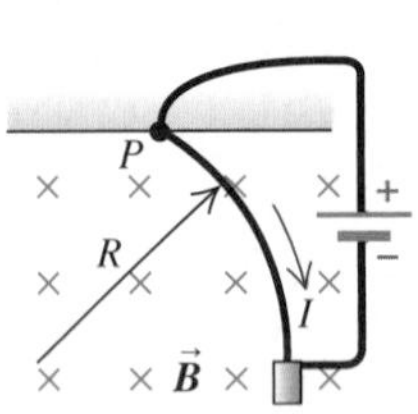

Figure 27.56 Problem 27.68.

27.69 Two positive ions having the same charge q, but different masses m_1 and m_2, are accelerated horizontally from rest through a potential difference V. They then enter a region where there is a uniform magnetic field $\vec{B}$ normal to the plane of the trajectory. a) Show that if the beam entered the magnetic field along the x-axis, the value of the y-coordinate for each ion at any time t is approximately

$$y = Bx^2\left(\frac{q}{8mV}\right)^{1/2}$$

provided y remains much smaller than x. b) Can this arrangement be used for isotope separation? Why or why not?

27.70 Mass Spectrograph. A mass spectrograph is used to measure the masses of ions, or to separate ions of different masses (Section 27.5). In one design for such an instrument, ions with mass m and charge q are accelerated through a potential difference V. They then enter a uniform magnetic field that is perpendicular to their velocity and are deflected in a semicircular path of radius R. A detector measures where the ions complete the semicircle and from this it is easy to calculate R. a) Derive the equation for calculating the mass of the ion from measurements of B, V, R, and q. b) What potential difference V is needed so that singly ionized ^{12}C atoms will have $R = 50.0\ \text{cm}$ in a 0.150-T magnetic field? c) Suppose the beam consists of a mixture of ^{12}C and ^{14}C ions. If V and B have the same values as in part (b), calculate the separation of these two isotopes at the detector. Do you think that this beam separation is sufficient for the two ions to be distinguished? (Make the assumption described in Problem 27.63 for the masses of the ions.)

27.71 A thin, uniform rod with negligible mass and length 0.200 m is attached to the floor by a frictionless hinge at point P (see Fig. 27.57). A horizontal spring with force constant $k = 4.80\ \text{N/m}$ connects the other end of the rod to a vertical wall. The rod is in a uniform magnetic field $B = 0.340\ \text{T}$ directed into the plane of the figure. There is current $I = 6.50\ \text{A}$ in the rod, in the direction shown. a) Calculate the torque due to the magnetic force on the rod, for an axis at P. Is it correct to take the total magnetic force to act at the center of gravity of the rod when calculating the torque? Explain. b) When the rod is in equilibrium and makes an angle of 53.0° with the floor, is the spring stretched or compressed? c) How much energy is stored in the spring when the rod is in equilibrium?

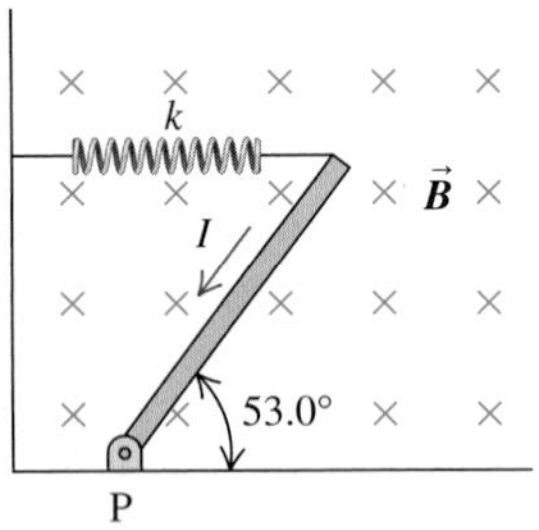

Figure 27.57 Problem 27.71.

27.72 The triangular loop of wire shown in Fig. 27.58 carries a current $I = 5.00\ \text{A}$ in the direction shown. The loop is in a uniform magnetic field that has magnitude $B = 3.00\ \text{T}$ and the same direction as the current in side PQ of the loop. a) Find the force exerted by the magnetic field on each side of the triangle. If the force is not zero, specify its direction. b) What is the net force on the loop?

c) The loop is pivoted about an axis that lies along side PR. Use the forces calculated in part (a) to calculate the torque on each side of the loop (see Problem 27.71). d) What is the magnitude of the net torque on the loop? Calculate the net torque from the torques calculated in part (c) and also from Eq. 27.28. Do these two results agree? e) Is the net torque directed to rotate point Q into the plane of the figure or out of the plane of the figure?

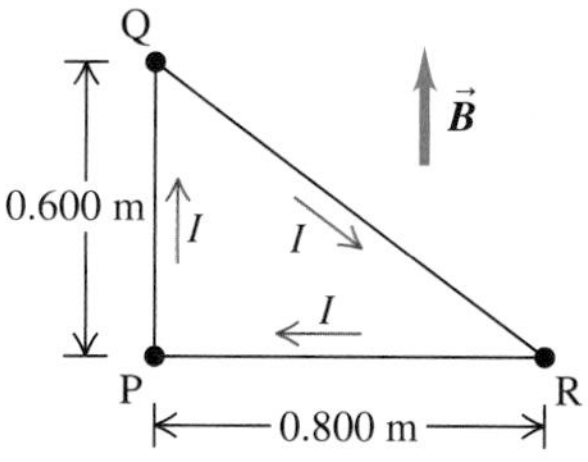

Figure 27.58 Problem 27.72.

27.73 The Haunted Flag. You have landed a job in the movie industry as a special effects engineer. The script of the movie to which you have been assigned calls for a square flag, which ordinarily hangs straight down from a horizontal flagpole, to swing up and hang at an angle of 37.0° from the vertical as though lifted by a ghostly wind. To create this effect, you set up a uniform, vertically upward 0.600-T magnetic field and run a wire around the entire perimeter of the flag. When you connect the wire to a voltage source, a current flows in the wire and the flag "levitates." The flag is 1.00 m on a side, uniform and weighs 16.0 N. Determine how much current must flow in the wire in order for the flag to hang at the desired angle.

27.74 A wire 25.0 cm long lies along the z-axis and carries a current of 9.00 A in the $+z$-direction. The magnetic field is uniform and has components $B_x = -0.242$ T, $B_y = -0.985$ T, and $B_z = -0.336$ T. a) Find the components of the magnetic force on the wire. b) What is the magnitude of the net magnetic force on the wire?

27.75 The rectangular loop of wire shown in Fig. 27.59 has a mass of 0.15 g per centimeter of length and is pivoted about side ab on a frictionless axis. The current in the wire is 8.2 A in the direction shown. Find the magnitude and direction of the magnetic field parallel to the y-axis that will cause the loop to swing up until its plane makes an angle of 30.0° with the yz-plane.

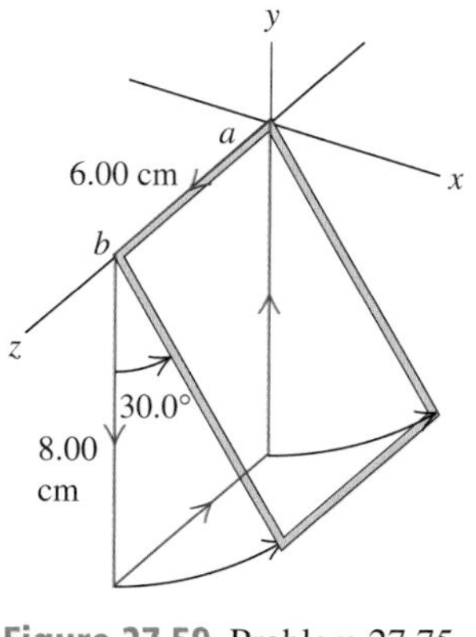

Figure 27.59 Problem 27.75.

27.76 The rectangular loop shown in Fig. 27.60 is pivoted about the y-axis and carries a current of 15.0 A in the direction indicated. a) If the loop is in a uniform magnetic field with magnitude 0.48 T in the $+x$-direction, find the magnitude and direction of the torque required to hold the loop in the position shown. b) Repeat part (a) for the case in which the field is in the $-z$-direction. c) For each of the above magnetic fields, what torque would be required if the loop were pivoted about an axis through its center, parallel to the y-axis?

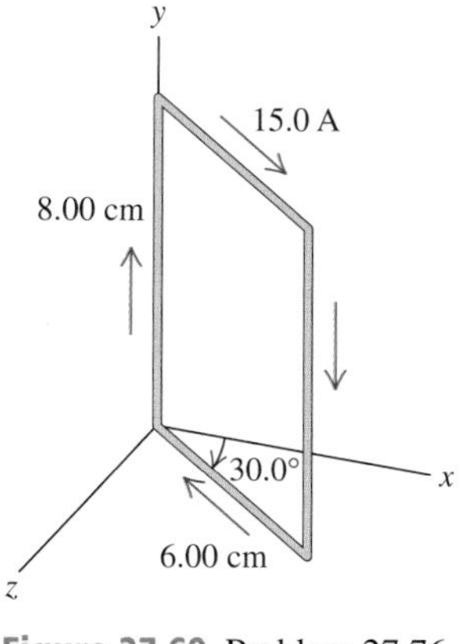

Figure 27.60 Problem 27.76.

27.77 A solenoid with N turns of wire, each of cross-sectional area A, is connected to a source of emf that maintains a constant current I in the wire. The solenoid is initially placed as shown in Fig. 27.32, with the solenoid axis at an angle ϕ to a uniform magnetic field $\vec{B}$. The solenoid is then released and is free to pivot about an axis perpendicular to the plane of Fig. 27.32. The moment of inertia of the solenoid about this axis is I_s. Determine the period of oscillation of the solenoid. Assume that the angle ϕ is much smaller than is shown in Fig. 27.32. (*Hint:* See Section 13.6.)

27.78 A thin wire of length L is made of an insulating material. The wire is bent to form a circular loop, and a positive charge q is distributed uniformly around the circumference of the loop. The loop is then set into rotation with angular speed ω around an axis perpendicular to the plane of the loop and passing through its center. If the loop is in a region where there is a uniform magnetic field $\vec{B}$ directed parallel to the plane of the loop, calculate the magnitude of the magnetic torque on the loop.

27.79 A Voice Coil. It was shown in Section 27.7 that the net force on a current loop in a *uniform* magnetic field is zero. The magnetic force on the voice coil of a loudspeaker (see Fig. 27.26) is nonzero because the magnetic field at the coil is not uniform. A voice coil in a loudspeaker has 50 turns of wire and a diameter of 1.56 cm, and the current in the coil is 0.950 A. Assume that the magnetic field at each point of the coil has a constant magnitude of 0.220 T and is directed at an angle of 60.0° outward from the normal to the plane of the coil (see Fig. 27.61). Let the axis of the coil be in the y-direction. The current in the coil is in the direction shown (counterclockwise as viewed from a point above the coil on the y-axis). Calculate the magnitude and direction of the net magnetic force on the coil.

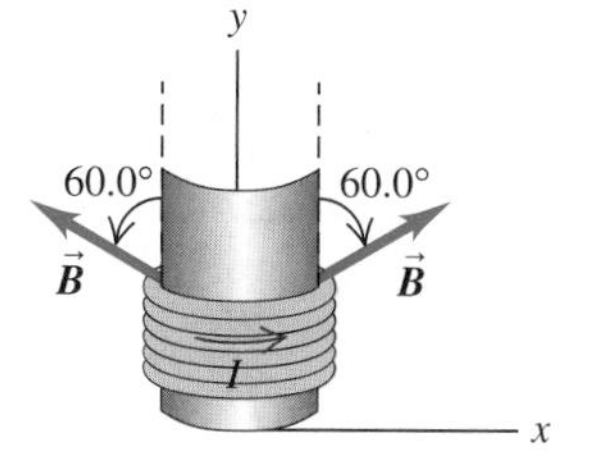

Figure 27.61 Problem 27.79.

27.80 Several external forces $\vec{F}_i$ are applied to a rigid body. With respect to an origin O, force $\vec{F}_1$ is applied at the point located at $\vec{r}_1$, force $\vec{F}_2$ at $\vec{r}_2$, and so on. Point P is located at $\vec{r}_p$. If the body is in translational equilibrium, so that $\sum \vec{F}_i = \mathbf{0}$, prove that the sum of the torques about point P equals the sum of torques about point O. (This justifies a statement made in Section 27.7: Because there is zero net magnetic force on a current loop in a uniform magnetic field, the magnetic torque has the same magnitude and direction for *any* choice of axis.)

27.81 Force on a Current Loop in a Non-Uniform Magnetic Field. It was shown in Section 27.7 that the net force on a current loop in a *uniform* magnetic field is zero. But what if $\vec{B}$ is *not* uniform? Figure 27.62 shows a square loop of wire that lies in the xy-plane. The loop has corners at (0, 0), (0, L), (L, 0), and (L, L), and carries a constant current I in the clockwise direction. The magnetic field has no x-component but has both y- and z-components: $\vec{B}$ =

$(B_0 z/L)\hat{\jmath} + (B_0 y/L)\hat{k}$, where B_0 is a positive constant. a) Sketch the magnetic field lines in the yz-plane. b) Find the magnitude and direction of the magnetic force exerted on each of the sides of the loop by integrating Eq. (27.20). c) Find the magnitude and direction of the net magnetic force on the loop.

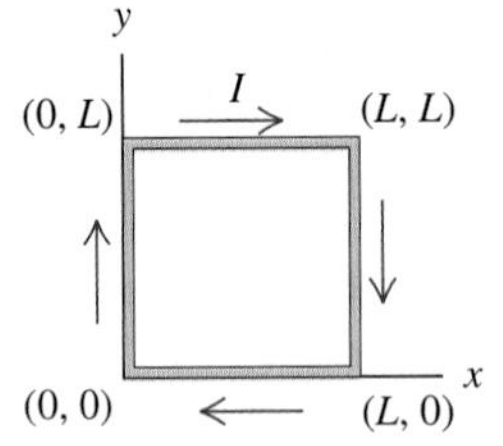

Figure 27.62 Problems 27.81 and 27.82.

27.82 Torque on a Current Loop in a Non-Uniform Magnetic Field. In Section 27.7 the expression for the torque on a current loop was derived assuming that the magnetic field $\vec{B}$ was uniform. But what if $\vec{B}$ is *not* uniform? Figure 27.62 shows a square loop of wire that lies in the xy-plane. The loop has corners at (0, 0), (0, L), (L, 0), and (L, L), and carries a constant current I in the clockwise direction. The magnetic field has no z-component but has both x- and y-components: $\vec{B} = (B_0 y/L)\hat{\imath} + (B_0 x/L)\hat{\jmath}$, where B_0 is a positive constant. a) Sketch the magnetic field lines in the xy-plane. b) Find the magnitude and direction of the magnetic force exerted on each of the sides of the loop by integrating Eq. (27.20). c) If the loop is free to rotate about the x-axis, find the magnitude and direction of the magnetic torque on the loop. d) Repeat part (c) for the case in which the loop is free to rotate about the y-axis. e) Is Eq. (27.26), $\vec{\tau} = \vec{\mu} \times \vec{B}$, an appropriate description of the torque on this loop? Why or why not?

27.83 An insulated wire with mass $m = 5.40 \times 10^{-5}$ kg is bent into the shape of an inverted U such that the horizontal part has a length $l = 15.0$ cm. The bent ends of the wire are partially immersed in two pools of mercury, with 2.5 cm of each end below the mercury's surface. The entire structure is in a region containing a uniform 0.00650-T magnetic field directed into the page (see Fig. 27.63). An electrical connection from the mercury pools is made through the ends of the wires. The mercury pools are connected to a 1.50-V battery and a switch S. When switch S is closed, the wire jumps 35.0 cm into the air, measured from its initial position. a) Determine the speed v of the wire as it leaves the mercury. b) Assuming that the current I through the wire was constant from the time the switch was closed until the wire left the mercury, determine I. c) Ignoring the resistance of the mercury and the circuit wires, determine the resistance of the moving wire.

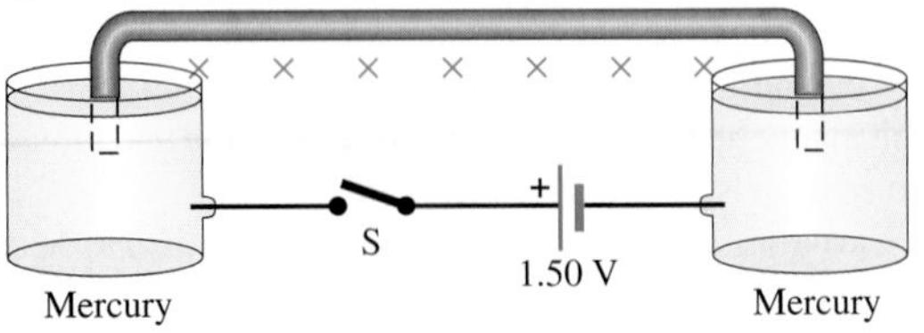

Figure 27.63 Problem 27.83.

27.84 Quark Model of the Neutron. The neutron is a particle with zero charge. Nonetheless, it has a nonzero magnetic moment with z-component $9.66 \times 10^{-27}\ \text{A}\cdot\text{m}^2$. This can be explained by the internal structure of the neutron. A substantial body of evidence indicates that a neutron is composed of three fundamental particles called *quarks*: an "up" (u) quark, of charge $+2e/3$, and two "down" (d) quarks, each of charge $-e/3$. The combination of the three quarks produces a net charge of $2e/3 - e/3 - e/3 = 0$. If the quarks are in motion, they can produce a nonzero magnetic moment. As a very simple model, suppose the u quark moves in a counterclockwise circular path and the d quarks move in a clockwise, circular path, all of radius r and all with the same speed v (see Fig. 27.64). a) Determine the current due to the circulation of the u quark. b) Determine the magnitude of the magnetic moment due to the circulating u quark. c) Determine the magnitude of the magnetic moment of the three-quark system. (Be careful to use the correct magnetic moment directions.) d) With what speed v must the quarks move if this model is to reproduce the magnetic moment of the neutron? Use $r = 1.20 \times 10^{-15}$ m (the radius of the neutron) for the radius of the orbits.

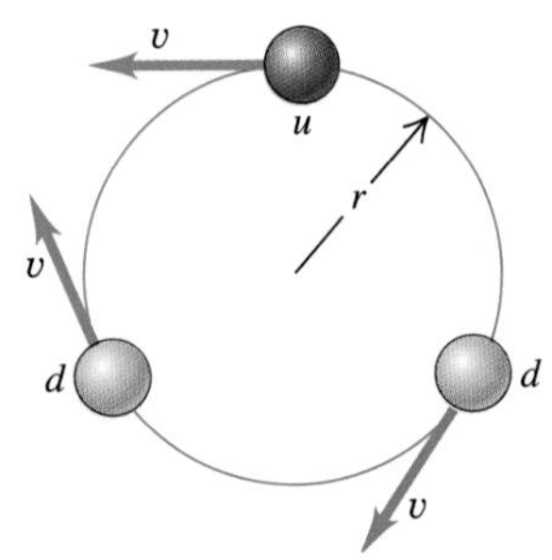

Figure 27.64 Problem 27.84.

27.85 A circular loop of wire with area A lies in the xy-plane. As viewed along the z-axis looking in the $-z$-direction toward the origin, a current I is circulating clockwise around the loop. The torque produced by an external magnetic field $\vec{B}$ is given by $\vec{\tau} = D(4\hat{\imath} - 3\hat{\jmath})$, where D is a positive constant, and for this orientation of the loop the magnetic potential energy $U = -\vec{\mu}\cdot\vec{B}$ is negative. The magnitude of the magnetic field is $B_0 = 13D/IA$. a) Determine the vector magnetic moment of the current loop. b) Determine the components B_x, B_y, and B_z of $\vec{B}$.

27.86 Derivation of Eq. (27.26) for a Circular Current Loop. A wire ring lies in the xy-plane with its center at the origin. The ring carries a counterclockwise current I (see Fig. 27.65). A uniform magnetic field $\vec{B}$ is in the $+x$-direction, $\vec{B} = B_x\hat{\imath}$. (The result is easily extended to $\vec{B}$ in an arbitrary direction.) a) In Fig. 27.65, show that the element $d\vec{l} = R\,d\theta(-\sin\theta\hat{\imath} + \cos\theta\hat{\jmath})$, and find $d\vec{F} = I\,d\vec{l} \times \vec{B}$. b) Integrate $d\vec{F}$ around the loop to show that the net force is zero. c) From part (a), find $d\vec{\tau} = \vec{r} \times d\vec{F}$, where $\vec{r} = R(\cos\theta\hat{\imath} + \sin\theta\hat{\jmath})$ is the vector from the center of the loop to the element $d\vec{l}$. (Note that $d\vec{l}$ is perpendicular to $\vec{r}$.) d) Integrate $d\vec{\tau}$ over the loop to find the total torque $\vec{\tau}$ on the loop. Show that the result can be written as $\vec{\tau} = \vec{\mu} \times \vec{B}$, where $\mu = IA$. (*Note:* $\int \cos^2 x\,dx = \frac{1}{2}x + \frac{1}{4}\sin 2x$, $\int \sin^2 x\,dx = \frac{1}{2}x - \frac{1}{4}\sin 2x$, and $\int \sin x \cos x\,dx = \frac{1}{2}\sin^2 x$.)

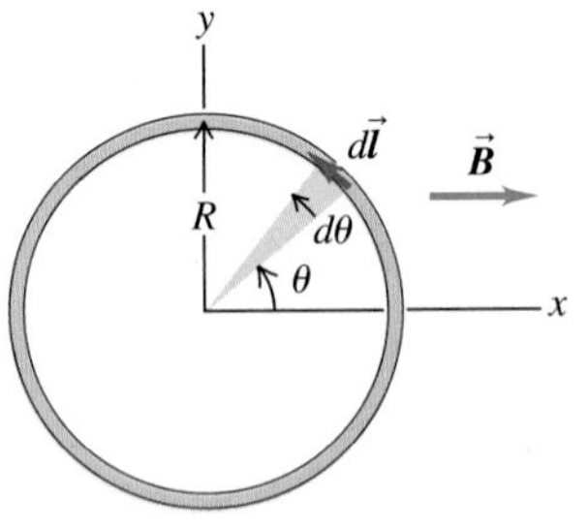

Figure 27.65 Problem 27.86.

27.87 Using Gauss's Law for Magnetism. In a certain region of space, the magnetic field $\vec{B}$ is not uniform. The magnetic field has both a z-component and a component that points radially away from or toward the z-axis. The z-component is given by $B_z(z) = \beta z$, where β is a positive constant. The radial component B_r depends only on r, the radial distance from the z-axis. a) Use Gauss's law for magnetism, Eq. (27.8), to find the radial component B_r as a function of r. (*Hint:* Try a cylindrical Gaussian surface of radius r concentric with the z-axis, with one end at $z = 0$ and the other at $z = L$.) b) Sketch the magnetic field lines.

27.88 A circular ring with area 4.45 cm^2 is carrying a current of 12.5 A. The ring is free to rotate about a diameter. The ring, initially at rest, is immersed in a region of uniform magnetic field given by $\vec{B} = (1.15 \times 10^{-2}\,\text{T})(12\hat{\imath} + 3\hat{\jmath} - 4\hat{k})$. The ring is positioned initially such that its magnetic moment is given by $\vec{\mu}_i = \mu(-0.800\hat{\imath} + 0.600\hat{\jmath})$, where μ is the (positive) magnitude of the magnetic moment. The ring is released and turns through an angle of 90.0°, at which point its magnetic moment is given by $\vec{\mu}_f = -\mu\hat{k}$. a) Determine the decrease in potential energy. b) If the moment of inertia of the ring about a diameter is $8.50 \times 10^{-7}\,\text{kg}\cdot\text{m}^2$, determine the angular speed of the ring as it passes through the second position.

Challenge Problems

27.89 A particle with charge 2.15 μC and mass 3.20×10^{-11} kg is initially traveling in the $+y$-direction with a speed $v_0 = 1.45 \times 10^5$ m/s. It then enters a region containing a uniform magnetic field that is directed into, and perpendicular to, the page in Fig. 27.66. The magnitude of the field is 0.420 T. The region extends a distance of 25.0 cm along the initial direction of travel; 75.0 cm from the point of entry into the magnetic field region is a wall. The length of the field-free region is thus 50.0 cm. When the charged particle enters the magnetic field, it follows a curved path whose radius of curvature is R. It then leaves the magnetic field after a time t_1, having been deflected a distance Δx_1. The particle then travels in the field-free region and strikes the wall after undergoing a total deflection Δx. a) Determine the radius R of the curved part of the path. b) Determine t_1, the time the particle spends in the magnetic field. c) Determine Δx_1, the horizontal deflection at the point of exit from the field. d) Determine Δx, the total horizontal deflection.

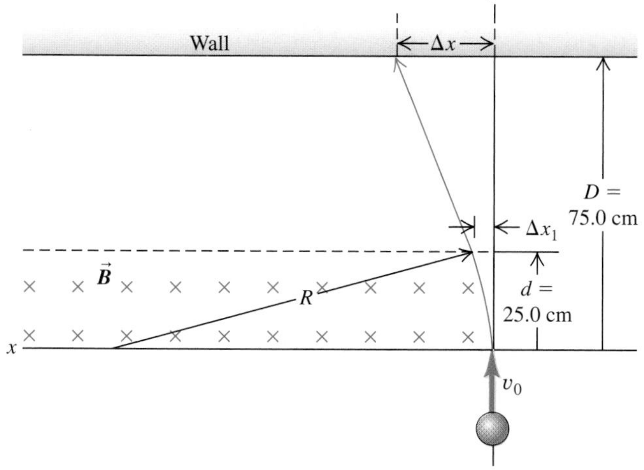

Figure 27.66 Challenge Problem 27.89.

27.90 The Electromagnetic Pump. Magnetic forces acting on conducting fluids provide a convenient means of pumping these fluids. For example, this method can be used to pump blood without the damage to the cells that can be caused by a mechanical pump. A horizontal tube with rectangular cross section (height h, width w) is placed at right angles to a uniform magnetic field with magnitude B so that a length l is in the field (see Fig. 27.67). The tube is filled with a conducting liquid and an electric current of density J is maintained in the third mutually perpendicular direction. a) Show that the difference of pressure between a point in the liquid on a vertical plane through ab and a point in the liquid on another vertical plane through cd, under conditions in which the liquid is prevented from flowing, is $\Delta p = JlB$. b) What current density is needed to provide a pressure difference of 1.00 atm between these two points if $B = 2.20$ T and $l = 35.0$ mm?

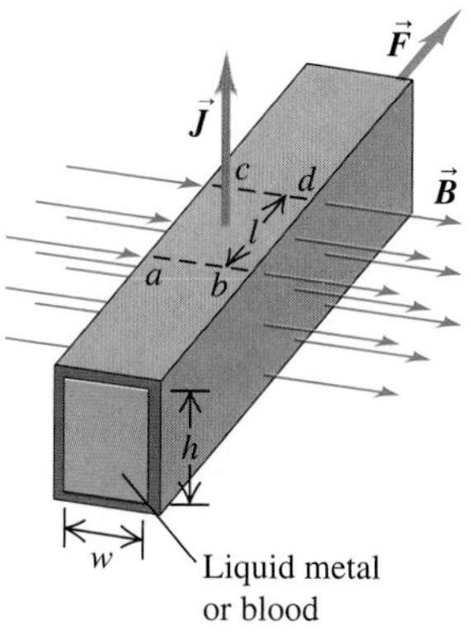

Figure 27.67 Challenge Problem 27.90.

27.91 A Cycloidal Path. A particle with mass m and positive charge q starts from rest at the origin shown in Fig. 27.68. There is a uniform electric field $\vec{E}$ in the $+y$-direction and a uniform magnetic field $\vec{B}$ directed out of the page. It is shown in more advanced books that the path is a *cycloid* whose radius of curvature at the top points is twice the y-coordinate at that level. a) Explain why the path has this general shape and why it is repetitive. b) Prove that the speed at any point is equal to $\sqrt{2qEy/m}$. (*Hint:* Use energy conservation.) c) Applying Newton's second law at the top point and taking as given that the radius of curvature here equals $2y$, prove that the speed at this point is $2E/B$.

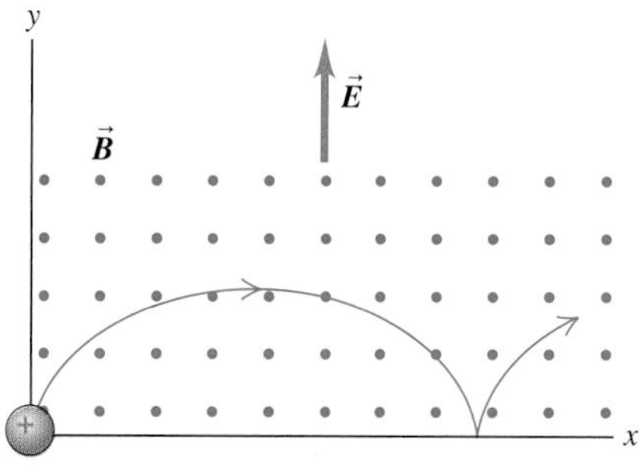

Figure 27.68 Challenge Problem 27.91.

28 SOURCES OF MAGNETIC FIELD

The medical diagnostic technique called magnetic resonance imaging (MRI) requires that the patient lie in a strong magnetic field. The first MRI devices used a solenoid to produce the field. Newer devices use two large coils, placed either above and below the patient or at the patient's head and feet. This "open" design provides a much less claustrophobic experience for the patient.

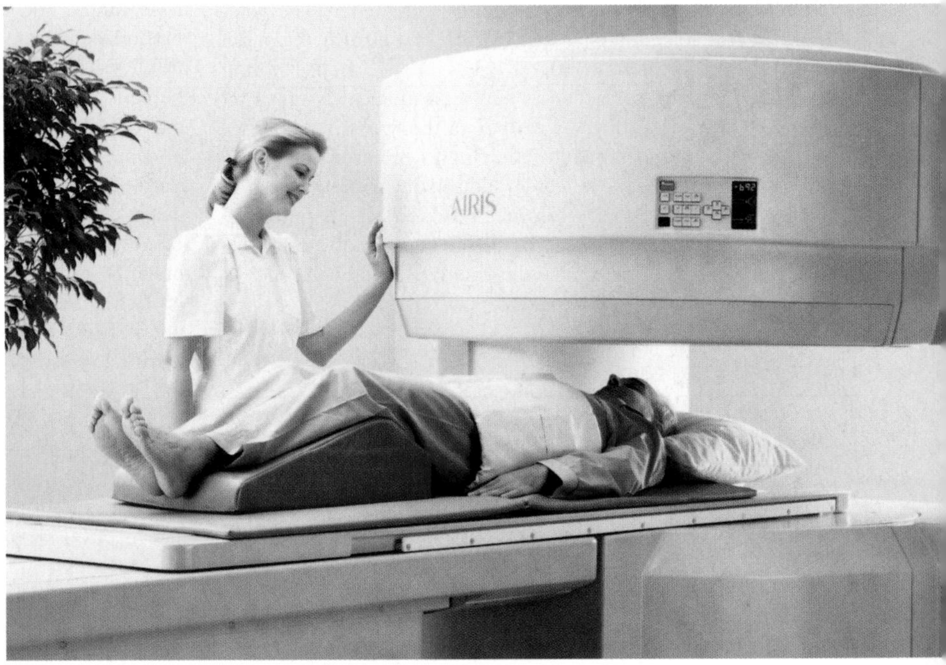

? **If the diameter of one of the MRI coils is increased without changing the current, does the magnetic field that it produces at its center increase, decrease, or stay the same?**

In Chapter 27 we studied the forces exerted on moving charges and on current-carrying conductors in a magnetic field. We didn't worry about how the magnetic field got there; we simply took its existence as a given fact. But how are magnetic fields *created?* We know that both permanent magnets and electric currents in electromagnets create magnetic fields. In this chapter we will study these sources of magnetic field in detail.

We've learned that a charge creates an electric field and that an electric field exerts a force on a charge. But a *magnetic* field exerts a force only on a *moving* charge. Is it also true that a charge *creates* a magnetic field only when the charge is moving? In a word, yes.

Our analysis will begin with the magnetic field created by a single moving point charge. We can use this analysis to determine the field created by a small segment of a current-carrying conductor. Once we can do that, we can in principle find the magnetic field produced by *any* shape of conductor.

Then we will introduce Ampere's law, which plays a role in magnetism analogous to that of Gauss's law in electrostatics. Ampere's law lets us exploit symmetry properties in relating magnetic fields to their sources.

Moving charged particles within atoms respond to magnetic fields and can also act as sources of magnetic field. We'll use these ideas to understand how certain magnetic materials can be used to intensify magnetic fields as well as why some materials such as iron act as permanent magnets.

28.1 | Magnetic Field of a Moving Charge

Let's start with the basics, the magnetic field of a single point charge q moving with a constant velocity $\vec{v}$. In practical applications, such as the coils shown in the photo that opens this chapter, magnetic fields are produced by tremendous numbers of charged particles moving together in a current. But once we understand how to calculate the magnetic field due to a single point charge, it's a small leap to calculate the field due to a current-carrying wire or collection of wires.

As we did for electric fields, we call the location of the moving charge at a given instant the **source point** and the point P where we want to find the field the **field point.** In Section 21.4 we found that at a field point a distance r from a point charge q, the magnitude of the *electric* field $\vec{E}$ caused by the charge is proportional to the charge magnitude $|q|$ and to $1/r^2$, and the direction of $\vec{E}$ (for positive q) is along the line from source point to field point. The corresponding relationship for the *magnetic* field $\vec{B}$ of a point charge q moving with constant velocity has some similarities and some interesting differences.

Experiments show that the magnitude of $\vec{B}$ is also proportional to $|q|$ and to $1/r^2$. But the *direction* of $\vec{B}$ is *not* along the line from source point to field point. Instead, $\vec{B}$ is perpendicular to the plane containing this line and the particle's velocity vector $\vec{v}$, as shown in Fig. 28.1. Furthermore, the field *magnitude* B is also proportional to the particle's speed v and to the sine of the angle ϕ. Thus the magnetic field magnitude at point P is given by

$$B = \frac{\mu_0}{4\pi}\frac{|q|v \sin\phi}{r^2} \tag{28.1}$$

where $\mu_0/4\pi$ is a proportionality constant (μ_0 is pronounced "mu-nought" or "mu-sub-zero"). The reason for writing the constant in this particular way will emerge shortly. We did something similar with Coulomb's law in Section 21.3.

We can incorporate both the magnitude and direction of $\vec{B}$ into a single vector equation using the vector product. To avoid having to say "the direction from the source q to the field point P" over and over, we introduce a *unit* vector $\hat{r}$ ("r-hat") that points from the source point to the field point. (We used $\hat{r}$ for the same

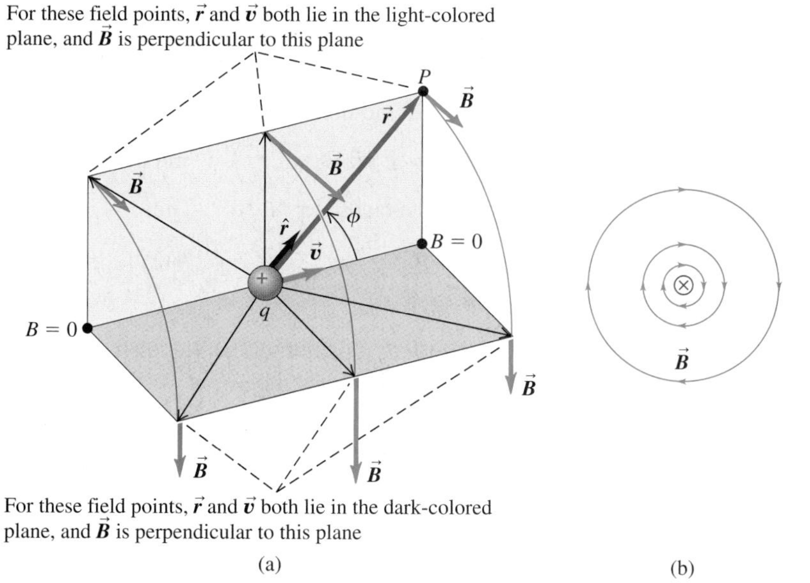

28.1 (a) Magnetic-field vectors due to a moving positive point charge q. At each point, $\vec{B}$ is perpendicular to the plane of $\vec{r}$ and $\vec{v}$, and its magnitude is proportional to the sine of the angle between them. (b) Magnetic field lines in a plane containing a moving positive charge. The $\times$ indicates that the charge is moving into the plane of the page.

purpose in Section 21.4). This unit vector is equal to the vector $\vec{r}$ from the source to the field point divided by its magnitude: $\hat{r} = \vec{r}/r$. Then the $\vec{B}$ field of a moving point charge is

$$\vec{B} = \frac{\mu_0}{4\pi}\frac{q\vec{v} \times \hat{r}}{r^2} \quad (28.2)$$

(magnetic field of a point charge with constant velocity)

Figure 28.1 shows the relation of $\hat{r}$ to P and also shows the magnetic field $\vec{B}$ at several points in the vicinity of the charge. At all points along a line through the charge parallel to the velocity $\vec{v}$, the field is zero because $\sin\phi = 0$ at all such points. At any distance r from q, $\vec{B}$ has its greatest magnitude at points lying in the plane perpendicular to $\vec{v}$ because at all such points, $\phi = 90°$ and $\sin\phi = 1$. If the charge q is negative, the directions of $\vec{B}$ are opposite to those shown in Fig. 28.1.

A point charge in motion also produces an *electric* field, with field lines that radiate outward from a positive charge. The *magnetic* field lines are completely different. The above discussion shows that for a point charge moving with velocity $\vec{v}$, the magnetic field lines are *circles* centered on the line of $\vec{v}$ and lying in planes perpendicular to this line. The field line directions for a positive charge are given by the following *right-hand rule,* one of several that we will encounter in this chapter for determining the direction of the magnetic field caused by different sources. Grasp the velocity vector $\vec{v}$ with your right hand so your right thumb points in the direction of $\vec{v}$; your fingers then curl around the line of $\vec{v}$ in the same sense as the magnetic field lines (assuming q is positive). Figure 28.1a shows parts of a few field lines; Fig. 28.1b shows some field lines in a plane through q, perpendicular to $\vec{v}$, as seen by looking in the direction of $\vec{v}$.

Equations (28.1) and (28.2) describe the $\vec{B}$ field of a point charge moving with *constant* velocity. If the charge *accelerates,* the field can be much more complicated. We won't need these more complicated results for our purposes. (The moving charged particles that make up a current in a wire accelerate at points where the wire bends and the direction of $\vec{v}$ changes. But because the magnitude v_d of the drift velocity in a conductor is typically very small, the acceleration v_d^2/r is also very small, and the effects of acceleration can be ignored.)

As we discussed in Section 27.2, the unit of B is one tesla (1 T):

$$1\ \text{T} = 1\ \text{N}\cdot\text{s/C}\cdot\text{m} = 1\ \text{N/A}\cdot\text{m}$$

Using this with Eq. (28.1) or (28.2), we find that the units of the constant μ_0 are

$$1\ \text{N}\cdot\text{s}^2/\text{C}^2 = 1\ \text{N/A}^2 = 1\ \text{Wb/A}\cdot\text{m} = 1\ \text{T}\cdot\text{m/A}$$

In SI units the numerical value of μ_0 is exactly $4\pi \times 10^{-7}$. Thus

$$\mu_0 = 4\pi \times 10^{-7}\ \text{N}\cdot\text{s}^2/\text{C}^2 = 4\pi \times 10^{-7}\ \text{Wb/A}\cdot\text{m}$$
$$= 4\pi \times 10^{-7}\ \text{T}\cdot\text{m/A} \quad (28.3)$$

It may seem incredible that μ_0 has *exactly* this numerical value! In fact this is a *defined* value that arises from the definition of the ampere, as we'll discuss in Section 28.4.

We mentioned in Section 21.3 that the constant $1/4\pi\epsilon_0$ in Coulomb's law is related to the speed of light c:

$$k = \frac{1}{4\pi\epsilon_0} = (10^{-7}\ \text{N}\cdot\text{s}^2/\text{C}^2)c^2$$

When we study electromagnetic waves in Chapter 32, we will find that their speed of propagation in vacuum, which is equal to the speed of light c, is given by

$$c^2 = \frac{1}{\epsilon_0 \mu_0} \tag{28.4}$$

If we solve the equation $k = 1/4\pi\epsilon_0$ for ϵ_0, substitute the resulting expression into Eq. (28.4), and solve for μ_0, we indeed get the value of μ_0 stated above. This discussion is a little premature, but it may give you a hint that electric and magnetic fields are intimately related to the nature of light.

Example 28.1 Forces between two moving protons

Two protons move parallel to the x-axis in opposite directions (Fig. 28.2) at the same speed v (small compared to the speed of light c.) At the instant shown, find the electric and magnetic forces on the upper proton and determine the ratio of their magnitudes.

SOLUTION

IDENTIFY and SET UP: We find the electric force on the upper proton using Coulomb's law. To find the magnetic force, we first use Eq. (28.2) to determine the magnetic field $\vec{B}$ caused by the lower proton at the location of the upper one. We then calculate the magnetic force on the upper proton using the magnetic force law, Eq. (27.2).

EXECUTE: From Coulomb's law, the magnitude of the electric force on the upper proton is

$$F_E = \frac{1}{4\pi\epsilon_0}\frac{q^2}{r^2}$$

The forces are repulsive, and the force on the upper proton is vertically upward (in the $+y$-direction).

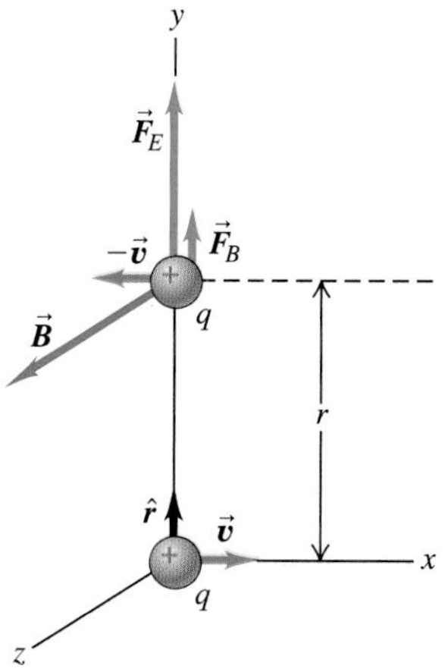

28.2 Electric and magnetic forces between two moving protons.

From the right-hand rule for the cross product $\vec{v} \times \hat{r}$ in Eq. (28.2), the $\vec{B}$-field due to the lower proton at the position of the upper proton is in the $+z$-direction (see Fig. 28.2). From Eq. (28.2), the magnitude of $\vec{B}$ is

$$B = \frac{\mu_0}{4\pi}\frac{qv}{r^2}$$

since $\phi = 90°$. Alternatively, from Eq. (28.2),

$$\vec{B} = \frac{\mu_0}{4\pi}\frac{q(v\hat{\imath}) \times \hat{\jmath}}{r^2} = \frac{\mu_0}{4\pi}\frac{qv}{r^2}\hat{k}$$

The velocity of the upper proton is $-\vec{v}$ and the magnetic force on it is $\vec{F} = q(-\vec{v}) \times \vec{B}$. Combining this with the expressions for $\vec{B}$, we find

$$F_B = \frac{\mu_0}{4\pi}\frac{q^2v^2}{r^2}$$

or

$$\vec{F}_B = q(-\vec{v}) \times \vec{B} = q(-v\hat{\imath}) \times \frac{\mu_0}{4\pi}\frac{qv}{r^2}\hat{k} = \frac{\mu_0}{4\pi}\frac{q^2v^2}{r^2}\hat{\jmath}$$

The magnetic interaction in this situation is also repulsive. The ratio of the magnitudes of the two forces is

$$\frac{F_B}{F_E} = \frac{\mu_0 q^2 v^2/4\pi r^2}{q^2/4\pi\epsilon_0 r^2} = \frac{\mu_0 v^2}{1/\epsilon_0} = \epsilon_0\mu_0 v^2$$

Using the relationship $\epsilon_0\mu_0 = 1/c^2$, Eq. (28.4), we can express our result very simply as

$$\frac{F_B}{F_E} = \frac{v^2}{c^2}$$

When v is small in comparison to c, the speed of light, the magnetic force is much smaller than the electric force.

EVALUATE: Note that it is essential to use the same frame of reference in this entire calculation. We have described the velocities and the fields as they appear to an observer who is stationary in the coordinate system of Fig. 28.2. In a coordinate system that moves with one of the charges, one of the velocities would be zero, so there would be *no* magnetic force. The explanation of this apparent paradox provided one of the paths that led to the special theory of relativity.

Test Your Understanding

If two protons are traveling parallel to each other in the *same* direction and at the same speed, is the magnetic force between them attractive or repulsive? What about the net force between them? (Assume that the protons' speed is far less than the speed of light.)

28.2 | Magnetic Field of a Current Element

Just as for the electric field, there is a **principle of superposition of magnetic fields: The total magnetic field caused by several moving charges is the vector sum of the fields caused by the individual charges.** We can use this principle with the results of Section 28.1 to find the magnetic field produced by a current in a conductor.

We begin by calculating the magnetic field caused by a short segment $d\vec{l}$ of a current-carrying conductor, as shown in Fig. 28.3a. The volume of the segment is $A\,dl$, where A is the cross-sectional area of the conductor. If there are n moving charged particles per unit volume, each of charge q, the total moving charge dQ in the segment is

$$dQ = nqA\,dl$$

The moving charges in this segment are equivalent to a single charge dQ, traveling with a velocity equal to the *drift* velocity $\vec{v}_{\mathrm{d}}$. (Magnetic fields due to the *random* motions of the charges will, on average, cancel out at every point.) From Eq. (28.1) the magnitude of the resulting field $d\vec{B}$ at any field point P is

$$dB = \frac{\mu_0}{4\pi}\frac{|dQ|v_{\mathrm{d}}\sin\phi}{r^2} = \frac{\mu_0}{4\pi}\frac{n|q|v_{\mathrm{d}}A\,dl\sin\phi}{r^2}$$

But from Eq. (25.2), $n|q|v_{\mathrm{d}}A$ equals the current I in the element. So

$$dB = \frac{\mu_0}{4\pi}\frac{I\,dl\sin\phi}{r^2} \tag{28.5}$$

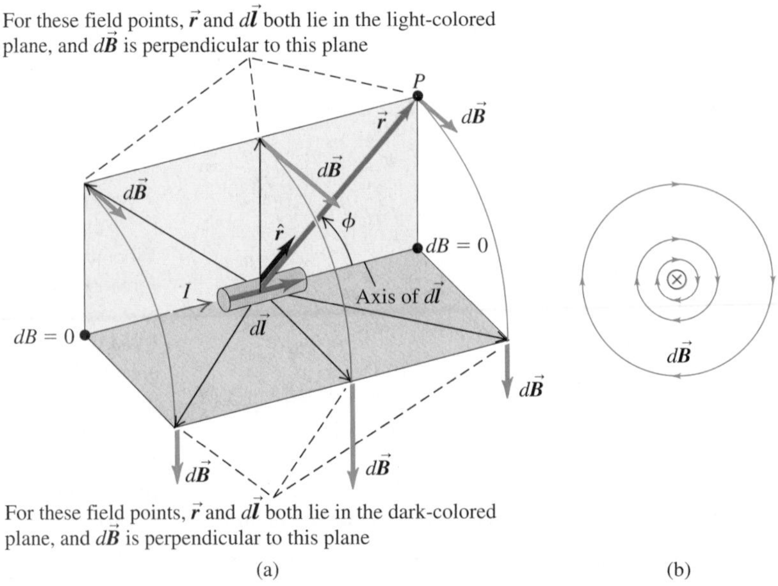

28.3 (a) Magnetic-field vectors due to a current element $d\vec{l}$. (b) Magnetic field lines in a plane containing the current element $d\vec{l}$. The $\times$ indicates that the current is directed into the plane of the page. Compare this figure to Fig. 28.1 for the field of a moving point charge.

In vector form, using the unit vector $\hat{r}$ as in Section 28.1, we have

$$d\vec{B} = \frac{\mu_0}{4\pi}\frac{I\,d\vec{l} \times \hat{r}}{r^2} \quad \text{(magnetic field of a current element)} \quad (28.6)$$

where $d\vec{l}$ is a vector with length dl, in the same direction as the current in the conductor.

Equations (28.5) and (28.6) are called the **law of Biot and Savart** (pronounced "Bee-oh" and "Suh-var"). We can use this law to find the total magnetic field $\vec{B}$ at any point in space due to the current in a complete circuit. To do this, we integrate Eq. (28.6) over all segments $d\vec{l}$ that carry current; symbolically,

$$\vec{B} = \frac{\mu_0}{4\pi}\int\frac{I\,d\vec{l} \times \hat{r}}{r^2} \quad (28.7)$$

In the following sections we will carry out this vector integration for several examples.

As Fig. 28.3a shows, the field vectors $d\vec{B}$ and the magnetic field lines of a current element are exactly like those set up by a positive charge dQ moving in the direction of the drift velocity $\vec{v}_d$. The field lines are circles in planes perpendicular to $d\vec{l}$ and centered on the line of $d\vec{l}$. Their directions are given by the same right-hand rule that we introduced for point charges in Section 28.1.

We can't verify Eq. (28.5) or (28.6) directly because we can never experiment with an isolated segment of a current-carrying circuit. What we measure experimentally is the *total* $\vec{B}$ for a complete circuit. But we can still verify these equations indirectly by calculating $\vec{B}$ for various current configurations using Eq. (28.7) and comparing the results with experimental measurements.

If matter is present in the space around a current-carrying conductor, the field at a field point P in its vicinity will have an additional contribution resulting from the *magnetization* of the material. We'll return to this point in Section 28.8. However, unless the material is iron or some other ferromagnetic material, the additional field is small and is usually negligible. Additional complications arise if time-varying electric or magnetic fields are present or if the material is a superconductor; we'll return to these topics later.

Problem-Solving Strategy: Magnetic Field Calculations

IDENTIFY *the relevant concepts:* The law of Biot and Savart allows you to calculate the magnetic field due to a current-carrying wire of any shape. The idea is to calculate the field due to a representative current element in the wire, then combine the contributions from all such elements to find the total field.

SET UP *the problem using the following steps:*
1. Make a diagram showing a representative current element and the point P at which the field is to be determined (the field point).
2. Draw the current element $d\vec{l}$, being careful to ensure that it points in the direction of the current.
3. Draw the unit vector $\hat{r}$. Note that it is always directed *from* the current element (the source point) to the field point P.
4. Identify the target variables. Usually they will be the magnitude and direction of the magnetic field $\vec{B}$.

EXECUTE *the solution as follows:*
1. Use Eq. (28.5) or (28.6) to express the magnetic field $d\vec{B}$ at P from the representative current element.
2. Add up all the $d\vec{B}$'s to find the total field at point P. In some situations the $d\vec{B}$'s at point P have the same direction for all the current elements; then the magnitude of the total $\vec{B}$ field is the sum of the magnitudes of the $d\vec{B}$'s. But often the $d\vec{B}$'s have different directions for different current elements. Then you have to set up a coordinate system and represent each $d\vec{B}$ in terms of its components.

The integral for the total $\vec{B}$ is then expressed in terms of an integral for each component.

3. Sometimes you can use the symmetry of the situation to prove that one component of $\vec{B}$ must vanish. Always be alert for ways to use symmetry to simplify the problem.
4. Look for ways to use the principle of superposition of magnetic fields. Later in this chapter we'll determine the fields produced by certain simple conductor shapes; if you encounter a conductor of a complex shape that can be represented as a combination of these simple shapes, you can use superposition to find the field of the complex shape. Examples include a rectangular loop and a semicircle with straight line segments on both sides.

EVALUATE *your answer:* Often your answer will be a mathematical expression for $\vec{B}$ as a function of the position of the field point. Check the answer by examining its behavior in as many limits as you can.

Example 28.2 Magnetic field of a current segment

A copper wire carries a steady current of 125 A to an electroplating tank. Find the magnetic field caused by a 1.0-cm segment of this wire at a point 1.2 m away from it, if the point is a) point P_1, straight out to the side of the segment; b) point P_2, on a line at 30° to the segment, as shown in Fig. 28.4.

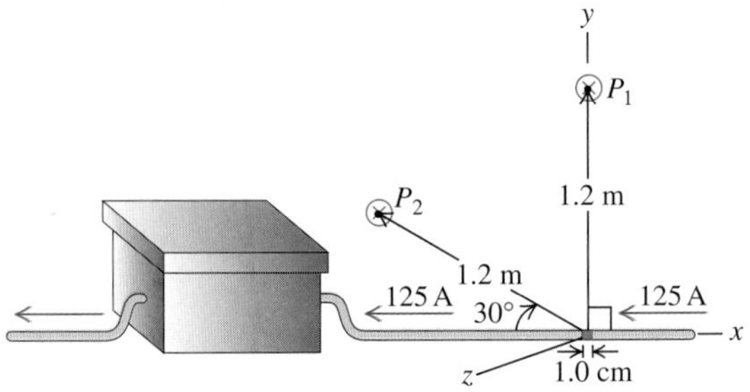

28.4 Finding the magnetic field at two points due to a 1.0-cm segment of current-carrying wire (not shown to scale).

SOLUTION

IDENTIFY: Although Eqs. (28.5) and (28.6) are strictly to be used with infinitesimal current elements only, we may use them here since the segment's 1.0-cm length is much smaller than the 1.2-m distance to the field point.

SET UP: The current element is shown in red in Fig. 28.4 and points in the $-x$-direction (the direction of the current). The unit vector $\hat{r}$ for each field point is directed from the current element toward that point: $\hat{r}$ is in the $+y$-direction for point P_1 and at an angle of 30° above the $-x$-direction for point P_2.

EXECUTE: a) From the right-hand rule, the direction of $\vec{B}$ at P_1 is *into* the xy-plane of Fig. 28.4. Or, using unit vectors, we note that $d\vec{l} = dl(-\hat{\imath})$. At point P_1, $\hat{r} = \hat{\jmath}$, so in Eq. (28.6),

$$d\vec{l} \times \hat{r} = dl(-\hat{\imath}) \times \hat{\jmath} = dl(-\hat{k})$$

The negative z-direction is *into* the plane.

To find the magnitude of $\vec{B}$, we use Eq. (28.5). At point P_1, the angle between $d\vec{l}$ and $\hat{r}$ is 90°, so

$$B = \frac{\mu_0}{4\pi}\frac{I\,dl\sin\phi}{r^2}$$
$$= (10^{-7}\,\mathrm{T\cdot m/A})\frac{(125\,\mathrm{A})(1.0\times10^{-2}\,\mathrm{m})(\sin 90°)}{(1.2\,\mathrm{m})^2}$$
$$= 8.7\times10^{-8}\,\mathrm{T}$$

b) At point P_2 the direction of $\vec{B}$ is again into the xy-plane of the figure. The angle between $d\vec{l}$ and $\hat{r}$ is 30°, and

$$B = (10^{-7}\,\mathrm{T\cdot m/A})\frac{(125\,\mathrm{A})(1.0\times10^{-2}\,\mathrm{m})(\sin 30°)}{(1.2\,\mathrm{m})^2}$$
$$= 4.3\times10^{-8}\,\mathrm{T}$$

EVALUATE: You can check our results for the direction of $\vec{B}$ by comparing them with Fig. 28.3. The xy-plane in Fig. 28.4 corresponds to the tan-colored plane in Fig. 28.3. However, in the present example the direction of the current and hence of $d\vec{l}$ is the reverse of the direction shown in Fig. 28.3, so the direction of the magnetic field is reversed as well. Hence the field at points in the xy-plane in Fig. 28.4 must point *into,* not out of, that plane. This is just what we concluded above.

Note that these magnetic-field magnitudes are very small; for comparison the magnetic field of the earth is of the order of 10^{-4} T. Note also that the values are not the *total* fields at points P_1 and P_2, but only the contributions from the short segment of conductor described.

Test Your Understanding

In Fig. 28.4, what are the magnitude and direction of the magnetic field caused by the 1.0-cm segment at a point 0.60 m directly below it (that is, on the negative

y-axis)? What are the magnitude and direction of the magnetic force that the segment exerts on a proton at that point moving at 4.0×10^3 m/s in the positive y-direction?

28.3 | Magnetic Field of a Straight Current-Carrying Conductor

An important application of the law of Biot and Savart is finding the magnetic field produced by a straight current-carrying conductor. This result is useful because straight conducting wires are found in essentially all electric and electronic devices. Figure 28.5 shows such a conductor with length $2a$ carrying a current I. We will find $\vec{B}$ at a point a distance x from the conductor on its perpendicular bisector.

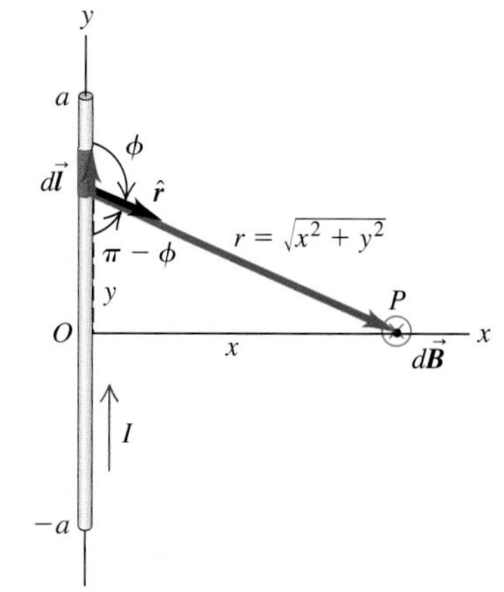

28.5 Magnetic field produced by a straight current-carrying conductor of length $2a$. At point P, $\vec{B}$ is directed into the plane of the page.

We first use the law of Biot and Savart, Eq. (28.5), to find the field $d\vec{B}$ caused by the element of conductor of length $dl = dy$ shown in Fig. 28.5. From the figure, $r = \sqrt{x^2 + y^2}$ and $\sin\phi = \sin(\pi - \phi) = x/\sqrt{x^2 + y^2}$. The right-hand rule for the vector product $d\vec{l} \times \hat{r}$ shows that the *direction* of $d\vec{B}$ is into the plane of the figure, perpendicular to the plane; furthermore, the directions of the $d\vec{B}$'s from *all* elements of the conductor are the same. Thus in integrating Eq. (28.7), we can just add the *magnitudes* of the $d\vec{B}$'s, a significant simplification.

Putting the pieces together, we find that the magnitude of the total $\vec{B}$ field is

$$B = \frac{\mu_0 I}{4\pi}\int_{-a}^{a}\frac{x\,dy}{(x^2 + y^2)^{3/2}}$$

We can integrate this by trigonometric substitution or by using an integral table. The final result is

$$B = \frac{\mu_0 I}{4\pi}\frac{2a}{x\sqrt{x^2 + a^2}} \qquad (28.8)$$

ActivPhysics ONLINE

13.1 Magnetic Field of a Wire

When the length $2a$ of the conductor is very great in comparison to its distance x from point P, we can consider it to be infinitely long. When a is much larger than x, $\sqrt{x^2 + a^2}$ is approximately equal to a; hence in the limit $a \to \infty$, Eq. (28.8) becomes

$$B = \frac{\mu_0 I}{2\pi x}$$

The physical situation has axial symmetry about the y-axis. Hence $\vec{B}$ must have the same *magnitude* at all points on a circle centered on the conductor and lying in a plane perpendicular to it, and the *direction* of $\vec{B}$ must be everywhere tangent to such a circle. Thus, at all points on a circle of radius r around the conductor, the magnitude B is

$$B = \frac{\mu_0 I}{2\pi r} \quad \text{(a long, straight, current-carrying conductor)} \qquad (28.9)$$

Part of the magnetic field around a long, straight, current-carrying conductor is shown in Fig. 28.6.

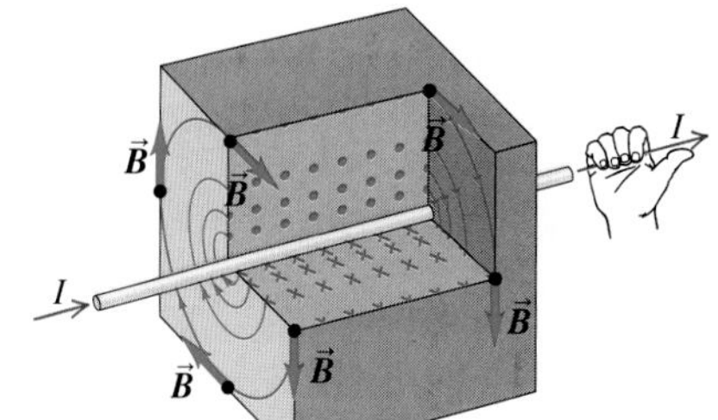

28.6 Magnetic field around a long, straight, current-carrying conductor. The field lines are circles, with directions determined by the right-hand rule.

The geometry of this problem is similar to that of Example 21.11 (Section 21.5), in which we solved the problem of the *electric* field caused by an infinite line of charge. The same integral appears in both problems, and the field magnitudes in both problems are proportional to $1/r$. But the lines of $\vec{B}$ in the magnetic problem have completely different shapes than the lines of $\vec{E}$ in the analogous electrical problem. Electric field lines radiate outward from a positive line charge distribution (inward for negative charges). By contrast, magnetic field lines *encircle* the current that acts as their source. Electric field lines due to charges begin and end at those charges, but magnetic field lines always form closed loops and *never* have end points, irrespective of the shape of the current-carrying conductor that sets up the field. As we discussed in Section 27.3, this is a consequence of Gauss's law for magnetism, which states that the total magnetic flux through *any* closed surface is always zero:

$$\oint \vec{B} \cdot d\vec{A} = 0 \qquad \text{(magnetic flux through any closed surface)} \qquad (28.10)$$

This implies that there are no isolated magnetic charges or magnetic monopoles. Any magnetic field line that enters a closed surface must also emerge from that surface.

Example 28.3 Magnetic field of a single wire

A long, straight conductor carries a current of 1.0 A. At what distance from the axis of the conductor is the magnetic field caused by the current equal in magnitude to the earth's magnetic field in Pittsburgh (about 0.5×10^{-4} T)?

SOLUTION

IDENTIFY and SET UP: The geometry is the same as that in Fig. 28.6, so we use Eq. (28.8). All of the quantities in this equation are known except the target variable, the distance r.

EXECUTE: We solve Eq. (28.8) for r and insert the appropriate numbers:

$$r = \frac{\mu_0 I}{2\pi B}$$

$$= \frac{(4\pi \times 10^{-7}\ \text{T}\cdot\text{m/A})(1.0\ \text{A})}{(2\pi)(0.5 \times 10^{-4}\ \text{T})} = 4 \times 10^{-3}\ \text{m} = 4\ \text{mm}$$

EVALUATE: Currents of an ampere or so are typical of those found in the wiring of home appliances. This example shows that the magnetic fields produced by these appliances are very weak even at points very close to the wire. At greater distances the field becomes even weaker; for example, at five times the distance ($r = 20$ mm $= 2$ cm $= 2 \times 10^{-2}$ m) the field is one-fifth as great ($B = 0.1 \times 10^{-4}$ T).

Example 28.4 Magnetic field of two wires

Figure 28.7a is an end view of two long, straight, parallel wires perpendicular to the xy-plane, each carrying a current I but in opposite directions. a) Find the magnitude and direction of $\vec{B}$ at points P_1, P_2, and P_3. b) Find the magnitude and direction of $\vec{B}$ at any point on the x-axis to the right of wire 2 in terms of the x-coordinate of the point.

SOLUTION

IDENTIFY and SET UP: We use Eq. (28.9) to find the magnitude of the fields $\vec{B}_1$ (due to wire 1) and $\vec{B}_2$ (due to wire 2) at any point. We find the directions of these fields using the right-hand rule. We then use the principle of superposition of magnetic fields to find the *total* magnetic field at the point in question: $\vec{B}_{total} = \vec{B}_1 + \vec{B}_2$.

EXECUTE: a) Point P_1 is closer to wire 1 (distance $2d$) than to wire 2 (distance $4d$), so at this point the field magnitude B_1 is greater than the magnitude B_2:

$$B_1 = \frac{\mu_0 I}{2\pi(2d)} = \frac{\mu_0 I}{4\pi d} \qquad B_2 = \frac{\mu_0 I}{2\pi(4d)} = \frac{\mu_0 I}{8\pi d}$$

The right-hand rule shows that $\vec{B}_1$ is in the negative y-direction and $\vec{B}_2$ is in the positive y-direction. Since B_1 is the larger magnitude, the total field $\vec{B}_{total} = \vec{B}_1 + \vec{B}_2$ is in the negative y-direction, with magnitude

$$B_{total} = B_1 - B_2 = \frac{\mu_0 I}{4\pi d} - \frac{\mu_0 I}{8\pi d} = \frac{\mu_0 I}{8\pi d} \qquad \text{(point } P_1\text{)}$$

At point P_2, a distance d from both wires, $\vec{B}_1$ and $\vec{B}_2$ are both in the positive y-direction, and both have the same magnitude:

$$B_1 = B_2 = \frac{\mu_0 I}{2\pi d}$$

so $\vec{B}_{total}$ is also in the positive y-direction and has magnitude

$$B_{total} = B_1 + B_2 = \frac{\mu_0 I}{\pi d} \qquad \text{(point } P_2\text{)}$$

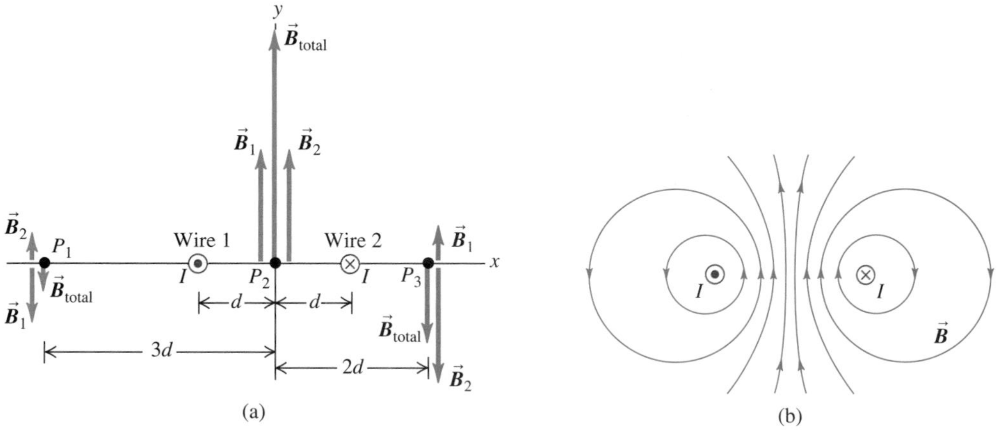

28.7 (a) Two long, straight conductors carrying equal currents in opposite directions. The conductors are seen end-on. (b) Map of the magnetic field produced by the two conductors. The field lines are closest together between the conductors, where the field is strongest.

Finally, at point P_3 the right-hand rule shows that $\vec{B}_1$ is in the positive y-direction and $\vec{B}_2$ is in the negative y-direction. This point is farther from wire 1 (distance $3d$) than from wire 2 (distance d), so B_1 is less than B_2:

$$B_1 = \frac{\mu_0 I}{2\pi(3d)} = \frac{\mu_0 I}{6\pi d} \qquad B_2 = \frac{\mu_0 I}{2\pi d}$$

The total field is in the negative y-direction, the same as $\vec{B}_2$, and has magnitude

$$B_{total} = B_2 - B_1 = \frac{\mu_0 I}{2\pi d} - \frac{\mu_0 I}{6\pi d} = \frac{\mu_0 I}{3\pi d} \qquad \text{(point } P_3\text{)}$$

You should be able to use the right-hand rule to verify for yourself the directions of $\vec{B}_1$ and $\vec{B}_2$ for each point.

The fields $\vec{B}_1$, $\vec{B}_2$, and $\vec{B}_{total}$ at each of the three points are shown in Fig. 28.7a. The same technique can be used to find $\vec{B}_{total}$ at any point; for points off the x-axis, caution must be taken in vector addition, since $\vec{B}_1$ and $\vec{B}_2$ need no longer be simply parallel or antiparallel (see Problems 28.56 and 28.57). Figure 28.7b shows some of the magnetic field lines due to this combination of wires.

b) At any point to the right of wire 2 (that is, for $x > d$), $\vec{B}_1$ and $\vec{B}_2$ are in the same directions as at P_3. As x increases, both $\vec{B}_1$ and $\vec{B}_2$ decrease in magnitude, so $\vec{B}_{total}$ must decrease as well. The magnitudes of the fields due to each wire are

$$B_1 = \frac{\mu_0 I}{2\pi(x+d)} \quad \text{and} \quad B_2 = \frac{\mu_0 I}{2\pi(x-d)}$$

At any field point to the right of wire 2, wire 2 is closer than wire 1, and so $B_2 > B_1$. Hence $\vec{B}_{total}$ is in the negative y-direction, the same as $\vec{B}_2$, and has magnitude

$$B_{total} = B_2 - B_1 = \frac{\mu_0 I}{2\pi(x-d)} - \frac{\mu_0 I}{2\pi(x+d)} = \frac{\mu_0 I d}{\pi(x^2 - d^2)}$$

where we combined the two terms using a common denominator.

EVALUATE: At points very far from the wires, so that x is much larger than d, the d^2 term in the denominator can be neglected, and

$$B_{total} = \frac{\mu_0 I d}{\pi x^2}$$

The magnetic-field magnitude for a single wire decreases with distance in proportion to $1/x$, as shown by Eq. (28.9); for two wires carrying opposite currents, $\vec{B}_1$ and $\vec{B}_2$ partially cancel each other, and so the magnitude of $\vec{B}_{total}$ decreases more rapidly, in proportion to $1/x^2$. This effect is used in communication systems such as telephone or computer networks. The wiring is arranged so that a conductor carrying a signal in one direction and the conductor carrying the return signal are side by side, as in Fig. 28.7a, or twisted around each other (Fig. 28.8). As a result, the magnetic field caused *outside* the conductors by these signals is greatly reduced and is less likely to exert unwanted forces on other information-carrying currents.

28.8 Computer cables, or cables for audio-video equipment, create little or no magnetic field. This is because within each cable, closely spaced wires carry current in both directions along the length of the cable. The magnetic fields from these opposing currents cancel each other.

Test Your Understanding

Suppose both conductors in Fig. 28.7 carried identical currents I in the direction out of the plane. What would be the magnetic field at point P_2?

28.4 | Force Between Parallel Conductors

In Example 28.4 (Section 28.3) we showed how to use the principle of superposition of magnetic fields to find the total field due to two long current-carrying conductors. Another important aspect of this configuration is the *interaction force* between the conductors. This force plays a role in many practical situations in which current-carrying wires are close to each other, and it also has fundamental significance in connection with the definition of the ampere. Figure 28.9 shows segments of two long, straight parallel conductors separated by a distance r and carrying currents I and I', respectively, in the same direction. Each conductor lies in the magnetic field set up by the other, so each experiences a force. The diagram shows some of the field lines set up by the current in the lower conductor.

From Eq. (28.9) the lower conductor produces a $\vec{B}$ field that, at the position of the upper conductor, has magnitude

$$B = \frac{\mu_0 I}{2\pi r}$$

From Eq. (27.19) the force that this field exerts on a length L of the upper conductor is $\vec{F} = I'\vec{L} \times \vec{B}$, where the vector $\vec{L}$ is in the direction of the current I' and has magnitude L. Since $\vec{B}$ is perpendicular to the length of the conductor and hence to $\vec{L}$, the magnitude of this force is

$$F = I'LB = \frac{\mu_0 I I' L}{2\pi r}$$

and the force *per unit length* F/L is

$$\frac{F}{L} = \frac{\mu_0 I I'}{2\pi r} \quad \text{(two long, parallel, current-carrying conductors)} \qquad (28.11)$$

Applying the right-hand rule to $\vec{F} = I'\vec{L} \times \vec{B}$ shows that the force on the upper conductor is directed *downward.*

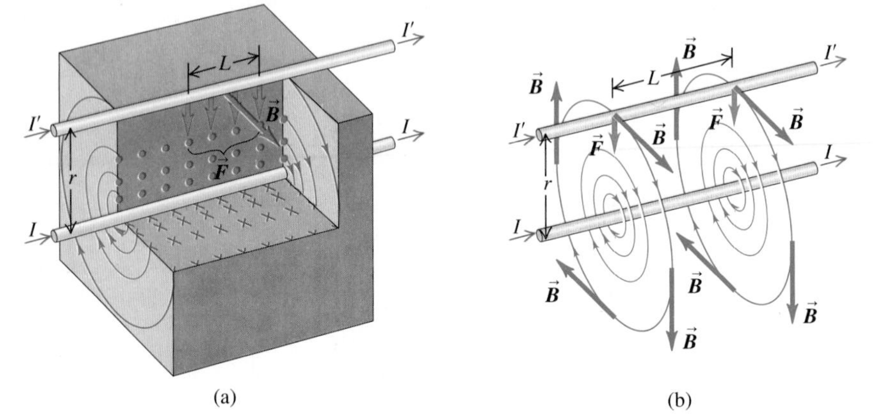

28.9 Parallel conductors carrying currents in the same direction attract each other. The diagrams show how the magnetic field $\vec{B}$ caused by the current in the lower conductor exerts a force $\vec{F}$ on the upper conductor.

The current in the upper conductor also sets up a field at the position of the lower one. Two successive applications of the right-hand rule for vector products (one to find the direction of the $\vec{B}$ field due to the upper conductor, as in Section 28.2, and one to find the direction of the force that this field exerts on the lower conductor, as in Section 27.6) show that the force on the lower conductor is *upward.* Thus *two parallel conductors carrying current in the same direction attract each other.* If the direction of either current is reversed, the forces also reverse. *Parallel conductors carrying currents in opposite directions repel each other.*

The attraction or repulsion between two straight, parallel, current-carrying conductors is the basis of the official SI definition of the **ampere:**

***One ampere* is that unvarying current that, if present in each of two parallel conductors of infinite length and one meter apart in empty space, causes each conductor to experience a force of exactly 2 × 10^{-7} newtons per meter of length.**

From Eq. (28.11) you can see that this definition of the ampere is what leads us to choose the value of $4\pi \times 10^{-7}\ \text{T}\cdot\text{m/A}$ for μ_0. It also forms the basis of the SI definition of the coulomb, which is the amount of charge transferred in one second by a current of one ampere.

This is an *operational definition;* it gives us an actual experimental procedure for measuring current and defining a unit of current. In principle we could use this definition to calibrate an ammeter, using only a meter stick and a spring balance. For high-precision standardization of the ampere, coils of wire are used instead of straight wires, and their separation is only a few centimeters. Even more precise measurements of the standardized ampere are possible using a version of the Hall effect (Section 27.8).

Mutual forces of attraction exist not only between *wires* carrying currents in the same direction, but also between the longitudinal elements of a single current-carrying conductor. If the conductor is a liquid or an ionized gas (a plasma), these forces result in a constriction of the conductor, as if its surface were acted on by an inward pressure. The constriction of the conductor is called the *pinch effect.* The high temperature produced by the pinch effect in a plasma has been used in one technique to bring about nuclear fusion.

Example 28.5 Forces between parallel wires

Two straight, parallel, superconducting wires 4.5 mm apart carry equal currents of 15,000 A in opposite directions. Should we worry about the mechanical strength of these wires?

SOLUTION

IDENTIFY: Whether or not we need to worry about the wires' mechanical strength depends on how much magnetic force each wire exerts on the other.

SET UP: Our target variable is the magnetic force per unit length of wire, which we find using Eq. (28.11).

EXECUTE: Because the currents are in opposite directions, the two conductors repel each other. From Eq. (28.11) the force per unit length is

$$\frac{F}{L} = \frac{\mu_0 I I'}{2\pi r} = \frac{(4\pi \times 10^{-7}\ \text{T}\cdot\text{m/A})(15{,}000\ \text{A})^2}{(2\pi)(4.5 \times 10^{-3}\ \text{m})}$$
$$= 1.0 \times 10^4\ \text{N/m}$$

EVALUATE: This is a large force, more than one ton per meter, so mechanical strength of the conductors and insulating materials are certainly a significant consideration. Currents and separations of this magnitude are used in superconducting electromagnets in particle accelerators, and mechanical stress analysis is a crucial part of the design process.

Conceptual Example 28.6

Mechanical stress in a solenoid

A *solenoid,* introduced in Section 27.7, is a wire wound into a helical coil. For a solenoid that carries a current I (Fig. 28.10), discuss the directions of the magnetic forces a) between adjacent turns of the coil; b) between opposite sides of the same turn.

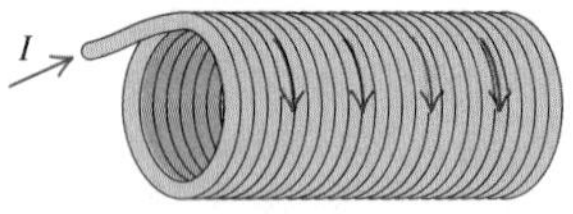

28.10 A current-carrying solenoid.

SOLUTION

a) Adjacent turns carry parallel currents and therefore attract one another, causing a *compressive* stress along the solenoid axis.
b) Opposite sides of the same turn carry current in opposite directions and thus repel each other. Therefore the coil experiences forces that pull radially outward, creating *tensile* stress in the conductors. These forces are proportional to the square of the current; as we mentioned above, they are an important consideration in electromagnet design. We'll discuss the nature of the *total* magnetic field produced by a solenoid in Section 28.7.

Test Your Understanding

A cable like those shown in Fig. 28.8 has two wires 4 mm apart, each carrying 1 A of current in opposite directions. Is the magnetic force between the wires attractive or repulsive? What is the force per unit length on either wire? Is this force noticeable in normal use?

28.5 | Magnetic Field of a Circular Current Loop

28.11 This electromagnet contains a current-carrying coil with numerous turns of wire. The resulting magnetic field is so strong that the electromagnet can pick up hundreds of steel cans at once.

If you look inside a doorbell, a transformer, an electric motor, or an electromagnet (Fig. 28.11), you will find coils of wire with a large number of turns, spaced so closely that each turn is very nearly a planar circular loop. A current in such a coil is used to establish a magnetic field. So it is worthwhile to derive an expression for the magnetic field produced by a single circular conducting loop carrying a current or by N closely spaced circular loops forming a coil. In Section 27.7 we considered the force and torque on such a current loop placed in an external magnetic field produced by other currents; we are now about to find the magnetic field produced by the loop itself.

Figure 28.12 shows a circular conductor with radius a that carries a current I. The current is led into and out of the loop through two long, straight wires side by side; the currents in these straight wires are in opposite directions, and their magnetic fields very nearly cancel each other (see Example 28.4 in Section 28.3).

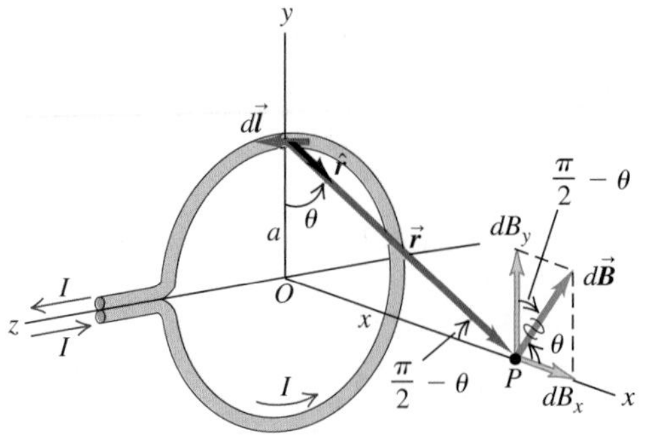

28.12 Magnetic field of a circular loop. The current in the segment $d\vec{l}$ causes the field $d\vec{B}$, which lies in the xy-plane. The currents in other $d\vec{l}$'s cause $d\vec{B}$'s with different components perpendicular to the x-axis; these components add to zero. The x-components of the $d\vec{B}$'s combine to give the total $\vec{B}$ field at point P.

We can use the law of Biot and Savart, Eq. (28.5) or (28.6), to find the magnetic field at a point P on the axis of the loop, at a distance x from the center. As the figure shows, $d\vec{l}$ and $\hat{r}$ are perpendicular, and the direction of the field $d\vec{B}$ caused by this particular element $d\vec{l}$ lies in the xy-plane. Since $r^2 = x^2 + a^2$, the magnitude dB of the field due to element $d\vec{l}$ is

$$dB = \frac{\mu_0 I}{4\pi} \frac{dl}{(x^2 + a^2)} \tag{28.12}$$

The components of the vector $d\vec{B}$ are

$$dB_x = dB \cos\theta = \frac{\mu_0 I}{4\pi} \frac{dl}{(x^2 + a^2)} \frac{a}{(x^2 + a^2)^{1/2}} \tag{28.13}$$

$$dB_y = dB \sin\theta = \frac{\mu_0 I}{4\pi} \frac{dl}{(x^2 + a^2)} \frac{x}{(x^2 + a^2)^{1/2}} \tag{28.14}$$

The situation has rotational symmetry about the x-axis, so there cannot be a component of the total field $\vec{B}$ perpendicular to this axis. For every element $d\vec{l}$ there is a corresponding element on the opposite side of the loop, with opposite direction. These two elements give equal contributions to the x-component of $d\vec{B}$, given by Eq. (28.13), but *opposite* components perpendicular to the x-axis. Thus all the perpendicular components cancel and only the x-components survive.

To obtain the x-component of the total field $\vec{B}$, we integrate Eq. (28.13), including all the $d\vec{l}$'s around the loop. Everything in this expression except dl is constant and can be taken outside the integral, and we have

ActivPhysics ONLINE

13.2 Magnetic Field of a Loop

$$B_x = \int \frac{\mu_0 I}{4\pi} \frac{a\,dl}{(x^2 + a^2)^{3/2}} = \frac{\mu_0 I a}{4\pi (x^2 + a^2)^{3/2}} \int dl$$

The integral of dl is just the circumference of the circle, $\int dl = 2\pi a$, and we finally get

$$B_x = \frac{\mu_0 I a^2}{2(x^2 + a^2)^{3/2}} \quad \text{(on the axis of a circular loop)} \tag{28.15}$$

Now suppose that instead of the single loop in Fig. 28.12 we have a coil consisting of N loops, all with the same radius. The loops are closely spaced so that the plane of each loop is essentially the same distance x from the field point P. Each loop contributes equally to the field, and the total field is N times the field of a single loop:

$$B_x = \frac{\mu_0 N I a^2}{2(x^2 + a^2)^{3/2}} \quad \text{(on the axis of } N \text{ circular loops)} \tag{28.16}$$

The factor N in Eq. (28.16) is the reason why coils of wire, not single loops, are used to produce strong magnetic fields; for a desired field strength, using a single loop might require a current I so great as to exceed the rating of the loop's wire.

Figure 28.13 shows a graph of B_x as a function of x. The maximum value of the field is at $x = 0$, the center of the loop or coil:

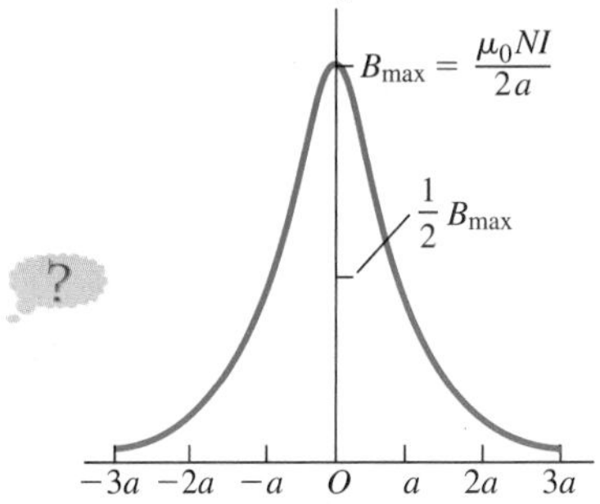

28.13 Graph of the magnetic field along the axis of a circular coil with N turns. When x is much larger than a, the field magnitude decreases approximately as $1/x^3$.

$$B_x = \frac{\mu_0 N I}{2a} \quad \text{(at the center of } N \text{ circular loops)} \tag{28.17}$$

As we go out along the axis, the field decreases in magnitude.

In Section 27.7 we defined the *magnetic dipole moment* μ (or *magnetic moment*) of a current-carrying loop to be equal to IA, where A is the cross-sectional area of the loop. If there are N loops, the total magnetic moment is NIA. The circular loop in Fig. 28.12 has area $A = \pi a^2$, so the magnetic moment of a single loop is $\mu = I\pi a^2$; for N loops, $\mu = NI\pi a^2$. Substituting these results into Eqs. (28.15) and (28.16), we find that both of these expressions can be written as

$$B_x = \frac{\mu_0 \mu}{2\pi(x^2 + a^2)^{3/2}} \qquad (28.18)$$

(on the axis of any number of circular loops)

We described a magnetic dipole in Section 27.7 in terms of its response to a magnetic field produced by currents outside the dipole. But a magnetic dipole is also a *source* of magnetic field; Eq. (28.18) describes the magnetic field *produced* by a magnetic dipole for points along the dipole axis. This field is directly proportional to the magnetic dipole moment μ. Note that the field along the x-axis is in the same direction as the vector magnetic moment $\vec{\mu}$; this is true on both the positive and negative x-axis.

CAUTION Equations (28.15), (28.16), and (28.18) are valid only on the *axis* of a loop or coil. Don't attempt to apply these equations at other points!

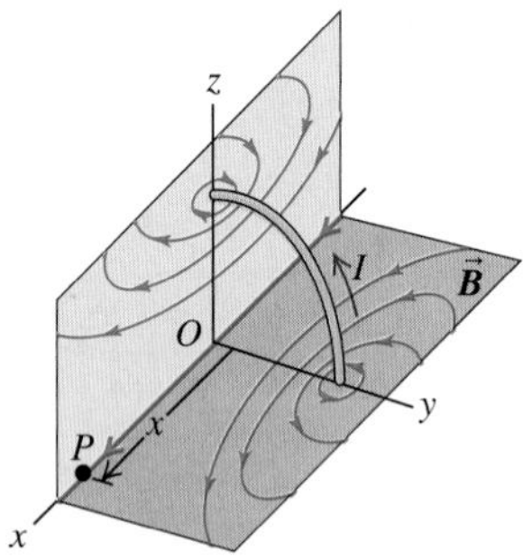

28.14 Magnetic field lines produced by the current in a circular loop. At points on the axis the $\vec{B}$ field has the same direction as the magnetic moment of the loop.

Some of the magnetic field lines surrounding a circular current loop (magnetic dipole), in planes through the axis, are shown in Fig. 28.14. The directions of the field lines are given by the same right-hand rule as for a long, straight conductor. Grab the conductor with your right hand, with your thumb in the direction of the current; your fingers curl around in the same direction as the field lines. The field lines for the circular current loop are closed curves that encircle the conductor; they are *not* circles, however.

Example 28.7 Magnetic field of a coil

A coil consisting of 100 circular loops with radius 0.60 m carries a current of 5.0 A. a) Find the magnetic field at a point along the axis of the coil, 0.80 m from the center. b) Along the axis, at what distance from the center is the field magnitude 1/8 as great as it is at the center?

SOLUTION

IDENTIFY and SET UP: We want the field on the axis of the coil, not necessarily at its center, so we use Eq. (28.16). We are given $N = 100$, $I = 5.0$ A, and $a = 0.60$ m. In part (a) our target variable is the magnetic field at a given value of the coordinate x; in part (b) the target variable is the value of x at which the field has 1/8 of the magnitude that it has at $x = 0$.

EXECUTE: a) Using $x = 0.80$ m, from Eq. (28.16) we have

$$B_x = \frac{(4\pi \times 10^{-7}\ \text{T}\cdot\text{m/A})(100)(5.0\ \text{A})(0.60\ \text{m})^2}{2[(0.80\ \text{m})^2 + (0.60\ \text{m})^2]^{3/2}}$$

$$= 1.1 \times 10^{-4}\ \text{T}$$

b) Considering Eq. (28.16), we want to find a value of x such that

$$\frac{1}{(x^2 + a^2)^{3/2}} = \frac{1}{8}\,\frac{1}{(0^2 + a^2)^{3/2}}$$

To solve this for x, we take the reciprocal of the whole thing, then take the 2/3 power of both sides; the result is

$$x = \pm\sqrt{3}a = \pm 1.04\ \text{m}$$

At a distance of about 1.7 radii from the center, the field has dropped off to 1/8 its value at the center.

EVALUATE: We can check our answer in part (a) by first finding the magnetic moment, then substituting the result into Eq. (28.18):

$$\mu = NI\pi a^2 = (100)(5.0\ \text{A})\pi(0.60\ \text{m})^2 = 5.7 \times 10^2\ \text{A}\cdot\text{m}^2$$

$$B_x = \frac{(4\pi \times 10^{-7}\ \text{T}\cdot\text{m/A})(5.7 \times 10^2\ \text{A}\cdot\text{m}^2)}{2\pi[(0.80\ \text{m})^2 + (0.60\ \text{m})^2]^{3/2}} = 1.1 \times 10^{-4}\ \text{T}$$

The magnetic moment μ is relatively large, yet this is a rather small field, comparable in magnitude to the earth's magnetic field. This example may give you some idea of the difficulty of producing a field of 1 T or more.

Test Your Understanding

Equations (28.13) and (28.14) give the x- and y-components of the magnetic field $d\vec{B}$ produced at point P by a segment $d\vec{l}$ that lies on the positive y-axis in Fig. 28.12 (at the top of the loop). The z-component of this $d\vec{B}$ is zero. What are the x-, y-, and z-components of the field $d\vec{B}$ produced at P by a segment $d\vec{l}$ on the negative y-axis? On the negative z-axis?

28.6 | Ampere's Law

So far our calculations of the magnetic field due to a current have involved finding the infinitesimal field $d\vec{B}$ due to a current element, then summing all the $d\vec{B}$'s to find the total field. This approach is directly analogous to our *electric*-field calculations in Chapter 21.

For the electric-field problem we found that in situations with a highly symmetric charge distribution, it was often easier to use Gauss's law to find $\vec{E}$. There is likewise a law that allows us to more easily find the *magnetic* fields caused by highly symmetric *current* distributions. But the law that allows us to do this, called *Ampere's law,* is rather different in character from Gauss's law.

Gauss's law for electric fields involves the flux of $\vec{E}$ through a closed surface; it states that this flux is equal to the total charge enclosed within the surface, divided by the constant ϵ_0. Thus this law relates electric fields and charge distributions. By contrast, Gauss's law for *magnetic* fields, Eq. (28.10), is *not* a relation between magnetic fields and current distributions; it states that the flux of $\vec{B}$ through *any* closed surface is always zero, whether or not there are currents within the surface. So Gauss's law for $\vec{B}$ can't be used to determine the magnetic field produced by a particular current distribution.

Ampere's law is formulated not in terms of magnetic flux, but rather in terms of the *line integral* of $\vec{B}$ around a closed path, denoted by

$$\oint \vec{B} \cdot d\vec{l}$$

We used line integrals to define work in Chapter 6 and to calculate electric potential in Chapter 23. To evaluate this integral, we divide the path into infinitesimal segments $d\vec{l}$, calculate the scalar product of $\vec{B} \cdot d\vec{l}$ for each segment, and sum these products. In general, $\vec{B}$ varies from point to point, and we must use the value of $\vec{B}$ at the location of each $d\vec{l}$. An alternative notation is $\oint B_{\parallel}\, dl$, where $B_{\parallel}$ is the component of $\vec{B}$ parallel to $d\vec{l}$ at each point. The circle on the integral sign indicates that this integral is always computed for a *closed* path, one whose beginning and end points are the same.

To introduce the basic idea of Ampere's law, let's consider again the magnetic field caused by a long, straight conductor carrying a current I. We found in Section 28.3 that the field at a distance r from the conductor has magnitude

$$B = \frac{\mu_0 I}{2\pi r}$$

and that the magnetic field lines are circles centered on the conductor. Let's take the line integral of $\vec{B}$ around one such circle with radius r, as in Fig. 28.15a. At every point on the circle, $\vec{B}$ and $d\vec{l}$ are parallel, and so $\vec{B} \cdot d\vec{l} = B\, dl$; since r is constant around the circle, B is constant as well. Alternatively, we can say that $B_{\parallel}$ is constant and equal to B at every point on the circle. Hence we can take B

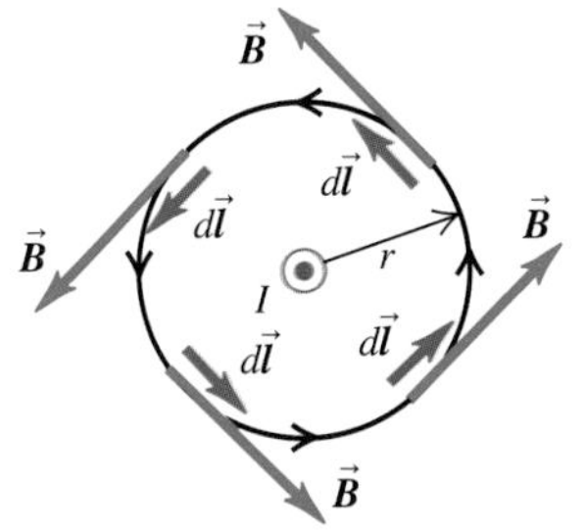

(a) Integration path is a circle centered on the conductor; integration goes around the circle counterclockwise

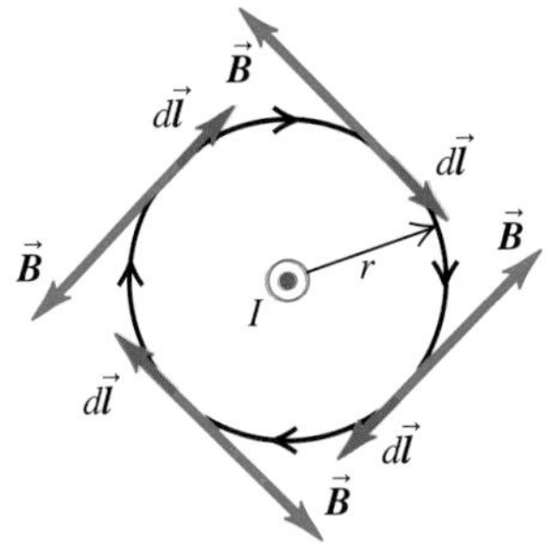

(b) Same integration path as in (a), but integration goes around the circle clockwise

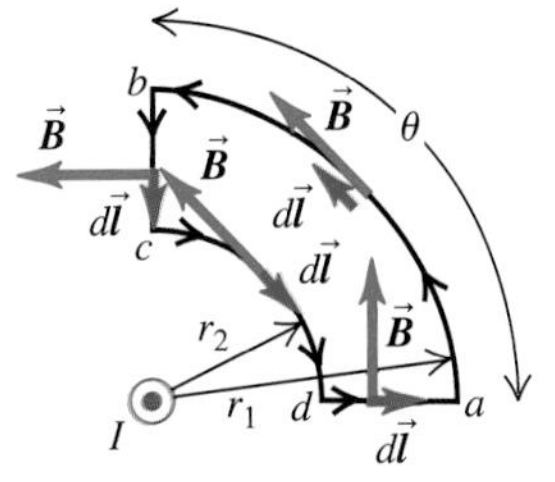

(c) An integration path that does not enclose the conductor

28.15 Some integration paths for the line integral of $\vec{B}$ in the vicinity of a long, straight conductor carrying current I *out* of the plane of the page (as indicated by the circle with a dot). The conductor is seen end on.

outside of the integral. The remaining integral $\oint dl$ is just the circumference of the circle, so

$$\oint \vec{B} \cdot d\vec{l} = \oint B_{\parallel}\, dl = B \oint dl = \frac{\mu_0 I}{2\pi r}(2\pi r) = \mu_0 I$$

The line integral is thus independent of the radius of the circle and is equal to μ_0 multiplied by the current passing through the area bounded by the circle.

In Fig. 28.15b the situation is the same, but the integration path now goes around the circle in the opposite direction. Now $\vec{B}$ and $d\vec{l}$ are antiparallel, so $\vec{B} \cdot d\vec{l} = -B\, dl$ and the line integral equals $-\mu_0 I$. We get the same result if the integration path is the same as in Fig. 28.15a, but the direction of the current is reversed. Thus the line integral $\oint \vec{B} \cdot d\vec{l}$ equals μ_0 multiplied by the current passing through the area bounded by the integration path, with a positive or negative sign depending on the direction of the current relative to the direction of integration.

There's a simple rule for the sign of the current; you won't be surprised to learn that it uses your right hand. Curl the fingers of your right hand around the integration path so that they curl in the direction of integration (that is, the direction that you use to evaluate $\oint \vec{B} \cdot d\vec{l}$). Then your right thumb indicates the positive current direction. Currents that pass through in this direction are positive; those in the opposite direction are negative. Using this rule, you should be able to convince yourself that the current is positive in Fig. 28.15a and negative in Fig. 28.15b. Here's another way to say the same thing: Looking at the surface bounded by the integration path, integrate counterclockwise around the path as in Fig. 28.15a. Currents moving toward you through the surface are positive, and those going away from you are negative.

An integration path that does *not* enclose the conductor is used in Fig. 28.15c. Along the circular arc ab of radius r_1, $\vec{B}$ and $d\vec{l}$ are parallel, and $B_{\parallel} = B_1 = \mu_0 I/2\pi r_1$; along the circular arc cd of radius r_2, $\vec{B}$ and $d\vec{l}$ are antiparallel and $B_{\parallel} = -B_2 = -\mu_0 I/2\pi r_2$. The $\vec{B}$ field is perpendicular to $d\vec{l}$ at each point on the straight sections bc and da, so $B_{\parallel} = 0$ and these sections contribute zero to the line integral. The total line integral is then

$$\oint \vec{B} \cdot d\vec{l} = \oint B_{\parallel}\, dl = B_1 \int_a^b dl + (0) \int_b^c dl + (-B_2) \int_c^d dl + (0) \int_d^a dl$$

$$= \frac{\mu_0 I}{2\pi r_1}(r_1 \theta) + 0 - \frac{\mu_0 I}{2\pi r_2}(r_2 \theta) + 0 = 0$$

The magnitude of $\vec{B}$ is greater on arc cd than on arc ab, but the arc length is less, so the contributions from the two arcs exactly cancel. Even though there is a magnetic field everywhere along the integration path, the line integral $\oint \vec{B} \cdot d\vec{l}$ is zero if there is no current passing through the area bounded by the path.

We can also derive these results for more general integration paths, such as the one in Fig. 28.16a. At the position of the line element $d\vec{l}$, the angle between $d\vec{l}$ and $\vec{B}$ is ϕ, and

$$\vec{B} \cdot d\vec{l} = B\, dl \cos\phi$$

From the figure, $dl \cos\phi = r\, d\theta$, where $d\theta$ is the angle subtended by $d\vec{l}$ at the position of the conductor and r is the distance of $d\vec{l}$ from the conductor. Thus

$$\oint \vec{B} \cdot d\vec{l} = \oint \frac{\mu_0 I}{2\pi r}(r\, d\theta) = \frac{\mu_0 I}{2\pi} \oint d\theta$$

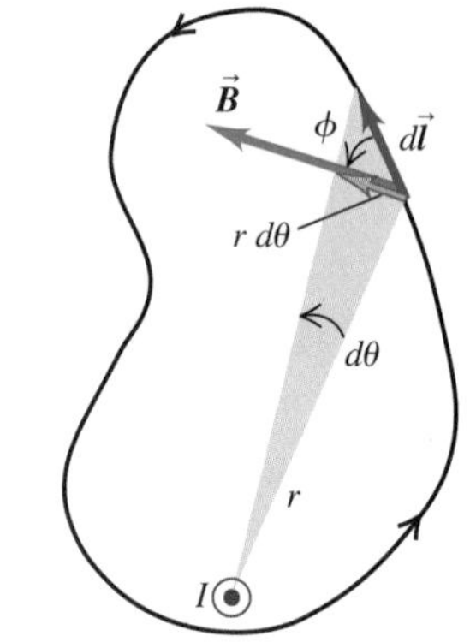

(a)

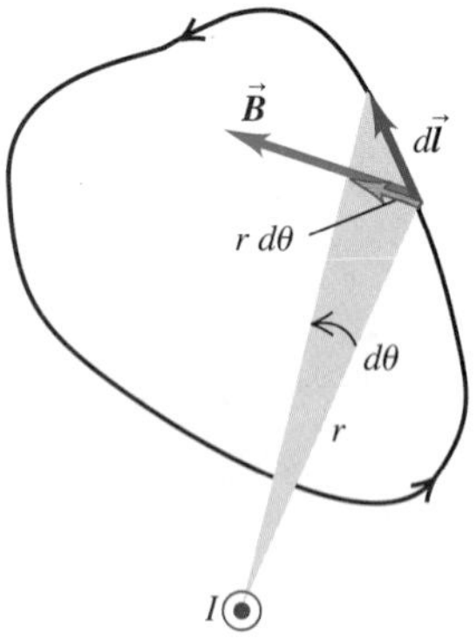

(b)

28.16 (a) A more general integration path for the line integral of $\vec{B}$ around a long, straight conductor carrying current I *out* of the plane of the page. The conductor is seen end-on. (b) A more general integration path that does not enclose the conductor.

But $\oint d\theta$ is just equal to 2π, the total angle swept out by the radial line from the conductor to $d\vec{l}$ during a complete trip around the path. So we get

$$\oint \vec{B} \cdot d\vec{l} = \mu_0 I \tag{28.19}$$

This result doesn't depend on the shape of the path or on the position of the wire inside it. If the current in the wire is opposite to that shown, the integral has the opposite sign. But if the path doesn't enclose the wire (Fig. 28.16b), then the net change in θ during the trip around the integration path is zero; $\oint d\theta$ is zero instead of 2π and the line integral is zero.

Equation (28.19) is almost, but not quite, the general statement of Ampere's law. To generalize it even further, suppose *several* long, straight conductors pass through the surface bounded by the integration path. The total magnetic field $\vec{B}$ at any point on the path is the vector sum of the fields produced by the individual conductors. Thus the line integral of the total $\vec{B}$ equals μ_0 times the *algebraic sum* of the currents. In calculating this sum, we use the sign rule for currents described above. If the integration path does not enclose a particular wire, the line integral of the $\vec{B}$ field of that wire is zero, because the angle θ for that wire sweeps through a net change of zero rather than 2π during the integration. Any conductors present that are not enclosed by a particular path may still contribute to the value of $\vec{B}$ at every point, but the *line integrals* of their fields around the path are zero.

Thus we can replace I in Eq. (28.19) with I_{encl}, the algebraic sum of the currents *enclosed* or *linked* by the integration path, with the sum evaluated by using the sign rule just described. Our statement of **Ampere's law** is then

$$\oint \vec{B} \cdot d\vec{l} = \mu_0 I_{encl} \quad \text{(Ampere's law)} \tag{28.20}$$

While we have derived Ampere's law only for the special case of the field of several long, straight parallel conductors, Eq. (28.20) is in fact valid for conductors and paths of *any* shape. The general derivation is no different in principle from what we have presented, but the geometry is more complicated.

If $\oint \vec{B} \cdot d\vec{l} = 0$, it *does not* necessarily mean that $\vec{B} = \mathbf{0}$ everywhere along the path, only that the total current through an area bounded by the path is zero. In Figs. 28.15c and 28.16b, the integration paths enclose no current at all; in Fig. 28.17 there are positive and negative currents of equal magnitude through the area enclosed by the path. In both cases, $I_{encl} = 0$ and the line integral is zero.

CAUTION In Chapter 23 we saw that the line integral of the electrostatic field $\vec{E}$ around any closed path is equal to zero; this is a statement that the electrostatic force $\vec{F} = q\vec{E}$ on a point charge q is conservative, so this force does zero work on a charge that moves around a closed path that returns to the starting

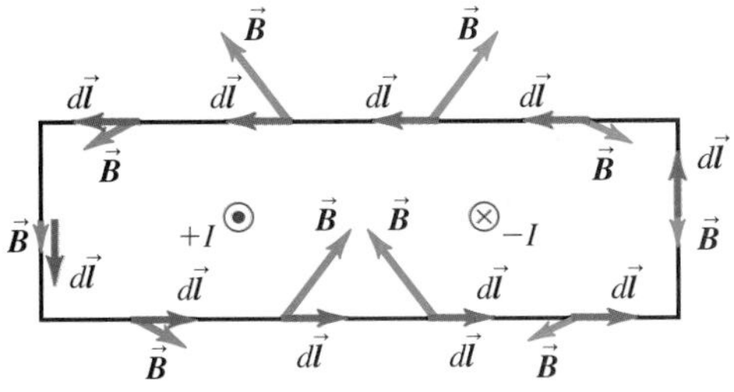

28.17 Two long, straight conductors carrying equal currents in opposite directions. The conductors are seen end-on, and the integration path is counterclockwise. The line integral $\oint \vec{B} \cdot d\vec{l}$ gets zero contribution from the upper and lower segments, a positive contribution from the left segment, and a negative contribution from the right segment; the net integral is zero.

point. You might think that the value of the line integral $\oint \vec{B} \cdot d\vec{l}$ is similarly related to the question of whether the *magnetic* force is conservative. This isn't the case at all. Remember that the magnetic force $\vec{F} = q\vec{v} \times \vec{B}$ on a moving charged particle is always *perpendicular* to $\vec{B}$, so $\oint \vec{B} \cdot d\vec{l}$ is *not* related to the work done by the magnetic force; as stated in Ampere's law, this integral is related only to the total current through a surface bounded by the integration path. In fact, the magnetic force on a moving charged particle is *not* conservative. A conservative force depends only on the position of the body on which the force is exerted, but the magnetic force on a moving charged particle also depends on the *velocity* of the particle.

ActivPhysics ONLINE

13.3 Magnetic Field of a Solenoid

In the form we have stated it, Ampere's law turns out to be valid only if the currents are steady and if no magnetic materials or time-varying electric fields are present. In Chapter 29 we will see how to generalize Ampere's law for time-varying fields.

Test Your Understanding

We learned in Section 27.3 that all magnetic field lines form closed loops. Use Ampere's law to explain why any such closed loop must enclose a current. By examining Fig. 27.11a, explain why a bar magnet must have currents in its interior.

28.7 | Applications of Ampere's Law

Ampere's law is useful when we can exploit the symmetry of a situation to evaluate the line integral of $\vec{B}$. Several examples are given below. The following problem-solving strategy is directly analogous to the strategy suggested in Section 22.4 for applications of Gauss's law; we suggest you review that strategy now and compare the two methods.

Problem-Solving Strategy

Ampere's Law

IDENTIFY *the relevant concepts:* Like Gauss's law for electricity, Ampere's law is always true but is most useful in situations where the magnetic field pattern is highly symmetrical. In such situations you can use Ampere's law to find a relationship between the magnetic field as a function of position and the current that generates the field.

SET UP *the problem using the following steps:*

1. Select the integration path you will use with Ampere's law. If you want to determine the magnetic field at a certain point, then the path must pass through that point. The integration path doesn't have to be any actual physical boundary. Usually it is a purely geometric curve; it may be in empty space, embedded in a solid body, or some of each. The integration path has to have enough *symmetry* to make evaluation of the integral possible. If the problem itself has cylindrical symmetry, the integration path will usually be a circle coaxial with the cylinder axis.
2. Determine the target variable(s). Usually this will be the magnitude of the $\vec{B}$ field as a function of position.

EXECUTE *the solution as follows:*

1. Carry out the integral $\oint \vec{B} \cdot d\vec{l}$ along your chosen integration path. If $\vec{B}$ is tangent to all or some portion of the integration path and has the same magnitude B at every point, then its line integral equals B multiplied by the length of that portion of the path. If $\vec{B}$ is perpendicular to some portion of the path, that portion makes no contribution to the integral.

2. In the integral $\oint \vec{B} \cdot d\vec{l}$, $\vec{B}$ is always the *total* magnetic field at each point on the path. This field can be caused partly by currents enclosed by the path and partly by currents outside the path. If *no* net current is enclosed by the path, the field at points on the path need not be zero, but the integral $\oint \vec{B} \cdot d\vec{l}$ is always zero.
3. Determine the current I_{encl} enclosed by the integration path. The sign of this current is given by a right-hand rule. Curl the fingers of your right hand so that they follow the integration path in the direction that you carry out the integration. Your right thumb then points in the direction of positive current. If $\vec{B}$ is tangent to the integration at all points along the path and I_{encl} is positive, then the direction of $\vec{B}$ is the same as the direction of the integration path; if instead I_{encl} is negative, $\vec{B}$ is in the direction opposite to that of the integration.
4. Use Ampere's law $\oint \vec{B} \cdot d\vec{l} = \mu_0 I$ to solve for the target variable.

EVALUATE *your answer:* If your result is an expression for the field magnitude as a function of position, you can check it by examining how the expression behaves in different limits.

Example 28.8 Field of a long, straight, current-carrying conductor

In Section 28.6 we derived Ampere's law using Eq. (28.9) for the field of a long, straight, current-carrying conductor. Reverse this process, and use Ampere's law to find the magnitude *and* direction of $\vec{B}$ for this situation.

SOLUTION

IDENTIFY and SET UP: We exploit the cylindrical symmetry of the situation by taking as our integration path a circle with radius r centered on the conductor and in a plane perpendicular to it, as in Fig. 28.15a (Section 28.6). At each point, $\vec{B}$ is tangent to this circle.

EXECUTE: With our choice of integration path, Ampere's law [Eq. (28.20)] becomes

$$\oint \vec{B} \cdot d\vec{l} = \oint B_{\parallel}\, dl = B(2\pi r) = \mu_0 I$$

Equation (28.9), $B = \mu_0 I/2\pi r$, follows immediately.

Ampere's law determines the direction of $\vec{B}$ as well as its magnitude. Since we go around the integration path in the counterclockwise direction, the positive direction for current is out of the plane of Fig. 28.15a; this is the same as the actual current direction in the figure, so I is positive and the integral $\oint \vec{B} \cdot d\vec{l}$ is also positive. Since the $d\vec{l}$'s run counterclockwise, the direction of $\vec{B}$ must be counterclockwise as well, as shown in Fig. 28.15a.

EVALUATE: Our results are consistent with those in Section 28.6, as they must be.

Example 28.9 Field inside a long cylindrical conductor

A cylindrical conductor with radius R carries a current I. The current is uniformly distributed over the cross-sectional area of the conductor. Find the magnetic field as a function of the distance r from the conductor axis for points both inside $(r < R)$ and outside $(r > R)$ the conductor.

SOLUTION

IDENTIFY and SET UP: To find the magnetic field *inside* the conductor, we take as our integration path a circle with radius $r < R$ as shown in Fig. 28.18. *Outside* the conductor, we again use a circle but with a radius $r > R$. In either case, the integration path takes advantage of the circular symmetry of the magnetic field pattern.

EXECUTE: Inside the conductor, $\vec{B}$ has the same magnitude at every point on the circular integration path and is tangent to the path. Thus the magnitude of the line integral is simply $B(2\pi r)$. If we use

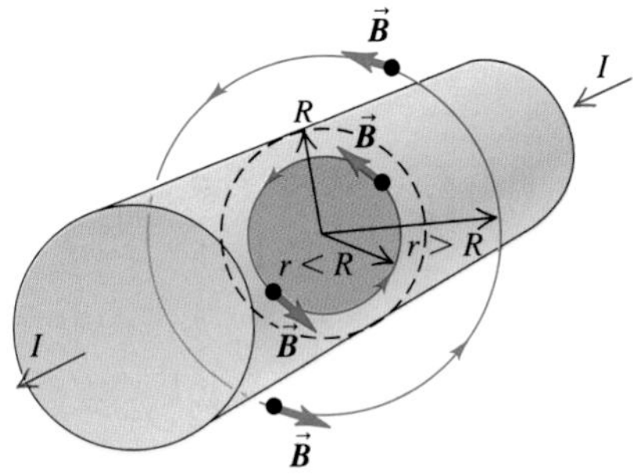

28.18 To find the magnetic field at radius $r < R$, we apply Ampere's law to the circle enclosing the red area. The current through the red area is $(r^2/R^2)I$. To find the magnetic field at radius $r > R$, we apply Ampere's law to the circle enclosing the entire conductor.

the right-hand rule for determining the sign of the current, the current through the red area enclosed by the path is positive; hence $\vec{B}$ points in the same direction as the integration path, as shown. To find the current I_{encl} enclosed by the path, note that the current density (current per unit area) is $J = I/\pi R^2$, so $I_{encl} = J(\pi r^2) = Ir^2/R^2$. Finally, Ampere's law gives

$$B(2\pi r) = \mu_0 \frac{Ir^2}{R^2}$$

$$B = \frac{\mu_0 I}{2\pi}\frac{r}{R^2} \quad \text{(inside the conductor, } r < R\text{)} \qquad (28.21)$$

For the circular integration path outside the conductor $(r > R)$, the same symmetry arguments apply and the magnitude of $\oint \vec{B} \cdot d\vec{l}$ is again $B(2\pi r)$. The right-hand rule gives the direction of $\vec{B}$ as shown in Fig. 28.18. For this path, $I_{encl} = I$, the total current in the conductor. Applying Ampere's law gives the same equation as in Example 28.8, with the same result for B:

$$B = \frac{\mu_0 I}{2\pi r} \quad \text{(outside the conductor, } r > R\text{)} \qquad (28.22)$$

Outside the conductor, the magnetic field is the same as that of a long, straight conductor carrying current I, independent of the radius R over which the current is distributed. Indeed, the magnetic field outside *any* cylindrically symmetric current distribution is the same as if the entire current were concentrated along the axis of the distribution. This is analogous to the results of Examples 22.5 and 22.9 (Section 22.4), in which we found that the *electric* field outside a spherically symmetric *charged* body is the same as though the entire charge were concentrated at the center.

EVALUATE: Note that at the surface of the conductor $(r = R)$, Eq. (28.21) for $r < R$ and Eq. (28.22) for $r > R$ agree (as they must). Figure 28.19 shows a graph of B as a function of r, both inside and outside the conductor.

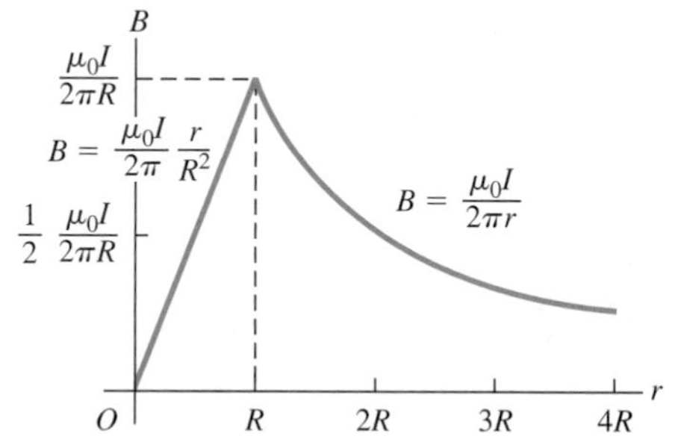

28.19 Magnitude of the magnetic field inside and outside a long, straight cylindrical conductor with radius R carrying a current I.

Example 28.10 Field of a solenoid

A solenoid, discussed in Example 28.6, consists of a helical winding of wire on a cylinder, usually circular in cross section. There can be hundreds or thousands of closely spaced turns, each of which can be regarded as a circular loop. There may be several layers of windings. For simplicity, Fig. 28.20 shows a solenoid with only a few turns. All turns carry the same current I, and the total $\vec{B}$ field at every point is the vector sum of the fields caused by the individual turns. The figure shows field lines in the xy- and xz-planes. We draw a set of field lines that are uniformly spaced at the center of the solenoid. Exact calculations show that for a long, closely wound solenoid, half of these field lines emerge from the ends and half "leak out" through the windings between the center and the end.

The field lines near the center of the solenoid are approximately parallel, indicating a nearly uniform $\vec{B}$; outside the solenoid, the field lines are spread apart, and the magnetic field is weak. If the solenoid is long in comparison with its cross-sectional diameter and the coils are tightly wound, the *internal* field near the midpoint of the solenoid's length is very nearly uniform over the cross section and parallel to the axis, and the *external* field near the midpoint is very small.

Use Ampere's law to find the field at or near the center of such a long solenoid. The solenoid has n turns of wire per unit length and carries a current I.

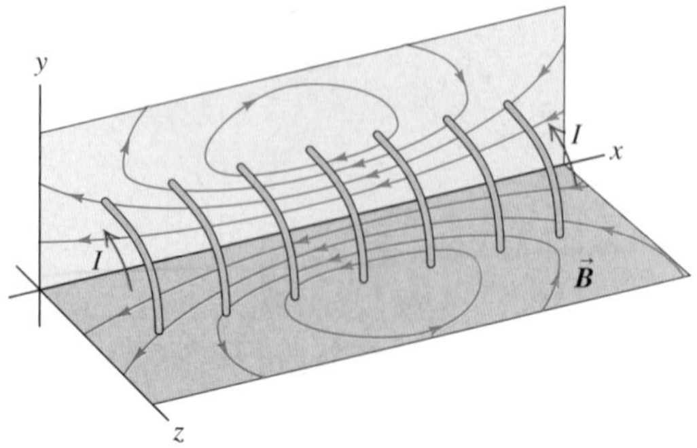

28.20 Magnetic field lines produced by the current in a solenoid. For clarity, only a few turns are shown.

SOLUTION

IDENTIFY and SET UP: We choose as our integration path the rectangle *abcd* in Fig. 28.21. Side *ab*, with length L, is parallel to the axis of the solenoid. Sides *bc* and *da* are taken to be very long so that side *cd* is far from the solenoid; then the field at side *cd* is negligibly small.

EXECUTE: By symmetry, the $\vec{B}$ field along side *ab* is parallel to this side and is constant. In carrying out the Ampere's-law integration, we go along side *ab* in the same direction as $\vec{B}$. So for this side, $B_\parallel = +B$ and

$$\int_a^b \vec{B} \cdot d\vec{l} = BL$$

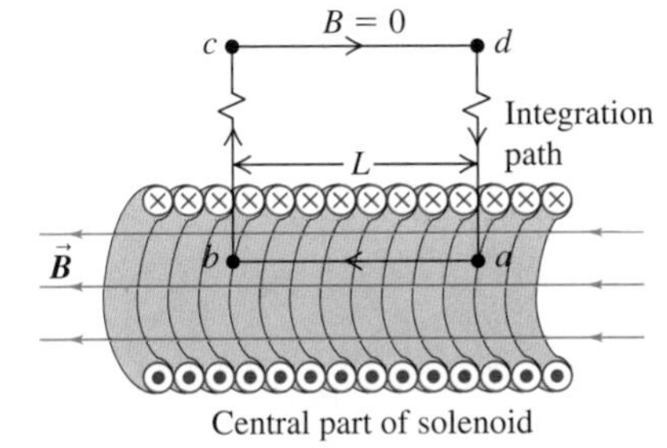

28.21 A section of a long, tightly wound solenoid centered on the x-axis, showing magnetic field lines in the xy- and xz-planes.

Along sides bc and da, $B_\parallel = 0$ because $\vec{B}$ is perpendicular to these sides; along side cd, $B_\parallel = 0$ because $\vec{B} = \mathbf{0}$. The integral $\oint \vec{B} \cdot d\vec{l}$ around the entire closed path therefore reduces to BL.

The number of turns in length L is nL. Each of these turns passes once through the rectangle $abcd$ and carries a current I, where I is the current in the windings. The total current enclosed by the rectangle is then $I_{encl} = nLI$. From Ampere's law, since the integral $\oint \vec{B} \cdot d\vec{l}$ is positive, I_{encl} must be positive as well; hence the current passing through the surface bounded by the integration path must be in the direction shown in Fig. 28.21. Ampere's law then gives the magnitude B:

$$BL = \mu_0 nLI$$

$$B = \mu_0 nI \quad \text{(solenoid)} \tag{28.23}$$

Side ab need not lie on the axis of the solenoid, so this calculation also proves that the field is uniform over the entire cross section at the center of the solenoid's length.

EVALUATE: Note that the *direction* of $\vec{B}$ inside the solenoid is in the same direction as the solenoid's vector magnetic moment $\vec{\mu}$. This is the same result that we found in Section 28.5 for a single current-carrying loop.

For points along the axis, the field is strongest at the center of the solenoid and drops off near the ends. For a solenoid that is very long in comparison to its diameter, the field at each end is exactly half as strong as the field at the center. For a short, fat solenoid the relationship is more complicated. Figure 28.22 shows a graph of B as a function of x for points on the axis of a short solenoid.

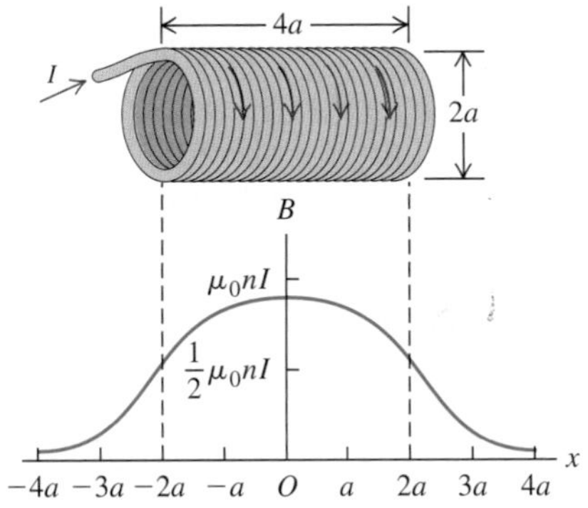

28.22 Magnitude of the magnetic field at points along the axis of a solenoid with length $4a$, equal to four times its radius a. The field magnitude at each end is about half its value at the center. (Compare with Fig. 28.13 for the field of N circular loops.)

Example 28.11 Field of a toroidal solenoid

Figure 28.23a shows a doughnut-shaped **toroidal solenoid,** also called a *toroid,* wound with N turns of wire carrying a current I. In a practical version the turns would be more closely spaced than they are in the figure. Find the magnetic field at all points.

SOLUTION

IDENTIFY and SET UP: The flow of current around the toroid's circumference produces a magnetic field component perpendicular to the plane of the figure, just as for the current loop discussed in Section 28.5. But if the coils are very tightly wound, we can consider them as circular loops that carry current between the inner and outer radii of the toroidal solenoid; the flow of current around the toroid's circumference is then negligible, and the perpendicular component of $\vec{B}$ is likewise negligible. In this idealized approximation the circular symmetry of the situation tells us that the magnetic field lines must be circles concentric with the axis of the toroid. To take advantage of this symmetry in finding the field, we choose circular integration paths for use with Ampere's law. Three such paths are shown as black lines in Fig. 28.23b.

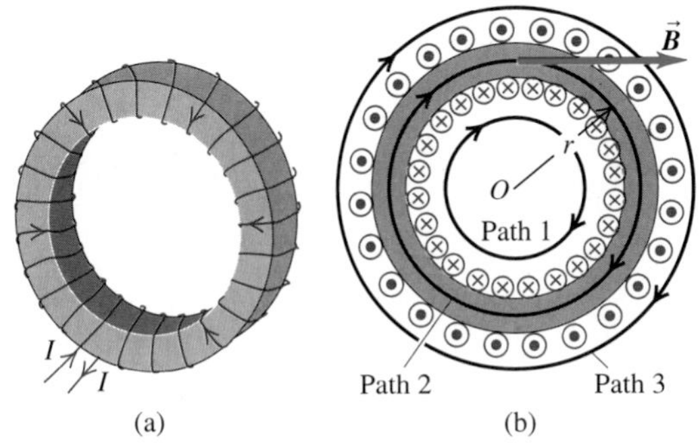

28.23 (a) A toroidal solenoid. For clarity, only a few turns of the winding are shown. (b) Integration paths (black circles) used to compute the magnetic field $\vec{B}$ set up by a current (shown as dots and crosses) in a toroidal solenoid. The field is very nearly zero at all points except those within the space enclosed by the windings.

EXECUTE: First consider integration path 1 in Fig. 28.23b. If the toroidal solenoid produces any field at all in this region, it must be *tangent* to the path at all points, and $\oint \vec{B} \cdot d\vec{l}$ will equal the product of B and the circumference $l = 2\pi r$ of the path. But the total current enclosed by the path is zero, so from Ampere's law the field $\vec{B}$ must be zero everywhere on this path.

Similarly, if the toroidal solenoid produces any field along path 3, it must also be tangent to the path at all points. Each turn of the winding passes *twice* through the area bounded by this path, carrying equal currents in opposite directions. The *net* current I_{encl} enclosed within this area is therefore zero, and hence $\vec{B} = \mathbf{0}$ at all points of the path. Conclusion: *The field of an idealized toroidal solenoid is confined completely to the space enclosed by the windings.* We can think of such an idealized toroidal solenoid as a tightly wound solenoid that has been bent into a circle.

Finally, we consider path 2, a circle with radius r. Again by symmetry we expect the $\vec{B}$ field to be tangent to the path, and $\oint \vec{B} \cdot d\vec{l}$ equals $2\pi rB$. Each turn of the winding passes *once* through the area bounded by path 2. The total current enclosed by the path is $I_{encl} = NI$, where N is the total number of turns in the winding; I_{encl} is positive for the clockwise direction of integration in Fig. 28.23b, so $\vec{B}$ is in the direction shown. Then, from Ampere's law,

$$2\pi rB = \mu_0 NI$$

$$B = \frac{\mu_0 NI}{2\pi r} \quad \text{(toroidal solenoid)} \qquad (28.24)$$

EVALUATE: The magnetic field is *not* uniform over a cross section of the core, because the radius r is larger at the outer side of the section than at the inner side. However, if the radial thickness of the core is small in comparison to r, the field varies only slightly across a section. In that case, considering that $2\pi r$ is the circumferential length of the toroid and that $N/2\pi r$ is the number of turns per unit length n, the field may be written as

$$B = \mu_0 nI$$

just as it is at the center of a long, *straight* solenoid.

In a real toroidal solenoid the turns are not precisely circular loops but rather segments of a bent helix. As a result, the field outside is not strictly zero. To estimate its magnitude, we imagine Fig. 28.23a as being roughly equivalent, for points outside the torus, to a circular loop with a single turn and radius r. Then we can use Eq. (28.17) to show that the field at the *center* of the torus is smaller than the field inside by approximately a factor of N/π.

The equations we have derived for the field in a closely wound straight or toroidal solenoid are strictly correct only for windings in *vacuum*. For most practical purposes, however, they can be used for windings in air or on a core of any nonmagnetic, nonsuperconducting material. In the next section we will show how they are modified if the core is a magnetic material.

Test Your Understanding

Consider a conducting wire that runs along the central axis of a hollow conducting cylinder. Such an arrangement, called a *coaxial cable,* has many applications in telecommunications. (See Fig. 22.20. The cable that connects a television set to a local cable provider is an example of a coaxial cable.) In such a cable a current I runs in one direction along the hollow conducting cylinder and is spread uniformly over the cylinder's cross-sectional area. An equal current runs in the opposite direction along the central wire. Use Ampere's law to explain why the currents produce no magnetic field outside the cable.

*28.8 | Magnetic Materials

In discussing how currents cause magnetic fields, we have assumed that the conductors are surrounded by vacuum. But the coils in transformers, motors, generators, and electromagnets nearly always have iron cores to increase the magnetic field and confine it to desired regions. Permanent magnets, magnetic recording tapes, and computer disks depend directly on the magnetic properties of materials; when you store information on a computer disk, you are actually setting up an array of microscopic permanent magnets on the disk. So it is worthwhile to examine some aspects of the magnetic properties of materials. After describing the atomic origins of magnetic properties, we will discuss three broad classes of magnetic behavior that occur in materials; these are called *paramagnetism, diamagnetism,* and *ferromagnetism.*

The Bohr Magneton

As we discussed briefly in Section 27.7, the atoms that make up all matter contain moving electrons, and these electrons form microscopic current loops that produce magnetic fields of their own. In many materials these currents are randomly oriented and cause no net magnetic field. But in some materials an external field (a field produced by currents outside the material) can cause these loops to become oriented preferentially with the field, so their magnetic fields *add* to the external field. We then say that the material is *magnetized.*

Let's look at how these microscopic currents come about. Figure 28.24 shows a primitive model of an electron in an atom. We picture the electron (mass m, charge $-e$) as moving in a circular orbit with radius r and speed v. This moving charge is equivalent to a current loop. In Section 27.7 we found that a current loop with area A and current I has a magnetic dipole moment μ given by $\mu = IA$; for the orbiting electron the area of the loop is $A = \pi r^2$. To find the current associated with the electron, we note that the orbital period T (the time for the electron to make one complete orbit) is the orbit circumference divided by the electron speed: $T = 2\pi r/v$. The equivalent current I is the total charge passing any point on the orbit per unit time, which is just the magnitude e of the electron charge divided by the orbital period T:

$$I = \frac{e}{T} = \frac{ev}{2\pi r}$$

The magnetic moment $\mu = IA$ is then

$$\mu = \frac{ev}{2\pi r}(\pi r^2) = \frac{evr}{2} \qquad (28.25)$$

It is useful to express μ in terms of the *angular momentum* L of the electron. For a particle moving in a circular path, the magnitude of angular momentum equals the magnitude of momentum mv multiplied by the radius r, that is, $L = mvr$ (Section 10.5). Comparing this with Eq. (28.25), we can write

$$\mu = \frac{e}{2m}L \qquad (28.26)$$

Equation (28.26) is useful in this discussion because atomic angular momentum is *quantized;* its component in a particular direction is always an integer multiple of $h/2\pi$, where h is a fundamental physical constant called *Planck's constant.* (We will discuss the quantization of angular momentum in more detail in Chapter 41.) The numerical value of h is

$$h = 6.626 \times 10^{-34}\ \text{J}\cdot\text{s}$$

The quantity $h/2\pi$ thus represents a fundamental unit of angular momentum in atomic systems, just as e is a fundamental unit of charge. Associated with the quantization of $\vec{L}$ is a fundamental uncertainty in the *direction* of $\vec{L}$ and therefore of $\vec{\mu}$. In the following discussion, when we speak of the magnitude of a magnetic moment, a more precise statement would be "maximum component in a given direction." Thus, to say that a magnetic moment $\vec{\mu}$ is aligned with a magnetic field $\vec{B}$ really means that $\vec{\mu}$ has its maximum possible component in the direction of $\vec{B}$; such components are always quantized.

Equation (28.26) shows that associated with the fundamental unit of angular momentum is a corresponding fundamental unit of magnetic moment. If $L = h/2\pi$, then

$$\mu = \frac{e}{2m}\left(\frac{h}{2\pi}\right) = \frac{eh}{4\pi m} \qquad (28.27)$$

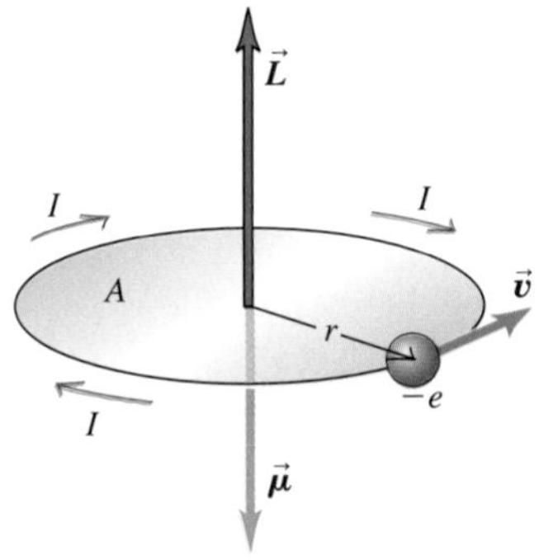

28.24 An electron moving with speed v in a circular orbit of radius r has an angular momentum $\vec{L}$ and an oppositely directed orbital magnetic dipole moment $\vec{\mu}$. It also has a spin angular momentum and an oppositely directed spin magnetic dipole moment.

This quantity is called the **Bohr magneton,** denoted by μ_B. Its numerical value is

$$\mu_B = 9.274 \times 10^{-24}\ \text{A}\cdot\text{m}^2 = 9.274 \times 10^{-24}\ \text{J/T}$$

You should verify that these two sets of units are consistent. The second set is useful when we compute the potential energy $U = -\vec{\mu}\cdot\vec{B}$ for a magnetic moment in a magnetic field.

Electrons also have an intrinsic angular momentum, called *spin,* that is not related to orbital motion but that can be pictured in a classical model as spinning on an axis. This angular momentum also has an associated magnetic moment, and its magnitude turns out to be almost exactly one Bohr magneton. (Effects having to do with quantization of the electromagnetic field cause the spin magnetic moment to be about 1.001 μ_B.)

Paramagnetism

In an atom, most of the various orbital and spin magnetic moments of the electrons add up to zero. However, in some cases the atom has a net magnetic moment that is of the order of μ_B. When such a material is placed in a magnetic field, the field exerts a torque on each magnetic moment, as given by Eq. (27.26): $\vec{\tau} = \vec{\mu} \times \vec{B}$. These torques tend to align the magnetic moments with the field, the position of minimum potential energy, as we discussed in Section 27.7. In this position, the directions of the current loops are such as to *add* to the externally applied magnetic field.

We saw in Section 28.5 that the $\vec{B}$ field produced by a current loop is proportional to the loop's magnetic dipole moment. In the same way, the additional $\vec{B}$ field produced by microscopic electron current loops is proportional to the total magnetic moment $\vec{\mu}_{\text{total}}$ per unit volume V in the material. We call this vector quantity the **magnetization** of the material, denoted by $\vec{M}$:

$$\vec{M} = \frac{\vec{\mu}_{\text{total}}}{V} \qquad (28.28)$$

The additional magnetic field due to magnetization of the material turns out to be equal simply to $\mu_0\vec{M}$, where μ_0 is the same constant that appears in the law of Biot and Savart and Ampere's law. When such a material completely surrounds a current-carrying conductor, the total magnetic field $\vec{B}$ in the material is

$$\vec{B} = \vec{B}_0 + \mu_0\vec{M} \qquad (28.29)$$

where $\vec{B}_0$ is the field caused by the current in the conductor.

To check that the units in Eq. (28.29) are consistent, note that magnetization $\vec{M}$ is magnetic moment per unit volume. The units of magnetic moment are current times area $(\text{A}\cdot\text{m}^2)$, so the units of magnetization are $(\text{A}\cdot\text{m}^2)/\text{m}^3 = \text{A/m}$. From Section 28.1, the units of the constant μ_0 are $\text{T}\cdot\text{m/A}$. So the units of $\mu_0\vec{M}$ are the same as the units of $\vec{B}$: $(\text{T}\cdot\text{m/A})(\text{A/m}) = \text{T}$.

A material showing the behavior just described is said to be **paramagnetic.** The result is that the magnetic field at any point in such a material is greater by a dimensionless factor K_m, called the **relative permeability** of the material, than it would be if the material were replaced by vacuum. The value of K_m is different for different materials; for common paramagnetic solids and liquids at room temperature, K_m typically ranges from 1.00001 to 1.003.

All of the equations in this chapter that relate magnetic fields to their sources can be adapted to the situation in which the current-carrying conductor is embedded in a paramagnetic material. All that need be done is to replace μ_0 by $K_m\mu_0$. This product is usually denoted as μ and is called the **permeability** of the material:

$$\mu = K_m\mu_0 \qquad (28.30)$$

CAUTION Equation (28.30) involves some really dangerous notation because we have also used μ for magnetic dipole moment. It's customary to use μ for both quantities, but beware; from now on, every time you see a μ, make sure you know whether it is permeability or magnetic moment. You can usually tell from the context.

The amount by which the relative permeability differs from unity is called the **magnetic susceptibility,** denoted by χ_m:

$$\chi_m = K_m - 1 \qquad (28.31)$$

Both K_m and χ_m are dimensionless quantities. Values of magnetic susceptibility for several materials are given in Table 28.1. For example, for aluminum, $\chi_m = 2.2 \times 10^{-5}$ and $K_m = 1.000022$. The first group of materials in the table are paramagnetic; we'll discuss the second group of materials, which are called *diamagnetic,* very shortly.

The tendency of atomic magnetic moments to align themselves parallel to the magnetic field (where the potential energy is minimum) is opposed by random thermal motion, which tends to randomize their orientations. For this reason, paramagnetic susceptibility always decreases with increasing temperature. In many cases it is inversely proportional to the absolute temperature T, and the magnetization M can be expressed as

$$M = C\frac{B}{T} \qquad (28.32)$$

This relation is called *Curie's law,* after its discoverer, Pierre Curie (1859–1906). The quantity C is a constant, different for different materials, called the *Curie constant.*

As we described in Section 27.7, a body with atomic magnetic dipoles is attracted to the poles of a magnet. In most paramagnetic substances this attraction is very weak due to thermal randomization of the atomic magnetic moments. That's why a magnet can't be used to pick up objects made of aluminum (a paramagnetic substance). But at very low temperatures the thermal effects are reduced, the magnetization increases in accordance with Curie's law, and the attractive forces are greater.

Table 28.1 Magnetic Susceptibilities of Paramagnetic and Diamagnetic Materials at $T = 20°C$

Material	$\chi_m = K_m - 1\ (\times 10^{-5})$
Paramagnetic	
Iron ammonium alum	66
Uranium	40
Platinum	26
Aluminum	2.2
Sodium	0.72
Oxygen gas	0.19
Diamagnetic	
Bismuth	−16.6
Mercury	−2.9
Silver	−2.6
Carbon (diamond)	−2.1
Lead	−1.8
Sodium chloride	−1.4
Copper	−1.0

Example 28.12 Magnetic dipoles in a paramagnetic material

Nitric oxide (NO) is a paramagnetic compound. Its molecules have a magnetic moment with a maximum component in any direction of about one Bohr magneton each. In a magnetic field with magnitude $B = 1.5$ T, compare the interaction energy of the magnetic moments with the field to the average translational kinetic energy of the molecules at a temperature of 300 K.

SOLUTION

IDENTIFY and SET UP: In Section 27.6 we derived the equation $U = -\vec{\mu}\cdot\vec{B}$ for the interaction energy of a magnetic moment $\vec{\mu}$ with a $\vec{B}$ field. From Section 18.3 the average translational kinetic energy of a molecule at temperature T is $K = \frac{3}{2}kT$, where k is the Boltzmann constant.

EXECUTE: We can write the interaction energy as $U = -(\mu \cos \phi)B$, where $\mu \cos \phi$ is the component of the magnetic moment $\vec{\mu}$ in the direction of the $\vec{B}$ field. In our case the maximum value of the component $\mu \cos \phi$ is about μ_B, so

$$|U|_{max} \approx \mu_B B = (9.27 \times 10^{-24}\ \text{J/T})(1.5\ \text{T})$$
$$= 1.4 \times 10^{-23}\ \text{J} = 8.7 \times 10^{-5}\ \text{eV}$$

The average translational kinetic energy K is

$$K = \frac{3}{2}kT = \frac{3}{2}(1.38 \times 10^{-23}\ \text{J/K})(300\ \text{K})$$
$$= 6.2 \times 10^{-21}\ \text{J} = 0.039\ \text{eV}$$

EVALUATE: At a temperature of 300 K the magnetic interaction energy is much *smaller* than the random kinetic energy, so we expect only a slight degree of alignment. This is why paramagnetic susceptibilities at ordinary temperature are usually very small.

Diamagnetism

In some materials the total magnetic moment of all the atomic current loops is zero when no magnetic field is present. But even these materials have magnetic effects because an external field alters electron motions within the atoms, causing additional current loops and induced magnetic dipoles comparable to the induced *electric* dipoles we studied in Section 24.5. In this case the additional field caused by these current loops is always *opposite* in direction to that of the external field. (This behavior is explained by Faraday's law of induction, which we will study in Chapter 29. An induced current always tends to cancel the field change that caused it.)

Such materials are said to be **diamagnetic.** They always have negative susceptibility, as shown in Table 28.1, and relative permeability K_m slightly *less* than unity, typically of the order of 0.99990 to 0.99999 for solids and liquids. Diamagnetic susceptibilities are very nearly temperature-independent.

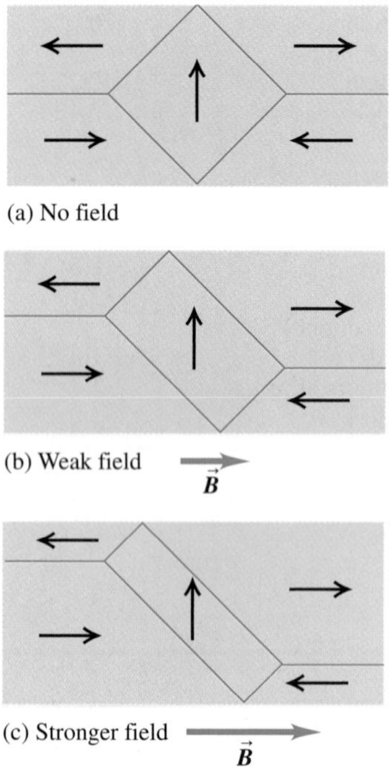

28.25 In this drawing adapted from a magnified photo, the arrows show the directions of magnetization in the domains of a single crystal of nickel. Domains that are magnetized in the direction of an applied magnetic field grow larger.

Ferromagnetism

There is a third class of materials, called **ferromagnetic** materials, which includes iron, nickel, cobalt, and many alloys containing these elements. In these materials, strong interactions between atomic magnetic moments cause them to line up parallel to each other in regions called **magnetic domains,** even when no external field is present. Figure 28.25 shows an example of magnetic domain structure. Within each domain, nearly all of the atomic magnetic moments are parallel.

When there is no externally applied field, the domain magnetizations are randomly oriented. But when a field $\vec{B}_0$ (caused by external currents) is present, the domains tend to orient themselves parallel to the field. The domain boundaries also shift; the domains that are magnetized in the field direction grow, and those that are magnetized in other directions shrink. Because the total magnetic moment of a domain may be many thousands of Bohr magnetons, the torques that tend to align the domains with an external field are much stronger than occur with paramagnetic materials. The relative permeability K_m is *much* larger than unity, typically of the order of 1,000 to 100,000. As a result, an object made of a ferromagnetic material such as iron is strongly magnetized by the field from a permanent magnet and is attracted to the magnet (see Fig. 27.36). A paramagnetic material such as aluminum is also attracted to a permanent magnet, but K_m for paramagnetic materials is so much smaller for such a material than for ferromagnetic materials that the attraction is very weak. Thus a magnet can pick up iron nails, but not aluminum cans (Fig. 28.26).

As the external field is increased, a point is eventually reached at which nearly *all* the magnetic moments in the ferromagnetic material are aligned parallel to the external field. This condition is called *saturation magnetization;* after it is reached, further increase in the external field causes no increase in magnetization or in the additional field caused by the magnetization.

Figure 28.27 shows a "magnetization curve," a graph of magnetization M as a function of external magnetic field B_0, for soft iron. An alternative description of this behavior is that K_m is not constant but decreases as B_0 increases. (Paramagnetic materials also show saturation at sufficiently strong fields. But the magnetic fields required are so large that departures from a linear relationship between M and B_0 in these materials can be observed only at very low temperatures, 1 K or so.)

For many ferromagnetic materials the relation of magnetization to external magnetic field is different when the external field is increasing from when it is decreasing. Figure 28.28a shows this relation for such a material. When the material is magnetized to saturation and then the external field is reduced to zero, some magnetization remains. This behavior is characteristic of permanent magnets, which retain most of their saturation magnetization when the magnetizing field is removed. To reduce the magnetization to zero requires a magnetic field in the reverse direction.

This behavior is called **hysteresis,** and the curves in Fig. 28.28 are called *hysteresis loops.* Magnetizing and demagnetizing a material that has hysteresis involves the dissipation of energy, and the temperature of the material increases during such a process.

Ferromagnetic materials are widely used in electromagnets, transformer cores, and motors and generators, in which it is desirable to have as large a magnetic field as possible for a given current. Because hysteresis dissipates energy, materials that are used in these applications should usually have as narrow a hysteresis loop as possible. Soft iron is often used; it has high permeability without appreciable hysteresis. For permanent magnets a broad hysteresis loop is usually desirable, with large zero-field magnetization and large reverse field needed to demagnetize. Many kinds of steel and many alloys, such as Alnico, are commonly

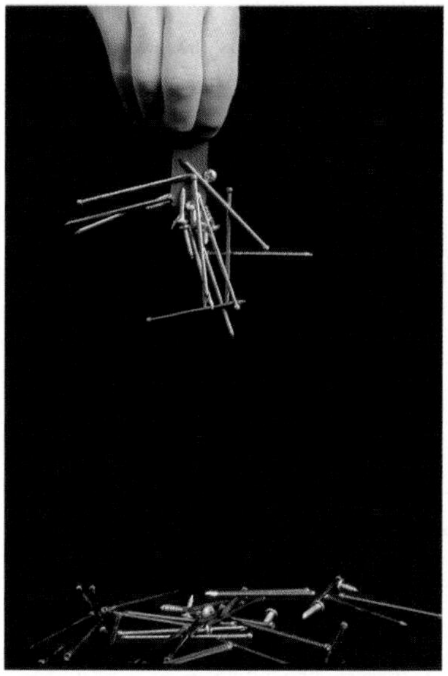

28.26 This permanent magnet can pick up screws and nails made of steel (a ferromagnetic material). Nails made of aluminum (which is paramagnetic) would also be attracted to the magnet, but with so little force that the magnet could not pick them up.

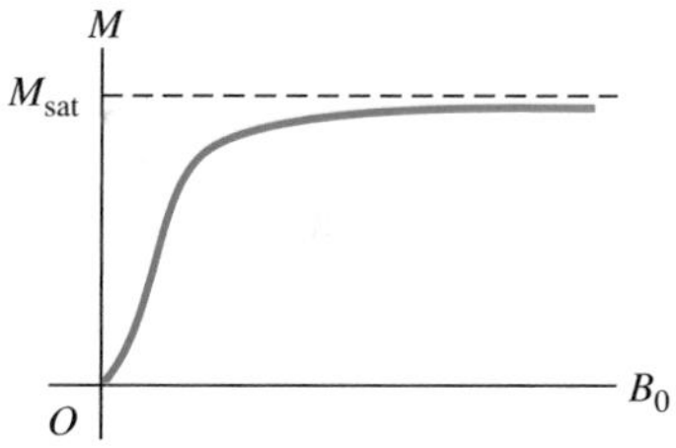

28.27 A magnetization curve for a ferromagnetic material. The magnetization M approaches its saturation value M_{sat} as the magnetic field B_0 (caused by external currents) becomes large.

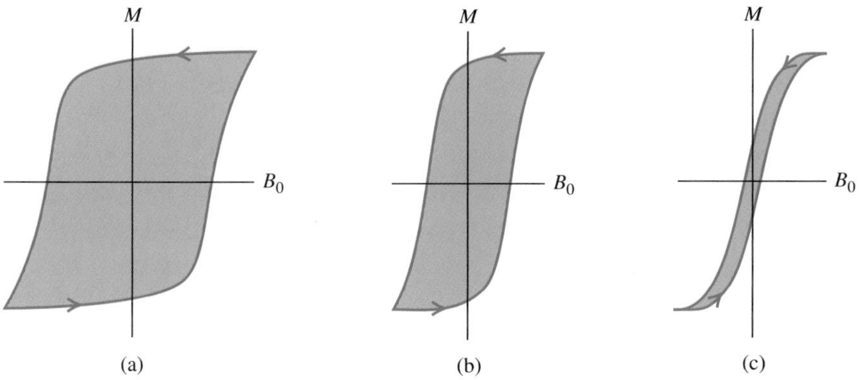

28.28 Hysteresis loops. The materials of both (a) and (b) remain strongly magnetized when B_0 is reduced to zero. Since (a) is also hard to demagnetize, it would be good for permanent magnets. Since (b) magnetizes and demagnetizes more easily, it could be used as a computer memory material. The material of (c) would be useful for transformers and other alternating-current devices where zero hysteresis would be optimal.

used for permanent magnets. The remaining magnetic field in such a material, after it has been magnetized to near saturation, is typically of the order of 1 T, corresponding to a remaining magnetization $M = B/\mu_0$ of about 800,000 A/m.

Example 28.13 A ferromagnetic material

A permanent magnet is made of a ferromagnetic material with a magnetization M of about 8×10^5 A/m. The magnet is in the shape of a cube of side 2 cm. a) Find the magnetic dipole moment of the magnet. b) Estimate the magnetic field due to the magnet at a point 10 cm from the magnet along its axis.

SOLUTION

IDENTIFY and SET UP: We find the magnetic dipole moment from the magnetization, which equals magnetic moment per unit volume. To estimate the magnetic field, we approximate the magnet as a current loop with the same magnetic moment and use the results of Section 28.5.

EXECUTE: a) The total magnetic moment is the magnetization multiplied by the volume:

$$\mu_{\text{total}} = MV = (8 \times 10^5 \text{ A/m})(2 \times 10^{-2} \text{ m})^3 = 6 \text{ A}\cdot\text{m}^2$$

b) We found in Section 28.5 that the magnetic field on the axis of a current loop with magnetic moment μ_{total} is given by Eq. (28.18),

$$B = \frac{\mu_0 \mu_{\text{total}}}{2\pi(x^2 + a^2)^{3/2}}$$

where x is the distance from the loop and a is its radius. We can use this same expression here, except that a refers to the size of the permanent magnet. Strictly speaking, there are complications because our magnet does not have the same geometry as a circular current loop. But because $x = 10$ cm is fairly large in comparison to the 2-cm size of the magnet, the term a^2 is negligible in comparison to x^2 and can be ignored. So

$$B \approx \frac{\mu_0 \mu_{\text{total}}}{2\pi x^3} = \frac{(4\pi \times 10^{-7} \text{ T}\cdot\text{m/A})(6 \text{ A}\cdot\text{m}^2)}{2\pi(0.1 \text{ m})^3}$$
$$= 1 \times 10^{-3} \text{ T} = 10 \text{ G}$$

which is about ten times stronger than the magnetic field of the earth. Such a magnet can easily deflect a compass needle.

EVALUATE: Note that we used μ_0, not the permeability μ of the magnetic material, in calculating B. The reason is that we are calculating B at a point *outside* the magnetic material. You would substitute permeability μ for μ_0 only if you were calculating B *inside* a material with relative permeability K_m, for which $\mu = K_m \mu_0$.

SUMMARY

The magnetic field $\vec{B}$ created by a charge q moving with velocity $\vec{v}$ depends on the distance r from the source point (the location of q) to the field point (where $\vec{B}$ is measured). The $\vec{B}$ field is perpendicular to $\vec{v}$ and to $\hat{r}$, the unit vector directed from the source point to the field point. The principle of superposition of magnetic fields states that the total $\vec{B}$ field produced by several moving charges is the vector sum of the fields produced by the individual charges. (See Example 28.1)

$$\vec{B} = \frac{\mu_0}{4\pi}\frac{q\vec{v} \times \hat{r}}{r^2} \quad (28.2)$$

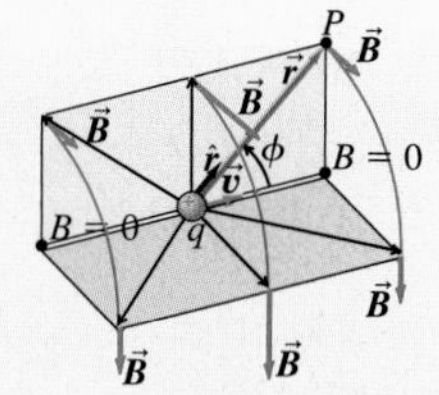

The law of Biot and Savart gives the magnetic field $d\vec{B}$ created by an element $d\vec{l}$ of a conductor carrying current I. The field $d\vec{B}$ is perpendicular to both $d\vec{l}$ and $\hat{r}$, the unit vector from the element to the field point. The $\vec{B}$ field created by a finite current-carrying conductor is the integral of $d\vec{B}$ over the length of the conductor. (See Example 28.2)

$$d\vec{B} = \frac{\mu_0}{4\pi}\frac{I\,d\vec{l} \times \hat{r}}{r^2} \quad (28.6)$$

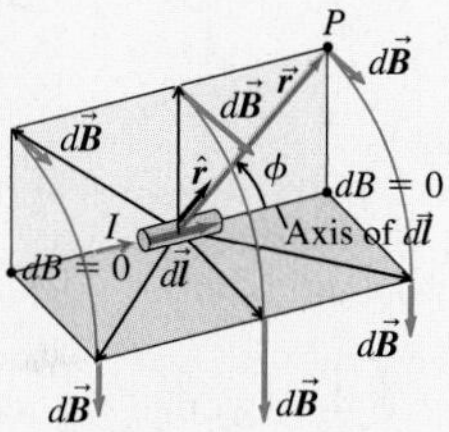

The magnetic field $\vec{B}$ at a distance r from a long, straight conductor carrying a current I has a magnitude that is inversely proportional to r. The magnetic field lines are circles coaxial with the wire, with directions given by the right-hand rule. (See Examples 28.3 and 28.4)

$$B = \frac{\mu_0 I}{2\pi r} \quad (28.9)$$

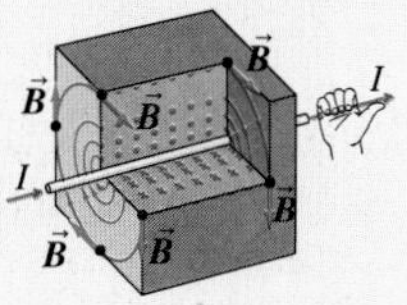

Two long, parallel, current-carrying conductors attract if the currents are in the same direction and repel if the currents are in opposite directions. The magnetic force per unit length between the conductors depends on their currents I and I' and their separation r. The definition of the ampere is based on this relation. (See Examples 28.5 and 28.6)

$$\frac{F}{L} = \frac{\mu_0 I I'}{2\pi r} \quad (28.11)$$

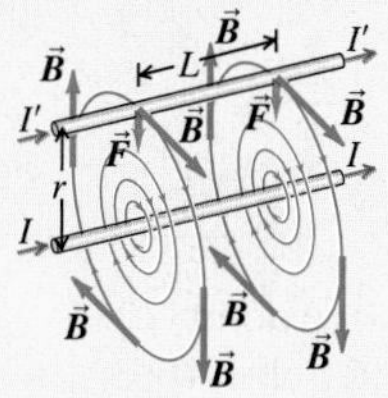

The law of Biot and Savart allows us to calculate the magnetic field produced along the axis of a circular conducting loop of radius a carrying current I. The field depends on the distance x along the axis from the center of the loop to the field point. If there are N loops, the field is multiplied by N. At the center of the loop, $x = 0$. (See Example 28.7)

$$B_x = \frac{\mu_0 I a^2}{2(x^2 + a^2)^{3/2}} \quad (28.15)$$

(circular loop)

$$B_x = \frac{\mu_0 N I}{2a} \quad (28.17)$$

(center of N circular loops)

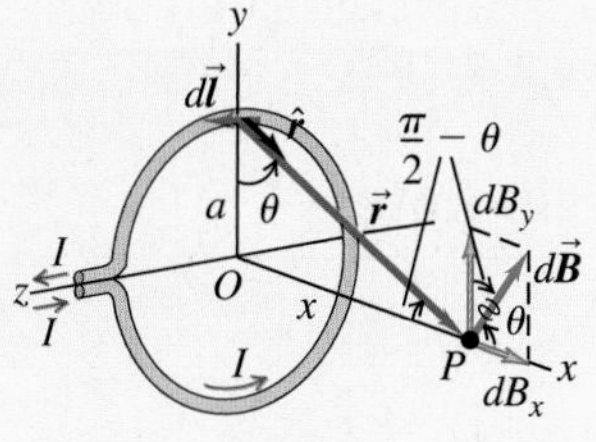

Ampere's law states that the line integral of $\vec{B}$ around any closed path equals μ_0 times the net current through the area enclosed by the path. The positive sense of current is determined by a right-hand rule. (See Examples 28.8 through 28.11)

$$\oint \vec{B} \cdot d\vec{l} = \mu_0 I_{encl} \quad (28.20)$$

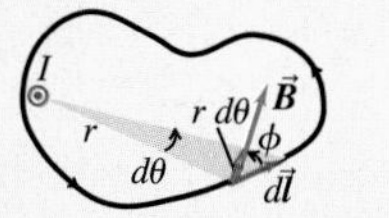

The following table lists magnetic fields caused by several current distributions. In each case the conductor is carrying current I.

Current Distribution	Point in Magnetic Field	Magnetic Field Magnitude
Long, straight conductor	Distance r from conductor	$B = \dfrac{\mu_0 I}{2\pi r}$
Circular loop of radius a	On axis of loop	$B = \dfrac{\mu_0 I a^2}{2(x^2 + a^2)^{3/2}}$
	At center of loop	$B = \dfrac{\mu_0 I}{2a}$ (for N loops, multiply these expressions by N)
Long cylindrical conductor of radius R	Inside conductor, $r < R$	$B = \dfrac{\mu_0 I}{2\pi}\dfrac{r}{R^2}$
	Outside conductor, $r > R$	$B = \dfrac{\mu_0 I}{2\pi r}$
Long, closely wound solenoid with n turns per unit length, near its midpoint	Inside solenoid, near center	$B = \mu_0 n I$
	Outside solenoid	$B \approx 0$
Tightly wound toroidal solenoid (toroid) with N turns	Within the space enclosed by the windings, distance r from symmetry axis	$B = \dfrac{\mu_0 N I}{2\pi r}$
	Outside the space enclosed by the windings	$B \approx 0$

When magnetic materials are present, the magnetization of the material causes an additional contribution to $\vec{B}$. For paramagnetic and diamagnetic materials, μ_0 is replaced in magnetic-field expressions by $\mu = K_m \mu_0$, where μ is the permeability of the material and K_m is its relative permeability. The magnetic susceptibility χ_m is defined as $\chi_m = K_m - 1$. Magnetic susceptibilities for paramagnetic materials are small positive quantities; those for diamagnetic materials are small negative quantities. For ferromagnetic materials, K_m is much larger than unity and is not constant. Some ferromagnetic materials are permanent magnets, retaining their magnetization even after the external magnetic field is removed. (See Examples 28.12 and 28.13)

Key Terms

source point, 1065
field point, 1065
principle of superposition of magnetic fields, 1068
law of Biot and Savart, 1069
ampere, 1075
Ampere's law, 1081
toroidal solenoid, 1085
Bohr magneton, 1088
magnetization, 1088
paramagnetic, 1088
relative permeability, 1088
permeability, 1089
magnetic susceptibility, 1089
diamagnetic, 1090
ferromagnetic, 1090
magnetic domain, 1090
hysteresis, 1091

Your Notes

Answer to Chapter Opening Question

From Eq. (28.17) the magnetic field at the center of N circular loops of radius a, each carrying current I, is $B_x = \mu_0 NI/2a$. Increasing the diameter and hence the radius a causes the magnetic field to decrease. A larger diameter means there are more moving charges to produce a magnetic field; it also means that the field point at the center of the coil is farther from those charges. In this case the latter of these factors is more important, so B_x decreases with increasing diameter.

Answers to Test Your Understanding Questions

Section 28.1 The situation is the same as shown in Fig. 28.2 except that the upper proton has velocity $\vec{v}$ rather than $-\vec{v}$. The magnetic field due to the lower proton is the same as shown in Fig. 28.2, but the direction of the magnetic force $\vec{F} = q\vec{v} \times \vec{B}$ on the upper proton is reversed. Hence the magnetic force is attractive. Since the speed v is small compared to c, the magnetic force is much smaller in magnitude than the repulsive electric force and the net force is still repulsive.

Section 28.2 The vector $d\vec{l}$ points in the negative x-direction and the unit vector $\hat{r}$ for this field points in the negative y-direction. The field is in the direction of $d\vec{l} \times \hat{r}$, which by the right-hand rule points out of the plane of Fig. 28.4 in the positive z-direction. The distance $r = 0.60$ m from the source point to the field point is half the 1.2-m distance in part (a) of Example 28.2, but the angle $\phi = 90°$ between $d\vec{l}$ and $\hat{r}$ is the same. The field magnitude B is proportional to $1/r^2$, so B has four times the value from part (a) of the example: $B = 4(8.7 \times 10^{-8}\,\text{T}) = 3.5 \times 10^{-7}\,\text{T}$. From $\vec{F} = q\vec{v} \times \vec{B}$ and the right-hand rule, the force on the proton is in the positive x-direction. Since $\vec{v}$ is perpendicular to $\vec{B}$, the force magnitude is $F = qvB = (1.6 \times 10^{-19}\,\text{C})(4.0 \times 10^3\,\text{m/s})(3.5 \times 10^{-7}\,\text{T})$.

Section 28.3 In this case the magnetic field $\vec{B}_1$ due to wire 1 is in the positive y-direction, as in Fig. 28.7, but the field $\vec{B}_2$ due to wire 2 is in the negative y-direction. Point P_2 is equidistant from the two wires, so $\vec{B}_1$ and $\vec{B}_2$ have the same magnitude. Hence the two fields cancel and the total field $\vec{B}_{\text{total}} = \vec{B}_1 + \vec{B}_2$ at P_2 is zero.

Section 28.4 The force is repulsive since the currents are in opposite directions. The force per unit length is $F/L = \mu_0 II'/2\pi r = (4\pi \times 10^{-7}\,\text{T}\cdot\text{m/A})(1\,\text{A})(1\,\text{A})/[2\pi(4 \times 10^{-3}\,\text{m})] = 5 \times 10^{-5}\,\text{N/m}$. This is so small that the force is totally negligible. (Contrast this result with Example 28.5, which describes a situation with much greater currents and forces.)

Section 28.5 For a segment on the negative y-axis (at the bottom of the loop in Fig. 28.12), $d\vec{l}$ is in the negative z-direction. The vectors $\vec{r}$ and $\hat{r}$ lie in the xy-plane and point from the bottom of the loop toward P. The field $d\vec{B}$ is in the direction of $d\vec{l} \times \hat{r}$, so it has a positive x-component, a negative y-component, and a zero z-component. Compared to the field due to a segment at the top of the loop, this $d\vec{B}$ has the same magnitude but is rotated by 180° around the x-axis:

$$dB_x = \frac{\mu_0}{4\pi}\frac{dl}{(x^2 + a^2)}\frac{a}{(x^2 + a^2)^{1/2}}$$

$$dB_y = -\frac{\mu_0}{4\pi}\frac{dl}{(x^2 + a^2)}\frac{x}{(x^2 + a^2)^{1/2}} \qquad dB_z = 0$$

For a segment on the negative z-axis (on the right side of the loop in Fig. 28.12), $d\vec{l}$ is in the positive y-direction. The vectors $\vec{r}$ and $\hat{r}$ (which point from the segment to P) lie in the xz-plane and point from the right side of the loop toward P. The field $d\vec{B}$ is in the direction of $d\vec{l} \times \hat{r}$, so it has a positive x-component, a zero y-component, and a negative z-component. Compared to the field due to a segment at the top of the loop, this $d\vec{B}$ has the same magnitude but is rotated by 90° around the x-axis:

$$dB_x = \frac{\mu_0}{4\pi}\frac{dl}{(x^2 + a^2)}\frac{a}{(x^2 + a^2)^{1/2}} \qquad dB_y = 0$$

$$dB_z = -\frac{\mu_0}{4\pi}\frac{dl}{(x^2 + a^2)}\frac{x}{(x^2 + a^2)^{1/2}}$$

Section 28.6 Imagine carrying out the integral $\oint \vec{B}\cdot d\vec{l}$ using an integration path that follows the closed loop of a magnetic field line in the direction of the field. At each point along the path the $\vec{B}$ field and the infinitesimal segment $d\vec{l}$ are both tangent to the path, so $\vec{B}\cdot d\vec{l}$ is positive at each point and the integral $\oint \vec{B}\cdot d\vec{l}$ must likewise be positive. (Figure 28.15a shows an integration path of this kind.) Since the integral is not zero, it follows from Ampere's law $\oint \vec{B}\cdot d\vec{l} = \mu_0 I_{\text{encl}}$, that there must be a current I_{encl} enclosed by the integration path and hence by the closed loop of the field line. Each closed field line in Fig. 27.11a encloses some empty space outside the magnet (where there cannot be any currents) and some of the interior of the magnet. Hence there must be currents inside the magnet (see Section 28.8).

Section 28.7 By symmetry, any $\vec{B}$ field outside the cable must circulate around the cable, with circular field lines like those surrounding the solid cylindrical conductor in Fig. 28.18. Choose an integration path like the one shown in Fig. 28.18 with radius $r > R$, so that the path completely encloses the cable. As in Example 28.9, the integral $\oint \vec{B}\cdot d\vec{l}$ for this path has magnitude $B(2\pi r)$. From Ampere's law this is equal to $\mu_0 I_{\text{encl}}$. The net enclosed current I_{encl} is zero because it includes two currents of equal magnitude but opposite direction: one in the central wire and one in the hollow cylinder. Hence $B(2\pi r) = 0$, and so $B = 0$ for any value of r outside the cable. (The field is nonzero *inside* the cable: see Exercise 28.32.)

Discussion Questions

Q28.1 A topic of current interest in physics research is the search (thus far unsuccessful) for an isolated magnetic pole, or magnetic *monopole*. If such an entity were found, how could it be recognized? What would its properties be?

Q28.2 Streams of charged particles emitted from the sun during periods of solar activity create a disturbance in the earth's magnetic field. How does this happen?

Q28.3 The text discussed the magnetic field of an infinitely long, straight conductor carrying a current. Of course, there is no such thing as an infinitely long *anything*. How do you decide whether a particular wire is long enough to be considered infinite?

Q28.4 Two parallel conductors carrying current in the same direction attract each other. If they are permitted to move toward each other, the forces of attraction do work. From where does the energy come? Does this contradict the assertion in Chapter 27 that magnetic forces on moving charges do no work? Explain.

Q28.5 Pairs of conductors carrying current into or out of the power-supply components of electronic equipment are sometimes twisted together to reduce magnetic field effects. Why does this help?

Q28.6 Suppose you have three long, parallel wires, arranged so that in cross section they are at the corners of an equilateral triangle. Is there any way to arrange the currents so that all three wires attract each other? So that all three wires repel each other? Explain.

Q28.7 In deriving the force on one of the long current-carrying conductors in Section 28.4, why did we use the magnetic field due to only one of the conductors? That is, why didn't we use the *total* magnetic field due to *both* conductors?

Q28.8 Two concentric, coplanar, circular loops of wire of different diameter carry currents in the same direction. Describe the nature of the forces exerted on the inner loop and on the outer loop.

Q28.9 A current was sent through a helical coil spring. The spring contracted, as though it had been compressed. Why?

Q28.10 What are the relative advantages and disadvantages of Ampere's law and the law of Biot and Savart for practical calculations of magnetic fields?

Q28.11 Magnetic field lines never have a beginning or an end. Use this to explain why it is reasonable for the field of a toroidal solenoid to be confined entirely to its interior, while a straight solenoid *must* have some field outside.

***Q28.12** Why should the permeability of a paramagnetic material be expected to decrease with increasing temperature?

***Q28.13** If a magnet is suspended over a container of liquid air, it attracts droplets to its poles. The droplets contain only liquid oxygen; even though nitrogen is the primary constituent of air, it is not attracted to the magnet. Explain what this tells you about the magnetic susceptibilities of oxygen and nitrogen, and explain why a magnet in ordinary, room-temperature air doesn't attract molecules of oxygen *gas* to its poles.

***Q28.14** What features of atomic structure determine whether an element is diamagnetic or paramagnetic? Explain.

***Q28.15** The magnetic susceptibility of paramagnetic materials is quite strongly temperature-dependent, but that of diamagnetic materials is nearly independent of temperature. Why the difference?

***Q28.16** A cylinder of iron is placed so that it is free to rotate around its axis. Initially the cylinder is at rest, and a magnetic field is applied to the cylinder so that it is magnetized in a direction parallel to its axis. If the direction of the external field is suddenly reversed, the direction of magnetization will also reverse and the cylinder will begin rotating around its axis. (This is called the *Einstein-de Haas Effect.*) Explain why the cylinder begins to rotate.

***Q28.17** The discussion of magnetic forces on current loops in Section 27.7 commented that no net force is exerted on a complete loop in a uniform magnetic field, only a torque. Yet magnetized materials, that contain atomic current loops, certainly *do* experience net forces in magnetic fields. How is this discrepancy resolved?

Exercises

Section 28.1 Magnetic Field of a Moving Charge

28.1 A $+6.00\text{-}\mu\text{C}$ point charge is moving at a constant 8.00×10^6 m/s in the $+y$-direction, relative to a reference frame. At the instant when the point charge is at the origin of this reference frame, what is the magnetic field vector $\vec{B}$ it produces at the following points: a) $x = 0.500$ m, $y = 0$, $z = 0$; b) $x = 0$, $y = -0.500$ m, $z = 0$; c) $x = 0$, $y = 0$, $z = +0.500$ m; d) $x = 0$, $y = -0.500$ m, $z = +0.500$ m?

28.2 Positive point charges $q = +8.00\ \mu\text{C}$ and $q' = +3.00\ \mu\text{C}$ are moving relative to an observer at point P, as shown in Fig. 28.29. The distance d is 0.120 m. When the two charges are at the locations shown in the figure, what are the magnitude and direction of the net magnetic field they produce at point P? Take $v = 4.50 \times 10^6$ m/s and $v' = 9.00 \times 10^6$ m/s.

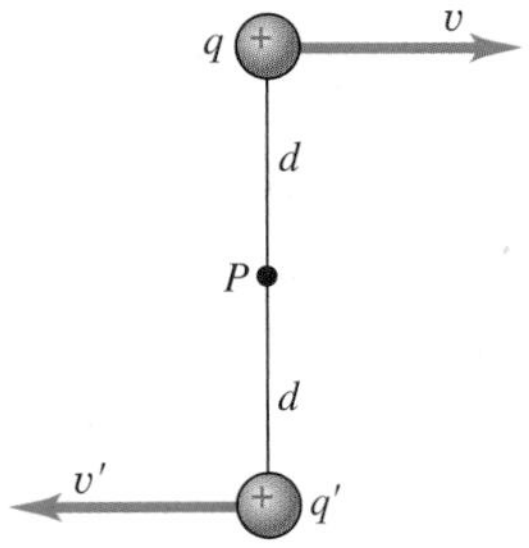

Figure 28.29 Exercises 28.2 and 28.6.

28.3 A $-4.80\text{-}\mu\text{C}$ charge is moving at a constant speed of 6.80×10^5 m/s in the $+x$-direction relative to a reference frame. At the instant when the point charge is at the origin, what is the magnetic field vector it produces at the following points: a) $x = 0.500$ m, $y = 0$, $z = 0$; b) $x = 0$, $y = 0.500$ m, $z = 0$; c) $x = 0.500$ m, $y = 0.500$ m, $z = 0$; d) $x = 0$, $y = 0$, $z = 0.500$ m?

28.4 a) In Exercise 28.2, what are the forces (magnitude and direction) that each charge exerts on the other? What are the Coulomb forces the charges exert on each other and what is the ratio of the magnitude of the Coulomb force to the magnitude of the magnetic force? b) If the direction of $\vec{v}'$ is reversed, so both charges are moving in the same direction, what are the magnitude and direction of the magnetic forces that the two charges exert on each other?

28.5 A pair of point charges, $q = +4.00\ \mu\text{C}$ and $q' = -1.50\ \mu\text{C}$, are moving in a reference frame as shown in Fig. 28.30. At this instant, what are the magnitude and direction of the magnetic field produced at the origin? Take $v = 2.00 \times 10^5$ m/s and $v' = 8.00 \times 10^5$ m/s.

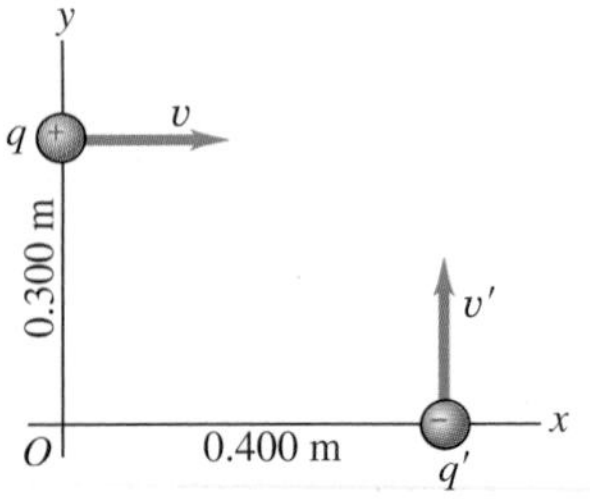

Figure 28.30 Exercise 28.5 and Problem 28.46.

28.6 Figure 28.29 shows two point charges, q and q', moving relative to an observer at point P. Suppose that the lower charge is actually *negative*, with $q' = -q$. a) Find the magnetic field (magnitude and direction) produced by the two charges at point P if i) $v' = v/2$; ii) $v' = v$; iii) $v' = 2v$. b) Find the direction of the magnetic force that q exerts on q', and find the direction of the magnetic force that q' exerts on q. c) If $v = v' = 3.00 \times 10^5$ m/s, what is the ratio of the magnitude of the magnetic force acting on each charge to that of the Coulomb force acting on each charge?

Section 28.2 Magnetic Field of a Current Element

28.7 a) In part (b) of Example 28.2 (Section 28.2), what is the unit vector $\hat{r}$ (expressed in terms of $\hat{i}$ and $\hat{j}$) that points from $d\vec{l}$ to point P_2? b) Compute the magnitude and direction of $d\vec{l} \times \hat{r}$. c) Use the result of part (b) and Eq. (28.6) to find the magnitude and direction of the magnetic field produced at point P_2 by the 1.0-cm segment of the wire.

28.8 A long, straight wire, carrying a current of 200 A, runs through a cubical wooden box, entering and leaving through holes in the centers of opposite faces (see Fig. 28.31). The length of each side of the box is 20.0 cm. Consider an element dl of the wire 0.100 cm long at the center of the box. Compute the magnitude dB of the magnetic field produced by this element at the points a, b, c, d, and e in Fig. 28.31. Points a, c, and d are at the centers of the faces of the cube; point b is at the midpoint of one edge; and point e is at a corner. Copy the figure and show the directions and relative magnitudes of the field vectors. (*Note:* Assume that dl is small in comparison to the distances from the current element to the points where the magnetic field is to be calculated.)

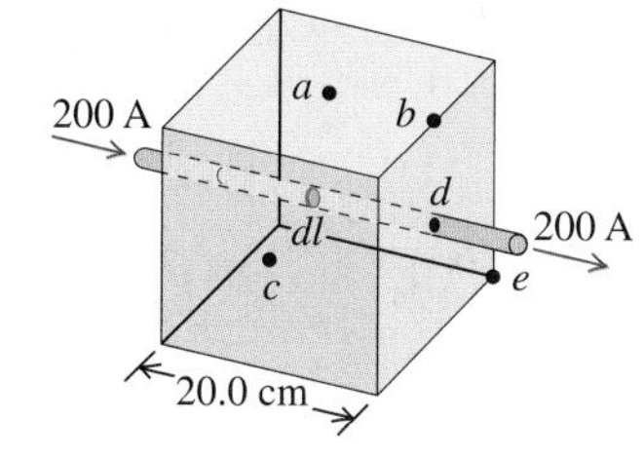

Figure 28.31 Exercise 28.8.

28.9 A long, straight wire lies along the z-axis and carries a 4.00-A current in the $+z$-direction. Find the magnetic field (magnitude and direction) produced at the following points by a 0.500-mm segment of the wire centered at the origin: a) $x = 2.00$ m, $y = 0$, $z = 0$; b) $x = 0$, $y = 2.00$ m, $z = 0$; c) $x = 2.00$ m, $y = 2.00$ m, $z = 0$; d) $x = 0$, $y = 0$, $z = 2.00$ m.

Section 28.3 Magnetic Field of a Straight Current-Carrying Conductor

28.10 In Fig. 28.7a (Example 28.4 of Section 28.3), what are the magnitude and direction of the net field produced by the wires on the x-axis at a) $x = d/2$? b) $x = -d/2$?

28.11 Two long, straight wires, one above the other, are separated by a distance $2a$ and are parallel to the x-axis. Let the $+y$-axis be in the plane of the wires in the direction from the lower wire to the upper wire. Each wire carries current I in the $+x$-direction. What are the magnitude and direction of the net magnetic field of the two wires at a point in the plane of the wires a) midway between them? b) at a distance a above the upper wire? c) at a distance a below the lower wire?

28.12 A long, straight wire lies along the y-axis and carries a current $I = 8.00$ A in the $-y$-direction (Fig. 28.32). In addition to the magnetic field due to the current in the wire, a uniform magnetic field $\vec{B}_0$ with magnitude 1.50×10^{-6} T is in the $+x$-direction What is the total field (magnitude and direction) at the following points in the xz-plane? a) $x = 0$, $z = 1.00$ m; b) $x = 1.00$ m, $z = 0$; c) $x = 0$, $z = -0.25$ m?

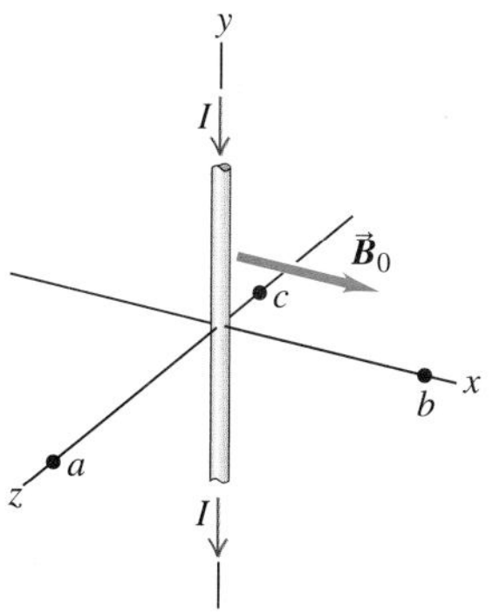

Figure 28.32 Exercise 28.12.

28.13 Provide the details of the derivation of Eq. (28.8) from the equation that precedes it.

28.14 You want to produce a magnetic field with a magnitude of 5.50×10^{-4} T at a distance of 0.040 m from a long, straight wire. a) What current is required to produce this field? b) With the current found in part (a), what is the magnitude of the field at a distance of 0.080 m from the wire and at 0.160 m?

28.15 Effect of Transmission Lines. Two hikers are reading a compass under an overhead transmission line that is 5.50 m above the ground and carries a current of 800 A in a horizontal direction from north to south. a) Find the magnitude and direction of the magnetic field at a point on the ground directly under the conductor. b) One hiker suggests they walk on another 50 m to avoid inaccurate compass readings caused by the current. Considering that the magnitude of the earth's field is of the order of 0.5×10^{-4} T, is the current really a problem?

28.16 Two long, straight, parallel wires, 10.0 cm apart carry equal 4.00-A currents in the same direction, as shown in Fig. 28.33. Find the magnitude and direction of the magnetic field at a) point P_1, midway between the wires; b) point P_2, 25.0 cm to the right of P_1; c) point P_3, 20.0 cm directly above P_1.

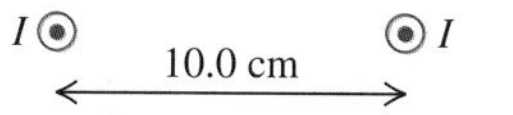

Figure 28.33 Exercise 28.16.

28.17 Two long, parallel transmission lines, 40.0 cm apart, carry 25.0-A and 75.0-A currents. Find all locations where the net magnetic field of the two wires is zero if these currents are in a) the same direction; b) the opposite direction.

28.18 Four, long, parallel power lines each carry 100-A currents. A cross-sectional diagram of these lines is a square, 20.0 cm on each side. For each case shown in Fig. 28.34, calculate the magnetic field at the center of the square.

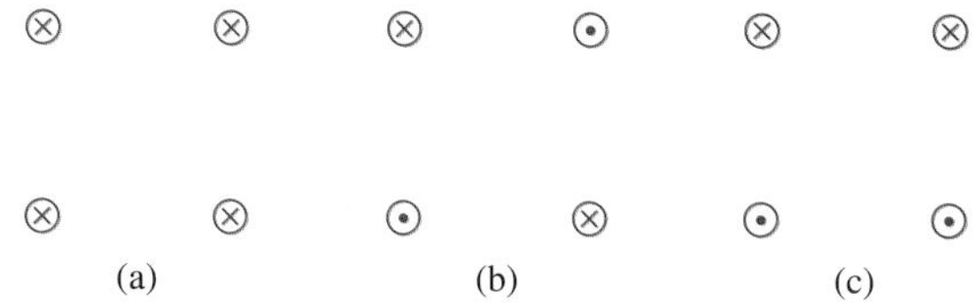

Figure 28.34 Exercise 28.18.

28.19 Four very long, current-carrying wires in the same plane intersect to form a square 40.0 cm on each side, as shown in Fig. 28.35. Find the magnitude and direction of the current I so that the magnetic field at the center of the square is zero.

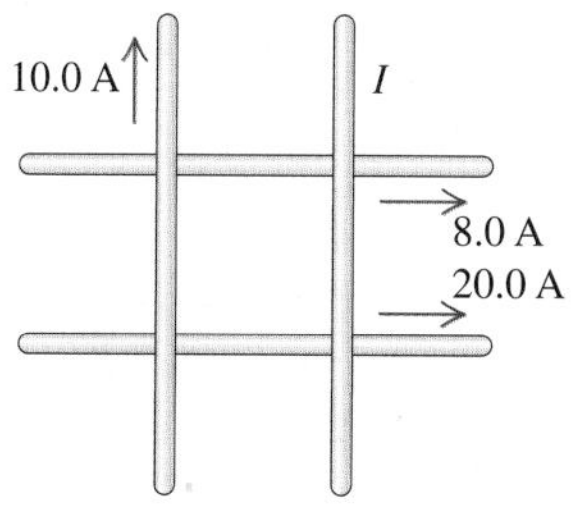

Figure 28.35 Exercise 28.19.

Section 28.4 Force between Parallel Conductors

28.20 Three parallel wires each carry current I in the directions shown in Fig. 28.36. If the separation between adjacent wires is d, calculate the magnitude and direction of the net magnetic force per unit length on each wire.

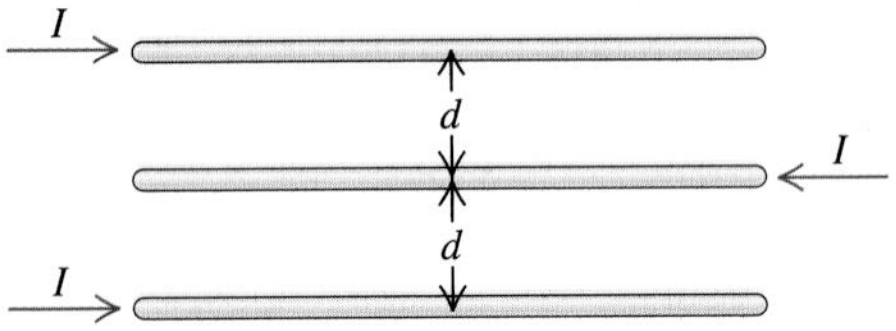

Figure 28.36 Exercise 28.20.

28.21 A long, horizontal wire AB rests on the surface of a table and carries a current I. Horizontal wire CD is vertically above wire AB, and is free to slide up and down on the two vertical metal guides C and D (Fig. 28.37). Wire CD is connected through the sliding contacts to another wire that also carries a current I, opposite in direction to the current in wire AB. The mass per unit length of the wire CD is λ. To what equilibrium height h will the wire CD rise, assuming that the magnetic force on it is due wholly to the current in the wire AB?

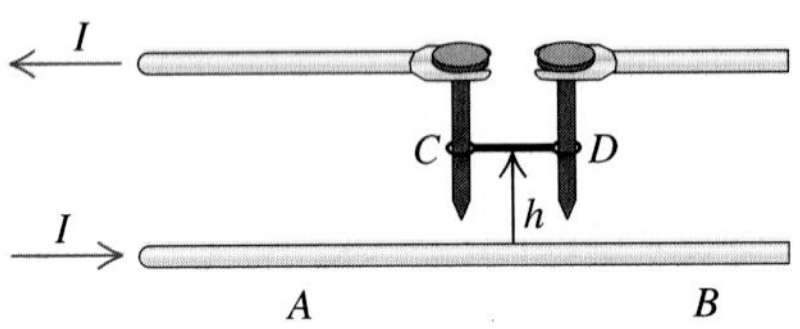

Figure 28.37 Exercise 28.21.

28.22 Two long, parallel wires are separated by a distance of 0.400 m (see Fig. 28.38). The currents I_1 and I_2 have the directions shown. a) Calculate the magnitude of the force exerted by each wire on a 1.20-m length of the other. Is the force attractive or repulsive? b) Each current is doubled, so that I_1 becomes 10.0 A and I_2 becomes 4.00 A. Now what is the magnitude of the force that each wire exerts on a 1.20-m length of the other?

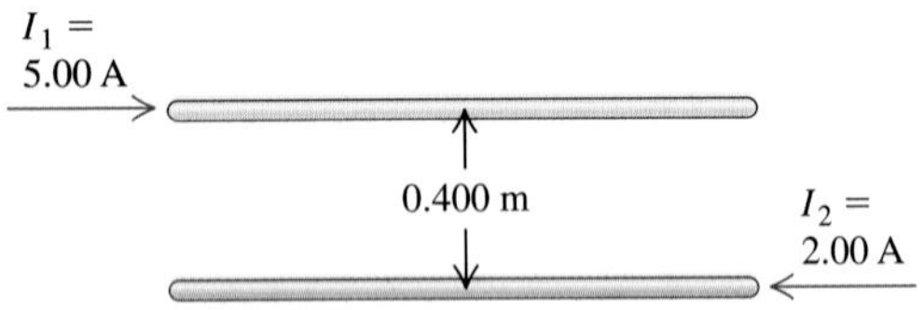

Figure 28.38 Exercise 28.22.

28.23 Two long, parallel wires are separated by a distance of 2.50 cm. The force per unit length that each wire exerts on the other is 4.00×10^{-5} N/m, and the wires repel each other. The current in one wire is 0.600 A. a) What is the current in the second wire? b) Are the two currents in the same direction or in opposite directions?

Section 28.5 Magnetic Field of a Circular Current Loop

28.24 Calculate the magnitude and direction of the magnetic field at point P due to the current in the semicircular section of wire shown in Fig. 28.39. (*Hint:* Does the current in the long, straight section of the wire produce any field at P?)

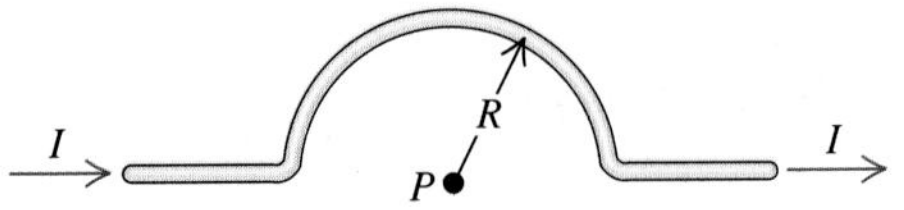

Figure 28.39 Exercise 28.24.

28.25 Calculate the magnitude of the magnetic field at point P of Fig. 28.40 in terms of R, I_1, and I_2. What does your expression give when $I_1 = I_2$?

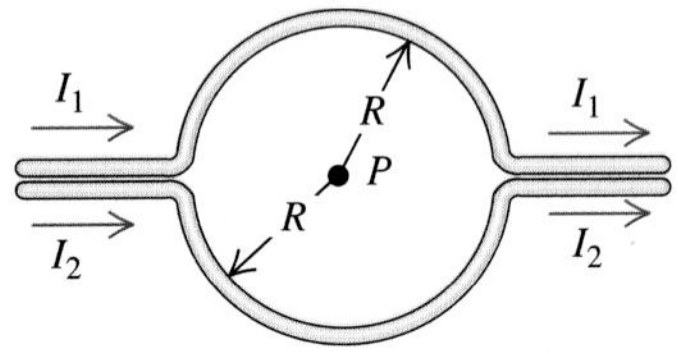

Figure 28.40 Exercise 28.25.

28.26 Consider the loop in Fig. 28.12. a) What is the direction of the magnetic field of the loop at points on the $-x$-axis? b) In Fig. 28.12, at point P the field points away from the loop. What is the direction of the field at point P if the direction of the current in the loop is reversed? c) A circular loop lies in the xy-plane, with the origin at the center of the loop. As you view the loop, the z-axis is directed out of the plane toward you. If the magnetic moment of the loop is in the $-z$-direction, is the current in the loop clockwise or counterclockwise?

28.27 A closely wound, circular coil with radius 2.40 cm has 800 turns. a) What must the current in the coil be if the magnetic field at the center of the coil is 0.0580 T? b) At what distance x from the center of the coil, on the axis of the coil, is the magnetic field $\frac{1}{2}$ its value at the center?

28.28 A closely wound circular coil with a diameter of 4.00 cm has 600 turns and carries a current 0.500 A. What is the magnitude of the magnetic field a) at the center of the coil? b) at a point on the axis of the coil 8.00 cm from its center?

28.29 A closely wound coil has a radius of 6.00 cm and carries a current of 2.50 A. How many turns must it have if, at a point on the coil axis 6.00 cm from the center of the coil, the magnetic field is 6.39×10^{-4} T?

Section 28.6 Ampere's Law

28.30 A closed curve encircles several conductors. The line integral $\oint \vec{B} \cdot d\vec{l}$ around this curve is 3.83×10^{-4} T·m. a) What is the net current in the conductors? b) If you were to integrate around the curve in the opposite direction, what would be the value of the line integral? Explain.

28.31 Figure 28.41 shows, in cross section, several conductors that carry currents through the plane of the figure. The currents have the magnitudes $I_1 = 4.0$ A, $I_2 = 6.0$ A, and $I_3 = 2.0$ A, and the directions shown. Four

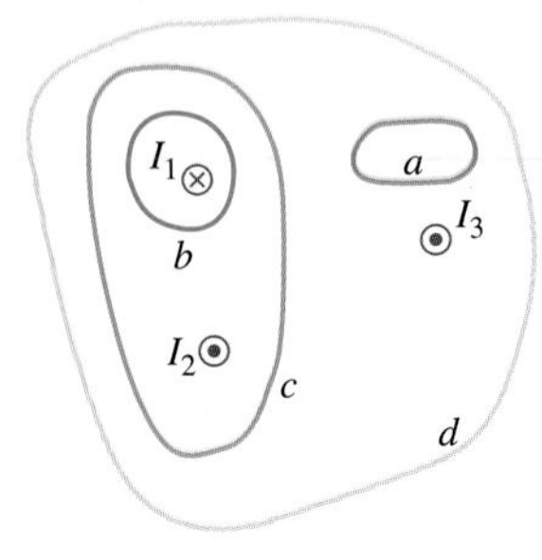

Figure 28.41 Exercise 28.31.

paths, labeled a through d, are shown. What is the line integral $\oint \vec{B} \cdot d\vec{l}$ for each path? Each integral involves going around the path in the counterclockwise direction. Explain your answers.

Section 28.7 Applications of Ampere's Law

28.32 Coaxial Cable. A solid conductor with radius a is supported by insulating disks on the axis of a conducting tube with inner radius b and outer radius c (see Fig. 28.42). The central conductor and tube carry equal currents I in opposite directions. The currents are distributed uniformly over the cross sections of each conductor. Derive an expression for the magnitude of the magnetic field a) at points outside the central, solid conductor, but inside the tube $(a < r < b)$; b) at points outside the tube $(r > c)$.

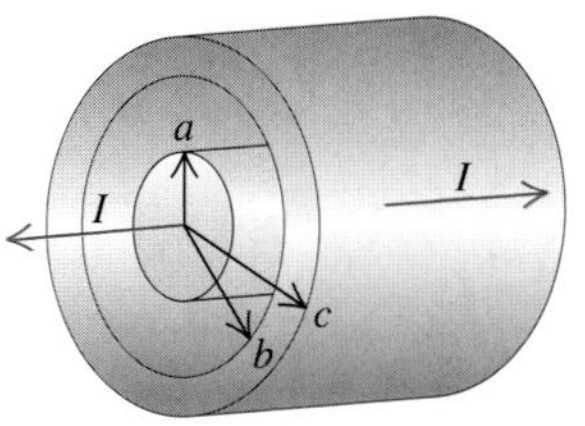

Figure 28.42 Exercises 28.32 and 28.33, and Problem 28.70.

28.33 Repeat Exercise 28.32 for the case in which the current in the central, solid conductor is I_1, that in the tube is I_2, and these currents are in the same direction rather than in opposite directions.

28.34 A 15.0-cm-long solenoid with radius 2.50 cm is closely wound with 600 turns of wire. The current in the windings is 8.00 A. Compute the magnetic field at a point near the center of the solenoid.

28.35 A solenoid is designed to produce a magnetic field of 0.0270 T at its center. It has radius 1.40 cm and length 40.0 cm, and the wire can carry a maximum current of 12.0 A. a) What minimum number of turns per unit length must the solenoid have? b) What total length of wire is required?

28.36 As a new electrical technician, you are designing a large solenoid to produce a uniform 0.150 T magnetic field near the center of the solenoid. You have enough wire for 4000 circular turns. This solenoid must be 1.40 m long and 20.0 cm in diameter. What current will you need to produce the necessary field?

28.37 A magnetic field of 37.2 T has been achieved at the MIT Francis Bitter National Magnetic Laboratory. Find the current needed to achieve such a field a) 2.00 cm from a long straight wire; b) at the center of a circular coil of radius 42.0 cm that has 100 turns; c) near the center of a solenoid with radius 2.40 cm, length 32.0 cm and 40,000 turns.

28.38 A toroidal solenoid (see Fig. 28.23) has inner radius $r_1 =$ 15.0 cm and outer radius $r_2 =$ 18.0 cm. The solenoid has 250 turns and carries a current of 8.50 A. What is the magnitude of the magnetic field at the following distances from the center of the torus: a) 12.0 cm; b) 16.0 cm? c) 20.0 cm?

28.39 A wooden ring whose mean diameter is 14.0 cm is wound with a closely spaced toroidal winding of 600 turns. Compute the magnitude of the magnetic field at the center of the cross section of the windings when the current in the windings is 0.650 A.

*Section 28.8 Magnetic Materials

***28.40** A toroidal solenoid with 400 turns of wire and a mean radius of 6.0 cm carries a current of 0.25 A. The relative permeability of the core is 80. a) What is the magnetic field in the core? b) What part of the magnetic field is due to atomic currents?

***28.41** A toroidal solenoid with 500 turns is wound on a ring with a mean radius of 2.90 cm. Find the current in the winding that is required to set up a magnetic field of 0.350 T in the ring a) if the ring is made of annealed iron $(K_m = 1400)$; b) if the ring is made of silicon steel $(K_m = 5200)$.

***28.42** The current in the windings of a toroidal solenoid is 2.400 A. There are 500 turns and the mean radius is 25.00 cm. The toroidal solenoid is filled with a magnetic material. The magnetic field inside the windings is found to be 1.940 T. Calculate a) the relative permeability; b) the magnetic susceptibility of the material that fills the toroid.

***28.43** A long solenoid with 60 turns of wire per centimeter carries a current of 0.15 A. The wire that makes up the solenoid is wrapped around a solid core of silicon steel $(K_m = 5200)$. (The wire of the solenoid is jacketed with an insulator so that none of the current flows into the core.) a) For a point inside the core, find the magnitudes of i) the magnetic field $\vec{B}_0$ due to the solenoid current, ii) the magnetization $\vec{M}$, and iii) the total magnetic field $\vec{B}$. b) In a sketch of the solenoid and core, show the directions of the vectors $\vec{B}$, $\vec{B}_0$, and $\vec{M}$ inside the core.

***28.44** Show that the units $A \cdot m^2$ and J/T for the Bohr magneton are equivalent.

***28.45 Curie's Law.** Experimental measurements of the magnetic susceptibility of iron ammonium alum are given below. Graph $1/\chi_m$ against Kelvin temperature. Does the material obey Curie's law? If so, what is the Curie constant?

T (°C)	χ_m
−258.15	129×10^{-4}
−173	19.4×10^{-4}
−73	9.7×10^{-4}
27	6.5×10^{-4}

Problems

28.46 A pair of point charges, $q = +8.00\ \mu C$ and $q' = -5.00\ \mu C$, are moving in a reference frame, as shown in Fig. 28.30, with speeds $v = 9.00 \times 10^4$ m/s and $v' = 6.50 \times 10^4$ m/s. When the point charges are at the locations shown in the figure, what magnetic force (magnitude and direction) does q' exert on q?

28.47 A long, straight wire carries a current of 2.50 A. An electron is traveling in the vicinity of the wire. At the instant when the electron is 4.50 cm from the wire and traveling with a speed of 6.00×10^4 m/s parallel to the wire in the direction opposite to the current, what are the magnitude and direction of the force that the magnetic field of the current exerts on the electron?

28.48 A long, straight wire carries a 25.0-A current. An electron is fired parallel to this wire with a velocity of 250 km/s in the same direction as the current, 2.00 cm from the wire. a) Find the magnitude and direction of the electron's initial acceleration. b) What should be the magnitude and direction of a uniform electric field that will allow the electron to continue to travel parallel to the wire? c) Is it necessary to include the effects of gravity? Justify your answer.

28.49 In Fig. 28.43, the battery branch of the circuit is very far from the two horizontal segments containing two resistors. These horizontal segments are separated by 5.00 cm, and their length is much greater than 5.00 cm. A proton is fired at 650 km/s from a point midway between the upper two horizontal segments of the circuit. The initial velocity of the proton is in the plane of the circuit and is directed toward the upper wire. Find the magnitude and direction of the initial magnetic force on the proton.

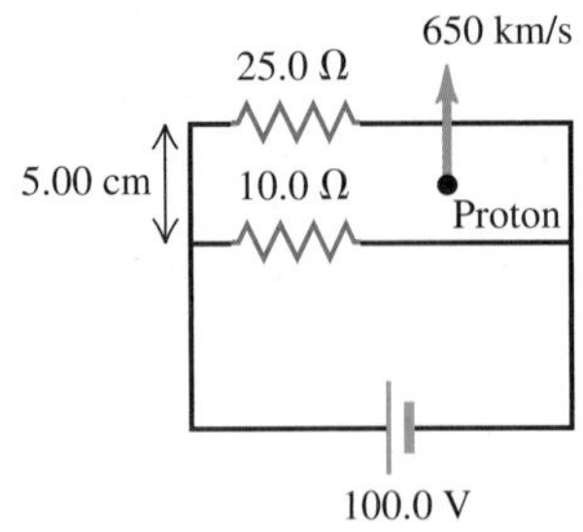

Figure 28.43 Problem 28.49.

28.50 Two identical, circular, wire loops 40.0 cm in diameter each carry a current of 1.50 A in the same direction. These loops are parallel to each other and are 25.0 cm apart. Line *ab* is normal to the plane of the loops and passes through their centers. A proton is fired at 2400 m/s perpendicular to line *ab* from a point midway between the centers of the loops. Find the magnitude and direction of the magnetic force these loops exert on the proton just after it is fired.

28.51 Two very long, straight wires carry currents as shown in Fig. 28.44. For each case shown, find all locations where the net magnetic field is zero.

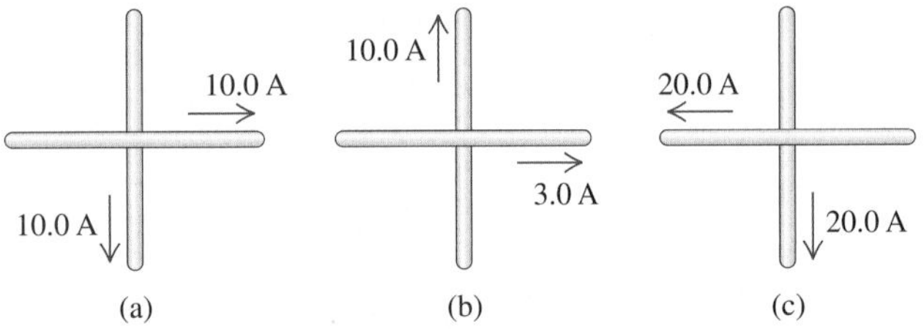

Figure 28.44 Problem 28.51.

28.52 A negative point charge $q = -7.20\text{ mC}$ is moving in a reference frame. When the point charge is at the origin, the magnetic field it produces at the point $x = 25.0\text{ cm}$, $y = 0$, $z = 0$ is $\vec{B} = (6.00\ \mu\text{T})\hat{\jmath}$, and its speed is 800 m/s. a) What are the x-, y-, and z-components of the velocity $\vec{v}_0$ of the charge? b) At this same instant, what is the magnitude of the magnetic field that the charge produces at the point $x = 0$, $y = 25.0\text{ cm}$, $z = 0$?

28.53 A neophyte magnet designer tells you that he can produce a magnetic field $\vec{B}$ in a vacuum that points everywhere in the x-direction and that increases in magnitude with increasing x. That is, $\vec{B} = B_0(x/a)\hat{\imath}$, where B_0 and a are constants with units of teslas and meters, respectively. Use Gauss's law for magnetic fields to show that this claim is *impossible*. (*Hint:* Use a Gaussian surface in the shape of a rectangular box, with edges parallel to the x-, y-, and z-axes.)

28.54 Three long, parallel wires are located as shown in Fig. 28.45. Wire 2 is at the origin, wire 1 is on the y-axis at $y = 3.00\text{ cm}$, and wire 3 is on the x-axis at $x = 4.00\text{ cm}$. The currents are $I_1 = 1.00\text{ A}$, $I_2 = 2.00\text{ A}$, and $I_3 = 4.00\text{ A}$. a) Copy Fig. 28.45 and show the directions of the magnetic fields of wires 2 and 3 in the xy-plane at the location of wire 1. b) Find the x- and y-components of the net magnetic field in the xy-plane at the location of wire 1 due to the currents in wires 2 and 3. c) Find the magnitude and direction of the net force exerted on a 1.00-cm section of wire 1 by the other two wires.

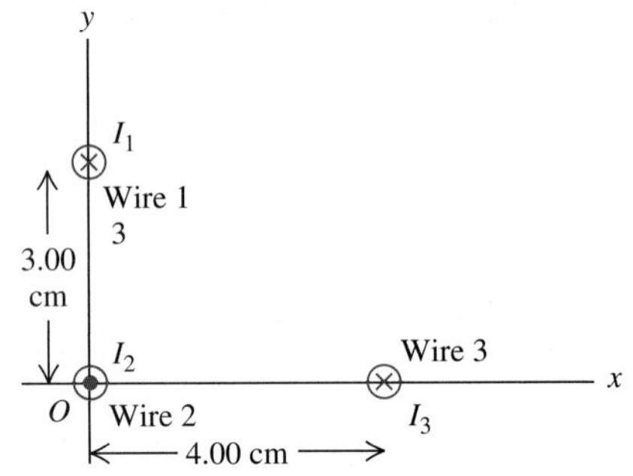

Figure 28.45 Problem 28.54.

28.55 Two long, straight, parallel wires are 1.00 m apart (Fig. 28.46). The upper wire carries a current I_1 of 6.00 A into the plane of the paper. a) What must the magnitude and direction of the current I_2 be for the net field at point P to be zero? b) Then, what are the magnitude and direction of the net field at Q? c) Then, what is the magnitude of the net field at S?

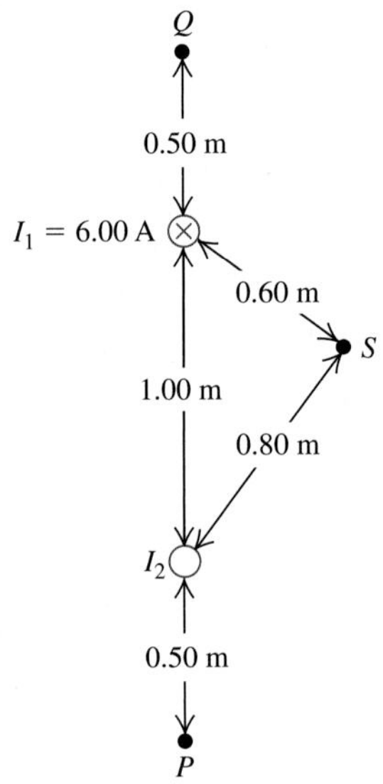

Figure 28.46 Problem 28.55.

28.56 Figure 28.47 shows an end view of two long, parallel wires perpendicular to the xy-plane, each carrying a current I, but in opposite directions. a) Copy the diagram, and draw vectors to show the $\vec{B}$ field of each wire and the net $\vec{B}$ field at point P. b) Derive the expression for the magnitude of $\vec{B}$ at any point on the x-axis in terms of the x-coordinate of the point. What is the direction of $\vec{B}$? c) Graph the magnitude of $\vec{B}$ at points on the x-axis. d) At what value of x is the magnitude of $\vec{B}$ a maximum? e) What is the magnitude of $\vec{B}$ when $x \gg a$?

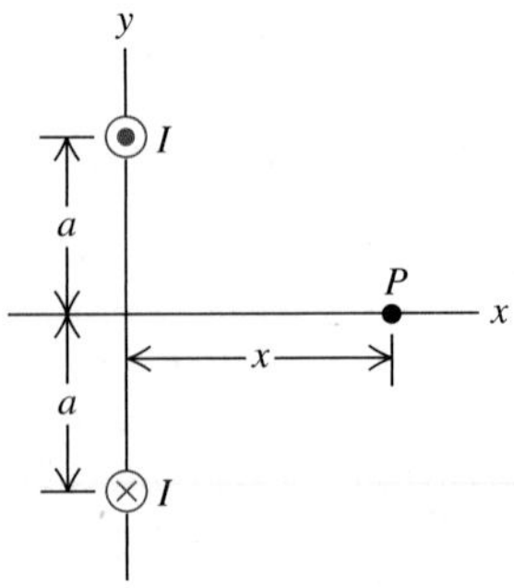

Figure 28.47 Problems 28.56, 28.57, and 28.58.

28.57 Repeat Problem 28.56, but with the current in both wires shown in Fig. 28.47 directed into the plane of the figure.

28.58 Refer to the situation in Problem 28.56. Suppose a third long, straight wire, parallel to the other two, passes through point P (see Fig. 28.47), and that each wire carries a current $I = 6.00\text{ A}$. Let $a = 40.0\text{ cm}$ and

$x = 60.0$ cm. Find the magnitude and direction of the force per unit length on the third wire, a) if the current in it is directed into the plane of the figure; b) if the current in it is directed out of the plane of the figure.

28.59 Two straight, 3.50-m-long wires, each with resistance 0.250 Ω, are lying parallel to each other on a smooth horizontal table. Their ends are connected by identical, non-conducting, light springs, each spring having an unstretched length of 1.00 cm. A wire of negligible resistance connects the wires at one end. When a switch is closed to connect a 45.0-V battery between the other ends of the wires, the wires move apart and come to rest 1.50 cm apart. Find the force constants of the springs.

28.60 Magnetic Current Meter. At your laboratory's location, the earth's magnetic field has magnitude 1.0×10^{-4} T and points north. A long, straight wire (wire A) runs in a north-south direction along the laboratory floor. You know that the wire carries current from south to north, but you do not know the magnitude of the current. To find out, you place two other long, straight wires on the floor parallel to wire A. Wire B is 5.0 cm to the east of wire A and wire C is 10.0 cm to the east of wire B. You connect sources of emf to wires B and C so that a constant current of 1.0 A flows from south to north in wire B, while an adjustable amount of current flows in wire C. Wires A and C are attached rigidly to the floor, while wire B is able to slide around. You find by experiment that wire B will indeed start to slide one way or the other unless there is a 3.0-A current in wire C that flows from south to north. If this is the case, wire B remains at rest. a) How much current is flowing in wire A? b) If the current in wire C were increased to 4.0 A, in which direction would wire B tend to slide?

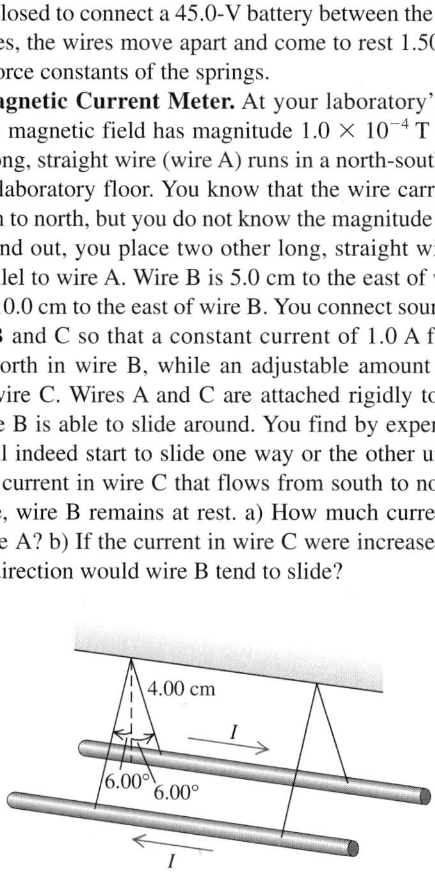

Figure 28.48 Problem 28.61.

28.61 Two long, parallel wires hang by 4.00-cm-long cords from a common axis (see Fig. 28.48). The wires have a mass per unit length of 0.0125 kg/m and carry the same current in opposite directions. What is the current in each wire if the cords hang at an angle of 6.00° with the vertical?

28.62 The long, straight wire *AB* shown in Fig. 28.49 carries a current of 14.0 A. The rectangular loop whose long edges are parallel to the wire carries a current of 5.00 A. Find the magnitude and direction of the net force exerted on the loop by the magnetic field of the wire.

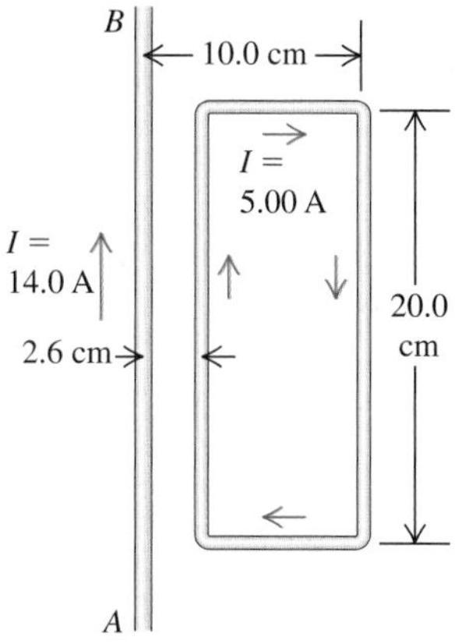

Figure 28.49 Problem 28.62.

28.63 A circular wire loop of radius a has N turns and carries a current I. A second loop with N' turns of radius a' carries current I' and is located on the axis of the first loop, a distance x from the center of the first loop. The second loop is tipped so that its axis is at an angle θ from the axis of the first loop. The distance x is large compared to both a and a'. a) Find the magnitude of the torque exerted on the second loop by the first loop. b) Find the potential energy for the second loop due to this interaction. c) What simplifications result from having x much larger than a? From having x much larger than a'?

28.64 The wire semicircles shown in Fig. 28.50 have radii a and b. Calculate the net magnetic field (magnitude and direction) that the current in the wires produces at point P.

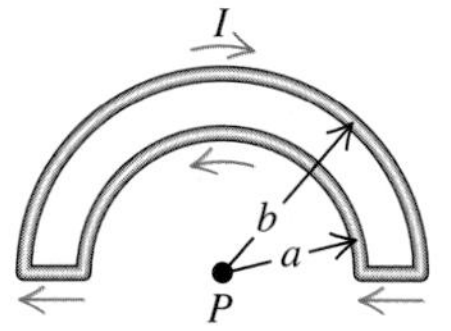

Figure 28.50 Problem 28.64.

28.65 Helmholtz Coils. Figure 28.51 is a sectional view of two circular coils with radius a, each wound with N turns of wire carrying a current I, circulating in the same direction in both coils. The coils are separated by a distance a equal to their radii. In this configuration the coils are called Helmholtz coils; they produce a very uniform magnetic field in the region between them. a) Derive the expression for the magnitude B of the magnetic field at a point on the axis a distance x to the right of point P, which is midway between the coils. b) Graph B versus x for $x = 0$ to $x = a/2$. Compare this graph to one for the magnetic field due to the right-hand coil alone. c) From part (a), obtain an expression for the magnitude of the magnetic field at point P. d) Calculate the magnitude of the magnetic field at P if $N = 300$ turns, $I = 6.00$ A, and $a = 8.00$ cm. e) Calculate dB/dx and d^2B/dx^2 at P $(x = 0)$. Discuss how your results show that the field is very uniform in the vicinity of P.

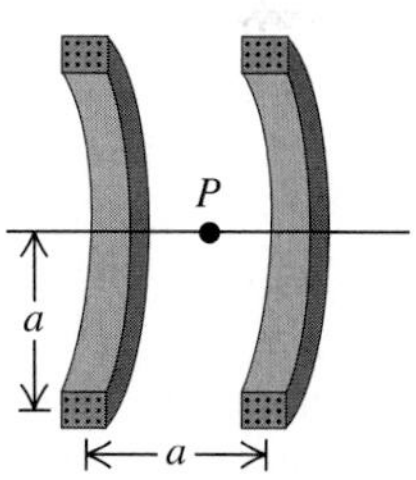

Figure 28.51 Problem 28.65.

28.66 Calculate the magnitude and direction of the magnetic field produced at point P shown in Fig. 28.52 by the current I in the rectangular wire loop. (Point P is at the center of the rectangle.) (*Hint:*

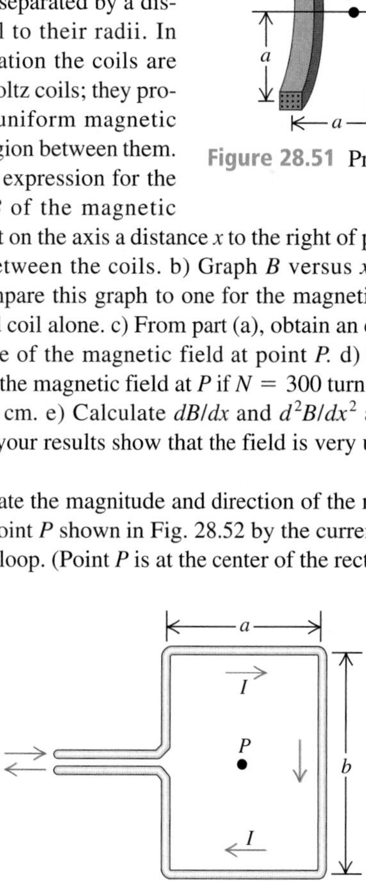

Figure 28.52 Problem 28.66.

The gap on the left-hand side where the wires enter and leave the rectangle is so small that this side of the rectangle can be taken to be a continuous wire with length b.)

28.67 The wire in Fig. 28.53 carries current I in the direction shown. The wire consists of a very long, straight section, a quarter-circle with radius R, and another long, straight section. What are the magnitude and direction of the net magnetic field at the center of curvature of the quarter-circle section (point P)?

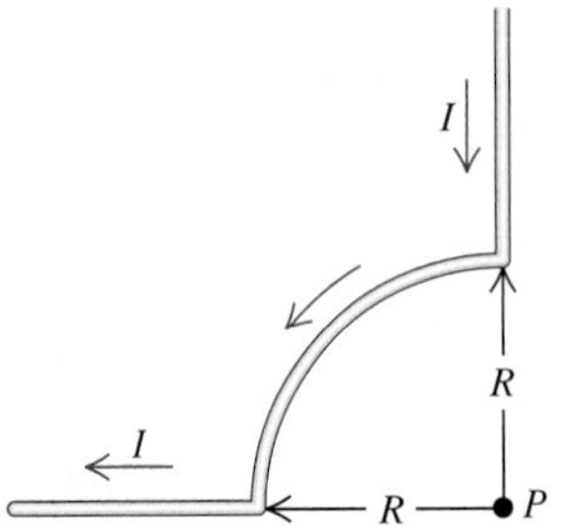

Figure 28.53 Problem 28.67.

28.68 The wire shown in Fig. 28.54 is infinitely long and carries a current I. Calculate the magnitude and direction of the magnetic field that this current produces at point P.

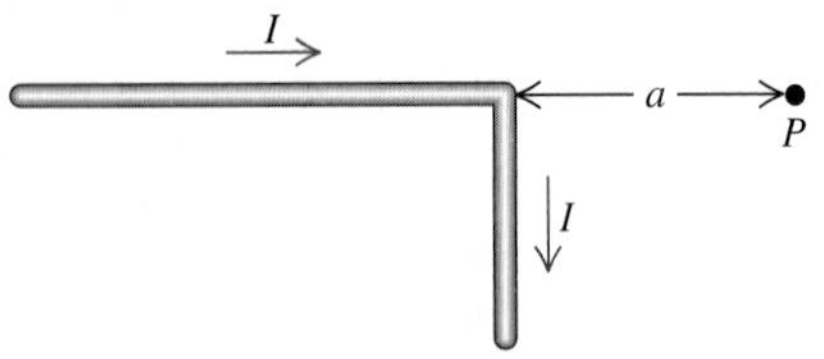

Figure 28.54 Problem 28.68.

28.69 A long, straight wire with a circular cross section of radius R carries a current I. Assume that the current density is not constant across the cross section of the wire, but rather, varies as $J = \alpha r$, where α is a constant. a) By the requirement that J integrated over the cross section of the wire gives the total current I, calculate the constant α in terms of I and R. b) Use Ampere's law to calculate the magnetic field $B(r)$ for i) $r \le R$; ii) $r \ge R$. Express your answers in terms of I.

28.70 a) For the coaxial cable of Exercise 28.32, derive an expression for the magnitude of the magnetic field at points inside the central solid conductor $(r < a)$. Compare your result when $r = a$ to the results of part (a) of Exercise 28.32 at that same point. b) For this coaxial cable derive an expression for the field within the tube $(b < r < c)$. Compare your result when $r = b$ to part (a) of Exercise 28.32 at that same point. Compare your result when $r = c$ to part (b) of Exercise 28.32 at that same point.

28.71 The electric field of an infinite line of positive charge is directed radially outward from the wire and can be calculated using Gauss's law for the electric field (see Example 22.6 in Section 22.4). Use Gauss's law for magnetism to show that the *magnetic* field of a straight, infinitely long, current-carrying *conductor* cannot have a radial component.

28.72 A conductor is made in the form of a hollow cylinder with inner and outer radii a and b, respectively. It carries a current I, uniformly distributed over its cross section. Derive expressions for the magnitude of the magnetic field in the regions a) $r < a$; b) $a < r < b$; c) $r > b$.

28.73 Knowing Magnetic Fields Inside and Out. You are given a hollow copper cylinder with inner radius a and outer radius $3a$. The cylinder's length is $200a$ and its electrical resistance to current flowing down its length is R. To test its suitability for use in a circuit, you connect the ends of the cylinder to a voltage source, causing a current I to flow down the length of the cylinder. The current is spread uniformly over the cylinder's cross section. You are interested in knowing the strength of the magnetic field that the current produces within the solid part of the cylinder, at a radius $2a$ from the cylinder axis. But since it's not easy to insert a magnetic-field probe into the solid metal, you decide instead to measure the field at a point outside the cylinder where the field should be as strong as at radius $2a$. At what distance from the axis of the cylinder should you place the probe?

28.74 A circular loop has radius R and carries current I_2 in a clockwise direction (see Fig. 28.55). The center of the loop is a distance D above a long, straight wire. What are the magnitude and direction of the current I_1 in the wire if the magnetic field at the center loop is zero?

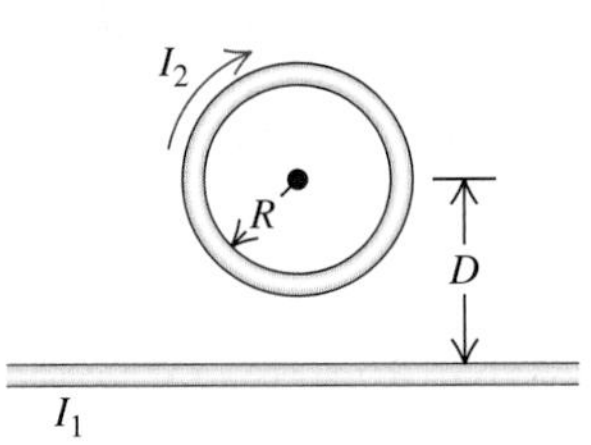

Figure 28.55 Problem 28.74.

28.75 A long, straight, solid cylinder, oriented with its axis in the z-direction, carries a current whose current density is $\vec{J}$. The current density, although symmetrical about the cylinder axis, is not constant but varies according to the relation

$$\vec{J} = \frac{2I_0}{\pi a^2}\left[1 - \left(\frac{r}{a}\right)^2\right]\hat{k} \quad \text{for } r \le a$$

$$= \mathbf{0} \quad \text{for } r \ge a$$

where a is the radius of the cylinder, r is the radial distance from the cylinder axis, and I_0 is a constant having units of amperes. a) Show that I_0 is the total current passing through the entire cross section of the wire. b) Using Ampere's law, derive an expression for the magnitude of the magnetic field $\vec{B}$ in the region $r \ge a$. c) Obtain an expression for the current I contained in a circular cross section of radius $r \le a$ and centered at the cylinder axis. d) Using Ampere's law, derive an expression for the magnitude of the magnetic field $\vec{B}$ in the region $r \le a$. How do your results in parts (b) and (d) compare for $r = a$?

28.76 A long, straight, solid cylinder, oriented with its axis in the z-direction, carries a current whose current density is $\vec{J}$. The current density, although symmetrical about the cylinder axis, is not constant and varies according to the relation

$$\vec{J} = \left(\frac{b}{r}\right)e^{(r-a)/\delta}\hat{k} \quad \text{for } r \le a$$

$$= \mathbf{0} \quad \text{for } r \ge a$$

where the radius of the cylinder is $a = 5.00$ cm, r is the radial distance from the cylinder axis, b is a constant equal to 600 A/m, and

δ is a constant equal to 2.50 cm. a) Let I_0 be the total current passing through the entire cross section of the wire. Obtain an expression for I_0 in terms of b, δ, and a. Evaluate your expression to obtain a numerical value for I_0. b) Using Ampere's law, derive an expression for the magnetic field $\vec{B}$ in the region $r \geq a$. Express your answer in terms of I_0 rather than b. c) Obtain an expression for the current I contained in a circular cross section of radius $r \leq a$ and centered at the cylinder axis. Express your answer in terms of I_0 rather than b. d) Using Ampere's law, derive an expression for the magnetic field $\vec{B}$ in the region $r \leq a$. e) Evaluate the magnitude of the magnetic field at $r = \delta$, $r = a$, and $r = 2a$.

28.77 Integrate B_x as given in Eq. (28.15) from $-\infty$ to $+\infty$. That is, calculate $\int_{-\infty}^{+\infty} B_x \, dx$. Explain the significance of your result.

28.78 In a region of space where there are no conduction or displacement currents, it is impossible to have a uniform magnetic field that abruptly drops to zero. To prove this statement, use the method of contradiction: Assume that such a case *is* possible, and then show that your assumption contradicts a law of nature. a) In the bottom half of a piece of paper, draw evenly spaced, horizontal lines representing a uniform magnetic field to your right. Use dashed lines to draw a rectangle *abcda* with horizontal side *ab* in the magnetic field region and horizontal side *cd* in the top half of your paper where $B = 0$. b) Show that integration around your rectangle contradicts Ampere's law.

28.79 An Infinite Current Sheet. Long, straight conductors with square cross sections and each carrying current I are laid side-by-side to form an infinite current sheet (see Fig. 28.56). The conductors lie in the xy-plane, are parallel to the y-axis and carry current in the $+y$-direction There are n conductors per unit length measured along the x-axis. a) What are the magnitude and direction of the magnetic field a distance a below the current sheet? b) What are the magnitude and direction of the magnetic field a distance a above the current sheet?

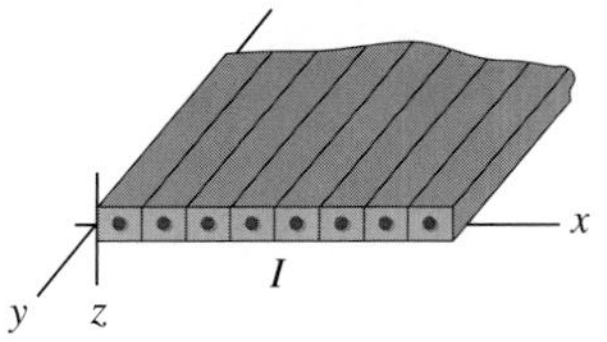

Figure 28.56 Problem 28.79.

28.80 Long, straight conductors with square cross section, each carrying current I, are laid side-by-side to form an infinite current sheet with current directed out of the plane of the page (see Fig. 28.57). A second infinite current sheet is a distance d below the first and is parallel to it. The second sheet carries current into the plane of the page. Each sheet has n conductors per unit length. (Refer to Problem 28.79.) Calculate the magnitude and direction of the net magnetic field at a) point P (above the upper sheet); b) point R (midway between the two sheets); c) point S (below the lower sheet).

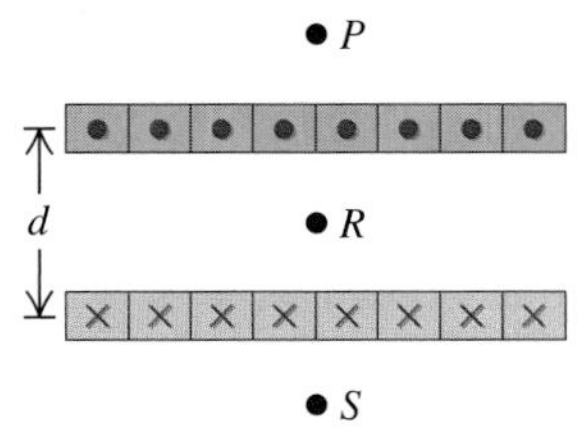

Figure 28.57 Problem 28.80.

***28.81** A piece of iron has magnetization $M = 6.50 \times 10^4$ A/m. Find the average magnetic dipole moment *per atom* in this piece of iron. Express your answer both in $\text{A} \cdot \text{m}^2$ and in Bohr magnetons. The density of iron is given in Table 14.1 and the atomic mass of iron (in grams per mole) is given in Appendix D. The chemical symbol for iron is Fe.

***28.82** a) In Section 27.7 we discussed how a magnetic dipole, such as a current loop or a magnetized object, can be attracted or repelled by a permanent magnet (see Figs. 27.34 and 27.36). Use this to explain why *either* pole of a magnet *attracts* both paramagnetic materials and (initially unmagnetized) ferromagnetic materials, but *repels* diamagnetic materials. b) The force that a magnet exerts on an object is directly proportional to the object's magnetic moment. A particular magnet is just strong enough to pick up a cube of annealed iron $(K_m = 1400)$ 2.00 cm on a side so that the iron sticks to one of the magnet's poles; that is, the magnet exerts an upward force on the iron cube equal to the cube's weight. If you tried to use this magnet to pick up a 2.00-cm cube of aluminum instead, what would be the upward force on the cube? How does this compare to the weight of the cube? Could the magnet pick up the cube? (*Hint:* You will need to use information from Tables 14.1 and 28.1.) c) If you tried to use the magnet to pick up a 2.00-cm cube of silver, what would be the magnitude and direction of the force on the cube? How does this magnitude compare to the weight of the cube? Would the effects of the magnetic force be noticeable?

Challenge Problems

28.83 Two long, straight conducting wires with linear mass density λ are suspended from cords so that they are each horizontal, parallel to each other, and a distance d apart. The back ends of the wires are connected to one another by a slack, low-resistance connecting wire. A charged capacitor (capacitance C) is now added to the system; the positive plate of the capacitor (initial charge $+Q_0$) is connected to the front end of one of the wires, and the negative plate of the capacitor (initial charge $-Q_0$) is connected to the front end of the other wire (see Fig. 28.58). Both of these connections are also made by slack, low-resistance wires. When the connection is made, the wires are pushed aside by the repulsive force between the wires, and each wire has an initial horizontal velocity of magnitude v_0. Assume that the time for the capacitor to discharge is negligible compared to the time it takes for any appreciable displacement in the position of the wires to occur. a) Show that the initial speed v_0 of either wire is given by

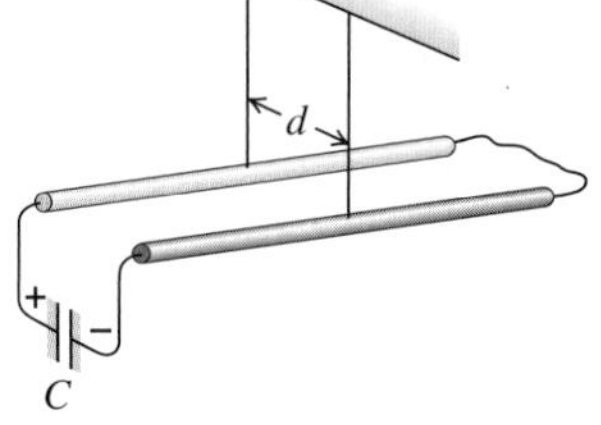

Figure 28.58 Challenge Problem 28.83.

$$v_0 = \frac{\mu_0 Q_0^2}{4\pi \lambda RCd}$$

where R is the total resistance of the circuit. b) Determine v_0 numerically if $\lambda = 4.50 \times 10^{-3}$ kg/m, $d = 3.00$ cm, $C = 2.50$ μF, and $R = 0.0480$ Ω, and if the capacitor was initially charged by connecting it to a 3.00-kV source. c) To what height h will each wire rise as a result of the circuit connection?

28.84 A wide, long, insulating belt has a uniform positive charge per unit area σ on its upper surface. Rollers at each end move the belt to the right at a constant speed v. Calculate the magnitude and direction of the magnetic field produced by the moving belt at a point just above its surface. (*Hint:* At points near the surface and far from its edges or ends, the moving belt can be considered to be an infinite current sheet like that in Problem 28.79.)

28.85 A Charged Dielectric Disk. A thin disk of dielectric material with radius a has a total charge $+Q$ distributed uniformly over its surface. It rotates n times per second about an axis perpendicular to the surface of the disk and passing through its center. Find the magnetic field at the center of the disk. (*Hint:* Divide the disk into concentric rings of infinitesimal width.)

28.86 A wire in the shape of a semicircle with radius a is oriented in the yz-plane with its center of curvature at the origin (see Fig. 28.59). If the current in the wire is I, calculate the magnetic field components produced at point P, a distance x out along the x-axis. (*Note:* Do not forget the contribution from the straight wire at the bottom of the semicircle that runs from $z = -a$ to $z = +a$. You may use the fact that the fields of the two antiparallel currents at $z > a$ cancel, but you must explain *why* they cancel.)

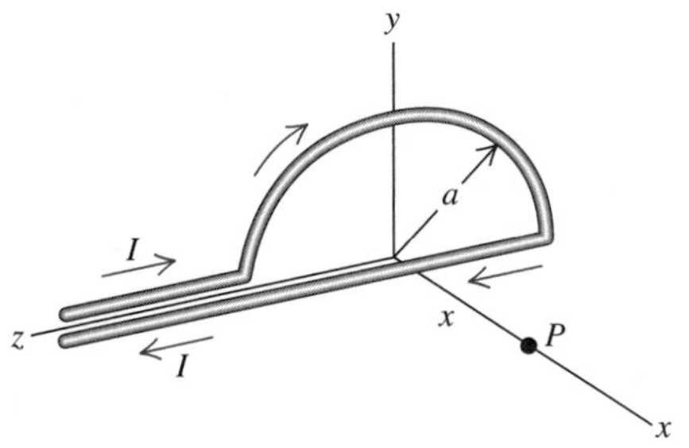

Figure 28.59 Challenge Problem 28.86.

ELECTROMAGNETIC INDUCTION 29

A credit card's number, expiration date, and cardholder name are coded into a magnetized pattern in a stripe on the back. When the card is "swiped" through a card reader, the moving stripe bathes the reader's circuitry in a varying magnetic field that induces currents in the circuits. These currents transmit the information in the stripe to the cardholder's bank.

? What would happen if the card were not swiped, but just sat motionless in the reader's slot?

Almost every modern device or machine, from a computer to a washing machine to a power drill, has electric circuits at its heart. We learned in Chapter 25 that an electromotive force (emf) is required for a current to flow in a circuit; in Chapters 25 and 26 we almost always took the source of emf to be a battery. But for the vast majority of electric devices that are used in industry and in the home (including any device that you plug into a wall socket), the source of emf is *not* a battery but an electrical generating station. Such a station produces electric energy by converting other forms of energy: gravitational potential energy at a hydroelectric plant, chemical energy in a coal- or oil-fired plant, nuclear energy at a nuclear plant. But how is this energy conversion done? In other words, what is the physics behind the production of almost all of our electric energy needs?

The answer is a phenomenon known as *electromagnetic induction:* If the magnetic flux through a circuit changes, an emf and a current are induced in the circuit. In a power-generating station, magnets move relative to coils of wire to produce a changing magnetic flux in the coils and hence an emf. Other key components of electric power systems, such as transformers, also depend on magnetically induced emfs. Indeed, thanks to its key role in electric power generation, electromagnetic induction is one of the foundations of our technological society.

The central principle of electromagnetic induction, and the keystone of this chapter, is *Faraday's law.* This law relates induced emf to changing magnetic flux in any loop, including a closed circuit. We also discuss Lenz's law, which helps us to predict the directions of induced emfs and currents. This chapter provides the principles we need to understand electrical energy-conversion devices such as motors, generators, and transformers.

Electromagnetic induction tells us that a time-varying magnetic field can act as a source of electric field. We will also see how a time-varying *electric* field can act as a source of *magnetic* field. These remarkable results form part of a neat package of formulas, called *Maxwell's equations,* that describe the behavior of electric and magnetic fields in *any* situation. Maxwell's equations pave the way toward an understanding of electromagnetic waves, the topic of Chapter 32.

29.1 | Induction Experiments

During the 1830s, several pioneering experiments with magnetically induced emf were carried out in England by Michael Faraday and in the United States by Joseph Henry (1797–1878), later the first director of the Smithsonian Institution. Figure 29.1 shows several examples. In Fig. 29.1a, a coil of wire is connected to a galvanometer. When the nearby magnet is stationary, the meter shows no current. This isn't surprising; there is no source of emf in the circuit. But when we *move* the magnet either toward or away from the coil, the meter shows current in the circuit, but *only* while the magnet is moving (Fig. 29.1b). If we keep the magnet stationary and move the coil, we again detect a current during the motion. We call this an **induced current,** and the corresponding emf required to cause this current is called an **induced emf.**

In Fig. 29.1c we replace the magnet with a second coil connected to a battery. When the second coil is stationary, there is no current in the first coil. However, when we move the second coil toward or away from the first or move the first toward or away from the second, there is current in the first coil, but again *only* while one coil is moving relative to the other.

Finally, using the two-coil setup in Fig. 29.1d, we keep both coils stationary and vary the current in the second coil, either by opening and closing the switch or by changing the resistance of the second coil with the switch closed (perhaps by changing the second coil's temperature). We find that as we open or close the switch, there is a momentary current pulse in the first circuit. When we vary the resistance (and thus the current) in the second coil, there is an induced current in the first circuit, but only while the current in the second circuit is changing.

To explore further the common elements in these observations, let's consider a more detailed series of experiments with the situation shown in Fig. 29.2. We connect a coil of wire to a galvanometer, then place the coil between the poles of an electromagnet whose magnetic field we can vary. Here's what we observe:

1. When there is no current in the electromagnet, so that $\vec{B} = \mathbf{0}$, the galvanometer shows no current.
2. When the electromagnet is turned on, there is a momentary current through the meter as $\vec{B}$ increases.
3. When $\vec{B}$ levels off at a steady value, the current drops to zero, no matter how large $\vec{B}$ is.
4. With the coil in a horizontal plane, we squeeze it so as to decrease the cross-sectional area of the coil. The meter detects current only *during* the defor-

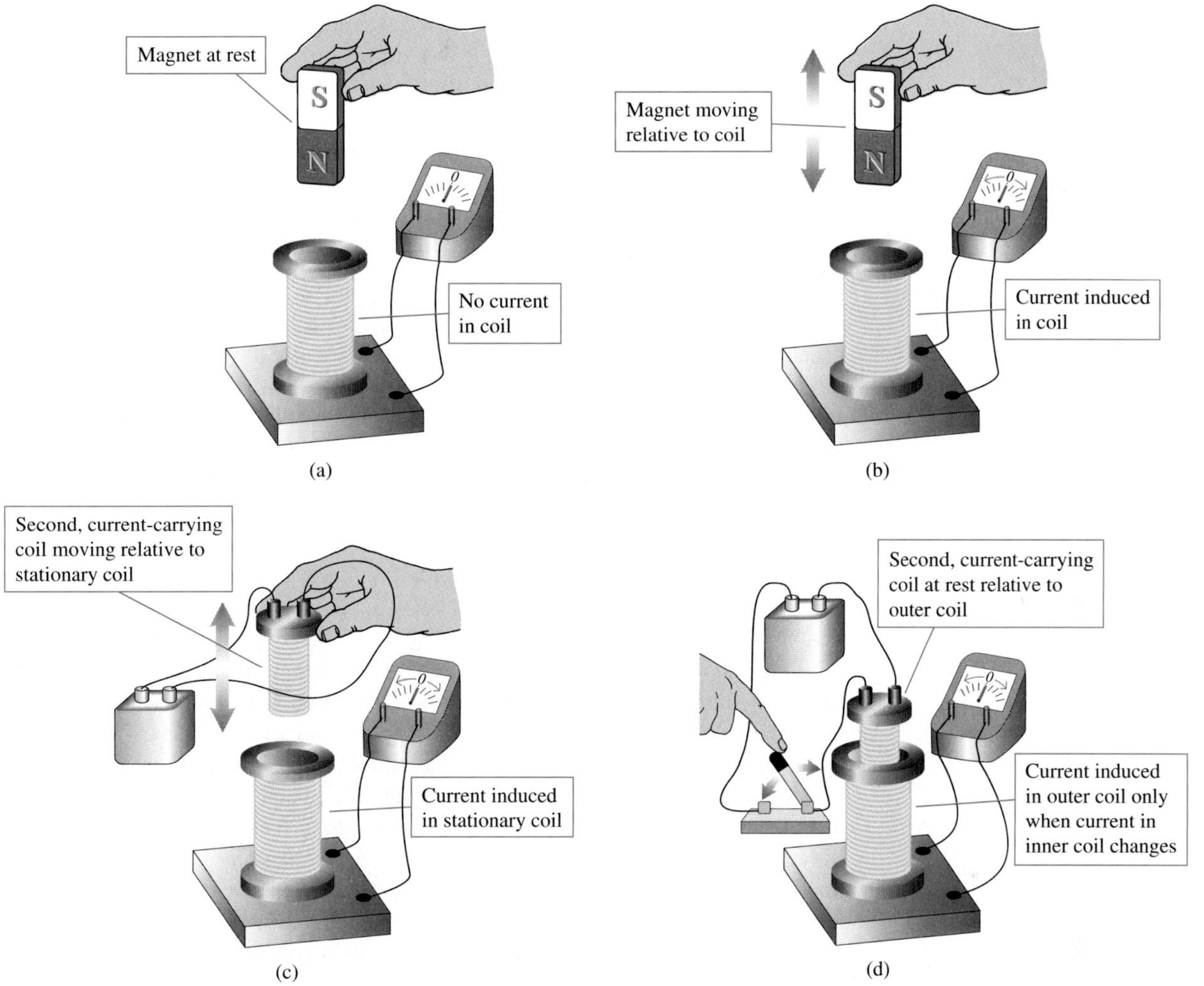

29.1 (a) A stationary magnet has no effect on a stationary coil of wire. A galvanometer connected to the coil shows zero current. (b) When the magnet and coil move relative to each other, a current is induced in the coil. The current is in one direction if the magnet moves down and the opposite direction if the magnet moves up. (c) We get the same effect as in (b) if we replace the magnet by a second coil carrying a constant current. (d) When the switch is opened or closed, the change in the inside coil's current induces a current in the outer coil.

mation, not before or after. When we increase the area to return the coil to its original shape, there is current in the opposite direction, but only while the area of the coil is changing.

5. If we rotate the coil a few degrees about a horizontal axis, the meter detects current during the rotation, in the same direction as when we decreased the area. When we rotate the coil back, there is a current in the opposite direction during this rotation.
6. If we jerk the coil out of the magnetic field, there is a current during the motion, in the same direction as when we decreased the area.
7. If we decrease the number of turns in the coil by unwinding one or more turns, there is a current during the unwinding, in the same direction as when we decreased the area. If we wind more turns onto the coil, there is a current in the opposite direction during the winding.

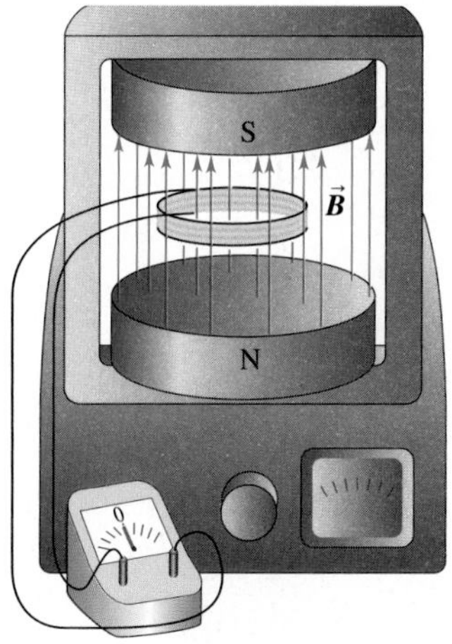

29.2 A coil in a magnetic field. When the $\vec{B}$ field is constant and the shape, location, and orientation of the coil do not change, no current is induced in the coil. A current is induced when any of these factors change.

8. When the magnet is turned off, there is a momentary current in the direction opposite to the current when it was turned on.
9. The faster we carry out any of these changes, the greater the current.
10. If all these experiments are repeated with a coil that has the same shape but different material and different resistance, the current in each case is inversely proportional to the total circuit resistance. This shows that the induced emfs that are causing the current do not depend on the material of the coil but only on its shape and the magnetic field.

The common element in all these experiments is changing *magnetic flux* Φ_B through the coil connected to the galvanometer. In each case the flux changes either because the magnetic field changes with time or because the coil is moving through a nonuniform magnetic field. Check back through the list to verify this statement. Faraday's law of induction, the subject of the next section, states that in all of these situations the induced emf is proportional to the *rate of change* of magnetic flux Φ_B through the coil. The *direction* of the induced emf depends on whether the flux is increasing or decreasing. If the flux is constant, there is no induced emf.

ActivPhysics Online 13.9 Electromagnetic Induction

Induced emfs are not mere laboratory curiosities but have a tremendous number of practical applications. If you are reading these words indoors, you are making use of induced emfs right now! At the power plant that supplies your neighborhood, an electric generator produces an emf by varying the magnetic flux through coils of wire. (In the next section we'll see in detail how this is done.) This emf supplies the voltage between the terminals of the wall sockets in your home, and this voltage supplies the power to your reading lamp. Indeed, any appliance that you plug into a wall socket makes use of induced emfs.

Magnetically induced emfs, just like the emfs discussed in Section 25.4, are always the result of the action of *nonelectrostatic* forces. When these forces are the result of additional electric fields induced by changing magnetic fields, we have to distinguish carefully between electric fields produced by charges (according to Coulomb's law) and those produced by changing magnetic fields. We'll denote these by $\vec{E}_c$ (where c stands for Coulomb or conservative) and $\vec{E}_n$ (where n stands for non-Coulomb or non-conservative), respectively. We'll return to this distinction later in this chapter and the next.

29.2 | Faraday's Law

The common element in all induction effects is changing magnetic flux through a circuit. Before stating the simple physical law that summarizes all of the kinds of experiments described in Section 29.1, let's first review the concept of magnetic flux Φ_B (which we introduced in Section 27.3). For an infinitesimal area element $d\vec{A}$ in a magnetic field $\vec{B}$ (Fig. 29.3), the magnetic flux $d\Phi_B$ through the area is

$$d\Phi_B = \vec{B} \cdot d\vec{A} = B_\perp \, dA = B \, dA \cos \phi.$$

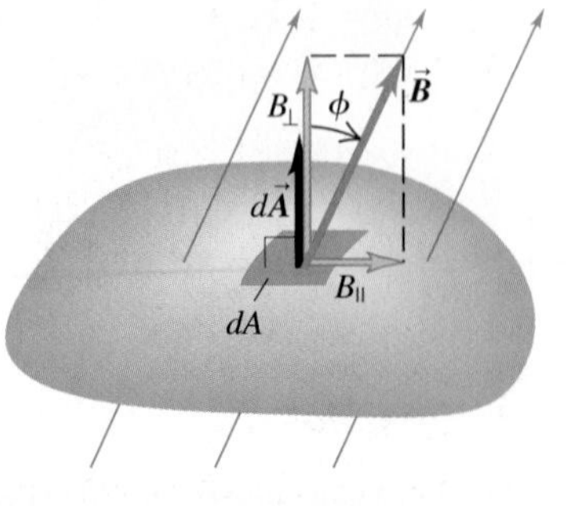

29.3 The magnetic flux through an area element dA is defined to be $d\Phi_B = B_\perp \, dA$.

where $B_\perp$ is the component of $\vec{B}$ perpendicular to the surface of the area element and ϕ is the angle between $\vec{B}$ and $d\vec{A}$. (As in Chapter 28, be careful to distinguish between two quantities named "phi," ϕ and Φ_B.) The total magnetic flux Φ_B through a finite area is the integral of this expression over the area:

$$\Phi_B = \int \vec{B} \cdot d\vec{A} = \int B \, dA \cos \phi \tag{29.1}$$

If $\vec{B}$ is uniform over a flat area $\vec{A}$, then

$$\Phi_B = \vec{B} \cdot \vec{A} = BA \cos \phi \qquad (29.2)$$

CAUTION In Eqs. (29.1) and (29.2) we have to be careful to define the direction of the vector area $d\vec{A}$ or $\vec{A}$ unambiguously. There are always two directions perpendicular to any given area, and the sign of the magnetic flux through the area depends on which one we choose to be positive. For example, in Fig. 29.3 we chose $d\vec{A}$ to point upward so ϕ is less than $90°$ and $\vec{B} \cdot d\vec{A}$ is positive. We could have chosen instead to have $d\vec{A}$ point downward, in which case ϕ would have been more than $90°$ and $\vec{B} \cdot d\vec{A}$ would have been negative. Either choice is equally good, but once we make a choice we must stick with it.

Faraday's law of induction states:

The induced emf in a closed loop equals the negative of the time rate of change of magnetic flux through the loop.

In symbols, Faraday's law is

$$\mathcal{E} = -\frac{d\Phi_B}{dt} \qquad \text{(Faraday's law of induction)} \qquad (29.3)$$

To understand the negative sign, we have to introduce a sign convention for the induced emf $\mathcal{E}$. But first let's look at a simple example of this law in action.

Example 29.1 Emf and current induced in a loop

The magnetic field between the poles of the electromagnet in Fig. 29.4 is uniform at any time, but its magnitude is increasing at the rate of 0.020 T/s. The area of the conducting loop in the field is 120 cm^2, and the total circuit resistance, including the meter and the resistor, is 5.0 Ω. a) Find the induced emf and the induced current in the circuit. b) If the loop is replaced by one made of an insulator, what effect does this have on the induced emf and induced current?

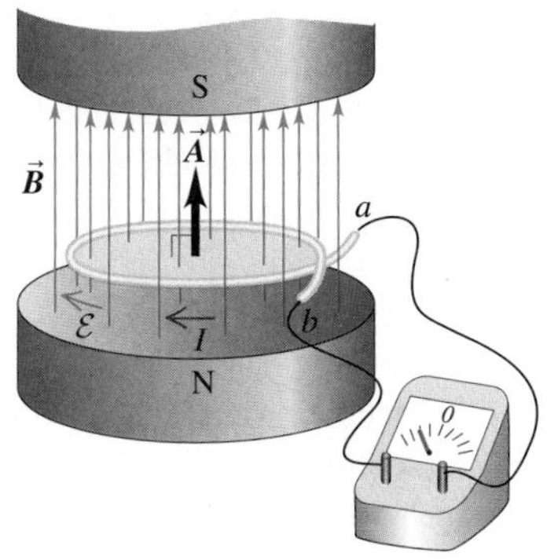

29.4 A stationary conducting loop in an increasing magnetic field.

SOLUTION

IDENTIFY: The magnetic flux through the loop changes as the magnetic field changes. Hence there will be an induced emf in the loop, and we can find its value (one of our target variables) using Faraday's law. We can determine the current produced in the loop by this emf (our other target variable) using the same techniques as in Chapter 25.

SET UP: We calculate the magnetic flux using Eq. (29.2), then use Faraday's law given by Eq. (29.3) to determine the resulting induced emf $\mathcal{E}$. Then we calculate the induced current produced by this emf using the relation $\mathcal{E} = IR$, where R is the total resistance of the circuit that includes the loop.

EXECUTE: a) The vector area of the loop is perpendicular to the plane of the loop; we choose it to be vertically upward. Then the vectors $\vec{A}$ and $\vec{B}$ are parallel. Since $\vec{B}$ is uniform, the magnetic flux through the loop is $\Phi_B = \vec{B}\cdot\vec{A} = BA\cos 0 = BA$. The area $A = 0.012\ \text{m}^2$ is constant, so the rate of change of magnetic flux is

$$\frac{d\Phi_B}{dt} = \frac{d(BA)}{dt} = \frac{dB}{dt}A = (0.020\ \text{T/s})(0.012\ \text{m}^2)$$
$$= 2.4 \times 10^{-4}\ \text{V} = 0.24\ \text{mV}$$

This, apart from a sign that we haven't discussed yet, is the induced emf $\mathcal{E}$. The corresponding induced current is

$$I = \frac{\mathcal{E}}{R} = \frac{2.4 \times 10^{-4}\ \text{V}}{5.0\ \Omega} = 4.8 \times 10^{-5}\ \text{A} = 0.048\ \text{mA}$$

b) By changing to a loop made of insulator, we've made the resistance of the loop very high. Faraday's law, Eq. (29.3), does not involve the resistance of the circuit in any way, so the induced *emf* does not change. But the *current* will be smaller, as given by the equation $I = \mathcal{E}/R$. If the loop is made of a perfect insulator with infinite resistance, the induced current is zero even though an emf is present. This situation is analogous to an isolated battery whose terminals aren't connected to anything: there is an emf present, but no current flows.

EVALUATE: It's worthwhile to verify unit consistency in this calculation. There are many ways to do this; one is to note that because of the magnetic force relation $\vec{F} = q\vec{v} \times \vec{B}$, the units of magnetic field are the units of force divided by the units of (charge times velocity): $1\ \text{T} = (1\ \text{N})/(1\ \text{C}\cdot\text{m/s})$. The units of magnetic flux can then be expressed as $(1\ \text{T})(1\ \text{m}^2) = 1\ \text{N}\cdot\text{s}\cdot\text{m/C}$, and the rate of change of magnetic flux as $1\ \text{N}\cdot\text{m/C} = 1\ \text{J/C} = 1\ \text{V}$. Thus the unit of $d\Phi_B/dt$ is the volt, as required by Eq. (29.3). Also recall that the unit of magnetic flux is the weber (Wb): $1\ \text{T}\cdot\text{m}^2 = 1\ \text{Wb}$, so $1\ \text{V} = 1\ \text{Wb/s}$.

Direction of Induced EMF

We can find the direction of an induced emf or current by using Eq. (29.3) together with some simple sign rules. Here's the procedure:

1. Define a positive direction for the vector area $\vec{A}$.
2. From the directions of $\vec{A}$ and the magnetic field $\vec{B}$, determine the sign of the magnetic flux Φ_B and its rate of change $d\Phi_B/dt$. Figure 29.5 shows several examples.
3. Determine the sign of the induced emf or current. If the flux is increasing, so $d\Phi_B/dt$ is positive, then the induced emf or current is negative; if the flux is decreasing, $d\Phi_B/dt$ is negative and the induced emf or current is positive.
4. Finally, determine the direction of the induced emf or current using your right hand. Curl the fingers of your right hand around the $\vec{A}$ vector, with your right thumb in the direction of $\vec{A}$. If the induced emf or current in the circuit is *positive,* it is in the same direction as your curled fingers; if the induced emf or current is *negative,* it is in the opposite direction.

In Example 29.1, in which $\vec{A}$ is upward, a positive $\mathcal{E}$ would be directed counterclockwise around the loop, as seen from above. Both $\vec{A}$ and $\vec{B}$ are upward in this example, so Φ_B is positive; the magnitude B is increasing, so $d\Phi_B/dt$ is positive. Hence by Eq. (29.3), $\mathcal{E}$ in Example 29.1 is *negative.* Its actual direction is thus *clockwise* around the loop, as shown in Fig. 29.4.

If the loop in Fig. 29.4 is a conductor, an induced current results from this emf; this current is also clockwise. This induced current produces an additional magnetic field through the loop, and the right-hand rule described in Section 28.6 shows that this field is *opposite* in direction to the increasing field produced by the electromagnet. This is an example of a general rule called *Lenz's law,* which says that any induction effect tends to oppose the change that caused it; in this case the change is the increase in the flux of the electromagnet's field through the loop. (We'll study this law in detail in the next section.)

You should check out the signs of the induced emfs and currents for the list of experiments cited at the end of Section 29.1. For example, when the loop in Fig. 29.2 is in a constant field and we tilt it or squeeze it to *decrease* the flux through it, the induced emf and current are counterclockwise, as seen from above.

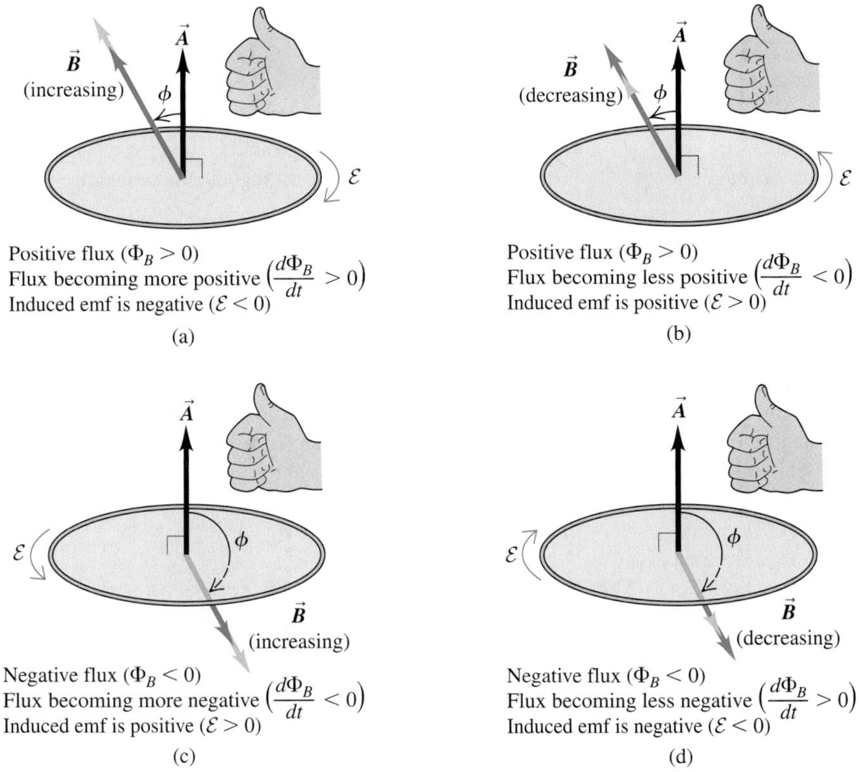

29.5 The magnetic flux is becoming (a) more positive, (b) less positive, (c) more negative, and (d) less negative. Therefore Φ_B is increasing in (a) and (d) and decreasing in (b) and (c). In (a) and (d) the emfs are negative (they are opposite to the direction of the curled fingers of your right hand when your right thumb points along $\vec{A}$). In (b) and (c) the emfs are positive (in the same direction as the curled fingers).

CAUTION Since magnetic flux plays a central role in Faraday's law, it's tempting to think that *flux* is the cause of induced emf and that an induced emf will appear in a circuit whenever there is a magnetic field in the region bordered by the circuit. But Eq. (29.3) shows that only a *change* in flux through a circuit, not flux itself, can induce an emf in a circuit. If the flux through a circuit has a constant value, whether positive, negative, or zero, there is no induced emf.

If we have a coil with N identical turns, and if the flux varies at the same rate through each turn, the *total* rate of change through all the turns is N times as large as for a single turn. If Φ_B is the flux through each turn, the total emf in a coil with N turns is

$$\mathcal{E} = -N\frac{d\Phi_B}{dt} \tag{29.4}$$

As we discussed in this chapter's introduction, induced emfs play an essential role in the generation of electric power for commercial use. Several of the following examples explore different methods of producing emfs by the motion of a conductor relative to a magnetic field, giving rise to a changing flux through a circuit.

Problem-Solving Strategy

Faraday's Law

IDENTIFY *the relevant concepts:* Faraday's law applies when there is a changing magnetic flux. To use the law, make sure you can identify an area through which there is a flux of magnetic field. This will usually be the area enclosed by a loop, usually made of a conducting material (though not always—see part (b) of Example 29.1). As always, identify the target variable(s).

SET UP *the problem using the following steps:*

1. Faraday's law relates the induced emf to the rate of change of magnetic flux. To calculate this rate of change, you first have to understand what is making the flux change. Is the conductor moving? Is it changing orientation? Is the magnetic field changing? Remember that it's not the flux itself that counts, but its *rate of change*.
2. Choose a direction for the area vector $\vec{A}$ or $d\vec{A}$. The direction must always be perpendicular to the plane of the area. Note that you always have two choices of direction. For instance, if the plane of the area is horizontal, $\vec{A}$ could point straight up or straight down. It's like choosing which direction is the positive one in a problem involving motion in a straight line; it doesn't matter which direction you choose, just so you use it consistently throughout the problem.

EXECUTE *the solution as follows:*

1. Calculate the magnetic flux using Eq. (29.2) if $\vec{B}$ is uniform over the area of the loop or Eq. (29.1) if it isn't uniform, being mindful of the direction you chose for the area vector.
2. Calculate the induced emf using Eq. (29.3) or (29.4). If your conductor has N turns in a coil, don't forget to multiply by N. Remember the sign rule for the positive direction of emf and use it consistently.
3. If the circuit resistance is known, you can calculate the magnitude of the induced current I using $\mathcal{E} = IR$.

EVALUATE *your answer:* Check your results for the proper units, and double-check that you have properly implemented the sign rules for calculating magnetic flux and induced emf.

Example 29.2

Magnitude and direction of an induced emf

A coil of wire containing 500 circular loops with radius 4.00 cm is placed between the poles of a large electromagnet, where the magnetic field is uniform and at an angle of 60° with the plane of the coil (Fig. 29.6a). The field decreases at a rate of 0.200 T/s. What are the magnitude and direction of the induced emf?

SOLUTION

IDENTIFY: Our target variable is the emf induced by a varying magnetic flux through the coil. The flux varies because the magnetic field decreases in amplitude.

SET UP: We choose the area vector $\vec{A}$ to be in the direction shown in Fig. 29.6a. With this choice, the geometry is very similar to Fig. 29.5b. That figure will help us determine the direction of the induced emf.

EXECUTE: The magnetic field is uniform over the loop, so we can calculate the flux using Eq. (29.2): $\Phi_B = BA\cos\phi$, where $\phi = 30°$. In this expression, the only quantity that changes with time is the magnitude B of the field.

CAUTION You may have been tempted to say that $\phi = 60°$ in this problem. If so, remember that ϕ is the angle between $\vec{A}$ and $\vec{B}$, *not* the angle between $\vec{B}$ and the plane of the loop (Fig. 29.6b).

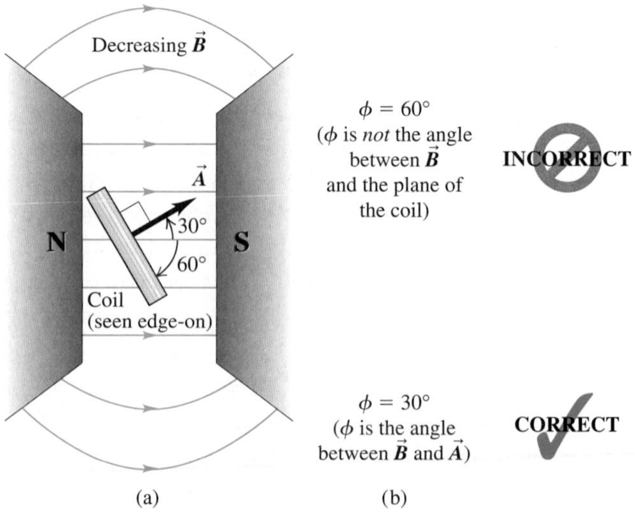

29.6 The magnitude of $\vec{B}$ is decreasing. With the direction of $\vec{A}$ shown, the flux through the coil is decreasing, and $\mathcal{E}$ is positive. This corresponds to clockwise emf and current as you look in from the left along the direction of $\vec{A}$. The additional $\vec{B}$ caused by the current tends to compensate for the decrease in flux. (b) Incorrect and correct choices of the angle ϕ.

The rate of change of the flux is $d\Phi_B/dt = (dB/dt)A\cos\phi$. In our problem, $dB/dt = -0.200\text{ T/s}$ and $A = \pi(0.0400\text{ m})^2 = 0.00503\text{ m}^2$, so

$$\begin{aligned}\frac{d\Phi_B}{dt} &= \frac{dB}{dt}A\cos 30^\circ\\ &= (-0.200\text{ T/s})(0.00503\text{ m}^2)(0.866)\\ &= -8.71\times10^{-4}\text{ T}\cdot\text{m}^2\text{/s} = -8.71\times10^{-4}\text{ Wb/s}\end{aligned}$$

From Eq. (29.4), the induced emf in the coil of $N = 500$ turns is

$$\begin{aligned}\mathcal{E} &= -N\frac{d\Phi_B}{dt}\\ &= -(500)(-8.71\times10^{-4}\text{ Wb/s}) = 0.435\text{ V}\end{aligned}$$

Note that the answer is positive. This means that when you point your right thumb in the direction of the area vector $\vec{A}$ (30° above the magnetic field $\vec{B}$), the positive direction for $\mathcal{E}$ is in the direction of the curled fingers of your right hand. Hence the emf in this example is in this same direction (compare Fig. 29.5b). If you were viewing the coil from the left side in Fig. 29.6a and looking in the direction of $\vec{A}$, the emf would be clockwise.

EVALUATE: If the ends of the wire are connected together, the direction of current in the coil is in the same direction as the emf; that is, clockwise as seen from the left side of the coil. A clockwise current gives added magnetic flux through the coil in the same direction as the flux from the electromagnet, and therefore tends to oppose the decrease in total flux. We'll see more examples of this in Section 29.3.

Conceptual Example 29.3 The search coil

One practical way to measure magnetic field strength uses a small, closely wound coil with N turns called a *search coil.* The coil, of area A, is initially held so that its area vector $\vec{A}$ is aligned with a magnetic field with magnitude B. The coil is then either quickly rotated a quarter-turn about a diameter or quickly pulled out of the field. Explain how this device can be used to measure the value of B.

SOLUTION

Initially, the flux through the coil is $\Phi_B = NBA$; when the coil is rotated or pulled from the field, the flux decreases rapidly from NBA to zero. While the flux is decreasing, there is a momentary induced emf, and a momentary induced current occurs in an external circuit connected to the coil. The rate of change of flux through the coil is proportional to the current, or rate of flow of charge, so it is easy to show that the *total* flux change is proportional to the total charge that flows around the circuit. We can build an instrument that measures this total charge, and from this we can compute B. We leave the details as a problem (see Exercise 29.3). Strictly speaking, this method gives only the *average* field over the area of the coil. But if the area is small, this average field is very nearly equal to the field at the center of the coil.

Example 29.4 Generator I: A simple alternator

Figure 29.7 shows a simple version of an *alternator,* a device that generates an emf. A rectangular loop is made to rotate with constant angular speed ω about the axis shown. The magnetic field $\vec{B}$ is uniform and constant. At time $t = 0$, $\phi = 0$. Determine the induced emf.

SOLUTION

IDENTIFY: Again the emf (our target variable) is produced by a varying magnetic flux. In this situation, however, the magnetic field $\vec{B}$ is constant; the flux changes because the direction of $\vec{A}$ changes as the loop rotates.

SET UP: Figure 29.7a shows the direction of the area vector $\vec{A}$. Note that as the loop rotates, the angle ϕ between $\vec{A}$ and $\vec{B}$ increases at a constant rate.

EXECUTE: Again the magnetic field is uniform over the loop, so the magnetic flux is easy to calculate. The rate of change of the angle ϕ is equal to ω, the angular speed of the loop, so we can write $\phi = \omega t$. Hence

$$\Phi_B = BA\cos\phi = BA\cos\omega t$$

The derivative of $\cos\omega t$ is $(d/dt)\cos\omega t = -\omega\sin\omega t$. Hence, by Faraday's law [Eq. (29.3)] the induced emf is

$$\mathcal{E} = -\frac{d\Phi_B}{dt} = \omega BA\sin\omega t$$

EVALUATE: The induced emf $\mathcal{E}$ varies sinusoidally with time (Fig. 29.7b). When the plane of the loop is perpendicular to $\vec{B}$ ($\phi = 0$ or 180°), Φ_B reaches its maximum and minimum values. At these times, its instantaneous rate of change is zero and $\mathcal{E}$ is zero. Also, $\mathcal{E}$ is greatest in absolute value when the plane of the loop is parallel to $\vec{B}$ ($\phi = 90°$ or 270°) and Φ_B is changing most rapidly. Finally, we note that the induced emf does not depend on

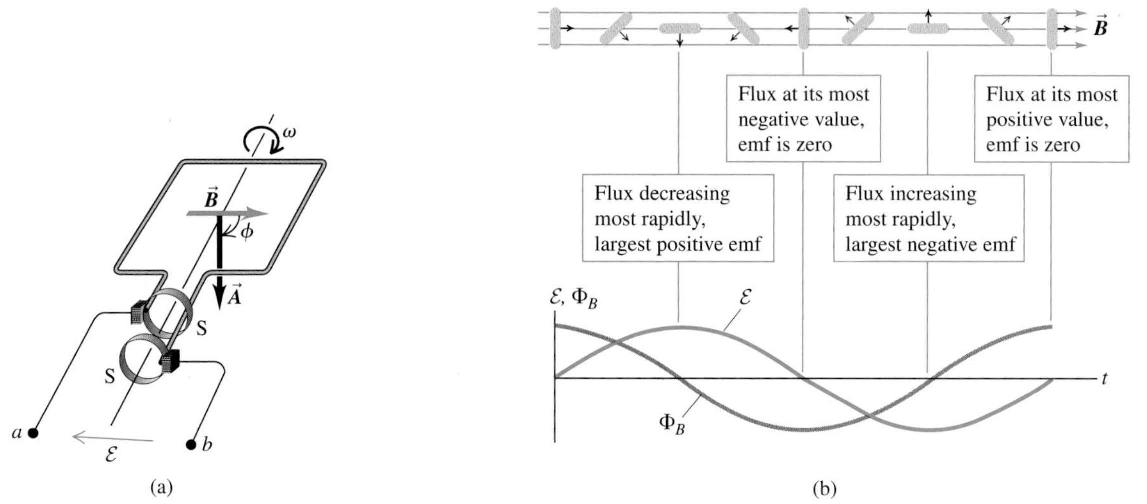

29.7 (a) Schematic diagram of an alternator. A conducting loop rotates in a magnetic field, producing an emf. Connections from each end of the loop to the external circuit are made by means of that end's slip ring S. The system is shown at the time when the angle $\phi = \omega t = 90°$. (b) Graph of the flux through the loop and the resulting emf at terminals *ab*, along with corresponding positions of the loop during one complete rotation. The emf is the negative slope of the flux graph: $\mathcal{E} = -d\Phi_B/dt$.

the *shape* of the loop, but only on its area. Because $\mathcal{E}$ is directly proportional to ω and B, some tachometers use the emf in a rotating coil to measure rotational speed. Other devices use an emf of this kind to measure magnetic field.

We can use the alternator as a source of emf in an external circuit by use of two *slip rings* S, which rotate with the loop, as shown in Fig. 29.7a. The rings slide against stationary contacts called *brushes,* which are connected to the output terminals a and b. Since the emf varies sinusoidally, the current that results in the circuit is an *alternating* current that also varies sinusoidally in magnitude and direction. An alternator is also called an *alternating-current* (ac) *generator* for this reason. The amplitude of the emf can be increased by increasing the rotation speed, the field magnitude, or the loop area or by using N loops instead of one, as in Eq. (29.4).

Alternators are used in automobiles to generate the currents in the ignition, the lights, and the entertainment system. The arrangement is a little different than in this example; rather than having a rotating loop in a magnetic field, the loop stays fixed and an electromagnet rotates. (The rotation is provided by a mechanical connection between the alternator and the engine.) But the result is the same; the flux through the loop varies sinusoidally, producing a sinusoidally varying emf. Larger alternators of this same type are used in electric power plants (Fig. 29.8).

29.8 A commercial alternator uses many loops of wire wound around a barrel-like structure called an armature. The armature and wire remain stationary while electromagnets rotate on a shaft (not shown) through the center of the armature. The resulting induced emf is far larger than would be possible with a single loop of wire.

Example 29.5 Generator II: A DC generator and back emf in a motor

The alternator in Example 29.4 produces a sinusoidally varying emf and hence an alternating current. We can use a similar scheme to make a *direct-current* (dc) *generator* that produces an emf that always has the same sign. A prototype dc generator is shown in Fig. 29.9a. The arrangement of split rings is called a *commutator;* it reverses the connections to the external circuit at angular positions where the emf reverses. The resulting emf is shown in Fig. 29.9b. Commercial dc generators have a large number of coils and commutator segments; this arrangement smooths out the bumps in the emf, so the terminal voltage is not only one-directional but also practically constant. This

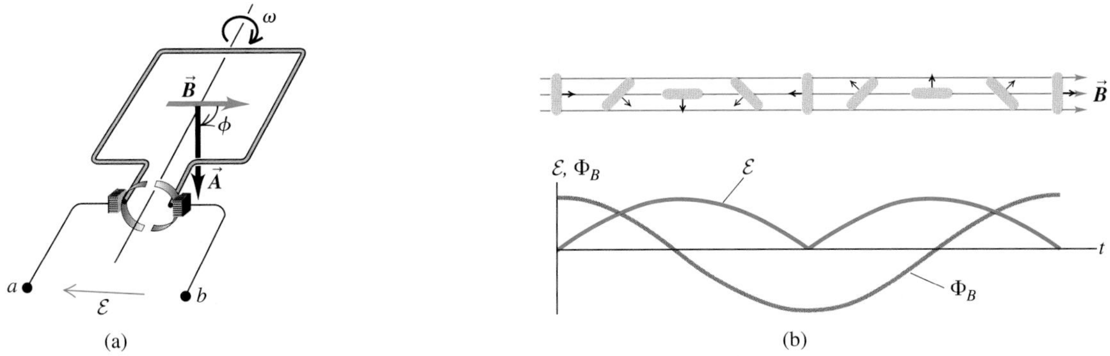

29.9 (a) Schematic diagram of a dc generator, using a split-ring commutator. The ring halves are attached to the loop and rotate with it. (b) Graph of the resulting induced emf at terminals *ab*. Compare to Fig. 29.7b.

brush-and-commutator arrangement is the same as that in the direct-current motor we discussed in Section 27.8. The motor's *back emf* is just the emf induced by the changing magnetic flux through its rotating coil. Consider a motor with a square coil 10.0 cm on a side, with 500 turns of wire. If the magnetic field has magnitude 0.200 T, at what rotation speed is the *average* back emf of the motor equal to 112 V?

SOLUTION

IDENTIFY: As far as the rotating loop is concerned, the situation is the same as in Example 29.4 except that we now have N turns of wire. Without the commutator, the emf would alternate between positive and negative values and have an average value of zero (Fig. 29.7b). But with the commutator added, the emf is never negative and its average value is positive (Fig. 29.9b). Using our result from Example 29.4, we'll determine an expression for this average value and solve that expression for the rotational speed ω (our target variable).

SET UP: The setup is the same as in Example 29.4.

EXECUTE: Comparing Figs. 29.7b and 29.9b shows that the back emf of the motor is just the absolute value of the emf found for an alternator in Example 29.4, multiplied by the number of turns N in the coil as in Eq. (29.4):

$$|\mathcal{E}| = N\omega BA|\sin \omega t|$$

To find the *average* back emf, we replace $|\sin \omega t|$ by its average value. The average value of the sine function is found by integrating $\sin \omega t$ over half a cycle, from $t = 0$ to $t = T/2 = \pi/\omega$, then dividing by the elapsed time π/ω. During this half cycle, the sine function is positive, so $|\sin \omega t| = \sin \omega t$, and we find

$$(|\sin \omega t|)_{\text{av}} = \frac{\int_0^{\pi/\omega} \sin \omega t\, dt}{\pi/\omega} = \frac{2}{\pi}$$

or about 0.64. The average back emf is then

$$\mathcal{E}_{\text{av}} = \frac{2N\omega BA}{\pi}$$

The back emf is proportional to the rotation speed ω, as was stated without proof in Section 27.8. Solving for ω, we obtain

$$\omega = \frac{\pi\mathcal{E}_{\text{av}}}{2NBA}$$

$$= \frac{\pi(112\text{ V})}{2(500)(0.200\text{ T})(0.100\text{ m})^2} = 176\text{ rad/s}$$

We used the relationships $1\text{ V} = 1\text{ Wb/s} = 1\text{ T}\cdot\text{m}^2\text{/s}$ from Example 29.1. We were able to add "radians" to the units of the answer because it is a dimensionless quantity, as we discussed in Chapter 9. The rotation speed can also be written as

$$\omega = 176\text{ rad/s}\frac{1\text{ rev}}{2\pi\text{ rad}}\frac{60\text{ s}}{1\text{ min}} = 1680\text{ rev/min}$$

EVALUATE: The average back emf is directly proportional to ω. Hence the slower the rotation speed, the less the back emf and the greater the possibility of burning out the motor, as we described in Example 27.12 (Section 27.8).

While we have used a very simple model of a generator in this and the previous example, the same principles apply to the operation of commercial generators.

Example 29.6 Generator III: The slidewire generator

Figure 29.10 shows a U-shaped conductor in a uniform magnetic field $\vec{B}$ perpendicular to the plane of the figure, directed *into* the page. We lay a metal rod with length L across the two arms of the conductor, forming a circuit, and move the rod to the right with constant velocity $\vec{v}$. This induces an emf and a current, which is why this device is called a *slidewire generator.* Find the magnitude and direction of the resulting induced emf.

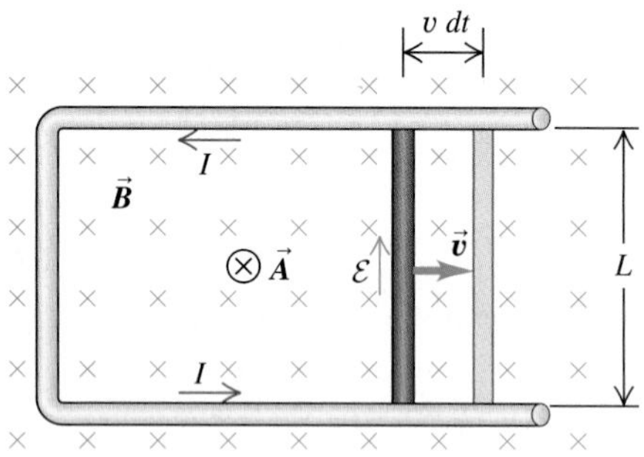

29.10 A slidewire generator. The magnetic field $\vec{B}$ and the vector area $\vec{A}$ are both directed into the figure. The increase in magnetic flux (caused by an increase in area) induces the emf and current shown.

SOLUTION

IDENTIFY: The magnetic flux changes because the area of the loop—bounded on the right by the moving rod—is increasing. Our target variable is the emf $\mathcal{E}$ induced in this expanding loop.

SET UP: The magnetic field is uniform over the area of the loop, so we can again calculate the magnetic flux using $\Phi_B = BA \cos \phi$. We choose the area vector $\vec{A}$ to point straight into the plane of the picture, in the same direction as $\vec{B}$. With this choice a positive emf will be one that is directed clockwise around the loop. (You can check this with the right-hand rule. Using your right hand, point your thumb into the page and curl your fingers as in Fig. 29.5.)

EXECUTE: Since $\vec{B}$ and $\vec{A}$ point in the same direction, the angle $\phi = 0$ and $\Phi_B = BA$. The magnetic field magnitude B is constant, so the induced emf is

$$\mathcal{E} = -\frac{d\Phi_B}{dt} = -B\frac{dA}{dt}$$

To calculate dA/dt, note that in a time dt the sliding rod moves a distance $v\,dt$ (Fig. 29.10) and the loop area increases by an amount $dA = Lv\,dt$. Hence the induced emf is

$$\mathcal{E} = -B\frac{Lv\,dt}{dt} = -BLv$$

The minus sign tells us that the emf is directed *counterclockwise* around the loop. The induced current is also counterclockwise, as shown in the figure.

EVALUATE: Note that the emf is constant if the velocity $\vec{v}$ of the rod is constant. In this case the slidewire generator acts as a *direct-current* generator. It's not a very practical device because the rod eventually moves beyond the U-shaped conductor and loses contact, after which the induced emf and current stop.

Example 29.7 Work and power in the slidewire generator

In the slidewire generator of Example 29.6, energy is dissipated in the circuit owing to its resistance. Let the resistance of the circuit (made up of the moving slidewire and the U-shaped conductor that connects the ends of the slidewire) at a given point in the slidewire's motion be R. Show that the rate at which energy is dissipated in the circuit is exactly equal to the rate at which work must be done to move the rod through the magnetic field.

SOLUTION

IDENTIFY: Our target variables are the *rates* at which energy is dissipated and at which work is done. This means that we'll be working with the concept of power (recall Section 6.4). Energy is dissipated in the circuit because there is resistance; to describe this we'll need the ideas of Section 25.5. It takes work to move the rod because there is an induced current flowing through it. The magnetic field exerts a force on this current-carrying rod, and whoever is pushing the rod has to do work against this force.

SET UP: We found the induced emf $\mathcal{E}$ in this circuit in Example 29.6. The current I in the circuit equals the absolute value of $\mathcal{E}$ divided by the resistance R, and the rate at which energy is dissipated in the rod is $P_{\text{dissipated}} = I^2R$. The magnetic force on the rod is

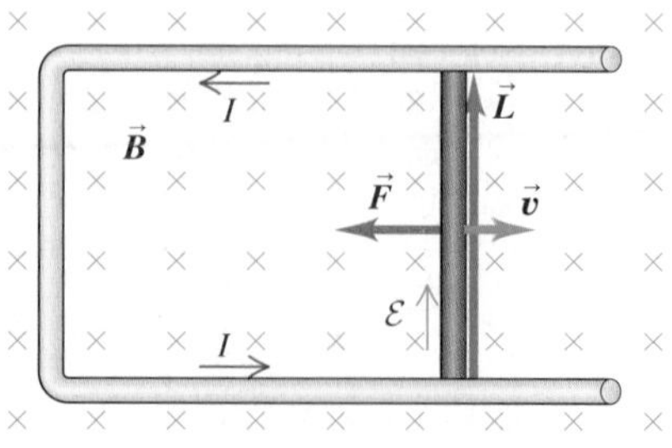

29.11 The magnetic force $\vec{F} = I\vec{L} \times \vec{B}$ that acts on the rod due to the induced current is to the left, opposite to $\vec{v}$.

$\vec{F} = I\vec{L} \times \vec{B}$; the vector $\vec{L}$ points along the rod in the direction of the current. Figure 29.11 shows that this force is opposite to the velocity of the rod, and so to maintain the motion a force of equal magnitude must be applied in the direction of the rod's motion (that is, in the direction of $\vec{v}$). The rate of doing work is equal to the product of the applied force and the speed of the rod: $P_{\text{applied}} = Fv$.

EXECUTE: First we'll calculate $P_{\text{dissipated}}$. From Example 29.6, $\mathcal{E} = -BLv$. Hence the current in the rod is

$$I = \frac{|\mathcal{E}|}{R} = \frac{BLv}{R}$$

and the rate of energy dissipation is

$$P_{\text{dissipated}} = I^2R = \left(\frac{BLv}{R}\right)^2 R = \frac{B^2L^2v^2}{R}$$

To calculate P_{applied}, we first calculate the magnitude of $\vec{F} = I\vec{L} \times \vec{B}$. Since $\vec{L}$ and $\vec{B}$ are perpendicular, this magnitude is

$$F = ILB = \frac{BLv}{R}LB = \frac{B^2L^2v}{R}$$

Hence the rate at which work is done by this applied force is

$$P_{\text{applied}} = Fv = \frac{B^2L^2v^2}{R}$$

EVALUATE: This is just equal to the rate at which energy is dissipated in the resistance.

CAUTION You might think that reversing the direction of $\vec{B}$ or of $\vec{v}$, might make it possible to have the magnetic force $\vec{F} = I\vec{L} \times \vec{B}$ be in the *same* direction as $\vec{v}$. This would be a pretty neat trick. Once the rod was moving, the changing magnetic flux would induce an emf and a current, and the magnetic force on the rod would make it move even faster, increasing the emf and current; this would go on until the rod was moving at tremendous speed and producing electric power at a prodigious rate. If this seems too good to be true, not to mention a violation of energy conservation, that's because it is. Reversing $\vec{B}$ also reverses the sign of the induced emf and current and hence the direction of $\vec{L}$, so the magnetic force still opposes the motion of the rod; a similar result holds true if we reverse $\vec{v}$. This behavior is part of Lenz's law, to be discussed in Section 29.3.

Example 29.7 shows that the slidewire generator doesn't produce electric energy out of nowhere; the energy is supplied by whatever body exerts the force that keeps the rod moving. All that the generator does is to *convert* that energy into a different form. The equality between the rate at which *mechanical* energy is supplied to a generator and the rate at which *electric* energy is generated holds for all types of generators. This is true in particular for the alternator described in Example 29.4; we leave the details as a problem. (We are neglecting the effects of friction in the bearings of an alternator or between the rod and the U-shaped conductor of a slidewire generator. If these are included, the conservation of energy demands that the energy lost to friction is not available for conversion to electric energy. In real generators the friction is kept to a minimum to keep the energy conversion process as efficient as possible.)

In Chapter 27 we stated that the magnetic force on moving charges can never do work. But you might think that the magnetic force $\vec{F} = I\vec{L} \times \vec{B}$ in Example 29.7 *is* doing (negative) work on the current-carrying rod as it moves, contradicting our earlier statement. In fact, the work done by the magnetic force is actually zero. The moving charges that make up the current in the rod in Fig. 29.11 have a vertical component of velocity, causing a horizontal component of force on these charges. As a result, there is a horizontal displacement of charge within the rod, the left side acquiring a net positive charge and the right side a net negative charge. The result is a horizontal component of electric field, perpendicular to the length of the rod (analogous to the Hall effect, described in Section 27.9). It is this field, in the direction of motion of the rod, that does work on the mobile charges in the rod and hence indirectly on the atoms making up the rod.

Test Your Understanding

In Example 29.2, what would be the magnitude and direction of the induced emf if the magnetic field *increased* at a rate of 0.300 T/s?

29.3 | Lenz's Law

Lenz's law is a convenient alternative method for determining the direction of an induced current or emf. Lenz's law is not an independent principle; it can be derived from Faraday's law. It always gives the same results as the sign rules we introduced in connection with Faraday's law, but it is often easier to use. Lenz's law also helps us gain intuitive understanding of various induction effects and of the role of energy conservation. H. F. E. Lenz (1804–1865) was a Russian scientist who duplicated independently many of the discoveries of Faraday and Henry. **Lenz's law** states:

The direction of any magnetic induction effect is such as to oppose the cause of the effect.

The "cause" may be changing flux through a stationary circuit due to a varying magnetic field, changing flux due to motion of the conductors that make up the circuit, or any combination. If the flux in a stationary circuit changes, as in Examples 29.1 and 29.2, the induced current sets up a magnetic field of its own. Within the area bounded by the circuit, this field is *opposite* to the original field if the original field is *increasing* but is in the *same* direction as the original field if the latter is *decreasing*. That is, the induced current opposes the *change in flux* through the circuit (*not* the flux itself).

If the flux change is due to motion of the conductors, as in Examples 29.3 through 29.7, the direction of the induced current in the moving conductor is such that the direction of the magnetic-field force on the conductor is opposite in direction to its motion. Thus the motion of the conductor, which caused the induced current, is opposed. We saw this explicitly for the slidewire generator in Example 29.7. In all these cases the induced current tries to preserve the *status quo* by opposing motion or a change of flux.

Lenz's law is also directly related to energy conservation. If the induced current in Example 29.7 were in the direction opposite to that given by Lenz's law, the magnetic force on the rod would accelerate it to ever-increasing speed with no external energy source, even though electric energy is being dissipated in the circuit. This would be a clear violation of energy conservation and doesn't happen in nature.

Conceptual Example 29.8 The slidewire generator, revisited

In Fig. 29.10, the induced current in the loop causes an additional magnetic field in the area bounded by the loop. The direction of the induced current is counterclockwise. From the discussion of Section 28.2, we see that the direction of the additional magnetic field caused by this current is *out of* the plane of the figure. Its direction is opposite that of the original magnetic field, so it tends to cancel the effect of that field. This is consistent with the prediction of Lenz's law.

Conceptual Example 29.9 Finding the direction of induced current

In Fig. 29.12 there is a uniform magnetic field $\vec{B}$ through the coil. The magnitude of the field is increasing, and the resulting induced emf causes an induced current. Use Lenz's law to determine the direction of the induced current.

SOLUTION

This situation is the same as in Example 29.1 (Section 29.2). By Lenz's law the induced current must produce a magnetic field $\vec{B}_{\text{induced}}$ inside the coil that is downward, opposing the change in flux. Using the right-hand rule we described in Section 28.5 for the direction of the magnetic field produced by a circular loop, $\vec{B}_{\text{induced}}$ will be in the desired direction if the induced current flows as shown in Fig. 29.12.

Figure 29.13 shows several applications of Lenz's law to the similar situation of a magnet moving near a conducting loop. In each of the four cases shown, the induced current produces a magnetic field of its own, in a direction that opposes the change in flux through the loop due to the magnet's motion.

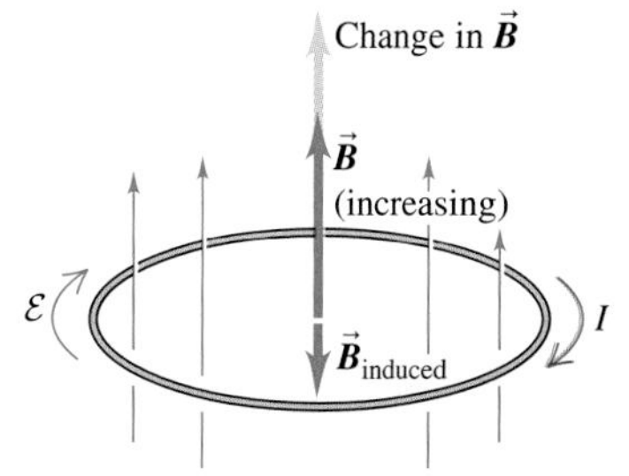

29.12 The induced current due to the change in $\vec{B}$ is clockwise, as seen from above the loop. The added field $\vec{B}_{\text{induced}}$ that it causes is downward, opposing the change in the upward field $\vec{B}$.

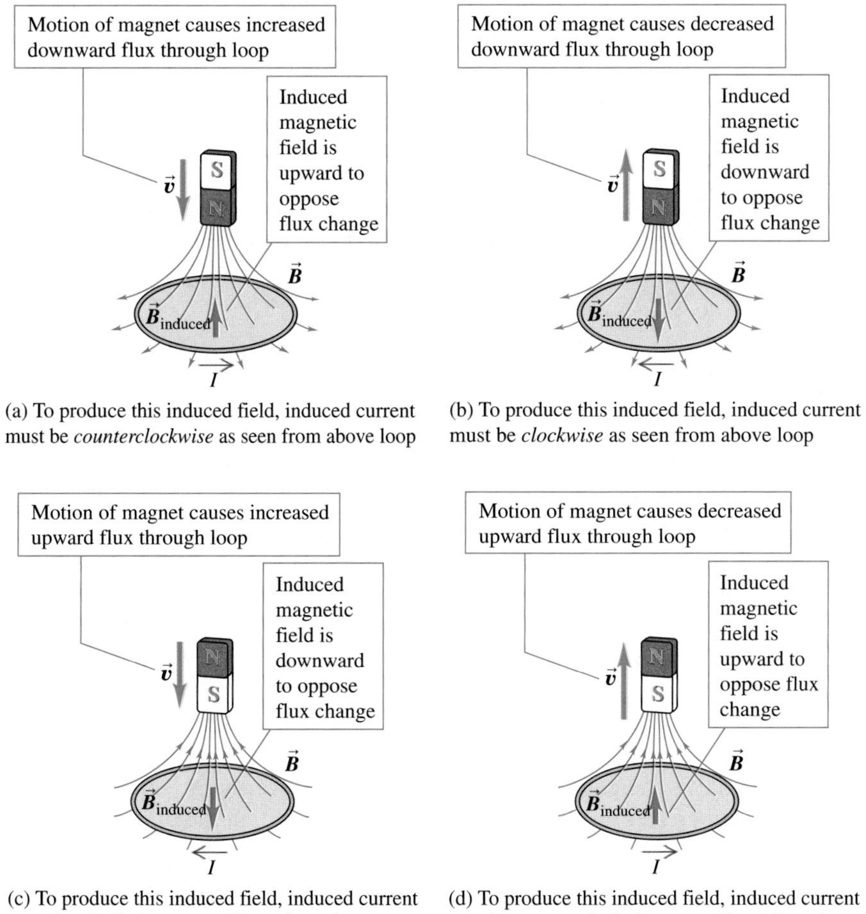

29.13 Directions of induced currents as a bar magnet moves along the axis of a conducting loop. If the bar magnet is stationary, there is no induced current.

Lenz's Law and the Response to Flux Changes

Since an induced current always opposes any change in magnetic flux through a circuit, how is it possible for the flux to change at all? The answer is that Lenz's law gives only the *direction* of an induced current; the *magnitude* of the current depends on the resistance of the circuit. The greater the circuit resistance, the less the induced current that appears to oppose any change in flux and the easier it is for a flux change to take effect. If the loop in Fig. 29.13 were made out of wood (an insulator), there would be almost no induced current in response to changes in the flux through the loop.

Conversely, the less the circuit resistance, the greater the induced current and the more difficult it is to change the flux through the circuit. If the loop in Fig. 29.13 is a good conductor, an induced current flows as long as the magnet moves relative to the loop. Once the magnet and loop are no longer in relative motion, the induced current very quickly decreases to zero because of the nonzero resistance in the loop.

The extreme case occurs when the resistance of the circuit is *zero.* Then the induced current in Fig. 29.13 will continue to flow even after the induced emf has disappeared; that is, even after the magnet has stopped moving relative to the loop. Thanks to this *persistent current,* it turns out that the flux through the loop is exactly the same as it was before the magnet started to move, so the flux through a loop of zero resistance *never* changes. Exotic materials called *superconductors* do indeed have zero resistance; we discuss these further in Section 29.8.

Test Your Understanding

Suppose the rod in Example 29.6 were moved to the *left* at constant velocity. Use Lenz's law to determine which direction the induced current would flow around the circuit in this case.

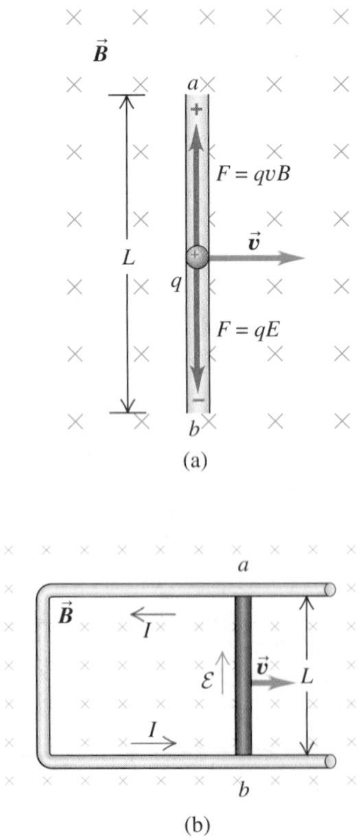

29.14 A conducting rod moving in a uniform magnetic field. (a) The rod, the velocity, and the field are mutually perpendicular. (b) Direction of induced current in the circuit.

29.4 | Motional Electromotive Force

We've seen several situations in which a conductor moves in a magnetic field, as in the generators discussed in Examples 29.4 through 29.7. We can gain additional insight into the origin of the induced emf in these situations by considering the magnetic forces on mobile charges in the conductor. Figure 29.14a shows the same moving rod that we discussed in Example 29.6, separated for the moment from the U-shaped conductor. The magnetic field $\vec{B}$ is uniform and directed into the page, and we move the rod to the right at a constant velocity $\vec{v}$. A charged particle q in the rod then experiences a magnetic force $\vec{F} = q\vec{v} \times \vec{B}$ with magnitude $F = |q|vB$. We'll assume in the following discussion that q is positive; in that case the direction of this force is upward along the rod, from b toward a.

This magnetic force causes the free charges in the rod to move, creating an excess of positive charge at the upper end a and negative charge at the lower end b. This in turn creates an electric field $\vec{E}$ within the rod, in the direction from a toward b (opposite to the magnetic force). Charge continues to accumulate at the ends of the rod until $\vec{E}$ becomes large enough for the downward electric force (with magnitude qE) to cancel exactly the *upward* magnetic force (with magnitude qvB). Then $qE = qvB$ and the charges are in equilibrium.

The magnitude of the potential difference $V_{ab} = V_a - V_b$ is equal to the electric field magnitude E multiplied by the length L of the rod. From the above discussion, $E = vB$, so

$$V_{ab} = EL = vBL \tag{29.5}$$

with point a at higher potential than point b.

Now suppose the moving rod slides along a stationary U-shaped conductor, forming a complete circuit (Fig. 29.14b). No *magnetic* force acts on the charges in the stationary U-shaped conductor, but the charge that was near points a and b redistributes itself along the stationary conductors, creating an *electric* field within them. This field establishes a current in the direction shown. The moving rod has become a source of electromotive force; within it, charge moves from lower to higher potential, and in the remainder of the circuit, charge moves from higher to lower potential. We call this emf a **motional electromotive force,** denoted by $\mathcal{E}$. From the above discussion, the magnitude of this emf is

ActivPhysics ONLINE

13.10 Motional emf

$$\mathcal{E} = vBL \qquad (29.6)$$

(motional emf; length and velocity perpendicular to $\vec{B}$ uniform)

corresponding to a force per unit charge of magnitude vB acting for a distance L along the moving rod. If the total circuit resistance of the U-shaped conductor and the sliding rod is R, the induced current I in the circuit is given by $vBL = IR$. This is the same result we obtained in Section 29.2 using Faraday's law, and indeed motional emf is a particular case of Faraday's law, one of the several examples described in Section 29.2.

The emf associated with the moving rod in Fig. 29.14 is analogous to that of a battery with its positive terminal at a and its negative terminal at b, although the origins of the two emfs are quite different. In each case a nonelectrostatic force acts on the charges in the device, in the direction from b to a, and the emf is the work per unit charge done by this force when a charge moves from b to a in the device. When the device is connected to an external circuit, the direction of current is from b to a in the device and from a to b in the external circuit. While we have discussed motional emf in terms of a closed circuit like that in Fig. 29.14b, a motional emf is also present in the isolated moving rod in Fig. 29.14a, in the same way that a battery has an emf even when it's not part of a circuit.

The direction of the induced emf in Fig. 29.14 can be deduced by using Lenz's law, even if (as in Fig. 29.14a) the conductor does not form a complete circuit. In this case we can mentally complete the circuit between the ends of the conductor and use Lenz's law to determine the direction of the current. From this we can deduce the polarity of the ends of the open-circuit conductor. The direction from the $-$end to the $+$end within the conductor is the direction the current would have if the circuit were complete.

You should verify that if we express v in meters per second, B in teslas, and L in meters, then $\mathcal{E}$ is in volts. (Recall that 1 V = 1 J/C.)

We can generalize the concept of motional emf for a conductor with *any* shape, moving in any magnetic field, uniform or not (assuming that the magnetic field at each point does not vary with time). For an element $d\vec{l}$ of conductor, the contribution $d\mathcal{E}$ to the emf is the magnitude dl multiplied by the component of $\vec{v} \times \vec{B}$ (the magnetic force per unit charge) parallel to $d\vec{l}$; that is,

$$d\mathcal{E} = (\vec{v} \times \vec{B}) \cdot d\vec{l}$$

For any closed conducting loop, the total emf is

$$\mathcal{E} = \oint (\vec{v} \times \vec{B}) \cdot d\vec{l} \qquad \text{(motional emf: closed conducting loop)} \qquad (29.7)$$

This expression looks very different from our original statement of Faraday's law, Eq. (29.3), which stated that $\mathcal{E} = -d\Phi_B/dt$. In fact, though, the two statements are equivalent. It can be shown that the rate of change of magnetic flux through a moving conducting loop is always given by the negative of the expression in Eq. (29.7). Thus this equation gives us an alternative formulation of Faraday's law. This alternative is often more convenient than the original one in problems with *moving* conductors. But when we have *stationary* conductors in changing magnetic fields, Eq. (29.7) *cannot* be used; in this case, $\mathcal{E} = -d\Phi_B/dt$ is the only correct way to express Faraday's law.

Example 29.10 Calculating motional emf

Suppose the length L in Fig. 29.14b is 0.10 m, the velocity v is 2.5 m/s, the total resistance of the loop is 0.030 Ω, and B is 0.60 T. Find $\mathcal{E}$, the induced current, and the force acting on the rod.

SOLUTION

IDENTIFY: The first target variable is the *motional* emf $\mathcal{E}$ due to the rod's motion. We'll find the current from the values of $\mathcal{E}$ and the resistance R. The force on the rod is actually a magnetic force exerted by $\vec{B}$ on the current in the rod.

SET UP: We'll use the motional emf expression developed in this section, the familiar relation $\mathcal{E} = IR$, and the formula $\vec{F} = I\vec{L} \times \vec{B}$ for the magnetic force on a current-carrying rod.

EXECUTE: From Eq. (29.6) the emf is

$$\mathcal{E} = vBL = (2.5\text{ m/s})(0.60\text{ T})(0.10\text{ m}) = 0.15\text{ V}$$

The resulting induced current in the loop is

$$I = \frac{\mathcal{E}}{R} = \frac{0.15\text{ V}}{0.030\ \Omega} = 5.0\text{ A}$$

The magnetic force on the rod carrying this current is directed *opposite* to the rod's motion. You can see this by applying the right-hand rule for vector products to the formula $\vec{F} = I\vec{L} \times \vec{B}$. Since $\vec{L}$ and $\vec{B}$ are perpendicular, this force has magnitude

$$F = ILB = (5.0\text{ A})(0.10\text{ m})(0.60\text{ T}) = 0.30\text{ N}$$

EVALUATE: We can check our answer for the direction of $\vec{F}$ by using Lenz's law. If we take the area vector $\vec{A}$ to point into the plane of the loop, the magnetic flux is positive and increasing as the rod moves to the right and increases the area of the loop. Lenz's law tells us that a force appears to oppose this increase in flux. Hence the force on the rod is to the left, opposite its motion.

Example 29.11 The Faraday disk dynamo

A conducting disk with radius R, shown in Fig. 29.15, lies in the xy-plane and rotates with constant angular velocity ω about the z-axis. The disk is in a uniform, constant $\vec{B}$ field parallel to the z-axis. Find the induced emf between the center and the rim of the disk.

SOLUTION

IDENTIFY: A motional emf is present because the conducting disk moves relative to the $\vec{B}$ field. The complication is that different parts of the disk move at different speeds v, depending on their distance from the rotation axis. We'll address this by considering small segments of the disk and adding (actually integrating) their contributions to determine our target variable, the emf between the center and the rim.

SET UP: Consider the small segment of the disk labeled by its velocity vector $\vec{v}$. The magnetic force per unit charge on this segment is $\vec{v} \times \vec{B}$, which points radially outward from the center of the disk (for the segment shown, in the negative y-direction). Hence the induced emf tends to make a current flow radially outward, which tells us that the moving conducting path to think about here is a straight line from the center to the rim. We can find the emf from each small disk segment along this line using the expression $d\mathcal{E} = (\vec{v} \times \vec{B}) \cdot d\vec{l}$, then integrate to find the total emf.

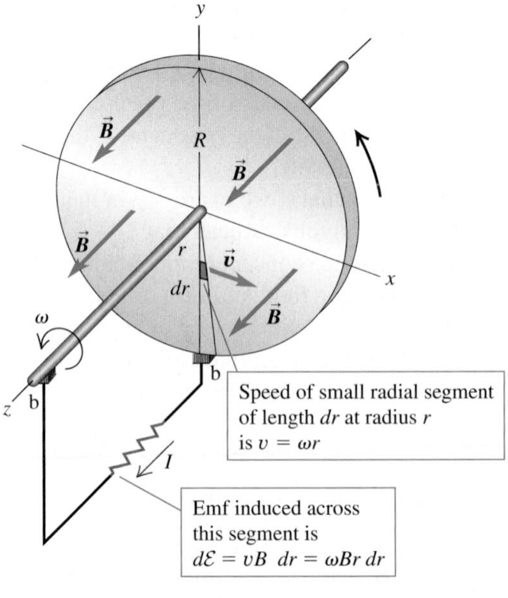

29.15 A conducting disk with radius R rotating at an angular speed ω in a magnetic field $\vec{B}$. The emf is induced along radial lines of the disk and is applied to an external circuit through the two sliding contacts labeled b.

CAUTION You might be tempted to use Eq. (29.5) and simply multiply vB times the length of the moving conducting path, which is just the radius R. That wouldn't be right, because v has different values at different points along the path.

EXECUTE: Let's consider the motional emf $d\mathcal{E}$ due to a small radial segment at a distance r from the rotation axis. The associated length vector $d\vec{l}$ (of length dr) points radially outward, in the same direction as $\vec{v} \times \vec{B}$. The vectors $\vec{v}$ and $\vec{B}$ are perpendicular, and the magnitude of $\vec{v}$ is $v = \omega r$. Hence the total emf between center and rim is the sum of all such contributions:

$$\mathcal{E} = \int_0^R \omega Br\, dr = \frac{1}{2}\omega BR^2$$

EVALUATE: We can use this device as a source of emf in a circuit by completing the circuit through stationary brushes (b in the figure) that contact the disk and its conducting shaft as shown. The emf in such a disk was studied by Faraday; the device is called *a Faraday disk dynamo* or a *homopolar generator.* Unlike the alternator in Example 29.4, the Faraday disk dynamo is a direct-current generator; it produces an emf that is constant in time. Can you use Lenz's law to show that for the direction of rotation in Fig. 29.15, the current in the external circuit must be in the direction shown?

Test Your Understanding

The earth's magnetic field points toward (magnetic) north. For simplicity, assume that the field has no vertical component (as is the case near the earth's equator). If you hold a metal rod in your hand and walk toward the east, how should you orient the rod to get the maximum motional emf between its ends? How should you hold it to get *zero* emf? In which direction should you walk so that the motional emf across the rod is zero no matter how the rod is oriented?

29.5 | Induced Electric Fields

When a conductor moves in a magnetic field, we can understand the induced emf on the basis of magnetic forces on charges in the conductor, as described in Section 29.4. But an induced emf also occurs when there is a changing flux through a stationary conductor. What is it that pushes the charges around the circuit in this type of situation?

As an example, let's consider the situation shown in Fig. 29.16. A long, thin solenoid with cross-sectional area A and n turns per unit length is encircled at its center by a circular conducting loop. The galvanometer G measures the current in

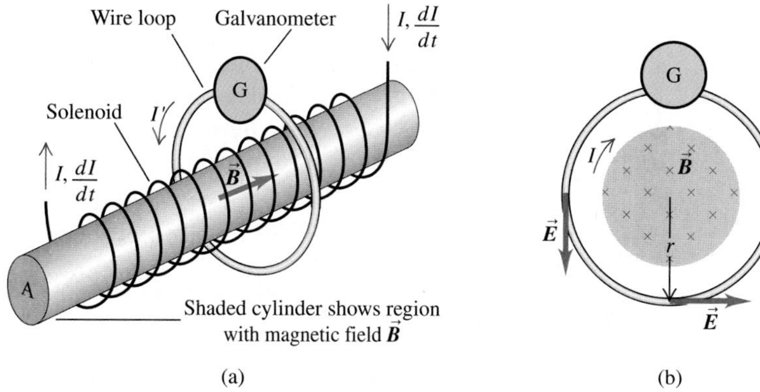

29.16 (a) The windings of a long solenoid carry a current I that is increasing at a rate dI/dt. The magnetic flux in the solenoid is increasing at a rate $d\Phi_B/dt$, and this changing flux passes through a wire loop. An emf $\mathcal{E} = -d\Phi_B/dt$ is induced in the loop, inducing a current I' that is measured by the galvanometer G. (b) Cross-sectional view.

the loop. A current I in the winding of the solenoid sets up a magnetic field $\vec{B}$ along the solenoid axis, as shown, with magnitude B as calculated in Example 28.10 (Section 28.7): $B = \mu_0 nI$, where n is the number of turns per unit length. If we neglect the small field outside the solenoid and take the area vector $\vec{A}$ to point in the same direction as $\vec{B}$, then the magnetic flux Φ_B through the loop is

$$\Phi_B = BA = \mu_0 nIA$$

When the solenoid current I changes with time, the magnetic flux Φ_B also changes, and according to Faraday's law the induced emf in the loop is given by

$$\mathcal{E} = -\frac{d\Phi_B}{dt} = -\mu_0 nA\frac{dI}{dt} \tag{29.8}$$

If the total resistance of the loop is R, the induced current in the loop, which we may call I', is $I' = \mathcal{E}/R$.

But what *force* makes the charges move around the loop? It can't be a magnetic force because the conductor isn't moving in a magnetic field and in fact isn't even *in* a magnetic field. We are forced to conclude that there has to be an **induced electric field** in the conductor *caused by the changing magnetic flux*. This may be a little jarring; we are accustomed to thinking about electric field as being caused by electric charges, and now we are saying that a changing magnetic field somehow acts as a source of electric field. Furthermore, it's a strange sort of electric field. When a charge q goes once around the loop, the total work done on it by the electric field must be equal to q times the emf $\mathcal{E}$. That is, the electric field in the loop *is not conservative,* as we used the term in Chapter 23, because the line integral of $\vec{E}$ around a closed path is not zero. Indeed, this line integral, representing the work done by the induced $\vec{E}$ field per unit charge, is equal to the induced emf $\mathcal{E}$:

$$\oint \vec{E}\cdot d\vec{l} = \mathcal{E} \tag{29.9}$$

From Faraday's law the emf $\mathcal{E}$ is also the negative of the rate of change of magnetic flux through the loop. Thus for this case we can restate Faraday's law as

$$\oint \vec{E}\cdot d\vec{l} = -\frac{d\Phi_B}{dt} \quad \text{(stationary integration path)} \tag{29.10}$$

Note that Faraday's law is always true in the form $\mathcal{E} = -d\Phi_B/dt$; the form given in Eq. (29.10) is valid *only* if the path around which we integrate is *stationary.*

As an example of a situation to which Eq. (29.10) can be applied, consider the stationary circular loop in Fig. 29.16b, which we take to have radius r. Because of cylindrical symmetry, the electric field $\vec{E}$ has the same magnitude at every point on the circle and is tangent to it at each point. (Symmetry would also permit the field to be *radial,* but then Gauss's law would require the presence of a net charge inside the circle, and there is none.) The line integral in Eq. (29.10) becomes simply the magnitude E times the circumference $2\pi r$ of the loop, $\oint \vec{E}\cdot d\vec{l} = 2\pi rE$, and Eq. (29.10) gives

$$E = \frac{1}{2\pi r}\left|\frac{d\Phi_B}{dt}\right| \tag{29.11}$$

The directions of $\vec{E}$ at several points on the loop are shown in Fig. 29.16b. We know that $\vec{E}$ has to have the direction shown when $\vec{B}$ in the solenoid is increasing, because $\oint \vec{E} \cdot d\vec{l}$ has to be negative when $d\Phi_B/dt$ is positive. The same approach can be used to find the induced electric field *inside* the solenoid when the solenoid $\vec{B}$ field is changing; we leave the details to you (see Exercise 29.28).

Now let's summarize what we've learned. Faraday's law, Eq. (29.3), is valid for two rather different situations. In one, an emf is induced by magnetic forces on charges when a conductor moves through a magnetic field. In the other, a time-varying magnetic field induces an electric field in a stationary conductor and hence induces an emf; in fact, the $\vec{E}$ field is induced even when no conductor is present. This $\vec{E}$ field differs from an electro*static* field in an important way. It is *nonconservative;* the line integral $\oint \vec{E} \cdot d\vec{l}$ around a closed path is not zero, and when a charge moves around a closed path, the field does a nonzero amount of work on it. It follows that for such a field the concept of *potential* has no meaning. We call such a field a **nonelectrostatic field.** In contrast, an electro*static* field is *always* conservative, as we discussed in Section 23.1, and always has an associated potential function. Despite this difference, the fundamental effect of *any* electric field is to exert a force $\vec{F} = q\vec{E}$ on a charge q. This relation is valid whether $\vec{E}$ is a conservative field produced by a charge distribution or a nonconservative field caused by changing magnetic flux.

So a changing magnetic field acts as a source of electric field of a sort that we *cannot* produce with any static charge distribution. This may seem strange, but it's the way nature behaves. What's more, we'll see in Section 29.7 that a changing *electric* field acts as a source of *magnetic* field. We'll explore this symmetry between the two fields in greater detail in our study of electromagnetic waves in Chapter 32.

If any doubt remains in your mind about the reality of magnetically induced electric fields, consider a few of the many practical applications (Fig. 29.17). In the playback head of a tape deck, currents are induced in a stationary coil as the variously magnetized regions of the tape move past it. Computer disk drives operate on the same principle. Pickups in electric guitars use currents induced in stationary pickup coils by the vibration of nearby ferromagnetic strings. Alternators in most cars use rotating magnets to induce currents in stationary coils. The list goes on and on; whether we realize it or not, magnetically induced electric fields play an important role in everyday life.

(a) (b) (c)

29.17 Applications of induced electric fields. (a) Data are stored on a computer hard disk in a pattern of magnetized areas on the surface of the disk. To read these data, a coil on a movable arm is placed next to the spinning disk. The coil experiences a changing magnetic flux, inducing a current whose characteristics depend on the pattern coded on the disk. (b) This hybrid automobile has both a gasoline engine and an electric motor. As the car comes to a halt, magnets in the spinning wheels move past stationary coils. The resulting induced current is used to recharge the car's batteries. (c) The rotating crankshaft of a piston-engine airplane spins a magnet, inducing an emf in an adjacent coil and generating the spark that ignites fuel in the engine cylinders. This keeps the engine running even if the airplane's other electrical systems fail.

Example 29.12 Induced electric fields

Suppose the long solenoid in Fig. 29.16a is wound with 500 turns per meter and the current in its windings is increasing at the rate of 100 A/s. The cross-sectional area of the solenoid is $4.0\ \text{cm}^2 = 4.0 \times 10^{-4}\ \text{m}^2$. a) Find the magnitude of the induced emf in the wire loop outside the solenoid. b) Find the magnitude of the induced electric field within the loop if its radius is 2.0 cm.

SOLUTION

IDENTIFY: As in Fig. 29.16b, the increasing magnetic field inside the solenoid causes a change in the magnetic flux through the wire loop and hence induces an electric field $\vec{E}$ around the loop. Our target variables are the induced emf $\mathcal{E}$ and the magnitude of $\vec{E}$.

SET UP: We use Eq. (29.8) to determine the emf. Determining the field magnitude E is simplified because the loop and the solenoid share the same central axis. Hence, by symmetry, the electric field is tangent to the loop and has the same magnitude all the way around its circumference. This makes it easy to find E from the emf $\mathcal{E}$ using Eq. (29.9).

EXECUTE: a) From Eq. (29.8), the induced emf is

$$\begin{aligned}\mathcal{E} &= -\frac{d\Phi_B}{dt} = -\mu_0 nA\frac{dI}{dt}\\ &= -(4\pi \times 10^{-7}\ \text{Wb/A}\cdot\text{m})(500\ \text{turns/m})\\ &\qquad (4.0 \times 10^{-4}\ \text{m}^2)(100\ \text{A/s})\\ &= -25 \times 10^{-6}\ \text{Wb/s} = -25 \times 10^{-6}\ \text{V} = -25\ \mu\text{V}\end{aligned}$$

b) By symmetry the line integral $\oint \vec{E}\cdot d\vec{l}$ has absolute value $2\pi rE$ (disregarding the direction in which we integrate around the loop). This is equal to the absolute value of the emf, so

$$E = \frac{|\mathcal{E}|}{2\pi r} = \frac{25 \times 10^{-6}\ \text{V}}{2\pi(2.0 \times 10^{-2}\ \text{m})} = 2.0 \times 10^{-4}\ \text{V/m}$$

EVALUATE: In Fig. 29.16b the magnetic flux *into* the plane of the figure is increasing. According to the right-hand rule for induced emf (illustrated in Fig. 29.5), a positive emf would be clockwise around the loop; the negative sign of $\mathcal{E}$ shows that the emf is in the counterclockwise direction. Can you also show this using Lenz's law?

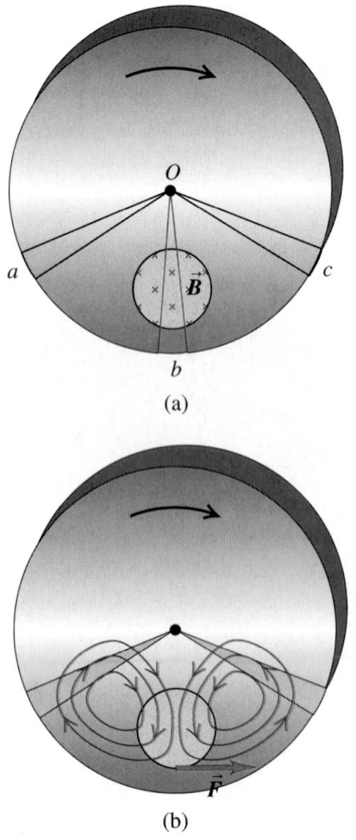

29.18 (a) A metallic disk rotating through a perpendicular magnetic field $\vec{B}$. (b) Eddy currents formed by the induced emf.

Test Your Understanding

If you wiggle a magnet back and forth in your hand, are you generating an electric field? If so, is this electric field conservative?

*29.6 | Eddy Currents

In the examples of induction effects that we have studied, the induced currents have been confined to well-defined paths in conductors and other components forming a circuit. However, many pieces of electrical equipment contain masses of metal moving in magnetic fields or located in changing magnetic fields. In situations like these we can have induced currents that circulate throughout the volume of a material. Because their flow patterns resemble swirling eddies in a river, we call these **eddy currents.**

As an example, consider a metallic disk rotating in a magnetic field perpendicular to the plane of the disk but confined to a limited portion of the disk's area, as shown in Fig. 29.18a. Sector Ob is moving across the field and has an emf induced in it. Sectors Oa and Oc are not in the field, but they provide return conducting paths for charges displaced along Ob to return from b to O. The result is a circulation of eddy currents in the disk, somewhat as sketched in Fig. 29.18b.

We can use Lenz's law to decide on the direction of the induced current in the neighborhood of sector Ob. This current must experience a magnetic force $\vec{F} = I\vec{L} \times \vec{B}$ that *opposes* the rotation of the disk, and so this force must be to the right in Fig. 29.18b. Since $\vec{B}$ is directed into the plane of the disk, the current and hence $\vec{L}$ have downward components. The return currents lie outside the field, so

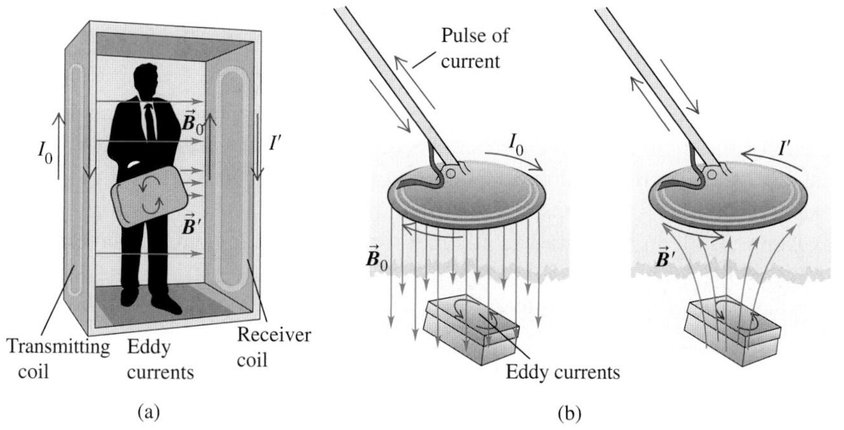

29.19 (a) A metal detector at an airport security checkpoint generates an alternating magnetic field $\vec{B}_0$. This induces eddy currents in a conducting object carried through the detector. The eddy currents in turn produce an alternating magnetic field $\vec{B}'$, and this field induces a current in the detector's receiver coil. (b) Portable metal detectors work on the same principle.

they do not experience magnetic forces. The interaction between the eddy currents and the field causes a braking action on the disk. Such effects can be used to stop the rotation of a circular saw quickly when the power is turned off. Some sensitive balances use this effect to damp out vibrations. Eddy current braking is used on some electrically powered rapid transit vehicles. Electromagnets mounted in the cars induce eddy currents in the rails; the resulting magnetic fields cause braking forces on the electromagnets and thus on the cars.

Eddy currents have many other practical uses. The shiny metal disk in the electric power company's meter outside your house rotates as a result of eddy currents. These currents are induced in the disk by magnetic fields caused by sinusoidally varying currents in a coil. In induction furnaces, eddy currents are used to heat materials in completely sealed containers for processes in which it is essential to avoid the slightest contamination of the materials. The metal detectors used at airport security checkpoints (Fig. 29.19a) operate by detecting eddy currents induced in metallic objects. Similar devices (Fig. 29.19b) are used to find buried treasure such as bottlecaps and lost pennies.

A particularly dramatic example of eddy currents in action is Jupiter's moon Io, which is slightly larger than the earth's moon (Fig. 29.20a). Io moves rapidly through Jupiter's intense magnetic field, and this sets up strong eddy currents within Io's interior. These currents dissipate energy at a rate of 10^{12} W, equivalent to setting off a one-kiloton nuclear weapon inside Io every four seconds! This dissipated energy helps to keep Io's interior hot and so helps to cause volcanic eruptions on its surface, like those in Fig. 29.20b. (Gravitational effects from Jupiter cause even more heating.)

Eddy currents also have undesirable effects. In an alternating-current transformer, coils wrapped around an iron core carry a sinusoidally varying current. The resulting eddy currents in the core waste energy through I^2R heating and themselves set up an unwanted opposing emf in the coils. To minimize these effects, the core is designed so that the paths for eddy currents are as narrow as possible. We'll describe how this is done when we discuss transformers in detail in Section 31.6.

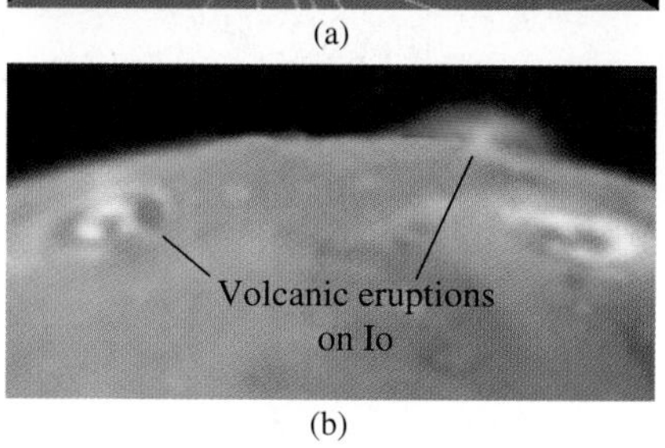

29.20 (a) As Jupiter's moon Io moves around it's orbit, the planet's powerful magnetic field induces eddy currents within Io. (b) Two simultaneous volcanic eruptions on Io, triggered in part by eddy current heating.

29.7 | Displacement Current and Maxwell's Equations

We have seen that a varying magnetic field gives rise to an induced electric field. In one of the more remarkable examples of the symmetry of nature, it turns out that a varying *electric* field gives rise to a *magnetic* field. This effect is of tremendous importance, for it turns out to explain the existence of radio waves, gamma rays, and visible light, as well as all other forms of electromagnetic waves.

To see the origin of the relationship between varying electric fields and magnetic fields, let's return to Ampere's law as given in Section 28.6, Eq. (28.20):

$$\oint \vec{B} \cdot d\vec{l} = \mu_0 I_{encl}$$

The problem with Ampere's law in this form is that it is *incomplete*. To see why, let's consider the process of charging a capacitor (Fig. 29.21). Conducting wires lead current i_C into one plate and out of the other; the charge Q increases, and the electric field $\vec{E}$ between the plates increases. The notation i_C indicates *conduction* current to distinguish it from another kind of current we are about to encounter, called *displacement* current i_D. We use lowercase i's and v's to denote instantaneous values of currents and potential differences, respectively, that may vary with time.

Let's apply Ampere's law to the circular path shown. The integral $\oint \vec{B} \cdot d\vec{l}$ around this path equals $\mu_0 I_{encl}$. For the plane circular area bounded by the circle, I_{encl} is just the current i_C in the left conductor. But the surface that bulges out to the right is bounded by the same circle, and the current through that surface is zero. So $\oint \vec{B} \cdot d\vec{l}$ is equal to $\mu_0 I_C$, and at the same time it is equal to zero! This is a clear contradiction.

But something else is happening on the bulged-out surface. As the capacitor charges, the electric field $\vec{E}$ and the electric *flux* Φ_E through the surface are increasing. We can determine their rates of change in terms of the charge and current. The instantaneous charge is $q = Cv$, where C is the capacitance and v is the instantaneous potential difference. For a parallel-plate capacitor, $C = \epsilon_0 A/d$, where A is the plate area and d is the spacing. The potential difference v between plates is $v = Ed$, where E is the electric field magnitude between plates. (We neglect fringing and assume that $\vec{E}$ is uniform in the region between the plates.) If this region is filled with a material with permittivity ϵ, we replace ϵ_0 by ϵ everywhere; we'll use ϵ in the following discussion.

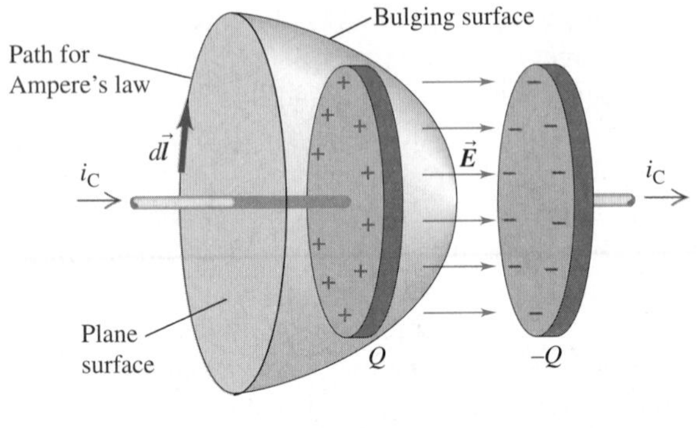

29.21 Parallel-plate capacitor being charged. The conduction current through the plane surface is i_C, but there is no conduction current through the surface that bulges out to pass between the plates. The two surfaces have a common boundary, so this difference in I_{encl} leads to an apparent contradiction in applying Ampere's law.

Substituting these expressions for C and v into $q = Cv$, we can express the capacitor charge q as

$$q = Cv = \frac{\epsilon A}{d}(Ed) = \epsilon EA = \epsilon \Phi_E \tag{29.12}$$

where $\Phi_E = EA$ is the electric flux through the surface.

As the capacitor charges, the rate of change of q is the conduction current, $i_C = dq/dt$. Taking the derivative of Eq. (29.12) with respect to time, we get

$$i_C = \frac{dq}{dt} = \epsilon \frac{d\Phi_E}{dt} \tag{29.13}$$

Now, stretching our imagination a little, we invent a fictitious current or pseudocurrent i_D in the region between the plates, defined as

$$i_D = \epsilon \frac{d\Phi_E}{dt} \quad \text{(displacement current)} \tag{29.14}$$

That is, we imagine that the changing flux through the curved surface in Fig. 29.21 is somehow equivalent, in Ampere's law, to a conduction current through that surface. We include this fictitious current, along with the real conduction current i_C, in Ampere's law:

$$\oint \vec{B} \cdot d\vec{l} = \mu_0 (i_C + i_D)_{\text{encl}} \quad \text{(generalized Ampere's law)} \tag{29.15}$$

Ampere's law in this form is obeyed no matter which surface we use in Fig. 29.21. For the flat surface, i_D is zero; for the curved surface, i_C is zero; and i_C for the flat surface equals i_D for the curved surface. Equation (29.15) remains valid in a magnetic material, provided that the magnetization is proportional to the external field and we replace μ_0 by μ.

The fictitious current i_D was invented in 1865 by the Scottish physicist James Clerk Maxwell (1831–1879), who called it **displacement current.** There is a corresponding *displacement current density* $j_D = i_D/A$; using $\Phi_E = EA$ and dividing Eq. (29.14) by A, we find

$$j_D = \epsilon \frac{dE}{dt} \tag{29.16}$$

We have pulled the concept out of thin air, as Maxwell did, but we see that it enables us to save Ampere's law in situations such as that in Fig. 29.21.

Another benefit of displacement current is that it lets us generalize Kirchhoff's junction rule, discussed in Section 26.2. Considering the left plate of the capacitor plate, we have conduction current into it but none out of it. But when we include the displacement current, we have conduction current coming in one side and an equal displacement current coming out the other side. With this generalized meaning of the term *current,* we can speak of current going *through* the capacitor.

You might well ask at this point whether displacement current has any real physical significance or whether it is just a ruse to satisfy Ampere's law and Kirchhoff's junction rule. Here's a fundamental experiment that helps to answer that question. We take a plane circular area between the capacitor plates, as shown

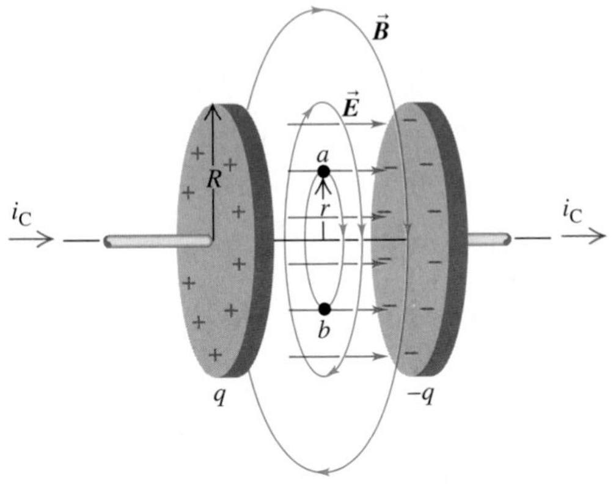

29.22 A capacitor being charged by a current i_C has a displacement current equal to i_C between the plates, with displacement-current density $j_D = \epsilon\, dE/dt$. This can be regarded as the source of the magnetic field between the plates.

in Fig. 29.22. If displacement current really plays the role in Ampere's law that we have claimed, then there ought to be a magnetic field in the region between the plates while the capacitor is charging. We can use our generalized Ampere's law, including displacement current, to predict what this field should be.

To be specific, let's picture round capacitor plates with radius R. To find the magnetic field at a point in the region between the plates at a distance r from the axis, we apply Ampere's law to a circle of radius r passing through the point, with $r < R$. This circle passes through points a and b in Fig. 29.22. The total current enclosed by the circle is j_D times its area, or $(i_D/\pi R^2)(\pi r^2)$. The integral $\oint \vec{B} \cdot d\vec{l}$ in Ampere's law is just B times the circumference $2\pi r$ of the circle, and because $i_D = i_C$ for the charging capacitor, Ampere's law becomes

$$\oint \vec{B} \cdot d\vec{l} = 2\pi r B = \mu_0 \frac{r^2}{R^2} i_C$$

or

$$B = \frac{\mu_0}{2\pi} \frac{r}{R^2} i_C \tag{29.17}$$

This result predicts that in the region between the plates $\vec{B}$ is zero at the axis and increases linearly with distance from the axis. A similar calculation shows that *outside* the region between the plates (that is, for $r > R$), $\vec{B}$ is the same as though the wire were continuous and the plates not present at all.

When we *measure* the magnetic field in this region, we find that it really is there and that it behaves just as Eq. (29.17) predicts. This confirms directly the role of displacement current as a source of magnetic field. It is now established beyond reasonable doubt that displacement current, far from being just an artifice, is a fundamental fact of nature. Maxwell's discovery was the bold step of an extraordinary genius.

Maxwell's Equations of Electromagnetism

We are now in a position to wrap up in a single package *all* of the relationships between electric and magnetic fields and their sources. This package consists of four equations, called **Maxwell's equations.** Maxwell did not discover all of these equations single-handedly (though he did develop the concept of displacement current). But he did put them together and recognized their significance, particularly in predicting the existence of electromagnetic waves.

For now we'll state Maxwell's equations in their simplest form, for the case in which we have charges and currents in otherwise empty space. In Chapter 32 we'll discuss how to modify these equations if a dielectric or a magnetic material is present.

Two of Maxwell's equations involve an integral of $\vec{E}$ or $\vec{B}$ over a closed surface. The first is simply Gauss's law for electric fields, Eq. (22.8), which states that the surface integral of $E_\perp$ over any closed surface equals $1/\epsilon_0$ times the total charge Q_{encl} enclosed within the surface:

$$\oint \vec{E} \cdot d\vec{A} = \frac{Q_{\text{encl}}}{\epsilon_0} \qquad (\text{Gauss's law for } \vec{E}) \qquad (29.18)$$

The second is the analogous relation for *magnetic* fields, Eq. (27.8), which states that the surface integral of $B_\perp$ over any closed surface is always zero:

$$\oint \vec{B} \cdot d\vec{A} = 0 \qquad (\text{Gauss's law for } \vec{B}) \qquad (29.19)$$

This statement means, among other things, that there are no magnetic monopoles (single magnetic charges) to act as sources of magnetic field.

The third equation is Ampere's law including displacement current. This states that both conduction current i_C and displacement current $\epsilon_0 d\Phi_E/dt$, where Φ_E is electric flux, act as sources of magnetic field:

$$\oint \vec{B} \cdot d\vec{l} = \mu_0 \left(i_C + \epsilon_0 \frac{d\Phi_E}{dt} \right)_{\text{encl}} \qquad (\text{Ampere's law}) \qquad (29.20)$$

The fourth and final equation is Faraday's law. It states that a changing magnetic field or magnetic flux induces an electric field:

$$\oint \vec{E} \cdot d\vec{l} = -\frac{d\Phi_B}{dt} \qquad (\text{Faraday's law}) \qquad (29.21)$$

If there is a changing magnetic flux, the line integral in Eq. (29.21) is not zero, which shows that the $\vec{E}$ field produced by a changing magnetic flux is not conservative. Recall that this line integral must be carried out over a *stationary* closed path.

It's worthwhile to look more carefully at the electric field $\vec{E}$ and its role in Maxwell's equations. In general, the total $\vec{E}$ field at a point in space can be the superposition of an electrostatic field $\vec{E}_c$ caused by a distribution of charges at rest and a magnetically induced, nonelectrostatic field $\vec{E}_n$. (The subscript c stands for Coulomb or conservative; the subscript n stands for non-Coulomb, nonelectrostatic, or nonconservative.) That is,

$$\vec{E} = \vec{E}_c + \vec{E}_n$$

The electrostatic part $\vec{E}_c$ is *always* conservative, so $\oint \vec{E}_c \cdot d\vec{l} = 0$. This conservative part of the field does not contribute to the integral in Faraday's law, so we can take $\vec{E}$ in Eq. (29.21) to be the *total* electric field $\vec{E}$, including both the part $\vec{E}_c$ due to charges and the magnetically induced part $\vec{E}_n$. Similarly, the nonconservative part $\vec{E}_n$ of the $\vec{E}$ field does not contribute to the integral in Gauss's law, because this part of the field is not caused by static charges. Hence $\oint \vec{E}_n \cdot d\vec{A}$ is always zero.

We conclude that in all the Maxwell equations, $\vec{E}$ is the *total* electric field; these equations don't distinguish between conservative and nonconservative fields.

There is a remarkable symmetry in Maxwell's four equations. In empty space where there is no charge, the first two equations (Eqs. (29.18) and (29.19)) are identical in form, one containing $\vec{E}$ and the other containing $\vec{B}$. When we compare the second two equations, Eq. (29.20) says that a changing electric flux creates a magnetic field, and Eq. (29.21) says that a changing magnetic flux creates an electric field. In empty space, where there is no conduction current, $i_C = 0$ and the two equations have the same form, apart from a numerical constant and a negative sign, with the roles of $\vec{E}$ and $\vec{B}$ exchanged in the two equations.

We can rewrite Eqs. (29.20) and (29.21) in a different but equivalent form by introducing the definitions of electric and magnetic flux, $\Phi_E = \int \vec{E} \cdot d\vec{A}$ and $\Phi_B = \int \vec{B} \cdot d\vec{A}$, respectively. In empty space, where there is no charge or conduction current, $i_C = 0$ and $Q_{encl} = 0$, and we have

$$\oint \vec{B} \cdot d\vec{l} = \epsilon_0 \mu_0 \frac{d}{dt} \int \vec{E} \cdot d\vec{A} \tag{29.22}$$

$$\oint \vec{E} \cdot d\vec{l} = -\frac{d}{dt} \int \vec{B} \cdot d\vec{A} \tag{29.23}$$

Again we notice the symmetry between the roles of $\vec{E}$ and $\vec{B}$ in these expressions.

The most remarkable feature of these equations is that a time-varying field of *either* kind induces a field of the other kind in neighboring regions of space. Maxwell recognized that these relationships predict the existence of electromagnetic disturbances consisting of time-varying electric and magnetic fields that travel or *propagate* from one region of space to another, even if no matter is present in the intervening space. Such disturbances, called *electromagnetic waves,* provide the physical basis for light, radio and television waves, infrared, ultraviolet, x rays, and the rest of the electromagnetic spectrum. We will return to this vitally important topic in Chapter 32.

Although it may not be obvious, *all* the basic relations between fields and their sources are contained in Maxwell's equations. We can derive Coulomb's law from Gauss's law, we can derive the law of Biot and Savart from Ampere's law, and so on. When we add the equation that defines the $\vec{E}$ and $\vec{B}$ fields in terms of the forces that they exert on a charge q, namely,

$$\vec{F} = q(\vec{E} + \vec{v} \times \vec{B}) \tag{29.24}$$

we have *all* the fundamental relations of electromagnetism!

Finally, we note that Maxwell's equations would have even greater symmetry between the $\vec{E}$ and $\vec{B}$ fields if single magnetic charges (magnetic monopoles) existed. The right side of Eq. (29.19) would contain the total *magnetic* charge enclosed by the surface, and the right side of Eq. (29.21) would include a magnetic monopole current term. Perhaps you can begin to see why some physicists wish that magnetic monopoles existed; they would help to perfect the mathematical poetry of Maxwell's equations.

The discovery that electromagnetism can be wrapped up so neatly and elegantly is a very satisfying one. In conciseness and generality, Maxwell's equations are in the same league with Newton's laws of motion and the laws of thermodynamics. Indeed, a major goal of science is learning how to express very broad and general relationships in a concise and compact form. Maxwell's synthesis of electromagnetism stands as a towering intellectual achievement, comparable to the Newtonian synthesis we described at the end of Section 12.5 and to the development of relativity and quantum mechanics in the twentieth century.

Test Your Understanding

Which of Maxwell's equations explains how a credit card reader works? Which one describes how a wire carrying a steady current generates a magnetic field?

*29.8 | Superconductivity

The most familiar property of a superconductor is the sudden disappearance of all electrical resistance when the material is cooled below a temperature called the *critical temperature,* denoted by T_c. We discussed this behavior and the circumstances of its discovery in Section 25.2. But superconductivity is far more than just the absence of measurable resistance. Superconductors also have extraordinary *magnetic* properties. We'll explore some of these properties in this section.

The first hint of unusual magnetic properties was the discovery that for any superconducting material the critical temperature T_c changes when the material is placed in an externally produced magnetic field $\vec{B}_0$. Figure 29.23 shows this dependence for mercury, the first element in which superconductivity was observed. As the external field magnitude B_0 increases, the superconducting transition occurs at lower and lower temperature. When B_0 is greater than 0.0412 T, *no* superconducting transition occurs. The minimum magnitude of magnetic field that is needed to eliminate superconductivity at a temperature below T_c is called the *critical field,* denoted by B_c.

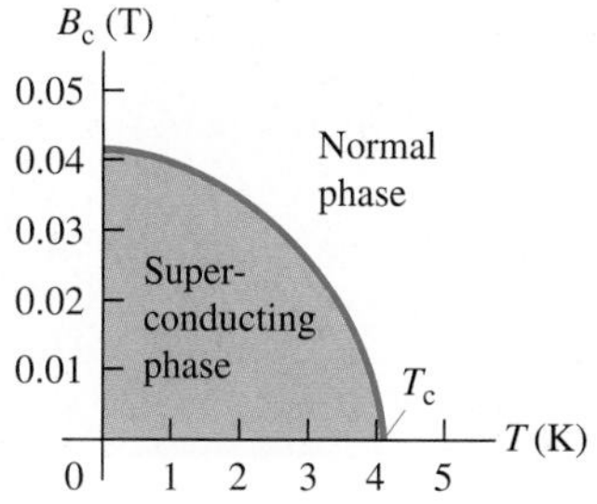

29.23 Phase diagram for pure mercury, showing the critical magnetic field B_c and its dependence on temperature. Superconductivity is impossible above the critical temperature T_c. The curves for other superconducting materials are similar but with different numerical values.

Another aspect of the magnetic behavior of superconductors appears if we place a homogeneous sphere of a superconducting material in a uniform applied magnetic field $\vec{B}_0$ at a temperature T greater than T_c. The material is then in the normal phase, not the superconducting phase. The field is as shown in Fig. 29.24a. Now we lower the temperature until the superconducting transition occurs. (We assume that the magnitude of $\vec{B}_0$ is not large enough to prevent the phase transition.) What happens to the field?

Measurements of the field outside the sphere show that the field lines become distorted as in Fig. 29.24b. There is no longer any field inside the material, except possibly in a very thin surface layer a hundred or so atoms thick. If a coil is wrapped around the sphere, the emf induced in the coil shows that during the superconducting

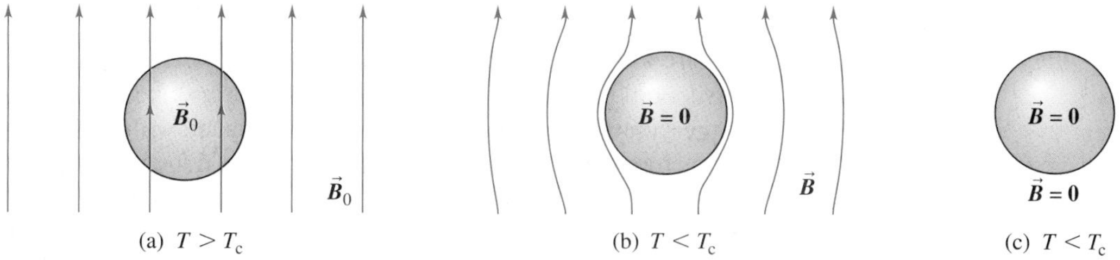

29.24 (a) When a superconducting material above the critical temperature is placed in an external magnetic field $\vec{B}_0$, the field inside the material is very nearly equal to $\vec{B}_0$. (b) As the material is cooled through the critical temperature T_c and becomes superconducting, the magnetic flux is expelled from the material, and the field inside it becomes zero. (c) When the external field is turned off with the material in the superconducting phase, the field is zero everywhere; there is no change of magnetic flux in the material.

transition the magnetic flux through the coil decreases from its initial value to zero; this is consistent with the absence of field inside the material. Finally, if the field is now turned off while the material is still in its superconducting phase, no emf is induced in the coil, and measurements show no field outside the sphere (Fig. 29.24c).

We conclude that during a superconducting transition in the presence of the field $\vec{B}_0$, all of the magnetic flux is expelled from the bulk of the sphere, and the magnetic flux Φ_B through the coil becomes zero. This expulsion of magnetic flux is called the *Meissner effect.* As shown in Fig. 29.24b, this expulsion crowds the magnetic field lines closer together to the side of the sphere, increasing $\vec{B}$ there.

29.25 A superconductor (the black slab) exerts a repulsive force on a magnet (the metallic cylinder), supporting the magnet in midair.

The diamagnetic nature of a superconductor has some interesting *mechanical* consequences. A paramagnetic or ferromagnetic material is attracted by a permanent magnet because the magnetic dipoles in the material align with the nonuniform magnetic field of the permanent magnet. (We discussed this in Section 27.7.) For a diamagnetic material the magnetization is in the opposite sense, and a diamagnetic material is *repelled* by a permanent magnet. By Newton's third law the magnet is also repelled by the diamagnetic material. Figure 29.25 shows the repulsion between a specimen of a high-temperature superconductor and a magnet; the magnet is supported ("levitated") by this repulsive magnetic force.

The behavior we have described is characteristic of what are called *type-I superconductors.* There is another class of superconducting materials called *type-II superconductors.* When such a material in the superconducting phase is placed in a magnetic field, the bulk of the material remains superconducting, but thin filaments of material, running parallel to the field, may return to the normal phase. Currents circulate around the boundaries of these filaments, and there *is* magnetic flux inside them. Type-II superconductors are used for electromagnets because they usually have much larger values of B_c than do type-I materials, permitting much larger magnetic fields without destroying the superconducting state. Type-II superconductors have *two* critical magnetic fields; the first, B_{c1}, is the field at which magnetic flux begins to enter the material, forming the filaments just described, and the second, B_{c2}, is the field at which the material becomes normal.

Many important and exciting applications of superconductors are under development. Superconducting electromagnets have been used in research laboratories for several years. Their advantages compared to conventional electromagnets include greater efficiency, compactness, and greater field magnitudes. Once a current is established in the coil of a superconducting electromagnet, no additional power input is required because there is no resistive energy loss. The coils can also be made more compact because there is no need to provide channels for the circulation of cooling fluids. Superconducting magnets routinely attain steady fields of the order of 10 T, much larger than the maximum fields that are available with ordinary electromagnets.

Superconductors are attractive for long-distance electric power transmission and for energy-conversion devices, including generators, motors, and transformers. Very sensitive measurements of magnetic fields can be made with superconducting quantum interference devices (SQUIDs), which can detect changes in magnetic flux of less than 10^{-14} Wb; these devices have applications in medicine, geology, and other fields. The number of potential uses for superconductors has increased greatly since the discovery in 1987 of high-temperature superconductors. These materials have critical temperatures that are above the temperature of liquid nitrogen (about 77 K) and so are comparatively easy to attain. Development of practical applications of superconductor science promises to be an exciting chapter in contemporary technology.

SUMMARY

Faraday's law states that the induced emf in a closed loop equals the negative of the time rate of change of magnetic flux through the loop. This relation is valid whether the flux change is caused by a changing magnetic field, motion of the loop, or both. (See Examples 29.1 through 29.7)

$$\mathcal{E} = -\frac{d\Phi_B}{dt} \qquad (29.3)$$

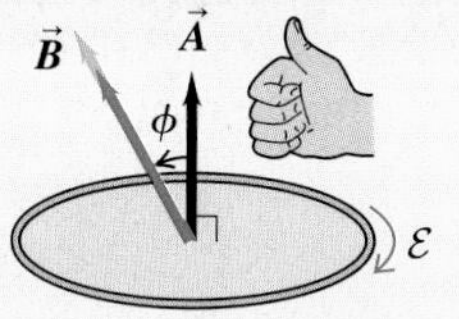

Lenz's law states that an induced current or emf always tends to oppose or cancel out the change that caused it. Lenz's law can be derived from Faraday's law, and is often easier to use. (See Examples 29.8 and 29.9)

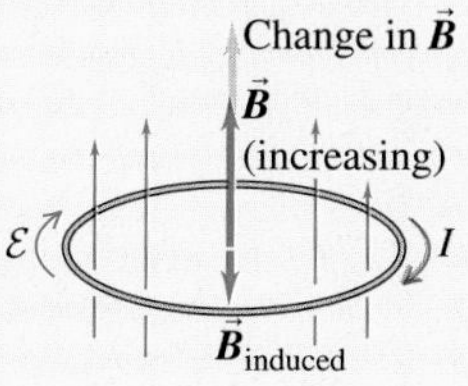

If a conductor moves in a magnetic field, a motional emf is induced. (See Examples 29.10 and 29.11)

$$\mathcal{E} = vBL \qquad (29.6)$$

(conductor with length L moves in uniform $\vec{B}$ field, $\vec{L}$ and $\vec{v}$ both perpendicular to $\vec{B}$ and to each other)

$$\mathcal{E} = \oint (\vec{v} \times \vec{B}) \cdot d\vec{l} \qquad (29.7)$$

(all or part of a closed loop moves in a $\vec{B}$ field)

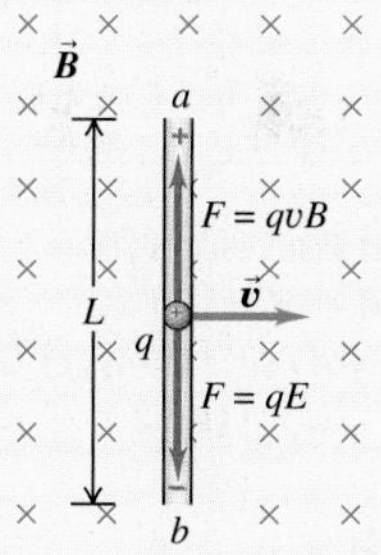

When an emf is induced by a changing magnetic flux through a stationary conductor, there is an induced electric field $\vec{E}$ of nonelectrostatic origin. This field is nonconservative and cannot be associated with a potential. (See Example 29.12)

$$\oint \vec{E} \cdot d\vec{l} = -\frac{d\Phi_B}{dt} \qquad (29.10)$$

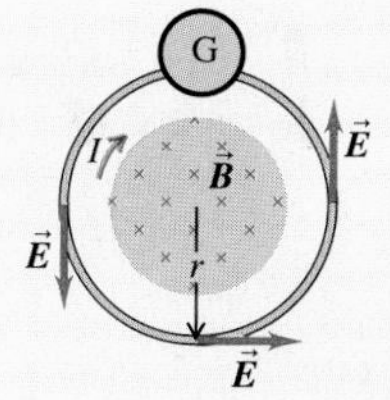

When a bulk piece of conducting material, such as a metal, is in a changing magnetic field or moves through a field, currents called eddy currents are induced in the volume of the material.

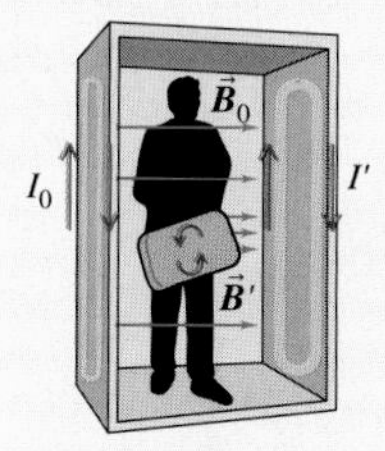

A time-varying electric field generates a displacement current i_D, which acts as a source of magnetic field in exactly the same way as conduction current.

$$i_D = \epsilon \frac{d\Phi_E}{dt} \qquad (29.14)$$

(displacement current)

The relationships between electric and magnetic fields and their sources can be stated compactly in four equations, called Maxwell's equations. Together they form a complete basis for the relation of $\vec{E}$ and $\vec{B}$ fields to their sources.

$$\oint \vec{E} \cdot d\vec{A} = \frac{Q_{\text{encl}}}{\epsilon_0} \quad (29.18)$$

(Gauss's law for $\vec{E}$ fields)

$$\oint \vec{B} \cdot d\vec{A} = 0 \quad (29.19)$$

(Gauss's law for $\vec{B}$ fields)

$$\oint \vec{B} \cdot d\vec{l} = \mu_0 \left(i_C + \epsilon_0 \frac{d\Phi_E}{dt} \right)_{\text{encl}} \quad (29.20)$$

(Ampere's law including displacement current)

$$\oint \vec{E} \cdot d\vec{l} = -\frac{d\Phi_B}{dt} \quad (29.21)$$

(Faraday's law)

Key Terms

induced current, 1106
induced emf, 1106
Faraday's law of induction, 1109
Lenz's law, 1118
motional electromotive force, 1121
induced electric field, 1124
nonelectrostatic field, 1125
eddy currents, 1126
displacement current, 1129
Maxwell's equations, 1130

Your Notes

Answer to Chapter Opening Question

To induce an emf in the card reader's circuits, the magnetized strip on the credit card must move past the circuits. If there is no motion, there is no induced emf or current and none of the credit card's information is read.

Answers to Test Your Understanding

Section 29.2 The direction of the induced emf would be reversed (counterclockwise as seen looking in the direction of $\vec{A}$) and the magnitude would be (0.300)/(0.200) times as large: $\mathcal{E} = -0.653$ V.
Section 29.3 The area of the loop and the magnetic flux are now decreasing. The induced current would flow clockwise around the circuit to generate a magnetic field to oppose this decrease.
Section 29.4 You will get the maximum motional emf if you hold the rod vertically, so that its length is perpendicular to both the magnetic field and the direction of motion. With this orientation, $\vec{L}$ is parallel to $\vec{v} \times \vec{B}$. If you hold the rod horizontally, $\vec{L}$ will be perpendicular to $\vec{v} \times \vec{B}$ and no emf will be induced. If you walk due north or south, $\vec{v} \times \vec{B} = \mathbf{0}$ and no emf will be induced for any orientation of the rod.
Section 29.5 Yes, because the magnetic field at a fixed position changes as you move the magnet. Such induced electric fields are *not* conservative.
Section 29.7 Faraday's law, Eq. (29.21), explains the operation of a credit card reader (see the photograph that opens the chapter). Ampere's law, Eq. (29.20), describes how currents of all kinds (both conduction currents and displacement currents) give rise to magnetic fields.

Discussion Questions

Q29.1 A sheet of copper is placed between the poles of an electromagnet with the magnetic field perpendicular to the sheet. When it is pulled out, a considerable force is required, and the force required increases with speed. Explain.
Q29.2 In Fig. 29.7, if the angular speed ω of the loop is doubled, then the frequency with which the induced current changes direction doubles, and the maximum emf also doubles. Why? Does the torque required to turn the loop change? Explain.
Q29.3 Two circular loops lie side-by-side in the same plane. One is connected to a source that supplies an increasing current; the other is a simple closed ring. Is the induced current in the ring in the same direction as that in the loop connected to the source, or opposite? What if the current in the first loop is decreasing? Explain.
Q29.4 A farmer claimed that the high-voltage transmission lines running parallel to his fence induced dangerously large voltages on the fence. Is this within the realm of possibility? Explain. (The lines carry alternating current that changes direction 120 times each second.)
Q29.5 A long, straight conductor passes through the center of a metal ring, perpendicular to its plane. If the current in the conductor increases, is a current induced in the ring? Explain.
Q29.6 A student asserted that if a permanent magnet is dropped down a vertical copper pipe, it eventually reaches a terminal velocity even if there is no air resistance. Why should this be? Or should it?
Q29.7 An airplane is in level flight over Antarctica, where the magnetic field of the earth is mostly directed upward, away from the ground (see Fig. 27.3). As viewed by a passenger facing toward the front of the plane, which wingtip, the left or the right, is at higher potential? Does your answer depend on the direction the plane is flying?
Q29.8 Consider the moving rod shown in Fig. 29.14a. If the charge q of charge carriers within the rod were negative rather than positive, how would this affect the directions of the electric and magnetic forces on these charge carriers? Would this affect the sign of the potential difference V_{ab} or the sign of the motional emf? Explain.
Q29.9 A square conducting loop is in a region of uniform, constant magnetic field. Can the loop be rotated about an axis along one side and no emf be induced in the loop? Discuss, in terms of the orientation of the rotation axis relative to the magnetic field direction.
Q29.10 Example 29.7 discusses the external force that must be applied to the slidewire to move it at constant speed. If there were a break in the left-hand end of the U-shaped conductor, how much force would be needed to move the slidewire at constant speed? As in the example, you can ignore friction.
Q29.11 In the situation shown in Fig. 29.16, would it be appropriate to ask how much *energy* an electron gains during a complete trip around the wire loop with current I'? Would it be appropriate to ask what *potential difference* the electron moves through during such a complete trip? Explain your answers.
Q29.12 A fellow student, who read this chapter in haste, writes Faraday's law [Eq. (29.21)] with $+d\Phi_B/dt$ on the right-hand side rather than $-d\Phi_B/dt$. If this were truly the form of Faraday's law, what would be the direction of the magnetic force on the sliding wire in Example 29.7 (Section 29.2)? Would energy be conserved? How does this argument tell you that the right-hand side of the equation must be $-d\Phi_B/dt$ rather than $+d\Phi_B/dt$?
Q29.13 Can one have a displacement current as well as a conduction current within a conductor? Explain.
Q29.14 Your physics study partner asks you to consider a parallel-plate capacitor that has a dielectric completely filling the volume between the plates. He then claims that Eqs. (29.13) and (29.14) show that the conduction current in the dielectric equals the displacement current in the dielectric. Do you agree? Explain.
Q29.15 Match the mathematical statements of Maxwell's equations as given in Section 29.7 to these verbal statements. a) Closed electric field lines are evidently produced only by changing magnetic flux. b) Closed magnetic field lines are produced both by the motion of electric charge and by changing electric flux. c) Electric field lines can start on positive charges and end on negative charges. d) Evidently there are no magnetic monopoles on which to start and end magnetic field lines.
Q29.16 If magnetic monopoles existed, the right-hand side of Eq. (29.21) would include a term proportional to the current of magnetic monopoles. Suppose a steady monopole current is moving in a long straight wire. Sketch the *electric* field lines that such a current would produce.

Q29.17 If magnetic monopoles existed, the right-hand side of Eq. (29.19) would be proportional to the total enclosed *magnetic* charge. Suppose an infinite line of magnetic monopoles were on the x-axis. Sketch the magnetic field lines that this line of monopoles would produce.

Exercises

Section 29.2 Faraday's Law

29.1 A closely wound, rectangular coil of 80 turns has dimensions of 25.0 cm $\times$ 40.0 cm. The plane of the coil is rotated, in 0.0600 s from a position where it makes an angle of 37.0° with a magnetic field of 1.10 T to a position perpendicular to the field. What is the average emf induced in the coil?

29.2 In a physics laboratory experiment, a coil with 200 turns enclosing an area of 12 cm^2 is rotated in 0.040 s from a position where its plane is perpendicular to the earth's magnetic field to one where its plane is parallel to the field. The earth's magnetic field at the lab location is 6.0×10^{-5} T. a) What is the total magnetic flux through the coil before it is rotated? After it is rotated? b) What is the average emf induced in the coil?

29.3 Search Coils and Credit Cards. a) Derive the equation relating the total charge Q that flows through a search coil (Conceptual Example 29.3 in Section 29.2) to the magnetic field magnitude B. The search coil has N turns, each with area A, and the flux through the coil is decreased from its initial maximum value to zero in a time Δt. The resistance of the coil is R, and the total charge is $Q = I\Delta t$, where I is the average current induced by the change in flux. b) In a credit card reader, the magnetic strip on the back of a credit card is rapidly "swiped" past a coil within the reader. Explain, using the same ideas that underlie the operation of a search coil, how the reader can decode the information stored in the pattern of magnetization on the strip. c) Is it necessary that the credit card be "swiped" through the reader at exactly the right speed? Why or why not?

29.4 The cross-sectional area of a closely wound search coil (Exercise 29.3) having 90 turns is 2.20 cm^2, and its resistance is 6.80 Ω. The coil is connected through leads of negligible resistance to a charge-measuring instrument with an internal resistance of 12.0 Ω. Find the quantity of charge displaced when the coil is pulled quickly out of a region, where $B = 2.05$ T, to a point where the magnetic field is zero. The plane of the coil, when in the field, makes an angle of 90° with the magnetic field.

29.5 A closely wound search coil (Exercise 29.3) has an area of 3.20 cm^2, 120 turns, and a resistance of 60.0 Ω. It is connected to a charge-measuring instrument whose resistance is 45.0 Ω. When the coil is rotated quickly from a position parallel to a uniform magnetic field to one perpendicular to the field, the instrument indicates a charge of 3.56×10^{-5} C. What is the magnitude of the field?

29.6 A coil 4.00 cm in radius, containing 500 turns, is placed in a uniform magnetic field that varies with time according to $B = (0.0120 \text{ T/s})t + (3.00 \times 10^{-5} \text{ T/s}^4)t^4$. The coil is connected to a 600-Ω resistor, and its plane is perpendicular to the magnetic field. You can ignore the resistance of the coil. a) Find the magnitude of the induced emf in the coil as a function of time. b) What is the current in the resistor at time $t = 5.00$ s?

29.7 A closely wound coil with N turns of cross-sectional area A lies in the xy-plane. The coil is placed in a uniform but time-dependent magnetic field $\vec{B} = B_z(t)\hat{k}$. Here $B_z(t)$ equals zero for $t \leq 0$, equals $B_0(1 - \cos(2\pi t/T))$ for $0 < t < T$, and equals zero for $t \geq T$; B_0 is a positive constant. a) Find the induced emf in the coil as a function of time. b) At what time(s) t, if any, in the range $0 < t < T$ is the induced emf equal to zero? c) At what time(s) t in the range $0 < t < T$ is the absolute value of the induced emf a maximum, and what is that maximum value? d) As seen by an observer on the positive z-axis looking back toward the coil, at what times is the induced emf in the clockwise direction? At what times is it in the counterclockwise direction?

29.8 A flat, circular, steel loop of radius 75 cm is at rest in a uniform magnetic field, as shown in an edge-on view in Fig. 29.26. The field is changing with time, according to $B(t) = (1.4 \text{ T})e^{-(0.057\text{s}^{-1})t}$. a) Find the emf induced in the loop as a function of time. b) When is the induced emf equal to $\frac{1}{10}$ of its initial value? c) Find the direction of the current induced in the loop, as viewed from above the loop.

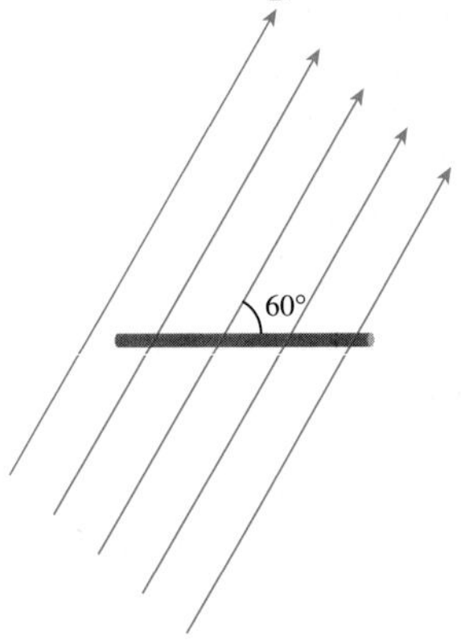

Figure 29.26 Exercise 29.8.

29.9 Shrinking Loop. A circular loop of flexible iron wire has an initial circumference of 165.0 cm, but its circumference is decreasing at a constant rate of 12.0 cm/s due to a tangential pull on the wire. The loop is in a constant, uniform magnetic field oriented perpendicular to the plane of the loop and with magnitude 0.500 T. a) Find the emf induced in the loop, at the instant when 9.0 s have passed. b) Find the direction of the induced current in the loop, as viewed looking along the direction of the magnetic field.

29.10 A circular loop of wire with radius of 12.0 cm and oriented in the horizontal xy-plane is located in a region of uniform magnetic field. A field of 1.5 T is directed along the positive z-direction, which is upward. a) If the loop is removed from the field region in a time interval of 2.0×10^{-3} s, find the average emf that will be induced in the wire loop during the extraction process. b) If the coil is viewed looking down on it from above, is the induced current in the loop clockwise or counterclockwise?

29.11 A square copper loop, 10.0 cm on a side, is located in a region of changing magnetic field. The direction of the magnetic field makes an angle of 37° with the plane of the loop. The time-changing field has the following time dependence: $B(t) = 0.10 \text{ T} + (1.00 \times 10^{-3} \text{ T/s})t$. Find the induced emf in the copper loop for times $t > 0$.

29.12 Back EMF. A motor with a brush-and-commutator arrangement, as described in Example 29.5 (Section 29.2) has a circular coil with radius 2.5 cm and 150 turns of wire. The magnetic field has magnitude 0.060 T and the coil rotates at 440 rev/min. a) What is the maximum emf induced in the coil? b) What is the average back emf?

29.13 If the rotation speed of the motor in Example 29.5 (Section 29.2) is increased to 3360 rev/min, what is the average back emf of the motor?

29.14 The armature of a small generator consists of a flat, square coil with 120 turns and sides with a length of 1.60 cm. The coil rotates in a magnetic field of 0.0750 T. What is the angular speed of the coil if the maximum emf produced is 24.0 mV?

Section 29.3 Lenz's Law

29.15 Redraw Figure 29.12 in Conceptual Example 29.9 (Section 29.3) for the case in which the upward magnetic field is *decreasing*.

29.16 A circular loop of wire is in a region of spatially uniform magnetic field, as shown in Fig. 29.27. The magnetic field is directed into the plane of the figure. Determine the direction (clockwise or counterclockwise) of the induced current in the loop when a) B is increasing; b) B is decreasing; c) B is constant with value B_0. Explain your reasoning.

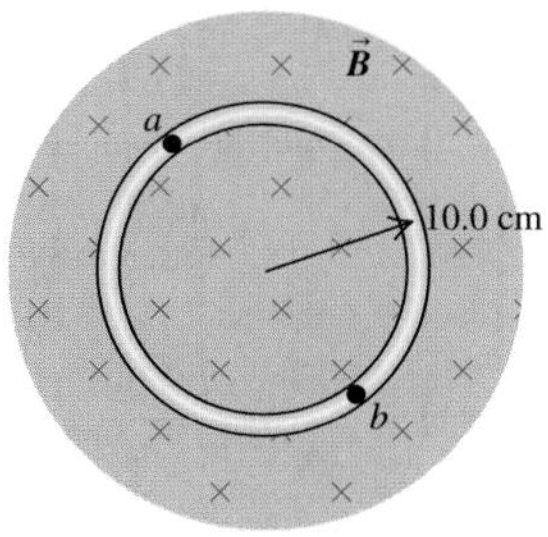

Figure 29.27 Exercises 29.16, 29.29, and Problem 29.48.

29.17 Using Lenz's law, determine the direction of the current in resistor ab of Fig. 29.28 when a) switch S is opened after having been closed for several minutes; b) coil B is brought closer to coil A with the switch closed; c) the resistance of R is decreased while the switch remains closed.

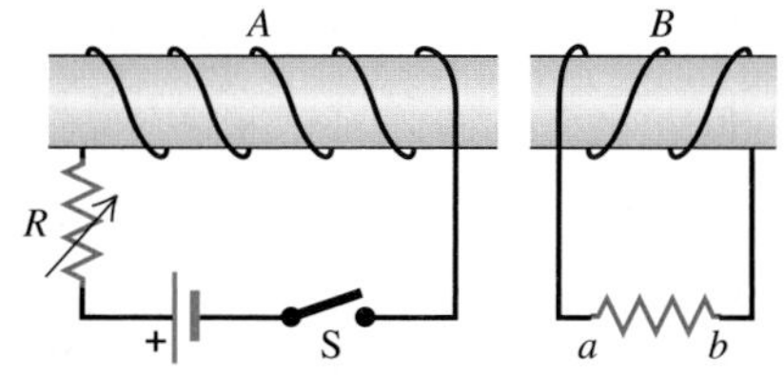

Figure 29.28 Exercise 29.17.

29.18 A cardboard tube is wrapped with two windings of insulated wire wound in opposite directions, as shown in Fig. 29.29. Terminals a and b of winding A may be connected to a battery through a reversing switch. State whether the induced current in the resistor R is from left to right or from right to left in the following circumstances: a) the current in winding A is from a to b and is increasing; b) the current in winding A is from b to a and is decreasing; c) the current in winding A is from b to a and is increasing.

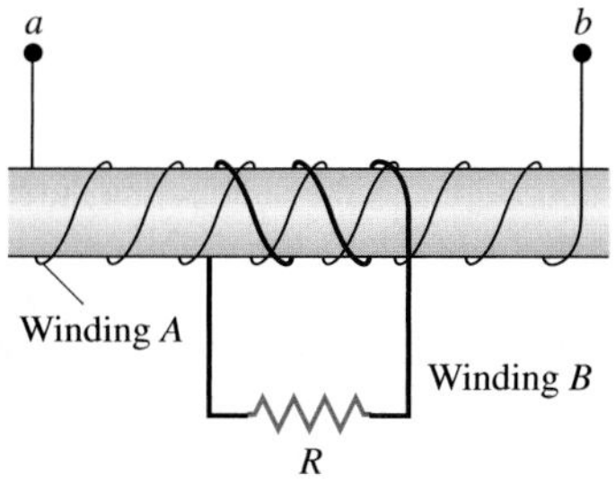

Figure 29.29 Exercise 29.18.

29.19 A small, circular ring is inside a larger loop that is connected to a battery and a switch, as shown in Fig. 29.30. Use Lenz's law to find the direction of the current induced in the small ring a) just after switch S is closed; b) after S has been closed a long time; c) just after S has been reopened after being closed a long time.

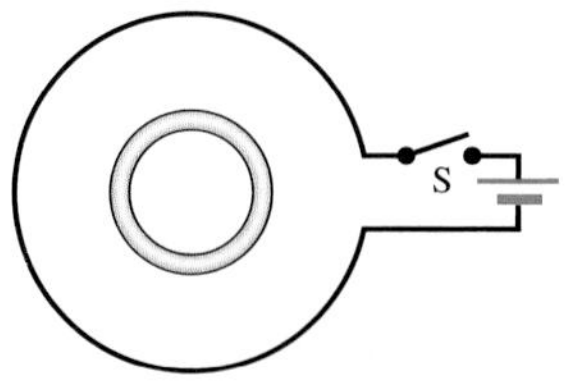

Figure 29.30 Exercise 29.19.

29.20 A 1.50-m-long metal bar is pulled to the right at a steady 5.0 m/s perpendicular to a uniform 0.750 T magnetic field. The bar rides on parallel metal rails connected through a 25.0 Ω resistor, as shown in Fig. 29.31, so the apparatus makes a complete circuit. You can ignore the resistance of the bar and the rails. a) Calculate the magnitude of the emf induced in the circuit. b) Find the direction of the current induced in the circuit i) using the magnetic force on the charges in the moving bar; ii) using Faraday's law; iii) using Lenz's law. c) Calculate the current through the resistor.

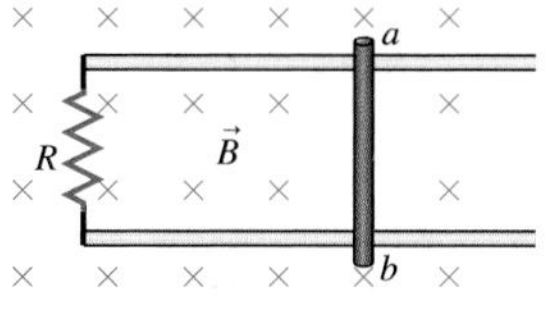

Figure 29.31 Exercise 29.20 and Problem 29.64.

Section 29.4 Motional Electromotive Force

29.21 For Eq. (29.6), show that if v is in meters per second, B in teslas, and L in meters, then the units of the right-hand side of the equation are joules per coulomb or volts (the correct SI units for $\mathcal{E}$).

29.22 In Fig. 29.32 a conducting rod with length $L = 30.0$ cm moves in a magnetic field $\vec{B}$ of magnitude 0.450 T directed into the plane of the figure. The rod moves with speed $v = 5.00$ m/s in the direction shown. a) What is the motional emf induced in the rod? b) What is the potential difference between the ends of the rod? c) Which point, a or b, is at higher potential? d) When the charges in the rod are in equilibrium, what are the magnitude and direction of the electric field within the rod? e) When the charges in the rod are in equilibrium, which point, a or b, has an excess of positive charge?

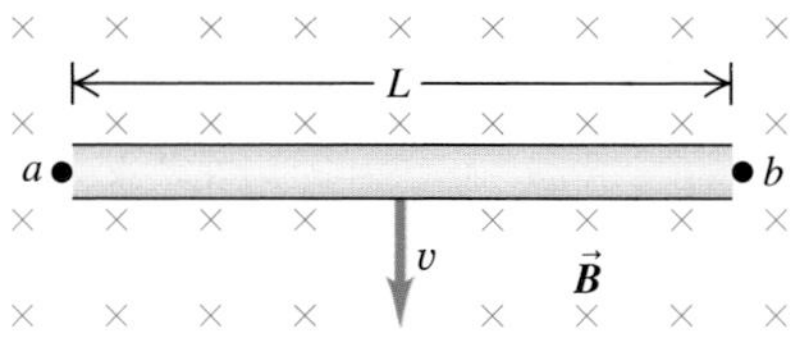

Figure 29.32 Exercise 29.22.

29.23 In Fig. 29.14b a rod with length $L = 85.0$ cm moves in a magnetic field with magnitude $B = 0.850$ T. The emf induced in the moving rod is 0.620 V. a) What is the speed of the rod? b) If the total circuit resistance is 0.750 Ω, what is the induced current? c) What force (magnitude and direction) does the field exert on the rod as a result of this current?

29.24 The conducting rod ab shown in Fig. 29.33 makes contact with metal rails ca and db. The apparatus is in a uniform magnetic field 0.800 T, perpendicular to the plane of the figure. a) Find the magnitude of the emf induced in the rod when it is moving toward the right with a speed 7.50 m/s. b) In what direction does the current flow in the rod? c) If the resistance of the circuit $abdc$ is 1.50 Ω (assumed to be constant), find the force (magnitude and direction) required to keep the rod moving to the right with a constant speed of 7.50 m/s. You can ignore friction. d) Compare the rate at which mechanical work is done by the force (Fv) with the rate at which thermal energy is developed in the circuit (I^2R).

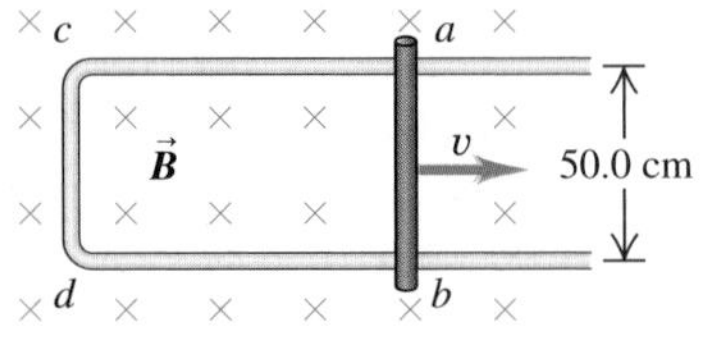

Figure 29.33 Exercise 29.24.

29.25 A square loop of wire with resistance R is moved at constant speed v across a uniform magnetic field confined to a square region whose sides are twice the length of those of the square loop (see Fig. 29.34). a) Graph the external force F needed to move the loop at constant speed as a function of the coordinate x from $x = -2L$ to $x = +2L$. (The coordinate x is measured from the center of the magnetic field region to the center of the loop. It is negative when the center of the loop is to the left of the center of the magnetic field region. Take positive force to be to the right.) b) Graph the induced current in the loop as a function of x. Take counterclockwise currents to be positive.

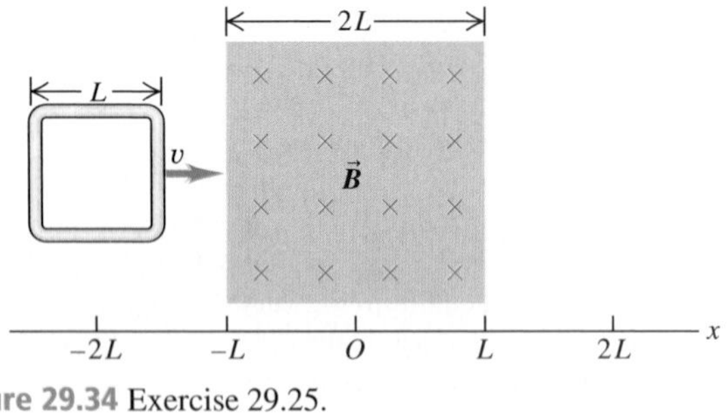

Figure 29.34 Exercise 29.25.

29.26 In Fig. 29.14a a 12.0-cm-long rod moves with constant speed 4.50 m/s in the direction shown. The induced emf is 0.450 V. a) What is the magnitude of the magnetic field? b) Which point is at higher potential, a or b?

Section 29.5 Induced Electric Fields

29.27 A long, thin solenoid has 900 turns per meter and radius 2.50 cm. The current in the solenoid is increasing at a uniform rate of 60.0 A/s. What is the magnitude of the induced electric field at a point near the center of the solenoid and a) 0.500 cm from the axis of the solenoid? b) 1.00 cm from the axis of the solenoid?

29.28 The magnetic field within a long, straight solenoid with a circular cross section and radius R is increasing at a rate of dB/dt. a) What is the rate of change of flux through a circle with radius r_1 inside the solenoid, normal to the axis of the solenoid, and with center on the solenoid axis? b) Find the magnitude of the induced electric field inside the solenoid, at a distance r_1 from its axis. Show the direction of this field in a diagram. c) What is the magnitude of the induced electric field *outside* the solenoid, at a distance r_2 from the axis? d) Graph the magnitude of the induced electric field as a function of the distance r from the axis from $r = 0$ to $r = 2R$. e) What is the magnitude of the induced emf in a circular turn of radius $R/2$ that has its center on the solenoid axis? f) What is the magnitude of the induced emf if the radius in part (e) is R? g) What is the induced emf if the radius in part (e) is $2R$?

29.29 The magnetic field $\vec{B}$ at all points within the colored circle shown in Fig. 29.27 has an initial magnitude of 0.750 T. (The circle could represent approximately the space inside a long, thin solenoid.) The magnetic field is directed into the plane of the diagram and is decreasing at the rate of −0.0350 T/s. a) What is the shape of the field lines of the induced electric field shown in Fig. 29.27, within the colored circle? b) What are the magnitude and direction of this field at any point on the circular conducting ring with radius 0.100 m? c) What is the current in the ring if its resistance is 4.00 Ω? d) What is the emf between points a and b on the ring? e) If the ring is cut at some point and the ends are separated slightly, what will be the emf between the ends?

29.30 A long, thin solenoid has 400 turns per meter and radius 1.10 cm. The current in the solenoid is increasing at a uniform rate di/dt. The induced electric field at a point near the center of the solenoid and 3.50 cm from its axis is 8.00×10^{-6} V/m. Calculate di/dt.

29.31 A +6.50-μC charge moves in a circular path centered on the center of the solenoid in Exercise 29.30. The circular path is 3.50 cm in radius and lies in a plane perpendicular to the axis of the solenoid. a) If the charge moves around the circular path in the direction of the induced electric field, how much work is done on the charge during one complete trip around the path? b) Explain how your answer to part (a) confirms that the electric field is nonconservative. c) How much work is done on the charge during one complete trip around the path if the path radius is 7.00 cm?

29.32 A long, straight solenoid with a cross-sectional area of 8.00 cm^2 is wound with 90 turns of wire per centimeter, and the windings carry a current of 0.350 A. A second winding of 12 turns encircles the solenoid at its center. The current in the solenoid is turned off such that the magnetic field of the solenoid becomes zero in 0.0400 s. What is the average induced emf in the second winding?

Section 29.7 Displacement Current and Maxwell's Equations

29.33 A dielectric of permittivity 3.5×10^{-11} F/m completely fills the volume between two capacitor plates. For $t > 0$ the electric flux through the dielectric is $(8.0 \times 10^3)\ \text{V} \cdot \text{m/s}^3)t^3$. The dielectric is ideal and nonmagnetic; the conduction current in the dielectric is zero. At what time does the displacement current in the dielectric equal 21 μA?

29.34 The electric flux through a certain area of a dielectric is $(8.76 \times 10^3)\ \text{V} \cdot \text{m/s}^4)t^4$. The displacement current through that area is 12.9 pA at time $t = 26.1$ ms. Calculate the dielectric constant for the dielectric.

29.35 A parallel-plate air-filled capacitor is being charged as in Fig. 29.22. The circular plates have radius 4.00 cm, and at a particular instant the conduction current in the wires is 0.280 A. a) What is the displacement current density j_D in the air space between the plates? b) What is the rate at which the electric field between the plates is changing? c) What is the induced magnetic field between the plates at a distance of 2.00 cm from the axis? d) at 1.00 cm from the axis?

29.36 Displacement Current in a Dielectric. Suppose that the parallel plates in Fig. 29.22 have an area of 3.00 cm^2 and are separated by a 2.50-mm-thick sheet of dielectric that completely fills the volume between the plates. The dielectric has dielectric constant 4.70. (You can ignore fringing effects.) At a certain instant, the potential difference between the plates is 120 V and the conduction current i_C equals 6.00 mA. At this instant, what is a) the charge q on each plate; b) the rate of change of charge on the plates; c) the displacement current in the dielectric?

29.37 In Fig. 29.22 the capacitor plates have area 5.00 cm^2 and separation 2.00 mm. The plates are in vacuum. The charging current i_C has a *constant* value of 1.80 mA. At $t = 0$ the charge on the plates is zero. a) Calculate the charge on the plates, the electric field between the plates, and the potential difference between the plates when $t = 0.500\ \mu s$. b) Calculate dE/dt, the time rate of change of the electric field between the plates. Does dE/dt vary in time? c) Calculate the displacement current density j_D between the plates, and from this the total displacement current i_D. How do i_C and i_D compare?

29.38 Displacement Current in a Wire. A long, straight, copper wire with a circular cross-sectional area of 2.1 mm^2 carries a current of 16 A. The resistivity of the material is $2.0 \times 10^{-8}\ \Omega \cdot m$. a) What is the uniform electric field in the material? b) If the current is changing at the rate of 4000 A/s, at what rate is the electric field in the material changing? c) What is the displacement current density in the material in part (b)? (*Hint:* Since K for copper is very close to 1, use $\epsilon = \epsilon_0$.) d) If the current is changing as in part (b), what is the magnitude of the magnetic field 6.0 cm from the center of the wire? Note that both the conduction current and the displacement current should be included in the calculation of B. Is the contribution from the displacement current significant?

*Section 29.8 Superconductivity

***29.39** A long, straight wire made of a type-I superconductor carries a constant current I along its length. Show that the current cannot be uniformly spread over the wire's cross section but instead must all be at the surface.

***29.40** A type-II superconductor in an external field between B_{c1} and B_{c2} has regions that contain magnetic flux and have resistance, and also has superconducting regions. What is the resistance of a long, thin cylinder of such material?

***29.41** At temperatures near absolute zero, B_c approaches 0.142 T for vanadium, a type-I superconductor. The normal phase of vanadium has a magnetic susceptibility close to zero. Consider a long, thin vanadium cylinder with its axis parallel to an external magnetic field $\vec{B}_0$ in the $+x$-direction. At points far from the ends of the cylinder, by symmetry, all the magnetic vectors are parallel to the x-axis. At temperatures near absolute zero, what are the resultant magnetic field $\vec{B}$ and the magnetization $\vec{M}$ inside and outside the cylinder (far from the ends) for a) $\vec{B}_0 = (0.130\ \text{T})\hat{\imath}$? b) $\vec{B}_0 = (0.260\ \text{T})\hat{\imath}$?

***29.42** The compound SiV_3 is a type-II superconductor. At temperatures near absolute zero the two critical fields are $B_{c1} = 55.0$ mT and $B_{c2} = 15.0$ T. The normal phase of SiV_3 has a magnetic susceptibility close to zero. A long, thin SiV_3 cylinder has its axis parallel to an external magnetic field $\vec{B}_0$ in the $+x$-direction. At points far from the ends of the cylinder, by symmetry, all the magnetic vectors are parallel to the x-axis. At a temperature near absolute zero the external magnetic field is slowly increased from zero. What are the resultant magnetic field $\vec{B}$ and the magnetization $\vec{M}$ inside the cylinder at points far from its ends a) just before the magnetic flux begins to penetrate the material; b) just after the material becomes completely normal?

Problems

29.43 A Changing Magnetic Field. You are testing a new data-acquisition system. This system allows you to record a graph of the current in a circuit as a function of time. As part of the test, you are using a circuit made up of a 4.00-cm-radius, 500-turn coil of copper wire connected in series to a 600-Ω resistor. Copper has resistivity $1.72 \times 10^{-8}\ \Omega \cdot m$, and the wire used for the coil has diameter 0.0300 mm. You place the coil on a table that is tilted 30.0° from the horizontal and that lies between the poles of an electromagnet. The electromagnet generates a vertically upward magnetic field that is zero for $t < 0$, equal to $(0.120\ \text{T})(1 - \cos \pi t)$ for $0 \le t \le 1.00$ s, and equal to 0.240 T for $t > 1.00$ s. a) Draw the graph that should be produced by your data-acquisition system. (This is a full-featured system, so the graph will include labels and numerical values on its axes.) b) If you were looking vertically downward at the coil, would the current be flowing clockwise or counterclockwise?

29.44 In the circuit shown in Fig. 29.35, the capacitor has capacitance $C = 20\ \mu F$ and is initially charged to 100 V with the polarity shown. The resistor R_0 has resistance 10 Ω. At time $t = 0$ the switch is closed. The small circuit is not connected in any way to the large one. The wire of the small circuit has a resistance of 1.0 Ω/m and contains 25 loops. The large circuit is a rectangle 2.0 m by 4.0 m, while the small one has dimensions $a = 10.0$ cm and $b = 20.0$ cm. The distance c is 5.0 cm. (The figure is not drawn to scale.) Both circuits are held stationary. Assume that only the wire nearest the small circuit produces an appreciable magnetic field through it. a) Find the current in the large circuit 200 μs after

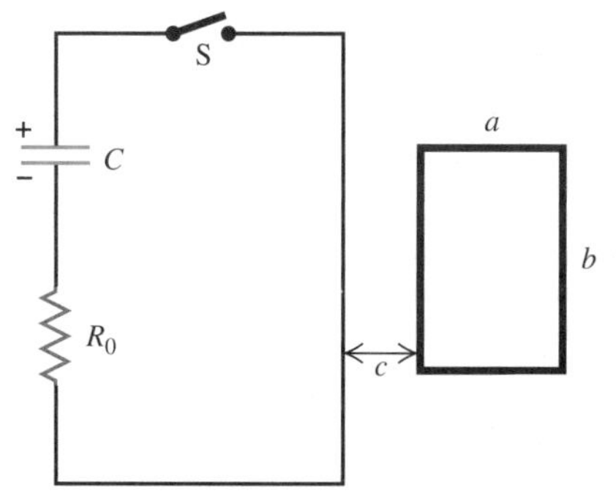

Figure 29.35 Problem 29.44.

closing S. b) Find the current in the small circuit 200 μs after closing S. (*Hint:* See Problem 29.54.) c) Find the direction of the current in the small circuit. d) Justify why we can ignore the magnetic field from all the wires of the large circuit except for the wire closest to the small circuit.

29.45 A circular coil of wire has a radius of 0.500 m, 20 turns, and a total resistance of 1.57 Ω. The coil lies in the xy-plane. The coil is in a uniform magnetic field $\vec{B}$ that is in the $-z$-direction, which is directed away from you as you view the coil. The magnitude B of the field depends on time as follows: increases at a constant rate from 0 at $t = 0$ to 0.800 T at $t = 0.500$ s; is constant at 0.800 T from $t = 0.500$ s to $t = 1.00$ s; decreases at a constant rate from 0.800 T at $t = 1.00$ s to 0 at $t = 2.00$ s. a) Graph B versus t for t from 0 to 2.00 s. b) Graph the current I induced in the coil versus t for t from 0 to 2.00 s. Let counterclockwise currents be positive and clockwise currents be negative. c) What is the maximum induced electric field magnitude in the coil during the 0- to 2.00-s time interval?

29.46 The rectangular loop in Example 29.4 (Section 29.2) is connected to an external circuit, forming a complete current path of resistance R. a) Show that the induced current is $i = (\omega BA/R)\sin\omega t$. b) Calculate the rate at which electrical energy is dissipated due to the resistance of the current path. c) Show that the magnetic dipole moment of the loop has magnitude $\mu = (\omega BA^2/R)\sin\omega t$. d) Show that the external torque required to maintain a constant ω is $\tau = (\omega B^2A^2/R)\sin^2\omega t$. e) Calculate the rate at which the external torque does work. How do the answers to parts (b) and (e) compare? Explain.

29.47 A circular wire loop of radius a and resistance R initially has a magnetic flux through it due to an external magnetic field. The external field then decreases to zero. A current is induced in the loop while the external field is changing; however, this current does not stop at the instant that the external field stops changing. The reason is that the current itself generates a magnetic field, which gives rise to a flux through the loop. If the current changes, the flux through the loop changes as well, and an induced emf appears in the loop to oppose the change. a) The magnetic field at the center of the loop of radius a produced by a current i in the loop is given by $B = \mu_0 i/2a$. If we use the crude approximation that the field has this same value at all points within the loop, what is the flux of this field through the loop? b) By using Faraday's law, Eq. (29.3), and the relationship $\mathcal{E} = iR$, show that after the external field has stopped changing, the current in the loop obeys the differential equation

$$\frac{di}{dt} = -\left(\frac{2R}{\pi\mu_0 a}\right)i$$

c) If the current has the value i_0 at $t = 0$, the instant that the external field stops changing, solve the equation in part (b) to find i as a function of time for $t > 0$. (*Hint:* In Section 26.4 we encountered a similar differential equation, Eq. (26.15), for the quantity q. This equation for i may be solved in the same way.) d) If the loop has radius $a = 50$ cm and resistance $R = 0.10\ \Omega$, how long after the external field stops changing will the current be equal to $0.010i_0$ (that is, $\frac{1}{100}$ of its initial value)? e) In solving the examples in this chapter, we ignored the effects described in this problem. Explain why this is a good approximation.

29.48 The circular loop of wire shown in Fig. 29.27 is in a spatially uniform magnetic field directed into the page. You measure the induced emf in the loop and for $0 < t < 5.00$ s you find that it is counterclockwise and has magnitude $\mathcal{E} = (2.00 \times 10^{-3}\ \text{V}) - (4.00 \times 10^{-4}\ \text{V/s})t$. a) If the magnitude of the magnetic field at $t = 0$ is 0.800 T, what is its magnitude at $t = 2.00$ s? b) Repeat part (a) if the emf has the same magnitude but is clockwise. c) In parts (a) and (b), is the flux through the loop larger or smaller at $t = 2.00$ s than at $t = 0$? Do your results agree with Lenz's law?

29.49 In Fig. 29.36, the loop is being pulled to the right at constant speed v. A constant current I flows in the long wire, in the direction shown. a) Calculate the magnitude of the net emf $\mathcal{E}$ induced in the loop. Do this two ways: (i) Using Faraday's law of induction. (*Hint:* See Problem 29.54.) (ii) By looking at the emf induced in each segment of the loop due to its motion. b) Find the direction (clockwise or counterclockwise) of the current induced in the loop. Do this two ways: (i) Using Lenz's law. (ii) Using the magnetic force on charges in the loop. c) Check your answer for the emf in part (a) in the following special cases to see if it is physically reasonable: (i) The loop is stationary. (ii) The loop is very thin, so $a \to 0$. (iii) The loop gets very far from the wire.

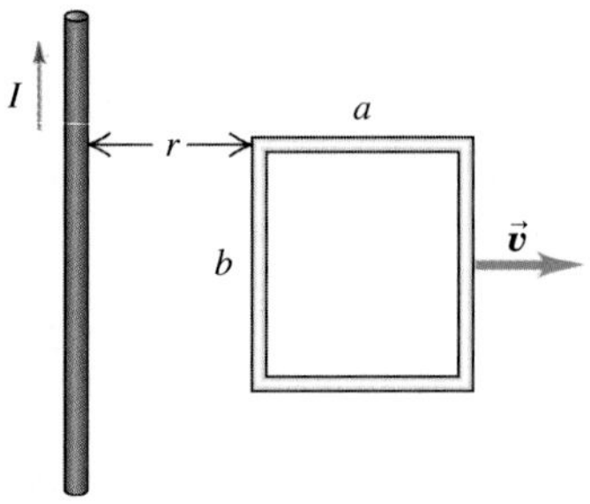

Figure 29.36 Problem 29.49.

29.50 Suppose the loop in Fig. 29.37 is a) rotated about the y-axis; b) rotated about the x-axis; c) rotated about an edge parallel to the z-axis. What is the maximum induced emf in each case if $A = 600\ \text{cm}^2$, $\omega = 35.0$ rad/s, and $B = 0.450$ T?

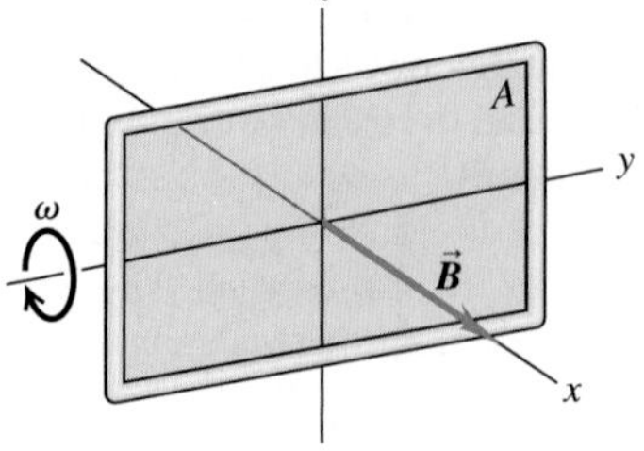

Figure 29.37 Problem 29.50.

29.51 As a new electrical engineer for the local power company, you are assigned the project of designing a generator of sinusoidal ac voltage with a maximum voltage of 120 V. Besides plenty of wire, you have two strong magnets that can produce a constant uniform magnetic field of 1.5 T over a square area of 10.0 cm on a side when they are 12.0 cm apart. The basic design should consist of a square coil turning in the uniform magnetic field. To have an acceptable coil resistance, the coil can have at most 400 loops. What is the minimum rotation rate (in rpm) of the coil so it will produce the required voltage?

29.52 Make a Generator? You are shipwrecked on a deserted tropical island. You have some electrical devices that you could operate using a generator but you have no magnets. The earth's magnetic field at your location is horizontal and equal to 8.0×10^{-5} T, and you decide to try to use this field for a generator by rotating a large circular coil of wire at a high rate. You need to produce an emf of 9.0 V and estimate that you can rotate the coil at 30 rpm by turning a crank handle. You also decide that to have an acceptable coil resistance, the maximum number of turns the coil can have is 2000. a) What area must the coil have? b) If the coil is circular, what is the maximum translational speed of a point on the coil as it rotates? Do you think this device is feasible? Explain.

29.53 A flexible circular loop 6.50 cm in diameter lies in a magnetic field with magnitude 0.950 T, directed into the plane of the page as shown in Fig. 29.38. The loop is pulled at the points indicated by the arrows, forming a loop of zero area in 0.250 s. a) Find the average induced emf in the circuit. b) What is the direction of the current in R, from a to b or from b to a? Explain your reasoning.

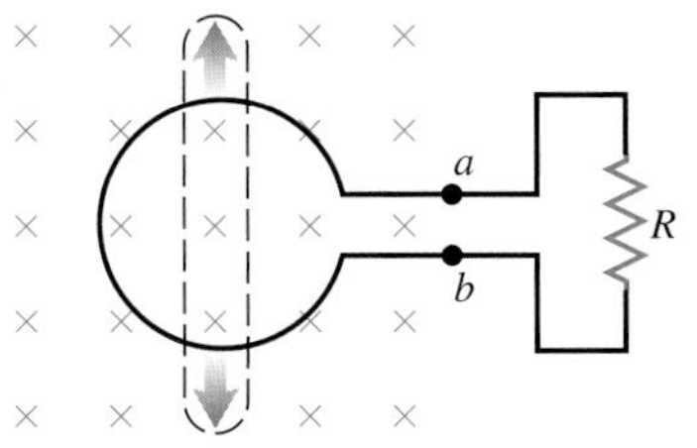

Figure 29.38 Problem 29.53.

29.54 The current in the long, straight wire AB shown in Fig. 29.39 is upward and is increasing steadily at a rate di/dt. a) At an instant when the current is i, what are the magnitude and direction of the field $\vec{B}$ at a distance r to the right of the wire? b) What is the flux $d\Phi_B$ through the narrow, shaded strip? c) What is the total flux through the loop? d) What is the induced emf in the loop? e) Evaluate the numerical value of the induced emf if $a = 12.0$ cm, $b = 36.0$ cm, $L = 24.0$ cm, and $di/dt = 9.60$ A/s.

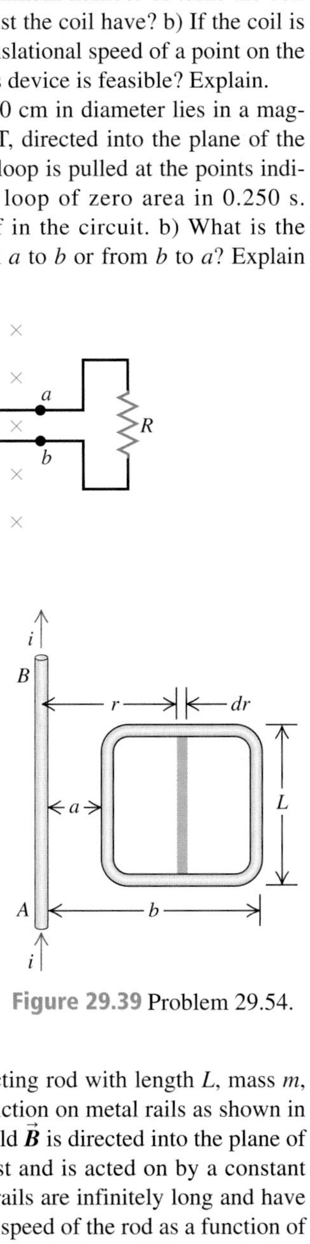

Figure 29.39 Problem 29.54.

29.55 Terminal Speed. A conducting rod with length L, mass m, and resistance R moves without friction on metal rails as shown in Fig. 29.10. A uniform magnetic field $\vec{B}$ is directed into the plane of the figure. The rod starts from rest and is acted on by a constant force $\vec{F}$ directed to the right. The rails are infinitely long and have negligible resistance. a) Graph the speed of the rod as a function of time. b) Find an expression for the terminal speed (the speed when the acceleration of the rod is zero).

29.56 Terminal Speed. A bar of length $L = 0.8$ m is free to slide without friction on horizontal rails, as shown in Fig. 29.40. There is a uniform magnetic field $B = 1.5$ T directed into the plane of the figure. At one end of the rails there is a battery with emf $\mathcal{E} = 12$ V and a switch. The bar has mass 0.90 kg and resistance 5.0 Ω, and all other resistance in the circuit can be ignored. The switch is closed at time $t = 0$. a) Sketch the speed of the bar as a function of time. b) Just after the switch is closed, what is the acceleration of the bar? c) What is the acceleration of the bar when its speed is 2.0 m/s? d) What is the terminal speed of the bar?

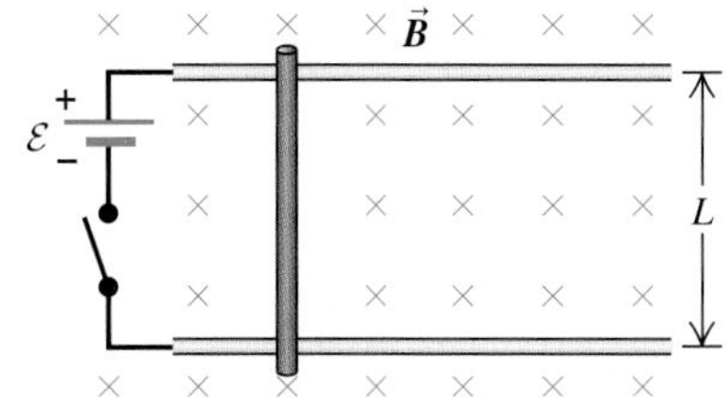

Figure 29.40 Problem 29.56.

29.57 Antenna emf. A satellite, orbiting the earth at the equator at an altitude of 400 km, has an antenna that can be modeled as a 2.0-m-long rod. The antenna is oriented perpendicular to the earth's surface. At the equator, the earth's magnetic field is essentially horizontal and has a value of 8.0×10^{-5} T; ignore any changes in B with altitude. Assuming the orbit is circular, determine the induced emf between the tips of the antenna.

29.58 EMF in a Bullet. At the equator, the earth's magnetic field is approximately horizontal, is directed toward the north, and has a value of 8×10^{-5} T. a) Estimate the emf induced between the top and bottom of a bullet shot horizontally at a target on the equator if the bullet is shot toward the east. Assume the bullet has a length of 1 cm and a diameter of 0.4 cm and is traveling at 300 m/s. Which is at higher potential, the top or bottom of the bullet? b) What is the emf if the bullet travels south? c) What is the emf induced between the front and back of the bullet for any horizontal velocity?

29.59 A very long, cylindrical wire of radius R carries a current I_0 uniformly distributed across the cross section of the wire. Calculate the magnetic flux through a rectangle that has one side of length W running down the center of the wire and another side of length R, as shown in Fig. 29.41 (see Problem 29.54).

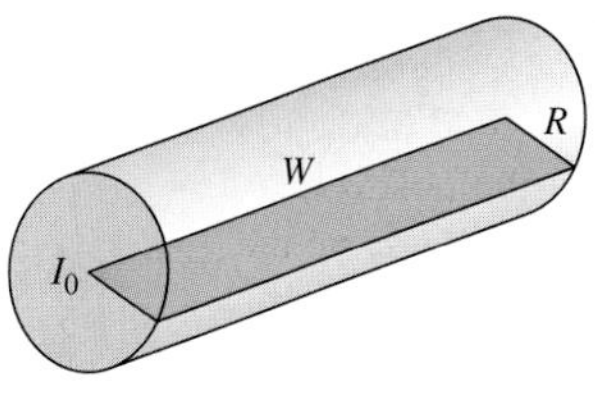

Figure 29.41 Problem 29.59.

29.60 A circular conducting ring with radius $r_0 = 0.0420$ m lies in the xy-plane in a region of uniform magnetic field $\vec{B} = B_0[1 - 3(t/t_0)^2 + 2(t/t_0)^3]\hat{k}$. In this expression, $t_0 = 0.0100$ s and is constant, t is time, $\hat{k}$ is the unit vector in the $+z$-direction, and

$B_0 = 0.0800$ T and is constant. At points a and b (see Fig. 29.42) there is a small gap in the ring with wires leading to an external circuit of resistance $R = 12.0\ \Omega$. There is no magnetic field at the location of the external circuit. a) Derive an expression, as a function of time, for the total magnetic flux Φ_B through the ring. b) Determine the emf induced in the ring at time $t = 5.00 \times 10^{-3}$ s. What is the polarity of the emf? c) Because of the internal resistance of the ring, the current through R at the time given in part (b) is only 3.00 mA. Determine the internal resistance of the ring. d) Determine the emf in the ring at a time $t = 1.21 \times 10^{-2}$ s. What is the polarity of the emf? e) Determine the time at which the current through R reverses its direction.

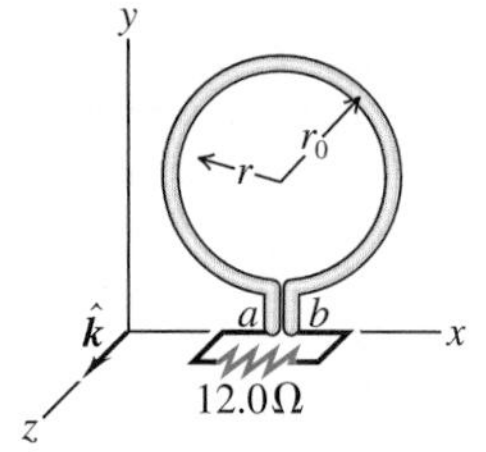

Figure 29.42 Problem 29.60.

29.61 The long, straight wire shown in Fig. 29.43a carries constant current I. A metal bar with length L is moving at constant velocity $\vec{v}$, as shown in the figure. Point a is a distance d from the wire. a) Calculate the emf induced in the bar. b) Which point, a or b, is at higher potential? c) If the bar is replaced by a rectangular wire loop of resistance R (see Fig. 29.43b) what is the magnitude of the current induced in the loop?

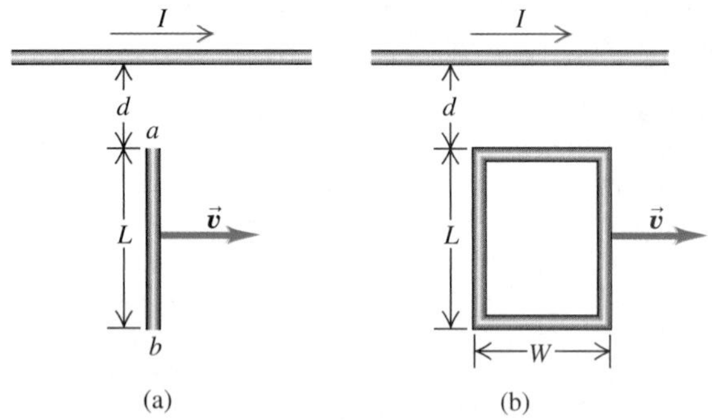

Figure 29.43 Problem 29.61.

29.62 The cube shown in Fig. 29.44, 50.0 cm on a side, is in a uniform magnetic field of 0.120 T, directed along the positive y-axis. Wires A, C, and D move in the directions indicated, each with a speed of 0.350 m/s. (Wire A moves parallel to the xy-plane, C moves at an angle of 45.0° below the xy-plane, and D moves parallel to the xz-plane.) What is the potential difference between the ends of each wire?

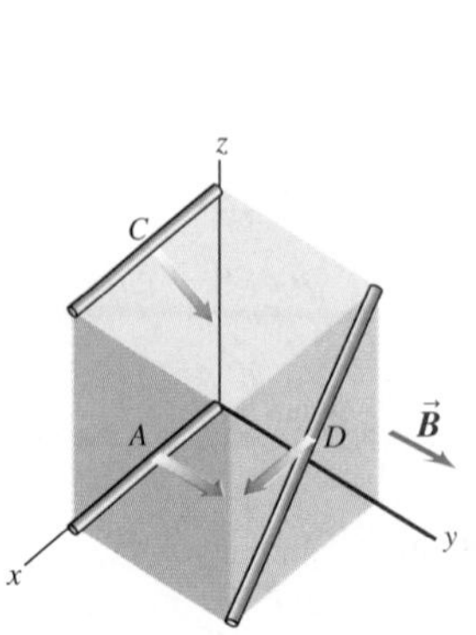

Figure 29.44 Problem 29.62.

29.63 A slender rod, 0.240 m long, rotates with an angular speed of 8.80 rad/s about an axis through one end and perpendicular to the rod. The plane of rotation of the rod is perpendicular to a uniform magnetic field with a magnitude of 0.650 T. a) What is the induced emf in the rod? b) What is the potential difference between its ends? c) Suppose instead the rod rotates at 8.80 rad/s about an axis through its center and perpendicular to the rod. In this case, what is the potential difference between the ends of the rod? Between the center of the rod and one end?

29.64 A Magnetic Exercise Machine. You have designed a new type of exercise machine with an extremely simple mechanism (see Fig. 29.31). A vertical bar of silver (chosen for its low resistivity and because it makes the machine look cool) with length $L = 3.0$ m is free to move left or right without friction on silver rails. The entire apparatus is placed in a horizontal uniform magnetic field of strength 0.25 T. When you push the bar to the left or right, the bar's motion sets up a current in the circuit that includes the bar. The resistance of the bar and the rails can be neglected. The magnetic field exerts a force on the current-carrying bar, and this force opposes the bar's motion. The health benefit is from the exercise that you do in working against this force. a) Your design goal is that the person doing the exercise is to do work at the rate of 25 watts when moving the bar at a steady 2.0 m/s. What should be the resistance R? b) You decide you want to be able to vary the power required from the person, to adapt the machine to the person's strength and fitness. If the power is to be increased to 50 W by altering R while leaving the other design parameters constant, should R be increased or decreased? Calculate the value of R for 50 W. c) When you start to construct a prototype machine, you find it is difficult to produce a 0.25-T magnetic field over such a large area. If you decrease the length of the bar to 0.20 m while leaving B, v, and R the same as in part (a), what will be the power required of the person?

29.65 A rectangular loop with width L and a slide wire with mass m are as shown in Fig. 29.45. A uniform magnetic field $\vec{B}$ is directed perpendicular to the plane of the loop into the plane of the figure. The slide wire is given an initial speed of v_0 and then released. There is no friction between the slide wire and the loop, and the resistance of the loop is negligible in comparison to the resistance R of the slide wire. a) Obtain an expression for F, the magnitude of the force exerted on the wire while it is moving at speed v. b) Show that the distance x that the wire moves before coming to rest is $x = mv_0R/a^2B^2$.

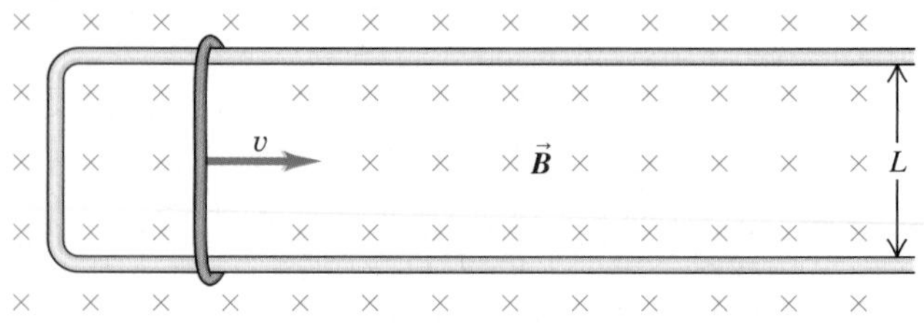

Figure 29.45 Problem 29.65.

29.66 A 25.0-cm-long metal rod lies in the xy-plane and makes an angle of 36.9° with the positive x-axis and an angle of 53.1° with the positive y-axis. The rod is moving in the $+x$-direction with a speed of 4.20 m/s. The rod is in a uniform magnetic field $\vec{B} = (0.120\ \text{T})\hat{\imath} - (0.220\ \text{T})\hat{\jmath} - (0.0900\ T)\hat{k}$. a) What is the magnitude of the emf induced in the rod? b) Indicate in a sketch which end of the rod is at higher potential.

29.67 The magnetic field $\vec{B}$, at all points within a circular region of radius R, is uniform in space and directed into the plane of the page shown in Fig. 29.46. (The region could be a cross section inside the windings of a long, straight solenoid.) If the magnetic field is increasing at a rate dB/dt, what are the magnitude and direction of the force on a stationary positive point charge q located at points a, b, and c? (Point a is a distance r above the center of the region, point b is a distance r to the right of the center, and point c is at the center of the region.)

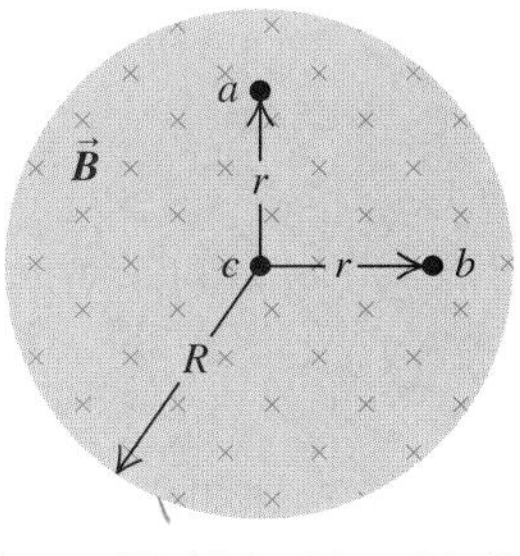

Figure 29.46 Problem 29.67.

29.68 It is impossible to have a uniform electric field that abruptly drops to zero in a region of space in which the magnetic field is constant and in which there are no electric charges. To prove this statement, use the method of contradiction: Assume that such a case *is* possible, then show that your assumption contradicts a law of nature. a) In the bottom half of a piece of paper, draw evenly, spaced horizontal lines representing a uniform electric field to your right. Use dashed lines to draw a rectangle *abcda* with horizontal side *ab* in the electric field region and horizontal side *cd* in the top half of your paper where $E = 0$. b) Show that integration around your rectangle contradicts Faraday's law, Eq. (29.21).

29.69 Falling Square Loop. A vertically oriented, square loop of copper wire falls from a region where the field $\vec{B}$ is horizontal, uniform, and perpendicular to the plane of the loop, into a region where the field is zero. The loop is released from rest and initially is entirely within the magnetic field region. Let the side length of the loop be s and let the diameter of the wire be d. The resistivity of copper is ρ_R and the density of copper is ρ_m. If the loop reaches its terminal speed while its upper segment is still in the magnetic field region, find an expression for the terminal speed.

29.70 In a region of space where there are no conduction or displacement currents, it is impossible to have a uniform magnetic field that abruptly drops to zero. To prove this statement, use the method of contradiction: Assume that such a case *is* possible, then show that your assumption contradicts a law of nature. a) In the bottom half of a piece of paper, draw evenly spaced horizontal lines representing a uniform magnetic field to your right. Use dashed lines to draw a rectangle *abcda* with horizontal side *ab* in the magnetic field region and horizontal side *cd* in the top half of your paper where $B = 0$. b) Show that integration around your rectangle contradicts Ampere's law, Eq. (29.15).

29.71 A capacitor has two parallel plates with area A separated by a distance d. The space between plates is filled with a material having dielectric constant K. The material is not a perfect insulator but has resistivity ρ. The capacitor is initially charged with charge of magnitude Q_0 on each plate, that gradually discharges by conduction through the dielectric. a) Calculate the conduction current density $j_C(t)$ in the dielectric. b) Show that at any instant the displacement current density in the dielectric is equal in magnitude to the conduction current density but opposite in direction, so the *total* current density is zero at every instant.

29.72 A rod of pure silicon (resistivity $\rho = 2300\ \Omega \cdot \text{m}$) is carrying a current. The electric field varies sinusoidally with time according to $E = E_0 \sin \omega t$, where $E_0 = 0.450$ V/m, $\omega = 2\pi f$, and the frequency $f = 120$ Hz. a) Find the magnitude of the maximum conduction current density in the wire. b) Assuming $\epsilon = \epsilon_0$, find the maximum displacement current density in the wire, and compare with the result of part (a). c) At what frequency f would the maximum conduction and displacement densities become equal if $\epsilon = \epsilon_0$ (which is not actually the case)? d) At the frequency determined in part (c), what is the relative *phase* of the conduction and displacement currents?

Challenge Problems

29.73 A square, conducting, wire loop of side L, total mass m, and total resistance R initially lies in the horizontal xy-plane, with corners at $(x, y, z) = (0, 0, 0)$, $(0, L, 0)$, $(L, 0, 0)$, and $(L, L, 0)$. There is a uniform, upward magnetic field $\vec{B} = B\hat{k}$ in the space within and around the loop. The side of the loop that extends from $(0, 0, 0)$ to $(L, 0, 0)$ is held in place on the x-axis; the rest of the loop is free to pivot around this axis. When the loop is released, it begins to rotate due to the gravitational torque. a) Find the *net* torque (magnitude and direction) that acts on the loop when it has rotated through an angle ϕ from its original orientation and is rotating downward at an angular speed ω. b) Find the angular acceleration of the loop at the instant described in part (a). c) Compared to the case with zero magnetic field, does it take the loop a longer or shorter time to rotate through 90°? Explain. d) Is mechanical energy conserved as the loop rotates downward? Explain.

29.74 A square conducting loop, 20.0 cm on a side, is placed in the same magnetic field as shown in Exercise 29.29. (See Fig. 29.47; the center of the square loop is at the center of the magnetic field region.) a) Copy Fig. 29.47, and draw vectors to show the directions and relative magnitudes of the induced electric field $\vec{E}$ at points a, b, and c. b) Prove that the component of $\vec{E}$ along the loop has the same value at every point of the loop and is equal to that of the ring shown in Fig. 29.27 (Exercise 29.29). c) What current is induced in the loop if its resistance is 1.90 Ω? d) What is the potential difference between points a and b?

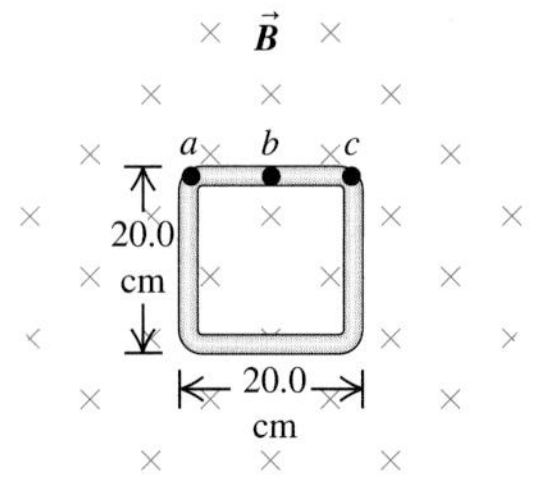

Figure 29.47 Challenge Problem 29.74.

29.75 A uniform, square conducting loop, 20.0 cm on a side, is placed in the same magnetic field as shown in Exercise 29.29, with side *ac* along a diameter and with point b at the center of the field (see Fig. 29.48). a) Copy Fig. 29.48, and draw vectors to show the direction and relative magnitude of the induced electric field $\vec{E}$ at the lettered points. b) What is the induced emf in side ac? c) What is the induced emf in the loop? d) What is the current in the loop if its resistance is 1.90 Ω? e) What is the potential difference between points a and c? Which is at higher potential?

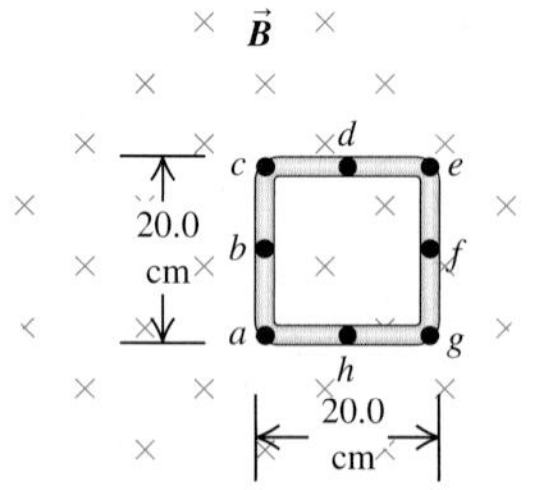

Figure 29.48 Challenge Problem 29.75.

29.76 A metal bar with length L, mass m, and resistance R is placed on frictionless, metal rails that are inclined at an angle ϕ above the horizontal. The rails have negligible resistance. A uniform magnetic field of magnitude B is directed downward as shown in Fig. 29.49. The bar is released from rest and slides down the rails. a) Is the direction of the current induced in the bar from a to b or from b to a? b) What is the terminal speed of the bar? c) What is the induced current in the bar when the terminal speed has been reached? d) After the terminal speed has been reached, at what rate is electrical energy being converted to thermal energy in the resistance of the bar? e) After the terminal speed has been reached, at what rate is work being done on the bar by gravity? Compare your answer to that in part (d).

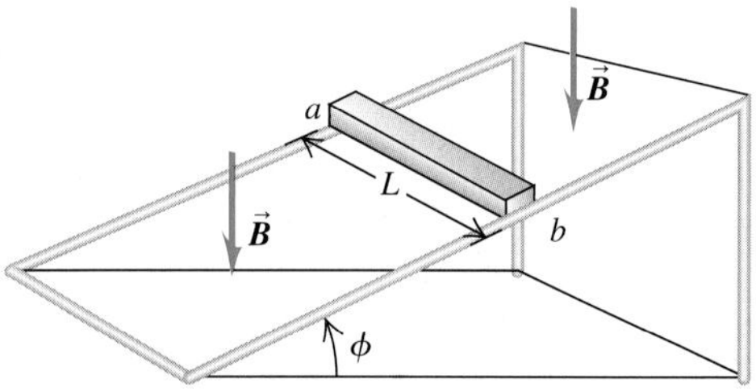

Figure 29.49 Challenge Problem 29.76.

29.77 Consider a uniform metallic disk rotating through a perpendicular magnetic field $\vec{B}$, as shown in Fig. 29.18a. The disk has mass m, radius R, and thickness t, is made of a material with resistivity ρ, and is rotating clockwise in Fig. 29.18a with angular speed ω. The magnetic field is directed into the plane of the disk. Suppose that the region to which the magnetic field is confined is not circular, as shown in Fig. 29.18a, but is a small square with sides of length L ($L \ll R$) centered a distance d from the point O (the center of the disk). The sides of this square are horizontal and vertical in Fig. 29.18a. a) Show that the current induced within the square is approximately equal to $I = \omega dBLt/\rho$. In which direction does this current flow? (*Hint:* Assume that the resistance to the current is confined to the region of the square. The current also encounters resistance as it flows outside the region to which the magnetic field is confined, as shown in Fig. 29.18b; however, this resistance is relatively small, since the current can flow through such a wide area. Recall Eq. (25.10) for resistance, given in Section 25.3.) b) Show that the induced current gives rise to a torque of approximate magnitude $\tau = \omega d^2 B^2 L^2 t/\rho$ that opposes the rotation of the disk (that is, a counterclockwise torque). c) What would be the magnitudes and directions of the induced current and torque if the direction of $\vec{B}$ were still into the plane of the disk but the disk rotated counterclockwise? What if the direction of $\vec{B}$ were out of the plane and the disk rotated counterclockwise?

INDUCTANCE 30

This electric car is being recharged by a 220-V line—but there is no physical contact between the recharging cable and the car's wires. Instead, mutual inductance is used: The alternating current in the recharging cable creates a time-varying magnetic field, which induces a current in the car's circuitry. This induced current recharges the car's batteries.

? **If there were a steady direct current in the recharging cable, but still no direct contact between the cable and the car's wires, would the cable recharge the electric vehicle's batteries?**

Take a length of copper wire and wrap it around a pencil to form a coil. If you put this coil in a circuit, does it behave any differently than a straight piece of wire? Remarkably, the answer is yes. In an ordinary gasoline-powered car, a coil of this kind makes it possible for the 12-volt car battery to provide thousands of volts to the spark plugs, which in turn makes it possible for the plugs to fire and make the engine run. Other coils of this type are used to keep fluorescent light fixtures shining. Larger coils placed under city streets are used to control the operation of traffic signals. All of these applications, and many others, involve the *induction* effects that we studied in Chapter 29.

A changing current in a coil induces an emf in an adjacent coil. The coupling between the coils is described by their *mutual inductance.* A changing current in a coil also induces an emf in that same coil. Such a coil is called an *inductor,* and the relationship of current to emf is described by the *inductance* (also called *self-inductance*) of the coil. If a coil is initially carrying a current, energy is released when the current decreases; this principle is used in automotive ignition systems. We'll find that this released energy was stored in the magnetic field caused by the current that was initially in the coil, and we'll look at some of the practical applications of magnetic-field energy.

We'll also take a first look at what happens when an inductor is part of a circuit. In Chapter 31 we'll go on to study how inductors behave in alternating-

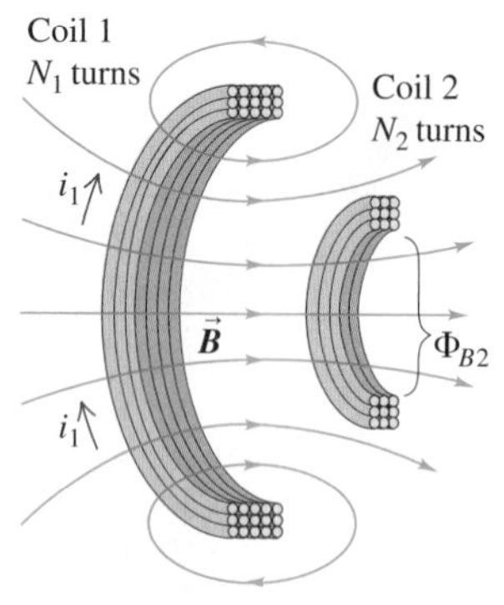

30.1 A current i_1 in coil 1 gives rise to a magnetic flux through coil 2. If i_1 changes, an emf is induced in coil 2; this is described in terms of mutual inductance.

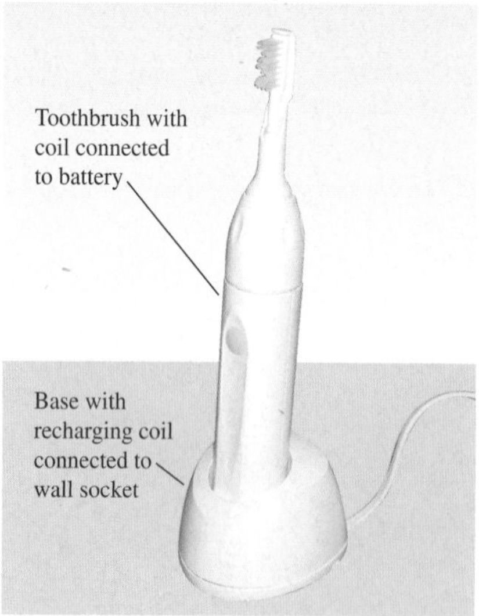

30.2 This electric toothbrush makes use of mutual inductance. The base contains a coil that is supplied with alternating current from a wall socket. This varying current induces an emf in a coil within the toothbrush itself, which is used to recharge the toothbrush battery. (A similar system is used to recharge the electric vehicle shown in the photograph that opens this chapter.)

current circuits; in that chapter we'll learn why inductors play an essential role in modern electronics, including communication systems, power supplies, and many other devices.

30.1 | Mutual Inductance

In Section 28.4 we considered the magnetic interaction between two wires carrying *steady* currents; the current in one wire causes a magnetic field, which exerts a force on the current in the second wire. But an additional interaction arises between two circuits when there is a *changing* current in one of the circuits. Consider two neighboring coils of wire, as in Fig. 30.1. A current flowing in coil 1 produces a magnetic field $\vec{B}$ and hence a magnetic flux through coil 2. If the current in coil 1 changes, the flux through coil 2 changes as well; according to Faraday's law, this induces an emf in coil 2. In this way, a change in the current in one circuit can induce a current in a second circuit.

Let's analyze the situation shown in Fig. 30.1 in more detail. We will use lowercase letters to represent quantities that vary with time; for example, a time-varying current is i, often with a subscript to identify the circuit. In Fig. 30.1 a current i_1 in coil 1 sets up a magnetic field (as indicated by the blue lines), and some of these field lines pass through coil 2. We denote the magnetic flux through *each* turn of coil 2, caused by the current i_1 in coil 1, as Φ_{B2}. (If the flux is different through different turns of the coil, then Φ_{B2} denotes the *average* flux.) The magnetic field is proportional to i_1, so Φ_{B2} is also proportional to i_1. When i_1 changes, Φ_{B2} changes; this changing flux induces an emf $\mathcal{E}_2$ in coil 2, given by

$$\mathcal{E}_2 = -N_2 \frac{d\Phi_{B2}}{dt} \tag{30.1}$$

We could represent the proportionality of Φ_{B2} and i_1 in the form $\Phi_{B2} = (\text{constant})i_1$, but instead it is more convenient to include the number of turns N_2 in the relation. Introducing a proportionality constant M_{21}, called the **mutual inductance** of the two coils, we write

$$N_2 \Phi_{B2} = M_{21} i_1 \tag{30.2}$$

where Φ_{B2} is the flux through a *single* turn of coil 2. From this,

$$N_2 \frac{d\Phi_{B2}}{dt} = M_{21} \frac{di_1}{dt}$$

and we can rewrite Eq. (30.1) as

$$\mathcal{E}_2 = -M_{21} \frac{di_1}{dt} \tag{30.3}$$

That is, a change in the current i_1 in coil 1 induces an emf in coil 2 that is directly proportional to the rate of change of i_1 (Fig. 30.2).

We may also write the definition of mutual inductance, Eq. (30.2), as

$$M_{21} = \frac{N_2 \Phi_{B2}}{i_1}$$

If the coils are in vacuum, the flux Φ_{B2} through each turn of coil 2 is directly proportional to the current i_1. Then the mutual inductance M_{21} is a constant that depends only on the geometry of the two coils (the size, shape, number of turns, and orientation of each coil and the separation between the coils). If a magnetic material is present, M_{21} also depends on the magnetic properties of the material. If the material has nonlinear magnetic properties, that is, if the relative permeability K_m (defined in Section 28.8) is not constant and magnetization is not proportional to magnetic field, then Φ_{B2} is no longer directly proportional to i_1. In that case the mutual inductance also depends on the value of i_1. In this discussion we will assume that any magnetic material present has constant K_m so that flux *is* directly proportional to current and M_{21} depends on geometry only.

We can repeat our discussion for the opposite case in which a changing current i_2 in coil 2 causes a changing flux Φ_{B1} and an emf $\mathcal{E}_1$ in coil 1. We might expect that the corresponding constant M_{12} would be different from M_{21} because in general the two coils are not identical and the flux through them is not the same. It turns out, however, that M_{12} is *always* equal to M_{21}, even when the two coils are not symmetric. We call this common value simply the mutual inductance, denoted by the symbol M without subscripts; it characterizes completely the induced-emf interaction of two coils. Then we can write

$$\mathcal{E}_2 = -M\frac{di_1}{dt} \quad \text{and} \quad \mathcal{E}_1 = -M\frac{di_2}{dt} \qquad \text{(mutually induced emfs)} \qquad (30.4)$$

where the mutual inductance M is

$$M = \frac{N_2\Phi_{B2}}{i_1} = \frac{N_1\Phi_{B1}}{i_2} \qquad \text{(mutual inductance)} \qquad (30.5)$$

The negative signs in Eq. (30.4) are a reflection of Lenz's law. The first equation says that a change in current in coil 1 causes a change in flux through coil 2, inducing an emf in coil 2 that opposes the flux change; in the second equation the roles of the two coils are interchanged.

CAUTION Note that only a *time-varying* current in a coil can induce an emf and hence a current in a second coil. Equations (30.4) show that the induced emf in each coil is directly proportional to the *rate of change* of the current in the other coil, not to the value of the current. A steady current in one coil, no matter how strong, cannot induce a current in a neighboring coil.

The SI unit of mutual inductance is called the **henry** (1 H), in honor of the American physicist Joseph Henry (1797–1878), one of the discoverers of electromagnetic induction. From Eq. (30.5), one henry is equal to *one weber per ampere.* Other equivalent units, obtained by using Eq. (30.4), are *one volt-second per ampere, one ohm-second,* or *one joule per ampere squared:*

$$1\ \text{H} = 1\ \text{Wb/A} = 1\ \text{V}\cdot\text{s/A} = 1\ \Omega\cdot\text{s} = 1\ \text{J/A}^2$$

Just as the farad is a rather large unit of capacitance (see Section 24.1), the henry is a rather large unit of mutual inductance. As Example 30.1 shows, typical values of mutual inductance can be in the millihenry (mH) or microhenry (μH) range.

Mutual inductance can be a nuisance in electric circuits, since variations in current in one circuit can induce unwanted emfs in other, nearby circuits. To minimize these effects, multiple-circuit systems must be designed so that M is as small as possible; for example, two coils would be placed far apart or with their planes perpendicular.

Happily, mutual inductance also has many useful applications. A *transformer,* used in alternating-current circuits to raise or lower voltages, is fundamentally no different from the two coils shown in Fig. 30.1. A time-varying alternating current in one coil of the transformer produces an alternating emf in the other coil; the value of M, which depends on the geometry of the coils, determines the amplitude of the induced emf in the second coil and hence the amplitude of the output voltage. (We'll describe transformers in more detail in Chapter 31 after we've discussed alternating current in greater depth.)

Example 30.1 Calculating mutual inductance

In one form of Tesla coil (a high-voltage generator that you may have seen in a science museum), a long solenoid with length l and cross-sectional area A is closely wound with N_1 turns of wire. A coil with N_2 turns surrounds it at its center (Fig. 30.3). Find the mutual inductance.

SOLUTION

IDENTIFY: Mutual inductance occurs in this situation because a current in one of the coils sets up a magnetic field that causes a flux through the other coil.

SET UP: We use Eq. (30.5) to determine the mutual inductance M. According to that equation, we need to know either (a) the flux Φ_{B2} through each turn of the outer coil due to a current i_1 in the solenoid or (b) the flux Φ_{B1} through each turn of the solenoid due to a current i_2 in the outer coil. We choose option (a) since from Example 28.10 (Section 28.7) we have a simple expression for the field at the center of a long current-carrying solenoid, given by Eq. (28.23). Note that we are not given a value for the current i_1 in the solenoid. This omission is not cause for alarm, however: the value of the mutual inductance doesn't depend on the value of the current, so the quantity i_1 should cancel out when we calculate M.

EXECUTE: From Example 28.10, a long solenoid carrying current i_1 produces a magnetic field $\vec{B}_1$ that points along the axis of the solenoid. The field magnitude B_1 is proportional to i_1 and to n_1, the number of turns per unit length:

$$B_1 = \mu_0 n_1 i_1 = \frac{\mu_0 N_1 i_1}{l}$$

The flux through a cross section of the solenoid equals $B_1 A$. Since a very long solenoid produces no magnetic field outside of its coil, this is also equal to the flux Φ_{B2} through each turn of the outer, surrounding coil, no matter what the cross-sectional area of the outer coil. From Eq. (30.5) the mutual inductance M is

$$M = \frac{N_2 \Phi_{B2}}{i_1} = \frac{N_2 B_1 A}{i_1} = \frac{N_2}{i_1}\frac{\mu_0 N_1 i_1}{l}A = \frac{\mu_0 A N_1 N_2}{l}$$

EVALUATE: The mutual inductance of any two coils is always proportional to the product $N_1 N_2$ of their numbers of turns. Notice that the mutual inductance M depends only on the geometry of the two coils, not on the current.

Here's a numerical example to give you an idea of magnitudes. Suppose $l = 0.50$ m, $A = 10\ \text{cm}^2 = 1.0 \times 10^{-3}\ \text{m}^2$, $N_1 = 1000$ turns, and $N_2 = 10$ turns. Then

$$M = \frac{(4\pi \times 10^{-7}\ \text{Wb/A}\cdot\text{m})(1.0 \times 10^{-3}\ \text{m}^2)(1000)(10)}{0.50\ \text{m}}$$
$$= 25 \times 10^{-6}\ \text{Wb/A} = 25 \times 10^{-6}\ \text{H} = 25\ \mu\text{H}$$

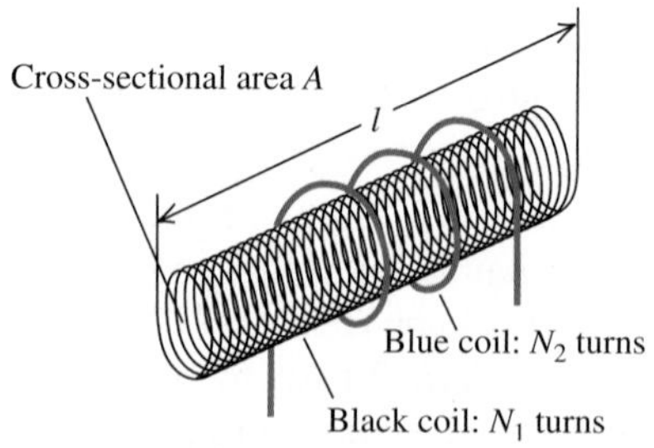

30.3 A long solenoid with cross-sectional area A and N_1 turns (shown in black) is surrounded at its center by a coil with N_2 turns (shown in blue).

Example 30.2 Emf due to mutual inductance

In Example 30.1, suppose the current i_2 in the outer, surrounding coil is given by $i_2 = (2.0 \times 10^6\ \text{A/s})t$ (currents in wires can indeed increase this rapidly for brief periods). a) At time $t = 3.0\ \mu\text{s}$, what average magnetic flux through each turn of the solenoid is caused by the current in the outer, surrounding coil? b) What is the induced emf in the solenoid?

SOLUTION

IDENTIFY: In Example 30.1 we found the mutual inductance by relating the current in the solenoid to the flux produced in the outer coil. In this example we are given the current in the outer coil and want to find the resulting flux in the solenoid. The key point is that the mutual inductance is the *same* in either case.

SET UP: Given the value of the mutual inductance $M = 25\ \mu\text{H}$ from Example 30.1, we use Eq. (30.5) to determine the flux Φ_{B1} through each turn of the solenoid caused by a given current i_2 in the outer coil. We then use Eq. (30.4) to determine the emf induced in the solenoid by the time variation of the outer coil's current.

EXECUTE: a) At time $t = 3.0\ \mu\text{s} = 3.0 \times 10^{-6}$ s, the current in the outer coil (coil 2) is $i_2 = (2.0 \times 10^6\ \text{A/s})(3.0 \times 10^{-6}\ \text{s}) = 6.0$ A. To find the average flux through each turn of the solenoid (coil 1), we solve Eq. (30.5) for Φ_{B1}:

$$\Phi_{B1} = \frac{Mi_2}{N_1} = \frac{(25 \times 10^{-6}\ \text{H})(6.0\ \text{A})}{1000} = 1.5 \times 10^{-7}\ \text{Wb}$$

Note that this is an *average* value; the flux can vary considerably between the center and the ends of the solenoid.

b) The induced emf $\mathcal{E}_1$ is given by Eq. (30.4):

$$\mathcal{E}_1 = -M\frac{di_2}{dt} = -(25 \times 10^{-6}\ \text{H})\frac{d}{dt}[(2.0 \times 10^6\ \text{A/s})t]$$

$$= -(25 \times 10^{-6}\ \text{H})(2.0 \times 10^6\ \text{A/s}) = -50\ \text{V}$$

EVALUATE: This is a substantial induced emf in response to a very rapid rate of change of current. In an operating Tesla coil, there is a high-frequency alternating current rather than a continuously increasing current as in this example; both di_2/dt and $\mathcal{E}_1$ alternate as well, with amplitudes that can be thousands of times larger than in this example.

Test Your Understanding

Consider the Tesla coil described in Example 30.1. If you make the solenoid out of twice as much wire, so that it has twice as many turns and is twice as long, how much larger is the mutual inductance? Using this double-size solenoid, how many turns would the outer coil need for the mutual inductance to be 1.0 H instead of 25 μH?

30.2 | Self-Inductance and Inductors

In our discussion of mutual inductance we considered two separate, independent circuits: a current in one circuit creates a magnetic field and this field gives rise to a flux through the second circuit. If the current in the first circuit changes, the flux through the second circuit changes and an emf is induced in the second circuit.

An important related effect occurs even if we consider only a *single* isolated circuit. When a current is present in a circuit, it sets up a magnetic field that causes a magnetic flux through the *same* circuit; this flux changes when the current changes. Thus any circuit that carries a varying current has an emf induced in it by the variation in *its own* magnetic field. Such an emf is called a **self-induced emf.** By Lenz's law, a self-induced emf always opposes the change in the current that caused the emf and so tends to make it more difficult for variations in current to occur. For this reason, self-induced emfs can be of great importance whenever there is a varying current.

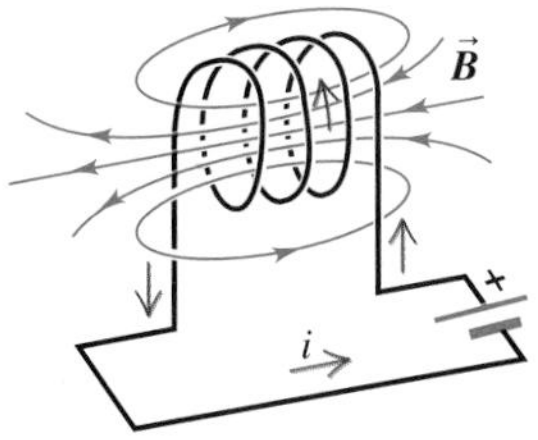

30.4 The current i in the circuit causes a magnetic field $\vec{B}$ in the coil and hence a flux through the coil. When the current i changes, the flux changes also and a self-induced emf appears in the circuit.

Self-induced emfs can occur in *any* circuit, since there is always some magnetic flux through the closed loop of a current-carrying circuit. But the effect is greatly enhanced if the circuit includes a coil with N turns of wire (Fig. 30.4). As a result of the current i, there is an average magnetic flux Φ_B through each turn of the coil. In analogy to Eq. (30.5) we define the **self-inductance** L of the circuit as

$$L = \frac{N\Phi_B}{i} \qquad \text{(self-inductance)} \qquad (30.6)$$

When there is no danger of confusion with mutual inductance, the self-inductance is called simply the **inductance.** Comparing Eqs. (30.5) and (30.6), we see that the units of self-inductance are the same as those of mutual inductance; the SI unit of self-inductance is one henry.

If the current i in the circuit changes, so does the flux Φ_B; from rearranging Eq. (30.6) and taking the derivative with respect to time, the rates of change are related by

$$N\frac{d\Phi_B}{dt} = L\frac{di}{dt}$$

From Faraday's law for a coil with N turns, Eq. (29.4), the self-induced emf is $\mathcal{E} = -N\, d\Phi_B/dt$, so it follows that

$$\mathcal{E} = -L\frac{di}{dt} \qquad \text{(self-induced emf)} \qquad (30.7)$$

The minus sign in Eq. (30.7) is a reflection of Lenz's law; it says that the self-induced emf in a circuit opposes any change in the current in that circuit. (Later in this section we'll explore in greater depth the significance of this minus sign.)

Equation (30.7) also states that the self-inductance of a circuit is the magnitude of the self-induced emf per unit rate of change of current. This relationship makes it possible to measure an unknown self-inductance in a relatively simple way: Change the current in the circuit at a known rate di/dt, measure the induced emf, and take the ratio to determine L.

A circuit device that is designed to have a particular inductance is called an **inductor,** or a *choke*. The usual circuit symbol for an inductor is

Like resistors and capacitors, inductors are among the indispensable circuit elements of modern electronics. Their purpose is to oppose any variations in the current through the circuit. An inductor in a direct-current circuit helps to maintain a steady current despite any fluctuations in the applied emf; in an alternating-current circuit, an inductor tends to suppress variations of the current that are more rapid than desired. In this chapter and the next we will explore the behavior and applications of inductors in circuits in more detail.

To understand the behavior of circuits containing inductors, we need to develop a general principle analogous to Kirchhoff's loop rule (discussed in Section 26.2). To apply that rule, we go around a conducting loop, measuring potential differences across successive circuit elements as we go. The algebraic sum of these differences around any closed loop must be zero because the electric field produced by charges distributed around the circuit is *conservative*. In Section 29.7 we denoted such a conservative field as $\vec{E}_c$.

When an inductor is included in the circuit, the situation changes. The magnetically induced electric field within the coils of the inductor is *not* conservative; as in Section 29.7, we'll denote it by $\vec{E}_n$. We need to think very carefully about the roles of the various fields. Let's assume we are dealing with an inductor whose coils have negligible resistance. Then a negligibly small electric field is required to make charge move through the coils, so the *total* electric field $\vec{E}_c + \vec{E}_n$ within the coils must be zero, even though neither field is individually zero. Because $\vec{E}_c$ is nonzero, we know there have to be accumulations of charge on the terminals of the inductor and the surfaces of its conductors, to produce this field.

Consider the circuit shown in Fig. 30.5; the box contains some combination of batteries and variable resistors that enables us to control the current i in the circuit. According to Faraday's law, Eq. (29.10), the line integral of $\vec{E}_n$ around the circuit is the negative of the rate of change of flux through the circuit, which in turn is given by Eq. (30.7). Combining these two relations, we get

$$\oint \vec{E}_n \cdot d\vec{l} = -L\frac{di}{dt}$$

where we integrate clockwise around the loop (the direction of the assumed current). But $\vec{E}_n$ is different from zero only within the inductor. Therefore the integral of $\vec{E}_n$ around the whole loop can be replaced by its integral only from a to b through the inductor; that is,

$$\int_a^b \vec{E}_n \cdot d\vec{l} = -L\frac{di}{dt}$$

Next, because $\vec{E}_c + \vec{E}_n = 0$ at each point within the inductor coils, we can rewrite this as

$$\int_a^b \vec{E}_c \cdot d\vec{l} = L\frac{di}{dt}$$

But this integral is just the potential V_{ab} of point a with respect to point b, so we finally obtain

$$V_{ab} = V_a - V_b = L\frac{di}{dt} \tag{30.8}$$

We conclude that there is a genuine potential difference between the terminals of the inductor, associated with conservative, electrostatic forces, despite the fact that the electric field associated with the magnetic induction effect is nonconservative. Thus we are justified in using Kirchhoff's loop rule to analyze circuits that include inductors. Equation (30.8) gives the potential difference across an inductor in a circuit.

CAUTION Note that the self-induced emf does not oppose the current i itself; rather, it opposes any *change* (di/dt) in the current. Thus the circuit behavior of an inductor is quite different from that of a resistor. Figure 30.6 compares the behavior of a resistor and an inductor and summarizes the sign relations.

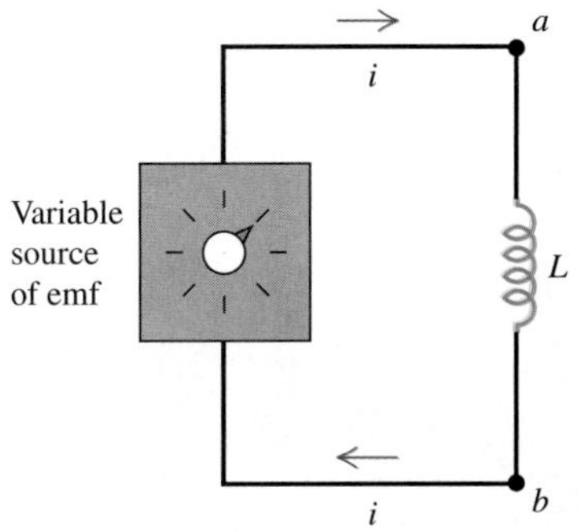

30.5 A circuit containing a source of emf and an inductor. The source is variable, so the current i and its rate of change di/dt can be varied.

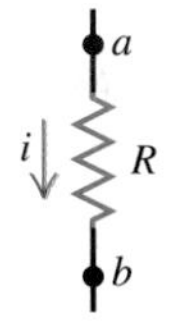

$V_{ab} = iR$

(a) Resistor with current i flowing from a to b: potential drops from a to b

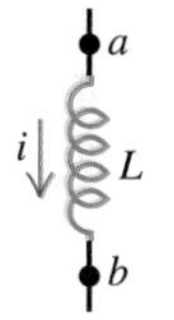

$V_{ab} = L\frac{di}{dt}$

(b) Inductor with current i flowing from a to b:
- If $di/dt > 0$: potential drops from a to b
- If $di/dt < 0$: potential increases from a to b
- If i is constant ($di/dt = 0$): no potential difference

30.6 (a) The potential difference across a resistor depends on the current. (b) The potential difference across an inductor depends on the rate of change of the current.

Because an inductor opposes changes in current, it plays an important role in fluorescent light fixtures (Fig. 30.7). In such fixtures, current flows from the wiring into the gas that fills the tube, ionizing the gas and causing it to glow. However, an ionized gas or *plasma* is a highly nonohmic conductor: the greater the

30.7 These fluorescent light tubes are wired in series with an inductor, or ballast, that helps to sustain the current flowing through the tubes.

current, the more highly ionized the plasma becomes and the lower its resistance. If a sufficiently large voltage is applied to the plasma, the current can grow enough that it damages the circuitry outside the fluorescent tube. To prevent this problem, an inductor or *magnetic ballast* is put in series with the fluorescent tube to keep the current from growing out of bounds.

The ballast also makes it possible for the fluorescent tube to work with the alternating voltage provided by household wiring. This voltage oscillates sinusoidally with a frequency of 60 Hz, so that it goes momentarily to zero 120 times per second. If there were no ballast, the plasma in the fluorescent tube would rapidly deionize when the voltage went to zero and the tube would shut off. With a ballast present, a self-induced emf sustains the current and keeps the tube lit. Magnetic ballasts are also used for this purpose in streetlights (which obtain their light from a glowing vapor of mercury or sodium atoms) and in neon lights. (In compact fluorescent lamps, the magnetic ballast is replaced by a more complicated scheme for regulating current. This scheme utilizes transistors, discussed in Chapter 42.)

The self-inductance of a circuit depends on its size, shape, and number of turns. For N turns close together, it is always proportional to N^2. It also depends on the magnetic properties of the material enclosed by the circuit. In the following examples we will assume that the circuit encloses only vacuum (or air, which from the standpoint of magnetism is essentially a vacuum). If, however, the flux is concentrated in a region containing a magnetic material with permeability μ, then in the expression for B we must replace μ_0 (the permeability of vacuum) by $\mu = K_m\mu_0$, as discussed in Section 28.8. If the material is diamagnetic or paramagnetic, this replacement makes very little difference since K_m is very close to 1. If the material is *ferromagnetic,* however, the difference is of crucial importance. A solenoid wound on a soft iron core having $K_m = 5000$ can have an inductance approximately 5000 times as great as that of the same solenoid with an air core. Ferromagnetic-core inductors are very widely used in a variety of electronic and electric-power applications.

An added complication is that with ferromagnetic materials the magnetization is in general not a linear function of magnetizing current, especially as saturation is approached. As a result, the inductance is not constant but can depend on current in a fairly complicated way. In our discussion we will ignore this complication and assume always that the inductance is constant. This is a reasonable assumption even for a ferromagnetic material if the magnetization remains well below the saturation level.

Because automobiles contain steel, a ferromagnetic material, driving an automobile over a coil causes an appreciable increase in the coil's inductance. This effect is used in traffic light sensors, which use a large, current-carrying coil embedded under the road surface near an intersection. The circuitry connected to the coil detects the inductance change as a car drives over. When a preprogrammed number of cars have passed over the coil, the light changes to green to allow the cars through the intersection.

Example 30.3 Calculating self-inductance

A toroidal solenoid with cross-sectional area A and mean radius r is closely wound with N turns of wire (Fig. 30.8). The toroid is wound on a nonmagnetic core. Determine its self-inductance L. Assume that B is uniform across a cross section (that is, neglect the variation of B with distance from the toroid axis).

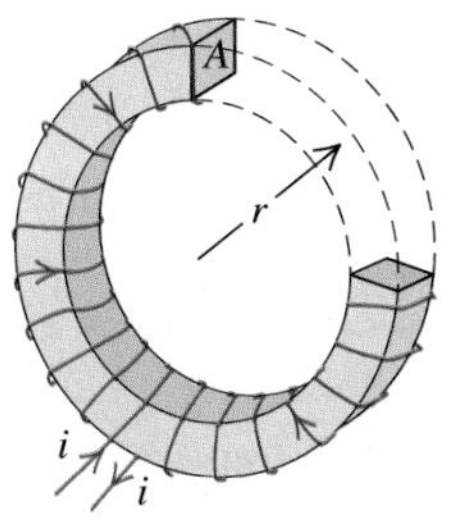

30.8 Determining the self-inductance of a closely wound toroidal solenoid. For clarity, only a few turns of the winding are shown. Part of the toroid has been cut away to show the cross-sectional area A and radius r.

SOLUTION

IDENTIFY and SET UP: The self-inductance of a coil can be found in one of two ways: either with Eq. (30.6), which requires knowing the flux Φ_B through each turn and the current i in the coil, or from Eq. (30.7), which requires knowing the self-induced emf $\mathcal{E}$ due to a given rate of change of current di/dt. We are not given any information about the emf, so we must use the first approach. We use the results of Example 28.11 (Section 28.7), in which we found the magnetic field in the interior of a toroidal solenoid.

EXECUTE: From Eq. (30.6), the self-inductance is $L = N\Phi_B/i$. From Example 28.11, the field magnitude at a distance r from the toroid axis is $B = \mu_0 Ni/2\pi r$. If we assume that the field has this magnitude over the entire cross-sectional area A, then the magnetic flux through the cross section is

$$\Phi_B = BA = \frac{\mu_0 NiA}{2\pi r}$$

The flux Φ_B is the same through each turn, and the self-inductance L is

$$L = \frac{N\Phi_B}{i} = \frac{\mu_0 N^2 A}{2\pi r}$$

(self-inductance of a toroidal solenoid)

EVALUATE: Suppose $N = 200$ turns, $A = 5.0\ \text{cm}^2 = 5.0 \times 10^{-4}\ \text{m}^2$, and $r = 0.10$ m; then

$$L = \frac{(4\pi \times 10^{-7}\ \text{Wb/A}\cdot\text{m})(200)^2(5.0 \times 10^{-4}\ \text{m}^2)}{2\pi(0.10\ \text{m})}$$
$$= 40 \times 10^{-6}\ \text{H} = 40\ \mu\text{H}$$

Later in this chapter we will use the expression $L = \mu_0 N^2 A/2\pi r$ for the inductance of a toroidal solenoid to help develop an expression for the energy stored in a magnetic field.

Example 30.4 Calculating self-induced emf

If the current in the toroidal solenoid in Example 30.3 increases uniformly from zero to 6.0 A in 3.0 μs, find the magnitude and direction of the self-induced emf.

SOLUTION

IDENTIFY and SET UP: Given the self-inductance L and the rate of change of the current di/dt, we calculate the emf using Eq. (30.7).

EXECUTE: The rate of change of the solenoid current is $di/dt = (6.0\ \text{A})/(3.0 \times 10^{-6}\ \text{s}) = 2.0 \times 10^6\ \text{A/s}$. From Eq. (30.7), the magnitude of the induced emf is

$$|\mathcal{E}| = L\left|\frac{di}{dt}\right| = (40 \times 10^{-6}\ \text{H})(2.0 \times 10^6\ \text{A/s}) = 80\ \text{V}$$

The current is increasing, so according to Lenz's law the direction of the emf is opposite to that of the current. This corresponds to the situation in Fig. 30.6b; the emf is in the direction from b to a, like a battery with a as the $(+)$ terminal and b the $(-)$ terminal, tending to oppose the current increase from the external circuit.

EVALUATE: This example shows that even a small inductance L can give rise to a substantial induced emf if the current changes rapidly.

Test Your Understanding

If the current in the toroidal solenoid of Example 30.3 decreases uniformly from 6.0 A to zero in 3.0 μs, how do the magnitude and direction of the induced emf compare to the results in Example 30.4?

30.3 | Magnetic-Field Energy

Establishing a current in an inductor requires an input of energy, and an inductor carrying a current has energy stored in it. Let's see how this comes about. In Fig. 30.5, an increasing current i in the inductor causes an emf $\mathcal{E}$ between its terminals, and a corresponding potential difference V_{ab} between the terminals of the source, with point a at higher potential than point b. Thus the source must be adding energy to the inductor, and the instantaneous power P (rate of transfer of energy into the inductor) is $P = V_{ab}i$.

We can calculate the total energy input U needed to establish a final current I in an inductor with inductance L if the initial current is zero. We assume that the inductor has zero resistance, so no energy is dissipated within the inductor. Let the current at some instant be i and let its rate of change be di/dt; the current is increasing, so $di/dt > 0$. The voltage between the terminals a and b of the inductor at this instant is $V_{ab} = L\,di/dt$, and the rate P at which energy is being delivered to the inductor (equal to the instantaneous power supplied by the external source) is

$$P = V_{ab}i = Li\frac{di}{dt}$$

The energy dU supplied to the inductor during an infinitesimal time interval dt is $dU = P\,dt$, so

$$dU = Li\,di$$

The total energy U supplied while the current increases from zero to a final value I is

$$U = L\int_0^I i\,di = \frac{1}{2}LI^2 \qquad \text{(energy stored in an inductor)} \qquad (30.9)$$

After the current has reached its final steady value I, $di/dt = 0$ and no more energy is input to the inductor. When there is no current, the stored energy U is zero; when the current is I, the energy is $\frac{1}{2}LI^2$.

When the current decreases from I to zero the inductor acts as a source that supplies a total amount of energy $\frac{1}{2}LI^2$ to the external circuit. If we interrupt the circuit suddenly by opening a switch or yanking a plug from a wall socket, the current decreases very rapidly, the induced emf is very large, and the energy may be dissipated in an arc across the switch contacts. This large emf is the electrical analog of the large force exerted by a car running into a brick wall and stopping very suddenly.

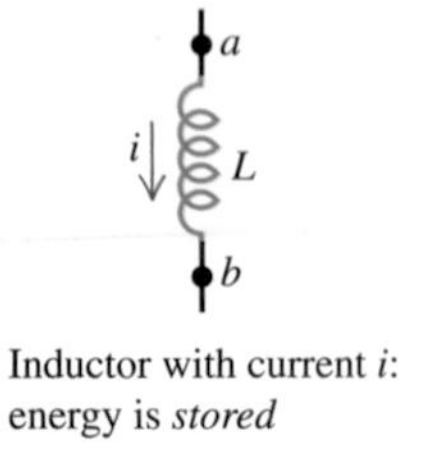

30.9 A resistor is a device in which energy is irrecoverably dissipated. By contrast, energy stored in a current-carrying inductor can be recovered when the current decreases to zero.

CAUTION It's important not to confuse the behavior of resistors and inductors where energy is concerned (Fig. 30.9). Energy flows into a resistor whenever a current passes through it, whether the current is steady or varying; this energy is dissipated in the form of heat. By contrast, energy flows into an ideal, zero-resistance inductor only when the current in the inductor *increases.* This energy is not dissipated; it is stored in the inductor and released when the current *decreases.* When a steady current flows through an inductor, there is no energy flow in or out.

The energy in an inductor is actually stored in the magnetic field within the coil, just as the energy of a capacitor is stored in the electric field between its plates. We can develop relations for magnetic-field energy analogous to those we obtained for electric-field energy in Section 24.3 [Eqs. (24.9) and (24.11)]. We will concentrate on one simple case, the ideal toroidal solenoid. This system has the advantage that its magnetic field is confined completely to a finite region of space within its core. As in Example 30.3, we assume that the cross-sectional area A is small enough that we can pretend that the magnetic field is uniform over the area. The volume V enclosed by the toroidal solenoid is approximately equal to the circumference $2\pi r$ multiplied by the area A: $V = 2\pi rA$. From Eq. (30.8) in Example 30.3, the self-inductance of the toroidal solenoid with vacuum within its coils is

$$L = \frac{\mu_0 N^2 A}{2\pi r}$$

From Eq. (30.9), the energy U stored in the toroidal solenoid when the current is I is

$$U = \frac{1}{2}LI^2 = \frac{1}{2}\frac{\mu_0 N^2 A}{2\pi r}I^2$$

The magnetic field and therefore this energy are localized in the volume $V = 2\pi rA$ enclosed by the windings. The energy *per unit volume,* or *magnetic energy density,* is $u = U/V$:

$$u = \frac{U}{2\pi rA} = \frac{1}{2}\mu_0 \frac{N^2 I^2}{(2\pi r)^2}$$

We can express this in terms of the magnitude B of the magnetic field inside the toroidal solenoid. From Eq. (28.24) in Example 28.11 (Section 28.7), this is

$$B = \frac{\mu_0 NI}{2\pi r}$$

and so

$$\frac{N^2 I^2}{(2\pi r)^2} = \frac{B^2}{\mu_0^2}$$

When we substitute this into the above equation for u, we finally find the expression for **magnetic energy density** in a vacuum:

$$u = \frac{B^2}{2\mu_0} \quad \text{(magnetic energy density in a vacuum)} \qquad (30.10)$$

This is the magnetic analog of the energy per unit volume in an *electric* field in a vacuum, $u = \frac{1}{2}\epsilon_0 E^2$, which we derived in Section 24.3.

When the material inside the toroid is not vacuum but a material with (constant) magnetic permeability $\mu = K_m \mu_0$, we replace μ_0 by μ in Eq. (30.10). The energy per unit volume in the magnetic field is then

$$u = \frac{B^2}{2\mu} \quad \text{(magnetic energy density in a material)} \qquad (30.11)$$

Although we have derived Eq. (30.11) only for one special situation, it turns out to be the correct expression for the energy per unit volume associated with *any* magnetic-field configuration in a material with constant permeability. For vac-

30.10 The energy required to fire an automobile spark plug is derived from magnetic-field energy stored in the ignition coil.

uum, Eq. (30.11) reduces to Eq. (30.10). We will use the expressions for electric-field and magnetic-field energy in Chapter 32 when we study the energy associated with electromagnetic waves.

Magnetic-field energy plays an important role in the ignition systems of gasoline-powered automobiles. A primary coil of about 250 turns is connected to the car's battery and produces a strong magnetic field. This coil is surrounded by a secondary coil with some 25,000 turns of very fine wire. When it is time for a spark plug to fire (see Fig. 20.4 in Section 20.3), the current to the primary coil is interrupted, the magnetic field quickly drops to zero, and an emf of tens of thousands of volts is induced in the secondary coil. The energy stored in the magnetic field thus goes into a powerful pulse of current that travels through the secondary coil to the spark plug, generating the spark that ignites the fuel–air mixture in the engine's cylinders (Fig. 30.10).

Example 30.5 Storing energy in an inductor

The electric-power industry would like to find efficient ways to store surplus energy generated during low-demand hours to help meet customer requirements during high-demand hours. Perhaps a large inductor can be used. What inductance would be needed to store 1.00 kWh of energy in a coil carrying a 200-A current?

SOLUTION

IDENTIFY and SET UP: We are given the required amount of stored energy U and the current I. We solve for the required inductance L using Eq. (30.9).

EXECUTE: We have $I = 200\text{ A}$ and $U = 1.00\text{ kWh} = (1.00 \times 10^3\text{ W})(3600\text{ s}) = 3.60 \times 10^6\text{ J}$. Solving Eq. (30.9) for L, we find

$$L = \frac{2U}{I^2} = \frac{2(3.60 \times 10^6\text{ J})}{(200\text{ A})^2} = 180\text{ H}$$

This is more than a *million* times greater than the self-inductance of the toroidal solenoid of Example 30.3 (Section 30.2).

EVALUATE: Conventional wires that are to carry 200 A would have to be of large diameter to keep the resistance low and avoid unacceptable energy losses due to I^2R heating. As a result, a 180-H inductor using conventional wire would be very large (room-size). A superconducting inductor could be much smaller, since the resistance of a superconductor is zero and much thinner wires could be used; one drawback is that the wires would have to be kept at low temperature to remain superconducting, and energy would have to be used to maintain this low temperature. As a result, this scheme is impractical with present technology.

Example 30.6 Magnetic energy density

In a proton accelerator used in elementary particle physics experiments, the trajectories of protons are controlled by bending magnets that produce a magnetic field of 6.6 T. What is the energy density in this field in the vacuum between the poles of such a magnet?

SOLUTION

IDENTIFY and SET UP: In a vacuum, $\mu = \mu_0$ and the energy density is given by Eq. (30.10).

EXECUTE: The energy density in the magnetic field is

$$u = \frac{B^2}{2\mu_0} = \frac{(6.6\text{ T})^2}{2(4\pi \times 10^{-7}\text{ T}\cdot\text{m/A})} = 1.73 \times 10^7\text{ J/m}^3$$

EVALUATE: As an interesting comparison, the heat of combustion of natural gas, expressed on an energy per unit volume basis, is about $3.8 \times 10^7\text{ J/m}^3$.

Test Your Understanding

If the current in a solenoid is reversed in direction while keeping the same magnitude, what effect is there on the magnetic field within the solenoid? What effect is there on the magnetic-field energy stored in the solenoid?

30.4 | The *R-L* Circuit

ActivPhysics ONLINE
14.1 The *RL* Circuit

Let's look at some examples of the circuit behavior of an inductor. One thing is clear already; an inductor in a circuit makes it difficult for rapid changes in current to occur, thanks to the effects of self-induced emf. Equation (30.7) shows that the greater the rate of change of current di/dt, the greater the self-induced emf and the greater the potential difference between the inductor terminals. This equation, together with Kirchhoff's rules (Section 26.2), gives us the principles we need to analyze circuits containing inductors.

Problem-Solving Strategy: Inductors in Circuits

IDENTIFY *the relevant concepts:* An inductor is just another circuit element, like a source of emf, a resistor, or a capacitor. One key difference is that when an inductor is included in a circuit, all the voltages, currents, and capacitor charges are in general functions of time, not constants as they have been in most of our previous circuit analysis. But Kirchhoff's rules, which we studied in Section 26.2, are still valid. When the voltages and currents vary with time, Kirchhoff's rules hold at each instant of time.

SET UP *the problem using the following steps:*

1. Follow the same procedure described in the Problem-Solving Strategy in Section 26.2. (Now would be an excellent time to review that strategy.) Draw a large circuit diagram and label all quantities, known and unknown. Apply the junction rule immediately at any junction.
2. Determine which quantities are the target variables.

EXECUTE *the solution as follows:*

1. As in the Problem-Solving Strategy in Section 26.2, apply Kirchhoff's loop rule to each loop in the circuit.
2. As in all circuit analysis, getting the correct sign for each potential difference is essential. (You should review the rules given in the Problem-Solving Strategy in Section 26.2.) To get the correct sign for the potential difference between the terminals of an inductor, remember Lenz's law and the sign rule described in Section 30.2 in conjunction with Eq. (30.7) and Fig. 30.6. In Kirchhoff's loop rule, when we go through an inductor in the *same* direction as the assumed current, we encounter a voltage *drop* equal to $L\,di/dt$, so the corresponding term in the loop equation is $-L\,di/dt$. When we go through an inductor in the *opposite* direction from the assumed current, the potential difference is reversed and the term to use in the loop equation is $+L\,di/dt$.
3. As always, solve for the target variables.

EVALUATE *your answer:* Check whether your answer is consistent with the way that inductors behave. If the current through an inductor is changing, your result should indicate that the potential difference across the inductor opposes the change. If not, you probably used an incorrect sign somewhere in your calculation.

Current Growth in an *R-L* Circuit

We can learn several basic things about inductor behavior by analyzing the circuit of Fig. 30.11. A circuit that includes both a resistor and an inductor, and possibly a source of emf, is called an ***R-L* circuit.** The inductor helps to prevent rapid changes in current, which can be useful if a steady current is required but the external source has a fluctuating emf. The resistor R may be a separate circuit element, or it may be the resistance of the inductor windings; every real-life inductor

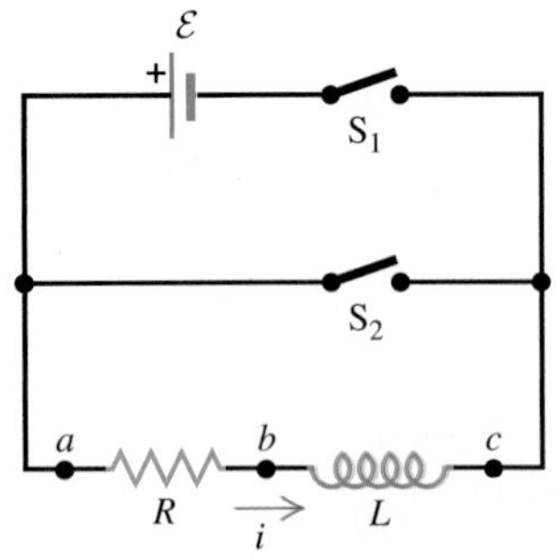

30.11 An *R-L* circuit. Closing switch S_1 connects the *R-L* combination in series with a source of emf $\mathcal{E}$. Closing switch S_2 while opening switch S_1 disconnects the combination from the source.

has some resistance unless it is made of superconducting wire. By closing switch S_1, we can connect the *R-L* combination to a source with constant emf $\mathcal{E}$. (We assume that the source has zero internal resistance, so the terminal voltage equals the emf.)

Suppose both switches are open to begin with, and then at some initial time $t = 0$ we close switch S_1. The current cannot change suddenly from zero to some final value, since di/dt and the induced emf in the inductor would both be infinite. Instead, the current begins to grow at a rate that depends only on the value of L in the circuit.

Let i be the current at some time t after switch S_1 is closed and let di/dt be its rate of change at that time. The potential difference v_{ab} across the resistor at that time is

$$v_{ab} = iR$$

and the potential difference v_{bc} across the inductor is

$$v_{bc} = L\frac{di}{dt}$$

Note that if the current is in the direction shown in Fig. 30.11 and is increasing, then both v_{ab} and v_{bc} are positive; a is at a higher potential than b, which in turn is at a higher potential than c. (Compare to Fig. 30.6b.) We apply Kirchhoff's loop rule, starting at the negative terminal and proceeding counterclockwise around the loop:

$$\mathcal{E} - iR - L\frac{di}{dt} = 0 \tag{30.12}$$

Solving this for di/dt, we find that the rate of increase of current is

$$\frac{di}{dt} = \frac{\mathcal{E} - iR}{L} = \frac{\mathcal{E}}{L} - \frac{R}{L}i \tag{30.13}$$

At the instant that switch S_1 is first closed, $i = 0$ and the potential drop across R is zero. The initial rate of change of current is

$$\left(\frac{di}{dt}\right)_{\text{initial}} = \frac{\mathcal{E}}{L}$$

As we would expect, the greater the inductance L, the more slowly the current increases.

As the current increases, the term $(R/L)i$ in Eq. (30.13) also increases, and the *rate* of increase of current given by Eq. (30.13) becomes smaller and smaller. This means that the current is approaching a final, steady-state value I. When the current reaches this value, its rate of increase is zero. Then Eq. (30.13) becomes

$$\left(\frac{di}{dt}\right)_{\text{final}} = 0 = \frac{\mathcal{E}}{L} - \frac{R}{L}I$$

and

$$I = \frac{\mathcal{E}}{R}$$

The *final* current I does not depend on the inductance L; it is the same as it would be if the resistance R alone were connected to the source with emf $\mathcal{E}$.

The behavior of the current as a function of time is shown by the graph of Fig. 30.12. To derive the equation for this curve (that is, an expression for current as a function of time), we proceed just as we did for the charging capacitor in Section 26.4. First we rearrange Eq. (30.13) to the form

$$\frac{di}{i - (\mathcal{E}/R)} = -\frac{R}{L}dt$$

This separates the variables, with i on the left side and t on the right. Then we integrate both sides, renaming the integration variables i' and t' so that we can use i and t as the upper limits. (The lower limit for each integral is zero, corresponding to zero current at the initial time $t = 0$.) We get

$$\int_0^i \frac{di'}{i' - (\mathcal{E}/R)} = -\int_0^t \frac{R}{L}dt'$$

$$\ln\left(\frac{i - (\mathcal{E}/R)}{-\mathcal{E}/R}\right) = -\frac{R}{L}t$$

Now we take exponentials of both sides and solve for i. We leave the details for you to work out; the final result is

$$i = \frac{\mathcal{E}}{R}\left(1 - e^{-(R/L)t}\right) \quad \text{(current in an } R\text{-}L \text{ circuit with emf)} \quad (30.14)$$

This is the equation of the curve in Fig. 30.12. Taking the derivative of Eq. (30.14), we find

$$\frac{di}{dt} = \frac{\mathcal{E}}{L}e^{-(R/L)t} \quad (30.15)$$

At time $t = 0$, $i = 0$ and $di/dt = \mathcal{E}/L$. As $t \to \infty$, $i \to \mathcal{E}/R$ and $di/dt \to 0$, as we predicted.

As Fig. 30.12 shows, the instantaneous current i first rises rapidly, then increases more slowly and approaches the final value $I = \mathcal{E}/R$ asymptotically. At a time equal to L/R the current has risen to $(1 - 1/e)$, or about 63%, of its final value. The quantity L/R is therefore a measure of how quickly the current builds toward its final value; this quantity is called the **time constant** for the circuit, denoted by τ:

$$\tau = \frac{L}{R} \quad \text{(time constant for an } R\text{-}L \text{ circuit)} \quad (30.16)$$

In a time equal to 2τ, the current reaches 86% of its final value; in 5τ, 99.3%; and in 10τ, 99.995%. (Compare the discussion in Section 26.4 of charging a capacitor of capacitance C that was in series with a resistor of resistance R; the time constant for that situation was the product RC.)

The graphs of i versus t have the same general shape for all values of L. For a given value of R, the time constant τ is greater for greater values of L. When L is small, the current rises rapidly to its final value; when L is large, it rises more slowly. For example, if $R = 100\ \Omega$ and $L = 10$ H,

$$\tau = \frac{L}{R} = \frac{10\text{ H}}{100\ \Omega} = 0.10\text{ s}$$

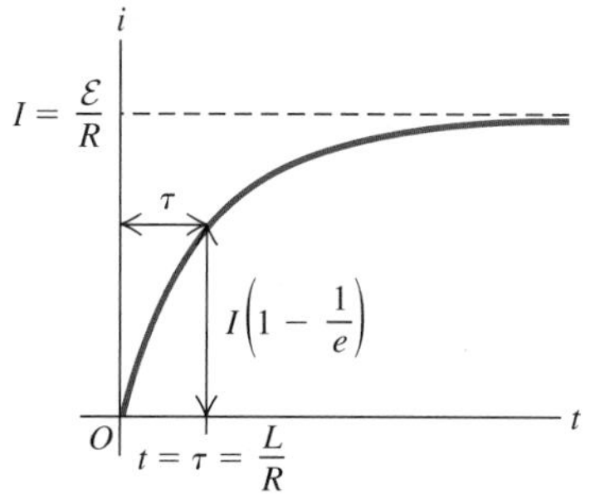

30.12 Graph of i versus t for growth of current in an R-L circuit with an emf in series. The final current is $I = \mathcal{E}/R$; after one time constant τ, the current is $1 - 1/e$ of this value.

and the current increases to about 63% of its final value in 0.10 s. (Recall that $1\ \text{H} = 1\ \Omega\cdot\text{s}$.) But if $L = 0.010\ \text{H}$, $\tau = 1.0 \times 10^{-4}\ \text{s} = 0.10\ \text{ms}$, and the rise is much more rapid.

Energy considerations offer us additional insight into the behavior of an *R-L* circuit. The instantaneous rate at which the source delivers energy to the circuit is $P = \mathcal{E}i$. The instantaneous rate at which energy is dissipated in the resistor is i^2R, and the rate at which energy is stored in the inductor is $iv_{bc} = Li\,di/dt$ (or, equivalently, $(d/dt)(\frac{1}{2}Li^2) = Li\,di/dt$). When we multiply Eq. (30.12) by i and rearrange, we find

$$\mathcal{E}i = i^2R + Li\frac{di}{dt} \tag{30.17}$$

Of the power $\mathcal{E}i$ supplied by the source, part (i^2R) is dissipated in the resistor and part $(Li\,di/dt)$ goes to store energy in the inductor. This discussion is completely analogous to our power analysis for a charging capacitor, given at the end of Section 26.4.

Example 30.7 Analyzing an *R-L* circuit

A sensitive electronic device of resistance 175 Ω is to be connected to a source of emf by a switch. The device is designed to operate with a current of 36 mA, but to avoid damage to the device, the current can rise to no more than 4.9 mA in the first 58 μs after the switch is closed. To protect the device, it is connected in series with an inductor as in Fig. 30.11; the switch in question is S_1. a) What emf must the source have? Assume negligible internal resistance. b) What inductance is required? c) What is the time constant?

SOLUTION

IDENTIFY and SET UP: Figure 30.12 shows that the final current is $I = \mathcal{E}/R$. Since the resistance is given, the emf is determined by the requirement that the final current is to be 36 mA. The other requirement is that the current be no more than $i = 4.9\ \text{mA}$ at $t = 58\ \mu\text{s}$; to satisfy this, we use Eq. (30.14) for the current as a function of time and solve for the inductance, which is the only unknown quantity. Equation (30.16) then tells us the time constant.

EXECUTE: a) Using $I = 36\ \text{mA} = 0.036\ \text{A}$ and $R = 175\ \Omega$ in the expression $I = \mathcal{E}/R$ for the final current and solving for the emf, we find

$$\mathcal{E} = IR = (0.036\ \text{A})(175\ \Omega) = 6.3\ \text{V}$$

b) To find the required inductance, we solve Eq. (30.14) for L. First we multiply through by $(-R/\mathcal{E})$ and then add 1 to both sides to obtain

$$1 - \frac{iR}{\mathcal{E}} = e^{-(R/L)t}$$

Then we take natural logs of both sides, solve for L, and insert the numbers:

$$L = \frac{-Rt}{\ln(1 - iR/\mathcal{E})}$$
$$= \frac{-(175\ \Omega)(58 \times 10^{-6}\ \text{s})}{\ln[1 - (4.9 \times 10^{-3}\ \text{A})(175\ \Omega)/(6.3\ \text{V})]} = 69\ \text{mH}$$

c) From Eq. (30.16),

$$\tau = \frac{L}{R} = \frac{69 \times 10^{-3}\ \text{H}}{175\ \Omega} = 3.9 \times 10^{-4}\ \text{s} = 390\ \mu\text{s}$$

EVALUATE: We note that 58 μs is much less than the time constant. In 58 μs the current builds up only from zero to 4.9 mA, a small fraction of its final value of 36 mA; after 390 μs the current equals $(1 - 1/e)$ of its final value, or about $(0.63)(36\ \text{mA}) = 23\ \text{mA}$.

Current Decay in an *R-L* Circuit

Now suppose switch S_1 in the circuit of Fig. 30.11 has been closed for a while and that the current has reached the value I_0. Resetting our stopwatch to redefine the initial time, we close switch S_2 at time $t = 0$, bypassing the battery. (At the same time we should open S_1 to save the battery from ruin.) The current through R and L does not instantaneously go to zero but decays smoothly, as shown in Fig. 30.13.

The Kirchhoff's-rule loop equation is obtained from Eq. (30.12) by simply omitting the $\mathcal{E}$ term. We challenge you to retrace the steps in the above analysis and show that the current i varies with time according to

$$i = I_0 e^{-(R/L)t} \tag{30.18}$$

where I_0 is the initial current at time $t = 0$. The time constant, $\tau = L/R$, is the time for current to decrease to $1/e$, or about 37%, of its original value. In time 2τ it has dropped to 13.5%, in time 5τ to 0.67%, and in 10τ to 0.0045%.

The energy that is needed to maintain the current during this decay is provided by the energy stored in the magnetic field of the inductor. The detailed energy analysis is simpler this time. In place of Eq. (30.17) we have

$$0 = i^2 R + Li\frac{di}{dt} \tag{30.19}$$

In this case, $Li\,di/dt$ is negative; Eq. (30.19) shows that the energy stored in the inductor *decreases* at a rate equal to the rate of dissipation of energy i^2R in the resistor.

This entire discussion should look familiar; the situation is very similar to that of a charging and discharging capacitor, analyzed in Section 26.4. It would be a good idea to compare that section with our discussion of the *R-L* circuit.

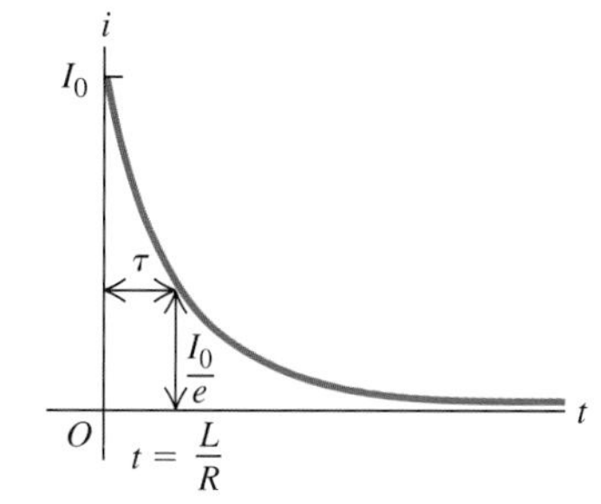

30.13 Graph of i versus t for decay of current in an *R-L* circuit. After one time constant τ, the current is $1/e$ of its initial value.

Example 30.8 Energy in an *R-L* circuit

When the current in an *R-L* circuit is decaying, what fraction of the original energy stored in the inductor has been dissipated after 2.3 time constants?

SOLUTION

IDENTIFY and SET UP: The current i at any time t for this situation is given by Eq. (30.18). The stored energy associated with this current is given by Eq. (30.9), $U = \frac{1}{2}Li^2$.

EXECUTE: From Eq. (30.18), the current i at any time t is

$$i = I_0 e^{-(R/L)t}$$

The energy U in the inductor at *any* time is obtained by substituting this expression into $U = \frac{1}{2}Li^2$. We obtain

$$U = \frac{1}{2}LI_0^2 e^{-2(R/L)t} = U_0 e^{-2(R/L)t}$$

where $U_0 = \frac{1}{2}LI_0^2$ is the energy at the initial time $t = 0$. When $t = 2.3\tau = 2.3L/R$, we have

$$U = U_0 e^{-2(2.3)} = U_0 e^{-4.6} = 0.010\,U_0$$

That is, only 0.010 or 1.0% of the energy initially stored in the inductor remains, so 99.0% has been dissipated in the resistor.

EVALUATE: To get a sense of what this result means, consider the *R-L* circuit we analyzed in Example 30.7, for which the time constant is 390 μs. With $L = 69$ mH $= 0.069$ H and an initial current $I_0 = 36$ mA $= 0.036$ A, the amount of energy in the inductor initially is $U_0 = \frac{1}{2}LI_0^2 = \frac{1}{2}(0.069\text{ H})(0.036\text{ A})^2 = 4.5 \times 10^{-5}$ J. Of this, 99.0% or 4.4×10^{-5} J is dissipated in $2.3(390\ \mu\text{s}) = 1.8 \times 10^{-3}$ s $= 1.8$ ms. In other words, this circuit can be powered off almost completely in 1.8 ms, and can be powered on in the same amount of time. The minimum time for a complete on-off cycle is therefore 3.6 ms. For many purposes, such as in fast switching networks for telecommunication, an even shorter cycle time is required. In such cases a smaller time constant $\tau = L/R$ is needed.

Test Your Understanding

In Fig. 30.11, what are the algebraic signs of the potential differences v_{ab} and v_{bc} when switch S_1 is closed and switch S_2 is open? When S_1 is open, S_2 is closed, and current is flowing in the direction shown?

30.5 | The *L-C* Circuit

A circuit containing an inductor and a capacitor shows an entirely new mode of behavior, characterized by *oscillating* current and charge. This is in sharp contrast to the *exponential* approach to a steady-state situation that we have seen with both *R-C* and *R-L* circuits. In the ***L-C* circuit** in Fig. 30.14 we charge the capacitor to a potential difference V_{m} and initial charge $Q = CV_{\mathrm{m}}$ on its left-hand plate, as shown in Fig. 30.14a, and then close the switch. What happens?

The capacitor begins to discharge through the inductor. Because of the induced emf in the inductor, the current cannot change instantaneously; it starts at zero and eventually builds up to a maximum value I_{m}. During this buildup the capacitor is discharging. At each instant the capacitor potential equals the induced emf, so as the capacitor discharges, the *rate of change* of current decreases. When the capacitor potential becomes zero, the induced emf is also zero, and the current has leveled off at its maximum value I_{m}. Figure 30.14b shows this situation; the capacitor has completely discharged. The potential difference between its terminals (and those of the inductor) has decreased to zero, and the current has reached its maximum value I_{m}.

During the discharge of the capacitor, the increasing current in the inductor has established a magnetic field in the space around it, and the energy that was initially stored in the capacitor's electric field is now stored in the inductor's magnetic field.

Although the capacitor is completely discharged in Fig. 30.14b, the current persists (it cannot change instantaneously), and the capacitor begins to charge with polarity opposite to that in the initial state. As the current decreases, the magnetic field also decreases, inducing an emf in the inductor in the *same* direction as the current; this slows down the decrease of the current. Eventually, the current and the magnetic field reach zero, and the capacitor has been charged in the sense *opposite* to its initial polarity (Fig. 30.14c), with potential difference $-V_{\mathrm{m}}$ and charge $-Q$ on its left-hand plate.

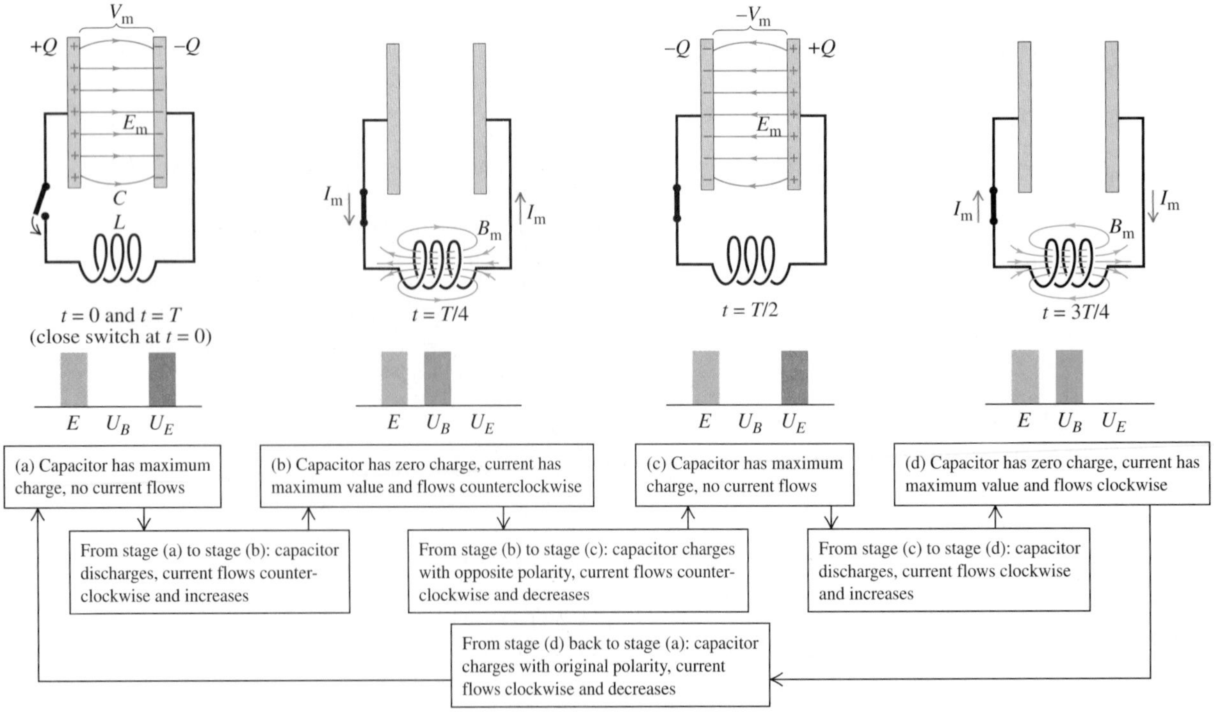

30.14 In an oscillating *L-C* circuit, the charge on the capacitor and the current through the inductor both vary sinusoidally with time. Energy is transferred between magnetic energy in the inductor (U_B) and electric energy in the capacitor (U_E). As in simple harmonic motion, the total energy E remains constant. (Compare Fig. 13.12 in Section 13.3.)

The process now repeats in the reverse direction; a little later, the capacitor has again discharged, and there is a current in the inductor in the opposite direction (Fig. 30.14d). Still later, the capacitor charge returns to its original value (Fig. 30.14a), and the whole process repeats. If there are no energy losses, the charges on the capacitor continue to oscillate back and forth indefinitely. This process is called an **electrical oscillation.**

From an energy standpoint the oscillations of an electrical circuit transfer energy from the capacitor's electric field to the inductor's magnetic field and back. The *total* energy associated with the circuit is constant. This is analogous to the transfer of energy in an oscillating mechanical system from potential energy to kinetic energy and back, with constant total energy. As we will see, this analogy goes much further.

To study the flow of charge in detail, we proceed just as we did for the *R-L* circuit. Figure 30.15 shows our definitions of q and i.

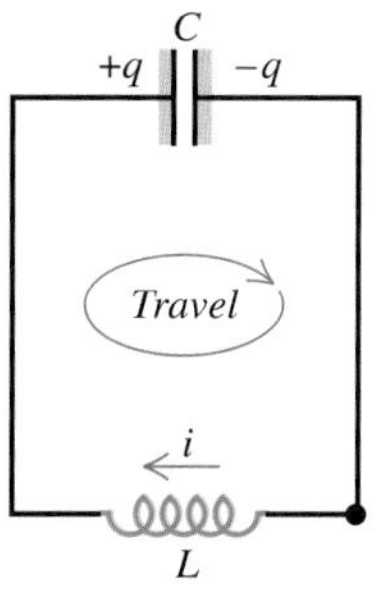

30.15 Applying Kirchhoff's loop rule to the *L-C* circuit. The direction of travel around the loop in the loop equation is shown. Just after the circuit is completed and the capacitor first begins to discharge, as in Fig. 30.14a, the current is negative (opposite to the direction shown).

CAUTION After examining Fig. 30.14, the positive direction for current in Fig. 30.15 may seem backward to you. In fact we've chosen this direction to simplify the relationship between current and capacitor charge. We define the current at each instant to be $i = dq/dt$, the rate of change of the charge on the left-hand capacitor plate. Hence if the capacitor is initially charged and begins to discharge as in Figs. 30.14a and 30.14b, then $dq/dt < 0$ and the initial current i is negative; the direction of the current is then opposite to the (positive) direction shown in Fig. 30.15.

ActivPhysics Online

14.2 The *RLC* Oscillator (Questions 1–6)

We apply Kirchhoff's loop rule to the circuit in Fig. 30.15. Starting at the lower-right corner of the circuit and adding voltages as we go clockwise around the loop, we obtain

$$-L\frac{di}{dt} - \frac{q}{C} = 0$$

Since $i = dq/dt$, it follows that $di/dt = d^2q/dt^2$. We substitute this expression into the above equation and divide by $-L$ to obtain

$$\frac{d^2q}{dt^2} + \frac{1}{LC}q = 0 \qquad (L\text{-}C \text{ circuit}) \qquad (30.20)$$

Equation (30.20) has exactly the same form as the equation we derived for simple harmonic motion in Section 13.2, Eq. (13.4). That equation is $d^2x/dt^2 = -(k/m)x$, or

$$\frac{d^2x}{dt^2} + \frac{k}{m}x = 0$$

(You should review Section 13.2 before going on with this discussion.) In the *L-C* circuit the capacitor charge q plays the role of the displacement x, and the current $i = dq/dt$ is analogous to the particle's velocity $v_x = dx/dt$. The inductance L is analogous to the mass m, and the reciprocal of the capacitance, $1/C$, is analogous to the force constant k.

Pursuing this analogy, we recall that the angular frequency $\omega = 2\pi f$ of the harmonic oscillator is equal to $(k/m)^{1/2}$, and the position is given as a function of time by Eq. (13.13),

$$x = A\cos(\omega t + \phi)$$

where the amplitude A and the phase angle ϕ depend on the initial conditions. In the analogous electrical situation the capacitor charge q is given by

$$q = Q\cos(\omega t + \phi) \tag{30.21}$$

and the angular frequency ω of oscillation is given by

$$\omega = \sqrt{\frac{1}{LC}} \tag{30.22}$$

(angular frequency of oscillation in an L-C circuit)

You should verify that Eq. (30.21) satisfies the loop equation, Eq. (30.20), when ω has the value given by Eq. (30.22). In doing this, you will find that the instantaneous current $i = dq/dt$ is given by

$$i = -\omega Q\sin(\omega t + \phi) \tag{30.23}$$

Thus the charge and current in an L-C circuit oscillate sinusoidally with time, with an angular frequency determined by the values of L and C. The ordinary frequency f, the number of cycles per second, is equal to $\omega/2\pi$ as always. The constants Q and ϕ in Eqs. (30.21) and (30.23) are determined by the initial conditions. If at time $t = 0$ the left-hand capacitor plate in Fig. 30.15 has its maximum charge Q and the current i is zero, then $\phi = 0$. If $q = 0$ at time $t = 0$, then $\phi = \pm\pi/2$ rad.

We can also analyze the L-C circuit using an energy approach. The analogy to simple harmonic motion is equally useful here. In the mechanical problem a body with mass m is attached to a spring with force constant k. Suppose we displace the body a distance A from its equilibrium position and release it from rest at time $t = 0$. The kinetic energy of the system at any later time is $\frac{1}{2}mv_x^2$, and its elastic potential energy is $\frac{1}{2}kx^2$. Because the system is conservative, the sum of these energies equals the initial energy of the system, $\frac{1}{2}kA^2$. We find the velocity v_x at any position x just as we did in Section 13.3, Eq. (13.22):

$$v_x = \pm\sqrt{\frac{k}{m}}\sqrt{A^2 - x^2} \tag{30.24}$$

The L-C circuit is also a conservative system. Again let Q be the maximum capacitor charge. The magnetic-field energy $\frac{1}{2}Li^2$ in the inductor at any time corresponds to the kinetic energy $\frac{1}{2}mv^2$ of the oscillating body, and the electric-field energy $q^2/2C$ in the capacitor corresponds to the elastic potential energy $\frac{1}{2}kx^2$ of the spring. The sum of these energies equals the total energy $Q^2/2C$ of the system:

$$\frac{1}{2}Li^2 + \frac{q^2}{2C} = \frac{Q^2}{2C} \tag{30.25}$$

The total energy in the L-C circuit is *constant;* it oscillates between the magnetic and the electric forms, just as the constant total mechanical energy in simple harmonic motion is constant and oscillates between the kinetic and potential forms.

Solving Eq. (30.25) for i, we find that when the charge on the capacitor is q, the current i is

$$i = \pm\sqrt{\frac{1}{LC}}\sqrt{Q^2 - q^2} \tag{30.26}$$

You can verify this equation by substituting q from Eq. (30.21) and i from Eq. (30.23). Comparing Eqs. (30.24) and (30.26), we see that current $i = dq/dt$

and charge q are related in the same way as are velocity $v_x = dx/dt$ and position x in the mechanical problem.

The analogies between simple harmonic motion and L-C circuit oscillations are summarized in Table 30.1. The striking parallel shown there between mechanical and electrical oscillations is one of many such examples in physics. This parallel is so close that we can solve complicated mechanical and acoustical problems by setting up analogous electrical circuits and measuring the currents and voltages that correspond to the mechanical and acoustical quantities to be determined. This is the basic principle of many analog computers. This analogy can be extended to *damped* oscillations, which we consider in the next section; in Chapter 31 we will extend the analogy further to include *forced* electrical oscillations, which occur in all alternating-current circuits.

Table 30.1 Oscillation of a Mass-Spring System Compared with Electrical Oscillation in an L-C Circuit

Mass-Spring System	Inductor-Capacitor Circuit
Kinetic energy $= \frac{1}{2}mv_x^2$	Magnetic energy $= \frac{1}{2}Li^2$
Potential energy $= \frac{1}{2}kx^2$	Electric energy $= q^2/2C$
$\frac{1}{2}mv_x^2 + \frac{1}{2}kx^2 = \frac{1}{2}kA^2$	$\frac{1}{2}Li^2 + q^2/2C = Q^2/2C$
$v_x = \pm\sqrt{k/m}\sqrt{A^2 - x^2}$	$i = \pm\sqrt{1/LC}\sqrt{Q^2 - q^2}$
$v_x = dx/dt$	$i = dq/dt$
$\omega = \sqrt{\dfrac{k}{m}}$	$\omega = \sqrt{\dfrac{1}{LC}}$
$x = A\cos(\omega t + \phi)$	$q = Q\cos(\omega t + \phi)$

Example 30.9 An oscillating circuit

A 300-V dc power supply is used to charge a 25-μF capacitor. After the capacitor is fully charged, it is disconnected from the power supply and connected across a 10-mH inductor. The resistance in the circuit is negligible.
a) Find the frequency and period of oscillation of the circuit.
b) Find the capacitor charge and the circuit current 1.2 ms after the inductor and capacitor are connected.

SOLUTION

IDENTIFY and SET UP: We are given the capacitance C and the inductance L, from which we can calculate the frequency and period using Eq. (30.22). We find the charge and current using Eqs. (30.21) and (30.23). Initially the capacitor is fully charged and the current is zero, as in Fig. 30.14a, so the phase angle is $\phi = 0$ [see the discussion that follows Eq. (30.23)].

EXECUTE: a) The natural *angular* frequency is

$$\omega = \sqrt{\frac{1}{LC}} = \sqrt{\frac{1}{(10 \times 10^{-3}\text{ H})(25 \times 10^{-6}\text{ F})}}$$
$$= 2.0 \times 10^3 \text{ rad/s}$$

The frequency f is $1/2\pi$ times this:

$$f = \frac{\omega}{2\pi} = \frac{2.0 \times 10^3 \text{ rad/s}}{2\pi \text{ rad/cycle}} = 320 \text{ Hz}$$

The period is the reciprocal of the frequency:

$$T = \frac{1}{f} = \frac{1}{320 \text{ Hz}} = 3.1 \times 10^{-3}\text{ s} = 3.1 \text{ ms}$$

b) Since the period of the oscillation is $T = 3.1$ ms, $t = 1.2$ ms equals $0.38T$; this corresponds to a situation intermediate between Fig. 30.14b ($t = T/4$) and Fig. 30.14c ($t = T/2$). Comparing those figures to Fig. 30.15, we expect the capacitor charge q to be negative (that is, there will be negative charge on the left-hand plate of the capacitor) and the current i to be negative as well (that is, current will be traveling in a counterclockwise direction).

To find the value of q, we use Eq. (30.21). The charge is maximum at $t = 0$, so $\phi = 0$ and $Q = C\mathcal{E} = (25 \times 10^{-6}\text{ F})(300\text{ V}) = 7.5 \times 10^{-3}$ C. The charge q at any time is

$$q = (7.5 \times 10^{-3}\text{ C})\cos\omega t$$

At time $t = 1.2 \times 10^{-3}$ s,

$$\omega t = (2.0 \times 10^3 \text{ rad/s})(1.2 \times 10^{-3} \text{ s}) = 2.4 \text{ rad}$$
$$q = (7.5 \times 10^{-3} \text{ C}) \cos(2.4 \text{ rad}) = -5.5 \times 10^{-3} \text{ C}$$

The current i at any time is

$$i = -\omega Q \sin \omega t$$

At time $t = 1.2 \times 10^{-3}$ s,

$$i = -(2.0 \times 10^3 \text{ rad/s})(7.5 \times 10^{-3} \text{ C}) \sin(2.4 \text{ rad}) = -10 \text{ A}$$

EVALUATE: Note that the signs of q and i are both negative, as we predicted.

Example 30.10 Energy in an oscillating circuit

Consider again the L-C circuit of Example 30.9. a) Find the magnetic energy and electric energy at $t = 0$. b) Find the magnetic energy and electric energy at $t = 1.2$ ms.

SOLUTION

IDENTIFY and SET UP: From Example 30.9 we know the values of the capacitor charge q and circuit current i for both of the times of interest. We use them to calculate the magnetic energy stored in the inductor, given by $U_B = \frac{1}{2}Li^2$, and the electric energy stored in the capacitor, given by $U_E = q^2/2C$.

EXECUTE: a) At $t = 0$ there is no current and $q = Q$. Hence there is no magnetic energy, and all the energy in the circuit is in the form of electric energy in the capacitor:

$$U_B = \frac{1}{2}Li^2 = 0 \qquad U_E = \frac{Q^2}{2C} = \frac{(7.5 \times 10^{-3} \text{ C})^2}{2(25 \times 10^{-6} \text{ F})} = 1.1 \text{ J}$$

b) As we mentioned in Example 30.9, $t = 1.2$ ms corresponds to a situation intermediate between Fig. 30.14b ($t = T/4$) and Fig. 30.14c ($t = T/2$). So we expect the energy to be part magnetic and part electric at this time. From Example 30.9, $i = -10$ A and $q = -5.5 \times 10^{-3}$ C, so

$$U_B = \frac{1}{2}Li^2 = \frac{1}{2}(10 \times 10^{-3} \text{ H})(-10 \text{ A})^2 = 0.5 \text{ J}$$

$$U_E = \frac{q^2}{2C} = \frac{(-5.5 \times 10^{-3} \text{ C})^2}{2(25 \times 10^{-6} \text{ F})} = 0.6 \text{ J}$$

EVALUATE: The magnetic and electric energies are the same at $t = 3T/8 = 0.375T$, exactly halfway between the situations in Figs. 30.14b and 30.14c. The time we are considering here is slightly later and U_B is slightly less than U_E, as we would expect. We emphasize that at *all* times, the *total* energy $E = U_B + U_E$ has the same value, 1.1 J. An L-C circuit without resistance is a conservative system; no energy is dissipated.

Test Your Understanding

One way to think about the energy stored in an L-C circuit is to say that the circuit elements do positive or negative work on the charges that move back and forth through the circuit. Between stages (a) and (b) in Fig. 30.14, does the capacitor do positive work or negative work on the charges? What kind of force (electric or magnetic) does the capacitor exert on the charges to do this work? During this process, does the inductor do positive or negative work on the charges, and what kind of force does it exert on the charges?

30.6 | The *L-R-C* Series Circuit

In our discussion of the L-C circuit we assumed that there was no *resistance* in the circuit. This is an idealization, of course; every real inductor has resistance in its windings, and there may also be resistance in the connecting wires. Because of resistance, the electromagnetic energy in the circuit is dissipated and converted to other forms, such as internal energy of the circuit materials. Resistance in an electric circuit is analogous to friction in a mechanical system.

Suppose an inductor with inductance L and a resistor of resistance R are connected in series across the terminals of a charged capacitor, forming a **series *L-R-C* circuit.** As before, the capacitor starts to discharge as soon as the circuit is completed. But because of i^2R losses in the resistor, the magnetic-field energy acquired by the inductor when the capacitor is completely discharged is *less* than the original electric-field energy of the capacitor. In the same way, the energy of the capacitor when the magnetic field has decreased to zero is still smaller, and so on.

If the resistance R is relatively small, the circuit still oscillates, but with **damped harmonic motion** (Fig. 30.16a), and we say that the circuit is **underdamped.** If we increase R, the oscillations die out more rapidly. When R reaches a certain value, the circuit no longer oscillates; it is **critically damped** (Fig. 30.16b). For still larger values of R the circuit is **overdamped** (Fig. 30.16c), and the capacitor charge approaches zero even more slowly. We used these same terms to describe the behavior of the analogous mechanical system, the damped harmonic oscillator, in Section 13.7.

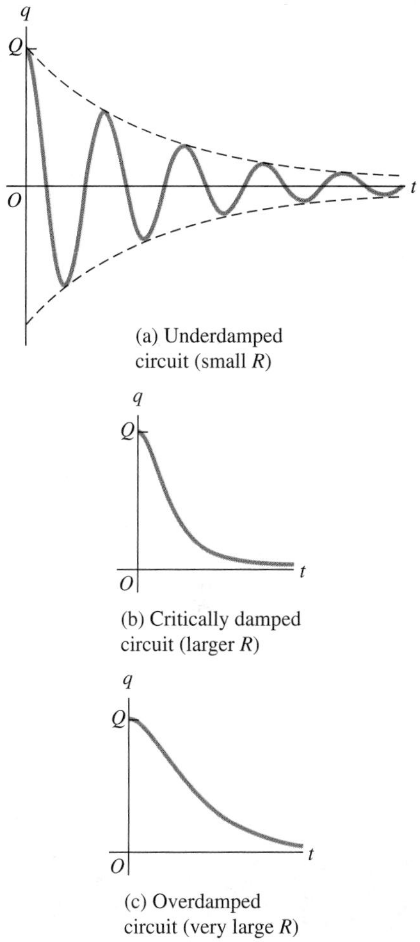

30.16 Graphs of capacitor charge as a function of time in an L-R-C series circuit with initial charge Q.

To analyze L-R-C circuit behavior in detail, we consider the circuit shown in Fig. 30.17. It is like the L-C circuit of Fig. 30.15 except for the added resistor R; we also show the source that charges the capacitor initially. The labeling of the positive senses of q and i are the same as for the L-C circuit.

First we close the switch in the upward position, connecting the capacitor to a source of emf $\mathcal{E}$ for a long enough time to ensure that the capacitor acquires its final charge $Q = C\mathcal{E}$ and any initial oscillations have died out. Then at time $t = 0$ we flip the switch to the downward position, removing the source from the circuit and placing the capacitor in series with the resistor and inductor. Note that the initial current is negative, opposite in direction to the direction of i shown in the figure.

To find how q and i vary with time, we apply Kirchhoff's loop rule. Starting at point a and going around the loop in the direction $abcda$, we obtain the equation

$$-iR - L\frac{di}{dt} - \frac{q}{C} = 0$$

Replacing i with dq/dt and rearranging, we get

$$\frac{d^2q}{dt^2} + \frac{R}{L}\frac{dq}{dt} + \frac{1}{LC}q = 0 \tag{30.27}$$

Note that when $R = 0$, this reduces to Eq. (30.20) for an L-C circuit.

There are general methods for obtaining solutions of Eq. (30.27). The form of the solution is different for the underdamped (small R) and overdamped (large R) cases. When R^2 is less than $4L/C$, the solution has the form

$$q = Ae^{-(R/2L)t}\cos\left(\sqrt{\frac{1}{LC} - \frac{R^2}{4L^2}}\,t + \phi\right) \tag{30.28}$$

where A and ϕ are constants. We invite you to take the first and second derivatives of this function and show by direct substitution that it does satisfy Eq. (30.27).

This solution corresponds to the *underdamped* behavior shown in Fig. 30.16a; the function represents a sinusoidal oscillation with an exponentially decaying amplitude. (Note that the exponential factor $e^{-(R/2L)t}$ is *not* the same as the factor $e^{-(R/L)t}$ that we encountered in describing the R-L circuit in Section 30.4.) When $R = 0$, Eq. (30.28) reduces to Eq. (30.21) for the oscillations in an L-C circuit. If R is not zero, the angular frequency of the oscillation is *less* than $1/(LC)^{1/2}$ because of the term containing R. The angular frequency ω' of the damped oscillations is given by

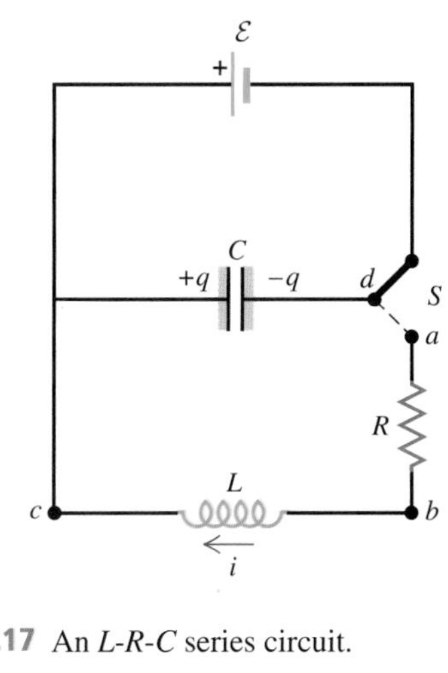

30.17 An L-R-C series circuit.

ActivPhysics ONLINE

14.2 The *RLC* Oscillator (Questions 7–10)

$$\omega' = \sqrt{\frac{1}{LC} - \frac{R^2}{4L^2}} \quad \text{(underdamped } L\text{-}R\text{-}C \text{ series circuit)} \tag{30.29}$$

When $R = 0$, this reduces to Eq. (30.22), $\omega = (1/LC)^{1/2}$. As R increases, ω' becomes smaller and smaller. When $R^2 = 4L/C$, the quantity under the radical becomes zero; the system no longer oscillates, and the case of *critical damping* (Fig. 30.16b) has been reached. For still larger values of R the system behaves as in Fig. 30.16c. In this case the circuit is *overdamped,* and q is given as a function of time by the sum of two decreasing exponential functions.

In the *underdamped* case the phase constant ϕ in the cosine function of Eq. (30.28) provides for the possibility of both an initial charge and an initial current at time $t = 0$, analogous to an underdamped harmonic oscillator given both an initial displacement and an initial velocity (see Exercise 30.38).

We emphasize once more that the behavior of the *L-R-C* circuit is completely analogous to that of the damped harmonic oscillator studied in Section 13.7. We invite you to verify, for example, that if you start with Eq. (13.41) and substitute L for m, $1/C$ for k, and R for the damping constant b, the result is Eq. (30.27). Similarly, the cross-over point between underdamping and overdamping occurs at $b^2 = 4km$ for the mechanical system and at $R^2 = 4L/C$ for the electrical one. Can you find still other aspects of this analogy?

The practical applications of the *L-R-C* series circuit emerge when we include a sinusoidally varying source of emf in the circuit. This is analogous to the *forced oscillations* that we discussed in Section 13.7, and there are analogous *resonance* effects. Such a circuit is called an *alternating-current (ac) circuit;* the analysis of ac circuits is the principal topic of the next chapter.

Example 30.11 An underdamped *L-R-C* series circuit

What resistance R is required (in terms of L and C) to give an *L-R-C* circuit a frequency that is one-half the undamped frequency?

SOLUTION

IDENTIFY and SET UP: The angular frequency of an underdamped *L-R-C* series circuit is given by Eq. (30.29); the angular frequency of an undamped *L-C* circuit is given by Eq. (30.22). We use these to solve for the target variable R.

EXECUTE: We want ω' given by Eq. (30.29) to be equal to one-half of ω given by Eq. (30.22):

$$\sqrt{\frac{1}{LC} - \frac{R^2}{4L^2}} = \frac{1}{2}\sqrt{\frac{1}{LC}}$$

When we square both sides and solve for R, we get

$$R = \sqrt{\frac{3L}{C}}$$

For example, adding 35 Ω to the circuit of Example 30.9 would reduce the frequency from 320 Hz to 160 Hz.

EVALUATE: The circuit becomes critically damped with no oscillations when $R = \sqrt{4L/C}$. Our result for R is smaller than that, as it should be; we want the circuit to be underdamped.

Test Your Understanding

The exponential factor $e^{-(R/2L)t}$ in Eq. (30.28) describes how the maximum value (amplitude) of the oscillating capacitor charge decreases with time. Suppose a 35-Ω resistor is added in series to the circuit of Example 30.9. One oscillation cycle after current first begins to flow, by what factor will the maximum capacitor charge have decreased from its initial value?

SUMMARY

When a changing current i_1 in one circuit causes a changing magnetic flux in a second circuit, an emf $\mathcal{E}_2$ is induced in the second circuit. Likewise, a changing current i_2 in the second circuit induces an emf $\mathcal{E}_1$ in the first circuit. The mutual inductance M depends on the geometry of the two coils and the material between them. If the circuits are coils of wire with N_1 and N_2 turns, respectively, M can be expressed in terms of the average flux Φ_{B2} through each turn of coil 2 that is caused by the current i_1 in coil 1, or in terms of the average flux Φ_{B1} through each turn of coil 1 that is caused by the current i_2 in coil 2. The SI unit of mutual inductance is the henry, abbreviated H. (See Examples 30.1 and 30.2)

$$\mathcal{E}_2 = -M\frac{di_1}{dt} \quad \text{and} \quad \mathcal{E}_1 = -M\frac{di_2}{dt} \tag{30.4}$$

$$M = \frac{N_2\Phi_{B2}}{i_1} = \frac{N_1\Phi_{B1}}{i_2} \tag{30.5}$$

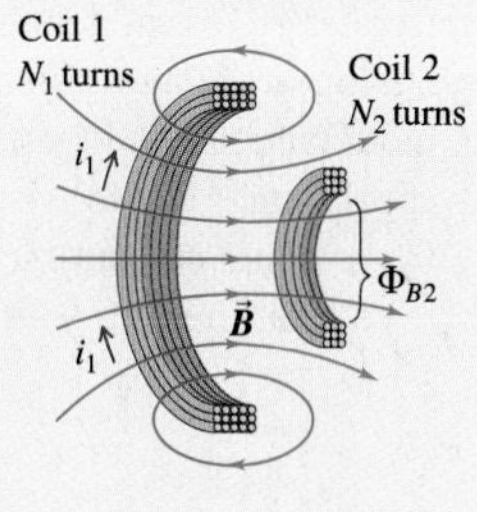

A changing current i in any circuit causes a self-induced emf $\mathcal{E}$. The inductance (or self-inductance) L depends on the geometry of the circuit and the material surrounding it. The inductance of a coil of N turns is related to the average flux Φ_B through each turn caused by the current i in the coil. An inductor is a circuit device, usually including a coil of wire, intended to have a substantial inductance. (See Examples 30.3 and 30.4)

$$\mathcal{E} = -L\frac{di}{dt} \tag{30.7}$$

$$L = \frac{N\Phi_B}{i} \tag{30.6}$$

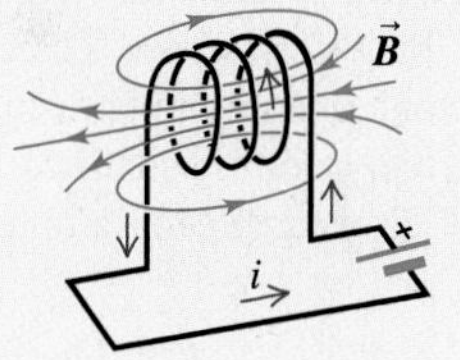

An inductor with inductance L carrying current I has energy U associated with the inductor's magnetic field. The magnetic energy density u (energy per unit volume) is proportional to the square of the magnetic field magnitude. (See Examples 30.5 and 30.6)

$$U = \frac{1}{2}LI^2 \tag{30.9}$$

$$u = \frac{B^2}{2\mu_0} \quad \text{(in a vacuum)} \tag{30.10}$$

$$u = \frac{B^2}{2\mu} \quad \text{(in a material with magnetic permeability } \mu\text{)} \tag{30.11}$$

In an R-L circuit containing a resistor R, an inductor L, and a source of emf, the growth and decay of current are exponential. The time constant τ is the time required for the current to approach within a fraction $1/e$ of its final value. (See Examples 30.7 and 30.8)

$$\tau = \frac{L}{R} \tag{30.16}$$

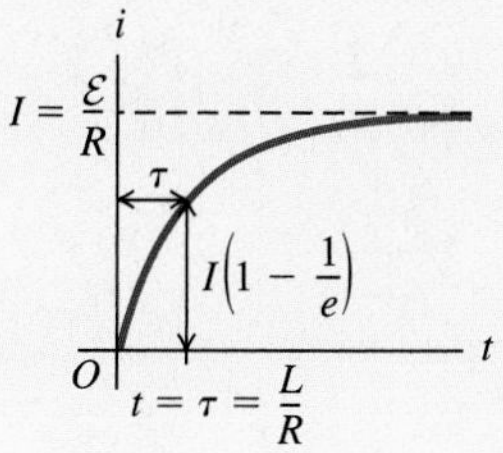

An *L-C* circuit, which contains inductance *L* and capacitance *C*, undergoes electrical oscillations with an angular frequency ω that depends on *L* and *C*. Such a circuit is analogous to a mechanical harmonic oscillator, with inductance *L* analogous to mass *m*, the reciprocal of capacitance $1/C$ to force constant *k*, charge *q* to displacement *x*, and current *i* to velocity v_x. (See Examples 30.9 and 30.10)

$$\omega = \sqrt{\frac{1}{LC}} \qquad (30.22)$$

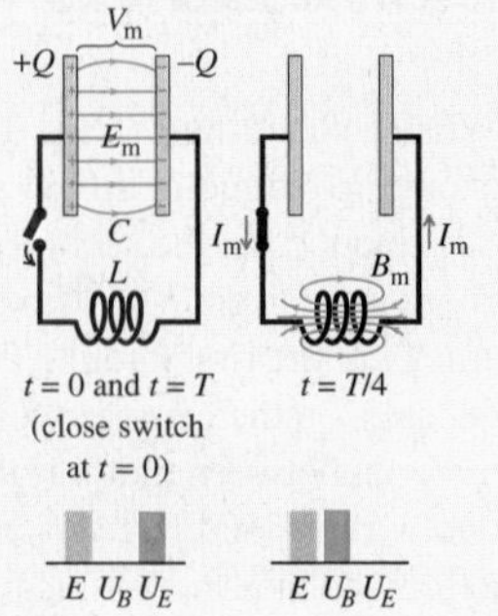

An *L-R-C* series circuit, which contains inductance, resistance, and capacitance, undergoes damped oscillations for sufficiently small resistance. The frequency ω' of damped oscillations depends on the values of *L*, *R*, and *C*. As *R* increases, the damping increases; if *R* is greater than a certain value the behavior becomes overdamped and no longer oscillates. The cross-over between underdamping and overdamping occurs when $R^2 = 4L/C$; when this condition is satisfied, the oscillations are critically damped. (See Example 30.11)

$$\omega' = \sqrt{\frac{1}{LC} - \frac{R^2}{4L^2}} \qquad (30.29)$$

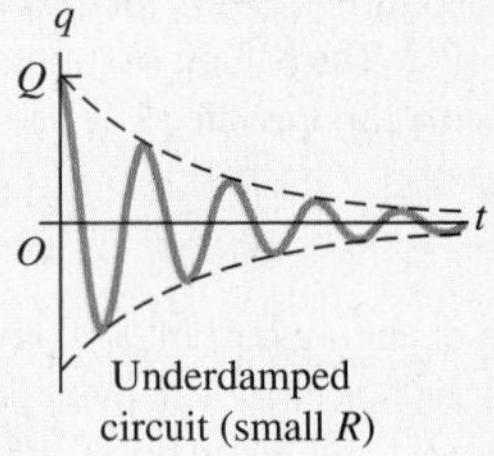

Key Terms

mutual inductance, 1148
henry, 1149
self-induced emf, 1151
inductance (self-inductance), 1152
inductor, 1152
magnetic energy density, 1157
***R-L* circuit, 1159**
time constant, 1161
***L-C* circuit, 1164**
electrical oscillation, 1165
series *L-R-C* circuit, 1169
damped harmonic motion, 1169
underdamped, 1169
critically damped, 1169
overdamped, 1169

Your Notes

Answer to Chapter Opening Question

No. For one circuit to induce a current in an adjacent circuit, the second circuit must experience a time-varying magnetic flux. But if the current in the first circuit is steady, it produces a steady magnetic field and hence a steady, unvarying flux in the second current. In this case no current would be induced in the second circuit.

Answers to Test Your Understanding Questions

Section 30.1 Doubling both the length of the solenoid (l) and the number of turns of wire in the solenoid (N_1) would have *no* effect on the mutual inductance M. Example 30.1 shows that M depends on the ratio of these quantities, which would remain unchanged. This is because the magnetic field produced by the solenoid depends on the number of turns *per unit length*, and the proposed change has no effect on this quantity. Increasing the value of M from 25 μH to 1.0 H $= 10^6\ \mu$H would therefore require increasing N_2, the number of turns in the outer coil, by a factor of $(10^6/25) = 4 \times 10^4$. Instead of 10 turns in the outer coil, we would need 4×10^5 or 400,000 turns! One henry is indeed a very large mutual inductance.

Section 30.2 The absolute value $|di/dt|$ of the rate of change of current is the same as in Example 30.4, so the magnitude of the self-induced emf is also the same: $|\mathcal{E}| = 80$ V. By Lenz's law the induced emf acts to oppose changes in the current. The emf therefore acts to sustain the decreasing current and is in the *same* direction as the current.

Section 30.3 Reversing the direction of the current reverses the direction of the magnetic field. However, it has no effect on the magnetic-field energy density, which is proportional to the square of the *magnitude* of the magnetic field. Hence there is no effect on the total energy stored in the solenoid. You can also see this from the expression $U = \frac{1}{2}LI^2$ for the energy stored in an inductor such as a solenoid: reversing the direction of the current changes I to $-I$, but has no effect on the square of the current.

Section 30.4 Recall that v_{ab} is the potential at a minus the potential at b, and similarly for v_{bc}. For either arrangement of the switches, current flows through the resistor from a to b. The upstream end of the resistor is always at the higher potential, so v_{ab} is positive. With S_1 closed and S_2 open, the current through the inductor flows from b to c and is increasing. The self-induced emf opposes this increase and is therefore directed from c toward b, which means that b is at the higher potential. Hence v_{bc} is positive. With S_1 open and S_2 closed, the inductor current again flows from b to c but is now decreasing. The self-induced emf is directed from b to c in an effort to sustain the decaying current, so c is at the higher potential and v_{bc} is negative.

Section 30.5 The capacitor loses energy between stages (a) and (b), so it does positive work on the charges. It does this by exerting an electric force that pushes current away from the positively charged left-hand capacitor plate and toward the negatively charged right-hand plate. At the same time, the inductor gains energy and does negative work on the moving charges. Although the inductor stores magnetic energy, the force that the inductor exerts is *electric*. This force comes about from the inductor's self-induced emf (see Section 30.2).

Section 30.6 This is an underdamped circuit because $R = 35\ \Omega$ is less than $\sqrt{4L/C} = \sqrt{4(10 \times 10^{-3}\text{ H})/(25 \times 10^{-6}\text{ F})} = 40\ \Omega$. From Example 30.11, the frequency of oscillation of this circuit is 160 Hz, so the oscillation period is $1/(160\text{ Hz}) = 6.3 \times 10^{-3}$ s. At this time the quantity $(R/2L)t = [(40\ \Omega)/2(10 \times 10^{-3}\text{ H})](6.3 \times 10^{-3}\text{ s}) = 12.5$, so the maximum capacitor charge has decreased to $e^{-12.5} = 3.7 \times 10^{-6}$ of its initial value. This oscillation decays *much* more rapidly than the one depicted in Fig. 30.16a.

Discussion Questions

Q30.1 In Fig. 30.1, if coil 2 is turned 90° so that its axis is vertical, does the mutual inductance increase or decrease? Explain.

Q30.2 The tightly wound toroidal solenoid is one of the few configurations for which it is easy to calculate self-inductance. What features of the toroidal solenoid give it this simplicity?

Q30.3 Two identical, closely wound, circular coils, each having self-inductance L, are placed next to each other, so they are coaxial and almost touching. If they are connected in series, what is the self-inductance of the combination? What if they are connected in parallel? Can they be connected so that the total inductance is zero? Explain.

Q30.4 Two closely wound circular coils have the same number of turns, but one has twice the radius of the other. How are the self-inductances of the two coils related? Explain your reasoning.

Q30.5 You are to make a resistor by winding a wire around a cylindrical form. To make the inductance as small as possible, it is proposed that you wind half the wire in one direction and the other half in the opposite direction. Would this achieve the desired result? Why or why not?

Q30.6 For the same magnetic field strength B, is the energy density larger in vacuum or in a magnetic material? Explain. Does Eq. (30.11) imply that for a long solenoid in which the current is I the energy stored is proportional to $1/\mu$? And does this mean that for the same current less energy is stored when the solenoid is filled with a ferromagnetic material rather than with air? Explain.

Q30.7 In Section 30.5, Kirchhoff's loop rule is applied to an L-C circuit where the capacitor is initially fully charged and the equation $-L\,di/dt - q/C = 0$ is derived. But as the capacitor starts to discharge, the current increases from zero. The equation says $L\,di/dt = -q/C$, so it says $L\,di/dt$ is negative. Explain how $L\,di/dt$ can be negative when the current is increasing.

Q30.8 In Section 30.5 the relation $i = dq/dt$ is used in deriving Eq. (30.20). But a flow of current corresponds to a decrease in the charge on the capacitor. Explain, therefore, why this is the correct equation to use in the derivation, rather than $i = -dq/dt$.

Q30.9 In the R-L circuit shown in Fig. 30.11, when the switch S_1 is closed, the potential v_{ab} changes suddenly and discontinuously, but the current does not. Explain why the voltage can change suddenly but the current can't.

Q30.10 In the R-L circuit shown in Fig. 30.11, is the current in the resistor always the same as that in the inductor? How do you know?

Q30.11 Suppose there is a steady current in an inductor. If you attempt to reduce the current to zero instantaneously by quickly opening a switch, an arc can appear at the switch contacts. Why? Is it physically possible to stop the current instantaneously? Explain.
Q30.12 In an *R-L-C* circuit, what criteria could be used to decide whether the system is overdamped or underdamped? For example, could we compare the maximum energy stored during one cycle to the energy dissipated during one cycle? Explain.

Exercises

Section 30.1 Mutual Inductance

30.1 Two coils have mutual inductance $M = 3.25 \times 10^{-4}$ H. The current i_1 in the first coil increases at a uniform rate of 830 A/s. a) What is the magnitude of the induced emf in the second coil? Is it constant? b) Suppose that the current described is in the second coil rather than the first. What is the magnitude of the induced emf in the first coil?
30.2 A toroidal solenoid has a mean radius r and a cross-sectional area A and is wound uniformly with N_1 turns. A second toroidal solenoid with N_2 turns is wound uniformly around the first. The two coils are wound in the same direction. What is their mutual inductance? (You can ignore the variation of the magnetic field across the cross section of the toroid.)
30.3 Two toroidal solenoids are wound around the same form so that the magnetic field of one passes through the turns of the other. Solenoid 1 has 700 turns and solenoid 2 has 400 turns. When the current in solenoid 1 is 6.52 A, the average flux through each turn of solenoid 2 is 0.0320 Wb. a) What is the mutual inductance of the pair of solenoids? b) When the current in solenoid 2 is 2.54 A, what is the average flux through each turn of solenoid 1?
30.4 Two coils are wound around the same cylindrical form, like the coils in Example 30.1. When the current in the first coil is decreasing at a rate of -0.242 A/s, the induced emf in the second coil has magnitude 1.65×10^{-3} V. a) What is the mutual inductance of the pair of coils? b) If the second coil has 25 turns, what is the flux through each turn when the current in the first coil equals 1.20 A? c) If the current in the second coil increases at a rate of 0.360 A/s, what is the magnitude of the induced emf in the first coil?
30.5 From Eq. (30.5), 1 H = 1 Wb/A and from Eq. (30.4), 1 H = 1 $\Omega \cdot$s. Show that these two definitions are equivalent.

Section 30.2 Self-Inductance and Inductors

30.6 When the current in a toroidal solenoid is changing at a rate of 0.0260 A/s, the magnitude of the induced emf is 12.6 mV. When the current equals 1.40 A, the average flux through each turn of the solenoid is 0.00285 Wb. How many turns does the solenoid have?

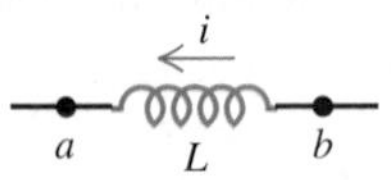

Figure 30.18 Exercises 30.7 and 30.10.

30.7 The inductor in Fig. 30.18 has inductance 0.260 H and carries a current in the direction shown that is decreasing at a uniform rate, $di/dt = -0.0180$ A/s. a) Find the self-induced emf. b) Which end of the inductor, a or b, is at a higher potential?
30.8 A toroidal solenoid has a cross-sectional area of 0.480 cm^2, a mean radius of 12.0 cm, and 1800 turns. The space inside the windings is filled with a ferromagnetic material that has a relative permeability 500. a) Calculate the inductance of the solenoid. (You can ignore the variation of the magnetic field across the cross section of the toroid.) b) If the material is removed from inside the solenoid and replaced by air, what is the inductance of the solenoid?
30.9 Inductance of a Solenoid. A long, straight solenoid has N turns, uniform cross-sectional area A, and length l. Derive an approximate expression for the inductance of the solenoid. Assume that the magnetic field is uniform inside the solenoid and zero outside. (Your answer is approximate because B is actually smaller at the ends than at the center. For this reason, your answer is actually an upper limit on the inductance.)
30.10 The inductor shown in Fig. 30.18 has inductance 0.260 H and carries a current in the direction shown. The current is changing at a constant rate. a) The potential between points a and b is $V_{ab} = 1.04$ V, with point a at higher potential. Is the current increasing or decreasing? b) If the current at $t = 0$ is 12.0 A, what is the current at $t = 2.00$ s?
30.11 At the instant when the current in an inductor is increasing at a rate of 0.0640 A/s, the magnitude of the self-induced emf is 0.0160 V. a) What is the inductance of the inductor? b) If the inductor is a solenoid with 400 turns, what is the average magnetic flux through each turn when the current is 0.720 A?

Section 30.3 Magnetic-Field Energy

30.12 An inductor used in a dc power supply has an inductance of 12.0 H and a resistance of 180 Ω. It carries a current of 0.300 A. a) What is the energy stored in the magnetic field? b) At what rate is thermal energy developed in the inductor? c) Does your answer to part (b) mean that the magnetic-field energy is decreasing with time? Explain.
30.13 An air-filled toroidal solenoid has a mean radius of 15.0 cm and a cross-sectional area of 5.00 cm^2. When the current is 12.0 A, the energy stored is 0.390 J. How many turns does the winding have?
30.14 It has been proposed to use large inductors as energy storage devices. a) How much electrical energy is converted to light and thermal energy by a 200-W light bulb in one day? b) If the amount of energy calculated in part (a) is stored in an inductor in which the current is 80.0 A, what is the inductance?
30.15 Starting from Eq. (30.9), derive in detail Eq. (30.11) for the energy density in a toroidal solenoid filled with a magnetic material.
30.16 A magnetic field with magnitude $B = 0.560$ T is uniform across a volume of 0.0290 m^3. Calculate the total magnetic energy in the volume if a) the volume is free space; b) the volume is filled with material with relative permeability 450.
30.17 It is proposed to store 1.00 kWh $= 3.60 \times 10^6$ J of electrical energy in a uniform magnetic field with magnitude 0.600 T. a) What volume (in vacuum) must the magnetic field occupy to store this amount of energy? b) If instead this amount of energy is to be stored in a volume (in vacuum) equivalent to a cube 40.0 cm on a side, what magnetic field is required?
30.18 An air-filled toroidal solenoid has 600 turns and a mean radius of 6.90 cm. Each winding has a cross-sectional area of

0.0350 cm^2. Assume that the magnetic field is uniform over the cross section of the windings. a) When the current in the solenoid is 2.50 A, what is the magnetic field within the solenoid? b) Use Eq. (30.10) to calculate the energy density in the solenoid. c) What is the total volume enclosed by the windings? d) Use the volume from part (c) and the energy density from part (b) to calculate the total energy stored in the solenoid. e) What is the inductance of the solenoid? f) Use Eq. (30.9) to calculate the energy stored in the solenoid. Compare your answer to what you obtained in part (d).

Section 30.4 The *R-L* Circuit

30.19 An inductor with an inductance of 2.50 H and a resistance of 8.00 Ω is connected to the terminals of a battery with an emf of 6.00 V and negligible internal resistance. Find a) the initial rate of increase of current in the circuit; b) the rate of increase of current at the instant when the current is 0.500 A; c) the current 0.250 s after the circuit is closed; d) the final steady-state current.

30.20 A 1.00-kΩ resistor is connected in series with a 10.0-mH inductor, a 30.0-V battery, and an open switch. At time $t = 0$, the switch is suddenly closed. a) What is the maximum current in this circuit and when does it occur? b) What are the voltage drops across the inductor and across the resistor 20.0 μs after the switch is closed? c) On a single set of axes, sketch the voltage across the resistor and the voltage across the inductor as functions of time. Also, sketch a graph of the current in the circuit as a function of time.

30.21 A 35.0-V battery with negligible internal resistance, a 50.0-Ω resistor, and a 1.25-mH inductor with negligible resistance are all connected in series with an open switch. The switch is suddenly closed. a) How long after closing the switch will the current through the inductor reach one-half of its maximum value? b) How long after closing the switch will the energy stored in the inductor reach one-half of its maximum value?

30.22 In Fig. 30.11, switch S_1 is closed while S_2 is kept open. The inductance is $L = 0.115$ H and the resistance is $R = 120\ \Omega$. a) When the current has reached its final value, the energy stored in the inductor is 0.260 J. What is the emf $\mathcal{E}$ of the battery? b) After the current has reached its final value, S_1 is opened and S_2 is closed. How much time does it take for the energy stored in the inductor to decrease to 0.130 J, half the original value?

30.23 In Fig. 30.11, suppose that $\mathcal{E} = 60.0$ V, $R = 240\ \Omega$, and $L = 0.160$ H. With switch S_2 open, switch S_1 is left closed until a constant current is established. Then S_2 is closed and S_1 opened, taking the battery out of the circuit. a) What is the initial current in the resistor, just after S_2 is closed and S_1 is opened? b) What is the current in the resistor at $t = 4.00 \times 10^{-4}$ s? c) What is the potential difference between points b and c at $t = 4.00 \times 10^{-4}$ s? Which point is at a higher potential? d) How long does it take the current to decrease to half its initial value?

30.24 In Fig. 30.11, suppose that $\mathcal{E} = 60.0$ V, $R = 240\ \Omega$, and $L = 0.160$ H. Initially, there is no current in the circuit. Switch S_2 is left open, and switch S_1 is closed. a) Just after S_1 is closed, what are the potential differences v_{ab} and v_{bc}? b) A long time (many time constants) after S_1 is closed, what are v_{ab} and v_{bc}? c) What are v_{ab} and v_{bc} at an intermediate time when $i = 0.150$ A?

30.25 Refer to Exercise 30.19. a) What is the power input to the inductor from the battery as a function of time if the circuit is completed at $t = 0$? b) What is the rate of dissipation of energy in the resistance of the inductor as a function of time? c) What is the rate at which the energy of the magnetic field in the inductor is increasing, as a function of time? d) Compare the results of (a), (b), and (c).

30.26 Write an equation corresponding to Eq. (30.13) for the current shown in Fig. 30.11 just after the switch S_2 is closed and S_1 is opened, if the initial current is I_0. Use integration methods to verify Eq. (30.18).

30.27 Show that L/R has units of time.

Section 30.5 The *L-C* Circuit

30.28 A Radio Tuning Circuit. The minimum capacitance of a variable capacitor in a radio is 4.18 pF. a) What is the inductance of a coil connected to this capacitor if the oscillation frequency of the *L-C* circuit is 1600×10^3 Hz, corresponding to one end of the AM radio broadcast band, when the capacitor is set to its minimum capacitance? b) The frequency at the other end of the broadcast band is 540×10^3 Hz. What is the maximum capacitance of the capacitor if the oscillation frequency is adjustable over the range of the broadcast band?

30.29 *L-C* Oscillations. A capacitor with capacitance 6.00×10^{-5} F is charged by connecting it to a 12.0-V battery. The capacitor is disconnected from the battery and connected across an inductor with $L = 1.50$ H. a) What are the angular frequency ω of the electrical oscillations and the period of these oscillations (the time for one oscillation)? b) What is the initial charge on the capacitor? c) How much energy is initially stored in the capacitor? d) What is the charge on the capacitor 0.0230 s after the connection to the inductor is made? Interpret the sign of your answer. e) At the time given in part (d), what is the current in the inductor? Interpret the sign of your answer. f) At the time given in part (d), how much electrical energy is stored in the capacitor and how much is stored in the inductor?

30.30 A 18.0-μF capacitor is placed across a 22.5-V battery for a few seconds and is then connected across a 12.0-mH inductor that has no appreciable resistance. a) After connecting the capacitor and inductor together, find the maximum current in the circuit. When the current is a maximum, what is the charge on the capacitor? b) How long after connecting the capacitor and inductor together does it take for the capacitor to be completely discharged for the first time? For the second time? c) Sketch graphs of the charge on the capacitor plates and the current through the inductor as functions of time.

30.31 When a voltmeter is placed across the plates of a capacitor, it reads 4.29 mV when the plates carry a charge of magnitude 150 nC. If the capacitor is now charged to 45.0 V and connected across a coil of negligible resistance, you observe that the current in the circuit oscillates with a period of 911 μs. Find the capacitance of the capacitor and the inductance of the coil.

30.32 In an *L-C* circuit, $L = 85.0$ mH and $C = 3.20$ μF. During the oscillations the maximum current in the inductor is 0.850 mA. a) What is the maximum charge on the capacitor? b) What is the magnitude of the charge on the capacitor at an instant when the current in the inductor has magnitude 0.500 mA?

30.33 In an L-C circuit, $L = 0.640$ H and $C = 3.60\ \mu$F. a) At the instant when the current in the inductor is changing at a rate of 2.80 A/s, what is the magnitude of the charge on the capacitor? b) When the charge on the capacitor is 8.50×10^{-6} C, what is the magnitude of the induced emf in the inductor?

30.34 An L-C circuit has a 0.400-H inductor and a 0.250-nF capacitor. During the current oscillations, the maximum current in the inductor is 1.5 A. a) What is the maximum energy stored in the capacitor at any time during the current oscillations? b) How many times per second does the capacitor contain the amount of energy found in part (a)?

30.35 Show that $\sqrt{LC}$ has units of time.

30.36 Show that the differential equation of Eq. (30.20) is satisfied by the function $q = Q\cos(\omega t + \phi)$, with ω given by $1/\sqrt{LC}$.

30.37 a) Using Eqs. (30.21) and (30.23) for an L-C circuit, write expressions for the energy stored in the capacitor as a function of time and for the energy stored in the inductor as a function of time. b) Using Eq. (30.22) and the trigonometric identity $\sin^2 x + \cos^2 x = 1$, show that the total energy in the L-C circuit is constant and equal to $Q^2/2C$.

Section 30.6 The *L-R-C* Series Circuit

30.38 a) Take first and second derivatives with respect to time of q given in Eq. (30.28), and show that it is a solution of Eq. (30.27). b) At $t = 0$ the switch shown in Fig. 30.17 is thrown so that it connects points d and a; at this time, $q = Q$ and $i = dq/dt = 0$. Show that the constants ϕ and A in Eq. (30.28) are given by

$$\tan\phi = -\frac{R}{2L\sqrt{(1/LC) - (R^2/4L^2)}} \quad \text{and} \quad A = \frac{Q}{\cos\phi}$$

30.39 a) In Eq. (13.41), substitute q for x, L for m, $1/C$ for k, and R for the damping constant b. Show that the result is Eq. (30.27). b) Make these same substitutions in Eq. (13.43) and show that Eq. (30.29) results. c) Make these same substitutions in Eq. (13.42) and show that Eq. (30.28) results.

30.40 Show that the quantity $\sqrt{L/C}$ has units of resistance (ohms).

30.41 An L-R-C circuit has $L = 0.285$ H, $C = 4.60 \times 10^{-4}$ F, and an angular frequency $\omega' = 1/\sqrt{6LC}$. What is the resistance R in the circuit?

30.42 An L-R-C circuit has $L = 0.450$ H, $C = 2.50 \times 10^{-5}$ F, and resistance R. a) What is the angular frequency of the circuit when $R = 0$? b) What value must R have to give a 5.0% decrease in angular frequency compared to the value calculated in part (a)?

Problems

30.43 A Differentiating Circuit. The current in a resistanceless inductor is caused to vary with time as shown in the graph of Fig. 30.19. a) Sketch the pattern that would be observed on the screen of an oscilloscope connected to the terminals of the inductor. (The oscilloscope spot sweeps horizontally across the screen at a constant speed, and its vertical deflection is proportional to the potential difference between the inductor terminals.) b) Explain why the inductor can be described as a "differentiating circuit."

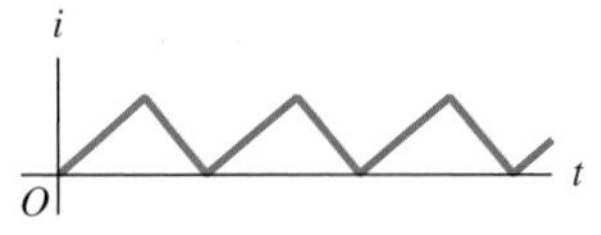

Figure 30.19 Problem 30.43.

30.44 A 0.250-H inductor carries a time-varying current given by the expression $i = (124\text{ mA})\cos[(240\pi/\text{s})t]$. a) Find an expression for the induced emf as a function of time. Graph the current and induced emf as functions of time for $t = 0$ to $t = \frac{1}{60}$ s. b) What is the maximum emf? What is the current when the induced emf is a maximum? c) What is the maximum current? What is the induced emf when the current is a maximum?

30.45 We previously ignored the variation of the magnetic field across the cross section of a toroidal solenoid. Let's now examine the validity of that approximation. A certain toroidal solenoid has a rectangular cross section (see Fig. 30.20). It has N uniformly spaced turns, with air inside. The magnetic field at a point inside the toroid is given by the equation derived in Example 28.11 (Section 28.7). *Do not* assume the field to be uniform over the cross section.

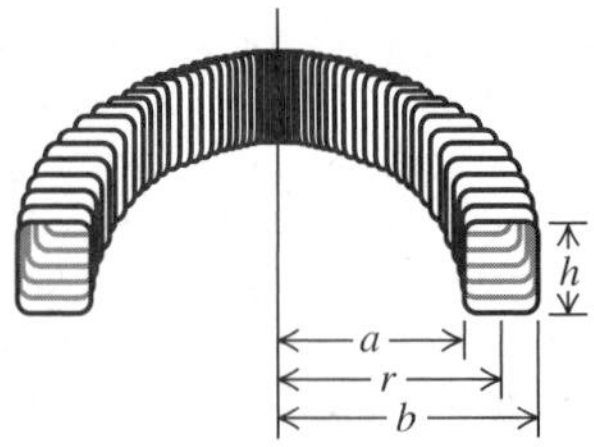

Figure 30.20 Problem 30.45.

a) Show that the magnetic flux through a cross section of the toroid is

$$\Phi_B = \frac{\mu_0 N i h}{2\pi}\ln\left(\frac{b}{a}\right)$$

b) Show that the inductance of the toroidal solenoid is given by

$$L = \frac{\mu_0 N^2 h}{2\pi}\ln\left(\frac{b}{a}\right)$$

c) The fraction b/a may be written as

$$\frac{b}{a} = \frac{a + b - a}{a} = 1 + \frac{b - a}{a}$$

Use the power series expansion $\ln(1 + z) = z + z^2/2 + \dots$, valid for $|z| < 1$, to show that when $b - a$ is much less than a, the inductance is approximately equal to

$$L = \frac{\mu_0 N^2 h(b - a)}{2\pi a}$$

Compare this result with the result given in Eq. (30.8), obtained in Example 30.3 (Section 30.2).

30.46 A solenoid has length l_1, radius r_1, and number of turns N_1. A second, smaller solenoid with length l_2, radius r_2 $(r_2 < r_1)$, and number of turns N_2 is placed at the center of the first solenoid, such that their axes coincide. Assume that the magnetic field of the first solenoid, at the location of the second, is uniform and has a magnitude given by the equation derived in Example 28.10 (Section 28.7). a) What is the mutual inductance of the pair of solenoids? b) If the current in the large solenoid is increasing at the rate di_1/dt, what is the magnitude of the emf induced in the small solenoid? c) If the current in the smaller solenoid is increasing at the rate di_2/dt, what is the magnitude of the emf induced in the larger solenoid?

30.47 The current in a coil of wire is initially zero but increases at a constant rate; after 12.0 s it is 48.0 A. The changing current induces an emf of magnitude 30.0 V in the coil. a) Determine the inductance of the coil. b) Determine the total magnetic flux through the coil when the current is 48.0 A. c) If the resistance of the coil is 60.0 Ω, determine the ratio of the rate at which energy is being stored in the magnetic field to the rate at which electrical energy is being dissipated by the resistance at the instant when the current is 48.0 A.

30.48 A coil has 400 turns and self-inductance 3.50 mH. The current in the coil varies with time according to $i = (680 \text{ mA}) \cos[\pi t/(0.0250 \text{ s})]$. a) What is the maximum emf induced in the coil? b) What is the maximum average flux through each turn of the coil? c) At $t = 0.0180$ s, what is the magnitude of the induced emf?

30.49 Inductors in Series and Parallel. You are given two inductors, one of self-inductance L_1 and the other of self-inductance L_2. a) You connect the two inductors in series and arrange them so that their mutual inductance is negligible. Show that the equivalent inductance of the combination is $L_{eq} = L_1 + L_2$. b) You now connect the two inductors in parallel, again arranging them so that their mutual inductance is negligible. Show that the equivalent inductance of the combination is $L_{eq} = (1/L_1 + 1/L_2)^{-1}$. (*Hint:* For either a series or a parallel combination, the potential difference across the combination is $L_{eq}(di/dt)$, where i is the current through the combination. For a parallel combination, i is the sum of the currents through the two inductors.)

30.50 A Coaxial Cable. A small solid conductor with radius a is supported by insulating, nonmagnetic disks on the axis of a thin-walled tube with inner radius b. The inner and outer conductors carry equal currents i in opposite directions. a) Use Ampere's law to find the magnetic field at any point in the volume between the conductors. b) Write the expression for the flux $d\Phi_B$ through a narrow strip of length l parallel to the axis, of width dr, at a distance r from the axis of the cable and lying in a plane containing the axis. c) Integrate your expression from part (b) over the volume between the two conductors to find the total flux produced by a current i in the central conductor. d) Show that the inductance of a length l of the cable is

$$L = l\frac{\mu_0}{2\pi} \ln\left(\frac{b}{a}\right)$$

e) Use Eq. (30.9) to calculate the energy stored in the magnetic field for a length l of the cable.

30.51 Consider the coaxial cable of Problem 30.50. The conductors carry equal currents i in opposite directions. a) Use Ampere's law to find the magnetic field at any point in the volume between the conductors. b) Use the energy density for a magnetic field, Eq. (30.10), to calculate the energy stored in a thin, cylindrical shell between the two conductors. Let the cylindrical shell have inner radius r, outer radius $r + dr$, and length l. c) Integrate your result in part (b) over the volume between the two conductors to find the total energy stored in the magnetic field for a length l of the cable. d) Use your result in part (c) and Eq. (30.9) to calculate the inductance L of a length l of the cable. Compare your result to L calculated in part (d) of Problem 30.50.

30.52 Consider the toroidal solenoid of Exercise 30.2. a) Derive an expression for the inductance L_1 when only the first coil is used and an expression for L_2 when only the second coil is used. b) Show that $M^2 = L_1L_2$.

30.53 Uniform electric and magnetic fields $\vec{E}$ and $\vec{B}$ occupy the same region of free space. If $E = 650$ V/m, what is B if the energy densities in the electric and magnetic fields are equal?

30.54 An inductor is connected to the terminals of a battery that has an emf of 12.0 V and negligible internal resistance. The current is 4.86 mA at 0.725 ms after the connection is completed. After a long time the current is 6.45 mA. What is a) the resistance R of the inductor? b) the inductance L of the inductor?

30.55 Continuation of Exercises 30.19 and 30.25. a) How much energy is stored in the magnetic field of the inductor one time constant after the battery has been connected? Compute this both by integrating the expression in Exercise 30.25(c) and by using Eq. (30.9), and compare the results. b) Integrate the expression obtained in Exercise 30.25(a) to find the *total* energy supplied by the battery during the time interval considered in part (a). c) Integrate the expression obtained in Exercise 30.25(b) to find the *total* energy dissipated in the resistance of the inductor during the same time period. d) Compare the results obtained in (a), (b), and (c).

30.56 Continuation of Exercise 30.23. a) What is the total energy initially stored in the inductor? b) At $t = 4.00 \times 10^{-4}$ s, at what rate is the energy stored in the inductor decreasing? c) At $t = 4.00 \times 10^{-4}$ s, at what rate is electrical energy being converted into thermal energy in the resistor? d) Obtain an expression for the rate at which electrical energy is being converted into thermal energy in the resistor as a function of time. Integrate this expression from $t = 0$ to $t = \infty$ to obtain the total electrical energy dissipated in the resistor. Compare your result to that of part (a).

30.57 The equation preceding Eq. (30.27) may be converted into an energy relation. Multiply both sides of this equation by $-i = -dq/dt$. The first term then becomes i^2R. Show that the second can be written $d(\frac{1}{2}Li^2)/dt$, and that the third can be written $d(q^2/2C)/dt$. What does the resulting equation say about energy conservation in the circuit?

30.58 At $t = 0$ in an L-C circuit, the capacitor has its maximum charge. a) Show that the electric and magnetic energies are equal at $t = 3T/8$, where T is the period of the oscillations. b) What is the next time after $t = 3T/8$ when the electric and magnetic energies again are equal?

30.59 An Electromagnetic Car Alarm. Your latest invention is a car alarm that produces sound at a particularly annoying frequency of 3500 Hz. To do this, the car alarm circuitry must produce an alternating electric current of the same frequency. That's why your design includes an inductor and capacitor in series. The maximum voltage across the capacitor is to be 12.0 V (the same voltage as the car battery). To produce a sufficiently loud sound, the capacitor must store 0.0160 J of energy. What values of capacitance and inductance should you choose for your car alarm circuit?

30.60 An L-C circuit consists of a 60.0-mH inductor and a 250-μF capacitor. The initial charge on the capacitor is 6.00 μC, and the initial current in the inductor is zero. a) What is the maximum voltage across the capacitor? b) What is the maximum current in the inductor? c) What is the maximum energy stored in the inductor? d) When the current in the inductor has half its maximum value, what is the charge on the capacitor and what is the energy stored in the inductor?

30.61 Solar Magnetic Energy. Magnetic fields within a sunspot can be as strong as 0.4 T. (By comparison, the earth's magnetic field is about 1/10,000 as strong.) Large sunspots can be as much as 25,000 km in radius. The material in a sunspot has a density of about 3×10^{-4} kg/m^3. If 100% of the magnetic field energy stored in a sunspot could be used to eject the sunspot's material away from the sun's surface, at what speed would that material be ejected? Compare to the sun's escape speed, which is about 6×10^5 m/s.

30.62 While studying a coil of unknown inductance and internal resistance, you connect it in series with a 25.0-V battery and a 150-Ω resistor. You then place an oscilloscope across one of these circuit elements and use the oscilloscope to measure the voltage across the circuit element as a function of time. The result is shown in Fig. 30.21. a) Across which circuit element (coil or resistor) is the oscilloscope connected? How do you know this? b) Find the inductance and the internal resistance of the coil. c) Carefully make a quantitative sketch showing the voltage versus time you would observe if you put the oscilloscope across the other circuit element (resistor or coil).

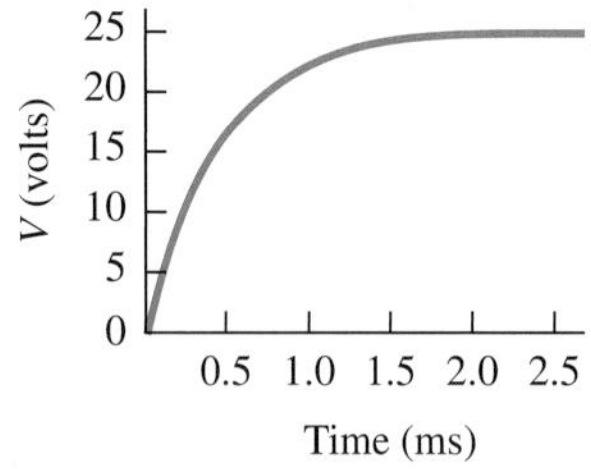

Figure 30.21 Problem 30.62.

30.63 In the lab, you are trying to find the inductance and internal resistance of a solenoid. You place it in series with a battery of negligible internal resistance, a 10.0-Ω resistor, and a switch. You then put an oscilloscope across one of these circuit elements, to measure the voltage across that circuit element as a function of time. You close the switch, and the oscilloscope shows voltage versus time as shown in Fig. 30.22. a) Across which circuit element (solenoid or resistor) is the oscilloscope connected? How do you know this? b) Why does the graph not approach zero as $t \to \infty$? c) What is the emf of the battery? d) Find the maximum current in the circuit. e) What are the internal resistance and self-inductance of the solenoid?

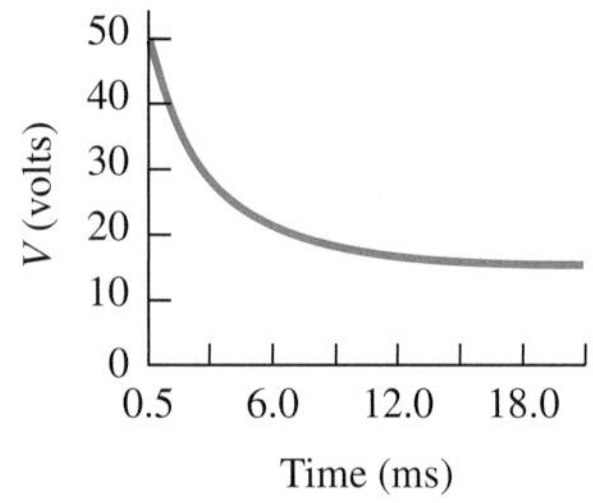

Figure 30.22 Problem 33.63.

30.64 In the circuit shown in Fig. 30.23, find the reading in each ammeter and voltmeter at the following times a) just after switch S is closed; b) after S has been closed a very long time.

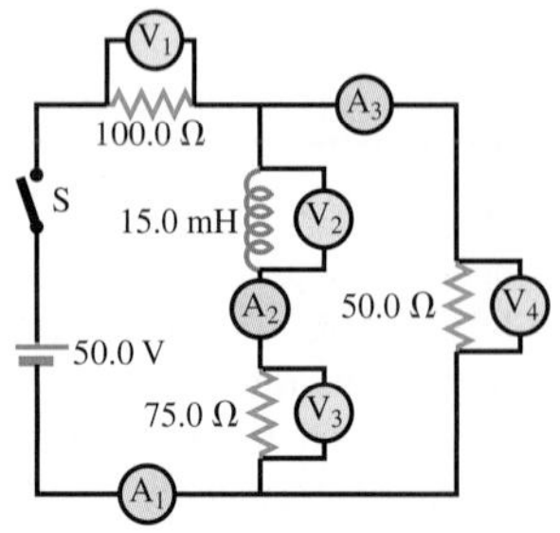

Figure 30.23 Problem 30.64.

30.65 In the circuit shown in Fig. 30.24, switch S is closed at time $t = 0$ with no charge initially on the capacitor. a) Find the reading of each ammeter and each voltmeter just after S is closed. b) Find the reading of each meter after a long time has elapsed. c) Find the maximum charge on the capacitor. d) Draw a qualitative graph of the reading of voltmeter V_2 as a function of time.

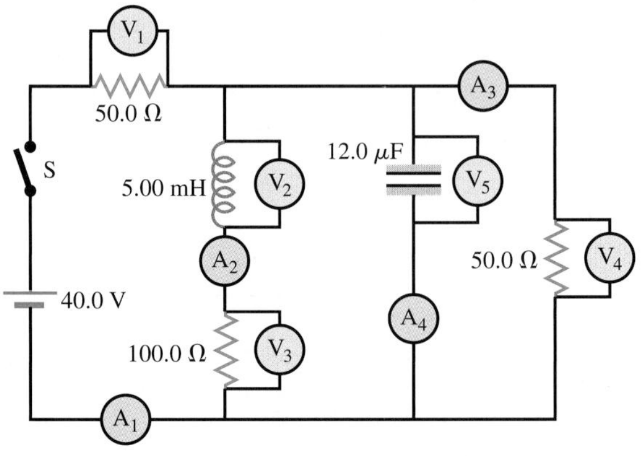

Figure 30.24 Problem 30.65.

30.66 In the circuit shown in Fig. 30.25, switch S is closed at time $t = 0$ with no charge initially on the capacitor. a) What is the reading of each meter just after S is closed? b) What does each meter read long after S is closed? c) What is the maximum charge on the capacitor and when does it occur?

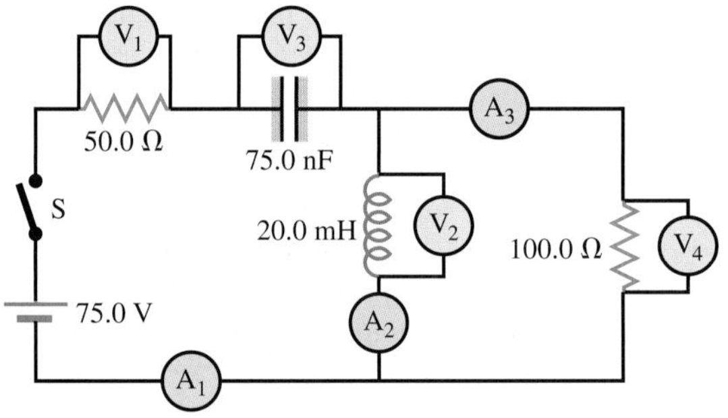

Figure 30.25 Problem 30.66.

30.67 In the circuit shown in Fig. 30.26, switch S is closed at time $t = 0$. a) Find the reading of each meter just after S is closed. b) What does each meter read long after S is closed?

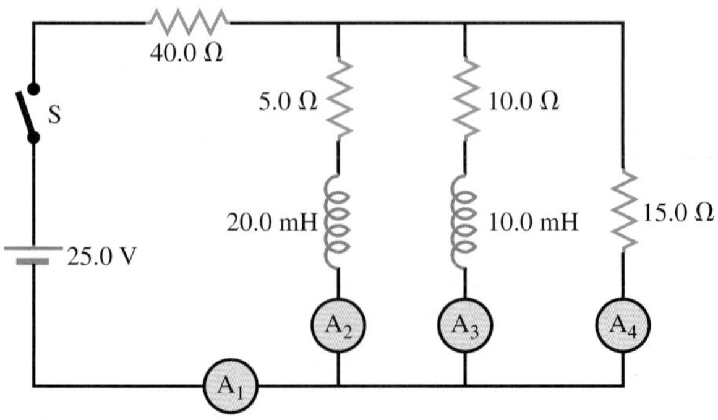

Figure 30.26 Problem 30.67.

30.68 In the circuit shown in Fig. 30.27, the capacitor is originally uncharged. The switch starts in the open position and is then flipped to position 1 for 0.500 s. It is then flipped to position 2 and left there. a) If the resistance r is very small, what is the upper limit for the amount of charge the capacitor could receive? b) Even if r is very small, how much electrical energy will be dissipated in it? c) Sketch a graph showing the reading of the ammeter as a function of time after the switch is in position 2, assuming that r is small.

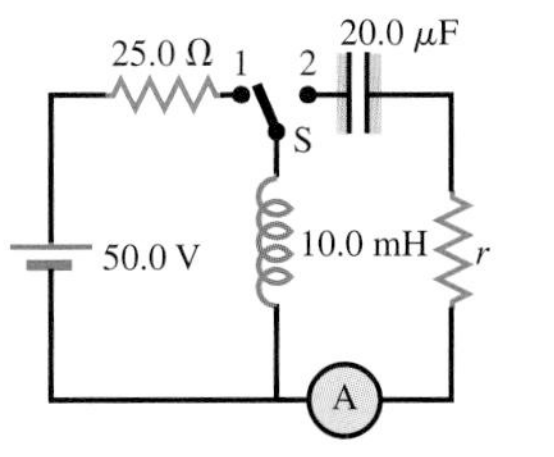

Figure 30.27 Problem 30.68.

30.69 In the circuit shown in Fig. 30.28, $\mathcal{E} = 60.0$ V, $R_1 = 40.0\ \Omega$, $R_2 = 25.0\ \Omega$, and $L = 0.300$ H. Switch S is closed at $t = 0$. Just after the switch is closed, a) What is the potential difference v_{ab} across the resistor R_1? b) Which point, a or b, is at a higher potential? c) What is the potential difference v_{cd} across the inductor L? d) Which point, c or d, is at a higher potential? The switch is left closed a long time and then opened. Just after the switch is opened, e) What is the potential difference v_{ab} across the resistor R_1? f) Which point, a or b, is at a higher potential? g) What is the potential difference v_{cd} across the inductor L? h) Which point, c or d, is at a higher potential?

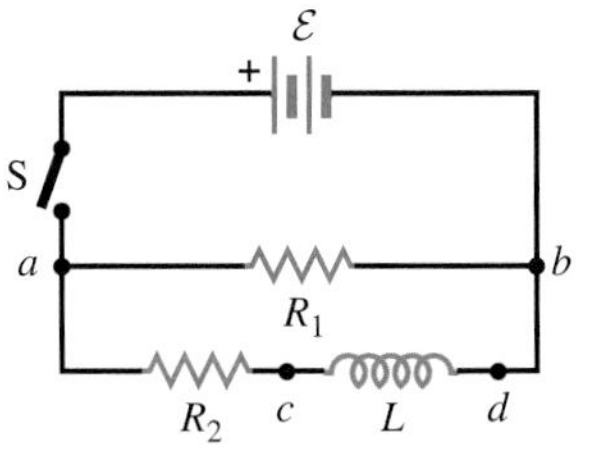

Figure 30.28 Problems 30.69, 30.70, and 30.75.

30.70 In the circuit shown in Fig. 30.28, $\mathcal{E} = 60.0$ V, $R_1 = 40.0\ \Omega$, $R_2 = 25.0\ \Omega$, and $L = 0.300$ H. a) Switch S is closed. At some time t afterward the current in the inductor is increasing at a rate of $di/dt = 50.0$ A/s. At this instant, what is the current i_1 through R_1 and the current i_2 through R_2? b) After the switch has been closed a long time, it is opened again. Just after it is opened, what is the current through R_1?

30.71 Consider the circuit shown in Fig. 30.29. Let $\mathcal{E} = 36.0$ V, $R_0 = 50.0\ \Omega$, $R = 150\ \Omega$, and $L = 4.00$ H. a) Switch S_1 is closed and switch S_2 is left open. Just after S_1 is closed, what are the current i_0 through R_0 and the potential differences v_{ac} and v_{cb}? b) After S_1 has been closed a long time (S_2 is still open) so that the current has reached its final steady value, what are i_0, v_{ac}, and v_{cb}? c) Find the expressions for i_0, v_{ac}, and v_{cb} as functions of the time t since S_1 was closed. Your results should agree with part (a) when $t = 0$ and with part (b) when $t \to \infty$. Graph i_0, v_{ac}, and v_{cb} versus time.

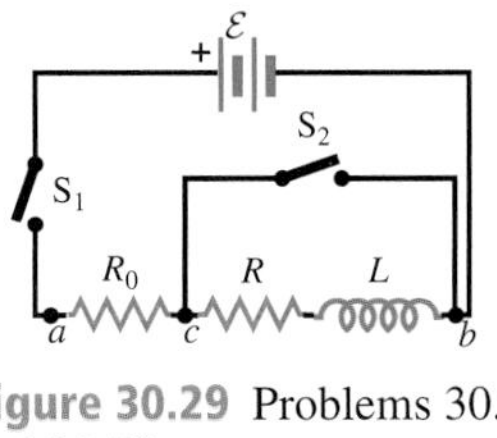

Figure 30.29 Problems 30.71 and 30.72.

30.72 After the current in the circuit of Fig. 30.29 has reached its final, steady value with switch S_1 closed and S_2 open, the switch S_2 is closed, thus short-circuiting the inductor. (Switch S_1 remains closed. See Problem 30.71 for numerical values of the circuit elements.) a) Just after S_2 is closed, what are v_{ac} and v_{cb}, and what are the currents through R_0, R, and S_2? b) A long time after S_2 is closed, what are v_{ac} and v_{cb}, and what are the currents through R_0, R, and S_2? c) Derive expressions for the currents through R_0, R, and S_2 as functions of the time t that has elapsed since S_2 was closed. Your results should agree with part (a) when $t = 0$ and with part (b) when $t \to \infty$. Graph these three currents versus time.

30.73 In the circuit shown in Fig. 30.30, the switch has been open for a long time and is suddenly closed. Neither the battery nor the inductors have any appreciable resistance. Review the results of Problem 30.49. a) What do the ammeter and voltmeter read just after closing S? b) What do the ammeter and the voltmeter read after S has been closed a very long time? c) What do the ammeter and the voltmeter read 0.115 ms after closing S?

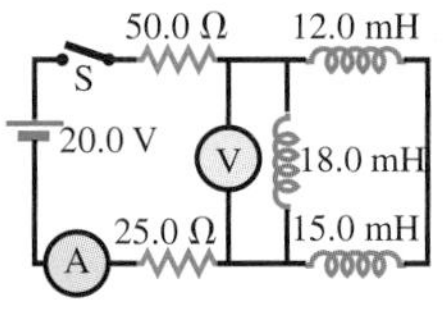

Figure 30.30 Problem 30.73.

30.74 In the circuit shown in Fig. 30.31, neither the battery nor the inductors have any appreciable resistance, the capacitors are initially uncharged, and the switch S has been in position 1 for a very long time. Review the results of Problem 30.49. a) What is the current in the circuit? b) The switch is now suddenly flipped to position 2. Find the maximum charge that each capacitor will receive, and how much time after the switch is flipped it will take them to acquire this charge.

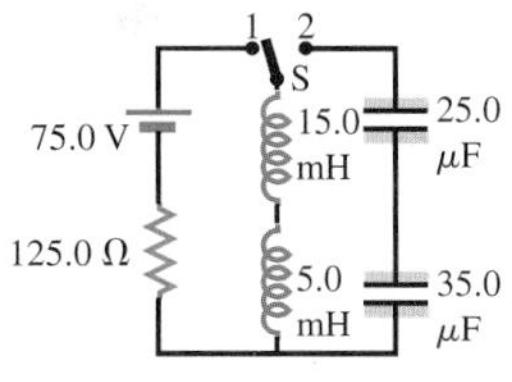

Figure 30.31 Problem 30.74.

30.75 Demonstrating Inductance. A common demonstration of inductance employs a circuit such as the one shown in Fig. 30.28. Switch S is closed, and the light bulb (represented by resistance R_1) just barely glows. After a period of time, switch S is opened, and the bulb lights up brightly for a short period of time. To understand this effect, think of an inductor as a device that imparts an "inertia" to the current, preventing a discontinuous change in the current through it. a) Derive, as explicit functions of time, expressions for i_1 (the current through the light bulb) and i_2 (the current through the inductor) after the switch S is closed. b) After a long period of time, the currents i_1 and i_2 reach their steady-state values. Obtain expressions for these steady-state currents. c) Switch S is now opened. Obtain an expression for the current through the inductor and light bulb as an explicit function of time. d) You have been asked to design a demonstration apparatus using the circuit shown in Fig. 30.28 with a 22.0-H inductor and a 40.0-W light bulb. You are to connect a resistor in series with the inductor, and R_2 represents the sum of that resistance plus the internal resistance of the inductor. When switch S is opened, a transient current is to be set up that starts at 0.600 A and is not to fall below 0.150 A until after 0.0800 s. For simplicity, assume that the resistance of the light bulb is constant and equals the resistance the bulb must have to dissipate 40.0 W at 120 V. Determine R_2 and $\mathcal{E}$ for the given design considerations. e) With the numerical values determined in part (d), what is the current through the light bulb just before the switch is opened? Does this result confirm the qualitative description of what is observed in the demonstration?

Challenge Problems

30.76 Two coils are wrapped around each other as shown in Fig. 30.3. The current travels in the same sense around each coil. One coil has self-inductance L_1, and the other coil has self-inductance L_2. The mutual inductance of the two coils is M. a) Show that if the two coils are connected in series, the equivalent inductance of the combination is $L_{eq} = L_1 + L_2 + 2M$. b) Show that if the two coils are connected in parallel, the equivalent inductance of the combination is

$$L_{eq} = \frac{L_1L_2 - M^2}{L_1 + L_2 - 2M}$$

(*Hint:* See the hint for Problem 30.49.)

30.77 Consider the circuit shown in Fig. 30.32. Switch S is closed at time $t = 0$, causing a current i_1 through the inductive branch and a current i_2 through the capacitive branch. The initial charge on the capacitor is zero, and the charge at time t is q_2. a) Derive expressions for i_1, i_2, and q_2 as functions of time. Express your answers in terms of $\mathcal{E}$, L, C, R_1, R_2, and t. For the remainder of the problem let the circuit elements have the following values: $\mathcal{E} = 48$ V, $L = 8.0$ H, $C = 20\ \mu$F, $R_1 = 25\ \Omega$, and $R_2 = 5000\ \Omega$. b) What is the initial current through the inductive branch? What is the initial current through the capacitive branch? c) What are the currents through the inductive and capacitive branches a long time after the switch has been closed? How long is a "long time"? Explain. d) At what time t_1 (accurate to two significant figures) will the currents i_1 and i_2 be equal? (*Hint:* You might consider using series expansions for the exponentials.) e) For the conditions given in part (d), determine i_1. f) The total current through the battery is $i = i_1 + i_2$. At what time t_2 (accurate to two significant figures) will i equal one-half of its final value? (*Hint:* The numerical work is greatly simplified if one makes suitable approximations. A sketch of i_1 and i_2 versus t may help you decide what approximations are valid.)

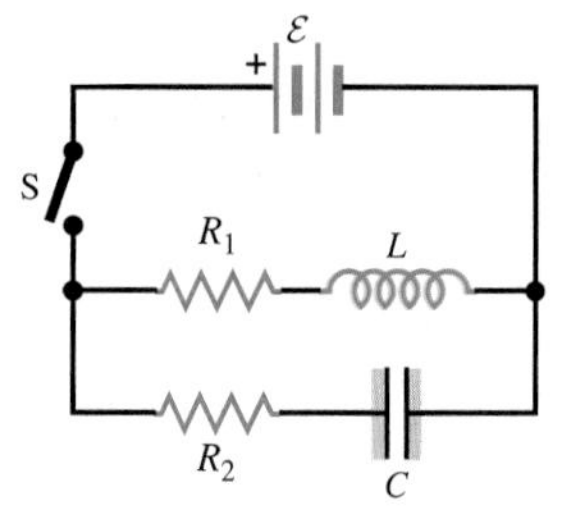

Figure 30.32 Challenge Problem 30.77.

30.78 Consider the circuit shown in Fig. 30.33. The circuit elements are as follows: $\mathcal{E} = 32.0$ V, $L = 0.640$ H, $C = 2.00\ \mu$F, and $R = 400\ \Omega$. At time $t = 0$, switch S is closed. The current through the inductor is i_1, the current through the capacitor branch is i_2, and the charge on the capacitor is q_2. a) Using Kirchhoff's rules, verify the circuit equations

$$R(i_1 + i_2) + L\left(\frac{di_1}{dt}\right) = \mathcal{E}$$

$$R(i_2 + i_2) + \frac{q_2}{C} = \mathcal{E}$$

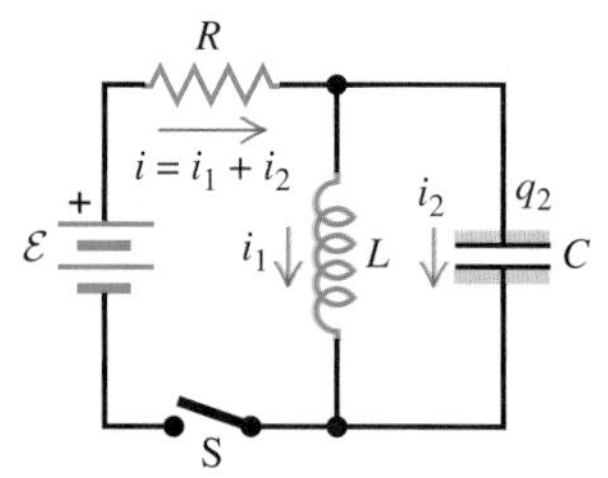

Figure 30.33 Challenge Problem 30.78.

b) What are the initial values of i_1, i_2, and q_2? c) Show by direct substitution that the following solutions for i_1 and q_2 satisfy the circuit equations from part (a). Also, show that they satisfy the initial conditions

$$i_1 = \left(\frac{\mathcal{E}}{R}\right)[1 - e^{-\beta t}(2\omega RC)^{-1}\sin(\omega t) + \cos(\omega t)]$$

$$q_2 = \left(\frac{\mathcal{E}}{\omega R}\right)e^{-\beta t}\sin(\omega t)$$

where $\beta = (2RC)^{-1}$ and $\omega = [(LC)^{-1} - (2RC)^{-2}]^{1/2}$. d) Determine the time t_1 at which i_2 first becomes zero.

30.79 A Volume Gauge. A tank containing a liquid has turns of wire wrapped around it, causing it to act as an inductor. The liquid content of the tank can be measured by using its inductance to determine the height of the liquid in the tank. The inductance of the tank changes from a value of L_0 corresponding to a relative permeability of 1 when the tank is empty to a value of L_f corresponding to a relative permeability of K_m (the relative permeability of the liquid) when the tank is full. The appropriate electronic circuitry can determine the inductance to five significant figures and thus the effective relative permeability of the combined air and liquid within the rectangular cavity of the tank. The four sides of the tank each have width W and height D (see Fig. 30.34). The height of the liquid in the tank is d. You can ignore any fringing effects and assume that the relative permeability of the material of which the tank is made can be ignored. a) Derive an expression for d as a function of L, the inductance corresponding to a certain fluid height, L_0, L_f, and D. b) What is the inductance (to five significant figures) for a tank $\frac{1}{4}$ full, $\frac{1}{2}$ full, $\frac{3}{4}$ full, and completely full if the tank contains liquid oxygen? Take $L_0 = 0.63000$ H. The magnetic susceptibility of liquid oxygen is $\chi_m = 1.52 \times 10^{-3}$. c) Repeat part (b) for mercury. The magnetic susceptibility of mercury is given in Table 28.1. d) For which material is this volume gauge more practical?

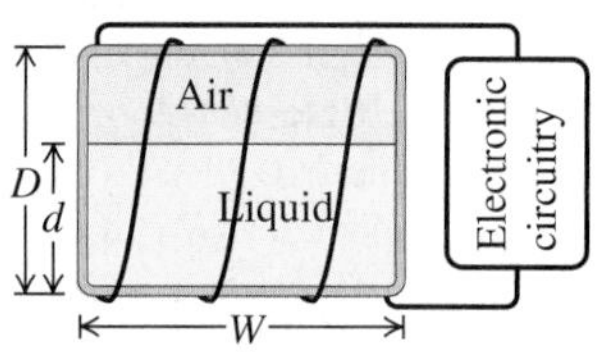

Figure 30.34 Challenge Problem 30.79.

ALTERNATING CURRENT 31

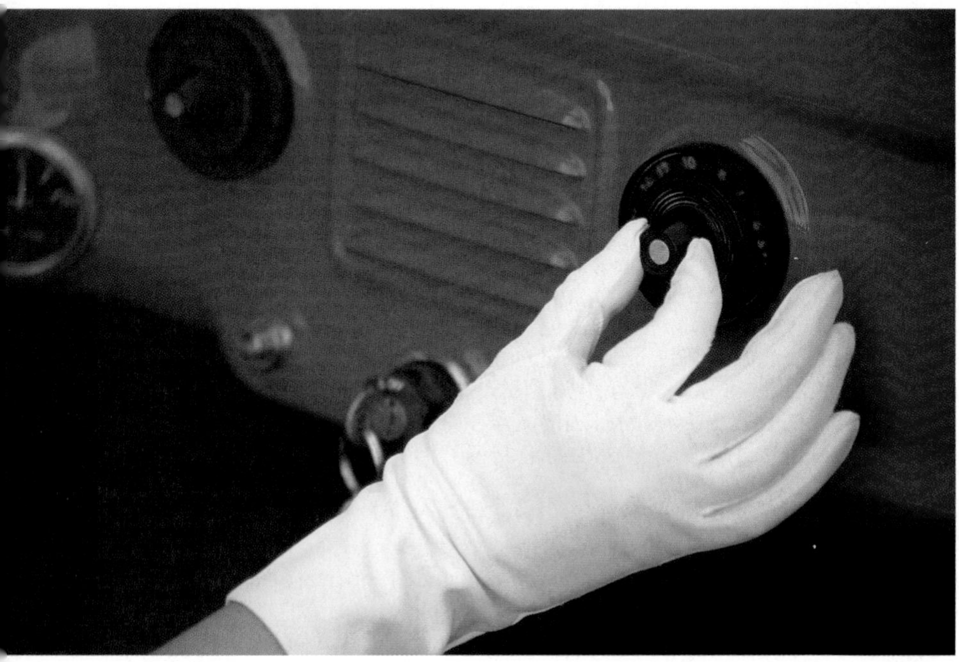

To tune the radio in this classic car, you turn a knob that adjusts a variable capacitance. This in turn adjusts the radio's resonant frequency, which depends on the capacitance as well as the radio's inductance. When the resonant frequency matches the frequency of the desired radio station, that station's signals come in loud and clear.

? **If a radio is tuned to a station at a frequency of 1000 kHz, does it also detect the transmissions from a station broadcasting at 600 kHz?**

During the 1880s in the United States there was a heated and acrimonious debate between two inventors over the best method of electric-power distribution. Thomas Edison favored direct current (dc), that is, steady current that does not vary with time. George Westinghouse favored **alternating current (ac),** with sinusoidally varying voltages and currents. He argued that transformers (which we will study in this chapter) can be used to step the voltage up and down with ac but not with dc; low voltages are safer for consumer use, but high voltages and correspondingly low currents are best for long-distance power transmission to minimize i^2R losses in the cables.

Eventually, Westinghouse prevailed, and most present-day household and industrial power-distribution systems operate with alternating current. Any appliance that you plug into a wall outlet uses ac, and many battery-powered devices such as radios and cordless telephones make use of the dc supplied by the battery to create or amplify alternating currents. Circuits in modern communication equipment, including pagers and television, also make extensive use of ac.

In this chapter we will learn how resistors, inductors, and capacitors behave in circuits with sinusoidally varying voltages and currents. Many of the principles that we found useful in Chapters 25, 28, and 30 are applicable, along with several new concepts related to the circuit behavior of inductors and capacitors. A key concept in this discussion is *resonance,* which we studied in Chapter 13 for mechanical systems.

31.1 | Phasors and Alternating Currents

To supply an alternating current to a circuit, a source of alternating emf or voltage is required. An example of such a source is a coil of wire rotating with constant angular velocity in a magnetic field, which we discussed in Example 29.4 (Section 29.2). This develops a sinusoidal alternating emf and is the prototype of the commercial alternating-current generator or *alternator* (see Fig. 29.8).

We use the term **ac source** for any device that supplies a sinusoidally varying voltage (potential difference) v or current i. The usual circuit-diagram symbol for an ac source is

A sinusoidal voltage might be described by a function such as

$$v = V \cos \omega t \tag{31.1}$$

In this expression, v (lowercase) is the *instantaneous* potential difference; V (uppercase) is the maximum potential difference, which we call the **voltage amplitude;** and ω is the *angular frequency,* equal to 2π times the frequency f.

In the United States and Canada, commercial electric-power distribution systems always use a frequency of $f = 60$ Hz, corresponding to $\omega = (2\pi \text{ rad})(60 \text{ s}^{-1}) = 377$ rad/s; in much of the rest of the world, $f = 50$ Hz ($\omega = 314$ rad/s) is used. Similarly, a sinusoidal current might be described as

$$i = I \cos \omega t \tag{31.2}$$

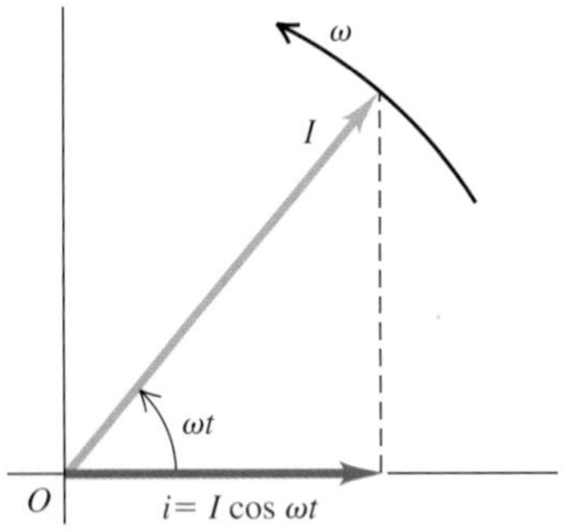

31.1 A phasor diagram. The projection of the rotating vector (phasor) onto the horizontal axis represents the instantaneous current.

where i (lowercase) is the instantaneous current and I (uppercase) is the maximum current or **current amplitude.**

To represent sinusoidally varying voltages and currents, we will use rotating vector diagrams similar to those we used in the study of simple harmonic motion in Section 13.2 (see Fig. 13.5). In these diagrams the instantaneous value of a quantity that varies sinusoidally with time is represented by the *projection* onto a horizontal axis of a vector with a length equal to the amplitude of the quantity. The vector rotates counterclockwise with constant angular speed ω. These rotating vectors are called **phasors,** and diagrams containing them are called **phasor diagrams.** Figure 31.1 shows a phasor diagram for the sinusoidal current described by Eq. (31.2). The projection of the phasor onto the horizontal axis at time t is $I \cos \omega t$; this is why we chose to use the cosine function rather than the sine in Eq. (31.2).

CAUTION A phasor is not a real physical quantity with a direction in space, such as velocity, momentum, or electric field. Rather, it is a *geometric* entity that helps us to describe and analyze physical quantities that vary sinusoidally with time. In Section 13.2 we used a single phasor to represent the position of a point mass undergoing simple harmonic motion. In this chapter we will use phasors to *add* sinusoidal voltages and currents. Combining sinusoidal quantities with phase differences then becomes a matter of vector addition. We will find a similar use for phasors in Chapters 35 and 36 in our study of interference effects with light.

How do we measure a sinusoidally varying current? In Section 26.3 we used a d'Arsonval galvanometer to measure steady currents. But if we pass a *sinusoidal* current through a d'Arsonval meter, the torque on the moving coil varies sinusoidally, with one direction half the time and the opposite direction the other half. The needle may wiggle a little if the frequency is low enough, but its average deflection is zero. Hence a d'Arsonval meter by itself isn't very useful for measuring alternating currents.

To get a measurable one-way current through the meter, we can use *diodes,* which we described in Section 25.3. A diode (or rectifier) is a device that conducts better in one direction than in the other; an ideal diode has zero resistance for one direction of current and infinite resistance for the other. One possible arrangement is shown in Fig. 31.2a. The current through the galvanometer G is always upward, regardless of the direction of the current from the ac source (i.e., which part of the cycle the source is in). The current through G is as shown by the graph in Fig. 31.2b. It pulsates but always has the same direction, and the average meter deflection is *not* zero. This arrangement of diodes is called a *full-wave rectifier circuit.*

The **rectified average current** I_{rav} is defined so that during any whole number of cycles, the total charge that flows is the same as though the current were constant with a value equal to I_{rav}. The notation I_{rav} and the name *rectified average* current emphasize that this is *not* the average of the original sinusoidal current. In Fig. 31.2b the total charge that flows in time t corresponds to the area under the curve of i versus t (recall that $i = dq/dt$, so q is the integral of t); this area must equal the rectangular area with height I_{rav}. We see that I_{rav} is less than the maximum current I; the two are related by

$$I_{rav} = \frac{2}{\pi}I = 0.637I \quad (31.3)$$

(rectified average value of a sinusoidal current)

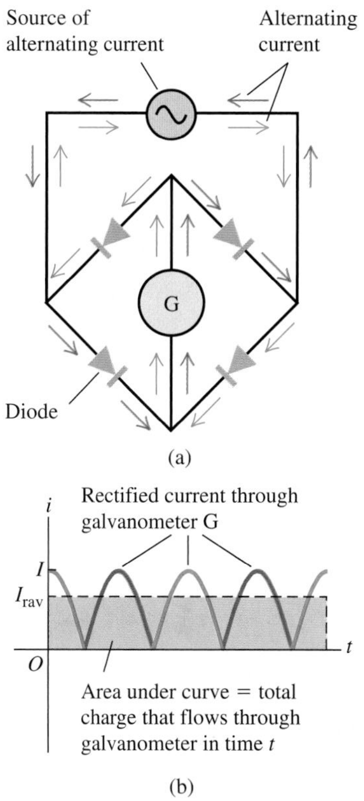

31.2 (a) A full-wave rectifier circuit. The arrow-bar symbol for a diode indicates the directions in which current can and cannot pass. (b) Graph of the full-wave rectified current and its average value, the rectified average current I_{rav}.

(The factor of $2/\pi$ is the average value of $|\cos \omega t|$ or of $|\sin \omega t|$; see Example 29.4 in Section 29.2.) The galvanometer deflection is proportional to I_{rav}. The galvanometer scale can be calibrated to read I, I_{rav}, or, most commonly, I_{rms} (discussed below).

A more useful way to describe a quantity that can be either positive or negative is the *root-mean-square (rms) value.* We used rms values in Section 18.3 in connection with the speeds of molecules in a gas. We *square* the instantaneous current i, take the *average* (mean) value of i^2, and finally take the *square root* of that average. This procedure defines the **root-mean-square current,** denoted as I_{rms}. Even when i is negative, i^2 is always positive, so I_{rms} is never zero (unless i is zero at every instant).

Here's how we obtain I_{rms}. If the instantaneous current is given by $i = I \cos \omega t$, then

$$i^2 = I^2 \cos^2 \omega t$$

Using a double-angle formula from trigonometry,

$$\cos^2 A = \frac{1}{2}(1 + \cos 2A)$$

we find

$$i^2 = I^2\frac{1}{2}(1 + \cos 2\omega t) = \frac{1}{2}I^2 + \frac{1}{2}I^2 \cos 2\omega t$$

The average of $\cos 2\omega t$ is zero because it is positive half the time and negative half the time. Thus the average of i^2 is simply $I^2/2$. The square root of this is I_{rms}:

$$I_{rms} = \frac{I}{\sqrt{2}} \quad \text{(root-mean-square value of a sinusoidal current)} \quad (31.4)$$

In the same way, the root-mean-square value of a sinusoidal voltage with amplitude (maximum value) V is

$$V_{rms} = \frac{V}{\sqrt{2}} \quad \text{(root-mean-square value of a sinusoidal voltage)} \qquad (31.5)$$

31.3 This wall socket delivers a root-mean-square voltage of 120 V. Sixty times per second, the instantaneous voltage across its terminals varies from $(\sqrt{2})(120\text{ V}) = 170\text{ V}$ to -170 V and back again.

We can convert a rectifying ammeter into a voltmeter by adding a series resistor, just as for the dc case discussed in Section 26.3. Meters used for ac voltage and current measurements are nearly always calibrated to read rms values, not maximum or rectified average. Voltages and currents in power distribution systems are always described in terms of their rms values. The usual household power supply, "120-volt ac," has an rms voltage of 120 V (Fig. 31.3). The voltage amplitude is

$$V = \sqrt{2}V_{rms} = \sqrt{2}(120\text{ V}) = 170\text{ V}$$

Example 31.1 Current in a personal computer

The plate on the back of a personal computer says that it draws 2.7 A from a 120-V, 60-Hz line. For this computer, what are a) the average current, b) the average of the square of the current, and c) the current amplitude?

SOLUTION

IDENTIFY and SET UP: This example is about alternating current. In parts (b) and (c) we use the idea that the root-mean-square current, given by Eq. (31.4), is the *square root* of the *mean* (average) of the *square* of the current.

EXECUTE: a) The average of *any* sinusoidal alternating current, over any whole number of cycles, is zero.

b) The current given is the rms value: $I_{rms} = 2.7$ A. The target variable $(i^2)_{av}$ is the *mean* of the *square* of the current. The rms current is the square root of this target variable, so

$$I_{rms} = \sqrt{(i^2)_{av}} \quad \text{or} \quad (i^2)_{av} = (I_{rms})^2 = (2.7\text{ A})^2 = 7.3\text{ A}^2$$

c) From Eq. (31.4) the current amplitude I is

$$I = \sqrt{2}I_{rms} = \sqrt{2}(2.7\text{ A}) = 3.8\text{ A}$$

EVALUATE: Why would we be interested in the average of the square of the current? Recall that the rate at which energy is dissipated in a resistor R equals i^2R. This rate varies if the current is alternating, so it is best described by its average value $(i^2)_{av}R = I_{rms}^2R$. We make use of this idea in Section 31.4.

Test Your Understanding

It is sometimes said that when you pay your home electric bill, you are "paying for electrons." Is this description accurate?

31.2 | Resistance and Reactance

In this section we will derive voltage–current relations for individual circuit elements carrying a sinusoidal current. We'll consider resistors, inductors, and capacitors.

Resistor in an ac Circuit

First let's consider a resistor with resistance R through which there is a sinusoidal current given by Eq. (31.2): $i = I\cos\omega t$. The positive direction of current is counterclockwise around the circuit, as in Fig. 31.4a. The current amplitude (maximum current) is I. From Ohm's law the instantaneous potential v_R of point a with respect to point b (that is, the instantaneous voltage across the resistor) is

$$v_R = iR = (IR)\cos\omega t \tag{31.6}$$

The maximum voltage V_R, the *voltage amplitude*, is the coefficient of the cosine function:

$$V_R = IR \qquad \text{(amplitude of voltage across a resistor, ac circuit)} \tag{31.7}$$

Hence we can also write

$$v_R = V_R\cos\omega t \tag{31.8}$$

The current i and voltage v_R are both proportional to $\cos\omega t$, so the current is *in phase* with the voltage. Equation (31.7) shows that the current and voltage amplitudes are related in the same way as in a dc circuit.

Figure 31.4b shows graphs of i and v_R as functions of time. The vertical scales for current and voltage are different, so the relative heights of the two curves are not significant. The corresponding phasor diagram is given in Fig. 31.4a. Because i and v_R are *in phase* and have the same frequency, the current and voltage phasors rotate together; they are parallel at each instant. Their projections on the horizontal axis represent the instantaneous current and voltage, respectively.

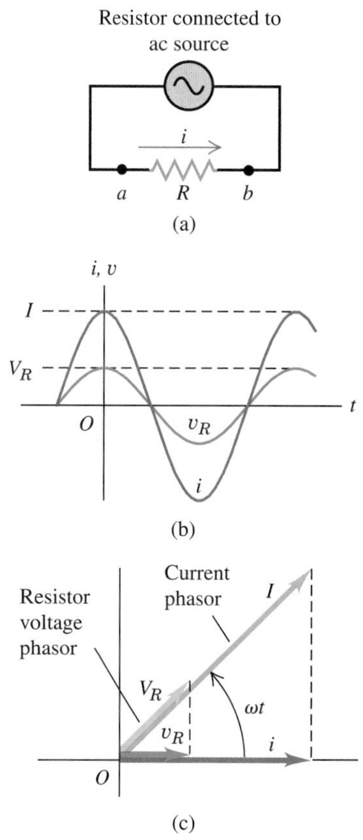

31.4 (a) Resistance R connected across an ac source. (b) Graphs of instantaneous voltage (blue) and current (purple). (c) Phasor diagram. As the voltage and current phasors rotate together in the counterclockwise direction, voltage and current remain in phase.

Inductor in an ac Circuit

Next, we replace the resistor in Fig. 31.4 with a pure inductor with self-inductance L and zero resistance (Fig. 31.5a). Again we assume that the current is $i = I\cos\omega t$, with the positive direction of current taken as counterclockwise around the circuit.

Although there is no resistance, there is a potential difference v_L between the inductor terminals a and b because the current varies with time, giving rise to a self-induced emf. The induced emf in the direction of i is given by Eq. (30.7), $\mathcal{E} = -L\,di/dt$; however, the voltage v_L is *not* simply equal to $\mathcal{E}$. To see why, notice that if the current in the inductor is in the positive (counterclockwise) direction

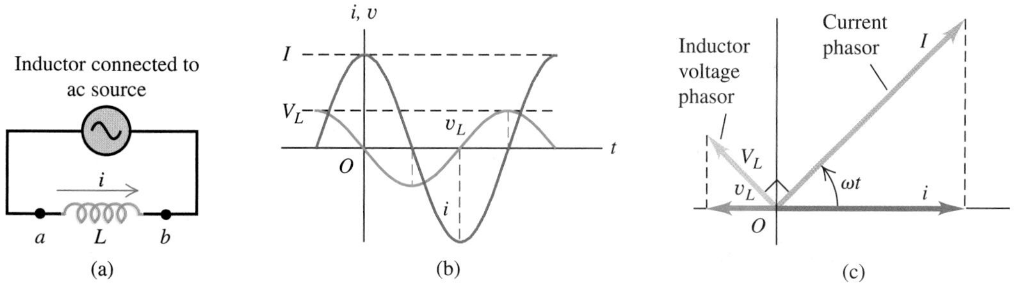

31.5 (a) Inductance L connected across an ac source. (b) Graphs of instantaneous voltage (blue) and current (purple). (c) Phasor diagram. As the voltage and current phasors rotate together in the counterclockwise direction, voltage *leads* current by 90°.

from a to b and is increasing, then di/dt is positive and the induced emf is directed to the left to oppose the increase in current; hence point a is at higher potential than is point b. Thus the potential of point a with respect to point b is positive and is given by $v_L = +L\, di/dt$, the *negative* of the induced emf. (You should convince yourself that this expression gives the correct sign of v_L in *all* cases, including i counterclockwise and decreasing, i clockwise and increasing, and i clockwise and decreasing; you should also review Section 30.2.) So we have

ActivPhysics ONLINE
14.3 AC Circuits: The Driven Oscillator (Questions 1–5)

$$v_L = L\frac{di}{dt} = L\frac{d}{dt}(I\cos\omega t) = -I\omega L\sin\omega t \qquad (31.9)$$

The voltage v_L across the inductor at any instant is proportional to the *rate of change* of the current. The points of maximum voltage on the graph correspond to maximum steepness of the current curve, and the points of zero voltage are the points where the current curve instantaneously levels off at its maximum and minimum values (Fig. 31.5b). The voltage and current are "out of step" or *out of phase* by a quarter-cycle. Since the voltage peaks occur a quarter-cycle earlier than the current peaks, we say that the voltage *leads* the current by 90°. The phasor diagram in Fig. 31.5c also shows this relationship; the voltage phasor is ahead of the current phasor by 90°.

We can also obtain this phase relationship by rewriting Eq. (31.9) using the identity $\cos(A + 90°) = -\sin A$:

$$v_L = I\omega L\cos(\omega t + 90°) \qquad (31.10)$$

This result shows that the voltage can be viewed as a cosine function with a "head start" of 90° relative to the current.

As we have done in Eq. (31.10), we will usually describe the phase of the *voltage* relative to the *current,* not the reverse. Thus if the current i in a circuit is

$$i = I\cos\omega t$$

and the voltage v of one point with respect to another is

$$v = V\cos(\omega t + \phi)$$

we call ϕ the **phase angle;** it gives the phase of the *voltage* relative to the *current.* For a pure resistor, $\phi = 0$, and for a pure inductor, $\phi = 90°$.

From Eq. (31.9) or (31.10) the amplitude V_L of the inductor voltage is

$$V_L = I\omega L \qquad (31.11)$$

We define the **inductive reactance** X_L of an inductor as

$$X_L = \omega L \qquad \text{(inductive reactance)} \qquad (31.12)$$

Using X_L, we can write Eq. (31.11) in a form similar to Eq. (31.7) for a resistor $(V_R = IR)$:

$$V_L = IX_L \quad \text{(amplitude of voltage across an inductor, ac circuit)} \qquad (31.13)$$

Because X_L is the ratio of a voltage and a current, its SI unit is the ohm, the same as for resistance.

CAUTION Keep in mind that Eq. (31.13) is a relation between the *amplitudes* of the oscillating voltage and current for the inductor in Fig. 31.5a. It does *not* say that the voltage at any instant is equal to the current at that instant multiplied by X_L. As Fig. 31.5b shows, the voltage and current are 90° out of phase. Voltage and current are only in phase for resistors, as in Eq. (31.6).

The inductive reactance X_L is really a description of the self-induced emf that opposes any change in the current through the inductor. From Eq. (31.13), for a given current amplitude I the voltage $v_L = +L\,di/dt$ across the inductor and the self-induced emf $\mathcal{E} = -L\,di/dt$ both have an amplitude V_L that is directly proportional to X_L. According to Eq. (31.12), the inductive reactance and self-induced emf increase with more rapid variation in current (that is, increasing angular frequency ω) and increasing inductance L.

If an oscillating voltage of a given amplitude V_L is applied across the inductor terminals, the resulting current will have a smaller amplitude I for larger values of X_L. Since X_L is proportional to frequency, a high-frequency voltage applied to the inductor gives only a small current, while a lower-frequency voltage of the same amplitude gives rise to a larger current. Inductors are used in some circuit applications, such as power supplies and radio-interference filters, to block high frequencies while permitting lower frequencies or dc to pass through. A circuit device that uses an inductor for this purpose is called a *low-pass filter* (see Problem 31.46).

Example 31.2 An inductor in an ac circuit

Suppose you want the current amplitude in a pure inductor in a radio receiver to be 250 μA when the voltage amplitude is 3.60 V at a frequency of 1.60 MHz (corresponding to the upper end of the AM broadcast band). a) What inductive reactance is needed? What inductance? b) If the voltage amplitude is kept constant, what will be the current amplitude through this inductor at 16.0 MHz? At 160 kHz?

SOLUTION

IDENTIFY: We are not told about any other elements of the circuit of which the inductor is part. Nor should we care about those other elements, since from the perspective of this example, all they do is provide the inductor with an oscillating voltage. Hence all of those other circuit elements are lumped into the ac source shown in Fig. 31.5a.

SET UP: We are given the current amplitude I and the voltage amplitude V. Our target variables in part (a) are the inductive reactance X_L at 1.60 MHz and the inductance L, which we find using Eqs. (31.13) and (31.12). Once we know L, we use these same two equations to find the inductive reactance and current amplitude at any other frequency.

EXECUTE: a) From Eq. (31.13),

$$X_L = \frac{V_L}{I} = \frac{3.60\ \text{V}}{250 \times 10^{-6}\ \text{A}} = 1.44 \times 10^4\ \Omega = 14.4\ \text{k}\Omega$$

From Eq. (31.12), with $\omega = 2\pi f$, we find

$$L = \frac{X_L}{2\pi f} = \frac{1.44 \times 10^4\ \Omega}{2\pi(1.60 \times 10^6\ \text{Hz})} = 1.43 \times 10^{-3}\ \text{H} = 1.43\ \text{mH}$$

b) Combining Eqs. (31.12) and (31.13), we find that the current amplitude is $I = V_L/X_L = V_L/\omega L = V_L/2\pi f L$. Thus the current amplitude is inversely proportional to the frequency f. Since $I = 250\ \mu\text{A}$ at $f = 1.60$ MHz, the current amplitude at 16.0 MHz (ten times the original frequency) will be one-tenth as great, or 25.0 μA; at 160 kHz = 0.160 MHz (one-tenth of the original frequency) the current amplitude is ten times as great, or 2500 μA = 2.50 mA.

EVALUATE: In general, the lower the frequency of an oscillating voltage applied across an inductor, the greater the amplitude of the oscillating current that results.

Capacitor in an ac Circuit

Finally, we connect a capacitor with capacitance C to the source, as in Fig. 31.6a, producing a current $i = I\cos\omega t$ through the capacitor. Again, the positive direction of current is counterclockwise around the circuit.

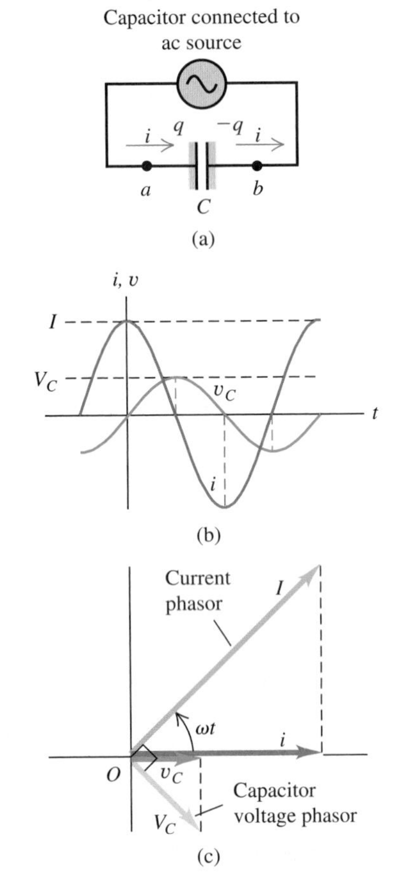

31.6 (a) Capacitor C connected across an ac source. (b) Graphs of instantaneous voltage (blue) and current (purple). (c) Phasor diagram. As the voltage and current phasors rotate together in the counterclockwise direction, voltage *lags* current by 90°.

CAUTION You may object that charge can't really move through the capacitor because its two plates are insulated from each other. True enough, but as the capacitor charges and discharges, there is at each instant a current i into one plate, an equal current out of the other plate, and an equal *displacement* current between the plates just as though the charge were being conducted through the capacitor. (You may want to review the discussion of displacement current in Section 29.7.) Thus we often speak about alternating current *through* a capacitor.

To find the instantaneous voltage v_C across the capacitor—that is, the potential of point a with respect to point b—we first let q denote the charge on the left-hand plate of the capacitor in Fig. 31.6a (so $-q$ is the charge on the right-hand plate). The current i is related to q by $i = dq/dt$; with this definition, positive current corresponds to an increasing charge on the left-hand capacitor plate. Then

$$i = \frac{dq}{dt} = I \cos \omega t$$

Integrating this, we get

$$q = \frac{I}{\omega} \sin \omega t \tag{31.14}$$

Also, from Eq. (24.1) the charge q equals the voltage v_C multiplied by the capacitance, $q = Cv_C$. Using this in Eq. (31.14), we find

$$v_C = \frac{I}{\omega C} \sin \omega t \tag{31.15}$$

The instantaneous current i is equal to the rate of change dq/dt of the capacitor charge q; since $q = Cv_C$, i is also proportional to the rate of change of voltage. (Compare to an inductor, for which the situation is reversed and v_L is proportional to the rate of change of i.) Figure 31.6b shows v_C and i as functions of t. Because $i = dq/dt = C\,dv_C/dt$, the current has its greatest magnitude when the v_C curve is rising or falling most steeply and is zero when the v_C curve instantaneously levels off at its maximum and minimum values.

The capacitor voltage and current are out of phase by a quarter-cycle. The peaks of voltage occur a quarter-cycle *after* the corresponding current peaks, and we say that the voltage *lags* the current by 90°. The phasor diagram in Fig. 31.6c shows this relationship; the voltage phasor is behind the current phasor by a quarter-cycle, or 90°.

We can also derive this phase difference by rewriting Eq. (31.15), using the identity $\cos(A - 90°) = \sin A$:

$$v_C = \frac{I}{\omega C} \cos(\omega t - 90°) \tag{31.16}$$

This corresponds to a phase angle $\phi = -90°$. This cosine function has a "late start" of 90° compared with the current $i = I \cos \omega t$.

Equations (31.15) and (31.16) show that the *maximum* voltage V_C (the voltage amplitude) is

$$V_C = \frac{I}{\omega C} \tag{31.17}$$

To put this expression in a form similar to Eq. (31.7) for a resistor, $V_R = IR$, we define a quantity X_C, called the **capacitive reactance** of the capacitor, as

$$X_C = \frac{1}{\omega C} \quad \text{(capacitive reactance)} \quad (31.18)$$

Then

$$V_C = IX_C \quad \text{(amplitude of voltage across a capacitor, ac circuit)} \quad (31.19)$$

The SI unit of X_C is the ohm, the same as for resistance and inductive reactance, because X_C is the ratio of a voltage and a current.

CAUTION Remember that Eq. (31.19) for a capacitor, like Eq. (31.13) for an inductor, is *not* a statement about the instantaneous values of voltage and current. The instantaneous values are actually 90° out of phase, as Fig. 31.6b shows. Rather, Eq. (31.19) relates the *amplitudes* of the voltage and current.

The capacitive reactance of a capacitor is inversely proportional both to the capacitance C and to the angular frequency ω; the greater the capacitance and the higher the frequency, the *smaller* the capacitive reactance X_C. Capacitors tend to pass high-frequency current and to block low-frequency currents and dc, just the opposite of inductors. A device that preferentially passes signals of high frequency is called a *high-pass filter* (see Problem 31.45).

Example 31.3 A resistor and a capacitor in an ac circuit

A 200-Ω resistor is connected in series with a 5.0-μF capacitor. The voltage across the resistor is $v_R = (1.20\text{ V})\cos(2500\text{ rad/s})t$. a) Derive an expression for the circuit current. b) Determine the capacitive reactance of the capacitor. c) Derive an expression for the voltage v_C across the capacitor.

SOLUTION

IDENTIFY and SET UP: We use the idea that circuit elements in series have the same current. We find the current through the resistor, and hence through the circuit as a whole, using Eq. (31.6). We use Eq. (31.18) to find the capacitive reactance X_C, Eq. (31.19) to find the voltage amplitude, and Eq. (31.16) to write an expression for the instantaneous voltage across the capacitor.

EXECUTE: a) Using $v_R = iR$, we find that the current i in the resistor, and hence the current through the circuit as a whole, is

$$i = \frac{v_R}{R} = \frac{(1.20\text{ V})\cos(2500\text{ rad/s})t}{200\ \Omega}$$
$$= (6.0 \times 10^{-3}\text{ A})\cos(2500\text{ rad/s})t$$

b) From Eq. (31.18), the capacitive reactance at $\omega = 2500$ rad/s is

$$X_C = \frac{1}{\omega C} = \frac{1}{(2500\text{ rad/s})(5.0 \times 10^{-6}\text{ F})} = 80\ \Omega$$

c) From Eq. (31.19), the amplitude V_C of the voltage across the capacitor is

$$V_C = IX_C = (6.0 \times 10^{-3}\text{ A})(80\ \Omega) = 0.48\text{ V}$$

The 80-Ω reactance of the capacitor is 40% of the resistor's 200-Ω resistance, so the value of V_C is 40% of V_R. The instantaneous capacitor voltage v_C is given by Eq. (31.16):

$$v_C = V_C\cos(\omega t - 90°)$$
$$= (0.48\text{ V})\cos[(2500\text{ rad/s})t - \pi/2\text{ rad}]$$

EVALUATE: Although the *current* through the capacitor is the same as through the resistor, the *voltages* across these two devices are different in both amplitude and phase. Note that in the expression for v_C we converted the 90° to $\pi/2$ rad so that all the angular quantities have the same units. In ac circuit analysis, phase angles are often given in degrees, so be careful to convert to radians when necessary.

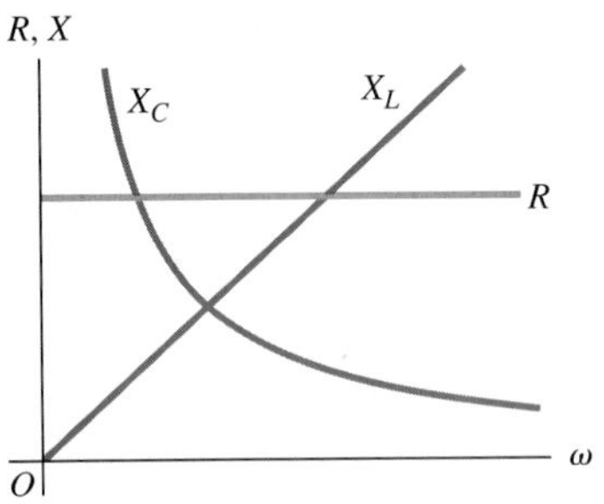

31.7 Graphs of R, X_L, and X_C as functions of angular frequency ω.

Table 31.1 Circuit Elements with Alternating Current

Circuit Element	Amplitude Relation	Circuit Quantity	Phase of v
Resistor	$V_R = IR$	R	In phase with i
Inductor	$V_L = IX_L$	$X_L = \omega L$	Leads i by 90°
Capacitor	$V_C = IX_C$	$X_C = 1/\omega C$	Lags i by 90°

Table 31.1 summarizes the relations of voltage and current amplitudes for the three circuit elements we have discussed. Note again that *instantaneous* voltage and current are proportional in a resistor, where there is zero phase difference between v_R and i (see Fig. 31.4b). The instantaneous voltage and current are *not* proportional in an inductor or capacitor, because there is a 90° phase difference in both cases (see Figs. 31.5b and 31.6b).

Figure 31.7 shows how the resistance of a resistor and the reactances of an inductor and a capacitor vary with angular frequency ω. Resistance R is independent of frequency, while the reactances X_L and X_C are not. If $\omega = 0$, corresponding to a dc circuit, there is *no* current through a capacitor because $X_C \rightarrow \infty$, and there is no inductive effect because $X_L = 0$. In the limit $\omega \rightarrow \infty$, X_L also approaches infinity, and the current through an inductor becomes vanishingly small; recall that the self-induced emf opposes rapid changes in current. In this same limit, X_C and the voltage across a capacitor both approach zero; the current changes direction so rapidly that no charge can build up on either plate.

Figure 31.8 shows an application of the above discussion to a loudspeaker system. Low-frequency sounds are produced by the *woofer,* which is a speaker with large diameter; the *tweeter,* a speaker of smaller diameter, produces high-frequency sounds. In order to route signals of different frequency to the appropriate speaker, the woofer and tweeter are connected in parallel across the amplifier output. The capacitor in the tweeter branch blocks the low-frequency components of sound but passes the higher frequencies; the inductor in the woofer branch does the opposite.

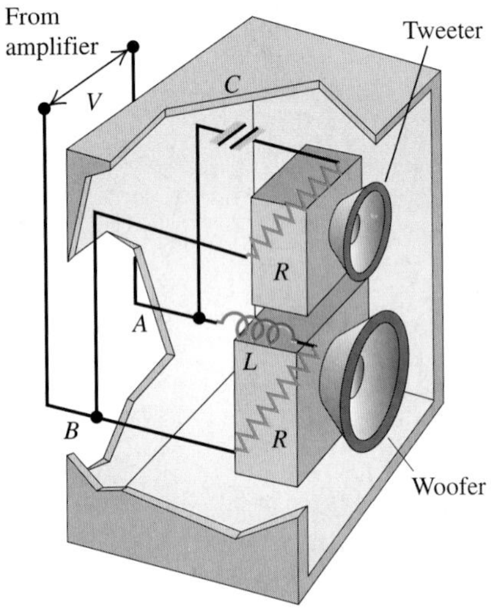

(a)

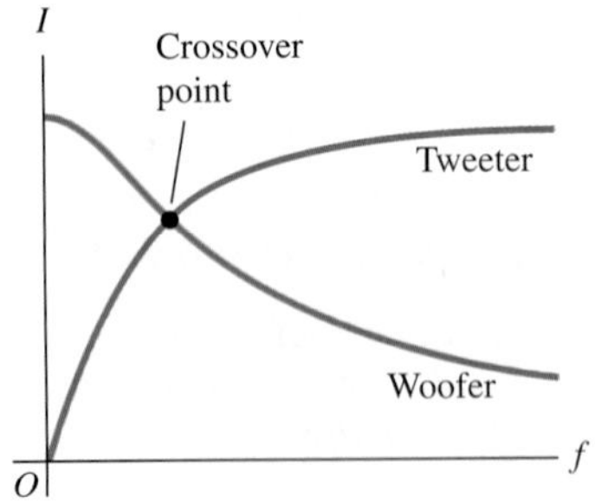

(b)

31.8 (a) A crossover network in a loudspeaker system. The two speakers are connected in parallel to the amplifier. (b) Graphs of current amplitude in the tweeter and woofer as functions of frequency for a given amplifier voltage amplitude. The point where the two curves cross is called the crossover point. The inductance and capacitance feed frequencies below the crossover point predominantly to the woofer, and frequencies above the crossover point to the tweeter.

Test Your Understanding

An oscillating voltage of fixed amplitude is applied across a circuit element. If the frequency of this voltage is increased, how does the amplitude of the current through the element change if it is a) a resistor, b) an inductor, or c) a capacitor?

31.3 | The *L-R-C* Series Circuit

Many ac circuits used in practical electronic systems involve resistance, inductive reactance, and capacitive reactance. A simple example is a series circuit containing a resistor, an inductor, a capacitor, and an ac source, as shown in Fig. 31.9a. (In Section 30.6 we considered the behavior of the current in an *L-R-C* series circuit *without* a source.)

To analyze this and similar circuits, we will use a phasor diagram that includes the voltage and current phasors for each of the components. In this circuit, because of Kirchhoff's loop rule, the instantaneous *total* voltage v_{ad} across all three components is equal to the source voltage at that instant. We will show that the phasor representing this total voltage is the *vector sum* of the phasors for the individual voltages. Complete phasor diagrams for this circuit are shown in Figs. 31.9a and 31.9b. These may appear complex, but we'll explain them one step at a time.

Let's assume that the source supplies a current i given by $i = I\cos\omega t$. Because the circuit elements are connected in series, the current at any instant is the same at every point in the circuit. Thus a *single phasor* I, with length proportional to the current amplitude, represents the current in *all* circuit elements.

As in Section 31.2, we use the symbols v_R, v_L, and v_C for the instantaneous voltages across R, L, and C, and the symbols V_R, V_L, and V_C for the maximum voltages. We denote the instantaneous and maximum *source* voltages by v and V. Then, in Fig. 31.9a, $v = v_{ad}$, $v_R = v_{ab}$, $v_L = v_{bc}$, and $v_C = v_{cd}$.

We have shown that the potential difference between the terminals of a resistor is *in phase* with the current in the resistor and that its maximum value V_R is given by Eq. (31.7):

$$V_R = IR$$

The phasor V_R in Fig. 31.9b, in phase with the current phasor I, represents the voltage across the resistor. Its projection onto the horizontal axis at any instant gives the instantaneous potential difference v_R.

The voltage across an inductor *leads* the current by 90°. Its voltage amplitude is given by Eq. (31.13):

$$V_L = IX_L$$

The phasor V_L in Fig. 31.9b represents the voltage across the inductor, and its projection onto the horizontal axis at any instant equals v_L.

The voltage across a capacitor *lags* the current by 90°. Its voltage amplitude is given by Eq. (31.19):

$$V_C = IX_C$$

The phasor V_C in Fig. 31.9b represents the voltage across the capacitor, and its projection onto the horizontal axis at any instant equals v_C.

The instantaneous potential difference v between terminals a and d is equal at every instant to the (algebraic) sum of the potential differences v_R, v_L, and v_C. That is, it equals the sum of the *projections* of the phasors V_R, V_L, and V_C. But the sum of the projections of these phasors is equal to the *projection* of their *vector sum*. So the vector sum V must be the phasor that represents the source voltage v and the instantaneous total voltage v_{ad} across the series of elements.

To form this vector sum, we first subtract the phasor V_C from the phasor V_L. (These two phasors always lie along the same line, with opposite directions.) This gives the phasor $V_L - V_C$. This is always at right angles to the phasor V_R, so from the Pythagorean theorem the magnitude of the phasor V is

$$V = \sqrt{V_R^2 + (V_L - V_C)^2} = \sqrt{(IR)^2 + (IX_L - IX_C)^2}$$

or

$$V = I\sqrt{R^2 + (X_L - X_C)^2} \tag{31.20}$$

We define the **impedance** Z of an ac circuit as the ratio of the voltage amplitude across the circuit to the current amplitude in the circuit. From Eq. (31.20) the impedance of the *L-R-C* series circuit is

$$Z = \sqrt{R^2 + (X_L - X_C)^2} \tag{31.21}$$

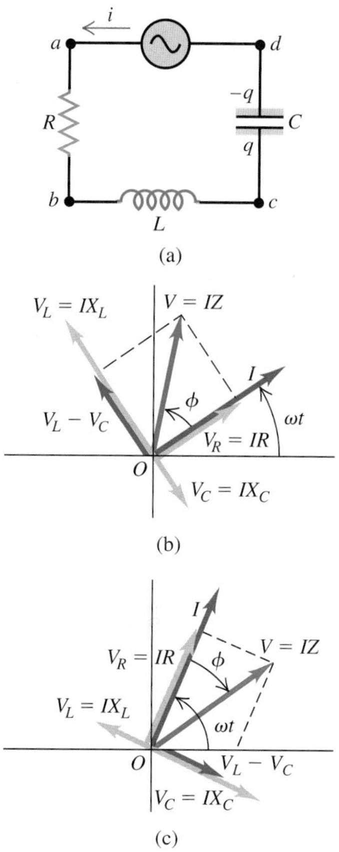

31.9 (a) An *L-R-C* series circuit with an ac source. (b) Phasor diagram for the case $X_L > X_C$. (c) Phasor diagram for the case $X_L < X_C$.

ActivPhysics Online

14.3 AC Circuits: The Driven Oscillator (Questions 6, 7, and 10)

31.10 This gas-filled glass sphere has an alternating voltage between its surface and the electrode at its center. The glowing streamers show the resulting alternating current that passes through the gas. When you touch the outside of the sphere, your fingertips and the inner surface of the sphere act as the plates of a capacitor, and the sphere and your body together form an *L-R-C* series circuit. The current (which is low enough to be harmless) is drawn to your fingers because the path through your body has a low impedance.

so we can rewrite Eq. (31.20) as

$$V = IZ \qquad \text{(amplitude of voltage across an ac circuit)} \qquad (31.22)$$

While Eq. (31.21) is valid only for an *L-R-C* series circuit, we can use Eq. (31.22) to define the impedance of *any* network of resistors, inductors, and capacitors as the ratio of the amplitude of the voltage across the network to the current amplitude. The SI unit of impedance is the ohm.

Equation (31.22) has a form similar to $V = IR$, with impedance Z in an ac circuit playing the role of resistance R in a dc circuit. Just as direct current tends to follow the path of least resistance, so alternating current tends to follow the path of lowest impedance (Fig. 31.10). Note, however, that impedance is actually a function of R, L, and C, as well as of the angular frequency ω. We can see this by substituting Eq. (31.12) for X_L and Eq. (31.18) for X_C into Eq. (31.21), giving the following complete expression for Z for a series circuit:

$$\begin{aligned} Z &= \sqrt{R^2 + (X_L - X_C)^2} \\ &= \sqrt{R^2 + [\omega L - (1/\omega C)]^2} \end{aligned} \qquad (31.23)$$

(impedance of an *L-R-C* series circuit)

Hence for a given amplitude V of the source voltage applied to the circuit, the amplitude $I = V/Z$ of the resulting current will be different at different frequencies. We'll explore this frequency dependence in detail in Section 31.5.

In the phasor diagram shown in Fig. 31.9b, the angle ϕ between the voltage and current phasors is the phase angle of the source voltage v with respect to the current i; that is, it is the angle by which the source voltage leads the current. From the diagram,

$$\tan\phi = \frac{V_L - V_C}{V_R} = \frac{I(X_L - X_C)}{IR} = \frac{X_L - X_C}{R}$$

$$\tan\phi = \frac{\omega L - 1/\omega C}{R} \qquad \text{(phase angle of an } L\text{-}R\text{-}C \text{ series circuit)} \qquad (31.24)$$

If the current is $i = I\cos\omega t$, then the source voltage v is

$$v = V\cos(\omega t + \phi) \qquad (31.25)$$

Figure 31.9b shows the behavior of a circuit in which $X_L > X_C$. Figure 31.9c shows the behavior when $X_L < X_C$; the voltage phasor V lies on the opposite side of the current phasor I and the voltage *lags* the current. In this case, $X_L - X_C$ is *negative*, $\tan\phi$ is negative, and ϕ is a negative angle between 0 and $-90°$. Since X_L and X_C depend on frequency, the phase angle ϕ depends on frequency as well. We'll examine the consequences of this in Section 31.5.

All of the expressions that we've developed for an *L-R-C* series circuit are still valid if one of the circuit elements is missing. If the resistor is missing, we set $R = 0$; if the inductor is missing, we set $L = 0$. But if the capacitor is missing, we set $C = \infty$, corresponding to the absence of any potential difference ($v_C = q/C = 0$) or any capacitive reactance ($X_C = 1/\omega C = 0$).

In this entire discussion we have described magnitudes of voltages and currents in terms of their *maximum* values, the voltage and current *amplitudes*. But we remarked at the end of Section 31.1 that these quantities are usually described in terms of rms values, not amplitudes. For any sinusoidally varying quantity the rms value is always $1/\sqrt{2}$ times the amplitude. All the relations between voltage and current that we have derived in this and the preceding sections are still valid if we use rms quantities throughout instead of amplitudes. For example, if we divide Eq. (31.22) by $\sqrt{2}$, we get

$$\frac{V}{\sqrt{2}} = \frac{I}{\sqrt{2}}Z$$

which we can rewrite as

$$V_{\mathrm{rms}} = I_{\mathrm{rms}}Z \tag{31.26}$$

We can translate Eqs. (31.7), (31.13), and (31.19) in exactly the same way.

We have considered only ac circuits in which an inductor, a resistor, and a capacitor are in series. You can do a similar analysis for a *parallel L-R-C* circuit; see Problem 31.50.

Finally, we remark that in this section we have been describing the *steady-state* condition of a circuit, the state that exists after the circuit has been connected to the source for a long time. When the source is first connected, there may be additional voltages and currents, called *transients,* whose nature depends on the time in the cycle when the circuit is initially completed. A detailed analysis of transients is beyond our scope. They always die out after a sufficiently long time, and they do not affect the steady-state behavior of the circuit. But they can cause dangerous and damaging surges in power lines, which is why delicate electronic systems such as computers are often provided with power-line surge protectors.

Problem-Solving Strategy: Alternating-Current Circuits

IDENTIFY *the relevant concepts:* All of the concepts that we used to analyze direct-current circuits also apply to alternating-current circuits. However, we must be careful to distinguish between the amplitudes of alternating currents and voltages and their instantaneous values. We must also keep in mind the distinctions between resistance (for resistors), reactance (for inductors or capacitors), and impedance (for composite circuits).

SET UP *the problem using the following steps:*

1. Draw a diagram of the circuit and label all known and unknown quantities.
2. Determine the target variables.

EXECUTE *the solution as follows:*

1. Use the relationships derived in Sections 31.2 and 31.3 to solve for the target variables, using the following hints.
2. In ac circuit problems it is nearly always easiest to work with angular frequency ω. If you are given the ordinary frequency f, expressed in Hz, convert it using the relation $\omega = 2\pi f$.
3. Keep in mind a few basic facts about phase relationships. For a resistor, voltage and current are always *in phase,* and the two corresponding phasors in a phasor diagram always have the same direction. For an inductor, the voltage always *leads* the current by 90° (i.e., $\phi = +90°$), and the voltage phasor is always turned 90° counterclockwise from the current phasor. For a capacitor, the voltage always *lags* the current by 90° (i.e., $\phi = -90°$), and the voltage phasor is always turned 90° clockwise from the current phasor.
4. Remember that with ac circuits, all voltages and currents are sinusoidal functions of time instead of being constant, but Kirchhoff's rules hold nonetheless at each instant. Thus, in a series circuit, the instantaneous current is the same in all circuit elements; in a parallel circuit, the instantaneous potential difference is the same across all circuit elements.
5. Inductive reactance, capacitive reactance, and impedance are analogous to resistance; each represents the ratio of voltage amplitude V to current amplitude I in a circuit element or combination of elements. Keep in mind, however, that

phase relations play an essential role. The effects of resistance and reactance have to be combined by *vector* addition of the corresponding voltage phasors, as in Figs. 31.9b and 31.9c. When you have several circuit elements in series, for example, you can't just *add* all the numerical values of resistance and reactance to get the impedance; that would ignore the phase relations.

EVALUATE *your answer:* When working with a series *L-R-C* circuit, you can check your results by comparing the values of the inductive reactance X_L and the capacitive reactance X_C. If $X_L > X_C$, then the voltage amplitude across the inductor is greater than that across the capacitor and the phase angle ϕ is positive (between 0 and 90°). If $X_L < X_C$, then the voltage amplitude across the inductor is less than that across the capacitor and the phase angle ϕ is negative between (0 and −90°).

Example 31.4 An *L-R-C* series circuit I

In the series circuit of Fig. 31.9a, suppose $R = 300\ \Omega$, $L = 60$ mH, $C = 0.50\ \mu$F, $V = 50$ V, and $\omega = 10{,}000$ rad/s. Find the reactances X_L and X_C, the impedance Z, the current amplitude I, the phase angle ϕ, and the voltage amplitude across each circuit element.

SOLUTION

IDENTIFY and SET UP: We use Eqs. (31.12) and (31.18) to determine the reactances and Eq. (31.23) to find the impedance. We then use Eq. (31.22) to find the current amplitude and Eq. (31.24) to calculate the phase angle. Given this information, the relationships in Table 31.1 tell us the voltage amplitudes.

EXECUTE: The inductive and capacitive reactances are

$$X_L = \omega L = (10{,}000\ \text{rad/s})(60\ \text{mH}) = 600\ \Omega$$

$$X_C = \frac{1}{\omega C} = \frac{1}{(10{,}000\ \text{rad/s})(0.50 \times 10^{-6}\ \text{F})} = 200\ \Omega$$

The impedance Z of the circuit is

$$Z = \sqrt{R^2 + (X_L - X_C)^2} = \sqrt{(300\ \Omega)^2 + (600\ \Omega - 200\ \Omega)^2}$$
$$= 500\ \Omega$$

With source voltage amplitude $V = 50$ V the current amplitude is

$$I = \frac{V}{Z} = \frac{50\ \text{V}}{500\ \Omega} = 0.10\ \text{A}$$

The phase angle ϕ is

$$\phi = \arctan\frac{X_L - X_C}{R} = \arctan\frac{400\ \Omega}{300\ \Omega} = 53^\circ$$

From Table 31.1, the voltage amplitudes V_R, V_L, and V_C across the resistor, inductor, and capacitor, respectively, are

$$V_R = IR = (0.10\ \text{A})(300\ \Omega) = 30\ \text{V}$$
$$V_L = IX_L = (0.10\ \text{A})(600\ \Omega) = 60\ \text{V}$$
$$V_C = IX_C = (0.10\ \text{A})(200\ \Omega) = 20\ \text{V}$$

EVALUATE: Note that $X_L > X_C$ and hence the voltage amplitude across the inductor is greater than across the capacitor and ϕ is negative. The value $\phi = -53^\circ$ means that the voltage *leads* the current by 53°; this is like the situation shown in Fig. 31.9b.

Note that the source voltage amplitude $V = 50$ V is *not* equal to the sum of the voltage amplitudes across the separate circuit elements. (That is, 50 V ≠ 30 V + 60 V + 20 V.) Make sure you understand why not!

Example 31.5 An *L-R-C* series circuit II

For the *L-R-C* series circuit described in Example 31.4, describe the time dependence of the instantaneous current and each instantaneous voltage.

SOLUTION

IDENTIFY and SET UP: If we describe the current using Eq. (31.2), the voltages are given by Eq. (31.8) for the resistor, Eq. (31.10) for the inductor, Eq. (31.16) for the capacitor, and Eq. (31.25) for the source.

EXECUTE: The current and all of the voltages oscillate with the same angular frequency, $\omega = 10{,}000$ rad/s, and hence with the same period, $2\pi/\omega = 2\pi/(10{,}000\ \text{rad/s}) = 6.3 \times 10^{-4}\ \text{s} = 0.63$ ms. Using Eq. (31.2), the current is

$$i = I\cos\omega t = (0.10\ \text{A})\cos(10{,}000\ \text{rad/s})t$$

this choice simply means that we choose $t = 0$ to be an instant when the current is maximum. The resistor voltage is *in phase* with the current, so

$$v_R = V_R\cos\omega t = (30\ \text{V})\cos(10{,}000\ \text{rad/s})t$$

The inductor voltage *leads* the current by 90°, so

$$v_L = V_L \cos(\omega t + 90°) = -V_L \sin \omega t$$
$$= -(60\text{ V}) \sin(10{,}000\text{ rad/s})t$$

The capacitor voltage *lags* the current by 90°, so

$$v_C = V_C \cos(\omega t - 90°) = V_c \sin \omega t$$
$$= (20\text{ V}) \sin(10{,}000\text{ rad/s})t$$

Finally, the source voltage (equal to the voltage across the entire combination of resistor, inductor, and capacitor) *leads* the current by $\phi = 53°$, so

$$v = V \cos(\omega t + \phi)$$
$$= (50\text{ V}) \cos\left[(10{,}000\text{ rad/s})t + \left(\frac{2\pi\text{ rad}}{360°}\right)(53°)\right]$$
$$= (50\text{ V}) \cos[(10{,}000\text{ rad/s})t + 0.93\text{ rad}]$$

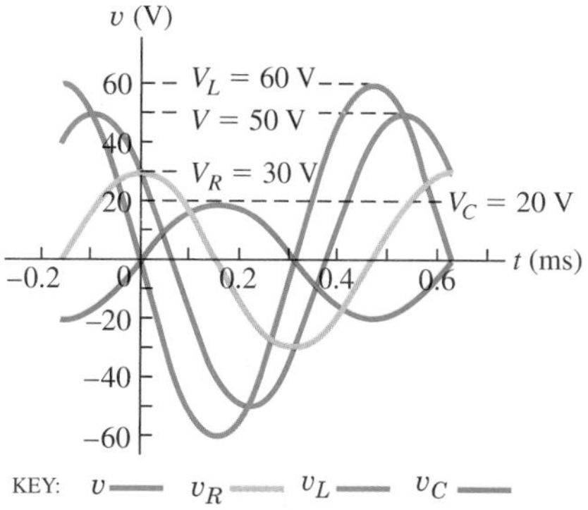

31.11 Graphs of the source voltage v, resistor voltage v_R, inductor voltage v_L, and capacitor voltage v_C as functions of time for the situation of Example 31.4. The current, which is not shown, is in phase with the resistor voltage.

EVALUATE: Figure 31.11 graphs the various voltages versus time. The inductor voltage has a larger amplitude than the capacitor voltage because $X_L > X_C$. While the source voltage amplitude V is not equal to the sum of the individual voltage amplitudes V_R, V_L, and V_C, the *instantaneous* source voltage v is always equal to the sum of the instantaneous voltages v_R, v_L, and v_C. You should verify this by measuring the values of the voltages shown in the graph at different values of the time t.

Test Your Understanding

For the situation described in Example 31.4, does the current amplitude increase or decrease if the inductor is removed from the circuit? If the inductor is kept but the capacitor is removed?

31.4 | Power in Alternating-Current Circuits

Alternating currents play a central role in systems for distributing, converting, and using electrical energy, so it's important to look at power relationships in ac circuits. For an ac circuit with instantaneous current i and current amplitude I, we'll consider an element of that circuit across which the instantaneous potential difference is v with voltage amplitude V. The instantaneous power p delivered to this circuit element is

$$p = vi$$

Let's first see what this means for individual circuit elements. We'll assume in each case that $i = I \cos \omega t$.

Suppose first that the circuit element is a *pure resistance* R, as in Fig. 31.4a; then $v = v_R$ and i are *in phase.* We obtain the graph representing p by multiplying the heights of the graphs of v and i in Fig. 31.4b at each instant. This graph is shown by the black curve in Fig. 31.12a. The product vi is always positive because v and i are always either both positive or both negative. Hence energy is supplied *to* the resistor at every instant for both directions of i, although the power is not constant.

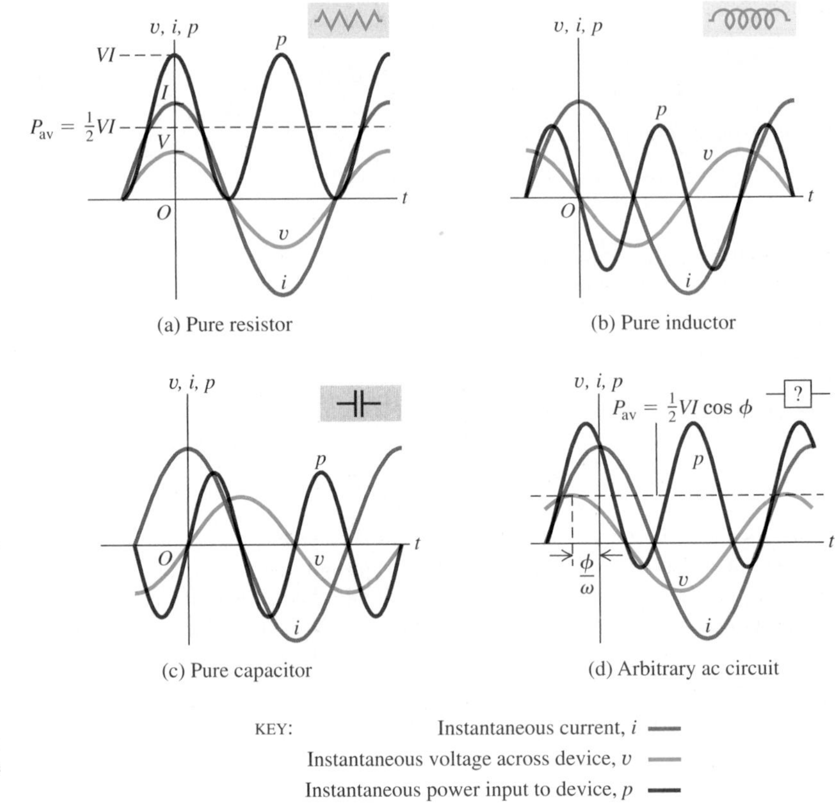

31.12 Graphs of voltage (blue), current (purple), and power (black) as functions of time for various circuits. (a) Instantaneous power input to a pure resistor. The average power is $\frac{1}{2}VI = V_{rms}I_{rms}$. (b) Instantaneous power input to a pure inductor. The average power is zero. (c) Instantaneous power input to a pure capacitor. The average power is zero. (d) Instantaneous power input to an arbitrary ac circuit. The average power is $\frac{1}{2}VI\cos\phi = V_{rms}I_{rms}\cos\phi$.

The power curve for a pure resistance is symmetrical about a value equal to one-half its maximum value VI, so the *average power* P_{av} is

$$P_{av} = \frac{1}{2}VI \qquad \text{(for a pure resistance)} \tag{31.27}$$

An equivalent expression is

$$P_{av} = \frac{V}{\sqrt{2}}\frac{I}{\sqrt{2}} = V_{rms}I_{rms} \qquad \text{(for a pure resistance)} \tag{31.28}$$

Also, $V_{rms} = I_{rms}R$, so we can express P_{av} by any of the equivalent forms

$$P_{av} = I_{rms}^2 R = \frac{V_{rms}^2}{R} = V_{rms}I_{rms} \qquad \text{(for a pure resistance)} \tag{31.29}$$

Note that the expressions in Eq. (31.29) have the same form as the corresponding relations for a dc circuit, Eq. (25.18). Also note that they are valid only for pure resistors, not for more complicated combinations of circuit elements.

Next we connect the source to a pure inductor L, as in Fig. 31.5a. The voltage $v = v_L$ leads the current i by 90°. When we multiply the curves of v and i, the product vi is *negative* during the half of the cycle when v and i have *opposite* signs. The power curve, shown in Fig. 31.12b, is symmetrical about the horizontal axis; it is positive half the time and negative the other half, and the average power is zero. When p is positive, energy is being supplied to set up the magnetic field in the inductor; when p is negative, the field is collapsing and the inductor is returning energy to the source. The net energy transfer over one cycle is zero.

Finally, we connect the source to a pure capacitor C, as in Fig. 31.6a. The voltage $v = v_C$ lags the current i by 90°. Figure 31.12c shows the power curve; the average power is again zero. Energy is supplied to charge the capacitor and is returned to the source when the capacitor discharges. The net energy transfer over one cycle is again zero.

In *any* ac circuit, with any combination of resistors, capacitors, and inductors, the voltage v across the entire circuit has some phase angle ϕ with respect to the current i. Then the instantaneous power p is given by

$$p = vi = [V\cos(\omega t + \phi)][I\cos\omega t] \tag{31.30}$$

The instantaneous power curve has the form shown in Fig. 31.12d. The area between the positive loops and the horizontal axis is greater than that between the negative loops and the horizontal axis, and the average power is positive.

We can derive from Eq. (31.30) an expression for the *average* power P_{av} by using the identity for the cosine of the sum of two angles:

$$\begin{aligned} p &= [V(\cos\omega t\cos\phi - \sin\omega t\sin\phi)][I\cos\omega t] \\ &= VI\cos\phi\cos^2\omega t - VI\sin\phi\cos\omega t\sin\omega t \end{aligned}$$

From the discussion in Section 31.1 that led to Eq. (31.4), we see that the average value of $\cos^2\omega t$ (over one cycle) is $\frac{1}{2}$. The average value of $\cos\omega t\sin\omega t$ is zero because this product is equal to $\frac{1}{2}\sin 2\omega t$, whose average over a cycle is zero. So the average power P_{av} is

$$P_{av} = \frac{1}{2}VI\cos\phi = V_{rms}I_{rms}\cos\phi \tag{31.31}$$

(average power into a general ac circuit)

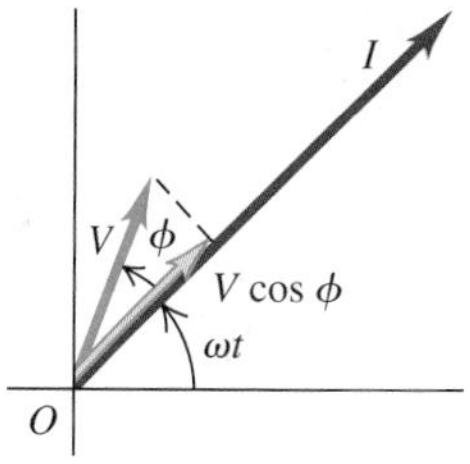

31.13 The average power is half the product of I and $V\cos\phi$, the component of V in phase with I.

When v and i are in phase, so $\phi = 0$, the average power equals $\frac{1}{2}VI = V_{rms}I_{rms}$; when v and i are 90° out of phase, the average power is zero. In the general case, when v has a phase angle ϕ with respect to i, the average power equals $\frac{1}{2}I$ multiplied by $V\cos\phi$, the component of the voltage phasor that is *in phase* with the current phasor. Figure 31.13 shows the general relationship of the current and voltage phasors. For the L-R-C series circuit, Figs. 31.9b and 31.9c show that $V\cos\phi$ equals the voltage amplitude V_R for the resistor; hence Eq. (31.31) is the average power dissipated in the resistor. On average there is no energy flow into or out of the inductor or capacitor, so none of P_{av} goes into either of these circuit elements.

The factor $\cos\phi$ is called the **power factor** of the circuit. For a pure resistance, $\phi = 0$, $\cos\phi = 1$, and $P_{av} = V_{rms}I_{rms}$. For a pure inductor or capacitor, $\phi = \pm 90°$, $\cos\phi = 0$, and $P_{av} = 0$. For an L-R-C series circuit the power factor is equal to R/Z; we leave the proof of this statement as a problem.

A low power factor (large angle ϕ of lag or lead) is usually undesirable in power circuits. The reason is that for a given potential difference, a large current is needed to supply a given amount of power. This results in large i^2R losses in the transmission lines. Your electric power company may charge a higher rate to a client with a low power factor. Many types of ac machinery draw a *lagging* current; that is, the current drawn by the machinery lags the applied voltage. Hence the voltage leads the current, so $\phi > 0$ and $\cos\phi < 1$. The power factor can be corrected toward the ideal value of 1 by connecting a capacitor in parallel with the load. The current drawn by the capacitor *leads* the voltage (that is, the voltage across the capacitor lags the current), which compensates for the lagging current in the other branch of the circuit. The capacitor itself absorbs no net power from the line.

Example 31.6 Power in a hair dryer

An electric hair dryer is rated at 1500 W at 120 V. The rated power of this hair dryer, or of any other ac device, is the *average* power drawn by the device, and the rated voltage is the *rms* voltage. Calculate a) the resistance, b) the rms current, and c) the maximum instantaneous power. Assume that the hair dryer is a pure resistance. (The hair dryer's heating element acts as a resistor.)

SOLUTION

IDENTIFY and SET UP: We are given the rms voltage V_{rms} and the average power P_{av}. Since the hair dryer is assumed to be a pure resistance, we solve Eq. (31.29) to determine the resistance R. We find the rms current from V_{rms} and P_{av} using Eq. (31.28), and find the maximum instantaneous power from Eq. (31.30).

EXECUTE: From Eq. (31.29), the resistance is

$$R = \frac{V_{rms}^{\ 2}}{P_{av}} = \frac{(120\ \text{V})^2}{1500\ \text{W}} = 9.6\ \Omega$$

b) From Eq. (31.28),

$$I_{rms} = \frac{P_{av}}{V_{rms}} = \frac{1500\ \text{W}}{120\ \text{V}} = 12.5\ \text{A}$$

c) For a pure resistance, the voltage and current are in phase and the phase angle ϕ is zero. Hence from Eq. (31.30), the instantaneous power is $p = VI\cos^2 \omega t$ and the maximum instantaneous power is $p_{max} = VI$. From Eq. (31.27), this is twice the average power P_{av}, so

$$p_{max} = VI = 2P_{av} = 2(1500\ \text{W}) = 3000\ \text{W}$$

EVALUATE: We can confirm our result in part (b) by using Eq. (31.7): $I_{rms} = V_{rms}/R = (120\ \text{V})/(9.6\ \Omega) = 12.5\ \text{A}$. Note that some manufacturers of stereo amplifiers state power outputs in terms of the peak value rather than the lower average value, to mislead the unwary consumer.

Example 31.7 Power in an *L-R-C* series circuit

For the *L-R-C* series circuit of Example 31.4, a) calculate the power factor; b) calculate the average power delivered to the entire circuit and to each circuit element.

SOLUTION

IDENTIFY and SET UP: We can use all of the results found in Example 31.4. The power factor is simply the cosine of the phase angle ϕ, and Eq. (31.31) allows us to find the average power delivered in terms of ϕ and the amplitudes of voltage and current.

EXECUTE: a) The power factor is $\cos\phi = \cos 53° = 0.60$.

b) From Eq. (31.31) the average power delivered to the circuit is

$$P_{av} = \frac{1}{2}VI\cos\phi = \frac{1}{2}(50\ \text{V})(0.10\ \text{A})(0.60) = 1.5\ \text{W}$$

EVALUATE: While P_{av} is the average power delivered to the *L-R-C* combination, all of this power is dissipated in the resistor. The average power delivered to a pure inductor or pure capacitor is always zero (see Figs. 31.12b and 31.12c).

Test Your Understanding

Figure 31.12d shows that during part of a cycle of oscillation, the instantaneous power delivered to the circuit is negative. This means that energy is being extracted from the circuit. Where is the energy extracted from, and where does it go?

31.5 | Resonance in Alternating-Current Circuits

?

Much of the practical importance of *L-R-C* series circuits arises from the way in which such circuits respond to sources of different angular frequency ω. For example, one type of tuning circuit used in radio receivers is simply an *L-R-C* series circuit. A radio signal of any given frequency produces a current of the same frequency in the receiver circuit, but the amplitude of the current is *greatest* if the signal frequency equals the particular frequency to which the receiver circuit is "tuned." This effect is called *resonance.* The circuit is designed so that signals at other than the tuned frequency produce currents that are too small to make an audible sound come out of the radio's speakers.

To see how an *L-R-C* series circuit can be used in this way, suppose we connect an ac source with constant voltage amplitude V but adjustable angular frequency ω across an *L-R-C* series circuit. The current that appears in the circuit has the same angular frequency as the source and a current amplitude $I = V/Z$, where Z is the impedance of the *L-R-C* series circuit. This impedance depends on the frequency, as Eq. (31.23) shows. Figure 31.14a shows graphs of R, X_L, X_C, and Z as functions of ω. We have used a logarithmic angular frequency scale so that we can cover a wide range of frequencies. As the frequency increases, X_L increases and X_C decreases; hence there is always one frequency at which X_L and X_C are equal and $X_L - X_C$ is zero. At this frequency the impedance $Z = \sqrt{R^2 + (X_L - X_C)^2}$ has its *smallest* value, equal simply to the resistance R.

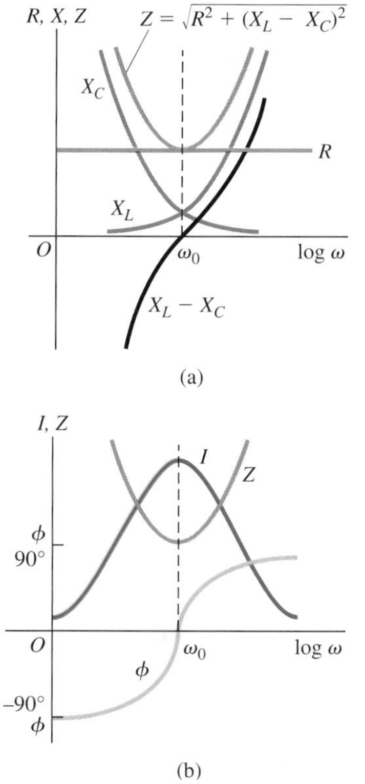

31.14 (a) Reactance, resistance, and impedance as functions of angular frequency. (The angular frequency scale is logarithmic.) (b) Impedance, current, and phase angle as functions of angular frequency. (The angular frequency scale is logarithmic.)

As we vary the angular frequency ω of the source, the current amplitude $I = V/Z$ varies as shown in Fig. 31.14b; the *maximum* value of I occurs at the frequency at which the impedance Z is *minimum.* This peaking of the current amplitude at a certain frequency is called **resonance.** The angular frequency ω_0 at which the resonance peak occurs is called the **resonance angular frequency.** This is the angular frequency at which the inductive and capacitive reactances are equal, so at resonance,

$$X_L = X_C \qquad \omega_0 L = \frac{1}{\omega_0 C} \qquad \omega_0 = \frac{1}{\sqrt{LC}} \tag{31.32}$$

(*L-R-C* series circuit at resonance)

Note that this is equal to the natural angular frequency of oscillation of an *L-C* circuit, which we derived in Section 30.5, Eq. (30.22). The **resonance frequency** f_0 is $\omega_0/2\pi$. This is the frequency at which the greatest current appears in the circuit for a given source voltage amplitude; in other words, f_0 is the frequency to which the circuit is "tuned."

It's instructive to look at what happens to the *voltages* in an *L-R-C* series circuit at resonance. The current at any instant is the same in L and C. The voltage across an inductor always *leads* the current by 90°, or 1/4 cycle, and the voltage across a capacitor always *lags* the current by 90°. Therefore the instantaneous voltages across L and C always differ in phase by 180°, or 1/2 cycle; they have opposite signs at each instant. At the resonance frequency, and *only* at the resonance frequency, $X_L = X_C$ and the voltage amplitudes $V_L = IX_L$ and $V_C = IX_C$ are *equal;* then the instantaneous voltages across L and C add to zero at each instant, and the *total* voltage v_{bd} across the *L-C* combination in Fig. 31.9a is exactly zero. The voltage across the resistor is then equal to the source voltage. So at the resonance frequency the circuit behaves as if the inductor and capacitor weren't there at all!

The *phase* of the voltage relative to the current is given by Eq. (31.24). At frequencies below resonance, X_C is greater than X_L; the capacitive reactance dominates, the voltage *lags* the current, and the phase angle ϕ is between zero and $-90°$. Above resonance, the inductive reactance dominates; the voltage *leads* the current, and the phase angle is between zero and $+90°$. This variation of ϕ with angular frequency is shown in Fig. 31.14b.

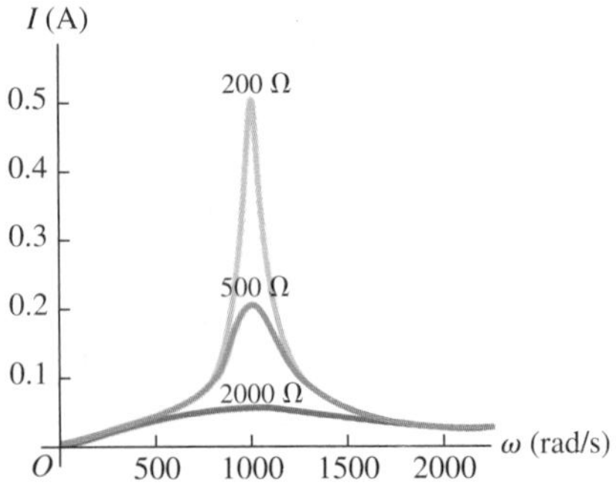

31.15 Graph of current amplitude I as a function of angular frequency ω for an L-R-C series circuit with $V = 100$ V, $L = 2.0$ H, $C = 0.50\ \mu$F, and $R = 500\ \Omega$ (the purple curve). The other curves show the relationship for different values of the resistance, $R = 2000\ \Omega$ (dark purple curve) and $R = 200\ \Omega$ (light purple curve).

14.3 The Driven Oscillator (Questions 8, 9, and 11)

If we can vary the inductance L or the capacitance C of a circuit, we can also vary the resonance frequency. This is exactly how a radio or television receiving set is "tuned" to receive a particular station. In the early days of radio this was accomplished by use of capacitors with movable metal plates whose overlap could be varied to change C. A more modern approach is to vary L by using a coil with a ferrite core that slides in or out.

In a series L-R-C circuit the impedance reaches its minimum value and the current its maximum value at the resonance frequency. The red curve in Fig. 31.15 is a graph of current as a function of frequency for such a circuit, with source voltage amplitude $V = 100$ V, $L = 2.0$ H, $C = 0.50\ \mu$F, and $R = 500\ \Omega$. This curve is called a *response curve* or a *resonance curve.* The resonance angular frequency is $\omega_0 = (LC)^{-1/2} = 1000$ rad/s. As we expect, the curve has a peak at this angular frequency.

The resonance frequency is determined by L and C; what happens when we change R? Figure 31.15 also shows graphs of I as a function of ω for $R = 200\ \Omega$ and for $R = 2000\ \Omega$. The curves are similar for frequencies far away from resonance, where the impedance is dominated by X_L or X_C. But near resonance, where X_L and X_C nearly cancel each other, the curve is higher and more sharply peaked for small values of R and broader and flatter for large values of R. At resonance, $Z = R$ and $I = V/R$, so the maximum height of the curve is inversely proportional to R.

The shape of the response curve is important in the design of radio and television receiving circuits. The sharply peaked curve is what makes it possible to discriminate between two stations broadcasting on adjacent frequency bands. But if the peak is *too* sharp, some of the information in the received signal is lost, such as the high-frequency sounds in music. The shape of the resonance curve is also related to the overdamped and underdamped oscillations that we described in Section 30.6. A sharply peaked resonance curve corresponds to a small value of R and a lightly damped oscillating system; a broad, flat curve goes with a large value of R and a heavily damped system.

In this section we have discussed resonance in an L-R-C *series* circuit. Resonance can also occur in an ac circuit in which the inductor, resistor, and capacitor are connected in *parallel.* We leave the details to you (see Problem 31.51).

Resonance phenomena occur not just in ac circuits, but in all areas of physics. We discussed examples of resonance in *mechanical* systems in Sections 13.8 and 16.5. The amplitude of a mechanical oscillation peaks when the driving-force frequency is close to a natural frequency of the system; this is analogous to the peaking of the current in an L-R-C series circuit. We suggest that you review the sections on mechanical resonance now, looking for the analogies. Other important examples of resonance occur in atomic and nuclear physics and in the study of fundamental particles (high-energy physics).

Example 31.8 Tuning a radio

The series circuit in Fig. 31.16 is similar to arrangements that are sometimes used in radio tuning circuits. This circuit is connected to the terminals of an ac source with a constant rms terminal voltage of 1.0 V and a variable frequency. Find a) the resonance frequency; b) the inductive reactance, the capacitive reactance, and the impedance at the resonance frequency; c) the rms current at resonance; and d) the rms voltage across each circuit element at resonance.

SOLUTION

IDENTIFY and SET UP: Equation (31.32) includes the formula for the resonance angular frequency ω_0, from which we find the resonance frequency f_0. We find the remaining target variables using the results of Sections 31.2 and 31.3.

EXECUTE: a) The resonance angular frequency is

$$\omega_0 = \frac{1}{\sqrt{LC}} = \frac{1}{\sqrt{(0.40 \times 10^{-3}\text{ H})(100 \times 10^{-12}\text{ F})}}$$
$$= 5.0 \times 10^6 \text{ rad/s}$$

The corresponding frequency $f_0 = \omega_0/2\pi$ is

$$f_0 = 8.0 \times 10^5 \text{ Hz} = 800 \text{ kHz}$$

This corresponds to the lower part of the AM radio band.
b) At this frequency,

$$X_L = \omega L = (5.0 \times 10^6 \text{ rad/s})(0.40 \times 10^{-3}\text{ H}) = 2000\ \Omega$$
$$X_C = \frac{1}{\omega C} = \frac{1}{(5.0 \times 10^6 \text{ rad/s})(100 \times 10^{-12}\text{ F})} = 2000\ \Omega$$

Since $X_L = X_C$ and $X_L - X_C = 0$, Eq. (31.23) shows that the impedance Z at resonance is equal to the resistance: $Z = R = 500\ \Omega$.
c) From Eq. (31.26) the rms current at resonance is

$$I_{\text{rms}} = \frac{V_{\text{rms}}}{Z} = \frac{V_{\text{rms}}}{R} = \frac{1.0\text{ V}}{500\ \Omega} = 0.0020\text{ A} = 2.0\text{ mA}$$

d) The rms potential difference across the resistor is

$$V_{R\text{-rms}} = I_{\text{rms}} R = (0.0020\text{ A})(500\ \Omega) = 1.0\text{ V}$$

The rms potential differences across the inductor and capacitor are, respectively,

$$V_{L\text{-rms}} = I_{\text{rms}} X_L = (0.0020\text{ A})(2000\ \Omega) = 4.0\text{ V}$$
$$V_{C\text{-rms}} = I_{\text{rms}} X_C = (0.0020\text{ A})(2000\ \Omega) = 4.0\text{ V}$$

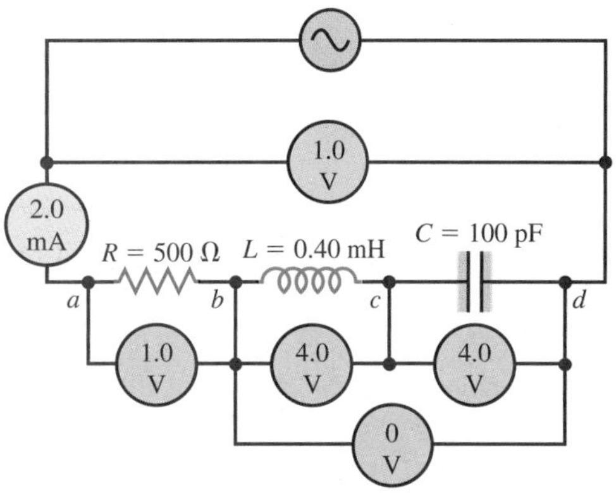

31.16 A radio tuning circuit at resonance. The circles denote rms current and voltages.

EVALUATE: The potential differences across the inductor and the capacitor have equal rms values and amplitudes, but are 180° out of phase and so add to zero at each instant. Note also that at resonance, $V_{R\text{-rms}}$ is equal to the source voltage V_{rms}, while in this example, $V_{L\text{-rms}}$ and $V_{C\text{-rms}}$ are both considerably *larger* than V_{rms}.

Test Your Understanding

How does the resonance frequency of an *L-R-C* series circuit change if the plates of the capacitor are brought closer together?

31.6 | Transformers

One of the great advantages of ac over dc for electric-power distribution is that it is much easier to step voltage levels up and down with ac than with dc. For long-distance power transmission it is desirable to use as high a voltage and as small a current as possible; this reduces i^2R losses in the transmission lines, and smaller wires can be used, saving on material costs. Present-day transmission lines routinely operate at rms voltages of the order of 500 kV. On the other hand, safety considerations and insulation requirements dictate relatively low voltages in generating equipment and in household and industrial power distribution. The standard voltage for household wiring is 120 V in the United States and Canada and 240 V in many other countries. The necessary voltage conversion is accomplished by the use of **transformers.**

Figure 31.17 shows an idealized transformer. The key components of the transformer are two coils or *windings,* electrically insulated from each other but wound on the same core. The core is typically made of a material, such as iron, with a very large relative permeability K_m. This keeps the magnetic field lines due to a current in one winding almost completely within the core. Hence almost all of these field lines pass through the other winding, maximizing the *mutual*

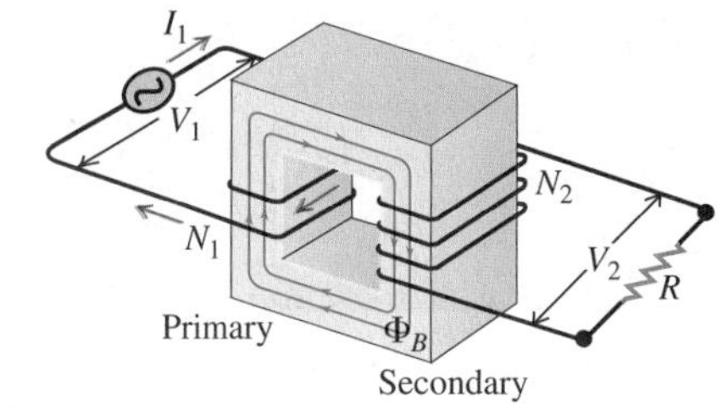

31.17 Schematic diagram of an idealized step-up transformer. The primary is connected to an ac source; the secondary is connected to a device with resistance R.

inductance of the two windings (see Section 30.1). The winding to which power is supplied is called the **primary;** the winding from which power is delivered is called the **secondary.** The circuit symbol for a transformer with an iron core, such as those used in power distribution systems, is

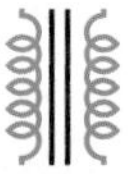

Here's how a transformer works. The ac source causes an alternating current in the primary, which sets up an alternating flux in the core; this induces an emf in each winding, in accordance with Faraday's law. The induced emf in the secondary gives rise to an alternating current in the secondary, and this delivers energy to the device to which the secondary is connected. All currents and emfs have the same frequency as the ac source.

Let's see how the voltage across the secondary can be made larger or smaller in amplitude than the voltage across the primary. We neglect the resistance of the windings and assume that all the magnetic field lines are confined to the iron core, so at any instant the magnetic flux Φ_B is the same in each turn of the primary and secondary windings. The primary winding has N_1 turns and the secondary winding has N_2 turns. When the magnetic flux changes because of changing currents in the two coils, the resulting induced emfs are

$$\mathcal{E}_1 = -N_1\frac{d\Phi_B}{dt} \quad \text{and} \quad \mathcal{E}_2 = -N_2\frac{d\Phi_B}{dt} \tag{31.33}$$

The flux *per turn* Φ_B is the same in both the primary and the secondary, so Eqs. (31.33) show that the induced emf *per turn* is the same in each. The ratio of the secondary emf $\mathcal{E}_2$ to the primary emf $\mathcal{E}_1$ is therefore equal at any instant to the ratio of secondary to primary turns:

$$\frac{\mathcal{E}_2}{\mathcal{E}_1} = \frac{N_2}{N_1} \tag{31.34}$$

Since $\mathcal{E}_1$ and $\mathcal{E}_2$ both oscillate with the same frequency as the ac source, Eq. (31.34) also gives the ratio of the amplitudes or of the rms values of the induced emfs. If the windings have zero resistance, the induced emfs $\mathcal{E}_1$ and $\mathcal{E}_2$ are equal to the terminal voltages across the primary and the secondary, respectively; hence

$$\frac{V_2}{V_1} = \frac{N_2}{N_1} \tag{31.35}$$

(terminal voltages of transformer primary and secondary)

where V_1 and V_2 are either the amplitudes or the rms values of the terminal voltages. By choosing the appropriate turns ratio N_2/N_1, we may obtain any desired secondary voltage from a given primary voltage. If $N_2 > N_1$, as in Fig. 31.17, then $V_2 > V_1$ and we have a *step-up* transformer; if $N_2 < N_1$, then $V_2 < V_1$ and we have a *step-down* transformer. At a power generating station, step-up transformers are used; the primary is connected to the power source and the secondary is connected to the transmission lines, giving the desired high voltage for transmission. Near the consumer, step-down transformers lower the voltage to a value suitable for use in home or industry (Fig. 31.18).

31.18 The cylindrical can near the top of this power pole is a step-down transformer. It converts the high-voltage ac in the power lines to low-voltage (120 V) ac, which is then distributed to the surrounding homes and businesses.

Even the relatively low voltage provided by a household wall socket is too large for many electronic devices, so a further step-down transformer is necessary. This is the role of an "ac adapter" (also called a "power cube" or "power adapter"), such as those used to run a portable stereo or recharge a mobile phone from line voltage. Such adapters contain a step-down transformer that converts line voltage to a smaller value, typically 3 to 12 volts, as well as diodes to convert alternating current to the direct current that small electronic devices require (Fig. 31.19).

If the secondary circuit is completed by a resistance R, then the amplitude or rms value of the current in the secondary circuit is $I_2 = V_2/R$. From energy considerations, the power delivered to the primary equals that taken out of the secondary (since there is no resistance in the windings), so

$$V_1 I_1 = V_2 I_2 \qquad \text{(currents in transformer primary and secondary)} \qquad (31.36)$$

We can combine Eqs. (31.35) and (31.36) and the relation $I_2 = V_2/R$ to eliminate V_2 and I_2; we obtain

$$\frac{V_1}{I_1} = \frac{R}{(N_2/N_1)^2} \qquad (31.37)$$

This shows that when the secondary circuit is completed through a resistance R, the result is the same as if the *source* had been connected directly to a resistance equal to R divided by the square of the turns ratio, $(N_2/N_1)^2$. In other words, the transformer "transforms" not only voltages and currents, but resistances as well. More generally, we can regard a transformer as "transforming" the *impedance* of the network to which the secondary circuit is completed.

Equation (31.37) has many practical consequences. The power supplied by a source to a resistor depends on the resistances of both the resistor and the source. It can be shown that the power transfer is greatest when the two resistances are *equal.* The same principle applies in both dc and ac circuits. When a high-impedance ac source must be connected to a low-impedance circuit, such as an audio amplifier connected to a loudspeaker, the source impedance can be *matched* to that of the circuit by use of a transformer with an appropriate turns ratio N_2/N_1.

Real transformers always have some energy losses. (That's why an ac adapter like the one shown in Fig. 31.19 feels warm to the touch after it's been in use for a while; the transformer is heated by the dissipated energy.) The windings have some resistance, leading to i^2R losses. There are also energy losses through hysteresis in the core (Section 28.8). Hysteresis losses are minimized by the use of soft iron with a narrow hysteresis loop.

Another important mechanism for energy loss in a transformer core involves eddy currents (Section 29.6). Consider a section AA through an iron transformer core (Fig. 31.20a). Since iron is a conductor, any such section can be pictured as several conducting circuits, one within the other (Fig. 31.20b). The flux through each of these circuits is continually changing, so eddy currents circulate in the entire volume of the core, with lines of flow that form planes perpendicular to the flux. These eddy currents are very undesirable; they waste energy through i^2R heating and themselves set up an opposing flux.

The effects of eddy currents can be minimized by the use of a *laminated* core, that is, one built up of thin sheets or laminae. The large electrical surface resistance of each lamina, due either to a natural coating of oxide or to an insulating varnish, effectively confines the eddy currents to individual laminae (Fig. 31.20c). The possible eddy-current paths are narrower, the induced emf in each path is smaller, and

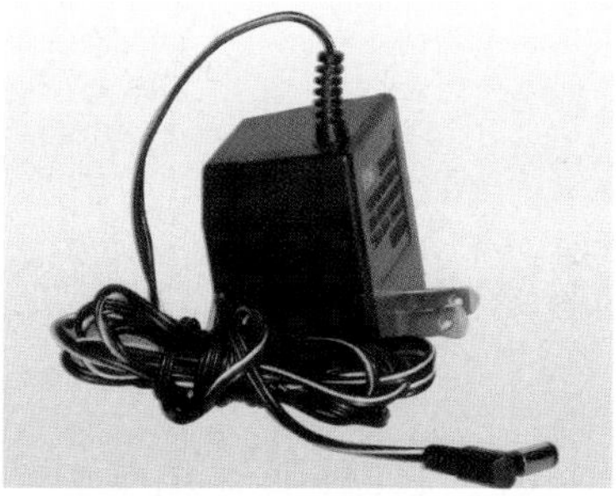

31.19 An ac adapter like this one converts household ac into low-voltage dc for use in electronic devices. It contains a step-down transformer to lower the voltage and diodes to rectify the output current (see Fig. 31.2).

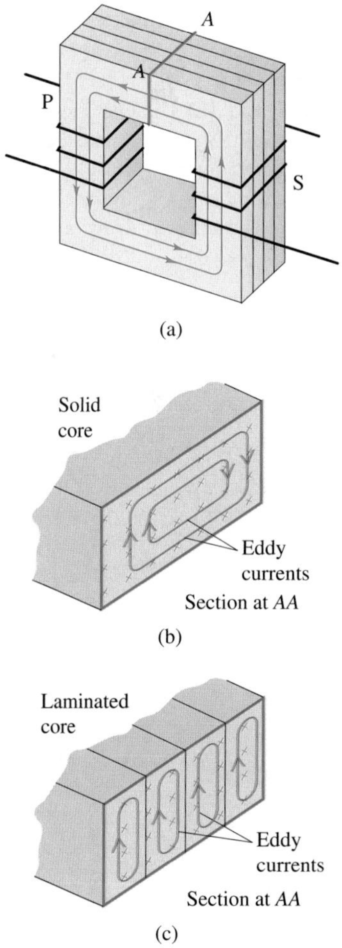

31.20 (a) Primary (P) and secondary (S) windings in a transformer. (b) Eddy currents in the iron core, shown in the cross section at AA. (c) Using a laminated core reduces the eddy currents.

the eddy currents are greatly reduced. The alternating magnetic field exerts forces on the current-carrying lamina that cause them to vibrate back and forth; this vibration causes the characteristic "hum" of an operating transformer. You can hear this same "hum" from the magnetic ballast of a fluorescent light fixture (see Section 30.2).

Thanks to the use of soft iron cores and lamination, transformer efficiencies are usually well over 90%; in large installations they may reach 99%.

Example 31.9 "Wake up and smell the (transformer)!"

A friend brings back from Europe a device that she claims to be the world's greatest coffeemaker. Unfortunately, it was designed to operate from a 240-V line to obtain the 960 W of power that it needs. a) What can she do to operate it at 120 V? b) What current will the coffeemaker draw from the 120-V line? c) What is the resistance of the coffeemaker? (The voltages are rms values.)

SOLUTION

IDENTIFY and SET UP: Our friend needs a step-up transformer to convert the 120-V ac available in the home to the 240-V ac that the coffeemaker requires. We use Eq. (31.35) to determine the transformer turns ratio N_2/N_1, the relationship $P_{av} = V_{rms}I_{rms}$ for a resistor to find the current draw, and Eq. (31.37) to calculate the resistance.

EXECUTE: a) To get $V_2 = 240$ V from $V_1 = 120$ V, the required turns ratio is $N_2/N_1 = V_2/V_1 = (240\text{ V})/(120\text{ V}) = 2$. That is, the secondary coil (connected to the coffeemaker) should have twice as many turns as the primary coil (connected to the 120-V line).

b) The rms current I_1 in the 120-V primary is found by using $P_{av} = V_1I_1$, where P_{av} is the average power drawn by the coffeemaker and hence the power supplied by the 120-V line. (We're assuming that there are no energy losses in the transformer.) Hence $I_1 = P_{av}/V_1 = (960\text{ W})/(120\text{ V}) = 8.0$ A. The secondary current is then $I_2 = P_{av}/V_2 = (960\text{ W})/(240\text{ V}) = 4.0$ A.

c) We have $V_1 = 120$ V, $I_1 = 8.0$ A, and $N_2/N_1 = 2$, so

$$\frac{V_1}{I_1} = \frac{120\text{ V}}{8.0\text{ A}} = 15\ \Omega$$

From Eq. (31.37),

$$R = 2^2(15\ \Omega) = 60\ \Omega$$

EVALUATE: As a check, $V_2/R = (240\text{ V})/(60\ \Omega) = 4.0\text{ A} = I_2$, the same value obtained previously. You can also check this result for R by using the expression $P_{av} = V_2^2/R$ for the power drawn by the coffeemaker.

Test Your Understanding

An ac adapter like the one shown in Fig. 31.19 has a transformer that converts 120-V ac into 9.0-V ac. (The diodes then convert this current into rectified ac.) What turns ratio is required? Which coil has more turns, the primary or the secondary?

SUMMARY

An alternator or ac source produces an emf that varies sinusoidally with time. A sinusoidal voltage or current can be represented by a phasor, a vector that rotates counterclockwise with constant angular velocity ω equal to the angular frequency of the sinusoidal quantity. Its projection on the horizontal axis at any instant represents the instantaneous value of the quantity.

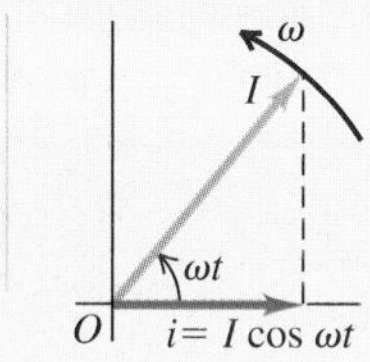

For a sinusoidal current, the rectified average and rms (root-mean-square) currents are proportional to the current amplitude I. Similarly, the rms value of a sinusoidal voltage is proportional to the voltage amplitude V. (See Example 31.1)

$$I_{rav} = \frac{2}{\pi} I = 0.637I \qquad (31.3)$$

$$I_{rms} = \frac{I}{\sqrt{2}} \qquad (31.4)$$

$$V_{rms} = \frac{V}{\sqrt{2}} \qquad (31.5)$$

In general, the instantaneous voltage between two points in an ac circuit is not in phase with the instantaneous current passing through those points. The quantity ϕ is called the phase angle of the voltage relative to the current.

$$i = I\cos\omega t \qquad (31.2)$$

$$v = V\cos(\omega t + \phi)$$

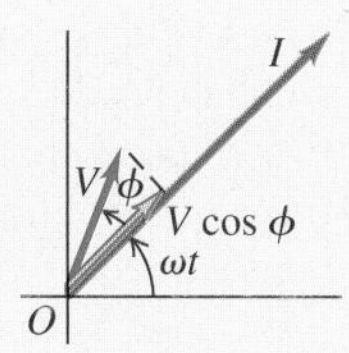

The voltage across a resistor R is in phase with the current. The voltage across an inductor L leads the current by 90° ($\phi = +90°$), while the voltage across a capacitor C lags the current by 90° ($\phi = -90°$). The voltage amplitude across each type of device is proportional to the current amplitude I. An inductor has inductive reactance $X_L = \omega L$ and a capacitor has capacitive reactance $X_C = 1/\omega C$. (See Examples 31.2 and 31.3)

$$V_R = IR \qquad (31.7)$$

$$V_L = IX_L \qquad (31.13)$$

$$V_C = IX_C \qquad (31.19)$$

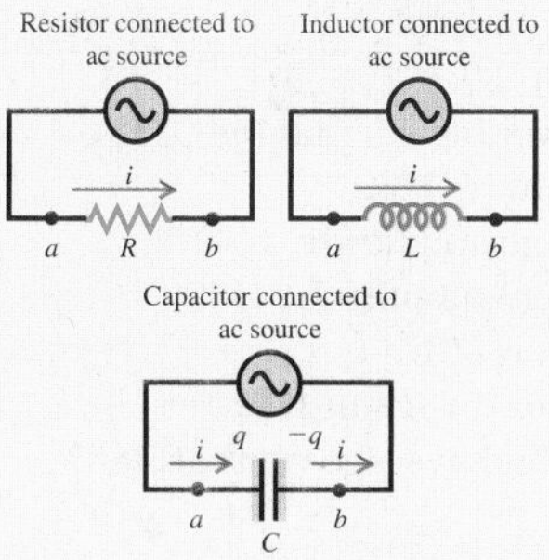

In a general ac circuit, the voltage and current amplitudes are related by the circuit impedance Z. In an L-R-C series circuit, the values of L, R, C, and the angular frequency ω determine the impedance and the phase angle ϕ of the voltage relative to the current. (See Examples 31.4 and 31.5)

$$V = IZ \qquad (31.22)$$

$$Z = \sqrt{R^2 + (X_L - X_C)^2}$$

$$= \sqrt{R^2 + [\omega L - (1/\omega C)]^2} \qquad (31.23)$$

$$\tan\phi = \frac{\omega L - 1/\omega C}{R} \qquad (31.24)$$

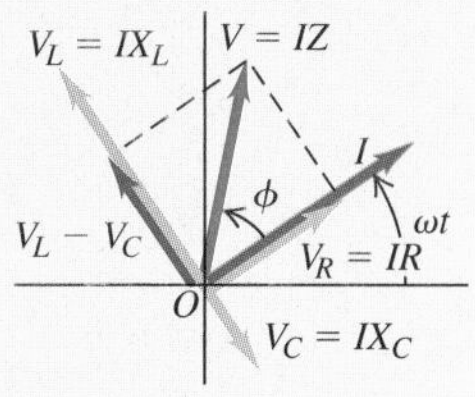

The average power input P_{av} to an ac circuit depends on the voltage and current amplitudes (or, equivalently, their rms values) and the phase angle ϕ of the voltage relative to the current. The quantity $\cos\phi$ is called the power factor. (See Examples 31.6 and 31.7)

$$P_{av} = \frac{1}{2}VI\cos\phi = V_{rms}I_{rms}\cos\phi \qquad (31.31)$$

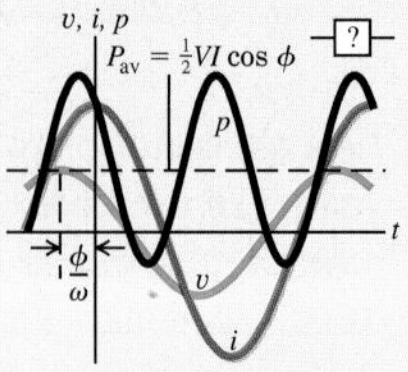

In an *L-R-C* series circuit, the current becomes maximum and the impedance becomes minimum at an angular frequency $\omega_0 = 1/\sqrt{LC}$ called the resonance angular frequency. This phenomenon is called resonance. At resonance the voltage and current are in phase, and the impedance Z is equal to the resistance R. (See Example 31.8)

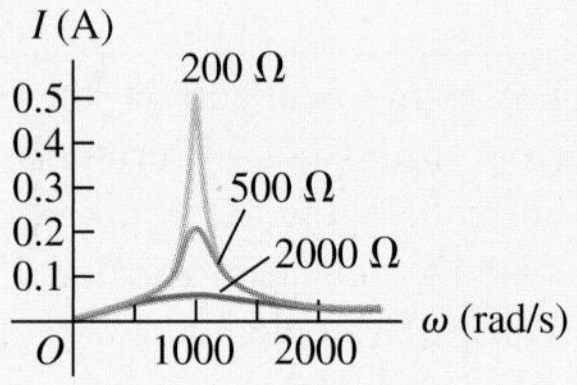

A transformer is used to transform the voltage and current levels in an ac circuit. In an ideal transformer with no energy losses, if the primary winding has N_1 turns and the secondary winding has N_2 turns, the amplitudes (or rms values) of the two voltages are related by Eq. (31.35). The amplitudes (or rms values) of the primary and secondary voltages and currents are related by Eq. (31.36). (See Example 31.9)

$$\frac{V_2}{V_1} = \frac{N_2}{N_1} \qquad (31.35)$$

$$V_1I_1 = V_2I_2 \qquad (31.36)$$

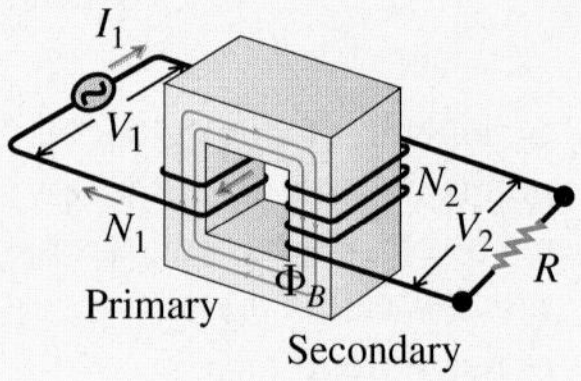

Key Terms

alternating current (ac), 1181
ac source, 1182
voltage amplitude, 1182
current amplitude, 1182
phasor, 1182
phasor diagram, 1182
rectified average current, 1183
root-mean-square (rms) current, 1183
phase angle, 1186
inductive reactance, 1186
capacitive reactance, 1189
impedance, 1191
power factor, 1197
resonance, 1199
resonance angular frequency, 1199
resonance frequency, 1199
transformer, 1201
primary, 1202
secondary, 1202

Your Notes

Answer to Chapter Opening Question

Yes. In fact, the radio simultaneously detects transmissions at *all* frequencies. However, a radio is an *L-R-C* series circuit, and at any given time it is tuned to have a resonance at just one frequency. Hence the response of the radio to that frequency is much greater than its response to any other frequency, which is why you hear only one broadcasting station through the radio's speaker. (You can sometimes hear a second station if its frequency is sufficiently close to the tuned frequency.)

Answers to Test Your Understanding Questions

Section 31.1 No. If an appliance is connected to an ac source (by plugging it into a wall socket), the electrons in the appliance wiring simply move back and forth in response to the source voltage. Thus, the electrons in the appliance stay in the appliance. What you pay for is the *energy* that is given to the electrons to make them oscillate. This energy is then absorbed by the appliance (in a toaster, for example, it goes into heating the toaster coils).

Section 31.2 a) For a resistor, $V_R = IR$, so $I = V_R/R$. The voltage amplitude V_R and resistance R do not change with frequency, so the current amplitude I remains constant. b) For an inductor, $V_L = IX_L = I\omega L$, so $I = V_L/\omega L$. The voltage amplitude V_L and inductance L are constant, so the current amplitude I decreases as the frequency increases. c) For a capacitor, $V_C = IX_C = I/\omega C$, so $I = V_C\omega C$. The voltage amplitude V_C and capacitance C are constant, so the current amplitude I increases as the frequency increases.

Section 31.3 Removing the inductor means that we set the inductive reactance X_L equal to zero. (This is equivalent to saying that there is zero potential difference $V_L = IX_L$ across the inductor.) Then the impedance becomes $Z = \sqrt{R^2 + (0 - X_C)^2} = \sqrt{(300\ \Omega)^2 + (-200\ \Omega)^2} = 360\ \Omega$, which is smaller than in Example 31.4. Hence the current amplitude $I = V/Z$ *increases* if we remove the inductor. If instead we remove the capacitor, we set the capacitive reactance X_C to zero and the impedance becomes $Z = \sqrt{R^2 + (X_L - 0)^2} = \sqrt{(300\ \Omega)^2 + (600\ \Omega)^2} = 670\ \Omega$. This is greater than in Example 31.4, so the current amplitude *decreases* if we remove the capacitor. At the angular frequency in question, the capacitor actually facilitates the flow of current. If you repeat Example 31.4 and this exercise with a sufficiently low angular frequency, say 2000 rad/s, you'll find that these results are reversed: the current is greater with the inductor removed and smaller with the capacitor removed.

Section 31.4 The energy cannot be extracted from the resistor, since energy is dissipated in a resistor cannot be recovered. Instead, the energy must be extracted from either the inductor (which stores magnetic-field energy) or the capacitor (which stores electric-field energy). Positive power means that energy is being transferred from the ac source to the circuit, so *negative* power implies that energy is being transferred back into the source.

Section 31.5 The capacitance C increases if the plate spacing is decreased (see Section 24.1). Hence the resonance frequency $f_0 = \omega_0/2\pi = 1/2\pi\sqrt{LC}$ decreases.

Section 31.6 We are given $V_1 = 120$ V and $V_2 = 9.0$ V, so from Eq. (31.35) the turns ratio is $N_2/N_1 = V_2/V_1 = (9.0\text{ V})/(120\text{ V}) = 0.075$. This number is less than 1, so there are more turns in the primary (N_1) than in the secondary (N_2): there is 0.075 turn in the secondary for each full turn in the primary, or one turn in the secondary for each $1/0.075 = 13$ turns in the primary.

Discussion Questions

Q31.1 Household electric power in most of western Europe is supplied at 240 V, rather than the 120 V that is standard in the United States and Canada. What are the advantages and disadvantages of each system?

Q31.2 The current in an ac power line changes direction 120 times per second, and its average value is zero. Explain how it is possible for power to be transmitted in such a system.

Q31.3 Equation (31.14) was derived by using the relation $i = dq/dt$ between the current and the charge on the capacitor. In Fig. 31.6a the positive counterclockwise current increases the charge on the capacitor. When the charge on the left plate is positive but decreasing in time, is $i = dq/dt$ still correct or should it be $i = -dq/dt$? Is $i = dq/dt$ still correct when the right-hand plate has positive charge that is increasing or decreasing in magnitude? Explain.

Q31.4 Fluorescent lights often use an inductor, called a ballast, to limit the current through the tubes. Why is it better to use an inductor rather than a resistor for this purpose?

Q31.5 Equation (31.9) says that $v_{ab} = L\,di/dt$ (Fig. 31.5a). Using Faraday's law, explain why point a is at higher potential than point b when i is in the direction shown in Fig. 31.5a and is increasing in magnitude. What about when i is counterclockwise and decreasing in magnitude, is $v_{ab} = L\,di/dt$ still correct, or should it be $v_{ab} = -L\,di/dt$? Is $v_{ab} = L\,di/dt$ still correct when i is clockwise and increasing or decreasing in magnitude? Explain.

Q31.6 In a series *L-R-C* circuit, can the instantaneous voltage across the capacitor exceed the source voltage at that same instant? Can this be true for the instantaneous voltage across the inductor? Across the resistor? Explain.

Q31.7 In a series *L-R-C* circuit, what are the phase angle ϕ and power factor $\cos\phi$ when the resistance is much smaller than the inductive or capacitive reactance and the circuit is operated far from resonance? Explain.

Q31.8 When a series *L-R-C* circuit is connected across a 120-V ac line, the voltage rating of the capacitor may be exceeded even if it is rated at 200 or 400 V. How can this be?

Q31.9 In Example 31.6 (Section 31.4), a hair dryer was treated as a pure resistance. But because there are coils in the heating element and in the motor that drives the blower fan, a hair dryer also has inductance. Qualitatively, does including an inductance increase or decrease the values of R, I_{rms}, and P?

Q31.10 A light bulb and a parallel-plate capacitor with air between the plates are connected in series to an ac source. What happens to the brightness of the bulb when a dielectric is inserted between the plates of the capacitor? Explain.

Q31.11 A coil of wire wrapped on a hollow tube and a light bulb are connected in series to an ac source. What happens to the brightness of the bulb when an iron rod is inserted in the tube?

Q31.12 A circuit consists of a lightbulb, a capacitor, and an inductor connected in series to an ac source. What happens to the brightness of the bulb when the inductor is removed? When the inductor is left in the circuit but the capacitor is removed? Explain.

Q31.13 A circuit consists of a lightbulb, a capacitor, and an inductor connected in series to an ac source. Is it possible for both the capacitor and the inductor to be removed and the brightness of the bulb to remain the same? Explain.

Q31.14 Can a transformer be used with dc? Explain. What happens if a transformer designed for 120-V ac is connected to a 120-V dc line?

Q31.15 Some electrical appliances operate equally well on ac or dc, and others work only on ac or only on dc. Give examples of each, and explain the differences.

Exercises

Section 31.1 Phasors and Alternating Currents

31.1 The voltage across the terminals of an ac power supply varies with time according to Eq. (31.1). The voltage amplitude is $V = 45.0$ V. What is a) the root-mean-square potential difference V_{rms}? b) the average potential difference V_{av} between the two terminals of the power supply?

31.2 A sinusoidal current $i = I\cos\omega t$ has an rms value $I_{rms} = 2.10$ A. a) What is the current amplitude? b) The current is passed through a full-wave rectifier circuit. What is the rectified average current? c) Which is larger, I_{rms} or I_{rav}? Explain, using graphs of i^2 and of the rectified current.

Section 31.2 Resistance and Reactance

31.3 A 5.00-H inductor with negligible resistance is connected across an ac source whose voltage amplitude is kept constant at 60.0 V but whose frequency can be varied. Find the current amplitude when the angular frequency is a) 100 rad/s; b) 1000 rad/s; c) 10,000 rad/s. d) Show the results of parts (a) through (c) in a plot of log I versus log ω.

31.4 A 2.20-μF capacitor is connected across the ac source of Exercise 31.3. Find the current amplitude when the angular frequency is a) 100 rad/s; b) 1000 rad/s; c) 10,000 rad/s. d) Show the results of parts (a) through (c) in a plot of log I versus log ω.

31.5 a) What is the reactance of a 3.00-H inductor at a frequency of 80.0 Hz? b) What is the inductance of an inductor whose reactance is 120 Ω at 80.0 Hz? c) What is the reactance of a 4.00-μF capacitor at a frequency of 80.0 Hz? d) What is the capacitance of a capacitor whose reactance is 120 Ω at 80.0 Hz?

31.6 a) Compute the reactance of a 0.450-H inductor at frequencies of 60.0 Hz and 600 Hz. b) Compute the reactance of a 2.50-μF capacitor at the same frequencies. c) At what frequency is the reactance of a 0.450-H inductor equal to that of a 2.50-μF capacitor?

31.7 Kitchen Capacitance. The wiring for a refrigerator contains a starter capacitor. A voltage of amplitude 170 V and frequency 60.0 Hz applied across the capacitor is to produce a current amplitude of 0.850 A through the capacitor. What capacitance C is required?

31.8 A Radio Inductor. You want the current amplitude through a 0.450-mH inductor (part of the circuitry for a radio receiver) to be 2.60 mA when a sinusoidal voltage with amplitude 12.0 V is applied across the inductor. What frequency is required?

31.9 A 150-Ω resistor is connected in series with a 0.250-H inductor. The voltage across the resistor is $v_R = (3.80\text{ V})\cos[(720\text{ rad/s})t]$. a) Derive an expression for the circuit current. b) Determine the inductive reactance of the inductor. c) Derive an expression for the voltage v_L across the inductor.

31.10 A 250-Ω resistor is connected in series with a 4.80-μF capacitor. The voltage across the capacitor is $v_C = (7.60\text{ V})\sin[(120\text{ rad/s})t]$. a) Determine the capacitive reactance of the capacitor. b) Derive an expression for the voltage v_R across the resistor.

Section 31.3 The *L-R-C* Series Circuit

31.11 Define the reactance X of an L-R-C circuit to be $X = X_L - X_C$. a) Show that $X = 0$ when the angular frequency ω of the current is equal to the resonance angular frequency ω_0. b) What is the sign of X when $\omega > \omega_0$? c) What is the sign of X when $\omega < \omega_0$? d) Graph X versus ω.

31.12 You have a 200-Ω resistor, a 0.400-H inductor, and a 6.00-μF capacitor. Suppose you take the resistor and inductor and make a series circuit with a voltage source that has voltage amplitude 30.0 V and an angular frequency of 250 rad/s. a) What is the impedance of the circuit? b) What is the current amplitude? c) What are the voltage amplitudes across the resistor and across the inductor? d) What is the phase angle ϕ of the source voltage with respect to the current? Does the source voltage lag or lead the current? e) Construct the phasor diagram.

31.13 Repeat Exercise 31.12 with the circuit consisting of only the resistor and the capacitor in series. For part (c), calculate the voltage amplitudes across the resistor and across the capacitor.

31.14 Repeat Exercise 31.12 with the circuit consisting of only the capacitor and the inductor in series. For part (c), calculate the voltage amplitudes across the capacitor and across the inductor.

31.15 a) For the R-L circuit of Exercise 31.12, graph v, v_R, and v_L versus t for $t = 0$ to $t = 50.0$ ms. The current is given by $i = I\cos\omega t$, so $v = V\cos(\omega t + \phi)$. b) What are v, v_R, and v_L at $t = 20.0$ ms? Compare $v_R + v_L$ to v at this instant. c) Repeat part (b) for $t = 40.0$ ms.

31.16 a) For the R-C circuit of Exercise 31.13, graph v, v_R, and v_C versus t for $t = 0$ to $t = 50.0$ ms. The current is given by $i = I\cos\omega t$, so $v = V\cos(\omega t + \phi)$. b) What are v, v_R, and v_C at $t = 20.0$ ms? Compare $v_R + v_C$ to v at this instant. c) Repeat part (b) for $t = 40.0$ ms.

31.17 The resistor, inductor, capacitor, and voltage source described in Exercise 31.12 are connected to form an L-R-C series circuit. a) What is the impedance of the circuit? b) What is the current amplitude? c) What is the phase angle of the source voltage with respect to the current? Does the source voltage lag or lead the current? d) What are the voltage amplitudes across the resistor, inductor, and capacitor? e) Explain how it is possible for the voltage amplitude across the capacitor to be greater than the voltage amplitude across the source.

31.18 a) For the *L-R-C* circuit of Exercise 31.17, graph v, v_R, v_L, and v_C versus t for $t = 0$ to $t = 50.0$ ms. The current is given by $i = I\cos\omega t$, so $v = V\cos(\omega t + \phi)$. b) What are v, v_R, v_L, and v_C at $t = 20.0$ ms? Compare $v_R + v_L + v_C$ to v at this instant. c) Repeat part (b) for $t = 40.0$ ms.

31.19 Analyzing an *L-R-C* Circuit. You have a 200-Ω resistor, a 0.400-H inductor, a 5.00-μF capacitor, and a variable-frequency ac source with an amplitude of 3.00 V. You connect all four elements together to form a series circuit. a) At what frequency will the current in the circuit be greatest? What will be the current amplitude at this frequency? b) What will be the current amplitude at an angular frequency of 400 rad/s? At this frequency, will the source voltage lead or lag the current?

31.20 The resistor, inductor, and capacitor described in Exercise 31.12 are connected with an ac source of voltage amplitude 30.0 V to form a series *L-R-C* circuit. a) Compute the impedance of the circuit at angular frequencies of 1000, 600, and 200 rad/s. b) Describe how the current amplitude varies as the frequency of the source is slowly reduced from 1000 rad/s to 200 rad/s. c) What is the phase angle of the source voltage with respect to the current when $\omega = 1000$ rad/s? d) Construct the phasor diagram when $\omega = 1000$ rad/s. e) Repeat parts (c) and (d) for $\omega = 200$ rad/s.

31.21 In an *L-R-C* series circuit, the rms voltage across the resistor is 30.0 V, across the capacitor it is 90.0 V and across the inductor it is 50.0 V. What is the rms voltage of the source?

Section 31.4 Power in Alternating-Current Circuits

31.22 In a series *L-R-C* circuit, the components have the following values: $L = 20.0$ mH, $C = 140$ nF, and $R = 350$ Ω. The generator has an rms voltage of 120 V and a frequency of 1.25 kHz. Determine a) the power supplied by the generator; b) the power dissipated in the resistor.

31.23 a) Show that for an *L-R-C* series circuit the power factor is equal to R/Z. (*Hint:* Use the phasor diagram; see Fig. 31.9b.) b) Show that for any ac circuit, not just one containing pure resistance only, the average power delivered by the voltage source is given by $P = I_{rms}{}^2R$.

31.24 An *L-R-C* series circuit is connected to a 120-Hz ac source that has $V_{rms} = 80.0$ V. The circuit has a resistance of 75.0 Ω and an impedance at this frequency of 105 Ω. What average power is delivered to the circuit by the source?

31.25 An *L-R-C* series circuit with $L = 0.120$ H, $R = 240$ Ω, and $C = 7.30$ μF carries an rms current of 0.450 A with a frequency of 400 Hz. a) What are the phase angle and power factor for this circuit? b) What is the impedance of the circuit? c) What is the rms voltage of the source? d) What average power is delivered by the source? e) What is the average rate at which electrical energy is converted to thermal energy in the resistor? f) What is the average rate at which electrical energy is dissipated (converted to other forms) in the capacitor? g) In the inductor?

31.26 A large 360-Ω, 5.20-H electromagnet coil is connected across the terminals of a source that has voltage amplitude 240 V and frequency 60.0 Hz. a) What is the power factor? b) What is the average power delivered by the source?

Section 31.5 Resonance in Alternating-Current Circuits

31.27 In an *L-R-C* series circuit, $R = 300$ Ω, $L = 0.400$ H, and $C = 6.00 \times 10^{-8}$ F. When the ac source operates at the resonance frequency of the circuit, the current amplitude is 0.500 A. a) What is the voltage amplitude of the source? b) What is the amplitude of the voltage across the resistor, across the inductor, and across the capacitor? c) What is the average power supplied by the source?

31.28 An *L-R-C* series circuit consists of a source with voltage amplitude 120 V and angular frequency 50.0 rad/s, a resistor with $R = 400$ Ω, an inductor with $L = 9.00$ H, and a capacitor with capacitance C. a) For what value of C will the current amplitude in the circuit be a maximum? b) When C has the value calculated in part (a), what is the amplitude of the voltage across the inductor?

31.29 In an *L-R-C* series circuit, $R = 150$ Ω, $L = 0.750$ H, and $C = 0.0180$ μF. The source has voltage amplitude $V = 150$ V and a frequency equal to the resonance frequency of the circuit. a) What is the power factor? b) What is the average power delivered by the source? c) The capacitor is replaced by one with $C = 0.0360$ μF and the source frequency is adjusted to the new resonance value. Then what is the average power delivered by the source?

31.30 In an *L-R-C* series circuit, $R = 400$ Ω, $L = 0.350$ H, and $C = 0.0120$ μF. a) What is the resonance angular frequency of the circuit? b) The capacitor can withstand a peak voltage of 550 V. If the voltage source operates at the resonance frequency, what maximum voltage amplitude can it have if the maximum capacitor voltage is not exceeded?

31.31 Consider an *L-R-C* series circuit composed of the inductor, resistor, and capacitor described in Exercise 31.12. The circuit is connected to an ac source with voltage amplitude 30.0 V. a) At what frequency (in Hz) is the circuit in resonance? b) Sketch the phasor diagram at the resonance frequency. c) What is the reading of each voltmeter shown in Fig. 31.21 when the source frequency equals the resonance frequency? The voltmeters are calibrated to read rms voltages. d) What is the resonance frequency if the resistance is reduced to 100 Ω? e) What is then the rms current at resonance?

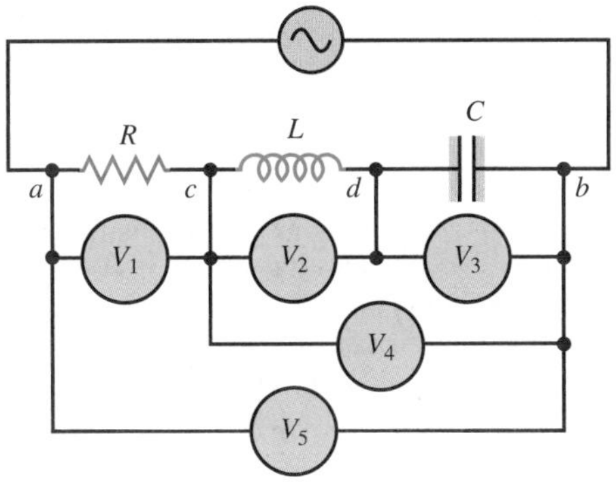

Figure 31.21 Exercise 31.31 and Problem 31.38.

31.32 In an *L-R-C* series circuit, $L = 0.280$ H and $C = 4.00$ μF. The voltage amplitude of the source is 120 V. a) What is the resonance angular frequency of the circuit? b) When the source operates at the resonance angular frequency, the current amplitude in the circuit is 1.70 A. What is the resistance R of the resistor? c) At the resonance angular frequency, what are the peak voltages across the inductor, the capacitor, and the resistor?

Section 31.6 Transformers

31.33 A Step-Down Transformer. A transformer connected to a 120-V (rms) ac line is to supply 12.0 V (rms) to a portable electronic device. The total equivalent resistance of the system is 5.00 Ω. a) What should the ratio of primary to secondary turns of the transformer be? b) What rms current must the secondary supply? c) What average power is delivered to the load? d) What resistance connected directly across the 120-V line would draw the same power as the transformer? Show that this is equal to 5.00 Ω times the square of the ratio of primary to secondary turns.

31.34 A Step-Up Transformer. A transformer connected to a 120-V (rms) ac line is to supply 13,000 V (rms) for a neon sign. To reduce shock hazard, a fuse is to be inserted in the primary circuit; the fuse is to blow when the rms current in the secondary circuit exceeds 8.50 mA. a) What is the ratio of secondary to primary turns of the transformer? b) What power must be supplied to the transformer when the rms secondary current is 8.50 mA? c) What current rating should the fuse in the primary circuit have?

31.35 The internal resistance of an ac source is 12.8 kΩ. a) What should the ratio of primary to secondary turns of a transformer be to match the source to a load with a resistance of 8.00 Ω? ("Matching" means that the effective load resistance equals the internal resistance of the source. See Exercise 31.33.) b) If the voltage amplitude of the source is 60.0 V, what is the voltage amplitude in the secondary circuit under open circuit conditions?

Problems

31.36 Figure 31.8a shows the crossover network in a loudspeaker system. One branch consists of a capacitor C and a resistor R in series (the tweeter). This branch is in parallel with a second branch (the woofer) that consists of an inductor L and a resistor R in series. The same source voltage with angular frequency ω is applied across each parallel branch. a) What is the impedance of the tweeter branch? b) What is the impedance of the woofer branch? c) Explain why the currents in the two branches are equal when the impedances of the branches are equal. d) Derive an expression for the frequency f that corresponds to the crossover point in Fig. 31.8b.

31.37 A coil has a resistance of 48.0 Ω. At a frequency of 80.0 Hz the voltage across the coil leads the current in it by 52.3°. Determine the inductance of the coil.

31.38 Five infinite-impedance voltmeters, calibrated to read rms values, are connected as shown in Fig 31.21. Let $R = 200\ \Omega$, $L = 0.400$ H, $C = 6.00\ \mu$F, and $V = 30.0$ V. What is the reading of each voltmeter if a) $\omega = 200$ rad/s; b) $\omega = 1000$ rad/s?

31.39 A sinusoidal current is given by $i = I \cos \omega t$. The full-wave rectified current is shown in Fig. 31.2b. a) Let t_1 and t_2 be the two smallest positive times at which the rectified current is zero. Express t_1 and t_2 in terms of ω. b) Find the area under the rectified i versus t curve between t_1 and t_2 by computing the integral $\int_{t_1}^{t_2} i\,dt$. Since $dq = i\,dt$, this area equals the charge that flows during the t_1 to t_2 time interval. c) Set the result in part (b) equal to $I_{rav}(t_2 - t_1)$ and calculate I_{rav} in terms of the current amplitude I. Compare your answer to Eq. (31.3).

31.40 A large electromagnet coil is connected to a 120-Hz ac source. The coil has resistance 400 Ω, and at this source frequency the coil has inductive reactance 250 Ω. a) What is the inductance of the coil? b) What must the rms voltage of the source be if the coil is to consume an average electrical power of 800 W?

31.41 A series circuit has an impedance of 60.0 Ω and a power factor of 0.720 at 50.0 Hz. The source voltage lags the current. a) What circuit element, an inductor or a capacitor, should be placed in series with the circuit to raise its power factor? b) What size element will raise the power factor to unity?

31.42 A circuit consists of a resistor and a capacitor in series with an ac source that supplies an rms voltage of 240 V. At the frequency of the source the reactance of the capacitor is 50.0 Ω. The rms current in the circuit is 3.00 A. What is the average power supplied by the source?

31.43 An LC circuit consists of an inductor coil with inductance $L = 5.00$ mH and a 20.0-μF capacitor. There is negligible resistance in the circuit. The circuit is driven by a voltage source with $v = V \cos \omega t$. If $V = 5.00$ mV and the frequency is twice the resonance frequency, determine a) the maximum charge on the capacitor; b) the maximum current in the circuit; c) the phase relationships between the voltages across the inductor, the capacitor and the source.

31.44 At a frequency ω_1 the reactance of a certain capacitor equals that of a certain inductor. a) If the frequency is changed to $\omega_2 = 2\omega_1$, what is the ratio of the reactance of the inductor to that of the capacitor? Which reactance is larger? b) If the frequency is changed to $\omega_3 = \omega_1/3$, what is the ratio of the reactance of the inductor to that of the capacitor? Which reactance is larger? c) If the capacitor and inductor are placed in series with a resistor of resistance R to form a series L-R-C circuit, what will be the resonance angular frequency of the circuit?

31.45 A High-Pass Filter. One application of L-R-C series circuits is to high-pass or low-pass filters, which filter out either the low- or high-frequency components of a signal. A high-pass filter is shown in Fig. 31.22, where the output voltage is taken across the L-R combination. (The L-R combination represents an inductive coil that also has resistance due to the large length of wire in the coil.) Derive an expression for V_{out}/V_s, the ratio of the output and source voltage amplitudes, as a function of the angular frequency ω of the source. Show that when ω is small, this ratio is proportional to ω and thus is small, and show that the ratio approaches unity in the limit of large frequency.

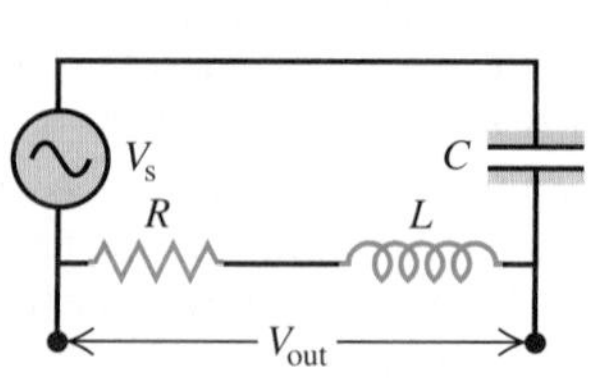

Figure 31.22 Problem 31.45.

31.46 A Low-Pass Filter. Figure 31.23 shows a low-pass filter (see Problem 31.45); the output voltage is taken across the capacitor in an L-R-C series circuit. Derive an expression for V_{out}/V_s, the ratio of the output and source voltage amplitudes, as a function of the angular frequency ω of the source. Show that when ω is large this ratio is proportional to ω^{-2} and thus is very small, and show that the ratio approaches unity in the limit of small frequency.

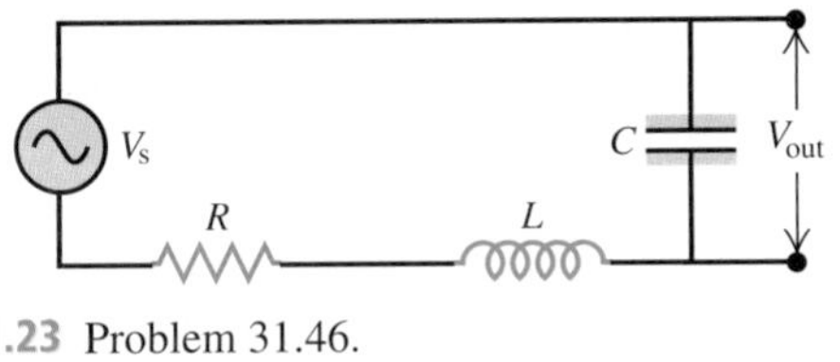

Figure 31.23 Problem 31.46.

31.47 An *L-R-C* series circuit is connected to an ac source of constant voltage amplitude V and variable angular frequency ω. a) Show that the current amplitude, as a function of ω, is

$$I = \frac{V}{\sqrt{R^2 + (\omega L - 1/\omega C)^2}}$$

b) Show that the average power dissipated in the resistor is

$$P = \frac{V^2 R/2}{R^2 + (\omega L - 1/\omega C)^2}$$

c) Show that I and P are *both* maximum when $\omega = 1/\sqrt{LC}$, that is, when the source frequency equals the resonance frequency of the circuit. d) Graph P as a function of ω for $V = 100$ V, $R = 200\ \Omega$, $L = 2.0$ H, and $C = 0.50\ \mu$F. Compare to the light purple curve in Fig. 31.15. Discuss the behavior of I and P in the limits $\omega = 0$ and $\omega \to \infty$.

31.48 An *L-R-C* series circuit is connected to an ac source of constant voltage amplitude V and variable angular frequency ω. Using the results of Problem 31.47, find an expression for a) the amplitude V_L of the voltage across the inductor as a function of ω; b) the amplitude V_C of the voltage across the capacitor as a function of ω. c) Graph V_L and V_C as functions of ω for $V = 100$ V, $R = 200\ \Omega$, $L = 2.0$ H, and $C = 0.50\ \mu$F. d) Discuss the behavior of V_L and V_C in the limits $\omega = 0$ and $\omega \to \infty$. For what value of ω is $V_L = V_C$? What is the significance of this value of ω?

31.49 An *L-R-C* series circuit is connected to an ac source of constant voltage amplitude V and variable angular frequency ω. a) Show that the time-averaged energy stored in the inductor is $U_B = \frac{1}{4}LI^2$ and the time-averaged energy stored in the capacitor is $U_E = \frac{1}{4}CV^2$. b) Use the results of Problems 31.47 and 31.48 to find expressions for U_B and U_E as functions of ω. c) Graph U_B and U_E as functions of ω for $V = 100$ V, $R = 200\ \Omega$, $L = 2.0$ H, and $C = 0.50\ \mu$F. d) Discuss the behavior of U_B and U_E in the limits $\omega = 0$ and $\omega \to \infty$. For what value of ω is $U_B = U_E$? What is the significance of this value of ω?

31.50 The *L-R-C* Parallel Circuit. A resistor, inductor, and capacitor are connected in parallel to an ac source with voltage amplitude V and angular frequency ω. Let the source voltage be given by $v = V\cos\omega t$. a) Show that the instantaneous voltages v_R, v_L, and v_C at any instant are each equal to v and that $i = i_R + i_L + i_C$, where i is the current through the source and i_R, i_L, and i_C are the currents through the resistor, the inductor, and the capacitor, respectively. b) What are the phases of i_R, i_L, and i_C with respect to v? Use current phasors to represent i, i_R, i_L, and i_C. In a phasor diagram, show the phases of these four currents with respect to v. c) Use the phasor diagram of part (b) to show that the current amplitude I for the current i through the source is given by $I = \sqrt{I_R^2 + (I_C - I_L)^2}$. d) Show that the result of part (c) can be written as $I = V/Z$, with $1/Z = \sqrt{1/R^2 + (\omega C - 1/\omega L)^2}$.

31.51 Parallel Resonance. The impedance of an *L-R-C* parallel circuit was derived in Problem 31.50. a) Show that at the resonance angular frequency $\omega_0 = 1/\sqrt{LC}$, $I_C = I_L$ and I is a *minimum*. b) Since I is a minimum at resonance, is it correct to say that the power delivered to the resistor is also a minimum at $\omega = \omega_0$? Explain. c) At resonance, what is the phase angle of the source current with respect to the source voltage? How does this compare to the phase angle for an *L-R-C series* circuit at resonance? d) Draw the circuit diagram for an *L-R-C* parallel circuit. Arrange the circuit elements in your diagram so that the resistor is closest to the ac source. Justify the following statement: When the angular frequency of the source is $\omega = \omega_0$, there is *no* current flowing between i) the part of the circuit that includes the source and the resistor and ii) the part that includes the inductor and capacitor, so you could cut the wires connecting these two parts of the circuit without affecting the currents. e) Is the statement in part (d) still valid if we consider that any real inductor or capacitor also has some resistance of its own? Explain.

31.52 A 400-Ω resistor and a 6.00-μF capacitor are connected in parallel to an ac generator that supplies an rms voltage of 220 V at an angular frequency of 360 rad/s. Use the results of Problem 31.50. Note that since there is no inductor in the circuit, the $1/\omega L$ term is not present in the expression for Z. Find a) the current amplitude in the resistor; b) the current amplitude in the capacitor; c) the phase angle of the source current with respect to the source voltage; d) the amplitude of the current through the generator. e) Does the source current lag or lead the source voltage?

31.53 An *L-R-C* parallel circuit is connected to an ac source of constant voltage amplitude V and variable angular frequency ω. a) Using the results of Problem 31.50, find expressions for the amplitudes I_R, I_L, and I_C of the currents through the resistor, inductor, and capacitor as functions of ω. b) Graph I_R, I_L, and I_C as functions of ω for $V = 100$ V, $R = 200\ \Omega$, $L = 2.0$ H, and $C = 0.50\ \mu$F. c) Discuss the behavior of I_L and I_C in the limits $\omega = 0$ and $\omega \to \infty$. Explain why I_L and I_C behave as they do in these limits. d) Calculate the resonance frequency (in Hz) of the circuit, and sketch the phasor diagram at the resonance frequency. e) At the resonance frequency, what is the current amplitude through the source? f) At the resonance frequency, what is the current amplitude through the resistor, through the inductor, and through the capacitor?

31.54 In Fig. 31.24, $R_1 = 60.0\ \Omega$, $R_2 = 40.0\ \Omega$, $L = 0.400$ H, $C = 5.00\ \mu$F, and $V_{rms} = 240$ V. What is the power supplied by the source in the limit that a) the frequency ω of the source is very large; b) ω is very small?

31.55 Finding an Unknown Inductance. Your boss gives you an inductor and asks you to measure its inductance. You have available a resistor, an ac voltmeter of high impedance, a capacitor and an ac source. Explain how you might use these to determine the inductance and cite any other piece of equipment you may need. Be

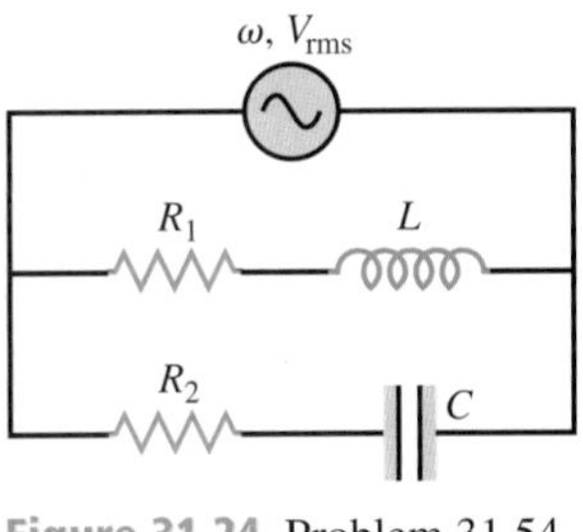

Figure 31.24 Problem 31.54.

sure to explain clearly how to use the equipment and what you need to measure to find the unknown inductance.

31.56 An L-R-C series circuit draws 220 W from a 120-V (rms), 50.0-Hz ac line. The power factor is 0.560, and the source voltage leads the current. a) What is the net resistance R of the circuit? b) Find the capacitance of the series capacitor that will result in a power factor of unity when it is added to the original circuit. c) What power will then be drawn from the supply line?

31.57 In an L-R-C series circuit the magnitude of the phase angle is 54.0°, with the source voltage lagging the current. The reactance of the capacitor is 350 Ω, and the resistor resistance is 180 Ω. The average power delivered by the source is 140 W. Find a) the reactance of the inductor; b) the rms current; c) the rms voltage of the source.

31.58 An L-R-C series circuit has $R = 500\ \Omega$, $L = 2.00$ H, $C = 0.500\ \mu$F, and $V = 100$ V. a) For $\omega = 800$ rad/s, calculate V_R, V_L, V_C, and ϕ. Using a single set of axes, graph v, v_R, v_L, and v_C as functions of time. Include two cycles of v on your graph. b) Repeat part (a) for $\omega = 1000$ rad/s. c) Repeat part (a) for $\omega = 1250$ rad/s.

31.59 In an L-R-C series circuit, the source has a voltage amplitude of 120 V, $R = 80.0\ \Omega$, and the reactance of the capacitor is 480 Ω. The voltage amplitude across the capacitor is 360 V. a) What is the current amplitude in the circuit? b) What is the impedance? c) What two values can the reactance of the inductor have? d) For which of the two values found in part (c) is the angular frequency less than the resonance angular frequency? Explain.

31.60 Designing an FM Radio Receiver. You enjoy listening to KONG-FM, which broadcasts at 94.1 MHz. You detest listening to KRUD-FM, which broadcasts at 94.0 MHz. You live the same distance from both stations and both transmitters are equally powerful, so both radio signals produce the same 1.0-V source voltage as measured at your house. Your goal is to design an L-R-C radio circuit with the following properties: i) It gives the maximum power response to the signal from KONG-FM; ii) the average power delivered to the resistor in response to KRUD-FM is 1.00% of the average power in response to KONG-FM. This limits the power received from the unwanted station, making it inaudible. You are required to use an inductor with $L = 1.00\ \mu$H. Find the capacitance C and resistance R that satisfy the design requirements.

31.61 The current in a certain circuit varies with time as shown in Fig. 31.25. Find the average current and the rms current in terms of I_0.

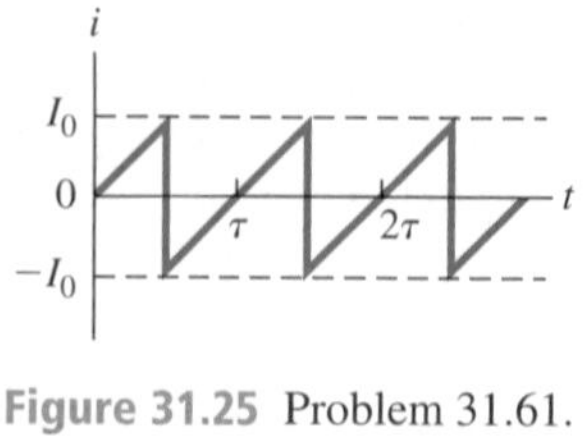

Figure 31.25 Problem 31.61.

31.62 The Resonance Width. Consider an L-R-C series circuit with a 1.80-H inductor, a 0.900-μF capacitor, and a 300-Ω resistor. The source has terminal rms voltage $V_{rms} = 60.0$ V and variable angular frequency ω. a) What is the resonance angular frequency ω_0 of the circuit? b) What is the rms current through the circuit at resonance, $I_{rms\text{-}0}$? c) For what two values of the angular frequency, ω_1 and ω_2, is the rms current half the resonance value? d) The quantity $|\omega_1 - \omega_2|$ defines the *resonance width.* Calculate $I_{rms\text{-}0}$ and the resonance width for $R = 300\ \Omega$, $30.0\ \Omega$, and $3.00\ \Omega$. Describe how your results compare to the discussion in Section 31.5.

31.63 The Resonance Width II. An L-R-C series circuit has inductance L, resistance R, and capacitance C. The ac source has voltage amplitude V and variable angular frequency ω. a) When ω equals the resonance angular frequency $\omega_0 = 1/\sqrt{LC}$, what is the amplitude I_0 of the current through the circuit? b) Suppose the angular frequency is only slightly different than ω_0. Then we can write $\omega = \omega_0 + \Delta\omega$, where $|\Delta\omega|$ is much less than ω_0. Show that the impedance is then approximately equal to $\sqrt{R^2 + 4L^2(\Delta\omega)}$. [*Hint:* Use the binomial expansion $(1 + z)^n = 1 + nz + n(n-1)z^2/2 + \ldots$, valid for the case $|z| < 1$.] c) Using the approximate expression for the impedance in part (b), find the two values of the angular frequency $\omega = \omega_0 + \Delta\omega$ for which the current amplitude is half the resonance value. Explain why these results are valid only if R is much less than $\sqrt{4L/3C}$. d) The difference between the two angular frequencies found in part (c) is the *resonance width.* Find an expression for the resonance width valid in the case $R \ll \sqrt{4L/3C}$. Use your result and the result of part (a) to explain how the shape of the curve in Fig. 31.15 changes as R is decreased. e) Determine I_0, the resonance angular frequency, and the resonance width for an L-R-C series circuit with $L = 2.50$ H, $C = 0.400\ \mu$F, and $V = 120$ V if i) $R = 15.0\ \Omega$; ii) $R = 1.50\ \Omega$.

31.64 A resistance R, capacitance C, and inductance L are connected in series to a voltage source with amplitude V and variable angular frequency ω. If $\omega = \omega_0$, the resonance angular frequency, find a) the maximum current in the resistor; b) the maximum voltage across the capacitor; c) the maximum voltage across the inductor; d) the maximum energy stored in the capacitor; and e) the maximum energy stored in the inductor. Give your answers in terms of R, C, L, and V.

31.65 Repeat Problem 31.64 for the case $\omega = \omega_0/2$.

31.66 Repeat Problem 31.64 for the case $\omega = 2\omega_0$.

31.67 In an L-R-C series circuit the current is given by $i = I \cos \omega t$. The voltage amplitudes for the resistor, inductor, and capacitor are V_R, V_L, and V_C. a) Show that the instantaneous power into the resistor is $p_R = V_R I \cos^2 \omega t = \frac{1}{2}V_R I(1 + \cos 2\omega t)$. What does this expression give for the average power into the resistor? b) Show that the instantaneous power into the inductor is $p_L = -V_L I \sin \omega t \cos \omega t = -\frac{1}{2}V_L I \sin 2\omega t$. What does this expression give for the average power into the inductor? c) Show that the instantaneous power into the capacitor is $p_C = V_C I \sin \omega t \cos \omega t = \frac{1}{2}V_C I \sin 2\omega t$. What does this expression give for the average power into the capacitor? d) The instantaneous power delivered by the source is shown in Section 31.4 to be $p = VI \cos \omega t (\cos \phi \cos \omega t - \sin \phi \sin \omega t)$. Show that $p_R + p_L + p_C$ equals p at each instant of time.

Challenge Problems

31.68 a) At what angular frequency is the voltage amplitude across the *resistor* in an L-R-C series circuit at maximum value? b) At what angular frequency is the voltage amplitude across the *induc-*

tor at maximum value? c) At what angular frequency is the voltage amplitude across the *capacitor* at maximum value? (You may want to refer to the results of Problem 31.48.)

31.69 Complex Numbers in a Circuit. The voltage across a circuit element in an ac circuit is not necessarily in phase with the current through that circuit element. Therefore the voltage amplitudes across the circuit elements in a branch in an ac circuit do not add algebraically. A method that is commonly employed to simplify the analysis of an ac circuit driven by a sinusoidal source is to represent the impedance Z as a *complex* number. The resistance R is taken to be the real part of the impedance, and the reactance $X = X_L - X_C$ is taken to be the imaginary part. Thus, for a branch containing a resistor, inductor, and capacitor in series, the complex impedance is $Z_{\text{cpx}} = R + iX$, where $i^2 = -1$. If the voltage amplitude across the branch is V_{cpx}, we define a *complex* current amplitude by $I_{\text{cpx}} = V_{\text{cpx}}/Z_{\text{cpx}}$. The *actual* current amplitude is the absolute value of the complex current amplitude, that is, $I = (I^*_{\text{cpx}} I_{\text{cpx}})^{1/2}$. The phase angle ϕ of the current with respect to the source voltage is given by $\tan\phi = \text{Im}(I_{\text{cpx}})/\text{Re}(I_{\text{cpx}})$. The voltage amplitudes $V_{R\text{-cpx}}$, $V_{L\text{-cpx}}$, and $V_{C\text{-cpx}}$ across the resistance, inductance, and capacitance, respectively, are found by multiplying I_{cpx} by R, iX_L, or $-iX_C$, respectively. Using the complex representation for the voltage amplitudes, the voltage across a branch is just the algebraic sum of the voltages across each circuit element; $V_{\text{cpx}} = V_{R\text{-cpx}} + V_{L\text{-cpx}} + V_{C\text{-cpx}}$. The actual value of any current amplitude or voltage amplitude is the absolute value of the corresponding complex quantity. Consider the series L-R-C circuit shown in Fig. 31.26. The values of the circuit elements, the source voltage amplitude, and the source angular frequency are as shown. Use the phasor diagram techniques presented in Section 31.1 to solve for a) the current amplitude; b) the phase angle ϕ of the current with respect to the source voltage. (Note that this angle is the negative of the phase angle defined in Fig. 31.9.) Now analyze the same circuit using the complex-number approach. c) Determine the complex impedance of the circuit, Z_{cpx}. Take the absolute value to obtain Z, the actual impedance of the circuit. d) Take the voltage amplitude of the source, V_{cpx}, to be real, and find the complex current amplitude I_{cpx}. Find the actual current amplitude by taking the absolute value of I_{cpx}. e) Find the phase angle ϕ of the current with respect to the source voltage by using the real and imaginary parts of I_{cpx}, as explained above. f) Find the complex representations of the voltages across the resistance, the inductance, and the capacitance. g) Adding the answers found in part (f), verify that the sum of these complex numbers is real and equal to 200 V, the voltage of the source.

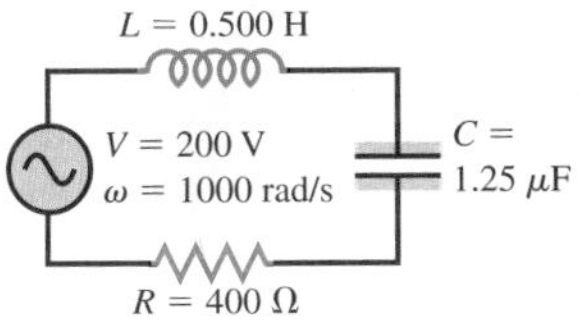

Figure 31.26 Challenge Problem 31.69.

32 ELECTROMAGNETIC WAVES

The cellular phone has revolutionized personal communications for millions of people around the world. Like any mobile radio, a cellular phone transmits and receives electromagnetic waves. Most cellular phone systems use wave frequencies between 806 and 902 MHz; PCS (personal communications system) phones use frequencies from 1850 to 1990 MHz.

? **Why do cellular phones sometime have poor reception inside steel-frame office buildings?**

What is light? This question has been asked by humans for countless centuries, but there was no answer until electricity and magnetism were unified into the single discipline of *electromagnetism,* as described by Maxwell's equations. These equations show that a time-varying magnetic field acts as a source of electric field and that a time-varying electric field acts as a source of magnetic field. These $\vec{E}$ and $\vec{B}$ fields can sustain each other, forming an *electromagnetic wave* that propagates through space. Visible light emitted by the glowing filament of a light bulb is one example of an electromagnetic wave; other kinds of electromagnetic waves are produced by sources such as TV and radio stations, microwave oscillators for ovens and radar, x-ray machines, and radioactive nuclei.

In this chapter we'll use Maxwell's equations as the theoretical basis for understanding electromagnetic waves. We'll find that these waves carry both energy and momentum. In sinusoidal electromagnetic waves, the $\vec{E}$ and $\vec{B}$ fields are sinusoidal functions of time and position, with a definite frequency and wavelength. The various types of electromagnetic waves—visible light, radio, x rays, and others—differ only in their frequency and wavelength. Our study of optics in the following chapters will be based in part on the electromagnetic nature of light.

Unlike waves on a string or sound waves in a fluid, electromagnetic waves do not require a material medium; the light that you see coming from the stars at night has traveled without difficulty across tens or hundreds of light years of (nearly) empty space. Nonetheless, electromagnetic waves and mechanical waves have much in common and are described in much the same language. Before reading further in this chapter, you should review the properties of mechanical waves as discussed in Chapters 15 and 16.

32.1 | Maxwell's Equations and Electromagnetic Waves

In the last several chapters we studied various aspects of electric and magnetic fields. We learned that when the fields don't vary with time, such as an electric field produced by charges at rest or the magnetic field of a steady current, we can analyze the electric and magnetic fields independently without considering interactions between them. But when the fields vary with time, they are no longer independent. Faraday's law (Section 29.2) tells us that a time-varying magnetic field acts as a source of electric field, as shown by induced emfs in inductors and transformers. Ampere's law, including the displacement current discovered by Maxwell (Section 29.7), shows that a time-varying electric field acts as a source of magnetic field. This mutual interaction between the two fields is summarized in Maxwell's equations, presented in Section 29.7.

Thus, when *either* an electric or a magnetic field is changing with time, a field of the other kind is induced in adjacent regions of space. We are led (as Maxwell was) to consider the possibility of an electromagnetic disturbance, consisting of time-varying electric and magnetic fields, that can propagate through space from one region to another, even when there is no matter in the intervening region. Such a disturbance, if it exists, will have the properties of a *wave,* and an appropriate term is **electromagnetic wave.**

Such waves do exist; radio and television transmission, light, x rays, and many other kinds of radiation are examples of electromagnetic waves. Our goal in this chapter is to see how such waves are explained by the principles of electromagnetism that we have studied thus far and to examine the properties of these waves.

32.1 James Clerk Maxwell (1831–1879) was the first person to truly understand the fundamental nature of light. He also made major contributions to thermodynamics, optics, astronomy, and color photography. Albert Einstein described Maxwell's accomplishments as "the most profound and the most fruitful that physics has experienced since the time of Newton."

As often happens in the development of science, the theoretical understanding of electromagnetic waves evolved along a considerably more devious path than the one just outlined. In the early days of electromagnetic theory (the early nineteenth century), two different units of electric charge were used, one for electrostatics and the other for magnetic phenomena involving currents. In the system of units used at that time, these two units of charge had different physical dimensions. Their *ratio* had units of velocity, and measurements showed that the ratio had a numerical value that was precisely equal to the speed of light, 3.00×10^8 m/s. At the time, physicists regarded this as an extraordinary coincidence and had no idea how to explain it.

In searching to understand this result, Maxwell (Fig. 32.1) proved in 1865 that an electromagnetic disturbance should propagate in free space with a speed equal to that of light and hence that light waves were likely to be electromagnetic in nature. At the same time, he discovered that the basic principles of electromagnetism can be expressed in terms of the four equations that we now call **Maxwell's equations,** which we discussed in Section 29.7. These four equations are (1) Gauss's law for electric fields; (2) Gauss's law for magnetic fields, showing the absence of magnetic monopoles; (3) Ampere's law, including displacement current; and (4) Faraday's law. Maxwell's equations are

$$\oint \vec{E} \cdot d\vec{A} = \frac{Q_{encl}}{\epsilon_0} \quad \text{(Gauss's law)} \tag{29.18}$$

$$\oint \vec{B} \cdot d\vec{A} = 0 \quad \text{(Gauss's law for magnetism)} \tag{29.19}$$

$$\oint \vec{B} \cdot d\vec{l} = \mu_0 \left(i_C + \epsilon_0 \frac{d\Phi_E}{dt} \right)_{encl} \quad \text{(Ampere's law)} \tag{29.20}$$

$$\oint \vec{E} \cdot d\vec{l} = -\frac{d\Phi_B}{dt} \quad \text{(Faraday's law)} \tag{29.21}$$

These equations apply to electric and magnetic fields *in vacuum.* If a material is present, the permittivity ϵ_0 and permeability μ_0 of free space are replaced by the permittivity ϵ and permeability μ of the material. If the values of ϵ and μ are different at different points in the regions of integration, then ϵ and μ have to be transferred to the left sides of Eqs. (29.18) and (29.20), respectively, and placed inside the integrals. The ϵ in Eq. (29.20) also has to be included in the integral that gives $d\Phi_E/dt$.

According to Maxwell's equations, a point charge at rest produces a static $\vec{E}$ field but no $\vec{B}$ field; a point charge moving with a constant velocity (Section 28.1) produces both $\vec{E}$ and $\vec{B}$ fields. Maxwell's equations can also be used to show that in order for a point charge to produce electromagnetic waves, the charge must *accelerate.* In fact, it's a general result of Maxwell's equations that *every* accelerated charge radiates electromagnetic energy. This is the reason for the shielding required around high-energy particle accelerators and high-voltage power supplies in TV sets.

One way in which a point charge can be made to emit electromagnetic waves is by making it oscillate in simple harmonic motion, so that it has an acceleration at almost every instant (the exception is when the charge is passing through its equilibrium position). Figure 32.2 shows some of the electric field lines produced by such an oscillating point charge. Field lines are *not* material objects, but you may nonetheless find it helpful to think of them as behaving somewhat like strings that extend from the point charge off to infinity. Oscillating the charge up and down makes waves that propagate outward from the charge along these "strings." Note that the charge does not emit waves equally in all directions; the waves are strongest at 90° to the axis of motion of the charge, while there are *no* waves along this axis. This is just what the "string" picture would lead you to conclude. There is also a *magnetic* disturbance that spreads outward from the charge; this is not shown in Fig. 32.2. Because the electric and magnetic disturbances spread or radiate away from the source, the name **electromagnetic radiation** is used interchangeably with the phrase "electromagnetic waves."

Electromagnetic waves with macroscopic wavelengths were first produced in the laboratory in 1887 by the German physicist Heinrich Hertz. As a source of waves, he used charges oscillating in *L-C* circuits of the sort discussed in Section 30.5; he detected the resulting electromagnetic waves with other circuits tuned to the same frequency. Hertz also produced electromagnetic *standing waves* and measured the distance between adjacent nodes (one half-wavelength) to determine the wavelength. Knowing the resonant frequency of his circuits, he then found the speed of the waves from the wavelength–frequency relation $v = \lambda f$. He established that their speed was the same as that of light; this verified Maxwell's theoretical prediction directly. The SI unit of frequency is named in honor of Hertz: One hertz (1 Hz) equals one cycle per second.

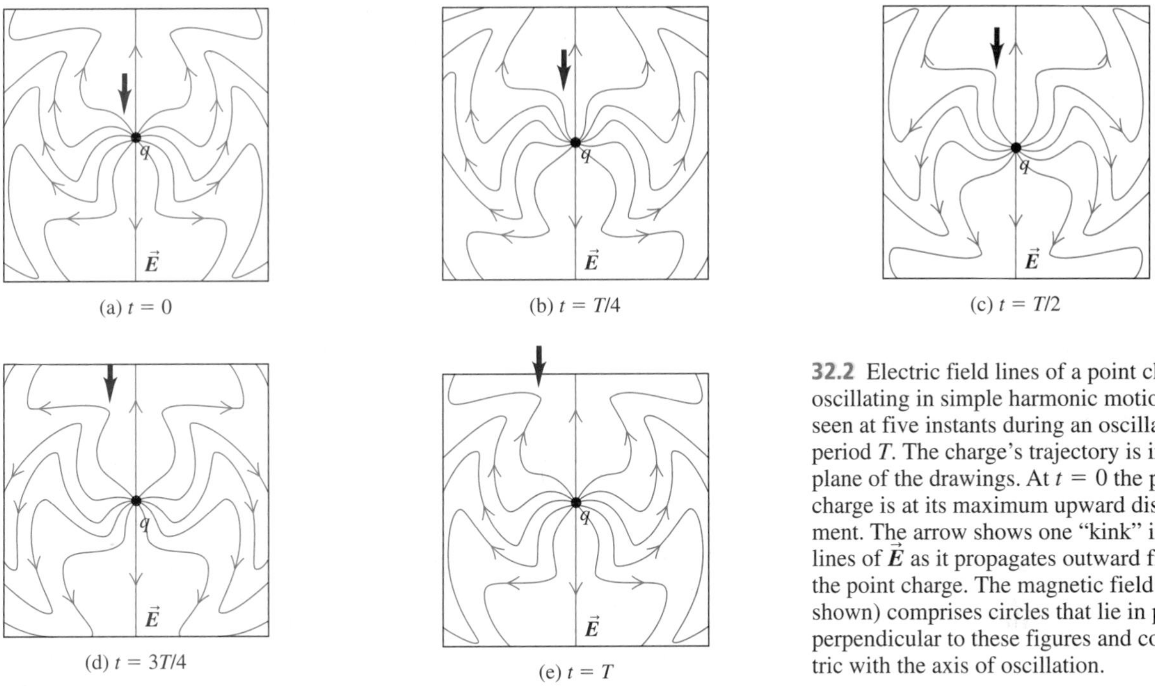

32.2 Electric field lines of a point charge oscillating in simple harmonic motion, seen at five instants during an oscillation period T. The charge's trajectory is in the plane of the drawings. At $t = 0$ the point charge is at its maximum upward displacement. The arrow shows one "kink" in the lines of $\vec{E}$ as it propagates outward from the point charge. The magnetic field (not shown) comprises circles that lie in planes perpendicular to these figures and concentric with the axis of oscillation.

The possible use of electromagnetic waves for long-distance communication does not seem to have occurred to Hertz. It remained to Marconi and others to make radio communication a familiar household experience. In a radio *transmitter,* electric charges are made to oscillate along the length of the conducting antenna, producing oscillating field disturbances like those shown in Fig. 32.2. Since many charges oscillate together in the antenna, the disturbances are much stronger than those of a single oscillating charge and can be detected at a much greater distance. In a radio *receiver* the antenna is also a conductor; the fields of the wave emanating from a distant transmitter exert forces on free charges within the receiver antenna, producing an oscillating current that is detected and amplified by the receiver circuitry.

For the remainder of this chapter our concern will be with electromagnetic waves themselves, not with the rather complex problem of how they are produced.

Test Your Understanding

Is it possible to have a purely electric wave propagate through empty space—that is, a wave made up of an electric field but no magnetic field? What about a purely magnetic wave, with a magnetic field but no electric field?

32.2 | Plane Electromagnetic Waves and the Speed of Light

We are now ready to develop the basic ideas of electromagnetic waves and their relation to the principles of electromagnetism. Our procedure will be to postulate a simple field configuration that has wavelike behavior. We'll assume an electric field $\vec{E}$ that has only a y-component and a magnetic field $\vec{B}$ with only a z-component, and

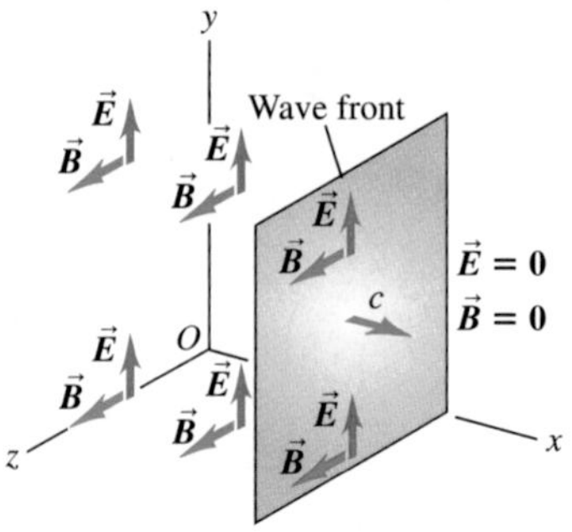

32.3 An electromagnetic wave front. The plane representing the wave front moves to the right with speed c. The $\vec{E}$ and $\vec{B}$ fields are uniform over the region behind the wave front but are zero everywhere in front of it.

we'll assume that both fields move together in the $+x$-direction with a speed c that is initially unknown. (As we go along, it will become clear why we choose $\vec{E}$ and $\vec{B}$ to be perpendicular to the direction of propagation as well as to each other.) Then we will test whether these fields are physically possible by asking whether they are consistent with Maxwell's equations, particularly Ampere's law and Faraday's law. We'll find that the answer is yes, provided that c has a particular value. We'll also show that the *wave equation,* which we encountered during our study of mechanical waves in Chapter 15, can be derived from Maxwell's equations.

A Simple Plane Electromagnetic Wave

Using an xyz-coordinate system (Fig. 32.3), we imagine that all space is divided into two regions by a plane perpendicular to the x-axis (parallel to the yz-plane). At every point to the left of this plane there are a uniform electric field $\vec{E}$ in the $+y$-direction and a uniform magnetic field $\vec{B}$ in the $+z$-direction, as shown. Furthermore, we suppose that the boundary plane, which we call the *wave front,* moves to the right in the $+x$-direction with a constant speed c, as yet unknown. Thus the $\vec{E}$ and $\vec{B}$ fields travel to the right into previously field-free regions with a definite speed. The situation, in short, describes a rudimentary electromagnetic wave. A wave such as this, in which at any instant the fields are uniform over any plane perpendicular to the direction of propagation, is called a **plane wave.** In the case shown in Fig. 32.3, the fields are zero for planes to the right of the wave front and have the same values on all planes to the left of the wave front; later we will consider more complex plane waves.

We won't concern ourselves with the problem of actually *producing* such a field configuration. Instead, we simply ask whether it is consistent with the laws of electromagnetism, that is, with Maxwell's equations. We'll consider each of these four equations in turn.

Let us first verify that our wave satisfies Maxwell's first and second equations, that is, Gauss's laws for electric and magnetic fields. To do this, we take as our Gaussian surface a rectangular box with sides parallel to the xy, xz, and yz coordinate planes (Fig. 32.4). The box encloses no electric charge, and you should be able to show that the total electric flux and magnetic flux through the box are both zero; this is true even if part of the box is in the region where $E = B = 0$. This would *not* be the case if $\vec{E}$ or $\vec{B}$ had an x-component, parallel to the direction of propagation. We leave the proof as a problem (see Problem 32.33). Thus to satisfy Maxwell's first and second equations, the electric and magnetic fields must be perpendicular to the direction of propagation; that is, the wave must be **transverse.**

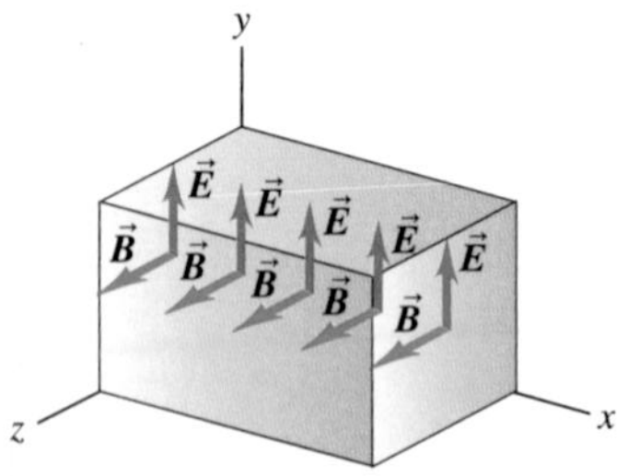

32.4 Gaussian surface for a plane electromagnetic wave. The total electric flux and the total magnetic flux through the surface are both zero.

The next of Maxwell's equations to be considered is Faraday's law:

$$\oint \vec{E} \cdot d\vec{l} = -\frac{d\Phi_B}{dt} \tag{32.1}$$

To test whether our wave satisfies Faraday's law, we apply this law to a rectangle *efgh* that is parallel to the xy-plane (Fig. 32.5a). As shown in Fig. 32.5b, a cross section in the xy-plane, this rectangle has height a and width Δx. At the time shown, the wave front has progressed partway through the rectangle, and $\vec{E}$ is zero along the side *ef.* In applying Faraday's law we take the vector area $d\vec{A}$ of rectangle *efgh* to be in the $+z$-direction. With this choice the right-hand rule requires

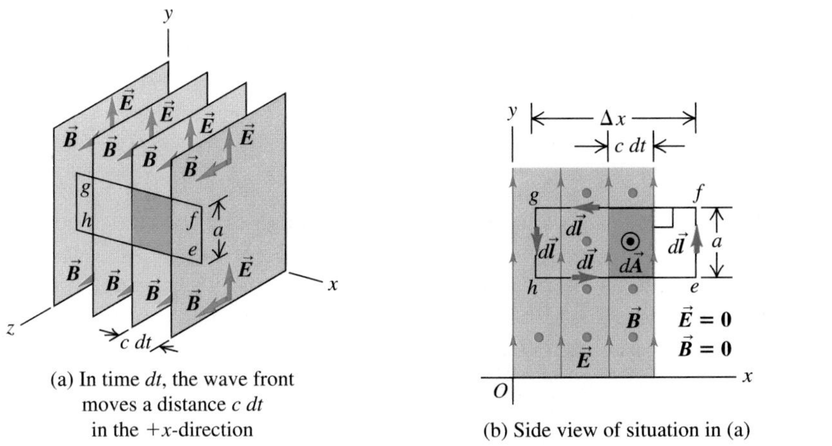

32.5 (a) Applying Faraday's law to a plane wave. (b) In a time dt, the magnetic flux through the rectangle in the xy-plane increases by an amount $d\Phi_B$. This increase equals the flux through the shaded rectangle with area $ac\,dt$; that is, $d\Phi_B = Bac\,dt$. Thus $d\Phi_B/dt = Bac$.

that we integrate $\vec{E}\cdot d\vec{l}$ *counterclockwise* around the rectangle. At every point on side *ef*, $\vec{E}$ is zero. At every point on sides *fg* and *he*, $\vec{E}$ is either zero or perpendicular to $d\vec{l}$. Only side *gh* contributes to the integral. On this side, $\vec{E}$ and $d\vec{l}$ are opposite, and we obtain

$$\oint \vec{E}\cdot d\vec{l} = -Ea \tag{32.2}$$

Hence, the left-hand side of Eq. (32.1) is nonzero.

To satisfy Faraday's law, Eq. (32.1), there must be a component of $\vec{B}$ in the z-direction (perpendicular to $\vec{E}$) so that there can be a nonzero magnetic flux Φ_B through the rectangle *efgh* and a nonzero derivative $d\Phi_B/dt$. Indeed, in our wave, $\vec{B}$ has *only* a z-component. We have assumed that this component is in the *positive* z-direction; let's see whether this assumption is consistent with Faraday's law. During a time interval dt the wave front moves a distance $c\,dt$ to the right in Fig. 32.5b, sweeping out an area $ac\,dt$ of the rectangle *efgh*. During this interval the magnetic flux Φ_B through the rectangle *efgh* increases by $d\Phi_B = B(ac\,dt)$, so the rate of change of magnetic flux is

$$\frac{d\Phi_B}{dt} = Bac \tag{32.3}$$

Now we substitute Eqs. (32.2) and (32.3) into Faraday's law, Eq. (32.1); we get

$$-Ea = -Bac$$

$$E = cB \qquad \text{(electromagnetic wave in vacuum)} \tag{32.4}$$

This shows that our wave is consistent with Faraday's law only if the wave speed c and the magnitudes of the perpendicular vectors $\vec{E}$ and $\vec{B}$ are related as in Eq. (32.4). Note that if we had assumed that $\vec{B}$ was in the *negative* z-direction, there would have been an additional minus sign in Eq. (32.4); since E, c, and B are all positive magnitudes, no solution would then have been possible. Furthermore, any component of $\vec{B}$ in the y-direction (parallel to $\vec{E}$) would not contribute to the changing magnetic flux Φ_B through the rectangle *efgh* (which is parallel to the xy plane) and so would not be part of the wave.

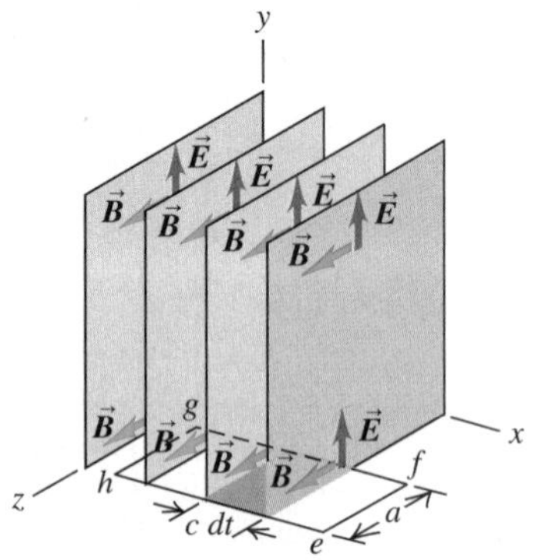

(a) In time dt, the wave front moves a distance $c\,dt$ in the $+x$-direction

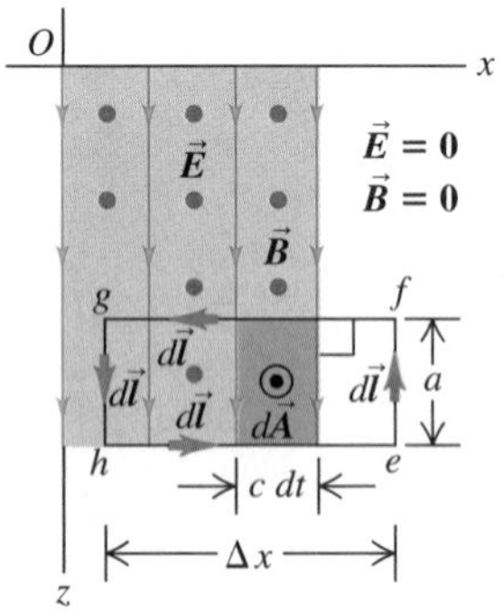

(b) Top view of situation in (a)

32.6 (a) Applying Ampere's law to a plane wave. (Compare to Fig. 32.5a.) (b) In a time dt, the electric flux through the rectangle in the xz-plane increases by an amount $d\Phi_E$. This increase equals the flux through the shaded rectangle with area $ac\,dt$; that is, $d\Phi_E = Eac\,dt$. Thus $d\Phi_E/dt = Eac$.

Finally, we carry out a similar calculation using Ampere's law, the remaining member of Maxwell's equations. There is no conduction current $(i_C = 0)$, so Ampere's law is

$$\oint \vec{B} \cdot d\vec{l} = \mu_0 \epsilon_0 \frac{d\Phi_E}{dt} \tag{32.5}$$

To check whether our wave is consistent with Ampere's law, we move our rectangle so that it lies in the xz-plane, as shown in Fig. 32.6, and we again look at the situation at a time when the wave front has traveled partway through the rectangle. We take the vector area $d\vec{A}$ in the $+y$-direction, and so the right-hand rule requires that we integrate $\vec{B} \cdot d\vec{l}$ counterclockwise around the rectangle. The $\vec{B}$ field is zero at every point along side ef, and at each point on sides fg and he it is either zero or perpendicular to $d\vec{l}$. Only side gh, where $\vec{B}$ and $d\vec{l}$ are parallel, contributes to the integral, and we find

$$\oint \vec{B} \cdot d\vec{l} = Ba \tag{32.6}$$

Hence, the left-hand side of Ampere's law, Eq. (32.5), is nonzero; the right-hand side must be nonzero as well. Thus $\vec{E}$ must have a y-component (perpendicular to $\vec{B}$) so that the electric flux Φ_E through the rectangle and the time derivative $d\Phi_E/dt$ can be nonzero. We come to the same conclusion that we inferred from Faraday's law: In an electromagnetic wave, $\vec{E}$ and $\vec{B}$ must be mutually perpendicular.

In a time interval dt the electric flux Φ_E through the rectangle increases by $d\Phi_E = E(ac\,dt)$. Since we chose $d\vec{A}$ to be in the $+y$-direction, this flux change is positive; the rate of change of electric field is

$$\frac{d\Phi_E}{dt} = Eac \tag{32.7}$$

Substituting Eqs. (32.6) and (32.7) into Ampere's law, Eq. (32.5), we find

$$Ba = \epsilon_0 \mu_0 Eac$$

$$B = \epsilon_0 \mu_0 cE \qquad \text{(electromagnetic wave in vacuum)} \tag{32.8}$$

Thus our assumed wave obeys Ampere's law only if B, c, and E are related as in Eq. (32.8).

Our electromagnetic wave must obey *both* Ampere's law and Faraday's law, so Eqs. (32.4) and (32.8) must both be satisfied. This can happen only if $\epsilon_0 \mu_0 c = 1/c$, or

$$c = \frac{1}{\sqrt{\epsilon_0 \mu_0}} \qquad \text{(speed of electromagnetic waves in vacuum)} \tag{32.9}$$

Inserting the numerical values of these quantities, we find

$$c = \frac{1}{\sqrt{(8.85 \times 10^{-12}\ \text{C}^2/\text{N}\cdot\text{m}^2)(4\pi \times 10^{-7}\ \text{N/A}^2)}}$$
$$= 3.00 \times 10^8\ \text{m/s}$$

Our assumed wave is consistent with all of Maxwell's equations, provided that the wave front moves with the speed given above, which you should recognize as the speed of light! Note that the *exact* value of c is defined to be 299,792,458 m/s; for our purposes, $c = 3.00 \times 10^8$ m/s is sufficiently accurate.

We chose a simple wave for our study in order to avoid mathematical complications, but this special case illustrates several important features of *all* electromagnetic waves:

1. The wave is *transverse;* both $\vec{E}$ and $\vec{B}$ are perpendicular to the direction of propagation of the wave. The electric and magnetic fields are also perpendicular to each other. The direction of propagation is the direction of the vector product $\vec{E} \times \vec{B}$.
2. There is a definite ratio between the magnitudes of $\vec{E}$ and $\vec{B}$: $E = cB$.
3. The wave travels in vacuum with a definite and unchanging speed.
4. Unlike mechanical waves, which need the oscillating particles of a medium such as water or air to transmit a wave, electromagnetic waves require no medium. What's "waving" in an electromagnetic wave are the electric and magnetic fields.

We can generalize this discussion to a more realistic situation. Suppose we have several wave fronts in the form of parallel planes perpendicular to the x-axis, all of which are moving to the right with speed c. Suppose that the $\vec{E}$ and $\vec{B}$ fields are the same at all points within a single region between two planes, but that the fields differ from region to region. The overall wave is a plane wave, but one in which the fields vary in steps along the x-axis. Such a wave could be constructed by superposing several of the simple step waves we have just discussed (shown in Fig. 32.3). This is possible because the $\vec{E}$ and $\vec{B}$ fields obey the superposition principle in waves just as in static situations: When two waves are superposed, the total $\vec{E}$ field at each point is the vector sum of the $\vec{E}$ fields of the individual waves, and similarly for the total $\vec{B}$ field.

We can extend the above development to show that a wave with fields that vary in steps is also consistent with Ampere's and Faraday's laws, provided that the wave fronts all move with the speed c given by Eq. (32.9). In the limit that we make the individual steps infinitesimally small, we have a wave in which the $\vec{E}$ and $\vec{B}$ fields at any instant vary *continuously* along the x-axis. The entire field pattern moves to the right with speed c. In Section 32.3 we will consider waves in which $\vec{E}$ and $\vec{B}$ are *sinusoidal* functions of x and t. Because at each point the magnitudes of $\vec{E}$ and $\vec{B}$ are related by $E = cB$, the periodic variations of the two fields in any periodic traveling wave must be *in phase.*

Electromagnetic waves have the property of **polarization.** In the above discussion the choice of the y-direction for $\vec{E}$ was arbitrary. We could just as well have specified the z-axis for $\vec{E}$; then $\vec{B}$ would have been in the $-y$-direction. A wave in which $\vec{E}$ is always parallel to a certain axis is said to be **linearly polarized** along that axis. More generally, *any* wave traveling in the x-direction can be represented as a superposition of waves linearly polarized in the y- and z-directions. We will study polarization in greater detail, with special emphasis on polarization of light, in Chapter 33.

*Derivation of the Wave Equation

Here is an alternative derivation of Eq. (32.9) for the speed of electromagnetic waves. It is more mathematical than our other treatment, but it includes a derivation of the wave equation for electromagnetic waves. This part of the section can be omitted without loss of continuity in the chapter.

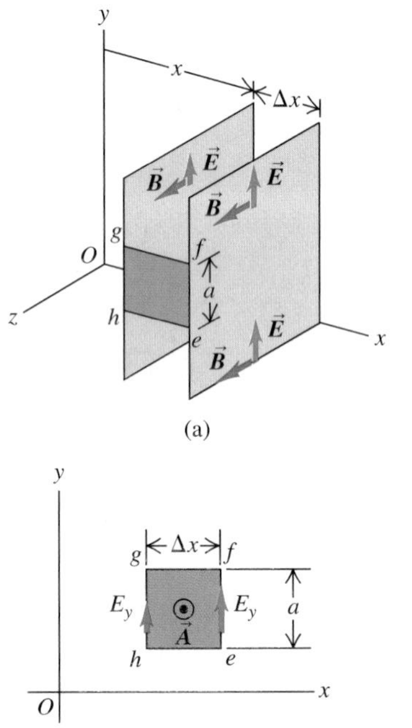

(a)

(b) Side view of situation in (a)

32.7 Faraday's law applied to a rectangle with height a and width Δx parallel to the xy-plane.

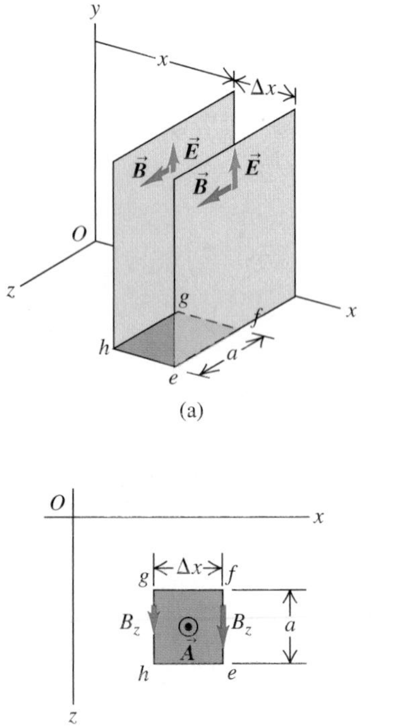

(a)

(b) Side view of situation in (a)

32.8 Ampere's law applied to a rectangle with height a and width Δx parallel to the xz-plane.

During our discussion of mechanical waves in Section 15.3, we showed that a function $y(x, t)$ that represents the displacement of any point in a mechanical wave traveling along the x-axis must satisfy a differential equation, Eq. (15.12):

$$\frac{\partial^2 y(x,t)}{\partial x^2} = \frac{1}{v^2}\frac{\partial^2 y(x,t)}{\partial t^2} \tag{32.10}$$

This equation is called the **wave equation,** and v is the speed of propagation of the wave.

To derive the corresponding equation for an electromagnetic wave, we again consider a plane wave. That is, we assume that at each instant, E_y and B_z are uniform over any plane perpendicular to the x-axis, the direction of propagation. But now we let E_y and B_z vary continuously as we go along the x-axis; then each is a function of x and t. We consider the values of E_y and B_z on two planes perpendicular to the x-axis, one at x and one at $x + \Delta x$.

Following the same procedure as previously, we apply Faraday's law to a rectangle lying parallel to the xy-plane, as in Fig. 32.7. This figure is similar to Fig. 32.5. Let the left end gh of the rectangle be at position x, and let the right end ef be at position $(x + \Delta x)$. At time t, the values of E_y on these two sides are $E_y(x, t)$ and $E_y(x + \Delta x, t)$, respectively. When we apply Faraday's law to this rectangle, we find that instead of $\oint \vec{E} \cdot d\vec{l} = -Ea$ as before, we have

$$\begin{aligned}\oint \vec{E} \cdot d\vec{l} &= -E_y(x,t)a + E_y(x+\Delta x, t)a \\ &= a[E_y(x+\Delta x,t) - E_y(x,t)]\end{aligned} \tag{32.11}$$

To find the magnetic flux Φ_B through this rectangle, we assume that Δx is small enough that B_z is nearly uniform over the rectangle. In that case, $\Phi_B = B_z(x, t)A = B_z(x, t)a\,\Delta x$, and

$$\frac{d\Phi_B}{dt} = \frac{\partial B_z(x,t)}{\partial t} a\,\Delta x$$

We use partial-derivative notation because B_z is a function of both x and t. When we substitute this expression and Eq. (32.11) into Faraday's law, Eq. (32.1), we get

$$a[E_y(x+\Delta x,t) - E_y(x,t)] = -\frac{\partial B_z}{\partial t} a\,\Delta x$$

$$\frac{E_y(x+\Delta x,t) - E_y(x,t)}{\Delta x} = -\frac{\partial B_z}{\partial t}$$

Finally, imagine shrinking the rectangle down to a sliver so that Δx approaches zero. When we take the limit of this equation as $\Delta x \to 0$, we get

$$\frac{\partial E_y(x,t)}{\partial x} = -\frac{\partial B_z(x,t)}{\partial t} \tag{32.12}$$

This equation shows that if there is a time-varying component B_z of magnetic field, there must also be a component E_y of electric field that varies with x, and conversely. We put this relation on the shelf for now; we'll return to it soon.

Next we apply Ampere's law to the rectangle shown in Fig. 32.8. The line integral $\oint \vec{B} \cdot d\vec{l}$ becomes

$$\oint \vec{B} \cdot d\vec{l} = -B_z(x+\Delta x,t)a + B_z(x,t)a \tag{32.13}$$

Again assuming that the rectangle is narrow, we approximate the electric flux Φ_E through it as $\Phi_E = E_y(x, t)\, A = E_y(x, t)a\, \Delta x$. The rate of change of Φ_E, which we need for Ampere's law, is then

$$\frac{d\Phi_E}{dt} = \frac{\partial E_y(x, t)}{\partial t} a\, \Delta x$$

Now substitute this expression and Eq. (32.13) into Ampere's law, Eq. (32.5):

$$-B_z(x + \Delta x, t)a + B_z(x, t)a = \epsilon_0 \mu_0 \frac{\partial E_y(x, t)}{\partial t} a\, \Delta x$$

Again we divide both sides by $a\, \Delta x$ and take the limit as $\Delta x \rightarrow 0$. We find

$$-\frac{\partial B_z(x, t)}{\partial x} = \epsilon_0 \mu_0 \frac{\partial E_y(x, t)}{\partial t} \tag{32.14}$$

Now comes the final step. We take the partial derivatives with respect to x of both sides of Eq. (32.12), and we take the partial derivatives with respect to t of both sides of Eq. (32.14). The results are

$$-\frac{\partial^2 E_y(x, t)}{\partial x^2} = \frac{\partial^2 B_z(x, t)}{\partial x \partial t}$$

$$-\frac{\partial^2 B_z(x, t)}{\partial x \partial t} = \epsilon_0 \mu_0 \frac{\partial^2 E_y(x, t)}{\partial t^2}$$

Combining these two equations to eliminate B_z, we finally find

$$\frac{\partial^2 E_y(x, t)}{\partial x^2} = \epsilon_0 \mu_0 \frac{\partial^2 E_y(x, t)}{\partial t^2} \tag{32.15}$$

(electromagnetic wave equation in vacuum)

This expression has the same form as the general wave equation, Eq. (32.10). Because the electric field E_y must satisfy this equation, it behaves as a wave with a pattern that travels through space with a definite speed. Furthermore, comparison of Eqs. (32.15) and (32.10) shows that the wave speed v is given by

$$\frac{1}{v^2} = \epsilon_0 \mu_0 \quad \text{or} \quad v = \frac{1}{\sqrt{\epsilon_0 \mu_0}}$$

This agrees with Eq. (32.9) for the speed c of electromagnetic waves.

We can show that B_z also must satisfy the same wave equation as E_y, Eq. (32.15). To prove this, we take the partial derivative of Eq. (32.12) with respect to t and the partial derivative of Eq. (32.14) with respect to x and combine the results. We leave this derivation as a problem (see Problem 32.35).

Test Your Understanding

In a wave of the sort depicted in Fig. 32.3, the magnitude of the electric field behind the wave front is 4.50×10^2 V/m. What is the magnitude of the magnetic field?

32.3 | Sinusoidal Electromagnetic Waves

ActivPhysics ONLINE
10.1 Properties of Mechanical Waves

Sinusoidal electromagnetic waves are directly analogous to sinusoidal transverse mechanical waves on a stretched string, which we studied in Section 15.3. In a sinusoidal electromagnetic wave, $\vec{E}$ and $\vec{B}$ at any point in space are sinusoidal

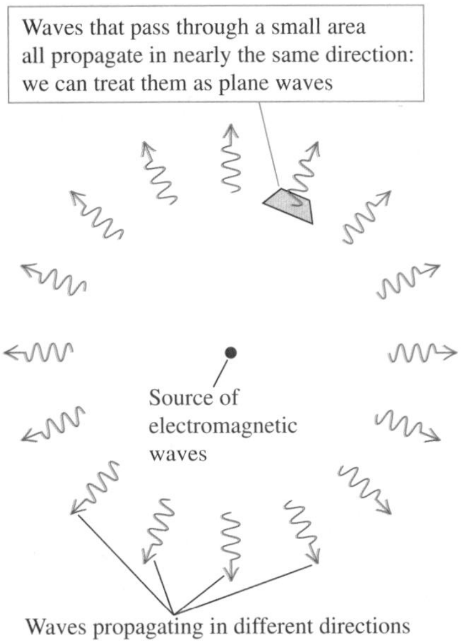

32.9 Waves passing through a small area at a sufficiently great distance from a source can be treated as plane waves.

functions of time, and at any instant of time the *spatial* variation of the fields is also sinusoidal.

Some sinusoidal electromagnetic waves are *plane waves;* they share with the waves described in Section 32.2 the property that at any instant the fields are uniform over any plane perpendicular to the direction of propagation. The entire pattern travels in the direction of propagation with speed c. The directions of $\vec{E}$ and $\vec{B}$ are perpendicular to the direction of propagation (and to each other), so the wave is *transverse.* Electromagnetic waves produced by an oscillating point charge, shown in Fig. 32.2, are an example of sinusoidal waves that are *not* plane waves. But if we restrict our observations to a relatively small region of space at a sufficiently great distance from the source, even these waves are well approximated by plane waves (Fig. 32.9). In the same way, the curved surface of the (nearly) spherical earth appears flat to us because of our small size relative to the earth's radius. In this section we'll restrict our discussion to plane waves.

The frequency f, the wavelength λ, and the speed of propagation c of any periodic wave are related by the usual wavelength–frequency relation $c = \lambda f$. If the frequency f is the power-line frequency of 60 Hz, the wavelength is

$$\lambda = \frac{c}{f} = \frac{3 \times 10^8 \text{ m/s}}{60 \text{ Hz}} = 5 \times 10^6 \text{ m} = 5000 \text{ km}$$

which is of the order of the earth's radius! For a wave with this frequency, even a distance of many miles includes only a small fraction of a wavelength. But if the frequency is 10^8 Hz (100 MHz), typical of commercial FM radio broadcasts, the wavelength is

$$\lambda = \frac{3 \times 10^8 \text{ m/s}}{10^8 \text{ Hz}} = 3 \text{ m}$$

and a moderate distance can include many complete waves.

Figure 32.10 shows a linearly polarized sinusoidal electromagnetic wave traveling in the $+x$-direction. The $\vec{E}$ and $\vec{B}$ vectors are shown for only a few points on the positive x-axis. Note that the electric and magnetic fields oscillate in phase: $\vec{E}$ is maximum where $\vec{B}$ is maximum and $\vec{E}$ is zero where $\vec{B}$ is zero. Note also that where $\vec{E}$ is in the $+y$-direction, $\vec{B}$ is in the $+z$-direction; where $\vec{E}$ is in the $-y$-direction, $\vec{B}$ is in the $-z$-direction. At all points the vector product $\vec{E} \times \vec{B}$ is in the direction in which the wave is propagating (the $+x$-direction). We mentioned this in Section 32.2 in the list of characteristics of electromagnetic waves.

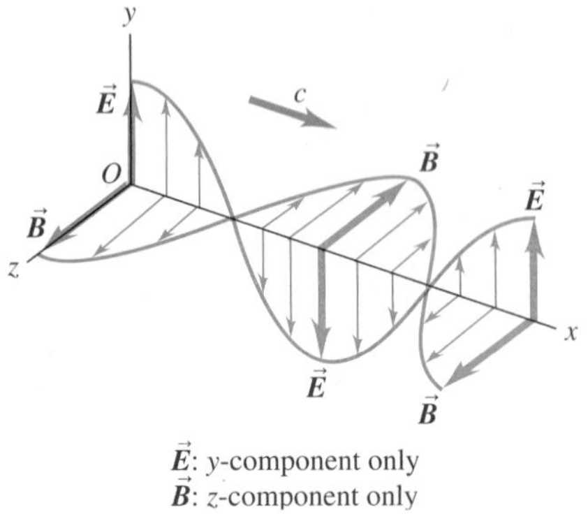

32.10 Representation of the electric and magnetic fields as functions of x for a linearly polarized sinusoidal plane electromagnetic wave. One wavelength of the wave is shown at time $t = 0$. The wave is traveling in the positive x-direction, the same as the direction of $\vec{E} \times \vec{B}$.

CAUTION Figure 32.10 may give you the erroneous impression that the electric and magnetic fields exist only along the x-axis. In fact, in a sinusoidal plane wave there are electric and magnetic fields at *all* points in space. Imagine a plane perpendicular to the x-axis (that is, parallel to the yz-plane) at a particular point, at a particular time; the fields have the same values at all points in that plane. The values are different on different planes.

We can describe electromagnetic waves by means of *wave functions,* just as we did in Section 15.3 for waves on a string. One form of the wave function for a transverse wave traveling in the $+x$-direction along a stretched string is Eq. (15.7):

$$y(x, t) = A \cos(kx - \omega t)$$

where $y(x, t)$ is the transverse displacement from its equilibrium position at time t of a point with coordinate x on the string. The quantity A is the maximum displacement, or *amplitude,* of the wave; ω is its *angular frequency,* equal to 2π times the frequency f; and k is the *wave number,* equal to $2\pi/\lambda$, where λ is the wavelength.

Let $E_y(x, t)$ and $B_z(x, t)$ represent the instantaneous values of the y-component of $\vec{E}$ and the z-component of $\vec{B}$, respectively, in Fig. 32.10, and let E_{max} and B_{max} represent the maximum values, or *amplitudes,* of these fields. The wave functions for the wave are then

$$E_y(x, t) = E_{max}\cos(kx - \omega t) \qquad B_z(x, t) = B_{max}\cos(kx - \omega t) \qquad (32.16)$$

(sinusoidal electromagnetic plane wave, propagating in $+x$-direction)

We can also write the wave functions in vector form:

$$\vec{E}(x, t) = \hat{\jmath}E_{max}\cos(kx - \omega t)$$
$$\vec{B}(x, t) = \hat{k}B_{max}\cos(kx - \omega t) \qquad (32.17)$$

CAUTION Note the two different k's, the unit vector $\hat{k}$ in the z-direction and the wave number k. Don't get these confused!

The sine curves in Fig. 32.10 represent instantaneous values of the electric and magnetic fields as functions of x at time $t = 0$, that is, $\vec{E}(x, t = 0)$ and $\vec{B}(x, t = 0)$. As time goes by, the wave travels to the right with speed c. Equations (32.16) and (32.17) show that at any point the sinusoidal oscillations of $\vec{E}$ and $\vec{B}$ are *in phase.* From Eq. (32.4) the amplitudes must be related by

$$E_{max} = cB_{max} \qquad \text{(electromagnetic wave in vacuum)} \qquad (32.18)$$

These amplitude and phase relations are also required for $E(x, t)$ and $B(x, t)$ to satisfy Eqs. (32.12) and (32.14), which came from Faraday's law and Ampere's law respectively. Can you verify this statement? (See Problem 32.34.)

Figure 32.11 shows the electric and magnetic fields of a wave traveling in the *negative* x-direction. At points where $\vec{E}$ is in the positive y-direction, $\vec{B}$ is in the *negative* z-direction; where $\vec{E}$ is in the negative y-direction, $\vec{B}$ is in the *positive* z-direction. The wave functions for this wave are

$$E_y(x, t) = E_{max}\cos(kx + \omega t) \qquad B_z(x, t) = -B_{max}\cos(kx + \omega t) \quad (32.19)$$

(sinusoidal electromagnetic plane wave, propagating in $-x$-direction)

As with the wave traveling in the $+x$-direction, at any point that the sinusoidal oscillations of the $\vec{E}$ and $\vec{B}$ fields are *in phase,* and the vector product $\vec{E} \times \vec{B}$ points in the direction of propagation.

The sinusoidal waves shown in Figs. 32.10 and 32.11 are both linearly polarized in the y-direction; the $\vec{E}$ field is always parallel to the y-axis. Example 32.1 concerns a wave that is linearly polarized in the z-direction.

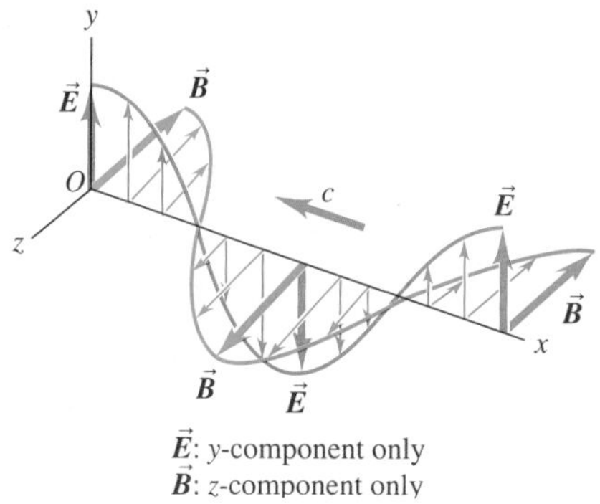

32.11 Representation of one wavelength of a linearly polarized sinusoidal plane electromagnetic wave traveling in the negative x-direction at $t = 0$. The fields are shown only for points along the x-axis. (Compare with Fig. 32.10.)

Problem-Solving Strategy — Electromagnetic Waves

IDENTIFY *the relevant concepts:* Many of the same ideas that apply to mechanical waves (discussed in Chapters 15 and 16) also apply to electromagnetic waves. The new feature is that the wave is described by two quantities, electric field $\vec{E}$ and magnetic field $\vec{B}$, instead of by a single quantity, such as the displacement of a string.

SET UP *the problem using the following steps:*

1. Draw a diagram showing the direction of wave propagation and the directions of $\vec{E}$ and $\vec{B}$.
2. Determine the target variables.

EXECUTE *the solution as follows:*

1. For problems involving electromagnetic waves, the best approach is to concentrate on basic relationships, such as the relation of $\vec{E}$ to $\vec{B}$ (both magnitude and direction), how the wave speed is determined, the transverse nature of the waves, and so on. Keep these relationships in mind when working through the mathematical details.
2. For sinusoidal electromagnetic waves, you need to use the language developed for sinusoidal mechanical waves in Chapters 15 and 16. Don't hesitate to go back and review that material, including the problem-solving strategies suggested in those chapters.
3. Keep in mind the basic relationships for periodic waves, $v = \lambda f$ and $\omega = vk$. For electromagnetic waves in vacuum, $v = c$. Be careful to distinguish between ordinary frequency f, usually expressed in hertz, and angular frequency $\omega = 2\pi f$, expressed in rad/s. Also remember that the wave number is $k = 2\pi/\lambda$.

EVALUATE *your answer:* Check that your result is reasonable. For electromagnetic waves in a vacuum, the magnitude of the magnetic field in teslas is much smaller (by a factor of 3.00×10^8) than the magnitude of the electric field in volts per meter. If your answer suggests otherwise, you probably made an error using the relationship $E = cB$. (We'll see later in this section that the relationship between E and B is different for electromagnetic waves in a material medium.)

Example 32.1 — Fields of a laser beam

A carbon dioxide laser emits a sinusoidal electromagnetic wave that travels in vacuum in the negative x-direction. The wavelength is 10.6 μm and the $\vec{E}$ field is parallel to the z-axis, with maximum magnitude of 1.5 MV/m. Write vector equations for $\vec{E}$ and $\vec{B}$ as functions of time and position.

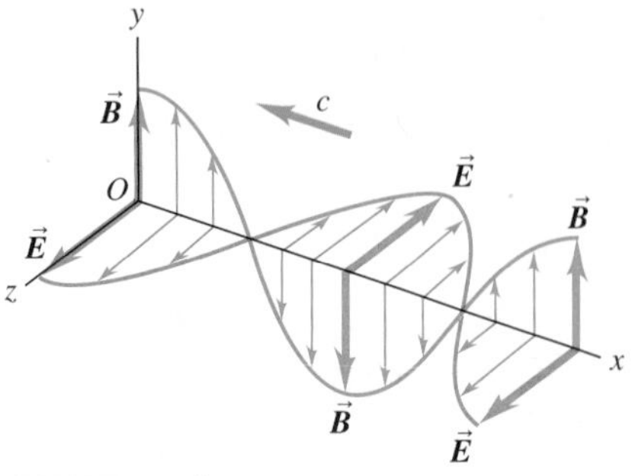

32.12 Representation of one wavelength of a sinusoidal plane electromagnetic wave traveling in the negative x-direction and linearly polarized along the z-axis. The fields are shown at $t = 0$ only for points along the x-axis. (Compare with Fig. 32.11, which shows a wave traveling in the same direction but linearly polarized along the y-axis.)

SOLUTION

IDENTIFY and SET UP: Equations (32.19) describe a wave traveling in the negative x-direction with $\vec{E}$ along the y-axis—that is, a wave that is linearly polarized along the y-axis. By contrast, the wave in this example is linearly polarized along the z-axis. At points where $\vec{E}$ is in the positive z-direction, $\vec{B}$ must be in the positive y-direction for the vector product $\vec{E} \times \vec{B}$ to be in the negative x-direction (the direction of propagation). Figure 32.12 shows a wave that satisfies these requirements.

EXECUTE: A possible pair of wave functions that describe the wave shown in Fig. 32.12 are

$$\vec{E}(x,t) = \hat{k}E_{max}\cos(kx + \omega t) \qquad \vec{B}(x,t) = \hat{\jmath}B_{max}\cos(kx + \omega t)$$

The plus sign in the arguments of the cosine functions indicates that the wave is propagating in the negative x-direction, as it should. Faraday's law requires that $E_{max} = cB_{max}$ [Eq. (32.18)], so

$$B_{max} = \frac{E_{max}}{c} = \frac{1.5 \times 10^6\ \text{V/m}}{3.0 \times 10^8\ \text{m/s}} = 5.0 \times 10^{-3}\ \text{T}$$

To check unit consistency, note that 1 V = 1 Wb/s and 1 Wb/m^2 = 1 T.

We have $\lambda = 10.6 \times 10^{-6}$ m, so the wave number and angular frequency are

$$k = 2\pi/\lambda = (2\pi \text{ rad})/(10.6 \times 10^{-6} \text{ m}) = 5.93 \times 10^5 \text{ rad/m}$$
$$\omega = ck = (3.00 \times 10^8 \text{ m/s})(5.93 \times 10^5 \text{ rad/m}) = 1.78 \times 10^{14} \text{ rad/s}$$

Substituting these values into the above wave functions, we get

$$\vec{E}(x,t) = \hat{k}(1.5 \times 10^6 \text{ V/m}) \cos[(1.78 \times 10^{14} \text{ rad/s})t + (5.93 \times 10^5 \text{ rad/m})x]$$
$$\vec{B}(x,t) = \hat{\jmath}(5.0 \times 10^{-3} \text{ T}) \cos[(1.78 \times 10^{14} \text{ rad/s})t + (5.93 \times 10^5 \text{ rad/m})x]$$

With these equations we can find the fields in the laser beam at any particular position and time by substituting specific values of x and t.

EVALUATE: As we expect, the magnitude B_{max} in teslas is much smaller than the magnitude E_{max} in volts per meter. To check the directions of $\vec{E}$ and $\vec{B}$, note that $\vec{E} \times \vec{B}$ is in the direction of $\hat{k} \times \hat{\jmath} = -\hat{\imath}$. This is as it should be for a wave that propagates in the negative x-direction.

Our expressions for $\vec{E}(x,t)$ and $\vec{B}(x,t)$ are not the only possible solutions. We could always add a phase ϕ to the arguments of the cosine function, so that $kx + \omega t$ would become $kx + \omega t + \phi$. To determine the value of ϕ we would need to know $\vec{E}$ and $\vec{B}$ either as functions of x at a given time t or as functions of t at a given coordinate x. However, the statement of the problem doesn't include this information.

Electromagnetic Waves in Matter

So far, our discussion of electromagnetic waves has been restricted to waves in *vacuum.* But electromagnetic waves can also travel in *matter;* think of light traveling through air, water, or glass. In this subsection we extend our analysis to electromagnetic waves in nonconducting materials, that is, *dielectrics.*

In a dielectric the wave speed is not the same as in vacuum, and we denote it by v instead of c. Faraday's law is unaltered, but in Eq. (32.4), derived from Faraday's law, the speed c is replaced by v. In Ampere's law the displacement current is given not by $\epsilon_0\, d\Phi_E/dt$, where Φ_E is the flux of $\vec{E}$ through a surface, but by $\epsilon\, d\Phi_E/dt = K\epsilon_0\, d\Phi_E/dt$, where K is the dielectric constant and ϵ is the permittivity of the dielectric. (We introduced these quantities in Section 24.4.) Also, the constant μ_0 in Ampere's law must be replaced by $\mu = K_m\mu_0$, where K_m is the relative permeability of the dielectric and μ is its permeability (see Section 28.8). Hence Eqs. (32.4) and (32.8) are replaced by

$$E = vB \quad \text{and} \quad B = \epsilon\mu vE \tag{32.20}$$

Following the same procedure as for waves in a vacuum, we find that the wave speed v is

$$v = \frac{1}{\sqrt{\epsilon\mu}} = \frac{1}{\sqrt{KK_m}}\frac{1}{\sqrt{\epsilon_0\mu_0}} = \frac{c}{\sqrt{KK_m}} \tag{32.21}$$

(speed of electromagnetic waves in a dielectric)

For most dielectrics the relative permeability K_m is very nearly equal to unity (except for insulating ferromagnetic materials). When $K_m \cong 1$,

$$v = \frac{1}{\sqrt{K}}\frac{1}{\sqrt{\epsilon_0\mu_0}} = \frac{c}{\sqrt{K}}$$

Because K is always greater than unity, the speed v of electromagnetic waves in a dielectric is always *less* than the speed c in vacuum by a factor of $1/\sqrt{K}$. The ratio of the speed c in vacuum to the speed v in a material is known in optics as the **index of refraction** n of the material. When $K_m \cong 1$,

$$\frac{c}{v} = n = \sqrt{KK_m} \cong \sqrt{K} \tag{32.22}$$

Usually, we can't use the values of K in Table 24.1 in this equation because those values are measured using *constant* electric fields. When the fields oscillate rapidly, there is usually not time for the re-orientating of electric dipoles that occurs with steady fields. Values of K with rapidly varying fields are usually much *smaller* than the values in the table. For example, K for water is 80.4 for steady fields but only about 1.8 in the frequency range of visible light. Thus the dielectric "constant" K is actually a function of frequency, called the *dielectric function* in more advanced treatments.

Example 32.2 Electromagnetic waves in different materials

a) While visiting a jewelry store one evening, you hold a diamond up to the light of a street lamp. The heated sodium vapor in the street lamp emits yellow light with a frequency of 5.09×10^{14} Hz. Find the wavelength in vacuum, the speed of wave propagation in diamond, and the wavelength in diamond. At this frequency, diamond has the properties $K = 5.84$ and $K_m = 1.00$. b) A radio wave with a frequency of 90.0 MHz (in the FM radio broadcast band) passes from vacuum into an insulating ferrite (a ferromagnetic material used in computer cables to suppress radio interference). Find the wavelength in vacuum, the speed of wave propagation in the ferrite, and the wavelength in the ferrite. At this frequency, the ferrite has the properties $K = 10.0$ and $K_m = 1000$.

SOLUTION

IDENTIFY and SET UP: In each case we find the wavelength in vacuum using $c = \lambda f$. The wave speed v is given in terms of c, the dielectric constant K, and the relative permeability K_m by Eq. (32.21). Once we know the value of v, we use $v = \lambda f$ to find the wavelength in the material in question.

EXECUTE: a) The wavelength in vacuum of the sodium light is

$$\lambda_{vacuum} = \frac{c}{f} = \frac{3.00 \times 10^8 \text{ m/s}}{5.09 \times 10^{14} \text{ Hz}} = 5.89 \times 10^{-7} \text{ m} = 589 \text{ nm}$$

The wave speed in diamond is

$$v_{diamond} = \frac{c}{\sqrt{KK_m}} = \frac{3.00 \times 10^8 \text{ m/s}}{\sqrt{(5.84)(1.00)}} = 1.24 \times 10^8 \text{ m/s}$$

This is about two-fifths of the speed in vacuum. The wavelength is proportional to the wave speed and so is reduced by the same factor:

$$\lambda_{diamond} = \frac{v_{diamond}}{f} = \frac{1.24 \times 10^8 \text{ m/s}}{5.09 \times 10^{14} \text{ Hz}} = 2.44 \times 10^{-7} \text{ m} = 244 \text{ nm}$$

b) Following the same steps as in part (a), we find that the wavelength in vacuum of the radio wave is

$$\lambda_{vacuum} = \frac{c}{f} = \frac{3.00 \times 10^8 \text{ m/s}}{90.0 \times 10^6 \text{ Hz}} = 3.33 \text{ m}$$

The wave speed in the ferrite is

$$v_{ferrite} = \frac{c}{\sqrt{KK_m}} = \frac{3.00 \times 10^8 \text{ m/s}}{\sqrt{(10.0)(1000)}} = 3.00 \times 10^6 \text{ m/s}$$

This is only 1% of the speed of light in a vacuum, so the wavelength is likewise 1% as large as the wavelength in vacuum:

$$\lambda_{ferrite} = \frac{v_{ferrite}}{f} = \frac{3.00 \times 10^9 \text{ m/s}}{90.0 \times 10^6 \text{ Hz}} = 3.33 \times 10^{-2} \text{ m} = 3.33 \text{ cm}$$

EVALUATE: The speed of light in transparent materials like diamond is typically between c and several percent of c. As our results in part (b) show, the speed of electromagnetic waves in dense materials like ferrite can be *far* slower than in vacuum.

Test Your Understanding

A sinusoidal electromagnetic wave is traveling in the negative z-direction. What are possible wave functions for $\vec{E}$ and $\vec{B}$ if the wave is linearly polarized in the x-direction?

32.4 | Energy and Momentum in Electromagnetic Waves

It is a familiar fact that energy is associated with electromagnetic waves; think of the energy in the sun's radiation. Practical applications of electromagnetic waves, such as microwave ovens, radio transmitters, and lasers for eye surgery, all make use of the energy that these waves carry. To understand how to utilize this energy, it's helpful to derive detailed relationships for the energy in an electromagnetic wave.

We begin with the expressions derived in Sections 24.3 and 30.3 for the **energy densities** in electric and magnetic fields; we suggest you review those derivations now. Equations (24.11) and (30.10) show that in a region of empty space where $\vec{E}$ and $\vec{B}$ fields are present, the total energy density u is given by

$$u = \frac{1}{2}\epsilon_0 E^2 + \frac{1}{2\mu_0}B^2 \tag{32.23}$$

where ϵ_0 and μ_0 are, respectively, the permittivity and permeability of free space. For electromagnetic waves in a vacuum, the magnitudes E and B are related by

$$B = \frac{E}{c} = \sqrt{\epsilon_0\mu_0}E \tag{32.24}$$

Combining Eqs. (32.23) and (32.24), we can also express the energy density u in a simple electromagnetic wave in vacuum as

$$u = \frac{1}{2}\epsilon_0 E^2 + \frac{1}{2\mu_0}(\sqrt{\epsilon_0\mu_0}E)^2 = \epsilon_0 E^2 \tag{32.25}$$

This shows that in a vacuum, the energy density associated with the $\vec{E}$ field in our simple wave is equal to the energy density of the $\vec{B}$ field. In general, the electric-field magnitude E is a function of position and time, as for the sinusoidal wave described by Eqs. (32.16); thus the energy density u of an electromagnetic wave, given by Eq. (32.25), also depends in general on position and time.

Electromagnetic Energy Flow and the Poynting Vector

Electromagnetic waves such as those we have described are *traveling* waves that transport energy from one region to another. For instance, in the wave described in Section 32.2 the $\vec{E}$ and $\vec{B}$ fields advance with time into regions where originally no fields were present and carry the energy density u with them as they advance. We can describe this energy transfer in terms of energy transferred *per unit time per unit cross-sectional area,* or *power per unit area,* for an area perpendicular to the direction of wave travel.

To see how the energy flow is related to the fields, consider a stationary plane, perpendicular to the x-axis, that coincides with the wave front at a certain time. In a time dt after this, the wave front moves a distance $dx = c\,dt$ to the right of the plane. Considering an area A on this stationary plane (Fig. 32.13), we note that the energy in the space to the right of this area must have passed through the area to reach the new location. The volume dV of the relevant region is the base area A times the length $c\,dt$, and the energy dU in this region is the energy density u times this volume:

$$dU = u\,dV = (\epsilon_0 E^2)(Ac\,dt)$$

This energy passes through the area A in time dt. The energy flow per unit time per unit area, which we will call S, is

$$S = \frac{1}{A}\frac{dU}{dt} = \epsilon_0 cE^2 \qquad \text{(in a vacuum)} \tag{32.26}$$

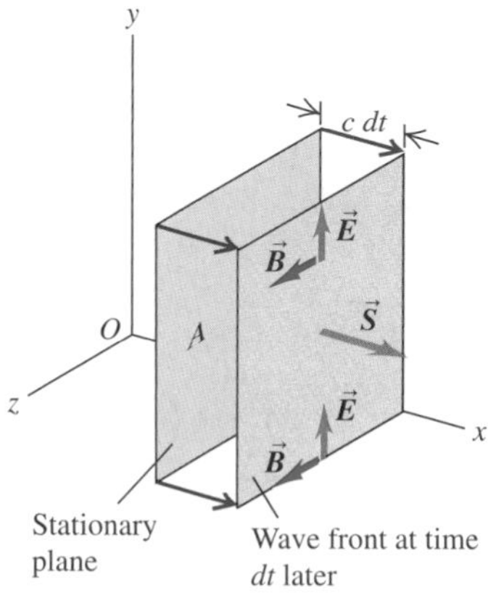

32.13 A wave front at a time dt after it passes through the stationary plane with area A. The volume between the plane and the wave front contains an amount of electromagnetic energy $uAc\,dt$.

32.14 These rooftop solar panels are tilted to be face-on to the sun—that is, face-on to the Poynting vector of electromagnetic waves from the sun, so that the panels can absorb the maximum amount of wave energy.

Using Eqs. (32.15) and (32.25), we can derive the alternate forms

$$S = \frac{\epsilon_0}{\sqrt{\epsilon_0 \mu_0}} E^2 = \sqrt{\frac{\epsilon_0}{\mu_0}} E^2 = \frac{EB}{\mu_0} \quad \text{(in a vacuum)} \quad (32.27)$$

The derivation of Eq. (32.27) from Eq. (32.26) is left as a problem (see Exercise 32.23). The units of S are energy per unit time per unit area, or power per unit area. The SI unit of S is 1 J/s·m^2 or 1 W/m^2.

We can define a *vector* quantity that describes both the magnitude and direction of the energy flow rate:

$$\vec{S} = \frac{1}{\mu_0} \vec{E} \times \vec{B} \quad \text{(Poynting vector in a vacuum)} \quad (32.28)$$

The vector $\vec{S}$ is called the **Poynting vector;** it was introduced by the British physicist John Poynting (1852–1914). Its direction is in the direction of propagation of the wave (Fig. 32.14). Since $\vec{E}$ are $\vec{B}$ are perpendicular, the magnitude of $\vec{S}$ is $S = EB/\mu_0$; from Eqs. (32.26) and (32.27) this is the energy flow per unit area and per unit time through a cross-sectional area perpendicular to the propagation direction. The total energy flow per unit time (power, P) out of any closed surface is the integral of $\vec{S}$ over the surface:

$$P = \oint \vec{S} \cdot d\vec{A}$$

For the sinusoidal waves studied in Section 32.3, as well as for other more complex waves, the electric and magnetic fields at any point vary with time, so the Poynting vector at any point is also a function of time. Because the frequencies of typical electromagnetic waves are very high, the time variation of the Poynting vector is so rapid that it's most appropriate to look at its *average* value. The magnitude of the average value of $\vec{S}$ at a point is called the **intensity** of the radiation at that point. The SI unit of intensity is the same as for S, 1 W/m^2 (watt per square meter).

Let's work out the intensity of the sinusoidal wave described by Eqs. (32.17). We first substitute $\vec{E}$ and $\vec{B}$ into Eq. (32.28):

$$\begin{aligned} \vec{S}(x,t) &= \frac{1}{\mu_0} \vec{E}(x,t) \times \vec{B}(x,t) \\ &= \frac{1}{\mu_0} [\hat{\jmath} E_{max} \cos(kx - \omega t)] \times [\hat{k} B_{max} \cos(kx - \omega t)] \end{aligned}$$

The vector product of the unit vectors is $\hat{\jmath} \times \hat{k} = \hat{\imath}$ and $\cos^2(kx - \omega t)$ is never negative, so $\vec{S}(x,t)$ always points in the positive x-direction (the direction of wave propagation). The x-component of the Poynting vector is

$$S_x(x,t) = \frac{E_{max} B_{max}}{\mu_0} \cos^2(kx - \omega t) = \frac{E_{max} B_{max}}{2\mu_0} [1 + \cos 2(kx - \omega t)]$$

The time average value of $\cos 2(kx - \omega t)$ is zero because at any point, it is positive during one half-cycle and negative during the other half. So the average value of the Poynting vector over a full cycle is $\vec{S}_{av} = \hat{\imath} S_{av}$, where

$$S_{av} = \frac{E_{max} B_{max}}{2\mu_0}$$

That is, the magnitude of the average value of $\vec{S}$ for a sinusoidal wave (the intensity I of the wave) is 1/2 the maximum value. By using the relations $E_{max} = B_{max}c$ and $\epsilon_0\mu_0 = 1/c^2$, we can express the intensity in several equivalent forms:

$$I = S_{av} = \frac{E_{max}B_{max}}{2\mu_0} = \frac{E_{max}{}^2}{2\mu_0 c} = \frac{1}{2}\sqrt{\frac{\epsilon_0}{\mu_0}}E_{max}{}^2 = \frac{1}{2}\epsilon_0 c E_{max}{}^2 \quad (32.29)$$

(intensity of a sinusoidal wave in a vacuum)

We invite you to verify that these expressions are all equivalent.

For a wave traveling in the $-x$-direction, represented by Eqs. (32.19), the Poynting vector is in the $-x$-direction at every point, but its magnitude is the same as for a wave traveling in the $+x$-direction. Verifying these statements is left to you (see Exercise 32.20).

CAUTION At any point x, the magnitude of the Poynting vector varies with time. Hence, the *instantaneous* rate at which electromagnetic energy in a sinusoidal plane wave arrives at a surface is not constant. This may seem to contradict everyday experience; the light from the sun, a light bulb, or the laser in a grocery-store scanner appears steady and unvarying in strength. In fact the Poynting vector from these sources *does* vary in time, but the variation isn't noticeable because the oscillation frequency is so high (around 5×10^{14} Hz for visible light). All that you sense is the *average* rate at which energy reaches your eye, which is why we commonly use intensity (the average value of S) to describe the strength of electromagnetic radiation.

Throughout this discussion we have considered only electromagnetic waves propagating in vacuum. If the waves are traveling in a dielectric medium, however, the expressions for energy density [Eq. (32.23)], the Poynting vector [Eq. (32.28)], and the intensity of a sinusoidal wave [Eq. (32.29)] must be modified. It turns out that the required modifications are quite simple: just replace ϵ_0 with the permittivity ϵ of the dielectric, replace μ_0 with the permeability μ of the dielectric, and replace c with the speed v of electromagnetic waves in the dielectric. Remarkably, the energy densities in the $\vec{E}$ and $\vec{B}$ fields are equal even in a dielectric.

Example 32.3 Energy in a nonsinusoidal wave

For the nonsinusoidal wave described in Section 32.2, suppose that $E = 100$ V/m $= 100$ N/C. Find the value of B, the energy density, and the rate of energy flow per unit area S.

SOLUTION

IDENTIFY and SET UP: In the wave described in Section 32.2, the electric and magnetic fields are uniform behind the wave front. Given the value of the magnitude E, we calculate the magnitude B using Eq. (32.4), the energy density u using Eq. (32.25), and the rate of energy flow per unit area S using Eq. (32.27). (Note that we cannot use Eq. (32.29), which applies to sinusoidal waves only.)

EXECUTE: From Eq. (32.4),

$$B = \frac{E}{c} = \frac{100 \text{ V/m}}{3.00 \times 10^8 \text{ m/s}} = 3.33 \times 10^{-7} \text{ T}$$

From Eq. (32.25),

$$u = \epsilon_0 E^2 = (8.85 \times 10^{-12} \text{ C}^2/\text{N}\cdot\text{m}^2)(100 \text{ N/C})^2$$
$$= 8.85 \times 10^{-8} \text{ N/m}^2 = 8.85 \times 10^{-8} \text{ J/m}^3$$

The magnitude of the Poynting vector is

$$S = \frac{EB}{\mu_0} = \frac{(100 \text{ V/m})(3.33 \times 10^{-7} \text{ T})}{4\pi \times 10^{-7} \text{ T}\cdot\text{m/A}}$$
$$= 26.5 \text{ V}\cdot\text{A/m}^2 = 26.5 \text{ W/m}^2$$

EVALUATE: We can check our result for S by using an alternative formula from Eq. (32.26):

$$\begin{aligned} S &= \epsilon_0 c E^2 \\ &= (8.85 \times 10^{-12}\ \mathrm{C^2/N \cdot m^2})(3.00 \times 10^8\ \mathrm{m/s})(100\ \mathrm{N/C})^2 \\ &= 26.5\ \mathrm{W/m^2} \end{aligned}$$

Since $\vec{E}$ and $\vec{B}$ have the same values at all points behind the wave front, the energy density u and Poynting vector magnitude S likewise have the same value everywhere behind the wave front. In front of the wave front, $\vec{E} = \mathbf{0}$ and $\vec{B} = \mathbf{0}$ and so $u = 0$ and $S = 0$; where there are no fields, there is no field energy.

Example 32.4 Energy in a sinusoidal wave

A radio station on the surface of the earth radiates a sinusoidal wave with an average total power of 50 kW (Fig. 32.15). Assuming that the transmitter radiates equally in all directions above the ground (which is unlikely in real situations), find the amplitudes E_{max} and B_{max} detected by a satellite at a distance of 100 km from the antenna.

SOLUTION

IDENTIFY: This is a sinusoidal wave, so we use the idea that the intensity is equal to the magnitude of the average value of the Poynting vector. We are not given the value of the intensity, but we *are* given the average total power of the transmitter. We use the idea that the intensity is the same as the average power per unit area.

SET UP: Figure 32.15 shows a hemisphere of radius 100 km centered on the transmitter. We divide the average power of the transmitter by the surface area of this hemisphere to find the intensity I at this distance from the transmitter. We then use Eq. (32.29) to find the electric-field magnitude and Eq. (32.4) to find the magnetic-field magnitude.

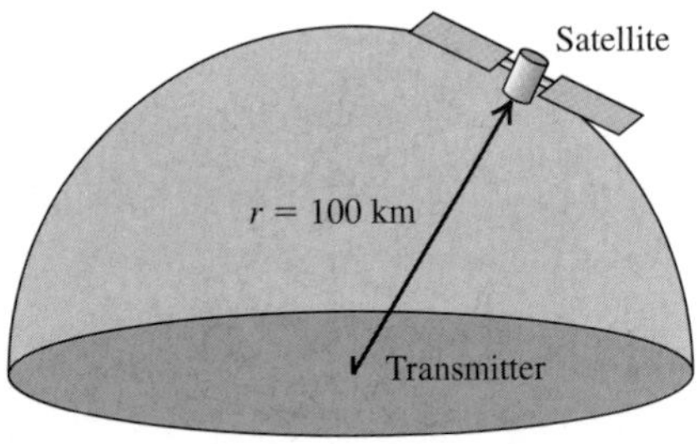

32.15 A radio station radiates waves into the hemisphere shown.

EXECUTE: The surface area of a hemisphere of radius $r = 100\ \mathrm{km} = 1.00 \times 10^5\ \mathrm{m}$ is

$$A = 2\pi R^2 = 2\pi(1.00 \times 10^5\ \mathrm{m})^2 = 6.28 \times 10^{10}\ \mathrm{m^2}$$

All the radiated power passes through this surface, so the average power per unit area (that is, the intensity) is

$$I = \frac{P}{A} = \frac{P}{2\pi R^2} = \frac{5.00 \times 10^4\ \mathrm{W}}{6.28 \times 10^{10}\ \mathrm{m^2}} = 7.96 \times 10^{-7}\ \mathrm{W/m^2}$$

From Eqs. (32.29), $I = S_{av} = E_{max}{}^2/2\mu_0 c$, so

$$\begin{aligned} E_{max} &= \sqrt{2\mu_0 c S_{av}} \\ &= \sqrt{2(4\pi \times 10^{-7}\ \mathrm{T \cdot m/A})(3.00 \times 10^8\ \mathrm{m/s})(7.96 \times 10^{-7}\ \mathrm{W/m^2})} \\ &= 2.45 \times 10^{-2}\ \mathrm{V/m} \end{aligned}$$

From Eq. (32.4),

$$B_{max} = \frac{E_{max}}{c} = 8.17 \times 10^{-11}\ \mathrm{T}$$

EVALUATE: Note that the magnitude of E_{max} is comparable to fields commonly seen in the laboratory, but B_{max} is extremely small in comparison to $\vec{B}$ fields we saw in previous chapters. For this reason, most detectors of electromagnetic radiation respond to the effect of the electric field, not the magnetic field. Loop radio antennas are an exception.

Electromagnetic Momentum Flow and Radiation Pressure

By using the observation that energy is required to establish electric and magnetic fields, we have shown that electromagnetic waves transport energy. It can also be shown that electromagnetic waves carry *momentum* p, with a corresponding momentum density (momentum dp per volume dV) of magnitude

$$\frac{dp}{dV} = \frac{EB}{\mu_0 c^2} = \frac{S}{c^2} \tag{32.30}$$

This momentum is a property of the field; it is not associated with the mass of a moving particle in the usual sense.

There is also a corresponding momentum flow rate. The volume dV occupied by an electromagnetic wave (speed c) that passes through an area A in time dt is $dV = Ac\,dt$. When we substitute this into Eq. (32.30) and rearrange, we find that the momentum flow rate per unit area is

$$\frac{1}{A}\frac{dp}{dt} = \frac{S}{c} = \frac{EB}{\mu_0 c} \quad \text{(flow rate of electromagnetic momentum)} \quad (32.31)$$

This is the momentum transferred per unit surface area per unit time. We obtain the *average* rate of momentum transfer per unit area by replacing S in Eq. (32.31) by $S_{av} = I$.

This momentum is responsible for the phenomenon of **radiation pressure.** When an electromagnetic wave is completely absorbed by a surface, the wave's momentum is also transferred to the surface. For simplicity we'll consider a surface perpendicular to the propagation direction. Using the ideas developed in Section 8.1, we see that the rate dp/dt at which momentum is transferred to the absorbing surface equals the *force* on the surface. The average force per unit area due to the wave, or *radiation pressure* p_{rad}, is the average value of dp/dt divided by the absorbing area A. (We use the subscript "rad" to distinguish pressure from momentum, for which the symbol p is also used.) From Eq. (32.31) the radiation pressure is

$$p_{rad} = \frac{S_{av}}{c} = \frac{I}{c} \quad \text{(radiation pressure, wave totally absorbed)} \quad (32.32)$$

If the wave is totally reflected, the momentum change is twice as great, and the pressure is

$$p_{rad} = \frac{2S_{av}}{c} = \frac{2I}{c} \quad \text{(radiation pressure, wave totally reflected} \quad (32.33)$$

For example, the value of I (or S_{av}) for direct sunlight, before it passes through the earth's atmosphere, is approximately 1.4 kW/m^2. From Eq. (32.32) the corresponding average pressure on a completely absorbing surface is

$$p_{rad} = \frac{I}{c} = \frac{1.4 \times 10^3\ \text{W/m}^2}{3.0 \times 10^8\ \text{m/s}} = 4.7 \times 10^{-6}\ \text{Pa}$$

From Eq. (32.33) the average pressure on a totally *reflecting* surface is twice this, $2I/c$ or 9.4×10^{-6} Pa. These are very small pressures, of the order of 10^{-10} atm, but they can be measured with sensitive instruments.

The radiation pressure of sunlight is much larger *inside* the sun than at the earth (see Problem 32.41). Inside stars that are much more massive and luminous than the sun, radiation pressure is so great that it substantially augments the gas pressure within the star and so helps to prevent the star from collapsing under its own gravity. In some cases the radiation pressure of a star actually pushes part of the star's material out into space (Fig. 32.16).

32.16 From time to time the unusual star at the center of this image undergoes a dramatic increase in brightness. When this happens, radiation pressure at the star's surface becomes so strong that part of its outer layers are ejected into space. Some of this ejected material appears as glowing blobs surrounding the star.

Example 32.5 Power and pressure from sunlight

An earth-orbiting satellite has solar-energy–collecting panels with a total area of 4.0 m^2 (Fig. 32.17). If the sun's radiation is perpendicular to the panels and is completely absorbed, find the average solar power absorbed and the average force associated with radiation pressure.

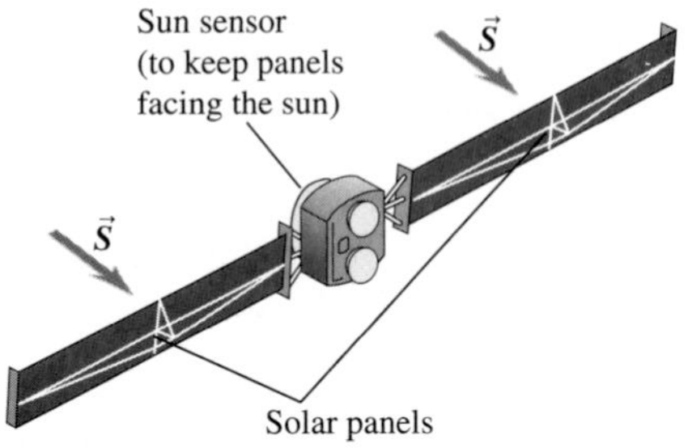

32.17 Solar panels on a satellite.

SOLUTION

IDENTIFY and SET UP: In the above discussion we calculated the intensity I (power per unit area) of sunlight as well as the radiation pressure p_{rad} (force per unit area) of sunlight on an absorbing surface. (We calculated these values for points above the atmosphere, which is where the satellite orbits.) Multiplying each value by the area of the solar panels gives the average power absorbed and the net radiation force on the panels.

EXECUTE: The intensity I (power per unit area) is $1.4 \times 10^3\ \text{W/m}^2$. Although the light from the sun is not a simple sinusoidal wave, we can still use the relationship that the average power P is the intensity I times the area A:

$$P = IA = (1.4 \times 10^3\ \text{W/m}^2)(4.0\ \text{m}^2)$$
$$= 5.6 \times 10^3\ \text{W} = 5.6\ \text{kW}$$

The radiation pressure of sunlight on an absorbing surface is $p_{\text{rad}} = 4.7 \times 10^{-6}\ \text{Pa} = 4.7 \times 10^{-6}\ \text{N/m}^2$. The total force F is the pressure p_{rad} times the area A:

$$F = p_{\text{rad}}A = (4.7 \times 10^{-6}\ \text{N/m}^2)(4.0\ \text{m}) = 1.9 \times 10^{-5}\ \text{N}$$

EVALUATE: The absorbed power is quite substantial. Part of it can be used to power the equipment aboard the satellite; the rest goes into heating the panels, either directly or due to inefficiencies in the photocells contained in the panels.

The total radiation force is comparable to the weight (on earth) of a single grain of salt. Over time, however, this small force can have a noticeable effect on the orbit of a satellite like that in Fig. 32.17, and so radiation pressure must be taken into account.

Test Your Understanding

Figure 32.10 shows one wavelength of a sinusoidal electromagnetic wave at time $t = 0$. For which values of x between $x = 0$ and $x = \lambda$ is the energy density at a maximum? For which values is it at a minimum? For which values of x is the magnitude of the instantaneous (not average) Poynting vector at a maximum? For which values is it at a minimum?

32.5 | Standing Electromagnetic Waves

Electromagnetic waves can be *reflected;* the surface of a conductor (like a polished sheet of metal) or of a dielectric (such as a sheet of glass) can serve as a reflector. The superposition principle holds for electromagnetic waves just as for electric and magnetic fields. The superposition of an incident wave and a reflected wave forms a **standing wave.** The situation is analogous to standing waves on a stretched string, discussed in Section 15.7; you should review that discussion.

Suppose a sheet of a perfect conductor (zero resistivity) is placed in the yz-plane of Fig. 32.18 and a linearly polarized electromagnetic wave, traveling in the negative x-direction, strikes it. As we discussed in Section 23.4, $\vec{E}$ cannot have a component parallel to the surface of a perfect conductor. Therefore in the present situation, $\vec{E}$ must be zero everywhere in the yz-plane. The electric field of the *incident* electro-

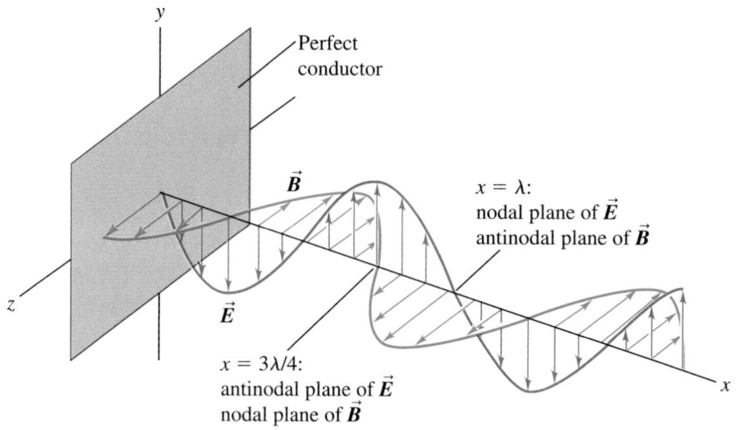

32.18 Representation of the electric and magnetic fields of a linearly polarized electromagnetic standing wave when $\omega t = 3\pi/4$ rad. In any plane perpendicular to the x-axis, E is maximum (an antinode) where B is zero (a node), and vice versa. As time elapses, the pattern does *not* move along the x-axis; instead, at every point the $\vec{E}$ and $\vec{B}$ vectors simply oscillate.

magnetic wave is *not* zero at all times in the yz-plane. But this incident wave induces oscillating currents on the surface of the conductor, and these currents give rise to an additional electric field. The *net* electric field, which is the vector sum of this field and the incident $\vec{E}$, *is* zero everywhere inside and on the surface of the conductor.

The currents induced on the surface of the conductor also produce a *reflected* wave that travels out from the plane in the $+x$-direction. Suppose the incident wave is described by the wave functions of Eqs. (32.19) (a sinusoidal wave traveling in the $-x$-direction) and the reflected wave by the negative of Eqs. (32.16) (a sinusoidal wave traveling in the $+x$-direction). We take the *negative* of the wave given by Eqs. (32.16) so that the incident and reflected electric fields cancel at $x = 0$ (the plane of the conductor, where the total electric field must be zero). The superposition principle states that the total $\vec{E}$ field at any point is the vector sum of the $\vec{E}$ fields of the incident and reflected waves, and similarly for the $\vec{B}$ field. Therefore the wave functions for the superposition of the two waves are

$$E_y(x, t) = E_{max}[\cos(kx + \omega t) - \cos(kx - \omega t)]$$
$$B_z(x, t) = B_{max}[-\cos(kx + \omega t) - \cos(kx - \omega t)]$$

We can expand and simplify these expressions, using the identities

$$\cos(A \pm B) = \cos A \cos B \mp \sin A \sin B$$

The results are

$$E_y(x, t) = -2E_{max} \sin kx \sin \omega t \qquad (32.34)$$
$$B_z(x, t) = -2B_{max} \cos kx \cos \omega t \qquad (32.35)$$

Equation (32.34) is analogous to Eq. (15.28) for a stretched string. We see that at $x = 0$ the electric field $E_y(x = 0, t)$ is *always* zero; this is required by the nature of the ideal conductor, which plays the same role as a fixed point at the end of a string. Furthermore, $E_y(x, t)$ is zero at *all* times at points in those planes perpendicular to the x-axis for which $\sin kx = 0$; that is, $kx = 0, \pi, 2\pi, \ldots$. Since $k = 2\pi/\lambda$, the positions of these planes are

$$x = 0, \frac{\lambda}{2}, \lambda, \frac{3\lambda}{2}, \ldots \text{ (nodal planes of } \vec{E}\text{)} \qquad (32.36)$$

These planes are called the **nodal planes** of the $\vec{E}$ field; they are the equivalent of the nodes, or nodal points, of a standing wave on a string. Midway between any two adjacent nodal planes is a plane on which $\sin kx = \pm 1$; on each such plane, the magnitude of $E(x, t)$ equals the maximum possible value of $2E_{max}$ twice per oscillation cycle. These are the **antinodal planes** of $\vec{E}$, corresponding to the antinodes of waves on a string.

The total magnetic field is zero at all times at points in planes on which $\cos kx = 0$. This occurs where

$$x = \frac{\lambda}{4}, \frac{3\lambda}{4}, \frac{5\lambda}{4}, \ldots \text{ (nodal planes of } \vec{B}) \tag{32.37}$$

These are the nodal planes of the $\vec{B}$ field; there is an antinodal plane of $\vec{B}$ midway between any two adjacent nodal planes.

Figure 32.18 shows a standing-wave pattern at one instant of time. The magnetic field is *not* zero at the conducting surface $(x = 0)$, and there is no reason it should be. The surface currents that must be present to make $\vec{E}$ exactly zero at the surface cause magnetic fields at the surface. The nodal planes of each field are separated by one half-wavelength. The nodal planes of one field are midway between those of the other; hence the nodes of $\vec{E}$ coincide with antinodes of $\vec{B}$, and conversely. Compare this situation to the distinction between pressure nodes and displacement nodes in Section 16.4.

The total electric field is a *sine* function of t, and the total magnetic field is a *cosine* function of t. The sinusoidal variations of the two fields are therefore 90° out of phase at each point. At times when $\sin \omega t = 0$, the electric field is zero *everywhere,* and the magnetic field is maximum. When $\cos \omega t = 0$, the magnetic field is zero everywhere, and the electric field is maximum. This is in contrast to a wave traveling in one direction, as described by Eqs. (32.16) or (32.19) separately, in which the sinusoidal variations of $\vec{E}$ and $\vec{B}$ at any particular point are *in phase.* It is interesting to check that Eqs. (32.34) and (32.35) satisfy the wave equation, Eq. (32.15). They also satisfy Eqs. (32.12) and (32.14) (the equivalents of Faraday's and Ampere's laws); we leave the proofs of these statements as problems.

Pursuing the stretched-string analogy, we may now insert a second conducting plane, parallel to the first and a distance L from it, along the $+x$-axis. This is analogous to a stretched string held at the points $x = 0$ and $x = L$. Both conducting planes must be nodal planes for $\vec{E}$; a standing wave can exist only when the second plane is placed at one of the positions where $E(x, t) = 0$. That is, for a standing wave to exist, L must be an integer multiple of $\lambda/2$. The wavelengths that satisfy this condition are

$$\lambda_n = \frac{2L}{n} \qquad (n = 1, 2, 3, \ldots) \tag{32.38}$$

The corresponding frequencies are

$$f_n = \frac{c}{\lambda_n} = n\frac{c}{2L} \qquad (n = 1, 2, 3, \ldots) \tag{32.39}$$

Thus there is a set of *normal modes,* each with a characteristic frequency, wave shape, and node pattern (Fig. 32.19). By measuring the node positions, we can measure the wavelength. If the frequency is known, the wave speed can be determined. This technique was first used by Hertz in the 1880s in his pioneering investigations of electromagnetic waves.

A laser has two mirrors; a standing wave is set up in the cavity between the mirrors. One of the mirrors has a small, partially transmitting aperture that allows waves to escape from this end of the laser.

32.19 A microwave oven sets up a standing electromagnetic wave with $\lambda = 12.2$ cm, a wavelength that is strongly absorbed by the water in food. Because the wave has nodes spaced $\lambda/2 = 6.1$ cm apart, the food must be rotated while cooking. Otherwise, the portion that lies at a node—where the electric-field amplitude is zero—will remain cold.

Conducting surfaces are not the only reflectors of electromagnetic waves. Reflections also occur at an interface between two insulating materials with different dielectric or magnetic properties. The mechanical analog is a junction of two strings with equal tension but different linear mass density. In general, a wave incident on such a boundary surface is partly transmitted into the second material and partly reflected back into the first. For example, light is transmitted through a glass window, but its surfaces also reflect light.

Example 32.6 Intensity in a standing wave

Calculate the intensity of the standing wave discussed in this section.

SOLUTION

IDENTIFY and SET UP: The intensity I of the wave is the average value S_{av} of the magnitude of the Poynting vector. We first find the instantaneous value of the Poynting vector, then average it over a whole number of cycles of the wave to determine I.

EXECUTE: Using the wave functions of Eqs. (32.34) and (32.35) in the expression for the Poynting vector $\vec{S}$, Eq. (32.28), we find

$$\begin{aligned}\vec{S}(x,t) &= \frac{1}{\mu_0}\vec{E}(x,t) \times \vec{B}(x,t)\\ &= \frac{1}{\mu_0}[-2\hat{\jmath}E_{max}\sin kx\cos\omega t] \times [-2\hat{k}B_{max}\cos kx\sin\omega t]\\ &= \hat{\imath}\frac{E_{max}B_{max}}{\mu_0}(2\sin kx\cos kx)(2\sin\omega t\cos\omega t)\\ &= \hat{\imath}S_x(x,t)\end{aligned}$$

Using the identity $\sin 2A = 2\sin A\cos A$, we can rewrite $S_x(x,t)$ as

$$S_x(x,t) = -\frac{E_{max}B_{max}\sin 2kx\sin 2\omega t}{\mu_0}$$

The average value of a sine function over any whole number of cycles is zero. Thus *the time average of* $\vec{S}$ *at any point is zero;* $I = S_{av} = 0$.

EVALUATE: This is just what we should expect. We formed our standing wave by superposing two waves with the same frequency and amplitude, traveling in opposite directions. All the energy transferred by one wave is completely cancelled by an equal amount transferred in the opposite direction by the other wave. When we use waves to transmit power, it is important to avoid reflections that give rise to standing waves.

Example 32.7 Standing waves in a cavity

Electromagnetic standing waves are set up in a cavity with two parallel, highly conducting walls separated by 1.50 cm. a) Calculate the longest wavelength and lowest frequency of electromagnetic standing waves between the walls. b) For this longest-wavelength standing wave, where in the cavity does $\vec{E}$ have maximum magnitude? Where is $\vec{E}$ zero? Where does $\vec{B}$ have maximum magnitude? Where is $\vec{B}$ zero?

SOLUTION

IDENTIFY and SET UP: The longest possible wavelength and lowest possible frequency correspond to the $n = 1$ mode in Eqs. (32.38) and (32.39). We use these equations to determine the values of λ and f. Equations (32.36) and (32.37) then tell us the locations of the nodal planes of $\vec{E}$ and $\vec{B}$; the antinodal planes of each field are midway between adjacent nodal planes.

EXECUTE: From Eq. (32.28), the $n = 1$ wavelength is

$$\lambda_1 = 2L = 2(1.50\text{ cm}) = 3.00\text{ cm}$$

The corresponding frequency is given by Eq. (32.29) with $n = 1$:

$$f_1 = \frac{c}{2L} = \frac{3.00 \times 10^8\text{ m/s}}{2(1.50 \times 10^{-2}\text{ m})} = 1.00 \times 10^{10}\text{ Hz} = 10\text{ GHz}$$

b) With $n = 1$ there is a single half-wavelength between the walls. The electric field has nodal planes ($\vec{E} = \mathbf{0}$) at the walls and an antinodal plane (where the maximum magnitude of $\vec{E}$ occurs) midway between them. The magnetic field has *antinodal* planes at the walls and a nodal plane midway between them.

EVALUATE: One application of standing waves of this kind is to produce an oscillating $\vec{E}$ field of definite frequency, which in turn is used to probe the behavior of a small sample of material placed inside the cavity. To subject the sample to the strongest possible field, the sample should be placed near the center of the cavity, at the antinode of $\vec{E}$.

Test Your Understanding

In the standing wave described in Example 32.7, is there any point where the energy density is zero at all times? If so, where? If not, why not?

32.6 | The Electromagnetic Spectrum

Electromagnetic waves cover an extremely broad spectrum of wavelength and frequency. This **electromagnetic spectrum** encompasses radio and TV transmission, visible light, infrared and ultraviolet radiation, x rays, and gamma rays. Electromagnetic waves have been detected with frequencies from at least 1 to 10^{24} Hz; the most commonly encountered portion of the spectrum is shown in Fig. 32.20, which gives approximate wavelength and frequency ranges for the various segments. Despite vast differences in their uses and means of production, these are all electromagnetic waves with the general characteristics described in preceding sections, including the propagation speed (in vacuum) $c = 299{,}792{,}458$ m/s. Electromagnetic waves may differ in frequency f and wavelength λ, but the relation $c = \lambda f$ in vacuum holds for each.

We can detect only a very small segment of this spectrum directly through our sense of sight. We call this range **visible light.** Its wavelengths range from about 400 to 700 nm (400 to 700×10^{-9} m), with corresponding frequencies from about 750 to 430 THz (7.5 to 4.3×10^{14} Hz). Different parts of the visible spectrum evoke in humans the sensations of different colors. Wavelengths for colors in the visible spectrum are given (very approximately) in Table 32.1.

Ordinary white light includes all visible wavelengths. However, by using special sources or filters, we can select a narrow band of wavelengths within a range of a few nm. Such light is approximately *monochromatic* (single-color) light. Absolutely monochromatic light with only a single wavelength is an unattainable idealization. When we use the expression "monochromatic light with $\lambda = 550$ nm" with reference to a laboratory experiment, we really mean a small

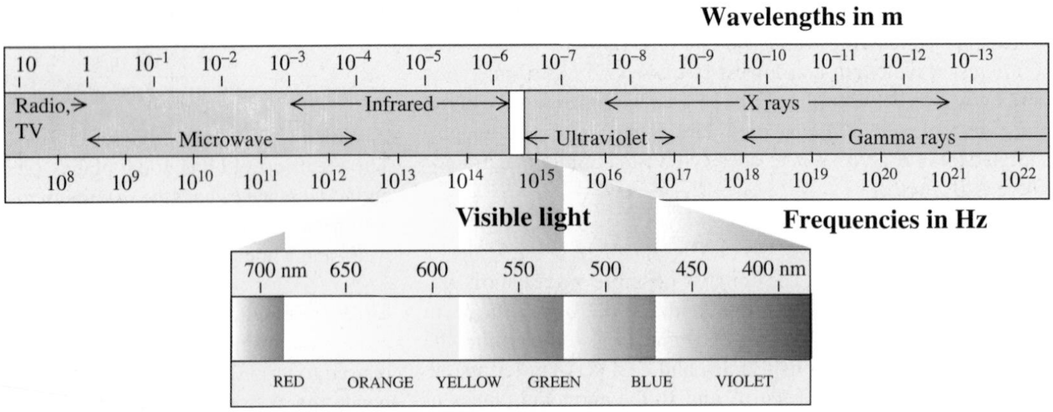

32.20 The electromagnetic spectrum. The frequencies and wavelengths found in nature extend over such a wide range that we have to use a logarithmic scale to show all important bands. The boundaries between bands are somewhat arbitrary.

band of wavelengths *around* 550 nm. Light from a *laser* is much more nearly monochromatic than is light obtainable in any other way.

Invisible forms of electromagnetic radiation are no less important than visible light. Our system of global communication, for example, depends on radio waves: AM radio uses waves with frequencies from 5.4×10^5 Hz to 1.6×10^6 Hz, while FM radio broadcasts are at frequencies from 8.8×10^7 Hz to 1.08×10^8 Hz. (Television broadcasts use frequencies that bracket the FM band.) Microwaves are also used for communication (for example, by cellular phones; see the photograph that opens this chapter) and for weather radar (at frequencies near 3×10^9 Hz). Many cameras have a device that emits a beam of infrared radiation; by analyzing the properties of the infrared radiation reflected from the subject, the camera determines the distance to the subject and automatically adjusts the focus. Ultraviolet radiation has shorter wavelengths than visible light; as we will learn in Chapter 36, this property allows it to be focused into very narrow beams for high-precision applications such as LASIK eye surgery. X rays are able to penetrate through flesh, which makes them invaluable in dentistry and medicine. The shortest-wavelength electromagnetic radiation, gamma rays, is produced in nature by radioactive materials (see Chapter 43). Gamma rays, which are very energetic, are used in medicine to destroy cancer cells.

Table 32.1 Wavelengths of Visible Light

400 to 440 nm	Violet
440 to 480 nm	Blue
480 to 560 nm	Green
560 to 590 nm	Yellow
590 to 630 nm	Orange
630 to 700 nm	Red

SUMMARY

Maxwell's equations predict the existence of electromagnetic waves that propagate in vacuum at the speed of light c. In a plane wave, $\vec{E}$ and $\vec{B}$ are uniform over any plane perpendicular to the propagation direction. Faraday's law and Ampere's law both give relationships between the magnitudes of $\vec{E}$ and $\vec{B}$; requiring both of these relationships to be satisfied gives an expression for c in terms of ϵ_0 and μ_0.

$$E = cB \tag{32.4}$$

$$B = \epsilon_0\mu_0 cE \tag{32.8}$$

$$c = \frac{1}{\sqrt{\epsilon_0\mu_0}} \tag{32.9}$$

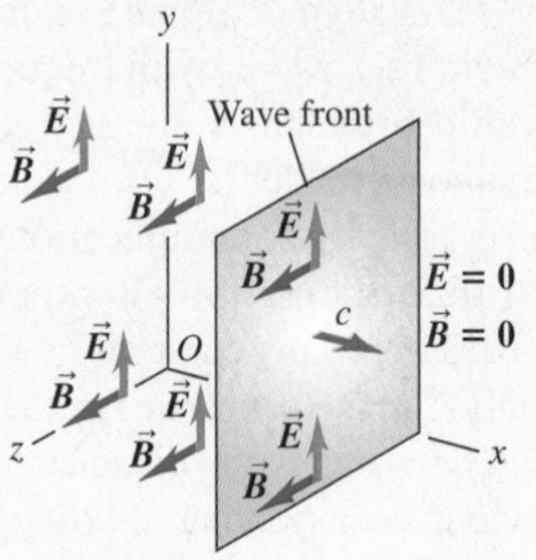

Electromagnetic waves are transverse; the $\vec{E}$ and $\vec{B}$ fields are perpendicular to the direction of propagation and to each other. The direction of propagation is the direction of $\vec{E} \times \vec{B}$. Equations (32.17) and (32.18) describe a sinusoidal plane electromagnetic wave traveling in vacuum in the $+x$-direction. (See Example 32.1)

$$\vec{E}(x,t) = \hat{\jmath}E_{max}\cos(kx - \omega t)$$
$$\vec{B}(x,t) = \hat{k}B_{max}\cos(kx - \omega t) \tag{32.17}$$

$$E_{max} = cB_{max} \tag{32.18}$$

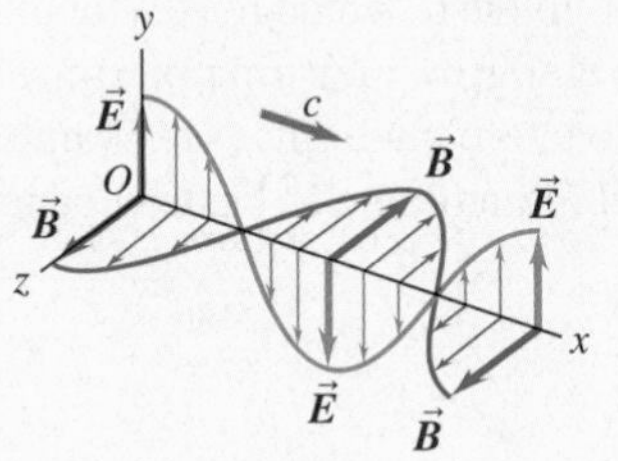

When an electromagnetic wave travels through a dielectric, the wave speed v is less than the speed of light in a vacuum c. (See Example 32.2)

$$v = \frac{1}{\sqrt{\epsilon\mu}} = \frac{1}{\sqrt{KK_m}}\frac{1}{\sqrt{\epsilon_0\mu_0}}$$
$$= \frac{c}{\sqrt{KK_m}} \tag{32.21}$$

The energy flow rate (power per unit area) in an electromagnetic wave in a vacuum is given by the Poynting vector $\vec{S}$. The magnitude of the time-averaged value of the Poynting vector is called the intensity I of the wave. Electromagnetic waves also carry momentum. When an electromagnetic wave strikes a surface, it exerts a radiation pressure p_{rad}. If the surface is perpendicular to the wave propagation direction and is totally absorbing, $p_{rad} = I/c$; if the surface is a perfect reflector, $p_{rad} = 2I/c$. (See Examples 32.3 through 32.5)

$$\vec{S} = \frac{1}{\mu_0}\vec{E} \times \vec{B} \tag{32.28}$$

$$I = S_{av} = \frac{E_{max}B_{max}}{2\mu_0} = \frac{E_{max}^2}{2\mu_0 c}$$
$$= \frac{1}{2}\sqrt{\frac{\epsilon_0}{\mu_0}}E_{max}^2$$
$$= \frac{1}{2}\epsilon_0 cE_{max}^2 \tag{32.29}$$

$$\frac{1}{A}\frac{dp}{dt} = \frac{S}{c} = \frac{EB}{\mu_0 c} \tag{32.31}$$

(flow rate of electromagnetic momentum)

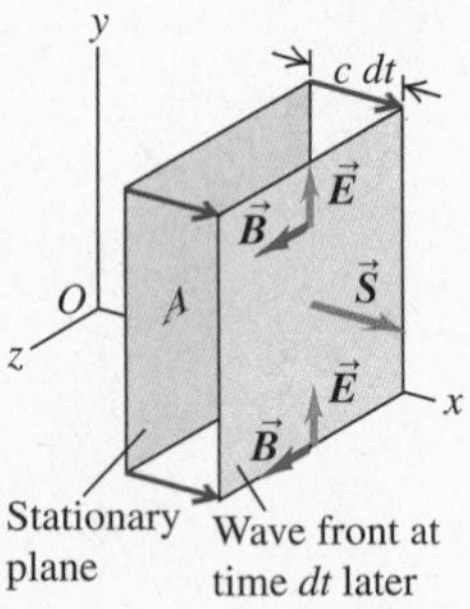

If a perfect reflecting surface is placed at $x = 0$, the incident and reflected waves form a standing wave. Nodal planes for $\vec{E}$ occur at $kx = 0, \pi, 2\pi, \ldots,$ and nodal planes for $\vec{B}$ at $kx = \pi/2, 3\pi/2, 5\pi/2, \ldots$. At each point, the sinusoidal variations of $\vec{E}$ and $\vec{B}$ with time are 90° out of phase. (See Examples 32.6 and 32.7)

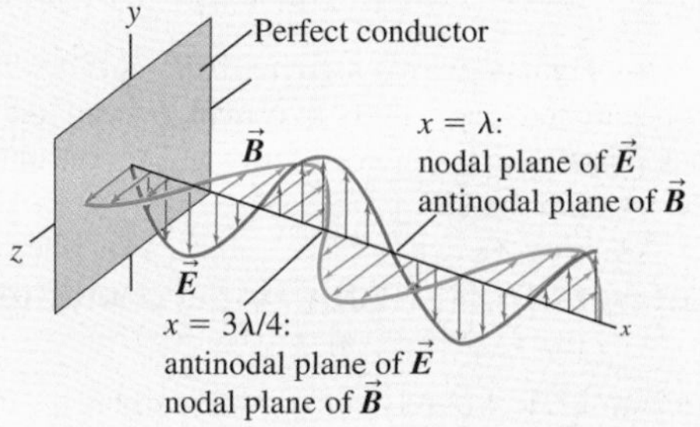

The electromagnetic spectrum covers a range of frequencies from at least 1 to 10^{24} Hz and a correspondingly broad range of wavelengths. Visible light constitutes only a very small part of this spectrum, with wavelengths of 400 to 700 nm.

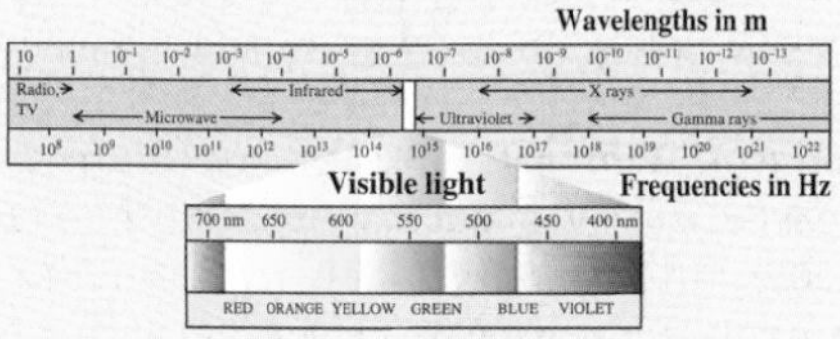

Key Terms

electromagnetic wave, 1215
Maxwell's equations, 1215
electromagnetic radiation, 1216
plane wave, 1218
transverse wave, 1218
polarization, 1221
linearly polarized, 1221
wave equation, 1222
index of refraction, 1227
energy density, 1229
Poynting vector, 1230
intensity, 1230
radiation pressure, 1233
standing wave, 1234
nodal plane, 1235
antinodal plane, 1236
electromagnetic spectrum, 1238
visible light, 1238

Your Notes

Answer to Chapter Opening Question

Electromagnetic waves arriving from outside the building cannot propagate into conductors like steel. Instead, the wave is reflected back outside, so less wave energy reaches the phone and the reception is compromised.

Answers to Test Your Understanding Questions

Section 32.1 A purely electric wave would have a varying electric field. Such a field necessarily generates a magnetic field through Ampere's law, Eq. (29.20), so a purely electric wave is impossible. In the same way, a purely magnetic wave is impossible: the varying magnetic field in such a wave would automatically give rise to an electric field through Faraday's law, Eq. (29.21).

Section 32.2 From Eq. (32.4), the electric field magnitude E and the magnetic field magnitude B are related by $E = cB$. Hence $B = E/c = (4.50 \times 10^2\ \text{V/m})/(3.00 \times 10^8\ \text{m/s}) = 1.50 \times 10^{-6}\ \text{V}\cdot\text{s/m}^2$. Note that 1 volt = 1 newton per coulomb ($1\ \text{V} = 1\ \text{N/C}$) and 1 coulomb per second = 1 ampere ($1\ \text{C/s} = 1\ \text{A}$), so the units of B are $\text{N}\cdot\text{s}/(\text{C}\cdot\text{m}^2) = \text{N}/(\text{A}\cdot\text{m})$. We saw in Section 27.2 that 1 tesla (1 T) is equal to $1\ \text{N}/(\text{A}\cdot\text{m})$. So just as we would expect, our result for B is in teslas: $B = 1.50 \times 10^{-6}\ \text{T}$. Because c in m/s is such a large number, the magnetic field magnitude in T is *much* smaller than the electric field magnitude in V/m.

Section 32.3 The cosine functions for $\vec{E}$ and $\vec{B}$ have the argument $kz + \omega t$. (Note that x is replaced by z because the propagation is along the z-axis. We use a plus sign because the wave propagates in the negative z-direction.) The direction of polarization tells us that $\vec{E}$ has only an x-component, so a possible wave function is $\vec{E}(z, t) = \hat{\imath}E_{max} \cos(kz + \omega t)$. The magnetic field $\vec{B}$ is perpendicular to both $\vec{E}$ and the direction of propagation, so it has only a y-component. It must also be in phase with $\vec{E}$, so B_y is equal to either $B_{max} \cos(kz + \omega t)$ or $-B_{max} \cos(kz + \omega t)$. To decide which is correct, recall that $\vec{E} \times \vec{B}$ is in the direction of propagation—that is, in the negative z-direction. Thus, if $\vec{E}$ is in the positive x-direction, $\vec{B}$ must be in the negative y-direction. Hence $B_y(z, t) = -B_{max} \cos(kz + \omega t)$ and $\vec{B}(z, t) = -\hat{\jmath}B_{max} \cos(kz + \omega t)$.

Section 32.4 Both the energy density u and the Poynting vector magnitude S are maximum where the $\vec{E}$ and $\vec{B}$ fields have their maximum magnitudes. (The direction of the fields doesn't matter.) From Fig. 32.10 this occurs at $x = 0$, $x = \lambda/2$, and $x = \lambda$. Both u and S have a minimum value of zero; that occurs where $\vec{E}$ and $\vec{B}$ are both zero. From Fig. 32.10, this occurs at $x = \lambda/4$ and $x = 3\lambda/4$.

Section 32.5 There are places where $\vec{E} = \mathbf{0}$ at all times (at the walls) and the electric energy density $\frac{1}{2}\epsilon_0 E^2$ is always zero. There are also places where $\vec{B} = \mathbf{0}$ at all times (on the plane midway between the walls) and the magnetic energy density $B^2/2\mu_0$ is always zero. However, there are *no* locations where both $\vec{E}$ and $\vec{B}$ are always zero. Hence the energy density at any point in the standing wave is always nonzero.

Discussion Questions

Q32.1 By measuring the electric and magnetic fields at a point in space where there is an electromagnetic wave, can you determine the direction from which the wave came? Explain.

Q32.2 According to Ampere's law, is it possible to have both a conduction current and a displacement current at the same time? Is it possible for the effects of the two kinds of current to cancel each other exactly so that *no* magnetic field is produced? Explain.

Q32.3 Give several examples of electromagnetic waves that are encountered in everyday life. How are they all alike? How do they differ?

Q32.4 Sometimes neon signs located near a powerful radio station are seen to glow faintly at night, even though they are not turned on. What is happening?

Q32.5 Is polarization a property of all electromagnetic waves, or is it unique to visible light? Can sound waves be polarized? What fundamental distinction in wave properties is involved? Explain.

Q32.6 Suppose that a positive point charge q is initially at rest on the x-axis, in the path of the electromagnetic plane wave described in Section 32.2. Will the charge move after the wave front reaches it? If not, why not? If the charge does move, describe its motion qualitatively. (Remember that $\vec{E}$ and $\vec{B}$ have the same value at all points behind the wave front.)

Q32.7 The light beam from a searchlight may have an electric-field magnitude of 1000 V/m, corresponding to a potential difference of 1500 V between the head and feet of a 1.5-m-tall person on whom the light shines. Does this cause the person to feel a strong electric shock? Why or why not?

Q32.8 The magnetic-field amplitude of the electromagnetic wave from the laser described in Example 32.1 (Section 32.3) is about 100 times greater than the earth's magnetic field. If you illuminate a compass with the light from this laser, would you expect the compass to deflect? Why or why not?

Q32.9 Most automobiles have vertical antennas for receiving radio broadcasts. Explain what this tells you about the direction of polarization of $\vec{E}$ in the radio waves used in broadcasting.

Q32.10 If a light beam carries momentum, should a person holding a flashlight feel a recoil analogous to the recoil of a rifle when it is fired? Why is this recoil not actually observed?

Q32.11 Does an electromagnetic *standing* wave have energy? Does it have momentum? Are your answers to these questions the same as for a *traveling* wave? Why or why not?

Q32.12 When driving on the upper level of the Bay Bridge, westbound from Oakland to San Francisco, you can easily pick up a number of radio stations on your car radio. But when driving eastbound on the lower level of the bridge, which has steel girders on either side to support the upper level, the radio reception is much worse. Why is there a difference?

Exercises

Section 32.2 Plane Electromagnetic Waves and the Speed of Light

32.1 a) How much time does it take light to travel from the moon to the earth, a distance of 384,000 km? b) Light from the star Sirius takes 8.61 years to reach the earth. What is the distance to Sirius in kilometers?

32.2 TV Ghosting. In a TV picture, ghost images are formed when the signal from the transmitter travels to the receiver both directly and indirectly after reflection from a building or other large metallic mass. In a 25-inch set, the ghost is about 1.0 cm to the right of the principal image if the reflected signal arrives 0.60 μs after the principal signal. In this case, what is the difference in path lengths for the two signals?

Section 32.3 Sinusoidal Electromagnetic Waves

32.3 A sinusoidal electromagnetic wave of frequency 6.10×10^{14} Hz travels in vacuum in the $+z$-direction. The $\vec{B}$ field is parallel to the y-axis and has amplitude 5.80×10^{-4} T. Write the vector equations for $\vec{E}(z, t)$ and $\vec{B}(z, t)$.

32.4 An electromagnetic wave of wavelength 435 nm is traveling in vacuum in the $-z$-direction. The electric field has amplitude 2.70×10^{-3} V/m and is parallel to the x-axis. What is a) the frequency? b) the magnetic-field amplitude? c) Write the vector equations for $\vec{E}(z, t)$ and $\vec{B}(z, t)$.

32.5 An electromagnetic wave has an electric field given by $\vec{E}(y, t) = -(3.10 \times 10^5 \text{ V/m})\hat{k} \sin[ky - (12.65 \times 10^{12} \text{ rad/s})t]$. a) In which direction is the wave traveling? b) What is the wavelength of the wave? c) Write the vector equation for $\vec{B}(y, t)$.

32.6 An electromagnetic wave has a magnetic field given by $\vec{B}(x, t) = (8.25 \times 10^{-9} \text{ T})\hat{j} \sin[(1.38 \times 10^4 \text{ rad/m})x + \omega t]$. a) In which direction is the wave traveling? b) What is the frequency f of the wave? c) Write the vector equation for $\vec{E}(x, t)$.

32.7 Radio station WCCO in Minneapolis broadcasts at a frequency of 830 kHz. At a point some distance from the transmitter, the magnetic-field amplitude of the electromagnetic wave from WCCO is 4.82×10^{-11} T. Calculate a) the wavelength; b) the wave number; c) the angular frequency; d) the electric-field amplitude.

32.8 The electric-field amplitude near a certain radio transmitter is 3.85×10^{-3} V/m. What is the amplitude of $\vec{B}$? How does this compare in magnitude with the earth's field?

32.9 An electromagnetic wave propagates in a dielectric material. At the frequency of the light, the dielectric constant of the material is 1.74 and the relative permeability is 1.23. If the magnetic-field amplitude is 3.80×10^{-9} T, what is the electric-field amplitude?

32.10 An electromagnetic wave with frequency 65.0 Hz travels in an insulating magnetic material that has dielectric constant 3.64 and relative permeability 5.18 at this frequency. The electric field has amplitude 7.20×10^{-3} V/m. a) What is the speed of propagation of the wave? b) What is the wavelength of the wave? c) What is the amplitude of the magnetic field? d) What is the intensity of the wave?

32.11 An electromagnetic wave with frequency 5.70×10^{14} Hz propagates with a speed of 2.17×10^8 m/s in a certain piece of glass. Find a) the wavelength of the wave in the glass; b) the wavelength of a wave of the same frequency propagating in air; c) the index of refraction n of the glass for an electromagnetic wave with this frequency; d) the dielectric constant for glass at this frequency, assuming that the relative permeability is unity.

32.12 An electromagnetic wave with frequency 38.0 MHz and wavelength 6.15 m travels in an insulating material that has K_m very nearly equal to unity. a) What is the speed of propagation of the wave? b) What is the dielectric constant of the insulating material at this frequency?

Section 32.4 Energy and Momentum in Electromagnetic Waves

32.13 Testing a Space Radio Transmitter. You are a NASA mission specialist on your first flight aboard the space shuttle. Thanks to your extensive training in physics, you have been assigned to evaluate the performance of a new radio transmitter on board the International Space Station (ISS). Perched on the shuttle's moveable arm, you aim a sensitive detector at the ISS, which is 2.5 km away. You find that the electric-field amplitude of the radio waves coming from the ISS transmitter is 0.090 V/m and that the frequency of the waves is 244 MHz. Find the following: a) the intensity of the radio wave at your location; b) the magnetic-field amplitude of the wave at your location; c) the total power output of the ISS radio transmitter. d) What assumptions, if any, did you make in your calculations?

32.14 A sinusoidal electromagnetic wave from a radio station passes perpendicularly through an open window that has area 0.500 m^2. At the window, the electric field of the wave has rms value 0.0200 V/m. How much energy does this wave carry through the window during a 30.0-s commercial?

32.15 A space probe 2.0×10^{10} m from a star measures the total intensity of electromagnetic radiation from the star to be 5.0×10^3 W/m^2. If the star radiates uniformly in all directions, what is its total average power output?

32.16 The intensity of a cylindrical laser beam is 0.800 W/m^2. The cross-sectional area of the beam is 3.0×10^{-4} m^2 and the intensity is uniform across the cross section of the beam. a) What is the average power output of the laser? b) What is the rms value of the electric field in the beam?

32.17 An intense light source radiates uniformly in all directions. At a distance of 5.0 m from the source the radiation pressure on a perfectly absorbing surface is 9.0×10^{-6} Pa. What is the total average power output of the source?

32.18 A sinusoidal electromagnetic wave emitted by a cellular phone has a wavelength of 35.4 cm and an electric-field amplitude of 5.40×10^{-2} V/m at a distance of 250 m from the antenna. Calculate a) the frequency of the wave; b) the magnetic-field amplitude; c) the intensity of the wave.

32.19 A monochromatic light source with power output 60.0 W radiates light of wavelength 700 nm uniformly in all directions. Calculate E_{max} and B_{max} for the 700-nm light at a distance of 5.00 m from the source.

32.20 For the electromagnetic wave represented by Eq. (32.19), show that the Poynting vector a) is in the same direction as the propagation of the wave; b) has average magnitude given by Eqs. (32.29).

32.21 If the intensity of direct sunlight at a point on the earth's surface is 0.78 kW/m^2, find a) the average momentum density (momentum per unit volume) in the sunlight; b) the average momentum flow rate in the sunlight.

32.22 In the 25-ft Space Simulator facility at NASA's Jet Propulsion Laboratory, a bank of overhead arc lamps can produce light of intensity 2500 W/m^2 at the floor of the facility. (This simulates the intensity of sunlight near the planet Venus.) Find the average radiation pressure (in pascals and in atmospheres) on a) a totally absorbing section of the floor; b) a totally reflecting section of the floor. c) Find the average momentum density (momentum per unit volume) in the light at the floor.

32.23 Verify that all the expressions in Eqs. (32.27) are equivalent to Eq. (32.26).

32.24 Consider each of the electric- and magnetic-field orientations given below. In each case, what is the direction of propagation of the wave? a) $\vec{E} = E\hat{\imath}$, $\vec{B} = -B\hat{\jmath}$; b) $\vec{E} = E\hat{\jmath}$, $\vec{B} = B\hat{\imath}$; c) $\vec{E} = -E\hat{k}$, $\vec{B} = -B\hat{\imath}$; d) $\vec{E} = E\hat{\imath}$, $\vec{B} = -B\hat{k}$.

32.25 A sinusoidal electromagnetic wave is propagating in a vacuum in the $+z$-direction. If at a particular instant and at a certain point in space the electric field is in the $+x$-direction and has magnitude 4.00 V/m, what are the magnitude and direction of the magnetic field of the wave at this same point in space and instant in time?

Section 32.5 Standing Electromagnetic Waves

32.26 An electromagnetic standing wave in air has frequency 75.0 MHz. a) What is the distance between nodal planes of the $\vec{E}$ field? b) What is the distance between a nodal plane of $\vec{E}$ and the closest nodal plane of $\vec{B}$?

32.27 An electromagnetic standing wave in a certain material has frequency 1.20×10^{10} Hz and speed of propagation 2.10×10^{8} m/s. a) What is the distance between a nodal plane of $\vec{B}$ and the closest antinodal plane of $\vec{B}$? b) What is the distance between an antinodal plane of $\vec{E}$ and the closest antinodal plane of $\vec{B}$? c) What is the distance between a nodal plane of $\vec{E}$ and the closest nodal plane of $\vec{B}$?

32.28 An electromagnetic standing wave in air of frequency 750 MHz is set up between two conducting planes 80.0 cm apart. At which positions between the planes could a point charge be placed at rest so that it would *remain* at rest? Explain.

32.29 A standing electromagnetic wave in a certain material has frequency 2.20×10^{10} Hz. The nodal planes of $\vec{B}$ are 3.55 mm apart. Find a) the wavelength of the wave in this material; b) the distance between adjacent nodal planes of the $\vec{E}$ field; c) the speed of propagation of the wave.

32.30 Show that the standing wave electric and magnetic fields for standing waves given by Eqs. (32.34) and (32.35) a) satisfy the wave equation, Eq. (32.15); b) satisfy Eqs. (32.12) and (32.14).

Section 32.6 The Electromagnetic Spectrum

32.31 For waves propagating in air, what is the wavelength in meters and nanometers of a) gamma rays of frequency 6.50×10^{21} Hz; b) visible light of frequency 5.75×10^{14} Hz?

32.32 For an electromagnetic wave propagating in air, determine the frequency of a wave with a wavelength of a) 5.0 km; b) 5.0 m; c) 5.0 μm; d) 5.0 nm.

Problems

32.33 Consider a plane electromagnetic wave such as that shown in Fig. 32.3, but in which $\vec{E}$ and $\vec{B}$ also have components in the x-direction (along the direction of wave propagation). Use Gauss's law for electric and magnetic fields to show that the components E_x and B_x must both be equal to zero so that the fields $\vec{E}$ and $\vec{B}$ are both transverse. (*Hint:* Use a Gaussian surface like that shown in Fig. 32.4. Of the two faces parallel to the yz-plane, choose one to be to the left of the wave front and the other to be to the right of the wave front.)

32.34 Consider a sinusoidal electromagnetic wave with fields $\vec{E} = E_{max}\hat{\jmath}\sin(kx - \omega t)$ and $\vec{B} = B_{max}\hat{k}\sin(kx - \omega t + \phi)$, with $-\pi \le \phi \le \pi$. Show that if $\vec{E}$ and $\vec{B}$ are to satisfy Eqs. (32.12) and (32.14), then $E_{max} = cB_{max}$ and $\phi = 0$. (The result $\phi = 0$ means the $\vec{E}$ and $\vec{B}$ fields oscillate in phase.)

32.35 Show that the *magnetic* field $B_z(x, t)$ in a plane electromagnetic wave propagating in the $+x$-direction must satisfy Eq. (32.15). (*Hint:* Take the partial derivative of Eq. (32.12) with respect to t and the partial derivative of Eq. (32.14) with respect to x. Then combine the results.)

32.36 For a sinusoidal electromagnetic wave in vacuum, such as that described by Eq. (32.16), show that the *average* energy density in the electric field is the same as that in the magnetic field.

32.37 A small mirror with area 5.00 cm^2 faces a monochromatic light source that is 3.20 m away. At the mirror the electric-field amplitude of the light from the source is 0.0280 V/m. a) How much energy is incident on the mirror in 1.00 s? b) What is the average radiation pressure exerted by the light on the mirror? c) What is the total radiant power output of the source if it is assumed to radiate uniformly in all directions?

32.38 A plane sinusoidal electromagnetic wave in air has a wavelength of 3.84 cm and an $\vec{E}$-field amplitude of 1.35 V/m. a) What is the frequency? b) What is the $\vec{B}$-field amplitude? c) What is the intensity? d) What average force does this radiation exert on a totally absorbing surface with area 0.240 m^2 perpendicular to the direction of propagation?

32.39 A small helium-neon laser emits red visible light with a power of 3.20 mW in a beam that has a diameter of 2.50 mm. a) What are the amplitudes of the electric and magnetic fields of the light? b) What are the average energy densities associated with the electric field and with the magnetic field? c) What is the total energy contained in a 1.00-m length of the beam?

32.40 The plane of a flat surface is perpendicular to the propagation direction of an electromagnetic wave of intensity I. The surface absorbs a fraction w of the incident intensity, where $0 \le w \le 1$, and reflects the rest. a) Show that the radiation pressure on the surface equals $(2 - w)I/c$. b) Show that this expression gives the correct results for a surface that is i) totally absorbing; ii) totally reflective. c) For an incident intensity of 1.40 kW/m^2, what is the radiation pressure for 90% absorption? For 90% reflection?

32.41 The sun emits energy in the form of electromagnetic waves at a rate of 3.9×10^{26} W. This energy is produced by nuclear reactions deep in the sun's interior. a) Find the intensity of electromagnetic radiation and the radiation pressure on an absorbing object at the surface of the sun (radius $r = R = 6.96 \times 10^5$ km) and at $r = R/2$, in the sun's interior. Ignore any scattering of the waves as they move radially outward from the center of the sun. Compare to the values given in Section 32.4 for sunlight just before it enters the earth's atmosphere. b) The gas pressure at the sun's surface is about 1.0×10^4 Pa; at $r = R/2$, the gas pressure is calculated from solar models to be about 4.7×10^{13} Pa. Comparing with your results in part (a), would you expect that radiation pressure is an important factor in determining the structure of the sun? Why or why not?

32.42 For a sinusoidal electromagnetic plane wave propagating in the $+x$-direction, the Poynting vector is $\vec{S}(x, t) = S(x, t)\hat{\imath}$, where $S(x, t) = (E_{max}B_{max}/2\mu_0)[1 - \cos 2(kx - \omega t)]$. a) Is $S(x, t)$ ever negative, corresponding to energy flowing in the direction *opposite* the direction of wave propagation? Explain why or why not. b) Figure 32.10 shows $\vec{E}$ and $\vec{B}$ for a plane electromagnetic wave at $t = 0$. Draw a copy of this figure, and on the copy draw vectors showing $\vec{S}$ for $t = 0$ at the following values of x: 0, $\lambda/8$, $\lambda/4$, $3\lambda/8$, $\lambda/2$, $5\lambda/8$, $3\lambda/4$, $7\lambda/8$, and λ.

32.43 A very long solenoid with n turns per unit length and radius a carries a current i that is increasing at a constant rate of di/dt. a) Calculate the magnetic field and induced electric field at a point inside the solenoid at a distance r from the solenoid axis. b) Compute the magnitude and direction of the Poynting vector $\vec{S}$ at this point. Show that $\vec{S}$ is directed inward toward the solenoid axis. c) Find the magnetic energy stored in a length l of the solenoid, and the rate at which that energy is increasing due to the increase in the current. d) Consider a cylindrical surface of radius a and length l that coincides with the coils of the solenoid. Integrate $\vec{S}$ over this surface to find the total rate at which electromagnetic energy is flowing into the solenoid through the solenoid walls. e) Compare the rate of change of magnetic-field energy from part (c) with the result of part (d). Discuss why the energy stored in a current-carrying solenoid can be thought of as entering through the cylindrical walls of the solenoid.

32.44 Consider the standing wave given by Eqs. (32.34) and (32.35). Let $\vec{E}$ be parallel to the y-axis and $\vec{B}$ be parallel to the z-axis; the equations then give the components of $\vec{E}$ and $\vec{B}$ along each axis. a) Plot the energy density as a function of x, $0 < x < \pi/k$, for the times $t = 0$, $\pi/4\omega$, $\pi/2\omega$, $3\pi/4\omega$, and π/ω. b) Find the direction of $\vec{S}$ in the regions $0 < x < \pi/2k$ and $\pi/2k < x < \pi/k$ at the times $t = \pi/4\omega$, $3\pi/4\omega$. c) Use your results in part (b) to explain the plots obtained in part (a).

32.45 A cylindrical conductor with a circular cross section has a radius a and a resistivity ρ and carries a constant current I. a) What are the magnitude and direction of the electric field vector $\vec{E}$ at a point just inside the wire at a distance a from the axis? b) What are the magnitude and direction of the magnetic field vector $\vec{B}$ at the same point? c) What are the magnitude and direction of the Poynting vector $\vec{S}$ at the same point? (The direction of $\vec{S}$ is the direction in which electromagnetic energy flows into or out of the conductor.) d) Use the result in part (c) to find the rate of flow of energy into the volume occupied by a length l of the conductor. (*Hint:* Integrate $\vec{S}$ over the surface of this volume.) Compare your result to the rate of generation of thermal energy in the same volume. Discuss why the energy dissipated in a current-carrying conductor, due to its resistance, can be thought of as entering through the cylindrical sides of the conductor.

32.46 A capacitor consists of two circular plates of radius r separated by a distance l. Neglecting fringing, show that while the capacitor is being charged, the rate at which energy flows into the space between the plates is equal to the rate at which the electrostatic energy stored in the capacitor increases. (*Hint:* Integrate $\vec{S}$ over the surface of the empty cylindrical volume between the plates.)

32.47 A circular loop of wire can be used as a radio antenna. If a 18.0-cm-diameter antenna is located 2.50 km from a 95.0-MHz source with a total power of 55.0 kW, what is the maximum emf induced in the loop? (Assume that the plane of the antenna loop is perpendicular to the direction of the radiation's magnetic field and that the source radiates uniformly in all directions.)

32.48 It has been proposed to place solar-power-collecting satellites in earth orbit. The power they collect would be beamed down to the earth as microwave radiation. For a microwave beam with a cross-sectional area of 36.0 m^2 and a total power of 2.80 kW at the earth's surface, what is the amplitude of the electric field of the beam at the earth's surface?

32.49 Flashlight to the Rescue. You are the sole crew member of the interplanetary spaceship *T:1339 Vorga,* which makes regular cargo runs between the earth and the mining colonies in the asteroid belt. You are working outside the ship one day while at a distance of 2.0 AU from the sun. [(1AU (astronomical unit) is the average distance from the earth to the sun, 149,600,000 km.)] Unfortunately, you lose contact with the ship's hull and begin to drift away into space. You use your spacesuit's rockets to try to push yourself back toward the ship, but they run out of fuel and stop working before you can return to the ship. You find yourself in an awkward position, floating 16.0 m from the spaceship with zero velocity relative to it. Fortunately, you are carrying a 200-W flashlight. You turn on the flashlight and use its beam as a "light rocket" to push yourself back toward the ship. a) If you, your spacesuit, and the flashlight have a combined mass of 150 kg, how long will it take you to get back to the ship? b) Is there another way you could use the flashlight to accomplish the same job of returning you to the ship?

32.50 The 19th-century inventor Nikola Tesla proposed to transmit electric power via sinusoidal electromagnetic waves. Suppose power is to be transmitted in a beam of cross-sectional area 100 m^2. What electric- and magnetic-field amplitudes are required to transmit an amount of power comparable to that handled by modern transmission lines (that carry voltages and currents of the order of 500 kV and 1000 A)?

32.51 Interplanetary space contains many small particles referred to as *interplanetary dust.* Radiation pressure from the sun sets a lower limit on the size of such dust particles. To see the origin of this limit, consider a spherical dust particle of radius R and mass density ρ. a) Write an expression for the gravitational force exerted on this particle by the sun (mass M) when the particle is a distance r from the sun. b) Let L represent the luminosity of the sun, equal to the rate at which it emits energy in electromagnetic radiation. Find the force exerted on the (totally absorbing) particle due to solar radiation pressure. The relevant area is the cross-sectional area of the particle, *not* the total surface area of the particle. As part of your answer, explain why this is so. c) The mass density of a typical interplanetary dust particle is about 3000 kg/m^3. Find the particle radius R such that the gravitational and radiation forces acting on the particle are equal in magnitude. The luminosity of the sun is 3.9×10^{26} W. Does your answer depend on the distance of the particle from the sun? Why or why not? d) Explain why dust particles with a radius less than that found in part (c) are unlikely to be found in the solar system. (*Hint:* Construct the ratio of the two force expressions found in parts (a) and (b).)

32.52 NASA is giving serious consideration to the concept of *solar sailing.* A solar sailcraft uses a large, low-mass sail and the energy and momentum of sunlight for propulsion. a) Should the sail be absorbing or reflective? Why? b) The total power output of the sun is 3.9×10^{26} W. How large a sail is necessary to propel a 10,000-kg spacecraft against the gravitational force of the sun? Express your result in square kilometers. c) Explain why your answer to part (b) is independent of the distance from the sun.

Challenge Problems

32.53 Electromagnetic radiation is emitted by accelerating charges. The rate at which energy is emitted from an accelerating charge that has charge q and acceleration a is given by

$$\frac{dE}{dt} = \frac{q^2a^2}{6\pi\epsilon_0 c^3}$$

where c is the speed of light. a) Verify that this equation is dimensionally correct. b) If a proton with a kinetic energy of 6.0 MeV is traveling in a particle accelerator in a circular orbit of radius 0.750 m, what fraction of its energy does it radiate per second? c) Consider an electron orbiting with the same speed and radius. What fraction of its energy does it radiate per second?

32.54 The Classical Hydrogen Atom. The electron in a hydrogen atom can be considered to be in a circular orbit with a radius of 0.0529 nm and a kinetic energy of 13.6 eV. If the electron behaved classically, how much energy would it radiate per second (see Challenge Problem 32.53)? What does this tell you about the use of classical physics in describing the atom?

32.55 Electromagnetic waves propagate much differently in *conductors* than they do in dielectrics or in vacuum. If the resistivity of the conductor is sufficiently low (that is, if it is a sufficiently good conductor), the oscillating electric field of the wave gives rise to an oscillating conduction current that is much larger than the displacement current. In this case, the wave equation for an electric field $\vec{E}(x, t) = E_y(x, t)\hat{\jmath}$ propagating in the $+x$-direction within a conductor is

$$\frac{\partial^2 E_y(x, t)}{\partial x^2} = \frac{\mu}{\rho}\frac{\partial E_y(x, t)}{\partial t}$$

where μ is the permeability of the conductor and ρ is its resistivity. a) A solution to this wave equation is

$$E_y(x, t) = E_{\max} e^{-k_C x} \sin(k_C x - \omega t)$$

where $k_C = \sqrt{\omega\mu/2\rho}$. Verify this by substituting $E_y(x, t)$ into the above wave equation. b) The exponential term shows that the electric field decreases in amplitude as it propagates. Explain why this happens. (*Hint:* The field does work to move charges within the conductor. The current of these moving charges causes i^2R heating within the conductor, raising its temperature. Where does the energy to do this come from?). c) Show that the electric-field amplitude decreases by a factor of $1/e$ in a distance $1/k_C = \sqrt{2\rho/\omega\mu}$, and calculate this distance for a radio wave with frequency $f = 1.0$ MHz in copper (resistivity $1.72 \times 10^{-8}\,\Omega\cdot$m, permeability $\mu = \mu_0$). Since this distance is so short, electromagnetic waves of this frequency can hardly propagate at all into copper. Instead, they are reflected at the surface of the metal. This is why radio waves cannot penetrate through copper or other metals, and why radio reception is poor inside a metal structure.

33 THE NATURE AND PROPAGATION OF LIGHT

You see a rainbow when you are facing falling rain with the sun behind you. The arc is formed by rays of sunlight that change direction three times: the rays refract, or bend, when they enter a raindrop; then reflect off the back of the raindrop; and finally refract again as they exit the raindrop. You see colors because different wavelengths refract through different angles, a phenomenon called dispersion.

? What causes the second, faint rainbow outside the primary one? Why are its colors in the reverse sequence from the primary rainbow?

Blue lakes, ochre deserts, green forests, and multicolored rainbows can be enjoyed by anyone who has eyes with which to see them. But by studying the branch of physics called **optics,** which deals with the behavior of light and other electromagnetic waves, we can reach a deeper appreciation of the visible world. A knowledge of the properties of light allows us to understand the blue color of the sky and the design of optical devices such as telescopes, microscopes, cameras, eyeglasses, and the human eye. The same basic principles of optics also lie at the heart of modern developments such as the laser, optical fibers, holograms, optical computers, and new techniques in medical imaging.

The importance of optics to physics, and to science and engineering in general, is so great that we will devote the next four chapters to its study. In this chapter we begin with a study of the laws of reflection and refraction and the concepts of dispersion, polarization, and scattering of light. Along the way we compare the various possible descriptions of light in terms of particles, rays, or waves, and we introduce Huygens' principle, an important link that connects the ray and wave viewpoints. In Chapter 34 we'll use the ray description of light to understand how mirrors and lenses work, and we'll see how mirrors and lenses are used in optical instruments such as cameras, microscopes, and telescopes. We'll explore the wave characteristics of light further in Chapters 35 and 36.

33.1 An electric heating element emits primarily infrared radiation. But if its temperature is high enough, it also emits a discernible amount of visible light.

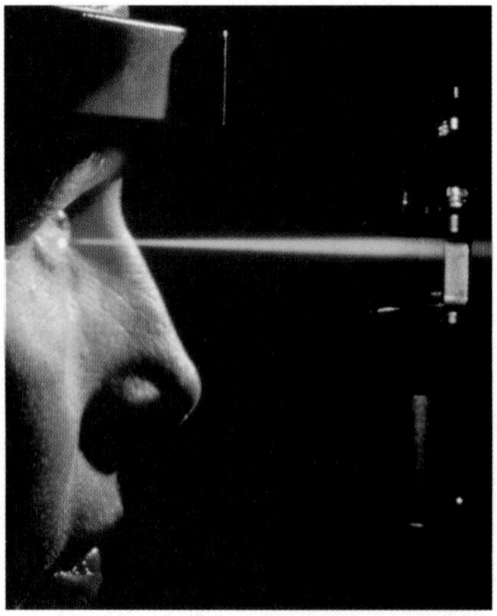

33.2 Ophthalmic surgeons use lasers for repairing detached retinas and for cauterizing blood vessels in retinopathy. Pulses of blue-green light from an argon laser are ideal for this purpose, since they pass harmlessly through the transparent part of the eye but are absorbed by red pigments in the retina.

33.1 | The Nature of Light

Until the time of Isaac Newton (1642–1727), most scientists thought that light consisted of streams of particles (called *corpuscles*) emitted by light sources. Galileo and others tried (unsuccessfully) to measure the speed of light. Around 1665, evidence of *wave* properties of light began to be discovered. By the early nineteenth century, evidence that light is a wave had grown very persuasive.

In 1873, James Clerk Maxwell predicted the existence of electromagnetic waves and calculated their speed of propagation, as we learned in Chapter 32. This development, along with the experimental work of Heinrich Hertz starting in 1887, showed conclusively that light is indeed an electromagnetic wave.

The wave picture of light is not the whole story, however. Several effects associated with emission and absorption of light reveal a particle aspect, in that the energy carried by light waves is packaged in discrete bundles called *photons* or *quanta.* These apparently contradictory wave and particle properties have been reconciled since 1930 with the development of quantum electrodynamics, a comprehensive theory that includes *both* wave and particle properties. The *propagation* of light is best described by a wave model, but understanding emission and absorption requires a particle approach.

The fundamental sources of all electromagnetic radiation are electric charges in accelerated motion. All bodies emit electromagnetic radiation as a result of thermal motion of their molecules; this radiation, called *thermal radiation,* is a mixture of different wavelengths. At sufficiently high temperatures, all matter emits enough visible light to be self-luminous; a very hot body appears "red hot" (Fig. 33.1) or "white hot." Thus hot matter in any form is a light source. Familiar examples are a candle flame, hot coals in a campfire, the coils in an electric room heater, and an incandescent lamp filament (which usually operates at a temperature of about 3000°C).

Light is also produced during electrical discharges through ionized gases. The bluish light of mercury-arc lamps, the orange-yellow of sodium-vapor lamps, and the various colors of "neon" signs are familiar. A variation of the mercury-arc lamp is the *fluorescent* lamp (Fig. 30.7). This light source uses a material called a *phosphor* to convert the ultraviolet radiation from a mercury arc into visible light. This conversion makes fluorescent lamps more efficient than incandescent lamps in converting electrical energy into light.

A light source that has attained prominence in the last forty years is the *laser.* In most light sources, light is emitted independently by different atoms within the source; in a laser, by contrast, atoms are induced to emit light in a cooperative, coherent fashion. The result is a very narrow beam of radiation that can be enormously intense and that is much more nearly *monochromatic,* or single-frequency, than light from any other source. Lasers are used by physicians for microsurgery, in CD players and computers to scan the information encoded on a compact disc or CD-ROM, in industry to cut through steel and to fuse high-melting-point materials, and in many other applications (Fig. 33.2).

No matter what its source, electromagnetic radiation travels in vacuum at the speed of light. The first demonstration that the speed of light is not infinite was made in 1676 by the Danish astronomer Ole Rømer, from observations of the motion of one of Jupiter's satellites. The first successful *terrestrial* measurement of the speed of light was made by the French scientist Armand Fizeau in 1849, using a reflected light beam interrupted by a notched rotating disk. More refined versions of this experiment were carried out later in the nineteenth century by Jean Foucault in France and by Albert A. Michelson in the United States.

From analysis of all measurements up to 1983, the most probable value for the speed of light at that time was

$$c = 2.99792458 \times 10^8 \text{ m/s}$$

As we explained in Section 1.3, the definition of the second based on the cesium clock is precise to within one part in 10 trillion (10^{13}). Up to 1983 the definition of the meter was much less precise, about four parts in a billion (10^9). Any attempt to measure the speed of light with greater precision foundered on this limitation. For this reason, in November 1983 the General Conference of Weights and Measures redefined the meter by *defining* the speed of light in vacuum to be precisely 299,792,458 m/s. One meter is now defined to be the distance traveled by light in a time of 1/299,792,458 s, with the second defined by the cesium clock.

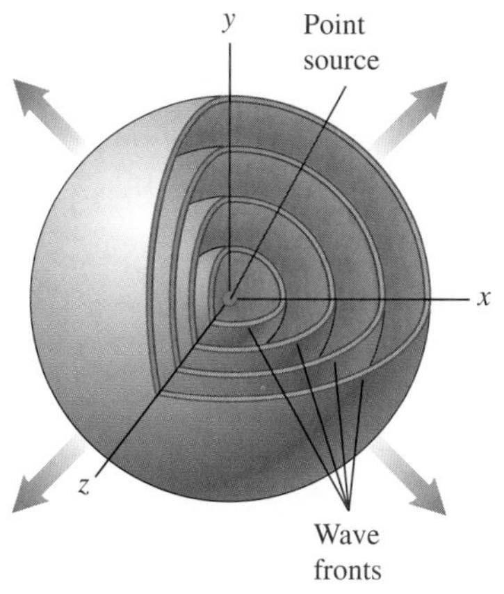

33.3 Spherical wave fronts spread out uniformly in all directions from a point source in a motionless medium, such as still air, that has the same properties in all regions and in all directions. Electromagnetic waves in a vacuum also spread out as shown here.

Waves, Wave Fronts, and Rays

We often use the concept of a **wave front** to describe wave propagation. We introduced this concept in Section 32.2 to describe the leading edge of a wave. More generally, we define a wave front as *the locus of all adjacent points at which the phase of vibration of a physical quantity associated with the wave is the same.* That is, at any instant, all points on a wave front are at the same part of the cycle of their variation.

When we drop a pebble into a calm pool, the expanding circles formed by the wave crests, as well as the circles formed by the wave troughs between them, are wave fronts. Similarly, when sound waves spread out in still air from a pointlike source, or when electromagnetic radiation spreads out from a pointlike emitter, any spherical surface that is concentric with the source is a wave front, as shown in Fig. 33.3. In diagrams of wave motion we usually draw only parts of a few wave fronts, often choosing consecutive wave fronts that have the same phase and thus are one wavelength apart, such as crests of water waves. Similarly, a diagram for sound waves might show only the "pressure crests," the surfaces over which the pressure is maximum, and a diagram for electromagnetic waves might show only the "crests" on which the electric or magnetic field is maximum.

We will often use diagrams that show the shapes of the wave fronts or their cross sections in some reference plane. For example, when electromagnetic waves are radiated by a small light source, we can represent the wave fronts as spherical surfaces concentric with the source or, as in Fig. 33.4a, by the circular intersections of these surfaces with the plane of the diagram. Far away from the source, where the radii of the spheres have become very large, a section of a spherical surface can be considered as a plane, and we have a *plane* wave like those discussed in Sections 32.2 and 32.3 (Fig, 33.4b).

To describe the directions in which light propagates, it's often convenient to represent a light wave by **rays** rather than by wave fronts. Rays were used to describe light long before its wave nature was firmly established. In a particle theory of light, rays are the paths of the particles. From the wave viewpoint *a ray is an imaginary line along the direction of travel of the wave.* In Fig. 33.4a the rays are the radii of the spherical wave fronts, and in Fig. 33.4b they are straight lines perpendicular to the wave fronts. When waves travel in a homogeneous isotropic material (a material with the same properties in all regions and in all directions), the rays are always straight lines normal to the wave fronts. At a boundary surface

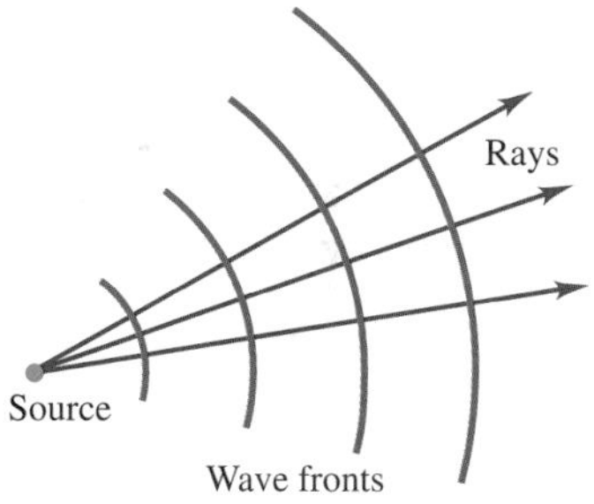

(a)

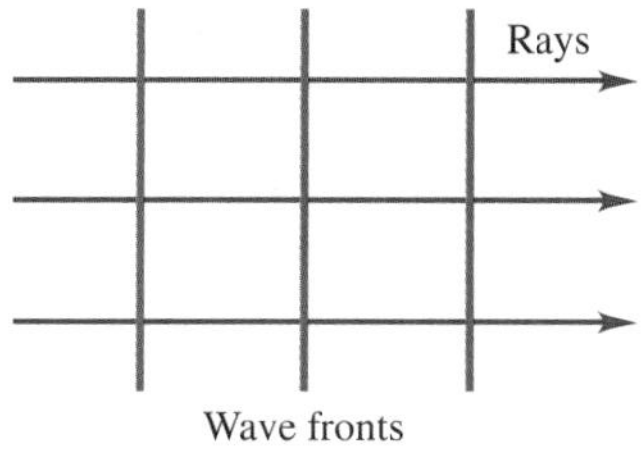

(b)

33.4 Wave fronts (blue) and rays (purple). (a) When the wave fronts are spherical, the rays radiate out from the center of the spheres. (b) When the wave fronts are planes, the rays are parallel.

between two materials, such as the surface of a glass plate in air, the wave speed and the direction of a ray may change, but the ray segments in the air and in the glass are straight lines.

The next several chapters will give you many opportunities to see the interplay of the ray, wave, and particle descriptions of light. The branch of optics for which the ray description is adequate is called **geometric optics;** the branch dealing specifically with wave behavior is called **physical optics.** This chapter and the following one are concerned mostly with geometric optics. In Chapters 35 and 36 we will study wave phenomena and physical optics.

Test Your Understanding

Some crystals are *not* isotropic: light travels through the crystal at a higher speed in some directions than in others. In a crystal in which light travels at the same speed in the x- and z-directions but at a faster speed in the y-direction, would the wave fronts in Fig. 33.3 still be spherical? If not, what would they look like?

33.2 | Reflection and Refraction

In this section we'll use the *ray* model of light to explore two of the most important aspects of light propagation: **reflection** and **refraction.** When a light wave strikes a smooth interface separating two transparent materials (such as air and glass or water and glass), the wave is in general partly *reflected* and partly *refracted* (transmitted) into the second material, as shown in Fig. 33.5a. For

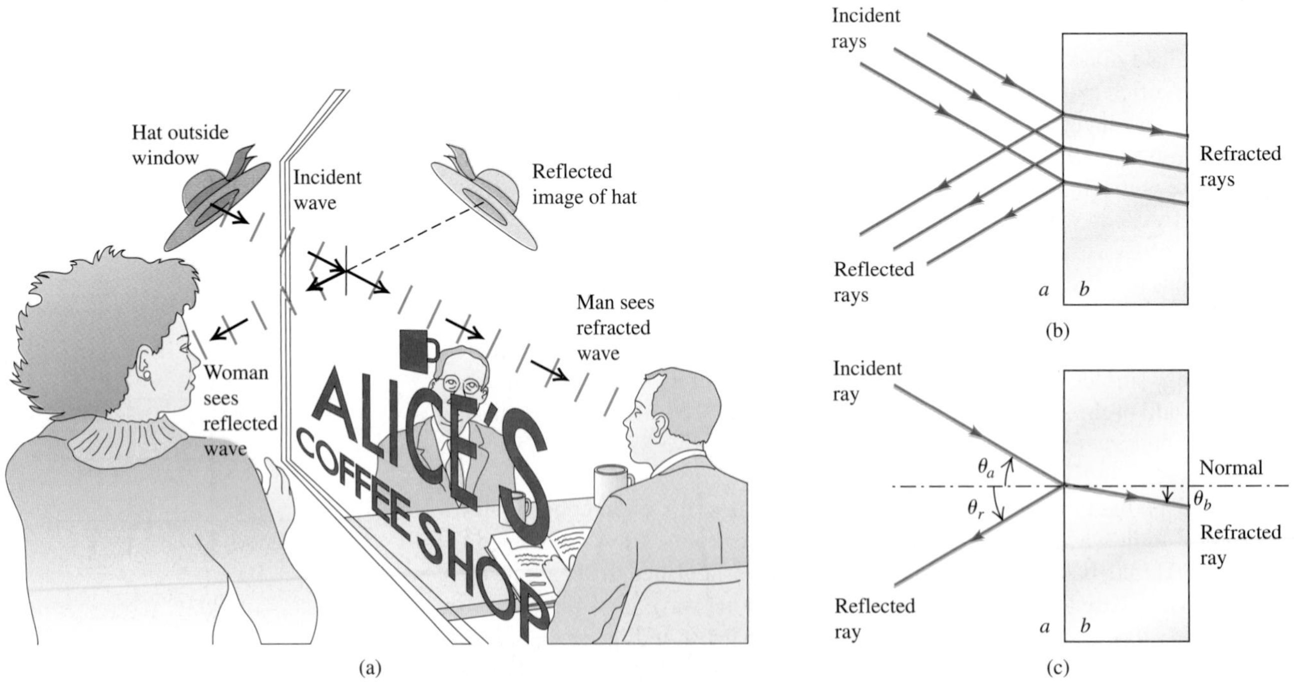

33.5 (a) A plane wave is in part reflected and in part refracted at the boundary between two media (in this case, air and glass). The light that reaches the inside of the coffee shop is refracted twice, once entering the glass and once exiting the glass. (b) The waves in the outside air and in glass in (a) are represented by rays. (c) For simplicity, only one example of an incident ray, a reflected ray, and a refracted ray is drawn. For the case shown here, material b has a greater index of refraction that material a $(n_b > n_a)$ and the angle θ_b is smaller than θ_a.

example, when you look into a restaurant window from the street, you see a reflection of the street scene, but a person inside the restaurant can look out through the window at the same scene as light reaches him by refraction.

The segments of plane waves shown in Fig. 33.5a can be represented by bundles of rays forming *beams* of light (Fig. 33.5b). For simplicity we often draw only one ray in each beam (Fig. 33.5c). Representing these waves in terms of rays is the basis of geometric optics. We begin our study with the behavior of an individual ray.

We describe the directions of the incident, reflected, and refracted (transmitted) rays at a smooth interface between two optical materials in terms of the angles they make with the *normal* (perpendicular) to the surface at the point of incidence, as shown in Fig. 33.5c. If the interface is rough, both the transmitted light and the reflected light are scattered in various directions, and there is no single angle of transmission or reflection. Reflection at a definite angle from a very smooth surface is called **specular reflection** (from the Latin word for "mirror"); scattered reflection from a rough surface is called **diffuse reflection.** This distinction is shown in Fig. 33.6. Both kinds of reflection can occur with either transparent materials or *opaque* materials that do not transmit light. The vast majority of objects in your environment (including clothing, plants, other people, and this book) are visible to you because they reflect light in a diffuse manner from their surfaces. Our primary concern, however, will be with specular reflection from a very smooth surface such as highly polished glass, plastic, or metal. Unless stated otherwise, when referring to "reflection" we will always mean *specular* reflection.

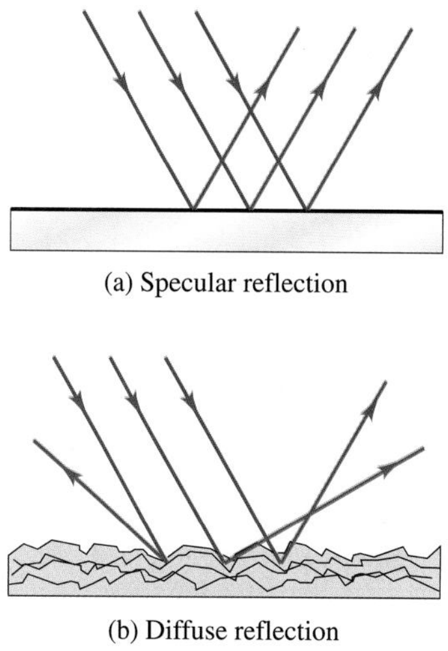

33.6 Two types of reflection.

The **index of refraction** of an optical material (also called the **refractive index**), denoted by n, plays a central role in geometric optics. It is the ratio of the speed of light c in vacuum to the speed v in the material:

$$n = \frac{c}{v} \quad \text{(index of refraction)} \tag{33.1}$$

Light always travels *more slowly* in a material than in vacuum, so the value of n in anything other than a vacuum is always greater than unity. For vacuum, $n = 1$. Since n is a ratio of two speeds, it is a pure number without units. (The relation of the value of n to the electric and magnetic properties of a material is described in Section 32.3.)

CAUTION Keep in mind that the wave speed v is *inversely* proportional to the index of refraction n. The greater the index of refraction in a material, the *slower* the wave speed in that material. Failure to remember this point can lead to serious confusion!

Laws of Reflection and Refraction

Experimental studies of the directions of the incident, reflected, and refracted rays at a smooth interface between two optical materials lead to the following conclusions:

1. **The incident, reflected, and refracted rays and the normal to the surface all lie in the same plane.** The plane of the three rays is perpendicular to the plane of the boundary surface between the two materials. We always

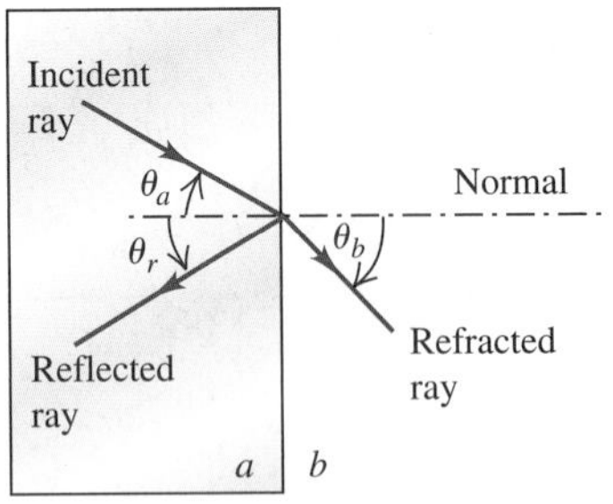

33.7 Reflection and refraction in the case where material b has a smaller index of refraction than material a, so $n_b < n_a$. The refracted angle θ_b is greater than the incident angle θ_a. Compare to Fig. 33.5c, which shows the situation for $n_b > n_a$.

draw ray diagrams so that the incident, reflected, and refracted rays are in the plane of the diagram.

2. **The angle of reflection θ_r is equal to the angle of incidence θ_a for all wavelengths and for any pair of materials.** That is, in Fig. 33.5c,

$$\theta_r = \theta_a \quad \text{(law of reflection)} \tag{33.2}$$

This relation, together with the observation that the incident and reflected rays and the normal all lie in the same plane, is called the **law of reflection.**

3. For monochromatic light and for a given pair of materials, a and b, on opposite sides of the interface, **the ratio of the sines of the angles θ_a and θ_b, where both angles are measured from the normal to the surface, is equal to the inverse ratio of the two indexes of refraction:**

$$\frac{\sin\theta_a}{\sin\theta_b} = \frac{n_b}{n_a} \tag{33.3}$$

or

$$n_a \sin\theta_a = n_b \sin\theta_b \quad \text{(law of refraction)} \tag{33.4}$$

This experimental result, together with the observation that the incident and refracted rays and the normal all lie in the same plane, is called the **law of refraction** or **Snell's law,** after the Dutch scientist Willebrord Snell (1591–1626). There is some doubt that Snell actually discovered it. The discovery that $n = c/v$ came much later.

ActivPhysics ONLINE
15.1 Reflection and Refraction
15.3 Refraction Applications

While these results were first observed experimentally, they can be derived theoretically from a wave description of light. We do this in Section 33.7.

Equations (33.3) and (33.4) show that when a ray passes from one material (a) into another material (b) having a larger index of refraction $(n_b > n_a)$ and hence a slower wave speed, the angle θ_b with the normal is *smaller* in the second material than the angle θ_a in the first; hence the ray is bent *toward* the normal (Fig. 33.5c). When the second material has a *smaller* index of refraction than the first material $(n_b < n_a)$ and hence a faster wave speed, the ray is bent *away from* the normal (Fig. 33.7). This explains why a partially submerged ruler or drinking straw appears bent; light rays coming from below the surface change in direction at the air–water interface, so the rays appear to be coming from a position above their actual point of origin (Fig. 33.8). A similar effect explains the appearance of the setting sun (Fig. 33.9).

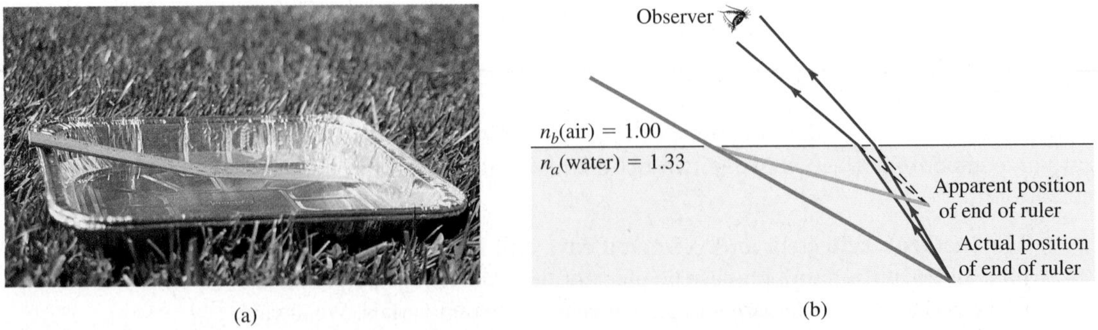

33.8 (a) This ruler is actually straight, but it appears to bend at the surface of the water. (b) Light rays from any submerged object bend away from the normal when they emerge into the air. As seen by an observer above the surface of the water, the object appears to be much closer to the surface than it actually is.

An important special case is refraction that occurs at an interface between vacuum, for which the index of refraction is unity by definition, and a material. When a ray passes from vacuum into a material (*b*), so that $n_a = 1$ and $n_b > 1$, the ray is always bent *toward* the normal. When a ray passes from a material into vacuum, so that $n_a > 1$ and $n_b = 1$, the ray is always bent *away from* the normal.

No matter what the materials on either side of the interface, the transmitted ray is not bent at all in the special case of *normal* incidence, in which the incident ray is perpendicular to the interface so that $\theta_a = 0$ and $\sin\theta_a = 0$. From Eq. (33.4) this means that θ_b is also equal to zero, so the transmitted ray is also perpendicular to the interface.

The laws of reflection and refraction apply regardless of which side of the interface the incident ray comes from. If a ray of light approaches the interface in Fig. 33.5c or Fig. 33.7 from the right rather than from the left, there are again reflected and refracted rays; these two rays, the incident ray, and the normal to the surface again lie in the same plane. Furthermore, the path of a refracted ray is *reversible;* it follows the same path when going from *b* to *a* as when going from *a* to *b*. (You can verify this using Eq. (33.4).) Since reflected and incident rays make the same angle with the normal, the path of a reflected ray is also reversible. That's why when you see someone's eyes in a mirror, they can also see you.

The *intensities* of the reflected and refracted rays depend on the angle of incidence, the two indexes of refraction, and the polarization (that is, the direction of the electric field vector) of the incident ray. The fraction reflected is smallest at normal incidence $(\theta_a = 0°)$, where it is about 4% for an air–glass interface. This fraction increases with increasing angle of incidence to 100% at grazing incidence, when $\theta_a = 90°$.

It's possible to use Maxwell's equations to predict the amplitude, intensity, phase, and polarization states of the reflected and refracted waves. Such an analysis is beyond our scope, however.

The index of refraction depends not only on the substance but also on the wavelength of the light. The dependence on wavelength is called *dispersion;* we will consider it in Section 33.4. Indexes of refraction for several solids and liquids are given in Table 33.1 for a particular wavelength of yellow light.

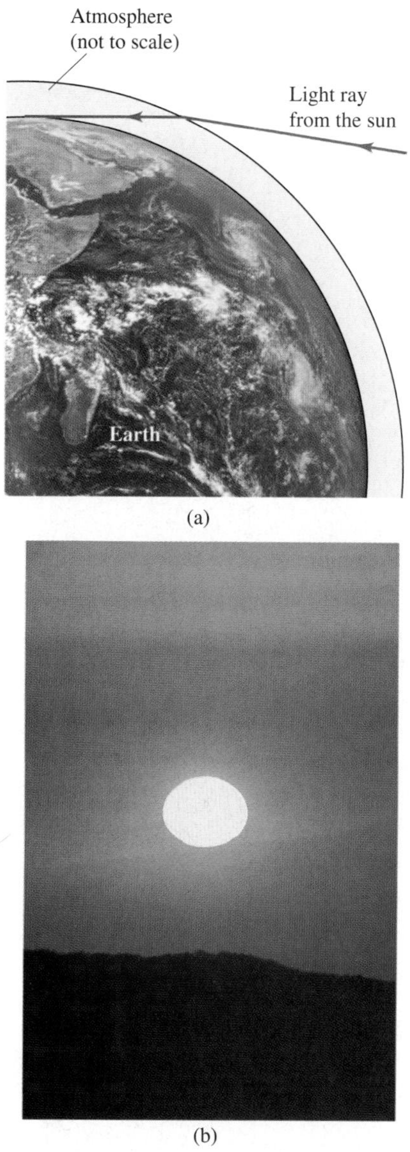

33.9 (a) The index of refraction of air is slightly greater than 1, so light rays from the setting sun bend downward when they enter our atmosphere. (The effect is exaggerated in this figure.) (b) Stronger refraction occurs for light coming from the lower limb of the sun (the part that appears closest to the horizon), which passes through denser air in the lower atmosphere. As a result, the setting sun appears flattened vertically. (See Problem 33.49.)

Table 33.1 Index of Refraction for Yellow Sodium Light $(\lambda_0 = 589\ \text{nm})$

Substance	Index of Refraction, n	Substance	Index of Refraction, n
Solids		Liquids at 20°C	
Ice (H_2O)	1.309	Methanol (CH_3OH)	1.329
Fluorite (CaF_2)	1.434	Water (H_2O)	1.333
Polystyrene	1.49	Ethanol (C_2H_5OH)	1.36
Rock salt (NaC1)	1.544	Carbon tetrachloride (CCl_4)	1.460
Quartz (SiO_2)	1.544		
Zircon ($ZrO_2 \cdot SiO_2$)	1.923	Turpentine	1.472
Diamond (C)	2.417	Glycerine	1.473
Fabulite ($SrTiO_3$)	2.409	Benzene	1.501
Rutile (TiO_2)	2.62	Carbon disulfide (CS_2)	1.628
Glasses (typical values)			
Crown	1.52		
Light flint	1.58		
Medium flint	1.62		
Dense flint	1.66		
Lanthanum flint	1.80		

The index of refraction of air at standard temperature and pressure is about 1.0003, and we will usually take it to be exactly unity. The index of refraction of a gas increases as its density increases. Most glasses used in optical instruments have indexes of refraction between about 1.5 and 2.0. A few substances have larger indexes; one example is diamond, with 2.417.

Index of Refraction and the Wave Aspects of Light

We have discussed how the direction of a light ray changes when it passes from one material to another material with a different index of refraction. It's also important to see what happens to the *wave* characteristics of the light when this happens.

First, the frequency f of the wave does not change when passing from one material to another. That is, the number of wave cycles arriving per unit time must equal the number leaving per unit time; this is a statement that the boundary surface cannot create or destroy waves.

Second, the wavelength λ of the wave *is* different in general in different materials. This is because in any material, $v = \lambda f$; since f is the same in any material as in vacuum and v is always less than the wave speed c in vacuum, λ is also correspondingly reduced. Thus the wavelength λ of light in a material is *less than* the wavelength λ_0 of the same light in vacuum. From the above discussion, $f = c/\lambda_0 = v/\lambda$. Combining this with Eq. (33.1), $n = c/v$, we find

$$\lambda = \frac{\lambda_0}{n} \quad \text{(wavelength of light in a material)} \qquad (33.5)$$

When a wave passes from one material into a second material with larger index of refraction, so that $n_b > n_a$, the wave speed decreases. The wavelength $\lambda_b = \lambda_0/n_b$ in the second material is then shorter than the wavelength $\lambda_a = \lambda_0/n_a$ in the first material. If instead the second material has a smaller index of refraction than the first material, so that $n_b < n_a$, then the wave speed increases. Then the wavelength λ_b in the second material is longer than the wavelength λ_a in the first material. This makes intuitive sense; the waves get "squeezed" (the wavelength gets shorter) if the wave speed decreases and get "stretched" (the wavelength gets longer) if the wave speed increases.

Problem-Solving Strategy

Reflection and Refraction

IDENTIFY *the relevant concepts:* You need to use the ideas of this section, called *geometric optics,* whenever light encounters a boundary between two different materials. In general, part of the light is reflected back into the first material and part is refracted into the second material. These ideas apply to electromagnetic radiation of all frequencies and wavelengths, not just visible light.

SET UP *the problem using the following steps:*

1. In geometric optics problems involving rays and angles, *always* start by drawing a large, neat diagram. Label all known angles and indexes of refraction.
2. Determine the target variables.

EXECUTE *the solution as follows:*

1. Apply the laws of reflection, Eq. (33.2), and refraction, Eq. (33.4). Remember to always measure the angles of incidence, reflection, and refraction from the *normal* to the surface where the reflection and refraction occur, *never* from the surface itself.
2. You will often have to use some simple geometry or trigonometry in working out angular relations. The sum of the interior angles in a triangle is 180°, an angle and its complement differ by 180°, and so on. Ask yourself, "What information am I given?", "What do I need to know in order to find this angle?", or "What other angles

or other quantities can I compute using the information given in the problem?"

3. Remember that the frequency of the light does not change when it moves from one material to another, but the wavelength changes in accordance with Eq. (33.5).

EVALUATE *your answer:* In problems that involve refraction, check that the direction of refraction makes sense. If the second material has a higher index of refraction than the first material, the refracted ray bends toward the normal and the refracted angle is less than the incident angle. If the first material has the higher index of refraction, the refracted ray bends away from the normal and the refracted angle is greater than the incident angle. Do your results agree with these rules?

Example 33.1 Reflection and refraction

In Fig. 33.10, material a is water and material b is a glass with index of refraction 1.52. If the incident ray makes an angle of 60° with the normal, find the directions of the reflected and refracted rays.

SOLUTION

IDENTIFY and SET UP: Figure 33.10 shows the rays and angles for this situation. The target variables are the reflected angle θ_r and the refracted angle θ_b. Since n_b is greater than n_a, the refracted angle must be smaller than the incident angle θ_a; this is shown in the figure.

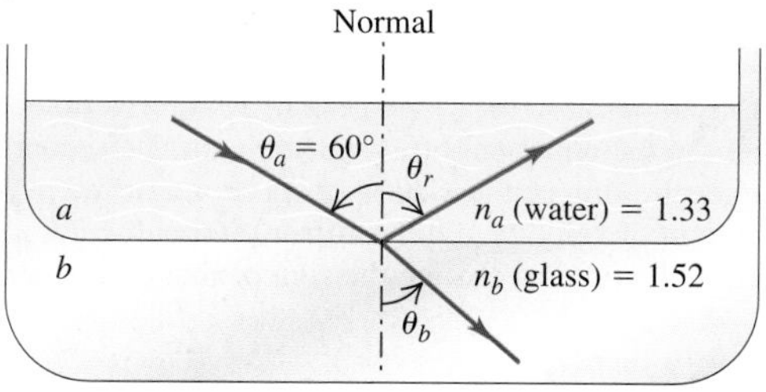

33.10 Reflection and refraction of light passing from water to glass.

EXECUTE: According to Eq. (33.2), the angle the reflected ray makes with the normal is the same as that of the incident ray, so $\theta_r = \theta_a = 60.0°$.

To find the direction of the refracted ray, we use Snell's law, Eq. (33.4), with $n_a = 1.33$, $n_b = 1.52$, and $\theta_a = 60.0°$. We find

$$n_a \sin\theta_a = n_b \sin\theta_b$$

$$\sin\theta_b = \frac{n_a}{n_b}\sin\theta_a = \frac{1.33}{1.52}\sin 60.0° = 0.758$$

$$\theta_b = 49.3°$$

EVALUATE: The second material has a larger refractive index than the first. Hence, the refracted ray is bent toward the normal as the wave slows down upon entering the second material, and $\theta_b < \theta_a$.

Example 33.2 Index of refraction in the eye

The wavelength of the red light from a helium-neon laser is 633 nm in air but 474 nm in the aqueous humor inside your eyeball. Calculate the index of refraction of the aqueous humor and the speed and frequency of the light in this substance.

SOLUTION

IDENTIFY and SET UP: The key ideas here are the definition of index of refraction, $n = c/v$; the relationship between index of refraction and wavelength, Eq. (33.5), and the relationship $v = \lambda f$ between wave speed, wavelength, and frequency.

EXECUTE: The index of refraction of air is very close to unity, so we assume that the wavelengths in air and vacuum are the same. Then the wavelength λ in the material is given by Eq. (33.5) with $\lambda_0 = 633$ nm:

$$\lambda = \frac{\lambda_0}{n} \qquad n = \frac{\lambda_0}{\lambda} = \frac{633\text{ nm}}{474\text{ nm}} = 1.34$$

This is about the same index of refraction as for water. Then $n = c/v$ gives

$$v = \frac{c}{n} = \frac{3.00 \times 10^8\text{ m/s}}{1.34} = 2.25 \times 10^8\text{ m/s}$$

Finally, from $v = \lambda f$,

$$f = \frac{v}{\lambda} = \frac{2.25 \times 10^8\text{ m/s}}{474 \times 10^{-9}\text{ m}} = 4.74 \times 10^{14}\text{ Hz}$$

EVALUATE: Note that while the speed and wavelength have different values in air and in the aqueous humor, the *frequency* in air, f_0, is the same as the frequency f in the aqueous humor:

$$f_0 = \frac{c}{\lambda_0} = \frac{3.00 \times 10^8\ \text{m/s}}{633 \times 10^{-9}\ \text{m}} = 4.74 \times 10^{14}\ \text{Hz}$$

This illustrates the general rule that when a light wave passes from one material into another, the wave frequency is unchanged.

Example 33.3 A twice-reflected ray

Two mirrors are perpendicular to each other. A ray traveling in a plane perpendicular to both mirrors is reflected from one mirror, then the other, as shown in Fig. 33.11. What is the ray's final direction relative to its original direction?

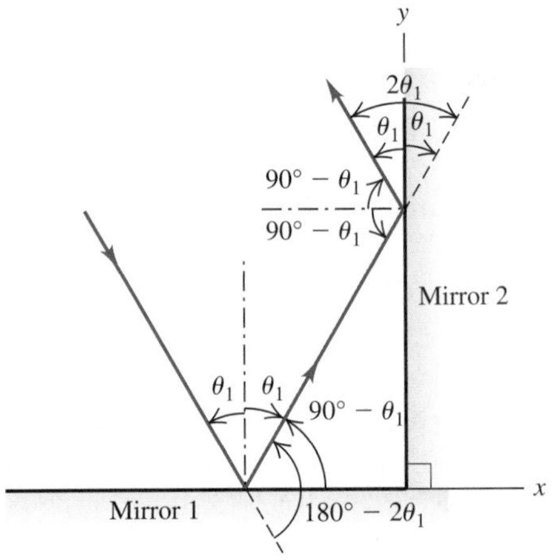

33.11 A ray moving in the xy-plane. The first reflection changes the sign of the y-component of its velocity, and the second reflection changes the sign of the x-component. For a different ray with a z-component of velocity, a third mirror (perpendicular to the two shown) could be used to change the sign of that component.

SOLUTION

IDENTIFY and SET UP: There are two reflections in this situation, so we must apply the law of reflection twice.

EXECUTE: For mirror 1 the angle of incidence is θ_1, and this equals the angle of reflection. The sum of interior angles in the triangle shown in the figure is 180°, so we see that the angles of incidence and reflection for mirror 2 are both $90° - \theta_1$. The total change in direction of the ray after both reflections is therefore $2(90° - \theta_1) + 2\theta_1 = 180°$. That is, the ray's final direction is opposite to its original direction.

EVALUATE: An alternative viewpoint is that specular reflection reverses the sign of the component of light velocity perpendicular to the surface but leaves the other components unchanged. We invite you to verify this in detail. You should also be able to use this result to show that when a ray of light is successively reflected by three mirrors forming a corner of a cube (a "corner reflector"), its final direction is again opposite to its original direction. This principle is widely used in tail-light lenses and highway signs to improve their night-time visibility. Apollo astronauts placed arrays of corner reflectors on the moon. By use of laser beams reflected from these arrays, the earth–moon distance has been measured to within 0.15 m.

Test Your Understanding

You are standing on the shore of a lake. You spot a tasty fish swimming some distance below the lake surface. If you want to spear the fish, should you aim the spear above, below, or directly at the apparent position of the fish? Where should you aim if instead you use a high-power laser to simultaneously kill and cook the fish?

ActivPhysics ONLINE

15.2 Total Internal Reflection

33.3 | Total Internal Reflection

We have described how light is partially reflected and partially transmitted at an interface between two materials with different indexes of refraction. Under certain circumstances, however, *all* of the light can be reflected back from the inter-

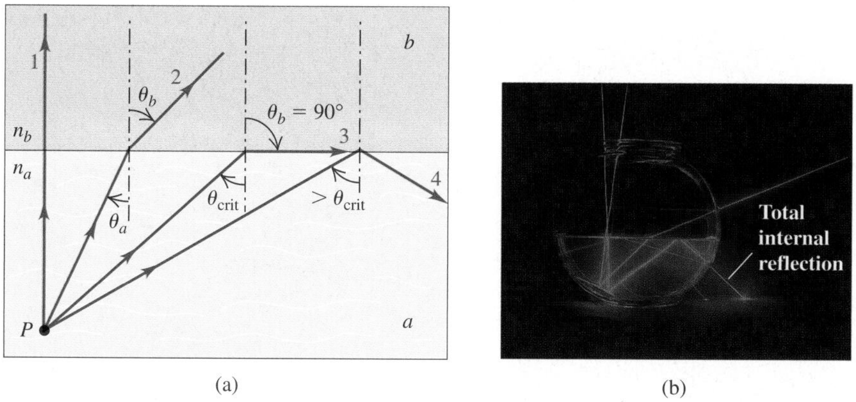

33.12 (a) Total internal reflection. The angle of incidence for which the angle of refraction is 90° is called the critical angle: this is the case for ray 3. The reflected portions of rays 1, 2, and 3 are omitted for clarity. (b) Rays of laser light enter the water in the fishbowl from above; they are reflected at the bottom by mirrors titled at slightly different angles. One ray undergoes total internal reflection at the air–water interface.

face, with none of it being transmitted, even though the second material is transparent. Figure 33.12a shows how this can occur. Several rays are shown radiating from a point source P in material a with index of refraction n_a. The rays strike the surface of a second material b with index n_b, where $n_a > n_b$. (For instance, materials a and b could be water and air, respectively.) From Snell's law of refraction,

$$\sin\theta_b = \frac{n_a}{n_b}\sin\theta_a$$

Because n_a/n_b is greater than unity, $\sin\theta_b$ is larger than $\sin\theta_a$; the ray is bent *away from* the normal. Thus there must be some value of θ_a *less than* 90° for which Snell's law gives $\sin\theta_b = 1$ and $\theta_b = 90°$. This is shown by ray 3 in the diagram, which emerges just grazing the surface at an angle of refraction of 90°. Compare the diagram in Fig. 33.12a to the photograph of light rays in Fig. 33.12b.

The angle of incidence for which the refracted ray emerges tangent to the surface is called the **critical angle,** denoted by θ_{crit}. (A more detailed analysis using Maxwell's equations shows that as the incident angle approaches the critical angle, the transmitted intensity approaches zero.) If the angle of incidence is *greater than* the critical angle, the sine of the angle of refraction, as computed by Snell's law, would have to be greater than unity, which is impossible. Beyond the critical angle, the ray *cannot* pass into the upper material; it is trapped in the lower material and is completely reflected at the boundary surface. This situation, called **total internal reflection,** occurs only when a ray is incident on the interface with a second material whose index of refraction is *smaller* than that of the material in which the ray is traveling.

We can find the critical angle for two given materials by setting $\theta_b = 90°$ $(\sin\theta_b = 1)$ in Snell's law. We then have

$$\sin\theta_{crit} = \frac{n_b}{n_a} \qquad \text{(critical angle for total internal reflection)} \qquad (33.6)$$

Total internal reflection will occur if the angle of incidence θ_a is greater than or equal to θ_{crit}.

As an example, for a glass–air surface with $n = 1.52$ for the glass,

$$\sin\theta_{crit} = \frac{1}{1.52} = 0.658 \qquad \theta_{crit} = 41.1°$$

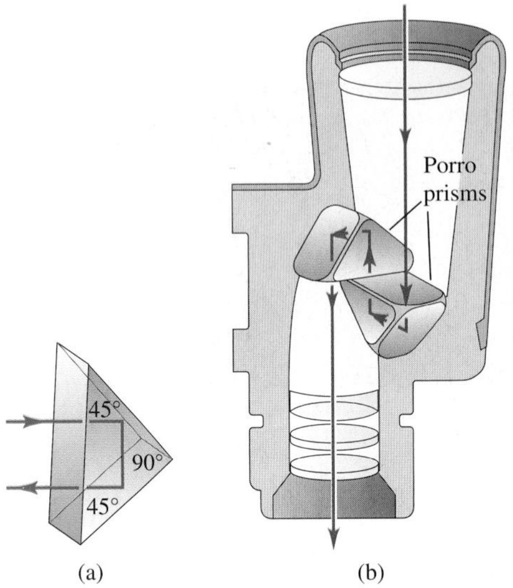

33.13 (a) Total internal reflection in a Porro prism. (b) A combination of two Porro prisms in binoculars.

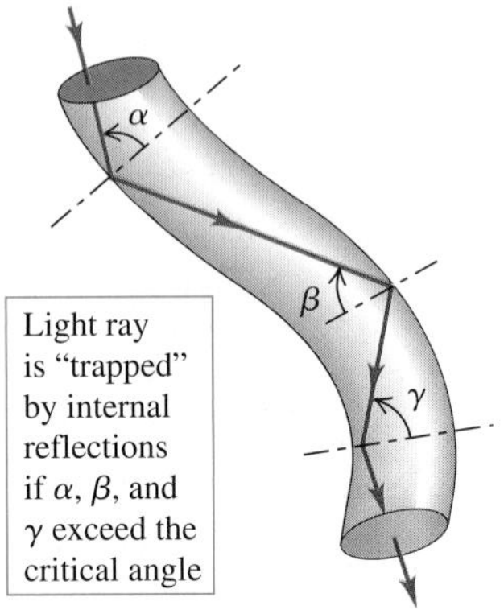

33.14 A transparent rod with refractive index greater than that of the surrounding material.

Light propagating within this glass will be *totally reflected* if it strikes the glass–air surface at an angle of 41.1° or greater. Because the critical angle is slightly less than 45°, it is possible to use a prism with angles of 45°−45°−90° as a totally reflecting surface. As reflectors, totally reflecting prisms have some advantages over metallic surfaces such as ordinary coated-glass mirrors. While no metallic surface reflects 100% of the light incident on it, light can be *totally* reflected by a prism. The reflecting properties of a prism have the additional advantages of being permanent and unaffected by tarnishing.

A 45°−45°−90° prism, used as in Fig. 33.13a, is called a *Porro* prism. Light enters and leaves at right angles to the hypotenuse and is totally reflected at each of the shorter faces. The total change of direction of the rays is 180°. Binoculars often use combinations of two Porro prisms, as in Fig. 33.13b.

The brilliance of diamond is due in large measure to its very high index of refraction $(n = 2.417)$ and correspondingly high critical angle. Light entering a cut diamond is totally internally reflected from facets on its back surface, then emerges from the front surface. "Imitation diamond" gems, such as cubic zirconia, are made from less expensive crystalline materials with comparable indexes of refraction.

33.15 Image transmission by a bundle of optical fibers.

When a beam of light enters at one end of a transparent rod (Fig. 33.14), the light can be totally reflected internally if the index of refraction of the rod is greater than that of the surrounding material. The light is "trapped" within the rod even if the rod is curved, provided that the curvature is not too great. Such a rod is sometimes called a *light pipe.* A bundle of fine glass or plastic fibers behaves in the same way and has the advantage of being flexible. A bundle may consist of thousands of individual fibers, each of the order of 0.002 to 0.01 mm in diameter. If the fibers are assembled in the bundle so that the relative positions of the ends are the same (or mirror images) at both ends, the bundle can transmit an image, as shown in Fig. 33.15.

Fiber-optic devices have found a wide range of medical applications in instruments called *endoscopes,* which can be inserted directly into the bronchial tubes, the bladder, the colon, and so on, for direct visual examination. A bundle of fibers can be enclosed in a hypodermic needle for study of tissues and blood vessels far beneath the skin.

Fiber optics also have applications in communication systems, in which they are used to transmit a modulated laser beam. The rate at which information can be transmitted by a wave (light, radio, or whatever) is proportional to the frequency. To see qualitatively why this is so, consider modulating (modifying) the wave by chopping off some of the wave crests. Suppose each crest represents a binary digit, with a chopped-off crest representing a zero and an unmodified crest representing a one. The number of binary digits we can transmit per unit time is thus proportional to the frequency of the wave. Infrared and visible-light waves have much higher frequency than do radio waves, so a modulated laser beam can transmit an enormous amount of information through a single fiber-optic cable.

Another advantage of optical fibers is that they can be made thinner than conventional copper wire, so more fibers can be bundled together in a cable of a given diameter. Hence more distinct signals (for instance, different phone lines) can be sent over the same cable. Because fiber-optic cables are electrical insulators, they are immune to electrical interference from lightning and other sources, and they don't allow unwanted currents between source and receiver. For these and other reasons, fiber-optic cables are playing an increasingly important role in long-distance telephone, television, and Internet communication.

Conceptual Example 33.4 A leaky periscope

A periscope for a submarine uses two totally reflecting 45°–45°–90° prisms with total internal reflection on the sides adjacent to the 45° angles. It springs a leak, and the bottom prism is covered with water. Explain why the periscope no longer works.

SOLUTION

The critical angle for water $(n_b = 1.33)$ on glass $(n_a = 1.52)$ is

$$\theta_{\text{crit}} = \arcsin \frac{1.33}{1.52} = 61.0^\circ$$

The 45° angle of incidence for a totally reflecting prism is *less than* the 61° critical angle, so total internal reflection does not occur at the glass–water boundary. Most of the light is transmitted into the water, and very little is reflected back into the prism.

Test Your Understanding

Consider light propagating in an optical fiber. Why doesn't the light undergo total internal reflection when it reaches the end of the fiber, preventing it from getting out?

*33.4 | Dispersion

Ordinary white light is a superposition of waves with wavelengths extending throughout the visible spectrum. The speed of light *in vacuum* is the same for all wavelengths, but the speed in a material substance is different for different wavelengths. Therefore the index of refraction of a material depends on wavelength. The dependence of wave speed and index of refraction on wavelength is called **dispersion.**

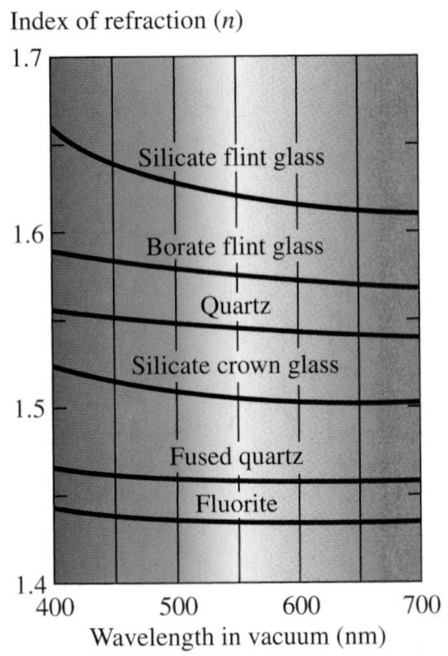

33.16 Variation of index of refraction n with wavelength for different transparent materials. The horizontal axis shows the wavelength λ_0 of the light *in vacuum;* the wavelength in the material is equal to $\lambda = \lambda_0/n$.

Figure 33.16 shows the variation of index of refraction n with wavelength for a few common optical materials. Note that the horizontal axis of this figure refers to the wavelength of the light *in vacuum,* λ_0; the wavelength in the material is given by Eq. (33.5), $\lambda = \lambda_0/n$. In most materials the value of n *decreases* with increasing wavelength and decreasing frequency, and thus n *increases* with decreasing wavelength and increasing frequency. In such a material, light of longer wavelength has greater speed than light of shorter wavelength.

Figure 33.17 shows a ray of white light incident on a prism. The deviation (change of direction) produced by the prism increases with increasing index of refraction and frequency and decreasing wavelength. Violet light is deviated most, and red is deviated least; other colors are in intermediate positions. When it comes out of the prism, the light is spread out into a fan-shaped beam, as shown. The light is said to be *dispersed* into a spectrum. The amount of dispersion depends on the *difference* between the refractive indexes for violet light and for red light. From Fig. 33.16 we can see that for a substance such as fluorite, the difference between the indexes for red and violet is small, and the dispersion will also be small. A better choice of material for a prism whose purpose is to produce a spectrum would be silicate flint glass, for which there is a larger difference in the value of n between red and violet.

As we mentioned in Section 33.3, the brilliance of diamond is due in part to its unusually large refractive index; another important factor is its large dispersion, which causes white light entering a diamond to emerge as a multicolored spectrum. Crystals of rutile and of strontium titanate, which can be produced synthetically, have about eight times the dispersion of diamond.

When you experience the beauty of a rainbow, as in the photograph that opens this chapter, you are seeing the combined effects of dispersion, refraction, and reflection. Sunlight comes from behind you, enters a water droplet, is (partially) reflected from the back surface of the droplet, and is refracted again upon exiting the droplet (Fig. 33.18a). A light ray that enters the middle of the raindrop is reflected straight back. All other rays exit the raindrop within an angle Δ of that middle ray, with many rays "piling up" at the angle Δ. What you see is a disk of light of angular radius Δ centered on the down-sun point (the point in the sky opposite the sun); due to the "piling up" of light rays, the disk is brightest around its rim, which we see as a rainbow. Because no light reaches your eye from angles greater than Δ, the sky looks dark outside the rainbow (see the photograph that

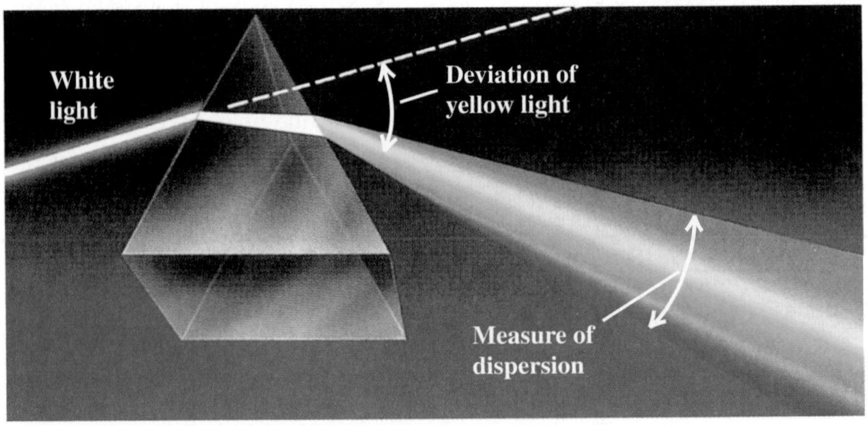

33.17 Dispersion of light by a prism. The band of colors is called a spectrum.

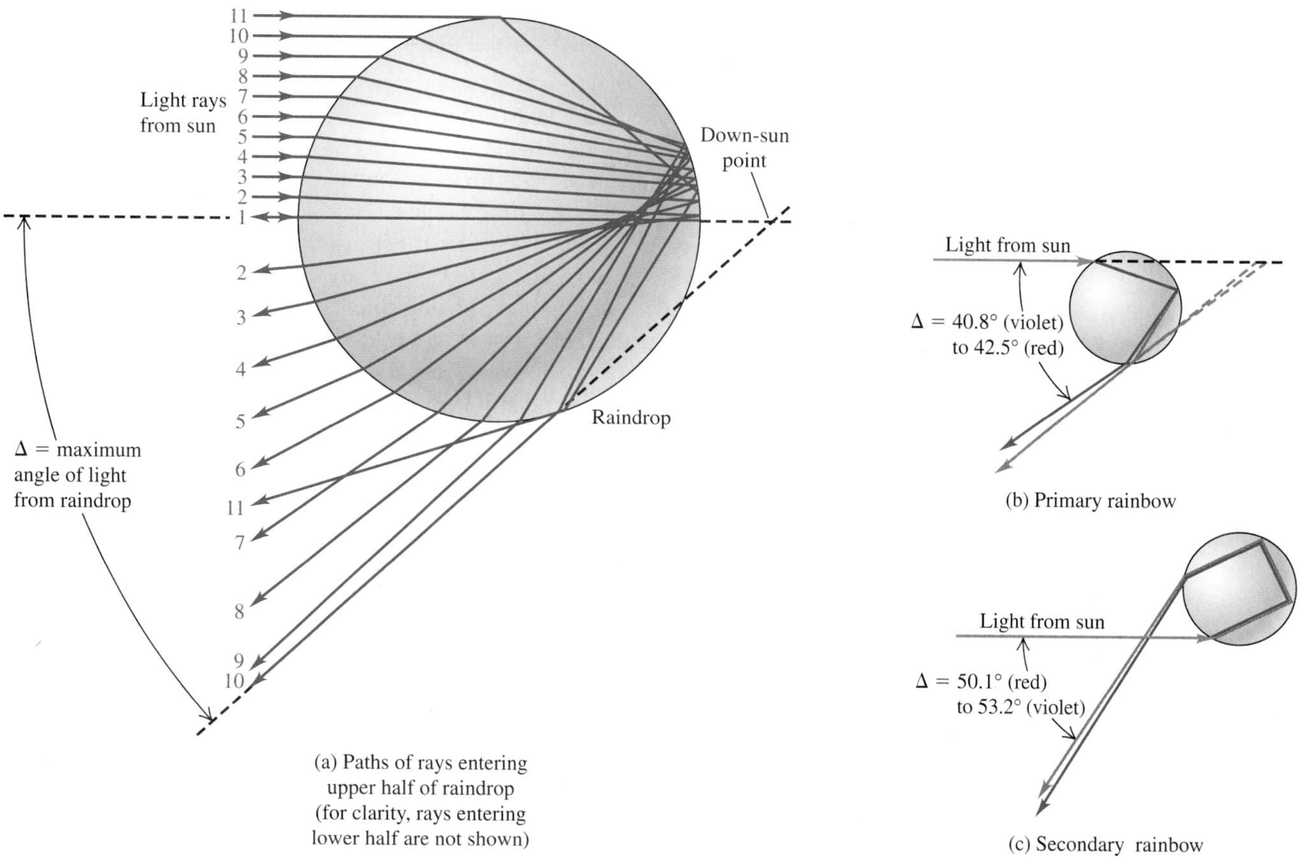

33.18 Rainbows are formed by refraction, reflection, and dispersion in water drops. (a) The paths of light rays entering the upper half of a raindrop. (The pattern of rays entering the lower half is the same, but flipped upside down.) All of the rays emerge within an angle Δ from the down-sun point. The angle Δ is the angular radius of the rainbow. (b) The colors of the rainbow are due to dispersion. Red appears on the *outside* of the primary rainbow. (c) Red appears on the *inside* of the fainter secondary rainbow.

opens this chapter). The value of the angle Δ depends on the index of refraction of the water that makes up the raindrops, which in turn depends on the wavelength (Fig. 33.18b). The bright disk of red light is slightly larger than that for orange light, which in turn is slightly larger than that for yellow light, and so on. As a result, you see the rainbow as a band of colors.

In many cases you can see a second, larger rainbow, such as the one shown in the photograph that opens this chapter. It is the result of dispersion, refraction, and *two* reflections from the back surface of the droplet (Fig. 33.18c). Each time a light ray hits the back surface, part of the light is refracted out of the drop (not shown in Fig. 33.18); after two such hits, relatively little light is left inside the drop, which is why the secondary rainbow is noticeably fainter than the primary rainbow. Just as a mirror held up to a book reverses the printed letters, so the second reflection reverses the sequence of colors in the secondary rainbow. You can see this effect in the photograph that opens this chapter.

?

33.5 | Polarization

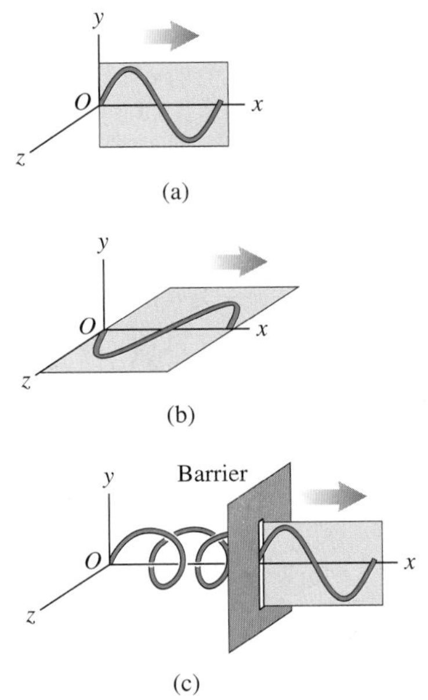

33.19 (a) Transverse wave on a string, polarized in the y-direction (b) Wave polarized in the z-direction. (c) A barrier with a frictionless vertical slot acts as a polarizing filter. It passes components polarized in the y-direction but blocks those polarized in the z-direction.

Polarization is a characteristic of all transverse waves. This chapter is about light, but to introduce some basic polarization concepts, let's go back to the transverse waves on a string that we studied in Chapter 15. For a string that in equilibrium lies along the x-axis, the displacements may be along the y-direction, as in Fig. 33.19a. In this case the string always lies in the xy-plane. But the displacements might instead be along the z-axis, as in Fig. 33.19b; then the string always lies in the xz-plane.

When a wave has only y-displacements, we say that it is **linearly polarized** in the y-direction; a wave with only z-displacements is linearly polarized in the z-direction. For mechanical waves we can build a **polarizing filter,** or **polarizer,** that permits only waves with a certain polarization direction to pass. In Fig. 33.19c the string can slide vertically in the slot without friction, but no horizontal motion is possible. This filter passes waves that are polarized in the y-direction but blocks those that are polarized in the z-direction.

This same language can be applied to electromagnetic waves, which also have polarization. As we learned in Chapter 32, an electromagnetic wave is a *transverse* wave; the fluctuating electric and magnetic fields are perpendicular to each other and to the direction of propagation. We always define the direction of polarization of an electromagnetic wave to be the direction of the *electric*-field vector $\vec{E}$, not the magnetic field, because many common electromagnetic-wave detectors respond to the electric forces on electrons in materials, not the magnetic forces. Thus the electromagnetic wave described by Eq. (32.17),

$$\vec{E}(x,t) = \hat{\jmath}E_{max}\cos(kx - \omega t)$$
$$\vec{B}(x,t) = \hat{k}B_{max}\cos(kx - \omega t)$$

is said to be polarized in the y-direction because the electric field has only a y-component.

CAUTION It's unfortunate that the same word "polarization" that is used to describe the direction of $\vec{E}$ in an electromagnetic wave is also used to describe the shifting of electric charge within a body, such as in response to a nearby charged body; we described this latter kind of polarization in Section 21.2 (see Fig. 21.7). You should remember that while these two concepts have the same name, they do not describe the same phenomenon.

ActivPhysics ONLINE

16.9 Polarization

Polarizing Filters

Waves emitted by a radio transmitter are usually linearly polarized. The vertical rod antennas that are used for cellular telephones in automobiles emit waves that, in a horizontal plane around the antenna, are polarized in the vertical direction (parallel to the antenna). Rooftop TV antennas have horizontal elements in the United States and vertical elements in Great Britain because the transmitted waves have different polarizations.

The situation is different for visible light. Light from ordinary sources, such as incandescent light bulbs and fluorescent light fixtures, is *not* polarized. The "antennas" that radiate light waves are the molecules that make up the sources. The waves emitted by any one molecule may be linearly polarized, like those from a radio antenna. But any actual light source contains a tremendous number of molecules with random orientations, so the emitted light is a random mixture of waves linearly polarized in all possible transverse directions. Such light is called **unpolarized light** or **natural light.** To create polarized light from unpolarized natural light requires a filter that is analogous to the slot for mechanical waves in Fig. 33.19c.

Polarizing filters for electromagnetic waves have different details of construction, depending on the wavelength. For microwaves with a wavelength of a few centimeters, a good polarizer is an array of closely spaced, parallel conducting wires that are insulated from each other. (Think of a barbecue grill with the outer metal ring replaced by an insulating one.) Electrons are free to move along the length of the conducting wires and will do so in response to a wave whose $\vec{E}$ field is parallel to the wires. The resulting currents in the wires dissipate energy by I^2R heating; the dissipated energy comes from the wave, so whatever wave passes through the grid is greatly reduced in amplitude. Waves with $\vec{E}$ oriented perpendicular to the wires pass through almost unaffected, since electrons cannot move through the air between the wires. Hence a wave that passes through such a filter will be predominantly polarized in the direction perpendicular to the wires.

The most common polarizing filter for visible light is a material known by the trade name Polaroid, widely used for sunglasses and polarizing filters for camera lenses. Developed originally by the American scientist Edwin H. Land, this material incorporates substances that have **dichroism,** a selective absorption in which one of the polarized components is absorbed much more strongly than the other (Fig. 33.20). A Polaroid filter transmits 80% or more of the intensity of a wave that is polarized parallel to a certain axis in the material, called the **polarizing axis,** but only 1% or less for waves that are polarized perpendicular to this axis. In one type of Polaroid filter, long-chain molecules within the filter are oriented with their axis perpendicular to the polarizing axis; these molecules preferentially absorb light that is polarized along their length, much like the conducting wires in a polarizing filter for microwaves.

An *ideal* polarizing filter (polarizer) passes 100% of the incident light that is polarized in the direction of the filter's polarizing axis but completely blocks all light that is polarized perpendicular to this axis. Such a device is an unattainable idealization, but the concept is useful in clarifying the basic ideas. In the following discussion we will assume that all polarizing filters are ideal. In

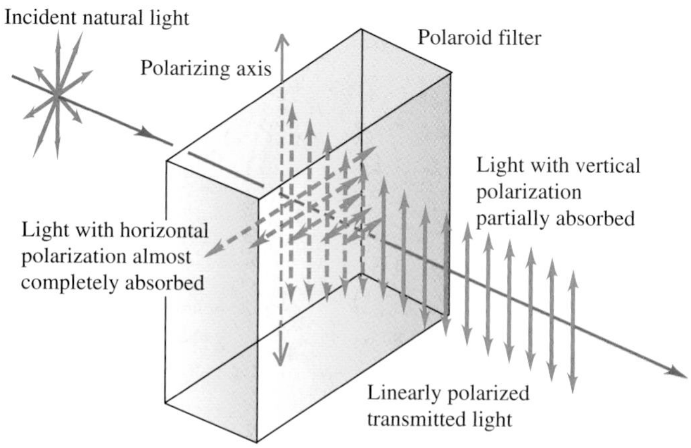

33.20 A Polaroid filter is illuminated by unpolarized natural light (shown by $\vec{E}$ vectors that point in all directions perpendicular to the direction of propagation). The components of $\vec{E}$ that are perpendicular to the polarizing axis are absorbed by the filter, and the transmitted light is linearly polarized (shown by $\vec{E}$ vectors along the polarization direction only).

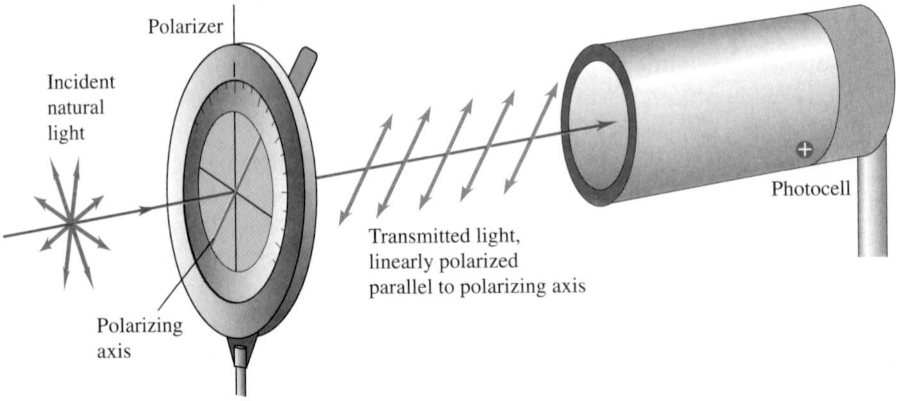

33.21 Unpolarized natural light is incident on the polarizing filter. The intensity of the transmitted linearly polarized light, measured by the photocell, is the same for all orientations of the polarizing filter. For an ideal polarizing filter, the transmitted intensity is half the incident intensity.

Fig. 33.21, unpolarized light is incident on a flat polarizing filter. The polarizing axis is represented by the blue line. The $\vec{E}$ vector of the incident wave can be represented in terms of components parallel and perpendicular to the polarizer axis; only the component of $\vec{E}$ parallel to the polarizing axis is transmitted. Hence the light emerging from the polarizer is linearly polarized parallel to the polarizing axis.

When unpolarized light is incident on an ideal polarizer as in Fig. 33.21, the intensity of the transmitted light is *exactly half* that of the incident unpolarized light, no matter how the polarizing axis is oriented. Here's why: We can resolve the $\vec{E}$ field of the incident wave into a component parallel to the polarizing axis and a component perpendicular to it. Because the incident light is a random mixture of all states of polarization, these two components are, on average, equal. The ideal polarizer transmits only the component that is parallel to the polarizing axis, so half the incident intensity is transmitted.

What happens when the linearly polarized light emerging from a polarizer passes through a second polarizer, as in Fig. 33.22? Consider the general case

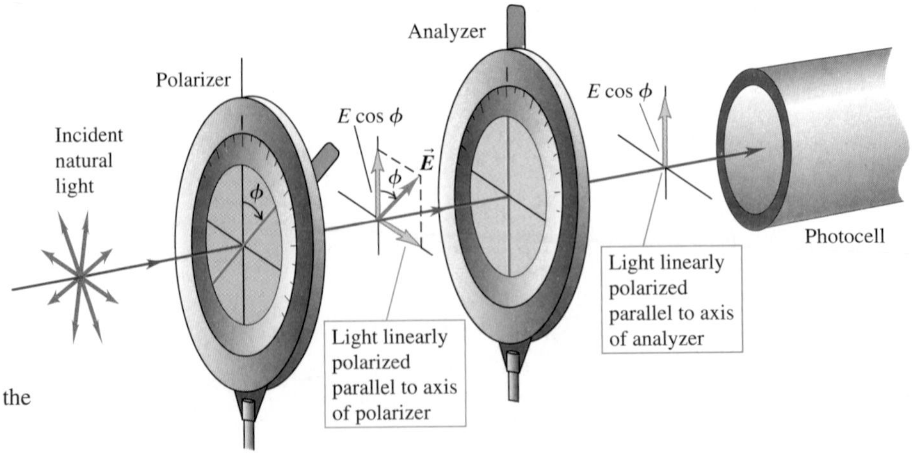

33.22 An ideal analyzer transmits only the component parallel to its transmission direction (that is, its polarizing axis).

in which the polarizing axis of the second polarizer, or *analyzer,* makes an angle ϕ with the polarizing axis of the first polarizer. We can resolve the linearly polarized light that is transmitted by the first polarizer into two components, as shown in Fig. 33.22, one parallel and the other perpendicular to the axis of the analyzer. Only the parallel component, with amplitude $E\cos\phi$, is transmitted by the analyzer. The transmitted intensity is greatest when $\phi = 0$, and it is zero when polarizer and analyzer are *crossed* so that $\phi = 90°$ (Fig. 33.23). To determine the direction of polarization of the light transmitted by the first polarizer, rotate the analyzer until the photocell in Fig. 33.22 measures zero intensity; the polarization axis of the first polarizer is then perpendicular to that of the analyzer.

To find the transmitted intensity at intermediate values of the angle ϕ, we recall from our energy discussion in Chapter 32 that the intensity of an electromagnetic wave is proportional to the *square* of the amplitude of the wave [see Eq. (32.29)]. The ratio of transmitted to incident *amplitude* is $\cos\phi$, so the ratio of transmitted to incident *intensity* is $\cos^2\phi$. Thus the intensity of the light transmitted through the analyzer is

$$I = I_{max}\cos^2\phi \qquad (33.7)$$

(Malus's law, polarized light passing through an analyzer)

where I_{max} is the maximum intensity of light transmitted (at $\phi = 0$) and I is the amount transmitted at angle ϕ. This relation, discovered experimentally by Etienne Louis Malus in 1809, is called **Malus's law.** Malus's law applies *only* if the incident light passing through the analyzer is already linearly polarized.

33.23 These photos show the view through Polaroid sunglasses whose polarizing axes are aligned ($\phi = 0$, top) and whose axes are perpendicular ($\phi = 90°$, bottom). The transmitted intensity is greatest when the axes are aligned; it is zero when the axes are perpendicular.

Problem-Solving Strategy — Linear Polarization

IDENTIFY *the relevant concepts:* Remember that in all electromagnetic waves, including light waves, the direction of the $\vec{E}$ field is the direction of polarization and is perpendicular to the propagation direction. When working with polarizers, you are really dealing with components of $\vec{E}$ parallel and perpendicular to the polarizing axis. Everything you know about components of vectors is applicable here.

SET UP *the problem using the following steps:*
1. Just as for problems in geometric optics, you should *always* start by drawing a large, neat diagram. Label all known angles, including the angles of any and all polarizing axes.
2. Determine the target variables.

EXECUTE *the solution as follows:*
1. Remember that a polarizer lets pass only electric field components parallel to its polarizing axis.
2. If the incident light is linearly polarized and has amplitude E and intensity I_{max}, the light that passes through an ideal polarizer has amplitude $E\cos\phi$ and intensity $I_{max}\cos^2\phi$, where ϕ is the angle between the incident polarization direction and the filter's polarizing axis.
3. Unpolarized light is a random mixture of all possible polarization states, so on the average it has equal components in any two perpendicular directions. When passed through an ideal polarizer, unpolarized light becomes linearly polarized light with half the incident intensity. Partially linearly polarized light is a superposition of linearly polarized and unpolarized light.
4. The intensity (average power per unit area) of a wave is proportional to the *square* of its amplitude. If you find that two waves differ in amplitude by a certain factor, their intensities differ by the square of that factor.

EVALUATE *your answer:* Check your answer for any obvious errors. If your results say that light emerging from a polarizer has greater intensity than the incident light, something's wrong: a polarizer can't add energy to a light wave.

Example 33.5 Two polarizers in combination

In Fig. 33.22 the incident unpolarized light has intensity I_0. Find the intensities transmitted by the first and second polarizers if the angle between the axes of the two filters is 30°.

SOLUTION

IDENTIFY and SET UP: The diagram has already been drawn for us in Fig. 33.22. We are given the intensity I_0 of the incident natural light and the angle $\phi = 30°$ between the polarizing axes. Our target variables are the intensities of the light emerging from the first polarizer and of the light emerging from the second polarizer.

EXECUTE: As we explained above, the intensity of the linearly polarized light transmitted by the first filter is $I_0/2$. According to Eq. (33.7) with $\phi = 30°$, the second filter reduces the intensity by a factor of $\cos^2 30° = \frac{3}{4}$. Thus the intensity transmitted by the second polarizer is

$$\left(\frac{I_0}{2}\right)\left(\frac{3}{4}\right) = \frac{3}{8} I_0$$

EVALUATE: Note that the intensity decreases after each passage through a polarizer. The only situation in which the transmitted intensity does *not* decrease is if the polarizer is ideal (so it absorbs none of the light that passes through it) and if the incident light is linearly polarized along the polarizing axis, so $\phi = 0$.

Polarization by Reflection

Unpolarized light can be polarized, either partially or totally, by *reflection.* In Fig. 33.24, unpolarized natural light is incident on a reflecting surface between two transparent optical materials; the plane containing the incident and reflected rays and the normal to the surface is called the **plane of incidence.** For most angles of incidence, waves for which the electric-field vector $\vec{E}$ is perpendicular to the plane of incidence (that is, parallel to the reflecting surface) are reflected more strongly than those for which $\vec{E}$ lies in this plane. In this case the reflected light is *partially polarized* in the direction perpendicular to the plane of incidence.

But at one particular angle of incidence, called the **polarizing angle** θ_p, the light for which $\vec{E}$ lies in the plane of incidence is *not reflected at all* but is completely refracted. At this same angle of incidence the light for which $\vec{E}$ is perpendicular to the plane of incidence is partially reflected and partially refracted. The *reflected*

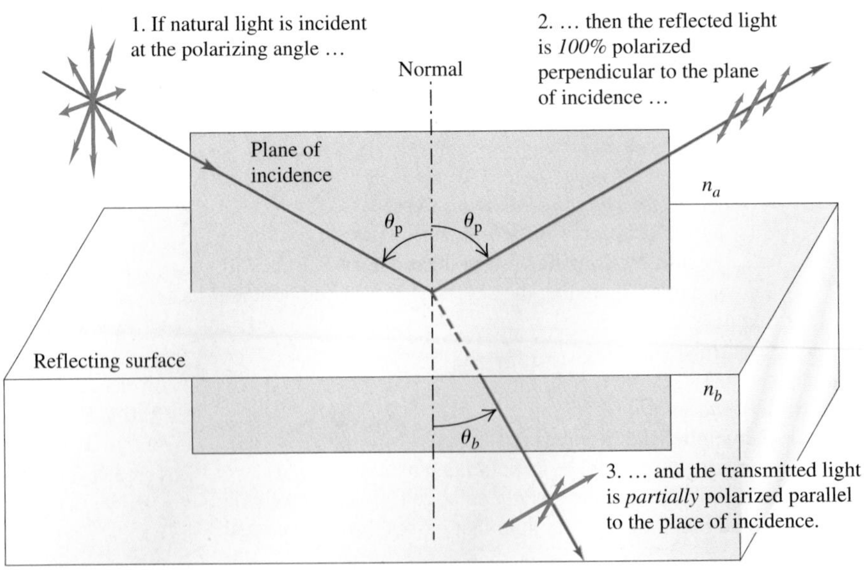

33.24 When light is incident on a reflecting surface at the polarizing angle, the reflected light is linearly polarized.

light is therefore *completely* polarized perpendicular to the plane of incidence, as shown in Fig. 33.24. The *refracted* light is *partially* polarized parallel to this plane; the refracted light is a mixture of the component parallel to the plane of incidence, all of which is refracted, and the remainder of the perpendicular component.

In 1812 the British scientist Sir David Brewster discovered that when the angle of incidence is equal to the polarizing angle θ_p, the reflected ray and the refracted ray are perpendicular to each other (Fig. 33.25). In this case the angle of refraction θ_b becomes the complement of θ_p, so $\theta_b = 90° - \theta_p$. From the law of refraction,

$$n_a \sin\theta_p = n_b \sin\theta_b$$

so we find

$$n_a \sin\theta_p = n_b \sin(90° - \theta_p) = n_b \cos\theta_p$$

$$\tan\theta_p = \frac{n_b}{n_a} \quad \text{(Brewster's law for the polarizing angle)} \quad (33.8)$$

This relation is known as **Brewster's law.** Although discovered experimentally, it can also be *derived* from a wave model using Maxwell's equations.

Polarization by reflection is the reason polarizing filters are widely used in sunglasses (Fig. 33.23). When sunlight is reflected from a horizontal surface, the plane of incidence is vertical, and the reflected light contains a preponderance of light that is polarized in the horizontal direction. When the reflection occurs at a smooth asphalt road surface or the surface of a lake, it causes unwanted glare. Vision can be improved by eliminating this glare. The manufacturer makes the polarizing axis of the lens material vertical, so very little of the horizontally polarized light reflected from the road is transmitted to the eyes. The glasses also reduce the overall intensity of the transmitted light to somewhat less than 50% of the intensity of the unpolarized incident light.

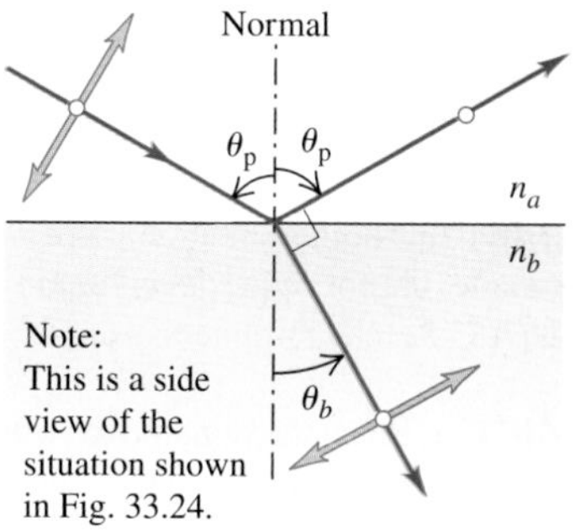

33.25 When light is incident at the polarizing angle, the reflected and refracted rays are perpendicular to each other. The open circles represent a component of $\vec{E}$ that is perpendicular to the plane of the figure (the plane of incidence) and parallel to the surface between the two materials.

Example 33.6 Reflection from a swimming pool's surface

Sunlight reflects off the smooth surface of an unoccupied swimming pool. a) At what angle of reflection is the light completely polarized? b) What is the corresponding angle of refraction for the light that is transmitted (refracted) into the water? c) At night an underwater floodlight is turned on in the pool. Repeat parts (a) and (b) for rays from the floodlight that strike the smooth surface from below.

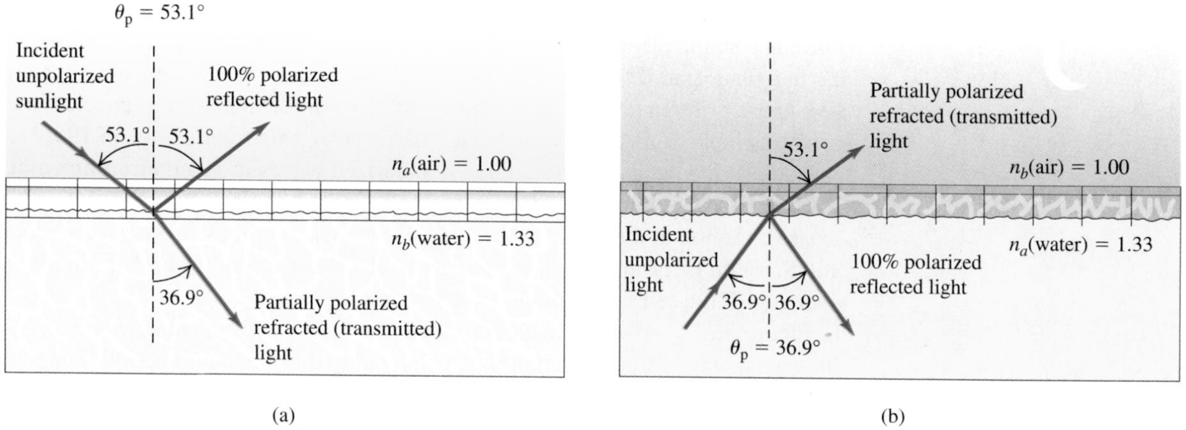

33.26 Light striking the water-air interface at the polarizing angle (a) from the air side; (b) from the water side.

SOLUTION

IDENTIFY and SET UP: In part (a) we're looking for the polarizing angle for light that is first in air, then in water; we find this with Brewster's law, Eq. (33.8). In part (b) we want the angle of the refracted light for this situation. In part (c) we again want the polarizing angle, but for light that is first in water, then in air. Again we use Eq. (33.8) to determine this angle.

EXECUTE: a) Since the light moves from air toward water, we have $n_a = 1.00$ (air) and $n_b = 1.33$ (water). From Eq. (33.8),

$$\theta_p = \arctan\frac{n_b}{n_a} = \arctan\frac{1.33}{1.00} = 53.1°$$

The angles are shown in Fig. 33.26a.

b) The incident light is at the polarizing angle, so the reflected and refracted rays are perpendicular; hence

$$\theta_p + \theta_b = 90°$$
$$\theta_b = 90° - 53.1° = 36.9°$$

c) Now the light is *first* in the water, then in the air, so $n_a = 1.33$ and $n_b = 1.00$. Again using Eq. (33.8), we have

$$\theta_p = \arctan\frac{1.00}{1.33} = 36.9°$$
$$\theta_b = 90° - 36.9° = 53.1°$$

The angles are shown in Fig. 33.26b.

EVALUATE: We can check our answer in part (b) using Snell's law, $n_a \sin\theta_a = n_b \sin\theta_b$, or

$$\sin\theta_b = \frac{n_a \sin\theta_p}{n_b} = \frac{1.00 \sin 53.1°}{1.33} = 0.600$$
$$\theta_b = 36.9°$$

Note that the two polarizing angles found in parts (a) and (c) add to 90°. This is *not* an accident; can you see why?

Circular and Elliptical Polarization

Light and other electromagnetic radiation can also have *circular* or *elliptical* polarization. To introduce these concepts, let's return once more to mechanical waves on a stretched string. In Fig. 33.19, suppose the two linearly polarized waves in parts (a) and (b) are in phase and have equal amplitude. When they are superposed, each point in the string has simultaneous *y*- and *z*-displacements of equal magnitude. A little thought shows that the resultant wave lies in a plane oriented at 45° to the *y*- and *z*-axes (i.e., in a plane making a 45° angle with the *xy*- and *xz*-planes). The amplitude of the resultant wave is larger by a factor of $\sqrt{2}$ than that of either component wave, and the resultant wave is linearly polarized.

But now suppose the two equal-amplitude waves differ in phase by a quarter-cycle. Then the resultant motion of each point corresponds to a superposition of two simple harmonic motions at right angles, with a quarter-cycle phase difference. The *y*-displacement at a point is greatest at times when the *z*-displacement is zero, and vice versa. The motion of the string as a whole then no longer takes place in a single plane. It can be shown that each point on the rope moves in a *circle* in a plane parallel to the *yz*-plane. Successive points on the rope have successive phase differences, and the overall motion of the string has the appearance of a rotating helix. This is shown to the left of the polarizing filter in Fig. 33.19c. This particular superposition of two linearly polarized waves is called **circular polarization.** By convention, the wave is said to be *right circularly polarized* when the sense of motion of a particle of the string, to an observer looking *backward* along the direction of propagation, is *clockwise;* the wave is said to be *left circularly polarized* if the sense of motion is the reverse.

Figure 33.27 shows the analogous situation for an electromagnetic wave. Two sinusoidal waves of equal amplitude, polarized in the *y*- and *z*-directions and with a quarter-cycle phase difference, are superposed. The result is a wave in which the $\vec{E}$ vector at each point has a constant magnitude but *rotates* around the direction of propagation. If the wave in Fig. 33.27 is propagating toward you, it is called a right circularly polarized electromagnetic wave.

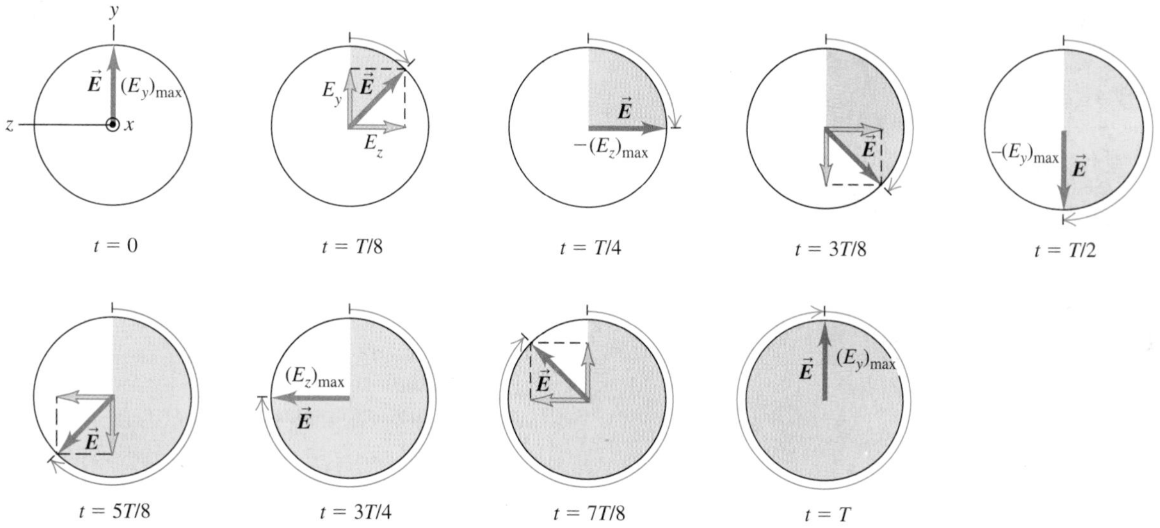

33.27 Circular polarization of an electromagnetic wave moving parallel to the x-axis. The y-component of $\vec{E}$ lags the z-component by a quarter-cycle. This phase difference results in right circular polarization if the wave is moving toward you (in the positive x-direction).

If the phase difference between the two component waves is something other than a quarter-cycle, or if the two component waves have different amplitudes, then each point on the string traces out not a circle but an *ellipse*. The resulting wave is said to be **elliptically polarized.**

For electromagnetic waves with radio frequencies, circular or elliptical polarization can be produced by using two antennas at right angles, fed from the same transmitter but with a phase-shifting network that introduces the appropriate phase difference. For light, the phase shift can be introduced by use of a material that exhibits *birefringence*—that is, has different indexes of refraction for different directions of polarization. A common example is calcite $(CaCO_3)$. When a calcite crystal is oriented appropriately in a beam of unpolarized light, its refractive index (for a wavelength in vacuum of 589 nm) is 1.658 for one direction of polarization and 1.486 for the perpendicular direction. When two waves with equal amplitude and with perpendicular directions of polarization enter such a material, they travel with different speeds. If they are in phase when they enter the material, then in general they are no longer in phase when they emerge. If the crystal is just thick enough to introduce a quarter-cycle phase difference, then the crystal converts linearly polarized light to circularly polarized light. Such a crystal is called a *quarter-wave plate.* Such a plate also converts circularly polarized light to linearly polarized light. Can you prove this? (See Problem 33.50.)

Photoelasticity

Some optical materials that are not normally birefringent become so when they are subjected to mechanical stress. This is the basis of the science of *photoelasticity.* Stresses in girders, boiler plates, gear teeth, and cathedral pillars can be analyzed by constructing a transparent model of the object, usually of a plastic material, subjecting it to stress, and examining it between a polarizer and an analyzer in the crossed position. Very complicated stress distributions can be studied

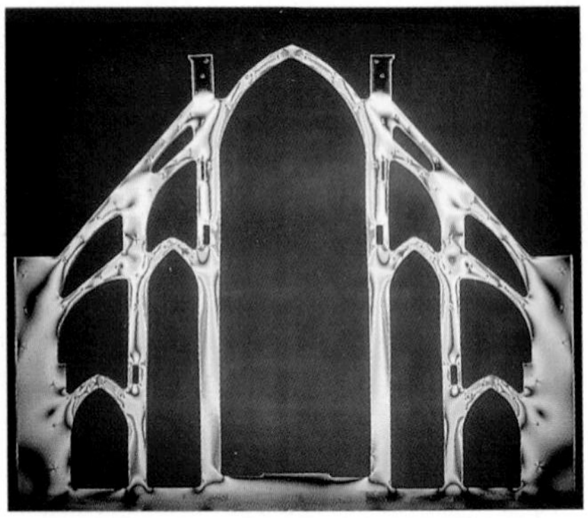

33.28 Photoelastic stress analysis of a model of a cross section of a Gothic cathedral. The masonry construction that was used for this kind of building had great strength in compression but very little in tension. Inadequate buttressing and high winds sometimes caused tensile stresses in normally compressed structural elements, leading to some spectacular collapses.

by these optical methods. Figure 33.28 is a photograph of a photoelastic model under stress.

Test Your Understanding

Why might a photographer use a Polaroid filter to take a photograph of a sunlit high-rise building?

*33.6 | Scattering of Light

The sky is blue. Sunsets are red. Skylight is partially polarized; that's why the sky looks darker from some angles than from others when it is viewed through Polaroid sunglasses. It turns out that one phenomenon is responsible for all of these effects.

When you look at the daytime sky, the light that you see is sunlight that has been absorbed and then re-radiated in a variety of directions. This process is called **scattering.** (If the earth had no atmosphere, the sky would appear as black in the daytime as it does at night, just as it does to an astronaut in space or on the moon; you would see the sun's light only if you looked directly at it, and the stars would be visible in the daytime.) Figure 33.29 shows some of the details of the scattering process. Sunlight, which is unpolarized, comes from the left along the x-axis and passes over an observer looking vertically upward along the y-axis. (We are viewing the situation from the side.) Consider the molecules of the earth's atmosphere located at point O. The electric field in the beam of sunlight sets the electric charges in these molecules into vibration. Since light is a transverse wave, the direction of the electric field in any component of the sunlight lies in the yz-plane, and the motion of the charges takes place in this plane. There is no field, and hence no motion of charges, in the direction of the x-axis.

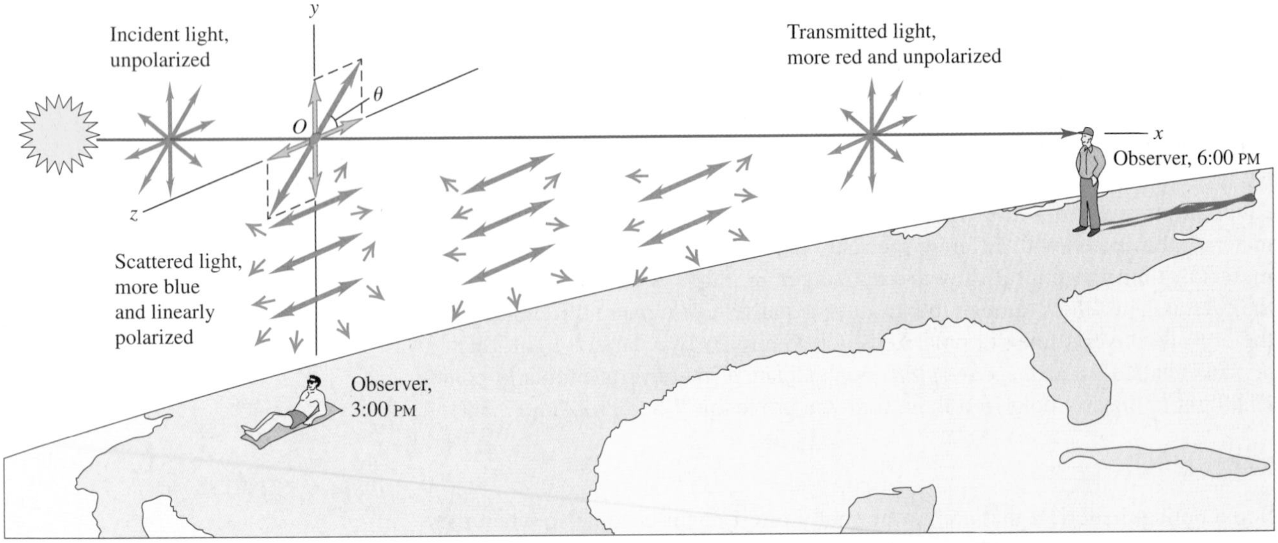

33.29 The sunbathing observer in the west sees sunlight that has been scattered by 90°. This scattered light is linearly polarized and contains predominantly light from the blue end of the spectrum. The white sunlight loses this blue light as it travels through the atmosphere, and the transmitted light seen by the observer in the east contains predominantly light from the red end of the spectrum.

An incident light wave with its $\vec{E}$ field at an angle θ with the z-axis sets the electric charges in the molecules vibrating along the line of $\vec{E}$, as indicated by the double-ended arrow through point O. We can resolve this vibration into two components, one along the y-axis and the other along the z-axis. Each component in the incident light produces the equivalent of two molecular "antennas," oscillating with the same frequency as the incident light and lying along the y- and z-axes.

We mentioned in Chapter 32 that an oscillating charge, like those in an antenna, does not radiate in the direction of its oscillation. (See Fig. 32.2 in Section 32.1.) Thus the "antenna" along the y-axis does not send any light to the observer directly below it, although it does emit light in other directions. Therefore the only light reaching this observer comes from the other molecular "antenna," corresponding to the oscillation of charge along the z-axis. This light is linearly polarized, with its electric field along the z-axis (parallel to the "antenna"). The red vectors on the y-axis below point O in Fig. 33.29 show the direction of polarization of the light reaching the observer.

As the original beam of sunlight passes though the atmosphere, its intensity decreases as its energy goes into the scattered light. Detailed analysis of the scattering process shows that the intensity of the light scattered from air molecules increases in proportion to the fourth power of the frequency (inversely to the fourth power of the wavelength). Thus the intensity ratio for the two ends of the visible spectrum is $(700 \text{ nm}/400 \text{ nm})^4 = 9.4$. Roughly speaking, scattered light contains nine times as much blue light as red, and that's why the sky is blue.

33.30 Clouds are white because they efficiently scatter sunlight of all wavelengths.

Clouds contain a high concentration of water droplets or ice crystals, which also scatter light. Because of this high concentration, light passing through the cloud has many more opportunities for scattering than does light passing through a clear sky. Thus light of *all* wavelengths is eventually scattered out of the cloud, so the cloud looks white (Fig. 33.30). Milk looks white for the same reason; the scattering is due to fat globules in the milk. If you dilute milk by mixing it with enough water, the concentration of fat globules will be so low that only blue light will be substantially scattered; the dilute solution will look blue, not white. (Nonfat milk, which also has a very low concentration of globules, looks somewhat bluish for this same reason.)

Near sunset, when sunlight has to travel a long distance through the earth's atmosphere, a substantial fraction of the blue light is removed by scattering. White light minus blue light looks yellow or red. This explains the yellow or red hue that we so often see from the setting sun (and that is seen by the observer at the far right of Fig. 33.29).

Because skylight is partially polarized, polarizers are useful in photography. The sky in a photograph can be darkened by orienting the polarizer axis to be perpendicular to the predominant direction of polarization of the scattered light. The most strongly polarized light comes from parts of the sky that are 90° away from the sun—for example, from directly overhead when the sun is on the horizon at sunrise or sunset.

33.7 | Huygens' Principle

The laws of reflection and refraction of light rays that we introduced in Section 33.2 were discovered experimentally long before the wave nature of light was firmly established. However, we can *derive* these laws from wave considerations and show that they are consistent with the wave nature of light. The same kind of analysis that we use here will be of central importance in Chapters 35 and 36 in our discussion of physical optics.

We begin with a principle called **Huygens' principle.** This principle, stated originally by the Dutch scientist Christiaan Huygens in 1678, is a geometrical method for finding, from the known shape of a wave front at some instant, the shape of the wave front at some later time. Huygens assumed that **every point of a wave front may be considered the source of secondary wavelets that spread out in all directions with a speed equal to the speed of propagation of the wave.** The new wave front at a later time is then found by constructing a surface *tangent* to the secondary wavelets or, as it is called, the *envelope* of the wavelets. All the results that we obtain from Huygens' principle can also be obtained from Maxwell's equations. Thus it is not an independent principle, but it is often very convenient for calculations with wave phenomena.

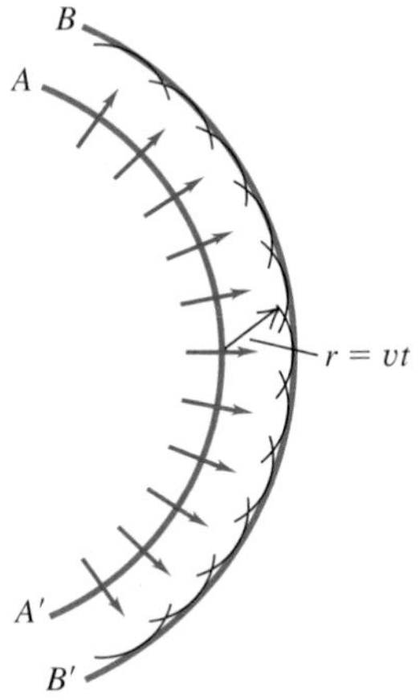

33.31 Applying Huygens' principle to wave front AA' to construct a new wave front BB'.

Huygens' principle is shown in Fig. 33.31. The original wave front AA' is traveling outward from a source, as indicated by the small arrows. We want to find the shape of the wave front after a time interval t. Let v be the speed of propagation of the wave; then in time t it travels a distance vt. We construct several circles (traces of spherical wavelets) with radius $r = vt$, centered at points along AA'. The trace of the envelope of these wavelets, which is the new wave front, is the curve BB'. We are assuming that the speed v is the same at all points and in all directions.

To derive the law of reflection from Huygens' principle, we consider a plane wave approaching a plane reflecting surface. In Fig. 33.32a the lines AA', OB', and NC' represent successive positions of a wave front approaching the surface MM'. Point A on the wave front AA' has just arrived at the reflecting surface. We can use Huygens' principle to find the position of the wave front after a time interval t. With points on AA' as centers, we draw several secondary wavelets with radius vt. The wavelets that originate near the upper end of AA' spread out unhindered, and their envelope gives the portion OB' of the new wave front. If the reflecting surface were not there, the wavelets originating near the lower end of AA' would similarly reach the positions shown by the broken circular arcs. Instead, these wavelets strike the reflecting surface.

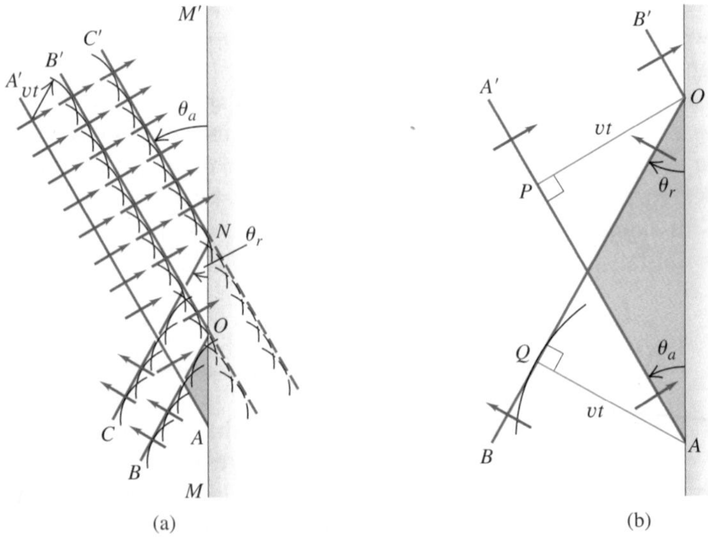

33.32 (a) Successive positions of a plane wave AA' as it is reflected from a plane surface. (b) Magnified portion of part (a).

The effect of the reflecting surface is to *change the direction* of travel of those wavelets that strike it, so the part of a wavelet that would have penetrated the surface actually lies to the left of it, as shown by the full lines. The first such wavelet is centered at point A; the envelope of all such reflected wavelets is the portion OB of the wave front. The trace of the entire wave front at this instant is the bent line BOB'. A similar construction gives the line CNC' for the wave front after another interval t.

From plane geometry the angle θ_a between the incident *wave front* and the *surface* is the same as that between the incident *ray* and the *normal* to the surface and is therefore the angle of incidence. Similarly, θ_r is the angle of reflection. To find the relation between these angles, we consider Fig. 33.32b. From O we draw $OP = vt$, perpendicular to AA'. Now OB, by construction, is tangent to a circle of radius vt with center at A. If we draw AQ from A to the point of tangency, the triangles APO and OQA are congruent because they are right triangles with the side AO in common and with $AQ = OP = vt$. The angle θ_a therefore equals the angle θ_r, and we have the law of reflection.

We can derive the law of *refraction* by a similar procedure. In Fig. 33.33 we consider a wave front, represented by line AA', for which point A has just arrived at the boundary surface SS' between two transparent materials a and b, with indexes of refraction n_a and n_b and wave speeds v_a and v_b. (The *reflected* waves are not shown in the figure; they proceed exactly as in Fig. 33.32.) We can apply Huygens' principle to find the position of the refracted wave fronts after a time t.

With points on AA' as centers, we draw several secondary wavelets. Those originating near the upper end of AA' travel with speed v_a and, after a time interval t, are spherical surfaces of radius $v_a t$. The wavelet originating at point A, however, is traveling in the second material b with speed v_b and at time t is a spherical surface of radius $v_b t$. The envelope of the wavelets from the original wave front is the plane whose trace is the bent line BOB'. A similar construction leads to the trace CPC' after a second interval t.

The angles θ_a and θ_b between the surface and the incident and refracted wave fronts are the angle of incidence and the angle of refraction, respectively. To find the relation between these angles, refer to Fig. 33.33b. Draw $OQ = v_a t$, perpendicular to AQ, and draw $AB = v_b t$, perpendicular to BO. From the right triangle AOQ,

$$\sin\theta_a = \frac{v_a t}{AO}$$

and from the right triangle AOB,

$$\sin\theta_b = \frac{v_b t}{AO}$$

Combining these, we find

$$\frac{\sin\theta_a}{\sin\theta_b} = \frac{v_a}{v_b} \tag{33.9}$$

We have defined the index of refraction n of a material as the ratio of the speed of light c in vacuum to its speed v in the material: $n_a = c/v_a$ and $n_b = c/v_b$. Thus

$$\frac{n_b}{n_a} = \frac{c/v_b}{c/v_a} = \frac{v_a}{v_b}$$

and we can rewrite Eq. (33.9) as

$$\frac{\sin\theta_a}{\sin\theta_b} = \frac{n_b}{n_a}$$

or

$$n_a \sin\theta_a = n_b \sin\theta_b$$

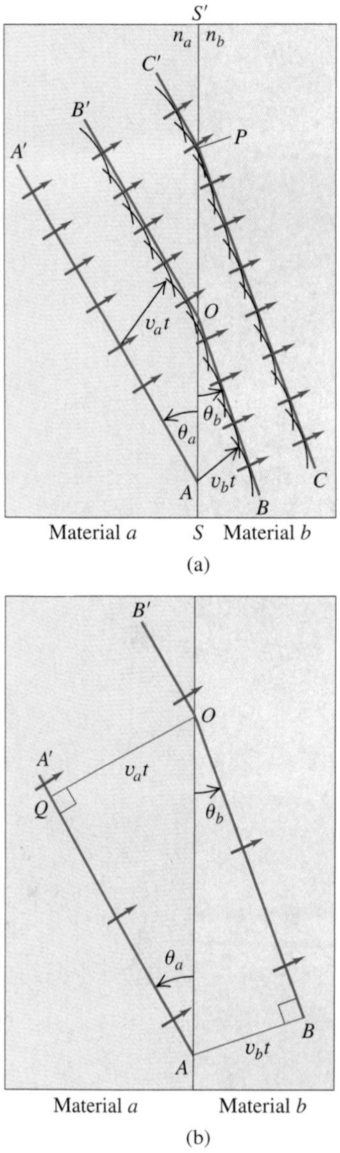

33.33 (a) Successive positions of a plane wave AA' as it is refracted at a plane surface. (b) Magnified portion of part (a). The case $v_b < v_a$ is shown.

which we recognize as Snell's law, Eq. (33.4). So we have derived Snell's law from a wave theory. Alternatively, we may choose to regard Snell's law as an experimental result that defines the index of refraction of a material; in that case this analysis helps to confirm the relationship $v = c/n$ for the speed of light in a material.

Mirages offer an interesting example of Huygens' principle in action. When the surface of pavement or desert sand is heated intensely by the sun, a hot, less dense, smaller-n layer of air forms near the surface. The speed of light is slightly greater in the hotter air near the ground, the Huygens wavelets have slightly larger radii, the wave fronts tilt slightly, and rays that were headed toward the surface with a large angle of incidence (near 90°) can be bent up as shown in Fig. 33.34. Light farther from the ground is bent less and travels nearly in a straight line. The observer sees the object in its natural position, with an inverted image below it, as though seen in a horizontal reflecting surface. Even when the turbulence of the heated air prevents a clear inverted image from being formed, the mind of the thirsty traveler can interpret the apparent reflecting surface as a sheet of water.

It is important to keep in mind that Maxwell's equations are the fundamental relations for electromagnetic wave propagation. But it is a remarkable fact that Huygens' principle anticipated Maxwell's analysis by two centuries. Maxwell provided the theoretical underpinning for Huygens' principle. Every point in an electromagnetic wave, with its time-varying electric and magnetic fields, acts as a source of the continuing wave, as predicted by Ampere's and Faraday's laws.

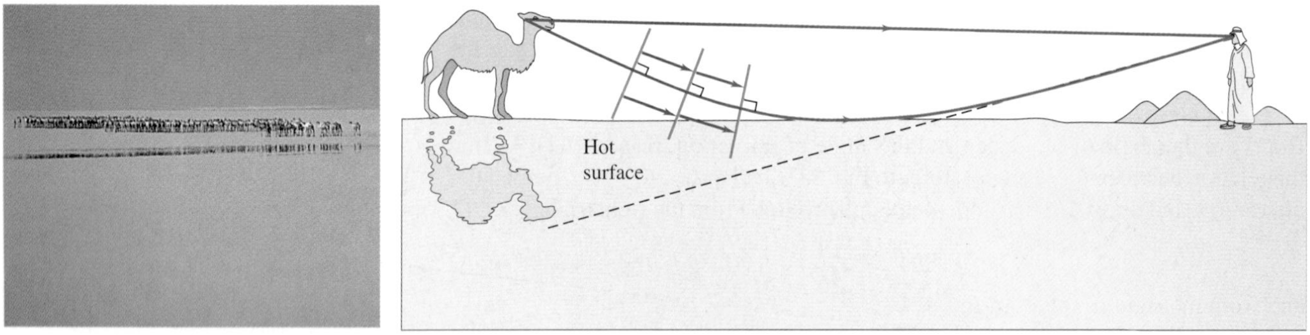

33.34 A mirage is observed because wavelets near the hot surface have slightly greater radii vt, gradually tilting the wave fronts and bending the rays.

Test Your Understanding

Explain how Fig. 33.33a shows that the wavelength decreases when a wave moves into a material with a higher index of refraction.

SUMMARY

Light is an electromagnetic wave. When emitted or absorbed, it also shows particle properties. It is emitted by accelerated electric charges. The speed of light is a fundamental physical constant.

A wave front is a surface of constant phase; wave fronts move with a speed equal to the propagation speed of the wave. A ray is a line along the direction of propagation, perpendicular to the wave fronts. Representation of light by rays is the basis of geometric optics.

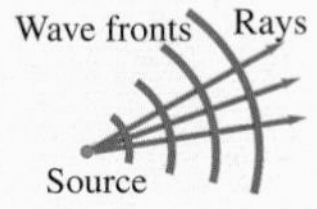

When light is transmitted from one material to another, the frequency of the light is unchanged, but the wavelength and wave speed can change. The index of refraction n of a material is the ratio of the speed of light in vacuum c to the speed v in the material. If λ_0 is the wavelength in a vacuum, the same wave has a shorter wavelength λ in a medium with index of refraction n. (See Example 33.2)

$$n = \frac{c}{v} \quad (33.1)$$

$$\lambda = \frac{\lambda_0}{v} \quad (33.5)$$

At a smooth interface between two optical materials, the incident, reflected, and refracted rays and the normal to the interface all lie in a single plane called the plane of incidence. The law of reflection states that the angles of incidence and reflection are equal. The law of refraction relates the angles of incidence and refraction to the indices of refraction of the materials. Angles of incidence, reflection, and refraction are always measured from the normal to the surface. (See Examples 33.1 and 33.3)

$$\theta_r = \theta_a \quad (33.2)$$

(law of reflection)

$$n_a \sin\theta_a = n_b \sin\theta_b \quad (33.4)$$

(law of refraction)

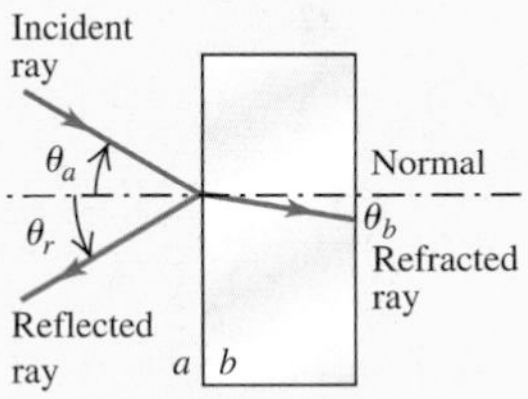

When a ray travels in a material of greater index of refraction n_a toward a material of smaller index n_b, total internal reflection occurs at the interface when the angle of incidence exceeds a critical angle θ_{crit}. (See Example 33.4)

$$\sin\theta_{crit} = \frac{n_b}{n_a} \quad (33.6)$$

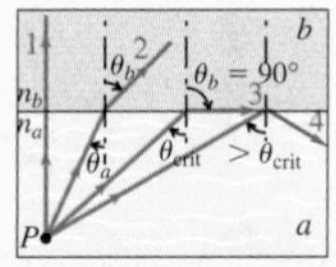

The variation of index of refraction n with wavelength λ is called dispersion. Usually, n decreases with increasing λ.

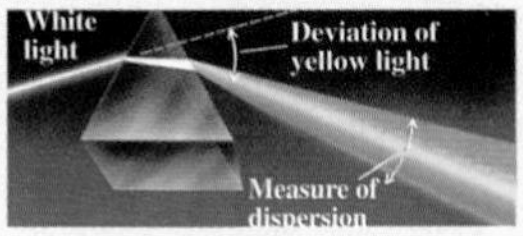

The direction of polarization of a linearly polarized electromagnetic wave is the direction of the $\vec{E}$ field. A polarizing filter passes waves that are linearly polarized along its polarizing axis and blocks waves polarized perpendicularly to that axis. When polarized light of intensity I_{max} is incident on a polarizing filter used as an analyzer, the intensity I of the light transmitted through the analyzer depends on the angle ϕ between the polarization direction of the incident light and the polarizing axis of the analyzer. (See Example 33.5)

$$I = I_{max}\cos^2\phi \quad (33.7)$$

(Malus's law)

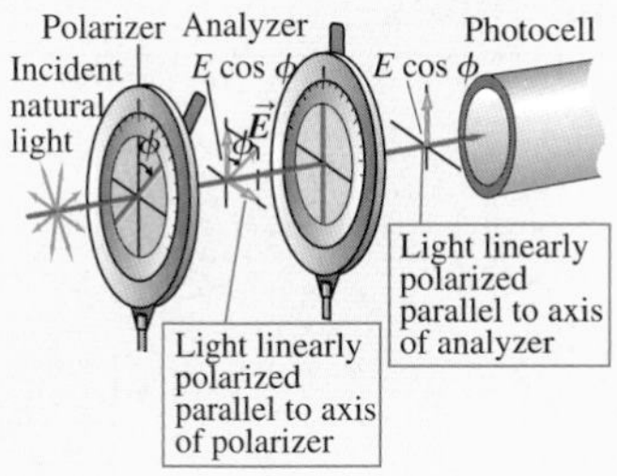

When unpolarized light strikes an interface between two materials, Brewster's law states that the reflected light is completely polarized perpendicular to the plane of incidence (parallel to the interface) if the angle of incidence equals the polarizing angle θ_p. (See Example 33.6)

$$\tan\theta_p = \frac{n_b}{n_a} \quad (33.8)$$

(Brewster's law)

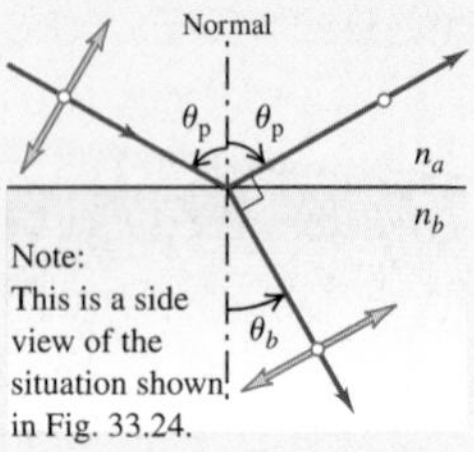

When two linearly polarized waves with a phase difference are superposed, the result is circularly or elliptically polarized light. In this case the $\vec{E}$ vector is not confined to a plane containing the direction of propagation, but rather describes circles or ellipses in planes perpendicular to the propagation direction.

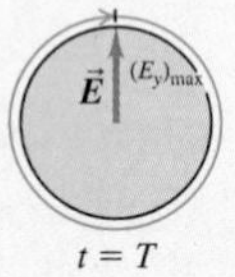

Light is scattered by air molecules. The scattered light is partially polarized.

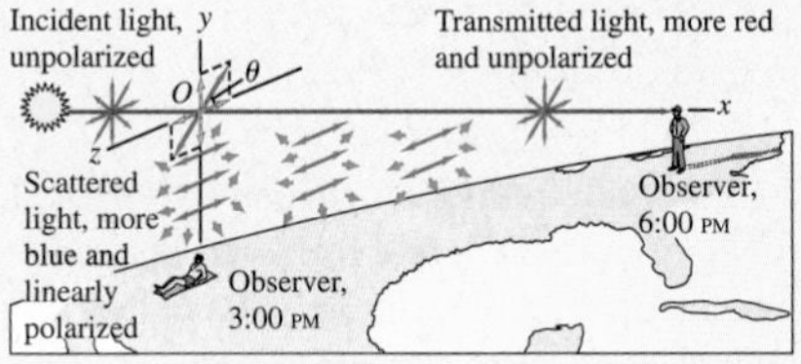

Huygens' principle states that if the position of a wave front at one instant is known, the position of the front at a later time can be constructed by imagining the front as a source of secondary wavelets. Huygens' principle can be used to derive the laws of reflection and refraction.

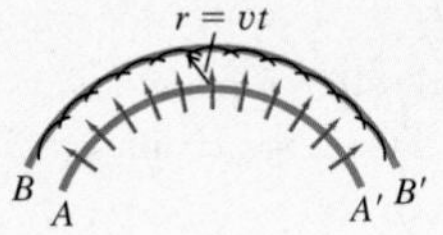

Key Terms

optics, 1247
wave front, 1249
ray, 1249
geometric optics, 1250
physical optics, 1250
reflection, 1250
refraction, 1250
specular reflection, 1251
diffuse reflection, 1251
index of refraction (refractive index), 1251
law of reflection, 1252
law of refraction (Snell's law), 1252
critical angle, 1257
total internal reflection, 1257
dispersion, 1259
linear polarization, 1262
polarizing filter (polarizer), 1262
unpolarized light (natural light), 1263
dichroism, 1263
polarizing axis, 1263
Malus's law, 1265
plane of incidence, 1266
polarizing angle, 1266
Brewster's law, 1267
circular polarization, 1268
elliptical polarization, 1269
scattering, 1270
Huygens' principle, 1272

Your Notes

Answer to Chapter Opening Question

The secondary rainbow is caused by light rays that reflect *twice* inside a raindrop before exiting the drop. The light rays flip over in the second reflection, which reverses the sequence of colors. See Fig. 33.18c.

Answers to Test Your Understanding Questions

Section 33.1 The wave fronts would not be spherical, since the waves do not propagate at equal speed in all directions. The waves go farther in the y-direction in a given amount of time than in the other directions, so the wave fronts are elongated in the y-direction.
Section 33.2 As shown in the figure, light rays coming from the fish bend away from the normal when they pass from the water $(n = 1.33)$ into the air $(n = 1.00)$. As a result, the fish appears to be higher in the water than it actually is. Hence you should aim a spear *below* the apparent position of the fish. If you use a laser beam, you should aim *at* the apparent position of the fish: the beam of laser light takes the same path from you to the fish as ordinary light takes from the fish to you (though in the opposite direction).

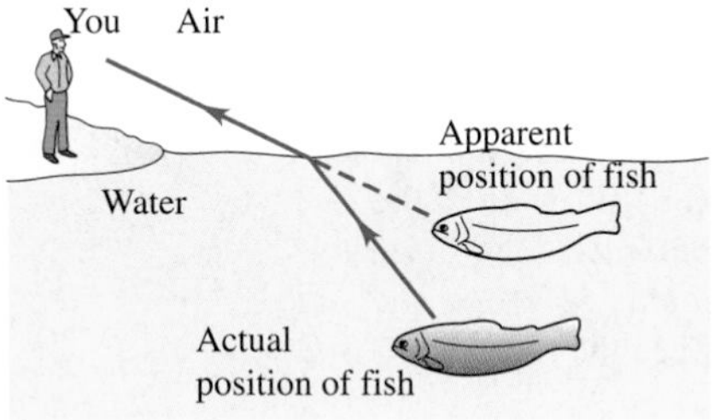

Section 33.3 The light strikes the end surface of the fiber at a small angle to the normal, as Fig. 33.14 shows. Hence total internal reflection, which requires a large angle of incidence (greater than the critical angle), cannot occur, and the light is able to escape the fiber.
Section 33.5 The sunlight reflected from the windows of the high-rise building is partially polarized. To minimize the reflections in the photograph, the photographer uses a Polaroid filter in front of the lens. The filter is oriented with its polarizing axis perpendicular to the dominant direction of polarization of the reflected light.
Section 33.7 If the time t required for a wave front to move from AA' to BB' is one period of the wave, then the distance between these wave front locations is one wavelength. Figure 33.33a shows that the spacing between successive wave fronts is less in material b, in which the index of refraction is greater and the wave speed is slower, than in material a. Hence the wavelength is less in material b. A more detailed analysis shows that the wavelength changes in accordance with Eq. (33.3).

Discussion Questions

Q33.1 Light requires about eight minutes to travel from the sun to the earth. Is it delayed appreciably by the earth's atmosphere? Explain.
Q33.2 Sunlight or starlight passing through the earth's atmosphere is always bent toward the vertical. Why? Does this mean that a star is not really where it appears to be? Explain.
Q33.3 The sun or moon usually appears flattened just before it sets. Is this related to refraction in the earth's atmosphere (mentioned in Question 33.2)? Explain.
Q33.4 A student claimed that, because of atmospheric refraction (see Question 33.2), the sun can be seen after it has set and that the day is therefore longer than it would be if the earth had no atmosphere. First, what does she mean by saying that the sun can be seen after it has set? Second, comment on the validity of her conclusion.
Q33.5 When hot air rises from a radiator or heating duct, objects behind it appear to shimmer or waver. What causes this?
Q33.6 Sometimes when looking at a window, you see two reflected images slightly displaced from each other. What causes this?
Q33.7 A ray of light in air strikes a glass surface. Is there a range of angles for which total reflection occurs? Explain.
Q33.8 When light is incident on an interface between two materials, the angle of the refracted ray depends on the wavelength, but the angle of the reflected ray does not. Why should this be?
Q33.9 A salesperson at a bargain counter claims that a certain pair of sunglasses has Polaroid filters; you suspect that the glasses are just tinted plastic. How could you find out for sure?
Q33.10 Does it make sense to talk about the polarization of a *longitudinal* wave, such as a sound wave? Why or why not?
Q33.11 How can you determine the direction of the polarizing axis of a single polarizer?
Q33.12 It has been proposed that automobile windshields and headlights should have polarizing filters to reduce the glare of oncoming lights during night driving. Would this work? How should the polarizing axes be arranged? What advantages would this scheme have? What disadvantages?
Q33.13 When a sheet of plastic food wrap is placed between two crossed polarizers, no light is transmitted. When the sheet is stretched in one direction, some light passes through the crossed polarizers. What is happening?
Q33.14 If you sit on the beach and look at the ocean through Polaroid sunglasses, the glasses help to reduce the glare from sunlight reflecting off the water. But if you lie on your side on the beach, there is little reduction in the glare. Explain why there is a difference.
Q33.15 When unpolarized light is incident on two crossed polarizers, no light is transmitted. A student asserted that if a third polarizer is inserted between the other two, some transmission will occur. Does this make sense? How can adding a third filter *increase* transmission?
Q33.16 In Fig. 33.29, since the light that is scattered out of the incident beam is polarized, why is the transmitted beam not also partially polarized?
Q33.17 You are sunbathing in the late afternoon when the sun is relatively low in the western sky. You are lying flat on your back, looking straight up through Polaroid sunglasses. To minimize the amount of sky light reaching your eyes, how should you lie: with your feet pointing north, east, south, west, or in some other direction? Explain your reasoning.

Q33.18 Light scattered from blue sky is strongly polarized because of the nature of the scattering process described in Section 33.6. But light scattered from white clouds is usually *not* polarized. Why not?

Q33.19 Atmospheric haze is due to water droplets or smoke particles ("smog"). Such haze reduces visibility by scattering light, so that the light from distant objects becomes randomized and images become indistinct. Explain why visibility through haze can be improved by wearing red-tinted sunglasses, which filter out blue light.

Q33.20 The explanation given in Section 33.6 for the color of the setting sun should apply equally well to the *rising* sun, since sunlight travels the same distance through the atmosphere to reach your eyes at either sunrise or sunset. Typically, however, sunsets are redder than sunrises. Why? (*Hint:* Particles of all kinds in the atmosphere contribute to scattering.)

Q33.21 Huygens' principle also applies to sound waves. During the day, the temperature of the atmosphere decreases with increasing altitude above the ground. But at night, when the ground cools, there is a layer of air just above the surface in which the temperature *increases* with altitude. Use this to explain why sound waves from distant sources can be heard more clearly at night than in the daytime. (*Hint:* The speed of sound increases with increasing temperature. Use the ideas displayed in Fig. 33.34 for light.)

Q33.22 Can water waves be reflected and refracted? Give examples. Does Huygens' principle apply to water waves? Explain.

Exercises

Section 33.2 Reflection and Refraction

33.1 A beam of light has a wavelength of 650 nm in vacuum. a) What is the speed of this light in a liquid whose index of refraction at this wavelength is 1.47? b) What is the wavelength of these waves in the liquid?

33.2 Light with a frequency of 5.80×10^{14} Hz travels in a block of glass that has an index of refraction of 1.52. What is the wavelength of the light a) in vacuum; b) in the glass?

33.3 A light beam travels at 1.94×10^8 m/s in quartz. The wavelength of the light in quartz is 355 nm. a) What is the index of refraction of quartz at this wavelength? b) If this same light travels through air, what is its wavelength there?

33.4 Light of a certain frequency has a wavelength of 438 nm in water. What is the wavelength of this light in benzene?

33.5 A parallel beam of light in air makes an angle of 47.5° with the surface of a glass plate having a refractive index of 1.66. a) What is the angle between the reflected part of the beam and the surface of the glass? b) What is the angle between the refracted beam and the surface of the glass?

33.6 Using a fast-pulsed laser and electronic timing circuitry, you find that light travels 2.50 m within a plastic rod in 11.5 ns. What is the refractive index of the plastic?

33.7 Light traveling in air is incident on the surface of a block of plastic at an angle of 62.7° to the normal and is bent so that it makes a 48.1° angle with the normal in the plastic. Find the speed of light in the plastic.

33.8 a) A tank containing methanol has walls 2.50 cm thick made of glass of refractive index 1.550. Light from the outside air strikes the glass at a 41.3° angle with the normal to the glass. Find the angle the light makes with the normal in the methanol. b) The tank is emptied and refilled with an unknown liquid. If light incident at the same angle as in part (a) enters the liquid in the tank at an angle of 20.2° from the normal, what is the refractive index of the unknown liquid?

33.9 a) Light passes through three parallel slabs of different thicknesses and refractive indices. The light is incident in the first slab and finally refracts into the third slab. Show that the middle slab has no effect on the final direction of the light. That is, show that the direction of the light in the third slab is the same as if the light had passed directly from the first slab into the third slab. b) Generalize this result to a stack of N slabs. What determines the final direction of the light in the last slab?

33.10 A horizontal, parallel-sided plate of glass having a refractive index of 1.52 is in contact with the surface of water in a tank. A ray coming from above in air makes an angle of incidence of 35.0° with the normal to the top surface of the glass. a) What angle does the ray refracted into the water make with the normal to the surface? b) What is the dependence of this angle on the refractive index of the glass?

33.11 A parallel beam of light is incident on a prism, as shown in Fig. 33.35. Part of the light is reflected from one face and part from another. Show that the angle θ between the two reflected beams is twice the angle A between the two reflecting surfaces.

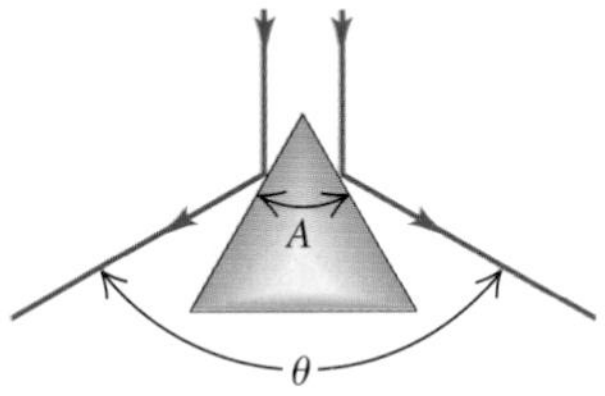

Figure 33.35 Exercise 33.11.

33.12 Prove that a ray of light reflected from a plane mirror rotates through an angle of 2θ when the mirror rotates through an angle θ about an axis perpendicular to the plane of incidence.

33.13 A ray of light is incident on a plane surface separating two sheets of glass with refractive indexes 1.70 and 1.58. The angle of incidence is 62.0°, and the ray originates in the glass with $n = 1.70$. Compute the angle of refraction.

33.14 In Example 33.1 (Section 33.2), the water-glass interface is horizontal. If instead this interface were tilted 15.0° above the horizontal, with the right side higher than the left side, what would be the angle from the vertical of the ray in the glass? (The ray in the water still makes an angle of 60.0° with the vertical.)

Section 33.3 Total Internal Reflection

33.15 The critical angle for total internal reflection at a liquid-air interface is 42.5°. a) If a ray of light traveling in the liquid has an angle of incidence at the interface of 35.0°, what angle does the refracted ray in the air make with the normal? b) If a ray of light traveling in air has an angle of incidence at the interface of 35.0°, what angle does the refracted ray in the liquid make with the normal?

33.16 At the very end of Wagner's series of operas *Ring of the Nibelung,* Brünnhilde takes the golden ring from the finger of the dead Siegfried and throws it into the Rhine, where it sinks to the bottom of the river. Assuming that the ring is small enough compared to the depth of the river to be treated as a point and that the Rhine is 10.0 m deep where the ring goes in, what is the area of the largest circle at the surface of the water over which light from the ring could escape from the water?

33.17 A ray of light is traveling in a glass cube that is totally immersed in water. You find that if the ray is incident on the glass-water interface at an angle to the normal greater than 48.7°, no light is refracted into the water. What is the refractive index of the glass?

33.18 Light is incident along the normal on face AB of a glass prism of refractive index 1.52, as shown in Fig. 33.36. Find the largest value the angle α can have without any light refracted out of the prism at face AC if: a) the prism is immersed in air; b) the prism is immersed in water.

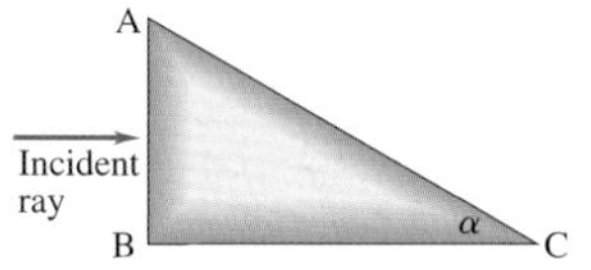

Figure 33.36 Exercise 33.18.

33.19 Define the index of refraction of a material for sound waves to be the ratio of the speed of sound in air to the speed of sound in the material. Snell's law then applies to the refraction of sound waves. The speed of a sound wave is 344 m/s in air and 1320 m/s in water. a) Which medium has the higher index of refraction for sound? b) What is the critical angle for a sound wave incident on the surface between air and water? c) For total internal reflection to occur, must the sound wave be traveling in the air or in the water?

33.20 A ray of light in diamond (index of refraction 2.42) is incident on an interface with air. What is the *largest* angle the ray can make with the normal and not be totally reflected back into the diamond?

Section 33.5 Polarization

33.21 A parallel beam of unpolarized light in air is incident at an angle of 54.5° (with respect to the normal) on a plane glass surface. The reflected beam is completely linearly polarized. a) What is the refractive index of the glass? b) What is the angle of refraction of the transmitted beam?

33.22 Light traveling in water strikes a glass plate at an angle of incidence of 53.0°; part of the beam is reflected and part is refracted. If the reflected and refracted portions make an angle of 90.0° with each other, what is the index of refraction of the glass?

33.23 Unpolarized light traveling in a liquid with refractive index n is incident on the surface of the liquid, above which there is air. If the light is incident on the surface at an angle of 31.2° with respect to the normal, the light reflected back into the liquid is completely polarized. a) What is the refractive index n of the liquid? b) What angle does the refracted light traveling in air make with the normal to the surface?

33.24 The refractive index of a certain glass is 1.66. For what incident angle is light reflected from the surface of this glass completely polarized if the glass is immersed in a) air? b) water?

33.25 Unpolarized light with intensity I_0 is incident on a polarizing filter. The emerging light strikes a second polarizing filter whose axis is at 41.0° to that of the first. Determine a) the intensity of the beam after it has passed through the second polarizer; b) its state of polarization.

33.26 A polarizer and an analyzer are oriented so that the maximum amount of light is transmitted. To what fraction of its maximum value is the intensity of the transmitted light reduced when the analyzer is rotated through a) 22.5°; b) 45.0°; c) 67.5°?

33.27 Unpolarized light of intensity 20.0 W/cm^2 is incident on two polarizing filters. The axis of the first filter is at an angle of 25.0° counterclockwise from the vertical (viewed in the direction the light is traveling) and the axis of the second filter is at 62.0° counterclockwise from the vertical. What is the intensity of the light after it has passed through the second polarizer?

33.28 Three polarizing filters are stacked, with the polarizing axis of the second and third filters at 23.0° and 62.0°, respectively, to that of the first. If unpolarized light is incident on the stack, the light has intensity 75.0 W/cm^2 after it passes through the stack. If the incident intensity is kept constant, what is the intensity of the light after it has passed through the stack if the second polarizer is removed?

33.29 Three Polarizing Filters. Three polarizing filters are stacked with the polarizing axes of the second and third at 45.0° and 90.0°, respectively, with that of the first. a) If unpolarized light of intensity I_0 is incident on the stack, find the intensity and state of polarization of light emerging from each filter. b) If the second filter is removed, what is the intensity of the light emerging from each remaining filter?

*Section 33.6 Scattering of Light

***33.30** A beam of light, after passing through the Polaroid disk P_1 shown in Fig. 33.37, traverses a cell containing a scattering medium. The cell is observed at right angles through another Polaroid disk P_2. Initially, the disks are oriented until the brightness of the scattered light from the cell as seen by the observer is a maximum. a) Disk P_2 is now rotated through 90°. Does the cell as viewed by the observer appear bright or dark? Explain. b) Disk P_1 is now rotated through 90°. Is the cell bright or dark? Explain. c) Disk P_2 is then restored to its original position. Is the cell bright or dark? Explain.

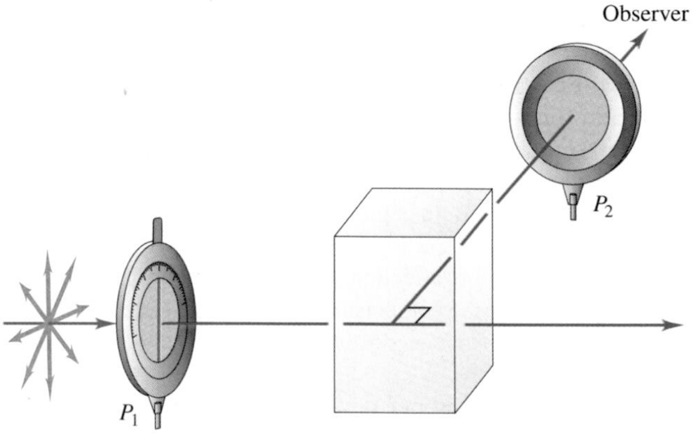

Figure 33.37 Exercise 33.30.

Problems

33.31 The Corner Reflector. An inside corner of a cube is lined with mirrors to make a corner reflector (Example 33.3 in Section 33.2). A ray of light is reflected successively from each of three mutually perpendicular mirrors; show that its final direction is always exactly opposite to its initial direction.

33.32 Sonogram of the Heart. Physicians use high-frequency ($f = 1$–5 MHz) sound waves, called ultrasound, to image internal organs. The speed of these ultrasound waves is 1480 m/s in muscle and 344 m/s in air (see Exercise 33.19). a) At what angle from the normal does an ultrasound beam enter the heart if it leaves the lungs at an angle of 9.73° from the normal to the heart wall? (Assume that the speed of sound in the lungs is 344 m/s.) b) What is the critical angle for sound waves in air incident on muscle?

33.33 Light traveling in a material of refractive index n_1 is incident at angle θ_1 with respect to the normal at the interface with a slab of material that has parallel faces and refractive index n_2. After the light passes through this material, it is refracted into a material with refractive index n_3 and in this third material it makes an angle of θ_3 with the normal. a) Find θ_3 in terms of θ_1 and the refractive indices of the materials. b) The ray in the third material is now reversed, so that it is incident on the n_3-to-n_2 interface with the angle θ_3 found in part (a). Show that when the light refracts into the material with refractive index n_1, the angle it makes with the normal is angle θ_1. This shows that the refracted ray is reversible. c) Are reflected rays also reversible? Explain.

33.34 In a physics lab, light with wavelength 490 nm travels in air from a laser to a photocell in 17.0 ns. When a slab of glass 0.840 m thick is placed in the light beam, with the beam incident along the normal to the parallel faces of the slab, it takes the light 21.2 ns to travel from the laser to the photocell. What is the wavelength of the light in the glass?

33.35 A ray of light is incident in air on a block of a transparent solid whose index of refraction is n. If $n = 1.38$, what is the *largest* angle of incidence θ_a for which total internal reflection will occur at the vertical face (point A shown in Fig. 33.38)?

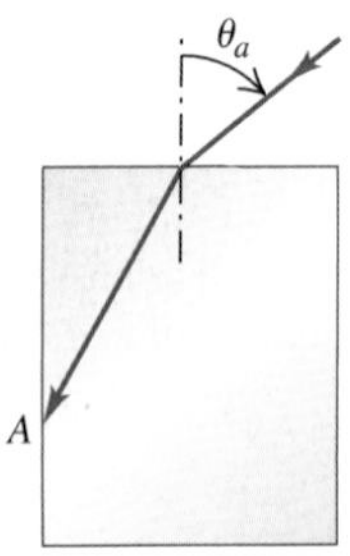

Figure 33.38 Problem 33.35.

33.36 A thin beam of light in air is incident on the surface of a lanthanum flint glass plate having a refractive index of 1.80. What is the angle of incidence θ_a with this plate for which the angle of refraction is $\theta_a/2$? Both angles are measured relative to the normal.

33.37 Use the law of reflection to show that specular reflection reverses the sign of the component of light velocity perpendicular to the reflecting surface, but leaves the other components unchanged.

33.38 A glass plate 2.50 mm thick, with an index of refraction of 1.40, is placed between a point source of light with wavelength 540 nm (in vacuum) and a screen. The distance from source to screen is 1.80 cm. How many wavelengths are there between the source and the screen?

33.39 Old photographic plates were made of glass with a light-sensitive emulsion on the front surface. This emulsion was somewhat transparent. When a bright point source is focused on the front of the plate, the developed photograph will show a halo around the image of the spot. If the glass plate is 3.10 mm thick and the halos have an inner radius of 5.34 mm, what is the index of refraction of the glass? (*Hint:* Light from the spot on the front surface is totally reflected at the back surface of the plate and comes back to the front surface.)

33.40 After a long day of driving you take a late-night swim in a motel swimming pool. When you go to your room, you realize that you have lost your room key in the pool. You borrow a powerful flashlight and walk around the pool, shining the light into it. The light shines on the key, which is lying on the bottom of the pool, when the flashlight is held 1.2 m above the water surface and is directed at the surface a horizontal distance of 1.5 m from the edge (see Fig. 33.39). If the water here is 4.0 m deep, how far is the key from the edge of the pool?

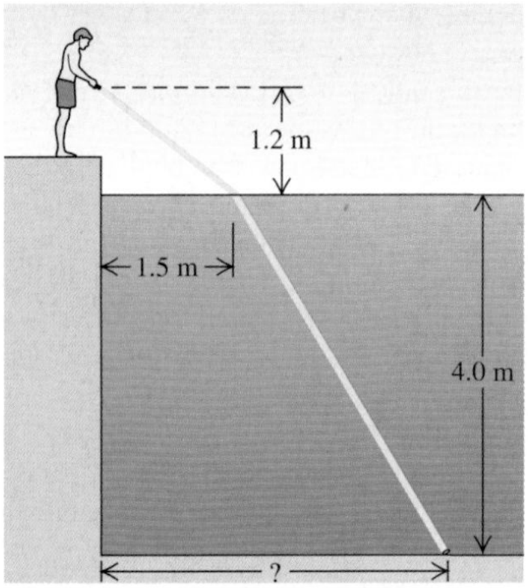

Figure 33.39 Problem 33.40.

33.41 You sight along the rim of a glass with vertical sides so that the top rim is lined up with the opposite edge of the bottom (see Fig. 33.40a). The glass is a thin-walled hollow cylinder 16.0 cm high with a top and bottom of the glass diameter of 8.0 cm. While you keep your eye in the same position, a friend fills the glass with a transparent liquid, and you then see a dime that is lying at the center of the bottom of the glass (see Fig. 33.40b). What is the index of refraction of the liquid?

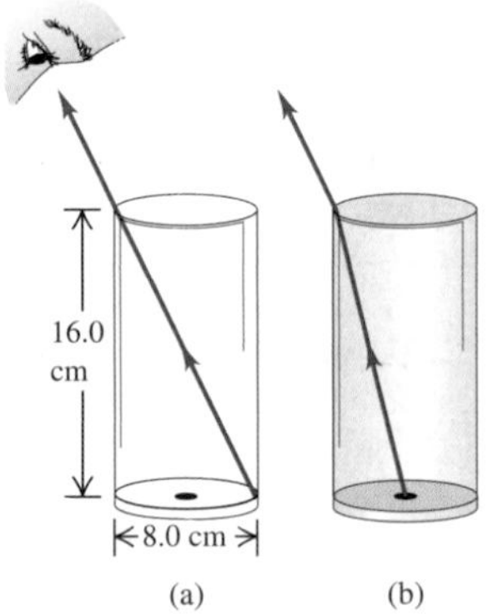

Figure 33.40 Problem 33.41.

33.42 A beaker with a mirrored bottom is filled with a liquid whose index of refraction is 1.63. A light beam strikes the top surface of the liquid at an angle of 42.5° from the normal. At what angle from the normal will the beam exit from the liquid after traveling down through the liquid, reflecting from the mirrored bottom, and returning to the surface?

33.43 A thin layer of ice $(n = 1.309)$ floats on the surface of water $(n = 1.333)$ in a bucket. A ray of light from the bottom of the bucket travels upward through the water. a) What is the largest angle with respect to the normal that the ray can make at the ice-water interface and still pass out into the air above the ice? b) What is this angle after the ice melts?

33.44 A 45°–45°–90° prism is immersed in water. A ray of light is incident normally on one of its shorter faces. What is the minimum index of refraction that the prism must have if this ray is to be totally reflected within the glass at the long face of the prism?

33.45 The prism shown in Fig. 33.41 has a refractive index of 1.66, and the angles A are 25.0°. Two light rays m and n are parallel as they enter the prism. What is the angle between them after they emerge?

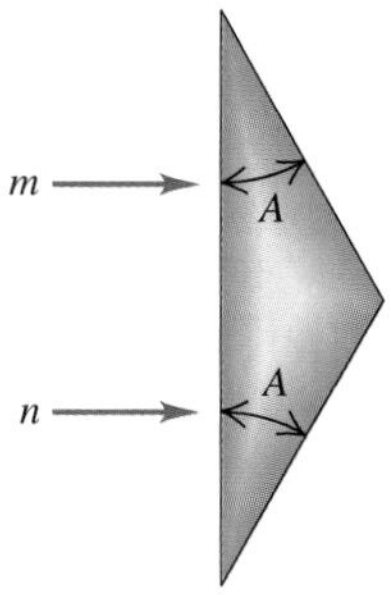

Figure 33.41 Problem 33.45.

33.46 Light is incident normally on the short face of a 30°–60°–90° prism (see Fig. 33.42). A drop of liquid is placed on the hypotenuse of the prism. If the index of the prism is 1.62, find the maximum index that the liquid may have if the light is to be totally reflected.

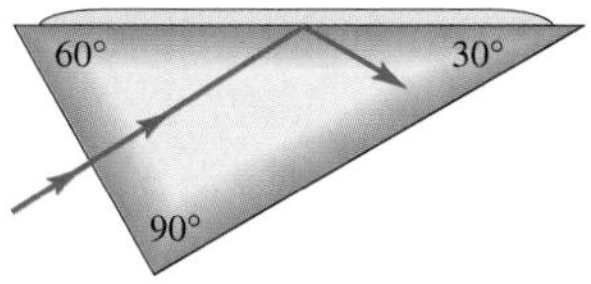

Figure 33.42 Problem 33.46.

33.47 A ray of light traveling in a block of glass $(n = 1.52)$ is incident on the top surface at an angle of 57.2° with respect to the normal. If a layer of oil is placed on the top surface of the glass, the ray is totally reflected. What is the maximum possible index of refraction of the oil?

33.48 Light is incident in air at an angle θ_a (see Fig. 33.43) on the upper surface of a transparent plate, the surfaces of the plate being plane and parallel to each other. a) Prove that $\theta_a = \theta_a'$. b) Show that this is true for any number of different parallel plates. c) Prove that the lateral displacement d of the emergent beam is given by the relation

$$d = t\frac{\sin(\theta_a - \theta_b')}{\cos\theta_b'}$$

where t is the thickness of the plate. d) A ray of light is incident at an angle of 66.0° on one surface of a glass plate 2.40 cm thick with an index of refraction 1.80. The medium on either side of the plate is air. Find the lateral displacement between the incident and emergent rays.

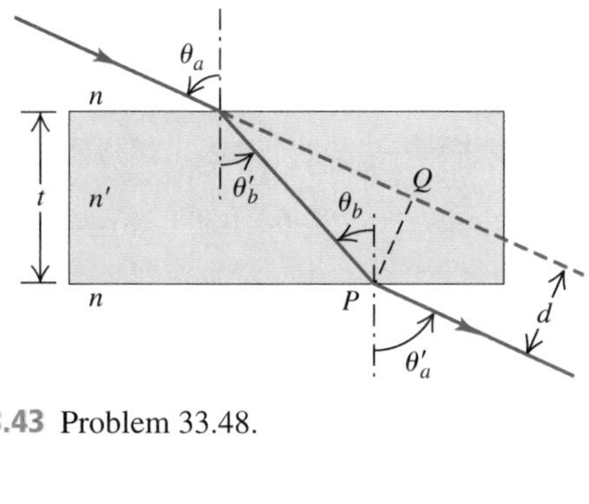

Figure 33.43 Problem 33.48.

33.49 When the sun is either rising or setting and appears to be just on the horizon, it is in fact *below* the horizon. The explanation for this seeming paradox is that light from the sun bends slightly when entering the earth's atmosphere, as shown in Fig. 33.44. Since our perception is based on the idea that light travels in straight lines, we perceive the light to be coming from an apparent position that is an angle δ above the sun's true position. a) Make the simplifying assumptions that the atmosphere has uniform density, and hence uniform index of refraction n, and extends to a height h above the earth's surface, at which point it abruptly stops. Show that the angle δ is given by

$$\delta = \arcsin\left(\frac{nR}{R + h}\right) - \arcsin\left(\frac{R}{R + h}\right)$$

where $R = 6378$ km is the radius of the earth. b) Calculate δ using $n = 1.0003$ and $h = 20$ km. How does this compare to the angular radius of the sun, which is about one quarter of a degree? (In actuality a light ray from the sun bends gradually, not abruptly, since the density and refractive index of the atmosphere change gradually with altitude.)

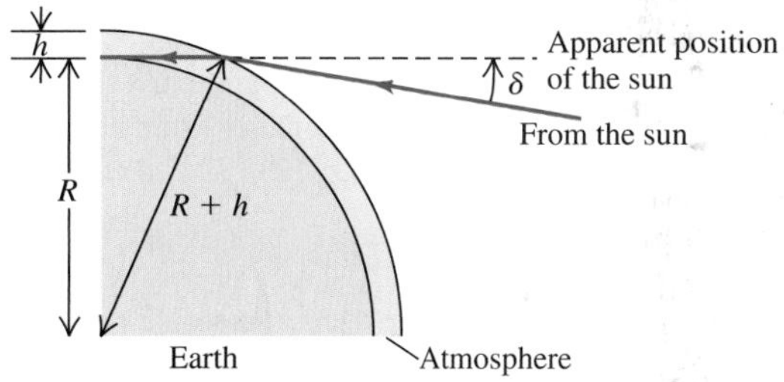

Figure 33.44 Problem 33.49.

33.50 A quarter-wave plate converts linearly polarized light to circularly polarized light. Prove that a quarter-wave plate also converts circularly polarized light to linearly polarized light.

33.51 Three polarizing filters are stacked, with the polarizing axes of the second and third at angles θ and 90°, respectively, with that of the first. Unpolarized light with intensity I_0 is incident on the stack. a) Derive an expression for the intensity of light transmitted through the stack as a function of I_0 and θ. b) For what value of θ does the maximum transmission occur?

33.52 Fermat's Principle of Least Time. A ray of light traveling with speed c leaves point 1 shown in Fig. 33.45 and is reflected to

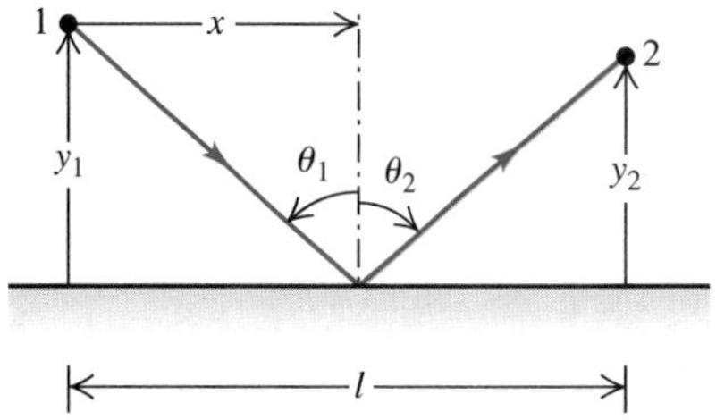

Figure 33.45 Problem 33.52.

point 2. The ray strikes the reflecting surface a horizontal distance x from point 1. a) Show that the time t required for the light to travel from 1 to 2 is

$$t = \frac{\sqrt{y_1^2 + x^2} + \sqrt{y_2^2 + (l - x)^2}}{c}$$

b) Take the derivative of t with respect to x. Set the derivative equal to zero to show that this time reaches its *minimum* value when $\theta_1 = \theta_2$, which is the law of reflection and corresponds to the actual path taken by the light. This is an example of Fermat's *principle of least time,* which states that among all possible paths between two points, the one actually taken by a ray of light is that for which the time of travel is a *minimum.* (In fact, there are some cases in which the time is a maximum rather than a minimum.)

33.53 A ray of light goes from point A in a medium in which the speed of light is v_1 to point B in a medium in which the speed is v_2 (Fig. 33.46). The ray strikes the interface a horizontal distance x to the right of point A. a) Show that the time required for the light to go from A to B is

$$t = \frac{\sqrt{h_1^2 + x^2}}{v_1} + \frac{\sqrt{h_2^2 + (l - x)^2}}{v_2}$$

b) Take the derivative of t with respect to x. Set this derivative equal to zero to show that this time reaches its *minimum* value when $n_1 \sin\theta_1 = n_2 \sin\theta_2$. This is Snell's law, and corresponds to the actual path taken by the light. This is another example of Fermat's principle of least time (see Problem 33.52).

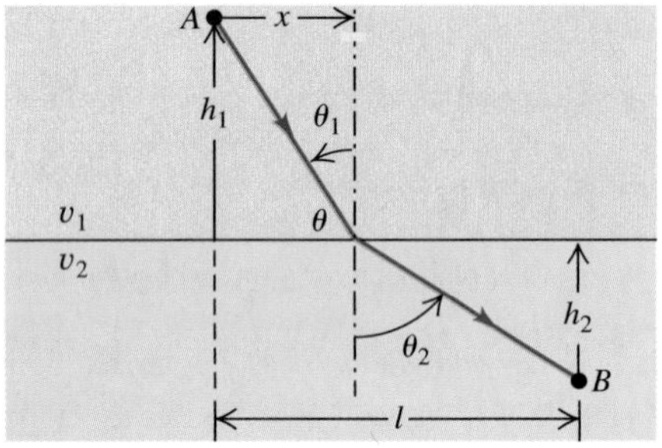

Figure 33.46 Problem 33.53.

33.54 A thin beam of white light is directed at a flat sheet of silicate flint glass at an angle of 20.0° to the surface of the sheet. Due to dispersion in the glass, the beam is spread out as shown in a spectrum in Fig. 33.47. The refractive index of silicate flint glass versus wavelength is graphed in Fig. 33.16. a) The rays a and b shown in Fig. 33.47 correspond to the extremes of the visible spectrum. Which corresponds to red and which to violet? Explain your reasoning. b) For what thickness d of the glass sheet will the spectrum be 1.0 mm wide, as shown (see Problem 33.48)?

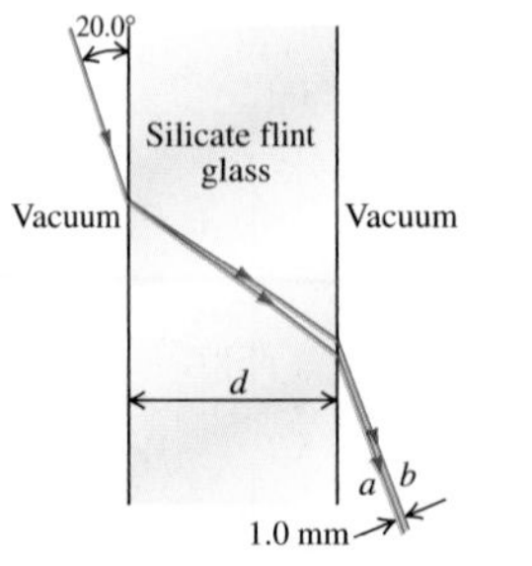

Figure 33.47 Problem 33.54.

33.55 Angle of Deviation. The incident angle θ_a shown in Fig. 33.48 is chosen so that the light passes symmetrically through the prism, that has refractive index n and apex angle A. a) Show that the angle of deviation δ (the angle between the initial and final directions of the ray) is given by

$$\sin\frac{A + \delta}{2} = n \sin\frac{A}{2}$$

(When the light passes through symmetrically, as shown, the angle of deviation is a minimum.) b) Use the result of part (a) to find the angle of deviation for a ray of light passing symmetrically through a prism having three equal angles ($A = 60.0°$) and $n = 1.52$. c) A certain glass has a refractive index of 1.61 for red light (700 nm) and 1.66 for violet light (400 nm). If both colors pass through symmetrically, as described in part (a), and if $A = 60.0°$, find the difference between the angles of deviation for the two colors.

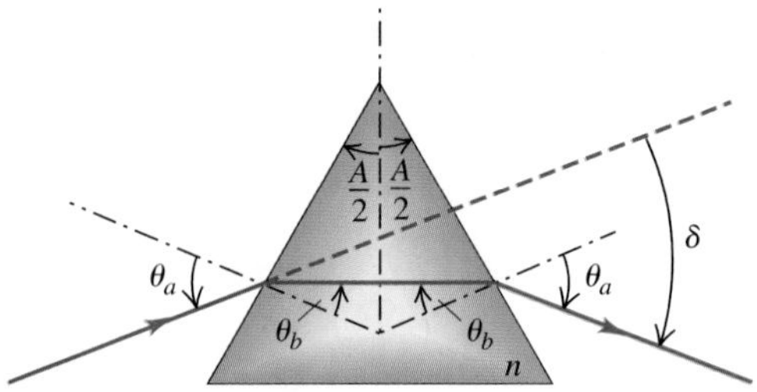

Figure 33.48 Problem 33.55.

33.56 Light traveling downward is incident on a horizontal film of thickness t, as shown in Fig. 33.49. The incident ray splits into two rays, A and B. Ray A reflects from the top of the film. Ray B reflects from the bottom of the film and then refracts back into the material that is above the film. If the film has parallel faces, show that rays A and B end up parallel to each other.

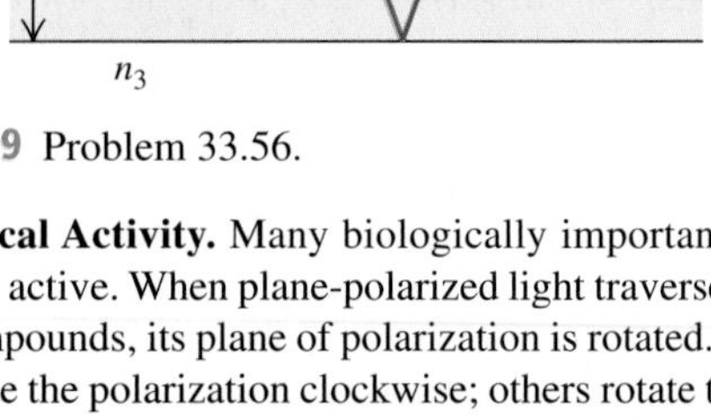

Figure 33.49 Problem 33.56.

33.57 Optical Activity. Many biologically important molecules are optically active. When plane-polarized light traverses a solution of these compounds, its plane of polarization is rotated. Some compounds rotate the polarization clockwise; others rotate the polarization counterclockwise. The amount of rotation depends on the amount of material in the light path. The following data give the amount of rotation through two amino acids over a path length of 100 cm. From this data, find the relationship between the concentration C (in grams per 100 mL) and the rotation of the polarization (in degrees) of each amino acid.

Rotation		Concentration
***l*-leucine**	***d*-glutamic acid**	**(g/100 mL)**
$-0.11°$	$0.124°$	1.0
$-0.22°$	$0.248°$	2.0
$-0.55°$	$0.620°$	5.0
$-1.10°$	$1.24°$	10.0
$-2.20°$	$2.48°$	20.0
$-5.50°$	$6.20°$	50.0
$-11.0°$	$12.4°$	100.0

33.58 A certain birefringent material has indexes of refraction n_1 and n_2 for the two perpendicular components of linearly polarized light passing through it. The corresponding wavelengths are $\lambda_1 = \lambda_0/n_1$ and λ_0/n_2, where λ_0 is the wavelength in vacuum. a) If the crystal is to function as a quarter-wave plate, the number of wavelengths of each component within the material must differ by $\frac{1}{4}$. Show that the minimum thickness for a quarter-wave plate is

$$d = \frac{\lambda_0}{4(n_1 - n_2)}$$

b) Find the minimum thickness of a quarter-wave plate made of siderite $(FeO \cdot CO_2)$ if the indexes of refraction are $n_1 = 1.875$ and $n_2 = 1.635$ and the wavelength in vacuum is $\lambda_0 = 589$ nm.

33.59 A beam of light traveling horizontally is made of an unpolarized component with intensity I_0 and a polarized component with intensity I_p. The plane of polarization of the polarized component is oriented at an angle of θ with respect to the vertical. The following data gives the intensity measured through a polarizer with an orientation of ϕ with respect to the vertical:

ϕ (°)	I_{total} (W/m²)	ϕ (°)	I_{total} (W/m²)
0	18.4	100	8.6
10	21.4	110	6.3
20	23.7	120	5.2
30	24.8	130	5.2
40	24.8	140	6.3
50	23.7	150	8.6
60	21.4	160	11.6
70	18.4	170	15.0
80	15.0	180	18.4
90	11.6		

a) What is the orientation of the polarized component? (That is, what is the angle θ?) b) What are the values of I_0 and I_p?

33.60 The polarization direction of linearly polarized light must be rotated by 90°. This is to be done by using N ideal polarizers. The intensity of the original linearly polarized light is I_0. a) Describe how the N filters should best be arranged to accomplish the desired rotation of the polarization direction, and show that the intensity of the light transmitted through the last filter is $I = I_0 \cos^{2N}(\pi/2N)$. b) If $N \gg 1$, use the information given in Appendix B to show that $I \cong I_0(1 - \pi^2/4N)$. Hence show that if N is very large, the final intensity is essentially the same as the initial intensity.

Challenge Problems

33.61 Consider two vibrations of equal amplitude and frequency but differing in phase, one along the x-axis,

$$x = a \sin(\omega t - \alpha)$$

and the other along the y-axis,

$$y = a \sin(\omega t - \beta)$$

These can be written as follows:

$$\frac{x}{a} = \sin \omega t \cos \alpha - \cos \omega t \sin \alpha \quad (1)$$

$$\frac{y}{a} = \sin \omega t \cos \beta - \cos \omega t \sin \beta \quad (2)$$

a) Multiply Eq. (1) by $\sin \beta$ and Eq. (2) by $\sin \alpha$, and then subtract the resulting equations. b) Multiply Eq. (1) by $\cos \beta$ and Eq. (2) by $\cos \alpha$, and then subtract the resulting equations. c) Square and add the results of parts (a) and (b). d) Derive the equation $x^2 + y^2 - 2xy \cos \delta = a^2 \sin^2 \delta$, where $\delta = \alpha - \beta$. e) Use the above result to justify each of the diagrams in Fig. 33.50. In the figure, the angle given is the phase difference between two simple harmonic motions of the same frequency and amplitude, one horizontal (along the x-axis) and the other vertical (along the y-axis). The figure thus shows the resultant motion from the superposition of the two perpendicular harmonic motions.

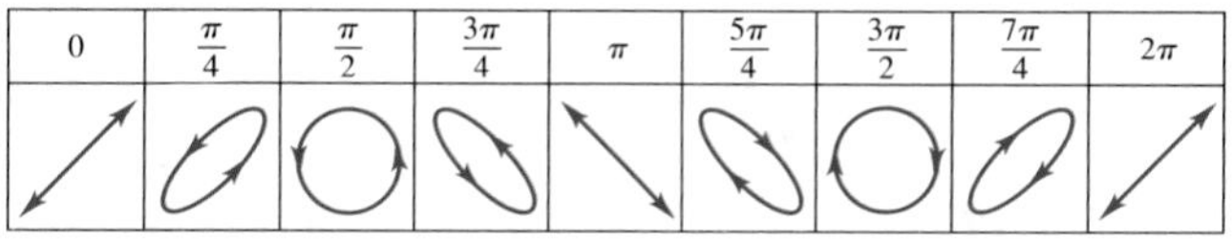

Figure 33.50 Challenge Problem 33.61.

33.62 A rainbow is produced by the reflection of sunlight by spherical drops of water in the air. Figure 33.51 shows a ray that refracts into a drop at point A, is reflected from the back surface of

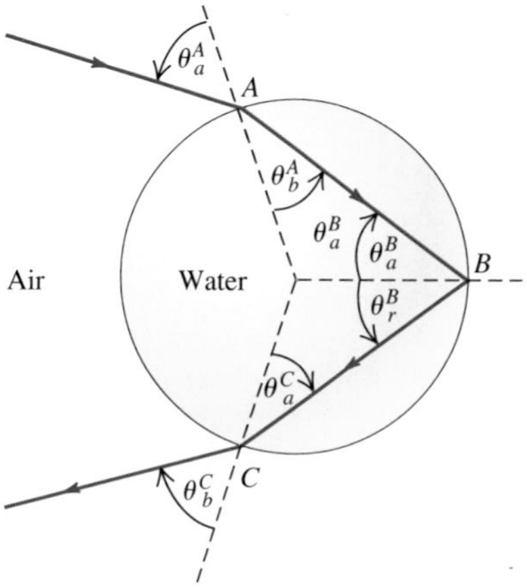

Figure 33.51 Exercise 33.62.

the drop at point B, and refracts back into the air at point C. The angles of incidence and refraction, θ_a and θ_b, are shown at points A and C and the angles of incidence and reflection, θ_a and θ_r, are shown at point B. a) Show that $\theta_a^B = \theta_b^A$, $\theta_a^C = \theta_b^A$, and $\theta_b^C = \theta_a^A$. b) Show that the angle in radians between the ray before it enters the drop at A and after it exits at C (the total angular deflection of the ray) is $\Delta = 2\theta_a^A - 4\theta_b^A + \pi$. (*Hint:* Find the angular deflections that occur at A, B, and C, and add to get Δ.) c) Use Snell's law to write Δ in terms of θ_a^A and n, the refractive index of the water in the drop. d) A rainbow will form when the angular deflection Δ is *stationary* in the incident angle θ_a^A, that is, when $d\Delta/d\theta_a^A = 0$. If this condition is satisfied, all the rays with incident angles close to θ_a^A will be sent back in the same direction, producing a bright zone in the sky. Let θ_1 be the value of θ_a^A for which this occurs. Show that $\cos^2\theta_1 = \frac{1}{3}(n^2 - 1)$. (*Hint:* You may find the derivative formula $d(\arcsin u(x))/dx = (1 - u^2)^{-1/2}(du/dx)$ helpful.) e) The index of refraction in water is 1.342 for violet light and 1.330 for red light. Use the results of parts (c) and (d) to find θ_1 and Δ for violet and red light. Do your results agree with the angles shown in Fig. 33.18b? When you view the rainbow, which color, red or violet, is higher above the horizon?

33.63 A *secondary rainbow* is formed when the incident light undergoes two internal reflections in a spherical drop of water as shown in Fig. 33.18c. (See Challenge Problem 33.62.) a) In terms of the incident angle θ_a^A and the refractive index n of the drop, what is the angular deflection Δ of the ray? That is, what is the angle between the ray before it enters the drop and after it exits? b) What is the incident angle θ_2 for which the derivative of Δ with respect to the incident angle θ_a^A is zero? c) The indexes of refraction for red and violet light in water are given in part (e) of Challenge Problem 33.62. Use the results of parts (a) and (b) to find θ_2 and Δ for violet and red light. Do your results agree with the angles shown in Fig. 33.18c? When you view a secondary rainbow, is red or violet highest above the horizon? Explain.

GEOMETRIC OPTICS

34

With a diameter of 8.2 m, this telescope mirror is one of the largest ever constructed for astronomy. Light from distant galaxies is collected by this immense "eye" and reflected from the polished concave surface, bringing the light to a focus. The mirror's surface is polished with a precision of just 8.5×10^{-9} m—equivalent to only a dozen or so atomic diameters—so the images are unusually sharp. This mirror is now in use at the European Southern Observatory in northern Chile.

? **Why do large telescopes use mirrors rather than lenses to collect light and form images?**

Your reflection in the bathroom mirror, the view of the moon through a telescope, the patterns seen in a kaleidoscope—all of these are examples of *images.* In each case the object that you're looking at appears to be in a different place than its actual position: your reflection is on the other side of the mirror, the moon appears to be much closer when seen through a telescope, and objects seen in a kaleidoscope seem to be in many places at the same time. In each case, light rays that come from a point on an object are deflected by reflection or refraction (or a combination of the two), so they converge toward or appear to diverge from a point called an *image point.* Our goal in this chapter is to see how this is done and to explore the different kinds of images that can be made with simple optical devices.

To understand images and image formation, all we need are the ray model of light, the laws of reflection and refraction, and some simple geometry and trigonometry. The key role played by geometry in our analysis is the reason for the name *geometric optics* that is given to the study of how light rays form images. We'll begin our analysis with one of the simplest of image-forming optical devices, a plane mirror. We'll go on to study how images are formed by curved mirrors, by refracting surfaces, and by thin lenses. Our results will lay the foundation for understanding many familiar optical instruments, including camera lenses, magnifiers, the human eye, microscopes, and telescopes.

34.1 | Reflection and Refraction at a Plane Surface

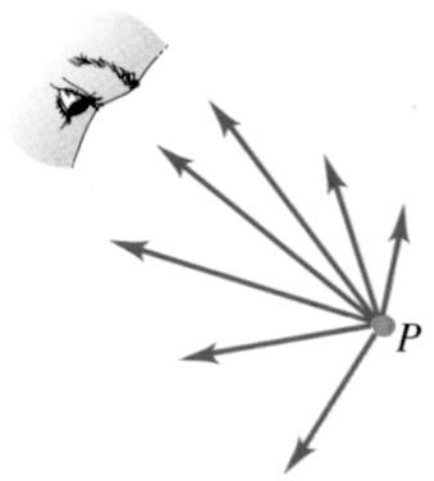

34.1 Light rays radiate from a point object P in all directions.

Before discussing what is meant by an image, we first need the concept of **object** as it is used in optics. By an *object* we mean anything from which light rays radiate. This light could be emitted by the object itself if it is *self-luminous,* like the glowing filament of a light bulb. Alternatively, the light could be emitted by another source (such as a lamp or the sun) and then reflected from the object; an example is the light you see coming from the pages of this book. Figure 34.1 shows light rays radiating in all directions from an object at a point P. For an observer to see this object directly, there must be no obstruction between the object and the observer's eyes. Note that light rays from the object reach the observer's left and right eyes at different angles; these differences are processed by the observer's brain to infer the *distance* from the observer to the object.

The object in Fig. 34.1 is a **point object** that has no physical extent. Real objects with length, width, and height are called **extended objects.** To start with, we'll consider only an idealized point object, since we can always think of an extended object as being made up of a very large number of point objects.

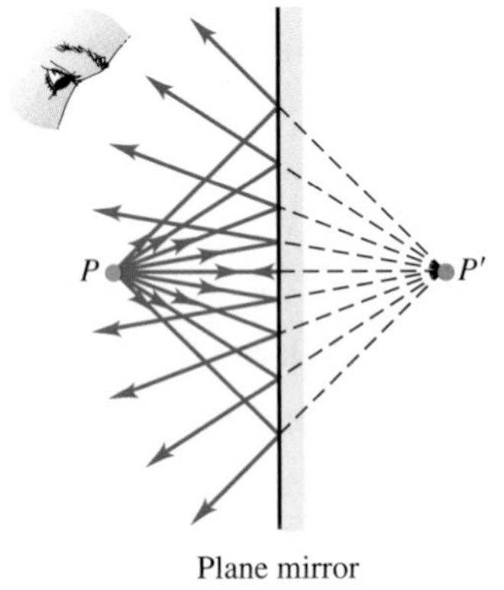

34.2 Light rays from the object at point P are reflected from a plane mirror. The reflected rays entering the eye look as though they had come from image point P'.

Suppose some of the rays from the object strike a smooth, plane reflecting surface (Fig. 34.2). This could be the surface of a material with a different index of refraction, which reflects part of the incident light, or a polished metal surface that reflects almost 100% of the light that strikes it. We will always draw the reflecting surface as a black line with a shaded area behind it, as in Fig. 34.2. Bathroom mirrors have a thin sheet of glass that lies in front of and protects the reflecting surface; we'll ignore the effects of this thin sheet (but see Problem 34.75).

According to the law of reflection, all rays striking the surface are reflected at an angle from the normal equal to the angle of incidence. Since the surface is plane, the normal is in the same direction at all points on the surface, and we have *specular* reflection. After the rays are reflected, their directions are the same as though they had come from point P'. We call point P an *object point* and point P' the corresponding *image point,* and we say that the reflecting surface forms an **image** of point P. An observer who can see only the rays reflected from the surface, and who doesn't know that he's seeing a reflection, *thinks* that the rays originate from the image point P'. The image point is therefore a convenient way to describe the directions of the various reflected rays, just as the object point P describes the directions of the rays arriving at the surface *before* reflection.

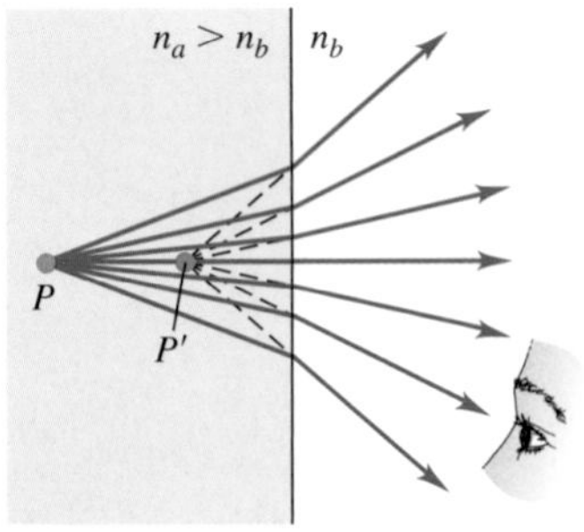

34.3 Light rays from the object at point P are refracted at the plane interface. The refracted rays entering the eye look as though they had come from image point P'. When $n_a > n_b$ as shown here, the image point P' is closer to the surface than the object point P. (The angles of incidence have been exaggerated for clarity.)

If the surface in Fig. 34.2 were *not* smooth, the reflection would be *diffuse,* and rays reflected from different parts of the surface would go in uncorrelated directions (see Fig. 33.6b). In this case there would not be a definite image point P' from which all reflected rays seem to emanate. You can't see your reflection in the surface of a tarnished piece of metal, because its surface is rough; polishing the metal smooths the surface so that specular reflection occurs and a reflected image becomes visible.

An image is also formed by a plane *refracting* surface, as shown in Fig. 34.3. Rays coming from point P are refracted at the interface between two optical materials. When the angles of incidence are small, the final directions of the rays after refraction are the same as though they had come from point P', as shown, and again we call P' an *image point.* In Section 33.2 we described how this effect makes underwater objects appear closer to the surface than they really are (see Fig. 33.8).

In both Figs. 34.2 and 34.3 the rays do not actually pass through the image point P'. Indeed, if the mirror in Fig. 34.2 is opaque, there is no light at all on its right side. If the outgoing rays don't actually pass through the image point, we call the image a **virtual image.** Later we will see cases in which the outgoing rays really *do* pass through an image point, and we will call the resulting image a **real image.** The images that are formed on a projection screen, on the photographic film in a camera, and on the retina of your eye are real images.

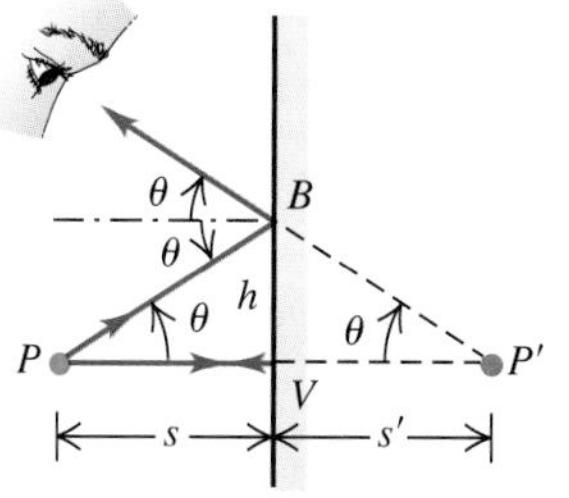

34.4 Construction for determining the location of the image formed by a plane mirror. The image point P' is as far behind the mirror as the object point P is in front of it. Since the light rays do not actually pass through P', it is a *virtual* image.

Image Formation by a Plane Mirror

Let's concentrate for now on images produced by *reflection;* we'll return to refraction later in the chapter. To find the precise location of the virtual image P' that a plane mirror forms of an object at P, we use the construction shown in Fig. 34.4. The figure shows two rays diverging from an object point P at a distance s to the left of a plane mirror. We call s the **object distance.** The ray PV is incident normally on the mirror (that is, it is perpendicular to the mirror surface), and it returns along its original path.

The ray PB makes an angle θ with PV. It strikes the mirror at an angle of incidence θ and is reflected at an equal angle with the normal. When we extend the two reflected rays backward, they intersect at point P', at a distance s' behind the mirror. We call s' the **image distance.** The line between P and P' is perpendicular to the mirror. The two triangles PVB and $P'VB$ are congruent, so P and P' are at equal distances from the mirror, and s and s' have equal magnitudes. The image point P' is located exactly opposite the object point P as far *behind* the mirror as the object point is from the front of the mirror.

ActivPhysics Online 15.4 Plane Mirrors

We can repeat the construction of Fig. 34.4 for each ray diverging from P. The directions of *all* the outgoing reflected rays are the same as though they had originated at point P', confirming that P' is the *image* of P. No matter where the observer is located, she will always see the image at the point P'.

Sign Rules

Before we go further, let's introduce some general sign rules. These may seem unnecessarily complicated for the simple case of an image formed by a plane mirror, but we want to state the rules in a form that will be applicable to *all* the situations we will encounter later. These will include image formation by a plane or spherical reflecting or refracting surface, or by a pair of refracting surfaces forming a lens. Here are the rules:

1. **Sign rule for the object distance:** When the object is on the same side of the reflecting or refracting surface as the incoming light, the object distance s is positive; otherwise, it is negative.
2. **Sign rule for the image distance:** When the image is on the same side of the reflecting or refracting surface as the outgoing light, the image distance s' is positive; otherwise, it is negative.
3. **Sign rule for the radius of curvature of a spherical surface:** When the center of curvature C is on the same side as the outgoing light, the radius of curvature is positive; otherwise, it is negative.

For a mirror the incoming and outgoing sides are always the same; for example, in Figs. 34.2 and 34.4 they are both the left side. For the refracting surface in Fig. 34.3 the incoming and outgoing sides are on the left and right sides, respectively, of the interface between the two materials. (Note that other textbooks may use different rules.)

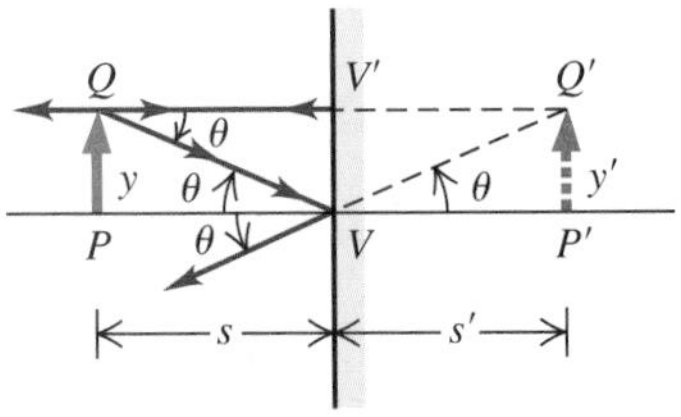

34.5 Construction for determining the height of an image formed by reflection at a plane reflecting surface. The image at P' (dashed arrow) is the same height as the object at P (solid arrow).

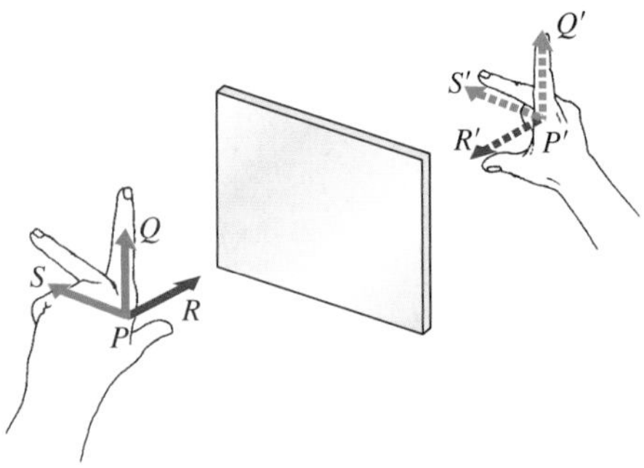

34.6 The image $S'P'Q'R'$ formed by a plane mirror is virtual, erect, and reversed. It is the same size as the object $SPQR$.

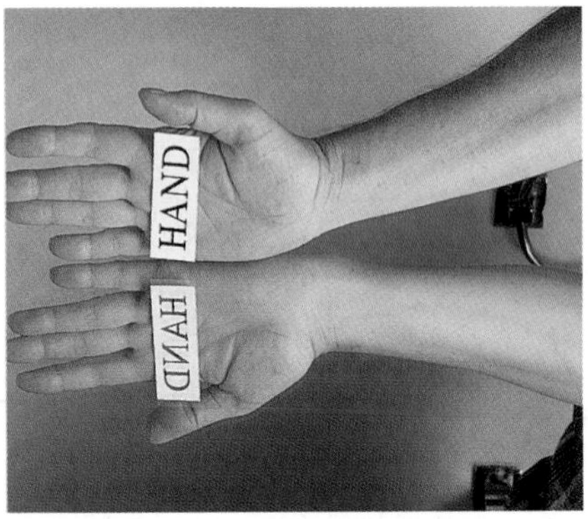

34.7 The image formed by a plane mirror is reversed; the image of a right hand is a left hand, and so on. (The hand is resting on a horizontal mirror.) Are images of the letters H and A reversed?

In Fig. 34.4 the object distance s is *positive* because the object point P is on the incoming side (the left side) of the reflecting surface. The image distance s' is *negative* because the image point P' is *not* on the outgoing side (the left side) of the surface. The object and image distances s and s' are related simply by

$$s = -s' \qquad \text{(plane mirror)} \tag{34.1}$$

For a plane reflecting or refracting surface, the radius of curvature is infinite and not a particularly interesting or useful quantity; in these cases we really don't need the third sign rule. But this rule will be of great importance when we study image formation by *curved* reflecting and refracting surfaces later in the chapter.

Image of an Extended Object—Plane Mirror

Next we consider an *extended* object with finite size. For simplicity we often consider an object that has only one dimension, like a slender arrow, oriented parallel to the reflecting surface; an example is the arrow PQ in Fig. 34.5. The distance from the head to the tail of an arrow oriented in this way is called its *height;* in Fig. 34.5 the height is y. The image formed by such an extended object is an extended image; to each point on the object, there corresponds a point on the image. Two of the rays from Q are shown; *all* the rays from Q appear to diverge from its image point Q' after reflection. The image of the arrow is the line $P'Q'$, with height y'. Other points of the object PQ have image points between P' and Q'. The triangles PQV and $P'Q'V$ are congruent, so the object PQ and image $P'Q'$ have the same size and orientation, and $y = y'$.

The ratio of image height to object height, y'/y, in *any* image-forming situation is called the **lateral magnification** m; that is,

$$m = \frac{y'}{y} \qquad \text{(lateral magnification)} \tag{34.2}$$

Thus for a plane mirror the lateral magnification m is unity. When you look at yourself in a plane mirror, your image is the same size as the real you.

In Fig. 34.5 the image arrow points in the *same* direction as the object arrow; we say that the image is **erect.** In this case, y and y' have the same sign, and the lateral magnification m is positive. The image formed by a plane mirror is always erect, so y and y' have both the same magnitude and the same sign; from Eq. (34.2) the lateral magnification of a plane mirror is always $m = +1$. Later we will encounter situations in which the image is **inverted**—that is, the image arrow points in the direction *opposite* to that of the object arrow. For an inverted image, y and y' have *opposite* signs, and the lateral magnification m is *negative.*

The object in Fig. 34.5 has only one dimension. Figure 34.6 shows a *three-dimensional* object and its three-dimensional virtual image formed by a plane mirror. The object and image are related in the same way as a left hand and a right hand.

CAUTION At this point, you may be asking, "Why does a plane mirror reverse images left and right but not top and bottom?" This question is quite misleading! As Fig. 34.6 shows, the up-down image $P'Q'$ and the left-right image $P'S'$ are parallel to their objects and are not reversed at all. Only the front-back image $P'R'$ is reversed relative to PR. Hence it's most correct to say that a plane mirror reverses *back to front.* To verify this object–image relationship, point your thumbs along PR and $P'R'$, your forefingers along PQ and $P'Q'$, and your middle fingers along *PS* and *P′S′*. When an object and its image are related in this way, the image is said to be **reversed;** this means that *only* the front-back dimension is reversed.

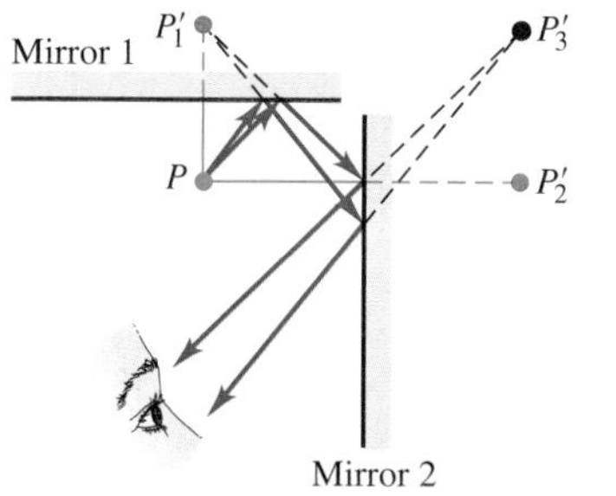

34.8 Images P_1' and P_2' are formed by a single reflection of each ray from the object at *P*. Image P_3', located by treating either of the other images as an object, is formed by a double reflection of each ray.

The reversed image of a three-dimensional object formed by a plane mirror is the same *size* as the object in all its dimensions. When the transverse dimensions of object and image are in the same direction, the image is erect. Thus a plane mirror always forms an erect but reversed image. Figure 34.7 illustrates this point.

An important property of all images formed by reflecting or refracting surfaces is that an *image* formed by one surface or optical device can serve as the *object* for a second surface or device. Figure 34.8 shows a simple example. Mirror 1 forms an image P_1' of the object point *P*, and mirror 2 forms another image P_2', each in the way we have just discussed. But in addition, the image P_1' formed by mirror 1 serves an object for mirror 2, which then forms an image of this object at point P_3' as shown. Similarly, mirror 1 uses the image P_2' formed by mirror 2 as an object and forms an image of it. We leave it to you to show that this image point is also at P_3'. The idea that an image formed by one device can act as the object for a second device is of great importance in geometric optics. We will use it later in this chapter to locate the image formed by two successive curved-surface refractions in a lens. This idea will help us to understand image formation by combinations of lenses, as in a microscope or a refracting telescope.

Test Your Understanding

If you walk directly toward a plane mirror at a speed v, at what speed does your image approach you?

34.2 | Reflection at a Spherical Surface

A plane mirror produces an image that is the same size as the object. But there are many applications for mirrors in which the image and object must be of different sizes. A magnifying mirror used when applying makeup gives an image that is *larger* than the object, and surveillance mirrors (used in stores to help spot shoplifters) give an image that is *smaller* than the object. There are also applications of mirrors in which a *real* image is desired, so light rays do indeed pass through the image point *P′*. A plane mirror by itself cannot perform any of these tasks. Instead, *curved* mirrors are used. (See the photograph that opens this chapter.)

We'll consider the special (and easily analyzed) case of image formation by a *spherical* mirror. Figure 34.9a shows a spherical mirror with radius of curvature *R*, with its concave side facing the incident light. The **center of curvature** of the surface (the center of the sphere of which the surface is a part) is at *C*, and the **vertex** of the mirror (the center of the mirror surface) is at *V*. The line *CV* is called the **optic axis.** Point *P* is an object point that lies on the optic axis; for the moment, we assume that the distance from *P* to *V* is greater than *R*.

Ray *PV*, passing through *C*, strikes the mirror normally and is reflected back on itself. Ray *PB*, at an angle α with the axis, strikes the mirror at *B*, where the angles of incidence and reflection are θ. The reflected ray intersects the axis at point *P′*. We will show shortly that *all* rays from *P* intersect the axis at the *same*

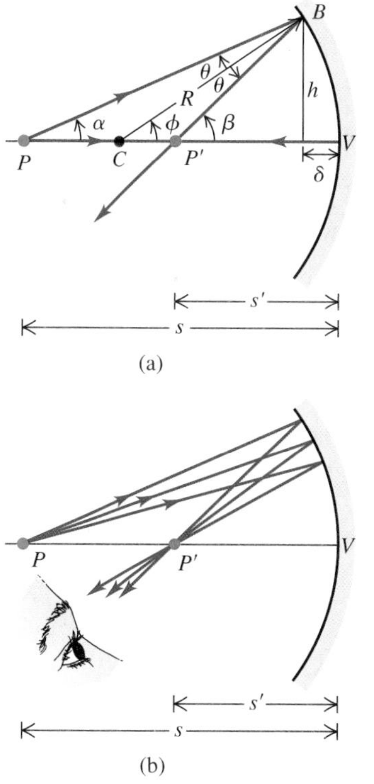

34.9 (a) Construction for finding the position of the image *P′* formed by a concave spherical mirror of a point object *P* on the mirror's optic axis. (b) If the angle α is small, *all* rays from *P* intersect at *P′*. The eye sees some of the outgoing rays and perceives them as having come from *P′*.

point P', as in Fig. 34.9b, provided that the angle α is small. Point P' is therefore the *image* of object point P. Unlike the reflected rays in Fig. 34.1, the reflected rays in Fig. 34.9b actually do intersect at point P', then diverge from P' *as if* they had originated at this point. Thus P' is a *real* image.

To see the usefulness of having a real image, suppose that the mirror is in a darkened room in which the only source of light is a self-luminous object at P. If you place a small piece of photographic film at P', all the rays of light coming from point P that reflect off the mirror will strike the same point P' on the film; when developed, the film will show a single bright spot, representing a sharply focused image of the object at point P. This principle is at the heart of most astronomical telescopes, which use large concave mirrors to make photographs of celestial objects. With a *plane* mirror like that in Fig. 34.2, placing a piece of film at the image point P' would be a waste of time; the light rays never actually pass through the image point, and the image can't be recorded on film. Real images are *essential* for photography.

Let's now find the location of the real image point P' in Fig. 34.9a and prove the assertion that all rays from P intersect at P' (provided that their angle with the optic axis is small). The object distance, measured from the vertex V, is s; the image distance, also measured from V, is s'. The signs of s, s', and the radius of curvature R are determined by the sign rules given in Section 34.1. The object point P is on the same side as the incident light, so according to the first sign rule, s is positive. The image point P' is on the same side as the reflected light, so according to the second sign rule, the image distance s' is also positive. The center of curvature C is on the same side as the reflected light, so according to the third sign rule, R, too, is positive; R is always positive when reflection occurs at the *concave* side of a surface.

We now use the following theorem from plane geometry: An exterior angle of a triangle equals the sum of the two opposite interior angles. Applying this theorem to triangles PBC and $P'BC$ in Fig. 34.9a, we have

$$\phi = \alpha + \theta \qquad \beta = \phi + \theta$$

Eliminating θ between these equations gives

$$\alpha + \beta = 2\phi \tag{34.3}$$

We may now compute the image distance s'. Let h represent the height of point B above the optic axis, and let δ represent the short distance from V to the foot of this vertical line. We now write expressions for the tangents of α, β, and ϕ, remembering that s, s', and R are all positive quantities:

$$\tan\alpha = \frac{h}{s - \delta} \qquad \tan\beta = \frac{h}{s' - \delta} \qquad \tan\phi = \frac{h}{R - \delta}$$

These trigonometric equations cannot be solved as simply as the corresponding algebraic equations for a plane mirror. However, *if the angle α is small,* the angles β and ϕ are also small. The tangent of an angle that is much less than one radian is nearly equal to the angle itself (measured in radians), so we can replace $\tan\alpha$ by α, and so on, in the equations above. Also, if α is small, we can neglect the distance δ compared with s', s, and R. So for small angles we have the following approximate relations:

$$\alpha = \frac{h}{s} \qquad \beta = \frac{h}{s'} \qquad \phi = \frac{h}{R}$$

Substituting these into Eq. (34.3) and dividing by h, we obtain a general relation among s, s', and R:

$$\frac{1}{s} + \frac{1}{s'} = \frac{2}{R} \quad \text{(object-image relation, spherical mirror)} \qquad (34.4)$$

This equation does not contain the angle α. Hence *all* rays from P that make sufficiently small angles with the axis intersect at P' after they are reflected; this verifies our earlier assertion. Such rays, nearly parallel to the axis and close to it, are called **paraxial rays.** (The term **paraxial approximation** is often used for the approximations we have just described.) Since all such reflected light rays converge on the image point, a concave mirror is also called a *converging mirror.*

Be sure you understand that Eq. (34.4), as well as many similar relations that we will derive later in this chapter and the next, is only *approximately* correct. It results from a calculation containing approximations, and it is valid only for paraxial rays. If we increase the angle α that a ray makes with the optic axis, the point P' where the ray intersects the optic axis moves somewhat closer to the vertex than for a paraxial ray. As a result, a spherical mirror, unlike a plane mirror, does not form a precise point image of a point object; the image is "smeared out." This property of a spherical mirror is called **spherical aberration.** The initially disappointing results from the Hubble Space Telescope when it was first placed in orbit in 1990 resulted in part from errors in the corrections for spherical aberration in its primary mirror (Fig. 34.10a). The performance of the telescope improved dramatically after the installation of corrective optics in 1993 (Fig. 34.10b).

If the radius of curvature becomes infinite ($R = \infty$), the mirror becomes *plane,* and Eq. (34.4) reduces to Eq. (34.1) for a plane reflecting surface.

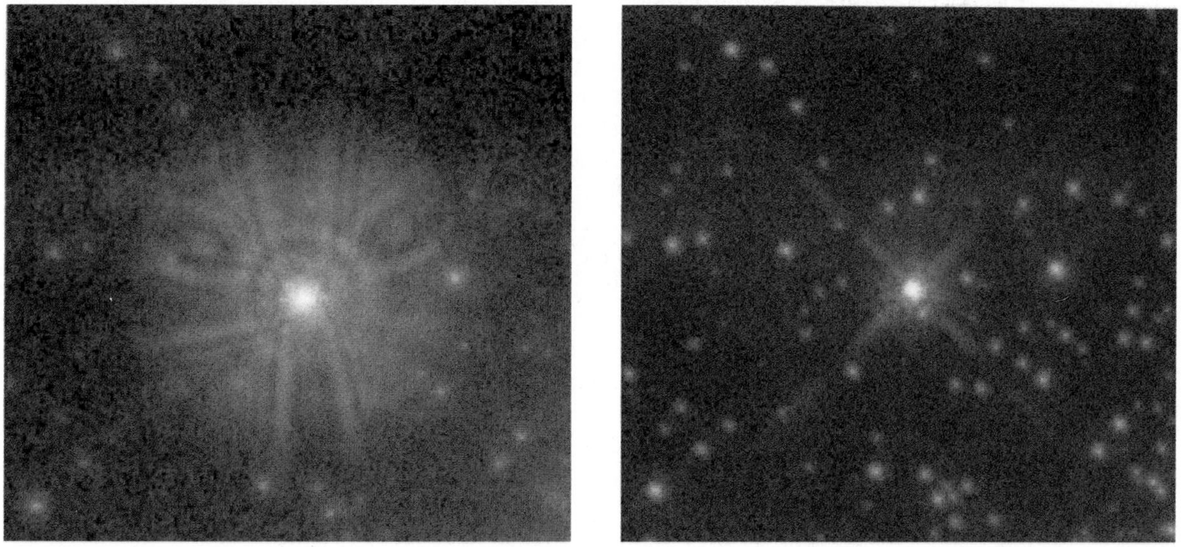

(a) (b)

34.10 (a) An image of a star made with the original Hubble Space Telescope (HST) after it was placed in orbit in 1990. The 2.4-m-diameter concave mirror was too shallow by about 1/50 the width of a human hair, leading to spherical aberration of the star's image. (b) After corrective optics were installed in 1993, the effects of spherical aberration were almost completely eliminated.

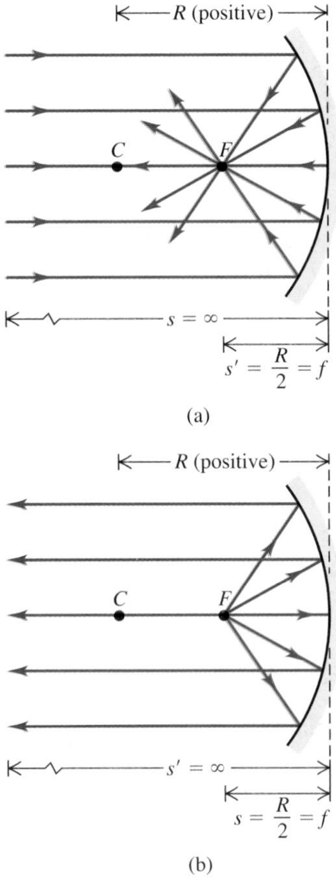

34.11 (a) Incident rays parallel to the axis of a concave mirror converge to the focal point F. (b) Rays diverging from the focal point F of a concave mirror are parallel to the axis after reflection. The angles are exaggerated for clarity.

Focal Point and Focal Length

When the object point P is very far from the spherical mirror ($s = \infty$), the incoming rays are parallel. (The star shown in Fig. 34.10 is an example of such a distant object.) From Eq. (34.4) the image distance s' in this case is given by

$$\frac{1}{\infty} + \frac{1}{s'} = \frac{2}{R} \qquad s' = \frac{R}{2}$$

The situation is shown in Fig. 34.11a. The beam of incident parallel rays converges, after reflection from the mirror, to a point F at a distance $R/2$ from the vertex of the mirror. The point F at which the incident parallel rays converge is called the **focal point;** we say that these rays are brought to a focus. The distance from the vertex to the focal point, denoted by f, is called the **focal length.** We see that f is related to the radius of curvature R by

$$f = \frac{R}{2} \qquad \text{(focal length of a spherical mirror)} \qquad (34.5)$$

The opposite situation is shown in Fig. 34.11b. Now the *object* is placed at the focal point F, so the object distance is $s = f = R/2$. The image distance s' is again given by Eq. (34.4):

$$\frac{2}{R} + \frac{1}{s'} = \frac{2}{R} \qquad \frac{1}{s'} = 0 \qquad s' = \infty$$

With the object at the focal point, the reflected rays in Fig. 34.11b are parallel to the optic axis; they meet only at a point infinitely far from the mirror, so the image is at infinity.

Thus the focal point F of a spherical mirror has the properties that (1) any incoming ray parallel to the optic axis is reflected through the focal point and (2) any incoming ray that passes through the focal point is reflected parallel to the optic axis. For spherical mirrors these statements are true only for paraxial rays. For parabolic mirrors these statements are *exactly* true; this is why parabolic mirrors are preferred for astronomical telescopes. Spherical or parabolic mirrors are used in flashlights and headlights to form the light from the bulb into a parallel beam. Some solar-power plants use an array of plane mirrors to simulate an approximately spherical concave mirror; light from the sun is collected by the mirrors and directed to the focal point, where a steam boiler is placed. (The concepts of focal point and focal length also apply to lenses, as we'll see in Section 34.4.)

We will usually express the relationship between object and image distances for a mirror, Eq. (34.4), in terms of the focal length f:

$$\frac{1}{s} + \frac{1}{s'} = \frac{1}{f} \qquad \text{(object-image relation, spherical mirror)} \qquad (34.6)$$

Image of an Extended Object–Spherical Mirror

Now suppose we have an object with *finite* size, represented by the arrow PQ in Fig. 34.12, perpendicular to the optic axis CV. The image of P formed by paraxial rays is at P'. The object distance for point Q is very nearly equal to that for point P, so the image $P'Q'$ is nearly straight and perpendicular to the axis. Note that the object and image arrows have different sizes, y and y', respectively, and

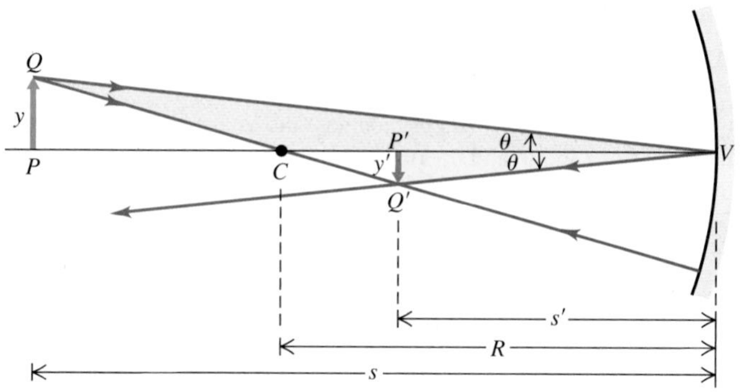

34.12 Construction for determining the position, orientation, and height of an image formed by a concave spherical mirror.

that they have opposite orientation. In Eq. (34.2) we defined the *lateral magnification m* as the ratio of image size y' to object size y:

$$m = \frac{y'}{y}$$

Because triangles PVQ and $P'VQ'$ in Fig. 34.12 are *similar,* we also have the relation $y/s = -y'/s'$. The negative sign is needed because object and image are on opposite sides of the optic axis; if y is positive, y' is negative. Therefore

$$m = \frac{y'}{y} = -\frac{s'}{s} \quad \text{(lateral magnification, spherical mirror)} \quad (34.7)$$

If m is positive, the image is erect in comparison to the object; if m is negative, the image is *inverted* relative to the object, as in Fig. 34.12. For a *plane* mirror, $s = -s'$, so $y' = y$ and $m = +1$; since m is positive, the image is erect, and since $|m| = 1$, the image is the same size as the object.

CAUTION Although the ratio of image size to object size is called the *magnification,* the image formed by a mirror or lens may be larger than, smaller than, or the same size as the object. If it is smaller, then the magnification is less than unity in absolute value: $|m| < 1$. The image formed by an astronomical telescope mirror or a camera lens is usually *much* smaller than the object. For example, the image of the bright star shown in Fig. 34.10 is just a few millimeters across, while the star itself is hundreds of thousands of kilometers in diameter.

Activ Online Physics

15.5 Spherical Mirrors: Ray Diagrams
15.6 Spherical Mirrors: The Mirror Equation
15.7 Spherical Mirrors: Linear Magnification
15.8 Spherical Mirrors: Problems

In our discussion of concave mirrors we have so far considered only objects that lie *outside* or at the focal point, so that the object distance s is greater than or equal to the (positive) focal length f. In this case the image point is on the same side of the mirror as the outgoing rays, and the image is real and inverted. If an object is placed *inside* the focal point of a concave mirror, so that $s < f$, the resulting image is *virtual* (that is, the image point is on the opposite side of the mirror from the object), *erect,* and *larger* than the object. Mirrors used when applying makeup (referred to at the beginning of this section) are concave mirrors; in use, the distance from the face to the mirror is less than the focal length, and an enlarged, erect image is seen. You can prove these statements about concave mirrors by applying Eqs. (34.6) and (34.7) (see Exercise 34.9). We'll also be able to verify these results later in this section, after we've learned some graphical methods for relating the positions and sizes of the object and the image.

Example 34.1 Image formation by a concave mirror I

A concave mirror forms an image, on a wall 3.00 m from the mirror, of the filament of a headlight lamp 10.0 cm in front of the mirror (Fig. 34.13). a) What are the radius of curvature and focal length of the mirror? b) What is the height of the image if the height of the object is 5.00 mm?

SOLUTION

IDENTIFY and SET UP: We are given the distances from the mirror to the object (s) and from the mirror to the image (s'). We use the object–image relation given by Eq. (34.6) to determine the focal length f, then find the radius of curvature R using Eq. (34.5). Equation (34.7) lets us calculate the image height y' from the distances s and s' and the object height y.

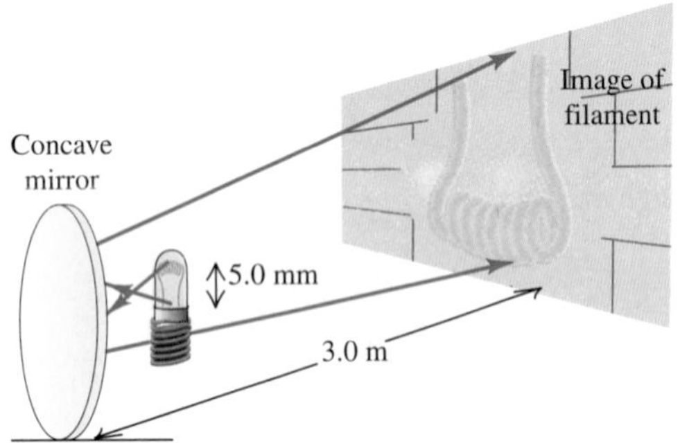

34.13 The concave mirror forms a real, enlarged, inverted image of the lamp filament.

EXECUTE: a) Both the object and the image are on the concave side of the mirror (the reflective side), so both object distance and image distance are positive; we have $s = 10.0$ cm and $s' = 300$ cm. From Eq. (34.4),

$$\frac{1}{10.0 \text{ cm}} + \frac{1}{300 \text{ cm}} = \frac{2}{R}$$

$$R = \frac{2}{0.100 \text{ cm}^{-1} + 3.33 \times 10^{-3} \text{ cm}^{-1}} = 19.4 \text{ cm}$$

The focal length of the mirror is $f = R/2 = 9.7$ cm.

b) From Eq. (34.7) the lateral magnification is

$$m = \frac{y'}{y} = -\frac{s'}{s} = -\frac{300 \text{ cm}}{10.0 \text{ cm}} = -30.0$$

Because m is negative, the image is inverted. The height of the image is 30.0 times the height of the object, or (30.0)(5.00 mm) = 150 mm.

EVALUATE: Note that the object is placed just outside the focal point ($s = 10.0$ cm compared to $f = 9.7$ cm). This is very similar to what is done in automobile headlights. By placing the filament close to the focal point, the concave mirror produces a beam of nearly parallel rays.

Conceptual Example 34.2 Image formation by a concave mirror II

In Example 34.1, suppose that the left half of the mirror's reflecting surface is covered with nonreflective soot. What effect will this have on the image of the filament?

SOLUTION

It would be natural to guess that the image would now show only half of the filament. But in fact the image will still show the *entire* filament. The explanation can be seen by examining Fig. 34.9b. Light rays coming from any object point P are reflected from *all* parts of the mirror and converge on the corresponding image point P'. If part of the mirror surface is made nonreflective or is removed altogether, the light rays from the remaining reflective surface still form an image of every part of the object.

The only effect of reducing the reflecting area is that the image becomes dimmer because less light energy reaches the image point. In our example the reflective area of the mirror is reduced by one half, and the image will be one half as bright. *Increasing* the reflective area makes the image brighter. To make reasonably bright images of distant stars, astronomical telescopes use mirrors that are up to several meters in diameter. The photograph that opens this chapter shows an example.

Convex Mirrors

In Fig. 34.14a the *convex* side of a spherical mirror faces the incident light. The center of curvature is on the side opposite to the outgoing rays; according to the third sign rule in Section 34.1, R is negative. Ray PB is reflected, with the angles

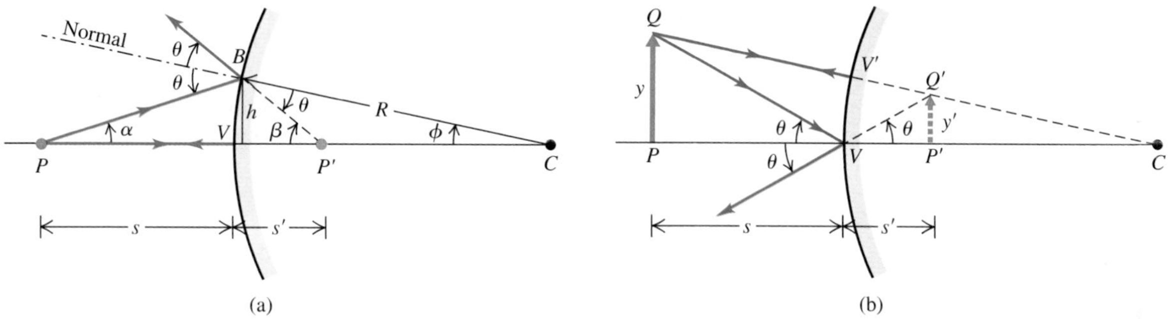

34.14 Constructions for finding (a) the position and (b) the magnification of the image formed by a convex mirror.

of incidence and reflection both equal to θ. The reflected ray, projected backward, intersects the axis at P'. As with a concave mirror, *all* rays from P that are reflected by the mirror diverge from the same point P', provided that the angle α is small. Therefore P' is the image of P. The object distance s is positive, the image distance s' is negative, and the radius of curvature R is *negative* for a *convex* mirror.

Figure 34.14b shows two rays diverging from the head of the arrow PQ and the virtual image $P'Q'$ of this arrow. The same procedure that we used for a concave mirror can be used to show that for a convex mirror,

$$\frac{1}{s} + \frac{1}{s'} = \frac{2}{R}$$

and the lateral magnification is

$$m = \frac{y'}{y} = -\frac{s'}{s}$$

These expressions are exactly the same as Eqs. (34.4) and (34.7) for a concave mirror; we leave the proof as a problem. Thus when we use our sign rules consistently, Eqs. (34.4) and (34.7) are valid for both concave and convex mirrors.

When R is negative (convex mirror), incoming rays that are parallel to the optic axis are not reflected through the focal point F. Instead, they diverge as though they had come from the point F at a distance f *behind* the mirror, as shown in Fig. 34.15a. In this case, f is the focal length, and F is called a *virtual focal point.* The corresponding image distance s' is negative, so both f and R are negative, and

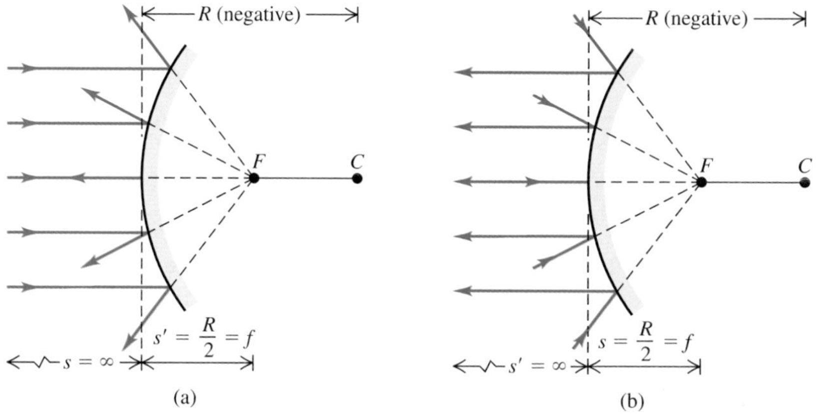

34.15 (a) Incident rays parallel to the axis of a convex mirror diverge as if from the virtual focal point F. (b) Rays aimed at the virtual focal point F of a convex mirror are parallel to the axis after reflection. The angles are exaggerated for clarity.

Eq. (34.5), $f = R/2$, holds for convex as well as concave mirrors. In Fig. 34.15b the incoming rays are converging as though they would meet at the virtual focal point F, and they are reflected parallel to the optic axis.

In summary, Eqs. (34.4) through (34.7), the basic relationships for image formation by a spherical mirror, are valid for both concave and convex mirrors, provided that we use the sign rules consistently.

Example 34.3

Santa's image problem

Santa checks himself for soot, using his reflection in a shiny silvered Christmas tree ornament 0.750 m away (Fig. 34.16a). The diameter of the ornament is 7.20 cm. Standard reference works state that he is a "right jolly old elf," so we estimate his height to be 1.6 m. Where and how tall is the image of Santa formed by the ornament? Is it erect or inverted?

SOLUTION

IDENTIFY: Santa is the object and the surface of the ornament closest to Santa acts as a convex mirror. The relationships between object distance, image distance, focal length, and magnification are the same as for concave mirrors, provided we use the sign rules consistently.

SET UP: Since the mirror is convex, its radius of curvature and focal length are negative. The object distance is $s = 0.750\text{ m} = 75.0\text{ cm}$ and Santa's height is $y = 1.65\text{ m}$. We use Eq. (34.6) to determine the image distance s', and then use Eq. (34.7) to find the lateral magnification m and hence the image height y'. The sign of y' tells us whether the image is erect or inverted.

EXECUTE: The radius of the convex mirror (half the diameter) is $R = -(7.20\text{ cm})/2 = -3.60\text{ cm}$ and the focal length is $f = R/2 = -1.80\text{ cm}$. From Eq. (34.6),

$$\frac{1}{s'} = \frac{1}{f} - \frac{1}{s} = \frac{1}{-1.80\text{ cm}} - \frac{1}{75.0\text{ cm}}$$

$$s' = -1.76\text{ cm}$$

Because s' is negative, the image is behind the mirror, that is, on the side opposite to the outgoing light (Fig. 34.16b), and it is virtual. The image is about halfway between the front surface of the ornament and its center.

The lateral magnification m is given by Eq. (34.7):

$$m = \frac{y'}{y} = -\frac{s'}{s} = -\frac{-1.76\text{ cm}}{75.0\text{ cm}} = 0.0234$$

Because m is positive, the image is erect. It is only about 0.0234 as tall as Santa himself:

$$y' = my = (0.0234)(1.6\text{ m}) = 3.8 \times 10^{-2}\text{ m} = 3.8\text{ cm}$$

EVALUATE: When the object distance s is positive, a convex mirror *always* forms an erect, virtual, diminished, reversed image. For this reason, convex mirrors are used for shoplifting surveillance in stores, at blind intersections, and as "wide-angle" rear-view mirrors for cars and trucks (including those that bear the legend "objects in mirror are closer than they appear").

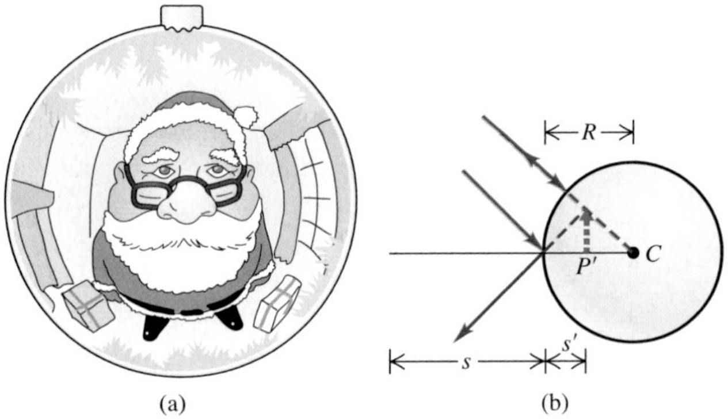

34.16 (a) The ornament forms a virtual, reduced, erect image of Santa. (b) Two of the rays forming the image. The angles are exaggerated for clarity.

Graphical Methods for Mirrors

In Examples 34.1 and 34.3, we used Eqs. (34.6) and (34.7) to find the position and size of the image formed by a mirror. We can also determine the properties of the image by a simple *graphical* method. This method consists of finding the point of intersection of a few particular rays that diverge from a point of the object (such as point Q in Fig. 34.17) and are reflected by the mirror. Then (neglecting aberra-

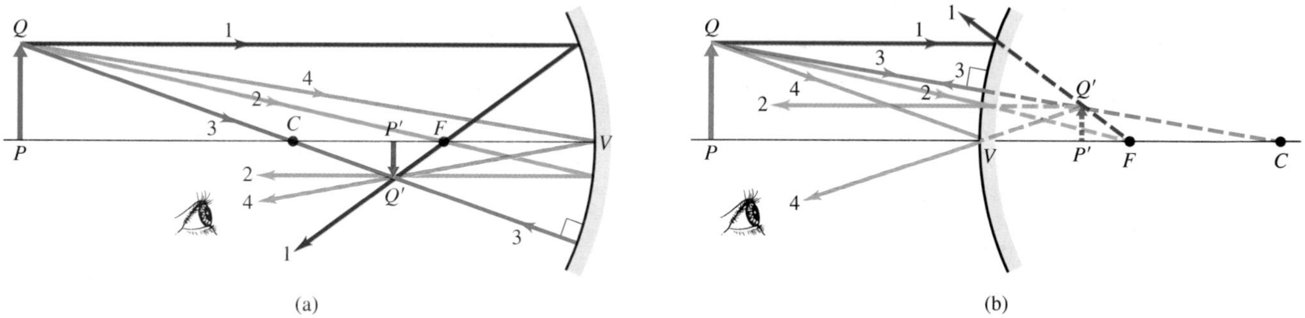

34.17 Principal-ray diagrams showing the graphical method of locating an image formed by (a) a concave mirror; (b) a convex mirror. The colors of the rays are for identification only; they do not refer to specific colors of light.

tions) *all* rays from this object point that strike the mirror will intersect at the same point. For this construction we always choose an object point that is *not* on the optic axis. Four rays that we can usually draw easily are shown in Fig. 34.17. These are called **principal rays.**

1. *A ray parallel to the axis,* after reflection, passes through the focal point F of a concave mirror or appears to come from the (virtual) focal point of a convex mirror.
2. *A ray through (or proceeding toward) the focal point F* is reflected parallel to the axis.
3. *A ray along the radius* through or away from the center of curvature C intersects the surface normally and is reflected back along its original path.
4. *A ray to the vertex V* is reflected forming equal angles with the optic axis.

Once we have found the position of the image point by means of the intersection of any two of these principal rays (1, 2, 3, 4), we can draw the path of any other ray from the object point to the same image point.

CAUTION Although we've emphasized the principal rays, in fact *any* ray from the object that strikes the mirror will pass through the image point (for a real image) or appear to originate from the image point (for a virtual image). Usually, you only need to draw the principal rays, because these are all you need to locate the image.

Problem-Solving Strategy

Image Formation by Mirrors

IDENTIFY *the relevant concepts:* There are two different and complementary ways to solve problems involving image formation by mirrors. One approach uses equations, while the other involves drawing a principal-ray diagram. A successful problem solution uses *both* approaches.

SET UP *the problem:* Determine the target variables. The three key quantities are the focal length, object distance, and image distance; typically you'll be given two of these and will have to determine the third.

EXECUTE *the solution as follows:*

1. The principal-ray diagram is to geometric optics what the free-body diagram is to mechanics. In any problem involving image formation by a mirror, *always* draw a principal-ray diagram first if you have enough information. (The same advice should be followed when dealing with lenses in the following sections.)
2. It is usually best to orient your diagrams consistently with the incoming rays traveling from left to right. Don't draw a lot of other rays at random; stick with the principal rays,

the ones you know something about. Use a ruler and measure distances carefully! A freehand sketch will *not* give good results.

3. If your principal rays don't converge at a real image point, you may have to extend them straight backward to locate a virtual image point, as in Fig. 34.17b. We recommend drawing the extensions with broken lines. Another useful aid is to color-code the different principal rays, as is done in Fig. 34.17.
4. Check your results using Eq. (34.6), $1/s + 1/s' = 1/f$, and the magnification equation, Eq. (34.7). The results you find using this equation must be consistent with your principal-ray diagram; if not, double-check both your calculations and your diagram.
5. Pay careful attention to signs on object and image distances, radii of curvature, and object and image heights. A negative sign on any of these quantities *always* has significance. Use the equations and the sign rules carefully and consistently, and they will tell you the truth! Note that the *same* sign rules (given in Section 34.1) work for all four cases in this chapter: reflection and refraction from plane and spherical surfaces.

EVALUATE *your answer:* You've already checked your results by using both diagrams and equations. But it always helps to take a look back and ask yourself, "Do these results make sense?"

Example 34.4 Concave mirror, different object distances

A concave mirror has a radius of curvature with absolute value 20 cm. Find graphically the image of an object in the form of an arrow perpendicular to the axis of the mirror at each of the following object distances: a) 30 cm, b) 20 cm, c) 10 cm, and d) 5 cm. Check the construction by *computing* the size and magnification of each image.

SOLUTION

IDENTIFY and SET UP: We are given the radius of curvature $R = 20$ cm (positive since the mirror is concave) and hence the focal length $f = R/2 = 10$ cm. In each case we are told the object distance s and are asked to find the image distance s' and the magnification $m = -s'/s$.

EXECUTE: The principal-ray diagrams for the four cases are shown in the four parts of Fig. 34.18. Study each of these diagrams carefully, comparing each numbered ray with the description above. Several points are worth noting. First, in (b) the object and image distances are equal. Ray 3 cannot be drawn in this case because a ray from Q through the center of curvature C does not strike the mirror. Ray 2 cannot be drawn in (c) because a ray from Q toward F also does not strike the mirror. In this case the outgoing rays are parallel, corresponding to an infinite image distance. In (d) the outgoing rays have no real intersection point; they must be extended backward to find the point from which they appear to diverge, that is, from the *virtual image point* Q'. The case shown in (d) illustrates the general observation that an object placed inside the focal point of a concave mirror produces a virtual image.

Measurements of the figures, with appropriate scaling, give the following approximate image distances: a) 15 cm; b) 20 cm; c) ∞ or $-\infty$ (because the outgoing rays are parallel and do not converge at any finite distance); d) -10 cm. To *compute* these distances, we use Eq. (34.6) with $f = 10$ cm:

a) $$\frac{1}{30\text{ cm}} + \frac{1}{s'} = \frac{1}{10\text{ cm}} \qquad s' = 15\text{ cm}$$

b) $$\frac{1}{20\text{ cm}} + \frac{1}{s'} = \frac{1}{10\text{ cm}} \qquad s' = 20\text{ cm}$$

c) $$\frac{1}{10\text{ cm}} + \frac{1}{s'} = \frac{1}{10\text{ cm}} \qquad s' = \infty\ (\text{or } -\infty)$$

d) $$\frac{1}{5\text{ cm}} + \frac{1}{s'} = \frac{1}{10\text{ cm}} \qquad s' = -10\text{ cm}$$

In (a) and (b) the image is real; in (d) it is virtual. In (c) the image is formed at infinity.

The lateral magnifications measured from the figures are approximately a) $-\frac{1}{2}$; b) -1; c) ∞ or $-\infty$ (because the image distance is infinite); d) $+2$. *Computing* the magnifications from Eq. (34.7), we find:

a) $$m = \frac{-15\text{ cm}}{30\text{ cm}} = -\frac{1}{2}$$

b) $$m = \frac{-20\text{ cm}}{20\text{ cm}} = -1$$

c) $$m = -\frac{\infty\text{ cm}}{10\text{ cm}} = -\infty\ (\text{or } +\infty)$$

d) $$m = -\frac{-10\text{ cm}}{5\text{ cm}} = +2$$

In (a) and (b) the image is inverted; in (d) it is erect.

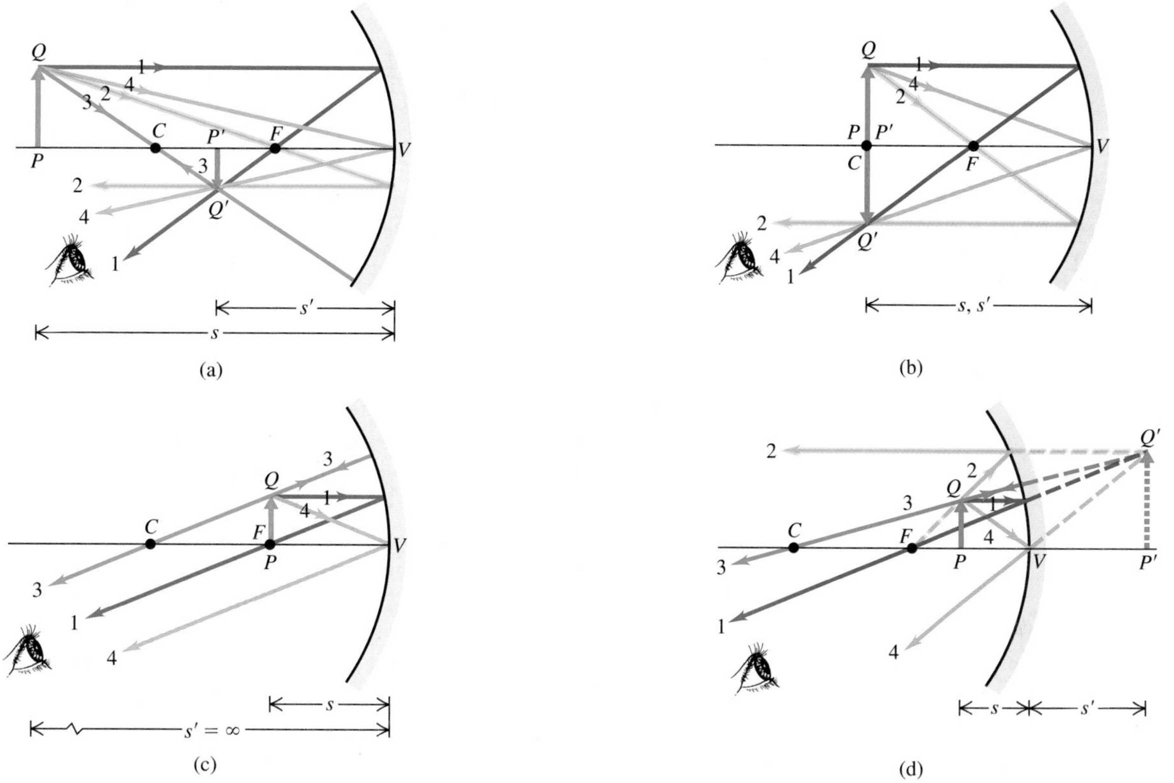

34.18 Using principal-ray diagrams to locate the image $P'Q'$ made by a concave mirror. In each case the object PQ is at a different distance from the mirror.

EVALUATE: Notice the trend as the object is moved closer to the mirror. When the object is far from the mirror, as in Fig. 34.18a, the image is smaller than the object, inverted, and real. As the object distance decreases, the image moves farther from the mirror and increases in size (Fig. 34.18b). When the object is at the focal point, the image is at infinity (Fig. 34.18c). If the object is moved inside the focal point, the image becomes larger than the object, erect, and virtual (Fig. 34.18d). You can test these conclusions by looking at objects reflected in the concave bowl of a metal spoon.

Test Your Understanding

A cosmetics mirror is designed so that your reflection appears right-side up and enlarged. Is the mirror concave or convex? To see an enlarged image, where should you place the mirror relative to your face?

34.3 | Refraction at a Spherical Surface

As we mentioned in Section 34.1, images can be formed by refraction as well as by reflection. To begin with, let's consider refraction at a spherical surface, that is, at a spherical interface between two optical materials with different indexes of refraction. This analysis is directly applicable to some real optical systems, such

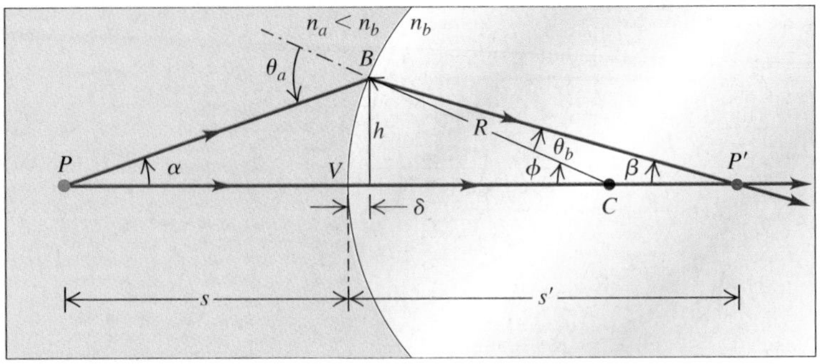

34.19 Construction for finding the position of the image point P' of a point object P formed by refraction at a spherical surface. The materials to the left and right of the interface have refractive lenses n_a and n_b, respectively. In the case shown here, $n_a < n_b$.

as the human eye. It also provides a stepping-stone for the analysis of lenses, which usually have *two* spherical (or nearly spherical) surfaces.

In Fig. 34.19 a spherical surface with radius R forms an interface between two materials with different indexes of refraction n_a and n_b. The surface forms an image P' of an object point P; we want to find how the object and image distances (s and s') are related. We will use the same sign rules that we used for spherical mirrors. The center of curvature C is on the outgoing side of the surface, so R is positive. Ray PV strikes the vertex V and is perpendicular to the surface (that is, to the plane that is tangent to the surface at the point of incidence V). It passes into the second material without deviation. Ray PB, making an angle α with the axis, is incident at an angle θ_a with the normal and is refracted at an angle θ_b. These rays intersect at P', a distance s' to the right of the vertex. The figure is drawn for the case $n_a < n_b$. The object and image distances are both positive.

We are going to prove that if the angle α is small, *all* rays from P intersect at the same point P', so P' is the *real image* of P. We use much the same approach as we did for spherical mirrors in Section 34.2. We again use the theorem that an exterior angle of a triangle equals the sum of the two opposite interior angles; applying this to the triangles PBC and $P'BC$ gives

$$\theta_a = \alpha + \phi \qquad \phi = \beta + \theta_b \tag{34.8}$$

From the law of refraction,

$$n_a \sin\theta_a = n_b \sin\theta_b$$

Also, the tangents of α, β, and ϕ are

$$\tan\alpha = \frac{h}{s+\delta} \qquad \tan\beta = \frac{h}{s'-\delta} \qquad \tan\phi = \frac{h}{R-\delta} \tag{34.9}$$

For paraxial rays, θ_a and θ_b are both small in comparison to a radian, and we may approximate both the sine and tangent of either of these angles by the angle itself (measured in radians). The law of refraction then gives

$$n_a\theta_a = n_b\theta_b$$

Combining this with the first of Eqs. (34.8), we obtain

$$\theta_b = \frac{n_a}{n_b}(\alpha + \phi)$$

When we substitute this into the second of Eqs. (34.8), we get

$$n_a\alpha + n_b\beta = (n_b - n_a)\phi \tag{34.10}$$

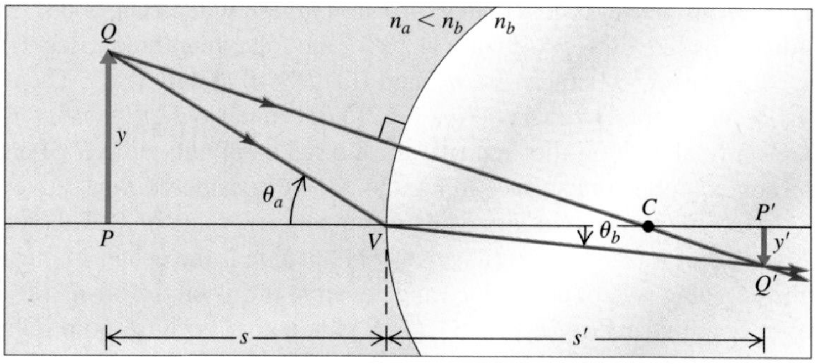

34.20 Construction for determining the height of an image formed by refraction at a spherical surface. In the case shown here, $n_a < n_b$.

Now we use the approximations $\tan \alpha = \alpha$, and so on, in Eqs. (34.9) and also neglect the small distance δ; those equations then become

$$\alpha = \frac{h}{s} \qquad \beta = \frac{h}{s'} \qquad \phi = \frac{h}{R}$$

Finally, we substitute these into Eq. (34.10) and divide out the common factor h. We obtain

$$\frac{n_a}{s} + \frac{n_b}{s'} = \frac{n_b - n_a}{R} \tag{34.11}$$

(object–image relation, spherical refracting surface)

This equation does not contain the angle α, so the image distance is the same for *all* paraxial rays emanating from P; this proves our assertion that P' is the image of P.

To obtain the lateral magnification m for this situation, we use the construction in Fig. 34.20. We draw two rays from point Q, one through the center of curvature C and the other incident at the vertex V. From the triangles PQV and $P'Q'V$,

$$\tan \theta_a = \frac{y}{s} \qquad \tan \theta_b = \frac{-y'}{s'}$$

and from the law of refraction,

$$n_a \sin \theta_a = n_b \sin \theta_b$$

For small angles,

$$\tan \theta_a = \sin \theta_a \qquad \tan \theta_b = \sin \theta_b$$

so finally

$$\frac{n_a y}{s} = -\frac{n_b y'}{s'}$$

or

$$m = \frac{y'}{y} = -\frac{n_a s'}{n_b s} \tag{34.12}$$

(lateral magnification, spherical refracting surface)

Equations (34.11) and (34.12) can be applied to both convex and concave refracting surfaces, provided that you use the sign rules consistently. It doesn't mat-

34.21 Light rays refract as they pass through the curved surfaces of these water droplets.

ter whether n_b is greater or less than n_a. To verify these statements, you should construct diagrams like Figs. 34.19 and 34.20 for the following three cases: (i) $R > 0$ and $n_a > n_b$, (ii) $R < 0$ and $n_a < n_b$, and (iii) $R < 0$ and $n_a > n_b$. Then in each case, use your diagram to again derive Eqs. (34.11) and (34.12).

Here's a final note on the sign rule for the radius of curvature R of a surface. For the convex reflecting surface in Fig. 34.14, we considered R negative, but the convex *refracting* surface in Fig. 34.19 has a *positive* value of R. This may seem inconsistent, but it isn't. The rule is that R is positive if the center of curvature C is on the outgoing side of the surface and negative if C is on the other side. For the convex reflecting surface in Fig. 34.14, R is negative because point C is to the right of the surface but outgoing rays are to the left. For the convex refracting surface in Fig. 34.19, R is positive because both C and the outgoing rays are to the right of the surface.

Refraction at a curved surface is one reason why gardeners avoid watering plants at midday. As sunlight enters a water drop resting on a leaf (Fig. 34.21), the light rays are refracted toward each other as in Figs. 34.19 and 34.20. The sunlight that strikes the leaf is therefore more concentrated and able to cause damage.

An important special case of a spherical refracting surface is a *plane* surface between two optical materials. This corresponds to setting $R = \infty$ in Eq. (34.11). In this case,

$$\frac{n_a}{s} + \frac{n_b}{s'} = 0 \qquad \text{(plane refracting surface)} \qquad (34.13)$$

To find the lateral magnification m for this case, we combine this equation with the general relation, Eq. (34.12), obtaining the simple result

$$m = 1$$

That is, the image formed by a *plane* refracting surface always has the same lateral size as the object, and it is always erect.

An example of image formation by a plane refracting surface is the appearance of a partly submerged drinking straw or canoe paddle. When viewed from some angles, the object appears to have a sharp bend at the water surface because the submerged part appears to be only about three quarters of its actual distance below the surface. (We commented on the appearance of a submerged object in Section 33.2; see Fig. 33.8.)

Example 34.5 Image formation by refraction I

A cylindrical glass rod in air (Fig. 34.22) has index of refraction 1.52. One end is ground to a hemispherical surface with radius $R = 2.00$ cm. a) Find the image distance of a small object on the axis of the rod, 8.00 cm to the left of the vertex. b) Find the lateral magnification.

SOLUTION

IDENTIFY and SET UP: Here material a is air $(n_a = 1.00)$ and material b is the glass of which the rod is made $(n_b = 1.52)$. We are given $s = 8.00$ cm; the radius of the spherical surface is positive $(R = +2.00\text{ cm})$ because the center of curvature is on the outgoing side of the surface. Our target variables are the image distance s' and the lateral magnification m.

EXECUTE: a) From Eq. (34.11),

$$\frac{1.00}{8.00\text{ cm}} + \frac{1.52}{s'} = \frac{1.52 - 1.00}{+2.00\text{ cm}}$$

$$s' = +11.3\text{ cm}$$

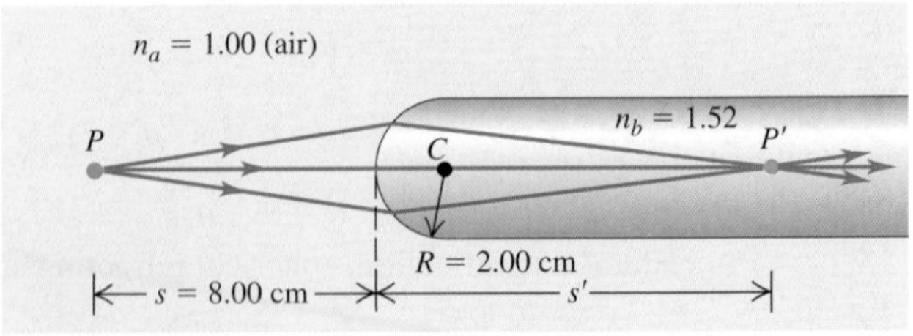

34.22 The glass rod in air forms a real image.

b) From Eq. (34.12),

$$m = -\frac{n_a s'}{n_b s} = -\frac{(1.00)(11.3\text{ cm})}{(1.52)(8.00\text{ cm})} = -0.929$$

EVALUATE: Because the image distance s' is positive, the image is formed 11.3 cm to the *right* of the vertex (on the outgoing side), as shown in Fig. 34.22. The value of m tells us that the image is somewhat smaller than the object, and it is inverted. If the object is an arrow 1.000 mm high, pointing upward, the image is an arrow 0.929 mm high, pointing downward.

Example 34.6 Image formation by refraction II

The glass rod in Example 34.5 is immersed in water (index of refraction $n = 1.33$), as shown in Fig. 34.23. The other quantities have the same values as before. Find the image distance and lateral magnification.

SOLUTION

IDENTIFY and SET UP: The situation is the same as in Example 34.5 except that now $n_a = 1.33$.

EXECUTE: From Eq. (34.11),

$$\frac{1.33}{8.00\text{ cm}} + \frac{1.52}{s'} = \frac{1.52 - 1.33}{+2.00\text{ cm}}$$

$$s' = -21.3\text{ cm}$$

The magnification in this case is

$$m = -\frac{(1.33)(-21.3\text{ cm})}{(1.52)(8.00\text{ cm})} = +2.33$$

The negative value of s' means that after the rays are refracted by the surface, they are not converging but *appear* to diverge from a point 21.3 cm to the *left* of the vertex. We saw a similar case in the reflection of light from a convex mirror; we call the point a *virtual image*. In this example the surface forms a virtual image 21.3 cm to the left of the vertex. The vertical image is erect (because m is positive) and 2.33 times as large as the object.

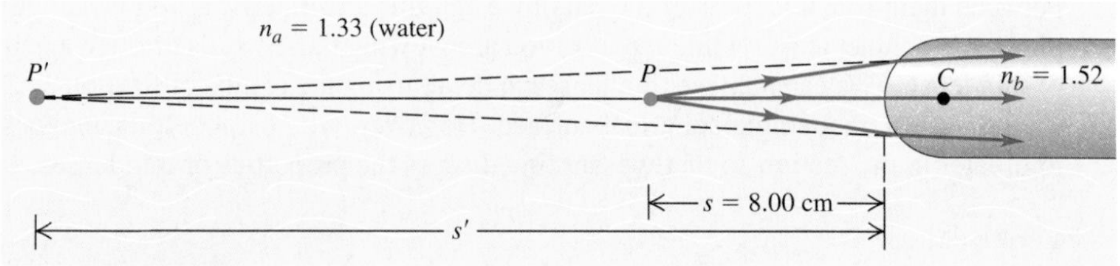

34.23 When immersed in water, the glass rod forms a virtual image.

Example 34.7 Apparent depth of a swimming pool

Swimming pool owners know that the pool always looks shallower than it really is and that it is important to identify the deep parts conspicuously so that people who can't swim won't jump into water that's over their heads. If a nonswimmer looks straight down into water that is actually 2.00 m (about 6 ft, 7 in.) deep, how deep does it appear to be?

SOLUTION

IDENTIFY: The surface of the water acts as a plane refracting surface.

SET UP: The situation is shown in Fig. 34.24. To determine the apparent depth of the pool, we imagine that there is an arrow PQ painted on the bottom of the pool. The refracting surface of the pool forms a virtual image $P'Q'$ of this arrow. We use Eq. (34.13) to find the depth of this arrow; it tells us the apparent depth of the pool.

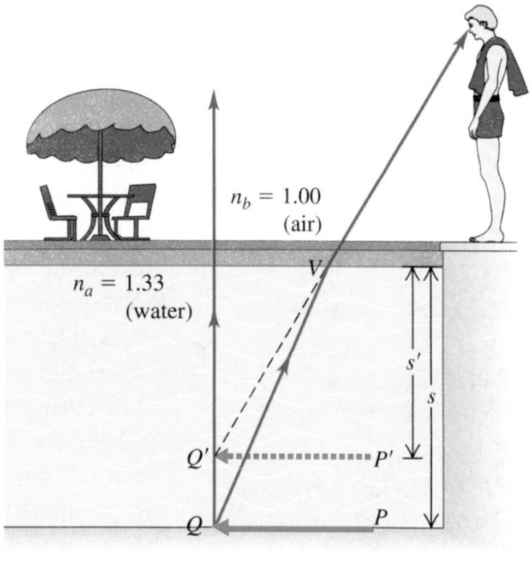

34.24 Arrow $P'Q'$ is the virtual image of the underwater arrow PQ. The angles of the ray with the vertical are exaggerated for clarity.

EXECUTE: The object distance is the actual depth of the pool, $s = 2.00\text{ m}$. Material a is the water $(n_a = 1.33)$ and material b is air $(n_b = 1.00)$. The position of the image is given by Eq. (34.13):

$$\frac{n_a}{s} + \frac{n_b}{s'} = \frac{1.33}{2.00\text{ m}} + \frac{1.00}{s'} = 0$$

$$s' = -1.50\text{ m}$$

The image distance is negative. From the sign rules in Section 34.1, this means that the image is virtual and on the incoming side of the refracting surface—that is, on the same side as the object. The apparent depth is 1.50 m (about 4 ft, 11 in.), or only three-quarters of the actual depth. A 6-ft nonswimmer who didn't allow for this effect would be in trouble.

EVALUATE: Recall that the lateral magnification for a plane refracting surface is $m = 1$. Hence the image $P'Q'$ of the arrow is the same *length* as the actual arrow PQ. Only its depth is different.

Test Your Understanding

The water droplets in Fig. 34.21 have radius of curvature R and index of refraction $n = 1.33$. Can they form an image of the sun on the leaf?

34.4 | Thin Lenses

The most familiar and widely used optical device (after the plane mirror) is the *lens.* A lens is an optical system with two refracting surfaces. The simplest lens has two *spherical* surfaces close enough together that we can neglect the distance between them (the thickness of the lens); we call this a **thin lens.** If you wear eyeglasses or contact lenses while reading, you are viewing these words through a pair of thin lenses. We can analyze thin lenses in detail using the results of Section 34.3 for refraction by a single spherical surface. However, we postpone this analysis until later in the section so that we can first discuss the properties of thin lenses.

Properties of a Lens

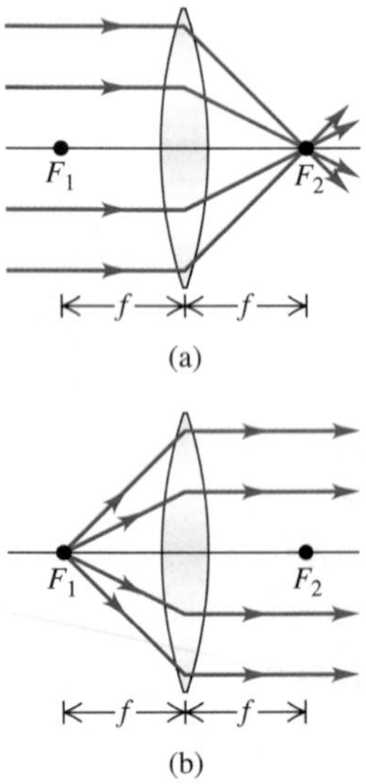

34.25 F_1 and F_2 are the first and second focal points of a converging thin lens. The numerical value of f is positive.

A lens of the shape shown in Fig. 34.25 has the property that when a beam of rays parallel to the axis passes through the lens, the rays converge to a point F_2 (Fig. 34.25a) and form a real image at that point. Such a lens is called a **converging lens.** Similarly, rays passing through point F_1 emerge from the lens as a beam of parallel rays (Fig. 34.25b). The points F_1 and F_2 are called the first and second *focal points,* and the distance f (measured from the center of the lens) is called the *focal length.* Note the similarities between the two focal points of a converging lens and the single focal point of a concave mirror (Fig. 34.11). As for a concave mirror, the focal length of a converging lens is defined to be a *positive* quantity, and such a lens is also called a *positive lens.*

The central horizontal line in Fig. 34.25 is called the *optic axis,* as with spherical mirrors. The centers of curvature of the two spherical surfaces lie on and define the optic axis. The two focal lengths in Fig. 34.25, both labeled f, *are always equal* for a thin lens, even when the two sides have different curvatures. We will derive this somewhat surprising result later in the section, when we derive the relationship of f to the index of refraction of the lens and the radii of curvature of its surfaces.

Like a concave mirror, a converging lens can form an image of an extended object. Figure 34.26 shows how to find the position and lateral magnification of an image made by a thin converging lens. Using the same notation and sign rules as before, we let s and s' be the object and image distances, respectively, and let y and

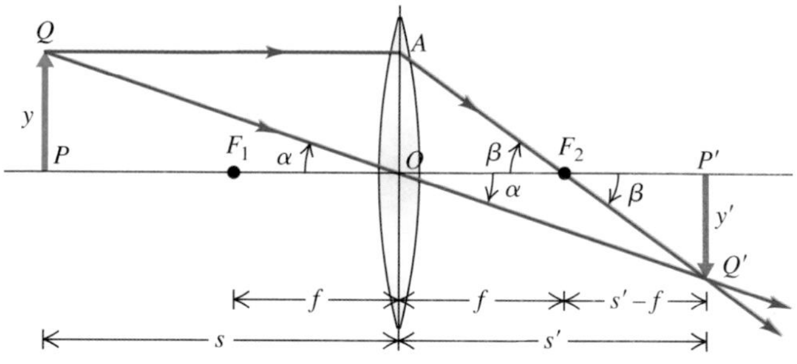

34.26 Construction used to find image position for a thin lens. To emphasize that the lens is assumed to be very thin, the ray QAQ' is shown as bent at the midplane of the lens rather than at the two surfaces and ray QOQ' is shown as a straight line.

y' be the object and image heights. Ray QA, parallel to the optic axis before refraction, passes through the second focal point F_2 after refraction. Ray QOQ' passes undeflected straight through the center of the lens because at the center the two surfaces are parallel and (we have assumed) very close together. There is refraction where the ray enters and leaves the material but no net change in direction.

The two angles labeled α in Fig. 34.26 are equal. Therefore the two right triangles PQO and $P'Q'O$ are *similar,* and ratios of corresponding sides are equal. Thus

$$\frac{y}{s} = -\frac{y'}{s'} \quad \text{or} \quad \frac{y'}{y} = -\frac{s'}{s} \qquad (34.14)$$

(The reason for the negative sign is that the image is below the optic axis and y' is negative.) Also, the two angles labeled β are equal, and the two right triangles OAF_2 and $P'Q'F_2$ are similar, so

$$\frac{y}{f} = -\frac{y'}{s' - f}$$

or

$$\frac{y'}{y} = -\frac{s' - f}{f} \qquad (34.15)$$

We now equate Eqs. (34.14) and (34.15), divide by s', and rearrange to obtain

$$\frac{1}{s} + \frac{1}{s'} = \frac{1}{f} \quad \text{(object–image relation, thin lens)} \qquad (34.16)$$

ActivPhysics Online

15.9 Thin-Lens Ray Diagrams
15.10 Converging Lens Problems
15.11 Diverging Lens Problems

This analysis also gives the lateral magnification $m = y'/y$ for the lens; from Eq. (34.14),

$$m = -\frac{s'}{s} \quad \text{(lateral magnification, thin lens)} \qquad (34.17)$$

The negative sign tells us that when s and s' are both positive, as in Fig. 34.26, the image is *inverted,* and y and y' have opposite signs.

Equations (34.16) and (34.17) are the basic equations for thin lenses. They are *exactly* the same as the corresponding equations for spherical mirrors, Eqs. (34.6) and (34.7). As we will see, the same sign rules that we used for spherical mirrors are also applicable to lenses. In particular, consider a lens with a positive focal length (a converging lens). When an object is outside the first focal point F_1 of this lens (that is, when $s > f$), the image distance s' is positive (that is, the image is on the same side as the outgoing rays); this image is real and inverted, as in Fig. 34.26. An object placed inside the first focal point of a converging lens, so that $s < f$, pro-

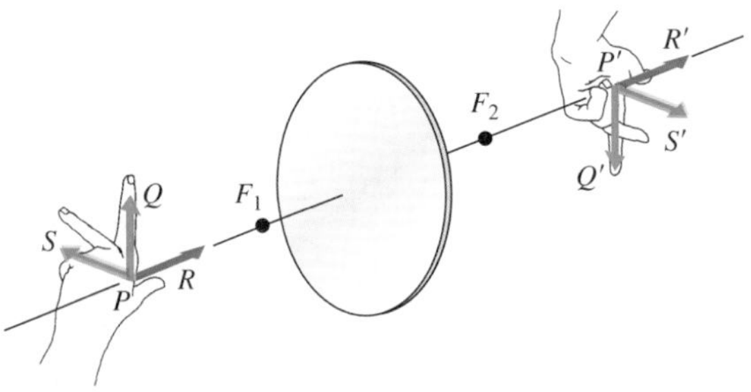

34.27 The image $S'P'Q'R'$ of a three-dimensional object $SPQR$ is *not* reversed by a lens.

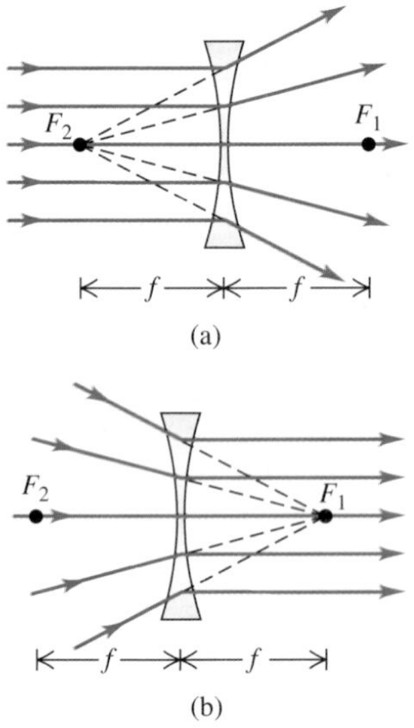

34.28 F_2 and F_1 are the second and first focal points of a diverging thin lens, respectively. The numerical value of f is negative.

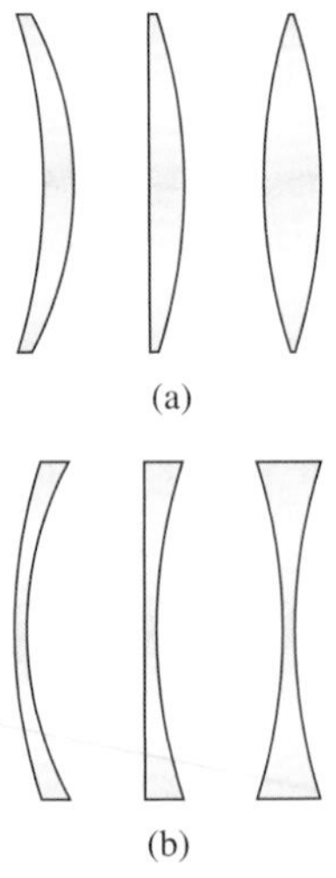

34.29 (a) Meniscus, plano-convex, and double-convex converging lenses. (b) Meniscus, plano-concave, and double-concave diverging lenses.

duces an image with a negative value of s'; this image is located on the same side of the lens as the object and is virtual, erect, and larger than the object. You can verify these statements algebraically using Eqs. (34.16) and (34.17); we'll also verify them in the next section, using graphical methods analogous to those introduced for mirrors in Section 34.2.

Figure 34.27 shows how a lens forms a three-dimensional image of a three-dimensional object. Point R is nearer the lens than point P. From Eq. (34.16), image point R' is farther from the lens than is image point P', and the image $P'R'$ points in the same direction as the object PR. Arrows $P'S'$ and $P'Q'$ are reversed relative to PS and PQ.

Let's compare Fig. 34.27 with Fig. 34.6, which shows the image formed by a plane *mirror.* We note that the image formed by the lens is inverted, but it is *not* reversed front to back along the optic axis. That is, if the object is a left hand, its image is also a left hand. You can verify this by pointing your left thumb along PR, your left forefinger along PQ, and your left middle finger along PS. Then rotate your hand 180°, using your thumb as an axis; this brings the fingers into coincidence with $P'Q'$ and $P'S'$. In other words, an *inverted* image is equivalent to an image that has been rotated by 180° about the lens axis.

So far we have been discussing *converging* lenses. Figure 34.28 shows a **diverging lens;** the beam of parallel rays incident on this lens *diverges* after refraction. The focal length of a diverging lens is a negative quantity, and the lens is also called a *negative lens.* The focal points of a negative lens are reversed, relative to those of a positive lens. The second focal point, F_2, of a negative lens is the point from which rays that are originally parallel to the axis *appear to diverge* after refraction, as in Fig. 34.28a. Incident rays converging toward the first focal point F_1, as in Fig. 34.28b, emerge from the lens parallel to its axis.

Equations (34.16) and (34.17) apply to *both* positive and negative lenses. Various types of lenses, both converging and diverging, are shown in Fig. 34.29. Here's an important observation: *Any lens that is thicker at its center than at its edges is a converging lens with positive f; and any lens that is thicker at its edges than at its center is a diverging lens with negative f* (provided that the lens has a greater index of refraction than the surrounding material). We can prove this using the *lensmaker's equation,* which it is our next task to derive.

The Lensmaker's Equation

We'll now derive Eq. (34.16) in more detail and at the same time derive the *lensmaker's equation,* which is a relationship between the focal length f, the index of refraction n of the lens, and the radii of curvature R_1 and R_2 of the lens surfaces.

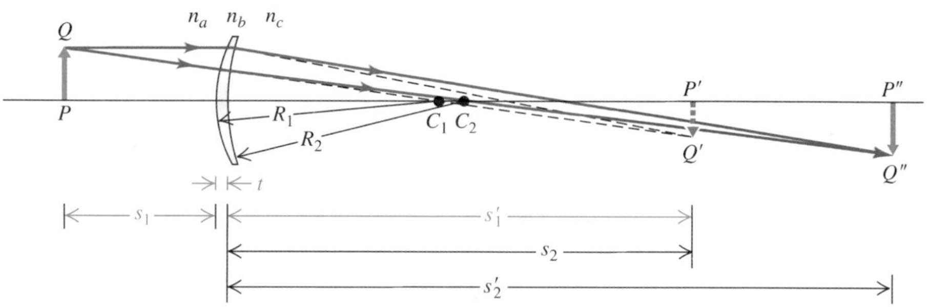

34.30 The image formed by the first surface of a lens serves as the object for the second surface. The distances s_1' and s_2 are taken to be equal; this is a good approximation if the lens thickness t is small.

We use the principle that an image formed by one reflecting or refracting surface can serve as the object for a second reflecting or refracting surface.

We begin with the somewhat more general problem of two spherical interfaces separating three materials with indexes of refraction n_a, n_b, and n_c, as shown in Fig. 34.30. The object and image distances for the first surface are s_1 and s_1', and those for the second surface are s_2 and s_2'. We assume that the lens is thin, so that the distance t between the two surfaces is small in comparison with the object and image distances and can therefore be neglected. This is usually the case with eyeglass lenses (Fig. 34.31). Then s_2 and s_1' have the same magnitude but opposite sign. For example, if the first image is on the outgoing side of the first surface, s_1' is positive. But when viewed as an object for the second surface, the first image is *not* on the incoming side of that surface. So we can say that $s_2 = -s_1'$.

We need to use the single-surface equation, Eq. (34.11), twice, once for each surface. The two resulting equations are

$$\frac{n_a}{s_1} + \frac{n_b}{s_1'} = \frac{n_b - n_a}{R_1}$$

$$\frac{n_b}{s_2} + \frac{n_c}{s_2'} = \frac{n_c - n_b}{R_2}$$

Ordinarily, the first and third materials are air or vacuum, so we set $n_a = n_c = 1$. The second index n_b is that of the lens, which we can call simply n. Substituting these values and the relation $s_2 = -s_1'$, we get

$$\frac{1}{s_1} + \frac{n}{s_1'} = \frac{n - 1}{R_1}$$

$$-\frac{n}{s_1'} + \frac{1}{s_2'} = \frac{1 - n}{R_2}$$

To get a relation between the initial object position s_1 and the final image position s_2', we add these two equations. This eliminates the term n/s_1', and we obtain

$$\frac{1}{s_1} + \frac{1}{s_2'} = (n - 1)\left(\frac{1}{R_1} - \frac{1}{R_2}\right)$$

Finally thinking of the lens as a single unit, we call the object distance simply s instead of s_1, and we call the final image distance s' instead of s_2'. Making these substitutions, we have

$$\frac{1}{s} + \frac{1}{s'} = (n - 1)\left(\frac{1}{R_1} - \frac{1}{R_2}\right) \qquad (34.18)$$

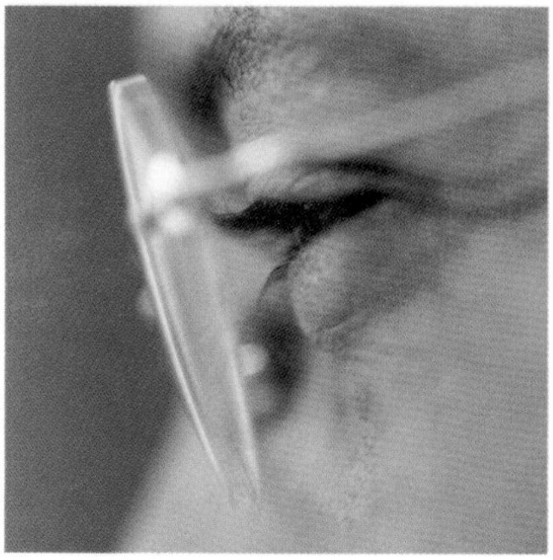

34.31 These eyeglass lenses satisfy the thin-lens approximation: their thickness is small compared to the object and image distances.

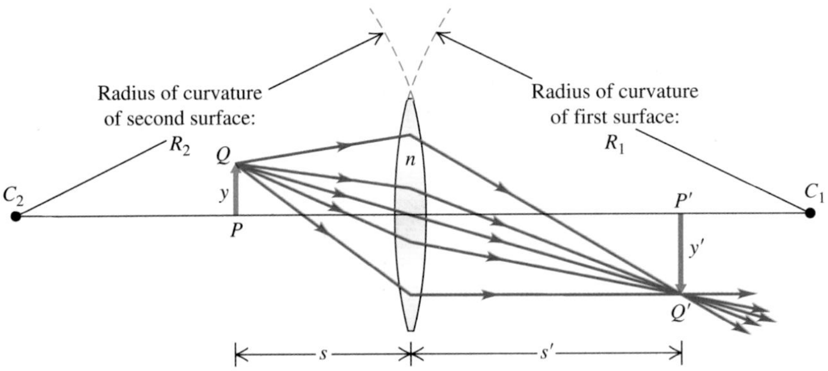

34.32 A thin lens. In this case, R_1 is positive (because the center of curvature C_1 is on the same side as the outgoing light) and R_2 is negative (because C_2 is on the side opposite the outgoing light). The focal length f is positive, and the lens is a converging lens. Here, s and s' are also positive, and m is therefore negative.

Now we compare this with the other thin-lens equation, Eq. (34.16). We see that the object and image distances s and s' appear in exactly the same places in both equations and that the focal length f is given by

$$\frac{1}{f} = (n - 1)\left(\frac{1}{R_1} - \frac{1}{R_2}\right) \quad \text{(lensmaker's equation for a thin lens)} \quad (34.19)$$

This is the **lensmaker's equation.** In the process of rederiving the relationship between object distance, image distance, and focal length for a thin lens, we have also derived an expression for the focal length f of a lens in terms of its index of refraction n and the radii of curvature R_1 and R_2 of its surfaces. This can be used to show that all the lenses in Fig. 34.29a are converging lenses with positive focal lengths and that all the lenses in Fig. 34.29b are diverging lenses with negative focal lengths (see Exercise 34.30).

We use all our previous sign rules from Section 34.1 with Eqs. (34.18) and (34.19). For example, in Fig. 34.32, s, s', and R_1 are positive, but R_2 is negative.

It is not hard to generalize Eq. (34.19) to the situation in which the lens is immersed in a material with an index of refraction greater than unity. We invite you to work out the lensmaker's equation for this more general situation.

We stress that the paraxial approximation is indeed an approximation! Rays that are at sufficiently large angles to the optic axis of a spherical lens will not be brought to the same focus as paraxial rays; this is the same problem of spherical aberration that plagues spherical *mirrors* (Section 34.2). To avoid this and other limitations of thin spherical lenses, lenses of more complicated shape are used in precision optical instruments.

Example 34.8 Determining the focal length of a lens

a) Suppose the absolute values of the radii of curvature of the lens surfaces in Fig. 34.32 are both equal to 10 cm and the index of refraction is $n = 1.52$. What is the focal length f of the lens? b) Suppose the lens in Fig. 34.28 also has $n = 1.52$, and the absolute values of the radii of curvature of its lens surfaces are also both equal to 10 cm. What is the focal length of this lens?

SOLUTION

IDENTIFY and SET UP: We use the lensmaker's equation, Eq. (34.19), to determine the focal length in each situation. Note that in part (a) the lens is convex on both sides (Fig. 34.32) while in part (a) it is concave on both sides (Fig. 34.28). We take account of these differences by paying careful attention to the signs of the radii of curvature R_1 and R_2.

EXECUTE: a) Figure 34.32 shows that the center of curvature of the first surface (C_1) is on the outgoing side of the lens, while the center of curvature of the second surface (C_2) is on the *incoming* side. Hence R_1 is positive but R_2 is negative: $R_1 = +10$ cm, $R_2 = -10$ cm. From Eq. (34.19),

$$\frac{1}{f} = (1.52 - 1)\left(\frac{1}{+10\text{ cm}} - \frac{1}{-10\text{ cm}}\right)$$

$$f = 9.6\text{ cm}$$

b) For a double-concave lens the center of curvature of the first surface is on the *incoming* side, while the center of curvature of the second surface is on the outgoing side. Hence R_1 is negative and R_2 is positive: $R_1 = -10$ cm, $R_2 = +10$ cm. Again using Eq. (34.19),

$$\frac{1}{f} = (1.52 - 1)\left(\frac{1}{-10\text{ cm}} - \frac{1}{+10\text{ cm}}\right)$$

$$f = -9.6\text{ cm}$$

EVALUATE: In part (a) the focal length is positive, so this is a converging lens; this makes sense since the lens is thicker at its center than at its edges. In part (b) the focal length is *negative,* so this is a *diverging* lens; this also makes sense, since the lens is thicker at its edges than at its center.

Graphical Methods for Lenses

We can determine the position and size of an image formed by a thin lens by using a graphical method very similar to the one we used in Section 34.2 for spherical mirrors. Again we draw a few special rays called *principal rays* that diverge from a point of the object that is *not* on the optic axis. The intersection of these rays, after they pass through the lens, determines the position and size of the image. In using this graphical method, we will consider the entire deviation of a ray as occurring at the midplane of the lens, as shown in Fig. 34.33. This is consistent with the assumption that the distance between the lens surfaces is negligible.

The three principal rays whose paths are usually easy to trace for lenses are shown in Fig. 34.33:

1. *A ray parallel to the axis* emerges from the lens in a direction that passes through the second focal point F_2 of a converging lens, or appears to come from the second focal point of a diverging lens.
2. *A ray through the center of the lens* is not appreciably deviated; at the center of the lens the two surfaces are parallel, so this ray emerges at essentially the same angle at which it enters and along essentially the same line.
3. *A ray through (or proceeding toward) the first focal point* F_1 emerges parallel to the axis.

When the image is real, the position of the image point is determined by the intersection of any two rays 1, 2, and 3 (Fig. 34.33a). When the image is virtual, we extend the diverging outgoing rays backward to their intersection point to find the image point (Fig. 34.33b).

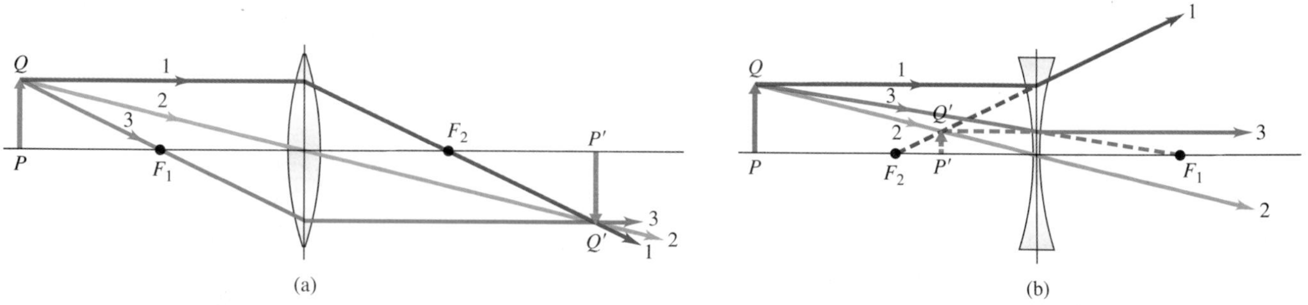

34.33 Principal-ray diagrams showing the graphical method of locating an image formed by a thin lens. (a) A converging lens; (b) a diverging lens. The colors of the rays are for identification only; they do not refer to specific colors of light.

CAUTION Keep in mind that *any* ray from the object that strikes the lens will pass through the image point (for a real image) or appear to originate from the image point (for a virtual image). (We made a similar comment about image formation by mirrors in Section 34.2.) We've emphasized the principal rays because they're the only ones you need to draw to locate the image.

Figure 34.34 shows principal-ray diagrams for a converging lens for several object distances. We suggest you study each of these diagrams very carefully, comparing each numbered ray with the above description.

Parts (a), (b), and (c) of Fig. 34.34 help explain what happens in focusing a camera. For a photograph to be in sharp focus, the film must be at the position of the real image made by the camera's lens. The image distance increases as the object is brought closer, so the film is moved farther behind the lens (i.e., the lens is moved farther in front of the film). In Fig. 34.34d the object is at the focal point; ray 3 can't be drawn because it doesn't pass through the lens. In Fig. 34.34e the object distance is less than the focal length. The outgoing rays are divergent, and the image is *virtual;* its position is located by extending the outgoing rays backward, so the image distance s' is negative. Note also that the image is erect and larger than the object. (We'll see the usefulness of this in Section 34.6.) Figure 34.34f corresponds to a *virtual object.* The incoming rays do not diverge from a real object, but are *converging* as though they would meet at the tip of the virtual object O on the right side; the object distance s is negative in this case. The image is real and is located between the lens and the second focal point. This situation can arise if the rays that strike the lens in Fig. 34.34f emerge from another converging lens (not shown) to the left of the figure.

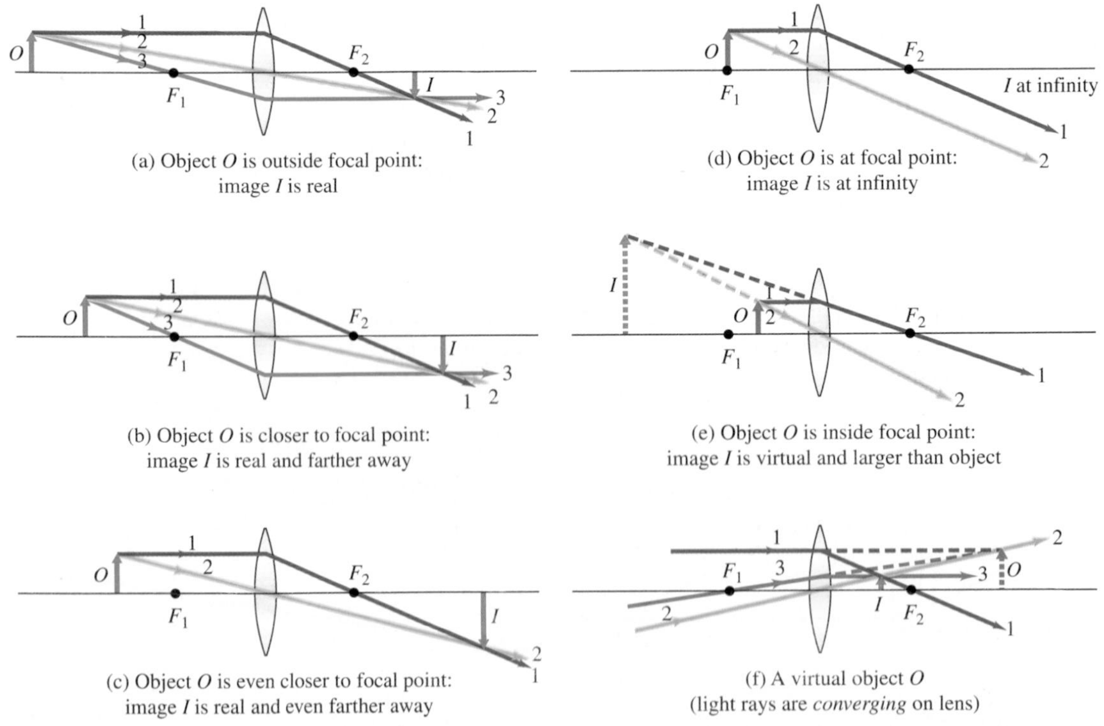

34.34 Formation of images by a thin converging lens for various object distances. The principal rays are numbered.

Problem-Solving Strategy: Image Formation by Thin Lenses

IDENTIFY *the relevant concepts:* The strategy outlined for mirrors in Section 34.2 is equally applicable here, and you should review it now. As for mirrors, you should solve problems involving image formation by lenses using *both* equations and a principal-ray diagram.

SET UP *the problem:* As always, determine the target variables.

EXECUTE *the solution as follows:*

1. Always begin with a principal-ray diagram if you have enough information. Orient your diagrams consistently so that light travels from left to right. Don't just sketch these diagrams; draw the rays with a ruler and measure the distances carefully.
2. Draw the principal rays so they bend at the midplane of the lens, as shown in Fig. 34.33. For a lens there are only three principal rays, compared to four for a mirror. Be sure to draw *all three* whenever possible. The intersection of any two determines the image, but if the third doesn't pass through the same intersection point, you know you have made a mistake. Redundancy can be useful in spotting errors.
3. If the outgoing principal rays don't converge at a real image point, the image is virtual. Then you have to extend the outgoing rays backward to find the virtual image point, which lies on the *incoming* side of the lens.
4. The same sign rules we have used for mirrors and single refracting surfaces (see Section 34.1) are also applicable for thin lenses. Be extremely careful to get your signs right and to interpret the signs of results correctly.
5. Use Eqs. (34.16) and (34.17) to confirm by calculation your graphical results for the image position and size. This gives an extremely useful consistency check.
6. The *image* from one lens or mirror may serve as the *object* for another. In that case, be careful in finding the object and image *distances* for this intermediate image; be sure you include the distance between the two elements (lenses and/or mirrors) correctly.

EVALUATE *your answer:* Cast a critical eye on your diagrams and calculations to make certain that your results are consistent.

Example 34.9 Image location and magnification with a converging lens

A converging lens has a focal length of 20 cm. Find graphically the image location for an object at each of the following distances from the lens: a) 50 cm; b) 20 cm; c) 15 cm; d) −40 cm. Determine the magnification in each case. Check your results by calculating the image position and magnification from Eqs. (34.16) and (34.17).

SOLUTION

IDENTIFY and SET UP: In each case we are given the focal length $f = 20$ cm and the value of the object distance s. Our target variables are the image distance s' and the magnification $m = -s'/s$.

EXECUTE: The appropriate principal-ray diagrams are shown in a) Fig. 34.34a, b) Fig. 34.34d, c) Fig. 34.34e, and d) Fig. 34.34f. The approximate image distances, from measurements of these diagrams, are 35 cm, $-\infty$, −40 cm, and 15 cm, and the approximate magnifications are $-\frac{2}{3}$, $+\infty$, and +3, and $+\frac{1}{3}$, respectively.

Calculating the image positions from Eq. (34.16), we find

a) $$\frac{1}{50\text{ cm}} + \frac{1}{s'} = \frac{1}{20\text{ cm}} \qquad s' = 33.3\text{ cm}$$

b) $$\frac{1}{20\text{ cm}} + \frac{1}{s'} = \frac{1}{20\text{ cm}} \qquad s' = \pm\infty$$

c) $$\frac{1}{15\text{ cm}} + \frac{1}{s'} = \frac{1}{20\text{ cm}} \qquad s' = -60\text{ cm}$$

d) $$\frac{1}{-40\text{ cm}} + \frac{1}{s'} = \frac{1}{20\text{ cm}} \qquad s' = 13.3\text{ cm}$$

The graphical results are fairly close to these except for part (c); the accuracy of the diagram in Fig. 34.34e is limited because the rays extended backward have nearly the same direction.

From Eq. (34.17) the magnifications are

a) $$m = -\frac{33.3\text{ cm}}{50\text{ cm}} = -\frac{2}{3}$$

b) $$m = -\frac{\pm\infty\text{ cm}}{20\text{ cm}} = \pm\infty$$

c) $$m = -\frac{-60\text{ cm}}{15\text{ cm}} = +4$$

d) $$m = -\frac{13.3\text{ cm}}{-40\text{ cm}} = +\frac{1}{3}$$

EVALUATE: Note that s' is positive in parts (a) and (d) but negative in part (c). This makes sense: the image is real in parts (a) and (d) but virtual in part (c). The light rays that emerge from the lens in part (b) are parallel and never converge, so the image can be regarded as being at either $+\infty$ or $-\infty$.

The values of magnification tell us that the image is inverted in part (a) and erect in parts (c) and (d), in accordance with the principal-ray diagrams. The infinite value of magnification in part (b) is another way of saying that the image is formed infinitely far away.

Example 34.10 Image formation by a diverging lens

You are given a thin diverging lens. You find that a beam of parallel rays spreads out after passing through the lens, as though all the rays came from a point 20.0 cm from the center of the lens. You want to use this lens to form an erect virtual image that is $\frac{1}{3}$ the height of the object. a) Where should the object be placed? b) Draw a principal-ray diagram.

SOLUTION

IDENTIFY and SET UP: The observation with parallel rays shows that the focal length is $f = -20$ cm. We want the lateral magnification to be $m = +\frac{1}{3}$ (positive because the image is to be erect). We use this to determine the ratio s'/s from Eq. (34.17), then determine the object distance s with Eq. (34.16).

EXECUTE: From Eq. (34.17), $m = +\frac{1}{3} = -s'/s$, so $s' = -s/3$. If we insert this result into Eq. (34.16), we find

$$\frac{1}{s} + \frac{1}{-s/3} = \frac{1}{-20.0\text{ cm}}$$

$$s = 40.0\text{ cm}$$

$$s' = -\frac{s}{3} = -\frac{40.0\text{ cm}}{3} = -13.3\text{ cm}$$

The image distance is negative, so the object and image are on the same side of the lens.

b) Figure 34.35 is a principal-ray diagram for this problem, with the rays numbered in the same way as in Fig. 34.33b.

EVALUATE: A diverging lens is often mounted in the front door of a home. It provides the occupant of the home with an erect, reduced image of anyone standing outside the door. The occupant can see the outside person's entire height and decide whether to let him or her in.

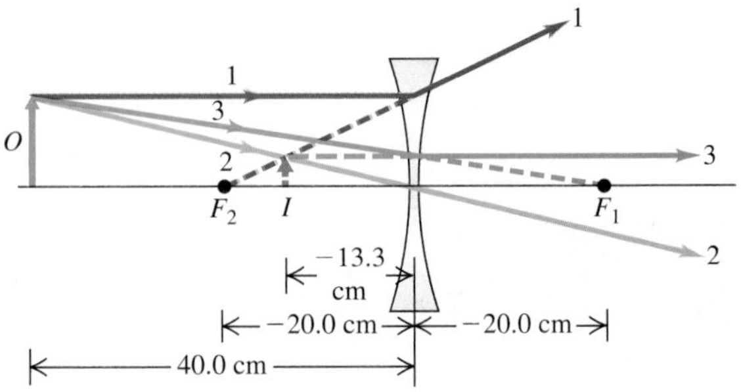

34.35 Principal-ray diagram for an image formed by a thin diverging lens.

Example 34.11 An image of an image

An object 8.0 cm high is placed 12.0 cm to the left of a converging lens of focal length 8.0 cm. A second converging lens of focal length 6.0 cm is placed 36.0 cm to the right of the first lens. Both lenses have the same optic axis. Find the position, size, and orientation of the image produced by the two lenses in combination. (Combinations of converging lenses are used in telescopes and microscopes, to be discussed in Section 34.7.)

SOLUTION

IDENTIFY and SET UP: The situation is shown in Fig. 34.36. The object O lies outside the first focal point F_1 of the first lens, so this lens produces a real image I. The light rays that strike the second lens diverge from this real image just as if I was a material object. Hence the *image* made by the *first* lens acts as an *object* for the *second* lens. Our goal is to determine the properties of the final image made by the second lens.

EXECUTE: In Fig. 34.36 we have drawn principal rays 1, 2, and 3 from the head of the object arrow O to find the position of the first image I and principal rays 1′, 2′, and 3′ from the head of the first image to find the position of the second image I' made by the second lens (even though rays 2′ and 3′ don't actually exist in this case). Note that the image is inverted *twice,* once by each lens, so the second image I' has the same orientation as the original object.

To *calculate* the position and size of the second image I', we must first find the position and size of the first image I. Applying Eq. (34.16), $1/s + 1/s' = 1/f$, to the first lens gives

$$\frac{1}{12.0\text{ cm}} + \frac{1}{s_1'} = \frac{1}{8.0\text{ cm}} \qquad s_1' = +24.0\text{ cm}$$

The first image I is 24.0 cm to the right of the first lens. The magnification is $m_1 = -(24.0\text{ cm})/(12.0\text{ cm}) = -2.00$, so the height of the first image is $(-2.0)(8.0\text{ cm}) = -16.0\text{ cm}$.

The first image is 36.0 cm − 24.0 cm = 12.0 cm to the left of the second lens, so the object distance for the second lens is +12.0 cm. Using Eq. (34.16) for the second lens gives the position of the second and final image:

$$\frac{1}{12.0\text{ cm}} + \frac{1}{s_2'} = \frac{1}{6.0\text{ cm}} \qquad s_2' = +12.0\text{ cm}$$

The final image is 12.0 cm to the right of the second lens and 48.0 cm to the right of the first lens. The magnification produced by the second converging lens is $m_2 = -(12.0\text{ cm})/(12.0\text{ cm}) = -1.0$.

EVALUATE: The value of m_2 means that the final image is just as large as the first image but has the opposite orientation. This is also shown in the principal-ray diagram.

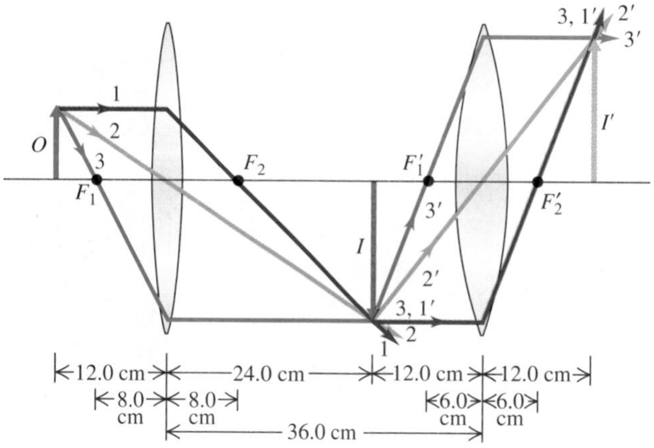

34.36 Principal-ray diagram for a combination of two converging lenses. The first lens makes a real image of the object. This real image acts as an object for the second lens.

Test Your Understanding

Suppose you cut away the lower two-thirds of the lens shown in Fig. 34.34a. Is an image I still formed at the same position? Why or why not?

34.5 | Cameras

The concept of *image,* which is so central to understanding simple mirror and lens systems, plays an equally important role in the analysis of optical instruments (also called *optical devices*). Among the most common of optical devices are cameras, which make an image of an object and record it either electronically or on film.

The basic elements of a **camera** are a light-tight box (*camera* is a Latin word meaning "a room or enclosure"), a converging lens, a shutter to open the lens for a prescribed length of time, and a light-sensitive recording medium—either photographic film or, in a digital camera, an electronic detector (Fig. 34.37). The

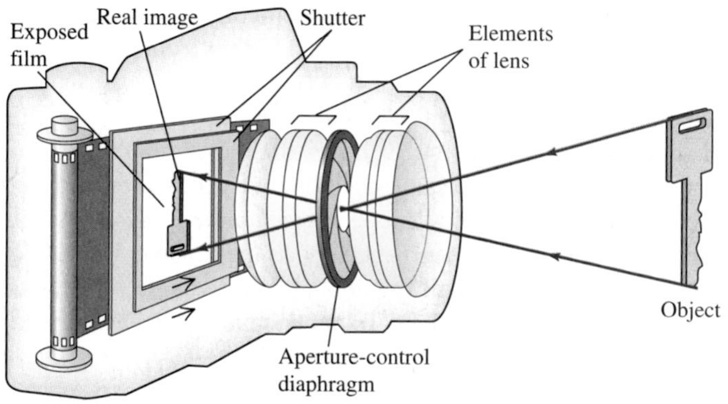

34.37 Key elements of a film camera. The lens forms a real, inverted image of the object in the plane of the film. The image is brought into focus by moving the lens forward or backward relative to the film. In a digital camera, the film is replaced by an electronic detector.

(a)

(b)

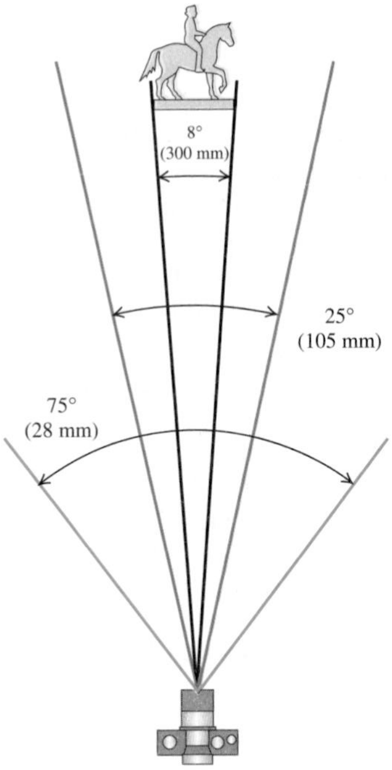

(d)

(c)

34.38 (a), (b), (c) Three photographs taken with the same camera from the same position in the Boston Public Garden using lenses with focal lengths $f = 28$ mm, 105 mm, and 300 mm. Increasing the focal length increases the image size proportionately. (d) The larger the value of f, the smaller the angle of view. The angles shown here are for a camera that uses 35-mm film (image area 24 mm $\times$ 36 mm) and refer to the angle of view along the diagonal dimension of the film.

lens forms an inverted real image on the recording medium of the object being photographed. High-quality camera lenses have several elements, permitting partial correction of various *aberrations,* including the dependence of index of refraction on wavelength and the limitations imposed by the paraxial approximation.

When the camera is in proper *focus,* the position of the recording medium coincides with the position of the real image formed by the lens. The resulting photograph will then be as sharp as possible. With a converging lens, the image distance increases as the object distance decreases (see Figs. 34.34a, 34.34b, and 34.34c, and the discussion in Section 34.4). Hence in "focusing" the camera, the lens is moved closer to the film for a distant object and farther from the film for a nearby object.

The choice of the focal length f for a camera lens depends on the film size and the desired angle of view. Figure 34.38 shows three photographs taken on 35-mm film with the same camera at the same position, but with lenses of different focal lengths. A lens of long focal length, called a *telephoto* lens, gives a small angle of view and a large image of a distant object (such as the statue in Fig. 34.38c); a lens of short focal length gives a small image and a wide angle of view (as in Fig. 34.38a) and is called a *wide-angle* lens. To understand this behavior, recall that the focal length is the distance from the lens to the image when the object is infinitely far away. In general, for *any* object distance, using a lens of longer focal length gives a greater image distance. This also increases the height of the image; as was discussed in Section 34.4, the ratio of the image height y' to the object height y (the *lateral magnification*) is equal in absolute value to the ratio of image distance s' to the object distance s [Eq. (34.17)]:

$$m = \frac{y'}{y} = -\frac{s'}{s}$$

With a lens of short focal length, the ratio s'/s is small, and a distant object gives only a small image. When a lens with a long focal length is used, the image of this same object may entirely cover the area of the film. Hence the longer the focal length, the narrower the angle of view (Fig. 34.38d).

For the film to record the image properly, the total light energy per unit area reaching the film (the "exposure") must fall within certain limits. This is controlled by the *shutter* and the *lens aperture.* The shutter controls the time interval during which light enters the lens. This is usually adjustable in steps corresponding to factors of about 2, often from 1 s to $\frac{1}{1000}$ s.

The intensity of light reaching the film is proportional to the area viewed by the camera lens and to the effective area of the lens. The size of the area that the lens "sees" is proportional to the square of the angle of view of the lens, and so is roughly proportional to $1/f^2$. The effective area of the lens is controlled by means of an adjustable lens aperture, or *diaphragm,* a nearly circular hole with variable diameter D; hence the effective area is proportional to D^2. Putting these factors together, we see that the intensity of light reaching the film with a particular lens is proportional to D^2/f^2. The light-gathering capability of a lens is commonly expressed by photographers in terms of the ratio f/D, called the ***f*-number** of the lens:

$$f\text{-number} = \frac{\text{Focal length}}{\text{Aperture diameter}} = \frac{f}{D} \qquad (34.20)$$

For example, a lens with a focal length $f = 50$ mm and an aperture diameter $D = 25$ mm is said to have an f-number of 2, or "an aperture of $f/2$." The light intensity reaching the film is *inversely* proportional to the square of the f-number.

For a lens with a variable-diameter aperture, increasing the diameter by a factor of $\sqrt{2}$ changes the f-number by $1/\sqrt{2}$ and increases the intensity at the film by a factor of 2. Adjustable apertures usually have scales labeled with successive numbers (often called *f-stops*) related by factors of $\sqrt{2}$, such as

$$f/2 \qquad f/2.8 \qquad f/4 \qquad f/5.6 \qquad f/8 \qquad f/11 \qquad f/16$$

and so on. The larger numbers represent smaller apertures and exposures, and each step corresponds to a factor of 2 in intensity (Fig. 34.39). The actual *exposure* (total amount of light reaching the film) is proportional both to the aperture area and the time of exposure. Thus $f/4$ and $\frac{1}{500}$ s, $f/5.6$ and $\frac{1}{250}$ s, and $f/8$ and $\frac{1}{125}$ s all correspond to the same exposure.

Many photographers use a *zoom lens,* which is not a single lens but a complex collection of several lens elements that give a continuously variable focal length,

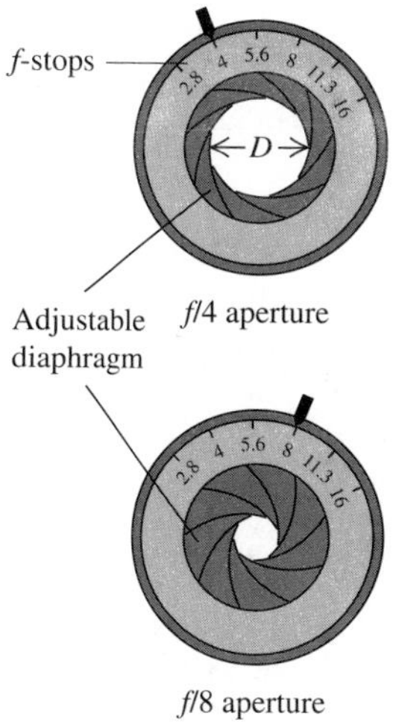

34.39 In a camera lens, larger f-numbers mean smaller aperture diameters.

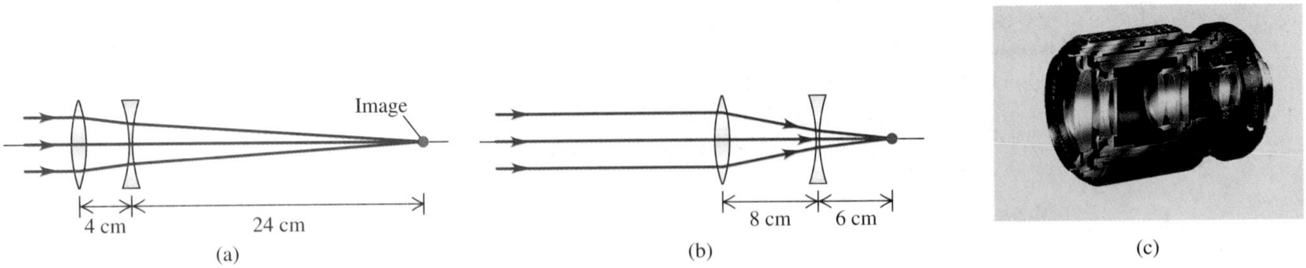

34.40 (a), (b) Principle of the zoom lens, using two elements with variable spacing. The effective focal length depends on the distance between the lens elements. (c) A typical zoom lens for a 35-mm camera, containing twelve elements arranged in four groups.

often over a range as great as 10 to 1. Figure 34.40a shows a simple system with variable focal length, and Fig. 34.40b shows a typical zoom lens for a 35-mm camera. Zoom lenses give a range of image sizes of a given object. It is an enormously complex problem in optical design to keep the image in focus and maintain a constant f-number while the focal length changes. When you vary the focal length of a typical zoom lens, two groups of elements move within the lens and a diaphragm opens and closes.

A *projector* for viewing slides, digital images, or motion pictures operates very much like a camera in reverse. In a movie projector, a lamp shines on the film, which acts as an object for the projection lens. The lens forms a real, enlarged, inverted image of the film on the projection screen. Because the image is inverted, the film goes through the projector upside down so that the image on the screen appears right-side up.

Example 34.12 Photographic exposures

A common telephoto lens for a 35-mm camera has a focal length of 200 mm and a range of f-stops from $f/5.6$ to $f/45$. a) What is the corresponding range of aperture diameters? b) What is the corresponding range of intensity of the image on the film?

SOLUTION

IDENTIFY and SET UP: We use Eq. (34.20) to relate the diameter D (the target variable) to the f-number and the focal length $f = 200$ mm. The intensity of the light reaching the film is proportional to D^2/f^2; since f is the same in each case, we conclude that the intensity in this case is proportional to D^2, the square of the aperture diameter.

EXECUTE: a) From Eq. (34.20) the range of diameters is from

$$D = \frac{f}{f\text{-number}} = \frac{200\ \text{mm}}{5.6} = 36\ \text{mm}$$

to

$$D = \frac{200\ \text{mm}}{45} = 4.4\ \text{mm}$$

b) Because the intensity is proportional to the square of the diameter, the ratio of the intensity at $f/5.6$ to the intensity at $f/45$ is

$$\left(\frac{36\ \text{mm}}{4.4\ \text{mm}}\right)^2 = \left(\frac{45}{5.6}\right)^2 = 65 \qquad (\text{about } 2^6)$$

EVALUATE: If the correct exposure time at $f/5.6$ is $\frac{1}{1000}$ s, then at $f/45$ it is $(65)\left(\frac{1}{1000}\ \text{s}\right) = \frac{1}{15}$ s to compensate for the lower intensity. This illustrates a general rule: the smaller the aperture and the larger the f-number, the longer the required exposure time. Nevertheless, many photographers prefer to use small apertures so that only the central part of the lens is used to make the image. This minimizes aberrations that occur near the edges of the lens and gives the sharpest possible images.

Test Your Understanding

When used with 35-mm film (image area 24 mm × 36 mm), a lens with $f = 50$ mm gives a 45° angle of view and is called a "normal lens." But when used with a portrait camera that uses 60 mm × 70 mm film, this same lens is called a wide-angle lens. Why?

34.6 | The Eye

The optical behavior of the eye is similar to that of a camera. The essential parts of the human eye, considered as an optical system, are shown in Fig. 34.41a. The eye is nearly spherical in shape and about 2.5 cm in diameter. The front portion is somewhat more sharply curved and is covered by a tough, transparent membrane called the *cornea.* The region behind the cornea contains a liquid called the *aqueous humor.* Next comes the *crystalline lens,* a capsule containing a fibrous jelly, hard at the center and progressively softer at the outer portions. The crystalline lens is held in place by ligaments that attach it to the ciliary muscle, which encircles it. Behind the lens, the eye is filled with a thin watery jelly called the *vitreous humor.* The indexes of refraction of both the aqueous humor and the vitreous humor are about 1.336, nearly equal to that of water. The crystalline lens, while not homogeneous, has an average index of 1.437. This is not very different from the indexes of the aqueous and vitreous humors. As a result, most of the refraction of light entering the eye occurs at the outer surface of the cornea.

Refraction at the cornea and the surfaces of the lens produces a *real image* of the object being viewed. This image is formed on the light-sensitive *retina,* lining the rear inner surface of the eye. The retina plays the same role as the film in a camera. The *rods* and *cones* in the retina act like an array of miniature photocells (Fig. 34.41b); they sense the image and transmit it via the *optic nerve* to the brain. Vision is most acute in a small central region called the *fovea centralis,* about 0.25 mm in diameter.

In front of the lens is the *iris.* It contains an aperture with variable diameter called the *pupil,* which opens and closes to adapt to changing light intensity. The receptors of the retina also have intensity adaptation mechanisms.

For an object to be seen sharply, the image must be formed exactly at the location of the retina. The eye adjusts to different object distances s by changing the focal length f of its lens; the lens-to-retina distance, corresponding to s', does not change.

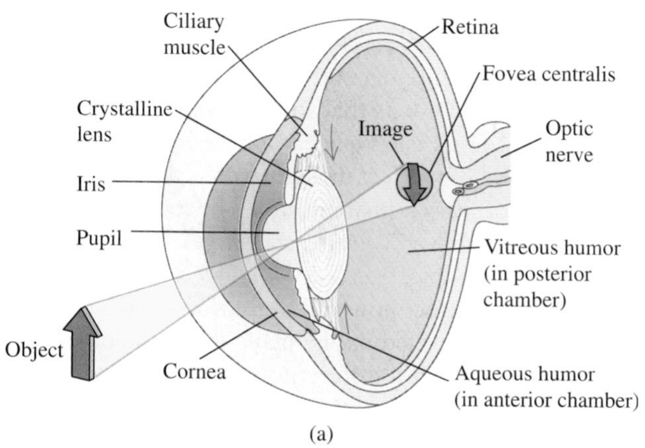

(a)

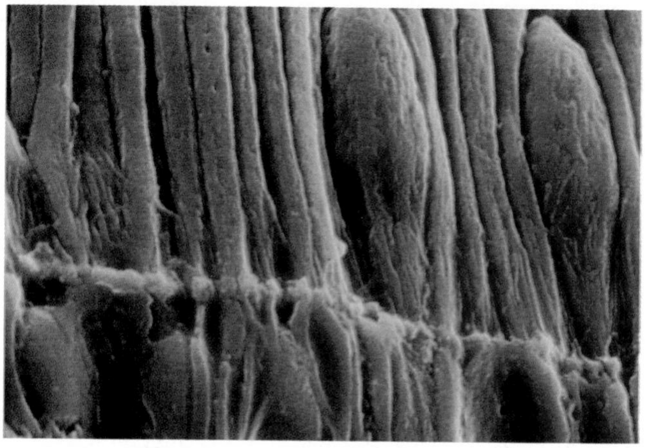

(b)

34.41 (a) The eye. The ciliary muscle contracts to change the focal length of the crystalline lens to image close objects sharply. (b) This magnified image shows rods and cones in the human eye. The rods are more sensitive to light than the cones, but only the cones are sensitive to differences in color. A typical human eye contains about 1.3×10^8 rods and about 7×10^6 cones.

(Contrast this with focusing a camera, in which the focal length is fixed and the lens-to-film distance is changed). For the normal eye, an object at infinity is sharply focused when the ciliary muscle is relaxed. To permit sharp imaging on the retina of closer objects, the tension in the ciliary muscle surrounding the lens increases, the ciliary muscle contracts, the lens bulges, and the radii of curvature of its surfaces decrease; this decreases the focal length. This process is called *accommodation.*

The extremes of the range over which distinct vision is possible are known as the *far point* and the *near point* of the eye. The far point of a normal eye is at infinity. The position of the near point depends on the amount by which the ciliary muscle can increase the curvature of the crystalline lens. The range of accommodation gradually diminishes with age because the crystalline lens grows throughout a person's life (it is about 50% larger at age 60 than at age 20) and the ciliary muscles are less able to distort a larger lens. For this reason, the near point gradually recedes as one grows older. This recession of the near point is called *presbyopia.* Table 34.1 shows the approximate position of the near point for an average person at various ages. For example, an average person 50 years of age cannot focus on an object that is closer than about 40 cm.

Table 34.1 Receding of Near Point with Age

Age (years)	Near Point (cm)
10	7
20	10
30	14
40	22
50	40
60	200

Several common defects of vision result from incorrect distance relations in the eye. A normal eye forms an image on the retina of an object at infinity when the eye is relaxed (Fig. 34.42a). In the *myopic* (nearsighted) eye, the eyeball is too long from front to back in comparison with the radius of curvature of the cornea (or the cornea is too sharply curved), and rays from an object at infinity are focused in front of the retina (Fig. 34.42b). The most distant object for which an image can be formed on the retina is then nearer than infinity. In the *hyperopic* (farsighted) eye, the eyeball is too short or the cornea is not curved enough, and the image of an infinitely distant object is behind the retina (Fig. 34.42c). The myopic eye produces *too much* convergence in a parallel bundle of rays for an image to be formed on the retina; the hyperopic eye, *not enough* convergence.

All of these defects can be corrected by the use of corrective lenses (eyeglasses or contact lenses). The near point of either a presbyopic or a hyperopic eye is *farther* from the eye than normal. To see clearly an object at normal reading distance (often assumed to be 25 cm), we need a lens that forms a virtual image of the object at or beyond the near point. This can be accomplished by a converging (positive) lens, as shown in Fig. 34.43. In effect the lens moves the object farther away from the eye to a point where a sharp retinal image can be formed. Similarly, correcting the myopic eye involves using a diverging (negative) lens to move the image closer to the eye than the actual object, as shown in Fig. 34.44.

Astigmatism is a different type of defect in which the surface of the cornea is not spherical but rather more sharply curved in one plane than in another. As a result, horizontal lines may be imaged in a different plane from vertical lines (Fig. 34.45a). Astigmatism may make it impossible, for example, to focus clearly on the horizontal and vertical bars of a window at the same time.

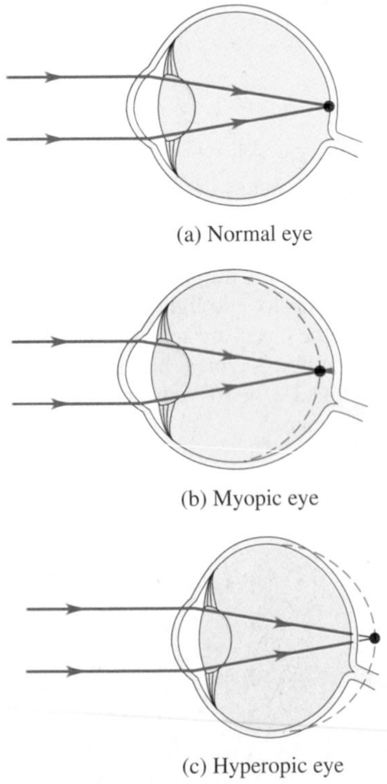
(a) Normal eye
(b) Myopic eye
(c) Hyperopic eye

34.42 Refractive errors for (a) a normal eye, (b) a myopic (nearsighted) eye, and (c) a hyperopic (farsighted) eye viewing a very distant object. The dashed blue curve indicates the required position of the retina.

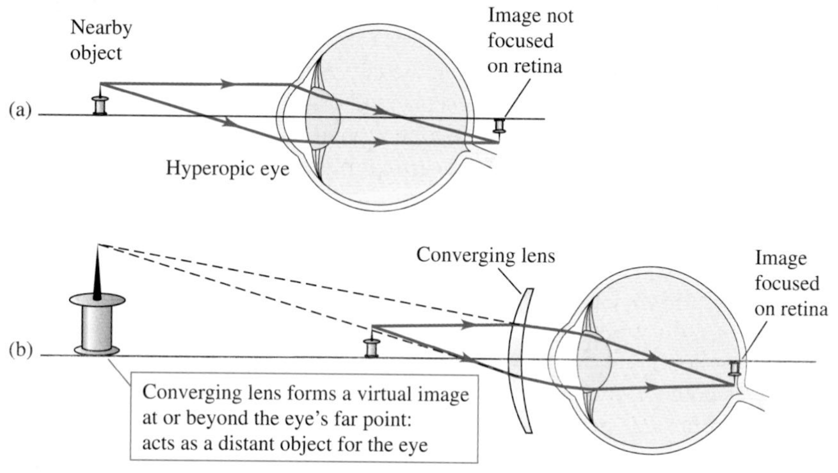

34.43 (a) An uncorrected hyperopic (farsighted) eye. (b) A positive (converging) lens gives the extra convergence needed for a hyperopic eye to focus the image on the retina.

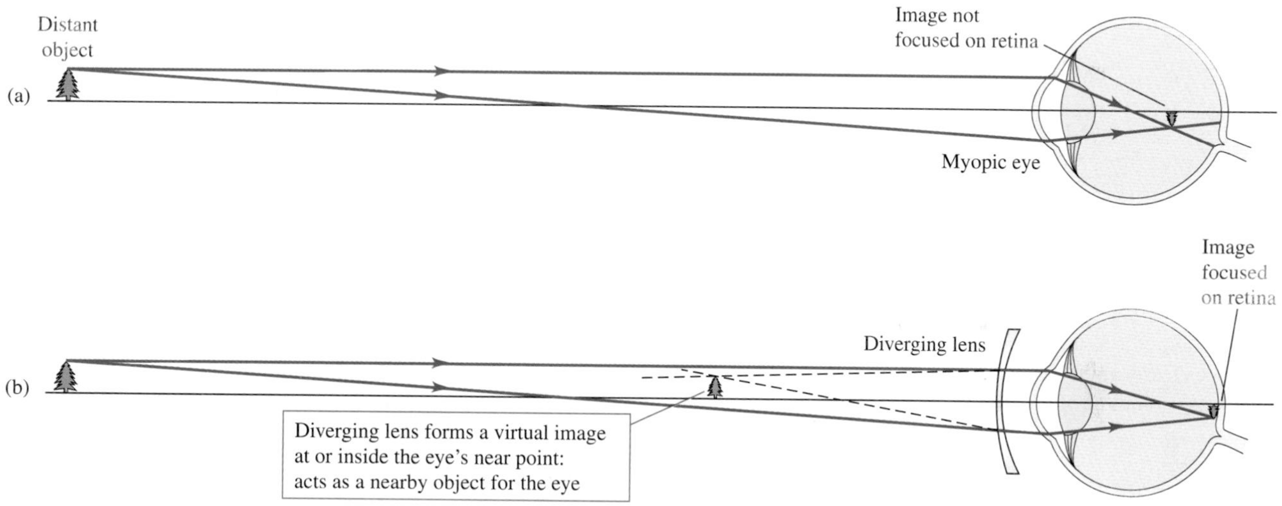

34.44 (a) An uncorrected myopic (nearsighted) eye. (b) A negative (diverging) lens spreads the rays farther apart to compensate for the excessive convergence of the myopic eye.

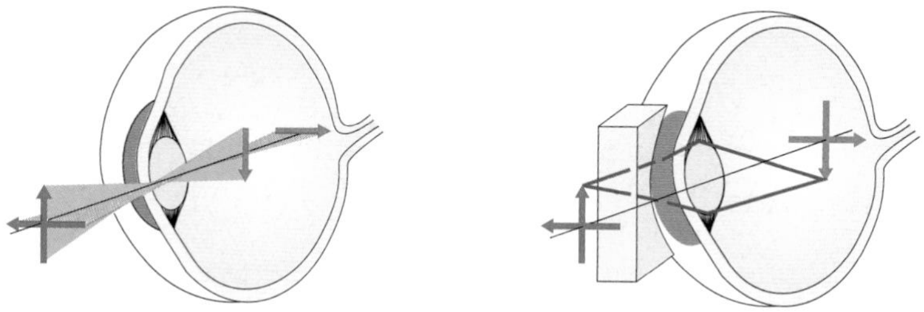

(a) Vertical lines are imaged in front of the retina

(b) A cylindrical lens corrects for astigmatism

34.45 (a) One type of astigmatic eye. (b) The diverging cylindrical lens with horizontal axis separates rays farther apart vertically (but not horizontally) so that the vertical line is also imaged on the retina.

Astigmatism can be corrected by use of a lens with a *cylindrical* surface. For example, suppose the curvature of the cornea in a horizontal plane is correct to focus rays from infinity on the retina but the curvature in the vertical plane is too great to form a sharp retinal image. When a cylindrical lens with its axis horizontal is placed before the eye, the rays in a horizontal plane are unaffected, but the additional divergence of the rays in a vertical plane causes these to be sharply imaged on the retina (Fig. 34.45b).

Lenses for vision correction are usually described in terms of the **power,** defined as the reciprocal of the focal length expressed in meters. The unit of power is the **diopter.** Thus a lens with $f = 0.50$ m has a power of 2.0 diopters, $f = -0.25$ m corresponds to -4.0 diopters, and so on. The numbers on a prescription for glasses are usually powers expressed in diopters. When the correction involves both astigmatism and myopia or hyperopia, there are three numbers: one for the spherical power, one for the cylindrical power, and an angle to describe the orientation of the cylinder axis.

An alternative approach for correcting many defects of vision is to reshape the cornea. This is often done using a procedure called *laser-assisted in situ keratomileusis,* or LASIK. An incision is made into the cornea and a flap of outer corneal tissue is folded back. A pulsed ultraviolet laser with a beam only 50 μm wide (about 1/200 the width of a human hair) is then used to vaporize away microscopic areas of the underlying tissue. The flap is then folded back into position, where it conforms to the new shape "carved" by the laser.

Example 34.13 Correcting for farsightedness

The near point of a certain hyperopic eye is 100 cm in front of the eye. To see clearly an object that is 25 cm in front of the eye, what contact lens is required?

SOLUTION

IDENTIFY and SET UP: We want the lens to form a virtual image of the object at the near point of the eye, 100 cm from it. That is, when $s = 25$ cm, we want s' to be 100 cm. We determine the required focal length of the contact lens using the object–image relation for a thin lens, Eq. (34.16).

EXECUTE: From Eq. (34.16),

$$\frac{1}{f} = \frac{1}{s} + \frac{1}{s'} = \frac{1}{+25\text{ cm}} + \frac{1}{-100\text{ cm}}$$

$$f = +33\text{ cm}$$

We need a converging lens with focal length $f = 33$ cm. The corresponding power is $1/(0.33\text{ m})$, or $+3.0$ diopters.

EVALUATE: In this example we used a contact lens to correct hyperopia. Had we used eyeglasses, we would have had to account for the separation between the eye and the eyeglass lens, and a somewhat different power would have been required (see Example 34.14).

Example 34.14 Correcting for nearsightedness

The far point of a certain myopic eye is 50 cm in front of the eye. To see clearly an object at infinity, what eyeglass lens is required? Assume that the lens is worn 2 cm in front of the eye.

SOLUTION

IDENTIFY and SET UP: The far point of a myopic eye is nearer than infinity. To see clearly objects beyond the far point, we need a lens that forms a virtual image of such objects no farther from the eye than the far point. Assume that the virtual image of the object at infinity is formed at the far point, 50 cm in front of the eye and 48 cm in front of the eyeglass lens. Then when $s = \infty$, we want s' to be -48 cm. As in Example 34.13, we use the values of s and s' to calculate the required focal length.

EXECUTE: From Eq. (34.16),

$$\frac{1}{f} = \frac{1}{s} + \frac{1}{s'} = \frac{1}{\infty} + \frac{1}{-48\text{ cm}}$$

$$f = -48\text{ cm}$$

We need a *diverging* lens with focal length -48 cm $= -0.48$ m. The power is -2.1 diopter.

EVALUATE: If a *contact* lens were used instead, we would need $f = -50$ cm and a power of -2.0 diopters. Can you see why?

Test Your Understanding

A certain eyeglass lens is thin at its center, even thinner at its top and bottom edges, and relatively thick at its left and right edges. Describe the defects in vision that this lens is intended to correct.

34.7 | The Magnifier

The apparent size of an object is determined by the size of its image on the retina. If the eye is unaided, this size depends upon the *angle* θ subtended by the object at the eye, called its **angular size** (Fig. 34.46a).

To look closely at a small object, such as an insect or a crystal, you bring it close to your eye, making the subtended angle and the retinal image as large as possible. But your eye cannot focus sharply on objects that are closer than the near point, so the angular size of an object is greatest (that is, it subtends the largest possible viewing angle) when it is placed at the near point. In the following discussion we will assume an average viewer for whom the near point is 25 cm from the eye.

A converging lens can be used to form a virtual image that is larger and farther from the eye than the object itself, as shown in Fig. 34.46b. Then the object can be moved closer to the eye, and the angular size of the image may be substantially larger than the angular size of the object at 25 cm without the lens. A lens used in this way is called a **magnifier,** otherwise known as a *magnifying glass* or a *simple magnifier.* The virtual image is most comfortable to view when it is placed at infinity, so that the ciliary muscle of the eye is relaxed; in the following discussion we assume that this is done.

In Fig. 34.46a the object is at the near point, where it subtends an angle θ at the eye. In Fig. 34.46b a magnifier in front of the eye forms an image at infinity, and the angle subtended at the magnifier is θ'. The usefulness of the magnifier is given

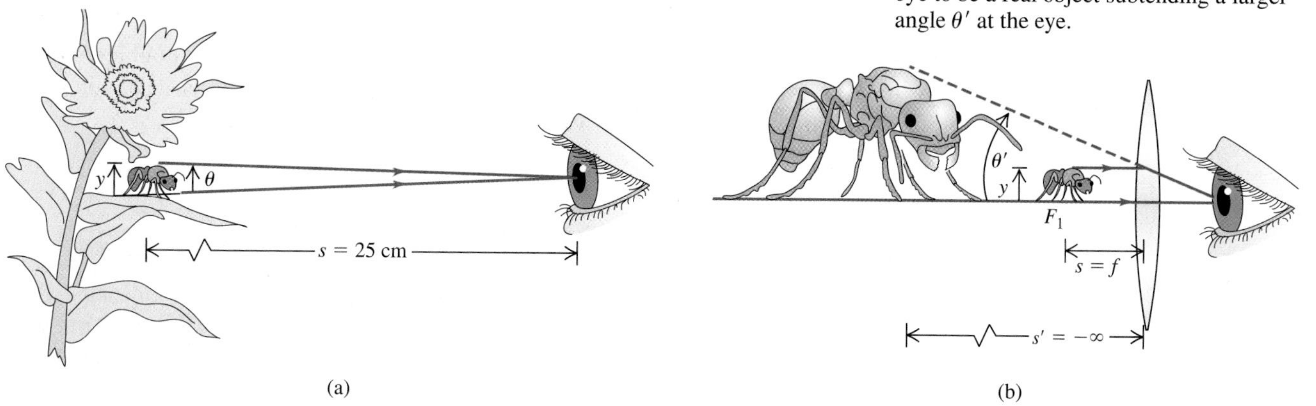

34.46 (a) The angular size θ is largest when the object is at the near point. (b) The magnifier gives a virtual image at infinity. This virtual image appears to the eye to be a real object subtending a larger angle θ' at the eye.

by the ratio of the angle θ' (with the magnifier) to the angle θ (without the magnifier). This ratio is called the **angular magnification** M:

$$M = \frac{\theta'}{\theta} \quad \text{(angular magnification)} \tag{34.21}$$

CAUTION Don't confuse the angular magnification M with the *lateral* magnification m. Angular magnification is the ratio of the *angular* size of an image to the angular size of the corresponding object; lateral magnification refers to the ratio of the *height* of an image to the height of the corresponding object. For the situation shown in Fig. 34.46b, the angular magnification is about 3×, since the insect subtends an angle about three times larger than that in Fig. 34.46a; hence the insect will look about three times larger to the eye. The *lateral* magnification $m = -s'/s$ in Fig. 34.46b is *infinite* because the virtual image is at infinity, but that doesn't mean that the insect looks infinitely large through the magnifier! (That's why we didn't attempt to draw an infinitely large ant in Fig. 34.46b.) When dealing with a magnifier, M is useful but m is not.

To find the value of M, we first assume that the angles are small enough that each angle (in radians) is equal to its sine and its tangent. Using Fig. 34.46a and drawing the ray in Fig. 34.46b that passes undeviated through the center of the lens, we find that θ and θ' (in radians) are

$$\theta = \frac{y}{25\text{ cm}} \qquad \theta' = \frac{y}{f}$$

Combining these expressions with Eq. (34.21), we find

$$M = \frac{\theta'}{\theta} = \frac{y/f}{y/25\text{ cm}} = \frac{25\text{ cm}}{f} \tag{34.22}$$

(angular magnification for a simple magnifier)

It may seem that we can make the angular magnification as large as we like by decreasing the focal length f. In fact, the aberrations of a simple double-convex lens set a limit to M of about 3× to 4×. If these aberrations are corrected, the angular magnification may be made as great as 20×. When greater magnification than this is needed, we usually use a compound microscope, discussed in the next section.

Test Your Understanding

You have two plastic lenses, one double-convex and the other double-concave, each with a focal length with absolute value 10.0 cm. Which lens can you use as a simple magnifier, and what is its angular magnification?

34.8 | Microscopes and Telescopes

Cameras, eyeglasses, and magnifiers use a single lens to form an image. Two important optical devices that use *two* lenses are the microscope and the telescope. In each device a primary lens, or *objective,* forms a real image, and a second lens, or *eyepiece,* is used as a magnifier to make an enlarged virtual image.

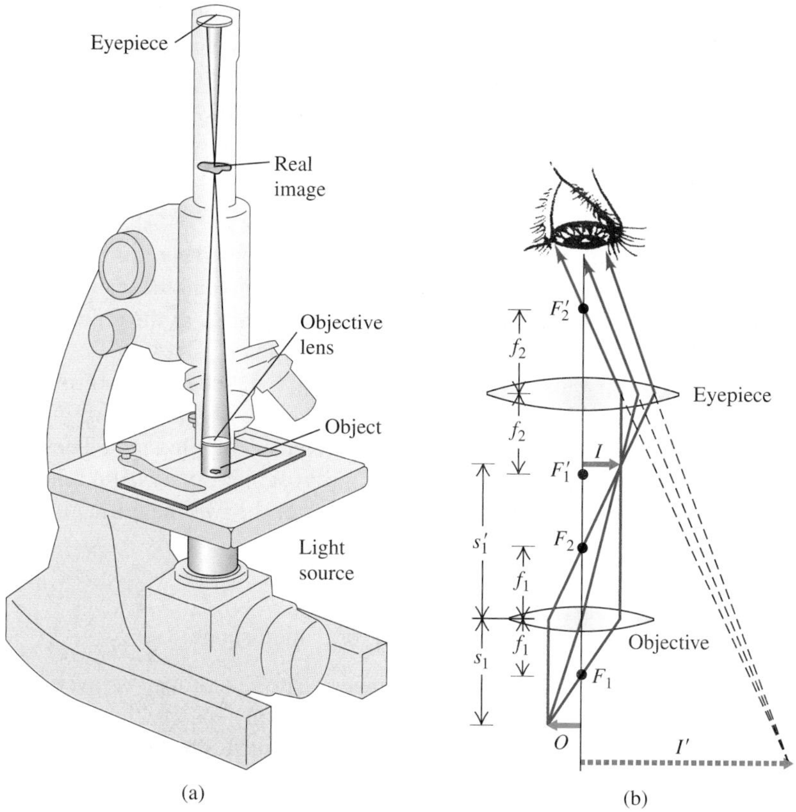

34.47 (a) Elements of a microscope. (b) The object O is placed just outside the first focal point of the objective (the distance s_1 has been exaggerated for clarity). The objective forms a real, inverted image I inside the first focal point F_1' of the eyepiece. The eyepiece uses that image as an object and forms an enlarged virtual image I' that remains inverted.

Microscopes

When we need greater magnification than we can get with a simple magnifier, the instrument that we usually use is the **microscope,** sometimes called a *compound microscope*. The essential elements of a microscope are shown in Fig. 34.47a. To analyze this system, we use the principle that an image formed by one optical element such as a lens or mirror can serve as the object for a second element. We used this principle in Section 34.4 when we derived the thin-lens equation by repeated application of the single-surface refraction equation; we used this principle again in Example 34.11 (Section 34.4), in which the image formed by a lens was used as the object of a second lens.

The object O to be viewed is placed just beyond the first focal point F_1 of the **objective,** a converging lens that forms a real and enlarged image I (Fig. 34.47b). In a properly designed instrument this image lies just inside the first focal point F_1' of a second converging lens called the **eyepiece** or *ocular*. (The reason why the image should lie just *inside* F_1' is left for you to discover; see Problem 34.108.) The eyepiece acts as a simple magnifier, as discussed in Section 34.7, and forms a final virtual image I' of I. The position of I' may be anywhere between the near and far points of the eye. Both the objective and the eyepiece of an actual microscope are highly corrected compound lenses with several optical elements, but for simplicity we show them here as simple thin lenses.

As for a simple magnifier, what matters when viewing through a microscope is the *angular* magnification M. The overall angular magnification of the compound microscope is the product of two factors. The first factor is the *lateral* magnifica-

Activ Physics ONLINE
15.12 Two-Lens Optical Systems

tion m_1 of the objective, which determines the linear size of the real image I; the second factor is the *angular* magnification M_2 of the eyepiece, which relates the angular size of the virtual image seen through the eyepiece to the angular size that the real image I would have if you viewed it *without* the eyepiece. The first of these factors is given by

$$m_1 = -\frac{s_1'}{s_1}$$

where s_1 and s_1' are the object and image distances, respectively, for the objective lens. Ordinarily, the object is very close to the focal point, and the resulting image distance s_1' is very great in comparison to the focal length f_1 of the objective lens. Thus s_1 is approximately equal to f_1, and we can write $m_1 = -s_1'/f_1$.

The real image I is close to the focal point F_1' of the eyepiece, so to find the eyepiece angular magnification, we can use Eq. (34.22): $M_2 = (25\text{ cm})/f_2$, where f_2 is the focal length of the eyepiece (considered as a simple lens). The overall angular magnification M of the compound microscope (apart from a negative sign, which is customarily ignored) is the product of the two magnifications:

$$M = m_1 M_2 = \frac{(25\text{ cm})s_1'}{f_1 f_2} \qquad (34.23)$$

(angular magnification for a microscope)

where s_1', f_1, and f_2 are measured in centimeters. The final image is inverted with respect to the object. Microscope manufacturers usually specify the values of m_1 and M_2 for microscope components rather than the focal lengths of the objective and eyepiece.

Equation (34.23) shows that the angular magnification of a microscope can be increased by using an objective of shorter focal length f_1, thereby increasing m_1 and the size of the real image I. Most optical microscopes have a rotating "turret" with three or more objectives of different focal lengths so that the same object can be viewed at different magnifications. The eyepiece should also have a short focal length f_2 to help to maximize the value of M.

Telescopes

The optical system of a **telescope** is similar to that of a compound microscope. In both instruments the image formed by an objective is viewed through an eyepiece. The key difference is that the telescope is used to view large objects at large distances and the microscope is used to view small objects close at hand. Another difference is that many telescopes use a curved mirror, not a lens, as an objective.

An *astronomical telescope* is shown in Fig. 34.48. Because this telescope uses a lens as an objective, it is called a *refracting telescope* or *refractor*. The objective lens forms a real, reduced image I of the object. This image is the object for the eyepiece lens, which forms an enlarged virtual image of I. Objects that are viewed with a telescope are usually so far away from the instrument that the first image I is formed very nearly at the second focal point of the objective lens. If the final image I' formed by the eyepiece is at infinity (for most comfortable viewing by a normal eye), the first image must also be at the first focal point of the eyepiece. The distance between objective and eyepiece, which is the length of the telescope, is therefore the *sum* of the focal lengths of objective and eyepiece, $f_1 + f_2$.

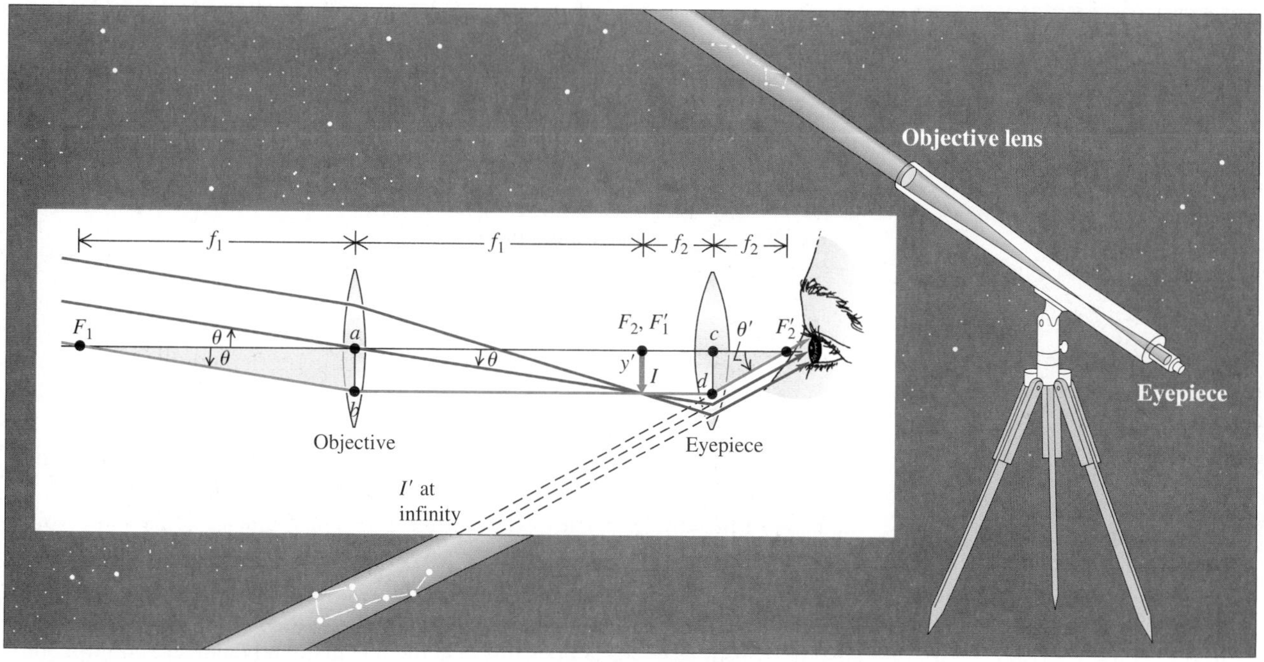

34.48 Optical system of an astronomical refracting telescope. The objective forms a real, inverted image I of the distant object (in this case, the Big Dipper) at its second focal point F_2; this point is also the first focal point F_1' of the eyepiece. The eyepiece uses that image as an object to form a magnified virtual image I at infinity that remains inverted.

The angular magnification M of a telescope is defined as the ratio of the angle subtended at the eye by the final image I' to the angle subtended at the (unaided) eye by the object. We can express this ratio in terms of the focal lengths of objective and eyepiece. In Fig. 34.48 the ray passing through F_1, the first focal point of the objective, and through F_2', the second focal point of the eyepiece, is shown as a darker color. The object (not shown) subtends an angle θ at the objective and would subtend essentially the same angle at the unaided eye. Also, since the observer's eye is placed just to the right of the focal point F_2', the angle subtended at the eye by the final image is very nearly equal to the angle θ'. Because bd is parallel to the optic axis, the distances ab and cd are equal to each other and also to the height y' of the real image I. Because the angles θ and θ' are small, they may be approximated by their tangents. From the right triangles F_1ab and $F_2'cd$,

ActivPhysics Online
15.3 The Telescope and Angular Magnification

$$\theta = \frac{-y'}{f_1} \qquad \theta' = \frac{y'}{f_2}$$

and the angular magnification M is

$$M = \frac{\theta'}{\theta} = -\frac{y'/f_2}{y'/f_1} = -\frac{f_1}{f_2} \tag{34.24}$$

(angular magnification for a telescope)

The angular magnification M of a telescope is equal to the ratio of the focal length of the objective to that of the eyepiece. The negative sign shows that the final image is inverted. Equation (34.24) shows that to achieve good angular magnification, a *tele-*

34.49 Built in 1897, this is the largest refracting telescope ever constructed. Its lens is one meter in diameter (D) and the length of the tube—essentially the focal length f—is 19.5 m. It is at the University of Chicago's Yerkes Observatory in Wisconsin.

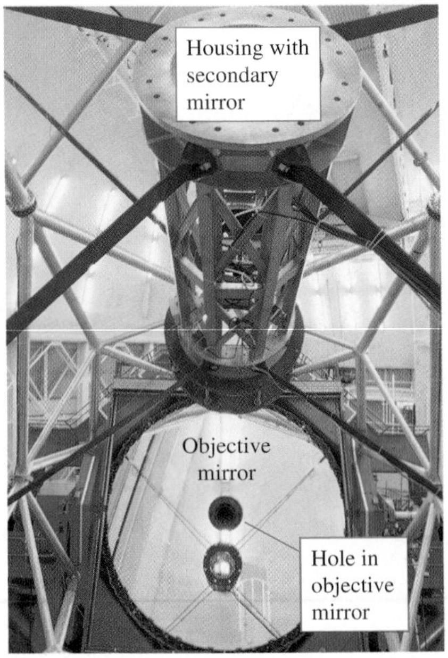

34.50 This photograph shows the interior of the Gemini North telescope atop Mauna Kea in Hawaii. Starlight is reflected from the 8-m diameter objective mirror, then reflected by a secondary mirror into the hole at the center of the objective.

scope should have a *long* objective focal length f_1. By contrast, Eq. (34.33) shows that a *microscope* should have a *short* objective focal length. However, a telescope objective with a long focal length should also have a large diameter D so that the f-number f_1/D will not be too large; as described in Section 34.5, a large f-number means a dim, low-intensity image (Fig. 34.49). Telescopes typically do not have interchangeable objectives; instead, the magnification is varied by using different eyepieces with different focal lengths f_2. Just as for a microscope, smaller values of f_2 give larger angular magnifications.

An inverted image is no particular disadvantage for astronomical observations. When we use a telescope or binoculars—essentially a pair of telescopes mounted side by side—to view objects on the earth, though, we want the image to be right-side up. In prism binoculars, this is accomplished by reflecting the light several times along the path from the objective to the eyepiece. The combined effect of the reflections is to flip the image both horizontally and vertically. Binoculars are usually described by two numbers separated by a multiplication sign, such as 7×50. The first number is the angular magnification M, and the second is the diameter of the objective lenses (in millimeters). The diameter helps to determine the light-gathering capacity of the objective lenses and thus the brightness of the image.

In the *reflecting telescope* (Fig. 34.50) the objective lens is replaced by a concave mirror. In large telescopes this scheme has many advantages, both theoretical and practical. Mirrors are inherently free of chromatic aberrations (dependence of focal length on wavelength), and spherical aberrations (associated with the paraxial approximation) are easier to correct than with a lens. The reflecting surface is sometimes parabolic rather than spherical. The material of the mirror need not be transparent, and it can be made more rigid than a lens, which has to be supported only at its edges.

The largest reflecting telescopes in the world, the Keck telescopes atop Mauna Kea in Hawaii, each have a mirror of overall diameter 10 m made up of 36 separate hexagonal reflecting elements. Lenses larger than 1 m in diameter are usually not practical.

One challenge in designing reflecting telescopes is that the image is formed in front of the objective mirror, in a region traversed by incoming rays. Figure 34.50 shows one way to meet this challenge: the light rays reflected from the objective are intercepted by a secondary mirror, which reflects the light into a hole in the objective. The image is formed some distance on the other side of the objective. (You can also see a hole of this kind at the center of the telescope mirror in the photograph that opens this chapter.)

When a telescope is used for photography, the eyepiece is removed, and either photographic film or an electronic detector is placed at the position of the real image formed by the objective. (Some long-focal-length "lenses" for photography are actually reflecting telescopes used in this way.) Most telescopes used for astronomical research are never used with an eyepiece.

Test Your Understanding

The photograph that opens this chapter shows a mirror with effective diameter $D = 8.12$ m. (This is smaller than the true outer diameter because of the hole at its center.) The f-number of this mirror is 13.4. What is the focal length of the mirror? Can you suggest another reason why it's helpful to use a secondary mirror?

SUMMARY

When rays diverge from an object point P and are reflected or refracted, the directions of the outgoing rays are the same as though they had diverged from a point P' called the image point. If they actually converge at P' and diverge again beyond it, P' is a real image of P; if they only appear to have diverged from P', it is a virtual image. Images can be either erect or inverted.

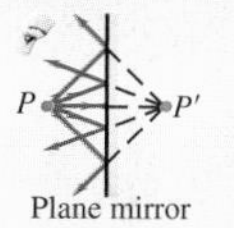

The lateral magnification m in any reflecting or refracting situation is defined as the ratio of image height y' to object height y. When m is positive, the image is erect; when m is negative, the image is inverted.

$$m = \frac{y'}{y} \quad (34.2)$$

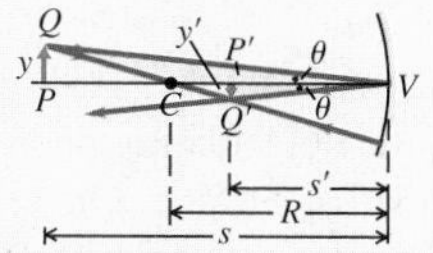

Object–image relations derived in this chapter are valid only for rays close to and nearly parallel to the optic axis; these are called paraxial rays. Nonparaxial rays do not converge precisely to an image point. This effect is called spherical aberration.

The focal point of a mirror is the point where parallel rays converge after reflection from a concave mirror, or the point from which they appear to diverge after reflection from a convex mirror. Rays diverging from the focal point of a concave mirror are parallel after reflection; and rays converging toward the focal point of a convex mirror are parallel after reflection. The distance from the focal point to the vertex is called the focal length, denoted as f. The focal points of a lens are defined similarly.

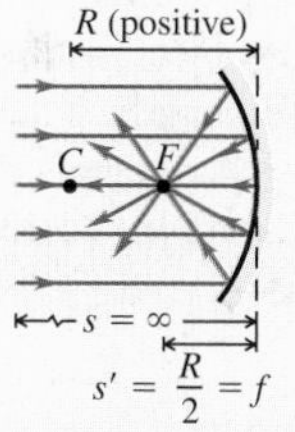

The formulas for object distance s and image distance s' for plane and spherical mirrors and single refracting surfaces are summarized in the following table. The equation for a plane surface can be obtained from the corresponding equation for a spherical surface by setting $R = \infty$. (See Examples 34.1 through 34.7)

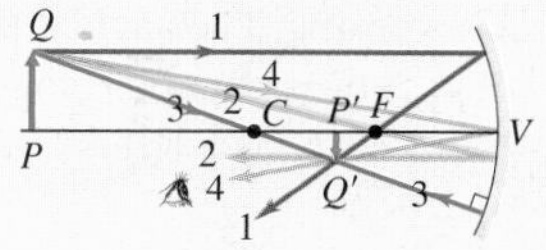

	Plane Mirror	Spherical Mirror	Plane Refracting Surface	Spherical Refracting Surface
Object and image distances	$\frac{1}{s} + \frac{1}{s'} = 0$	$\frac{1}{s} + \frac{1}{s'} = \frac{2}{R} = \frac{1}{f}$	$\frac{n_a}{s} + \frac{n_b}{s'} = 0$	$\frac{n_a}{s} + \frac{n_b}{s'} = \frac{n_b - n_a}{R}$
Lateral magnification	$m = -\frac{s'}{s} = 1$	$m = -\frac{s'}{s}$	$m = -\frac{n_a s'}{n_b s} = 1$	$m = -\frac{n_a s'}{n_b s}$

The object–image relation, given by Eq. (34.16), is the same for a thin lens as for a spherical mirror. Equation (34.19), the lensmaker's equation, relates the focal length of a lens to its index of refraction and the radii of curvature of its surfaces. (See Examples 34.8 through 34.11)

$$\frac{1}{s} + \frac{1}{s'} = \frac{1}{f} \quad (34.16)$$

$$\frac{1}{f} = (n - 1)\left(\frac{1}{R_1} - \frac{1}{R_2}\right) \quad (34.19)$$

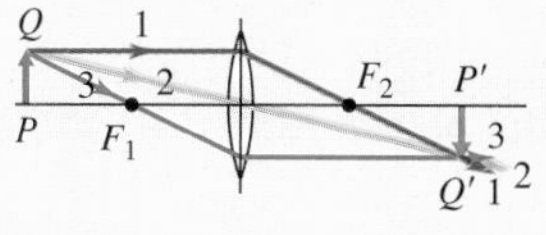

The following sign rules are used with all plane and spherical reflecting and refracting surfaces.

- $s > 0$ when the object is on the incoming side of the surface (a real object); $s < 0$ otherwise.
- $s' > 0$ when the image is on the outgoing side of the surface (a real image); $s' < 0$ otherwise.
- $R > 0$ when the center of curvature is on the outgoing side of the surface; $R < 0$ otherwise.
- $m > 0$ when the image is erect; $m < 0$ when inverted.

A camera forms a real, inverted, reduced image of the object being photographed on a light-sensitive surface. The amount of light striking this surface is controlled by the shutter speed and the aperture. The intensity of this light is inversely proportional to the square of the *f*-number of the lens. (See Example 34.12)

$$f\text{-number} = \frac{\text{Focal length}}{\text{Aperture diameter}} = \frac{f}{D} \qquad (34.20)$$

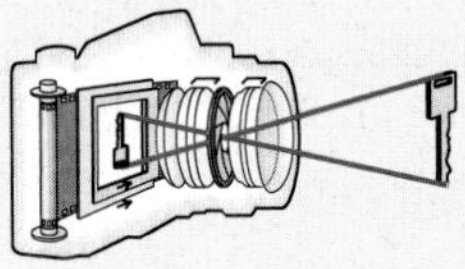

In the eye, refraction at the surface of the cornea forms a real image on the retina. Adjustment for various object distances is made by squeezing the lens, thereby making it bulge and decreasing its focal length. A nearsighted eye is too long for its lens; a farsighted eye is too short. The power of a corrective lens, in diopters, is the reciprocal of the focal length in meters. (See Examples 34.13 and 34.14)

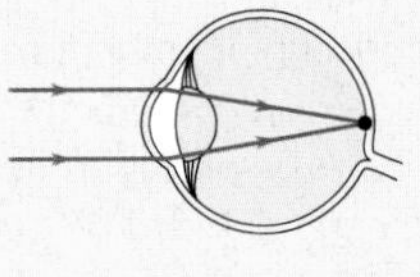

The simple magnifier creates a virtual image whose angular size θ' is larger than the angular size θ of the object itself at a distance of 25 cm, the nominal closest distance for comfortable viewing. The angular magnification M of a simple magnifier is the ratio of the angular size of the virtual image to that of the object at this distance.

$$M = \frac{\theta'}{\theta} = \frac{25\text{ cm}}{f} \qquad (34.22)$$

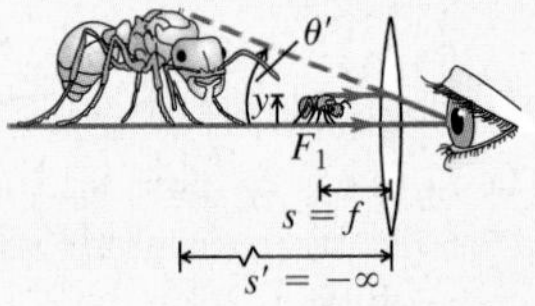

In a compound microscope, the objective lens forms a first image in the barrel of the instrument, and the eyepiece forms a final virtual image, often at infinity, of the first image. The telescope operates on the same principle, but the object is far away. In a reflecting telescope, the objective lens is replaced by a concave mirror, which eliminates chromatic aberrations.

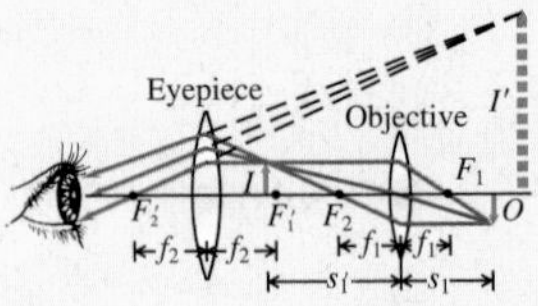

Key Terms

object, 1286
point object, 1286
extended object, 1286
image, 1286
virtual image, 1287
real image, 1287
object distance, 1287
image distance, 1287
lateral magnification, 1288
erect image, 1288
inverted image, 1288
reversed image, 1289
center of curvature, 1289
vertex, 1289
optic axis, 1289
paraxial rays, 1291
paraxial approximation, 1291
spherical aberration, 1291
focal point, 1292
focal length, 1292
principal rays, 1297
thin lens, 1304
converging lens, 1304
diverging lens, 1306
lensmaker's equation, 1308
camera, 1313
f-number, 1315
power, 1320
diopter, 1320
angular size, 1321
magnifier, 1321
angular magnification, 1322
microscope, 1323
objective, 1323
eyepiece, 1323
telescope, 1324

Answer to Chapter Opening Question

Lenses are plagued by chromatic aberrations. By contrast, a mirror is intrinsically free of chromatic aberrations. Furthermore, its shape can be maintained by supports all along its back surface.

Answers to Test Your Understanding Questions

Section 34.1 When you are a distance s from the mirror, your image is a distance s on the other side of the mirror and the distance from you to your image is $2s$. As you move toward the mirror, the distance $2s$ changes at twice the rate of the distance s, so your image moves toward you at speed $2v$.

Section 34.2 A convex mirror always produces an erect image, but that image is smaller than the object (see Fig. 34.14b). Hence a concave mirror must be used. The image will be erect and enlarged only if the distance from the object (your face) to the mirror is less than the focal length of the mirror, as in Fig. 34.18d.

Section 34.3 The sun is very far away, so the object distance is essentially infinite: $s = \infty$ and $1/s = 0$. Material a is air $(n_a = 1.00)$ and material b is water, so the image position s' is given by

$$\frac{n_a}{s} + \frac{n_b}{s'} = \frac{n_b - n_a}{R} \quad \text{or} \quad 0 + \frac{1.33}{s'} = \frac{1.33 - 1.00}{R}$$

$$s' = \frac{1.33}{0.33}R = 4.0R$$

The image would be formed 4.0 drop radii from the front surface of the drop. But since each drop is only a part of a complete sphere, the distance from the front to the back of the drop is less than $2R$. Thus the rays of sunlight never reach the image point, and the drops do not form an image of the sun on the leaf. While the rays are not focused to a point, they are nonetheless concentrated and can cause damage to the leaf.

Section 34.4 The key idea is that *all* rays from the object that enter the lens are refracted to form the image, not just the principal rays. With the lens cut away in the fashion described, principal rays 2 and 3 do not strike the lens and hence do not contribute to the image. But all of the rays that strike the remaining portion of the lens come to a focus to form the same image I. The only difference is that since there are fewer rays, the image is dimmer. (Compare Conceptual Example 34.2 in Section 34.2.)

Section 34.5 The larger film used with a portrait camera means that the angle of view is increased for a given focal length. Individual objects make images of the same size on either type of film; when a larger piece of film is used, more images fit on to the film and the field of view is wider.

Section 34.6 This lens is designed to correct for a type of astigmatism. Along the vertical axis, the lens is configured as a converging lens; along the horizontal axis, the lens is configured as a diverging lens. Hence the eye is hyperopic (see Fig. 34.43) for objects that are oriented vertically but myopic for objects that are oriented horizontally (see Fig. 34.44). Without correction, the eye focuses vertical objects behind the retina but horizontal objects in front of the retina.

Section 34.7 To form a virtual image as in Fig. 34.46, a converging lens (positive focal length) is required. So the double-convex lens is the one to use (see Fig. 34.29 in Section 34.4). From Eq. (34.22), the angular magnification with this lens (focal length $f = +10$ cm) is

$$M = \frac{25\text{ cm}}{10\text{ cm}} = 2.5 \quad \text{or} \quad 2.5\times$$

Section 34.8 The f-number is equal to the ratio f/D, so $f = 13.4D = 13.4(8.12\text{ m}) = 109$ m. Were it not for the secondary mirror, astronomers would have to locate a detector 109 m (357 ft) in front of the objective mirror, which is not a very convenient location. By using a secondary mirror, the image path can be folded on itself so that the overall dimensions of the observatory can be dramatically reduced.

Discussion Questions

Q34.1 Can you see a real image by looking backward along the direction from which the rays come? Can you see a virtual image? Explain. Can you tell by looking whether an image is real or virtual? If not, how *can* the two be distinguished?

Q34.2 For the situation shown in Fig. 34.3, is the image distance s' positive or negative? Is the image real or virtual? Explain your answers.

Q34.3 The laws of optics also apply to electromagnetic waves invisible to the eye. A satellite TV dish is used to detect radio waves coming from orbiting satellites. Why is a curved reflecting surface (a "dish") used? The dish is always concave, never convex; why? The actual radio receiver is placed on an arm and suspended in front of the dish. How far in front of the dish should it be placed?

Q34.4 Explain why the focal length of a *plane* mirror is infinite, and explain what it means for the focal point to be at infinity.

Q34.5 If a spherical mirror is immersed in water, does its focal length change? Explain.

Q34.6 For what range of object positions does a concave spherical mirror form a real image? What about a convex spherical mirror?

Q34.7 When a room has mirrors on two opposite walls, an infinite series of reflections can be seen. Discuss this phenomenon in terms of images. Why do the distant images appear fainter?

Q34.8 For a spherical mirror, if $s = f$, then $s' = \infty$, and the lateral magnification m is infinite. Does this make sense? If so, what does it mean?

Q34.9 A concave mirror (sometimes surrounded by lights) is often used as an aid for applying cosmetics to the face. Why is such a mirror always concave rather than convex? What considerations determine its radius of curvature?

Q34.10 A student claims that she can start a fire on a sunny day using just the sun's rays and a concave mirror. How is this done? Is the concept of image relevant? Can she do the same thing with a convex mirror? Explain.

Q34.11 A person looks at his reflection in the concave side of a shiny spoon. Is it right side up or inverted? Does it matter how far his face is from the spoon? What if he looks in the convex side? (Try this yourself!)

Q34.12 In Example 34.4 (Section 34.2), there appears to be an ambiguity for the case $s = 10$ cm, as to whether s' is $+\infty$ or $-\infty$

and as to whether the image is erect or inverted. How is this resolved? Or is it?

Q34.13 Suppose that in the situation of Example 34.7 of Section 34.3 (see Fig. 34.24) a vertical arrow 2.00 m tall is painted on the side of the pool beneath the water line. According to the calculations in the example, this arrow would appear to the person shown in Fig. 34.24 to be 1.50 m long. But the discussion following Eq. (34.13) states that the magnification for a plane refracting surface is $m = +1$, which suggests that the arrow would appear to the person to be 2.00 m long. How can you resolve this apparent contradiction?

Q34.14 There have been reports of round fishbowls starting fires by focusing the sun's rays coming in a window. Is this possible? Explain.

Q34.15 How could you very quickly make an approximate measurement of the focal length of a converging lens? Could the same method be applied if you wished to use a diverging lens? Explain.

Q34.16 The focal length of a simple lens depends on the color (wavelength) of light passing through it. Why? Is it possible for a lens to have a positive focal length for some colors and negative for others? Explain.

Q34.17 When a converging lens is immersed in water, does its focal length increase or decrease in comparison with the value in air? Explain.

Q34.18 A spherical air bubble in water can function as a lens. Is it a converging or diverging lens? How is its focal length related to its radius?

Q34.19 Can an image formed by one reflecting or refracting surface serve as an object for a second reflection or refraction? Does it matter whether the first image is real or virtual? Explain.

Q34.20 If a piece of photographic film is placed at the location of a real image, the film will record the image. Can this be done with a virtual image? How might one record a virtual image?

Q34.21 According to the discussion in Section 34.2, light rays are reversible. Are the formulas in the table in this chapter's summary still valid if object and image are interchanged? What does reversibility imply with respect to the *forms* of the various formulas?

Q34.22 While lost in the mountains, a person who is very far-sighted in one eye and barely farsighted in the other made a crude emergency telescope from the two lenses of his eyeglasses. How did he do this?

Q34.23 You can't see clearly underwater with the naked eye, but you *can* if you wear a face mask or goggles (with air between your eyes and the mask or goggles). Why is there a difference? Could you instead wear eyeglasses (with water between your eyes and the eyeglasses) in order to see underwater? If so, should the lenses be converging or diverging? Explain.

Q34.24 You are marooned on a desert island and want to use your eyeglasses to start a fire. Can this be done if you are nearsighted? If you are farsighted? Explain.

Exercises

Section 34.1 Reflection and Refraction at a Plane Surface

34.1 As shown in Fig. 34.8, mirror 1 uses the image P_2' formed by mirror 2 as an object and forms an image of it. Show that this image is at point P_3' in the figure.

34.2 The image of a tree just covers the length of a plane mirror 4.00 cm tall when the mirror is held 35.0 cm from the eye. The tree is 28.0 m from the mirror. What is its height?

34.3 A candle 4.85 cm tall is 39.2 cm to the left of a plane mirror. Where is the image formed by the mirror, and what is the height of this image?

Section 34.2 Reflection at a Spherical Surface

34.4 A concave mirror has a radius of curvature of 34.0 cm. a) What is its focal length? b) If the mirror is immersed in water (refractive index 1.33), what is its focal length?

34.5 An object 0.600 cm tall is placed 16.5 cm to the left of the vertex of a concave spherical mirror having a radius of curvature of 22.0 cm. a) Draw a principal-ray diagram showing formation of the image. b) Determine the position, size, orientation, and nature (real or virtual) of the image.

34.6 Repeat Exercise 34.5 for the case in which the mirror is convex.

34.7 The diameter of Mars is 6794 km, and its minimum distance from the earth is 5.58×10^7 km. When Mars is at this distance, find the diameter of the image of Mars formed by a spherical, concave, telescope mirror with a focal length of 1.75 m.

34.8 An object is 24.0 cm from the center of a silvered spherical glass Christmas tree ornament 6.00 cm in diameter. What are the position and magnification of its image?

34.9 a) Show that Eq. (34.6) can be written as $s' = sf/(s - f)$ and hence that the lateral magnification, given by Eq. (34.7), can be expressed as $m = f/(f - s)$. b) Consider a concave mirror, which has $f > 0$, and let $s > f$ (that is, let the object lie outside the focal point). Show that the image distance s' is *positive,* and explain why this means that the image is *real.* Also show that the image is *inverted.* c) Show that if the object distance s is greater than $2f$, the image is smaller than the object, and that if s is between f and $2f$, the image is larger than the object. d) For a concave mirror, show that if $0 < s < f$ (that is, if the object lies inside the focal point), the image distance s' is *negative.* Explain why this means that the image is *virtual.* Show also that the image is *erect* and larger than the object.

34.10 Consider a convex mirror, which has $f < 0$, so that $f = -|f|$. Use the results of part (a) of Exercise 34.9 to show that for *any* positive object distance s, the image distance s' is negative and the image is virtual, erect, and smaller than the object.

34.11 a) Using the formulas for s' and m obtained in part (a) of Exercise 34.9, graph s' as a function of s for the case $f > 0$ (a concave mirror). b) For what values of s is s' positive, so that the image is real? c) For what values of s is s' negative, so that the image is virtual? d) Where is the image if the object is just outside the focal point (s slightly greater than f)? e) If the object is just inside the focal point (s slightly less than f)? f) Where is the image if the object is at infinity? g) Where is the image if the object is next to the mirror $(s = 0)$? h) Graph m as a function of s for the case of a concave mirror. i) For which values of s is the image erect? j) For what values of s is the image inverted? k) For which values of s is the image larger than the object? l) For which values of s is the image smaller than the object? m) What happens to the size of the image when the object is placed at the focal point?

34.12 Using the formulas for s' and m obtained in part (a) of Exercise 34.9, graph s' as a function of s, and graph m as a function of

s, for the case $f < 0$ (a convex mirror), so that $f = -|f|$. a) For which values of s is s' positive? b) For what values of s is s' negative? c) Where is the image if the object is at infinity? d) Where is the image if the object is next to the mirror $(s = 0)$? For which values of s is the image e) erect; f) inverted; g) larger than the object; h) smaller than the object?

34.13 An object 9.00 mm tall is placed 12.0 cm to the left of the vertex of a convex spherical mirror whose radius of curvature has a magnitude of 20.0 cm. a) Draw a principal-ray diagram showing formation of the image. b) Determine the position, size, orientation, and nature (real or virtual) of the image.

34.14 A spherical, concave, shaving mirror has a radius of curvature of 32.0 cm. a) What is the magnification of a person's face when it is 12.0 cm to the left of the vertex of the mirror? b) Where is the image? Is the image real or virtual? c) Draw a principal-ray diagram showing formation of the image.

Section 34.3 Refraction at a Spherical Surface

34.15 A speck of dirt is embedded 3.50 cm below the surface of a sheet of ice $(n = 1.309)$. What is its apparent depth when viewed at normal incidence?

34.16 A tank whose bottom is a mirror is filled with water to a depth of 20.0 cm. A small fish floats motionless 7.0 cm under the surface of the water. a) What is the apparent depth of the fish when viewed at normal incidence? b) What is the apparent depth of the image of the fish when viewed at normal incidence?

34.17 Equations (34.11) and (34.12) were derived in the text for the case in which R is positive and $n_a < n_b$ (see Figs. 34.19 and 34.20). a) Carry through the derivation of these two equations for the case $R > 0$ and $n_a > n_b$. b) Carry through the derivation for $R < 0$ and $n_a < n_b$.

34.18 The left end of a long glass rod 6.00 cm in diameter has a convex hemispherical surface 3.00 cm in radius. The refractive index of the glass is 1.60. Determine the position of the image if an object is placed in air on the axis of the rod at the following distances to the left of the vertex of the curved end: a) infinitely far, b) 12.0 cm, c) 2.00 cm.

34.19 The rod of Exercise 34.18 is immersed in a liquid. An object 90.0 cm from the vertex of the left end of the rod and on its axis is imaged at a point 1.60 m inside the rod. What is the refractive index of the liquid?

34.20 The left end of a long glass rod 8.00 cm in diameter, with an index of refraction 1.60, is ground and polished to a convex hemispherical surface with a radius of 4.00 cm. An object in the form of an arrow 1.50 mm tall, at right angles to the axis of the rod, is located on the axis 24.0 cm to the left of the vertex of the convex surface. Find the position and height of the image of the arrow formed by paraxial rays incident on the convex surface. Is the image erect or inverted?

34.21 Repeat Exercise 34.20 for the case in which the end of the rod is ground to a *concave* hemispherical surface with radius 4.00 cm.

34.22 A Spherical Fish Bowl. A small tropical fish is at the center of a water-filled spherical fish bowl 28.0 cm in diameter. a) Find the apparent position and magnification of the fish to an observer outside the bowl. The effect of the thin walls of the bowl may be ignored. b) A friend advised the owner of the bowl to keep it out of direct sunlight to avoid blinding the fish, which might swim into the focal point of the parallel rays from the sun. Is the focal point actually within the bowl?

Section 34.4 Thin Lenses

34.23 A converging lens has a focal length of 14.0 cm. For an object to the left of the lens, at distances of 18.0 cm and 7.00 cm, determine a) the image position; b) the magnification; c) whether the image is real or virtual; d) whether the image is erect or inverted. Draw a principal-ray diagram in each case.

34.24 A lens forms an image of an object. The object is 16.0 cm from the lens. The image is 12.0 cm from the lens on the same side as the object. a) What is the focal length of the lens? Is the lens converging or diverging? b) If the object is 8.50 mm tall, how tall is the image? Is it erect or inverted? c) Draw a principal-ray diagram.

34.25 A converging lens with a focal length of 7.00 cm forms an image of a 4.00-mm-tall real object that is to the left of the lens. The image is 1.30 cm tall and erect. Where are the object and image located? Is the image real or virtual?

34.26 A converging lens with a focal length of 90.0 cm forms an image of a 3.20-cm-tall real object that is to the left of the lens. The image is 4.50 cm tall and inverted. Where are the object and image located in relation to the lens? Is the image real or virtual?

34.27 A diverging meniscus lens (see Fig. 34.29b) with a refractive index of 1.48 has spherical surfaces whose radii are 5.00 cm and 3.50 cm. What is the position of the image if an object is placed 18.0 cm to the left of the lens?

34.28 A photographic slide is to the left of a lens. The lens projects an image of the slide onto a wall 6.00 m to the right of the slide. The image is 80.0 times the size of the slide. a) How far is the slide from the lens? b) Is the image erect or inverted? c) What is the focal length of the lens? Is the lens converging or diverging?

34.29 Sketch the various possible thin lenses that can be obtained by combining two surfaces whose radii of curvature are 4.00 cm and 8.00 cm in absolute magnitude. Which are converging and which are diverging? Find the focal length of each if the surfaces are made of glass with index of refraction 1.60.

34.30 Six lenses in air are shown in Fig. 34.29. Each lens is made of a material with index of refraction $n > 1$. Considering each lens individually, imagine that light enters the lens from the left. Show that the three lenses shown in Fig. 34.29a have *positive* focal lengths and hence are *converging* lenses. In addition, show that the three lenses in Fig. 34.29b have *negative* focal lengths and hence are *diverging* lenses.

34.31 Exercises 34.9 through 34.12 deal with spherical mirrors. a) Show that the equations for s' and m derived in part (a) of Exercise 34.9 also apply to a thin lens. b) A concave mirror is used in Exercises 34.9 and 34.11. Repeat these exercises for a converging lens. Are there any differences in the results when the mirror is replaced by a lens? Explain. c) A convex mirror is used in Exercises 34.10 and 34.12. Repeat these exercises for a diverging lens. Are there any differences in the results when the mirror is replaced by a lens? Explain.

34.32 A converging lens with a focal length of 12.0 cm forms a virtual image 8.00 mm tall, 17.0 cm to the right of the lens. Determine the position and size of the object. Is the image erect or inverted? Are the object and image on the same side or opposite sides of the lens? Draw a principal-ray diagram for this situation.

34.33 Repeat Exercise 34.32 for the case in which the lens is diverging, with a focal length of -48.0 cm.

34.34 An object is 16.0 cm to the left of a lens. The lens forms an image 36.0 cm to the right of the lens. a) What is the focal length of the lens? Is the lens converging or diverging? b) If the object is 8.00 mm tall, how tall is the image? Is it erect or inverted? c) Draw a principal-ray diagram.

Section 34.5 Cameras

34.35 Choosing a Camera Lens. The picture size on ordinary 35-mm camera film is 24 mm $\times$ 36 mm. Focal lengths of lenses available for 35-mm cameras typically include 28, 35, 50 (the "normal" lens), 85, 100, 135, 200, and 300 mm, among others. Which of these lenses should be used to photograph the following objects, assuming that the object is to fill most of the picture area? a) A building 240 m tall and 160 m wide at a distance of 600 m. b) A mobile home 9.6 m in length at a distance of 40.0 m.

34.36 When a camera is focused, the lens is moved away from or toward the film. If you take a picture of your friend, who is standing 3.90 m from the lens, using a camera with a lens with a 85-mm focal length, how far from the film is the lens? Will the whole image of your friend, who is 175 cm tall, fit on film that is 24 $\times$ 36 mm?

34.37 A camera lens has a focal length of 200 mm. How far from the lens should the subject for the photo be if the lens is 20.4 cm from the film?

34.38 A photographer takes a photograph of a Boeing 747 airliner (length 70.7 m) when it is flying directly overhead at an altitude of 9.50 km. The lens has a focal length of 5.00 m. How long is the image of the airliner on the film?

34.39 Figure 34.38 shows photographs of the same scene taken with the same camera with lenses of different focal length. If the object is 200 m from the lens, what is the magnitude of the lateral magnification for a lens of focal length a) 28 mm? b) 105 mm? c) 300 mm?

34.40 Zoom Lens. Consider the simple model of the zoom lens shown in Fig. 34.40a. The converging lens has focal length $f_1 = 12$ cm and the diverging lens has focal length $f_2 = -12$ cm. The lenses are separated by 4 cm as shown in Fig. 34.40a. a) For a distant object, where is the image of the converging lens? b) The image of the converging lens serves as the object for the diverging lens. What is the object distance for the diverging lens? c) Where is the final image? Compare your answer to Fig. 34.40a. d) Repeat parts (a), (b), and (c) for the situation shown in Fig. 34.40b, in which the lenses are separated by 8 cm.

34.41 The focal length of an *f*/4 camera lens is 300 mm. a) What is the aperture diameter of the lens? b) If the correct exposure of a certain scene is $\frac{1}{250}$ s at *f*/4, what is the correct exposure at *f*/8?

34.42 Camera A has a lens with an aperture diameter of 8.00 mm. It photographs an object using the correct exposure time of $\frac{1}{30}$ s. What exposure time should be used with camera B in photographing the same object with the same film if this camera has a lens with an aperture diameter of 23.1 mm?

34.43 Photography. A 35-mm camera has a standard lens with focal length 50 mm and can focus on objects between 45 cm and infinity. a) Is the lens for such a camera a concave or a convex lens? b) The camera is focused by rotating the lens, which moves it on the camera body and changes its distance from the film. In what range of distances between the lens and the film plane must the lens move to focus properly over the 45 cm to infinity range?

34.44 The dimensions of the picture on a 35-mm color slide are 24 mm $\times$ 36 mm. An image of the slide is projected on a screen 9.00 m from the projector lens. The focal length of the projector lens is 150 mm. a) How far is the slide from the lens? b) What are the dimensions of the image on the screen?

Section 34.6 The Eye

34.45 a) Where is the near point of an eye for which a contact lens with a power of $+2.75$ diopters is prescribed? b) Where is the far point of an eye for which a contact lens with a power of -1.30 diopters is prescribed for distant vision?

34.46 Curvature of the Cornea. In a simplified model of the human eye, the aqueous and vitreous humors and the lens all have a refractive index of 1.40, and all the refraction occurs at the cornea, whose vertex is 2.60 cm from the retina. What should be the radius of curvature of the cornea such that the image of an object 40.0 cm from the cornea's vertex is focused on the retina?

34.47 Corrective Lenses. Determine the power of the corrective contact lenses required by a) a hyperopic eye whose near point is at 60.0 cm; b) a myopic eye whose far point is at 60.0 cm.

Section 34.7 The Magnifier

34.48 A thin lens with a focal length of 6.00 cm is used as a simple magnifier. a) What angular magnification is obtainable with the lens if the object is at the focal point? b) When an object is examined through the lens, how close can it be brought to the lens? Assume that the image viewed by the eye is at the near point, 25.0 cm from the eye, and that the lens is very close to the eye.

34.49 The focal length of a simple magnifier is 8.00 cm. Assume the magnifier to be a thin lens placed very close to the eye. a) How far in front of the magnifier should an object be placed if the image is formed at the observer's near point, 25.0 cm in front of her eye? b) If the object is 1.00 mm high, what is the height of its image formed by the magnifier?

34.50 You want to view an insect 2.00 mm in length through a magnifier. If the insect is to be at the focal point of the magnifier, what focal length will give the image of the insect an angular size of 0.025 radian?

34.51 You are examining a flea with a magnifying lens that has focal length 4.00 cm. If the image of the flea is 6.50 times the size of the flea, how far is the flea from the lens? Where, relative to the lens, is the image?

Section 34.8 Microscopes and Telescopes

34.52 Resolution of a Microscope. The image formed by a microscope objective with a focal length of 5.00 mm is 160 mm from its second focal point. The eyepiece has a focal length of 26.0 mm. a) What is the angular magnification of the microscope? b) The unaided eye can distinguish two points at its near point as separate if they are about 0.10 mm apart. What is the minimum separation that can be resolved with this microscope?

34.53 The focal length of the eyepiece of a certain microscope is 18.0 mm. The focal length of the objective is 8.00 mm. The distance between objective and eyepiece is 19.7 cm. The final image formed by the eyepiece is at infinity. Treat all lenses as thin. a) What is the distance from the objective to the object being viewed? b) What is the magnitude of the linear magnification produced by the objective? c) What is the overall angular magnification of the microscope?

34.54 A certain microscope is provided with objectives that have focal lengths of 16 mm, 4 mm, and 1.9 mm and with eyepieces that have angular magnifications of 5× and 10×. Each objective forms an image 120 mm beyond its second focal point. Determine a) the largest overall angular magnification obtainable; b) the least overall angular magnification obtainable.

34.55 A telescope is constructed from two lenses with focal lengths of 95.0 cm and 15.0 cm, the 95.0-cm lens being used as the objective. Both the object being viewed and the final image are at infinity. a) Find the angular magnification for the telescope. b) Find the height of the image formed by the objective of a building 60.0 m tall, 3.00 km away. c) What is the angular size of the final image as viewed by an eye very close to the eyepiece?

34.56 The eyepiece of a refracting telescope (see Fig. 34.48) has a focal length of 9.00 cm. The distance between objective and eyepiece is 1.80 m, and the final image is at infinity. What is the angular magnification of the telescope?

34.57 The Yerkes refracting telescope of the University of Chicago has an objective 1.02 m in diameter with an *f*-number of 19.0. (This is the largest-diameter refracting telescope in the world.) What is its focal length?

34.58 As viewed from the earth, Jupiter subtends an angle of approximately 0.014°. What is the diameter of the image of Jupiter produced by the objective of the Lick Observatory refracting telescope of focal length 18 m?

34.59 A reflecting telescope (see Fig. 34.51a) is to be made by using a spherical mirror with a radius of curvature of 1.30 m and an eyepiece with a focal length of 1.10 cm. The final image is at infinity. a) What should the distance between the eyepiece and the mirror vertex be if the object is taken to be at infinity? b) What will the angular magnification be?

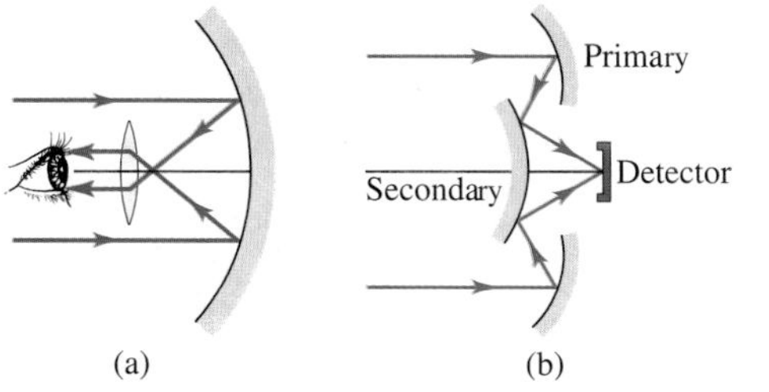

Figure 34.51 Exercises 34.59, 34.60, and Problem 34.112.

34.60 A reflecting telescope uses an optical system similar to that shown in Fig. 34.51b (Cassegrain system). The image of a distant galaxy is focused on the detector through a hole in the large (primary) mirror. The primary mirror has a focal length of 1.3 m, the distance between the vertexes of the two mirrors is 75 cm, and the distance from the vertex of the primary mirror to the detector is 12 cm. Should the smaller mirror (the secondary) be concave or convex? What should its radius of curvature be?

Problems

34.61 If you run away from a plane mirror at 2.40 m/s, at what speed does your image move away from you?

34.62 An object is placed between two plane mirrors arranged at right angles to each other at a distance d_1 from the surface of one mirror and a distance d_2 from the other. a) How many images are formed? Show the location of the images in a diagram. b) Draw the paths of rays from the object to the eye of an observer.

34.63 What is the size of the smallest vertical plane mirror in which a woman of height h can see her full-length image?

34.64 A luminous object is 4.00 m from a wall. You are to use a concave mirror to project an image of the object on the wall, with the image 2.25 times the size of the object. How far should the mirror be from the wall? What should its radius of curvature be?

34.65 A concave mirror is to form an image of the filament of a headlight lamp on a screen 8.00 m from the mirror. The filament is 6.00 mm tall, and the image is to be 36.0 cm tall. a) How far in front of the vertex of the mirror should the filament be placed? b) What should be the radius of curvature of the mirror?

34.66 Rear-View Mirror. A mirror on the passenger side of your car is convex and has a radius of curvature with magnitude 18.0 cm. a) Another car is seen in this side mirror and is 13.0 m behind the mirror. If this car is 1.5 m tall, what is the height of the image? b) The mirror has a warning attached that objects viewed in it are closer than they appear. Why is this so?

34.67 Virtual Object. If the light incident from the left onto a convex mirror does not diverge from an object point but instead converges toward a point at a (negative) distance s to the right of the mirror, this point is called a *virtual object.* a) For a convex mirror having a radius of curvature of 24.0 cm, for what range of virtual-object positions is a real image formed? b) What is the orientation of this real image? c) Draw a principal-ray diagram showing formation of such an image.

34.68 Equations (34.6) and (34.7) were derived in the text for the case of a concave mirror. Carry out a similar derivation for a convex mirror, and show that the same equations result if you use the sign convention established in the text.

34.69 Suppose the lamp filament shown in Example 34.1 (Section 34.2) is moved to a position 8.0 cm in front of the mirror. a) Where is the image located now? Is it real or virtual? b) What is the height of the image? Is it erect or inverted? c) In Example 34.1, the filament is 10.0 cm in front of the mirror, and an image of the filament is formed on a wall 3.00 m from the mirror. If the filament is 8.0 cm from the mirror, can a wall be placed so that an image is formed on it? If so, where should the wall be placed? If not, why not?

34.70 Where must you place an object in front of a concave mirror with radius R so that the image is erect and $2\frac{1}{2}$ times the size of the object? Where is the image?

34.71 A Three-Dimensional Image. The *longitudinal* magnification is defined as $m' = ds'/ds$. It relates the longitudinal dimension of a small object to the longitudinal dimension of its image. a) Show that for a spherical mirror, $m' = -m^2$. What is the significance of the fact that m' is *always* negative? b) A wire frame in the form of a small cube 1.00 mm on a side is placed with its center on the axis of a concave mirror with radius of curvature 150.0 cm. The sides of the cube are all either parallel or perpendicular to the axis. The cube face toward the mirror is 200.0 cm to the left of the mirror vertex. Find i) the location of the image of this face and of the opposite face of the cube; ii) the lateral and longitudinal magnifications; iii) the shape and dimensions of each of the six faces of the image.

34.72 Refer to Problem 34.71. Show that the longitudinal magnification m' for refraction at a spherical surface is given by

$$m' = -\frac{n_b}{n_a}m^2$$

34.73 As you sit in your parked car, you glance in the convex side mirror (radius of curvature 125 cm) and notice a jogger approaching. If the jogger is running at a speed of 2.50 m/s, how fast is her image moving when she is a) 10.0 m away; b) 2.0 m away?

34.74 A microscope is focused on the upper surface of a glass plate. A second plate is then placed over the first. To focus on the bottom surface of the second plate, the microscope must be raised 0.780 mm. To focus on the upper surface, it must be raised another 2.50 mm. Find the index of refraction of the second plate.

34.75 The primary reflecting surface of a bathroom mirror lies behind a thin sheet of glass. This sheet protects the reflecting surface and keeps it from tarnishing. Because some light reflects from the front surface of the glass, and because light refracts upon entering or leaving the glass, the formation of images is somewhat more complex than is shown in Fig. 34.2. Consider a bathroom mirror with a glass sheet of thickness d and index of refraction n. Your face is located a distance h from the front surface of the glass sheet. a) Find the position of the image of your face that you see due to reflection from the front surface of the glass. b) Find the position of the image of your face that you see due to reflection from the mirror surface behind the glass. Take into account the effects of refraction at the glass-air interface. c) How far apart are the images in parts (a) and (b)?

34.76 A Glass Rod. Both ends of a glass rod with an index of refraction 1.60 are ground and polished to convex hemispherical surfaces. The radius of curvature at the left end is 6.00 cm, and the radius of curvature at the right end is 12.0 cm. The length of the rod between vertexes is 40.0 cm. The object for the surface at the left end is an arrow that lies 23.0 cm to the left of the vertex of this surface. The arrow is 1.50 mm tall and at right angles to the axis. a) What constitutes the object for the surface at the right end of the rod? b) What is the object distance for this surface? c) Is the object for this surface real or virtual? (*Hint:* See Problem 34.67.) d) What is the position of the final image? e) Is the final image real or virtual? Is it erect or inverted with respect to the original object? f) What is the height of the final image?

34.77 The rod in Problem 34.76 is shortened to a distance of 25.0 cm between its vertexes; the curvatures of its ends remain the same. As in Problem 34.76, the object for the surface at the left end is an arrow that lies 23.0 cm to the left of the vertex of this surface. The arrow is 1.50 mm tall and at right angles to the axis. a) What is the object distance for the surface at the right end of the rod? b) Is the object for this surface real or virtual? c) What is the position of the final image? d) Is the final image real or virtual? Is it erect or inverted with respect to the original object? e) What is the height of the final image?

34.78 A layer of benzene $(n = 1.50)$ 2.60 cm deep floats on water $(n = 1.33)$ that is 6.50 cm deep. What is the apparent distance from the upper benzene surface to the bottom of the water layer when it is viewed at normal incidence?

34.79 What should be the index of refraction of a transparent sphere in order for paraxial rays from an infinitely distant object to be brought to a focus at the vertex of the surface opposite the point of incidence?

34.80 A transparent rod 50.0 cm long and with a refractive index of 1.60 is cut flat at the right end and rounded to a hemispherical surface with a 15.0-cm radius at the left end. An object is placed on the axis of the rod 12.0 cm to the left of the vertex of the hemispherical end. a) What is the position of the final image? b) What is its magnification?

34.81 A narrow beam of parallel rays enters a solid glass sphere in a radial direction. At what point outside the sphere are these rays brought to a focus? The radius of the sphere is 4.00 cm, and its index of refraction is 1.80.

34.82 A transparent rod 30.0 cm long is cut flat at one end and rounded to a hemispherical surface of radius 10.0 cm at the other end. A small object is embedded within the rod along its axis and halfway between its ends, 15.0 cm from the flat end and 15.0 cm from the vertex of the curved end. When viewed from the flat end of the rod, the apparent depth of the object is 9.50 cm from the flat end. What is its apparent depth when viewed from the curved end?

34.83 A solid glass hemisphere of radius 12.0 cm and index of refraction $n = 1.50$ is placed with its flat face downward on a table. A parallel beam of light with a circular cross section 3.80 mm in diameter travels straight down and enters the hemisphere at the center of its curved surface. a) What is the diameter of the circle of light formed on the table? b) How does your result depend on the radius of the hemisphere?

34.84 A thick-walled wine goblet sitting on a table can be considered to be a hollow glass sphere with an outer radius of 4.00 cm and an inner radius of 3.40 cm. The index of refraction of the goblet glass is 1.50. a) A beam of parallel light rays enters the side of the empty goblet along a horizontal radius. Where, if anywhere, will an image be formed? b) The goblet is filled with white wine $(n = 1.37)$. Where is the image formed?

34.85 A thin-walled glass sphere with radius R is filled with water. An object is placed a distance $3R$ from the surface of the sphere. Determine the position of the final image. The effect of the glass wall may be ignored. Use $n = \frac{4}{3}$ for water.

34.86 For refraction at a spherical surface, the first focal length f is defined as the value of s corresponding to $s' = \infty$, as shown in Fig. 34.52a. The second focal length f' is defined as the value

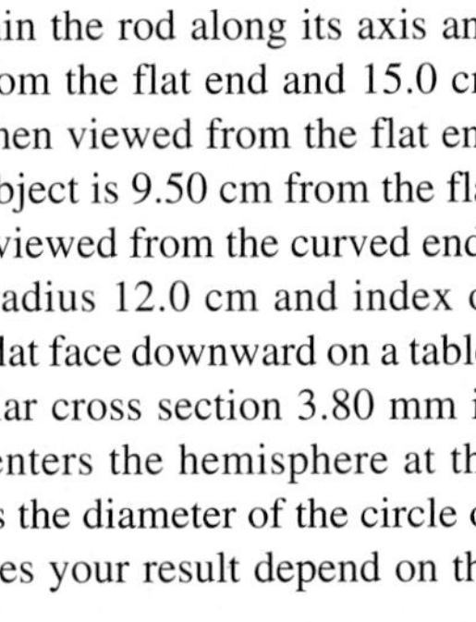

Figure 34.52 Problem 34.86.

of s' when $s = \infty$, as shown in Fig. 34.52b. a) Prove that $n_a/n_b = f/f'$. b) Prove that the general relation between object and image distance is

$$\frac{f}{s} + \frac{f'}{s'} = 1$$

34.87 An object to the left of a lens is imaged by the lens on a screen 30.0 cm to the right of the lens. When the lens is moved 4.00 cm to the right, the screen must be moved 4.00 cm to the left to refocus the image. Determine the focal length of the lens.

34.88 A glass rod with a refractive index of 1.55 is ground and polished at both ends to hemispherical surfaces with radii of 6.00 cm. When an object is placed on the axis of the rod, 25.0 cm to the left of the left-hand end, the final image is formed 65.0 cm to the right of the right-hand end. What is the length of the rod measured between the vertexes of the two hemispherical surfaces?

34.89 Two thin lenses with a focal length of magnitude 12.0 cm, the first converging and the second diverging, are placed 9.00 cm apart. An object 2.50 mm tall is placed 20.0 cm to the left of the first (converging) lens. a) How far from this first lens is the final image formed? b) Is the final image real or virtual? c) What is the height of the final image? Is the final image erect or inverted?

34.90 The radii of curvature of the surfaces of a thin converging meniscus lens are $R_1 = +12.0$ cm and $R_2 = +28.0$ cm. The index of refraction is 1.60. a) Compute the position and size of the image of an object in the form of an arrow 5.00 mm tall, perpendicular to the lens axis, 45.0 cm to the left of the lens. b) A second converging lens with the same focal length is placed 3.15 m to the right of the first. Find the position and size of the final image. Is the final image erect or inverted with respect to the original object? c) Repeat part (b) except with the second lens 45.0 cm to the right of the first.

34.91 Three thin lenses, each with a focal length of 40.0 cm, are aligned on a common axis; adjacent lenses are separated by 52.0 cm. Find the position of the image of a small object on the axis, 80.0 cm to the left of the first lens.

34.92 An object is placed 18.0 cm from a screen. a) At what two points between object and screen may a converging lens with a 3.00-cm focal length be placed to obtain an image on the screen? b) What is the magnification of the image for each position of the lens?

34.93 A convex mirror and a concave mirror are placed on the same optic axis, separated by a distance $L = 0.600$ m. The radius of curvature of each mirror has a magnitude of 0.360 m. A light source is located a distance x from the concave mirror, as shown in Fig. 34.53. a) What distance x will result in the rays from the source returning to the source after reflecting first from the convex mirror and then from the concave mirror? b) Repeat part (a), but now let the rays reflect first from the concave mirror and then from the convex one.

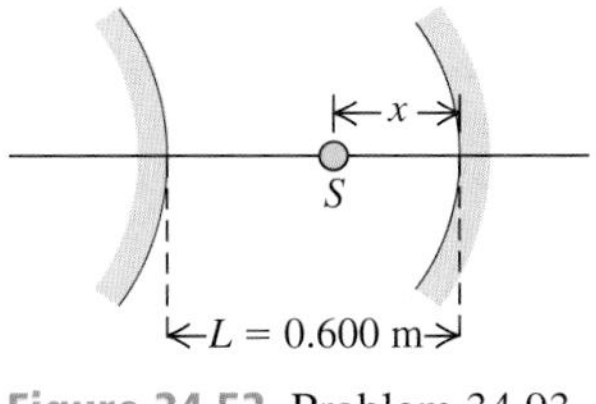

Figure 34.53 Problem 34.93.

34.94 As shown in Fig. 34.54 the candle is at the center of curvature of the concave mirror, whose focal length is 10.0 cm. The converging lens has a focal length of 32.0 cm and is 85.0 cm to the right of the candle. The candle is viewed looking through the lens from the right. The lens forms two images of the candle. The first is formed by light passing directly through the lens. The second image is formed from the light that goes from the candle to the mirror, is reflected, and then passes through the lens. a) For each of these two images, draw a principal-ray diagram that locates the image. b) For each image, answer the following questions: i) Where is the image? ii) Is the image real or virtual? iii) Is the image erect or inverted with respect to the original object?

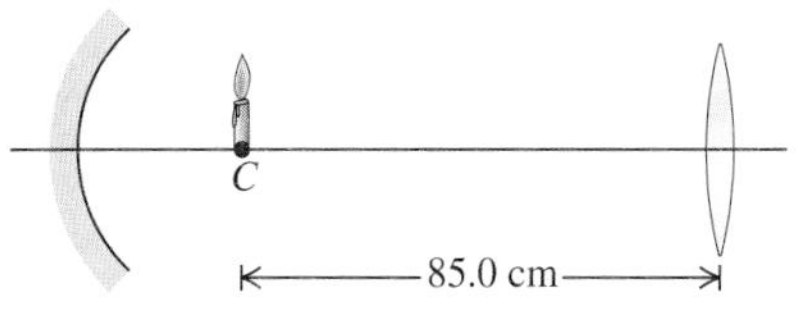

Figure 34.54 Problem 34.94.

34.95 An eyepiece consists of two converging thin lenses, each with a focal length 12.0 cm, separated by a distance of 4.00 cm. Where are the first and second focal points of the eyepiece?

34.96 Two Lenses in Contact. a) Prove that when two thin lenses with focal lengths f_1 and f_2 are placed *in contact*, the focal length f of the combination is given by the relation

$$\frac{1}{f} = \frac{1}{f_1} + \frac{1}{f_2}$$

b) A converging meniscus lens (see Fig. 34.29a) has index of refraction of 1.55 and radii of curvature for its surfaces of 4.50 cm and 9.00 cm. The concave surface is placed upward and filled with carbon tetrachloride (CCl_4), which has $n = 1.46$. What is the focal length of the CCl_4-glass combination?

34.97 Rays from a lens are converging toward a point image P located to the right of the lens. What thickness t of glass with an index of refraction 1.60 must be interposed between the lens and P for the image to be formed at P', located 0.30 cm to the right of P? The locations of the piece of glass and of points P and P' are shown in Fig. 34.55.

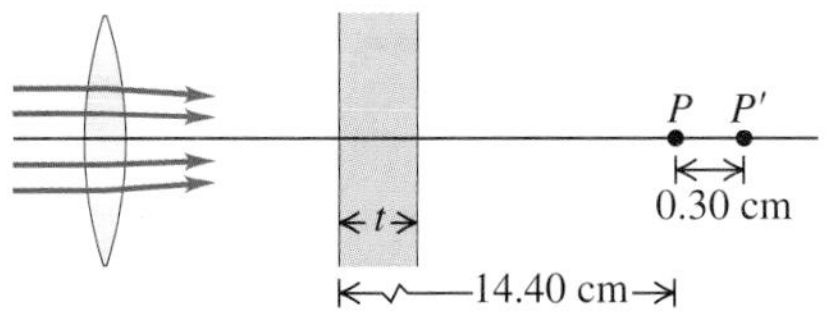

Figure 34.55 Problem 34.97.

34.98 A Lens in a Liquid. A lens obeys Snell's law, bending light rays at each surface an amount determined by the index of refraction of the lens and the index of the medium in which the lens is located. a) Equation (34.19) assumes that the lens is surrounded by air. Consider instead a thin lens immersed in a liquid with refractive index n_{liq}. Prove that the focal length f' is then given by Eq. (34.19) with n replaced by n/n_{liq}. b) A thin lens with index n has focal length f in vacuum. Use the result of part (a) to show that when this lens is immersed in a liquid of index n_{liq}, it will have a new focal length given by

$$f' = \left[\frac{n_{liq}(n-1)}{n - n_{liq}}\right] f$$

34.99 When an object is placed at the proper distance to the left of a converging lens, the image is focused on a screen 30.0 cm to the right of the lens. A diverging lens is now placed 15.0 cm to the right of the converging lens, and it is found that the screen must be moved 19.2 cm farther to the right to obtain a sharp image. What is the focal length of the diverging lens?

34.100 A convex spherical mirror with a focal length of magnitude 24.0 cm is placed 20.0 cm to the left of a plane mirror. An object 0.250 cm tall is placed midway between the surface of the plane mirror and the vertex of the spherical mirror. The spherical mirror forms multiple images of the object. Where are the two images of the object formed by the spherical mirror that are closest to the spherical mirror, and how tall is each image?

34.101 A glass plate 3.50 cm thick, with an index of refraction 1.55 and plane parallel faces, is held with its faces horizontal and its lower face 6.00 cm above a printed page. Find the position of the image of the page formed by rays making a small angle with the normal to the plate.

34.102 A symmetric double-convex thin lens made of glass with an index of refraction 1.52 has a focal length in air of 40.0 cm. The lens is sealed into an opening in the left-hand end of a tank filled with water. At the right-hand end of the tank, opposite the lens, is a plane mirror 90.0 cm from the lens. The index of refraction of the water is $\frac{4}{3}$. a) Find the position of the image formed by the lens-water-mirror system of a small object outside the tank on the lens axis and 70.0 cm to the left of the lens. b) Is the image real or virtual? c) Is it erect or inverted? d) If the object has a height of 4.00 mm, what is the height of the image?

34.103 Your camera has a lens with a 35.0-mm focal length and film that is 36.0 mm wide. In taking a picture of an airplane 22.7 m in length you find that the image of the airplane fills only $\frac{3}{4}$ of the width of the film. a) How far are you from the airplane? b) How close should you stand if you want to fill the width of the film with the airplane's image?

34.104 Resolution of a Camera. The *resolution* of a camera lens can be defined as the maximum number of lines per millimeter in the image that can barely be distinguished as separate lines. A certain lens has a focal length of 35.0 mm and resolution of 120 lines/mm. What is the minimum separation of two lines in an object 25.0 m away if they are to be visible in the image as separate lines?

34.105 Angle of View. The angle of view for a camera and lens is defined as the angle subtended by the widest object for which the width of the image equals the diagonal dimension d of the film. a) For a distant object for which the image distance is approximately equal to the focal length of the camera lens, show that the angle of view θ is given by $\theta = 2\arctan(d/2f)$. b) The film for a 35-mm camera is 24 mm by 36 mm. Use the result of part (a) to calculate the angle of view for focal lengths of 28 mm, 105 mm, and 300 mm. Compare your results to the angles of view shown in Fig. 34.38d.

34.106 A camera with a 90-mm focal length lens is focused on an object 1.30 m from the lens. To refocus on an object 6.50 m from the lens, by how much must the distance between the lens and the film be changed? To refocus on the more distant object, is the lens moved toward or away from the film?

34.107 The derivation of the expression for angular magnification, Eq. (34.22), assumed a near point of 25 cm. In fact, the near point changes with age as shown in Table 34.1. In order to achieve an angular magnification of 2.0×, what focal length should be used by a person of a) age 10? b) age 30? c) age 60? d) If the lens that gives $M = 2.0$ for a 10-year-old is used by a 60-year-old, what angular magnification will the older viewer obtain? e) Does your answer in (d) mean that older viewers are able to see more highly magnified images than younger viewers? Explain.

34.108 Angular Magnification. In deriving Eq. (34.22) for the angular magnification of a magnifier, we assumed that the object is placed at the focal point of the magnifier so that the virtual image is formed at infinity. Suppose instead that the object is placed so that the virtual image appears at an average viewer's near point of 25 cm, the closest point at which the viewer can bring an object into focus. a) Where should the object be placed to achieve this? Give your answer in terms of the magnifier focal length f. b) What angle θ' will an object of height y subtend at the position found in part (a)? c) Find the angular magnification M with the object at the position found in part (a). The angle θ is the same as in Fig. 34.46a, since it refers to viewing the object *without* the magnifier. d) For a convex lens with $f = +10.0$ cm, what is the value of M with the object at the position found in part (a)? How many times greater is M in this case than in the case where the image is formed at infinity? e) In the description of a compound microscope in Section 34.8, it is stated that in a properly designed instrument, the real image formed by the objective lies *just inside* the first focal point F_1' of the eyepiece. What advantages are gained by having the image formed by the objective be just inside F_1', as opposed to precisely at F_1'? What happens if the image formed by the objective is *just outside* F_1'?

34.109 In one form of cataract surgery the person's natural lens, which has become cloudy, is replaced by an artificial lens. The refracting properties of the replacement lens can be chosen so that the person's eye focuses on distant objects. But there is no accommodation, and glasses or contact lenses are needed for close vision. What is the power, in diopters, of the corrective contact lenses that will enable a person who has had such surgery to focus on the page of a book at a distance of 24 cm?

34.110 A Nearsighted Eye. A certain very nearsighted person cannot focus on anything farther than 36.0 cm from the eye. Consider the simplified model of the eye described in Exercise 34.46. If the radius of curvature of the cornea is 0.75 cm when the eye is focusing on an object 36.0 cm from the cornea vertex and the indexes of refraction are as described in Exercise 34.46, what is the distance from the cornea vertex to the retina? What does this tell you about the shape of the nearsighted eye?

34.111 Focal Length of a Zoom Lens. Figure 34.56 shows a simple version of a zoom lens. The converging lens has focal length f_1, and the diverging lens has focal length $f_2 = -|f_2|$. The two lenses are separated by a variable distance d that is always less than f_1. Also, the magnitude of the focal length of the diverging lens satisfies the inequality $|f_2| > (f_1 - d)$. To determine the effective focal length of the combination lens, consider a bundle of parallel rays of radius r_0 entering the converging lens. a) Show that the radius of the ray bundle decreases to $r_0' = r_0(f_1 - d)/f_1$ at the point that it enters the diverging lens. b) Show that the final image I' is formed a distance $s_2' = |f_2|(f_1 - d)/(|f_2| - f_1 + d)$ to the right of the diverging lens. c) If the rays that emerge from the diverging lens and reach the final image point are extended backward to the left of the diverging lens, they will eventually expand to the original radius r_0 at some point Q. The distance from the final image I' to the point Q is the *effective focal length* f of the lens combination; if the combination were replaced by a single lens of focal length f placed at Q, parallel rays would still be brought to a focus at I'. Show that the effective focal length is given by $f = f_1|f_2|/(|f_2| - f_1 + d)$. d) If $f_1 = 12.0$ cm, $f_2 = -18.0$ cm, and the separation d is adjustable between zero and 4.0 cm, find the maximum and minimum focal lengths of the combination. What value of d gives $f = 30.0$ cm?

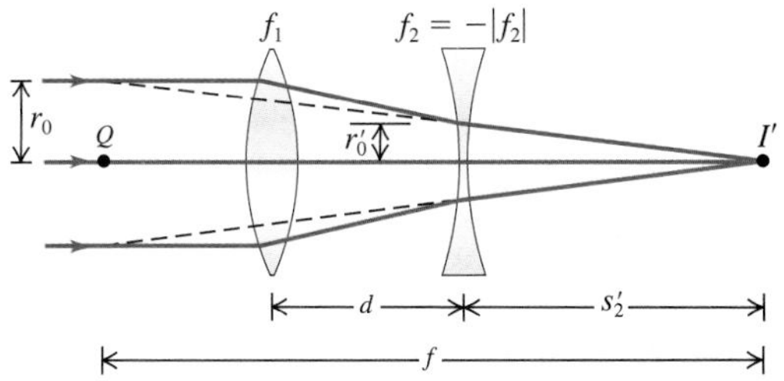

Figure 34.56 Problem 34.111.

34.112 A certain reflecting telescope, constructed as shown in Fig. 34.51a, has a spherical mirror with a radius of curvature of 96.0 cm and an eyepiece with a focal length of 1.20 cm. If the angular magnification has a magnitude of 36 and the object is at infinity, find the position of the eyepiece and the position and nature (real or virtual) of the final image. (*Note:* $|M|$ is *not* equal to $|f_1/f_2|$, so the image formed by the eyepiece is *not* at infinity.)

34.113 A microscope with an objective of focal length 8.00 mm and an eyepiece of focal length 7.50 cm is used to project an image on a screen 2.00 m from the eyepiece. Let the image distance of the objective be 18.0 cm. a) What is the lateral magnification of the image? b) What is the distance between the objective and the eyepiece?

34.114 The Galilean Telescope. Figure 34.57 is a diagram of a *Galilean telescope,* or *opera glass,* with both the object and its final image at infinity. The image I serves as a virtual object for the eyepiece. The final image is virtual and erect. a) Prove that the angular magnification is $M = -f_1/f_2$. b) A Galilean telescope is to be constructed with the same objective lens as in Exercise 34.55. What focal length should the eyepiece have if this telescope is to have the same magnitude of angular magnification as the one in Exercise 34.55? c) Compare the lengths of the telescopes.

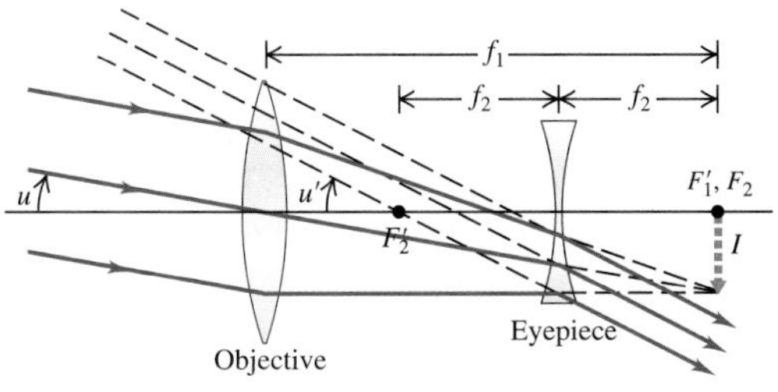

Figure 34.57 Problem 34.114.

Challenge Problems

34.115 An Object at an Angle. A 16.0-cm-long pencil is placed at a 45.0° angle, with its center 15.0 cm above the optic axis and 45.0 cm from a lens with a 20.0-cm focal length as shown in Fig. 34.58. (Note that the figure is not drawn to scale.) Assume that the diameter of the lens is large enough for the paraxial approximation to be valid. a) Where is the image of the pencil? (Give the location of the images of the points A, B, and C on the object, which are located at the eraser, point, and center of the pencil, respectively.) b) What is the length of the image (that is, the distance between the images of points A and B)? c) Show the orientation of the image in a sketch.

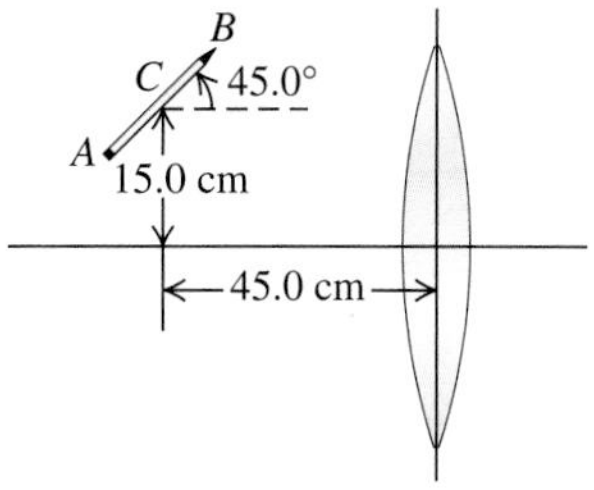

Figure 34.58 Problem 33.115.

34.116 *Spherical aberration* is a blurring of the image formed by a spherical mirror. It occurs because parallel rays striking the mirror far from the optic axis are focused at a different point than are rays near the axis. This problem is usually minimized by using only the center of a spherical mirror. a) Show that for a spherical concave mirror, the focus moves toward the mirror as the parallel rays move toward the outer edge of the mirror. (*Hint:* Derive an analytic expression for the distance from the vertex to the focus of the ray for a particular parallel ray. This expression should be in terms of i) the radius of curvature R of the mirror and ii) the angle θ between the incident ray and a line connecting the center of curvature of the mirror with the point where the ray strikes the mirror.) b) What value of θ produces a 2% change in the location of the focus, compared to the location for θ very close to zero?

34.117 a) For a lens with focal length f, find the smallest distance possible between the object and its real image. b) Graph the distance between the object and real image as a function of the distance of the object from the lens. Does your graph agree with the result you found in part (a)?

34.118 Two mirrors are placed together as shown in Fig. 34.59. a) Show that a point source in front of these mirrors and its two images lie on a circle. b) Find the center of the circle. c) In a diagram, show where an observer should stand so as to be able to see both images.

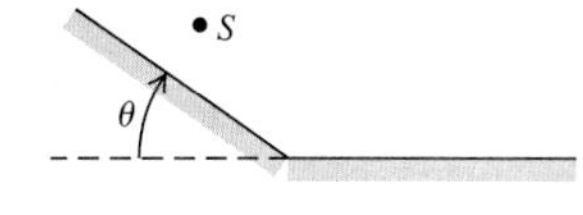

Figure 34.59 Challenge Problem 34.118.

34.119 People with normal vision cannot focus their eyes underwater if they aren't wearing a face mask or goggles and there is water in contact with their eyes (see Question Q34.23). a) Why not? b) With the simplified model of the eye described in Exercise 34.46, what corrective lens (specified by focal length as measured in air) would be needed to enable a person underwater to focus an infinitely distant object? (Be careful—the focal length of a lens underwater is *not* the same as in air! See Problem 34.98. Assume that the corrective lens has a refractive index of 1.62 and that the lens is used in eyeglasses, not goggles, so there is water on both sides of the lens. Assume that the eyeglasses are 2.00 cm in front of the eye.)

35 INTERFERENCE

Oil and water are both colorless, but a thin film of oil floating on water shows vibrant colors. The colors are caused by interference between light waves reflecting from the upper and lower surfaces of the oil film. For a given thickness of the film, the two waves reinforce each other strongly only if their wavelength—and hence their color—has a particular value.

? **How thin must an oil film be to show a rainbow of colors?**

An ugly black oil spot on the pavement can become a thing of beauty after a rain, when the oil reflects a rainbow of colors. Multicolored reflections can also be seen from the surfaces of soap bubbles and compact discs. These familiar sights give us a hint that there are aspects of light that we haven't yet explored.

In our discussion of lenses, mirrors, and optical instruments we used the model of *geometric optics,* in which we represent light as *rays,* straight lines that are bent at a reflecting or refracting surface. But many aspects of the behavior of light *can't* be understood on the basis of rays. We have already learned that light is fundamentally a *wave,* and in some situations we have to consider its wave properties explicitly. If two or more light waves of the same frequency overlap at a point, the total effect depends on the *phases* of the waves as well as their amplitudes. The resulting patterns of light are a result of the *wave* nature of light and cannot be understood on the basis of rays. Optical effects that depend on the wave nature of light are grouped under the heading **physical optics.**

In this chapter we'll look at *interference* phenomena that occur when two waves combine. The colors seen in oil films and soap bubbles are a result of interference between light reflected from the front and back surfaces of a thin film of oil or soap solution. Effects that occur when *many* sources of waves are present are called *diffraction* phenomena; we'll study these in Chapter 36. In that chapter we'll see that diffraction effects occur whenever a wave passes through an aperture or around an obstacle. They are important in practical applications of physical optics such as diffraction gratings, x-ray diffraction, and holography.

While our primary concern is with light, interference and diffraction can occur with waves of *any* kind. As we go along, we'll point out applications to other types of waves such as sound and water waves.

35.1 | Interference and Coherent Sources

As we discussed in Chapter 15, the term **interference** refers to any situation in which two or more waves overlap in space. When this occurs, the total wave at any point at any instant of time is governed by the **principle of superposition,** which we introduced in Section 15.6 in the context of waves on a string. This principle also applies to electromagnetic waves and is the most important principle in all of physical optics, so make sure you understand it well. The principle of superposition states: **When two or more waves overlap, the resultant displacement at any point and at any instant is found by adding the instantaneous displacements that would be produced at the point by the individual waves if each were present alone.** (In some special physical situations, such as electromagnetic waves propagating in a crystal, this principle may not apply. A discussion of these is beyond our scope.)

We use the term *displacement* in a general sense. With waves on the surface of a liquid, we mean the actual displacement of the surface above or below its normal level. With sound waves, the term refers to the excess or deficiency of pressure. For electromagnetic waves, we usually mean a specific component of electric or magnetic field.

We have already discussed one important case of interference, in which two identical waves propagating in opposite directions combine to produce a *standing wave.* We saw this in Chapters 15 and 16 for transverse waves on a string and for longitudinal waves in a fluid filling a pipe; we described the same phenomenon for electromagnetic waves in Section 32.5. In all of these cases the waves propagated along only a single axis: along a string, along the length of a fluid-filled pipe, or along the propagation direction of an electromagnetic plane wave. But light waves can (and do) travel in two or three dimensions, as can any kind of wave that propagates in a two- or three-dimensional medium. In this section we'll see what happens when we combine waves that spread out in two or three dimensions from a pair of identical wave sources.

Interference effects are most easily seen when we combine *sinusoidal* waves with a single frequency f and wavelength λ. Figure 35.1 shows a "snapshot" or "freeze-frame" of a *single* source S_1 of sinusoidal waves and some of the wave fronts produced by this source. The figure shows only the wave fronts corresponding to wave *crests,* so the spacing between successive wave fronts is one wavelength. The material surrounding S_1 is uniform, so the wave speed is the same in all directions, and there is no refraction (and hence no bending of the wave fronts). If the waves are two-dimensional, like waves on the surface of a liquid, the circles in Fig. 35.1 represent circular wave fronts; if the waves propagate in three dimensions, the circles represent spherical wave fronts spreading away from S_1.

In optics, sinusoidal waves are characteristic of **monochromatic light** (light of a single color). While it's fairly easy to make water waves or sound waves of a single frequency, common sources of light *do not* emit monochromatic (single-

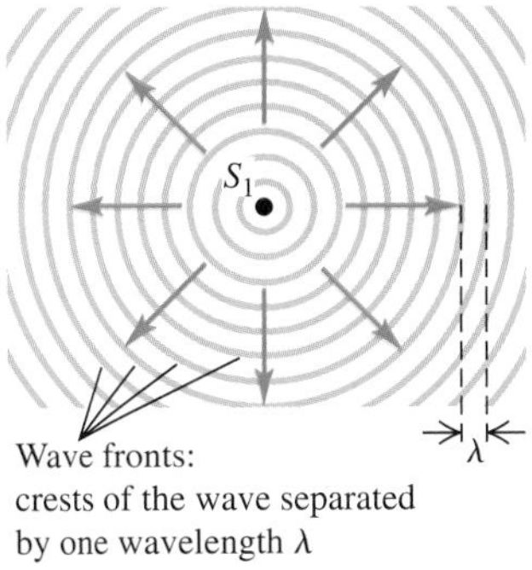

35.1 A "snapshot" of sinusoidal waves of frequency f and wavelength λ spreading out from source S_1 in all directions. The wave fronts move outward at the wave speed v, equal to the product of f and λ.

frequency) light. For example, incandescent light bulbs and flames emit a continuous distribution of wavelengths. However, there are several ways to produce *approximately* monochromatic light. For example, some filters block all but a very narrow range of wavelengths. By far the most nearly monochromatic source that is available at present is the *laser.* The familiar helium-neon laser, which is inexpensive and readily available, emits red light at 632.8 nm with a wavelength range of the order of ± 0.000001 nm, or about one part in 10^9. As we analyze interference and diffraction effects in this chapter and the next, we will assume that we are working with monochromatic waves (unless we explicitly state otherwise).

Constructive and Destructive Interference

Two identical sources of monochromatic waves, S_1 and S_2, are shown in Fig. 35.2a. The two sources produce waves of the same amplitude and the same wavelength λ. In addition, the two sources are permanently *in phase;* they vibrate in unison. They might be two synchronized agitators in a ripple tank, two loudspeakers driven by the same amplifier, two radio antennas powered by the same transmitter, or two small holes or slits in an opaque screen, illuminated by the same monochromatic light source. We will see that if there were not a constant phase relationship between the two sources, the phenomena we are about to discuss would not occur. Two monochromatic sources of the same frequency and with any definite, constant phase relation (not necessarily in phase) are said to be **coherent.** We also use the term *coherent waves* (or, for light waves, *coherent light*) to refer to the waves emitted by two such sources.

If the waves emitted by the two sources are *transverse,* like electromagnetic waves, then we will also assume that the wave disturbances produced by both sources have the same *polarization* (that is, they lie along the same line). For example, the sources S_1 and S_2 in Fig. 35.2a could be two radio antennas in the form of long rods oriented parallel to the z-axis (perpendicular to the plane of the figure); at any point in the xy-plane the waves produced by both antennas have $\vec{E}$ fields with only a z-component. Then we need only a single scalar function to describe each wave; this makes the analysis much easier.

We position the two sources of equal amplitude, equal wavelength, and (if the waves are transverse) the same polarization along the y-axis in Fig. 35.2a, equidistant from the origin. Consider a point a on the x-axis. From symmetry the two distances from S_1 to a and from S_2 to a are *equal;* waves from the two sources thus require equal times to travel to a. Hence waves that leave S_1 and S_2 in phase arrive at a in phase. The two waves add, and the total amplitude at a is *twice* the amplitude of each individual wave. This is true for *any* point on the x-axis.

35.2 (a) A "snapshot" of sinusoidal waves spreading out from two coherent sources S_1 and S_2. In this case the distance between the sources is 4λ. Constructive interference occurs at point a (equidistant from the two sources) and (b) at point b. (c) Destructive interference occurs at point c.

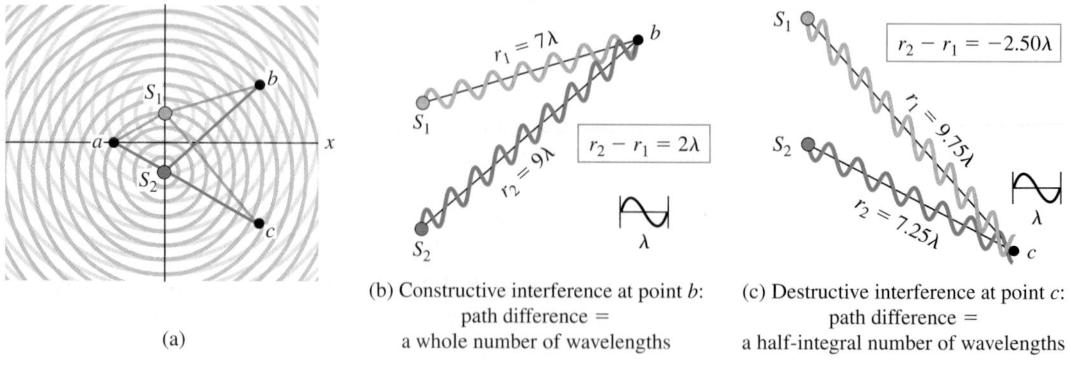

Similarly, the distance from S_2 to point b is exactly two wavelengths *greater* than the distance from S_1 to b. A wave crest from S_1 arrives at b exactly two cycles earlier than a crest emitted at the same time from S_2, and again the two waves arrive in phase. As at point a, the total amplitude is the sum of the amplitudes of the waves from S_1 and S_2.

In general, when waves from two or more sources arrive at a point *in phase,* the amplitude of the resultant wave is the *sum* of the amplitudes of the individual waves; the individual waves reinforce each other. This is called **constructive interference** (Fig. 35.2b). Let the distance from S_1 to any point P be r_1, and let the distance from S_2 to P be r_2. For constructive interference to occur at P, the path difference $r_2 - r_1$ for the two sources must be an integral multiple of the wavelength λ:

$$r_2 - r_1 = m\lambda \qquad (m = 0, \pm1, \pm2, \pm3, \ldots) \qquad (35.1)$$

(constructive interference, sources in phase)

In Fig. 35.2a, points a and b satisfy Eq. (35.1) with $m = 0$ and $m = +2$, respectively.

Something different occurs at point c in Fig. 35.2a. At this point, the path difference $r_2 - r_1 = -2.50\lambda$, which is a *half-integral* number of wavelengths. Waves from the two sources arrive at point c exactly a half-cycle out of phase. A crest of one wave arrives at the same time as a crest in the opposite direction (a "trough") of the other wave (Fig. 35.2c). The resultant amplitude is the *difference* between the two individual amplitudes. If the individual amplitudes are equal, then the total amplitude is *zero!* This cancellation or partial cancellation of the individual waves is called **destructive interference.** The condition for destructive interference in the situation shown in Fig. 35.2a is

$$r_2 - r_1 = \left(m + \frac{1}{2}\right)\lambda \qquad (m = 0, \pm1, \pm2, \pm3, \ldots) \qquad (35.2)$$

(destructive interference, sources in phase)

The path difference at point c in Fig. 35.2a satisfies Eq. (35.2) with $m = -3$.

Figure 35.3 shows the same situation as in Fig. 35.2a, but with red curves that denote all points on which *constructive* interference occurs. On each curve, the path difference $r_2 - r_1$ is equal to an integer m times the wavelength, as in Eq. (35.1). These curves are called **antinodal curves.** They are directly analogous to *antinodes* in the standing-wave patterns described in Chapters 15 and 16 and Section 32.5. In a standing wave formed by interference between waves propagating in opposite directions, the antinodes are points at which the amplitude is maximum; likewise, the wave amplitude in the situation of Fig. 35.3 is maximum along the antinodal curves. Not shown in Fig. 35.3 are the **nodal curves,** which are the curves denoting points on which *destructive* interference occurs in accordance with Eq. (35.2); these are analogous to the *nodes* in a standing wave pattern. A nodal curve lies between each two adjacent antinodal curves in Fig. 35.3; one such curve, corresponding to $r_2 - r_1 = -2.50\lambda$, passes through point c.

In some cases, such as two loudspeakers or two radio-transmitter antennas, the interference pattern is three-dimensional. Think of rotating the color curves of Fig. 35.3 around the y-axis; then maximum constructive interference occurs at all points on the resulting surfaces of revolution.

CAUTION The interference patterns in Figs. 35.2a and 35.3 are *not* standing waves, though they have some similarities to the standing-wave patterns described in Chapters 15 and 16 and Section 32.5. In a standing wave, the inter-

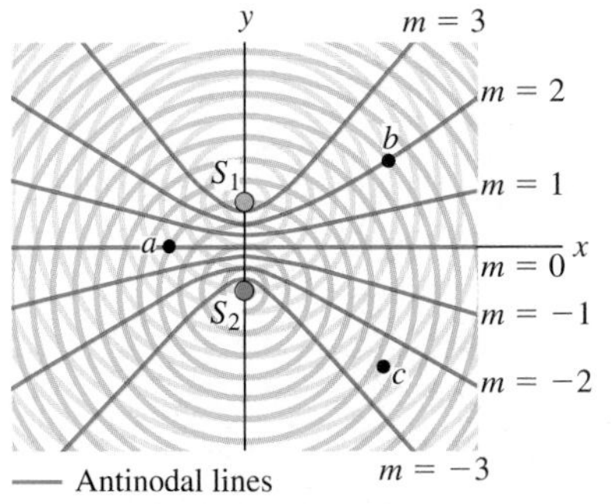

35.3 The same as Fig. 35.2a, but with red antinodal curves (curves of maximum amplitude) superimposed. All points on each curve satisfy Eq. (35.1) with the value of m shown. The nodal curves (not shown) lie between each adjacent pair of antinodal curves.

ference is between two waves propagating in opposite directions; a stationary pattern of antinodes and nodes appears, and there is *no* net energy flow in either direction (the energy in the wave is left "standing"). In the situations shown in Figs. 35.2a and 35.3, there is likewise a stationary pattern of antinodal and nodal curves, but there is a net flow of energy *outward* from the two sources. From the energy standpoint, all that interference does is to "channel" the energy flow so that it is greatest along the antinodal curves and least along the nodal curves.

For Eqs. (35.1) and (35.2) to hold, the two sources must have the same wavelength and must *always* be in phase. These conditions are rather easy to satisfy for sound waves (see Example 16.15 in Section 16.6). But with *light* waves there is no practical way to achieve a constant phase relationship (coherence) with two independent sources. This is because of the way light is emitted. In ordinary light sources, atoms gain excess energy by thermal agitation or by impact with accelerated electrons. An atom that is "excited" in such a way begins to radiate energy and continues until it has lost all the energy it can, typically in a time of the order of 10^{-8} s. The many atoms in a source ordinarily radiate in an unsynchronized and random phase relationship, and the light that is emitted from *two* such sources has no definite phase relation.

However, the light from a single source can be split so that parts of it emerge from two or more regions of space, forming two or more *secondary sources*. Then any random phase change in the source affects these secondary sources equally and does not change their *relative* phase.

The distinguishing feature of light from a *laser* is that the emission of light from many atoms is synchronized in frequency and phase. As a result, the random phase changes mentioned above occur much less frequently. Definite phase relations are preserved over correspondingly much greater lengths in the beam, and laser light is much more coherent than ordinary light.

Test Your Understanding

On a copy of Fig. 35.3, draw the nodal curves that lie between the antinodal curves shown in the figure. To which value of m in Eq. (35.2) does each nodal curve correspond?

35.4 The concepts of constructive interference and destructive interference apply to these water waves as well as to light waves or sound waves.

35.2 | Two-Source Interference of Light

The interference pattern produced by two coherent sources of *water* waves of the same wavelength can be readily seen in a ripple tank with a shallow layer of water (Fig. 35.4). This pattern is not directly visible when the interference is between *light* waves, since light traveling in a uniform medium cannot be seen. (A shaft of afternoon sunlight in a room is made visible by scattering from airborne dust particles.)

One of the earliest quantitative experiments to reveal the interference of light from two sources was performed in 1800 by the English scientist Thomas Young. We will refer back to this experiment several times in this and later chapters, so it's important to understand it in detail. Young's apparatus is shown in perspective in Fig. 35.5a. A light source (not shown) emits monochromatic light; however, this light is not suitable for use in an interference experiment because emissions from different parts of an ordinary source are not synchronized. To remedy this, the light is directed at a screen with a narrow slit S_0, 1 μm or so wide. The light emerging from the slit originated from only a small region of the light source; thus slit S_0 behaves more nearly like the idealized source shown in Fig. 35.1. (In modern versions of the experiment, a laser is used as a source of coherent light, and the slit S_0 isn't needed.) The light from slit S_0 falls on a screen with two other narrow slits S_1 and S_2, each 1 μm or so wide and a few tens or hundreds of micrometers apart. Cylindrical wave fronts spread out from slit S_0 and reach slits S_1 and S_2 *in phase* because they travel equal distances from S_0. The waves *emerging* from slits S_1 and S_2 are therefore also always in phase, so S_1 and S_2 are *coherent* sources. The interference of waves from S_1 and S_2 produces a pattern in space like that to the right of the sources in Figs. 35.2a and 35.3.

ActivPhysics Online

16.1 Two-Source Interference: Introduction
16.2 Two-Source Interference: Qualitative Questions
16.3 Two-Source Interference: Problems

To visualize the interference pattern, a screen is placed so that the light from S_1 and S_2 falls on it (Fig. 35.5b). The screen will be most brightly illuminated at points P, where the light waves from the slits interfere constructively, and will be darkest at points where the interference is destructive.

To simplify the analysis of Young's experiment, we assume that the distance R from the slits to the screen is so large in comparison to the distance d between the slits that the lines from S_1 and S_2 to P are very nearly parallel, as in Fig. 35.5c. This is usually the case for experiments with light; the slit separation is typically a few millimeters, while the screen may be a meter or more away. The difference in path length is then given by

$$r_2 - r_1 = d \sin\theta \tag{35.3}$$

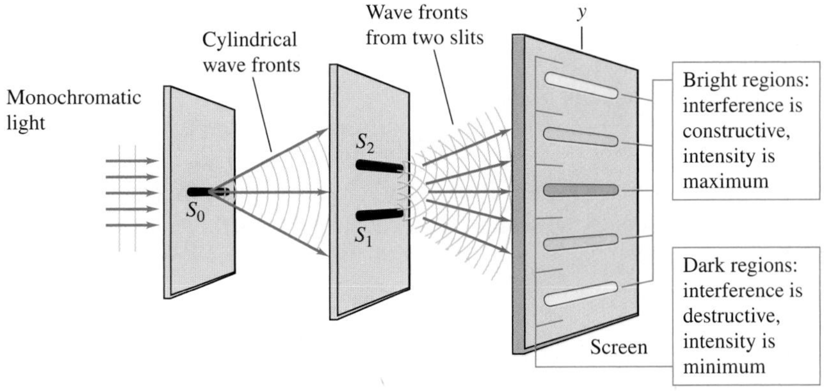

(a)

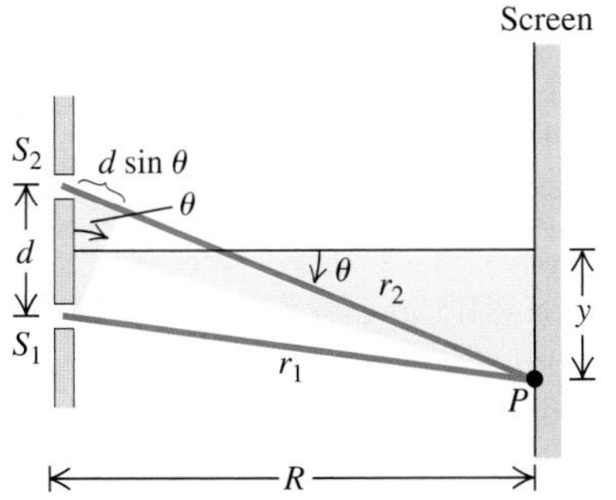

(b) Slits S_1 and S_2 are horizontal and seen from the side in cross section

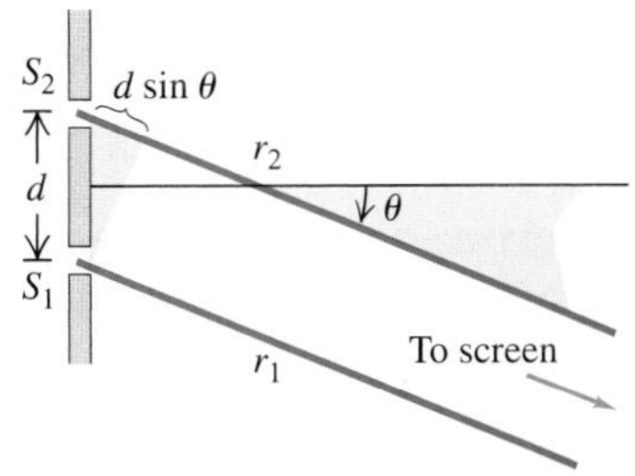

(c) Same as (b), but with screen very far from slits (R much greater than d)

35.5 (a) Young's experiment to show interference of light passing through two slits. A pattern of bright and dark areas appears on the screen (see Fig. 35.6). (b) Geometrical analysis of Young's experiment. For the case shown, $r_2 > r_1$ and both y and θ are positive. If point P is on the other side of the screen's center, $r_2 < r_1$ and both y and θ are negative. (c) Approximate geometry when the distance R to the screen is much larger than the distance d between the slits.

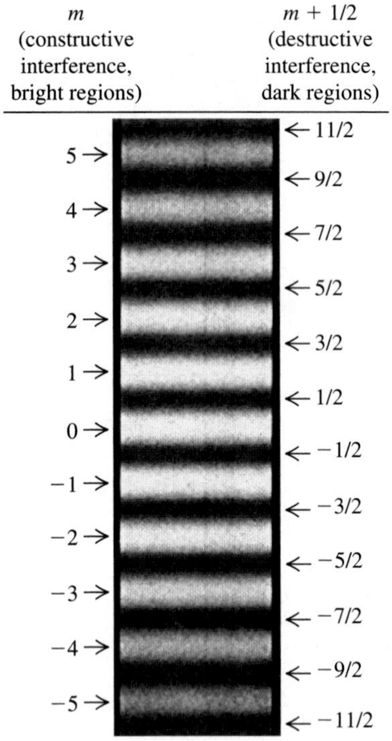

35.6 Photograph of interference fringes produced on a screen in Young's double-slit experiment.

where θ is the angle between a line from slits to screen (shown in blue in Fig. 35.5c) and the normal to the plane of the slits (shown as a thin black line).

We found in Section 35.1 that constructive interference (reinforcement) occurs at points where the path difference $d \sin\theta$ is an integral number of wavelengths, $m\lambda$, where $m = 0, \pm1, \pm2, \pm3, \ldots$. So the bright regions on the screen occur at angles θ for which

$$d \sin\theta = m\lambda \qquad (m = 0, \pm1, \pm2, \ldots) \qquad (35.4)$$

(constructive interference, two slits)

Similarly, destructive interference (cancellation) occurs, forming dark regions on the screen, at points for which the path difference is a half-integral number of wavelengths, $(m + \frac{1}{2})\lambda$:

$$d \sin\theta = \left(m + \frac{1}{2}\right)\lambda \qquad (m = 0, \pm1, \pm2, \ldots) \qquad (35.5)$$

(destructive interference, two slits)

Thus the pattern on the screen of Figs. 35.5a and 35.5b is a succession of bright and dark bands, or **interference fringes,** parallel to the slits S_1 and S_2. A photograph of such a pattern is shown in Fig. 35.6. The center of the pattern is a bright band corresponding to $m = 0$ in Eq. (35.4); this point on the screen is equidistant from the two slits.

We can derive an expression for the positions of the centers of the bright bands on the screen. In Fig. 35.5b, y is measured from the center of the pattern, corresponding to the distance from the center of Fig. 35.6. Let y_m be the distance from the center of the pattern $(\theta = 0)$ to the center of the mth bright band. Let θ_m be the corresponding value of θ; then

$$y_m = R \tan\theta_m$$

In experiments such as this, the distances y_m are often much smaller than the distance R from the slits to the screen. Hence θ_m is very small, $\tan\theta_m$ is very nearly equal to $\sin\theta_m$, and

$$y_m = R \sin\theta_m$$

Combining this with Eq. (35.4), we find that *for small angles only,*

$$y_m = R\frac{m\lambda}{d} \qquad \text{(constructive interference in Young's experiment)} \qquad (35.6)$$

We can measure R and d, as well as the positions y_m of the bright fringes, so this experiment provides a direct measurement of the wavelength λ. Young's experiment was in fact the first direct measurement of wavelengths of light.

CAUTION While Eqs. (35.4) and (35.5) are valid at any angle, Eq. (35.6) is valid only for *small* angles. It can be used *only* if the distance R from slits to screen is much greater than the slit separation d and if R is much greater than the distance y_m from the center of the interference pattern to the mth bright fringe.

The distance between adjacent bright bands in the pattern is *inversely* proportional to the distance d between the slits. The closer together the slits are, the more the pattern spreads out. When the slits are far apart, the bands in the pattern are closer together.

While we have described the experiment that Young performed with visible light, the results given in Eqs. (35.4) and (35.5) are valid for *any* type of wave, provided that the resultant wave from two coherent sources is detected at a point that is far away in comparison to the separation d.

Example 35.1 Two-slit interference

In a two-slit interference experiment, the slits are 0.20 mm apart, and the screen is at a distance of 1.0 m. The third bright fringe (not counting the central bright fringe straight ahead from the slits) is found to be displaced 7.5 mm from the central fringe. Find the wavelength of the light used.

SOLUTION

IDENTIFY and SET UP: The third bright fringe corresponds to $m = 3$ in Eqs. (35.4) and (35.6), as well as to the bright fringe labeled $m = 3$ in Fig. 35.6. To determine the value of the target variable λ, we may use Eq. (35.6) since $R = 1.0$ m is much greater than either $d = 0.20$ mm or $y_3 = 7.5$ mm.

EXECUTE: Solving Eq. (35.6) for λ, we find

$$\lambda = \frac{y_m d}{mR} = \frac{(7.5 \times 10^{-3}\ \text{m})(0.20 \times 10^{-3}\ \text{m})}{(3)(1.0\ \text{m})}$$
$$= 500 \times 10^{-9}\ \text{m} = 500\ \text{nm}$$

EVALUATE: This bright fringe could also correspond to $m = -3$; can you show that this gives the same result for λ?

Example 35.2 Broadcast pattern of a radio station

A radio station operating at a frequency of 1500 kHz $= 1.5 \times 10^6$ Hz (near the top end of the AM broadcast band) has two identical vertical dipole antennas spaced 400 m apart, oscillating in phase. At distances much greater than 400 m, in what directions is the intensity greatest in the resulting radiation pattern? (This is not just a hypothetical problem. It is often desirable to beam most of the radiated energy from a radio transmitter in particular directions rather than uniformly in all directions. Pairs or rows of antennas are often used to produce the desired radiation pattern.)

SOLUTION

IDENTIFY and SET UP: The two antennas, seen in Fig. 35.7, correspond to sources S_1 and S_2 in Fig. 35.5. The wavelength is $\lambda = c/f = 200$ m. Since the resultant wave is detected at distances much greater than $d = 400$ m, we may use Eq. (35.4) to give the directions of the intensity *maxima,* the values of θ for which the path difference is zero or a whole number of wavelengths.

EXECUTE: From Eq. (35.4) with $m = 0, \pm1$, and ±2, the intensity maxima are given by

$$\sin\theta = \frac{m\lambda}{d} = \frac{m(200\ \text{m})}{400\ \text{m}} = \frac{m}{2} \qquad \theta = 0, \pm30°, \pm90°$$

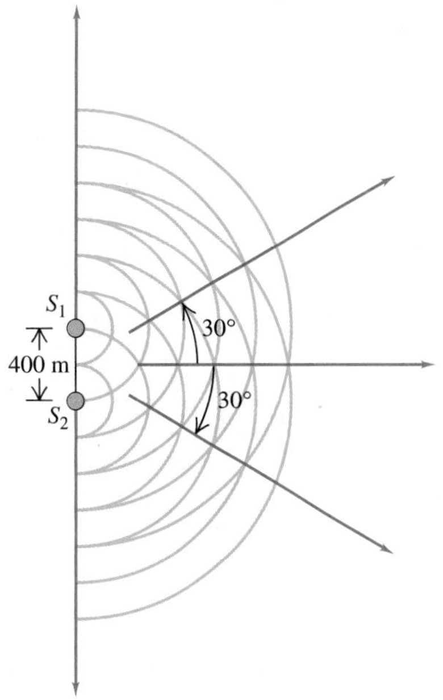

35.7 Two radio antennas broadcasting in phase. The red arrows indicate the directions of maximum intensity. The waves that are emitted to the left of the two sources are not shown.

In this example, values of m greater than 2 or less than -2 give values of $\sin\theta$ greater than 1 or less than -1, which is impossible. There is *no* direction for which the path difference is three or more wavelengths. Thus values of m of ± 3 and beyond have no physical meaning in this example.

EVALUATE: We can check our result by calculating the angles for *minimum* intensity (destructive interference). There should be one intensity minimum between each pair of intensity maxima, just as in the interference pattern shown in Fig. 35.6. The angles of the intensity minima are given by Eq. (35.5) with $m = -2, -1, 0,$ and 1:

$$\sin\theta = \frac{(m + \frac{1}{2})\lambda}{d} = \frac{m + \frac{1}{2}}{2} \qquad \theta = \pm 14.5°, \pm 48.6°$$

(Other values of m have no physical significance in this example.) Note that these angles are intermediate between the angles for intensity maxima, as they should be. Note also that since the angles are not small, the angles for the minima are *not* exactly halfway between the angles for the maxima.

Test Your Understanding

You shine a tunable laser (whose wavelength can be adjusted by turning a knob) on a pair of closely spaced slits. The light emerging from the two slits produces an interference pattern on a screen like that shown in Fig. 35.6. If you adjust the wavelength so that the laser light changes from red to blue, how will the spacing between bright fringes change?

35.3 | Intensity in Interference Patterns

In Section 35.2 we found the positions of maximum and minimum intensity in a two-source interference pattern. Let's now see how to find the intensity at *any* point in the pattern. To do this, we have to combine the two sinusoidally varying fields (from the two sources) at a point P in the radiation pattern, taking proper account of the phase difference of the two waves at point P, which results from the path difference. The intensity is then proportional to the square of the resultant electric-field amplitude, as we learned in Chapter 32.

To calculate the intensity, we will assume that the two sinusoidal functions (corresponding to waves from the two sources) have equal amplitude E and that the $\vec{E}$ fields lie along the same line (have the same polarization). This assumes that the sources are identical and neglects the slight amplitude difference caused by the unequal path lengths (the amplitude decreases with increasing distance from the source). From Eq. (32.29), each source by itself would give an intensity $\frac{1}{2}\epsilon_0 c E^2$ at point P. If the two sources are in phase, then the waves that arrive at P differ in phase by an amount proportional to the difference in their path lengths, $(r_2 - r_1)$. If the phase angle between these arriving waves is ϕ, then we can use the following expressions for the two electric fields superposed at P:

$$E_1(t) = E\cos(\omega t + \phi)$$

$$E_2(t) = E\cos\omega t$$

Here is our program. The superposition of the two fields at P is a sinusoidal function with some amplitude E_P that depends on E and the phase difference ϕ. First we'll work on finding the amplitude E_P if E and ϕ are known. Then we'll find the intensity I of the resultant wave, which is proportional to E_P^2. Finally, we'll relate the phase difference ϕ to the path difference, which is determined by the geometry of the situation.

Amplitude in Two-Source Interference

To add the two sinusoidal functions with a phase difference, we use the same *phasor* representation that we used for simple harmonic motion (Section 13.2) and for voltages and currents in ac circuits (Section 31.1). We suggest that you review these sections so that phasors are fresh in your mind. Each sinusoidal function is represented by a rotating vector (phasor) whose projection on the horizontal axis at any instant represents the instantaneous value of the sinusoidal function.

In Fig. 35.8, E_1 is the horizontal component of the phasor representing the wave from source S_1, and E_2 is the horizontal component of the phasor for the wave from S_2. As shown in the diagram, both phasors have the same magnitude E, but E_1 is *ahead* of E_2 in phase by an angle ϕ. Both phasors rotate counterclockwise with constant angular velocity ω, and the sum of the projections on the horizontal axis at any time gives the instantaneous value of the total E field at point P. Thus the amplitude E_P of the resultant sinusoidal wave at P is the magnitude of the dark red phasor in the diagram (labeled E_P); this is the *vector sum* of the other two phasors. To find E_P, we use the law of cosines and the trigonometric identity $\cos(\pi - \phi) = -\cos\phi$:

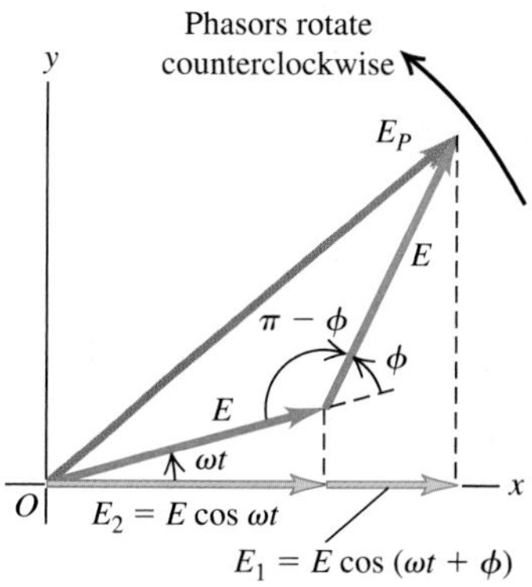

35.8 Phasor diagram for the superposition at a point P of two waves of equal amplitude E with a phase difference ϕ. The resultant wave has amplitude E_P, equal to the magnitude of the phasor shown in brown.

$$E_P^2 = E^2 + E^2 - 2E^2\cos(\pi - \phi)$$
$$= E^2 + E^2 + 2E^2\cos\phi$$

Then, using the identity $1 + \cos\phi = 2\cos^2(\phi/2)$, we obtain

$$E_P^2 = 2E^2(1 + \cos\phi) = 4E^2\cos^2\left(\frac{\phi}{2}\right)$$

$$E_P = 2E\left|\cos\frac{\phi}{2}\right| \quad \text{(amplitude in two-source interference)} \quad (35.7)$$

You can also obtain this result algebraically without using phasors (see Problem 35.46).

When the two waves are in phase, $\phi = 0$ and $E_P = 2E$. When they are exactly a half-cycle out of phase, $\phi = \pi$ rad $= 180°$, $\cos(\phi/2) = \cos(\pi/2) = 0$, and $E_P = 0$. Thus the superposition of two sinusoidal waves with the same frequency and amplitude but with a phase difference yields a sinusoidal wave with the same frequency and an amplitude between zero and twice the individual amplitudes, depending on the phase difference.

Intensity in Two-Source Interference

To obtain the intensity I at point P, we recall from Section 32.4 that I is equal to the average magnitude of the Poynting vector, S_{av}. For a sinusoidal wave with electric-field amplitude E_P, this is given by Eq. (32.29) with E_{max} replaced by E_P. Thus we can express the intensity in any of the following equivalent forms:

$$I = S_{av} = \frac{E_P^2}{2\mu_0 c} = \frac{1}{2}\sqrt{\frac{\epsilon_0}{\mu_0}}E_P^2 = \frac{1}{2}\epsilon_0 c E_P^2 \quad (35.8)$$

The essential content of these expressions is that I is proportional to E_P^2. When we substitute Eq. (35.7) into the last expression in Eq. (35.8), we get

$$I = \frac{1}{2}\epsilon_0 c E_P^2 = 2\epsilon_0 c E^2\cos^2\frac{\phi}{2} \quad (35.9)$$

In particular, the *maximum* intensity I_0, which occurs at points where the phase difference is zero $(\phi = 0)$, is

$$I_0 = 2\epsilon_0 c E^2$$

Note that the maximum intensity I_0 is *four times* (not twice) as great as the intensity $\frac{1}{2}\epsilon_0 c E^2$ from each individual source.

Substituting the expression for I_0 into Eq. (35.9), we can express the intensity I at any point very simply in terms of the maximum intensity I_0:

$$I = I_0 \cos^2 \frac{\phi}{2} \quad \text{(intensity in two-source interference)} \quad (35.10)$$

Note that if we average this over all possible phase differences, the result is $I_0/2 = \epsilon_0 c E^2$ (the average of $\cos^2(\phi/2)$ is $\frac{1}{2}$). This is just twice the intensity from each individual source, as we should expect. The total energy output from the two sources isn't changed by the interference effects, but the energy is redistributed (as we mentioned in Section 35.1). For some phase angles the intensity is four times as great as for an individual source, but for others it is zero. It averages out.

Phase Difference and Path Difference

Now we have to find how the phase difference ϕ between the two fields at point P is related to the geometry of the situation. We know that ϕ is proportional to the difference in path length from the two sources to point P. When the path difference is one wavelength, the phase difference is one cycle, and $\phi = 2\pi \text{ rad} = 360°$. When the path difference is $\lambda/2$, $\phi = \pi \text{ rad} = 180°$, and so on. That is, the ratio of the phase difference ϕ to 2π is equal to the ratio of the path difference $r_2 - r_1$ to λ:

$$\frac{\phi}{2\pi} = \frac{r_2 - r_1}{\lambda}$$

Thus a path difference $(r_2 - r_1)$ causes a phase difference given by

$$\phi = \frac{2\pi}{\lambda}(r_2 - r_1) = k(r_2 - r_1) \quad (35.11)$$

(phase difference related to path difference)

where $k = 2\pi/\lambda$ is the *wave number* introduced in Section 15.3.

If the material in the space between the sources and P is anything other than vacuum, we must use the wavelength *in the material* in Eq. (35.11). If the material has index of refraction n, then

$$\lambda = \frac{\lambda_0}{n} \quad \text{and} \quad k = nk_0 \quad (35.12)$$

where λ_0 and k_0 are the wavelength and the wave number, respectively, in vacuum.

Finally, if the point P is far away from the sources in comparison to their separation d, the path difference is given by Eq. (35.3):

$$r_2 - r_1 = d \sin\theta$$

Combining this with Eq. (35.11), we find

$$\phi = k(r_2 - r_1) = kd\sin\theta = \frac{2\pi d}{\lambda}\sin\theta \quad (35.13)$$

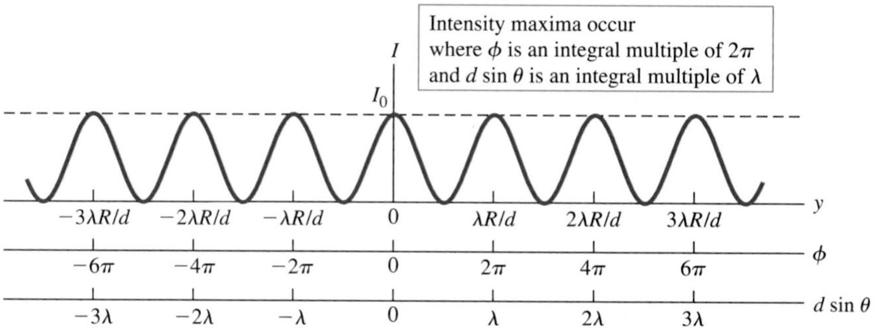

35.9 Intensity distribution in the interference pattern from two identical slits. The three scales show the distance y of a point in the pattern from the center $(y = 0)$, the phase difference ϕ between the two waves at each point in the pattern, and the path difference $d \sin\theta$ expressed as a multiple of the wavelength λ.

When we substitute this into Eq. (35.10), we find

$$I = I_0 \cos^2\left(\frac{1}{2}kd \sin\theta\right) = I_0 \cos^2\left(\frac{\pi d}{\lambda} \sin\theta\right) \qquad (35.14)$$

(intensity far from two sources)

The directions of *maximum* intensity occur when the cosine has the values ± 1, that is, when

$$\frac{\pi d}{\lambda} \sin\theta = m\pi \qquad (m = 0, \pm 1, \pm 2, \ldots)$$

or

$$d \sin\theta = m\lambda$$

in agreement with Eq. (35.4). We leave it as an exercise to show that Eq. (35.5) for the zero-intensity directions can also be derived from Eq. (35.14).

As we noted in Section 35.2, in experiments with light we visualize the interference pattern due to two slits by using a screen placed at a distance R from the slits. We can describe positions on the screen with the coordinate y; the positions of the bright fringes are given by Eq. (35.6), where ordinarily $y \ll R$. In that case, $\sin\theta$ is approximately equal to y/R, and we obtain the following expressions for the intensity at *any* point on the screen as a function of y:

$$I = I_0 \cos^2\left(\frac{kdy}{2R}\right) = I_0 \cos^2\left(\frac{\pi dy}{\lambda R}\right) \qquad (35.15)$$

(intensity in two-slit interference)

Figure 35.9 shows a graph of Eq. (35.15); we can compare this with the photographically recorded pattern of Fig. 35.6. The peaks in Fig. 35.9 all have the same intensity, while those in Fig. 35.6 fade off as we go away from the center. We'll explore the reasons for this variation in peak intensity in Chapter 36.

Example 35.3 A directional transmitting antenna array

Suppose the two identical radio antennas in Fig. 35.7 are moved to be only 10.0 m apart and the frequency of the radiated waves is increased to $f = 60.0$ MHz. The intensity at a distance of 700 m in the $+x$-direction (corresponding to $\theta = 0$ in Fig. 35.5) is $I_0 = 0.020\ \text{W/m}^2$. a) What is the intensity in the direction $\theta = 4.0°$? b) In what direction near $\theta = 0$ is the intensity $I_0/2$? c) In what directions is the intensity zero?

SOLUTION

IDENTIFY: This problem involves the intensity distribution as a function of *direction*—that is, as a function of angle. (In other problems we are concerned with the intensity as a function of *position* on a screen, as in the interference pattern shown in Fig. 35.6.)

SET UP: Because the 700-m distance from the antennas to the point where the intensity is measured is much greater than the distance between the antennas $(d = 10.0\text{ m})$, the amplitudes of the waves from the two antennas are very nearly equal. Hence we can use Eq. (35.14) to relate intensity I and angle θ.

EXECUTE: To use Eq. (35.14), we must first find the wavelength λ using the relation $c = \lambda f$:

$$\lambda = \frac{c}{f} = \frac{3.00 \times 10^8\text{ m/s}}{60.0 \times 10^6\text{ s}^{-1}} = 5.00\text{ m}$$

The spacing $d = 10.0$ m between the antennas is just twice the wavelength. Equation (35.14) then becomes

$$I = I_0 \cos^2\left(\frac{\pi d}{\lambda}\sin\theta\right)$$
$$= (0.020\text{ W/m}^2)\cos^2\left[\frac{\pi(10.0\text{ m})}{(5.00\text{ m})}\sin\theta\right]$$
$$= (0.020\text{ W/m}^2)\cos^2((2.00\pi\text{ rad})\sin\theta)$$

a) When $\theta = 4.0°$,

$$I = (0.020\text{ W/m}^2)\cos^2((2.00\pi\text{ rad})\sin 4.0°)$$
$$= 0.016\text{ W/m}^2$$

This is about 82% of the intensity at $\theta = 0$.

b) The intensity I equals $I_0/2$ when the cosine in Eq. (35.14) has the value $\pm 1/\sqrt{2}$. This occurs when $2.00\pi \sin\theta = \pm\pi/4$ rad, so that $\sin\theta = \pm(1/8.00) = \pm 0.125$ and $\theta = \pm 7.2°$.

c) The intensity is zero when $\cos((2.00\pi\text{ rad})\sin\theta) = 0$. This occurs when $2.00\pi\sin\theta = \pm\pi/2, \pm 3\pi/2, \pm 5\pi/2, \ldots$, or $\sin\theta = \pm 0.250, \pm 0.750, \pm 1.25, \ldots$. Values of $\sin\theta$ greater than 1 have no meaning, and we find

$$\theta = \pm 14.5°, \pm 48.6°$$

EVALUATE: The condition in part (b) that $I = I_0/2$, so that $(2.00\pi\text{ rad})\sin\theta = \pm\pi/4$ rad, is also satisfied when $\sin\theta = \pm 0.375, \pm 0.625$, or ± 0.875 so that $\theta = \pm 22.0°, \pm 38.7°$, or $\pm 61.0°$. (Can you verify this?) It would be incorrect to include these angles in the solution, however, because the problem asked for the angle *near* $\theta = 0$ at which $I = I_0/2$. These additional values of θ aren't the ones we're looking for.

Test Your Understanding

In a two-slit interference pattern with maximum intensity I_0, what is the intensity at a point P on the screen that is closer to one slit than the other by 7/4 of a wavelength?

35.4 | Interference in Thin Films

You often see bright bands of color when light reflects from a soap bubble or from a thin layer of oil floating on water (see the photograph that opens this chapter). These are the results of interference. Light waves are reflected from the front and back surfaces of such thin films, and constructive interference between the two reflected waves (with different path lengths) occurs in different places for different wavelengths. The situation is shown schematically in Fig. 35.10. Light shining on the upper surface of a thin film with thickness t is partly reflected at the upper surface (path *abc*). Light *transmitted* through the upper surface is partly reflected at the lower surface (path *abdef*). The two reflected waves come together at point P on the retina of the eye. Depending on the phase relationship, they may interfere constructively or destructively. Different colors have different wavelengths, so the interference may be constructive for some colors and destructive for others. That's why we see colored rings or fringes. The complex shapes of the colored rings in the photograph that opens this chapter result from variations in the thickness of the oil film.

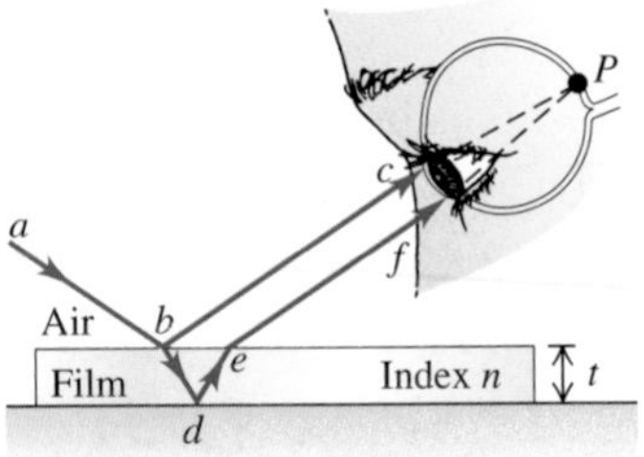

35.10 Interference between light waves reflected from the upper and lower surfaces of a thin film. The angles have been exaggerated for clarity.

Here's an example involving *monochromatic* light reflected from two nearly parallel surfaces at nearly normal incidence. Figure 35.11 shows two plates of glass separated by a thin wedge, or film, of air. We want to consider interference between the two light waves reflected from the surfaces adjacent to the air wedge, as shown. (Reflections also occur at the top surface of the upper plate and the bottom surface of the lower plate; to keep our discussion simple, we won't include these.) The situation is the same as in Fig. 35.10 except that the film (wedge) thickness is not uniform. The path difference between the two waves is just twice the thickness t of the air wedge at each point. At points where $2t$ is an integer number of wavelengths, we expect to see constructive interference and a bright area; where it is a half-integer number of wavelengths, we expect to see destructive interference and a dark area. Along the line where the plates are in contact, there is practically *no* path difference, and we expect a bright area.

When we carry out the experiment, the bright and dark fringes appear, but they are interchanged! Along the line where the plates are in contact, we find a *dark* fringe, not a bright one. This suggests that one or the other of the reflected waves has undergone a half-cycle phase shift during its reflection. In that case the two waves that are reflected at the line of contact are a half-cycle out of phase even though they have the same path length.

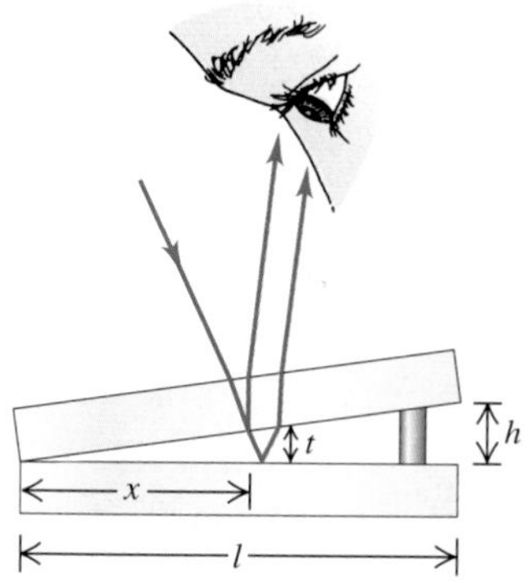

35.11 Interference between light waves reflected from the two sides of an air wedge separating two glass plates. The angles and the thickness of the air wedge have been exaggerated for clarity; in the text we assume that the light strikes the upper plate at normal incidence and that the distances h and t are much less than l.

In fact, this phase shift can be predicted from Maxwell's equations and the electromagnetic nature of light. The details of the derivation are beyond our scope, but here is the result. Suppose a light wave with electric-field amplitude E_i is traveling in an optical material with index of refraction n_a. It strikes, at normal incidence, an interface with another optical material with index n_b. The amplitude E_r of the wave reflected from the interface is proportional to the amplitude E_i of the incident wave and is given by

$$E_\mathrm{r} = \frac{n_a - n_b}{n_a + n_b} E_\mathrm{i} \qquad \text{(normal incidence)} \qquad (35.16)$$

This result shows that the incident and reflected amplitudes have the same sign when n_a is larger than n_b and opposite sign when n_b is larger than n_a. We can distinguish three cases, as shown in Fig. 35.12:

Figure 35.12a: When $n_a > n_b$, light travels more slowly in the first medium than in the second. In this case, E_r and E_i have the same sign, and the phase shift of the reflected wave relative to the incident wave is zero. This is analogous to reflection of a transverse mechanical wave on a heavy rope at a point where it is tied to a lighter rope or a ring that can move vertically without friction.

Figure 35.12b: When $n_a = n_b$, the amplitude E_r of the reflected wave is zero. The incident light wave can't "see" the interface, and there *is no* reflected wave.

Figure 35.12c: When $n_a < n_b$, light travels more slowly in the second material than in the first. In this case, E_r and E_i have opposite signs, and the phase shift of the reflected wave relative to the incident wave is π rad ($180°$ or a half-cycle). This is analogous to reflection (with inversion) of a transverse mechanical wave on a light rope at a point where it is tied to a heavier rope or a rigid support.

Let's compare with the situation of Fig. 35.11. For the wave reflected from the upper surface of the air wedge, n_a (glass) is greater than n_b, so this wave has zero phase shift. For the wave reflected from the lower surface, n_a (air) is less than n_b (glass), so this wave has a half-cycle phase shift. Waves that are reflected from the line of contact have no path difference to give additional phase shifts, and they interfere destructively; this is what we observe. You can use the above principle to show that for normal incidence, the wave reflected at point b in Fig. 35.10 is shifted by a half-cycle, while the wave reflected at d is not (if there is air below the film).

We can summarize this discussion mathematically. If the film has thickness t, the light is at normal incidence and has wavelength λ in the film; if neither or both

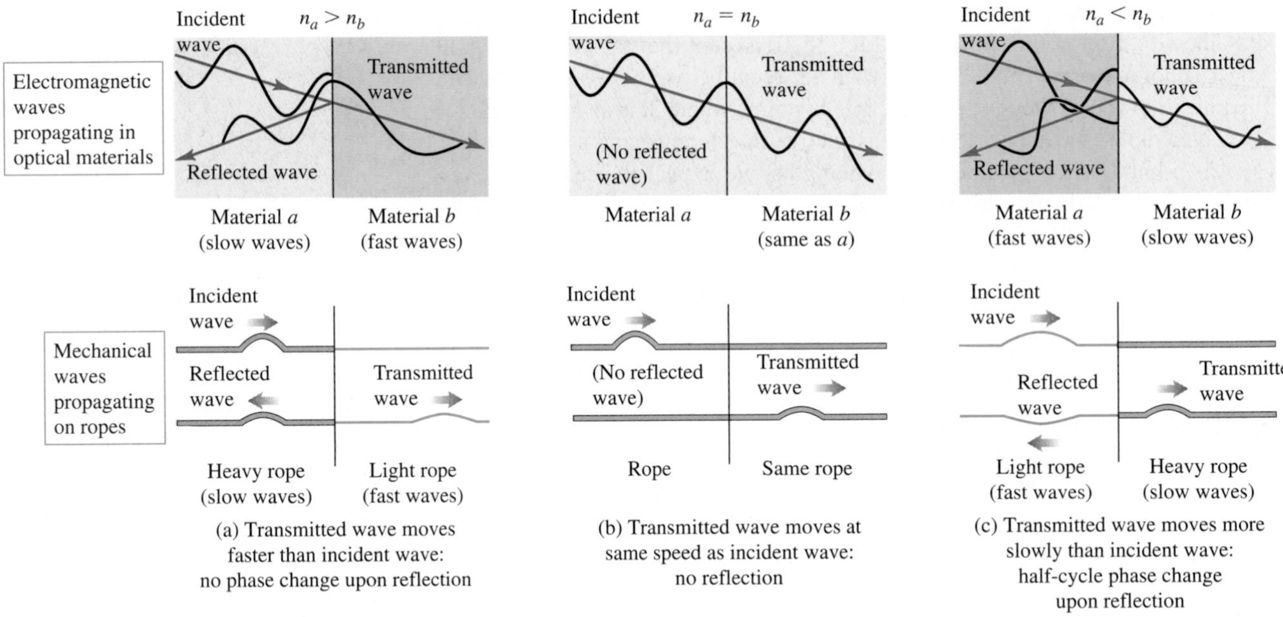

35.12 Upper figures: electromagnetic waves striking an interface between optical materials at normal incidence (shown as a small angle for clarity). Lower figures: mechanical wave pulses on ropes.

of the reflected waves from the two surfaces have a half-cycle reflection phase shift, the condition for constructive interference is

$$2t = m\lambda \qquad (m = 0, 1, 2, \dots) \tag{35.17}$$

(constructive reflection from thin film, no relative phase shift)

However, when *one* of the two waves has a half-cycle reflection phase shift, this equation is the condition for *destructive* interference.

Similarly, if neither wave or both have a half-cycle phase shift, the condition for *destructive* interference in the reflected waves is

$$2t = \left(m + \frac{1}{2}\right)\lambda \qquad (m = 0, 1, 2, \dots) \tag{35.18}$$

(destructive reflection from thin film, no relative phase shift)

However, if one wave has a half-cycle phase shift, this is the condition for *constructive* interference.

Problem-Solving Strategy: Interference in Thin Films

IDENTIFY *the relevant concepts:* Problems with thin films involve interference of two waves, one reflected from the film's front surface and one reflected from the back surface. Typically you will be asked to relate the wavelength, the index of refraction of the film, and the dimensions of the film.

SET UP *the problem using the following steps:*

1. Make a drawing showing the geometry of the thin film. Your drawing should also depict the materials that adjoin the film; their properties determine whether one or both of the reflected waves have a half-cycle phase shift.
2. Determine the target variable.

EXECUTE *the solution as follows:*

1. Apply the rule for phase changes to each reflected wave. There is a half-cycle phase shift when $n_b > n_a$, and none when $n_b < n_a$.
2. If neither reflected wave undergoes a phase shift, or if both reflected waves do, you can apply Eqs. (35.17) and (35.18) as they are written. If only one of the reflected waves undergoes a phase shift, the two equations are reversed: Eq. (35.17) is then the condition for destructive interference and Eq. (35.18) is the condition for constructive interference.
3. Solve the resulting interference equation for the target variable. If the film consists of anything other than vacuum, be sure to use the wavelength of light *in the film* in your calculations. If the film is anything except vacuum, this is smaller than the wavelength in vacuum by a factor of n. (For air, $n = 1.000$ to four-figure precision.)
4. If you are asked about the wave that is transmitted through the film, keep in mind that *minimum* intensity in the *reflected* wave corresponds to *maximum transmitted* intensity, and vice versa.

EVALUATE *your answer:* You can interpret your results by examining what would happen if the wavelength were changed or if the film had a different thickness.

Example 35.4 Thin-film interference I

Suppose the two glass plates in Fig. 35.11 are two microscope slides 10.0 cm long. At one end they are in contact; at the other end they are separated by a piece of paper 0.0200 mm thick. What is the spacing of the interference fringes seen by reflection? Is the fringe at the line of contact bright or dark? Assume monochromatic light with a wavelength in air of $\lambda = \lambda_0 = 500$ nm.

SOLUTION

IDENTIFY: We'll consider only interference between the light reflected from the upper and lower surfaces of the air wedge between the slides.

SET UP: Figure 35.11 depicts the situation. Light travels more slowly in the glass of the microscope slides than it does in air. Hence the wave reflected from the upper surface of the air wedge has no phase shift (see Fig. 35.12a), while the wave reflected from the lower surface has a half-cycle phase shift (see Fig. 35.12c).

EXECUTE: Since only one of the reflected waves undergoes a phase shift, the condition for *destructive* interference (a dark fringe) is Eq. (35.17):

$$2t = m\lambda_0 \qquad (m = 0, 1, 2, \ldots)$$

From similar triangles in Fig. 35.11 the thickness t of the air wedge at each point is proportional to the distance x from the line of contact:

$$\frac{t}{x} = \frac{h}{l}$$

Combining this with Eq. (35.17), we find

$$\frac{2xh}{l} = m\lambda_0$$

$$x = m\frac{l\lambda_0}{2h} = m\frac{(0.100\text{ m})(500 \times 10^{-9}\text{ m})}{(2)(0.0200 \times 10^{-3}\text{ m})} = m(1.25\text{ mm})$$

Successive dark fringes, corresponding to successive integer values of m, are spaced 1.25 mm apart. Substituting $m = 0$ into this equation gives $x = 0$, corresponding to the line of contact between the two slides (at the left-hand side of Fig. 35.11). Hence there is a dark fringe at the line of contact.

EVALUATE: Our result shows that the fringe spacing is proportional to the wavelength of the light used; the fringes would be farther apart with red light (larger λ_0) than with blue light (smaller λ_0). If we use white light, the reflected light at any point is a mixture of wavelengths for which constructive interference occurs; the wavelengths that interfere destructively are weak or absent in the reflected light. (This same effect explains the colors seen when an oil film on water is illuminated by white light, as in the photograph that opens this chapter.)

We've ignored interference between the light reflected from the upper and lower surfaces of the *glass* plate on top of the air wedge. This is acceptable if the glass plate is very much thicker than a wavelength of light. Example 35.7 shows one reason why interference effects are usually unimportant with a *thick* film.

Example 35.5 Thin-film interference II

In Example 35.4, suppose the glass plates have $n = 1.52$ and the space between plates contains water $(n = 1.33)$ instead of air. What happens now?

SOLUTION

IDENTIFY and SET UP: The index of refraction of the water film is still less than that of the glass on either side of the film, so the phase shifts are the same as in Example 35.4. The only difference is that the wavelength in water is different than in air.

EXECUTE: In the film of water $(n = 1.33)$, the wavelength is

$$\lambda = \frac{\lambda_0}{n} = \frac{500\text{ nm}}{1.33} = 376\text{ nm}$$

When we replace λ_0 by λ in the expression from Example 35.4 for the position x of the mth dark fringe, we find that the fringe spacing is reduced by the same factor of 1.33 and is equal to 0.940 mm. Note that there is still a dark fringe at the line of contact.

EVALUATE: Can you see that to return to the same fringe spacing as in Example 35.4, the dimension h in Fig. 35.11 would have to be reduced to $(0.0200\text{ mm})/1.33 = 0.0150\text{ mm}$? This shows that what matters in thin-film interference is the *ratio* between the wavelength and the thickness of the film. (You can see this by considering Eqs. (35.17) and (35.18).)

Example 35.6 Thin-film interference III

Suppose the upper of the two plates in Example 35.4 is a plastic material with $n = 1.40$, the wedge is filled with a silicone grease having $n = 1.50$, and the bottom plate is a dense flint glass with $n = 1.60$. What happens now?

SOLUTION

IDENTIFY and SET UP: The geometry is still as shown in Fig. 35.12, but now half-cycle phase shifts occur at *both* surfaces of the wedge of grease (see Fig. 35.12c).

EXECUTE: The positions of the dark fringes are now given by Eq. (35.18), with λ equal to the wavelength in the silicone grease: $\lambda = \lambda_0/n = (500\text{ nm})/1.50 = 333\text{ nm}$. You can readily show that the fringe spacing is 0.833 mm. Note that the two reflected waves from the line of contact are in phase (they both undergo the same phase shift), so the line of contact is at a *bright* fringe.

EVALUATE: What would happen if you carefully removed the upper microscope slide so that the grease wedge retained its shape? There would still be half-cycle phase changes at the upper and lower surfaces of the wedge, so the pattern of fringes would be the same as with the upper slide present.

Example 35.7 Thick films versus thin films

We have emphasized interference effects in rather *thin* films. Suppose the film in Fig. 35.10 is a layer of water $(n = 1.33)$ 1.00 cm thick on top of a glass surface with $n = 1.52$. What will you see if this rather *thick* film is illuminated from above?

SOLUTION

IDENTIFY and SET UP: As in Example 35.6, half-cycle phase changes occur at both the top and bottom surfaces of the film.

EXECUTE: Equation (35.17) gives the condition for *constructive* interference in the reflected waves. Hence the wavelengths in *air* that undergo constructive interference are given by

$$\lambda_0 = n\lambda = \frac{2nt}{m}$$

where n is the index of refraction of water.

Because $t = 1.00$ cm is large in comparison to the wavelengths of visible light, the values of m in this equation will be large. For instance, using $m = 40{,}000$ in the first equation tells us that light will be strongly reflected if it has a wavelength in air of

$$\lambda_0 = \frac{2nt}{m} = \frac{2(1.33)(1.00 \times 10^{-2}\text{ m})}{40{,}000}$$
$$= 6.67 \times 10^{-7}\text{ m} = 667\text{ nm}$$

corresponding to orange-red light. You might be tempted to conclude that this thick film would appear strongly orange-red; however, there will also be strong reflection of a slightly longer wavelength corresponding to $m = 39{,}999$ and of a slightly shorter wavelength corresponding to $m = 40{,}001$. These wavelengths differ from each other by only one part in 40,000, or $(667\text{ nm})/(40{,}000) = 0.017\text{ nm}$. This is a much smaller wavelength difference than the eye can detect. Hence no one wavelength appears to be emphasized more than any other in the light that is reflected from a *thick* film. If you shine white light on this thick film, the reflected light appears white.

EVALUATE: It's useful to compare this situation to a *thin* film that is no more than a few wavelengths in thickness. Then the integer m in Eqs. (35.17) and (35.18) will be small, the difference between adjacent strongly reflected wavelengths will be substantial, and the preferential reflection of certain wavelengths (as in the thin oil film shown in the chapter-opening paragraph) will be apparent to the eye.

?

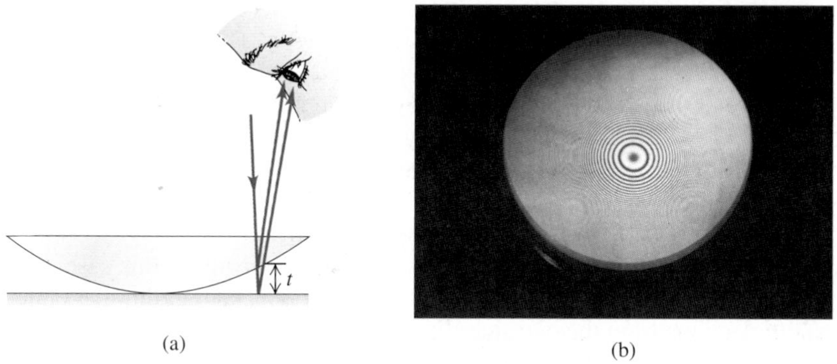

35.13 (a) Air film between a convex lens and a plane surface. The thickness of the film t increases from zero as we move out from the center, giving a series of alternating dark and bright rings for monochromatic light. (b) A photograph of Newton's rings. What do you suppose causes the fainter sets of rings in the photograph?

Newton's Rings

Figure 35.13a shows the convex surface of a lens in contact with a plane glass plate. A thin film of air is formed between the two surfaces. When you view the setup with monochromatic light, you see circular interference fringes (Fig. 35.13b). These were studied by Newton and are called **Newton's rings.**

We can use interference fringes to compare the surfaces of two optical parts by placing the two in contact and observing the interference fringes. Figure 35.14 is a photograph made during the grinding of a telescope objective lens. The lower, larger-diameter, thicker disk is the correctly shaped master, and the smaller, upper disk is the lens under test. The "contour lines" are Newton's interference fringes; each one indicates an additional distance between the specimen and the master of one half-wavelength. At 10 lines from the center spot the distance between the two surfaces is 5 wavelengths, or about 0.003 mm. This isn't very good; high-quality lenses are routinely ground with a precision of less than one wavelength. The surface of the primary mirror of the Hubble Space Telescope was ground to a precision of better than $\frac{1}{50}$ wavelength. Unfortunately, it was ground to incorrect specifications, creating one of the most precise errors in the history of optical technology (see Section 34.2).

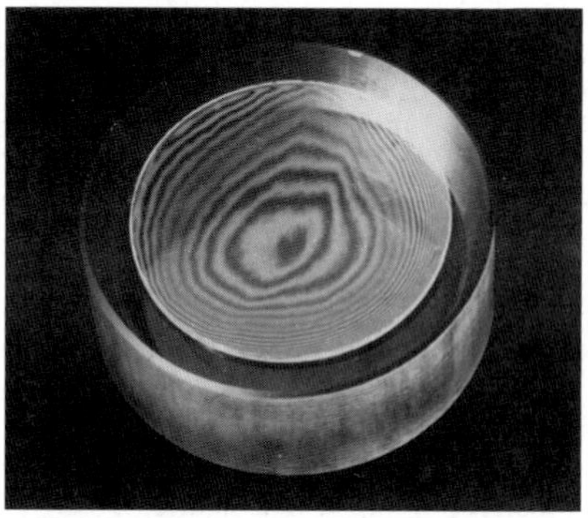

35.14 The surface of a telescope objective under inspection during manufacture.

Nonreflective and Reflective Coatings

Nonreflective coatings for lens surfaces make use of thin-film interference. A thin layer or film of hard transparent material with an index of refraction smaller than that of the glass is deposited on the lens surface, as in Fig. 35.15. Light is reflected from both surfaces of the layer. In both reflections the light is reflected from a medium of greater index than that in which it is traveling, so the same phase change occurs in both reflections. If the film thickness is a quarter (one-fourth) of the wavelength *in the film* (assuming normal incidence), the total path difference is a half-wavelength. Light reflected from the first surface is then a half-cycle out of phase with light reflected from the second, and there is destructive interference.

The thickness of the nonreflective coating can be a quarter-wavelength for only one particular wavelength. This is usually chosen in the central yellow-green portion of the spectrum ($\lambda = 550\text{ nm}$), where the eye is most sensitive. Then there is somewhat more reflection at both longer (red) and shorter (blue) wavelengths, and the reflected light has a purple hue. The overall reflection from a lens or prism surface can be reduced in this way from 4–5% to less than 1%. This treatment is particularly important in eliminating stray reflected light in highly

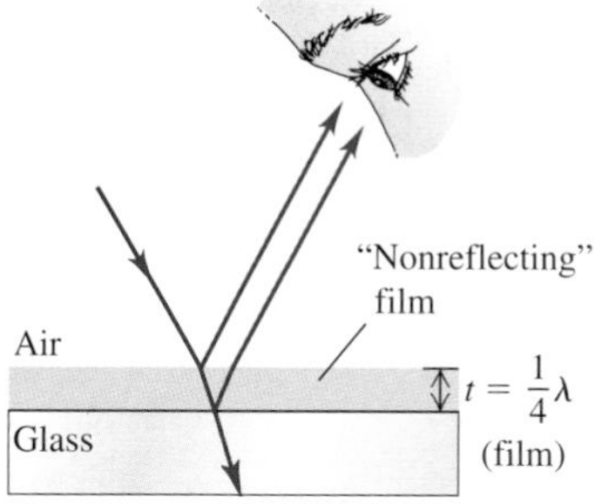

35.15 A nonreflective coating has an index of refraction intermediate between those of glass and air. Destructive interference results when the film thickness is one-fourth of the wavelength in the film.

corrected photographic lenses with many individual pieces of glass and many air–glass surfaces. It also increases the net amount of light that is *transmitted* through the lens, since light that is not reflected will be transmitted. The same principle is used to minimize reflection from silicon photovoltaic solar cells $(n = 3.5)$ by use of a thin surface layer of silicon monoxide $(\mathrm{SiO}, n = 1.45)$; this helps to increase the amount of light that actually reaches the solar cells.

If a quarter-wavelength thickness of a material with an index of refraction *greater* than that of glass is deposited on glass, then the reflectivity is *increased,* and the deposited material is called a **reflective coating.** In this case there is a half-cycle phase shift at the air-film interface but none at the film-glass interface, and reflections from the two sides of the film interfere constructively. For example, a coating with refractive index 2.5 causes 38% of the incident energy to be reflected, compared with 4% or so with no coating. By use of multiple-layer coatings, it is possible to achieve nearly 100% transmission or reflection for particular wavelengths. Some practical applications of these coatings are for color separation in color television cameras and for infrared "heat reflectors" in motion-picture projectors, solar cells, and astronauts' visors. Reflective coatings occur in nature on the scales of herring and other silvery fish; this gives these fish their characteristic shiny appearance (see Problem 35.54).

Example 35.8 A nonreflective coating

A commonly used lens coating material is magnesium fluoride, MgF_2, with $n = 1.38$. What thickness should a nonreflective coating have for 550-nm light if it is applied to glass with $n = 1.52$?

SOLUTION

IDENTIFY and SET UP: This coating is of the sort depicted in Fig. 35.15. The thickness must be one-quarter of the wavelength in the coating.

EXECUTE: The wavelength of yellow-green light in air is $\lambda_0 = 550$ nm, so its wavelength in the MgF_2 coating is

$$\lambda = \frac{\lambda_0}{n} = \frac{550\text{ nm}}{1.38} = 400\text{ nm}$$

To be a nonreflective film, the coating should have a thickness of one-quarter λ, or 100 nm. This is a very thin film, no more than a few hundred molecules thick.

EVALUATE: Note that such a coating becomes *reflective* if its thickness is equal to one-*half* of a wavelength; then light reflected from the lower surface of the coating travels one wavelength farther than light reflected from the upper surface, so the two waves are in phase and interfere constructively. This occurs for light with a wavelength in MgF_2 of 200 nm and a wavelength in air of $(200\text{ nm})(1.38) = 276$ nm. This is an ultraviolet wavelength (see Section 32.6), so designers of optical lenses with nonreflective coatings need not worry about enhanced reflection of this kind.

Test Your Understanding

When a convex lens on a plane surface is illuminated as shown in Fig. 35.13a, the center of the pattern is black (Fig. 35.13b). Why?

35.5 | The Michelson Interferometer

An important experimental device that uses interference is the **Michelson interferometer.** A century ago, it helped to provide one of the key experimental underpinnings of the theory of relativity. More recently, Michelson interferometers have been used to make precise measurements of wavelengths and of very small

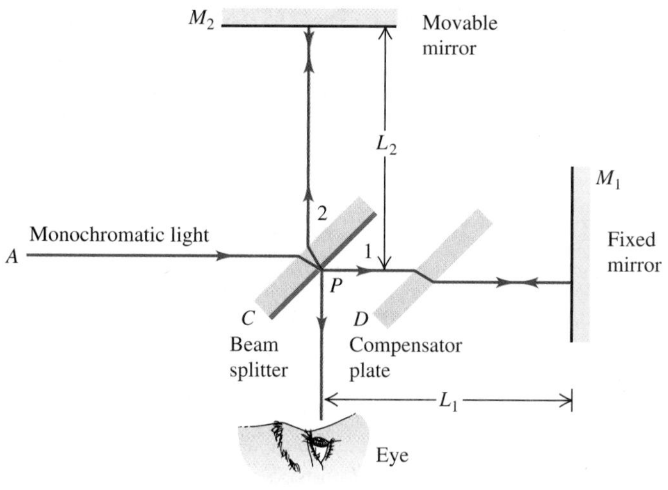

35.16 A schematic Michelson interferometer. The observer sees an interference pattern that results from the difference in path lengths for rays 1 and 2.

distances, such as the minute changes in thickness of an axon when a nerve impulse propagates along its length. Like the Young two-slit experiment, a Michelson interferometer takes monochromatic light from a single source and divides it into two waves that follow different paths. In Young's experiment, this is done by sending part of the light through one slit and part through another; in a Michelson interferometer a device called a *beam splitter* is used. Interference occurs in both experiments when the two light waves are recombined.

The principal components of a Michelson interferometer are shown schematically in Fig. 35.16. A ray of light from a monochromatic source A strikes the beam splitter C, which is a glass plate with a thin coating of silver on its right side. Part of the light (ray 1) passes through the silvered surface and the compensator plate D and is reflected from mirror M_1. It then returns through D and is reflected from the silvered surface of C to the observer. The remainder of the light (ray 2) is reflected from the silvered surface at point P to the mirror M_2 and back through C to the observer's eye. The purpose of the compensator plate D is to ensure that rays 1 and 2 pass through the same thickness of glass; plate D is cut from the same piece of glass as plate C, so their thicknesses are identical to within a fraction of a wavelength.

The whole apparatus in Fig. 35.16 is mounted on a very rigid frame, and the position of mirror M_2 can be adjusted with a fine, very accurate micrometer screw. If the distances L_1 and L_2 are exactly equal and the mirrors M_1 and M_2 are exactly at right angles, the virtual image of M_1 formed by reflection at the silvered surface of plate C coincides with mirror M_2. If L_1 and L_2 are *not* exactly equal, the image of M_1 is displaced slightly from M_2; and if the mirrors are not exactly perpendicular, the image of M_1 makes a slight angle with M_2. Then the mirror M_2 and the virtual image of M_1 play the same roles as the two surfaces of a wedge-shaped thin film (Section 35.4), and light reflected from these surfaces forms the same sort of interference fringes.

Suppose the angle between mirror M_2 and the virtual image of M_1 is just great enough that five or six vertical fringes are present in the field of view. If we now move the mirror M_2 slowly either backward or forward a distance $\lambda/2$, the difference in path length between rays 1 and 2 changes by λ, and each fringe moves to the left or right a distance equal to the fringe spacing. If we observe the fringe positions through a telescope with a crosshair eyepiece and m fringes cross the crosshairs when we move the mirror a distance y, then

$$y = m\frac{\lambda}{2} \quad \text{or} \quad \lambda = \frac{2y}{m} \tag{35.19}$$

If m is several thousand, the distance y is large enough that it can be measured with good accuracy, and we can obtain an accurate value for the wavelength λ. Alternatively, if the wavelength is known, a distance y can be measured by simply counting fringes when M_2 is moved by this distance. In this way, distances that are comparable to a wavelength of light can be measured with relative ease.

The original application of the Michelson interferometer was to the historic **Michelson-Morley experiment.** Before the electromagnetic theory of light and Einstein's special theory of relativity became established, most physicists believed that the propagation of light waves occurred in a medium called the **ether,** which was believed to permeate all space. In 1887 the American scientists Albert Michelson and Edward Morley used the Michelson interferometer in an attempt to detect the motion of the earth through the ether. Suppose the interferometer in Fig. 35.16 is moving from left to right relative to the ether. According to the ether theory, this would lead to changes in the speed of light in the portions of the path shown as horizontal lines in the figure. There would be fringe shifts relative to the positions that the fringes would have if the instrument were at rest in the ether. Then when the entire instrument was rotated 90°, the other portions of the paths would be similarly affected, giving a fringe shift in the opposite direction.

Michelson and Morley expected that the motion of the earth through the ether would cause a fringe shift of about four-tenths of a fringe when the instrument was rotated. The shift that was actually observed was less than a hundredth of a fringe and, within the limits of experimental uncertainty, appeared to be exactly zero. Despite its orbital motion around the sun, the earth appeared to be *at rest* relative to the ether. This negative result baffled physicists until Einstein developed the special theory of relativity in 1905. Einstein postulated that the speed of a light wave in vacuum has the same magnitude c relative to *all* inertial reference frames, no matter what their velocity may be relative to each other. The presumed ether then plays no role, and the concept of an ether has been abandoned.

The theory of relativity is a well-established cornerstone of modern physics, and we will study it in detail in Chapter 37. In retrospect, the Michelson-Morley experiment gives strong experimental support to the special theory of relativity, and it is often called the most significant "negative-result" experiment ever performed.

Test Your Understanding

A Michelson interferometer is used with light of wavelength 605.78 nm. If the observer views the interference pattern through a telescope with a crosshair eyepiece, how many fringes pass the crosshairs when mirror M_2 moves exactly one centimeter?

SUMMARY

Monochromatic light is light with a single frequency. Coherence is a definite, unchanging phase relationship between two waves. The overlap of waves from two coherent sources of monochromatic light forms an interference pattern. The principle of superposition states that the total wave disturbance at any point is the sum of the disturbances from the separate waves.

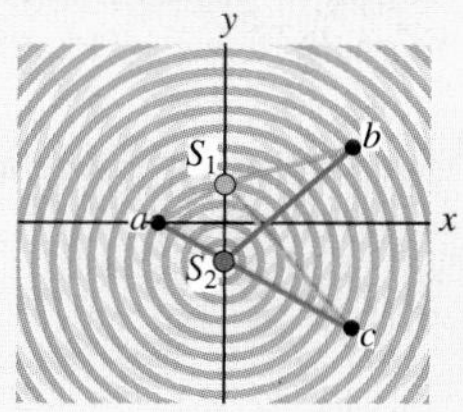

When two sources are in phase, constructive interference occurs at points where the difference in path length from the two sources is zero or an integer number of wavelengths; destructive interference occurs at points where the path difference is a half-integer number of wavelengths. If two sources separated by a distance d are both very far from a point P, and the line from the sources to P makes an angle θ with the line perpendicular to the line of the sources, then the condition for constructive interference at P is Eq. (35.4). The condition for destructive interference is Eq. (35.5). When θ is very small, the position y_m of the mth bright fringe on a screen located a distance R from the sources is given by Eq. (35.6). (See Examples 35.1 and 35.2)

$$d \sin \theta = m\lambda \quad (m = 0, \pm 1, \pm 2, \dots) \tag{35.4}$$

$$d \sin \theta = \left(m + \frac{1}{2}\right)\lambda \quad (m = 0, \pm 1, \pm 2, \dots) \tag{35.5}$$

$$y_m = R\frac{m\lambda}{d} \tag{35.6}$$

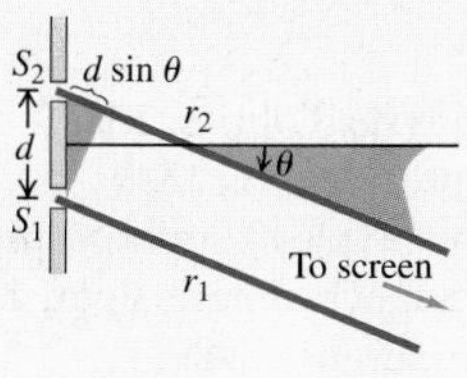

When two sinusoidal waves with equal amplitude E and phase difference ϕ are superimposed, the resultant amplitude E_P and intensity I are given by Eqs. (35.7) and (35.10) respectively. If the two sources emit in phase, the phase difference ϕ at a point P (located a distance r_1 from source 1 and a distance r_2 from source 2) is directly proportional to the difference in path length $r_2 - r_1$. (See Example 35.3)

$$E_P = 2E\left|\cos\frac{\phi}{2}\right| \tag{35.7}$$

$$I = I_0 \cos^2 \frac{\phi}{2} \tag{35.10}$$

$$\phi = \frac{2\pi}{\lambda}(r_2 - r_1) = k(r_2 - r_1) \tag{35.11}$$

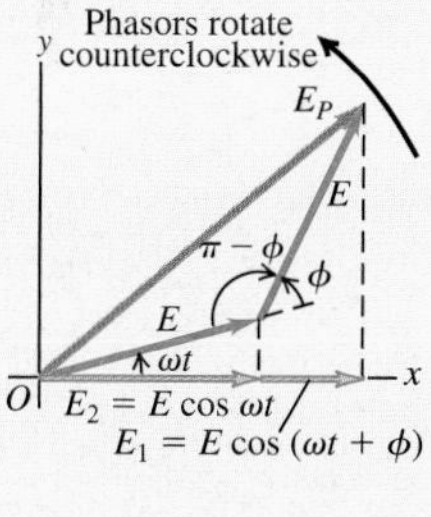

When light is reflected from both sides of a thin film of thickness t and no phase shift occurs at either surface, constructive interference between the reflected waves occurs when $2t$ is equal to an integral number of wavelengths. If a half-cycle phase shift occurs at one surface, this is the condition for destructive interference. A half-cycle phase shift occurs during reflection whenever the index of refraction in the second material is greater than that in the first. (See Examples 35.4 through 35.8)

$$2t = m\lambda \quad (m = 0, 1, 2, \dots) \tag{35.17}$$

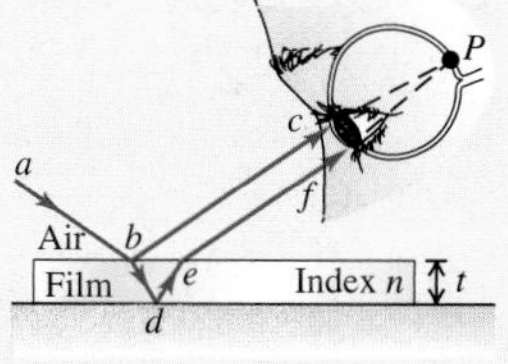

The Michelson interferometer uses a monochromatic light source and can be used for high-precision measurements of wavelengths. Its original purpose was to detect motion of the earth relative to a hypothetical ether, the supposed medium for electromagnetic waves. The ether has never been detected, and the concept has been abandoned; the speed of light is the same relative to all observers. This is part of the foundation of the special theory of relativity.

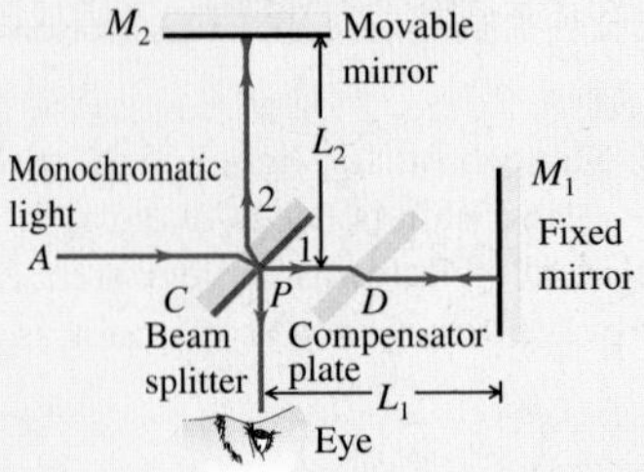

Key Terms

physical optics, 1338
interference, 1339
principle of superposition, 1339
monochromatic light, 1339
coherent, 1340
constructive interference, 1341
destructive interference, 1341
antinodal curves, 1341
nodal curves, 1341
interference fringes, 1344
Newton's rings, 1355
nonreflective coating, 1355
reflective coating, 1356
Michelson interferometer, 1356
Michelson-Morley experiment, 1358
ether, 1358

Your Notes

Answer to Chapter Opening Question

To display the kinds of colors shown in the chapter-opening photograph, the oil film can be no more than a few wavelengths of light in thickness (that is, a few micrometers). This is explained in Example 35.7.

Answers to Test Your Understanding Questions

Section 35.1 The nodal curves have the same shape as the antinodal curves but lie between them. From top to bottom, the nodal curves correspond to $m = 2, 1, 0, -1, -2$, and -3 in Eq. (35.2). Note that the $m = -3$ nodal curve passes through point c.
Section 35.2 Blue light has a shorter wavelength than red light (see Section 32.6). Equation (35.6) tells us that the distance y_m from the center of the pattern to the mth bright fringe is proportional to the wavelength λ. Hence all of the fringes will move toward the center of the pattern as the wavelength decreases, and the spacing between fringes will decrease.
Section 35.3 For this point the *path* difference is $r_2 - r_1 = d\sin\theta = \pm 7\lambda/4$, so the *phase* difference is $\phi = (\pi d/\lambda)\sin\theta = (\pi/\lambda)(\pm 7\lambda/4) = \pm 7\pi/4$. The intensity at this point is therefore $I = I_0 \cos^2(\phi/2) = I_0 \cos^2(\pm 7\pi/4) = I_0(1/\sqrt{2})^2 = I_0/2$. You can also conclude this from examining Fig. 35.9, which shows that $I = I_0/2$ at $d\sin\theta = \pm 7\lambda/4$.
Section 35.4 Light reflected from the glass–air interface at the bottom of the convex lens undergoes no phase shift (see Fig. 35.12a). By contrast, light reflected at the interface between the air and the plane surface undergoes a half-cycle phase shift (see Fig. 35.12c). The center of the interference pattern comes from the region where the lens and plane surface are almost touching, so there is essentially zero path difference between the two reflected light waves and hence no additional phase difference. Hence, the two waves are a half-cycle out of phase, so they interfere destructively and we see a dark central spot in the pattern.
Section 35.5 Equation (35.19) gives the number m of fringes that cross the crosshairs when mirror M_2 moves a distance y. Solving for m,

$$m = \frac{2y}{\lambda} = \frac{2(1.0000 \times 10^{-2}\ \text{m})}{605.78 \times 10^{-9}\ \text{m}} = 33{,}015$$

Another way to look at this result is that *one* fringe moves past the crosshairs when mirror M_2 moves exactly 1/33015 cm, or 0.30289 μm (about 1/300 the width of a human hair). This demonstrates how the Michelson interferometer can be used to measure very small displacements.

Discussion Questions

Q35.1 A two-slit interference experiment is set up, and the fringes are displayed on a screen. Then the whole apparatus is immersed in the nearest swimming pool. How does the fringe pattern change?
Q35.2 Could an experiment similar to Young's two-slit experiment be performed with sound? How might this be carried out? Does it matter that sound waves are longitudinal and electromagnetic waves are transverse? Explain.
Q35.3 You place a bright flashlight inside a shoe box and cut very narrow slits a few micrometers apart in one end. When the light from the two slits falls on a screen in a dark room, will interference fringes be observed? What if the flashlight is replaced by a laser? Explain your answers.
Q35.4 Would the headlights of a distant car form a two-source interference pattern? If so, how might it be observed? If not, why not?
Q35.5 The two sources S_1 and S_2 shown in Fig. 35.3 emit waves of the same wavelength λ and are in phase with each other. Suppose S_1 is a weaker source, so that the waves emitted by S_1 have half the amplitude of the waves emitted by S_2. How would this affect the positions of the antinodal lines and nodal lines? Would there be total reinforcement at points on the antinodal curves? Would there be total cancellation at points on the nodal curves? Explain your answers.
Q35.6 Could the Young two-slit interference experiment be performed with gamma rays? If not, why not? If so, discuss differences in the experimental design compared to the experiment with visible light.
Q35.7 Coherent red light illuminates two narrow slits that are 25 cm apart. Will a two-slit interference pattern be observed when the light from the slits falls on a screen? Explain.
Q35.8 Coherent light with wavelength λ falls on two narrow slits separated by a distance d. If d is less than some minimum value, no dark fringes are observed. Explain. In terms of λ, what is this minimum value of d?
Q35.9 A fellow student, who values memorizing equations above understanding them, combines Eqs. (35.4) and (35.13) to "prove" that ϕ can *only* equal $2\pi m$. How would you explain to this student that ϕ can have values other than $2\pi m$?
Q35.10 If the monochromatic light shown in Fig. 35.5a were replaced by white light, would a two-slit interference pattern be seen on the screen? Explain.
Q35.11 In using the superposition principle to calculate intensities in interference patterns, could you add the intensities of the waves instead of their amplitudes? Explain.
Q35.12 A glass windowpane with a thin film of water on it reflects less than when it is perfectly dry. Why?
Q35.13 A *very* thin soap film ($n = 1.33$), whose thickness is much less than a wavelength of visible light, looks black; it appears to reflect no light at all. Why? By contrast, an equally thin layer of soapy water ($n = 1.33$) on glass ($n = 1.50$) appears quite shiny. Why is there a difference?
Q35.14 Interference can occur in thin films. Why is it important that the films be *thin?* Why don't you get these effects with a relatively *thick* film? Where should you put the dividing line between "thin" and "thick"? Explain your reasoning.
Q35.15 If we shine white light on an air wedge like that shown in Fig. 35.11, the colors that are weak in the light *reflected* from any point along the wedge are strong in the light *transmitted* through the wedge. Explain why this should be so.
Q35.16 Monochromatic light is directed at normal incidence on a thin film. There is destructive interference for the reflected light, so

the intensity of the reflected light is very low. What happened to the energy of the incident light?

Q35.17 When a thin oil film spreads out on a puddle of water, the thinnest part of the film looks dark in the resulting interference pattern. What does this tell you about the relative magnitudes of the refractive indices of oil and water?

Exercises

Section 35.1 Interference and Coherent Sources

35.1 Consider Fig. 35.3, which could represent interference between water waves in a ripple tank. Pick at least three points on the antinodal curve labeled "$m = 3$," and make measurements from the figure to show that Eq. (35.1) is indeed satisfied. Explain what measurements you made and how you measured the wavelength λ.

35.2 Figure 35.3 shows the wave pattern produced by two identical, coherent sources emitting waves with wavelength λ and separated by a distance $d = 4\lambda$. a) Explain why the positive y-axis above S_1 constitutes an antinodal curve with $m = +4$ and why the negative y-axis below S_2 constitutes an antinodal curve with $m = -4$. b) Draw the wave pattern produced when the separation between the sources is reduced to 3λ. In your drawing, sketch all antinodal curves, that is, the curves on which $r_2 - r_1 = m\lambda$. Label each curve by its value of m. c) In general, what determines the maximum (most positive) and minimum (most negative) values of the integer m that labels the antinodal lines? d) Suppose the separation between the sources is increased to $7\frac{1}{2}\lambda$. How many antinodal curves will there be? To what values of m do they correspond? Explain your reasoning. (You should not have to make a drawing to answer these questions.)

35.3 Two coherent sources A and B of radio waves are 5.00 m apart. Each source emits waves with wavelength 6.00 m. Consider points along the line between the two sources. At what distances, if any, from A is the interference a) constructive; b) destructive?

35.4 Radio Interference. Two radio antennas A and B radiate in phase. Antenna B is 120 m to the right of antenna A. Consider point Q along the extension of the line connecting the antennas, a horizontal distance of 40 m to the right of antenna B. The frequency, and hence the wavelength, of the emitted waves can be varied. a) What is the longest wavelength for which there will be destructive interference at point Q? b) What is the longest wavelength for which there will be constructive interference at point Q?

35.5 A radio transmitting station operating at a frequency of 120 MHz has two identical antennas that radiate in phase. Antenna B is 9.00 m to the right of antenna A. Consider point P between the antennas and along the line connecting them, a horizontal distance x to the right of antenna A. For what values of x will constructive interference occur at point P?

35.6 Two light sources can be adjusted to emit monochromatic light of any visible wavelength. The two sources are coherent, 2.04 μm apart, and in line with an observer, so that one source is 2.04 μm farther from the observer than the other. a) For what visible wavelengths (400 to 700 nm) will the observer see the brightest light, owing to constructive interference? b) How would your answers to part (a) be affected if the two sources were not in line with the observer, but were still arranged so that one source is 2.04 μm farther away from the observer than the other?

35.7 In part (a) of Exercise 35.6, for what visible wavelengths will there be destructive interference at the location of the observer?

Section 35.2 Two-Source Interference of Light

35.8 An FM radio station has a frequency of 107.9 MHz and uses two identical antennas mounted at the same elevation, 12.0 m apart. The antennas radiate in phase. The resulting radiation pattern has a maximum intensity along a horizontal line perpendicular to the line joining the antennas and midway between them. Assume that the intensity is observed at distances from the antennas that are much larger than 12.0 m. a) At which other angles (measured from the line of maximum intensity) is the intensity maximum? b) At which angles is it zero?

35.9 Coherent light from a sodium-vapor lamp is passed through a filter that blocks everything except for light of a single wavelength. It then falls on two slits separated by 0.460 mm. In the resulting interference pattern on a screen 2.20 m away, adjacent bright fringes are separated by 2.82 mm. What is the wavelength?

35.10 Young's experiment is performed with light from excited helium atoms $(\lambda = 502 \text{ nm})$. Fringes are measured carefully on a screen 1.20 m away from the double slit, and the center of the twentieth fringe (not counting the central bright fringe) is found to be 10.6 mm from the center of the central bright fringe. What is the separation of the two slits?

35.11 Two slits spaced 0.450 mm apart are placed 75.0 cm from a screen. What is the distance between the second and third dark lines of the interference pattern on the screen when the slits are illuminated with coherent light with a wavelength of 500 nm?

35.12 Coherent light with wavelength 400 nm passes through two very narrow slits that are separated by 0.200 mm and the interference pattern is observed on a screen 4.00 m from the slits. a) What is the width (in mm) of the central interference maximum? b) What is the width of the first-order bright fringe?

35.13 Coherent light with wavelength 600 nm passes through two very narrow slits and the interference pattern is observed on a screen 3.00 m from the slits. The first-order bright fringe is at 4.84 mm from the center of the central bright fringe. For what wavelength of light will the first-order dark fringe be observed at this same point on the screen?

35.14 Coherent light that contains two wavelengths, 660 nm (red) and 470 nm (blue), passes through two narrow slits separated by 0.300 mm and the interference pattern is observed on a screen 5.00 m from the slits. What is the distance on the screen between the first-order bright fringe for each wavelength?

35.15 Two very narrow slits are spaced 1.80 μm apart and are placed 35.0 cm from a screen. What is the distance between the first and second dark lines of the interference pattern when the slits are illuminated with coherent light with $\lambda = 550$ nm? (*Hint:* The angle θ in Eq. (35.5) is *not* small.)

35.16 A sheet of blackened photographic film has two very narrow scratches in it. When you shine laser light of wavelength 624 nm on it, you find that the first dark fringe occurs at an angle of $\pm 11.0°$ on either side of the central bright line. a) How far apart are these scratches? b) Find the total number of bright fringes and the total number of dark fringes that the film will produce with this laser.

35.17 White light illuminates two thin slits that are 0.100 mm apart. Calculate the angular width of the first full-color visible spectrum on either side of the central bright line. (*Note:* The wavelength range for visible light is 400 nm to 700 nm.)

35.18 Coherent light with wavelength 450 nm falls on a double slit. On a screen 1.80 m away the distance between dark fringes is 4.20 mm. What is the separation of the slits?

Section 35.3 Intensity in Interference Patterns

35.19 Coherent light with wavelength 500 nm passes through narrow slits separated by 0.340 mm. At a distance from the slits large compared to their separation, what is the phase difference (in radians) in the light from the two slits at an angle of 23.0° from the centerline?

35.20 Coherent sources A and B emit electromagnetic waves with wavelength 2.00 cm. Point P is 4.86 m from A and 5.24 m from B. What is the phase difference at P between these two waves?

35.21 In a two-slit interference pattern, the intensity at the peak of the central maximum is I_0. a) At a point in the pattern where the phase difference between the waves from the two slits is 60.0°, what is the intensity? b) What is the path difference for 480 nm light from the two slits at a point where the phase angle is 60.0°?

35.22 Consider two antennas separated by 9.00 m, as described in Exercise 35.5. The two antennas radiate in phase at 15.0 MHz, and the resulting interference pattern is observed at distances from the antennas that are much greater than 9.00 m. a) In which directions is the intensity maximum, considering points in a horizontal plane? b) Calling the maximum intensity in part (a) I_0, determine in terms of I_0 the intensity in directions making angles of 30°, 45°, 60°, and 90° to the direction of maximum intensity.

35.23 Two slits spaced 0.260 mm apart are placed 0.700 m from a screen and illuminated by coherent light with wavelength of 660 nm. The intensity at the center of the central maximum $(\theta = 0°)$ is I_0. a) What is the distance on the screen from the center of the central maximum to the first minimum? b) What is the distance on the screen from the center of the central maximum to the point where the intensity has fallen to $I_0/2$?

35.24 Consider two antennas separated by 9.00 m that radiate in phase at 120 MHz, as described in Exercise 35.5. A receiver placed 150 m from both antennas measures an intensity I_0. The receiver is moved so that it is 1.8 m closer to one antenna than to the other. a) What is the phase difference ϕ between the two radio waves produced by this path difference? b) In terms of I_0, what is the intensity measured by the receiver at its new position?

35.25 Two slits spaced 0.130 mm apart are placed 90.0 cm from a screen and illuminated by coherent light with wavelength 550 nm. The intensity at the center of the central maximum $(\theta = 0°)$ is 4.00×10^{-6} W/m^2. a) What is the distance on the screen from the center of the central maximum to the first minimum? b) What is the intensity at a point midway between the center of the central maximum and the first *minimum?*

35.26 Show that Eq. (35.14) gives zero-intensity directions that agree with Eq. (35.5).

Section 35.4 Interference in Thin Films

35.27 Two rectangular pieces of plane glass are laid one upon the other on a table. A thin strip of paper is placed between them at one edge so that a very thin wedge of air is formed. The plates are illuminated at normal incidence by 546-nm light from a mercury-vapor lamp. Interference fringes are formed, with 15.0 fringes per centimeter. Find the angle of the wedge.

35.28 A plate of glass 9.00 cm long is placed in contact with a second plate and is held at a small angle with it by a metal strip 0.0800 mm thick placed under one end. The space between the plates is filled with air. The glass is illuminated from above with light having a wavelength in air of 656 nm. How many interference fringes are observed per centimeter in the reflected light?

35.29 What is the thinnest film of a coating with $n = 1.42$ on glass $(n = 1.52)$ for which destructive interference of the red component (650 nm) of an incident white light beam in air can take place by reflection?

35.30 A thin film of polystyrene is used as a nonreflecting coating for fabulite (see Table 33.1). What is the minimum thickness of the film required? Assume that the wavelength of the light in air is 480 nm.

35.31 An oil film covers the surface of a small pond. The refractive index of the oil is greater than that of water. At one point on the film, the film has the smallest nonzero thickness for which there will be destructive interference in the reflected light when infrared radiation with wavelength 800 nm is incident normal to the film. When this film is viewed at normal incidence at this same point, for what visible wavelengths, if any, will there be constructive interference? (Visible light has wavelengths between 400 nm and 700 nm.)

35.32 Light with wavelength 648 nm in air is incident perpendicularly from air on a film 8.76 μm thick and with refractive index 1.35. Part of the light is reflected from the first surface of the film, and part enters the film and is reflected back at the second surface, where the film is again in contact with air. a) How many waves are contained along the path of this second part of the light in its round-trip through the film? b) What is the phase difference between these two parts of the light as they leave the film?

35.33 Compact Disc Player. A compact disc (CD) is read from the bottom by a semiconductor laser with wavelength 790 nm passing through a plastic substrate of refractive index 1.8. When the beam encounters a pit, part of the beam is reflected from the pit and part from the flat region between the pits, so these two beams interfere with each other (see Fig. 35.17). What must the minimum pit depth be so that the part of the beam reflected from a pit cancels the part of the beam reflected from the flat region? (It is this cancel-

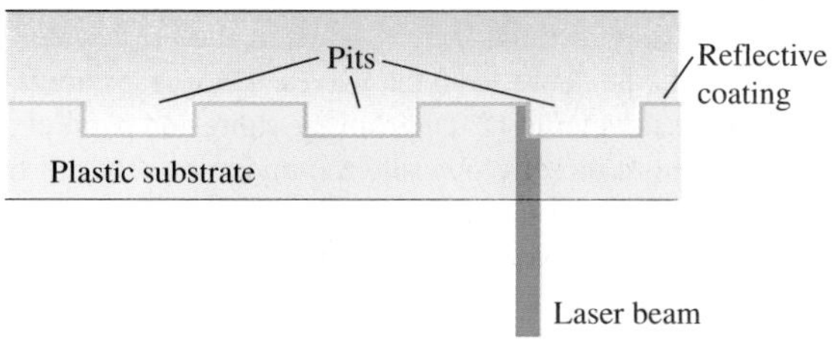

Figure 35.17 Exercise 35.33.

lation which allows the player to recognize the beginning and end of a pit. For a fuller explanation of the physics behind CD technology, see the article "The Compact Disc Digital Audio System," by Thomas D. Rossing, in the Dec. 1987 issue of *The Physics Teacher.*)

35.34 What is the thinnest soap film (excluding the case of zero thickness) that appears black when illuminated with light with wavelength 480 nm? The index of refraction of the film is 1.33, and there is air on both sides of the film.

35.35 The walls of a soap bubble have about the same index of refraction as that of plain water, $n = 1.33$. There is air both inside and outside the bubble. a) What wavelength (in air) of visible light is most strongly reflected from a point on a soap bubble where its wall is 290 nm thick? To what color does this correspond (see Fig. 32.20 and Table 32.1)? b) Repeat part (a) for a wall thickness of 340 nm.

35.36 A plastic film with index of refraction 1.85 is put on the surface of a car window to increase the reflectivity and thus to keep the interior of the car cooler. The window glass has index of refraction 1.52. a) What minimum thickness is required if light with wavelength 550 nm in air reflected from the two sides of the film is to interfere constructively? b) It is found to be difficult to manufacture and install coatings as thin as calculated in part (a). What is the next greatest thickness for which there will also be constructive interference?

Section 35.5 The Michelson Interferometer

35.37 How far must the mirror M_2 (see Fig. 35.16) of the Michelson interferometer be moved so that 1800 fringes of He-Ne laser light ($\lambda = 633$ nm) move across a line in the field of view?

35.38 Jan first uses a Michelson interferometer with the 606-nm light from a krypton-86 lamp. He displaces the movable mirror away from him, counting 818 fringes moving across a line in his field of view. Then Linda replaces the krypton lamp with filtered 502-nm light from a helium lamp and displaces the movable mirror toward her. She also counts 818 fringes, but they move across the line in her field of view opposite to the direction they moved for Jan. Assume that both Jan and Linda counted to 818 correctly. a) What distance did each person move the mirror? b) What is the resultant displacement of the mirror?

Problems

35.39 If the entire apparatus of Exercise 35.11 (slits, screen, and space in between) is immersed in water, what then is the distance between the second and third dark lines?

35.40 While walking along the shore at your beachfront home, you notice that there are two narrow gaps in the breakwater, the wall that protects the shore from the waves. These gaps are 9.0 m apart and the breakwater is 12.0 m from the shore and parallel to it. You go to the shore directly opposite the midpoint between these gaps. As you walk along the shore, the first point where no waves reach you is 1.7 m from your starting point. Out beyond the breakwater you observe that there are ten wave crests in 18 s. How far apart are the wave crests?

35.41 Two identical audio speakers, connected to the same amplifier, produce monochromatic sound waves with a frequency that can be varied between 300 and 600 Hz. The speed of sound is 340 m/s. You find that where you are standing, you hear minimum-intensity sound. a) Explain why you hear minimum-intensity sound. b) If one of the speakers is moved 39.8 cm toward you, the sound you hear has maximum intensity. What is the frequency of the sound? c) How much closer to you from the position in part (b) must the speaker be moved to the next position where you hear maximum intensity?

35.42 Two radio antennas radiating in phase are located at points A and B, 200 m apart (see Fig. 35.18). The radio waves have a frequency of 5.80 MHz. A radio receiver is moved out from point B along a line perpendicular to the line connecting A and B (line BC shown in Fig. 35.18). At what distances from B will there be *destructive* interference? (*Note:* The distance of the receiver from the sources is not large in comparison to the separation of the sources, so Eq. (35.5) does not apply.)

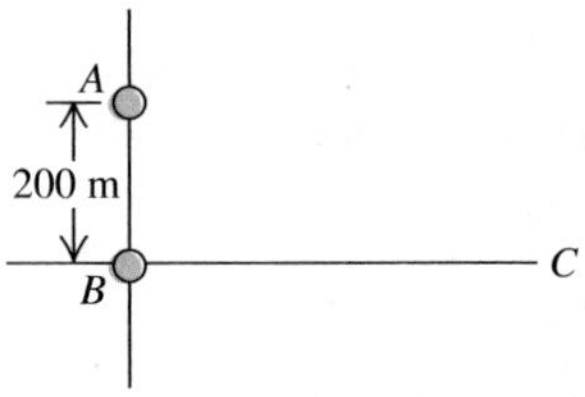

Figure 35.18 Problem 35.42.

35.43 Two speakers, A and B, are 3.36 m apart and each one is emitting sound with wavelength 0.800 m. Speaker A is $\frac{1}{6}$ of a period behind B in phase. The sound from the speakers is detected at distances much greater than 3.36 m. Find all the angles relative to the centerline at which sound waves in front of the speakers will completely cancel. (*Hint:* Consider what happens on both sides of the centerline separately.)

35.44 Two sound speakers, 42.2 cm apart, are driven at 1570 Hz by a single oscillator. However, because the cables to these speakers have different lengths, the signal reaches one of them 0.159 ms before the other. The speed of sound in the air is 330 m/s. The sound from the speakers is detected at distances from the speakers much greater than 42.2 cm. Find the directions, measured relative to the centerline between the speakers, at which the sound will have maximum intensity. (*Hint:* Consider what happens on both sides of the centerline separately.)

35.45 a) In Fig. 35.3, suppose source S_2 is *not* in phase with S_1, but instead is *out* of phase by $\frac{1}{2}$ cycle. In this situation, Eq. (35.1) is the condition for *destructive* interference, and Eq. (35.2) is the condition for *constructive* interference. Explain why this is so. b) Suppose S_2 *leads* S_1 by a phase angle ϕ; that is, if the displacement of source S_1 is given by $x_1(t) = A\cos\omega t$, then the displacement of source S_2 is $x_2(t) = A\cos(\omega t + \phi)$. (In the situation of part (a), $\phi = \pi$.) Find expressions for the values of the path difference $r_2 - r_1$ that correspond to constructive interference and to destructive interference.

35.46 The electric fields received at point P from two identical, coherent wave sources are $E_1(t) = E\cos(\omega t + \phi)$ and $E_2(t) = E\cos\omega t$. a) Use one of the trigonometric identities in Appendix B to show that the resultant wave is $E_P(t) = 2E\cos(\phi/2)\cos(\omega t + \phi/2)$. b) Show that the amplitude of this resultant wave is given by Eq. (35.7). c) Use the result of part (a) to show that at an interference maximum, the amplitude of the resultant wave is in phase with the original waves $E_1(t)$ and $E_2(t)$. d) Use the result of part (a) to show that near an interference minimum, the resultant wave is approximately $\frac{1}{4}$ cycle out of phase with either of the original waves. e) Show that the *instantaneous* Poynting vector at point P has magnitude $S = 4\epsilon_0 cE^2 \cos^2(\phi/2)\cos^2(\omega t + \phi/2)$, and that the *time-averaged* Poynting vector is given by Eq. (35.9).

35.47 Let the two sources S_1 and S_2 shown in Fig. 35.3 be located at $y = d$ and $y = -d$, respectively. a) Rewrite Eq. (35.1) in terms of the x- and y-coordinates of a point P in Fig. 35.3 at which constructive interference occurs. b) Your expression in part (a) is the equation for the antinodal curves shown in Fig. 35.3. Show that these curves are hyperbolas. (*Hint:* You may want to review the definition of a hyperbola in analytic geometry.) c) Repeat part (a) for Eq. (35.2), that describes points at which *destructive* interference occurs, and show that the *nodal* curves (not shown in Fig. 35.3) are also hyperbolas.

35.48 Consider a two-slit interference experiment in which the two slits are of different widths. As measured on a distant screen, the amplitude of the wave from the first slit is E, while the amplitude of the wave from the second slit is $2E$. a) Show that the intensity at any point in the interference pattern is

$$I = I_0\left(\frac{5}{9} + \frac{4}{9}\cos\phi\right)$$

where ϕ is the phase difference between the two waves as measured at a particular point on the screen and I_0 is the maximum intensity in the pattern. b) Graph I versus ϕ (like Fig. 35.9). What is the minimum value of the intensity, and for which values of ϕ does it occur?

35.49 A thin uniform film of refractive index 1.750 is placed on a sheet of glass of refractive index 1.50. At room temperature (20.0°C), this film is just thick enough for light with wavelength 582.4 nm reflected off the top of the film to be canceled by light reflected from the top of the glass. After the glass is placed in an oven and slowly heated to 170°C, you find that the film cancels reflected light with wavelength 588.5 nm. What is the coefficient of linear expansion of the film? (Ignore any changes in the refractive index of the film due to the temperature change.)

35.50 Red light with wavelength 700 nm is passed through a two-slit apparatus. At the same time, monochromatic visible light with another wavelength passes through the same apparatus. As a result, most of the pattern that appears on the screen is a mixture of two colors; however, the center of the third bright fringe $(m = 3)$ of the red light appears pure red, with none of the other color. What are the possible wavelengths of the second type of visible light? Do you need to know the slit spacing to answer this question? Why or why not?

35.51 Consider a two-slit interference pattern, for which the intensity distribution is given by Eq. (35.14). Let θ_m be the angular position of the mth bright fringe, where the intensity is I_0. Assume that θ_m is small, so that $\sin\theta_m \cong \theta_m$. Let θ_m^+ and θ_m^- be the two angles on either side of θ_m for which $I = \frac{1}{2}I_0$. The quantity $\Delta\theta_m = |\theta_m^+ - \theta_m^-|$ is the half-width of the mth fringe. Calculate $\Delta\theta_m$. How does $\Delta\theta_m$ depend on m?

35.52 White light reflects at normal incidence from the top and bottom surface of a glass plate $(n = 1.52)$. There is air above and below the plate. Constructive interference is observed for light whose wavelength in air is 477.0 nm. What is the thickness of the plate if the next longer wavelength for which there is constructive interference is 540.6 nm?

35.53 A source S of monochromatic light and a detector D are both located in air a distance h above a horizontal plane sheet of glass, and are separated by a horizontal distance x. Waves reaching D directly from S interfere with waves that reflect off the glass. The distance x is small compared to h so that the reflection is at close to normal incidence. a) Show that the condition for constructive interference is $\sqrt{x^2 + 4h^2} - x = (m + \frac{1}{2})\lambda$, and the condition for destructive interference is $\sqrt{x^2 + 4h^2} - x = m\lambda$. (*Hint:* Take into account the phase change on reflection.) b) Let $h = 24$ cm and $x = 14$ cm. What is the longest wavelength for which there will be constructive interference?

35.54 Reflective Coatings and Herring. Herring and related fish have a brilliant silvery appearance that camouflages them while swimming in a sunlit ocean. The silveriness is due to *platelets* attached to the surfaces of these fish. Each platelet is made up of several alternating layers of crystalline guanine $(n = 1.80)$ and of cytoplasm ($n = 1.333$, the same as water), with a guanine layer on the outside in contact with the surrounding water (Fig. 35.19). In one typical platelet, the guanine layers are 74 nm thick and the cytoplasm layers are 100 nm thick. a) For light striking the platelet surface at normal incidence, for which vacuum wavelengths of visible light will all of the reflections R_1, R_2, R_3, R_4, and R_5, shown in Fig. 35.19, be approximately in phase? If white light is shone on this platelet, what color will be most strongly reflected (see Fig. 32.20)? The surface of a herring has very many platelets side by side with layers of different thickness, so that *all* visible wavelengths are reflected. b) Explain why such a "stack" of layers is more reflective than a single layer of guanine with cytoplasm underneath. (A stack of five guanine layers separated by cytoplasm layers reflects over 80% of incident light at the wavelength for which it is "tuned.") c) The color that is most strongly reflected from a platelet depends on the angle at which it is viewed. Explain why this should be so. (You can see these changes in color by examining a herring from different angles. Most of the platelets on

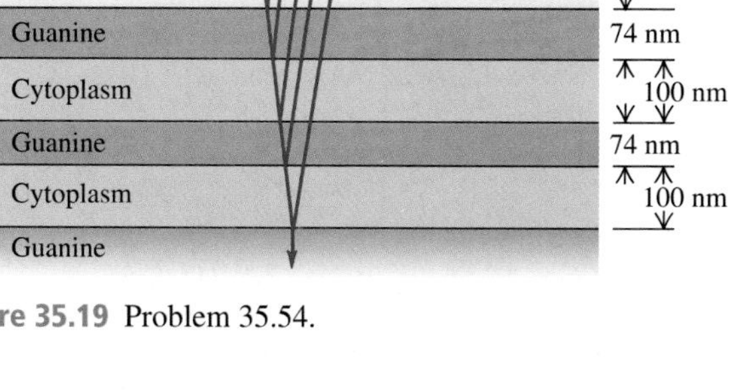

Figure 35.19 Problem 35.54.

these fish are oriented in the same way, so that they are vertical when the fish is swimming.)

35.55 A glass plate ($n = 1.53$) that is 0.485 μm thick and surrounded by air is illuminated by a beam of white light normal to the plate. a) What wavelengths (in air) within the limits of the visible spectrum ($\lambda = 400$ to 700 nm) are intensified in the reflected beam? b) What wavelengths within the visible spectrum are intensified in the transmitted light?

35.56 An oil tanker spills a large amount of oil ($n = 1.45$) into the sea ($n = 1.33$). a) If you look down onto the oil spill from overhead, what predominant wavelength of light do you see at a point where the oil is 380 nm thick? What color is the light? (*Hint:* See Table 32.1.) b) In the water under the slick, what visible wavelength (as measured in air) is predominant in the transmitted light at the same place in the slick as in part (a)?

35.57 The radius of curvature of the convex surface of a plano-convex lens is 95.2 cm. The lens is placed convex side down on a perfectly flat glass plate that is illuminated from above with red light having a wavelength of 580 nm. Find the diameter of the second bright ring in the interference pattern.

35.58 Newton's rings can be seen when a plano-convex lens is placed on a flat glass surface (Problem 35.57). For a particular lens with an index of refraction of $n = 1.50$ and a glass plate with an index of $n = 1.80$, the diameter of the third bright ring is 0.850 mm. If water ($n = 1.33$) now fills the space between the lens and the plate, what is the new diameter of this ring?

35.59 In a Young's two-slit experiment a piece of glass with an index of refraction n and a thickness L is placed in front of the upper slit. a) Describe qualitatively what happens to the interference pattern. b) Derive an expression for the intensity I of the light at points on a screen as a function of n, L, and θ. Here θ is the usual angle measured from the center of the two slits. That is, determine the equation analogous to Eq. (35.14) but that also involves L and n for the glass plate. c) From your result in part (b) derive an expression for the values of θ that locate the maxima in the interference pattern. [That is, derive an equation analogous to Eq. (35.4).]

35.60 An initially evacuated tube is placed in one arm of a Michelson interferometer. As a gas is slowly added to the tube, 48 fringes are seen to cross the telescope crosshairs. If the length of the tube is 5.00 cm and the light source is a mercury-vapor lamp (546 nm), what is the index of refraction of the final quantity of gas within the tube? (*Note:* For gases it is convenient to give the value of $n - 1$ rather than n itself, since n is very close to unity.)

Challenge Problems

35.61 The index of refraction of a glass rod is 1.48 at $T = 20.0°\text{C}$ and varies linearly with temperature, with a coefficient of $2.50 \times 10^{-5}/\text{C}°$. The coefficient of linear expansion of the glass is $5.00 \times 10^{-6}/\text{C}°$. At 20.0°C the length of the rod is 3.00 cm. A Michelson interferometer has this glass rod in one arm, and the rod is being heated so that its temperature increases at a rate of 5.00 C°/min. The light source has wavelength $\lambda = 589$ nm, and the rod initially is at $T = 20.0°\text{C}$. How many fringes cross the field of view each minute?

35.62 Figure 35.20 shows an interferometer known as *Fresnel's biprism.* The magnitude of the prism angle A is extremely small. a) If S_0 is a very narrow source slit, show that the separation of the two virtual coherent sources S_1 and S_2 is given by $d = 2aA(n - 1)$, where n is the index of refraction of the material of the prism. b) Calculate the spacing of the fringes of green light with wavelength 500 nm on a screen 2.00 m from the biprism. Take $a = 0.200$ m, $A = 3.50$ mrad, and $n = 1.50$.

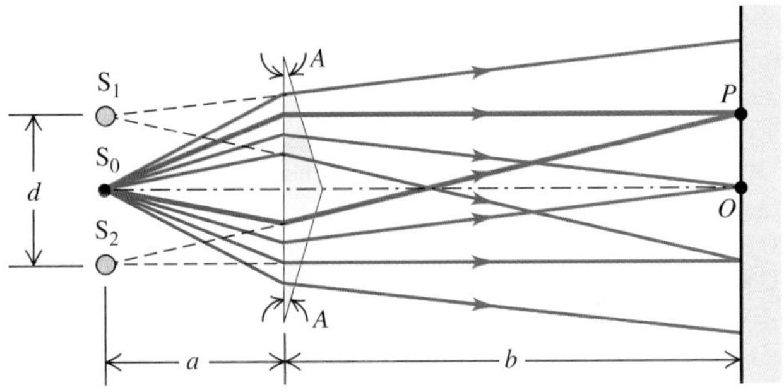

Figure 35.20 Challenge Problem 35.62.

36 DIFFRACTION

The brilliant blue colors of the tropical butterfly *Morpho peleides* are a result of diffraction—the interference of light waves reflected from millions of microscopic scales in the butterfly's wings. When viewed at the right angle, these scales strongly reflect blue light. The flashes of light from the flapping wings of a *Morpho* may momentarily dazzle predators such as lizards and birds.

? Why does the color of a butterfly's wing depend on the angle at which it is viewed?

Everyone is used to the idea that sound bends around corners. If sound didn't behave this way, you couldn't hear a police siren that's out of sight around a corner or the speech of a person whose back is turned to you. What may surprise you (and certainly surprised many scientists of the early nineteenth century) is that *light* can bend around corners as well. When light from a point source falls on a straight edge and casts a shadow, the edge of the shadow is never perfectly sharp. Some light appears in the area that we expect to be in the shadow, and we find alternating bright and dark fringes in the illuminated area. In general, light emerging from apertures doesn't behave precisely according to the predictions of the straight-line ray model of geometric optics.

The reason for these effects is that light, like sound, has wave characteristics. In Chapter 35 we studied the interference patterns that can arise when two light waves are combined. In this chapter we'll investigate interference effects due to combining *many* light waves. Such effects are referred to as *diffraction.* We'll find that the behavior of waves after they pass through an aperture is an example of diffraction; each infinitesimal part of the aperture acts as a source of waves, and the resulting pattern of light and dark is a result of interference among the waves emanating from these sources.

Light emerging from arrays of apertures also forms patterns whose character depends on the color of the light and the size and spacing of the apertures. Examples of this effect include the colors of iridescent butterflies and the "rainbow" you see reflected from the surface of a compact disc. We'll explore similar effects with x rays that are used to study the atomic structure of solids and liquids. Finally, we'll look at the physics of a *hologram,* a special kind of interference pattern recorded on photographic film and reproduced. When properly illuminated, it forms a three-dimensional image of the original object.

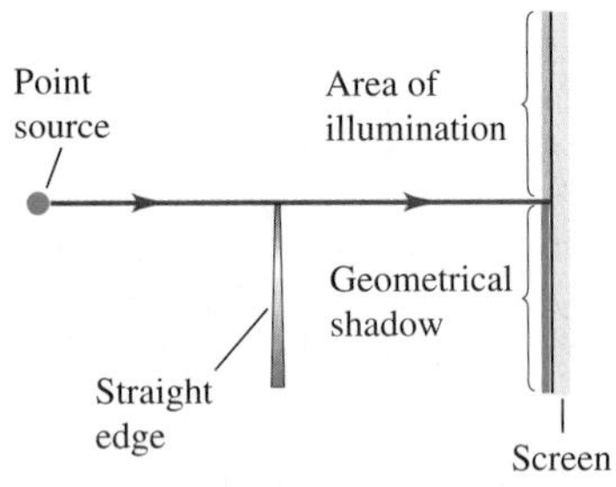

36.1 Geometric optics predicts that a straight edge should give a shadow with a sharp boundary and an area of relatively uniform illumination above it.

36.1 | Fresnel and Fraunhofer Diffraction

According to geometric optics, when an opaque object is placed between a point light source and a screen, as in Fig. 36.1, the shadow of the object forms a perfectly sharp line. No light at all strikes the screen at points within the shadow, and the area outside the shadow is illuminated nearly uniformly. But as we saw in Chapter 35, the *wave* nature of light causes effects that can't be understood with the simple model of geometric optics. An important class of such effects occurs when light strikes a barrier that has an aperture or an edge. The interference patterns formed in such a situation are grouped under the heading **diffraction.**

An example of diffraction is shown in Fig. 36.2. The photograph in Fig. 36.2a was made by placing a razor blade halfway between a pinhole, illuminated by monochromatic light, and a photographic film. The film recorded the shadow cast by the blade. Figure 36.2b is an enlargement of a region near the shadow of the right edge of the blade. The position of the *geometrical* shadow line is indicated by arrows. The area outside the geometric shadow is bordered by alternating bright and dark bands. There is some light in the shadow region, although this is not very visible in the photograph. The first bright band in Fig. 36.2b, just to the right of the geometrical shadow, is considerably brighter than in the region of uniform illumination to the extreme right. This simple experiment gives us some idea of the richness and complexity of what might seem to be a simple idea, the casting of a shadow by an opaque object.

We don't often observe diffraction patterns such as Fig. 36.2 in everyday life because most ordinary light sources are not monochromatic and are not point sources. If we use a white frosted light bulb instead of a point source in Fig. 36.1, each wavelength of the light from every point of the bulb forms its own diffraction pattern, but the patterns overlap to such an extent that we can't see any individual pattern.

36.2 (a) Actual shadow of a razor blade illuminated by monochromatic light from a point source. (b) Enlarged shadow of the straight edge. The arrows show the position of the *geometric* shadow.

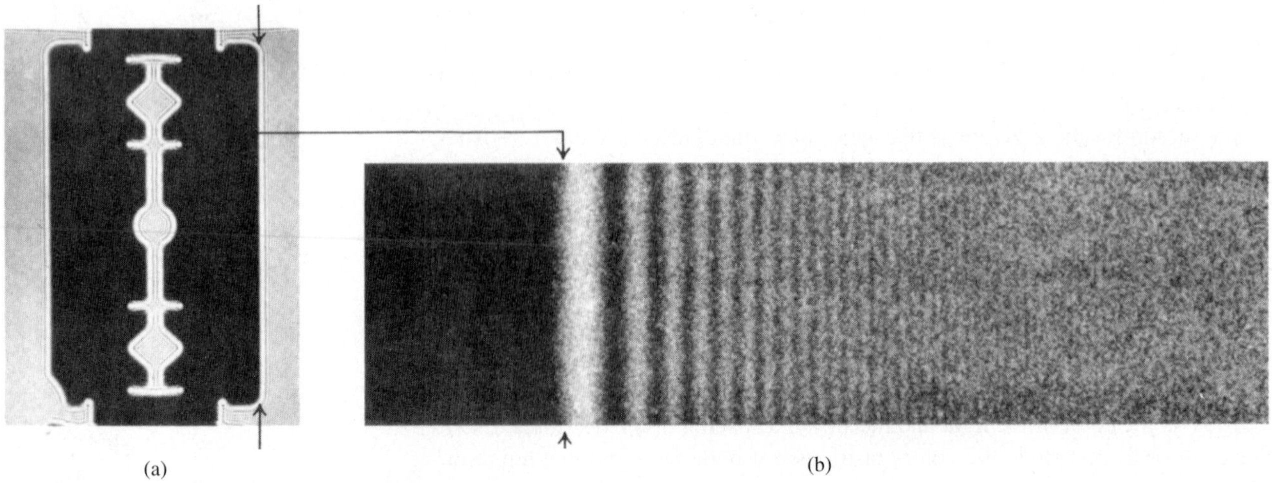

Diffraction patterns can be analyzed by use of Huygens' principle (Section 33.7). Let's review that principle briefly. Every point of a wave front can be considered the source of secondary wavelets that spread out in all directions with a speed equal to the speed of propagation of the wave. The position of the wave front at any later time is the *envelope* of the secondary waves at that time. To find the resultant displacement at any point, we combine all the individual displacements produced by these secondary waves, using the superposition principle and taking into account their amplitudes and relative phases.

In Fig. 36.1, both the point source and the screen are relatively close to the obstacle forming the diffraction pattern. This situation is described as *near-field diffraction* or **Fresnel diffraction,** pronounced "Freh-nell" (after the French scientist Augustin Jean Fresnel, 1788–1827). If the source, obstacle, and screen are far enough away that all lines from the source to the obstacle can be considered parallel and all lines from the obstacle to a point in the pattern can be considered parallel, the phenomenon is called *far-field diffraction* or **Fraunhofer diffraction** (after the German physicist Joseph von Fraunhofer, 1787–1826). We will restrict the following discussion to Fraunhofer diffraction, which is usually simpler to analyze in detail than Fresnel diffraction.

Diffraction is sometimes described as "the bending of light around an obstacle." But the process that causes diffraction is present in the propagation of *every* wave. When part of the wave is cut off by some obstacle, we observe diffraction effects that result from interference of the remaining parts of the wave fronts. Optical instruments typically use only a limited portion of a wave; for example, a telescope uses only the part of a wave that is admitted by its objective lens or mirror. Thus diffraction plays a role in nearly all optical phenomena.

Finally, we emphasize that there is no fundamental distinction between *interference* and *diffraction.* In Chapter 35 we used the term *interference* for effects involving waves from a small number of sources, usually two. *Diffraction* usually involves a *continuous* distribution of Huygens' wavelets across the area of an aperture, or a very large number of sources or apertures. But both categories of phenomena are governed by the same basic physics of superposition and Huygens' principle.

Test Your Understanding

Can sound waves undergo diffraction around an edge?

36.2 | Diffraction from a Single Slit

ActivPhysics Online 16.6 Single-Slit Diffraction

In this section we'll discuss the diffraction pattern formed by plane-wave (parallel-ray) monochromatic light when it emerges from a long, narrow slit, as shown in Fig. 36.3. We call the narrow dimension the *width,* even though in this figure it is a vertical dimension.

According to geometric optics, the transmitted beam should have the same cross section as the slit, as in Fig. 36.3a. What is *actually* observed is the pattern shown in Fig. 36.3b. The beam spreads out vertically after passing through the slit. The diffraction pattern consists of a central bright band, which may be much broader than the width of the slit, bordered by alternating dark and bright bands with rapidly decreasing intensity. About 85% of the power in the transmitted beam is in the central bright band, whose width is found to be *inversely* proportional to the width of the slit. In general, the smaller the width of the slit, the

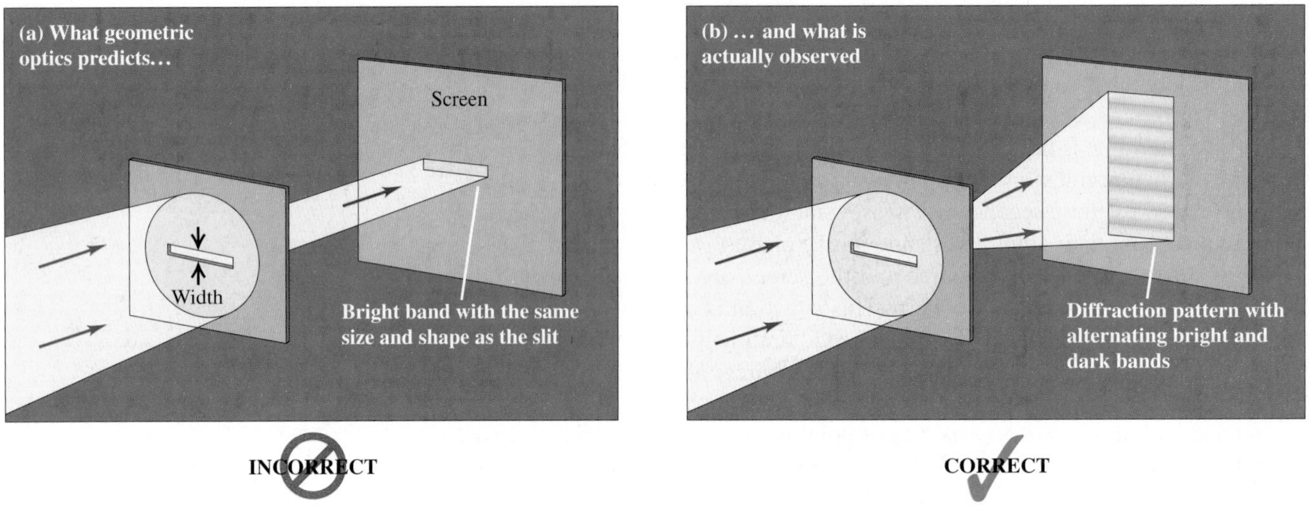

36.3 (a) The "shadow" of a horizontal slit as incorrectly predicted by geometric optics. (b) A horizontal slit actually produces a diffraction pattern. The slit width has been greatly exaggerated.

broader the entire diffraction pattern. (The *horizontal* spreading of the beam in Fig. 36.3b is negligible because the horizontal dimension of the slit is relatively large.) You can easily observe a similar diffraction pattern by looking at a point source, such as a distant street light, through a narrow slit formed between your two thumbs held in front of your eye; the retina of your eye corresponds to the screen.

Figure 36.4 shows a side view of the same setup; the long sides of the slit are perpendicular to the figure, and plane waves are incident on the slit from the left. According to Huygens' principle, each element of area of the slit opening can be considered as a source of secondary waves. In particular, imagine dividing the slit into several narrow strips of equal width, parallel to the long edges and perpendicular to the page. Two such strips are shown in Fig. 36.4a. Cylindrical secondary wavelets, shown in cross section, spread out from each strip.

In Fig. 36.4b a screen is placed to the right of the slit. We can calculate the resultant intensity at a point P on the screen by adding the contributions from the individual wavelets, taking proper account of their various phases and amplitudes. It's easiest to do this calculation if we assume that the screen is far enough away that all the rays from various parts of the slit to a particular point P on the screen are parallel, as in Fig. 36.4c. An equivalent situation is Fig. 36.4d, in which the rays to the lens are parallel and the lens forms a reduced image of the same pattern that would be formed on an infinitely distant screen without the lens. We

36.4 Diffraction by a single rectangular slit. The long sides of the slit are perpendicular to the figure. Plane waves are incident on the slit. (a) Dividing the slit into strips. (b) Fresnel (near-field) diffraction. (c) Fraunhofer (far-field) diffraction. (d) Using a converging cylindrical lens to obtain a Fraunhofer diffraction pattern on a nearby screen.

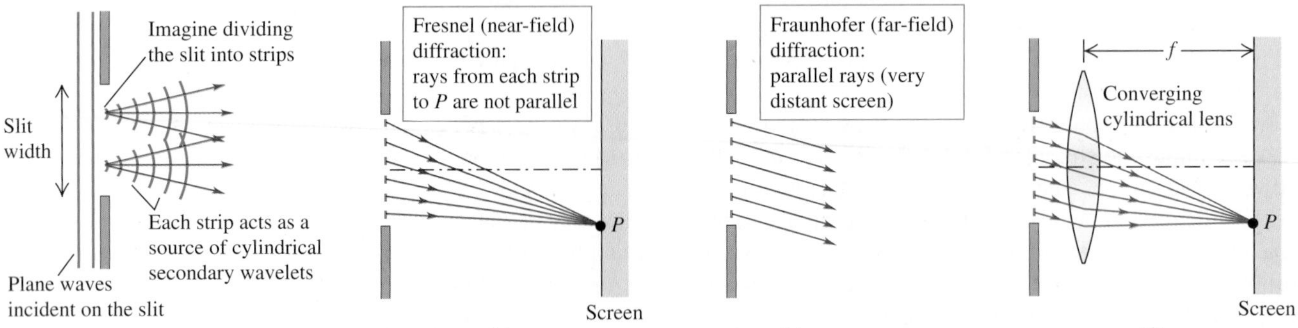

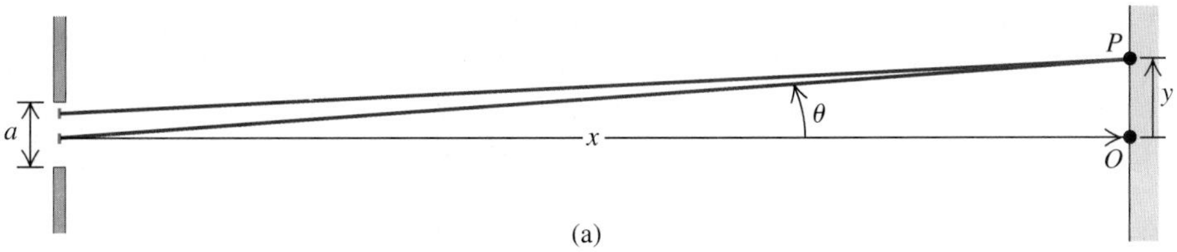

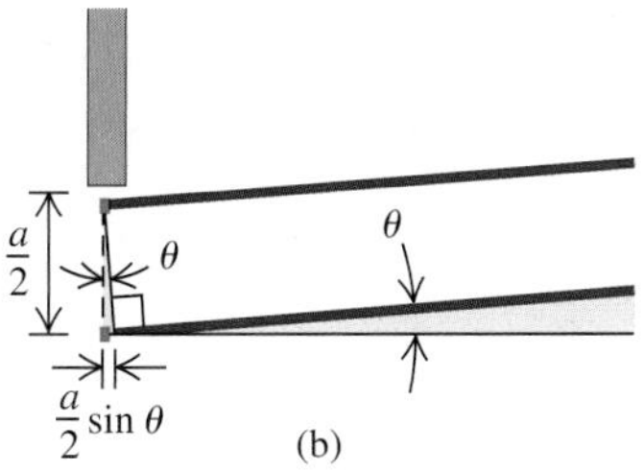

36.5 Side view of a horizontal slit. (a) When the distance x to the screen is much greater than the slit width a, the rays from a distance $a/2$ apart may be considered parallel. (b) Enlarged view of half of the slit. The ray from the middle of the slit travels a distance $(a/2)\sin\theta$ farther to point P than does the ray from the top edge of the slit.

might expect that the various light paths through the lens would introduce additional phase shifts, but in fact it can be shown that all the paths have *equal* phase shifts, so this is not a problem.

The situation of Fig. 36.4b is Fresnel diffraction; those in Figs. 36.4c and 36.4d, where the outgoing rays are considered parallel, are Fraunhofer diffraction. We can derive quite simply the most important characteristics of the Fraunhofer diffraction pattern from a single slit. First consider two narrow strips, one just below the top edge of the drawing of the slit and one at its center, shown in end view in Fig. 36.5. The difference in path length to point P is $(a/2)\sin\theta$, where a is the slit width and θ is the angle between the perpendicular to the slit and a line from the center of the slit to P. Suppose this path difference happens to be equal to $\lambda/2$; then light from these two strips arrives at point P with a half-cycle phase difference, and cancellation occurs.

Similarly, light from two strips immediately *below* the two in the figure also arrives at P a half-cycle out of phase. In fact, the light from *every* strip in the top half of the slit cancels out the light from a corresponding strip in the bottom half. The result is complete cancellation at P for the combined light from the entire slit, giving a dark fringe in the interference pattern. That is, a dark fringe occurs whenever

$$\frac{a}{2}\sin\theta = \pm\frac{\lambda}{2} \quad \text{or} \quad \sin\theta = \pm\frac{\lambda}{a} \tag{36.1}$$

The plus-or-minus $(\pm)$ sign in Eq. (36.1) says that there are symmetrical dark fringes above and below point O in Fig. 36.5a. The upper fringe $(\theta > 0)$ occurs at a point P where light from the bottom half of the slit travels $\lambda/2$ farther to P than does light from the top half; the lower fringe $(\theta < 0)$ occurs where light from the *top* half travels $\lambda/2$ farther than light from the *bottom* half.

We may also divide the screen into quarters, sixths, and so on, and use the above argument to show that a dark fringe occurs whenever $\sin\theta = \pm 2\lambda/a$, $\pm 3\lambda/a$, and so on. Thus the condition for a *dark* fringe is

$$\sin\theta = \frac{m\lambda}{a} \qquad (m = \pm 1, \pm 2, \pm 3, \ldots) \tag{36.2}$$

(dark fringes in single-slit diffraction)

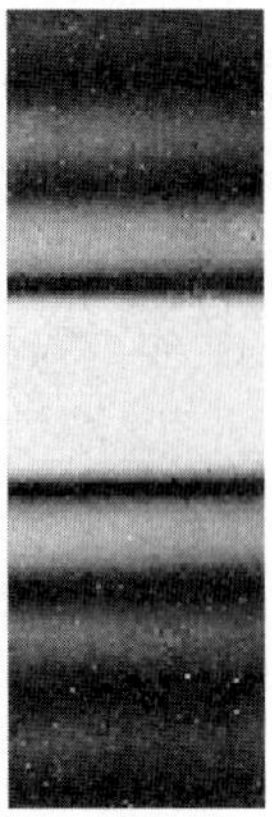

36.6 Photograph of the Fraunhofer diffraction pattern of a single horizontal slit.

For example, if the slit width is equal to ten wavelengths ($a = 10\lambda$), dark fringes occur at $\sin\theta = \pm\frac{1}{10}, \pm\frac{2}{10}, \pm\frac{3}{10}, \ldots$. Between the dark fringes are bright fringes. We also note that $\sin\theta = 0$ corresponds to a *bright* band; in this case, light from the entire slit arrives at P in phase. Thus it would be wrong to put $m = 0$ in Eq. (36.2). The central bright fringe is wider than the other bright fringes, as Fig. 36.3 shows. In the small-angle approximation that we will use below, it is exactly *twice* as wide.

With light, the wavelength λ is of the order of 500 nm $= 5 \times 10^{-7}$ m. This is often much smaller than the slit width a; a typical slit width is 10^{-2} cm $= 10^{-4}$ m. Therefore the values of θ in Eq. (36.2) are often so small that the approximation $\sin\theta \approx \theta$ (where θ is in radians) is a very good one. In that case we can rewrite this equation as

$$\theta = \frac{m\lambda}{a} \qquad (m = \pm1, \pm2, \pm3, \ldots) \qquad \text{(for small angles } \theta\text{)}$$

where θ is in *radians*. Also, if the distance from slit to screen is x, as in Fig. 36.5a, and the vertical distance of the mth dark band from the center of the pattern is y_m, then $\tan\theta = y_m/x$. For small θ we may also approximate $\tan\theta$ by θ (in radians), and we then find

$$y_m = x\frac{m\lambda}{a} \qquad (\text{for } y_m \ll x) \qquad (36.3)$$

Figure 36.6 is a photograph of a single-slit diffraction pattern with the $m = \pm1, \pm2$, and ±3 minima labeled.

CAUTION This equation has the same form as the equation for the two-slit pattern, Eq. (35.6), except that in Eq. (36.3) we use x rather than R for the distance to the screen. But Eq. (36.3) gives the positions of the *dark* fringes in a *single-slit* pattern rather than the *bright* fringes in a *double-slit* pattern. Also, $m = 0$ is *not* a dark fringe. Be careful!

Example 36.1 Single-slit diffraction

You pass 633-nm laser light through a narrow slit and observe the diffraction pattern on a screen 6.0 m away. You find that the distance on the screen between the centers of the first minima outside the central bright fringe is 32 mm. How wide is the slit?

SOLUTION

IDENTIFY and SET UP: The distances between points on the screen are much smaller than the distance from the slit to the screen, so the angle θ shown in Fig. 36.5a is very small. Hence we can use the approximate relation of Eq. (36.3) to solve for the slit width a (the target variable).

EXECUTE: The first minimum corresponds to $m = 1$ in Eq. (36.3). The distance y_1 from the central maximum to the first minimum on either side is half the distance between the two first minima, so $y_1 = (32\text{ mm})/2$. Substituting these values and solving for a, we find

$$a = \frac{x\lambda}{y_1} = \frac{(6.0\text{ m})(633 \times 10^{-9}\text{ m})}{(32 \times 10^{-3}\text{ m})/2}$$
$$= 2.4 \times 10^{-4}\text{ m} = 0.24\text{ mm}$$

EVALUATE: The angle θ is small only if the wavelength is small compared to the slit width. Since $\lambda = 633\text{ nm} = 6.33 \times 10^{-7}$ m and we have found $a = 0.24\text{ mm} = 2.4 \times 10^{-4}$ m, our result is consistent with this: the wavelength is $(6.33 \times 10^{-7}\text{ m})/(2.4 \times 10^{-4}\text{ m}) = 0.0026$ as large as the slit width.

Can you show that the distance between the *second* minima on the two sides is $2(32\text{ mm}) = 64$ mm, and so on?

Test Your Understanding

At what angle would the first diffraction minimum occur if the slit in Example 36.1 were used with infrared radiation of wavelength 0.18 mm? (You would need an infrared-sensitive detector to observe the diffraction pattern.) Would it be appropriate to use the approximations $\tan\theta \approx \sin\theta \approx \theta$ that lead to Eq. (36.3)?

36.3 | Intensity in the Single-Slit Pattern

We can derive an expression for the intensity distribution for the single-slit diffraction pattern by the same phasor-addition method that we used in Section 35.3 to obtain Eqs. (35.10) and (35.14) for the two-slit interference pattern. We again imagine a plane wave front at the slit subdivided into a large number of strips. We superpose the contributions of the Huygens wavelets from all the strips at a point P on a distant screen at an angle θ from the normal to the slit plane. To do this, we use a phasor to represent the sinusoidally varying $\vec{E}$ field from each individual strip. The magnitude of the vector sum of the phasors at each point P is the amplitude E_P of the total $\vec{E}$ field at that point. The intensity at P is proportional to E_P^2.

At the point O shown in Figure 36.5a, corresponding to the center of the pattern where $\theta = 0$, there are negligible path differences for $x \gg a$; the phasors are all essentially *in phase* (that is, have the same direction). In Fig. 36.7a we draw the phasors at time $t = 0$ and denote the resultant amplitude at O by E_0. In this illustration we have divided the slit into 14 strips.

Now consider wavelets arriving from different strips at point P in Fig. 36.5a, at an angle θ from point O. Because of the differences in path length, there are now phase differences between wavelets coming from adjacent strips; the corresponding phasor diagram is shown in Fig. 36.7b. The vector sum of the phasors is now part of the perimeter of a many-sided polygon, and E_P, the amplitude of the resultant electric field at P, is the *chord*. The angle β is the total phase difference between the wave from the top strip of Fig. 36.5a and the wave from the bottom strip; that is, β is the phase of the wave at P from the top strip with respect to the wave received at P from the bottom strip.

We may imagine dividing the slit into narrower and narrower strips. In the limit that there is an infinite number of infinitesimally narrow strips, the curved trail of phasors becomes an *arc of a circle* (Fig. 36.7c), with arc length equal to the length E_0 in Fig. 36.7a. The center C of this arc is found by constructing perpendiculars at A and B. From the relationship between arc length, radius, and angle, the radius of the arc is E_0/β; the amplitude E_P of the resultant electric field at P is equal to the chord AB, which is $2(E_0/\beta)\sin(\beta/2)$. (Note that β *must* be in radians!) We then have

$$E_P = E_0\frac{\sin(\beta/2)}{\beta/2} \quad \text{(amplitude in single-slit diffraction)} \tag{36.4}$$

The intensity at each point on the screen is proportional to the square of the amplitude given by Eq. (36.4). If I_0 is the intensity in the straight-ahead direction where $\theta = 0$ and $\beta = 0$, then the intensity I at any point is

$$I = I_0\left[\frac{\sin(\beta/2)}{\beta/2}\right]^2 \quad \text{(intensity in single-slit diffraction)} \tag{36.5}$$

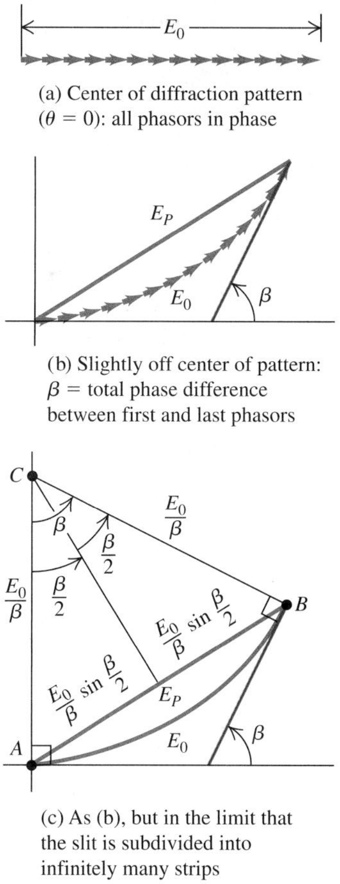

(a) Center of diffraction pattern ($\theta = 0$): all phasors in phase

(b) Slightly off center of pattern: β = total phase difference between first and last phasors

(c) As (b), but in the limit that the slit is subdivided into infinitely many strips

36.7 Phasor diagrams used to find the amplitude of the $\vec{E}$ field in single-slit diffraction. Each phasor represents the $\vec{E}$ field from a single strip within the slit.

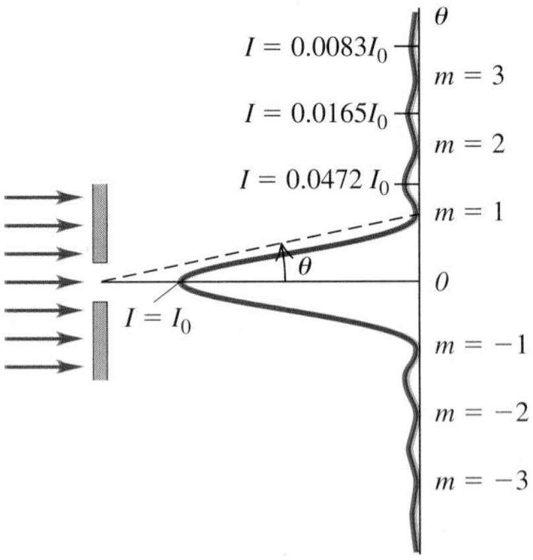

(a)

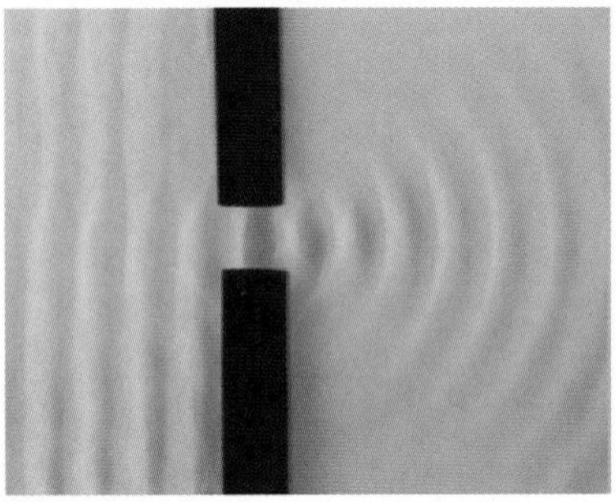

(b)

36.8 (a) Intensity versus angle in single-slit diffraction. The values of m label intensity minima given by Eq. (36.8). Most of the wave power goes into the central intensity peak (between the $m = 1$ and $m = -1$ intensity minima). (b) These water waves passing through a small aperture behave exactly like light waves in single-slit diffraction. Only the diffracted waves within the central intensity peak are visible; the waves at larger angles are too faint to see.

We can express the phase difference β in terms of geometric quantities, as we did for the two-slit pattern. From Eq. (35.11) the phase difference is $2\pi/\lambda$ times the path difference. Figure 36.5 shows that the path difference between the ray from the top of the slit and the ray from the middle of the slit is $(a/2)\sin\theta$. The path difference between the rays from the top of the slit and the bottom of the slit is twice this, so

$$\beta = \frac{2\pi}{\lambda} a \sin\theta \qquad (36.6)$$

and Eq. (36.5) becomes

$$I = I_0 \left\{ \frac{\sin[\pi a(\sin\theta)/\lambda]}{\pi a(\sin\theta)/\lambda} \right\}^2 \qquad \text{(intensity in single-slit diffraction)} \qquad (36.7)$$

This equation expresses the intensity directly in terms of the angle θ. In many calculations it is easier first to calculate the phase angle β, using Eq. (36.6), and then to use Eq. (36.5).

Equation (36.7) is plotted in Fig. 36.8a. Note that the central intensity peak is much larger than any of the others. This means that most of the power in the wave remains within an angle θ from the perpendicular to the slit, where $\sin\theta = \lambda/a$ (the first diffraction minimum). You can see this easily in Fig. 36.8b, which is a photograph of water waves undergoing single-slit diffraction. Note also that the peak intensities in Fig. 36.8a decrease rapidly as we go away from the center of the pattern. (Compare Fig. 36.6, which shows a single-slit diffraction pattern for light.)

The dark fringes in the pattern are the places where $I = 0$. These occur at points for which the numerator of Eq. (36.5) is zero so that β is a multiple of 2π. From Eq. (36.6) this corresponds to

$$\frac{a\sin\theta}{\lambda} = m \qquad (m = \pm1, \pm2, \dots)$$

$$\sin\theta = \frac{m\lambda}{a} \qquad (m = \pm1, \pm2, \dots) \qquad (36.8)$$

This agrees with our previous result, Eq. (36.2). Note again that $\beta = 0$ (corresponding to $\theta = 0$) is *not* a minimum. Equation (36.5) is indeterminate at $\beta = 0$, but we can evaluate the limit as $\beta \to 0$ using L'Hôpital's rule. We find that at $\beta = 0$, $I = I_0$, as we should expect.

Intensity Maxima in the Single-Slit Pattern

We can also use Eq. (36.5) to calculate the positions of the peaks, or *intensity maxima,* and the intensities at these peaks. This is not quite as simple as it may appear. We might expect the peaks to occur where the sine function reaches the value ±1, namely, where $\beta = \pm\pi, \pm3\pi, \pm5\pi$, or in general,

$$\beta \approx \pm(2m+1)\pi \qquad (m = 0, 1, 2, \dots) \qquad (36.9)$$

This is *approximately* correct, but because of the factor $(\beta/2)^2$ in the denominator of Eq. (36.5), the maxima don't occur precisely at these points. When we take the derivative of Eq. (36.5) with respect to β and set it equal to zero to try to find the maxima and minima, we get a transcendental equation that has to be solved numerically. In fact there is *no* maximum near $\beta = \pm\pi$. The first maxima on either side of the central maximum, near $\beta = \pm 3\pi$, actually occur at $\pm 2.860\pi$. The second side maxima, near $\beta = \pm 5\pi$, are actually at $\pm 4.918\pi$, and so on. The error in Eq. (36.9) vanishes in the limit of large m, that is, for intensity maxima far from the center of the pattern.

To find the intensities at the side maxima, we substitute these values of β back into Eq. (36.5). Using the approximate expression in Eq. (36.9), we get

$$I_m \approx \frac{I_0}{\left(m + \frac{1}{2}\right)^2 \pi^2} \qquad (36.10)$$

where I_m is the intensity of the mth side maximum and I_0 is the intensity of the central maximum. Equation (36.10) gives the series of intensities

$$0.0450I_0 \qquad 0.0162I_0 \qquad 0.0083I_0$$

and so on. As we have pointed out, this equation is only approximately correct. The actual intensities of the side maxima turn out to be

$$0.0472I_0 \qquad 0.0165I_0 \qquad 0.0083I_0 \qquad \ldots$$

Note that the intensities of the side maxima decrease very rapidly, as Fig. 36.8a also shows. Even the first side maxima have less than 5% of the intensity of the central maximum.

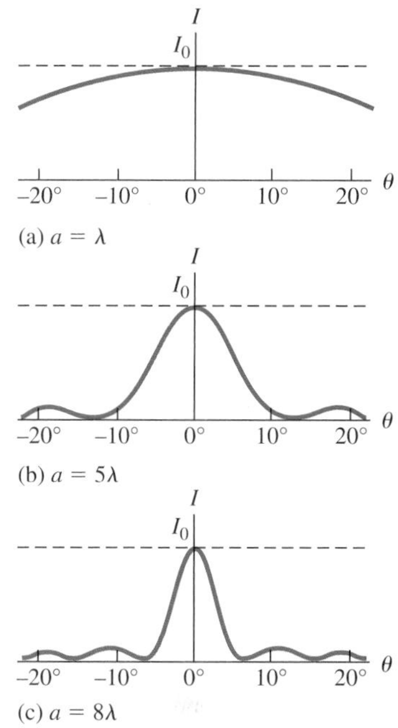

36.9 (a) When the slit width a is less than or equal to the wavelength λ, the central maximum of the diffraction pattern is spread out from $+90°$ to $-90°$. (b), (c) The angular width of the central maximum decreases when the ratio a/λ is increased, either by increasing the slit width or by decreasing the wavelength.

Width of the Single-Slit Pattern

For small angles the angular spread of the diffraction pattern is inversely proportional to the slit width a or, more precisely, to the ratio of a to the wavelength λ. Figure 36.9 shows graphs of intensity I as a function of the angle θ for various values of the ratio a/λ.

With light waves, the wavelength λ is often much smaller than the slit width a, and the values of θ in Eqs. (36.6) and (36.7) are so small that the approximation $\sin\theta = \theta$ is very good. With this approximation the position θ_1 of the first minimum beside the central maximum, corresponding to $\beta/2 = \pi$, is, from Eq. (36.7),

$$\theta_1 = \frac{\lambda}{a} \qquad (36.11)$$

This characterizes the width (angular spread) of the central maximum, and we see that it is *inversely* proportional to the slit width a. When the small-angle approximation is valid, the central maximum is exactly twice as wide as each side maximum. When a is of the order of a centimeter or more, θ_1 is so small that we can consider practically all the light to be concentrated at the geometrical focus. But when a is less than λ, the central maximum spreads over 180°, and the fringe pattern is not seen at all.

36.10 The sound waves used in speech have a long wavelength (about 1 m) and can easily bend around this instructor's head. By contrast, light waves have very short wavelengths and undergo very little diffraction. Hence you can't *see* around his head!

It's important to keep in mind that diffraction occurs for *all* kinds of waves, not just light. Sound waves undergo diffraction when they pass through a slit or aperture such as an ordinary doorway. The sound waves used in speech have wavelengths of about a meter or greater, and a typical doorway is less than 1 m wide; in this situation, a is less than λ, and the central intensity maximum extends over 180°. This is why the sounds coming through an open doorway can easily be heard by an eavesdropper hiding out of sight around the corner. In the same way, sound waves can bend around the head of an instructor who faces the blackboard while lecturing (Fig. 36.10). By contrast, there is essentially no diffraction of visible light through such a doorway because the width a is very much greater than the wavelength λ (of order 5×10^{-7} m). You can *hear* around corners because typical sound waves have relatively long wavelengths; you cannot *see* around corners because the wavelength of visible light is very short.

Example 36.2 Single-slit diffraction: Intensity I

a) In a single-slit diffraction pattern, what is the intensity at a point where the total phase difference between wavelets from the top and bottom of the slit is 66 rad? b) If this point is 7.0° away from the central maximum, how many wavelengths wide is the slit?

SOLUTION

IDENTIFY and SET UP: The total phase difference between wavelets from the two edges of the slit is the quantity we called β in Fig. 36.7c. Given $\beta = 66$ rad, we use Eq. (36.5) to find the intensity I at the point in question and use Eq. (36.6) to find the slit width a in terms of the wavelength λ.

EXECUTE: a) Since $\beta = 66$ rad, $\beta/2 = 33$ rad and Eq. (36.5) becomes

$$I = I_0\left[\frac{\sin(33\text{ rad})}{33\text{ rad}}\right]^2 = (9.2 \times 10^{-4})I_0$$

b) We solve Eq. (36.6) for a:

$$a = \frac{\beta\lambda}{2\pi \sin\theta} = \frac{(66\text{ rad})\lambda}{(2\pi\text{ rad})\sin 7.0°} = 86\lambda$$

For example, for 550-nm light, the slit width a is $(86)(550\text{ nm}) = 4.7 \times 10^{-5}\text{ m} = 0.047$ mm, or roughly $\frac{1}{20}$ mm.

EVALUATE: To what point in the diffraction pattern does this value of β correspond? To find out, note that $\beta = 66\text{ rad} = 21\pi$. Comparing to Eq. (36.9) shows that this is approximately equal to the value of β at the *tenth* side maximum, well beyond the range shown in Fig. 36.8a (which shows only the first three side maxima). The intensity is very much less than the intensity I_0 at the central maximum. (The *actual* position of this maximum is at $\beta = 65.91\text{ rad} = 20.98\pi$, or approximately midway between the minima at $\beta = 20\pi$ and $\beta = 22\pi$.)

Example 36.3 Single-slit diffraction: Intensity II

In the experiment described in Example 36.1 (Section 36.2), what is the intensity at a point on the screen 3.0 mm from the center of the pattern? The intensity at the center of the pattern is I_0.

SOLUTION

IDENTIFY and SET UP: Unlike in Example 36.2, we are not given the value of the phase difference β at the point in question. Instead, we use geometry to determine the angle θ for our point, then use Eq. (36.7) to calculate the intensity I (our target variable).

EXECUTE: Referring to Fig. 36.5a, we have $y = 3.0$ mm and $x = 6.0$ m, so $\tan\theta = y/x = (3.0 \times 10^{-3}\text{ m})/(6.0\text{ m}) = 5.0 \times 10^{-4}$; since this is so small, the values of $\tan\theta$, $\sin\theta$, and θ (in radians) are all nearly the same. Then, using Eq. (36.7), we have

$$\frac{\pi a \sin\theta}{\lambda} = \frac{\pi(2.4 \times 10^{-4}\text{ m})(5.0 \times 10^{-4})}{6.33 \times 10^{-7}\text{ m}} = 0.60$$

$$I = I_0\left(\frac{\sin 0.60}{0.60}\right)^2 = 0.89I_0$$

EVALUATE: Examining Fig. 36.8a shows that an intensity this large can occur only within the central intensity maximum. This checks out; from Example 36.1, the first intensity minimum ($m = 1$ in Fig. 36.8a) is $(32\text{ mm})/2 = 16$ mm from the center of the pattern, so the point in question here does, indeed, lie within the central maximum.

Test Your Understanding

For what slit width would there be an intensity minimum at the point on the screen in Example 38.3?

36.4 | Multiple Slits

In Sections 35.2 and 35.3 we analyzed interference from two point sources or from two very narrow slits; in this analysis we ignored effects due to the finite (that is, nonzero) slit width. In Sections 36.2 and 36.3 we considered the diffraction effects that occur when light passes through a single slit of finite width. Additional interesting effects occur when we have two slits with finite width or when there are several very narrow slits.

Two Slits of Finite Width

Let's take another look at the two-slit pattern in the more realistic case in which the slits have finite width. If the slits are narrow in comparison to the wavelength, we can assume that light from each slit spreads out uniformly in all directions to the right of the slit. We used this assumption in Section 36.2 to calculate the interference pattern described by Eq. (36.10) or (36.15), consisting of a series of equally spaced, equally intense maxima. However, when the slits have finite width, the peaks in the two-slit interference pattern are modulated by the single-slit diffraction pattern characteristic of the width of each slit.

Figure 36.11a shows the intensity in a single-slit diffraction pattern with slit width a. The *diffraction minima* are labeled by the integer $m_d = \pm 1, \pm 2, \ldots$ ("d" for "diffraction"). Figure 36.11b shows the pattern formed by two very narrow slits with distance d between slits, where d is four times as great as the single-slit width a in Fig. 36.11a; that is, $d = 4a$. The *interference maxima* are labeled by the integer $m_i = 0, \pm 1, \pm 2, \ldots$ ("i" for "interference"). We note that the spacing between adjacent minima in the single-slit pattern is four times as great as in the two-slit pattern. Now suppose we widen each of the narrow slits to the same width a as that of the single slit in Fig. 36.11a. Figure 36.11c shows the pattern from two slits with width a, separated by a distance (between centers) $d = 4a$. The effect of the finite width of the slits is to superimpose the two patterns, that is, to multiply the two intensities at each point. The two-slit peaks are in the same positions as before, but their intensities are modulated by the single-slit pattern, which acts as an "envelope" for the intensity function. The expression for the intensity shown in Fig. 36.11c is proportional to the product of the two-slit and single-slit expressions, Eqs. (35.10) and (36.5):

$$I = I_0 \cos^2 \frac{\phi}{2} \left[\frac{\sin(\beta/2)}{\beta/2} \right]^2 \quad \text{(two slits of finite width)} \qquad (36.12)$$

where, as before,

$$\phi = \frac{2\pi d}{\lambda} \sin \theta \qquad \beta = \frac{2\pi a}{\lambda} \sin \theta$$

Note that in Fig. 36.11c, every fourth interference maximum at the sides is *missing* because these interference maxima $(m_i = \pm 4, \pm 8, \ldots)$ coincide with diffraction minima $(m_d = \pm 1, \pm 2, \ldots)$. This can also be seen in Fig. 36.11d, which is a photograph of an actual pattern with $d = 4a$. You should be able to convince yourself that there will be "missing" maxima whenever d is an integer multiple of a.

Figures 36.11c and 36.11d show that as you move away from the central bright maximum of the two-slit pattern, the intensity of the maxima decreases. This is a result of the single-slit modulating pattern shown in Fig. 36.11a; mathematically,

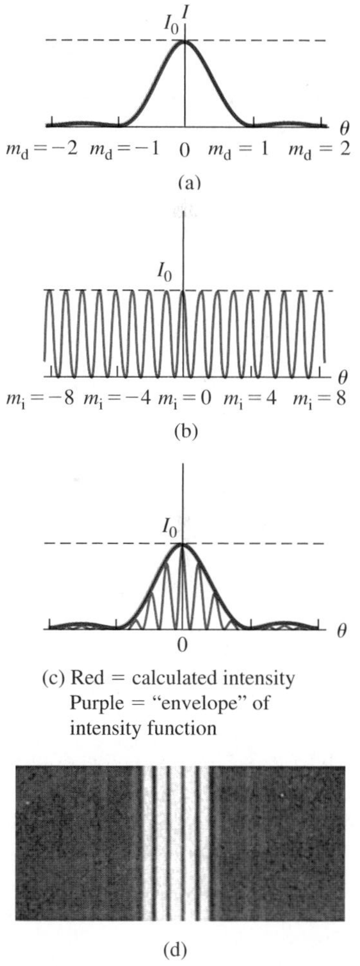

36.11 (a) Single-slit diffraction pattern for a slit of width a. (b) Double-slit interference pattern for narrow slits whose separation d is four times the width of the slit in (a). (c) Calculated double-slit pattern for $d = 4a$. (d) Actual photograph of the pattern calculated in (c). For $d = 4a$, every fourth interference maximum at the sides is missing.

the decrease in intensity arises from the factor $(\beta/2)^2$ in the denominator of Eq. (36.12). This decrease in intensity can also be seen in Fig. 35.6 (Section 35.2). The narrower the slits, the broader the single-slit pattern (as in Fig. 36.9) and the slower the decrease in intensity from one interference maximum to the next.

Shall we call the pattern in Fig. 36.11d *interference* or *diffraction?* It's really both, since it results from superposition of waves coming from various parts of the two apertures. There is no truly fundamental distinction between interference and diffraction.

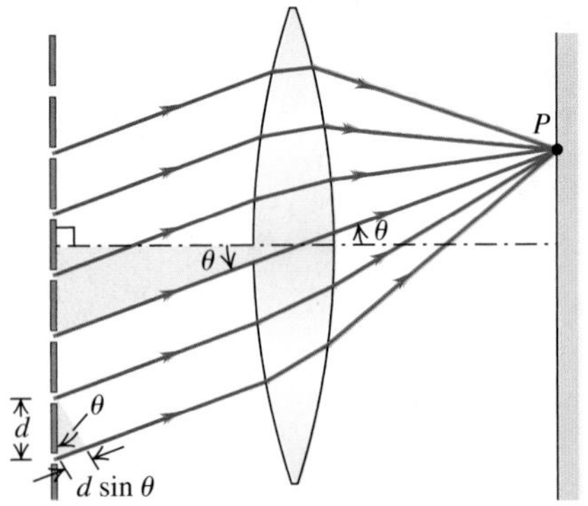

36.12 In multiple-slit diffraction, rays from every slit arrive in phase, giving a sharp maximum if the path difference between adjacent slits is a whole number of wavelengths. Here a lens is used to give a Fraunhofer pattern on a nearby screen, as in Fig. 36.4d.

Several Slits

Next let's consider patterns produced by *several* very narrow slits. As we will see, systems of narrow slits are of tremendous practical importance in *spectroscopy,* the determination of the particular wavelengths of light coming from a source. Assume that each slit is narrow in comparison to the wavelength, so its diffraction pattern spreads out nearly uniformly. Figure 36.12 shows an array of eight narrow slits, with distance d between adjacent slits. Constructive interference occurs for rays at angle θ to the normal that arrive at point P with a path difference between adjacent slits equal to an integer number of wavelengths,

$$d \sin\theta = m\lambda \qquad (m = 0, \pm1, \pm2, \ldots)$$

This means that reinforcement occurs when the phase difference ϕ at P for light from adjacent slits is an integer multiple of 2π. That is, the maxima in the pattern occur at the *same* positions as for *two* slits with the same spacing. To this extent the pattern resembles the two-slit pattern.

But what happens *between* the maxima? In the two-slit pattern, there is exactly one intensity minimum located midway between each pair of maxima, corresponding to angles for which the phase difference between waves from the two sources is π, 3π, 5π, and so on. In the eight-slit pattern these are also minima because the light from adjacent slits cancels out in pairs, corresponding to the phasor diagram in Fig. 36.13a. But these are not the only minima in the eight-slit pattern. For example, when the phase difference ϕ from adjacent sources is $\pi/4$, the phasor diagram is as shown in Fig. 36.13b; the total (resultant) phasor is zero, and the intensity is zero. When $\phi = \pi/2$, we get the phasor diagram of Fig. 36.13c, and again both the total phasor and the intensity are zero. More generally, the intensity with eight slits is zero whenever ϕ is an integer multiple of $\pi/4$, *except* when ϕ is a multiple of 2π. Thus there are seven minima for every maximum.

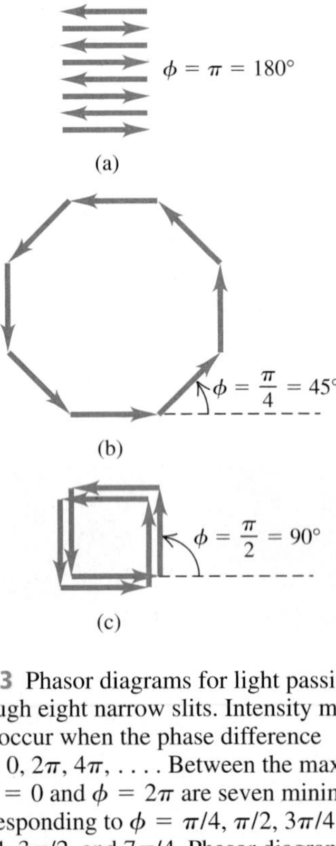

36.13 Phasor diagrams for light passing through eight narrow slits. Intensity maxima occur when the phase difference $\phi = 0, 2\pi, 4\pi, \ldots$. Between the maxima at $\phi = 0$ and $\phi = 2\pi$ are seven minima, corresponding to $\phi = \pi/4, \pi/2, 3\pi/4, \pi, 5\pi/4, 3\pi/2$, and $7\pi/4$. Phasor diagrams are shown for (a) $\phi = \pi$, (b) $\phi = \pi/4$, (c) $\phi = \pi/2$. Can you draw phasor diagrams for the other minima?

Detailed calculation shows that the eight-slit pattern is as shown in Fig. 36.14b. The large maxima, called *principal maxima,* are in the same positions as for the two-slit pattern of Fig. 36.14a but are much narrower. If the phase difference ϕ between adjacent slits is slightly different from a multiple of 2π, the waves from slits 1 and 2 will be only a little out of phase; however, the phase difference between slits 1 and 3 will be greater, that between slits 1 and 4 will be greater still, and so on. This leads to a partial cancellation for angles that are only slightly different from the angle for a maximum, giving the narrow maxima in Fig. 36.14b. The maxima are even narrower with sixteen slits (Fig. 36.14c).

You should show that when there are N slits, there are $(N - 1)$ minima between each pair of principal maxima and a minimum occurs whenever ϕ is an integral multiple of $2\pi/N$ (except when ϕ is an integral multiple of 2π, which gives a principal maximum). There are small *secondary* intensity maxima between the minima; these become smaller in comparison to the principal maxima as N increases. The greater the value of N, the narrower the principal maxima

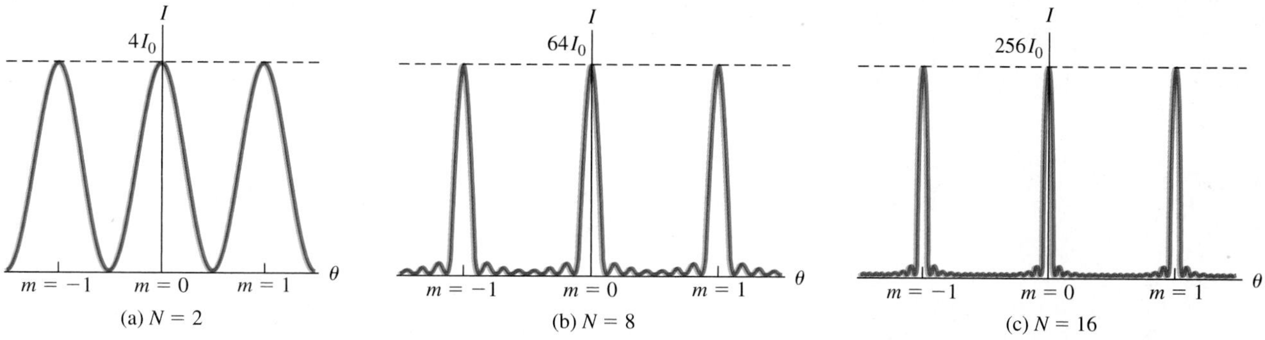

36.14 Interference patterns for N equally spaced very narrow slits. (a) Two slits. One minimum occurs between adjacent maxima. (b) Eight slits. Seven minima occur between adjacent pairs of the narrower large maxima. (c) Sixteen slits. The vertical scales are different for each graph; I_0 is the maximum intensity for a single slit, and the maximum intensity for N slits is N^2I_0. The width of each peak is proportional to $1/N$.

become. From an energy standpoint the total power in the entire pattern is proportional to N. The height of each principal maximum is proportional to N^2, so from energy conservation the width of each principal maximum must be proportional to $1/N$. As we will see in the next section, the narrowness of the principal maxima in a multiple-slit pattern is of great practical importance in physics and astronomy.

Test Your Understanding

Suppose two slits, each of width a, are separated by a distance $d = 2.5a$. Are there any missing maxima in the interference pattern produced by these slits? If so, which are missing? If not, why not?

36.5 | The Diffraction Grating

ActivPhysics ONLINE

16.4 The Grating: Introduction and Qualitative Questions

16.5 The Grating: Problems

We have just seen that increasing the number of slits in an interference experiment (while keeping the spacing of adjacent slits constant) gives interference patterns in which the maxima are in the same positions, but progressively sharper and narrower, than with two slits. Because these maxima are so sharp, their angular position, and hence the wavelength, can be measured to very high precision. As we will see, this effect has many important applications.

An array of a large number of parallel slits, all with the same width a and spaced equal distances d between centers, is called a **diffraction grating.** The first one was constructed by Fraunhofer using fine wires. Gratings can be made by using a diamond point to scratch many equally spaced grooves on a glass or metal surface, or by photographic reduction of a pattern of black and white stripes on paper. For a grating, what we have been calling *slits* are often called *rulings* or *lines.*

In Fig. 36.15, GG' is a cross section of a *transmission grating;* the slits are perpendicular to the plane of the page, and an interference pattern is formed by the light that is transmitted through the slits. The diagram shows only six slits; an actual grating may contain several thousand. The spacing d between centers of adjacent slits is called the *grating spacing.* A plane monochromatic wave is incident normally on the grating from the left side. We assume far-field (Fraunhofer) conditions; that is, the pattern is formed on a screen that is far enough away that all rays emerging from the grating and going to a particular point on the screen can be considered to be parallel.

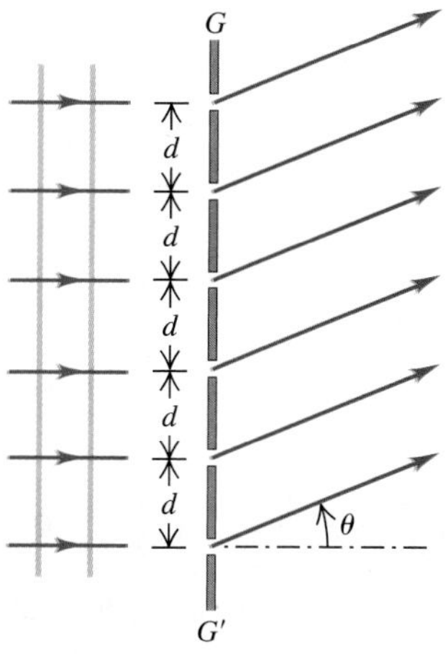

36.15 A portion of a transmission diffraction grating. The separation between the centers of adjacent slits is d.

We found in Section 36.4 that the principal intensity maxima with multiple slits occur in the same directions as for the two-slit pattern. These are the direc-

36.16 Microscopic pits on the surface of this compact disc act as a reflection grating, splitting white light into its component colors.

tions for which the path difference for adjacent slits is an integer number of wavelengths. So the positions of the maxima are once again given by

$$d \sin \theta = m\lambda \qquad (m = 0, \pm 1, \pm 2, \pm 3, \ldots) \qquad (36.13)$$
(intensity maxima, multiple slits)

The intensity patterns for two, eight, and sixteen slits displayed in Fig. 36.14 show the progressive increase in sharpness of the maxima as the number of slits increases.

When a grating containing hundreds or thousands of slits is illuminated by a beam of parallel rays of monochromatic light, the pattern is a series of very sharp lines at angles determined by Eq. (36.13). The $m = \pm 1$ lines are called the *first-order lines,* the $m = \pm 2$ lines the *second-order lines,* and so on. If the grating is illuminated by white light with a continuous distribution of wavelengths, each value of m corresponds to a continuous spectrum in the pattern. The angle for each wavelength is determined by Eq. (36.13); for a given value of m, long wavelengths (the red end of the spectrum) lie at larger angles (that is, are deviated more from the straight-ahead direction) than do the shorter wavelengths at the violet end of the spectrum.

As Eq. (36.13) shows, the sines of the deviation angles of the maxima are proportional to the ratio λ/d. For substantial deviation to occur, the grating spacing d should be of the same order of magnitude as the wavelength λ. Gratings for use with visible light (λ from 400 to 700 nm) usually have about 1000 slits per millimeter; the value of d is the *reciprocal* of the number of slits per unit length, so d is of the order of $\frac{1}{1000}$ mm $=$ 1000 nm.

?

In a *reflection grating,* the array of equally spaced slits shown in Fig. 36.15 is replaced by an array of equally spaced ridges or grooves on a reflective screen. The reflected light has maximum intensity at angles where the phase difference between light waves reflected from adjacent ridges or grooves is an integral multiple of 2π. If light of wavelength λ is incident normally on a reflection grating with a spacing d between adjacent ridges or grooves, the *reflected* angles at which intensity maxima occur are given by Eq. (36.13). The iridescent colors of certain butterflies arise from microscopic ridges on the butterfly's wings that form a reflection grating (see the photograph that opens this chapter). When the wings are viewed from different angles, corresponding to varying θ in Eq. (36.13), the wavelength and color that are predominantly reflected to the viewer's eye vary as well.

The rainbow-colored reflections that you see from the surface of a compact disc, CD-ROM, or DVD are a reflection-grating effect (Fig. 36.16). The "grooves" are tiny pits 0.1 μm deep in the surface of the disc, with a uniform radial spacing of $d = 1.60\ \mu\text{m} = 1600$ nm. Information is coded on the CD by varying the *length* of the pits; the reflection-grating aspect of the disc is merely an aesthetic side benefit.

Example 36.4 Width of a grating spectrum

The wavelengths of the visible spectrum are approximately 400 nm (violet) to 700 nm (red). a) Find the angular width of the first-order visible spectrum produced by a plane grating with 600 slits per millimeter when white light falls normally on the grating. b) Do the first-order and second-order spectra overlap? What about the second-order and third-order spectra? Do your answers depend on the grating spacing?

SOLUTION

IDENTIFY and SET UP: The first-order spectrum corresponds to $m = 1$ in Eq. (36.13). We use this equation to find the angular deviation for 400-nm violet light and 700-nm red light; the difference between these is the angular width, our target variable in part (a). Using the same technique for $m = 2$ and $m = 3$ tells us the maximum and minimum angular deviation for these orders.

EXECUTE: a) The grating spacing d is

$$d = \frac{1}{600 \text{ slits/mm}} = 1.67 \times 10^{-6} \text{ m}$$

From Eq. (36.13), with $m = 1$, the angular deviation θ_v of the violet light (400 nm or 400×10^{-9} m) is

$$\sin\theta_v = \frac{400 \times 10^{-9} \text{ m}}{1.67 \times 10^{-6} \text{ m}} = 0.240$$

$$\theta_v = 13.9°$$

The angular deviation θ_r of the red light (700 nm) is

$$\sin\theta_r = \frac{700 \times 10^{-9} \text{ m}}{1.67 \times 10^{-6} \text{ m}} = 0.419$$

$$\theta_r = 24.8°$$

So the angular width of the first-order visible spectrum is

$$24.8° - 13.9° = 10.9°$$

b) From Eq. (36.13), with a grating spacing of d the angular deviation θ_{vm} of the 400-nm violet light in the mth order spectrum is given by

$$\sin\theta_{vm} = \frac{m(400 \times 10^{-9} \text{ m})}{d}$$

$$= \frac{4.00 \times 10^{-7} \text{ m}}{d}(m = 1) \qquad \frac{8.00 \times 10^{-7} \text{ m}}{d}(m = 2)$$

$$\frac{1.20 \times 10^{-6} \text{ m}}{d}(m = 3)$$

Similarly, the angular deviation θ_{rm} of the 700-nm red light in the mth order is given by

$$\sin\theta_{rm} = \frac{m(700 \times 10^{-9} \text{ m})}{d}$$

$$= \frac{7.00 \times 10^{-7} \text{ m}}{d}(m = 1) \qquad \frac{1.40 \times 10^{-6} \text{ m}}{d}(m = 2)$$

$$\frac{2.10 \times 10^{-6} \text{ m}}{d}(m = 3)$$

The greater the value of $\sin\theta$, the greater the value of θ (for angles between zero and 90°). Hence our results show for any value of the grating spacing d, the largest angle (at the red end) of the $m = 1$ spectrum is always less than the smallest angle (at the violet end) of the $m = 2$ spectrum, so the first and second orders *never* overlap. By contrast, the largest (red) angle of the $m = 2$ spectrum is always greater than the smallest (violet) angle of the $m = 3$ spectrum, so the second and third orders *always* overlap.

EVALUATE: The fundamental reason why the first-order and second-order visible spectra don't overlap is that the human eye is sensitive to only a narrow range of wavelengths. Can you show that if the eye could detect wavelengths from 400 nm to 900 nm (in the near-infrared range), the first and second orders *would* overlap?

Grating Spectrographs

Diffraction gratings are widely used to measure the spectrum of light emitted by a source, a process called *spectroscopy* or *spectrometry.* Light incident on a grating of known spacing is dispersed into a spectrum. The angles of deviation of the maxima are then measured, and Eq. (36.13) is used to compute the wavelength. By using a grating with many slits, very sharp maxima are produced, and the angle of deviation (and hence the wavelength) can be measured very precisely.

An important application of this technique is to astronomy. As light generated within the sun passes through the sun's atmosphere, certain wavelengths are selectively absorbed. The result is that the spectrum of sunlight produced by a diffraction grating has dark *absorption lines* (Fig. 36.17). Experiments in the laboratory show that different types of atoms and ions absorb light at different wavelengths. By comparing these laboratory results with the wavelengths of absorption lines in the spectrum of sunlight, astronomers can deduce the chemical composition of the sun's atmosphere. The same technique is used to make chemical assays of galaxies that are millions of light years away.

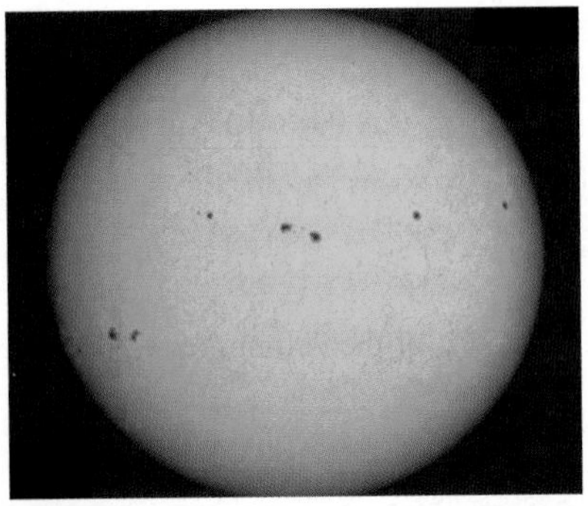

(a)

(b)

36.17 (a) A visible-light photograph of the sun. (b) Sunlight is dispersed into a spectrum by a diffraction grating. Specific wavelengths are absorbed as sunlight passes through the sun's atmosphere, leaving dark lines in the spectrum.

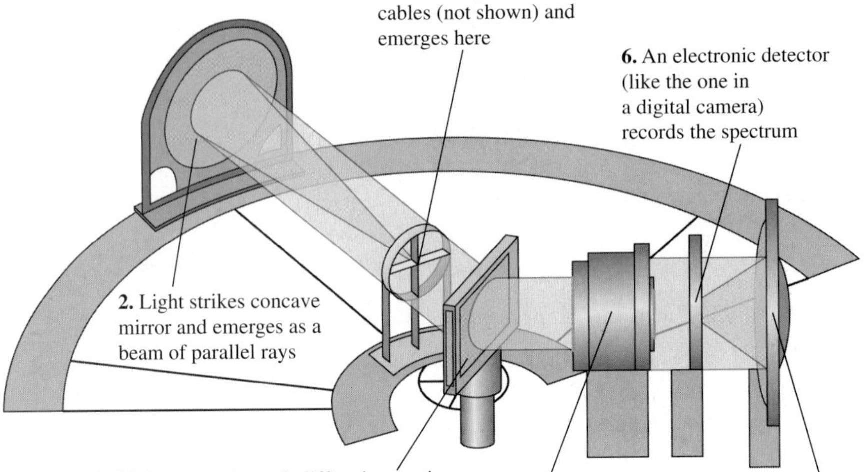

36.18 A schematic diagram of a diffraction-grating spectrograph for use in astronomy. Note that the light does not strike the grating normal to its surface, so the intensity maxima are given by a somewhat different expression than Eq. (36.13). (See Problem 36.64).

Figure 36.18 shows one design for a *grating spectrograph* used in astronomy. A transmission grating is used in the figure; in other setups, a reflection grating is used. In older designs a prism was used rather than a grating, and a spectrum was formed by dispersion (Section 33.4) rather than diffraction. However, there is no simple relationship between wavelength and angle of deviation for a prism, prisms absorb some of the light that passes through them, and they are less effective for many nonvisible wavelengths that are important in astronomy. For these and other reasons, gratings are preferred in precision applications.

Resolution of a Grating Spectrograph

In spectroscopy it is often important to distinguish slightly differing wavelengths. The minimum wavelength difference $\Delta\lambda$ that can be distinguished by a spectrograph is described by the **chromatic resolving power** R, defined as

$$R = \frac{\lambda}{\Delta\lambda} \qquad \text{(chromatic resolving power)} \qquad (36.14)$$

As an example, when sodium atoms are heated, they emit strongly at the yellow wavelengths 589.00 nm and 589.59 nm. A spectrograph that can barely distinguish these two lines in the spectrum of sodium light (called the *sodium doublet*) has a chromatic resolving power $R = (589.00\text{ nm})/(0.59\text{ nm}) = 1000$. (You can see these wavelengths when boiling water on a gas range. If the water boils over onto the flame, dissolved sodium from table salt emits a burst of yellow light.)

We can derive an expression for the resolving power of a diffraction grating used in a spectrograph. Two different wavelengths give diffraction maxima at slightly different angles. As a reasonable (though arbitrary) criterion, let's assume that we can distinguish them as two separate peaks if the maximum of one coincides with the first minimum of the other.

From our discussion in Section 36.4 the mth-order maximum occurs when the phase difference ϕ for adjacent slits is $\phi = 2\pi m$. The first minimum beside that maximum occurs when $\phi = 2\pi m + 2\pi/N$, where N is the number of slits. Now ϕ is also given by $\phi = (2\pi d \sin\theta)/\lambda$, so the angular interval $d\theta$ corresponding to a small increment $d\phi$ in the phase shift can be obtained from the differential of this equation:

$$d\phi = \frac{2\pi d \cos\theta \, d\theta}{\lambda}$$

When $d\phi = 2\pi/N$, this corresponds to the angular interval $d\theta$ between a maximum and the first adjacent minimum. Thus $d\theta$ is given by

$$\frac{2\pi}{N} = \frac{2\pi d \cos\theta \, d\theta}{\lambda} \quad \text{or} \quad d\cos\theta \, d\theta = \frac{\lambda}{N}$$

CAUTION Don't confuse the spacing d with the differential "d" in the angular interval $d\theta$ or in the phase shift increment $d\phi$!

Now we need to find the angular spacing $d\theta$ between maxima for two slightly different wavelengths. This is easy; we have $d \sin\theta = m\lambda$, so the differential of this equation gives

$$d\cos\theta \, d\theta = m \, d\lambda$$

According to our criterion, the limit or resolution is reached when these two angular spacings are equal. Equating the two expressions for the quantity $(d\cos\theta \, d\theta)$, we find

$$\frac{\lambda}{N} = m \, d\lambda \quad \text{and} \quad \frac{\lambda}{d\lambda} = Nm$$

If $\Delta\lambda$ is small, we can replace $d\lambda$ by $\Delta\lambda$, and the resolving power R is given simply by

$$R = \frac{\lambda}{\Delta\lambda} = Nm \tag{36.15}$$

The greater the number of slits N, the better the resolution; also, the higher the order m of the diffraction-pattern maximum that we use, the better the resolution.

Test Your Understanding

What minimum number of slits would be required in a grating to resolve the sodium doublet in the first order? In the fourth order?

36.6 | X-Ray Diffraction

X rays were discovered by Wilhelm Röntgen (1845–1923) in 1895, and early experiments suggested that they were electromagnetic waves with wavelengths of the order of 10^{-10} m. At about the same time, the idea began to emerge that in a crystalline solid the atoms are arranged in a regular repeating pattern, with spacing between adjacent atoms also of the order of 10^{-10} m. Putting these two ideas

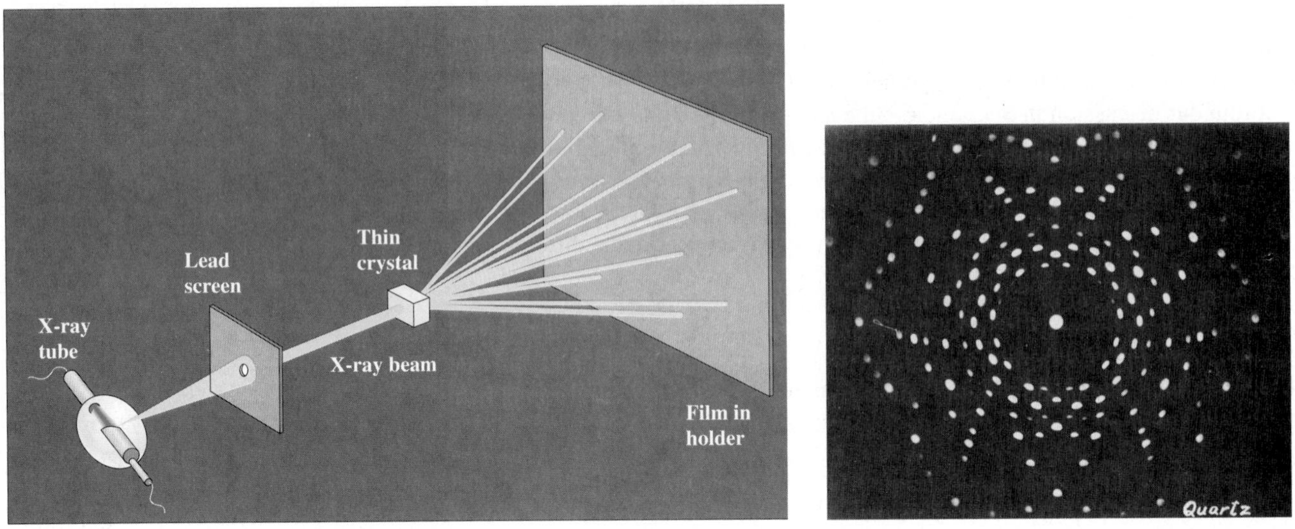

(a) (b)

36.19 (a) In an x-ray diffraction experiment, most x rays pass straight through the crystal. But some x rays are scattered and form an interference pattern that exposes the film in a pattern related to the atomic arrangement in the crystal. (b) Diffraction pattern (or *Laue pattern*) formed by directing a beam of x rays at a thin section of quartz crystal.

together, Max von Laue (1879–1960) proposed in 1912 that a crystal might serve as a kind of three-dimensional diffraction grating for x rays. That is, a beam of x rays might be scattered (that is, absorbed and re-emitted) by the individual atoms in a crystal, and the scattered waves might interfere just like waves from a diffraction grating.

The first **x-ray diffraction** experiments were performed in 1912 by Friederich, Knipping, and von Laue, using the experimental setup sketched in Fig. 36.19a. The scattered x rays *did* form an interference pattern, which they recorded on photographic film. Figure 36.19b is a photograph of such a pattern. These experiments verified that x rays *are* waves, or at least have wavelike properties, and also that the atoms in a crystal *are* arranged in a regular pattern (Fig. 36.20). Since that time, x-ray diffraction has proved to be an invaluable research tool, both for measuring x-ray wavelengths and for studying the structure of crystals and complex molecules.

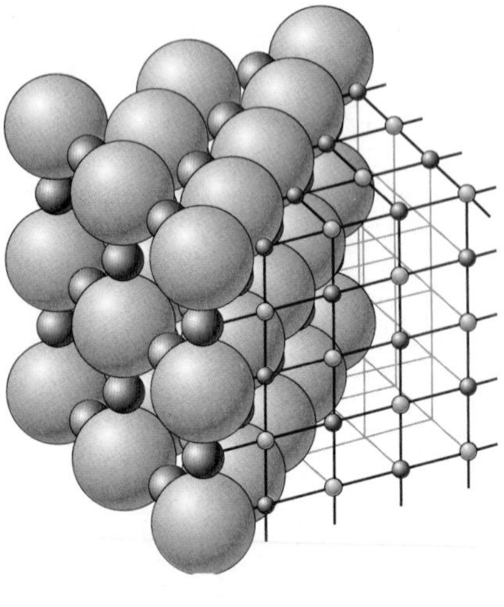

36.20 Model of the arrangement of ions in a crystal of NaCl (table salt). Black spheres are sodium (Na); red spheres are chlorine (Cl). The spacing of adjacent atoms is 0.282 nm. (The electron clouds of the atoms actually overlap slightly.)

To introduce the basic ideas, we consider first a two-dimensional scattering situation, as shown in Fig. 36.21a, in which a plane wave is incident on a rectangular array of scattering centers. The situation might be a ripple tank with an array of small posts, 3-cm microwaves striking an array of small conducting spheres, or x rays incident on an array of atoms. In the case of electromagnetic waves, the wave induces an oscillating electric dipole moment in each scatterer. These dipoles act like little antennas, emitting scattered waves. The resulting interference pattern is the superposition of all these scattered waves. The situation is different from that with a diffraction grating, in which the waves from all the slits are emitted *in phase* (for a plane wave at normal incidence). Here the scattered waves are *not* all in phase because their distances from the *source* are different. To compute the interference pattern, we have to consider the *total* path differences for the scattered waves, including the distances from source to scatterer and from scatterer to observer.

As Fig. 36.21b shows, the path length from source to observer is the same for all the scatterers in a single row if the two angles θ_a and θ_r are equal. Scattered radiation from *adjacent* rows is *also* in phase if the path difference for adjacent rows is an integer number of wavelengths. Figure 36.21c shows that this path difference is

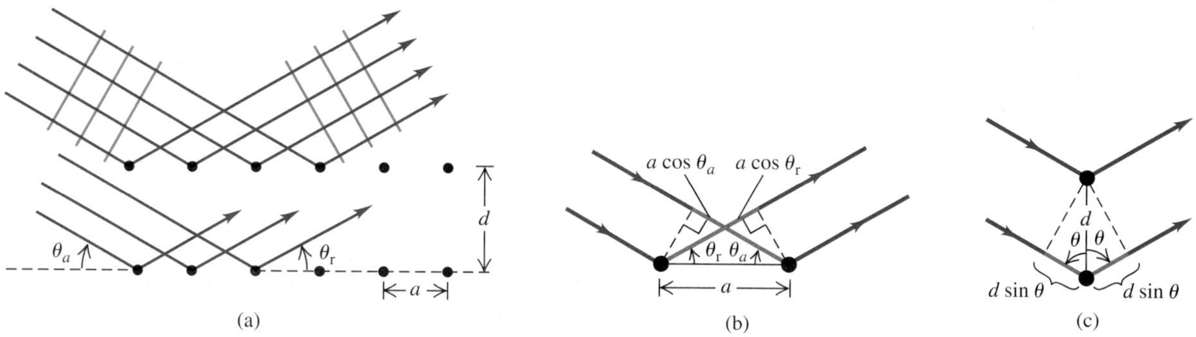

36.21 (a) Scattering of waves from a rectangular array. (b) Interference of waves scattered from adjacent atoms in a row is constructive when $a\cos\theta_a = a\cos\theta_r$, that is, when the angle of incidence θ_a equals the angle of reflection θ_r. Both angles are measured from the *surface* of the crystal, not from its normal. (c) Interference from adjacent rows is also constructive when the path difference $2d\sin\theta$ equals an integral number of wavelengths, as in Eq. (35.16).

$2d\sin\theta$, where θ is the common value of θ_a and θ_r. Therefore the conditions for radiation from the *entire array* to reach the observer in phase are (1) the angle of incidence must equal the angle of scattering and (2) the path difference for adjacent rows must equal $m\lambda$, where m is an integer. We can express the second condition as

$$2d\sin\theta = m\lambda \qquad (m = 1, 2, 3, \ldots) \qquad (36.16)$$

(Bragg condition for constructive interference from an array)

CAUTION In this equation the angle θ is measured with respect to the *surface* of the crystal, rather than with respect to the *normal* to the plane of an array of slits or a grating. Also, note that the path difference in Eq. (36.16) is $2d\sin\theta$, not $d\sin\theta$ as in Eq. (36.13) for a diffraction grating.

In directions for which Eq. (36.16) is satisfied, we see a strong maximum in the interference pattern. We can describe this interference in terms of *reflections* of the wave from the horizontal rows of scatterers in Fig. 36.21a. Strong reflection (constructive interference) occurs at angles such that the incident and scattered angles are equal and Eq. (36.16) is satisfied.

We can extend this discussion to a three-dimensional array by considering *planes* of scatterers instead of *rows*. Figure 36.22 shows two different sets of parallel planes that pass through all the scatterers. Waves from all the scatterers in a given plane interfere constructively if the angles of incidence and scattering are equal. There is also constructive interference between planes when Eq. (36.16) is satisfied, where d is now the distance between adjacent planes. Because there are many different sets of parallel planes, there are also many values of d and many sets of angles that give constructive interference for the whole crystal lattice. This phenomenon is called **Bragg reflection,** and Eq. (36.16) is called the **Bragg condition,** in honor of Sir William Bragg and his son Laurence Bragg, two pioneers in x-ray analysis.

CAUTION While we are using the term *reflection,* remember that we are dealing with an *interference* effect. In fact, the reflections from various planes are closely analogous to interference effects in thin films (Section 35.4).

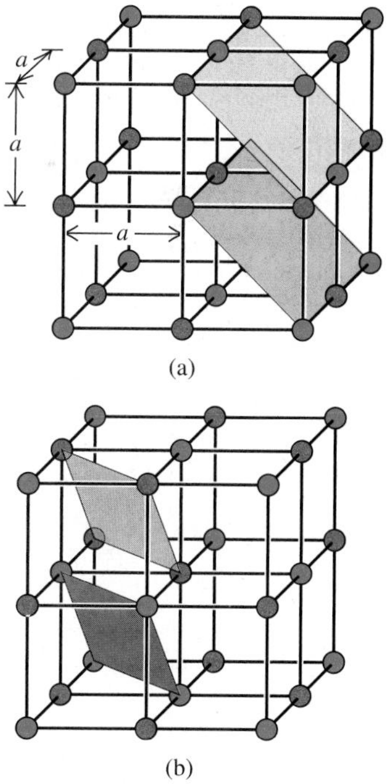

36.22 A cubic crystal and two different families of crystal planes. The spacing of the planes in (a) is $d = a/\sqrt{2}$; that of the planes in (b) is $a/\sqrt{3}$. There are also three sets of planes parallel to the cube faces, with spacing a.

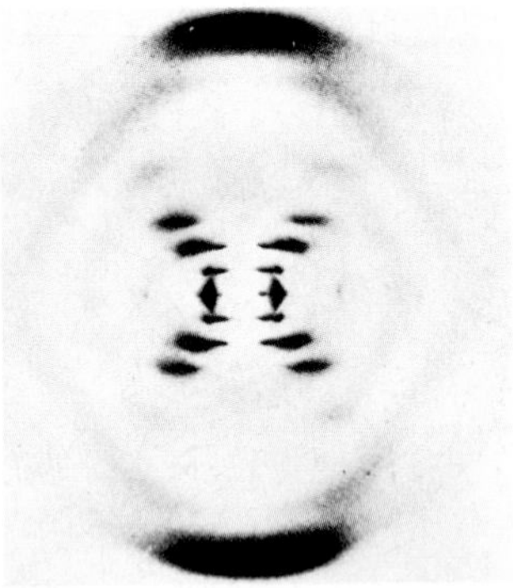

36.23 The British scientist Rosalind Franklin made this groundbreaking x-ray diffraction image of DNA in 1953. The dark bands arranged in a cross provided the first evidence of the helical structure of the DNA molecule.

As Fig. 36.19b shows, in x-ray diffraction there is nearly complete cancellation in all but certain very specific directions in which constructive interference occurs and forms bright spots. Such a pattern is usually called an x-ray *diffraction* pattern, although *interference* pattern might be more appropriate.

We can determine the wavelength of x rays by examining the diffraction pattern for a crystal of known structure and known spacing between atoms, just as we determined wavelengths of visible light by measuring patterns from slits or gratings. (The spacing between atoms in simple crystals of known structure, such as sodium chloride, can be found from the density of the crystal and Avogadro's number.) Then, once we know the x-ray wavelength, we can use x-ray diffraction to explore the structure and determine the spacing between atoms in crystals with unknown structure.

X-ray diffraction is by far the most important experimental tool in the investigation of crystal structure of solids. X-ray diffraction also plays an important role in studies of the structures of liquids and of organic molecules. It has been one of the chief experimental techniques in working out the double-helix structure of DNA (Fig. 36.23) and subsequent advances in molecular genetics.

Example 36.5 X-ray diffraction

You direct a beam of x rays with wavelength 0.154 nm at certain planes of a silicon crystal. As you increase the angle of incidence from zero, you find the first strong interference maximum from these planes when the beam makes an angle of 34.5° with the planes. a) How far apart are the planes? b) Will you find other interference maxima from these planes at larger angles?

SOLUTION

IDENTIFY and SET UP: In part (a) we use the Bragg condition, Eq. (36.16), to relate the wavelength λ and the angle θ for the $m = 1$ interference maximum (both of which are given) to the spacing d between planes (which is the target variable). Given the value of d, we use the Bragg condition again in part (b) to find the values of θ for interference maxima corresponding to other values of m.

EXECUTE: a) We solve the Bragg equation, Eq. (36.16), for d and set $m = 1$:

$$d = \frac{m\lambda}{2\sin\theta} = \frac{(1)(0.154\text{ nm})}{2\sin 34.5^\circ} = 0.136\text{ nm}$$

This is the distance between adjacent planes.

b) To calculate other angles, we solve Eq. (36.16) for $\sin\theta$:

$$\sin\theta = \frac{m\lambda}{2d} = m\frac{0.154\text{ nm}}{2(0.136\text{ nm})} = m(0.566)$$

Values of m of 2 or greater give values of $\sin\theta$ greater than unity, which is impossible. Hence there are *no* other angles for interference maxima for this particular set of crystal planes.

EVALUATE: Our result in part (b) shows that there *would* be a second interference maximum if the quantity $\lambda/2d$ were equal to 0.500 or less. This would be the case if the wavelength of the x rays were less than $2d = 0.272$ nm. How short would the wavelength need to be to have *three* interference maxima?

Test Your Understanding

An engineer is designing an array of metal spheres to reflect a radio beam coming from a weather radar antenna. If the radio waves have a wavelength of 10.0 cm and the beam is to strike the surface of the array at a 45.0° angle, what should be the spacing between adjacent planes of spheres to maximize the reflected wave?

ActivPhysics ONLINE

16.7 Circular Hole Diffraction

16.8 Resolving Power

36.7 | Circular Apertures and Resolving Power

We have studied in detail the diffraction patterns formed by long, thin slits or arrays of slits. But an aperture of *any* shape forms a diffraction pattern. The dif-

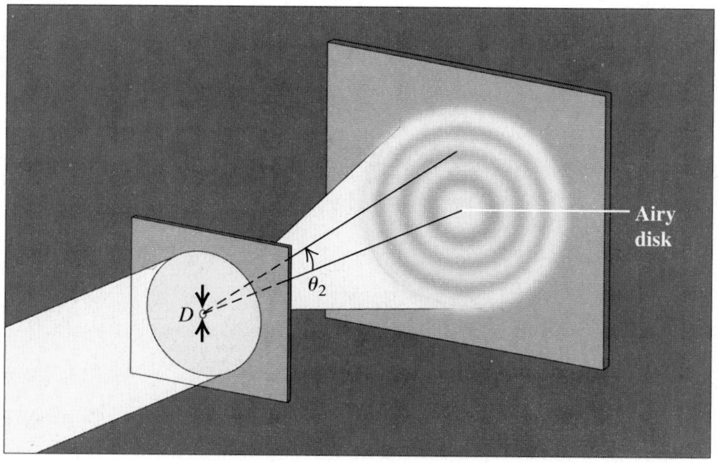

36.24 Diffraction pattern formed by a circular aperture of diameter D. The pattern consists of a central bright spot and alternating dark and bright rings. The angular radius θ_2 of the second dark ring is shown. (This diagram is not drawn to scale.)

fraction pattern formed by a *circular* aperture is of special interest because of its role in limiting how well an optical instrument can resolve fine details. In principle, we could compute the intensity at any point P in the diffraction pattern by dividing the area of the aperture into small elements, finding the resulting wave amplitude and phase at P, and then integrating over the aperture area to find the resultant amplitude and intensity at P. In practice, the integration cannot be carried out in terms of elementary functions. We will simply *describe* the pattern and quote a few relevant numbers.

The diffraction pattern formed by a circular aperture consists of a central bright spot surrounded by a series of bright and dark rings, as shown in Fig. 36.24. We can describe the pattern in terms of the angle θ, representing the angular radius of each ring. If the aperture diameter is D and the wavelength is λ, the angular radius θ_1 of the first *dark* ring is given by

$$\sin\theta_1 = 1.22\frac{\lambda}{D} \quad \text{(diffraction by a circular aperture)} \quad (36.17)$$

The angular radii of the next two dark rings are given by

$$\sin\theta_2 = 2.23\frac{\lambda}{D} \qquad \sin\theta_3 = 3.24\frac{\lambda}{D} \quad (36.18)$$

Between these are bright rings with angular radii given by

$$\sin\theta = 1.63\frac{\lambda}{D} \qquad 2.68\frac{\lambda}{D} \qquad 3.70\frac{\lambda}{D} \quad (36.19)$$

and so on. The central bright spot is called the **Airy disk,** in honor of Sir George Airy (1801–1892), Astronomer Royal of England, who first derived the expression for the intensity in the pattern. The angular radius of the Airy disk is that of the first dark ring, given by Eq. (36.17).

The intensities in the bright rings drop off very quickly with increasing angle. When D is much larger than the wavelength λ, the usual case for optical instruments, the peak intensity in the first ring is only 1.7% of the value at the center of the Airy disk, and the peak intensity of the second ring is only 0.4%. Most (85%) of the light energy falls within the Airy disk. Figure 36.25 shows a diffraction pattern from a circular aperture 1.0 mm in diameter.

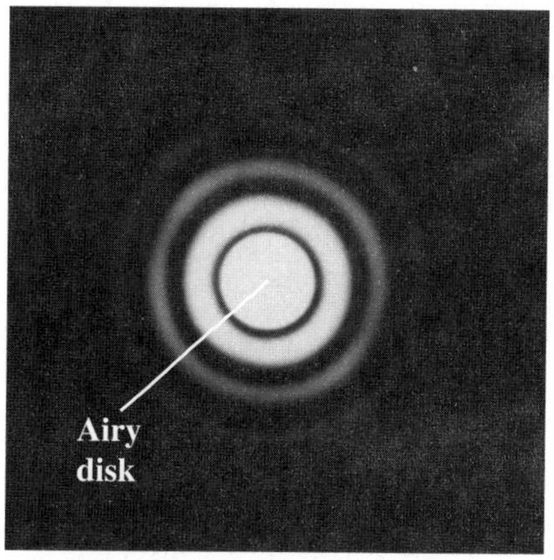

36.25 Diffraction pattern formed by a circular aperture 1.0 mm in diameter. The Airy disk is overexposed so that other rings may be seen.

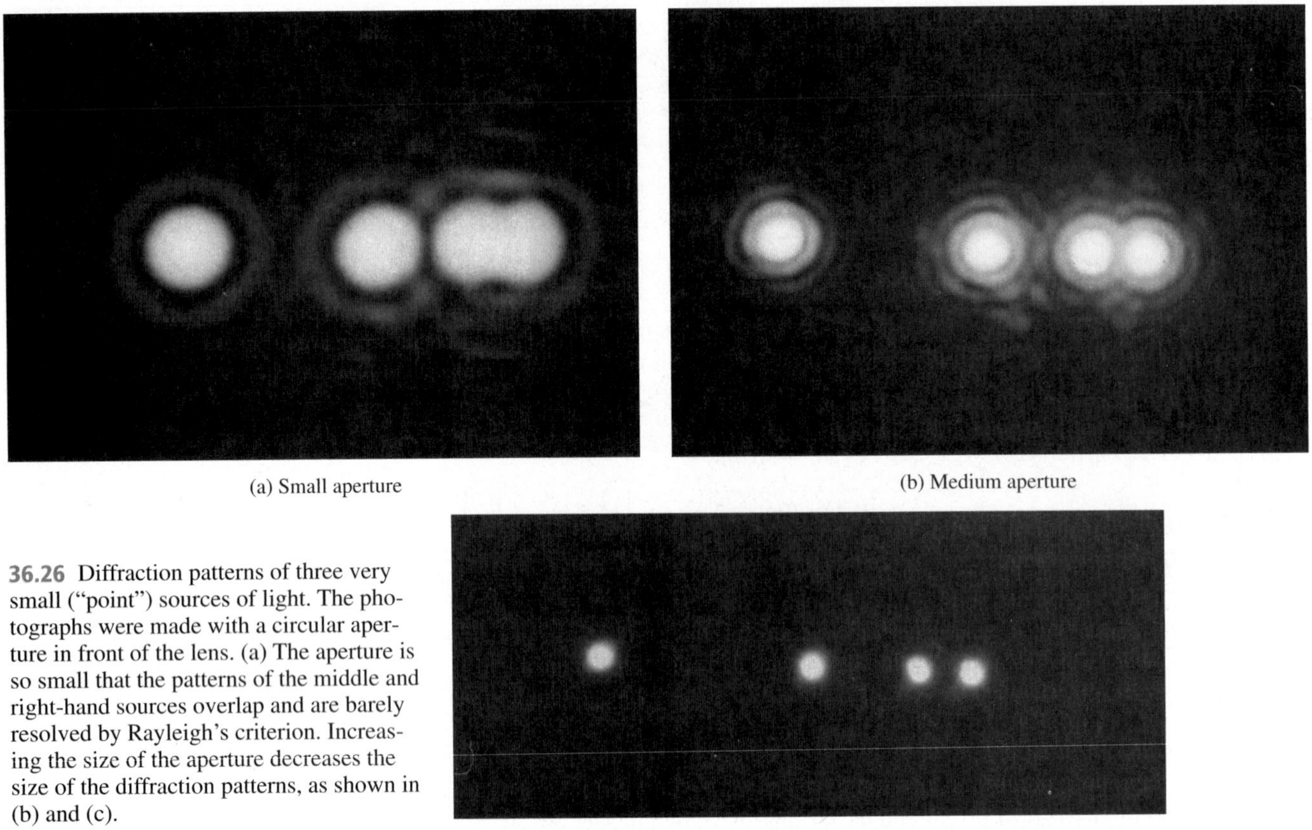

(a) Small aperture

(b) Medium aperture

(c) Large aperture

36.26 Diffraction patterns of three very small ("point") sources of light. The photographs were made with a circular aperture in front of the lens. (a) The aperture is so small that the patterns of the middle and right-hand sources overlap and are barely resolved by Rayleigh's criterion. Increasing the size of the aperture decreases the size of the diffraction patterns, as shown in (b) and (c).

Diffraction has far-reaching implications for image formation by lenses and mirrors. In our study of optical instruments in Chapter 34 we assumed that a lens with focal length f focuses a parallel beam (plane wave) to a *point* at a distance f from the lens. This assumption ignored diffraction effects. We now see that what we get is not a point but the diffraction pattern just described. If we have two point objects, their images are not two points but two diffraction patterns. When the objects are close together, their diffraction patterns overlap; if they are close enough, their patterns overlap almost completely and cannot be distinguished. The effect is shown in Fig. 36.26, which presents the patterns for three very small "point" sources of light. In Fig. 36.26a the image of the left-hand source is well separated from the other two, but the images of the middle and right-hand sources have merged. In Fig. 36.26b, with a larger aperture diameter and hence smaller Airy disks, the middle and right-hand images are better resolved. In Fig. 36.26c, with a still larger aperture, they are well resolved.

A widely used criterion for resolution of two point objects, proposed by the English physicist Lord Rayleigh (1842–1919) and called **Rayleigh's criterion,** is that the objects are just barely resolved (that is, distinguishable) if the center of one diffraction pattern coincides with the first minimum of the other. In that case the angular separation of the image centers is given by Eq. (36.17). The angular separation of the *objects* is the same as that of the *images* made by a telescope, microscope, or other optical device. So two point objects are barely resolved, according to Rayleigh's criterion, when their angular separation is given by Eq. (36.17).

The minimum separation of two objects that can just be resolved by an optical instrument is called the **limit of resolution** of the instrument. The smaller the limit of resolution, the greater the *resolution,* or **resolving power,** of the instrument. Diffraction sets the ultimate limits on resolution of lenses. *Geometric* optics may make it seem that we can make images as large as we like. Eventually, though, we always reach a point at which the image becomes larger but does not gain in detail. The images in Fig. 36.26 would not become sharper with further enlargement.

CAUTION Be careful not to confuse the resolving power of an optical instrument with the *chromatic* resolving power of a grating (described in Section 36.5). Resolving power refers to the ability to distinguish the images of objects that appear close to each other, either when looking through an optical instrument or in a photograph made with the instrument. Chromatic resolving power describes how well different wavelengths can be distinguished in a spectrum formed by a diffraction grating.

Rayleigh's criterion combined with Eq. (36.17) shows that resolution (resolving power) improves with larger diameter; it also improves with shorter wavelengths. Ultraviolet microscopes have higher resolution than visible-light microscopes. In electron microscopes the resolution is limited by the wavelengths associated with the electrons, which have wavelike aspects (to be discussed further in Chapter 39). These wavelengths can be made 100,000 times smaller than wavelengths of visible light, with a corresponding gain in resolution. One reason for building very large telescopes is to increase the aperture diameter and thus minimize diffraction effects.

Diffraction is an important consideration for satellite "dishes," parabolic reflectors designed to receive satellite transmission. Satellite dishes have to be able to pick up transmissions from two satellites that are only a few degrees apart, transmitting at the same frequency; the need to resolve two such transmissions determines the minimum diameter of the dish. As higher frequencies are used, the needed diameter decreases. For example, when two satellites 5.0° apart broadcast 7.5-cm microwaves, the minimum dish diameter to resolve them (by Rayleigh's criterion) is about 1.0 m.

The effective diameter of a telescope can be increased in some cases by using arrays of smaller telescopes. The Very Large Array (VLA) is a collection of 27 radio telescopes that can be spread out in a Y-shaped arrangement 36 km across (Fig. 36.27). Hence the effective aperture diameter is 36 km, giving the VLA a limit of resolution of less than 3×10^{-7} rad. This is comparable, in the optical realm, to being able to read the bottom line of an eye chart 7 km away! NASA has proposed placing a number of infrared telescopes in space to form an array 6000 km across. The superb resolution of this array would allow us to see continents on planets many light years away.

36.27 Each of the 27 radio telescopes of the Very Large Array in New Mexico is 25 m in diameter. Spreading the telescopes out and combining their signals electronically gives the array the resolving power of a single telescope 36 km in diameter.

Example 36.6 Resolving power of a camera lens

A camera lens with focal length $f = 50$ mm and maximum aperture $f/2$ forms an image of an object 9.0 m away. a) If the resolution is limited by diffraction, what is the minimum distance between two points on the object that are barely resolved, and what is the corresponding distance between image points? b) How does the situation change if the lens is "stopped down" to $f/16$? Assume that $\lambda = 500$ nm in both cases.

SOLUTION

IDENTIFY: This example uses ideas from this section as well as Sections 34.4 (in which we discussed image formation by a lens) and 34.5 (in which the idea of *f*-number was introduced).

SET UP: From Eq. (34.20) the *f*-number of a lens is its focal length f divided by the aperture diameter D. We use the information provided to determine D, then use Eq. (36.17) to find the angular separation θ between two barely resolved points on the object. We then use the geometry of image formation by a lens (Section 34.4) to determine the distance between those points and the distance between the corresponding image points.

EXECUTE: a) The aperture diameter is $D = f/(f\text{-number}) = (50\text{ mm})/2 = 25\text{ mm} = 25 \times 10^{-3}\text{ m}$. From Eq. (36.17) the angular separation θ of two object points that are barely resolved is given by

$$\theta \approx \sin\theta = 1.22\frac{\lambda}{D} = 1.22\frac{500 \times 10^{-9}\text{ m}}{25 \times 10^{-3}\text{ m}}$$
$$= 2.4 \times 10^{-5}\text{ rad}$$

Let y be the separation of the object points, and let y' be the separation of the corresponding image points. We know from our thin-lens analysis in Section 34.4 that, apart from sign, $y/s = y'/s'$. Thus the angular separations of the object points and the corresponding image points are both equal to θ. Because the object distance s is much greater than the focal length $f = 50$ mm, the image distance s' is approximately equal to f. Thus

$$\frac{y}{9.0\text{ m}} = 2.4 \times 10^{-5} \qquad y = 2.2 \times 10^{-4}\text{ m} = 0.22\text{ mm}$$

$$\frac{y'}{50\text{ mm}} = 2.4 \times 10^{-5} \qquad y' = 1.2 \times 10^{-3}\text{ mm}$$
$$= 0.0012\text{ mm} \approx \frac{1}{800}\text{ mm}$$

b) The aperture diameter is now $(50\text{ mm})/16$, or one-eighth as large as before. The angular separation between barely resolved points is eight times as great, and the values of y and y' are also eight times as great as before:

$$y = 1.8\text{ mm} \qquad y' = 0.0096\text{ mm} = \frac{1}{100}\text{ mm}$$

Only the best camera lenses can approach this resolving power.

EVALUATE: Many photographers always use the smallest possible aperture for maximum sharpness, since lens aberrations cause light rays that are far from the optic axis to converge to a different image point than do rays near the axis. Photographers should be aware that diffraction effects become more significant at small apertures. One cause of fuzzy images has to be balanced against another.

Test Your Understanding

The Hubble Space Telescope (Fig. 1.1b), which has an objective mirror 2.4 m in diameter, has the same resolving power for visible light as the VLA (Fig. 36.27) has for 40-GHz radio waves. Since resolving power is worse for smaller apertures, how can this be?

*36.8 | Holography

Holography is a technique for recording and reproducing an image of an object through the use of interference effects. Unlike the two-dimensional images recorded by an ordinary photograph or television system, a holographic image is truly three-dimensional. Such an image can be viewed from different directions to reveal different sides and from various distances to reveal changing perspective. If you had never seen a hologram, you wouldn't believe it was possible!

The basic procedure for making a hologram is shown in Fig. 36.28a. We illuminate the object to be holographed with monochromatic light, and we place a photographic film so that it is struck by scattered light from the object and also by direct light from the source. In practice, the light source must be a laser, for reasons we will discuss later. Interference between the direct and scattered light leads to the formation and recording of a complex interference pattern on the film.

To form the images, we simply project light through the developed film, as shown in Fig. 36.28b. Two images are formed, a virtual image on the side of the film nearer the source and a real image on the opposite side.

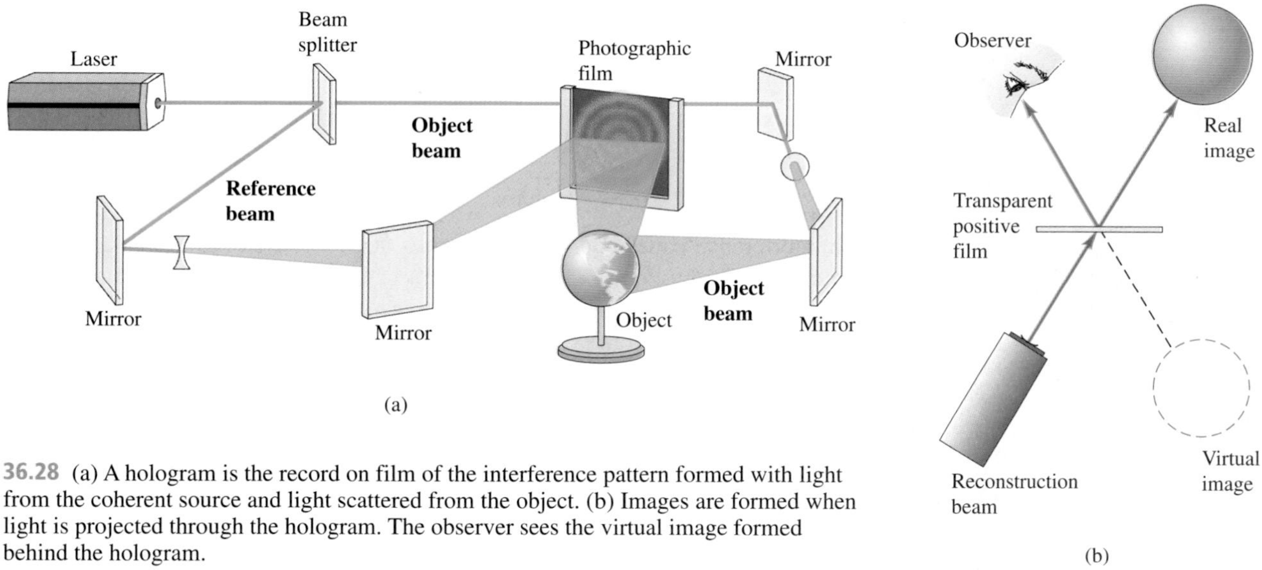

36.28 (a) A hologram is the record on film of the interference pattern formed with light from the coherent source and light scattered from the object. (b) Images are formed when light is projected through the hologram. The observer sees the virtual image formed behind the hologram.

A complete analysis of holography is beyond our scope, but we can gain some insight into the process by looking at how a single point is holographed and imaged. Consider the interference pattern that is formed on a sheet of photographic negative film by the superposition of an incident plane wave and a spherical wave, as shown in Fig. 36.29a. The spherical wave originates at a point source P at a distance b_0 from the film; P may in fact be a small object that scatters part of the incident plane wave. We assume that the two waves are monochromatic and coherent and that the phase relation is such that constructive interference occurs at point O on the diagram. Then constructive interference will *also* occur at any point Q on the film that is farther from P than O is by an integer number of wavelengths. That is, if $b_m - b_0 = m\lambda$, where m is an integer, then constructive interference occurs. The points where this condition is satisfied form circles on the film centered at O, with radii r_m given by

$$b_m - b_0 = \sqrt{b_0^2 + r_m^2} - b_0 = m\lambda \qquad (m = 1, 2, 3, \ldots) \qquad (36.20)$$

36.29 (a) Constructive interference of the plane and spherical waves occurs in the plane of the film at every point Q for which the distance b_m from P is greater than the distance b_0 from P to O by an integral number of wavelengths $m\lambda$. For the point Q shown, $m = 2$. (b) When a plane wave strikes a transparent positive print of the developed film, the diffracted wave consists of a wave converging to P' and then diverging again and a diverging wave that appears to originate at P. These waves form the real and virtual images, respectively.

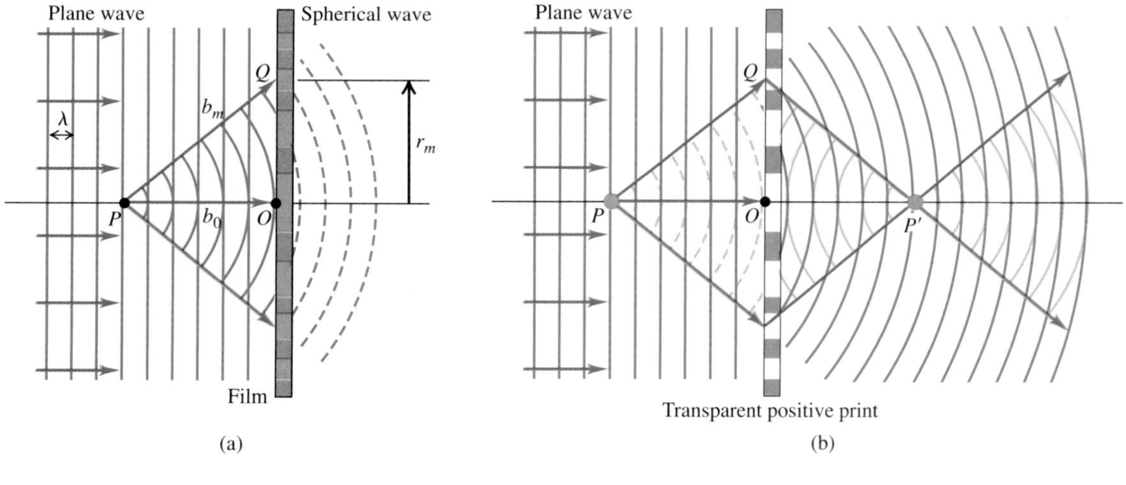

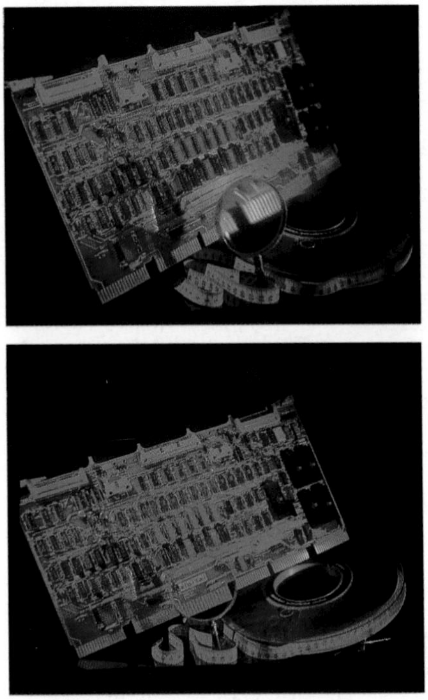

36.30 Two views of the same hologram seen from different angles.

Solving this for r_m^2, we find

$$r_m^2 = \lambda(2mb_0 + m^2\lambda)$$

Ordinarily, b_0 is very much larger than λ, so we neglect the second term in parentheses and obtain

$$r_m = \sqrt{2m\lambda b_0} \qquad (m = 1, 2, 3, \ldots) \tag{36.21}$$

The interference pattern consists of a series of concentric bright circular fringes with radii given by Eq. (36.21). Between these bright fringes are dark fringes.

Now we develop the film and make a transparent positive print, so the bright-fringe areas have the greatest transparency on the film. Then we illuminate it with monochromatic plane-wave light of the same wavelength λ that we used initially. In Fig. 36.29b, consider a point P' at a distance b_0 along the axis from the film. The centers of successive bright fringes differ in their distances from P' by an integer number of wavelengths, and therefore a strong *maximum* in the diffracted wave occurs at P'. That is, light converges to P' and then diverges from it on the opposite side. Therefore P' is a *real image* of point P.

This is not the entire diffracted wave, however. The interference of the wavelets that spread out from all the transparent areas forms a second spherical wave that is diverging rather than converging. When this wave is traced back behind the film in Fig. 36.29b, it appears to be spreading out from point P. Thus the total diffracted wave from the hologram is a superposition of a spherical wave converging to form a real image at P' and a spherical wave that diverges as though it had come from the virtual image point P.

Because of the principle of superposition for waves, what is true for the imaging of a single point is also true for the imaging of any number of points. The film records the superposed interference pattern from the various points, and when light is projected through the film, the various image points are reproduced simultaneously. Thus the images of an extended object can be recorded and reproduced just as for a single point object. Figure 36.30 shows photographs of a holographic image from two different angles, showing the changing perspective in this three-dimensional image.

In making a hologram, we have to overcome several practical problems. First, the light used must be *coherent* over distances that are large in comparison to the dimensions of the object and its distance from the film. Ordinary light sources *do not* satisfy this requirement, for reasons that we discussed in Section 35.1. Therefore laser light is essential for making a hologram. (Ordinary white light can be used for *viewing* certain types of hologram, such as those used on credit cards.) Second, extreme mechanical stability is needed. If any relative motion of source, object, or film occurs during exposure, even by as much as a quarter of a wavelength, the interference pattern on the film is blurred enough to prevent satisfactory image formation. These obstacles are not insurmountable, however, and holography has become important in research, entertainment, and a wide variety of technological applications.

SUMMARY

Diffraction occurs when light passes through an aperture or around an edge. When the source and the observer are so far away from the obstructing surface that the outgoing rays can be considered parallel, it is called Fraunhofer diffraction. When the source or the observer is relatively close to the obstructing surface, it is Fresnel diffraction.

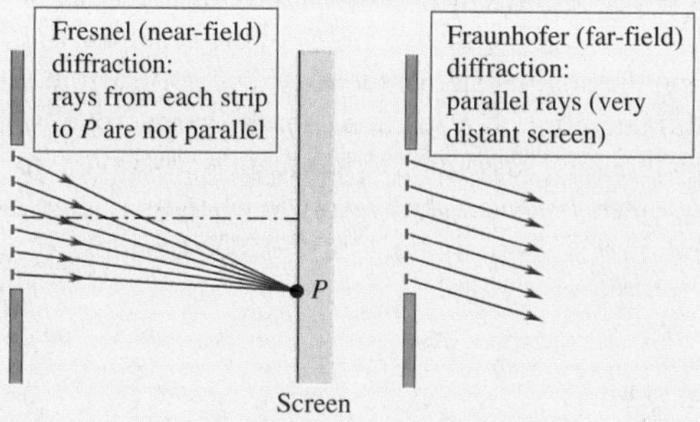

Monochromatic light sent through a narrow slit of width a produces a diffraction pattern on a distant screen. Equation (36.2) gives the condition for destructive interference (a dark fringe) at a point P in the pattern at angle θ. Equation (36.7) gives the intensity in the pattern as a function of θ. (See Examples 36.1 through 36.3)

$$\sin\theta = \frac{m\lambda}{a} \quad (m = \pm 1, \pm 2, \dots) \tag{36.2}$$

$$I = I_0\left\{\frac{\sin[\pi a(\sin\theta)/\lambda]}{\pi a(\sin\theta)/\lambda}\right\}^2 \tag{36.7}$$

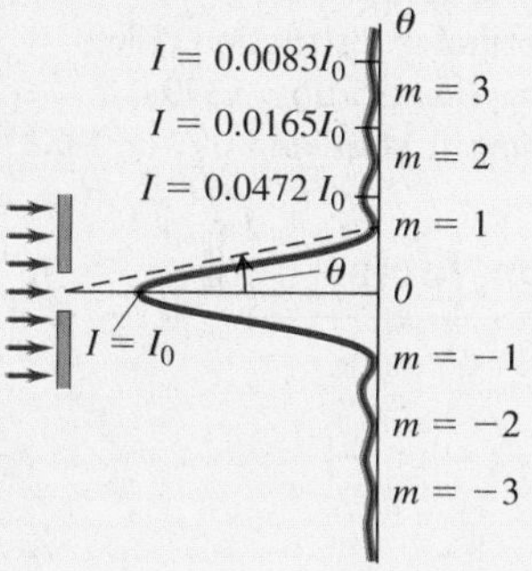

A diffraction grating consists of a large number of thin parallel slits, spaced a distance d apart. The condition for maximum intensity in the interference pattern is the same as for the two-source pattern, but the maxima for the grating are very sharp and narrow. (See Example 36.4)

$$d\sin\theta = m\lambda \quad (m = 0, \pm 1, \pm 2, \pm 3, \dots) \tag{36.13}$$

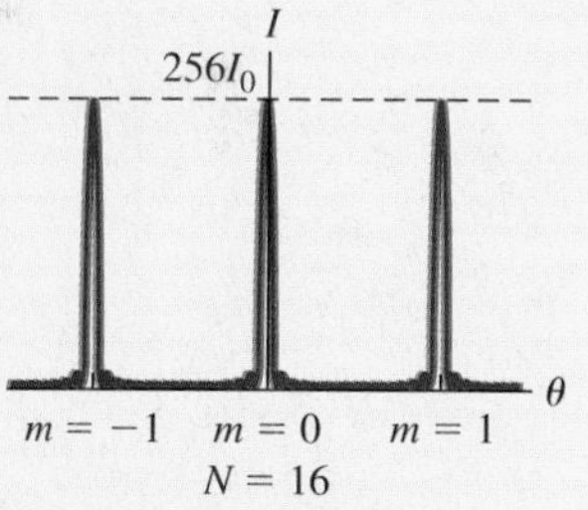

A crystal serves as a three-dimensional diffraction grating for x rays with wavelengths of the same order of magnitude as the spacing between atoms in the crystal. For a set of crystal planes spaced a distance d apart, constructive interference occurs when the angles of incidence and scattering (measured from the crystal planes) are equal and when the Bragg condition [Eq. (36.16)] is satisfied. (See Example 36.5)

$$2d\sin\theta = m\lambda \quad (m = 1, 2, 3, \dots) \tag{36.16}$$

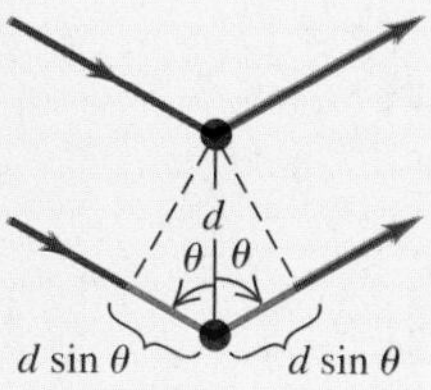

The diffraction pattern from a circular aperture of diameter D consists of a central bright spot, called the Airy disk, and a series of concentric dark and bright rings. Equation (36.17) gives the angular radius θ_1 of the first dark ring, equal to the angular size of the Airy disk. Diffraction sets the ultimate limit on resolution (image sharpness) of optical instruments. According to Rayleigh's criterion, two point objects are just barely resolved when their angular separation θ is given by Eq. (36.17). (See Example 36.6)

$$\sin\theta_1 = 1.22\frac{\lambda}{D} \qquad (36.17)$$

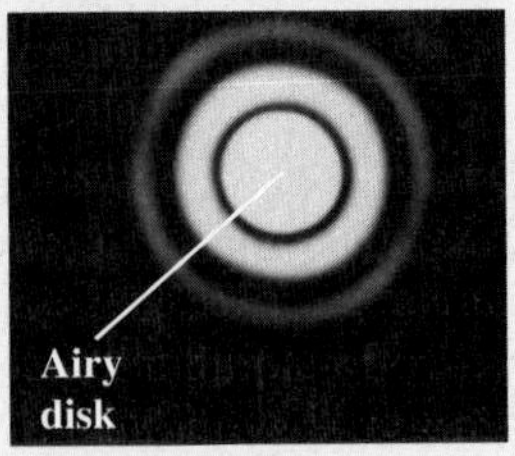

Key Terms

diffraction, 1368
Fresnel diffraction, 1369
Fraunhofer diffraction, 1369
diffraction grating, 1379
chromatic resolving power, 1382
x-ray diffraction, 1384
Bragg reflection, 1385
Bragg condition, 1385
Airy disk, 1387
Rayleigh's criterion, 1388
limit of resolution, 1389
resolving power, 1389
holography, 1390

Your Notes

Answer to Chapter Opening Question

The ridges of a butterfly wing act like a reflection grating. For each wavelength there is a certain set of angles at which there is a maximum in the reflected intensity. If you shine white light on the wing and change the angle at which you view the wing, different wavelengths will be reflected most strongly into your eye and the apparent color of the wing will change.

Answers to Test Your Understanding Questions

Section 36.1 Yes. When you hear the voice of someone standing around a corner, you are hearing sound waves that underwent diffraction. If there were no diffraction of sound, you could hear sounds only from objects that were in plain view.

Section 36.2 With $\lambda = 0.18$ mm and $a = 0.24$ mm, from Eq. (36.2) the first $(m = 1)$ diffraction minimum occurs when $\sin\theta = \lambda/a = (0.18\text{ mm})/(0.24\text{ mm}) = 0.75$, corresponding to $\theta = 49° = 0.85$ rad. In this case the wavelength is *not* small compared to the slit width, so $\tan\theta = 1.1$ and $\sin\theta = 0.75$ are not very close to the value of θ in radians. Hence the approximations cannot be used.

Section 36.3 From Eq. (36.8), an intensity minimum is given by $\sin\theta = m\lambda/a$, so $a = m\lambda/(\sin\theta) = m(6.33 \times 10^{-7}\text{ m})/(5.0 \times 10^{-4}) = m(1.3 \times 10^{-3}\text{ m}) = m(1.3\text{ mm})$. The point in question will lie at the first intensity minimum $(m = 1)$ if $a = 1.3$ mm, at the second intensity minimum $(m = 2)$ if $a = 2(1.3\text{ mm}) = 2.6$ mm, and so on.

Section 36.4 A "missing maximum" satisfies both $d\sin\theta = m_i\lambda$ (the condition for an interference maximum) and $a\sin\theta = m_d\lambda$ (the condition for a diffraction minimum). Substituting $d = 2.5a$, we can combine these two conditions into the relation $m_i = 2.5m_d$. This is satisfied for $m_i = 5$ and $m_d = 2$ (the fifth interference maximum is missing because it coincides with the second diffraction minimum), $m_i = 10$ and $m_d = 4$ (the tenth interference maximum is missing because it coincides with the fourth diffraction minimum), and so on.

Section 36.5 As described in the text, the resolving power needed is $R = 1000$. In the first order $(m = 1)$ we need 1000 slits, but in the fourth order we need only 250 slits. (These numbers are only approximate because of the arbitrary nature of our criterion for resolution and because real gratings always have slight imperfections in the shapes and spacings of the slits.)

Section 36.6 The same principle applies as for x-ray diffraction: the reflected wave has maximum intensity if the Bragg condition, Eq. (36.16), is satisfied. From this condition, the spacing between planes is $d = m\lambda/(2\sin\theta) = m(10.0\text{ cm})/(2\sin 45.0°) = m(7.07\text{ cm})$. Thus any plane spacing that is a multiple of 7.07 cm will do the job.

Section 36.7 Equation (36.17) shows that the angular radius of the Airy disk (which determines the resolving power) depends on the *ratio* of wavelength λ to angular diameter D. Visible light has a much shorter wavelength than radio waves: $\lambda = 500\text{ nm} = 5 \times 10^{-7}\text{ m}$ for green light (see Table 32.1 in Section 32.6) versus $\lambda = c/f = (3.00 \times 10^8\text{ m/s})/(40 \times 10^9\text{ Hz}) = 7.5 \times 10^{-3}\text{ m}$ for 40-GHz radio waves. Hence a visible-light telescope can have a much smaller diameter than the VLA and still achieve the same resolving power. (You can show that the ratio λ/D is the *same* for the Hubble Space Telescope as for the VLA.) Conversely, a radio telescope must be far larger than a visible-light telescope to produce equally sharp images.

Discussion Questions

Q36.1 You use a lens of diameter D and light of wavelength λ and frequency f to form an image of two closely spaced and distant objects. Which of the following will increase the resolving power? a) Use a lens with a smaller diameter. b) Use light of larger frequency. c) Use light of larger wavelength. In each case justify your answer.

Q36.2 Light of wavelength λ and frequency f passes through a single slit of width a. The diffraction pattern is observed on a screen a distance x from the slit. Which of the following will *decrease* the width of the central maximum? a) Decrease the slit width. b) Decrease the frequency f of the light. c) Decrease the wavelength λ of the light. d) Decrease the distance x of the screen from the slit. In each case justify your answer.

Q36.3 Describe what happens to the single-slit diffraction pattern when the width of the slit is less than the wavelength of the wave.

Q36.4 The sound waves used predominantly in human speech have wavelengths in the range from 1.0 to 3.0 meters. Using the ideas of diffraction, explain how it is possible to hear a person's voice even when they are facing away from you.

Q36.5 In a diffraction experiment with waves of wavelength λ, there will be *no* intensity minima (that is, no dark fringes) if the slit width is small enough. What is the maximum slit width for which this occurs? Explain your answer.

Q36.6 A rainbow ordinarily shows a range of colors (see Section 33.4). But if the water droplets that form the rainbow are small enough, the rainbow will appear white. Explain why, using diffraction ideas. How small do you think the raindrops would have to be for this to occur?

Q36.7 Some loudspeaker horns for outdoor concerts (at which the entire audience is seated on the ground) are wider vertically than horizontally. Use diffraction ideas to explain why this is more efficient at spreading the sound uniformly over the audience than either a square speaker horn or a horn that is wider horizontally than vertically. Would this still be the case if the audience was seated at different elevations, as in an amphitheater? Why or why not?

Q36.8 Figure 31.8 (Section 31.2) shows a loudspeaker system. Low-frequency sounds are produced by the *woofer,* which is a speaker with large diameter; the *tweeter,* a speaker of smaller diameter, produces high-frequency sounds. Use diffraction ideas to explain why the tweeter is more effective for distributing high-frequency sounds uniformly over a room than is the woofer.

Q36.9 Information is stored on an audio compact disc, CD-ROM, or DVD disc in a series of pits on the disc. These pits are scanned by a laser beam. An important limitation on the amount of information that can be stored on such a disc is the width of the laser beam. Explain why this should be, and explain how using a shorter-wavelength laser allows more information to be stored on a disc of the same size.

Q36.10 With which kind of light can the Hubble Space Telescope see finer detail in a distant astronomical object: red, blue, or ultraviolet? Explain your answer.

Q36.11 A typical telescope used by amateur astronomers has a mirror 20 cm in diameter. With such a telescope (and a filter to cut the intensity of sunlight to a safe level for viewing), fine details can be seen on the surface of the sun. Explain why a *radio* telescope would have to be *much* larger to "see" comparable details on the sun.

Q36.12 Could x-ray diffraction effects with crystals be observed by using visible light instead of x rays? Why or why not?

Q36.13 Why is a diffraction grating better than a two-slit setup for measuring wavelengths of light?

Q36.14 One sometimes sees rows of evenly spaced radio antenna towers. A student remarked that these act like diffraction gratings. What did she mean? Why would one want them to act like a diffraction grating?

Q36.15 If a hologram is made using 600-nm light and then viewed with 500-nm light, how will the images look compared to those observed when viewed with 600-nm light? Explain.

Q36.16 A hologram is made using 600-nm light and is then viewed by using white light from an incandescent bulb. What will be seen? Explain.

Q36.17 Ordinary photographic film reverses black and white, in the sense that the most brightly illuminated areas become blackest upon development (hence the term *negative*). Suppose a hologram negative is viewed directly, without making a positive transparency. How will the resulting images differ from those obtained with the positive? Explain.

Exercises

Section 36.2 Diffraction from a Single Slit

36.1 Monochromatic light from a distant source is incident on a slit 0.750 mm wide. On a screen 2.00 m away, the distance from the central maximum of the diffraction pattern to the first minimum is measured to be 1.35 mm. Calculate the wavelength of the light.

36.2 Parallel rays of green mercury light with a wavelength of 546 nm pass through a slit covering a lens with a focal length of 60.0 cm. In the focal plane of the lens the distance from the central maximum to the first minimum is 10.2 mm. What is the width of the slit?

36.3 In the situation described in Example 36.1 (Section 36.2), what is the angle θ of the first minimum (dark fringe)?

36.4 Light of wavelength 633 nm from a distant source is incident on a slit 0.750 mm wide, and the resulting diffraction pattern is observed on a screen 3.50 m away. What is the distance between the two dark fringes on either side of the central bright fringe?

36.5 Diffraction occurs for all types of waves, including sound waves. High-frequency sound from a distant source with wavelength 9.00 cm passes through a narrow slit 12.0 cm wide. A microphone is placed 40.0 cm directly in front of the center of the slit, corresponding to point O in Fig. 36.5a. The microphone is then moved in a direction perpendicular to the line from the center of the slit to point O. At what distances from O will the intensity detected by the microphone be zero?

36.6 Monochromatic light of wavelength 580 nm passes through a single slit and the diffraction pattern is observed on a screen. Both the source and screen are far enough from the slit for Fraunhofer diffraction to apply. a) If the first diffraction minima are at $\pm 90.0°$, so the central maximum completely fills the screen, what is the width of the slit? b) For the width of the slit as calculated in part (a), what is the ratio of the intensity at $\theta = 45.0°$ to the intensity at $\theta = 0$?

36.7 Doorway Diffraction. Sound of frequency 1250 Hz leaves a room through a 1.00 m wide doorway (see Exercise 36.5). At which angles relative to the centerline perpendicular to the doorway will someone outside the room hear no sound? Use 344 m/s for the speed of sound in air and assume that the source and listener are both far enough from the doorway for Fraunhofer diffraction to apply. You can ignore effects of reflections.

36.8 Light waves, for which the electric field is given by $E_y(x, t) = E_{max} \sin[(1.20 \times 10^7 \text{ m}^{-1})x - \omega t]$, pass through a slit and produce the first dark bands at $\pm 28.6°$ from the center of the diffraction pattern. a) What is the frequency of this light? b) How wide is the slit? c) At which angles will other dark bands occur?

36.9 Parallel rays of light with wavelength 620 nm pass through a slit covering a lens with a focal length of 40.0 cm. The diffraction pattern is observed in the focal plane of the lens and the distance from the center of the central maximum to the first minimum is 36.5 cm. What is the width of the slit? (*Note:* The angle that locates the first minimum is *not* small.)

36.10 Monochromatic electromagnetic radiation with wavelength λ from a distant source passes through a slit. The diffraction pattern is observed on a screen 2.50 m from the slit. If the width of the central maximum is 6.00 mm, what is the slit width a if the wavelength is a) 500 nm (visible light)? b) 50.0 μm (infrared radiation)? c) 0.500 nm (x rays)?

36.11 Red light of wavelength 633 nm from a helium-neon laser passes through a slit 0.350 mm wide. The diffraction pattern is observed on a screen 3.00 m away. Define the width of a bright fringe as the distance between the minima on either side. a) What is the width of the central bright fringe? b) What is the width of the first bright fringe on either side of the central one?

Section 36.3 Intensity in the Single-Slit Pattern

36.12 Monochromatic light of wavelength $\lambda = 620$ nm from a distant source passes through a slit 0.450 mm wide. The diffraction pattern is observed on a screen 3.00 m from the slit. In terms of the intensity I_0 at the peak of the central maximum, what is the intensity of the light at the screen the following distances from the center of the central maximum: a) 1.00 mm; b) 3.00 mm; c) 5.00 mm?

36.13 A slit 0.240 mm wide is illuminated by parallel light rays of wavelength 540 nm. The diffraction pattern is observed on a screen that is 3.00 m from the slit. The intensity at the center of the central maximum $(\theta = 0°)$ is 6.00×10^{-6} W/m^2. a) What is the distance on the screen from the center of the central maximum to the first minimum? b) What is the intensity at a point on the screen midway between the center of the central maximum and the first minimum?

36.14 A diffraction pattern is formed by passing parallel rays of 600-nm light through a slit 0.150 mm wide. What is the phase angle β (the phase difference between wavelets from the top and bottom of the slit) at a) the center of the central maximum; b) the second minimum out from the central maximum; c) 7.0° from the central maximum?

36.15 In a single-slit diffraction pattern produced by electromagnetic radiation from a distant source passing through a slit 0.320 mm wide, the phase angle β (the phase difference between wavelets from the top and bottom of the slit) is $\pi/2$ rad at an angle of 0.24° from the central maximum. What is the wavelength of the electromagnetic radiation?

36.16 Consider a single-slit diffraction experiment in which the amplitude of the wave at point O in Fig. 36.5a is E_0. For each of the following cases, draw a phasor diagram like that in Fig. 36.7b and determine *graphically* the amplitude of the wave at the point in question. (*Hint:* Use Eq. (36.6) to determine the value of β for each case.) Compute the intensity and compare to Eq. (36.5). a) $\sin\theta = \lambda/2a$; b) $\sin\theta = \lambda/a$; c) $\sin\theta = 3\lambda/2a$.

36.17 A single-slit diffraction pattern is formed by monochromatic electromagnetic radiation from a distant source passing through a slit 0.105 mm wide. At the point in the pattern 3.25° from the center of the central maximum, the total phase difference between wavelets from the top and bottom of the slit is 56.0 rad. a) What is the wavelength of the radiation? b) What is the intensity at this point, if the intensity at the center of the central maximum is I_0?

Section 36.4 Multiple Slits

36.18 Diffraction in an Interference Pattern. Consider the interference pattern produced by two parallel slits with width a and separation d. Let $d = 4a$. a) Ignoring diffraction effects due to the slit width, at what angles θ from the central maximum will the next five maxima in the two-slit interference pattern occur? (Your answer will be in terms of d and the wavelength λ of the light.) b) Now include the effects of diffraction. If the intensity at $\theta = 0$ is I_0, what is the intensity at each of the angles calculated in part (a)? Compare your results to Fig. 36.11c.

36.19 Number of Fringes in a Diffraction Maximum. In Fig. 36.11c the central diffraction maximum contains exactly seven interference fringes, and in this case $d/a = 4$. a) What must the ratio d/a be if the central maximum contains exactly five fringes? b) In the case considered in part (a), how many fringes are contained within the first diffraction maximum on one side of the central maximum?

36.20 An interference pattern is produced by eight parallel and equally spaced, narrow slits. There is an interference minimum when the phase difference ϕ between light from adjacent slits is $\pi/4$. The phasor diagram is given in Fig. 36.13b. For which pairs of slits is there totally destructive interference?

36.21 An interference pattern is produced by light of wavelength 580 nm from a distant source incident on two identical parallel slits separated by a distance (between centers) of 0.530 mm. a) If the slits are very narrow, what would be the angular positions of the first-order and second-order, two-slit, interference maxima? b) Let the slits have width 0.320 mm. In terms of the intensity I_0 at the center of the central maximum, what is the intensity at each of the angular positions in part (a)?

36.22 An interference pattern is produced by light of wavelength 620 nm from a distant source incident on two parallel slits of width 0.280 mm and separated by a distance (between centers) of $d = 0.800$ mm. In terms of the intensity I_0 at the center of the central maximum, what is the intensity at the following angular positions θ in the pattern: a) 0.125 mrad; a) 0.250 mrad; a) 0.300 mrad? ($1 \text{ mrad} = 10^{-3} \text{ rad}$)

36.23 An interference pattern is produced by four parallel and equally spaced narrow slits. By drawing appropriate phasor diagrams, show that there is an interference minimum when the phase difference ϕ from adjacent slits is i) $\pi/2$; ii) π; iii) $3\pi/2$. In each case, for which pairs of slits is there totally destructive interference?

36.24 An interference pattern is produced by two identical parallel slits of width a and separation (between centers) $d = 3a$. Which interference maxima m_i will be missing in the pattern?

36.25 An interference pattern is produced by monochromatic light of wavelength 630 nm from a distant source incident on two identical, parallel slits separated by a distance (between centers) of 0.840 mm. a) It is found that the $m_i = 3$ interference maximum is missing from the pattern because it coincides with the first diffraction minimum. What is the width of each slit? b) Which other interference maxima should also be missing from the pattern? c) If the wavelength is decreased from 630 nm to 420 nm, which interference maxima will be missing from the pattern?

36.26 Laser light of wavelength 500.0 nm illuminates two identical slits, producing an interference pattern on a screen 90.0 cm from the slits. The bright bands are 1.00 cm apart, and the third bright bands on either side of the central maximum are missing in the pattern. Find the width and the separation of the two slits.

Section 36.5 The Diffraction Grating

36.27 If a diffraction grating produces its third-order bright band at an angle of 78.4° for light of wavelength 681 nm, find a) the number of slits per centimeter for the grating; b) the angular location of the first-order and second-order bright bands. c) Will there be a fourth-order bright band? Explain.

36.28 Monochromatic light is at normal incidence on a plane transmission grating. The first-order maximum in the interference pattern is at an angle of 8.94°. What is the angular position of the fourth-order maximum?

36.29 Visible light passes through a diffraction grating that has 900 slits/cm and the interference pattern is observed on a screen that is 2.50 m from the grating. a) Is the angular position of the first-order spectrum small enough for $\sin\theta \approx \theta$ to be a good approximation? b) In the first-order spectrum, maxima for two different wavelengths are separated on the screen by 3.00 mm. What is the difference in these wavelengths?

36.30 Identifying Isotopes by Spectra. Different isotopes of the same element emit light at slightly different wavelengths. A wavelength in the emission spectrum of a hydrogen atom is 656.45 nm; for deuterium, the corresponding wavelength is 656.27 nm. a) What minimum number of slits is required to resolve these two wavelengths in second order? b) If the grating has 500.00 slits/mm, find the angles and angular separation of these two wavelengths in the second order.

36.31 Measuring Wavelengths with a CD. A laser beam of wavelength $\lambda = 632.8$ nm shines at normal incidence on the reflective side of a compact disc. The tracks of tiny pits in which information is coded onto the CD are 1.60 μm apart. For what angles of reflection (measured from the normal) will the intensity of light be maximum?

36.32 a) What is the wavelength of light that is deviated in the first order through an angle of 13.5° by a transmission grating having 5000 slits/cm? b) What is the second-order deviation of this wavelength? Assume normal incidence.

36.33 Plane monochromatic waves with wavelength 520 nm are incident normally on a plane transmission grating having 350 slits/m. Find the angles of deviation in the first, second, and third orders.

36.34 The wavelength range of the visible spectrum is approximately 400 nm to 700 nm. White light falls at normal incidence on a diffraction grating that has 350 slits/mm. Find the angular width of the visible spectrum in a) the first order; b) the third order. (*Note:* An advantage of working in higher orders is the greater angular spread and better resolution. A disadvantage is the overlapping of different orders, as shown in Example 36.4 in Section 36.5.)

36.35 A plane transmission grating is ruled with 4000 slits/cm. Assume normal incidence. The α and β lines emitted by atomic hydrogen have wavelengths 656 nm and 486 nm, respectively. Compute the angular separation in degrees between these lines in a) the first-order spectrum; b) the second-order spectrum.

36.36 The light from an iron arc includes many different wavelengths. Two of these are at $\lambda = 587.9782$ nm and $\lambda = 587.8002$ nm. You wish to resolve these spectral lines in first order using a grating 1.20 cm in length. What minimum number of slits per centimeter must the grating have?

Section 36.6 X-Ray Diffraction

36.37 X rays of wavelength 0.0850 nm are scattered from the atoms of a crystal. The second-order maximum in the Bragg reflection occurs when the angle θ in Fig. 36.21 is 21.5°. What is the spacing between adjacent atomic planes in the crystal?

36.38 Monochromatic x rays are incident on a crystal for which the spacing of the atomic planes is 0.440 nm. The first-order maximum in the Bragg reflection occurs when the incident and reflected x rays make an angle of 39.4° with the crystal planes. What is the wavelength of the x rays?

Section 36.7 Circular Apertures and Resolving Power

36.39 Due to blurring caused by atmospheric distortion, the best resolution that can be obtained by a normal, earth-based, visible-light telescope is about 0.3 arcsecond (there are 60 arcminutes in a degree and 60 arcseconds in an arcminute). a) Using Rayleigh's criterion, calculate the diameter of an earth-based telescope that gives this resolution with 550-nm light. b) Increasing the telescope diameter beyond the value found in part (a) will increase the light-gathering power of the telescope, allowing more distant and dimmer astronomical objects to be studied, but will not improve the resolution. In what ways are the Keck telescopes (each of 10-m diameter) atop Mauna Kea in Hawaii superior to the Hale Telescope (5-m diameter) on Palomar Mountain in California? In what ways are they *not* superior? Explain.

36.40 If you can read the bottom row of your doctor's eye chart, your eye has a resolving power of one arcminute, equal to $\frac{1}{60}$ degree. If this resolving power is diffraction-limited, to what effective diameter of your eye's optical system does this correspond? Use Rayleigh's criterion and assume $\lambda = 550$ nm.

36.41 Two satellites at an altitude of 1200 km are separated by 28 km. If they broadcast 3.6-cm microwaves, what minimum receiving dish diameter is needed to resolve (by Rayleigh's criterion) the two transmissions?

36.42 The VLBA can resolve (by Rayleigh's criterion) signals from sources separated by 1.0×10^{-8} rad. If the effective diameter of the receiver is 8000 km, what is the wavelength of the signal?

36.43 Monochromatic light with wavelength 620 nm passes through a circular aperture with diameter 7.4 μm. The resulting diffraction pattern is observed on a screen that is 4.5 m from the aperture. What is the diameter of the Airy disk on the screen?

36.44 Figure 36.31 shows two point sources of light, a and b, that are 25.0 m from lens L and 8.00 mm apart. The two sources produce images at c that are just resolved according to Rayleigh's criterion. The focal length of the lens is 180 mm. What is the diameter of the diffraction circles at c? (Note that the figure is not to scale.)

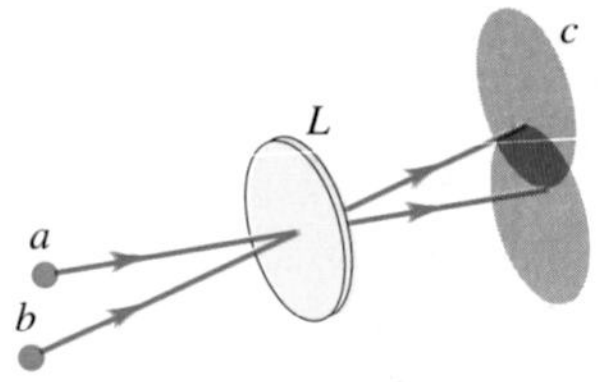

Figure 36.31 Exercise 36.44.

36.45 Observing Jupiter. You are asked to design a space telescope for earth orbit. When Jupiter is 5.93×10^8 km away (its closest approach to the earth), the telescope is to resolve, by Rayleigh's criterion, features on Jupiter that are 250 km apart. What minimum diameter mirror is required? Assume a wavelength of 500 nm.

36.46 A converging lens 7.20 cm in diameter has a focal length of 300 mm. If the resolution is diffraction-limited, how far away can an object be if points on it 4.00 mm apart are to be resolved (according to Rayleigh's criterion)? Use $\lambda = 550$ nm.

36.47 Hubble vs. Arecibo. The Hubble Space Telescope has an aperture of 2.4 m and focuses visible light (400 nm to 700 nm). The Arecibo radio telescope in Puerto Rico is 305 m (1000 ft) in diameter (it is built in a mountain valley) and focuses radio waves of wavelength 75 cm. a) Under optimal viewing conditions, what is the smallest crater that each of these telescopes could resolve on our moon? b) If the Hubble telescope were to be converted to surveillance use, what is the highest orbit above the surface of the earth it could have and still be able to resolve the license plate (not the letters, just the plate) of a car on the ground? Assume optimal viewing conditions, so that the resolution is diffraction limited.

36.48 Searching for Starspots. The Hale telescope on Palomar Mountain, California has a mirror 200 in. (5.08 m) in diameter and it focuses visible light. Given that a large sunspot is about 10,000 mi in diameter, what is the most distant star on which this telescope could resolve a sunspot to see if other stars have them? (Assume optimal viewing conditions, so that the resolution is diffraction limited.) Are there any stars this close to us, besides our Sun?

36.49 Searching for Planets. The Keck telescope, on Mauna Kea, Hawaii has a 10.0-m diameter mirror. Could this telescope resolve Jupiter-sized planets about our nearest star, Alpha Centauri, that is 4.28 light-years away?

Problems

36.50 Thermal Effects. A metal sheet has a thin slit in it. When light of wavelength 500 nm shines through the slit, the central bright region on a screen 8.00 m away is 10.0 cm wide. When the temperature of the sheet is increased by 80.0 C°, the width of the central bright region changes by 0.0020 cm. a) Does the central bright region get wider or narrower when the sheet is heated? b) What is the coefficient of linear expansion for the metal sheet?

36.51 Consider a single-slit diffraction pattern. The center of the central maximum, where the intensity is I_0, is located at $\theta = 0$. a) Let θ_+ and θ_- be the two angles on either side of $\theta = 0$ for which $I = \frac{1}{2}I_0$. $\Delta\theta = |\theta_+ - \theta_-|$ is called the *full width at half maximum,* or *FWHM,* of the central diffraction maximum. Solve for $\Delta\theta$ when the ratio between the slit width a and wavelength λ is i) $a/\lambda = 2$; ii) $a/\lambda = 5$; iii) $a/\lambda = 10$. (*Hint:* Your equation for θ_+ or θ_- cannot be solved analytically. You must use trial and error or solve it graphically.) b) The width of the central maximum can alternatively be defined as $2\theta_0$, where θ_0 is the angle that locates the minimum on one side of the central maximum. Calculate $2\theta_0$ for each case considered in part (a), and compare to $\Delta\theta$.

36.52 Suppose the entire apparatus (slit, screen, and space in between) in Exercise 36.4 is immersed in water $(n = 1.33)$. Then what is the distance between the two dark fringes?

36.53 Measuring Refractive Index. A thin slit illuminated by light of frequency f produces its first dark band at $\pm 38.2°$ in air. When the entire apparatus (slit, screen and space in between) is immersed in an unknown transparent liquid, the slit's first dark bands occur instead at $\pm 17.4°$. Find the refractive index of the liquid.

36.54 Grating Design. Your boss asks you to design a diffraction grating which will disperse the first-order visible spectrum through an angular range of 15.0° (see Example 36.4). a) What must the number of slits per centimeter be for this grating? b) At what angles will the first-order visible spectrum begin and end?

36.55 A slit 0.360 mm wide is illuminated by parallel rays of light that have a wavelength of 540 nm. The diffraction pattern is observed on a screen that is 1.20 m from the slit. The intensity at the center of the central maximum $(\theta = 0°)$ is I_0. a) What is the distance on the screen from the center of the central maximum to the first minimum? b) What is the distance on the screen from the center of the central maximum to the point where the intensity has fallen to $I_0/2$? (See Problem 36.51a for a hint about how to solve for the phase angle β.)

36.56 The intensity of light in the Fraunhofer diffraction pattern of a single slit is

$$I = I_0\left(\frac{\sin\gamma}{\gamma}\right)^2$$

where

$$\gamma = \frac{\pi a \sin\theta}{\lambda}$$

a) Show that the equation for the values of γ at which I is a maximum is $\tan\gamma = \gamma$. b) Determine the three smallest positive values of γ that are solutions of this equation. (*Hint:* You can use a trial-and-error procedure. Guess a value of γ and adjust your guess to bring $\tan\gamma$ closer to γ. A graphical solution of the equation is very helpful in locating the solutions approximately, to get good initial guesses.)

36.57 Angular Width of a Principal Maximum. Consider N evenly spaced, narrow slits. Use the small-angle approximation $\sin\theta = \theta$ (for θ in radians) to prove the following: For an intensity maximum that occurs at an angle θ, the intensity minima immediately adjacent to this maximum are at angles $\theta + \lambda/Nd$ and $\theta - \lambda/Nd$, so that the angular width of the principal maximum is $2\lambda/Nd$. This is proportional to $1/N$, as we concluded in Section 36.4 on the basis of energy conservation.

36.58 Calculating the Intensity for Three Slits. An opaque sheet has three narrow, parallel slits width a distance d between adjacent slits. Plane waves of light with wavelength λ are incident on the sheet, and the light that passes through the slits produces a Fraunhofer pattern on a distant screen. a) Let E_0 be the electric field amplitude of the wave reaching the screen from each individual slit, and let ϕ be the phase difference between waves from adjacent slits arriving at a point P on the screen. Draw a phasor diagram for the *total* wave at P, and show that the amplitude of this total wave is $E_P = E_0(1 + 2\cos\phi)$. (*Hint:* The three phasors of length E_0 and the total phasor of length E_P form a trapezoid.) b) Calculate the intensity measured on the screen as a function of d, λ, and the angle θ (the angle between the straight-ahead direction and a line extending from the center slit to point P.) Graph the intensity as a function of ϕ. c) Show that the intensity function calculated in part (b) has the following properties: i) The intensity at the principal maximum is nine times greater than the intensity from a single slit. ii) The locations of the principal maxima are given by $d\sin\theta = m\lambda$, where $m = 0, \pm 1, \pm 2, \ldots$, corresponding to values of ϕ that are integral multiples of 2π, just as in the two-slit pattern. iii) There are $(3 - 1) = 2$ minima between each pair of principal maxima. iv) The minima are located at $d\sin\theta = \pm\lambda/3, \pm 2\lambda/3, \pm 4\lambda/3, \pm 5\lambda/3, \ldots$, corresponding to values of ϕ that are integral multiples of $2\pi/3$ but not integral multiples of 2π. v) Between each pair of principal maxima there is a *secondary* maximum at which the intensity is only $\frac{1}{9}$ the value at a principal maximum. Show that these conclusions are in agreement with the general statements about N slits made in Section 36.5. d) Draw the phasor diagram for the following cases: i) $\phi = 2\pi/3$; ii) $\phi = \pi$; iii) $\phi = 4\pi/3$; iv) $\phi = 2\pi$. State which of these cases are intensity maxima, which are minima, and which are neither.

36.59 Phasor Diagram for Eight Slits. An interference pattern is produced by eight equally spaced, narrow slits. Figure 36.13 shows phasor diagrams for the cases in which the phase difference ϕ between light from adjacent slits is $\phi = \pi$, $\phi = \pi/4$, and $\phi = \pi/2$. Each of these cases gives an intensity minimum. The caption for Fig. 36.13 also claims that a minimum occurs for

$\phi = 3\pi/4$, $\phi = 5\pi/4$, $\phi = 3\pi/2$, and $\phi = 7\pi/4$. a) Draw the phasor diagram for each of these four cases, and explain why each diagram proves that there is in fact a minimum. (*Note:* You may find it helpful to use a different colored pencil for each slit!) b) For each of the four cases $\phi = 3\pi/4$, $\phi = 5\pi/4$, $\phi = 3\pi/2$, and $\phi = 7\pi/4$, for which pairs of slits is there totally destructive interference?

36.60 An interference pattern is produced by six equally spaced, narrow slits. a) Draw the phasor diagram for the case in which the phase difference between light from adjacent slits is zero. In this case, if the electric field amplitude due to each individual slit is E, what is the resultant amplitude due to all six slits? If the intensity of light through each slit is I, what is the resultant intensity? b) Repeat part (a) for the case in which $\phi = 2\pi$. c) For what values of ϕ between 0 and 2π is there an interference minimum? For each ϕ draw the phasor diagram.

36.61 At the end of Section 36.4, the following statements were made about an array of N slits. Explain, using phasor diagrams, why each statement is true. a) A minimum occurs whenever ϕ is an integral multiple of $2\pi/N$, except when ϕ is an integral multiple of 2π (which gives a principal maximum). b) There are $(N - 1)$ minima between each pair of principal maxima.

36.62 In Eq. (36.12), consider the case in which $d = a$. In a sketch, show that in this case the two slits reduce to a single slit with width $2a$. Then show that Eq. (36.12) reduces to Eq. (36.5) with slit width $2a$.

36.63 What is the longest wavelength that can be observed in the third order for a transmission grating having 6500 slits/cm? Assume normal incidence.

36.64 a) Figure 36.15 shows plane waves of light incident *normally* on a diffraction grating. If instead the light strikes the grating at an angle of incidence θ' (measured from the normal), show that the condition for an intensity maximum is *not* Eq. (36.13), but rather

$$d(\sin\theta + \sin\theta') = m\lambda \quad (m = 0, \pm1, \pm2, \pm3, \dots)$$

b) For the grating described in Example 36.4 (Section 36.5), with 600 slits per millimeter, find the angles of the maxima corresponding to $m = 0, 1$, and -1 with red light $(\lambda = 650\text{ nm})$ for the cases $\theta' = 0$ (normal incidence) and $\theta' = 20.0°$.

36.65 A diffraction grating has 650 slits/mm. What is the highest order that contains the entire visible spectrum? (The wavelength range of the visible spectrum is approximately 400 nm to 700 nm.)

36.66 X-Ray Diffraction of Salt. X rays with a wavelength of 0.125 nm are scattered from a cubic array (of a sodium chloride crystal), for which the spacing of adjacent atoms is $a = 0.282$ nm. a) If diffraction from planes parallel to a cube face is considered, at what angles θ of the incoming beam relative to the crystal planes will maxima be observed? b) Repeat part (a) for diffraction produced by the planes shown in Fig. 36.22b, which are separated by $a/\sqrt{2}$.

36.67 Phased-Array Radar. In one common type of radar installation, a rotating antenna sweeps a radio beam around the sky. But in a *phased-array* radar system, the antennas remain stationary and the beam is swept electronically. To see how this is done, consider an array of N antennas that are arranged along the horizontal x-axis at $x = 0, \pm d, \pm 2d, \dots, \pm(N-1)d/2$. (The number N is odd.) Each antenna emits radiation uniformly in all directions in the horizontal xy-plane. The antennas all emit radiation coherently, with the same amplitude E_0 and the same wavelength λ. The relative phase δ of the emission from adjacent antennas can be varied, however. If the antenna at $x = 0$ emits a signal that is given by $E_0 \cos \omega t$, as measured at a point next to the antenna, the antenna at $x = d$ emits a signal given by $E_0 \cos(\omega t + \delta)$, as measured at a point next to that antenna. The corresponding quantity for the antenna at $x = -d$ is $E_0 \cos(\omega t - \delta)$; for the antennas at $x = \pm 2d$, it is $E_0 \cos(\omega t \pm 2\delta)$; and so on. a) If $\delta = 0$, the interference pattern at a distance from the antennas is large compared to d and has a principal maximum at $\theta = 0$ (that is, in the $+y$-direction, perpendicular to the line of the antennas). Show that if $d < \lambda$, this is the *only* principal interference maximum in the angular range $-90° < \theta < 90°$. Hence this principal maximum describes a beam emitted in the direction $\theta = 0$. As described in Section 36.4, if N is large, the beam will have a large intensity and be quite narrow. b) If $\delta \neq 0$, show that the principal intensity maximum described in part (a) is located at

$$\theta = \arcsin\left(\frac{\delta\lambda}{2\pi d}\right)$$

where δ is measured in radians. Thus by varying δ from positive to negative values and back again, which can easily be done electronically, the beam can be made to sweep back and forth around $\theta = 0$. c) A weather radar unit to be installed on an airplane emits radio waves at 8800 MHz. The unit uses 15 antennas in an array 28.0 cm long (from the antenna at one end of the array to the antenna at the other end). What must the maximum and minimum values of δ be (that is, the most positive and most negative values) if the radar beam is to sweep 45° to the left or right of the airplane's direction of flight? Give your answer in radians.

36.68 Resolution of Radio and Optical Telescopes. Diffraction occurs for circular antennas and for circular mirrors, just as for circular apertures. What is the minimum angular separation θ of two sources of electromagnetic radiation that can be resolved (according to Rayleigh's criterion) by a) the 8.3-m-diameter Subaru reflecting telescope atop Mauna Kea, Hawaii, for a wavelength of 500 nm; b) the 305-m-diameter radio telescope at Arecibo, Puerto Rico, for a wavelength of 21.0 cm? (This shows that radio telescopes must be much larger than optical ones to achieve the same diffraction-limited resolution.)

36.69 A telescope is used to observe two distant point sources 2.50 m apart with light of wavelength 600 nm. The objective of the telescope is covered with a slit of width 0.350 mm. What is the maximum distance in meters at which the two sources may be distinguished if the resolution is diffraction limited and Rayleigh's criterion is used?

36.70 *Quasars,* an abbreviation for *quasi-stellar radio sources,* are distant objects that look like stars through a telescope but that emit far more electromagnetic radiation than an entire normal galaxy of stars. An example is the bright object below and to the left of center in Fig. 36.32; the other elongated objects in this image are normal galaxies. The leading model for the structure of a quasar is a galaxy with a supermassive black hole at its center. In this model, the radiation is emitted by interstellar gas and dust within the galaxy as this material falls toward the black hole. The radiation is thought to emanate from a region just a few light years in diameter. (The diffuse glow surrounding the bright quasar shown in Fig.

36.32 is thought to be this quasar's host galaxy.) To investigate this model of quasars and to study other exotic astronomical objects, the Russian Space Agency plans to place a radio telescope in an orbit that extends to 77,000 km from the earth. When the signals from this telescope are combined with signals from the ground-based telescopes of the VLBA, the resolution will be that of a single radio telescope 77,000 km in diameter. What is the size of the smallest detail that this arrangement could resolve in the quasar 3C 405, which is 7.2×10^8 light years from earth, using radio waves at a frequency of 1665 MHz? (*Hint:* Use Rayleigh's criterion.) Give your answer in light years and in kilometers.

Figure 36.32 Problem 36.70.

36.71 An astronaut in orbit can just resolve two point sources on the earth that are 75.0 m apart. Assume that the resolution is diffraction-limited and use Rayleigh's criterion. What is the astronaut's altitude above the earth? Treat her eye as a circular aperture with a diameter of 4.00 mm (the diameter of her pupil), and take the wavelength of the light to be 500 nm.

36.72 Observing Planets Beyond Our Solar System. NASA is considering a project called *Planet Imager* that would give astronomers the ability to see details on planets orbiting other stars. Using the same principle as the Very Long Baseline Array (see Section 36.7), *Planet Imager* will use an array of infrared telescopes spread over thousands of kilometers of space. (Visible light would give even better resolution. Unfortunately, at visible wavelengths, stars are so bright that a planet would be lost in the glare. This is less of a problem at infrared wavelengths.) a) If *Planet Imager* has an effective diameter of 6000 km and observes infrared radiation at a wavelength of 10 μm, what is the greatest distance at which it would be able to observe details as small as 250 km across (about the size of the greater Los Angeles area) on a planet? Give your answer in light years (see Appendix E). (*Hint:* Use Rayleigh's criterion.) b) For comparison, consider the resolution of a single infrared telescope in space with a diameter of 1.0 m and that observes 10-μm radiation. What is the size of the smallest details that such a telescope could resolve at the distance of the nearest star to the sun, Proxima Centauri, which is 4.22 light years distant? How does this compare to the diameter of the earth $(1.27 \times 10^4 \text{ km})$? To the average distance from the earth to the sun $(1.50 \times 10^8 \text{ km})$? Would a single telescope of this kind be able to detect the presence of a planet like the earth, in an orbit the size of the earth's orbit, around *any* other star? Explain. c) Suppose *Planet Imager* is used to observe a planet orbiting the star 70 Virginis, that lies 59 light years from our solar system. A planet (though not an earthlike one) has in fact been detected orbiting this star, not by imaging it directly but by observing the slight "wobble" of the star as both it and the planet orbit their common center of mass. What is the size of the smallest details that *Planet Imager* could hope to resolve on the planet of 70 Virginis? How does this compare to the diameter of the planet, assumed to be comparable to that of Jupiter $(1.38 \times 10^5 \text{ km})$? (Although the planet of 70 Virginis is thought to be at least 6.6 times more massive than Jupiter, its radius is probably not too different from that of Jupiter. The reason is that such large planets are thought to be composed primarily of gases, not rocky material, and hence can be greatly compressed by the mutual gravitational attraction of different parts of the planet.)

Challenge Problems

36.73 It is possible to calculate the intensity in the single-slit Fraunhofer diffraction pattern *without* using the phasor method of Section 36.3. Let y' represent the position of a point within the slit of width a in Fig. 36.5a, with $y' = 0$ at the center of the slit so that the slit extends from $y' = -a/2$ to $y' = a/2$. We imagine dividing the slit up into infinitesimal strips of width dy', each of which acts as a source of secondary wavelets. a) The amplitude of the total wave at the point O on the distant screen in Fig. 36.5a is E_0. Explain why the amplitude of the wavelet from each infinitesimal strip within the slit is $E_0(dy'/a)$, so that the electric field of the wavelet a distance x from the infinitesimal strip is $dE = E_0(dy'/a)\sin(kx - \omega t)$. b) Explain why the wavelet from each strip as detected at point P in Fig. 36.5a can be expressed as

$$dE = E_0\frac{dy'}{a}\sin[k(D - y'\sin\theta) - \omega t]$$

where D is the distance from the center of the slit to point P and $k = 2\pi/\lambda$. c) By integrating the contributions dE from all parts of the slit, show that the total wave detected at point P is

$$E = E_0\sin(kD - \omega t)\frac{\sin[ka(\sin\theta)/2]}{ka(\sin\theta)/2}$$

$$= E_0\sin(kD - \omega t)\frac{\sin[\pi a(\sin\theta)/\lambda]}{\pi a(\sin\theta)/\lambda}$$

(The trigonometric identities in Appendix B will be useful.) Show that at $\theta = 0$, corresponding to point O in Fig. 36.5a, the wave is $E = E_0\sin(kD - \omega t)$ and has amplitude E_0, as stated in part (a).

d) Use the result of part (c) to show that if the intensity at point O is I_0, then the intensity at a point P is given by Eq. (36.7).

36.74 Intensity Pattern of *N* Slits. a) Consider an arrangement of N slits with a distance d between adjacent slits. The slits emit coherently and in phase at wavelength λ. Show that at a time t, the electric field at a distant point P is

$$E_P(t) = E_0 \cos(kR-\omega t) + E_0 \cos(kR - \omega t + \phi) + E_0 \cos(kR - \omega t + 2\phi) + \cdots + E_0 \cos(kR - \omega t + (N-1)\phi)$$

where E_0 is the amplitude at P of the electric field due to an individual slit, $\phi = (2\pi d \sin\theta)/\lambda$, θ is the angle of the rays reaching P (as measured from the perpendicular bisector of the slit arrangement), and R is the distance from P to the most distant slit. In this problem, assume that R is much larger than d. b) To carry out the sum in part (a), it is convenient to use the complex-number relationship

$$e^{iz} = \cos z + i \sin z$$

where $i = \sqrt{-1}$. In this expression, $\cos z$ is the *real part* of the complex number e^{iz}, and $\sin z$ is its *imaginary part.* Show that the electric field $E_P(t)$ is equal to the real part of the complex quantity

$$\sum_{n=0}^{N-1} E_0 e^{i(kR-\omega t+n\phi)}$$

c) Using the properties of the exponential function that $e^A e^B = e^{(A+B)}$ and $(e^A)^n = e^{nA}$, show that the sum in part (b) can be written as

$$E_0\left(\frac{e^{iN\phi} - 1}{e^{i\phi} - 1}\right)e^{i(kR-\omega t)} = E_0\left(\frac{e^{iN\phi/2} - e^{-iN\phi/2}}{e^{i\phi/2} - e^{-i\phi/2}}\right)e^{i[kR-\omega t+(N-1)\phi/2]}$$

Then, using the relationship $e^{iz} = \cos z + i \sin z$, show that the (real) electric field at point P is

$$E_p(t) = \left[E_0 \frac{\sin(N\phi/2)}{\sin(\phi/2)}\right] \cos[kR - \omega t + (N-1)\phi/2]$$

The quantity in the first square brackets in this expression is the amplitude of the electric field at P. d) Use the result for the electric field amplitude in part (c) to show that the intensity at an angle θ is

$$I = I_0\left[\frac{\sin(N\phi/2)}{\sin(\phi/2)}\right]^2$$

where I_0 is the maximum intensity for an individual slit. e) Check the result in part (d) for the case $N = 2$. It will help to recall that $\sin 2A = 2 \sin A \cos A$. Explain why your result differs from Eq. (35.10), the expression for the intensity in two-source interference, by a factor of 4. (*Hint:* Is I_0 defined in the same way in both expressions?)

36.75 Intensity Pattern of *N* Slits, continued. Part (d) of the previous problem gives an expression for the intensity in the interference pattern of N identical slits. Use this result to verify the following statements. a) The maximum intensity in the pattern is $N^2 I_0$. b) The principal maximum at the center of the pattern extends from $\phi = -2\pi/N$ to $\phi = 2\pi/N$, so its width is inversely proportional to $1/N$. c) A minimum occurs whenever ϕ is an integral multiple of $2\pi/N$, except when ϕ is an integral multiple of 2π (which gives a principal maximum). d) There are $(N - 1)$ minima between each pair of principal maxima. e) Halfway between two principal maxima, the intensity can be no greater than I_0; that is, it can be no greater than $1/N^2$ times the intensity at a principal maximum.

RELATIVITY 37

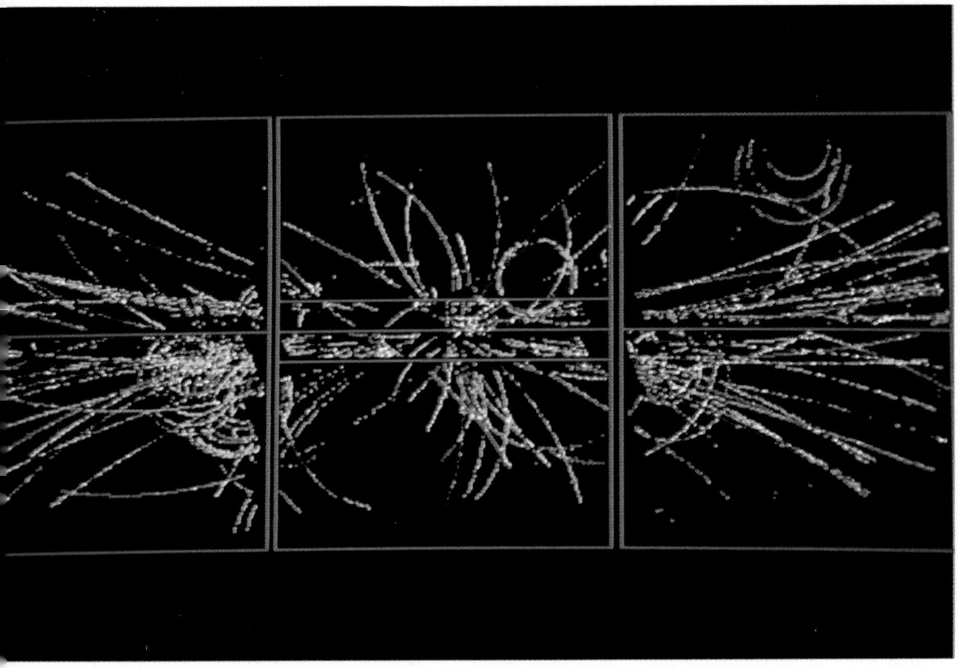

One of the many remarkable predictions of the special theory of relativity is that energy can be converted into mass, and vice versa. This image of a collision between a proton and an antiproton (essentially a proton with a negative charge) provides dramatic confirmation of this idea. The kinetic energies of the proton and antiproton—both of which were moving at nearly the speed of light—were converted into a spray of new particles, each of which is shown by a colored trail.

? **The speed of light is an upper limit on the speed of any particle. Does that mean there is an upper limit on the kinetic energy of a particle?**

When the year 1905 began, Albert Einstein was an unknown 25-year-old clerk in the Swiss patent office. By the end of that amazing year he had published three papers of extraordinary importance. One was an analysis of Brownian motion; a second (for which he was awarded the Nobel Prize) was on the photoelectric effect. In the third, Einstein introduced his **special theory of relativity,** proposing drastic revisions in the Newtonian concepts of space and time.

The special theory of relativity has made wide-ranging changes in our understanding of nature, but Einstein based it on just two simple postulates. One states that the laws of physics are the same in all inertial frames of reference; the other states that the speed of light in vacuum is the same in all inertial frames. These innocent-sounding propositions have far-reaching implications. Here are three: (1) Events that are simultaneous for one observer may not be simultaneous for another. (2) When two observers moving relative to each other measure a time interval or a length, they may not get the same results. (3) For the conservation principles for momentum and energy to be valid in all inertial systems, Newton's second law and the equations for momentum and kinetic energy have to be revised.

Relativity has important consequences in *all* areas of physics, including electromagnetism, atomic and nuclear physics, and high-energy physics. Although many of the results derived in this chapter may run counter to your intuition, the theory is in solid agreement with experimental observations.

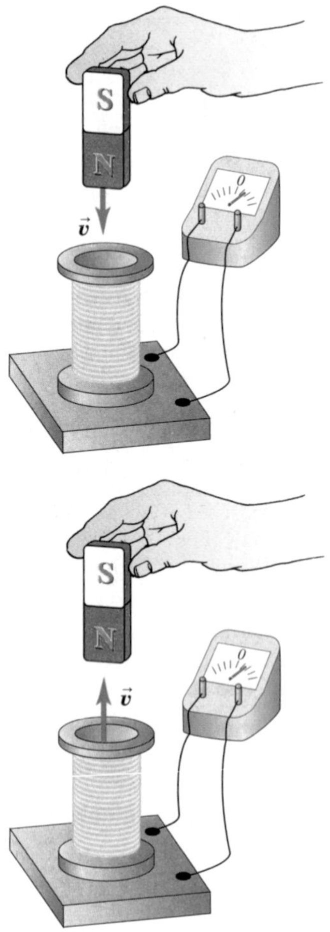

37.1 The same emf is induced in the coil whether (a) the magnet moves relative to the coil or (b) the coil moves relative to the magnet.

37.1 | Invariance of Physical Laws

Let's take a look at the two postulates that make up the special theory of relativity. Both postulates describe what is seen by an observer in an *inertial frame of reference,* which we introduced in Section 4.2. The theory is "special" in the sense that it applies to observers in such special reference frames.

Einstein's First Postulate

Einstein's first postulate, called the **principle of relativity,** states: **The laws of physics are the same in every inertial frame of reference.** If the laws differed, that difference could distinguish one inertial frame from the others or make one frame somehow more "correct" than another. Here are two examples. Suppose you watch two children playing catch with a ball while the three of you are aboard a train moving with constant velocity. Your observations of the motion *of the ball,* no matter how carefully done, can't tell you how fast (or whether) the train is moving. This is because Newton's laws of motion are the same in every inertial frame.

Another example is the electromotive force (emf) induced in a coil of wire by a nearby moving permanent magnet. In the frame of reference in which the *coil* is stationary (Fig. 37.1a), the moving magnet causes a change of magnetic flux through the coil, and this induces an emf. In a different frame of reference in which the *magnet* is stationary (Fig. 37.1b), the motion of the coil through a magnetic field induces the emf. According to the principle of relativity, both of these frames of reference are equally valid. Hence the same emf must be induced in both situations shown in Fig. 37.1. As we saw in Chapter 29, this is indeed the case, so Faraday's law is consistent with the principle of relativity. Indeed, *all* of the laws of electromagnetism are the same in every inertial frame of reference.

Equally significant is the prediction of the speed of electromagnetic radiation, derived from Maxwell's equations (Section 32.2). According to this analysis, light and all other electromagnetic waves travel in vacuum with a constant speed, now defined to equal exactly 299,792,458 m/s. (We often use the approximate value $c = 3.00 \times 10^8$ m/s, which is within one part in 1000 of the exact value.) As we will see, the speed of light in vacuum plays a central role in the theory of relativity.

Einstein's Second Postulate

During the nineteenth century, most physicists believed that light traveled through a hypothetical medium called the *ether,* just as sound waves travel through air. If so, the speed of light measured by observers would depend on their motion relative to the ether and would therefore be different in different directions. The Michelson-Morley experiment, described in Section 35.5, was an effort to detect motion of the earth relative to the ether. Einstein's conceptual leap was to recognize that if Maxwell's equations are valid in all inertial frames, then the speed of light in vacuum should also be the same in all frames and in all directions. In fact, Michelson and Morley detected *no* ether motion across the earth, and the ether concept has been discarded. Although Einstein may not have known about this negative result, it supported his bold hypothesis of the constancy of the speed of light in vacuum.

Einstein's second postulate states: **The speed of light in vacuum is the same in all inertial frames of reference and is independent of the motion of the**

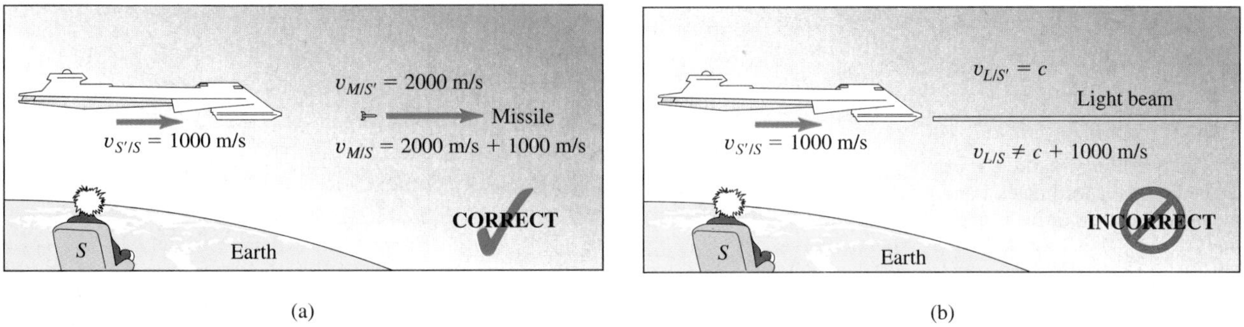

37.2 A spaceship (S') moves with speed $v_{S'/S} = 1000$ m/s relative to an observer on earth (S). (a) A missile (M) is fired with speed $v_{M/S'} = 2000$ m/s relative to the spaceship. Newtonian mechanics tells us correctly that the missile moves with speed $v_{M/S} = 3000$ m/s relative to the observer on earth. (b) Newtonian mechanics also tells us that the light beam (L) emitted by the spaceship moves at a speed greater than c relative to the earth. This is incorrect because it contradicts Einstein's second postulate.

source. Let's think about what this means. Suppose two observers measure the speed of light in vacuum. One is at rest with respect to the light source, and the other is moving away from it. Both are in inertial frames of reference. According to the principle of relativity, the two observers must obtain the same result, despite the fact that one is moving with respect to the other.

If this seems too easy, consider the following situation. A spacecraft moving past the earth at 1000 m/s fires a missile straight ahead with a speed of 2000 m/s (relative to the spacecraft) (Fig. 37.2). What is the missile's speed relative to the earth? Simple, you say; this is an elementary problem in relative velocity (Section 3.5). The correct answer, according to Newtonian mechanics, is 3000 m/s. But now suppose the spacecraft turns on a searchlight, pointing in the same direction in which the missile was fired. An observer on the spacecraft measures the speed of light emitted by the searchlight and obtains the value c. According to Einstein's second postulate, the motion of the light after it has left the source cannot depend on the motion of the source. So the observer on earth who measures the speed of this same light must also obtain the value c, *not* $c + 1000$ m/s. This result contradicts our elementary notion of relative velocities, and it may not appear to agree with common sense. But "common sense" is intuition based on everyday experience, and this does not usually include measurements of the speed of light.

Einstein's second postulate immediately implies the following result: *It is impossible for an inertial observer to travel at c, the speed of light in vacuum.* We can prove this by showing that travel at c implies a logical contradiction. Suppose that the spacecraft S' in Fig. 37.2b is moving at the speed of light relative to an observer on the earth, so that $v_{S'/S} = c$. If the spacecraft turns on a headlight, the second postulate now asserts that the earth observer S measures the headlight beam to be also moving at c. Thus this observer measures that the headlight beam and the spacecraft move together and are always at the same point in space. But Einstein's second postulate also asserts that the headlight beam moves at a speed c relative to the spacecraft, so they *cannot* be at the same point in space. This contradictory result can be avoided only if it is impossible for an inertial observer, such as a passenger on the spacecraft, to move at c. As we go through our discussion of relativity, you may find yourself asking the question Einstein asked himself as a 16-year-old student, "What would I see if I were traveling at the speed of light?" Einstein realized only years later that his question's basic flaw was that he could *not* travel at c.

Let's restate this argument symbolically, using two inertial frames of reference, labeled S for the observer on earth and S' for the moving spacecraft, as shown in Fig. 37.3. To keep things as simple as possible, we have omitted the z-axes. The x-axes of the two frames lie along the same line, but the origin O' of frame S'

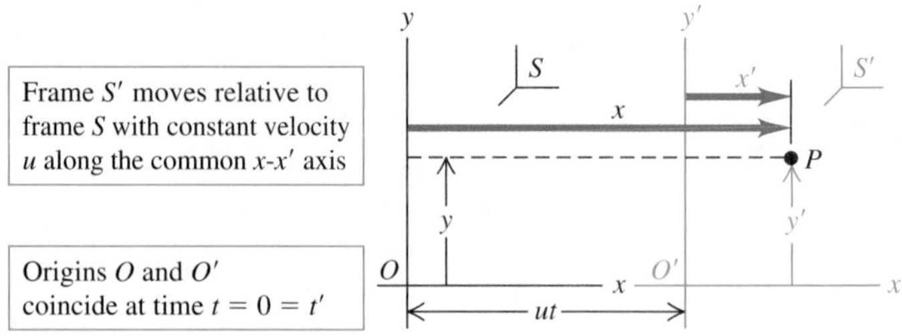

37.3 The position of particle P can be described by the coordinates x and y in frame of reference S or by x' and y' in frame S'.

moves relative to the origin O of frame S with constant velocity u along the common x-x'-axis. We on earth set our clocks so that the two origins coincide at time $t = 0$, so their separation at a later time t is ut.

CAUTION Many of the equations derived in this chapter are true *only* if you define your inertial reference frames as stated in the preceding paragraph. For instance, the positive x-direction must be the direction in which the origin O' moves relative to the origin O. In Fig. 37.3 this direction is to the right; if instead O' moves to the left relative to O, you must define the positive x-direction to be to the left.

Now think about how we describe the motion of a particle P. This might be an exploratory vehicle launched from the spacecraft or a pulse of light from a laser. We can describe the *position* of this particle by using the earth coordinates (x, y, z) in S or the spacecraft coordinates (x', y', z') in S'. Figure 37.3 shows that these are simply related by

$$x = x' + ut \qquad y = y' \qquad z = z' \qquad (37.1)$$
(Galilean coordinate transformation)

These equations, based on the familiar Newtonian notions of space and time, are called the **Galilean coordinate transformation.**

If particle P moves in the x-direction, its instantaneous velocity v_x as measured by an observer stationary in S is $v_x = dx/dt$. Its velocity v'_x as measured by an observer stationary in S' is $v'_x = dx'/dt$. We can derive a relation between v_x and v'_x by taking the derivative with respect to t of the first of Eqs. (37.1):

$$\frac{dx}{dt} = \frac{dx'}{dt} + u$$

Now dx/dt is the velocity v_x measured in S, and dx'/dt is the velocity v'_x measured in S', so we get the *Galilean velocity transformation* for one-dimensional motion:

$$v_x = v'_x + u \quad \text{(Galilean velocity transformation)} \qquad (37.2)$$

Although the notation differs, this result agrees with our discussion of relative velocities in Section 3.5.

Now here's the fundamental problem. Applied to the speed of light in vacuum, Eq. (37.2) says that $c = c' + u$. Einstein's second postulate, supported subsequently by a wealth of experimental evidence, says that $c = c'$. This is a genuine inconsistency, not an illusion, and it demands resolution. If we accept this postulate, we are forced to conclude that Eqs. (37.1) and (37.2) *cannot* be precisely correct, despite our convincing derivation. These equations have to be modified to bring them into harmony with this principle.

The resolution involves some very fundamental modifications in our kinematic concepts. The first idea to be changed is the seemingly obvious assumption that the observers in frames S and S' use the same *time scale,* formally stated as $t = t'$. Alas, we are about to show that this everyday assumption cannot be correct; the two observers *must* have different time scales. We must define the velocity v' in frame S' as $v' = dx'/dt'$, not as dx'/dt; the two quantities are not the same. The difficulty lies in the concept of *simultaneity,* which is our next topic. A careful analysis of simultaneity will help us develop the appropriate modifications of our notions about space and time.

Test Your Understanding

As a high-speed spaceship flies past you, it fires a strobe light that sends out a pulse of light in all directions. An observer aboard the spaceship measures a spherical wave front that spreads away from the spaceship with the same speed c in all directions. What is the shape of the wave front that *you* measure? Does the wave front that you measure travel at the same speed in all directions? Is the wave front centered on the spaceship?

37.2 | Relativity of Simultaneity

Measuring times and time intervals involves the concept of **simultaneity.** In a given frame of reference, an **event** is an occurrence that has a definite position and time. When you say that you awoke at seven o'clock, you mean that two events (your awakening and your clock showing 7:00) occurred *simultaneously.* The fundamental problem in measuring time intervals is this: In general, two events that are simultaneous in one frame of reference are *not* simultaneous in a second frame that is moving relative to the first, even if both are inertial frames.

This may seem to be contrary to common sense. To illustrate the point, here is a version of one of Einstein's *thought experiments*—mental experiments that follow concepts to their logical conclusions. Imagine a train moving with a speed comparable to c, with uniform velocity (Fig. 37.4). Two lightning bolts strike a passenger car, one near each end. Each bolt leaves a mark on the car and one on the ground at the instant the bolt hits. The points on the ground are labeled A and B in the figure, and the corresponding points on the car are A' and B'. Stanley is stationary on the ground at O, midway between A and B. Mavis is moving with the train at O' in the middle of the passenger car, midway between A' and B'. Both Stanley and Mavis see both light flashes emitted from the points where the lightning strikes.

Suppose the two wave fronts from the lightning strikes reach Stanley at O simultaneously. He knows that he is the same distance from B and A, so Stanley concludes that the two bolts struck B and A simultaneously. Mavis agrees that the two wave fronts reached Stanley at the same time, but she disagrees that the flashes were emitted simultaneously.

Stanley and Mavis agree that the two wave fronts do not reach Mavis at the same time. Mavis at O' is moving to the right with the train, so she runs into the wave front from B' *before* the wave front from A' catches up to her. However, because she is in the middle of the passenger car equidistant from A' and B', her observation is that both wave fronts took the same time to reach her because both

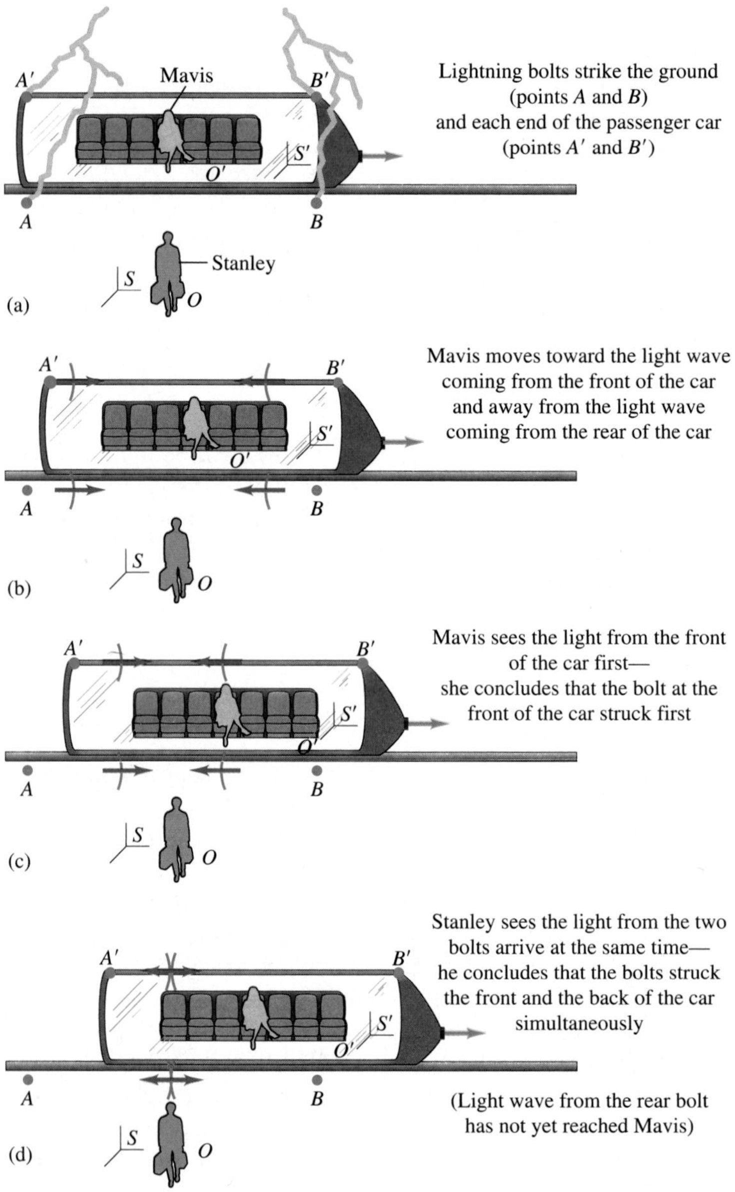

37.4 A thought experiment in simultaneity.

moved the same distance at the same speed c. (Recall that the speed of each wave front with respect to *either* observer is c.) Thus she concludes that the lightning bolt at B' struck *before* the one at A'. Stanley at O measures the two events to be simultaneous, but Mavis at O' does not! *Whether or not two events at different x-axis locations are simultaneous depends on the state of motion of the observer.*

You may want to argue that in this example the lightning bolts really *are* simultaneous and that if Mavis at O' could communicate with the distant points without the time delay caused by the finite speed of light, she would realize this. But that would be erroneous; the finite speed of information transmission is not the real issue. If O' is midway between A' and B', then in her frame of reference the time for a signal to travel from A' to O' is the same as that from B' to O'. Two signals arrive simultaneously at O' only if they were emitted simultaneously at A'

and B'. In this example they *do not* arrive simultaneously at O', and so Mavis must conclude that the events at A' and B' were *not* simultaneous.

Furthermore, there is no basis for saying that Stanley is right and Mavis is wrong, or vice versa. According to the principle of relativity, no inertial frame of reference is more correct than any other in the formulation of physical laws. Each observer is correct *in his or her own frame of reference.* In other words, simultaneity is not an absolute concept. Whether two events are simultaneous depends on the frame of reference. As we mentioned at the beginning of this section, simultaneity plays an essential role in measuring time intervals. It follows that *the time interval between two events may be different in different frames of reference.* So our next task is to learn how to compare time intervals in different frames of reference.

Test Your Understanding

Stanley, who works for the rail system shown in Fig. 37.4, has carefully synchronized the clocks at all of the rail stations. At the moment that Stanley measures all of the clocks striking noon, Mavis is on a high-speed passenger car traveling from Ogdenville toward North Haverbrook. According to Mavis, when the Ogdenville clock strikes noon, what time is it in North Haverbrook: noon, before noon, or afternoon?

37.3 | Relativity of Time Intervals

ActivPhysics Online

17.1 Relativity of Time

To derive a quantitative relation between time intervals in different coordinate systems, let's consider another thought experiment. As before, a frame of reference S' moves along the common x-x' axis with constant speed u relative to a frame S. As discussed in Section 37.1, u must be less than the speed of light c. Mavis, who is riding along with frame S', measures the time interval between two events that occur at the *same* point in space. Event 1 is when a flash of light from a light source leaves O'. Event 2 is when the flash returns to O', having been reflected from a mirror a distance d away, as shown in Fig. 37.5a. We label the time interval Δt_0, using the subscript zero as a reminder that the apparatus is at rest, with zero velocity, in frame S'. The flash of light moves a total distance $2d$, so the time interval is

$$\Delta t_0 = \frac{2d}{c} \tag{37.3}$$

The round-trip time measured by Stanley in frame S is a different interval Δt; in his frame of reference the two events occur at *different* points in space. During the

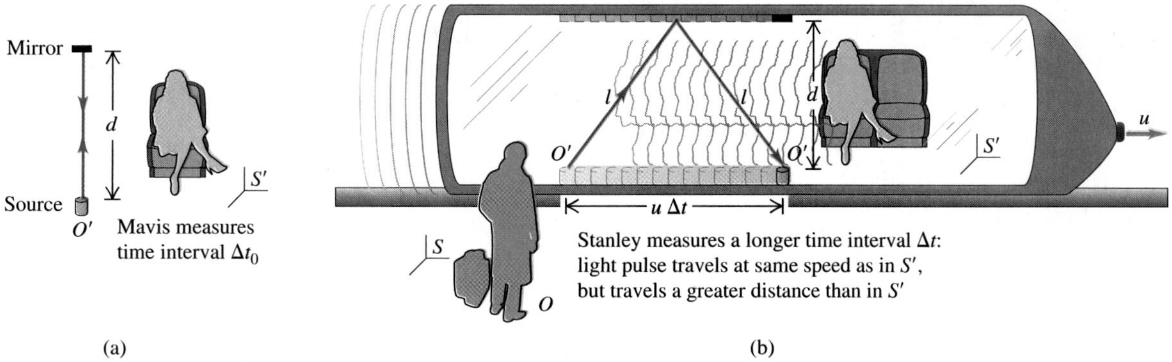

37.5 (a) Mavis, in frame of reference S', observes a light pulse emitted from a source at O' and reflected back along the same line. (b) Stanley, in frame of reference S, sees the same light pulse follow a diagonal path. The positions of O' at the times of departure and return of the pulse are shown.

time Δt, the source moves relative to S a distance $u\,\Delta t$ (Fig. 37.5b). In S' the round-trip distance is $2d$ perpendicular to the relative velocity, but the round-trip distance in S is the longer distance $2l$, where

$$l = \sqrt{d^2 + \left(\frac{u\,\Delta t}{2}\right)^2}$$

In writing this expression, we have assumed that both observers measure the same distance d. We will justify this assumption in the next section. The speed of light is the same for both observers, so the round-trip time measured in S is

$$\Delta t = \frac{2l}{c} = \frac{2}{c}\sqrt{d^2 + \left(\frac{u\,\Delta t}{2}\right)^2} \tag{37.4}$$

We would like to have a relation between Δt and Δt_0 that is independent of d. To get this, we solve Eq. (37.3) for d and substitute the result into Eq. (37.4), obtaining

$$\Delta t = \frac{2}{c}\sqrt{\left(\frac{c\,\Delta t_0}{2}\right)^2 + \left(\frac{u\,\Delta t}{2}\right)^2} \tag{37.5}$$

Now we square this and solve for Δt; the result is

$$\Delta t = \frac{\Delta t_0}{\sqrt{1 - u^2/c^2}}$$

We may generalize this important result. In a particular frame of reference, suppose that two events occur at the same point in space. The time interval between these events, as measured by an observer at rest in this same frame (which we call the *rest frame* of this observer), is Δt_0. Then an observer in a second frame moving with constant speed u relative to the rest frame will measure the time interval to be Δt, where

$$\Delta t = \frac{\Delta t_0}{\sqrt{1 - u^2/c^2}} \quad \text{(time dilation)} \tag{37.6}$$

37.6 This image shows an exploding star, called a *supernova,* within a distant galaxy. The brightness of a typical supernova decays at a certain rate. But supernovae that are moving away from us at a substantial fraction of the speed of light decay more slowly, in accordance with Eq. (37.6): the decaying supernova is a moving "clock" that runs slow.

We recall that no inertial observer can travel at $u = c$ and note that $\sqrt{1 - u^2/c^2}$ is imaginary for $u > c$. Thus Eq. (37.6) gives sensible results only when $u < c$. The denominator of Eq. (37.6) is always smaller than 1, so Δt is always *larger* than Δt_0. Thus we call this effect **time dilation.**

Think of an old-fashioned pendulum clock that has one second between ticks, as measured by Mavis in the clock's rest frame; this is Δt_0. If the clock's rest frame is moving relative to Stanley, he measures a time between ticks Δt that is longer than one second. In brief, *observers measure any clock to run slow if it moves relative to them* (Fig. 37.6). Note that this conclusion is a direct result of the fact that the speed of light in vacuum is the same in both frames of reference.

The quantity $1/\sqrt{1 - u^2/c^2}$ in Eq. (37.6) appears so often in relativity that it is given its own symbol γ (gamma):

$$\gamma = \frac{1}{\sqrt{1 - u^2/c^2}} \tag{37.7}$$

In terms of this symbol, we can express the time dilation formula, Eq. (37.6), as

$$\Delta t = \gamma\,\Delta t_0 \quad \text{(time dilation)} \tag{37.8}$$

As a further simplification, u/c is sometimes given the symbol β (beta); then $\gamma = 1/\sqrt{1 - \beta^2}$.

Figure 37.7 shows a graph of γ as a function of the relative speed u of two frames of reference. When u is very small compared to c, u^2/c^2 is much smaller than 1 and γ is very nearly *equal* to 1. In that limit, Eqs. (37.6) and (37.8) approach the Newtonian relation $\Delta t = \Delta t_0$, corresponding to the same time interval in all frames of reference.

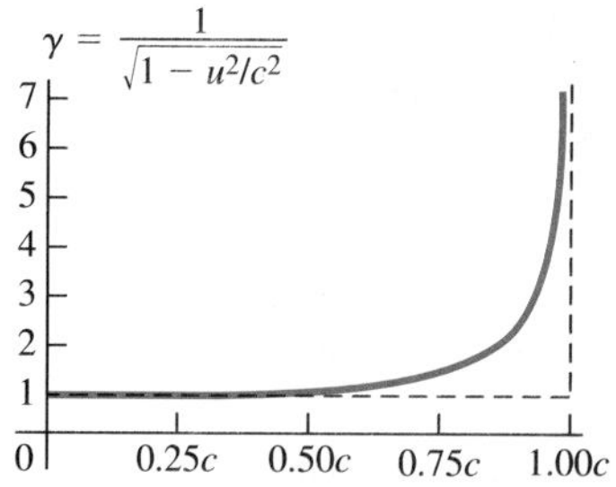

37.7 The quantity $\gamma = 1/\sqrt{1 - u^2/c^2}$ as a function of the relative speed u of two frames of reference.

Proper Time

There is only one frame of reference in which a clock is at rest, and there are infinitely many in which it is moving. Therefore the time interval measured between two events (such as two ticks of the clock) that occur at the same point in a particular frame is a more fundamental quantity than the interval between events at different points. We use the term **proper time** to describe the time interval Δt_0 between two events that occur *at the same point.*

CAUTION It is important to note that the time interval Δt in Eq. (37.6) involves events that occur *at different space points* in the frame of reference S. Note also that any differences between Δt and the proper time Δt_0, are *not* caused by differences in the times required for light to travel from those space points to an observer at rest in S. We assume that our observer is able to correct for differences in light transit times, just as an astronomer who's observing the sun understands that an event seen now on earth actually occurred 500 s ago on the sun's surface. Alternatively, we can use *two* observers, one stationary at the location of the first event and the other at the second, each with his or her own clock. We can synchronize these two clocks without difficulty, as long as they are at rest in the same frame of reference. For example, we could send a light pulse simultaneously to the two clocks from a point midway between them. When the pulses arrive, the observers set their clocks to a prearranged time. (But note that clocks that are synchronized in one frame of reference *are not* in general synchronized in any other frame.)

In thought experiments, it's often helpful to imagine many observers with synchronized clocks at rest at various points in a particular frame of reference. We can picture a frame of reference as a coordinate grid with lots of synchronized clocks distributed around it, as suggested by Fig. 37.8. Only when a clock is moving relative to a given frame of reference do we have to watch for ambiguities of synchronization or simultaneity.

Throughout this chapter we will frequently use phrases like "Stanley *observes* that Mavis passes the point $x = 5.00$ m, $y = 0$, $z = 0$ at time 2.00 s." This means that Stanley is using a grid of clocks in his frame of reference, like the grid shown in Fig. 37.8, to record the time of an event. We could restate the phrase as "When Mavis passes the point at $x = 5.00$ m, $y = 0$, $z = 0$, the clock at that location in Stanley's frame of reference reads 2.00 s." We will avoid using phrases like "Stanley *sees* that Mavis is a certain point at a certain time," because there is a time delay for light to travel to Stanley's eye from the position of an event.

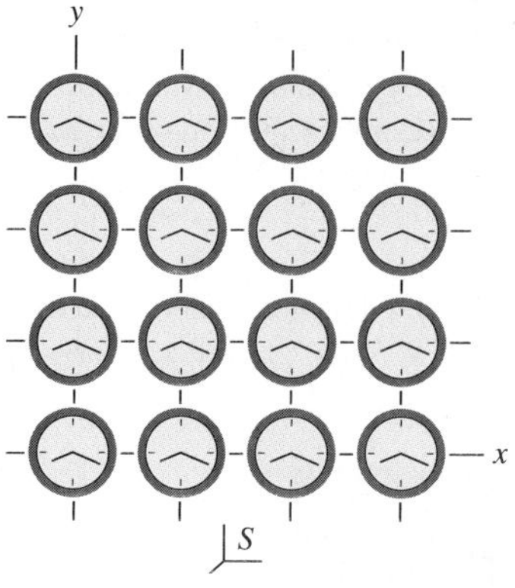

37.8 A frame of reference pictured as a coordinate system with a grid of synchronized clocks. The grid is actually three-dimensional: there are identical planes of clocks that are all parallel to the page, in front of and behind it, connected by grid lines perpendicular to the page.

Problem-Solving Strategy

Time Dilation

IDENTIFY *the relevant concepts:*
The concept of time dilation is used whenever we compare the time interval between events as measured by observers in different inertial frames of reference.

SET UP *the problem using the following steps:*

1. To describe a time interval, you must first decide what two events define the beginning and the end of the interval. You must also identify the two frames of reference in which the time interval is measured.
2. Determine what the target variable is.

EXECUTE *the solution as follows:*

1. In many problems involving time dilation, the time interval between events as measured in one frame of reference is the *proper* time Δt_0. The proper time is the time interval between two events in a frame of reference in which the two events occur at the same point in space. The dilated time Δt is the longer time interval between the same two events as measured in a frame of reference that has a speed u relative to the first frame. The two events occur at different points as measured in the second frame. You will need to decide in which frame the time interval is Δt_0, and in which frame it is Δt.
2. Use Eq. (37.6) or (37.8) to relate Δt_0 and Δt, and then solve for the target variable.

EVALUATE *your answer:*
Note that Δt is never smaller than Δt_0, and u is never greater than c. If your results suggest otherwise, you need to rethink your calculation.

Example 37.1

Time dilation at 0.990*c*

High-energy subatomic particles coming from space interact with atoms in the earth's upper atmosphere, producing unstable particles called *muons*. A muon decays with a mean lifetime of 2.20×10^{-6} s as measured in a frame of reference in which it is at rest. If a muon is moving at $0.990c$ (about 2.97×10^8 m/s) relative to the earth, what will you (an observer on earth) measure its mean lifetime to be?

SOLUTION

IDENTIFY and SET UP: Let S be your frame of reference on earth, and let S' be the muon's frame of reference. The two events in question are the production of the muon and its subsequent decay; the target variable is the interval between these events as measured in S.

EXECUTE: The time interval between the two events as measured in S', 2.20×10^{-6} s, is a *proper* time, since the two events occur at the same position relative to the muon. Hence $\Delta t_0 = 2.20 \times 10^{-6}$ s. The muon moves relative to the earth between the two events, so the two events occur at different positions as measured in S and the time interval in that frame is Δt (the target variable). From Eq. (37.6),

$$\Delta t = \frac{\Delta t_0}{\sqrt{1 - u^2/c^2}} = \frac{2.20 \times 10^{-6}\ \text{s}}{\sqrt{1 - (0.990)^2}} = 15.6 \times 10^{-6}\ \text{s}.$$

EVALUATE: Our result predicts that the mean lifetime of the muon in the earth frame (Δt) is about seven times longer than in the muon's frame (Δt_0). This prediction has been verified experimentally; indeed, it was the first experimental confirmation of the time dilation formula, Eq. (37.6).

Example 37.2

Time dilation at jetliner speeds

An airplane flies from San Francisco to New York (about 4800 km, or 4.80×10^6 m) at a steady speed of 300 m/s (about 670 mi/h). How much time does the trip take, as measured by an observer on the ground? By an observer in the plane?

SOLUTION

IDENTIFY and SET UP: The two events are the airplane departing San Francisco and arriving at New York. The target variables are the time intervals between these events as measured in the frame of reference of the ground S and in the frame of reference of the airplane S'.

EXECUTE: The two events occur at different positions (San Francisco and New York) as measured in S, so the time interval measured by ground observers corresponds to Δt in Eq. (37.6). To find it, we simply divide the distance by the speed:

$$\Delta t = \frac{4.80 \times 10^6\ \text{m}}{300\ \text{m/s}} = 1.60 \times 10^4\ \text{s} \quad (\text{about } 4\tfrac{1}{2}\ \text{hours})$$

In the airplane's frame S', San Francisco and New York passing under the plane occur at the same point (the position of the plane). The time interval in the airplane is a proper time, corresponding to Δt_0 in Eq. (37.6). We have

$$\frac{u^2}{c^2} = \frac{(300\ \text{m/s})^2}{(3.00 \times 10^8\ \text{m/s})^2} = 1.00 \times 10^{-12}$$

From Eq. (37.6),

$$\Delta t_0 = (1.60 \times 10^4\ \text{s})\sqrt{1 - 1.00 \times 10^{-12}}$$

The radical can't be evaluated with adequate precision with an ordinary calculator. But we can approximate it using the binomial theorem (Appendix B):

$$(1 - 1.00 \times 10^{-12})^{1/2} = 1 - \left(\frac{1}{2}\right)(1.00 \times 10^{-12}) + \cdots$$

The remaining terms are of the order of 10^{-24} or smaller and can be discarded. The approximate result for Δt_0 is

$$\Delta t_0 = (1.60 \times 10^4\ \text{s})(1 - 0.50 \times 10^{-12})$$

The proper time Δt_0, measured in the airplane, is very slightly less (by less than one part in 10^{12}) than the time measured on the ground.

EVALUATE: We don't notice such effects in everyday life. But we mentioned in Section 1.3 that present-day atomic clocks can attain a precision of about one part in 10^{13}. A cesium clock traveling a long distance in an airliner has been used to measure this effect and thereby verify Eq. (37.6) even at speeds much less than c.

Example 37.3 Just when is it proper?

Mavis boards a spaceship, then zips past Stanley on earth at a relative speed of $0.600c$. At the instant she passes, both start timers. a) At the instant when Stanley measures that Mavis has traveled 9.00×10^7 m past him, what does Mavis's timer read? b) At the instant when Mavis reads 0.400 s on her timer, what does Stanley read on his?

SOLUTION

IDENTIFY AND SET UP: Let S be Stanley's frame of reference, and let S' be Mavis's frame of reference. The two events of interest in part (a) are when Mavis passes Stanley and when Stanley measures Mavis as having traveled a distance of 9.00×10^7 m; the target variables are the time intervals between these two events as measured in S and in S'. The two events in part (b) are when Mavis passes Stanley and when Mavis measures an elapsed time of 0.400 s; the target variable is the time interval between these two events as measured in S. As we will see, understanding this example hinges on understanding the difference between these two pairs of events.

EXECUTE: a) The two events, Mavis passing the earth and Mavis reaching $x = 9.00 \times 10^7$ m as measured by Stanley, occur at different positions in Stanley's frame but at the same position in Mavis's frame. Hence the time interval in Stanley's frame S is Δt, while the time interval in Mavis's frame S' is the proper time Δt_0. As measured by Stanley, Mavis moves at $0.600c = 0.600(3.00 \times 10^8\ \text{m/s}) = 1.80 \times 10^8$ m/s and travels the 9.00×10^7 m in a time $\Delta t = (9.00 \times 10^7\ \text{m})/(1.80 \times 10^8\ \text{m/s}) = 0.500$ s. From Eq. (37.6), Mavis's timer reads an elapsed time of

$$\Delta t_0 = \Delta t\sqrt{1 - u^2/c^2} = 0.500\ \text{s}\sqrt{1 - (0.600)^2} = 0.400\ \text{s}$$

b) It is tempting—but wrong—to answer that Stanley's timer reads 0.500 s. We are now considering a *different* pair of events, the starting and the reading of Stanley's timer, that both occur at the same point in Stanley's earth frame S. These two events occur at different positions in Mavis's frame S', so the time interval of 0.400 s that she measures between these events is equal to Δt. (In her frame, Stanley passes her at time zero and is a distance behind her of $(1.80 \times 10^8\ \text{m/s})(0.400\ \text{s}) = 7.20 \times 10^7$ m at time 0.400 s.) The time on Stanley's timer is now the proper time:

$$\Delta t_0 = \Delta t\sqrt{1 - u^2/c^2} = 0.400\ \text{s}\sqrt{1 - (0.600)^2} = 0.320\ \text{s}$$

EVALUATE: If the difference between 0.500 s and 0.320 s still troubles you, consider the following: Stanley, taking proper account of the transit time of a signal from $x = 9.00 \times 10^7$ m, says that Mavis passed that point and his timer read 0.500 s at the same instant. But Mavis says that those two events occurred at different positions and were *not* simultaneous—she passed the point at the instant his timer read 0.320 s. This example shows the relativity of simultaneity.

The Twin Paradox

Equations (37.6) and (37.8) for time dilation suggest an apparent paradox called the **twin paradox.** Consider identical twin astronauts named Eartha and Astrid. Eartha remains on earth while her twin Astrid takes off on a high-speed trip through the galaxy. Because of time dilation, Eartha sees Astrid's heartbeat and all other life processes proceeding more slowly than her own. Thus to Eartha, Astrid ages more slowly; when Astrid returns to earth she is younger (has aged less) than Eartha.

Now here is the paradox: All inertial frames are equivalent. Can't Astrid make exactly the same arguments to conclude that Eartha is in fact the younger? Then each twin measures the other to be younger when they're back together, and that's a paradox.

To resolve the paradox, we recognize that the twins are *not* identical in all respects. While Eartha remains in an approximately inertial frame at all times, Astrid must accelerate with respect to that inertial frame during parts of her trip in order to leave, turn around, and return to earth. Eartha's reference frame is always approximately inertial; Astrid's is often far from inertial. Thus there is a real physical difference between the circumstances of the two twins. Careful analysis shows that Eartha is correct; when Astrid returns, she *is* younger than Eartha.

Test Your Understanding

Samir (who is standing on the ground) starts his stopwatch at the instant that Maria flies past him in her spaceship at a speed of $0.600c$. At the same instant, Maria starts her stopwatch. How much time has elapsed in Maria's frame at the instant that Samir's stopwatch reads 16.0 s? How much time has elapsed in Samir's frame at the instant that Maria's stopwatch reads 20.0 s?

37.4 | Relativity of Length

ActivPhysics ONLINE

17.2 Relativity of Length

Not only does the time interval between two events depend on the observer's frame of reference, but the *distance* between two points may also depend on the observer's frame of reference. The concept of simultaneity is involved. Suppose you want to measure the length of a moving car. One way is to have two assistants make marks on the pavement at the positions of the front and rear bumpers. Then you measure the distance between the marks. But your assistants have to make their marks *at the same time.* If one marks the position of the front bumper at one time and the other marks the position of the rear bumper half a second later, you won't get the car's true length. Since we've learned that simultaneity isn't an absolute concept, we have to proceed with caution.

Lengths Parallel to the Relative Motion

To develop a relation between lengths that are measured parallel to the direction of motion in various coordinate systems, we consider another thought experiment. We attach a light source to one end of a ruler and a mirror to the other end. The ruler is at rest in reference frame S', and its length in this frame is l_0 (Fig. 37.9a). Then the time Δt_0 required for a light pulse to make the round trip from source to mirror and back is

$$\Delta t_0 = \frac{2l_0}{c} \qquad (37.9)$$

This is a proper time interval because departure and return occur at the same point in S'.

In reference frame S the ruler is moving to the right with speed u during this travel of the light pulse (Fig. 37.9b). The length of the ruler in S is l, and the time of travel from source to mirror, as measured in S, is Δt_1. During this interval the

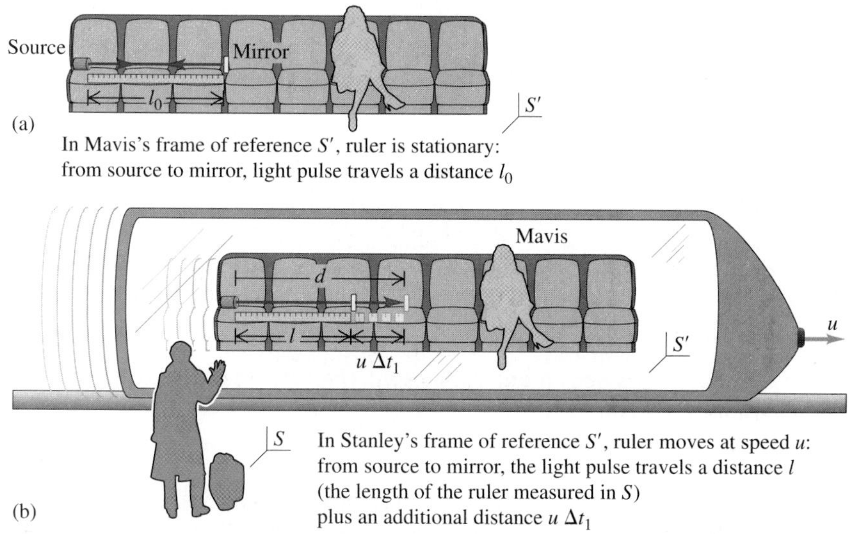

37.9 (a) A ruler is at rest in Mavis's frame S'. A light pulse is emitted from a source at one end of the ruler, reflected by a mirror at the other end, and returned to the source position. (b) Motion of the light pulse as measured in Stanley's frame S.

ruler, with source and mirror attached, moves a distance $u\,\Delta t_1$. The total length of path d from source to mirror is not l, but rather

$$d = l + u\,\Delta t_1 \tag{37.10}$$

The light pulse travels with speed c, so it is also true that

$$d = c\,\Delta t_1 \tag{37.11}$$

Combining Eqs. (37.10) and (37.11) to eliminate d, we find

$$c\,\Delta t_1 = l + u\,\Delta t_1$$

or

$$\Delta t_1 = \frac{l}{c - u} \tag{37.12}$$

(Dividing the distance l by $c - u$ does *not* mean that light travels with speed $c - u$, but rather that the distance the pulse travels in S is greater than l.)

In the same way we can show that the time Δt_2 for the return trip from mirror to source is

$$\Delta t_2 = \frac{l}{c + u} \tag{37.13}$$

The *total* time $\Delta t = \Delta t_1 + \Delta t_2$ for the round trip, as measured in S, is

$$\Delta t = \frac{l}{c - u} + \frac{l}{c + u} = \frac{2l}{c(1 - u^2/c^2)} \tag{37.14}$$

We also know that Δt and Δt_0 are related by Eq. (37.6), because Δt_0 is a proper time in S'. Thus Eq. (37.9) for the round-trip time in the rest frame S' of the ruler becomes

$$\Delta t\sqrt{1 - \frac{u^2}{c^2}} = \frac{2l_0}{c} \tag{37.15}$$

Finally, combining Eqs. (37.14) and (37.15) to eliminate Δt and simplifying, we obtain

$$l = l_0\sqrt{1 - \frac{u^2}{c^2}} = \frac{l_0}{\gamma} \quad \text{(length contraction)} \tag{37.16}$$

(We have used the quantity $\gamma = 1/\sqrt{1 - u^2/c^2}$ defined in Eq. (37.7).) Thus the length l measured in S, in which the ruler is moving, is *shorter* than the length l_0 measured in its rest frame S'.

CAUTION This is *not* an optical illusion! The ruler really is shorter in reference frame S than it is in S'.

A length measured in the frame in which the body is at rest (the rest frame of the body) is called a **proper length;** thus l_0 is a proper length in S', and the length measured in any other frame moving relative to S' is *less than* l_0. This effect is called **length contraction.**

When u is very small in comparison to c, γ approaches 1. Thus in the limit of small speeds we approach the Newtonian relation $l = l_0$. This and the corresponding result for time dilation show that Eqs. (37.1), the Galilean coordinate transformation, are usually sufficiently accurate for relative speeds much smaller than c.

Lengths Perpendicular to the Relative Motion

We have derived Eq. (37.16) for lengths measured in the direction *parallel* to the relative motion of the two frames of reference. Lengths that are measured *perpendicular* to the direction of motion are *not* contracted. To prove this, consider two identical meter sticks. One stick is at rest in frame S and lies along the positive y-axis with one end at O, the origin of S. The other is at rest in frame S' and lies along the positive y'-axis with one end at O', the origin of S'. Frame S' moves in the positive x-direction relative to frame S. Observers Stanley and Mavis, at rest in S and S' respectively, station themselves at the 50-cm mark of their sticks. At the instant the two origins coincide, the two sticks lie along the same line. At this instant, Mavis makes a mark on Stanley's stick at the point that coincides with her own 50-cm mark, and Stanley does the same to Mavis's stick.

Suppose for the sake of argument that Stanley sees Mavis's stick as longer than his own. Then the mark Stanley makes on her stick is *below* its center. In that case, Mavis will think Stanley's stick has become shorter, since half of its length coincides with *less* than half her stick's length. So Mavis sees moving sticks getting shorter and Stanley sees them getting longer. But this implies an asymmetry between the two frames that contradicts the basic postulate of relativity that tells us all inertial frames are equivalent. We conclude that consistency with the postulates of relativity requires that both observers see the rulers as having the same length, even though to each observer one of them is stationary and the other is moving (see Fig. 37.10). So *there is no length contraction perpendicular to the direction of relative motion of the coordinate systems.* We used this result in our derivation of Eq. (37.6) in assuming that the distance d is the same in both frames of reference.

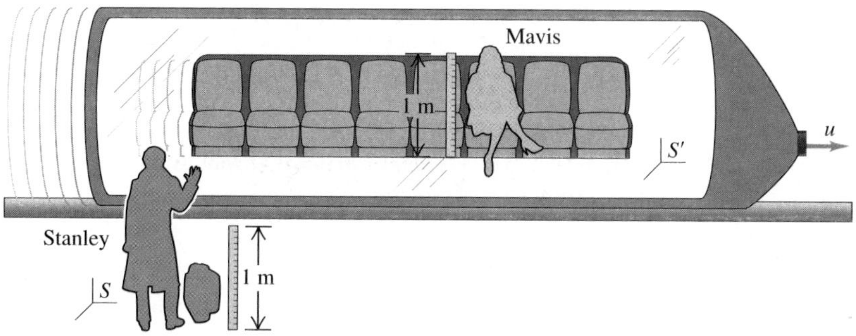

37.10 The meter sticks are perpendicular to the relative velocity. For any value of u, both Stanley and Mavis measure either meter stick to have a length of one meter.

For example, suppose a moving rod of length l_0 makes an angle θ_0 with the direction of relative motion (the x-axis) as measured in its rest frame. Its length component in that frame parallel to the motion, $l_0 \cos \theta_0$, is contracted to $(l_0 \cos \theta_0)/\gamma$. However, its length component perpendicular to the motion, $l_0 \sin \theta_0$, remains the same.

Problem-Solving Strategy — Length Contraction

IDENTIFY *the relevant concepts:* The concept of length contraction is used whenever we compare the length of an object as measured by observers in different inertial frames of reference.

SET UP *the problem using the following steps:*

1. Decide what defines the length in question. If the problem statement describes an object such as a ruler, it is just the distance between the ends of the object. If, however, the problem is about a distance between two points with no object between them, it can help to envision a ruler or rod that extends from one point to the other.
2. Determine what the target variable is.

EXECUTE *the solution as follows:*

1. Determine the reference frame in which the object in question is at rest. In this frame, the length of the object is its proper length l_0. In a second reference frame moving at speed u relative to the first frame, the object has contracted length l.
2. Keep in mind that length contraction occurs only for lengths parallel to the direction of relative motion of the two frames. Any length that is perpendicular to the relative motion is the same in both frames.
3. Use Eq. (37.16) to relate l and l_0, and then solve for the target variable.

EVALUATE *your answer:*
Check that your answers make sense: l is never larger than l_0, and u is never greater than c.

Example 37.4 — How long is the spaceship?

A spaceship flies past earth at a speed of $0.990c$. A crew member on board the spaceship measures its length, obtaining the value 400 m. What length do observers measure on earth?

SOLUTION

IDENTIFY and SET UP: The length in question is the distance between the front and the back of the spaceship. This length is along the direction of relative motion (Fig. 37.11), so there will be

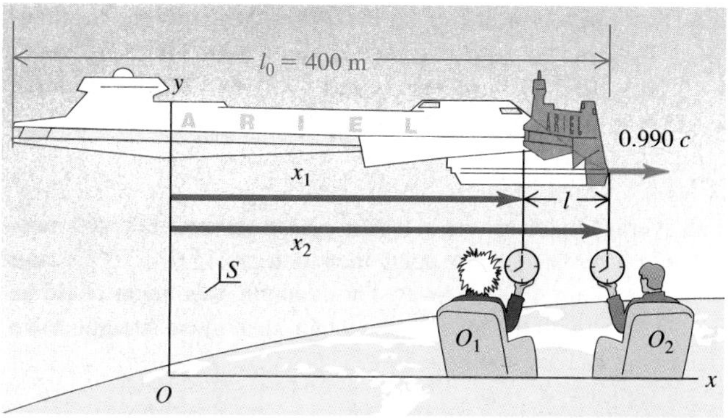

37.11 The two observers on earth (frame S) must measure x_2 and x_1 simultaneously to correctly obtain the length $l = x_2 - x_1$ in their frame of reference.

length contraction as measured in one of the frames of reference. Our target variable is the length measured in the earth frame.

EXECUTE: The 400-m length of the spaceship is the *proper* length l_0 because it is measured in the frame in which the spaceship is at rest. We want to find the length l measured by observers on earth. From Eq. (37.16),

$$l = l_0\sqrt{1 - \frac{u^2}{c^2}} = (400\text{ m})\sqrt{1 - (0.990)^2}$$
$$= 56.4\text{ m}$$

EVALUATE: This answer makes sense: the spaceship is shorter in a frame in which it is in motion than in a frame in which it is at rest. To measure the length l, two observers with synchronized clocks could measure the positions of the two ends of the spaceship simultaneously in the earth's reference frame, as shown in Fig. 37.11. (These two measurements will *not* appear simultaneous to an observer in the spaceship.)

Example 37.5 How far apart are the observers?

The two observers mentioned in Example 37.4 are 56.4 m apart on the earth. How far apart does the spaceship crew measure them to be?

SOLUTION

IDENTIFY and SET UP: The answer is *not* 400 m, because the length in question in this example is *not* the length of the spaceship. Rather, it is the distance between the two observers O_1 and O_2 shown in Fig. 37.11. The target variable is the distance between these two observers as measured in the spaceship frame.

EXECUTE: Think of a 56.4-m-long telephone cable stretching from one observer to the other. This cable is at rest in the earth frame, and so 56.4 m is the proper length l_0. Relative to the crew members, the earth zooms by at $u = 0.990c$, so Eq. (37.16) gives

$$l = l_0\sqrt{1 - \frac{u^2}{c^2}} = (56.4\text{ m})\sqrt{1 - (0.990)^2}$$
$$= 7.96\text{ m}$$

EVALUATE: This answer does *not* say that the crew measures their spaceship to be both 400 m long and 7.96 m long. The observers on earth measure the spaceship to have a contracted length of 56.4 m because they are 56.4 m apart when they measure the positions of the ends at what they measure to be the same instant. (Viewed from the spaceship frame, the observers do not measure those positions simultaneously.) The crew then measure the 56.4-m proper length to be contracted to 7.96 m. The key point is that two different measurements are being made in Example 37.4 (in which the earth observers measure the distance between the ends of the spaceship) and in this example (in which the spaceship crew measures the distance between the earth observers).

Conceptual Example 37.6 Moving with a muon

As was stated in Example 37.1, a muon has, on average, a proper lifetime of 2.20×10^{-6} s and a dilated lifetime of 15.6×10^{-6} s in a frame in which its speed is $0.990c$. Multiplying constant speed by time to find distance gives $0.990(3.00 \times 10^8\text{ m/s}) \times (2.20 \times 10^{-6}\text{ s}) = 653\text{ m}$ and $0.990(3.00 \times 10^8\text{ m/s}) \times (15.6 \times 10^{-6}\text{ s}) = 4630\text{ m}$. Interpret these two distances.

SOLUTION

If an average muon moves at $0.990c$ past observers, they will measure it to be created at one point, then to decay 15.6×10^{-6} s later at another point 4630 m away. For example, this muon could be created level with the top of a mountain, then move straight down to decay at its base 4630 m below.

However, an observer moving with an average muon will say that it traveled only 653 m because it existed for only 2.20×10^{-6} s. To show that this answer is completely consistent, consider the mountain. The 4630-m distance is its height, a proper length in the direction of motion. Relative to the observer traveling with this muon, the mountain moves up at $0.990c$ with the 4630-m length contracted to

$$l = l_0\sqrt{1 - \frac{u^2}{c^2}} = (4630\text{ m})\sqrt{1 - (0.990)^2}$$
$$= 653\text{ m}$$

Thus we see that length contraction is consistent with time dilation.

Finally, let's think a little about the visual appearance of a moving three-dimensional body. If we could see the positions of all points of the body simultaneously, it would appear to shrink only in the direction of motion. But we *don't* see all the points simultaneously; light from points farther from us takes longer to reach us than does light from points near to us, so we see the farther points at the positions they had at earlier times.

Suppose we have a rectangular rod with its faces parallel to the coordinate planes. When we look end-on at the center of the closest face of such a rod at rest, we see only that face. (See the center rod in computer-generated Fig. 37.12a.) But when that rod is moving past us toward the right at relativistic speed, we may also see its left side because of the earlier-time effect just described. That is, we can see some points that we couldn't see when the rod was at rest because the rod moves out of the way of the light rays from those points to us. Conversely, some light that can get to us when the rod is at rest is blocked by the moving rod. Because of all this, the rods in Figs. 37.12b and 37.12c appear rotated and distorted.

(a) Array at rest

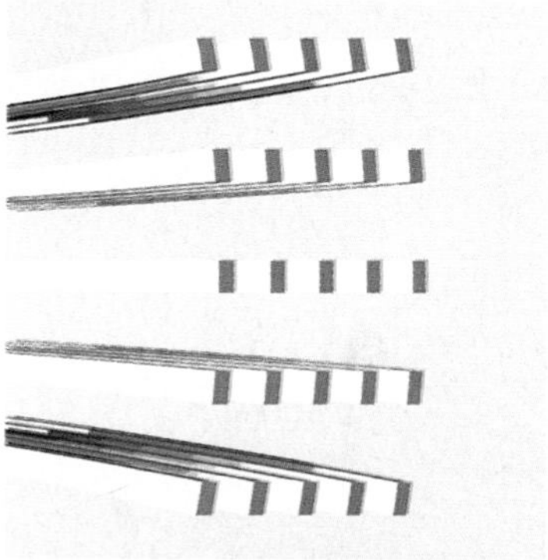
(b) Array moving to the right at $0.2c$

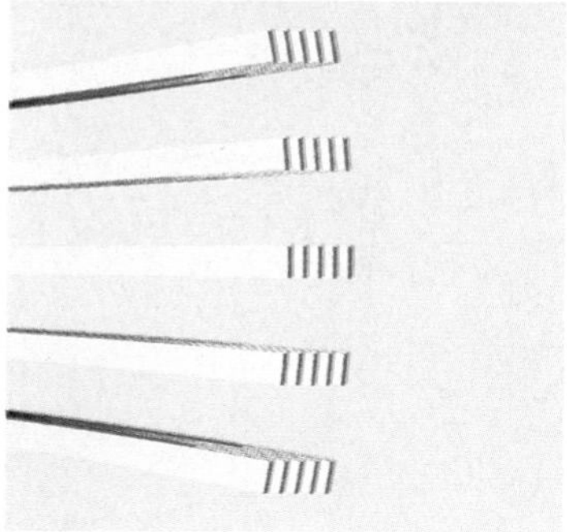
(c) Array moving to the right at $0.9c$

37.12 Computer simulation of the appearance of an array of 25 rods with square cross section. The center rod is viewed end-on. (The simulation ignores color changes in the array caused by the Doppler effect.)

Test Your Understanding

Melinda holds a cube with sides of length 15.0 cm. The edges of the cube are parallel to the x-, y-, and z-axes. If Santiago flies past her in the negative y-direction at a speed of $0.800c$, what will he measure for the dimensions of the cube?

37.5 | The Lorentz Transformations

In Section 37.1 we discussed the Galilean coordinate transformation equations, Eqs. (37.1). They relate the coordinates (x, y, z) of a point in frame of reference S to the coordinates (x', y', z') of the point in a second frame S'. The second frame moves with constant speed u relative to S in the positive direction along the common x-x' axis. This transformation also assumes that the time scale is the same in the two frames of reference, as expressed by the additional relation $t = t'$. This Galilean transformation, as we have seen, is valid only in the limit when u approaches zero. We are now ready to derive more general transformations that are consistent with the principle of relativity. The more general relations are called the **Lorentz transformations.**

Our first question is this: When an event occurs at point (x, y, z) at time t, as observed in a frame of reference S, what are the coordinates (x', y', z') and time t' of the event as observed in a second frame S' moving relative to S with constant speed u in the $+x$-direction?

To derive the coordinate transformation, we refer to Fig. 37.13, which is the same as Fig. 37.3. As before, we assume that the origins coincide at the initial time $t = 0 = t'$. Then in S the distance from O to O' at time t is still ut. The coor-

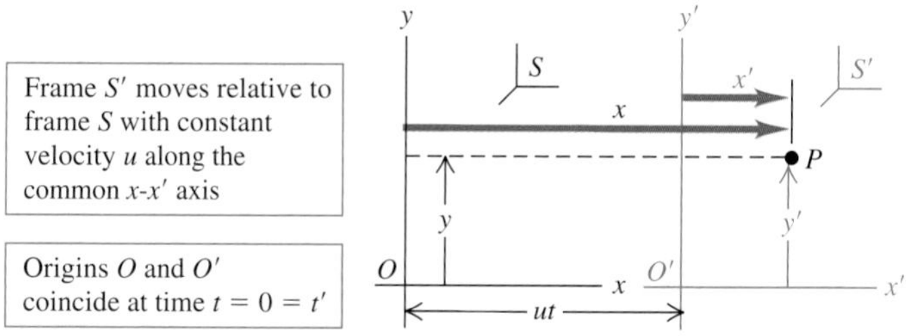

37.13 As measured in frame of reference S, x' is contracted to x'/γ, so $x = ut + x'/\gamma$ and $x' = \gamma(x - ut)$.

dinate x' is a *proper length* in S', so in S it is contracted by the factor $1/\gamma = \sqrt{1 - u^2/c^2}$, as in Eq. (37.16). Thus the distance x from O to P, as seen in S, is not simply $x = ut + x'$, as in the Galilean coordinate transformation, but

$$x = ut + x'\sqrt{1 - \frac{u^2}{c^2}} \tag{37.17}$$

Solving this equation for x', we obtain

$$x' = \frac{x - ut}{\sqrt{1 - u^2/c^2}} \tag{37.18}$$

Equation (37.18) is part of the Lorentz coordinate transformation; another part is the equation giving t' in terms of x and t. To obtain this, we note that the principle of relativity requires that the *form* of the transformation from S to S' be identical to that from S' to S. The only difference is a change in the sign of the relative velocity component u. Thus from Eq. (37.17) it must be true that

$$x' = -ut' + x\sqrt{1 - \frac{u^2}{c^2}} \tag{37.19}$$

We now equate Eqs. (37.18) and (37.19) to eliminate x'. This gives us an equation for t' in terms of x and t. We leave the algebraic details for you to work out; the result is

$$t' = \frac{t - ux/c^2}{\sqrt{1 - u^2/c^2}} \tag{37.20}$$

As we discussed previously, lengths perpendicular to the direction of relative motion are not affected by the motion, so $y' = y$ and $z' = z$.

Collecting all these transformation equations, we have

$$\begin{aligned} x' &= \frac{x - ut}{\sqrt{1 - u^2/c^2}} = \gamma(x - ut) \\ y' &= y \\ z' &= z \\ t' &= \frac{t - ux/c^2}{\sqrt{1 - u^2/c^2}} = \gamma(t - ux/c^2) \end{aligned} \quad \text{(Lorentz coordinate transformation)} \tag{37.21}$$

These equations are the *Lorentz coordinate transformation,* the relativistic generalization of the Galilean coordinate transformation, Eqs. (37.1) and $t = t'$. For values of u that approach zero, the radicals in the denominators and γ approach 1, and the ux/c^2 term approaches zero. In this limit, Eqs. (37.21) become identical to Eqs. (37.1) along with $t = t'$. In general, though, both the coordinates and time of an event in one frame depend on its coordinates and time in another frame. *Space and time have become intertwined; we can no longer say that length and time have absolute meanings independent of the frame of reference.* For this reason, we refer to time and the three dimensions of space collectively as a four-dimensional entity called **spacetime,** and we call (x, y, z, t) together the **spacetime coordinates** of an event.

The Lorentz Velocity Transformation

We can use Eqs. (37.21) to derive the relativistic generalization of the Galilean velocity transformation, Eq. (37.2). We consider only one-dimensional motion along the x-axis and use the term "velocity" as being short for the "x-component of the velocity." Suppose that in a time dt a particle moves a distance dx, as measured in frame S. We obtain the corresponding distance dx' and time dt' in S' by taking differentials of Eqs. (37.22):

$$dx' = \gamma(dx - u\,dt)$$

$$dt' = \gamma(dt - u\,dx/c^2)$$

We divide the first equation by the second and then divide the numerator and denominator of the result by dt to obtain

$$\frac{dx'}{dt'} = \frac{\dfrac{dx}{dt} - u}{1 - \dfrac{u}{c^2}\dfrac{dx}{dt}}$$

Now dx/dt is the velocity v_x in S, and dx'/dt' is the velocity v_x' in S', so we finally obtain the relativistic generalization

$$v_x' = \frac{v_x - u}{1 - uv_x/c^2} \quad \text{(Lorentz velocity transformation)} \qquad (37.22)$$

When u and v_x are much smaller than c, the denominator in Eq. (37.22) approaches 1, and we approach the nonrelativistic result $v_x' = v_x - u$. The opposite extreme is the case $v_x = c$; then we find

$$v_x' = \frac{c - u}{1 - uc/c^2} = \frac{c(1 - u/c)}{1 - u/c} = c$$

This says that anything moving with velocity $v_x = c$ measured in S also has velocity $v_x' = c$ measured in S', despite the relative motion of the two frames. So Eq. (37.22) is consistent with Einstein's postulate that the speed of light in vacuum is the same in all inertial frames of reference.

The principle of relativity tells us there is no fundamental distinction between the two frames S and S'. Thus the expression for v_x in terms of v_x' must have the same form as Eq. (37.22), with v_x changed to v_x', and vice versa, and the sign of u reversed. Carrying out these operations with Eq. (37.22), we find

$$v_x = \frac{v_x' + u}{1 + uv_x'/c^2} \quad \text{(Lorentz velocity transformation)} \qquad (37.23)$$

This can also be obtained algebraically by solving Eq. (37.22) for v_x. Both Eqs. (37.22) and (37.23) are *Lorentz velocity transformations* for one-dimensional motion.

CAUTION Keep in mind that the Lorentz transformation equations given by Eqs. (37.21), (37.22), and (37.23) assume that frame S' is moving in the positive x-direction with velocity u relative to frame S. You should always set up your coordinate system to follow this convention.

When u is less than c, the Lorentz velocity transformations show us that a body moving with a speed less than c in one frame of reference always has a speed less than c in *every other* frame of reference. This is one reason for concluding that no material body may travel with a speed equal to or greater than that of light in vacuum, relative to *any* inertial frame of reference. The relativistic generalizations of energy and momentum, which we will explore later, give further support to this hypothesis.

Problem-Solving Strategy: Lorentz Transformations

IDENTIFY *the relevant concepts:* The Lorentz *coordinate* transformation tells you how to relate the spacetime coordinates of an event in one inertial frame of reference to the spacetime coordinates of the same event in a second inertial frame. The Lorentz *velocity* transformation relates the velocity of an object in one inertial frame to its velocity in a second inertial frame.

SET UP *the problem using the following steps:*

1. Determine what the target variable is.
2. Define the two inertial frames S and S'. Remember that S' moves relative to S at a constant velocity u in the $+x$-direction.
3. If the coordinate transformation equations are needed, make a list of spacetime coordinates in the two frames, such as x_1, x'_1, t_1, t'_1, and so on. Label carefully which of these you know and which you don't.
4. In velocity-transformation problems, clearly identify the velocities u (the relative velocity of the two frames of reference), v_x (the velocity of the object relative to S), and v'_x (the velocity of the object relative to S').

EXECUTE *the solution as follows:*

1. In a coordinate-transformation problem, use Eqs. (37.21) to solve for the spacetime coordinates of the event as measured in S' in terms of the corresponding values in S. (If you need to solve for the spacetime coordinates in S in terms of the corresponding values in S', you can easily convert the expressions in Eqs. (37.21): replace all of the primed quantities with unprimed ones, and vice versa, and replace u with $-u$.)
2. In a velocity-transformation problem, use either Eq. (37.22) or Eq. (37.23), as appropriate, to solve for the target variable.

EVALUATE *your answer:* Don't be discouraged if some of your results don't seem to make sense or if they disagree with "common sense." It takes time to develop intuition about relativity; you'll gain it with experience. (One result that would definitely be in error is a speed greater than c.)

Example 37.7 Was it received before it was sent?

Winning an interstellar race, Mavis pilots her spaceship across a finish line in space at a speed of $0.600c$ relative to that line. A "hooray" message is sent from the back of her ship (event 2) at the instant (in her frame of reference) that the front of her ship crosses the line (event 1). She measures the length of her ship to be 300 m. Stanley is at the finish line and is at rest relative to it. When and where does he measure events 1 and 2 to occur?

SOLUTION

IDENTIFY and SET UP: This example involves the Lorentz coordinate transformation. Our derivation of this transformation assumes that the origins of frames S and S' coincide at $t = 0 = t'$. Thus for simplicity we fix the origin of S at the finish line and the origin of S' at the front of the spaceship so that Stanley and Mavis measure event 1 to be at $x = 0 = x'$ and $t = 0 = t'$.

Mavis in S' measures her spaceship to be 300 m long, so she has the "hooray" sent from 300 m behind her spaceship's front at the instant she measures the front to cross the finish line. That is, she measures event 2 at $x' = -300$ m and $t' = 0$.

Our target variables are the coordinate x and time t of event 2 that Stanley measures in S.

EXECUTE: To most easily solve for the target variables, we modify the first and last of Eqs. (37.21) to give x and t as functions of x' and t'. We do so by using the principle of relativity in the same way that we obtained Eq. (37.23) from Eq. (37.22). We remove the primes from x' and t', add primes to x and t, and replace each u with $-u$. The results are

$$x = \gamma(x' + ut') \quad \text{and} \quad t = \gamma(t' + ux'/c^2)$$

From Eq. (37.7), $\gamma = 1.25$ for $u = 0.600c = 1.80 \times 10^8$ m/s. We also substitute $x' = -300$ m, $t' = 0$, $c = 3.00 \times 10^8$ m/s, and $u = 1.80 \times 10^8$ m/s in the equations for x and t to find $x = -375$ m at $t = -7.50 \times 10^{-7}$ s $= -0.750\ \mu$s for event 2.

EVALUATE: Mavis says that the events are simultaneous, but Stanley disagrees. In fact, he says that the "hooray" was sent *before* Mavis crossed the finish line. This does not mean that the effect preceded the cause. The fastest that Mavis can send a signal the length of her ship is $300 \text{ m}/3.00 \times 10^8 \text{ m/s} = 1.00\ \mu\text{s}$. She cannot send a signal from the front at the instant it crosses the finish line that would cause a "hooray" to be broadcast from the back at the same instant. She would have to send that signal from the front at least $1.00\ \mu\text{s}$ before then, so she had to slightly anticipate her success.

Note that relativity *is* consistent. In his frame, Stanley measures Mavis's ship to be $l_0/\gamma = 300 \text{ m}/1.25 = 240 \text{ m}$ long with its back at $x = -375$ m at $t = -0.750\ \mu\text{s} = -7.50 \times 10^{-7}$ s when the "hooray" is sent. At that instant he thus measures the front of her 240-m-long ship to be a distance of $(375 - 240) \text{ m} = 135 \text{ m}$ from the finish line. However, since $(1.80 \times 10^8 \text{ m/s})(7.50 \times 10^{-7} \text{ s}) = 135$ m, the front does cross the line at $t = 0$.

Example 37.8 Relative velocities

a) A spaceship moving away from the earth with speed $0.900c$ fires a robot space probe in the same direction as its motion, with speed $0.700c$ relative to the spaceship. What is the probe's velocity relative to the earth? b) A scoutship tries to catch up with the spaceship by traveling at $0.950c$ relative to the earth. What is the velocity of the scoutship relative to the spaceship?

SOLUTION

IDENTIFY and SET UP: This example uses the Lorentz velocity transformation. Let the earth's frame of reference be S, and let the spaceship's frame of reference be S' (Fig. 37.14). The relative velocity of the two frames is $u = 0.700c$. The target variable in part (a) is the velocity of the probe relative to S; the target variable in part (b) is the velocity of the scoutship relative to S'.

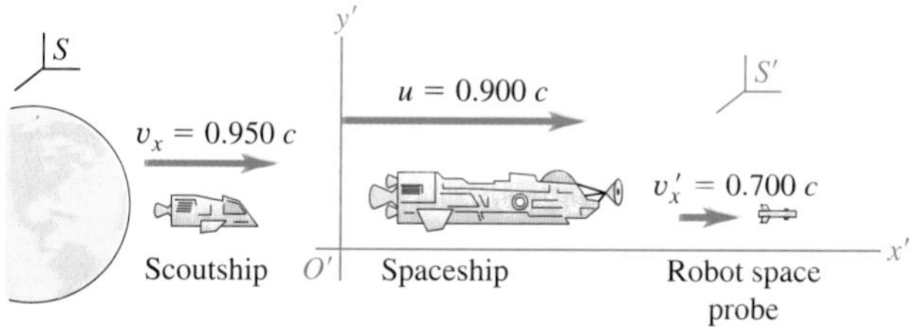

37.14 The spaceship, robot space probe, and scoutship.

EXECUTE: a) We are given the velocity of the probe relative to the spaceship, $v'_x = 0.700c$. We use Eq. (37.23) to determine its velocity v_x relative to the earth:

$$v_x = \frac{v'_x + u}{1 + uv'_x/c^2} = \frac{0.700c + 0.900c}{1 + (0.900c)(0.700c)/c^2} = 0.982c$$

b) We are given the velocity of the scoutship relative to the earth, $v_x = 0.950c$. We use Eq. (37.22) to determine its velocity v'_x relative to the spaceship:

$$v'_x = \frac{v_x - u}{1 - uv_x/c^2} = \frac{0.950c - 0.900c}{1 - (0.950c)(0.900c)/c^2} = 0.345c$$

EVALUATE: It's instructive to compare our results to what we would have obtained had we used the Galilean velocity transformation formula, Eq. (37.2). In part (a) we would have found the probe's velocity relative to the earth to be $v_x = v'_x + u = 0.700c + 0.900c = 1.600c$. This value is greater than the speed of light and so must be incorrect. In part (b) we would have found that the scoutship's velocity relative to the spaceship to be $v'_x = v_x - u = 0.950c - 0.900c = 0.050c$; the relativistically correct value, $v'_x = 0.345c$, is almost seven times greater than the incorrect Galilean value.

Test Your Understanding

In Example 37.8, what is the velocity of the robot space probe relative to the scoutship?

*37.6 | The Doppler Effect for Electromagnetic Waves

An additional important consequence of relativistic kinematics is the Doppler effect for electromagnetic waves. In our previous discussion of the Doppler effect (Section 16.8) we quoted without proof the formula, Eq. (16.30), for the frequency shift that results from motion of a source of electromagnetic waves relative to an observer. We can now derive that result.

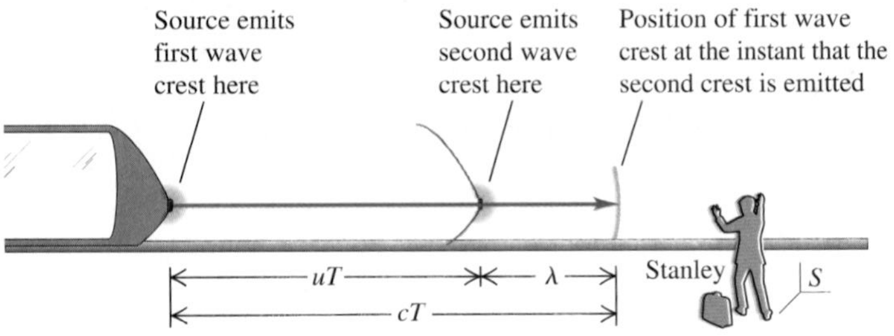

37.15 The Doppler effect for light. A light source moving at speed u relative to Stanley emits a wave crest, then travels a distance uT toward an observer and emits the next crest. In Stanley's reference frame S, the second crest is a distance λ behind the first crest.

Here's a statement of the problem. A source of light is moving with constant speed u toward Stanley, who is stationary in an inertial frame (Fig. 37.15). As measured in its rest frame, the source emits light waves with frequency f_0 and period $T_0 = 1/f_0$. What is the frequency f of these waves as received by Stanley?

Let T be the time interval between *emission* of successive wave crests as observed in Stanley's reference frame. Note that this is *not* the interval between the *arrival* of successive crests at his position, because the crests are emitted at different points in Stanley's frame. In measuring only the frequency f he receives, he does not take into account the difference in transit times for successive crests. Therefore the frequency he receives is *not* $1/T$. What is the equation for f?

During a time T the crests ahead of the source move a distance cT, and the source moves a shorter distance uT in the same direction. The distance λ between successive crests—that is, the wavelength—is thus $\lambda = (c - u)T$, as measured in Stanley's frame. The frequency that he measures is c/λ. Therefore

$$f = \frac{c}{(c - u)T} \tag{37.24}$$

So far we have followed a pattern similar to that for the Doppler effect for sound from a moving source (Section 16.8). In that discussion our next step was to equate T to the time T_0 between emissions of successive wave crests by the source. However, due to time dilation it is *not* relativistically correct to equate T to T_0. The time T_0 is measured in the rest frame of the source, so it is a proper time. From Eq. (37.6), T_0 and T are related by

$$T = \frac{T_0}{\sqrt{1 - u^2/c^2}} = \frac{cT_0}{\sqrt{c^2 - u^2}}$$

or, since $T_0 = 1/f_0$,

$$\frac{1}{T} = \frac{\sqrt{c^2 - u^2}}{cT_0} = \frac{\sqrt{c^2 - u^2}}{c} f_0$$

Remember, $1/T$ is not equal to f. We must substitute this expression for $1/T$ into Eq. (37.24) to find f:

$$f = \frac{c}{c - u} \frac{\sqrt{c^2 - u^2}}{c} f_0$$

Using $c^2 - u^2 = (c - u)(c + u)$ gives

$$f = \sqrt{\frac{c + u}{c - u}}\, f_0 \quad \text{(Doppler effect, electromagnetic waves)} \tag{37.25}$$

This shows that when the source moves *toward* the observer, the observed frequency f is *greater* than the emitted frequency f_0. The difference $f - f_0 = \Delta f$ is called the Doppler frequency shift. When u/c is much smaller than 1, the fractional shift $\Delta f/f$ is also small and is approximately equal to u/c:

$$\frac{\Delta f}{f} = \frac{u}{c}$$

When the source moves *away from* the observer, we change the sign of u in Eq. (37.25) to get

$$f = \sqrt{\frac{c-u}{c+u}}\,f_0 \qquad (37.26)$$

This agrees with Eq. (16.30), which we quoted previously, with minor notation changes.

With light, unlike sound, there is no distinction between motion of source and motion of observer; only the *relative* velocity of the two is significant. The last four paragraphs of Section 16.8 discuss several practical applications of the Doppler effect with light and other electromagnetic radiation; we suggest you review those paragraphs now. Figure 37.16 shows one common application.

37.16 This handheld radar gun emits a radio beam of frequency f_0, which in the frame of reference of an approaching car has a higher frequency f given by Eq. (37.25). The reflected beam also has frequency f in the car's frame, but has an even higher frequency f' in the police officer's frame. The radar gun calculates the car's speed by comparing the frequencies of the emitted beam and the doubly Doppler-shifted reflected beam. (Compare Example 16.20 in Section 16.8.)

Example 37.9 The Doppler effect

A number of galaxies have supermassive black holes at their centers (see Section 12.8). As material swirls around such a black hole, it is heated, becomes ionized, and generates strong magnetic fields. The resulting magnetic forces steer some of the material into high-speed jets that blast out of the galaxy and into intergalactic space (Fig. 37.17). The blue light we observe from the jet in Fig. 37.17 has a frequency of 6.66×10^{14} Hz, but in the frame of reference of the jet material the light has a frequency of 5.55×10^{13} Hz (in the infrared region of the electromagnetic spectrum). At what speed is the jet material moving toward us?

SOLUTION

IDENTIFY and SET UP: The frequency we observe is $f = 6.66 \times 10^{14}$ Hz, and the frequency in the frame of the source is $f_0 = 5.55 \times 10^{13}$ Hz. Since $f > f_0$, the source is approaching us and therefore we use Eq. (37.25) to find the target variable u.

EXECUTE: We need to solve Eq. (37.25) for u. That takes a little algebra; we'll leave it as an exercise for you to show that the result is

$$u = \frac{(f/f_0)^2 - 1}{(f/f_0)^2 + 1}c$$

We have $f/f_0 = (6.66 \times 10^{14}\ \text{Hz})/(5.55 \times 10^{13}\ \text{Hz}) = 12.0$, so we find

$$u = \frac{(12.0)^2 - 1}{(12.0)^2 + 1}c = 0.986c$$

EVALUATE:
Because the frequency shift is quite substantial, it would have been erroneous to use the approximate expression $\Delta f/f = u/c$. Had you tried to do so, you would have found $u = c\,\Delta f/f_0 = c(6.66 \times 10^{14}\ \text{Hz} - 5.55 \times 10^{13}\ \text{Hz})/(5.55 \times 10^{13}\ \text{Hz}) = 11.0c$. This result cannot be correct, because the jet material cannot travel faster than light.

37.17 This image shows a fast-moving jet 5000 light years in length emanating from the center of the galaxy M87. The light from the jet is emitted by fast-moving electrons spiraling around magnetic field lines (see Fig. 27.16).

Test Your Understanding

Galaxies with supermassive black holes typically have *two* jets moving in opposite directions. If a jet is moving *away* from us at $0.986c$ and emits light with the same frequency (as measured in the reference frame of the jet material) as the jet described in Example 37.9, what frequency do we measure on earth?

37.7 | Relativistic Momentum

Newton's laws of motion have the same form in all inertial frames of reference. When we use transformations to change from one inertial frame to another, the laws should be *invariant* (unchanging). But we have just learned that the principle of relativity forces us to replace the Galilean transformations with the more general Lorentz transformations. As we will see, this requires corresponding generalizations in the laws of motion and the definitions of momentum and energy.

The principle of conservation of momentum states that *when two bodies interact, the total momentum is constant,* provided that the net external force acting on the bodies in an inertial reference frame is zero (for example, if they form an isolated system, interacting only with each other). If conservation of momentum is a valid physical law, it must be valid in *all* inertial frames of reference. Now, here's the problem: Suppose we look at a collision in one inertial coordinate system S and find that momentum is conserved. Then we use the Lorentz transformation to obtain the velocities in a second inertial system S'. We find that if we use the Newtonian definition of momentum ($\vec{p} = m\vec{v}$), momentum is *not* conserved in the second system! If we are convinced that the principle of relativity and the Lorentz transformation are correct, the only way to save momentum conservation is to generalize the *definition* of momentum.

We won't derive the correct relativistic generalization of momentum, but here is the result. Suppose we measure the mass of a particle to be m when it is at rest relative to us: we often call m the **rest mass.** We will use the term *material particle* for a particle that has a nonzero rest mass. When such a particle has a velocity $\vec{v}$, its **relativistic momentum** $\vec{p}$ is

$$\vec{p} = \frac{m\vec{v}}{\sqrt{1 - v^2/c^2}} \quad \text{(relativistic momentum)} \qquad (37.27)$$

When the particle's speed v is much less than c, this is approximately equal to the Newtonian expression $\vec{p} = m\vec{v}$, but in general the momentum is greater in magnitude than mv (Fig. 37.18). In fact, as v approaches c, the momentum approaches infinity.

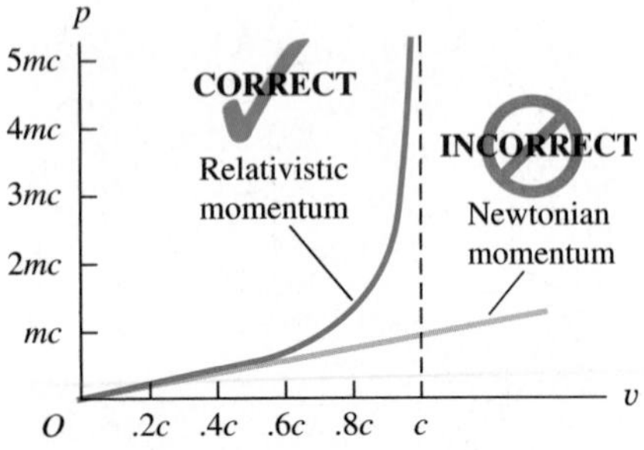

37.18 Graph of the magnitude of the momentum of a particle of rest mass m as a function of speed v. Also shown is the Newtonian prediction, which gives correct results only at speeds much less than c.

Newton's Second Law

What about the relativistic generalization of Newton's second law? In Newtonian mechanics the most general form of the second law is

$$\vec{F} = \frac{d\vec{p}}{dt} \qquad (37.28)$$

That is, the net force $\vec{F}$ on a particle equals the time rate of change of its momentum. Experiments show that this result is still valid in relativistic mechanics, provided that we use the relativistic momentum given by Eq. (37.27). That is, the relativistically correct generalization of Newton's second law is

$$\vec{F} = \frac{d}{dt}\frac{m\vec{v}}{\sqrt{1 - v^2/c^2}} \tag{37.29}$$

Because momentum is no longer directly proportional to velocity, the rate of change of momentum is no longer directly proportional to the acceleration. As a result, *constant force does not cause constant acceleration.* For example, when the net force and the velocity are both along the *x*-axis, Eq. (37.29) gives

$$F = \frac{m}{(1 - v^2/c^2)^{3/2}}\,a \qquad (\vec{F} \text{ and } \vec{v} \text{ along the same line}) \tag{37.30}$$

where a is the acceleration, also along the *x*-axis. Solving Eq. (37.30) for the acceleration a gives

$$a = \frac{F}{m}\left(1 - \frac{v^2}{c^2}\right)^{3/2}$$

We see that as a particle's speed increases, the acceleration caused by a given force continuously *decreases*. As the speed approaches c, the acceleration approaches zero, no matter how great a force is applied. Thus it is impossible to accelerate a particle with nonzero rest mass to a speed equal to or greater than c. The speed of light in vacuum is sometimes called "the ultimate speed."

Equation (37.27) for relativistic momentum is sometimes interpreted to mean that a rapidly moving particle undergoes an increase in mass. If the mass at zero velocity (the rest mass) is denoted by m, then the "relativistic mass" m_{rel} is given by

$$m_{\text{rel}} = \frac{m}{\sqrt{1 - v^2/c^2}}$$

Indeed, when we consider the motion of a system of particles (such as rapidly moving ideal gas molecules in a stationary container), the total rest mass of the system is the sum of the relativistic masses of the particles, not the sum of their rest masses. Also with relativistic mass, the famous equation $E = mc^2$ can be applied to all types of energy, not just most types.

However, if blindly applied, the concept of relativistic mass has its pitfalls. As Eq. (37.29) shows, the relativistic generalization of Newton's second law is *not* $\vec{F} = m_{\text{rel}}\vec{a}$, and we will show in Section 37.8 that the relativistic kinetic energy of a particle is *not* $K = \frac{1}{2}m_{\text{rel}}v^2$. The use of relativistic mass has its supporters and detractors, some quite strong in their opinions. We will mostly deal with individual particles, so we will sidestep the controversy and use Eq. (37.27) as the generalized definition of momentum with m as a constant for each particle, independent of its state of motion.

We will use the abbreviation

$$\gamma = 1/\sqrt{1 - v^2/c^2}$$

When we used this abbreviation in Section 37.3 with u in place of v, u was the relative speed of two coordinate systems. Here v is the speed of a particle in a particular coordinate system, that is, the speed of the particle's *rest frame* with respect to that system. In terms of γ, Eqs. (37.27) and (37.30) become

$$\vec{p} = \gamma m\vec{v} \quad \text{(relativistic momentum)} \tag{37.31}$$

$$F = \gamma^3 ma \quad (\vec{F} \text{ and } \vec{v} \text{ along the same line}) \tag{37.32}$$

In linear accelerators (used in medicine as well as nuclear and elementary-particle physics) the net force $\vec{F}$ and the velocity $\vec{v}$ of the accelerated particle are along the same straight line. But for much of the path in most circular accelerators the particle moves in uniform circular motion at constant speed v. Then the net force and velocity are perpendicular, so the force can do no work on the particle and the kinetic energy and speed remain constant. Thus the denominator in Eq. (37.29) is constant, and we obtain

$$F = \frac{m}{(1 - v^2/c^2)^{1/2}}a = \gamma ma \quad (\vec{F} \text{ and } \vec{v} \text{ perpendicular}) \tag{37.33}$$

Recall from Section 3.4 that if the particle moves in a circle, the net force and acceleration are directed inward along the radius r, and $a = v^2/r$.

What about the general case in which $\vec{F}$ and $\vec{v}$ are neither along the same line nor perpendicular? Then we can resolve the net force $\vec{F}$ at any instant into components parallel to and perpendicular to $\vec{v}$. The resulting acceleration will have corresponding components obtained from Eqs. (37.32) and (37.33). Because of the different γ^3 and γ factors, the acceleration components will not be proportional to the net force components. That is, *unless the net force on a relativistic particle is either along the same line as the particle's velocity or perpendicular to it, the net force and acceleration vectors are not parallel.*

Example 37.10 Relativistic dynamics of an electron

An electron (rest mass 9.11×10^{-31} kg, charge -1.60×10^{-19} C) is moving opposite to an electric field of magnitude $E = 5.00 \times 10^5$ N/C. All other forces are negligible in comparison to the electric field force. a) Find the magnitudes of momentum and of acceleration at the instants when $v = 0.010c$, $0.90c$, and $0.99c$. b) Find the corresponding accelerations if a net force of the same magnitude is perpendicular to the velocity.

SOLUTION

IDENTIFY: In addition to the expressions from this section for relativistic momentum and acceleration, we need the relationship between electric force and electric field from Chapter 21.

SET UP: In part (a) we use Eq. (37.31) to determine the magnitude of momentum and Eq. (37.32) to determine the magnitude of acceleration due to a force along the same line as the velocity. In part (b) the force is perpendicular to the velocity, so we use Eq. (37.33) to determine the magnitude of acceleration.

EXECUTE: (a) To find both the magnitude of momentum and the magnitude of acceleration, we need the values of $\gamma = \sqrt{1 - v^2/c^2}$ for each of the three speeds. We find $\gamma = 1.00$, 2.29, and 7.09. The values of the momentum magnitude p are

$$\begin{aligned}
p_1 &= \gamma_1 m v_1 \\
&= (1.00)(9.11 \times 10^{-31}\ \text{kg})(0.010)(3.00 \times 10^8\ \text{m/s}) \\
&= 2.7 \times 10^{-24}\ \text{kg}\cdot\text{m/s at } v_1 = 0.010c \\
p_2 &= (2.29)(9.11 \times 10^{-31}\ \text{kg})(0.90)(3.00 \times 10^8\ \text{m/s}) \\
&= 5.6 \times 10^{-22}\ \text{kg}\cdot\text{m/s at } v_2 = 0.90c \\
p_3 &= (7.09)(9.11 \times 10^{-31}\ \text{kg})(0.99)(3.00 \times 10^8\ \text{m/s}) \\
&= 1.9 \times 10^{-21}\ \text{kg}\cdot\text{m/s at } v_3 = 0.99c
\end{aligned}$$

From Chapter 21, the magnitude of the force on the electron is

$$\begin{aligned}
F &= |q|E = (1.60 \times 10^{-19}\ \text{C})(5.00 \times 10^5\ \text{N/C}) \\
&= 8.00 \times 10^{-14}\ \text{N}
\end{aligned}$$

From Eq. (37.32), $a = F/\gamma^3 m$. When $v = 0.010c$ and $\gamma = 1.00$,

$$a_1 = \frac{8.00 \times 10^{-14}\ \text{N}}{(1.00)^3(9.11 \times 10^{-31}\ \text{kg})} = 8.8 \times 10^{16}\ \text{m/s}^2$$

The accelerations at the two higher speeds are smaller by factors of γ^3:

$$a_2 = 7.3 \times 10^{15}\ \text{m/s}^2 \qquad a_3 = 2.5 \times 10^{14}\ \text{m/s}^2$$

These last two accelerations are only 8.3% and 0.28%, respectively, of the values predicted by nonrelativistic mechanics.

b) From Eq. (37.33), $a = F/\gamma m$ if $\vec{F}$ and $\vec{v}$ are perpendicular. When $v = 0.010c$ and $\gamma = 1.00$,

$$a_1 = \frac{8.00 \times 10^{-14}\ \text{N}}{(1.00)(9.11 \times 10^{-31}\ \text{kg})} = 8.8 \times 10^{16}\ \text{m/s}^2$$

The accelerations at the two higher speeds are smaller by a factor of γ.

$$a_2 = 3.8 \times 10^{16}\ \text{m/s}^2 \qquad a_3 = 1.2 \times 10^{16}\ \text{m/s}^2$$

These accelerations are larger than the corresponding ones in part (a) by factors of γ^2.

EVALUATE: Our results in part (a) show that at higher speeds, the relativistic values of momentum differ more and more from the nonrelativistic values computed using $p = mv$. Note that the momentum at $0.99c$ is more than three times as great as at $0.90c$ because of the increase in the factor γ.

Our results also show that the acceleration drops off very quickly as v approaches c. At the Stanford Linear Accelerator Center, an essentially constant electric force is used to accelerate electrons to a speed only slightly less than c. If the acceleration were constant as predicted by Newtonian mechanics, this speed would be attained after the electrons had traveled a mere 1.5 cm. In fact, because of the decrease of acceleration with speed, a path length of 3 km is needed.

Test Your Understanding

At what speed would a particle have double the momentum predicted by Newtonian mechanics? Does the answer depend on the mass of the particle?

37.8 | Relativistic Work and Energy

When we developed the relationship between work and kinetic energy in Chapter 6, we used Newton's laws of motion. When we generalize these laws according to the principle of relativity, we need a corresponding generalization of the equation for kinetic energy.

We use the work-energy principle, beginning with the definition of work. When the net force and displacement are in the same direction, the work done by that force is $W = \int F\,dx$. We substitute the expression for F from Eq. (37.30), the applicable relativistic version of Newton's second law. In moving a particle of rest mass m from point x_1 to point x_2,

$$W = \int_{x_1}^{x_2} F\,dx = \int_{x_1}^{x_2} \frac{ma\,dx}{(1 - v^2/c^2)^{3/2}} \tag{37.34}$$

To derive the generalized expression for kinetic energy K as a function of speed v, we would like to convert this to an integral on v. But first we remember that the kinetic energy of a particle equals the net work done on it in moving it from rest to the speed v: $K = W$. Thus we let the speeds be zero at point x_1 and v at point x_2. So as not to confuse the variable of integration with the final speed, we change v to v_x in Eq. (37.34). That is, v_x is the varying x-component of the velocity of the particle as the net force accelerates it from rest to a speed v. We also realize that dx and dv_x are the infinitesimal changes in x and v_x, respectively, in the time interval dt. Because $v_x = dx/dt$ and $a = dv_x/dt$, we can rewrite $a\,dx$ in Eq. (37.34) as

$$a\,dx = \frac{dv_x}{dt}dx = dx\frac{dv_x}{dt} = \frac{dx}{dt}dv_x = v_x\,dv_x$$

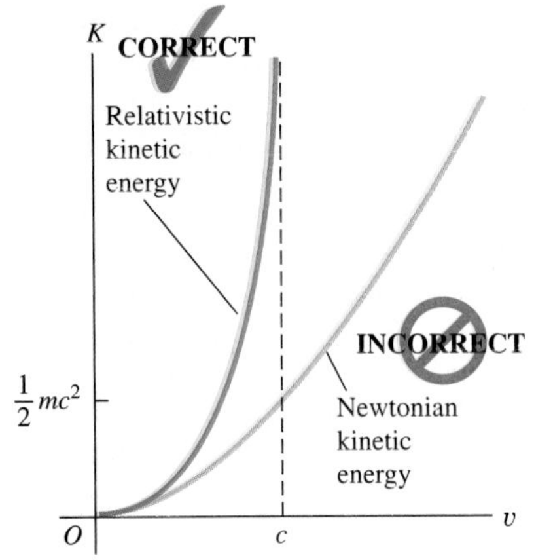

37.19 Graph of the kinetic energy of a particle of rest mass m as a function of speed v. Also shown is the Newtonian prediction, which gives correct results only at speeds much less than c.

Making these substitutions gives us

$$K = W = \int_0^v \frac{mv_x\, dv_x}{(1 - v_x^2/c^2)^{3/2}} \tag{37.35}$$

We can evaluate this integral by a simple change of variable; the final result is

$$K = \frac{mc^2}{\sqrt{1 - v^2/c^2}} - mc^2 = (\gamma - 1)mc^2 \tag{37.36}$$

(relativistic kinetic energy)

As v approaches c, the kinetic energy approaches infinity. If Eq. (37.36) is correct, it must also approach the Newtonian expression $K = \frac{1}{2}mv^2$ when v is much smaller than c (see Fig. 37.19). To verify this, we expand the radical, using the binomial theorem in the form

$$(1 + x)^n = 1 + nx + n(n - 1)x^2/2 + \cdots$$

In our case, $n = -1/2$ and $x = -v^2/c^2$, and we get

$$\gamma = \left(1 - \frac{v^2}{c^2}\right)^{-1/2} = 1 + \frac{1}{2}\frac{v^2}{c^2} + \frac{3}{8}\frac{v^4}{c^4} + \cdots$$

Combining this with $K = (\gamma - 1)mc^2$, we find

$$K = \left(1 + \frac{1}{2}\frac{v^2}{c^2} + \frac{3}{8}\frac{v^4}{c^4} + \cdots - 1\right)mc^2$$

$$= \frac{1}{2}mv^2 + \frac{3}{8}\frac{mv^4}{c^2} + \cdots \tag{37.37}$$

When v is much smaller than c, all the terms in the series except the first are negligibly small, and we obtain the Newtonian expression $\frac{1}{2}mv^2$.

Rest Energy

Equation (37.36) for the kinetic energy, the energy due to the motion of the particle, includes an energy term $mc^2/\sqrt{1 - v^2/c^2}$ that depends on the motion and a second energy term mc^2 that is independent of the motion. It seems that the kinetic energy of a particle is the difference between some **total energy** E and an energy mc^2 that it has even when it is at rest. Thus we can rewrite Eq. (37.36) as

$$E = K + mc^2 = \frac{mc^2}{\sqrt{1 - v^2/c^2}} = \gamma mc^2 \tag{37.38}$$

(total energy of a particle)

For a particle at rest $(K = 0)$, we see that $E = mc^2$. The energy mc^2 associated with rest mass m rather than motion is called the **rest energy** of the particle.

There is in fact direct experimental evidence that rest energy really does exist. The simplest example is the decay of a neutral *pion*. This is an unstable subatomic particle of rest mass m_π; when it decays, it disappears and electromagnetic radiation appears. If a neutral pion has no kinetic energy before its decay, the total energy of the radiation after its decay is found to equal exactly $m_\pi c^2$. In many other fundamental particle transformations the sum of the rest masses of the particles changes. In every case there is a corresponding energy change, consistent with the assumption of a rest energy mc^2 associated with a rest mass m.

Historically, the principles of conservation of mass and of energy developed quite independently. The theory of relativity shows that they are actually two special cases of a single broader conservation principle, the *principle of conservation of mass and energy.* In some physical phenomena, neither the sum of the rest masses of the particles nor the total energy other than rest energy is separately conserved, but there is a more general conservation principle: In an isolated system, when the sum of the rest masses changes, there is always a change in $1/c^2$ times the total energy other than the rest energy. This change is equal in magnitude but opposite in sign to the change in the sum of the rest masses.

This more general mass-energy conservation law is the fundamental principle involved in the generation of power through nuclear reactions. When a uranium nucleus undergoes fission in a nuclear reactor, the sum of the rest masses of the resulting fragments is *less than* the rest mass of the parent nucleus. An amount of energy is released that equals the mass decrease multiplied by c^2. Most of this energy can be used to produce steam to operate turbines for electric power generators (Fig. 37.20).

37.20 Although the control room of a nuclear power plant is very complex, the physical principle whereby such a plant operates is a simple one: Part of the rest energy of atomic nuclei is converted to thermal energy, which in turn is used to produce steam to drive electric generators.

We can also relate the total energy E of a particle (kinetic energy plus rest energy) directly to its momentum by combining Eq. (37.27) for relativistic momentum and Eq. (37.38) for total energy to eliminate the particle's velocity. The simplest procedure is to rewrite these equations in the following forms:

$$\left(\frac{E}{mc^2}\right)^2 = \frac{1}{1 - v^2/c^2} \quad \text{and} \quad \left(\frac{p}{mc}\right)^2 = \frac{v^2/c^2}{1 - v^2/c^2}$$

Subtracting the second of these from the first and rearranging, we find

$$E^2 = (mc^2)^2 + (pc)^2 \quad \text{(total energy, rest energy, and momentum)} \tag{37.39}$$

Again we see that for a particle at rest $(p = 0)$, $E = mc^2$.

Equation (37.39) also suggests that a particle may have energy and momentum even when it has no rest mass. In such a case, $m = 0$ and

$$E = pc \quad \text{(zero rest mass)} \tag{37.40}$$

In fact, zero rest mass particles do exist. Such particles always travel at the speed of light in vacuum. One example is the *photon,* the quantum of electromagnetic radiation (to be discussed in Chapter 38). Photons are emitted and absorbed during changes of state of an atomic or nuclear system when the energy and momentum of the system change.

Example 37.11 Energetic electrons

a) Find the rest energy of an electron $(m = 9.109 \times 10^{-31}\ \text{kg}$, $q = -e = -1.602 \times 10^{-19}\ \text{C})$ in joules and in electron volts. b) Find the speed of an electron that has been accelerated by an electric field, from rest, through a potential increase of 20.0 kV (typical of TV picture tubes) or of 5.00 MV (a high-voltage x-ray machine).

SOLUTION

IDENTIFY and SET UP: We use the relationship $E = mc^2$ to find the rest energy and Eq. (37.38) to find the speed that gives the stated total energy. We also use the concepts of electric potential energy from Chapter 23.

EXECUTE: a) The rest energy is

$$mc^2 = (9.109 \times 10^{-31}\ \text{kg})(2.998 \times 10^8\ \text{m/s})^2$$
$$= 8.187 \times 10^{-14}\ \text{J}$$

From the definition of the electron volt in Section 23.2, $1\ \text{eV} = 1.602 \times 10^{-19}\ \text{J}$. Using this, we find

$$mc^2 = (8.187 \times 10^{-14}\ \text{J})\frac{1\ \text{eV}}{1.602 \times 10^{-19}\ \text{J}}$$
$$= 5.11 \times 10^5\ \text{eV} = 0.511\ \text{MeV}$$

b) In calculations such as this, it is often convenient to work with the quantity γ defined from the modified Eq. (37.7):

$$\gamma = \frac{1}{\sqrt{1 - v^2/c^2}}$$

Solving this for v, we get

$$v = c\sqrt{1 - (1/\gamma)^2}$$

The total energy E of the accelerated electron is the sum of its rest energy mc^2 and the kinetic energy eV_{ba} that it gains from the work done on it by the electric field in moving from point a to point b:

$$E = \gamma mc^2 = mc^2 + eV_{ba}$$

or

$$\gamma = 1 + \frac{eV_{ba}}{mc^2}$$

An electron accelerated through a potential increase of $V_{ba} = 20.0\ \text{kV}$ gains an amount of energy 20.0 keV, so for this electron we have

$$\gamma = 1 + \frac{20.0 \times 10^3\ \text{eV}}{0.511 \times 10^6\ \text{eV}}$$
$$= 1.039$$

and

$$v = c\sqrt{1 - (1/1.039)^2} = 0.272c$$
$$= 8.15 \times 10^7\ \text{m/s}$$

Repeating the calculation for $V_{ba} = 5.00\ \text{MV}$, we find $eV_{ba}/mc^2 = 9.78$, $\gamma = 10.78$, and $v = 0.996c$.

EVALUATE: These results make sense: With $V_{ba} = 20.0\ \text{kV}$, the added kinetic energy of 20.0 keV is less than 4% of the rest energy of 0.511 MeV, and the final speed is about one-fourth of the speed of light. With $V_{ba} = 5.00\ \text{MV}$, the added kinetic energy of 5.00 MeV is much larger than the rest energy and the speed is close to c.

CAUTION The electron that accelerated from rest through a potential increase of 5.00 MV had a kinetic energy of 5.00 MeV. By convention we call such an electron a "5.00-MeV electron." A 5.00-MeV electron has a rest energy of 0.511 MeV (as do all electrons), a kinetic energy of 5.00 MeV, and a total energy of 5.51 MeV. Be careful not to mix these energies up.

Example 37.12 A relativistic collision

Two protons (each with $M = 1.67 \times 10^{-27}\ \text{kg}$) are initially moving with equal speeds in opposite directions. They continue to exist after a head-on collision that also produces a neutral pion ($m = 2.40 \times 10^{-28}\ \text{kg}$). If the protons and the pion are at rest after the collision, find the initial speed of the protons. Energy is conserved in the collision.

SOLUTION

IDENTIFY and SET UP: Since energy is conserved, we equate the (unknown) total energy of the two protons before the collision to the rest energy of the two protons and the pion after the collision. We then use Eq. (37.38) to solve for the speed of each proton.

EXECUTE: The energy of each proton before the collision is γMc^2. By conservation of energy,

$$2(\gamma Mc^2) = 2(Mc^2) + mc^2$$

$$\gamma = 1 + \frac{m}{2M} = 1 + \frac{2.40 \times 10^{-28}\ \text{kg}}{2(1.67 \times 10^{-27}\ \text{kg})} = 1.072$$

so

$$v = c\sqrt{1 - (1/\gamma)^2} = 0.360c$$

EVALUATE: The initial kinetic energy of each proton is $(\gamma - 1)Mc^2 = 0.072Mc^2$. The rest energy of a proton is 938 MeV, so the kinetic energy is $(0.072)(938\ \text{MeV}) = 67.5\ \text{MeV}$. (These are "67.5-MeV protons.") You can verify that the rest energy of the pion is twice this, or 135 MeV. All the kinetic energy "lost" in this completely inelastic collision is transformed into the rest energy of the pion.

Test Your Understanding

Each of the colliding protons in Example 37.12 is accelerated from rest to the collision speed in a distance of 40 m. What constant force is required to do this? Is the acceleration constant?

37.9 | Newtonian Mechanics and Relativity

The sweeping changes required by the principle of relativity go to the very roots of Newtonian mechanics, including the concepts of length and time, the equations of motion, and the conservation principles. Thus it may appear that we have destroyed the foundations on which Newtonian mechanics is built. In one sense this is true, yet the Newtonian formulation is still accurate whenever speeds are small in comparison with the speed of light in vacuum. In such cases, time dilation, length contraction, and the modifications of the laws of motion are so small that they are unobservable. In fact, every one of the principles of Newtonian mechanics survives as a special case of the more general relativistic formulation.

The laws of Newtonian mechanics are not *wrong;* they are *incomplete.* They are a limiting case of relativistic mechanics. They are *approximately* correct when all speeds are small in comparison to c, and they become exactly correct in the limit when all speeds approach zero. Thus relativity does not completely destroy the laws of Newtonian mechanics but *generalizes* them. Newton's laws rest on a very solid base of experimental evidence, and it would be very strange to advance a new theory that is inconsistent with this evidence. This is a common pattern in the development of physical theory. Whenever a new theory is in partial conflict with an older, established theory, the new must yield the same predictions as the old in areas in which the old theory is supported by experimental evidence. Every new physical theory must pass this test, called the **correspondence principle.**

At this point we may ask whether the special theory of relativity gives the final word on mechanics or whether *further* generalizations are possible or necessary. For example, inertial frames have occupied a privileged position in our discussion. Can the principle of relativity be extended to noninertial frames as well?

Here's an example that illustrates some implications of this question. A student decides to go over Niagara Falls while enclosed in a large wooden box. During her free fall she can float through the air inside the box. She doesn't fall to the floor because both she and the box are in free fall with a downward acceleration of 9.8 m/s^2. But an alternative interpretation, from her point of view, is that she doesn't fall to the floor because her gravitational interaction with the earth has suddenly been turned off. As long as she remains in the box and it remains in free fall, she cannot tell whether she is indeed in free fall or whether the gravitational interaction has vanished.

A similar problem appears in a space station in orbit around the earth. Objects in the space station *seem* to be weightless, but without looking outside the station there is no way to determine whether gravity has been turned off or whether the station and all its contents are accelerating toward the center of the earth. Figure 37.21 makes a similar point for a spaceship that is not in free fall but may be accelerating relative to an inertial frame or be at rest on the earth's surface.

These considerations form the basis of Einstein's **general theory of relativity.** If we cannot distinguish experimentally between a uniform gravitational field at a particular location and a uniformly accelerated reference frame, then there cannot be any real distinction between the two. Pursuing this concept, we may try to rep-

37.21 (a) An astronaut drops his watch in the spaceship. (b) In gravity-free space the floor accelerates upward at $a = g$ and hits the watch. (c) On the earth's surface the watch accelerates downward at $a = g$ and hits the floor. Without information from outside the spaceship, the astronaut cannot distinguish situation (b) from situation (c).

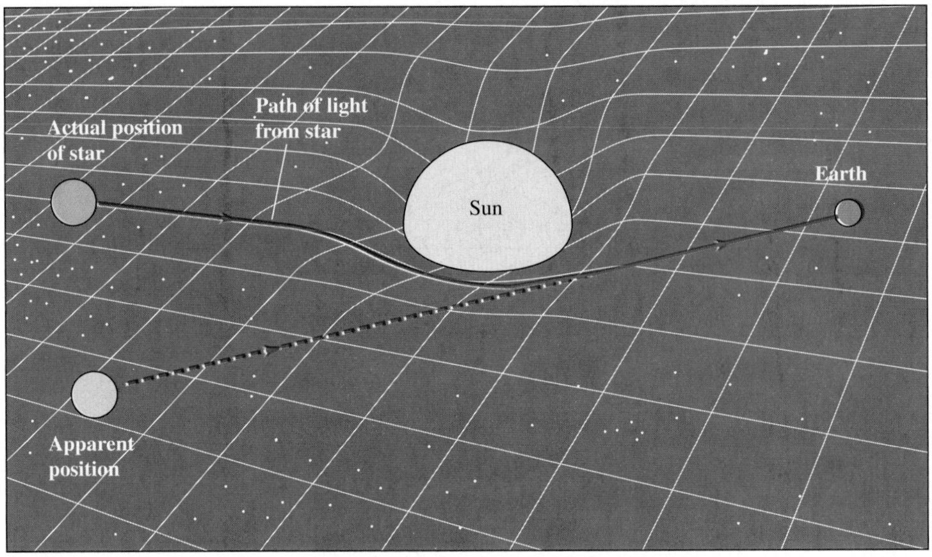

37.22 A two-dimensional representation of curved space. We imagine the space (a plane) as being distorted as shown by a massive object (the sun). Light from a distant star (solid line) follows the distorted surface on its way to the earth. The dashed line shows the direction from which the light *appears* to be coming. The effect is greatly exaggerated; for the sun, the maximum deviation is only 0.00048°.

resent *any* gravitational field in terms of special characteristics of the coordinate system. This turns out to require even more sweeping revisions of our space-time concepts than did the special theory of relativity. In the general theory of relativity the geometric properties of space are in general non-Euclidean (Fig. 37.22).

The general theory of relativity has passed several experimental tests, including three proposed by Einstein. One test has to do with understanding the rotation of the axes of the planet Mercury's elliptical orbit, called the *precession of the perihelion.* (The perihelion is the point of closest approach to the sun.) A second test concerns the apparent bending of light rays from distant stars when they pass near the sun. The third test is the *gravitational red shift,* the increase in wavelength of light proceeding outward from a massive source. Some details of the general theory are more difficult to test, but this theory has played a central role in cosmological investigations of the structure of the universe, the formation and evolution of stars, black holes, and related matters. In recent years the theory has also been confirmed by several purely *terrestrial* experiments.

SUMMARY

All of the fundamental laws of physics have the same form in all inertial frames of reference. The speed of light in vacuum is the same in all inertial frames and is independent of the motion of the source. Simultaneity is not an absolute concept; events that are simultaneous in one frame are not necessarily simultaneous in a second frame moving relative to the first.

Mavis sees the light from the front of the car first—she concludes that the bolt at the front of the car struck first

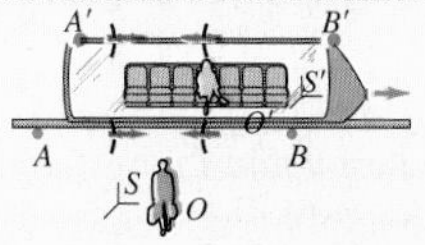

If two events occur at the same space point in a particular frame of reference, the time interval Δt_0 between the events as measured in that frame is called a proper time interval. If this frame moves with constant velocity u relative to a second frame, the time interval Δt between the events as observed in the second frame is longer than Δt_0. This effect is called time dilation. (Examples 37.1 through 37.3)

$$\Delta t = \frac{\Delta t_0}{\sqrt{1 - u^2/c^2}} = \gamma\,\Delta t_0 \quad (37.6), (37.8)$$

$$\gamma = \frac{1}{\sqrt{1 - u^2/c^2}} \quad (37.7)$$

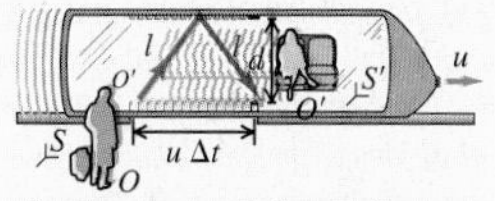

Stanley measures a longer time interval Δt: light pulse travels at same speed as in S', but travels a greater distance than in S'

If two points are at rest in a particular frame of reference, the distance l_0 between the points as measured in that frame is called a proper length. If this frame moves with constant velocity u relative to a second frame and the distances are measured parallel to the motion, the distance l between the points as measured in the second frame is shorter than l_0. This effect is called length contraction. (See Examples 37.4 through 37.6)

$$l = l_0\sqrt{1 - u^2/c^2} = \frac{l_0}{\gamma} \quad (37.16)$$

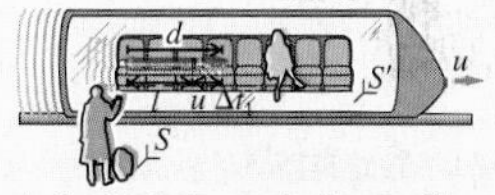

In Stanley's frame of reference S', ruler moves at speed u: from source to mirror, the light pulse travels a distance l (the length of the ruler measured in S) plus an additional distance $u\,\Delta t_1$

The Lorentz coordinate transformation relates the coordinates and time of an event in an inertial frame S to the coordinates and time of the same event as observed in a second inertial frame S' moving at velocity u relative to the first. For one-dimensional motion, a particle's velocities v_x in S and v_x' in S' are related by the Lorentz velocity transformation. (See Examples 37.7 and 37.8)

$$x' = \frac{x - ut}{\sqrt{1 - u^2/c^2}} = \gamma(x - ut)$$

$$y' = y \qquad z' = z$$

$$t' = \frac{t - ux/c^2}{\sqrt{1 - u^2/c^2}} = \gamma(t - ux/c^2) \quad (37.21)$$

$$v_x' = \frac{v_x - u}{1 - uv_x/c^2} \quad (37.22)$$

$$v_x = \frac{v_x' + u}{1 + uv_x'/c^2} \quad (37.23)$$

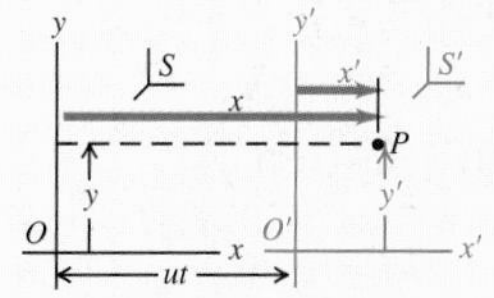

Origins O and O' coincide at time $t = 0 = t'$

The Doppler effect is the frequency shift in light from a source due to the relative motion of source and observer. For a source moving toward the observer with speed u, Eq. (37.25) gives the received frequency f in terms of the emitted frequency f_0. (See Example 37.9)

$$f = \sqrt{\frac{c + u}{c - u}}\,f_0 \quad (37.25)$$

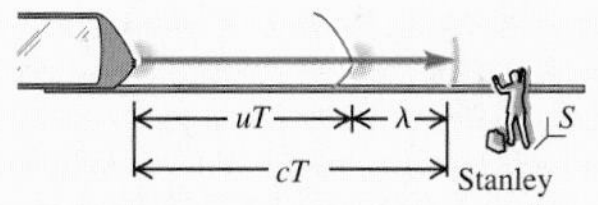

For a particle of rest mass m moving with velocity $\vec{v}$, the relativistic momentum $\vec{p}$ is given by Eq. (37.27) or (37.31) and the relativistic kinetic energy K is given by Eq. (37.36). The total energy E is the sum of the kinetic energy and the rest energy mc^2. The total energy can also be expressed in terms of the magnitude of momentum p and rest mass m. (See Examples 37.10 through 37.12)

$$\vec{p} = \frac{m\vec{v}}{\sqrt{1 - v^2/c^2}} = \gamma m\vec{v} \qquad (37.27), (37.31)$$

$$K = \frac{mc^2}{\sqrt{1 - v^2/c^2}} - mc^2 = (\gamma - 1)mc^2 \qquad (37.36)$$

$$E = K + mc^2 = \frac{mc^2}{\sqrt{1 - v^2/c^2}} = \gamma mc^2 \qquad (37.38)$$

$$E^2 = (mc^2)^2 + (pc)^2 \qquad (37.39)$$

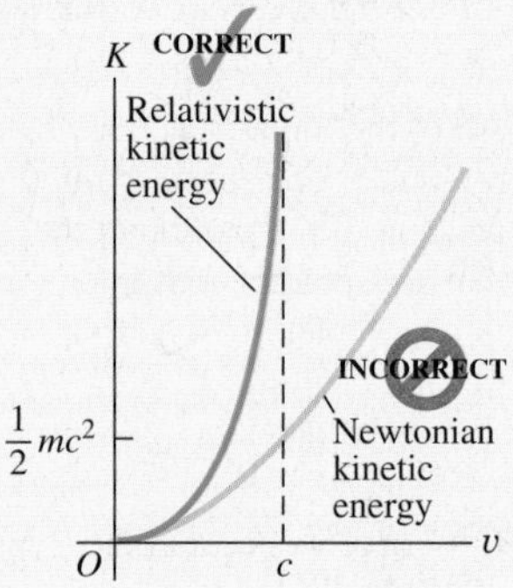

The special theory of relativity is a generalization of Newtonian mechanics. All the principles of Newtonian mechanics are present as limiting cases when all the speeds are small compared to c. Further generalization to include noninertial frames of reference and their relation to gravitational fields leads to the general theory of relativity.

Key Terms

special theory of relativity, 1403
principle of relativity, 1404
Galilean coordinate transformation, 1406
simultaneity, 1407
event, 1407
time dilation, 1410
proper time, 1411
twin paradox, 1413
proper length, 1416
length contraction, 1416
Lorentz transformations, 1419
spacetime, 1420
spacetime coordinates, 1420
rest mass, 1426
relativistic momentum, 1426
total energy, 1430
rest energy, 1430
correspondence principle, 1433
general theory of relativity, 1433

Your Notes

Answer to Chapter Opening Question

While the speed of light c is the ultimate "speed limit" for any particle, there is *no* upper limit on a particle's kinetic energy (see Fig. 37.19). As the speed approaches c, a small increase in speed corresponds to a large increase in kinetic energy.

Answers to Test Your Understanding Questions

Section 37.1 You, too, will measure a spherical wave front that expands at the same speed c in all directions. This is a consequence of Einstein's second postulate. The wave front that you measure is *not* centered on the current position of the spaceship; rather, it is centered on the point P where the spaceship was located at the instant that it emitted the light pulse. For example, suppose the spaceship is moving at speed $c/2$. When your watch shows that a time t has elapsed since the pulse of light was emitted, your measurements will show that the wave front is a sphere of radius ct centered on P and that the spaceship is a distance $ct/2$ from P.

Section 37.2 In Mavis's frame of reference, the two events (the Ogdenville clock striking noon and the North Haverbrook clock striking noon) are not simultaneous. Figure 37.4 shows that the event toward the front of the monorail car occurs first. Since the monorail car is moving toward North Haverbrook, that clock struck noon before the one on Ogdenville. So, according to Mavis, it is after noon in North Haverbrook.

Section 37.3 With $u = 0.600c$, $\gamma = 1/\sqrt{1 - u^2/c^2} = 1/\sqrt{1 - (0.600)^2} = 1.25$. For the first question, the two events are when Maria passes Samir and when Samir's stopwatch reads 16.0 s. These events occur at the same position in Samir's frame, so the time interval in his frame is the proper time: $\Delta t_0 = 1.00$ s. The time interval between these two events in Maria's frame is $\Delta t = \gamma\,\Delta t_0 = (1.25)(16.0\text{ s}) = 20.0$ s, so Maria concludes that Samir's stopwatch is running slow (it ticked off only 16.0 s while her stopwatch ticked off 20.0 s). The second question is about a *different* pair of events: when Maria passes Samir, and when *Maria's* stopwatch reads 20.0 s. These two events occur at the same position in *Maria's* frame, so the time interval in *her* proper time: $\Delta t_0 = 1.25$. The time interval between *these* two events in Samir's frame is $\Delta t = \gamma\,\Delta t_0 = (1.25)(20.0\text{ s}) = 25.0$ s. Hence Samir concludes that Maria's stopwatch is running slow (it ticked off only 20.0 s while his stopwatch ticked off 25.0 s). There is no contradiction, because the two observers are using different pairs of events. *Both* observers conclude that a moving stopwatch runs slow. This is consistent with the principle of relativity (Section 37.1), which states that the laws of physics are the same in all inertial frames of reference.

Section 37.4 The cube is at rest in Melinda's frame of reference, so its dimensions as measured in her frame are all proper lengths. Santiago measures a contracted length along the y-direction (the direction of relative motion): $l = l_0\sqrt{1 - u^2/c^2} = (15.0\text{ cm})\sqrt{1 - (0.800)^2} = 9.00$ cm. There is no length contraction along directions perpendicular to the relative motion, so Santiago measures both the x- and z-dimensions of the cube to be 15.0 cm. In Santiago's frame of reference, the cube is not cubical!

Section 37.5 To determine the velocity of the robot space probe relative to the scoutship, we let S be the frame of reference of the earth (as in Fig. 37.14) and let S' be the frame of reference of the *scoutship* (we want the velocity of the robot space probe relative to this frame). The velocity of S' with respect to S is therefore $u = 0.950c$ (the velocity of the scoutship relative to the earth). From Example 37.8 we know the velocity of the robot space probe with respect to S, $v_x = 0.980c$, and our target variable is the velocity of the robot space probe with respect to S', v'_x. Using Eq. (37.22), we find

$$v'_x = \frac{v_x - u}{1 - uv_x/c^2} = \frac{0.980c - 0.950c}{1 - (0.950c)(0.980c)/c^2} = 0.43c$$

Had you used the Galilean velocity transformation, you would have gotten a very different (and incorrect!) answer, $(0.980c) - 0.950c = 0.030c$.

Section 37.6 Since the source is moving away from us, we use Eq. (37.26) to determine f from $f_0 = 5.55 \times 10^{13}$ Hz and $u = 0.986c$. We find

$$f = \sqrt{\frac{c - u}{c + u}}f_0 = \sqrt{\frac{1 - u/c}{1 + u/c}}f_0 = \sqrt{\frac{1 - 0.986}{1 + 0.986}}(5.55 \times 10^{13}\text{ Hz})$$
$$= 4.66 \times 10^{12}\text{ Hz}$$

This is in the far-infrared part of the electromagnetic spectrum (see Section 32.6).

Section 37.7 This occurs when the relativistic value $p = \gamma mv$ is twice the nonrelativistic value $p = mv$, so $\gamma mv = 2mv$ or $\gamma = 1/\sqrt{1 - v^2/c^2} = 2$. Solving for v, we find $v = c\sqrt{1 - (1/\gamma)^2} = c\sqrt{1 - (1/2)^2} = c\sqrt{3}/2 = 0.866c$.

Section 37.8 A constant force F that acts over a distance s does work $W = Fs$, which is converted into kinetic energy K. From Eq. (37.36), $K = (\gamma - 1)Mc^2 = Fs$, so $F = (\gamma - 1)Mc^2/s = (0.072)(938\text{ MeV})(1.602 \times 10^{-13}\text{ J/MeV})/(40\text{ m}) = 2.7 \times 10^{-13}$ N. The acceleration is *not* constant; it decreases as the speed increases (see the discussion of Eq. (37.30) in Section 37.7).

Discussion Questions

Q37.1 You are standing on a train platform watching a high-speed train pass by. A light inside one of the train cars is turned on and then a little later it is turned off. a) Who can measure the proper time interval for the duration of the light, you or a passenger on the train? b) Who can measure the proper length of the train car, you or a passenger on the train? c) Who can measure the proper length of a sign attached to a post on the train platform, you or a passenger on the train? In each case explain your answer.

Q37.2 If simultaneity is not an absolute concept, does that mean that we must discard the concept of causality? If event A is to *cause* event B, A must occur first. Is it possible that in some frames, A appears to be the cause of B, and in others, B appears to be the cause of A? Explain.

Q37.3 A new book purports to refute the special theory of relativity. The author begins with a premise that might be paraphrased as follows: "Either two events occur at the same time, or they don't. That's just common sense." In the light of our discussion of the relative nature of simultaneity, how would you respond to this?

Q37.4 What do you think would be different in everyday life if the speed of light were 10 m/s instead of 3.00×10^8 m/s?

Q37.5 The average life span in the United States is about 70 years. Does this mean that it is impossible for an average person to travel a distance greater than 70 light years away from the earth? (A light year is the distance light travels in a year.) Explain.

Q37.6 A physicist claims that it is impossible to define what is meant by a *rigid body* in a relativistically correct way. Why should this be?

Q37.7 Two events occur at the same space point in a particular inertial frame of reference and are simultaneous in that frame. Is it possible that they may not be simultaneous in a different inertial frame? Explain.

Q37.8 As Mavis flies past Stanley at $0.80c$, she holds a meter stick that is aligned parallel to her flight path. She says that her meter stick is 1.00 m in length, but Stanley says it is $(1.00 \text{ m})\sqrt{1-(0.80)^2} = 0.60$ m in length. Which observer is correct?

Q37.9 The theory of relativity sets an upper limit on the speed that a particle can have. Are there also limits on the energy and momentum of a particle? Explain.

Q37.10 When a monochromatic light source moves toward an observer, its wavelength appears to be shorter than the value measured when the source is at rest. Does this contradict the hypothesis that the speed of light is the same for all observers? Explain.

Q37.11 The speed of light relative to still water is 2.25×10^8 m/s. If the water is moving past us, the speed of light we measure depends on the speed of the water. Do these facts violate Einstein's second postulate? Explain.

Q37.12 A student asserts that a material particle must always have a speed less than that of light, and a massless particle must always move at exactly the speed of light. Is she correct? If so, how do massless particles such as photons and neutrinos acquire this speed? Can't they start from rest and accelerate? Explain.

Q37.13 In principle, does a hot gas have more mass than the same gas when it is cold? Explain. In practice, would this be a measurable effect? Explain.

Q37.14 Why do you think the development of Newtonian mechanics preceded the more refined relativistic mechanics by so many years?

Exercises

Section 37.2 Relativity of Simultaneity

37.1 Suppose the two lightning bolts shown in Fig. 37.4a are simultaneous to an observer on the train. Show that they are *not* simultaneous to an observer on the ground. Which lightning strike does the ground observer measure to come first?

Section 37.3 Relativity of Time Intervals

37.2 The positive muon (μ^+), an unstable particle, lives on average 2.20×10^{-6} s (measured in its own frame of reference) before decaying. a) If such a particle is moving, with respect to the laboratory, with a speed of $0.900c$, what average lifetime is measured in the laboratory? b) What average distance, measured in the laboratory, does the particle move before decaying?

37.3 Everyday Time Dilation. Two atomic clocks are carefully synchronized. One remains in New York, and the other is loaded on an airliner that travels at an average speed of 250 m/s and then returns to New York. When the plane returns, the elapsed time on the clock that stayed behind is 4.00 h. By how much will the reading of the two clocks differ, and which clock will show the shorter elapsed time? (*Hint:* Since $u \ll c$, you can simplify $\sqrt{1-u^2/c^2}$ by a binomial expansion.)

37.4 A spaceship flies past Mars with a speed of $0.978c$ relative to the surface of the planet. When the spaceship is directly overhead at an altitude of 1200 km, a very bright signal light on the Martian surface blinks on and then off. An observer on Mars measures that the signal light was on for 82.4 μs. What is the duration of the light pulse measured by the pilot of the spaceship?

37.5 The negative pion (π^-) is an unstable particle with an average lifetime of 2.60×10^{-8} s (measured in the rest frame of the pion). a) If the pion is made to travel at very high speed relative to a laboratory, its average lifetime is measured in the laboratory to be 4.20×10^{-7} s. Calculate the speed of the pion expressed as a fraction of c. b) What distance, measured in the laboratory, does the pion travel during its average lifetime?

37.6 As you pilot your space utility vehicle at a constant speed toward the moon, a race pilot flies past you in her spaceracer at a constant speed of $0.800c$ relative to you. At the instant the spaceracer passes you, both of you start timers at zero. a) At the instant when you measure that the spaceracer has traveled 1.20×10^8 m past you, what does the race pilot read on her timer? b) When the race pilot reads the value calculated in part (a) on her timer, what does she measure to be your distance from her? c) At the instant when the race pilot reads the value calculated in part (a) on her timer, what do you read on yours?

37.7 A spacecraft flies away from the earth with a speed of 4.80×10^6 m/s relative to the earth and then returns at the same speed. The spacecraft carries an atomic clock that has been carefully synchronized with an identical clock that remains at rest on earth. The spacecraft returns to its starting point 365 days (1 year) later, as measured by the clock that remained on earth. What is the difference in the elapsed times on the two clocks, measured in hours? Which clock, the one in the spacecraft or the one on earth, shows the smallest elapsed time?

37.8 An alien spacecraft is flying overhead at a great distance as you stand in your backyard. You see its searchlight blink on for 0.190 s. The first officer on the spacecraft measures that the searchlight is on for 12.0 ms. a) Which of these two measured times is the proper time? b) What is the speed of the spacecraft relative to the earth expressed as a fraction of the speed of light c?

Section 37.4 Relativity of Length

37.9 A muon is created 55.0 km above the surface of the earth (as measured in the earth's frame). The average lifetime of a muon, measured in its own rest frame, is 2.20×10^{-6} s; the muon we are considering has this lifetime. In the frame of the muon the earth is moving toward the muon with a speed of $0.9860c$. a) In the muon's frame, what is its initial height above the surface of the earth? b) In the muon's frame, how much closer does the earth get during the lifetime of the muon? What fraction is this of the muon's original height as measured in the muon's frame? c) In the earth's frame, what is the lifetime of the muon? In the earth's frame, how far does the muon travel during its lifetime? What fraction is this of the muon's original height in the earth's frame?

37.10 An unstable particle is created in the upper atmosphere from a cosmic ray and travels straight down toward the surface of the earth with a speed of $0.99540c$ relative to the earth. A scientist at rest on the earth's surface measures that the particle is created at an altitude of 45.0 km. a) As measured by the scientist, how much time does it take the particle to travel the 45.0 km to the surface of the earth? b) Use the length contraction formula to calculate the distance from where the particle is created to the surface of the earth as measured in the particle's frame. c) In the particle's frame, how much time does it take the particle to travel from where it is created to the surface of the earth? Calculate this time both by the time dilation formula and also from the distance calculated in part (b). Do the two results agree?

37.11 As measured by an observer on the earth, a spacecraft runway on earth has a length of 3600 m. a) What is the length of the runway as measured by a pilot of a spacecraft flying past at a speed of 4.00×10^7 m/s relative to the earth? b) An observer on earth measures the time interval from when the spacecraft is directly over one end of the runway until it is directly over the other end. What result does she get? c) The pilot of the spacecraft measures the time it takes him to travel from one end of the runway to the other end. What value does he get?

37.12 A meter stick moves past you at great speed. Its motion relative to you is parallel to its long axis. If you measure the length of the moving meter stick to be 1.00 ft ($1 \text{ ft} = 0.3048 \text{ m}$), for example by comparing it to a one-foot ruler that is at rest relative to you, at what speed is the meter stick moving relative to you?

37.13 A spacecraft of the Trade Federation flies past the planet Coruscant at a speed of $0.600c$. A scientist on Coruscant measures the length of the moving spacecraft to be 74.0 m. The spacecraft later lands on Coruscant, and the same scientist measures the length of the now stationary spacecraft. What value does she get?

Section 37.5 The Lorentz Transformations

37.14 Solve Eq. (37.21) to obtain x and t in terms of x' and t', and show that the resulting transformation has the same form as the original one except for a change of sign for u.

37.15 An observer in frame S' is moving to the right ($+x$-direction) at speed $u = 0.600c$ away from a stationary observer in frame S. The observer in S' measures the speed v' of a particle moving to the right away from her. What speed v does the observer in S measure for the particle if a) $v' = 0.400c$? b) $v' = 0.900c$? c) $v' = 0.990c$?

37.16 Space pilot Mavis zips past Stanley at a constant speed relative to him of $0.800c$. Mavis and Stanley start timers at zero when the front of Mavis's ship is directly above Stanley. When Mavis reads 5.00 s on her timer, she turns on a bright light under the front of her spaceship. a) Use the Lorentz coordinate transformation derived in Exercise 37.14 and Example 37.7 (Section 37.5) to calculate x and t as measured by Stanley for the event of turning on the light. b) Use the time dilation formula, Eq. (37.6), to calculate the time interval between the two events (the front of the spaceship passing overhead and turning on the light) as measured by Stanley. Compare to the value of t you calculated in part (a). c) Multiply the time interval by Mavis's speed, both as measured by Stanley, to calculate the distance she has traveled as measured by him when the light turns on. Compare to the value of x you calculated in part (a).

37.17 Show in detail the derivation of Eq. (37.20) from Eqs. (37.18) and (37.19).

37.18 Show the details of the derivation of Eq. (37.23) from (37.22).

37.19 A pursuit spacecraft from the planet Tatooine is attempting to catch up with a Trade Federation cruiser. As measured by an observer on Tatooine, the cruiser is traveling away from the planet with a speed of $0.600c$. The pursuit ship is traveling at a speed of $0.800c$ relative to Tatooine, in the same direction as the cruiser. What is the speed of the cruiser relative to the pursuit ship?

37.20 Two particles in a high-energy accelerator experiment are approaching each other head-on, each with a speed $0.9520c$ as measured in the laboratory. What is the magnitude of the velocity of one particle relative to the other?

37.21 Two particles are created in a high-energy accelerator and move off in opposite directions. The speed of one particle, as measured in the laboratory, is $0.650c$, and the speed of each particle relative to the other is $0.950c$. What is the speed of the second particle, as measured in the laboratory?

37.22 An enemy spaceship is moving toward your starfighter with a speed, as measured in your frame, of $0.400c$. The enemy ship fires a missile toward you at a speed of $0.700c$ relative to the enemy ship (see Fig. 37.23). a) What is the speed of the missile relative to you? Express your answer in terms of the speed of light. b) If you measure that the enemy ship is 8.00×10^6 km away from you when the missile is fired, how much time, measured in your frame, will it take the missile to reach you?

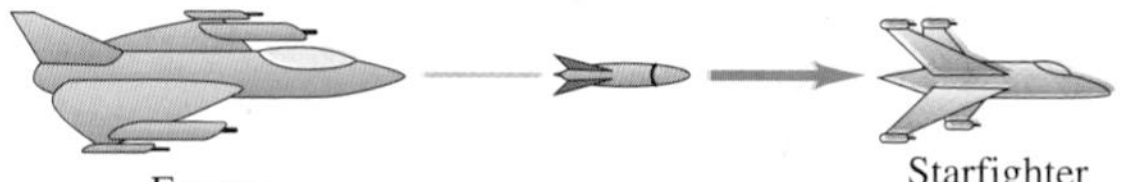

Figure 37.23 Exercise 37.22.

37.23 An imperial spaceship, moving at high speed relative to the planet Arrakis, fires a rocket toward the planet with a speed of $0.920c$ relative to the spaceship. An observer on Arrakis measures that the rocket is approaching with a speed of $0.360c$. What is the speed of the spaceship relative to Arrakis? Is the spaceship moving toward or away from Arrakis?

*Section 37.6 The Doppler Effect for Electromagnetic Waves

***37.24** In terms of c, what relative velocity u between a source of electromagnetic waves and an observer produces a) a 2.0% decrease in frequency? b) an increase by a factor of four of the frequency of the observed light?

***37.25 Tell It to the Judge.** a) How fast must you be approaching a red traffic light ($\lambda = 675$ nm) for it to appear yellow ($\lambda = 575$ nm)? Express your answer in terms of the speed of light. b) If you used this as an excuse for not getting a ticket for running a red light, how much of a fine would you get for speeding? Assume that the fine is \$1.00 for each kilometer per hour that your speed exceeds the posted limit of 90 km/h.

***37.26** Show that when the source of electromagnetic waves moves away from us at $0.600c$, the frequency we measure is half the value measured in the rest frame of the source.

Section 37.7 Relativistic Momentum

37.27 a) A particle with mass m moves along a straight line under the action of a force F directed along the same line. Evaluate the derivative in Eq. (37.29) to show that the acceleration $a = dv/dt$ of the particle is given by $a = (F/m)(1 - v^2/c^2)^{3/2}$. b) Evaluate the derivative in Eq. (37.29) to find the expression for the magnitude of the acceleration in terms of F, m, and v/c if the force is perpendicular to the velocity.

37.28 Relativistic Baseball. Calculate the magnitude of the force required to give a 0.145-kg baseball an acceleration $a = 1.00\ \text{m/s}^2$ in the direction of the baseball's initial velocity when this velocity has a magnitude of a) 10.0 m/s; b) $0.900c$; c) $0.990c$. d) Repeat parts (a), (b), and (c) if the force and acceleration are perpendicular to the velocity.

37.29 a) At what speed is the momentum of a particle twice as great as the result obtained from the nonrelativistic expression mv? Express your answer in terms of the speed of light. b) A force is applied to a particle along its direction of motion. At what speed is the magnitude of force required to produce a given acceleration twice as great as the force required to produce the same acceleration when the particle is at rest? Express your answer in terms of the speed of light.

37.30 a) At what speed does the momentum of a particle differ by 1.0% from the value obtained using the nonrelativistic expression mv? b) Is the correct relativistic value greater or less than that obtained from the nonrelativistic expression?

Section 37.8 Relativistic Work and Energy

37.31 What is the speed of a particle whose kinetic energy is equal to a) its rest energy; b) five times its rest energy?

37.32 Annihilation. In proton-antiproton annihilation a proton and an antiproton (a negatively charged proton) collide and disappear, producing electromagnetic radiation. If each particle has a mass of 1.67×10^{-27} kg and they are at rest just before the annihilation, find the total energy of the radiation. Give your answers in joules and in electron volts.

37.33 What is the speed of a particle if its kinetic energy is 2.0% larger than $\frac{1}{2}mv^2$? (*Hint:* Use Eq. (37.37).)

37.34 a) How much work must be done on a particle with mass m to accelerate it a) from rest to a speed of $0.090c$? b) from a speed $0.900c$ to a speed $0.990c$? (Express the answers in terms of mc^2.) c) How do your answers in parts (a) and (b) compare?

37.35 a) By what percentage does your rest mass increase when you climb 30 m to the top of a ten-story building? Are you aware of this increase? Explain. b) By how many grams does the mass of a 12.0-g spring with force constant 200 N/cm change when you compress it by 6.0 cm? Does the mass increase or decrease? Would you notice the change in mass if you were holding the spring? Explain.

37.36 A 60.0-kg person is standing at rest on level ground. How fast would she have to run to a) double her total energy? b) increase her total energy by a factor of ten?

37.37 A proton (rest mass 1.67×10^{-27} kg) has total energy that is 4.00 times its rest energy. What is a) the kinetic energy of the proton? b) the magnitude of the momentum of the proton? c) the speed of the proton?

37.38 How much work must be done on a proton to accelerate it from rest to a) $0.100c$? b) $0.500c$? c) $0.900c$?

37.39 A particle has rest mass 6.64×10^{-27} kg and momentum $2.10 \times 10^{-18}\ \text{kg}\cdot\text{m/s}$. a) What is the total energy (kinetic plus rest energy) of the particle? b) What is the kinetic energy of the particle? c) What is the ratio of the kinetic energy to the rest energy of the particle?

37.40 Starting from Eq. (37.39), show that in the classical limit $(pc \ll mc^2)$ the energy approaches the classical kinetic energy $\frac{1}{2}mv^2$ plus the rest mass energy mc^2.

37.41 Compute the kinetic energy of a proton (mass 1.67×10^{-27} kg) using both the nonrelativistic and relativistic expressions, and compute the ratio of the two results (relativistic divided by nonrelativistic) for speeds of a) 8.00×10^7 m/s; b) 2.85×10^8 m/s.

37.42 A ψ ("psi") particle has mass 5.52×10^{-27} kg. Compute the rest energy of the ψ in MeV.

37.43 a) Through what potential difference does an electron have to be accelerated, starting from rest, to achieve a speed of $0.980c$? b) What is the kinetic energy of the electron at this speed? Express your answer in J and in eV.

37.44 Creating a Particle. Two protons (each with rest mass $M = 1.67 \times 10^{-27}$ kg) are initially moving with equal speeds in opposite directions. The protons continue to exist after a collision that also produces an η^0 particle (Chapter 44). The rest mass of the η^0 is $m = 9.75 \times 10^{-28}$ kg. a) If the two protons and the η^0 are all at rest after the collision, find the initial speed of the protons, expressed as a fraction of the speed of light. b) What is the kinetic energy of each proton? Express your answer in MeV. c) What is the rest energy of the η^0, expressed in MeV? d) Discuss the relationship between the answers to parts (b) and (c).

37.45 An 0.420-MeV electron in an x-ray tube strikes an anode. When it arrives at the anode, what is a) its kinetic energy in eV? b) its total energy in eV? c) its speed? d) What is the speed of the electron, calculated classically?

37.46 Energy of Fusion. In a hypothetical nuclear-fusion reactor, two deuterium nuclei combine or "fuse" to form one helium nucleus. The mass of a deuterium nucleus, expressed in atomic mass units (u), is 2.0136 u; that of a helium nucleus is 4.0015 u $(1\ \text{u} = 1.6605402 \times 10^{-27}\ \text{kg})$. a) How much energy is released when 1.0 kg of deuterium undergoes fusion? b) The annual consumption of electrical energy in the United States is of the order of 1.0×10^{19} J. How much deuterium must react to produce this much energy?

37.47 The sun produces energy by nuclear fusion reactions, in which matter is converted into energy. By measuring the amount of energy we receive from the sun, we know that it is producing energy at a rate of 3.8×10^{26} W. a) How many kilograms of matter does the sun lose each second? Approximately how many tons of matter is this? b) At this rate, how long would it take the sun to use up all its mass?

37.48 A very small object with mass 2.00×10^{-12} kg is accelerated from rest to a speed of $0.920c$ by a constant force of 4.20×10^4 N. a) If the nonrelativistic form of Newton's second law is used, $F = ma$, what distance has the object traveled by the time it reaches this speed? b) Using the correct relativistic treatment of Section 37.8, what distance has the object traveled by the time it has reached this speed? c) What is the magnitude of the acceleration of the particle when its speed is i) $0.100c$? ii) $0.462c$? iii) $0.920c$?

Problems

37.49 After being produced in a collision between elementary particles, a positive pion (π^+) must travel down a 1.20-km-long tube to reach an experimental area. A π^+ particle has an average lifetime (measured in its rest frame) of 2.60×10^{-8} s; the π^+ we are considering has this lifetime. a) How fast must the π^+ travel if it is not to decay before it reaches the end of the tube? (Since u will be very close to c, write $u = (1 - \Delta)c$ and give your answer in terms of Δ rather than u.) b) The π^+ has a rest energy of 139.6 MeV. What is the total energy of the π^+ at the speed calculated in part (a)?

37.50 A cube of metal with sides of length a sits at rest in a frame S with one edge parallel to the x-axis. Therefore, in S the cube has volume a^3. Frame S' moves along the x-axis with a speed u. As measured by an observer in frame S', what is the volume of the metal cube?

37.51 The starships of the Solar Federation are marked with the symbol of the Federation, a circle, while starships of the Denebian Empire are marked with the Empire's symbol, an ellipse whose major axis is 1.40 times its minor axis ($a = 1.40b$ in Fig. 37.24). How fast, relative to an observer, does a Federation ship have to travel for its markings to be confused with those of an Empire ship?

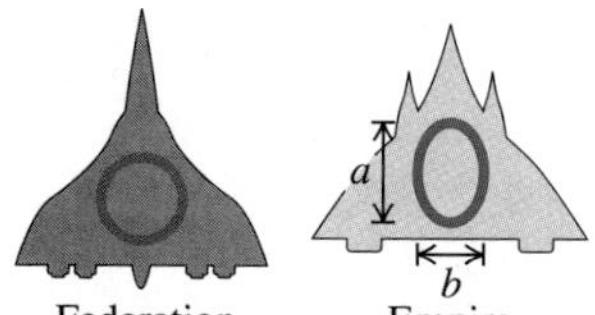

Figure 37.24 Problem 37.51.

37.52 A space probe is sent to the vicinity of the star Capella, which is 42.2 light years from the earth. (A light year is the distance light travels in a year.) The probe travels with a speed of $0.9910c$. An astronaut recruit on board is 19 years old when the probe leaves the earth. What is her biological age when the probe reaches Capella?

37.53 A particle is said to be *extremely relativistic* when its kinetic energy is much larger than its rest energy. a) What is the speed of a particle (expressed as a fraction of c) such that the total energy is ten times the rest energy? b) What is the percentage difference between the left and right sides of Eq. (37.39) if $(mc^2)^2$ is neglected for a particle with the speed calculated in part (a)?

37.54 A proton (rest mass 1.67×10^{-27} kg) is initially moving at $0.900c$ in the $+x$-direction, relative to the reference frame of a physics laboratory. Then a constant force $\vec{F} = (-3.00 \times 10^{-12}\text{ N})\hat{\imath} + (5.00 \times 10^{-12}\text{ N})\hat{\jmath}$ is applied to the proton. a) What are the x- and y-components of the acceleration of the proton just after the force is applied? b) What is the angle between $\vec{F}$ and the acceleration you calculated in part (a)?

37.55 The Superconducting Supercollider (SSC). The U.S. was planning to build (in Texas) the largest accelerator in the world, the Superconducting Super Collider, or SSC for short. This machine would accelerate protons to kinetic energies of 20 TeV in an underground ring approximately 17 miles in diameter. (For good discussions of this accelerator, along with justification for it and a discussion of some of the physics problems it would be used to investigate, see "The SSC: A Machine for the Nineties" by Sheldon Glashow and Leon Lederman in the March 1985 issue of *Physics Today* and "The Superconducting Supercollider" by J.D. Jackson, M. Tigner, and S. Wojcicki in the April 1986 issue of *Scientific American*. This project was canceled by Congress in 1993.) a) What speed v would the protons reach in the SSC? (Since v is very close to c, write $v = (1 - \Delta)c$ and give your answer in terms of Δ.) b) Find the relativistic mass, m_{rel}, of the accelerated protons in terms of their rest mass.

37.56 A nuclear bomb containing 8.00 kg of plutonium explodes. The sum of the rest masses of the products of the explosion is less than the original rest mass by one part in 10^4. a) How much energy is released in the explosion? b) If the explosion takes place in 4.00 μs, what is the average power developed by the bomb? c) What mass of water could the released energy lift to a height of 1.00 km?

37.57 By what minimum amount does the mass of 4.00 kg of ice increase when the ice melts at 0°C to form water at 0°C?

37.58 A photon with energy E is emitted by an atom with mass m, which recoils in the opposite direction. a) Assuming that the motion of the atom can be treated nonrelativistically, compute the recoil speed of the atom. b) From the result of part (a), show that the recoil speed is much smaller than c whenever E is much smaller than the rest energy mc^2 of the atom.

37.59 Čerenkov Radiation. The Russian physicist P. A. Čerenkov discovered that a charged particle traveling in a solid with a speed exceeding the speed of light in that material radiates electromagnetic radiation. (This is analogous to the sonic boom produced by an aircraft moving faster than the speed of sound in air; see Section 16.9. Čerenkov shared the 1958 Nobel Prize for this discovery.) What is the minimum kinetic energy (in electron volts) that an electron must have while traveling inside a slab of crown glass $(n = 1.52)$ in order to create this Čerenkov radiation?

37.60 Albert in Wonderland. Einstein and Lorentz, being avid tennis players, play a fast-paced game on a court where they stand 20.0 m from each other. Being very skilled players, they play without a net. The tennis ball has mass 0.0580 kg. You can ignore gravity and assume that the ball travels parallel to the ground as it travels between the two players. Unless otherwise specified, all measurements are made by the two men. a) Lorentz serves the ball at 80.0 m/s. What is the ball's kinetic energy? b) Einstein slams a return at 1.80×10^8 m/s. What is the ball's kinetic energy? c) During Einstein's return of the ball in part (a), a white rabbit runs beside the court in the direction from Einstein to Lorentz. The rabbit has a speed of 2.20×10^8 m/s relative to the two men. What is the speed of the rabbit relative to the ball? d) What does the rabbit measure as the distance from Einstein to Lorentz? e) How much time does it take for the rabbit to run 20.0 m, according to the players? f) The white rabbit carries a pocket watch. He uses this watch to measure the time (as he sees it) for the distance from Einstein to Lorentz to pass by under him. What time does he measure?

37.61 Frame S' has an x-component of velocity u relative to frame S and at $t = t' = 0$ the two frames coincide (see Fig. 37.3). A light pulse with a spherical wave front is emitted at the origin of S' at time $t' = 0$. Its distance x' from the origin after a time t' is given by $x'^2 = c^2t'^2$. Use the Lorentz coordinate transformation to transform this equation to an equation in x and t, and show that the result is $x^2 = c^2t^2$; that is, the motion appears exactly the same in the frame of reference S as it does in S'; the wave front is observed to be spherical in both frames.

37.62 An astronaut, her rocket pack, and her spacesuit have a combined rest mass of 90.0 kg and a speed of 3.00×10^4 m/s. a) What is the difference between her correct relativistic kinetic energy and her Newtonian kinetic energy, $\frac{1}{2}mv^2$? Note that you cannot simply calculate both values and subtract, as the precision of most calculators is insufficient. Instead, use Eq. (37.37). b) What fraction of the Newtonian kinetic energy is this difference?

37.63 A particle with mass m accelerated from rest by a constant force F will, according to Newtonian mechanics, continue to accelerate without bound. That is, as $t \rightarrow \infty, v \rightarrow \infty$. Show that according to relativistic mechanics, the particle's speed approaches c as $t \rightarrow \infty$. [*Note:* A useful integral is $\int(1 - x^2)^{-3/2}\,dx = x/\sqrt{1 - x^2}$.]

37.64 Two events are observed in a frame of reference S to occur at the same space point, the second occurring 1.80 s after the first. In a second frame S' moving relative to S, the second event is observed to occur 2.35 s after the first. What is the difference between the positions of the two events as measured in S'?

37.65 Two events observed in a frame of reference S have positions and times given by (x_1, t_1) and (x_2, t_2), respectively. a) Frame S' moves along the x-axis just fast enough that the two events occur at the same position in S'. Show that in S', the time interval $\Delta t'$ between the two events is given by

$$\Delta t' = \sqrt{(\Delta t)^2 - \left(\frac{\Delta x}{c}\right)^2}$$

where $\Delta x = x_2 - x_1$ and $\Delta t = t_2 - t_1$. Hence show that if $\Delta x > c\,\Delta t$, there is *no* frame S' in which the two events occur at the same point. The interval $\Delta t'$ is sometimes called the *proper time interval* for the events. Is this term appropriate? b) Show that if $\Delta x > c\,\Delta t$, there is a different frame of reference S' in which the two events occur *simultaneously*. Find the distance between the two events in S'; express your answer in terms of Δx, Δt, and c. This distance is sometimes called a *proper length*. Is this term appropriate? c) Two events are observed in a frame of reference S' to occur simultaneously at points separated by a distance of 2.50 m. In a second frame S moving relative to S' along the line joining the two points in S', the two events appear to be separated by 5.00 m. What is the time interval between the events as measured in S? [*Hint:* Apply the result obtained in part (b).]

37.66 Tachyons. Tachyons are hypothetical particles that can travel faster than light. Consider two observers, Sebulba (at rest in frame S, which uses coordinates x, y, z, and t) and Watto (at rest in frame S', which uses coordinates x', y', z', and t'). Watto moves relative to Sebulba with speed $v = 0.600c$. At the instant they pass each other, they both set their clocks to zero. a) At time $t_1 = 100$ s in Sebulba's frame, Sebulba transmits a tachyon message to Watto. At this instant, how far away is Watto from Sebulba (as measured in Sebulba's frame)? b) The speed of the tachyons relative to Sebulba is $u = 4.00c$. What is the time t_2 (in Sebulba's frame) at which Watto receives the message? c) Watto is displeased with the message that Sebulba sends him, and with zero response time sends Sebulba a very unpleasant tachyon message in return. The tachyons have the same speed as those in part (b), so that their speed relative to Watto is $v' = 4.00c$. What is the velocity (magnitude and direction) *relative to Sebulba* of Watto's return message? d) What is the time t_3 (in Sebulba's frame) at which Sebulba receives Watto's message? e) The message that Watto sent overloaded the circuits of Sebulba's transmitter, converting it into a pile of useless metal and making it impossible for Selbulba to send tachyon signals. How do t_1 (the time when Sebulba sent the original signal) and t_3 (the time at which Sebulba's transmitter was destroyed) compare? Discuss your result and the possible existence of tachyons.

***37.67** One of the wavelengths of light emitted by hydrogen atoms under normal laboratory conditions is $\lambda = 656.3$ nm, in the red portion of the electromagnetic spectrum. In the light emitted from a distant galaxy this same spectral line is observed to be Doppler-shifted to $\lambda = 953.4$ nm, in the infrared portion of the spectrum. How fast are the emitting atoms moving relative to the earth? Are they approaching the earth or receding from it?

***37.68 Measuring Speed by Radar.** A baseball coach uses a radar device to measure the speed of an approaching pitched baseball. This device sends out electromagnetic waves with frequency f_0 and then measures the shift in frequency Δf of the waves reflected from the moving baseball. If the fractional frequency shift produced by a baseball is $\Delta f/f_0 = 2.86 \times 10^{-7}$, what is the baseball's speed in km/h?

37.69 A particle with mass m moves at speed v. Construct a right triangle in which one of the angles is α, where $\sin\alpha = v/c$. If the length of the side of the triangle adjacent to α equals the rest energy mc^2, show that a) the length of the hypotenuse equals the total energy; b) the length of the side opposite α equals c times the relativistic momentum. c) Describe a simple graphical procedure for finding the kinetic energy K.

***37.70** A spaceship moving at constant speed u relative to us broadcasts a radio signal at constant frequency f_0. As the spaceship approaches us, we receive a higher frequency f; after it has passed, we receive a lower frequency. a) As the spaceship passes by, so it is instantaneously neither moving toward or away from us, show that the frequency we receive is not f_0, and derive an expression for the frequency we do receive. Is the frequency we receive larger or smaller than f_0? (*Hint:* In this case, successive wave crests move the same distance to the observer and so they have the same transit time. Thus f equals $1/T$. Use the time dilation formula to relate the periods in the stationary and moving frames.) b) A spaceship emits electromagnetic waves of frequency $f_0 = 345$ MHz as measured in a frame moving with the ship. The spaceship is moving at a constant speed $0.758c$ relative to us. What frequency f do we receive when the spaceship is approaching us? When it is moving away? In each case what is the shift in frequency, $f - f_0$? c) Use the result of part (a) to calculate the frequency f and frequency shift $f - f_0$ we receive at the instant that the ship passes by us. How does the shift in frequency calculated here compare to the shifts calculated in part (b)?

***37.71 The Pole and Barn Paradox.** Suppose a *very* fast runner ($v = 0.600c$) holding a long, horizontal pole, runs through a barn open at both ends. The length of the pole (in its rest frame) is 6.00 m and the length of the barn (in *its* rest frame) is 5.00 m. In the barn's reference frame, the pole will undergo length contraction and can all fit inside the barn at the same time. But in the runner's reference frame, the *barn* will undergo length contraction and the entire pole can *never* be entirely within the barn at any time! Explain the resolution of this paradox.

37.72 The French physicist Armand Fizeau was the first to measure the speed of light accurately. He also found experimentally that the speed, relative to the lab frame, of light traveling in a tank of water that is itself moving at a speed V relative to the lab frame is

$$v = (c/n) + kV$$

where $n = 1.333$ is the index of refraction of water. Fizeau called k the dragging coefficient and obtained an experimental value of $k = 0.44$. What value of k do you calculate from relativistic transformations?

Challenge Problems

37.73 Lorentz Transformation for Acceleration. Using a method analogous to the one in the text to find the Lorentz transformation formula for velocity, we can find the Lorentz transformation for *acceleration*. Let frame S' have a constant x-component of velocity u relative to frame S. An object moves relative to frame S along the x-axis with instantaneous velocity v_x and instantaneous acceleration a_x. a) Show that its instantaneous acceleration in frame S' is

$$a'_x = a_x\left(1 - \frac{u^2}{c^2}\right)^{3/2}\left(1 - \frac{uv_x}{c^2}\right)^{-3}$$

[*Hint:* Express the acceleration in S' as $a'_x = dv_x'/dt'$. Then use Eq. (37.21) to express dt' in terms of dt and dx, and use Eq. (37.22) to express dv'_x in terms of u and dv_x. The velocity of the object in S is $v_x = dx/dt$.)] b) Show that the acceleration in frame S can be expressed as

$$a_x = a'_x\left(1 - \frac{u^2}{c^2}\right)^{3/2}\left(1 + \frac{uv'_x}{c^2}\right)^{-3}$$

where $v'_x = dx'/dt'$ is the velocity of the object in frame S'.

37.74 A Realistic Version of the Twin Paradox. A rocket ship leaves the earth on January 1, 2100. Stella, one of a pair of twins born in the year 2075, pilots the rocket (reference frame S'); the other twin, Terra, stays on the earth (reference frame S). The rocket ship has an acceleration of constant magnitude g in its own reference frame (this makes the pilot feel at home, since it simulates the earth's gravity). The path of the rocket ship is a straight line in the $+x$-direction in frame S. a) Using the results of Challenge Problem 37.73, show that in Terra's earth frame S, the rocket's acceleration is

$$\frac{du}{dt} = g\left(1 - \frac{u^2}{c^2}\right)^{3/2}$$

where u is the rocket's instantaneous velocity in frame S. b) Write the result of part (a) in the form $dt = f(u)\,du$, where $f(u)$ is a function of u, and integrate both sides. (*Hint:* Use the integral given in Problem 37.63.) Show that in Terra's frame, the time when Stella attains a velocity v_{1x} is

$$t_1 = \frac{v_{1x}}{g\sqrt{1 - v_{1x}^2/c^2}}$$

c) Use the time dilation formula to relate dt and dt' (infinitesimal time intervals measured in frames S and S', respectively). Combine this result with the result of part (a) and integrate as in part (b) to show the following: When Stella attains a velocity v_{1x} relative to Terra, the time t'_1 that has elapsed in frame S' is

$$t'_1 = \frac{c}{g}\,\text{arctanh}\left(\frac{v_{1x}}{c}\right)$$

Here arctanh is the inverse hyperbolic tangent. (*Hint:* Use the integral given in Challenge Problem 5.124.) d) Combine the results of parts (b) and (c) to find t_1 in terms of t'_1, g, and c alone. e) Stella accelerates in a straight-line path for five years (by her clock), slows down at the same rate for five years, turns around, accelerates for five years, slows down for five years, and lands back on the earth. According to Stella's clock, the date is January 1, 2120. What is the date according to Terra's clock?

***37.75 Determining the Masses of Stars.** Many of the stars in the sky are actually *binary stars,* in which two stars orbit about their common center of mass. If the orbital speeds of the stars are high enough, the motion of the stars can be detected by the Doppler shifts of the light that they emit. Stars for which this is the case are called *spectroscopic binary stars.* Figure 37.25 shows the simplest case of a spectroscopic binary star: two identical stars, each with mass m, orbiting their center of mass in a circle of radius R. The plane of the stars' orbits is edge-on to the line of sight of an observer on the earth. a) The light produced by heated hydrogen gas in a laboratory on the earth has a frequency of 4.568110×10^{14} Hz. In the light received from the stars by a telescope on the earth, hydrogen light is observed to vary in frequency between 4.567710×10^{14} Hz and 4.568910×10^{14} Hz. Determine whether the binary star system as a whole is moving toward or away from the earth, the speed of this motion, and the orbital speeds of the stars. (*Hint:* The speeds involved are much less than c, so you may use the approximate result $\Delta f/f = u/c$ given in Section 37.6.) b) The light from each star in the binary system varies from its maximum frequency to its minimum frequency and back

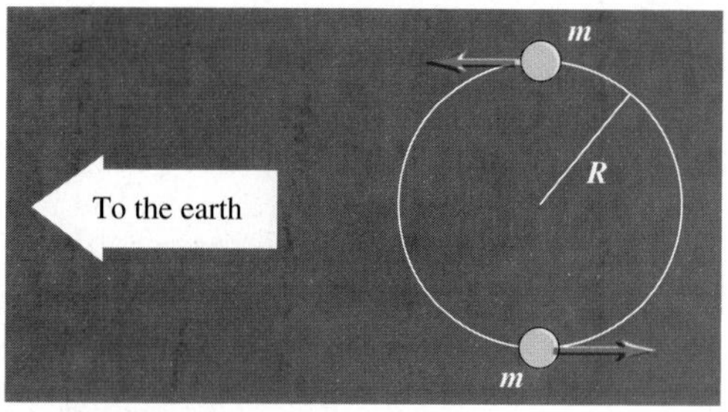

Figure 37.25 Challenge Problem 37.75.

again in 11.0 days. Determine the orbital radius R and the mass m of each star. Give your answer for m in kilograms and as a multiple of the mass of the sun, 1.99×10^{30} kg. Compare the value of R to the distance from the earth to the sun, 1.50×10^{11} m. (This technique is actually used in astronomy to determine the masses of stars. In practice, the problem is more complicated because the two stars in a binary system are usually not identical, the orbits are usually not circular, and the plane of the orbits is usually tilted with respect to the line of sight from the earth.)

37.76 Relativity and the Wave Equation. a) Consider the Galilean transformation along the x-direction: $x' = x - vt$ and $t' = t$. In frame S the wave equation for electromagnetic waves in a vacuum is

$$\frac{\partial^2 E(x,t)}{\partial x^2} - \frac{1}{c^2}\frac{\partial^2 E(x,t)}{\partial t^2} = 0$$

where E represents the electric field in the wave. Show that by using the Galilean transformation the wave equation in frame S' is found to be

$$\left(1 - \frac{v^2}{c^2}\right)\frac{\partial^2 E(x',t')}{\partial x'^2} + \frac{2v}{c^2}\frac{\partial^2 E(x',t')}{\partial x' \partial t'} - \frac{1}{c^2}\frac{\partial^2 E(x',t')}{\partial t'^2} = 0$$

This has a different form than the wave equation in S. Hence the Galilean transformation *violates* the first relativity postulate that all physical laws have the same form in all inertial reference frames. (*Hint:* Express the derivatives $\partial/\partial x$ and $\partial/\partial t$ in terms of $\partial/\partial x'$ and $\partial/\partial t'$ by use of the chain rule.) b) Repeat the analysis of part (a), but use the Lorentz coordinate transformations, Eqs. (37.21), and show that in frame S' the wave equation has the same form as in frame S:

$$\frac{\partial^2 E(x',t')}{\partial x'^2} - \frac{1}{c^2}\frac{\partial^2 E(x',t')}{\partial t'^2} = 0$$

Explain why this shows that the speed of light in vacuum is c in both frames S and S'.

37.77 Kaon Production. In high-energy physics, new particles can be created by collisions of fast-moving projectile particles with stationary particles. Some of the kinetic energy of the incident particle is used to create the mass of the new particle. A proton-proton collision can result in the creation of a negative kaon (K^-) and a positive kaon (K^+):

$$p + p \rightarrow p + p + K^- + K^+$$

a) Calculate the minimum kinetic energy of the incident proton that will allow this reaction to occur if the second (target) proton is initially at rest. The rest energy of each kaon is 493.7 MeV, and the rest energy of each proton is 938.3 MeV. (*Hint:* It is useful here to work in the frame in which the total momentum is zero. See Problem 8.100, but note that here the Lorentz transformation must be used to relate the velocities in the laboratory frame to those in the zero-total-momentum frame.) b) How does this calculated minimum kinetic energy compare with the total rest mass energy of the created kaons? c) Suppose that instead the two protons are both in motion with velocities of equal magnitude and opposite direction. Find the minimum combined kinetic energy of the two protons that will allow the reaction to occur. How does this calculated minimum kinetic energy compare with the total rest mass energy of the created kaons? (This example shows that when colliding beams of particles are used instead of a stationary target, the energy requirements for producing new particles are reduced substantially.)

PHOTONS, ELECTRONS, AND ATOMS 38

Though the glowing gas cloud of the Eagle Nebula is 6500 light years away, we can tell that it contains mostly hydrogen. The key is the particular red color of the gas, which is emitted by excited hydrogen atoms and by no other element. In their quest to understand the spectrum of hydrogen and other atoms, 20th-century physicists revolutionized our picture of what light is and what atoms are.

? The red light from the Eagle Nebula has a wavelength of 656.3 nm. What happens within a hydrogen atom to emit light of this wavelength?

In Chapter 32 we saw how Maxwell, Hertz, and others established firmly that light is an electromagnetic wave. Interference, diffraction, and polarization, discussed in Chapters 35 and 36, further demonstrate this *wave nature* of light.

When we look more closely at the emission, absorption, and scattering of electromagnetic radiation, however, we discover a completely different aspect of light. We find that the energy of an electromagnetic wave is *quantized;* it is emitted and absorbed in particle-like packages of definite energy, called *photons* or *quanta.* The energy of a single photon is proportional to the frequency of the radiation.

The internal energy of atoms is also quantized. For a given kind of individual atom the energy can't have just any value; only discrete values called *energy levels* are possible.

The basic ideas of photons and energy levels take us a long way toward understanding a wide variety of otherwise puzzling observations. Among these are the unique sets of wavelengths emitted and absorbed by gaseous elements, the emission of electrons from an illuminated surface, the operation of lasers, and the production and scattering of x rays. Our studies of photons and energy levels will take us to the threshold of *quantum mechanics,* which involves some radical changes in our views of the nature of electromagnetic radiation and of matter itself.

38.1 | Emission and Absorption of Light

How is light produced? In Chapter 32 we discussed how Heinrich Hertz produced electromagnetic waves using oscillations in a resonant L-C circuit similar to those we studied in Chapter 30. He used frequencies of the order of 10^8 Hz, but visible light has frequencies of the order of 10^{15} Hz, far higher than any frequency that can be attained with conventional electronic circuits. At the end of the nineteenth century, some physicists speculated that waves in this frequency range might be produced by oscillating electric charges within individual atoms. However, their speculations were unable to give explanations for some key experimental data. Among the great challenges facing physicists in 1900 were how to explain *line spectra,* the *photoelectric effect,* and the nature of *x rays.* We'll describe each of these in turn.

Line Spectra

We can use a prism or a diffraction grating to separate the various wavelengths in a beam of light into a spectrum. If the light source is a hot solid (such as a light-bulb filament) or liquid, the spectrum is *continuous;* light of all wavelengths is present (Fig. 38.1a). But if the source is a gas carrying an electric discharge (as in a neon sign) or a volatile salt heated in a flame (as when table salt is thrown into a campfire), only a few colors appear, in the form of isolated sharp parallel lines (Fig. 38.1b). (Each "line" is an image of the spectrograph slit, deviated through an angle that depends on the wavelength of the light forming that image; see Section 36.5.) A spectrum of this sort is called a **line spectrum.** Each line corresponds to a definite wavelength and frequency.

It was discovered early in the nineteenth century that each element in its gaseous state has a unique set of wavelengths in its line spectrum. The spectrum of hydrogen always contains a certain set of wavelengths; sodium produces a different set, iron still another, and so on. Scientists find the use of spectra to identify elements and compounds to be an invaluable tool (see the photograph that opens this chapter). For instance, astronomers have detected the spectra from more than 100 different molecules in interstellar space, including some that are not found naturally on earth. The characteristic spectrum of an atom was presumably related to its internal structure, but attempts to understand this relation solely on the basis

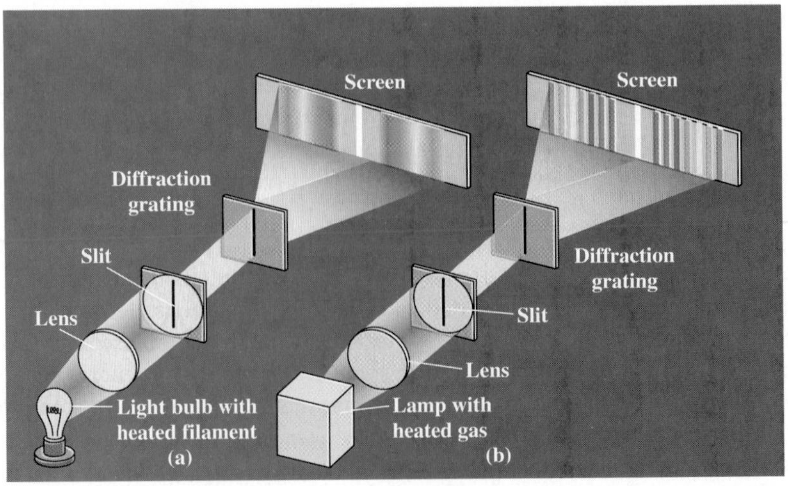

38.1 (a) Continuous spectrum produced by a glowing light-bulb filament. (b) Line spectrum emitted by a lamp containing a heated gas.

of *classical* mechanics and electrodynamics—the physics summarized in Newton's three laws and Maxwell's four equations—were not successful.

ActivPhysics Online
17.3 Photoelectric Effect

Photoelectric Effect

There were also mysteries associated with *absorption* of light. In 1887, during his electromagnetic-wave experiments, Hertz discovered the *photoelectric effect.* When light struck a metal surface, some electrons near the surface absorbed enough energy to overcome the attraction of the positive ions in the metal and escape into the surrounding space. Detailed investigation of this effect revealed some puzzling features that couldn't be understood on the basis of classical optics. We will discuss these in the next section.

X Rays

Other unsolved problems in the emission and absorption of radiation centered on the production and scattering of *x rays,* discovered in 1895. These rays were produced in high-voltage electric discharge tubes, but no one understood how or why they were produced or what determined their wavelengths (which are much shorter those of visible light). Even worse, when x rays collided with matter, the scattered rays sometimes had longer wavelengths than the original ray. This is analogous to a beam of blue light striking a mirror and reflecting back as red!

Photons and Energy Levels

All these phenomena (and several others) pointed forcefully to the conclusion that classical optics, successful though it was in explaining lenses, mirrors, interference, and polarization, had its limitations. We now understand that all these phenomena result from the *quantum* nature of radiation. Electromagnetic radiation, along with its *wave* nature, has properties resembling those of *particles*. In particular, the energy in an electromagnetic wave is always emitted and absorbed in packages called *photons* or *quanta,* with energy proportional to the frequency of the radiation.

The two common threads that are woven through this chapter are the quantization of electromagnetic radiation and the existence of discrete energy levels in atoms. In the remainder of this chapter we will show how these two concepts contribute to understanding the phenomena mentioned above. We are not yet ready for a comprehensive theory of atomic structure; that will come in Chapters 40 and 41. But we'll look at the Bohr model of the hydrogen atom, one attempt to predict atomic energy levels on the basis of atomic structure.

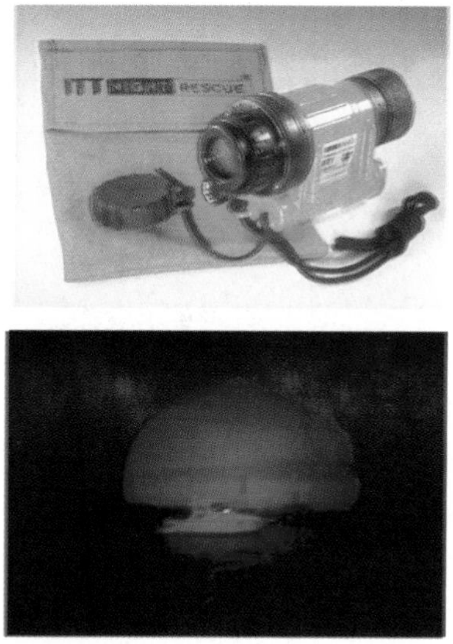

38.2 A night vision scope makes use of the photoelectric effect. Photons entering the scope strike a plate, ejecting electrons that pass through a thin disk in which there are millions of tiny channels. The current through each channel is amplified electronically, then directed toward a screen that glows when hit by electrons. The image formed on the screen (shown below), which is a combination of these millions of glowing spots, is thousands of times brighter than the naked-eye view.

38.2 | The Photoelectric Effect

The **photoelectric effect** is the emission of electrons when light strikes a surface. This effect has numerous practical applications (Fig. 38.2). To escape from the surface, the electron must absorb enough energy from the incident radiation to overcome the attraction of positive ions in the material of the surface. This attraction causes a potential-energy barrier that normally confines the electrons inside the material. Think of this barrier as being like a rounded curb separating a flat street from a raised sidewalk. The curb will keep a slowly moving soccer ball in the street. But if the ball is kicked hard enough, it can roll up onto the sidewalk, with the work done against the gravitational attraction (the gain in gravitational potential energy) equal to its loss in kinetic energy.

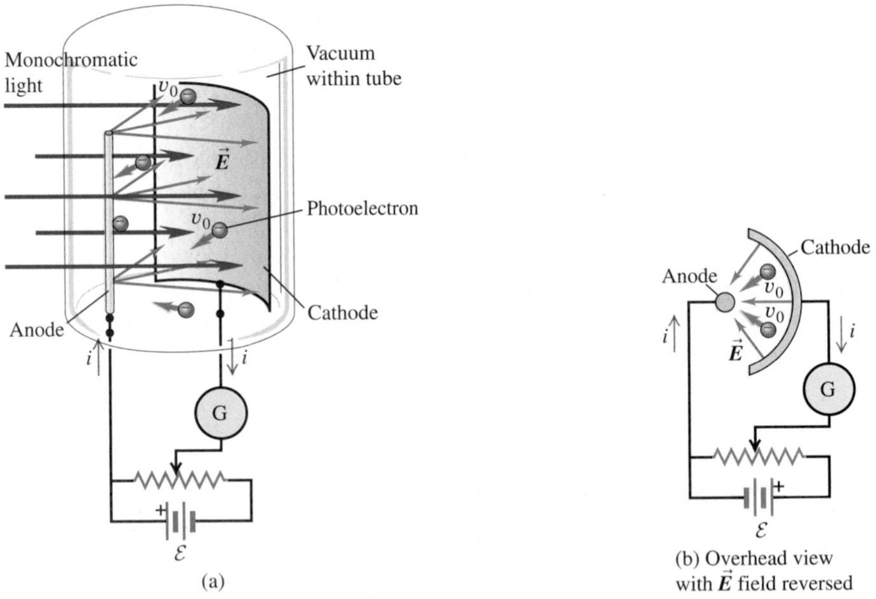

38.3 Demonstrating the photoelectric effect. (a) Light shining on the cathode emits electrons, which are pushed toward the anode by the electric-field force. (b) Even when the direction of the $\vec{E}$ field is reversed, some electrons still reach the anode. The stopping potential V_0 is the minimum absolute value of the reverse potential difference that gives zero current.

The photoelectric effect was first observed in 1887 by Heinrich Hertz, who noticed that a spark would jump more readily between two electrically charged spheres when their surfaces were illuminated by the light from another spark. Light shining on the surfaces somehow facilitated the escape of what we now know to be electrons. This idea in itself was not revolutionary. The existence of the surface potential-energy barrier was already known. In 1883, Thomas Edison had discovered *thermionic emission,* in which the escape energy is supplied by heating the material to a very high temperature, liberating electrons by a process analogous to boiling a liquid. The *minimum* amount of energy an individual electron has to gain to escape from a particular surface is called the **work function** for that surface, denoted by ϕ. However, the surfaces that Hertz used were not at the high temperatures needed for thermionic emission.

The photoelectric effect was investigated in detail by the German physicists Wilhelm Hallwachs and Philipp Lenard during the years 1886–1900; their results were quite unexpected. We will describe their work in terms of a more modern phototube (Fig. 38.3). Two conducting electrodes, the anode and the cathode, are enclosed in an evacuated glass tube. The battery or other source of potential difference creates an electric field in the direction from anode to cathode in Fig. 38.3a. Light (indicated by the magenta arrows) falling on the surface of the cathode causes a current in the external circuit; the current is measured by the galvanometer (G). Hallwachs and Lenard studied how this *photocurrent* varies with voltage and with the frequency and intensity of the light.

After the discovery of the electron in 1897, it became clear that the light causes electrons to be emitted from the cathode. Because of their negative charge $-e$, the emitted *photoelectrons* are then pushed toward the anode by the electric field. A high vacuum with residual pressure of 0.01 Pa (10^{-7} atm) or less is needed to minimize collisions of the electrons with gas molecules.

Hallwachs and Lenard found that when monochromatic light fell on the cathode, *no* photoelectrons at all were emitted unless the frequency of the light was greater than some minimum value called the **threshold frequency.** This minimum frequency depends on the material of the cathode. For most metals the threshold frequency is in the ultraviolet (corresponding to wavelengths λ between 200 and 300 nm), but for potassium and cesium oxides it is in the visible spectrum (λ between 400 and 700 nm).

When the frequency f is *greater* than the threshold frequency, some electrons are emitted from the cathode with substantial initial speeds. This can be shown by reversing the polarity of the battery (Fig. 38.3b) so that the electric-field force on the electrons is back toward the cathode. If the magnitude of the field is not too great, the highest-energy emitted electrons still reach the anode and there is still a current. We can determine the *maximum* kinetic energy of the emitted electrons by making the potential of the anode relative to the cathode, V_{AC}, just negative enough so that the current stops. This occurs for $V_{AC} = -V_0$, where V_0 is called the **stopping potential.** As an electron moves from the cathode to the anode, the potential decreases by V_0 and negative work $-eV_0$ is done on the (negatively charged) electron; the most energetic electron leaves the cathode with kinetic energy $K_{max} = \frac{1}{2}mv_{max}{}^2$ and has zero kinetic energy at the anode. Using the work-energy theorem, we have

$$W_{tot} = -eV_0 = \Delta K = 0 - K_{max}$$

$$K_{max} = \frac{1}{2}mv_{max}{}^2 = eV_0 \qquad (38.1)$$

(maximum kinetic energy of photoelectrons)

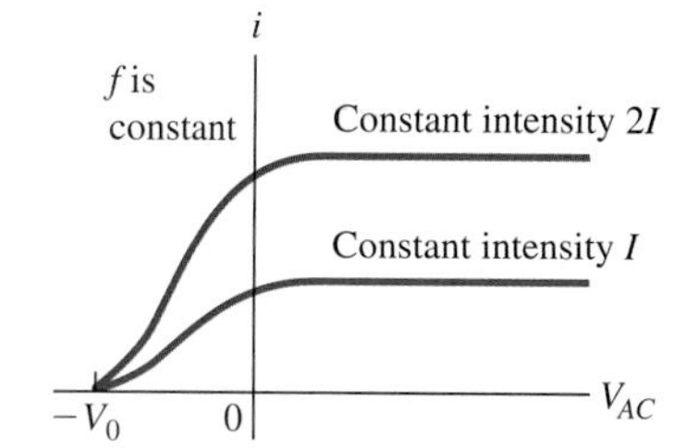

38.4 Photocurrent i for a constant light frequency f as a function of the potential V_{AC} of the anode with respect to the cathode. The stopping potential V_0 is independent of the light intensity, but the photocurrent for large positive V_{AC} is directly proportional to the intensity.

Hence by measuring the stopping potential V_0, we can determine the maximum kinetic energy with which electrons leave the cathode. (We are ignoring any effects due to differences in the materials of the cathode and anode.)

Figure 38.4 shows graphs of photocurrent as a function of potential difference V_{AC} for light of constant frequency and two different intensities. When V_{AC} is sufficiently large and positive, the curves level off, showing that *all* the emitted electrons are being collected by the anode. The reverse potential difference $-V_0$ needed to reduce the current to zero is shown.

If the intensity of light is increased while its frequency is kept the same, the current levels off at a higher value, showing that more electrons are being emitted per time. But the stopping potential V_0 is found to be the same.

Figure 38.5 shows current as a function of potential difference for two different frequencies, with the same intensity in each case. We see that when the frequency of the incident monochromatic light is increased, the stopping potential V_0 increases. In fact, V_0 turns out to be a linear function of the frequency f.

These results are hard to understand on the basis of classical physics. When the intensity (average energy per unit area per unit time) increases, electrons should be able to gain more energy, increasing the stopping potential V_0. But V_0 was found *not* to depend on intensity. Also, classical physics offers no explanation for the threshold frequency. In Section 32.4 we found that the intensity of an electromagnetic wave such as light does not depend on frequency, so an electron should be able to acquire its needed escape energy from light of any frequency. Thus there should not be a threshold frequency f_0. Finally, we would expect it to take a while for an electron to collect enough energy from extremely faint light. But experiment also shows that electrons are emitted as soon as *any* light with $f \geq f_0$ hits the surface.

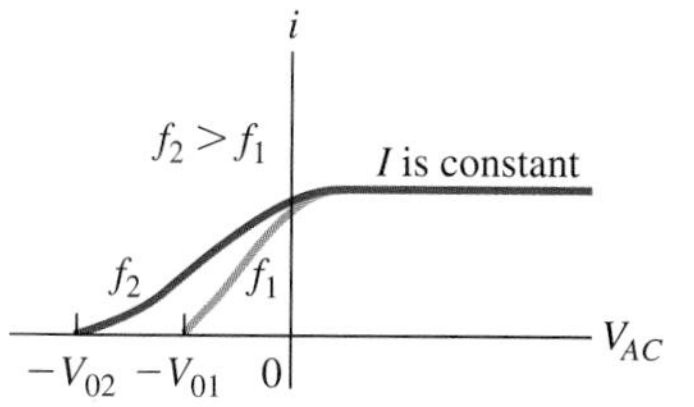

38.5 Photocurrent i for two different light frequencies f_1 and f_2 as functions of the potential V_{AC} of the anode with respect to the cathode. The stopping potential V_0 (and therefore the maximum kinetic energy of the photoelectrons) increases linearly with current.

The correct analysis of the photoelectric effect was developed by Albert Einstein in 1905. Building on an assumption made five years earlier by Max Planck (to be discussed in Section 38.8), Einstein postulated that a beam of light consists of small packages of energy called **photons** or *quanta*. The energy E of a photon is equal to a constant h times its frequency f. From $f = c/\lambda$ for electromagnetic waves in vacuum we have

$$E = hf = \frac{hc}{\lambda} \quad \text{(energy of a photon)} \qquad (38.2)$$

where h is a universal constant called **Planck's constant.** The numerical value of this constant, to the accuracy known at present, is

$$h = 6.62606876(52) \times 10^{-34}\ \text{J}\cdot\text{s}$$

A photon arriving at the surface is absorbed by an electron. This energy transfer is an all-or-nothing process, in contrast to the continuous transfer of energy in the classical theory; the electron gets all the photon's energy or none at all. If this energy is greater than the work function ϕ, the electron may escape from the surface. Greater intensity at a particular frequency means a proportionally greater number of photons per second absorbed, and thus a proportionally greater number of electrons emitted per second and the proportionally greater current seen in Fig. 38.4.

Recall that ϕ is the *minimum* energy needed to remove an electron from the surface. Einstein applied conservation of energy to find that the *maximum* kinetic energy $K_{max} = \frac{1}{2}mv_{max}^2$ for an emitted electron is the energy hf gained from a photon minus the work function ϕ:

$$K_{max} = \frac{1}{2}mv_{max}^2 = hf - \phi \qquad (38.3)$$

Substituting $K_{max} = eV_0$ from Eq. (38.1), we find

$$eV_0 = hf - \phi \quad \text{(photoelectric effect)} \qquad (38.4)$$

We can measure the stopping potential V_0 for each of several values of frequency f for a given cathode material. A graph of V_0 as a function of f turns out to be a straight line, verifying Eq. (38.4), and from such a graph we can determine both the work function ϕ for the material and the value of the quantity h/e. (We do this graphical analysis in Example 38.3.) After the electron charge $-e$ was measured by Robert Millikan in 1909, Planck's constant h could also be determined from these measurements.

Electron energies and work functions are usually expressed in electron volts (eV), defined in Section 23.2. To four significant figures,

$$1\ \text{eV} = 1.602 \times 10^{-19}\ \text{J}$$

To this accuracy, Planck's constant is

$$h = 6.626 \times 10^{-34}\ \text{J}\cdot\text{s} = 4.136 \times 10^{-15}\ \text{eV}\cdot\text{s}$$

Table 38.1 lists a few typical work functions of elements. These values are approximate because they are very sensitive to surface impurities. The greater the work function, the higher the minimum frequency needed to emit photoelectrons.

We have discussed photons mostly in the context of light. However, the quantization concept applies to *all* regions of the electromagnetic spectrum, including radio waves, x rays, and so on. A photon of any electromagnetic radiation with frequency f and wavelength λ has energy E given by Eq. (38.2). Furthermore, according to the special theory of relativity, every particle that has energy must

Table 38.1 Work Functions of Several Elements

Element	Work Function (eV)
Aluminum	4.3
Carbon	5.0
Copper	4.7
Gold	5.1
Nickel	5.1
Silicon	4.8
Silver	4.3
Sodium	2.7

also have momentum, even if it has no rest mass. Photons have zero rest mass. As we saw in Eq. (37.40), a photon with energy E has momentum with magnitude p given by $E = pc$. Thus the wavelength λ of a photon and the magnitude of its momentum p are related simply by

$$p = \frac{E}{c} = \frac{hf}{c} = \frac{h}{\lambda} \quad \text{(momentum of a photon)} \qquad (38.5)$$

The direction of the photon's momentum is simply the direction in which the electromagnetic wave is moving.

Problem-Solving Strategy — Photons

IDENTIFY *the relevant concepts:* The energy and momentum of an individual photon are proportional to the frequency and inversely proportional to the wavelength. These concepts, along with the idea of stopping potential in the photoelectric effect, allow you to solve almost any problem involving photons.

SET UP *the problem:* Decide what your target variable is. It could be the photon's wavelength λ, frequency f, energy E, or momentum p. If the problem involves the photoelectric effect, the target variable could be the maximum kinetic energy of photoelectrons K_{max}, the stopping potential V_0, or the work function ϕ.

EXECUTE *the solution as follows:*

1. Use Eqs. (38.2) and (38.5) to relate the energy and momentum of a photon to its wavelength and frequency. If the problem involves the photoelectric effect, use Eqs. (38.1), (38.3), and (38.4) to relate the photon frequency, stopping potential, work function, and maximum photoelectron kinetic energy.
2. The electron volt is an important and convenient unit. We used it in Chapter 37, and we will use it even more in this and the next three chapters. One electron volt (eV) is the amount of kinetic energy gained by an electron when it moves freely through an increase of potential of one volt: $1\ \text{eV} = 1.602 \times 10^{-19}\ \text{J}$. If the photon energy E is given in electron volts, use $h = 4.136 \times 10^{-15}\ \text{eV} \cdot \text{s}$; if E is in joules, use $h = 6.626 \times 10^{-34}\ \text{J} \cdot \text{s}$.

EVALUATE *your answer:* In problems involving photons, the numbers are in such unfamiliar ranges that common sense may not help you identify whether your calculation is off by a large factor. It helps to remember that a visible-light photon with $\lambda = 600\ \text{nm}$ and $f = 5 \times 10^{14}\ \text{Hz}$ has an energy E of about 2 eV, or about $3 \times 10^{-19}\ \text{J}$.

Example 38.1 — FM radio photons

Radio station WQED in Pittsburgh broadcasts at 89.3 MHz with a radiated power of 43.0 kW. a) What is the magnitude of the momentum of each photon? b) How many photons does WQED emit each second?

SOLUTION

IDENTIFY and SET UP: We are given the frequency of the radiation, so we can use Eq. (38.5) directly to find the magnitude of each photon's momentum in part (a). In part (b), note that "radiated power" means the energy emitted per second and that Eq. (38.2) gives the energy per photon. We can combine these to calculate the number of photons emitted per second.

EXECUTE: a) From Eq. (38.5), the magnitude of momentum of each photon is

$$p = \frac{hf}{c} = \frac{(6.626 \times 10^{-34}\ \text{J} \cdot \text{s})(89.3 \times 10^6\ \text{Hz})}{3.00 \times 10^8\ \text{m/s}}$$
$$= 1.97 \times 10^{-34}\ \text{kg} \cdot \text{m/s}$$

(Recall that $1\ \text{J} = 1\ \text{kg} \cdot \text{m}^2/\text{s}^2$ and that $1\ \text{Hz} = 1\ \text{s}^{-1}$.)

b) From Eq. (38.2), the energy of each photon is

$$E = pc = (1.97 \times 10^{-34}\ \text{kg} \cdot \text{m/s})(3.00 \times 10^8\ \text{m/s})$$
$$= 5.92 \times 10^{-26}\ \text{J}$$

The station emits 43.0×10^3 joules each second, so the rate at which photons is emitted is

$$\frac{(43.0 \times 10^3\ \text{J/s})}{5.92 \times 10^{-26}\ \text{J/photon}} = 7.26 \times 10^{29}\ \text{photon/s}$$

EVALUATE: The result in part (a) is very small, about the magnitude of the momentum an electron would have if it crawled along at a speed of one meter per hour. The photon energy E calculated in part (b) is also very small, equal to 3.69×10^{-7} eV. A visible-light photon has about 10^7 times as much energy. This makes sense: the energy of a photon is proportional to its frequency, and the frequencies of visible light are about 10^7 times greater than those used in FM radio. (You can also check the value of the photon energy by using $E = hf$.)

Our result in part (b) shows that a huge number of photons leave the station each second, each of which has an infinitesimal amount of energy. Hence the discreteness of the photons isn't noticed, and the radiated energy appears to be a continuous flow.

Example 38.2 A photoelectric experiment

While conducting a photoelectric-effect experiment with light of a certain frequency, you find that a reverse potential difference of 1.25 V is required to reduce the current to zero. Find a) the maximum kinetic energy; b) the maximum speed of the emitted photoelectrons.

SOLUTION

IDENTIFY and SET UP: The value of 1.25 V is the stopping potential V_0. We can find the maximum photoelectron kinetic energy K_{max} by using Eq. (38.1); given this value, we can find the maximum photoelectron speed.

EXECUTE: a) From Eq. (38.1),

$$K_{max} = eV_0 = (1.60 \times 10^{-19}\ \text{C})(1.25\ \text{V}) = 2.00 \times 10^{-19}\ \text{J}$$

Recall that 1 V = 1 J/C. In terms of electron volts,

$$K_{max} = eV_0 = e(1.25\ \text{V}) = 1.25\ \text{eV}$$

since the electron volt (eV) is the magnitude of the electron charge e times one volt (1 V).

b) From $K_{max} = \frac{1}{2}mv_{max}{}^2$ we get

$$v_{max} = \sqrt{\frac{2K_{max}}{m}} = \sqrt{\frac{2(2.00 \times 10^{-19}\ \text{J})}{9.11 \times 10^{-31}\ \text{kg}}} = 6.63 \times 10^5\ \text{m/s}$$

EVALUATE: The value of V_{max} is about 1/500 the speed of light c, so we are justified in using the nonrelativistic expression for kinetic energy. An equivalent justification is that the electron's 1.25-eV kinetic energy is much smaller than its rest energy $mc^2 = 0.511$ MeV.

Example 38.3 Determining ϕ and h experimentally

For a certain cathode material in a photoelectric-effect experiment, you measure a stopping potential of 1.0 V for light of wavelength 600 nm, 2.0 V for 400 nm, and 3.0 V for 300 nm. Determine the work function for this material and the value of Planck's constant.

SOLUTION

IDENTIFY and SET UP: According to Eq. (38.4), a graph of stopping potential V_0 versus frequency f should be a straight line. Such a graph is completely determined by its slope and the value at which it intercepts the vertical axis; we will use these to determine the values of the target variables ϕ and h.

EXECUTE: We rewrite Eq. (38.4) as

$$V_0 = \frac{h}{e}f - \frac{\phi}{e}$$

In this form we see that the slope of the line is h/e and the intercept on the vertical axis (corresponding to $f = 0$) is at $-\phi/e$. The frequencies, obtained from $f = c/\lambda$ and $c = 3.00 \times 10^8$ m/s, are 0.50×10^{15} Hz, 0.75×10^{15} Hz, and 1.0×10^{15} Hz, respectively. The graph is shown in Fig. 38.6. From it we find

$$-\frac{\phi}{e} = \text{vertical intercept} = -1.0\ \text{V}$$

$$\phi = 1.0\ \text{eV} = 1.6 \times 10^{-19}\ \text{J}$$

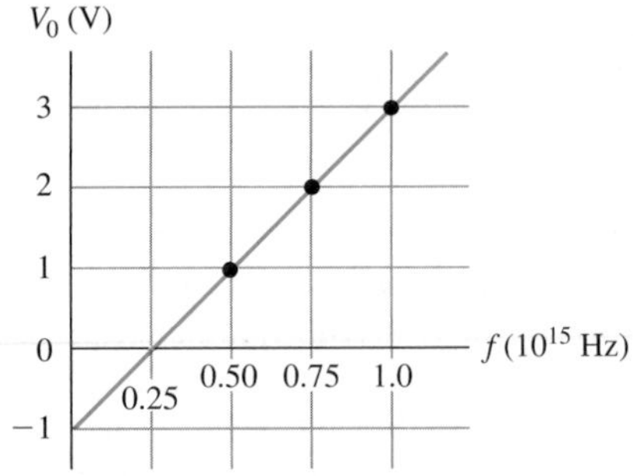

38.6 Stopping potential as a function of frequency. For a different cathode material having a different work function, the line would be displaced up or down but would have the same slope, equal to h/e within experimental error.

and

$$\text{Slope} = \frac{\Delta V_0}{\Delta f} = \frac{3.0\text{ V} - (-1.0\text{ V})}{1.00 \times 10^{15}\text{ s}^{-1} - 0} = 4.0 \times 10^{-15}\text{ J}\cdot\text{s/C}$$

$$h = \text{slope} \times e = (4.0 \times 10^{-15}\text{ J}\cdot\text{s/C})(1.60 \times 10^{-19}\text{ C})$$
$$= 6.4 \times 10^{-34}\text{ J}\cdot\text{s}$$

EVALUATE: This experimental value differs by about 3% from the accepted value. The small value of ϕ tells us that the cathode surface is not composed solely of one of the elements in Table 38.1.

Test Your Understanding

Silicon films become better electrical conductors when illuminated by photons with energies of 1.14 eV or greater, an effect called *photoconductivity*. What range of wavelengths causes photoconductivity in silicon films?

38.3 | Atomic Line Spectra and Energy Levels

ActivPhysics Online
18.2 Spectroscopy

The origin of *line spectra,* described in Section 38.1, can be understood in general terms on the basis of two central ideas. One is the photon concept; the other is the concept of *energy levels* of atoms. These two ideas were combined by the Danish physicist Niels Bohr in 1913.

Bohr's hypothesis represented a bold breakaway from nineteenth-century ideas. His reasoning went like this. The line spectrum of an element results from the emission of photons with specific energies from the atoms of that element. During the emission of a photon, the internal energy of the atom changes by an amount equal to the energy of the photon. Therefore, said Bohr, each atom must be able to exist only with certain specific values of internal energy. Each atom has a set of possible **energy levels.** An atom can have an amount of internal energy equal to any one of these levels, but it *cannot* have an energy *intermediate* between two levels. All isolated atoms of a given element have the same set of energy levels, but atoms of different elements have different sets. In electric discharge tubes atoms are raised, or *excited,* to higher energy levels mainly through inelastic collisions with electrons.

According to Bohr, an atom can make a *transition* from one energy level to a lower level by emitting a photon with energy equal to the energy *difference* between the initial and final levels (Fig. 38.7). If E_i is the initial energy of the atom before such a transition, E_f is its final energy after the transition, and the photon's energy is $hf = hc/\lambda$, then conservation of energy gives

$$hf = \frac{hc}{\lambda} = E_i - E_f \qquad \text{(energy of emitted photon)} \qquad (38.6)$$

For example, when a krypton atom emits a photon of orange light with wavelength $\lambda = 606$ nm, the corresponding photon energy is

$$E = \frac{hc}{\lambda} = \frac{(6.63 \times 10^{-34}\text{ J}\cdot\text{s})(3.00 \times 10^{8}\text{ m/s})}{606 \times 10^{-9}\text{ m}}$$
$$= 3.28 \times 10^{-19}\text{ J} = 2.05\text{ eV}$$

This photon is emitted during a transition like that in Fig. 38.7 between two levels of the atom that differ in energy by 2.05 eV.

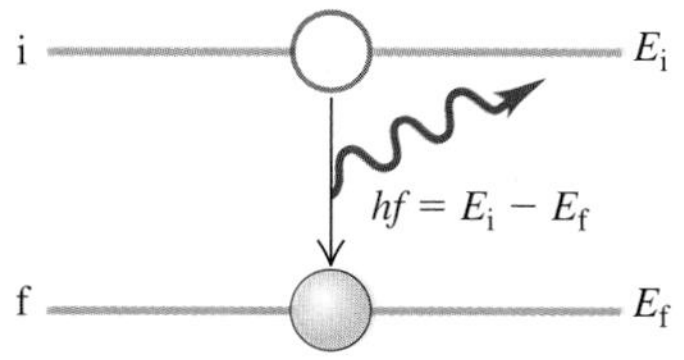

38.7 An atom drops from an initial level i to a lower-energy final level f by emitting a photon with energy equal to $E_i - E_f$.

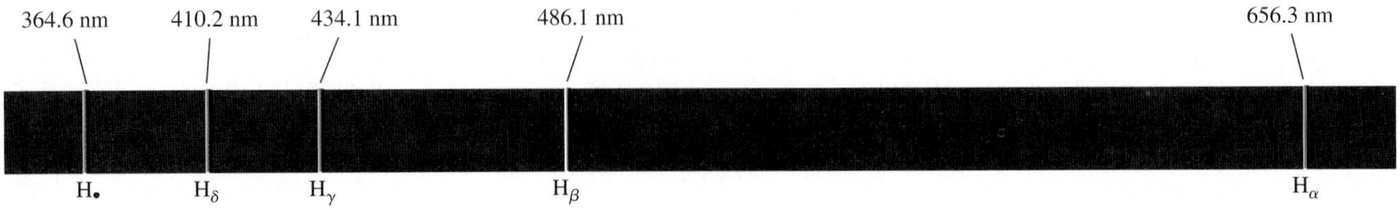

38.8 The Balmer series of spectral lines for atomic hydrogen. The H_α, H_β, H_γ, and H_δ lines are all in the visible portion of the spectrum; all other Balmer-series lines are in the ultraviolet.

The Hydrogen Spectrum

By 1913 the spectrum of hydrogen, the simplest and least massive atom, had been studied intensively. In an electric discharge tube, atomic hydrogen emits the series of lines shown in Fig. 38.8. The visible line with longest wavelength, or lowest frequency, is in the red and is called H_α; the next line, in the blue-green, is called H_β; and so on. In 1885 the Swiss teacher Johann Balmer (1825–1898) found (by trial and error) a formula that gives the wavelengths of these lines, which are now called the *Balmer series*. We may write Balmer's formula as

$$\frac{1}{\lambda} = R\left(\frac{1}{2^2} - \frac{1}{n^2}\right) \tag{38.7}$$

where λ is the wavelength, R is a constant called the **Rydberg constant** (chosen to make Eq. (38.7) fit the measured wavelengths), and n may have the integer values 3, 4, 5, ... If λ is in meters, the numerical value of R is

$$R = 1.097 \times 10^7\ \text{m}^{-1}$$

Letting $n = 3$ in Eq. (38.7), we obtain the wavelength of the H_α line:

$$\frac{1}{\lambda} = (1.097 \times 10^7\ \text{m}^{-1})\left(\frac{1}{4} - \frac{1}{9}\right) \quad \text{or} \quad \lambda = 656.3\ \text{nm}$$

For $n = 4$ we obtain the wavelength of the H_β line, and so on. For $n = \infty$ we obtain the *smallest* wavelength in the series, $\lambda = 364.6$ nm.

Balmer's formula has a very direct relation to Bohr's hypothesis about energy levels. Using the relation $E = hc/\lambda$, we can find the *photon energies* corresponding to the wavelengths of the Balmer series. Multiplying Eq. (38.7) by hc, we find

$$E = \frac{hc}{\lambda} = hcR\left(\frac{1}{2^2} - \frac{1}{n^2}\right) = \frac{hcR}{2^2} - \frac{hcR}{n^2} \tag{38.8}$$

Equations (38.6) and (38.8) for the photon's energy agree most simply if we identify $-hcR/n^2$ as the initial energy E_i of the atom and $-hcR/2^2$ as its final energy E_f, in a transition in which a photon with energy $E_i - E_f$ is emitted. The energies of the levels are negative because we choose the potential energy to be zero when the electron and nucleus are infinitely far apart. The Balmer (and other) series suggest that the hydrogen atom has a series of energy levels, which we may call E_n, given by

$$E_n = -\frac{hcR}{n^2} \qquad n = 1, 2, 3, 4, \ldots \tag{38.9}$$

(energy levels of the hydrogen atom)

Each wavelength in the Balmer series corresponds to a transition from a level with n equal to 3 or greater to the level with $n = 2$.

ActivPhysics ONLINE

18.1 The Bohr Model

The numerical value of the product hcR is

$$hcR = (6.626 \times 10^{-34}\ \mathrm{J \cdot s})(2.998 \times 10^{8}\ \mathrm{m/s})(1.097 \times 10^{7}\ \mathrm{m^{-1}})$$
$$= 2.179 \times 10^{-18}\ \mathrm{J} = 13.60\ \mathrm{eV}$$

Thus the magnitudes of the energy levels given by Eq. (38.9) are -13.60 eV, -3.40 eV, -1.51 eV, -0.85 eV, ...

Other spectral series for hydrogen have been discovered. These are known, after their discoverers, as the Lyman, Paschen, Brackett, and Pfund series. Their wavelengths can be represented by formulas similar to Balmer's formula:

Lyman series $$\frac{1}{\lambda} = R\left(\frac{1}{1^2} - \frac{1}{n^2}\right) \quad n = 2, 3, 4, \ldots$$

Paschen series $$\frac{1}{\lambda} = R\left(\frac{1}{3^2} - \frac{1}{n^2}\right) \quad n = 4, 5, 6, \ldots$$

Brackett series $$\frac{1}{\lambda} = R\left(\frac{1}{4^2} - \frac{1}{n^2}\right) \quad n = 5, 6, 7, \ldots$$

Pfund series $$\frac{1}{\lambda} = R\left(\frac{1}{5^2} - \frac{1}{n^2}\right) \quad n = 6, 7, 8, \ldots$$

The Lyman series is in the ultraviolet, and the Paschen, Brackett, and Pfund series are in the infrared. We see that the Balmer series fits into the scheme between the Lyman and Paschen series.

Thus *all* the spectral series of hydrogen can be understood on the basis of Bohr's picture of transitions from one energy level (and corresponding electron orbit) to another, with the energy levels given by Eq. (38.9) with $n = 1, 2, 3, \ldots$. For the Lyman series the final level is always $n = 1$; for the Paschen series it is $n = 3$; and so on. The relation of the various spectral series to the energy levels and to electron orbits is shown in Fig. 38.9. Taken together, these spectral series give very strong support to Bohr's picture of energy levels in atoms.

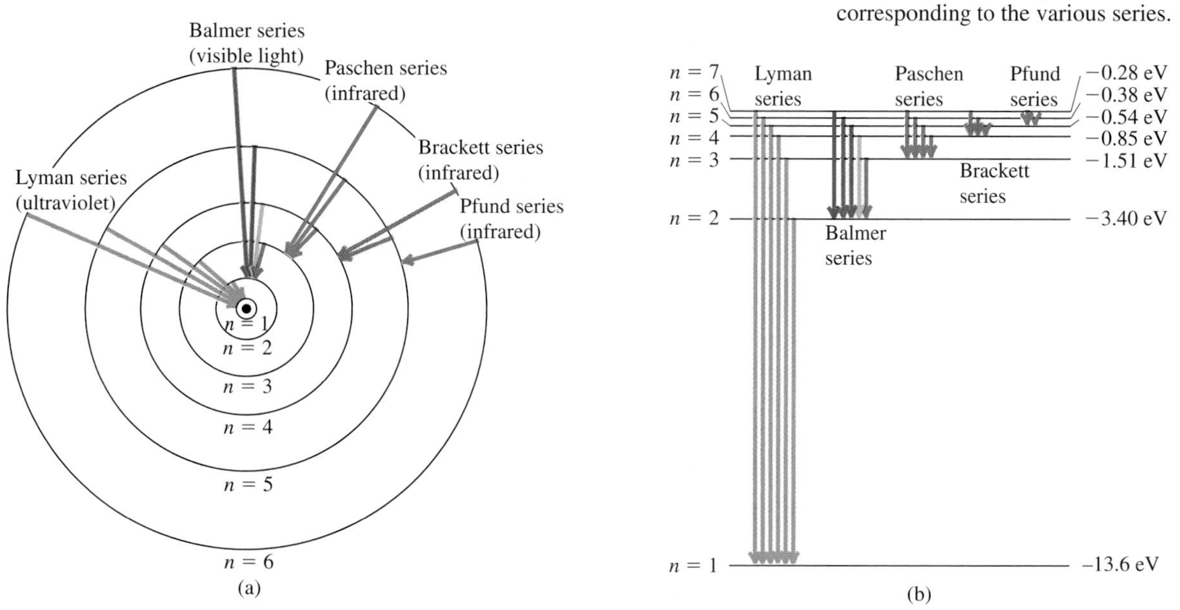

38.9 a) "Permitted" orbits of an electron in the Bohr model of a hydrogen atom (not to scale). Arrows indicate the transitions responsible for some of the lines of the various series. (b) Energy-level diagram for hydrogen, showing some transitions corresponding to the various series.

CAUTION The lines of a spectrum, such as the hydrogen spectrum shown in Fig. 38.8, are *not* all produced by a single atom. The sample of hydrogen gas that produced the spectrum in Fig. 38.8 contained a large number of atoms; these were excited in an electric discharge tube to various energy levels. The spectrum of the gas shows the light emitted from all the different transitions that occurred in different atoms of the sample.

We haven't yet discussed any way to *predict* what the energy levels for a particular atom should be. Neither have we shown how to derive Eq. (38.9) from fundamental theory or to relate the Rydberg constant R to other fundamental constants. We'll return to these problems later.

The Franck-Hertz Experiment

In 1914, James Franck and Gustav Hertz found even more direct experimental evidence for the existence of atomic energy levels. Franck and Hertz studied the motion of electrons through mercury vapor under the action of an electric field. They found that when the electron kinetic energy was 4.9 eV or greater, the vapor emitted ultraviolet light of wavelength 0.25 μm. Suppose mercury atoms have an energy level 4.9 eV above the lowest energy level. An atom can be raised to this level by collision with an electron; it later decays back to the lowest energy level by emitting a photon. According to Eq. (38.2), the wavelength of the photon should be

$$\lambda = \frac{hc}{E} = \frac{(4.136 \times 10^{-15}\ \text{eV}\cdot\text{s})(3.00 \times 10^{8}\ \text{m/s})}{4.9\ \text{eV}}$$
$$= 2.5 \times 10^{-7}\ \text{m} = 0.25\ \mu\text{m}$$

This is equal to the measured wavelength, confirming the existence of this energy level of the mercury atom. Similar experiments with other atoms yield the same kind of evidence for atomic energy levels.

Energy Levels

Only a few atoms and ions (such as hydrogen, singly ionized helium, doubly ionized lithium) have spectra whose wavelengths can be represented by a simple formula such as Balmer's. But it is *always* possible to analyze the more complicated spectra of other elements in terms of transitions among various energy levels and to deduce the numerical values of these levels from the measured spectrum wavelengths.

Every atom has a lowest energy level that includes the *minimum* internal energy state that the atom can have. This is called the *ground-state level,* or **ground level,** and all higher levels are called **excited levels.** A photon corresponding to a particular spectrum line is emitted when an atom makes a transition from a state in an excited level to a state in a lower excited level or the ground level.

Some energy levels for sodium are shown in Fig. 38.10. You may have noticed the yellow-orange light emitted by sodium vapor street lights. Sodium atoms emit this characteristic yellow-orange light with wavelengths 589.0 and 589.6 nm when they make transitions from the two closely spaced levels labeled *lowest excited levels* to the ground level.

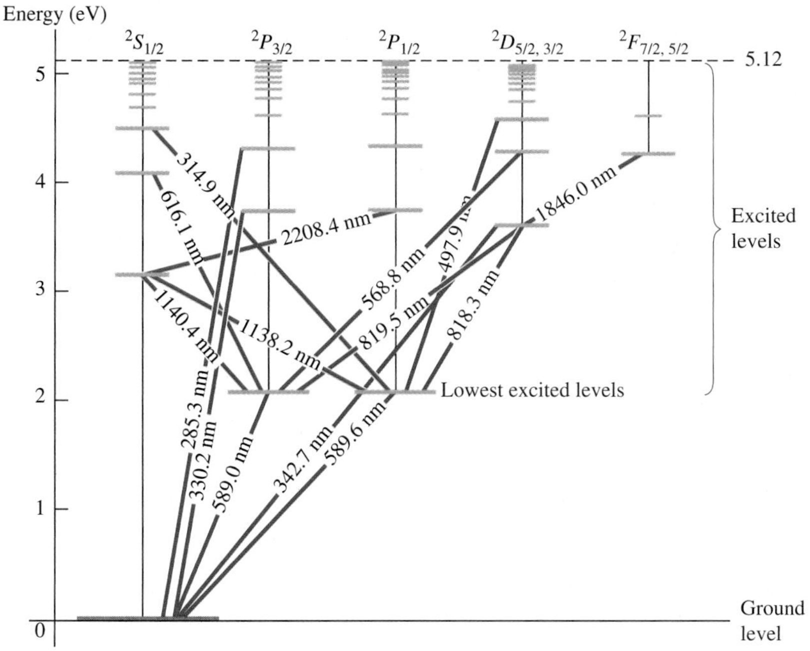

38.10 Energy levels of the sodium atom relative to the ground level. Numbers on the lines between levels are wavelengths. The column labels, such as $^2S_{1/2}$, refer to some quantum states of the valence electron (to be discussed in Chapter 41).

A sodium atom in the ground level can also *absorb* a photon with wavelength 589.0 or 589.6 nm. To demonstrate this process, we pass a beam of light from a sodium-vapor lamp through a bulb containing sodium vapor. The atoms in the vapor absorb the 589.0-nm or 589.6-nm photons from the beam, reaching the lowest excited levels; after a short time they return to the ground level, emitting photons in all directions and causing the sodium vapor to glow with the characteristic yellow light. The average time spent in an excited level is called the *lifetime* of the level; for the lowest excited levels of the sodium atom, the lifetime is about 1.6×10^{-8} s.

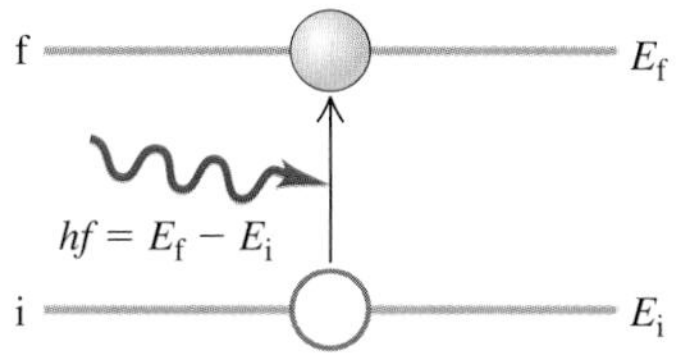

38.11 An atom is raised from an initial level i to a higher-energy final level f by absorbing a photon with energy equal to $E_f - E_i$. (Compare with Fig. 38.7.)

More generally, a photon *emitted* when an atom makes a transition from an excited level to a lower level can also be *absorbed* by a similar atom that is initially in the lower level (Fig. 38.11). If we pass white (continuous-spectrum) light through a gas and look at the *transmitted* light with a spectrometer, we find a series of dark lines corresponding to the wavelengths that have been absorbed, as shown in Fig. 38.12. This is called an **absorption spectrum.**

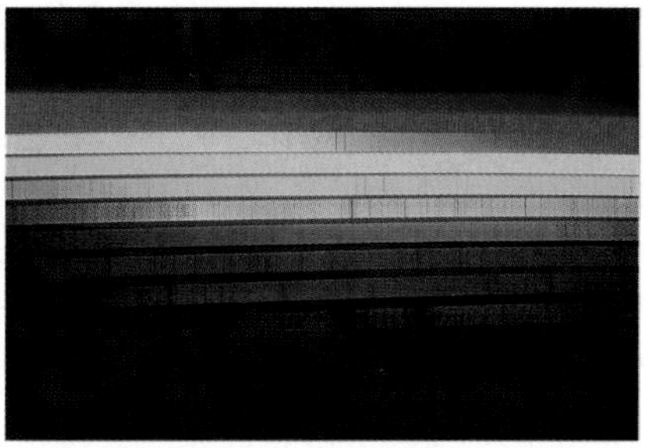

38.12 The absorption spectrum of the sun. The relatively cool solar atmosphere selectively absorbs photons from the sun's glowing surface, producing dark absorption lines. The wavelengths of the dark lines indicate what kinds of atoms are present in the solar atmosphere.

A related phenomenon is *fluorescence.* An atom absorbs a photon (often in the ultraviolet region) to reach an excited level and then drops back to the ground level in steps, emitting two or more photons with smaller energy and longer wavelength. For example, the electric discharge in a fluorescent tube causes the mercury vapor in the tube to emit ultraviolet radiation. This radiation is absorbed by the atoms of the coating on the inside of the tube. The coating atoms then re-emit light in the longer-wavelength visible portion of the spectrum. Fluorescent lamps are more efficient than incandescent lamps in converting electrical energy to visible light because they do not waste as much energy producing (invisible) infrared photons.

The Bohr hypothesis established the relation of wavelengths to energy levels, but it provided no general principles for *predicting* the energy levels of a particular atom. Bohr provided a partial analysis for the hydrogen atom; we will discuss this in Section 38.5. A more general understanding of atomic structure and energy levels rests on the concepts of *quantum mechanics,* which we will introduce in Chapters 39 and 40. Quantum mechanics provides all the principles needed to calculate energy levels from fundamental theory. Unfortunately, for many-electron atoms the calculations are so complex that they can be carried out only approximately.

Example 38.4 Emission and absorption spectra

A hypothetical atom has three energy levels: the ground level and 1.00 eV and 3.00 eV above the ground level. a) Find the frequencies and wavelengths of the spectral lines that this atom can emit when excited. b) What wavelengths can be absorbed by this atom if it is initially in its ground level?

SOLUTION

IDENTIFY and SET UP: We use the idea that energy is conserved when a photon is emitted or absorbed. Hence the photon energy is equal to the difference between the energies of the higher level and the lower level involved in the transition.

EXECUTE: a) Figure 38.13a shows an energy-level diagram. The possible photon energies, corresponding to the transitions shown, are 1.00 eV, 2.00 eV, and 3.00 eV. For 1.00 eV we have, from Eq. (38.2),

$$f = \frac{E}{h} = \frac{1.00\text{ eV}}{4.136 \times 10^{-15}\text{ eV}\cdot\text{s}} = 2.42 \times 10^{14}\text{ Hz}$$

For 2.00 eV and 3.00 eV, $f = 4.84 \times 10^{14}$ Hz and 7.25×10^{14} Hz, respectively. We can find the wavelengths using $\lambda = c/f$. For 1.00 eV,

$$\lambda = \frac{c}{f} = \frac{3.00 \times 10^{8}\text{ m/s}}{2.42 \times 10^{14}\text{ Hz}} = 1.24 \times 10^{-6}\text{ m} = 1240\text{ nm}$$

in the infrared region of the spectrum. For 2.00 eV and 3.00 eV the wavelengths are 620 nm (red) and 414 nm (violet), respectively (Fig. 38.13b).

b) From the atom's ground level, only a 1.00-eV or 3.00-eV photon can be absorbed; a 2.00-eV photon cannot be because there is no energy level 2.00 eV above the ground level. As we have calculated, the wavelengths of 1.00-eV and 3.00-eV photons are 1240 nm and 414 nm, respectively. Passing light from a hot solid through a cool gas of these atoms would result in a continuous spectrum with dark absorption lines at 1240 nm and 414 nm.

EVALUATE: Note that if a gas of these atoms were at a high enough temperature, collisions would excite a number of atoms into the 1.00-eV energy level. Such excited atoms *can* absorb 2.00-eV photons, and an absorption line at 620 nm would appear in the spectrum. Thus the observed spectrum of a given substance depends on its energy levels and its temperature.

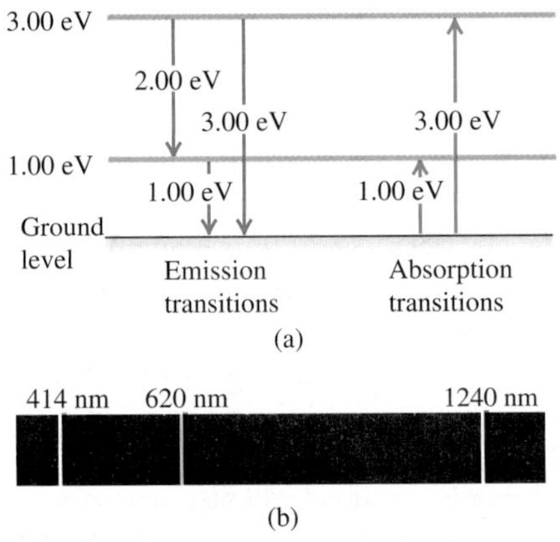

38.13 (a) Energy-level diagram, showing the possible transitions for emission from excited levels and for absorption from the ground level. (b) Emission spectrum of this hypothetical atom.

Test Your Understanding

Would it be possible to excite the hypothetical atom in Example 38.4 from the ground level to the 1.00-eV level with a 1.50-eV photon?

38.4 | The Nuclear Atom

Before we can make further progress in relating the energy levels of an atom to its internal structure, we need to have a better idea of what the inside of an atom is like. We know that atoms are much smaller than the wavelengths of visible light, so there is no hope of actually *seeing* an atom using that light. But we can still describe how the mass and electric charge are distributed throughout the volume of the atom.

Here's where things stood in 1910. J. J. Thomson had discovered the electron and measured its charge-to-mass ratio (e/m) in 1897; and by 1909, Millikan had completed his first measurements of the electron charge $-e$. These and other experiments showed that almost all the mass of an atom had to be associated with the *positive* charge, not with the electrons. It was also known that the overall size of atoms is of the order of 10^{-10} m and that all atoms except hydrogen contain more than one electron. What was *not* known was how the mass and charge were distributed within the atom. Thomson had proposed a model in which the atom consisted of a sphere of positive charge, of the order of 10^{-10} m in diameter, with the electrons embedded in it like raisins in a more or less spherical muffin.

38.14 Born in New Zealand, Ernest Rutherford (1871–1937) spent his professional life in England and Canada. Before carrying out the experiments that established the existence of atomic nuclei, he shared (with Frederick Soddy) the 1908 Nobel Prize in chemistry for showing that radioactivity results from the disintegration of atoms.

The first experiments designed to probe the interior structure of the atom were the **Rutherford scattering experiments,** carried out in 1910–1911 by Ernest Rutherford (Fig. 38.14) and two of his students, Hans Geiger and Ernest Marsden, at the University of Manchester in England. These experiments consisted of projecting charged particles at thin foils of the elements under study and observing the deflections of the particles. The particle accelerators that are now in common use in laboratories had not yet been invented, and Rutherford's projectiles were *alpha particles* emitted from naturally radioactive elements. Alpha particles are identical with the nuclei of most helium atoms: two protons and two neutrons bound together. They are ejected from unstable nuclei with speeds of the order of 10^7 m/s, and they can travel several centimeters through air or 0.1 mm or so through solid matter before they are brought to rest by collisions.

ActivPhysics Online
19.1 Particle Scattering

Rutherford's experimental setup is shown schematically in Fig. 38.15. A radioactive substance at the left emits alpha particles. Thick lead screens stop all particles except those in a narrow beam defined by small holes. The beam then passes through a target consisting of a thin gold, silver, or copper foil and strikes screens coated with zinc sulfide, similar in principle to the screen of a TV picture tube. A momentary flash, or *scintillation,* can be seen on the screen whenever it is struck by an alpha particle. Rutherford and his students counted the numbers of particles deflected through various angles.

Think of the atoms of the target material as being packed together like marbles in a box. An alpha particle can pass through a thin sheet of metal foil, so the alpha particle must be able actually to pass through the interiors of atoms. The *total*

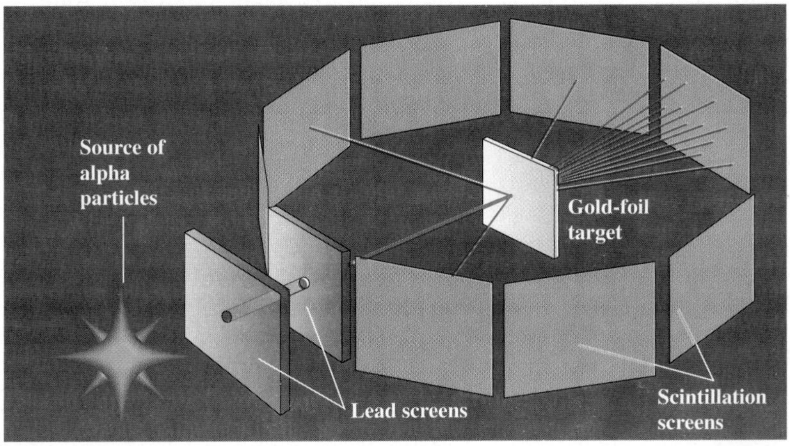

38.15 The scattering of alpha particles by a thin metal foil. The source of alpha particles is a radioactive element such as radium. The two lead screens with small holes form a narrow beam of alpha particles, which are scattered by the gold foil. The directions of the scattered particles are determined from the flashes on the scintillation screens.

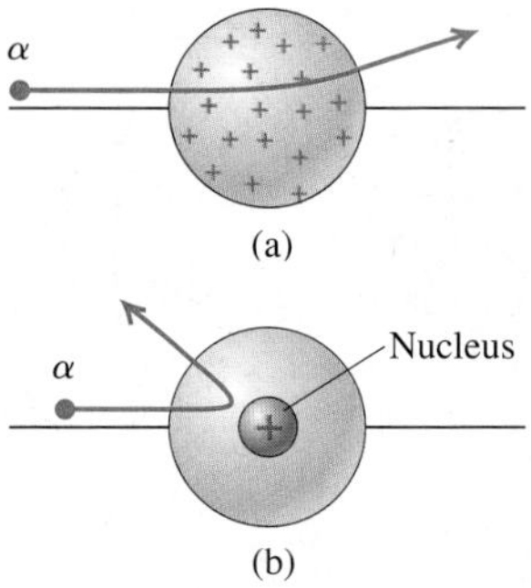

38.16 (a) Thomson's model of the atom: An alpha particle is scattered through only a small angle. (b) Rutherford's model of the atom: An alpha particle can be scattered by the compact, positively charged nucleus (not drawn to scale).

electric charge of the atom is zero, so outside the atom there is little force on the alpha particle. Within the atom there are electrical forces caused by the electrons and by the positive charge. But the mass of an alpha particle is about 7300 times that of an electron. Momentum considerations show that the alpha particle can be scattered only a very small amount by its interaction with the much lighter electrons. It's like throwing a pebble through a swarm of mosquitoes; the mosquitoes don't deflect the pebble very much. Only interactions with the *positive* charge, which is tied to almost all of the mass of the atom, can deflect the alpha particle appreciably.

In the Thomson model, the positive charge and the negative electrons are distributed through the whole atom. Hence the electric field inside the atom should be quite small, and the electric force on an alpha particle that enters the atom should be quite weak. The maximum deflection to be expected is then only a few degrees (Fig. 38.16a).

The results were very different from this and were totally unexpected. Some alpha particles were scattered by nearly 180°, that is, almost straight backward (Fig. 38.16b). Rutherford later wrote:

> It was quite the most incredible event that ever happened to me in my life. It was almost as incredible as if you had fired a 15-inch shell at a piece of tissue paper and it came back and hit you.

Back to the drawing board! Suppose the positive charge, instead of being distributed through a sphere with atomic dimensions (of the order of 10^{-10} m), is all concentrated in a much *smaller* space. Then it would act like a point charge down to much smaller distances. The maximum electric field repelling the alpha particle would be much larger, and the amazing large-angle scattering would be possible. Rutherford called this concentration of positive charge the **nucleus.** He again computed the numbers of particles expected to be scattered through various angles. Within the accuracy of his experiments, the computed and measured results agreed, down to distances of the order of 10^{-14} m. His experiments therefore established that the atom does have a nucleus, a very small, very dense structure, no larger than 10^{-14} m in diameter. The nucleus occupies only about 10^{-12} of the total volume of the atom or less, but it contains *all* the positive charge and at least 99.95% of the total mass of the atom.

Figure 38.17 shows a computer simulation of the scattering of 5.0-MeV alpha particles from a gold nucleus of radius 7.0×10^{-15} m (the actual value) and from a nucleus with a hypothetical radius ten times this great. In the second case there is *no* large-angle scattering. The presence of large-angle scattering in Rutherford's experiments thus attested to the small size of the nucleus.

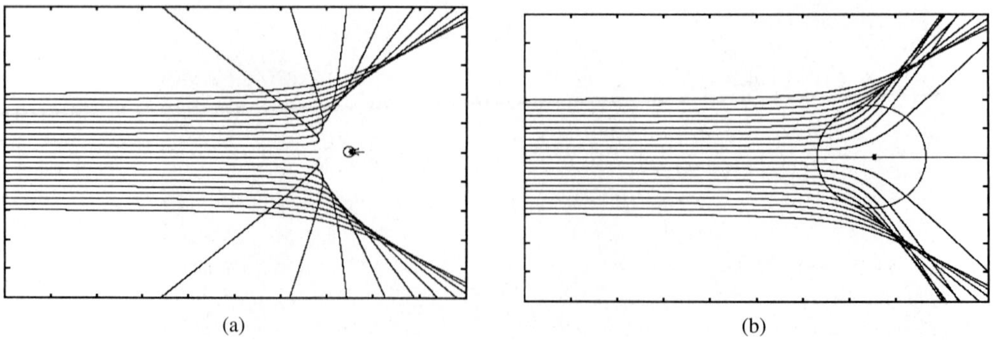

38.17 Computer simulation of scattering of 5.0-MeV alpha particles from a gold nucleus. Each curve shows a possible alpha-particle trajectory. (a) A gold nucleus with radius 7.0×10^{-15} m (its actual size) gives large-angle scattering. (b) A nucleus with ten times the radius of that in (a) shows *no* large-angle scattering.

Example 38.5 A Rutherford experiment

An alpha particle is aimed directly at a gold nucleus. An alpha particle has two protons and a charge of $2e = 2(1.60 \times 10^{-19}\text{ C})$, while a gold nucleus has 79 protons and a charge of $79e = 79(1.60 \times 10^{-19}\text{ C})$. What minimum initial kinetic energy must the alpha particle have to approach within 5.0×10^{-14} m of the center of the gold nucleus? Assume that the gold nucleus, which has about 50 times the rest mass of an alpha particle, remains at rest.

SOLUTION

IDENTIFY and SET UP: Let point 1 be very far from the gold nucleus (the initial position of the alpha particle) and let point 2 be 5.0×10^{-14} m from the center of the gold nucleus. Our target variable is the kinetic energy of the alpha particle at point 1 that allows it to reach point 2. To find this we use the law of conservation of energy and Eq. (23.9) for electric potential energy, $U = qq_0/4\pi\epsilon_0 r$.

EXECUTE: The kinetic energy at point 1 is K_1 (the target variable), and the kinetic energy at point 2 (where the alpha particle comes momentarily to rest) is $K_2 = 0$. At point 1 the alpha particle and gold nucleus are so far apart that we can regard their separation r as infinite, so from Eq. (23.9) the electric potential energy at point 1 is $U_1 = 0$. At point 2, the potential is

$$U_2 = \frac{1}{4\pi\epsilon_0}\frac{qq_0}{r}$$

$$= (9.0 \times 10^9\text{ N}\cdot\text{m}^2/\text{C}^2)\frac{(2)(79)(1.60 \times 10^{-19}\text{ C})^2}{5.0 \times 10^{-14}\text{ M}}$$

$$= 7.3 \times 10^{-13}\text{ J} = 4.6 \times 10^6\text{ eV} = 4.6\text{ MeV}$$

By energy conservation $K_1 + U_1 = K_2 + U_2$, so $K_1 = K_2 + U_2 - U_1 = 0 + 4.6\text{ MeV} - 0 = 4.6$ MeV. Thus, to approach to 5.0×10^{-14} m, the alpha particle must have 4.6 MeV when it is far away from the nucleus; to approach even closer, even greater kinetic energy is needed. In fact, alpha particles emitted from naturally occurring radioactive elements typically have energies in the range 4 to 6 MeV. For example, the common isotope of radium, ^{226}Ra, emits an alpha particle with energy 4.78 MeV.

EVALUATE: Was it valid to assume that the gold nucleus remains at rest? To find out, note that when the alpha particle stops momentarily, all of its initial momentum has been transferred to the gold nucleus. An alpha particle has a mass $m_\alpha = 6.64 \times 10^{-27}$ kg; if its initial kinetic energy $K_1 = \frac{1}{2}mv_1^2$ is 7.3×10^{-13} J, you can show that its initial speed is $v_1 = 1.5 \times 10^7$ m/s and its initial momentum is $p_1 = m_\alpha v_1 = 9.8 \times 10^{-20}$ kg·m/s. A gold nucleus (mass $m_{Au} = 3.27 \times 10^{-25}$ kg) with this much momentum has a much slower speed $v_{Au} = 3.0 \times 10^5$ m/s and a kinetic energy $K_{Au} = \frac{1}{2}mv_{Au}^2 = 1.5 \times 10^{-14}\text{ J} = 0.092$ MeV. This *recoil kinetic energy* of the gold nucleus is a small fraction of the total energy in this situation, so we are justified in neglecting it.

Test Your Understanding

Suppose you repeated Rutherford's scattering experiment using a thin sheet of solid hydrogen in place of the gold foil. (Hydrogen is a solid at temperatures below 14.0 K.) Would you expect to observe large-angle scattering of alpha particles?

38.5 | The Bohr Model

At the same time (1913) that Bohr established the relationship between spectral wavelengths and energy levels, he also proposed a model of the hydrogen atom. He developed his ideas while working in Rutherford's laboratory. Using this model, now known as the **Bohr model,** he was able to *calculate* the energy levels of hydrogen and obtain agreement with values determined from spectra.

Rutherford's discovery of the atomic nucleus raised a serious question. What kept the negatively charged electrons at relatively large distances $(\sim 10^{-10}\text{ m})$ away from the very small $(\sim 10^{-14}\text{ m})$, positively charged nucleus despite their electrostatic attraction? Rutherford suggested that perhaps the electrons *revolve* in orbits about the nucleus, just as the planets revolve around the sun.

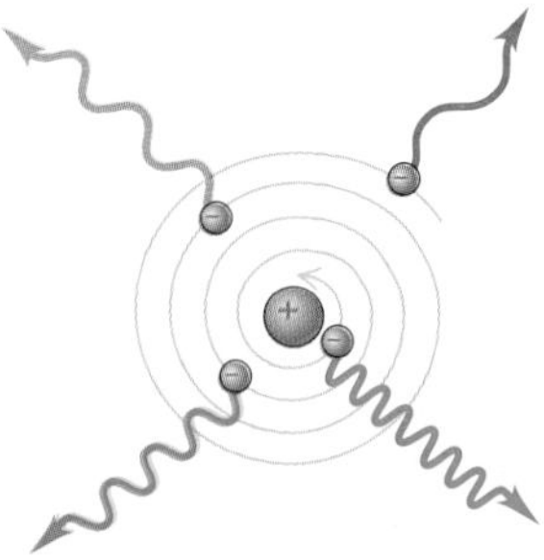

38.18 Classical physics predicts that an orbiting electron should continuously radiate electromagnetic waves and spiral into the nucleus. Because the electron's angular speed varies with its radius, classical physics predicts that the frequency of the emitted radiation should change continuously.

But according to classical electromagnetic theory, any accelerating electric charge (either oscillating or revolving) radiates electromagnetic waves. An example is radiation from an oscillating point charge, shown in Fig. 32.2 (Section 32.1). The energy of an orbiting electron should therefore decrease continuously, its orbit should become smaller and smaller, and it should spiral rapidly into the nucleus (Fig. 38.18). Even worse, according to classical theory the *frequency* of the electromagnetic waves emitted should equal the frequency of revolution. As the electrons radiated energy, their angular speeds would change continuously, and they would emit a *continuous* spectrum (a mixture of all frequencies), not the *line* spectrum actually observed.

Stable Orbits

ActivPhysics ONLINE

18.1 The Bohr Model

To solve this problem, Bohr made a revolutionary proposal. He postulated that an electron in an atom can move around the nucleus in certain circular *stable orbits, without* emitting radiation, contrary to the predictions of classical electromagnetic theory. According to Bohr, there is a definite energy associated with each stable orbit, and an atom radiates energy only when it makes a transition from one of these orbits to another. The energy is radiated in the form of a photon with energy and frequency given by Eq. (38.6), $hf = E_i - E_f$.

As a result of a rather complicated argument that related the angular frequency of the light emitted to the angular speeds of the electron in highly excited energy levels, Bohr found that the magnitude of the electron's angular momentum is *quantized,* that this magnitude for the electron must be an integral multiple of $h/2\pi$. (Note that because $1\ \text{J} = 1\ \text{kg}\cdot\text{m}^2/\text{s}^2$, the SI units of Planck's constant h, $\text{J}\cdot\text{s}$, are the same as the SI units of angular momentum, usually written as $\text{kg}\cdot\text{m}^2/\text{s}$.) From Section 10.5, Eq. (10.28), the magnitude of the angular momentum is $L = mvr$ for a particle with mass m moving with speed v in a circle of radius r (thus with $\phi = 90°$). So Bohr's argument led to

$$L = mvr = n\frac{h}{2\pi}$$

where $n = 1, 2, 3, \ldots$. Each value of n corresponds to a permitted value of the orbit radius, which we denote from now on by r_n, and a corresponding speed v_n. The value of n for each orbit is called the **principal quantum number** for the orbit. With this notation the above equation becomes

$$L_n = mv_n r_n = n\frac{h}{2\pi} \qquad \text{(quantization of angular momentum)} \qquad (38.10)$$

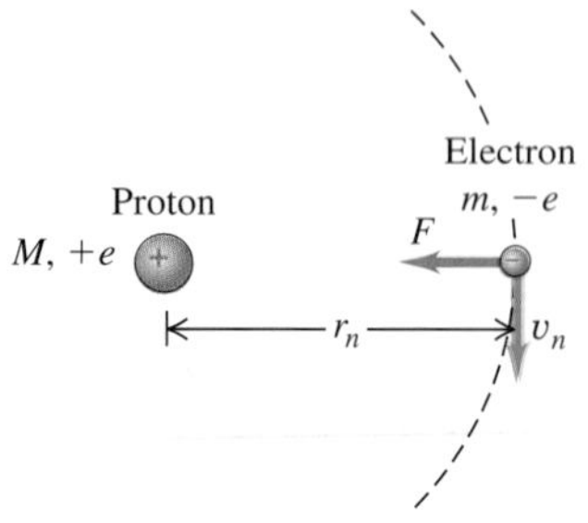

38.19 Bohr model of the hydrogen atom. The proton is assumed to be stationary; the electron revolves in a circle of radius r_n with speed v_n. The electrostatic attraction provides the needed centripetal acceleration.

Now let's consider a model of the hydrogen atom (Fig. 38.19) that is Newtonian in spirit but incorporates this quantization assumption. This atom consists of a single electron with mass m and charge $-e$, revolving around a single proton with charge $+e$. The proton is nearly 2000 times as massive as the electron, so we have been assuming that the proton does not move. We learned in Section 5.4 that when a particle with mass m moves with speed v_n in a circular orbit with radius r_n, its radially inward acceleration is v_n^2/r_n. According to Newton's second law, a radially inward net force with magnitude $F = mv_n^2/r_n$ is needed to cause this acceleration. We discussed in Section 12.4 how the gravitational attraction provided that force for satellite orbits. In hydrogen the force F is provided by the electrical attraction between the positive proton and the negative electron. From Coulomb's law (Eq. 21.2),

$$F = \frac{1}{4\pi\epsilon_0}\frac{e^2}{r_n^2}$$

so Newton's second law states that

$$\frac{1}{4\pi\epsilon_0}\frac{e^2}{r_n^2} = \frac{mv_n^2}{r_n} \qquad (38.11)$$

When we solve Eqs. (38.10) and (38.11) simultaneously for r_n and v_n, we get

$$r_n = \epsilon_0\frac{n^2h^2}{\pi me^2} \quad \text{(orbit radii in the Bohr model)} \qquad (38.12)$$

$$v_n = \frac{1}{\epsilon_0}\frac{e^2}{2nh} \quad \text{(orbital speeds in the Bohr model)} \qquad (38.13)$$

Equation (38.12) shows that the orbit radius r_n is proportional to n^2; the smallest orbit radius corresponds to $n = 1$. We'll denote this minimum radius, called the *Bohr radius,* as a_0:

$$a_0 = \epsilon_0\frac{h^2}{\pi me^2} \qquad (38.14)$$

Then Eq. (38.12) can be written as

$$r_n = n^2a_0 \qquad (38.15)$$

The permitted orbits have radii a_0, $4a_0$, $9a_0$, and so on.

The numerical values of the quantities on the right side of Eq. (38.14) are given in Appendix F. Using these values, we find that the radius a_0 of the smallest Bohr orbit is

$$a_0 = \frac{(8.854 \times 10^{-12}\ \text{C}^2/\text{N}\cdot\text{m}^2)(6.626 \times 10^{-34}\ \text{J}\cdot\text{s})^2}{(3.142)(9.109 \times 10^{-31}\ \text{kg})(1.602 \times 10^{-19}\ \text{C})^2}$$
$$= 5.29 \times 10^{-11}\ \text{m}$$

This result, giving an atomic diameter of about 10^{-10} m $= 0.1$ nm, is consistent with atomic dimensions estimated by other methods.

We can use Eq. (38.13) to find the orbital *speed* of the electron. We leave this calculation as an exercise; the result is that for the $n = 1$ state, $v_1 = 2.19 \times 10^6$ m/s. This, the greatest possible speed of the electron in the hydrogen atom, is less than 1% of the speed of light, showing that relativistic considerations aren't significant.

Energy Levels

We can use Eqs. (38.13) and (38.12) to find the kinetic and potential energies K_n and U_n when the electron is in the orbit with quantum number n:

$$K_n = \frac{1}{2}mv_n^2 = \frac{1}{\epsilon_0^2}\frac{me^4}{8n^2h^2} \qquad (38.16)$$

$$U_n = -\frac{1}{4\pi\epsilon_0}\frac{e^2}{r_n} = -\frac{1}{\epsilon_0^2}\frac{me^4}{4n^2h^2} \qquad (38.17)$$

The total energy E_n is the sum of the kinetic and potential energies:

$$E_n = K_n + U_n = -\frac{1}{\epsilon_0^2}\frac{me^4}{8n^2h^2} \qquad (38.18)$$

The potential energy has a negative sign because we have taken the potential energy to be zero when the electron is infinitely far from the nucleus. We are interested only in energy *differences,* so the reference position doesn't matter.

The orbits and energy levels are depicted in Fig. 38.9 (Section 38.3). The possible energy levels of the atom are labeled by values of the quantum number n. For each value of n there are corresponding values of orbit radius r_n, speed v_n, angular momentum $L_n = nh/2\pi$, and total energy E_n. The energy of the atom is least when $n = 1$ and E_n has its most negative value. This is the *ground level* of the atom; it is the level with the smallest orbit, with radius a_0. For $n = 2, 3, \ldots$, the absolute value of E_n is smaller and the energy is progressively larger (less negative). The orbit radius increases as n^2, as shown by Eqs. (38.12) and (38.15), while the speed decreases as $1/n$, as shown by Eq. (38.13).

Comparing the expression for E_n in Eq. (38.18) with Eq. (38.9) (deduced from the measured spectrum of hydrogen), we see that they agree only if the coefficients are equal:

$$hcR = \frac{1}{\epsilon_0^2}\frac{me^4}{8h^2} \quad \text{or} \quad R = \frac{me^4}{8\epsilon_0^2h^3c} \qquad (38.19)$$

This equation therefore shows us how to *calculate* the value of the Rydberg constant from the fundamental physical constants m, c, e, h, and ϵ_0, all of which can be determined quite independently of the Bohr theory. When we substitute the numerical values of these quantities, we obtain the value $R = 1.097 \times 10^7\ \text{m}^{-1}$. To four significant figures, this is the value determined from wavelength measurements. This agreement provides very strong and direct confirmation of Bohr's theory. You should substitute numerical values into Eq. (38.19) and compute the value of R to confirm these statements.

The *ionization energy* of the hydrogen atom is the energy required to remove the electron completely. Ionization corresponds to a transition from the ground level $(n = 1)$ to an infinitely large orbit radius $(n = \infty)$ and thus equals $-E_1$. Substituting the constants from Appendix F into Eq. (38.18) gives an ionization energy of 13.606 eV. The ionization energy can also be measured directly; the result is 13.60 eV. These two values agree within 0.1%.

Example 38.6 Exploring the Bohr model

Find the kinetic, potential, and total energies of the hydrogen atom in the first excited level, and find the wavelength of the photon emitted in the transition from the first excited level to the ground level.

SOLUTION

IDENTIFY and SET UP: We use Eqs. (38.16), (38.17), and (38.18) to find the energies of the atom. To find the photon wavelength λ we use Eq. (38.6), which relates the energy and wavelength of an emitted photon to the energies of the initial and final levels of the atom.

EXECUTE: We could evaluate Eqs. (38.16), (38.17), and (38.18) for the nth level by substituting the values of m, e, ϵ_0, and h. But we can simplify the calculation by comparing with Eq. (38.19), which shows that the constant that appears in Eqs. (38.14), (38.15), and (38.16) is equal to hcR and is experimentally equal to 13.60 eV:

$$\frac{me^4}{8\epsilon_0^2h^2} = hcR = 13.60\ \text{eV}$$

Using this expression, we can rewrite Eqs. (38.16), (38.17), and (38.18) as

$$K_n = \frac{13.60\text{ eV}}{n^2} \qquad U_n = \frac{-27.20\text{ eV}}{n^2} \qquad E_n = \frac{-13.60\text{ eV}}{n^2}$$

The first excited level is the $n = 2$ level, which has $K_2 = 3.40$ eV, $U_2 = -6.80$ eV, and $E_2 = -3.40$ eV.

The ground level is the $n = 1$ level, for which the energy is $E_1 = -13.60$ eV. The energy of the emitted photon is $E_2 - E_1 = -3.40\text{ eV} - (-13.60\text{ eV}) = 10.20$ eV. The photon energy equals hc/λ, so we find

$$\lambda = \frac{hc}{E_2 - E_1} = \frac{(4.136 \times 10^{-15}\text{ eV}\cdot\text{s})(3.00 \times 10^8\text{ m/s})}{10.2\text{ eV}}$$
$$= 1.22 \times 10^{-7}\text{ m} = 122\text{ nm}$$

This is the wavelength of the "Lyman alpha" line, the longest-wavelength line in the Lyman series of ultraviolet lines in the hydrogen spectrum (Fig. 38.9).

EVALUATE: Our results show that the total mechanical energy is negative and equal to one-half the potential energy. Remarkably, we found the same energy relationship for satellite orbits in Section 12.4. The reason for the similarity between the two situations is that both the electrostatic force and the gravitational force are inversely proportional to $1/r^2$.

Reduced Mass

The values of the Rydberg constant R and the ionization energy of hydrogen predicted by Bohr's analysis are within 0.1% of the measured values. The agreement would be even better if we had not assumed that the nucleus (a proton) remains at rest. Rather, the proton and electron both revolve in circular orbits about their common *center of mass* (Section 8.5), as shown in Fig. 38.20. It turns out that we can take this motion into account very simply by using in Bohr's equations not the electron rest mass m but a quantity called the **reduced mass** m_r of the system. For a system composed of two bodies of masses m_1 and m_2, the reduced mass is defined as

$$m_r = \frac{m_1 m_2}{m_1 + m_2}$$

For ordinary hydrogen we let m_1 equal m and m_2 equal the proton mass, $m_p = 1836.2m$. Thus the proton-electron system of ordinary hydrogen has a reduced mass of

$$m_r = \frac{m(1836.2m)}{m + 1836.2m} = 0.99946m$$

When this value is used instead of the electron mass m in the Bohr equations, the predicted values agree very well with the measured values.

In an atom of deuterium, also called *heavy hydrogen,* the nucleus is not a single proton but a proton and a neutron bound together to form a composite body called the *deuteron.* The reduced mass of the deuterium atom turns out to be $0.99973m$. Equations (38.6) and (38.18) (with m replaced by m_r) show that all energies and frequencies are proportional to m_r, while all wavelengths are inversely proportional to m_r. Thus the spectral wavelengths of deuterium should be those of hydrogen divided by $(0.99973m)/(0.99946m) = 1.00027$. This is a small effect but well within the precision of modern spectrometers. This small wavelength shift led the American scientist Harold Urey to the discovery of deuterium in 1932, an achievement that earned him the 1934 Nobel Prize in chemistry.

The concept of reduced mass has other applications. A *positron* has the same rest mass as an electron but a charge $+e$. A positronium "atom" (Fig. 38.21) con-

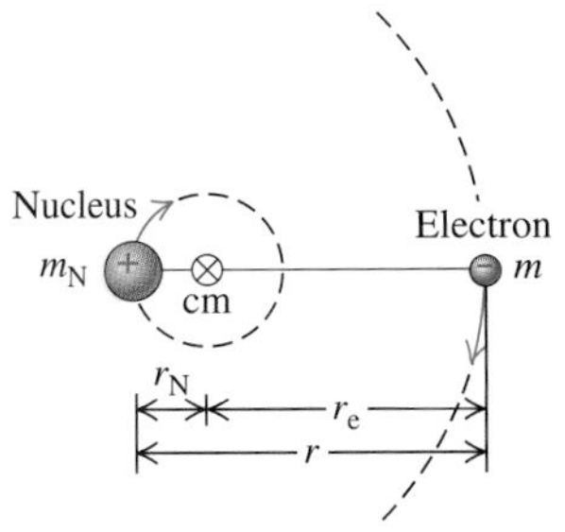

38.20 The nucleus and the electron revolve around their common center of mass. The distance r_N has been exaggerated for clarity; for ordinary hydrogen it actually equals $r_e/1836.2$.

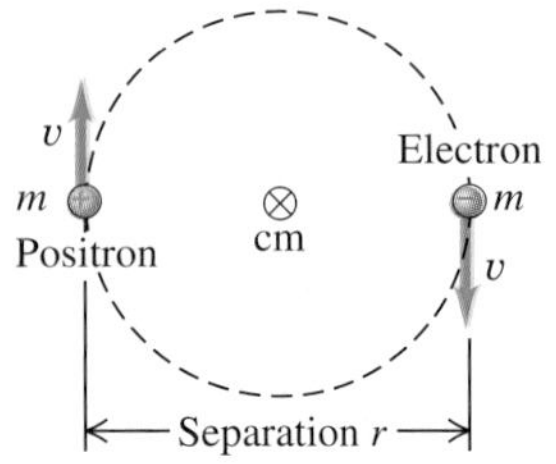

38.21 Applying the Bohr model to positronium. The electron and the positron revolve about their common center of mass, which is located midway between them because they have equal mass. Their separation is twice either orbit radius.

sists of an electron and a positron, each with mass m, in orbit around their common center of mass. This structure lasts only about 10^{-6} s before the two particles annihilate one another and disappear, but this is enough time to study the positronium spectrum. The reduced mass is $m/2$, so the energy levels and photon frequencies have exactly half the values for the simple Bohr model with infinite proton mass. The existence of positronium atoms was confirmed by observation of the corresponding spectrum lines.

Hydrogenlike Atoms

We can extend the Bohr model to other one-electron atoms, such as singly ionized helium (He^+), doubly ionized lithium (Li^{2+}), and so on. Such atoms are called *hydrogenlike* atoms. In such atoms, the nuclear charge is not e but Ze, where Z is the *atomic number*, equal to the number of protons in the nucleus. The effect in the previous analysis is to replace e^2 everywhere by Ze^2. In particular, the orbit radii r_n given by Eq. (38.12) become smaller by a factor of Z, and the energy levels E_n given by Eq. (38.18) are multiplied by Z^2. We invite you to verify these statements. The reduced-mass correction in these cases is even less than 0.1% because the nuclei are more massive than the single proton of ordinary hydrogen. Figure 38.22 compares the energy levels for H and for He^+, which has $Z = 2$.

Atoms of the *alkali* metals have one electron outside a core consisting of the nucleus and the inner electrons, with net core charge $+e$. These atoms are approximately hydrogenlike, especially in excited levels. Even trapped electrons and electron vacancies in semiconducting solids act somewhat like hydrogen atoms.

Although the Bohr model predicted the energy levels of the hydrogen atom correctly, it raised as many questions as it answered. It combined elements of classical physics with new postulates that were inconsistent with classical ideas. The model provided no insight into what happens *during* a transition from one orbit to another; the angular speeds of the electron motion were not in general the angular frequencies of the emitted radiation, a result that is contrary to classical electrodynamics. Attempts to extend the model to atoms with two or more electrons were not successful. An electron moving in one of Bohr's circular orbits forms a current loop and should produce a magnetic moment. However, a hydrogen atom in

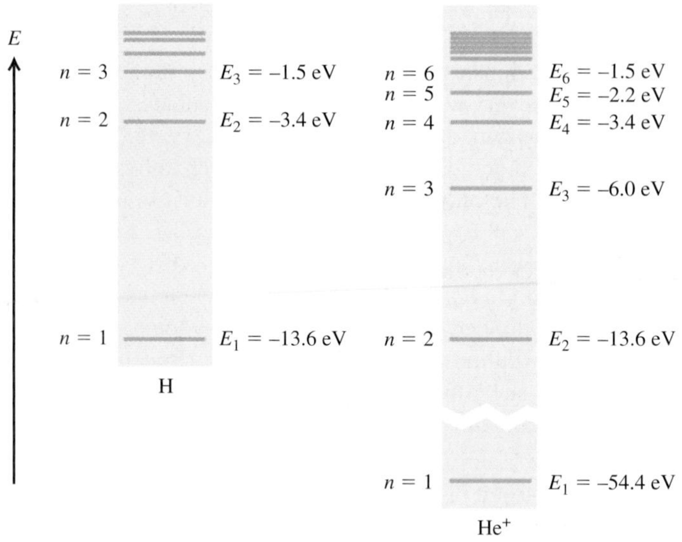

38.22 Energy levels of H and He^+. The energy expression, Eq. (38.18), is multiplied by $Z^2 = 4$ for He^+, so the energy of an He^+ ion with a given n is almost exactly four times that of an H atom with the same n. (There are small differences of the order of 0.05% because the reduced masses are slightly different.)

its ground level has *no* magnetic moment due to orbital motion. In Chapters 39 and 40 we will find that an even more radical departure from classical concepts was needed before the understanding of atomic structure could progress further.

18.3 The Laser

Test Your Understanding

Many of the wavelengths in the spectrum of light emitted by excited He^+ ions are almost the same as the wavelengths emitted by excited H atoms. Explain why.

38.6 | The Laser

The **laser** is a light source that produces a beam of highly coherent and very nearly monochromatic light as a result of cooperative emission from many atoms. The name *laser* is an acronym for "light amplification by stimulated emission of radiation." We can understand the principles of laser operation on the basis of photons and atomic energy levels. During the discussion we'll also introduce two new concepts: *stimulated emission* and *population inversion.*

If an atom has an excited level an energy E above the ground level, the atom in its ground level can absorb a photon with frequency f given by $E = hf$. This process is shown schematically in Fig. 38.23a, which shows a gas in a transparent container. Three atoms A each absorb a photon, reaching an excited level and being denoted as A*. Some time later, the excited atoms return to the ground level by each emitting a photon with the same frequency as the one originally absorbed. This process is called *spontaneous emission;* the direction and phase of the emitted photons are random (Fig. 38.23b).

In **stimulated emission** (Fig. 38.23c), each incident photon encounters a previously excited atom. A kind of resonance effect induces each atom to emit a second photon with the same frequency, direction, phase, and polarization as the incident photon, which is not changed by the process. For each atom there is one photon before a stimulated emission and two photons after—thus the name *light amplification.* Because the two photons have the same phase, they emerge together as *coherent* radiation. The laser makes use of stimulated emission to produce a beam consisting of a large number of such coherent photons.

To discuss stimulated emission from atoms in excited levels, we need to know something about how many atoms are in each of the various energy levels. First, we need to make the distinction between the terms *energy level* and *state.* A system may have more than one way to attain a given energy level; each different way is a different **state.** For instance, there are two ways of putting an ideal unstretched spring in a given energy level. Remembering that $U = \frac{1}{2}kx^2$, we could compress the spring by $x = -b$ or we could stretch it by $x = +b$ to get the same $U = \frac{1}{2}kb^2$. The Bohr model had only one state in each energy level, but we will find in Section 41.4 that the hydrogen atom actually has two states in its −13.60-eV ground level, eight states in its −3.40-eV first exited level, and so on.

The Maxwell-Boltzmann distribution function (Section 18.5) determines the number of atoms in a given state in a gas. The function tells us that when the gas is in thermal equilibrium at absolute temperature T, the number n_i of atoms in a state with energy E_i equals $Ae^{-E_i/kT}$, where k is Boltzmann's constant and A is another constant determined by the total number of atoms in the gas. (In Section 18.5, E was the kinetic energy $\frac{1}{2}mv^2$ of a gas molecule; here we're talking about the internal energy of an atom.) Because of the negative exponent, fewer atoms are in higher-energy states, as we should expect. If E_g is a ground-state

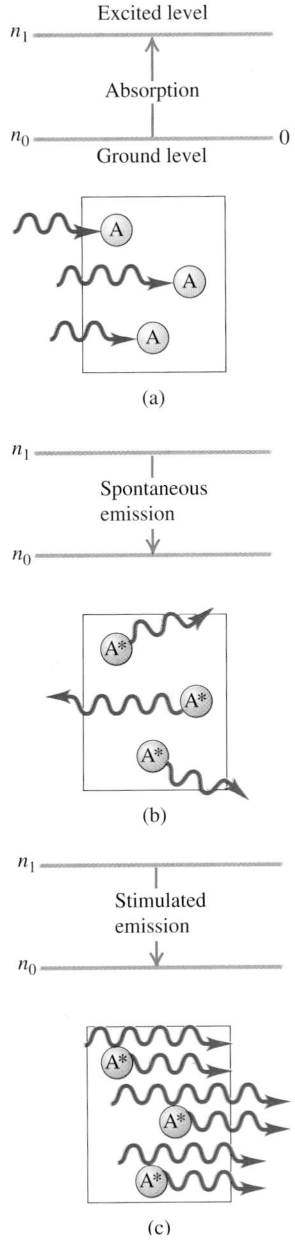

38.23 Three interaction processes between an atom and electromagnetic waves. In part (a), A denotes an atom in its ground level; in parts (b) and (c), A* denotes an atom in an excited level.

energy and E_{ex} is the energy of an excited state, then the ratio of numbers of atoms in the two states is

$$\frac{n_{ex}}{n_g} = \frac{Ae^{-E_{ex}/kT}}{Ae^{-E_g/kT}} = e^{-(E_{ex}-E_g)/kT} \quad (38.21)$$

For example, suppose $E_{ex} - E_g = 2.0\ \text{eV} = 3.2 \times 10^{-19}\ \text{J}$, the energy of a 620-nm visible-light photon. At $T = 3000\ \text{K}$ (the temperature of the filament in an incandescent light bulb),

$$\frac{E_{ex}-E_g}{kT} = \frac{3.2 \times 10^{-19}\ \text{J}}{(1.38 \times 10^{-23}\ \text{J/K})(3000\ \text{K})} = 7.73$$

and

$$e^{-(E_{ex}-E_g)/kT} = e^{-7.73} = 0.00044$$

That is, the fraction of atoms in a state 2.0 eV above a ground state is extremely small, even at this high temperature. The point is that at any reasonable temperature there aren't enough atoms in excited states for any appreciable amount of stimulated emission from these states to occur. Rather, absorption is much more probable.

We could try to enhance the number of atoms in excited states by sending through the container a beam of radiation with frequency $f = E/h$ corresponding to the energy difference $E = E_{ex} - E_g$. Some of the atoms absorb photons of energy E and are raised to the excited state, and the population ratio n_{ex}/n_g momentarily increases. But because n_g is originally so much larger than n_{ex}, an enormously intense beam of light would be required to momentarily increase n_{ex} to a value comparable to n_g. The rate at which energy is *absorbed* from the beam by the n_g ground-state atoms far outweighs the rate at which energy is added to the beam by stimulated emission from the relatively rare (n_{ex}) excited atoms.

We need to create a non-equilibrium situation in which the number of atoms in a higher-energy state is greater than the number in a lower-energy state. Such a situation is called a **population inversion.** Then the rate of energy radiation by stimulated emission can *exceed* the rate of absorption, and the system acts as a net *source* of radiation with photon energy E. Furthermore, because the photons are the result of stimulated emission, they all have the same frequency, phase, polarization, and direction of propagation. The resulting radiation is therefore very much more coherent than light from ordinary sources, in which the emissions of individual atoms are not coordinated. This coherent emission is exactly what happens in a laser.

The necessary population inversion can be achieved in a variety of ways. A familiar example is the helium-neon laser, a common and inexpensive laser that is available in many undergraduate laboratories. A mixture of helium and neon, typically at a pressure of the order of 10^2 Pa $(10^{-3}\ \text{atm})$, is sealed in a glass enclosure provided with two electrodes. When a sufficiently high voltage is applied, an electric discharge occurs. Collisions between ionized atoms and electrons carrying the discharge current excite atoms to various energy states.

Figure 38.24a shows an energy-level diagram for the system. The labels for the various energy levels, such as $1s$, $3p$, and $5s$, refer to states of the electrons in the levels. (We'll discuss this notation in Chapter 41; we don't need detailed knowledge of the nature of the states to understand laser action.) Because of restrictions imposed by conservation of angular momentum, a helium atom with an electron excited to a $2s$ state cannot return to a ground $(1s)$ state by emitting a 20.61-eV

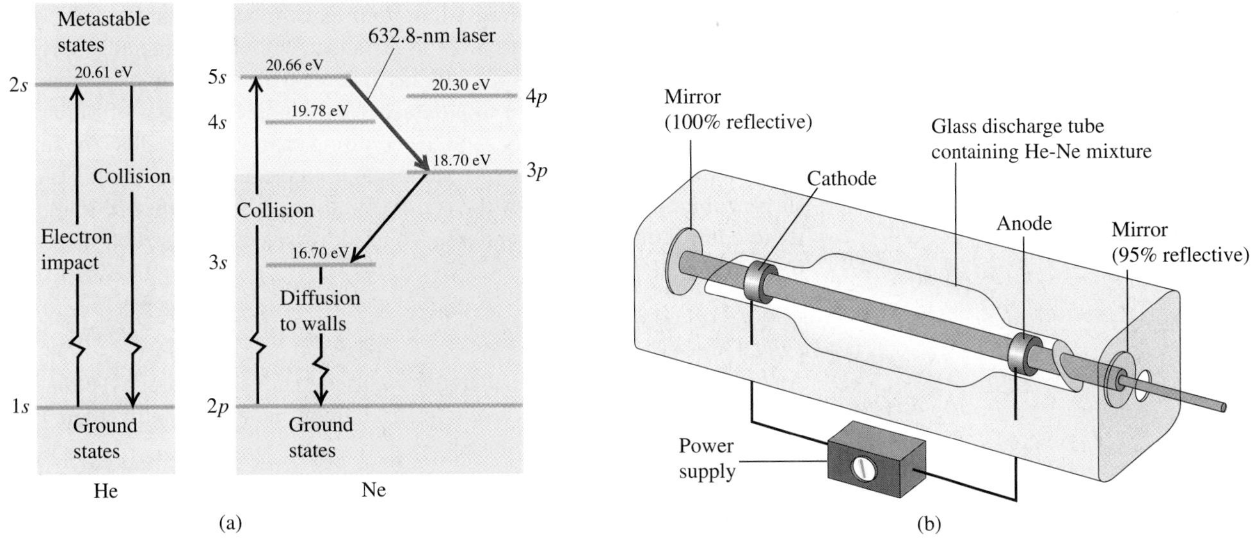

38.24 (a) Energy-level diagram for a helium-neon laser. (b) Operation of a helium-neon laser. The light emitted by electrons returning to lower-energy states is reflected between mirrors, so it continues to stimulate emission of more coherent light. One mirror is partially transmitting and allows the high-intensity light beam to escape.

photon, as you might expect it to do. Such a state, in which single-photon emission is impossible, has an unusually long lifetime and is called a **metastable state.** Helium atoms "pile up" in the metastable 2s states, creating a population inversion relative to the ground states.

However, excited helium atoms *can* lose energy by energy-exchange collisions with neon atoms that are initially in a ground state. A helium atom, excited to a 2s state 20.61 eV above its ground states and possessing a little additional kinetic energy, can collide with a neon atom in a ground state, exciting the *neon* atom to a 5s excited state at 20.66 eV and dropping the *helium* atom back to a ground state. Thus we have the necessary mechanism for a population inversion in neon, with more neon atoms per 5s state than per 3p state.

Stimulated emissions during transitions from a 5s to a 3p state then result in the emission of highly coherent light at 632.8 nm, as shown in the diagram. In practice the beam is sent back and forth through the gas many times by a pair of parallel mirrors (Fig. 38.24b), so as to stimulate emission from as many excited atoms as possible. One of the mirrors is partially transparent, so a portion of the beam emerges as an external beam. The net result of all these processes is a beam of light that can be quite intense, has very parallel rays, is highly monochromatic, and is spatially *coherent* at all points within a given cross section.

Other types of lasers use different processes to achieve the necessary population inversion. In a semiconductor laser the inversion is obtained by driving electrons and holes to a p-n junction (to be discussed in Section 42.7) with a steady electric field. In one type of *chemical* laser a chemical reaction forms molecules in metastable excited states. In a carbon dioxide gas dynamic laser the population inversion results from rapid expansion of the gas. A *maser* (acronym for "microwave amplification by stimulated emission of radiation") operates on the basis of population inversions in molecules, using closely spaced energy levels. The corresponding emitted radiation is in the microwave range. Maser action even occurs naturally in interstellar molecular clouds.

Lasers have found a wide variety of practical applications. A high-intensity laser beam can drill a very small hole in a diamond for use as a die in drawing very small-diameter wire. Surveyors often use lasers, especially in situations requiring great precision such as drilling a long tunnel from both ends; the laser beam has very parallel rays that can travel long distances without appreciable spreading.

Lasers are widely used in medicine. A laser with a very narrow intense beam can be used in the treatment of a detached retina; a short burst of radiation damages a small area of the retina, and the resulting scar tissue "welds" the retina back to the membrane from which it has become detached. Laser beams are used in surgery; blood vessels cut by the beam tend to seal themselves off, making it easier to control bleeding. Lasers are also used for selective destruction of tissue, as in the removal of tumors.

Test Your Understanding

An ordinary neon light fixture (like those used in advertising signs) emits red light with the same 632.8-nm wavelength as a helium-neon laser. Why, then, isn't such a light fixture classified as a laser?

38.7 | X-Ray Production and Scattering

The production and scattering of x rays provide additional examples of the quantum nature of electromagnetic radiation. X rays are produced when rapidly moving electrons that have been accelerated through a potential difference of the order of 10^3 to 10^6 V strike a metal target. They were first produced in 1895 by Wilhelm Röntgen (1845–1923), using an apparatus similar in principle to the setup shown in Fig. 38.25. Electrons are "boiled off" from the heated cathode by thermionic emission and are accelerated toward the anode (the target) by a large potential difference V_{AC}. The bulb is evacuated (residual pressure 10^{-7} atm or less), so the electrons can travel from the cathode to the anode without colliding with air molecules. When V_{AC} is a few thousand volts or more, a very penetrating radiation is emitted from the anode surface.

X-Ray Photons

Because they are emitted by accelerated charges, it is clear that x rays are electromagnetic waves. Like light, x rays are governed by quantum relations in their interaction with matter. Thus we can talk about x-ray photons or quanta, and the energy of an x-ray photon is related to its frequency and wavelength in the same way as for photons of light, $E = hf = hc/\lambda$. Typical x-ray wavelengths are 0.001 to 1 nm (10^{-12} to 10^{-9} m). X-ray wavelengths can be measured quite precisely by crystal diffraction techniques, which we discussed in Section 36.6.

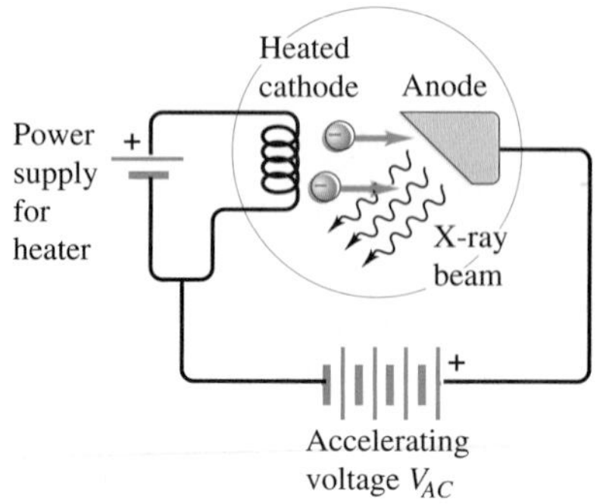

38.25 An apparatus used to produce x rays, similar to Röntgen's 1895 apparatus. Electrons are emitted thermionically from the heated cathode and are accelerated toward the anode; when they strike it, x rays are produced.

X-ray emission is the inverse of the photoelectric effect. In photoelectric emission there is a transformation of the energy of a photon into the kinetic energy of an electron; in x-ray production there is a transformation of the kinetic energy of an electron into the energy of a photon. The energy relationships are similar. In x-ray production we usually neglect the work function of the target and the initial kinetic energy of the "boiled-off" electrons because they are very small in comparison to the other energies.

Two distinct processes are involved in x-ray emission. In the first process, some electrons are slowed down or stopped by the target, and part or all of their kinetic energy is converted directly to a continuous spectrum of photons, including x rays. This process is called *bremsstrahlung* (German for "braking radiation"). Classical physics is completely unable to explain why the x rays that are emitted in a bremsstrahlung process have a maximum frequency f_{max} and a corresponding minimum wavelength λ_{min}, much less predict their values. With quantum concepts it is easy. An electron has charge $-e$ and gains kinetic energy eV_{AC} when accelerated through a potential increase V_{AC}. The most energetic photon (highest frequency and shortest wavelength) is produced when all the electron's kinetic energy goes to produce one photon; that is,

$$eV_{AC} = hf_{max}\frac{hc}{\lambda_{min}} \quad \text{(bremsstrahlung limits)} \tag{38.22}$$

Note that the maximum frequency and minimum wavelength in the bremsstrahlung process do *not* depend on the target material.

The second process gives peaks in the x-ray spectrum at characteristic frequencies and wavelengths that *do* depend on the target material. Other electrons, if they have enough kinetic energy, can transfer that energy partly or completely to individual atoms within the target. These atoms are left in excited levels; when they decay back to their ground levels, they may emit x-ray photons. Since each element has a unique set of atomic energy levels, each also has a characteristic x-ray spectrum. The energy levels associated with x rays are rather different in character from those associated with visible spectra. They involve vacancies in the inner electron configurations of complex atoms. The energy differences between these levels can be hundreds or thousands of electron volts, rather than a few electron volts as is typical for optical spectra. We will return to energy levels and spectra associated with x rays in Section 41.5.

Example 38.7 Producing x rays

Electrons in an x-ray tube are accelerated by a potential difference of 10.0 kV. If an electron produces one photon on impact with the target, what is the minimum wavelength of the resulting x rays? Answer using both SI units and electron volts.

SOLUTION

IDENTIFY and SET UP: To produce an x-ray photon with minimum wavelength and hence maximum energy, all of the kinetic energy of an electron must go into producing a single x-ray photon. We use Eq. (38.22) to determine the wavelength in this situation.

EXECUTE: From Eq. (38.22), using SI units we find

$$\lambda_{min} = \frac{hc}{eV_{AC}} = \frac{(6.626 \times 10^{-34}\ \text{J}\cdot\text{s})(3.00 \times 10^8\ \text{m/s})}{(1.602 \times 10^{-19}\ \text{C})(10.0 \times 10^3\ \text{V})}$$
$$= 1.24 \times 10^{-10}\ \text{m} = 0.124\ \text{nm}$$

Using electron volts, we have

$$\lambda_{min} = \frac{hc}{eV_{AC}} = \frac{(4.136 \times 10^{-15}\ \text{eV}\cdot\text{s})(3.00 \times 10^8\ \text{m/s})}{e(10.0 \times 10^3\ \text{V})}$$
$$= 1.24 \times 10^{-10}\ \text{m} = 0.124\ \text{nm}$$

Note that the "e" in the unit eV is canceled by the "e" for the magnitude of the electron charge, because the electron volt (eV) is the magnitude of the electron charge e times one volt (1 V).

EVALUATE: To check our result, recall from Example 38.6 (Section 38.5) that a photon of energy 10.2 eV has a wavelength of 122 nm. In the present example the electron energy, and therefore the energy of the x-ray photon, is 10.0×10^3 eV = 10.0 keV (about 10^3 times greater than in Example 38.6) and the wavelength is about 10^{-3} times as great as in Example 38.6. This makes sense, since wavelength and photon energy are inversely proportional.

Compton Scattering

A phenomenon called **Compton scattering,** first explained in 1923 by the American physicist Arthur H. Compton, provides additional direct confirmation of the quantum nature of x rays. When x rays strike matter, some of the radiation is *scattered,* just as visible light falling on a rough surface undergoes diffuse reflection.

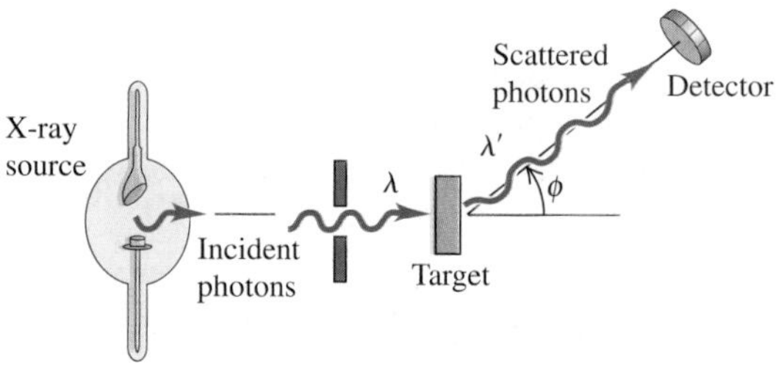

38.26 A Compton-effect experiment.

Compton and others discovered that some of the scattered radiation has smaller frequency (longer wavelength) than the incident radiation and that the change in wavelength depends on the angle through which the radiation is scattered. Specifically, if the scattered radiation emerges at an angle ϕ with respect to the incident direction (Fig. 38.26) and if λ and λ' are the wavelengths of the incident and scattered radiation, respectively, we find that

ActivPhysics Online

17.4 Compton Scattering

$$\lambda' - \lambda = \frac{h}{mc}(1 - \cos\phi) \qquad \text{(Compton scattering)} \qquad (38.23)$$

where m is the electron rest mass. The quantity h/mc that appears in Eq. (38.23) has units of length. Its numerical value is

$$\frac{h}{mc} = \frac{6.626 \times 10^{-34}\ \text{J}\cdot\text{s}}{(9.109 \times 10^{-31}\ \text{kg})(2.998 \times 10^{8}\ \text{m/s})} = 2.426 \times 10^{-12}\ \text{m}$$

We cannot explain Compton scattering using classical electromagnetic theory, which predicts that the scattered wave has the same wavelength as the incident wave. However the quantum theory provides a beautifully clear explanation. We imagine the scattering process as a collision of two *particles,* the incident photon and an electron that is initially at rest, as in Fig. 38.27a. The incident photon disappears, giving part of its energy and momentum to the electron, which recoils as a result of this impact. The remainder goes to a new scattered photon that therefore has less energy, smaller frequency, and longer wavelength than the incident one (Fig. 38.27b).

We can derive Eq. (38.23) from the principles of conservation of energy and momentum. We outline the derivation below; we invite you to fill in the details (see Exercise 38.41). The electron recoil energy may be in the relativistic range, so we have to use the relativistic energy-momentum relations, Eqs. (37.39) and (37.40). The incident photon has momentum $\vec{p}$, with magnitude p, and energy pc. The scattered photon has momentum $\vec{p}\,'$, with magnitude p', and energy $p'c$. The electron is initially at rest, so its initial momentum is zero and its initial energy is its rest energy mc^2. The final electron momentum $\vec{P}_e$ has momentum P_e, and the final electron energy is given by $E_e = (mc^2)^2 + (P_e c)^2$. Then energy conservation gives us the relation

$$pc + mc^2 = p'c + E_e$$

Rearranging,

$$(pc - p'c + mc^2)^2 = E_e^2 = (mc^2)^2 + (P_e c)^2 \qquad (38.24)$$

We can eliminate the electron momentum $\vec{P}_e$ from Eq. (38.24) by using momentum conservation (Fig. 38.27c):

$$\vec{p} = \vec{p}\,' + \vec{P}_e$$

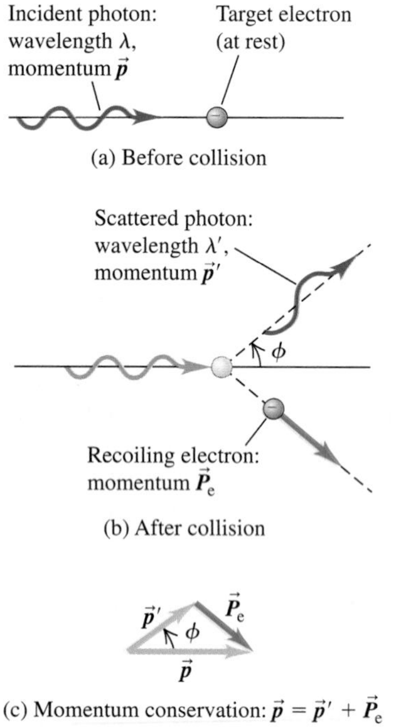

38.27 Schematic diagram of Compton scattering. (a) The target electron is at rest. (b) The angle between the directions of the scattered photon and incident photon (and hence between their momenta $\vec{p}\,'$ and $\vec{p}$) is ϕ. (c) Vector diagram showing the conservation of momentum in the collision.

or

$$\vec{P}_e = \vec{p} - \vec{p}\,' \tag{38.25}$$

By taking the scalar product of each side of Eq. (38.25) with itself or by using the law of cosines with the vector diagram in Fig. 38.27c, we find

$$P_e^2 = p^2 + p'^2 - 2pp' \cos\phi \tag{38.26}$$

We now substitute this expression for P_e^2 into Eq. (38.24) and multiply out the left side. We divide out a common factor c^2; several terms cancel, and when the resulting equation is divided through by (pp'), the result is

$$\frac{mc}{p'} - \frac{mc}{p} = 1 - \cos\phi \tag{38.27}$$

Finally, we substitute $p' = h/\lambda'$ and $p = h/\lambda$, then multiply by h/mc to obtain Eq. (38.23).

When the wavelengths of x rays scattered at a certain angle are measured, the curve of intensity per unit wavelength as a function of wavelength has two peaks (Fig. 38.28). The longer-wavelength peak represents Compton scattering. The shorter-wavelength peak, labeled λ_0, is at the wavelength of the incident x rays and corresponds to x-ray scattering from tightly bound electrons. In such scattering processes the entire atom must recoil, so the m in Eq. (38.23) is the mass of the entire atom rather than of a single electron. The resulting wavelength shifts are negligible.

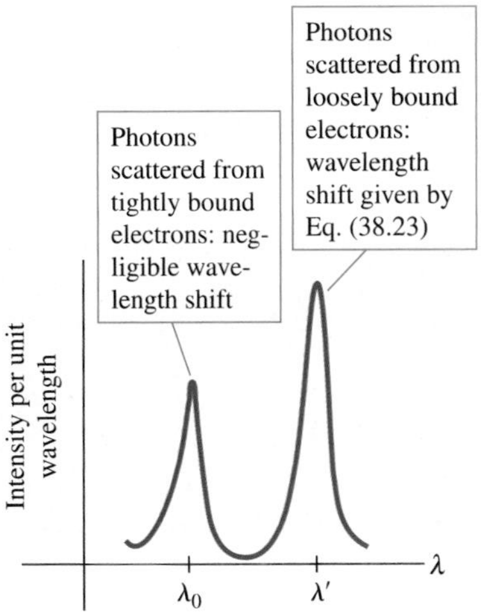

38.28 Intensity as a function of wavelength for photons scattered at an angle of 135° in a Compton scattering experiment.

Example 38.8 Compton scattering

You use the x-ray photons in Example 38.7 (with $\lambda = 0.124$ nm) in a Compton scattering experiment. a) At what angle is the wavelength of the scattered x rays 1.0% longer than that of the incident x rays? b) At what angle is it 0.050% greater?

SOLUTION

IDENTIFY and SET UP: In each case our target variable is the angle ϕ shown in Fig. 38.27b. We solve for it using Eq. (38.23).

EXECUTE: a) In Eq. (38.23) we want $\Delta\lambda = \lambda' - \lambda$ to be 1.0% of 0.124 nm. That is, $\Delta\lambda = 0.00124\text{ nm} = 1.24 \times 10^{-12}$ m. Using the value $h/mc = 2.426 \times 10^{-12}$ m, we find

$$\Delta\lambda = \frac{h}{mc}(1 - \cos\phi)$$

$$\cos\phi = 1 - \frac{\Delta\lambda}{(h/mc)} = 1 - \frac{1.24 \times 10^{-12}\text{ m}}{2.426 \times 10^{-12}\text{ m}} = 0.4889$$

$$\phi = 60.7^\circ$$

b) For $\Delta\lambda$ to be 0.050% of 0.124 nm, or 6.2×10^{-14} m,

$$\cos\phi = 1 - \frac{6.2 \times 10^{-14}\text{ m}}{2.426 \times 10^{-12}\text{ m}} = 0.9744$$

$$\phi = 13.0^\circ$$

EVALUATE: Comparing our results for parts (a) and (b) shows that smaller angles give smaller wavelength shifts. Thus in a grazing collision the photon energy loss and the electron recoil energy are smaller than when the scattering angle is larger. This is just what we would expect for an elastic collision, whether between a photon and an electron or between two billiard balls.

X rays have many practical applications in medicine and industry. Because x-ray photons are of such high energy, they can penetrate several centimeters of solid matter and so can be used to visualize the interiors of materials that are opaque to ordinary light, such as broken bones or defects in structural steel. The

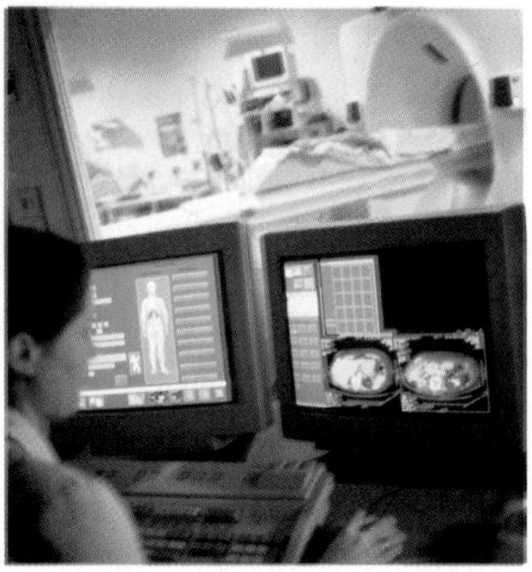

38.29 This radiologist is operating a CT scanner (seen through the window) from a separate room to avoid repeated exposure to x rays.

object to be visualized is placed between an x-ray source and a large sheet of photographic film; the darkening of the film is proportional to the radiation exposure. A crack or air bubble allows greater transmission and shows as a dark area. Bones appear lighter than the surrounding soft tissue because they contain greater proportions of elements with higher atomic number (and greater absorption); in the soft tissue the light elements carbon, hydrogen, and oxygen predominate.

A widely used and vastly improved x-ray technique is *computed tomography;* the corresponding instrument is called a *CT scanner.* The x-ray source produces a thin, fan-shaped beam that is detected on the opposite side of the subject by an array of several hundred detectors in a line. Each detector measures absorption along a thin line through the subject. The entire apparatus is rotated around the subject in the plane of the beam during a few seconds. The changing photon-counting rates of the detectors are recorded digitally; a computer processes this information and reconstructs a picture of absorption over an entire cross section of the subject (Fig. 38.29). Differences as small as 1% can be detected with CT scans, and tumors and other anomalies that are much too small to be seen with older x-ray techniques can be detected.

X rays cause damage to living tissues. As x-ray photons are absorbed in tissues, their energy breaks molecular bonds and creates highly reactive free radicals (such as neutral H and OH), which in turn can disturb the molecular structure of proteins and especially genetic material. Young and rapidly growing cells are particularly susceptible; thus x rays are useful for selective destruction of cancer cells. Conversely, however, a cell may be damaged by radiation but survive, continue dividing, and produce generations of defective cells; thus x rays can *cause* cancer.

Even when the organism itself shows no apparent damage, excessive radiation exposure can cause changes in the organism's reproductive system that will affect its offspring. A careful assessment of the balance between risks and benefits of radiation exposure is essential in each individual case.

Test Your Understanding

If you used visible-light photons in the experiment shown in Fig. 38.26, would the photons undergo a wavelength shift in the scattering? Is it possible to detect the shift with the human eye?

38.8 | Continuous Spectra

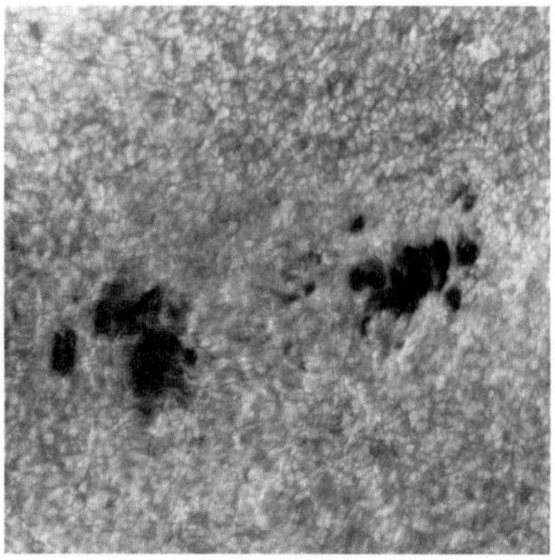

38.30 This close-up view of the sun's surface shows two dark sunspots. Their temperature is about 4000 K, while the surrounding solar material is at $T =$ 5800 K. From the Stefan-Boltzmann law, the intensity from a given area of sunspot is only $(4000\ \text{K}/5800\ \text{K})^4 = 0.23$ as great as the intensity from the same area of the surrounding material—which is why sunspots appear dark.

Line spectra are emitted by matter in the gaseous state, in which the atoms are so far apart that interactions between them are negligible and each atom behaves as an isolated system. Hot matter in condensed states (solid or liquid) nearly always emits radiation with a *continuous* distribution of wavelengths rather than a line spectrum. An ideal surface that absorbs all wavelengths of electromagnetic radiation incident upon it is also the best possible emitter of electromagnetic radiation at any wavelength. Such an ideal surface is called a *blackbody,* and the continuous-spectrum radiation that it emits is called **blackbody radiation.** By 1900 this radiation had been studied extensively, and several characteristics had been established.

First, the total intensity I (the average rate of radiation of energy per unit surface area or average power per area) emitted from the surface of an ideal radiator is proportional to the fourth power of the absolute temperature (Fig. 38.30). We studied this relationship in Section 17.7 during our study of heat transfer mechanisms. This total intensity I emitted at absolute temperature T is given by the **Stefan-Boltzmann law:**

$$I = \sigma T^4 \qquad \text{(Stefan-Boltzmann law for a blackbody)} \qquad (38.28)$$

where σ is a fundamental physical constant called the *Stefan-Boltzmann constant.* In SI units,

$$\sigma = 5.670400(40) \times 10^{-8} \frac{\text{W}}{\text{m}^2 \cdot \text{K}^4}$$

Second, the intensity is not uniformly distributed over all wavelengths. Its distribution can be measured and described by the intensity per wavelength interval $I(\lambda)$, called the *spectral emittance.* Thus $I(\lambda)\,d\lambda$ is the intensity corresponding to wavelengths in the interval from λ to $\lambda + d\lambda$. The *total* intensity I, given by Eq. (38.28), is the *integral* of the distribution function $I(\lambda)$ over all wavelengths, which equals the area under the $I(\lambda)$ versus λ curve:

$$I = \int_0^\infty I(\lambda)\,d\lambda \qquad (38.29)$$

CAUTION Although we use the symbol $I(\lambda)$ for spectral emittance, keep in mind that spectral emittance is *not* the same thing as intensity I. Intensity is power per unit area, with units W/m^2; spectral emittance is power per unit area per unit wavelength interval, with units W/m^3.

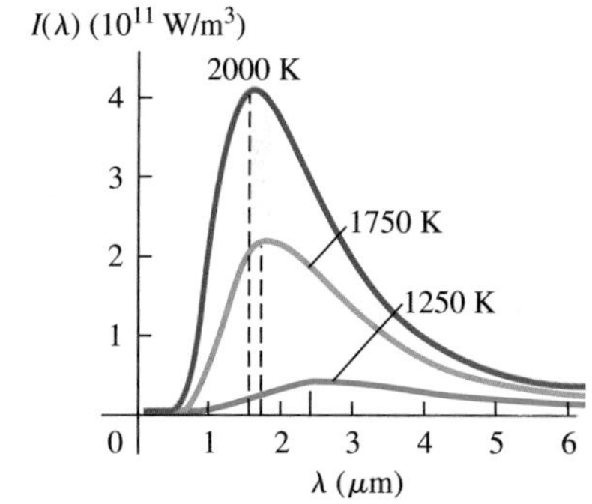

38.31 Spectral emittance $I(\lambda)$ for radiation from a blackbody at three different temperatures. The dashed vertical lines show the value of λ_m in Eq. (38.30) for each temperature. As the temperature increases, the peak grown larger and shifts to shorter wavelengths.

Measured spectral emittances $I(\lambda)$ for three different temperatures are shown in Fig. 38.31. Each has a peak wavelength λ_m at which the emitted intensity per wavelength interval is largest. Experiment shows that λ_m is inversely proportional to T, so their product is constant. This result is called the **Wien displacement law.** The experimental value of the constant is $2.90 \times 10^{-3}\ \text{m} \cdot \text{K}$:

$$\lambda_\text{m} T = 2.90 \times 10^{-3}\ \text{m} \cdot \text{K} \qquad \text{(Wien displacement law)} \qquad (38.30)$$

As the temperature rises, the peak of $I(\lambda)$ becomes higher and shifts to shorter wavelengths. A body that glows yellow is hotter and brighter than one that glows red; yellow light has shorter wavelengths than red light. Finally, experiments show that the *shape* of the distribution function is the same for all temperatures; we can make a curve for one temperature fit any other temperature by simply changing the scales on the graph.

During the last decade of the nineteenth century, many attempts were made to *derive* these empirical results from basic principles. In one attempt, the English physicist Lord Rayleigh considered light enclosed in a rectangular box with perfectly reflecting sides. Such a box has a series of possible *normal modes* for electromagnetic waves, as discussed in Section 32.5. It seemed reasonable to assume that the distribution of energy among the various modes was given by the equipartition principle (Section 18.4), which had been successfully used in the analysis of heat capacities. A small hole in the box would behave as an ideal blackbody radiator.

Including both the electric- and magnetic-field energies, Rayleigh assumed that the total energy of each normal mode was equal to kT. Then by computing the *number* of normal modes corresponding to a wavelength interval $d\lambda$, Rayleigh could predict the distribution of wavelengths in the radiation within the box. Finally, he could compute the intensity distribution $I(\lambda)$ of the radiation emerging from a small hole in the box. His result was quite simple:

$$I(\lambda) = \frac{2\pi ckT}{\lambda^4} \tag{38.31}$$

At large wavelengths this formula agrees quite well with the experimental results shown in Fig. 38.30, but there is serious disagreement at small wavelengths. The experimental curve falls toward zero at small λ, but Rayleigh's curve goes the opposite direction, approaching infinity as $1/\lambda^4$, a result that was called in Rayleigh's time the "ultraviolet catastrophe." Even worse, the integral of Eq. (38.31) over all λ is infinite, indicating an infinitely large *total* radiated intensity. Clearly, something is wrong.

Finally, in 1900, the German physicist Max Planck succeeded in deriving a function, now called the **Planck radiation law,** that agreed very well with experimental intensity distribution curves. In his derivation he made what seemed at the time to be a crazy assumption. Planck assumed that electromagnetic oscillators (electrons) in the walls of Rayleigh's box vibrating at a frequency f could have only certain values of energy equal to nhf, where $n = 0, 1, 2, 3, \ldots$ and h turned out to be the constant that now bears Planck's name. These oscillators were in equilibrium with the electromagnetic waves in the box. His assumption gave quantized energy levels and was in sharp contrast to Rayleigh's point of view, which was that each normal mode could have any amount of energy.

Planck was not comfortable with this *quantum hypothesis;* he regarded it as a calculational trick rather than a fundamental principle. In a letter to a friend, he called it "an act of desperation" into which he was forced because "a theoretical explanation had to be found at any cost, whatever the price." But five years later, Einstein identified the energy change hf between levels as the energy of a photon to explain the photoelectric effect (Section 38.2), and other evidence quickly mounted. By 1915 there was little doubt about the validity of the quantum concept and the existence of photons. By discussing atomic spectra *before* continuous spectra, we have departed from the historical order of things. The credit for inventing the concept of quantization of energy levels goes to Planck, even though he didn't believe it at first.

We won't go into the details of Planck's derivation of the intensity distribution. Here is his result:

$$I(\lambda) = \frac{2\pi hc^2}{\lambda^5(e^{hc/\lambda kT} - 1)} \quad \text{(Planck radiation law)} \tag{38.32}$$

where h is Planck's constant, c is the speed of light, k is Boltzmann's constant, T is the *absolute* temperature, and λ is the wavelength. This function turns out to agree well with experimental intensity curves such as those in Fig. 38.31.

The Planck radiation law also contains the Wien displacement law and the Stefan-Boltzmann law as consequences. To derive the Wien law, we take the derivative of Eq. (38.32) and set it equal to zero to find the value of λ at which $I(\lambda)$ is maximum. We leave the details as a problem (see Exercise 38.49); the result is

$$\lambda_m = \frac{hc}{4.965kT} \tag{38.33}$$

To obtain this result, you have to solve the equation

$$5 - x = 5e^{-x} \tag{38.34}$$

The root of this equation, found by trial and error or more sophisticated means, is 4.965 to four significant figures. You should evaluate the constant $hc/4.965k$ and show that it agrees with the experimental value of $2.90 \times 10^{-3}\ \text{m}\cdot\text{K}$ given in Eq. (38.30).

We can obtain the Stefan-Boltzmann law for a blackbody by integrating Eq. (38.32) over all λ to find the *total* radiated intensity (see Problem 38.77). This is not a simple integral; the result is

$$I = \int_0^\infty I(\lambda)\,d\lambda = \frac{2\pi^5 k^4}{15c^2h^3}T^4 = \sigma T^4 \qquad (38.35)$$

in agreement with Eq. (38.28). This result also shows that the constant σ in that law can be expressed as a combination of other fundamental constants:

$$\sigma = \frac{2\pi^5 k^4}{15c^2h^3} \qquad (38.36)$$

You should substitute the values of k, c, and h from Appendix F and verify that you obtain the Stefan-Boltzmann constant $\sigma = 5.6705 \times 10^{-8}\ \text{W/m}^2\cdot\text{K}^4$.

The general form of Eq. (38.33) is what we should expect from kinetic theory. If photon energies are typically of the order of kT, as suggested by the equipartition theorem, then for a typical photon we would expect

$$E \approx kT \approx \frac{hc}{\lambda} \quad \text{and} \quad \lambda \approx \frac{hc}{kT} \qquad (38.37)$$

Indeed, a photon with wavelength given by Eq. (38.33) has an energy $E = 4.965kT$.

The Planck radiation law, Eq. (38.32), looks so different from the unsuccessful Rayleigh expression, Eq. (38.31), that it may seem unlikely that they would agree at large values of λ. But when λ is large, the exponent in the denominator of Eq. (38.32) is very small. We can then use the approximation $e^x \approx 1 + x$ (for $x \ll 1$). You should verify that when this is done, the result approaches Eq. (38.31), showing that the two expressions *do* agree in the limit of very large λ. We also note that the Rayleigh expression does not contain h. At very long wavelengths and correspondingly very small photon energies, quantum effects become unimportant.

Example 38.9 Light from the sun

The surface of the sun has a temperature of approximately 5800 K. To a good approximation we may treat it as a blackbody. a) What is the peak-intensity wavelength λ_m? b) What is the total radiated power per unit area?

SOLUTION

IDENTIFY and SET UP: We use the Wien displacement law [Eq. (38.30)] to determine the value of λ_m, and the Stefan-Boltzmann law [Eq. (38.28)] to determine I, the total radiated power per unit area.

EXECUTE: a) From Eq. (38.30),

$$\lambda_m = \frac{2.90 \times 10^{-3}\ \text{m}\cdot\text{K}}{T} = \frac{2.90 \times 10^{-3}\ \text{m}\cdot\text{K}}{5800\ \text{K}}$$
$$= 0.500 \times 10^{-6}\ \text{m} = 500\ \text{nm}$$

b) From Eq. (38.28),

$$I = \sigma T^4 = (5.67 \times 10^{-8}\ \text{W/m}^2\cdot\text{K}^4)(5800\ \text{K})^4$$
$$= 6.42 \times 10^7\ \text{W/m}^2 = 64.2\ \text{MW/m}^2$$

EVALUATE: The wavelength found in part (b) is near the middle of the visible spectrum. This is not a surprising result: the human eye evolved to take maximum advantage of natural light.

The enormous value of I found in part (b) is the intensity at the *surface* of the sun. When the radiated power reaches the earth, the intensity drops to about 1.4 kW/m^2 because the power spreads out over the much larger area of a sphere with a radius equal to that of the earth's orbit.

Example 38.10 A slice of sunlight

Find the power per unit area radiated from the surface of the sun in the wavelength range 600.0 to 605.0 nm.

SOLUTION

IDENTIFY and SET UP: For an exact result we should integrate Eq. (38.32) between the limits 600.0 and 605.0 nm, finding the area under the $I(\lambda)$ curve between these limits. This integral can't be evaluated in terms of familiar functions, so we *approximate* the area by the height of the curve at the median wavelength $\lambda = 602.5$ nm multiplied by the width of the interval ($\Delta\lambda = 5.0$ nm).

EXECUTE: To obtain the height of the $I(\lambda)$ curve at $\lambda = 602.5\ \text{nm} = 6.025 \times 10^{-7}$ m, we first evaluate the quantity $hc/\lambda kT$ for this value of λ and for $T = 5800$ K (the temperature of the solar surface; see Example 38.9). We then substitute the result into Eq. (38.32):

$$\frac{hc}{\lambda kT} = \frac{(6.626 \times 10^{-34}\ \text{J}\cdot\text{s})(2.998 \times 10^{8}\ \text{m/s})}{(6.025 \times 10^{-7}\ \text{m})(1.381 \times 10^{-23}\ \text{J/K})(5800\ \text{K})}$$

$$= 4.116$$

$$I(\lambda) = \frac{2\pi(6.626 \times 10^{-34}\ \text{J}\cdot\text{s})(2.998 \times 10^{8}\ \text{m/s})^2}{(6.025 \times 10^{-7}\ \text{m})^5(e^{4.116} - 1)}$$

$$= 7.81 \times 10^{13}\ \text{W/m}^3$$

The intensity in the 5.0-nm range from 600.0 to 605.0 nm is then approximately

$$I(\lambda)\Delta\lambda = (7.81 \times 10^{13}\ \text{W/m}^3)(5.0 \times 10^{-9}\ \text{m})$$

$$= 3.9 \times 10^{5}\ \text{W/m}^2 = 0.39\ \text{MW/m}^2$$

EVALUATE: We see from this result and from part (b) of Example 38.9 (in which we found the power radiated per unit area at *all* wavelengths) that about 0.6% of the total power radiated from the sun is in the wavelength range from 600 to 605 nm. Had we found that more power is radiated in this range than the total calculated in Example 38.9, something would have been very wrong with our calculations!

Test Your Understanding

Does a blackbody at 2000 K emit x rays? Does it emit radio waves?

38.9 | Wave-Particle Duality

We have studied many examples of the behavior of light and other electromagnetic radiation. Some, including the interference and diffraction effects described in Chapters 35 and 36, demonstrate conclusively the *wave* nature of light. Others, the subject of the present chapter, point with equal force to the *particle* nature of light. At first glance these two aspects seem to be in direct conflict. How can light be a wave and a particle at the same time?

We can find the answer to this apparent wave-particle conflict in the **principle of complementarity,** first stated by Niels Bohr in 1928. The wave descriptions and the particle descriptions are complementary. That is, we need both descriptions to complete our model of nature, but we will never need to use both descriptions at the same time to describe a single part of an occurrence.

Let's start by considering again the diffraction pattern for a single slit, which we analyzed in Sections 36.2 and 36.3. Instead of recording the pattern on photographic film, we use a detector called a *photomultiplier* that can actually detect individual photons. Using the setup shown in Fig. 38.32, we place the photomultiplier at various positions for equal time intervals, count photons at each position, and plot out the intensity distribution.

We find that, on average, the distribution of photons agrees with our predictions from Section 36.3. At points corresponding to the maxima of the pattern, we count many photons; at minimum points, we count almost none, and so on. The graph of the counts at various points gives us the same diffraction pattern that we predicted with Eq. (36.7).

But suppose we now reduce the intensity to the point at which only a few photons per second pass through the slit. With only a few photons we can't expect to get the smooth diffraction curve that we found with very large numbers. In fact, there is no way to predict exactly where any individual photon will go. To reconcile the wave and particle aspects of this pattern, we have to regard the pattern as a *statistical* distribution that tells us how many photons, on average, go various places, or the *probability* for an individual photon to land in each of several places. But we can't predict exactly where an individual photon will go.

Now let's take a brief look at a quantum interpretation of a *two-slit* optical interference pattern, which we studied in Section 35.2. We can again trace out the pattern using a photomultiplier and a counter. We reduce the light intensity to a level of a few photons per second (Fig 38.33). Again we can't predict exactly where an individual photon will go; the interference pattern is a *statistical distribution.*

How does the principle of complementarity apply to these diffraction and interference experiments? The wave description, not the particle description, explains the single- and double-slit patterns. But the particle description, not the wave description, explains why the photomultiplier measures the pattern to be built up by discrete packages of energy. The two descriptions complete our understanding of the results.

In these experiments, light goes through a slit or slits, and photons are detected at various positions on a screen. Suppose we ask the following question: "We just detected a photon at a certain position; when the photon went through the slit, how did it know which way to go?" The problem with the question is that it is asked in terms of a *particle* description. The wave nature of light, not the particle nature, determines the distribution of photons. Asking that question is trying to force a particle description (the photon being somehow directed which way to go) on a wave phenomenon (the formation of the pattern).

We know from experiments that sending electromagnetic waves through slits gives interference patterns and that these patterns are built up by individual photons. But if we didn't know the experimental results, how would we predict when to apply the wave description and when to apply the particle description? We need a theory that includes both these descriptions and predicts as well as explains both types of behavior. Such a comprehensive theory is called *quantum electrodynamics* (QED). In this theory the concept of energy levels of an atomic system is extended to electromagnetic fields. Just as an atom exists only in certain definite energy states, so the electromagnetic field has certain well-defined energy states, corresponding to the presence of various numbers of photons with assorted energies, momenta, and polarizations. QED came to full flower 50 years after Planck's quantum hypothesis of 1900 gave quantum mechanics its conceptual birth.

In the following chapters we will find that particles such as electrons also have a dual wave-particle personality. One of the great achievements of quantum mechanics has been to reconcile these apparently incompatible aspects of behavior of photons, electrons, and other constituents of matter.

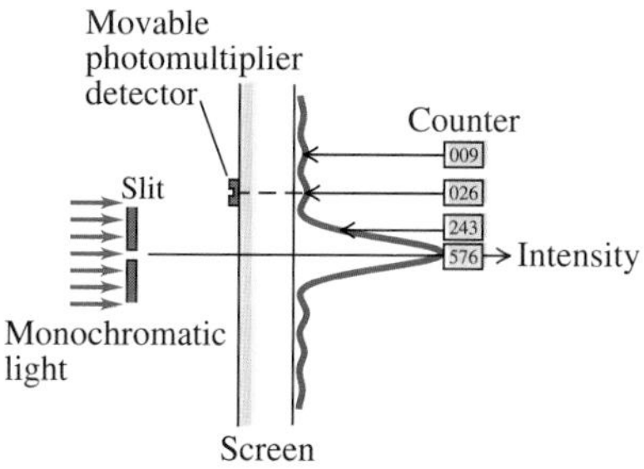

38.32 Single-slit diffraction pattern observed with a movable photomultiplier. The curve shows the intensity distribution predicted by the wave picture. The photon distribution is shown by the numbers of photons counted at various positions.

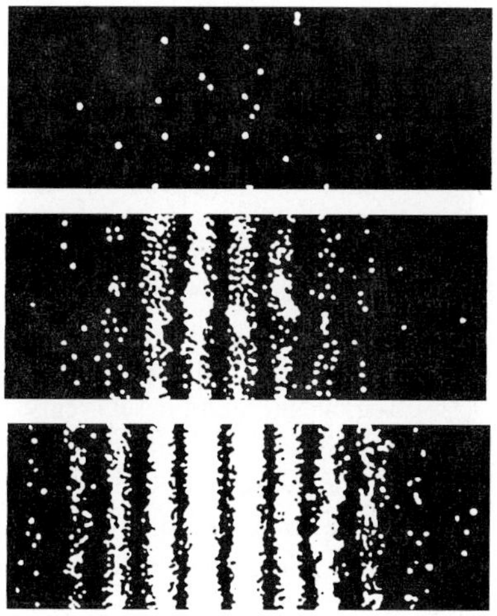

38.33 These images record the positions where individual photons in a two-slit interference experiment strike the screen. The top image shows the pattern after 21 photons have reached the screen; the middle image, after 1000 photons; and the bottom image, after 10,000 photons. As more photons reach the screen, a recognizable interference pattern appears.

SUMMARY

The energy in an electromagnetic wave is carried in units called photons. The energy E of one photon is proportional to the wave frequency f and inversely proportional to the wavelength λ, and is proportional to a universal quantity h called Planck's constant. The momentum of a photon has magnitude E/c. (See Example 38.1)

$$E = hf = \frac{hc}{\lambda} \quad (38.2)$$

$$p = \frac{E}{c} = \frac{hf}{c} = \frac{b}{\lambda} \quad (38.5)$$

In the photoelectric effect, a surface can eject an electron by absorbing a photon whose energy hf is greater than or equal to the work function ϕ of the material. The stopping potential V_0 is the voltage required to stop a current of ejected electrons from reaching an anode. (See Examples 38.2 and 38.3)

$$eV_0 = hf - \phi \quad (38.4)$$

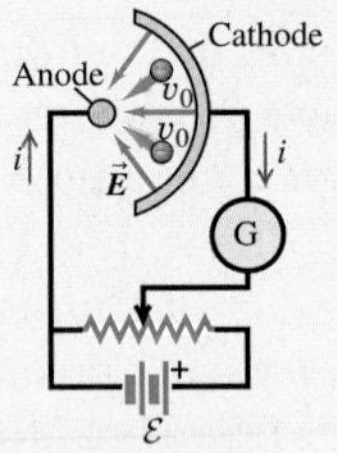

When an atom makes a transition from an energy level E_i to a lower level E_f, the energy of the emitted photon is equal to $E_i - E_f$. The energy levels of the hydrogen atom are given by Eq. (38.9), where R is the Rydberg constant. All of the observed spectral series of hydrogen can be understood in terms of these levels. (See Example 38.4)

$$hf = \frac{hc}{\lambda} = E_i - E_f \quad (38.6)$$

$$E_n = -\frac{hcR}{n^2} = -\frac{13.60\ \text{eV}}{n^2}$$

$$n = 1, 2, 3, \ldots \quad (38.9)$$

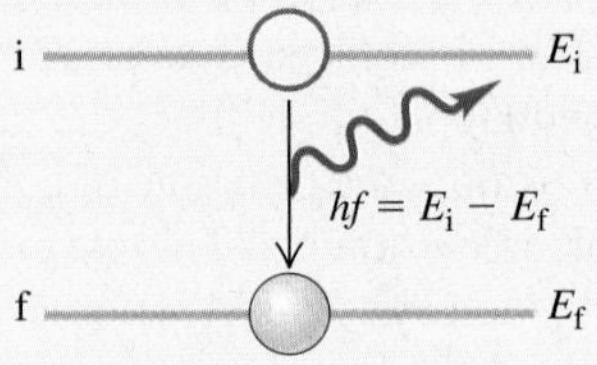

The Rutherford scattering experiments show that at the center of an atom is a dense nucleus, much smaller than the overall size of the atom but containing all of the positive charge and most of the mass. (See Example 38.5)

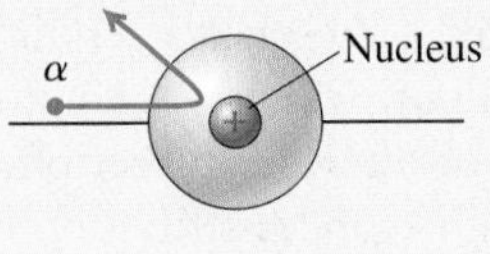

In the Bohr model of the hydrogen atom, the permitted values of angular momentum are integral multiples of $h/2\pi$. The integer multiplier n is called the principal quantum number for the level. The orbital radii are proportional to n^2 and the orbital speeds are proportional to $1/n$. (See Example 38.6)

$$L_n = mv_n r_n = n\frac{h}{2\pi} \quad (38.10)$$

$$n = 1, 2, 3, \ldots$$

$$r_n = \epsilon_0 \frac{n^2 h^2}{\pi m e^2} = n^2 a_0$$

$$= n^2 (5.29 \times 10^{-11}\ \text{m}) \quad (38.12)$$

$$v_n = \frac{1}{\epsilon_0}\frac{e^2}{2nh} = \frac{2.19 \times 10^6\ \text{m/s}}{n} \quad (38.13)$$

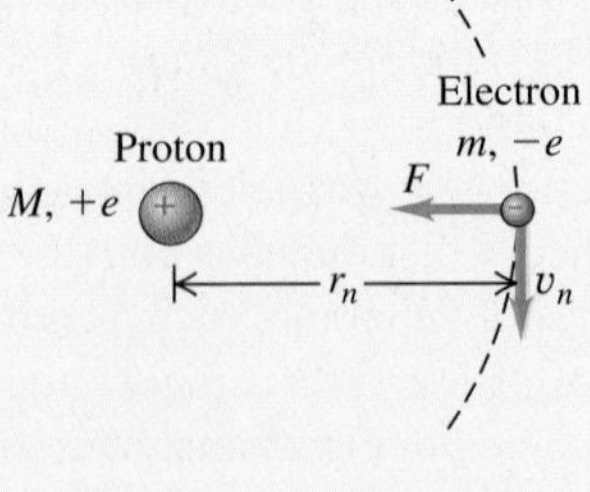

The laser operates on the principle of stimulated emission, by which many photons with identical wavelength and phase are emitted. Laser operation requires a non-equilibrium condition called a population inversion, in which more atoms are in a higher-energy state than are in a lower-energy state.

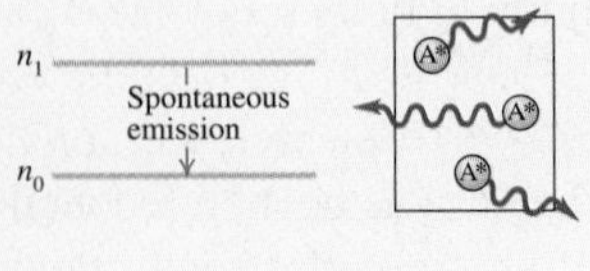

X rays can be produced by electron impact on a target. If electrons are accelerated through a potential increase V_{AC}, the maximum frequency and minimum wavelength that they can produce is given by Eq. (38.22). (See Example 38.7)

$$eV_{AC} = hf_{\max} = \frac{hc}{\lambda_{\min}} \qquad (38.22)$$

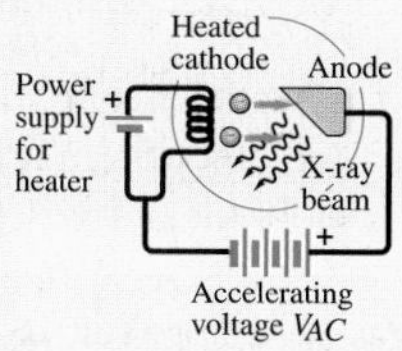

Compton scattering is scattering of x-ray photons by electrons. For free electrons (mass m), the wavelengths of incident and scattered photons are related to the photon scattering angle ϕ by Eq. (38.23). (See Example 38.8)

$$\lambda' - \lambda = \frac{h}{mc}(1 - \cos\phi) \qquad (38.23)$$

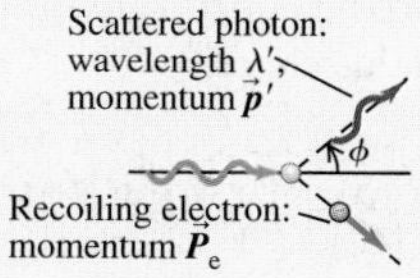

The total radiated intensity (average power radiated per area) from a blackbody surface is proportional to the fourth power of the absolute temperature T. The quantity $\sigma = 5.67 \times 10^{-8}\ \mathrm{W/m^2 \cdot K^4}$ is called the Stefan-Boltzmann constant. The wavelength λ_m at which a blackbody radiates most strongly is inversely proportional to T. The Planck radiation law gives the spectral emittance $I(\lambda)$ (intensity per wavelength interval in blackbody radiation. (See Examples 38.9 and 38.10)

$$I = \sigma T^4 \qquad (38.28)$$

(Stefan-Boltzmann law)

$$\lambda_m T = 2.90 \times 10^{-3}\ \mathrm{m \cdot K} \qquad (38.30)$$

(Wien displacement law)

$$I(\lambda) = \frac{2\pi hc^2}{\lambda^5 (e^{hc/\lambda kT} - 1)} \qquad (38.32)$$

(Planck radiation law)

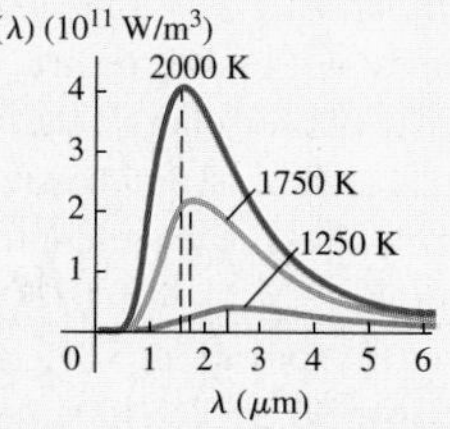

Electromagnetic radiation behaves as both waves and particles. A comprehensive theory must include both of these aspects of its behavior.

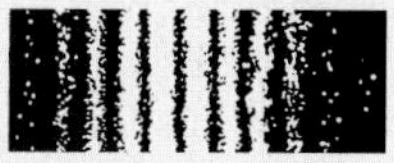

Key Terms

line spectrum, 1446
photoelectric effect, 1447
work function, 1448
threshold frequency, 1448
stopping potential, 1449
photon, 1450
Planck's constant, 1450
energy level, 1453
Rydberg constant, 1454
ground level, 1456
excited level, 1456
absorption spectrum, 1457
Rutherford scattering experiments, 1459
nucleus, 1460
Bohr model, 1461
principal quantum number, 1462
reduced mass, 1465
laser, 1467
stimulated emission, 1467
state, 1467
population inversion, 1468
metastable state, 1469
Compton scattering, 1471
blackbody radiation, 1474
Stefan-Boltzmann law, 1475
Wien displacement law, 1475
Planck radiation law, 1476
principle of complementarity, 1478

Your Notes

Answer to Chapter Opening Question

A photon of wavelength 656.3 nm (the H_α line in Fig. 38.8) is emitted when the electron in a hydrogen atom drops from the $n = 3$ energy level to the $n = 2$ level. The fact that we observe this tells us that the hydrogen gas in the Eagle Nebula is very highly excited. (If the gas were not excited, all of the atoms would be in the lowest energy level and could not emit photons.) The gas is excited by high-energy ultraviolet photons coming from the hot, luminous stars to the upper right of the Eagle Nebula in the photograph that opens this chapter.

Answers to Test Your Understanding Questions

Section 38.2 From Eq. (38.2), a photon of energy $E = 1.14$ eV has wavelength $\lambda = hc/E = (4.136 \times 10^{-15}\ \text{eV})(3.00 \times 10^8\ \text{m/s})/(1.14\ \text{eV}) = 1.09 \times 10^{-6}\ \text{m} = 1090$ nm. This is in the infrared part of the spectrum. Since wavelength is inversely proportional to photon energy, the *minimum* photon energy of 1.14 eV corresponds to the *maximum* wavelength that causes photoconductivity in silicon. Thus the wavelength must be 1090 nm or less. Photons of visible light (λ from 400 to 700 nm) will produce the effect, but radio waves (λ from about 3 m to about 1 km) will not.

Section 38.3 No. For the atom to absorb a photon, the photon energy must be precisely equal to the energy difference between the initial and final energy levels of the atom (Fig. 38.11).

Section 38.4 The nucleus of a hydrogen atom is a proton (mass 1.67×10^{-27} kg), which has only about one-fourth of the mass of an alpha particle $(6.64 \times 10^{-27}\ \text{kg})$. Because the alpha particle is more massive, it won't bounce back in even a head-on collision with a proton. It's like a bowling ball colliding with a ping-pong ball at rest (see Fig. 8.19b). Thus there would be no large-angle scattering in this case. In Rutherford's experiment, by contrast, there *was* large-angle scattering because a gold nucleus is more massive than an alpha particle. The analogy there is a ping-pong ball hitting a bowling ball at rest (Fig. 8.19a).

Section 38.5 Figure 38.22 shows that many of the energy levels of He^+ are the same as those of H. Hence photons emitted during transitions between corresponding pairs of levels in He^+ and H have the same energy E and the same wavelength $\lambda = hc/E$. For example, a H atom that drops from the $n = 2$ level to the $n = 1$ level emits a photon of energy 10.20 eV and wavelength 122 nm (see Example 38.6); a He^+ ion emits a photon of the same energy and wavelength when it drops from the $n = 4$ level to the $n = 2$ level.

Section 38.6 In a neon light fixture, a large potential difference is applied between the ends of a neon-filled glass tube. This ionizes some of the neon atoms, allowing a current of electrons to flow through the gas. Some of the neon atoms are struck by fast-moving electrons, exciting the atoms to the 5*s* level shown in Fig. 38.24a. From this level the atoms transition spontaneously to the 3*p* level and emit 632.8-nm photons in the process. The photons escape out of the sides of the glass tube. This process is *spontaneous* emission, as depicted in Fig. 38.23a. No population inversion occurs and the photons are not trapped by mirrors as shown in Fig. 38.24b, so there is no stimulated emission. Hence there is no laser action.

Section 38.7 Equation (38.23) shows that the wavelength shift $\Delta\lambda = \lambda' - \lambda$ depends only on the photon scattering angle ϕ, not on the wavelength of the incident photon. So a visible-light photon scattered through an angle ϕ undergoes the same wavelength shift as an x-ray photon. Equation (38.23) also shows that this shift is of the order of $h/mc = 2.426 \times 10^{-12}\ \text{m} = 0.002426$ nm. This is a few percent of the wavelength of x rays (see Example 38.8), so the effect is noticeable in x-ray scattering. However, h/mc is a tiny fraction of the wavelength of visible light (between 400 and 700 nm). The human eye cannot distinguish such miniscule differences in wavelength (that is, differences in color).

Section 38.8 The Planck radiation law, Eq. (38.32), shows that an ideal blackbody emits radiation at *all* wavelengths: the spectral emittance $I(\lambda)$ is equal to zero only for $\lambda = 0$ and in the limit $\lambda \to \infty$. So a blackbody at 2000 K does indeed emit both x rays and radio waves. However, Fig. 38.31 shows that the spectral emittance for this temperature is very low for wavelengths much less than 1 μm (including x rays) and for wavelengths much longer than a few μm (including radio waves). Hence such a blackbody emits very little in the way of x rays or radio waves.

Discussion Questions

Q38.1 In what ways do photons resemble other particles such as electrons? In what ways do they differ? Do photons have mass? Do they have electric charge? Can they be accelerated? What mechanical properties do they have?

Q38.2 There is a certain probability that a single electron may simultaneously absorb *two* identical photons from a high-intensity laser. How would such an occurrence affect the threshold frequency and the equations of Section 38.2? Explain.

Q38.3 Would you expect effects due to the photon nature of light to be generally more important at the low-frequency end of the electromagnetic spectrum (radio waves) or at the high-frequency end (x rays and gamma rays)? Why?

Q38.4 Most black-and-white photographic film (with the exception of some special-purpose films) is less sensitive to red light than blue light, and has almost no sensitivity to infrared. How can these properties be understood on the basis of photons?

Q38.5 Human skin is relatively insensitive to visible light, but ultraviolet radiation can cause severe burns. Does this have anything to do with photon energies? Explain.

Q38.6 Explain why Fig. 38.4 shows that most photoelectrons have kinetic energies less than $hf - \phi$, and also explain how these smaller kinetic energies occur.

Q38.7 Figure 38.5 shows that in a photoelectric-effect experiment, the photocurrent i for large positive values of V_{AC} has the same value no matter what the light frequency f (provided that f is greater than the threshold frequency f_0). Explain why.

Q38.8 On a graph of K_{max} versus f for the photoelectric effect, what is the physical significance of a) the slope? b) the intercept with the horizontal (f) axis? c) the intercept with the vertical (K_{max}) axis?

Q38.9 The materials called *phosphors* that coat the inside of a fluorescent lamp convert ultraviolet radiation (from the mercury-vapor discharge inside the tube) into visible light. Could one also make a phosphor that converts visible light to ultraviolet? Explain.

Q38.10 In a photoelectric effect experiment, which of the following will increase the maximum kinetic energy of the photoelectrons: a) use light of greater intensity; b) use light of greater frequency; c) use light of greater wavelength; d) use a metal surface with a larger work function. In each case justify your answer.

Q38.11 The total internal energy (kinetic plus potential) of the hydrogen atom is negative. What significance does this have?

Q38.12 Galaxies tend to be strong emitters of Lyman-α photons (from the $n = 2$ to $n = 1$ transition in atomic hydrogen). But the intergalactic medium—the very thin gas between the galaxies—tends to *absorb Lyman-α* photons. What can you infer from these observations about the temperature in these two environments? Explain.

Q38.13 A doubly ionized lithium atom (Li^{++}) is one that has had two of its three electrons removed. The energy levels of the remaining single-electron ion are closely related to those of the hydrogen atom. The nuclear charge for lithium is $+3e$ instead of just $+e$. How are the energy levels related to those of hydrogen? How is the *radius* of the ion in the ground level related to that of the hydrogen atom? Explain.

Q38.14 The emission of a photon by an isolated atom is a recoil process in which momentum is conserved. Thus Eq. (38.6) should include a recoil kinetic energy K_r for the atom. Why is this energy negligible in that equation?

Q38.15 How might the energy levels of an atom be measured directly, that is, without recourse to analysis of spectra?

Q38.16 Elements in the gaseous state emit line spectra with well-defined wavelengths. But hot solid bodies always emit a continuous spectrum, that is, a continuous smear of wavelengths. Can you account for this difference?

Q38.17 As a body is heated to a very high temperature and becomes self-luminous, the apparent color of the emitted radiation shifts from red to yellow and finally to blue as the temperature increases. Why does the color shift? What other changes in the character of the radiation occur?

Q38.18 A photon of frequency f undergoes Compton scattering from an electron at rest and scatters through an angle θ. The frequency of the scattered photon is f'. How is f' related to f? Does your answer depend on θ? Explain.

Q38.19 Can Compton scattering occur with protons as well as electrons? For example, suppose a beam of x rays is directed at a target of liquid hydrogen. (Recall that the nucleus of hydrogen consists of a single proton.) Compared to Compton scattering with electrons, what similarities and differences would you expect? Explain.

Q38.20 Why must engineers and scientists shield against x-ray production in high-voltage equipment?

Exercises

Section 38.2 The Photoelectric Effect

38.1 A photon of green light has a wavelength of 520 nm. Find the photon's frequency, magnitude of momentum, and energy. Express the energy both in joules and electron volts.

38.2 A laser used to weld detached retinas emits light with a wavelength of 652 nm in pulses that are 20.0 ms in duration. The average power during each pulse is 0.600 W. a) How much energy is in each pulse in joules? In electron volts? b) What is the energy of one photon in joules? In electron volts? c) How many photons are in each pulse?

38.3 An excited nucleus emits a gamma-ray photon with an energy of 2.45 MeV. a) What is the photon frequency? b) What is the photon wavelength? c) How does the wavelength compare with a typical nuclear diameter of 10^{-14} m?

38.4 The predominant wavelength emitted by an ultraviolet lamp is 248 nm. If the total power emitted at this wavelength is 12.0 W, how many photons are emitted per second?

38.5 A clean nickel surface is exposed to light of wavelength 235 nm. What is the maximum speed of the photoelectrons emitted from this surface? Use Table 38.1.

38.6 The photoelectric threshold wavelength of a tungsten surface is 272 nm. Calculate the maximum kinetic energy of the electrons ejected from this tungsten surface by ultraviolet radiation of frequency 1.45×10^{15} Hz. Express the answer in electron volts.

38.7 A 75-W light source consumes 75 W of electrical power. Assume all this energy goes into emitted light of wavelength 600 nm. a) Calculate the frequency of the emitted light. b) How many photons per second does the source emit? c) Are the answers to parts (a) and (b) the same? Is the frequency of the light the same thing as the number of photons emitted per second? Explain.

38.8 What would the minimum work function for a metal have to be for visible light (400 nm to 700 nm) to eject photoelectrons?

38.9 a) A proton is moving at a speed much less than the speed of light. It has kinetic energy K_1 and momentum p_1. If the momentum of the proton is doubled, so $p_2 = 2p_1$, how is its new kinetic energy K_2 related to K_1? b) A photon with energy E_1 has momentum p_1. If another photon has momentum p_2 that is twice p_1, how is the energy E_2 of the second photon related to E_1?

38.10 When ultraviolet light with a wavelength of 400 nm falls on a certain metal surface, the maximum kinetic energy of the emitted photoelectrons is 1.10 eV. What is the maximum kinetic energy of the photoelectrons when light of wavelength 300 nm falls on the same surface?

38.11 When ultraviolet light with a wavelength of 254 nm falls upon a clean copper surface, the stopping potential necessary to stop emission of photoelectrons is 0.181 V. a) What is the photoelectric threshold wavelength for this copper surface? b) What is the work function for this surface, and how does your calculated value compare with that given in Table 38.1?

38.12 The photoelectric work function of potassium is 2.3 eV. If light having a wavelength of 250 nm falls on potassium, find a) the stopping potential in volts; b) the kinetic energy in electron volts of the most energetic electrons ejected; c) the speeds of these electrons.

38.13 A photon has momentum of magnitude 8.24×10^{-28} kg$\cdot$m/s. a) What is the energy of this photon? Give your answer in joules and in electron volts. b) What is the wavelength of this photon? In what region of the electromagnetic spectrum does it lay?

38.14 a) In the photoelectric effect, what is the relation between the threshold frequency f_0 and the work function ϕ? b) The threshold wavelength for producing photoelectrons from a metal surface is 372 nm. What is the work function for this surface, in eV?

Section 38.3 Atomic Line Spectra and Energy Levels

38.15 a) An atom initially in an energy level with $E = -6.52$ eV absorbs a photon that has wavelength 860 nm. What is the internal energy of the atom after it absorbs the photon? b) An atom initially in an energy level with $E = -2.68$ eV emits a photon that has wavelength 420 nm. What is the internal energy of the atom after it emits the photon?

38.16 The energy-level scheme for the hypothetical one-electron element Searsium is shown in Fig. 38.34. The potential energy is taken to be zero for an electron at an infinite distance from the nucleus. a) How much energy (in electron volts) does it take to ionize an electron from the ground level? b) An 18-eV photon is absorbed by a Searsium atom in its ground level. As the atom returns to its ground level, what possible energies can the emitted photons have? Assume that there can be transitions between all pairs of levels. c) What will happen if a photon with an energy of 8 eV strikes a Searsium atom in its ground level? Why? d) Photons emitted in the Searsium transitions $n = 3 \rightarrow n = 2$ and $n = 3 \rightarrow n = 1$ will eject photoelectrons from an unknown metal, but the photon emitted from the transition $n = 4 \rightarrow n = 3$ will not. What are the limits (maximum and minimum possible values) of the work function of the metal?

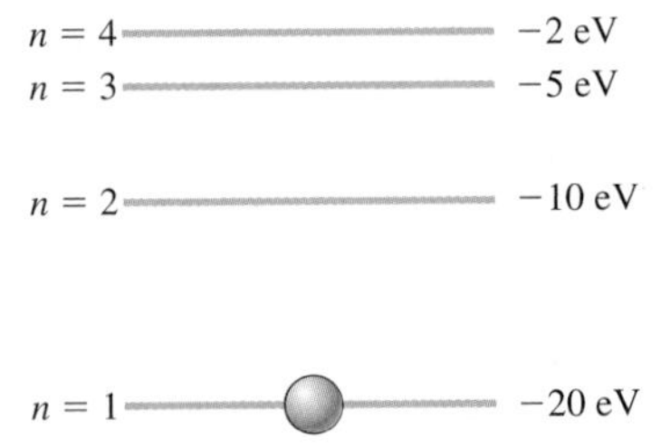

Figure 38.34 Exercise 38.16.

38.17 Use Balmer's formula to calculate a) the wavelength, b) the frequency, and c) the photon energy for the H_γ line of the Balmer series for hydrogen.

38.18 Find the longest and shortest wavelengths in the Lyman and Paschen series for hydrogen. In what region of the electromagnetic spectrum does each series lay?

38.19 Using the data shown in Fig. 38.10, determine the *difference* in energy (in electron volts) between the two lowest excited levels of the sodium atom.

Section 38.4 The Nuclear Atom

38.20 A 4.78-MeV alpha particle from a radium ^{226}Ra decay makes a head-on collision with a uranium nucleus. A uranium nucleus has 92 protons. a) What is the distance of closest approach of the alpha particle to the center of the nucleus? Assume that the uranium nucleus remains at rest and that the distance of closest approach is much larger than the radius of the uranium nucleus. b) What is the force on the alpha particle at the instant when it is at the distance of closest approach?

38.21 A beam of alpha particles is incident on a target of lead. A particular alpha particle comes in "head-on" to a particular lead nucleus and stops 6.50×10^{-14} m away from the center of the nucleus. (This point is well outside the nucleus). Assume that the lead nucleus, which has 82 protons, remains at rest. The mass of the alpha particle is 6.64×10^{-27} kg. a) Calculate the electrostatic potential energy at the instant that the alpha particle stops. Express your result in joules and in MeV. b) What initial kinetic energy (in joules and in MeV) did the alpha particle have? c) What was the initial speed of the alpha particle?

Section 38.5 The Bohr Model

38.22 A hydrogen atom initially in the ground level absorbs a photon, which excites it to the $n = 4$ level. Determine the wavelength and frequency of the photon.

38.23 A triply ionized beryllium ion, Be^{3+} (a beryllium atom with three electrons removed), behaves very much like a hydrogen atom except that the nuclear charge is four times as great. a) What is the ground-level energy of Be^{3+}? How does this compare to the ground-level energy of the hydrogen atom? b) What is the ionization energy of Be^{3+}? How does this compare to the ionization energy of the hydrogen atom? c) For the hydrogen atom the wavelength of the photon emitted in the $n = 2$ to $n = 1$ transition is 122 nm (see Example 38.6 of Section 38.5). What is the wavelength of the photon emitted when a Be^{3+} ion undergoes this transition? d) For a given value of n, how does the radius of an orbit in Be^{3+} compare to that for hydrogen?

38.24 a) What is the angular momentum L of the electron in a hydrogen atom, with respect to an origin at the nucleus, when the atom is in its lowest energy level? b) Repeat part (a) for the ground level of He^{+}. Compare to the answer in part (a).

38.25 A hydrogen atom is in a state with energy -1.51 eV. In the Bohr model, what is the angular momentum of the electron in the atom, with respect to an axis at the nucleus?

38.26 A hydrogen atom undergoes a transition from the $n = 5$ to the $n = 2$ state. a) What are the energy and wavelength of the photon that is emitted? b) If the angular momentum is conserved and if the Bohr model is used to describe the atom, what must the angular momentum be of the photon that is emitted? (As we will see in Chapter 41, the modern quantum-mechanical description of the hydrogen atom gives a different result.)

38.27 a) Using the Bohr model, calculate the speed of the electron in a hydrogen atom in the $n = 1$, 2, and 3 levels. b) Calculate the orbital period in each of these levels. c) The average lifetime of the first excited level of a hydrogen atom is 1.0×10^{-8} s. In the Bohr model, how many orbits does an electron in the $n = 2$ level complete before returning to the ground level?

38.28 According to the Bohr model (with the nucleus assumed to be at rest), the Rydberg constant R is equal to $me^4/8\epsilon_0^2 h^3 c$. a) Calculate R in m^{-1} and compare with the experimental value. b) Calculate the energy (in electron volts) of a photon whose wavelength equals R^{-1}. (This quantity is known as the *Rydberg energy*.) c) To account for the motion of the nucleus, replace the mass of the electron m with the reduced mass m_r for ordinary hydrogen and determine the change in R that this produces.

Section 38.6 The Laser

38.29 A large number of neon atoms are in thermal equilibrium. What is the ratio of the number of atoms in a $5s$ state to the number in a $3p$ state at a) 300 K; b) 600 K; c) 1200 K? The energies of these states are shown in Fig. 38.24a. d) At any of these temperatures, the rate at which a neon gas will spontaneously emit 632.8-nm radiation is quite low. Explain why.

38.30 Figure 38.10 shows the energy levels of the sodium atom. The two lowest excited levels are shown in columns labeled $^2P_{3/2}$ and $^2P_{1/2}$. Find the ratio of the number of atoms in a $^2P_{3/2}$ state to the number in a $^2P_{1/2}$ state for a sodium gas in thermal equilibrium at 500 K. In which state are more atoms found?

38.31 How many photons per second are emitted by a 7.50-mW CO_2 laser that has a wavelength of 10.6 μm?

38.32 Using the information shown in Fig. 38.24a, compute the energy difference for the $5s$ to $3p$ transition in neon. Express your result in electron volts and in joules. Compute the wavelength of a photon having this energy, and compare your result with the observed wavelength of the laser light.

Section 38.7 X-Ray Production and Scattering

38.33 Protons are accelerated from rest by a potential difference of 4.00 kV and strike a metal target. If a proton produces one photon on impact, what is the minimum wavelength of the resulting x rays? How does your answer compare to the minimum wavelength if 4.00-keV electrons are used instead? Why do x-ray tubes use electrons rather than protons to produce x rays?

38.34 a) What is the minimum potential difference between the filament and the target of an x-ray tube if the tube is to produce x rays with a wavelength of 0.150 nm? b) What is the shortest wavelength produced in an x-ray tube operated at 30.0 kV?

38.35 The cathode-ray tubes that generated the picture in early color televisions were sources of x rays. If the acceleration voltage in a television tube is 15.0 kV, what are the shortest wavelength x rays produced by the television? (Modern televisions contain shielding to stop these x rays.)

38.36 A beam of x rays with wavelength 0.0500 nm is Compton-scattered by the electrons in a sample. At what angle from the incident beam should you look to find x rays with a wavelength of a) 0.0542 nm; b) 0.0521 nm; c) 0.0500 nm?

38.37 X rays with initial wavelength 0.0665 nm undergo Compton scattering. What is the largest wavelength found in the scattered x rays? At which scattering angle is this wavelength observed?

38.38 X rays are produced in a tube operating at 18.0 kV. After emerging from the tube, x rays with the minimum wavelength produced strike a target and are Compton-scattered through an angle of 45.0°. a) What is the original x-ray wavelength? b) What is the wavelength of the scattered x rays? c) What is the energy of the scattered x rays (in electron volts)?

38.39 Consider Compton scattering with visible light. A photon with wavelength 500 nm scatters backward ($\theta = 180°$) from a free electron initially at rest. What is the fractional shift in wavelength, $\Delta\lambda/\lambda$, for the photon?

38.40 A photon scatters in the backward direction ($\theta = 180°$) from a free proton that is initially at rest. What must the wavelength of the incident photon be if it is to undergo a 10.0% change in wavelength as a result of the scattering?

38.41 Complete the derivation of the Compton-scattering formula, Eq. (38.23), following the outline given in Eqs. (38.24) through (38.27).

Section 38.8 Continuous Spectra

38.42 The shortest visible wavelength is about 400 nm. What is the temperature of an ideal radiator whose spectral emittance peaks at this wavelength?

38.43 Radiation has been detected from space that is characteristic of an ideal radiator at $T = 2.728$ K. (This radiation is a relic of the "Big Bang" at the beginning of the universe.) For this temperature, at what wavelength does the Planck distribution peak? In what part of the electromagnetic spectrum is this wavelength?

38.44 Determine λ_m, the wavelength at the peak of the Planck distribution, and the corresponding frequency f, at the following Kelvin temperatures: a) 3.00 K; b) 300 K; c) 3000 K.

38.45 A 100-W incandescent light bulb has a cylindrical tungsten filament 30.0 cm long, 0.40 mm in diameter and with an emissivity of 0.26. a) What is the temperature of the filament? b) For what wavelength does the spectral emittance of the bulb peak? Incandescent light bulbs are not very efficient sources of visible light. Explain why this is so.

38.46 Blue Supergiants. A typical "blue supergiant" star (the type that explode and leave behind black holes) has a surface temperature of 30,000 K and a visual luminosity 100,000 times that of our sun. Our sun radiates at the rate of 3.86×10^{26} W. (Visual luminosity is the total power radiated at visible wavelengths.) a) Assuming that this star behaves like an ideal black body, what is the principal wavelength it radiates? Is this light visible? Use your answer to explain why these stars are blue. b) If we assume that the power radiated by the star is also 100,000 times that of our sun, what is the radius of this star? Compare its size to that of our sun, that has a radius of 6.96×10^5 km. c) Is it really correct to say that the visual luminosity is proportional to the total power radiated? Explain.

38.47 Show that for large values of λ the Planck distribution, Eq. (38.32), agrees with the Rayleigh distribution, Eq. (38.31).

38.48 Light is emitted from an ideal blackbody that has temperature T. Use the approximate method of Example 38.10 (Section 38.8) to calculate the percent of the total spectral emittance that is in the wavelength range 500.0 nm to 501.0 nm for a) $T = 300$ K; b) $T = 2000$ K; c) $T = 6000$ K. d) How do the answers to parts (a), (b), and (c) compare? What does this tell you about the change in the intensity distribution with temperature?

38.49 a) Show that the maximum in the Planck distribution, Eq. (38.32), occurs at a wavelength λ_m given by $\lambda_m = hc/4.965kT$

(Eq. 38.33). As discussed in the text, 4.965 is the root of Eq. (38.34). b) Evaluate the constants in the expression derived in part (a) to show that $\lambda_m T$ has the numerical value given in the Wien displacement law, Eq. (38.30).

38.50 An ideal radiator radiates with a total intensity of $I = 6.94\ \mathrm{MW/m^2}$. At what wavelength does the spectral emittance $I(\lambda)$ peak?

Problems

38.51 Exposing Photographic Film. The light-sensitive compound on most photographic films is silver bromide, AgBr. A film is "exposed" when the light energy absorbed dissociates this molecule into its atoms. (The actual process is more complex, but the quantitative result does not differ greatly.) The energy of dissociation of AgBr is 1.00×10^5 J/mol. For a photon that is just able to dissociate a molecule of silver bromide, find a) the photon energy in electron volts, b) the wavelength of the photon, and c) the frequency of the photon. d) What is the energy in electron volts of a photon having a frequency of 100 MHz? e) Light from a firefly can expose photographic film, but the radiation from an FM station broadcasting 50,000 W at 100 MHz cannot. Explain why this is so.

38.52 An atom with mass m emits a photon of wavelength λ. a) What is the recoil speed of the atom? b) What is the kinetic energy K of the recoiling atom? c) Find the ratio K/E, where E is the energy of the emitted photon. If this ratio is much less than unity, the recoil of the atom can be neglected in the emission process. Is the recoil of the atom more important for small or large atomic masses? For long or short wavelengths? d) Calculate K (in electron volts) and K/E for a hydrogen atom (mass 1.67×10^{-27} kg) that emits an ultraviolet photon of energy 10.2 eV. Is recoil an important consideration in this emission process?

38.53 The directions of spontaneous emission of photons from a source of radiation are random. According to the wave theory, the intensity of radiation from a point source varies inversely as the square of the distance from the source. Show that the number of photons from a point source passing out through a unit area is also given by an inverse-square law.

38.54 The photoelectric work functions for particular samples of certain metals are as follows: cesium, 2.1 eV; copper, 4.7 eV; potassium, 2.3 eV; and zinc, 4.3 eV. a) What is the threshold wavelength for each metal surface? b) Which of these metals could *not* emit photoelectrons when irradiated with visible light (400-700 nm)?

38.55 When a certain photoelectric surface is illuminated with light of different wavelengths, the following stopping potentials are observed:

Wavelength (nm)	Stopping potential (V)
366	1.48
405	1.15
436	0.93
492	0.62
546	0.36
579	0.24

Plot the stopping potential on the vertical axis against the frequency of the light on the horizontal axis. Determine a) the threshold frequency; b) the threshold wavelength; c) the photoelectric work function of the material (in electron volts); d) the value of Planck's constant h (assuming that the value of e is known).

38.56 a) If the average frequency emitted by a 200-W light bulb is 5.00×10^{14} Hz, and 10.0% of the input power is emitted as visible light, approximately how many visible-light photons are emitted per second? b) At what distance would this correspond to 1.00×10^{11} visible-light photons per square centimeter per second if the light is emitted uniformly in all directions?

38.57 a) The wavelength of light incident on a metal surface is reduced from λ_1 to λ_2. (Both λ_1 and λ_2 are less than the threshold wavelength for the surface.) When the wavelength is reduced in this way, what is the change in the stopping potential for photoelectrons emitted from the surface? b) Evaluate the change in stopping potential for $\lambda_1 = 295$ nm and $\lambda_2 = 265$ nm.

38.58 Consider a hydrogenlike atom with nuclear charge Ze. a) For what value of Z (rounded to the nearest integer value) is the Bohr speed of the electron in the ground level equal to 10.0% of the speed of light? b) For what value of Z (rounded to the nearest integer value) is the ionization energy of the ground level equal to 1.0% of the rest energy of the electron?

38.59 An incident x-ray photon of wavelength 0.0900 nm is scattered in the backward direction from a free electron that is initially at rest. a) What is the magnitude of the momentum of the scattered photon? b) What is the kinetic energy of the electron after the photon is scattered?

38.60 An incident x-ray photon is scattered from a free electron that is initially at rest. The photon is scattered straight back at an angle of 180° from its initial direction. The wavelength of the scattered photon is 0.0830 nm. a) What is the wavelength of the incident photon? b) What is the magnitude of the momentum of the electron after the collision? c) What is the kinetic energy of the electron after the collision?

38.61 The negative muon has a charge equal to that of an electron but a mass that is 207 times as great. Consider a hydrogen-like atom consisting of a proton and a muon. a) What is the reduced mass of the atom? b) What is the ground-level energy (in electron volts)? c) What is the wavelength of the radiation emitted in the transition from the $n = 2$ level to the $n = 1$ level?

38.62 An unknown element has a spectrum for absorption from its ground level with lines at 2.0, 5.0, and 9.0 eV. Its ionization energy is 10.0 eV. a) Draw an energy-level diagram for this element. b) If a 9.0-eV photon is absorbed, what energies can the subsequently emitted photons have?

38.63 A sample of hydrogen atoms is irradiated with light with a wavelength of 85.5 nm, and electrons are observed leaving the gas. a) If each hydrogen atom were initially in its ground level, what would be the maximum kinetic energy in electron volts of these photoelectrons? b) A few electrons are detected with energies as much as 10.2 eV greater than the maximum kinetic energy calculated in part (b). How can this be?

38.64 Bohr Orbits of a Satellite. A 20.0-kg satellite circles the earth once every 2.00 h in an orbit having a radius of 8060 km. a) Assuming that Bohr's angular-momentum result $(L = nh/2\pi)$

applies to satellites just as it does to an electron in the hydrogen atom, find the quantum number n of the orbit of the satellite. b) Show from Bohr's angular momentum result and Newton's law of gravitation that the radius of an earth-satellite orbit is directly proportional to the square of the quantum number, $r = kn^2$, where k is the constant of proportionality. c) Using the result from part (b), find the distance between the orbit of the satellite in this problem and its next "allowed" orbit. (Calculate a numerical value.) d) Comment on the possibility of observing the separation of the two adjacent orbits. e) Do quantized and classical orbits correspond for this satellite? Which is the "correct" method for calculating the orbits?

38.65 A nonrelativistic particle with mass m is held in circular orbit around the origin by an attractive force $F(r) = -Dr$, where D is a positive constant. Use the Bohr-model idea that only certain values of the angular momentum are allowed to answer the following questions. a) What are the allowed values of the orbital radius? b) What are the allowed energies? (Let the potential energy be zero when $r = 0$.) c) If the particle is excited, it can decay back down to the ground level by emitting one or more photons. What are the possible photon energies? d) Describe a possible physical situation to which this problem corresponds.

38.66 The mass of a deuteron is 3.34359×10^{-27} kg, the mass of a proton is 1.672623×10^{-27} kg, and the mass of an electron is 9.10939×10^{-31} kg. a) Calculate the reduced mass of a deuterium atom and express your result in terms of the electron mass m. b) Consider the $n = 2$ to $n = 1$ transition in atomic hydrogen. Calculate the difference in the wavelength of light emitted in this transition by a hydrogen atom and by a deuterium atom.

38.67 a) What is the least amount of energy in electron volts that must be given to a hydrogen atom initially in its ground level so that it can emit the H_α line in the Balmer series? b) How many different possibilities of spectral-line emissions are there for this atom when the electron starts in the $n = 3$ level and eventually ends up in the ground level? Calculate the wavelength of the emitted photon in each case.

38.68 A large number of hydrogen atoms are in thermal equilibrium. Let n_2/n_1 be the ratio of the number of atoms in an $n = 2$ excited state to the number of atoms in an $n = 1$ ground state. At what temperature is n_2/n_1 equal to a) 10^{-12}; b) 10^{-8}; c) 10^{-4}? d) Like the sun, other stars have continuous spectra with dark absorption lines (see Fig. 38.12). The absorption takes place in the star's atmosphere, which in all stars is composed primarily of hydrogen. Explain why the Balmer absorption lines are relatively weak in stars with low atmospheric temperatures such as the sun (atmosphere temperature $= 5800$ K) but strong in stars with higher atmospheric temperatures.

38.69 When a photon is emitted by an atom, the atom must recoil to conserve momentum. This means that the photon and the recoiling atom share the transition energy. a) For an atom with mass m, calculate the correction $\Delta\lambda$ due to recoil to the wavelength of an emitted photon. Let λ be the wavelength of the photon if recoil is not taken into consideration. (*Hint:* The correction is very small, as Problem 38.52 suggests, so $|\Delta\lambda|/\lambda \ll 1$. Use this fact to obtain an approximate but very accurate expression for $\Delta\lambda$.) b) Evaluate the correction for a hydrogen atom in which an electron in the nth level returns to the ground level. How does the answer depend on n?

38.70 A photon with a wavelength of 0.1800 nm is Compton-scattered through an angle of 180°. a) What is the wavelength of the scattered photon? b) How much energy is given to the electron? c) What is the recoil speed of the electron? Is it necessary to use the relativistic kinetic-energy relationship?

38.71 a) Calculate the maximum increase in photon wavelength that can occur during Compton scattering. b) What is the energy (in electron volts) of the smallest-energy x-ray photon for which Compton scattering could result in doubling the original wavelength?

38.72 An x-ray tube is operating at voltage V and current I. a) If only a fraction p of the electric power supplied is converted into x rays, at what rate is energy being delivered to the target? b) If the target has mass m and specific heat capacity c (in J/kg·K), at what average rate would its temperature rise if there were no thermal losses? c) Evaluate your results from parts (a) and (b) for an x-ray tube operating at 18.0 kV and 60.0 mA that converts 1.0% of the electric power into x rays. Assume that the 0.250-kg target is made of lead ($c = 130$ J/kg·K). d) What must the physical properties of a practical target material be? What would be some suitable target elements?

38.73 A photon with wavelength 0.1100 nm collides with a free electron that is initially at rest. After the collision the wavelength is 0.1132 nm. a) What is the kinetic energy of the electron after the collision? What is its speed? b) If the electron is suddenly stopped (for example, in a solid target), all of its kinetic energy is used to create a photon. What is the wavelength of this photon?

38.74 a) Derive an expression for the total shift in photon wavelength after two successive Compton scatterings from electrons at rest. The photon is scattered by an angle θ_1 in the first scattering and by θ_2 in the second. b) In general, is the total shift in wavelength produced by two successive scatterings of an angle $\theta/2$ the same as by a single scattering of θ? If not, are there any specific values of θ, other than $\theta = 0°$, for which the total shifts are the same? c) Use the result of part (a) to calculate the total wavelength shift produced by two successive Compton scatterings of 30.0° each. Express your answer in terms of h/mc. d) What is the wavelength shift produced by a single Compton scattering of 60.0°? Compare to the answer in part (c).

38.75 Nuclear fusion reactions at the center of the sun produce gamma-ray photons with energies of order 1 MeV (10^6 eV). By contrast, what we see emanating from the sun's surface are visible-light photons with wavelengths of order 500 nm. A simple model that explains this difference in wavelength is that a photon undergoes Compton scattering many times—in fact, about 10^{26} times, as suggested by models of the solar interior—as it travels from the center of the sun to its surface. a) Estimate the increase in wavelength of a photon in an average Compton-scattering event. b) Find the angle in degrees through which the photon is scattered in the scattering event described in part (a). (*Hint:* A useful approximation is $\cos\phi \approx 1 - \phi^2/2$, which is valid for $\phi \ll 1$. Note that ϕ is in radians in this expression.) c) It is estimated that a photon takes about 10^6 years to travel from the core to the surface of the sun. Find the average distance that light can travel within the interior of the sun without being scattered. (This distance is roughly equivalent to how far you could see if you were inside the sun and could survive the extreme temperatures there. As your answer shows, the interior of the sun is *very* opaque.)

38.76 An x-ray photon is scattered from a free electron (mass m) at rest. The wavelength of the scattered photon is λ', and the final speed of the struck electron is v. a) What was the initial wavelength λ of the photon? Express your answer in terms of λ', v, and m. (*Hint:* Use the relativistic expression for the electron kinetic energy.) b) Through what angle ϕ is the photon scattered? Express your answer in terms of λ, λ', and m. c) Evaluate your results in parts (a) and (b) for a wavelength of 5.10×10^{-3} nm for the scattered photon and a final electron speed of 1.80×10^8 m/s. Give ϕ in degrees.

38.77 a) Write the Planck distribution law in terms of the frequency f rather than the wavelength λ, to obtain $I(f)$. b) Show that

$$\int_0^\infty I(\lambda)\,d\lambda = \frac{2\pi^5 k^4}{15c^2h^3}T^4$$

where $I(\lambda)$ is the Planck distribution formula of Eq. (38.32). (*Hint:* Change the integration variable from λ to f.) You will need to use the following tabulated integral:

$$\int_0^\infty \frac{x^3}{e^{\alpha x} - 1}\,dx = \frac{1}{240}\left(\frac{2\pi}{\alpha}\right)^4$$

c) The result of (b) is I, and has the form of the Stefan-Boltzmann law, $I = \sigma T^4$ (Eq. 38.28). Evaluate the constants in (b) to show that σ has the value given in Section 38.8.

38.78 An Ideal Blackbody. A large cavity with a very small hole and maintained at a temperature T is a good approximation to an ideal radiator or blackbody. Radiation can pass into or out of the cavity only through the hole. The cavity is a perfect absorber, since any radiation incident on the hole becomes trapped inside the cavity. Such a cavity at 200°C has a hole with area 4.00 mm^2. How long does it take for the cavity to radiate 100 J of energy through the hole?

Challenge Problems

38.79 a) Show that in the Bohr model, the frequency of revolution of an electron in its circular orbit around a stationary hydrogen nucleus is $f = me^4/4\epsilon_0^2 n^3 h^3$. b) In classical physics, the frequency of revolution of the electron is equal to the frequency of the radiation that it emits. Show that when n is very large, the frequency of revolution does indeed equal the radiated frequency calculated from Eq. (38.6) for a transition from $n_1 = n + 1$ to $n_2 = n$. (This illustrates Bohr's *correspondence principle,* which is often used as a check on quantum calculations. When n is small, quantum physics gives results that are very different from those of classical physics. When n is large, the differences are not significant, and the two methods then "correspond." In fact, when Bohr first tackled the hydrogen atom problem, he sought to determine f as a function of n such that it would correspond to classical results for large n.)

38.80 Consider a beam of monochromatic light with intensity I incident on a perfectly absorbing surface oriented perpendicular to the beam. Use the photon concept to show that the radiation pressure exerted by the light on the surface is given by I/c.

38.81 Consider Compton scattering of a photon by a *moving* electron. Before the collision the photon has wavelength λ and is moving in the $+x$-direction, and the electron is moving in the $-x$-direction with total energy E (including its rest energy mc^2). The photon and electron collide head-on. After the collision, both are moving in the $-x$-direction (that is, the photon has been scattered by 180°). a) Derive an expression for the wavelength λ' of the scattered photon. Show that if $E \gg mc^2$, where m is the rest mass of the electron, your result reduces to

$$\lambda' = \frac{hc}{E}\left(1 + \frac{m^2c^4\lambda}{4hcE}\right)$$

b) A beam of infrared radiation from a CO_2 laser ($\lambda = 10.6$ mm) collides head-on with a beam of electrons, each of total energy $E = 10.0$ GeV ($1\text{ GeV} = 10^9$ eV). Calculate the wavelength λ' of the scattered photons, assuming a 180° scattering angle. c) What kind of scattered photons are these (infrared, microwave, ultraviolet, etc.)? Can you think of an application of this effect?

39 THE WAVE NATURE OF PARTICLES

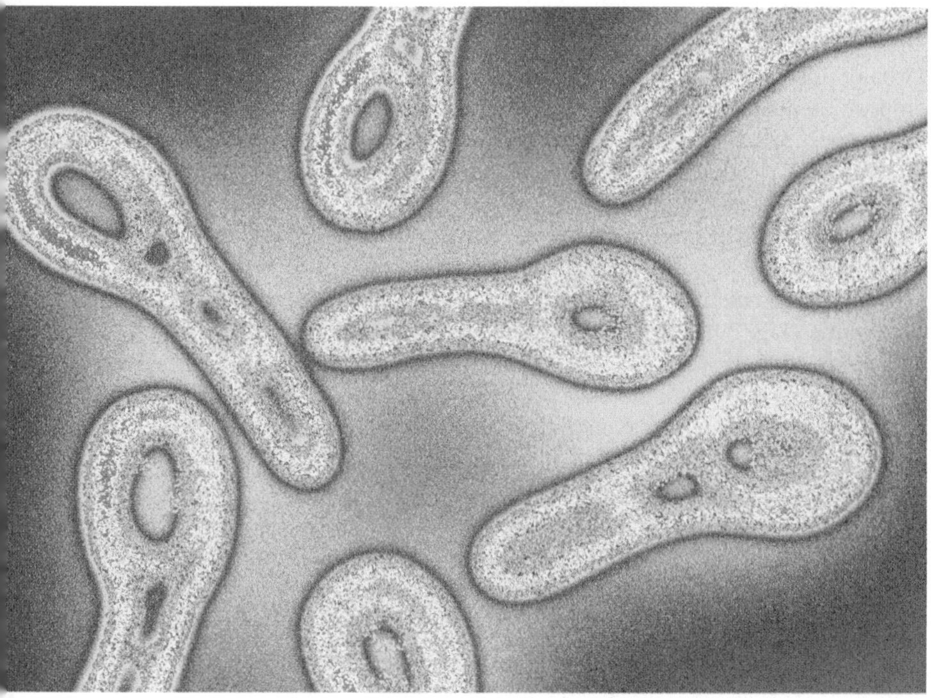

This image of the bacteria that carry diphtheria (shown magnified 60,000 times) was made with a transmission electron microscope. Electrons accelerated through tens of thousands of volts act like waves with wavelengths of a fraction of a nanometer. Aiming a beam of such electrons at the bacteria allows us to see far finer details than would be possible with a beam of visible light, which has wavelengths of 400 to 700 nm.

? A television picture tube uses a fast-moving beam of electrons to create an image on the screen. Can you see the wave aspect of these electrons when you look at the screen?

In preceding chapters we've seen one aspect of nature's wave-particle duality: Light and other electromagnetic radiation sometimes act like waves and sometimes like particles. Interference and diffraction demonstrate wave behavior, while emission and absorption of photons demonstrate the particle behavior.

A complete theory should also be able to *predict,* on theoretical grounds, the energy levels of any particular atom. The 1913 Bohr model of the hydrogen atom was a step in this direction. But it combined classical principles with new ideas that were inconsistent with classical theory, and it raised as many questions as it answered. More drastic departures from classical concepts were needed.

A successful drastic departure is *quantum mechanics,* a theory that began to emerge in the 1920s. Besides waves that sometimes act like particles, quantum mechanics extends the concept of wave-particle duality to include particles that sometimes show *wavelike* behavior. In these situations a particle is modeled as an inherently spread-out entity that can't be described as a point with a perfectly definite position and velocity.

In this chapter we will explore the wavelike nature of matter and some of its applications. We'll also introduce the *Schrödinger equation,* which is as fundamental to quantum mechanics as Newton's laws are to mechanics or as Maxwell's equations are to electromagnetism. In Chapter 40 we'll see how to use this equation to find the possible energy levels of a quantum-mechanical system.

Quantum mechanics is the key to understanding atoms and molecules, including their structure, spectra, chemical behavior, and many other properties. It has the happy effect of restoring unity and symmetry to our description of both particles and radiation. We will learn about quantum mechanics and use its results throughout the remainder of this book.

39.1 | De Broglie Waves

A major advance in the understanding of atomic structure began in 1924 with a proposition made by a French physicist and nobleman, Prince Louis de Broglie (pronounced "de broy"). His reasoning, freely paraphrased, went like this: Nature loves symmetry. Light is dualistic in nature, behaving in some situations like waves and in others like particles. If nature is symmetric, this duality should also hold for matter. Electrons and protons, which we usually think of as *particles,* may in some situations behave like *waves.*

39.1 Louis-Victor de Broglie, the 7th Duke de Broglie (1892–1987), broke with family tradition by choosing a career in physics rather than as a diplomat. His revolutionary proposal that particles have wave characteristics—for which de Broglie won the 1929 Nobel Prize in physics—was published in his doctoral thesis.

If a particle acts like a wave, it should have a wavelength and a frequency. De Broglie (Fig. 39.1) postulated that a free particle with rest mass m, moving with nonrelativistic speed v, should have a wavelength λ related to its momentum $p = mv$ in exactly the same way as for a photon, as expressed by Eq. (38.5): $\lambda = h/p$. The **de Broglie wavelength** of a particle is then

$$\lambda = \frac{h}{p} = \frac{h}{mv} \quad \text{(de Broglie wavelength of a particle)} \qquad (39.1)$$

where h is Planck's constant. If the particle's speed is an appreciable fraction of the speed of light c, we use Eq. (37.27) to replace mv in Eq. (39.1) with $\gamma mv = mv/\sqrt{1 - v^2/c^2}$. The frequency f, according to de Broglie, is also related to the particle's energy E in the same way as for a photon, namely,

$$E = hf \qquad (39.2)$$

Thus the relations of wavelength to momentum and of frequency to energy, in de Broglie's hypothesis, are exactly the same for particles as for photons.

CAUTION The relationship $E = hf$ must be carefully applied to nonzero-rest-mass particles such as electrons and protons. Unlike a photon, they do *not* travel at speed c, so neither the equation $f = c/\lambda$ nor $E = pc$ applies to them!

To appreciate the enormous significance of de Broglie's proposal, we have to realize that at the time there was no direct experimental evidence that particles have wave characteristics. It is one thing to suggest a new hypothesis to explain experimental observations; it is quite another to propose such a radical departure from established concepts on theoretical grounds alone. But it was clear that a radical idea was needed. The dual nature of electromagnetic radiation had led to adoption of the photon concept, also a radical idea. The limited success in understanding atomic structure indicated that a similar revolution was needed in the mechanics of particles.

De Broglie's hypothesis was the beginning of that revolution. Within a few years after 1924 it was developed by Heisenberg, Schrödinger, Dirac, Born, and many others into a detailed theory called **quantum mechanics.** This development was well under way even before direct experimental evidence for the wave properties of particles was found.

Quantum mechanics involves sweeping revisions of our fundamental concepts of the description of matter. A particle is not a geometric point but an entity that is spread out in space. The spatial distribution of a particle is defined by a function called a **wave function,** which is closely analogous to the wave functions we used for mechanical waves in Chapter 15 and for electromagnetic waves in Chapter 32. The wave function for a *free* particle with definite energy has a recurring wave pattern with definite wavelength and frequency. The wave and particle aspects are not inconsistent; the *principle of complementarity,* which we discussed in Section 38.8, tells us that we need both the particle model and the wave model for a complete description of nature.

In the Bohr model, we pictured the energy levels of the hydrogen atom in terms of definite electron orbits, as shown in Fig. 38.9. This is an oversimplification and should not be taken literally. But the most important idea in Bohr's theory was the existence of discrete energy levels and their relation to the frequencies of emitted photons. The new quantum mechanics still assigns only certain allowed energy states to an atom, but with a more general description of the electron motion in terms of wave functions. In the hydrogen atom the energy levels predicted by quantum mechanics turn out to be the same as those given by Bohr's theory. In more complicated atoms, for which the Bohr theory does not work, the quantum-mechanical picture is in excellent agreement with observation.

The de Broglie wave hypothesis has an interesting relation to the Bohr model. We can use Eq. (39.1) to obtain the Bohr quantum condition that the angular momentum $L = mvr$ must be an integer multiple of Planck's constant h. The method is analogous to determining the normal-mode frequencies of standing waves. We discussed this problem in Sections 15.8 and 32.5; the central idea was to satisfy the **boundary conditions** for the waves. For example, for standing waves on a string that is fixed at both ends, the ends are always nodes, and there will be additional nodes along the string for all but the fundamental mode. For the boundary conditions to be satisfied, the total length of the string must equal some *integral* number of half-wavelengths.

A standing wave on a string transmits no energy, and electrons in Bohr's orbits radiate no energy. So think of an electron as a standing wave fitted around a circle in one of the Bohr orbits. For the wave to "come out even" and join onto itself smoothly, the circumference of this circle must include some *whole number* of wavelengths, as suggested by Fig. 39.2. For an orbit with radius r and circumference $2\pi r$, we must have $2\pi r = n\lambda$, where $n = 1, 2, 3, \ldots$. According to the de Broglie relation, Eq. (39.1), the wavelength λ of a particle with rest mass m, moving with nonrelativistic speed v, is $\lambda = h/mv$. Combining $2\pi r = n\lambda$ and $\lambda = h/mv$, we find $2\pi r = nh/mv$, or

$$mvr = n\frac{h}{2\pi} \qquad (39.3)$$

We recognize Eq. (39.3) as being identical to Eq. (38.10), Bohr's result that the magnitude of the angular momentum $L = mvr$ must equal an integer n times $h/2\pi$. Thus a wave-mechanical picture leads naturally to the quantization of the electron's angular momentum.

To be sure, the idea of fitting a standing wave around a circular orbit is a rather vague notion. But the agreement of Eq. (39.3) with Bohr's result is much too remarkable to be a coincidence. It strongly suggests that the wave properties of electrons do indeed have something to do with atomic structure.

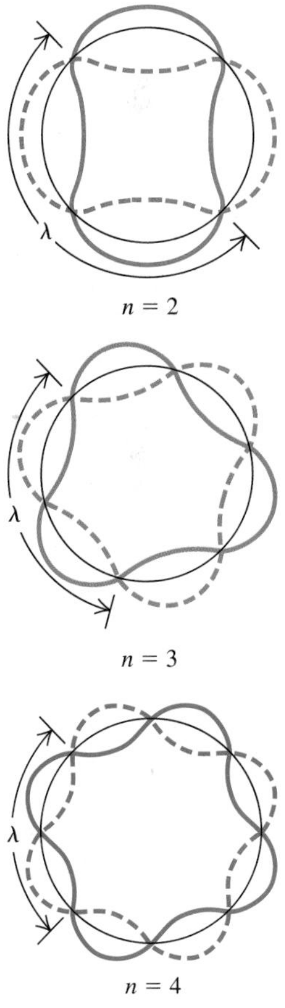

39.2 Diagrams showing the idea of fitting a standing wave around a circular orbit. For the wave to join onto itself smoothly, the circumference of the orbit must be an integral number n of wavelengths. Examples are shown for $n = 2$, $n = 3$, and $n = 4$.

Later we will learn how wave functions for specific systems are determined by solution of a wave equation called the Schrödinger equation. Boundary conditions play a central role in finding solutions of this equation and thus in determining possible energy levels, values of angular momentum, and other properties.

Problem-Solving Strategy: Particles and Waves

IDENTIFY *the relevant concepts:* Jarring as it may seem, both particles and light have wavelike properties as well as particle-like properties. The wavelength of a particle is inversely proportional to the momentum and the frequency is proportional to the energy. These concepts will help you solve many problems involving the wave character of matter.

SET UP *the problem:* As always, determine the target variable and decide which equations you will use to calculate it.

EXECUTE *the solution as follows:*

1. Use Eq. (39.1) to relate the momentum p to the wavelength λ, and use Eq. (39.2) to relate the energy E to the frequency f.
2. Nonrelativistic kinetic energy may be expressed as either $K = \frac{1}{2}mv^2$ or (because $p = mv$) $K = p^2/2m$. The latter form is often useful in calculations involving the de Broglie wavelength.
3. Be careful to use consistent units. Lengths, such as wavelengths, are always in meters if you use the other quantities consistently in SI units. If you want a length in nanometers or some other units, don't forget to convert. As in Chapter 38, energies may be expressed in either joules or electron volts. Depending on your choice, you can use either $h = 6.626 \times 10^{-34}\ \text{J}\cdot\text{s}$ or $h = 4.136 \times 10^{-15}\ \text{eV}\cdot\text{s}$.

EVALUATE *your answer:* To check your numerical results, it helps to remember some typical orders of magnitude for quantities on the atomic scale:

Size of an atom: 10^{-10} m

Mass of an atom: 10^{-26} kg

Mass of an electron: 10^{-30} kg

Energy magnitude of an atomic state: 1 to 10 eV (10^{-19} to 10^{-18} J) (but some interaction energies are much smaller)

Speed of an electron in the Bohr model of a hydrogen atom: 10^6 m/s

Electron charge magnitude: 10^{-19} C

kT at room temperature: $\frac{1}{40}$ eV

You may want to add items to this list. These approximate values will also help you in Chapter 43, in which we will deal with magnitudes that are characteristic of *nuclear* structure rather than atomic structure; these are often different by factors of 10^4 to 10^6.

Example 39.1 Energy of a thermal neutron

Find the speed and kinetic energy of a neutron ($m = 1.675 \times 10^{-27}$ kg) that has a de Broglie wavelength $\lambda = 0.200$ nm, approximately the atomic spacing in many crystals. Compare the energy with the average translational kinetic energy of a gas molecule at room temperature ($T = 20°\text{C} = 293$ K).

SOLUTION

IDENTIFY and SET UP: Our target variables are the speed v and kinetic energy K. We determine v using Eq. (39.1), then calculate the kinetic energy using $K = \frac{1}{2}mv^2$. To find the average kinetic energy of a molecule at temperature T, we use the results of our discussion of kinetic theory in Section 18.3.

EXECUTE: From Eq. (39.1),

$$v = \frac{h}{\lambda m} = \frac{6.626 \times 10^{-34}\ \text{J}\cdot\text{s}}{(0.200 \times 10^{-9}\ \text{m})(1.675 \times 10^{-27}\ \text{kg})}$$
$$= 1.98 \times 10^3\ \text{m/s}$$

The kinetic energy is

$$K = \frac{1}{2}mv^2 = \frac{1}{2}(1.675 \times 10^{-27}\ \text{kg})(1.98 \times 10^3\ \text{m/s})^2$$
$$= 3.28 \times 10^{-21}\ \text{J} = 0.0204\ \text{eV}$$

The average translational kinetic energy of a molecule of an ideal gas is given by Eq. (18.16):

$$\frac{1}{2}m(v^2)_{av} = \frac{3}{2}kT = \frac{3}{2}(1.38 \times 10^{-23}\ \text{J/K})(293\ \text{K})$$

$$= 6.07 \times 10^{-21}\ \text{J} = 0.0397\ \text{eV}$$

The two energies are comparable in magnitude. In fact, a neutron with kinetic energy in this range is called a *thermal neutron.* Diffraction of thermal neutrons, which we'll discuss in the next section, is used to study crystal and molecular structure in the same way as x-ray diffraction. Neutron diffraction has proved to be especially useful in the study of large organic molecules.

EVALUATE: Note that the calculated neutron speed is much less than the speed of light. This justifies our use of the nonrelativistic form of Eq. (39.1).

Test Your Understanding

A proton has a slightly smaller mass than a neutron. Compared to the neutron described in Example 39.1, would a proton of the same wavelength have more or less kinetic energy?

39.2 | Electron Diffraction

De Broglie's wave hypothesis, radical though it seemed, almost immediately received experimental confirmation. The first direct evidence involved a diffraction experiment with electrons that was analogous to the x-ray diffraction experiments that we described in Section 36.6. In those experiments, atoms in a crystal act as a three-dimensional diffraction grating for x rays. An x-ray beam is strongly reflected when it strikes a crystal at an angle that gives constructive interference among the waves scattered from the various atoms in the crystal. These interference effects demonstrate the *wave* nature of x rays.

In 1927, Clinton Davisson and Lester Germer, working at the Bell Telephone Laboratories, were studying the surface of a piece of nickel by directing a beam of *electrons* at the surface and observing how many electrons bounced off at various angles. Figure 39.3 shows an experimental setup like theirs. The specimen was *polycrystalline:* like many ordinary metals, it consisted of many microscopic crystals bonded together with random orientations. The experimenters expected that even the smoothest surface attainable would still look rough to an electron and that the electron beam would be diffusely reflected, with a smooth distribution of intensity as a function of the angle θ.

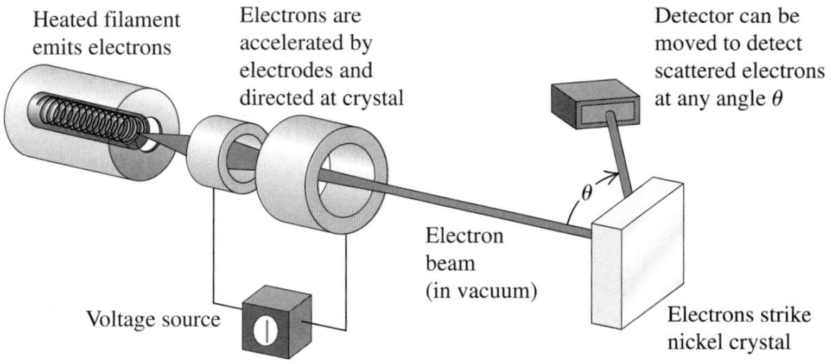

39.3 An apparatus similar to that used by Davisson and Germer to discover electron diffraction.

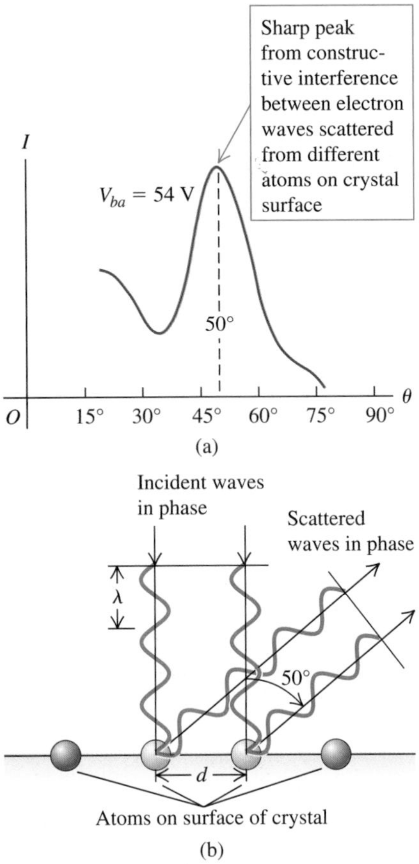

39.4 (a) A graph of intensity of the scattered electron beam shown in Fig. 39.3 as a function of the scattering angle θ. (b) Constructive interference of electron waves scattered from two adjacent atoms occurs when $d \sin\theta = m\lambda$. In the case shown here, $\theta = 50°$ and $m = 1$.

During the experiment an accident occurred that permitted air to enter the vacuum chamber, and an oxide film formed on the metal surface. To remove this film, Davisson and Germer baked the specimen in a high-temperature oven, almost hot enough to melt it. Unknown to them, this had the effect of creating large single-crystal regions with crystal planes that were continuous over the width of the electron beam.

When the observations were repeated, the results were quite different. Strong maxima in the intensity of the reflected electron beam occurred at specific angles (Fig. 39.4a), in contrast to the smooth variation of intensity with angle that Davisson and Germer had observed before the accident. The angular positions of the maxima depended on the accelerating voltage V_{ba} used to produce the electron beam. Davisson and Germer were familiar with de Broglie's hypothesis, and they noticed the similarity of this behavior to x-ray diffraction. This was not the effect they had been looking for, but they immediately recognized that the electron beam was being *diffracted.* They had discovered a very direct experimental confirmation of the wave hypothesis.

Davisson and Germer could determine the speeds of the electrons from the accelerating voltage, so they could compute the de Broglie wavelength from Eq. (39.1). The electrons were scattered primarily by the planes of atoms at the surface of the crystal. Atoms in a surface plane are arranged in rows, with a distance d that can be measured by x-ray diffraction techniques. These rows act like a reflecting diffraction grating; the angles at which strong reflection occurs are the same as for a grating with center-to-center distance d between its slits (Fig. 39.4b). From Eq. (36.13) the angles of maximum reflection are given by

$$d \sin\theta = m\lambda \qquad (m = 1, 2, 3, \ldots) \tag{39.4}$$

where θ is the angle shown in Fig. 39.3. The angles predicted by this equation, using the de Broglie wavelength, were found to agree with the observed values (Fig. 39.4a). Thus the accidental discovery of **electron diffraction** was the first direct evidence confirming de Broglie's hypothesis.

The de Broglie wavelength of a nonrelativistic particle is $\lambda = h/p = h/mv$. We can also express λ in terms of the particle's kinetic energy. For example, consider an electron freely accelerated from rest at point a to point b through a potential increase $V_b - V_a = V_{ba}$. The work done on the electron eV_{ba} equals its kinetic energy K. Using $K = p^2/2m$, we have

$$eV_{ba} = \frac{p^2}{2m} \qquad p = \sqrt{2meV_{ba}}$$

and the de Broglie wavelength of the electron is

$$\lambda = \frac{h}{p} = \frac{h}{\sqrt{2meV_{ba}}} \qquad \text{(de Broglie wavelength of an electron)} \tag{39.5}$$

Example 39.2 An electron-diffraction experiment

In a particular electron-diffraction experiment using an accelerating voltage of 54 V, an intensity maximum occurs when the angle θ in Fig. 39.3 is 50° (see Fig. 39.4a). The initial kinetic energy of the electrons is negligible. The rows of atoms have been found by x-ray diffraction to have a separation $d = 2.15 \times 10^{-10}\ \text{m} = 0.215\ \text{nm}$. Find the electron wavelength.

SOLUTION

IDENTIFY and SET UP: We are given the accelerating voltage $V_{ba} = 54$ V, so we can determine the wavelength λ using Eq. (39.5). We are also given the atomic spacing d and the angle θ for an intensity maximum, so we can also determine λ using Eq. (39.4).

EXECUTE: From Eq. (39.5), the electron wavelength is

$$\lambda = \frac{6.626 \times 10^{-34}\ \text{J}\cdot\text{s}}{\sqrt{2(9.109 \times 10^{-31}\ \text{kg})(1.602 \times 10^{-19}\ \text{C})(54\ \text{V})}} = 1.7 \times 10^{-10}\ \text{m}$$

Alternatively, using Eq. (39.4) and assuming $m = 1$,

$$\lambda = d \sin\theta = (2.15 \times 10^{-10}\ \text{m}) \sin 50^\circ = 1.7 \times 10^{-10}\ \text{m}$$

EVALUATE: The two numbers agree within the accuracy of the experimental results, which gives us an excellent check on our calculations. Note that this electron wavelength is less than the spacing between the atoms.

In 1928, just a year after the Davisson-Germer discovery, the English physicist G. P. Thomson carried out electron diffraction experiments using a thin polycrystalline metallic foil as a target. Debye and Sherrer had used a similar technique several years earlier to study x-ray diffraction from polycrystalline specimens. Because of the random orientations of the individual microscopic crystals in his foil, the diffraction pattern consisted of intensity maxima forming rings around the direction of the incident beam. Thomson's results again confirmed the de Broglie relationship. Figure 39.5 shows both x-ray and electron diffraction patterns for a polycrystalline aluminum foil. (G. P. Thomson was the son of J. J. Thomson, who 31 years earlier had performed the definitive experiment to establish the *particle* nature of electrons.)

Additional experiments were soon carried out in many laboratories. In Germany, Estermann and Stern demonstrated diffraction of alpha particles. Diffraction experiments have also been performed with various ions and low-energy neutrons (see Example 39.1). Thus the wave nature of particles, so strange in 1924, became firmly established in the years that followed.

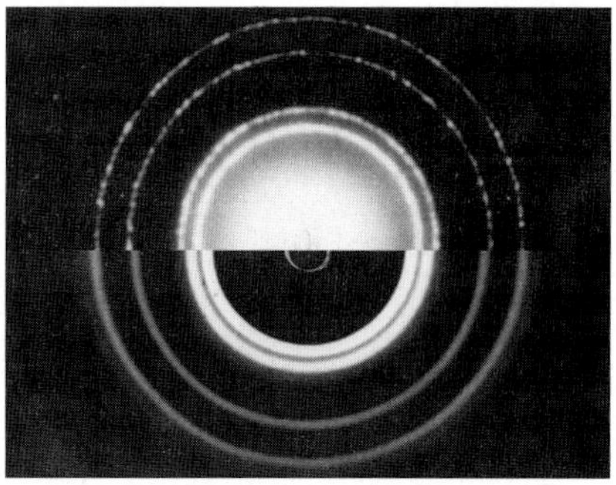

39.5 X-ray and electron diffraction. The upper half of the photo shows the diffraction pattern for 71-pm x rays passing through aluminum foil. The lower half, with a different scale, shows the diffraction pattern for 600-eV electrons from aluminum.

ActivPhysics ONLINE

17.5 Electron Interference

Test Your Understanding

Example 39.2 shows that to give electrons a wavelength of 1.7×10^{-10} m, they must be accelerated from rest through a voltage of 54 V and so acquire a kinetic energy of 54 eV. Does a photon of this same energy also have a wavelength of 1.7×10^{-10} m?

39.3 | Probability and Uncertainty

The discovery of the dual wave-particle nature of matter has forced us to reevaluate the kinematic language we use to describe the position and motion of a particle. In classical Newtonian mechanics we think of a particle as a point. We can describe its location and state of motion at any instant with three spatial coordinates and three components of velocity. But in general such a specific description is not possible. When we look on a small enough scale, there are fundamental limitations on the precision with which we can determine the position and velocity of a particle. Many aspects of a particle's behavior can be stated only in terms of *probabilities.*

ActivPhysics ONLINE
17.6 Uncertainty Principle

Single-Slit Diffraction

To try to get some insight into the nature of the problem, let's review the single-slit diffraction of light described in Section 36.2. Suppose the wavelength λ is much less than the slit width a. Then most (85%) of the light in the diffraction pattern is concentrated in the central maximum, bounded on either side by the first intensity minimum. We use θ_1 to denote the angle between the central maximum and the first minimum. Using Eq. (36.2) with $m = 1$, we find that θ_1 is given by $\sin\theta_1 = \lambda/a$. Since we assume $\lambda \ll a$, it follows that θ_1 is very small, $\sin\theta_1$ is very nearly equal to θ_1 (in radians), and

$$\theta_1 = \frac{\lambda}{a} \tag{39.6}$$

Now we perform the same experiment again, but using a beam of *electrons* instead of a beam of monochromatic light (Fig. 39.6). We have to do the experiment in vacuum (10^{-7} atm or less) so that the electrons don't bounce off air molecules. We can produce the electron beam with a setup that is similar in principle to the electron gun in a cathode-ray tube. This produces a narrow beam of electrons that all have very nearly the same direction and speed and therefore also the same de Broglie wavelength.

The result of this experiment, recorded on photographic film or by use of more sophisticated detectors, is a diffraction pattern identical to the one shown in Fig. 36.6. This pattern gives us additional direct evidence of the *wave* nature of electrons. About 85% of the electrons strike the film within the central maximum; the remainder strike the film within the subsidiary maxima on both sides.

If we believe that electrons are waves, the wave behavior in this experiment isn't surprising. But if we try to interpret it in terms of *particles,* we run into very serious problems. First, the electrons don't all follow the same path, even though they all have the same initial state of motion. In fact, we can't predict the exact trajectory of any individual electron from knowledge of its initial state. The best we can do is to say that *most* of the electrons go to a certain region, *fewer* go to other regions, and so on. That is, we can describe only the *probability* that an individual electron will

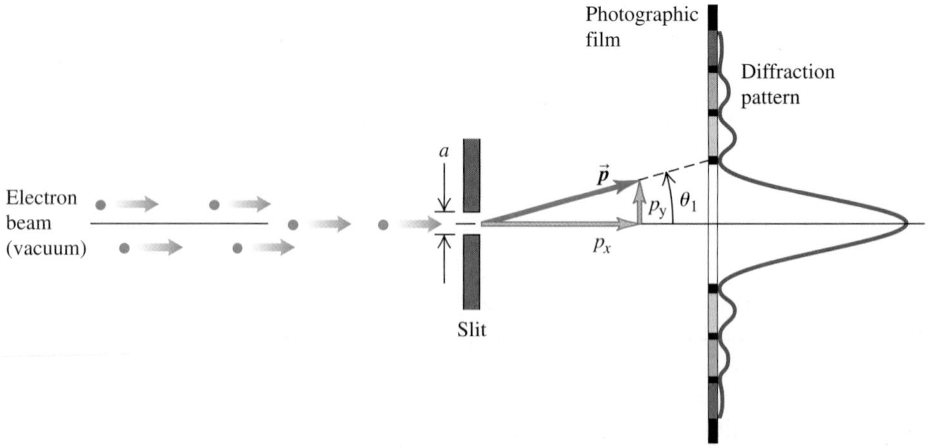

39.6 An electron diffraction experiment. The graph at the right shows the degree of exposure of the film, which in any region is proportional to the number of electrons striking that region. The components of momentum of an electron striking the outer edge of the central maximum, at angle θ_1, are shown.

strike each of various areas on the film. This fundamental indeterminacy has no counterpart in Newtonian mechanics, in which the motion of a particle can always be well predicted if we know the initial position and motion with sufficient accuracy.

Second, there are fundamental *uncertainties* in both the position and the momentum of an individual particle, and these two uncertainties are related inseparably. To clarify this point, let's go back to Fig. 39.6. An electron that strikes the film at the outer edge of the central maximum, at angle θ_1, must have a component of momentum p_y in the y-direction, as well as a component p_x in the x-direction, despite the fact that initially the beam was directed along the x-axis. From the geometry of the situation the two components are related by $p_y/p_x = \tan\theta_1$. Since θ_1 is small, we may use the approximation $\tan\theta_1 = \theta_1$, and

$$p_y = p_x\theta_1 \tag{39.7}$$

Substituting Eq. (39.6), $\theta_1 = \lambda/a$, we have

$$p_y = p_x\frac{\lambda}{a} \tag{39.8}$$

For the 85% of the electrons that strike the film within the central maximum (that is, at angles between $-\lambda/a$ and $+\lambda/a$), we see that the y-component of momentum is spread out over a range from $-p_x\lambda/a$ to $+p_x\lambda/a$. Now let's consider *all* the electrons that pass through the slit and strike the film. Again, they may hit above or below the center of the pattern, so their component p_y may be positive or negative. However the symmetry of the diffraction pattern shows us the average value $(p_y)_{\text{av}} = 0$. There will be an *uncertainty* Δp_y in the y-component of momentum at least as great as $p_x\lambda/a$. That is,

$$\Delta p_y \geq p_x\frac{\lambda}{a} \tag{39.9}$$

The narrower the slit width a, the broader is the diffraction pattern and the greater is the uncertainty in the y-component of momentum p_y.

The electron wavelength λ is related to the momentum $p_x = mv_x$ by the de Broglie relation, Eq. (39.1), which we can rewrite as $\lambda = h/p_x$. Using this relation in Eq. (39.9) and simplifying, we find

$$\begin{aligned}\Delta p_y &\geq p_x\frac{h}{p_x a} = \frac{h}{a}\\ \Delta p_y a &\geq h\end{aligned} \tag{39.10}$$

What does this result mean? The slit width a represents an uncertainty in the y-component of the *position* of an electron as it passes through the slit. We don't know exactly *where* in the slit each particle passes through. So both the y-position and the y-component of momentum have uncertainties, and the two uncertainties are related by Eq. (39.10). We can reduce the *momentum* uncertainty Δp_y only by reducing the width of the diffraction pattern. To do this, we have to increase the slit width a, which increases the *position* uncertainty. Conversely, when we *decrease* the position uncertainty by narrowing the slit, the diffraction pattern broadens and the corresponding momentum uncertainty *increases.*

You may protest that it doesn't seem to be consistent with common sense for a particle not to have a definite position and momentum. We reply that what we call *common sense* is based on familiarity gained through experience. Our usual experience includes very little contact with the microscopic behavior of particles. Sometimes we have to accept conclusions that violate our intuition when we are dealing with areas that are far removed from everyday experience.

The Uncertainty Principle

In more general discussions of uncertainty relations, the uncertainty of a quantity is usually described in terms of the statistical concept of *standard deviation,* which is a measure of the spread or dispersion of a set of numbers around their average value. Suppose we now begin to describe uncertainties in this way [neither Δp_y nor a in Eq. (39.10) is a standard deviation]. If a coordinate x has an uncertainty Δx and if the corresponding momentum component p_x has an uncertainty Δp_x, then those standard-deviation uncertainties are found to be related in general by the inequality

$$\Delta x \Delta p_x \geq \hbar \tag{39.11}$$

(Heisenberg uncertainty principle for position and momentum)

In this expression the quantity $\hbar$ (pronounced "h-bar") is Planck's constant divided by 2π:

$$\hbar = \frac{h}{2\pi} = 1.054571596(82) \times 10^{-34}\ \mathrm{J \cdot s} \tag{39.12}$$

We will use this quantity frequently to avoid writing a lot of factors of 2π in later equations.

Equation (39.11) is one form of the **Heisenberg uncertainty principle,** first discovered by the German physicist Werner Heisenberg (1901–1976). It states that, in general, neither the position nor the momentum of a particle can be determined with arbitrarily great precision, as classical physics would predict. Instead, the uncertainties in the two quantities play complementary roles, as we have described. Figure 39.7 shows the relationship between the two uncertainties.

It is tempting to suppose that we could get greater precision by using more sophisticated detectors of position and momentum. This turns out not to be possible. To detect a particle, the detector must *interact* with it, and this interaction unavoidably changes the state of motion of the particle, introducing uncertainty about its original state. For example, if we were to bounce shorter-wavelength photons off a particle to better locate its position, the larger photon momentum h/λ would make the particle recoil more, giving us greater uncertainty in its momentum. A more detailed analysis of such hypothetical experiments shows that the uncertainties we have described are fundamental and intrinsic. They *cannot* be circumvented *even in principle* by any experimental technique, no matter how sophisticated.

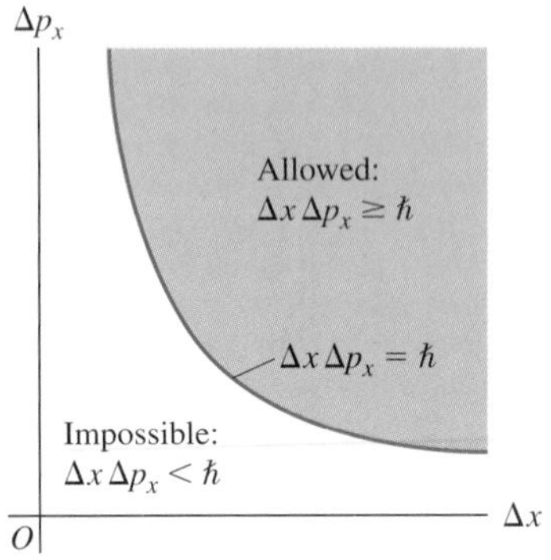

39.7 The Heisenberg uncertainty principle for position and momentum components. It is impossible for the product $\Delta x\, \Delta p_x$ to be less than $\hbar = h/2\pi$.

There is nothing special about the x-axis. In a three-dimensional situation with coordinates (x, y, z) there is an uncertainty relation for each coordinate and its corresponding momentum component: $\Delta x\, \Delta p_x \geq \hbar$, $\Delta y\, \Delta p_y \geq \hbar$, and $\Delta z\, \Delta p_z \geq \hbar$. However, the uncertainty in one coordinate is *not* related to the uncertainty in a different component of momentum. For example, Δx is not related directly to Δp_y.

For a particle moving along a radius, we can replace x in Eq. (39.11) with r, giving $\Delta r\, \Delta p_r \geq \hbar$. In the Bohr model, an electron moves in a circle of exact radius r, giving $\Delta r = 0$ and $\Delta p_r = 0$. Thus the Bohr model violates the Heisenberg uncertainty principle. We'll give a more correct description of atomic structure in Chapter 41; happily, it turns out that the energy-level predictions of the Bohr model *are* correct.

Uncertainty in Energy

There is also an uncertainty principle for *energy.* It turns out that the energy of a system also has inherent uncertainty. The uncertainty ΔE depends on the *time interval* Δt during which the system remains in the given state. The relation is

$$\Delta E\,\Delta t \geq \hbar \quad (39.13)$$

(Heisenberg uncertainty principle for energy and time interval)

A system that remains in a metastable state for a very long time (large Δt) can have a very well-defined energy (small ΔE), but if it remains in a state for only a short time (small Δt), the uncertainty in energy must be correspondingly greater (large ΔE). Figure 39.8 illustrates this idea.

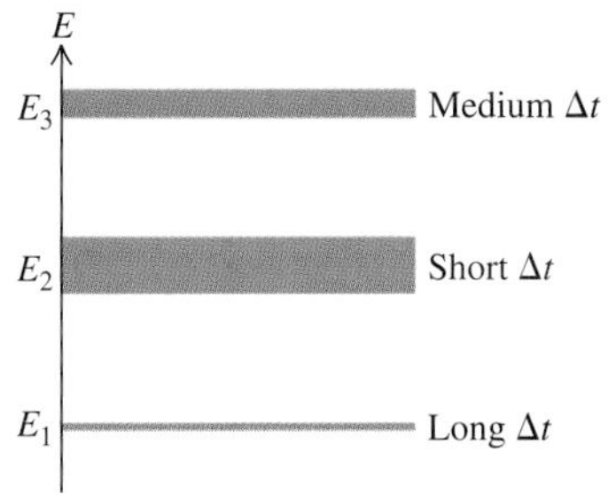

39.8 The longer the lifetime Δt of a state, the smaller is its spread in energy (shown by the width of the energy levels).

Example 39.3 The uncertainty principle: momentum and position

An electron is confined within a region of width 1.0×10^{-10} m. a) Estimate the minimum uncertainty in the x-component of the electron's momentum. b) If the electron has momentum with magnitude equal to the uncertainty found in part (a), what is its kinetic energy? Express the result in joules and in electron volts.

SOLUTION

IDENTIFY and SET UP: We know that the electron is *somewhere* within the region to which it is confined. Since we don't know exactly where within this region the electron is, the width of the region is the position uncertainty: $\Delta x = 1.0 \times 10^{-10}$ m. (You could argue that the uncertainty should actually be half the width of the region. All we want is an estimate, however, so we won't worry about factors of 2.) We then find the momentum uncertainty Δp_x using Eq. (39.11), and we find the kinetic energy using $K = p^2/2m$.

EXECUTE: a) For a given value of Δx, there is minimum uncertainty in momentum when the product $\Delta x\,\Delta p_x$ in Eq. (39.11) is equal to $\hbar$. Using the value of $\hbar$ from Eq. (39.12),

$$(\Delta p_x)_{\text{min}} = \frac{\hbar}{\Delta x} = \frac{1.055 \times 10^{-34}\ \text{J}\cdot\text{s}}{1.0 \times 10^{-10}\ \text{m}} = 1.1 \times 10^{-24}\ \text{kg}\cdot\text{m/s}$$

b) An electron with this magnitude of momentum has kinetic energy

$$K = \frac{p^2}{2m} = \frac{(1.1 \times 10^{-24}\ \text{kg}\cdot\text{m/s})^2}{2(9.11 \times 10^{-31}\ \text{kg})}$$
$$= 6.1 \times 10^{-19}\ \text{J} = 3.8\ \text{eV}$$

EVALUATE: The region is roughly the same width as an atom, and the energy is of the same order of magnitude as typical electron energies in atoms. This is a very rough calculation, but it is reassuring that the energy has a reasonable order of magnitude.

In a type of radioactive decay called *beta-minus decay,* an electron is emitted from a nucleus. Although it seems reasonable to assume that the electron was confined within the nucleus before the decay, that would require the position of the electron to be known to within an uncertainty of $\Delta x = 10^{-14}$ m or so. But this would give the confined electron a value of Δp_x that is 10^4 times greater than for the electron in this example and an expected kinetic energy so large that we would need to use relativistic equations to calculate it. This high energy is one reason to believe that there are no electrons confined in nuclei. In beta-minus decay the electron is actually *produced* within the nucleus when a neutron changes into a proton.

Example 39.4 The uncertainty principle: energy and time

A sodium atom is in one of the states labeled "Lowest excited levels" in Fig. 38.10 (Section 38.3). It remains in that state for an average time of 1.6×10^{-8} s before it makes a transition back to a ground state, emitting a photon with wavelength 589.0 nm and energy 2.105 eV. What is the uncertainty in energy of that excited state? What is the wavelength spread of the corresponding spectrum line?

SOLUTION

IDENTIFY and SET UP: The average time that the atom spends in this excited state is equal to Δt in Eq. (39.13). We find the uncertainty in energy by replacing the $\geq$ sign in Eq. (39.13) with an equals sign and solving for ΔE.

EXECUTE: From Eq. (39.13),

$$\Delta E = \frac{\hbar}{\Delta t} = \frac{1.055 \times 10^{-34}\ \mathrm{J \cdot s}}{1.6 \times 10^{-8}\ \mathrm{s}}$$
$$= 6.6 \times 10^{-27}\ \mathrm{J} = 4.1 \times 10^{-8}\ \mathrm{eV}$$

The atom remains an indefinitely long time in the ground state, so there is *no* fundamental uncertainty there. The fractional uncertainty of the photon energy is

$$\frac{4.1 \times 10^{-8}\ \mathrm{eV}}{2.105\ \mathrm{eV}} = 1.95 \times 10^{-8}$$

The corresponding spread in wavelength, or "width," of the spectrum line is approximately

$$\Delta\lambda = (1.95 \times 10^{-8})(589.0\ \mathrm{nm}) = 0.000011\ \mathrm{nm}$$

EVALUATE: The irreducible uncertainty $\Delta\lambda$ that we have calculated is called the *natural line width* of this particular spectral line. Though very small, it is within the limits of resolution of present-day spectrometers. Ordinarily, the natural line width is much smaller than the line width from other causes such as the Doppler effect and collisions among the rapidly moving atoms.

Two-Slit Interference

Now let's take a brief look at a quantum interpretation of a *two-slit* optical interference pattern. We studied these patterns in detail for light in Sections 35.2 and 35.3, and in Section 38.9 we discussed their interpretation in terms of the probability that photons strike various regions of the screen where the pattern is formed.

In view of our discussion of the wave properties of electrons, it is natural to ask what happens when we do a two-slit interference experiment with electrons. The answer is: exactly the same thing as we saw in Section 38.9 with photons! We can again use photographic film (Fig. 39.9) or electronic detectors to trace out the interference pattern, as we did with photons. The principle of complementarity, introduced in Section 38.9, again tells us that we cannot simultaneously attempt to apply the wave model and the particle model to describe a single part of this experiment. Thus we *cannot* predict exactly where in the pattern (a wave phenomenon) any individual electron (a particle) will land. We can't even ask which slit an individual electron passed through in building up the two-slit interference pattern. If we do determine the slits that the electrons pass through by scattering

39.9 (a) Formation of an interference pattern for electrons incident on two slits, (b) after 28 electrons, after 1000 electrons, and after 10,000 electrons.

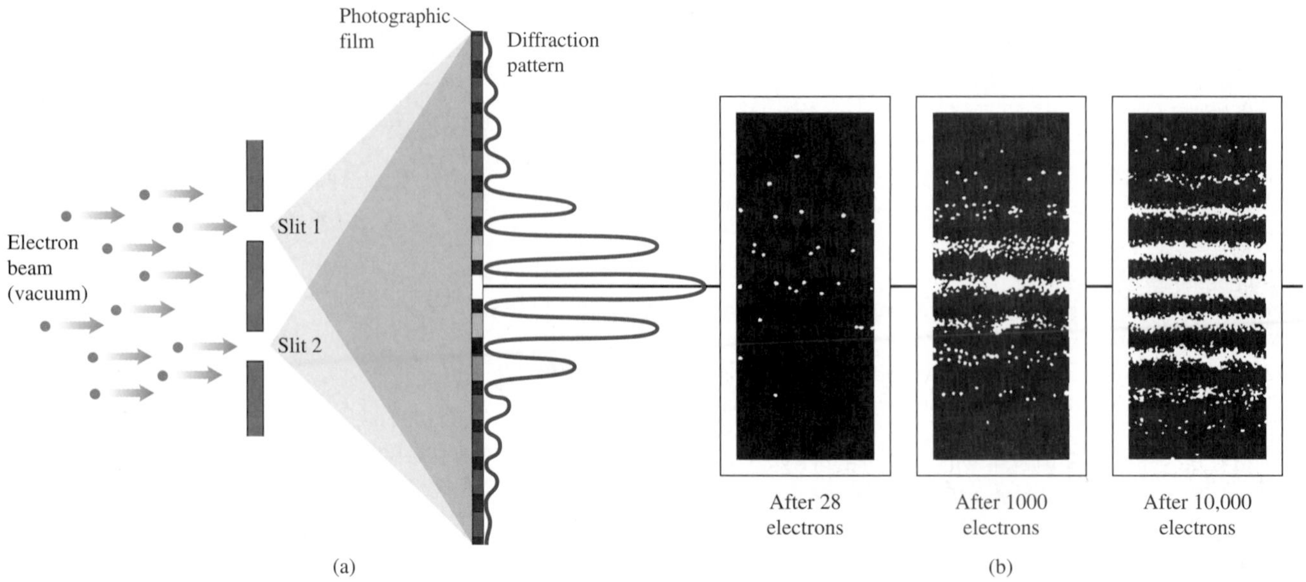

photons off them, the electrons recoil, and the two-slit interference pattern is not built up.

Test Your Understanding

Absolute zero ($T = 0$ K) is the temperature at which molecular motion is at an absolute minimum (see Sections 17.3 and 20.6). Use the Heisenberg uncertainty principle to explain why there must be *some* molecular motion even at $T = 0$ K.

39.4 | The Electron Microscope

The **electron microscope** offers an important and interesting example of the interplay of wave and particle properties of electrons. An electron beam can be used to form an image of an object in much the same way as a light beam. A ray of light can be bent by reflection or refraction, and an electron trajectory can be bent by an electric or magnetic field. Rays of light diverging from a point on an object can be brought to convergence by a converging lens or concave mirror, and electrons diverging from a small region can be brought to convergence by electric and/or magnetic fields.

The analogy between light rays and electrons goes deeper. The *ray* model of geometric optics is an approximate representation of the more general *wave* model. Geometric optics (ray optics) is valid whenever interference and diffraction effects can be neglected. Similarly, the model of an electron as a point particle following a line trajectory is an approximate description of the actual behavior of the electron; this model is useful when we can neglect effects associated with the wave nature of electrons.

How is an electron microscope superior to an optical microscope? The *resolution* of an optical microscope is limited by diffraction effects, as we discussed in Section 36.7. Using wavelengths around 500 nm, an optical microscope can't resolve objects smaller than a few hundred nanometers, no matter how carefully its lenses are made. The resolution of an electron microscope is similarly limited by the wavelengths of the electrons, but these wavelengths may be many thousands of times *smaller* than wavelengths of visible light. As a result, the useful magnification of an electron microscope can be thousands of times as great as that of an optical microscope.

Note that the ability of the electron microscope to form a magnified image *does not* depend on the wave properties of electrons. Within the limitations of the Heisenberg uncertainty principle, we can compute the electron trajectories by treating them as classical charged particles under the action of electric- and magnetic-field forces (in analogy to ray optics). Only when we talk about *resolution* do the wave properties become important.

Example 39.5 An electron microscope

The nonrelativistic electron beam in an electron microscope is formed by a setup similar to the electron gun used in the Davisson-Germer experiment (Fig. 39.3). What accelerating voltage is needed to produce electrons with wavelength 10 pm = 0.010 nm (roughly 50,000 times smaller than typical visible-light wavelengths)? The initial kinetic energy of the electrons is negligible.

SOLUTION

IDENTIFY and SET UP: The accelerating voltage (our target variable) is the quantity V_{ba} in Eq. (39.5). We are given the values of all the other quantities in that equation, so we can solve for V_{ba}.

EXECUTE: Solving Eq. (39.5) for V_{ba} and inserting the appropriate numbers, we find

$$V_{ba} = \frac{h^2}{2me\lambda^2}$$
$$= \frac{(6.626 \times 10^{-34}\ \mathrm{J \cdot s})^2}{2(9.109 \times 10^{-31}\ \mathrm{kg})(1.602 \times 10^{-19}\ \mathrm{C})(10 \times 10^{-12}\ \mathrm{m})^2}$$
$$= 1.5 \times 10^4\ \mathrm{V} = 15{,}000\ \mathrm{V}$$

EVALUATE: This is approximately equal to the accelerating voltage for the electron beam in a TV picture tube. This example shows, incidentally, that the sharpness of a TV picture is *not* limited by electron diffraction effects. Also note that this 15-kV voltage will increase the kinetic energy of the electrons from a relatively small value to 15 keV. Since electrons have a rest energy of 0.511 MeV = 511 keV, we can accurately describe these 15-keV electrons as nonrelativistic.

The Transmission Electron Microscope

Except within their electron guns, most practical electron microscopes use magnetic fields rather than electric fields as "lenses" for focusing the beam. A common setup for *transmission electron microscopes* includes three such lenses in a compound-microscope arrangement, as shown in Fig. 39.10. Electrons are emitted from a hot cathode and accelerated by a potential difference, typically 10 to 100 kV. The electrons pass through a condensing lens and are formed into a parallel beam before passing through the specimen or object to be viewed. The specimen to be viewed is very thin, typically 10 to 100 nm, so the electrons are not slowed appreciably as they pass through. The objective lens then forms an intermediate image of this object, and the projection lens produces a final real image of that image. The objective and projection lenses play the roles of the objective and eyepiece lenses, respectively, of a compound optical microscope (Section 34.8). The final image is recorded on photographic film or projected onto a fluorescent screen for viewing or photographing. The entire apparatus, including the specimen, must be enclosed in a vacuum container, just as with the cathode-ray tube; otherwise, electrons would scatter off air molecules and muddle the image. The image that opens this chapter was made with a transmission electron microscope.

We might think that when the electron wavelength is 0.01 nm (as in Example 39.5), the resolution would also be about 0.01 nm. In fact, it is seldom better than 0.5 nm, in part because the focal length of a magnetic lens depends on the electron speed, which is never exactly the same for all electrons in the beam.

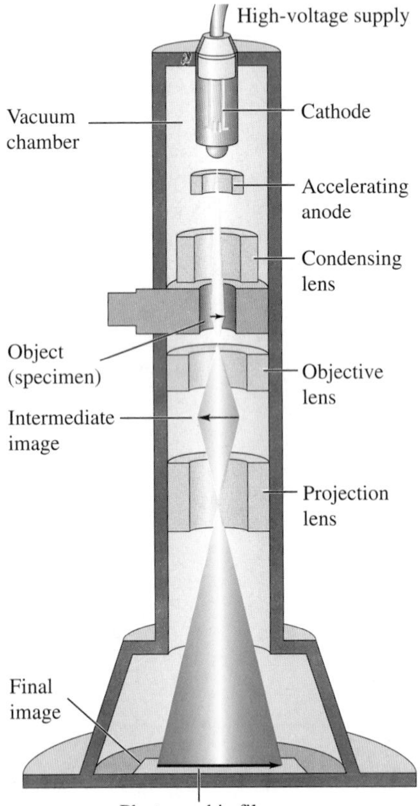

39.10 Schematic diagram of a transmission electron microscope. For an example of an image made with such a microscope, see the first page of this chapter.

The Scanning Electron Microscope

An important variation is the *scanning electron microscope.* The electron beam is focused to a very fine line and is swept across the specimen, just as the electron beam in a TV picture tube traces out the picture. As the beam scans the specimen, electrons are knocked off and are collected by a collecting anode that is kept at a potential a few hundred volts positive with respect to the specimen. The current in the collecting anode is amplified and used to modulate the electron beam in a cathode-ray tube, which is swept in synchronization with the microscope beam. Thus the cathode-ray tube traces out a greatly magnified image of the specimen. This scheme has several advantages. The specimen can be thick because the beam does not need to pass through it. Also, the knock-off electron production depends on the *angle* at which the beam strikes the surface. Thus scanning electron micrographs have an appearance that is much more three-dimensional than conventional visible-light micrographs. The resolution is typically of the order of 10 nm, still much finer than the best optical microscopes. Figure 39.11 shows an image made with a scanning electron microscope.

39.5 | Wave Functions and the Schrödinger Equation

We have now seen persuasive evidence that on an atomic or subatomic scale, a particle such as an electron cannot be described simply as a point. Instead we use a *wave function* to describe the state of a particle. Let's describe more specifically the kinematic language we must use to replace the classical scheme of describing a particle by its coordinates and velocity components.

Our new scheme for describing the state of a particle has a lot in common with the language of classical wave motion. In Chapter 15, we described transverse waves on a string by specifying the position of each point in the string at each instant of time by means of a *wave function* (Section 15.3). If y represents the displacement from equilibrium, at time t, of a point on the string at a distance x from the origin, then the function $y(x, t)$ represents the displacement of any point x at any time t. Once we know the wave function for a particular wave motion, we know everything there is to know about the motion. We can find the position and velocity of any point on the string at any time, and so on. We worked out specific forms for these functions for *sinusoidal* waves, in which each particle undergoes simple harmonic motion.

We followed a similar pattern for sound waves in Chapter 16. The wave function $p(x, t)$ for a wave traveling along the x-direction represented the pressure variation at any point x at any time t. We used this language once more in Section 32.3, where we used *two* wave functions to describe the electric and magnetic fields of *electromagnetic* waves at any point in space at any time.

Thus it is natural to use a wave function as the central element of our new language. The symbol customarily used for this wave function is Ψ or ψ. In general, Ψ is a function of all the space coordinates and time, whereas ψ is a function of the space coordinates only—*not* of time. Just as the wave function $y(x, t)$ for mechanical waves on a string provides a complete description of the motion, so the wave function $\Psi(x, y, z, t)$ for a particle contains all the information that can be known about the particle. The mathematical theory of quantum mechanics describes how to use $\Psi(x, y, z, t)$ to determine the average values of the particle's position, velocity, momentum, energy, and angular momentum. The required techniques are beyond the scope of this discussion, but they are well established and well supported by experimental results.

CAUTION Unlike for mechanical waves on a string or sound waves in air, the wave function for a particle is *not* a mechanical wave that needs some material medium in order to propagate. The wave function describes the particle, but we cannot define the function itself in terms of anything material. We can only describe how it is related to physically observable effects.

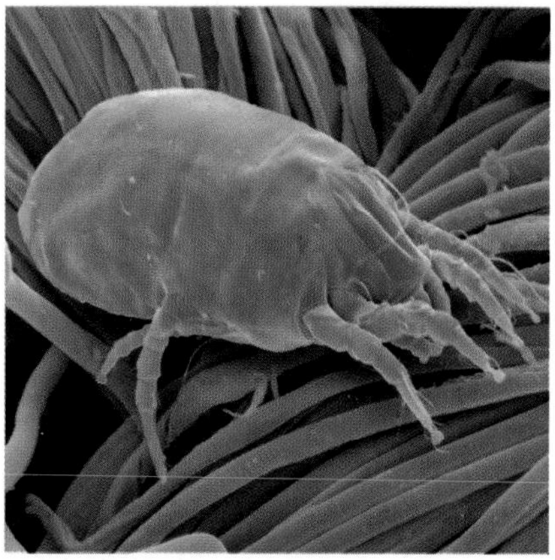

39.11 This image from a scanning electron microscope shows a dust mite (genus *Dermatophagoides*) on the threads of a piece of fabric. Colors have been added to the image. The dust mite is shown at a magnification of 1200×.

Interpretation of the Wave Function

The wave function describes the distribution of a particle in space, just as the wave functions for an electromagnetic wave describe the distribution of the electric and magnetic fields. When we worked out interference and diffraction patterns in Chapters 35 and 36, we found that the intensity I of the radiation at any point in a pattern is proportional to the square of the electric-field magnitude—that is, to E^2. In the photon interpretation of interference and diffraction (Section 38.9), the intensity at each point is proportional to the number of photons strik-

39.12 In 1926, the German physicist Max Born (1882–1970) devised the interpretation that $|\Psi|^2\,dV$ is the probability of finding a particle within a given volume dV. He also coined the term "quantum mechanics" (in the original German, *Quantenmechanik*). For his contributions, Born shared (with Walther Bothe) the 1954 Nobel Prize in physics.

ing around that point or, alternatively, to the *probability* that any individual photon will strike around the point. Thus the square of the electric-field magnitude at each point is proportional to the probability of finding a photon around that point.

In exactly the same way, the square of the wave function of a particle at each point tells us about the probability of finding the particle around that point. More precisely, we should say the square of the *absolute value* of the wave function, $|\Psi|^2$. This is necessary because, as we'll see later, Ψ may be a *complex* quantity with real and imaginary parts. (The imaginary part of the function is a real function multiplied by the imaginary number $i = \sqrt{-1}$.)

For a particle moving in three dimensions, the quantity $|\Psi(x, y, z, t)|^2\,dV$ is the probability that the particle will be found at time t within a volume dV around the point (x, y, z). The particle is most likely to be found in regions where $|\Psi|^2$ is large, and so on. This interpretation, first made by the German physicist Max Born (Fig. 39.12), requires that the wave function Ψ be *normalized*. That is, the integral of $|\Psi|^2\,dV$ over all space must equal exactly 1. In other words, the probability is exactly 1, or 100%, that the particle is *somewhere* in the universe.

CAUTION Note that $|\Psi(x, y, z, t)|^2$ itself is *not* a probability. Rather, $|\Psi(x, y, z, t)|^2\,dV$ is the probability of finding the particle within a volume dV around the point (x, y, z) at time t. If the volume is made smaller, it becomes less likely that the particle will be found within that volume, so the probability decreases. A better name for $|\Psi(x, y, z, t)|^2$ is the **probability distribution function,** since it describes how the probability of finding the particle at different locations is distributed over space.

Stationary States

In general, the value of $|\Psi(x, y, z, t)|^2$ at a particular point varies with time. This makes sense: as an electron in a television tube flies from the cathode toward the screen, the place it is most likely to be found changes with time. But if the particle is in a state of a definite energy, such as an atomic electron in an atom in a definite energy level, the value of $|\Psi|^2$ at each point is *independent* of time. Because the particle's probability distribution in such a state doesn't change with time, a state with a definite energy is called a **stationary state.** Such states are of tremendous importance in quantum mechanics. For example, for each definite-energy, stationary state in a hydrogen atom, there is a specific wave function. It is possible for an atom to be in a state that is not identical to any one of these stationary-state wave functions and that does not have a definite energy. Nevertheless, the wave function for any state can always be written as a combination of stationary-state wave functions. We saw something similar in Section 15.8: any standing wave on a string, no matter how complex, can be expressed as a combination of the normal-mode vibrations of the string. We will devote most of Chapter 40 to finding the stationary-state wave functions for different physical situations.

Why *is* $|\Psi|^2$ independent of time if the particle is in a state of definite energy? That is, why are such states stationary? To answer this question, we first note the following result from quantum mechanics: for a particle in a state of definite energy E, the time-dependent wave function $\Psi(x, y, z, t)$ can be written as a product of a time-*independent* function $\psi(x, y, z)$ and a simple function of time:

$$\Psi(x, y, z, t) = \psi(x, y, z)\, e^{-iEt/\hbar} \tag{39.14}$$

(time-dependent wave function for a stationary state)

The exponential function in Eq. (39.14) is defined by *Euler's formula*, which states that for any angle θ,

$$e^{i\theta} = \cos\theta + i\sin\theta \quad \text{and} \quad e^{-i\theta} = \cos\theta - i\sin\theta \tag{39.15}$$

Thus Eq. (39.14) shows that the wave function for any stationary state is a *complex* function.

Let's now look at the probability distribution function $|\Psi|^2$ obtained from Eq. (39.14). Note that $|\Psi|^2$ is the product of Ψ and its *complex conjugate* Ψ^*. To find the complex conjugate of a complex number, simply replace all i with $-i$. For example, the complex conjugate of $c = a + ib$, where a and b are real, is $c^* = a - ib$, so $|c|^2 = c^*c = (a + ib)(a - ib) = a^2 + b^2$ (recall that $i^2 = -1$). The complex conjugate of Eq. (39.14) is

$$\Psi^*(x, y, z, t) = \psi^*(x, y, z)\, e^{+iEt/\hbar} \tag{39.16}$$

Hence

$$\begin{aligned} |\Psi(x, y, z, t)|^2 &= \Psi^*(x, y, z, t)\Psi(x, y, z, t) \\ &= \psi^*(x, y, z)\psi(x, y, z)\, e^{+iEt/\hbar}\, e^{-iEt/\hbar} \\ &= \psi^*(x, y, z)\psi(x, y, z)e^0 = |\psi(x, y, z)|^2 \end{aligned} \tag{39.17}$$

Since $|\psi(x, y, z)|^2$ does not depend on time, Eq. (39.17) shows that the same must be true for the probability distribution function $|\Psi(x, y, z, t)|^2$. This justifies the term *stationary state* for a state of definite energy.

The Schrödinger Equation

We have stressed the importance of stationary states in describing quantum-mechanical systems. As Eq. (39.14) shows, to describe a stationary state we must know its spatial wave function $\psi(x, y, z)$ and its energy E. To determine these values we use a tool developed in 1926 by the Austrian physicist Erwin Schrödinger (Fig. 39.13) and known as the *Schrödinger equation*. The Schrödinger equation plays the same central role in quantum mechanics as do Newton's laws in mechanics and Maxwell's equations in electromagnetism. Our understanding of every quantum-mechanical system, including atoms, molecules, atomic nuclei, and electrons in solids, is based on the solutions to this equation for that system.

We can't *derive* the Schrödinger equation from other principles; it is a new principle of its own. But we can show how it is related to the de Broglie relations, and we can make it seem plausible.

The simplest form of the Schrödinger equation is for a particle of mass m that moves in one dimension only, parallel to the x-axis, so that the spatial wave function ψ is a function only of x. We assume that the particle moves in the presence of a conservative force that has only an x-component, so there is a corresponding

39.13 Erwin Schrödinger (1887–1961) developed the equation that bears his name in 1926, an accomplishment for which he shared (with the British physicist P. A. M. Dirac) the 1933 Nobel Prize in physics. In later life he did research into the philosophy and history of science.

potential energy $U(x)$. The **Schrödinger equation** for such a particle with a definite energy E is

$$-\frac{\hbar^2}{2m}\frac{d^2\psi(x)}{dx^2} + U(x)\psi(x) = E\psi(x) \tag{39.18}$$

(one-dimensional Schrödinger equation)

In this equation E is a constant. How do we know that this equation is correct? Because it works. Predictions made using this equation agree with experimental results. In Chapter 40 we'll apply Eq. (39.18) to several physical situations, each with a different form of the function $U(x)$.

As an example, let's consider a *free* particle that experiences no force at all. If there is no force, $U(x)$ is independent of x; for simplicity we choose $U(x) = 0$. If such a free particle is moving in the $+x$-direction with momentum of magnitude p, its kinetic energy (and hence its total energy) is $E = p^2/2m$. Such a particle is in a state of definite energy (a stationary state).

From the de Broglie equations, the particle has a definite wavelength $\lambda = h/p$ and a definite frequency $f = E/h$. By analogy to the traveling mechanical waves we studied in Section 15.3, we write the wave function for the particle as

$$\Psi(x, t) = A\cos(kx - \omega t) + B\sin(kx - \omega t) \tag{39.19}$$

where A and B are constants. As we did for mechanical waves in Chapter 15 and electromagnetic waves in Chapter 32, we use the wave number $k = 2\pi/\lambda$ and the angular frequency $\omega = 2\pi f$ in Eq. (39.19). These are related to the momentum and energy by

$$k = \frac{2\pi}{\lambda} = \frac{2\pi}{h}\frac{h}{\lambda} = \frac{p}{\hbar} \qquad \omega = 2\pi f = \frac{2\pi}{h}hf = \frac{E}{\hbar} \tag{39.20}$$

The wave function in Eq. (39.19) doesn't seem to be of the stationary-state form given in Eq. (39.14). We can cast it into this form, however, if we let $B = iA$. Then

$$\begin{aligned}\Psi(x, t) &= A\cos(kx - \omega t) + iA\sin(kx - \omega t)\\ &= A[\cos(kx - \omega t) + i\sin(kx - \omega t)]\\ &= Ae^{i(kx-\omega t)} = Ae^{ikx}e^{-i\omega t}\end{aligned} \tag{39.21}$$

In Eq. (39.21) we have again used Euler's formula, which states that for any angle θ,

$$e^{i\theta} = \cos\theta + i\sin\theta \quad \text{and} \quad e^{-i\theta} = \cos\theta - i\sin\theta \tag{39.22}$$

Comparing Eq. (39.21) and Eq. (39.14), we see that our wave function does indeed represent a stationary state with energy $E = \hbar\omega$ and spatial wave function $\psi(x) = Ae^{ikx}$. If we substitute $\psi(x)$ into the left-hand side of the Schrödinger equation, Eq. (39.18), we obtain

$$\begin{aligned}-\frac{\hbar^2}{2m}\frac{d^2(Ae^{ikx})}{dx^2} + (0)(Ae^{ikx}) &= -\frac{\hbar^2}{2m}(ik)^2Ae^{ikx} = \frac{\hbar^2k^2}{2m}Ae^{ikx}\\ &= \frac{p^2}{2m}\psi(x)\end{aligned} \tag{39.23}$$

(We used the result that the second derivative of e^{ax}, where a is a constant, is a^2e^{ax}.) The right-hand side of Eq. (39.18) is $E\psi(x)$. Since $E = p^2/2m$ from Eq. (39.18), we see from Eq. (39.23) that the free-particle wave function in Eq. (39.21) satisfies the Schrödinger equation.

If k is positive in Eq. (39.21), the wave function represents a free particle moving in the positive x-direction. If k is negative, the momentum and hence the motion is in the negative x-direction. (With a negative value of k, the wavelength is $\lambda = 2\pi/|k|$.)

For a free particle, there is no restriction on the value of p and hence no restriction on the value of the energy $E = p^2/2m$. We will find in Chapter 40 that if $U(x)$ is not a constant, then solutions of the Schrödinger equation are possible only for certain values of E. These values represent the allowed *energy levels* of the system described by $U(x)$. This discovery is of the utmost importance. Before the development of the Schrödinger equation, there was no way to predict energy levels from any fundamental theory except the Bohr model, which had very limited success.

There is also a version of the Schrödinger equation that includes time dependence. It is needed to study states that are *not* stationary and for which the probability distribution function $|\Psi(x, y, z, t)|^2$ depends on time. However, we don't need that version to calculate the energy levels and wave functions of stationary states. If we do need a time-dependent wave function for a stationary state of energy E, we simply use Eq. (39.14). The time-dependent Schrödinger equation is essential for studying the details of *transitions* between states. In more advanced courses you may study the quantum mechanics of time-dependent phenomena such as photon emission and absorption and the lifetimes of states.

Example 39.6 The Schrödinger equation

Consider the wave function $\psi(x) = A_1e^{ikx} + A_2e^{-ikx}$, where k is positive. Is this a valid stationary-state wave function for a free particle? What is the energy?

SOLUTION

IDENTIFY and SET UP: A valid stationary-state wave function for a free particle must satisfy the Schrödinger equation, Eq. (39.18), with $U(x) = 0$. To test the given function $\psi(x)$, we simply substitute it into the left-hand side of the equation. If the result is a constant times $\psi(x)$, then the wave function is indeed a solution and the constant is equal to the energy E.

EXECUTE: Substituting $\psi(x) = A_1e^{ikx} + A_2e^{-ikx}$ and $U(x) = 0$ into Eq. (39.18), we obtain

$$\begin{aligned} -\frac{\hbar^2}{2m}\frac{d^2\psi(x)}{dx^2} &= -\frac{\hbar^2}{2m}\frac{d^2(A_1e^{ikx} + A_2e^{-ikx})}{dx^2} \\ &= -\frac{\hbar^2}{2m}[(ik)^2A_1e^{ikx} + (-ik)^2A_2e^{ikx}] \\ &= \frac{\hbar^2k^2}{2m}(A_1e^{ikx} + A_2e^{-ikx}) = \frac{\hbar^2k^2}{2m}\psi(x) \end{aligned}$$

We see that the left-hand side of the Schrödinger equation is indeed equal to a constant times $\psi(x)$, so this $\psi(x)$ is indeed a valid stationary-state wave function for a free particle. The constant is just equal to the energy: $E = \hbar^2k^2/2m$.

EVALUATE: Note that $\psi(x)$ is a *superposition* of two different wave functions: one for a particle with a given magnitude of momentum $p = \hbar k$ moving in the positive x-direction and one for a particle with the same magnitude of momentum moving in the negative x-direction. While the combined wave function represents a stationary state with a definite energy, this state does *not* have a definite momentum. We will see in Chapter 40 that such a wave function can represent a *standing wave*, and we will explore the situations in which such standing matter waves can arise.

Wave Packets

The wave function for a free particle given by Eq. (39.21) has a definite momentum p in the x-direction. For such a state, there is *no* uncertainty in momentum: $\Delta p_x = 0$. The Heisenberg uncertainty principle, Eq. (39.11), says that $\Delta x \, \Delta p_x \geq \hbar$. If Δp_x is zero, then Δx must be infinite. The price we pay for knowing the particle's momentum precisely is that we have no idea *where* the particle is! We can show this by calculating the probability distribution function $|\Psi(x,t)|^2$:

$$
\begin{aligned}
|\Psi(x,t)|^2 &= \Psi^*(x,t)\,\Psi(x,t) = (A^* e^{-ikx} e^{+i\omega t})(A e^{ikx} e^{-i\omega t}) \\
&= A^* A e^0 = |A|^2
\end{aligned}
$$

The probability distribution function doesn't depend on time, as we would expect because this is a stationary state (a state of definite energy). It also doesn't depend on position, which says that we are equally likely to find the particle *anywhere* in space! Mathematically, this situation occurs because the spatial wave function $\psi(x) = Ae^{ikx} = A\cos kx + iA\sin kx$ is a sinusoidal function that extends all the way from $x = -\infty$ to $x = +\infty$ with the same amplitude A. (This also means that the wave function can't be normalized: the integral of $|\Psi(x,t)|^2$ over all space would be infinite for any value of A.)

In practical situations, we always have *some* idea where a particle is. To describe this situation, we need a wave function that is more localized in space. We can create one by superimposing two or more sinusoidal functions, as we did in Example 39.6. In that example we added two wave functions with opposite wave numbers k and $-k$; to make a localized wave function, we must make a different choice.

To keep things simple, we'll imagine wave functions that depend on only one spatial coordinate (x), and we'll consider them at one instant of time (say, $t = 0$). Our wave functions are then functions only of x, so we'll denote them as ψ. For example, let's add together two different stationary-state wave functions for a free particle as given by Eq. (39.21), but with slightly different values k_1 and k_2 of the wave number. At $t = 0$ the time factors $e^{-i\omega_1 t}$ and $e^{-i\omega_2 t}$ are both equal to $e^0 = 1$, so the wave function at $t = 0$ is

$$
\begin{aligned}
\psi(x) &= A_1 e^{ik_1 x} + A_2 e^{ik_2 x} \\
&= [A_1\cos(k_1 x) + iA_1\sin(k_1 x)] + [A_2\cos(k_2 x) + iA_2\sin(k_2 x)] \\
&= [A_1\cos(k_1 x) + A_2\cos(k_2 x)] + i[A_1\sin(k_1 x) + A_2\sin(k_2 x)]
\end{aligned}
\tag{39.24}
$$

We have used Euler's formula, Eq. (39.22), to separate the real and imaginary parts of $\psi(x)$.

Figure 39.14a is a graph of the real parts of the individual wave functions for the case $A_2 = -A_1$; Fig. 39.14b graphs the real part of the combined wave function $\psi(x)$ given by Eq. (39.24). We saw something very similar to Fig. 39.14b in our discussion of beats in Section 16.7: when we superimposed two sinusoidal waves with slightly different frequencies (Fig. 16.25), the result was a wave with a lumpy character that the individual waves did not possess. In the same way, a particle represented by the lumpy wave function given in Eq. (39.24) is more likely to be found in some regions than in others; that is, it is *localized*. However,

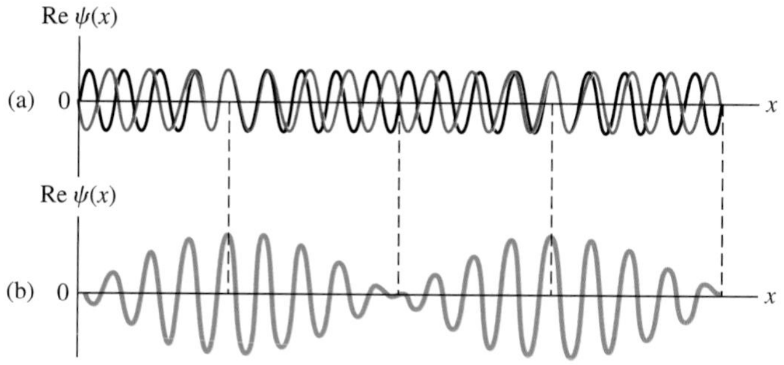

39.14 (a) Two sinusoidal waves with slightly different wave numbers, shown at one instant of time. (b) The superposition of these waves has a wave number equal to the average of the two individual wave numbers. The amplitude varies, giving the total wave a lumpy character not possessed by either individual wave.

the particle's momentum no longer has a definite value because we began with two different wave numbers. This agrees with the Heisenberg uncertainty principle: by decreasing the uncertainty in the particle's position, we have increased the uncertainty in its momentum.

It's not hard to imagine superposing two additional sinusoidal waves with different wave numbers and amplitudes so as to reinforce alternate lumps in Fig. 39.14b and cancel out the in-between ones. Finally, if we superpose waves with a very large number of different wave numbers, we can construct a wave with only one lump (Fig. 39.15). Then, finally, we have something that begins to look like both a particle and a wave. It is a particle in the sense that it is localized in space; if we look from a distance, it may look like a point. But it also has a periodic structure that is characteristic of a wave.

Such a wave pulse is called a **wave packet.** We can represent such a superposition by an expression such as

$$\psi(x) = \int_{-\infty}^{\infty} A(k) e^{ikx} dk \tag{39.25}$$

This integral represents a superposition of a very large number of waves, each with a different wave number and each with an amplitude $A(k)$ that depends on k.

There is an important relation between the two functions $\psi(x)$ and $A(k)$. This relation is shown qualitatively in Fig. 39.16. If the function $A(k)$ is sharply peaked, as in Fig. 39.16a, we are superposing only a narrow range of wave numbers. The resulting wave pulse is then relatively broad (Fig. 39.16b). But if we use a wider range of wave numbers, so that the function $A(k)$ is broader (Fig. 39.16c), then the wave pulse is more narrowly localized (Fig. 39.16d).

What we are seeing is the uncertainty principle in action. A narrow range of k means a narrow range of $p_x = \hbar k$ and thus a small Δp_x; the result is a relatively large Δx. A broad range of k corresponds to a large Δp_x, and the resulting Δx is

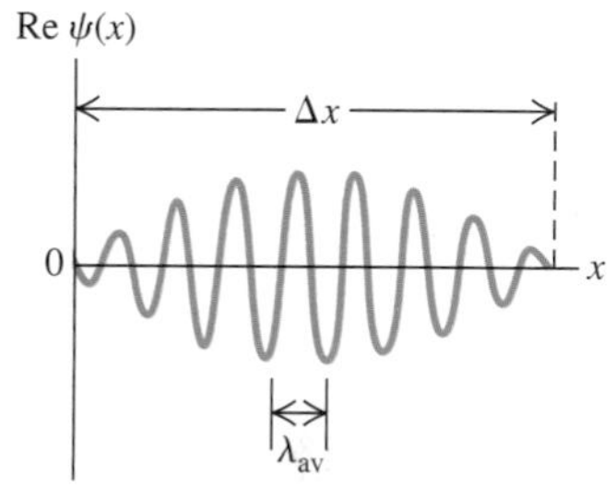

39.15 Superimposing a large number of sinusoidal waves with different wave numbers and appropriate amplitudes can produce a wave pulse with a wavelength $\lambda_{av} = 2\pi/k_{av}$ that is localized in a region of space of length Δx. Is the pulse a particle or a wave?

ActivPhysics ONLINE

17.7 Wave Packets

39.16 (a), (b) A narrow function $A(k)$ gives a wave function $\psi(x)$ with a broad spatial extent. (c), (d) A broad function $A(k)$ gives a wave function $\psi(x)$ with a narrow spatial extent.

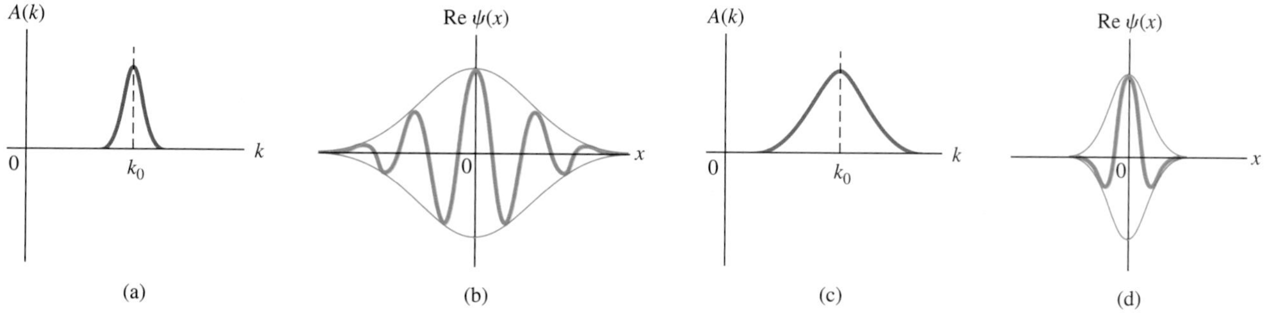

smaller. Thus we see that the uncertainty principle $\Delta x \, \Delta p_x \geq \hbar$ is an inevitable consequence of the de Broglie relation and the properties of integrals such as Eq. (39.25). These integrals are called *Fourier integrals;* they are a generalization of the concept of Fourier series, which we mentioned in Section 15.8. In both cases we are representing a complex wave form as a superposition of sinusoidal functions. With Fourier series we use a sequence of frequencies (or values of k) that are integer multiples of some basic value, while with Fourier integrals we superpose functions with a *continuous* distribution of values of k.

Test Your Understanding

Does a wave packet given by Eq. (39.25) represent a stationary state?

SUMMARY

Electrons and other particles have wave properties. The wavelength of the wave depends on the particle's momentum in the same way as for photons. The state of a particle is described not by its coordinates and velocity components but rather by a wave function, which in general is a function of the three space coordinates and time. (See Example 39.1)

$$\lambda = \frac{h}{p} = \frac{h}{mv} \quad (39.1)$$

Diffraction of an electron beam from the surface of a metallic crystal provided the first direct confirmation of the wave nature of particles. If a nonrelativistic electron has been accelerated from rest through a potential difference V_{ba}, its wavelength is given by Eq. (39.5). Electron microscopes use the very small wavelengths of fast-moving electrons to make images with resolution thousands of times finer than is possible with visible-light microscopes. (See Examples 39.2 and 39.5)

$$\lambda = \frac{h}{p} = \frac{h}{\sqrt{2meV_{ba}}} \quad (39.5)$$

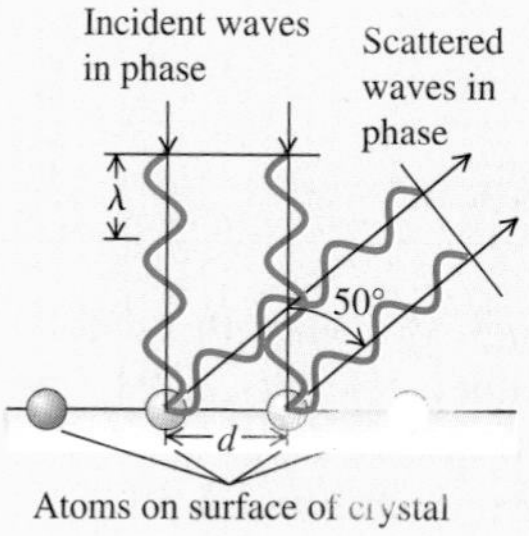

It is impossible to make precise determinations of a coordinate of a particle and of the corresponding momentum component at the same time. The precision of such measurements for the x-components is limited by the Heisenberg uncertainty principle, Eq. (39.11); there are corresponding relations for the y- and z-components. The uncertainty ΔE in the energy of a state that is occupied for a time Δt is given by Eq. (39.13). In these expressions, $\hbar = h/2\pi$. (See Examples 39.3 and 39.4)

$$\Delta x\, \Delta p_x \geq \hbar \quad (39.11)$$

(Heisenberg uncertainty principle for position and momentum)

$$\Delta E\, \Delta t \geq \hbar \quad (39.13)$$

(Heisenberg uncertainty principle for energy and time)

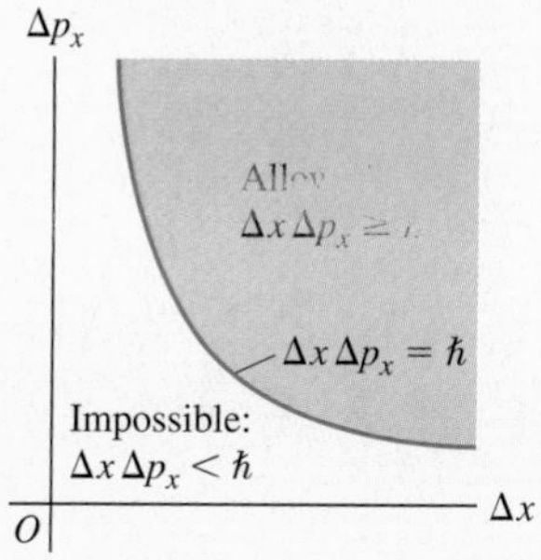

The wave function $\Psi(x, y, z, t)$ for a particle contains all of the information about that particle. The quantity $|\Psi(x, y, z, t)|^2$, called the probability distribution function, determines the relative probability of finding a particle near a given position at a given time. If the particle is in a state of definite energy, called a stationary state, $\Psi(x, y, z, t)$ is a product of a function ψ that depends only on spatial coordinates and a function $e^{-iEt/\hbar}$ that depends only on time. For a stationary state, the probability distribution function is independent of time.

$$\Psi(x, y, z, t) = \psi(x, y, z)\, e^{-iEt/\hbar} \quad (39.14)$$

(stationary state)

For a particle that moves in one dimension in the presence of a potential energy function $U(x)$, the wave function for a stationary state of energy E satisfies the Schrödinger equation. More complex wave functions can be made by superposing stationary-state wave functions. These can represent particles that are localized in a certain region and still have wave properties, giving it both particle and wave aspects. (See Example 39.6)

$$-\frac{\hbar^2}{2m}\frac{d^2\psi(x)}{dx^2} + U(x)\psi(x) = E\psi(x) \quad (39.18)$$

(one-dimensional Schrödinger equation)

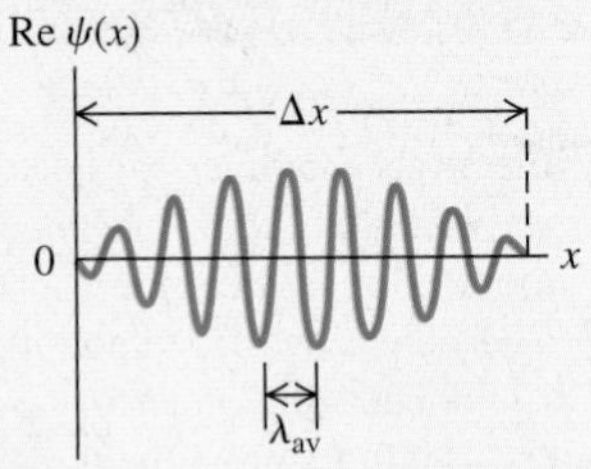

Key Terms

de Broglie wavelength, 1490
quantum mechanics, 1490
wave function, 1491
boundary condition, 1491
electron diffraction, 1494
Heisenberg uncertainty principle, 1498
electron microscope, 1501
probability distribution function, 1504
stationary state, 1504
Schrödinger equation, 1506
wave packet, 1509

Your Notes

Answer to Chapter Opening Question ?

From Eq. (39.4), the wavelength of electrons accelerated through a potential difference V_{ba} is $\lambda = h/\sqrt{2meV_{ba}}$. In a television tube $V_{ba} = 20\text{ kV} = 2.0 \times 10^4\text{ V}$, so we find $\lambda = 8.7 \times 10^{-12}\text{ m} = 0.0087\text{ nm}$. This is far too small to see, and far smaller than the width of the electron beam.

Answers to Test Your Understanding Questions

Section 39.1 From Example 39.1, the speed of a particle is $v = h/\lambda m$ and the kinetic energy is $K = \frac{1}{2}mv^2 = (m/2)(h/\lambda m)^2 = h^2/2\lambda^2 m$. This shows that for a given wavelength, the kinetic energy is inversely proportional to the mass. Hence the proton, with a smaller mass, has more kinetic energy than the neutron.

Section 39.2 The energy of a photon is $E = hf$, and the frequency of a photon is $f = c/\lambda$. Hence $E = hc/\lambda$ and $\lambda = hc/E = (4.136 \times 10^{-15}\text{ eV}\cdot\text{s}) = (2.998 \times 10^8\text{ m/s})/(54\text{ eV}) = 2.3 \times 10^{-8}\text{ m}$. This is more than 100 times greater than the wavelength of an electron of the same energy. While both photons and electrons have wavelike properties, they have different relationships between their energy and momentum and hence between their frequency and wavelength.

Section 39.3 In Newtonian physics, a state of no molecular motion would mean that every molecule was at rest (that is, had exactly zero momentum) at a fixed position. If this were the case, there would be zero uncertainty in each molecule's momentum as well as zero uncertainty in its position. But the Heisenberg uncertainty principle, Eq. (39.14), tells us that the product of these two uncertainties has a minimum value $\hbar$ and hence can *never* be zero. Thus there must be some molecular motion even at absolute zero.

Section 39.5 Equation (39.25) represents a superposition of wave functions with different values of wave number k and hence different values of energy $E = \hbar^2 k^2/2m$. The state that this combined wave function represents is not a state of definite energy, and therefore not a stationary state. Another way to see this is to note that the time-dependent wave function $\Psi(x, t)$ would include a factor $e^{-iEt/\hbar}$ inside the integral in Eq. (39.25), with a different value of E for each value of k. This wave function therefore has a very complicated time dependence, and the probability distribution function $|\Psi(x, t)|^2$ does depend on time.

Discussion Questions

Q39.1 If a proton and an electron have the same speed, which has the longer de Broglie wavelength? Explain.

Q39.2 If a proton and an electron have the same kinetic energy, which has the longer de Broglie wavelength? Explain.

Q39.3 Does a photon have a de Broglie wavelength? If so, how is it related to the wavelength of the associated electromagnetic wave? Explain.

Q39.4 A stone is dropped from the top of a building. You can ignore the effects of air resistance. As the stone falls, does its de Broglie wavelength increase, decrease or stay the same? Explain.

Q39.5 You have been asked to design a magnet system to steer a beam of 54-V electrons like those described in Example 39.2 (Section 39.2). The goal is to be able to direct the electron beam to a specific target location with an accuracy of ± 1.0 mm. In your design, do you need to take the wave nature of electrons into account? Explain.

Q39.6 The upper half of the electron diffraction pattern shown in Fig. 39.6 is the mirror image of the lower half of the pattern. Would it be correct to say that the upper half of the pattern is caused by electrons that pass through the upper half of the slit? Explain.

Q39.7 A particular electron in the experimental setup shown in Fig. 39.6 lands on the photographic film a distance x above the center of the pattern. Given the value of x, is it possible to calculate the precise trajectory that the electron followed? Explain.

Q39.8 Does the uncertainty principle have anything to do with marksmanship? That is, is the accuracy with which a bullet can be aimed at a target limited by the uncertainty principle? Explain.

Q39.9 What happens to the uncertainty in the y-component of the momentum of a particle as Δy approaches zero? As Δz approaches zero? Explain.

Q39.10 Suppose a two-slit interference experiment is carried out using an electron beam. Would the same interference pattern result if one slit at a time is uncovered instead of both at once? If not, why not? Does not each electron go through one slit or the other? Or does every electron go through both slits? Discuss the latter possibility in light of the principle of complementarity.

Q39.11 Equation (39.13) states that the energy of a system can have uncertainty. Does this mean that the principle of conservation of energy is no longer valid? Explain.

Q39.12 Laser light results from transitions from long-lived metastable states. Why is it more monochromatic than ordinary light?

Q39.13 Could an electron-diffraction experiment be carried out using three or four slits? Using a grating with many slits? What sort of results would you expect with a grating? Would the uncertainty principle be violated? Explain.

Q39.14 As the lower half of Fig. 39.5 shows, the diffraction pattern made by electrons that pass through an aluminum foil is a series of concentric rings. But if the aluminum foil is replaced by a single crystal of aluminum, only certain points on these rings appear in the pattern. Explain.

Q39.15 Why can an electron microscope have greater magnification than an ordinary microscope?

Q39.16 If quantum mechanics replaces the language of Newtonian mechanics, why do we not have to use wave functions to describe the motion of macroscopic bodies such as baseballs and cars?

Q39.17 A student remarks that the relation of ray optics to the more general wave picture is analogous to the relation of Newtonian mechanics, with well-defined particle trajectories, to quantum mechanics. Comment on this remark.

Q39.18 When you check the air pressure in a tire, a little air always escapes; the process of making the measurement changes the quantity being measured. Think of other examples of measurements that change or disturb the quantity being measured.

Q39.19 If a particle is in a stationary state, does that mean that the particle is not moving? If a particle moves in empty space with constant momentum $\vec{p}$ and hence constant energy $E = p^2/2m$, is it in a stationary state? Explain your answers.

Q39.20 Some lasers emit light in pulses that are only 10^{-12} s in duration. The length of such a pulse is $(3 \times 10^8 \text{ m/s})(10^{-12} \text{ s}) = 3 \times 10^{-4} \text{ m} = 0.3$ mm. Can pulsed laser light be as monochromatic as light from a laser that emits a steady, continuous beam? Explain.

Exercises

Section 39.1 De Broglie Waves

39.1 An electron has a de Broglie wavelength of 2.80×10^{-10} m. Determine a) the magnitude of its momentum; b) its kinetic energy (in joules and in electron volts).

39.2 Wavelength of an Alpha Particle. An alpha particle $(m = 6.64 \times 10^{-27} \text{ kg})$ emitted in the radioactive decay of uranium-238 has an energy of 4.20 MeV. What is its de Broglie wavelength?

39.3 a) An electron moves with a speed of 4.70×10^6 m/s. What is its de Broglie wavelength? b) A proton moves with the same speed. Determine its de Broglie wavelength.

39.4 For crystal diffraction experiments (discussed in Section 39.2), wavelengths on the order of 0.20 nm are often appropriate. Find the energy in electron volts for a particle with this wavelength if the particle is a) a photon; b) an electron; c) an alpha particle $(m = 6.64 \times 10^{-27} \text{ kg})$.

39.5 In the Bohr model of the hydrogen atom, what is the de Broglie wavelength for the electron when it is in a) the $n = 1$ level? b) the $n = 4$ level? In each case, compare the de Broglie wavelength to the circumference $2\pi r_n$ of the orbit.

39.6 a) A nonrelativistic free particle with mass m has kinetic energy K. Derive an expression for the de Broglie wavelength of the particle in terms of m and K. b) What is the de Broglie wavelength of an 800-eV electron?

39.7 Wavelength of a Bullet. Calculate the de Broglie wavelength of a 5.00-g bullet that is moving at 340 m/s. Will the bullet exhibit wavelike properties?

39.8 What is the de Broglie wavelength for an electron with speed a) $v = 0.480c$; b) $v = 0.960c$? (*Hint:* Use the correct relativistic expression for linear momentum if necessary.)

39.9 a) If a photon and an electron each have the same energy of 20.0 eV, find the wavelength of each. b) If a photon and an electron each have the same wavelength of 250 nm, find the energy of each. c) You want to study an organic molecule that is about 250 nm long using either a photon or an electron microscope. Approximately what wavelength should you use, and which probe, the electron or the photon, is likely to damage the molecule the least?

39.10 If a proton and an electron have the same de Broglie wavelength, will they have the same kinetic energy? If so, show why. If not, find the ratio of the proton's kinetic energy to that of the electron.

Section 39.2 Electron Diffraction

39.11 a) Approximately how fast should an electron move so it has a wavelength that makes it useful to measure the distance between adjacent atoms in typical crystals (about 0.10 nm)? b) What is the kinetic energy of the electron in part (a)? c) What would be the energy of a photon of the same wavelength as the electron in part (b)? d) Which would make a more effective probe of small-scale structures, electrons or photons? Why?

39.12 Through what potential difference must electrons be accelerated so they will have a) the same wavelength as an x ray of wavelength 0.150 nm? b) the same energy as the x ray in part (a)?

39.13 A beam of neutrons that all have the same energy scatters from the atoms that have a spacing of 0.0910 nm in the surface plane of a crystal. The $m = 1$ intensity maximum occurs when the angle θ in Fig. 39.3 is 28.6°. What is the kinetic energy in eV of each neutron in the beam?

39.14 A beam of 188-eV electrons is directed at normal incidence onto a crystal surface as shown in Fig. 39.4b. The $m = 2$ intensity maximum occurs at an angle $\theta = 60.6°$. a) What is the spacing between adjacent atoms on the surface? b) At what other angle or angles is there an intensity maximum? c) For what electron energy (in eV) would the $m = 1$ intensity maximum occur at $\theta = 60.6°$? For this energy, is there an $m = 2$ intensity maximum? Explain.

39.15 A beam of 840-eV alpha particles $(m = 6.64 \times 10^{-27} \text{ kg})$ scatters from the atoms from the atoms that have spacing of 0.0834 nm in the surface plane of a crystal. At what angle θ in Fig. 39.3 does the $m = 1$ intensity maximum occur?

39.16 A CD-ROM is used instead of a crystal in an electron-diffraction experiment like that shown in Fig. 39.3. The surface of the CD-ROM has tracks of tiny pits with a uniform spacing of 1.60 μm. a) If the speed of the electrons is 1.26×10^4 m/s, at which values of θ will the $m = 1$ and $m = 2$ intensity maxima appear? b) The scattered electrons in these maxima strike at normal incidence a piece of photographic film that is 50.0 cm from the CD-ROM. What is the spacing on the film between these maxima?

Section 39.3 Probability and Uncertainty

39.17 By extremely careful measurement, you determine the x-coordinate of a car's center of mass with an uncertainty of only 1.00 μm. The car has a mass of 1200 kg. a) What is the minimum uncertainty in the x-component of the velocity of the car's center of mass as prescribed by the Heisenberg uncertainty principle? b) Does the uncertainty principle impose a practical limit on our ability to make simultaneous measurements of the positions and velocities of ordinary objects like cars, books, and people? Explain.

39.18 a) The uncertainty in the y-component of a proton's position is 2.0×10^{-12} m. What is the minimum uncertainty in a simultaneous measurement of the y-component of the proton's velocity? b) The uncertainty in the z-component of an electron's velocity is 0.250 m/s. What is the minimum uncertainty in a simultaneous measurement of the z-coordinate of the electron?

39.19 A scientist has devised a new method of isolating individual particles. He claims that this method enables him to detect simultaneously the position of a particle along an axis with a standard deviation of 0.12 nm and its momentum component along this axis with a standard deviation of 3.0×10^{-25} kg·m/s. Use the Heisenberg uncertainty principle to evaluate the validity of this claim.

39.20 a) The x-coordinate of an electron is measured with an uncertainty of 0.20 mm. What is the x-component of the electron's velocity, v_x, if the minimum percentage uncertainty in a simultaneous measurement of v_x is 1.0%? b) Repeat part (a) for a proton.

39.21 The spacing between adjacent atoms on the surface of a nickel crystal (see Fig. 39.4b) is 0.215 nm. If the line of atoms shown in Fig. 39.4b lays along the x-direction, we can regard the uncertainty in the x-coordinate of each atom as approximately half the spacing. The mass of a single nickel atom is 9.75×10^{-26} kg. a) Estimate the minimum uncertainty in the x-component of momentum of a nickel atom in the crystal. b) If the magnitude of the momentum of a nickel atom is equal to the uncertainty found in part (a), what is its kinetic energy? Express the result in joules and in electron volts. c) If every atom in a 1.00-kg nickel crystal had the kinetic energy found in part (b), what would be the combined kinetic energy (in joules) of all the atoms? (The motion of each atom is in a random direction, so the center of mass of the crystal has zero momentum on average.) d) If all of this kinetic energy could be converted to gravitational potential energy, how high would the nickel crystal rise? e) Use the Heisenberg uncertainty principle to explain why the energy conversion described in part (d) is impossible.

39.22 Particle Lifetime. The unstable W^+ particle has a rest energy of 80.41 GeV ($1 \text{ GeV} = 10^9 \text{ eV}$) and an uncertainty in rest energy of 2.06 GeV. Estimate the lifetime of the W^+ particle.

39.23 The ψ ("psi") particle has a rest energy of 3097 MeV ($1 \text{ MeV} = 10^6 \text{ eV}$). The ψ is unstable with a lifetime of 7.6×10^{-21} s. Estimate the uncertainty in rest energy of the ψ particle. Express your answer in MeV and as a fraction of the rest energy of the particle.

39.24 An atom in a metastable state has a lifetime of 5.2 ms. What is the uncertainty in energy of the metastable state?

39.25 An unstable particle produced in a high-energy collision has a mass that is 4.50 times that of the proton and an uncertainty in mass that is 14.5% of the particle's mass. Using the relationship $E = mc^2$ between rest mass and energy, estimate the lifetime of the particle.

Section 39.4 The Electron Microscope

39.26 a) In an electron microscope, what accelerating voltage is needed to produce electrons with wavelength 0.0600 nm? b) If protons are used instead of electrons, what accelerating voltage is needed to produce protons with wavelength 0.0600 nm? (*Hint:* In each case the initial kinetic energy is negligible.)

39.27 a) What is the de Broglie wavelength of an electron that has been accelerated from rest through a potential increase of 800 V? b) What is the de Broglie wavelength of a proton accelerated from rest through a potential decrease of 800 V?

Section 39.5 Wave Functions and the Schrödinger Equation

39.28 Compute $|\Psi|^2$ for $\Psi = \psi \sin \omega t$, where ψ is time independent and ω is a real constant. Is this a wave function for a stationary state? Why or why not?

39.29 Consider a wave function given by $\psi(x) = A \sin kx$, where $k = 2\pi/\lambda$ and A is a real constant. a) For what values of x is there the highest probability of finding the particle described by this wave function? Explain. b) For which values of x is the probability *zero?* Explain.

39.30 A particle is described by a wave function $\psi(x) = Ae^{-\alpha x^2}$, where A and α are real, positive constants. If the value of α is increased, what effect does this have on a) the particle's uncertainty in position? b) the particle's uncertainty in momentum? Explain your answers.

39.31 Consider the complex-valued function $f(x, y) = (x - iy)/(x + iy)$. Calculate $|f|^2$.

39.32 Particle A is described by the wave function $\psi(x, y, z)$. Particle B is described by the wave function $\psi(x, y, z)e^{i\phi}$, where ϕ is a real constant. How does the probability of finding particle A within a volume dV around a certain point in space compare with the probability of finding particle B within this same volume?

39.33 At points within the region $0 \le x \le L$, $0 \le y \le L$, $0 \le z \le L$, a particle is described by the normalized wave function $\psi(x, y, z) = (2/L)^{3/2} \sin(\pi x/L) \sin(\pi y/L) \sin(\pi z/L)$. Outside this region, $\psi(x, y, z) = 0$. Calculate the probability that the particle will be found within a cube of side $0.01L$ a) centered on the point $(L/4, L/4, L/4)$; b) centered on the point $(L/2, L/2, L/2)$. [*Hint:* Assume that $\psi(x, y, z)$ is uniform over the volume of such a small cube.]

39.34 Linear Combinations of Wave Functions. Let ψ_1 and ψ_2 be two solutions of Eq. (39.18) with the same energy E. Show that $\psi = B\psi_1 + C\psi_2$ is also a solution with energy E, for any values of the constants B and C.

39.35 Let ψ_1 and ψ_2 be two solutions of Eq. (39.18) with energies E_1 and E_2, respectively, where $E_1 \neq E_2$. Is $\psi = A\psi_1 + B\psi_2$, where A and B are nonzero constants, a solution to Eq. (39.18)? Explain your answer.

Problems

39.36 A beam of 40-eV electrons traveling in the $+x$-direction passes through a slit that is parallel to the y-axis and 5.0 μm wide. The diffraction pattern is recorded on a screen 2.5 m from the slit. a) What is the de Broglie wavelength of the electrons? b) How much time does it take the electrons to travel from the slit to the screen? c) Use the width of the central diffraction pattern to calculate the uncertainty in the y-component of momentum of an electron just after it has passed through the slit. d) Use the result of part (c) and the Heisenberg uncertainty principle (Eq. 39.11 for y) to estimate the minimum uncertainty in the y-coordinate of an electron just after it has passed through the slit. Compare your result to the width of the slit.

39.37 a) What is the energy of a photon that has wavelength 0.10 μm? b) Through approximately what potential difference must electrons be accelerated so that they will exhibit wave nature in passing through a pinhole 0.10 μm in diameter? What is the speed of these electrons? c) If protons rather than electrons were used, through what potential difference would protons have to be accelerated so they would exhibit wave nature in passing through this pinhole? What would be the speed of these protons?

39.38 Electrons go through a single slit 150 nm wide and strike a screen 24.0 cm away. You find that at angles of $\pm 20.0°$ from the center of the diffraction pattern, no electrons hit the screen but electrons hit at all points closer to the center. a) How fast were these electrons moving when they went through the slit? b) What will be the next larger angles at which no electrons hit the screen?

39.39 A beam of photons with wavelength 150 nm and a beam of electrons having the same energy as the photons go through the same slit. You observe the diffraction pattern on a distant screen and measure that the photons produce their first dark band at

$\pm 25.0°$ from the centerline. a) How wide is the slit? b) At what angles relative to the centerline will no electrons be detected?

39.40 Coherent light is passed through two narrow slits whose separation is 40.0 μm. The second-order bright fringe in the interference pattern is located at an angle of 0.0300 rad. If electrons are used instead of light, what must the kinetic energy (in eV) of the electrons be if they are to produce an interference pattern for which the second-order maximum is also at 0.0300 rad?

39.41 What is the de Broglie wavelength of a red blood cell, with mass 1.00×10^{-11} g, that is moving with a speed of 0.400 cm/s? Do we need to be concerned with the wave nature of the blood cells when we describe the flow of blood in the body?

39.42 High-speed electrons are used to probe the interior structure of the atomic nucleus. For such electrons the expression $\lambda = h/p$ still holds, but we must use the relativistic expression for momentum, $p = mv/\sqrt{1 - v^2/c^2}$. a) Show that the speed of an electron that has de Broglie wavelength λ is

$$v = \frac{c}{\sqrt{1 + \left(\frac{mc\lambda}{h}\right)^2}}$$

b) The quantity h/mc equals 2.426×10^{-12} m. (As we saw in Section 38.7, this same quantity appears in Eq. (38.23), the expression for Compton scattering of photons by electrons.) If λ is small compared to h/mc, the denominator in the expression found in part (a) is close to unity and the speed v is very close to c. In this case it is convenient to write $v = (1 - \Delta)c$ and express the speed of the electron in terms of Δ rather than v. Find an expression for Δ valid when $\lambda \ll h/mc$. [*Hint:* Use the binomial expansion $(1 + z)^n = 1 + nz + n(n-1)z^2/2 + \cdots$, valid for the case $|z| < 1$.] c) How fast must an electron move for its de Broglie wavelength to be 1.00×10^{-15} m, comparable to the size of a proton? Express your answer in the form $v = (1 - \Delta)c$, and state the value of Δ.

39.43 a) What is the de Broglie wavelength of an electron accelerated from rest through a potential increase of 125 V? b) What is the de Broglie wavelength of an alpha particle ($q = +2e$, $m = 6.64 \times 10^{-27}$ kg) accelerated from rest through a potential drop of 125 V?

39.44 For relativistic particles the de Broglie relation $\lambda = h/p$ still holds, but the magnitude of momentum p is related to the total energy E by $E^2 = (pc)^2 + (mc^2)^2$ (Eq. 37.39). The kinetic energy is $K = E - mc^2$. a) Show that the de Broglie wavelength of a particle of kinetic energy K and rest mass m is

$$\lambda = \frac{hc}{\sqrt{K(K + 2mc^2)}}$$

b) Find approximate expressions for λ as a function of K in the special cases i) $K \ll mc^2$ (nonrelativistic limit) and ii) $K \gg mc^2$ (extreme relativistic limit). c) Calculate the de Broglie wavelength for a proton with kinetic energy 7.00 GeV. d) Repeat part (c) for an electron with kinetic energy 25.0 MeV.

39.45 a) A particle with mass m has kinetic energy equal to three times its rest energy. What is the de Broglie wavelength of this particle? (*Hint:* You must use the relativistic expression for kinetic energy. See Problem 39.44.) b) Determine the numerical value of the kinetic energy (in MeV) and the wavelength (in meters) if the particle in part (a) is i) an electron; ii) a proton.

39.46 Suppose that the uncertainty of position of an electron is equal to the radius of the $n = 1$ Bohr orbit for hydrogen. Calculate the simultaneous minimum uncertainty of the corresponding momentum component, and compare this with the magnitude of the momentum of the electron in the $n = 1$ Bohr orbit. Discuss your results.

39.47 Suppose that the uncertainty in position of a particle in a particular direction is 40% of its de Broglie wavelength. Show that in this case the simultaneous minimum uncertainty in its momentum component along the same direction is 40% of its momentum.

39.48 Proton Energy in a Nucleus. The radii of atomic nuclei are of the order of 5.0×10^{-15} m. a) Estimate the minimum uncertainty in the momentum of a proton if it is confined within a nucleus. b) Take this uncertainty in momentum to be an estimate of the magnitude of the momentum. Use the relativistic relationship between energy and momentum, Eq. (37.39), to obtain an estimate of the kinetic energy of a proton confined within a nucleus. c) For a proton to remain bound within a nucleus, what must the magnitude of the (negative) potential energy for a proton be within the nucleus? Give your answer in eV and in MeV. Compare to the potential energy for an electron in a hydrogen atom, that has a magnitude of a few tens of eV. (This shows why the interaction that binds the nucleus together is called the "strong nuclear force.")

39.49 Electron Energy in a Nucleus. The radii of atomic nuclei are of the order of 5.0×10^{-15} m. a) Estimate the minimum uncertainty in the momentum of an electron if it is confined within a nucleus. b) Take this uncertainty in momentum to be an estimate of the magnitude of the momentum. Use the relativistic relationship between energy and momentum, Eq. (37.39), to obtain an estimate of the kinetic energy of an electron confined within a nucleus. c) Compare the energy calculated in part (b) to the magnitude of the Coulomb potential energy of a proton and an electron separated by 5.0×10^{-15} m. On the basis of your result, could there be electrons within the nucleus? (*Note:* It is interesting to compare this result to that of Problem 39.48.)

39.50 In a TV picture tube the accelerating voltage is 15.0 kV, and the electron beam passes through an aperture 0.50 mm in diameter to a screen 0.300 m away. a) What is the uncertainty in position of the point where the electrons strike the screen? b) Does this uncertainty affect the clarity of the picture significantly? (Use nonrelativistic expressions for the motion of the electrons. This is fairly accurate and is certainly adequate for obtaining an estimate of uncertainty effects.)

39.51 The neutral pion (π^0) is an unstable particle produced in high-energy particle collisions. Its mass is about 264 times that of the electron, and it exists for an average lifetime of 8.4×10^{-17} s before decaying into two gamma-ray photons. Using the relationship $E = mc^2$ between rest mass and energy, find the uncertainty in the mass of the particle and express it as a fraction of the mass.

39.52 a) The average kinetic energy of a molecule in a gas at absolute temperature T is $\frac{3}{2}kT$. Find the de Broglie wavelength of a molecule with mass m with this (nonrelativistic) kinetic energy. b) A volume V of the gas contains N molecules. Explain why the average distance between atoms is approximately $(V/N)^{1/3}$. c) The classical description of an ideal gas is based on the premise that molecules can be regarded as particles. This premise will break down if the average de Broglie wavelength of a molecule is greater than the average distance between molecules, since in this case the wave nature of the molecules cannot be ignored. If a gas at temperature T is comprised of N molecules, each with mass m, find an approximate expression for the minimum volume at which it becomes necessary to use a wave description of the molecules. d) Evaluate the volume in part (c) for a gas consisting of 1.00 mol of oxygen molecules and one atmosphere pressure $(\text{mass } 2.66 \times 10^{-26} \text{ kg})$ at room temperature (293 K). How does this compare to the volume that such a gas actually occupies, as found using the ideal-gas equation? Is the wave nature of oxygen molecules important at room temperature and ordinary pressures? e) The conduction electrons within a metal can be regarded as a gas. Consider the metal silver, that has density $1.05 \times 10^4 \text{ kg/m}^3$. A single silver atom has mass 1.79×10^{-25} kg, and there is about one conduction electron per atom. Evaluate the volume in part (c) for a gas made up of all the conduction electrons in 1.00 kg of silver at 293 K. How does this compare with the actual volume of 1.00 kg of silver? In describing the behavior of conduction electrons in silver, is it necessary to account for their wave nature? (In Chapter 42 we will study the behavior of electrons in solids in greater depth.)

39.53 Doorway Diffraction. If your wavelength were 1.0 m, you would undergo considerable diffraction in moving through a doorway. a) What must your speed be for you to have this wavelength? (Assume that your mass is 60.0 kg.) b) At the speed calculated in part (a), how many years would it take you to move 0.80 m (one step)? Will you notice diffraction effects as you walk through doorways?

39.54 Atomic Spectra Uncertainties. A certain atom has an energy level 2.58 eV above the ground level. Once excited to this level, the atom remains in this level for 1.64×10^{-7} s (on average) before emitting a photon and returning to the ground level. a) What is the energy of the photon (in electron volts)? What is its wavelength (in nanometers)? b) What is the smallest possible uncertainty in energy of the photon? Give your answer in electron volts. c) Show that $|\Delta E/E| = |\Delta\lambda/\lambda|$ if $|\Delta\lambda/\lambda| \ll 1$. Use this to calculate the magnitude of the smallest possible uncertainty in the wavelength of the photon. Give your answer in nanometers.

39.55 In another universe the value of Planck's constant is $6.63 \times 10^{-22} \text{ J} \cdot \text{s}$. Assume that the physical laws and all other physical constants are the same as in our universe. In the other universe, an atom is in an excited state 4.50 eV above the ground state. The lifetime of this excited state (the average time the electron stays in this state) is 2.24×10^{-3} s. What is the minimum uncertainty (in eV) in the energy of the photon emitted when the atom makes the transition from this excited state to the ground state?

39.56 For x rays with wavelength 0.0300 nm, the $m = 1$ intensity maximum for a crystal occurs when the angle θ in Fig. 36.21c is 35.8°. At what angle θ does the $m = 1$ maximum occur when a beam of 4.50-keV electrons is used instead? Assume that the electrons also scatter from the atoms in the surface plane of this same crystal.

39.57 Electron diffraction can also take place when there is interference between electron waves that scatter from atoms on the surface of a crystal and waves that scatter from atoms in the next plane below the surface, a distance d from the surface (see Fig. 36.21c). a) Find an equation for the angles θ at which there is an intensity maximum for electron waves of wavelength λ. b) The spacing between crystal planes in a certain metal is 0.091 nm. If 71.0-eV electrons are used, find the angle at which there is an intensity maximum due to interference between scattered waves from adjacent crystal planes. The angle is measured as shown in Fig. 36.21c. c) The actual angle of the intensity maximum is slightly different from your result in part (b). The reason is the work function ϕ of the metal (see Section 38.2), which changes the electron potential energy by $-e\phi$ when it moves from vacuum into the metal. If the effect of the work function is taken into account, is the angle of the intensity maximum greater than or less than the value found in part (b)? Explain.

39.58 Zero-Point Energy. Consider a particle with mass m moving in a potential $U = \frac{1}{2}kx^2$, as in a mass-spring system. The total energy of the particle is $E = p^2/2m + \frac{1}{2}kx^2$. Assume that p and x are approximately related by the Heisenberg uncertainty principle, $px \approx h$. a) Calculate the minimum possible value of the energy E, and the value of x that gives this minimum E. This lowest possible energy, which is not zero, is called the *zero-point energy.* b) For the x calculated in part (a), what is the ratio of the kinetic to the potential energy of the particle?

39.59 A particle with mass m moves in a potential $U(x) = A|x|$, where A is a positive constant. In a simplified picture, quarks (the constituents of protons, neutrons, and other particles, as will be described in Chapter 44) have a potential energy of interaction of approximately this form, where x represents the separation between a pair of quarks. Because $U(x) \to \infty$ as $x \to \infty$, it's not possible to separate quarks from each other (a phenomenon called *quark confinement*). a) Classically, what is the force acting on this particle as a function of x? b) Using the uncertainty principle as in Problem 39.58, determine approximately the zero-point energy of the particle.

39.60 The discussion in Section 39.5 shows that the wave function $\Psi = \psi e^{-i\omega t}$ is a stationary state, where ψ is time independent and ω is a real (not complex) constant. Consider the wave function $\Psi = \psi_1 e^{-i\omega_1 t} + \psi_2 e^{-i\omega_2 t}$, where ψ_1 and ψ_2 are different time-independent functions and ω_1 and ω_2 are different real constants. Assume that ψ_1 and ψ_2 are real-valued functions, so that $\psi_1{}^* = \psi_1$ and $\psi_2{}^* = \psi_2$. Is this Ψ a wave function for a stationary state? Why or why not?

39.61 The Time-Dependent Schrödinger Equation. Equation (39.18) is the time-independent Schrödinger equation in one dimension. The time-dependent Schrödinger equation is

$$-\frac{\hbar^2}{2m}\frac{\partial^2\Psi(x,t)}{\partial x^2} + U(x)\Psi(x,t) = i\hbar\frac{\partial\Psi(x,t)}{\partial t}$$

If $\psi(x)$ is a solution to Eq. (39.18) with energy E, show that the time-dependent function $\Psi(x,t) = \psi(x)e^{-i\omega t}$ is a solution to the time-dependent Schrödinger equation if ω is chosen appropriately. What is the value of ω that makes Ψ a solution?

39.62 Time-Dependent Wave Function for a Free Particle. One example of a time-dependent wave function is that of a free particle [one for which $U(x) = 0$ for all x] of energy E and x-component of momentum p. From the de Broglie relations (Section 39.1), such a particle has associated with it a frequency $f = E/h$ and a wavelength $\lambda = h/p$. A reasonable first guess for the time-dependent wave function for such a particle is $\Psi(x,t) = A\cos(kx - \omega t)$, where A is a constant, $\omega = 2\pi f$ is the angular frequency, and $k = 2\pi/\lambda$ is the wave number. This is the same function we used to describe a mechanical wave [see Eq. (15.7)] or an electromagnetic wave propagating in the x-direction [see Eq. (32.16)]. a) Show that $\omega = E/\hbar$, $k = p/\hbar$, and $\omega = \hbar k^2/2m$. (*Hint:* The energy is purely kinetic, so $E = p^2/2m$.) b) To check this guess for the time-dependent wave function, substitute $\Psi(x, t) = A\cos(kx - \omega t)$ into the time-dependent Schrödinger equation (Problem 39.61) with $U(x) = 0$ (so the particle is free). Show that this guess for $\Psi(x, t)$ does *not* satisfy this equation, and so is not a suitable wave function for a free particle. c) Use the procedure described in part (b) to show that a second guess, $\Psi(x, t) = A\sin(kx - \omega t)$, is also not a suitable wave function for a free particle. d) Consider a combination of the functions proposed in parts (b) and (c):

$$\Psi(x, t) = A\cos(kx - \omega t) + B\sin(kx - \omega t)$$

By using the procedure described in part (b), show that this wave function is a solution to the time-dependent Schrödinger equation with $U(x) = 0$, but only if $B = iA$. (*Hint:* To satisfy the time-dependent Schrödinger equation for all x and t, the coefficients of $\cos(kx - \omega t)$ on both sides of the equation must be equal. The same is true for the coefficients of $\sin(kx - \omega t)$ on both sides of the equation.) This is an example of the general result that time-dependent wave functions always have both a real part and an imaginary part.

39.63 Imagine another universe in which the value of Planck's constant is 0.0663 J·s, but in which the physical laws and all other physical constants are the same as in our universe. In this universe, two physics students are playing catch. They are 12 m apart, and one throws a 0.25-kg ball directly toward the other with a speed of 6.0 m/s. a) What is the uncertainty in the ball's horizontal momentum, in a direction perpendicular to that in which it is being thrown, if the student throwing the ball knows that it is located within a cube with volume 125 cm^3 at the time she throws it? b) By what horizontal distance could the ball miss the second student?

39.64 A particle is described by the normalized wave function $\psi(x, y, z) = Axe^{-\alpha x^2}e^{-\beta y^2}e^{-\gamma z^2}$, where A, α, β, and γ are all real, positive constants. The probability that the particle will be found in the infinitesimal volume $dx\,dy\,dz$ centered at the point (x_0, y_0, z_0) is $|\psi(x_0, y_0, z_0)|^2\,dx\,dy\,dz$. a) At what value of x_0 is the particle most likely to be found? b) Are there values of x_0 for which the probability of the particle being found is zero? If so, at what x_0?

39.65 A particle is described by the normalized wave function $\psi(x, y, z) = Ae^{-\alpha(x^2+y^2+z^2)}$, where A and α are real, positive constants. a) Determine the probability of finding the particle at a distance between r and $r + dr$ from the origin. (*Hint:* See Problem 39.64. Consider a spherical shell centered on the origin with inner radius r and thickness dr.) b) For what value of r does the probability in part (a) have its maximum value? Is this the same value of r for which $|\psi(x, y, z)|^2$ is a maximum? Explain any differences.

39.66 Consider the wave packet defined by

$$\psi(x) = \int_0^\infty B(k)\cos kx\,dk$$

Let $B(k) = e^{-\alpha^2 k^2}$. a) The function $B(k)$ has its maximum value at $k = 0$. Let k_h be the value of k at which $B(k)$ has fallen to half its maximum value, and define the width of $B(k)$ as $w_k = k_h$. In terms of α, what is w_k? b) Use integral tables to evaluate the integral that gives $\psi(x)$. For what value of x is $\psi(x)$ maximum? c) Define the width of $\psi(x)$ as $w_x = x_h$, where x_h is the positive value of x where $\psi(x)$ has fallen to half its maximum value. Calculate w_x in terms of α. d) The momentum p is equal to $hk/2\pi$, so the width of B in momentum is $w_p = hw_k/2\pi$. Calculate the product $w_p w_x$ and compare to the Heisenberg uncertainty principle.

39.67 a) Using the integral in Problem 39.66, determine the wave function $\psi(x)$ for a function $B(k)$ given by

$$B(k) = \begin{cases} 0 & k < 0 \\ 1/k_0 & 0 \le k \le k_0 \\ 0 & k > k_0 \end{cases}$$

This represents an equal combination of all wave numbers between 0 and k_0. Thus $\psi(x)$ represents a particle with average wave number $k_0/2$, with a total spread or uncertainty in wave number of k_0. We will call this spread the *width* w_k of $B(k)$, so $w_k = k_0$. b) Graph $B(k)$ versus k and $\psi(x)$ versus x for the case $k_0 = 2\pi/L$, where L is a length. Locate the point where $\psi(x)$ has its maximum value and label this point on your graph. Locate the two points closest to this maximum (one on each side of it) where $\psi(x) = 0$, and define the distance along the x-axis between these two points as w_x, the width of $\psi(x)$. Indicate the distance w_x on your graph. What is the value of w_x if $k_0 = 2\pi/L$? c) Repeat part (b) for the case $k_0 = \pi/L$. d) The momentum p is equal to $hk/2\pi$, so the width of B in momentum is $w_p = hw_k/2\pi$. Calculate the product $w_p w_x$ for each of the cases $k_0 = 2\pi/L$ and $k_0 = \pi/L$. Discuss your results in light of the Heisenberg uncertainty principle.

Challenge Problems

39.68 The wave nature of particles results in the quantum mechanical situation that a particle confined in a box can only assume wavelengths that result in standing waves in the box, with nodes at the box walls. a) Show that an electron confined in a one-dimensional box of length L will have energy levels given by

$$E_n = \frac{n^2h^2}{8mL^2}$$

(*Hint:* Recall that the relation between the de Broglie wavelength and the speed of a nonrelativistic particle is $mv = h/\lambda$. The energy of the particle is $\frac{1}{2}mv^2$.) b) If a hydrogen atom is modeled as a one-dimensional box with length equal to the Bohr radius, what is the energy (in electron volts) of the lowest energy level of the electron?

39.69 You have entered a contest in which the contestants drop a marble with mass 20.0 g from the roof of a building onto a small target 25.0 m below. From uncertainty considerations, what is the typical distance by which you will miss the target, given that you aim with the highest possible precision? (*Hint:* The uncertainty Δx_f in the x-coordinate of the marble when it reaches the ground comes in part from the uncertainty Δx_i in the x-coordinate initially and in part from the initial uncertainty in v_x. The latter gives rise to an uncertainty Δv_x in the horizontal motion of the marble as it falls. The values of Δx_i and Δv_x are related by the uncertainty principle. A small Δx_i gives rise to a large Δv_x, and vice versa. Find the value of Δx_i that gives the smallest total uncertainty in x at the ground. Ignore any effects of air resistance.)

40 QUANTUM MECHANICS

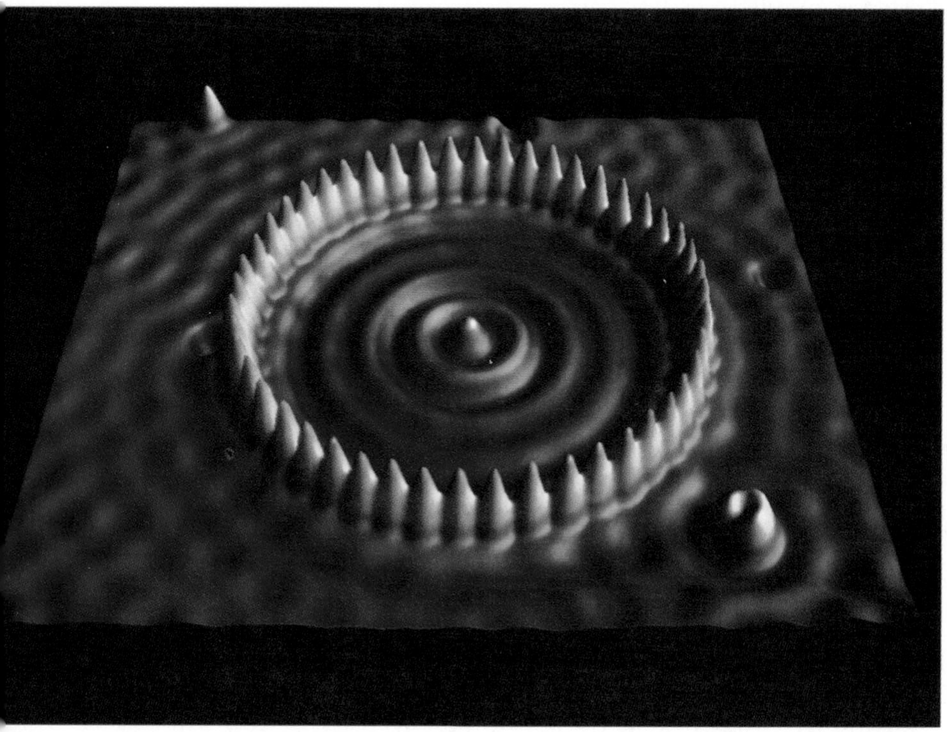

This STM (scanning tunneling microscope) image of a copper surface is graphic evidence of the wave behavior of electrons. The "elevation" at each point in the image indicates the density of electrons at that point. Forty-eight iron atoms (shown as yellow peaks) have been placed in a circle on the surface. The electrons within the circle form a clear standing-wave pattern.

? **The wavelength of the electron standing wave has nearly the same value at all points inside the circle of iron atoms shown in the image. What does this tell you about the potential energy function inside the circle?**

In the preceding chapter we found that particles sometimes behave like waves and that they can be described by wave functions. Now we're ready for a systematic analysis of particles in bound states (such as electrons in atoms), including finding their possible wave functions and energy levels.

Our analysis involves funding solutions of the Schrödinger equation, which we introduced in Chapter 39. For any quantum-mechanical system, the wave functions for any state of definite energy must be solutions of the Schrödinger equation for that system. As we will see, solving the Schrödinger equation automatically gives the possible energy levels for the system. We'll discuss several simple one-dimensional applications of the Schrödinger equation.

Besides energies, solving the Schrödinger equation gives us the probabilities of finding a particle in various regions. One surprising result is that there is a nonzero probability that microscopic particles will pass through thin barriers, even though such a process is forbidden by Newtonian mechanics.

Finally, we'll generalize the Schrödinger equation to three dimensions. This will pave the way for describing the wave functions for the hydrogen atom in Chapter 41. The hydrogen-atom wave functions in turn form the foundation for our analysis of more complex atoms, of the periodic table of the elements, of x-ray energy levels, and of other properties of atoms.

40.1 | Particle in a Box

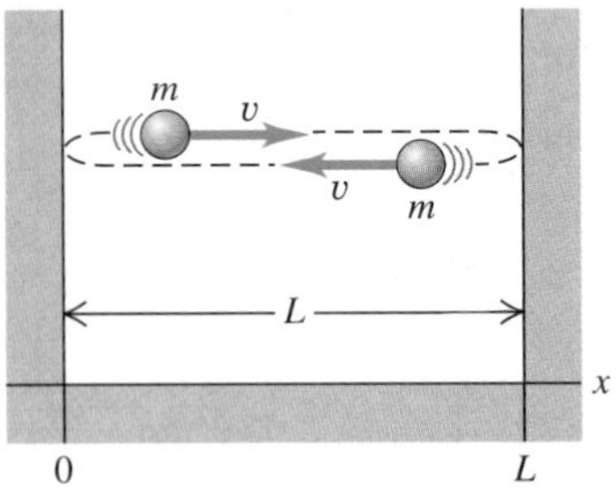

40.1 The Newtonian view of a particle in a box. A particle with mass m moves along a straight line at constant speed, bouncing between two rigid walls that are a distance L apart.

In this chapter we want to learn how to use the Schrödinger equation to determine the possible energy levels and the corresponding wave functions for various systems. As in Chapter 39, we refer to the states of definite energy as *stationary states.* For simplicity, we begin with problems in which the particle can move only along the x-axis. From Section 39.5, the Schrödinger equation for such one-dimensional problems is

$$-\frac{\hbar^2}{2m}\frac{d^2\psi(x)}{dx^2} + U(x)\psi(x) = E\psi(x) \tag{40.1}$$

(one-dimensional Schrödinger equation)

This equation is really a statement that the kinetic energy [represented by the term $-(\hbar^2/2m)d^2\psi(x)/dx^2$] plus the potential energy [represented by the term $U(x)\psi(x)$] is equal to the total energy [represented by the term $E\psi(x)$]. The fundamental problem is then the following: For a given potential energy function $U(x)$, what are the possible stationary-state wave functions $\psi(x)$, and what are the corresponding energies E?

In Section 39.5 we solved this problem for the case $U(x) = 0$, corresponding to a *free* particle. The allowed wave functions and corresponding energies are

$$\psi(x) = Ae^{ikx} \qquad E = \frac{\hbar^2k^2}{2m} \qquad \text{(free particle)} \tag{40.2}$$

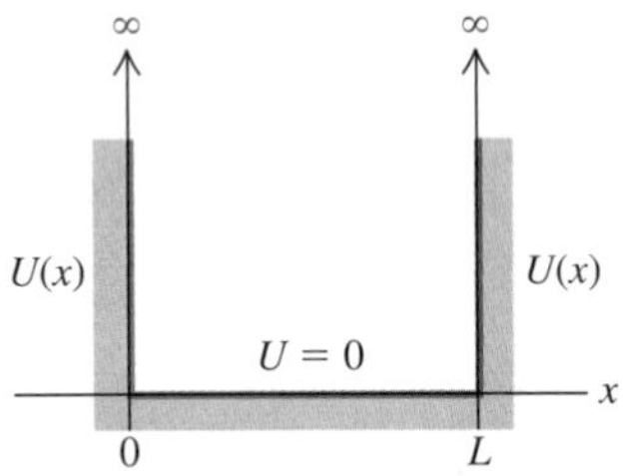

40.2 The potential energy function for a particle in a box. The potential energy U is zero in the interval $0 \le x \le L$ and is infinite everywhere outside this interval.

The wave number k is equal to $2\pi/\lambda$, where λ is the wavelength. We found that k can have any real value, so the energy E of a free particle can have any value from zero to infinity. Furthermore, the particle can be found with equal probability at any value of x from $-\infty$ to $+\infty$.

Now let's look at a simple model in which a particle is *bound* so that it cannot escape to infinity, but rather is confined to a restricted region of space. Our system consists of a particle confined between two rigid walls separated by a distance L (Fig. 40.1). The motion is purely one-dimensional, with the particle moving along the x-axis only and the walls at $x = 0$ and $x = L$. The potential energy corresponding to the rigid walls is infinite, so the particle cannot escape; between the walls, the potential energy is zero. This situation is often described as a **"particle in a box."** This model might represent an electron that is free to move within a long, straight molecule or along a very thin wire. The potential energy function is graphed in Fig. 40.2.

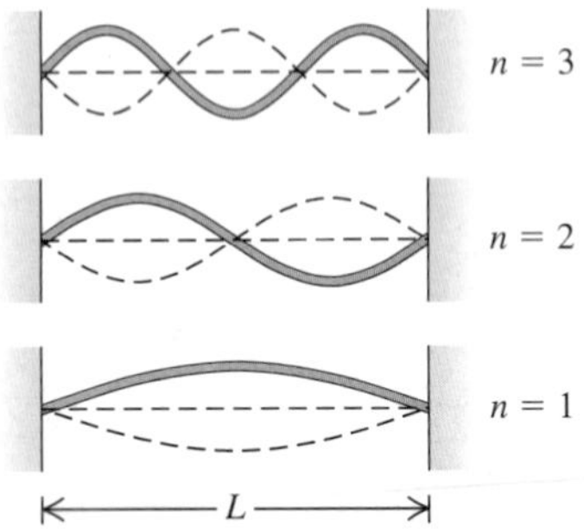

40.3 Normal modes of vibration for a string with length L, held at both ends. Each end is a node, and there may be $n - 1$ additional nodes. Thus the length is an integral number of half-wavelengths; $L = n\lambda_n/2$.

To solve the Schrödinger equation for this system, we begin with some restrictions on the particle's wave function $\psi(x)$. Because the particle is confined to the region $0 \le x \le L$, we expect $\psi(x)$ to be zero outside that region. This agrees with the Schrödinger equation: if the term $U(x)\psi(x)$ in Eq. (40.1) is to be finite, $\psi(x)$ must be zero where $U(x)$ is infinite.

Furthermore, $\psi(x)$ must be a *continuous* function to be a mathematically well-behaved solution to the Schrödinger equation. If it is, then $\psi(x)$ must be zero at the region's boundary, $x = 0$ and $x = L$. These two conditions serve as *boundary conditions* for the problem. They should look familiar, because they are the same conditions that we used to find the normal modes of a vibrating string in Section 15.8 (Fig. 40.3); you should review that discussion.

An additional condition is that to calculate the second derivative $d^2\psi(x)/dx^2$ in Eq. (40.1), the *first* derivative $d\psi(x)/dx$ must also be continuous except at points where the potential energy becomes infinite (as it does at the walls of the box). This is analogous to the requirement that a vibrating string, like those shown in Fig. 40.3, cannot have a kink in it (which would correspond to a discontinuity in the first derivative of the wave function) except at the ends of the string.

We now solve for the wave functions in the region $0 \le x \le L$ subject to the above conditions. In this region $U(x) = 0$, so the wave function in this region must satisfy

$$-\frac{\hbar^2}{2m}\frac{d^2\psi(x)}{dx^2} = E\psi(x) \qquad \text{(particle in a box)} \qquad (40.3)$$

ActivPhysics ONLINE
20.2 Particle in a Box

Equation (40.3) is the *same* Schrödinger equation as for a free particle, so it is tempting to conclude that the wave functions and energies are given by Eq. (40.2). It is true that $\psi(x) = Ae^{ikx}$ satisfies the Schrödinger equation with $U(x) = 0$, is continuous, and has a continuous first derivative $d\psi(x)/dx = ikAe^{ikx}$. However, this wave function does *not* satisfy the boundary conditions that $\psi(x)$ must be zero at $x = 0$ and $x = L$: at $x = 0$ the wave function in Eq. (40.2) is equal to $Ae^0 = A$, and at $x = L$ it is equal to Ae^{ikL}. (These would be equal to zero if $A = 0$, but then the wave function would be zero and there would be no particle at all!)

The way out of this dilemma is to recall Example 39.6 (Section 39.5), in which we found that a more general stationary-state solution to the Schrödinger equation with $U(x) = 0$ is

$$\psi(x) = A_1e^{ikx} + A_2e^{-ikx} \qquad (40.4)$$

This wave function is a superposition of two waves: one traveling in the $+x$-direction of amplitude A_1, and one traveling in the $-x$-direction with the same wave number but amplitude A_2. This is analogous to the situation for standing waves on a string (Fig. 40.3), which we can regard as the superposition of two sinusoidal waves propagating in opposite directions (see Section 15.7). The energy that corresponds to Eq. (40.4) is $E = \hbar^2k^2/2m$, just as for a single wave.

To see whether the wave function given by Eq. (40.4) can satisfy the boundary conditions, let's first rewrite it in terms of sines and cosines using Euler's formula, Eq. (39.15):

$$\begin{aligned}\psi(x) &= A_1(\cos kx + i\sin kx) + A_2[\cos(-kx) + i\sin(-kx)]\\ &= A_1(\cos kx + i\sin kx) + A_2(\cos kx - i\sin kx)\\ &= (A_1 + A_2)\cos kx + i(A_1 - A_2)\sin kx \qquad (40.5)\end{aligned}$$

At $x = 0$ this is equal to $\psi(0) = A_1 + A_2$, which must equal zero if we are to satisfy the boundary condition at that point. Hence $A_2 = -A_1$, and Eq. (40.5) becomes

$$\psi(x) = 2iA_1 \sin kx = C\sin kx \qquad (40.6)$$

We have simplified the expression by introducing the constant $C = 2iA_1$. (We'll come back to this constant later.) We can also satisfy the second boundary condition that $\psi = 0$ at $x = L$ by choosing values of k such that $kL = n\pi$ ($n = 1, 2, 3, \ldots$). Hence Eq. (40.6) does give the stationary-state wave functions for a particle in a box in the region $0 \le x \le L$. (Outside this region, $\psi(x) = 0$.) The possible values of k and the wavelength $\lambda = 2\pi/k$ are

$$k = \frac{n\pi}{L} \quad \text{and} \quad \lambda = \frac{2\pi}{k} = \frac{2L}{n} \qquad (n = 1, 2, 3, \ldots) \qquad (40.7)$$

Just as for the string in Fig. 40.3, the length L of the region is an integral number of half-wavelengths.

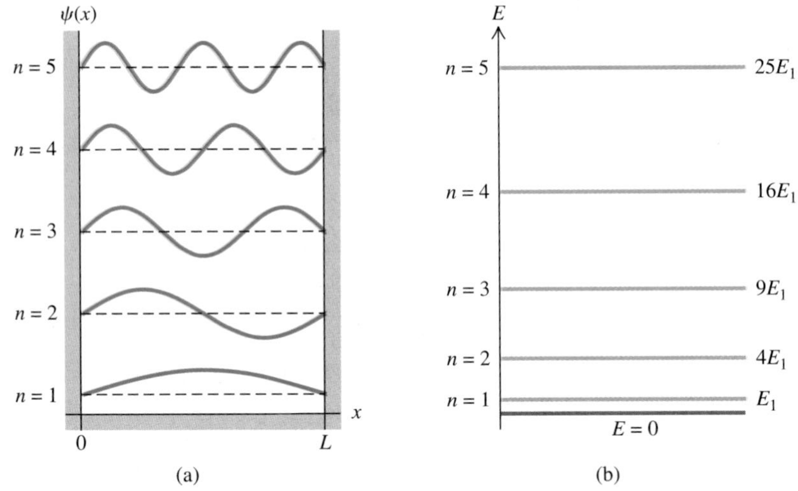

40.4 (a) Wave functions for a particle in a box, with $n = 1, 2, 3, 4$, and 5. Each of the horizontal dashed lines represents $\psi = 0$ for the respective wave function. The five graphs have been displaced vertically for clarity, as in Fig. 40.3. (b) Energy-level diagram for a particle in a box. Each energy is n^2E_1, where E_1 is the ground-level energy.

The possible energy levels are given by $E = \hbar^2k^2/2m = p^2/2m$, where $p = \hbar k = (h/2\pi)(2\pi/\lambda) = h/\lambda$ is the magnitude of momentum of a free particle with wave number k and wavelength λ. This makes sense, since inside the region $0 \leq x \leq L$ the potential energy is zero and the energy is all kinetic. For each value of n there are corresponding values of p, λ, and E; let's call them p_n, λ_n, and E_n. Putting the pieces together, we get

$$p_n = \frac{h}{\lambda_n} = \frac{nh}{2L} \tag{40.8}$$

and

$$E_n = \frac{p_n^2}{2m} = \frac{n^2h^2}{8mL^2} = \frac{n^2\pi^2\hbar^2}{2mL^2} \qquad (n = 1, 2, 3 \ldots) \tag{40.9}$$

(energy levels, particle in a box)

These are the possible energy levels for a particle in a box. Each energy level has its own value of the quantum number n and a corresponding wave function, which we denote by ψ_n. When we replace k in Eq. (40.6) by $n\pi/L$ from Eq. (40.7), we find

$$\psi_n(x) = C \sin\frac{n\pi x}{L} \qquad (n = 1, 2, 3 \ldots) \tag{40.10}$$

Figure 40.4a shows graphs of the wave functions $\psi_n(x)$ for $n = 1, 2, 3, 4$, and 5. Note that these functions look identical to those for a standing wave on a string (Fig. 40.3). Figure 40.4b shows the energy-level diagram for this system. The levels of successively higher energies, proportional to n^2, are spaced farther and farther apart. There are an infinite number of levels because the walls are perfectly rigid; even a particle of infinitely great kinetic energy is confined within the box.

CAUTION Note that the energy of a particle in a box *cannot* be zero. Equation (40.9) shows that $E = 0$ would require $n = 0$, but substituting $n = 0$ into Eq. (40.10) gives a zero wave function. Since a particle is described by a nonzero wave function, this means that there cannot be a particle with $E = 0$.

Example 40.1 Electron in an atom-size box

Find the lowest energy level for a particle in a box if the particle is an electron in a box 5.0×10^{-10} m across, or a little bigger than an atom.

SOLUTION

IDENTIFY and SET UP: The lowest energy level (that is, the ground level) corresponds to $n = 1$ in Eq. (40.9).

EXECUTE: From Eq. (40.9),

$$E_1 = \frac{h^2}{8mL^2} = \frac{(6.626 \times 10^{-34}\ \mathrm{J \cdot s})^2}{8(9.109 \times 10^{-31}\ \mathrm{kg})(5.0 \times 10^{-10}\ \mathrm{m})^2}$$
$$= 2.4 \times 10^{-19}\ \mathrm{J} = 1.5\ \mathrm{eV}$$

EVALUATE: A particle trapped in a box is rather different from an electron bound in an atom, but it is reassuring that this energy is of the same order of magnitude as actual atomic energy levels.

You should be able to show that replacing the electron with a proton or a neutron $(m = 1.67 \times 10^{-27}\ \mathrm{kg})$ in a box the width of a medium-sized nucleus $(L = 1.1 \times 10^{-14}\ \mathrm{m})$ gives $E_1 = 1.7$ MeV. This shows us that the energies for particles in the nucleus are about a million times larger than the energies of electrons in atoms, giving us a clue as to why each nuclear fission and fusion reaction provides so much more energy than each chemical reaction.

If you repeat this calculation for a billiard ball $(m = 0.2\ \mathrm{kg})$ bouncing back and forth between the cushions of a frictionless, perfectly elastic billiard table $(L = 1.5\ \mathrm{m})$, you'll find the separation between energy levels to be as small as 4×10^{-67} J (Exercise 40.1). This negligible value shows that quantum effects won't have much effect on your billiard game.

Probability and Normalization

Several additional points need to be discussed. First is the *probability* interpretation of the wave function ψ, which we discussed in Section 39.5. In our one-dimensional situation the quantity $|\psi(x)|^2\, dx$ is proportional to the probability that the particle will be found within a small interval dx about x. For a particle in a box,

$$|\psi(x)|^2\, dx = C^2 \sin^2 \frac{n\pi x}{L} dx$$

Both $\psi(x)$ and $|\psi(x)|^2$ are plotted in Fig. 40.5 for $n = 1$, 2, and 3. We note that not all positions are equally likely. This is in contrast to the situation in classical mechanics, in which all positions between $x = 0$ and $x = L$ are equally likely. We see in Fig. 40.5b that $|\psi(x)|^2 = 0$ at some points, so the probability is zero of finding the particle exactly at these points. Don't let that bother you; the uncertainty principle has already shown us that we can't measure position exactly. The particle is localized only to be somewhere between $x = 0$ and $x = L$, not to be exactly at some particular value of x.

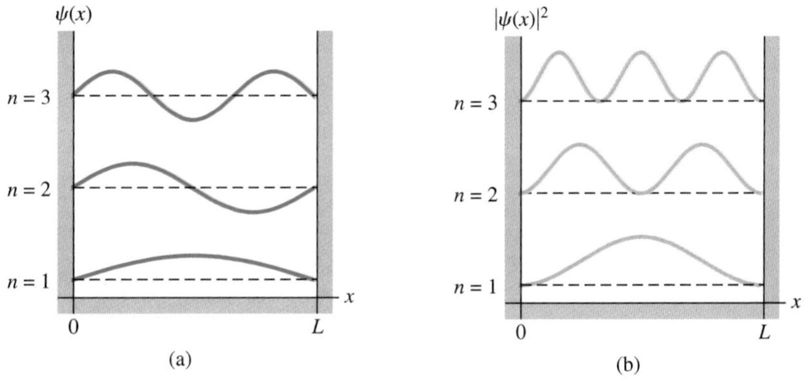

40.5 Graphs of (a) $\psi(x)$ and (b) $|\psi(x)|^2$ for the first three wave functions $(n = 1, 2, 3)$ for a particle in a box. The horizontal dashed lines represent $\psi(x) = 0$ and $|\psi(x)|^2 = 0$ for each of the three levels. The value of $|\psi(x)|^2\, dx$ at each point is the probability of finding the particle in a small interval dx about the point.

We know that the particle *must* be somewhere in the universe, that is, between $x = -\infty$ and $x = +\infty$. So the *sum* of the probabilities for all the dx's everywhere (the *total* probability of finding the particle) must equal 1. That is,

$$\int_{-\infty}^{\infty} |\psi(x)|^2\, dx = 1 \qquad \text{(normalization condition)} \qquad (40.11)$$

A wave function is said to be *normalized* if it has a constant such as C in Eq. (40.10) that is calculated to make the total probability equal 1 in Eq. (40.11). The process of determining the constant is called **normalization,** as was mentioned in Section 39.5. Why bother with normalization? Because for a normalized wave function, $|\psi(x)|^2\, dx$ is not merely proportional to, but *equals,* the probability of finding the particle in dx. Recall from Section 39.5 that we call $|\psi(x)|^2$ the *probability distribution function.*

Now let's normalize the wave function $\psi(x)$ given by Eq. (40.10) for the particle in a box. Since $\psi(x)$ is zero except between $x = 0$ and $x = L$, Eq. (40.11) becomes

$$\int_0^L C^2 \sin^2 \frac{n\pi x}{L}\, dx = 1 \qquad (40.12)$$

You can evaluate this integral using the trigonometric identity $\sin^2\theta = \frac{1}{2}(1 - \cos 2\theta)$; the result is $C^2L/2$. Thus our probability interpretation of the wave function demands that $C^2L/2 = 1$, or $C = (2/L)^{1/2}$; the constant C is *not* arbitrary. (This is in contrast to the classical vibrating string problem, in which C represents an amplitude that depends on initial conditions.) Thus the normalized stationary-state wave functions for a particle in a box are

$$\psi_n(x) = \sqrt{\frac{2}{L}} \sin \frac{n\pi x}{L} \qquad (n = 1, 2, 3, \dots) \qquad \text{(particle in a box)} \quad (40.13)$$

Next, let's check whether our results for the particle in a box are consistent with the uncertainty principle. We can write the position of the particle as $x = L/2 \pm L/2$, so we can estimate the uncertainty in position as $\Delta x \approx L/2$. From Eq. (40.8) the magnitude of momentum in state n is $p_n = nh/2L$. A reasonable estimate of the momentum uncertainty is the difference in momentum of two levels that differ by 1 in their n values, that is, $\Delta p_x \approx h/2L$. Then the product $\Delta x\, \Delta p_x$ is

$$\Delta x\, \Delta p_x = \frac{h}{4}$$

This is consistent with the uncertainty principle, Eq. (39.11), $\Delta x\, \Delta p_x \geq \hbar = h/2\pi$, in which the uncertainties were more precisely defined as standard deviations.

Example 40.2 A nonsinusoidal wave function?

a) Show that $\psi(x) = Ax + B$, where A and B are constants, is a solution of the Schrödinger equation for an $E = 0$ energy level of a particle in a box. b) What constraints do the boundary conditions at $x = 0$ and $x = L$ place on the constants A and B?

SOLUTION

IDENTIFY and SET UP: In part (a) we substitute $\psi(x)$ into the Schrödinger equation for a particle in a box, Eq. (40.3), to determine whether it is a solution. In part (b) we demand that $\psi(x)$ be equal to zero at $x = 0$ and at $x = L$.

EXECUTE: a) From Eq. (40.3), the Schrödinger equation for this situation is $-(\hbar^2/2m)\, d^2\psi(x)/dx^2 = E\psi(x)$. Differentiating $\psi(x) = Ax + B$ twice with respect to x gives $d^2\psi(x)/dx^2 = 0$, so the left side of the equation is zero. Since we assume $E = 0$, the right side of the equation is also zero. Since $0 = 0$, $\psi(x) = Ax + B$ is a solution of this Schrödinger equation for $E = 0$. Note that both $\psi(x)$ and its derivative $d\psi(x)/dx = A$ are continuous functions, as they must be.
b) Applying the boundary condition at $x = 0$ gives $\psi(0) = A = 0$, so the constant A must be zero and $\psi(x) = Bx$. Applying the boundary condition at $x = L$ gives $\psi(L) = BL = 0$, so the constant B must also be zero. Thus the wave function equals zero inside the box as well as outside the box, and there is *zero* probability of finding the particle anywhere with this wave function. Hence $\psi(x) = Ax + B$ is not a physically valid wave function.

EVALUATE: The moral of this story is that a wave function for a particle must satisfy *both* the Schrödinger equation *and* the boundary conditions, and be nonzero in some region in space.

Time Dependence

Finally, we note that the wave functions $\psi_n(x)$ in Eq. (40.13) depend only on the *spatial* coordinate x. From Section 39.5, the full wave function also depends on *time*. Equation (39.14) shows that if $\psi(x)$ is the wave function for a state of definite energy E, the full time-dependent wave function $\Psi(x, t)$ is

$$\Psi(x,t) = \psi(x)\, e^{-iEt/\hbar} \quad (40.14)$$

(time-dependent wave function for a stationary state)

From Eq. (40.12), it follows that the time-dependent wave functions for a particle in a box are

$$\Psi(x,t) = \sqrt{\frac{2}{L}} \sin\left(\frac{n\pi x}{L}\right) e^{-iE_n t/\hbar} \qquad (n = 1, 2, 3, \dots) \quad (40.15)$$

In this expression the energies E_n are given by Eq. (40.8).

The absolute value of $e^{-iEt/\hbar}$ is unity: $|e^{-iEt/\hbar}|^2 = e^{+iEt/\hbar}e^{-iEt/\hbar} = e^0 = 1$. Hence $|\Psi(x,t)|^2$ is equal to $|\psi(x)|^2$, and the probability distribution function does not depend on time. We saw in Section 39.5 that this is the origin of the term *stationary state* for a state of definite energy. So in calculations with states that have definite energy, we are justified in omitting the time factor. We'll be concerned mostly with stationary states in our applications of quantum mechanics.

Test Your Understanding

If a particle in a box is in the nth energy level, what is the average value of its x-component of momentum p_x?

40.2 | Potential Wells

A **potential well** is a potential-energy function $U(x)$ that has a minimum. We introduced this term in Section 7.5, and we also used it in our discussion of periodic motion in Chapter 13. In Newtonian mechanics a particle trapped in a potential well can vibrate back and forth with periodic motion. Our first application of the Schrödinger equation, the particle in a box, involved a rudimentary potential well with a function $U(x)$ that is zero within a certain interval and infinite everywhere else. This function corresponds to a few situations found in nature, as mentioned in Section 40.1, but the correspondence is only approximate.

A potential well that serves as a better approximation to several actual physical situations is a well with straight sides but *finite* height. Figure 40.6 shows a

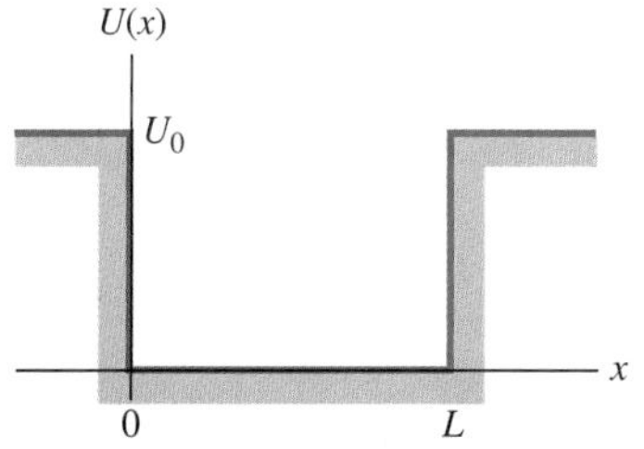

40.6 A square-well potential. The potential-energy function $U(x)$ is zero in the interval $0 \le x \le L$ and has the constant value U_0 everywhere outside this interval.

ActivPhysics ONLINE

20.1 Potential Energy Diagrams

potential-energy function that is zero in the interval $0 \le x \le L$ and has the value U_0 outside this interval. This function is often called a **square-well potential.** It could serve as a simple model of an electron within a metallic sheet with thickness L, moving perpendicular to the surfaces of the sheet. The electron can move freely inside the metal but has to climb a potential-energy barrier with height U_0 to escape from either surface of the metal. The energy U_0 is related to the *work function* that we discussed in Section 38.2 in connection with the photoelectric effect. In a three-dimensional application a spherical version of a potential well can be used to approximately describe the motions of protons and neutrons within a nucleus.

In Newtonian mechanics the particle is trapped (localized) in a well if the total energy E is less than U_0. In quantum mechanics, such a trapped state is often called a **bound state.** All states are bound for an infinitely deep well, but if E is greater than U_0 for a finite well, the particle is *not* bound.

For a finite square well, let's consider bound-state solutions of the Schrödinger equation, corresponding to $E < U_0$. The easiest approach is to consider separately the regions where $U = 0$ and where $U = U_0$. Where $U = 0$, the Schrödinger equation reduces to Eq. (40.3). Rearranging this equation, we find

$$\frac{d^2\psi(x)}{dx^2} = -\frac{2mE}{\hbar^2}\psi(x) \tag{40.16}$$

From Eq. (40.5), we can express the solutions of this equation as combinations of $\cos kx$ and $\sin kx$, where $E = \hbar^2k^2/2m$ and so $k = \sqrt{2mE}/\hbar$. Hence inside the square well $(0 \le x \le L)$ we have

$$\psi(x) = A\cos\frac{\sqrt{2mE}}{\hbar}x + B\sin\frac{\sqrt{2mE}}{\hbar}x \qquad \text{(inside the well)} \tag{40.17}$$

where A and B are constants. So far, this looks a lot like the particle-in-a-box analysis in Section 40.1.

In the regions $x < 0$ and $x > L$ we use Eq. (40.1) with $U = U_0$. Rearranging, we get

$$\frac{d^2\psi(x)}{dx^2} = \frac{2m(U_0 - E)}{\hbar^2}\psi(x) \tag{40.18}$$

The quantity $U_0 - E$ is positive, so the solutions of this equation are exponential. Using κ (the Greek letter "kappa") as positive in the abbreviation $\kappa = [2m(U_0 - E)]^{1/2}/\hbar$, we can write the solutions as

$$\psi(x) = Ce^{\kappa x} + De^{-\kappa x} \qquad \text{(outside the well)} \tag{40.19}$$

where C and D are constants with different values in the $x < 0$ and $x > L$ regions.

We see that the bound-state wave functions for this system are sinusoidal inside the well and exponential outside it. We have to use the positive exponent in the $x < 0$ region and the negative exponent in the $x > L$ region. That is, $D = 0$ for $x < 0$ and $C = 0$ for $x > L$. Without this choice of constants, ψ would approach infinity as $|x|$ approached infinity, and the normalization condition, Eq. (40.11), could not be satisfied. We also have to match the wave functions so that they satisfy the boundary conditions that we mentioned in Section 40.1: $\psi(x)$ and $d\psi(x)/dx$ must be continuous at the boundary points ($x = 0$ and $x = L$). If the wave function $\psi(x)$ or the slope $d\psi(x)/dx$ were to change discontinuously at a point, the second derivative $d^2\psi(x)/dx^2$ would be *infinite* at that point. But that would violate the Schrödinger equation, which says that at every point, $d^2\psi(x)/dx^2$ is proportional to $U - E$. In our situation, $U - E$ is finite everywhere, so $d^2\psi(x)/dx^2$ must also be finite everywhere.

Comparing Finite and Infinite Square Wells

Matching the sinusoidal and exponential functions at the boundary points so that they join smoothly is possible only for certain specific values of the total energy E, so this requirement determines the possible energy levels of the finite square well. There is no simple formula for the energy levels as there was for the infinitely deep well. Finding the levels is a fairly complex mathematical problem that requires solving a transcendental equation by numerical approximation; we won't go into the details. Figure 40.7 shows the general shape of a possible wave function. The most striking features of this wave function are the "exponential tails" that extend outside the well into regions that are forbidden by Newtonian mechanics (because in those regions the particle would have negative kinetic energy). We see that there is some probability for finding the particle *outside* the potential well, despite the fact that in classical mechanics this is impossible. This penetration into classically forbidden regions is a quantum effect that has no classical analog for particles. We will discuss an amazing result of this effect in Section 40.3.

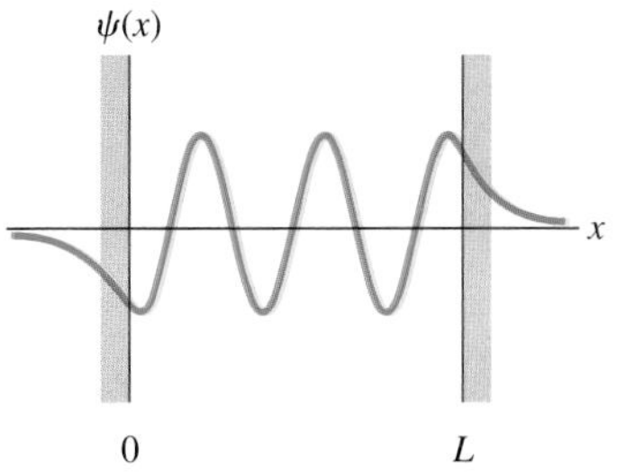

40.7 A possible wave function for a particle in a finite potential well. The function is sinusoidal inside the well $(0 \le x \le L)$ and exponential outside it. It approaches zero asymptotically at large $|x|$. The functions must join smoothly at $x = 0$ and $x = L$; the wave function and its derivative must be continuous.

Example 40.3 Well versus box

Show that Eq. (40.19), $\psi(x) = Ce^{\kappa x} + De^{-\kappa x}$, is consistent with the corresponding particle-in-a-box wave function.

SOLUTION

IDENTIFY and SET UP: Equation (40.19) gives the possible wave functions outside the potential well. Since the wave function outside the box (an infinite potential well) equals zero, we must show that $\psi(x) = Ce^{\kappa x} + De^{-\kappa x}$ approaches zero for both $x > 0$ and $x > L$ as the depth of the potential well (U_0) approaches infinity.

EXECUTE: The quantity κ in Eq. (40.19) depends on U_0: $\kappa = [2m(U_0 - E)]^{1/2}/\hbar$. As U_0 approaches infinity, κ also approaches infinity. In the region $x < 0$, the wave function is $\psi(x) = Ce^{\kappa x}$; as $\kappa \to \infty$, $\kappa x \to -\infty$ and $e^{\kappa x} \to 0$, so the wave function approaches zero for all $x < 0$. In the region $x > L$, the wave function is $\psi(x) = De^{-\kappa x}$; as $\kappa \to \infty$, $-\kappa x \to -\infty$ and $e^{-\kappa x} \to 0$, so the wave function also approaches zero for all $x > L$.

EVALUATE: In Section 40.1 we *assumed* $\psi(x) = 0$ outside the box. This analysis validates our assumption.

In our study of mechanics, we often considered limiting cases to check our results (see Examples 5.11 and 5.13 in Section 5.2). This example shows that this technique is also useful for problems in quantum mechanics, as it is for problems in all branches of physics.

Let's continue our comparison of the finite-depth potential well with the infinitely deep well. First, because the wave functions for the finite well aren't zero at $x = 0$ and $x = L$, the wavelength of the sinusoidal part of each wave function is *longer* than it would be with an infinite well. From $p = h/\lambda$ this increase in λ corresponds to a reduced magnitude of momentum and therefore a reduced energy. Thus each energy level, including the ground level, is *lower* for a finite well than for an infinitely deep well with the same width.

Second, a well with finite depth U_0 has only a *finite* number of bound states and corresponding energy levels, compared to the *infinite* number for an infinitely deep well. How many levels there are depends on the magnitude of U_0 in comparison with the ground-level energy for the infinite well, which we call E_∞. From Eq. (40.9),

$$E_\infty = \frac{\pi^2\hbar^2}{2mL^2} \tag{40.20}$$

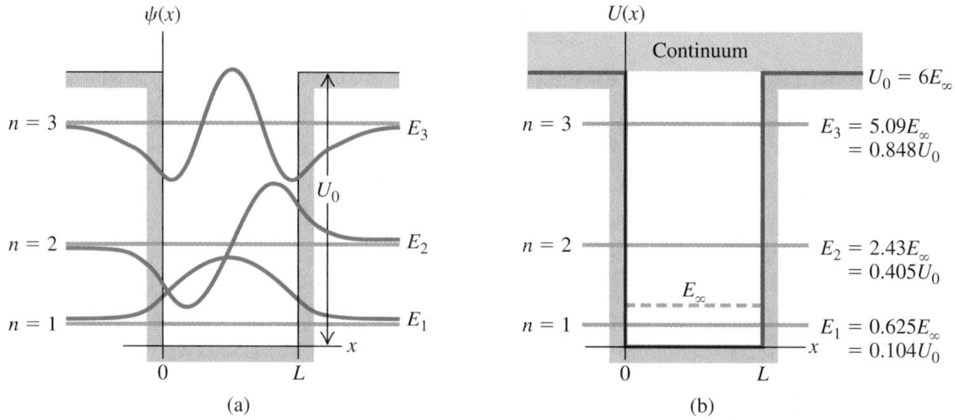

40.8 (a) Wave functions for the three bound states for a particle in a finite potential well with depth U_0, for the case $U_0 = 6E_\infty$. The horizontal brown line for each wave function corresponds to $\psi = 0$; the vertical placement of these lines indicates the energy of each bound state. (b) Energy-level diagram for this system. The energies are expressed both as multiples of E_∞ and as fractions of U_0. All energies greater than U_0 are possible; states with $E > U_0$ form a continuum.

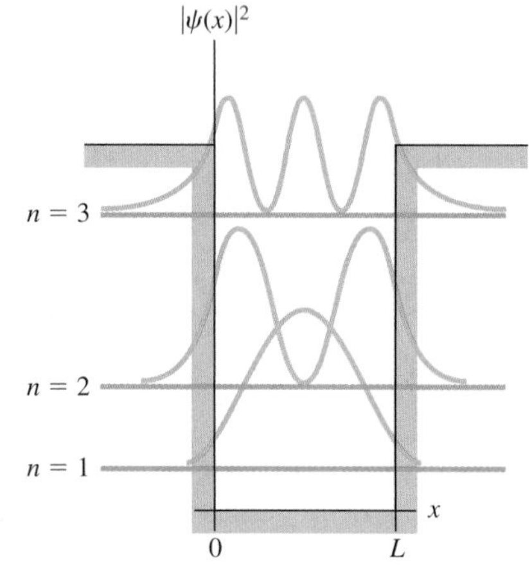

40.9 Probability distribution functions $|\psi(x)|^2$ for the square-well wave functions shown in Fig. 40.8. The horizontal brown line for each wave function corresponds to $|\psi|^2 = 0$.

ActivPhysics Online 20.3 Potential Wells

When U_0 is much *larger* than E_∞ (a very deep well), there are many bound states, and the energies of the lowest few are nearly the same as the energies for the infinitely deep well. When U_0 is only a few times as large as E_∞, there are only a few bound states. (But there is always at least one bound state, no matter how shallow the well.) As with the infinitely deep well, there is *no* state with $E = 0$; such a state would violate the uncertainty principle.

Figure 40.8 shows the particular case in which $U_0 = 6E_\infty$; for this case there are three bound states. The energy levels are expressed both as fractions of the well depth U_0 and as multiples of E_∞. Note that if the well were infinitely deep, the lowest three levels, as given by Eq. (40.9), would be E_∞, $4E_\infty$, and $9E_\infty$. The wave functions for the three bound states are also shown.

It turns out that when U_0 is less than E_∞, there is only one bound state. In the limit when U_0 is *much smaller* than E_∞ (a very shallow or very narrow well), the energy of this single state is approximately $E = 0.68U_0$.

Figure 40.9 shows graphs of the probability distributions, that is, the values of $|\psi|^2$, for the wave functions shown in Fig. 40.8a. As with the infinite well, not all positions are equally likely. We have already commented on the possibility of finding the particle outside the well, in the classically forbidden regions.

There are also states for which E is *greater than* U_0. In this case the particle is not bound but is free to move through all values of x. Thus *any* energy E greater than U_0 is possible. These free-particle states thus form a *continuum* rather than a discrete set of states with definite energy levels. The free-particle wave functions are sinusoidal both inside and outside the well. The wavelength is shorter inside the well than outside, corresponding to greater kinetic energy in the well than outside it.

The square-well potential described in this section has a number of practical applications. We mentioned earlier the example of an electron inside a metallic sheet. A three-dimensional version, in which U is zero inside a spherical region with radius R and has the value U_0 outside, provides a simple model to represent the interaction of a neutron with a nucleus in neutron-scattering experiments. In this context the model is called the *crystal-ball model* of the nucleus, because neutrons interacting with such a potential are scattered in a way that's analogous to scattering of light by a crystal ball.

Example 40.4 Electron in a square well

An electron is trapped in a square well with a width of 0.50 nm (comparable to a few atomic diameters). a) Find the ground-level energy if the well is infinitely deep. b) If the actual well depth is six times the ground-level energy found in part (a), find the energy levels. c) If the atom makes a transition from a state with energy E_2 to a state with energy E_1 by emitting a photon, find the wavelength of the photon. In what region of the electromagnetic spectrum does the photon lie? d) If the electron is initially in its ground level and absorbs a photon, what is the minimum energy the photon must have to liberate the electron from the well? In what region of the spectrum does the photon lie?

SOLUTION

IDENTIFY and SET UP: We use Eq. (40.20) to find the ground-level energy E_∞ for an infinitely deep well. The energies for a square well with $U_0 = 6E_\infty$ are shown in Fig. 40.8b. The energy of the photon emitted or absorbed in a transition is equal to the difference in energy between two states involved in the transition; the photon wavelength is given by the relation $E = hf = hc/\lambda$ [Eq. (38.8)].

EXECUTE: a) From Eq. (40.20),

$$E_\infty = \frac{\pi^2\hbar^2}{2mL^2} = \frac{\pi^2(1.055 \times 10^{-34}\ \text{J}\cdot\text{s})^2}{2(9.11 \times 10^{-31}\ \text{kg})(0.50 \times 10^{-9}\ \text{m})^2}$$
$$= 2.4 \times 10^{-19}\ \text{J} = 1.5\ \text{eV}$$

b) We are given $U_0 = 6E_\infty$, so $U_0 = 6(1.5\ \text{eV}) = 9.0\ \text{eV}$. We can simply read off the energy levels from Fig. 40.8b. Our answers are

$$E_1 = 0.104U_0 = 0.104(9.0\ \text{eV}) = 0.94\ \text{eV}$$
$$E_2 = 0.405U_0 = 0.405(9.0\ \text{eV}) = 3.6\ \text{eV}$$
$$E_3 = 0.848U_0 = 0.848(9.0\ \text{eV}) = 7.6\ \text{eV}$$

Alternatively,

$$E_1 = 0.625E_\infty = 0.625(1.5\ \text{eV}) = 0.94\ \text{eV}$$
$$E_2 = 2.43E_\infty = 0.243(1.5\ \text{eV}) = 3.6\ \text{eV}$$
$$E_3 = 5.09E_\infty = 5.09(1.5\ \text{eV}) = 7.6\ \text{eV}$$

c) The photon energy is

$$E_2 - E_1 = 3.6\ \text{eV} - 0.94\ \text{eV} = 2.7\ \text{eV}$$

We determine the wavelength from $E = hf = hc/\lambda$:

$$\lambda = \frac{hc}{E} = \frac{(4.136 \times 10^{-15}\ \text{eV}\cdot\text{s})(3.00 \times 10^8\ \text{m/s})}{2.7\ \text{eV}} = 460\ \text{nm}$$

This photon is in the blue region of the visible spectrum.

d) We see from Fig. 40.8b that the minimum energy needed to lift the electron out of the well from its $n = 1$ ground level is the well depth $(U_0 = 9.0\ \text{eV})$ minus the electron's initial energy $(E_1 = 0.94\ \text{eV})$, or 8.1 eV. Since 8.1 eV is three times the 2.7-eV photon energy of part (c), the corresponding photon wavelength is one-third of 460 nm, or 150 nm, which is in the ultraviolet region of the spectrum.

EVALUATE: To check our results for the energies of the three bound states, note that if the well had been infinitely deep, the energies would have been

$$E_1 = E_\infty = 1.5\ \text{eV} \qquad E_2 = 4E_\infty = 6.0\ \text{eV}$$
$$E_3 = 9E_\infty = 13.5\ \text{eV}$$

The energies we found in part (b) are less than these values. This makes sense: as we mentioned earlier, the finite depth of the well lowers the energy levels compared to the levels for an infinitely deep well.

Test Your Understanding

Suppose that the width of the finite potential well shown in Fig. 40.8 is reduced by one-half. How must the value of U_0 change so that there are still just three bound energy levels whose energies are the fractions of U_0 shown in Fig. 40.8b? What will be the effect on the energies of these levels?

40.3 | Potential Barriers and Tunneling

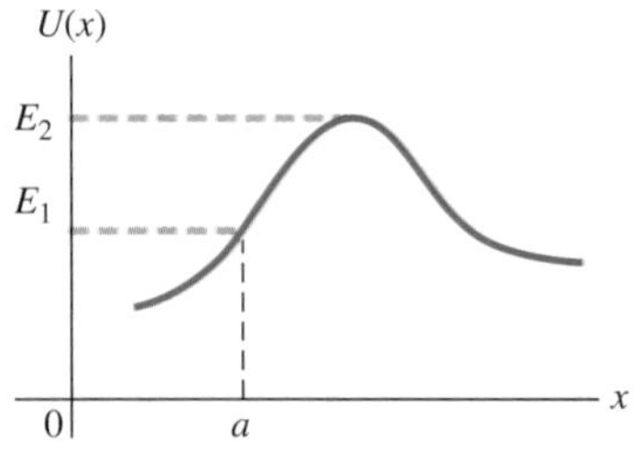

40.10 A potential-energy barrier. According to Newtonian mechanics, if the total energy is E_1, a particle that is on the left side of the barrier can go no farther than $x = a$. If the total energy is greater than E_2, the particle can pass the barrier.

A **potential barrier** is the opposite of a potential well; it is a potential-energy function with a *maximum.* Figure 40.10 shows an example. In Newtonian mechanics, if the total energy is E_1, a particle that is initially to the left of the barrier must remain to the left of the point $x = a$. If it were to move to the right of this point, the potential energy U would be greater than the total energy E. Because $K = E - U$, the kinetic energy would be negative, and this is impossible since a negative value of $K = \frac{1}{2}mv^2$ would require negative mass or imaginary speed.

If the total energy is greater than E_2, the particle can get past the barrier. A roller-coaster car can surmount a hill if it has enough kinetic energy at the bottom. If it doesn't, it stops partway up and then rolls back down.

Quantum mechanics supplies us with a peculiar and very interesting phenomenon in connection with potential barriers. A particle that encounters such a barrier is not necessarily turned back; there is some probability for it to emerge on the other side, even when it doesn't have enough kinetic energy to surmount the barrier according to Newtonian mechanics. This penetration of a barrier is called **tunneling.** It's a natural name; if you dig a tunnel through a hill, you don't have to go over the top. However, in quantum-mechanical tunneling, the particle does not actually bore a hole through the barrier and loses no energy in the tunneling process.

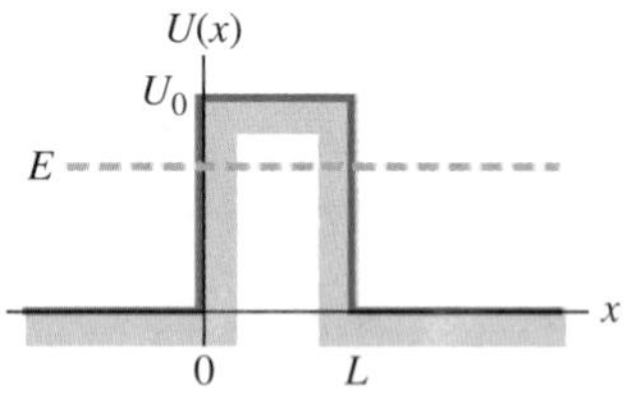

40.11 A potential-energy barrier with width L and height U_0. According to Newtonian mechanics, if the total energy E is less than U_0, a particle cannot pass over this barrier but is confined to the side where it starts.

ActivPhysics ONLINE

20.4 Potential Barriers

To understand how tunneling can occur, let's look at the potential-energy function $U(x)$ shown in Fig. 40.11. It's like Fig. 40.6 turned upside-down; the potential energy is zero everywhere except in the range $0 \le x \le L$, where it has the value U_0. This might represent a simple model for an electron and two slabs of metal separated by an air gap of thickness L. We let the potential energy be zero within either slab but equal to U_0 in the gap between them.

Let's consider solutions of the Schrödinger equation for this potential-energy function for the case in which E is less than U_0. We can use our results from Section 40.2. In the regions $x < 0$ and $x > L$ the solution is sinusoidal and given by Eq. (40.17). Within the barrier $(0 \le x \le L)$ the solution is exponential, as in Eq. (40.19). Just as with the finite potential well, the functions have to join smoothly at the boundary points $x = 0$ and $x = L$. That is, the function and its derivative have to be continuous at these points.

These requirements lead to a wave function like the one shown in Fig. 40.12. The function is *not* zero inside the barrier (the region forbidden by Newtonian mechanics). Even more remarkable, a particle that is initially to the *left* of the barrier has some probability of being found to the *right* of the barrier. How great a probability? That depends on the width L of the barrier and the particle's energy E (all kinetic) in comparison with the barrier height U_0. The probability T that the particle gets through the barrier is proportional to the square of the ratio of the

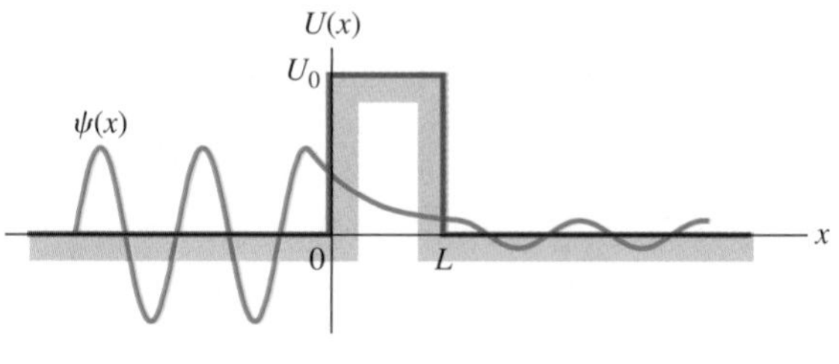

40.12 A possible wave function for a particle tunneling through the potential-energy barrier shown in Fig. 40.11. The function is exponential within the barrier $(0 \le x \le L)$ and sinusoidal outside the barrier. The functions must join smoothly at $x = 0$ and $x = L$; the function and its derivative (slope) must be continuous.

amplitudes of sinusoidal wave functions on the two sides of the barrier. These amplitudes are determined by matching wave functions and their derivatives at the boundary points, a fairly involved mathematical problem. When T is much smaller than unity, it is given approximately by

$$T = Ge^{-2\kappa L} \quad \text{where} \quad G = 16\frac{E}{U_0}\left(1 - \frac{E}{U_0}\right) \quad \text{and} \quad \kappa = \frac{\sqrt{2m(U_0 - E)}}{\hbar}$$

(probability of tunneling) (40.21)

The probability decreases rapidly with increasing barrier width L. It also depends critically on the energy difference $U_0 - E$, representing the additional kinetic energy the particle would need to be able to climb over the barrier in a Newtonian analysis.

Example 40.5 Tunneling through a barrier

A 2.0-eV electron encounters a barrier with height 5.0 eV. What is the probability that it will tunnel through the barrier if the barrier width is a) 1.00 nm; b) 0.50 nm (roughly ten and five atomic diameters, respectively)?

SOLUTION

IDENTIFY and SET UP: Our target variable is the tunneling probability T in Eq. (40.21), which we evaluate for the given values $E = 2.0\text{ eV}$ (electron energy), $U = 5.0\text{ eV}$ (barrier height), $m = 9.11 \times 10^{-31}\text{ kg}$ (mass of the electron), and $L = 1.00\text{ nm}$ or 0.50 nm (barrier width).

EXECUTE: First we evaluate G and κ in Eq. (40.21), using $E = K = 2.0\text{ eV}$:

$$G = 16\frac{2.0\text{ eV}}{5.0\text{ eV}}\left(1 - \frac{2.0\text{ eV}}{5.0\text{ eV}}\right) = 3.8$$

$$U_0 - E = 5.0\text{ eV} - 2.0\text{ eV} = 3.0\text{ eV} = 4.8 \times 10^{-19}\text{ J}$$

$$\kappa = \frac{\sqrt{2(9.11 \times 10^{-31}\text{ kg})(4.8 \times 10^{-19}\text{ J})}}{1.055 \times 10^{-34}\text{ J}\cdot\text{s}} = 8.9 \times 10^9\text{ m}^{-1}$$

(Note that the eV energy units canceled in calculating G, but we had to convert eV to J = kg$\cdot$m^2/s^2 to find κ in m^{-1}.)

a) When $L = 1.00\text{ nm} = 1.00 \times 10^{-9}\text{m}$, $2\kappa L = 2(8.9 \times 10^9\text{ m}^{-1}) \times (1.00 \times 10^{-9}\text{ m}) = 17.8$, and

$$T = Ge^{-2\kappa L} = 3.8e^{-17.8} = 7.1 \times 10^{-8}$$

b) When $L = 0.50\text{ nm}$, half of 1.00 nm, $2\kappa L$ is half of 17.8, or 8.9, and

$$T = 3.8e^{-8.9} = 5.2 \times 10^{-4}$$

EVALUATE: Halving the width of this barrier increases the tunneling probability by a factor of nearly ten thousand.

Applications of Tunneling

Tunneling is significant in many areas of physics, including some with considerable practical importance. For example, when you twist two copper wires together or close the contacts of a switch, current passes from one conductor to the other despite a thin layer of nonconducting copper oxide between them. The electrons tunnel through this thin insulating layer. The *tunnel diode* is a semiconductor device in which electrons tunnel through a potential barrier. The current can be switched on and off very quickly (within a few picoseconds) by varying the height of the barrier. The *Josephson junction* consists of two superconductors separated by an oxide layer a few atoms (1 to 2 nm) thick. Electron pairs in the superconductors can tunnel through the barrier layer, giving such a device unusual circuit properties. Josephson junctions are useful for establishing precise voltage standards and measuring tiny magnetic fields.

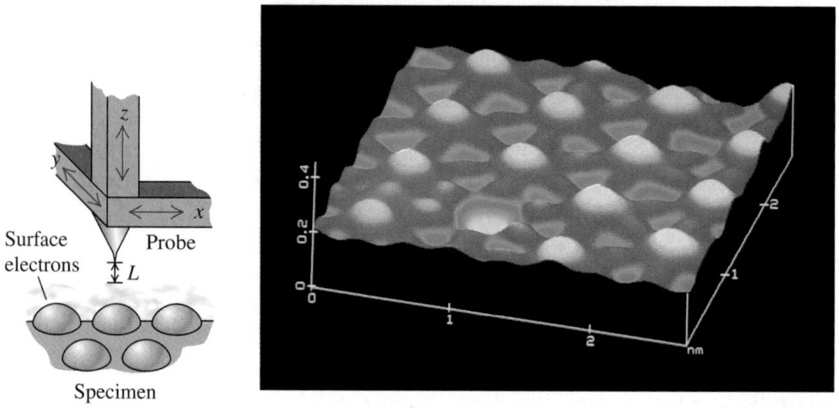

(a) (b)

40.13 (a) Schematic diagram of the probe of a scanning tunneling microscope (STM). As the sharp conducting probe is scanned across the surface in the x- and y-directions, it is also moved in the z-direction to maintain a constant tunneling current. The changing position of the probe is recorded and used to construct an image of the surface. (b) This colored STM image shows sodium atoms (pink) absorbed on a platinum surface. The yellow spot indicates a missing atom.

The *scanning tunneling microscope* (STM) uses electron tunneling to create images of surfaces down to the scale of individual atoms. An extremely sharp conducting needle is brought very close to the surface, within 1 nm or so (Fig. 40.13a). When the needle is at a positive potential with respect to the surface, electrons can tunnel through the surface potential-energy barrier and reach the needle. As Example 40.5 shows, the tunneling probability and hence the tunneling current depend strongly on the width L of the barrier (the distance between the surface and the needle tip). In one mode of operation the needle is scanned across the surface and at the same time is moved perpendicular to the surface to maintain a constant tunneling current. The needle motion is recorded, and after many parallel scans, an image of the surface can be reconstructed. Extremely precise control of needle motion, including isolation from vibration, is essential. Figure 40.13b shows an image of iodine atoms on the surface of a platinum crystal.

Tunneling is of great importance in nuclear physics. A fusion reaction can occur when two nuclei tunnel through the barrier caused by their electrical repulsion and approach each other closely enough for the attractive nuclear force to cause them to fuse. Fusion reactions occur in the cores of stars, including the sun; without tunneling, the sun wouldn't shine. The emission of alpha particles from unstable nuclei also involves tunneling. An alpha particle at the surface of a nucleus encounters a potential barrier that results from the combined effect of the attractive nuclear force and the electrical repulsion of the remaining part of the nucleus (Fig. 40.14). The alpha particle tunnels through this barrier. Because the tunneling probability depends so critically on the barrier height and width, the lifetimes of alpha-emitting nuclei vary over an extremely wide range. We'll return to alpha decay in Chapter 43.

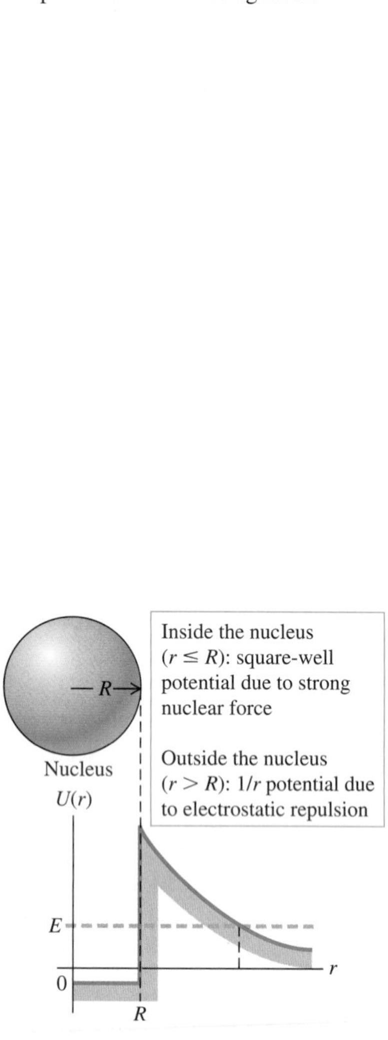

40.14 Approximate potential-energy function for an alpha particle interacting with a nucleus of radius R. If an alpha particle inside the nucleus has energy E greater than zero, it can tunnel through the barrier and escape from the nucleus.

Test Your Understanding

Is it possible for a particle undergoing tunneling to be found *within* the barrier rather than on either side of it?

40.4 | The Harmonic Oscillator

In Newtonian mechanics a **harmonic oscillator** is a particle with mass m acted on by a conservative force component $F_x = -k'x$. (In this discussion we will use k' for the *force constant* to minimize confusion with the wave number $k = 2\pi/\lambda$.) This force component is proportional to the particle's displacement x from its equilibrium position, $x = 0$. The corresponding potential-energy function is $U = \frac{1}{2}k'x^2$ (Fig. 40.15). When the particle is displaced from equilibrium, it undergoes sinusoidal motion with angular frequency $\omega = (k'/m)^{1/2}$. We studied this system in detail in Chapter 13; you should review that discussion.

A quantum-mechanical analysis of the harmonic oscillator using the Schrödinger equation is an interesting and useful study. The solutions provide insight into vibrations of molecules, the quantum theory of heat capacities, atomic vibrations in crystalline solids, and many other situations.

Before we get into the details, let's make an enlightened guess about the energy levels. The energy E of a *photon* is related to its angular frequency ω by $E = hf = (h/2\pi)(\omega/2\pi) = \hbar\omega$. The harmonic oscillator has a characteristic angular frequency ω, at least in Newtonian mechanics. So a reasonable guess would be that in the quantum-mechanical analysis the energy levels of a harmonic oscillator would be multiples of the quantity

$$\hbar\omega = \hbar\sqrt{\frac{k'}{m}}$$

These are the energy levels that Planck assumed in deriving his radiation law (Section 38.8). It was a good assumption; the energy levels are in fact half-integer $(\frac{1}{2}, \frac{3}{2}, \frac{5}{2}, \ldots)$ multiples of $\hbar\omega$.

For the harmonic oscillator we write $\frac{1}{2}k'x^2$ in place of U in the one-dimensional Schrödinger equation, Eq. (40.1):

$$-\frac{\hbar^2}{2m}\frac{d^2\psi}{dx^2} + \frac{1}{2}k'x^2\psi = E\psi \qquad (40.22)$$

(Schrödinger equation for the harmonic oscillator)

The solutions of this equation are wave functions for the physically possible states of the system.

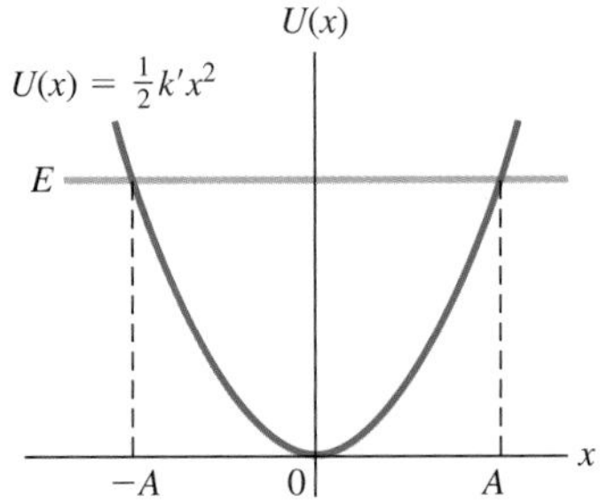

40.15 Potential energy function for the harmonic oscillator. In Newtonian mechanics the amplitude A is related to the total energy E by $E = \frac{1}{2}k'A^2$, and the particle is restricted to the range from $x = -A$ to $x = A$.

Wave Functions, Boundary Conditions, and Energy Levels

In the discussion of square-well potentials in Section 40.2 we found that the energy levels are determined by boundary conditions at the walls of the well. However, the harmonic-oscillator potential has no walls as such; what, then, are the appropriate boundary conditions? Classically, $|x|$ cannot be greater than the amplitude A, which is the maximum displacement from equilibrium. Quantum mechanics does allow some penetration into classically forbidden regions, but the probability decreases as that penetration increases. Thus the wave functions must approach zero as $|x|$ grows large.

Satisfying the requirement that $\psi(x) \to 0$ as $|x| \to \infty$ is not as trivial as it may seem. Suppose we compute solutions of Eq. (40.22) *numerically.* We start at some point x, choosing values of $\psi(x)$ and $d\psi(x)/dx$ at that point. Using the Schrödinger equation to evaluate $d^2\psi(x)/dx^2$ at the point, we can compute $\psi(x)$ and $d\psi(x)/dx$ at a neighboring point $x + \Delta x$, and so on. By using a computer to iterate this process many times with sufficiently small steps Δx, we can compute the wave function with great precision. There is no guarantee, however, that a function obtained in this way will approach zero at large x. To see what kinds of trouble we can get into, let's rewrite Eq. (40.22) in the form

$$\frac{d^2\psi(x)}{dx^2} = \frac{2m}{\hbar^2}\left(\frac{1}{2}k'x^2 - E\right)\psi(x) \qquad (40.23)$$

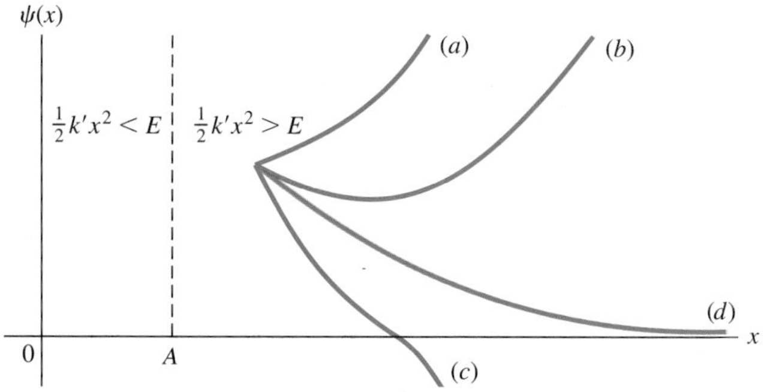

40.16 Possible behaviors of harmonic-oscillator wave functions in the region $\frac{1}{2}k'x^2 > E$. In this region, $\psi(x)$ and $d^2\psi(x)/dx^2$ have the same sign. The curve is concave upward when $d^2\psi(x)/dx^2$ is positive and concave downward when $d^2\psi(x)/dx^2$ is negative. Only curve (d), which approaches the x-axis asymptotically for large x, is an acceptable wave function for this system.

In this form, the equation shows that when x is large enough (either positive or negative) to make the quantity $(\frac{1}{2}k'x^2 - E)$ positive, the function $\psi(x)$ and its second derivative $d^2\psi(x)/dx^2$ have the same sign.

The second derivative of $\psi(x)$ is the rate of change of the *slope* of $\psi(x)$. Consider a point with $x > A$ for which $\frac{1}{2}k'x^2 - E > 0$. If $\psi(x)$ is positive, so is $d^2\psi(x)/dx^2$ and the function is *concave upward.* Figure 40.16 shows four possible kinds of behavior beginning at a point $x > A$. If the slope is initially positive, the function curves upward more and more steeply (curve a) and goes to infinity. If the slope is initially negative at the point, there are three possibilities. If the slope changes too quickly (curve b), the curve bends up and again goes to infinity. If the slope doesn't change quickly enough, the curve bends down and crosses the x-axis. After it has crossed, $\psi(x)$ and $d^2\psi(x)/dx^2$ are both negative (curve c) and the concave-downward curve heads for *negative* infinity. Between these infinities is the chance that the curve may bend just enough to glide in asymptotically to the x-axis (curve d). In this case, $\psi(x)$, $d\psi(x)/dx$, and $d^2\psi(x)/dx^2$ all approach zero at large x; in this possibility lies the only hope of satisfying the boundary condition that $\psi(x) \rightarrow 0$ as $|x| \rightarrow \infty$, and it occurs only for certain very special values of the constant E.

This qualitative discussion offers some insight into how the boundary conditions for this problem determine the possible energy levels. Equation (40.22) can also be solved exactly. The solutions, although not encountered in elementary calculus courses, are well known to mathematicians; they are called *Hermite functions.* Each one is an exponential function multiplied by a polynomial in x. The state with lowest energy (the ground level) has the wave function

$$\psi(x) = Ce^{-\sqrt{mk'}\,x^2/2\hbar} \tag{40.24}$$

The constant C is chosen to normalize the function, that is, to make $\int_{-\infty}^{\infty} |\psi|^2\,dx = 1$. You can find C using, from integral tables,

$$\int_{-\infty}^{\infty} e^{-a^2x^2}\,dx = \frac{\sqrt{\pi}}{a}$$

The corresponding energy, which we'll call E_0, is the ground-state energy; it is

$$E_0 = \frac{1}{2}\hbar\omega = \frac{1}{2}\hbar\sqrt{\frac{k'}{m}} \tag{40.25}$$

You may not believe that Eq. (40.24) really *is* a solution of Eq. (40.22) (the Schrödinger equation for the harmonic oscillator) with the energy given by Eq. (40.25). We invite you to calculate the second derivative of Eq. (40.24), substitute it into Eq. (40.22), and verify that it really is a solution if Eq. (40.25) gives E_0 (Exercise 40.28). It's a little messy, but the result is satisfying and worth the effort.

Further analysis of the Schrödinger equation for the harmonic oscillator shows that it can be written in the form

$$\frac{d^2\psi(x)}{dx^2} = \frac{2m}{\hbar^2}\left(\frac{1}{2}k'x^2 - \left(n + \frac{1}{2}\right)\hbar\sqrt{\frac{k'}{m}}\right)\psi(x)$$

Comparing this complicated expression to Eq. (40.23) gives us a pleasant surprise: The energy levels, which we'll call E_n, are given by the simple formula

$$E_n = \left(n + \frac{1}{2}\right)\hbar\sqrt{\frac{k'}{m}} = \left(n + \frac{1}{2}\right)\hbar\omega \qquad (n = 0, 1, 2, \dots) \qquad (40.26)$$

(energy levels, harmonic oscillator)

where n is the quantum number identifying each state and energy level. Note that the ground level of energy $E_0 = \frac{1}{2}\hbar\omega$ is denoted by $n = 0$, *not* $n = 1$.

Equation (40.26) confirms our guess that the energy levels are multiples of $\hbar\omega$. Adjacent energy levels are separated by a constant interval of $\hbar\omega = hf$, as Planck assumed in 1900. There are infinitely many levels; this shouldn't be surprising, because we are dealing with an infinitely deep potential well. As $|x|$ increases, $U = \frac{1}{2}k'x^2$ increases without bound.

Figure 40.17 shows the lowest six energy levels and the potential-energy function $U(x)$. For each level n, the value of $|x|$ at which the horizontal line representing the total energy E_n intersects $U(x)$ gives the amplitude A_n of the corresponding Newtonian oscillator.

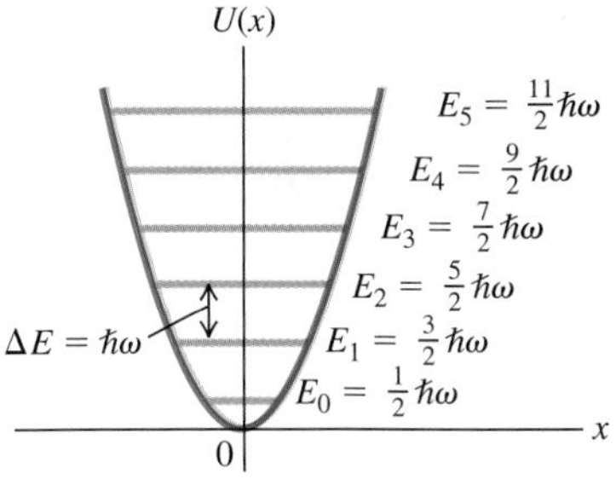

40.17 Energy levels for the harmonic oscillator. The spacing between any two adjacent levels is $\Delta E = \hbar\omega$. The energy of the ground level is $E_0 = \frac{1}{2}\hbar\omega$.

Example 40.6 Vibration in a sodium crystal

A sodium atom of mass 3.82×10^{-26} kg vibrates with simple harmonic motion in a crystal. The potential energy increases by 0.0075 eV when the atom is displaced 0.014 nm from its equilibrium position. a) Find the angular frequency, according to Newtonian mechanics. b) Find the spacing of adjacent energy levels in electron volts. c) If an atom emits a photon during a transition from one vibrational level to the next lower level, what is the wavelength of the emitted photon? In what region of the electromagnetic spectrum does it lie?

SOLUTION

IDENTIFY and SET UP: To find the angular frequency $\omega = (k'/m)^{1/2}$, we use the information given and the expression $U(x) = \frac{1}{2}k'x^2$ for the potential energy. The spacing between adjacent energy levels is given by Eq. (40.26). We calculate the wavelength of the emitted photon using the same ideas as in Example 40.4 (Section 40.2).

EXECUTE: We don't know k', but we can find it from the fact that $U = 0.0075\ \text{eV} = 1.2 \times 10^{-21}$ J when $x = 0.014 \times 10^{-9}$ m. We have

$$U = \frac{1}{2}k'x^2$$

$$k' = \frac{2U}{x^2} = \frac{2(1.2 \times 10^{-21}\ \text{J})}{(0.014 \times 10^{-9}\ \text{m})^2} = 12.2\ \text{N/m}$$

The angular frequency is

$$\omega = \sqrt{\frac{k'}{m}} = \sqrt{\frac{12.2\ \text{N/m}}{3.82 \times 10^{-26}\ \text{kg}}} = 1.79 \times 10^{13}\ \text{rad/s}$$

b) From Eq. (40.26) and Fig. 40.17 the spacing between adjacent energy levels is

$$\hbar\omega = (1.054 \times 10^{-34}\ \text{J}\cdot\text{s})(1.79 \times 10^{13}\ \text{s}^{-1})$$
$$= 1.88 \times 10^{-21}\ \text{J} = 0.0118\ \text{eV}$$

c) From $E = hc/\lambda$ the corresponding wavelength is

$$\lambda = \frac{hc}{E} = \frac{(4.136 \times 10^{-15}\ \text{eV}\cdot\text{s})(3.00 \times 10^{8}\ \text{m/s})}{0.0118\ \text{eV}}$$
$$= 1.05 \times 10^{-4}\ \text{m} = 1.05\ \mu\text{m}$$

This photon is in the infrared region of the spectrum.

EVALUATE: This example shows us that the force constants for interatomic forces are on the order of a few N/m, not very different from the values for household springs or spring-based toys such as a Slinky. It also suggests that physicists can learn about the vibrations of atoms by measuring the radiation that they emit as they transition from one vibrational state to a lower state. We will explore this idea further in Chapter 42.

Quantum and Newtonian Oscillators

Figure 40.18 shows the first four harmonic-oscillator wave functions. Each graph also shows the amplitude A of a Newtonian harmonic oscillator with the same energy, that is, the value of A determined from

$$\frac{1}{2}k'A^2 = \left(n + \frac{1}{2}\right)\hbar\omega \tag{40.27}$$

In each case there is some penetration of the wave function into the regions $|x| > A$ that are forbidden by Newtonian mechanics. This is similar to the effect that we noted with the particle in a finite square well.

CAUTION Be very careful with your algebraic symbols. This Newtonian amplitude A (in units of m) is *not* the same as the quantum-mechanical normalizing constant A (in $\text{m}^{-1/2}$) that is used in several wave functions in this chapter. Also the angular frequency ω (in rad/s) of the harmonic oscillator is *not* the same as the angular frequency ω (also in rad/s) obtained from the de Broglie relation $E = \hbar\omega$ (which is why we expressed Eq. (40.14) in terms of $2\pi E/h = E/\hbar$ rather than in terms of ω). Finally, we have distinguished the force constant k' (in N/m or kg/s^2) from the wave number k (in rad/m). Neither of these is the same as κ (in m^{-1}) used in Eqs. (40.19) and (40.21) or the kinetic energy K (in J or eV).

Figure 40.19 shows the probability distributions $|\psi(x)|^2$ for these same states. Each graph also shows the probability distributions determined from Newtonian analysis, in which the probability of finding the particle near a randomly chosen point is inversely proportional to its speed at that point. If we average out the wiggles in the quantum-mechanical probability curves, the results for $n > 0$ resemble the Newtonian predictions. This agreement improves with increasing n; Fig. 40.20 shows the classical and quantum-mechanical probability functions for

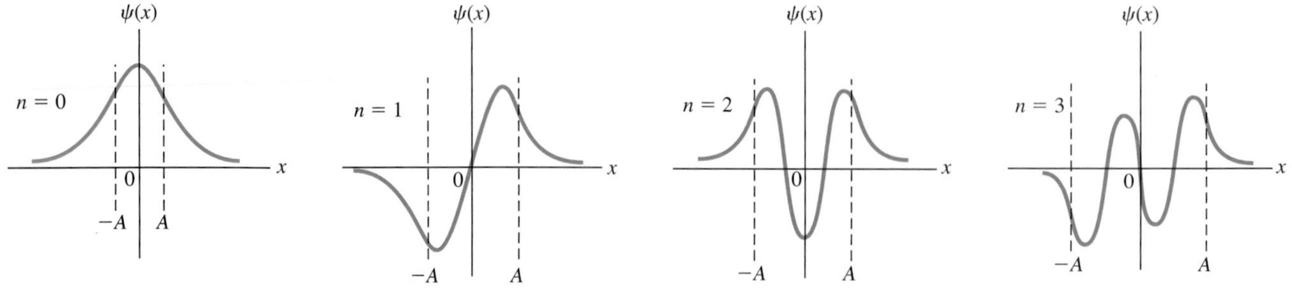

40.18 The first four wave functions for the harmonic oscillator. The amplitude A of a Newtonian oscillator with the same total energy is shown for each. Each wave function penetrates somewhat into the classically forbidden regions $|x| > A$. The total number of finite maxima and minima for each function is $n + 1$, one more than the quantum number.

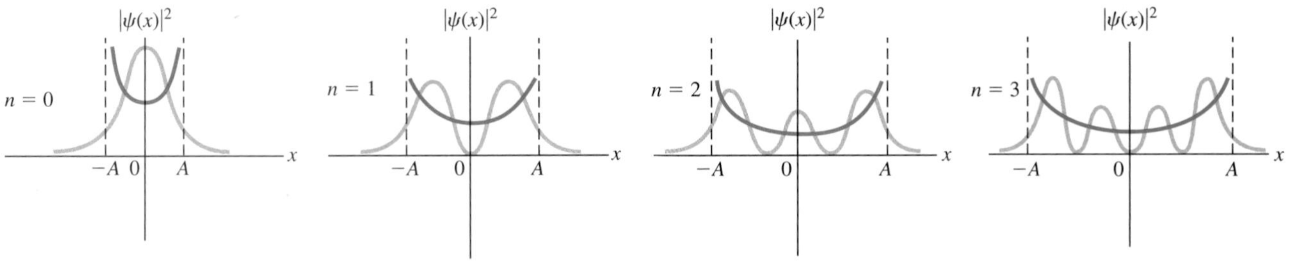

40.19 Probability distribution functions $|\psi|^2$ for the harmonic-oscillator wave functions shown in Fig. 40.18. The amplitude A of the Newtonian motion with the same energy is shown for each. The blue lines show the corresponding probability distributions for the Newtonian motion. As n increases, the averaged-out quantum-mechanical functions resemble the Newtonian curves more and more.

$n = 10$. Notice that the spacing between zeros of $|\psi(x)|^2$ in Fig. 40.20 increases with increasing distance from $x = 0$. This makes sense from the Newtonian perspective: as a particle moves away from $x = 0$, its kinetic energy K and the magnitude p of its momentum both decrease. Thinking quantum-mechanically, this means that the wavelength $\lambda = h/p$ increases, so the spacing between zeros of $\psi(x)$ (and hence of $|\psi(x)|^2$) also increases.

In the Newtonian analysis of the harmonic oscillator the minimum energy is zero, with the particle at rest at its equilibrium position. This is not possible in quantum mechanics; no solution of the Schrödinger equation has $E = 0$ and satisfies the boundary conditions. Furthermore, if there were such a state, it would violate the uncertainty principle because there would be no uncertainty in either position or momentum. Indeed, the energy must be at least $\frac{1}{2}\hbar\omega$ for the system to conform to the uncertainty principle. To see qualitatively why this is so, consider a Newtonian oscillator with total energy $\frac{1}{2}\hbar\omega$. We can find the amplitude A and the maximum velocity just as we did in Section 13.3. The appropriate relations are Eqs. (13.21) for the total energy and (13.23) for the maximum speed. Setting $E = \frac{1}{2}\hbar\omega$, we find

$$E = \frac{1}{2}k'A^2 = \frac{1}{2}\hbar\omega = \frac{1}{2}\hbar\sqrt{\frac{k'}{m}} \qquad A = \sqrt{\frac{\hbar}{\sqrt{k'm}}}$$

$$p_{\text{max}} = mv_{\text{max}} = m\sqrt{\frac{k'}{m}}A = m\sqrt{\frac{k'}{m}}\sqrt{\frac{\hbar}{\sqrt{k'm}}} = \sqrt{\hbar\sqrt{k'm}}$$

Finally, if we assume that A represents the uncertainty Δx in position and p_{max} is the corresponding uncertainty Δp_x in momentum, then the product of the two uncertainties is

$$\Delta x\,\Delta p_x = \sqrt{\frac{\hbar}{\sqrt{k'm}}}\sqrt{\hbar\sqrt{k'm}} = \hbar$$

So the product equals the minimum value allowed by Eq. (39.11), $\Delta x\,\Delta p_x \geq \hbar$, and thus satisfies the uncertainty principle. If the energy had been less than $\frac{1}{2}\hbar\omega$,

ActivPhysics Online

20.1.6 Potential Energy Diagrams, Question 6

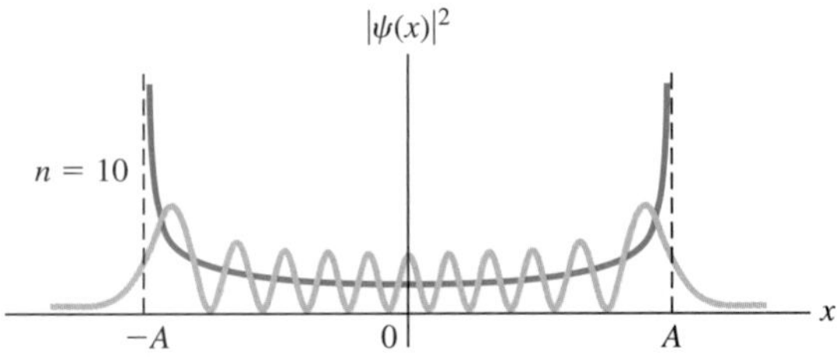

40.20 Newtonian and quantum-mechanical probability distribution functions for a harmonic oscillator for the state $n = 10$. The Newtonian amplitude A is also shown.

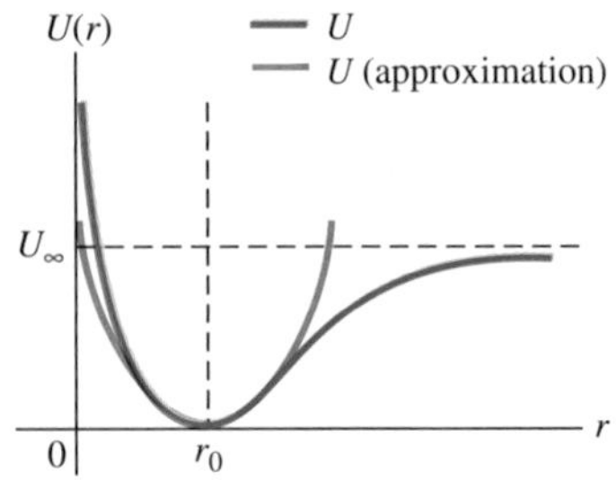

40.21 A potential-energy function describing the interaction of two atoms in a diatomic molecule. The equilibrium position is at $r = r_0$. When r is near r_0, the curve is approximately parabolic (as shown by the bright red curve), and the motion is approximately simple harmonic. The potential energy needed to dissociate the molecule is U_∞.

the product $\Delta x\, \Delta p_x$ would have been less than $\hbar$, and the uncertainty principle would have been violated.

Even when a potential-energy function isn't precisely parabolic in shape, we may be able to approximate it by the harmonic-oscillator potential for sufficiently small displacements from equilibrium. Figure 40.21 shows a typical potential-energy function for an interatomic force in a molecule. At large separations it levels off, corresponding to the absence of force at great distances. But it is approximately parabolic near the minimum point (the equilibrium position of the atoms). Near equilibrium the molecular vibration is approximately simple harmonic with energy levels given by Eq. (40.26), as we assumed in Example 40.6.

Test Your Understanding

If a harmonic oscillator is excited from the $n = N$ energy level to the $n = N + 1$ energy level, by what factor does the amplitude of the corresponding Newtonian harmonic oscillator increase?

40.5 | Three-Dimensional Problems

We have discussed the Schrödinger equation and its applications only for *one-dimensional* problems, the analog of a Newtonian particle moving along a straight line. The straight-line model is adequate for some applications, but to understand atomic structure, we need a three-dimensional generalization.

It's not difficult to guess what the three-dimensional Schrödinger equation should look like. First, the wave function ψ is a function of all three space coordinates (x, y, z). In general, the potential-energy function also depends on all three coordinates and can be written as $U(x, y, z)$. Next, note that the term $-(\hbar^2/2m)d^2\psi/dx^2$ in the one-dimensional Schrödinger equation, Eq. (40.1) is related to the kinetic energy of the particle in the state described by the wave function ψ. For example, if we insert into this term the wave function $\psi(x) = Ae^{ikx}$ for a free particle with magnitude of momentum $p = \hbar k$ and kinetic energy $K = p^2/2m$, we obtain $-(\hbar^2/2m)(ik)^2Ae^{ikx} = (\hbar^2k^2/2m)Ae^{ikx} = (p^2/2m)\psi(x) = K\psi(x)$. If the particle's momentum has three components (p_x, p_y, p_z), then the kinetic energy is

$$K = \frac{p_x^2}{2m} + \frac{p_y^2}{2m} + \frac{p_z^2}{2m} \tag{40.28}$$

These observations, taken together, suggest that the correct generalization of the Schrödinger equation to three dimensions is

$$-\frac{\hbar^2}{2m}\left(\frac{\partial^2\psi(x, y, z)}{\partial x^2} + \frac{\partial^2\psi(x, y, z)}{\partial y^2} + \frac{\partial^2\psi(x, y, z)}{\partial z^2}\right) + U(x, y, z)\psi(x, y, z) = E\psi(x, y, z) \tag{40.29}$$

(three-dimensional Schrödinger equation)

We use partial derivatives since ψ is a function of x, y, and z. As in one dimension, E is constant for a state of particular energy.

We won't pretend that we have *derived* Eq. (40.29). Like the one-dimensional version, this equation has to be tested by comparison of its predictions with experimental results. As we will see in later chapters, Eq. (40.29) passes this test with flying colors, so we are confident that it *is* the correct equation.

In many practical problems, in atomic structure and elsewhere, the potential-energy function is *spherically symmetric;* it depends only on the distance $r = (x^2 + y^2 + z^2)^{1/2}$ from the origin of coordinates. To take advantage of this symmetry, we use *spherical coordinates* (r, θ, ϕ) (Fig. 40.22) instead of Cartesian coordinates (x, y, z). Then a spherically symmetric potential-energy function is a function only of r, not of θ or ϕ, so $U = U(r)$. This fact turns out to simplify greatly the problem of finding solutions of the Schrödinger equation, even though the derivatives in Eq. (40.29) are considerably more complex when expressed in terms of spherical coordinates. Be careful; many math texts exchange the angles θ and ϕ shown in Fig. 40.22.

For the hydrogen atom the potential-energy function $U(r)$ is the familiar Coulomb's-law function:

$$U(r) = -\frac{1}{4\pi\epsilon_0}\frac{e^2}{r} \tag{40.30}$$

We will find that for *all* spherically symmetric potential-energy functions $U(r)$, each possible wave function can be expressed as a product of three functions: one a function only of r, one only of θ, and one only of ϕ. Furthermore, the functions of θ and ϕ are *the same* for *every* spherically symmetric potential-energy function. This result is directly related to the problem of finding the possible values of *angular momentum* for the various states. We'll discuss these matters more in the next chapter.

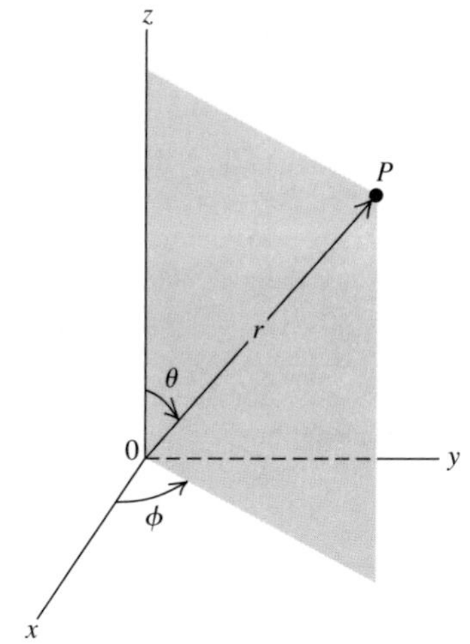

40.22 The position of a point P can be described by the rectangular coordinates (x, y, z) or by the spherical coordinates (r, θ, ϕ). In quantum mechanics, spherical coordinates are particularly useful in problems in which the potential energy depends only on r, the distance from the origin.

SUMMARY

To be a solution of the Schrödinger equation, the wave function $\psi(x)$ and its derivative $d\psi(x)/dx$ must be continuous everywhere, except where the potential-energy function $U(x)$ has an infinite discontinuity. Wave functions are usually normalized so that the total probability for finding the particle somewhere is unity.

$$\int_{-\infty}^{\infty} |\psi(x)|^2 \, dx = 1 \qquad (40.11)$$

(normalization condition)

The energy levels for a particle of mass m in a box (an infinitely deep square potential well) with width L are given by Eq. (40.9). The corresponding normalized wave functions of the particle are given by Eq. (40.13). (See Examples 40.1 and 40.2)

$$E_n = \frac{p_n^2}{2m} = \frac{n^2h^2}{8mL^2} = \frac{n^2\pi^2\hbar^2}{2mL^2} \qquad (40.9)$$

$(n = 1, 2, 3, \dots)$

$$\psi_n(x) = \sqrt{\frac{2}{L}} \sin\frac{n\pi x}{L} \qquad (40.13)$$

$(n = 1, 2, 3, \dots)$

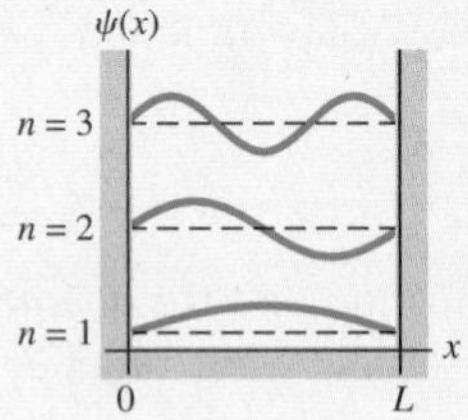

In a potential well with finite depth U_0, the energy levels are lower than those for an infinitely deep well with the same width, and the number of energy levels corresponding to bound states is finite. The levels are obtained by matching wave functions at the well walls to satisfy the continuity of $\psi(x)$ and $d\psi(x)/dx$. (See Examples 40.3 and 40.4)

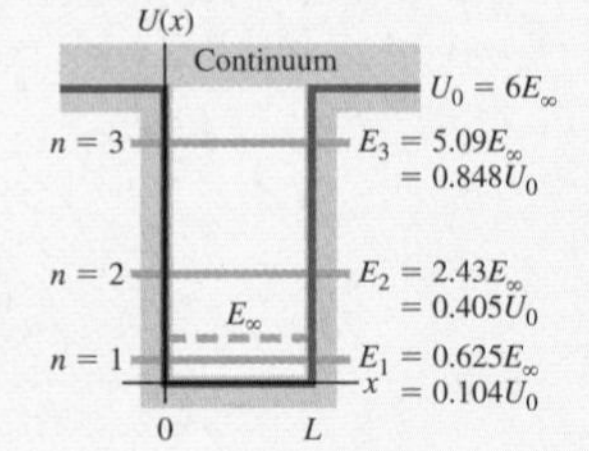

There is a certain probability that a particle will penetrate a potential energy barrier although its initial kinetic energy is less than the barrier height. This process is called tunneling. (See Example 40.5)

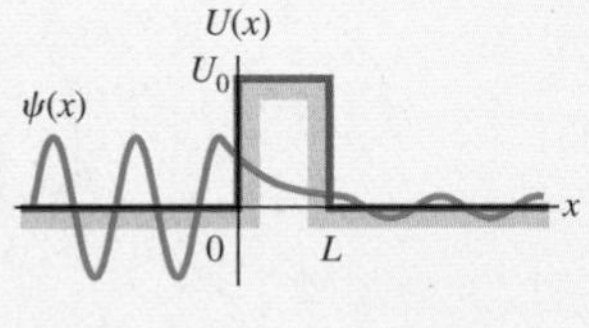

The energy levels for the harmonic oscillator, for which $U(x) = \frac{1}{2}k'x^2$, are given by Eq. (40.26). The spacing between any two adjacent levels is $\hbar\omega$, where $\omega = \sqrt{k'/m}$ is the oscillation angular frequency of the corresponding Newtonian harmonic oscillator. (See Example 40.6)

$$E_n = \left(n + \frac{1}{2}\right)\hbar\omega = \left(n + \frac{1}{2}\right)\hbar\sqrt{\frac{k'}{m}} \qquad (40.26)$$

$(n = 0, 1, 2, 3, \dots)$

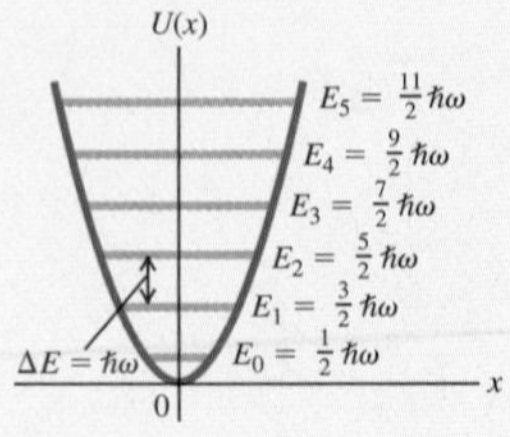

The Schrödinger equation for three-dimensional problems is given by Eq. (40.29).	$$-\frac{\hbar^2}{2m}\left(\frac{\partial^2\psi(x,y,z)}{\partial x^2}+\frac{\partial^2\psi(x,y,z)}{\partial y^2}+\frac{\partial^2\psi(x,y,z)}{\partial z^2}\right)+U(x,y,z)\psi(x,y,z) = E\psi\,(x,y,z) \quad (40.29)$$

Key Terms

particle in a box, 1520
normalization, 1524
potential well, 1525
square-well potential, 1526
bound state, 1526
potential barrier, 1530
tunneling, 1530
harmonic oscillator, 1532

Your Notes

Answer to Chapter Opening Question

The wavelength of a stationary-state wave function varies with position if the potential energy varies with position. One example is a harmonic oscillator wave function, as discussed in Section 40.4. But if the potential energy is independent of position, as is the case inside the "box" described in Section 40.1, the wavelength is also independent of position. Hence the potential energy has almost the same value at all points inside the circle of iron atoms shown in the image that opens this chapter.

Answers to Test Your Understanding Questions

Section 40.1 Our derivation of the stationary-state wave functions for a particle in a box shows that they are superpositions of waves propagating in opposite directions, just like a standing wave on a string. One wave has momentum in the positive x-direction, while the other wave has an equal magnitude of momentum in the negative x-direction. The *total* x-component of momentum is zero.

Section 40.2 The energy levels are arranged as shown in Fig. 40.8b if $U_0 = 6E_\infty$, where $E_\infty = \pi^2\hbar^2/2mL^2$ is the ground-level energy of an infinite well. If the well width L is reduced to $\frac{1}{2}$ of its initial value, E_∞ increases by a factor of 4 and so U_0 must also increase by a factor of 4. The energies E_1, E_2, and E_3 shown in Fig. 40.8b are all specific fractions of U_0, so they will also increase by a factor of 4.

Section 40.3 Yes. Figure 40.12 shows a possible wave function $\psi(x)$ for tunneling; since $\psi(x)$ is not zero within the barrier $(0 \le x \le L)$, there is some probability that the particle can be found there.

Section 40.4 From Eq. (40.26), the energies of the two levels are $E_N = (N + \frac{1}{2})\hbar\omega$ and $E_{N+1} = (N + 1 + \frac{1}{2})\hbar\omega = (N + \frac{3}{2})\hbar\omega$. The amplitude A of a Newtonian harmonic oscillator is related to the energy E by $E = \frac{1}{2}k'A^2$, so $A = \sqrt{2E/k'}$. Hence the ratio of the amplitudes is $A_{N+1}/A_N = (\sqrt{2E_{N+1}/k'})/(\sqrt{2E_N/k'}) = \sqrt{E_{N+1}/E_N}$, where $E_N = (N + \frac{1}{2})\hbar\omega$ and $E_{N+1} = (N + 1 + \frac{1}{2})\hbar\omega = (N + \frac{3}{2})\hbar\omega$ are the energies of the two levels. Substituting, $A_{N+1}/A_N = \sqrt{(N + 3/2)\hbar\omega/(N + 1/2)\hbar\omega} = \sqrt{(N + 3/2)/(N + 1/2)}$. For example, if the oscillator is excited from the $n = 2$ level to the $n = 3$ level, $N = 2$ and the Newtonian amplitude increases by a factor $A_3/A_2 = \sqrt{(2 + 3/2)/(2 + 1/2)} = \sqrt{7/5}$.

Discussion Questions

Q40.1 For the particle in a box, we chose $k = n\pi/L$ with $n = 1, 2, 3, \ldots$ to fit the boundary condition that $\psi = 0$ at $x = L$. However, $n = 0$ and $n = -1, -2, -3, \ldots$ also satisfy that boundary condition. Why didn't we also choose those values of n?

Q40.2 If ψ is normalized, what is the physical significance of the area under a graph of $|\psi|^2$ versus x between x_1 and x_2? What is the total area under the graph of $|\psi|^2$ when all x are included? Explain.

Q40.3 For a particle in a box, what would the probability distribution function $|\psi|^2$ look like if the particle behaved as a classical (Newtonian) particle? Do the actual probability distributions approach this classical form when n is very large? Explain.

Q40.4 In Chapter 15 we represented a standing wave as a superposition of two waves traveling in opposite directions. Can the wave functions for a particle in a box also be thought of as a combination of two traveling waves? Why or why not? What physical interpretation does this representation have? Explain.

Q40.5 A particle in a box is in the ground level. What is the probability of finding the particle in the right half of the box? (Refer to Fig. 40.5, but don't evaluate an integral.) Is the answer the same if the particle is in an excited level? Explain.

Q40.6 The wave functions for a particle in a box (Fig. 40.5a) are zero at certain points. Does this mean that the particle can't move past one of these points? Explain.

Q40.7 For a particle confined to an infinite square well, is it correct to say that each state of definite energy is also a state of definite wavelength? Is it also a state of definite momentum? Explain. (*Hint:* Remember that momentum is a vector.)

Q40.8 For a particle in a finite potential well, is it correct to say that each bound state of definite energy is also a state of definite wavelength? Is it a state of definite momentum? Explain.

Q40.9 Equation 40.15 gives the time-dependent wave functions for a particle in a box. How can a wave function that has both a real and an imaginary part describe the position (a real number) of a particle?

Q40.10 In Fig. 40.5b, the probability function is zero at the points $x = 0$ and $x = L$, the "walls" of the box. Does this mean that the particle never strikes the walls? Explain.

Q40.11 In Fig. 40.7, the second derivative $d^2\psi/dx^2$ for x slightly less than zero has the opposite sign to the second derivative for x slightly greater than zero. Use the properties of the Schrödinger equation, Eq. (40.1), to explain why this should be.

Q40.12 Compare the wave functions for the first three energy levels for a particle in a box of width L (Fig. 40.5a) to the corresponding wave functions for a finite potential well of the same width (Fig. 40.8a). How does the wavelength in the interval $0 \le x \le L$ for the $n = 1$ level of the particle in a box compare to the corresponding wavelength for the $n = 1$ level of the finite potential well? Use this to explain why E_1 is less than E_∞ in the situation depicted in Fig. 40.8b.

Q40.13 It is stated in Section 40.2 that a finite potential well always has at least one bound level, no matter how shallow the well. Does this mean that as $U_0 \to 0$, $E_1 \to 0$? Does this violate the Heisenberg uncertainty principle? Explain.

Q40.14 Figure 40.8a shows that the greater the energy of a bound state for a finite potential well, the more the wave function extends outside the well (into the intervals $x < 0$ and $x > L$). Explain why this happens.

Q40.15 In classical (Newtonian) mechanics, the total energy E of a particle can never be less than the potential energy U because the kinetic energy K cannot be negative. Yet in barrier tunneling (Section 40.3) a particle passes through regions where E is less than U. Is this a contradiction? Explain.

Q40.16 In Example 40.5, halving L increased T by a factor of nearly 10,000. Keeping L constant, would halving the barrier height U_0 have the same effect? The initial kinetic energy of the particle is kept fixed at 2.0 eV. Explain.

Q40.17 Qualitatively, how would you expect the probability for a particle to tunnel through a potential barrier to depend on the height of the barrier? Explain.

Q40.18 The wave function shown in Fig. 40.12 is nonzero for both $x < 0$ and $x > L$. Does this mean that the particle splits into two parts when it strikes the barrier, with one part tunneling through the barrier and the other part bouncing off the barrier? Explain.

Q40.19 The probability distributions for the harmonic oscillator wave functions (see Figs. 40.19 and 40.20) begin to resemble the classical (Newtonian) probability distribution when the quantum number n becomes large. Would the distributions become the same as in the classical case in the limit of very large n? Explain.

Q40.20 For a particle in a harmonic oscillator potential, what is the probability of finding the particle at any $x \geq 0$ if the particle is in the ground level? (Refer to Fig. 40.19, but do not evaluate an integral.) Is the answer the same if the particle is in an excited level? Explain.

Q40.21 Compare the allowed energy levels for the hydrogen atom, the particle in a box, and the harmonic oscillator. What are the values of the quantum number n for the ground level and the second excited level of each system?

Q40.22 In the hydrogen atom the potential-energy function depends only on distance r from the nucleus, not on direction. That is, it is spherically symmetric. Would you expect all the corresponding wave functions for the electron in the hydrogen atom to be spherically symmetric? Explain.

Q40.23 The hydrogen atom, particle in a box, and harmonic oscillator are all bound systems with discrete energy levels. Why are the bound-state energies E_n negative for the hydrogen atom but positive for the other two systems?

Exercises

Section 40.1 Particle in a Box

40.1 Ground-Level Billiards. a) Find the lowest energy level for a particle in a box if the particle is a billiard ball $(m = 0.20 \text{ kg})$ and the box has a width of 1.5 m, the size of a billiard table. (Assume that the billiard ball slides without friction rather than rolls. That is, ignore the *rotational* kinetic energy.) b) Since the energy in part (a) is all kinetic, to what speed does this correspond? How much time would it take at this speed for the ball to move from one side of the table to the other? c) What is the difference in energy between the $n = 2$ and $n = 1$ levels? d) Are quantum-mechanical effects important for the game of billiards?

40.2 A proton is in a box of width L. What must the width of the box be for the ground-level energy to be 5.0 MeV, a typical value for the energy with which the particles in a nucleus are bound? Compare your result to the size of a nucleus, that is on the order of 10^{-14} m.

40.3 It takes 3.0 eV of energy to excite an electron in a box from the ground level to the first excited level. What is the width L of the box?

40.4 When a hydrogen atom undergoes a transition from the $n = 2$ to the $n = 1$ level, a photon with $\lambda = 122$ nm is emitted. a) If the atom is modeled as an electron in a one-dimensional box, what is the width of the box in order for the $n = 2$ to $n = 1$ transition to correspond to emission of a photon of this energy? b) For a box with the width calculated in part (a), what is the ground state energy? How does this correspond to the ground state energy of a hydrogen atom? c) Do you think a one-dimensional box is a good model for a hydrogen atom? Explain. (*Hint:* Compare the spacing between adjacent energy levels as a function of n.)

40.5 A particle with mass $m = 5.00$ g is in a box with width L. a) Classically, what is the kinetic energy of the particle if it has speed 1.0 cm/s? b) What is L if the ground state energy of the particle equals the kinetic energy calculated in part (a)?

40.6 Recall that $|\psi|^2\,dx$ is the probability of finding the particle that has normalized wave function $\psi(x)$ in the interval x to $x + dx$. Consider a particle in a box with rigid walls at $x = 0$ and $x = L$. Let the particle be in the ground level and use ψ_n as given in Eq. (40.13). a) For which values of x, if any, in the range from 0 to L is the probability of finding the particle zero? b) For which values of x is the probability largest? c) In parts (a) and (b) are your answers consistent with Fig. 40.8? Explain.

40.7 Repeat Exercise 40.6 for the particle in the first excited level.

40.8 a) Find the excitation energy from the ground level to the third excited level for an electron confined to a box that has a width of 0.125 nm. b) The electron makes a transition from the $n = 1$ to $n = 4$ level by absorbing a photon. Calculate the wavelength of this photon.

40.9 An electron is in a box of width 3.0×10^{-10} m. What are the de Broglie wavelength and the magnitude of the momentum of the electron if it is in a) the $n = 1$ level; b) the $n = 2$ level; c) the $n = 3$ level? In each case how does the wavelength compare to the width of the box?

40.10 a) Show that $\psi = A \sin kx$ is a solution to Eq. (40.3) if $k = \sqrt{2mE}/\hbar$. b) Explain why this is an acceptable wave function for a particle in a box with rigid walls at $x = 0$ and $x = L$ only if k is an integer multiple of π/L.

40.11 a) Repeat Exercise 40.10 for $\psi = A \cos kx$. b) Explain why this cannot be an acceptable wave function for a particle in a box with rigid walls at $x = 0$ and $x = L$ no matter what the value of k.

40.12 Let ψ be a solution of Eq. (40.1) with energy E. Show that the function $\psi' = A\psi$ is also a solution with the same E, for any values of the constant A. (This result shows that normalizing a wave function (Section 40.1) does not change the energy associated with the wave function.)

Section 40.2 Potential Wells

40.13 a) Show that $\psi = A \sin kx$, where k is a constant, is *not* a solution of Eq. (40.1) for $U = U_0$ and $E < U_0$. b) Is this ψ a solution for $E > U_0$?

40.14 An electron is moving past the square well shown in Fig. 40.6. The electron has energy $E = 3U_0$. What is the ratio of the de Broglie wavelength of the electron in the region $x > L$ to the wavelength for $0 < x < L$?

40.15 An electron is bound in a square well of depth $U_0 = 6E_\infty$. What is the width of the well if its ground state energy is 2.00 eV?

40.16 An electron is bound in a square well of width 1.50 nm and depth $U_0 = 6E_\infty$. If the electron is initially in the ground level and absorbs a photon, what maximum wavelength can the photon have and still liberate the electron from the well?

40.17 A proton is bound in a square well of width 4.0 fm = 4.0×10^{-15} m. The depth of the well is six times the ground level energy E_∞ of the corresponding infinite well. If the proton makes a transition from the level with energy E_1 to the level with energy E_3 by absorbing a photon, find the wavelength of the photon.

40.18 An electron is bound in a square well with a depth equal to six times the ground-level energy E_∞ of an infinite well of the same width. When the electron makes a transition from the level with energy E_2 to the level with energy E_1, it emits a photon of wavelength 455 nm. Determine the width of the well.

40.19 Calculate $d^2\psi/dx^2$ for the wave function of Eq. (40.17), and show that the function is a solution of Eq. (40.16).

40.20 Calculate $d^2\psi/dx^2$ for the wave function of Eq. (40.19), and show that for any values of C and D it is a solution of Eq. (40.1) for $x \leq 0$ and $x \geq L$ when $U = U_0$ and $E < U_0$.

Section 40.3 Potential Barriers and Tunneling

40.21 An electron with initial kinetic energy 6.0 eV encounters a barrier with height 11.0 eV. What is the probability of tunneling if the width of the barrier is a) 0.80 nm; b) 0.40 nm?

40.22 An electron with initial kinetic energy 5.0 eV encounters a barrier with height U_0 and width 0.60 nm. What is the transmission coefficient if a) $U_0 = 7.0$ eV; b) $U_0 = 9.0$ eV; c) $U_0 = 13.0$ eV?

40.23 An electron is moving past the square barrier shown in Fig. 40.11, but the energy of the electron is *greater* than the barrier height. If $E = 2U_0$, what is the ratio of the de Broglie wavelength of the electron in the region $x > L$ to the wavelength for $0 < x < L$?

40.24 An electron with initial kinetic energy 12.0 eV encounters a barrier of height 15.0 eV. What is the width of the barrier if the probability of tunneling is 2.5%?

40.25 a) An electron with initial kinetic energy 32 eV encounters a square barrier with height 41 eV and width 0.25 nm. What is the probability that the electron will tunnel through the barrier? b) A proton with the same kinetic energy encounters the same barrier. What is the probability that the proton will tunnel through the barrier?

40.26 Alpha Decay. In a simple model for a radioactive nucleus, an alpha particle ($m = 6.64 \times 10^{-27}$ kg) is trapped by a square barrier that has width 2.0 fm and height 30.0 MeV. a) What is the tunneling probability when the alpha particle encounters the barrier if its kinetic energy is 1.0 MeV below the top of the barrier (see Fig. 40.23)? b) What is the tunneling probability if the energy of the alpha particle is 10.0 MeV below the top of the barrier?

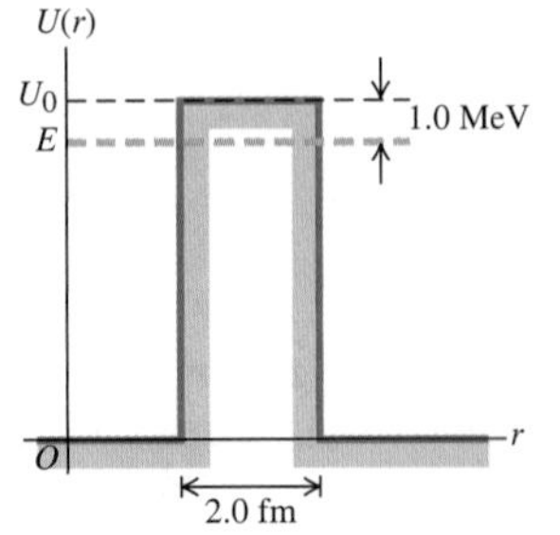

Figure 40.23 Exercise 40.26.

Section 40.4 The Harmonic Oscillator

40.27 A wooden block with mass 0.250 kg is oscillating on the end of a spring that has force constant 110 N/m. Calculate the ground-level energy and the energy separation between adjacent levels. Express your results in joules and in electron volts. Are quantum effects important?

40.28 Show that $\psi(x)$ given by Eq. (40.24) is a solution to Eq. (40.22) with energy $E_0 = \hbar\omega/2$.

40.29 An atom with mass 9.4×10^{-26} kg vibrates with simple harmonic motion in a crystal lattice. If the atom absorbs a photon with wavelength 525 μm when it makes a transition from the ground level to the first excited level, what is the force constant?

40.30 The ground state energy of a harmonic oscillator is 5.60 eV. If the oscillator undergoes a transition from its $n = 3$ to $n = 2$ level by emitting a photon, what is the wavelength of the photon?

40.31 A diatomic molecule is modeled as a harmonic oscillator with mass 5.60×10^{-26} kg and force constant 12.0 N/m. a) If the oscillator undergoes a transition from the $n = 4$ to $n = 3$ level by emitting a photon, what is the wavelength of this photon? b) What is the ground state energy (in eV) of this oscillator?

40.32 For the ground-level harmonic oscillator wave function $\psi(x)$ given in Eq. (40.24), $|\psi|^2$ has a maximum at $x = 0$. a) Compute the ratio of $|\psi|^2$ at $x = +A$ to $|\psi|^2$ at $x = 0$, where A is given by Eq. (40.27) with $n = 0$ for the ground level. b) Compute the ratio of $|\psi|^2$ at $x = +2A$ to $|\psi|^2$ at $x = 0$. In each case is your result consistent with what is shown in Fig. 40.19?

40.33 In Section 40.4 it is shown that for the ground level of a harmonic oscillator, $\Delta x \, \Delta p_x = \hbar$. Do a similar analysis for an excited level that has quantum number n. How does the uncertainty product $\Delta x \, \Delta p_x$ depend on n?

Problems

40.34 A particle is in the ground level of a box that extends from $x = 0$ to $x = L$. a) What is the probability of finding the particle in the region between 0 and $L/4$? Calculate this by integrating $|\psi(x)|^2 \, dx$, where ψ is normalized, from $x = 0$ to $x = L/4$. b) What is the probability of finding the particle in the region $x = L/4$ to $x = L/2$? c) How do the results of parts (a) and (b) compare? Explain. d) Add the probabilities calculated in parts (a) and (b). e) Are your results in parts (a), (b) and (d) consistent with Fig. 40.5b? Explain.

40.35 What is the probability of finding a particle in a box of length L in the region between $x = L/4$ and $x = 3L/4$ when the particle is in a) the ground level; b) the first excited level? (*Hint:* Integrate $|\psi(x)|^2 \, dx$, where ψ is normalized, between $L/4$ and $3L/4$.) c) Are your results in parts (a) and (b) consistent with Fig. 40.5b? Explain.

40.36 Consider a particle in a box with rigid walls at $x = 0$ and $x = L$. Let the particle be in the ground level. Calculate the probability $|\psi|^2 \, dx$ that the particle will be found in the interval x to $x + dx$ for a) $x = L/4$; b) $x = L/2$; c) $x = 3L/4$.

40.37 Repeat Problem 40.36 for a particle in the first excited level.

40.38 Let ΔE_n be the energy difference between the adjacent energy levels E_n and E_{n+1} for a particle in a box. The ratio $R_n = \Delta E_n/E_n$ compares the energy of a level to the energy separation of the next higher energy level. a) For what value of n is R_n largest, and what is this largest R_n? b) What does R_n approach as n

becomes very large? How does this result compare to the classical value for this quantity?

40.39 Photon in a Dye Laser. An electron in a long, organic molecule used in a dye laser behaves approximately like a particle in a box with width 4.18 nm. What is the wavelength of the photon emitted when the electron undergoes a transition a) from the first excited level to the ground level? b) from the second excited level to the first excited level?

40.40 Consider a particle in a box that extends from $x = 0$ to $x = L$. The particle is in its ground level. a) What is $d\psi/dx$ for $x > L$? b) Calculate $d\psi/dx$ for $0 < x < L$ and from this result obtain $d\psi/dx$ as $x \to L$. c) Is $d\psi/dx$ continuous at $x = L$? Use Eq. (40.1) to show that $d\psi/dx$ need not be continuous at points where U becomes infinite. [*Hint:* Solve Eq. (40.1) for $d^2\psi/dx^2$.]

40.41 It is stated in Section 40.1 that $d\psi/dx$ need not be continuous at points where U becomes infinite (see Problem 40.40). For a particle in a box with rigid walls at $x = 0$ and $x = L$, what is $d\psi/dx$ for x slightly greater than zero if a) the particle is in the $n = 1$ level? b) the particle is in the $n = 2$ level? What is $d\psi/dx$ for x slightly less than L if c) the particle is in the $n = 1$ level? d) the particle is in the $n = 2$ level? e) For which of these levels does the slope of the wave function near the walls have the greater magnitude? Do your results agree with Fig. 40.5a?

40.42 A particle is confined within a box with perfectly rigid walls at $x = 0$ and $x = L$. Although the magnitude of the instantaneous force exerted on the particle by the walls is infinite and the time over which it acts is zero, the impulse (that involves a product of force and time) is both finite and quantized. Show that the impulse exerted by the wall at $x = 0$ is $(nh/L)\hat{\imath}$ and that the impulse exerted by the wall at $x = L$ is $-(nh/L)\hat{\imath}$. (*Hint:* You may wish to review Section 8.1.)

40.43 A fellow student proposes that a possible wave function for a free particle with mass m [one for which the potential-energy function $U(x)$ is zero] is

$$\psi(x) = \begin{cases} e^{+\kappa x} & x < 0 \\ e^{-\kappa x} & x \geq 0 \end{cases}$$

where κ is a positive constant. a) Graph this proposed wave function. b) Show that the proposed wave function satisfies the Schrödinger equation for $x < 0$ if the energy is $E = -\hbar^2\kappa^2/2m$, that is, if the energy of the particle is *negative.* c) Show that the proposed wave function also satisfies the Schrödinger equation for $x \geq 0$ with the same energy as in part (b). d) Explain why the proposed wave function is nonetheless *not* an acceptable solution of the Schrödinger equation for a free particle. (*Hint:* What is the behavior of the function at $x = 0$?) It is in fact impossible for a free particle [one for which $U(x) = 0$] to have an energy less than zero.

40.44 Figure 40.8a shows several bound-state wave functions for a particle in a finite potential well. a) By inspection of the wave functions, verify that ψ and its second derivative $d^2\psi/dx^2$ have opposite signs at each point inside the well $(0 \leq x \leq L)$, but have the same sign at each point outside the well $(x < 0$ and $x > L)$. b) Use the time-independent Schrödinger equation, Eq. (40.1), to explain the observations in part (a) about ψ and $d^2\psi/dx^2$.

40.45 a) For the finite potential well of Fig. 40.6, what relations among the constants A and B of Eq. (40.17) and C and D of Eq. (40.19) are obtained by applying the boundary condition that ψ be continuous at $x = 0$ and at $x = L$? b) What relations among A, B, C, and D are obtained by applying the boundary condition that $d\psi/dx$ be continuous at $x = 0$ and at $x = L$?

40.46 An electron with initial kinetic energy 5.5 eV encounters a square potential barrier with height 10.0 eV. What is the width of the barrier if the electron has a 0.10% probability of tunneling through the barrier?

40.47 A particle with mass m and total energy E tunnels through a square barrier of height U_0 and width L. When the transmission coefficient is *not* much smaller than unity, it is given by

$$T = \left[1 + \frac{(U_0 \sinh \kappa L)^2}{4E(U_0 - E)}\right]^{-1}$$

where $\sinh \kappa L = (e^{\kappa L} - e^{-\kappa L})/2$ is the hyperbolic sine of κL. a) Show that if $\kappa L \gg 1$, this expression for T approaches Eq. (40.21). b) Explain why the restriction $\kappa L \gg 1$ in part (a) implies either that the barrier is relatively wide or that the energy E is relatively small compared to U_0. c) Show that as the particle's incident kinetic energy E approaches the barrier height U_0, T approaches $[1 + (kL/2)^2]^{-1}$, where $k = \sqrt{2mE}/\hbar$ is the wave number of the incident particle. (*Hint:* If $|z| \ll 1$, $\sinh z \approx z$.)

40.48 A particle with mass m and total energy E tunnels through a square barrier of height U_0 and width L. Show that the wave number k of the tunneling particle is imaginary when the particle is within the barrier. That is, show that $k = i\kappa$, where κ is as given in Eq. (40.21).

40.49 For small amplitudes of oscillation the motion of a pendulum is simple harmonic. For a pendulum with a period of 0.500 s, find the ground-level energy and the energy difference between adjacent energy levels. Express your results in joules and in electron volts. Are these values detectable?

40.50 A harmonic oscillator consists of a 0.020-kg mass on a spring. Its frequency is 1.50 Hz, and the mass has a speed of 0.360 m/s as it passes the equilibrium position. a) What is the value of the quantum number n for its energy level? b) What is the difference in energy between the levels E_n and E_{n+1}? Is this difference detectable?

40.51 a) Show that $\psi(x) = Cxe^{-m\omega x^2/2\hbar}$, where C is a normalization constant, is a solution to Eq. (40.22) with energy $E_1 = \frac{3}{2}\hbar\omega$. b) Sketch $\psi(x)$ as a function of x. Which of the graphs in Fig. 40.18 does your sketch resemble most closely?

40.52 The wave function for the first excited level of a harmonic oscillator is given by $\psi(x) = Cxe^{-m\omega x^2/2\hbar}$, where C is a normalization constant (Problem 40.51). a) At which values of x is $|\psi|^2$ a maximum? b) At which values of x is $|\psi|^2$ zero? Compare your results to the corresponding graph in Fig. 40.19.

40.53 A Three-Dimensional Isotropic Harmonic Oscillator. An isotropic harmonic oscillator has the potential energy function $U(x, y, z) = \frac{1}{2}k'(x^2 + y^2 + z^2)$. (*Isotropic* means that the force constant k' is the same in all three coordinate directions.) a) Show that for this potential, a solution to Eq. (40.29) is given by $\psi = \psi_{n_x}(x)\psi_{n_y}(y)\psi_{n_z}(z)$. In this expression, $\psi_{n_x}(x)$ is a solution to the one-dimensional harmonic oscillator Schrödinger equation,

Eq. (40.22), with energy $E_{n_x} = (n_x + \frac{1}{2})\hbar\omega$. The functions $\psi_{n_y}(y)$ and $\psi_{n_z}(z)$ are analogous one-dimensional wave functions for oscillations in the y- and z-directions. Find the energy associated with this ψ. b) From your results in part (a) what are the ground level and first excited-level energies of the three-dimensional isotropic oscillator? c) Show that there is only one state (one set of quantum numbers n_x, n_y, and n_z) for the ground level but three states for the first excited level.

40.54 Three-Dimensional Anisotropic Harmonic Oscillator. An oscillator has the potential energy function $U(x, y, z) = \frac{1}{2}k_1'(x^2 + y^2) + \frac{1}{2}k_2'z^2$, where $k_1' > k_2'$. This oscillator is called *anisotropic* because the force constant is not the same in all three coordinate directions. a) Find a general expression for the energy levels of the oscillator (see Problem 40.53). b) From your results in part (a), what are the ground level and first excited-level energies of this oscillator? c) How many states (different sets of quantum numbers n_x, n_y, and n_z) are there for the ground level and for the first excited level? Compare to part (c) of Problem 40.53.

Challenge Problems

40.55 Section 40.1 considered a box with walls at $x = 0$ and $x = L$. Now consider a box with width L but centered at $x = 0$, so that it extends from $x = -L/2$ to $x = +L/2$ (see Fig. 40.24). Note that this box is symmetric about $x = 0$. a) Consider possible wave functions of the form $\psi(x) = A \sin kx$. Apply the boundary conditions at the wall to obtain the allowed energy levels. b) Another set of possible wave functions are functions of the form $\psi(x) = A \cos kx$. Apply the boundary conditions at the wall to obtain the allowed energy levels. c) Compare the energies obtained in parts (a) and (b) to the set of energies given in Eq. (40.9). d) An odd function f satisfies the condition $f(x) = -f(-x)$, and an even function g satisfies $g(x) = g(-x)$. Of the wave functions from parts (a) and (b), which are even and which are odd?

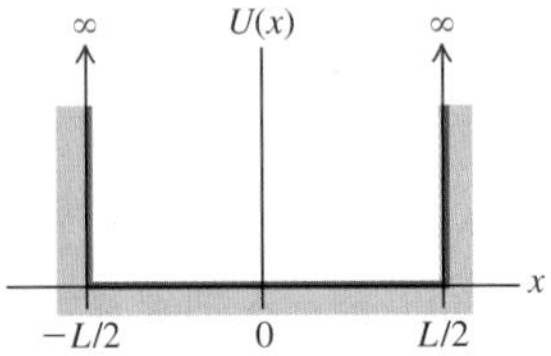

Figure 40.24 Challenge Problem 40.55.

40.56 Consider a potential well defined as follows: $U(x) = \infty$ for $x < 0$, $U(x) = 0$ for $0 < x < L$, and $U(x) = U_0 > 0$ for $x > L$ (see Fig. 40.25). Consider a particle with mass m and kinetic energy $E < U_0$ that is trapped in the well. a) The boundary condition at the infinite wall $(x = 0)$ is $\psi(0) = 0$. What must the form of the function $\psi(x)$ for $0 < x < L$ be in order to satisfy both the Schrödinger equation and this boundary condition? b) The wave function must remain finite as $x \to \infty$. What must the form of the function $\psi(x)$ for $x > L$ be in order to satisfy both the Schrödinger equation and this boundary condition at infinity? c) Impose the boundary conditions that ψ and $d\psi/dx$ are continuous at $x = L$. Show that the energies of the allowed levels are obtained from solutions of the equation $k \cot kL = -\kappa$, where $k = \sqrt{2mE}/\hbar$ and $\kappa = \sqrt{2m(U_0 - E)}/\hbar$.

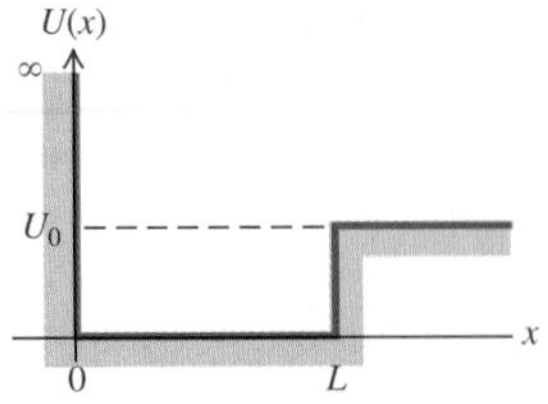

Figure 40.25 Challenge Problem 40.56.

40.57 The WKB Approximation. It can be a challenge to solve the Schrödinger equation for the bound-state energy levels of an arbitrary potential well. An alternative approach that can yield good approximate results for the energy levels is the *WKB approximation* (named for the physicists Gregor Wentzel, Hendrik Kramers, and Léon Brillouin, who pioneered its application to quantum mechanics). The WKB approximation begins from three physical statements: i) According to de Broglie, the magnitude of momentum p of a quantum-mechanical particle is $p = h/\lambda$. ii) The magnitude of momentum is related to the kinetic energy K by the relation $K = p^2/2m$. iii) If there are no nonconservative forces, then in Newtonian mechanics the energy E for a particle is constant and equal at each point to the sum of the kinetic and potential energies at that point: $E = K + U(x)$, where x is the coordinate. a) Combine these three relations to show that the wavelength of the particle at a coordinate x can be written as

$$\lambda(x) = \frac{h}{\sqrt{2m[E - U(x)]}}$$

Thus we envision a quantum-mechanical particle in a potential well $U(x)$ as being like a free particle, but with a wavelength $\lambda(x)$ that is a function of position. b) When the particle moves into a region of increasing potential energy, what happens to its wavelength? c) At a point where $E = U(x)$, Newtonian mechanics says that the particle has zero kinetic energy and must be instantaneously at rest. Such a point is called a *classical turning point,* since this is where a Newtonian particle must stop its motion and reverse direction. As an example, an object oscillating in simple harmonic motion with amplitude A moves back and forth between the points $x = -A$ and $x = +A$; each of these is a classical turning point, since there the potential energy $\frac{1}{2}k'x^2$ equals the total energy $\frac{1}{2}k'A^2$. In the WKB expression for $\lambda(x)$, what is the wavelength at a classical turning point? d) For a particle in a box with length L, the walls of the box are classical turning points (see Fig. 40.1). Furthermore, the number of wavelengths that fit within the box must be a half-integer (see Fig. 40.31545-1), so that $L = (n/2)\lambda$ and hence $L/\lambda = n/2$, where $n = 1, 2, 3, \ldots$. [Note that this is a restatement of Eq. (40.7).] The WKB scheme for finding the allowed bound-state energy levels of an *arbitrary* potential well is an extension of these observations. It demands that for an allowed energy E, there must be a half-integer number of wavelengths between the classical turning points for that energy. Since the wavelength in the WKB approximation is not a constant but depends on x, the number of wavelengths between the classical turning points a and b for a given value of the energy is the integral of $1/\lambda(x)$ between those points:

$$\int_a^b \frac{dx}{\lambda(x)} = \frac{n}{2} \qquad n = 1, 2, 3, \ldots$$

Using the expression for $\lambda(x)$ you found in part (a), show that the *WKB condition for an allowed bound-state energy* can be written as

$$\int_a^b \sqrt{2m[E - U(x)]}\, dx = \frac{nh}{2} \qquad n = 1, 2, 3, \ldots$$

e) As a check on the expression in part (d), apply it to a particle in a box with walls at $x = 0$ and $x = L$. Evaluate the integral and show that the allowed energy levels according to the WKB approximation are the same as those given by Eq. (40.9). (*Hint:* Since the walls of the box are infinitely high, the points $x = 0$ and $x = L$ are classical turning points for *any* energy E. Inside the box, the potential energy is zero.) f) For the finite square well shown in Fig. 40.6, show that the WKB expression given in part (d) predicts the *same* bound-state energies as for an infinite square well of the same width. (*Hint:* Assume $E < U_0$. Then the classical turning points are at $x = 0$ and $x = L$.) This shows that the WKB approximation does a poor job when the potential energy function changes discontinuously, as for a finite potential well. In the next two problems we consider situations in which the potential energy function changes gradually and the WKB approximation is much more useful.

40.58 The WKB approximation (Challenge Problem 40.57) can be used to calculate the energy levels for a harmonic oscillator. In this approximation, the energy levels are the solutions to the equation

$$\int_a^b \sqrt{2m[E - U(x)]}\,dx = \frac{nh}{2} \qquad n = 1, 2, 3, \ldots$$

Here E is the energy, $U(x)$ is the potential energy function, and $x = a$ and $x = b$ are the classical turning points (the points at which E is equal to the potential energy, so the Newtonian kinetic energy would be zero). a) Determine the classical turning points for a harmonic oscillator with energy E and force constant k'. b) Carry out the integral in the WKB approximation and show that the energy levels in this approximation are $E_n = \hbar\omega$, where $\omega = \sqrt{k'/m}$ and $n = 1, 2, 3, \ldots$. (*Hint:* Recall that $\hbar = h/2\pi$. A useful standard integral is

$$\int \sqrt{A^2 - x^2}\,dx = \frac{1}{2}\left[x\sqrt{A^2 - x^2} + A^2 \arcsin\left(\frac{x}{|A|}\right)\right]$$

where arcsin denotes the inverse sine function. Note that the integrand is even, so the integral from $-x$ to x is equal to twice the integral from 0 to x.) c) How do the approximate energy levels found in part (b) compare with the true energy levels given by Eq. (40.26)? Does the WKB approximation give an underestimate or an overestimate of the energy levels?

40.59 Protons, neutrons, and many other particles are made of more fundamental particles called *quarks* and *antiquarks* (the antimatter equivalent of quarks). A quark and an antiquark can form a bound state with a variety of different energy levels, each of which corresponds to a different particle observed in the laboratory. As an example, the ψ particle is low-energy bound state of a so-called charm quark and its antiquark, with a rest energy of 3097 MeV; the $\psi(2S)$ particle is an excited state of this same quark-antiquark combination, with a rest energy of 3686 MeV. A simplified representation of the potential energy of interaction between a quark and an antiquark is $U(x) = A|x|$ where A is a positive constant and x represents the distance between the quark and the antiquark. You can use the WKB approximation (Challenge Problem 40.57) to determine the bound-state energy levels for this potential energy function. In the WKB approximation, the energy levels are the solutions to the equation

$$\int_a^b \sqrt{2m[E - U(x)]}\,dx = \frac{nh}{2} \qquad n = 1, 2, 3, \ldots$$

Here E is the energy, $U(x)$ is the potential energy function, and $x = a$ and $x = b$ are the classical turning points (the points at which E is equal to the potential energy, so the Newtonian kinetic energy would be zero). a) Determine the classical turning points for the potential $U(x) = A|x|$ and for an energy E. b) Carry out the above integral and show that the allowed energy levels in the WKB approximation are given by

$$E_n = \frac{1}{2m}\left(\frac{3mAh}{4}\right)^{2/3} n^{2/3} \qquad n = 1, 2, 3, \ldots$$

(*Hint:* The integrand is even, so the integral from $-x$ to x is equal to twice the integral from 0 to x.) c) Does the difference in energy between successive levels increase, decrease, or remain the same as n increases? How does this compare to the behavior of the energy levels for the harmonic oscillator? For the particle in a box? Can you suggest a simple rule that relates the difference in energy between successive levels to the shape of the potential energy function?

41 ATOMIC STRUCTURE

Why is seawater so salty? That is, why does salt dissolve so readily in water? The answer lies in the electronic structure of the sodium and chlorine atoms that make up salt molecules. Sodium can easily lose an electron to form an Na^+ ion, just as chlorine can easily form a Cl^- ion by gaining an electron. These ions are attracted to the polar ends of water molecules, holding the ions in solution.

? **If a sodium atom can easily lose one electron, why can't it just as easily lose two, three, or more electrons?**

Some physicists claim that all of chemistry is contained in the Schrödinger equation. This is somewhat of an exaggeration, but this equation can teach us a great deal about the chemical behavior of elements and the nature of chemical bonds. It provides insight into the periodic table of the elements and the microscopic basis of magnetism.

We can learn a great deal about the structure and properties of *all* atoms from the solutions to the Schrödinger equation for the hydrogen atom. These solutions have quantized values of angular momentum; we don't need to make a separate statement about quantization as we did with the Bohr model. We label the states with a set of quantum numbers, which we'll use later with many-electron atoms as well. We'll find that the electron also has an intrinsic "spin" angular momentum in addition to the orbital angular momentum associated with its motion.

We'll also encounter the exclusion principle, a kind of microscopic zoning ordinance that is the key to understanding many-electron atoms. This principle says that no two electrons in an atom can have the same quantum-mechanical state. Finally, we'll use the principles of this chapter to explain the characteristic x-ray spectra of atoms.

41.1 | The Hydrogen Atom

Let's continue the discussion of the hydrogen atom that we began in Chapter 38. In the Bohr model, electrons move in circular orbits like Newtonian particles, but with quantized values of angular momentum. While this model gave the correct energy levels of the hydrogen atom, as deduced from spectra, it had many conceptual difficulties. It mixed classical physics with new and seemingly contradictory concepts. It provided no insight into the process by which photons are emitted and absorbed. It could not be generalized to atoms with more than one electron. It predicted the wrong magnetic properties for the hydrogen atom. And perhaps most important, its picture of the electron as a localized point particle was inconsistent with the more general view we developed in Chapters 39 and 40.

Now let's see how to apply the Schrödinger equation to find the wave functions for stationary states (states of definite energy) of the hydrogen atom. As in Section 38.5, we include the motion of the nucleus by simply replacing the electron mass m with the reduced mass m_r. We discussed the three-dimensional version of the Schrödinger equation in Section 40.5. The hydrogen-atom problem is best formulated in spherical coordinates (r, θ, ϕ), shown in Fig. 40.22; the potential energy is then simply

$$U(r) = -\frac{1}{4\pi\epsilon_0}\frac{e^2}{r} \tag{41.1}$$

The Schrödinger equation with this potential-energy function can be solved exactly; the solutions are combinations of familiar functions. Without going into a lot of detail, we can describe the most important features of the procedure and the results.

First, the solutions are obtained by a method called *separation of variables,* in which we express the wave function $\psi(r, \theta, \phi)$ as a product of three functions, each one a function of only one of the three coordinates:

$$\psi(r, \theta, \phi) = R(r)\Theta(\theta)\Phi(\phi) \tag{41.2}$$

That is, the function $R(r)$ depends *only* on r, $\Theta(\theta)$ depends *only* on θ, and $\Phi(\phi)$ depends *only* on ϕ. When we substitute Eq. (41.2) into the Schrödinger equation, we get three separate equations, each containing only one of the coordinates. This is an enormous simplification; it reduces the problem of solving a fairly complex *partial* differential equation with three independent variables to the much simpler problem of solving three separate *ordinary* differential equations with one independent variable each.

The physically acceptable solutions of these three equations are determined by *boundary conditions.* The radial function $R(r)$ must approach zero at large r, because we are describing *bound states* of the electron that are localized near the nucleus. This is analogous to the requirement that the harmonic-oscillator wave functions (Section 40.4) must approach zero at large x. The angular functions $\Theta(\theta)$ and $\Phi(\phi)$ must be *periodic.* For example, (r, θ, ϕ) and $(r, \theta, \phi + 2\pi)$ describe the same point, so $\Phi(\phi + 2\pi)$ must equal $\Phi(\phi)$. Also, the angular functions must be *finite* for all relevant values of the angles. For example, there are solutions of the Θ equation that become infinite at $\theta = 0$ and $\theta = \pi$; these are unacceptable, since $\psi(r, \theta, \phi)$ must be normalizable.

The radial functions $R(r)$ turn out to be an exponential function $e^{-\alpha r}$ (where α is positive) multiplied by a polynomial in r. The functions $\Theta(\theta)$ are polynomials containing various powers of $\sin\theta$ and $\cos\theta$, and the functions $\Phi(\phi)$ are simply proportional to $e^{im_l\phi}$, where $i = \sqrt{-1}$ and m_l is an integer that may be positive, zero, or negative.

In the process of finding solutions that satisfy the boundary conditions, we also find the corresponding energy levels. Their energies, which we denote by E_n $(n = 1, 2, 3, \ldots)$, turn out to be *identical* to those from the Bohr model, as given by Eq. (38.18), with the electron rest mass m replaced by the reduced mass m_r. Rewriting that equation using $\hbar$, we have

$$E_n = -\frac{1}{(4\pi\epsilon_0)^2}\frac{m_r e^4}{2n^2\hbar^2} = -\frac{13.60\ \text{eV}}{n^2} \qquad (41.3)$$

(energy levels of hydrogen)

As in Section 38.5, we call n the **principal quantum number** for the level of energy E_n.

The result that Eq. (41.3) can be obtained from the Schrödinger equation is an important confirmation that this approach is valid. The Schrödinger analysis is quite different from the Bohr model, both formally and conceptually, yet both yield the same energy-level scheme, a scheme that agrees with the energies determined from spectra. As we will see, the Schrödinger analysis can explain many more aspects of the hydrogen atom than can the Bohr model.

Quantization of Orbital Angular Momentum

The solutions that satisfy the boundary conditions mentioned above also have quantized values of *orbital angular momentum.* That is, only certain discrete values of the magnitude and components of orbital angular momentum are permitted. In discussing the Bohr model in Section 38.5, we mentioned that quantization of angular momentum was a result with little fundamental justification. With the Schrödinger equation it appears automatically!

The possible values of the magnitude L of orbital angular momentum $\vec{L}$ are determined by the requirement that the $\Theta(\theta)$ function must be finite at $\theta = 0$ and $\theta = \pi$. In a level with energy E_n and principal quantum number n, the possible values of L are

$$L = \sqrt{l(l+1)}\hbar \qquad (l = 0, 1, 2, \ldots, n-1) \qquad (41.4)$$

(magnitude of orbital angular momentum)

The *orbital angular-momentum quantum number l* is called the **orbital quantum number** for short. In the Bohr model, each energy level corresponded to a single value of angular momentum. Equation (41.4) shows that in fact there are n different possible values of L for the nth energy level.

An interesting feature of Eq. (41.4) is that the orbital angular momentum is *zero* for $l = 0$ states. This result disagrees with the Bohr model, in which the electron always moved in a circle of definite radius and L was never zero. The $l = 0$ wave functions ψ depend only on r; the $\Theta(\theta)$ and $\Phi(\phi)$ functions for these states are constants. Thus the wave functions for $l = 0$ states are spherically symmetric; there is nothing in their probability distribution $|\psi|^2$ to favor one direction over any other, and there is no orbital angular momentum.

The permitted values of the *component* of $\vec{L}$ in a given direction, say the z-component L_z, are determined by the requirement that the $\Phi(\phi)$ function must equal $\Phi(\phi + 2\pi)$. The possible values of L_z are

$$L_z = m_l\hbar \qquad (m_l = 0, \pm 1, \pm 2, \ldots, \pm l) \quad \text{(components of orbital angular momentum)} \tag{41.5}$$

We see that m_l can be zero or a positive or negative integer up to, but no larger in magnitude than, l. That is, $|m_l| \le l$. For example, if $l = 1$, m_l can equal 1, 0, or -1. For reasons that will emerge later, we call m_l the *orbital magnetic quantum number* or **magnetic quantum number** for short.

The component L_z can never be quite as large as L (unless both are zero). For example, when $l = 2$, the largest possible value of m_l is also 2; then Eqs. (41.4) and (41.5) give

$$L = \sqrt{2(2+1)}\hbar = \sqrt{6}\hbar = 2.45\hbar$$
$$L_z = 2\hbar$$

Figure 41.1 shows the situation. The minimum value of the angle θ_L between the vector $\vec{L}$ and the z-axis is given by

$$\theta_L = \arccos\frac{L_z}{L} = \arccos\frac{2}{2.45} = 35.3°$$

That $|L_z|$ is always less than L is also required by the uncertainty principle. Suppose we could know the precise *direction* of the orbital angular momentum vector. Then we could let that be the direction of the z-axis, and L_z would equal L. This corresponds to a particle moving in the xy-plane only, in which case the z-component of the linear momentum $\vec{p}$ would be zero with no uncertainty Δp_z. Then the uncertainty principle $\Delta z\,\Delta p_z \ge \hbar$ requires infinite uncertainty Δz in the coordinate z. This is impossible for a localized state; we conclude that we can't know the direction of $\vec{L}$ precisely. Thus, as we've already stated, the component of $\vec{L}$ in a given direction can never be quite as large as its magnitude L. Also, if we can't know the direction of $\vec{L}$ precisely, we can't determine the components

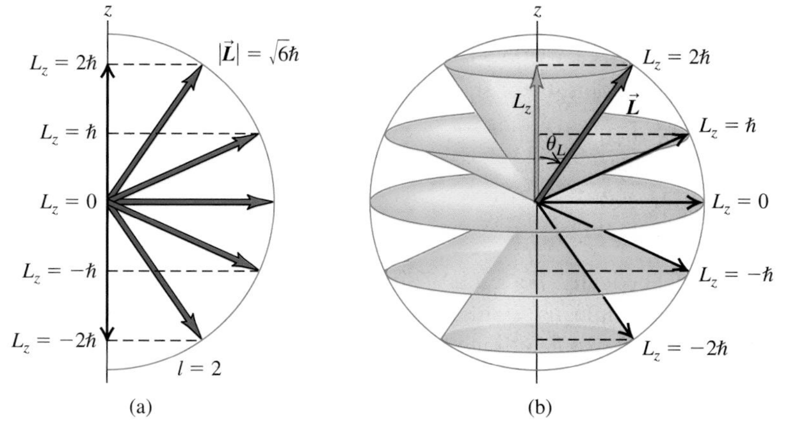

41.1 (a) When $l = 2$, the magnitude of $\vec{L}$ is $\sqrt{6}\hbar = 2.45\hbar$, but the direction of $\vec{L}$ is not definite. In this semiclassical vector picture, $\vec{L}$ makes an angle of 35.3° with the z-axis when the z-component has its maximum value of $2\hbar$. (b) Cones of the possible directions of $\vec{L}$.

L_x and L_y precisely. Thus we show *cones* of possible directions for $\vec{L}$ in Fig. 41.1b.

You may wonder why we have singled out the z-axis for special attention. There's no fundamental reason for this; the atom certainly doesn't care what coordinate system we use. The point is that we can't determine all three components of orbital angular momentum with certainty, so we arbitrarily pick one as the component we want to measure. When we discuss interactions of the atom with a magnetic field, we will consistently choose the positive z-axis to be in the direction of $\vec{B}$.

41.2 The energy for an orbiting satellite depends on the average distance between the satellite and the center of the earth. It does *not* depend on whether the orbit is circular (with a large orbital angular momentum L) or elliptical (in which case L is smaller). In the same way, the energy of a hydrogen atom does not depend on the orbital angular momentum.

Quantum Number Notation

The wave functions for the hydrogen atom are determined by the values of three quantum numbers n, l, and m_l. The energy E_n is determined by the principal quantum number n according to Eq. (41.3). The magnitude of orbital angular momentum is determined by the orbital quantum number l, as in Eq. (41.4). The component of orbital angular momentum in a specified axis direction (customarily the z-axis) is determined by the magnetic quantum number m_l, as in Eq. (41.5). The energy does not depend on the values of l or m_l (Fig. 41.2), so for each energy level E_n given by Eq. (41.3), there is more than one distinct state having the same energy but different quantum numbers. The existence of more than one distinct state with the same energy is called **degeneracy;** it has no counterpart in the Bohr model.

States with various values of the orbital quantum number l are often labeled with letters, according to the following scheme:

$l = 0$: *s* states
$l = 1$: *p* states
$l = 2$: *d* states
$l = 3$: *f* states
$l = 4$: *g* states

and so on alphabetically. This seemingly irrational choice of the letters *s*, *p*, *d*, and *f* originated in the early days of spectroscopy and has no fundamental significance. In an important form of *spectroscopic notation* that we'll use often, a state with $n = 2$ and $l = 1$ is called a 2*p* state; a state with $n = 4$ and $l = 0$ is a 4*s* state, and so on. Only *s* states $(l = 0)$ are spherically symmetric.

Here's another bit of notation. The radial extent of the wave functions increases with the principal quantum number n, and we can speak of a region of space associated with a particular value of n as a **shell.** Especially in discussions of many-electron atoms, these shells are given letters:

$n = 1$: *K* shell
$n = 2$: *L* shell
$n = 3$: *M* shell
$n = 4$: *N* shell

and so on alphabetically. For each n, different values of l correspond to different *subshells.* For example, the *L* shell $(n = 2)$ contains the 2*s* and 2*p* subshells.

Table 41.1 shows some of the possible combinations of the quantum numbers n, l, and m_l for hydrogen-atom wave functions. The spectroscopic notation and the shell notation for each are also shown.

Table 41.1 Quantum States of the Hydrogen Atom

n	l	m_l	Spectroscopic Notation	Shell
1	0	0	1s	K
2	0	0	2s	L
2	1	$-1, 0, 1$	2p	
3	0	0	3s	M
3	1	$-1, 0, 1$	3p	
3	2	$-2, -1, 0, 1, 2$	3d	
4	0	0	4s	N
and so on.				

Problem-Solving Strategy: Atomic Structure

IDENTIFY *the relevant concepts:* Many problems in atomic structure can be solved just by reference to the quantum numbers that describe the total energy E, the magnitude of the orbital angular momentum $\vec{L}$, the z-component of $\vec{L}$, and other properties of an atom.

SET UP *the problem:* Decide what the target variable is and collect the equations that you will use to determine its value. These may include Eqs. (41.3), (41.4), and (41.5).

EXECUTE *the solution as follows:*

1. Be sure you know the possible values of the quantum numbers n, l, and m_l for the hydrogen atom. They are all integers; n is always greater than zero, l can be zero or positive up to $n - 1$, and m_l can range from $-l$ to l. Be able to count the number of (n, l, m_l) states in each shell $(K, L, M,$ and so on$)$ and subshell $(3s, 3p, 3d,$ and so on$)$. You should know how to construct, not simply memorize, Table 41.1.
2. Solve for the target variable.

EVALUATE *your answer:* When checking numerical values, it's helpful to be familiar with some typical magnitudes. For example, the electric potential energy of a proton and electron that are -0.10 nm apart (typical of atomic dimensions) is about -15 eV. Visible light (emitted in a number of transitions between states of the hydrogen atom) has wavelengths around 500 nm and frequencies around 5×10^{14} Hz. You'll find other typical magnitudes in the problem-solving strategy for Section 39.1.

Example 41.1 Counting hydrogen states

How many distinct (n, l, m_l) states of the hydrogen atom with $n = 3$ are there? Find the energy of these states.

SOLUTION

IDENTIFY and SET UP: To determine the number of states, we use the idea that l can have integer values from 0 to $n - 1$ and that m_l can have values from $-l$ to l. The energy of any particular state is given by Eq. (41.3).

EXECUTE: When $n = 3$, l can be 0, 1, or 2. When $l = 0$, m_l can be only 0 (1 state). When $l = 1$, m_l can be -1, 0, or 1 (3 states). When $l = 2$, m_l can be -2, -1, 0, 1, or 2 (5 states). The total number of (n, l, m_l) states with $n = 3$ is therefore $1 + 3 + 5 = 9$. (In Section 41.3 we'll find that the total number of $n = 3$ states is in fact twice this, or 18, because of electron spin.)

The energies of these states are all the same because the energy depends only on n. According to Eq. (41.3), with $n = 3$,

$$E_3 = \frac{-13.60\text{ eV}}{3^2} = -1.51\text{ eV}$$

EVALUATE: It's useful to remember that the ground level of hydrogen has $n = 1$ and $E_1 = -13.6$ eV. These excited states have a higher energy, as they must.

Example 41.2

Angular momentum in an excited level of hydrogen

Consider the $n = 4$ states of hydrogen. a) What is the maximum magnitude L of the orbital angular momentum? b) What is the maximum value of L_z? c) What is the minimum angle between $\vec{L}$ and the z-axis? Give your answers to parts (a) and (b) in terms of $\hbar$.

SOLUTION

IDENTIFY and SET UP: We use Eq. (41.4) in part (a) to determine the maximum value of L, then use Eq. (41.5) in part (b) to determine the maximum value of L_z. The minimum angle between $\vec{L}$ and the z-axis occurs when L_z is maximum (so that $\vec{L}$ is most nearly aligned with the positive z-axis).

EXECUTE: a) When $n = 4$, the maximum value of the orbital angular-momentum quantum number l is $(n - 1) = (4 - 1) = 3$; from Eq. (41.4),

$$L = \sqrt{3(3 + 1)}\hbar = \sqrt{12}\hbar = 3.464\hbar$$

b) For $l = 3$ the maximum value of the magnetic quantum number m_l is 3; from Eq. (41.5),

$$L_z = 3\hbar$$

c) The *minimum* allowed angle between $\vec{L}$ and the z-axis corresponds to the *maximum* allowed values of L_z and m_l (Fig. 41.1b shows an $l = 2$ example). For the state with $l = 3$ and $m_l = 3$,

$$\theta_{\text{min}} = \arccos\frac{(L_z)_{\text{max}}}{L} = \arccos\frac{3\hbar}{3.464\hbar} = 30.0^\circ$$

EVALUATE: As a check, you can verify that the angle is greater than 30.0° for all states with smaller values of l.

Electron Probability Distributions

Rather than picturing the electron as a point particle moving in a precise circle, the Schrödinger equation gives a *probability distribution* surrounding the nucleus. Because the hydrogen-atom probability distributions are three-dimensional, they are harder to visualize than the two-dimensional circular orbits of the Bohr model. It's helpful to look at the *radial probability distribution function* $P(r)$, that is, the probability per radial length for the electron to be found at various distances from the proton. From Section 39.5 the probability for finding the electron in a small volume element dV is $|\psi|^2\,dV$. (We assume that ψ is normalized—that is, that the integral of $|\psi|^2\,dV$ over all space equals unity so that there is 100% probability of finding the electron somewhere in the universe.) Let's take as our volume element a thin spherical shell with inner radius r and outer radius $r + dr$. The volume dV of this shell is approximately its area $4\pi r^2$ multiplied by its thickness dr:

$$dV = 4\pi r^2\,dr \qquad (41.6)$$

We denote by $P(r)\,dr$ the probability of finding the particle within the radial range dr; then

$$P(r)\,dr = |\psi|^2\,dV = |\psi|^2 4\pi r^2\,dr \qquad (41.7)$$

(probability that the electron is between r and $r + dr$)

For wave functions that depend on θ and ϕ as well as r, we use the value of $|\psi|^2$ averaged over all angles in Eq. (41.7).

Figure 41.3 shows graphs of $P(r)$ for several hydrogen-atom wave functions. The r scales are labeled in multiples of a, the smallest distance between the electron and the nucleus in the Bohr model,

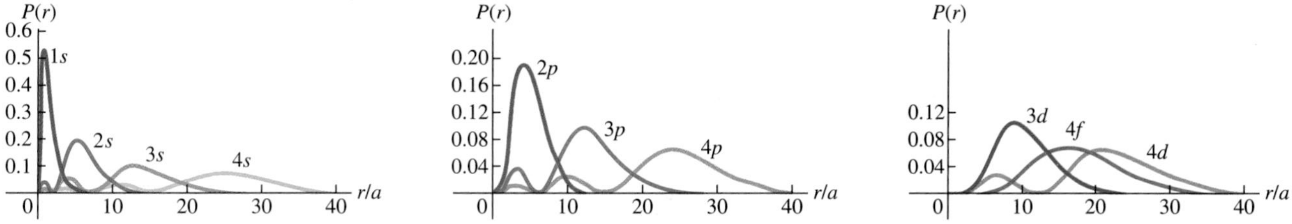

41.3 Radial probability distribution functions $P(r)$ for several hydrogen-atom wave functions, plotted as functions of the ratio r/a [see Eq. (41.8)]. For each function, the number of maxima is $(n - l)$. The curves for which $l = n - 1$ have only one maximum, located at $r = n^2a$.

$$a = \frac{\epsilon_0 h^2}{\pi m_r e^2} = \frac{4\pi\epsilon_0 \hbar^2}{m_r e^2} = 5.29 \times 10^{-11}\ \text{m} \tag{41.8}$$
(smallest r, Bohr model)

Just as for a particle in a box (Section 40.1), there are some positions where the probability is zero. But again, the uncertainty principle tells us not to worry; we can't localize the electron exactly anyway. Note that for the states having the largest possible l for each n (such as $1s$, $2p$, $3d$ and $4f$ states), $P(r)$ has a single maximum at n^2a. For these states, the electron is most likely to be found at the distance from the nucleus that is predicted by the Bohr model, $r = n^2a$.

Figure 41.3 shows *radial* probability distribution functions $P(r) = 4\pi r^2|\psi|^2$, which indicate the relative probability of finding the electron within a thin spherical shell of radius r. By contrast, Figs. 41.4 and 41.5 show the *three-dimensional* probability distribution functions $|\psi|^2$, which indicate the relative probability of finding the electron within a small box at a given position. The darker the blue "cloud," the greater the value of $|\psi|^2$. Figure 41.4 shows cross sections of the spherically symmetric probability clouds for the lowest three s subshells, for

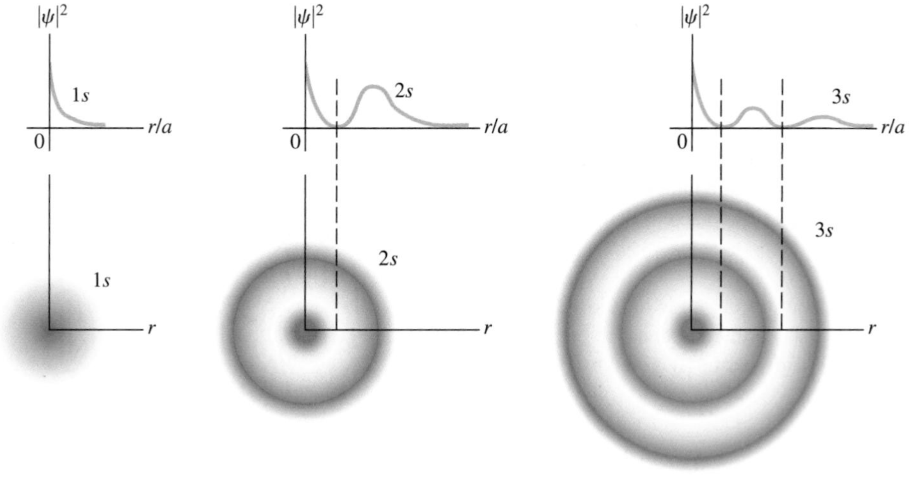

41.4 Three-dimensional probability distribution functions $|\psi|^2$ for the spherically symmetric $1s$, $2s$, and $3s$ hydrogen-atom wave functions.

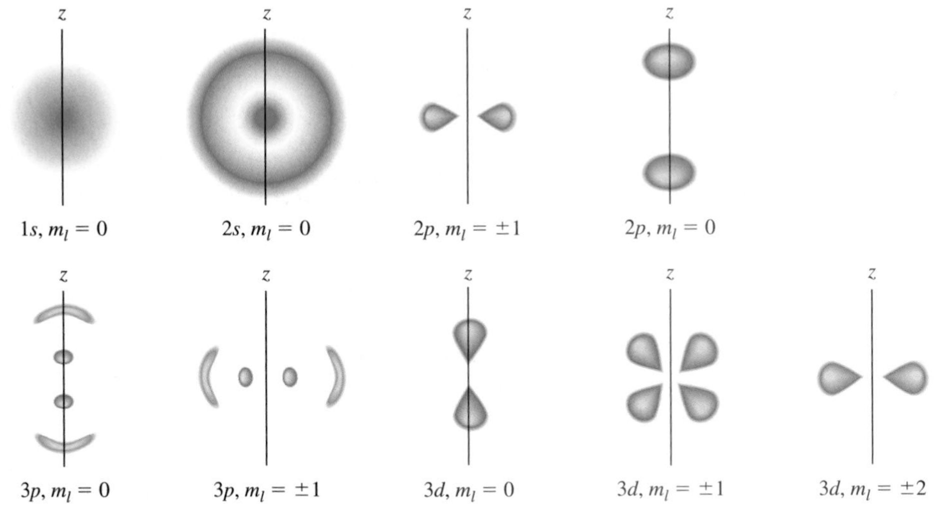

41.5 Cross sections of three-dimensional probability distributions for a few quantum states of the hydrogen atom. They are not to the same scale. Mentally rotate each drawing about the z-axis to obtain the three-dimensional representation of $|\psi|^2$. For example, the $2p$, $m_l = \pm 1$ probability distribution looks like a fuzzy doughnut.

which $|\psi|^2$ depends only on the radial coordinate r. Figure 41.5 shows cross sections of the clouds for other electron states for which $|\psi|^2$ depends on both r and θ. In *any* stationary state of the hydrogen atom, $|\psi|^2$ is independent of ϕ.

Example 41.3 A hydrogen wave function

The ground-state wave function for hydrogen (a $1s$ state) is

$$\psi_{1s}(r) = \frac{1}{\sqrt{\pi a^3}} e^{-r/a}$$

a) Verify that this function is normalized. b) What is the probability that the electron will be found at a distance less than a from the nucleus?

SOLUTION

IDENTIFY and SET UP: In part (a) we carry out the integral $\int |\psi|^2\, dV$ over all space; if it is equal to 1, the wave function is normalized. In part (b) we carry out the same integral over a spherical volume that extends from the origin (the nucleus) out to a distance a from the nucleus.

EXECUTE: Since the wave function depends only on the radial coordinate r, we can choose our volume elements to be spherical shells of radius r, thickness dr, and volume dV given by Eq. (41.6). We then have

$$\int_{\text{all space}} |\psi_{1s}|^2\, dV = \int_0^\infty \frac{1}{\pi a^3} e^{-2r/a}(4\pi r^2\, dr) = \frac{4}{a^3}\int_0^\infty r^2 e^{-2r/a}\, dr$$

The following indefinite integral can be found in a table of integrals or by integrating by parts:

$$\int r^2 e^{-2r/a}\, dr = \left(-\frac{ar^2}{2} - \frac{a^2 r}{2} - \frac{a^3}{4}\right)e^{-2r/a}$$

Evaluating this between the limits $r = 0$ and $r = \infty$ is simple; it is zero at $r = \infty$ because of the exponential factor, and at $r = 0$ only the last term in the parentheses survives. Thus the value of the integral is $a^3/4$. Putting it all together, we find

$$\int_0^\infty |\psi_{1s}|^2\, dV = \frac{4}{a^3}\int_0^\infty r^2 e^{-2r/a}\, dr = \frac{4}{a^3}\frac{a^3}{4} = 1$$

The wave function *is* normalized.

b) To find the probability P that the electron is found within $r < a$, we carry out the same integration but with the limits 0 and a. We'll leave the details as an exercise. From the upper limit we get $-5e^{-2}a^3/4$; the final result is

$$P = \int_0^a |\psi_{1s}|^2 4\pi r^2\, dr = \frac{4}{a^3}\left(-\frac{5a^3 e^{-2}}{4} + \frac{a^3}{4}\right)$$
$$= 1 - 5e^{-2} = 0.323$$

EVALUATE: Our results tell us that in a ground state we expect to find the electron at a distance from the nucleus less than a about $\frac{1}{3}$ of the time and at a greater distance about $\frac{2}{3}$ of the time. It's hard to tell, but in Fig. 41.3, about $\frac{2}{3}$ of the area under the $1s$ curve is at distances greater than a (that is, $r/a > 1$).

Two generalizations that we discussed with the Bohr model in Section 38.5 are equally valid in the Schrödinger analysis. First, if the "atom" is not composed of a single proton and a single electron, the reduced mass m_r of the system will give changes in Eqs. (41.3) and (41.8) that are substantial with some exotic systems. One example is *positronium,* composed of a positron and an electron (Fig. 38.21). Second, our analysis is applicable to single-electron ions, such as He^+, Li^{2+}, and so on. For such ions we replace e^2 by Ze^2 in Eqs. (41.3) and (41.8), where Z is the number of protons (the **atomic number**). This replacement will far overshadow the small change in the reduced mass m_r in these equations.

The Schrödinger analysis of the hydrogen atom is a lot more complex, both conceptually and mathematically, than the Newtonian analysis of planetary motion or the semiclassical Bohr model. It deals with probabilities rather than certainties, and it predicts discrete rather than continuous values of energy and angular momentum. But this analysis enables us to understand phenomena for which classical mechanics and electromagnetism are inadequate. Added complexity is the price we pay for expanded understanding.

Test Your Understanding

In a stationary state with magnetic quantum number m_l, the function $\Phi(\phi)$ in Eq. (41.2) is proportional to $e^{im_l\phi}$. Use this fact to explain why the stationary-state probability distributions for a hydrogen atom, like those shown in Fig. 41.5, are independent of ϕ (i.e., are rotationally symmetric around the z-axis).

41.2 | The Zeeman Effect

The **Zeeman effect** is the splitting of atomic energy levels and the associated spectrum lines when the atoms are placed in a magnetic field (Fig. 41.6). This effect confirms experimentally the quantization of angular momentum. The discussion in this section, which assumes that the only angular momentum is the *orbital* angular momentum of a single electron, also shows why we call m_l the magnetic quantum number.

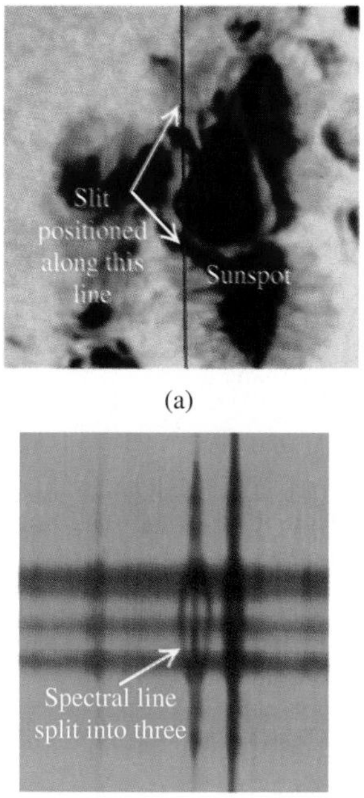

41.6 Magnetic effects on the spectrum of sunlight. (a) The slit of a spectrograph is positioned along the black line crossing a portion of a sunspot. (b) The 0.4-T magnetic field in the sunspot (a thousand times greater than the earth's field) splits the middle spectral line into three lines.

Atoms contain charges in motion, so it should not be surprising that magnetic forces cause changes in that motion and in the energy levels. As early as the middle of the nineteenth century, physicists speculated that the sources of visible light might be vibrating electric charge on an atomic scale. In 1896 the Dutch physicist Pieter Zeeman was the first to show that in the presence of a magnetic field, some spectrum lines were split into groups of closely spaced lines (Fig. 41.7). This effect now bears his name.

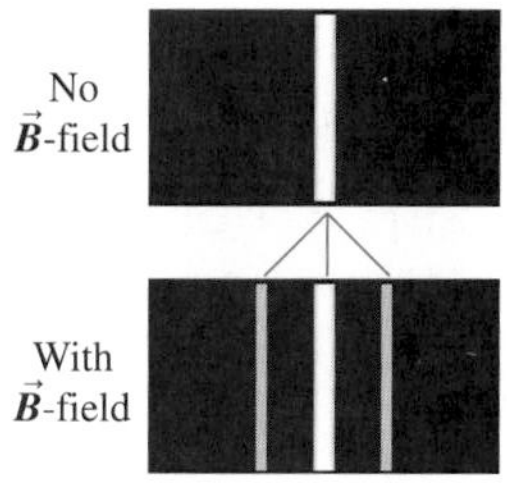

41.7 The normal Zeeman effect. When a source of radiation is placed in a magnetic field, the interactions of orbital magnetic moments with the field split individual spectral lines into sets of three lines (compare with Fig. 41.6b).

Let's begin our analysis of the Zeeman effect by reviewing the concept of *magnetic dipole moment* or *magnetic moment,* introduced in Section 27.7. A plane current loop with vector area $\vec{A}$ carrying current I has a magnetic moment $\vec{\mu}$ given by

$$\vec{\mu} = I\vec{A} \tag{41.9}$$

When a magnetic dipole of moment $\vec{\mu}$ is placed in a magnetic field $\vec{B}$, the field exerts a torque $\vec{\tau} = \vec{\mu} \times \vec{B}$ on the dipole. The potential energy U associated with this interaction is given by Eq. (27.27):

$$U = \vec{\mu} \cdot \vec{B} \tag{41.10}$$

Now let's use Eqs. (41.9) and (41.10) and the Bohr model to look at the interaction of a hydrogen atom with a magnetic field. The orbiting electron (speed v) is equivalent to a current loop with radius r and area πr^2. The average current I is the average charge per unit time that passes a point of the orbit. This is equal to the charge magnitude e divided by the time T for one revolution, given by $T = 2\pi r/v$. Thus $I = ev/2\pi r$, and from Eq. (41.9) the magnitude μ of the magnetic moment is

$$\mu = IA = \frac{ev}{2\pi r}\pi r^2 = \frac{evr}{2} \tag{41.11}$$

We can also express this in terms of the magnitude L of the orbital angular momentum. From Eq. (10.28), $L = mvr$ for a particle in a circular orbit, so Eq. (41.11) becomes

$$\mu = \frac{e}{2m}L \tag{41.12}$$

The ratio of the magnitude of $\vec{\boldsymbol{\mu}}$ to the magnitude of $\vec{\boldsymbol{L}}$ is $\mu/L = e/2m$ and is called the *gyromagnetic ratio.*

In the Bohr model, $L = nh/2\pi = n\hbar$, where $n = 1, 2, \ldots$. For an $n = 1$ state (a ground state), Eq. (41.12) becomes $\mu = (e/2m)\hbar$. This quantity is a natural unit for magnetic moment; it is called one **Bohr magneton,** denoted by μ_B:

$$\mu_B = \frac{e\hbar}{2m} \quad \text{(definition of the Bohr magneton)} \tag{41.13}$$

Evaluating Eq. (41.13) gives

$$\mu_B = 5.788 \times 10^{-5}\ \text{eV/T} = 9.274 \times 10^{-24}\ \text{J/T} \quad \text{or} \quad \text{A}\cdot\text{m}^2$$

Note that the units J/T and $\text{A}\cdot\text{m}^2$ are equivalent. We defined this quantity previously in (optional) Section 28.8.

While the Bohr model suggests that the orbital motion of an atomic electron gives rise to a magnetic moment, this model does *not* give correct predictions about magnetic interactions. As an example, the Bohr model predicts that an electron in a hydrogen-atom ground state has an orbital magnetic moment equal in magnitude to μ_B. But the Schrödinger picture tells us that such a ground-state electron is in an s state with zero angular momentum, so the orbital magnetic moment must be *zero!* To get the correct results, we must describe the states by using Schrödinger wave functions.

It turns out that in the Schrödinger formulation, electrons have the same ratio of μ to L (gyromagnetic ratio) as in the Bohr model, namely, $e/2m$. Suppose the magnetic field $\vec{\boldsymbol{B}}$ is directed along the $+z$-axis. From Eq. (41.10) the interaction energy U of the atom's magnetic moment with the field is

$$U = -\mu_z B \tag{41.14}$$

where μ_z is the z-component of the vector $\vec{\boldsymbol{\mu}}$.

Now we use Eq. (41.12) to find μ_z, recalling that e is the *magnitude* of the electron charge and that the actual charge is $-e$. Because the electron charge is negative, the orbital angular momentum and magnetic moment vectors are opposite. We find

$$\mu_z = -\frac{e}{2m}L_z \tag{41.15}$$

For the Schrödinger wave functions, $L_z = m_l\hbar$, with $m_l = 0, \pm 1, \pm 2, \ldots, \pm l$, so

$$\mu_z = -\frac{e}{2m}L_z = -m_l\frac{e\hbar}{2m} \tag{41.16}$$

CAUTION Be careful not to confuse the electron mass m with the magnetic quantum number m_l.

Finally, we can express the interaction energy, Eq. (41.14), as

$$U = -\mu_z B = m_l \frac{e\hbar}{2m} B \qquad (m_l = 0, \pm 1, \pm 2, \ldots, \pm l) \qquad (41.17)$$

(orbital magnetic interaction energy)

In terms of the Bohr magneton $\mu_B = e\hbar/2m$,

$$U = m_l \mu_B B \qquad \text{(orbital magnetic interaction energy)} \qquad (41.18)$$

The effect of the magnetic field is to shift the energy of each orbital state by an amount U. The interaction energy U depends on the value of m_l because m_l determines the orientation of the orbital magnetic moment relative to the magnetic field. This dependence is the reason m_l is called the magnetic quantum number.

Because the values of m_l range from $-l$ to $+l$ in steps of one, an energy level with a particular value of the orbital quantum number l contains $(2l + 1)$ different orbital states. Without a magnetic field these states all have the same energy; that is, they are degenerate. The magnetic field removes this degeneracy. In the presence of a magnetic field they are split into $2l + 1$ distinct energy levels; adjacent levels differ in energy by $(e\hbar/2m)B = \mu_B B$. The effect on the energy levels of hydrogen is shown in Fig. 41.8. Spectral lines corresponding to transitions from one set of levels to another set are correspondingly split and appear as a series of three closely spaced spectral lines replacing a single line. As the following example shows, the splitting of spectral lines is quite small because the value of $\mu_B B$ is small even for substantial magnetic fields.

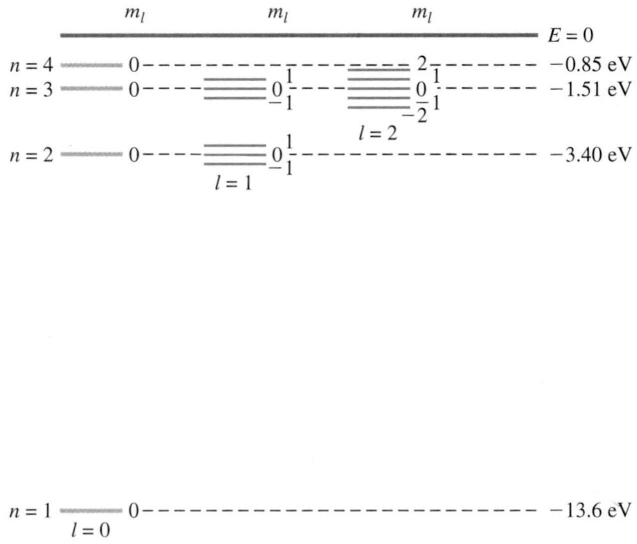

41.8 Energy-level diagram for hydrogen, showing the splitting of energy levels resulting from the interaction of the electron's orbital magnetic moment with an external magnetic field. Values of m_l are shown adjacent to the various levels. Relative magnitudes of splittings are exaggerated for clarity. The $n = 4$ splittings are not shown; can you draw them in?

Example 41.4 An atom in a magnetic field

An atom in a state with $l = 1$ emits a photon with wavelength 600.000 nm as it decays to a state with $l = 0$. If the atom is placed in a magnetic field with magnitude $B = 2.00$ T, determine the shifts in the energy levels and in the wavelength resulting from the interaction of the magnetic field and the atom's orbital magnetic moment.

SOLUTION

IDENTIFY and SET UP: We use Eq. (41.17) or (41.18) to determine the shifts in the energy levels. The relationship $E = hc/\lambda$ between the energy and wavelength of a photon then lets us calculate the wavelengths emitted during transitions from the $l = 1$ states to the $l = 0$ state.

EXECUTE: The energy of a 600-nm photon is

$$E = \frac{hc}{\lambda} = \frac{(4.14 \times 10^{-15}\,\text{eV}\cdot\text{s})(3.00 \times 10^8\,\text{m/s})}{600 \times 10^{-9}\,\text{m}} = 2.07\,\text{eV}$$

Hence, if there is no external magnetic field, the difference in energy between the $l = 0$ and $l = 1$ levels is 2.07 eV.

With a 2.00-T field present, Eq. (41.18) shows that there is no shift of the $l = 0$ state (which has $m_l = 0$). For the $l = 1$ states, the splitting of levels is given by

$$U = m_l \mu_B B = m_l(5.788 \times 10^{-5}\,\text{eV/T})(2.00\,\text{T})$$
$$= m_l(1.16 \times 10^{-4}\,\text{eV}) = m_l(1.85 \times 10^{-23}\,\text{J})$$

When $l = 1$, the possible values of m_l are -1, 0, and $+1$, and the three corresponding levels are separated by equal intervals of 1.16×10^{-4} eV. This is a small fraction of the 2.07-eV photon energy:

$$(1.16 \times 10^{-4}\,\text{eV})/(2.07\,\text{eV}) = 5.60 \times 10^{-5}$$

The corresponding *wavelength* shifts are approximately $(5.60 \times 10^{-5})(600\,\text{nm}) = 0.034$ nm. The original 600.000-nm line is split into a triplet with wavelengths 599.966, 600.000 and 600.034 nm.

EVALUATE: Even though 2.00 T would be a strong field in most laboratories, the wavelength splittings are extremely small. Nonetheless, modern spectrographs have more than enough chromatic resolving power to measure these splittings (see Section 36.5).

Selection Rules

Figure 41.9 shows what happens to a set of d states $(l = 2)$ as the magnetic field increases. The five states, $m_l = -2, -1, 0, 1$, and 2, are degenerate (have the same energy) with zero field, but the increasing field spreads the states out and removes their degeneracy. Figure 41.10 shows the splittings of the 3d and 2p states. Equal energy differences $(e\hbar/2m)B = \mu_B B$ separate adjacent levels. In the absence of a magnetic field, a transition from a 3d to a 2p state would yield a single spectral line with photon energy $E_i - E_f$. With the levels split as shown, it might seem that there are five possible photon energies.

In fact, there are only three possibilities. Not all combinations of initial and final levels are possible, because of a restriction associated with conservation of angular momentum. The photon ordinarily carries off one unit $(\hbar)$ of angular momentum, which leads to the requirements that in a transition l must change by 1 and m_l must change by 0 or ± 1. These requirements are called **selection rules.** Transitions that obey these rules are called *allowed transitions;* those that don't are *forbidden transitions.* The allowed transitions are shown by solid arrows in Fig. 41.10. You should count the possible transition energies to convince yourself that the nine solid arrows give only three possible energies; the zero-field value $E_i - E_f$, and that value plus or minus $\Delta E = (e\hbar/2m)B = \mu_B B$. The corresponding spectral lines are shown.

What we have described is called the *normal* Zeeman effect. It is based entirely on the orbital angular momentum of the electron. In fact, there's nothing particularly *normal* about it. It leaves out a very important consideration: the *spin* angular momentum, the subject of the next section.

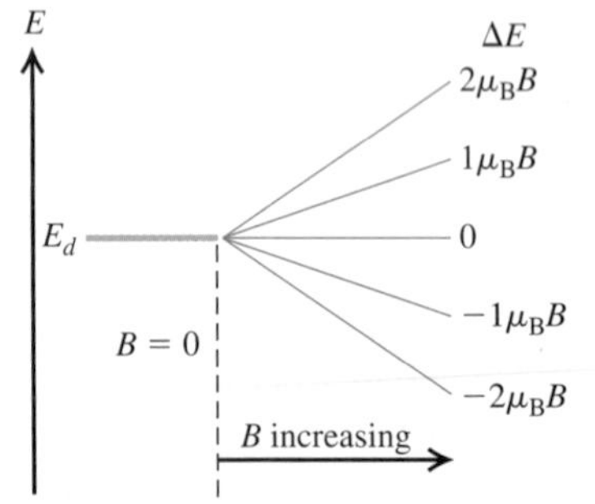

41.9 Splitting of the energy levels of a d state $(l = 2)$ caused by an external magnetic field, assuming only an orbital magnetic moment.

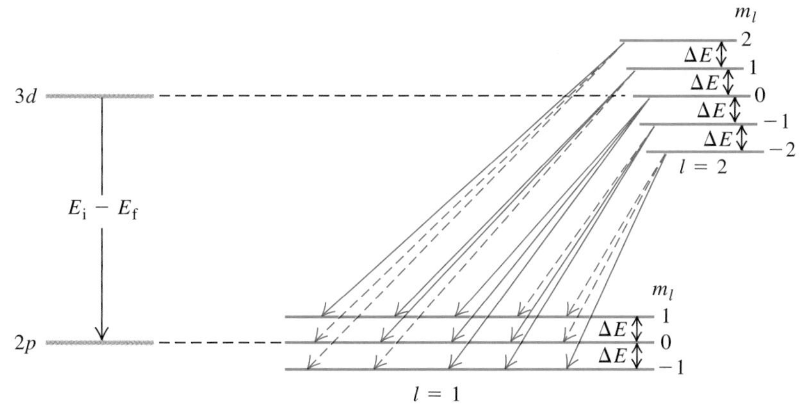

41.10 The cause of the normal Zeeman effect. The magnetic field splits the levels, but selection rules allow transitions with only three different energy changes, giving three different photon frequencies and wavelengths. Solid lines show allowed transitions; dashed lines show forbidden transitions.

Test Your Understanding

In this section we assumed that the magnetic field points in the positive z-direction. How would the results be affected if the magnetic field pointed in the positive x-direction?

41.3 | Electron Spin

Despite the success of the Schrödinger equation in predicting the energy levels of the hydrogen atom, several experimental observations indicate that it doesn't tell the whole story of the behavior of electrons in atoms. First, spectroscopists have found magnetic-field splitting into other than the three lines we've explained, sometimes unequally spaced. Before this effect was understood, it was called the *anomalous* Zeeman effect to distinguish it from the "normal" effect discussed in the preceding section. Both kinds of splittings are shown in Fig. 41.11.

Second, some energy levels show splittings that resemble the Zeeman effect even when there is *no* external magnetic field. For example, when the lines in the hydrogen spectrum are examined with a high-resolution spectrograph, some lines are found to consist of sets of closely spaced lines called *multiplets.* Similarly, the orange-yellow line of sodium, corresponding to the transition $4p \rightarrow 3s$ of the outer electron, is found to be a doublet ($\lambda = 589.0, 589.6$ nm), suggesting that the $4p$ level might in fact be two closely spaced levels. The Schrödinger equation in its original form didn't predict any of this.

41.11 Illustrations of the normal and anomalous Zeeman effects for two elements, zinc and sodium. The brackets under each illustration show the "normal" splitting predicted by neglecting the effect of electron spin.

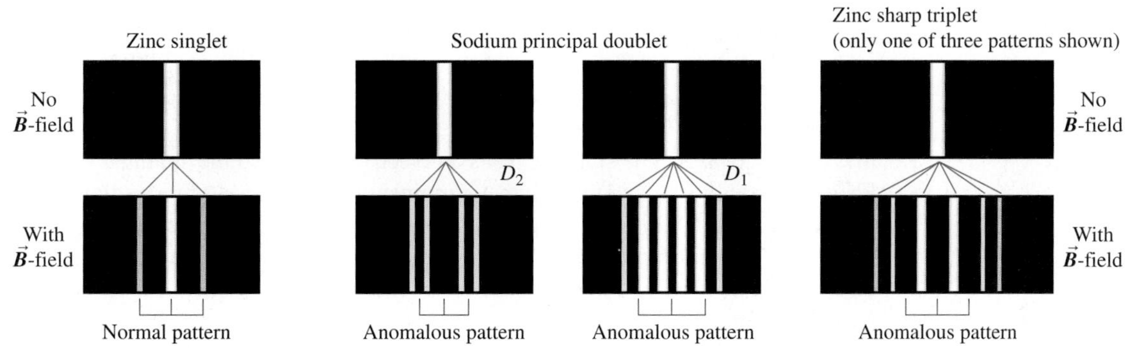

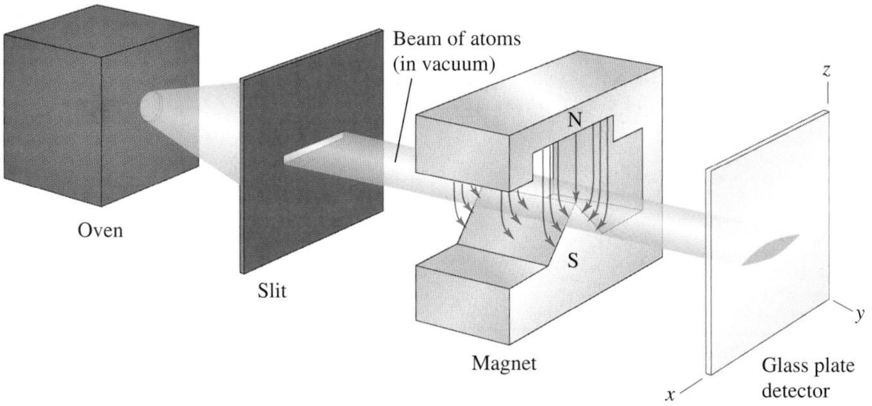

41.12 The Stern-Gerlach experiment. A beam of atoms is directed parallel to the y-axis. The specially shaped magnet poles produce a strongly non-uniform magnetic field that exerts a net force on the atomic magnetic dipoles. The atoms are deflected upward or downward according to the orientation of their magnetic moment.

Similar anomalies appeared in 1922 in atomic-beam experiments performed in Germany by Otto Stern and Walter Gerlach. When they passed a beam of neutral atoms through a non-uniform magnetic field (Fig. 41.12), atoms were deflected according to the orientation of their magnetic moments with respect to the field. These experiments demonstrated the quantization of angular momentum in a very direct way. If there were only orbital angular momentum, the deflections would split the beam into an odd number $(2l + 1)$ of different components. However, some atomic beams were split into an *even* number of components. If we use a different symbol j for an angular momentum quantum number, setting $2j + 1$ equal to an even number gives $j = \frac{1}{2}, \frac{3}{2}, \frac{5}{2}, \ldots$, suggesting a half-integer angular momentum. This can't be understood on the basis of the Bohr model and similar pictures of atomic structure.

In 1925, two graduate students in the Netherlands, Samuel Goudsmidt and George Uhlenbeck, proposed that the electron might have some additional motion. Using a semiclassical model, they suggested that the electron might behave like a spinning sphere of charge instead of a particle. If so, it would have an additional *spin* angular momentum and magnetic moment. If these were quantized in much the same way as *orbital* angular momentum and magnetic moment, they might help to explain the observed energy-level anomalies.

To introduce the concept of **electron spin,** let's start with an analogy. The earth travels in a nearly circular orbit around the sun, and at the same time it *rotates* on its axis. Each motion has its associated angular momentum, which we call the *orbital* and *spin* angular momentum, respectively. The total angular momentum of the earth is the vector sum of the two. If we were to model the earth as a single point, it would have no moment of inertia about its spin axis and thus no spin angular momentum. But when our model includes the finite size of the earth, spin angular momentum becomes possible.

In the Bohr model, suppose the electron is not just a point charge, but a small spinning sphere moving in orbit. Then the electron has not only orbital angular momentum but also spin angular momentum associated with the rotation of its mass about its axis. The sphere carries an electric charge, so the spinning motion leads to current loops and to a magnetic moment, as we discussed in Section 27.7. In a magnetic field, the *spin* magnetic moment has an interaction energy in addition to that of the *orbital* magnetic moment (the normal Zeeman-effect interaction discussed in Section 41.2). We should see additional Zeeman shifts due to the spin magnetic moment.

As we mentioned above, such shifts *are* indeed observed in precise spectroscopic analysis. This and a variety of other experimental evidence have shown conclusively that the electron *does* have spin angular momentum and a spin magnetic moment that do not depend on its orbital motion but are intrinsic to the electron itself.

Spin Quantum Numbers

Like orbital angular momentum, the spin angular momentum of an electron (denoted by $\vec{S}$) is found to be *quantized.* Suppose we have some apparatus that measures a particular component of $\vec{S}$, say the z-component S_z. We find that the only possible values are

$$S_z = \pm\frac{1}{2}\hbar \qquad \text{(components of spin angular momentum)} \qquad (41.19)$$

This relation is reminiscent of the expression $L_z = m_l\hbar$ for the z-component of orbital angular momentum, except that $|S_z|$ is *one-half* of $\hbar$ instead of an *integer* multiple. Equation (41.19) also suggests that the magnitude S of the spin angular momentum is given by an expression analogous to Eq. (41.4) with the orbital quantum number l replaced by the **spin quantum number** $s = \frac{1}{2}$:

$$S = \sqrt{\frac{1}{2}\left(\frac{1}{2}+1\right)}\hbar = \sqrt{\frac{3}{4}}\hbar \qquad (41.20)$$

(magnitude of spin angular momentum)

The electron is often called a "spin-$\frac{1}{2}$ particle."

The Bohr model is an oversimplified picture of electron behavior. In quantum mechanics, in which the Bohr orbits are superseded by probability distributions $|\psi|^2$, we can't really *picture* electron spin. If we visualize a probability distribution as a cloud surrounding the nucleus, then we can imagine many tiny spin arrows distributed throughout the cloud, either all with components in the $+z$-direction or all with components in the $-z$-direction. But don't take this picture too seriously.

To label completely the state of the electron in a hydrogen atom, we now need a fourth quantum number m_s, to specify the electron spin orientation. For an electron we give m_s the values $+\frac{1}{2}$ or $-\frac{1}{2}$ to agree with Eq. (41.19):

$$S_z = m_s\hbar \left(m_s = \pm\frac{1}{2}\right) \qquad \text{(allowed values of } m_s \text{ and } S_z \text{ for an electron)} \qquad (41.21)$$

The spin angular momentum vector $\vec{S}$ can have only two orientations in space relative to the z-axis: "*spin up*" with a z-component of $+\frac{1}{2}\hbar$ and "*spin down*" with a z-component of $-\frac{1}{2}\hbar$.

The z-component of the associated spin magnetic moment (μ_z) turns out to be related to S_z by

$$\mu_z = -(2.00232)\frac{e}{2m}S_z \qquad (41.22)$$

where $-e$ and m are (as usual) the charge and mass of the electron. When the atom is placed in a magnetic field, the interaction energy $-\vec{\boldsymbol{\mu}}\cdot\vec{\boldsymbol{B}}$ of the spin magnetic dipole moment with the field causes further splittings in energy levels and in the corresponding spectrum lines.

Equation (41.22) shows that the gyromagnetic ratio for electron spin is approximately *twice* as great as the value $e/2m$ for *orbital* angular momentum and magnetic dipole moment. This result has no classical analog. But in 1928, Paul Dirac developed a relativistic generalization of the Schrödinger equation for electrons. His equation gave a spin gyromagnetic ratio of exactly $2(e/2m)$. It took another two decades to develop the area of physics called *quantum electrodynamics,* abbreviated QED, that predicts the value we've given to "only" six significant figures as 2.00232. In fact, QED now predicts a value that agrees with a recent measurement of 2.0023193043737(82), making QED the most precise theory in all science.

Example 41.5 Spin magnetic interaction energy

Calculate the interaction energy for an electron in an $l = 0$ state (with no orbital magnetic moment) in a magnetic field with magnitude 2.00 T.

SOLUTION

IDENTIFY and SET UP: The interaction energy is $U = -\vec{\mu}\cdot\vec{B}$ as in Eq. (41.10), but now the magnetic moment is due to the electron spin $\vec{S}$. Just as in Section 41.2, we take $\vec{B}$ to be in the positive z-direction so that the interaction energy can be written in the form $U = -\mu_z B$ [Eq. (41.14)]. The z-component of the magnetic moment is given in terms of S_z by Eq. (41.22), and S_z is given by Eq. (41.19).

EXECUTE: Combining Eqs. (41.14), (41.19), and (41.22), we have

$$U = -\left[-(2.00232)\left(\frac{e}{2m}\right)\left(\pm\frac{1}{2}\hbar\right)B\right]$$
$$= \pm\frac{1}{2}(2.00232)\left(\frac{e\hbar}{2m}\right)B = \pm(1.00116)\mu_B B$$
$$= \pm(1.00116)(9.274\times10^{-24}\ \text{J/T})(2.00\ \text{T})$$
$$= \pm1.86\times10^{-23}\ \text{J} = \pm1.16\times10^{-5}\ \text{eV}$$

The positive value of U corresponds to $S_z = +\frac{1}{2}\hbar$ (spin up) and the negative value of U to $S_z = -\frac{1}{2}\hbar$ (spin down).

EVALUATE: Let's check the *sign* of our results. If the electron is spin down, the spin angular momentum $\vec{S}$ points generally opposite to $\vec{B}$, and hence the magnetic moment $\vec{\mu}$ (which is opposite to $\vec{S}$ because the electron charge is negative) points generally parallel to $\vec{B}$. From Eq. (41.10), $U = -\vec{\mu}\cdot\vec{B}$, the interaction energy is negative if $\vec{\mu}$ and $\vec{B}$ are parallel. Our results show that U is indeed negative in this case. In a similar way, we can argue that U must be positive for a spin-up electron, which is just what we calculated.

The red lines in Fig. 41.13 show how increasing the magnetic field magnitude affects the interaction energies for the two different spin states. The quantity 5.795×10^{-5} eV/T is just $(1.00116)\mu_B$. The graphs are straight lines because, as Eq. (41.14) shows, U is proportional to B.

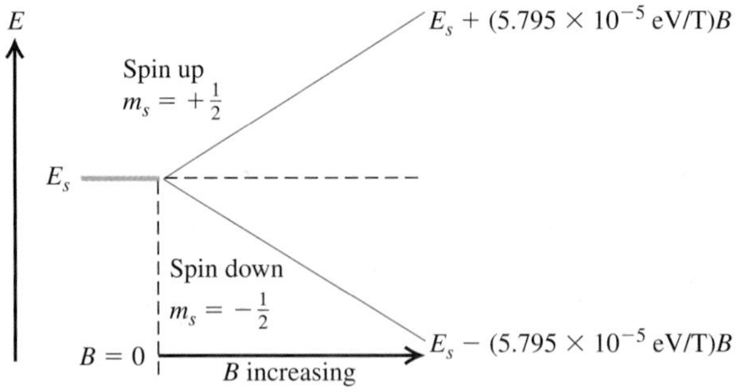

41.13 An $l = 0$ level of a single electron is split by interaction of the spin magnetic moment with a magnetic field.

Spin-Orbit Coupling

As has already been mentioned, the spin magnetic dipole moment also gives splitting of energy levels even when there is *no* external field. One cause involves the orbital motion of the electron. In the Bohr model, observers moving with the electron would see the positively charged nucleus revolving around them (just as to earthbound observers the sun seems to be orbiting the earth). This apparent motion of charge causes a magnetic field at the location of the electron, as measured in the electron's moving frame of reference. The resulting interaction with the spin magnetic moment causes a twofold splitting of this level, corresponding to the two possible orientations of electron spin.

Discussions based on the Bohr model can't be taken too seriously, but a similar result can be derived from the Schrödinger equation. The interaction energy U can be expressed in terms of the scalar product of the angular-momentum vectors $\vec{L}$ and $\vec{S}$. This effect is called **spin-orbit coupling;** it is responsible for the small energy difference between the two closely spaced lowest excited levels of sodium shown in Fig. 38.10 and for the corresponding doublet (589.0, 589.6 nm) in the spectrum of sodium.

Example 41.6 An effective magnetic field

Calculate the effective magnetic field experienced by the electron in the $3p$ levels of the sodium atom.

SOLUTION

IDENTIFY: The two lines in the sodium doublet result from transitions from the two $3p$ levels, which are split by spin-orbit coupling, to the $3s$ level, which is *not* split because it has $L = 0$. We picture the spin-orbit coupling as an interaction between the electron spin magnetic moment and an effective magnetic field due to the nucleus. Hence this example is like Example 41.5 in reverse: there we were given B and found the difference between the energies of the two spin states; here we use the energy difference to find B.

SET UP: The difference in energy between the two $3p$ levels is equal to the difference in energy between the two photons of the sodium doublet. We use this relationship and the results of Example 41.5 to determine our target variable, the effective magnetic field strength B.

EXECUTE: The energies of the two photons are

$$E = \frac{hc}{\lambda} = \frac{(4.136 \times 10^{-15}\ \text{eV}\cdot\text{s})(2.998 \times 10^{8}\ \text{m/s})}{589.0 \times 10^{-9}\ \text{m}} = 2.1052\ \text{eV}$$

and

$$E = \frac{hc}{\lambda} = \frac{(4.136 \times 10^{-15}\ \text{eV}\cdot\text{s})(2.998 \times 10^{8}\ \text{m/s})}{589.6 \times 10^{-9}\ \text{m}} = 2.1031\ \text{eV}$$

Hence the energy difference between the two $3p$ levels is

$$2.1052\ \text{eV} - 2.1031\ \text{eV} = 0.0021\ \text{eV} = 3.4 \times 10^{-22}\ \text{J}$$

The spin-orbit interaction raises one level and lowers the other, each by 1.7×10^{-22} J, half this difference. From Example 41.5, the amount each state is raised or lowered is $|U| = (1.00116)\mu_B B$, so

$$B = \left|\frac{U}{(1.00116)\mu_B}\right| = \frac{1.7 \times 10^{-22}\ \text{J}}{9.28 \times 10^{-24}\ \text{J/T}} = 18\ \text{T}$$

EVALUATE: This result tells us that the effective magnetic field experienced by an atomic electron is *very* strong. To produce a steady field of this magnitude in the laboratory requires state-of-the-art electromagnets.

The orbital and spin angular momenta ($\vec{L}$ and $\vec{S}$, respectively) can combine in various ways. We define the total angular momentum $\vec{J}$ as

$$\vec{J} = \vec{L} + \vec{S} \tag{41.23}$$

The possible values of the magnitude J are given in terms of a quantum number j by

$$J = \sqrt{j(j+1)}\hbar \tag{41.24}$$

We can then have states in which $j = \left|l \pm \frac{1}{2}\right|$. The $l + \frac{1}{2}$ states correspond to the case in which the vectors $\vec{L}$ and $\vec{S}$ have parallel z-components; while for the $l - \frac{1}{2}$ states, $\vec{L}$ and $\vec{S}$ have antiparallel z-components. For example, when $l = 1$, j can be

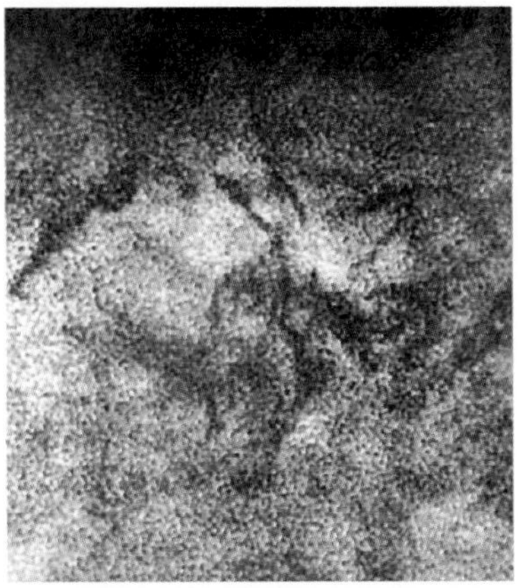

41.14 This false-color image shows 21-cm radio emission from a large cloud of interstellar hydrogen about 7000 light years from the earth. The dark areas within the cloud are regions where hydrogen is at such a low temperature—just a few kelvins—that it cannot emit even low-energy 21-cm photons, but absorbs them instead.

$\frac{1}{2}$ or $\frac{3}{2}$. In another spectroscopic notation these p states are labeled $^2P_{1/2}$ and $^2P_{3/2}$, respectively. The superscript is the number of possible spin orientations, the letter P (now capitalized) indicates states with $l = 1$, and the subscript is the value of j. We used this scheme to label the energy levels of the sodium atom in Fig. 38.10.

The various line splittings resulting from magnetic interactions are collectively called *fine structure.* There are also additional, much smaller splittings associated with the fact that the *nucleus* of the atom has a magnetic dipole moment that interacts with the orbital and/or spin magnetic dipole moments of the electrons. These effects are called *hyperfine structure.* For example, the ground level of hydrogen is split into two states, separated by only 5.9×10^{-6} eV. The photon that is emitted in the transitions between these states has a wavelength of 21 cm. Radio astronomers use this wavelength to map interstellar clouds of hydrogen gas that are too cold to emit visible light (Fig. 41.14).

Test Your Understanding

If all the electrons in your body were somehow flipped to the spin-up orientation, what would be the net z-component of your body's spin angular momentum? Would you notice this? (*Hint:* Electrons make up about 1/4000 of your body's mass.)

41.4 | Many-Electron Atoms and the Exclusion Principle

So far, our analysis of atomic structure has concentrated on the hydrogen atom. That's natural; neutral hydrogen, with only one electron, is the simplest atom. If we can't understand hydrogen, we certainly can't understand anything more complex. But now let's move to many-electron atoms.

In general, an atom in its normal (electrically neutral) state has Z electrons and Z protons. Recall from Section 41.1 that we call Z the *atomic number.* The total electric charge of such an atom is exactly zero because the neutron has no charge while the proton and electron charges have the same magnitude but opposite sign.

We can apply the Schrödinger equation to this general atom. However, the complexity of the analysis increases very rapidly with increasing Z. Each of the Z electrons interacts not only with the nucleus but also with every other electron. The wave functions and the potential energy are functions of $3Z$ coordinates, and the equation contains second derivatives with respect to all of them. The mathematical problem of finding solutions of such equations is so complex that it has not been solved exactly even for the neutral helium atom, which has only two electrons.

Fortunately, various approximation schemes are available. The simplest approximation is to ignore all interactions between electrons and consider each electron as moving under the action only of the nucleus (considered to be a point charge). In this approximation the wave function for each electron is a function like those for the hydrogen atom, specified by four quantum numbers (n, l, m_l, m_s); the nuclear charge is Ze instead of e. This requires replacement of every factor of e^2 in the wave functions and the energy levels by Ze^2. In particular, the energy levels are given by Eq. (41.3) with e^4 replaced by Z^2e^4:

$$E_n = -\frac{1}{(4\pi\epsilon_0)^2}\frac{m_r Z^2 e^4}{2n^2\hbar^2} = -\frac{Z^2}{n^2}(13.6\ \text{eV}) \qquad (41.25)$$

This approximation is fairly drastic; when there are many electrons, their interactions with each other are as important as the interaction of each with the nucleus. So this model isn't very useful for quantitative predictions.

The Central-Field Approximation

A less drastic and more useful approximation is to think of all the electrons together as making up a charge cloud that is, on average, *spherically symmetric.* We can then think of each individual electron as moving in the total electric field due to the nucleus and this averaged-out cloud of all the other electrons. There is a corresponding spherically symmetric potential-energy function $U(r)$. This picture is called the **central-field approximation;** it provides a useful starting point for the understanding of atomic structure. If you are disappointed that we have to make approximations at such an early stage in our discussion, keep in mind that we are dealing with problems that initially defied all attempts at analysis, with or without approximations.

In the central-field approximation we can again deal with one-electron wave functions. The Schrödinger equation differs from the equation for hydrogen only in that the $1/r$ potential-energy function is replaced by a different function $U(r)$. But it turns out that $U(r)$ does not enter the differential equations for $\Theta(\theta)$ and $\Phi(\phi)$, so those angular functions are exactly the same as for hydrogen, and the orbital angular-momentum *states* are also the same as before. The quantum numbers l, m_l, and m_s have the same meaning as before, and the magnitude and z-component of the orbital angular momentum are again given by Eqs. (41.4) and (41.5).

The radial wave functions and probabilities are different than for hydrogen because of the change in $U(r)$, so the energy levels are no longer given by Eq. (41.3). We can still label a state using the four quantum numbers (n, l, m_l, m_s). In general, the energy of a state now depends on both n and l, rather than just on n as with hydrogen. The restrictions on values of the quantum numbers are the same as before:

$$n \geq 1 \quad 0 \leq l \leq n-1 \quad |m_l| \leq l, m_s = \pm\frac{1}{2} \qquad (41.26)$$

(allowed values of quantum numbers)

The Exclusion Principle

To understand the structure of many-electron atoms, we need an additional principle, the *exclusion principle.* To see why this principle is needed, let's consider the lowest-energy state or *ground state* of a many-electron atom. In the one-electron states of the central-field model, there is a lowest-energy state (corresponding to an $n = 1$ state of hydrogen). We might expect that in the ground state of a complex atom, *all* the electrons should be in this lowest state. If so, then we should see only gradual changes in physical and chemical properties when we look at the behavior of atoms with increasing numbers of electrons (Z).

Such gradual changes are *not* what is observed. Instead, properties of elements vary widely from one to the next, with each element having its own distinct personality. For example, the elements fluorine, neon, and sodium have 9, 10, and 11 electrons, respectively, per atom. Fluorine $(Z = 9)$ is a *halogen;* it tends strongly to form compounds in which each fluorine atom acquires an extra electron. Sodium $(Z = 11)$ is an *alkali metal;* it forms compounds in which each sodium atom *loses* an electron. Neon $(Z = 10)$ is an *noble gas,* forming no compounds

41.15 The key to understanding the periodic table of the elements was the discovery by Wolfgang Pauli (1900–1958) of the exclusion principle. Pauli received the 1945 Nobel Prize in physics for his accomplishment. This photo shows Pauli (on the left) and Niels Bohr watching a toy top spinning on the floor.

at all. Such observations show that in the ground state of a complex atom the electrons *cannot* all be in the lowest-energy states. But why not?

The key to this puzzle, discovered by the Austrian physicist Wolfgang Pauli (Fig. 41.15) in 1925, is called the **exclusion principle.** This principle states that **no two electrons can occupy the same quantum-mechanical state** in a given system. That is, **no two electrons in an atom can have the same values of all four quantum numbers** (n, l, m_l, m_s). Each quantum state corresponds to a certain distribution of the electron "cloud" in space. Therefore the principle also says, in effect, that no more than two electrons with opposite values of the quantum number m_s can occupy the same region of space. We shouldn't take this last statement too seriously because the electron probability functions don't have sharp, definite boundaries. But the exclusion principle limits the amount that electron wave functions can overlap. Think of it as the quantum-mechanical analog of a university rule that allows only one student per desk.

CAUTION Don't confuse the exclusion principle with the electric repulsion between electrons. While both effects tend to keep electrons within an atom separated from each other, they are very different in character. Two electrons can always be pushed closer together by adding energy to combat electric repulsion; in contrast, *nothing* can overcome the exclusion principle and force two electrons into the same quantum-mechanical state.

Table 41.2 lists some of the sets of quantum numbers for electron states in an atom. It's similar to Table 41.1 (Section 41.1), but we've added the number of states in each subshell and shell. Because of the exclusion principle, the "number of states" is the same as the *maximum* number of electrons that can be found in those states. For each state, m_s can be either $+\frac{1}{2}$ or $-\frac{1}{2}$.

As with the hydrogen wave functions, different states correspond to different spatial distributions; electrons with larger values of n are concentrated at larger distances from the nucleus. Figure 41.3 (Section 41.1) shows this effect. When an atom has more than two electrons, they can't all huddle down in the low-energy $n = 1$ states nearest to the nucleus because there are only two of these states; the exclusion principle forbids multiple occupancy of a state. Some electrons are forced into states farther away, with higher energies. Each value of n corresponds roughly to a region of space around the nucleus in the form of a spherical *shell.* Hence we speak of the K shell as the region that is occupied by the electrons in the

Table 41.2 Quantum States of Electrons in the First Four Shells

n	l	m_l	Spectroscopic Notation	Number of States		Shell
1	0	0	1*s*	2		*K*
2	0	0	2*s*	2	8	*L*
2	1	−1, 0, 1	2*p*	6		
3	0	0	3*s*	2	18	*M*
3	1	−1, 0, 1	3*p*	6		
3	2	−2, −1, 0, 1, 2	3*d*	10		
4	0	0	4*s*	2	32	*N*
4	1	−1, 0, 1	4*p*	6		
4	2	−2, −1, 0, 1, 2	4*d*	10		
4	3	−3, −2, −1, 0, 1, 2, 3	4*f*	14		

$n = 1$ states, the L shell as the region of the $n = 2$ states, and so on. States with the same n but different l form *subshells,* such as the $3p$ subshell.

The Periodic Table

We can use the exclusion principle to derive the most important features of the structure and chemical behavior of multielectron atoms, including the periodic table of the elements. Let's imagine constructing a neutral atom by starting with a bare nucleus with Z protons and adding Z electrons, one by one. To obtain the ground state of the atom as a whole, we fill the lowest-energy electron states (those closest to the nucleus, with the smallest values of n and l) first, and we use successively higher states until all the electrons are in place. The chemical properties of an atom are determined principally by interactions involving the outermost, or *valence,* electrons, so we particularly want to learn how these electrons are arranged.

Let's look at the ground-state electron configurations for the first few atoms (in order of increasing Z). For hydrogen the ground state is $1s$; the single electron is in a state $n = 1$, $l = 0$, $m_l = 0$, and $m_s = \pm\frac{1}{2}$. In the helium atom $(Z = 2)$, *both* electrons are in $1s$ states, with opposite spins; one has $m_s = -\frac{1}{2}$ and the other has $m_s = +\frac{1}{2}$. We denote the helium ground state as $1s^2$. (The superscript 2 is not an exponent; the notation $1s^2$ tells us that there are two electrons in the $1s$ subshell. Also, the superscript 1 is understood, as in $2s$.) For helium the K shell is completely filled, and all others are empty. Helium is a noble gas; it has no tendency to gain or lose an electron, and it forms no compounds.

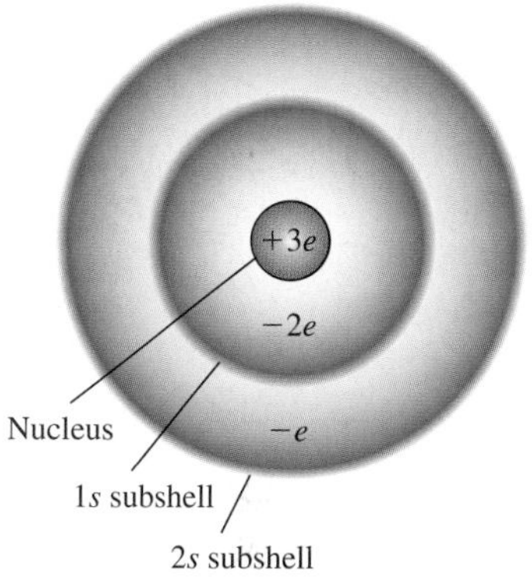

41.16 Schematic representation of the charge distribution in a lithium atom. The nucleus has a charge of $+3e$; the two $1s$ electrons are closer to the nucleus than is the $2s$ electron, which moves in a field roughly equal to that of a point charge of $+3e - 2e = +e$.

Lithium $(Z = 3)$ has three electrons. In its ground state, two are in $1s$ states and one is in a $2s$ state, so we denote the lithium ground state as $1s^2 2s$. On average, the $2s$ electron is considerably farther from the nucleus than are the $1s$ electrons (Fig. 41.16). According to Gauss's law, the *net* charge Q_{encl} attracting the $2s$ electron is nearer to $+e$ than to the value $+3e$ it would have without the two $1s$ electrons present. As a result, the $2s$ electron is loosely bound; only 5.4 eV is required to remove it, compared with the 30.6 eV given by Eq. (41.25) with $Z = 3$ and $n = 2$. In chemical behavior, lithium is an *alkali metal.* It forms ionic compounds in which each lithium atom loses an electron and has a valence of $+1$.

Next is beryllium $(Z = 4)$; its ground-state configuration is $1s^2 2s^2$, with its two valence electrons filling the s subshell of the L shell. Beryllium is the first of the *alkaline-earth* elements, forming ionic compounds in which the valence of the atoms is $+2$.

Table 41.3 shows the ground-state electron configurations of the first 30 elements. The L shell can hold eight electrons. At $Z = 10$, both the K and L shells are filled, and there are no electrons in the M shell. We expect this to be a particularly stable configuration, with little tendency to gain or lose electrons. This element is neon, a noble gas with no known compounds. The next element after neon is sodium $(Z = 11)$, with filled K and L shells and one electron in the M shell. Its "noble-gas-plus-one-electron" structure resembles that of lithium; both are alkali metals. The element *before* neon is fluorine, with $Z = 9$. It has a vacancy in the L shell and has an affinity for an extra electron to fill the shell. Fluorine forms ionic compounds in which it has a valence of -1. This behavior is characteristic of the *halogens* (fluorine, chlorine, bromine, iodine, and astatine), all of which have "noble-gas-minus-one" configurations.

Proceeding down the list, we can understand the regularities in chemical behavior displayed by the **periodic table of the elements** (Appendix D) on the basis of electron configurations. The similarity of elements in each *group* (verti-

Table 41.3 Ground-State Electron Configurations

Element	Symbol	Atomic Number (Z)	Electron Configuration
Hydrogen	H	1	$1s$
Helium	He	2	$1s^2$
Lithium	Li	3	$1s^22s$
Beryllium	Be	4	$1s^22s^2$
Boron	B	5	$1s^22s^22p$
Carbon	C	6	$1s^22s^22p^2$
Nitrogen	N	7	$1s^22s^22p^3$
Oxygen	O	8	$1s^22s^22p^4$
Fluorine	F	9	$1s^22s^22p^5$
Neon	Ne	10	$1s^22s^22p^6$
Sodium	Na	11	$1s^22s^22p^63s$
Magnesium	Mg	12	$1s^22s^22p^63s^2$
Aluminum	Al	13	$1s^22s^22p^63s^23p$
Silicon	Si	14	$1s^22s^22p^63s^23p^2$
Phosphorus	P	15	$1s^22s^22p^63s^23p^3$
Sulfur	S	16	$1s^22s^22p^63s^23p^4$
Chlorine	Cl	17	$1s^22s^22p^63s^23p^5$
Argon	Ar	18	$1s^22s^22p^63s^23p^6$
Potassium	K	19	$1s^22s^22p^63s^23p^64s$
Calcium	Ca	20	$1s^22s^22p^63s^23p^64s^2$
Scandium	Sc	21	$1s^22s^22p^63s^23p^64s^23d$
Titanium	Ti	22	$1s^22s^22p^63s^23p^64s^23d^2$
Vanadium	V	23	$1s^22s^22p^63s^23p^64s^23d^3$
Chromium	Cr	24	$1s^22s^22p^63s^23p^64s3d^5$
Manganese	Mn	25	$1s^22s^22p^63s^23p^64s^23d^5$
Iron	Fe	26	$1s^22s^22p^63s^23p^64s^23d^6$
Cobalt	Co	27	$1s^22s^22p^63s^23p^64s^23d^7$
Nickel	Ni	28	$1s^22s^22p^63s^23p^64s^23d^8$
Copper	Cu	29	$1s^22s^22p^63s^23p^64s3d^{10}$
Zinc	Zn	30	$1s^22s^22p^63s^23p^64s^23d^{10}$

cal column) of the periodic table is the result of similarity in outer electron configuration. All the noble gases (helium, neon, argon, krypton, xenon, and radon) have filled-shell or filled-shell plus filled p subshell configurations. All the alkali metals (lithium, sodium, potassium, rubidium, cesium, and francium) have "noble-gas-plus-one" configurations. All the alkaline earth metals (beryllium, magnesium, calcium, strontium, barium, and radium) have "noble-gas-plus-two" configurations, and all the halogens (fluorine, chlorine, bromine, iodine, and astatine) have "noble-gas-minus-one" structures.

A slight complication occurs with the M and N shells because the $3d$ and $4s$ subshell levels ($n = 3, l = 2$, and $n = 4, l = 0$, respectively) have similar energies. (We'll discuss in the next subsection why this happens.) Argon $(Z = 18)$ has all the $1s$, $2s$, $2p$, $3s$, and $3p$ subshells filled, but in potassium $(Z = 19)$ the additional electron goes into a $4s$ energy state rather than a $3d$ state (because the $4s$ state has slightly lower energy).

The next several elements have one or two electrons in the $4s$ subshell and increasing numbers in the $3d$ subshell. These elements are all metals with rather similar chemical and physical properties; they form the first *transition series,* starting with scandium $(Z = 21)$ and ending with zinc $(Z = 30)$, for which all the $3d$ and $4s$ subshells are filled.

Something similar happens with $Z = 57$ through $Z = 71$, which have one or two electrons in the $6s$ subshell but only partially filled $4f$ and $5d$ subshells. These are the *rare earth* elements; they all have very similar physical and chemical properties. Yet another such series, called the *actinide* series, starts with $Z = 91$.

Screening

We have mentioned that in the central-field picture, the energy levels depend on l as well as n. Let's take sodium $(Z = 11)$ as an example. If ten of its electrons fill its K and L shells, the energies of some of the states for the remaining electron are found experimentally to be

$3s$ states:	-5.138 eV,
$3p$ states:	-3.035 eV,
$3d$ states:	-1.521 eV,
$4s$ states:	-1.947 eV.

The $3s$ states are the lowest (most negative); one is the ground state for the eleventh electron in sodium. The energy of the $3d$ states is quite close to the energy of the $n = 3$ state in hydrogen. The surprise is that the $4s$ state energy is 0.426 eV *below* the $3d$ state, even though the $4s$ state has larger n.

We can understand these results using Gauss's law and the radial probability distribution. For any spherically symmetric charge distribution the electric-field magnitude at a distance r from the center is $(\frac{1}{4}\pi\epsilon_0)Q_{encl}/r^2$, where Q_{encl} is the total charge enclosed within a sphere with radius r. Mentally remove the outer (valence) electron atom from a sodium atom. What you have left is a spherically symmetric collection of 10 electrons (filling the K and L shells) and 11 protons, so $Q_{encl} = -10e + 11e = +e$. If the eleventh electron is completely outside this collection of charges, it is attracted by an effective charge of $+e$, not $+11e$.

This effect is called **screening;** the 10 electrons *screen* 10 of the 11 protons, leaving an effective net charge of $+e$. In general, an electron that spends all its time completely outside a positive charge $Z_{eff}e$ has energy levels given by the hydrogen expression with e^2 replaced by $Z_{eff}e^2$. From Eq. (41.25) this is

$$E_n = -\frac{Z_{eff}^{\ 2}}{n^2}(13.6\text{ eV}) \qquad \text{(energy levels with screening)} \qquad (41.27)$$

If the eleventh electron in the sodium atom is completely outside the remaining charge distribution, then $Z_{eff} = 1$.

CAUTION Equations (41.3), (41.25), and (41.27) all give values of E_n in terms of $(13.6\text{ eV})/n^2$, but they don't apply in general to the same atoms. Equation (41.3) is *only* for hydrogen, Eq. (41.25) is only for the case in which there is no interaction with any other electron (and is thus accurate only when the atom has just one electron), and Eq. (41.27) is useful when one electron is screened from the nucleus by other electrons.

Now let's use the radial probability functions shown in Fig. 41.3 to explain why the energy of a sodium $3d$ state is approximately the same as the $n = 3$ value of hydrogen, -1.51 eV. The distribution for the $3d$ state (for which l has the maximum value $n - 1$) has one peak, and its most probable radius is *outside* the positions of the electrons with $n = 1$ or 2. (Those electrons also are pulled closer to the nucleus than in hydrogen because they are less effectively screened from the positive charge $11e$ of the nucleus.) Thus in sodium a $3d$ electron spends most of its time well outside the $n = 1$ and $n = 2$ states (the K and L shells). The 10 electrons in these shells screen almost all but one of the 11 protons, leaving a net charge of about $Z_{\text{eff}}e = (1)e$. Then, from Eq. (41.27), the corresponding energy is approximately $-(1)^2(13.6\text{ eV})/3^2 = -1.51\text{ eV}$. This approximation is very close to the experimental value of -1.521 eV.

Looking again at Fig. 41.3, we see that the radial probability density for the $3p$ state (for which $l = n - 2$) has two peaks and that for the $3s$ state $(l = n - 3)$ has three peaks. For sodium the first small peak in the $3p$ distribution gives a $3p$ electron a higher probability (compared to the $3d$ state) of being *inside* the charge distributions for the electrons in the $n = 2$ states. That is, a $3p$ electron is less completely screened from the nucleus than is a $3d$ electron because it spends some of its time within the filled K and L shells. Thus for the $3p$ electrons, Z_{eff} is greater than unity. From Eq. (41.27) the $3p$ energy is lower (more negative) than the $3d$ energy of -1.521 eV. The actual value is -3.035 eV. A $3s$ electron spends even more time within the inner electron shells than a $3p$ electron does, giving an even larger Z_{eff} and an even more negative energy.

Example 41.7 Determining Z_{eff} experimentally

The measured energy of a $3s$ state of sodium is -5.138 eV. Calculate the value of Z_{eff}.

SOLUTION

IDENTIFY and SET UP: For this state we are given $n = 3$ and $E_n = -5.138$ eV, so we can use Eq. (41.27) to determine Z_{eff}.

EXECUTE: Solving Eq. (41.27) for Z_{eff}, we have

$$Z_{\text{eff}}^{\ 2} = -\frac{n^2 E_n}{13.6\text{ eV}} = -\frac{3^2(-5.138\text{ eV})}{13.6\text{ eV}} = 3.40$$

$$Z_{\text{eff}} = 1.84$$

EVALUATE: The effective charge attracting a $3s$ electron is $1.84e$. Sodium's 11 protons are screened by an average of $11 - 1.84 = 9.16$ electrons instead of 10 electrons because of the penetration of the inner (K and L) shells by the $3s$ electron.

Each alkali metal (lithium, sodium, potassium, rubidium, and cesium) has one more electron than the corresponding noble gas (helium, neon, argon, krypton, and xenon). This extra electron is mostly outside the other electrons in the filled shells and subshells. Therefore all the alkali metals behave similarly to sodium.

Example 41.8 Energies for a valence electron

The valence electron in potassium has a $4s$ ground state. Calculate the approximate energy of the state having the smallest Z_{eff}, and discuss the relative energies of the $4s$, $4p$, $4d$, and $4f$ states.

SOLUTION

IDENTIFY and SET UP: The state with the smallest Z_{eff} is the one for which the nucleus is most effectively screened by the other electrons. We use this idea along with Eq. (41.27) to determine the energy of this state.

EXECUTE: A $4f$ state has $n = 4$ and $l = 3 = 4 - 1$. Thus it is the state of largest orbital angular momentum for $n = 4$, and thus the state in which the electron spends the most time outside the electron charge clouds of the inner filled shells and subshells. This makes Z_{eff} for a $4f$ state close to unity. Equation (41.27) then gives

$$E_4 = -\frac{Z_{\text{eff}}^{\ 2}}{n^2}(13.6\text{ eV}) = -\frac{1^2}{4^2}(13.6\text{ eV}) = -0.85\text{ eV}$$

This approximation agrees with the measured energy to the precision given.

An electron in a $4d$ state spends a bit more time within the inner shells, and its energy is therefore a bit more negative (measured to be -0.94 eV). For the same reason, a $4p$ state has an even lower energy (measured to be −2.73 eV), and a 4*s* state has the lowest energy (measured to be −4.339 eV).

EVALUATE: We can extend this analysis to the singly ionized alkaline earth elements: Be^+, Mg^+, Ca^+, Sr^+, and Ba^+. For any allowed value of *n*, the highest-*l* state $(l = n - 1)$ of the one remaining outer electron sees an effective charge of almost $+2e$, so for these states, $Z_{eff} = 2$. A 3*d* state for Mg^+, for example, has an energy of about $-2^2(13.6\ \text{eV})/3^2 = -6.0\ \text{eV}$.

Test Your Understanding

If electrons did not obey the exclusion principle, what would be the ground-state electron configuration of sodium? Compared to a real sodium atom, would it be easier or more difficult to remove the first electron from sodium in this case?

41.5 | X-Ray Spectra

X-ray spectra provide yet another example of the richness and power of the Schrödinger equation and of the model of atomic structure that we derived from it in the preceding section. In Section 38.7 we discussed x-ray production on the basis of the photon concept. With the development of x-ray diffraction techniques (Section 36.6) by von Laue, Bragg, and others, beginning in 1912, it became possible to measure x-ray wavelengths quite precisely (to within 0.1% or less).

Detailed studies of x-ray spectra showed a continuous spectrum of wavelengths (Fig. 41.17), with *minimum* wavelength (corresponding to *maximum* frequency and photon energy) determined by the accelerating voltage V_{AC} in the x-ray tube, according to the relation derived in Section 38.7 for *bremsstrahlung* processes:

$$\lambda_{min} = \frac{hc}{eV_{AC}} \tag{41.28}$$

This continuous-spectrum radiation is nearly independent of the target material in the x-ray tube.

Depending on the accelerating voltage and the target element, we may find sharp peaks superimposed on this continuous spectrum, as in Fig. 41.18. These peaks are at different wavelengths for different elements; they form what is called a *characteristic x-ray spectrum* for each target element. In 1913, the British scientist H. G. J. Moseley studied these spectra in detail using x-ray diffraction techniques. He found that the most intense short-wavelength line in the characteristic

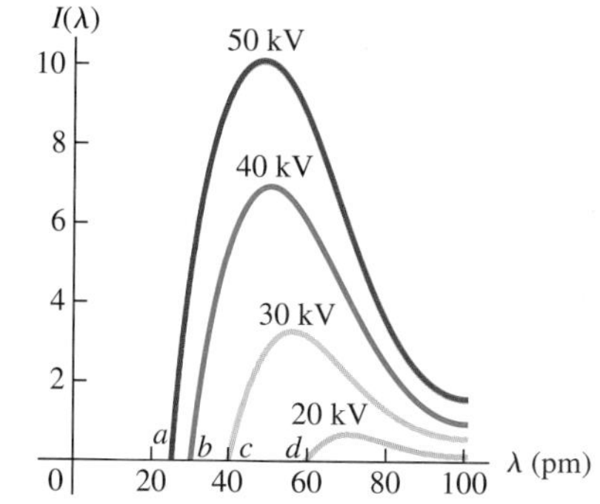

41.17 Graphs of x-ray intensity per unit wavelength as a function of wavelength, showing a continuous x-ray spectrum, for various accelerating voltages with a tungsten target. The minimum wavelength for each voltage is shown by the points *a*, *b*, *c*, and *d*.

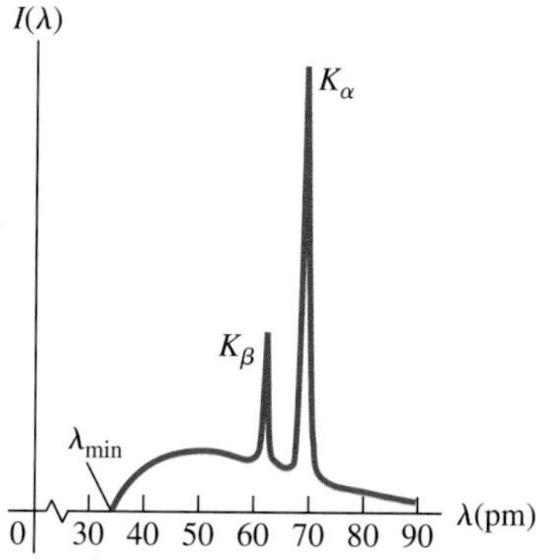

41.18 Graph of intensity per unit wavelength as a function of wavelength for x rays produced with an accelerating voltage of 35 kV and a molybdenum target. The curve is a smooth function similar to Fig. 41.17, but with two sharp spikes corresponding to part of the characteristic x-ray spectrum for molybdenum.

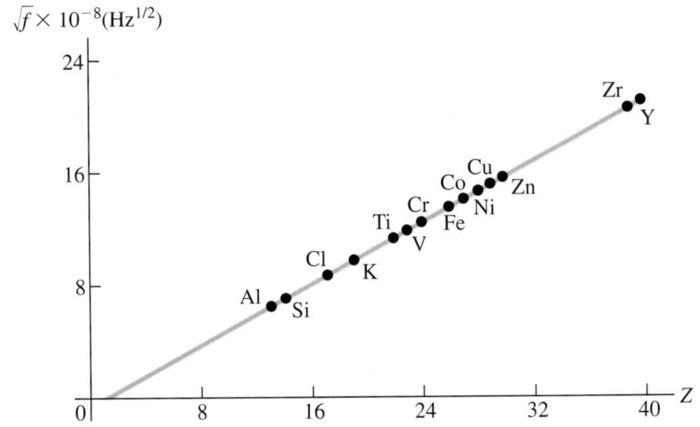

41.19 The square root of Moseley's measured frequencies of the K_α line for 14 elements. The graph of $\sqrt{f}$ versus Z is a straight line with an intercept at $Z = 1$, confirming Moseley's law, Eq. (41.29).

x-ray spectrum from a particular target element, called the K_α line, varied smoothly with that element's atomic number Z (Fig. 41.19). This is in sharp contrast to optical spectra, in which elements with adjacent Z values have spectra that often bear no resemblance to each other.

Moseley found that the relationship could be expressed in terms of x-ray frequencies f by a simple formula called *Moseley's law:*

$$f = (2.48 \times 10^{15}\ \text{Hz})(Z - 1)^2 \qquad \text{(Moseley's law)} \qquad (41.29)$$

Moseley's discovery of this formula is comparable to Balmer's discovery of the empirical formula, Eq. (38.8), for the wavelengths in the spectrum of hydrogen.

But Moseley went far beyond this empirical relationship; he showed how characteristic x-ray spectra could be understood on the basis of energy levels of atoms in the target. His analysis was based on the Bohr model, published in the same year. We will recast it somewhat, using the ideas of atomic structure discussed in Section 41.4. First recall that the *outer* electrons of an atom are responsible for optical spectra. Their excited states are usually only a few electron volts above their ground state. In transitions from excited states to the ground state, they usually emit photons in or near the visible region.

Characteristic x rays, by contrast, are emitted in transitions involving the *inner* shells of a complex atom. We mentioned these briefly in Section 38.7. In an x-ray tube the electrons may strike the target with enough energy to knock electrons out of the inner shells of the target atoms. These inner electrons are much closer to the nucleus than are the electrons in the outer shells; they are much more tightly bound, and hundreds or thousands of electron volts may be required to remove them.

Suppose one electron is knocked out of the K shell. This process leaves a vacancy, which we'll call a *hole.* (One electron remains in the K shell.) The hole can then be filled by an electron falling in from one of the outer shells, such as the $L, M, N, \ldots$ shell. This transition is accompanied by a decrease in the energy of the atom (because *less* energy would be needed to remove an electron from an $L, M, N, \ldots$ shell), and an x-ray photon is emitted with energy equal to this decrease. Each state has definite energy, so the emitted x rays have definite wavelengths; the emitted spectrum is a *line spectrum.*

We can estimate the energy and frequency of K_α x-ray photons using the concept of *screening.* A K_α x-ray photon is emitted when an electron in the L shell

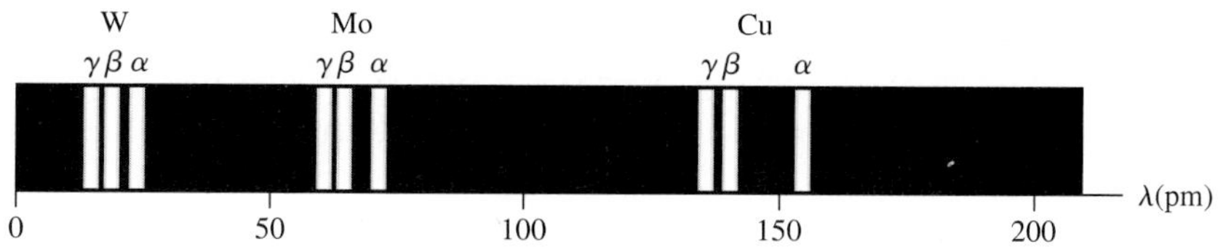

41.20 Wavelengths of the K_α, K_β, and K_γ lines of tungsten (W), molybdenum (Mo), and copper (Cu).

$(n = 2)$ drops down to fill a hole in the K shell $(n = 1)$. As the electron drops down, it is attracted by the Z protons in the nucleus screened by the one remaining electron in the K shell. We therefore approximate the energy by Eq. (41.27), with $Z_{\text{eff}} = Z - 1$, $n_{\text{i}} = 2$, and $n_{\text{f}} = 1$. The energy before the transition is

$$E_{\text{i}} \approx -(Z-1)^2(13.6\text{ eV})/2^2 = -(Z-1)^2(3.4\text{ eV})$$

and the energy after the transition is

$$E_{\text{f}} \approx -(Z-1)^2(13.6\text{ eV})/1^2 = -(Z-1)^2(13.6\text{ eV})$$

The energy of the K_α x-ray photon is $E_{K\alpha} = E_{\text{i}} - E_{\text{f}} \approx (Z-1)^2(-3.4\text{ eV} + 13.6\text{ eV})$. That is,

$$E_{K\alpha} \approx (Z-1)^2(10.2\text{ eV}) \qquad (41.30)$$

The frequency of any photon is its energy divided by Planck's constant,

$$f = \frac{E}{h} \approx \frac{(Z-1)^2 10.2\text{ eV}}{4.136 \times 10^{-15}\text{ eV}\cdot\text{s}} = (2.47 \times 10^{15}\text{ Hz})(Z-1)^2$$

This relation agrees almost exactly with Moseley's experimental law, Eq. (41.29). Indeed, considering the approximations we have made, the agreement is better than we have a right to expect. But our calculation does show how Moseley's law can be understood on the bases of screening and transitions between energy levels.

The hole in the K shell may also be filled by an electron falling from the M or N shell, assuming that these are occupied. If so, the x-ray spectrum of a large group of atoms of a single element shows a series, named the K series, of three lines, called the K_α, K_β, and K_γ lines. These three lines result from transitions in which the K-shell hole is filled by an L, M, or N electron, respectively. Figure 41.20 shows the K series for tungsten $(Z = 74)$, molybdenum $(Z = 42)$, and copper $(Z = 29)$.

There are other series of x-ray lines, called the L, M, and N series that are produced after the ejection of electrons from the L, M, and N shells rather than the K shell. Electrons in these outer shells are farther away from the nucleus and are not held as tightly as are those in the K shell. Their removal requires less energy, and the x-ray photons that are emitted when these vacancies are filled have lower energy than those in the K series.

Example 41.9 Chemical analysis by x-ray emission

You measure the K_α wavelength for an unknown element, obtaining the value 0.0709 nm. What is the element?

SOLUTION

IDENTIFY and SET UP: We use the relationship $f = c/\lambda$ to calculate the frequency for the K_α line, and then use Moseley's law,

Eq. (41.29), to find the corresponding value of the atomic number Z. We then consult the periodic table (Appendix D) to determine which element has this atomic number.

EXECUTE: The frequency is

$$f = \frac{c}{\lambda} = \frac{3.00 \times 10^8 \text{ m/s}}{0.0709 \times 10^{-9} \text{ m}} = 4.23 \times 10^{18} \text{ Hz}$$

Solving Moseley's law for Z,

$$Z = 1 + \sqrt{\frac{f}{2.48 \times 10^{15} \text{ Hz}}} = 1 + \sqrt{\frac{4.23 \times 10^{18} \text{ Hz}}{2.48 \times 10^{15} \text{ Hz}}} = 42.3$$

We know that Z has to be an integer; we conclude that $Z = 42$, corresponding to the element molybdenum.

EVALUATE: If you're worried that our calculation did not give an integer for Z, remember that Moseley's law is an empirical relationship. There are slight variations from one atom to another due to differences in the structure of the electron shells. Nonetheless, this example suggests the power of Moseley's law.

Niels Bohr commented that it was Moseley's observations, not the alpha-particle scattering experiments of Rutherford, Geiger, and Marsden (Section 38.4), which truly convinced physicists that the atom consists of a positive nucleus surrounded by electrons in motion. Unlike Bohr or Rutherford, Moseley did not receive a Nobel Prize for his important work; these awards are given only to living scientists, and Moseley was killed in combat during the First World War.

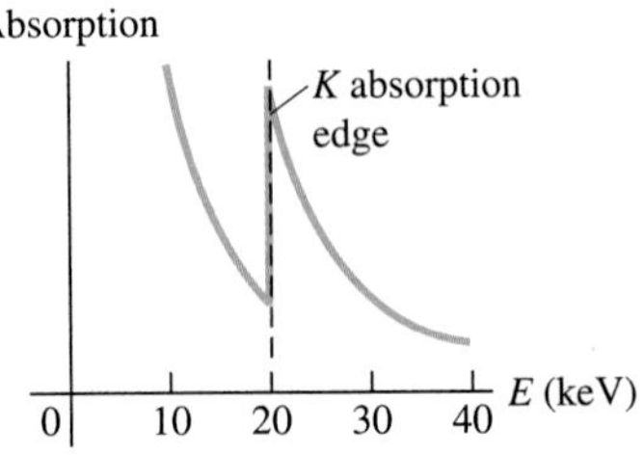

41.21 When a beam of x rays is passed through a slab of molybdenum, the extent to which the beam is absorbed depends on the energy E of the x-ray photons. A sharp increase in absorption occurs at the K absorption edge at 20 keV. Photons with energies above this value can excite an electron from the K shell into an empty state.

We can also observe x-ray *absorption* spectra. Unlike optical spectra, the absorption wavelengths are usually not the same as those for emission, especially in many-electron atoms, and do not give simple line spectra. For example, the K_α emission line results from a transition from the L shell to a hole in the K shell. The reverse transition doesn't occur in atoms with $Z \geq 10$ because in the atom's ground state, there is no vacancy in the L shell. To be absorbed, a photon must have enough energy to move an electron to an empty state. Since empty states are only a few electron volts in energy below the free-electron continuum, the minimum absorption energies in many-electron atoms are about the same as the minimum energies that are needed to remove an electron from its shell. Experimentally, if we gradually increase the accelerating voltage and hence the maximum photon energy, we observe sudden increases in absorption when we reach these minimum energies. These sudden jumps of absorption are called *absorption edges* (Fig. 41.21).

Characteristic x-ray spectra provide a very useful analytical tool. Satellite-borne x-ray spectrometers are used to study x-ray emission lines from highly excited atoms in distant astronomical sources. X-ray spectra are also used in air-pollution monitoring and in studies of the abundance of various elements in rocks.

Test Your Understanding

Explain why atomic hydrogen has an absorption edge at $E = 13.6$ eV.

SUMMARY

The Schrödinger equation for the hydrogen atom gives the same energy levels as the Bohr model. If the nucleus has charge Ze, there is an additional factor of Z^2 in the numerator of Eq. (41.3). The Schrödinger equation also shows that the possible magnitudes L of orbital angular momentum are given by Eq. (41.4), and that the possible values of the z-component of orbital angular momentum are given by Eq. (41.5) (See Example 41.1 and 41.2)

$$E_n = -\frac{1}{(4\pi\epsilon_0)^2}\frac{m_r e^4}{2n^2\hbar^2} = -\frac{13.60\text{ eV}}{n^2} \quad (41.3)$$

(energy levels of hydrogen)

$$L = \sqrt{l(l+1)}\hbar \quad (41.4)$$

$$(l = 0, 1, 2, \ldots, n-1)$$

$$L_z = m_l\hbar \quad (41.5)$$

$$(m_l = 0, \pm1, \pm2, \ldots, \pm l)$$

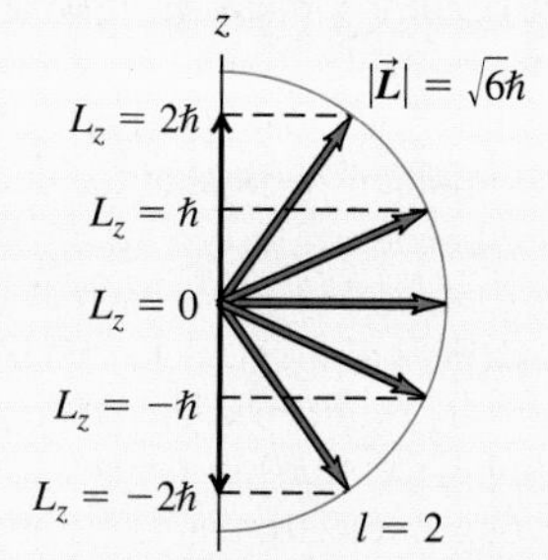

The probability that an atomic electron is between r and $r + dr$ from the nucleus is $P(r)\,dr$, given by Eq. (41.7). Atomic distances are often measured in units of a, the smallest distance between the electron and the nucleus in the Bohr model. (See Example 41.3)

$$P(r)\,dr = |\psi|^2\,dV = |\psi|^2\,4\pi r^2\,dr \quad (41.7)$$

$$a = \frac{\epsilon_0 h^2}{\pi m_r e^2} = \frac{4\pi\epsilon_0\hbar^2}{m_r e^2}$$

$$= 0.529 \times 10^{-10}\text{ m} \quad (41.8)$$

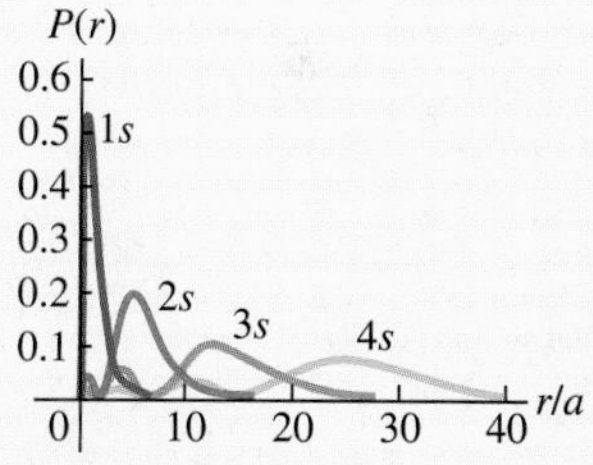

The interaction energy of an electron (mass m) with magnetic quantum number m_l in a magnetic field $\vec{B}$ along the $+z$-direction is given by Eq. (43.17) or (43.18), where $\mu_B = e\hbar/2m$ is called the Bohr magneton. (See Example 41.4)

$$U = -\mu_z B = m_l\frac{e\hbar}{2m}B = m_l\mu_B B$$

$$(m_l = 0, \pm1, \pm2, \ldots, \pm l) \quad (43.17), (43.18)$$

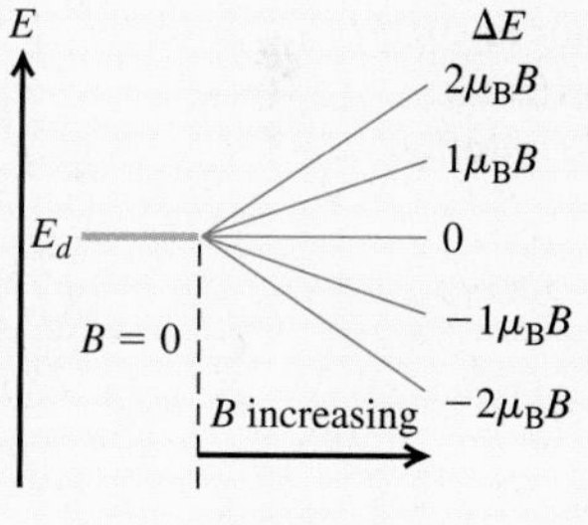

An electron has an intrinsic spin angular momentum of magnitude S, given by Eq. (41.20). The possible values of the z-component of the spin angular momentum are $S_z = m_s\hbar$, where $m_s = \pm\frac{1}{2}$. (See Examples 41.5 and 41.6)

$$S = \sqrt{\frac{1}{2}\left(\frac{1}{2}+1\right)}\hbar = \sqrt{\frac{3}{4}}\hbar \quad (41.20)$$

$$S_z = \pm\frac{1}{2}\hbar \quad (41.19)$$

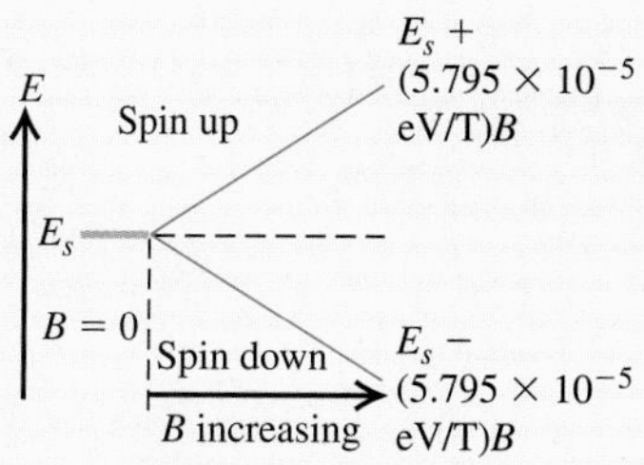

In a hydrogen atom, the quantum numbers n, l, m_l, and m_s of the electron have certain allowed values given by Eq. (41.26). In a many-electron atom, the quantum numbers for each electron are the same, but the energy levels depend on both n and l because of screening, the partial cancellation of the field of the nucleus by the inner electrons. The idea of screening is related to the central-field approximation, in which each electron moves in the electric field of the nucleus and of the averaged-out, spherically symmetric charge distribution of all the remaining electrons. If the effective (screened) charge attracting an electron is $Z_{eff}e$, the energies of the levels are given approximately by Eq. (41.27). (See Examples 41.7 and 41.8)

$$n \geq 1 \qquad 0 \leq l \leq n-1$$

$$|m_l| \leq l \quad m_s = \pm\frac{1}{2} \qquad (41.26)$$

$$E_n = -\frac{Z_{eff}^2}{n^2}(13.6\text{ eV}) \qquad (41.27)$$

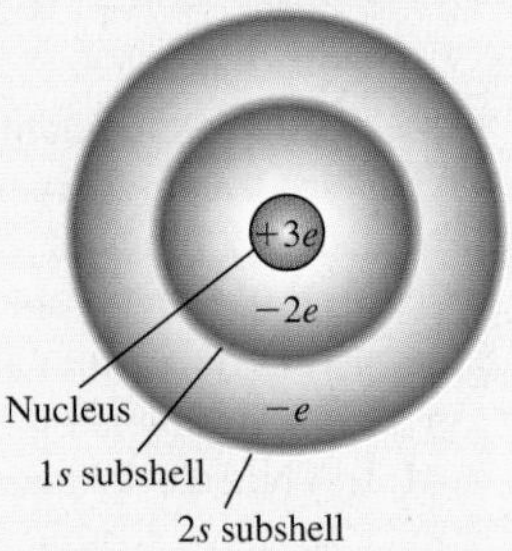

Moseley's law states that the frequency of a K_α x ray from a target with atomic number Z is given by Eq. (41.29). Characteristic x-ray spectra result from transitions to a hole in an inner energy level of an atom. (See Example 41.9)

$$f = (2.48 \times 10^{15}\text{ Hz})(Z-1)^2 \qquad (41.29)$$

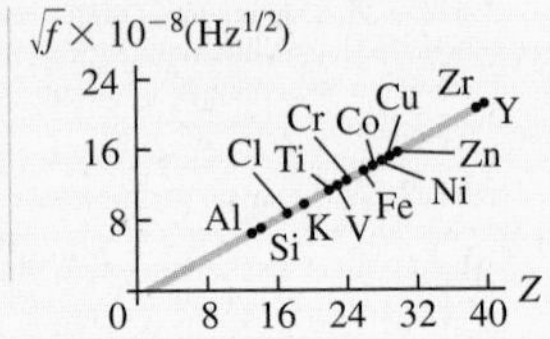

Key Terms

principal quantum number, 1550
orbital quantum number, 1550
magnetic quantum number, 1551
degeneracy, 1552
shell, 1552
atomic number, 1557
Zeeman effect, 1557
Bohr magneton, 1558
selection rules, 1560
electron spin, 1562
spin quantum number, 1563
spin-orbit coupling, 1565
central-field approximation, 1567
exclusion principle, 1568
periodic table of the elements, 1569
screening, 1571

Your Notes

Answer to Chapter Opening Question

The entry for sodium in Table 41.3 (Section 41.4) shows that sodium has one electron in the 3*s* level, which has relatively high energy and hence is easily removed from the atom. The other electrons are in the $n = 2$ or $n = 1$ levels, which have substantially lower energies. Hence removing more than one electron from a sodium atom requires much more energy than is available in chemical reactions at ordinary temperatures.

Answers to Test Your Understanding Questions

Section 41.1 If $\Phi(\phi) = Ae^{im_l\phi}$, where A is a constant, then $|\Phi(\phi)|^2 = \Phi(\phi)^*\Phi(\phi) = (A^*e^{-im_l\phi})(Ae^{im_l\phi}) = |A|^2$. In this case the probability distribution function is $|\psi(r, \theta, \phi)|^2 = [R(r)^*\Theta(\theta)^*\Phi(\phi)^*][R(r)\Theta(\theta)\Phi(\phi)] = |R(r)|^2|\Theta(\theta)|^2 |\Phi(\phi)|^2 = |R(r)|^2|\Theta(\theta)|^2|A|^2$. Hence $|\psi|^2$ is independent of ϕ and the probability distribution is symmetric around the z-axis.

Section 41.2 There would be no change, because all that matters is the component of the electron's orbital magnetic moment along the direction of $\vec{B}$. We called this μ_z in Eq. (41.14) because we *defined* the positive z-axis to be in the direction of $\vec{B}$. In reality, the names of the axes are entirely arbitrary.

Section 41.3 Let's use the mass of an average adult, about 75 kg. The total mass of electrons in this person's body is $(75\text{ kg})/4000 = 0.02$ kg, and one electron has a mass of 9.11×10^{-31} kg. Hence the total number of electrons in this person's body is $(0.02\text{ kg})/(9.11 \times 10^{-31}\text{ kg/electron}) = 2 \times 10^{28}$. If each electron has $S_z = \frac{1}{2}\hbar$ (spin up), the total spin angular momentum in the person's body is $(2 \times 10^{28})(\frac{1}{2}\hbar) = 10^{28}\hbar = 10^{28}(1.055 \times 10^{-34}\text{ J}\cdot\text{s})$, or about $10^{-6}\text{ J}\cdot\text{s}$.

To decide whether this would be noticeable, recall from Chapter 10 that the angular momentum of a rigid body of moment of inertia I is $I\omega$. In Example 10.13 (Section 10.6) we estimated the moment of inertia of a person to be 2 or 3 $\text{kg}\cdot\text{m}^2$. Thus the angular speed corresponding to an angular momentum of $10^{-6}\text{ J}\cdot\text{s}$ is about $(10^{-6}\text{ J}\cdot\text{s})/(2.5\text{ kg}\cdot\text{m}^2) = 4 \times 10^{-7}$ rad/s. This is equivalent to one revolution in six months! Such a minimal amount of rotation would be unnoticeable.

Section 41.4 If there were no exclusion principle, all 11 electrons in the sodium atom would be in the level of lowest energy (the 1*s* level) and the configuration would be $1s^{11}$. Consequently, it would be more difficult to remove the first electron. (In a real sodium atom the valence electron is in a screened 3*s* state, which has a comparatively high energy.)

Section 41.5 The energy of the ground level of hydrogen is -13.6 eV, so a photon of 13.6 eV or greater can be absorbed by hydrogen and excite the electron out of the atom (i.e., ionize the atom). Although a 13.6-eV photon is in the ultraviolet part of the spectrum, the principle is the same as for x-ray absorption edges in atoms of larger atomic number.

Discussion Questions

Q41.1 Do gravitational forces play a significant role in atomic structure? Explain.

Q41.2 What are the most significant differences between the Bohr model of the hydrogen atom and the Schrödinger analysis? What are the similarities?

Q41.3 Why is the analysis of the helium atom much more complex than that of the hydrogen atom, either in a Bohr type of model or using the Schrödinger equation?

Q41.4 The Stern-Gerlach experiment is always performed with beams of *neutral* atoms. Wouldn't it be easier to form beams using *ionized* atoms? Why won't this work?

Q41.5 In the Stern-Gerlach experiment, why is it essential for the magnetic field to be *inhomogeneous* (that is, nonuniform)?

Q41.6 In the ground state of the helium atom one electron must have "spin down" and the other "spin up." Why?

Q41.7 An electron in a hydrogen atom is in an *s* level and the atom is in a magnetic field $\vec{B} = B\hat{k}$. Explain why the "spin up" state $(m_s = +\frac{1}{2})$ has a higher energy than the "spin down" state $(m_s = -\frac{1}{2})$.

Q41.8 The central-field approximation is more accurate for alkali metals than for transition metals such as iron, nickel, or copper. Why?

Q41.9 Table 41.3 shows that for the ground state of the potassium atom, the outermost electron is in a 4*s* state. What does this tell you about the relative energies of the 3*d* and 4*s* levels for this atom? Explain.

Q41.10 A student asserted that any filled shell must have zero total angular momentum and hence must be spherically symmetric. Do you believe this is true? What about a filled *subshell?* Explain.

Q41.11 Why do the transition elements $(Z = 21 \text{ to } 30)$ all have similar chemical properties?

Q41.12 Use Table 41.3 to help determine the ground-state electron configuration of the neutral gallium atom (Ga), as well as the ions Ga^+ and Ga^-. Gallium has an atomic number of 31.

Q41.13 On the basis of the Pauli exclusion principle, the structure of the periodic table of the elements shows that there must be a fourth quantum number in addition to n, l, and m_l. Explain.

Q41.14 The ionization energies of the alkali metals (that is, the smallest energy required to remove one outer electron when the atom is in its ground state) are about 4 or 5 eV, while those of the noble gases are in the range from 11 to 25 eV. Why is there a difference?

Q41.15 The energy required to remove the 3*s* electron from a sodium atom in its ground state is about 5 eV. Would you expect the energy required to remove an additional electron to be about the same, or more, or less? Why?

Q41.16 The nucleus of a gold atom contains 79 protons. How does the energy required to remove a 1*s* electron completely from a gold atom compare with the energy required to remove the electron from the ground level in a hydrogen atom? In what region of the electromagnetic spectrum would a photon with this energy for each of these two atoms lay?

Q41.17 What is the basic distinction between x-ray energy levels and ordinary energy levels?

Q41.18 An atom in its ground level absorbs a photon with energy equal to the K absorption edge. Does absorbing this photon ionize this atom? Explain.

Q41.19 Can a hydrogen atom emit x rays? If so, how? If not, why not?

Exercises

Section 41.1 The Hydrogen Atom

41.1 The orbital angular momentum of an electron has a magnitude of 4.716×10^{-34} kg·m²/s. What is the angular-momentum quantum number l for this electron?

41.2 Consider states with angular-momentum quantum number $l = 2$. a) In units of $\hbar$, what is the largest possible value of L_z? b) In units of $\hbar$, what is the value of L? Which is larger, L or the maximum possible L_z? c) For each allowed value of L_z, what angle does the vector $\vec{L}$ make with the $+z$-axis? How does the minimum angle for $l = 2$ compare to the minimum angle for $l = 3$ calculated in Example 41.2 (Section 41.1)?

41.3 Calculate, in units of $\hbar$, the magnitude of the maximum orbital angular momentum for an electron in a hydrogen atom for states with a principal quantum number of 2, 20, and 200. Compare each with the value of $n\hbar$ postulated in the Bohr model. What trend do you see?

41.4 a) Make a chart showing all the possible sets of quantum numbers l and m_l for the states of the electron in the hydrogen atom when $n = 5$. How many combinations are there? b) What are the energies of these states?

41.5 The Problem-Solving Strategy in Section 41.1 claims that the electric potential energy of a proton and an electron 0.10 nm apart has magnitude 15 eV. Verify this claim.

41.6 a) What is the probability that an electron in the 1s state of a hydrogen atom will be found at a distance less than $a/2$ from the nucleus? b) Use the results of part (a) and of Example 41.3 (Section 41.1) to calculate the probability that the electron will be found at distances between $a/2$ and a from the nucleus.

41.7 a) For the wave function $\psi(r, \theta, \phi) = R(r)\Theta(\theta)\Phi(\phi)$ with $\Phi(\phi) = Ae^{im_l\phi}$, show that $|\psi|^2$ is independent of ϕ. b) What value must A have if $\Phi(\phi)$ is to satisfy the normalization condition $\int_0^{2\pi}|\Phi(\phi)|^2\,d\phi = 1$?

41.8 For ordinary hydrogen, the reduced mass of the electron and proton is $m_r = 0.99946m$, where m is the electron mass (Section 38.5). For each of the following cases, find the numerical coefficient of $-1/n^2$ in Eq. (41.3) and the energy of the photon emitted in a transition from the $n = 2$ level to the $n = 1$ level. a) A hydrogen atom in which the nucleus is taken to be infinitely massive, so $m_r = m$. b) Positronium (Section 38.5), for which $m_r = m/2$ exactly. c) Muonium, an atom composed of a muon and a proton (see Problem 38.61), for which $m_r = 185.8m$.

41.9 Find the numerical value of a in Eq. (41.8) for a) a hydrogen atom in which the nucleus is taken to be infinitely massive, so $m_r = m$; b) positronium (Section 38.5), for which $m_r = m/2$ exactly; c) muonium, an atom composed of a muon and a proton (see Problem 38.61), for which $m_r = 185.8m$. In part (a), a becomes a_0, the Bohr radius.

41.10 Show that $\Phi(\phi) = e^{im_l\phi} = \Phi(\phi + 2\pi)$ (that is, show that $\Phi(\phi)$ is periodic with period 2π) if and only if m_l is restricted to the values $0, \pm1, \pm2, \ldots$. (*Hint:* Euler's formula states that $e^{i\phi} = \cos\phi + i\sin\phi$.)

41.11 In Example 41.3 (Section 41.1), fill in the missing details that show that $P = 1 - 5e^{-2}$.

Section 41.2 The Zeeman Effect

41.12 A hydrogen atom is in a d state. In the absence of an external magnetic field the states with different m_l have (approximately) the same energy. Consider the interaction of the magnetic field with the atom's orbital magnetic dipole moment. a) Calculate the splitting in electron volts of the m_l levels when the atom is put in a 0.400-T magnetic field that is in the $+z$-direction. b) Which m_l level will have the lowest energy? c) Draw an energy-level diagram that shows the d levels with and without the external magnetic field.

41.13 A hydrogen atom in the 5g state is placed in a magnetic field of 0.600 T that is in the z-direction. a) Into how many levels is this state split by the interaction of the atom's orbital magnetic dipole moment with the magnetic field? b) What is the energy separation between adjacent levels? c) What is the energy separation between the level of lowest energy and the level of highest energy?

41.14 A hydrogen atom undergoes a transition from a 2p state to the 1s ground state. In the absence of a magnetic field, the energy of the photon emitted is 122 nm. The atom is then placed in a strong magnetic field in the z-direction. Ignore spin effects; consider only the interaction of the magnetic field with the atom's orbital magnetic moment. a) How many different photon wavelengths are observed for the $2p \rightarrow 1s$ transition? What are the m_l values for the initial and final states for the transition that leads to each photon wavelength? b) One observed wavelength is exactly the same with the magnetic field as without. What are the initial and final m_l values for the transition that produces a photon of this wavelength? c) One observed wavelength with the field is larger than the wavelength without the field. What are the initial and final m_l values for the transition that produces a photon of this wavelength? d) Repeat part (c), for the wavelength that is shorter than the wavelength in the absence of the field.

41.15 A hydrogen atom in a 3p state is placed in a uniform external magnetic field $\vec{B}$. Consider the interaction of the magnetic field with the atom's orbital magnetic dipole moment. a) What field magnitude B is required to split the 3p state into multiple levels with an energy difference of 2.71×10^{-5} eV between adjacent levels? b) How many levels will there be?

Section 41.3 Electron Spin

41.16 A hydrogen atom in the $n = 1$, $m_s = -\frac{1}{2}$ state is placed in a magnetic field with a magnitude of 0.480 T in the $+z$-direction. a) Find the magnetic interaction energy (in electron volts) of the electron with the field. b) Is there any orbital magnetic dipole moment interaction for this state? Explain. Can there be an orbital magnetic dipole moment interaction for $n \neq 1$?

41.17 Calculate the energy difference between the $m_s = \frac{1}{2}$ ("spin up") and $m_s = -\frac{1}{2}$ ("spin down") levels of a hydrogen atom in the

1*s* state when it is placed in a 1.45-T magnetic field in the *negative* *z*-direction. Which level, $m_s = \frac{1}{2}$ or $m_s = -\frac{1}{2}$, has the lower energy?

41.18 List the different possible combinations of l and j for a hydrogen atom in the $n = 3$ level.

41.19 A hydrogen atom in a particular orbital angular momentum state is found to have j quantum numbers $\frac{7}{2}$ and $\frac{9}{2}$. What is the letter that labels the value of l for the state?

41.20 The hyperfine interaction in a hydrogen atom between the magnetic dipole moment of the proton and the spin magnetic dipole moment of the electron splits the ground level into two levels separated by 5.9×10^{-6} eV. a) Calculate the wavelength and frequency of the photon emitted when the atom makes a transition between these states and compare your answer to the value given at the end of Section 41.3. In what part of the electromagnetic spectrum does this lay? Such photons are emitted by cold hydrogen clouds in interstellar space; by detecting these photons, astronomers can learn about the number and density of such clouds. b) Calculate the effective magnetic field experienced by the electron in these states (see Fig. 41.3). Compare your result to the effective magnetic field due to the spin-orbit coupling calculated in Example 41.6.

41.21 Classical Electron Spin. a) If you treat an electron as a classical spherical object with a radius of 1.0×10^{-17} m, what angular speed is necessary to produce a spin angular momentum of magnitude $\sqrt{\frac{3}{4}}\hbar$? b) Use $v = r\omega$ and the result of part (a) to calculate the speed v of a point at the electron's equator. What does your result suggest about the validity of this model?

Section 41.4 Many-Electron Atoms and the Exclusion Principle

41.22 For magnesium, the first ionization potential is 7.6 eV. The second ionization potential (additional energy required to remove a second electron) is almost twice this, 15 eV, and the third ionization potential is much larger, about 80 eV. How can these numbers be understood?

41.23 The 5*s* electron in rubidium (Rb) sees an effective charge of 2.771*e*. Calculate the ionization energy of this electron.

41.24 For germanium (Ge, $Z = 32$), make a list of the number of electrons in each subshell (1*s*, 2*s*, 2*p*, . . .). Use the allowed values of the quantum numbers along with the exclusion principle; do *not* refer to Table 41.3.

41.25 Make a list of the four quantum numbers n, l, m_l, and m_s for each of the ten electrons in the ground state of the neon atom. Do *not* refer to Tables 41.2 or 41.3.

41.26 The energies of the 4*s*, 4*p*, and 4*d* levels of potassium are given in Example 41.8 (Section 41.4). Calculate Z_{eff} for each state. What trend does your results show? How can you explain this trend?

41.27 a) The doubly charged ion N^{2+} is formed by removing two electrons from a nitrogen atom. What is the ground-state electron configuration for the N^{2+} ion? b) Estimate the energy of the least strongly bound level in the L shell of N^{2+}. c) The doubly charged ion P^{2+} is formed by removing two electrons from a phosphorus atom. What is the ground-state electron configuration for the P^{2+} ion? d) Estimate the energy of the least strongly bound level in the M shell of P^{2+}.

41.28 a) The energy of the 2*s* state of lithium is -5.391 eV. Calculate the value of Z_{eff} for this state. b) The energy of the 4*s* state of potassium is -4.339 eV. Calculate the value of Z_{eff} for this state. c) Compare Z_{eff} for the 2*s* state of lithium, the 3*s* state of sodium (Example 41.7 in Section 41.4), and the 4*s* state of potassium. What trend do you see? How can you explain this trend?

41.29 Estimate the energy of the highest-l state for a) the L shell of Be^+; b) the N shell of Ca^+.

Section 41.5 X-Ray Spectra

41.30 A K_α x ray emitted from a sample has an energy of 7.46 keV. Of which element is the sample made?

41.31 Calculate the frequency, energy (in keV), and wavelength of the K_α x ray for the elements a) calcium (Ca, $Z = 20$); b) cobalt (Co, $Z = 27$); c) cadmium (Cd, $Z = 48$).

Problems

41.32 For a hydrogen atom, the probability $P(r)$ of finding the electron within a spherical shell with inner radius r and outer radius $r + dr$ is given by Eq. (41.7). For a hydrogen atom in the 1*s* ground state, at what value of r does $P(r)$ have its maximum value? How does your result compare to the distance between the electron and the nucleus for the $n = 1$ state in the Bohr model (Eq. 41.8)?

41.33 Consider a hydrogen atom in the 1*s* state. a) For what value of r is the potential energy $U(r)$ equal to the total energy E? Express your answer in terms of a. This value of r is called the *classical turning point,* since this is where a Newtonian particle would stop its motion and reverse direction. b) For r greater than the classical turning point, $U(r) > E$. Classically, the particle cannot be in this region since the kinetic energy cannot be negative. Calculate the probability of the electron being found in this classically forbidden region.

41.34 Rydberg Atoms. Rydberg atoms are atoms whose outermost electron is in an excited state with a *very* large principal quantum number. Rydberg atoms have been produced in the laboratory and detected in interstellar space. a) Why do all neutral Rydberg atoms with the same n value have essentially the same ionization energy, independent of the total number of electrons in the atom? b) What is the ionization energy for a Rydberg atom with a principal quantum number of 350? What is the radius in the Bohr model of the Rydberg electron's orbit? c) Repeat part (b) for $n = 650$.

41.35 The wave function for a hydrogen atom in the 2*s* state is

$$\psi_{2s}(r) = \frac{1}{\sqrt{32\pi a^3}}\left(2 - \frac{r}{a}\right)e^{-r/2a}$$

a) Verify that this function is normalized. b) In the Bohr model, the distance between the electron and the nucleus in the $n = 2$ state is exactly 4*a*. Calculate the probability that an electron in the 2*s* state will be found at a distance less than 4*a* from the nucleus.

41.36 The normalized wave function for a hydrogen atom in the 2*s* state is given in Problem 41.35. a) For a hydrogen atom in the 2*s* state, at what value of r is $P(r)$ maximum? How does your result compare to 4*a*, the distance between the electron and the nucleus in the $n = 2$ state of the Bohr model? b) At what value of r (other than

$r = 0$ or $r = \infty$) is $P(r)$ equal to zero, so that the probability of finding the electron at that separation from the nucleus is zero? Compare your result to Fig. 41.4.

41.37 a) For an excited state of hydrogen, show that the smallest angle that the orbital angular momentum vector $\vec{L}$ can have with the z-axis is

$$(\theta_L)_{\min} = \arccos\left(\frac{n-1}{\sqrt{n(n-1)}}\right)$$

b) What is the corresponding expression for $(\theta_L)_{\max}$, the largest possible angle between $\vec{L}$ and the z-axis?

41.38 a) If the value of L_z is known, we cannot know either L_x or L_y precisely. But we *can* know the value of the quantity $\sqrt{L_x^2 + L_y^2}$. Write an expression for this quantity in terms of l, m_l, and $\hbar$. b) What is the meaning of $\sqrt{L_x^2 + L_y^2}$? c) For a state of nonzero orbital angular momentum, find the maximum and minimum values of $\sqrt{L_x^2 + L_y^2}$. Explain your results.

41.39 The normalized radial wavefunction (Eq. 41.2) for the $2p$ state of the hydrogen atom is $R_{2p} = (1/\sqrt{24a^5})re^{-r/2a}$. After averaging over the angular variables, the radial probability function becomes $P(r)dr = (R_{2p})^2r^2dr$. At what value of r is $P(r)$ for the $2p$ state a maximum? Compare your results to the radius of the $n = 2$ state in the Bohr model.

41.40 Stern-Gerlach Experiment. In a Stern-Gerlach experiment, the deflecting force on the atom is $F_z = -\mu_z(dB_z/dz)$, where μ_z is given by Eq. 41.22 and dB_z/dz is the magnetic-field gradient. In a particular experiment the magnetic field region is 50.0 cm long; assume the magnetic-field gradient is constant in this region. A beam of silver atoms enters the magnetic field with a speed of 525 m/s. What value of dB_z/dz is required to give a separation of 1.0 mm between the two spin components as they exit the field? (*Note*: The magnetic dipole moment of silver is the same as that for hydrogen, since its valence electron is in a $l = 0$ state.)

41.41 Consider the transition from a $3d$ to a $2p$ state of hydrogen in an external magnetic field. Assume that the effects of electron spin can be ignored (which is not actually the case) so that the magnetic field interacts only with the orbital angular momentum. Identify each allowed transition by the m_l values of the initial and final states. For each of these allowed transitions, determine the shift of the transition energy from the zero-field value and show that there are three different transition energies.

41.42 A hydrogen atom makes a transition from a $n = 3$ state to a $n = 2$ state (the Balmer H_α line) while in a magnetic field in the $+z$-direction and with magnitude 1.40 T. a) If the magnetic quantum number is $m_l = 2$ in the initial ($n = 3$) state and $m_l = 1$ in the final ($n = 2$) state, by how much is each energy level shifted from the zero-field value? b) By how much is the wavelength of the H_α line shifted from the zero-field value? Is the wavelength increased or decreased? Disregard the effect of electron spin. [*Hint:* Use the result of Problem 39.54(c).]

41.43 A large number of hydrogen atoms in $1s$ states are placed in an external magnetic field that is in the $+z$-direction. Assume that the atoms are in thermal equilibrium at room temperature, $T = 300$ K. According to the Boltzmann distribution (Section 38.6), what is the ratio of the number of atoms in the $m_s = \frac{1}{2}$ state to the number in the $m_s = -\frac{1}{2}$ when the magnetic field magnitude is a) 5.00×10^{-5} T (approximately the earth's field); b) 0.500 T; c) 5.00 T?

41.44 Effective Magnetic Field. An electron in a hydrogen atom is in the $2p$ state. In a simple model of the atom, assume that the electron circles the proton in an orbit with radius r equal to the Bohr-model radius for $n = 2$. Assume that the speed v of the orbiting electron can be calculated by setting $L = mvr$ and taking L to have the quantum-mechanical value for a $2p$ state. In the frame of the electron, the proton orbits with radius r and speed v. Model the orbiting proton as a circular current loop and calculate the magnetic field it produces at the location of the electron.

41.45 Weird Universe. In another universe, the electron is a spin-3/2 rather than a spin-1/2 particle, but all other physics are the same as in our universe. In this universe, a) what are the atomic numbers of the lightest two inert gases? b) What is the ground-state electron configuration of sodium?

41.46 For an ion with nuclear charge Z and a single electron, the electric potential energy is $-Ze^2/4\pi\epsilon_0 r$ and the expression for the energies of the states and for the normalized wave functions are obtained from those for hydrogen by replacing e^2 by Ze^2. Consider the N^{6+} ion, with seven protons and one electron. a) What is the ground-state energy in electron volts? b) What is the ionization energy, the energy required to remove the electron from the N^{6+} ion if it is initially in the ground state? c) What is the distance a [given for hydrogen by Eq. (41.8)] for this ion? d) What is the wavelength of the photon emitted when the N^{6+} ion makes a transition from the $n = 2$ state to the $n = 1$ ground state?

41.47 A hydrogen atom in an $n = 2$, $l = 1$, $m_l = -1$ state emits a photon when it decays to an $n = 1$, $l = 0$, $m_l = 0$ ground state. a) In the absence of an external magnetic field, what is the wavelength of this photon? b) If the atom is in a magnetic field in the $+z$-direction and with a magnitude of 2.20 T, what is the shift in the wavelength of the photon from the zero-field value? Does the magnetic field increase or decrease the wavelength? Disregard the effect of electron spin. [*Hint:* Use the result of Problem 39.54(c).]

41.48 A lithium atom has three electrons, and the $^2S_{1/2}$ ground-state electron configuration is $1s^22s$. The $1s^22p$ excited state is split into two closely spaced levels, $^2P_{3/2}$ and $^2P_{1/2}$, by the spin-orbit interaction (Example 41.6 in Section 41.3). A photon with wavelength 67.09608 μm is emitted in the $^2P_{3/2} \rightarrow {}^2S_{1/2}$ transition, and a photon with wavelength 67.09761 μm is emitted in the $^2P_{1/2} \rightarrow {}^2S_{1/2}$ transition. Calculate the effective magnetic field seen by the electron in the $1s^22p$ state of the lithium atom. How does your result compare to that for the $3p$ level of sodium found in Example 41.6?

41.49 Estimate the minimum and maximum wavelengths of the characteristic x rays emitted by a) vanadium ($Z = 23$); b) rhenium ($Z = 45$). Discuss any approximations that you make.

41.50 Electron Spin Resonance. Electrons in the lower of two spin states in a magnetic field can absorb a photon of the right frequency and move to the higher state. a) Find the magnetic field magnitude B required for this transition in a hydrogen atom with $n = 1$ and $l = 0$ to be induced by microwaves with wavelength λ. b) Calculate the value of B for a wavelength of 3.50 cm.

41.51 a) Show that the total number of atomic states (including different spin states) in a shell of principal quantum number n is

$2n^2$. [*Hint:* The sum of the first N integers $1 + 2 + 3 + \cdots + N$ is given by $N(N + 1)/2$.] b) Which shell has 50 states?

Challenge Problems

41.52 Consider a simple model of the helium atom in which two electrons, each with mass m, move around the nucleus (charge $+2e$) in the same circular orbit. Each electron has orbital angular momentum $\hbar$ (that is, the orbit is the smallest-radius Bohr orbit), and the two electrons are always on opposite sides of the nucleus. Ignore the effects of spin. a) Determine the radius of the orbit and the orbital speed of each electron. [*Hint:* Follow the procedure used in Section 38.5 to derive Eqs. (38.12) and (38.13). Each electron experiences an attractive force from the nucleus and a repulsive force from the other electron.] b) What is the total kinetic energy of the electrons? c) What is the potential energy of the system (the nucleus and the two electrons)? d) In this model, how much energy is required to remove both electrons to infinity? How does this compare to the experimental value of 79.0 eV?

41.53 Each of $2N$ electrons (mass m) is free to move along the x-axis. The potential-energy function for each electron is $U(x) = \frac{1}{2}k'x^2$, where k' is a positive constant. The electric and magnetic interactions between electrons can be ignored. Use the exclusion principle to show that the minimum energy of the system of $2N$ electrons is $\hbar N^2\sqrt{k'/m}$. (*Hint:* See Section 40.4 and the hint given in Problem 41.51.)

41.54 Repeat the calculation of Problem 41.33 for a one-electron ion with nuclear charge Z. (See Problem 41.46.) How does the probability of the electron being found in the classically forbidden region depend on Z?

42 MOLECULES AND CONDENSED MATTER

The burning of fossil fuels has caused an 18% increase in the concentration of atmospheric carbon dioxide (CO_2) since continuous measurements began in 1958. (The yearly variations are due to increased intake of CO_2 by plants in spring and summer.) Because CO_2 molecules absorb infrared radiation emitted by the earth's surface, an increase in their concentration prevents the earth from radiating excess heat into space and leads to a rise in global temperatures. This is one cause of the effect called global warming.

? **What aspects of the CO_2 molecule enable it to absorb infrared radiation?**

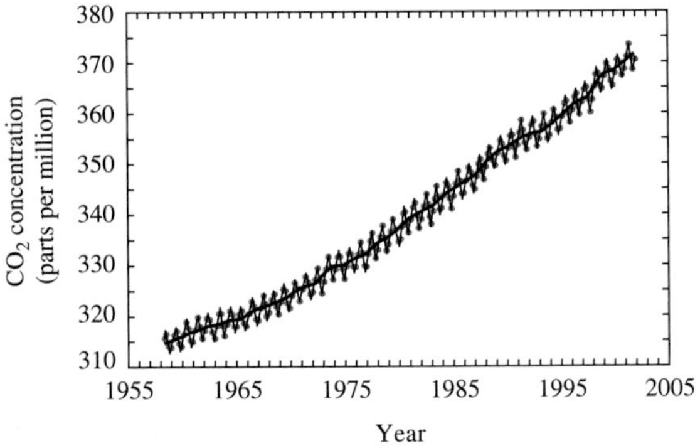

In Chapter 41 we discussed the structure and properties of isolated atoms. But such atoms are the exception; usually, we find atoms combined to form molecules or more extended structures we call condensed matter (liquid or solid). It's the attractive forces between atoms, called molecular bonds, that causes them to combine. In this chapter we'll study several kinds of bonds as well as the energy levels and spectra associated with diatomic molecules. We will see that just as atoms have quantized energies determined by the quantum-mechanical state of their electrons, so molecules have quantized energies determined by their rotational and vibrational states.

The same physical principles behind molecular bonds also apply to the study of condensed matter, in which various types of bonding occur. We'll explore the concept of energy bands and see how it helps understanding the properties of solids. Then we'll look more closely at the properties of a special class of solids called semiconductors. Devices using semiconductors are found in every radio, TV, pocket calculator, and computer used today; they have revolutionized the entire field of electronics during the past half-century.

42.1 | Types of Molecular Bonds

We can use our discussion of atomic structure in Chapter 41 as a basis for exploring the nature of *molecular bonds,* the interactions that hold atoms together to form stable structures such as molecules and solids.

Ionic Bonds

The **ionic bond** is an interaction between oppositely charged *ionized* atoms. The most familiar example is sodium chloride (NaCl), in which the sodium atom gives its one 3*s* electron to the chlorine atom, filling the vacancy in the 3*p* subshell of chlorine.

Let's look at the energy balance in this transaction. Removing the 3*s* electron from a neutral sodium atom requires 5.138 eV of energy; this is called the *ionization energy* or *ionization potential* of sodium. The neutral chlorine atom can attract an extra electron into the vacancy in the 3*p* subshell, where it is incompletely screened by the other electrons and therefore is attracted to the nucleus. This state has 3.613 eV lower energy than a neutral chlorine atom and a distant free electron; 3.613 eV is the magnitude of the *electron affinity* of chlorine. Thus creating the well-separated Na^+ and Cl^- ions requires a net investment of only $5.138\text{ eV} - 3.613\text{ eV} = 1.525\text{ eV}$. When the two oppositely charged ions are brought together by their mutual attraction, the magnitude of their negative potential energy is determined by how closely they can approach each other. This in turn is limited by the exclusion principle, which forbids extensive overlap of the electron clouds of the two ions. As the distance decreases, the exclusion principle distorts the charge clouds, so the ions no longer interact like point charges and the interaction eventually becomes repulsive (Fig. 42.1).

The minimum electrical potential energy for NaCl turns out to be -5.7 eV at a separation of 0.24 nm. The net energy released in creating the ions and letting them come together to the equilibrium separation of 0.24 nm is $5.7\text{ eV} - 1.525\text{ eV} = 4.2\text{ eV}$. Thus, if the kinetic energy of the ions is neglected, 4.2 eV is the *binding energy* of the NaCl molecule, the energy that is needed to dissociate the molecule into separate neutral atoms.

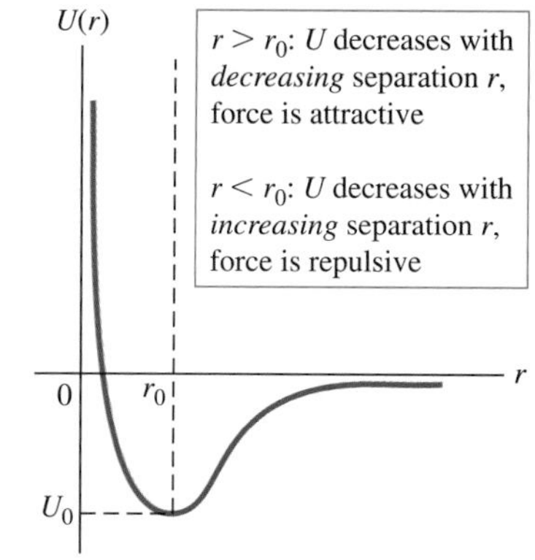

42.1 When the separation r between two oppositely charged ions is large, the potential energy $U(r)$ is proportional to $1/r$ as for point charges and the force is attractive. As r decreases, the charge clouds of the two atoms overlap and the force becomes less attractive. If r is less than the equilibrium separation r_0, the force is repulsive.

Example 42.1 Electric potential energy of the NaCl molecule

Find the electric potential energy of Na^+ and Cl^- ions separated by 0.24 nm if they can be treated as point charges.

SOLUTION

IDENTIFY and SET UP: We use Eq. (23.9) from Section 23.1, which tells us that the electric potential energy of two point charges q and q_0 separated by a distance r is $U = qq_0/4\pi\epsilon_0 r$.

EXECUTE: The two charges are $q = +e$ (for Na^+) and $q_0 = -e$ (for Cl^-), and the separation is $r = 0.24\text{ nm} = 0.24 \times 10^{-9}\text{ m}$. From Eq. (23.9),

$$U = -\frac{1}{4\pi\epsilon_0}\frac{e^2}{r_0} = -(9.0 \times 10^9\ \text{N}\cdot\text{m}^2/\text{C}^2)\frac{(1.6 \times 10^{-19}\ \text{C})^2}{0.24 \times 10^{-9}\ \text{m}}$$
$$= -9.6 \times 10^{-19}\ \text{J} = -6.0\ \text{eV}$$

EVALUATE: This result agrees fairly well with the observed value of -5.7 eV. The difference is because at the equilibrium separation the ions don't behave exactly like point charges; at this separation the electron clouds of the two ions are overlapping.

Covalent Bonds

Ionic bonds are interactions between charge distributions that are nearly spherically symmetric; hence they are not highly directional. They can involve more than one electron per atom. The alkaline earth elements form ionic compounds in which an atom loses *two* electrons; an example is $Mg^{2+}(Cl^-)_2$. Loss of more than two electrons is relatively rare; instead, a different kind of bond comes into operation.

The **covalent bond** is characterized by a more egalitarian participation of the two atoms than occurs with the ionic bond. The simplest covalent bond is found in the hydrogen molecule, a structure containing two protons and two electrons. This bond is shown schematically in Fig. 42.2. As the separate atoms (Fig. 42.2a) come together, the electron wave functions are distorted and become more concentrated in the region between the two protons (Fig. 42.2b). The net attraction of the electrons for each proton more than balances the repulsion of the two protons and of the two electrons.

The attractive interaction is then supplied by a *pair* of electrons, one contributed by each atom, with charge clouds that are concentrated primarily in the region between the two atoms. The energy of the covalent bond in the hydrogen molecule H_2 is -4.48 eV.

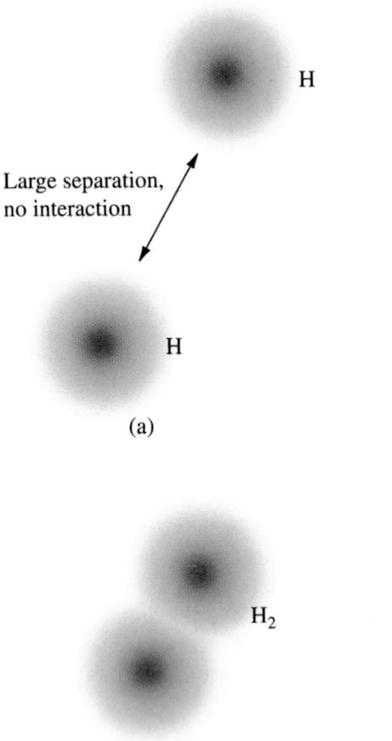

42.2 Covalent bond in a hydrogen molecule. (a) Two separate hydrogen atoms. (b) The covalent bond; the charge clouds for the two electrons with opposite spins are concentrated in the region between the nuclei (protons).

As we saw in Chapter 41, the exclusion principle permits two electrons to occupy the same region of space (that is, to be in the same spatial quantum state) only when they have opposite spins. When the spins are parallel, the exclusion principle forbids the molecular state that would be most favorable from energy considerations (with both electrons in the region between atoms). Opposite spins are an essential requirement for a covalent bond, and no more than two electrons can participate in such a bond.

However, an atom with several electrons in its outermost shell can form several covalent bonds. The bonding of carbon and hydrogen atoms, of central importance in organic chemistry, is an example. In the *methane* molecule (CH_4) the carbon atom is at the center of a regular tetrahedron, with a hydrogen atom at each corner. The carbon atom has four electrons in its L shell, and each of these four electrons forms a covalent bond with one of the four hydrogen atoms (Fig. 42.3). Similar patterns occur in more complex organic molecules.

Because of the role played by the exclusion principle, covalent bonds are highly directional. In the methane molecule the wave function for each of carbon's four valence electrons is a combination of the 2s and 2p wave functions called a *hybrid wave function.* The probability distribution for each one has a lobe protruding toward a corner of a tetrahedron. This symmetrical arrangement minimizes the overlap of wave functions for the electron pairs, minimizing their repulsive potential energy.

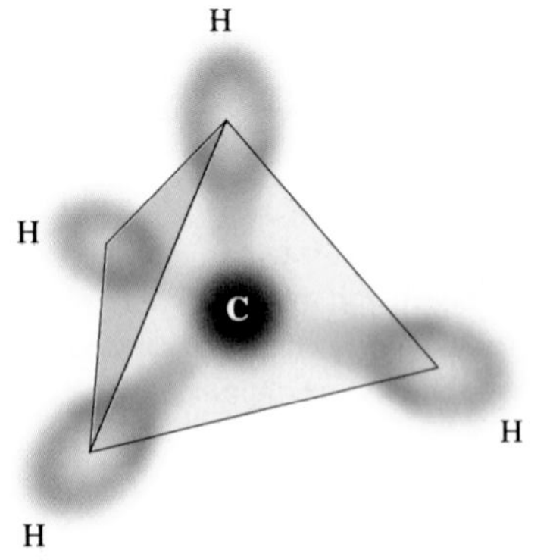

42.3 Schematic diagram of the methane (CH_4) molecule. The carbon atom is at the center of a regular tetrahedron and forms four covalent bonds with the hydrogen atoms at the corners. Each covalent bond includes two electrons with opposite spins, forming a charge cloud that is concentrated between the carbon atom and a hydrogen atom.

Ionic and covalent bonds represent two extremes in molecular bonding, but there is no sharp division between the two types. Often there is a *partial* transfer of one or more electrons from one atom to another. As a result, many molecules that have dissimilar atoms have electric dipole moments, that is, a preponderance of positive charge at one end and of negative charge at the other. Such molecules are called *polar* molecules. Water molecules have large electric dipole moments; these are responsible for the exceptionally large dielectric constant of liquid water (see Sections 24.4 and 24.5).

van der Waals Bonds

Ionic and covalent bonds, with typical bond energies of 1 to 5 eV, are called *strong bonds.* There are also two types of weaker bonds. One of these, the **van der Waals bond,** is an interaction between the electric dipole moments of atoms or molecules; typical energies are 0.1 eV or less. The bonding of water molecules in the liquid and solid state results partly from dipole–dipole interactions.

No atom has a permanent electric dipole moment, nor do many molecules. However, fluctuating charge distributions can lead to fluctuating dipole moments;

these in turn can induce dipole moments in neighboring structures. Overall, the resulting dipole–dipole interaction is attractive, giving a weak bonding of atoms or molecules. The interaction potential energy drops off very quickly with distance r between molecules, usually as $1/r^6$. The liquefaction and solidification of the inert gases and of molecules such as H_2, O_2, and N_2 is due to induced-dipole van der Waals interactions. Not much thermal-agitation energy is needed to break these weak bonds, so such substances usually exist in the liquid and solid states only at very low temperatures.

Hydrogen Bonds

In another type of weak bond, the **hydrogen bond,** a proton (H^+ ion) gets between two atoms, polarizing them and attracting them by means of the induced dipoles. This bond is unique to hydrogen-containing compounds because only hydrogen has a singly ionized state with no remaining electron cloud; the hydrogen ion is a bare proton, much smaller than any other singly ionized atom. The bond energy is usually less than 0.5 eV. The hydrogen bond plays an essential role in many organic molecules, including the cross-linking of polymer chains such as polyethylene and cross-link bonding between the two strands of the double-helix DNA molecule. Hydrogen bonding also plays a role in the structure of ice.

All these bond types hold the atoms together in *solids* as well as in molecules. Indeed, a solid is in many respects a giant molecule. Still another type of bonding, the *metallic bond,* comes into play in the structure of metallic solids. We'll return to this subject in Section 42.3.

Test Your Understanding

If electrons obeyed the exclusion principle but did *not* have spin, how many electrons could participate in a covalent bond?

42.2 | Molecular Spectra

Molecules have energy levels that are associated with rotational motion of a molecule as a whole and with vibrational motion of the atoms relative to each other. Just as transitions between energy levels in atoms lead to atomic spectra, transitions between rotational and vibrational levels in molecules lead to *molecular spectra.*

Rotational Energy Levels

In this discussion we'll concentrate mostly on *diatomic* molecules, to keep things as simple as possible. In Fig. 42.4 we picture a diatomic molecule as a rigid dumbbell (two point masses m_1 and m_2 separated by a constant distance r_0) that can *rotate* about axes through its center of mass, perpendicular to the line joining them. What are the energy levels associated with this motion?

We showed in Section 10.5 that when a rigid body rotates with angular speed ω about a perpendicular axis through its center of mass, the magnitude L of its angular momentum is given by Eq. (10.31), $L = I\omega$, where I is its moment of inertia about that symmetry axis. Its kinetic energy is given by Eq. (9.17), $K = \frac{1}{2}I\omega^2$. Combining these two equations, we find $K = L^2/2I$. There is no potential energy U, so the kinetic energy K is equal to the total mechanical energy E:

$$E = \frac{L^2}{2I} \qquad (42.1)$$

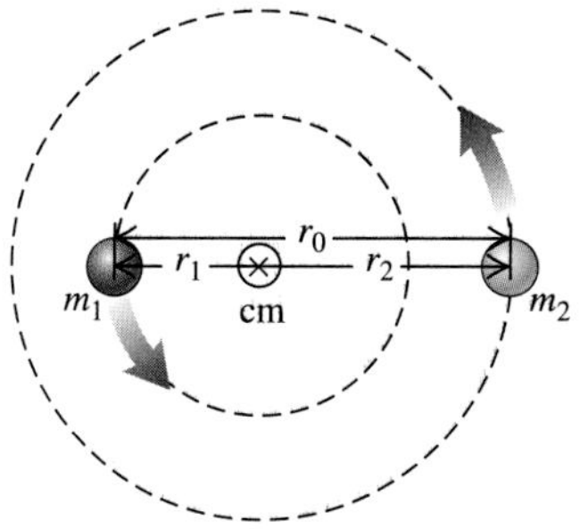

42.4 A diatomic molecule modeled as two point masses m_1 and m_2 separated by a distance r_0. The distances of the masses from the center of mass are r_1 and r_2, where $r_1 + r_2 = r_0$.

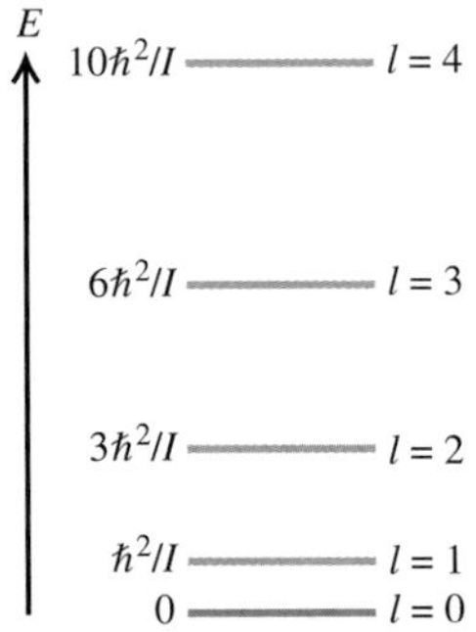

42.5 The ground level and first four excited rotational energy levels for a diatomic molecule. The levels are not equally spaced.

Zero potential energy U means no dependence of U on θ or ϕ. But the potential-energy function U in the hydrogen atom also has no dependence on θ or ϕ. Thus the angular solutions to the Schrödinger equation for rigid-body rotation are the same as for the hydrogen atom, and the angular momentum is quantized in the same way. As in Eq. (41.4),

$$L^2 = l(l+1)\hbar^2 \qquad (l = 0, 1, 2\ldots) \tag{42.2}$$

Combining Eqs. (42.1) and (42.2), we obtain the *rotational energy levels:*

$$E_l = l(l+1)\frac{\hbar^2}{2I} \qquad (l = 0, 1, 2, \ldots) \tag{42.3}$$

(rotational energy levels, diatomic molecule)

Figure 42.5 is an energy-level diagram showing these rotational levels. The ground level has zero quantum number l and zero energy E, corresponding to zero angular momentum (no rotation). The spacing of adjacent levels increases with increasing l.

We can express the moment of inertia I in Eqs. (42.1) and (42.3) in terms of the *reduced mass* m_r of the molecule:

$$m_r = \frac{m_1 m_2}{m_1 + m_2} \tag{42.4}$$

We introduced this quantity in Section 38.5 to accommodate the finite nuclear mass of the hydrogen atom. In Fig. 42.4 the distances r_1 and r_2 are the distances from the center of mass to the nuclei of the atoms. By definition of the center of mass, $m_1 r_1 = m_2 r_2$, and the figure also shows that $r_0 = r_1 + r_2$. Solving these equations for r_1 and r_2, we find

$$r_1 = \frac{m_2}{m_1 + m_2} r_0 \qquad r_2 = \frac{m_1}{m_1 + m_2} r_0 \tag{42.5}$$

The moment of inertia is $I = m_1 r_1^2 + m_2 r_2^2$; substituting Eq. (42.5), we find

$$I = m_1 \frac{m_2^2}{(m_1 + m_2)^2} r_0^2 + m_2 \frac{m_1^2}{(m_1 + m_2)^2} r_0^2 = \frac{m_1 m_2}{m_1 + m_2} r_0^2$$

or

$$I = m_r r_0^2 \qquad \text{(moment of inertia of a diatomic molecule)} \tag{42.6}$$

The reduced mass enables us to reduce this two-body problem to an equivalent one-body problem (a particle of mass m_r moving around a circle with radius r_0), just as we did with the hydrogen atom. Indeed, the only difference between this problem and the hydrogen atom is the difference in the radial forces. To conserve angular momentum, the allowed transitions are determined by the same selection rule as for the hydrogen atom: in allowed transitions, l must change by exactly one unit.

Example 42.2 Rotational spectrum of carbon monoxide

The two nuclei in the carbon monoxide (CO) molecule are 0.1128 nm apart. The mass of the most common carbon atom is exactly 12 u, or 1.993×10^{-26} kg. The mass of the most common oxygen atom is 15.995 u $= 2.656 \times 10^{-26}$ kg. a) Find the energies of the lowest three rotational energy levels. Express your results in electron volts. b) Find the wavelength of the photon emitted in the transition from the $l = 2$ to the $l = 1$ level.

SOLUTION

IDENTIFY and SET UP: We are given the distance r_0 between the atoms and their masses m_1 and m_2. We calculate the reduced mass m_r using Eq. (42.4), the moment of inertia I of the molecule using Eq. (42.6), and the energies of the levels using Eq. (42.3). The energy E of the emitted photon is equal to the difference in energy between the $l = 2$ and $l = 1$ levels; we determine the wavelength using the relationship $E = hc/\lambda$ for a photon.

EXECUTE: a) Using Eq. (42.4), we find that the reduced mass is $m_r = 1.139 \times 10^{-26}$ kg. From Eq. (42.6),

$$I = m_r r_0^2 = (1.139 \times 10^{-26}\ \text{kg})(0.1128 \times 10^{-9}\ \text{m})^2$$
$$= 1.449 \times 10^{-46}\ \text{kg} \cdot \text{m}^2$$

The rotational levels are given by Eq. (42.3):

$$E_l = l(l+1)\frac{\hbar^2}{2I} = l(l+1)\frac{(1.0546 \times 10^{-34}\ \text{J} \cdot \text{s})^2}{2(1.449 \times 10^{-46}\ \text{kg} \cdot \text{m}^2)}$$
$$= l(l+1)(3.838 \times 10^{-23}\ \text{J}) = l(l+1)0.2395\ \text{meV}$$

Substituting $l = 0, 1, 2$, we find

$$E_0 = 0 \qquad E_1 = 0.479\ \text{meV} \qquad E_2 = 1.437\ \text{meV}$$

b) The photon energy is

$$E = E_2 - E_1 = 0.958\ \text{meV}$$

The photon wavelength is

$$\lambda = \frac{hc}{E} = \frac{(4.136 \times 10^{-15}\ \text{eV} \cdot \text{s})(3.00 \times 10^8\ \text{m/s})}{0.958 \times 10^{-3}\ \text{eV}}$$
$$= 1.29 \times 10^{-3}\ \text{m} = 1.29\ \text{mm}$$

EVALUATE: The difference between the first two rotational energy levels of CO is very small (about $\frac{1}{2}$ meV, where 1 meV $= 10^{-3}$ eV) compared to the difference between atomic energy levels (typically a few eV) associated with optical spectra. Hence a photon emitted by a CO molecule in a transition from the $l = 2$ to the $l = 1$ level has very low energy and a very long wavelength compared to visible light. Indeed, the wavelength calculated in part (b) is in the microwave part of the spectrum. Photon wavelengths for rotational transitions in other molecules are also long, falling within the microwave and far infrared regions of the spectrum.

In this example we were given the equilibrium separation between the atoms, also called the *bond length,* and used it to calculate one of the wavelengths emitted by excited CO molecules. In actual experiments, scientists work this problem backward: by measuring the microwave emissions of a sample of diatomic molecules, they determine the moment of inertia of the molecule and hence the bond length.

Rotational levels of water molecules play an important role in the operation of microwave ovens. The magnetron oscillator in such an oven generates microwaves with a frequency of 2450 MHz; the corresponding photon energy is 1.01×10^{-5} eV. The water molecule has an $l = 1$ rotational level 1.01×10^{-5} eV above the nonrotating $(l = 0)$ ground level, so the microwaves are strongly absorbed. Energy transfer from the rotating water molecules to their surroundings then heats up the substance.

Vibrational Energy Levels

Molecules are never completely rigid. In a more realistic model of a diatomic molecule we represent the connection between atoms not as a rigid rod but as a *spring* (Fig. 42.6). Then in addition to rotating, the atoms of the molecule can *vibrate* about their equilibrium positions along the line joining them. For small oscillations the restoring force can be taken as proportional to the displacement from the equilibrium separation r_0 (like a spring that obeys Hooke's law with a force constant k'), and the system is a harmonic oscillator. We discussed the quantum-mechanical harmonic oscillator in Section 40.4. The energy levels are given by Eq. (40.25), with the mass m replaced with the reduced mass m_r:

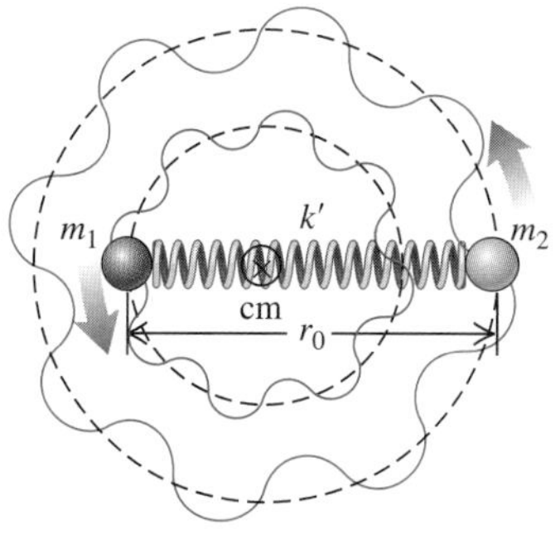

42.6 A diatomic molecule modeled as two point masses m_1 and m_2 connected by a spring with force constant k'.

$$E_n = \left(n + \frac{1}{2}\right)\hbar\omega = \left(n + \frac{1}{2}\right)\hbar\sqrt{\frac{k'}{m_r}} \qquad (n = 0, 1, 2, \dots) \qquad (42.7)$$

(vibrational energy levels of a diatomic molecule)

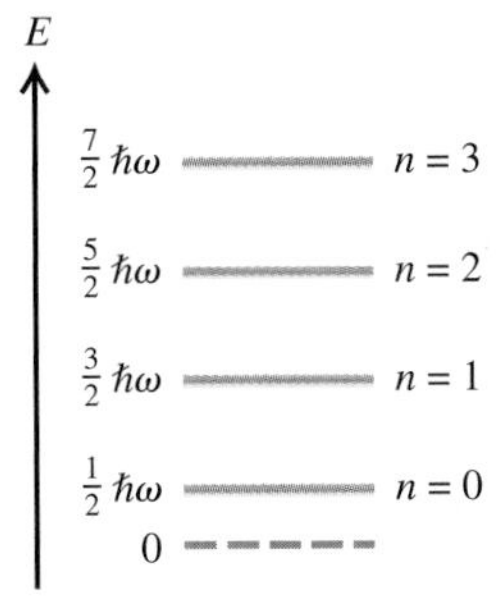

42.7 The ground level and first three excited vibrational levels for a diatomic molecule, assuming small oscillations. The levels are equally spaced, with spacing $\Delta E = \hbar\omega$.

This represents a series of levels equally spaced in energy, with an energy separation of

$$\Delta E = \hbar\omega = \hbar\sqrt{\frac{k'}{m_r}} \tag{42.8}$$

Figure 42.7 is an energy-level diagram showing these vibrational levels.

CAUTION We're again using k' for the force constant, this time to minimize confusion with Boltzmann's constant k, the gas constant per molecule (introduced in Section 18.3). Besides the quantities k and k', we also use the absolute temperature unit 1 K = 1 kelvin.

Example 42.3 Force constant of carbon monoxide

For the carbon monoxide molecule of Example 42.2, the spacing of vibrational energy levels is found to be $\Delta E = 0.2690$ eV. Find the force constant k' for the interatomic force.

SOLUTION

IDENTIFY and SET UP: We are given the value of the energy separation ΔE between adjacent vibrational energy levels, so we solve Eq. (42.8) for the force constant k' (our target variable).

EXECUTE: From Eq. (42.8),

$$k' = m_r\left(\frac{\Delta E}{\hbar}\right)^2 = (1.139 \times 10^{-26}\ \text{kg})\left(\frac{0.2690\ \text{eV}}{6.582 \times 10^{-16}\ \text{eV}\cdot\text{s}}\right)^2$$
$$= 1902\ \text{N/m}$$

EVALUATE: This corresponds to a fairly loose spring; to stretch a macroscopic spring with this force constant by 1.0 cm would require a pull of 19 N (about 4 lb). Force constants for diatomic molecules are typically about 100 to 2000 N/m.

Visible-light photons have energies between 1.77 eV and 3.10 eV. The 0.269-eV energy difference in Example 42.3 corresponds to a photon in the infrared region of the spectrum, though closer to the visible region than the photon in the *rotational* transition in Example 42.2. Vibrational energy differences, while usually much smaller than those that produce atomic spectra, are usually much *larger* than the rotational energy differences.

When we include *both* rotational and vibrational energies, the energy levels for our diatomic molecule are

$$E_{nl} = l(l+1)\frac{\hbar^2}{2I} + \left(n + \frac{1}{2}\right)\hbar\sqrt{\frac{k'}{m_r}} \tag{42.9}$$

The energy-level diagram is shown in Fig. 42.8. For each value of n there are many values of l, forming a series of closely spaced levels.

The arrows in Fig. 42.8 show several possible transitions in which a molecule goes from a level with $n = 2$ to a level with $n = 1$ by emitting a photon. Because the photon carries angular momentum, these molecular transitions obey a *selection rule* (Section 41.3): the angular momentum quantum number l must either increase

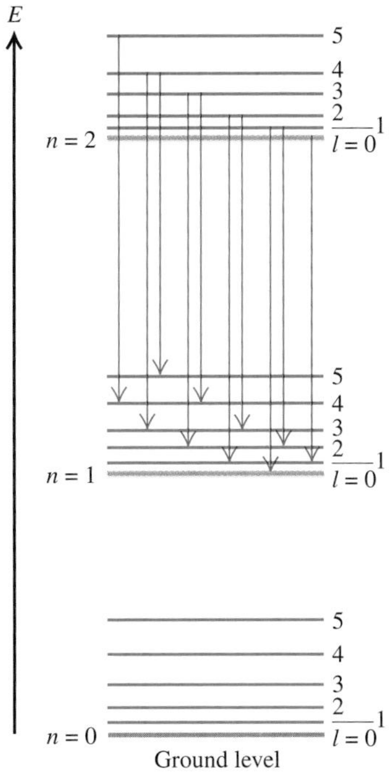

42.8 Energy-level diagram for vibrational and rotational energy levels of a diatomic molecule. For each vibrational level (n) there is a series of more closely spaced rotational levels (l). Several transitions corresponding to a single band in a band spectrum are shown. These transitions obey the selection rule $\Delta l = \pm 1$.

by 1 or decrease by 1: $\Delta l = \pm 1$. An additional selection rule states that if the vibrational level changes, the vibrational quantum number n in Eq. (42.9) must increase by 1 $(\Delta n = 1)$ if a photon is absorbed or decrease by 1 $(\Delta n = -1)$ if a photon is emitted. For example, Fig. 42.8 shows that a molecule in the $n = 2$, $l = 4$ level can emit a photon and drop into the $n = 1$, $l = 5$ level $(\Delta n = -1, \Delta l = +1)$ or the $n = 1$, $l = 3$ level $(\Delta n = -1, \Delta l = -1)$, but is forbidden from making a $\Delta n = -1, \Delta l = 0$ transition into the $n = 1, l = 4$ level.

Example 42.4 Vibration-rotation spectrum of carbon monoxide

Again consider the CO molecule of Examples 42.2 and 42.3. Find the photon wavelength that is emitted when there is a change in its vibrational energy, and its rotational energy is a) initially zero and b) finally zero.

SOLUTION

IDENTIFY and SET UP: In either case the energy of the emitted photon equals the difference between the initial and final energies of the molecule. Since a photon is emitted as the vibrational energy changes, the vibrational quantum number decreases by 1. For the case in part (a) the initial value of l is zero (no rotational energy), so the selection rule $\Delta l = \pm 1$ tells us that the *final* value of l is 1. For the case in part (a) the *final* value of l is zero, so the same selection rule tells us that the *initial* value of l is 1. Hence in part (a) the photon energy E is equal to the difference $\hbar\omega$ between adjacent vibrational energy levels *minus* the amount of rotational energy that the molecule *gains,* while in part (b) E is equal to $\hbar\omega$ *plus* the amount of rotational energy that the molecules *loses*. We use the results of Examples 42.2 and 42.3 to tell us the rotational and vibrational energy changes respectively, and we use $E = hc/\lambda$ to determine the wavelengths (our target variables).

EXECUTE: a) From Example 42.2 the energy difference between the $l = 0$ and $l = 1$ rotational levels of CO is 0.479 meV = 0.000479 meV, and from Example 42.3 the energy difference between adjacent vibrational levels is 0.2690 eV. Hence the energy of this emitted photon is 0.2690 eV − 0.000479 eV = 0.2685 eV, and the photon wavelength is

$$\lambda = \frac{hc}{E} = \frac{(4.136 \times 10^{-16}\,\text{eV}\cdot\text{s})(2.998 \times 10^{8}\,\text{m/s})}{0.2685\,\text{eV}}$$
$$= 4.618 \times 10^{-6}\,\text{m} = 4.618\,\mu\text{m}$$

b) Now the photon has 0.2690 eV from the decrease in vibrational energy and 0.000479 eV from the decrease in the rotational energy for a total of $E = 0.2695$ eV for the photon. The photon's wavelength is then

$$\lambda = \frac{hc}{E} = \frac{(4.136 \times 10^{-16}\ \text{eV}\cdot\text{s})(2.998 \times 10^{8}\ \text{m/s})}{0.2695\ \text{eV}}$$
$$= 4.601 \times 10^{-6}\ \text{m} = 4.601\ \mu\text{m}$$

EVALUATE: In part (b) the molecule loses more energy than it does in part (a), so the emitted photon must have a greater energy and a shorter wavelength. This is just what our results show.

Transitions between states with various pairs of n values give different series of spectrum lines, and the resulting spectrum has a series of *bands.* Each band corresponds to a particular vibrational transition, and each individual line in a band represents a particular rotational transition, with the selection rule $\Delta l = \pm 1$. A typical *band spectrum* is shown in Fig. 42.9. All molecules can have excited states of the *electrons* in addition to the rotational and vibrational states that we have described. In general, these lie at higher energies than the rotational and vibrational states, and there is no simple rule relating them. When there is a transition between electronic states, the $\Delta n = \pm 1$ selection rule for the vibrational levels no longer holds.

We can apply these same principles to more complex molecules. A molecule with three or more atoms has several different kinds or *modes* of vibratory motion. Each mode has its own set of energy levels, related to its frequency by Eq. (42.7). In nearly all cases the associated radiation lies in the infrared region of the electromagnetic spectrum. One example is the carbon dioxide (CO_2) molecule, which is the fourth most abundant constituent of the earth's atmosphere (after nitrogen, oxygen, and argon). The fact that CO_2 can *absorb* infrared radiation and go into an excited state explains why this substance is implicated in global warming (see the image that opens this chapter).

Infrared spectroscopy has proved to be an extremely valuable analytical tool. It provides information about the strength, rigidity, and length of molecular bonds and the structure of complex molecules. Also, because every molecule (like every atom) has its characteristic spectrum, infrared spectroscopy can be used to identify unknown compounds.

Test Your Understanding

Figure 42.5 shows that the rotational energy levels of a diatomic molecule are not evenly spaced. Prove that the energy difference between levels $l - 1$ and l is proportional to l, so that the difference increases with increasing energy.

42.9 A typical molecular band spectrum.

42.3 | Structure of Solids

The term *condensed matter* includes both solids and liquids. In both states, the interactions between atoms or molecules are strong enough to give the material a definite volume that changes relatively little with applied stress. In condensed matter, adjacent atoms attract one another until their outer electron charge clouds begin to overlap significantly. Thus the distances between adjacent atoms in condensed matter are about the same as the diameters of the atoms themselves, typically 0.1 to 0.5 nm. Also, when we speak of the distances between atoms, we mean the center-to-center (nucleus-to-nucleus) distances.

Ordinarily, we think of a liquid as a material that can flow and of a solid as a material with a definite shape. However, if you heat a horizontal glass rod in the flame of a burner, you'll find that the rod begins to sag (flow) more and more easily as its temperature rises. Glass has no definite transition from solid to liquid, and no definite melting point. On this basis, we can consider glass at room temperature as being an extremely viscous liquid. Tar and butter show similar behavior.

What is the microscopic difference between materials like glass or butter and solids like ice or copper, which do have definite melting points? Ice and copper are examples of *crystalline solids* in which the atoms have *long-range order,* a recurring pattern of atomic positions that extends over many atoms. This pattern is called the *crystal structure.* In contrast, glass at room temperature is an example of an *amorphous* solid, one that has no long-range order, but only *short-range order* (correlations between neighboring atoms or molecules). Liquids also have only short-range order. The boundaries between crystalline solid, amorphous solid, and liquid may be sometimes blurred. Some solids, crystalline when perfect, can form with so many imperfections in their structure that they have almost no long-range order. Conversely, some liquids have a fairly high degree of long-range order; a familiar example is the type of *liquid crystal* illustrated in Fig. 42.10.

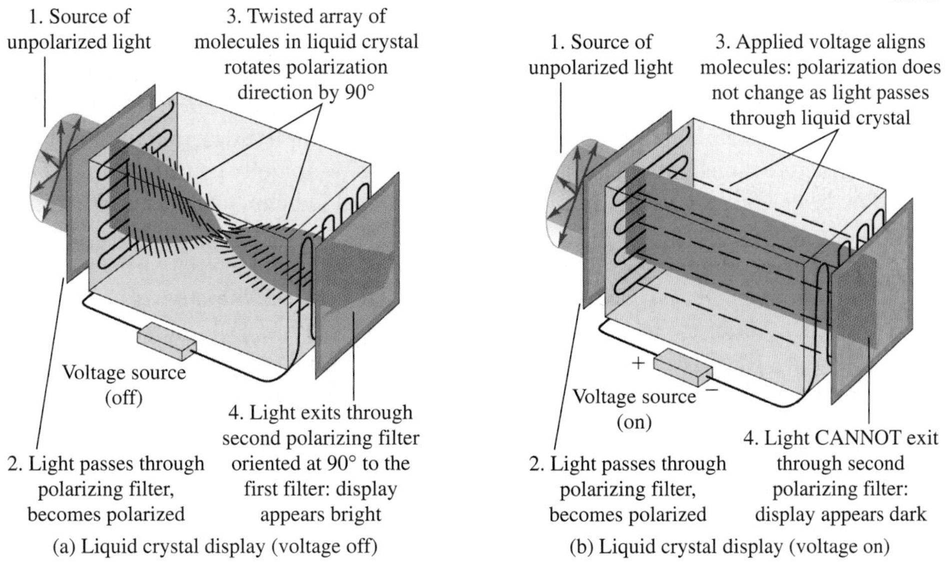

(a) Liquid crystal display (voltage off)

(b) Liquid crystal display (voltage on)

42.10 A simple version of a liquid-crystal display (LCD). A thin layer of liquid crystal—an organic compound whose cylindrical molecules tend to line up parallel to each other—is confined between two parallel glass plates and placed between two crossed polarizing filters. (a) The molecules tend to align with fine scratches on the glass plates, twisting them so that the light can pass through the second polarizing filter. (b) Each glass plate also has a pattern of fine electrodes. When a voltage is applied between the plates, the molecules align with the electric field and the light does not pass through the second polarizing filter. Hence switching the voltage on changes the display from light to dark.

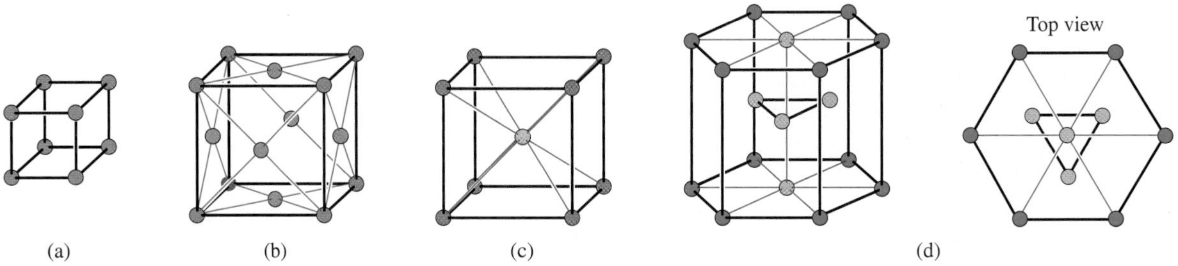

42.11 Portions of some common types of crystal lattices. (a) Simple cubic (sc); (b) face-centered cubic (fcc); (c) body-centered cubic (bcc); (d) hexagonal close-packed (hcp) in two views.

Nearly everything we know about crystal structure has been learned from diffraction experiments, initially with x rays, later with electrons and neutrons. A typical distance between atoms is of the order of 0.1 nm. You can show that 12.4-keV x rays, 150-eV electrons, and 0.0818-eV neutrons all have wavelengths $\lambda = 0.1$ nm.

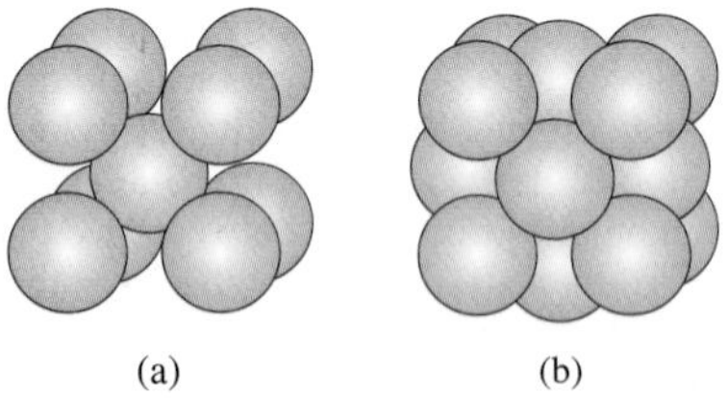

42.12 (a) The bcc *structure* is composed of a bcc *lattice* with a basis of one atom for each lattice point. (b) The fcc *structure* is composed of an fcc *lattice* with a basis of one atom for each lattice point. These structures repeat precisely to make up perfect crystals.

Crystal Lattices

A *crystal lattice* is a repeating pattern of mathematical points that extends throughout space. There are 14 general types of such patterns; Fig. 42.11 shows small portions of a few common examples. The *simple cubic lattice* (sc) has a lattice point at each corner of a cubic array (Fig. 42.11a). The *face-centered cubic lattice* (fcc) is like the simple cubic but with an additional lattice point at the center of each cube face (Fig. 42.11b). The *body-centered cubic lattice* (bcc) is like the simple cubic but with an additional point at the center of each cube (Fig. 42.11c). The *hexagonal close-packed lattice* has layers of lattice points in hexagonal patterns, each hexagon made up of six equilateral triangles (Fig. 42.11d).

CAUTION Figure 42.11 shows just enough lattice points to easily visualize the pattern; the lattice, a mathematical abstraction, extends throughout space. Thus the lattice points shown repeat endlessly in all directions.

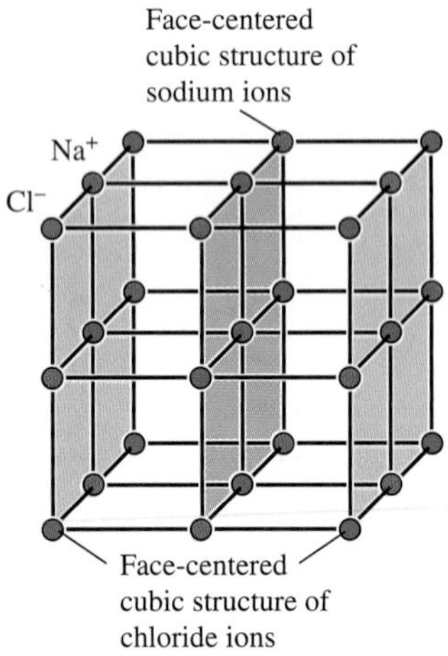

42.13 Representation of part of the sodium chloride crystal structure. The distances between ions are exaggerated.

In a crystal structure, a single atom or a group of atoms is associated with *each* lattice point. The group may contain the same or different kinds of atoms. This atom or group of atoms is called a *basis.* Thus a complete description of a crystal structure includes both the lattice and the basis. We initially consider *perfect crystals,* or *ideal single crystals,* in which the crystal structure extends uninterrupted throughout space.

The bcc and fcc structures are two common simple crystal structures. The alkali metals have a bcc structure, that is, a bcc lattice with a basis of one atom at each lattice point. Each atom in a bcc structure has eight nearest neighbors (Fig. 42.12a). The elements Al, Ca, Cu, Ag, and Au have an fcc structure, that is, an fcc lattice with a basis of one atom at each lattice point. Each atom in an fcc structure has 12 nearest neighbors (Fig. 42.12b).

Figure 42.13 shows a representation of the structure of sodium chloride (NaCl, ordinary salt). It may look like a simple cubic structure, but it isn't. The sodium and chloride ions each form an fcc structure, so we can think roughly of the sodium chloride structure as being composed of two interpenetrating fcc structures. More correctly, the sodium chloride crystal structure of Fig. 42.13 has a fcc lattice with one chloride ion at each lattice point and one sodium ion half a cube length above it. That is, its basis consists of one chloride and one sodium ion.

Another example is the *diamond structure;* it's called that because it is the crystal structure of carbon in the diamond form (Fig. 42.14). It's also the crystal structure of silicon, germanium, and gray tin (all four are Group IV elements in the periodic table). The diamond lattice is fcc; the basis consists of one atom at each lattice point and a second *identical* atom displaced a quarter of a cube length in each of the three cube-edge directions. Figure 42.15 will help you visualize this. The shaded volume in Fig. 42.15 shows the bottom right front eighth of the basic cube; the four atoms at alternate corners of this cube are at the corners of a regular tetrahedron, and there is an additional atom at the center. Thus each atom in the diamond structure is at the center of a regular tetrahedron with four nearest-neighbor atoms at the corners.

In the diamond structure, both the purple and green spheres in Fig. 42.15 represent *identical* atoms—for example, both carbon or both silicon. In the cubic zinc sulfide structure, the purple spheres represent one type of atom and the green spheres represent a *different* type. For example, in zinc sulfide (ZnS) each zinc atom (purple in Fig. 42.15) is at the center of a regular tetrahedron with four sulfur atoms (green in Fig. 42.15) at its corners, and vice versa. Gallium arsenide (GaAs) and similar compounds have this same structure.

42.14 The crystal structure of diamond is similar to that of many other, less highly coveted materials. What makes diamond unique are its beautiful optical properties. A large index of refraction makes diamond sparkle due to total internal reflection, and a large amount of dispersion turns white light into a rainbow of colors.

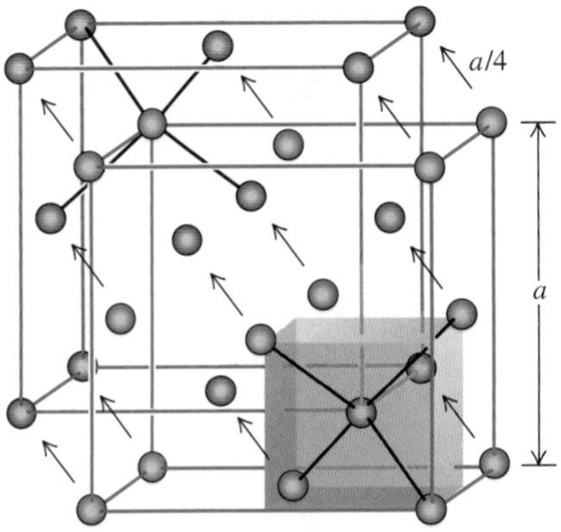

42.15 (a) The diamond structure, shown as two interpenetrating face-centered cubic structures with distances between atoms exaggerated. Relative to the corresponding green atom, each purple atom is shifted up, back, and to the left by a distance $a/4$.

Bonding in Solids

The forces that are responsible for the regular arrangement of atoms in a crystal are the same as those involved in molecular bonds, plus one additional type. Not surprisingly, *ionic* and *covalent* molecular bonds are found in ionic and covalent crystals, respectively. The most familiar *ionic crystals* are the alkali halides, such as ordinary salt (NaCl). The positive sodium ions and the negative chloride ions occupy alternate positions in a cubic arrangement (Fig. 42.13). The attractive forces are the familiar Coulomb's-law forces between charged particles. These forces have no preferred direction, and the arrangement in which the material crystallizes is partly determined by the relative size of the two ions. Such a structure is *stable* in the sense that it has lower total energy than the separated ions (see Example 42.5). The negative potential energies of pairs of opposite charges are greater in absolute value than the positive energies of pairs of like charges because the pairs of unlike charges are closer together, on average.

Example 42.5 Potential energy of an ionic crystal

Consider a fictitious one-dimensional ionic crystal consisting of a very large number of alternating positive and negative ions with charges e and $-e$, with equal spacing a along a line. Prove that the total interaction potential energy is negative.

SOLUTION

IDENTIFY and SET UP: We treat each ion as a point charge and consider the electric potential energy U of each pair of charges. The total potential energy of the system is the sum of the values of U for every possible pair.

EXECUTE: Let's pick an ion somewhere in the middle of the string and add up the potential energies of its interactions with all the ions to one side of it. We get the series

$$\sum U = -\frac{e^2}{4\pi\epsilon_0}\frac{1}{a} + \frac{e^2}{4\pi\epsilon_0}\frac{1}{2a} - \frac{e^2}{4\pi\epsilon_0}\frac{1}{3a} + \cdots$$

$$= -\frac{e^2}{4\pi\epsilon_0 a}\left(1 - \frac{1}{2} + \frac{1}{3} - \frac{1}{4} + \cdots\right)$$

You may notice the resemblance of the series in parentheses to the Taylor series for $\ln(1+x)$:

$$\ln(1+x) = x - \frac{x^2}{2} + \frac{x^3}{3} - \frac{x^4}{4} + \cdots$$

When $x = 1$, we have the series in the parentheses, and

$$\sum U = -\frac{e^2}{4\pi\epsilon_0 a}\ln 2$$

This is certainly a negative quantity. The atoms on the other side of the ion that we're considering make an equal contribution to the potential energy. And if we include the potential energies of all pairs of atoms, the sum is certainly negative.

EVALUATE: We conclude that this structure has lower energy than the zero electric potential energy that is obtained when all the ions are infinitely separated from each other.

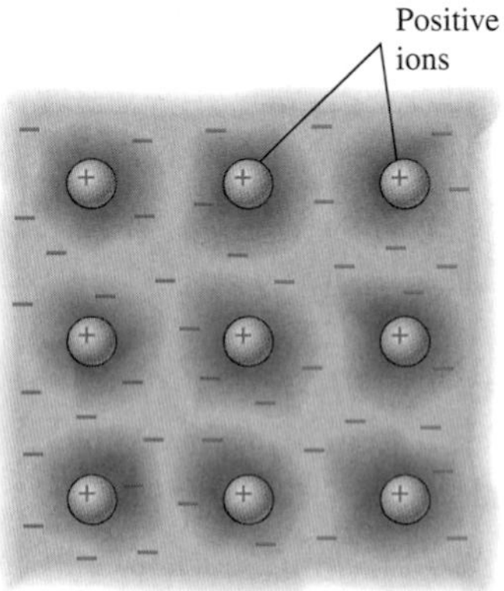

42.16 In a metallic solid, one or more electrons are detached from each atom and are free to wander around the crystal, forming an "electron gas." The wave functions for these electrons extend over many atoms. The positive ions vibrate around fixed locations in the crystal.

Carbon, silicon, germanium, and tin in the diamond structure are simple examples of *covalent crystals.* These elements are in Group IV of the periodic table, meaning that each atom has four electrons in its outermost shell. Each atom forms a covalent bond with each of four adjacent atoms at the corners of a tetrahedron (Fig. 42.15). These bonds are strongly directional because of the asymmetrical electron distributions dictated by the exclusion principle, and the result is the tetrahedral diamond structure.

A third crystal type, less directly related to the chemical bond than are ionic or covalent crystals, is the **metallic crystal.** In this structure, one or more of the outermost electrons in each atom become detached from the parent atom (leaving a positive ion) and are free to move through the crystal. These electrons are not localized near the individual ions. The corresponding electron wave functions extend over many atoms.

Thus we can picture a metallic crystal as an array of positive ions immersed in a sea of freed electrons whose attraction for the positive ions holds the crystal together (Fig. 42.16). These electrons also give metals their high electrical and thermal conductivities. This sea of electrons has many of the properties of a gas, and indeed we speak of the *electron-gas model* of metallic solids. The simplest version of this model is the *free-electron model,* which ignores interactions with the ions completely (except at the surface). We'll return to this model in Section 42.5.

In a metallic crystal the freed electrons are not localized but are shared among *many* atoms. This gives a bonding that is neither localized nor strongly directional. The crystal structure is determined primarily by considerations of *close packing,* that is, the maximum number of atoms that can fit into a given volume. The two most common metallic crystal lattices, the face-centered cubic and the hexagonal close-packed, are shown in Figs. 42.11b and 42.11d. In structures composed of these lattices with a basis of one atom, each atom has 12 nearest neighbors.

As we mentioned in Section 42.1, van der Waals interactions and hydrogen bonding also play a role in the structure of some solids. In polyethylene and similar polymers, covalent bonding of atoms forms long-chain molecules, and hydrogen bonding forms cross-links between adjacent chains. In solid water, both van der Waals forces and hydrogen bonds are significant in determining the crystal structures of ice.

Our discussion has centered on perfect crystals, or ideal single crystals. Real crystals show a variety of departures from this idealized structure. Materials are often *polycrystalline,* composed of many small single crystals bonded together at *grain boundaries.* Within a single crystal, *interstitial* atoms may occur in places where they do not belong, and there may be *vacancies,* positions that should be occupied by an atom but are not. A point defect of particular interest in semiconductors, which we will discuss in Section 42.6, is the *substitutional impurity,* a foreign atom replacing a regular atom (for example, arsenic in a silicon crystal).

There are several basic types of extended defects called *dislocations.* One type is the *edge dislocation,* shown schematically in Fig. 42.17, in which one plane of atoms slips relative to another. The mechanical properties of metallic crystals are influenced strongly by the presence of dislocations. The ductility and malleability of some metals depend on the presence of dislocations that can move through the crystal during plastic deformations. Solid-state physicists often point out that the biggest extended defect of all, present in *all* real crystals, is the surface of the material with its dangling bonds and abrupt change in potential energy.

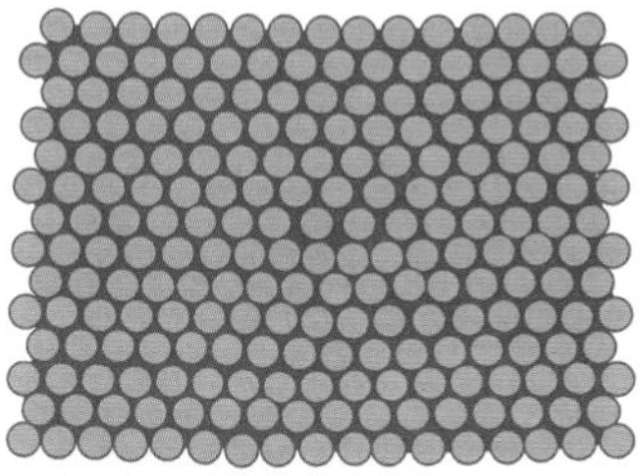

42.17 An edge dislocation in two dimensions. The irregularity is seen most easily by viewing the figure from various directions at a grazing angle with the page. In three dimensions an edge dislocation would look like an extra plane of atoms slipped partway into the crystal.

Test Your Understanding

In a sodium chloride crystal (Fig. 42.13), each sodium ion is surrounded by six chloride ions. Discuss why sodium chloride has this structure rather than a diamond structure like that of zinc sulfide (Fig. 42.15).

42.4 | Energy Bands

The **energy band** concept, introduced in 1928 (Fig. 42.18), is a great help in understanding several properties of solids. To introduce the idea, suppose we have a large number N of identical atoms, far enough apart that their interactions are negligible. Every atom has the same energy-level diagram. We can draw an energy-level diagram for the *entire system.* It looks just like the diagram for a single atom, but the exclusion principle, applied to the entire system, permits each state to be occupied by N electrons instead of just one.

42.18 The concept of energy bands was first developed by the Swiss American physicist Felix Bloch (1905–1983) in his doctoral thesis. Our modern understanding of electrical conductivity stems from that landmark work. Bloch's work in nuclear physics brought him (along with Edward Purcell) the 1952 Nobel Prize in physics.

Now we begin to push the atoms uniformly closer together. Because of the electrical interactions and the exclusion principle, the wave functions begin to distort, especially those of the outer, or *valence,* electrons. The corresponding energies also shift, some upward and some downward, by varying amounts, as the valence electron wave functions become less localized and extend over more and more atoms. Thus the valence states that formerly gave the *system* a state with a sharp energy level that could accommodate N electrons now give a *band* containing N closely spaced levels (Fig. 42.19). Ordinarily, N is very large, somewhere

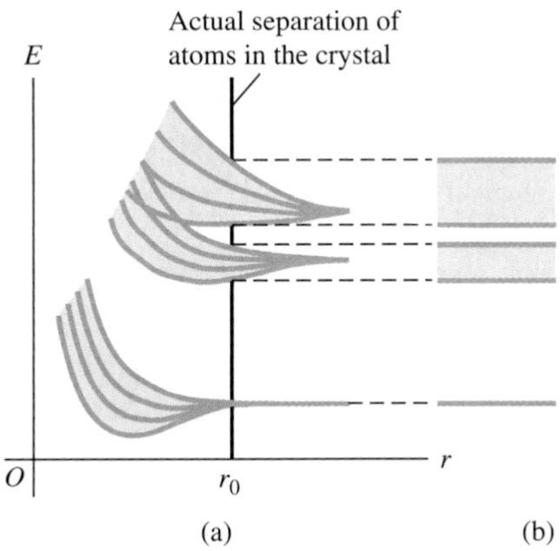

42.19 Origin of energy bands in a solid. (a) As the distance r between atoms decreases, the energy levels spread into bands. The vertical line at r_0 shows the actual atomic spacing in the crystal. (b) Symbolic representation of energy bands.

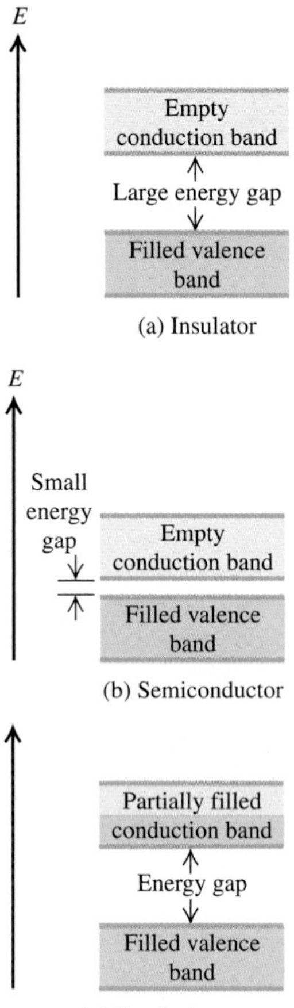

42.20 Three types of energy-band structure. (a) In an insulator at absolute zero, there are no electrons in the conduction band. (b) A semiconductor has the same structure as an insulator but a smaller energy gap between the valence and conduction bands. (c) A conductor has a partially filled conduction band.

near the order of Avogadro's number (10^{24}), so we can accurately treat the levels as forming a *continuous* distribution of energies within a band. Between adjacent energy bands are gaps or forbidden regions where there are *no* allowed energy levels. The inner electrons in an atom are affected much less by nearby atoms than are the valence electrons, and their energy levels remain relatively sharp.

Insulators, Semiconductors, and Conductors

The nature of the energy bands determines whether the material is an electrical insulator, a semiconductor, or a conductor. In particular, what matters are the extent to which the states in each band are occupied and the spacing, or *energy gap,* between adjacent bands. A crucial factor is the exclusion principle (Section 41.4), which states that only one electron can occupy a given quantum-mechanical state.

In an *insulator* at absolute zero temperature, the highest band that is completely filled, called the **valence band,** is also the highest band that has *any* electrons in it. The next higher band, called the **conduction band,** is completely empty; there are no electrons in its states (Fig. 42.20a). Imagine what happens if an electric field is applied to a material of this kind. To move in response to the field, an electron would have to go into a different quantum state with a slightly different energy. It can't do that, however, because all the neighboring states are already occupied. The only way such an electron can move is to jump across the energy gap into the conduction band, where there are plenty of nearby unoccupied states. At any temperature above absolute zero there is some probability this jump can happen, because an electron can gain energy from thermal motion. In an insulator, however, the energy gap between the valence and conduction bands can be 5 eV or more, and that much thermal energy is not ordinarily available. Hence little or no current flows in response to an applied electric field, and the electric conductivity (Section 25.2) is low. The thermal conductivity (Section 17.7), which also depends on mobile electrons, is likewise low.

We saw in Section 24.4 that an insulator becomes a conductor if it is subjected to a large enough electric field; this is called *dielectric breakdown.* If the electric field is of order 10^{10} V/m, there is a potential difference of a few volts over a distance comparable to atomic sizes. In this case the field can do enough work on a valence electron to boost it across the energy gap and into the conduction band. (In practice dielectric breakdown occurs for fields much less than 10^{10} V/m, because imperfections in the structure of an insulator provide some more accessible energy states *within* the energy gap.)

As in an insulator, a *semiconductor* at absolute zero has an empty conduction band above the full valence band. The difference is that in a semiconductor the energy gap between these bands is relatively small and electrons can more readily jump into the conduction band (Fig. 42.20b). As the temperature of a semiconductor increases, the population in the conduction band increases very rapidly, as does the electric conductivity. For example, in a semiconductor near room temperature with an energy gap of 1 eV, the number of conduction electrons doubles when the temperature rises by just 10°C. We will use the concept of energy bands to explore semiconductors in more depth in Section 42.6.

In a *conductor* such as a metal, there are electrons in the conduction band even at absolute zero (Fig. 42.20c). The metal sodium is an example. An analysis of the atomic energy-level diagram for sodium (Fig. 38.10) shows that for an isolated sodium atom, the six lowest excited states (all $3p$ states) are about 2.1 eV above the two $3s$ ground states. In solid sodium, however, the atoms are so close together that the $3s$ and $3p$ *bands* spread out and overlap into a single band. Each sodium atom contributes one electron to the band, leaving an Na^+ ion behind. Each atom also contributes eight *states* to that band (two $3s$, six $3p$), so the band is only one-eighth occupied. We call this structure a *conduction* band because it is only partially occupied. Electrons near the top of the filled portion of the band have many adjacent unoccupied states available, and they can easily gain or lose small amounts of energy in response to an applied electric field. Therefore these electrons are mobile, giving solid sodium its high electrical and thermal conductivity. A similar description applies to other conducting materials.

Example 42.6 Photoconductivity in germanium

Even at room temperature, pure germanium has an almost completely filled valence band separated by a gap of 0.67 eV from an almost completely empty conduction band. It is a poor electrical conductor, but its conductivity increases substantially when it is irradiated with electromagnetic waves of certain maximum wavelength. What maximum wavelength is appropriate?

SOLUTION

IDENTIFY and SET UP: The conductivity of a semiconductor increases greatly when electrons are excited from the valence band into the conduction band. In this example, the excitation occurs when an electron absorbs a photon of with an energy of at least E_{min}. From the relation $E = hc/\lambda$ for photons, the *maximum* wavelength corresponds to the *minimum* photon energy.

EXECUTE: An electron at the top of the valence band can absorb a photon with energy of 0.67 eV (no less) and move to the bottom of the conduction band, where it is a mobile charge. Thus the maximum wavelength is

$$\lambda_{max} = \frac{hc}{E_{min}} = \frac{(4.136 \times 10^{-15}\ \text{eV}\cdot\text{s})(3.00 \times 10^{8}\ \text{m/s})}{0.67\ \text{eV}}$$
$$= 1.9 \times 10^{-6}\ \text{m} = 1.9\ \mu\text{m} = 1900\ \text{nm}$$

EVALUATE: This wavelength is in the infrared part of the spectrum, so visible-light photons (which have shorter wavelength) will also induce conductivity in germanium. As we will see in Section 42.7, semiconductor crystals are widely used as photocells as well as for many other applications.

Test Your Understanding

Pure sodium (Na) is a good conductor of electricity because the $3s$ and $3p$ atomic bands overlap to form a partially filled conduction band. By contrast, the ionic sodium chloride (NaCl) crystal is an insulator. Explain why. (*Hint:* In NaCl, the atomic bands do not overlap.)

42.5 | Free-Electron Model of Metals

Studying the energy states of electrons in metals can give us a lot of insight into their electrical and magnetic properties, the electron contributions to heat capacities, and other behavior. As we discussed in Section 42.3, one of the distinguishing features of a metal is that one or more valence electrons are detached from their home atom and can move freely within the metal, with wave functions that extend over many atoms.

The **free-electron model** assumes that these electrons are completely free inside the material, that they don't interact at all with the ions or with each other, but that there are infinite potential-energy barriers at the surfaces. The wave functions and energy levels are then the three-dimensional versions of those for the particle in a box that we analyzed in Section 40.1 in one dimension. Suppose the

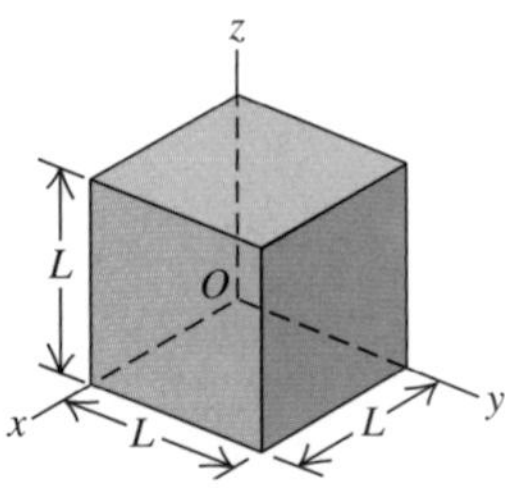

42.21 A cubical box with rigid walls and side length L. This is the three-dimensional version of the infinite square well discussed in Section 40.1. The energy levels for a particle in this box are given by Eq. (42.11).

box is a cube with side length L (Fig. 42.21). Then the possible wave functions, analogous to Eq. (42.10), are

$$\psi(x, y, z) = A \sin\frac{n_x \pi x}{L} \sin\frac{n_y \pi y}{L} \sin\frac{n_z \pi z}{L} \qquad (42.10)$$

where (n_x, n_y, n_z) is a set of three positive-integer quantum numbers that identify the state. We invite you to verify that these functions are zero at the surfaces of the cube, satisfying the boundary conditions. You can also substitute Eq. (42.10) into the three-dimensional Schrödinger equation, Eq. (40.28), with $U = 0$ to show that the energies of the states are

$$E = \frac{(n_x^2 + n_y^2 + n_z^2)\pi^2\hbar^2}{2mL^2} \qquad (42.11)$$

This equation is the three-dimensional analog of Eq. (40.9) for the energy levels of a particle in a box.

Density of States

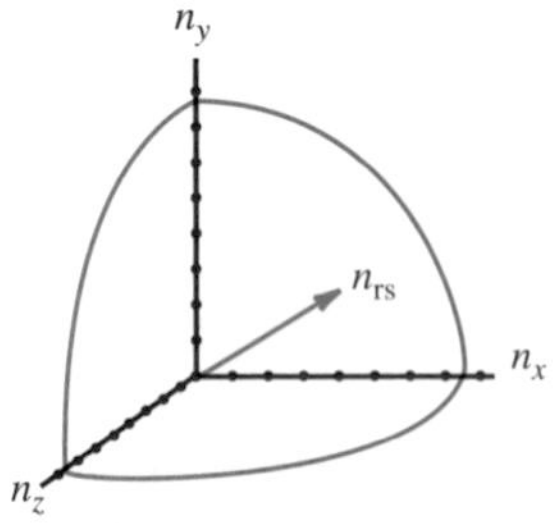

42.22 The allowed values of n_x, n_y, and n_z are positive integers for the electron states in the free-electron gas model. Including spin, there are two states for each unit volume in n space.

Later we'll need to know the *number dn* of quantum states that have energies in a given range dE. The number of states per unit energy range dn/dE is called the **density of states,** denoted by $g(E)$. We'll begin by working out an expression for $g(E)$. Think of a three-dimensional space with coordinates (n_x, n_y, n_z) (Fig. 42.22). The radius n_{rs} of a sphere centered at the origin in that space is given by $n_{rs}{}^2 = n_x^2 + n_y^2 + n_z^2$. Each point with integer coordinates in that space represents one spatial quantum state. Thus each point corresponds to one unit of volume in the space, and the total number of points with integer coordinates inside a sphere equals the volume of the sphere, $\frac{4}{3}\pi n_{rs}{}^3$. Because all our n's are positive, we must take only one *octant* of the sphere, with $\frac{1}{8}$ the total volume, or $(\frac{1}{8})(\frac{4}{3}\pi n_{rs}{}^3) = \frac{1}{6}\pi n_{rs}{}^3$. The particles are electrons, so each point corresponds to *two* states with opposite spin components $(m_s = \pm\frac{1}{2})$, and the total number n of electron states corresponding to points inside the octant is twice $\frac{1}{6}\pi n_{rs}{}^3$, or

$$n = \frac{\pi n_{rs}{}^3}{3} \qquad (42.12)$$

The energy E of states at the surface of the sphere can be expressed in terms of n_{rs}. Equation (42.11) becomes

$$E = \frac{n_{rs}{}^2\pi^2\hbar^2}{2mL^2} \qquad (42.13)$$

We can combine Eqs. (42.12) and (42.13) to get a relation between E and n that doesn't contain n_{rs}. We'll leave the details as an exercise (Exercise 42.24); the result is

$$n = \frac{(2m)^{3/2}VE^{3/2}}{3\pi^2\hbar^3} \qquad (42.14)$$

where $V = L^3$ is the volume of the box. Equation (42.14) gives the total number of states with energies of E or less.

To get the number of states dn in an energy interval dE, we take differentials of both sides of Eq. (42.14). We get

$$dn = \frac{(2m)^{3/2}VE^{1/2}}{2\pi^2\hbar^3}\,dE \qquad (42.15)$$

The density of states $g(E)$ is equal to dn/dE, so from Eq. (42.15) we get

$$g(E) = \frac{(2m)^{3/2}V}{2\pi^2\hbar^3}E^{1/2} \quad \text{(density of states, free-electron model)} \quad (42.16)$$

Fermi-Dirac Distribution

Now we need to know how the electrons are distributed among the various quantum states at any given temperature. The *Maxwell-Boltzmann distribution* states that the average number of particles in a state of energy E is proportional to $e^{-E/kT}$ (see Section 38.6). However, it wouldn't be right to use the Maxwell-Boltzmann distribution, for two very important reasons. The first reason is the *exclusion principle.* At absolute zero the Maxwell-Boltzmann function predicts that *all* the electrons would go into the two ground states of the system, with $n_x = n_y = n_z = 1$ and $m_s = \pm\frac{1}{2}$. But the exclusion principle allows only one electron in each state. At absolute zero the electrons can fill up the lowest *available* states, but there's not enough room for *all* of them to go into the lowest states. Thus a reasonable guess as to the shape of the distribution would be Fig. 42.23. At absolute zero temperature the states are filled up to some value E_{F0}, and all states above this value are empty.

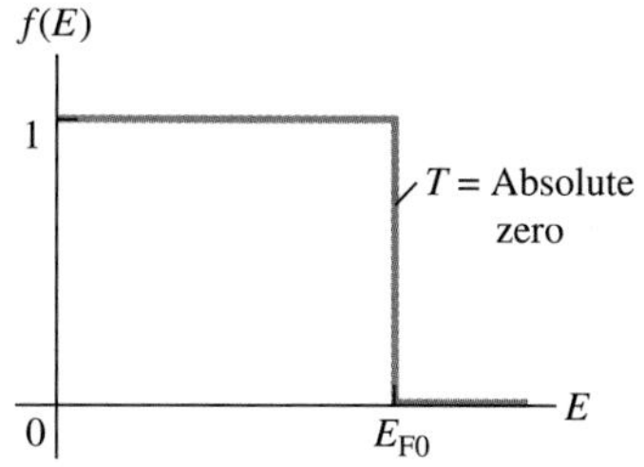

42.23 The probability distribution for occupation of free-electron energy states at absolute zero. All states are occupied (occupation probability 1) up to E_{F0}, and all states above E_{F0} are empty (occupation probability zero).

The second reason we can't use the Maxwell-Boltzmann distribution is more subtle. That distribution assumes that we are dealing with *distinguishable* particles. It might seem that we could put a tag on each electron and know which is which. But overlapping electrons in a system such as a metal are *indistinguishable.* Suppose we have two electrons; a state in which the first is in an energy level E_1 and the second is in level E_2 is not distinguishable from a state in which the two electrons are reversed, because we can't tell which electron is which.

The statistical distribution function that emerges from the exclusion principle and the indistinguishability requirement is called (after its inventors) the **Fermi-Dirac distribution.** Because of the exclusion principle, the probability that a particular state with energy E is occupied by an electron is the same as $f(E)$, the fraction of states with that energy that are occupied:

$$f(E) = \frac{1}{e^{(E-E_F)/kT} + 1} \quad \text{(Fermi-Dirac distribution)} \quad (42.17)$$

The energy E_F is called the **Fermi energy** or the *Fermi level;* we'll discuss its significance below. We use E_{F0} for its value at absolute zero $(T = 0)$ and E_F for other temperatures. We can accurately let $E_F = E_{F0}$ for metals because the Fermi energy does not change much with temperature for solid conductors. However, it is not safe to assume that $E_F = E_{F0}$ for semiconductors, in which the Fermi energy usually does change with temperature.

Figure 42.24 shows graphs of Eq. (42.17) for several temperatures. The trend of this function as kT approaches zero confirms our guess. When $E = E_F$, the exponent is zero and $f(E_F) = \frac{1}{2}$. That is, the probability is $\frac{1}{2}$ that a state at the Fermi energy contains an electron. Alternatively, at $E = E_F$, half the states are filled (and half are empty).

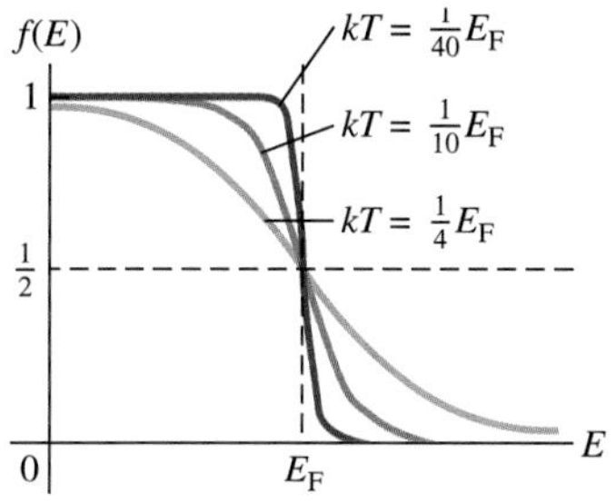

42.24 Graphs of the Fermi-Dirac distribution function for various values of kT, assuming that the Fermi energy E_F is independent of the temperature T. As T increases, more and more of the electrons are excited to states with energy $E > E_F$.

For $E < E_F$ the exponent is negative, and $f(E) > \frac{1}{2}$. For $E > E_F$ the exponent is positive, and $f(E) < \frac{1}{2}$. The shape depends on the ratio E_F/kT. At $T \ll E_F/k$ this ratio is very large. Then for $E < E_F$ the curve very quickly approaches 1, and for $E > E_F$ it quickly approaches zero. When T is larger, the changes are more gradual. When T is zero, all the states up to the Fermi level E_{F0} are filled, and all states above that level are empty (Fig. 42.23).

Example 42.7 Probabilities in the free-electron model

For free electrons in a solid, at what energy is the probability that a particular state is occupied equal to a) 0.01; b) 0.99?

SOLUTION

IDENTIFY and SET UP: Equation (42.17) tells us the occupation probability $f(E)$ for a given energy E. If we solve this equation for E, we get an expression for the energy that corresponds to a given occupation probability—which is just what we need to solve this problem.

EXECUTE: Solving Eq. (42.17) for E, we get

$$E = E_F + kT \ln\left(\frac{1}{f(E)} - 1\right)$$

a) When $f(E) = 0.01$,

$$E = E_F + kT \ln\left(\frac{1}{0.01} - 1\right) = E_F + 4.6kT$$

A state $4.6kT$ above the Fermi level is occupied only 1% of the time.

b) When $f(E) = 0.99$,

$$E = E_F + kT \ln\left(\frac{1}{0.99} - 1\right) = E_F - 4.6kT$$

A state $4.6kT$ below the Fermi level is occupied 99% of the time.

EVALUATE: At very low temperatures, $4.6kT$ becomes very small. Then levels even slightly below E_F are nearly always full, and levels even slightly above E_F are nearly always empty (see Fig. 42.24). In general, if the probability is P that a state with an energy ΔE *above* E_F is occupied, then the probability is $1 - P$ that a state ΔE *below* E_F is occupied. We leave the proof as a problem (Problem 42.46).

Equation (42.17) gives the probability that any specific state with energy E is occupied at a temperature T. To get the actual number of electrons in any energy range dE, we have to multiply this probability by the number dn of states in that range $g(E)\,dE$. Thus the number dN of electrons with energies in the range dE is

$$dN = g(E)f(E)\,dE = \frac{(2m)^{3/2}VE^{1/2}}{2\pi^2\hbar^3}\frac{1}{e^{(E-E_F)/kT} + 1}\,dE \qquad (42.18)$$

The Fermi energy E_F is determined by the total number N of electrons; at any temperature the electron states are filled up to a point at which all electrons are accommodated. At absolute zero there is a simple relation between E_{F0} and N. *All* states below E_{F0} are filled; in Eq. (42.14) we set n equal to the total number of electrons N and E to the Fermi energy at absolute zero E_{F0}:

$$N = \frac{(2m)^{3/2}VE_{F0}{}^{3/2}}{3\pi^2\hbar^3} \qquad (42.19)$$

Solving for E_{F0}, we get

$$E_{F0} = \frac{3^{2/3}\pi^{4/3}\hbar^2}{2m}\left(\frac{N}{V}\right)^{2/3} \qquad (42.20)$$

The quantity N/V is the number of free electrons per unit volume. It is called the *electron concentration* and is usually denoted by n.

If we replace N/V with n, Eq. (42.20) becomes

$$E_{F0} = \frac{3^{2/3}\pi^{4/3}\hbar^2 n^{2/3}}{2m} \qquad (42.21)$$

CAUTION Don't confuse the electron concentration n with any quantum number n. Furthermore, the number of states n is *not* in general the same as the total number of electrons N.

Example 42.8 The Fermi energy in copper

At low temperatures, copper has a free-electron concentration of $8.45 \times 10^{28}\ \text{m}^{-3}$. Using the free-electron model, find the Fermi energy for solid copper, and find the speed of an electron with a kinetic energy equal to the Fermi energy.

SOLUTION

IDENTIFY and SET UP: Because copper is a solid conductor, its Fermi energy changes very little with temperature and we can safely use the expression for the Fermi energy at absolute zero, Eq. (42.21). We determine the speed v_F that corresponds to kinetic energy E_F using the familiar nonrelativistic formula for kinetic energy.

EXECUTE: Using the given value of n,

$$E_F = \frac{3^{2/3}\pi^{4/3}(1.055 \times 10^{-34}\ \text{J}\cdot\text{s})^2(8.45 \times 10^{28}\ \text{m}^{-3})^{2/3}}{2(9.11 \times 10^{-31}\ \text{kg})}$$
$$= 1.126 \times 10^{-18}\ \text{J} = 7.03\ \text{eV}$$

To find the corresponding speed v_F, we use $E_F = \frac{1}{2}mv_F^2$:

$$v_F = \sqrt{\frac{2E_F}{m}} = \sqrt{\frac{2(1.126 \times 10^{-18}\ \text{J})}{9.11 \times 10^{-31}\ \text{kg}}} = 1.57 \times 10^6\ \text{m/s}$$

EVALUATE: Typical Fermi energies for metals range from 1.6 to 14 eV; our value is within this range, as it should be.

The quantity v_F is called the *Fermi speed*. The calculated value of v_F is far less than the speed of light $c = 3.00 \times 10^8$ m/s, which justifies our use of nonrelativistic formulas. Typical Fermi speeds for metals range from 0.8 to 2.2×10^6 m/s.

Our results show that the Fermi energy is much *larger* than kT at ordinary temperatures, so it is a good approximation to take almost all the states below E_F as completely full and almost all those above E_F as completely empty (see Fig. 42.23).

We can also use Eq. (42.16) to find $g(E)$ if E and V are known. You can show that if $E = 7$ eV and $V = 1\ \text{cm}^3$, $g(E)$ is about 2×10^{22} states/eV. This huge number shows why we were justified in treating n and E as continuous variables in our density-of-states derivation.

We can calculate the *average* free-electron energy in a metal at absolute zero by using the same ideas that we used to find E_{F0}. From Eq. (42.18) the number dN of electrons with energies in the range dE is $g(E)f(E)\,dE$. The energy of these electrons is $E\,dN = Eg(E)f(E)\,dE$. At absolute zero we substitute $f(E) = 1$ from $E = 0$ to $E = E_{F0}$ and $f(E) = 0$ for all other energies. Therefore the total energy E_{tot} of all the N electrons is

$$E_{tot} = \int_0^{E_{F0}} Eg(E)(1)\,dE + \int_{E_{F0}}^{\infty} Eg(E)(0)\,dE = \int_0^{E_{F0}} Eg(E)\,dE$$

The simplest way to evaluate this expression is to compare Eqs. (42.16) and (42.20), noting that

$$g(E) = \frac{3NE^{1/2}}{2E_{F0}^{3/2}}$$

Substituting this expression into the integral and using $E_{av} = E_{tot}/N$, we get

$$E_{av} = \frac{3}{2E_{F0}^{3/2}}\int_0^{E_{F0}} E^{3/2}\,dE = \frac{3}{5}E_{F0} \qquad (42.22)$$

That is, at absolute zero the average free-electron energy equals $\frac{3}{5}$ of the corresponding Fermi energy.

Example 42.9 Free-electron gas versus ideal gas

a) Find the average energy of the free electrons in copper at absolute zero (Example 42.8). b) If the electrons behaved as an ideal gas (Section 18.3) at room temperature, what would be their average kinetic energy? What would be the speed of an electron with this kinetic energy?

SOLUTION

IDENTIFY and SET UP: We use Eq. (42.22) to determine the average kinetic energy of free electrons. If the electrons behaved as a classical ideal gas with temperature T, from Eq. (18.16) the average kinetic energy per electron would be $E_{av} = \frac{3}{2}kT$; we use this relationship to determine the energy and the speed asked for in part (b).

EXECUTE: a) From Example 42.8, the Fermi energy in copper is 1.126×10^{-18} J = 7.03 eV. According to Eq. (42.22), the average energy is $\frac{3}{5}$ of this, or 6.76×10^{-19} J = 4.22 eV.
b) If the electrons behaved like an ideal gas at room temperature ($T = 20°C = 293$ K), the average kinetic energy per electron would be

$$E_{av} = \frac{3}{2}kT = \frac{3}{2}(1.381 \times 10^{-23}\ \text{J/K})(293\ \text{K})$$
$$= 6.07 \times 10^{-21}\ \text{J} = 0.0379\ \text{eV}$$

The speed of an electron with this kinetic energy would be

$$v = \sqrt{\frac{2E_{av}}{m}} = \sqrt{\frac{2(6.07 \times 10^{-21}\ \text{J})}{9.11 \times 10^{-31}\ \text{kg}}} = 1.15 \times 10^5\ \text{m/s}$$

EVALUATE: The average energy predicted by the ideal-gas model is less than 1% of the value given (correctly) by the free-electron model, and the corresponding speed in the ideal-gas model is only a small fraction of the characteristic Fermi speed given by the free-electron model (see Example 42.8). Thus temperature plays a *very* small role in determining the properties of electrons in metals; instead, their average energies are determined almost entirely by the exclusion principle.

A similar analysis allows us to determine the contributions of electrons to the heat capacities of a solid metal. If there is one conduction electron per atom, the principle of equipartition of energy (Section 18.4) would predict that the kinetic energies of these electrons contribute $3R/2$ to the molar heat capacity at constant volume C_V. But when kT is much smaller than E_F, which is usually the situation in metals, only those few electrons near the Fermi level can find empty states and change energy appreciably when the temperature changes. The number of such electrons is proportional to kT/E_F, so we expect that the electron molar heat capacity at constant volume is proportional to $(kT/E_F)(3R/2) = (3kT/2E_F)R$. A more detailed analysis shows that the actual electron contribution to C_V for a solid metal is $(\pi^2 kT/2E_F)R$, not far from our prediction. You can verify that if $T = 293$ K and $E_F = 7.03$ eV, the electron contribution to C_V is $0.018R$, which is only 1.2% of the (incorrect) $3R/2$ prediction of the equipartition principle. Because the electronic contribution is so small, the overall heat capacity of most solid metals is due primarily to vibration of the atoms in the crystal structure (see Fig. 18.18 in Section 18.4).

Test Your Understanding

An ideal gas obeys the relation $pV = nRT$ (Section 18.1). That is, for a given volume V and a number of moles n, as the temperature T decreases the pressure p decreases proportionately and tends to zero as T approaches absolute zero. Is the same true of the free-electron gas in a solid metal?

42.6 | Semiconductors

A **semiconductor** has an electrical resistivity that is intermediate between those of good conductors and of good insulators. The tremendous importance of semiconductors in present-day electronics stems in part from the fact that their electrical properties are very sensitive to very small concentrations of impurities. We'll discuss the basic concepts using the semiconductor elements silicon (Si) and germanium (Ge) as examples.

Silicon and germanium are in Group IV of the periodic table. Both have four electrons in the outermost atomic subshells ($3s^2 3p^2$ for silicon, $4s^2 4p^2$ for germanium), and both crystallize in the covalently bonded diamond structure discussed in Section 42.2 (Fig. 42.15). Because all four of the outer electrons are involved in the bonding, at absolute zero the band structure (Section 42.4) has a completely empty conduction band (Fig. 42.20b). As we discussed in Section 42.4, at very low temperatures electrons cannot jump from the filled valence band into the conduction band, so silicon and germanium are insulators at such temperatures. This distribution makes these materials insulators at very low temperatures; their electrons have no nearby states available into which they can move in response to an applied electric field.

However, the energy gap E_g between the valence and conduction bands is small in comparison to the gap of 5 eV or more for many insulators; room temperature values are 1.12 eV for silicon and only 0.67 eV for germanium. Thus even at room temperature a substantial number of electrons can gain enough energy to jump the gap to the conduction band, where they are dissociated from their parent atoms and are free to move about the crystal. The number of these electrons increases rapidly with temperature.

Example 42.10 Jumping the energy gap

Consider a material with the band structure described above, with its Fermi energy in the middle of the gap (Fig. 42.25). Find the probability that a state at the bottom of the conduction band is occupied at a temperature of 300 K if the band gap is a) 0.200 eV; b) 1.00 eV; c) 5.00 eV. Repeat the calculations for a temperature of 310 K.

SOLUTION

IDENTIFY and SET UP: The Fermi-Dirac distribution function gives us the probability that a state of energy E is occupied at temperature T. Figure 42.25 shows that the state of interest at the bottom of the conduction band has energy $E = E_F + E_g/2$, so $E - E_F = E_g/2$.

EXECUTE: a) When $E_g = 0.200$ eV,

$$\frac{E - E_F}{kT} = \frac{0.100 \text{ eV}}{(8.617 \times 10^{-5} \text{ eV/K})(300 \text{ K})} = 3.87$$

$$f(E) = \frac{1}{e^{3.87} + 1} = 0.0205$$

With $E_g = 0.200$ eV and $T = 310$ K, the exponent is 3.74 and $f(E) = 0.0231$, a 13% increase for a temperature rise of only 10 K.

b) When $E_g = 1.00$ eV, both exponents are five times as large as before, 19.3 and 18.7; the values of $f(E)$ are 4.0×10^{-9} and 7.4×10^{-9}. In this case the probability nearly doubles with a temperature rise of 10 K.

c) When $E_g = 5.0$ eV, the exponents are 96.7 and 93.6, the values of $f(E)$ are 1.0×10^{-42} and 2.3×10^{-41}. The probability increases by a factor of 23 for a 10-K temperature rise, but it is still extremely small. Pure diamond, with a 5.47-eV band gap, has essentially no electrons in the conduction band and is an excellent insulator.

EVALUATE: This example illustrates two important points. First, the probability of finding an electron in a state at the bottom of the conduction band is extremely sensitive to the width of the band gap. When the gap is 0.20 eV the chance is about 2%, but when it is 1.00 eV, the chance is a few in a thousand million, and for a band gap of 5.0 eV it is essentially zero. Second, for any given band gap the probability is very temperature dependent, more so for large gaps than for small.

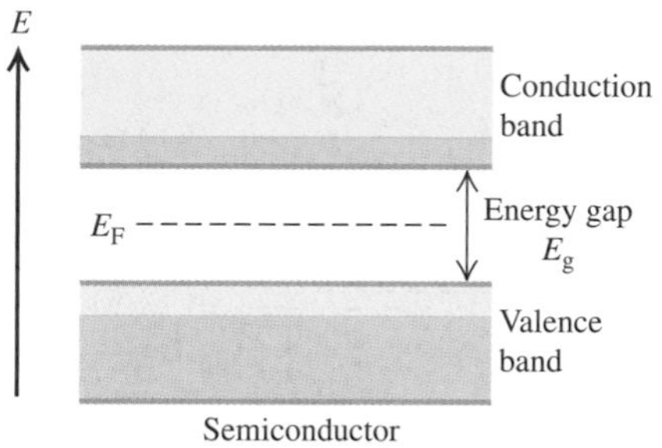

42.25 Band structure of a semiconductor. At absolute zero a completely filled valence band is separated by a narrow energy gap E_g of 1 eV or so from a completely empty conduction band. At ordinary temperatures, a number of electrons are excited to the conduction band.

In principle, we could continue the calculation in Example 42.10 to find the actual density $n = N/V$ of electrons in the conduction band at any temperature. To do this, we would have to evaluate the integral $\int g(E)f(E)\, dE$ from the bottom of the conduction band to its top. First we would need to know the density of states function $g(E)$. It wouldn't be correct to use Eq. (42.16) because the energy-level structure and the density of states for real solids are more complex than those for the simple free-electron model. However, there are theoretical methods for predicting what $g(E)$ should be near the bottom of the conduction band, and such calculations have been carried out. Once we know n, we can *begin* to determine the resistivity of the material (and its temperature dependence) using the analysis of Section 25.2, which you may want to review. But next we'll see that the electrons in the conduction band don't tell the whole story about conduction in semiconductors.

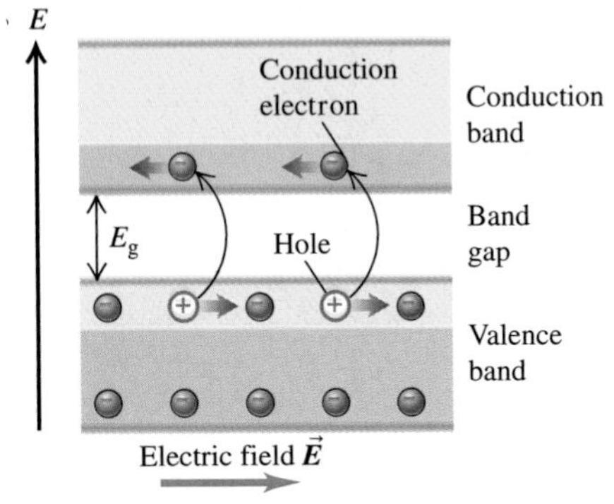

42.26 Motion of electrons in the conduction band and of holes in the valence band of a semiconductor under the action of an applied electric field $\vec{E}$.

Holes

When an electron is removed from a covalent bond, it leaves a vacancy behind. An electron from a neighboring atom can move into this vacancy, leaving the neighbor with the vacancy. In this way the vacancy, called a **hole,** can travel through the material and serve as an additional current carrier. It's like describing the motion of a bubble in a liquid. In a pure, or *intrinsic,* semiconductor, valence-band holes and conduction-band electrons are always present in equal numbers. When an electric field is applied, they move in opposite directions (Fig. 42.26). Thus a hole in the valence band behaves like a positively charged particle, even though the moving charges in that band are electrons. The conductivity that we just described for a pure semiconductor is called *intrinsic conductivity.* Another kind of conductivity, to be discussed in the next subsection, is due to impurities.

An analogy helps to picture conduction in an intrinsic semiconductor. The valence band at absolute zero is like a floor of a parking garage that's filled bumper to bumper with cars (which represent electrons). No cars can move because there is nowhere for them to go. But if one car is moved to the vacant floor above, it can move freely, just as electrons can move freely in the conduction band. Also, the empty space that it leaves permits cars to move on the nearly filled floor, thereby moving the empty space just as holes move in the normally filled valence band.

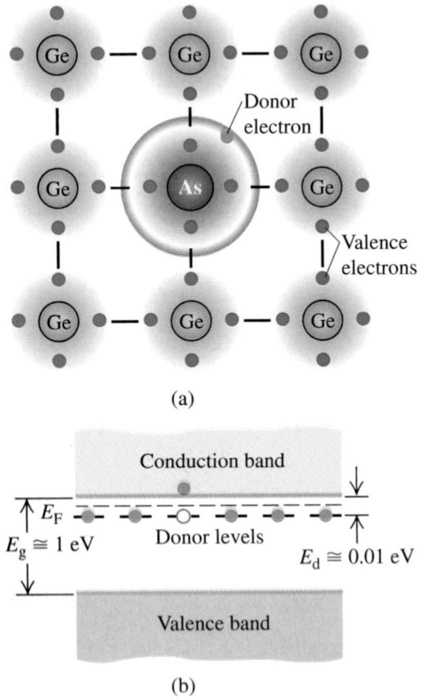

42.27 (a) A donor (n-type) impurity atom has a fifth valence electron that does not participate in the covalent bonding and is very loosely bound. (b) Energy-band diagram for an n-type semiconductor at a low temperature. One donor electron has been excited from the donor levels into the conduction band.

Impurities

Suppose we mix into melted germanium ($Z = 32$) a small amount of arsenic ($Z = 33$), the next element after germanium in the periodic table. This deliberate addition of impurity elements is called *doping.* Arsenic is in Group V; it has *five* valence electrons. When one of these electrons is removed, the remaining electron structure is essentially identical to that of germanium. The only difference is that it is smaller; the arsenic nucleus has a charge of $+33e$ rather than $+32e$, and it pulls the electrons in a little more. An arsenic atom can comfortably take the place of a germanium atom as a substitutional impurity. Four of its five valence electrons form the necessary nearest-neighbor covalent bonds.

The fifth valence electron is very loosely bound (Fig. 42.27a); it doesn't participate in the covalent bonds, and it is screened from the nuclear charge of $+33e$ by the 32 electrons, leaving a net effective charge of about $+e$. We might guess that the binding energy would be of the same order of magnitude as the energy of the $n = 4$ level in hydrogen, that is, $(\frac{1}{4})^2(13.6\text{ eV}) = 0.85$ eV. In fact, it is much smaller than this, only about 0.01 eV, because the electron probability distribution actually extends over many atomic diameters and the polarization of intervening atoms provides additional screening.

The energy level of this fifth electron corresponds in the band picture to an isolated energy level lying in the gap, about 0.01 eV below the bottom of the conduction band (Fig. 42.27b). This level is called a *donor level,* and the impurity atom that is responsible for it is simply called a *donor.* All Group V elements, including N, P, As, Sb, and Bi, can serve as donors. At room temperature, kT is about 0.025 eV. This is substantially greater than 0.01 eV, so at ordinary temperatures, most electrons can gain enough energy to jump from donor levels into the conduction band, where they are free to wander through the material. The remaining ionized donor stays at its site in the structure and does not participate in conduction.

Example 42.10 shows that at ordinary temperatures and with a band gap of 1.0 eV, only a very small fraction (of the order of 10^{-9}) of the states at the bottom of the conduction band in a pure semiconductor contain electrons to participate in intrinsic conductivity. Thus we expect the conductivity of such a semiconductor to be about 10^{-9} as great as that of good metallic conductors, and measurements bear out this prediction. However, a concentration of donors as small as one part in 10^8 can increase the conductivity so drastically that conduction due to impurities becomes by far the dominant mechanism. In this case the conductivity is due almost entirely to *negative* charge (electron) motion. We call the material an ***n*-type semiconductor,** with n-type impurities.

Adding atoms of an element in Group III (B, Al, Ga, In, Tl), with only *three* valence electrons, has an analogous effect. An example is gallium $(Z = 31)$; as a substitutional impurity in germanium, the gallium atom would like to form four covalent bonds, but it has only three outer electrons. It can, however, steal an electron from a neighboring germanium atom to complete the required four covalent bonds (Fig. 42.28a). The resulting atom has the same electron configuration as Ge but is somewhat larger because gallium's nuclear charge is smaller, $+31e$ instead of $+32e$.

This theft leaves the neighboring atom with a *hole,* or missing electron. The hole acts as a positive charge that can move through the crystal just as with intrinsic conductivity. The stolen electron is bound to the gallium atom in a level called an *acceptor level* about 0.01 eV above the top of the valence band (Fig. 42.28b). The gallium atom, called an *acceptor,* thus accepts an electron to complete its desire for four covalent bonds. This extra electron gives the previously neutral gallium atom a net charge of $-e$. The resulting gallium ion is *not* free to move. In a semiconductor that is doped with acceptors, we consider the conductivity to be almost entirely due to *positive* charge (hole) motion. We call the material a ***p*-type semiconductor,** with p-type impurities. Some semiconductors are doped with *both* n- and p-type impurities. Such materials are called *compensated* semiconductors.

CAUTION Saying that a material is a p-type semiconductor does *not* mean that the material has a positive charge; ordinarily, it would be neutral. Rather, it means that its *majority carriers* of current are positive holes (and therefore its *minority carriers* are negative electrons). The same idea holds for an n-type semiconductor; ordinarily, it will *not* have a negative charge, but its majority carriers are negative electrons.

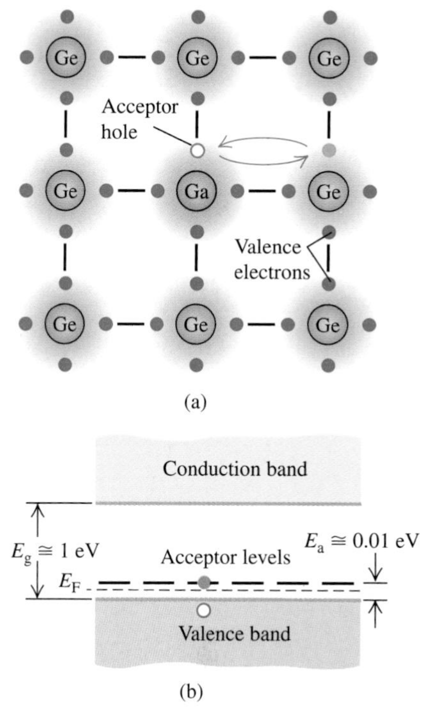

42.28 (a) An acceptor (p-type) impurity atom has only three valence electrons, so it can borrow an electron from a neighboring atom. The resulting hole is free to move about the crystal. (b) Energy-band diagram for a p-type semiconductor at a low temperature. One acceptor level has accepted an electron from the valence band, leaving a hole behind.

We can verify the assertion that the current in n and p semiconductors really *is* carried by electrons and holes, respectively, by using the Hall effect (optional Section 27.9). The sign of the Hall emf is opposite in the two cases. Hall-effect devices constructed from semiconductor materials are used in probes to measure magnetic fields and the currents that cause those fields.

Test Your Understanding

Would there be any advantage to adding n-type or p-type impurities to copper? Why or why not?

42.7 | Semiconductor Devices

Semiconductor devices play an indispensable role in contemporary electronics. In the early days of radio and television, transmitting and receiving equipment relied on vacuum tubes, but these have been almost completely replaced in the last four decades by solid-state devices, including transistors, diodes, integrated circuits, and other semiconductor devices. The only surviving vacuum tubes in consumer electronics are the picture tubes in most TV receivers and many computer monitors, but even these are gradually being replaced by flat-screen displays.

One simple semiconductor device is the *photocell*. When a thin slab of semiconductor is irradiated with an electromagnetic wave whose photons have at least as much energy as the band gap between the valence and conduction bands, an electron in the valence band can absorb a photon and jump to the conduction band, where it and the hole it left behind contribute to the conductivity (see Example 42.6 in Section 42.5). The conductivity therefore increases with wave intensity. Detectors for charged particles operate on the same principle. An external circuit applies a voltage across a semiconductor. An energetic charged particle passing through the semiconductor collides inelastically with valence electrons, exciting them from the valence to the conduction band and creating pairs of holes and conduction electrons. The conductivity increases momentarily, causing a pulse of current in the external circuit. Solid-state detectors are widely used in nuclear and high-energy physics research; without them, some of the most recent discoveries could not have been made.

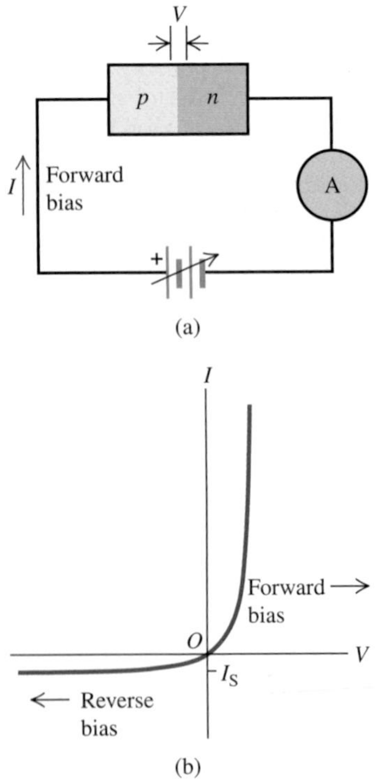

42.29 (a) A semiconductor p-n junction in a circuit. (b) Graph showing the asymmetric current-voltage relationship. The curve is described by Eq. (42.23).

The p-n Junction

In many semiconductor devices the essential principle is the fact that the conductivity of the material is controlled by impurity concentrations, which can be varied within wide limits from one region of a device to another. An example is the **p-n junction** at the boundary between one region of a semiconductor with p-type impurities and another region containing n-type impurities. One way of fabricating a p-n junction is to deposit some n-type material on the *very* clean surface of some p-type material. (We can't just stick p- and n-type pieces together and expect the junction to work properly, because of the impossibility of matching their surfaces at the atomic level.)

When a p-n junction is connected to an external circuit, as in Fig. 42.29a, and the potential difference $V_p - V_n = V$ across the junction is varied, the current I varies as shown in Fig. 42.29b. In striking contrast to the symmetrical behavior of resistors that obey Ohm's law and give a straight line on an I-V graph, a p-n junction conducts much more readily in the direction from p to n than the reverse. Such a (mostly) one-way device is called a **diode rectifier.** Later we'll discuss a simple model of p-n junction behavior that predicts a current–voltage relation in the form

$$I = I_S(e^{eV/kT} - 1) \qquad \text{(current through a } p\text{-}n \text{ junction)} \qquad (42.23)$$

In the exponent, e is the quantum of charge, k is Boltzmann's constant, and T is absolute temperature.

CAUTION In $e^{eV/kT}$ the base of the exponent also uses the symbol e, standing for the base of the natural logarithms, 2.71828 . . . . This e is quite different from $e = 1.602 \times 10^{-19}$ C in the exponent.

Equation (42.23) is valid for both positive and negative values of V; note that V and I always have the same sign. As V becomes very negative, I approaches the value $-I_S$. The magnitude I_S (always positive) is called the *saturation current.*

Example 42.11 Is a *p-n* junction diode always a one-way device?

At a temperature of 290 K, a certain *p-n* junction diode has a saturation current $I_S = 0.500$ mA. Find the current at this temperature when the voltage is 1.00 mV, -1.00 mV, 100 mV, and -100 mV.

SOLUTION

IDENTIFY and SET UP: We are given the values of the saturation current I_S and the temperature T, so we can use Eq. (42.23) to determine the current I for various values of the voltage V.

EXECUTE: At $T = 290$ K, $kT = 0.0250$ eV $= 25.0$ meV. When $V = 1.00$ mV, $eV/kT = e(1.00\text{ mV})/(25.0\text{ meV}) = 0.0400$. From Eq. (42.23) the current is

$$I = (0.500\text{ mA})(e^{0.0400} - 1) = 0.0204\text{ mA}$$

When $V = -1.00$ mV,

$$I = (0.500\text{ mA})(e^{-0.0400} - 1) = -0.0196\text{ mA}$$

The values of I for the other two voltages are obtained in the same way; when $V = 100$ mV, $I = 26.8$ mA; and when $V = -100$ mV, $I = -0.491$ mA. We summarize the data in the following table, also calculating the resistance $R = V/I$.

V (mV)	I (mA)	R (Ω)
+1.00	+0.0204	49.0
−1.00	−0.0196	51.0
+100	+26.8	3.73
−100	−0.491	204

EVALUATE: Note that at $|V| = 1.00$ mV the current has nearly the same magnitude for both directions. That is, when $|V| \ll kT/e$ (near the origin of Fig. 42.29b), the curve approaches a straight line, and this junction diode acts more like a 50.0-Ω resistor than like a rectifier. However, as the voltage increases, the directional asymmetry becomes more and more pronounced. At $|V| = 0.100$ V the negative current is nearly equal to the saturation value and has a magnitude that is less than 2% of the positive current.

Currents Through a *p-n* Junction

We can understand the behavior of a *p-n* junction diode qualitatively on the basis of the mechanisms for conductivity in the two regions. Suppose, as in Fig. 42.29a, you connect the positive terminal of the battery to the p region and the negative terminal to the n region. Then the p region is at higher potential than the n, corresponding to positive V in Eq. (42.23), and the resulting electric field is in the direction p to n. This is called the *forward* direction, and the positive potential difference is called *forward bias.* Holes, plentiful in the p region, flow easily across the junction into the n region, and free electrons, plentiful in the n region, easily flow into the p region; these movements of charge constitute a *forward* current. Connecting the battery with the opposite polarity gives *reverse bias,* and the field tends to push electrons from p to n and holes from n to p. But there are very few free electrons in the p region and very few holes in the n region. As a result, the current in the *reverse* direction is much smaller than that with the same potential difference in the forward direction.

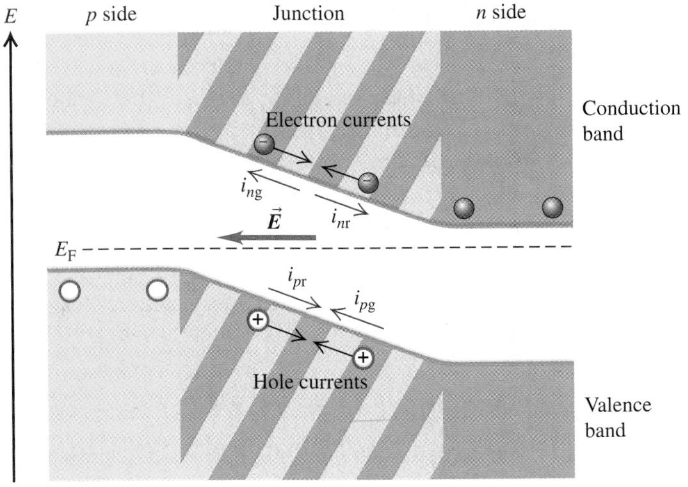

42.30 A *p*-*n* junction in equilibrium, with no externally applied field or potential difference. The generation and recombination currents exactly balance. The Fermi energy E_F is the same on both sides of the junction. There is an excess of positive charge on the *n* side and of negative charge on the *p* side, resulting in an electric field $\vec{E}$ in the direction shown. (The energy *E* increases when the electric potential decreases because electrons have a negative charge.)

Suppose you have a box with a barrier separating the left and right sides: You fill the left side with oxygen gas and the right side with nitrogen gas. What happens if the barrier leaks? Oxygen diffuses to the right, and nitrogen diffuses to the left. A similar diffusion occurs across a *p*-*n* junction. First consider the equilibrium situation with no applied voltage (Fig. 42.30). The many holes in the *p* region act like a hole gas that diffuses across the junction into the *n* region. Once there, the holes recombine with some of the many free electrons. Similarly, electrons diffuse from the *n* region to the *p*-region and fall into some of the many holes there. The hole and electron diffusion currents lead to a net positive charge in the *n* region and a net negative charge in the *p* region, causing an electric field in the direction from *n* to *p* at the junction. The potential energy associated with this field raises the electron energy levels in the *p* region relative to the same levels in the *n* region.

There are four currents across the junction, as shown. The diffusion processes lead to *recombination currents* of holes and electrons, labeled i_{pr} and i_{nr} in Fig. 42.30. At the same time, electron–hole pairs are generated in the junction region by thermal excitation. The electric field described above sweeps these electrons and holes out of the junction; electrons are swept opposite the field to the *n* side, and holes are swept in the same direction as the field to the *p* side. The corresponding currents, called *generation currents,* are labeled i_{pg} and i_{ng}. At equilibrium the magnitudes of the generation and recombination currents are equal:

$$|i_{pg}| = |i_{pr}| \quad \text{and} \quad |i_{ng}| = |i_{nr}| \tag{42.24}$$

At thermal equilibrium the Fermi energy is the same at each point across the junction.

Now we apply a forward bias, that is, a positive potential difference *V* across the junction. A forward bias *decreases* the electric field in the junction region. It also decreases the difference between the energy levels on the *p* and *n* sides (Fig. 42.31) by an amount $\Delta E = -eV$. It becomes easier for the electrons in the *n* region to climb the potential-energy hill and diffuse into the *p* region and for the holes in the *p* region to diffuse into the *n* region. This effect increases both recombination currents by the Maxwell-Boltzmann factor $e^{-\Delta E/kT} = e^{eV/kT}$. (We don't have to use the Fermi-Dirac distribution because most of the available states for the

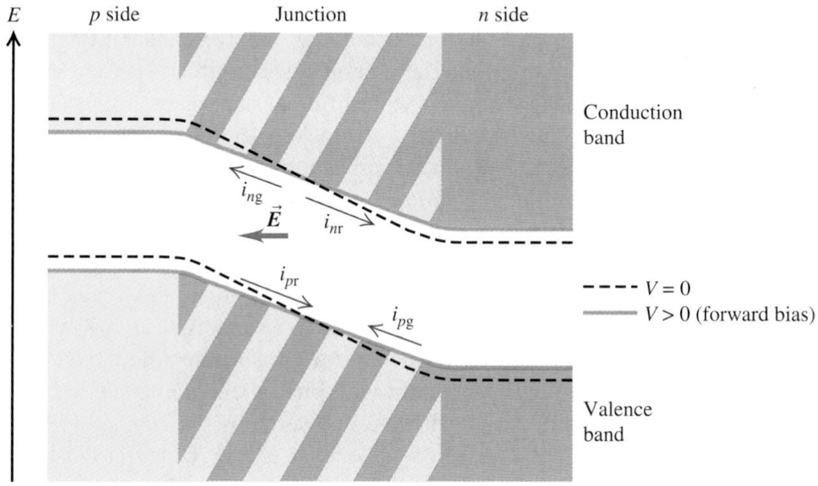

42.31 A *p*-*n* junction under forward-bias conditions. The potential difference between *p* and *n* regions is reduced, as is the electric field within the junction. The recombination currents increase but the generation currents are nearly constant, causing a net current from left to right.

diffusing electrons and holes are empty, so the exclusion principle has little effect.) The generation currents don't change appreciably, so the net hole current is

$$\begin{aligned} i_{p\text{tot}} &= i_{p\text{r}} - |i_{pg}| \\ &= |i_{pg}|e^{eV/kT} - |i_{pg}| \\ &= |i_{pg}|(e^{eV/kT} - 1) \end{aligned} \tag{42.25}$$

The net electron current $i_{n\text{tot}}$ is given by a similar expression, so the total current $I = i_{p\text{tot}} + i_{n\text{tot}}$ is

$$I = I_S(e^{eV/kT} - 1) \tag{42.26}$$

in agreement with Eq. (42.23). This entire discussion can be repeated for reverse bias (negative V and I) with the same result. Therefore Eq. (42.23) is valid for both positive and negative values.

Several effects make the behavior of practical *p*-*n* junction diodes more complex than this simple analysis predicts. One effect, *avalanche breakdown,* occurs under large reverse bias. The electric field in the junction is so great that the carriers can gain enough energy between collisions to create electron–hole pairs during inelastic collisions. The electrons and holes then gain energy and collide to form more pairs, and so on. (A similar effect occurs in dielectric breakdown in insulators, discussed in Section 42.4.)

A second type of breakdown begins when the reverse bias becomes large enough that the top of the valence band in the *p* region is just higher in energy than the bottom of the conduction band in the *n* region (Fig. 42.32). If the junction region is thin enough, the probability becomes large that electrons can *tunnel* from the valence band of the *p* region to the conduction band of the *n* region. This process is called *Zener breakdown.* It occurs in Zener diodes, which are widely used for voltage regulation and protection against voltage surges.

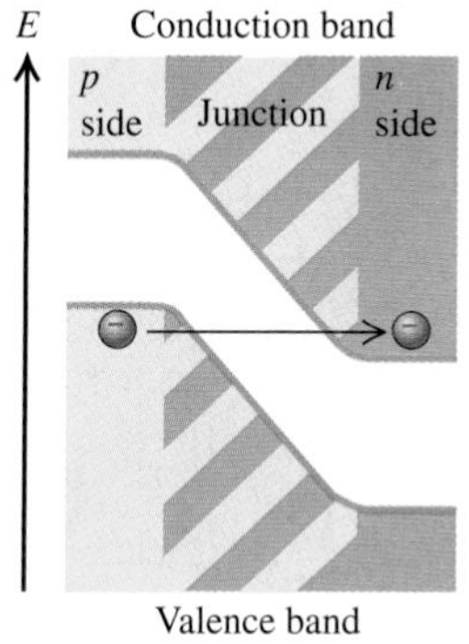

42.32 Under reverse-bias conditions the potential-energy difference between the *p* and *n* sides of a junction is greater than at equilibrium. If this difference is great enough, the bottom of the conduction band on the *n* side may actually be below the top of the valence band on the *p* side. If the junction is thin enough, electrons can tunnel from the valence band to the conduction band; this process is called Zener breakdown.

Semiconductor Devices and Light

A *light-emitting diode (LED)* is a *p*-*n* junction diode that emits light. When the junction is forward biased, many holes are pushed from their *p* region to the junction region, and many electrons are pushed from their *n* region to the junction region. In the junction region the electrons fall into holes (recombine). In recombining, the electron can emit a photon with energy approximately equal to the band gap. This energy (and therefore the photon wavelength and the color of the light) can be varied by using materials with different band gaps. Light-emitting diodes are widely used for digital displays in clocks, electronic equipment, automobile instrument panels, and many other applications.

The reverse process is called the *photovoltaic effect.* Here the material absorbs photons, and electron–hole pairs are created. Pairs that are created in the *p-n* junction, or close enough to migrate to it without recombining, are separated by the electric field we described above that sweeps the electrons to the *n* side and the holes to the *p* side. We can connect this device to an external circuit, where it becomes a source of emf and power. Such a device is often called a *solar cell,* although sunlight isn't required. *Any* light with photon energies greater than the band gap will do. You might have a calculator powered by such cells. Production of low-cost photovoltaic cells for large-scale solar energy conversion is a very active field of research. The same basic physics is used in charge-coupled device (CCD) image detectors, digital cameras, and camcorders.

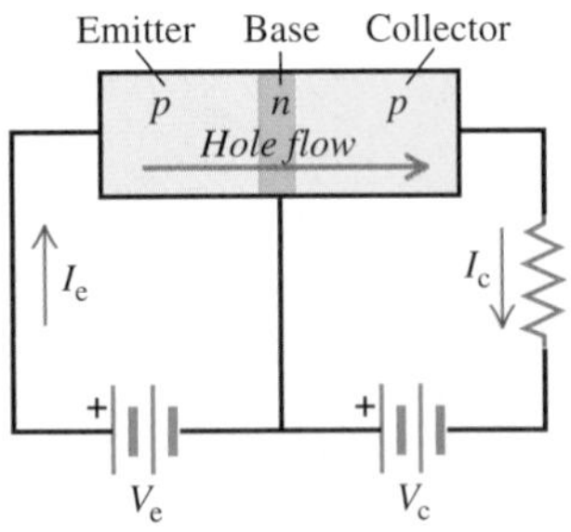

42.33 Schematic diagram of a *p-n-p* transistor and circuit. When $V_e = 0$, the current in the collector circuit is very small. When a potential V_e is applied between emitter and base, holes travel from emitter to base, as shown; when V_c is sufficiently large, most of the holes continue into the collector.

Transistors

A *bipolar junction transistor* includes two *p-n* junctions in a "sandwich" configuration, which may be either *p-n-p* or *n-p-n*. Such a *p-n-p* transistor is shown in Fig. 42.33. The three regions are called the emitter, base, and collector, as shown. When there is no current in the left loop of the circuit, there is only a very small current through the resistor R because the voltage across the base-collector junction is in the reverse direction. But when a forward bias is applied between emitter and base, as shown, most of the holes traveling from emitter to base travel *through* the base (which is typically both narrow and lightly doped) to the second junction, where they come under the influence of the collector-to-base potential difference and flow on through the collector to give an increased current to the resistor.

In this way the current in the collector circuit is *controlled* by the current in the emitter circuit. Furthermore, V_c may be considerably larger than V_e, so the *power* dissipated in R may be much larger than the power supplied to the emitter circuit by the battery V_e. Thus the device functions as a *power amplifier.* If the potential drop across R is greater than V_e, it may also be a voltage amplifier.

In this configuration the *base* is the common element between the "input" and "output" sides of the circuit. Another widely used arrangement is the *common-emitter* circuit, shown in Fig. 42.34. In this circuit the current in the collector side of the circuit is much larger than that in the base side, and the result is current amplification.

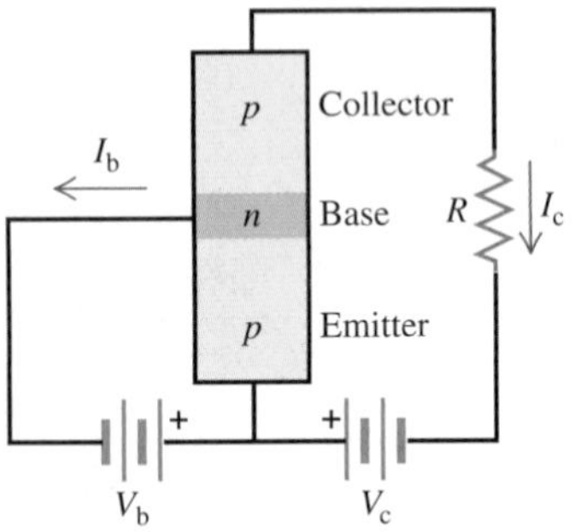

42.34 A common-emitter circuit. When $V_b = 0$, I_c is very small, and most of the voltage V_c appears across the base-collector junction. As V_b increases, the base-collector potential decreases, and more holes can diffuse into the collector; thus I_c increases. Ordinarily, I_c is much larger than I_b.

The *field-effect transistor* (Fig. 42.35) is an important type. In one variation a slab of *p*-type silicon is made with two *n*-type regions on the top, called the *source* and the *drain;* a metallic conductor is fastened to each. A third electrode called the *gate is* separated from the slab, source, and drain by an insulating layer of SiO_2. When there is no charge on the gate and a potential difference of either polarity is applied between the source and the drain, there is very little current because one of the *p-n* junctions is reverse biased.

Now we place a positive charge on the gate. With dimensions of the order of 10^{-6} m, it takes little charge to provide a substantial electric field. Thus there is very little current into or out of the gate. There aren't many free electrons in the *p*-type material, but there are some, and the effect of the field is to attract them

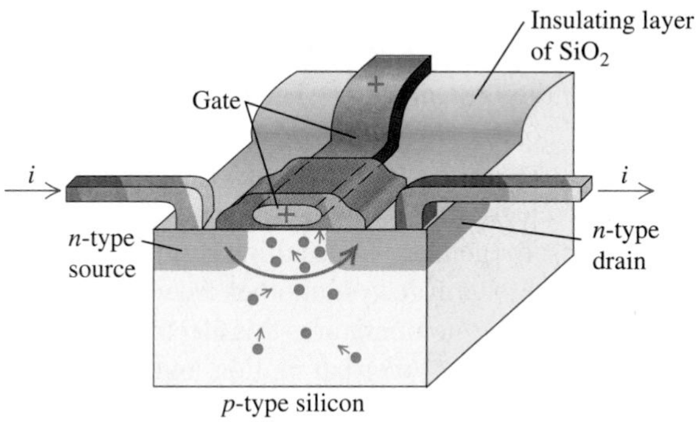

42.35 A field-effect transistor. The current from source to drain is controlled by the potential difference between the source and the drain and by the charge on the gate; no current flows through the gate.

toward the positive gate. The resulting greatly enhanced concentration of electrons near the gate (and between the two junctions) permits current to flow between the source and the drain. The current is very sensitive to the gate charge and potential, and the device functions as an amplifier. The device just described is called an *enhancement-type MOSFET* (metal-oxide-semiconductor field-effect transistor).

Integrated Circuits

A further refinement in semiconductor technology is the *integrated circuit.* By successively depositing layers of material and etching patterns to define current paths, we can combine the functions of several MOSFETs, capacitors, and resistors on a single square of semiconductor material that may be only a few millimeters on a side. An elaboration of this idea leads to *large-scale integrated circuits* and *very-large-scale integration* (VLSI). The resulting integrated circuit chips are the heart of all pocket calculators and present-day computers, large and small. An example is shown in Fig. 42.36.

The first semiconductor devices were invented in 1947. Since then, they have completely revolutionized the electronics industry through miniaturization, reliability, speed, energy usage, and cost. They have found applications in communications, computer systems, control systems, and many other areas. In transforming these areas, they have changed, and continue to change, human civilization itself.

42.36 An integrated circuit chip the size of your thumb can contain millions of transistors.

Test Your Understanding

Suppose a negative charge is placed on the gate of the MOSFET shown in Fig. 42.35. Will a substantial current flow between the source and the drain?

42.8 | Superconductivity

Superconductivity is the complete disappearance of all electrical resistance at low temperatures. We described this property at the end of Section 25.2 and the magnetic properties of type I and type II superconductors in Section 29.8. In this section we'll relate superconductivity to the structure and energy-band model of a solid.

Although superconductivity was discovered in 1911, it was not well understood on a theoretical basis until 1957. In that year, the American physicists John Bardeen, Leon Cooper, and Robert Schrieffer published the theory of superconductivity, now called the BCS theory, that was to earn them the Nobel Prize in physics in 1972. (It was Bardeen's second Nobel Prize; he shared his first for his work on the development of the transistor.) The key to the BCS theory is an interaction between *pairs* of conduction electrons, called *Cooper pairs,* caused by an interaction with the positive ions of the crystal. Here's a rough qualitative picture of what happens. A free electron exerts attractive forces on nearby positive ions, pulling them slightly closer together. The resulting slight concentration of positive charge then exerts an attractive force on another free electron with momentum opposite to the first. At ordinary temperatures this electron-pair interaction is very small in comparison to energies of thermal motion, but at very low temperatures it becomes significant.

Bound together this way, the pairs of electrons cannot *individually* gain or lose very small amounts of energy, as they would ordinarily be able to do in a partly filled conduction band. Their pairing gives an energy gap in the allowed electron quantum levels, and at low temperatures there is not enough collision energy to jump this gap. Therefore the electrons can move freely through the crystal without any energy exchange through collisions, that is, with zero resistance.

Researchers have not yet reached a consensus on whether some modification of the BCS theory can explain the properties of the high-T_C superconductors that have been discovered since 1986. There *is* evidence for pairing, but possibly of holes rather than electrons. Furthermore, the original pairing mechanism of the BCS theory seems too weak to explain the high transition temperatures and critical fields of these new superconductors.

SUMMARY

The principal types of molecular bonds are ionic, covalent, van der Waals, and hydrogen bonds. In a diatomic molecule the rotational energy levels are given by Eq. (42.3), where I is the moment of inertia of the molecule and m_r is its reduced mass. The vibrational energy levels are given by Eq. (42.7), where k' is the effective force constant of the interatomic force.
(See Examples 42.1 through 42.4)

$$E_l = l(l+1)\frac{\hbar^2}{2I} \qquad (l = 0, 1, 2, \dots) \qquad (42.3)$$

$$I = m_r r_0^2 \qquad (42.6)$$

$$m_r = \frac{m_1 m_2}{m_1 + m_2} \qquad (42.4)$$

$$E_n = \left(n + \frac{1}{2}\right)\hbar\omega = \left(n + \frac{1}{2}\right)\hbar\sqrt{\frac{k'}{m_r}} \qquad (42.7)$$

$$(n = 0, 1, 2, \dots)$$

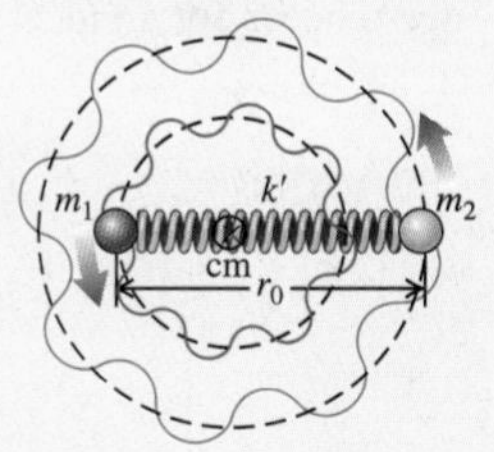

Interatomic bonds in solids are of the same types as in molecules plus one additional type, the metallic bond. Associating the basis with each lattice point gives the crystal structure. (See Example 42.5)

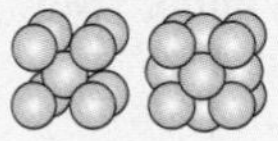

When atoms are bound together in condensed matter, their outer energy levels spread out into bands. At absolute zero, insulators and conductors have a completely filled valence band separated by an energy gap from an empty conduction band. Conductors, including metals, have partially filled conduction bands.
(See Example 42.6)

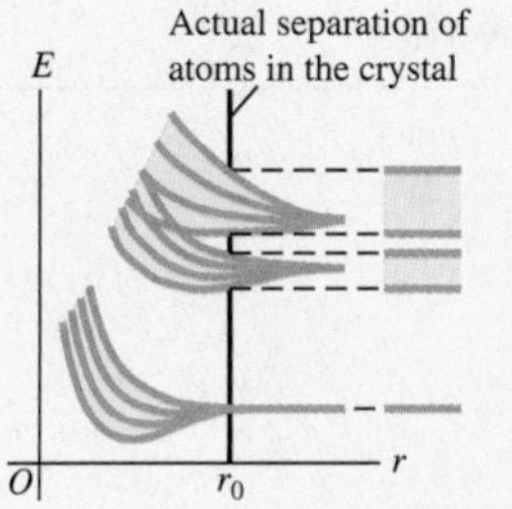

In the free-electron model of the behavior of conductors, the electrons are treated as completely free particles within the conductor. In this model the density of states is given by Eq. (42.16). The probability that an energy state of energy E is occupied is given by the Fermi-Dirac distribution, Eq. (42.17), which is a consequence of the exclusion principle. In Eq. (42.17), E_F is the Fermi energy.
(See Examples 42.7 through 42.9)

$$g(E) = \frac{(2m)^{3/2}V}{2\pi^2\hbar^3}E^{1/2} \qquad (42.16)$$

$$f(E) = \frac{1}{e^{(E-E_F)/kT} + 1} \qquad (42.17)$$

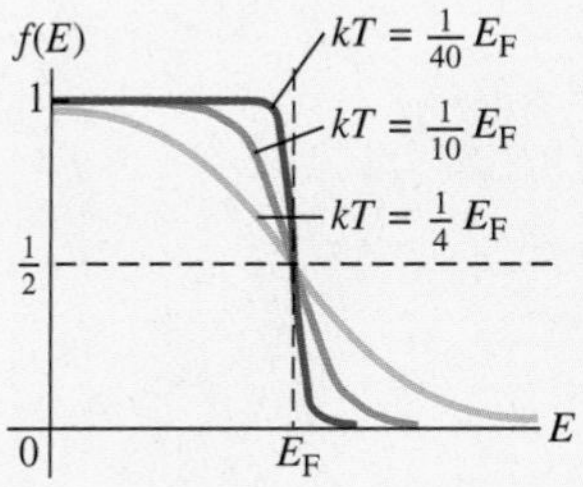

A semiconductor has an energy gap of about 1 eV between its valence and conduction bands. Its electrical properties may be drastically changed by the addition of small concentrations of donor impurities, giving an n-type semiconductor, or acceptor impurities, giving a p-type semiconductor. (See Example 42.10)

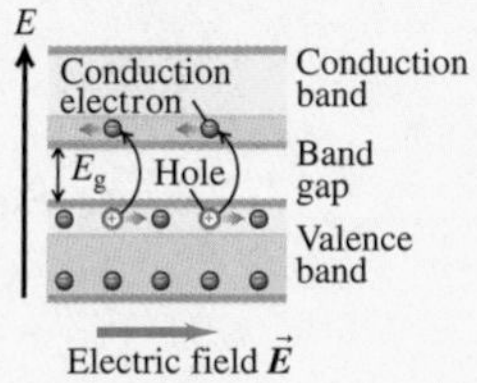

Many semiconductor devices, including diodes, transistors, and integrated circuits, use one or more p-n junctions. The current–voltage relation for an ideal p-n junction diode is given by Eq. (42.23). (See Example 42.11)

$$I = I_S(e^{eV/kT} - 1) \quad (42.23)$$

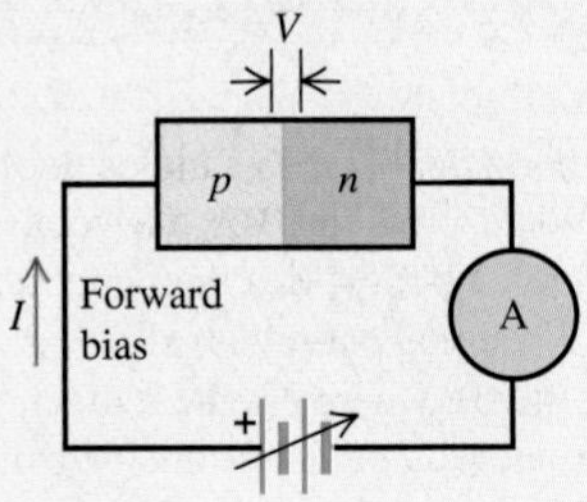

Key Terms

ionic bond, 1585
covalent bond, 1586
van der Waals bond, 1586
hydrogen bond, 1587
metallic crystal, 1596
energy band, 1597
valence band, 1598
conduction band, 1598
free-electron model, 1599
density of states, 1600
Fermi-Dirac distribution, 1601
Fermi energy, 1601
semiconductor, 1604
hole, 1606
***n*-type semiconductor, 1607**
***p*-type semiconductor, 1607**
***p-n* junction, 1608**
diode rectifier, 1608

Your Notes

Answer to Chapter Opening Question

As for a diatomic molecule, the energy levels of a CO_2 molecule have different rotational and vibrational quantum numbers. A CO_2 molecule can make a transition from one rotational-vibrational state to a higher-energy state by absorbing a photon. As is the case for most molecules, the energy associated with these transitions corresponds to a photon wavelength in the infrared part of the spectrum.

Answers to Test Your Understanding Questions

Section 42.1 The exclusion principle states that only one electron can be in a given state. Real electrons have spin, so two electrons (one spin-up, one spin-down) can be in a given *spatial* state and hence two can participate in a given covalent bond between two atoms. If electrons obeyed the exclusion principle but did not have spin, that state of an electron would be completely described by its spatial distribution and only *one* electron could participate in a covalent bond. (We will learn in Chapter 44 that this situation is wholly imaginary: there are subatomic particles without spin, but they do *not* obey the exclusion principle.)
Section 42.2 From Eq. (42.3), the $l - 1$ level has energy $E_{l-1} = (l-1)[(l-1)+1]\hbar^2/2I = l(l-1)\hbar^2/2I$ and the l level has energy $E_l = l(l+1)\hbar^2/2I$. The energy difference is $\Delta E = E_l - E_{l+1} = [l(l+1) - l(l-1)]\hbar^2/2I = l(\hbar^2/I)$, which is proportional to l. Hence the energy difference between the $l = 3$ and $l = 4$ levels is twice the difference between the $l = 1$ and $l = 2$ levels ($4\hbar^2/I$ versus $2\hbar^2/I$).
Section 42.3 If sodium chloride had the structure shown in Fig. 42.15, each positive sodium ion would be surrounded by only four negative chloride ions. The more negative charges there are near each positive charge, the more negative the potential energy and the more stable the crystal. Thus sodium chloride "prefers" the interpenetrating fcc structure shown in Fig. 42.13 because it has a lower potential energy.
Section 42.4 In sodium chloride the Na^+ and Cl^- ions both have noble gas electron configurations corresponding to completely filled bands. Since the bands do not overlap, there must be a gap between the filled bands and the empty bands above them. This is the same band structure shown in Fig. 42.20a, so NaCl is an insulator.
Section 42.5 The kinetic-molecular model of an ideal gas (Section 18.3) shows that the gas pressure is proportional to the average translational kinetic energy E_{av} of the particles that make up the gas. In a classical ideal gas, E_{av} is directly proportional to the average temperature T, so the pressure decreases as T decreases. In a free-electron gas, the average kinetic energy per electron is *not* related simply to T; as Example 42.9 shows, for the free-electron gas in a metal, E_{av} is almost completely a consequence of the exclusion principle at room temperature and colder. Hence the pressure of a free-electron gas in a solid metal does *not* change appreciably between room temperature and absolute zero.
Section 42.6 No. Pure copper is already an excellent conductor since it has a partially filled conduction band (Fig. 42.20c). Furthermore, copper forms a metallic crystal (Fig. 42.16) as opposed to the covalent crystals of silicon or germanium, so the scheme of using an impurity to donate or accept an electron does not work for copper. In fact, adding impurities to copper *decreases* the conductivity because an impurity tends to scatter electrons, impeding the flow of current.
Section 42.7 A negative charge on the gate will repel, not attract, electrons in the *p*-type silicon. Hence the electron concentration in the region between the two *p-n* junctions will be made even smaller. With so few charge carriers present in this region, very little current will flow between the source and the drain.

Discussion Questions

Q42.1 Ionic bonds result from the electrical attraction of oppositely charged particles. Are other types of molecular bonds also electrical in nature, or is some other interaction involved? Explain.
Q42.2 In ionic bonds, an electron is transferred from one atom to another and thus no longer "belongs" to the atom from which it came. Are there similar transfers of ownership of electrons with other types of molecular bonds? Explain.
Q42.3 Van der Waals bonds occur in many molecules, but hydrogen bonds occur only with materials containing hydrogen. Why is this type of bond unique to hydrogen?
Q42.4 The bonding of gallium arsenide (GaAs) is said to be 31% ionic and 69% covalent. Explain.
Q42.5 The H_2^+ molecule consists of two hydrogen nuclei and a single electron. What kind of molecular bond do you think holds this molecule together? Explain.
Q42.6 The moment of inertia for an axis through the center of mass of a diatomic molecule calculated from the wavelength emitted in an $l = 19 \rightarrow l = 18$ transition is different from the moment of inertia calculated from the wavelength of the photon emitted in an $l = 1 \rightarrow l = 0$ transition. Explain this difference. Which transition corresponds to the larger moment of inertia?
Q42.7 Analysis of the photon absorption spectrum of a diatomic molecule shows that the vibrational energy levels for small values of n are very nearly equally spaced but the levels for large n are not equally spaced. Discuss the reason for this observation. Would you expect the adjacent levels to move closer together or farther apart as n increases? Explain.
Q42.8 Discuss the differences between the rotational and vibrational energy levels of the deuterium ("heavy hydrogen") molecule D_2 and those of the ordinary hydrogen molecule H_2. A deuterium atom has twice the mass of an ordinary hydrogen atom.
Q42.9 Various organic molecules have been discovered in interstellar space. Why were these discoveries made with radio telescopes rather than optical telescopes?
Q42.10 The air you are breathing contains primarily nitrogen (N_2) and oxygen (O_2). Many of these molecules are in excited rotational energy levels ($l = 1, 2, 3, \ldots$), but almost all of them are in the vibrational ground level ($n = 0$). Explain this difference between the rotational and vibrational behavior of the molecules.
Q42.11 In what ways do atoms in a diatomic molecule behave as though they were held together by a spring? In what ways is this a poor description of the interaction between the atoms?

Q42.12 What factors determine whether a material is a conductor of electricity or an insulator? Explain.

Q42.13 Ionic crystals are often transparent, while metallic crystals are always opaque. Why?

Q42.14 Speeds of molecules in a gas vary with temperature, while speeds of electrons in the conduction band of a metal are nearly independent of temperature. Why are these behaviors so different?

Q42.15 Use the band model to explain how it is possible for some materials to undergo a semiconductor-to-metal transition as the temperature or pressure varies.

Q42.16 An isolated zinc atom has a ground-state electron configuration of filled $1s$, $2s$, $2p$, $3s$, $3p$, and $4s$ subshells. How can zinc be a conductor if its valence subshell is full?

Q42.17 Why are materials that are good thermal conductors also good electrical conductors? What kinds of problems does this pose for the design of appliances such as clothes irons and electric heaters? Are there materials that do not follow this general rule?

Q42.18 What is the essential characteristic for an element to serve as a donor impurity in a semiconductor such as Si or Ge? For it to serve as an acceptor impurity? Explain.

Q42.19 There are several methods for removing electrons from the surface of a semiconductor. Can holes be removed from the surface? Explain.

Q42.20 A student asserts that silicon and germanium become good insulators at very low temperatures and good conductors at very high temperatures. Do you agree? Explain your reasoning.

Q42.21 The electric conductivities of most metals decrease gradually with increasing temperature, but the intrinsic conductivity of semiconductors always *increases* rapidly with increasing temperature. What causes the difference?

Q42.22 How could you make compensated silicon that has twice as many acceptors as donors?

Q42.23 For electronic devices such as amplifiers, what are some advantages of transistors compared to vacuum tubes? What are some disadvantages? Are there any situations in which vacuum tubes *cannot* be replaced by solid-state devices? Explain your reasoning.

Q42.24 Why does tunneling limit the miniaturization of MOSFETs?

Q42.25 The saturation current I_S for a *p-n* junction [Eq. (42.23)] depends strongly on temperature. Explain why.

Exercises

Section 42.1 Types of Molecular Bonds

42.1 We know from Chapter 18 that the average kinetic energy of an ideal gas atom or molecule at Kelvin temperature T is $\frac{3}{2}kT$. For what value of T does this energy correspond to a) the bond energy of the van der Waals bond in He_2 $(7.9 \times 10^{-4}\ \text{eV})$; b) the bond energy of the covalent bond in H_2 (4.48 eV)? c) The kinetic energy in a collision between molecules can go into dissociating one or both molecules, provided the kinetic energy is greater than the bond energy. At room temperature (300 K), is it likely that He_2 molecules will remain intact after a collision? What about H_2 molecules? Explain.

42.2 An Ionic Bond. a) Calculate the electric potential energy for a K^+ ion and a Br^- ion separated by a distance of 0.29 nm, the equilibrium separation in the KBr molecule. Treat the ions as point charges. b) The ionization energy of the potassium atom is 4.3 eV. Atomic bromine has an electron affinity of 3.5 eV. Use these data and the results of part (a) to estimate the binding energy of the KBr molecule. Do you expect the actual binding energy to be larger or smaller than your estimate? Explain your reasoning.

Section 42.2 Molecular Spectra

42.3 In Example 42.2, the moment of inertia for CO was calculated using Eq. (42.6). a) In CO, how far is each atom from the center of mass of the molecule? b) Use $I = m_1 r_1^2 + m_2 r_2^2$ to calculate the moment of inertia of CO about an axis through the center of mass and perpendicular to the line joining the centers of the two atoms. Does your result agree with the value obtained in Example 42.2?

42.4 The water molecule has a $l = 1$ rotational level 1.01×10^{-5} eV above the $l = 0$ ground level (Section 42.2). Calculate the wavelength and frequency of the photon absorbed by water when it undergoes a rotational level transition from $l = 0$ to $l = 1$. How does your calculated frequency compare to the value given in Section 42.2 for the magnetron oscillator frequency for a microwave oven?

42.5 The rotational energy levels of CO are calculated in Example 42.2. If the energy of the rotating molecule is described by the classical expression $K = \frac{1}{2}I\omega^2$, for the $l = 1$ level what is a) the angular speed of the rotating molecule; b) the linear speed of each atom (use the result of Exercise 42.3); c) the rotational period (the time for one rotation)?

42.6 The vibrational energy levels of CO are specified in Example 42.3. a) If the energy of the vibrating molecule is described by the classical expression $E = \frac{1}{2}m_r v^2 + \frac{1}{2}k'x^2$, what is the maximum speed v_{max} for the molecule in its ground state? b) Use the value of ω for CO to calculate the vibrational period (the time for one vibration) for the ground state. c) Compare your answer in part (b) to the rotational period for the $l = 1$ state of CO that is calculated in Exercise 42.5. Which motion is faster, rotation or vibration?

42.7 A lithium atom has mass 1.17×10^{-26} kg, and a hydrogen atom has mass 1.67×10^{-27} kg. The equilibrium separation between the two nuclei in the LiH molecule is 0.159 nm. a) What is the difference in energy between the $l = 3$ and $l = 4$ rotational levels? b) What is the wavelength of the photon emitted in a transition from the $l = 4$ to $l = 3$ level?

42.8 Two atoms of cesium (Cs) can form a Cs_2 molecule. The equilibrium distance between the nuclei in a Cs_2 molecule is 0.447 nm. Calculate the moment of inertia about an axis through the center of mass of the two nuclei and perpendicular to the line joining them. The mass of a cesium atom is 2.21×10^{-25} kg.

42.9 a) Show that the energy difference between rotational levels with angular momentum quantum numbers l and $l - 1$ is $l\hbar^2/I$. b) In terms of l, $\hbar$, and I, what is the frequency of the photon emitted in the pure rotation transition $l \rightarrow l - 1$?

42.10 If a sodium chloride (NaCl) molecule could undergo an $n \rightarrow n - 1$ vibrational transition with no change in rotational quantum number, a photon with wavelength 20.0 μm would be emitted. The mass of a sodium atom is 3.82×10^{-26} kg, and the mass of a chlorine atom is 5.81×10^{-26} kg. Calculate the force constant k' for the interatomic force in NaCl.

42.11 The rotational energy levels of the CO molecule are calculated in Example 42.1 and the vibrational level energy differences are given in Example 42.3 (Section 42.2). The vibration-rotation energies are given by Eq. (42.9). Calculate the wavelength of the photon absorbed by CO in each of the following vibration-rotation transitions: a) $n = 0$, $l = 1 \rightarrow n = 1$, $l = 2$; b) $n = 0$, $l = 2 \rightarrow n = 1, l = 1$; c) $n = 0, l = 3 \rightarrow n = 1, l = 2$.

42.12 The force constant for the O_2 molecule is 1180 N/m. The mass of an oxygen atom is 2.66×10^{-26} kg. What is the energy separation in electron volts between adjacent vibrational levels of the molecule?

42.13 The vibration frequency for the molecule HF is 1.24×10^{14} Hz. The mass of a hydrogen atom is 1.67×10^{-27} kg, and the mass of a fluorine atom is 3.15×10^{-26} kg. a) What is the force constant k' for the interatomic force? b) What is the spacing between adjacent vibrational energy levels in joules and in electron volts? c) What is the wavelength of a photon of energy equal to the energy difference between two adjacent vibrational levels? In what region of the spectrum does it lay?

Section 42.3 Structure of Solids

42.14 Calculate the wavelength of a) a 6.20-keV x ray; b) a 37.6-eV electron; c) a 0.0205-eV neutron.

42.15 Density of NaCl. The spacing of adjacent atoms in a crystal of sodium chloride is 0.282 nm. The mass of a sodium atom is 3.82×10^{-26} kg, and the mass of a chlorine atom is 5.89×10^{-26} kg. Calculate the density of sodium chloride.

42.16 Potassium bromide, KBr, has a density of 2.75×10^3 kg/m^3 and the same crystal structure as NaCl. The mass of a potassium atom is 6.49×10^{-26} kg, and the mass of a bromine atom is 1.33×10^{-25} kg. a) Calculate the average spacing between adjacent atoms in a KBr crystal. b) How does the value calculated in part (a) compare with the spacing in NaCl (Exercise 42.15)? Is the relation between the two values qualitatively what you would expect? Explain.

Section 42.4 Energy Bands

42.17 The gap between valence and conduction bands in silicon is 1.12 eV. A nickel nucleus in an excited state emits a gamma-ray photon with wavelength 9.31×10^{-4} nm. How many electrons can be excited from the top of the valence band to the bottom of the conduction band by the absorption of this gamma ray?

42.18 The gap between valence and conduction bands in diamond is 5.47 eV. a) What is the maximum wavelength of a photon that can excite an electron from the top of the valence band into the conduction band? In what region of the electromagnetic spectrum does this photon lay? b) Explain why pure diamond is transparent and colorless. c) Most gem diamonds have a yellow color. Explain how impurities in the diamond can cause this color.

42.19 Consider a photocell that uses pure silicon rather than germanium (Example 42.6, Section 42.4). a) If the gap between valence and conduction bands in silicon is 1.12 eV, what maximum wavelength can be detected by a silicon photocell? In what region of the electromagnetic spectrum does this photon lie? b) Explain why pure silicon is opaque.

Section 42.5 Free-Electron Model of Metals

42.20 Calculate v_{rms} for free electrons with average kinetic energy $\frac{3}{2}kT$ at a temperature of 300 K. How does your result compare to the speed of an electron with a kinetic energy equal to the Fermi energy of copper, calculated in Example 42.9 (Section 42.5). Why is there such a difference between these speeds?

42.21 a) Show that the wave function ψ given in Eq. (42.10) is a solution of the three-dimensional Schrödinger equation [Eq. (40.29)] with the energy as given by Eq. (42.11). b) What are the energies of the ground level and the lowest two excited levels? What is the degeneracy of each of these levels? (Include the factor of 2 in the degeneracy that is due to the two possible spin states.)

42.22 What is the value of the constant A in Eq. (42.10) that makes $\psi(x, y, z)$ normalized?

42.23 Calculate the density of states $g(E)$ for the free-electron model of a metal if $E = 7.0$ eV and $V = 1.0$ cm^3. Express your answer in units of states per electron volt.

42.24 Supply the details in the derivation of Eq. (42.14) from Eqs. (42.13) and (42.12).

42.25 Calculate n_{rs} from Eq. (42.13) for $E = 7.0$ eV and $L = 1.0$ cm.

42.26 The Fermi energy of sodium is 3.23 eV. a) Find the average energy E_{av} of the electrons at absolute zero. b) What is the speed of an electron that has energy E_{av}? c) At what Kelvin temperature T is kT equal to E_{F}? (This is called the *Fermi temperature* for the metal. It is approximately the temperature at which molecules in a classical ideal gas would have the same kinetic energy as the fastest-moving electron in the metal.)

42.27 Silver has a Fermi energy of 5.48 eV. Calculate the electron contribution to the molar heat capacity at constant volume of silver, C_V, at 300 K. Express your result a) as a multiple of R; b) as a fraction of the actual value for silver, $C_V = 25.3$ J/mol·K. c) Is the value of C_V due principally to the electrons? If not, to what is it due? (*Hint:* See Section 18.4.)

Section 42.6 Semiconductors

42.28 Pure germanium has a band gap of 0.67 eV. The Fermi energy is in the middle of the gap. a) For temperatures of 250 K, 300 K, and 350 K, calculate the probability $f(E)$ that a state at the bottom of the conduction band is occupied. b) For each temperature in part (a), calculate the probability that a state at the top of the valence band is empty.

42.29 Germanium has a band gap of 0.67 eV. Doping with arsenic adds donor levels in the gap 0.01 eV below the bottom of the conduction band. At a temperature of 300 K, the probability is 4.4×10^{-4} that an electron state is occupied at the bottom of the conduction band. Where is the Fermi level relative to the conduction band in this case?

Section 42.7 Semiconductor Devices

42.30 A *p-n* junction has a saturation current of 3.60 mA. a) At a temperature of 300 K, what voltage is needed to produce a positive current of 40.0 mA? b) For a voltage equal to the negative of the value calculated in part (a), what is the negative current?

42.31 a) A forward-bias voltage of 15.0 mV produces a positive current of 9.25 mA through a *p-n* junction at 300 K. What does the positive current become if the forward-bias voltage is reduced to 10.0 mV? b) For reverse-bias voltages of −15.0 mV and −10.0 mV, what is the reverse-bias negative current?

Problems

42.32 When a diatomic molecule undergoes a transition from the $l = 2$ to the $l = 1$ rotational state, a photon with wavelength 63.8 μm is emitted. What is the moment of inertia of the molecule for an axis through its center of mass and perpendicular to the line connecting the nuclei?

42.33 a) The equilibrium separation of the two nuclei in a NaCl molecule is 0.24 nm. If the molecule is modeled as charges $+e$ and $-e$ separated by 0.24 nm, what is the electric dipole moment of the molecule (see Section 21.7)? b) The measured electric dipole moment of an NaCl molecule is 3.0×10^{-29} C·m. If this dipole moment arises from point charges $+q$ and $-q$ separated by 0.24 nm, what is q? c) A definition of the *fractional ionic character* of the bond is q/e. If the sodium atom has charge $+e$ and the chlorine atom has charge $-e$, the fractional ionic character would be equal to 1. What is the actual fractional ionic character for the bond in NaCl? d) The equilibrium distance between nuclei in the hydrogen iodide (HI) molecule is 0.16 nm, and the measured electric dipole moment of the molecule is 1.5×10^{-30} C·m. What is the fractional ionic character for the bond in HI? How does your answer compare to that for NaCl calculated in part (c)? Discuss reasons for the difference in these results.

42.34 The binding energy of a potassium chloride molecule (KCl) is 4.43 eV. The ionization energy of a potassium atom is 4.3 eV, and the electron affinity of chlorine is 3.6 eV. Use these data to estimate the equilibrium separation between the two atoms in the KCl molecule. Explain why your result is only an estimate and not a precise value.

42.35 a) For the sodium chloride molecule (NaCl) discussed at the beginning of Section 42.1, what is the maximum separation of the ions for stability if they may be regarded as point charges? That is, what is the largest separation for which the energy of an Na^+ ion and a Cl^- ion, calculated in this model, is lower than the energy of the two separate atoms Na and Cl? b) Calculate this distance for the potassium bromide molecule, described in Exercise 42.2.

42.36 The rotational spectrum of HCl contains the following wavelengths (among others): 60.4 μm, 69.0 μm, 80.4 μm, 96.4 μm, and 120.4 μm. Use this spectrum to find the moment of inertia of the HCl molecule about an axis through the center of mass and perpendicular to the line joining the two nuclei.

42.37 a) Use the result of Problem 42.36 to calculate the equilibrium separation of the atoms in an HCl molecule. The mass of a chlorine atom is 5.81×10^{-26} kg, and the mass of a hydrogen atom is 1.67×10^{-27} kg. b) The value of l changes by ± 1 in rotational transitions. What is the value of l for the upper level of the transition that gives rise to each of the wavelengths listed in Problem 42.36? c) What is the longest-wavelength line in the rotational spectrum of HCl? d) Calculate the wavelengths of the emitted light for the corresponding transitions in the deuterium chloride (DCl) molecule. In this molecule the hydrogen atom in HCl is replaced by an atom of deuterium, an isotope of hydrogen with a mass of 3.34×10^{-27} kg. Assume that the equilibrium separation between the atoms is the same as for HCl.

42.38 When a NaF molecule makes a transition from the $l = 3$ to $l = 2$ rotational level with no change in vibrational quantum number or electronic state, a photon with wavelength 3.83 mm is emitted. A sodium atom has mass 3.82×10^{-26} kg, and a fluorine atom has mass 3.15×10^{-26} kg. Calculate the equilibrium separation between the nuclei in a NaF molecule. How does your answer compare with the value for NaCl given in Section 42.1? Is this result reasonable? Explain.

42.39 Consider a gas of diatomic molecules (moment of inertia I) at an absolute temperature T. If E_g is a ground-state energy and E_{ex} is the energy of an excited state, then the Boltzmann distribution (Section 38.6) predicts that the ratio of the numbers of molecules in the two states is

$$\frac{n_{ex}}{n_g} = e^{-(E_{ex}-E_g)/kT}$$

a) Explain why the ratio of the number of molecules in the lth rotational energy *level* to the number of molecules in the ground $(l = 0)$ rotational level is

$$\frac{n_l}{n_0} = (2l + 1)e^{-[l(l+1)\hbar^2]/2IkT}$$

(*Hint:* For each value of l, how many states are there with different values of m_l?) b) Determine the ratio n_l/n_0 for a gas of CO molecules at 300 K for the cases i) $l = 1$; ii) $l = 2$; iii) $l = 10$; iv) $l = 20$; v) $l = 50$. The moment of inertia of the CO molecule is given in Example 42.2 (Section 42.2). c) Your results in part (b) show that as l is increased, the ratio n_l/n_0 first increases, then decreases. Explain why.

42.40 Our galaxy contains numerous *molecular clouds,* regions many light years in extent in which the density is high enough and the temperature low enough for atoms to form into molecules. Most of the molecules are H_2, but a small fraction of the molecules are carbon monoxide (CO). Such a molecular cloud in the constellation Orion is shown in Fig. 42.37. The left-hand image was made with an ordinary visible-light telescope; the right-hand image shows the

Figure 42.37 Problem 42.40.

molecular cloud in Orion as imaged with a radio telescope tuned to a wavelength emitted by CO in a rotational transition. The different colors in the radio image indicate regions of the cloud that are moving either toward us (blue) or away from us (red) relative to the motion of the cloud as a whole, as determined by the Doppler shift of the radiation. (Since a molecular cloud has about 10,000 hydrogen molecules for each CO molecule, it might seem more reasonable to tune a radio telescope to emissions from H_2 than to emissions from CO. Unfortunately, it turns out that the H_2 molecules in molecular clouds do not radiate in either the radio or visible portions of the electromagnetic spectrum.) a) Using the data in Example 42.2 (Section 42.2), calculate the energy and wavelength of the photon emitted by a CO molecule in an $l = 1 \rightarrow l = 0$ rotational transition. b) As a rule, molecules in a gas at temperature T will be found in a certain excited rotational energy level provided the energy of that level is no greater than kT (see Problem 42.39). Use this rule to explain why astronomers can detect radiation from CO in molecular clouds even though the typical temperature of a molecular cloud is a very low 20 K.

42.41 Spectral Lines from Isotopes. The equilibrium separation for NaCl is 0.2361 nm. The mass of a sodium atom is 3.8176×10^{-26} kg. Chlorine has two stable isotopes, ^{35}Cl and ^{37}Cl, that have different masses but identical chemical properties. The atomic mass of ^{35}Cl is 5.8068×10^{-26} kg, and the atomic mass of ^{37}Cl is 6.1384×10^{-26} kg. a) Calculate the wavelength of the photon emitted in the $l = 2 \rightarrow l = 1$ and $l = 1 \rightarrow l = 0$ transitions for $Na^{35}Cl$. b) Repeat part (a) for $Na^{37}Cl$. What are the differences in the wavelengths for the two isotopes?

42.42 When an OH molecule undergoes a transition from the $n = 0$ to the $n = 1$ vibrational level, its internal vibrational energy increases by 0.463 eV. Calculate the frequency of vibration and the force constant for the interatomic force. (The mass of an oxygen atom is 2.66×10^{-26} kg, and the mass of a hydrogen atom is 1.67×10^{-27} kg.)

42.43 The force constant for the internuclear force in a hydrogen molecule (H_2) is $k' = 576$ N/m. A hydrogen atom has mass 1.67×10^{-27} kg. Calculate the zero-point vibrational energy for H_2 (that is, the vibrational energy the molecule has in the $n = 0$ ground vibrational level). How does this energy compare in magnitude with the H_2 bond energy of -4.48 eV?

42.44 Suppose the hydrogen atom in HF (see Exercise 42.13) is replaced by an atom of deuterium, an isotope of hydrogen with a mass of 3.34×10^{-27} kg. The force constant is determined by the electron configuration, so it is the same as for the normal HF molecule. a) What is the vibrational frequency of this molecule? b) What wavelength of light corresponds to the energy difference between the $n = 1$ and $n = 0$ levels? In what region of the spectrum does this wavelength lay?

42.45 The hydrogen iodide (HI) molecule has equilibrium separation 0.160 nm and vibrational frequency 6.93×10^{13} Hz. The mass of a hydrogen atom is 1.67×10^{-27} kg, and the mass of an iodine atom is 2.11×10^{-25} kg. a) Calculate the moment of inertia of HI about a perpendicular axis through its center of mass. b) Calculate the wavelength of the photon emitted in each of the following vibration-rotation transitions: i) $n = 1, l = 1 \rightarrow n = 0, l = 0$; ii) $n = 1, l = 2 \rightarrow n = 0, l = 1$; iii) $n = 2, l = 2 \rightarrow n = 1, l = 3$.

42.46 Prove the following statement: For free electrons in a solid, if a state that is at an energy ΔE above E_F has probability P of being occupied, then the probability is $1 - P$ that a state at an energy ΔE below E_F is occupied.

42.47 Compute the Fermi energy of potassium by making the simple approximation that each atom contributes one free electron. The density of potassium is 851 kg/m^3, and the mass of a single potassium atom is 6.49×10^{-26} kg.

42.48 Neutron Star. The mass density of a typical neutron star is about 7.0×10^{17} kg/m^3 (Problem 9.98). In a neutron star, the energy levels are occupied by neutrons instead of electrons. Calculate the zero-temperature Fermi energy E_{F0} and the Fermi temperature (Exercise 42.26) for a neutron star of this density.

42.49 Metallic lithium has a bcc crystal structure. Each unit cell is a cube of side length $a = 0.35$ nm. a) For a bcc lattice, what is the number of atoms per unit volume? Give your answer in terms of a. (*Hint*: How many atoms are there per unit cell?) b) Use the result of part (a) to calculate the zero-temperature Fermi energy E_{F0} for metallic lithium. Assume there is one free electron per atom.

42.50 The one-dimensional calculation of Example 42.5 (Section 42.3) can be extended to three dimensions. For the three-dimensional fcc NaCl lattice, the result for the potential energy of a pair of Na^+ and Cl^- ions due to the electrostatic interaction with all of the ions in the crystal is $U = -\alpha e^2/4\pi\epsilon_0 r$, where $\alpha = 1.75$ is the *Madelung constant*. Another contribution to the potential energy is a repulsive interaction at small ionic separation r due to overlap of the electron clouds. This contribution can be represented by A/r^8, where A is a positive constant, so the expression for the total potential energy is

$$U_{tot} = -\frac{\alpha e^2}{4\pi\epsilon_0 r} + \frac{A}{r^8}$$

a) Let r_0 be the value of the ionic separation r for which U_{tot} is a minimum. Use this definition to find an equation that relates r_0 and A, and use this to write U_{tot} in terms of r_0. For NaCl, $r_0 = 0.281$ nm. Obtain a numerical value (in electron volts) of U_{tot} for NaCl. b) The quantity $-U_{tot}$ is the energy required to remove a Na^+ ion and a Cl^- ion from the crystal. To form a pair of neutral atoms from this pair of ions involves a release of 5.14 eV (the ionization energy of Na) and the expenditure of 3.61 eV (the electron affinity of Cl). Use the result of part (a) to calculate the energy required to remove a pair of neutral Na and Cl atoms from the crystal. The experimental value for this quantity is 6.39 eV; how well does your calculation agree?

42.51 Consider a system of N free electrons within a volume V. Even at absolute zero, such a system exerts a pressure p on its surroundings due to the motion of the electrons. To calculate this pressure, imagine that the volume increases by a small amount dV. The electrons will do an amount of work pdV on their surroundings, which means that the total energy E_{tot} of the electrons will change by an amount $dE_{tot} = -pdV$. Hence $p = -dE_{tot}/dV$. a) Show that the pressure of the electrons at absolute zero is

$$p = \frac{3^{2/3}\pi^{4/3}\hbar^2}{5m}\left(\frac{N}{V}\right)^{5/3}$$

b) Evaluate this pressure for copper, which has a free-electron concentration of 8.45×10^{28} m^{-3}. Express your result in pascals and in atmospheres. c) The pressure you found in part (b) is extremely high. Why, then, don't the electrons in a piece of copper simply explode out of the metal?

42.52 When the pressure p on a material increases by an amount Δp, the volume of the material will change from V to $V + \Delta V$, where ΔV is negative. The *bulk modulus B* of the material is defined to be the ratio of the pressure change Δp to the absolute value $|\Delta V/V|$ of the fractional volume change. The greater the bulk modulus, the greater the pressure increase required for a given fractional volume change, and the more incompressible the material (Section 11.4). Since $\Delta V < 0$, the bulk modulus can be written as $B = -\Delta p/(\Delta V/V_0)$. In the limit that the pressure and volume changes are very small, this becomes

$$B = -V\frac{dp}{dV}$$

a) Use the result of Problem 42.51 to show that the bulk modulus for a system of N free electrons in a volume V at low temperatures is $B = \frac{5}{3}p$. [*Hint:* The quantity p in the expression $B = -V(dp/dV)$ is the *external* pressure on the system. Can you explain why this is equal to the *internal* pressure of the system itself, as found in Problem 42.51?] b) Evaluate the bulk modulus for the electrons in copper, which has a free-electron concentration of $8.45 \times 10^{28}\ \text{m}^{-3}$. Express your result in pascals. c) The actual bulk modulus of copper is 1.4×10^{11} Pa. Based on your result in part (b), what fraction of this is due to the free electrons in copper? (This result shows that the free electrons in a metal play a major role in making the metal resistant to compression.) What do you think is responsible for the remaining fraction of the bulk modulus?

42.53 In the discussion of free electrons in Section 42.5, we assumed that we could ignore the effects of relativity. This is not a safe assumption if the Fermi energy is more than about $\frac{1}{100}\ mc^2$ (that is, more than about 1% of the rest energy of an electron). a) Assume that the Fermi energy at absolute zero, as given by Eq. (42.20), is equal to $\frac{1}{100}\ mc^2$. Show that the electron concentration is

$$\frac{N}{V} = \frac{2^{3/2}m^3c^3}{3000\pi^2\hbar^3}$$

and determine the numerical value of N/V. b) Is it a good approximation to ignore relativistic effects for electrons in a metal such as copper, for which the electron concentration is $8.45 \times 10^{28}\ \text{m}^{-3}$? Explain. c) A *white dwarf star* is what is left behind by a star like the sun after it has ceased to produce energy by nuclear reactions. (Our own sun will become a white dwarf in another 6×10^9 years or so.) A typical white dwarf has mass 2×10^{30} kg (comparable to the sun) and a radius of 6000 km (comparable to that of the earth). The gravitational attraction of different parts of the white dwarf for each other tends to compress the star; what prevents it from compressing is the pressure of free electrons within the star (see Problem 42.51). Estimate the electron concentration within a typical white dwarf using the following assumptions: i) the white dwarf is made of carbon, which has a mass per atom of 1.99×10^{-26} kg; ii) all six of the electrons from each carbon atom are able to move freely throughout the star. d) Is it a good approximation to ignore relativistic effects in the structure of a white dwarf star? Explain.

Challenge Problems

42.54 a) Consider the hydrogen molecule (H_2) to be a simple harmonic oscillator with an equilibrium spacing of 0.074 nm, and estimate the vibrational energy-level spacing for H_2. The mass of a hydrogen atom is 1.67×10^{-27} kg. (*Hint:* Estimate the force constant by equating the change in Coulomb repulsion of the protons, when the atoms move slightly closer together than r_0, to the "spring" force. That is, assume that the chemical binding force remains approximately constant as r is decreased slightly from r_0.) b) Use the results of part (a) to calculate the vibrational energy level spacing for the deuterium molecule, D_2. Assume that the spring constant is the same for H_2 and D_2. The mass of a deuterium atom is 3.34×10^{-27} kg.

42.55 Van der Waals bonds arise from the interaction between two permanent or induced electric dipole moments in a pair of atoms or molecules. a) Consider two identical dipoles, each consisting of charges $+q$ and $-q$ separated by a distance d and oriented as shown in Fig. 42.38a. Calculate the electric potential energy, expressed in terms of the electric dipole moment $p = qd$, for the situation where $r \gg d$. Is the interaction attractive or repulsive, and how does this potential energy vary with r, the separation between the centers of the two dipoles? b) Repeat part (a) for the orientation of the dipoles shown in Fig. 42.38b. The dipole interaction is more complicated when we have to average over the relative orientations of the two dipoles due to thermal motion or when the dipoles are induced rather than permanent.

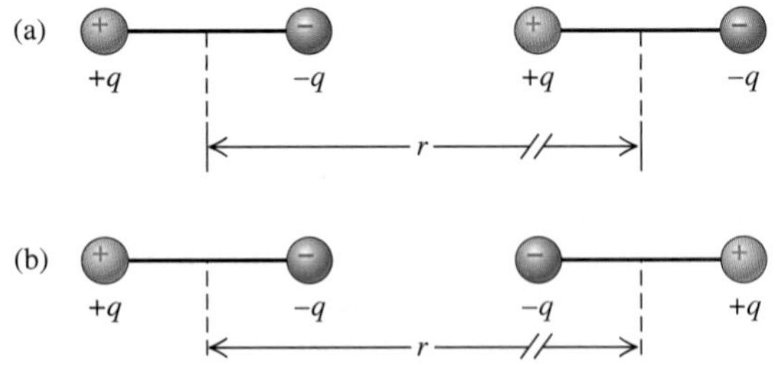

Fig. 42.38 Challenge Problem 42.55.

NUCLEAR PHYSICS 43

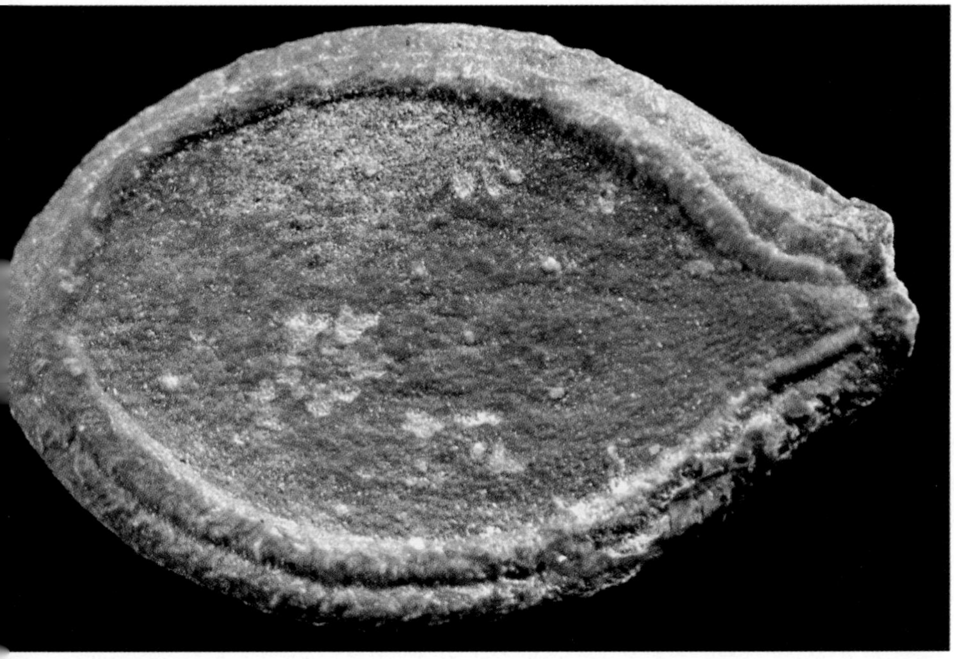

Using radioactive dating techniques, scientists have learned that this domesticated squash seed—one of several found in a cave in Mexico—is 8,000–10,000 years old. This discovery shows that agriculture began in the Americas at least as early as it did in Europe and Asia. This is just one example of how nuclear physics has become an essential tool for archaeology.

? How do physicists use radioactive dating to determine the age of biological specimens?

During the past century, applications of nuclear physics have had enormous effects on humankind, some beneficial, some catastrophic. Many people have strong opinions about applications such as bombs and reactors. Ideally, those opinions should be based on understanding, not on prejudice or emotion, and we hope this chapter will help you to reach that ideal.

Every atom contains at its center an extremely dense, positively charged *nucleus,* which is much smaller than the overall size of the atom but contains most of its total mass. We will look at several important general properties of nuclei and of the nuclear force that holds them together. The stability or instability of a particular nucleus is determined by the competition between the attractive nuclear force among the protons and neutrons and the repulsive electrical interactions among the protons. Unstable nuclei *decay,* transforming themselves spontaneously into other structures, by a variety of decay processes. Structure-altering nuclear reactions can also be induced by impact on a nucleus of a particle or another nucleus. Two classes of reactions of special interest are *fission* and *fusion.* We could not survive without the 3.90×10^{26}-watt output of one nearby fusion reactor, our sun.

43.1 | Properties of Nuclei

As we described in Section 38.4, Rutherford found that the nucleus is tens of thousands of times smaller in radius than the atom itself. Since Rutherford's initial experiments, many additional scattering experiments have been performed, using high-energy protons, electrons, and neutrons as well as alpha particles (helium-4 nuclei). These experiments show that we can model a nucleus as a sphere with a radius R that depends on the total number of *nucleons* (neutrons and protons) in the nucleus. This number is called the **nucleon number,** A. The radii of most nuclei are represented quite well by the equation

$$R = R_0 A^{1/3} \quad \text{(radius of a nucleus)} \tag{43.1}$$

where R_0 is an experimentally determined constant:

$$R_0 = 1.2 \times 10^{-15}\ \text{m} = 1.2\ \text{fm}$$

The nucleon number A in Eq. (43.1) is also called the **mass number** because it is the nearest whole number to the mass of the nucleus measured in unified atomic mass units (u). (The proton mass and the neutron mass are both approximately 1 u.) The best current conversion factor is

$$1\ \text{u} = 1.66053873(13) \times 10^{-27}\ \text{kg}$$

In Section 43.2 we'll discuss the masses of nuclei in more detail. Note that when we speak of the masses of nuclei and particles, we mean their *rest* masses.

Nuclear Density

The volume V of a sphere is equal to $4\pi R^3/3$, so Eq. (43.1) shows that the *volume* of a nucleus is proportional to A. Dividing A (the approximate mass in u) by the volume gives us the approximate density and cancels out A. Thus *all nuclei have approximately the same density.* This fact is of crucial importance in understanding nuclear structure.

Example 43.1 Calculating nuclear properties

The most common kind of iron nucleus has a mass number of 56. Find the radius, approximate mass, and approximate density of the nucleus.

SOLUTION

IDENTIFY and SET UP: We use Eq. (43.1) to determine the radius of the nucleus. The mass of the nucleus in atomic mass units is approximately equal to the mass number, and the density is the mass divided by the volume.

EXECUTE: The radius is

$$R = R_0 A^{1/3} = (1.2 \times 10^{-15}\ \text{m})(56)^{1/3}$$
$$= 4.6 \times 10^{-15}\ \text{m} = 4.6\ \text{fm}$$

Since $A = 56$, the mass of the nucleus is approximately 56 u, or

$$m \approx (56)(1.66 \times 10^{-27}\ \text{kg}) = 9.3 \times 10^{-26}\ \text{kg}$$

The volume is

$$V = \frac{4}{3}\pi R^3 = \frac{4}{3}\pi(4.6 \times 10^{-15}\ \text{m})^3 = 4.1 \times 10^{-43}\ \text{m}^3$$

and the density ρ is approximately

$$\rho = \frac{m}{V} \approx \frac{9.3 \times 10^{-26}\ \text{kg}}{4.1 \times 10^{-43}\ \text{m}^3} = 2.3 \times 10^{17}\ \text{kg/m}^3$$

EVALUATE: The density of solid iron is about 7000 kg/m^3, so we see that the nucleus is over 10^{13} times as dense as the bulk material. Densities of this magnitude are also found in *neutron stars,* which are similar to gigantic nuclei made almost entirely of neutrons. A 1-cm cube of material with this density would have a mass of 2.3×10^{11} kg, or 230 million metric tons!

Nuclides and Isotopes

The basic building blocks of the nucleus are the proton and the neutron. In a neutral atom, the nucleus is surrounded by one electron for every proton in the nucleus. We introduced these particles in Section 21.2; we recount the discovery of the neutron in Chapter 44. The masses of these particles are

Proton: $m_p = 1.007276\ \text{u} = 1.672622 \times 10^{-27}\ \text{kg}$

Neutron: $m_n = 1.008665\ \text{u} = 1.674927 \times 10^{-27}\ \text{kg}$

Electron: $m_e = 0.000548580\ \text{u} = 9.10938 \times 10^{-31}\ \text{kg}$

The number of protons in a nucleus is the **atomic number** Z. The number of neutrons is the **neutron number** N. The nucleon number or mass number A is the sum of the number of protons Z and the number of neutrons N,

$$A = Z + N \tag{43.2}$$

A single nuclear species having specific values of both Z and N is called a **nuclide.** Table 43.1 lists values of A, Z, and N for a few nuclides. The electron structure of an atom, which is responsible for its chemical properties, is determined by the charge Ze of the nucleus. The table shows some nuclides that have the same Z but different N. These nuclides are called **isotopes** of that element; they have different masses because they have different numbers of neutrons in their nuclei. A familiar example is chlorine (Cl, $Z = 17$). About 76% of chlorine nuclei have $N = 18$; the other 24% have $N = 20$. Different isotopes of an element usually have slightly different physical properties such as melting and boiling temperatures and diffusion rates. The two common isotopes of uranium with $A = 235$ and 238 are usually separated industrially by taking advantage of the

Table 43.1 Compositions of Some Common Nuclides

Nucleus	Mass Number (Total Number of Nucleons), A	Atomic Number (Number of Protons), Z	Neutron Number, $N = A - Z$
$^{1}_{1}H$	1	1	0
$^{2}_{1}D$	2	1	1
$^{4}_{2}He$	4	2	2
$^{6}_{3}Li$	6	3	3
$^{7}_{3}Li$	7	3	4
$^{9}_{4}Be$	9	4	5
$^{10}_{5}B$	10	5	5
$^{11}_{5}B$	11	5	6
$^{12}_{6}C$	12	6	6
$^{13}_{6}C$	13	6	7
$^{14}_{7}N$	14	7	7
$^{16}_{8}O$	16	8	8
$^{23}_{11}Na$	23	11	12
$^{65}_{29}Cu$	65	29	36
$^{200}_{80}Hg$	200	80	120
$^{235}_{92}U$	235	92	143
$^{238}_{92}U$	238	92	146

Table 43.2 Neutral Atomic Masses for Some Light Nuclides

Element and Isotope	Atomic Number Z	Neutron Number N	Atomic Mass (u)	Mass Number A
Hydrogen ($^{1}_{1}H$)	1	0	1.007825	1
Deuterium ($^{2}_{1}H$)	1	1	2.014102	2
Tritium ($^{3}_{1}H$)	1	2	3.016049	3
Helium ($^{3}_{2}He$)	2	1	3.016029	3
Helium ($^{4}_{2}He$)	2	2	4.002603	4
Lithium ($^{6}_{3}Li$)	3	3	6.015122	6
Lithium ($^{7}_{3}Li$)	3	4	7.016004	7
Beryllium ($^{9}_{4}Be$)	4	5	9.012182	9
Boron ($^{10}_{5}B$)	5	5	10.012937	10
Boron ($^{11}_{5}B$)	5	6	11.009305	11
Carbon ($^{12}_{6}C$)	6	6	12.000000	12
Carbon ($^{13}_{6}C$)	6	7	13.003355	13
Nitrogen ($^{14}_{7}N$)	7	7	14.003074	14
Nitrogen ($^{15}_{7}N$)	7	8	15.000109	15
Oxygen ($^{16}_{8}O$)	8	8	15.994915	16
Oxygen ($^{17}_{8}O$)	8	9	16.999132	17
Oxygen ($^{18}_{8}O$)	8	10	17.999160	18

Source: A. H. Wapstra and G. Audi, *Nuclear Physics* **A595,** 4 (1995).

different diffusion rates of gaseous uranium hexafluoride (UF_6) containing the two isotopes.

Table 43.1 also shows the usual notation for individual nuclides: the symbol of the element, with a pre-subscript equal to Z and a pre-superscript equal to the mass number A. The general format for an element El is $^{A}_{Z}El$. The isotopes of chlorine mentioned above, with $A = 35$ and 37, are written $^{35}_{17}Cl$ and $^{37}_{17}Cl$ and pronounced "chlorine-35" and "chlorine-37," respectively. This name of the element determines the atomic number Z, so the pre-subscript Z is sometimes omitted, as in ^{35}Cl.

Table 43.2 gives the masses of some common atoms, including their electrons. This table gives masses of *neutral* atoms (with Z electrons) rather than masses of *bare* nuclei, because it is much more difficult to measure masses of bare nuclei with high precision. The mass of a neutral carbon-12 atom is exactly 12 u; that's how the unified atomic mass unit is defined. The masses of other atoms are *approximately* equal to A atomic mass units, as we stated earlier. You may notice that the atomic masses are *less* than the sum of the masses of their parts (the Z protons, the Z electrons, and the N neutrons). We'll explain this very important mass difference in the next section.

Nuclear Spins and Magnetic Moments

Like electrons, protons and neutrons are also spin-$\frac{1}{2}$ particles with spin angular momentum given by the same equations as in Section 41.3. The magnitude of the spin angular momentum $\vec{S}$ of a nucleon is

$$S = \sqrt{\frac{1}{2}\left(\frac{1}{2} + 1\right)}\hbar = \sqrt{\frac{3}{4}}\hbar \tag{43.3}$$

and the z-component is

$$S_z = \pm\frac{1}{2}\hbar \tag{43.4}$$

In addition to the spin angular momentum of the nucleons, there may be *orbital* angular momentum associated with their motions within the nucleus. The orbital angular momentum of the nucleons is quantized in the same way as that of electrons in atoms.

The *total* angular momentum $\vec{J}$ of the nucleus is the vector sum of the individual spin and orbital angular momenta of all the nucleons. It has magnitude

$$J = \sqrt{j(j+1)}\hbar \tag{43.5}$$

and z-component

$$J_z = m_j\hbar \qquad (m_j = -j, -j+1, \ldots, j-1, j) \tag{43.6}$$

When the total number of nucleons A is *even*, j is an integer; when it is *odd*, j is a half-integer. All nuclides for which both Z and N are even have $J = 0$, which suggests that pairing of particles with opposite spin components may be an important consideration in nuclear structure. The total nuclear angular momentum quantum number j is usually called the *nuclear spin*, even though in general it refers to a combination of the orbital and spin angular momenta of the nucleons that make up the nucleus.

Associated with nuclear angular momentum is a *magnetic moment*. When we discussed *electron* magnetic moments in Section 41.2, we introduced the Bohr magneton $\mu_B = e\hbar/2m_e$ as a natural unit of magnetic moment. We found that the magnitude of the z-component of the electron-spin magnetic moment is almost exactly equal to μ_B. That is, $|\mu_{sz}|_{\text{electron}} \approx \mu_B$. In discussing *nuclear* magnetic moments, we can define an analogous quantity, the **nuclear magneton** μ_n:

$$\mu_n = \frac{e\hbar}{2m_p} = 5.05079 \times 10^{-27}\ \text{J/T} = 3.15245 \times 10^{-8}\ \text{eV/T} \tag{43.7}$$

(nuclear magneton)

where m_p is the proton mass. Because the proton mass m_p is 1836 times larger than the electron mass m_e, the nuclear magneton μ_n is 1836 times smaller than the Bohr magneton μ_B.

We might expect the magnitude of the z-component of the spin magnetic moment of the proton to be approximately μ_n. Instead, it turns out to be

$$|\mu_{sz}|_{\text{proton}} = 2.7928\mu_n \tag{43.8}$$

Even more surprising, the neutron, which has no charge, has a corresponding magnitude of

$$|\mu_{sz}|_{\text{neutron}} = 1.9130\mu_n \tag{43.9}$$

The proton has a positive charge; as expected, its spin magnetic moment $\vec{\mu}$ is parallel to its spin angular momentum $\vec{S}$. However, $\vec{\mu}$ and $\vec{S}$ are opposite for a neutron, as would be expected for a negative charge distribution. These *anomalous* magnetic moments arise because the proton and neutron aren't really fundamental particles, but are made of simpler particles called *quarks*. We'll discuss quarks in some detail in the next chapter.

The magnetic moment of an entire nucleus is typically a few nuclear magnetons. When a nucleus is placed in an external magnetic field $\vec{B}$, there is an interaction energy $U = -\vec{\mu} \cdot \vec{B} = -\mu_z B$ just as with atomic magnetic moments. The components of the magnetic moment in the direction of the field μ_z are quantized, so a series of energy levels results from this interaction.

Example 43.2 Proton spin flips

Protons are placed in a magnetic field in the z-direction with magnitude 2.30 T. a) What is the energy difference between a state with the z-component of proton spin angular momentum parallel to the field and one with the component antiparallel to the field? b) A proton can make a transition from one of these states to the other by emitting or absorbing a photon with energy equal to the energy difference of the two states. Find the frequency and wavelength of such a photon.

SOLUTION

IDENTIFY and SET UP: The magnetic field $\vec{B}$ is assumed to be in the positive z-direction, and the magnetic moment $\vec{\mu}$ of the proton is in the same direction as its spin. If the z-component of spin is aligned with $\vec{B}$, μ_z is equal to the positive value given in Eq. (43.8); if the z-component of spin is opposite $\vec{B}$, then μ_z is the negative of this value. The interaction energy in either case is $U = -\mu_z B$, and the energy difference [our target variable in part (a)] is the difference between the values of U for the two spin orientations. We find the photon frequency and wavelength using the relationships $E = hf = hc/\lambda$.

EXECUTE: a) When the z-component of $\vec{S}$ (and $\vec{\mu}$) is parallel to the field, the interaction energy is

$$U = -|\mu_z|B = -(2.7928)(3.152 \times 10^{-8}\ \text{eV/T})(2.30\ \text{T})$$
$$= -2.025 \times 10^{-7}\ \text{eV}$$

When the components are antiparallel to the field, the energy is $+2.025 \times 10^{-7}$ eV, and the energy *difference* between the two states is

$$\Delta E = 2(2.025 \times 10^{-7}\ \text{eV}) = 4.05 \times 10^{-7}\ \text{eV}$$

b) The corresponding photon frequency and wavelength are

$$f = \frac{\Delta E}{h} = \frac{4.05 \times 10^{-7}\ \text{eV}}{4.136 \times 10^{-15}\ \text{eV}\cdot\text{s}} = 9.79 \times 10^{7}\ \text{Hz} = 97.9\ \text{MHz}$$

$$\lambda = \frac{c}{f} = \frac{3.00 \times 10^{8}\ \text{m/s}}{9.79 \times 10^{7}\ \text{s}^{-1}} = 3.06\ \text{m}$$

EVALUATE: This frequency is in the middle of the FM radio band. When a hydrogen specimen is placed in a 2.30-T magnetic field and then irradiated with radiation of this frequency, the proton *spin flips* can be detected by the absorption of energy from the radiation.

Spin-flip experiments of the sort referred to in Example 43.2 are called *nuclear magnetic resonance* (NMR). They have been carried out with many different nuclides. Frequencies and magnetic fields can be measured very precisely, so this technique permits precise measurements of nuclear magnetic moments. An elaboration of this basic idea leads to *magnetic resonance imaging* (MRI), a noninvasive imaging technique that discriminates among various body tissues on the basis of the differing environments of protons in the tissues. The principles of MRI are shown in Fig. 43.1.

The magnetic moment of a nucleus is also the *source* of a magnetic field. In an atom the interaction of an electron's magnetic moment with the field of the nucleus's magnetic moment causes additional splittings in atomic energy levels and spectra. We called this effect *hyperfine structure* in Section 41.3. Measurements of the hyperfine structure may be used to directly determine the nuclear spin.

Test Your Understanding

By what factor must the mass number of a nucleus increase to double its volume? To double its radius?

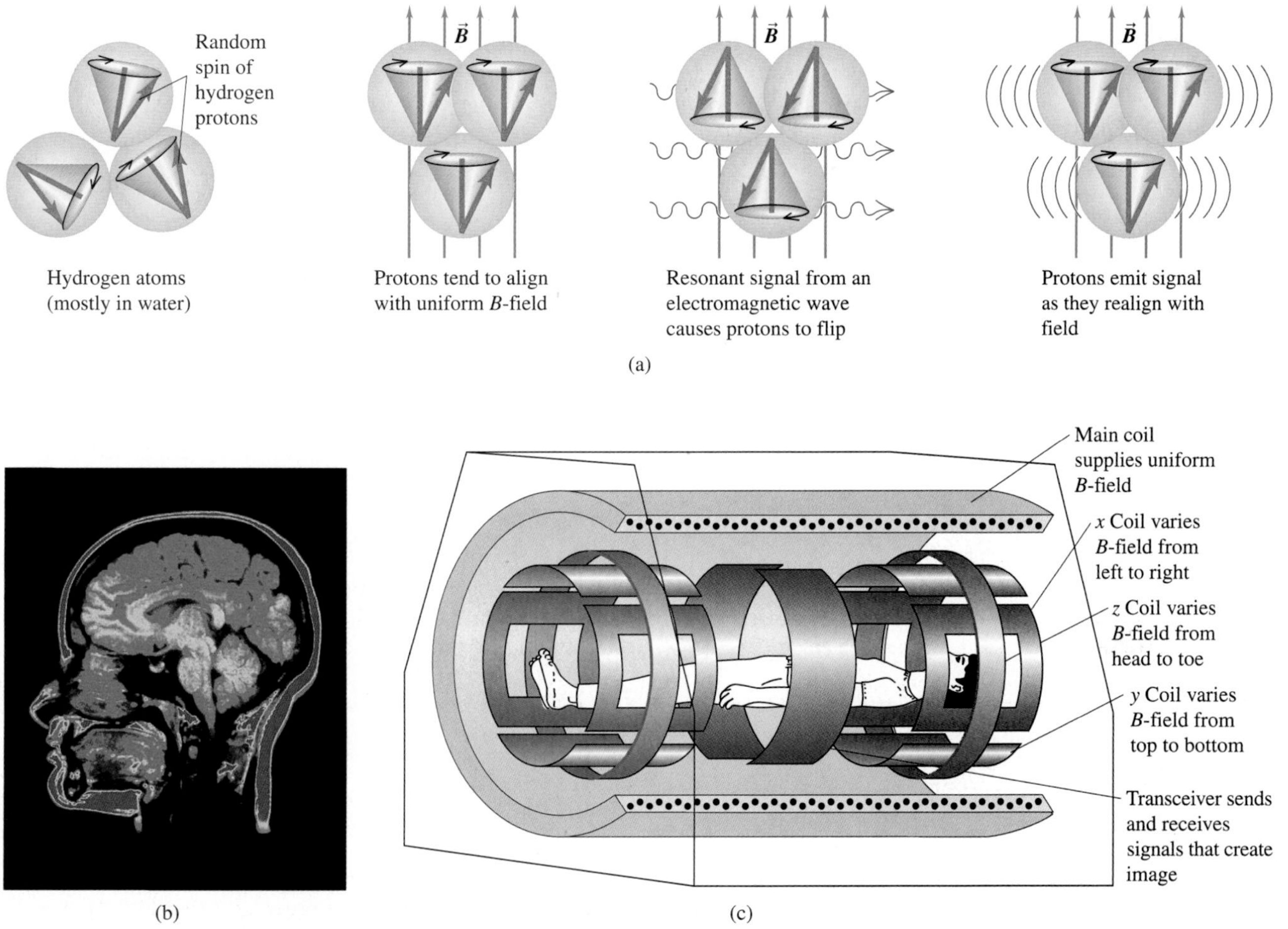

43.1 Magnetic resonance imaging (MRI). (a) Protons, the nuclei of hydrogen atoms in the tissue under study, normally have random spin orientations. In the presence of a strong magnetic field, they become aligned with a component parallel to the field. A brief radio signal flips the spins; as their components reorient parallel to the field, they emit signals that are picked up by sensitive detectors. The differing magnetic environment in various regions permits reconstruction of an image showing the types of tissue present. (b) A color-enhanced MRI image showing a cross section through a patient's head. (c) An electromagnet used for MRI.

43.2 | Nuclear Binding and Nuclear Structure

Because energy must be added to a nucleus to separate it into its individual protons and neutrons, the total rest energy E_0 of the separated nucleons is greater than the rest energy of the nucleus. The energy that must be added to separate the nucleons is called the **binding energy** E_B; it is the magnitude of the energy by which the nucleons are bound together. Thus the rest energy of the nucleus is $E_0 - E_B$. Using the equivalence of rest mass and energy (Section 37.8), we see that the total mass of the nucleons is always greater than the mass of the nucleus by an amount E_B/c^2 called the *mass defect*. The binding energy for a nucleus containing Z protons and N neutrons is defined as

$$E_B = (ZM_H + Nm_n - {}^{A}_{Z}M)c^2 \qquad \text{(nuclear binding energy)} \qquad (43.10)$$

where ${}^{A}_{Z}M$ is the mass of the *neutral* atom containing the nucleus, the quantity in the parentheses is the mass defect, and $c^2 = 931.5\ \text{MeV/u}$. Note that Eq. (43.10) does not include Zm_p, the mass of Z protons. Rather, it contains ZM_H, the mass of Z protons and Z electrons combined as Z neutral ${}^{1}_{1}\text{H}$ atoms, to balance the Z electrons included in ${}^{A}_{Z}M$, the mass of the neutral atom.

ActivPhysics ONLINE
19.2 Nuclear Binding Energy

The simplest nucleus is that of hydrogen, a single proton. Next comes the nucleus of ${}^{2}_{1}H$, the isotope of hydrogen with mass number 2, usually called *deuterium.* Its nucleus consists of a proton and a neutron bound together to form a particle called the *deuteron.* By using values from Table 43.2 in Eq. (43.10), the binding energy of the deuteron is

$$E_B = (1.007825 \text{ u} + 1.008665 \text{ u} - 2.014102 \text{ u})(931.5 \text{ MeV/u})$$
$$= 2.224 \text{ MeV}$$

This much energy would be required to pull the deuteron apart into a proton and a neutron. An important measure of how tightly a nucleus is bound is the *binding energy per nucleon, E_B/A.* At $(2.224 \text{ MeV})(2 \text{ nucleons}) = 1.112$ MeV per nucleon, ${}^{2}_{1}H$ has the smallest binding energy per nucleon of all nuclides.

Problem-Solving Strategy: Nuclear Properties

IDENTIFY *the relevant concepts:* The key properties of any nucleus include the mass, radius, binding energy, mass defect, binding energy per nucleon, and angular momentum.

SET UP *the problem:* Once you have identified your target variables, collect the equations you need to solve the problem. A relatively small number of equations from this section and Section 43.1 are all you need, but make sure that you understand their meanings.

EXECUTE *the solution:* Solve the equations for the target variables. When doing energy calculations involving the binding energy and binding energy per nucleon, note that mass tables almost always list the masses of *neutral* atoms, including their full complement of electrons. To compensate for this, use the mass M_H of a ${}^{1}_{1}H$ atom in Eq. (43.10) rather than the mass of a bare proton. The binding energies of the electrons in the neutral atoms are much smaller and tend to cancel in the subtraction in Eq. (43.10), so we won't worry about them. As Eq. (43.10) shows, binding energy calculations often involve subtracting two nearly equal quantities. To get enough precision in the difference, you often have to carry seven, eight, or nine significant figures, if that many are available. If not, you may have to be content with an approximate result.

EVALUATE *your answer:* Familiarity with some typical numerical magnitudes is very useful when checking your results. Many of these magnitudes are quite different in nuclear physics than in atomic physics. Protons and neutrons are about 1840 times as massive as electrons. The radius of a nucleus is of the order 10^{-15} m; the electric potential energy of two protons is of the order 10^{-13} J or 1 MeV. Thus typical nuclear interaction energies are of the order of a few MeV, rather than a few eV as with atoms. The typical binding energy per nucleon is about 1% of the rest energy of a nucleon. For comparison, the ionization energy of the hydrogen atom is only 0.003% of the electron's rest energy.

Angular momentum is of the same order of magnitude in both nuclei and atoms because it is determined by the value of $\hbar$. Magnetic moments of nuclei, however, are about 1000 times *smaller* than those of electrons in atoms because the nuclei are so much more massive than electrons.

Example 43.3 The most strongly bound nuclide

Because it has the highest binding energy per nucleon of all nuclides, ${}^{62}_{28}Ni$ may be described as the most strongly bound. Its neutral atomic mass is 61.928349 u. Find its mass defect, its total binding energy, and its binding energy per nucleon.

SOLUTION

IDENTIFY and SET UP: The mass defect (the difference between the mass of the nucleus and the combined mass of its constituent nucleons) is the quantity in parentheses in Eq. (43.10). The binding energy E_B is this quantity multiplied by c^2, and the binding energy per nucleon is E_B divided by the mass number A.

EXECUTE: We use $Z = 28$, $M_H = 1.007825$ u, $N = A - Z = 62 - 38 = 34$, $m_n = 1.008665$ u, and ${}^{A}_{Z}M = 61.928349$ u in the parentheses of Eq. (43.10) to find a mass defect of 0.585361 u. Then

$$E_B = (0.585361\ \text{u})(931.5\ \text{MeV/u}) = 545.3\ \text{MeV}$$

It would require a minimum of 545.3 MeV to pull a $^{62}_{28}Ni$ nucleus completely apart into 62 separate nucleons. The binding energy *per nucleon* is 1/62 of this, or 8.795 MeV per nucleon.

EVALUATE: The mass defect of $^{62}_{28}Ni$ is about 1% of the mass of the atom: $(0.585361\ \text{u})/(61.928349\ \text{u}) = 0.00945223 = 0.945223\%$. Since the mass of the atom is almost the same as the mass of the nucleus, this means that the binding energy (mass defect times c^2) is about 1% of the rest energy of the nucleus, and the binding energy per nucleon is about 1% of the rest energy of a nucleon. This result is as expected. Note that the mass defect is more than half the mass of a neutron or a proton, which suggests how tightly bound nuclei are.

Nearly all stable nuclides, from the lightest to the most massive, have binding energies in the range of 7 to 9 MeV per nucleon. Figure 43.2 is a graph of binding energy per nucleon as a function of the mass number A. Note the spike at $A = 4$, showing the unusually large binding energy per nucleon of the ^4_2He nucleus (alpha particle) relative to its neighbors. To explain this curve, we must consider the interactions among the nucleons.

The Nuclear Force

The force that binds protons and neutrons together in the nucleus, despite the electrical repulsion of the protons, is an example of the *strong interaction* that we mentioned in Section 5.5. In the context of nuclear structure, this interaction is called the *nuclear force*. Here are some of its characteristics. First, it does not depend on charge; neutrons as well as protons are bound, and the binding is the same for both. Second, it has short range, of the order of nuclear dimensions, that is, 10^{-15} m. (Otherwise, the nucleus would grow by pulling in additional protons and neutrons.) But within its range, the nuclear force is much stronger than electrical forces; otherwise, the nucleus could never be stable. It would be nice if we could write a simple equation like Newton's law of gravitation or Coulomb's law for this force, but physicists have yet to fully determine its dependence on the

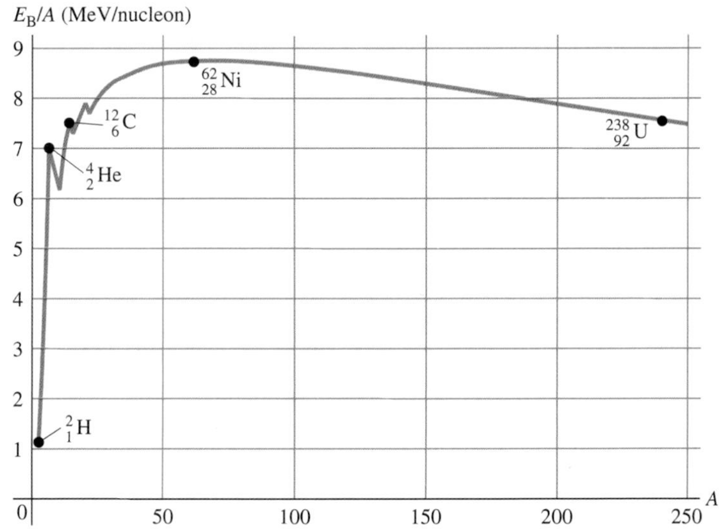

43.2 Approximate binding energy per nucleon as a function of mass number A (the total number of nucleons) for stable nuclides. The curve reaches a peak of about 8.8 MeV/nucleon at $A = 62$, corresponding to the element nickel. The spike at $A = 4$ shows the unusual stability of the ^4_2He structure.

separation r. Third, the nearly constant density of nuclear matter and the nearly constant binding energy per nucleon of larger nuclides show that a particular nucleon cannot interact simultaneously with *all* the other nucleons in a nucleus, but only with those few in its immediate vicinity. This is different from electrical forces; *every* proton in the nucleus repels every other one. This limited number of interactions is called *saturation;* it is analogous to covalent bonding in molecules and solids. Finally, the nuclear force favors binding of *pairs* of protons or neutrons with opposite spins and of *pairs of pairs,* that is, a pair of protons and a pair of neutrons, each pair having opposite spins. Hence the alpha particle (two protons and two neutrons) is an exceptionally stable nucleus for its mass number. We'll see other evidence for pairing effects in nuclei in the next subsection. (In Section 42.8 we described an analogous pairing that binds opposite-spin electrons in Cooper pairs in the BCS theory of superconductivity.)

The analysis of nuclear structure is more complex than the analysis of many-electron atoms. Two different kinds of interactions are involved (electrical and nuclear), and the nuclear force is not yet completely understood. Even so, we can gain some insight into nuclear structure by the use of simple models. We'll discuss briefly two rather different but successful models, the *liquid-drop model* and the *shell model.*

The Liquid-Drop Model

The **liquid-drop model,** first proposed in 1928 by the Russian physicist George Gamow and later expanded on by Niels Bohr, is suggested by the observation that all nuclei have nearly the same density. The individual nucleons are analogous to molecules of a liquid, held together by short-range interactions and surface-tension effects. We can use this simple picture to derive a formula for the estimated total binding energy of a nucleus. We'll include five contributions:

1. We've remarked that nuclear forces show *saturation;* an individual nucleon interacts only with a few of its nearest neighbors. This effect gives a binding energy term that is proportional to the number of nucleons. We write this term as C_1A, where C_1 is an experimentally determined constant.
2. The nucleons on the surface of the nucleus are less tightly bound than those in the interior because they have no neighbors outside the surface. This decrease in the binding energy gives a *negative* energy term proportional to the surface area $4\pi R^2$. Because R is proportional to $A^{1/3}$, this term is proportional to $A^{2/3}$; we write it as $-C_2A^{2/3}$, where C_2 is another constant.
3. Every one of the Z protons repels every one of the $(Z - 1)$ other protons. The total repulsive electric potential energy is proportional to $Z(Z - 1)$ and inversely proportional to the radius R and thus to $A^{1/3}$. This energy term is negative because the nucleons are less tightly bound than they would be without the electrical repulsion. We write this correction as $-C_3Z(Z - 1)/A^{1/3}$.
4. To be in a stable, low-energy state, the nucleus must have a balance between the energies associated with the neutrons and with the protons. This means that N is close to Z for small A and N is greater than Z (but not too much greater) for larger A. We need a negative energy term corresponding to the difference $|N - Z|$. The best agreement with observed binding energies is obtained if this term is proportional to $(N - Z)^2/A$. If we use

$N = A - Z$ to express this energy in terms of A and Z, this correction is $-C_4(A - 2Z)^2/A$.

5. Finally, the nuclear force favors *pairing* of protons and of neutrons. This energy term is positive (more binding) if both Z and N are even, negative (less binding) if both Z and N are odd, and zero otherwise. The best fit to the data occurs with the form $\pm C_5 A^{-4/3}$ for this term.

The total estimated binding energy E_B is the sum of these five terms:

$$E_B = C_1 A - C_2 A^{2/3} - C_3 \frac{Z(Z-1)}{A^{1/3}} - C_4 \frac{(A-2Z)^2}{A} \pm C_5 A^{-4/3} \quad (43.11)$$

(nuclear binding energy)

The constants C_1, C_2, C_3, C_4, and C_5, chosen to make this formula best fit the observed binding energies of nuclides, are

$$C_1 = 15.75 \text{ MeV}$$
$$C_2 = 17.80 \text{ MeV}$$
$$C_3 = 0.7100 \text{ MeV}$$
$$C_4 = 23.69 \text{ MeV}$$
$$C_5 = 39 \text{ MeV}$$

The constant C_1 is the binding energy per nucleon due to the saturated nuclear force. This energy is almost 16 MeV per nucleon, about double the *total* binding energy per nucleon in most nuclides.

If we estimate the binding energy E_B using Eq. (43.11), we can solve Eq. (43.10) to use it to estimate the mass of any neutral atom:

$$^{A}_{Z}M = ZM_H + Nm_n - E_B/c^2 \qquad \text{(semi-empirical mass formula)} \quad (43.12)$$

Equation (43.12) is called the *semi-empirical mass formula.* The name is apt; it is *empirical* in the sense that the C's have to be determined empirically (experimentally), yet it does have a sound theoretical basis.

Example 43.4 Estimating the binding energy and mass

Consider the nuclide $^{62}_{28}\text{Ni}$ of Example 43.3. a) Calculate the five terms in the binding energy and the total estimated binding energy. b) Find its neutral atomic mass using the semi-empirical mass formula.

SOLUTION

IDENTIFY and SET UP: We use Eq. (43.11) to calculate the individual terms in the binding energy as well as the total binding energy, and we use Eq. (43.12) to find the neutral atomic mass.

EXECUTE: Substituting $Z = 28$, $A = 62$, and $N = 34$ into Eq. (43.11), the individual terms are

1. $C_1 A = (15.75 \text{ MeV})(62) = 976.5 \text{ MeV}$

2. $-C_2 A^{2/3} = -(17.80 \text{ MeV})(62)^{2/3} = -278.8 \text{ MeV}$

3. $-C_3 \frac{Z(Z-1)}{A^{1/3}} = -(0.7100 \text{ MeV}) \frac{(28)(27)}{(62)^{1/3}}$
$= -135.6 \text{ MeV}$

4. $-C_4 \frac{(A-2Z)^2}{A} = -(23.69 \text{ MeV}) \frac{(62-56)^2}{62}$
$= -13.8 \text{ MeV}$

5. $+C_5 A^{-4/3} = (39 \text{ MeV})(62)^{-4/3} = 0.2 \text{ MeV}$

The pairing correction (term 5) is positive because both Z and N are even. Note that this is by far the smallest of all the terms. The total estimated binding energy is the sum of these five terms, or 548.5 MeV.

b) Now we use the $E_B = 548.5$ MeV value in Eq. (43.12) to find

$$M = 28(1.007825\text{ u}) + 34(1.008665\text{ u}) - \frac{548.5\text{ MeV}}{931.5\text{ MeV/u}}$$
$$= 61.925\text{ u}$$

This calculated mass is only about 0.005% smaller than the measured value of 61.928349 u.

EVALUATE: The binding energy of ${}^{62}_{28}\text{Ni}$ calculated in part (a) is only about 0.6% larger than the true value of 545.3 MeV found in Example 43.3, and the mass calculated in part (b) is only about 0.005% smaller than the measured value of 61.928349 u. These results show how accurate the semi-empirical mass formula can be.

The liquid-drop model and the mass formula derived from it are quite successful in correlating nuclear masses, and we will see later that they are a great help in understanding decay processes of unstable nuclides. Some other aspects of nuclei, such as angular momentum and excited states, are better approached with different models.

The Shell Model

The **shell model** of nuclear structure is analogous to the central-field approximation in atomic physics (Section 41.4). We picture each nucleon as moving in a potential that represents the averaged-out effect of all the other nucleons. This may not seem to be a very promising approach; the nuclear force is very strong, very short-range, and therefore strongly distance-dependent. However, in some respects, this model turns out to work fairly well.

The potential-energy function for the nuclear force is the same for protons as for neutrons. A reasonable assumption as to the shape of this function is shown in Fig. 43.3a. This function is a three-dimensional version of the square well we discussed in Section 40.2. The corners are somewhat rounded because the nucleus doesn't have a sharply defined surface. For protons there is an additional potential energy associated with electrical repulsion. We consider each proton to interact with a sphere of uniform charge density, with radius R and total charge $(Z - 1)e$. Figure 43.3b shows the nuclear, electric, and total potential energies for a proton as functions of the distance r from the center of the nucleus.

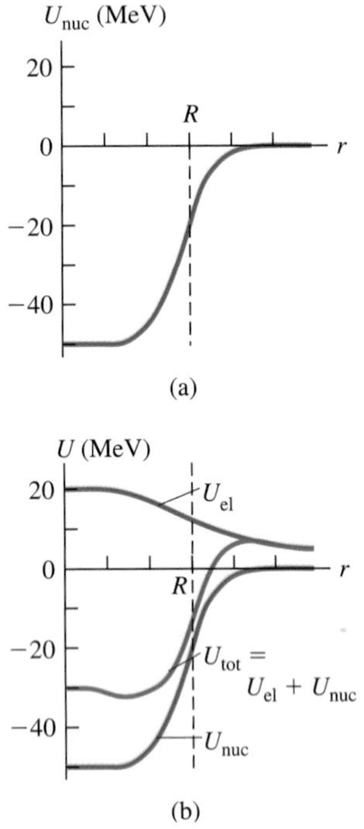

43.3 Approximate potential-energy functions for a nucleon in a nucleus. The approximate nuclear radius is R. (a) The potential energy U_{nuc} due to the nuclear force. It is the same for protons and neutrons; for neutrons, it is the *total* potential energy. (b) The total potential energy U_{tot} for a proton is the sum of the nuclear (U_{nuc}) and electric (U_{el}) potential energies.

In principle, we could solve the Schrödinger equation for a proton or neutron moving in such a potential. For any spherically symmetric potential energy, the angular-momentum states are the same as for the electrons in the central-field approximation in atomic physics. In particular, we can use the concept of *filled shells and subshells* and their relation to stability. In atomic structure we found that the values $Z = 2, 10, 18, 36, 54$, and 86 (the atomic numbers of the noble gases) correspond to particularly stable electron arrangements.

A comparable effect occurs in nuclear structure. The numbers are different because the potential-energy function is different and the nuclear spin–orbit interaction is much stronger and of opposite sign than in atoms, so the subshells fill up in a different order from that for electrons in an atom. It is found that when the number of neutrons *or* the number of protons is 2, 8, 20, 28, 50, 82, or 126, the resulting structure is unusually stable, that is, has an unusually great binding energy. (Nuclides with $Z = 126$ have not been observed in nature.) These numbers are called *magic numbers.* Nuclides in which Z is a magic number tend to have an above-average number of stable isotopes. There are several nuclides for which both Z and N are magic, including

$$ {}^{4}_{2}\text{He} \quad {}^{16}_{8}\text{O} \quad {}^{40}_{20}\text{Ca} \quad {}^{48}_{20}\text{Ca} \quad \text{and} \quad {}^{208}_{82}\text{Pb} $$

All these nuclides have substantially larger binding energy per nucleon than do nuclides with neighboring values of N or Z. They also all have zero nuclear spin. The magic numbers correspond to filled-shell or -subshell configurations of nucleon energy levels with a relatively large jump in energy to the next allowed level.

Test Your Understanding

Helium-4 (^4_2He) has a binding energy of 28.3 MeV, but Eq. (43.11) gives $E_B = 23.4$ MeV, which is in error by 17%. Why does Eq. (43.11) give better results for more massive nuclides, as in Example 43.4?

43.3 | Nuclear Stability and Radioactivity

Among about 2500 known nuclides, fewer than 300 are stable. The others are unstable structures that decay to form other nuclides by emitting particles and electromagnetic radiation, a process called **radioactivity.** The time scale of these decay processes ranges from a small fraction of a microsecond to billions of years. The *stable* nuclides are shown by dots on the graph in Fig. 43.4, where the neutron

ActivPhysics ONLINE
19.2 Nuclear Binding Energy

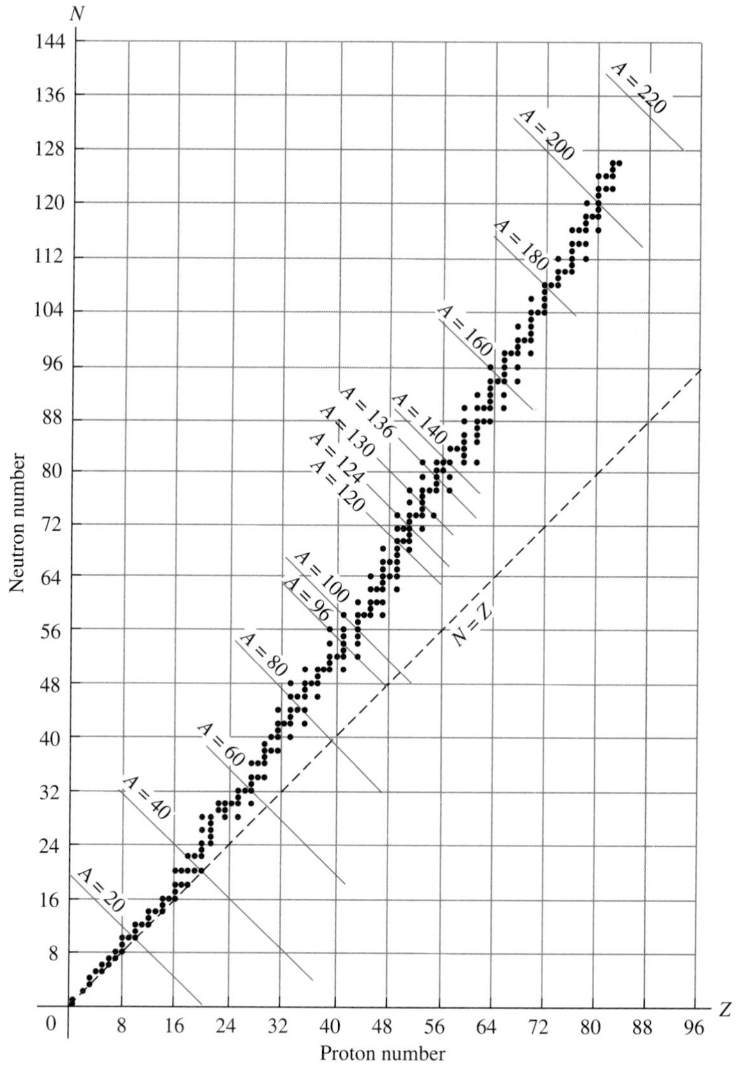

43.4 Segrè chart showing neutron number and proton number for stable nuclides. As the mass number A increases, stable nuclides have a greater number of neutrons relative to protons.

number N and proton number (or atomic number) Z for each nuclide are plotted. Such a chart is called a *Segrè chart,* after its inventor, the Italian-American physicist Emilio Segrè (1905–1989).

Each blue line perpendicular to the line $N = Z$ represents a specific value of the mass number $A = Z + N$. Most lines of constant A pass through only one or two stable nuclides; that is, there is usually a very narrow range of stability for a given mass number. The lines at $A = 20$, $A = 40$, $A = 60$, and $A = 80$ are examples. In four cases these lines pass through *three* stable nuclides—namely, at $A = 96$, 124, 130, and 136.

Our four stable nuclides have both odd Z and odd N:

$$^{2}_{1}\text{H} \qquad ^{6}_{3}\text{Li} \qquad ^{10}_{5}\text{B} \qquad ^{14}_{7}\text{N}$$

These are called *odd-odd nuclides.* The absence of other odd-odd nuclides shows the influence of pairing. Also, there is *no* stable nuclide with $A = 5$ or $A = 8$. The doubly magic $^{4}_{2}\text{He}$ nucleus, with a pair of protons and a pair of neutrons, has no interest in accepting a fifth particle into its structure. Collections of eight nucleons decay to smaller nuclides, with a $^{8}_{4}\text{Be}$ nucleus immediately splitting into two $^{4}_{2}\text{He}$ nuclei.

The points on the Segrè chart representing stable nuclides define a rather narrow stability region. For low mass numbers, the numbers of protons and neutrons are approximately equal, $N \approx Z$. The ratio N/Z increases gradually with A, up to about 1.6 at large mass numbers, because of the increasing influence of the electrical repulsion of the protons. Points to the right of the stability region represent nuclides that have too many protons relative to neutrons to be stable. In these cases, repulsion wins, and the nucleus comes apart. To the left are nuclides with too many neutrons relative to protons. In these cases the energy associated with the neutrons is out of balance with that associated with the protons, and the nuclides decay in a process that converts neutrons to protons. The graph also shows that no nuclide with $A > 209$ or $Z > 83$ is stable. A nucleus is unstable if it is too big. We also note that there is no stable nuclide with $Z = 43$ (technetium) or 61 (promethium). Figure 43.5, a three-dimensional version of the Segrè chart, shows the "valley of stability" for light nuclides (up to $Z = 22$).

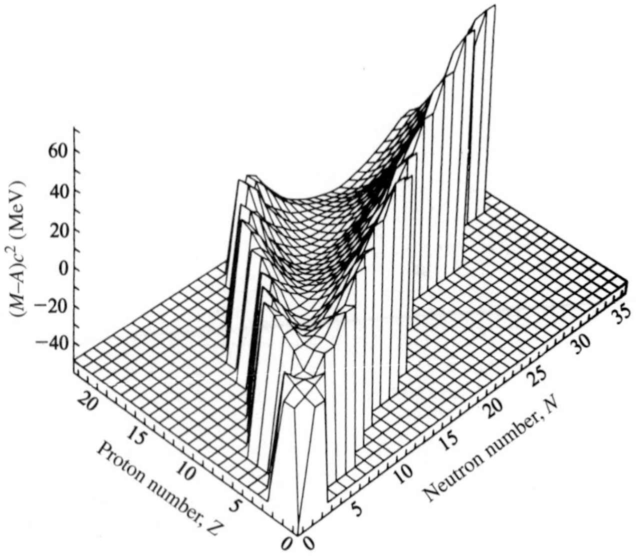

43.5 A three-dimensional Segrè chart for light nuclides up to $Z = 22$ (titanium). The quantity plotted on the third axis is $(M - A)c^2$, where M is the nuclide mass expressed in u. This quantity is related to the binding energy by a different constant for each nuclide. The lower points on the raised surface represent especially stable nuclides.

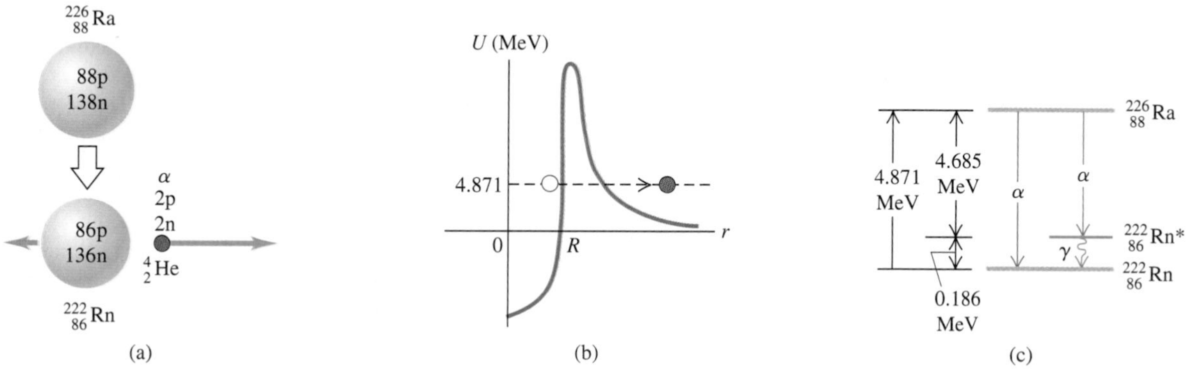

43.6 (a) The nuclide $^{226}_{88}\text{Ra}$ decays by alpha emission to $^{222}_{86}\text{Rn}$. (b) Potential-energy curve for an α particle and $^{222}_{86}\text{Rn}$ nucleus. The α particle tunnels through the potential-energy barrier. (c) Energy-level diagram for the system, showing the excited level $^{222}_{86}\text{Rn}^*$ at an energy 0.186 MeV above the ground level. The system can decay from this level to the ground level $^{222}_{86}\text{Rn}$ by emission of a γ photon with energy 0.186 MeV.

Alpha Decay

Nearly 90% of the 2500 known nuclides are *radioactive;* they are not stable but decay into other nuclides. When unstable nuclides decay into different nuclides, they usually emit alpha (α) or beta (β) particles. An **alpha particle** is a ^4He nucleus, two protons and two neutrons bound together, with total spin zero. Alpha emission occurs principally with nuclei that are too large to be stable. When a nucleus emits an alpha particle, its N and Z values each decrease by two and A decreases by four, moving it closer to stable territory on the Segrè chart.

A familiar example of an alpha emitter is radium, $^{226}_{88}\text{Ra}$ (Fig. 43.6a). The speed of the emitted alpha particle, determined from the curvature of its path in a transverse magnetic field, is about 1.52×10^7 m/s. This speed, although large, is only 5% of the speed of light, so we can use the nonrelativistic kinetic-energy expression $K = \frac{1}{2}mv^2$:

$$K = \frac{1}{2}(6.64 \times 10^{-27}\ \text{kg})(1.52 \times 10^7\ \text{m/s})^2 = 7.7 \times 10^{-13}\ \text{J} = 4.8\ \text{MeV}$$

Alpha particles are always emitted with definite kinetic energies, determined by conservation of momentum and energy. Because of their charge and mass, alpha particles can travel only several centimeters in air, or a few tenths or hundredths of a millimeter through solids, before they are brought to rest by collisions.

Some nuclei can spontaneously decay by emission of α particles because energy is released in their alpha decay. You can use conservation of mass-energy to show that *alpha decay is possible whenever the mass of the original neutral atom is greater than the sum of the masses of the final neutral atom and the neutral helium-4 atom.* In alpha decay, the α particle tunnels through a potential-energy barrier, as shown in Fig. 43.6b. You may want to review the discussion of tunneling in Section 40.3.

Example 43.5 Alpha decay of radium

You are given the following neutral atomic masses:

$^{226}_{88}\text{Ra}$	226.025403 u
$^{222}_{86}\text{Rn}$	222.017571 u

Show that alpha emission is energetically possible and that the calculated kinetic energy of the emitted α particle agrees with the experimentally measured value of 4.78 MeV.

SOLUTION

IDENTIFY and SET UP: Alpha emission is possible if the mass of the $^{226}_{88}\text{Ra}$ atom is greater than the sum of the atomic masses of $^{222}_{86}\text{Rn}$ and $^{4}_{2}\text{He}$. The mass difference between the initial radium atom and the final radon and helium atoms corresponds (through $E = mc^2$) to the energy E released in the decay. Because momentum is conserved as well as energy, *both* the alpha particle and the $^{222}_{86}Rn$ atom are in motion after the decay; we will have to account for this fact in determining the kinetic energy of the alpha particle.

EXECUTE: From Table 43.2, the mass of the $^{4}_{2}He$ atom is 4.002603 u. The difference in mass between the original nucleus and the decay products is

$$226.025403\text{ u} - (222.017571\text{ u} + 4.002603\text{ u}) = +0.005229\text{ u}$$

Since this is positive, alpha decay is energetically possible.

The energy equivalent of 0.005229 u is

$$E = (0.005229\text{ u})(931.5\text{ MeV/u}) = 4.871\text{ MeV}$$

Thus we expect the decay products to emerge with total kinetic energy 4.871 MeV. Momentum is also conserved; if the parent nucleus is at rest, the daughter and the α particle have momenta of equal magnitude p but opposite direction. Kinetic energy is $K = p^2/2m$: since p is the same for the two particles, the kinetic energy divides inversely as their masses. The α particle gets $222/(222 + 4)$ of the total, or 4.78 MeV, equal to the observed α-particle energy.

EVALUATE: An excellent way to check your results is to verify that the alpha particle and the $^{222}_{86}Rn$ nucleus produced in the decay have the same magnitude of momentum $p = mv$. You can calculate the speed v of each of the decay products from its respective kinetic energy. You'll find that the alpha particle moves at a sprightly $0.0506c = 1.52 \times 10^7$ m/s; if momentum is conserved, you should find that the $^{222}_{86}Rn$ nucleus moves 4/222 as fast. Does it?

Beta Decay

There are three different simple types of *beta decay: beta-minus, beta-plus,* and *electron capture.* A **beta-minus particle** (β^-) is an electron. It's not obvious how a nucleus can emit an electron if there aren't any electrons in the nucleus. Emission of a β^- involves *transformation* of a neutron into a proton, an electron, and a third particle called an *antineutrino.* In fact, if you freed a neutron from a nucleus, it would decay into a proton, an electron, and an antineutrino in an average time of about 15 minutes.

Beta particles can be identified and their speeds can be measured with techniques that are similar to the Thomson experiments we described in Section 27.5. The speeds of beta particles range up to 0.9995 of the speed of light, so their motion is highly relativistic. They are emitted with a continuous spectrum of energies. This would not be possible if the only two particles were the β^- and the recoiling nucleus, since energy and momentum conservation would then require a definite speed for the β^-. Thus there must be a *third* particle involved. From conservation of charge, it must be neutral, and from conservation of angular momentum, it must be a spin-$\frac{1}{2}$ particle.

This third particle is an antineutrino, the *antiparticle* of a **neutrino.** The symbol for a neutrino is ν (the Greek letter "nu"). Both the neutrino and the antineutrino have zero charge and zero (or very small) mass and therefore produce very little observable effect when passing through matter. Both evaded detection until 1953, when Frederick Reines and Clyde Cowan succeeded in observing the antineutrino directly. We now know that there are at least three varieties of neutrinos, each with its corresponding antineutrino; one is associated with beta decay and the other two are associated with the decay of two unstable particles, the muon and the tau particle. We'll discuss these particles in more detail in Chapter 44. The antineutrino that is emitted in β^- decay is denoted as $\overline{\nu}_e$. The basic process of β^- decay is

$$\text{n} \rightarrow \text{p} + \beta^- + \overline{\nu}_e \tag{43.13}$$

Beta-minus decay usually occurs with nuclides for which the neutron-to-proton ratio N/Z is too large for stability. In β^- decay, N decreases by one, Z increases by one, and A doesn't change. You can use conservation of mass-energy to show that *beta-minus decay can occur whenever the neutral atomic mass of the original atom is larger than that of the final atom.*

Example 43.6 Why cobalt-60 is a beta-minus emitter

The nuclide ${}^{60}_{27}\text{Co}$, an odd-odd unstable nucleus, is used in medical applications of radiation. Show that it is unstable relative to β^- decay. The following masses are given:

${}^{60}_{27}\text{Co}$:	59.933822 u
${}^{60}_{28}\text{Ni}$:	59.930791 u

SOLUTION

IDENTIFY and SET UP: The technique for solving this problem has just two steps: (1) decide which nuclide will result if the original nuclide decay undergoes β^- decay, and (2) check whether the natural atomic mass of the original atom is greater than that of the final atom.

EXECUTE: The original nuclide is ${}^{60}_{27}\text{Co}$. In β^- decay, Z increases by one from 27 to 28 and A remains at 60, so the final nuclide is ${}^{60}_{28}\text{Ni}$. Its mass is less than that of ${}^{60}_{27}\text{Co}$ by 0.003031 u, so β^- decay *can* occur.

EVALUATE: As with alpha decay, this beta-minus decay obeys the conservation laws for both momentum and energy. But with three decay products in β^- decay—the ${}^{60}_{28}\text{Ni}$ nucleus, the electron, and the antineutrino—the energy can be shared in many different ways that are consistent with the conservation laws. It's impossible to predict precisely how the energy will be shared for the decay of a particular ${}^{60}_{27}\text{Co}$ nucleus. By contrast, in alpha decay there are just two decay products, and their energies and momenta are determined uniquely (see Example 43.5).

We have noted that β^- decay occurs with nuclides that have too large a neutron-to-proton ratio N/Z. Nuclides for which N/Z is too *small* for stability can emit a *positron,* the electron's antiparticle, which is identical to the electron but with positive charge. (We mentioned the positron in connection with positronium in Section 38.5 and will discuss it in more detail in Chapter 44.) The basic process, called *beta-plus decay* (β^+), is

$$\text{p} \rightarrow \text{n} + \beta^+ + \nu_\text{e} \tag{43.14}$$

where β^+ is a positron and ν_e is the electron neutrino. *Beta-plus decay can occur whenever the neutral atomic mass of the original atom is at least two electron masses larger than that of the final atom;* you can show this using conservation of mass-energy.

The third type of beta decay is *electron capture.* There are a few nuclides for which β^+ emission is not energetically possible but in which an orbital electron (usually in the K shell) can combine with a proton in the nucleus to form a neutron and a neutrino. The neutron remains in the nucleus and the neutrino is emitted. The basic process is

$$\text{p} + \beta^- \rightarrow \text{n} + \nu_\text{e} \tag{43.15}$$

You can use conservation of mass-energy to show that *electron capture can occur whenever the neutral atomic mass of the original atom is larger than that of the final atom.* In all types of beta decay, A remains constant. However, in beta-plus decay and electron capture, N increases by one and Z decreases by one as the neutron–proton ratio increases toward a more stable value. The reaction of Eq. (43.15) also helps to explain the formation of a neutron star, mentioned in Example 43.1.

CAUTION The beta-decay reactions given by Eqs. (43.13), (43.14), and (43.15) occur *within* a nucleus. Although the decay of a neutron outside the nucleus proceeds through the reaction of Eq. (43.13), the reaction of Eq. (43.14) is forbidden by conservation of mass-energy for a proton outside the nucleus. The reaction of Eq. (43.15) can occur outside the nucleus only with the addition of some extra energy, as in a collision.

Example 43.7

Why cobalt-57 is not a beta-plus emitter

The nuclide $^{57}_{27}Co$, an odd-even unstable nucleus, is often used as a source of radiation in a nuclear process called the *Mössbauer effect.* Show that this nuclide is stable relative to β^+ decay but can decay by electron capture. The following masses are given:

$^{57}_{27}Co$:	56.936296 u
$^{57}_{26}Fe$:	56.935399 u

SOLUTION

IDENTIFY and SET UP: As in Example 43.6, we first decide what nuclide results from β^+ decay or electron capture, then compare the mass of the neutral $^{57}_{27}Co$ atom to the mass of the final atom.

EXECUTE: The original nuclide is $^{57}_{27}Co$. In β^+ decay and electron capture, Z decreases by one from 27 to 26, and A remains at 57. Thus the final nuclide is $^{57}_{26}Fe$. Its mass is less than that of $^{57}_{27}Co$ by 0.000897 u, a value smaller than 0.001097 u (two electron masses), so β^+ decay *cannot* occur. However, the mass of the original atom is greater than the mass of the final atom, so electron capture *can* occur. In Section 43.4 we'll see how to relate the probability that electron capture will occur to the *half-life* of this nuclide.

EVALUATE: In electron capture there are just two decay products: the final nucleus and the emitted neutrino. Unlike in β^- decay (Example 43.6) but like in alpha decay (Example 43.5), the decay products of electron capture have unique energies and momenta.

Gamma Decay

The energy of internal motion of a nucleus is quantized. A typical nucleus has a set of allowed energy levels, including a *ground state* (state of lowest energy) and several *excited states.* Because of the great strength of nuclear interactions, excitation energies of nuclei are typically of the order of 1 MeV, compared with a few eV for atomic energy levels. In ordinary physical and chemical transformations the nucleus always remains in its ground state. When a nucleus is placed in an excited state, either by bombardment with high-energy particles or by a radioactive transformation, it can decay to the ground state by emission of one or more photons called **gamma rays** or *gamma-ray photons,* with typical energies of 10 keV to 5 MeV. This process is called *gamma* (γ) *decay.* For example, alpha particles emitted from ^{226}Ra have two possible kinetic energies, either 4.784 MeV or 4.602 MeV. Including the recoil energy of the resulting ^{222}Rn nucleus, these correspond to a total released energy of 4.871 MeV or 4.685 MeV, respectively. When an alpha particle with the smaller energy is emitted, the ^{222}Rn nucleus is left in an excited state. It then decays to its ground state by emitting a gamma-ray photon with energy

$$(4.871 - 4.685)\text{ MeV} = 0.186\text{ MeV}$$

A photon with this energy is observed during this decay (Fig. 43.6c).

CAUTION In both α and β decay, the Z value of a nucleus changes and the nucleus of one element becomes the nucleus of a different element. In γ decay, the element does *not* change; the nucleus merely goes from an excited state to a less excited state.

Natural Radioactivity

Many radioactive elements occur in nature. For example, you are very slightly radioactive because of unstable nuclides such as carbon-14 and potassium-40 that are present throughout your body. The study of natural radioactivity began in 1896, one year after Röntgen discovered x rays. Henri Becquerel discovered a radiation from uranium salts that seemed similar to x rays. Intensive investigation in the following two decades by Marie and Pierre Curie, Ernest Rutherford, and many others revealed that the emissions consist of positively and negatively charged particles and neutral rays; they were given the names *alpha, beta,* and *gamma* because of their differing penetration characteristics.

The decaying nucleus is usually called the *parent nucleus;* the resulting nucleus is the *daughter nucleus.* When a radioactive nucleus decays, the daughter nucleus may also be unstable. In this case a *series* of successive decays occurs until a stable configuration is reached. Several such series are found in nature. The most abundant radioactive nuclide found on earth is the uranium isotope ^{238}U, which undergoes a series of 14 decays, including eight α emissions and six β^- emissions, terminating at a stable isotope of lead, ^{206}Pb (Fig. 43.7).

Radioactive decay series can be represented on a Segrè chart, as in Fig. 43.8. The neutron number N is plotted vertically, and the atomic number Z is plotted horizontally. In alpha emission, both N and Z decrease by two. In β^- emission, N

43.7 Earthquakes are caused in part by the radioactive decay of ^{238}U in the earth's interior. These decays release energy that helps to keep the interior molten. The earth's solid crust is able to slide over this molten material. When two adjacent sections of crust slide past each other, an earthquake results.

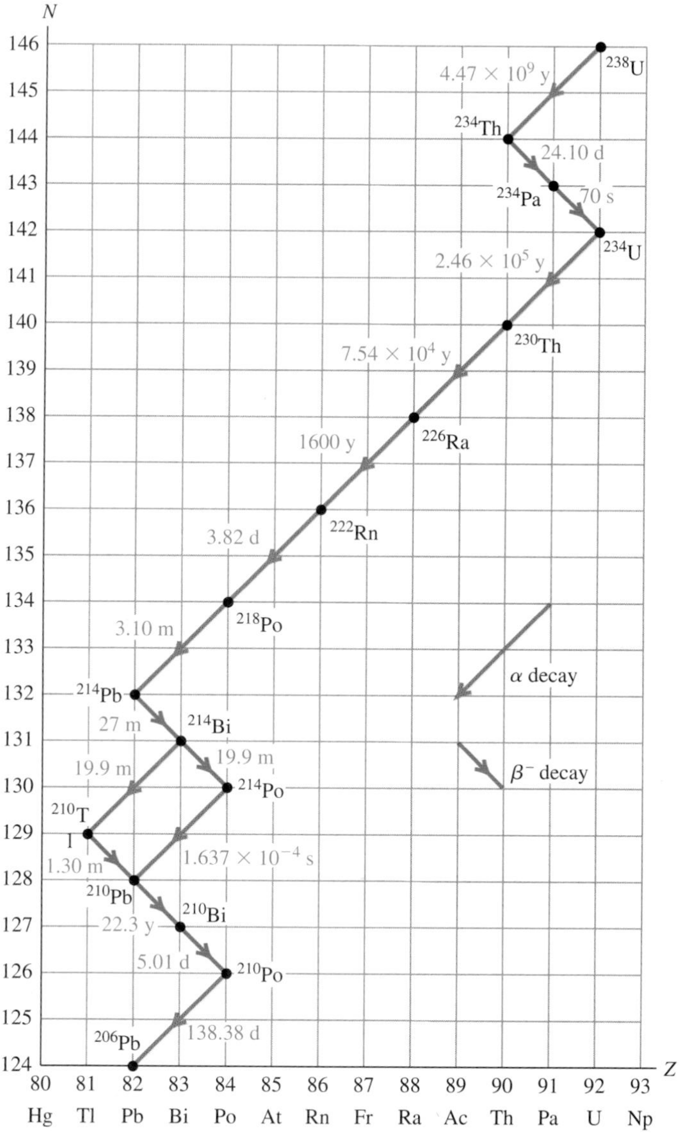

43.8 Segrè chart showing the uranium ^{238}U decay series, terminating with the stable nuclide ^{206}Pb. The times are half-lives (discussed in the next section), given in years (y), days (d), hours (h), minutes (m), or seconds (s).

decreases by one and Z increases by one. The decays can also be represented in equation form; the first two decays in the series are written as

$$^{238}\text{U} \rightarrow {}^{234}\text{Th} + \alpha$$

$$^{234}\text{Th} \rightarrow {}^{234}\text{Pa} + \beta^{-} + \overline{\nu}_{\text{e}}$$

or more briefly as

$$^{238}\text{U} \xrightarrow{\alpha} {}^{234}\text{Th}$$

$$^{234}\text{Th} \xrightarrow{\beta^{-}} {}^{234}\text{Pa}$$

In the second process, the beta decay leaves the daughter nucleus ^{234}Pa in an excited state, from which it decays to the ground state by emitting a gamma-ray photon. An excited state is denoted by an asterisk, so we can represent the γ emission as

$$^{234}\text{Pa}^{*} \rightarrow {}^{234}\text{Pa} + \gamma$$

or

$$^{234}\text{Pa}^{*} \xrightarrow{\gamma} {}^{234}\text{Pa}$$

An interesting feature of the ^{238}U decay series is the branching that occurs at ^{214}Bi. This nuclide decays to ^{210}Pb by emission of an α and a β^{-}, which can occur in either order. We also note that the series includes unstable isotopes of several elements that also have stable isotopes, including thallium (Tl), lead (Pb), and bismuth (Bi). The unstable isotopes of these elements that occur in the ^{238}U series all have too many neutrons to be stable.

Many other decay series are known. Two of these occur in nature, one starting with the uncommon isotope ^{235}U and ending with ^{207}Pb, the other starting with thorium (^{232}Th) and ending with ^{208}Pb.

Test Your Understanding

Use conservation of mass-energy to explain why it is impossible for electron capture to occur in the $^{1}_{1}$H atom.

43.4 | Activities and Half-Lives

Suppose you need to dispose of some radioactive waste that contains a certain number of a particular radioactive nuclide. If no more are produced, that number decreases in a simple manner as the nuclei decay. This decrease is a statistical process; there is no way to predict when any individual nucleus will decay. No change in physical or chemical environment, such as chemical reactions or heating or cooling, greatly affects most decay rates. The rate varies over an extremely wide range for different nuclides.

Let $N(t)$ be the (very large) number of radioactive nuclei in a sample at time t, and let $dN(t)$ be the (negative) change in that number during a short time interval dt. (We'll use $N(t)$ to minimize confusion with the neutron number N.) The number of decays during the interval dt is $-dN(t)$. The rate of change of $N(t)$ is the negative quantity $dN(t)/dt$; thus $-dN(t)/dt$ is called the *decay rate* or the **activity** of the specimen. The larger the number of nuclei in the specimen, the more nuclei decay during any time interval. That is, the activity is directly proportional to $N(t)$; it equals a constant λ multiplied by $N(t)$:

$$-\frac{dN(t)}{dt} = \lambda N(t) \tag{43.16}$$

The constant λ is called the **decay constant,** and it has different values for different nuclides. A large value of λ corresponds to rapid decay; a small value corresponds to slower decay. Solving Eq. (43.16) for λ shows us that λ is the ratio of the number of decays per time to the number of remaining radioactive nuclei; λ can then be interpreted as the *probability per time* that any individual nucleus will decay.

The situation is reminiscent of a discharging capacitor, which we studied in Section 26.4. Equation (43.16) has the same form as the negative of Eq. (26.15), with q and $1/RC$ replaced by $N(t)$ and λ. Then we can make the same substitutions in Eq. (26.16), with the initial number of nuclei $N(0) = N_0$, to find the exponential function:

$$N(t) = N_0 e^{-\lambda t} \quad \text{(number of remaining nuclei)} \tag{43.17}$$

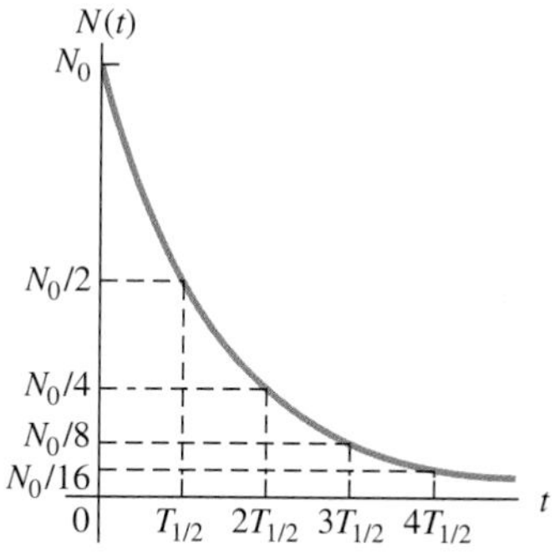

43.9 The number of nuclei in a sample of a radioactive element as a function of time. The sample's activity has an exponential decay curve with the same shape.

Figure 43.9 is a graph of this function, showing the number of remaining nuclei $N(t)$ as a function of time.

The **half-life** $T_{1/2}$ is the time required for the number of radioactive nuclei to decrease to one-half the original number N_0. Then half of the remaining radioactive nuclei decay during a second interval $T_{1/2}$, and so on. The numbers remaining after successive half-lives are $N_0/2$, $N_0/4$, $N_0/8$, . . .

To get the relation between the half-life $T_{1/2}$ and the decay constant λ, we set $N(t)/N_0 = 1/2$ and $t = T_{1/2}$ in Eq. (43.17), obtaining

$$\frac{1}{2} = e^{-\lambda T_{1/2}}$$

We take logarithms of both sides and solve for $T_{1/2}$:

$$T_{1/2} = \frac{\ln 2}{\lambda} = \frac{0.693}{\lambda} \tag{43.18}$$

The mean lifetime T_{mean}, generally called the *lifetime,* of a nucleus or unstable particle is proportional to the half-life $T_{1/2}$:

$$T_{\text{mean}} = \frac{1}{\lambda} = \frac{T_{1/2}}{\ln 2} = \frac{T_{1/2}}{0.693} \tag{43.19}$$

(lifetime T_{mean}, decay constant λ, and half-life $T_{1/2}$)

In particle physics the life of an unstable particle is usually described by the lifetime, not the half-life.

Because the activity $-dN(t)/dt$ at any time equals $\lambda N(t)$, Eq. (43.17) tells us that the activity also depends on time as $e^{-\lambda t}$. Thus the graph of activity versus time has the same shape as Fig. 43.9. Also, after successive half-lives, the activity is one-half, one-fourth, one-eighth, and so on of the original activity.

19.4 Radioactivity

CAUTION It is sometimes implied that any radioactive sample will be safe after a half-life has passed. That's wrong. If your radioactive waste initially has ten times too much activity for safety, it is not safe after one half-life, when it still has five times too much. Even after three half-lives it still has 25% more activity than is safe. The number of radioactive nuclei and the activity approach zero only as t approaches infinity.

A common unit of activity is the **curie,** abbreviated Ci, which is defined to be 3.70×10^{10} decays per second. This is approximately equal to the activity of one gram of radium. The SI unit of activity is the *becquerel,* abbreviated Bq. One becquerel is one decay per second, so

$$1 \text{ Ci} = 3.70 \times 10^{10} \text{ Bq} = 3.70 \times 10^{10} \text{ decays/s}$$

Example 43.8

Activity of ^{57}Co

The radioactive isotope ^{57}Co decays by electron capture with a half-life of 272 days. a) Find the decay constant and the lifetime. b) If you have a radiation source containing ^{57}Co, with activity 2.00 μCi, how many radioactive nuclei does it contain? c) What will be the activity of your source after one year?

SOLUTION

IDENTIFY and SET UP: We determine the decay constant λ and lifetime T_{mean} from the half-life $T_{1/2}$ using Eq. (43.19). Once we have found λ, we calculate the number of nuclei $N(t)$ from the activity (which is the same as the decay rate $-dN(t)/dt$) using Eq. (43.16). We then use Eq. (43.17) to find the number of nuclei remaining after one year, and from this value find the activity after one year by using Eq. (43.16) once again.

EXECUTE: a) To simplify the units, we convert the half-life to seconds:

$$T_{1/2} = (272 \text{ days})(86{,}400 \text{ s/day}) = 2.35 \times 10^7 \text{ s}$$

From Eq. (43.19) the lifetime is

$$T_{\text{mean}} = \frac{T_{1/2}}{\ln 2} = \frac{2.35 \times 10^7 \text{ s}}{0.693} = 3.39 \times 10^7 \text{ s}$$

The decay constant is

$$\lambda = \frac{1}{T_{\text{mean}}} = 2.95 \times 10^{-8} \text{ s}^{-1}$$

b) The activity $-dN(t)/dt$ is given as 2.00 μCi, so

$$-\frac{dN(t)}{dt} = 2.00 \ \mu\text{Ci} = (2.00 \times 10^{-6})(3.70 \times 10^{10} \text{ s}^{-1})$$
$$= 7.40 \times 10^4 \text{ decays/s}$$

From Eq. (43.16) this is equal to $\lambda N(t)$, so we find

$$N(t) = -\frac{dN(t)/dt}{\lambda} = \frac{7.40 \times 10^4 \text{ s}^{-1}}{2.95 \times 10^{-8} \text{ s}^{-1}} = 2.51 \times 10^{12} \text{ nuclei}$$

If you feel we're being too cavalier about the "units" decays and nuclei, you can use decays/(nucleus · s) as the unit for λ.

c) From Eq. (43.17) the number $N(t)$ of nuclei remaining after one year $(3.156 \times 10^7 \text{ s})$ is

$$N(t) = N_0 e^{-\lambda t} = N_0 e^{-(2.95\times10^{-8} \text{ s}^{-1})(3.156\times10^7 \text{ s})}$$
$$= 0.394 N_0$$

The number of nuclei has decreased to 0.394 of the original number. Equation (43.16) says that the activity is proportional to the number of nuclei, so the activity has decreased by this same factor to $(0.394)(2.00 \ \mu\text{Ci}) = 0.788 \ \mu\text{Ci}$.

EVALUATE: The number of nuclei found in part (b) is equivalent to 4.17×10^{-12} mol, with a mass of 2.38×10^{-10} g. This is a far smaller mass than even the most sensitive balance can measure.

After one 272-day half-life, the number of ^{57}Co nuclei has decreased to $N_0/2$; after $2(272 \text{ d}) = 544$ d, it has decreased to $N_0/2^2 = N_0/4$. This result agrees with our answer to part (c), which says that after 365 d the number of nuclei is between $N_0/2$ and $N_0/4$.

Radioactive Dating

An interesting application of radioactivity is the dating of archeological and geological specimens by measuring the concentration of radioactive isotopes. The most familiar example is *carbon dating*. The unstable isotope ^{14}C, produced during nuclear reactions in the atmosphere that result from cosmic-ray bombardment, gives a small proportion of ^{14}C in the CO_2 in the atmosphere. Plants that obtain their carbon from this source contain the same proportion of ^{14}C as the atmosphere. When a plant dies, it stops taking in carbon, and its ^{14}C β^- decays to ^{14}N with a half-life of 5730 years. By measuring the proportion of ^{14}C in the remains, we can determine how long ago the organism died.

One difficulty with radiocarbon dating is that the ^{14}C concentration in the atmosphere changes over long time intervals. Corrections can be made on the basis of other data such as measurements of tree rings that show annual growth cycles. Similar radioactive techniques are used with other isotopes for dating geologic specimens. Some rocks, for example, contain the unstable potassium isotope ^{40}K, a beta emitter that decays to the stable nuclide ^{40}Ar with a half-life of 2.4×10^8 y. The age of the rock can be determined by comparing the concentrations of ^{40}K and ^{40}Ar.

Example 43.9 Radiocarbon dating

Before 1900 the activity per mass of atmospheric carbon due to the presence of ^{14}C averaged about 0.255 Bq per gram of carbon. a) What fraction of carbon atoms were ^{14}C? b) In analyzing an archeological specimen containing 500 mg of carbon, you observe 174 decays in one hour. What is the age of the specimen, assuming that its activity per mass of carbon when it died was that average value of the air?

SOLUTION

IDENTIFY and SET UP: In part (a) we determine the number of ^{14}C atoms $N(t)$ from the activity $-dN(t)/dt$ using Eq. (43.16). We find the total number of carbon atoms in 500 mg by using the molar mass of carbon (12.011 g/mol, given in Appendix D), and we use the result to calculate the fraction of carbon atoms that are ^{14}C. The activity decays at the same rate as the number of ^{14}C nuclei; we use this fact in conjunction with Eq. (43.17) to solve for the age t of the specimen.

EXECUTE: a) To use Eq. (43.16), we must first find λ from Eq. (43.18):

$$T_{1/2} = 5730\text{ y} = (5730\text{ y})(3.156 \times 10^7\text{ s/y}) = 1.808 \times 10^{11}\text{ s}$$

$$\lambda = \frac{\ln 2}{T_{1/2}} = \frac{0.693}{1.808 \times 10^{11}\text{ s}} = 3.83 \times 10^{-12}\text{ s}^{-1}$$

Alternatively,

$$\lambda = \frac{0.693}{5730\text{ y}} = 1.209 \times 10^{-4}\text{ y}^{-1}$$

Then, from Eq. (43.16),

$$N(t) = \frac{-dN/dt}{\lambda} = \frac{0.255\text{ s}^{-1}}{3.83 \times 10^{-12}\text{ s}^{-1}} = 6.66 \times 10^{10}\text{ atoms}$$

The *total* number of C atoms in one gram (1/12.011 mol) is $(1/12.011)(6.022 \times 10^{23}) = 5.01 \times 10^{22}$. The ratio of ^{14}C atoms to all C atoms is

$$\frac{6.66 \times 10^{10}}{5.01 \times 10^{22}} = 1.33 \times 10^{-12}$$

Only four carbon atoms in every 3 million million are ^{14}C.

b) Assuming that the activity per gram of carbon in the specimen when it died was $0.255\text{ Bq/g} = (0.255\text{ s}^{-1}\cdot\text{g}^{-1})(3600\text{ s/h}) = 918\text{ h}^{-1}\cdot\text{g}^{-1}$, the activity of 500 mg of carbon then was $(0.500\text{ g}) \times (918\text{ h}^{-1}\cdot\text{g}^{-1}) = 459\text{ h}^{-1}$. The observed activity now, at time t later, is 174 h^{-1}. Since the activity is proportional to the number of radioactive nuclei, the activity ratio $174/459 = 0.379$ equals the number ratio $N(t)/N_0$.

Now we solve Eq. (43.17) for t and insert values for $N(t)/N_0$ and λ:

$$t = \frac{\ln(N(t)/N_0)}{-\lambda} = \frac{\ln 0.379}{-1.209 \times 10^{-4}\text{ y}^{-1}} = 8020\text{ y}$$

EVALUATE: After 8020 y the ^{14}C activity has decreased from 459 to 174 decays per hour. The specimen died and stopped taking CO_2 out of the air about 8000 years ago.

A serious health hazard in some areas is the accumulation in houses of ^{222}Rn, an inert, colorless, odorless radioactive gas. Looking at the ^{238}U decay chain (Fig. 43.8), we see that the half-life of ^{222}Rn is 3.82 days. If so, why not just move out of the house for a while and let it decay away? The answer is that ^{222}Rn is continuously being *produced* by the decay of ^{226}Ra, which is found in minute quantities in the rocks and soil on which houses are built. It's a dynamic equilibrium situation, in which the rate of production equals the rate of decay. The reason why ^{222}Rn is a bigger hazard than the other elements in the ^{238}U decay series is that it's a gas. During its short half-life of 3.82 days it can migrate from the soil into your house. If a ^{222}Rn nucleus decays in your lungs, it emits a damaging α particle and its daughter nucleus ^{218}Po, which is *not* chemically inert and is likely to stay in your lungs until it decays, emits another damaging α particle and so on down the ^{238}U decay series.

How much of a hazard is radon? Although reports indicate values as high as 3500 pCi/L, the average activity per volume in the air inside American homes due to ^{222}Rn is about 1.5 pCi/L (over a thousand decays each second in an average-sized room). *If* your environment has this level of activity, it has been estimated that a lifetime exposure would reduce your life expectancy by about 40 days. For comparison, smoking one pack of cigarettes per day reduces life expectancy by 6 years, and it is estimated that the average emission from all the nuclear power plants in the world reduces life expectancy by anywhere from 0.01 day to 5 days. These figures include catastrophies such as the 1986 nuclear reactor disaster at Chernobyl, for which the *local* effect on life expectancy is much greater.

Test Your Understanding

Which sample contains a greater number of nuclei: a 5.00-μCi sample of ^{240}Pu (half-life 6560 y) or a 4.45-μCi sample of ^{243}Am (half-life 7370 y)?

43.5 | Biological Effects of Radiation

The above discussion of radon introduced the interaction of radiation with living organisms, a topic of vital interest and importance. Under *radiation* we include radioactivity (alpha, beta, gamma, and neutrons) and electromagnetic radiation such as x rays. As these particles pass through matter, they lose energy, breaking molecular bonds and creating ions—hence the term *ionizing radiation.* Charged particles interact directly with the electrons in the material. X rays and γ rays interact by the photoelectric effect, in which an electron absorbs a photon and breaks loose from its site, or by Compton scattering (Section 38.7). Neutrons cause ionization indirectly through collisions with nuclei or absorption by nuclei with subsequent radioactive decay of the resulting nuclei.

These interactions are extremely complex. It is well known that excessive exposure to radiation, including sunlight, x rays, and all the nuclear radiations, can destroy tissues. In mild cases it results in a burn, as with common sunburn. Greater exposure can cause very severe illness or death by a variety of mechanisms, including massive destruction of tissue cells, alterations of genetic material, and destruction of the components in bone marrow that produce red blood cells.

Calculating Radiation Doses

Radiation dosimetry is the quantitative description of the effect of radiation on living tissue. The *absorbed dose* of radiation is defined as the energy delivered to the tissue per unit mass. The SI unit of absorbed dose, the joule per kiligram, is called the *gray* (Gy); 1 Gy = 1 J/kg. Another unit, in more common use at present, is the *rad,* defined as 0.01 J/kg:

$$1 \text{ rad} = 0.01 \text{ J/kg} = 0.01 \text{ Gy}$$

Absorbed dose by itself is not an adequate measure of biological effect because equal energies of different kinds of radiation cause different extents of biological effect. This variation is described by a numerical factor called the **relative bio-**

logical effectiveness (RBE), also called the *quality factor* (QF), of each specific radiation. X rays with 200 keV of energy are defined to have an RBE of unity, and the effects of other radiations can be compared experimentally. Table 43.3 shows approximate values of RBE for several radiations. All these values depend somewhat on the kind of tissue in which the radiation is absorbed and on the energy of the radiation.

The biological effect is described by the product of the absorbed dose and the RBE of the radiation; this quantity is called the *biologically equivalent dose,* or simply the equivalent dose. The SI unit of equivalent dose for humans is the Sievert (Sv):

$$\text{Equivalent dose (Sv)} = \text{RBE} \times \text{absorbed dose (Gy)} \qquad (43.20)$$

A more common unit, corresponding to the rad, is the rem (röntgen equivalent for man):

$$\text{Equivalent dose (rem)} = \text{RBE} \times \text{absorbed dose (rad)} \qquad (43.21)$$

Thus the unit of the RBE is 1 Sv/Gy or 1 rem/rad, and 1 rem = 0.01 Sv.

Table 43.3 Relative Biological Effectiveness (RBE) for Several Types of Radiation

Radiation	RBE (Sv/Gy or rem/rad)
X rays and γ rays	1
Electrons	1.0–1.5
Slow neutrons	3–5
Protons	10
α particles	20
Heavy ions	20

Example 43.10 A medical x-ray exam

During a diagnostic x-ray examination a 1.2-kg portion of a broken leg receives an equivalent dose of 0.40 mSv. a) What is the equivalent dose in mrem? b) What is the absorbed dose in mrad and mGy? c) If the x-ray energy is 50 keV, how many x-ray photons are absorbed?

SOLUTION

IDENTIFY and SET UP: In part (a) we use the conversion factor 1 rem = 0.01 Sv for equivalent dose. Table 43.3 gives the RBE for x rays; we use this value in part (b) to determine the absorbed dose using Eqs. (43.20) and (43.21). Finally, in part (c) we use the mass and the definition of absorbed dose to find the total energy absorbed and the total number of photons absorbed.

EXECUTE: a) The equivalent dose in mrem is

$$\frac{0.40\ \text{mSv}}{0.01\ \text{Sv/rem}} = 40\ \text{mrem}$$

b) For x rays, RBE = 1 rem/rad or 1 Sv/Gy, so the absorbed dose is

$$\frac{40\ \text{mrem}}{1\ \text{rem/rad}} = 40\ \text{mrad}$$

$$\frac{0.40\ \text{mSv}}{1\ \text{Sv/Gy}} = 0.40\ \text{mGy} = 4.0 \times 10^{-4}\ \text{J/kg}$$

c) The total energy absorbed is

$$(4.0 \times 10^{-4}\ \text{J/kg})(1.2\ \text{kg}) = 4.8 \times 10^{-4}\ \text{J} = 3.0 \times 10^{15}\ \text{eV}$$

The number of x-ray photons is

$$\frac{3.0 \times 10^{15}\ \text{eV}}{5.0 \times 10^{4}\ \text{eV/photon}} = 6.0 \times 10^{10}\ \text{photons}$$

EVALUATE: The absorbed dose is relatively large because x rays have a low RBE. If the ionizing radiation had been a beam of α particles, for which RBE = 20, the absorbed dose needed for an equivalent dose of 0.40 mSv would be 0.020 mGy, corresponding to a total absorbed energy of 2.4×10^{-5} J.

Radiation Hazards

Here are a few numbers for perspective. To convert from Sv to rem, simply multiply by 100. An ordinary chest x-ray exam delivers about 0.20 to 0.40 mSv to about 5 kg of tissue. Radiation exposure from cosmic rays and natural radioactivity in soil, building materials, and so on is of the order of 1.0 mSv per year at sea level and twice that at an elevation of 1500 m (5000 ft). A whole-body dose of up to about 0.20 Sv causes no immediately detectable effect. A short-term whole-body

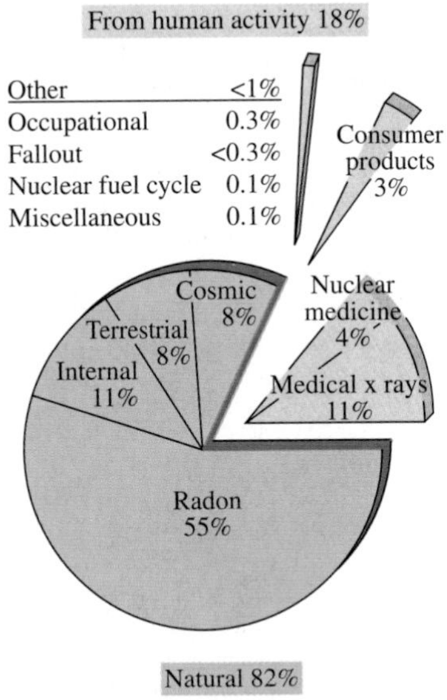

43.10 Contribution of various sources to the total average radiation exposure in the U.S. population, expressed as percentages of the total.

dose of 5 Sv or more usually causes death within a few days or weeks. A localized dose of 100 Sv causes complete destruction of the exposed tissues.

The long-term hazards of radiation exposure in causing various cancers and genetic defects have been widely publicized, and the question of whether there is any "safe" level of radiation exposure has been hotly debated. U.S. government regulations are based on a maximum *yearly* exposure, from all except natural resources, of 2 to 5 mSv. Workers with occupational exposure to radiation are permitted 50 mSv per year. Recent studies suggest that these limits are too high and that even extremely small exposures carry hazards, but it is very difficult to gather reliable statistics on the effects of low dosages. It has become clear that any use of x rays for medical diagnosis should be preceded by a very careful estimation of the relation of risk to possible benefit.

Another sharply debated question is that of radiation hazards from nuclear power plants. The radiation level from these plants is *not* negligible. However, to make a meaningful evaluation of hazards, we must compare these levels with the alternatives, such as coal-powered plants. The health hazards of coal smoke are serious and well documented, and the natural radioactivity in the smoke from a coal-fired power plant is believed to be roughly 100 times as great as that from a properly operating nuclear plant with equal capacity. But the comparison is not this simple; the possibility of a nuclear accident and the very serious problem of safe disposal of radioactive waste from nuclear plants must also be considered. It is clearly impossible to eliminate *all* hazards to health. Our goal should be to try to take a rational approach to the problem of *minimizing* the hazard from all sources. Figure 43.10 shows one estimate of the various sources of radiation exposure for the U.S. population. Ionizing radiation is a two-edged sword; it poses very serious health hazards, yet it also provides many benefits to humanity, including the diagnosis and treatments of disease and a wide variety of analytical techniques.

Beneficial Uses of Radiation

Radiation is widely used in medicine for intentional selective destruction of tissue such as tumors. The hazards are considerable, but if the disease would be fatal without treatment, any hazard may be preferable. Artificially produced isotopes are often used as sources. Such isotopes have several advantages over naturally radioactive isotopes. They may have shorter half-lives and correspondingly greater activity. Isotopes can be chosen that emit the type and energy of radiation desired. Some artificial isotopes have been replaced by photon and electron beams from linear accelerators.

Nuclear medicine is an expanding field of application. Radioactive isotopes have virtually the same electron configurations and resulting chemical behavior as stable isotopes of the same element. But the location and concentration of radioactive isotopes can easily be detected by measurements of the radiation they emit. A familiar example is the use of radioactive iodine for thyroid studies. Nearly all the iodine ingested is either eliminated or stored in the thyroid, and the body's chemical reactions do not discriminate between the unstable isotope ^{131}I and the stable isotope ^{127}I. A minute quantity of ^{131}I is fed or injected into the patient, and the speed with which it becomes concentrated in the thyroid provides a measure of thyroid function. The half-life is 8.02 days, so there are no long-lasting radiation hazards. By use of more sophisticated scanning detectors, one can also obtain a "picture" of the thyroid, which shows enlargement and other abnormalities. This procedure, a type of *autoradiography,* is comparable to photographing the glowing filament of an incandescent light bulb by using the light emitted by the filament itself. If this process discovers cancerous thyroid nodules, they can be destroyed by much larger quantities of ^{131}I.

Another useful nuclide for nuclear medicine is technetium-99 (^{99}Tc), which is formed in an excited state by the β^- decay of molybdenum (^{99}Mo). The technetium then decays to its ground state by emitting a γ-ray photon with energy 143 KeV. The half-life is 6.01 hours, unusually long for γ emission. (The ground state of ^{99}Tc is also unstable, with a half-life of 2.11×10^5 y; it decays by β^- emission to the stable ruthenium nuclide ^{99}Ru.) The chemistry of technetium is such that it can readily be attached to organic molecules that are taken up by various organs of the body. A small quantity of such technetium-bearing molecules is injected into a patient, and a scanning detector or *gamma camera* is used to produce an image, or *scintigram,* that reveals which parts of the body take up these γ-emitting molecules. This technique, in which ^{99}Tc acts as a radioactive *tracer,* plays an important role in locating cancers, embolisms, and other pathologies (Fig. 43.11).

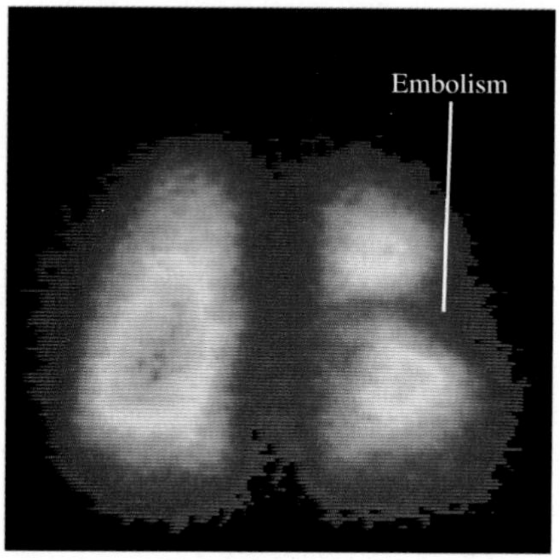

43.11 This colored scintigram shows where a chemical containing radioactive ^{99}Tc was taken up by a patient's lungs. The orange color in the lung on the left indicates strong γ-ray emission by the ^{99}Tc, which shows that the chemical was able to pass into this lung through the bloodstream. The lung on the right shows weaker emission, indicating the presence of an embolism (a blood clot or other obstruction in an artery) that is restricting the flow of blood to this lung.

Tracer techniques have many other applications. Tritium (^3H), a hydrogen isotope, is used to tag molecules in complex organic reactions; radioactive tags on pesticide molecules, for example, can be used to trace their passage through food chains. In the world of machinery, radioactive iron can be used to study piston-ring wear. Laundry detergent manufacturers have even tested the effectiveness of their products using radioactive dirt.

Many direct effects of radiation are also useful, such as strengthening of polymers by cross-linking, sterilizing surgical tools, dispersion of unwanted static electricity in the air, and intentional ionization of air in smoke detectors. Gamma rays are also being used to sterilize and preserve some food products.

Test Your Understanding

Why is ^{99}Tc a better choice for medical studies than radioactive nuclides that undergo α decay?

43.6 | Nuclear Reactions

In the preceding sections we studied the decay of unstable nuclei, especially spontaneous emission of an α or β particle, sometimes followed by γ emission. Nothing was done to initiate this decay, and nothing could be done to control it. This section examines some *nuclear reactions,* rearrangements of nuclear components that result from a bombardment by a particle rather than a spontaneous natural process. Rutherford suggested in 1919 that a massive particle with sufficient kinetic energy might be able to penetrate a nucleus. The result would be either a new nucleus with greater atomic number and mass number or a decay of the original nucleus. Rutherford bombarded nitrogen (^{14}N) with α particles and obtained an oxygen (^{17}O) nucleus and a proton:

$$^4_2He + {}^{14}_7N \rightarrow {}^{17}_8O + {}^1_1H \qquad (43.22)$$

Rutherford used alpha particles from naturally radioactive sources. In Chapter 44 we'll describe some of the particle accelerators that are used nowadays to initiate nuclear reactions.

Nuclear reactions are subject to several *conservation laws.* The classical conservation principles for charge, momentum, angular momentum, and energy (including rest energies) are obeyed in all nuclear reactions. An additional conservation law, not anticipated by classical physics, is conservation of the total number of nucleons. The numbers of protons and neutrons need not be conserved separately; we have seen that in β decay, neutrons and protons change into one another. We'll study the basis of the conservation of nucleon number in Chapter 44.

When two nuclei interact, charge conservation requires that the sum of the initial atomic numbers must equal the sum of the final atomic numbers. Because of conservation of nucleon number, the sum of the initial mass numbers must also equal the sum of the final mass numbers. In general, these are *not* elastic collisions, and, correspondingly, the total initial mass does *not* equal the total final mass.

Reaction Energy

The difference between the masses before and after the reaction corresponds to the **reaction energy,** according to the mass-energy relation $E = mc^2$. If initial particles A and B interact to produce final particles C and D, the reaction energy Q is defined as

$$Q = (M_A + M_B - M_C - M_D)c^2 \qquad \text{(reaction energy)} \qquad (43.23)$$

To balance the electrons, we use the neutral atomic masses in Eq. (43.23). That is, we use the mass of ${}^1_1\text{H}$ for a proton, ${}^2_1\text{H}$ for a deuteron, ${}^4_2\text{He}$ for an α particle, and so on. When Q is positive, the total mass decreases and the total kinetic energy increases. Such a reaction is called an *exoergic reaction.* When Q is negative, the mass increases and the kinetic energy decreases, and the reaction is called an *endoergic reaction.* The terms *exothermal* and *endothermal,* borrowed from chemistry, are also used. In an endoergic reaction the reaction cannot occur at all unless the initial kinetic energy in the center of mass reference frame is at least as great as $|Q|$. That is, there is a **threshold energy,** the minimum kinetic energy to make an endoergic reaction go.

Example 43.11 An exoergic reaction

When lithium (${}^7\text{Li}$) is bombarded by a proton, two alpha particles (${}^4\text{He}$) are produced. Find the reaction energy.

SOLUTION

IDENTIFY and SET UP: We determine the reaction energy using Eq. (43.23). The required masses are given in Table 43.2.

EXECUTE: The reaction can be written

$${}^1_1\text{H} + {}^7_3\text{Li} \rightarrow {}^4_2\text{He} + {}^4_2\text{He}$$

Here are the initial and final masses:

A: ${}^1_1\text{H}$	1.007825 u	C: ${}^4_2\text{He}$	4.002603 u
B: ${}^7_3\text{Li}$	7.016004 u	D: ${}^4_2\text{He}$	4.002603 u
	8.023829 u		8.005206 u

We see that

$$M_A + M_B - M_C - M_D = 0.018623 \text{ u}$$

Then Eq. (43.23) gives a reaction energy of

$$Q = (0.018623 \text{ u})(931.5 \text{ MeV/u}) = 17.35 \text{ MeV}$$

EVALUATE: This is an exoergic reaction; the final total kinetic energy of the two separating alpha particles is 17.35 MeV *greater* than the initial total kinetic energy of the proton and the lithium nucleus.

Example 43.12 An endoergic reaction

Calculate the reaction energy for the nuclear reaction represented by Eq. (43.22).

SOLUTION

IDENTIFY and SET UP: As in Example 43.11, we use masses from Table 43.2 in Eq. (43.23).

EXECUTE: The masses of the various particles are

A: ${}^4_2\text{He}$	4.002603 u	C: ${}^{17}_{8}\text{O}$	16.999132 u
B: ${}^{14}_{7}\text{N}$	14.003074 u	D: ${}^1_1\text{H}$	1.007825 u
	18.005677 u		18.006957 u

We see that the mass increases by 0.001280 u, and the corresponding reaction energy is

$$Q = (-0.001280\ \text{u})(931.5\ \text{MeV/u}) = -1.192\ \text{MeV}$$

EVALUATE: This answer means that in the center-of-mass system, that is, in a head-on collision with zero total momentum, the minimum total initial kinetic energy for this reaction to occur is 1.192 MeV.

Ordinarily, the endoergic reaction of Example 43.12 would be produced by bombarding stationary ${}^{14}\text{N}$ nuclei with alpha particles from an accelerator. In this case an alpha's kinetic energy must be *greater than* 1.191 MeV. If all the alpha's kinetic energy went solely to increasing the rest energy, the final kinetic energy would be zero, and momentum would not be conserved. When a particle with mass m and kinetic energy K collides with a stationary particle with mass M, the total kinetic energy K_{cm} in the center-of-mass coordinate system (the energy available to cause reactions) is

$$K_{\text{cm}} = \frac{M}{M + m}K \tag{43.24}$$

This expression assumes that the kinetic energies of the particles and nuclei are much less than their rest energies. We leave the derivation of Eq. (43.24) as a problem. In the present example, $K_{\text{cm}} = (14.00/18.01)K$, so K must be at least $(18.01/14.00) \times (1.191\ \text{MeV}) = 1.532\ \text{MeV}$.

For a charged particle such as a proton or an α particle to penetrate the nucleus of another atom and cause a reaction, it must usually have enough initial kinetic energy to overcome the potential-energy barrier caused by the repulsive electrostatic forces. In the reaction of Example 43.11, if we treat the proton and the ${}^7\text{Li}$ nucleus as spherically symmetric charges with radii given by Eq. (43.1), their centers will be 3.5×10^{-15} m apart when they touch. The repulsive potential energy of the proton (charge $+e$) and the ${}^7\text{Li}$ nucleus (charge $+3e$) at this separation r is

$$U = \frac{1}{4\pi\epsilon_0}\frac{(e)(3e)}{r} = (9.0 \times 10^9\ \text{N}\cdot\text{m}^2/\text{C}^2)\frac{(3)(1.6 \times 10^{-19}\ \text{C})^2}{3.5 \times 10^{-15}\ \text{m}}$$
$$= 2.0 \times 10^{-13}\ \text{J} = 1.2\ \text{MeV}$$

Even though the reaction is exoergic, the proton must have a minimum kinetic energy of about 1.2 MeV for the reaction to occur, unless the proton *tunnels* through the barrier (see Section 40.3).

Neutron Absorption

Absorption of *neutrons* by nuclei forms an important class of nuclear reactions. Heavy nuclei bombarded by neutrons in a nuclear reaction can undergo a series of neutron absorptions alternating with beta decays, in which the mass number A increases by as much as 25. Some of the *transuranic elements,* elements having Z larger than 92, are produced in this way. These elements have not been found in nature. Many transuranic elements, having Z possibly as high as 118, have been identified.

The analytical technique of *neutron activation analysis* uses similar reactions. When bombarded by neutrons, many stable nuclides absorb a neutron to become unstable and then undergo β^- decay. The energies of the β^- and γ emissions depend on the unstable nuclide and provide a means of identifying it and the original stable nuclide. Quantities of elements that are far too small for conventional chemical analysis can be detected in this way.

Test Your Understanding

If the reaction described in Example 43.11 is exoergic, why doesn't it happen naturally when a sample of solid lithium is placed in a flask of hydrogen gas?

43.7 | Nuclear Fission

Nuclear fission is a decay process in which an unstable nucleus splits into two fragments of comparable mass. Fission was discovered in 1938 through the experiments of Otto Hahn and Fritz Strassman in Germany. Pursuing earlier work by Fermi, they bombarded uranium $(Z = 92)$ with neutrons. The resulting radiation did not coincide with that of any known radioactive nuclide. Urged on by their colleague Lise Meitner, they used meticulous chemical analysis to reach the astonishing but inescapable conclusion that they had found a radioactive isotope of barium $(Z = 56)$. Later, radioactive krypton $(Z = 36)$ was also found. Meitner and Otto Frisch correctly interpreted these results as showing that uranium nuclei were splitting into two massive fragments called *fission fragments.* Two or three free neutrons usually appear along with the fission fragments and, very occasionally, a light nuclide such as ^3H.

Both the common isotope (99.3%) ^{238}U and the uncommon isotope (0.7%) ^{235}U (as well as several other nuclides) can be easily split by neutron bombardment: ^{235}U by slow neutrons but ^{238}U only by neutrons with a minimum of about 1 MeV of energy. Fission resulting from neutron absorption is called *induced fission.* Some nuclides can also undergo *spontaneous fission* without initial neutron absorption, but this is quite rare. When ^{235}U absorbs a neutron, the resulting nuclide $^{236}U^*$ is in a highly excited state and splits into two fragments almost instantaneously. Strictly speaking, it is $^{236}U^*$, not ^{235}U, that undergoes fission, but it's usual to speak of the fission of ^{235}U.

Over 100 different nuclides, representing more than 20 different elements, have been found among the fission products. Figure 43.12 shows the distribution of mass numbers for fission fragments from the fission of ^{235}U. Most of the fragments have mass numbers from 90 to 100 and from 135 to 145; fission into two fragments with nearly equal mass is unlikely.

You should check the following two typical fission reactions for conservation of nucleon number and charge:

$$^{235}_{92}U + ^{1}_{0}n \rightarrow ^{236}_{92}U^* \rightarrow ^{144}_{56}Ba + ^{89}_{36}Kr + 3^{1}_{0}n$$

$$^{235}_{92}U + ^{1}_{0}n \rightarrow ^{236}_{92}U^* \rightarrow ^{140}_{54}Xe + ^{94}_{38}Sr + 2^{1}_{0}n$$

The total kinetic energy of the fission fragments is enormous, about 200 MeV (compared to typical α and β energies of a few MeV). The reason for this is that nuclides at the high end of the mass spectrum (near $A = 240$) are less tightly bound than those nearer the middle ($A = 90$ to 145). Referring to Fig. 43.2, we see that the average binding energy per nucleon is about 7.6 MeV at $A = 240$ but about 8.5 MeV at $A = 120$. Therefore a rough estimate of the expected *increase*

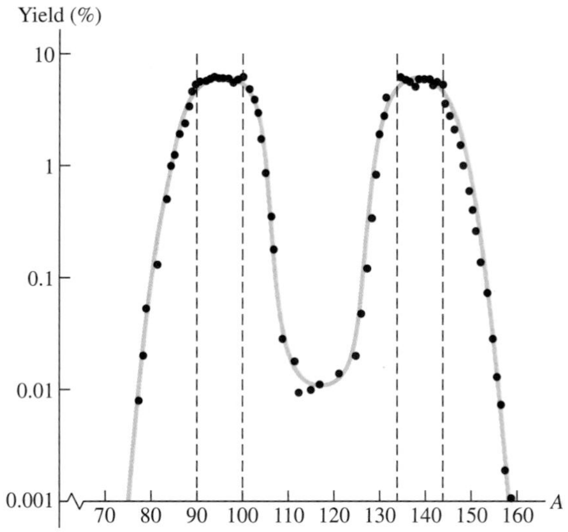

43.12 Mass distribution of fission fragments from the fission of $^{236}U^*$ (an excited state of ^{236}U), which is produced when ^{235}U absorbs a neutron. The vertical scale is logarithmic.

in binding energy during fission is about 8.5 MeV − 7.6 MeV = 0.9 MeV per nucleon, or a total of $(235)(0.9\text{ MeV}) \approx 200\text{ MeV}$.

CAUTION It may seem to be a violation of conservation of energy to have an increase in both the binding energy and the kinetic energy during a fission reaction. But relative to the total rest energy E_0 of the separated nucleons, the rest energy of the nucleus is E_0 *minus* E_B. Thus an *increase* in binding energy corresponds to a *decrease* in rest energy as rest energy is converted to the kinetic energy of the fission fragments.

Fission fragments always have too many neutrons to be stable. We noted in Section 43.3 that the neutron–proton ratio (N/Z) for stable nuclides is about 1 for light nuclides but almost 1.6 for the heaviest nuclides because of the increasing influence of the electrical repulsion of the protons. The N/Z value for stable nuclides is about 1.3 at $A = 100$ and 1.4 at $A = 150$. The fragments have about the same N/Z as ^{235}U, about 1.55. They usually respond to this surplus of neutrons by undergoing a series of β^- decays (each of which increases Z by one and decreases N by one) until a stable value of N/Z is reached. A typical example is

$$^{140}_{54}\text{Xe} \xrightarrow{\beta^-} {}^{140}_{55}\text{Cs} \xrightarrow{\beta^-} {}^{140}_{56}\text{Ba} \xrightarrow{\beta^-} {}^{140}_{57}\text{La} \xrightarrow{\beta^-} {}^{140}_{58}\text{Ce}$$

The nuclide ^{140}Ce is stable. This series of β^- decays produces, on average, about 15 MeV of additional kinetic energy. The neutron excess of fission fragments also explains why two or three free neutrons are released during the fission.

Fission appears to set an upper limit on the production of transuranic nuclei, mentioned in Section 43.6, that are relatively stable. There are theoretical reasons to expect that nuclei near $Z = 114$, $N = 184$ or 196, might be stable with respect to spontaneous fission. In the shell model (Section 43.2), these numbers correspond to filled shells and subshells in the nuclear energy-level structure. Such *superheavy nuclei* would still be unstable with respect to alpha emission, but they might live long enough to be identified. As of this writing, there is evidence that nuclei with $Z = 114$ have been produced in the laboratory.

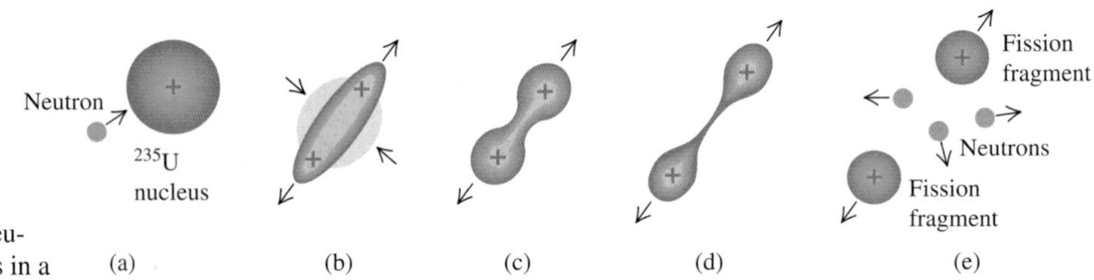

43.13 (a) A ^{235}U nucleus absorbs a neutron. (b) The resulting ^{236}U* nucleus is in a highly excited state and oscillates strongly. (c) A neck develops, and electrical repulsion pushes the two lobes apart. (d) The two lobes separate, forming fission fragments. (e) Neutrons are emitted from the fission fragments at the time of fission or occasionally a few seconds later.

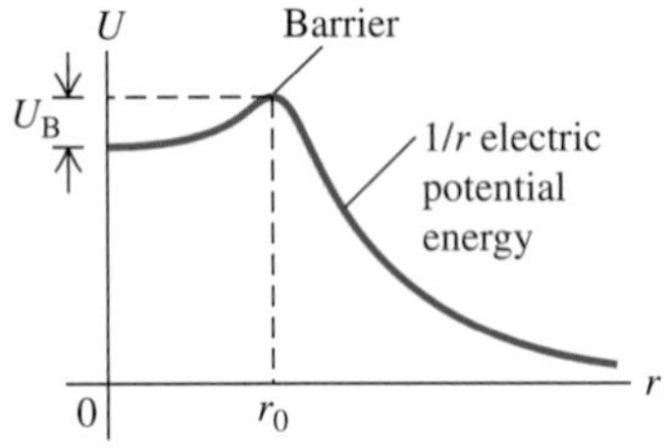

43.14 Hypothetical potential-energy function for two fission fragments in a fissionable nucleus. At distances r beyond the range of the nuclear force, the potential energy varies approximately as $1/r$. Fission occurs if there is an excitation energy greater than U_B or an appreciable probability for tunneling through the potential-energy barrier.

Liquid-Drop Model

We can understand fission qualitatively on the basis of the liquid-drop model of the nucleus (Section 43.2). The process is shown in Fig. 43.13 in terms of an electrically charged liquid drop. These sketches shouldn't be taken too literally, but they may help to develop your intuition about fission. A ^{235}U nucleus absorbs a neutron (Fig. 43.13a), becoming a ^{236}U* nucleus with excess energy (Fig. 43.13b). This excess energy causes violent oscillations (Fig. 43.13c), during which a neck between two lobes develops. The electrical repulsion of these two lobes stretches the neck farther (Fig. 43.13d), and finally two smaller drops are formed (Fig. 43.13e) that move rapidly apart.

This qualitative picture has been developed into a more quantitative theory to explain why some nuclei undergo fission and others don't. Figure 43.14 shows a hypothetical potential-energy function for two possible fission fragments. If neutron absorption results in an excitation energy greater than the energy barrier height U_B, fission occurs immediately. Even when there isn't quite enough energy to surmount the barrier, fission can take place by quantum-mechanical *tunneling,* discussed in Section 40.3. In principle, many stable heavy nuclei can fission by tunneling. But the probability depends very critically on the height and width of the barrier. For most nuclei this process is so unlikely that it is never observed.

Chain Reactions

Fission of a uranium nucleus, triggered by neutron bombardment, releases other neutrons that can trigger more fissions, suggesting the possibility of a **chain reaction** (Fig. 43.15). The chain reaction may be made to proceed slowly and in a controlled manner in a nuclear reactor or explosively in a bomb. The energy release in a nuclear chain reaction is enormous, far greater than that in any chemical reaction. (In a sense, *fire* is a chemical chain reaction.) For example, when uranium is "burned" to uranium dioxide in the chemical reaction

$$U + O_2 \rightarrow UO_2$$

the heat of combustion is about 4500 J/g. Expressed as energy per atom, this is about 11 eV per atom. By contrast, fission liberates about 200 MeV per atom, nearly 20 million times as much energy.

Nuclear Reactors

A *nuclear reactor* is a system in which a controlled nuclear chain reaction is used to liberate energy. In a nuclear power plant, this energy is used to generate steam, which operates a turbine and turns an electrical generator.

On average, each fission of a ^{235}U nucleus produces about 2.5 free neutrons, so 40% of the neutrons are needed to sustain a chain reaction. A ^{235}U nucleus is

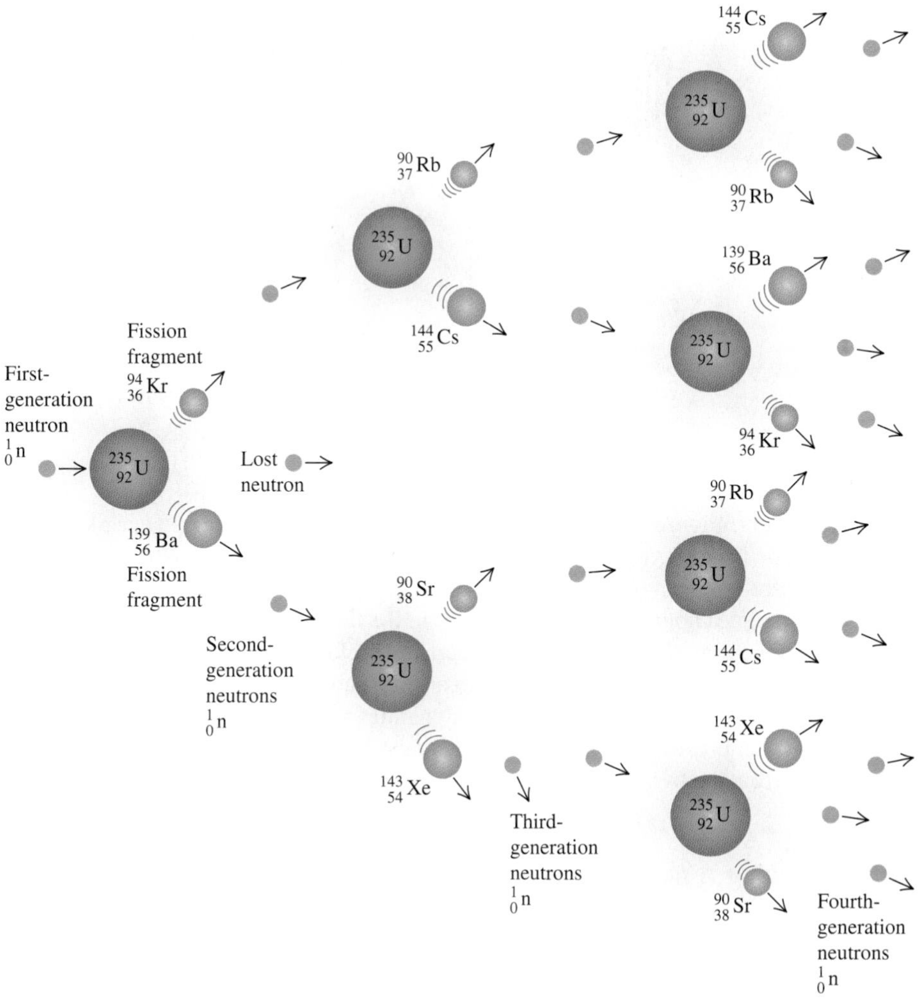

43.15 Schematic diagram of a nuclear fission chain reaction.

much more likely to absorb a low-energy neutron (less than 1 eV) than one of the higher-energy neutrons (1 MeV or so) that are liberated during fission. In a nuclear reactor the higher-energy neutrons are slowed down by collisions with nuclei in the surrounding material, called the *moderator,* so they are much more likely to cause further fissions. In nuclear power plants, the moderator is often water, occasionally graphite. The *rate* of the reaction is controlled by inserting or withdrawing *control rods* made of elements (such as boron or cadmium) whose nuclei *absorb* neutrons without undergoing any additional reaction. The isotope ^{238}U can also absorb neutrons, leading to $^{239}U^*$, but not with high enough probability for it to sustain a chain reaction by itself. Thus uranium that is used in reactors is often "enriched" by increasing the proportion of ^{235}U above the natural value of 0.7%, typically to 3% or so, by isotope-separation processing.

The most familiar application of nuclear reactors is for the generation of electric power. As was noted above, the fission energy appears as kinetic energy of the fission fragments, and its immediate result is to increase the internal energy of the fuel elements and the surrounding moderator. This increase in internal energy is transferred as heat to generate steam to drive turbines, which spin the electrical

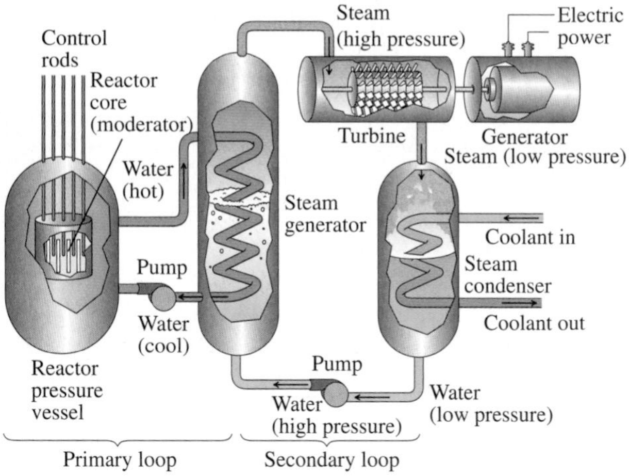

43.16 Schematic diagram of a nuclear power plant.

generators. Figure 43.16 is a schematic diagram of a nuclear power plant. The energetic fission fragments heat the water surrounding the reactor core. The steam generator is a heat exchanger that takes heat from this highly radioactive water and generates nonradioactive steam to run the turbines.

A typical nuclear plant has an electric-generating capacity of 1000 MW (or 10^9 W). The turbines are heat engines and are subject to the efficiency limitations imposed by the second law of thermodynamics, discussed in Chapter 20. In modern nuclear plants the overall efficiency is about one-third, so 3000 MW of thermal power from the fission reaction is needed to generate 1000 MW of electrical power.

Example 43.13 Uranium consumption in a nuclear reactor

What mass of ^{235}U has to undergo fission each day to provide 3000 MW of thermal power?

SOLUTION

IDENTIFY and SET UP: The critical bit of information is that fission liberates about 200 MeV per atom. We use this fact along with the mass of the ^{235}U atom to determine the required amount of uranium.

EXECUTE: Each second, we need 3000 MJ or 3000×10^6 J. Each fission provides 200 MeV, which is

$$(200\ \text{MeV})(1.6 \times 10^{-13}\ \text{J/MeV}) = 3.2 \times 10^{-11}\ \text{J}$$

The number of fissions needed each second is

$$\frac{3000 \times 10^6\ \text{J}}{3.2 \times 10^{-11}\ \text{J}} = 9.4 \times 10^{19}$$

Each ^{235}U atom has a mass of $(235\ \text{u})(1.66 \times 10^{-27}\ \text{kg/u}) = 3.9 \times 10^{-25}$ kg, so the mass of ^{235}U needed each second is

$$(9.4 \times 10^{19})(3.9 \times 10^{-25}\ \text{kg}) = 3.7 \times 10^{-5}\ \text{kg} = 37\ \mu\text{g}$$

In one day (86,400 s), the total consumption of ^{235}U is

$$(3.7 \times 10^{-5}\ \text{kg/s})(86{,}400\ \text{s}) = 3.2\ \text{kg}$$

EVALUATE: For comparison, a 1000-MW coal-fired power plant burns 10,600 tons (about ten million kg) of coal per day!

Nuclear fission reactors have many other practical uses. Among these are the production of artificial radioactive isotopes for medical and other research, production of high-intensity neutron beams for research in nuclear structure, and production of fissionable nuclides such as ^{239}Pu from the common isotope ^{238}U. The last is the function of *breeder reactors,* which can produce more fuel than they use.

We mentioned above that about 15 MeV of the energy released after fission of a ^{235}U nucleus comes from the β^- decays of the fission fragments. This fact poses a serious problem with respect to control and safety of reactors. Even after the chain reaction has been completely stopped by insertion of control rods into the core, heat continues to be evolved by the β^- decays, which cannot be stopped. For a 3000-MW reactor this heat power is initially very large, about 200 MW. In the event of total loss of cooling water, this power is more than enough to cause a catastrophic meltdown of the reactor core and possible penetration of the containment vessel. The difficulty in achieving a "cold shutdown" following an accident at the Three Mile Island nuclear power plant in Pennsylvania in March 1979 was a result of the continued evolution of heat due to β^- decays.

The catastrophe of April 26, 1986, at Chernobyl reactor No. 4 in Ukraine resulted from a combination of an inherently unstable design and several human errors committed during a test of the emergency core cooling system. Too many control rods were withdrawn to compensate for a decrease in power caused by a buildup of neutron absorbers such as ^{135}Xe. The power level rose from 1% of normal to 100 times normal in 4 seconds; a steam explosion ruptured pipes in the core cooling system and blew the heavy concrete cover off the reactor. The graphite moderator caught fire and burned for several days, and there was a meltdown of the core. The total activity of the radioactive material released into the atmosphere has been estimated as about 10^8 Ci.

Test Your Understanding

The fission of ^{235}U can be triggered by the absorption of a slow neutron by a nucleus. Give one reason why a slow *proton* cannot be used to trigger ^{235}U fission.

43.8 | Nuclear Fusion

ActivPhysics Online 19.3 Fusion

In a **nuclear fusion** reaction, two or more small light nuclei come together, or *fuse,* to form a larger nucleus. Fusion reactions release energy for the same reason as fission reactions; the binding energy per nucleon after the reaction is greater than before. Referring to Fig. 43.2, we see that the binding energy per nucleon increases with A up to about $A = 60$, so fusion of nearly any two light nuclei to make a nucleus with A less than 60 is likely to be an exoergic reaction. In comparison to fission, we are moving toward the peak of this curve from the opposite side. Another way to express the energy relations is that the total mass of the products is less than that of the initial particles.

Here are three examples of energy-liberating fusion reactions, written in terms of the neutral atoms:

$$^1_1\text{H} + {}^1_1\text{H} \rightarrow {}^2_1\text{H} + \beta^+ + \nu_e$$

$$^2_1\text{H} + {}^1_1\text{H} \rightarrow {}^3_2\text{He} + \gamma$$

$$^3_2\text{He} + {}^3_2\text{He} \rightarrow {}^4_2\text{He} + {}^1_1\text{H} + {}^1_1\text{H}$$

In the first reaction, two protons combine to form a deuteron (^{2}H), with the emission of a positron (β^+) and an electron neutrino. In the second, a proton and a deuteron combine to form the nucleus of the light isotope of helium, ^{3}He, with the emission of a gamma ray. Now double the first two reactions to provide the two ^{3}He nuclei that fuse in the third reaction to form an alpha particle (^{4}He) and two protons. Together the reactions make up the process called the *proton-proton chain* (Fig. 43.17).

The net effect of the chain is the conversion of four protons into one α particle, two positrons, two electron neutrinos, and two γ's. We can calculate the energy

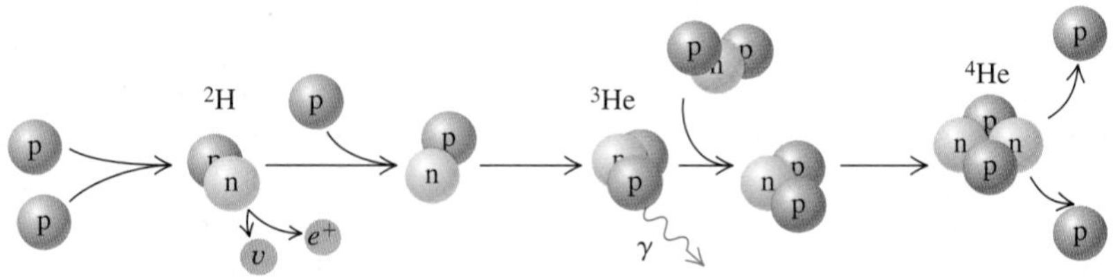

43.17 The proton-proton chain.

43.18 The energy released as starlight comes from fusion reactions deep within a star's interior. When a star is first formed and for most of its life, it converts the hydrogen in its core into helium. As a star ages, the core temperature can become high enough for additional fusion reactions that convert helium into carbon, oxygen, and other elements.

release from this part of the process: The mass of an α particle plus two positrons is the mass of neutral ^{4}He, the neutrinos have zero (or negligible) mass, and the gammas have zero mass.

Mass of four protons	4.029106 u
Mass of ^{4}He	4.002603 u
Mass difference and energy release	0.026503 u and 24.69 MeV

The two positrons that are produced during the first step of the proton-proton chain collide with two electrons; mutual annihilation of the four particles takes place, and their rest energy is converted into $4(0.511\text{ MeV}) = 2.044\text{ MeV}$ of gamma radiation. Thus the total energy release is $(24.69 + 2.044)\text{ MeV} = 26.73\text{ MeV}$. The proton-proton chain takes place in the interior of the sun and other stars (Fig. 43.18). Each gram of the sun's mass contains about 4.5×10^{23} protons. If all of these protons were fused into helium, the energy released would be about 130,000 kWh. If the sun were to continue to radiate at its present rate, it would take about 75×10^9 years to exhaust its supply of protons. As we will see below, fusion reactions can only occur at extremely high temperatures; in the sun, these temperatures are found only deep within the interior. Hence the sun cannot fuse *all* of its protons, and can do so for a total of only about 10×10^9 years in total. The present age of the solar system (including the sun) is 4.6×10^9 years, so the sun is about halfway through its available store of protons.

Example 43.14 A fusion reaction

Two deuterons fuse to form a *triton* (a nucleus of tritium, or ^{3}H) and a proton. How much energy is liberated?

SOLUTION

IDENTIFY and SET UP: This is a nuclear reaction of the type discussed in Section 43.6. We find the energy released using Eq. (43.23).

EXECUTE: Adding one electron to each particle makes each a neutral atom; we find their masses in Table 43.2. Substituting into Eq. (43.23), we find

$$\begin{aligned} Q &= [2(2.014102\text{ u}) - 3.016049\text{ u} - 1.007825\text{ u}] \\ &\quad \times (931.5\text{ MeV/u}) \\ &= 4.03\text{ MeV} \end{aligned}$$

EVALUATE: Thus 4.03 MeV is released in the reaction; the triton and proton together have 4.03 MeV more kinetic energy than the two deuterons had together.

Achieving Fusion

For two nuclei to undergo fusion, they must come together to within the range of the nuclear force, typically of the order of 2×10^{-15} m. To do this, they must overcome the electrical repulsion of their positive charges. For two protons at this distance, the corresponding potential energy is about 1.2×10^{-13} J or 0.7 MeV;

this represents the total initial *kinetic* energy that the fusion nuclei must have—for example, 0.6×10^{-13} J each in a head-on collision.

Atoms have this much energy only at extremely high temperatures. The discussion of Section 18.3 showed that the average translational kinetic energy of a gas molecule at temperature T is $\frac{3}{2}kT$, where k is Boltzmann's constant. The temperature at which this is equal to $E = 0.6 \times 10^{-13}$ J is determined by the relation

$$E = \frac{3}{2}kT$$

$$T = \frac{2E}{3k} = \frac{2(0.6 \times 10^{-13}\ \text{J})}{3(1.38 \times 10^{-23}\ \text{J/K})} = 3 \times 10^{9}\ \text{K}$$

Fusion reactions are possible at lower temperatures because the Maxwell-Boltzmann distribution function (Section 18.5) gives a small fraction of protons with kinetic energies much higher than the average value. The proton-proton reaction occurs at "only" 1.5×10^{7} K at the center of the sun, making it an extremely low-probability process; but that's why the sun is expected to last so long. At these temperatures the fusion reactions are called *thermonuclear* reactions.

Intensive efforts are under way to achieve controlled fusion reactions, which potentially represent an enormous new resource of energy. At the temperatures mentioned, light atoms are fully ionized, and the resulting state of matter is called a *plasma.* In one kind of experiment using *magnetic confinement,* a plasma is heated to extremely high temperature by an electrical discharge, while being contained by appropriately shaped magnetic fields. In another, using *inertial confinement,* pellets of the material to be fused are heated by a high-intensity laser beam. Some of the reactions being studied are

$$^{2}_{1}\text{H} + {}^{2}_{1}\text{H} \rightarrow {}^{3}_{1}\text{H} + {}^{1}_{1}\text{H} + 4.0\ \text{MeV} \qquad (1)$$

$$^{3}_{1}\text{H} + {}^{2}_{1}\text{H} \rightarrow {}^{4}_{2}\text{He} + {}^{1}_{0}\text{n} + 17.6\ \text{MeV} \qquad (2)$$

$$^{2}_{1}\text{H} + {}^{2}_{1}\text{H} \rightarrow {}^{3}_{2}\text{He} + {}^{1}_{0}\text{n} + 3.3\ \text{MeV} \qquad (3)$$

$$^{3}_{2}\text{He} + {}^{2}_{1}\text{H} \rightarrow {}^{4}_{2}\text{He} + {}^{1}_{1}\text{H} + 18.3\ \text{MeV} \qquad (4)$$

We described the first reaction in Example 43.14; two deuterons fuse to form a triton and a proton. In the second, a triton combines with another deuteron to form an alpha particle and a neutron. The result of both of these reactions together is the conversion of three deuterons into an alpha particle, a proton, and a neutron, with the liberation of 21.6 MeV of energy. Reactions (3) and (4) together achieve the same conversion. In a plasma that contains deuterons, the two pairs of reactions occur with roughly equal probability. As yet, no one has succeeded in producing these reactions under controlled conditions in such a way as to yield a net surplus of usable energy.

Methods of achieving fusion that don't require high temperatures are also being studied; these are called *cold fusion.* One scheme that does work uses an unusual hydrogen molecule ion. The usual H_2^+ ion consists of two protons bound by one shared electron; the nuclear spacing is about 0.1 nm. If the protons are replaced by a deuteron (^{2}H) and a triton (^{3}H) and the electron by a *muon,* which is 208 times as massive as the electron, the spacing is made smaller by a factor of 208. The probability then becomes appreciable for the two nuclei to tunnel through the thin repulsive potential-energy barrier and fuse in reaction (2) above. The prospect of making this process, called *muon-catalyzed fusion,* into a practical energy source is still distant.

Test Your Understanding

Are all fusion reactions exoergic?

SUMMARY

A nucleus is composed of A nucleons (Z protons and N neutrons). All nuclei have about the same density. The radius of a nucleus with mass number A is given approximately by Eq. (43.1). A single nuclear species is called a nuclide. Isotopes are nuclides of the same element that have different numbers of neutrons. Nuclear masses are measured in atomic mass units. Nucleons have angular momentum and a magnetic moment. (See Examples 43.1 and 43.2)

$$R = R_0 A^{1/3} \quad (43.1)$$

$$(R_0 = 1.2 \times 10^{-15}\ \text{m})$$

The mass of a nucleus is always less than the mass of the protons and neutrons within it. The mass difference multiplied by c^2 gives the binding energy E_B. The binding energy for a given nuclide is determined by the nuclear force, which is short-range and favors pairs of particles, and by the electric repulsion between protons. A nucleus is unstable if A or Z is too large or if the ratio N/Z is wrong. Two widely used models of the nucleus are the liquid-drop model and the shell model; the latter is analogous to the central-field approximation for atomic structure.
(See Examples 43.3 and 43.4)

$$E_B = (ZM_H + Nm_n - {}^{A}_{Z}M)c^2 \quad (43.10)$$

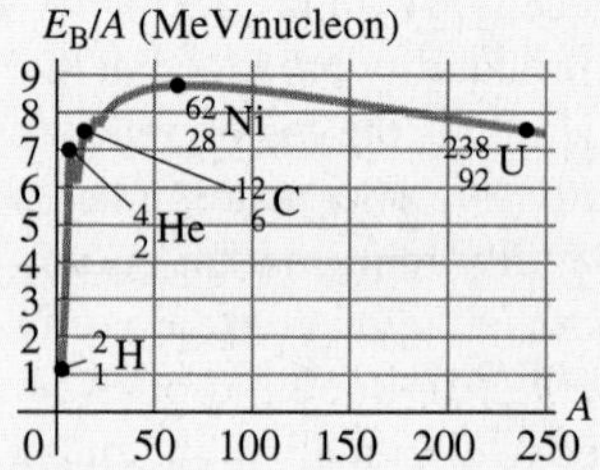

Unstable nuclides usually emit an alpha particle (a $^{4}_{2}$He nucleus) or a beta particle (an electron) in the process of changing to another nuclide, sometimes followed by a gamma-ray photon. The rate of decay of an unstable nucleus is described by the decay constant λ, the half-life $T_{1/2}$, or the lifetime T_{mean}. If the number of nuclei at time $t = 0$ is N_0 and no more are produced, the number at time t is given by Eq. (43.17)
(See Examples 43.5 through 43.9)

$$N(t) = N_0 e^{-\lambda t} \quad (43.17)$$

$$T_{\text{mean}} = \frac{1}{\lambda} = \frac{T_{1/2}}{\ln 2} = \frac{T_{1/2}}{0.693} \quad (43.19)$$

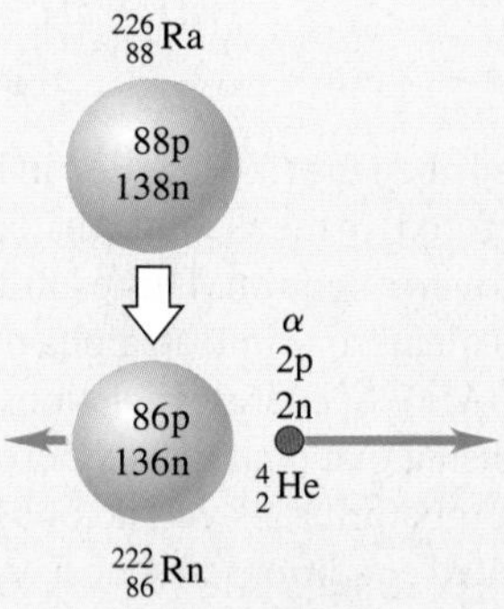

The biological effect of any radiation depends on the product of the energy absorbed per unit mass and the relative biological effectiveness (RBE), which is different for different radiations. (See Example 43.10)

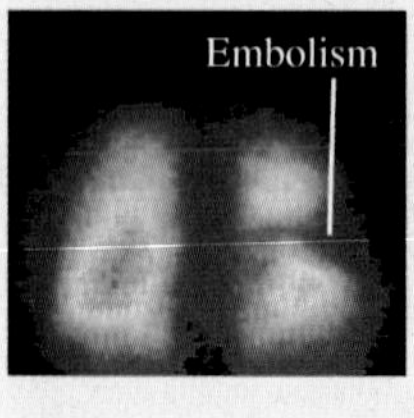

In a nuclear reaction, two nuclei or particles collide to produce two new nuclei or particles. Reactions can be exoergic or endoergic. Several conservation laws, including charge, energy, momentum, angular momentum, and nucleon number, are obeyed. Energy is released by the fission of a heavy nucleus into two lighter, always unstable, nuclei. Energy is also released by the fusion of two light nuclei into a heavier nucleus. (See Examples 43.11 through 43.14)

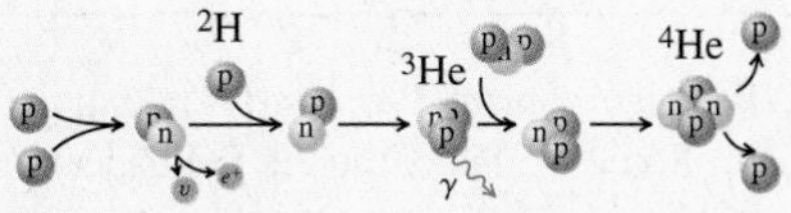

Key Terms

nucleon number, 1624
mass number, 1624
atomic number, 1625
neutron number, 1625
nuclide, 1625
isotope, 1625
nuclear magneton, 1627
binding energy, 1629
liquid-drop model, 1632
shell model, 1634
radioactivity, 1635
alpha particle, 1637
beta-minus particle, 1638
neutrino, 1638
gamma ray, 1640
activity, 1642
decay constant, 1643
half-life, 1643
curie, 1644
relative biological effectiveness (RBE), 1646
reaction energy, 1650
threshold energy, 1650
nuclear fission, 1652
chain reaction, 1654
nuclear fusion, 1657

Your Notes

Answer to Chapter Opening Question

When an organism dies, it stops taking in carbon from atmospheric CO_2. Some of this carbon is radioactive ^{14}C, which decays with a half-life of 5730 years. By measuring the proportion of ^{14}C that remains in the specimen, it is possible to determine how long ago the organism died. (See Section 43.4.)

Answers to Test Your Understanding Questions

Section 43.1 The radius R is proportional to the cube root of the mass number A, while the volume is proportional to R^3 and hence to $(A^{1/3})^3 = A$. Therefore, doubling the volume requires increasing the mass number by a factor of 2, while doubling the radius implies increasing both the volume and the mass number by a factor of $2^3 = 8$.

Section 43.2 For very light nuclides, the total number of nucleons is so small that it's not very accurate to describe the nucleus as a spherical liquid drop. For more massive nuclides, the liquid-drop model is more realistic and Eqs. (43.11) and (43.12), both of which are based on this model, give more accurate results.

Section 43.3 If electron capture were to occur in the ^1_1H atom, the single atomic electron would combine with the single proton of the nucleus to form a neutron as in Eq. (43.15). However, from Section 43.2 the mass of the ^1_1H atom is 1.007825 u while the neutron has a *greater* mass, 1.008665 u. Hence this process is impossible because it would require a net input of energy.

Section 43.4 The activity $-dN(t)/dt$ of a sample is the product of the number of nuclei in the sample $N(t)$ and the decay constant $\lambda = (\ln 2)/T_{1/2}$. Hence $N(t) = (-dN(t)/dt)T_{1/2}/(\ln 2)$. Taking the ratio of this expression for ^{240}Pu to this same expression for ^{243}Am, the factors of ln 2 cancel and we get

$$\frac{N_{Pu}}{N_{Am}} = \frac{(-dN_{Pu}/dt)T_{1/2\text{-}Pu}}{(-dN_{Am}/dt)T_{1/2\text{-}Am}} = \frac{(5.00\ \mu\text{Ci})(6560\ \text{y})}{(4.45\ \mu\text{Ci})(7370\ \text{y})} = 1.00$$

The two samples contain *equal* numbers of nuclei. The ^{243}Am sample has a longer half-life and hence a slower decay rate, so it has a lower activity than the ^{240}Pu sample.

Section 43.5 We saw in Section 43.3 that alpha particles can travel only a very short distance before they are stopped. By contrast, γ-ray photons are very penetrating, so they can easily pass through bones and tissue and reach detectors outside the body.

Section 43.6 The reaction $^1_1H + ^7_3Li \rightarrow ^4_2He + ^4_2He$ is a *nuclear* reaction, which can take place only if a proton (a hydrogen nucleus) comes into contact with a lithium nucleus. If the hydrogen is in atomic form, the interaction between its electron cloud and the electron cloud of a lithium atom keeps the two nuclei from getting close to each other. Even if isolated protons are used, they must be fired at the lithium atoms with enough kinetic energy to overcome the electric repulsion between the protons and the lithium nuclei. The statement that the reaction is exoergic means that more energy is released by the reaction than had to be put in to make the reaction occur.

Section 43.7 Because the neutron has no electric charge, it experiences no electric repulsion from a ^{235}U nucleus. Hence a slow-moving neutron can approach and enter a ^{235}U nucleus, thereby providing the excitation needed to trigger fission. By contrast, a slow-moving *proton* (charge $+e$) feels a strong electric repulsion from a ^{235}U nucleus (charge $+92e$). It never gets close to the nucleus, so it cannot trigger fission.

Section 43.8 Fusion reactions between sufficiently light nuclei are exoergic because the binding energy per nucleon E_B/A increases. If the nuclei are too massive, however, E_B/A decreases and fusion is *endoergic* (i.e., it takes in energy rather than releasing it). As an example, imagine fusing together two nuclei of $A = 100$ to make a single nucleus with $A = 200$. From Fig. 43.2, E_B/A is more than 8.5 MeV for the $A = 100$ nuclei but is less than 8 MeV for the $A = 200$ nucleus. Such a fusion reaction is possible, but requires a substantial input of energy.

Discussion Questions

Q43.1 Neutrons have a magnetic dipole moment and can undergo spin flips by absorbing electromagnetic radiation. Why, then, are protons rather than neutrons used in MRI of body tissues? (See Fig. 43.1.)

Q43.2 How can you be sure that nuclei are not made of protons and electrons, rather than protons and neutrons?

Q43.3 Why aren't the masses of all nuclei integer multiples of the mass of a single nucleon?

Q43.4 Can you tell from the value of the mass number A whether to use a plus value, a minus value, or zero for the fifth term of Eq. (43.11)? Explain.

Q43.5 The binding energy per nucleon for most nuclides doesn't vary much (see Fig. 43.2). Is there similar consistency in the *atomic* energy of atoms, on an "energy per electron" basis? If so, why? If not, why not?

Q43.6 Heavy unstable nuclei usually decay by emitting an α or β particle. Why do they not usually emit a single proton or neutron?

Q43.7 The only two stable nuclides with more protons than neutrons are ^1_1H and ^3_2He. Why is $Z > N$ so uncommon?

Q43.8 Since lead is a stable element, why doesn't the ^{238}U decay series shown in Fig. 43.8 stop at lead, ^{214}Pb?

Q43.9 In the ^{238}U decay chain shown in Fig. 43.8, some nuclides in the chain are found much more abundantly in nature than others, even though every ^{238}U nucleus goes through every step in the chain before finally becoming ^{206}Pb. Why don't the intermediate nuclides all have the same abundance?

Q43.10 Compared to α particles with the same energy, β particles can much more easily penetrate through matter. Why is this?

Q43.11 Some decays that are not possible from a nucleus in its ground state are possible if the nucleus is in an excited state. Explain why.

Q43.12 If $^A_Z\text{El}_i$ represents the initial nuclide, what is the decay process or processes if the final nuclide is a) $_{Z+1}{}^A\text{El}_f$? b) $^{A-4}_{Z-2}\text{El}_f$? c) $_{Z-1}{}^A\text{El}_f$?

Q43.13 In a nuclear decay equation, why can we represent an electron as $_{-1}^{0}\beta^-$? What are the equivalent representations for a positron, a neutrino, and an antineutrino?

Q43.14 Why is the alpha, beta, or gamma decay of an unstable nucleus unaffected by the *chemical* situation of the atom, such as the nature of the molecule or solid in which it is bound? The chemical situation of the atom can, however, have an effect on the half-life in electron capture. Why is this?

Q43.15 In the process of *internal conversion,* a nucleus decays from an excited state to a ground state by giving the excitation energy directly to an atomic electron rather than emitting a gamma-ray photon. Why can this process also produce x-ray photons?

Q43.16 In Example 43.9 (Section 43.4), the activity of atmospheric carbon *before* 1900 was given. Discuss why this activity may have changed since 1900.

Q43.17 The most common radium isotope found on earth, ^{226}Ra, has a half-life of about 1600 years. If the earth was formed well over 10^9 years ago, why is there any radium left now?

Q43.18 Fission reactions only occur for nuclei with large nucleon numbers, while exoergic fusion reactions only occur for nuclei with small nucleon numbers. Why is this?

Q43.19 In the interiors of some stars, fusion reactions take place in which three $^4_2$He combine to form a $^{12}_6$C nucleus. These reactions can only take place if the interior temperature of the star is substantially higher than that of the sun. Why does the temperature play a role?

Q43.20 An *antiproton* has the same rest energy as a proton but an opposite charge $-e$. What are the relative directions of $\vec{\mu}$ and $\vec{S}$ for an antiproton?

Exercises

Section 43.1 Properties of Nuclei

43.1 How many protons and how many neutrons are there in a nucleus of the most common isotope of a) silicon, $^{28}_{14}$Si; b) rubidium, $^{85}_{37}$Rb; c) thallium, $^{205}_{81}$Tl?

43.2 Consider the three nuclei of Exercise 43.1. Estimate a) the radius, b) the surface area, and c) the volume of each nucleus. Determine d) the mass density (in kg/m^3) and e) the nucleon density (in nucleons per cubic meter) for each nucleus. Assume that the mass of each nucleus is A atomic mass units.

43.3 Hydrogen atoms are placed in an external magnetic field. The protons can make transitions between states in which the nuclear spin component is parallel and antiparallel to the field by absorbing or emitting a photon. What magnetic field magnitude is required for this transition to be induced by photons with frequency 22.7 MHz?

43.4 Neutrons are placed in a magnetic field with magnitude 2.30 T. a) What is the energy difference between the states with the nuclear spin angular momentum components parallel and antiparallel to the field? Which state is lower in energy, the one with its spin component parallel to the field or the one with its spin component antiparallel to the field? How do your results compare with the energy states for a proton in the same field (Example 43.2 in Section 43.1)? b) The neutrons can make transitions from one of these states to the other by emitting or absorbing a photon with energy equal to the energy difference of the two states. Find the frequency and wavelength of such a photon.

43.5 Hydrogen atoms are placed in an external 1.65-T magnetic field. a) The *protons* can make transitions between states where the nuclear spin component is parallel and antiparallel to the field by absorbing or emitting a photon. Which state lays lower in energy, the state with the nuclear spin component parallel or antiparallel to the field? What are the frequency and wavelength of the photon? In which region of the electromagnetic spectrum does it lay? b) The *electrons* can make transitions between states where the electron spin component is parallel and antiparallel to the field by absorbing or emitting a photon. Which state lay lower in energy, the state with the electron spin component parallel or antiparallel to the field? What are the frequency and wavelength of the photon? In which region of the electromagnetic spectrum does it lay?

Section 43.2 Nuclear Binding and Nuclear Structure

43.6 a) Show that the ionization energy of the hydrogen atom (13.6 eV) is 0.0027% of an electron's rest energy. b) We found in Example 43.3 (Section 43.2) that the binding energy per nucleon for $^{62}_{28}$Ni is 8.795 MeV. What fraction is this of the rest energy of a proton?

43.7 What is the maximum wavelength of a γ ray that could break a deuteron into a proton and a neutron? (This process is called photodisintegration.)

43.8 Calculate the mass defect, the binding energy (in MeV), and the binding energy per nucleon of a) the nitrogen nucleus, $^{14}_7$N; b) the helium nucleus, $^4_2$He. c) How do the results of parts (a) and (b) compare?

43.9 The most common isotope of boron is $^{11}_5$B. a) Determine the total binding energy of $^{11}_5$B from Table 43.2 in Section 43.1. b) Calculate this binding energy from Eq. (43.11). (Why is the fifth term zero?) Compare to the result you obtained in part (a). What is the percent difference? Compare the accuracy of Eq. (43.11) for $^{11}_5$B to its accuracy for $^{62}_{28}$Ni (Example 43.4 in Section 43.2).

43.10 The most common isotope of copper is $^{63}_{29}$Cu. The measured mass of the neutral atom is 62.929601 u. a) From the measured mass, determine the mass defect, and use it to find the total binding energy and the binding energy per nucleon. b) Calculate the binding energy from Eq. (43.11). (Why is the fifth term zero?) Compare to the result you obtained in part (a). What is the percent difference? What do you conclude about the accuracy of Eq. (43.11)?

43.11 What are the six known elements for which Z is a magic number? Discuss what properties these elements have as a consequence of their special values of Z.

43.12 The most common isotope of uranium, $^{238}_{92}$U, has atomic mass 238.050783 u. Calculate a) the mass defect; b) the binding energy (in MeV); c) the binding energy per nucleon.

Section 43.3 Nuclear Stability and Radioactivity

43.13 What nuclide is produced in the following radioactive decays? a) α decay of $^{239}_{94}$Pu; b) β^- decay of $^{24}_{11}$Na; c) β^+ decay of $^{15}_8$O.

43.14 a) Is the decay $n \rightarrow p + \beta^- + \bar{\nu}_e$ energetically possible? If not, explain why not. If so, calculate the total energy released. b) Is the decay $p \rightarrow n + \beta^+ + \nu_e$ energetically possible? If not, explain why not. If so, calculate the total energy released.

43.15 Calculate the mass defect for the decay of $^8_4$Be into two alpha particles, $^8_4\text{Be} \rightarrow {}^4_2\text{He} + {}^4_2\text{He}$. The mass of $^8_4$Be is 8.005305 u.

43.16 What particle (α particle, electron, or positron) is emitted in the following radioactive decays? a) ${}^{27}_{14}Si \rightarrow {}^{27}_{13}Al$; b) ${}^{238}_{92}U \rightarrow {}^{234}_{90}Th$; c) ${}^{74}_{33}As \rightarrow {}^{74}_{34}Se$.

43.17 The atomic mass of ${}^{14}C$ is 14.003242 u. Show that the β^- decay of ${}^{14}C$ is energetically possible, and calculate the energy released in the decay.

43.18 a) Calculate the energy released by the electron-capture decay of ${}^{57}_{27}Co$ (Example 43.7 in Section 43.3). b) A negligible amount of this energy goes to the resulting ${}^{57}_{26}Fe$ atom as kinetic energy. About 90% of the time, the ${}^{57}_{26}Fe$ nucleus emits two successive gamma-ray photons after the electron-capture process, of energies 0.122 MeV and 0.014 MeV, respectively, in decaying to its ground state. What is the energy of the neutrino emitted in this case?

43.19 Tritium Decay. Tritium, ${}^{3}_{1}H$, is an unstable isotope of hydrogen produced in nuclear reactors. a) Show that tritium must be unstable with respect to β^- decay. b) Determine the total kinetic energy of the decay products.

Section 43.4 Activities and Half-Lives

43.20 Radioactive isotopes used in cancer therapy have a "shelf-life", just like pharmaceuticals used in chemotherapy. Just after it has been manufactured in a nuclear reactor, the activity of a sample of ${}^{60}Co$ is 5000 Ci. When its activity falls below 3500 Ci, it is considered to be too weak a source to use in treatment. You work in the radiology department of a large hospital. One of these ${}^{60}Co$ sources in your inventory was manufactured on October 6, 2001. It is now April 6, 2004. Is the source still usable? The half-life of ${}^{60}Co$ is 5.271 years.

43.21 A 12.0-g sample of carbon from living matter decays at the rate of 180.0 decays/minute due to the radioactive ${}^{14}C$ in it. What will be the decay rate of this sample in a) 1000 years; b) 50,000 years?

43.22 The isotope ${}^{90}Sr$ undergoes β^- decay with a half-life of 28 years. a) What nucleus is produced by this decay? b) If a nuclear power plant is contaminated with ${}^{90}Sr$, how long will it take for the radiation level to decrease to 1.0% of its initial value?

43.23 Tritium (${}^{3}H$) undergoes β^- decay with a half-life of 12.3 years. It is also highly toxic to living things. a) What nucleus is produced in the β^- decay of tritium? b) Suppose some tritium gas is released into the atmosphere in a nuclear power plant accident. How long will it take for 90.0% of the tritium to become nonradioactive?

43.24 As a health physicist, you are being consulted about a spill in a radiochemistry lab. The isotope spilled was 500 μCi of ${}^{131}Ba$, which has a half-life of 12 days. a) What mass of ${}^{131}Ba$ was spilled? b) Your recommendation is to clear the lab until the radiation level is down to 1.00 μCi. How long will the lab have to be closed?

43.25 Measurements on a certain isotope tell you that the decay rate decreases from 8318 decays/min to 3091 decays/min in 4.00 days. What is the half-life of this isotope?

43.26 The isotope ${}^{226}Ra$ undergoes α decay with a half-life of 1620 years. What is the activity of 1.00 g of ${}^{226}Ra$? Express your answer in Bq and in Ci.

43.27 The ratio of ${}^{14}C$ to ${}^{12}C$ in living matter is measured to be ${}^{14}C/{}^{12}C = 1.3 \times 10^{-12}$ at the present time. A 12.0-g sample of carbon produces 180 decays/min due to the small amount of ${}^{14}C$ in it. From this information, calculate the half-life of ${}^{14}C$.

43.28 An unstable isotope of cobalt, ${}^{60}Co$, has one more neutron in its nucleus than does the stable isotope ${}^{59}Co$ and is a β^- emitter with a half-life of 5.27 y. This isotope is used in medicine. A certain radiation source in a hospital contains 0.0360 mg of ${}^{60}Co$. a) What is the decay constant for this isotope? b) How many atoms are in the source? c) How many decays occur per second? d) What is the activity of the source in curies? How does this compare with the activity of an equal mass of radium (${}^{226}Ra$, half-life 1600 y)?

43.29 The radioactive nuclide ${}^{199}Pt$ has a half-life of 30.8 minutes. A sample is prepared that has an initial activity of 7.56×10^{11} Bq. a) How many ${}^{199}Pt$ nuclei are initially present in the sample? b) How many are present after 30.8 minutes? What is the activity at this time? c) Repeat part (b) for a time 92.4 minutes after the sample is first prepared.

43.30 Radiocarbon Dating. A sample from timbers at an archeological site containing 500 g of carbon provides 3070 decays per minute. What is the age of the sample?

43.31 The unstable isotope ${}^{40}K$ is used for dating of rock samples. Its half-life is 1.28×10^9 y. a) How many decays occur per second in a sample containing 1.63×10^{-6} g of ${}^{40}K$? b) What is the activity of the sample in curies?

43.32 If you are of average mass, about 360 million nuclei in your body undergo radioactive decay each day. Express your activity in curies.

43.33 The most common (99.3%) isotope of uranium, ${}^{238}U$, has a half-life of 4.47×10^9 y, decaying to ${}^{234}Th$ by α emission. a) What is the decay constant? b) What mass of uranium would be required for an activity of 12.0 μCi? c) How many α particles are emitted per second by 60.0 g of ${}^{238}U$?

Section 43.5 Biological Effects of Radiation

43.34 A person exposed to fast neutrons receives a radiation dose of 200 rem on part of his hand, affecting 25 g of tissue. The RBE of these neutrons is 10. a) How many rad did he receive? b) How many joules of energy did this person receive? c) Suppose the person received the same rad dosage, but from beta rays with a RBE of 1.0 instead of neutrons. How many rem would he have received?

43.35 A nuclear chemist receives an accidental radiation dose of 5.0 Gy from slow neutrons (RBE = 4.0). What does did she receive in rad, rem, and J/kg?

43.36 In an industrial accident a 65-kg person receives a lethal whole-body equivalent dose of 5.4 Sv from x rays. a) What is the equivalent dose in rem? b) What is the absorbed dose in rad? c) What is the total energy absorbed by the person's body? How does this amount of energy compare to the amount of energy required to raise the temperature of 65 kg of water 0.010°C?

43.37 A 50-kg person accidentally ingests 0.35 Ci of tritium. a) Assume that the tritium spreads uniformly over the body and that each decay leads on the average to the absorption of 5.0 keV of energy from the electrons emitted in the decay. The half-life of tritium is 12.3 y, and the RBE of the electrons is 1.0. Calculate the absorbed dose in rad and the equivalent dose in rem during one week. b) The β^- decay of tritium releases more than 5.0 keV of energy (see Exercise 43.19). Why is the average energy absorbed less than the total energy released in the decay?

43.38 In an experimental radiation therapy, an equivalent dose of 0.900 rem is given by 0.800-MeV protons to a localized area of tissue with mass 0.150 kg. a) What is the absorbed dose in rad? b) How many protons are absorbed by the tissue? c) How many α particles of the same energy of 0.800 MeV are required to deliver the same equivalent dose of 0.900 rem?

43.39 In a diagnostic x-ray procedure, 6.50×10^{10} photons are absorbed by tissue with mass 0.600 kg. The x-ray wavelength is 0.0200 nm. a) What is the total energy absorbed by the tissue? b) What is the equivalent dose in rem?

43.40 A person ingests an amount of a radioactive source with a very long lifetime and activity 0.72 μCi. The radioactive material lodges in the lungs, where all of the 4.0-MeV α particles emitted are absorbed within a 0.50-kg mass of tissue. Calculate the absorbed dose and the equivalent dose for one year.

Sections 43.6 Nuclear Reactions, 43.7 Nuclear Fission, and 43.8 Nuclear Fusion

43.41 Consider the nuclear reaction

$$^{2}_{1}\text{H} + ^{9}_{4}\text{Be} \rightarrow \text{X} + ^{4}_{2}\text{He}$$

where X is a nuclide. a) What are the values of Z and A for the nuclide X? b) How much energy is liberated? c) Estimate the threshold energy for this reaction.

43.42 Energy from Nuclear Fusion. Calculate the energy released in the fusion reaction

$$^{3}_{2}\text{He} + ^{2}_{1}\text{H} \rightarrow ^{4}_{2}\text{He} + ^{1}_{1}\text{H}$$

43.43 Consider the nuclear reaction

$$^{2}_{1}\text{H} + ^{14}_{7}\text{N} \rightarrow \text{X} + ^{10}_{5}\text{B}$$

where X is a nuclide. a) What are Z and A for the nuclide X? b) Calculate the reaction energy Q (in MeV). c) If the $^{2}_{1}$H nucleus is incident on a stationary $^{14}_{7}$N nucleus, what minimum kinetic energy must it have for the reaction to occur?

43.44 Heat of Fission. In the fission of one ^{238}U nucleus, 200 MeV of energy is released. Express this energy in joules per mole, and compare with typical heats of combustion, which are on the order of 1.0×10^5 J/mol.

43.45 A $^{235}_{92}$U nucleus at rest absorbs a low-energy neutron. What is the internal excitation energy of the $^{236}_{92}$U* nucleus that is produced? The atomic masses of the neutral atoms in their nuclear ground states are 235.043923 u for $^{235}_{92}$U and 236.045562 u for $^{236}_{92}$U.

43.46 Consider the nuclear reaction

$$^{4}_{2}\text{He} + ^{7}_{3}\text{Li} \rightarrow \text{X} + ^{1}_{0}\text{n}$$

where X is a nuclide. a) What are Z and A for the nuclide X? b) Is energy absorbed or liberated? How much?

43.47 The second reaction in the proton-proton chain (see Fig. 43.16) produces a $^{3}_{2}$He nucleus. A $^{3}_{2}$He nucleus produced in this way can combine with a $^{4}_{2}$He nucleus:

$$^{3}_{2}\text{He} + ^{4}_{2}\text{He} \rightarrow ^{7}_{4}\text{Be} + \gamma$$

Calculate the energy liberated in this process. (This is shared between the energy of the photon and the recoil kinetic energy of the beryllium nucleus.) The mass of a $^{7}_{4}$Be atom is 7.016929 u.

43.48 Consider the nuclear reaction

$$^{28}_{14}\text{Si} + \gamma \rightarrow ^{24}_{12}\text{Mg} + \text{X}$$

where X is a nuclide. a) What are Z and A for the nuclide X? b) Ignoring the effects of recoil, what minimum energy must the photon have for this reaction to occur? The mass of a $^{28}_{14}$Si atom is 27.976927 u, and the mass of a $^{24}_{12}$Mg atom is 23.985042 u.

Problems

43.49 Use conservation of mass-energy to show that the energy released in alpha decay is positive whenever the mass of the original neutral atom is greater than the sum of the masses of the final neutral atom and the neutral ^{4}He atom. (*Hint:* Let the parent nucleus have atomic number Z and nucleon number A. First write the reaction in terms of the nuclei and particles involved, and then add Z electron masses to both sides of the reaction and allot them as needed to arrive at neutral atoms.)

43.50 Use conservation of mass-energy to show that the energy released in β^- decay is positive whenever the neutral atomic mass of the original atom is larger than that of the final atom. (See the *Hint* in Problem 43.49.)

43.51 Use conservation of mass-energy to show that the energy released in β^+ decay is positive whenever the neutral atomic mass of the original atom is at least two electron masses larger than that of the final atom. (See the *Hint* in Problem 43.49.)

43.52 Use conservation of mass-energy to show that the energy released in electron capture is positive whenever the neutral atomic mass of the original atom is larger than that of the final atom. (See the *Hint* in Problem 43.49, except add $Z - 1$ electron masses to both sides of the reaction.)

43.53 The atomic mass of $^{25}_{12}$Mg is 24.985837 u, and the atomic mass of $^{25}_{13}$Al is 24.990429 u. a) Which of these nuclei will decay into the other? b) What type of decay will occur? Explain how you determined this. c) How much energy (in MeV) is released in the decay?

43.54 The polonium isotope $^{210}_{84}$Po has atomic mass 209.982857 u. Other atomic masses are $^{206}_{82}$Pb, 205.974449 u; $^{209}_{83}$Bi, 208.980383 u; $^{210}_{83}$Bi, 209.984105 u; $^{209}_{84}$Po, 208.982416 u; and $^{210}_{85}$At, 209.987131 u. a) Show that the alpha decay of $^{210}_{84}$Po is energetically possible, and find the energy of the emitted α particle. b) Is $^{210}_{84}$Po energetically stable with respect to emission of a proton? Why or why not? c) Is $^{210}_{84}$Po energetically stable with respect to emission of a neutron? Why or why not? d) Is $^{210}_{84}$Po energetically stable with respect to β^- decay? Why or why not? e) Is $^{210}_{84}$Po energetically stable with respect to β^+ decay? Why or why not?

43.55 The experimentally determined mass of the neutral ^{24}Na atom is 23.990963 u. Calculate the mass from the semi-empirical mass formula (Eq. 43.12). What is the percent error of the result as compared to the experimental value? What percent error is made if the E_B term is ignored entirely?

43.56 Thorium $^{230}_{90}$Th decays to radium $^{226}_{88}$Ra by α emission. The masses of the neutral atoms are 230.033127 u for $^{230}_{90}$Th and 226.025403 u for $^{226}_{88}$Ra. If the parent thorium nucleus is at rest, what is the kinetic energy of the emitted α particle? (Be sure to account for the recoil of the daughter nucleus.)

43.57 Gold $^{198}_{79}$Au undergoes β^- decay to an excited state of $^{198}_{80}$Hg. If the excited state decays by emission of a γ photon with energy 0.412 MeV, what is the maximum kinetic energy of the electron emitted in the decay? This maximum occurs when the antineutrino has negligible energy. (The recoil energy of the $^{198}_{80}$Hg nucleus can be ignored. The masses of the neutral atoms in their ground states are 197.968225 u for $^{198}_{79}$Au and 197.966752 u for $^{198}_{80}$Hg.)

43.58 Calculate the mass defect for the β^+ decay of $^{11}_{6}$C. Is this decay energetically possible? Why or why not? The atomic mass of $^{11}_{6}$C is 11.011434 u.

43.59 Calculate the mass defect for the β^+ decay of $^{13}_{7}$N. Is this decay energetically possible? Why or why not? The atomic mass of $^{13}_{7}$N is 13.005739 u.

43.60 The results of activity measurements on a radioactive sample are given below. a) Find the half-life. b) How many radioactive nuclei were present in the sample at $t = 0$? c) How many were present after 7.0 h?

Time (h)	Decays/s
0	20,000
0.5	14,800
1.0	11,000
1.5	8,130
2.0	6,020
2.5	4,460
3.0	3,300
4.0	1,810
5.0	1,000
6.0	550
7.0	300

43.61 If $A(t)$ is the activity of a sample at some time t and A_0 is the activity at $t = 0$, show that $A(t) = A_0 e^{-\lambda t}$.

43.62 Show that Eq. (43.17) may be written as $N(t) = N_0(\frac{1}{2})^n$, where $n = t/T_{1/2}$ is the number of half-lives that have elapsed since $t = 0$. (This expression is valid even if n is not a whole number.)

43.63 Measurements indicate that 27.83% of all rubidium atoms currently on the earth are the radioactive ^{87}Rb isotope. The rest are the stable ^{85}Rb isotope. The half-life of ^{87}Rb is 4.75×10^{10} y. Assuming that no rubidium atoms have been formed since, what percentage of rubidium atoms were ^{87}Rb when our solar system was formed 4.6×10^9 y ago?

43.64 A 70.0-kg person experiences a whole-body exposure to α radiation with energy 4.77 MeV. A total of 6.25×10^{12} α particles are absorbed. a) What is the absorbed dose in rad? b) What is the equivalent dose in rem? c) If the source is 0.0320 g of ^{226}Ra (half-life 1600 y) somewhere in the body, what is the activity of this source? d) If all the alpha particles produced are absorbed, what time is required for this dose to be delivered?

43.65 A ^{60}Co source with activity 2.6×10^{-4} Ci is imbedded in a tumor that has mass 0.500 kg. The source emits γ photons with average energy 1.25 MeV. Half the photons are absorbed in the tumor, and half escape. a) What energy is delivered to the tumor per second? b) What absorbed dose (in rad) is delivered per second? c) What equivalent dose (in rem) is delivered per second if the RBE for these γ rays is 0.70? d) What exposure time is required for an equivalent dose of 200 rem?

43.66 The nucleus $^{15}_{8}$O has a half-life of 122.2 s; $^{19}_{8}$O has a half-life of 26.9 s. If at some time a sample contains equal amounts of $^{15}_{8}$O and $^{19}_{8}$O, what is the ratio of $^{15}_{8}$O to $^{19}_{8}$O a) after 4.0 minutes? b) after 15.0 minutes?

43.67 A bone fragment found in a cave believed to have been inhabited by early humans contains 0.21 times as much ^{14}C as an equal amount of carbon in the atmosphere when the organism containing the bone died. (See Example 43.9 in Section 43.4.) Find the approximate age of the fragment.

43.68 A patient is given a 4.2 μCi dose of a radioactive isotope with a half-life of 2.0 h. Assuming that the entire dose remains in the body, how much time must elapse before the activity of the radioactive isotope is 8.5 counts/minute, which is about three times the normal background of 3 counts/minute?

43.69 Consider the fusion reaction

$$^{2}_{1}\text{H} + ^{2}_{1}\text{H} \rightarrow ^{3}_{2}\text{He} + ^{1}_{0}\text{n}$$

a) Estimate the barrier energy by calculating the repulsive electrostatic potential energy of the two $^{2}_{1}$H nuclei when they touch. b) Compute the energy liberated in this reaction in MeV and in joules. c) Compute the energy liberated *per mole* of deuterium, remembering that the gas is diatomic, and compare with the heat of combustion of hydrogen, about 2.9×10^5 J/mol.

43.70 Separation Energy. a) The proton separation energy is the minimum amount of energy that must be supplied to a nucleus to remove a proton. Calculate the proton separation energy for $^{16}_{8}$O. b) The neutron separation energy is the minimum amount of energy that must be supplied to a nucleus to remove a neutron. Calculate the neutron separation energy for $^{16}_{8}$O. The atomic mass of $^{15}_{8}$O is 15.003065 u. c) Compare the answers to parts (a) and (b). Which takes less energy to remove from $^{16}_{8}$O, a proton or a neutron?

43.71 Natural Body Radioactivity. On average, 0.21% of the mass of the human body is potassium, of which 0.012% is radioactive ^{40}K with a half-life of 1.28×10^9 y. From each decay, an average of 0.50 MeV of β and γ ray radiation is absorbed by the body. Take the RBE of the β radiation to be 1.0. Calculate the absorbed dose in rad and the equivalent dose in rem for 50 y from the potassium in the body.

43.72 In the 1986 disaster at the Chernobyl reactor in the Soviet Union (now Ukraine), about $\frac{1}{8}$ of the ^{137}Cs present in the reactor was released. The isotope ^{137}Cs has a half-life for β decay of 30.07 y and decays with the emission of a total of 1.17 MeV of energy per decay. Of this, 0.51 MeV goes to the emitted electron and the remaining 0.66 MeV to a γ ray. The radioactive ^{137}Cs is absorbed by plants which are eaten by livestock and humans. How many ^{137}Cs atoms would need to be present in each kilogram of body tissue if an equivalent dose for one week is 3.5 Sv? Assume that all of the energy from the decay is deposited in that 1.0 kg of tissue and that the RBE of the electrons is 1.5.

43.73 a) Prove that when a particle with mass m and kinetic energy K collides with a stationary particle with mass M, the total kinetic energy K_{cm} in the center-of-mass coordinate system (the energy available to cause reactions) is

$$K_{\text{cm}} = \frac{M}{M + m} K$$

Assume that the kinetic energies of the particles and nuclei are much less than their rest energies. b) If K_{th} is the minimum, or threshold, kinetic energy to cause an endoergic reaction to occur in the situation of part (a), show that

$$K_{th} = -\frac{M + m}{M}Q$$

43.74 A $^{186}_{76}Os$ nucleus at rest decays by the emission of a 2.76-MeV α particle. Calculate the atomic mass of the daughter nuclide produced by this decay, assuming that it is produced in its ground state. The atomic mass of $^{186}_{76}Os$ is 185.953838 u.

43.75 Calculate the energy released in the fission reaction

$$^{235}_{92}U + ^{1}_{0}n \rightarrow ^{140}_{54}Xe + ^{94}_{38}Sr + 2^{1}_{0}n$$

You can ignore the initial kinetic energy of the absorbed neutron. The atomic masses are $^{235}_{92}U$, 235.043923 u; $^{140}_{54}Xe$, 139.921636 u; and $^{94}_{38}Sr$, 93.915360 u.

Challenge Problems

43.76 The results of activity measurements on a mixed sample of radioactive elements are shown below. a) How many different nuclides are present in the mixture? b) What are their half-lives? c) How many nuclei of each type are initially present in the sample? d) How many of each type are present at t = 5.0 h?

Time (h)	Decays/s
0	7500
0.5	4120
1.0	2570
1.5	1790
2.0	1350
2.5	1070
3.0	872
4.0	596
5.0	414
6.0	288
7.0	201
8.0	140
9.0	98
10.0	68
12.0	33

43.77 In an experiment, the isotope ^{128}I is created by the irradiation of ^{127}I with a beam of neutrons that creates 1.5×10^6 ^{128}I nuclei per second. Initially no ^{128}I nuclei are present. The half-life of ^{128}I is 25 minutes. a) Graph the number of ^{128}I nuclei present as a function of time. b) What is the activity of the sample 1, 10, 25, 50, 75, and 180 minutes after irradiation begin? c) What is the maximum number of ^{128}I atoms that can be created in the sample after it is irradiated for a long time? (This steady-state situation is called *saturation.*) d) What is the maximum activity that can be produced?

43.78 Industrial Radioactivity. Radioisotopes are used in a variety of manufacturing and testing techniques. Wear measurements can be made using the following method. An automobile engine is produced using piston rings with a total mass of 100 g, that includes 9.4 μCi of ^{59}Fe whose half-life is 45 days. The engine is test run for 1000 hours, after which the oil is drained and its activity is measured. If the activity of the engine oil is 84 decays/s, how much mass was worn from the piston rings per hour of operation?

44 PARTICLE PHYSICS AND COSMOLOGY

Where did we come from? This false-color map of microwave radiation from the entire sky provides an answer. The radiation was emitted 380,000 years after the Big Bang, before any stars or galaxies existed. Regions shown in blue were slightly cooler and denser than their surroundings; within these regions formed galaxies, including the Milky Way galaxy of which our solar system is part.

? How does the existence of microwave radiation from the sky provide evidence that the universe began with a Big Bang?

What is the world made of? What are the most fundamental constituents of matter? Philosophers and scientists have been asking these questions for at least 2500 years. We still don't have anything that could be called a final answer, but we've come a long way. This final chapter is a progress report on where we are now and where we hope to go.

The chapter title, "Particle Physics and Cosmology," may seem strange. Fundamental particles are the *smallest* things in the universe, and cosmology deals with the *biggest* thing there is—the universe itself. Nonetheless, we'll see in this chapter that physics on the most microscopic scale plays an essential role in determining the nature of the universe on the largest scale.

Fundamental particles, we'll find, are not permanent entities; they can be created and destroyed. The development of high-energy particle accelerators and associated detectors has played an essential role in our emerging understanding of particles. We can classify particles and their interactions in several ways in terms of conservation laws and symmetries, some of which are absolute and others of which are obeyed only in certain kinds of interactions. We'll try in this final chapter to convey some idea of the remaining frontiers in this vital and exciting area of fundamental research. We'll conclude by discussing our present understanding of the nature and evolution of the universe as a whole.

44.1 | Fundamental Particles—A History

The idea that the world is made of *fundamental particles* has a long history. In about 400 B.C. the Greek philosophers Democritus and Leucippus suggested that matter is made of indivisible particles that they called *atoms,* a word derived from *a-* (not) and *tomos* (cut or divided). This idea lay dormant until about 1804, when the English scientist John Dalton (1766–1844), often called the father of modern chemistry, discovered that many chemical phenomena could be explained if atoms of each element are the basic, indivisible building blocks of matter.

The Electron and the Proton

Toward the end of the nineteenth century it became clear that atoms are *not* indivisible. The existence of characteristic atomic spectra of elements suggested that atoms have internal structure, and J. J. Thomson's discovery of the *electron* in 1897 showed that atoms could be taken apart into charged particles. The hydrogen nucleus was identified as a *proton,* and in 1911 the sizes of nuclei were measured by Rutherford's experiments. Quantum mechanics, including the Schrödinger equation, blossomed during the next 15 years. Scientists were on their way to understanding the principles that underlie atomic structure, although many details remained to be worked out.

The Photon

Einstein explained the photoelectric effect in 1905 by assuming that the energy of electromagnetic waves is quantized; that is, it comes in little bundles called *photons* with energy $E = hf$. Atoms and nuclei can emit (create) and absorb (destroy) photons. Considered as particles, photons have zero charge and zero rest mass. (Note that any discussions of a particle's mass in this chapter will refer to its rest mass.) In particle physics, a photon is denoted by the symbol γ (the Greek letter "gamma").

The Neutron

The discovery of the neutron was an important milestone. In 1930, the German physicists Walther Bothe and Herbert Becker observed that when beryllium, boron, or lithium was bombarded by α particles from radioactive polonium, the target material emitted a radiation that had much greater penetrating power than the original α particles. Experiments by the English physicist James Chadwick in 1932 showed that the emitted particles were electrically neutral, with mass approximately equal to that of the proton. Chadwick christened these particles *neutrons.* A typical reaction of the type studied by Bothe and Becker, using a beryllium target, is

$$ {}^{4}_{2}\mathrm{He} + {}^{9}_{4}\mathrm{Be} \rightarrow {}^{12}_{6}\mathrm{C} + {}^{1}_{0}\mathrm{n} \qquad (44.1) $$

It was difficult to detect neutrons directly because they have no charge. Hence they produce little ionization when they pass through matter and they are not deflected by electric or magnetic fields. Neutrons almost always interact only with nuclei; they can be slowed down during scattering, and they can penetrate the nucleus. Slow neutrons can be detected by means of a nuclear reaction in which a neutron is absorbed and an α particle is emitted. An example is

$$ {}^{1}_{0}\mathrm{n} + {}^{10}_{5}\mathrm{B} \rightarrow {}^{7}_{3}\mathrm{Li} + {}^{4}_{2}\mathrm{He} \qquad (44.2) $$

The ejected α particle is easy to detect because it is charged. Later experiments showed that neutrons, like protons and electrons, are spin-$\frac{1}{2}$ particles (see Section 43.1).

The neutron was a welcome discovery because it cleared up a mystery about the composition of the nucleus. Before 1930 the mass of a nucleus was thought to be due only to protons, but no one understood why the charge-to-mass ratio was not the same for all nuclides. It soon became clear that all nuclides (except ^1_1H) contain both protons and neutrons. In fact, the proton, the neutron, and the electron are the building blocks of atoms. One might think that would be the end of the story. On the contrary, it is barely the beginning. These are not the only particles, and they can do more than build atoms.

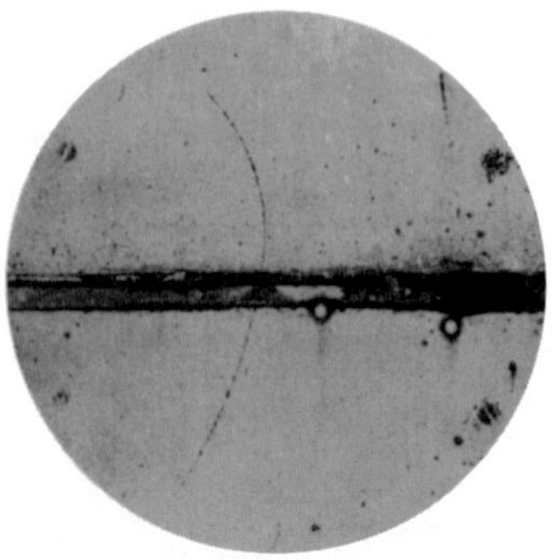

44.1 Photograph of the cloud-chamber track made by the first positron ever identified. The photograph was made by Carl D. Anderson in 1932. The lead plate is 6 mm thick.

The Positron

The positive electron, or *positron,* was discovered by the American physicist Carl D. Anderson in 1932, during an investigation of particles bombarding the earth from space. Figure 44.1 shows a historic photograph made with a *cloud chamber,* an instrument that is used to visualize the tracks of charged particles. The chamber contained a supercooled vapor; ions created by the passage of charged particles through the vapor served as nucleation centers, and liquid droplets formed around them, making a visible track.

The cloud chamber in Fig. 44.1 is in a magnetic field directed into the plane of the photograph. The particle has passed through a thin lead plate (which extends from left to right in the figure) that lies within the chamber. The curvature of the track is greater above the plate than below it, showing that the speed was less above the plate than below it. Therefore the particle had to be moving upward; it could not have gained energy passing through the lead. The thickness and curvature of the track suggested that its mass and the magnitude of its charge equaled those of the electron. But the directions of the magnetic field and the velocity in the magnetic force equation $\vec{F} = q\vec{v} \times \vec{B}$ showed that the particle had *positive* charge. Anderson christened this particle the *positron.*

To theorists, the appearance of the positron was a welcome development. In 1928, the English physicist Paul Dirac had developed a relativistic generalization of the Schrödinger equation for the electron. In Section 41.3 we discussed how Dirac's generalization helped to explain the spin magnetic moment of the electron.

One of the puzzling features of the Dirac equation was that for a *free* electron it predicted not only a continuum of energy states greater than its rest energy m_ec^2, as should be expected, but also a continuum of *negative* energy states *less than* $-m_ec^2$ (Fig. 44.2a). That posed a problem. What was to prevent an electron from emitting a photon with energy $2m_ec^2$ or greater and hopping from a positive state to a negative state? It wasn't clear what these negative-energy states meant, and there was no obvious way to get rid of them. Dirac's ingenious interpretation was that all the negative-energy states were filled with electrons, and that these electrons were for some reason unobservable. The exclusion principle would then forbid a transition to a state that was already occupied.

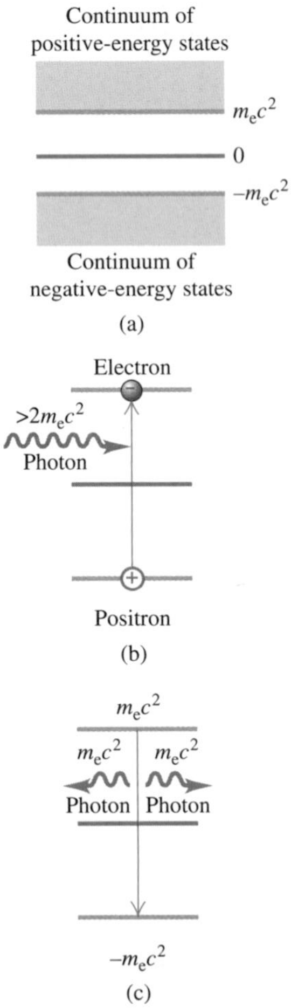

44.2 (a) Energy states for a free electron predicted by the Dirac equation. (b) Raising an electron from an $E < 0$ state to an $E > 0$ state corresponds to electron–positron pair production. (c) An electron dropping from an $E > 0$ state to a vacant $E < 0$ state corresponds to electron–positron pair annihilation.

A vacancy in a negative-energy state would act like a positive charge, just as a hole in the valence band of a semiconductor (Section 42.6) acts like a positive charge. Initially, Dirac tried to argue that such vacancies were protons. But after Anderson's discovery it became clear that the vacancies were observed physically as *positrons.* Furthermore, the Dirac energy-state picture provides a mechanism for the *creation* of positrons. When an electron in a negative-energy state absorbs a photon with energy greater than $2m_ec^2$, it goes to a positive state (Fig. 44.2b), in which it becomes observable. The vacancy that it leaves behind is observed as a positron; the result is the creation of an electron–positron pair. Similarly, when an electron in a positive-energy state falls into a vacancy, both the electron and the vacancy (that is, the positron) disappear, and photons are emitted (Fig. 44.2c). Thus the Dirac theory leads naturally to the conclusion that, like photons, *electrons can be created and destroyed.* While photons can be created and destroyed singly, electrons can be produced or destroyed only in electron–positron pairs or in association with other particles. We'll see why later.

In 1949, the American physicist Richard Feynman showed that a positron could be described mathematically as an electron traveling backward in time. His reformulation of the Dirac theory eliminated difficult calculations involving the infinite sea of negative-energy states and put electrons and positrons on the same footing. But the creation and destruction of electron–positron pairs remain. The Dirac theory is inherently a *many-particle* theory, and it provides the beginning of a theoretical framework for creation and destruction of all fundamental particles.

Experiment and theory tell us that the masses of the positron and electron are identical, and that their charges are equal in magnitude but opposite in sign. The positron's spin angular momentum $\vec{S}$ and magnetic moment $\vec{\mu}$ are parallel; they are opposite for the electron. However, $\vec{S}$ and $\vec{\mu}$ have the same magnitude for both particles because they have the same spin. We use the term **antiparticle** for a particle that is related to another particle as the positron is to the electron. Each kind of particle has a corresponding antiparticle. For a few kinds of particles (necessarily all neutral) the particle and antiparticle are identical, and we can say that they are their own antiparticles. The photon is an example; there is no way to distinguish a photon from an antiphoton. We'll use the standard symbols e^- for the electron and e^+ for the positron, and the generic term *electron* will often include both electrons and positrons. Other antiparticles are often denoted by a bar over the particle's symbol; for example, an antiproton is $\bar{p}$. We'll see several other examples of antiparticles later.

Positrons do not occur in ordinary matter. Electron–positron pairs are produced during high-energy collisions of charged particles or γ rays with matter, an example of the process called e^+e^- *pair production* (Fig. 44.3). Electric charge is conserved, and enough energy E must be available to account for the rest energy $2m_ec^2$ of the two particles. The minimum energy for electron–positron pair production is

$$E_{min} = 2m_ec^2 = 2(9.109 \times 10^{-31}\ \text{kg})(2.998 \times 10^8\ \text{m/s})^2$$

$$= 1.637 \times 10^{-13}\ \text{J} = 1.022\ \text{MeV}$$

The inverse process, e^+e^- *pair annihilation,* occurs when a positron and an electron collide. Both particles disappear, and two (or occasionally three) photons can appear, with total energy of at least $2m_ec^2 = 1.022$ MeV. Decay into a *single* photon is impossible because such a process could not conserve both energy and momentum. It's easiest to analyze e^+e^- annihilation processes in the frame of reference called the *center-of-momentum system,* in which the total momentum is zero. It is the relativistic generalization of the center-of-mass system that we discussed in Section 8.5.

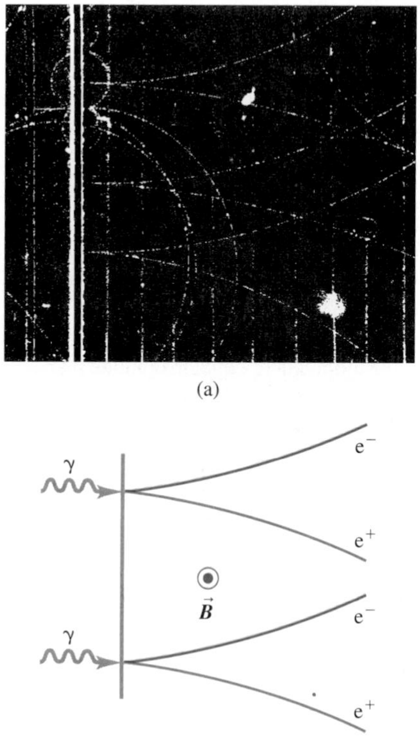

44.3 (a) Photograph of bubble-chamber tracks of electron–positron pairs that are produced when 300-MeV photons strike a lead sheet. A magnetic field directed out of the photograph made the electrons and positrons curve in opposite directions. (b) Diagram showing the pair-production process for two of the photons.

Example 44.1 Pair annihilation

An electron and positron are very far apart and moving toward each other with the same speed. They collide head-on, annihilating each other and producing two photons. Find the energies, wavelengths, and frequencies of the two photons if the initial kinetic energies of the e^- and e^+ are a) both negligibly small; b) both 5.000 MeV. The rest energy of an electron is 0.511 MeV.

SOLUTION

IDENTIFY: As in the ordinary collisions we studied in Chapter 8, in this process both momentum and energy are conserved.

SET UP: Because the electron and positron are initially very far apart, the initial electric potential energy is zero and the initial energy is all kinetic. The final energy is the sum of the energies of the two photons. Before the collision the momenta of the electron and positron have the same magnitude and opposite directions, so the total momentum is zero and we are in the center-of-momentum system; the total momentum of the two photons after the collision must likewise be zero. We solve for the energy E of each photon using conservation of energy, conservation of momentum, and the relationship $E = pc$ for each photon (Section 38.2). We then calculate the wavelengths and frequencies using the relations $E = hc/\lambda = hf$.

EXECUTE: For momentum to be conserved, the two photons must have momenta of equal magnitude p and opposite directions. Since $E = pc = hc/\lambda = hf$, the two photons must also have the same energy E, wavelength λ, and frequency f.

Before the collision the energy of each electron is $K + mc^2$, where K is its kinetic energy and $mc^2 = 0.511$ MeV. The equation for conservation of energy is therefore

$$(K + mc^2) + (K + mc^2) = E + E$$

Hence the energy of each photon is $E = K + mc^2$.

a) If the electron kinetic energy K is negligible compared to its rest energy mc^2, then each photon has energy $E = mc^2 = 0.511$ MeV. The wavelength and frequency of each photon are

$$\lambda = \frac{hc}{E} = \frac{(4.136 \times 10^{-15}\ \text{eV}\cdot\text{s})(3.00 \times 10^{8}\ \text{m/s})}{0.511 \times 10^{6}\ \text{eV}}$$
$$= 2.43 \times 10^{-12}\ \text{m} = 2.43\ \text{pm}$$

$$f = \frac{E}{h} = \frac{0.511 \times 10^{6}\ \text{eV}}{4.136 \times 10^{-15}\ \text{eV}\cdot\text{s}} = 1.24 \times 10^{20}\ \text{Hz}$$

b) Now $K = 5.000$ MeV for each electron, so each photon has energy $E = 5.000\ \text{MeV} + 0.511\ \text{MeV} = 5.511$ MeV. The wavelength and frequency of each photon are 0.2250 pm and 1.333×10^{21} Hz, respectively.

EVALUATE: As a check on our results, recall from Section 38.2 that a visible-light photon with an energy of 2 eV has wavelength 600 nm and frequency 5×10^{14} Hz. Each photon in part (a) has an energy that is about 2.5×10^{5} times greater; as expected, the photon's wavelength is smaller and its frequency larger than those for a visible-light photon by the same factor. You can check the results for part (b) in the same way.

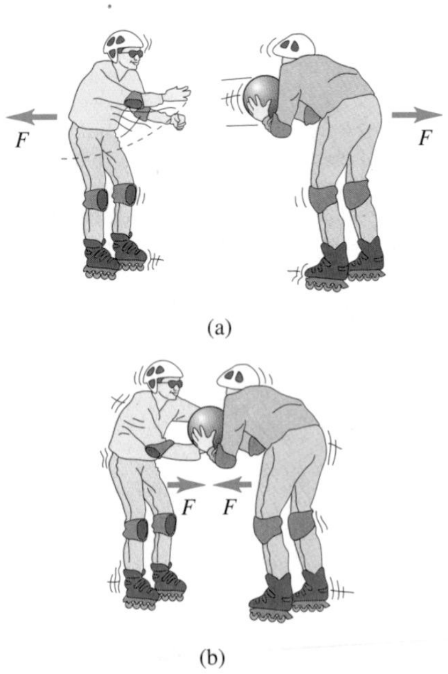

44.4 (a) Two skaters exert repulsive forces on each other by tossing a ball back and forth. (b) Two skaters exert attractive forces on each other when one tries to grab the ball out of the other's hands.

Positrons also occur in the decay of some unstable nuclei, in which they are called beta-plus particles (β^{+}). We discussed β^{+} decay in Section 43.3.

It's often convenient to represent particle masses in terms of the equivalent rest energy using $m = E/c^2$. Then typical mass units are MeV/c^2; for example, $m = 0.511$ MeV/c^2 for an electron or positron. We'll use these units frequently in the following discussions.

Particles as Force Mediators

In classical physics we describe the interaction of charged particles in terms of Coulomb's-law forces. In quantum mechanics we can describe this interaction in terms of emission and absorption of photons. Two electrons repeal each other as one emits a photon and the other absorbs it, just as two skaters can push one another apart by tossing a large ball back and forth between them (Fig. 44.4a). For an electron and a proton, in which the charges are opposite and the force is attractive, we imagine the skaters trying to grab the ball away from one other (Fig. 44.4b). The electromagnetic interaction between two charged particles is *mediated* or transmitted by photons.

If charged-particle interactions are mediated by photons, where does the energy to create the photons come from? Recall from our discussion of the uncertainty principle (Section 39.3) that a state that exists for a short time Δt has an uncertainty ΔE in its energy such that

$$\Delta E\, \Delta t \geq \hbar \qquad (44.3)$$

This uncertainty permits the creation of a photon with energy ΔE, provided that it lives no longer than the time Δt given by Eq. (44.3). A photon that can exist for a short time because of this energy uncertainty is called a *virtual photon.* It's as though there were an energy bank; you can borrow energy, provided that you pay it back within the time limit. According to Eq. (44.3), the more you borrow, the sooner you have to pay it back. Later we'll discuss other virtual particles that live for a short time on "borrowed" energy.

Mesons

Is there a particle that mediates the *nuclear* force? By the mid-1930s the nuclear force between two nucleons (neutrons or protons) appeared to be described by a potential energy $U(r)$ with the general form

$$U(r) = -f^2\frac{e^{-r/r_0}}{r} \qquad \text{(nuclear potential energy)} \qquad (44.4)$$

The constant f characterizes the strength of the interaction, and r_0 describes its range. Figure 44.5 shows a graph of the absolute value of this function and compares it with the function f^2/r, which would be analogous to the *electric* interaction of two protons:

$$U(r) = \frac{1}{4\pi\epsilon_0}\frac{e^2}{r} \qquad \text{(electric potential energy)} \qquad (44.5)$$

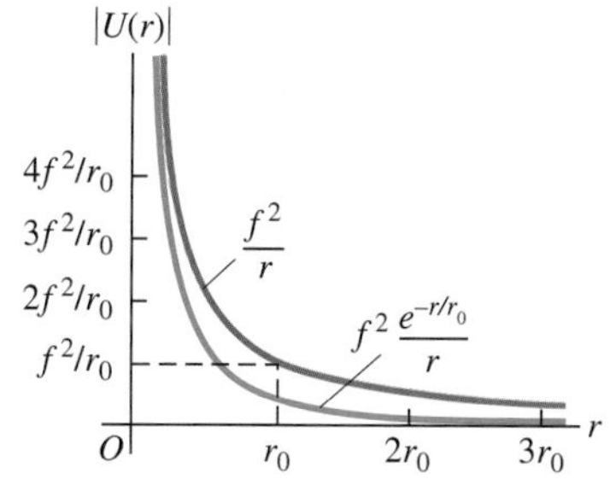

44.5 Graph of the magnitude of the Yukawa potential-energy function for nuclear forces, $|U(r)| = f^2e^{-r/r_0}/r$. The function $U(r) = f^2/r$, proportional to the potential energy for Coulomb's law, is also shown. The two functions are similar at small r, but the Yukawa potential energy drops off much more quickly at large r.

In 1935 the Japanese physicist Hideki Yukawa suggested that a hypothetical particle that he called a **meson** might act as a mediator of the nuclear force. He showed that the range of the force was related to the mass of the particle. His argument went like this: The particle must live for a time Δt long enough to travel a distance comparable to the range of the nuclear force. This range was known from the sizes of nuclei and other information to be of the order of $r_0 = 1.5 \times 10^{-15}$ m $= 1.5$ fm. If we assume that the particle's speed is comparable to c, its lifetime Δt must be of the order of

$$\Delta t = \frac{r_0}{c} = \frac{1.5 \times 10^{-15}\ \text{m}}{3.0 \times 10^8\ \text{m/s}} = 5.0 \times 10^{-24}\ \text{s}$$

From Eq. (44.3) the minimum necessary uncertainty ΔE in energy is

$$\Delta E = \frac{\hbar}{\Delta t} = \frac{1.05 \times 10^{-34}\ \text{J} \cdot \text{s}}{5.0 \times 10^{-24}\ \text{s}} = 2.1 \times 10^{-11}\ \text{J} = 130\ \text{MeV}$$

The mass equivalent Δm of this energy is

$$\Delta m = \frac{\Delta E}{c^2} = \frac{2.1 \times 10^{-11}\ \text{J}}{(3.00 \times 10^8\ \text{m/s})^2} = 2.3 \times 10^{-28}\ \text{kg} = 130\ \text{MeV}/c^2$$

This is about 250 times the electron mass, and Yukawa postulated that a particle with this mass serves as the messenger for the nuclear force. This was a courageous act; at the time, there was not a shred of experimental evidence that such a particle existed.

A year later, Carl Anderson and his colleague Seth Neddermeyer discovered in cosmic radiation two new particles, now called **muons.** The μ^- has charge equal to that of the electron, and its antiparticle the μ^+ has a positive charge with equal magnitude. The two particles have equal mass, about 207 times the electron mass.

But it soon became clear that muons were *not* Yukawa's particles because they interacted with nuclei only very weakly.

In 1947 a family of three particles, called π *mesons* or **pions,** were discovered. Their charges are $+e$, $-e$, and zero, and their masses are about 270 times the electron mass. The pions interact strongly with nuclei, and they *are* the particles predicted by Yukawa. Other, heavier mesons, the ω and ρ, evidently also act as shorter-range messengers of the nuclear force. The complexity of this explanation suggests that it has simpler underpinnings; these involve the quarks and gluons that we'll discuss in Section 44.4. Before discussing mesons further, we'll describe some particle accelerators and detectors to see how mesons and other particles are created in a controlled fashion and observed.

Test Your Understanding

As a virtual photon is being exchanged between a proton and an electron, it can change into a virtual electron–positron pair and then change back into a virtual photon. This effect, called *vacuum polarization,* has a small but measurable effect on the electromagnetic interaction between the proton and the electron. Over what range of distances between the proton and the electron does vacuum polarization have noticeable effects?

44.2 | Particle Accelerators and Detectors

Early nuclear physicists used alpha and beta particles from naturally occurring radioactive elements for their experiments, but they were restricted in energy to the few MeV that are available in such random decays. Present-day particle accelerators can produce precisely controlled beams of particles, from electrons and positrons up to heavy ions, with a wide range of energies. These beams have three main uses. First, high-energy particles can collide to produce new particles, just as a collision of an electron and a positron can produce photons. Second, a high-energy particle has a short de Broglie wavelength and so can probe the small-scale interior structure of other particles, just as electron microscopes (Section 39.4) can give better resolution than optical microscopes. Third, they can be used to produce nuclear reactions of scientific or medical use.

Linear Accelerators

Particle accelerators use electric and magnetic fields to accelerate and guide beams of charged particles. A *linear accelerator* (linac) accelerates particles in a straight line. The earliest examples were J. J. Thomson's cathode-ray tubes and William Coolidge's x-ray tubes. More sophisticated linacs use a series of electrodes with gaps to give the beam of particles a series of boosts. Most present-day high-energy linear accelerators use a traveling electromagnetic wave; the charged particles "ride" the wave in more or less the way that a surfer rides an incoming ocean wave. In the highest-energy linac in the world today, at the Stanford Linear Accelerator Center (SLAC), electrons and positrons are accelerated to 50 GeV in a tube 3 km long. At this energy their de Broglie wavelengths are 0.025 fm, much smaller than the size of a proton or a neutron.

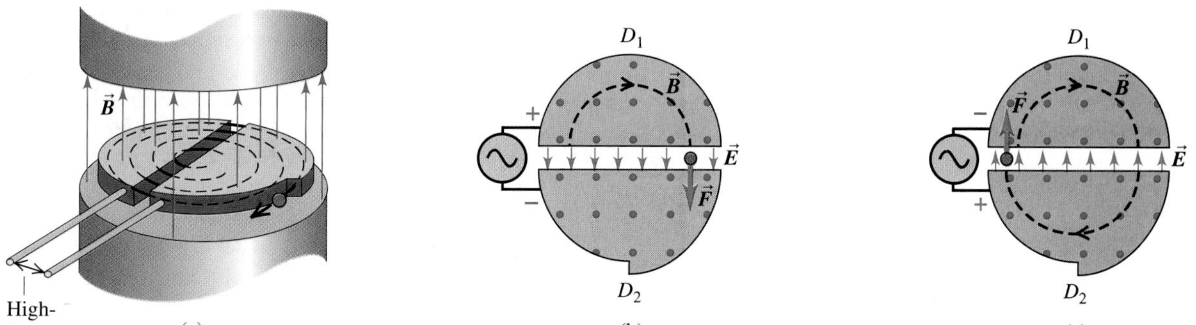

44.6 (a) Schematic diagram of a cyclotron. (b) As the positive particle reaches the gap, it is accelerated by the electric-field force, and the next semicircular orbit has a larger radius. (c) By the time the particle reaches the gap again, the dee voltage has reversed and the particle is again accelerated.

The Cyclotron

Many accelerators use magnets to deflect the charged particles into circular paths. The first of these was the *cyclotron,* invented in 1931 by E. O. Lawrence and M. Stanley Livingston at the University of California. In the cyclotron, shown schematically in Fig. 44.6a, particles with mass m and charge q move inside a vacuum chamber in a uniform magnetic field $\vec{B}$ that is perpendicular to the plane of their paths. In Section 27.4 we showed that in such a field, a particle with speed v moves in a circular path with radius r given by

$$r = \frac{mv}{|q|B} \tag{44.6}$$

and with angular speed (angular frequency) ω given by

$$\omega = \frac{v}{r} = \frac{|q|B}{m} \tag{44.7}$$

An alternating potential difference is applied between the two hollow electrodes D_1 and D_2 (called *dees*), creating an electric field in the gap between them. The polarity of the potential difference and electric field is changed precisely twice each revolution (Fig. 44.6b and 44.6c), so that the particles get a push each time they cross the gap. The pushes increase their speed and kinetic energy, boosting them into paths of larger radius. The maximum speed v_{max} and kinetic energy K_{max} are determined by the radius R of the largest possible path. Solving Eq. (44.6) for v, we find $v = |q|Br/m$ and $v_{max} = |q|BR/m$. Assuming nonrelativistic speeds, we have

$$K_{max} = \frac{1}{2}mv_{max}{}^2 = \frac{q^2B^2R^2}{2m} \tag{44.8}$$

Example 44.2 A proton cyclotron

One cyclotron built during the 1930s has a path of maximum radius 0.500 m and a magnetic field of magnitude 1.50 T. If it is used to accelerate protons, a) calculate the frequency of the alternating voltage applied to the dees; b) find the maximum particle energy.

SOLUTION

IDENTIFY and SET UP: The angular frequency ω of the cyclotron must equal the angular frequency of the proton orbital motion, given by Eq. (44.7). The target variable in part (a) is the ordinary frequency $f = \omega/2\pi$. The maximum kinetic energy of the protons is given by Eq. (44.8).

EXECUTE: a) For protons, $q = 1.60 \times 10^{-19}$ C and $m = 1.67 \times 10^{-27}$ kg. Using Eq. (44.7), we have

$$f = \frac{\omega}{2\pi} = \frac{|q|B}{2\pi m} = \frac{(1.60 \times 10^{-19}\ \text{C})(1.50\ \text{T})}{2\pi(1.67 \times 10^{-27}\ \text{kg})}$$
$$= 2.3 \times 10^{7}\ \text{Hz} = 23\ \text{MHz}$$

b) From Eq. (44.8) the maximum kinetic energy is

$$K_{\text{max}} = \frac{(1.60 \times 10^{-19}\ \text{C})^2(1.50\ \text{T})^2(0.50\ \text{m})^2}{2(1.67 \times 10^{-27}\ \text{kg})}$$
$$= 4.3 \times 10^{-12}\ \text{J} = 2.7 \times 10^{7}\ \text{eV} = 27\ \text{MeV}$$

This energy, which is much larger than energies that are available from natural radioactivity, can cause a variety of interesting nuclear reactions.

EVALUATE: From Eq. (44.6) or (44.7), the proton speed is $v = 7.2 \times 10^7$ m/s, which is about 25% of the speed of light. At such speeds, relativistic effects are beginning to become important. Since we ignored these effects in our calculation, the above result for f and K_{max} are in error by a few percent; this is why we kept only two significant figures.

The maximum energy that can be attained with a cyclotron is limited by relativistic effects. The relativistic version of Eq. (44.7) is

$$\omega = \frac{|q|B}{m}\sqrt{1 - v^2/c^2}$$

As the particles speed up, their angular frequency ω *decreases,* and their motion gets out of phase with the alternating dee voltage. In the *synchrocyclotron* the particles are accelerated in bursts. For each burst, the frequency of the alternating voltage is decreased as the particles speed up, maintaining the correct phase relation with the particles' motion.

Another limitation of the cyclotron is the difficulty of building very large electromagnets. The largest synchrocyclotron ever built has a vacuum chamber that is about 8 m in diameter and accelerates protons to energies of about 600 MeV.

The Synchrotron

To attain higher energies, another type of machine, called the *synchrotron,* is more practical. Particles move in a vacuum chamber in the form of a thin doughnut, called the *accelerating ring.* The particle beam is bent to follow the ring by a series of electromagnets placed around the ring. As the particles speed up, the magnetic field is increased so that the particles retrace the same trajectory over and over. The Tevatron at the Fermi National Accelerator Laboratory (Fermilab) in Batavia, Illinois, is currently the highest-energy accelerator in the world; it can accelerate protons to a maximum energy of nearly 1 TeV $(10^{12}\ \text{eV})$. The accelerating ring is 2 km in diameter and uses superconducting electromagnets (Fig. 44.7). In each machine cycle, a few seconds long, the Tevatron accelerates approximately 10^{13} protons.

As we pointed out in Section 38.5, accelerated charges radiate electromagnetic energy. Examples include the transmissions from radio and TV antennas. In an accelerator in which the particles move in curved paths, this radiation is often called *synchrotron radiation.* High-energy accelerators are typically constructed underground to provide protection from this radiation. From the accelerator standpoint, synchrotron radiation is undesirable since the energy given to an accelerated particle is radiated right back out. It can be minimized by making the accelerator radius r large so that the centripetal acceleration v^2/r is small. On the

(a) (b)

44.7 The Tevatron, the 1-TeV (1000-GeV) accelerator at the Fermi National Accelerator Laboratory in Illinois, is the highest-energy accelerator in the world. (a) An aerial view of the main accelerator ring. (b) A section of the main tunnel. The original upper ring of conventional magnets (red and blue) and the newer lower ring of superconducting magnets can be seen. The conventional ring is used to inject protons and antiprotons into the superconducting ring. The proton and antiproton beams travel in opposite directions as they are accelerated to 0.980 TeV, then collide at interaction points, making 1.96 TeV available to create new particles.

positive side, synchrotron radiation is used as a source of well-controlled high-frequency electromagnetic waves.

Available Energy

When a beam of high-energy particles collides with a stationary target, not all the kinetic energy of the incident particles is *available* to form new particle states. Because momentum must be conserved, the particles emerging from the collision must have some net motion and thus some kinetic energy. The discussion following Example 43.12 (Section 43.6) presented a nonrelativistic example of this principle. The maximum available energy is the kinetic energy in the frame of reference in which the total momentum is zero. We called this the *center-of-momentum system* in Section 44.1. In this system the total kinetic energy after the collision can be zero, so that the maximum amount of the initial kinetic energy becomes available to cause the reaction being studied.

Consider the *laboratory system,* in which a target particle with mass M is initially at rest and is bombarded by a particle with mass m and total energy (including rest energy) E_m. The total available energy E_a in the center-of-momentum system (including rest energies of all the particles) can be shown to be given by

$$E_a^2 = 2Mc^2E_m + (Mc^2)^2 + (mc^2)^2 \qquad \text{(available energy)} \qquad (44.9)$$

When the masses of the target and projectile particles are equal, this can be simplified to

$$E_a^2 = 2mc^2(E_m + mc^2) \qquad \text{(available energy, equal masses)} \qquad (44.10)$$

In the *extreme-relativistic range,* in which the kinetic energy of the bombarding particle is much larger than its rest energy, available energy is a very severe limitation. If the beam and target particles have equal masses and E_m is much greater than mc^2, we can neglect the second term in the parentheses in Eq (44.10). Then E_a is

$$E_a = \sqrt{2mc^2E_m} \qquad \text{(available energy, equal masses, } E_m \gg mc^2\text{)} \qquad (44.11)$$

Example 44.3

Threshold energy for pion production

A proton (rest energy 938 MeV) with kinetic energy K collides with another proton at rest. Both protons survive the collision, but in addition a neutral pion (π^0, rest energy 135 MeV) is produced. What is the threshold energy (minimum value of K) needed for this process?

SOLUTION

IDENTIFY and SET UP: The final state includes the two original protons (mass m) and the pion (mass m_π). The threshold energy corresponds to the minimum-energy case in which all three particles are at rest in the center-of-momentum system. The total available energy in this system must be at least their total rest energy, $2mc^2 + m_\pi c^2$. We use this in Eq. (44.10) and solve for the total energy E_m of the bombarding proton; the kinetic energy K (our target variable) is the total energy minus the proton rest energy mc^2.

EXECUTE: We substitute $E_a = 2mc^2 + m_\pi c^2$ into Eq. (44.10), simplify, and solve for E_m:

$$4m^2c^4 + 4mm_\pi c^4 + m_\pi^2 c^4 = 2mc^2 E_m + 2(mc^2)^2$$

$$E_m = mc^2 + m_\pi c^2\left(2 + \frac{m_\pi}{2m}\right)$$

The first term in the expression for E_m is the rest energy of the bombarding proton, and the remaining terms give its kinetic energy. We see that the kinetic energy must be somewhat greater than twice the rest energy of the pion we want to create. Using $mc^2 = 938$ MeV and $m_\pi c^2 = 135$ MeV in this expression, we find $m_\pi/2m = 0.072$ and

$$E_m = mc^2 + (135\text{ MeV})(2 + 0.072) = mc^2 + 280\text{ MeV}$$

To create a pion with rest energy 135 MeV, a bombarding proton needs a kinetic energy of at least 280 MeV.

EVALUATE: You should compare this result with the result of Example 37.12 (Section 37.8), which required only 67.5 MeV of kinetic energy for each proton in a head-on collision. We discuss the energy advantage of such collisions in the next subsection.

Example 44.4 Increasing the available energy

a) The Fermilab accelerator was originally designed for a proton beam energy of 800 GeV $(800 \times 10^9\text{ eV})$ on a stationary target. Find the available energy in a proton–proton collision. b) If the proton beam energy is increased to 980 GeV, what is the available energy?

SOLUTION

IDENTIFY and SET UP: Our target variable is the available energy E_a. In both parts (a) and (b) the two particles are identical and the beam energy E_m is much larger than the proton rest energy $mc^2 = 938\text{ MeV} = 0.938\text{ GeV}$, so we can safely use Eq. (44.11).

EXECUTE: a) When $E_m = 800$ GeV,

$$E_a = \sqrt{2(0.938\text{ GeV})(800\text{ GeV})} = 38.7\text{ GeV}$$

b) When $E_m = 980$ GeV,

$$E_a = \sqrt{2(0.938\text{ GeV})(980\text{ GeV})} = 42.9\text{ GeV}$$

EVALUATE: With a stationary proton target, increasing the proton beam energy by 200 GeV increases the available energy by only 4.2 GeV! This shows a major limitation of experiments in which one of the colliding particles is initially at rest. Below we describe how physicists can overcome these limitations.

Colliding Beams

The limitation illustrated by Example 44.4 is circumvented in *colliding-beam* experiments. In these experiments there is no stationary target; instead, beams of particles moving in opposite directions are tightly focused onto one another so that head-on collisions can occur. Usually the two colliding particles have momenta of equal magnitude and opposite direction, so the total momentum is zero. Hence the laboratory system is also the center-of-momentum system, and the available energy is maximized. If one beam contains particles and the other beam contains antiparticles (for instance, electrons and positrons or protons and antiprotons), the available energy E_a is the *total* energy of the two colliding particles.

The highest-energy colliding beams currently available are those at the Tevatron at Fermilab (Fig. 44.7). A beam of 980-GeV (0.980-TeV) protons circulates around a ring in the opposite direction from a beam of 0.980-TeV antiprotons. Magnets are used to steer the beams into each other at various points around the ring. The available energy in the resulting head-on collisions is $2(0.980\text{ TeV}) = 1.96$ TeV. These colliding beams were used in the discovery of the top quark, to be discussed in Section 44.4.

Other laboratories use colliding beams of electrons and positrons. A common design uses *storage rings,* in which bunches of electrons and positrons are kept circulating in opposite directions around a circular tunnel, giving them many opportunities to interact. At present the largest e^+e^- storage ring is the Large Electron-Positron Collider (LEP), situated in an underground tunnel 27 km (about 17 miles) in circumference at the European Laboratory for Particle Physics (CERN) in Geneva, Switzerland. The total available energy E_a is 180 GeV. Experiments using a lower-energy e^+e^- storage ring at the Stanford Linear Accelerator Center (SLAC) revealed the properties of the charm quark and were used to discover the tau lepton, a particle that is closely related to the electron and the muon. (In Section 44.4 we will see how physicists systematize their knowledge of these and many other particles.)

Even higher-energy colliding beams will be available with the Large Hadron Collider (LHC) at CERN, which will use the same tunnel as the LEP but will accelerate colliding beams of protons to 7 TeV each, giving an available energy of 14 TeV. It will also be possible to collide electrons from the LEP with protons from the LHC, giving an available energy of 1.5 TeV. The LHC is scheduled to go into operation in 2007.

Detectors

Ordinarily, we can't see or feel individual subatomic particles or photons. How, then, do we measure their properties? A wide variety of devices have been designed. Many detectors use the ionization caused by charged particles as they move through a gas, liquid, or solid. The ions along the particle's path act as nucleation centers for droplets of liquid in the supersaturated vapor of a cloud chamber (Fig. 44.1) or cause small volumes of vapor in the superheated liquid of a bubble chamber (Fig. 44.3a). In a semiconducting solid the ionization can take the form of electron–hole pairs. We discussed their detection in Section 42.7. *Wire chambers* contain arrays of closely spaced wires that detect the ions. The charge collected and time information from each wire are processed using computers to reconstruct the particle trajectories (Fig. 44.8; see also the photograph that opens Chapter 37).

44.8 This computer-generated image shows a typical result of a proton–antiproton collision in the Tevatron recorded by the Collider Detector. Particles produced in the collision include pions, kaons, muons, and others, each indicated by a different color.

Cosmic Ray Experiments

Large numbers of particles called *cosmic rays* continually bombard the earth from sources both within and beyond our galaxy. These particles consist mostly of neutrinos, protons, and heavier nuclei, with energies ranging from less than 1 MeV to more than 10^{20} eV. The earth's atmosphere and magnetic field protect us from much of this radiation. However, this means that cosmic-ray experimentation often must be carried out above all or most of the atmosphere by means of rockets or high-altitude balloons.

In contrast, neutrino detectors are buried below the earth's surface in tunnels or mines or submerged deep in the ocean. This is done to screen out all other types of particles so that only neutrinos, which interact only very weakly with matter, reach the detector. It would take a light-year's thickness of lead to absorb a sizable fraction of a beam of neutrinos. Thus neutrino detectors consist of huge amounts of matter: the Super-Kamiokande detector looks for flashes of light produced when a neutrino interacts in a tank containing 5×10^7 kg of water (see Section 44.5).

Cosmic rays were important in early particle physics, and their study currently brings us important information about the rest of the universe. Although cosmic rays provide a source of high-energy particles that does not depend on expensive accelerators, most particle physicists use accelerators because the high-energy cosmic-ray particles they want are too few and too random.

Test Your Understanding

In a colliding-beam experiment, a 90-GeV electron collides head-on with a 90-GeV positron. The electron and the positron annihilate each other, forming a single virtual photon that then transforms into other particles. Does the virtual photon obey the same relationship $E = pc$ as real photons do?

44.3 | Particles and Interactions

44.9 The ties that bind us together originate in the fundamental interactions of nature. The nuclei within our bodies are held together by the strong interaction. The electromagnetic interaction binds nuclei and electrons together to form atoms, binds atoms together to form molecules, and binds molecules together to form us.

We have mentioned the array of subatomic particles that were known as of 1947: photons, electrons, positrons, protons, neutrons, muons, and pions. Since then, literally hundreds of additional particles have been discovered in accelerator experiments. The vast majority of known particles are *unstable* and decay spontaneously into other particles. Particles of all kinds, whether stable or unstable, can be created or destroyed in interactions between particles. Each such interaction involves the exchange of virtual particles, which exist on borrowed energy allowed by the uncertainty principle.

Although the world of subatomic particles and their interactions is complex, some key results bring order and simplicity to the seeming chaos. One key simplification is that there are only four fundamental types of interactions, each mediated or transmitted by the exchange of certain characteristic virtual particles. Furthermore, not all particles respond to all four kinds of interaction. In this section we will examine the fundamental interactions more closely and see how physicists classify particles in terms of the ways in which they interact.

Four Forces and Their Mediating Particles

In Section 5.5 we first described the four fundamental types of forces or interactions (Fig. 44.9). They are, in order of decreasing strength,

1. the strong interaction,
2. the electromagnetic interaction,
3. the weak interaction, and
4. the gravitational interaction.

The *electromagnetic* and *gravitational* interactions are familiar from classical physics. Both are characterized by a $1/r^2$ dependence on distance. In this scheme, the mediating particles for both interactions have mass zero and are stable as ordinary particles. The mediating particle for the electromagnetic interaction is the familiar photon, which has spin 1. (That means its spin quantum number is $s = 1$, so the magnitude of its spin angular momentum is $S = \sqrt{s(s+1)}\hbar = \sqrt{2}\hbar$.) That particle for the gravitational force is the spin-2 *graviton* $(s = 2, S = \sqrt{s(s+1)}\hbar = \sqrt{6}\hbar)$. The graviton has not yet been observed experimentally because the gravitational force is very much weaker than the electromagnetic force. For example, the gravitational attraction of two protons is smaller than their electrical repulsion by a factor of about 10^{36}. The gravitational force is of primary importance in the structure of stars and the large-scale behavior of the universe, but it is not believed to play a significant role in particle interactions at the energies that are currently attainable.

The other two forces are less familiar. One, usually called the *strong interaction,* is responsible for the nuclear force and also for the production of pions and several other particles in high-energy collisions. At the most fundamental level, the mediating particle for the strong interaction is called a *gluon.* However, the force between nucleons is more easily described in terms of mesons as the mediating particles. We'll discuss the spin-1, massless gluon in Section 44.4.

Equation (44.4) is a possible potential-energy function for the nuclear force. The strength of the interaction is described by the constant f^2, which has units of energy times distance. A better basis for comparison with other forces is the dimensionless ratio $f^2/\hbar c$, called the *coupling constant* for the interaction. (We invite you to verify that this ratio is a pure number and so must have the same value in all systems of units.) The observed behavior of nuclear forces suggests that $f^2/\hbar c \approx 1$. The dimensionless coupling constant for *electromagnetic* interactions is

$$\frac{1}{4\pi\epsilon_0}\frac{e^2}{\hbar c} = 7.297 \times 10^{-3} = \frac{1}{137.0} \tag{44.12}$$

Thus the strong interaction is roughly 100 times as strong as the electromagnetic interaction; however, it drops off with distance more quickly than $1/r^2$.

The fourth interaction is called the *weak* interaction. It is responsible for beta decay, such as the conversion of a neutron into a proton, an electron, and an antineutrino. It is also responsible for the decay of many unstable particles (pions into muons, muons into electrons, and so on). Its mediating particles are the short-lived particles W^+, W^-, and Z^0. The existence of these particles was confirmed in 1983 in experiments at CERN, for which Carlo Rubbia and Simon van der Meer were awarded the Nobel Prize in 1984. The $W^{\pm}$ and Z^0 have spin 1 like the photon and the gluon, but they are *not* massless. In fact, they have enormous masses, 80.4 GeV/c^2 for the W's and 91.2 GeV/c^2 for the Z^0. With such massive mediating particles the weak interaction has a much shorter range than the strong interaction. It also lives up to its name by being weaker than the strong interaction by a factor of about 10^9.

Table 44.1 compares the main features of these four fundamental interactions.

Table 44.1 Four Fundamental Interactions

Interaction	Relative Strength	Range	Mediating Particle: Name	Mass	Charge	Spin
Strong	1	Short (~1 fm)	Gluon	0	0	1
Electromagnetic	$\frac{1}{137}$	Long ($1/r^2$)	Photon	0	0	1
Weak	10^{-9}	Short (~0.001 fm)	$W^{\pm}$, Z^0	80.4, 91.2 GeV/c^2	$\pm e$, 0	1
Gravitational	10^{-38}	Long ($1/r^2$)	Graviton	0	0	2

More Particles

In Section 44.1 we mentioned the discoveries of muons in 1937 and of pions in 1947. The electric charges of the muons and the charged pions have the same magnitude e as the electron charge. The positive muon μ^+ is the antiparticle of the negative muon μ^-. Each has spin $\frac{1}{2}$, like the electron, and a mass of about $207m_e = 106\ \text{MeV}/c^2$. Muons are unstable; each decays with a lifetime of 2.2×10^{-6} s into an electron of the same sign, a neutrino, and an antineutrino.

There are three kinds of pions, all with spin 0; they have *no* spin angular momentum. The π^+ and π^- have masses of $273m_e = 140\ \text{MeV}/c^2$. They are unstable; each $\pi^\pm$ decays with a lifetime of 2.6×10^{-8} s into a muon of the same sign along with a neutrino for the π^+ and an antineutrino for the π^-. The π^0 is somewhat less massive, $264m_e = 135\ \text{MeV}/c^2$, and it decays with a lifetime of 8.4×10^{-17} s into two photons. The π^+ and π^- are antiparticles of one another, while the π^0 is its own antiparticle. (That is, there is no distinction between particle and antiparticle for the π^0.)

The existence of the *antiproton* $\overline{\text{p}}$ had been suspected ever since the discovery of the positron. The $\overline{\text{p}}$ was found in 1955, when proton–antiproton $(\text{p}\overline{\text{p}})$ pairs were created by use of a beam of 6-GeV protons from the Bevatron at the University of California, Berkeley. The *antineutron* $\overline{\text{n}}$ was found soon afterward. After 1960, as higher-energy accelerators and more sophisticated detectors were developed, a veritable blizzard of new unstable particles were identified. To describe and classify them, we need a small blizzard of new terms.

Initially, particles were classified by mass into three categories: (1) leptons ("light ones" such as electrons); (2) mesons ("intermediate ones" such as pions); and (3) baryons ("heavy ones" such as nucleons and more massive particles). But this scheme has been superseded by a more useful one in which particles are classified in terms of their *interactions.* For instance, *hadrons* (which include mesons and baryons) have strong interactions, and *leptons* do not.

In the following discussion we will also distinguish between **fermions,** which have half-integer spins, and **bosons,** which have zero or integer spins. Fermions obey the exclusion principle, on which the Fermi-Dirac distribution function (Section 42.5) is based. Bosons do not obey the exclusion principle and have a different distribution function, the Bose-Einstein distribution.

Leptons

The **leptons,** which do not have strong interactions, include six particles; the electron (e^-) and its neutrino (ν_e), the muon (μ^-) and its neutrino (ν_μ), and the tau particle (τ^-) and its neutrino (ν_τ). Each of the six particles has a distinct antiparticle. All leptons have spin and thus are fermions. The family of leptons is shown in Table 44.2. The taus have mass $3478m_e = 1777\ \text{MeV}/c^2$. Taus and muons are

Table 44.2 The Leptons

Particle Name	Symbol	Anti-Particle	Mass (MeV/c^2)	L_e	L_μ	L_τ	Lifetime (s)	Principal Decay Modes
Electron	e^-	e^+	0.511	+1	0	0	Stable	
Electron neutrino	ν_e	$\overline{\nu}_e$	$<3 \times 10^{-6}$	+1	0	0	Stable	
Muon	μ^-	μ^+	105.7	0	+1	0	2.20×10^{-6}	$e^-\overline{\nu}_e\nu_\mu$
Muon neutrino	ν_μ	$\overline{\nu}_\mu$	<0.19	0	+1	0	Stable	
Tau	τ^-	τ^+	1777	0	0	+1	2.9×10^{-13}	$\mu^-\overline{\nu}_\mu\nu_\tau$ or $e^-\overline{\nu}_e\nu_\tau$
Tau neutrino	ν_τ	$\overline{\nu}_\tau$	<18.2	0	0	+1	Stable	

unstable; a τ^- often decays into a μ^- plus a tau neutrino and a muon antineutrino, and a μ^- decays into a electron plus a muon neutrino and an electron antineutrino. They have relatively long lifetimes because their decays are mediated by the weak interaction. Despite their zero charge, a neutrino is distinct from an antineutrino; the spin angular momentum of a neutrino has a component that is opposite its linear momentum, while for an antineutrino that component is parallel to its linear momentum. Because neutrinos are so elusive, physicists have only been able to place upper limits on the rest masses of the ν_e, the ν_μ, and the ν_τ. Until recently, it was thought that the rest masses of the neutrinos were actually zero; compelling evidence now indicates that they have small but nonzero masses. We'll return to this point and its implications later.

Leptons obey a *conservation principle.* Corresponding to the three pairs of leptons are three lepton numbers L_e, L_μ, and L_τ. The electron e^- and the electron neutrino ν_e are assigned $L_e = 1$, and their antiparticles e^+ and $\bar{\nu}_e$ are given $L_e = -1$. Corresponding assignments of L_μ and L_τ are made for the μ and τ particles and their neutrinos. **In all interactions, each lepton number is separately conserved.** For example, in the decay of the μ^-, the lepton numbers are

$$\begin{array}{ccccccc} \mu^- & \rightarrow & e^- & + & \bar{\nu}_e & + & \nu_\mu \\ L_\mu = 1 & & L_e = 1 & & L_e = -1 & & L_\mu = 1 \end{array}$$

These conservation principles have no counterpart in classical physics.

Example 44.5 Lepton number conservation

Check conservation of lepton numbers for the following decay schemes:

a) $\mu^+ \rightarrow e^+ + \nu_e + \bar{\nu}_\mu$

b) $\pi^- \rightarrow \mu^- + \bar{\nu}_\mu$

c) $\pi^0 \rightarrow \mu^- + e^+ + \nu_e$

SOLUTION

IDENTIFY and SET UP: Checking for lepton number conservation is straightforward: Use Table 44.2 to find the lepton numbers L_e, L_μ, and L_τ for each particle, then compare the total value of each lepton number before the process and after the process.

EXECUTE: In each of the three decays, there are no τ particles nor any τ neutrinos, so $L_\tau = 0$ both before and after each decay; L_τ is therefore conserved.

a) The initial lepton numbers are $L_e = 0$ and $L_\mu = -1$. The final values are $L_e = -1 + 1 + 0 = 0$ and $L_\mu = 0 + 0 + (-1) = -1$. All lepton numbers are conserved.

b) All lepton numbers are initially zero (the π^- is *not* a lepton). The final values are $L_\mu = 1 + (-1) = 0$ and $L_e = 0$. Again, all lepton numbers are conserved.

c) The initial L's are again all zero; in the final state, $L_e = 0 + (-1) + 1 = 0$ and $L_\mu = 1 + 0 + 0 = 1$. Thus L_μ is *not* conserved.

EVALUATE: Decays (a) and (b), which are consistent with the conservation of lepton number, are observed in nature. Decay (c), which violates the conservation of L_μ, has *never* been observed. Physicists used these and may other experimental results to deduce the principle that all three lepton numbers must be conserved.

Hadrons

Hadrons, the strongly interacting particles, are a more complex family than leptons. Each hadron has an antiparticle, often denoted with an overbar, as with the antiproton $\bar{p}$. There are two subclasses of hadrons: *mesons* and *baryons.*

Table 44.3 Some Hadrons and Their Properties

Particle	Mass (MeV/c^2)	Charge Ratio Q/e	Spin	Baryon Number B	Strangeness S	Mean Lifetime (s)	Typical Decay Modes	Quark Content
Mesons								
π^0	135.0	0	0	0	0	8.4×10^{-17}	$\gamma\gamma$	$u\bar{u}, d\bar{d}$
π^+	139.6	+1	0	0	0	2.60×10^{-8}	$\mu^+\nu_\mu$	$u\bar{d}$
π^-	139.6	−1	0	0	0	2.60×10^{-8}	$\mu^-\bar{\nu}_\mu$	$\bar{u}d$
K^+	493.7	+1	0	0	+1	1.24×10^{-8}	$\mu^+\nu_\mu$	$u\bar{s}$
K^-	493.7	−1	0	0	−1	1.24×10^{-8}	$\mu^-\bar{\nu}_\mu$	$\bar{u}s$
η^0	547.3	0	0	0	0	$\approx 10^{-18}$	$\gamma\gamma$	$u\bar{u}, d\bar{d}, s\bar{s}$
Baryons								
p	938.3	+1	$\frac{1}{2}$	1	0	Stable	–	*uud*
n	939.6	0	$\frac{1}{2}$	1	0	886	$pe^-\bar{\nu}_e$	*udd*
Λ^0	1116	0	$\frac{1}{2}$	1	−1	2.63×10^{-10}	$p\pi^-$ or $n\pi^0$	*uds*
Σ^+	1189	+1	$\frac{1}{2}$	1	−1	8.02×10^{-11}	$p\pi^0$ or $n\pi^+$	*uus*
Σ^0	1193	0	$\frac{1}{2}$	1	−1	7.4×10^{-20}	$\Lambda^0\gamma$	*uds*
Σ^-	1197	−1	$\frac{1}{2}$	1	−1	1.48×10^{-10}	$n\pi^-$	*dds*
Ξ^0	1315	0	$\frac{1}{2}$	1	−2	2.90×10^{-10}	$\Lambda^0\pi^0$	*uss*
Ξ^-	1321	−1	$\frac{1}{2}$	1	−2	1.64×10^{-10}	$\Lambda^0\pi^-$	*dss*
Δ^{++}	1232	+2	$\frac{3}{2}$	1	0	$\approx 10^{-23}$	$p\pi^+$	*uuu*
Ω^-	1672	−1	$\frac{3}{2}$	1	−3	8.2×10^{-11}	Λ^0K^-	*sss*
Λ_c^+	2285	+1	$\frac{1}{2}$	1	0	2.0×10^{-13}	$pK^-\pi^+$	*udc*

Table 44.3 shows some of the many hadrons that are currently known. (We'll discuss what is meant by *strangeness* and *quark content* later in this section and in the next one.)

Mesons include the pions that have already been mentioned, K mesons or *kaons,* η mesons, and others that we will discuss later. Mesons have spin 0 or 1 and therefore are all bosons. There are no stable mesons; all can and do decay to less massive particles, obeying all the conservation laws for such decays.

Baryons include the nucleons and several particles called *hyperons,* including the Λ, Σ, Ξ, and Ω. These resemble the nucleons but are more massive. Baryons have half-integer spin, and therefore all are fermions. The only stable baryon is the proton; a free neutron decays to a proton, and the hyperons decay to other hyperons or to nucleons by various processes. Baryons obey the principle of *conservation of baryon number,* analogous to conservation of lepton numbers, again with no counterpart in classical physics. We assign a baryon number $B = 1$ to each baryon (p, n, Λ, Σ, and so on) and $B = -1$ to each antibaryon ($\bar{p}$, $\bar{n}$, $\bar{\Lambda}$, $\bar{\Sigma}$, and so on). **In all interactions, the total baryon number is conserved.** This principle is the reason why the mass number A was conserved in all the nuclear reactions that we studied in Chapter 43.

Example 44.6

Baryon number conservation

Which of the following reactions obey the principle of conservation of baryons?

a) $n + p \rightarrow n + p + p + \bar{p}$

b) $n + p \rightarrow n + p + \bar{p}$

SOLUTION

IDENTIFY and SET UP: This example is similar to Example 44.5: we need only compare the total baryon number before the process to the total baryon number after the process.

EXECUTE: In both reactions the initial baryon number is $1 + 1 = 2$.
a) The final baryon number is $1 + 1 + 1 + (-1) = 2$. Baryon number is conserved.
b) The final baryon number for this reaction is $1 + 1 + (-1) = 1$. Baryon number is *not* conserved.

EVALUATE: Reaction (a) can and does occur, provided that enough energy is available in the n + p collision. Reaction (b), which violates the principle of conservation, has never been observed. Can you show that this reaction also violates the law of conservation of electric charge?

Example 44.7 Antiproton creation

Antiprotons can be produced by bombarding a stationary proton target (liquid hydrogen) with a beam of protons. Find the minimum beam energy that is needed for this reaction to occur.

SOLUTION

IDENTIFY and SET UP: The reaction must satisfy the conservation of baryon number, of charge, and of energy. Since the target and bombarding protons are of equal mass and the target is at rest, we determine the minimum energy E_m of the bombarding proton using Eq. (44.10)

EXECUTE: Conservation of charge and conservation of baryon number forbid the creation of an antiproton by itself; it must be created as part of a proton–antiproton pair. The appropriate reaction is

$$\mathrm{p} + \mathrm{p} \rightarrow \mathrm{p} + \mathrm{p} + \mathrm{p} + \overline{\mathrm{p}}$$

For this reaction to occur, the minimum available energy E_a in Eq. (44.10) is the final rest energy $4mc^2$. With that substitution, Eq. (44.10) gives

$$(4mc^2)^2 = 2mc^2(E_m + mc^2)$$
$$E_m = 7mc^2$$

EVALUATE: The energy E_m of the bombarding particle includes its rest energy mc^2, so its minimum *kinetic* energy must be $6mc^2 = 6(938\ \mathrm{MeV}) = 5.63\ \mathrm{GeV}$.

The search for the antiproton was one of the principal reasons for the construction of the Bevatron at the University of California, Berkeley, with beam energy of 6 GeV. The search was successful; in 1959, Emilio Segrè and Owen Chamberlain were awarded the Nobel Prize for its discovery.

Strangeness

The K mesons and the Λ and Σ hyperons were discovered during the late 1950s. Because of their unusual behavior they were called *strange particles.* They were produced in high-energy collisions such as $\pi^- + \mathrm{p}$, and a K meson and a hyperon were always produced *together.* The relatively high rate of production of these particles suggested that it was a *strong*-interaction process, but their relatively long lifetimes suggested that their decay was a *weak*-interaction process. The K^0 appeared to have *two* lifetimes, one about 9×10^{-11} s and another nearly 600 times longer. Where the K mesons strongly interacting hadrons or not?

The search for the answer to this questions led physicists to introduce a new quantity called **strangeness.** The hyperons Λ^0 and $\Sigma^{\pm,0}$ were assigned a strangeness quantum number $S = -1$, and the associated K^0 and K^+ mesons were assigned $S = +1$. The corresponding antiparticles had opposite strangeness, $S = +1$ for $\overline{\Lambda}^0$ and $\overline{\Sigma}^{\pm,0}$ and $S = -1$ for $\overline{\mathrm{K}}^0$ and K^-. Then strangeness was *conserved* in production processes such as

$$\mathrm{p} + \pi^- \rightarrow \Sigma^- + \mathrm{K}^+$$
$$\mathrm{p} + \pi^- \rightarrow \Lambda^0 + \mathrm{K}^0$$

The process

$$\mathrm{p} + \pi^- \rightarrow \mathrm{p} + \mathrm{K}^-$$

does not conserve strangeness, and it doesn't occur.

When strange particles decay individually, strangeness is usually *not* conserved. Typical processes include

$$\Sigma^+ \rightarrow n + \pi^+$$
$$\Lambda^0 \rightarrow p + \pi^-$$
$$K^- \rightarrow \pi^+ + \pi^- + \pi^-$$

In each of these decays, the initial strangeness is 1 or -1, and the final value is zero. All observations of these particles are consistent with the conclusion that *strangeness is conserved in strong interactions but it can change by zero or one unit in weak interactions.* There is no counterpart to the strangeness quantum number in classical physics.

CAUTION Take care not to confuse the symbol S for strangeness with the identical symbol for the magnitude of the spin angular momentum.

ActivPhysics ONLINE

19.5 Particle Physics

Conservation Laws

The decay of strange particles provides our first example of a conditional conservation law, one that is obeyed in some interactions and not in others. Let's review *all* the conservation laws we know about and see what conclusions we can draw from them.

Several conservation laws are obeyed in *all* interactions. These include the familiar conservation laws; energy, momentum, angular momentum, and electric charge. These are called *absolute conservation laws.* Baryon number and the three lepton numbers are also conserved in all interactions. Strangeness is conserved in strong and electromagnetic interactions but *not* in all weak interactions.

Two other quantities, which are conserved in some but not all interactions, are useful in classifying particles and their interactions. One is *isospin,* a quantity that is used to describe the charge independence of the strong interactions. The other is *parity,* which describes the comparative behavior of two systems that are mirror images of each other. Isospin is conserved in strong interactions, which are charge-independent, but not in electromagnetic or weak interactions. (The electromagnetic interaction is certainly *not* charge-independent.) Parity is conserved in strong and electromagnetic interactions but not in weak ones. The Chinese-American physicists T.D. Lee and C.N. Yang received the Nobel Prize in 1957 for laying the theoretical foundations for nonconservation of parity in weak interactions.

This discussion shows that conservation laws provide another basis for classifying particles and their interactions. Each conservation law is also associated with a *symmetry* property of the system. A familiar example is angular momentum. If a system is in an environment that has spherical symmetry, there can be no torque acting on it because the direction of the torque would violate the symmetry. In such a system, total angular momentum is *conserved.* When a conservation law is violated, the interaction is often described as a *symmetry-breaking interaction.*

Test Your Understanding

From conservation of energy, a particle of mass m and rest energy mc^2 can decay only if the decay products have a total mass less than m. (The remaining energy goes into the kinetic energy of the decay products.) Why, then, can't a proton decay into less massive mesons?

44.4 | Quarks and the Eightfold Way

The leptons form a fairly neat package: three particles and three neutrinos, each with its antiparticle, and a conservation law relating their numbers. Physicists believe that leptons are genuinely fundamental particles. The hadron family, by comparison, is a mess. Table 44.3 contains only a sample of well over 100 hadrons that have been discovered since 1960, and it has become clear that these particles *do not* represent the most fundamental level of the structure of matter.

Our present understanding of the structure of hadrons is based on a proposal made initially in 1964 by the American physicist Murray Gell-Mann and his collaborators. In this proposal, hadrons are not fundamental particles but are composite structures whose constituents are spin-$\frac{1}{2}$ fermions called **quarks.** (The name is found in the line "Three quarks for Muster Mark!" from *Finnegan's Wake* by James Joyce.) Each baryon is composed of three quarks (qqq), each antibaryon of three antiquarks ($\bar{q}\,\bar{q}\,\bar{q}$), and each meson of a quark–antiquark pair ($q\bar{q}$). Table 44.3 of the previous section gives the quark content of many hadrons. No other compositions seem to be necessary. This scheme requires that quarks have electric charges with magnitudes $\frac{1}{3}$ and $\frac{2}{3}$ of the electron charge e, which had previously been thought to be the smallest unit of charge. Each quark also has a fractional value $\frac{1}{3}$ for its baryon number B, and each antiquark has a baryon-number value $-\frac{1}{3}$. In a meson, a quark and antiquark combine with net baryon number 0 and can have their spin angular momentum components parallel to form a spin-1 meson or antiparallel to form a spin-0 meson. Similarly, the three quarks in a baryon combine with net baryon number 1 and can form a spin-$\frac{1}{2}$ baryon or a spin-$\frac{3}{2}$ baryon.

The Three Original Quarks

The first (1964) quark theory included three types (called *flavors*) of quarks, labeled u (up), d (down), and s (strange). Their principal properties are listed in Table 44.4. The corresponding antiquarks $\bar{u}$, $\bar{d}$, and $\bar{s}$ have opposite values of Q, B, and S. Protons, neutrons, π and K mesons, and several hyperons can be constructed from these three quarks. For example, the proton quark content is uud. Checking Table 44.4, we see that the values of Q/e add to 1 and that the values of the baryon number B also add to 1, as we should expect. The neutron is udd, with total $Q = 0$ and $B = 1$. The π^+ meson is $u\bar{d}$, with $Q/e = 1$ and $B = 0$, and the K^+ meson is $u\bar{s}$. Checking the values of S for the quark content, we see that the proton, neutron, and π^+ have strangeness 0 and that the K^+ has strangeness 1, in agreement with Table 44.3. The antiproton is $\bar{p} = \bar{u}\,\bar{u}\,\bar{d}$, the negative pion is $\pi^- = \bar{u}d$, and so on. The quark content can also be used to explain dynamic prop-

Table 44.4 Properties of the Three Original Quarks

Symbol	Q/e	Spin	Baryon Number, B	Strangeness, S	Charm, C	Bottomness, B'	Topness, T
u	$\frac{2}{3}$	$\frac{1}{2}$	$\frac{1}{3}$	0	0	0	0
d	$-\frac{1}{3}$	$\frac{1}{2}$	$\frac{1}{3}$	0	0	0	0
s	$-\frac{1}{3}$	$\frac{1}{2}$	$\frac{1}{3}$	-1	0	0	0

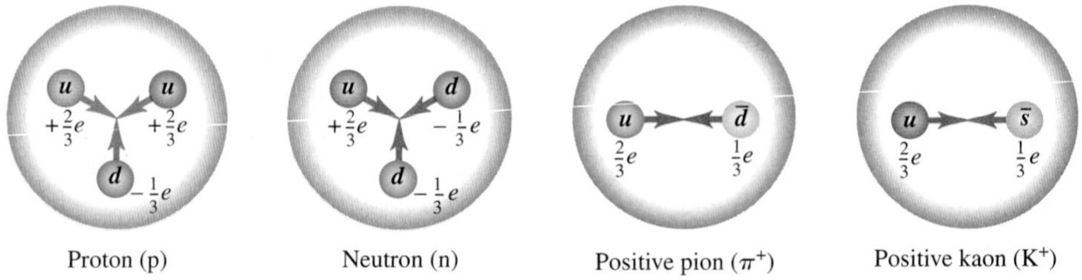

44.10 Quark content of four different hadrons. The various color combinations that are needed for color neutrality are not shown.

erties of hadrons such as their excited states and magnetic moments. Figure 44.10 shows the quark content of two baryons and two mesons.

Example 44.8 Determining the quark content of baryons

Given that they contain only $\boldsymbol{u}$, $\boldsymbol{d}$, $\boldsymbol{s}$, $\bar{\boldsymbol{u}}$, $\bar{\boldsymbol{d}}$, and/or $\bar{s}$, find the quark content of a) Σ^+; b) $\overline{\Lambda^0}$. The Σ^+ and Λ^0 particles are both baryons with strangeness $S = -1$.

SOLUTION

IDENTIFY and SET UP: We use the quark properties given in Table 44.4. The total charge of each baryon is the sum of the individual quark charges, and similarly for the baryon number and strangeness.

EXECUTE: Baryons contain three quarks; if $S = -1$, one and only one of the three must be an s quark.

a) The Σ^+ has $Q/e = +1$, so the other two quarks must both be $\boldsymbol{u}$. Hence the quark content of Σ^+ is $\boldsymbol{uus}$.

b) First we find the quark content of the Λ^0. For zero total charge, the other two quarks must be $\boldsymbol{u}$ and $\boldsymbol{d}$, so the quark content of the Λ^0 is $\boldsymbol{uds}$. The $\overline{\Lambda^0}$ is the antiparticle of the Λ^0, so its quark content is $\bar{\boldsymbol{u}}\,\bar{\boldsymbol{d}}\,\bar{s}$.

EVALUATE: Note that while the Λ^0 and $\overline{\Lambda^0}$ are both electrically neutral, and both have the same mass, they are different particles: Λ^0 has $B = 1$ and $S = -1$, while $\overline{\Lambda^0}$ has $B = -1$ and $S = 1$.

What caused physicists to suspect that hadrons were made up of something smaller? The magnetic moment of the neutron (Section 43.1) was one of the first reasons. In Section 27.7 we learned that a magnetic moment results from a circulating current (a motion of electric charge). But the neutron has *no* charge, or, to be more accurate, no *total* charge. It could be made up of smaller particles whose charges add to zero. The quantum motion of these particles within the neutron would then give its surprising magnetic moment. To verify this hypothesis by "seeing" inside a neutron, we need a probe with a wavelength that is much less than the neutron's size of about a femtometer. This probe should not be affected by the strong interaction, so that it won't interact with the neutron as a whole but will penetrate into it and interact electromagnetically with these supposed smaller charged particles. A probe with these properties is an electron with energy above 10 GeV. In experiments carried out at SLAC, such electrons were scattered from neutrons and protons to help show that nucleons are indeed made up of fractionally charged, spin-$\frac{1}{2}$ pointlike particles.

The Eightfold Way

Symmetry considerations play a very prominent role in particle theory. Here are two examples. Consider the eight spin-$\frac{1}{2}$ baryons we've mentioned: the familiar p and n, the strange Λ^0, Σ^+, Σ^0, and Σ^-; and the doubly strange Ξ^0 and Ξ^-. For each we plot the value of strangeness S versus the value of charge Q in Fig. 44.11.

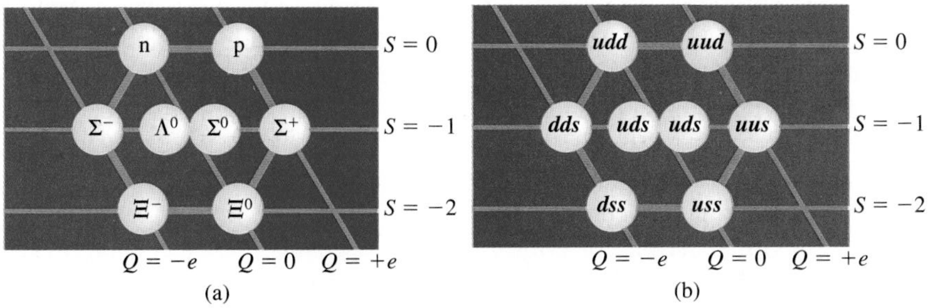

44.11 (a) Plot of S and Q values for spin-$\frac{1}{2}$ baryons showing the symmetry pattern of the eightfold way. (b) Quark content of each spin-$\frac{1}{2}$ baryon. The quark contents of the Σ^0 and Λ^0 are the same; the Σ^0 is an excited state of the Λ^0 and can decay into it by photon emission.

The result is a hexagonal pattern. A similar plot for the nine spin-0 mesons (six shown in Table 44.3 plus three others not included in that table) is shown in Fig. 44.12; the particles fall in exactly the same hexagonal pattern! In each plot, all the particles have masses that are within about $\pm 200\ \text{MeV}/c^2$ of the median mass value of that plot, with variations due to differences in quark masses and internal potential energies.

The symmetries that lead to these and similar patterns are collectively called the **eightfold way.** They were discovered in 1961 by Murray Gell-Mann and independently by Yu'val Ne'eman. (The name is a slightly irreverent reference to the Noble Eightfold Path, a set of principles for right living in Buddhism.) A similar pattern for the spin-$\frac{3}{2}$ baryons contains *ten* particles, arranged in a triangular pattern like pins in a bowling alley. When this pattern was first discovered, one of the particles was missing. But Gell-Mann gave it a name anyway (Ω^-), predicted the properties it should have, and told experimenters what they should look for. Three years later, the particle was found during an experiment at Brookhaven National Laboratory, a spectacular success for Gell-Mann's theory. The whole series of events is reminiscent of the way in which Mendeleev used gaps in the periodic table of the elements to predict properties of undiscovered elements and to guide chemists in their search for these elements.

What binds quarks to one another? The attractive interactions among quarks are mediated by massless spin-1 bosons called **gluons** in much the same way that photons mediate the electromagnetic interaction or that pions mediated the nucleon–nucleon force in the old Yukawa theory.

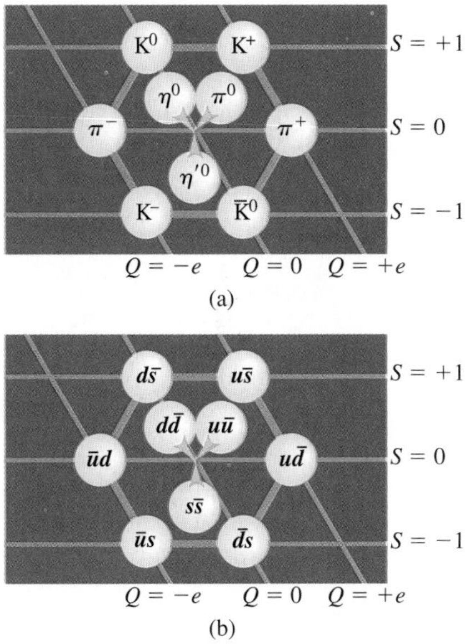

44.12 (a) Plot of S and Q values for nine spin-0 mesons, showing the symmetry pattern of the eightfold way. Each particle is on the opposite side of the hexagon from its antiparticle; each of the three particles in the center is its own antiparticle.
(b) Quark content of each spin-0 meson. The particles in the center are different mixtures of the three quark–antiquark pairs shown.

Color

Quarks, having spin $\frac{1}{2}$, are fermions and so are subject to the exclusion principle. This would seem to forbid a baryon having two or three quarks with the same flavor and same spin component. To avoid this difficulty, it is assumed that each quark comes in three varieties, which are whimsically called *colors.* Red, green, and blue are the usual choices. The exclusion principle applies separately to each color. A baryon always contains one red, one green, and one blue quark, so the baryon itself has no net color. Each gluon has a color–anticolor combination (for example, blue–antired) that allows it to transmit color when exchanged, and color is conserved during emission and absorption of a gluon by a quark. The gluon-exchange process changes the colors of the quarks in such a way that there is always one quark of each color in every baryon. The color of an individual quark changes continually as gluons are exchanged.

Similar processes occur in mesons such as pions. The quark–antiquark pairs of mesons have canceling color and anticolor (for example, blue and antiblue), so mesons also have no net color. Suppose a pion initially consists of a blue quark and

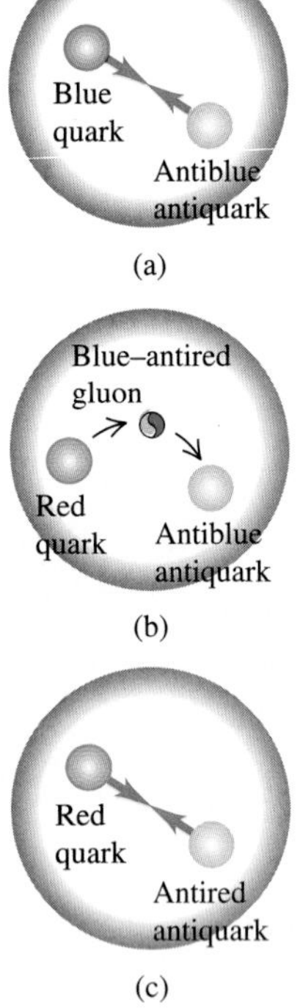

44.13 (a) A pion containing a blue quark and an antiblue antiquark. (b) The blue quark emits a blue–antired gluon, changing to a red quark. (c) The gluon is absorbed by the antiblue antiquark, which becomes an antired antiquark. The pion now consists of a red–antired quark–antiquark pair. The actual quantum state of the pion is an equal superposition of red–antired, green–antigreen, and blue–antiblue pairs.

an antiblue antiquark. The blue quark can become a red quark by emitting a blue–antired virtual gluon. The gluon is then absorbed by the antiblue antiquark, converting it to an antired antiquark (Fig. 44.13). Color is conserved in each emission and absorption, but a blue–antiblue pair has become a red–antired pair. Such changes occur continually, so we have to think of a pion as a superposition of three quantum states, blue–antiblue, green–antigreen, and red–antired. On a larger scale, the strong interaction between nucleons was described in Section 44.3 as due to the exchange of virtual mesons. In terms of quarks and gluons, these mediating virtual mesons are quark–antiquark systems bound together by the exchange of gluons.

The theory of strong interactions is known as *quantum chromodynamics* (QCD). No one has yet been able to isolate an individual quark for study. Most QCD theories contain phenomena associated with the binding of quarks that make it impossible to obtain a free quark. An impressive body of experimental evidence supports the correctness of the quark structure of hadrons and the belief that quantum chromodynamics is the key to understanding the strong interactions.

Three More Quarks

Before the tau particles were discovered, there were four known leptons. This fact, together with some puzzling decay rates, led to the speculation that there might be a fourth quark flavor. This quark is labeled ***c*** (the *charmed* quark); it has $Q/e = \frac{2}{3}$, $B = \frac{1}{3}$, $S = 0$, and a new quantum number **charm** $C = +1$. This was confirmed in 1974 by the observation at both SLAC and the Brookhaven National Laboratory of a meson, now named ψ, with mass 3097 MeV/c^2. This meson was found to have several decay modes, decaying into e^+e^-, $\mu^+\mu^-$, or hadrons. The mean lifetime was found to be about 10^{-20} s. These results are consistent with ψ being a spin-1 $c\bar{c}$ system. Almost immediately after this, similar mesons of greater mass were observed and identified as excited states of the $c\bar{c}$ system. A few years later, individual mesons with a nonzero net charm quantum number were also observed. These mesons, D^0 $(c\bar{u})$ and D^+ $(c\bar{d})$, and their excited states are now firmly established. A charmed baryon, Λ_c^+ (udc), has also been observed.

In 1977 a meson with mass 9460 MeV/c^2, called upsilon (Υ) was discovered at Brookhaven. Because it had properties similar to ψ, it was conjectured that the meson was really the bound system of a new quark, ***b*** (the *bottom* quark), and its antiquark, $\bar{b}$. The bottom quark has the value 1 of a new quantum number B' (not to be confused with baryon number B) called *bottomness.* Excited states of the Υ were soon observed, and the B^+ $(\bar{b}u)$ and B^0 $(\bar{b}d)$ mesons are now well established.

With five flavors of quarks (u, d, s, c, and b) and the six flavors of leptons (e, μ, τ, ν_e, ν_μ, and ν_τ) it was an appealing conjecture that nature is symmetrical in its building blocks and that therefore there should be a *sixth* quark. This quark, labeled ***t*** (top), would have $Q/e = \frac{2}{3}$, $B = \frac{1}{3}$, and a new quantum number, $T = 1$. In 1995, groups using two different detectors at Fermilab's Tevatron announced the discovery of the top quark. The groups collided 0.9-TeV protons with 0.9-TeV antiprotons, but even with 1.8 TeV of available energy, a top–antitop $(t\bar{t})$ pair was detected in fewer than two of every 10^{11} collisions! Table 44.5 lists some properties of the six quarks. Each has a corresponding antiquark with opposite values of Q, B, S, C, B', and T.

Test Your Understanding

Is it possible to have a baryon with charge $Q = +e$ and strangeness $S = -2$? Why or why not?

Table 44.5 Properties of Quarks

Symbol	Q/e	Spin	Baryon Number, B	Strange-ness, S	Charm, C	Bottom-ness, B'	Topness, T
u	$\frac{2}{3}$	$\frac{1}{2}$	$\frac{1}{3}$	0	0	0	0
d	$-\frac{1}{3}$	$\frac{1}{2}$	$\frac{1}{3}$	0	0	0	0
s	$-\frac{1}{3}$	$\frac{1}{2}$	$\frac{1}{3}$	-1	0	0	0
c	$\frac{2}{3}$	$\frac{1}{2}$	$\frac{1}{3}$	0	$+1$	0	0
b	$-\frac{1}{3}$	$\frac{1}{2}$	$\frac{1}{3}$	0	0	$+1$	0
t	$\frac{2}{3}$	$\frac{1}{2}$	$\frac{1}{3}$	0	0	0	$+1$

44.5 | The Standard Model and Beyond

The particles and interactions that we've discussed in this chapter provide a reasonably comprehensive picture of the fundamental building blocks of nature. There is enough confidence in the basic correctness of this picture that it is called the **standard model.**

The standard model includes three families of particles: (1) the six leptons, which have no strong interactions; (2) the six quarks, from which all hadrons are made; and (3) the particles that mediate the various interactions. These mediators are gluons for the strong interaction among quarks, photons for the electromagnetic interaction, the $W^{\pm}$ and Z^0 particles for the weak interaction, and the graviton for the gravitational interaction.

Electroweak Unification

Theoretical physicists have long dreamed of combining all the interactions of nature into a single unified theory. As a first step, Einstein spent much of his later life trying to develop a field theory that would unify gravitation and electromagnetism. He was only partly successful.

Between 1961 and 1967, Sheldon Glashow, Abdus Salam, and Steven Weinberg developed a theory that unifies the weak and electromagnetic forces. One outcome of their **electroweak theory** is a prediction of the weak-force mediator particles, the Z^0 and $W^{\pm}$ bosons, including their masses. The basic idea is that the mass difference between photons (zero mass) and the weak bosons ($\approx 100\ \text{GeV}/c^2$) makes the electromagnetic and weak interactions behave quite differently at low energies. At sufficiently high energies (well above 100 GeV), however, the distinction disappears, and the two merge into a single interaction. This prediction was verified experimentally in 1983 by two experimental groups working at the $p\overline{p}$ collider at CERN, as was mentioned in Section 44.3. The weak bosons were found, again with the help provided by the theoretical description, and their observed masses agreed with the predictions of the electroweak theory, a wonderful convergence of theory and experiment. The electroweak theory and quantum chromodynamics form the backbone of the standard model. Glashow, Salam, and Weinberg received the Nobel Prize in 1979.

A remaining difficulty in the electroweak theory is that photons are massless but the weak bosons are very massive. To account for the broken symmetry among these interaction mediators, a particle called the Higgs boson has been proposed. Its mass is expected to be less than $1\ \text{TeV}/c^2$, but to produce it in the laboratory may require a much greater available energy. The search for the Higgs particle (or particles) will be part of the mission of the Large Hadron Collider at CERN.

Grand Unified Theories

Perhaps at sufficiently high energies the strong interaction and the electroweak interaction have a convergence similar to that between the electromagnetic and weak interactions. If so, they can be unified to give a comprehensive theory of strong, weak, and electromagnetic interactions. Such schemes, called **grand unified theories** (GUTs), lean heavily on symmetry considerations, and they are still speculative.

One interesting feature of some grand unified theories is that they predict the decay of the proton (in violation of conservation of baryon number), with an estimated lifetime of more than 10^{28} years. (For comparison the age of the universe is estimated to be of the order of 10^{10} years.) With a lifetime of 10^{28} years, six metric tons of protons would be expected to have only one decay per day, so huge amounts of material must be examined. Some of the neutrino detectors that we mentioned in Section 44.2 originally looked for, and failed to find, evidence of proton decay. Nevertheless, experimental work continues, with current estimates setting the proton lifetime well over 10^{30} years. Some GUTs also predict the existence of magnetic monopoles, which we mentioned in Chapter 27. At present there is no confirmed experimental evidence that magnetic monopoles exist, but the search goes on.

44.14 This photo shows the interior of the Super-Kamiokande neutrino detector in Japan. When in operation, the detector is filled with 5×10^7 kg of water. A neutrino passing through the detector can produce a faint flash of light, which is detected by the 13,000 photomultiplier tubes lining the detector walls. Data from this detector were the first to indicate that neutrinos have mass.

In the standard model, the neutrinos have zero mass. Nonzero values are controversial because experiments to determine neutrino masses are difficult both to perform and to analyze. In most GUTs the neutrinos *must* have nonzero masses. If neutrinos do have mass, transitions called *neutrino oscillations* can occur, in which one type of neutrino $(\nu_e, \nu_\mu$ or $\nu_\tau)$ changes into another type. In 1998, scientists using the Super-Kamiokande neutrino detector in Japan (Fig. 44.14) reported the discovery of oscillations between muon neutrinos and tau neutrinos. Subsequent measurements at the Sudbury Neutrino Observatory in Canada have confirmed the existence of neutrino oscillations. This discovery is the first evidence for exciting new physics beyond that predicted by the standard model.

The discovery of neutrino oscillations has cleared up a long-standing mystery about the sun. Since the 1960s, physicists have been using sensitive detectors to look for electron neutrinos produced by nuclear fusion reactions in the sun's core (see Section 43.8). However, the observed flux of solar electron neutrinos is only one-third of the predicted value. The resolution of this mystery was provided in 2002 by physicists using the Sudbury Neutrino Observatory, which can detect neutrinos of all three flavors. The results showed that the combined flux of solar neutrinos of *all* flavors is equal to the theoretical prediction for the flux of *electron* neutrinos. The explanation is that the sun is indeed producing electron neutrinos at the rate predicted by theory, but that two-thirds of these electron neutrinos are transformed into muon or tau neutrinos as they pass through and interact with the material of the sun.

Supersymmetric Theories and TOEs

The ultimate dream of theorists is to unify all four fundamental interactions, adding gravitation to the strong and electroweak interactions that are included in GUTs. Such a unified theory is whimsically called a Theory of Everything (TOE). It turns out that an essential ingredient of such theories is a space-time continuum with more than four dimensions. The additional dimensions are "rolled up" into extremely tiny structures that we ordinarily do not notice. Depending on the scale of these structures, it may be possible for the next generation of particle accelerators to reveal the presence of extra dimensions.

Another ingredient of many theories is *supersymmetry,* which gives every boson and fermion a "superpartner" of the other spin type. For example, the proposed supersymmetric partner of the spin-$\frac{1}{2}$ electron is a spin-0 particle called the *selectron,* and that of the spin-1 photon is a spin-$\frac{1}{2}$ *photino.* As yet, no superpartner particles have been discovered, perhaps because they are too massive to be produced by the present generation of accelerators. Within a few years, new data from the Large Hadron Collider and other proposed accelerators will help us decide whether these intriguing theories have merit or whether nature is even more exotic than we imagine.

Test Your Understanding

One aspect of the standard model is that a $\boldsymbol{d}$ quark can transform into a $\boldsymbol{u}$ quark, an electron, and an antineutrino by means of the weak interaction. What is the connection between this process and nuclear β^- decay (Section 43.3)?

44.6 | The Expanding Universe

In the last two sections of this chapter we'll explore briefly the connections between the early history of the universe and the interactions of fundamental particles. It is remarkable that there are such close ties between physics on the smallest scale that we've explored experimentally (the range of the weak interaction, of the order of 10^{-18} m) and physics on the largest scale (the universe itself, of the order of at least 10^{26} m).

Gravitational interactions play an essential role in the large-scale behavior of the universe. One of the great achievements of Newtonian mechanics, including the law of gravitation, was the understanding it brought to the motion of planets in the solar system. Astronomical evidence, such as observations of the motions of binary stars around their common center of mass, shows that gravitational interactions also operate in larger astronomical systems, including stars, galaxies, and nebulae.

Until early in the twentieth century it was usually assumed that the universe was *static;* stars might move relative to each other, but there was not thought to be any overall expansion or contraction. But stars have gravitational attractions. If everything is initially sitting still in the universe, why doesn't gravity just pull it all together into one big clump? Newton himself recognized the seriousness of this troubling question.

Measurements that were begun in 1912 by Vesto Slipher at Lowell Observatory in Arizona, and continued in the 1920s by Edwin Hubble with the help of Milton Humason at Mount Wilson in California, indicated that the universe is *not* static. The motions of galaxies relative to the earth can be measured by observing the shifts in the wavelengths of their spectra. For distant galaxies these shifts are always toward longer wavelength, so they appear to be receding from us and from each other. Astronomers first assumed that these were Doppler shifts and used a relation between the wavelength λ_0 of light measured now from a source receding at speed v and the wavelength λ_S measured in the rest frame of the source when it was emitted. We can derive this relation by inverting Eq. (37.25) for the Doppler effect, making subscript changes, and using $\lambda = c/f$; the result is

$$\lambda_0 = \lambda_S \sqrt{\frac{c + v}{c - v}} \tag{44.13}$$

Wavelengths from receding sources are always shifted toward longer wavelengths; this increase in λ is called the **redshift.** We can solve Eq. (44.13) for v. We leave the details as a problem (Exercise 44.36); the result is

$$v = \frac{(\lambda_0/\lambda_S)^2 - 1}{(\lambda_0/\lambda_S)^2 + 1}c \tag{44.14}$$

CAUTION Equations (44.13) and (44.14) are from the *special* theory of relativity and refer to the Doppler effect. As we'll see, the redshift from *distant* galaxies is caused by an effect that is explained by the *general* theory of relativity and is *not* a Doppler shift. However, as the ratio v/c and the fractional wavelength change $(\lambda_0 - \lambda_S)/\lambda_S$ become small, the general theory's equations approach Eqs. (44.13) and (44.14), and those equations may be used.

Example 44.9 Recession speed of a galaxy

The spectral lines of various elements are detected in light from a galaxy in the constellation Ursa Major. An ultraviolet line from singly ionized calcium ($\lambda_S = 393$ nm) is observed with a wavelength $\lambda_0 = 414$ nm, redshifted into the visible portion of the spectrum. At what speed is this galaxy receding from the earth?

SOLUTION

IDENTIFY and SET UP: We are given the wavelength at which the light is emitted and the wavelength that we detect on earth. We use these in Eq. (44.14) to determine the recession speed v of the galaxy, provided the fractional wavelength shift is not too great.

EXECUTE: First we calculate $\lambda_0/\lambda_S = (414\text{ nm})/(393\text{ nm}) = 1.053$. This is only a 5.3% increase, so we can use Eq. (44.14) with reasonable accuracy:

$$v = \frac{(1.053)^2 - 1}{(1.053)^2 + 1}c = 0.0516c = 1.55 \times 10^7 \text{ m/s}$$

EVALUATE: The galaxy is receding from the earth at 5.16% of the speed of light. Note that rather than doing this calculation, astronomers often simply state the recession speed in terms of the *redshift,* defined as $z = (\lambda_0 - \lambda_S)/\lambda_S = (\lambda_0/\lambda_S) - 1$. Thus this galaxy has redshift $z = 0.053$.

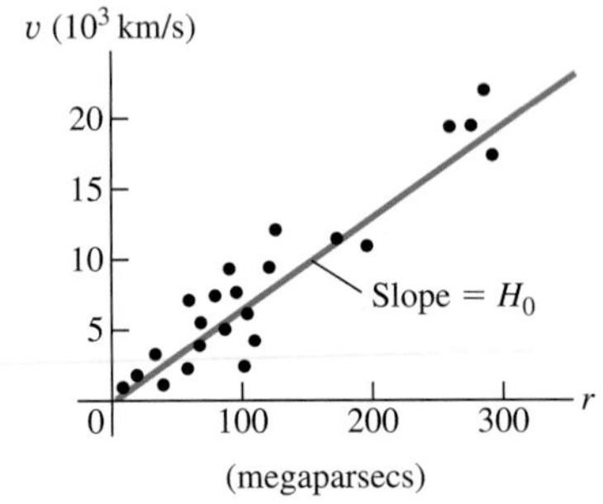

44.15 Graph of recession speed versus distance for several galaxies. The best-fit straight line illustrates Hubble's law. The slope of the line is the Hubble constant, H_0.

The Hubble Law

Analysis of redshifts from many distant galaxies led Edwin Hubble to a remarkable conclusion: The speed of recession v of a galaxy is proportional to its distance r from us (Fig. 44.15). This relation is now called the **Hubble law;** its symbolic statement is

$$v = H_0 r \tag{44.15}$$

where H_0 is an experimental quantity commonly called the *Hubble constant,* since at any given time it is constant over space. Determining H_0 has been a key goal of the Hubble Space Telescope, which can measure distances to galaxies with unprecedented accuracy. The current best value is $2.3 \times 10^{-18}\text{ s}^{-1}$, with an uncertainty of 15%.

Astronomical distances are often measured in *parsecs* (pc); one parsec is the distance at which there is a one-arcsecond $(1/3600°)$ angular separation between two objects 1.50×10^{11} m apart (the average distance from the earth to the sun).

A distance of 1 pc is equal to 3.26 *light years* (ly), where 1 ly $= 9.46 \times 10^{15}$ m is the distance that light travels in one year. The Hubble constant is then commonly expressed in the mixed units (km/s)/Mpc (kilometers per second per megaparsec), where 1 Mpc $= 10^6$ pc:

$$H_0 = (2.3 \times 10^{-18}\ \text{s}^{-1})\left(\frac{9.46 \times 10^{15}\ \text{km}}{1\ \text{ly}}\right)\left(\frac{3.26\ \text{ly}}{1\ \text{pc}}\right)\left(\frac{10^6\ \text{pc}}{1\ \text{Mpc}}\right) = 71\ \frac{\text{km/s}}{\text{Mpc}}$$

Example 44.10 Determining distance with the Hubble law

Find the distance of the galaxy in Example 44.9 from the earth, according to the Hubble law.

SOLUTION

IDENTIFY and SET UP: We solve Eq. (44.15) for the distance r, then substitute the value of the recession speed v from Example 44.9.

EXECUTE: Using $H_0 = 71\ (\text{km/s})/\text{Mpc} = 7.1 \times 10^4\ (\text{m/s})/\text{Mpc}$,

$$r = \frac{v}{H_0} = \frac{1.55 \times 10^7\ \text{m/s}}{7.1 \times 10^4\ (\text{m/s})/\text{Mpc}}$$
$$= 220\ \text{Mpc} = 2.2 \times 10^8\ \text{pc} = 7.1 \times 10^8\ \text{ly} = 6.7 \times 10^{24}\ \text{m}$$

EVALUATE: A distance of 220 million parsecs (710 million light years) is truly stupendous, but many galaxies lie much farther away. In contrast, the closest known galaxy to ours is the Sagittarius Dwarf, only 80,000 light years from us—so close that it is held in gravitational orbit around our galaxy. To appreciate the immensity of even this relatively small distance, consider that our farthest-ranging unmanned spacecraft have traveled only about 0.001 ly from our planet.

Another aspect of Hubble's observations was that, *in all directions,* distant galaxies appeared to be receding from us. There is no particular reason to think that our galaxy is at the very center of the universe; if we lived in some other galaxy, every distant galaxy would still seem to be moving away. That is, at any given time, *the universe looks more or less the same, no matter where in the universe we are.* This important idea is called the **cosmological principle.** There are local fluctuations in density, but on average, the universe looks the same from all locations. Thus the Hubble constant is constant in space although not necessarily constant in time, and the laws of physics are the same everywhere.

The Big Bang

The Hubble law suggests that at some time in the past, all the matter in the universe was far more concentrated than it is today. It was then blown apart in an immense explosion called the **Big Bang,** giving all observable matter more or less the velocities that we observe today. When did this happen? According to the Hubble law, matter at a distance r away from us is traveling with speed $v = H_0 r$. The time t needed to travel a distance r is

$$t = \frac{r}{v} = \frac{r}{H_0 r} = \frac{1}{H_0} = 4.3 \times 10^{17}\ \text{s} = 1.4 \times 10^{10}\ \text{y}$$

By this hypothesis the Big Bang occurred about 14 billion years ago. It assumes that all speeds are *constant* after the Big Bang; that is, it neglects any change in the expansion rate due to gravitational attraction or other effects. We'll return to this point later. For now, however, notice that the age of the earth determined from radioactive dating (Section 43.4) is 4.6 billion (4.6×10^9) years. It's encouraging that our simple hypothesis tells us that the universe is older than the earth!

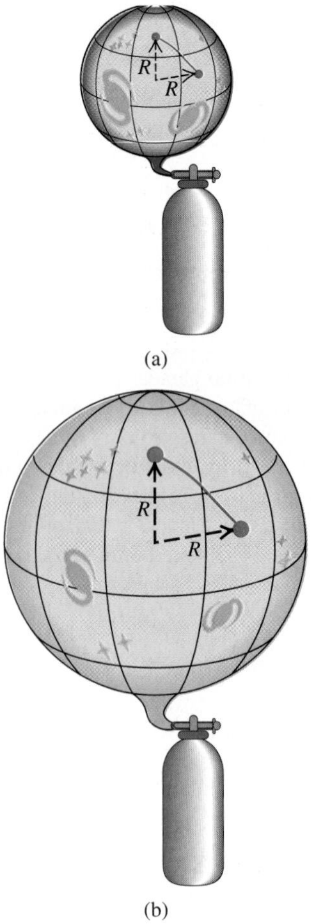

44.16 (a) Points (representing positions of astronomical objects) on the surface of a spherical balloon are described by their latitude and longitude coordinates. (b) The radius R of the balloon has increased. The coordinates of the points are the same, but the distance between them has increased. The rate of recession for any two points is proportional to the distance between them.

Expanding Space

The general theory of relativity takes a radically different view of the expansion just described. According to this theory, the increased wavelength is *not* caused by a Doppler shift as the universe expands into a previously empty void. Rather, the increase comes from *the expansion of space itself* and everything in intergalactic space, including the wavelengths of light traveling to us from distant sources. This is not an easy concept to grasp, and if you haven't encountered it before, it may sound like doubletalk.

Here's an analogy that may help to develop some intuition on this point. Imagine we are all bugs crawling around on a horizontal surface. We can't leave the surface, and we can see in any direction along the surface, but not up or down. We are then living in a two-dimensional world; some writers have called such a world *flatland.* If the surface is a plane, we can locate our position with two Cartesian coordinates (x, y). If the plane extends indefinitely in both the x- and y-directions we described our space as having *infinite* extent, or as being *unbounded.* No matter how far we go, we never reach an edge or a boundary.

An alternative habitat for us bugs would be the surface of a sphere with radius R. The space would still seem infinite in the sense that we could crawl forever and never reach an edge or a boundary. Yet in this case the space is *finite* or *bounded.* To describe the location of a point in this space, we could still use two coordinates: latitude and longitude, or the spherical coordinates θ and ϕ shown in Fig. 40.22.

Now suppose the spherical surface is that of a balloon (Fig. 44.16). As we inflate the balloon more and more, increasing the radius R, the coordinates of a point don't change, yet the distance between any two points gets larger and larger. Furthermore, as R increases, the *rate of change* of distance between two points (their recession speed) is proportional to their distance apart. *The recession speed is proportional to the distance,* just as with the Hubble law. For example, the distance from Pittsburgh to Miami is twice as great as the distance from Pittsburgh to Boston. If the earth were to begin to swell, Miami would recede from Pittsburgh twice as fast as Boston would.

We see that although the quantity R isn't one of the two coordinates giving the position of a point on the balloon's surface, it nevertheless plays an essential role in any discussion of distance. It is the radius of curvature of our two-dimensional space, and it is also a varying *scale factor* that changes as this two-dimensional universe expands.

Generalizing this picture to three dimensions isn't so easy. We have to think of our three-dimensional space as being embedded in a space with four or more dimensions, just as we visualized the two-dimensional spherical flatland as being embedded in a three-dimensional Cartesian space. Our real three-space is *not Cartesian;* to describe its characteristics in any small region requires at least one additional parameter, the curvature of space, which is analogous to the radius of the sphere. In a sense, this scale factor, which we'll continue to call R, describes the *size* of the universe, just as the radius of the sphere described the size of our two-dimensional spherical universe. We'll return later to the question of whether the universe is bounded or unbounded.

Any length that is measured in intergalactic space is proportional to R, so the wavelength of light traveling to us from a distant galaxy increases along with every other dimension as the universe expands. That is,

$$\frac{\lambda_0}{\lambda} = \frac{R_0}{R} \tag{44.16}$$

The zero subscripts refer to the values of the wavelength and scale factor *now,* just as H_0 is the current value of the Hubble constant. The quantities λ and R without subscripts are the values at *any* time, past, present, or future. In the situation described in Example 44.9, we have $\lambda_0 = 414$ nm and $\lambda = \lambda_S = 393$ nm, so Eq. (44.16) gives $R_0/R = 1.053$. That is, the scale factor *now* (R_0) is 5.3% larger than it was 710 million years ago when the light was emitted from that galaxy in Ursa Major. This increase of wavelength with time as the scale factor increases in our expanding universe is called the *cosmological redshift.* The farther away an object is, the longer its light takes to get to us and the greater the change in R and λ. The current largest measured wavelength ratio for galaxies is about 7, meaning that the volume of space itself is about $7^3 \approx 340$ times larger than it was when the light was emitted. Do *not* attempt to substitute $\lambda_0/\lambda_S = 7$ into Eq. (44.14) to find the recession speed; that equation is accurate only for small cosmological redshifts and $v \ll c$. The actual value of v depends on the density of the universe, the value of H_0, and the expansion history of the universe.

Here's a surprise for you: If the distance from us in the Hubble law is large enough, then the speed of recession will be greater than the speed of light! This does *not* violate the special theory of relativity because the recession speed is *not* caused by the motion of the astronomical object relative to some coordinates in its region of space. Rather, we can have $v > c$ when two sets of coordinates move apart fast enough as space itself expands. In other words, there are objects whose coordinates have been moving away from our coordinates so fast that light from them hasn't had enough time in the entire history of the universe to reach us. What we see is just the *observable* universe; we have no direct evidence about what lies beyond its horizon.

CAUTION We have spoken of the expansion of space with time, but in an attempt to find the real meaning of the Big Bang, we can also extrapolate backward to the initial time when all of space was concentrated in a very small region, or perhaps a single point. This region (or point) contained the entire universe at that time. It's important to understand that the Big Bang was not an expansion *in* space; it was an expansion *of* space itself. As space expanded, objects in intergalactic space expanded along with it.

Critical Density

We've mentioned that the law of gravitation isn't consistent with a static universe. We need to look at the role of gravity in an *expanding* universe. Gravitational attractions should slow the initial expansion, but by how much? If these attractions are strong enough, the universe should expand more and more slowly, eventually stop, and then begin to contract, perhaps all the way down to what's been called a *Big Crunch.* On the other hand, if gravitational forces are much weaker, they slow the expansion only a little, and the universe should continue to expand forever.

The situation is analogous to the problem of escape speed of a projectile launched from the earth. We studied this problem in Example 12.5 (Section 12.3); now would be an excellent time to review that discussion. The total energy $E = K + U$ when a projectile of mass m and speed v is at a distance r from the center of the earth (mass m_E) is

$$E = \frac{1}{2}mv^2 - \frac{Gmm_E}{r}$$

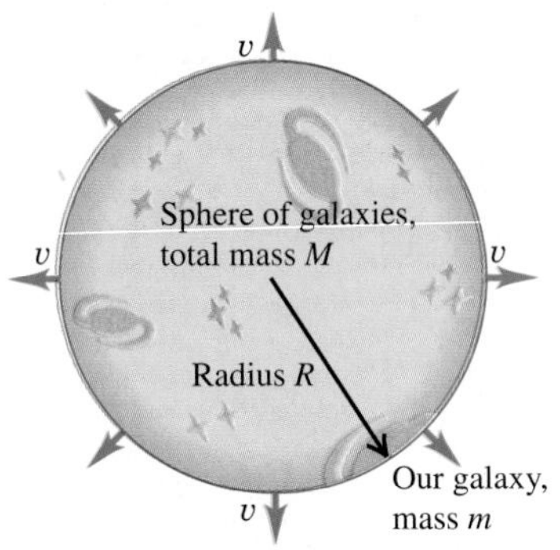

44.17 An imaginary sphere of galaxies. The net gravitational force exerted on our galaxy (at the surface of the sphere) by the other galaxies is the same as if all of their mass were concentrated at the center of the sphere.

If E is positive, the projectile has enough kinetic energy to move infinitely far from the earth $(r \to \infty)$ and have some kinetic energy left over. If E is negative, the kinetic energy $K = \frac{1}{2}mv^2$ becomes zero and the projectile stops when $r = -Gmm_E/E$. In that case, no greater value of r is possible, and the projectile can't escape the earth's gravity.

We can carry out a similar analysis for the universe. Whether the universe continues to expand indefinitely should depend on the average *density* of matter. If matter is relatively dense, there is a lot of gravitational attraction to slow and eventually stop the expansion and make the universe contract again. If not, the expansion should continue indefinitely. We can derive an expression for the *critical density* ρ_c needed to just barely stop the expansion.

Here's a calculation based on Newtonian mechanics; it isn't relativistically correct, but it illustrates the idea. Consider a large sphere with radius R, containing many galaxies (Fig. 44.17), with total mass M. Suppose our own galaxy has mass m and is located at the surface of this sphere. According to the cosmological principle, the average distribution of matter within the sphere is uniform. The total gravitational force on our galaxy is just the force due to the mass M inside the sphere. The force on our galaxy and potential energy U due to this spherically symmetric distribution are the same as though m and M were both points, so $U = -GmM/R$, just as in Section 12.3. The net force from all the uniform distribution of mass *outside* the sphere is zero, so we'll ignore it.

The total energy E (kinetic plus potential) for our galaxy is

$$E = \frac{1}{2}mv^2 - \frac{GmM}{R} \tag{44.17}$$

If E is *positive,* our galaxy has enough energy to escape from the gravitational attraction of the mass M inside the sphere; in this case the universe should keep expanding forever. If E is negative, our galaxy cannot escape and the universe should eventually pull back together. The crossover between these two cases occurs when $E = 0$, so that

$$\frac{1}{2}mv^2 = \frac{GmM}{R} \tag{44.18}$$

The total mass M inside the sphere is the volume $4\pi R^3/3$ times the density ρ_c:

$$M = \frac{4}{3}\pi R^3 \rho_c$$

We'll assume that the speed v of our galaxy relative to the center of the sphere is given by the Hubble law: $v = H_0 R$. Substituting these expressions for M and v into Eq. (44.18), we get

$$\frac{1}{2}m(H_0 R)^2 = \frac{Gm}{R}\left(\frac{4}{3}\pi R^3 \rho_c\right)$$

or

$$\rho_c = \frac{3H_0^2}{8\pi G} \quad \text{(critical density of the universe)} \tag{44.19}$$

This is the *critical density.* If the average density is less than ρ_c, the universe should continue to expand indefinitely; if it is greater, the universe should eventually stop expanding and begin to contract.

Putting numbers into Eq. (44.19), we find

$$\rho_c = \frac{3(2.3 \times 10^{-18}\ \text{s}^{-1})^2}{8\pi(6.67 \times 10^{-11}\ \text{kg}\cdot\text{m}^2/\text{s}^2)} = 6.3 \times 10^{-27}\ \text{kg/m}^3$$

The mass of a hydrogen atom is 1.67×10^{-27} kg, so this density if equivalent to about four hydrogen atoms per cubic meter.

Dark Matter, Dark Energy, and the Accelerating Universe

Astronomers have made extensive studies of the average density of matter in the universe. One way to do so is to count the number of galaxies in a patch of sky. Based on the mass of an average star and the number of stars in an average galaxy, this effort gives an estimate of the average density of *luminous* matter in the universe—that is, matter that emits electromagnetic radiation. (You are made of luminous matter, because you emit infrared radiation as a consequence of your temperature; see Sections 17.7 and 38.8.) It's also necessary to take into account other luminous matter within a galaxy, including the tenuous gas and dust between the stars.

Another technique is to study the motions of galaxies within clusters of galaxies (Fig. 44.18). The motions are so slow that we can't actually see galaxies changing positions within a cluster. However, observations show that different galaxies within a cluster have somewhat different redshifts, which indicates that the galaxies are moving relative to the center of mass of the cluster. The speeds of these motions are related to the gravitational force exerted on each galaxy by the other members of the cluster, which in turn depends on the total mass of the cluster. By measuring these speeds, astronomers can determine the average density of *all* kinds of matter within the cluster, whether or not the matter emits electromagnetic radiation.

44.18 The bright spots in this image are not stars, but entire galaxies. They are part of a cluster of galaxies about 1000 Mpc (330 Mly) away. When the galaxies emitted the light used to make this image, the scale factor of the universe was about 25% smaller than it is now.

Observations using these and other techniques show that the average density of *all* matter in the universe is 27% of the critical density, but the average density of *luminous* matter is only 4% of the critical density. In other words, most of the matter in the universe is not luminous: it does not emit electromagnetic radiation of *any* kind. At present, the nature of this **dark matter** remains an outstanding mystery. Some proposed candidates for dark matter are WIMPs (weakly interacting massive particles, which are hypothetical subatomic particles far more massive than those produced in accelerator experiments) and MACHOs (massive compact halo objects, which include objects such as black holes that might form "halos" around galaxies). Whatever the true nature of dark matter, it is by far the dominant form of matter in the universe. For every kilogram of the ordinary matter that has been our subject for most of this book—including electrons, protons, atoms, molecules, blocks on inclined planes, planets, and stars—there are almost *six* kilograms of dark matter.

Since the average density of matter in the universe is less than the critical density, it might seem fair to conclude that the universe will continue to expand indefinitely, and that gravitational attraction between matter in different parts of the universe should slow the expansion down (albeit not enough to stop it). One way to test this prediction is to examine the redshifts of extremely distant objects. When astronomers look at a galaxy 10^9 light years away, the light they receive has been in transit for 10^9 years, so they are seeing 10^9 years into the past. If the expansion of the universe has been slowing down, the expansion must have been more rapid in the distant past. Thus we would expect very distant galaxies to have *greater* redshifts than predicted by the Hubble law, Eq. (44.15).

Only in the last decade has it become possible to accurately measure both the distances and the redshifts of extremely distant galaxies. The results have been totally surprising: very distant galaxies actually have *smaller* redshifts than predicted by the Hubble law! The implication is that the expansion of the universe was slower in the past than it is now, so the expansion has been *speeding up* rather than slowing down.

If gravitational attraction should make the expansion slow down, why is it speeding up instead? The explanation generally accepted by astronomers and physicists is that space is suffused with a kind of energy that has no gravitational effect and emits no electromagnetic radiation, but rather acts as a kind of "antigravity" that produces a universal *repulsion.* This invisible, immaterial energy is called **dark energy.** As the name suggests, the nature of dark energy is poorly understood but is the subject of very active research.

Observations show that the *energy* density of dark energy (measured in, say, joules per cubic meter) is 73% of the critical density times c^2; that is, it is equal to $0.73\rho_c c^2$. As described above, the average density of matter of all kinds is 27% of the critical density. From the Einstein relation $E = mc^2$, the average *energy* density of matter in the universe is therefore $0.27\rho_c c^2$. Because the energy density of dark energy is nearly three times greater than that of matter, the expansion of the universe will continue to accelerate. This expansion will never stop, and the universe will never contract.

If we account for energy of *all* kinds, the average energy density of the universe is equal to $0.73\rho_c c^2 + 0.27\rho_c c^2 = 1.00\rho_c c^2$. Of this, 73% is the mysterious dark energy, 23% is the no less mysterious dark matter, and a mere 4% is the well-understood ordinary matter. How little we know about the contents of our universe!

What is the significance of the result that within observational error, the average energy density of the universe is equal to $\rho_c c^2$? It tells us that the universe is infinite and unbounded, but just barely so. If the average energy density were even slightly larger than $\rho_c c^2$, the universe would be finite like the surface of the balloon depicted in Fig. 44.16. As of this writing, the observational error in the average energy density is still large enough that we can't be totally sure that the universe *is* unbounded. Improving these measurements will be an important task for physicists and astronomers in the years ahead.

Test Your Understanding

A fellow student asks you, "If the universe is expanding, what is it expanding into?" How should you respond to this question?

44.7 | The Beginning of Time

What an odd title for the very last section of a book! We will describe in general terms some of the current theories about the very early history of the universe and their relation to fundamental particle interactions. We'll find that an astonishing amount happened in the very first second. A lot of loose ends will be left untied, and many questions will be left unanswered. This is, after all, one of the frontiers of theoretical physics, and continually brings new theories, new discoveries, and new questions.

Temperatures

The early universe was extremely dense and extremely hot, and the average particle energies were extremely large, all many orders of magnitude beyond anything that exists in the present universe. We can compare particle energy E and absolute temperature T using the equipartition principle

$$E = \frac{3}{2}kT \tag{44.20}$$

where k is Boltzmann's constant, which we'll often express in eV/K:

$$k = 8.617 \times 10^{-5}\ \text{eV/K}$$

Thus we can replace Eq. (44.20) by $E \approx (10^{-4}\ \text{eV/K})T = (10^{-13}\ \text{GeV/K})T$ when we're discussing orders of magnitude.

Example 44.11 Temperature and energy

a) What is the average kinetic energy in electron volts of particles at room temperature $(T = 290\ \text{K})$ and at the surface of the sun $(T = 5800\ \text{K})$? b) What approximate temperature corresponds to the ionization energy of the hydrogen atom, to the rest energy of the electron, and to the rest energy of the proton?

SOLUTION

IDENTIFY and SET UP: We use Eq. (44.20) in each part to determine the target variable T.

EXECUTE: a) From Eq. (44.20),

$$E = \frac{3}{2}kT = \frac{3}{2}(8.617 \times 10^{-5}\ \text{eV/K})(290\ \text{K})$$
$$= 0.0375\ \text{eV}$$

The temperature at the sun's surface is larger by a factor of $(5800\ \text{K})/(290\ \text{K}) = 20$, so the average kinetic energy there is $20(0.0375\ \text{eV}) = 0.75\ \text{eV}$.

b) The ionization energy of hydrogen is 13.6 eV. Using the approximation $E \approx (10^{-4}\ \text{eV/K})T$, we have

$$T \approx \frac{E}{10^{-4}\ \text{eV/K}} = \frac{13.6\ \text{eV}}{10^{-4}\ \text{eV/K}}$$
$$\approx 10^5\ \text{K}$$

The rest energies of the electron and proton are 0.511 MeV and 938 MeV, respectively. Repeating the calculation for these values gives temperatures of 10^{10} K, corresponding to the electron rest energy, and 10^{13} K, corresponding to the proton rest energy.

EVALUATE: Temperatures in excess of 10^5 K are found in the interior of the sun, so the hydrogen there is predominantly ionized. Temperatures of 10^{10} K or 10^{13} K are not found anywhere in the solar system; as we will see, temperatures were high in the very early universe.

Uncoupling of Interactions

The evolution of the universe has been characterized by a continual increase of the scale factor R, which we can think of very roughly as characterizing the *size* of the universe, and by a corresponding decrease in average density. As the total gravitational potential energy increased during expansion, there were corresponding *decreases* in temperature and average particle energy. As this happened, the basic interactions became progressively uncoupled.

To understand the uncouplings, recall that the unification of the electromagnetic and weak interactions occurs at energies that are large enough that the differences in mass among the various spin-1 bosons that mediate the interactions become insignificant by comparison. The electromagnetic interaction is mediated by the massless photon, and the weak interaction is mediated by the weak bosons $W^{\pm}$ and Z^0 with masses of the order of 100 GeV/c^2. At energies much *less* than 100 GeV the two interactions seem quite different, but at energies much *greater* than 100 GeV they become part of a single interaction.

The grand unified theories (GUTs) provide a similar behavior for the strong interaction. It becomes unified with the electroweak interaction at energies of the order of 10^{14} GeV, but at lower energies the two appear quite distinct. One of the reasons GUTs are still very speculative is that there is no way to do controlled experiments in this energy range, which is larger by a factor of 10^{11} than energies available with any current accelerator.

Finally, at sufficiently high energies and short distances, it is assumed that gravitation becomes unified with the other three interactions. The distance at which this happens is thought to be of the order of 10^{-35} m. This distance, called the *Planck length* l_P, is determined by the speed of light c and the fundamental constants of quantum mechanics and gravitation, h and G, respectively. The Planck length l_P is defined as

$$l_P = \sqrt{\frac{\hbar G}{c^3}} = 1.616 \times 10^{-35}\ \text{m} \tag{44.21}$$

You should verify that this combination of constants does indeed have units of length. The *Planck time* $t_P = l_P/c$ is the time required for light to travel a distance l_P:

$$t_P = \frac{l_P}{c} = \sqrt{\frac{\hbar G}{c^5}} = 0.539 \times 10^{-43}\ \text{s} \tag{44.22}$$

If we mentally go backward in time, we have to stop when we reach $t = 10^{-43}$ s because we have no adequate theory that unifies all four interactions. So as yet we have no way of knowing what happened or how the universe behaved at times earlier than the Planck time or when its size was less than the Planck length.

The Standard Model of the History of the Universe

The description that follows is called the *standard model* of the history of the universe. The title may be slightly optimistic, but it does indicate that there are substantial areas of theory that rest on solid experimental foundations and are quite generally accepted. The figure on pages 1704–1705 is a graphical description of this history, with the characteristic sizes, particle energies, and temperatures at various times. Referring to this chart frequently will help you to understand the following discussion.

In this standard model, the temperature of the universe at time $t = 10^{-43}$ s (the Planck time) was about 10^{32} K, and the average energy per particle was approximately

$$E \approx (10^{-13}\ \text{GeV/K})(10^{32}\ \text{K}) = 10^{19}\ \text{GeV}$$

In a totally unified theory this is about the energy below which gravity begins to behave as a separate interaction. This time therefore marked the transition from any proposed TOE to the GUT period.

During the GUT period, roughly $t = 10^{-43}$ to 10^{-35} s, the strong and electroweak forces were still unified, and the universe consisted of a soup of quarks and leptons transforming into each other so freely that there was no distinction between the two families of particles. Other, much more massive particles may also have been freely created and destroyed. One important characteristic of GUTs is that at sufficiently high energies, baryon number is not conserved. (We mentioned earlier the proposed decay of the proton, which has not yet been observed.) Thus by the end of the GUT period the numbers of quarks and antiquarks may have been unequal. This point has important implications; we'll return to it at the end of the section.

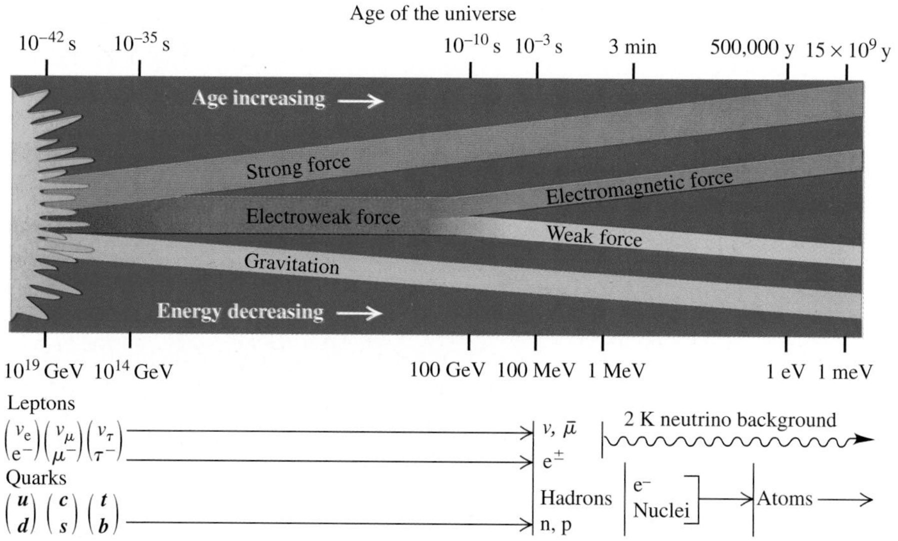

44.19 Schematic diagram showing the times and energies at which the various interactions are thought to have uncoupled. The energy scale is backward because the average energy decreased as the age of the universe increased.

By $t = 10^{-35}$ s the temperature had decreased to about 10^{27} K and the average energy to about 10^{14} GeV. At this energy the strong force separated from the electroweak force (Fig. 44.19), and baryon number and lepton numbers began to be separately conserved. In some models, called *inflationary models,* this separation of the strong force was analogous to a *phase change,* such as boiling of a liquid, with an associated heat of vaporization. Think of it as being similar to boiling a heavy nucleus, pulling the particles apart beyond the short range of the nuclear force. As a result, the inflationary models predict that there was a very rapid expansion. In one model, the scale factor R increased by a factor of 10^{50} in 10^{-32} s.

At $t = 10^{-32}$ s the universe was a mixture of quarks, leptons, and the mediating bosons (gluons, photons, and the weak bosons $W^{\pm}$ and Z^0). It continued to expand and cool from the inflationary period to $t = 10^{-6}$ s, when the temperature was about 10^{13} K and typical energies were about 1 GeV (comparable to the rest energy of a nucleon; see Example 44.11). At this time the quarks began to bind together to form nucleons and antinucleons. Also there were still enough photons of sufficient energy to produce nucleon–antinucleon pairs to balance the process of nucleon–antinucleon annihilation. However, by about $t = 10^{-2}$ s, most photon energies fell well below the threshold energy for such pair production. There was a slight excess of nucleons over antinucleons; as a result, virtually all of the antinucleons and most of the nucleons annihilated one another. A similar equilibrium occurred later between the production of electron–positron pairs from photons and the annihilation of such pairs. At about $t = 14$ s the average energy dropped to around 1 MeV, below the threshold for e^+e^- pair production. After pair production ceased, virtually all of the remaining positrons were annihilated, leaving the universe with many more protons and electrons than the antiparticles of each.

Up until about $t = 1$ s, neutrons and neutrinos could be produced in the endoergic reaction

$$e^- + p \rightarrow n + \nu_e$$

After this time, most electrons no longer had enough energy for this reaction. The average neutrino energy also decreased, and as the universe expanded, equilibrium reactions that involved *absorption* of neutrinos (which occurred with decreasing probability) became inoperative. At this time, in effect, the flux of neutrinos and

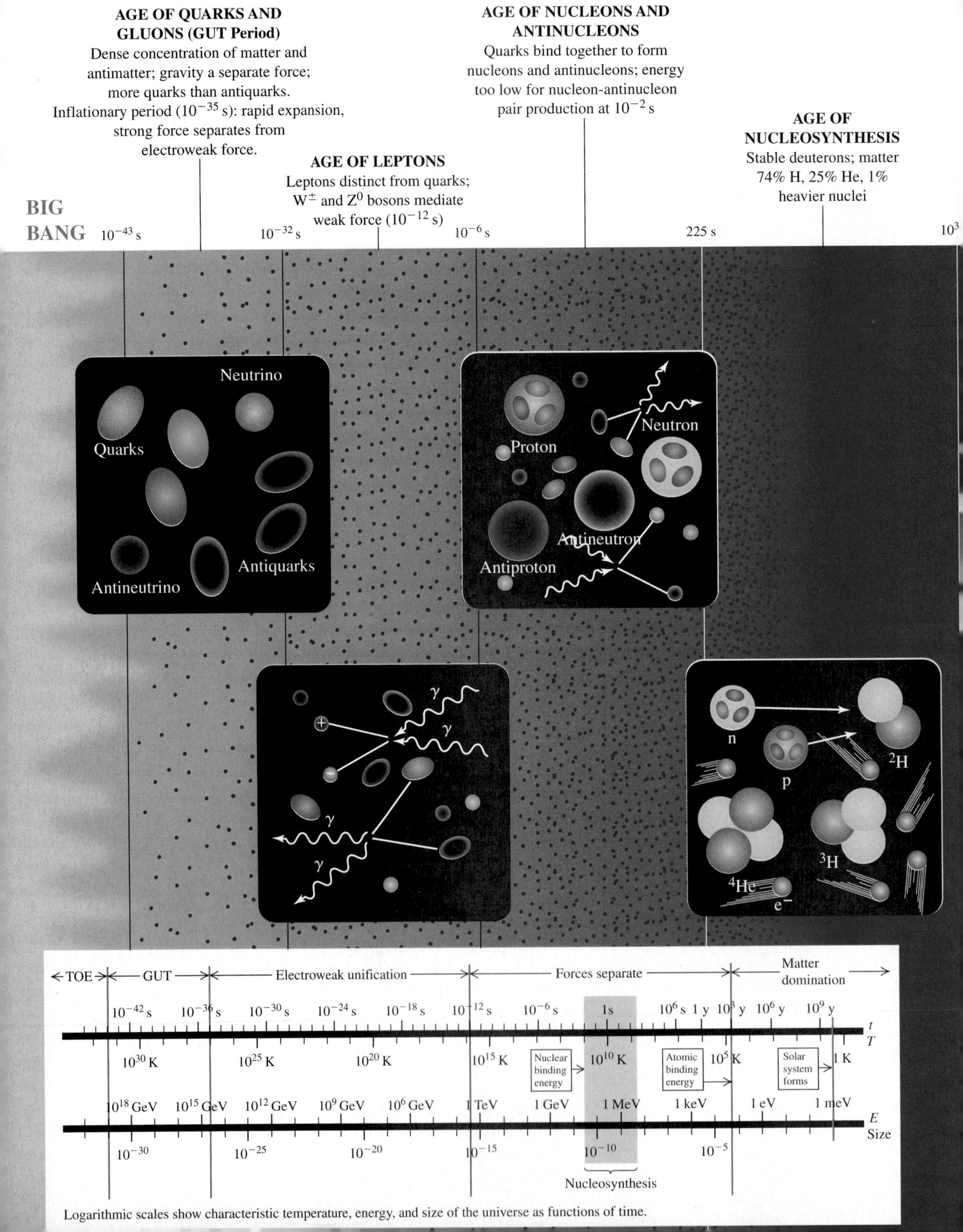

AGE OF QUARKS AND GLUONS (GUT Period)
Dense concentration of matter and antimatter; gravity a separate force; more quarks than antiquarks. Inflationary period (10^{-35} s): rapid expansion, strong force separates from electroweak force.
AGE OF LEPTONS
Leptons distinct from quarks; $W^{\pm}$ and Z^0 bosons mediate weak force (10^{-12} s)
AGE OF NUCLEONS AND ANTINUCLEONS
Quarks bind together to form nucleons and antinucleons; energy too low for nucleon-antinucleon pair production at 10^{-2} s
AGE OF NUCLEOSYNTHESIS
Stable deuterons; matter 74% H, 25% He, 1% heavier nuclei
BIG BANG
10^{-43} s
10^{-32} s
10^{-6} s
225 s
10^3
Neutrino
Quarks
Antiquarks
Antineutrino
Proton
Neutron
Antineutron
Antiproton
γ
n
p
^{2}H
^{3}H
^{4}He
e^-
TOE
GUT
Electroweak unification
Forces separate
Matter domination
10^{-42} s
10^{-36} s
10^{-30} s
10^{-24} s
10^{-18} s
10^{-12} s
10^{-6} s
1s
10^6 s
1 y
10^3 y
10^6 y
10^9 y
t
T
10^{30} K
10^{25} K
10^{20} K
10^{15} K
Nuclear binding energy
10^{10} K
Atomic binding energy
10^5 K
Solar system forms
1 K
10^{18} GeV
10^{15} GeV
10^{12} GeV
10^9 GeV
10^6 GeV
1 TeV
1 GeV
1 MeV
1 keV
1 eV
1 meV
E
Size
10^{-30}
10^{-25}
10^{-20}
10^{-15}
10^{-10}
10^{-5}
Nucleosynthesis
Logarithmic scales show characteristic temperature, energy, and size of the universe as functions of time.

A Brief History of the Universe

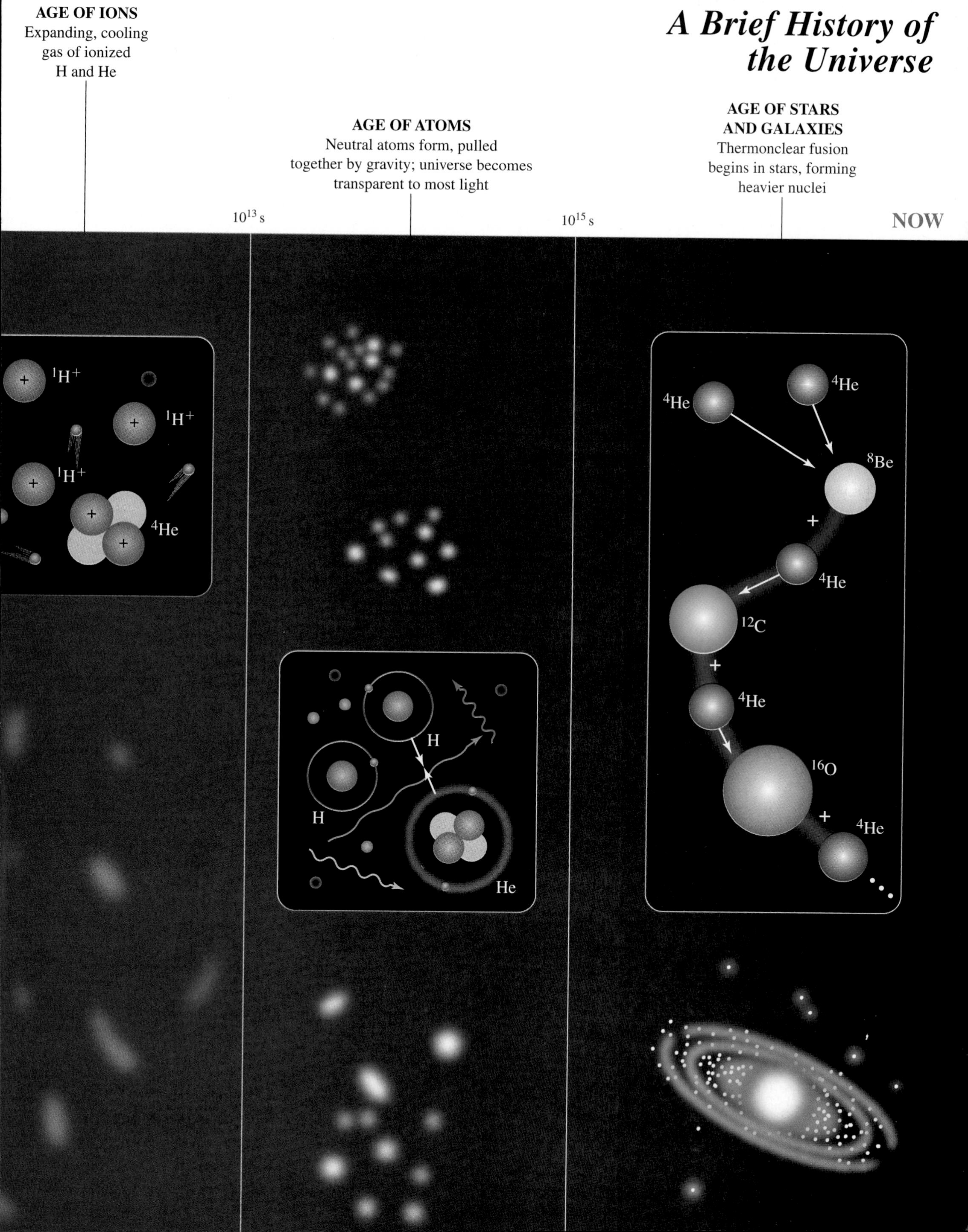

antineutrinos throughout the universe uncoupled from the rest of the universe. Because of the extraordinarily low probability for neutrino absorption, most of this flux is still present today, although cooled greatly by expansion. The standard model of the universe predicts a present neutrino temperature of about 2 K, but no one has yet been able to carry out an experiment to test this prediction.

Nucleosynthesis

At about $t = 1$ s, the ratio of protons to neutrons was determined by the Boltzmann distribution factor $e^{-\Delta E/kT}$, where ΔE is the difference between the neutron and proton rest energies, $\Delta E = 1.294$ MeV. At a temperature of about 10^{10} K, this distribution factor gives about 4.5 times as many protons as neutrons. However, as we have discussed, free neutrons (with a lifetime of 887 s) decay spontaneously to protons. This decay caused the proton–neutron ratio to increase until about $t = 225$ s. At this time, the temperature was about 10^9 K, and the average energy was well below 2 MeV.

This energy distribution was critical because the binding energy of the *deuteron* (a neutron and a proton bound together) is 2.22 MeV (see Section 43.2). A neutron bound in a deuteron does not decay spontaneously. As the average energy decreased, a proton and a neutron could combine to form a deuteron, and there were fewer and fewer photons with 2.22 MeV or more of energy to dissociate the deuterons again. Therefore the combining of protons and neutrons into deuterons halted the decay of free neutrons.

The formation of deuterons starting at about $t = 225$ s marked the beginning of the period of formation of nuclei, or *nucleosynthesis*. At this time, there were about seven protons for each neutron. The deuteron (^2H) can absorb a neutron and form a triton (^3H), or it can absorb a proton and form ^{3}He. Then ^{3}H can absorb a proton, and ^{3}He can absorb a neutron, each yielding ^{4}He (the alpha particle). A few ^{7}Li nuclei may also have been formed by fusion of ^{3}H and ^{4}He nuclei. According to the theory, essentially all the ^{1}H and ^{4}He in the present universe was formed at this time. But then the building of nuclei almost ground to a halt. The reason is that *no* nuclide with mass number $A = 5$ has a half-life greater than 10^{-21} s. Alpha particles simply do not permanently absorb neutrons or protons. The nuclide ^{8}Be that is formed by fusion of two ^{4}He nuclei is unstable, with an extremely short half-life, about 7×10^{-17} s. Note also that at this time, the average energy was still much too large for electrons to be bound to nuclei; there were not yet any atoms.

Conceptual Example 44.12 The relative abundance of hydrogen and helium in the universe

Nearly all of the protons and neutrons in the seven-to-one ratio at $t = 225$ s either formed ^{4}He or remained as ^{1}H. After this time, what was the relative abundance of ^{1}H and ^{4}He by mass?

SOLUTION

The ^{4}He nucleus contains two protons and two neutrons. For every two neutrons we have 14 protons. The two neutrons and two of the 14 protons make up one ^{4}He nucleus, leaving 12 protons (^{1}H nuclei). So at this time there were 12 ^{1}H nuclei for every ^{4}He nucleus. The masses of ^{1}H and ^{4}He are about 1 u and 4 u, respectively, so there were 12 u of ^{1}H for every 4 u of ^{4}He. Therefore the mix, by mass, was 75% ^{1}H and 25% ^{4}He. This result agrees very well with estimates of the present H-He ratio in the universe, an important confirmation of this part of the theory.

Further nucleosynthesis did not occur until very much later, well after the time $t = 10^{13}$ s (about 380,000 y). At that time, the temperature was about 3000 K, and the average energy was a few tenths of an electron volt. Because the ionization energies of hydrogen and helium *atoms* are 13.6 eV and 24.5 eV, respectively, almost all the hydrogen and helium was electrically neutral (not ionized). With the electrical repulsions of the nuclei canceled out, gravitational attraction could slowly pull the neutral atoms together to form clouds of gas and eventually stars. Thermonuclear reactions in stars are believed to have produced all of the more massive nuclei. In Section 43.8 we discussed one cycle of thermonuclear reactions in which ^{1}H becomes ^{4}He; this cycle is one of the sources of the energy radiated by stars.

As a star uses up its hydrogen, the inward gravitational pressure exceeds the outward radiation and gas pressure, and the star's core begins to contract. As it does so, the gravitational potential energy decreases, and the kinetic energies of the star's atoms increase. For stars with sufficient mass, there are both enough energy and sufficient density to begin another process, *helium fusion.* First two ^{4}He nuclei fuse to form ^{8}Be. The extremely short lifetime of this unstable nuclide is compensated by the density of the nuclear core and by an usually large probability for absorption of another ^{4}He nucleus with a specific energy, a kind of resonance effect. Thus a reasonable fraction of the ^{8}Be nuclei fuse with ^{4}He to form the stable nuclide ^{12}C. The net result is the fusion of three ^{4}He nuclei to form one ^{12}C, the *triple-alpha process.* Then successive fusions with ^{4}He give ^{16}O, ^{20}Ne, and ^{24}Mg. All these reactions are exoergic. They release energy to heat up the star, and ^{12}C and ^{16}O can fuse to form elements with higher and higher atomic number.

For nuclides that can be created in this manner, the binding energy per nucleon peaks at mass number $A = 56$ with the nuclide ^{56}Fe, so exoergic fusion reactions stop with Fe. But successive neutron captures followed by beta decays can continue the synthesis of more massive nuclei. If the star is massive enough, it may eventually explode as a *supernova,* sending out into space the heavy elements that were produced by the earlier processes (Fig. 44.20; see also Fig. 37.6). In space, the debris and other interstellar matter can gravitationally bunch together to form a new generation of stars and planets. Our own sun is one such "second-generation" star. This means that the sun's planets and everything on them (including you) contain matter that was long ago blasted into space by an exploding supernova.

44.20 The Veil Nebula in the constellation Cygnus is a remnant of a supernova explosion that occurred more than 20,000 years ago. The gas ejected from the supernova is still moving very rapidly. Collisions between this fast-moving gas and the tenuous material of interstellar space excites the gas and causes it to glow. The portion of the nebula shown here is about 40 ly (12 pc) in length.

Background Radiation

In 1965, Arno Penzias and Robert Wilson, working at Bell Telephone Laboratories in New Jersey on satellite communications, turned a microwave antenna skyward and found a background signal that had no apparent preferred direction. (This signal produces about 1% of the "hash" you see on a TV screen when you turn to an unused channel.) Further research has shown that the radiation that is received has a frequency spectrum that fits Planck's blackbody radiation law, Eq. (38.32) (see Section 38.8). The wavelength of peak intensity is 1.063 mm (in the microwave region of the spectrum), with a corresponding absolute temperature $T = 2.725$ K. Penzias and Wilson contacted physicists at nearby Princeton University who had begun the design of an antenna to search for radiation that was a remnant from the early evolution of the universe. We mentioned above that neutral atoms began to form at about $t = 380{,}000$ years when the temperature was 3000 K. With far fewer charged particles present than previously, the universe became transparent at this time to electromagnetic radiation of long wavelength. The 3000-K blackbody radiation therefore survived, cooling to its present 2.725-K temperature as the universe expanded. The *cosmic background radiation* is among the most clear-cut experimental confirmations of the Big Bang theory. The image that opens this chapter shows a modern map of the cosmic background radiation.

Example 44.13 Expansion of the universe

By approximately what factor has the universe expanded since $t = 380{,}000$ y?

SOLUTION

IDENTIFY and SET UP: As the universe has expanded, all intergalactic wavelengths have expanded with it. Furthermore, the Wien displacement law, Eq. (38.30), relates the peak wavelength in blackbody radiation to the temperature. Given the temperatures of the cosmic background radiation today and at $t = 380{,}000$ y, we can determine the factor by which wavelengths have changed and hence determine the factor by which the universe has expanded.

EXECUTE: We can rewrite Eq. (38.30) to show that the peak wavelength λ_m in blackbody radiation is inversely proportional to absolute temperature:

$$\lambda_m = \frac{2.90 \times 10^{-3}\ \text{m}\cdot\text{K}}{T}$$

As the universe expands, all intergalactic wavelengths (including λ_m) increase in proportion to the scale factor R. The temperature decreased by a factor of $(3000\ \text{K})/(2.7\ \text{K})$, so λ_m and the scale factor must have *increased* by this factor. Thus, between $t = 380{,}000$ y and the present, the universe expanded by a factor of $(3000\ \text{K})/(2.7\ \text{K})$, or about 1100.

EVALUATE: Our results show that since $t = 380{,}000$ y, any particular intergalactic volume has increased by a factor $(1100)^3 = 1.3 \times 10^9$. They also show that when the cosmic background radiation was emitted, its peak wavelength was $(1.063\ \text{mm})/1100 = 967$ nm, in the infrared region of the spectrum.

Matter and Antimatter

One of the most remarkable features of our universe is the asymmetry between matter and antimatter. One might think that the universe should have equal numbers of protons and antiprotons and of electrons and positrons, but this doesn't appear to be the case. There is no evidence for the existence of substantial amounts of antimatter (matter composed of antiprotons, antineutrons, and positrons) anywhere in the universe. Theories of the early universe must explain this imbalance.

We've mentioned that most GUTs include violation of conservation of baryon number at energies at which the strong and electroweak interactions have converged. If particle–antiparticle symmetry is also violated, we have a mechanism for making more quarks than antiquarks, more leptons than antileptons, and eventually more matter than antimatter. One serious problem is that any asymmetry that is created in this way during the GUT era might be wiped out by the electroweak interaction after the end of the GUT era. If so, there must be some mechanism that creates particle–antiparticle asymmetry at a much *later* time. The problem of the matter–antimatter asymmetry is still very much an open one.

We hope that this qualitative discussion has given you a hint of the close connections between particle physics and cosmology. There are still many unanswered questions in this active field of research. Is the energy density of the universe precisely equal to $\rho_c c^2$, or are there small but important differences? What is dark energy? Has the density of dark energy remained constant over the history of the universe, or has the density changed? What is dark matter, and why is it so difficult to detect other than by its gravitational influence? What happened during the first 10^{-43} s after the Big Bang? Can we see evidence that the strong and electroweak interactions undergo a grand unification at high energies? The search for the answers to these and many other questions about our physical world continues to be one of the most exciting adventures of the human mind.

Test Your Understanding

Why do we not detect photons emitted earlier than $t = 380{,}000$ y?

SUMMARY

Each particle has an antiparticle; some particles are their own antiparticles. Particles can be created and destroyed, some of them (including electrons and positrons) only in pairs or in conjunction with other particles and antiparticles.

Particles serve as mediators for the fundamental interactions. The photon is the mediator of the electromagnetic interaction. Yukawa proposed the existence of mesons to mediate the nuclear interaction. Mediating particles that can exist only because of the uncertainty principle for energy are called virtual particles. (See Example 44.1)

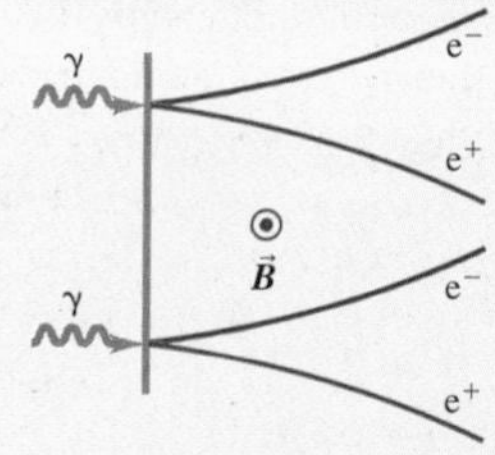

Cyclotrons, synchrotrons, and linear accelerators are used to accelerate charged particles to high energies for experiments with particle interactions. Only part of the beam energy is available to cause reactions with targets at rest. This problem is avoided in colliding-beam experiments. (See Examples 44.2 through 44.4)

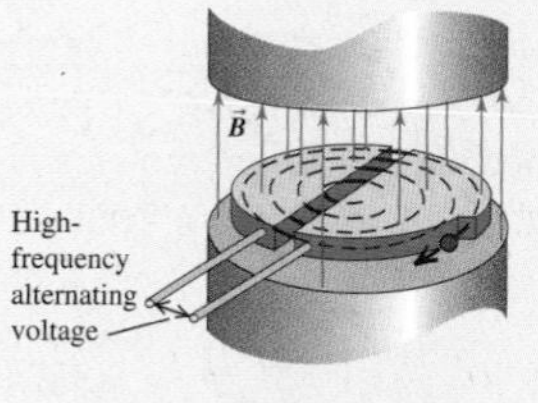

Four fundamental interactions are found in nature: the strong, electromagnetic, weak, and gravitational interactions. Particles can be described in terms of their interactions and of quantities that are conserved in all or some of the interactions.

Fermions have half-integer spins; bosons have integer spins. Leptons, which are fermions, have no strong interactions. Strongly interacting particles are called hadrons. They include mesons, which are always bosons, and baryons, which are always fermions. There are conservation laws for three different lepton numbers and for baryon number. Additional quantum numbers, including strangeness and charm, are conserved in some interactions and not in others. (See Examples 44.5 through 44.7)

Hadrons are composed of quarks. There are thought to be six types of quarks. The interaction between quarks is mediated by gluons. Quarks and gluons have an additional attribute called color. (See Example 44.8)

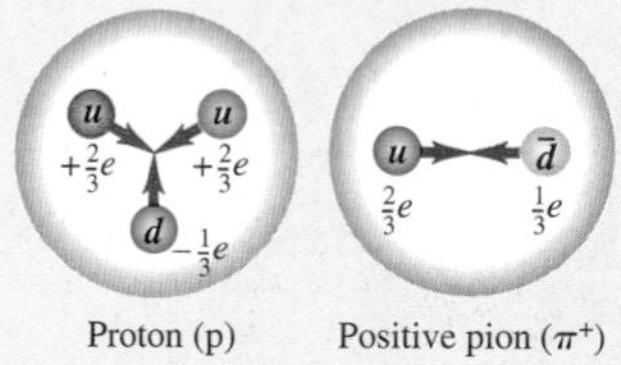

Symmetry considerations play a central role in all fundamental-particle theories. The electromagnetic and weak interactions become unified at high energies into the electroweak interaction. In grand unified theories the strong interaction is also unified with these interactions, but at much higher energies.

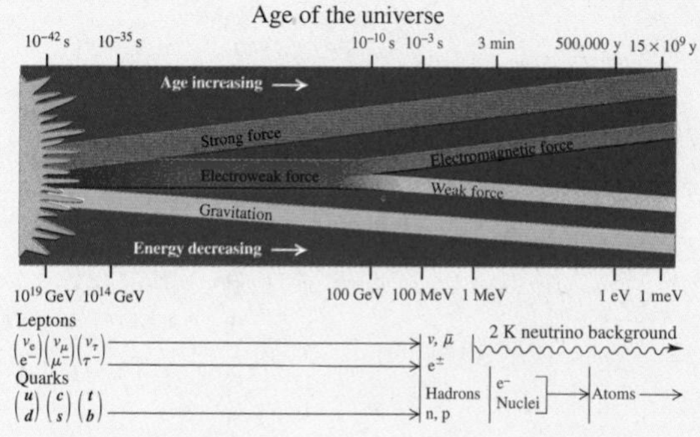

The Hubble law shows that galaxies are receding from each other and that the universe is expanding. Observations show that the rate of expansion is accelerating due to the presence of dark energy, which makes up 73% of the energy in the universe. Only 4% of the energy in the universe is in the form of ordinary matter; the remaining 23% is dark matter, whose nature is poorly understood. (See Examples 44.9 and 44.10)

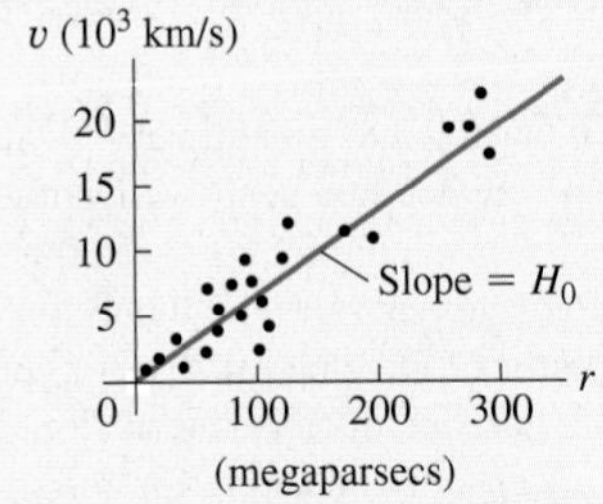

In the standard model of the universe, a Big Bang gave rise to the first fundamental particles. They eventually formed into the lightest atoms as the universe expanded and cooled. The cosmic background radiation is a relic of the time when these atoms formed. The heavier elements were manufactured much later by fusion reactions inside stars. (See Examples 44.11 and 44.12)

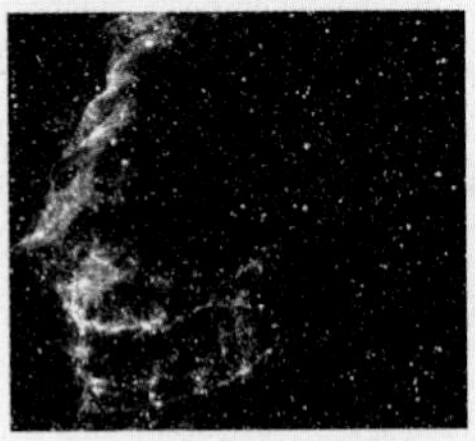

Key Terms

antiparticle, 1671
meson, 1673
muon, 1673
pion, 1674
fermion, 1682
boson, 1682
lepton, 1682
hadron, 1683
baryon, 1684
strangeness, 1685
quark, 1687
eightfold way, 1689
gluon, 1689
charm, 1690
standard model, 1691
electroweak theory, 1691
grand unified theory, 1692
redshift, 1694
Hubble law, 1694
cosmological principle, 1695
Big Bang, 1695
dark matter, 1699
dark energy, 1700

Your Notes

Answer to Chapter Opening Question

The cosmic background radiation is evidence that the universe was once at a much higher temperature than it is today, which must have been the case if the universe began with a Big Bang.

Answers to Test Your Understanding Questions

Section 44.1 An electron–positron pair has a rest energy of 2(0.511 Mev) $= 1.022$ MeV. This is about 1/125 the rest energy of a pion, so from Eq. (44.3) such a virtual pair can last for a time Δt that is about 125 times longer than the lifetime of a virtual pion. In this time the virtual pair can travel about 125 times farther than a virtual pion, or $125(1.5 \times 10^{-15}\ \text{m}) = 1.9 \times 10^{-13}$ m. This is the range of a virtual electron–positron pair; vacuum polarization becomes noticeable at smaller distances. At such close distances, the electric force between the electron and proton is slightly different from the $1/r^2$ force we first studied in Chapter 21.

Section 44.2 In a head-on collision between an electron and a positron of equal energy, the net momentum is zero. Since both momentum and energy are conserved in the collision, the virtual photon also has momentum $p = 0$ but has energy $E = 90\ \text{GeV} + 90\ \text{GeV} = 180\ \text{GeV}$. Hence the relationship $E = pc$ is definitely *not* true for this virtual photon. In fact, it is a hallmark of virtual particles of all kinds that they need not have the same relationship between energy and momentum that real particles do.

Section 44.3 Mesons all have baryon number $B = 0$, while a proton has $B = 1$. The decay of a proton into one or more mesons would require that baryon number *not* be conserved. No violation of this conservation principle has ever been observed, so the proposed decay is impossible. (As discussed in Section 44.5, no evidence has been found suggesting that the proton decays in *any* fashion.)

Section 44.4 Only the s quark, with $S = -1$, has nonzero strangeness. For a baryon to have $S = -2$, it must have two s quarks and one quark of a different flavor. Since each s quark has charge $-\frac{1}{3}e$, the nonstrange quark must have charge $+\frac{4}{3}e$ to make the net charge equal to $+e$. But *no* quark has charge $+\frac{4}{3}e$, so the proposed baryon is impossible.

Section 44.5 If a d quark in a neutron (quark content udd) undergoes the process $d \rightarrow u + e^- + \bar{\nu}_e$, the remaining baryon has quark content uud and hence is a proton (see Fig. 44.11). An electron is the same as a β^- particle, so the net result is beta-minus decay: $n \rightarrow p + \beta^- + \bar{\nu}_e$. Thus, in the standard model, beta-minus decay of a nucleus is a process whereby a quark within the nucleus undergoes a change in flavor from d to u.

Section 44.6 The question is based on the misconception that all of the matter and energy in the universe is localized and is expanding into emptiness, like a skyrocket exploding in empty space. In fact, the universe is infinite; it has no edges, so there is nothing "outside" it and it isn't "expanding into" anything. The expansion of the universe simply means that the coordinates of space are stretching out with the passage of time. In a sense, the infinite universe is becoming more infinite.

Section 44.7 Prior to $t = 380{,}000$ y the temperature was so high that atoms could not form, so free electrons and protons were plentiful. These charged particles are very effective at scattering photons, so light could not propagate very far and the universe was opaque. The oldest photons that we can detect date from the time $t = 380{,}000$ y when atoms formed and the universe became transparent.

Discussion Questions

Q44.1 Is it possible that some parts of the universe contain antimatter whose atoms have nuclei made of antiprotons and antineutrons, surrounded by positrons? How could we detect this condition without actually going there? Can we detect these antiatoms by identifying the light they emit as composed of antiphotons? Explain. What problems might arise if we actually *did* go there?

Q44.2 What are the similarities between a neutrino and a photon? What are the differences?

Q44.3 When they were first discovered during the 1930s and 1940s, there was confusion as to the identities of the pions and the muons. In retrospect, what are the similarities? What are the most significant differences?

Q44.4 The gravitational force between two electrons is weaker than the electrical force by the order of 10^{-40}. Yet the gravitational interactions of matter were observed and analyzed long before electrical interactions were understood. Why?

Q44.5 When a π^0 decays to two photons, what happens to the quarks of which it was made?

Q44.6 Why can't an electron decay into two photons? Into two neutrinos?

Q44.7 According to the standard model of the fundamental particles, what are the similarities between baryons and leptons? What are the most important differences?

Q44.8 According to the standard model of the fundamental particles, what are the similarities between quarks and leptons? What are the most important differences?

Q44.9 What are the main advantages of colliding-beam accelerators compared with those using stationary targets? What are the main disadvantages?

Q44.10 Does the weak force act on hadrons, or do they experience only the strong force? Explain.

Q44.11 Does the universe have a center? Explain.

Q44.12 Assume that the universe has an edge. Placing yourself at that edge in a thought experiment, explain why this assumption violates the cosmological principle.

Q44.13 Explain why the cosmological principle requires that H_0 must have the same value everywhere in space, but does not require that it be constant in time.

Exercises

Section 44.1 Fundamental Particles—A History

44.1 An electron and a positron are moving toward each other and each have speed $0.500c$ in the lab frame. a) What is the kinetic energy of each particle? b) The e^+ and e^- meet head-on and annihilate. What is the energy of each photon that is produced? c) What is the wavelength of each photon? How does the wavelength compare to the photon wavelength when the initial kinetic energy of the e^+ and e^- is negligibly small (Example 44.1)?

44.2 In Example 44.1 (Section 44.1), what is the speed of a positron that has kinetic energy 5.00 MeV? Express your result in terms of the speed of light.

44.3 A neutral pion at rest decays into two photons. Find the approximate energy, frequency, and wavelength of each photon. In which part of the electromagnetic spectrum does each photon lie? (Use the pion mass given in terms of the electron mass in Section 44.1.)

44.4 Two equal-energy photons collide head-on and annihilate one another, producing a $\mu^+\mu^-$ pair. The muon mass is given in terms of the electron mass in Section 44.1. a) Calculate the maximum wavelength of the photons for this to occur. If the photons have this wavelength, describe the motion of the μ^+ and μ^- immediately after they are produced. b) What happens if the photons have shorter wavelength than this maximum but still annihilate and produce a $\mu^+\mu^-$ pair?

44.5 A positive pion at rest decays into a positive muon and a neutrino. a) Approximately how much energy is released in the decay? (Assume the neutrino has zero rest mass. Use the muon and pion masses given in terms of the electron mass given in Section 44.1.) b) Why can a positive muon not decay into a positive pion?

44.6 a) A proton and an antiproton annihilate, producing two photons. Find the energy, frequency, and wavelength of each photon emitted in the center-of-momentum reference frame a) if the initial kinetic energies of the proton and antiproton are negligible; b) if each particle has an initial kinetic energy of 830 MeV.

44.7 "Maximum Power, Scotty!" The starship *Enterprise,* of television and movie fame, is powered by the controlled combination of matter and antimatter. If the entire 400-kg antimatter fuel supply of the *Enterprise* combines with matter, how much energy is released?

44.8 Calculate the reaction energy Q for the reaction given in Eq. (44.1). Is the reaction exoergic or endoergic?

44.9 Calculate the reaction energy Q for the neutron absorption reaction given in Eq. (44.2). Is the reaction exoergic or endoergic?

Section 44.2 Particle Accelerators and Detectors

44.10 In the head-on collision of an electron and a positron of equal energy, the total available energy is 100 GeV. a) What is the total energy of the electron or positron? b) If an electron with the total energy calculated in part (a) is incident on a stationary positron, what is the available energy?

44.11 Deuterons in a cyclotron travel in a circle with radius 32.0 cm just before emerging from the dees. The frequency of the applied alternating voltage is 9.00 MHz. Find a) the magnetic field; b) the kinetic energy and speed of the deuterons upon emergence.

44.12 The magnetic field in a cyclotron that accelerates protons is 1.30 T. a) How many times per second should the potential across the dees reverse? (This is twice the frequency of the circulating protons.) b) The maximum radius of the cyclotron is 0.250 m. What is the maximum speed of the proton? c) Through what potential difference would the proton have to be accelerated from rest to give it the same speed as calculated in part (b)?

44.13 a) A high-energy beam of alpha particles collides with a stationary helium gas target. What must the total energy of a beam particle be if the available energy in the collision is 16.0 GeV? b) What must the energy of each beam be to produce the same available energy in a colliding-beam experiment?

44.14 a) What is the speed of a proton that has total energy 1000 GeV? b) What is the angular frequency ω of a proton with the speed calculated in part (a) in a magnetic field of 4.00 T? Use both the nonrelativistic Eq. (44.7) and the correct relativistic expression, and compare the results.

44.15 In Example 44.4 (Section 44.2), it was shown that a proton beam with an 800-GeV beam energy gives an available energy of 38.7 GeV for collisions with a stationary proton target. a) You are asked to design an upgrade of the accelerator that will double the available energy in stationary-target collisions. What beam energy is required? b) In a colliding-beam experiment, what total energy of each beam is needed to give an available energy of $2(38.7\text{ GeV}) = 77.4\text{ GeV}$?

44.16 Calculate the threshold kinetic energy for the production of a η^0 particle in the collision of a proton beam with a stationary proton target: $\text{p} + \text{p} \rightarrow \text{p} + \text{p} + \eta^0$. The rest energy of the η^0 is 547.3 MeV (Table 44.3).

Section 44.3 Particles and Interactions

44.17 What is the mass in kg of the Z^0? What is the ratio of the mass of the Z^0 to the mass of the proton?

44.18 Table 44.3 shows that a Σ^0 decays into a Λ^0 and a photon. a) Calculate the energy of the photon emitted in this decay, if the Λ^0 is at rest. b) What is the magnitude of the momentum of the photon? Is it reasonable to ignore the final momentum and kinetic energy of the Λ^0? Explain.

44.19 If a Σ^+ at rest decays into a proton and a π^0, what is the total kinetic energy of the decay products?

44.20 If a muon at rest decays into an electron and two neutrinos, what is the total kinetic energy of the decay products? Assume that the neutrinos have zero rest mass.

44.21 In which of the following decays are the three lepton numbers conserved? In each case, explain your reasoning.

a) $\mu^- \rightarrow \text{e}^- + \nu_\text{e} + \bar{\nu}_\mu$ c) $\pi^+ \rightarrow \text{e}^+ + \gamma$

b) $\tau^- \rightarrow \text{e}^- + \bar{\nu}_\text{e} + \nu_\tau$ d) $\text{n} \rightarrow \text{p} + \text{e}^- + \bar{\nu}_\text{e}$

44.22 In which of the following reactions or decays is the baryon number conserved? In each case, explain your reasoning.

a) $\text{n} \rightarrow \text{p} + \text{e}^- + \bar{\nu}_\text{e}$ c) $\text{p} \rightarrow \pi^+ + \pi^0$

b) $\text{p} + \text{n} \rightarrow \text{p} + \pi^0$ d) $\text{p} + \text{p} \rightarrow \text{p} + \text{p} + \pi^0$

44.23 In which of the following reactions or decays is strangeness conserved? In each case, explain your reasoning.

a) $\text{K}^+ \rightarrow \mu^+ + \nu_\mu$ c) $\text{K}^+ + \text{K}^- \rightarrow \pi^0 + \pi^0$

b) $\text{n} + \text{K}^+ \rightarrow \text{p} + \pi^0$ d) $\text{p} + \text{K}^- \rightarrow \Lambda^0 + \pi^0$

44.24 a) Show that the coupling constant for the electromagnetic interaction, $e^2/4\pi\epsilon_0\hbar c$, is dimensionless and has the numerical value 1/137.0. b) Show that in the Bohr model (Section 38.5), the orbital speed of an electron in the $n = 1$ orbit is equal to c times the coupling constant $e^2/4\pi\epsilon_0\hbar c$.

44.25 Show that the nuclear force coupling constant $f^2/\hbar c$ is dimensionless.

Section 44.4 Quarks and the Eightfold Way

44.26 Nine of the spin-$\frac{3}{2}$ baryons are four Δ particles, each with mass 1232 MeV/c^2, strangeness 0, and charges $+2e$, $+e$, 0, and $-e$; three Σ^* particles, each with mass 1385 MeV/c^2, strangeness -1, and charges $+e$, 0, and $-e$; and two Ξ^* particles, each with mass 1530 MeV/c^2, strangeness -2, and charges 0 and $-e$. a) Place these particles on a plot of S versus Q. Deduce the Q and S values of the tenth spin-$\frac{3}{2}$ baryon, the Ω^- particle, and place it on your diagram. Also label the particles with their masses. The mass of the Ω^- is 1672 MeV/c^2; is this value consistent with your diagram? b) Deduce the three-quark combinations (of $\boldsymbol{u}$, $\boldsymbol{d}$, and $\boldsymbol{s}$) that comprise each of these ten particles. Redraw the plot of S versus Q from part (a) with each particle labeled by its quark content. What regularities do you see?

44.27 Determine the electric charge, baryon number, strangeness, and charm quantum numbers for the following quark combinations a) $\boldsymbol{uds}$; b) $\boldsymbol{c\bar{u}}$; c) $\boldsymbol{ddd}$; d) $\boldsymbol{d\bar{c}}$. Explain your reasoning.

44.28 Given that each particle contains only combinations of $\boldsymbol{u}$, $\boldsymbol{d}$, $\boldsymbol{s}$, $\bar{\boldsymbol{u}}$, $\bar{\boldsymbol{d}}$, and $\bar{\boldsymbol{s}}$, use the method of Example 44.8 (Section 44.4) to deduce the quark content of a) a particle with charge $+e$, baryon number 0, and strangeness $+1$; b) a particle with charge $+e$, baryon number -1, and strangeness $+1$; c) a particle with charge 0, baryon number $+1$, and strangeness -2.

44.29 The quark content of the neutron is $\boldsymbol{udd}$. a) What is the quark content of the antineutron? Explain your reasoning. b) Is the neutron its own antiparticle? Why or why not? c) The quark content of the ψ is $\boldsymbol{c\bar{c}}$. Is the ψ its own antiparticle? Explain your reasoning.

44.30 What is the total kinetic energy of the decay products when an upsilon particle at rest decays to $\tau^+ + \tau^-$?

44.31 The Test Your Understanding question at the end of Section 44.5 discusses β^- decay in terms of change in quark flavor. Make a similar analysis for β^+ decay.

Section 44.6 The Expanding Universe

44.32 Redshift Parameter. The definition of the redshift parameter z is given in Example 44.9. a) Show that Eq. (44.13) can be written as $1 + z = ([1 + \beta]/[1 - \beta])^{1/2}$, where $\beta = v/c$. b) The observed redshift parameter for a certain galaxy is $z = 0.500$. Find the speed of the galaxy relative to the earth, if the redshift is due to the Doppler shift. c) Use the Hubble law to find the distance of this galaxy from the earth.

44.33 A galaxy in the constellation Pisces is 5210 Mly from the earth. a) Use the Hubble law to calculate the speed at which this galaxy is receding from earth. b) What redshifted ratio λ_0/λ_s is expected for light from this galaxy?

44.34 a) According to the Hubble law, what is the distance r from us for galaxies that are receding from us with a speed c? b) Explain why the distance calculated in part (a) is the size of our observable universe (ignoring any slowing of the expansion of the universe due to gravitational attraction).

44.35 The spectrum of the sodium atom is detected in the light from a distant galaxy. a) If the 590.0-nm line is redshifted to 658.5 nm, at what speed is the galaxy receding from the earth? b) Use the Hubble law to calculate the distance of the galaxy from the earth.

44.36 Derive Eq. (44.14) from (44.13).

Section 44.7 The Beginning of Time

44.37 Calculate the energy released in each of the following reactions: a) $\mathrm{p} + {}^2\mathrm{H} \rightarrow {}^3\mathrm{He}$; b) $\mathrm{n} + {}^3\mathrm{He} \rightarrow {}^4\mathrm{He}$.

44.38 Calculate the energy (in MeV) released in the triple-alpha process $3\,{}^4\mathrm{He} \rightarrow {}^{12}\mathrm{C}$.

44.39 Calculate the reaction energy Q (in MeV) for the reaction

$$\mathrm{e}^- + \mathrm{p} \rightarrow \mathrm{n} + \nu_\mathrm{e}$$

Is this reaction endoergic or exoergic?

44.40 Calculate the reaction energy Q (in MeV) for the nucleosynthesis reaction

$${}^{12}_{6}\mathrm{C} + {}^{4}_{2}\mathrm{He} \rightarrow {}^{16}_{8}\mathrm{O}$$

Is this reaction endoergic or exoergic?

44.41 The 2.728 K blackbody radiation has its peak wavelength at 1.062 mm. What was the peak wavelength at $t = 700{,}000$ y when the temperature was 3000 K?

44.42 a) Show that the expression for the Planck length, $\sqrt{\hbar G/c^3}$, has dimensions of length. b) Evaluate the numerical value of $\sqrt{\hbar G/c^3}$, and verify the value given in Eq. (44.21).

Problems

44.43 In the LHC, each proton will be accelerated to a kinetic energy of 7.0 TeV. a) In the colliding beams, what is the available energy E_a in a collision? b) In a fixed-target experiment in which a beam of protons is incident on a stationary proton target, what must the total energy (in TeV) of the particles in the beam be to produce the same available energy as in part (a)?

44.44 A proton and an antiproton collide head-on with equal kinetic energies. In the center of momentum frame, two γ rays with wavelengths of 0.780 fm are produced. Calculate the kinetic energy of the incident proton.

44.45 An electron with kinetic energy K collides with another electron at rest. Both electrons survive the collision, and a neutral pion (π^0) is produced. What is the threshold energy (minimum value of K) needed for this reaction?

44.46 Calculate the threshold kinetic energy for the reaction

$$\pi^- + \mathrm{p} \rightarrow \Sigma^0 + \mathrm{K}^0$$

if a π^- beam is incident on a stationary proton target. The K^0 has a mass of 497.7 MeV/c^2.

44.47 Calculate the threshold kinetic energy for the reaction

$$\mathrm{K}^- + \mathrm{p} \rightarrow \Lambda^0 + \mathrm{K}^+ + \mathrm{K}^-$$

if a K^- beam is incident on a stationary proton target.

44.48 An η^0 meson at rest decays into three π mesons. a) What are the allowed combinations of π^0, π^+, and π^- as decay products? b) Find the total kinetic energy of the π mesons.

44.49 a) What energy is released when a negative pion (π^-) at rest decays to stable products (see Tables 44.2 and 44.3 in Section 44.3)? b) Why do the antineutrinos and neutrino carry away most of this energy?

44.50 The ϕ meson has mass 1019.4 MeV/c^2 and a measured energy width of 4.4 MeV/c^2. Using the uncertainty principle, Eq. (39.13), estimate the lifetime of the ϕ meson.

44.51 A ϕ meson (Problem 44.50) at rest decays via $\phi \rightarrow \mathrm{K}^+ + \mathrm{K}^-$. It has strangeness 0. a) Find the kinetic energy of the K^+ meson. (Assume that the two decay products share kinetic energy equally, since their masses are equal.) b) Suggest a reason why the decay $\phi \rightarrow \mathrm{K}^+ + \mathrm{K}^- + \pi^0$ has not been observed. c) Suggest reasons why the decays $\phi \rightarrow \mathrm{K}^+ + \pi^-$ and $\phi \rightarrow \mathrm{K}^+ + \mu^-$ have not been observed.

44.52 Each reaction listed below is missing a single particle. Calculate the baryon number, charge, strangeness, and the three lepton numbers (where appropriate) of the missing particle, and from this identify the particle. a) $\mathrm{p} + \mathrm{p} \rightarrow \mathrm{p} + \Lambda^0 + ?$ b) $\mathrm{K}^- + \mathrm{n} \rightarrow \Lambda^0 + ?$ c) $\mathrm{p} + \overline{\mathrm{p}} \rightarrow \mathrm{n} + ?$ d) $\overline{\nu}_\mu + \mathrm{p} \rightarrow \mathrm{n} + ?$

44.53 Estimate the energy width (energy uncertainty) of the ψ if its mean lifetime is 7.6×10^{-21} s. What fraction is this of its rest energy?

44.54 Proton Decay. Proton decay is a feature of some grand unification theories. One possible decay could be $\mathrm{p}^+ \rightarrow \mathrm{e}^+ + \pi^0$, which violates both baryon and lepton number conservation, so the proton lifetime is expected to be very long. Suppose the proton half-life were 1.0×10^{18} y. a) Calculate the energy deposited per kilogram of body tissue (in rad) due to the decay of the protons in your body in one year. Model your body as consisting entirely of water. Only the two protons in the hydrogen atoms in each H_2O molecule would decay in the manner shown; do you see why? Assume that the π^0 decays into two γ rays, that the positron annihilates with an electron, and that all the energy produced in the primary decay and these secondary decays remains in your body. b) Calculate the equivalent dose (in rem) assuming a RBE of 1.0 for all the radiation products and compare with the 0.1 rem due to the natural background and the 5.0 rem guideline for industrial workers. Based on your calculation, can the proton lifetime be as short as 1.0×10^{18} y?

44.55 A Ξ^- particle at rest decays into a Λ^0 and a π^-. a) Find the total kinetic energy of the decay products. b) What fraction of the energy is carried off by each particle? (For simplicity, use nonrelativistic momentum and kinetic-energy expressions.)

44.56 Consider the spherical balloon model of a two-dimensional expanding universe (see Fig. 44.16 in Section 44.6). The shortest distance between two points on the surface, measured along the surface, is the arc length r, where $r = R\theta$. As the balloon expands, its radius R increases, but the angle θ between the two points remains constant. a) Explain why, at any given time, $(dR/dt)/R$ is the same for all points on the balloon. b) Show that $v = dr/dt$ is directly proportional to r at any instant. c) From your answer to part (b), what is the expression for the Hubble constant H_0 in terms of R and dR/dt? d) The expression for H_0 you found in part (c) is constant in space. How would R have to depend on time for H_0 to be constant in time? e) Is your answer to part (d) consistent with the gravitational attraction of matter in the universe?

44.57 Suppose all the conditions are the same as in Problem 44.56, except that $v = dr/dt$ is constant for a given θ, rather than H_0 being constant in time. If so, show that the Hubble constant is $H_0 = 1/t$ and, hence, that the current value is $1/T$, where T is the age of the universe.

Challenge Problems

44.58 Consider a collision in which a stationary particle with mass M is bombarded by a particle with mass m, speed v_0, and total energy (including rest energy) E_m. a) Use the Lorentz transformation to write the velocities v_m and v_M of particles m and M in terms of the speed v_{cm} of the center of momentum. b) Use the fact that the total momentum in the center-of-momentum frame is zero to obtain an expression for v_{cm} in terms of m, M, and v_0. c) Combine the results of parts (a) and (b) to obtain Eq. (44.9) for the total energy in the center-of-momentum frame.

44.59 A Λ^0 hyperon at rest decays into a neutron and a π^0. a) Find the total kinetic energy of the decay products. b) What fraction of the total kinetic energy is carried off by each particle? (*Hint:* Use the relativistic expressions for momentum and kinetic energy.)

APPENDIX A

The International System of Units

The Système International d'Unités, abbreviated SI, is the system developed by the General Conference on Weights and Measures and adopted by nearly all the industrial nations of the world. The following material is adapted from B. N. Taylor, ed., National Institute of Standards and Technology Spec. Pub. 811 (U.S. Govt. Printing Office, Washington, DC, 1995). See also **http://physics.nist.gov/cuu**

Quantity	Name of unit	Symbol	
	SI base units		
length	meter	m	
mass	kilogram	kg	
time	second	s	
electric current	ampere	A	
thermodynamic temperature	kelvin	K	
amount of substance	mole	mol	
luminous intensity	candela	cd	
	SI derived units		**Equivalent units**
area	square meter	m^2	
volume	cubic meter	m^3	
frequency	hertz	Hz	s^{-1}
mass density (density)	kilogram per cubic meter	kg/m^3	
speed, velocity	meter per second	m/s	
angular velocity	radian per second	rad/s	
acceleration	meter per second squared	m/s^2	
angular acceleration	radian per second squared	rad/s^2	
force	newton	N	$kg \cdot m/s^2$
pressure (mechanical stress)	pascal	Pa	N/m^2
kinematic viscosity	square meter per second	m^2/s	
dynamic viscosity	newton-second per square meter	$N \cdot s/m^2$	
work, energy, quantity of heat	joule	J	$N \cdot m$
power	watt	W	J/s
quantity of electricity	coulomb	C	$A \cdot s$
potential difference, electromotive force	volt	V	J/C, W/A
electric field strength	volt per meter	V/m	N/C
electric resistance	ohm	Ω	V/A
capacitance	farad	F	$A \cdot s/V$
magnetic flux	weber	Wb	$V \cdot s$
inductance	henry	H	$V \cdot s/A$
magnetic flux density	tesla	T	Wb/m^2
magnetic field strength	ampere per meter	A/m	
magnetomotive force	ampere	A	
luminous flux	lumen	lm	$cd \cdot sr$
luminance	candela per square meter	cd/m^2	
illuminance	lux	lx	lm/m^2
wave number	1 per meter	m^{-1}	
entropy	joule per kelvin	J/K	
specific heat capacity	joule per kilogram-kelvin	$J/kg \cdot K$	
thermal conductivity	watt per meter-kelvin	$W/m \cdot K$	

Quantity	Name of unit	Symbol	Equivalent units
radiant intensity	watt per steradian	W/sr	
activity (of a radioactive source)	becquerel	Bq	s^{-1}
radiation dose	gray	Gy	J/kg
radiation dose equivalent	sievert	Sv	J/kg
	SI supplementary units		
plane angle	radian	rad	
solid angle	steradian	sr	

DEFINITIONS OF SI UNITS

meter (m) The *meter* is the length equal to the distance traveled by light, in vacuum, in a time of 1/299,792,458 second.

kilogram (kg) The *kilogram* is the unit of mass; it is equal to the mass of the international prototype of the kilogram. (The international prototype of the kilogram is a particular cylinder of platinum-iridium alloy that is preserved in a vault at Sèvres, France, by the International Bureau of Weights and Measures.)

second (s) The *second* is the duration of 9,192,631,770 periods of the radiation corresponding to the transition between the two hyperfine levels of the ground state of the cesium-133 atom.

ampere (A) The *ampere* is that constant current that, if maintained in two straight parallel conductors of infinite length, of negligible circular cross section, and placed 1 meter apart in vacuum, would produce between these conductors a force equal to 2×10^{-7} newton per meter of length.

kelvin (K) The *kelvin,* unit of thermodynamic temperature, is the fraction 1/273.16 of the thermodynamic temperature of the triple point of water.

ohm (Ω) The *ohm* is the electric resistance between two points of a conductor when a constant difference of potential of 1 volt, applied between these two points, produces in this conductor a current of 1 ampere, this conductor not being the source of any electromotive force.

coulomb (C) The *coulomb* is the quantity of electricity transported in 1 second by a current of 1 ampere.

candela (cd) The *candela* is the luminous intensity, in a given direction, of a source that emits monochromatic radiation of frequency 540×10^{12} hertz and that has a radiant intensity in that direction of 1/683 watt per steradian.

mole (mol) The *mole* is the amount of substance of a system that contains as many elementary entities as there are carbon atoms in 0.012 kg of carbon 12. The elementary entities must be specified and may be atoms, molecules, ions, electrons, other particles, or specified groups of such particles.

newton (N) The *newton* is that force that gives to a mass of 1 kilogram an acceleration of 1 meter per second per second.

joule (J) The *joule* is the work done when the point of application of a constant force of 1 newton is displaced a distance of 1 meter in the direction of the force.

watt (W) The *watt* is the power that gives rise to the production of energy at the rate of 1 joule per second.

volt (V) The *volt* is the difference of electric potential between two points of a conducting wire carrying a constant current of 1 ampere, when the power dissipated between these points is equal to 1 watt.

weber (Wb) The *weber* is the magnetic flux that, linking a circuit of one turn, produces in it an electromotive force of 1 volt as it is reduced to zero at a uniform rate in 1 second.

lumen (lm) The *lumen* is the luminous flux emitted in a solid angle of 1 steradian by a uniform point source having an intensity of 1 candela.

farad (F) The *farad* is the capacitance of a capacitor between the plates of which there appears a difference of potential of 1 volt when it is charged by a quantity of electricity equal to 1 coulomb.

henry (H) The *henry* is the inductance of a closed circuit in which an electromotive force of 1 volt is produced when the electric current in the circuit varies uniformly at a rate of 1 ampere per second.

radian (rad) The *radian* is the plane angle between two radii of a circle that cut off on the circumference an arc equal in length to the radius.

steradian (sr) The *steradian* is the solid angle that, having its vertex in the center of a sphere, cuts off an area of the surface of the sphere equal to that of a square with sides of length equal to the radius of the sphere.

SI Prefixes The names of multiples and submultiples of SI units may be formed by application of the prefixes listed in Appendix F.

APPENDIX B

Useful Mathematical Relations

ALGEBRA

$a^{-x} = \dfrac{1}{a^x}$ $\quad a^{(x+y)} = a^x a^y$ $\quad a^{(x-y)} = \dfrac{a^x}{a^y}$

Logarithms: If $\log a = x$, then $a = 10^x$. $\quad \log a + \log b = \log(ab)$ $\quad \log a - \log b = \log(a/b)$ $\quad \log(a^n) = n \log a$

If $\ln a = x$, then $a = e^x$. $\quad \ln a + \ln b = \ln(ab)$ $\quad \ln a - \ln b = \ln(a/b)$ $\quad \ln(a^n) = n \ln a$

Quadratic formula: If $ax^2 + bx + c = 0$, $\quad x = \dfrac{-b \pm \sqrt{b^2 - 4ac}}{2a}$.

BINOMIAL THEOREM

$$(a+b)^n = a^n + na^{n-1}b + \frac{n(n-1)a^{n-2}b^2}{2!} + \frac{n(n-1)(n-2)a^{n-3}b^3}{3!} + \cdots$$

TRIGONOMETRY

In the right triangle ABC, $x^2 + y^2 = r^2$.

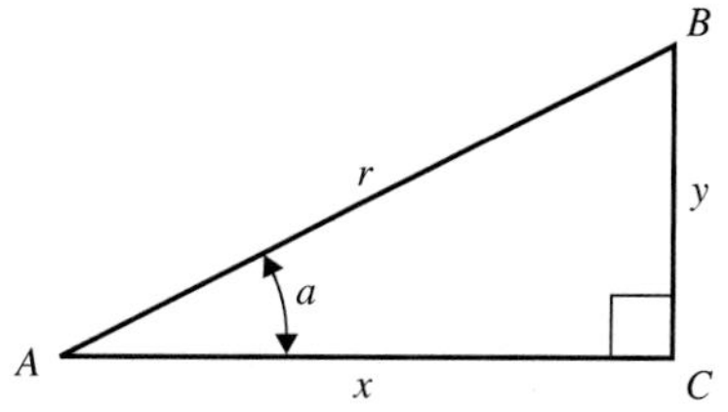

Definitions of the trigonometric functions: $\sin a = y/r$ $\quad \cos a = x/r$ $\quad \tan a = y/x$

Identities: $\sin^2 a + \cos^2 a = 1$ $\qquad \tan a = \dfrac{\sin a}{\cos a}$

$\sin 2a = 2 \sin a \cos a$ $\qquad \cos 2a = \cos^2 a - \sin^2 a = 2\cos^2 a - 1 = 1 - 2\sin^2 a$

$\sin \tfrac{1}{2}a = \sqrt{\dfrac{1 - \cos a}{2}}$ $\qquad \cos \tfrac{1}{2}a = \sqrt{\dfrac{1 + \cos a}{2}}$

$\sin(-a) = -\sin a$ $\qquad \sin(a \pm b) = \sin a \cos b \pm \cos a \sin b$

$\cos(-a) = \cos a$ $\qquad \cos(a \pm b) = \cos a \cos b \mp \sin a \sin b$

$\sin(a \pm \pi/2) = \pm \cos a$ $\qquad \sin a + \sin b = 2 \sin \tfrac{1}{2}(a+b) \cos \tfrac{1}{2}(a-b)$

$\cos(a \pm \pi/2) = \mp \sin a$ $\qquad \cos a + \cos b = 2 \cos \tfrac{1}{2}(a+b) \cos \tfrac{1}{2}(a-b)$

GEOMETRY

Circumference of circle of radius r: $\quad C = 2\pi r$

Area of circle of radius r: $\quad A = \pi r^2$

Volume of sphere of radius r: $\quad V = 4\pi r^3/3$

Surface area of sphere of radius r: $\quad A = 4\pi r^2$

Volume of cylinder of radius r and height h: $V = \pi r^2 h$

CALCULUS

Derivatives:

$$\frac{d}{dx}x^n = nx^{n-1}$$

$$\frac{d}{dx}\sin ax = a\cos ax$$

$$\frac{d}{dx}\cos ax = -a\sin ax$$

$$\frac{d}{dx}e^{ax} = ae^{ax}$$

$$\frac{d}{dx}\ln ax = \frac{1}{x}$$

Integrals:

$$\int x^n\,dx = \frac{x^{n+1}}{n+1} \quad (n \neq -1)$$

$$\int \frac{dx}{x} = \ln x$$

$$\int \sin ax\,dx = -\frac{1}{a}\cos ax$$

$$\int \cos ax\,dx = \frac{1}{a}\sin ax$$

$$\int e^{ax}dx = \frac{1}{a}e^{ax}$$

$$\int \frac{dx}{\sqrt{a^2 - x^2}} = \arcsin\frac{x}{a}$$

$$\int \frac{dx}{\sqrt{x^2 + a^2}} = \ln\left(x + \sqrt{x^2 + a^2}\right)$$

$$\int \frac{dx}{x^2 + a^2} = \frac{1}{a}\arctan\frac{x}{a}$$

$$\int \frac{dx}{(x^2 + a^2)^{3/2}} = \frac{1}{a^2}\frac{x}{\sqrt{x^2 + a^2}}$$

$$\int \frac{x\,dx}{(x^2 + a^2)^{3/2}} = -\frac{1}{\sqrt{x^2 + a^2}}$$

Power series (convergent for range of x shown):

$$(1 + x)^n = 1 + nx + \frac{n(n-1)x^2}{2!} + \frac{n(n-1)(n-2)}{3!}x^3 + \cdots \quad (|x| < 1)$$

$$\sin x = x - \frac{x^3}{3!} + \frac{x^5}{5!} - \frac{x^7}{7!} + \cdots \quad (\text{all } x)$$

$$\cos x = 1 - \frac{x^2}{2!} + \frac{x^4}{4!} - \frac{x^6}{6!} + \cdots \quad (\text{all } x)$$

$$\tan x = x + \frac{x^3}{3} + \frac{2x^2}{15} + \frac{17x^7}{315} + \cdots \quad (|x| < \pi/2)$$

$$e^x = 1 + x + \frac{x^2}{2!} + \frac{x^3}{3!} + \cdots \quad (\text{all } x)$$

$$\ln(1 + x) = x - \frac{x^2}{2} + \frac{x^3}{3} - \frac{x^4}{4} + \cdots \quad (|x| < 1)$$

APPENDIX C

The Greek Alphabet

Name	Capital	Lowercase	Name	Capital	Lowercase
Alpha	Α	α	Nu	Ν	ν
Beta	Β	β	Xi	Ξ	ξ
Gamma	Γ	γ	Omicron	Ο	o
Delta	Δ	δ	Pi	Π	π
Epsilon	Ε	ϵ	Rho	Ρ	ρ
Zeta	Ζ	ζ	Sigma	Σ	σ
Eta	Η	η	Tau	Τ	τ
Theta	Θ	θ	Upsilon	Υ	υ
Iota	Ι	ι	Phi	Φ	ϕ
Kappa	Κ	κ	Chi	Χ	χ
Lambda	Λ	λ	Psi	Ψ	ψ
Mu	Μ	μ	Omega	Ω	ω

APPENDIX D

Periodic Table of the Elements

Group / Period	1	2	3	4	5	6	7	8	9	10	11	12	13	14	15	16	17	18
1	1 **H** 1.008																	2 **He** 4.003
2	3 **Li** 6.941	4 **Be** 9.012											5 **B** 10.811	6 **C** 12.011	7 **N** 14.007	8 **O** 15.999	9 **F** 18.998	10 **Ne** 20.180
3	11 **Na** 22.990	12 **Mg** 24.305											13 **Al** 26.982	14 **Si** 28.086	15 **P** 30.974	16 **S** 32.065	17 **Cl** 35.453	18 **Ar** 39.948
4	19 **K** 39.098	20 **Ca** 40.078	21 **Sc** 44.956	22 **Ti** 47.867	23 **V** 50.942	24 **Cr** 51.996	25 **Mn** 54.938	26 **Fe** 55.845	27 **Co** 58.933	28 **Ni** 58.693	29 **Cu** 63.546	30 **Zn** 65.409	31 **Ga** 69.723	32 **Ge** 72.64	33 **As** 74.922	34 **Se** 78.96	35 **Br** 79.904	36 **Kr** 83.80
5	37 **Rb** 85.468	38 **Sr** 87.62	39 **Y** 88.906	40 **Zr** 91.224	41 **Nb** 92.906	42 **Mo** 95.94	43 **Tc** (98)	44 **Ru** 101.07	45 **Rh** 102.905	46 **Pd** 106.42	47 **Ag** 107.868	48 **Cd** 112.411	49 **In** 114.818	50 **Sn** 118.710	51 **Sb** 121.760	52 **Te** 127.60	53 **I** 126.904	54 **Xe** 131.293
6	55 **Cs** 132.905	56 **Ba** 137.327	71 **Lu** 174.967	72 **Hf** 178.49	73 **Ta** 180.948	74 **W** 183.84	75 **Re** 186.207	76 **Os** 190.23	77 **Ir** 192.217	78 **Pt** 195.078	79 **Au** 196.967	80 **Hg** 200.59	81 **Tl** 204.383	82 **Pb** 207.2	83 **Bi** 208.980	84 **Po** (209)	85 **At** (210)	86 **Rn** (222)
7	87 **Fr** (223)	88 **Ra** (226)	103 **Lr** (262)	104 **Rf** (261)	105 **Db** (262)	106 **Sg** (266)	107 **Bh** (264)	108 **Hs** (269)	109 **Mt** (268)	110 **Ds** (271)	111 **Uuu** (272)	112 **Uub** (285)	113 **Uut**	114 **Uuq** (289)	115 **Uup**	116 **Uuh**	117 **Uus**	118 **Uuo**

Lanthanoids	57 **La** 138.905	58 **Ce** 140.116	59 **Pr** 140.908	60 **Nd** 144.24	61 **Pm** (145)	62 **Sm** 150.36	63 **Eu** 151.964	64 **Gd** 157.25	65 **Tb** 158.925	66 **Dy** 162.500	67 **Ho** 164.930	68 **Er** 167.259	69 **Tm** 168.934	70 **Yb** 173.04
Actinoids	89 **Ac** (227)	90 **Th** (232)	91 **Pa** (231)	92 **U** (238)	93 **Np** (237)	94 **Pu** (244)	95 **Am** (243)	96 **Cm** (247)	97 **Bk** (247)	98 **Cf** (251)	99 **Es** (252)	100 **Fm** (257)	101 **Md** (258)	102 **No** (259)

For each element the average atomic mass of the mixture of isotopes occurring in nature is shown. For elements having no stable isotope, the approximate atomic mass of the longest-lived isotope is shown in parentheses. For elements that have been predicted but not yet detected, no atomic mass is given. All atomic masses are expressed in atomic mass units ($1\ u = 1.66053873(13) \times 10^{-27}$ kg), equivalent to grams per mole (g/mol).

APPENDIX E

Unit Conversion Factors

LENGTH
1 m = 100 cm = 1000 mm = 10^6 μm = 10^9 nm
1 km = 1000 m = 0.6214 mi
1 m = 3.281 ft = 39.37 in.
1 cm = 0.3937 in.
1 in. = 2.540 cm
1 ft = 30.48 cm
1 yd = 91.44 cm
1 mi = 5280 ft = 1.609 km
1 Å = 10^{-10} m = 10^{-8} cm = 10^{-1} nm
1 nautical mile = 6080 ft
1 light year = 9.461×10^{15} m

AREA
1 cm^2 = 0.155 $in.^2$
1 m^2 = 10^4 cm^2 = 10.76 ft^2
1 $in.^2$ = 6.452 cm^2
1 ft = 144 $in.^2$ = 0.0929 m^2

VOLUME
1 liter = 1000 cm^3 = 10^{-3} m^3 = 0.03531 ft^3 = 61.02 $in.^3$
1 ft^3 = 0.02832 m^3 = 28.32 liters = 7.477 gallons
1 gallon = 3.788 liters

TIME
1 min = 60 s
1 h = 3600 s
1 d = 86,400 s
1 y = 365.24 d = 3.156×10^7 s

ANGLE
1 rad = 57.30° = 180°/π
1° = 0.01745 rad = π/180 rad
1 revolution = 360° = 2π rad
1 rev/min (rpm) = 0.1047 rad/s

SPEED
1 m/s = 3.281 ft/s
1 ft/s = 0.3048 m/s
1 mi/min = 60 mi/h = 88 ft/s
1 km/h = 0.2778 m/s = 0.6214 mi/h
1 mi/h = 1.466 ft/s = 0.4470 m/s = 1.609 km/h
1 furlong/fortnight = 1.662×10^{-4} m/s

ACCELERATION
1 m/s^2 = 100 cm/s^2 = 3.281 ft/s^2
1 cm/s^2 = 0.01 m/s^2 = 0.03281 ft/s^2
1 ft/s^2 = 0.3048 m/s^2 = 30.48 cm/s^2
1 mi/h · s = 1.467 ft/s^2

MASS
1 kg = 10^3 g = 0.0685 slug
1 g = 6.85×10^{-5} slug
1 slug = 14.59 kg
1 u = 1.661×10^{-27} kg
1 kg has a weight of 2.205 lb when g = 9.80 m/s^2

FORCE
1 N = 10^5 dyn = 0.2248 lb
1 lb = 4.448 N = 4.448×10^5 dyn

PRESSURE
1 Pa = 1 N/m^2 = 1.450×10^{-4} lb/$in.^2$ = 0.209 lb/ft^2
1 bar = 10^5 Pa
1 lb/$in.^2$ = 6895 Pa
1 lb/ft^2 = 47.88 Pa
1 atm = 1.013×10^5 Pa = 1.013 bar
= 14.7 lb/$in.^2$ = 2117 lb/ft^2
1 mm Hg = 1 torr = 133.3 Pa

ENERGY
1 J = 10^7 ergs = 0.239 cal
1 cal = 4.186 J (based on 15° calorie)
1 ft · lb = 1.356 J
1 Btu = 1055 J = 252 cal = 778 ft · lb
1 eV = 1.602×10^{-19} J
1 kWh = 3.600×10^6 J

MASS–ENERGY EQUIVALENCE
1 kg $\leftrightarrow$ 8.988×10^{16} J
1 u $\leftrightarrow$ 931.5 MeV
1 eV $\leftrightarrow$ 1.074×10^{-9} u

POWER
1 W = 1 J/s
1 hp = 746 W = 550 ft · lb/s
1 Btu/h = 0.293 W

APPENDIX F

Numerical Constants

*FUNDAMENTAL PHYSICAL CONSTANTS**

Name	*Symbol*	*Value*
Speed of light	c	2.99792458×10^8 m/s
Magnitude of charge of electron	e	$1.602176462(63) \times 10^{-19}$ C
Gravitational constant	G	$6.673(10) \times 10^{-11}$ N · m²/kg²
Planck's constant	h	$6.62606876(52) \times 10^{-34}$ J · s
Boltzmann constant	k	$1.3806503(24) \times 10^{-23}$ J/K
Avogadro's number	N_A	$6.02214199(47) \times 10^{23}$ molecules/mol
Gas constant	R	8.314472(15) J/mol · K
Mass of electron	m_e	$9.10938188(72) \times 10^{-31}$ kg
Mass of proton	m_p	$1.67262158(13) \times 10^{-27}$ kg
Mass of neutron	m_n	$1.67492716(13) \times 10^{-27}$ kg
Permeability of free space	μ_0	$4\pi \times 10^{-7}$ Wb/A · m
Permittivity of free space	$\epsilon_0 = 1/\mu_0 c^2$	$8.854187817 \ldots \times 10^{-12}$ $C^2/N \cdot m^2$
	$1/4\pi\epsilon_0$	$8.987551787 \ldots \times 10^9$ $N \cdot m^2/C^2$

*OTHER USEFUL CONSTANTS**

Mechanical equivalent of heat		4.186 J/cal (15° calorie)
Standard atmospheric pressure	1 atm	1.01325×10^5 Pa
Absolute zero	0 K	−273.15°C
Electron volt	1 eV	$1.602176462(63) \times 10^{-19}$ J
Atomic mass unit	1 u	$1.66053873(13) \times 10^{-27}$ kg
Electron rest energy	$m_e c^2$	0.510998902(21) MeV
Volume of ideal gas (0°C and 1 atm)		22.413996(39) liter/mol
Acceleration due to gravity (standard)	g	9.80665 m/s²

*Source: National Institute of Standards and Technology (**http://physics.nist.gov/cuu**). Numbers in parentheses show the uncertainty in the final digits of the main number; for example, the number 1.6454(21) means 1.6454 ± 0.0021. Values shown without uncertainties are exact.

ASTRONOMICAL DATA†

Body	***Mass (kg)***	***Radius (m)***	***Orbit radius (m)***	***Orbit period***
Sun	1.99×10^{30}	6.96×10^{8}	—	—
Moon	7.35×10^{22}	1.74×10^{6}	3.84×10^{8}	27.3 d
Mercury	3.30×10^{23}	2.44×10^{6}	5.79×10^{10}	88.0 d
Venus	4.87×10^{24}	6.05×10^{6}	1.08×10^{11}	224.7 d
Earth	5.97×10^{24}	6.38×10^{6}	1.50×10^{11}	365.3 d
Mars	6.42×10^{23}	3.40×10^{6}	2.28×10^{11}	687.0 d
Jupiter	1.90×10^{27}	6.91×10^{7}	7.78×10^{11}	11.86 y
Saturn	5.68×10^{26}	6.03×10^{7}	1.43×10^{12}	29.45 y
Uranus	8.68×10^{25}	2.56×10^{7}	2.87×10^{12}	84.02 y
Neptune	1.02×10^{26}	2.48×10^{7}	4.50×10^{12}	164.8 y
Pluto	1.31×10^{22}	1.15×10^{6}	5.91×10^{12}	247.9 y

†Source: NASA Jet Propulsion Laboratory Solar System Dynamics Group (**http://ssd.jpl.nasa.gov**), and P. Kenneth Seidelmann, ed., ***Explanatory Supplement to the Astronomical Almanac*** (University Science Books, Mill Valley, CA, 1992), pp. 704–706. For each body, "radius" is its radius at its equator and "orbit radius" is its average distance from the sun (for the planets) or from the earth (for the moon).

PREFIXES FOR POWERS OF 10

Power of ten	***Prefix***	***Abbreviation***	***Pronunciation***
10^{-24}	yocto-	y	*yoc*-toe
10^{-21}	zepto-	z	*zep*-toe
10^{-18}	atto-	a	*at*-toe
10^{-15}	femto-	f	*fem*-toe
10^{-12}	pico-	p	*pee*-koe
10^{-9}	nano-	n	*nan*-oe
10^{-6}	micro-	μ	*my*-crow
10^{-3}	milli-	m	*mil*-i
10^{-2}	centi-	c	*cen*-ti
10^{3}	kilo-	k	*kil*-oe
10^{6}	mega-	M	*meg*-a
10^{9}	giga-	G	*jig*-a or *gig*-a
10^{12}	tera-	T	*ter*-a
10^{15}	peta-	P	*pet*-a
10^{18}	exa-	E	*ex*-a
10^{21}	zetta-	Z	*zet*-a
10^{24}	yotta-	Y	*yot*-a

Examples:

1 femtometer = 1 fm = 10^{-15} m
1 picosecond = 1 ps = 10^{-12} s
1 nanocoulomb = 1 nC = 10^{-9} C
1 microkelvin = 1 μK = 10^{-6} K

1 millivolt = 1 mV = 10^{-3} V
1 kilopascal = 1 kPa = 10^{3} Pa
1 megawatt = 1 MW = 10^{6} W
1 gigahertz = 1 GHz = 10^{9} Hz

Answers to Odd-Numbered Problems

Chapter 1

1.1 1.61 km
1.3 3.34×10^3 ns
1.5 5.36 L
1.7 a) 2330 km/h b) 648 m/s
1.9 35.3 mi/gal
1.11 9.0 cm
1.13 a) $1.1 \times 10^{-3}\%$ b) no
1.15 a) 0.1% b) 0.008% c) 0.03%
1.17 a) $2.8 \pm 0.3\ \text{cm}^3$ b) 170 ± 20
1.19 ten thousand
1.21 10^6
1.23 10^5 (with a full head of hair)
1.25 about $\$10^8$
1.29 $\$9 \times 10^{14}$; about \$3.50 million per person
1.31 7.8 m, 38° north of east
1.33 144 m, 41° south of west
1.35 $A_x = 7.22$ m, $A_y = 9.58$ m; $B_x = 11.5$ m, $B_y = -9.64$ m; $C_x = -3.00$ m, $C_y = -5.20$ m
1.37 1190 N; 13.4° above forward direction
1.39 a) 11.1 m, 77.6° b) 11.1 m, 77.6° c) 28.5 m, 202.3° d) 28.5 m, 22.3°
1.41 5.06 km, 20.2° north of west
1.43 a) 2.48 cm, 18.3° b) 4.10 cm, 83.7° c) 4.10 cm, 263.7°
1.45 $\vec{A} = (7.2\text{ m})\hat{\imath} + (9.6\text{ m})\hat{\jmath}$, $\vec{B} = (11.5\text{ m})\hat{\imath} - (9.6\text{ m})\hat{\jmath}$, $\vec{C} = (-3.0\text{ m})\hat{\imath} - (5.2\text{ m})\hat{\jmath}$
1.47 a) $A = 5.00$, $B = 5.39$ b) $-1.00\hat{\imath} + 5.00\hat{\jmath}$ c) 5.10, 101.3°
1.51 a) 14.00 b) 58.7°
1.53 a) out of the page b) into the page
1.55 $-23.0\hat{k}$; 23.0
1.57 a) 640 acres/mi² b) 43,560 ft²/acre c) 3.26×10^5 gal/acre-ft
1.59 a) 7.04×10^{-10} s b) 5.11×10^{12} cycle/h c) 2.1×10^{26} d) 4.6×10^4 s
1.61 a) 0.5 kg b) 6×10^{-14} g c) 0.3 g
1.63 About 1 dentist per 1,000 population
1.65 90.2 N, 256° counterclockwise from $+x$-axis
1.67 b) $A_x = 3.03$ cm, $A_y = 8.10$ cm c) 8.65 cm, 69.5° measured in the sense from the $+x$-axis toward the $+y$-axis
1.69 144 m, 41° south of west
1.71 b) 1.45 km
1.73 a) (87, 258) b) 136 pixels, 25° below straight left
1.75 380 km, 28.8° south of east
1.77 160 N, 13° below horizontal
1.79 B = 255 m, C = 70 m
1.83 b) 90°
1.85 a) $A = 5.39$, $B = 4.36$ b) $-5.00\hat{\imath} + 2.00\hat{\jmath} + 7.00\hat{k}$ c) 8.83; yes
1.89 a) 54.7° b) 35.3°
1.91 $C_x = 8.0$, $C_y = 6.1$
1.93 b) 72.2
1.95 38.5 yd, 24.6° right of downfield
1.97 a) 76 ly b) 129°

Chapter 2

2.1 a) 197 m/s b) 169 m/s
2.3 1 h 10 min
2.5 286 s; fast: 1770 m; slow: 1570 m
2.7 a) 14.0 m/s b) 11.4 m/s c) case (a)
2.9 a) 12.0 m/s b) i) 0 m/s ii) 15.0 m/s iii) 12.0 m/s c) 13.3 s
2.11 a) (in m/s²) 0, 1.0, 2.0, 2.0, 3.0, 1.5, 1.5, 0; no; yes b) 2.5 m/s², 1.5 m/s², 0
2.13 a) near 5 s b) 30 s to 40 s c) 0 d) -1.7 m/s²
2.15 a) 50.0 cm, 2.00 cm/s, -0.125 cm/s² b) 16.0 s c) 32.0 s d) 6.20 s, 1.22 cm/s; 25.8 s, -1.22 cm/s; 36.4 s, -2.55 cm/s
2.17 a) about 10 m/s²
2.21 a) 5.0 m/s b) 1.43 m/s²
2.23 1.70 m
2.25 a) 1.7 m/s² b) 12 s c) 240 m
2.27 a) speeding up: 6.67 ft/s²; slowing down: -26.5 ft/s² b) 90.4 mph; a_x not constant c) 3.32 s
2.31 a) 0, 6.3 m/s², -11.2 m/s² b) 100 m, 230 m, 320 m
2.33 a) 1.80×10^4 m/s b) 0.957 c) 6 h 11 min
2.35 b) 1 s, 3 s d) 2 s e) 3 s f) 1 s
2.37 b) $d/4$
2.39 a) 2.94 m/s b) 0.599 s
2.41 a) $t = \sqrt{2d/g}$ b) 0.190 s
2.43 a) 5.56 m/s, downward b) 9.80 m/s², downward c) 2.16 s d) 16.1 m/s
2.45 a) 25.6 m/s b) 31.6 m c) 15.2 m/s
2.47 a) 249 m/s² b) 25.4 c) 101 m d) no
2.49 a) 7.5 m b) 180 m c) 20 m
2.51 a) $x(t) = (0.250\text{ m/s}^3)t^3 - (0.0100\text{ m/s}^4)t^4$, $v_x(t) = (0.750\text{ m/s}^3)t^2 - (0.0400\text{ m/s}^4)t^3$ b) 39.1 m/s
2.53 a) 30.0 cm/s
2.55 b) 0.627 s, 1.60 s c) negative at 0.627 s, positive at 1.60 s d) 1.11 s e) 2.45 m f) 2.00 s, 0 s
2.57 a) 82 km/h b) 31 km/h
2.59 a) 3.5 m/s² b) 0 c) 1.5 m/s²
2.61 brake pedal; no
2.63 a) 464 m/s b) 2.99×10^4 m/s c) 7.48
2.65 50.0 m
2.67 4.6 m/s²
2.69 a) 6.17 s b) 24.8 m c) $v_{\text{truck}} = 13.0$ m/s, $v_{\text{auto}} = 21.0$ m/s
2.71 a) 17 m/s b) 1.6 s
2.73 a) 15.9 s b) 393 m c) 29.5 m/s
2.75 a) -4.00 m/s b) 12.0 m/s
2.77 28.6 m
2.79 a) no b) yes, 14.4 m/s; not physically attainable
2.81 a) 13.3 m b) 1.65 s
2.83 a) 7.59 m/s b) 5.14 m c) 1.60 s
2.85 a) 7.7 m/s b) 0.78 s c) 0.59 s d) 1.3 m
2.87 270 m
2.89 a) 20.5 m/s b) yes
2.91 a) 92.2 m b) 75.1 m/s
2.93 a) A b) 2.27 s, 5.73 s c) 1.00 s, 4.33 s d) 2.67 s
2.95 a) 9.55 s, 47.8 m b) 1.62 m/s d) 8.38 m/s e) no f) 3.69 m/s, 21.7 s, 80.0 m
2.97 a) 8.18 m/s b) i) 0.411 m ii) 1.15 km c) 9.80 m/s d) 4.90 m/s

Chapter 3

3.1 a) $(v_{\text{av}})_x = 1.4$ m/s, $(v_{\text{av}})_y = -1.3$ m/s b) 1.9 m/s, $-43°$
3.3 a) 7.1 cm/s, 45° b) 5.0 cm/s, 90°; 7.1 cm/s, 45°; 11 cm/s, 27°
3.5 b) $(a_{\text{av}})_x = -8.67$ m/s², $(a_{\text{av}})_y = -2.33$ m/s² c) 8.98 m/s², 195°
3.7 b) $\vec{v} = \alpha\hat{\imath} - 2\beta t\hat{\jmath}$, $\vec{a} = -2\beta\hat{\jmath}$ c) $v = 5.4$ m/s, $-63°$; $a = 2.4$ m/s², $-90°$ d) speeding up and turning right
3.9 a) 0.600 m b) 0.385 m c) $v_x = 1.10$ m/s, $v_y = -3.43$ m/s; $v = 3.60$ m/s, 72.2° below the horizontal
3.11 2.86 m
3.13 a) 30.6 m/s b) 36.3 m/s
3.15 a) 1.08 s b) 6.18 m, 4.51 m; 11.5 m, 5.74 m; 16.8 m, 4.51 m c) 11.7 m/s, +24.8°; 10.6 m/s, 0°; 11.7 m/s, $-24.8°$ d) parallel: -4.11 m/s², 0, 4.11 m/s²; perpendicular: 8.90 m/s², 9.80 m/s², 8.90 m/s²
3.17 a) 1.4 km b) 8.5 km
3.19 a) 0.682 s, 2.99 s b) 24.0 m/s, 11.3 m/s; 24.0 m/s, -11.3 m/s c) 30.0 m/s, $-36.9°$
3.21 a) 1.5 m b) -0.89 m/s
3.23 a) 13.6 m b) 34.6 m/s c) 103 m
3.25 a) 296 m b) 176 m c) 198 m d) horizontal: 15 m/s; vertical: 58.8 m/s e) horizontal: 15 m/s; vertical: 78.8 m/s
3.27 795 m
3.29 a) 0.034 m/s² = $0.0034g$ b) 1.4 h
3.31 a) 3.07 s b) 1.68 s
3.33 a) 3.50 m/s², upward b) 3.50 m/s², downward c) 12.6 s
3.35 b) no c) where the car is farthest from the center of the ellipse
3.37 a) 14 s b) 70 s
3.39 0.36 m/s, 38° west of south
3.41 a) 4.7 m/s, 25° south of east b) 190 s c) 380 m
3.43 b) -7.1 m/s, -42 m/s c) 43 m/s, 9.5° west of south
3.45 a) $t = 0$ b) $t = 0$ c) $t = 0$, $x = 0$, $y = 15.0$ m; $t = 5.21$ s, $x = 6.25$ m, $y = 1.44$ m d) 6.41 m, 5.21 s
3.47 a) 1.0 m/s² b) 5.4° c) 2.3 m/s d) 31 s
3.49 22 m/s
3.51 40 m/s
3.53 274 m
3.55 a) 42.8 m/s b) 42.0 m
3.57 $4\sqrt{2}D$
3.59 c) less than 45°
3.61 b) 15°, 75°
3.63 a) 17.8 m/s b) in river, 28.4 m from the near bank
3.65 a) 81.6 m b) in cart c) 245 m d) 53.1°
3.67 a) 49 m/s b) 50 m
3.69 a) 2000 m b) 2180 m
3.71 a) 2.23 m b) 3.84 m c) 8.65 m/s d) 3.09 m, 0.62 m
3.73 c) $(18/25)gT^2$ d) $(3/2)gT^2$
3.77 b) $v_x = R\omega(1 - \cos\omega t)$, $v_y = R\omega\sin\omega t$, $a_x = R\omega^2\sin\omega t$, $a_y = R\omega^2\cos\omega t$ c) $t = 0$, $2\pi/\omega, 4\pi/\omega, \ldots$; $x = 0, 2\pi R, 4\pi R, \ldots$; $y = 0$; $a = R\omega^2$ in the $+y$-direction d) no
3.79 30 km
3.81 a) 44.7 km/h, 26.6° west of south b) 10.5° north of west
3.83 a) 0.782 s b) 7.67 m/s c) 5.17 m/s d) 1.04 m
3.85 7.39 m/s, 12.4° north of east
3.87 a) 80 m b) 1.6×10^{-3} c) overall effect is to reduce the radius
3.89 a) $\left(\dfrac{2v_0^2}{g}\right)[\tan(\theta + \phi) - \tan\theta]\dfrac{\cos^2(\theta + \phi)}{\cos\theta}$ b) $\dfrac{\pi}{4} - \dfrac{\theta}{2}$
3.91 $\Delta t = 0.5$ s: 9.589 m/s², 118.6°; $\Delta t = 0.1$ s: 9.983 m/s², 95.73°; $\Delta t = 0.05$ s: 9.996 m/s², 92.86°
3.93 a) 1.5 km/h b) 3.5 km/h

Chapter 4

4.1 a) 0 b) 90° c) 180°
4.3 7.1 N to the right, 7.1 N downward
4.5 494 N, 31.7°
4.7 2.2 m/s²
4.9 16.0 kg
4.11 a) 3.13 m, 3.13 m/s b) 21.9 m, 6.25 m/s
4.13 a) $\Sigma\vec{F} = \mathbf{0}$
4.15 2.94×10^3 N
4.17 a) 4.49 kg b) 4.49 kg, 8.13 N
4.19 825 N, blocks
4.23 7.4×10^{-23} m/s²
4.25 no
4.27 142 N

4.31 1840 N, 135°
4.33 a) 17 N, 90° clockwise from $+x$-direction b) 840 N
4.35 a) 4.8 m/s b) 16 m/s^2 c) 2360 N
4.37 b) 5.83 m/s^2
4.39 a) 2.50 m/s^2 b) 10.0 N c) to the right; F d) 25.0 N
4.41 a) 4.4 m b) 300 m/s c) i) 2.7×10^4 N ii) 9.0×10^3 N
4.43 a) $4m\vec{a}$ b) $3m\vec{a}$ c) $2m\vec{a}$ d) $-m\vec{a}$ e) same magnitudes but opposite directions
4.45 a) 2.93 m/s^2 b) 11.1 m/s^2
4.47 b) 5.9×10^2 N c) 2.2×10^3 N
4.49 b) 3.53 m/s^2 upward c) 120 N d) 93.3 N
4.51 b) $2Mg/3$ c) 5/9
4.53 $(1.7 \times 10^4\ \mathrm{N})\hat{\imath} - (3.4 \times 10^3)\hat{k}$
4.55 $(k_1 t/m)\hat{\imath} + (k_2 t^4/4m)\hat{\jmath}$
4.57 $\vec{r} = \left(\frac{k_1}{2m}t^2 + \frac{k_2 k_3}{120m^2}t^5\right)\hat{\imath} + \left(\frac{k_3}{6m}t^3\right)\hat{\jmath}$

$\vec{v} = \left(\frac{k_1}{m}t + \frac{k_2 k_3}{24m^2}t^4\right)\hat{\imath} + \left(\frac{k_3}{2m}t^2\right)\hat{\jmath}$

Chapter 5

5.1 a) 25.0 N b) 50.0 N
5.3 a) 2.54×10^3 N b) 1.01°
5.5 yes
5.7 a) 5.23×10^4 N b) 3.36×10^4 N
5.9 523 N
5.11 a) 84.9 N b) 60.0 N, 60.0 N
5.13 a) $w \sin\alpha$ b) $2w \sin\alpha$ c) $w \cos\alpha$ for each block
5.15 b) 2.96 m/s^2 c) 191 N; more than the bricks, less than the counterweight
5.17 b) 2.50 m/s^2 c) 1.37 kg d) $T < mg$
5.19 a) 1.78 m/s^2, downward b) 2.14 m/s^2, upward c) yes
5.23 a) 22 N b) 3.1 m
5.25 a) 7.06 N b) 8.14 N c) 1.17 N, 2.25 N
5.27 a) pulled down b) 379 N
5.29 11 times farther
5.31 a) $\mu_k(m_A + m_B)g$ b) $\mu_k m_A g$
5.33 3.82 m/s^2
5.35 a) 0.218 m/s b) 11.7 N
5.37 a) $\mu_k mg/(\cos\theta - \mu_k \sin\theta)$ b) $1/(\tan\theta) = \mu_s$
5.39 b) 8.75 N c) 30.8 N d) 1.54 m/s^2
5.41 a) 0.44 kg/m b) 42 m/s
5.43 26.0 m/s
5.45 a) 75° b) no
5.47 20.7°
5.49 a) 1.49 rev/min b) 0.918 rev/min
5.51 a) 138 km/h b) 3580 N
5.53 2.43 m/s
5.55 a) rope making 60° angle b) 6400 N
5.57 $mg \tan\alpha$
5.59 $F[M + m(1 - x/L)]/(M + m)$
5.61 a) 12.0 N b) 15.0 N
5.63 a) 16.9 N b) 10.2 N
5.65 a) 25.3 s; 5.72×10^5 N b) 4170 m
5.67 5.9 m/s
5.69 2.8 m/s^2
5.71 a) plunge into hole b) 16 m/s
5.73 $\mu_s/(1 + \mu_s)$
5.75 a) 66 N, northward b) 59 N, southward
5.79 3.09×10^3 N
5.81 2.52 N
5.83 a) 12.9 kg b) 47.2 in left-hand cord, 101 N in right-hand cord
5.85 $a_1 = 2m_2 g/(4m_1 + m_2)$, $a_2 = m_2 g/(4m_1 + m_2)$
5.87 1.46 m above the floor
5.89 g/μ_s
5.91 b) 0.450
5.93 a) $(F\cos\alpha)/(m_1 + m_2)$
5.95 a) 21 m/s b) 8.5 m/s
5.97 a) 19 m/s b) 50 m behind you; yes
5.99 a) move up b) remains constant c) remains constant d) stop
5.101 a) 6.00 m/s^2 b) 3.80 m/s^2 c) 7.36 m/s d) 8.18 m/s e) 7.78 m, 6.29 m/s, 1.38 m/s^2 f) 3.14 s
5.103 1/3
5.105 a) $v_y(t) = v_t + (v_0 - v_t)e^{-kt/m}$ b) $v_y(t) = v_t$
5.107 a) 0.015; 0.36 N·s^2/m^2 b) 29 m/s c) ratio is $(\sin\beta - 0.015\cos\beta)^{1/2}$
5.109 a) 120 N b) 3.79 m/s
5.111 b) 0.28 c) no
5.113 a) right b) 120 m
5.115 a) 81.1° b) no c) bead rides at bottom of loop ($\beta = 0$)
5.119 $T_{max} = 2\pi\sqrt{\left(\frac{h\tan\beta}{g}\right)\left(\frac{\sin\beta + \mu_s\cos\beta}{\cos\beta - \mu_s\sin\beta}\right)}$

$T_{min} = 2\pi\sqrt{\left(\frac{h\tan\beta}{g}\right)\left(\frac{\sin\beta - \mu_s\cos\beta}{\cos\beta + \mu_s\sin\beta}\right)}$

5.121 $(M + m)g\tan\alpha$
5.123 a) $F = \frac{\mu_k w}{\cos\theta + \mu_k \sin\theta}$ c) $\theta = \tan^{-1}\mu_k = 14.0°$
5.125 a) $a_3 = g\frac{(-4m_1m_2 + m_2m_3 + m_3m_1)}{(4m_1m_2 + m_2m_3 + m_3m_1)}$ b) $a_B = -a_3$ c) $a_1 = g\frac{(4m_1m_2 - 3m_2m_3 + m_3m_1)}{(4m_1m_2 + m_2m_3 + m_3m_1)}$ d) $a_2 = g\frac{(4m_1m_2 + m_2m_3 - 3m_3m_1)}{(4m_1m_2 + m_2m_3 + m_3m_1)}$ e) $T_A = \frac{1}{2}T_C$ f) $T_C = \frac{8gm_1m_2m_3}{(4m_1m_2 + m_2m_3 + m_3m_1)}$ g) $a_1 = a_2 = a_3 = a_B = 0$, $T_C = 2m_2g$, $T_A = m_2g$; yes
5.127 $\cos^2\beta$

Chapter 6

6.1 a) 3.60 J b) −0.900 J c) 2.70 J
6.3 300 J
6.5 a) 99.2 N b) 387 J c) −387 J d) zero; zero e) zero
6.7 2.6×10^9 J
6.9 a) (i) zero (ii) zero b) (i) zero (ii) −25.1 J
6.11 a) 4.3×10^3 J b) 40 km/h
6.13 $(2gh(1 + \mu_k/\tan\alpha))^{1/2}$
6.15 c) 17.1 m/s
6.17 74.2 J
6.19 a) $-(15/16)K_1$ b) no
6.21 16.8 cm
6.23 a) −28.4 J b) 15.3 m/s c) no
6.25 a) 4.96 m/s b) $a = 1.43$ m/s^2, $v = 4.96$ m/s
6.27 a) $v_0^2/2\mu_k g$ b) 51.3 m
6.29 a) 48.0 N, 64.0 N b) 0.360 J, 0.640 J
6.31 a) 2.8 m/s b) 3.5 m/s
6.33 8.5 cm
6.35 a) 1.76 b) 0.67 m/s
6.37 a) 4.0 J b) zero c) −1.0 J d) 3.0 J e) −1.0 J
6.39 a) 2.83 m/s b) 2.40 m/s
6.41 a) 5.65 cm b) no; 0.57 J
6.43 3.6×10^5 J; 100 m/s
6.45 743 W = 0.995 hp
6.47 0.23
6.49 28
6.51 8.1×10^6 N
6.53 b) 9 c) 108 W
6.55 0.20 W
6.57 a) 532 J b) −315 J c) zero d) −203 J e) 14.7 J f) 1.21 m/s
6.59 a) $1/\sin\alpha$ b) $W_{in} = W_{out}$
6.61 a) 2.59×10^{12} J b) 4800 J
6.63 2.59 m
6.65 11,000 m/s; 1.2×10^{12} J
6.67 a) 4.00 N b) 32.0 N c) −15.0 J
6.69 a) 0.15 N b) 9.4 N c) 0.44 J
6.71 a) 2.56 m/s b) 5.28 N c) 19.7 J
6.73 a) −910 J b) 3.17×10^3 J
6.75 1.0×10^5 N/m
6.77 1.1 m from where the spring is released
6.79 a) 1.02×10^4 N/m, 8.16 m
6.81 a) 0.600 m b) 1.50 m/s
6.83 0.786
6.85 1.5 m
6.87 a) 1.10×10^5 J b) 1.30×10^5 J c) 3.99 kW
6.89 3.6 h
6.91 1.30×10^3 m^3/s
6.93 a) 1.26×10^5 J b) 1.46 W
6.95 a) 2.4 MW b) 61 MW c) 6.0 MW
6.97 a) 513 W b) 355 W c) 52.1 W
6.99 a) 358 N b) 47.2 hp c) 4.06 hp d) 2.03%
6.101 a) $\frac{1}{6}Mv^2$ b) 6.1 m/s c) 3.9 m/s d) $K_{ball} = 0.40$ J, $K_{spring} = 0.60$ J
6.103 a) 2.0×10^5 J b) 2.8×10^5 J c) 2.8×10^5 J d) 5 km/h

Chapter 7

7.1 3.45×10^6 J
7.5 a) 24.0 m/s b) 24.0 m/s c) part (b)
7.9 2.8 m/s
7.11 −5400 J
7.13 a) 880 J b) −157 J c) 471 J d) 253 J e) $a = 3.16$ m/s^2, $v = 7.11$ m/s, $\Delta K = 253$ J
7.15 a) 80.0 J b) 5.00 J
7.17 a) 6.32 cm b) 12 cm
7.19 1.72 m
7.21 a) 0.500 m/s b) 0.500 m
7.23 a) 23.6 m/s b) 69.2 m/s^2, upward
7.25 a) -1.2×10^2 J b) $+1.2 \times 10^2$ J c) zero d) conservative
7.27 a) zero b) −0.10 J c) +0.10 J d) conservative; $U(x) = \frac{1}{3}\alpha x^3$
7.29 a) −3.6 J b) −3.6 J c) −7.2 J d) nonconservative
7.31 a) −59 J b) −42 J c) −59 J d) nonconservative
7.33 $-6C_6/x^7$, attractive
7.35 $\vec{F} = -(2kx + k'y)\hat{\imath} - (k'x + 2ky)\hat{\jmath}$
7.37 a) $F(r) = (12a/r^{13}) - (6b/r^7)$ b) $(2a/b)^{1/6}$, stable c) $b^2/4a$ d) $a = 6.68 \times 10^{-138}$ J·m^{12}, $b = 6.41 \times 10^{-78}$ J·m^6
7.39 a) 229 N b) 823 J
7.41 a) no b) yes, \$150
7.43 0.41
7.45 a) 15.9 J b) 4.0 J c) 3.0 J
7.47 a) 20.0 m from left-hand edge of horizontal section b) −78.4 J
7.49 a) 22.2 m/s b) 16.4 m c) no
7.51 0.602 m
7.53 15.5 m/s
7.55 4.4 m/s
7.57 a) 919 N/m, 39.8 m b) 17.0 m/s^2
7.59 a) 7.00 m/s b) 2.94 N
7.61 a) $mg(1 - h/d)$ b) 4.4×10^2 N c) $\sqrt{2gh(1 - y/d)}$
7.63 48.2°
7.65 a) 0.392 b) −0.832 J
7.67 a) $U(x) = (30.0\ \mathrm{N/m})x^2 + (6.00\ \mathrm{N/m^2})x^3$ b) 7.85 m/s
7.69 7.01 m/s
7.71 a) $m(g + a)^2/2gh$ b) $2gh/(g + a)$
7.73 119 J
7.75 a) 3.87 m/s b) 0.10 m
7.77 a) 3.1 m/s b) 1.4 m c) no
7.79 a) 0.053 gallons b) 19
7.81 3.3×10^{14} J = 9.2×10^7 kWh
7.83 a) −50.6 J b) −67.5 J c) nonconservative
7.85 a) no b) $x_0 = F/k$ d) no e) $3F/k$, $-F/k$ f) $v_{max} = 2F/\sqrt{mk}$ at $x = x_0 = F/k$
7.87 b) $v(x) = \left[\frac{2\alpha}{mx_0^2}\left(\left(\frac{x_0}{x}\right) - \left(\frac{x_0}{x}\right)^2\right)\right]^{1/2}$ c) $x = 2x_0$, $v = \sqrt{\alpha/2mx_0^2}$ d) zero e) $v(x) = \left[\frac{2\alpha}{mx_0^2}\left(\left(\frac{x_0}{x}\right) - \left(\frac{x_0}{x}\right)^2 - \frac{2}{9}\right)\right]^{1/2}$ f) first case: x_0, ∞; second case: $3x_0/2$, $3x_0$

Chapter 8

8.1 a) 1.20×10^5 kg·m/s b) i) 60.0 m/s ii) 26.8 m/s
8.3 b) baseball, 0.525 c) woman, 0.643
8.5 0.256 kg·m/s, $-y$-direction
8.7 562 N, no
8.9 a) 10.8 m/s to the right b) 0.75 m/s to the left
8.11 a) $(18.8\ \text{kg}\cdot\text{m/s})\hat{\imath}$ b) $(-3.55 \times 10^{-3}\ \text{kg}\cdot\text{m/s})\hat{\jmath}$ c) $(7.50 \times 10^3\ \text{N})\hat{\imath}$ d) $(13.0\hat{\imath} - 0.73\hat{\jmath})$ kg·m/s, $(89.3\hat{\imath} - 5.0\hat{\jmath})$ m/s
8.13 a) $At_2 + (B/3)t_2^3$ b) $(A/m)t_2 + (B/3m)t_2^3$
8.15 a) 5.68×10^{-2} m/s b) 0.103 m/s
8.17 a) 4.66 m/s in the direction opposite the defender's original motion b) -6580 J
8.19 a) 3.60 m/s b) 8.64 J
8.21 a) $v_B = (m_A/m_B)v_A$
8.23 2.11 m/s
8.25 525 N
8.27 a) $v_{A2} = 29.3$ m/s, $v_{B2} = 20.7$ m/s b) $0.196 = 19.6\%$
8.29 a) 6.44 m/s, eastbound b) 2.49 m/s c) -2.81×10^5 J, -1.38×10^5 J; part (a)
8.31 0.0300 J; converted to other forms of energy
8.33 35.3 km/h, 19.3° west of south
8.35 229 m/s
8.37 a) Sam: 9.67 m/s; Abigail: 2.26 m/s b) 640 J
8.39 0.150-kg glider: 3.2 m/s to the left; 0.300-kg glider: 0.20 m/s to the left
8.43 a) $9.00m$ b) 3.00×10^6 m/s
8.45 7.42×10^8 m from the center of the sun; outside
8.47 a) 0.30 kg b) $(2.0\ \text{kg}\cdot\text{m/s})\hat{\imath}$ c) $(6.7\ \text{m/s})\hat{\imath}$
8.49 a) 1.25 kg b) $(1.5\ \text{m/s}^3)t\hat{\imath}$ c) $(5.6\ \text{N})\hat{\imath}$
8.51 a) 80.0 N b) yes
8.53 75.0 kg
8.55 a) 0.442 b) 800 m/s c) 530 m/s
8.57 a) 7.2×10^{-66} b) 0.223
8.59 $(0.521\hat{\imath} + 7.81\hat{\jmath})$ m/s
8.61 15
8.63 30° east of north, 30° east of south
8.65 a) 5.00 m/s, east b) 5.71 m/s, east c) 3.78 m/s, east
8.67 a) Nash, 1.68 b) Packard, 0.933 c) F_P is larger, 0.933 d) F_N is larger, 1.68
8.69 23.2 cm
8.71 a) 25.8 m/s, 35.5° b) no
8.73 $R/4$
8.75 65.5 m/s
8.77 a) 0.222 b) -291 J c) 0.784 J
8.79 b) $M = m$ c) zero
8.81 a) 9.35 m/s b) 3.29 m/s
8.83 b) $\frac{1}{2}Mv_{\text{cm}}^2$
8.85 a) 3.56 m/s b) 5.22 m/s c) 4.67 m/s
8.87 0.00544%
8.89 1.61×10^{-22} kg·m/s, to the left
8.91 A: 13.6 m/s; B: 6.34 m/s, 65.0°
8.93 a) $(L/2)\cos(\alpha/2)$, along axis from apex b) $(L/3)$, along bisector from bottom c) $L/\sqrt{8}$ along bisector d) $L/\sqrt{12}$ from each side
8.95 0.400 m/s
8.97 a) 1.40 kg: 14.3 m/s; 0.28 kg: 71.6 m/s b) 347 m
8.99 222 m/s, 1.01×10^3 m/s; $v_{\text{Kr}} = 1.5v_{\text{Ba}}$
8.101 a) zero b) 1 d) 0.87 m f) 0.089 m
8.103 a) yes b) no; kinetic energy decreases by 4.8×10^3 J
8.105 a) $1.37v_{\text{ex}}$ b) $1.18v_{\text{ex}}$ c) $2.38v_{\text{ex}}$ d) 2.94 km/s
8.107 a)

$$v = \begin{cases} (2400\ \text{m/s})\ln\left[\dfrac{1}{1 - t/(120\ \text{s})}\right], & 0 \le t \le 90\ \text{s} \\ 3.33\ \text{km/s}, & t \ge 90\ \text{s} \end{cases}$$

b) $(20\ \text{m/s}^2)/[1 - t/(120\ \text{s})]$ for $0 \le t \le 90$ s c) 6000 N vs. 735 N on earth
8.109 b) $2L/3$
8.111 a) $l^2\lambda g/32$ b) $l^2\lambda g/32$

Chapter 9

9.1 a) 0.600 rad = 34.4° b) 6.27 cm c) 1.05 m
9.3 a) 42 rad/s² b) 74 rad/s
9.5 a) $\omega_z(t) = (0.400\ \text{rad/s}) + (0.0360\ \text{rad/s}^3)t^2$ b) 0.400 rad/s c) $\omega_z = 1.30$ rad/s, $\omega_{\text{av}-z} = 0.700$ rad/s
9.7 a) $\alpha_z(t) = 2b - 6ct$ b) $b/3c$
9.9 a) 300 rpm b) 75.0 s; 312 rev
9.11 a) -1.25 rev/s², 23.3 rev b) 2.67 s
9.13 a) 24.0 s b) 68.8 rev
9.15 10.5 rad/s
9.17 9.00 rev
9.19 a) 540 rad b) 12.3 s c) -8.17 rad/s²
9.21 a) 47.1 m/s b) 47.3 m/s
9.23 a) $a_{\text{rad}} = 18.0$ m/s² b) $v = 3.00$ m/s, $a_{\text{rad}} = 18.0$ m/s²
9.25 a) 0.180 m/s², 0, 0.180 m/s² b) 0.180 m/s², 0.377 m/s², 0.418 m/s² c) 0.180 m/s², 0.754 m/s², 0.775 m/s²
9.27 10.7 cm; no
9.29 a) 0.831 m/s b) 109 m/s²
9.31 a) 2.29 b) 1.51 c) 15.7 m/s, 1.06×10^3 m/s² = $108g$
9.33 2.99 cm
9.35 $(M/12 + m/2)L^2$
9.37 a) 0.0640 kg·m² b) 0.0320 kg·m² c) 0.0320 kg·m²
9.39 8.52 kg·m²
9.41 0.193 kg·m²
9.43 b) $K = \pi^2 I\omega^2/1800$
9.45 0.600 kg·m²
9.47 7.35×10^4 J
9.49 a) $K = 2\pi^2 I/T^2$ b) $dK/dt = (-4\pi^2 I/T^3)(dT/dt)$ c) 70 J d) -0.56 J/s
9.51 823 J
9.53 an axis parallel to a diameter and $(2/\sqrt{15})R$ from the center of the sphere
9.55 $\frac{1}{3}M(a^2 + b^2)$
9.57 $\frac{1}{3}M(L^2 - 3Lh + 3h^2)$
9.59 $\frac{1}{3}ML^2$
9.63 a) 840 rpm b) 75 mph c) 60 mph
9.65 a) $(6.40\ \text{rad/s}^2)t - (1.50\ \text{rad/s}^3)t^2$ b) $6.40\ \text{rad/s}^2 - (3.00\ \text{rad/s}^3)t$ c) $\omega_{\text{max}} = 6.83$ rad/s at $t = 2.13$ s
9.67 a) 35.0 km/h = 9.72 m/s b) 8.51 J c) 652 rad/s
9.69 a) 211 rev/min b) 800 W
9.71 a) 1.70 m/s b) 84.8 rad/s
9.73 b) 2.00 m/s² d) 0.208 kg·m²
9.77 a) 7.36 m b) 327 m/s²
9.79 a) 2.14×10^{29} J b) 2.66×10^{33} J
9.81 a) $Mb^2/6$ b) 182 J
9.83 a) -0.784 J b) 5.42 rad/s c) 5.42 m/s d) particle speed = 4.43 m/s
9.85 $\sqrt{(2gd)(m_B - \mu_k m_A)/(m_A + m_B + I/R^2)}$
9.87 $\sqrt{(g/R)(1 - \cos\beta)}$
9.89 a) 2.25×10^{-3} kg·m² b) 3.40 m/s c) 4.95 m/s
9.91 7.23 m
9.93 a) $(247/512)MR^2$ b) $(383/512)MR^2$
9.95 b) $\frac{1}{4}M(R_1^2 + R_2^2)$
9.97 a) $\frac{3}{5}MR^2$ b) larger
9.99 b) 5.97×10^{24} kg c) $0.334MR^2$
9.101 a) $s = r_0\theta + \beta\theta^2/2$

$$\text{b) } \theta = \left(\frac{1}{\beta}\right)\left[\sqrt{r_0^2 + 2\beta vt} - r_0\right]$$

$$\text{c) } \omega_z = \frac{v}{\sqrt{r_0^2 + 2\beta vt}},\ \alpha_z = -\frac{\beta v^2}{(r_0^2 + 2\beta vt)^{3/2}};$$

no d) $r_0 = 2.50$ cm, $\beta = 0.247\ \mu$m/rad, 2.13×10^4 rev

Chapter 10

10.1 a) 40.0 N·m, out of the page b) 34.6 N·m, out of the page c) 20.0 N·m, out of the page d) 17.3 N·m, into the page e) zero f) zero
10.3 2.50 N·m, counterclockwise
10.5 c) $(-1.05\ \text{N}\cdot\text{m})\hat{k}$
10.7 0.0524 N·m
10.9 1.2 m/s
10.11 a) $g(M + 3m)/(1 + 2m/M)$ b) less c) no effect
10.13 0.482
10.15 7.47 N
10.17 a) 7.5 N in horizontal part, 18.2 N in hanging part b) 0.0160 kg·m²
10.19 a) 0.882 N b) 0.553 s c) 33.9 rad/s
10.21 a) 1/3 b) 2/7 c) 2/5 d) 5/13
10.23 a) 0.613 b) no, requires $\mu_s = 0.858$ c) no slipping
10.25 11.7 m
10.27 a) 0.309 rad/s b) 100 J c) 6.67 W
10.29 a) 0.38 N·m b) 160 rad c) 59 J d) 59 J
10.31 b) 65.6 N
10.33 a) 358 N·m b) 1.79×10^3 N c) 83.8 m/s
10.35 a) 115 kg·m²/s, into the page b) 125 kg·m²/s², out of the page
10.37 4.71×10^{-6} kg·m²/s
10.39 b) 7.00 rad/s c) 1.03×10^{-2} J d) 1.03×10^{-2} J
10.41 0.60 rev
10.43 a) 1.38 rad/s b) 1080 J before, 499 J after
10.45 a) 0.120 rad/s b) 3.20×10^{-4} J; work done by bug
10.49 a) 36.8 min b) 1.10×10^5 N·m
10.51 5.4×10^{22} N·m
10.53 a) 0.955 kg·m² b) -0.0800 N·m c) 104 rev
10.55 b) 4500 W d) 2600 W
10.57 0.675 s
10.59 a) $L/4$ from the end with the clay b) $(9g/8L)\sin\theta$ c) $(3g/2L)\sin\theta$
10.61 1200 N
10.63 a) 1.12 m/s² b) 14.0 N
10.65 $a = F/2M$, $f = F/2$
10.67 a) 0.957 m
10.69 a) $(27R - 17r)/10$ b) $(5R - 3r)/2$
10.71 a) 1.76 N b) 123 rad/s² c) 9.80 m/s² d) T would be unchanged, α and a would double
10.73 a) 23 m c) 16 m; 0.7 times smaller
10.75 a) 36.5 m, 28.0 m/s b) no; total energy is same but less of it is rotational kinetic energy
10.77 a) 2.07 m/s b) 0.50 rev/s c) 1.50 rev/s
10.79 a) 9.34 m/s b) 18.7 m/s c) zero d) 5.60 m
10.81 3.4 m
10.83 $g/3$
10.85 1.87 m
10.87 a) $6v/19L$ b) 3/19
10.89 a) 5.46 rad/s b) 3.17 cm c) 1.01×10^3 m/s
10.91 3200 J
10.93 0.30 rad/s, clockwise
10.97 12.7 rad/s
10.101 a) $a = +\mu_k g$, $\alpha = -2\mu_k g/R$ b) $\omega_0^2 R^2/18\mu_k g$ c) $-M\omega_0^2 R^2/6$
10.103 a) $mv_1^2 r_1^2/r^3$ b) $(mv_1^2/2)[(r_1/r_2)^2 - 1]$ c) same

Chapter 11

11.1 0.387 m from center of 1.00-kg sphere
11.5 5450 N
11.7 1000 N, 1.20 m from end where 400-N force is applied
11.11 a) 1920 N b) 1140 N
11.13 a) $T = 2.60w$; $F_{\text{pivot}} = 3.28w$ at 37.6° b) $T = 4.10w$; $F_{\text{pivot}} = 5.38w$ at 48.8°
11.15 140 N by each hinge
11.17 246 N; 0.34 m behind front feet
11.19 $T_{\text{left}} = 270$ N, $T_{\text{right}} = 304$ N, $\theta = 40°$
11.21 a) 0.800 m b) clockwise c) 0.800 m, clockwise
11.23 1.4 mm
11.25 2.0×10^{11} Pa
11.27 a) upper: 3.1×10^{-3}; lower: 2.0×10^{-3} b) upper: 1.6 mm; lower: 0.98 mm
11.29 9.1×10^6 N
11.31 a) 3.33×10^6 Pa b) 1.33×10^5 N
11.33 4.8×10^9 Pa, 2.1×10^{-10} Pa^{-1}
11.35 1/9

11.37 3.41×10^7 Pa
11.39 10.2 m/s^2
11.43 wing force: 7300 N upward; tail force: 600 N downward
11.45 a) 590 N b) neutral
11.49 120 N to the right, 160 N upward
11.51 a) fore: 37 N; hind: 58 N b) fore: 53 N; hind: 55 N
11.53 a) $\tau = Fl \sin\phi$ b) 12.6 N·m with either method of calculation
11.55 a) $V = mg + w$, $H = T = (w + mg/4)\cot\theta$ b) 9.50×10^2 N c) 4.00°
11.57 760 N
11.59 a) 2700 N b) 19
11.61 a) 4.90 m b) 60 N
11.63 a) 300 N upward, 225 N·m clockwise b) $T = 132$ N; $F_{pivot} = 194$ N, 71.2° above horizontal
11.65 a) 60.0 N b) 53.1°
11.67 person above: 590 N; person below: 1370 N; above
11.69 a) $w_{max} = T_{max}hD/L\sqrt{h^2 + D^2}$
11.71 a) 7140 N; tall walls b) 7900 N
11.73 a) 268 N b) 232 N c) 366 N
11.75 a) A: 0.424 N; B: 1.47 N; C: 0.424 N b) 0.848 N
11.77 a) tips at 27°, slips at 31°; the bale tips before it slips b) tips at 27°, slips at 22°; the bale slips before it tips
11.79 a) $F_A = 80$ N, $F_B = 870$ N b) 1.92 m
11.81 a) 3.7 kN, 2.0 kN vertically upward
11.83 a) 5.4 mm b) 4.2 mm
11.85 a) 0.70 m from wire *A* b) 0.45 m from wire *B*
11.87 a) 1.63 m b) brass: 2.00×10^8 Pa; nickel: 4.00×10^8 Pa c) brass: 2.2×10^{-3}; nickel: 1.9×10^{-3}
11.89 a) 0.36 mm b) 0.045 mm c) 0.33 mm
11.91 a) $(F\cos^2\theta)/A$ b) $(F\sin 2\theta)/2A$ c) 0 d) 45°
11.93 a) 600 N b) 13.5 kN c) slide: $\mu_s w/(\sin\theta - \mu_s\cos\theta)$; tip: $w/[(1/9)\cos\theta + 2\sin\theta]$; 66°
11.95 the lesser of $h^2/L + L/2$ and L
11.97 $[(A^2x/F) - k_O V_O]/V_S$
11.99 a) 0.662 mm b) 2.20×10^{-2} J c) 8.33×10^{-3} J d) -3.04×10^{-2} J e) 3.04×10^{-2} J

Chapter 12

12.1 2.18
12.3 F_{12}
12.5 a) 2.59×10^8 m
12.7 2.4×10^{-3} N
12.9 a) 6.30×10^{20} N, toward sun b) 4.77×10^{20} N, 24.6° toward earth from sun c) 2.37×10^{20} N, toward sun
12.11 1.6×10^{-3} N, to the left
12.13 $2.1 \times 10^{-9} \text{ m/s}^2$, down
12.15 1.38×10^7 m
12.17 a) 0.37 m/s^2 b) 1700 kg/m^3
12.19 610 N; 83% of weight at surface
12.21 5.98×10^{24} kg
12.23 0.83 m/s; yes
12.25 a) 5.02×10^3 m/s b) 6.06×10^4 m/s
12.27 a) 175 min b) 3.71 m/s^2
12.29 7.46×10^3 m/s
12.31 4.7 m/s; yes
12.33 a) 8.3×10^4 m/s b) 1.3×10^6 s
12.35 a) 4.45×10^{12} m, 4.55×10^{12} m b) 248 y
12.39 a) $-GMm/\sqrt{a^2 + x^2}$ c) $GMmx/(a^2 + x^2)^{3/2}$ toward the center of the ring e) $-GMm/a$, zero
12.41 a) 53 N b) 52 N
12.43 a) 4.3×10^{37} kg $= 2.1 \times 10^7 M_{sun}$ b) no c) 6.3×10^{10} m; yes
12.45 1.39×10^{-9}
12.47 a) 2.20×10^{-10} N, 163° from $+x$-direction toward $+y$-direction b) $x = 0$, $y = 1.39$ m
12.49 a) 3.40 N, 0.61° b) 1.31×10^9 J
12.51 3.83×10^8 m
12.53 a) 3.58×10^7 m
12.55 1.8×10^2 m/s
12.57 a) 1.39×10^7 m b) 3.59×10^7 m
12.59 $0.01R_E = 64$ km
12.61 0.28%
12.63 6.06×10^3 km/h
12.65 $\sqrt{2Gm_E h/(R_E^2 + hR_E)}$
12.67 a) 13.7 km/s b) 13.3 km/s c) 13.2 km/s
12.69 a) i) 2.8 y ii) 6.1 y b) 4.90×10^8 km c) 4.22×10^8 km
12.71 a) $GM^2/4R^2$ b) $\sqrt{GM/4R}$, $4\pi\sqrt{R^3/GM}$ c) $GM^2/4R$
12.73 6.8×10^4 m/s
12.75 a) 7.91×10^3 s b) 1.53 c) 8.43×10^3 m/s, 5.51×10^3 m/s d) 2.41×10^3 m/s, 3.26×10^3 m/s; perigee
12.77 3.22×10^9 J
12.79 1.4×10^{14} m
12.81 $2\pi GmM/L^2$, toward center of semicircle
12.83 $GMm/(x^2 + xL)$
12.85 b) 22 min c) 2.5×10^8 s = 2900 d = 7.9 y
12.87 a) 2.1 kN b) no
12.89 perpendicular: $GMm/(a\sqrt{L^2 + a^2})$; parallel: zero

Chapter 13

13.1 a) 4.54×10^{-3} s, 1.38×10^3 rad/s b) 1.14×10^{-3} s, 5.53×10^3 rad/s
13.3 5.53×10^3 rad/s, 1.14×10^{-3} s
13.5 0.0500 s
13.7 a) 0.167 s b) 37.7 rad/s c) 8.44×10^{-2} kg
13.9 a) 0.375 s b) 2.66 Hz c) 16.7 rad/s
13.11 a) $x = (3.0 \text{ mm})\cos[2\pi(440 \text{ Hz})t]$ b) 8.3 m/s, $2.3 \times 10^4 \text{ m/s}^2$ c) jerk $= (6.3 \times 10^7 \text{ m/s}^3)\sin[2\pi(440 \text{ Hz})t]$, $6.3 \times 10^7 \text{ m/s}^3$
13.13 a) 0.383 m b) 1.02 rad = 58.5° c) $x = (0.383 \text{ m})\cos[(12.2 \text{ rad/s})t + 1.02 \text{ rad}]$
13.15 0.353 m
13.17 2.09 s
13.19 a) 1.51 s b) 26.0 N/m c) 0.308 m/s d) 1.92 N e) −0.0125 m; 0.303 m/s; 0.216 m/s^2
13.23 a) $A/\sqrt{2}$, $\omega A/\sqrt{2}$ b) four times; $\pi/2\omega$ c) 3/4 kinetic, 1/4 potential
13.25 a) 5.13 m/s^2, 0.961 m/s b) -2.57 m/s^2, 0.833 m/s c) 0.137 s d) v, a can be found with energy approach; t cannot
13.27 a) 2.84×10^{-2} J b) 0.0138 m c) 0.615 m/s
13.29 0.509 m/s
13.31 a) 5.31×10^3 N/m b) 0.695 s
13.33 3.97 cm
13.35 a) 2.7×10^{-8} kg·m² b) 4.3×10^{-6} N·m/rad
13.37 5.12×10^{-2} kg·m²
13.41 2.60 s
13.43 2.00 m
13.45 10.7 m/s^2
13.49 0.58 s
13.51 a) 0.393 Hz b) 1.73 kg/s
13.53 a) A b) magnitude $= bA/2m$, in $-x$-direction; slope is negative c) $a(0) = A\left(\dfrac{b^2}{2m^2} - \dfrac{k}{m}\right)$; if $b < \sqrt{2mk}$, $a(0) < 0$; if $b = \sqrt{2mk}$, $a(0) = 0$; if $b > \sqrt{2mk}$, $a(0) > 0$
13.55 a) kg/s c) i) $5.0F_{max}/k$ ii) $2.5F_{max}/k$
13.57 a) $6.72 \times 10^3 \text{ m/s}^2$ b) 3.02 kN c) 18.3 m/s, 75.6 J d) 17.6 kW e) 12.1 kN, 36.7 m/s, 302 J, 141 kW
13.59 a) all unchanged b) 1/4 as large c) halved d) $1/\sqrt{5}$ as great e) U: unchanged; K: 1/5 as large
13.61 a) 1.68 s b) 0.090 m c) 0.143
13.63 $\mu_s(m + M)g/k$
13.65 a) $A_1/\sqrt{2}$; $T_1/\sqrt{2}$ b) A_1; $T_1/\sqrt{2}$
13.69 a) 0.150 m/s b) 0.112 m/s^2, downward c) 0.700 s d) 4.38 m
13.71 a) 2.6 m/s b) 0.21 m c) 0.49 s
13.73 9.08×10^{24} kg
13.75 1.17 s
13.77 $t_1 = T/4$
13.83 d) $\sqrt{A/\pi^2 m R_0^3}$
13.85 $2\pi\sqrt{3M/2k}$
13.87 2.74 s; smaller
13.89 0.88 m
13.91 a) 3.97 m
13.93 c) 0.430 m
13.95 a) spring 1: 0.350 m; spring 2: 0.250 m b) 0.702 s
13.97 a) $\Delta T/T = -\frac{1}{2}(\Delta g/g)$ b) 9.7791 m/s^2
13.99 a) $T/T_0 = \sqrt{(L^2 + 3y^2)/(L^2 + 2yL)}$ b) $y = 2L/3$
13.101 a) $kl_0/(k - m\omega'^2)$

Chapter 14

14.1 $w = 41.8$ N; no
14.3 $7.02 \times 10^3 \text{ kg/m}^3$; yes
14.5 1.6
14.7 9.91 m
14.9 a) 706 Pa b) 3160 Pa
14.11 a) 2.52×10^6 Pa b) 1.78×10^5 N
14.13 a) 1.07×10^5 Pa b) 1.03×10^5 Pa c) 1.03×10^5 Pa d) 5.33×10^3 Pa
14.15 816 cm^3
14.17 2.3×10^5 N
14.19 4.14 m
14.21 0.562 m^3
14.23 a) $\rho < \rho_{fluid}$ c) submerged: ρ/ρ_{fluid}; above: $(\rho_{fluid} - \rho)/\rho_{fluid}$ d) 32%
14.25 a) 116 Pa b) 921 Pa c) 0.822 kg, 822 kg/m^3
14.27 a) larger for B b) same c) larger for B
14.29 9.6 m/s
14.31 a) 17.0 m/s b) 0.317 m
14.33 28.4 m/s
14.35 1.47×10^5 Pa
14.37 500 N downward
14.39 0.41 cm
14.41 2.25×10^5 Pa
14.43 a) 1.10×10^8 Pa b) 1080 kg/m^3, 5%
14.45 2.61×10^4 N·m
14.47 $(p - p_0)VR^2/Gmd$
14.49 a) $12{,}700 \text{ kg/m}^3$, 3140 kg/m^3
14.51 a) 1470 Pa b) 13.9 cm
14.53 9.8×10^6 kg, yes
14.55 a) 30% b) 70%
14.57 $4.66 \times 10^{-4} \text{ m}^3$, 5.27 kg
14.59 a) $1.10 \times 10^4 \text{ m}^3$ b) 112 kN
14.61 a) 0.107 m b) 2.42 s
14.63 0.0958 kg
14.65 33.5 N
14.67 b) 12.2 N c) 11.8 N
14.69 b) $2.52 \times 10^{-4} \text{ m}^3$, 0.124
14.71 risen by 5.57×10^{-4} m
14.73 a) $1 - \rho_B/\rho_L$ b) $(\rho_L - \rho_B)L/(\rho_L - \rho_W)$ c) 4.60 cm
14.75 a) la/g b) $\omega^2 l^2/2g$
14.79 7.0 N·m
14.81 13.1 cm
14.83 133 m/s
14.85 $3h_1$
14.87 a) $r = r_0(1 + 2gy/v_0^2)^{-1/4}$ b) 1.10 m
14.89 a) 80.4 N
14.91 a) $\sqrt{2gh}$ b) $(p_a/\rho g) - h$

Chapter 15

15.1 a) 1.2 m/s b) 0.31 m c) a): unchanged; b): 0.15 m
15.3 a) 0.439 m b) 5.25×10^6 Hz
15.5 a) 17.2 m, 0.0172 m b) 74.0 m, 0.0740 m
15.7 a) $f = 25.0$ Hz, $T = 0.0400$ s, $k = 19.6$ rad/m b) $y(x, t) = (0.0700 \text{ m})\cos 2\pi\left(\dfrac{x}{0.320 \text{ m}} + \dfrac{t}{0.0400 \text{ s}}\right)$ c) +0.0495 m d) 0.0050 s
15.9 c) $-x$-direction for both d) $v_y(x, t) = \omega A\cos(kx + \omega t)$, $a_y(x, t) = -\omega^2 A\sin(kx + \omega t)$
15.11 a) 4.0 mm b) 0.040 s c) 0.14 m, 3.6 m/s d) 0.24 m, 6.0 m/s e) no
15.13 b) $+x$-direction
15.15 a) 16.3 m/s b) 0.136 m c) both increase by $\sqrt{2}$

15.17 0.390 s
15.19 a) 10.0 m/s b) 0.250 m c) $y(x, t) = (3.00\text{ cm})\cos[(8.00\pi\text{ rad/m})x - (80.0\pi\text{ rad/s})t]$ d) 1890 m/s^2 e) yes
15.23 a) 0.050 W/m^2 b) 2.2×10^4 J
15.25 707 W
15.31 a) $(1.33\text{ m})n$, $n = 0, 1, 2, 3, \ldots$ b) $(1.33\text{ m})(n + 1/2)$, $n = 0, 1, 2, 3, \ldots$
15.33 a) $y(x, t) = (0.850\text{ cm})\sin[2\pi x/(0.300\text{ m})] \times \sin[2\pi t/(0.0750\text{ s})]$ b) 4.00 m/s c) 0.688 cm
15.37 a) 96.0 m/s b) 461 N c) 1.13 m/s, 426 m/s^2
15.39 b) 2.80 cm c) 277 cm d) 185 cm, 0.126 s, 1470 cm/s e) 280 cm/s f) $y(x, t) = (5.60\text{ cm})\sin[(0.0907\text{ rad/cm})x] \times \sin([133\text{ rad/s}]t)$
15.41 a) $y(x, t) = (4.60\text{ mm})\sin[(6.98\text{ rad/m})x] \times \sin([742\text{ rad/s}]t)$ b) 3rd c) 39.4 Hz
15.43 a) 45.0 cm b) no
15.45 a) 3.00 m, 16.0 Hz b) 1.00 m, 48.0 Hz c) 0.750 m, 64.0 Hz
15.47 a) 20.0 Hz, 126 rad/s, 3.49 rad/m b) $y(x, t) = (2.50\text{ mm}) \times \cos[(3.49\text{ rad/m})x - (126\text{ rad/s})t]$ c) $y(t) = (2.50\text{ mm})\cos[(126\text{ rad/s})t]$ d) $y(t) = (2.50\text{ mm}) \times \cos[3\pi/2 - (126\text{ rad/s})t]$ e) 0.314 m/s f) -2.50 mm, 0
15.49 a) $(7L/2)\sqrt{\mu_1/F}$ b) no
15.51 a) $(2\pi A/\lambda)\sqrt{FL/M}$ b) increase by factor of 4
15.53 a) $4\pi^2 F\Delta x/\lambda^2$
15.55 a) 1, 0; 2, +; 3, −; 4, 0; 5, −; 6, + b) 1, −; 2, +; 3, −; 4, +; 5, −; 6; 0 c) (a): answers would reverse sign; (b): no change
15.59 c) C/B
15.61 b) k decreases by factor of $2\sqrt{2}$; ω decreases by factor of $\sqrt{2}$
15.63 a) 7.07 cm b) 400.0 W
15.65 $\alpha = (v_1^2 - v_2^2)\rho/Y\Delta T$
15.67 $n(0.800\text{ Hz})$, $n = 1, 2, 3, \ldots$
15.69 c) yes
15.71 c) $2A$, $2A\omega$, $2A\omega^2$
15.73 a) 20.0 m/s b) four times per second
15.75 a) 0, L b) 0, $L/2$, L d) no
15.77 a) 148 N b) 26%
15.79 b) $(1/2)\mu\omega^2 A^2 \sin^2(kx - \omega t)$ e) $(1/2)Fk^2A^2\sin^2(kx - \omega t)$
15.81 $\pi/\omega\sqrt{2}$
15.83 a) 99.4 N c) -4.25 Hz, falls

Chapter 16

16.1 a) 0.344 m b) 1.2×10^{-5} m c) 6.9 m, 50 Hz
16.3 a) 7.78 Pa b) 77.8 Pa c) 778 Pa
16.5 a) 1.33×10^{10} Pa b) 9.47×10^{10} Pa
16.7 90.8 m
16.9 a) -82°C b) no
16.11 0.208 s
16.13 $Y/900$
16.15 a) 9.44×10^{-11} m; 0.434 m b) 5.66×10^{-9} m; 0.101 m c) air; $A_{air}/A_{water} = 60.0$
16.17 a) 1.94 Pa b) 4.58×10^{-3} W/m^2 c) 96.6 dB
16.19 4.36×10^{-12} W/m^2, 6.40 dB
16.21 14.0 dB
16.23 a) 20.0
16.25 a) *fundamental:* displacement node at 0.60 m, pressure nodes at 0 and 1.20 m; *first overtone:* displacement nodes at 0.30 m and 0.90 m, pressure nodes at 0, 0.60 m, and 1.20 m; *second overtone:* displacement nodes at 0.20 m, 0.60 m, and 1.00 m, pressure nodes at 0, 0.40 m, 0.80 m, and 1.20 m b) *fundamental:* displacement node at 0, pressure node at 1.20 m; *first overtone:* displacement nodes at 0 and 0.80 m, pressure nodes at 0.40 m and 1.20 m; *second overtone:* displacement nodes at 0, 0.48 m, and 0.96 m, pressure nodes at 0.24 m, 0.72 m, and 1.20 m
16.27 506 Hz, 1520 Hz, 2530 Hz
16.29 a) 614 Hz b) 1.23 kHz
16.31 a) 172 Hz b) 86 Hz
16.33 0.125 m
16.35 destructive
16.37 a) 4 beats/s b) 3.0×10^{-8} m, 0
16.39 267 Hz
16.41 a) 0.798 m b) 0.922 m c) 431 Hz d) 373 Hz
16.43 a) 2000 Hz b) 1500 Hz
16.45 a) 302 Hz b) 228 Hz
16.49 a) 0.0114 Pa b) 7.51×10^{-9} m c) 63.0 m
16.51 flute harmonic $3N$ ($N = 1, 3, \ldots$) resonates with string harmonic $4N$
16.53 a) stopped b) $n = 7$, $n = 9$ c) 43.9 cm
16.55 a) $v/2L$, v/L, $3v/2L$
16.57 a) 375 m/s b) 1.39 c) 0.8 cm
16.59 $n(77.3\text{ Hz})$, $n = 1, 2, 3, \ldots$
16.61 1.27
16.63 a) 548 Hz b) 652 Hz
16.65 a) $I = 2\pi^2\sqrt{\rho B}f^2(\Delta R)^2$ b) $P = 8\pi^3\sqrt{\rho B}f^2R^2(\Delta R)^2$ c) $A = (R/d)\Delta R$, $p_{max} = 2\pi\sqrt{\rho B}(Rf/d)\Delta R$, $I = 2\pi^2\sqrt{\rho B}(fR/d)^2(\Delta R)^2$
16.67 a) 6.74 cm b) 147 Hz
16.69 b) 2.0 m/s
16.71 c) 56.8 m/s
16.73 a) greater, 109 Hz b) greater, 218 Hz
16.77 a) 180° b) A: 3.98×10^{-6} W/m^2, 66.0 dB; B: 5.31×10^{-7} W/m^2, 57.2 dB c) 1.60×10^{-6} W/m^2, 62.0 dB

Chapter 17

17.1 a) -81.0°F b) 134.1°F c) 88.0°F
17.3 140.2°F
17.5 a) 104.4°F, yes b) 54°F
17.7 B
17.9 a) 216.5 K b) 325.9 K c) 205.4 K
17.11 27.23 K
17.13 444 mm Hg
17.15 a) -282°C b) no, 4.76×10^4 Pa
17.17 0.39 m
17.19 a) 1.9014 cm b) 1.8964 cm
17.21 2.3×10^{-5} (C°)$^{-1}$
17.23 11 L
17.25 1.7×10^{-5} (C°)$^{-1}$
17.27 a) 1.431 cm^2 b) 1.437 cm^2
17.29 0.008 cm
17.31 a) 3.2×10^{-5} (C°)$^{-1}$ b) 2.5×10^9 Pa
17.33 a) 38 J b) 4.5×10^4 J
17.35 0.526 C°
17.37 3.03×10^5 J
17.39 5.79×10^5 J
17.41 a) 2.51×10^3 J/kg·K b) overestimate
17.43 a) 114 C° b) 6.35 C°
17.45 a) 215 J/kg·K b) water c) too small
17.47 8 min
17.49 3.64×10^4 J $= 8.69 \times 10^3$ cal $= 34.5$ Btu
17.51 2.39×10^4 Btu/h $= 7.01 \times 10^3$ W
17.53 357 m/s
17.55 3.45 L
17.57 27.5°C
17.59 0.0940 kg
17.61 2.10 kg
17.63 190 g
17.65 a) 222 K/m b) 10.7 W c) 73.3°C
17.67 a) -5.8°C b) 11 W/m^2
17.69 7.1×10^2 Btu $= 7.5 \times 10^5$ J
17.71 105.5°C
17.73 90.2°C
17.75 167 W
17.77 2.1 cm^2
17.79 a) 35.2°M b) 39.6 C°
17.81 b) 5.0×10^{-3} cm^3, -23 kg/m^3
17.83 37.5°C
17.85 415°C
17.87 23.0 cm, 7.0 cm
17.89 b) 1.9×10^8 Pa
17.91 a) 87°C b) -80°C
17.93 20.2°C
17.95 a) 54.3
17.97 a) 83.6 J b) 1.86 J/mol·K c) 5.60 J/mol·K
17.99 a) 2.7×10^7 J b) 6.89 K c) 19.3 K
17.101 a) 4.7×10^{-3} kg b) yes
17.103 a) 86.1°C b) no ice, 0.130 kg liquid water, no steam
17.105 a) 100°C b) 0.0214 kg steam, 0.219 kg liquid water
17.107 a) 100°C b) no ice, 0.419 kg liquid water, 0.0264 kg steam c) 0°C, 0.154 kg ice, 0.408 kg liquid water, no steam
17.109 a) 94 W b) 1.3
17.111 2.9
17.113 c) 6.0×10^5 s (about 170 h) d) 1.5×10^{10} s (about 500 y); no
17.115 305 min
17.117 5.82×10^{-3} kg
17.119 a) 69.6°C
17.121 a) $H = 4\pi kab\Delta T/(b - a)$ b) $T(r) = \dfrac{[T_1 b(r - a) + T_2 a(b - r)]}{r(b - a)}$ c) $H = 2\pi kL\Delta T/\ln(b/a)$ d) $T(r) = \dfrac{[T_1\ln(r/a) + T_2\ln(b/r)]}{\ln(b/a)}$
17.123 b) uniform 0°C d) 3.14×10^3 C°/m e) 121 W f) 0 g) 1.1×10^{-4} m^2/s h) -11 C°/s i) 9.1 s j) decrease k) -7.7 C°/s
17.125 20.42 K

Chapter 18

18.1 a) 56.2 mol b) 6.81×10^6 Pa $= 67.2$ atm
18.3 0.959 atm
18.5 a) 3×10^{27} molecules b) 3×10^{19} molecules/cm^3
18.7 503.0°C
18.9 3.36×10^5 Pa
18.11 0.159 L
18.13 1.05 atm
18.15 a) 70.2°C b) yes
18.17 850 m
18.19 density at sea level 1.2% larger
18.21 2.28×10^4 Pa
18.23 21.4 cm^3
18.25 a) 8.2×10^{-17} atm b) no
18.27 55.6 mol, 3.35×10^{25} molecules
18.29 a) 9.00×10^{-5} m^3 b) 3.10×10^{-10} m c) about the same
18.31 b) 1.004
18.33 a) could be true b) could be true c) not true d) must be true e) could be true
18.35 a) 1.9×10^6 m/s; no, 0.64% of c b) 7.3×10^{10} K
18.37 a) 6.21×10^{-21} J b) 2.34×10^5 m^2/s^2 c) 484 m/s d) 2.57×10^{-23} kg·m/s e) 1.24×10^{-19} N f) 1.24×10^{-17} Pa g) 8.15×10^{21} h) 2.45×10^{22}
18.39 3800°C
18.41 a) 1380 J/kg·K
18.43 a) 1560 J b) 935 J
18.45 a) 987 K b) 438 K c) 110 K
18.49 2500 m
18.51 solid and vapor, with no liquid
18.53 a) 6.9×10^{-3} L b) yes
18.55 0.213 kg
18.57 0.00123 g
18.59 1.92 atm
18.61 a) 31 b) 8.41×10^3 N c) 7.80×10^3 N
18.63 a) 26.2 m/s b) 16.1 m/s, 5.44 m/s c) 1.74 m
18.65 5.0×10^{27}
18.67 b) 303 mol/m^3 c) van der Waals
18.69 b) r_2 c) $r_1 = R_0/2^{1/6}$, $r_2 = R_0$, $r_1/r_2 = 2^{-1/6}$ d) U_0
18.71 a) 517 m/s b) 299 m/s
18.73 b) 1.40×10^5 K, 1.01×10^4 K c) 6.37×10^3 K, 4.59×10^2 K
18.75 a) 1.24×10^{-14} kg b) 4.16×10^{11} c) 2.95×10^{-6} m, no
18.77 a) $2R$ b) less
18.79 CO_2: 20.79 J/mol·K, 0.270; SO_2: 24.94 J/mol·K, 0.205; H_2S, 24.94 J/mol·K, 0.039
18.83 b) $0.0420N$ c) $(2.94 \times 10^{-21})N$ d) $0.0297N$, $(2.08 \times 10^{-21})N$ e) $0.0595N$, $(4.15 \times 10^{-21})N$
18.85 42.6%

18.87 a) 4.5×10^{11} m b) 703 m/s, 6.4×10^8 s (about 20 y), no c) 1.4×10^{-14} Pa d) about 650 m/s; evaporate f) 2×10^5 K; no
18.89 d) $T_c = 8a/27Rb$, $(V/n)_c = 3b$ e) $p_c = a/27b^2$ f) 8/3 g) 3.28, 3.44, 4.35

Chapter 19

19.1 b) 1330 J
19.3 b) -1.49×10^4 J
19.5 b) 0
19.7 b) 0 for first process, -4.00×10^4 J for second process
19.9 327 J
19.11 113°C
19.13 a) -1.15×10^5 J b) 2.55×10^5 J, out of gas c) no
19.15 a) 16.4 min b) 139 m/s = 501 km/h
19.17 a) absorbs b) 7200 J c) liberates 7200 J
19.19 a) yes b) no c) negative
19.21 a) 948 K b) 900 K
19.23 a) 3120 J b) 1870 J c) 5540 J
19.25 into gas
19.27 a) -605 J b) 0 c) yes, 605 J, liberates
19.29 $C_V = 65.5$ J/mol·K, $C_p = 73.8$ J/mol·K
19.31 a) 553 J b) 454 J
19.33 b) 495°C, 29.1 atm
19.35 decreases
19.37 b) 224 J c) $Q = 0$ d) -224 J
19.39 a) 4.50×10^4 Pa b) 345 J c) 0.818
19.41 -2.45×10^5 J; out of gas
19.43 a) 45.0 J b) liberates 65.0 J c) $Q_{ad} = 23.0$ J, $Q_{db} = 22.0$ J
19.45 a) $W_{ab} = 0$, $W_{bc} = p_c(V_c - V_a)$, $W_{ad} = p_a(V_c - V_a)$, $W_{dc} = 0$ b) $Q_{ab} = U_b - U_a$, $Q_{bc} = U_c - U_b + p_c(V_c - V_a)$, $Q_{ad} = U_d - U_a + p_a(V_c - V_a)$, $Q_{dc} = U_c - U_d$ c) $Q_{abc} = U_c - U_a + p_c(V_c - V_a)$, $W_{abc} = p_c(V_d - V_a)$; $Q_{adc} = U_c - U_a + p_a(V_c - V_a)$, $W_{adc} = p_a(V_d - V_a)$; *abc*, *abc*
19.47 -2460 J
19.49 b) -2.40×10^4 J
19.51 -0.226 m^3
19.53 a) 4.32×10^{-4} m^3 b) 648 J c) 7.15×10^5 J d) 7.14×10^5 J e) no substantial difference
19.55 3.4×10^5 J/kg
19.57 b) 11.9 C°
19.59 a) 0.173 m b) 206°C c) 7.46×10^4 J
19.61 a) $Q = 300$ J, $\Delta U = 0$ b) $Q = 0$, $\Delta U = -300$ J c) $Q = 750$ J, $\Delta U = 450$ J
19.63 a) $W = 738$ J, $Q = 2588$ J, $\Delta U = 1850$ J b) $W = 0$, $Q = -1850$ J, $\Delta U = -1850$ J c) 0
19.65 a) $W = -187$ J, $Q = -654$ J, $\Delta U = -467$ J b) $W = 113$ J, $Q = 0$, $\Delta U = -113$ J c) $W = 0$, $Q = 580$ J, $\Delta U = 580$ J
19.67 b) 3.63×10^3 J c) 0 d) 3.63×10^3 J e) 0.114 m^3
19.69 a) $W = nRT\ln[(V_2 - nb)/(V_1 - nb)] + an^2(1/V_2 - 1/V_1)$ b) i) 2.80×10^3 J ii) 3.11×10^3 J c) 310 J; larger for ideal gas

Chapter 20

20.1 a) 6500 J b) 0.34 = 34%
20.3 a) 0.23 = 23% b) 12,400 J c) 0.350 g d) 222 kW = 298 hp
20.5 a) 0.25 = 25% b) 970 MW
20.7 a) 453.2°C b) 1.99×10^6 Pa
20.9 a) 1.62×10^4 J b) 5.02×10^4 J
20.11 a) 767 W b) 7.27
20.13 a) 215 J b) 378 K c) 0.391 = 39.1%
20.15 a) 3.09×10^7 J b) 2.49×10^6 J
20.17 a) -665 J, refrigerator b) -190 J, refrigerator c) 285 J, engine
20.19 3790 cycles
20.21 4.89×10^3 J
20.23 b) $e \to 1$, $K = 0$; $e \to 0$, $K \to \infty$
20.25 a) irreversible b) 7.7 J/K
20.27 47.4 J/K
20.29 -644 J/K
20.31 a) 109 J/K b) N_2: 72.8 J/K; Ag: 102 J/K; Hg: 86.6 J/K
20.33 6.74 J/K
20.35 a) no b) 18.3 J/K c) 18.3 J/K
20.37 a) -73°C b) 3.34×10^4 cycles
20.39 a) $p_1 = 1.00$ atm, $V_1 = 8.62 \times 10^{-3}$ m^3; $p_2 = 2.00$ atm, $V_2 = 8.62 \times 10^{-3}$ m^3; $p_3 = 1.00$ atm, $V_3 = 1.41 \times 10^{-2}$ m^3 b) $1 \to 2$: $Q = 2183$ J, $W = 0$, $\Delta U = 2183$ J; $2 \to 3$: $Q = 0$, $W = 786$ J, $\Delta U = -786$ J; $3 \to 1$: $Q = -1960$ J, $W = -559$ J, $\Delta U = -1400$ J c) 227 J d) 227 J e) $e = 10.4\%$, $e_{\text{Carnot}} = 50.0\%$
20.41 a) 13.2 J b) 13.2 J
20.43 b) 45°C c) 58.8% d) 206 J e) 2.57×10^6 Pa
20.45 10.5%
20.47 a) 7.36×10^5 J b) 9.33×10^5 J c) 1.98×10^5 J d) 3.73
20.49 a) $a \to b$: $Q = -nRT_1 \ln r$, $W = -nRT_1 \ln r$, $\Delta U = 0$; $b \to c$: $Q = nC_V(T_2 - T_1)$, $W = 0$, $\Delta U = nC_V(T_2 - T_1)$; $c \to d$: $Q = nRT_2 \ln r$, $W = nRT_2 \ln r$, $\Delta U = 0$; $d \to a$: $Q = -nC_V(T_2 - T_1)$, $W = 0$, $\Delta U = -nC_V(T_2 - T_1)$ c) $1 - T_1/T_2$, same, no
20.51 b) 14 m/s = 50 km/h
20.53 a) 122 J, 78 J b) 5.10×10^{-4} m^3 c) $p_b = 2.32 \times 10^6$ Pa, $V_b = 4.81 \times 10^{-5}$ m^3, $T_b = 771$ K; $p_c = 4.00 \times 10^6$ Pa, $V_c = 4.81 \times 10^{-5}$ m^3, $T_c = 1333$ K; $p_d = 1.47 \times 10^5$ Pa, $V_d = 5.10 \times 10^{-4}$ m^3, $T_d = 518$ K d) $e_{\text{Carnot}} = 77.5\%$, $e = 61.1\%$
20.55 b) 6.20×10^4 J c) 3.42×10^4 J d) before: 6.20×10^4 J; after 3.42×10^4 J
20.57 10.5 J/K
20.61 a) 150 J/K b) -120 J/K c) 30 J/K d) irreversible
20.63 a) $e = 1 - \dfrac{1}{\gamma}\left[\dfrac{(T_c/T_a)^{\gamma} r^{\gamma(1-\gamma)} - 1}{(T_c/T_a) - r^{\gamma-1}}\right]$ b) 70.8%

Chapter 21

21.1 a) 2.00×10^{10} b) 8.58×10^{-13}
21.3 2×10^{28} electrons, 3×10^9 C
21.5 -1.74×10^5 C
21.7 a) 7.42×10^{-7} C on each sphere b) 3.71×10^{-7} C on one, 1.48×10^{-6} C on the other
21.9 5.08 m
21.11 +0.750 nC
21.13 charge on the left exerts a 0.172-N force to the right on the right-hand charge, charge on the right exerts a 0.172-N force to the left on the left-hand charge
21.15 1.8×10^{-4} N, $+x$-direction
21.17 $x = -0.144$ m
21.19 2.58×10^{-6} N, $-y$-direction
21.21 b) $F_x = 0$, $F_y = +2kqQa/(x^2 + a^2)^{3/2}$ c) $2kqQ/a^2$, $+y$-direction
21.23 b) $kq^2(1 + 2\sqrt{2})/2L^2$
21.25 23.5 N/C, to the left
21.27 a) -21.9 μC b) 1.02×10^{-7} N/C
21.29 a) 1.13×10^{-4} N/C, upward b) 5.24×10^{-22} N, upward
21.31 a) 364 N/C b) no, 2.73 μm downward
21.33 1.79×10^6 m/s
21.35 a) $mg = 8.93 \times 10^{-30}$ N; $F_e = 1.60 \times 10^{-15}$ N; yes b) 1.63×10^{-16} kg $= 1.79 \times 10^{14}\, m_e$ c) no
21.37 a) $-\hat{j}$ b) $(\hat{\imath} + \hat{j})/\sqrt{2}$ c) $-0.390\hat{\imath} + 0.921\hat{j}$
21.39 a) 6.33×10^5 m/s b) 1.59×10^4 m/s
21.41 a) 0 b) $E_x = -2kq(x^2 + a^2)/(x^2 - a^2)^2$ for $x < -a$; $E_x = -4kqax/(x^2 - a^2)^2$ for $-a < x < +a$; $E_x = +2kq(x^2 + a^2)/(x^2 - a^2)^2$ for $x > +a$
21.43 a) i) 574 N/C, $+x$-direction ii) 268 N/C, $-x$-direction iii) 404 N/C, $-x$-direction b) i) 9.20×10^{-17} N, $-x$-direction ii) 4.30×10^{-17} N, $+x$-direction iii) 6.48×10^{-17} N, $+x$-direction
21.45 a) $E_x = E_y = E = 0$ b) $E_x = +2.66 \times 10^3$ N/C, $E_y = 0$, $E = 2.66 \times 10^3$ N/C, $+x$-direction c) $E_x = +129$ N/C, $E_y = -510$ N/C, $E = 526$ N/C, 284° counterclockwise from $+x$-direction d) $E_x = 0$, $E_y = E = +1.38 \times 10^3$ N/C, $+y$-direction
21.47 a) $E_x = -4.79 \times 10^3$ N/C, $E_y = 0$, $E = 4.79 \times 10^3$ N/C, $-x$-direction b) $E_x = +2.13 \times 10^3$ N/C, $E_y = 0$, $E = 2.13 \times 10^3$ N/C, $+x$-direction c) $E_x = -129$ N/C, $E_y = -164$ N/C, $E = 209$ N/C, 232° counterclockwise from $+x$-direction d) $E_x = -1.04 \times 10^3$ N/C, $E_y = 0$, $E = 1.04 \times 10^3$ N/C, $-x$-direction
21.53 226 N/C, toward sheet
21.59 a) 1.4×10^{-11} C·m, from q_1 toward q_2 b) 860 N/C
21.61 a) $+8.0 \times 10^{-24}$ J b) 0.39 K
21.67 a) 1680 N, from +5.00 μC charge toward -5.00 μC charge b) 22.3 N·m, clockwise
21.69 a) $(1/2)\sqrt{qQ/m\pi^3\epsilon_0 a^3}$, b) accelerates along y-axis away from origin
21.71 b) 2.80×10^{-6}°C c) 39.5°
21.73 a) 2.09×10^{21} N b) 5.90×10^{23} m/s^2 c) no
21.75 a) $6kq^2/L^2$, away from vacant corner b) $(3kq^2/2L^2)(1 + 2\sqrt{2})$, toward center of square
21.77 a) $(2kq/x^2)[1 - (1 + a^2/x^2)^{-3/2}]$, $-x$-direction b) $3kqa^2/x^4$
21.79 a) 480 C b) 8.3×10^{13} N, repulsive; 8.3×10^{13} m/s^2
21.81 a) 3.5×10^{20} b) 1.6 C; 2.4×10^{10} N
21.83 a) $(mv_0^2 \sin^2\alpha)/2eE$ b) $(mv_0^2 \sin 2\alpha)/eE$ c) $h_{\max} = 0.418$ m, $d = 2.89$ m
21.85 +73.1 nC
21.87 a) $E_x = \dfrac{kQ}{x\sqrt{x^2 + a^2}}$, $E_y = \dfrac{-kQ}{a}\left(\dfrac{1}{x} - \dfrac{1}{\sqrt{x^2 + a^2}}\right)$ b) $F_x = \dfrac{-kqQ}{x\sqrt{x^2 + a^2}}$, $F_y = \dfrac{+kqQ}{a}\left(\dfrac{1}{x} - \dfrac{1}{\sqrt{x^2 + a^2}}\right)$
21.89 a) $-(7850 \text{ N/C})\hat{\imath}$ b) smaller c) 18 cm
21.91 a) $+(0.89 \text{ N/C})\hat{\imath}$ c) smaller d) i) 1.2% ii) 4.5%
21.93 a) $F = \dfrac{qQ}{2\pi\epsilon_0 a}\left(\dfrac{1}{y} - \dfrac{1}{\sqrt{a^2 + y^2}}\right)$, $-x$-direction b) $F = \dfrac{qQ}{4\pi\epsilon_0 a}\left(\dfrac{1}{x - a} + \dfrac{1}{x + a} - \dfrac{2}{x}\right)$, $+x$-direction
21.95 $E_x = E_y = 2Q/4\pi^2\epsilon_0 a^2$
21.97 a) 1.43×10^3 kg/C b) 1.15×10^{10}
21.99 a) 5.65×10^8 N/C, to the left b) 1.69×10^9 N/C, to the right c) 2.28×10^9 N/C, to the right d) 5.65×10^8 N/C, to the right
21.101 $\vec{\boldsymbol{E}} = \dfrac{\sigma}{2\epsilon_0}\left[-\left(\dfrac{x}{|x|}\right)\hat{\imath} + \left(\dfrac{z}{|z|}\hat{k}\right)\right]$
21.103 b) $q_1 < 0$, $q_2 > 0$ c) 0.844 μC d) 56.2 N
21.105 a) $\dfrac{kQ}{L}\left[\dfrac{1}{x + (a/2)} - \dfrac{1}{x + L + (a/2)}\right]$

Chapter 22

22.1 a) 1.75 N·m^2/C b) no c) i) 0 ii) 90°
22.3 a) S_1: $-CL^2$ S_2: $-DL^2$ S_3: CL^2 S_4: DL^2 S_5: $-BL^2$ S_6: BL^2 b) 0
22.5 a) 2.71×10^5 N·m^2/C b) 2.71×10^5 N·m^2/C c) 5.42×10^5 N·m^2/C
22.7 a) -4.07×10^5 N·m^2/C b) 6.91 nC c) no
22.9 b) no

22.11 a) $1.81 \times 10^5\ \text{N}\cdot\text{m}^2/\text{C}$ b) no change
22.13 3.19 nC
22.15 a) 1.62 m
22.17 2.04×10^{10}
22.19 a) 6.47×10^5 N/C, $+y$-direction
b) 7.2×10^4 N/C, $-y$-direction
22.21 8.06×10^5 N/C, toward negatively charged sphere
22.23 10.2 nC
22.25 between: $E = \sigma/\epsilon_0$; outside: $E = 0$
22.27 a) $\lambda = 2\pi R\sigma$ b) $\sigma R/r\epsilon_0$
22.29 a) yes; $+Q$ b) no c) yes d) no; no e) no; yes; no
22.31 b) $q/24\epsilon_0$
22.33 a) -5.98×10^{-10} C
22.35 0.097 m above center of sphere
22.37 a) $0 < r < R$, $E = 0$; $R < r < 2R$, $E = Q/4\pi\epsilon_0 r^2$, radially outward; $r > 2R$, $E = 2Q/4\pi\epsilon_0 r^2$, radially outward
22.39 a) i) 0 ii) 0 iii) $q/2\pi\epsilon_0 r^2$, radially outward iv) 0 v) $3q/2\pi\epsilon_0 r^2$, radially outward b) i) 0 ii) $+2q$ iii) $-2q$ iv) $+6q$
22.41 a) i) 0 ii) 0 iii) $q/2\pi\epsilon_0 r^2$, radially outward iv) 0 v) $q/2\pi\epsilon_0 r^2$, radially inward b) i) 0 ii) $+2q$ iii) $-2q$ iv) $-2q$
22.43 a) $Qq/4\pi\epsilon_0 r^2$, toward the center of the shell b) 0
22.45 a) $\lambda/2\pi\epsilon_0 r$, radially outward b) $\lambda/2\pi\epsilon_0 r$, radially outward d) inner: $-\lambda$; outer: $+\lambda$
22.47 a) i) $\alpha/2\pi\epsilon_0 r$, radially outward ii) 0 iii) 0 b) i) $-\alpha$ ii) 0
22.51 $d = R/2$
22.53 a) $1.0 \times 10^{32}\ \text{m/s}^2$ b) $4.1 \times 10^{32}\ \text{m/s}^2$ c) $2.1 \times 10^{32}\ \text{m/s}^2$ d) 0
22.55 b) $|x| \le d$: $\vec{E} = (\rho_0 x^3/3\epsilon_0 d^2)\hat{\imath}$; $|x| \ge d$, $\vec{E} = (\rho_0 d/3\epsilon_0)(x/|x|)\hat{\imath}$
22.57 c) $E(r) = \dfrac{Q}{\pi\epsilon_0 R^2}\left(\dfrac{r}{R} - \dfrac{3r^2}{4R^2}\right)$
e) $E_{\text{max}} = Q/3\pi\epsilon_0 R^2$ at $r = 2R/3$
22.61 b) $\rho\vec{b}/3\epsilon_0$
22.63 a) $-(Q/16\pi\epsilon_0 R^2)\hat{\imath}$ b) $(Q/72\pi\epsilon_0 R^2)\hat{\imath}$ c) 0 d) $(5Q/18\pi\epsilon_0 R^2)\hat{\imath}$
22.65 a) $\alpha = 8Q/5\pi R^3$ b) $r \le R/2$: $E(r) = 8Qr/15\pi\epsilon_0 R^3$; $R/2 \le r \le R$: $E(r) = (Q/60\pi\epsilon_0 r^2) \times [64(r/R)^3 - 48(r/R)^4 - 1]$; $r \ge R$: $E(r) = Q/4\pi\epsilon_0 r^2$ c) 4/15
e) $2\pi\sqrt{15m\pi\epsilon_0 R^3/8Qe}$ f) no

Chapter 23

23.1 -0.356 J
23.3 a) 12.5 m/s b) 0.323 m
23.5 a) 0.198 J b) i) 26.6 m/s ii) 36.7 m/s iii) 37.6 m/s
23.7 a) -3.60×10^{-7} J b) $x = 0.0743$ m
23.9 $-q/2$
23.11 1.44×10^{-18} J
23.13 7.42 m/s; faster
23.15 B: larger; C: smaller; D: same
23.17 a) 0 b) $+7.50 \times 10^{-4}$ J c) -2.06×10^{-3} J
23.19 a) 2.50 mm b) 7.49 mm
23.21 a) -737 V b) -704 V c) $+8.2 \times 10^{-8}$ J
23.23 b) 0 d) 0
23.25 b) $V = \dfrac{q}{4\pi\epsilon_0}\left(\dfrac{1}{|x|} - \dfrac{2}{|x - a|}\right)$ d) $x = -a$, $x = a/3$ e) $-q/4\pi\epsilon_0 x$
23.27 1.02×10^7 m/s
23.29 a) b b) 800 V/m c) -4.8×10^{-5} J
23.31 a) increase of 156 V b) decrease of 182 V
23.33 a) oscillatory b) 1.67×10^7 m/s
23.35 a) 8.00 kV/m b) 1.92×10^{-5} N c) 8.64×10^{-7} J d) -8.64×10^{-7} J
23.37 a) 1.58 mm b) $2.66 \times 10^{-5}\ \text{C/m}^2$
23.39 b) -20 nC c) no
23.41 a) $E_x = -Ay + 2Bx$, $E_y = -Ax - C$, $E_z = 0$ b) $x = -C/A$, $y = -2BC/A^2$, any value of z
23.43 a) $\vec{E} = -C\hat{k}$ for $0 < z < d$, $\vec{E} = \mathbf{0}$ otherwise b) parallel plates at $z = 0$ (charge density $-C\epsilon_0$) and $z = d$ (charge density $= +C\epsilon_0$)
23.45 a) 7.63×10^{-10} C
23.49 a) -2.15×10^{-5} J b) -2.83 kV c) 3.54×10^4 V/m
23.51 a) $7.85 \times 10^4\ \text{V/m}^{4/3}$ b) $\vec{E} = (-1.05 \times 10^5\ \text{V/m}^{4/3})x^{1/3}\hat{\imath}$ c) $\vec{F} = (3.13 \times 10^{-15}\ \text{N})\hat{\imath}$
23.53 a) $-1.46q^2/\pi\epsilon_0 d$
23.55 a) -8.62×10^{-18} J b) 2.87×10^{-11} m
23.57 a) i) $V = (\lambda/2\pi\epsilon_0)\ln(b/a)$ ii) $V = (\lambda/2\pi\epsilon_0)\ln(b/r)$ iii) $V = 0$ d) $(\lambda/2\pi\epsilon_0)\ln(b/a)$
23.59 a) 1.76×10^{-16} N, downward b) $1.93 \times 10^{14}\ \text{m/s}^2$, downward c) 8.24 mm d) 15.4° e) 4.12 cm
23.61 a) 9.71×10^4 V/m b) 3.03×10^{-11} C
23.63 a) $(\sigma/2\epsilon_0)(\sqrt{x^2 + R^2} - x)$
23.65 a) $r \le R$: $V = (\lambda/4\pi\epsilon_0)[1 - (r/R)^2]$; $r \ge R$, $V = -(\lambda/2\pi\epsilon_0)\ln(r/R)$
23.67 $Q/4\pi\epsilon_0\sqrt{x^2 + a^2}$
23.69 a) 0 b) 0.135 V; S_4 c) 0.225 V; S_5
23.71 a) $Q/8\pi\epsilon_0 R$ b) (i) center (ii) surface
23.73 b) yes c) no
23.75 $Q/8\pi\epsilon_0 R$
23.77 0
23.79 a) $(Q/4\pi\epsilon_0 a)\ln[1 + (a/x)]$ b) $(Q/4\pi\epsilon_0 a)\ln\left[(a/y) + \sqrt{1 + (a/y)^2}\right]$ c) in (a), $Q/4\pi\epsilon_0 x$; in (b), $Q/4\pi\epsilon_0 y$
23.81 a) 1/3 b) 3
23.83 a) $E = Q_1/4\pi\epsilon_0 R_1^2$, $V = Q_1/4\pi\epsilon_0 R_1$ b) sphere 1: $Q_1R_1/(R_1 + R_2)$; sphere 2: $Q_1R_2/(R_1 + R_2)$ c) $V = Q_1/4\pi\epsilon_0(R_1 + R_2)$ for either sphere d) sphere 1: $E = Q_1/4\pi\epsilon_0 R_1(R_1 + R_2)$; sphere 2: $E = Q_1/4\pi\epsilon_0 R_2(R_1 + R_2)$
23.85 a) 7.6×10^6 m/s b) 7.3×10^6 m/s c) 2.3×10^9 K; 6.4×10^9 K
23.87 1.01×10^{-12} m, 1.11×10^{-13} m, 2.54×10^{-14} m
23.89 c) 3 electrons, 0.507 μm
23.91 a) 700 m/s c) -1.89 J d) no e) 4.76 cm f) 9.61 J, yes, 980 m/s

Chapter 24

24.1 $+1.82 \times 10^{-4}$ C on one plate, -1.82×10^{-4} C on the other
24.3 a) 604 V b) 90.8 cm^2 c) 1.84×10^6 V/m d) 16.3 $\mu\text{C/m}^2$
24.5 a) 120 μC b) 60 μC c) 480 μC
24.7 2.8 mm; yes
24.9 a) 4.35×10^{-12} F b) 2.30 V
24.11 a) 66 pF/m b) inner: $+64$ pC; outer: -64 pC
24.13 a) 89.5 pF b) 6.08×10^3 V/m c) 4.47×10^3 V/m d) no
24.15 a) $Q_1 = Q_2 = 22.4\ \mu$C, $Q_3 = 44.8\ \mu$C, $Q_4 = 67.2\ \mu$C b) $V_1 = V_2 = 5.60$ V, $V_3 = 11.2$ V, $V_4 = 16.8$ V c) 11.2 V
24.17 a) $Q_1 = 156\ \mu$C, $Q_2 = 260\ \mu$C b) $V_1 = V_2 = 52$ V
24.21 57 μF
24.23 $V_2 = 50$ V, $V_3 = 70$ V
24.25 0.0283 J/m^3
24.27 19.6 J
24.29 a) $Q^2x/2\epsilon_0 A$ b) $(Q^2/2\epsilon_0 A)dx$ c) $Q^2/2\epsilon_0 A$
24.31 a) 3.12×10^{15} electrons b) halve plate area or double plate separation
24.33 a) 1.60 nC b) 8.0
24.35 a) $1.64 \times 10^{-4}\ \text{J/m}^3$ b) $8.83 \times 10^{-5}\ \text{J/m}^3$ c) no
24.37 a) $U_{\text{parallel}} = 4U_{\text{series}}$ b) $Q_{\text{parallel}} = 2Q_{\text{series}}$ c) $E_{\text{parallel}} = 2E_{\text{series}}$
24.39 a) $6.20 \times 10^{-7}\ \text{C/m}^2$ b) 1.28
24.41 0.0135 m^2
24.43 a) $2.3 \times 10^{-11}\ \text{C}^2/\text{N}\cdot\text{m}^2$ b) 40 kV c) $\sigma = 4.6 \times 10^{-4}\ \text{C/m}^2$, $\sigma_i = 2.8 \times 10^{-4}\ \text{C/m}^2$
24.45 a) 10.1 V b) 2.25
24.47 a) before: 3.00 V; after: 0.600 V b) before: 1500 V/m; after: 300 V/m
24.49 a) $Q/K\epsilon_0 A$ b) $Qd/K\epsilon_0 A$ c) $K\epsilon_0 A/d$
24.51 a) 2.4×10^{-11} F b) 2.9×10^{-10} C c) 1.3×10^3 V/m d) 1.7×10^{-9} J
24.53 a) 421 J b) 5.39×10^{-2} F
24.57 a) 4.00 μF: 2.64×10^{-3} C, 660 V; 6.00 μF: 3.96×10^{-3} C, 660 V b) 4.00 μF: 5.28×10^{-4} C, 132 V; 6.00 μF: 7.92×10^{-4} C, 132 V
24.59 a) 2.5 μF b) $Q_1 = 5.5 \times 10^{-4}$ C, $V_1 = 66$ V; $Q_2 = 3.7 \times 10^{-4}$ C, $V_2 = 88$ V; $Q_3 = 1.8 \times 10^{-4}$ C, $V_3 = 44$ V; $Q_4 = 1.8 \times 10^{-4}$ C, $V_4 = 44$ V; $Q_5 = 5.5 \times 10^{-4}$ C, $V_5 = 66$ V
24.61 a) 76 μC b) 1.4×10^{-3} J c) 11 V d) 1.2×10^{-3} J
24.63 a) 2.3 μF b) C_1: 9.7×10^{-4} C; C_2: 6.4×10^{-4} C c) 47 V
24:65 a) 3.91 b) 22.8 V
24.67 c) 710 μF
24.69 a) $Q^2r^2/32\pi^2\epsilon_0 R^6$ b) $Q^2/32\pi^2\epsilon_0 r^4$ c) $3Q^2/20\pi\epsilon_0 R$
24.73 a) 1.0×10^7 V/m b) 0.052 V, outer c) 1.4×10^{-15} J
24.75 b) 14.0 μF c) 72.0 μF: 505 μC, 7.02 V; 28.0 μF: 259 μC, 9.24 V; 21.0 μF: 246 μC, 11.7 V; 18.0 μF: 229 μC, 12.7 V; 27.0 μF: 276 μC, 10.2 V; 6.0 μF: 14.9 μC, 2.49 V
24.77 a) $(\epsilon_0 L/D)[L + (K - 1)x]$
24.79 b) 2.38×10^{-9} F

Chapter 25

25.1 3.9×10^4 C
25.3 a) 110 min b) 440 min
25.5 6.00×10^{-4} m/s
25.7 5.86×10^{28} free electrons/m^3; same
25.9 a) 330 C b) 41 A
25.11 0.125 Ω
25.13 a) 5.16×10^{-3} V/m b) 2.70×10^{-3} V/m
25.15 15 g
25.17 a) 820 A b) 0.36 mm
25.19 $1.53 \times 10^{-8}\ \Omega$
25.21 0.205 mm
25.23 a) 11 A b) 3.1 V c) 0.28 Ω
25.25 a) 1.25 V/m b) $2.84 \times 10^{-8}\ \Omega\cdot\text{m}$
25.27 a) 99.54 Ω b) 0.0158 Ω
25.29 a) $4.67 \times 10^{-8}\ \Omega$ b) $6.72 \times 10^{-4}\ \Omega$
25.31 a) 27.4 V b) 1.23×10^7 J
25.33 a) 0 b) 5.0 V c) 5.0 V
25.35 $\mathcal{E} = 3.08$ V, $r = 0.067\ \Omega$, $R = 1.80\ \Omega$
25.37 a) 1.41 A, from a toward b b) -13.7 V c) -1.0 V
25.39 b) yes c) 3.88 Ω
25.41 a) 0.101 Ω b) 0.221 Ω c) 0.0126 Ω
25.43 520 W
25.45 a) EJ b) ρJ^2 c) E^2/ρ
25.47 a) 2.6×10^6 J b) 0.063 L c) 5.8×10^3 s
25.49 a) 24 W b) 4 W c) 20 W
25.51 a) 26.7 Ω b) 4.50 A c) 454 W d) greater
25.53 a) $3.65 \times 10^{-8}\ \Omega\cdot\text{m}$ b) 172 A c) 2.58×10^{-3} m/s
25.55 0.060 Ω
25.57 a) 2.5 mA b) 2.1×10^{-5} V/m c) 8.6×10^{-5} V/m d) 1.8×10^{-4} V
25.59 a) $\rho h/\pi r_1 r_2$
25.63 a) 0.057 Ω b) $3.3 \times 10^{-8}\ \Omega\cdot\text{m}$ c) 0.86 mm d) 0.0024 Ω e) $1.1 \times 10^{-3}\ (\text{C}°)^{-1}$
25.65 a) 0.20 Ω b) 8.7 V
25.67 a) 1000 Ω b) 100 V c) 10 W
25.69 1.4 A
25.71 a) $I_A(1 + R_A/(r + R))$ b) 0.043 Ω
25.73 b) 8 gauge c) 106 W d) \$19

25.75 a) 0.40 A b) 1.6 W c) 4.8 W d) 3.2 W

25.77 a) $|q|/m = a/E$ b) $|q|/m = aL/V_{bc}$ c) c d) 3.5×10^8 m/s^2

25.79 a) 14.4 V b) 2.6×10^6 J c) 4.3×10^5 J d) 0.96 Ω e) 1.7×10^6 J f) 4.3×10^5 J

25.81 b) $a = 8.0 \times 10^{-5}\ \Omega \cdot \text{m} \cdot \text{K}^n$, $n = 0.15$ c) $4.3 \times 10^{-5}\ \Omega \cdot$m, $3.2 \times 10^{-5}\ \Omega \cdot$m

25.83 a) $R = (\rho_0 L/A)(1 - 1/e)$, V_0/R b) $(V_0/L)(1 - 1/e)e^{-x/L}$ c) $V_0(e^{-x/L} - 1/e)/(1 - 1/e)$

Chapter 26

26.1 a) 12.3 Ω b) 19.5 A c) 32-Ω resistor: 7.5 A; 20-Ω resistor: 12.0 A

26.3 a) true b) false c) false d) true e) false f) false g) false h) true

26.5 a) 0.80 Ω b) 1.60-Ω resistor: 17.5 A; 2.40-Ω resistor, 11.7 A; 4.80-Ω resistor: 5.8 A c) 35.0 A d) 28.0 V for each e) 1.60-Ω resistor: 490 W; 2.40-Ω resistor: 327 W; 4.80-Ω resistor: 163 W f) least resistance

26.7 a) 270 V b) 1.6 W

26.9 $R_{eq} = 3.00\ \Omega$; 1.00-Ω resistor: 12.0 A; 3.00-Ω resistor: 12.0 A; 7.00-Ω resistor: 4.0 A; 5.00-Ω resistor: 4.0 A

26.11 a) $I_1 = 1.50$ A, $I_2 = I_3 = I_4 = 0.50$ A b) $P_1 = 10.1$ W, $P_2 = P_3 = P_4 = 1.12$ W c) $I_1 = 1.33$ A, $I_2 = I_3 = 0.667$ A d) $P_1 = 8.00$ W, $P_2 = P_3 = 2.00$ W e) R_2 and R_3 are brighter, R_1 is dimmer

26.13 18.0 V; 3.00 A

26.15 a) 0.100 A for each b) 400-Ω bulb: 4.00 W; 800-Ω bulb: 8.00 W; total: 12.0 W c) 400-Ω bulb: 0.300 A; 800-Ω bulb: 0.150 A d) 400-Ω bulb: 36.0 W; 800-Ω bulb: 18.0 W; total: 54.0 W e) in series, 800-Ω bulb is brighter; in parallel, 400-Ω bulb is brighter and total light output is greater

26.17 1010 s

26.19 a) 2.00 A b) 5.00 Ω c) 42.0 V d) 3.50 A

26.21 a) 8.00 A b) $\mathcal{E}_1 = 36.0$ V, $\mathcal{E}_2 = 54.0$ V c) 9.00 Ω

26.23 a) 1.60 A, 1.40 A, 0.20 A b) 10.4 V

26.25 4.0 W

26.27 a) 6.5 A in each 1-Ω resistor, 0 in R b) 1.0 Ω c) 0

26.29 12.9 Ω

26.31 c) 3.34 V

26.33 218 Ω

26.37 a) 0.112 mA b) 0.582 ms

26.39 a) 0, 2.70×10^{-4} C, 4.42×10^{-4} C, 6.21×10^{-4} C, 7.44×10^{-4} C b) 6.70×10^{-5} A, 4.27×10^{-5} A, 2.72×10^{-5} A, 1.11×10^{-5} A, 8.19×10^{-9} A

26.41 190 μC

26.43 a) 1.33×10^{-4} C b) $V_R = 9.12$ V, $V_C = 8.88$ V c) $V_R = V_C = 8.88$ V d) 6.75×10^{-5} C

26.45 900 W

26.47 a) 6.0 A, 720 W b) 3.5 A, 420 W

26.49 a) two in parallel in series with two in parallel; two in series in parallel with two in series b) 0.60 W

26.51 1.25 Ω

26.53 54 W

26.55 a) 18.7 Ω b) 7.5 Ω

26.57 $I_1 = 0.848$ A, $I_2 = 2.14$ A, $I_3 = 0.171$ A

26.59 2.00-Ω resistor: 5.21 A; 4.00-Ω resistor: 1.11 A; 5.00-Ω resistor: 6.32 A

26.61 a) 0.222 V b) 0.464 A

26.63 12.7 V

26.65 a) 186 V, upper terminal + b) 3.00 A, from − to + terminal c) 20.0 Ω

26.67 a) $P_1 + P_2$ b) $P_1P_2/(P_1 + P_2)$

26.69 a) −12.0 V b) 1.71 A c) 4.20 Ω

26.71 a) −6.0 V b) b c) +6.0 V d) 54 μC

26.73 $R_1 = 2960\ \Omega$, $R_2 = 12.0$ kΩ, $R_3 = 135$ kΩ; $R_{overall} = 3000\ \Omega$ (3.00-V range), 15.0 kΩ (15.0-V range), 150 kΩ (150-V range)

26.75 18.5 kΩ

26.77 b) 1897 Ω

26.79 a) 224-Ω resistor: 24.8 V; 589-Ω resistor: 65.2 V b) 3.87 kΩ c) 62.6 V d) no

26.81 a) 8.24 mA toward negative plate b) 612 μC

26.83 1.7 MΩ, $C = 3.7$ μF

26.85 a) 19.4 s = 31.4τ b) yes

26.87 a) $C\mathcal{E}^2$ b) $C\mathcal{E}^2/2$ c) $C\mathcal{E}^2/2$ d) 1/2

26.89 a) 0.300 A, 0.500 A, 0.200 A b) 0.541 A, 0.077 A, −0.464 A c) −0.287 A, 0.192 A, 0.479 A d) 0.046 A, 0.231 A, 0.185 A f) 0.184 A, 0.577 A, 0.393 A

26.93 b) at least four c) $R_T = 3.2$ MΩ, $\beta = 4.0 \times 10^{-3}$ d) 3.4×10^{-4} e) 0.88

Chapter 27

27.1 a) $(-6.68 \times 10^{-4}\ \text{N})\hat{k}$ b) $(+6.68 \times 10^{-4}\ \text{N})\hat{\imath} + (7.27 \times 10^{-4}\ \text{N})\hat{\jmath}$

27.3 positive

27.5 a: $-qvB\hat{k}$; b: $+qvB\hat{\jmath}$; c: $\mathbf{0}$; d: $(-qvB/\sqrt{2})\hat{\jmath}$; e: $(-qvB/\sqrt{2})(\hat{\jmath} + \hat{k})$

27.7 9.47×10^6 m/s

27.9 a) $B_x = -0.175$ T, $B_z = -0.256$ T b) yes, B_y c) zero; 90.0°

27.11 a) 3.05×10^{-3} Wb b) 1.83×10^{-3} Wb c) 0

27.13 a) −0.0108 Wb b) no

27.15 a) 1.60×10^{-4} T, into page b) 1.11×10^{-7} s

27.17 a) 1.2×10^7 m/s b) 0.10 T

27.19 7.93×10^{-10} N, south

27.21 a) 8.35×10^5 m/s b) 2.62×10^{-8} s c) 7.26 kV

27.23 a) 107 T b) no

27.25 a) no b) 1.40 cm

27.27 8.38×10^{-4} T

27.29 a) 0.10 T b) 2.1×10^4 m/s to 3.2×10^5 m/s

27.31 1.29×10^{-25} kg, 78

27.33 a) mg/IL b) due west

27.35 9.7 A

27.37 a) $(0.023\ \text{N})\hat{k}$ b) $(0.020\ \text{N})\hat{\jmath}$ c) 0 d) $(-0.0098\ \text{N})\hat{\jmath}$ e) $(-0.013\ \text{N})\hat{\jmath} + (-0.026\ \text{N})\hat{k}$

27.39 a) 817 V b) 113 m/s^2

27.41 a) $-ILB\hat{\jmath}$ b) yes

27.43 a) the normal to the plane of the coil is perpendicular to $\vec{B}$; 0.13 N·m b) the normal to the plane is 45° from the direction of $\vec{B}$

27.45 a) 1.5×10^{-16} s b) 1.1 mA c) 9.3×10^{-24} A·m^2

27.47 −2.42 J

27.49 a) 1.13 A b) 3.69 A c) 98.2 V d) 362 W

27.51 a) 4.7 mm/s b) 4.5×10^{-3} V/m in the $+z$-direction c) 53 μV

27.53 F_2/qv_1 in the $-y$-direction b) $F_2/\sqrt{2}$

27.55 7.00 V/m, horizontal and perpendicular to velocity

27.57 a) 8.9×10^{-13} J $= 5.5 \times 10^6$ eV b) 7.7×10^{-8} s c) 1.2 T d) same as in (a)

27.59 4.24×10^3

27.61 a) -1.98×10^{-6} C b) $(9.69 \times 10^{13}\ \text{m/s}^2)(4\hat{\imath} + 3\hat{\jmath})$ c) $R = 5.69$ cm d) 1.47×10^7 Hz e) $(R, 0, 1.72\ \text{m})$

27.63 1.6 mm

27.65 a) ab: $(-4.24\ \text{N})\hat{k}$; bc: $(-4.24\ \text{N})\hat{\jmath}$; cd: $(+4.24\ \text{N})(\hat{\jmath} + \hat{k})$; de: $(-4.24\ \text{N})\hat{\jmath}$; ef: 0 b) $(-4.24\ \text{N})\hat{\jmath}$

27.67 $(Mg \tan\theta)/LB$, right to left

27.71 a) 0.0442 N·m, clockwise b) stretched c) 7.98×10^{-3} J

27.73 10.0 A

27.75 0.0242 T, $+y$-direction

27.77 $2\pi\sqrt{\dfrac{I_S}{NIAB}}$

27.79 0.444 N, $-y$-direction

27.81 b) side $(0, 0)$ to $(0, L)$: $(B_0IL/2)\hat{\imath}$; side $(0, L)$ to (L, L): $-B_0IL\hat{\jmath}$; side (L, L) to $(L, 0)$: $(-B_0IL/2)\hat{\imath}$; side $(L, 0)$ to $(0, 0)$: $\mathbf{0}$ c) $-B_0IL\hat{\jmath}$

27.83 a) 2.52 m/s b) 7.60 A c) 0.197 Ω

27.85 a) $\vec{\mu} = -IA\hat{k}$ b) $B_x = 3D/IA$, $B_y = 4D/IA$, $B_z = -12D/IA$

27.87 $-\beta r/2$

27.89 a) 5.14 m b) 1.72×10^{-6} s c) 6.09 mm d) 3.04 cm

Chapter 28

28.1 a) $(-1.92 \times 10^{-5}\ \text{T})\hat{k}$ b) 0 c) $(1.92 \times 10^{-5}\ \text{T})\hat{\imath}$ d) $(6.79 \times 10^{-6}\ \text{T})\hat{\imath}$

28.3 a) zero b) $(-1.31 \times 10^{-6}\ \text{T})\hat{k}$ c) $(-4.62 \times 10^{-7}\ \text{T})\hat{k}$ d) $(1.31 \times 10^{-6}\ \text{T})\hat{\jmath}$

28.5 1.64×10^{-6} T, into page

28.7 a) $-0.866\hat{\imath} + 0.500\hat{\jmath}$ b) $(-5.00 \times 10^{-3}\ \text{m})\hat{k}$ c) $(-4.34 \times 10^{-8}\ \text{T})\hat{k}$

28.9 a) $(5.00 \times 10^{-11}\ \text{T})\hat{\jmath}$ b) $(-5.00 \times 10^{-11}\ \text{T})\hat{\imath}$ c) $(-1.77 \times 10^{-11}\ \text{T})(\hat{\imath} - \hat{\jmath})$ d) $\mathbf{0}$

28.11 a) 0 b) $2\mu_0I/3\pi a$, $+z$-direction c) $2\mu_0I/3\pi a$, $-z$-direction

28.15 a) 2.91×10^{-5} T, east b) yes

28.17 a) between the wires, 30.0 cm from wire carrying 75.0 A b) 20.0 cm from wire carrying 25.0 A and 60.0 cm from wire carrying 75.0 A

28.19 2.0 A, toward bottom of page

28.21 $\mu_0I^2/2\pi\lambda g$

28.23 a) 8.33 A b) opposite

28.25 $\mu_0|I_1 - I_2|/4R$, zero

28.27 a) 2.77 A b) 0.0184 m

28.29 69

28.31 a: 0 b: -5.0×10^{-6} T·m c: $+2.5 \times 10^{-6}$ T·m d: $+5.0 \times 10^{-6}$ T·m

28.33 a) $\mu_0I_1/2\pi r$ b) $\mu_0(I_1 + I_2)/2\pi r$

28.35 a) 1790 turns/m b) 63.0 m

28.37 a) 3.72×10^6 A b) 2.49×10^5 A c) 237 A

28.39 1.11×10^{-3} T

28.41 a) 0.0725 A b) 0.0195 A

28.43 a) i) 1.1×10^{-3} T ii) 4.7×10^6 A/m iii) 5.9 T

28.45 yes, $C = 1.55 \times 10^5$ K·A/T·m

28.47 1.07×10^{-19} N, toward wire

28.49 5.00×10^{-18} N, to the right

28.51 a) along line passing through intersection of wires and with slope −1.00 b) along line passing through intersection of wires and with slope +0.333 c) along line passing through intersection of wires and with slope +1.00

28.55 a) 2.00 A, out of paper b) 2.13×10^{-6} T, to the right c) 2.06×10^{-6} T

28.57 b) $\mu_0Ix/\pi(a^2 + x^2)$, $-y$-direction d) $x = \pm a$ e) $\mu_0I/\pi x$

28.59 37.8 N/m

28.61 23.2 A

28.63 a) $\mu_0\pi NN'II'a^2a'^2(\sin\theta)/2x^3$ b) $-\mu_0\pi NN'II'a^2a'^2(\cos\theta)/2x^3$

28.65 a) $(\mu_0NIa^2/2)[((x + a/2)^2 + a^2)^{-3/2} + ((x - a/2)^2 + a^2)^{-3/2}]$ c) $(\mu_0NI/a)(4/5)^{3/2}$ d) 0.0202 T e) 0, 0

28.67 $\mu_0I/8R$, out of paper

28.69 a) $3I/2\pi R^3$ b) i) $\mu_0Ir^2/2\pi R^3$ ii) $\mu_0I/2\pi r$

28.73 $16a/3$

28.75 b) $\mu_0I_0/2\pi r$ c) $(I_0r^2/a^2)(2 - r^2/a^2)$ d) $(\mu_0I_0r/2\pi a^2)(2 - r^2/a^2)$

28.77 μ_0I

28.79 a) $\mu_0nI/2$, $+x$-direction b) $\mu_0nI/2$, $-x$-direction

28.81 7.73×10^{-25} J/T $= 0.0833\ \mu_B$

28.83 b) 0.347 m/s c) 6.15 mm

28.85 μ_0Qn/a

Chapter 29

29.1 58.4 V

29.3 a) $Q = NBA/R$ c) no

29.5 0.0973 T

29.7 a) $\mathcal{E} = -(2\pi NB_0A/T)\sin(2\pi t/T)$ for $0 \le t \le T$, $\mathcal{E} = 0$ for $t \le 0$ and $t \ge T$ b) $t = T/2$ c) $t = T/4$ and $t = 3T/4$; $|\mathcal{E}_{max}| = 2\pi NB_0A/T$ d) clockwise for $0 < t < T/2$, counterclockwise for $T/2 < t < T$

29.9 a) 5.44 mV b) clockwise

29.11 6.02×10^{-6} V

29.13 224 V

29.17 a) a to b b) b to a c) b to a

29.19 a) clockwise b) zero c) counterclockwise
29.23 a) 0.858 m/s b) 0.827 A c) 0.597 N, opposite $\vec{v}$ (to the left)
29.27 a) 1.70×10^{-4} V/m b) 3.39×10^{-4} V/m
29.29 a) concentric circles b) 1.75×10^{-3} V/m, clockwise c) 2.75×10^{-4} A d) 5.50×10^{-4} V e) 1.10×10^{-3} V
29.31 a) 1.14×10^{-11} J c) 1.14×10^{-11} J
29.33 5.0 s
29.35 a) 55.7 A/m^2 b) 6.29×10^{12} V/m·s c) 7.00×10^{-7} T d) 3.50×10^{-7} T
29.37 a) 0.900 nC, 2.03×10^5 V/m, 407 V b) 4.07×10^{11} V/m·s; no c) 3.60 A/m^2, $i_D = 1.80$ mA, $i_D = i_C$
29.41 a) inside: $\vec{B} = 0$, $\vec{M} = -(1.03 \times 10^5 \text{ A/m})\hat{\imath}$; outside: $\vec{B} = (+0.130 \text{ T})\hat{\imath}$, $\vec{M} = 0$ b) $\vec{B} = (+0.260 \text{ T})\hat{\imath}$, $\vec{M} = 0$ both inside and outside
29.43 b) clockwise
29.45 c) 0.400 V/m
29.47 a) $\mu_0 i\pi a/2$ c) $i = i_0 \exp(-2Rt/\mu_0\pi a)$ d) 45 μs
29.49 a) $\mu_0 Iabv/2\pi r(a + r)$ b) clockwise
29.51 191 rpm
29.53 a) 0.0126 V b) a to b
29.55 b) FR/B^2L^2
29.57 1.2 V
29.59 $\mu_0 IW/4\pi$
29.61 a) $(\mu_0 Iv/2\pi)\ln((L + d)/d)$ b) a c) 0
29.63 a) 0.165 V b) 0.165 V c) zero; 0.0412 V
29.65 a) B^2a^2v/R
29.67 a: $(qr/2)(dB/dt)$, to the left; b: $(qr/2)(dB/dt)$, upward; c: 0
29.69 $16\rho_m\rho_R g/B^2$
29.71 a) $(Q_0/AK\rho\epsilon_0)\exp(-t/K\epsilon_0\rho)$
29.73 a) $\tau = (mgL/2)\cos\phi - (\omega B^2L^4/R)\sin^2\phi$; in $-x$-direction b) $\alpha = (6g/5L)\cos\phi - (12\omega B^2L^2/5mR)\sin^2\phi$ c) longer d) no
29.75 b) 0 c) 1.40×10^{-3} V d) 7.37×10^{-4} A e) 3.50×10^{-4} V, a
29.77 a) radially outward c) in both cases, magnitudes of I and τ are unchanged: in first case, I radially inward and τ clockwise; in second case, I radially outward and τ clockwise

Chapter 30

30.1 a) 0.270 V, yes b) 0.270 V
30.3 a) 1.96 H b) 7.12×10^{-3} Wb
30.7 a) 4.68 mV b) a
30.9 $\mu_0 N^2A/l$
30.11 a) 0.250 H b) 4.50×10^{-4} Wb
30.13 2850
30.17 a) 25.1 m^3 b) 11.9 T
30.19 a) 2.40 A/s b) 0.800 A/s c) 0.413 A d) 0.750 A
30.21 a) 17.3 μs b) 30.7 μs
30.23 a) 0.250 A b) 0.137 A c) 32.9 V, c d) 4.62×10^{-4} s
30.25 a) $(4.50 \text{ W})[(1 - \exp(-(3.20 \text{ s}^{-1})t)]$ b) $(4.50 \text{ W})[(1 - \exp(-(3.20 \text{ s}^{-1})t]^2$ c) $(4.50 \text{ W})[\exp(-(3.20 \text{ s}^{-1})t) - \exp(-(6.40 \text{ s}^{-1})t)]$
30.29 a) 105 rad/s, 59.6 ms b) 7.20×10^{-4} C c) 4.32×10^{-3} J d) -543 μC e) -49.9 mA f) 2.45×10^{-3} J in the capacitor, 1.87×10^{-3} J in the inductor
30.31 a) 35.0 μF; 0.601 mH
30.33 a) 6.45 μC b) 2.36 V
30.37 a) $U_C = (Q^2/2C)\cos^2(\omega t + \phi)$, $U_L = (Q^2/2C)\sin^2(\omega t + \phi)$
30.41 45.4 Ω
30.47 a) 7.50 H b) 360 Wb c) 0.0104
30.51 a) $\mu_0 i/2\pi r$ b) $(\mu_0 i^2 l/4\pi r)\,dr$ c) $(\mu_0 i^2 l/4\pi)\ln(b/a)$
30.53 2.17×10^{-6} T
30.55 a) 0.281 J b) 0.517 J c) 0.236 J
30.59 222 μF; 9.31 μH
30.61 2×10^4 m/s
30.63 a) solenoid c) 50 V d) 3.5 A e) 4.3 Ω; 43 mH
30.65 a) $V_1 = 40.0$ V; $A_1 = A_4 = 0.80$A; all others read zero b) $V_1 = 24.0$ V, $V_2 = 0$, $V_3 = V_4 = V_5 = 16.0$ V; $A_1 = 0.48$ A, $A_2 = 0.16$ A, $A_3 = 0.32$ A, $A_4 = 0$ c) 192 μC
30.67 a) $A_1 = A_4 = 0.45$ A, $A_2 = A_3 = 0$ b) $A_1 = 0.58$ A, $A_2 = 0.32$ A, $A_3 = 0.16$ A, $A_4 = 0.11$ A
30.69 a) 60.0 V b) a c) 60.0 V d) c e) -96.0 V f) b g) -156 V h) d
30.71 a) $i_0 = 0$, $v_{ac} = 0$, $v_{cb} = 36.0$ V b) $i_0 = 0.180$ A, $v_{ac} = 9.00$ V, $v_{cb} = 27.0$ V c) $i_0 = (0.180 \text{ A})[1 - \exp(-(50.0 \text{ s}^{-1})t)]$, $v_{ac} = (9.00 \text{ V})[1 - \exp(-(50.0 \text{ s}^{-1})t)]$, $v_{cb} = 27.0 \text{ V} + (9.00 \text{ V})\exp(-(50.0 \text{ s}^{-1})t)$
30.73 a) 0; 20 V b) 0.267 A; 0 c) 0.147 A; 9.0 V
30.75 a) $i_1 = \mathcal{E}/R_1$, $i_2 = (\mathcal{E}/R_2)[1 - \exp(-R_2t/L)]$ b) $i_1 = \mathcal{E}/R_1$, $i_2 = \mathcal{E}/R_2$ c) $i = (\mathcal{E}/R_2)\exp(-(R_1 + R_2)t/L)$ d) $R_2 = 21.2$ Ω, $\mathcal{E} = 12.7$ V e) 0.0354 A
30.77 a) $i_1 = (\mathcal{E}/R_1)[1 - \exp(-R_1t/L)]$, $i_2 = (\mathcal{E}/R_2)\exp(-t/R_2C)$, $q_2 = C\mathcal{E}[1 - \exp(-t/R_2C)]$ b) 0, 9.6 mA c) 1.9 A, 0 d) 1.6 ms e) 9.4 mA f) 0.22 s
30.79 a) $d = [(L - L_0)/(L_f - L_0)]D$ b) 0.63024 H, 0.63048 H, 0.63072 H, 0.63096 H c) 0.63000 H, 0.62999 H, 0.62999 H, 0.62998 H

Chapter 31

31.1 a) 31.8 V b) 0
31.3 a) 0.120 A b) 1.20×10^{-2} A c) 1.20×10^{-3} A
31.5 a) 1.51 kΩ b) 0.239 H c) 497 Ω d) 16.6 μF
31.7 13.3 μF
31.9 a) $i = (0.0253 \text{ A})\cos[(720 \text{ rad/s})t]$ b) 180 Ω c) $v_L = (-4.56 \text{ V})\sin[(720 \text{ rad/s})t]$
31.11 b) positive c) negative
31.13 a) 696 Ω b) 0.0431 A c) $V_R = 8.62$ V, $V_C = 28.7$ V d) $-73.3°$
31.15 b) $v = 20.5$ V, $v_R = 7.6$ V, $v_L = 12.9$ V c) $v = -15.2$ V, $v_R = -22.5$ V, $v_L = 7.3$ V
31.17 a) 601 Ω b) 49.9 mA c) $-70.6°$, lags d) $V_R = 9.98$ V, $V_L = 4.99$ V, $V_C = 33.3$ V
31.19 a) 113 Hz; 15.0 mA b) 7.61 mA; lag
31.21 50.0 V
31.25 a) $+45.8°$, 0.697 b) 344 Ω c) 155 V d) 48.6 W e) 48.6 W f) 0 g) 0
31.27 a) 150 V b) 150 V, 1290 V, 1290 V c) 37.5 W
31.29 a) 1.00 b) 75.0 W c) 75.0 W
31.31 a) 103 Hz c) $V_1 = 21.2$ V, $V_2 = 27.4$ V, $V_3 = 27.4$ V, $V_4 = 0$, $V_5 = 21.2$ V d) 103 Hz e) 0.212 A
31.33 a) 10 b) 2.40 A c) 28.8 W d) 500 Ω
31.35 a) 40 b) 1.50 V
31.37 0.124 H
31.39 a) $t_1 = \pi/2\omega$, $t_2 = 3\pi/2\omega$ b) $2I/\omega$ c) $I_{rav} = 2I/\pi$
31.41 a) inductor b) 0.133 H
31.43 a) 33.3 nC b) 0.211 mA c) source and inductor voltages in phase; capacitor voltage lags by 180°
31.45 $\sqrt{(R^2 + \omega^2L^2)/[R^2 + (\omega L - 1/\omega C)^2]}$
31.49 b) $U_B = LV^2/4[R^2 + (\omega L - 1/\omega C)^2]$, $U_E = V^2/4\omega^2C[R^2 + (\omega L - 1/\omega C)^2]$ d) $\omega = 0$: $U_B = 0$, $U_E = CV^2/4$; $\omega \to \infty$: both U_B and $U_E \to 0$; $U_B = U_E$ at $\omega = \omega_0 = 1/\sqrt{LC}$
31.53 a) $I_R = V/R$, $I_L = V/\omega L$, $I_C = \omega CV$ c) $\omega = 0$: $I_L \to \infty$, $I_C = 0$; $\omega \to \infty$, $I_L = 0$, $I_C \to \infty$ d) 159 Hz e) 0.50 A f) $I_R = 0.50$ A, $I_L = I_C = 0.050$ A
31.57 a) 102 Ω b) 0.882 A c) 270 V
31.59 a) 0.750 A b) 160 Ω c) 619 Ω, 341 Ω d) 341 Ω
31.61 $i_{av} = 0$, $i_{rms} = I_0/\sqrt{3}$
31.63 a) $I_0 = V/R$ c) $\omega_1 = \omega_0 + \sqrt{3}R/2L$, $\omega_2 = \omega_0 - \sqrt{3}R/2L$ d) $\omega_1 - \omega_2 = \sqrt{3}R/L$ e) i) $I_0 = 8.00$ A, $\omega_0 = 1000$ rad/s, $\omega_1 - \omega_2 = 10.4$ rad/s ii) $I_0 = 80.0$ A, $\omega_0 = 1000$ rad/s, $\omega_1 - \omega_2 = 1.04$ rad/s
31.65 a) $V/\sqrt{R^2 + 9L/4C}$ b) $[2V/\sqrt{R^2 + 9L/4C}]\sqrt{L/C}$ c) $[V/2\sqrt{R^2 + 9L/4C}]\sqrt{L/C}$ d) $2LV^2/(R^2 + 9L/4C)$ e) $LV^2/2(R^2 + 9L/4C)$
31.67 a) $V_RI/2$ b) 0 c) 0
31.69 a) 0.400 A b) 36.9° c) $Z_{cpx} = (400 \text{ Ω}) - i(300 \text{ Ω})$, $Z = 500$ Ω d) $I_{cpx} = (0.320 \text{ A}) + i(0.240 \text{ A})$, $I = 0.400$ A e) 36.9° f) $V_{Rcpx} = (128 \text{ V}) + i(96 \text{ V})$, $V_{Lcpx} = (-120 \text{ V}) + i(160 \text{ V})$, $V_{Ccpx} = (192 \text{ V}) - i(256 \text{ V})$

Chapter 32

32.1 a) 1.28 s b) 8.15×10^{13} km
32.3 $\vec{E}(z, t) = (1.74 \times 10^5 \text{ V/m})\hat{\imath} \times \cos[(1.28 \times 10^7 \text{ rad/m})z - (3.83 \times 10^{15} \text{ rad/s})t]$, $\vec{B}(z, t) = (5.80 \times 10^{-4} \text{ T})\hat{\jmath} \times \cos[(1.28 \times 10^7 \text{ rad/m})z - (3.83 \times 10^{15} \text{ rad/s})t]$
32.5 a) $+y$-direction b) 7.11×10^{-4} m c) $\vec{B}(y, t) = (-1.03 \times 10^{-3} \text{ T})\hat{\imath} \times \sin[(8.84 \times 10^3 \text{ rad/m})y - (2.65 \times 10^{12} \text{ rad/s})t]$
32.7 a) 361 m b) 0.0174 rad/m c) 5.22×10^6 rad/s d) 0.0144 V/m
32.9 0.779 V/m
32.11 a) 381 nm b) 526 nm c) 1.38 d) 1.91
32.13 a) 1.1×10^{-5} W/m^2 b) 3.0×10^{-10} T c) 840 W; assuming isotropic transmission
32.15 2.5×10^{25} J
32.17 8.5×10^5 W
32.19 $E_{max} = 12.0$ V/m, $B_{max} = 4.00 \times 10^{-8}$ T
32.21 a) 8.68×10^{-15} kg/m^2·s b) 2.60×10^{-6} kg/m·s^2
32.25 1.33×10^{-8} T, $+y$-direction
32.27 a) 4.38 mm b) 4.38 mm c) 4.38 mm
32.29 a) 7.10 mm b) 3.55 mm c) 1.56×10^8 m/s
32.31 a) 4.61×10^{-14} m $= 4.61 \times 10^{-5}$ nm b) 5.21×10^{-7} m $= 521$ nm
32.37 a) 5.20×10^{-10} J b) 6.94×10^{-15} Pa c) 1.34×10^{-4} W
32.39 a) $E_{max} = 701$ V/m, $B_{max} = 2.34 \times 10^{-6}$ T b) $u_E = u_B = 1.09 \times 10^{-6}$ J/m^3 c) 1.07×10^{-11} J
32.41 a) $r = R$: $I = 6.4 \times 10^7$ W/m^2, $p_{rad} = 0.21$ Pa; $r = R/2$: $I = 2.6 \times 10^8$ W/m^2, $p_{rad} = 0.85$ Pa b) no
32.43 a) $B = \mu_0 ni$, $E = (\mu_0 nr/2)(di/dt)$ b) $S = (\mu_0 n^2 ri/2)(di/dt)$, radially inward c) $U_B = (1/2)\mu_0 n^2\pi a^2 li^2$, $dU_B/dt = (\mu_0 n^2\pi a^2 li)(di/dt)$ d) $(\mu_0 n^2\pi a^2 li)(di/dt)$
32.45 a) $I\rho/\pi a^2$, in direction of current b) $\mu_0 I/2\pi a$, counterclockwise if current out of page, clockwise if current into page c) $I^2\rho/2\pi^2a^3$, radially inward d) $I^2\rho l/\pi a^2 = I^2R$
32.47 0.0368 V
32.49 a) 23.6 h b) throw it
32.51 a) $4\pi R^3\rho\, GM/3r^2$ b) $LR^2/4r^2c$ c) 1.9×10^{-7} m, independent of r
32.53 b) 1.4×10^{-11} s^{-1} c) 2.6×10^{-8} s^{-1}
32.55 c) 0.066 mm

Chapter 33

33.1 a) 2.04×10^8 m/s b) 442 nm
33.3 a) 1.55 b) 549 nm
33.5 a) 47.5° b) 66.0°
33.7 2.51×10^8 m/s
33.13 71.8°
33.15 a) 58.1° b) 22.8°
33.17 1.77
33.19 a) air b) 15.1° c) air
33.21 a) 1.40 b) 35.5°
33.23 a) 1.65 b) 58.8°
33.25 a) $0.285I_0$ b) linearly polarized along axis of

second filter
33.27 6.38 W/m^2
33.29 a) first: $I = I_0/2$, linearly polarized along axis of first filter; second: $I = 0.25I_0$, linearly polarized along axis of second filter; third: $I = 0.125I_0$, linearly polarized along axis of third filter b) first: $I_0/2$; third: zero
33.33 a) $\sin\theta_3 = (n_1 \sin\theta_1)/n_3$ c) yes
33.35 72.0°
33.39 1.53
33.41 1.8
33.43 a) 48.6° b) 48.6°
33.45 39.1°
33.47 1.28
33.49 b) 0.22°
33.51 a) $I = (1/8)I_0 \sin^2 2\theta$ b) 45°
33.55 b) 38.9° c) 5.0°
33.57 *l*-leucine: $-(0.11° \text{ 100 mL/g})C$; *d*-glutamic acid, $(0.124° \text{ 100 mL/g})C$
33.59 a) 35° b) 10.1 W/m^2, 19.9 W/m^2
33.63 a) $\Delta = 2\theta_a^A - 6 \arcsin\left(\frac{1}{n}\sin\theta_a^A\right) + 2\pi$ b) $\cos^2\theta_2 = (n^2 - 1)/8$ c) red: $\theta_2 = 71.9°$, $\Delta = 230.1°$; violet: $\theta_2 = 71.6°$, $\Delta = 233.2°$; violet

Chapter 34

34.3 39.2 cm to right of mirror; 4.85 cm
34.5 b) 33.0 cm to left of vertex, 1.20 cm tall, inverted, real
34.7 0.213 mm
34.11 b) $s > f$ c) $s < f$ d) s' at ∞ e) s' at $-\infty$ f) $s' = f$ g) $s' = 0$ i) $s < f$ j) $s > f$ k) $0 < s < 2f$ l) $s > 2f$ m) $m \to \infty$
34.13 b) 5.46 cm to right of vertex, 4.09 mm tall, erect, virtual
34.15 2.67 cm
34.19 1.52
34.21 8.35 cm to left of vertex, 0.326 mm, erect
34.23 $s = 18.0$ cm: a) 63.0 cm to right of lens b) −3.50 c) real d) inverted, $s = 7.00$ cm: a) 14.0 cm to left of lens b) +2.00 c) virtual d) erect
34.25 4.85 cm to left of lens; image: 15.8 cm to left of lens; virtual
34.27 10.3 cm to left of lens
34.29 $f = \pm 4.44$ cm, ± 13.3 cm
34.33 object is 26.3 cm from the lens and has height 1.24 cm; image is erect; same side
34.35 a) 85 mm b) 135 mm
34.37 10.2 m
34.39 a) 1.40×10^{-4} b) 5.25×10^{-4} c) 1.50×10^{-3}
34.41 a) 75.0 mm b) (1/62.5) s
34.43 a) convex b) 50 mm to 56 mm
34.45 a) 80.0 cm b) 76.9 cm
34.47 a) +2.33 diopters b) −1.67 diopters
34.49 a) 6.06 cm b) 4.12 nm
34.51 flea: 3.38 cm from lens; image: 22.0 cm from lens, on same side as flea
34.53 a) 8.37 mm b) 21.4 c) 297
34.55 a) −6.33 b) 1.90 cm c) 0.126 rad = 7.22°
34.57 19.4 m
34.59 a) 66.1 cm b) −59.1
34.61 4.80 m/s
34.63 $h/2$
34.65 a) 13.3 cm b) 26.2 cm
34.67 a) $-12.0 \text{ cm} < s < 0$ b) erect
34.69 a) 46.2 cm from mirror, on opposite side of mirror; virtual b) 2.88 cm, erect c) no
34.71 b) i) 120.00 cm from mirror, 119.96 cm from mirror ii) $m = -0.600$, $m' = -0.360$ iii) faces perpendicular to axis: squares with side 0.600 mm; faces parallel to axis: rectangles with sides of length 0.360 mm (parallel to axis) and 0.600 mm (perpendicular to axis)
34.73 a) 8.65 mm/s b) 14.2 cm/s
34.75 a) distance h behind front surface of glass b) distance $(h + 2d/n)$ behind front surface of glass c) $2d/n$
34.77 a) −3.3 cm b) virtual c) 1.9 cm to right of vertex at right end of rod d) real, inverted e) 1.05 mm
34.79 2.00
34.81 0.50 cm to right of vertex opposite where rays enter
34.83 a) 2.53 mm
34.85 outside sphere, $3R$ from vertex opposite the object
34.87 10.6 cm
34.89 a) 19.0 cm to left b) virtual c) 5.00 mm, erect
34.91 134 cm to left of object
34.93 a) 0.24 m b) 0.24 m
34.95 4.8 cm to left of first lens, 4.8 cm to right of second lens
34.97 0.80 cm
34.99 −26.7 cm
34.101 1.24 cm above page
34.103 a) 29.5 m b) 22.1 m
34.105 b) 75°, 23°, 8.2°
34.107 a) 3.5 cm b) 7 cm c) 100 cm d) 57 e) no
34.109 4.17 diopters
34.111 d) $f_{max} = 36.0$ cm, $f_{min} = 21.6$ cm; $d = 1.2$ cm
34.113 a) 552 b) 25.8 cm
34.115 a) The distances from the lens and the distance from the axis for each point are *A*: (33.0 cm, 6.10 cm); *B*: (40.7 cm, 21.4 cm); *C*: (36.0 cm, 12.0 cm) b) 17.1 cm
34.117 a) $4f$
34.119 b) 1.34 cm

Chapter 35

35.3 a) 2.50 m b) 1.00 m, 4.00 m
35.5 0.75 m, 2.00 m, 3.25 m, 4.50 m, 5.75 m, 7.00 m, 8.25 m
35.7 453 nm, 583 nm
35.9 590 nm
35.11 0.83 mm
35.13 1200 nm
35.15 12.6 cm
35.17 0.172°
35.19 1670 rad
35.21 a) $0.750I_0$ b) 80 nm
35.23 a) 0.888 mm b) 0.444 mm
35.25 a) 1.90 mm b) 2.00×10^{-6} W/m^2
35.27 0.0235°
35.29 114 nm
35.31 533 nm
35.33 0.11 μm
35.35 a) 514 nm; green b) 603 nm; orange
35.37 0.570 mm
35.39 0.63 mm
35.41 b) 427 Hz c) 79.6 cm
35.43 toward A: 9.13°, 23.4°, 39.4°, 60.8°; toward B: 4.55°, 18.5°, 33.7°, 52.5°
35.45 b) constructive: $r_2 - r_1 = (m + \phi/2\pi)\lambda$, $m = 0, \pm 1, \pm 2, \ldots$; destructive: $r_2 - r_1 = (m + \frac{1}{2} + \phi/2\pi)\lambda$, $m = 0, \pm 1, \pm 2, \ldots$
35.47 a) $\sqrt{x^2 + (y+d)^2} - \sqrt{x^2 + (y-d)^2} = m\lambda$ c) $\sqrt{x^2 + (y+d)^2} - \sqrt{x^2 + (y-d)^2} = \left(m + \frac{1}{2}\right)\lambda$
35.49 6.8×10^{-5} $(C°)^{-1}$
35.51 $\lambda/2d$, independent of m
35.53 b) 72 cm
35.55 a) 424 nm, 594 nm b) 495 nm
35.57 1.82 mm
35.59 a) pattern moves down the screen b) $I = I_0 \cos^2[(\pi/\lambda)(d \sin\theta + (n-1)L)]$ c) $d \sin\theta = m\lambda - (n-1)L$
35.61 14.0

Chapter 36

36.1 506 nm
36.3 0.15°
36.5 ±45.4 cm
36.7 ±16.0°, ±33.4°, ±55.6°
36.9 0.920 μm
36.11 a) 10.8 mm b) 5.4 mm
36.13 a) 6.75 mm b) 2.43×10^{-6} W/m^2
36.15 5.4 μm
36.17 a) 668 nm b) $9.36 \times 10^{-5} I_0$
36.19 a) 3 b) 2
36.21 a) ±0.0627°, ±0.125° b) $0.249I_0$, $0.0256I_0$
36.23 cases (i), (iii): slits 1 and 3 and slits 2 and 4; case (ii): slits 1 and 2 and slits 3 and 4
36.25 a) 0.280 mm b) $m = 6, 9, 12, \ldots$ c) $m = 3, 6, 9, 12, \ldots$
36.27 a) 4790 b) 19.0°, 40.7° c) no
36.29 a) yes b) 13.3 nm
36.31 23.3°, 52.3°
36.33 10.5°, 21.3°, 33.1°
36.35 a) 4.00° b) 8.77°
36.37 0.232 nm
36.39 a) 0.461 m
36.41 1.9 m
36.43 92 cm
36.45 1.45 m
36.47 a) Hubble: 77 m; Arecibo: 1.1×10^6 m b) 1500 km
36.49 no
36.51 a) i) 25.6° ii) 10.2° iii) 5.1° b) i) 60.0° ii) 23.1° iii) 11.5°
36.53 2.07
36.55 a) 1.80 mm b) 0.798 mm
36.59 b) for $3\pi/2$, any two slits separated by one other slit; for the other cases, any two slits separated by three other slits
36.63 513 nm
36.65 second order
36.67 c) ±2.6 rad
36.69 1.46 km
36.71 492 km

Chapter 37

37.1 Flash at AA'
37.3 5.00×10^{-9} s, clock on airliner
37.5 a) $0.998c$ b) 126 m
37.7 1.12 h, clock on spacecraft
37.9 a) 9.17 km b) 0.651 km, 0.0710 c) 1.32×10^{-5} s, 3.90 km, 0.0710
37.11 a) 3.57 km b) 9.00×10^{-5} s c) 8.92×10^{-5} s
37.13 92.5 m
37.15 a) $0.806c$ b) $0.974c$ c) $0.997c$
37.19 $0.385c$
37.21 $0.784c$
37.23 $0.837c$, away
37.25 a) $0.159c$ b) $\$1.72 \times 10^8$
37.27 b) $a = (F/m)(1 - v^2/c^2)^{1/2}$
37.29 a) $(\sqrt{3}/2)\,c = 0.866c$ b) $c\sqrt{1 - \left(\frac{1}{2}\right)^{2/3}} = 0.608c$
37.31 a) $(\sqrt{3}/2)\,c = 0.866c$ b) $\sqrt{35/36}c = 0.986c$
37.33 $0.16c$
37.35 a) 3.3×10^{-13}%; no b) 4.0×10^{-16} kg; increases; no
37.37 a) 4.50×10^{-10} J b) 1.94×10^{-18} kg·m/s c) $0.968c$
37.39 a) 8.68×10^{-10} J b) 2.71×10^{-10} J c) 0.453
37.41 a) nonrelativistic: 5.34×10^{-12} J; relativistic: 5.65×10^{-12} J; 1.06 b) nonrelativistic: 6.78×10^{-11} J; relativistic: 3.31×10^{-10} J; 4.88
37.43 a) 2.06×10^6 V b) 3.30×10^{-13} J; 2.06 MeV
37.45 a) 4.20×10^5 eV b) 9.31×10^5 eV c) 2.51×10^8 m/s d) 3.84×10^8 m/s
37.47 a) 4.2×10^9 kg/s; 4.6×10^6 tons b) 1.5×10^{13} y
37.49 a) $\Delta = 2.11 \times 10^{-5}$ b) 2.15×10^4 MeV
37.51 $0.700c$
37.53 a) $0.995c$ b) 1.0%
37.55 a) 1.0×10^{-9} b) $(2.2 \times 10^4)m$
37.57 1.48×10^{-11} kg
37.59 1.68×10^5 eV
37.65 b) $\Delta x' = \sqrt{(\Delta x)^2 - (c\Delta t)^2}$ c) 1.44×10^{-8} s
37.67 $0.357c$, receding

37.75 a) toward the earth at 13.1 km/s, 39.4 km/s
b) 5.96×10^9 m,
5.54×10^{29} kg $= 0.278 \times$ (mass of the sun)
37.77 a) 2494 MeV b) 2.526 times larger
c) 987.4 MeV, the same

Chapter 38

38.1 5.77×10^{14} Hz, 1.27×10^{-27} kg·m/s, 3.82×10^{-19} J = 2.38 eV
38.3 a) 5.92×10^{20} Hz b) 5.06×10^{-13} m
38.5 2.5×10^5 m/s
38.7 a) 5.0×10^{14} Hz b) 2.3×10^{20} photons/s c) no
38.9 a) $K_2 = 4K_1$ b) $E_2 = 2E_1$
38.11 a) 264 nm b) 4.70 eV
38.13 a) 2.47×10^{-19} J = 1.54 eV b) 804 nm, infrared
38.15 a) −5.08 eV b) −5.63 eV
38.17 a) 434.1 nm b) 6.906×10^{14} Hz
c) 4.576×10^{-19} J = 2.856 eV
38.19 0.002 eV
38.21 a) 5.82×10^{-13} J = 3.63 MeV
b) 5.82×10^{-13} J = 3.63 MeV
c) 1.32×10^7 m/s
38.23 a) −218 eV, 16 times greater b) 218 eV, 16 times greater c) 7.63 nm d) 1/4 as large
38.25 $3\hbar = 3.16 \times 10^{-34}$ kg m²/s
38.27 a) $v_1 = 2.19 \times 10^6$ m/s, $v_2 = 1.09 \times 10^6$ m/s, $v_3 = 7.29 \times 10^5$ m/s b) $T_1 = 1.52 \times 10^{-16}$ s, $T_2 = 1.22 \times 10^{-15}$ s, $T_3 = 4.10 \times 10^{-15}$ s
c) 8.22×10^6
38.29 a) 1.2×10^{-33} b) 3.4×10^{-17} c) 5.9×10^{-9}
38.31 4.00×10^{17}
38.33 0.310 nm, the same
38.35 0.0827 nm
38.37 0.0714 nm, 180°
38.39 9.70×10^{-6}
38.43 1.06 mm, microwave
38.45 a) 2.06×10^4 K b) 141 nm
38.51 a) 1.04 eV b) 1.20 μm c) 2.51×10^{14} Hz
d) 4.14×10^{-7} eV
38.55 a) 4.59×10^{14} Hz b) 653 nm c) 1.89 eV
d) 6.59×10^{-34} J·s
38.57 a) $hc(\lambda_1 - \lambda_2)/e\lambda_1\lambda_2$ b) 0.476 V
38.59 a) 6.99×10^{-24} kg·m/s b) 705 eV
38.61 a) 1.69×10^{-28} kg b) −2.53 keV
c) 0.653 nm
38.63 a) 0.90 eV
38.65 a) $r_n = (n^2h^2/4\pi^2mD)^{1/4}$
b) $E_n = (nh/2\pi)\sqrt{D/m}$
c) integer multiples of $(h/2\pi)\sqrt{D/m}$
38.67 a) 12.09 eV b) 3 possibilities: 3 → 2 (656 nm), 3 → 1 (103 nm), 2 → 1 (122 nm)
38.69 a) $\Delta\lambda = h/2mc$ b) 6.61×10^{-16} m for any n
38.71 a) 4.85×10^{-12} m b) 255 keV
38.73 a) 5.10×10^{-17} J = 319 eV, 1.06×10^7 m/s
b) 3.89 nm
38.75 a) 5×10^{-33} m b) $(4 \times 10^{-9})°$ c) 0.1 mm
38.81 a) $\lambda' = \dfrac{\lambda(E - Pc) + 2hc}{E + Pc}$, where
$Pc = \sqrt{E^2 - m^2c^2}$ b) 7.04×10^{-15} m
c) gamma rays

Chapter 39

39.1 a) 2.37×10^{-24} kg·m/s
b) 3.07×10^{-18} J = 19.2 eV
39.3 a) 1.55×10^{-10} m b) 8.44×10^{-14} m
39.5 a) 3.32×10^{-10} m b) 1.33×10^{-9} m
39.7 3.90×10^{-34} m; no
39.9 a) photon: 62.0 nm; electron: 0.274 nm
b) photon: 4.96 eV; electron: 2.41×10^{-5} eV
c) 250 nm; electron
39.11 a) 7.3×10^6 m/s b) 150 eV c) 12 keV
d) electrons
39.13 0.431 eV
39.15 0.340°
39.17 a) 8.79×10^{-32} m/s b) no
39.19 claim is not valid
39.21 a) 9.81×10^{-25} kg·m/s
b) 4.94×10^{-24} J = 3.08×10^{-5} eV
c) 50.6 J d) 5.16 m
39.23 0.087 MeV = $(2.8 \times 10^{-5})E$
39.25 1.08×10^{-24} s
39.27 a) 4.34×10^{-11} m b) 1.01×10^{-12} m
39.29 a) $x = (2n + 1)(\lambda/4)$, n an integer
b) $x = n\lambda/2$, n an integer
39.31 1
39.33 a) 1.0×10^{-6} b) 8.0×10^{-6}
39.37 a) 12 eV b) 1.5×10^{-4} V; 7.3×10^3 m/s
c) 8.2×10^{-8} V; 4.0 m/s
39.39 a) 355 nm b) ±0.0688°; ±0.138°; ±0.206°, . . .
39.41 1.66×10^{-17} m; no
39.43 a) 1.10×10^{-10} m b) 9.09×10^{-13} m
39.45 a) $(1/\sqrt{15})(h/mc)$ b) i) 1.53 MeV, 6.26×10^{-13} m ii) 2.81×10^3 MeV, 3.41×10^{-16} m
39.49 a) 2.1×10^{-20} kg·m/s b) 39 MeV
c) 0.29 MeV; no
39.51 1.4×10^{-35} kg = $5.8 \times 10^{-8} m_{\text{pion}}$
39.53 a) 1.1×10^{-35} m/s b) 2.3×10^{27} y, no
39.55 0.294 eV
39.57 a) $2d \sin\theta = m\lambda$, $m = 1, 2, 3, \ldots$ b) 53.1°
c) less
39.59 a) $-A|x|/x$, $x \neq 0$ b) $(3/2)(\hbar^2A^2/m)^{1/3}$
39.61 $\omega = E/\hbar$
39.63 a) 0.21 kg·m/s b) 1.7 m
39.65 a) $4\pi A^2r^2e^{-2\alpha r^2}dr$ b) $1/\sqrt{2\alpha}$, no
39.67 a) $\psi(x) = (\sin k_0x)/k_0x$ b) L c) $2L$ d) h, h
39.69 2.2×10^{-16} m

Chapter 40

40.1 a) 1.2×10^{-67} J b) 1.1×10^{-33} m/s, 1.4×10^{33} s c) 3.7×10^{-67} J d) no
40.3 0.61 nm
40.5 a) 2.5×10^{-7} J b) 6.6×10^{-30} m
40.7 a) 0, $L/2$, L b) $L/4$, $3L/4$ c) yes
40.9 a) 6.0×10^{-10} m, 1.1×10^{-24} kg·m/s
b) 3.0×10^{-10} m, 2.2×10^{-24} kg·m/s
c) 2.0×10^{-10} m, 3.3×10^{-24} kg·m/s
40.15 3.43×10^{-10} m
40.17 2.2×10^{-14} m
40.21 a) 4.3×10^{-8} b) 4.2×10^{-4}
40.23 $1/\sqrt{2}$
40.25 a) 1.3×10^{-3} b) 10^{-143}
40.27 1.11×10^{-33} J = 6.90×10^{-15} eV; 2.21×10^{-33} J = 1.38×10^{-14} eV
40.29 1.2 N/m
40.31 a) 129 μm b) 4.82 meV
40.33 $\Delta x\Delta p = (2n + 1)\hbar$
40.35 a) $(1/2) + (1/\pi)$ b) 1/2 c) yes
40.37 a) $2dx/L$ b) 0 c) $2dx/L$
40.39 a) 19.2 μm b) 11.5 μm
40.41 a) $\sqrt{2\pi^2/L^3}$ b) $\sqrt{8\pi^2/L^3}$ c) $-\sqrt{2\pi^2/L^3}$
d) $\sqrt{8\pi^2/L^3}$ e) $n = 2$, yes
40.45 a) $B = C$,
$A \sin\dfrac{\sqrt{2mE}}{\hbar}L + B\cos\dfrac{\sqrt{2mE}}{\hbar}L = De^{-\kappa L}$
b) $\dfrac{\sqrt{2mE}}{\hbar}A = \kappa C$,
$\dfrac{\sqrt{2mE}}{\hbar}\left(A\cos\dfrac{\sqrt{2mE}}{\hbar}L - B\sin\dfrac{\sqrt{2mE}}{\hbar}L\right) = -\kappa De^{-\kappa L}$
40.49 6.63×10^{-34} J = 4.14×10^{-15} eV, 1.33×10^{-33} J = 8.27×10^{-15} eV, no
40.53 a) $(n_x + n_y + n_z + (3/2))\hbar\omega$ b) $(3/2)\hbar\omega$, $(5/2)\hbar\omega$
40.55 a) $E_n = n^2h^2/8mL^2$, $n = 2, 4, 6, \ldots$
b) $E_n = n^2h^2/8mL^2$, $n = 1, 3, 5, \ldots$
c) same d) odd in part (a), even in part (b)
40.57 b) increases c) infinite
40.59 a) $-E/A$, $+E/A$ c) decrease

Chapter 41

41.1 $l = 4$
41.3 $1.414\hbar$, $19.49\hbar$, $199.5\hbar$
41.7 b) $1/\sqrt{2\pi}$
41.9 a) 5.29×10^{-11} m b) 1.06×10^{-10} m
c) 2.85×10^{-13} m
41.13 a) 9 b) 3.47×10^{-5} eV c) 2.78×10^{-4} eV
41.15 a) 0.468 T b) 3
41.17 1.68×10^{-4} eV; $m_s = 1/2$
41.19 g
41.21 a) 2.5×10^{30} rad/s b) 2.5×10^{13} m/s, not valid
41.23 4.18 eV
41.25 see Tables 41.2 and 41.3
41.27 a) $1s^22s^22p$ b) −30.6 eV c) $1s^22s^22p^63s^23p$
d) −13.6 eV
41.29 a) −13.6 eV b) −3.4 eV
41.31 a) 8.95×10^{17} Hz, 3.70 keV, 0.335 nm
b) 1.68×10^{18} Hz, 6.93 keV, 0.179 nm
c) 5.48×10^{18} Hz, 22.7 keV, 0.0547 nm
41.33 a) $2a$ b) 0.238
41.35 b) 0.176
41.37 b) $\arccos[-\sqrt{1 - (1/n)}]$
41.39 $4a$
41.41 $2 \to 1, 1 \to 0, 0 \to -1, e\hbar B/2m$; $1 \to 1, 0 \to 0, -1 \to -1, 0$; $0 \to 1, -1 \to 0, -2 \to -1, -e\hbar B/2m$
41.43 a) $1 - 2.2 \times 10^{-7}$ b) 0.998 c) 0.978
41.45 a) 4, 20 b) $1s^42s^42p^3$
41.47 a) 122 nm b) 1.52×10^{-3} nm; increases
41.49 a) 0.188 nm, 0.250 nm b) 0.0471 nm, 0.0624 nm
41.51 b) O shell

Chapter 42

42.1 a) 6.1 K b) 3.47×10^4 K
42.3 a) carbon: 0.0644 nm; oxygen: 0.0484 nm
b) 1.45×10^{-46} kg·m²; yes
42.5 a) 1.03×10^{12} rad/s b) carbon: 66.3 m/s; oxygen: 49.8 m/s c) 6.11×10^{-12} s
42.7 a) 1.20×10^{-21} J = 7.52×10^{-3} eV
b) 0.165 mm
42.9 b) $\hbar l/2\pi I$
42.11 a) 4.593 μm b) 4.626 μm c) 4.634 μm
42.13 a) 963 N/m
b) 8.22×10^{-20} J = 0.513 eV c) 2.42 μm, infrared
42.15 2.16×10^3 kg/m³
42.17 1.19×10^6
42.19 a) 1.11 μm, infrared
42.21 b) ground: $E = 3\pi^2\hbar^2/2mL^2$, 2; first: $E = 6\pi^2\hbar^2/2mL^2$, 6; second: $E = 9\pi^2\hbar^2/2mL^2$, 6
42.23 1.5×10^{22} states/eV
42.25 4.3×10^7
42.27 a) $0.0233R$ b) 7.65×10^{-3} c) no; the ions
42.29 0.20 eV below the band
42.31 a) 5.56 mA b) −5.18 mA, −3.77 mA
42.33 a) 3.8×10^{-29} C·m b) 1.3×10^{-19} C
c) 0.78 d) 0.059
42.35 a) 0.96 nm b) 1.8 nm
42.37 a) 0.129 nm b) 8, 7, 6, 5, 4 c) 484 μm
d) 118 μm, 135 μm, 157 μm, 189 μm, 236 μm
42.39 b) i) 2.94 ii) 4.73 iii) 7.58 iv) 0.837
v) 5.5×10^{-9}
42.41 a) 1.147 cm, 2.293 cm b) 1.172 cm, 2.344 cm; 0.025 cm, 0.050 cm
42.43 4.38×10^{-20} J = 0.273 eV
42.45 a) 4.24×10^{-47} kg·m² b) i) 4.30 μm
ii) 4.28 μm iii) 4.40 μm
42.47 2.03 eV
42.49 a) 4.66×10^{28} atoms/m³ b) 4.7 eV
42.51 b) 3.80×10^{10} Pa = 3.75×10^5 atm
42.53 a) 1.66×10^{33} m^{-3} b) yes c) 7×10^{35} m^{-3}
d) no
42.55 a) $-p^2/2\pi\epsilon_0r^3$ b) $+p^2/2\pi\epsilon_0r^3$

Chapter 43

43.1 a) $Z = 14$, $N = 14$ b) $Z = 37$, $N = 48$
c) $Z = 81$, $N = 124$
43.3 0.533 T
43.5 a) parallel, 70.3 MHz, 4.27 m, radio
b) antiparallel, 46.2 GHz, 6.48 mm, microwave
43.7 5.575×10^{-13} m
43.9 a) 76.21 MeV b) 76.67 MeV c) 0.6%

43.11 He, O, Ca, Ni, Sn, Pb
43.13 a) $^{235}_{92}U$ b) $^{24}_{12}Mg$ c) $^{15}_{7}N$
43.15 -9.9×10^{-5} u
43.17 156 keV
43.19 b) 19 keV
43.21 a) 159 decays/min b) 0.43 decays/min
43.23 a) $^{3}_{2}He$ b) 40.9 y
43.25 2.80 days
43.27 5730 y
43.29 a) 2.02×10^{15} b) 1.01×10^{15}, 3.78×10^{11} Bq c) 2.52×10^{14}, 9.45×10^{10} Bq
43.31 a) 0.421 Bq b) 1.14×10^{-11} Ci
43.33 a) $4.91 \times 10^{-18}\ s^{-1}$ b) 35.7 g c) 7.46×10^{5}
43.35 500 rad, 2000 rem, 5.0 J/kg
43.37 a) 12.5 rad, 12.5 rem b) the antineutrinos are not absorbed
43.39 a) 6.46×10^{-4} J b) 0.108 rem
43.41 a) $Z = 3$, $A = 7$ b) 7.151 MeV c) 1.4 MeV
43.43 a) $Z = 3$, $A = 6$ b) -10.14 MeV c) 11.60 MeV
43.45 6.545 MeV
43.47 1.586 MeV
43.53 a) $^{25}_{12}Al$ to $^{25}_{12}Mg$ b) β^{+} or electron capture c) 3.255 MeV or 4.277 MeV
43.55 23.9858 u, 0.021%, 0.9%
43.57 0.960 MeV
43.59 1.287×10^{-3} u
43.63 29.2%
43.65 a) 9.6×10^{-7} J b) 1.9×10^{-4} rad c) 1.3×10^{-4} rem d) 1.5×10^{6} s = 17 days
43.67 1.3×10^{4} y
43.69 a) 0.48 MeV b) 3.270 MeV = 5.239×10^{-13} J c) 3.155×10^{11} J/mol
43.71 0.82 rad, 0.82 rem
43.75 185 MeV
43.77 b) 4.1×10^{4} Bq, 3.6×10^{5} Bq, 7.5×10^{5} Bq, 1.1×10^{6} Bq, 1.3×10^{6} Bq, 1.5×10^{6} Bq c) 3.2×10^{9} d) 1.5×10^{6} Bq

Chapter 44

44.1 a) 1.27×10^{-14} J b) 9.46×10^{-14} J c) 2.10 pm; smaller
44.3 69 MeV, 1.7×10^{22} Hz, 1.8×10^{-14} m, gamma ray
44.5 a) 32 MeV
44.7 7.2×10^{19} J
44.9 2.79 MeV; exoergic
44.11 a) 1.18 T b) 3.41 MeV, 1.81×10^{7} m/s
44.13 a) 30.6 GeV b) 8.0 GeV
44.15 a) 3.2 TeV b) 38.7 GeV
44.17 1.63×10^{-25} kg; 97.2
44.19 116 MeV
44.21 a) no b) yes c) no d) yes
44.23 a) no b) no c) yes d) yes
44.27 a) 0, 1, -1, 0 b) 0, 0, 0, 1 c) $-e$, $+1$, 0, 0 d) $-e$, 0, 0, -1
44.29 a) $\bar{u}\bar{d}\bar{d}$ b) no c) yes
44.33 a) 1.04×10^{8} m/s b) 1.44
44.35 a) 3.28×10^{7} m/s b) 1640 Mly
44.37 a) 5.494 MeV b) 20.58 MeV
44.39 -783 keV, endoergic
44.41 0.966 μm
44.43 a) 14.0 TeV b) 1.0×10^{5} TeV
44.45 18.1 GeV
44.47 1265 MeV
44.49 a) 139.1 MeV b) momentum conservation
44.51 a) 16.0 MeV
44.53 87 keV, 2.8×10^{-5}
44.55 a) 65 MeV b) 89% by the π^{-}, 11% by the Λ^{0}
44.59 a) 41 MeV b) 14% by the neutron, 86% by the π^{0}

PHOTO CREDITS

About the Author Hugh D. Young; John P. Surey

Chapter 1 Opener: Fermilab/Photo Researchers, Inc.; 1.1a: John G. Ross/Photo Researchers, Inc.; 1.1b: NASA; 1.3a: R. Williams (STScI), the HDF-S Team, and NASA; 1.3b: SOHO (ESA & NASA); 1.3c: Courtesy of NASA/JPL/Caltech; 1.3d: Ryan McVay/PhotoDisc/Getty; 1.3e: Chad Baker/PhotoDisc/Getty Images; 1.3f: Courtesy of Veeco, Courtesy of Purdue University; 1.4: Pearson Addison Wesley, San Francisco, California; 1.5: ND-Viollet/Roger Viollet/Liaison Agency, Inc.

Chapter 2 Opener: NASA/Dryden Flight Research Center; 2.4: Tim Davis/Photo Researchers, Inc.; 2.5: Michael Newman/PhotoEdit; 2.21: Richard Megna/Fundamental Photographs; 2.26: Digital Vision/Getty Images

Chapter 3 Opener: Royalty-free/Digital Vision/Getty Images; 3.7: Stephen Dalton/Animals Animals/Earth Scenes; 3.13: Dean Abramson/Stock Boston; 3.17: Richard Megna/Fundamental Photographs; 3.19: Richard Megna/Fundamental Photographs; 3.31: Eagles Aerobatic Flight Team

Chapter 4 Opener: Jess Stoc/Stone/Getty Images; 4.9: Wayne Eastep/Tony Stone Images; 4.14: Peter Beattie/Liaison Agency, Inc.; 4.17: James H. Robinson/Animals Animals/Earth Scenes; 4.25: Steve Mason/PhotoDisc, Inc.; U4.1: John McDonough/Time Inc. Magazines/Sports Illustrated; U4.2: Mark M. Lawrence/The Stock Market; U4.3a: John McDonough/Time Inc. Magazines/Sports Illustrated

Chapter 5 Opener: David Tipling/Tony Stone Images; 5.12: NASA/STS110; 5.17: David Madison/Tony Stone Images; 5.27: Didier Klein Vandystadt/Allsport Photography (USA), Inc.; 5.36: Courtesy of AOPA; 5.39a: NASA Headquarters; 5.39b: Helen Hansma, University of California, Santa Barbara; 5.39c: Frank Whitney/The Image Bank; 5.39d: Anglo-Australian Observatory, Photograph by David Malin

Chapter 6 Opener: ©Bohemian Nomad Picturemakers/CORBIS; 6.10: Michael Newman/PhotoEdit; 6.21a: Courtesy of Jeffrey H. Whitesell/Airliners of America; 6.21b: Courtesy of Jeffrey H. Whitesell/Airliners of America

Chapter 7 Opener: Allsport/Clive Brunskil/IOC/Olympic Museum Collections; 7.2: Dennis O'Clair/Tony Stone Images; 7.12: Rudi Von Briel/PhotoEdit; 7.15: Paul A. Souders/Corbis; 7.21: Aldo Torelli/Stone/Getty Images

Chapter 8 Opener: Jules Frazier/PhotoDisc/Getty Images; 8.2: Jim Cummins/Getty Images; 8.4: Mike Powell/Allsport Photography; 8.15: Tony Freeman/PhotoEdit; 8.18: David Leah/Tony Stone Images; 8.25: Richard Megna/Fundamental Photographs; 8.27: John V.A.F. Neal/Photo Researchers, Inc.; 8.29: NASA Headquarters

Chapter 9 Opener: Stuart Dee/The Image Bank; 9.3: Gorcher/Gorcher; 9.6: Blair Seitz/Photo Researchers, Inc.; 9.20: ©Jose Azel/Aurora/PNI; 9.23: NASA/Johnson Space Center

Chapter 10 Opener: Ray Massey/Stone/Getty Images; 10.7: John A. Rizzo/PhotoDisc, Inc.; 10.12: ©The Harold E. Edgerton 1992 Trust, courtesy of Palm Press, Inc.; 10.15: Tom Raymond/Stone/Getty Images; 10.18: Lori Adamski Peek/Tony Stone Images; 10.29: Gerard Lacz/Natural History Photographic Agency

Chapter 11 Opener: Chris Sattlberger/The Image Bank/Getty Images; 11.2: Larry Dale Gordon/Getty Images; 11.11a: Doug Menuez/Digital Vision/Getty Images; 11.11b: Lawrence M. Sawyer/PhotoDisc/Getty Images; 11.11c: David Buffinton/PhotoDisc/Getty Images

Chapter 12 Opener: NASA and Hubble Heritage Team (STScI); 12.3: Courtesy of NASA/JPL/Caltech; 12.6: George Hall/Corbis; 12.12: NASA; 12.15: STS-90 Crew, NASA; 12.16a: NASA/STS-108; 12.16b: STS-108 Crew, NASA; 12.17: Jet Propulsion Laboratory/NASA Headquarters; 12.20: Smithsonian Astrophysical Observatory; 12.26: NASA/Johnson Space Center; 12.29: A.M. Ghez, M. Morris, E.E. Becklin, A. Tanner, and T. Kremenek, UCLA

Chapter 13 Opener: James Stevenson/Photo Researchers, Inc.; 13.6: American Diagnostic Corporation; 13.24: Philip Habib/The Image Bank/Getty Images; 13.28a: AP/Wide World Photos; 13.28b: AP/Wide World Photos

Chapter 14 Opener: Nicholas Pinturas/Getty Images; 14.1: Nicholas Pinturas/Getty Images; 14.6: Courtesy of Cenco; 14.9: Photodisc Green/Getty Images; 14.15: Robert B. Suter, Vassar College; 14.19: From Sears, Zemansky, and Young, *College Physics*, 7th ed., ©1991 by Addison-Wesley Publishing Company. Reprinted by permission of Pearson Addison Wesley, San Francisco, California; 14.20: Cordelia Molloy/Photo Researchers, Inc.; 14.27: Paul Chesley/Tony Stone Images; 14.30: The Harold E. Edgerton 1992 Trust, courtesy of Palm Press, Inc.

Chapter 15 Opener: U.S. Geological Survey, Denver; 15.4: David Parker/Photo Researchers, Inc.; 15.9: R. Dolan/Camerique/H. Armstrong Roberts; 15.15: Reproduced from *PSSC Physics,* 2nd ed. (1965), D.C. Heath & Company with Educational Development Center, Inc., Newton Massachusets; 15.20a: Richard Megna/ Fundamental Photographs; 15.20b: Richard Megna/ Fundamental Photographs; 15.20c: Richard Megna/ Fundamental Photographs; 15.20d: Richard Megna/ Fundamental Photographs; 15.22: Blair Seitz/Photo Researchers, Inc.; 15.24: National Optical Astronomy Observatories

Chapter 16 Opener: Royalty-free/PhotoDisc/Getty Images; 16.9: Bernard Benoit/Photo Researchers, Inc.; 16.10: Royalty-free/Digital Vision/Getty Images; 16.16: John Bacchus/Pearson Education Corporate Digital Archive; 16.21: Martin Bough/Fundamental Photographs; 16.26: Raytheon Aircraft Co.; 16.29: Dennis O'Clair/Tony Stone Images; 16.36: Leonard M. Weinstein/NASA/Langley Research Center; 16.37: Robert A. Hoover/NASA/Dryden Flight Research Center

Chapter 17 Opener: Akira Fujii; 17.4: Photodisc Green/Getty Images; 17.5: Courtesy of Cenco; 17.10: Jim Ross/NASA/Dryden Flight Research Center; 17.12: Marshall Henrichs/Marshall Henrichs; 17.15: Paul Seheult/Corbis; 17.16: Adam Hart-Davis/Photo Researchers, Inc.; 17.17: Richard Megna/Fundamental Photographs; 17.19: Bill Bachmann/PhotoEdit; 17.21: Lockheed Martin Missiles & Space Company, Inc., Russ Underwood, photographer; 17.25: Nancy Rogers/Nancy Rogers; 17.26: Dr, Arthur Tucker/Photo Researchers, Inc.

Chapter 18 Opener: Stone/Getty Images; 18.8: Park Scientific Instruments; 18.11: Izzy Schwartz/PhotoDisc/Getty Images; 18.13: Keith Brofsky/PhotoDisc, Inc.; 18.22: PhotoDisc, Inc.; 18.25: Royal Observatory, Edinburgh, Anglo-Australian Obsevatory; 18.28: Anglo-Australian Observatory, Photograph by David Malin

Chapter 19 Opener: Richard A. Cooke III/Tony Stone Images; 19.1: John P. Surey; 19.2a: PhotoDisc/StockTrek/Getty Images; 19.2b: AP/Wide World Photos; 19.6: Stephen Frisch/Stock Boston; 19.11: John Kelly/Tony Stone Images; 19.15: Mark Adams Photography/Liaison Agency, Inc.; 19.20: Beaura Katherine Ringrose; 19.31: Thomas Eisner and Daniel Aneshansley

Chapter 20 Opener: Myrleen Ferguson/PhotoEdit; 20.2: Mark Richards/PhotoEdit; 20.13: Mark Wagner/Tony Stone Images; 20.17: Paul Silverman/Fundamental Photographs

Chapter 21 Opener: Definitive Stock; 21.3: The Cavendish Laboratory, University of Cambridge, Cambridge, England; 21.28B: Tony Craddock/Photo Researchers, Inc.

Chapter 22 Opener: Department of Energy/Photo Researchers, Inc.; 22.10: Lawrence Lawry/Getty Images; 22.11: AKG London Ltd.; 22.20: C. Sherburne/PhotoLink/Getty Images; 22.29: Russ Kinne/Comstock

Chapter 23 Opener: Keith Kent/Photo Researchers, Inc.; 23.8: Courtesy of NASA/JPL/Caltech; 23.10: ©Lester V. Berman/CORBIS; 23.12: Fermilab Visual Media Services; 23.17: Hulton Archive/Keystone/Getty Images

Chapter 24 Opener: Richard Olivier/Corbis; 24.5: Paul Silverman/Fundamental Photographs; 24.9: David Young-Wolff/PhotoEdit; 24.15: Stanford Linear Accelerator/Science Photo Library/Photo Researchers, Inc.

Chapter 25 Opener: Don Farrall/Getty Images; 25.4: Fundamental Photographs; 25.5: PhotoDisc/Getty Images; 25.12: Martial Colomb/Photodisc/Getty Images; 25.15: Richard Megna/Fundamental Photographs

Chapter 26 Opener: Nick Koudis/PhotoDisc, Inc.; 26.2: Mario Nette/SPS/Vandystadt/Photo Researchers, Inc.; 26.5: John P. Surey; 26.13: John P. Surey; 26.19: Mehau Kulyk/Photo Researchers, Inc.; 26.25: TEK Image/Photo Researchers, Inc.

Chapter 27 Opener: Paul McCormick/The Image Bank/Getty Images; 27.15: Courtesy of Cenco; 27.38: Dr. Jeremy Burgess/Photo Researchers, Inc.

Chapter 28 Opener: Hitachi Medical Systems America, Inc.; 28.8: Definitive Stock; 28.11: George Haling/Photo Researchers, Inc.; 28.26: Bill Aron/PhotoEdit

Chapter 29 Opener: Shaun Egan/Tony Stone Images; 29.8: Maximilian Stock Ltd./Photo Researcher, Inc.; 29.17a: Steve Horrell/Photo Researchers, Inc.; 29.17b: Courtesy Toyota Motor Corporation; 29.17c: ©2003, Cirrus Design Corp.; 29.20a: Courtesy of NASA/JPL/Caltech; 29.20b: Courtesy of NASA/JPL/Caltech; 29.25: Ken Gatherum/Boeing Computer Services

Chapter 30 Opener: Spencer Grant/PhotoEdit; 30.2: Thompson Steele, Inc.; 30.7: Photodisc/Getty Images; 30.10: John Walsh/Photo Researchers, Inc.

Chapter 31 Opener John P. Surey/Caroline J. Robillard; 31.10: John P. Surey; 31.18: Lawrence Migdale/Photo Researchers, Inc.

Chapter 32 Opener: Definitive Stock; 32.1: Corbis; 32.14: Volker Steger/Photo Researchers, Inc.; 32.19: Tony Freeman/PhotoEdit

Chapter 33 Opener: Royalty-free/Digital Vision/Getty Images; 33.1: Michael Newman/PhotoEdit; 33.2: National Institutes of Health/Photo Researchers, Inc.; 33.8: Roger A. Freedman; 33.9a: NASA/WMAP Science Team; 33.9b: Spencer Grant/PhotoEdit; 33.12: Copyright The Exploratorium, photograph by Susan Schwartzenberg; 33.15: Barry Blanchard/Barry Blanchard; 33.28: Sepp Seitz/Woodfin Camp & Associates; 33.30: Camerique/H. Armstrong Roberts; 33.34: K. Nomachi/Photo Researchers, Inc.

Chapter 34 Opener: European Southern Observatory; 34.7: Marshall Henrichs; 34.10a: NASA Headquarters; 34.10b: NASA Headquarters; 34.21: PhotoLink/Getty Images; 34.31: Hoby Finn/PhotoDisc/Getty Images; 34.38a: Marshall Henrichs/Marshall Henrichs; 34.38b: Marshall Henrichs/Marshall Henrichs; 34.38c: Marshall Henrichs/Marshall Henrichs; 34.40: THK Photo Products, Inc.; 34.41: Photo Lennart Nilsson/Albert Bonniers Forlag AB; 34.49: Yerkes Observatory; 34.50: NOAO/AURA/NSF

Chapter 35 Opener: Peter Aprahamian/Photo Researchers, Inc.; 35.13: Bausch & Lomb Incorporated; 35.14: Bausch & Lomb Incorporated

Chapter 36 Opener: Tom Boyden Photography; 36.2: From Sears, Zemansky, and Young, *College Physics,* 7th Ed., ©1991 by Addison Wesley Publishing Company. Reprinted by permission of Addison Wesley Longman, Reading, Massachusetts; 36.6: From Sears, Zemansky, and Young, *College Physics,* 7th Ed., ©1991 by Addison Wesley Publishing Company. Reprinted by permission of Pearson Addison Wesley, San Francisco, California; 36.8: Richard Megna/Fundamental Photographs; 36.10: Dane Hornsleth; 36.16: Steve Percival/Photo Researchers, Inc.; 36.17a: NOAO/AURA/NSF; 36.17b: Physics Dept., Imperial College/Photo Researchers, Inc.; 36.18: S. Barden, R. Robles (NOAO/AURA/NSF; 36.19: Estate of Bertram Eugene Warren; 36.23: Science Photo Library/Photo Researchers, Inc.; 36.25: From Sears, Zemansky, and Young, *College Physics,* 7th Ed., ©1991 by Addison Wesley Publishing Company;. Reprinted by permission of Pearson Addison Wesley, San Francisco, California; 36.26b: From Sears, Zemansky, and Young, *College Physics,* 7th Ed., ©1991 by Addison Wesley Publishing Company. Reprinted by permission of Pearson Addison Wesley, San Francisco, California; 36.26c: From Sears, Zemansky, and Young, *College Physics,* 7th Ed., ©1991 by Addison Wesley Publishing Company. Reprinted by permission of Pearson Addison Wesley, San Francisco, California; 36.26: From Sears, Zemansky, and Young, *College Physics,* 7th Ed., ©1991 by Addison Wesley Publishing Company. Reprinted by permission of Pearson Addison Wesley, San Francisco, California; 36.27: Joel Gordon Photography; 36.30: Ronald R. Erickson/Media Interface, Ltd.

Chapter 37 Opener: Cern/Photo Researchers, Inc.; 37.6: NASA/STScl-1999-19; 37.16: Adam Hart-Davis/Photo Researchers, Inc.; 37.17: NASA and The Hubble Heritage Team (STScI/AURA); 37.20: Royalty-free/Digital Vision/Getty Images

Chapter 38 Opener: Bill Schoening/NOAO/AURA/NSF; 38.2: ITT Industries Night Vision; 38.12: National Optical Astronomy Observatories; 38.14: American Institute of Physics/Emilio Segré Visual Archives; 38.29: Michael Donne/Photo Researchers, Inc.; 38.30: W. Livingston/National Solar Observatory; 38.33: From Huggins, *Physics 1,* ©1968 by W.A. Benjamin, Inc. Reprinted by permission of Pearson Addison Wesley, San Francisco, California

Chapter 39 Opener: Alfred Pasieka/Photo Researchers, Inc.; 39.1: AIP Meggers Gallery of Nobel Laureates; 39.5: Reproduced from *PSSC Physics,* 2nd ed. (1965), D.C. Heath & Company with Educational Development Center, Inc., Newton Massachusetts; 39.11: Andrew Syred/Photo Researchers; 39.12: American Institute of Physics/Niels Bohr Library; 39.13: UPI/Corbis

Chapter 40 Opener: Courtesy of International Business Machines Corporation, Unauthorized use not permitted; 40.13: Digital Instruments

Chapter 41 Opener: Warren Bolster/Tony Stone Images; 41.2: Courtesy of NASA/JPL/Caltech; 41.6a: National Optical Astronomy Observatories; 41.6b: National Optical Astronomy Observatories; 41.14: Steven J. Gibson and A. Russell Taylor, University of Calgary; 41.15: American Institute of Physics/Emilio Segré Visual Archives

Chapter 42 Opener: Royalty-free/Digital Vision/Getty Images; 42.14: Definitive Stock; 42.18: Stanford University, courtesy AIP Emilio Segré Visual Archives; 42.36: Royalty-free/Getty Images

Chapter 43 Opener: Photo by Carl Hensen, courtesy of Bruce D. Smith, Department of Anthropology, National Museum of Natural History, Smithsonian Institution. ©Smithsonian Institution; 43.1: Science Photo Library/Photo Researchers, Inc.; 43.7: Lloyd Cluff/Corbis; 43.11: Dept. of Nuclear Medicine, Charing Cross Hospital/Photo Researchers, Inc.; 43.18: Anglo-Australian Observatory/Royal Observatory Edinburgh, photograph made from UK Schmidt plates by David Malin

Chapter 44 Opener: NASA/WMAP /Science Team; 44.1: Carl D. Anderson/Ernest Orlando, Lawrence Berkeley National Laboratory; 44.3: Ernest Orlando, Lawrence Berkeley National Laboratory; 44.7a: Fermilab Visual Media Services; 44.7b: Fermilab Visual Media Services; 44.9: Mark R. Lowenstine; 44.14: Institute for Cosmic Ray Research, University of Tokyo; 44.18: European Southern Observatory; 44.20: N.A. Sharp, REU program/NOAO/AURA/NSF; SummArt: N.A. Sharp, REU program/NOAO/AURA/NSF

INDEX

For users of the three-volume edition, pages 1–791 (Chapters 1–20) are in Volume 1; pages 792–1444 (Chapters 21–37) are in Volume 2; and pages 1403–1714 (Chapters 37–44) are in Volume 3.

Pages 1445–1714 (Chapters 38–44) are not in the Standard Edition of this textbook.

Note: Pages in **bold** locate figures; pages in *italic* locate tables.

D

E

G

O

P

Q

S

APPENDIX F

Numerical Constants

*FUNDAMENTAL PHYSICAL CONSTANTS**

Name	*Symbol*	*Value*
Speed of light	c	2.99792458×10^{8} m/s
Magnitude of charge of electron	e	$1.602176462(63) \times 10^{-19}$ C
Gravitational constant	G	$6.673(10) \times 10^{-11}$ N · m^2/kg^2
Planck's constant	h	$6.62606876(52) \times 10^{-34}$ J · s
Boltzmann constant	k	$1.3806503(24) \times 10^{-23}$ J/K
Avogadro's number	N_A	$6.02214199(47) \times 10^{23}$ molecules/mol
Gas constant	R	8.314472(15) J/mol · K
Mass of electron	m_e	$9.10938188(72) \times 10^{-31}$ kg
Mass of proton	m_p	$1.67262158(13) \times 10^{-27}$ kg
Mass of neutron	m_n	$1.67492716(13) \times 10^{-27}$ kg
Permeability of free space	μ_0	$4\pi \times 10^{-7}$ Wb/A · m
Permittivity of free space	$\epsilon_0 = 1/\mu_0 c^2$	$8.854187817 \ldots \times 10^{-12}$ C^2/N · m^2
	$1/4\pi\epsilon_0$	$8.987551787 \ldots \times 10^{9}$ N · m^2/C^2

*OTHER USEFUL CONSTANTS**

Mechanical equivalent of heat		4.186 J/cal (15° calorie)
Standard atmospheric pressure	1 atm	1.01325×10^{5} Pa
Absolute zero	0 K	−273.15°C
Electron volt	1 eV	$1.602176462(63) \times 10^{-19}$ J
Atomic mass unit	1 u	$1.66053873(13) \times 10^{-27}$ kg
Electron rest energy	$m_e c^2$	0.510998902(21) MeV
Volume of ideal gas (0°C and 1 atm)		22.413996(39) liter/mol
Acceleration due to gravity (standard)	g	9.80665 m/s^2

*Source: National Institute of Standards and Technology (**http://physics.nist.gov/cuu**). Numbers in parentheses show the uncertainty in the final digits of the main number; for example, the number 1.6454(21) means 1.6454 ± 0.0021. Values shown without uncertainties are exact.